HANDBOOK OF
PHYSICAL PROPERTIES
OF
ORGANIC CHEMICALS

HANDBOOK OF
PHYSICAL
PROPERTIES
OF
ORGANIC
CHEMICALS

Edited by

Philip H. Howard and William M. Meylan

Environmental Sciences Center
Syracuse Research Corporation

Associate Editors

Julie Funk, Michelle Pepling, Gloria W. Sage,
Jay Tunkel, Amy Hueber, and Dallas Aronson

LEWIS PUBLISHERS

Boca Raton   New York   London   Tokyo

Acquiring Editor:	Joel Stein
Marketing Manager:	Greg Daurelle
Direct Marketing Manager:	Arline Massey
Cover design:	Jonathan Pennell
Manufacturing:	Sheri Schwartz

Library of Congress Cataloging-in-Publication Data

Catalog record is available from the Library of Congress

This book contains information obtained from authentic and highly regarded sources. Reprinted material is quoted with permission, and sources are indicated. A wide variety of references are listed. Reasonable efforts have been made to publish reliable data and information, but the author and the publisher cannot assume responsibility for the validity of all materials or for the consequences of their use.

No claim to original U.S. Government works
International Standard Book Number 1-56670-227-5
Printed in the United States of America 1 2 3 4 5 6 7 8 9 0
Printed on acid-free paper

Preface

The Handbook of Physical Properties of Organic Chemicals contains data of physical and chemical properties for approximately 13,000 organic chemicals of environmental, pharmaceutical, or commercial interest. Each chemical record includes chemical structure, chemical formula, molecular weight, melting point/freezing point, boiling point, water solubility, octanol-water partition coefficient, vapor pressure, acid dissociation constant, Henry's Law constant, and atmospheric hydroxyl radial reaction rate constant. This handbook will be especially useful to scientists in the chemical, environmental, pharmaceutical, biochemical, toxicological, and related sciences as a comprehensive source of evaluated data on the basic physical/chemical properties of organic compounds used by these disciplines.

This manual is a text version of the PHYSPROP© Database available from Syracuse Research Corporation (http://esc.syrres.com). The chemicals for the book were selected from the PHYSPROP© database. Due to space constraints, all records from the database could not be put into book format. Originally, PHYSPROP© was created from several thousand chemicals related to various U.S. EPA lists, in particular the National Library of Medicine's (NLM) Hazardous Substances Data Bank (HSDB©). PHYSPROP© is currently, and continues to be, updated with organic chemicals selected for inclusion in these EPA lists. A large number of chemicals in PHYSPROP© has been added from other Syracuse Research Corporation (SRC) databases of reliable, experimental values; these data have then been used to develop structure/property estimation methods (e.g., estimation of octanol/water partition coefficient [Meylan and Howard, 1995]). These include databases of octanol-water partition coefficients, water solubilities, Henry's Law constants, and hydroxyl radical reaction rate constants.

Many experimental (measured) physical property values in PHYSPROP© were located using SRC's Environmental Fate Data Bases (EFDB©) (Howard et al., 1982, 1986). The DATALOG© and CHEMFATE© files in EFDB© were especially useful in identifying physical/chemical property sources of data. Physical property values were prudently selected by SRC scientists from the many sources identified in SRC's DATALOG© file. Whenever possible, experimental values were selected from primary literature sources or a database of evaluated values; in the case of vapor pressures, some values were interpolated or extrapolated from the original experimentally determined data using a temperature-dependent equation. When an experimental or extrapolated value could not be found or determined, values were estimated using SRC's estimation programs (see http://esc.syrres.com).

Explanation of format

Chemical data records in the Handbook of Physical Properties of Organic Chemicals are ordered by the Chemical Abstract Service (CAS) Registry Number. The figure below illustrates a typical chemical record.

CAS #:	000050-00-0			FORMALDEHYDE
Formula:	CH_2O			
Mol Weight:	30.03			
MP (deg C):	-92	FP (deg C):		
BP (deg C):	-21			
BP pressure (mm Hg):				

Property/Value	Units	Temp	Data Type	Reference
Wsol 4.00e+005	mg/L	20	EXP	PICKRELL,JA ET AL. (1983)
logP 0.35			EXP	HANSCH,C & LEO,AJ (1985)
VP 3.89e+003	mm Hg	25	EXT	BOUBLIK,T ET AL. (1984)
DC	pKa			
HL 3.37e-007	atm m3/mol	25	EXP	BETTERTON,EA & HOFFMAN,MR (1988)
OH 9.70e-012	cm3/molc sec	25	EXP	ATKINSON,R (1989)

Abbreviations in fields to the left of the structure are as follows:

Mol Weight   molecular weight
MP   melting point
FP   freezing point
BP   boiling point

Abbreviations in the bottom half of the form are as follows:

Temp   temperature in °C
Wsol   water solubility
log P   log octanol-water partition coefficient
DC   acid dissociation constant (pKa)
HL   Henry's Law constant
OH   atmospheric hydroxyl radical reaction rate constant

Under the "Data Type" heading, three abbreviations are used as follows: experimentally measured or derived data (EXP), estimated data (EST), and data extrapolated out of range (EXT). Extrapolated values were determined only for vapor pressure. Extrapolated vapor pressures were calculated at 25°C from an Antoine's or similar equation, which was derived from an experimental data range above or below 25°C. When the experimental data range includes 25°C (interpolated value), the calculated value is considered experimentally measured (EXP).

The "Reference" heading refers to the source of the property value. References appear in abbreviated form; see Appendix III for full reference citations. For most estimated data types, the reference refers to the journal article that describes the estimation methodology.

How to use the handbook

This book is divided into four sections. The main section of the book consists of the chemical data records in CAS Registry Number order. The next two sections are indexes, Appendix I and Appendix II. The final section is an alphabetical list of full author references (Appendix III).

Appendix I is ordered by molecular formula. The molecular formula is ordered by the Hill system (number of carbons, number of hydrogens, and then alphabetical by element). For example, C_2H_5NO would precede C_2H_5O and C_2H_5O would precede C_2H_6. Because there will often be more than one CAS Registry Number for a molecular formula (e.g., o-, m-, and p-dichlorobenzene all have a molecular formula of $C_6H_4CL_2$), the chemical name and CAS Registry Number are also given. After finding the appropriate molecular formula and chemical name, the CAS Registry Number can be recorded; at this point, one may go to the main section, look for the indicated record and obtain the physical/chemical information needed.

Sample molecular formula index:

C6H12O6	003458-28-4	D-MANNOSE
C6H12O6	007660-25-5	BETA-FRUCTOPYRANOSE (D)
C6H12O7	000526-95-4	D-GLUCONIC ACID
C6H12S	001569-69-3	CYCLOHEXANETHIOL
C6H13Br	000111-25-1	1-BROMOHEXANE
C6H13Br	003377-86-4	2-BROMOHEXANE
C6H13Cl	000544-10-5	1-CHLOROHEXANE
C6H13Cl	000638-28-8	2-CHLOROHEXANE
C6H13Cl3Si	000928-65-4	HEXYLTRICHLOROSILANE
C6H13F	000373-14-8	1-FLUOROHEXANE
C6H13I	000638-45-9	1-IODOHEXANE
C6H13N	000108-91-8	CYCLOHEXANAMINE
C6H13N	000109-05-7	2-METHYLPIPERIDINE

Appendix II is ordered by chemical name. The chemical name index is in capital letters and has been alphabetically ordered. Because the length of the names are limited to 48 characters, some chemical names have been abbreviated and/or truncated. For example, IPR is short for isopropyl, METO is short for methoxy, NO_2 is short for nitro, and so on. Prefixes commonly used in organic chemistry which are not normally considered part of the name for alphabetizing purposes (e.g., o- ortho, m- meta, p- para, alpha, beta, N-, cis-, trans-, sec-), as well as numbers, have not been considered for alphabetical order. Other prefixes which normally are considered part of the names (e.g., di-, tri-, tetra-, iso-, cyclo-) are used for alphabetically positioning. For example, 2,4-Dinitrotoluene is under D and tert-Butyl alcohol is under B. After finding the appropriate chemical name, the CAS Registry Number can be recorded; at this point, one may go to the main section, look for the indicated record, and obtain the physical/chemical information needed.

Sample chemical name index:

PERFLUIDONE	037924-13-3
PERFLUOROCYCLOBUTANE	000115-25-3
PERFLUOROCYCLOHEXANE	000355-68-0
PERFLUORO-2,9-DIMETHYLDECANE	103188-55-2
PERFLUORO-2,7-DIMETHYLOCTANE	003021-63-4
PERFLUOROHEPTANE	000335-57-9
PERFLUORO-N-HEXANE	000355-42-0
PERFLUOROMETHYLCYCLOHEXANE	000355-02-2
PERICYAZINE	002622-26-6
CIS-PERMETHRIN	061949-76-6
PERMETHRIN	052645-53-1
TRANS-PERMETHRIN	061949-77-7
PEROXIDE, BIS(1-METHYL-1-PHENYLETHYL)	000080-43-3
PEROXIDE-2-BUTANONE	001338-23-4

Appendix III is an alphabetical list of full reference citations.

Substance identification

CAS Registry Number: This number is assigned by the American Chemical Society's Chemical Abstracts Services as a unique identifier.

Chemical Name: At the top of the form is the chemical name. The length of the names are assigned 48 characters; therefore, many names are truncated. An attempt was made to select the one relatively short name that would be most recognizable. For users who do not have the CAS Registry Number, books or databases that can be searched by the chemical name are recommended (e.g., *Dictionary of Chemical Names and Synonyms,* Howard and Neal (1992)).

Structure: The chemical structures have been drawn using ISIS™/DRAW software. For ease of reading, bond angles have been altered and some terminal functional groups have not been labeled, e.g., CH_3. Also, not all hydrogens are labeled.

Molecular Formula: The formula is in Hill notation, which is given as the number of carbons followed by the number of hydrogens followed by any other elements in alphabetic order. It is automatically assigned from the structure drawn using ISIS™/DRAW software.

Chemical and physical properties

Boiling Point: The boiling point or boiling point range is given, along with the pressure at which it was obtained. Units are in °C and mmHg, respectively. Where pressure is not reported, it is accepted that the value is at 760 mmHg. Values were collected primarily from standard reference sources such as the *CRC Handbook of Chemistry and Physics* (various annual editions through 1996), *The Merck Index* (9th, 10th, and 11th editions), the *Aldrich Catalog* (various annual editions through 1996), *Kirk-Othmer Encyclopedia of Chemical Technology* (3rd and 4th editions), *The Pesticide Manual* (various editions, e.g., Tomlin, 1994), the *Herbicide Handbook* (5th and 6th editions, e.g., WSSA, 1983), and other handbook compilations such as Riddick et al. (1986). For this property, no specific reference is cited.

Melting Point: The melting point or melting point range is recorded. In some instances, the temperature at which decomposition occurs is given in lieu of the melting point. Values were obtained from a variety of reliable sources such as the *CRC Handbook of Chemistry and Physics* (various annual editions through 1996),

The Merck Index (9th, 10th, and 11th editions), the *Aldrich Catalog* (various annual editions through 1996), *Kirk-Othmer Encyclopedia of Chemical Technology* (3rd and 4th editions), *The Pesticide Manual* (various editions, e.g., Tomlin, 1994), the *Herbicide Handbook* (5th and 6th editions, e.g., WSSA, 1983), and other handbook compilations such as Riddick et al. (1986). For this property, no specific reference is cited.

Freezing Point: The freezing point is recorded when available. However, the melting point was found to be more common than the freezing point. Values were collected primarily from standard reference sources such as the *CRC Handbook of Chemistry and Physics* (various annual editions through 1996), *The Merck Index* (9th, 10th, and 11th editions), the *Aldrich Catalog* (various annual editions through 1996), *Kirk-Othmer Encyclopedia of Chemical Technology* (3rd and 4th editions), *The Pesticide Manual* (various editions, e.g., Tomlin, 1994), the *Herbicide Handbook* (5th and 6th editions, e.g., WSSA, 1983), and other handbook compilations such as Riddick et al. (1986). For this property, no specific reference is cited.

Molecular Weight: The molecular weight to two decimal points is given. It was automatically assigned from the structure drawn using ISIS™/DRAW software.

Water Solubility: Water solubility is a physical property which determines the extent to which a compound will dissolve in water. The values are reported in units of mg/L (ppm) at a temperature at or as close as possible to 25°C. Fate processes that are, or can be, affected by water solubility include photolysis, hydrolysis, oxidation, and washout from the atmosphere by rain or fog. The water solubility of a chemical provides considerable insight into the fate and transport of a chemical in the environment. Chemicals that are highly water soluble have a tendency to remain dissolved and not partition to soil or sediment or bioconcentrate in aquatic organisms. They are also less likely to volatilize from water (depending upon the vapor pressure), more likely to washout from the atmosphere with rain or fog, and generally more likely to biodegrade. Low water soluble chemicals follow the opposite trend; they partition to soil or sediments, bioconcentrate in aquatic organisms, volatilize more readily from water, are less likely to be stripped from the atmosphere, and are less likely to biodegrade.

Water solubilities were initially collected primarily from references identified in SRC's DATALOG© file. Values from the AQUASOL dATABASE™ of the University of Arizona (Yalkowsky and Dannenfelser, 1992) were also obtained. Many pesticide water solubilities were taken from the U.S. Department of Agriculture's Pesticide Properties Database (Wauchope et al., 1991) and various editions of *The Pesticide Manual* (e.g., Tomlin, 1994) and the *Herbicide Handbook* (e.g., WSSA, 1983).

For those compounds still lacking an experimental value, water solubilities were estimated using recommended regression equations (Lyman et al., 1982) or with SRC's WSKOWWIN© software. This computer program uses the log octanol-water partition coefficient of a compound to estimate the water solubility. The program methodology is described in a journal article (Meylan et al., 1996); the same description appears in two reports prepared for the U.S. EPA (Meylan and Howard, 1994a, 1994b).

Log Octanol/Water Partition Coefficient: The octanol/water partition coefficient (log P) is the ratio of a chemical's concentration in the octanol phase versus its concentration in the water phase for a two phase system at equilibrium. It is a unitless quantity. The octanol/water partition coefficient has been shown to correlate well with bioconcentration factors in aquatic organisms (Lyman et al., 1982; Veith et al., 1979) and adsorption to soil or sediment (Lyman et al., 1982; Meylan et al., 1992). In general, the log P values pertain to compounds that exist predominantly in a nonionized form; values for compounds considered "ion pairs" such as sodium salts, quaternary ammonium compounds, and pyridinium compounds are exceptions. The zwitterionic log P values for amino acids are also used.

Experimental log P values were collected primarily from the large data compilations of Hansch and Leo (1981, 1985; MedChem log P database), Hansch et al. (1995), and Sangster (1993, 1994). The Hansch compilations identify recommended log P values with a "starlist" designation; with rare exception, values having a "starlist" designation were selected. The Sangster compilations also have "recommended" values many of which were selected. Other experimental log P values were identified through papers in SRC's DATALOG© file and recent journal articles.

Nearly all chemicals that do not have experimental log P values were estimated with SRC's estimation software. The LOGKOW© methodology is described in a journal article (Meylan and Howard, 1995). Occasionally, chemical octanol/water partition coefficients were not calculated because a necessary fragment constant for the chemical was not available.

Vapor Pressure: Vapor pressure is the pressure exerted by a vapor in equilibrium with its liquid or solid phase. Vapor pressure was reported in mmHg at or as close as possible to 25°C. The vapor pressure of a chemical provides considerable insight into the transport and partitioning of a chemical in the environment and in commercial settings. The volatility of the pure chemical is dependent upon the vapor pressure, and volatilization from water is dependent upon the vapor pressure and water solubility. The form in which a chemical will be found in the atmosphere is dependent upon the vapor pressure; chemicals with a vapor pressure less than 10^{-8} mmHg will be mostly associated with particulate matter (Bidleman, 1988; Eisenreich, 1981).

Reliable compilation sources such as Boublik et al. (1984) or Daubert and Danner (1989, 1991) were checked initially for experimental values. SRC's DATALOG© file was then searched to identify other possible sources of experimental values. An attempt was made to identify any values that were measured near 25°C or that had a derived Antoine or similar equation based on values near 25°C. In some instances, the vapor pressure was extrapolated to 25°C using Antoine or other constants. In cases where the extrapolated value was a supercooled liquid at 25°C, a correction was calculated using an equation from Bidleman (1988).

Most estimated vapor pressures were estimated with SRC's MPBPVP© software. This computer program uses the boiling point of a compound's to estimate the vapor pressure. It estimates vapor pressure by the Antoine method and the modified Grain method. The modified Grain method is a revision and significant improvement of the modified Watson method. Estimated vapor pressures are referenced as Neely and Blau (1985), the source from which the MPBPVP© software is an application.

Acid Dissociation Constant: The acid dissociation constant as its negative log (pKa) is given for chemicals that are likely to dissociate at pHs commonly found in the environment (between 5 and 9). Some chemical classes where dissociation is important include phenols, carboxylic acids, sulfonic acids, phosphates, and aliphatic and aromatic amines. The pKa determines the percent of a chemical in dissociated and undissociated form as a function of pH. For example, for a weak acid with a pKa of 4.75, the percent of the acid dissociated at one and two pH units above the pKa is:

1% dissociated at pH 2.75
10% dissociated at pH 3.75
50% dissociated at pH 4.75
90% dissociated at pH 5.75
99% dissociated at pH 6.75

The degree of dissociation affects such processes as photolysis (absorption spectra of chemicals that dissociate can be sensitive to pH), evaporation from water (ionic forms are nonvolatile), soil or sediment adsorption (ionic forces may have significant effects), bioconcentration in fish, and toxicity. Evaluated dissociation constants were preferentially obtained from evaluated sources such as Serjeant and Dempsey (1979) or Perrin (1972). SRC's DATALOG file was then searched to identify other possible sources of values. At present, there are only approximately 1,000 compounds that have experimental pKa values. Of these, 890 experimental values were collected from a database shared from the EPA (Hilal et al., 1994). Very few estimated pKa values presently appear in the handbook due to the lengthy calculation time required for estimation using pKalc (PALLAS© Software, CompuDrug Chemistry, Ltd 1994) or SPARC© (Hilal et al., 1994).

Henry's Law Constant: The Henry's Law constant, H, can be defined as the air/water partition coefficient at low concentrations of a chemical in water. A nondimensional H relates the chemical concentration in the gas phase to its concentration in the water phase. Henry's Law constants in the handbook are reported in units of atm-m^3/mole. H provides an indication of the partition between air and water at equilibrium and also is used to calculate the rate of evaporation from water. Henry's Law constants can be directly measured, calculated from the experimental water solubility and vapor pressure (dividing the vapor pressure in atm by the water solubility in mole/m^3 to give H in atm-m^3/mole), or estimated from the structure. Some critical review data of Henry's Law constants are available, e.g., Mackay and Shiu (1981).

SRC's DATALOG file was searched to identify possible sources of Henry's Law constants. When available, reliably measured experimental values were selected. In the absence of direct experimental values, Henry's Law constants were estimated by one of two different methods. When possible, Henry's Law constants were estimated from reliable experimental vapor pressure *and* water solubility data (referenced as VP/WSOL); Meylan and Howard (1991) have demonstrated that dividing the vapor pressure by the water solubility yields excellent Henry's Law constant values. If a reliable vapor pressure *or* water solubility was not available, Henry's Law constants were estimated with SRC's HENRY© software. This computer program uses bond or group fragments to calculate the estimated value. A journal article describes the methodology (Meylan and Howard, 1991), which is an update of the Hine and Mookerjee (1975) methodology.

Hydroxyl radical reaction rate constant: The hydroxyl radical reaction rate constant gives the gas phase reaction rate of organic chemicals with OH radicals in the atmosphere. For most organic chemicals in the vapor phase, reaction with photochemically generated hydroxyl radicals is the most important degradation process in the troposphere. It is expressed in units of cm^3/mole sec in the handbook.

Most experimental values were taken from evaluated data compilations (Atkinson, 1989, 1994; Kwok and Atkinson, 1995). Nearly all estimated OH rate constants were calculated with SRC's AOP© software using a fragment constant method. The AOP program is described in a journal article (Meylan and Howard, 1993); it uses the methodology detailed in several other journal articles (Atkinson, 1987, 1988; Kwok and Atkinson, 1995).

CHEMICAL DATA RECORDS

FORMALDEHYDE

CAS #: 000050-00-0			
Formula: CH_2O			
Mol Weight: 30.03			
MP (deg C): -92	FP (deg C):		
BP (deg C): -21			
BP pressure (mm Hg):			

Property/Value	Units	Temp	Data Type	Reference
WS 4.00E+005	mg/L	20	EXP	PICKRELL,JA ET AL. (1983)
logP 0.35			EXP	HANSCH,C & LEO,AJ (1985)
VP 3.89E+003	mm Hg	25	EXT	BOUBLIK,T ET AL. (1984)
DC 13.27	pKa	25	EXP	SERJEANT,EP & DEMPSEY,B (1979)
HL 3.37E-007	atm m3/mol	25	EXP	BETTERTON,EA & HOFFMAN,MR (1988)
OH 9.70E-012	cm3/molc sec	25	EXP	ATKINSON,R (1989)

DEXAMETHASONE

CAS #: 000050-02-2			
Formula: $C_{22}H_{29}FO_5$			
Mol Weight: 392.47			
MP (deg C): 236	FP (deg C):		
BP (deg C):			
BP pressure (mm Hg):			

Property/Value	Units	Temp	Data Type	Reference
WS 8.90E+001	mg/L	25	EXP	YALKOWSKY,SH & DANNENFELSER,RM (1992)
logP 1.83			EXP	HANSCH,C & LEO,AJ (1985)
VP 8.86E-014	mm Hg	25	EST	NEELY,WB & BLAU,GE (1985)
DC	pKa			
HL 7.15E-008	atm m3/mol	25	EST	MEYLAN,WM & HOWARD,PH (1991)
OH 6.87E-011	cm3/molc sec	25	EST	MEYLAN,WM & HOWARD,PH (1993)

HYDROCORTISONE ACETATE

CAS #: 000050-03-3			
Formula: $C_{23}H_{32}O_6$			
Mol Weight: 404.51			
MP (deg C): 223 dec	FP (deg C):		
BP (deg C):			
BP pressure (mm Hg):			

Property/Value	Units	Temp	Data Type	Reference
WS 1.41E+001	mg/L	25	EXP	YALKOWSKY,SH & DANNENFELSER,RM (1992)
logP 2.30			EXP	HANSCH,C & LEO,AJ (1985)
VP 2.56E-013	mm Hg	25	EST	NEELY,WB & BLAU,GE (1985)
DC	pKa			
HL 6.24E-012	atm m3/mol	25	EST	MEYLAN,WM & HOWARD,PH (1991)
OH 1.09E-010	cm3/molc sec	25	EST	MEYLAN,WM & HOWARD,PH (1993)

CORTISONE ACETATE

CAS #: 000050-04-4			
Formula: $C_{23}H_{30}O_6$			
Mol Weight: 402.49			
MP (deg C): 222	FP (deg C):		
BP (deg C):			
BP pressure (mm Hg):			

Property/Value	Units	Temp	Data Type	Reference
WS 2.00E+001	mg/L	25	EXP	YALKOWSKY,SH & DANNENFELSER,RM (1992)
logP 2.10			EXP	HANSCH,C & LEO,AJ (1985)
VP 1.10E-012	mm Hg	25	EST	NEELY,WB & BLAU,GE (1985)
DC	pKa			
HL 6.51E-014	atm m3/mol	25	EST	MEYLAN,WM & HOWARD,PH (1991)
OH 1.05E-010	cm3/molc sec	25	EST	MEYLAN,WM & HOWARD,PH (1993)

PHENOBARBITAL

CAS #: 000050-06-6			
Formula: $C_{12}H_{12}N_2O_3$			
Mol Weight: 232.24			
MP (deg C): 176.5	FP (deg C):		
BP (deg C):			
BP pressure (mm Hg):			

Property/Value	Units	Temp	Data Type	Reference
WS 1.11E+003	mg/L	25	EXP	YALKOWSKY,SH & DANNENFELSER,RM (1992)
logP 1.47			EXP	HANSCH,C & LEO,AJ (1985)
VP 1.98E-012	mm Hg	25	EST	NEELY,WB & BLAU,GE (1985)
DC 7.30	pKa		EST	HANSCH,C & LEO,AJ (1985)
HL 3.80E-016	atm m3/mol	25	EST	MEYLAN,WM & HOWARD,PH (1991)
OH 3.20E-011	cm3/molc sec	25	EST	MEYLAN,WM & HOWARD,PH (1993)

MITOMYCIN C

CAS #: 000050-07-7			
Formula: $C_{15}H_{18}N_4O_5$			
Mol Weight: 334.33			
MP (deg C): >360	FP (deg C):		
BP (deg C): 534			
BP pressure (mm Hg):			

Property/Value	Units	Temp	Data Type	Reference
WS 8.43E+003	mg/L	25	EST	MEYLAN,WM ET AL. (1996)
logP -0.40			EXP	HANSCH,C & LEO,AJ (1995)
VP 6.78E-010	mm Hg	25	EST	NEELY,WB & BLAU,GE (1985)
DC 10.85	pKa		EXP	HANSCH,C & LEO,AJ (1995)
HL 1.14E-024	atm m3/mol	25	EST	MEYLAN,WM & HOWARD,PH (1991)
OH 6.70E-011	cm3/molc sec	25	EST	MEYLAN,WM & HOWARD,PH (1993)

METHARBITAL

CAS #: 000050-11-3			
Formula: $C_9H_{14}N_2O_3$			
Mol Weight: 198.22			
MP (deg C): 115	FP (deg C):		
BP (deg C):			
BP pressure (mm Hg):			

Property/Value	Units	Temp	Data Type	Reference
WS 1.98E+003	mg/L	25	EXP	YALKOWSKY,SH & DANNENFELSER,RM (1992)
logP 1.15			EXP	HANSCH,C & LEO,AJ (1985)
VP 7.07E-010	mm Hg	25	EST	NEELY,WB & BLAU,GE (1985)
DC 8.01	pKa	25	EXP	HANSCH,C & LEO,AJ (1985)
HL 7.92E-013	atm m3/mol	25	EST	MEYLAN,WM & HOWARD,PH (1991)
OH 1.02E-011	cm3/molc sec	25	EST	MEYLAN,WM & HOWARD,PH (1993)

MEPHENYTOIN

CAS #: 000050-12-4			
Formula: $C_{12}H_{22}N_2O_2$			
Mol Weight: 226.32			
MP (deg C): 135	FP (deg C):		
BP (deg C):			
BP pressure (mm Hg):			

Property/Value	Units	Temp	Data Type	Reference
WS 1.27E+003	mg/L	25	EST	MEYLAN,WM ET AL. (1996)
logP 1.69			EXP	SANGSTER,J (1994)
VP 3.80E-008	mm Hg	25	EST	NEELY,WB & BLAU,GE (1985)
DC	pKa			
HL 4.90E-010	atm m3/mol	25	EST	MEYLAN,WM & HOWARD,PH (1991)
OH 1.07E-011	cm3/molc sec	25	EST	MEYLAN,WM & HOWARD,PH (1993)

CAS #: 000050-13-5				MEPERIDINE

Formula: $C_{15}H_{22}ClNO_2$

Mol Weight: 283.80

MP (deg C): 187.5 FP (deg C):

BP (deg C):

BP pressure (mm Hg):

Property/Value	Units	Temp	Data Type	Reference
WS 6.55E+003	mg/L	25	EXP	ROY,SD & FLYNN,GL (1988)
logP 3.03			EST	MEYLAN,WM & HOWARD,PH (1995)
VP 6.89E-005	mm Hg	25	EST	NEELY,WB & BLAU,GE (1985)
DC	pKa			
HL 9.61E-009	atm m3/mol	25	EST	MEYLAN,WM & HOWARD,PH (1991)
OH 9.74E-011	cm3/molc sec	25	EST	MEYLAN,WM & HOWARD,PH (1993)

CAS #: 000050-18-0				CYCLOPHOSPHAMIDE

Formula: $C_7H_{15}Cl_2N_2O_2P$

Mol Weight: 261.09

MP (deg C): 51.5 FP (deg C):

BP (deg C):

BP pressure (mm Hg):

Property/Value	Units	Temp	Data Type	Reference
WS 4.00E+004	mg/L	20	EXP	IARC (1975)
logP 0.63			EXP	HANSCH,C & LEO,AJ (1985)
VP 4.45E-005	mm Hg	25	EST	NEELY,WB & BLAU,GE (1985)
DC	pKa			
HL 1.40E-011	atm m3/mol	25	EST	MEYLAN,WM & HOWARD,PH (1991)
OH 7.03E-011	cm3/molc sec	25	EST	MEYLAN,WM & HOWARD,PH (1993)

CAS #: 000050-21-5				LACTIC ACID

Formula: $C_3H_6O_3$

Mol Weight: 90.08

MP (deg C): 16.8 FP (deg C):

BP (deg C): 122

BP pressure (mm Hg): 1.50E+001

Property/Value	Units	Temp	Data Type	Reference
WS 1.00E+006	mg/L	25	EST	MEYLAN,WM ET AL. (1996)
logP -0.72			EXP	HANSCH,C & LEO,AJ (1985)
VP 8.13E-002	mm Hg	25	EXP	DAUBERT,TE & DANNER,RP (1989)
DC 3.86	pKa	20	EXP	KORTUM,G ET AL (1961)
HL 1.13E-007	atm m3/mol	25	EST	MEYLAN,WM & HOWARD,PH (1991)
OH 5.92E-012	cm3/molc sec	25	EST	MEYLAN,WM & HOWARD,PH (1993)

CAS #: 000050-22-6				CORTICOSTERONE

Formula: $C_{21}H_{30}O_4$

Mol Weight: 346.47

MP (deg C): 181 FP (deg C):

BP (deg C):

BP pressure (mm Hg):

Property/Value	Units	Temp	Data Type	Reference
WS 1.43E+002	mg/L	25	EST	MEYLAN,WM ET AL. (1996)
logP 1.94			EXP	HANSCH,C & LEO,AJ (1985)
VP 4.74E-012	mm Hg	25	EST	NEELY,WB & BLAU,GE (1985)
DC	pKa			
HL 3.70E-010	atm m3/mol	25	EST	MEYLAN,WM & HOWARD,PH (1991)
OH 1.12E-010	cm3/molc sec	25	EST	MEYLAN,WM & HOWARD,PH (1993)

CAS #: 000050-23-7				HYDROCORTISONE

Formula: $C_{21}H_{30}O_5$

Mol Weight: 362.47

MP (deg C): 162.5 FP (deg C):

BP (deg C):

BP pressure (mm Hg):

Property/Value	Units	Temp	Data Type	Reference
WS 3.20E+002	mg/L	25	EXP	YALKOWSKY,SH & DANNENFELSER,RM (1992)
logP 1.61			EXP	HANSCH,C & LEO,AJ (1985)
VP 1.35E-013	mm Hg	25	EST	NEELY,WB & BLAU,GE (1985)
DC	pKa			
HL 5.77E-008	atm m3/mol	25	EST	MEYLAN,WM & HOWARD,PH (1991)
OH 1.10E-010	cm3/molc sec	25	EST	MEYLAN,WM & HOWARD,PH (1993)

CAS #: 000050-24-8				PREDNISOLONE

Formula: $C_{21}H_{28}O_5$

Mol Weight: 360.45

MP (deg C): 235 FP (deg C):

BP (deg C):

BP pressure (mm Hg):

Property/Value	Units	Temp	Data Type	Reference
WS 2.23E+002	mg/L	25	EXP	YALKOWSKY,SH & DANNENFELSER,RM (1992)
logP 1.62			EXP	HANSCH,C & LEO,AJ (1985)
VP 1.18E-013	mm Hg	25	EST	NEELY,WB & BLAU,GE (1985)
DC	pKa			
HL 2.71E-008	atm m3/mol	25	EST	MEYLAN,WM & HOWARD,PH (1991)
OH 6.97E-011	cm3/molc sec	25	EST	MEYLAN,WM & HOWARD,PH (1993)

CAS #: 000050-27-1				ESTRIOL

Formula: $C_{18}H_{24}O_3$

Mol Weight: 288.39

MP (deg C): 126 FP (deg C):

BP (deg C):

BP pressure (mm Hg):

Property/Value	Units	Temp	Data Type	Reference
WS 4.41E+002	mg/L	25	EST	MEYLAN,WM ET AL. (1996)
logP 2.45			EXP	HANSCH,C ET AL. (1995)
VP 1.97E-010	mm Hg	25	EST	NEELY,WB & BLAU,GE (1985)
DC	pKa			
HL 1.33E-012	atm m3/mol	25	EST	MEYLAN,WM & HOWARD,PH (1991)
OH 1.30E-010	cm3/molc sec	25	EST	MEYLAN,WM & HOWARD,PH (1993)

CAS #: 000050-28-2				ESTRADIOL

Formula: $C_{18}H_{24}O_2$

Mol Weight: 272.39

MP (deg C): 176 FP (deg C):

BP (deg C):

BP pressure (mm Hg):

Property/Value	Units	Temp	Data Type	Reference
WS 3.60E+000	mg/L	27	EXP	YALKOWSKY,SH & DANNENFELSER,RM (1992)
logP 4.01			EXP	HANSCH,C ET AL. (1995)
VP 1.26E-008	mm Hg	25	EST	NEELY,WB & BLAU,GE (1985)
DC	pKa			
HL 3.64E-011	atm m3/mol	25	EST	MEYLAN,WM & HOWARD,PH (1991)
OH 1.23E-010	cm3/molc sec	25	EST	MEYLAN,WM & HOWARD,PH (1993)

CAS #: 000050-29-3				P,P'-DDT

Formula: $C_{14}H_9Cl_5$

Mol Weight: 354.49

MP (deg C): 109 FP (deg C):

BP (deg C):

BP pressure (mm Hg):

Property/ Value	Units	Temp	Data Type	Reference
WS 2.50E-002	mg/L	25	EXP	BIGGAR,JW & RIGGS,RI (1974)
logP 6.91			EXP	HANSCH,C ET AL. (1995)
VP 1.60E-007	mm Hg	20	EXP	BIDLEMAN,TF & FOREMAN,WT (1987)
DC	pKa			
HL 8.30E-006	atm m3/mol	20	EST	VP/WSOL
OH 3.62E-012	cm3/molc sec	25	EST	MEYLAN,WM & HOWARD,PH (1993)

CAS #: 000050-30-6				2,6-DICHLOROBENZOIC ACID

Formula: $C_7H_4Cl_2O_2$

Mol Weight: 191.01

MP (deg C): 144 FP (deg C):

BP (deg C):

BP pressure (mm Hg):

Property/ Value	Units	Temp	Data Type	Reference
WS 1.41E+004	mg/L		EXP	YALKOWSKY,SH & DANNENFELSER,RM (1992)
logP 2.23			EXP	HANSCH,C ET AL. (1995)
VP 4.47E-004	mm Hg	25	EST	NEELY,WB & BLAU,GE (1985)
DC 1.59	pKa	25	EXP	SERJEANT,EP & DEMPSEY,B (1979)
HL 5.95E-008	atm m3/mol	25	EST	MEYLAN,WM & HOWARD,PH (1991)
OH 8.77E-013	cm3/molc sec	25	EST	MEYLAN,WM & HOWARD,PH (1993)

CAS #: 000050-31-7				2,3,6-TRICHLOROBENZOIC ACID

Formula: $C_7H_3Cl_3O_2$

Mol Weight: 225.46

MP (deg C): 125.5 FP (deg C):

BP (deg C):

BP pressure (mm Hg):

Property/ Value	Units	Temp	Data Type	Reference
WS 7.70E+003	mg/L	22	EXP	MARTIN,H & WORTHING,CR (1977)
logP 2.71			EST	MEYLAN,WM & HOWARD,PH (1995)
VP 5.50E-004	mm Hg	25	EXP	FREAR,DS (1976)
DC 1.50	pKa	25	EXP	WEBER,JB (1972); APPROXIMATION
HL 2.12E-008	atm m3/mol	25	EST	VP/WSOL
OH 1.43E-013	cm3/molc sec	25	EST	ATKINSON,R (1988)

CAS #: 000050-32-8				BENZO(A)PYRENE

Formula: $C_{20}H_{12}$

Mol Weight: 252.32

MP (deg C): 179.2 FP (deg C):

BP (deg C): 311

BP pressure (mm Hg): 1.00E+001

Property/ Value	Units	Temp	Data Type	Reference
WS 1.62E-003	mg/L	25	EXP	MAY,WE ET AL. (1983)
logP 5.97			EXP	HANSCH,C & LEO,AJ (1985)
VP 5.49E-009	mm Hg	25	EXT	MURRAY,JJ ET AL. (1974)
DC	pKa			
HL 2.45E-006	atm m3/mol	25	EXP	SOUTHWORTH,GR (1979A)
OH 1.50E-010	cm3/molc sec	25	EST	MEYLAN,WM & HOWARD,PH (1993)

CAS #: 000050-33-9				PHENYLBUTAZONE

Formula: $C_{19}H_{20}N_2O_2$

Mol Weight: 308.38

MP (deg C): 105 FP (deg C):

BP (deg C):

BP pressure (mm Hg):

Property/ Value	Units	Temp	Data Type	Reference
WS 7.00E+002	mg/L	23	EXP	MERCK INDEX (1996)
logP 3.16			EXP	SANGSTER,J (1994)
VP 2.97E-010	mm Hg	25	EST	NEELY,WB & BLAU,GE (1985)
DC 4.50	pKa		EXP	SERJEANT,EP & DEMPSEY,B (1979)
HL 6.56E-009	atm m3/mol	25	EST	MEYLAN,WM & HOWARD,PH (1991)
OH 3.31E-011	cm3/molc sec	25	EST	MEYLAN,WM & HOWARD,PH (1993)

CAS #: 000050-35-1				THALIDOMIDE

Formula: $C_{13}H_{10}N_2O_4$

Mol Weight: 258.24

MP (deg C): 269.5 FP (deg C):

BP (deg C):

BP pressure (mm Hg):

Property/ Value	Units	Temp	Data Type	Reference
WS 5.45E+002	mg/L	25	EST	MEYLAN,WM ET AL. (1996)
logP 0.33			EXP	HANSCH,C & LEO,AJ (1985)
VP 1.80E-014	mm Hg	25	EST	NEELY,WB & BLAU,GE (1985)
DC	pKa			
HL 1.29E-016	atm m3/mol	25	EST	MEYLAN,WM & HOWARD,PH (1991)
OH 3.92E-011	cm3/molc sec	25	EST	MEYLAN,WM & HOWARD,PH (1993)

CAS #: 000050-36-2				COCAINE

Formula: $C_{17}H_{21}NO_4$

Mol Weight: 303.36

MP (deg C): 98 FP (deg C):

BP (deg C):

BP pressure (mm Hg):

Property/ Value	Units	Temp	Data Type	Reference
WS 1.80E+003	mg/L	22	EXP	YALKOWSKY,SH & DANNENFELSER,RM (1992)
logP 2.30			EXP	HANSCH,C ET AL. (1995)
VP 1.91E-007	mm Hg	25	EXP	LAWERENCE,AH ET AL. (1984)
DC	pKa			
HL 4.24E-011	atm m3/mol	25	EST	VP/WSOL
OH 6.18E-011	cm3/molc sec	25	EST	MEYLAN,WM & HOWARD,PH (1993)

CAS #: 000050-37-3				LSD

Formula: $C_{20}H_{25}N_3O$

Mol Weight: 323.44

MP (deg C): 82.5 FP (deg C):

BP (deg C):

BP pressure (mm Hg):

Property/ Value	Units	Temp	Data Type	Reference
WS 2.10E+000	mg/L	25	EST	MEYLAN,WM ET AL. (1996)
logP 2.95			EXP	HANSCH,C & LEO,AJ (1985)
VP 9.03E-010	mm Hg	25	EST	NEELY,WB & BLAU,GE (1985)
DC	pKa			
HL 1.53E-016	atm m3/mol	25	EST	MEYLAN,WM & HOWARD,PH (1991)
OH 4.16E-010	cm3/molc sec	25	EST	MEYLAN,WM & HOWARD,PH (1993)

CAS #: 000050-44-2 — 6-PURINETHIOL HYDRATE

Formula: $C_5H_4N_4S$

Mol Weight: 152.18

MP (deg C): 313 dec

FP (deg C):

BP (deg C):

BP pressure (mm Hg):

Property/Value	Units	Temp	Data Type	Reference
WS 6.85E+004	mg/L		EXP	YALKOWSKY,SH & DANNENFELSER,RM (1992)
logP 0.01			EXP	HANSCH,C & LEO,AJ (1985)
VP 1.43E-006	mm Hg	25	EST	NEELY,WB & BLAU,GE (1985)
DC	pKa			
HL 7.49E-010	atm m3/mol	25	EST	MEYLAN,WM & HOWARD,PH (1991)
OH 2.00E-010	cm3/molc sec	25	EST	MEYLAN,WM & HOWARD,PH (1993)

CAS #: 000050-47-5 — DESIPRAMINE

Formula: $C_{18}H_{22}N_2$

Mol Weight: 266.39

MP (deg C):

FP (deg C):

BP (deg C): 172-174

BP pressure (mm Hg): 2.00E-002

Property/Value	Units	Temp	Data Type	Reference
WS 5.86E+001	mg/L	24	EXP	YALKOWSKY,SH & DANNENFELSER,RM (1992)
logP 4.90			EXP	HANSCH,C & LEO,AJ (1985)
VP 1.56E-006	mm Hg	25	EST	NEELY,WB & BLAU,GE (1985)
DC	pKa			
HL 4.60E-009	atm m3/mol	25	EST	MEYLAN,WM & HOWARD,PH (1991)
OH 2.89E-010	cm3/molc sec	25	EST	MEYLAN,WM & HOWARD,PH (1993)

CAS #: 000050-48-6 — AMITRYPTYLINE

Formula: $C_{20}H_{23}N$

Mol Weight: 277.41

MP (deg C): 196.5

FP (deg C):

BP (deg C):

BP pressure (mm Hg):

Property/Value	Units	Temp	Data Type	Reference
WS 9.71E+000	mg/L	24	EXP	YALKOWSKY,SH & DANNENFELSER,RM (1992)
logP 5.04			EXP	HANSCH,C & LEO,AJ (1985)
VP 1.84E-006	mm Hg	25	EST	NEELY,WB & BLAU,GE (1985)
DC	pKa			
HL 6.85E-008	atm m3/mol	25	EST	MEYLAN,WM & HOWARD,PH (1991)
OH 1.79E-010	cm3/molc sec	25	EST	MEYLAN,WM & HOWARD,PH (1993)

CAS #: 000050-49-7 — IMIPRAMINE

Formula: $C_{19}H_{24}N_2$

Mol Weight: 280.42

MP (deg C): 174.5

FP (deg C):

BP (deg C): 160

BP pressure (mm Hg): 1.00E-001

Property/Value	Units	Temp	Data Type	Reference
WS 1.82E+001	mg/L	24	EXP	YALKOWSKY,SH & DANNENFELSER,RM (1992)
logP 4.80			EXP	HANSCH,C & LEO,AJ (1985)
VP 1.53E-006	mm Hg	25	EST	NEELY,WB & BLAU,GE (1985)
DC	pKa			
HL 1.01E-008	atm m3/mol	25	EST	MEYLAN,WM & HOWARD,PH (1991)
OH 2.94E-010	cm3/molc sec	25	EST	MEYLAN,WM & HOWARD,PH (1993)

CAS #: 000050-52-2 — THIORIDAZINE

Formula: $C_{21}H_{26}N_2S_2$

Mol Weight: 370.58

MP (deg C): 73

FP (deg C):

BP (deg C): 230

BP pressure (mm Hg): 2.00E-002

Property/Value	Units	Temp	Data Type	Reference
WS 3.36E-002	mg/L	25	EST	MEYLAN,WM ET AL. (1996)
logP 5.90			EXP	HANSCH,C & LEO,AJ (1985)
VP 1.42E-009	mm Hg	25	EST	NEELY,WB & BLAU,GE (1985)
DC	pKa			
HL 1.50E-011	atm m3/mol	25	EST	MEYLAN,WM & HOWARD,PH (1991)
OH 3.23E-010	cm3/molc sec	25	EST	MEYLAN,WM & HOWARD,PH (1993)

CAS #: 000050-53-3 — CHLORPROMAZINE

Formula: $C_{17}H_{19}ClN_2S$

Mol Weight: 318.87

MP (deg C):

FP (deg C):

BP (deg C): 200-205

BP pressure (mm Hg): 8.00E-001

Property/Value	Units	Temp	Data Type	Reference
WS 2.55E+000	mg/L	24	EXP	YALKOWSKY,SH & DANNENFELSER,RM (1992)
logP 5.19			EXP	HANSCH,C ET AL. (1995)
VP 1.77E-007	mm Hg	25	EST	NEELY,WB & BLAU,GE (1985)
DC 9.30	pKa	25	EXP	HANSCH,C & LEO,AJ (1985)
HL 3.95E-011	atm m3/mol	25	EST	MEYLAN,WM & HOWARD,PH (1991)
OH 1.67E-010	cm3/molc sec	25	EST	MEYLAN,WM & HOWARD,PH (1993)

CAS #: 000050-55-5 — RESERPINE

Formula: $C_{33}H_{40}N_2O_9$

Mol Weight: 608.69

MP (deg C): 264.5 dec

FP (deg C):

BP (deg C):

BP pressure (mm Hg):

Property/Value	Units	Temp	Data Type	Reference
WS 7.30E+001	mg/L	30	EXP	YALKOWSKY,SH & DANNENFELSER,RM (1992)
logP 3.32			EST	MEYLAN,WM & HOWARD,PH (1995)
VP 4.51E-016	mm Hg	25	EST	NEELY,WB & BLAU,GE (1985)
DC	pKa			
HL 5.36E-023	atm m3/mol	25	EST	MEYLAN,WM & HOWARD,PH (1991)
OH 3.76E-010	cm3/molc sec	25	EST	MEYLAN,WM & HOWARD,PH (1993)

CAS #: 000050-65-7 — NICLOSAMIDE

Formula: $C_{13}H_8Cl_2N_2O_4$

Mol Weight: 327.13

MP (deg C): 230

FP (deg C):

BP (deg C):

BP pressure (mm Hg):

Property/Value	Units	Temp	Data Type	Reference
WS 6.50E+000	mg/L		EXP	TOMLIN,C (1994)
logP 4.56			EST	MEYLAN,WM & HOWARD,PH (1995)
VP 1.49E-011	mm Hg	20	EXP	TOMLIN,C (1994)
DC	pKa			
HL 9.90E-013	atm m3/mol	20	EST	VP/WSOL
OH 9.46E-012	cm3/molc sec		EST	MEYLAN,WM & HOWARD,PH (1993)

CAS #: 000050-67-9				5-HYDROXYTRYPTAMINE

Formula: $C_{10}H_{12}N_2O$

Mol Weight: 176.22

MP (deg C): 167.5   FP (deg C):

BP (deg C):

BP pressure (mm Hg):

Property/Value	Units	Temp	Data Type	Reference
WS 2.00E+004	mg/L	27	EXP	HORN,AS (1981)
logP 0.21			EXP	HANSCH,C & LEO,AJ (1985)
VP 1.91E-006	mm Hg	25	EST	NEELY,WB & BLAU,GE (1985)
DC 9.97	pKa		EXP	HANSCH,C & LEO,AJ (1985)
HL 1.39E-014	atm m3/mol	25	EST	MEYLAN,WM & HOWARD,PH (1991)
OH 2.33E-010	cm3/molc sec	25	EST	MEYLAN,WM & HOWARD,PH (1993)

CAS #: 000050-69-1				RIBOSE

Formula: $C_5H_{10}O_5$

Mol Weight: 150.13

MP (deg C): 87   FP (deg C):

BP (deg C):

BP pressure (mm Hg):

Property/Value	Units	Temp	Data Type	Reference
WS 1.00E+006	mg/L	25	EST	MEYLAN,WM ET AL. (1996)
logP -2.32			EXP	HANSCH,C & LEO,AJ (1985)
VP 8.55E-007	mm Hg	25	EST	NEELY,WB & BLAU,GE (1985)
DC	pKa			
HL 2.00E-013	atm m3/mol	25	EST	MEYLAN,WM & HOWARD,PH (1991)
OH 7.61E-011	cm3/molc sec	25	EST	MEYLAN,WM & HOWARD,PH (1993)

CAS #: 000050-70-4				SORBITOL

Formula: $C_6H_{14}O_6$

Mol Weight: 182.17

MP (deg C): 111   FP (deg C):

BP (deg C):

BP pressure (mm Hg):

Property/Value	Units	Temp	Data Type	Reference
WS 2.75E+006	mg/L	30	EXP	MULLIN,JW (1972)
logP -2.20			EXP	SANGSTER,J (1994)
VP 4.92E-009	mm Hg	25	EST	NEELY,WB & BLAU,GE (1985)
DC 13.57	pKa		EXP	BENSON,FR (1978)
HL 7.26E-013	atm m3/mol	25	EST	MEYLAN,WM & HOWARD,PH (1991)
OH 5.00E-011	cm3/molc sec	25	EST	MEYLAN,WM & HOWARD,PH (1993)

CAS #: 000050-71-5				ALLOXANE

Formula: $C_4H_2N_2O_4$

Mol Weight: 142.07

MP (deg C): 256 dec   FP (deg C):

BP (deg C):

BP pressure (mm Hg):

Property/Value	Units	Temp	Data Type	Reference
WS 8.00E+001	mg/L		EXP	YALKOWSKY,SH & DANNENFELSER,RM (1992)
logP -3.30			EST	MEYLAN,WM & HOWARD,PH (1995)
VP 2.14E-010	mm Hg	25	EST	NEELY,WB & BLAU,GE (1985)
DC	pKa			
HL 2.74E-015	atm m3/mol	25	EST	MEYLAN,WM & HOWARD,PH (1991)
OH 2.00E-012	cm3/molc sec	25	EST	MEYLAN,WM & HOWARD,PH (1993)

CAS #: 000050-78-2				ACETYLSALICYLIC ACID

Formula: $C_9H_8O_4$

Mol Weight: 180.16

MP (deg C): 136   FP (deg C):

BP (deg C):

BP pressure (mm Hg):

Property/Value	Units	Temp	Data Type	Reference
WS 4.60E+003	mg/L	25	EXP	YALKOWSKY,SH & DANNENFELSER,RM (1992)
logP 1.19			EXP	HANSCH,C & LEO,AJ (1985)
VP 1.84E-004	mm Hg	25	EST	NEELY,WB & BLAU,GE (1985)
DC 3.49	pKa	25	EXP	MERCK INDEX (1983)
HL 1.30E-009	atm m3/mol	25	EST	MEYLAN,WM & HOWARD,PH (1991)
OH 8.10E-013	cm3/molc sec	25	EST	ATKINSON,R (1987)

CAS #: 000050-79-3				2,5-DICHLOROBENZOIC ACID

Formula: $C_7H_4Cl_2O_2$

Mol Weight: 191.01

MP (deg C): 154.4   FP (deg C):

BP (deg C): 301

BP pressure (mm Hg):

Property/Value	Units	Temp	Data Type	Reference
WS 1.90E+002	mg/L	25	EST	MEYLAN,WM ET AL. (1996)
logP 2.82			EXP	DA,YZ ET AL. (1992)
VP 4.47E-004	mm Hg	25	EST	NEELY,WB & BLAU,GE (1985)
DC 2.47	pKa	25	EXP	SERJEANT,EP & DEMPSEY,B (1979)
HL 5.95E-008	atm m3/mol	25	EST	MEYLAN,WM & HOWARD,PH (1991)
OH 6.68E-013	cm3/molc sec	25	EST	MEYLAN,WM & HOWARD,PH (1993)

CAS #: 000050-81-7				ASCORBIC ACID

Formula: $C_6H_8O_6$

Mol Weight: 176.13

MP (deg C): 191   FP (deg C):

BP (deg C):

BP pressure (mm Hg):

Property/Value	Units	Temp	Data Type	Reference
WS 1.00E+006	mg/L	25	EST	MEYLAN,WM ET AL. (1996)
logP -1.64			EXP	HANSCH,C ET AL. (1995)
VP 2.49E-010	mm Hg	25	EST	NEELY,WB & BLAU,GE (1985)
DC 4.70	pKa	10	EXP	KORTUM,G ET AL (1961)
HL 4.07E-008	atm m3/mol	25	EST	MEYLAN,WM & HOWARD,PH (1991)
OH 5.64E-011	cm3/molc sec	25	EST	MEYLAN,WM & HOWARD,PH (1993)

CAS #: 000050-82-8				2,4,5-TRICHLOROBENZOIC ACID

Formula: $C_7H_3Cl_3O_2$

Mol Weight: 225.46

MP (deg C): 166.5   FP (deg C):

BP (deg C):

BP pressure (mm Hg):

Property/Value	Units	Temp	Data Type	Reference
WS 8.10E+002	mg/L	25	EXP	SHIU,WY ET AL. (1990)
logP 3.47			EST	MEYLAN,WM & HOWARD,PH (1995)
VP 8.27E-005	mm Hg	25	EST	NEELY,WB & BLAU,GE (1985)
DC	pKa			
HL 4.41E-008	atm m3/mol	25	EST	MEYLAN,WM & HOWARD,PH (1991)
OH 6.24E-013	cm3/molc sec	25	EST	MEYLAN,WM & HOWARD,PH (1993)

000050-84-0 — 2,4-DICHLOROBENZOIC ACID

CAS #: 000050-84-0

Formula:	$C_7H_4Cl_2O_2$
Mol Weight:	191.01
MP (deg C): 164.2	FP (deg C):
BP (deg C):	
BP pressure (mm Hg):	

Property/Value	Units	Temp	Data Type	Reference
WS 4.80E+002	mg/L		EXP	YALKOWSKY,SH & DANNENFELSER,RM (1992)
logP 2.82			EST	MEYLAN,WM & HOWARD,PH (1995)
VP 4.47E-004	mm Hg	25	EST	NEELY,WB & BLAU,GE (1985)
DC 2.68	pKa	25	EXP	SERJEANT,EP & DEMPSEY,B (1979)
HL 5.95E-008	atm m3/mol	25	EST	MEYLAN,WM & HOWARD,PH (1991)
OH 8.77E-013	cm3/molc sec	25	EST	MEYLAN,WM & HOWARD,PH (1993)

000050-85-1 — 4-METHYLSALICYLIC ACID

CAS #: 000050-85-1

Formula:	$C_8H_8O_3$
Mol Weight:	152.15
MP (deg C): 177	FP (deg C):
BP (deg C):	
BP pressure (mm Hg):	

Property/Value	Units	Temp	Data Type	Reference
WS 1.01E+004	mg/L	100	EXP	YALKOWSKY,SH & DANNENFELSER,RM (1992)
logP 2.99			EXP	HANSCH,C & LEO,AJ (1985)
VP 5.14E-005	mm Hg	25	EST	NEELY,WB & BLAU,GE (1985)
DC 3.40	pKa	25	EXP	SERJEANT,EP & DEMPSEY,B (1979)
HL 1.57E-008	atm m3/mol	25	EST	MEYLAN,WM & HOWARD,PH (1991)
OH 3.30E-011	cm3/molc sec	25	EST	MEYLAN,WM & HOWARD,PH (1993)

000050-86-2 — BENZOIC ACID, 4-(ACETYLAMINO)-2-HYDROXY

CAS #: 000050-86-2

Formula:	$C_9H_9NO_4$
Mol Weight:	195.18
MP (deg C):	FP (deg C):
BP (deg C):	
BP pressure (mm Hg):	

Property/Value	Units	Temp	Data Type	Reference
WS 1.92E+003	mg/L	25	EST	MEYLAN,WM ET AL. (1996)
logP 1.62			EXP	LAZNICEK,M ET AL. (1987)
VP 1.13E-008	mm Hg	25	EST	NEELY,WB & BLAU,GE (1985)
DC	pKa			
HL 1.63E-014	atm m3/mol	25	EST	MEYLAN,WM & HOWARD,PH (1991)
OH 7.93E-011	cm3/molc sec	25	EST	MEYLAN,WM & HOWARD,PH (1993)

000050-89-5 — THYMIDINE

CAS #: 000050-89-5

Formula:	$C_{10}H_{14}N_2O_5$
Mol Weight:	242.23
MP (deg C): 185	FP (deg C):
BP (deg C):	
BP pressure (mm Hg):	

Property/Value	Units	Temp	Data Type	Reference
WS 7.96E+003	mg/L	25	EST	MEYLAN,WM ET AL. (1996)
logP -0.93			EXP	SANGSTER,J (1993)
VP 4.56E-014	mm Hg	25	EST	NEELY,WB & BLAU,GE (1985)
DC	pKa			
HL 1.21E-017	atm m3/mol	25	EST	MEYLAN,WM & HOWARD,PH (1991)
OH 9.61E-011	cm3/molc sec	25	EST	MEYLAN,WM & HOWARD,PH (1993)

000050-91-9 — FLOXURIDINE

CAS #: 000050-91-9

Formula:	$C_9H_{11}FN_2O_5$
Mol Weight:	246.20
MP (deg C): 150.5	FP (deg C):
BP (deg C):	
BP pressure (mm Hg):	

Property/Value	Units	Temp	Data Type	Reference
WS 1.19E+004	mg/L	25	EST	MEYLAN,WM ET AL. (1996)
logP -1.16			EXP	HANSCH,C & LEO,AJ (1985)
VP 1.13E-013	mm Hg	25	EST	NEELY,WB & BLAU,GE (1985)
DC	pKa			
HL 1.48E-017	atm m3/mol	25	EST	MEYLAN,WM & HOWARD,PH (1991)
OH 8.15E-011	cm3/molc sec	25	EST	MEYLAN,WM & HOWARD,PH (1993)

000050-99-7 — GLUCOSE

CAS #: 000050-99-7

Formula:	$C_6H_{12}O_6$
Mol Weight:	180.16
MP (deg C): 83	FP (deg C):
BP (deg C):	
BP pressure (mm Hg):	

Property/Value	Units	Temp	Data Type	Reference
WS 1.20E+006	mg/L	30	EXP	MULLIN,JW (1972)
logP -1.88			EXP	HANSCH,C ET AL. (1995)
VP 1.82E-008	mm Hg	25	EST	NEELY,WB & BLAU,GE (1985)
DC 12.92	pKa	0	EXP	KORTUM,G ET AL (1961)
HL 5.88E-011	atm m3/mol	25	EST	MEYLAN,WM & HOWARD,PH (1991)
OH 8.49E-011	cm3/molc sec	25	EST	MEYLAN,WM & HOWARD,PH (1993)

000051-03-6 — PIPERONYL BUTOXIDE

CAS #: 000051-03-6

Formula:	$C_{19}H_{30}O_5$
Mol Weight:	338.45
MP (deg C):	FP (deg C):
BP (deg C): 180	
BP pressure (mm Hg): 1.00E+000	

Property/Value	Units	Temp	Data Type	Reference
WS 6.38E-001	mg/L	25	EST	MEYLAN,WM ET AL. (1996)
logP 4.75			EXP	TOMLIN,C (1994)
VP 2.60E-007	mm Hg	25	EST	NEELY,WB & BLAU,GE (1985)
DC	pKa			
HL 8.89E-011	atm m3/mol	25	EST	MEYLAN,WM & HOWARD,PH (1991)
OH 1.08E-010	cm3/molc sec	25	EST	MEYLAN,WM & HOWARD,PH (1993)

000051-05-8 — PROCAINE HYDROCHLORIDE

CAS #: 000051-05-8

Formula:	$C_{13}H_{21}ClN_2O_2$
Mol Weight:	272.78
MP (deg C): 155.5	FP (deg C):
BP (deg C):	
BP pressure (mm Hg):	

Property/Value	Units	Temp	Data Type	Reference
WS 9.40E+005	mg/L	25	EST	MEYLAN,WM ET AL. (1996)
logP -0.84			EST	MEYLAN,WM & HOWARD,PH (1995)
VP 7.78E-010	mm Hg	25	EST	NEELY,WB & BLAU,GE (1985)
DC	pKa			
HL 1.32E-018	atm m3/mol	25	EST	MEYLAN,WM & HOWARD,PH (1991)
OH 9.64E-011	cm3/molc sec	25	EST	MEYLAN,WM & HOWARD,PH (1993)

PROCAINAMIDE

CAS #: 000051-06-9				
Formula: $C_{13}H_{21}N_3O$				
Mol Weight: 235.33				
MP (deg C):		FP (deg C):		
BP (deg C):				
BP pressure (mm Hg):				

Property/Value	Units	Temp	Data Type	Reference
WS 5.05E+003	mg/L	25	EST	MEYLAN,WM ET AL. (1996)
logP 0.88			EXP	HANSCH,C & LEO,AJ (1985)
VP 9.57E-008	mm Hg	25	EST	NEELY,WB & BLAU,GE (1985)
DC	pKa			
HL 1.99E-015	atm m3/mol	25	EST	MEYLAN,WM & HOWARD,PH (1991)
OH 2.03E-010	cm3/molc sec	25	EST	MEYLAN,WM & HOWARD,PH (1993)

NIALAMIDE

CAS #: 000051-12-7				
Formula: $C_{16}H_{18}N_4O_2$				
Mol Weight: 298.35				
MP (deg C): 152.5		FP (deg C):		
BP (deg C):				
BP pressure (mm Hg):				

Property/Value	Units	Temp	Data Type	Reference
WS 2.27E+003	mg/L	25	EST	MEYLAN,WM ET AL. (1996)
logP 0.87			EXP	HANSCH,C & LEO,AJ (1985)
VP 8.75E-012	mm Hg	25	EST	NEELY,WB & BLAU,GE (1985)
DC	pKa			
HL 1.51E-018	atm m3/mol	25	EST	MEYLAN,WM & HOWARD,PH (1991)
OH 1.18E-010	cm3/molc sec	25	EST	MEYLAN,WM & HOWARD,PH (1993)

SESAMEX

CAS #: 000051-14-9				
Formula: $C_{15}H_{22}O_6$				
Mol Weight: 298.34				
MP (deg C): 122-123		FP (deg C):		
BP (deg C): 137-141				
BP pressure (mm Hg): 1.00E+000				

Property/Value	Units	Temp	Data Type	Reference
WS 1.55E+002	mg/L	25	EST	MEYLAN,WM ET AL. (1996)
logP 2.23			EST	MEYLAN,WM & HOWARD,PH (1995)
VP 5.65E-006	mm Hg	25	EST	NEELY,WB & BLAU,GE (1985)
DC	pKa			
HL 1.85E-012	atm m3/mol	25	EST	MEYLAN,WM & HOWARD,PH (1991)
OH 2.44E-010	cm3/molc sec	25	EST	MEYLAN,WM & HOWARD,PH (1993)

BENZIMIDAZOLE

CAS #: 000051-17-2				
Formula: $C_7H_6N_2$				
Mol Weight: 118.14				
MP (deg C): 170.5		FP (deg C):		
BP (deg C): 360 +				
BP pressure (mm Hg):				

Property/Value	Units	Temp	Data Type	Reference
WS 2.01E+003	mg/L	20	EXP	PEARLMAN,RS ET AL. (1984)
logP 1.32			EXP	HANSCH,C ET AL. (1995)
VP 7.64E-005	mm Hg	25	EST	NEELY,WB & BLAU,GE (1985)
DC 5.30	pKa	25	EXP	KORTUM,G ET AL (1961)
HL 3.67E-007	atm m3/mol	25	EST	MEYLAN,WM & HOWARD,PH (1991)
OH 3.60E-011	cm3/molc sec	25	EST	MEYLAN,WM & HOWARD,PH (1993)

2,4,6-TRIS(1-AZIRIDINYL)-1,3,5-TRIAZINE

CAS #: 000051-18-3				
Formula: $C_9H_{12}N_6$				
Mol Weight: 204.24				
MP (deg C):		FP (deg C):		
BP (deg C):				
BP pressure (mm Hg):				

Property/Value	Units	Temp	Data Type	Reference
WS 2.69E+002	mg/L	25	EST	MEYLAN,WM ET AL. (1996)
logP 2.56			EST	MEYLAN,WM & HOWARD,PH (1995)
VP 3.61E-005	mm Hg	25	EST	NEELY,WB & BLAU,GE (1985)
DC	pKa			
HL 1.48E-009	atm m3/mol	25	EST	MEYLAN,WM & HOWARD,PH (1991)
OH 1.94E-012	cm3/molc sec	25	EST	MEYLAN,WM & HOWARD,PH (1993)

5-BROMOURACIL

CAS #: 000051-20-7				
Formula: $C_4H_3BrN_2O_2$				
Mol Weight: 190.99				
MP (deg C): 293		FP (deg C):		
BP (deg C):				
BP pressure (mm Hg):				

Property/Value	Units	Temp	Data Type	Reference
WS 8.15E+002	mg/L		EXP	GUNTHER,FA ET AL. (1968)
logP -0.21			EXP	HANSCH,C & LEO,AJ (1985)
VP 2.27E-007	mm Hg	25	EST	NEELY,WB & BLAU,GE (1985)
DC	pKa			
HL 1.73E-011	atm m3/mol	25	EST	MEYLAN,WM & HOWARD,PH (1991)
OH 6.74E-012	cm3/molc sec	25	EST	MEYLAN,WM & HOWARD,PH (1993)

5-FLUOROURACIL

CAS #: 000051-21-8				
Formula: $C_4H_3FN_2O_2$				
Mol Weight: 130.08				
MP (deg C): 283		FP (deg C):		
BP (deg C):				
BP pressure (mm Hg):				

Property/Value	Units	Temp	Data Type	Reference
WS 2.59E+004	mg/L	25	EST	MEYLAN,WM ET AL. (1996)
logP -0.89			EXP	HANSCH,C & LEO,AJ (1985)
VP 2.68E-006	mm Hg	25	EST	NEELY,WB & BLAU,GE (1985)
DC	pKa			
HL 1.66E-010	atm m3/mol	25	EST	MEYLAN,WM & HOWARD,PH (1991)
OH 5.83E-012	cm3/molc sec	25	EST	MEYLAN,WM & HOWARD,PH (1993)

2,4-DINITROPHENOL

CAS #: 000051-28-5				
Formula: $C_6H_4N_2O_5$				
Mol Weight: 184.11				
MP (deg C): 112-114		FP (deg C):		
BP (deg C):				
BP pressure (mm Hg):				

Property/Value	Units	Temp	Data Type	Reference
WS 2.79E+003	mg/L	20	EXP	SCHWARZENBACH,RP ET AL. (1988)
logP 1.67			EXP	HANSCH,C ET AL. (1995)
VP 5.10E-003	mm Hg	20	EXP	SCHWARZENBACH,RP ET AL. (1988)
DC 4.09	pKa	25	EXP	PEARCE,PJ & SIMKINS,RJJ (1968)
HL 4.43E-007	atm m3/mol	20	EST	VP/WSOL
OH 5.76E-013	cm3/molc sec	25	EST	MEYLAN,WM & HOWARD,PH (1993)

CAS #: 000051-30-9 — 1,2-BENZENEDIOL, 4-[1-HYDROXY-2-[(1-METHYLETHYL)

Formula: $C_{11}H_{18}ClNO_3$

Mol Weight: 247.72

MP (deg C): | FP (deg C):

BP (deg C):

BP pressure (mm Hg):

Property/Value	Units	Temp	Data Type	Reference
WS 1.00E+006	mg/L	25	EST	MEYLAN,WM ET AL. (1996)
logP -2.69			EXP	SANGSTER,J (1994)
VP 3.84E-013	mm Hg	25	EST	NEELY,WB & BLAU,GE (1985)
DC	pKa			
HL 9.51E-026	atm m3/mol	25	EST	MEYLAN,WM & HOWARD,PH (1991)
OH 9.23E-011	cm3/molc sec	25	EST	MEYLAN,WM & HOWARD,PH (1993)

CAS #: 000051-34-3 — SCOPOLAMINE

Formula: $C_{17}H_{21}NO_4$

Mol Weight: 303.36

MP (deg C): 59 | FP (deg C):

BP (deg C):

BP pressure (mm Hg):

Property/Value	Units	Temp	Data Type	Reference
WS 1.00E+005	mg/L		EXP	YALKOWSKY,SH & DANNENFELSER,RM (1992)
logP 0.98			EXP	SANGSTER,J (1994)
VP 7.18E-009	mm Hg	25	EST	NEELY,WB & BLAU,GE (1985)
DC 7.75	pKa	25	EXP	SANGSTER,J (1994)
HL 3.36E-016	atm m3/mol	25	EST	MEYLAN,WM & HOWARD,PH (1991)
OH 6.48E-011	cm3/molc sec	25	EST	MEYLAN,WM & HOWARD,PH (1993)

CAS #: 000051-35-4 — HYDROXYPROLINE

Formula: $C_5H_9NO_3$

Mol Weight: 131.13

MP (deg C): 274 | FP (deg C):

BP (deg C):

BP pressure (mm Hg):

Property/Value	Units	Temp	Data Type	Reference
WS 3.95E+005	mg/L	25	EST	MEYLAN,WM ET AL. (1996)
logP -3.17			EXP	HANSCH,C & LEO,AJ (1985)
VP 1.52E-009	mm Hg	25	EST	NEELY,WB & BLAU,GE (1985)
DC 1.92	pKa	25	EXP	KORTUM,G ET AL (1961)
HL 7.01E-014	atm m3/mol	25	EST	MEYLAN,WM & HOWARD,PH (1991)
OH 9.49E-011	cm3/molc sec	25	EST	MEYLAN,WM & HOWARD,PH (1993)

CAS #: 000051-36-5 — 3,5-DICHLOROBENZOIC ACID

Formula: $C_7H_4Cl_2O_2$

Mol Weight: 191.01

MP (deg C): 188 | FP (deg C):

BP (deg C):

BP pressure (mm Hg):

Property/Value	Units	Temp	Data Type	Reference
WS 1.47E+002	mg/L		EXP	YALKOWSKY,SH & DANNENFELSER,RM (1992)
logP 3.00			EXP	HANSCH,C & LEO,AJ (1985)
VP 4.47E-004	mm Hg	25	EST	NEELY,WB & BLAU,GE (1985)
DC 3.54	pKa	22	EXP	SERJEANT,EP & DEMPSEY,B (1979)
HL 5.95E-008	atm m3/mol	25	EST	MEYLAN,WM & HOWARD,PH (1991)
OH 7.83E-013	cm3/molc sec	25	EST	MEYLAN,WM & HOWARD,PH (1993)

CAS #: 000051-41-2 — (-)-NORADRENALINE

Formula: $C_8H_{11}NO_3$

Mol Weight: 169.18

MP (deg C): 214 dec | FP (deg C):

BP (deg C):

BP pressure (mm Hg):

Property/Value	Units	Temp	Data Type	Reference
WS 1.00E+006	mg/L	25	EST	MEYLAN,WM ET AL. (1996)
logP -1.24			EXP	HANSCH,C ET AL. (1995)
VP 8.19E-007	mm Hg	25	EST	NEELY,WB & BLAU,GE (1985)
DC 8.58	pKa		EXP	PERRIN,DD (1965)
HL 3.21E-019	atm m3/mol	25	EST	MEYLAN,WM & HOWARD,PH (1991)
OH 9.49E-011	cm3/molc sec	25	EST	MEYLAN,WM & HOWARD,PH (1993)

CAS #: 000051-43-4 — EPINEPHRINE

Formula: $C_9H_{13}NO_3$

Mol Weight: 183.21

MP (deg C): 157 | FP (deg C):

BP (deg C):

BP pressure (mm Hg):

Property/Value	Units	Temp	Data Type	Reference
WS 1.80E+002	mg/L	20	EXP	YALKOWSKY,SH & DANNENFELSER,RM (1992)
logP -1.37			EXP	HANSCH,C & LEO,AJ (1985)
VP 7.37E-007	mm Hg	25	EST	NEELY,WB & BLAU,GE (1985)
DC 8.59	pKa	25	EXP	HANSCH,C & LEO,AJ (1985)
HL 7.06E-019	atm m3/mol	25	EST	MEYLAN,WM & HOWARD,PH (1991)
OH 1.38E-010	cm3/molc sec	25	EST	MEYLAN,WM & HOWARD,PH (1993)

CAS #: 000051-44-5 — 3,4-DICHLOROBENZOIC ACID

Formula: $C_7H_4Cl_2O_2$

Mol Weight: 191.01

MP (deg C): 350 dec | FP (deg C):

BP (deg C):

BP pressure (mm Hg):

Property/Value	Units	Temp	Data Type	Reference
WS 6.11E+001	mg/L		EXP	YALKOWSKY,SH & DANNENFELSER,RM (1992)
logP 3.25			EXP	DA,YZ ET AL. (1992)
VP 4.47E-004	mm Hg	25	EST	NEELY,WB & BLAU,GE (1985)
DC 3.64	pKa	25	EXP	SERJEANT,EP & DEMPSEY,B (1979)
HL 5.95E-008	atm m3/mol	25	EST	MEYLAN,WM & HOWARD,PH (1991)
OH 6.68E-013	cm3/molc sec	25	EST	MEYLAN,WM & HOWARD,PH (1993)

CAS #: 000051-45-6 — 1H-IMIDAZOLE-4-ETHANAMINE

Formula: $C_5H_9N_3$

Mol Weight: 111.15

MP (deg C): 83-84 | FP (deg C):

BP (deg C): 209-210

BP pressure (mm Hg): 1.80E+001

Property/Value	Units	Temp	Data Type	Reference
WS 1.00E+006	mg/L	25	EST	MEYLAN,WM ET AL. (1996)
logP -0.70			EXP	SANGSTER,J (1993)
VP 2.22E-004	mm Hg	25	EST	NEELY,WB & BLAU,GE (1985)
DC 9.80	pKa	25	EXP	PERRIN,DD (1965)
HL 5.66E-010	atm m3/mol	25	EST	MEYLAN,WM & HOWARD,PH (1991)
OH 1.22E-010	cm3/molc sec	25	EST	MEYLAN,WM & HOWARD,PH (1993)

CAS #: 000051-48-9 — THYROXINE

Formula: $C_{15}H_{11}I_4NO_4$
Mol Weight: 776.88
MP (deg C): 235.5
FP (deg C):
BP (deg C):
BP pressure (mm Hg):

Property/Value	Units	Temp	Data Type	Reference
WS 1.05E-004	mg/L	25	EST	MEYLAN,WM ET AL. (1996)
logP 4.12			EST	MEYLAN,WM & HOWARD,PH (1995)
VP 1.24E-017	mm Hg	25	EST	NEELY,WB & BLAU,GE (1985)
DC	pKa			
HL 7.91E-019	atm m3/mol	25	EST	MEYLAN,WM & HOWARD,PH (1991)
OH 4.30E-011	cm3/molc sec	25	EST	MEYLAN,WM & HOWARD,PH (1993)

CAS #: 000051-55-8 — ATROPINE

Formula: $C_{17}H_{23}NO_3$
Mol Weight: 289.38
MP (deg C): 191-194
FP (deg C):
BP (deg C):
BP pressure (mm Hg):

Property/Value	Units	Temp	Data Type	Reference
WS 2.20E+003	mg/L		EXP	DEHN,WM (1917)
logP 1.83			EXP	HANSCH,C & LEO,AJ (1985)
VP 1.95E-008	mm Hg	25	EST	NEELY,WB & BLAU,GE (1985)
DC	pKa			
HL 2.73E-013	atm m3/mol	25	EST	MEYLAN,WM & HOWARD,PH (1991)
OH 6.98E-011	cm3/molc sec	25	EST	MEYLAN,WM & HOWARD,PH (1993)

CAS #: 000051-56-9 — HOMATROPINE HYDROBROMIDE

Formula: $C_{16}H_{22}BrNO_3$
Mol Weight: 356.27
MP (deg C): 250 dec
FP (deg C):
BP (deg C):
BP pressure (mm Hg):

Property/Value	Units	Temp	Data Type	Reference
WS 1.49E+005	mg/L	25	EXP	STEPHEN,H & STEPHEN,T (1963)
logP -0.65			EST	MEYLAN,WM & HOWARD,PH (1995)
VP 1.09E-014	mm Hg	25	EST	NEELY,WB & BLAU,GE (1985)
DC	pKa			
HL 4.40E-017	atm m3/mol	25	EST	MEYLAN,WM & HOWARD,PH (1991)
OH 6.01E-011	cm3/molc sec	25	EST	MEYLAN,WM & HOWARD,PH (1993)

CAS #: 000051-61-6 — DOPAMINE

Formula: $C_8H_{11}NO_2$
Mol Weight: 153.18
MP (deg C): 128
FP (deg C):
BP (deg C): 227
BP pressure (mm Hg): 2.30E+001

Property/Value	Units	Temp	Data Type	Reference
WS 1.00E+006	mg/L	25	EST	MEYLAN,WM ET AL. (1996)
logP -0.98			EXP	HANSCH,C ET AL. (1995)
VP 2.57E-005	mm Hg	25	EST	NEELY,WB & BLAU,GE (1985)
DC 8.93	pKa		EXP	PERRIN,DD (1965)
HL 8.79E-015	atm m3/mol	25	EST	MEYLAN,WM & HOWARD,PH (1991)
OH 9.01E-011	cm3/molc sec	25	EST	MEYLAN,WM & HOWARD,PH (1993)

CAS #: 000051-66-1 — P-METHOXYACETANILIDE

Formula: $C_9H_{11}NO_2$
Mol Weight: 165.19
MP (deg C): 138
FP (deg C):
BP (deg C): 335
BP pressure (mm Hg):

Property/Value	Units	Temp	Data Type	Reference
WS 1.50E+004	mg/L	25	EXP	SUZUKI,T (1991)
logP 1.03			EXP	HANSCH,C & LEO,AJ (1985)
VP 6.59E-005	mm Hg	25	EST	NEELY,WB & BLAU,GE (1985)
DC	pKa			
HL 3.65E-010	atm m3/mol	25	EST	MEYLAN,WM & HOWARD,PH (1991)
OH 1.22E-011	cm3/molc sec	25	EST	MEYLAN,WM & HOWARD,PH (1993)

CAS #: 000051-67-2 — 4-(2-AMINOETHYL)PHENOL

Formula: $C_8H_{11}NO$
Mol Weight: 137.18
MP (deg C): 164-165
FP (deg C):
BP (deg C): 205-207
BP pressure (mm Hg): 2.50E+001

Property/Value	Units	Temp	Data Type	Reference
WS 1.04E+004	mg/L	15	EXP	YALKOWSKY,SH & DANNENFELSER,RM (1992)
logP 0.86			EST	MEYLAN,WM & HOWARD,PH (1995)
VP 3.17E-003	mm Hg	25	EST	NEELY,WB & BLAU,GE (1985)
DC 9.77	pKa		EXP	PERRIN,DD (1965)
HL 8.45E-011	atm m3/mol	25	EST	MEYLAN,WM & HOWARD,PH (1991)
OH 7.36E-011	cm3/molc sec	25	EST	MEYLAN,WM & HOWARD,PH (1993)

CAS #: 000051-75-2 — 2-CHLORO-N-(2-CHLOROETHYL)-N-METHYLETHANAMINE

Formula: $C_5H_{11}Cl_2N$
Mol Weight: 156.06
MP (deg C): 287 dec
FP (deg C):
BP (deg C): 87
BP pressure (mm Hg):

Property/Value	Units	Temp	Data Type	Reference
WS 2.29E+005	mg/L	25	EST	MEYLAN,WM ET AL. (1996)
logP 0.91			EXP	HANSCH,C & LEO,AJ (1985) @ pH=7.4
VP 6.51E+001	mm Hg	25	EST	NEELY,WB & BLAU,GE (1985)
DC 6.76	pKa	0	EXP	PERRIN,DD (1965)
HL 8.48E-008	atm m3/mol	25	EST	MEYLAN,WM & HOWARD,PH (1991)
OH 8.39E-012	cm3/molc sec	25	EST	MEYLAN,WM & HOWARD,PH (1993)

CAS #: 000051-79-6 — O-ETHYL CARBAMATE (URETHANE)

Formula: $C_3H_7NO_2$
Mol Weight: 89.09
MP (deg C): 236-240 de
FP (deg C):
BP (deg C): 182-184
BP pressure (mm Hg):

Property/Value	Units	Temp	Data Type	Reference
WS 4.78E+005	mg/L	25	EXP	SUZUKI,T (1991)
logP -0.15			EXP	HANSCH,C & LEO,AJ (1985)
VP 2.62E-001	mm Hg	25	EXT	PERRY,RH & GREEN,D (1984)
DC	pKa			
HL 5.25E-008	atm m3/mol	25	EST	MEYLAN,WM & HOWARD,PH (1991)
OH 3.66E-012	cm3/molc sec	25	EST	MEYLAN,WM & HOWARD,PH (1993)

ETHYLTRIMETHYLAMMONIUM IODIDE

CAS #: 000051-93-4

Formula: $C_5H_{14}IN$

Mol Weight: 215.08

MP (deg C): 245 dec

FP (deg C):

BP (deg C):

BP pressure (mm Hg):

Property/Value	Units	Temp	Data Type	Reference
WS 1.00E+006	mg/L	25	EST	MEYLAN,WM ET AL. (1996)
logP -3.14			EXP	KIMURA,R ET AL. (1991)
VP 2.73E-006	mm Hg	25	EST	NEELY,WB & BLAU,GE (1985)
DC	pKa			
HL 3.50E-012	atm m3/mol	25	EST	MEYLAN,WM & HOWARD,PH (1991)
OH 1.26E-011	cm3/molc sec	25	EST	MEYLAN,WM & HOWARD,PH (1993)

SPIRONOLACTONE

CAS #: 000052-01-7

Formula: $C_{24}H_{32}O_4S$

Mol Weight: 416.58

MP (deg C): 134-135

FP (deg C):

BP (deg C):

BP pressure (mm Hg):

Property/Value	Units	Temp	Data Type	Reference
WS 2.20E+001	mg/L	25	EXP	YALKOWSKY,SH & DANNENFELSER,RM (1992)
logP 2.78			EXP	HANSCH,C & LEO,AJ (1985)
VP 3.24E-011	mm Hg	25	EST	NEELY,WB & BLAU,GE (1985)
DC	pKa			
HL 1.14E-010	atm m3/mol	25	EST	MEYLAN,WM & HOWARD,PH (1991)
OH 1.03E-010	cm3/molc sec	25	EST	MEYLAN,WM & HOWARD,PH (1993)

PREDNISOLONE ACETATE

CAS #: 000052-21-1

Formula: $C_{23}H_{30}O_6$

Mol Weight: 402.49

MP (deg C):

FP (deg C):

BP (deg C):

BP pressure (mm Hg):

Property/Value	Units	Temp	Data Type	Reference
WS 2.64E+001	mg/L	25	EST	MEYLAN,WM ET AL. (1996)
logP 2.40			EXP	HANSCH,C & LEO,AJ (1985)
VP 2.24E-013	mm Hg	25	EST	NEELY,WB & BLAU,GE (1985)
DC	pKa			
HL 2.93E-012	atm m3/mol	25	EST	MEYLAN,WM & HOWARD,PH (1991)
OH 6.83E-011	cm3/molc sec	25	EST	MEYLAN,WM & HOWARD,PH (1993)

THIOTEPA

CAS #: 000052-24-4

Formula: $C_6H_{12}N_3PS$

Mol Weight: 189.22

MP (deg C): 51.5

FP (deg C):

BP (deg C):

BP pressure (mm Hg):

Property/Value	Units	Temp	Data Type	Reference
WS 1.75E+004	mg/L	25	EST	MEYLAN,WM ET AL. (1996)
logP 0.53			EXP	HANSCH,C & LEO,AJ (1985)
VP 8.45E-003	mm Hg	25	EST	NEELY,WB & BLAU,GE (1985)
DC	pKa			
HL 2.83E-010	atm m3/mol	25	EST	MEYLAN,WM & HOWARD,PH (1991)
OH 7.23E-011	cm3/molc sec	25	EST	MEYLAN,WM & HOWARD,PH (1993)

CODEINE PHOSPHATE

CAS #: 000052-28-8

Formula: $C_{18}H_{24}NO_7P$

Mol Weight: 397.37

MP (deg C): 220-235 de

FP (deg C):

BP (deg C):

BP pressure (mm Hg):

Property/Value	Units	Temp	Data Type	Reference
WS 3.10E+005	mg/L	25	EXP	STEPHEN,H & STEPHEN,T (1963)
logP 1.14			EXP	HANSCH,C & LEO,AJ (1985)
VP 2.06E-011	mm Hg	25	EST	NEELY,WB & BLAU,GE (1985)
DC 8.10	pKa	25	EXP	HANSCH,C & LEO,AJ (1985)
HL 4.50E-031	atm m3/mol	25	EST	MEYLAN,WM & HOWARD,PH (1991)
OH 1.70E-010	cm3/molc sec	25	EST	MEYLAN,WM & HOWARD,PH (1993)

CYCLOBARBITAL

CAS #: 000052-31-3

Formula: $C_{12}H_{16}N_2O_3$

Mol Weight: 236.27

MP (deg C): 171-174

FP (deg C):

BP (deg C):

BP pressure (mm Hg):

Property/Value	Units	Temp	Data Type	Reference
WS 8.27E+003	mg/L	25	EXP	YALKOWSKY,SH & DANNENFELSER,RM (1992)
logP 1.77			EXP	HANSCH,C ET AL. (1995)
VP 2.95E-012	mm Hg	25	EST	NEELY,WB & BLAU,GE (1985)
DC	pKa			
HL 5.13E-013	atm m3/mol	25	EST	MEYLAN,WM & HOWARD,PH (1991)
OH 9.78E-011	cm3/molc sec	25	EST	MEYLAN,WM & HOWARD,PH (1993)

ALDOSTERONE

CAS #: 000052-39-1

Formula: $C_{21}H_{28}O_5$

Mol Weight: 360.45

MP (deg C): 200 dec

FP (deg C):

BP (deg C):

BP pressure (mm Hg):

Property/Value	Units	Temp	Data Type	Reference
WS 5.12E+001	mg/L	37	EXP	YALKOWSKY,SH & DANNENFELSER,RM (1992)
logP 1.08			EXP	HANSCH,C & LEO,AJ (1985)
VP 1.25E-012	mm Hg	25	EST	NEELY,WB & BLAU,GE (1985)
DC	pKa			
HL 1.81E-013	atm m3/mol	25	EST	MEYLAN,WM & HOWARD,PH (1991)
OH 1.43E-010	cm3/molc sec	25	EST	MEYLAN,WM & HOWARD,PH (1993)

ALLOBARBITAL

CAS #: 000052-43-7

Formula: $C_{10}H_{12}N_2O_3$

Mol Weight: 208.22

MP (deg C): 171-173

FP (deg C):

BP (deg C):

BP pressure (mm Hg):

Property/Value	Units	Temp	Data Type	Reference
WS 2.25E+003	mg/L	25	EXP	YALKOWSKY,SH & DANNENFELSER,RM (1992)
logP 1.05			EXP	HANSCH,C & LEO,AJ (1985)
VP 3.42E-011	mm Hg	25	EST	NEELY,WB & BLAU,GE (1985)
DC 7.77	pKa	25	EXP	KORTUM,G ET AL (1961)
HL 3.53E-013	atm m3/mol	25	EST	MEYLAN,WM & HOWARD,PH (1991)
OH 6.19E-011	cm3/molc sec	25	EST	MEYLAN,WM & HOWARD,PH (1993)

CAS #: 000052-44-8				2,4-DIAZASPIRO[5.5]UNDECANE-1,3,5-TRIONE

Formula: $C_9H_{12}N_2O_3$

Mol Weight: 196.21

MP (deg C):   FP (deg C):

BP (deg C): 167

BP pressure (mm Hg): 5.00E-001

Property/ Value	Units	Temp	Data Type	Reference
WS 7.64E+003	mg/L	25	EST	MEYLAN,WM ET AL. (1996)
logP 0.91			EXP	PRANKERD,RJ & MCKEOWN,RH (1992)
VP 4.07E-011	mm Hg	25	EST	NEELY,WB & BLAU,GE (1985)
DC	pKa			
HL 2.11E-013	atm m3/mol	25	EST	MEYLAN,WM & HOWARD,PH (1991)
OH 1.52E-011	cm3/molc sec	25	EST	MEYLAN,WM & HOWARD,PH (1993)

CAS #: 000052-52-8				CYCLOLEUCINE

Formula: $C_6H_{11}NO_2$

Mol Weight: 129.16

MP (deg C): 330   FP (deg C):

BP (deg C):

BP pressure (mm Hg):

Property/ Value	Units	Temp	Data Type	Reference
WS 6.98E+004	mg/L	25	EST	MEYLAN,WM ET AL. (1996)
logP -2.28			EXP	TSAI,RS ET AL. (1991)
VP 2.63E-009	mm Hg	25	EST	NEELY,WB & BLAU,GE (1985)
DC	pKa			
HL 1.54E-009	atm m3/mol	25	EST	MEYLAN,WM & HOWARD,PH (1991)
OH 2.60E-011	cm3/molc sec	25	EST	MEYLAN,WM & HOWARD,PH (1993)

CAS #: 000052-53-9				VERAPAMIL

Formula: $C_{27}H_{38}N_2O_4$

Mol Weight: 454.61

MP (deg C):   FP (deg C):

BP (deg C): 243-246

BP pressure (mm Hg): 1.00E-002

Property/ Value	Units	Temp	Data Type	Reference
WS 4.47E+000	mg/L	25	EST	MEYLAN,WM ET AL. (1996)
logP 3.83			EXP	HANSCH,C ET AL. (1995)
VP 2.79E-011	mm Hg	25	EST	NEELY,WB & BLAU,GE (1985)
DC	pKa			
HL 8.79E-015	atm m3/mol	25	EST	MEYLAN,WM & HOWARD,PH (1991)
OH 1.86E-010	cm3/molc sec	25	EST	MEYLAN,WM & HOWARD,PH (1993)

CAS #: 000052-67-5				PENCILLAMINE

Formula: $C_5H_{11}NO_2S$

Mol Weight: 149.21

MP (deg C): 202-206   FP (deg C):

BP (deg C):

BP pressure (mm Hg):

Property/ Value	Units	Temp	Data Type	Reference
WS 1.11E+005	mg/L	20	EXP	YALKOWSKY,SH & DANNENFELSER,RM (1992)
logP -1.78			EXP	HANSCH,C & LEO,AJ (1985)
VP 6.27E-007	mm Hg	25	EST	NEELY,WB & BLAU,GE (1985)
DC	pKa			
HL 9.33E-011	atm m3/mol	25	EST	MEYLAN,WM & HOWARD,PH (1991)
OH 7.10E-011	cm3/molc sec	25	EST	MEYLAN,WM & HOWARD,PH (1993)

CAS #: 000052-68-6				TRICHLORFON

Formula: $C_4H_8Cl_3O_4P$

Mol Weight: 257.44

MP (deg C): 83-84   FP (deg C):

BP (deg C): 100

BP pressure (mm Hg): 1.00E-001

Property/ Value	Units	Temp	Data Type	Reference
WS 1.54E+005	mg/L	25	EXP	FREED,VH ET AL. (1977)
logP 0.51			EXP	HANSCH,C & LEO,AJ (1985)
VP 7.80E-006	mm Hg	20	EXP	FREED,VH ET AL. (1977)
DC	pKa			
HL 1.70E-011	atm m3/mol	20	EST	VP/WSOL
OH 6.36E-012	cm3/molc sec	25	EST	ATKINSON, R (1988)

CAS #: 000052-85-7				FAMPHUR

Formula: $C_{10}H_{16}NO_5PS_2$

Mol Weight: 325.34

MP (deg C): 52.5-53.5   FP (deg C):

BP (deg C):

BP pressure (mm Hg):

Property/ Value	Units	Temp	Data Type	Reference
WS 1.09E+002	mg/L	25	EST	MEYLAN,WM ET AL. (1996)
logP 2.23			EXP	HANSCH,C & LEO,AJ (1985)
VP 1.36E-006	mm Hg	25	EST	NEELY,WB & BLAU,GE (1985)
DC	pKa			
HL 1.61E-008	atm m3/mol	25	EST	MEYLAN,WM & HOWARD,PH (1991)
OH 6.19E-011	cm3/molc sec	25	EST	MEYLAN,WM & HOWARD,PH (1993)

CAS #: 000052-86-8				HALOPERIDOL

Formula: $C_{21}H_{23}ClFNO_2$

Mol Weight: 375.87

MP (deg C): 293   FP (deg C):

BP (deg C):

BP pressure (mm Hg):

Property/ Value	Units	Temp	Data Type	Reference
WS 5.93E+001	mg/L	25	EST	MEYLAN,WM ET AL. (1996)
logP 3.36			EXP	SANGSTER,J (1994)
VP 4.85E-011	mm Hg	25	EST	NEELY,WB & BLAU,GE (1985)
DC	pKa			
HL 2.26E-014	atm m3/mol	25	EST	MEYLAN,WM & HOWARD,PH (1991)
OH 1.16E-010	cm3/molc sec	25	EST	MEYLAN,WM & HOWARD,PH (1993)

CAS #: 000052-90-4				CYSTEINE

Formula: $C_3H_7NO_2S$

Mol Weight: 121.16

MP (deg C): 213-214 de   FP (deg C):

BP (deg C):

BP pressure (mm Hg):

Property/ Value	Units	Temp	Data Type	Reference
WS 1.13E+005	mg/L	25	EST	MEYLAN,WM ET AL. (1996)
logP -2.49			EXP	HANSCH,C & LEO,AJ (1985)
VP 2.07E-006	mm Hg	25	EST	NEELY,WB & BLAU,GE (1985)
DC	pKa			
HL 5.30E-011	atm m3/mol	25	EST	MEYLAN,WM & HOWARD,PH (1991)
OH 7.96E-011	cm3/molc sec	25	EST	MEYLAN,WM & HOWARD,PH (1993)

CAS #: 000053-03-2				PREDNISONE

Formula: $C_{21}H_{26}O_5$

Mol Weight: 358.44

MP (deg C): 233-235 de   FP (deg C):

BP (deg C):

BP pressure (mm Hg):

Property/ Value	Units	Temp	Data Type	Reference
WS 3.12E+002	mg/L	25	EST	MEYLAN,WM ET AL. (1996)
logP 1.46			EXP	HANSCH,C & LEO,AJ (1985)
VP 5.09E-013	mm Hg	25	EST	NEELY,WB & BLAU,GE (1985)
DC	pKa			
HL 2.83E-010	atm m3/mol	25	EST	MEYLAN,WM & HOWARD,PH (1991)
OH 6.60E-011	cm3/molc sec	25	EST	MEYLAN,WM & HOWARD,PH (1993)

CAS #: 000053-06-5				CORTISONE

Formula: $C_{21}H_{28}O_5$

Mol Weight: 360.45

MP (deg C): 220-224   FP (deg C).

BP (deg C):

BP pressure (mm Hg):

Property/ Value	Units	Temp	Data Type	Reference
WS 2.80E+002	mg/L	25	EXP	YALKOWSKY,SH & DANNENFELSER,RM (1992)
logP 1.47			EXP	HANSCH,C & LEO,AJ (1985)
VP 5.80E-013	mm Hg	25	EST	NEELY,WB & BLAU,GE (1985)
DC	pKa			
HL 6.03E-010	atm m3/mol	25	EST	MEYLAN,WM & HOWARD,PH (1991)
OH 1.07E-010	cm3/molc sec	25	EST	MEYLAN,WM & HOWARD,PH (1993)

CAS #: 000053-16-7				ESTRONE

Formula: $C_{18}H_{22}O_2$

Mol Weight: 270.37

MP (deg C): 251-254   FP (deg C):

BP (deg C): 154

BP pressure (mm Hg):

Property/ Value	Units	Temp	Data Type	Reference
WS 8.00E-001	mg/L	25	EXP	YALKOWSKY,SH & DANNENFELSER,RM (1992)
logP 3.13			EXP	HANSCH,C ET AL. (1995)
VP 1.42E-007	mm Hg	25	EST	NEELY,WB & BLAU,GE (1985)
DC	pKa			
HL 3.80E-010	atm m3/mol	25	EST	MEYLAN,WM & HOWARD,PH (1991)
OH 1.26E-010	cm3/molc sec	25	EST	MEYLAN,WM & HOWARD,PH (1993)

CAS #: 000053-19-0				O,P'-DDD

Formula: $C_{14}H_{10}Cl_4$

Mol Weight: 320.05

MP (deg C): 76-78   FP (deg C):

BP (deg C):

BP pressure (mm Hg):

Property/ Value	Units	Temp	Data Type	Reference
WS 1.00E-001	mg/L	25	EXP	BIGGAR,JW & RIGGS,RI (1974)
logP 5.87			EST	MEYLAN,WM & HOWARD,PH (1995)
VP 1.94E-006	mm Hg	30	EXP	SUNTIO,LR ET AL. (1988)
DC	pKa			
HL 8.17E-006	atm m3/mol	25	EST	VP/WSOL
OH 4.34E-012	cm3/molc sec	25	EST	MEYLAN,WM & HOWARD,PH (1993)

CAS #: 000053-21-4				COCAINE HYDROCHLORIDE

Formula: $C_{17}H_{22}ClNO_4$

Mol Weight: 339.82

MP (deg C): 290 dec   FP (deg C):

BP (deg C):

BP pressure (mm Hg):

Property/ Value	Units	Temp	Data Type	Reference
WS 7.14E+005	mg/L	25	EXP	STEPHEN,H & STEPHEN,T (1963)
logP 0.10			EST	MEYLAN,WM & HOWARD,PH (1995)
VP 4.30E-011	mm Hg	25	EST	NEELY,WB & BLAU,GE (1985)
DC	pKa			
HL 2.07E-018	atm m3/mol	25	EST	MEYLAN,WM & HOWARD,PH (1991)
OH 5.48E-011	cm3/molc sec	25	EST	MEYLAN,WM & HOWARD,PH (1993)

CAS #: 000053-41-8				ANDROSTERONE

Formula: $C_{19}H_{30}O_2$

Mol Weight: 290.45

MP (deg C): 283 dec   FP (deg C):

BP (deg C):

BP pressure (mm Hg):

Property/ Value	Units	Temp	Data Type	Reference
WS 1.20E+001	mg/L	23	EXP	YALKOWSKY,SH & DANNENFELSER,RM (1992)
logP 3.69			EXP	HANSCH,C ET AL. (1995)
VP 2.79E-008	mm Hg	25	EST	NEELY,WB & BLAU,GE (1985)
DC	pKa			
HL 6.37E-009	atm m3/mol	25	EST	MEYLAN,WM & HOWARD,PH (1991)
OH 4.15E-011	cm3/molc sec	25	EST	MEYLAN,WM & HOWARD,PH (1993)

CAS #: 000053-43-0				PRASTERONE

Formula: $C_{19}H_{28}O_2$

Mol Weight: 288.43

MP (deg C): 140-141   FP (deg C):

BP (deg C):

BP pressure (mm Hg):

Property/ Value	Units	Temp	Data Type	Reference
WS 6.35E+001	mg/L	25	EXP	YALKOWSKY,SH & DANNENFELSER,RM (1992)
logP 3.23			EXP	HANSCH,C ET AL. (1995)
VP 2.23E-008	mm Hg	25	EST	NEELY,WB & BLAU,GE (1985)
DC	pKa			
HL 6.62E-009	atm m3/mol	25	EST	MEYLAN,WM & HOWARD,PH (1991)
OH 1.24E-010	cm3/molc sec	25	EST	MEYLAN,WM & HOWARD,PH (1993)

CAS #: 000053-60-1				PROMAZINE HYDROCHLORIDE

Formula: $C_{17}H_{21}ClN_2S$

Mol Weight: 320.89

MP (deg C):   FP (deg C):

BP (deg C): 203-210

BP pressure (mm Hg): 3.00E-001

Property/ Value	Units	Temp	Data Type	Reference
WS 8.60E+000	mg/L	25	EXP	BAILEY,GW ET AL. (1968)
logP 2.94			EXP	BROWN,DS & FLAGG,EW (1981)
VP 7.89E-007	mm Hg	25	EST	NEELY,WB & BLAU,GE (1985)
DC	pKa			
HL 4.98E-010	atm m3/mol	25	EST	MEYLAN,WM & HOWARD,PH (1991)
OH 2.71E-010	cm3/molc sec	25	EST	MEYLAN,WM & HOWARD,PH (1993)

CAS #: 000053-70-3				DIBENZ(A,H)ANTHRACENE

Formula:	$C_{22}H_{14}$
Mol Weight:	278.36
MP (deg C):	269	FP (deg C):
BP (deg C):	197.5
BP pressure (mm Hg):	

Property/Value	Units	Temp	Data Type	Reference
WS 2.49E-006	mg/L	25	EXP	HASSETT,JJ ET AL. (1980)
logP 6.75			EXP	SANGSTER,J (1993)
VP 1.00E-010	mm Hg	20	EST	CALLAHAN,MA ET AL. (1979A)
DC	pKa			
HL 1.15E-004	atm m3/mol	20	EST	VP/WSOL
OH 1.50E-010	cm3/molc sec	25	EST	MEYLAN,WM & HOWARD,PH (1993)

CAS #: 000053-79-2				PUROMYCIN

Formula:	$C_{22}H_{29}N_7O_5$
Mol Weight:	471.52
MP (deg C):	175.5-177	FP (deg C):
BP (deg C):	
BP pressure (mm Hg):	

Property/Value	Units	Temp	Data Type	Reference
WS 5.03E+001	mg/L	25	EST	MEYLAN,WM ET AL. (1996)
logP 0.03			EXP	HANSCH,C & LEO,AJ (1985)
VP 1.03E-022	mm Hg	25	EST	NEELY,WB & BLAU,GE (1985)
DC	pKa			
HL 5.94E-031	atm m3/mol	25	EST	MEYLAN,WM & HOWARD,PH (1991)
OH 3.24E-010	cm3/molc sec	25	EST	MEYLAN,WM & HOWARD,PH (1993)

CAS #: 000053-86-1				INDOMETHACIN

Formula:	$C_{19}H_{16}ClNO_4$
Mol Weight:	357.80
MP (deg C):	155-162	FP (deg C):
BP (deg C):	117.9
BP pressure (mm Hg):	

Property/Value	Units	Temp	Data Type	Reference
WS 9.37E-001	mg/L	25	EXP	YALKOWSKY,SH & DANNENFELSER,RM (1992)
logP 4.27			EXP	HANSCH,C ET AL. (1995) @ pH=2
VP 9.89E-011	mm Hg	25	EST	NEELY,WB & BLAU,GE (1985)
DC 4.50	pKa		EXP	BUDAVARI,S ET AL. (1989)
HL 3.13E-014	atm m3/mol	25	EST	MEYLAN,WM & HOWARD,PH (1991)
OH 2.03E-010	cm3/molc sec	25	EST	MEYLAN,WM & HOWARD,PH (1993)

CAS #: 000053-96-3				N-2-FLUORENYLACETAMIDE

Formula:	$C_{15}H_{13}NO$
Mol Weight:	223.28
MP (deg C):	192-196	FP (deg C):
BP (deg C):	303
BP pressure (mm Hg):	

Property/Value	Units	Temp	Data Type	Reference
WS 5.53E+000	mg/L	25	EST	MEYLAN,WM ET AL. (1996)
logP 3.12			EST	MEYLAN,WM & HOWARD,PH (1995)
VP 9.44E-008	mm Hg	25	EST	NEELY,WB & BLAU,GE (1985)
DC	pKa			
HL 1.92E-010	atm m3/mol	25	EST	MEYLAN,WM & HOWARD,PH (1991)
OH 4.04E-011	cm3/molc sec	25	EST	MEYLAN,WM & HOWARD,PH (1993)

CAS #: 000054-04-6				MESCALINE

Formula:	$C_{11}H_{17}NO_3$
Mol Weight:	211.26
MP (deg C):	180	FP (deg C):
BP (deg C):	35-36
BP pressure (mm Hg):	1.20E+001

Property/Value	Units	Temp	Data Type	Reference
WS 8.41E+004	mg/L	25	EST	MEYLAN,WM ET AL. (1996)
logP 0.78			EXP	HANSCH,C & LEO,AJ (1985)
VP 3.22E-004	mm Hg	25	EST	NEELY,WB & BLAU,GE (1985)
DC	pKa			
HL 1.68E-010	atm m3/mol	25	EST	MEYLAN,WM & HOWARD,PH (1991)
OH 2.35E-010	cm3/molc sec	25	EST	MEYLAN,WM & HOWARD,PH (1993)

CAS #: 000054-05-7				CHLOROQUINE

Formula:	$C_{18}H_{26}ClN_3$
Mol Weight:	319.88
MP (deg C):	288-290 de	FP (deg C):
BP (deg C):	
BP pressure (mm Hg):	

Property/Value	Units	Temp	Data Type	Reference
WS 1.06E+001	mg/L	25	EST	MEYLAN,WM ET AL. (1996)
logP 4.63			EXP	HANSCH,C & LEO,AJ (1985)
VP 1.92E-007	mm Hg	25	EST	NEELY,WB & BLAU,GE (1985)
DC	pKa			
HL 1.07E-012	atm m3/mol	25	EST	MEYLAN,WM & HOWARD,PH (1991)
OH 1.59E-010	cm3/molc sec	25	EST	MEYLAN,WM & HOWARD,PH (1993)

CAS #: 000054-11-5				NICOTINE

Formula:	$C_{10}H_{14}N_2$
Mol Weight:	162.24
MP (deg C):	159-162	FP (deg C):
BP (deg C):	247
BP pressure (mm Hg):	7.45E+002

Property/Value	Units	Temp	Data Type	Reference
WS 1.00E+006	mg/L		EXP	SEIDELL,A (1941)
logP 1.17			EXP	HANSCH,C & LEO,AJ (1985)
VP 3.80E-002	mm Hg	25	EXT	BOUBLIK,T ET AL. (1984)
DC 6.16	pKa	15	EXP	MERCK INDEX (1983)
HL 2.81E-009	atm m3/mol	25	EST	MEYLAN,WM & HOWARD,PH (1991)
OH 9.21E-011	cm3/molc sec	25	EST	MEYLAN,WM & HOWARD,PH (1993)

CAS #: 000054-12-6				DL-TRYPTOPHAN

Formula:	$C_{11}H_{12}N_2O_2$
Mol Weight:	204.23
MP (deg C):	230-231 de	FP (deg C):
BP (deg C):	
BP pressure (mm Hg):	

Property/Value	Units	Temp	Data Type	Reference
WS 2.85E+003	mg/L	25	EST	MEYLAN,WM ET AL. (1996)
logP -1.05			EXP	HANSCH,C ET AL. (1995)
VP 2.12E-009	mm Hg	25	EST	NEELY,WB & BLAU,GE (1985)
DC 7.38	pKa	25	EXP	KORTUM,G ET AL (1961)
HL 1.98E-014	atm m3/mol	25	EST	MEYLAN,WM & HOWARD,PH (1991)
OH 2.39E-010	cm3/molc sec	25	EST	MEYLAN,WM & HOWARD,PH (1993)

CAS #: 000054-20-6 — TRIFLUOROMETHYLURACIL

Formula: $C_5H_3F_3N_2O_2$

Mol Weight: 180.09

MP (deg C): 316

FP (deg C):

BP (deg C):

BP pressure (mm Hg):

Property/Value	Units	Temp	Data Type	Reference
WS 2.49E+003	mg/L	25	EST	MEYLAN,WM ET AL. (1996)
logP 0.04			EXP	HANSCH,C & LEO,AJ (1985)
VP 2.30E-006	mm Hg	25	EST	NEELY,WB & BLAU,GE (1985)
DC	pKa			
HL 1.07E-009	atm m3/mol	25	EST	MEYLAN,WM & HOWARD,PH (1991)
OH 1.39E-011	cm3/molc sec	25	EST	MEYLAN,WM & HOWARD,PH (1993)

CAS #: 000054-21-7 — SODIUM SALICYLATE

Formula: $C_7H_5NaO_3$

Mol Weight: 160.11

MP (deg C): 213 dec

FP (deg C):

BP (deg C):

BP pressure (mm Hg):

Property/Value	Units	Temp	Data Type	Reference
WS 1.19E+005	mg/L	25	EXP	CHESSELLS,M ET AL. (1992)
logP -1.43			EXP	HANSCH,C & LEO,AJ (1985)
VP 3.68E-011	mm Hg	25	EST	NEELY,WB & BLAU,GE (1985)
DC	pKa			
HL	atm m3/mol			
OH 3.05E-011	cm3/molc sec	25	EST	MEYLAN,WM & HOWARD,PH (1993)

CAS #: 000054-25-1 — 6-AZAURIDINE

Formula: $C_8H_{11}N_3O_6$

Mol Weight: 245.19

MP (deg C): 146

FP (deg C):

BP (deg C):

BP pressure (mm Hg):

Property/Value	Units	Temp	Data Type	Reference
WS 8.28E+004	mg/L	25	EST	MEYLAN,WM ET AL. (1996)
logP -2.14			EXP	HANSCH,C & LEO,AJ (1985)
VP 6.82E-016	mm Hg	25	EST	NEELY,WB & BLAU,GE (1985)
DC	pKa			
HL 6.23E-018	atm m3/mol	25	EST	MEYLAN,WM & HOWARD,PH (1991)
OH 1.00E-010	cm3/molc sec	25	EST	MEYLAN,WM & HOWARD,PH (1993)

CAS #: 000054-31-9 — FUROSEMIDE

Formula: $C_{12}H_{11}ClN_2O_5S$

Mol Weight: 330.75

MP (deg C): 295

FP (deg C):

BP (deg C):

BP pressure (mm Hg):

Property/Value	Units	Temp	Data Type	Reference
WS 7.31E+001	mg/L	30	EXP	YALKOWSKY,SH & DANNENFELSER,RM (1992)
logP 2.03			EXP	SANGSTER,J (1993)
VP 3.23E-010	mm Hg	25	EST	NEELY,WB & BLAU,GE (1985)
DC	pKa			
HL 3.94E-016	atm m3/mol	25	EST	MEYLAN,WM & HOWARD,PH (1991)
OH 1.12E-010	cm3/molc sec	25	EST	MEYLAN,WM & HOWARD,PH (1993)

CAS #: 000054-32-0 — MOXISYLYTE

Formula: $C_{16}H_{25}NO_3$

Mol Weight: 279.38

MP (deg C): 144.5

FP (deg C):

BP (deg C):

BP pressure (mm Hg):

Property/Value	Units	Temp	Data Type	Reference
WS 3.23E+002	mg/L	25	EST	MEYLAN,WM ET AL. (1996)
logP 3.17			EXP	DALLET ET AL. (1985)
VP 2.78E-005	mm Hg	25	EST	NEELY,WB & BLAU,GE (1985)
DC	pKa			
HL 5.42E-009	atm m3/mol	25	EST	MEYLAN,WM & HOWARD,PH (1991)
OH 1.58E-010	cm3/molc sec	25	EST	MEYLAN,WM & HOWARD,PH (1993)

CAS #: 000054-42-2 — URIDINE, 2'-DEOXY-5-IODO-

Formula: $C_9H_{11}IN_2O_5$

Mol Weight: 354.10

MP (deg C): 263

FP (deg C):

BP (deg C):

BP pressure (mm Hg):

Property/Value	Units	Temp	Data Type	Reference
WS 1.89E+003	mg/L	25	EST	MEYLAN,WM ET AL. (1996)
logP -0.96			EXP	NARURKAR,MM & MITRA,AK (1988)
VP 1.07E-015	mm Hg	25	EST	NEELY,WB & BLAU,GE (1985)
DC	pKa			
HL 9.71E-019	atm m3/mol	25	EST	MEYLAN,WM & HOWARD,PH (1991)
OH 8.25E-011	cm3/molc sec	25	EST	MEYLAN,WM & HOWARD,PH (1993)

CAS #: 000054-47-7 — PYRIDOXAL-5-PHOSPHATE

Formula: $C_8H_{10}NO_6P$

Mol Weight: 247.15

MP (deg C): 255

FP (deg C):

BP (deg C):

BP pressure (mm Hg):

Property/Value	Units	Temp	Data Type	Reference
WS 1.19E+004	mg/L	25	EST	MEYLAN,WM ET AL. (1996)
logP 0.37			EST	MEYLAN,WM & HOWARD,PH (1995)
VP 3.28E-010	mm Hg	25	EST	NEELY,WB & BLAU,GE (1985)
DC	pKa			
HL 9.26E-020	atm m3/mol	25	EST	MEYLAN,WM & HOWARD,PH (1991)
OH 4.29E-011	cm3/molc sec	25	EST	MEYLAN,WM & HOWARD,PH (1993)

CAS #: 000054-49-9 — METARAMINOL

Formula: $C_9H_{13}NO_2$

Mol Weight: 167.21

MP (deg C): 107.5

FP (deg C):

BP (deg C): 218

BP pressure (mm Hg):

Property/Value	Units	Temp	Data Type	Reference
WS 1.00E+006	mg/L	25	EST	MEYLAN,WM ET AL. (1996)
logP -0.27			EXP	HANSCH,C & LEO,AJ (1985)
VP 1.67E-005	mm Hg	25	EST	NEELY,WB & BLAU,GE (1985)
DC	pKa			
HL 4.10E-015	atm m3/mol	25	EST	MEYLAN,WM & HOWARD,PH (1991)
OH 1.31E-010	cm3/molc sec	25	EST	MEYLAN,WM & HOWARD,PH (1993)

000054-71-7 — PILOCARPINE, MONOHYDROCHLORIDE

Formula: $C_{11}H_{17}ClN_2O_2$

Mol Weight: 244.72

MP (deg C): 204-205

FP (deg C):

BP (deg C):

BP pressure (mm Hg):

Property/Value	Units	Temp	Data Type	Reference
WS 1.00E+006	mg/L	25	EXP	SEIDELL,A (1941)
logP -3.03			EST	MEYLAN,WM & HOWARD,PH (1995)
VP 5.14E-007	mm Hg	25	EST	NEELY,WB & BLAU,GE (1985)
DC	pKa			
HL 2.88E-013	atm m3/mol	25	EST	MEYLAN,WM & HOWARD,PH (1991)
OH 1.07E-010	cm3/molc sec	25	EST	MEYLAN,WM & HOWARD,PH (1993)

000054-80-8 — PRONETHALOL

Formula: $C_{15}H_{19}NO$

Mol Weight: 229.32

MP (deg C): 108

FP (deg C):

BP (deg C):

BP pressure (mm Hg):

Property/Value	Units	Temp	Data Type	Reference
WS 8.57E+002	mg/L	25	EST	MEYLAN,WM ET AL. (1996)
logP 3.00			EXP	SANGSTER,J (1993)
VP 3.08E-007	mm Hg	25	EST	NEELY,WB & BLAU,GE (1985)
DC	pKa			
HL 1.12E-011	atm m3/mol	25	EST	MEYLAN,WM & HOWARD,PH (1991)
OH 1.52E-010	cm3/molc sec	25	EST	MEYLAN,WM & HOWARD,PH (1993)

000054-85-3 — ISONIAZID

Formula: $C_6H_7N_3O$

Mol Weight: 137.14

MP (deg C): 171.4

FP (deg C):

BP (deg C):

BP pressure (mm Hg):

Property/Value	Units	Temp	Data Type	Reference
WS 1.67E+004	mg/L	25	EST	MEYLAN,WM ET AL. (1996)
logP -0.70			EXP	HANSCH,C ET AL. (1995)
VP 4.64E-005	mm Hg	25	EST	NEELY,WB & BLAU,GE (1985)
DC 1.82	pKa	20	EXP	PERRIN,DD (1965)
HL 1.21E-014	atm m3/mol	25	EST	MEYLAN,WM & HOWARD,PH (1991)
OH 5.84E-012	cm3/molc sec	25	EST	MEYLAN,WM & HOWARD,PH (1993)

000054-91-1 — PIPOBROMAN

Formula: $C_{10}H_{16}Br_2N_2O_2$

Mol Weight: 356.07

MP (deg C): > 300

FP (deg C):

BP (deg C):

BP pressure (mm Hg):

Property/Value	Units	Temp	Data Type	Reference
WS 2.49E+003	mg/L	25	EST	MEYLAN,WM ET AL. (1996)
logP 0.42			EXP	HANSCH,C ET AL. (1995)
VP 1.78E-007	mm Hg	25	EST	NEELY,WB & BLAU,GE (1985)
DC	pKa			
HL 8.57E-013	atm m3/mol	25	EST	MEYLAN,WM & HOWARD,PH (1991)
OH 5.15E-011	cm3/molc sec	25	EST	MEYLAN,WM & HOWARD,PH (1993)

000054-92-2 — IPRONIAZID

Formula: $C_9H_{13}N_3O$

Mol Weight: 179.22

MP (deg C): 150 dec

FP (deg C):

BP (deg C):

BP pressure (mm Hg):

Property/Value	Units	Temp	Data Type	Reference
WS 2.68E+004	mg/L	25	EST	MEYLAN,WM ET AL. (1996)
logP 0.37			EXP	HANSCH,C ET AL. (1995)
VP 1.36E-005	mm Hg	25	EST	NEELY,WB & BLAU,GE (1985)
DC	pKa			
HL 4.68E-014	atm m3/mol	25	EST	MEYLAN,WM & HOWARD,PH (1991)
OH 8.72E-011	cm3/molc sec	25	EST	MEYLAN,WM & HOWARD,PH (1993)

000054-95-5 — PENTYLENETETRAZOLE

Formula: $C_6H_{10}N_4$

Mol Weight: 138.17

MP (deg C): 57-60

FP (deg C):

BP (deg C):

BP pressure (mm Hg):

Property/Value	Units	Temp	Data Type	Reference
WS 3.17E+003	mg/L	25	EST	MEYLAN,WM ET AL. (1996)
logP 0.14			EXP	HANSCH,C & LEO,AJ (1985)
VP 2.41E-002	mm Hg	25	EST	NEELY,WB & BLAU,GE (1985)
DC	pKa			
HL 1.48E-005	atm m3/mol	25	EST	MEYLAN,WM & HOWARD,PH (1991)
OH 6.79E-012	cm3/molc sec	25	EST	MEYLAN,WM & HOWARD,PH (1993)

000055-18-5 — N-NITROSODIETHYLAMINE

Formula: $C_4H_{10}N_2O$

Mol Weight: 102.14

MP (deg C): -11--9 EST

FP (deg C):

BP (deg C): 175-177

BP pressure (mm Hg):

Property/Value	Units	Temp	Data Type	Reference
WS 4.48E+004	mg/L	25	EST	MEYLAN,WM ET AL. (1996)
logP 0.48			EXP	HANSCH,C & LEO,AJ (1985)
VP 8.60E-001	mm Hg	20	EXT	KLEIN,RG (1982)
DC	pKa			
HL 3.63E-006	atm m3/mol	25	EXP	MIRVISH,SS ET AL. (1976)
OH 1.71E-011	cm3/molc sec	25	EST	MEYLAN,WM & HOWARD,PH (1993)

000055-21-0 — BENZAMIDE

Formula: C_7H_7NO

Mol Weight: 121.14

MP (deg C): 208 dec

FP (deg C):

BP (deg C): 288

BP pressure (mm Hg):

Property/Value	Units	Temp	Data Type	Reference
WS 1.35E+004	mg/L	25	EXP	SEIDELL,A (1941)
logP 0.64			EXP	HANSCH,C & LEO,AJ (1985)
VP 9.39E-004	mm Hg	25	EST	NEELY,WB & BLAU,GE (1985)
DC	pKa			
HL 2.21E-009	atm m3/mol	25	EST	MEYLAN,WM & HOWARD,PH (1991)
OH 1.33E-011	cm3/molc sec	25	EST	MEYLAN,WM & HOWARD,PH (1993)

000055-22-1 — ISONICOTINIC ACID

Formula: $C_6H_5NO_2$
Mol Weight: 123.11
MP (deg C): 315
FP (deg C):
BP (deg C):
BP pressure (mm Hg):

Property/Value	Units	Temp	Data Type	Reference
WS 5.20E+003	mg/L	20	EXP	MERCK INDEX (1976)
logP 0.32			EXP	HANSCH,C ET AL. (1995)
VP 7.89E-003	mm Hg	25	EST	NEELY,WB & BLAU,GE (1985)
DC 4.90	pKa		EXP	SERJEANT,EP & DEMPSEY,B (1979)
HL 1.42E-010	atm m3/mol	25	EST	MEYLAN,WM & HOWARD,PH (1991)
OH 6.57E-013	cm3/molc sec	25	EST	MEYLAN,WM & HOWARD,PH (1993)

000055-27-6 — NORADRENALINE (HCL)

Formula: $C_8H_{12}ClNO_3$
Mol Weight: 205.64
MP (deg C): 292 dec
FP (deg C):
BP (deg C):
BP pressure (mm Hg):

Property/Value	Units	Temp	Data Type	Reference
WS 1.00E+006	mg/L	25	EST	MEYLAN,WM ET AL. (1996)
logP -4.12			EST	MEYLAN,WM & HOWARD,PH (1995)
VP 3.56E-012	mm Hg	25	EST	NEELY,WB & BLAU,GE (1985)
DC	pKa			
HL 2.98E-026	atm m3/mol	25	EST	MEYLAN,WM & HOWARD,PH (1991)
OH 7.39E-011	cm3/molc sec	25	EST	MEYLAN,WM & HOWARD,PH (1993)

000055-38-9 — FENTHION

Formula: $C_{10}H_{15}O_3PS_2$
Mol Weight: 278.33
MP (deg C): 7.5
FP (deg C):
BP (deg C): 87
BP pressure (mm Hg): 1.00E-002

Property/Value	Units	Temp	Data Type	Reference
WS 7.50E+000	mg/L	20	EXP	BOWMAN,BT & SANS,WW (1983)
logP 4.09			EXP	HANSCH,C & LEO,AJ (1985)
VP 3.00E-005	mm Hg	20	EXP	WORTHING,CR & WALKER,SB (1987)
DC	pKa			
HL 1.46E-006	atm m3/mol	20	EST	VP/WSOL
OH 7.60E-011	cm3/molc sec	25	EST	MEYLAN,WM & HOWARD,PH (1993)

000055-56-1 — CHLORHEXIDINE

Formula: $C_{22}H_{30}Cl_2N_{10}$
Mol Weight: 505.46
MP (deg C): 140
FP (deg C):
BP (deg C):
BP pressure (mm Hg):

Property/Value	Units	Temp	Data Type	Reference
WS 4.20E+001	mg/L		EXP	GEYER,H ET AL. (1981)
logP 0.08			EXP	HANSCH,C & LEO,AJ (1985)
VP 1.92E-016	mm Hg	25	EST	NEELY,WB & BLAU,GE (1985)
DC 10.78	pKa	25	EXP	HANSCH,C & LEO,AJ (1985)
HL 1.13E-030	atm m3/mol	25	EST	MEYLAN,WM & HOWARD,PH (1991)
OH 3.04E-010	cm3/molc sec	25	EST	MEYLAN,WM & HOWARD,PH (1993)

000055-63-0 — 1,2,3-PROPANETRIOL TRINITRATE

Formula: $C_3H_5N_3O_9$
Mol Weight: 227.09
MP (deg C): 13.5
FP (deg C):
BP (deg C): 125
BP pressure (mm Hg): 2.00E+000

Property/Value	Units	Temp	Data Type	Reference
WS 1.38E+003	mg/L	20	EXP	SEIDELL,A (1941)
logP 1.62			EXP	HANSCH,C & LEO,AJ (1985)
VP 2.00E-004	mm Hg	20	EXP	KEMP,MD ET AL. (1957)
DC	pKa			
HL 4.33E-008	atm m3/mol	20	EST	VP/WSOL
OH 1.20E-013	cm3/molc sec	25	EST	MEYLAN,WM & HOWARD,PH (1993)

000055-80-1 — 4-DIMETHYLAMINO-3'-METHYLAZOBENZENE

Formula: $C_{15}H_{17}N_3$
Mol Weight: 239.32
MP (deg C): 138-139
FP (deg C):
BP (deg C):
BP pressure (mm Hg):

Property/Value	Units	Temp	Data Type	Reference
WS 4.00E-001	mg/L	25	EXP	TAKAGISHI,T ET AL. (1969)
logP 4.84			EST	MEYLAN,WM & HOWARD,PH (1995)
VP 3.52E-005	mm Hg	25	EST	NEELY,WB & BLAU,GE (1985)
DC	pKa			
HL 2.58E-007	atm m3/mol	25	EST	MEYLAN,WM & HOWARD,PH (1991)
OH 1.50E-010	cm3/molc sec	25	EST	MEYLAN,WM & HOWARD,PH (1993)

000055-98-1 — MYLERLAN

Formula: $C_6H_{14}O_6S_2$
Mol Weight: 246.30
MP (deg C): 287 dec
FP (deg C):
BP (deg C):
BP pressure (mm Hg):

Property/Value	Units	Temp	Data Type	Reference
WS 6.90E+004	mg/L	25	EST	MEYLAN,WM ET AL. (1996)
logP -0.52			EXP	HANSCH,C & LEO,AJ (1985)
VP 6.65E-006	mm Hg	25	EST	NEELY,WB & BLAU,GE (1985)
DC	pKa			
HL 5.16E-010	atm m3/mol	25	EST	MEYLAN,WM & HOWARD,PH (1991)
OH 4.71E-012	cm3/molc sec	25	EST	MEYLAN,WM & HOWARD,PH (1993)

000056-04-2 — METACIL

Formula: $C_5H_6N_2OS$
Mol Weight: 142.18
MP (deg C): 320-325 de
FP (deg C):
BP (deg C):
BP pressure (mm Hg):

Property/Value	Units	Temp	Data Type	Reference
WS 5.33E+002	mg/L	25	EXP	YALKOWSKY,SH & DANNENFELSER,RM (1992)
logP 1.45			EST	MEYLAN,WM & HOWARD,PH (1995)
VP 1.08E-003	mm Hg	25	EST	NEELY,WB & BLAU,GE (1985)
DC	pKa			
HL 1.26E-009	atm m3/mol	25	EST	MEYLAN,WM & HOWARD,PH (1991)
OH 5.99E-011	cm3/molc sec	25	EST	MEYLAN,WM & HOWARD,PH (1993)

000056-12-2 — 4-AMINOBUTYRIC ACID

Formula: $C_4H_9NO_2$
Mol Weight: 103.12
MP (deg C): 202
FP (deg C):
BP (deg C):
BP pressure (mm Hg):

Property/Value	Units	Temp	Data Type	Reference
WS 1.30E+006	mg/L	25	EXP	YALKOWSKY,SH & DANNENFELSER,RM (1992)
logP -3.17			EXP	HANSCH,C & LEO,AJ (1985)
VP 1.98E-008	mm Hg	25	EST	NEELY,WB & BLAU,GE (1985)
DC 4.05	pKa	15	EXP	KORTUM,G ET AL (1961)
HL 9.93E-011	atm m3/mol	25	EST	MEYLAN,WM & HOWARD,PH (1991)
OH 3.45E-011	cm3/molc sec	25	EST	MEYLAN,WM & HOWARD,PH (1993)

000056-23-5 — CARBON TETRACHLORIDE

Formula: CCl_4
Mol Weight: 153.82
MP (deg C): 285-287 de
FP (deg C):
BP (deg C): 76.7
BP pressure (mm Hg):

Property/Value	Units	Temp	Data Type	Reference
WS 7.93E+002	mg/L	25	EXP	HORVATH,AL (1982)
logP 2.83			EXP	HANSCH,C & LEO,AJ (1985)
VP 1.15E+002	mm Hg	25	EXP	BOUBLIK,T ET AL. (1984)
DC	pKa			
HL 2.76E-002	atm m3/mol	25	EXP	LEIGHTON,DTJR & CALO,JM (1981)
OH 1.20E-016	cm3/molc sec	25	EXP	ATKINSON,R (1989)

000056-29-1 — HEXABARITAL

Formula: $C_{12}H_{16}N_2O_3$
Mol Weight: 236.27
MP (deg C): 256 dec
FP (deg C):
BP (deg C):
BP pressure (mm Hg):

Property/Value	Units	Temp	Data Type	Reference
WS 4.35E+002	mg/L	25	EXP	YALKOWSKY,SH & DANNENFELSER,RM (1992)
logP 1.98			EXP	SANGSTER,J (1994)
VP 3.11E-011	mm Hg	25	EST	NEELY,WB & BLAU,GE (1985)
DC	pKa			
HL 8.48E-013	atm m3/mol	25	EST	MEYLAN,WM & HOWARD,PH (1991)
OH 9.51E-011	cm3/molc sec	25	EST	MEYLAN,WM & HOWARD,PH (1993)

000056-33-7 — 1,1,3,3-TETRAMETHYL-1,3-DIPHENYLDISILOXANE

Formula: $C_{16}H_{22}OSi_2$
Mol Weight: 286.53
MP (deg C): 290-292 de
FP (deg C):
BP (deg C): 155-158
BP pressure (mm Hg): 1.30E+001

Property/Value	Units	Temp	Data Type	Reference
WS 1.76E-003	mg/L	25	EST	MEYLAN,WM ET AL. (1996)
logP 7.20			EST	MEYLAN,WM & HOWARD,PH (1995)
VP 2.02E-004	mm Hg	25	EST	NEELY,WB & BLAU,GE (1985)
DC	pKa			
HL 2.52E-005	atm m3/mol	25	EST	MEYLAN,WM & HOWARD,PH (1991)
OH 4.50E-012	cm3/molc sec	25	EST	MEYLAN,WM & HOWARD,PH (1993)

000056-34-8 — TETRAETHYL AMMONIUM CHLORIDE

Formula: $C_8H_{20}ClN$
Mol Weight: 165.71
MP (deg C): 360 dec
FP (deg C):
BP (deg C):
BP pressure (mm Hg):

Property/Value	Units	Temp	Data Type	Reference
WS 1.00E+006	mg/L	25	EXP	PEDDLE,CJ & TURNER,WES (1913)
logP -3.15			EXP	KIMURA,R ET AL. (1991)
VP 3.08E-006	mm Hg	25	EST	NEELY,WB & BLAU,GE (1985)
DC	pKa			
HL 1.30E-011	atm m3/mol	25	EST	MEYLAN,WM & HOWARD,PH (1991)
OH 3.54E-011	cm3/molc sec	25	EST	MEYLAN,WM & HOWARD,PH (1993)

000056-38-2 — PARATHION

Formula: $C_{10}H_{14}NO_5PS$
Mol Weight: 291.26
MP (deg C): 360 dec
FP (deg C):
BP (deg C): 375
BP pressure (mm Hg):

Property/Value	Units	Temp	Data Type	Reference
WS 6.54E+000	mg/L	25	EXP	FELSOT,A & DAHM,PA (1979)
logP 3.83			EXP	HANSCH,C & LEO,AJ (1985)
VP 9.65E-006	mm Hg	25	EXP	SPENCER,WF & CLIATH,MM (1983)
DC	pKa			
HL 5.65E-007	atm m3/mol	25	EST	VP/WSOL
OH 9.03E-011	cm3/molc sec	25	EST	MEYLAN,WM & HOWARD,PH (1993)

000056-40-6 — GLYCINE

Formula: $C_2H_5NO_2$
Mol Weight: 75.07
MP (deg C): 244 dec
FP (deg C):
BP (deg C):
BP pressure (mm Hg):

Property/Value	Units	Temp	Data Type	Reference
WS 2.49E+005	mg/L	25	EXP	YALKOWSKY,SH & DANNENFELSER,RM (1992)
logP -3.21			EXP	HANSCH,C & LEO,AJ (1985)
VP 1.28E-007	mm Hg	25	EST	NEELY,WB & BLAU,GE (1985)
DC 2.37	pKa	20	EXP	KORTUM,G ET AL (1961)
HL 1.12E-009	atm m3/mol	25	EST	MEYLAN,WM & HOWARD,PH (1991)
OH 2.80E-011	cm3/molc sec	25	EST	MEYLAN,WM & HOWARD,PH (1993)

000056-41-7 — ALANINE

Formula: $C_3H_7NO_2$
Mol Weight: 89.09
MP (deg C): -87.9
FP (deg C):
BP (deg C): 68
BP pressure (mm Hg):

Property/Value	Units	Temp	Data Type	Reference
WS 1.64E+005	mg/L	25	EXP	YALKOWSKY,SH & DANNENFELSER,RM (1992)
logP -2.96			EXP	HANSCH,C & LEO,AJ (1985)
VP 1.05E-007	mm Hg	25	EST	NEELY,WB & BLAU,GE (1985)
DC 2.34	pKa	25	EXP	KORTUM,G ET AL (1961)
HL 1.49E-009	atm m3/mol	25	EST	MEYLAN,WM & HOWARD,PH (1991)
OH 3.52E-011	cm3/molc sec	25	EST	MEYLAN,WM & HOWARD,PH (1993)

CAS #: 000056-45-1				SERINE

Formula: $C_3H_7NO_3$

Mol Weight: 105.09

MP (deg C): -187.6 **FP (deg C):**

BP (deg C):

BP pressure (mm Hg): 1.00E+001

Property/Value	Units	Temp	Data Type	Reference
WS 4.25E+005	mg/L	25	EXP	YALKOWSKY,SH & DANNENFELSER,RM (1992)
logP -3.07			EXP	HANSCH,C & LEO,AJ (1985)
VP 4.02E-008	mm Hg	25	EST	NEELY,WB & BLAU,GE (1985)
DC 2.21	pKa	25	EXP	KORTUM,G ET AL (1961)
HL 5.46E-014	atm m3/mol	25	EST	MEYLAN,WM & HOWARD,PH (1991)
OH 4.23E-011	cm3/molc sec	25	EST	MEYLAN,WM & HOWARD,PH (1993)

CAS #: 000056-47-3				DEOXYCORTICOSTERONE ACETATE

Formula: $C_{23}H_{32}O_4$

Mol Weight: 372.51

MP (deg C): -136 **FP (deg C):**

BP (deg C): -51.6

BP pressure (mm Hg):

Property/Value	Units	Temp	Data Type	Reference
WS 1.06E+001	mg/L	25	EST	MEYLAN,WM ET AL. (1996)
logP 3.08			EXP	HANSCH,C & LEO,AJ (1985)
VP 1.32E-008	mm Hg	25	EST	NEELY,WB & BLAU,GE (1985)
DC	pKa			
HL 1.09E-009	atm m3/mol	25	EST	MEYLAN,WM & HOWARD,PH (1991)
OH 1.03E-010	cm3/molc sec	25	EST	MEYLAN,WM & HOWARD,PH (1993)

CAS #: 000056-49-5				3-METHYLCHOLANTHRENE

Formula: $C_{21}H_{16}$

Mol Weight: 268.36

MP (deg C): 2.5 **FP (deg C):**

BP (deg C): 109

BP pressure (mm Hg): 1.50E+001

Property/Value	Units	Temp	Data Type	Reference
WS 2.90E-003	mg/L	25	EXP	MACKAY,D & SHIU,WY (1977)
logP 6.42			EXP	HANSCH,C & LEO,AJ (1985)
VP 3.30E-008	mm Hg	25	EST	NEELY,WB & BLAU,GE (1985)
DC	pKa			
HL 3.01E+000	atm m3/mol	25	EST	MEYLAN,WM & HOWARD,PH (1991)
OH 2.02E-010	cm3/molc sec	25	EST	MEYLAN,WM & HOWARD,PH (1993)

CAS #: 000056-53-1				DES

Formula: $C_{18}H_{20}O_2$

Mol Weight: 268.36

MP (deg C): 94 **FP (deg C):**

BP (deg C): 172

BP pressure (mm Hg):

Property/Value	Units	Temp	Data Type	Reference
WS 3.32E+000	mg/L	25	EST	MEYLAN,WM ET AL. (1996)
logP 5.07			EXP	HANSCH,C & LEO,AJ (1985)
VP 1.41E-008	mm Hg	25	EST	NEELY,WB & BLAU,GE (1985)
DC	pKa			
HL 5.80E-012	atm m3/mol	25	EST	MEYLAN,WM & HOWARD,PH (1991)
OH 1.79E-010	cm3/molc sec	25	EST	MEYLAN,WM & HOWARD,PH (1993)

CAS #: 000056-54-2				QUINIDINE

Formula: $C_{20}H_{24}N_2O_2$

Mol Weight: 324.43

MP (deg C): 15.4 **FP (deg C):**

BP (deg C): 130

BP pressure (mm Hg):

Property/Value	Units	Temp	Data Type	Reference
WS 1.40E+002	mg/L	25	EXP	YALKOWSKY,SH & DANNENFELSER,RM (1992)
logP 3.44			EXP	HANSCH,C & LEO,AJ (1985)
VP 1.10E-010	mm Hg	25	EST	NEELY,WB & BLAU,GE (1985)
DC 8.00	pKa	25	EXP	HANSCH,C & LEO,AJ (1985)
HL 8.58E-016	atm m3/mol	25	EST	MEYLAN,WM & HOWARD,PH (1991)
OH 1.75E-010	cm3/molc sec	25	EST	MEYLAN,WM & HOWARD,PH (1993)

CAS #: 000056-55-3				BENZ(A)ANTHRACENE

Formula: $C_{18}H_{12}$

Mol Weight: 228.30

MP (deg C): -138.3 **FP (deg C):**

BP (deg C): -11.7

BP pressure (mm Hg):

Property/Value	Units	Temp	Data Type	Reference
WS 9.40E-003	mg/L	25	EXP	MAY,WE ET AL. (1978)
logP 5.76			EXP	WANG,L ET AL. (1986)
VP 3.05E-008	mm Hg	25	EXT	PUPP,C ET AL. (1974)
DC	pKa			
HL 8.02E-006	atm m3/mol	25	EXP	SOUTHWORTH,GR (1979A)
OH 8.00E-011	cm3/molc sec	25	EST	MEYLAN,WM & HOWARD,PH (1993)

CAS #: 000056-57-5				4-NITROQUINOLINE OXIDE

Formula: $C_9H_6N_2O_3$

Mol Weight: 190.16

MP (deg C): -130.5 **FP (deg C):**

BP (deg C): 52.6

BP pressure (mm Hg):

Property/Value	Units	Temp	Data Type	Reference
WS 2.34E+003	mg/L	25	EST	MEYLAN,WM ET AL. (1986)
logP 1.09			EXP	HANSCH,C & LEO,AJ (1985)
VP 2.60E-006	mm Hg	25	EST	NEELY,WB & BLAU,GE (1985)
DC	pKa			
HL 2.72E-014	atm m3/mol	25	EST	MEYLAN,WM & HOWARD,PH (1991)
OH 1.80E-012	cm3/molc sec	25	EST	MEYLAN,WM & HOWARD,PH (1993)

CAS #: 000056-72-4				COUMAPHOS

Formula: $C_{14}H_{16}ClO_5PS$

Mol Weight: 362.77

MP (deg C): -117 **FP (deg C):**

BP (deg C): -24.9

BP pressure (mm Hg):

Property/Value	Units	Temp	Data Type	Reference
WS 1.50E+000	mg/L	20	EXP	MELNIKOV,NN (1971)
logP 4.47			EST	MEYLAN,WM & HOWARD,PH (1995)
VP 9.44E-008	mm Hg	25	EST	NEELY,WB & BLAU,GE (1985)
DC	pKa			
HL 1.09E-007	atm m3/mol	25	EST	MEYLAN,WM & HOWARD,PH (1991)
OH 1.03E-010	cm3/molc sec	25	EST	MEYLAN,WM & HOWARD,PH (1993)

CHLORAMPHENICOL

CAS #: 000056-75-7

Formula: $C_{11}H_{12}Cl_2N_2O_5$

Mol Weight: 323.13

MP (deg C): -157.4

FP (deg C):

BP (deg C): -40.7

BP pressure (mm Hg):

Property/Value	Units	Temp	Data Type	Reference
WS 3.75E+003	mg/L	25	EXP	YALKOWSKY,SH & DANNENFELSER,RM (1992)
logP 1.14			EXP	HANSCH,C & LEO,AJ (1985)
VP 1.73E-012	mm Hg	25	EST	NEELY,WB & BLAU,GE (1985)
DC	pKa			
HL 2.29E-018	atm m3/mol	25	EST	MEYLAN,WM & HOWARD,PH (1991)
OH 3.10E-011	cm3/molc sec	25	EST	MEYLAN,WM & HOWARD,PH (1993)

GLYCEROL

CAS #: 000056-81-5

Formula: $C_3H_8O_3$

Mol Weight: 92.10

MP (deg C): -93

FP (deg C):

BP (deg C): 41

BP pressure (mm Hg):

Property/Value	Units	Temp	Data Type	Reference
WS 1.00E+006	mg/L		EXP	RIDDICK,JA ET AL. (1986)
logP -1.76			EXP	HANSCH,C & LEO,AJ (1985)
VP 1.68E-004	mm Hg	25	EXP	DAUBERT,TE & DANNER,RP (1989)
DC 14.40	pKa		EXP	SERJEANT,EP & DEMPSEY,B (1979)
HL 1.73E-008	atm m3/mol	25	EXP	HINE,J & MOOKERJEE,PK (1975)
OH 1.78E-011	cm3/molc sec	25	EST	MEYLAN,WM & HOWARD,PH (1993)

ASPARTIC ACID

CAS #: 000056-84-8

Formula: $C_4H_7NO_4$

Mol Weight: 133.10

MP (deg C): -110.1

FP (deg C):

BP (deg C): 25

BP pressure (mm Hg):

Property/Value	Units	Temp	Data Type	Reference
WS 5.02E+003	mg/L	25	EXP	YALKOWSKY,SH & DANNENFELSER,RM (1992)
logP -3.89			EXP	CHMELIK,J ET AL. (1991)
VP 1.30E-006	mm Hg	25	EST	NEELY,WB & BLAU,GE (1985)
DC 2.01	pKa	0	EXP	KORTUM,G ET AL (1961)
HL 1.11E-014	atm m3/mol	25	EST	MEYLAN,WM & HOWARD,PH (1991)
OH 3.95E-011	cm3/molc sec	25	EST	MEYLAN,WM & HOWARD,PH (1993)

GLUTAMINE

CAS #: 000056-85-9

Formula: $C_5H_{10}N_2O_3$

Mol Weight: 146.15

MP (deg C): -5.7

FP (deg C):

BP (deg C): 105

BP pressure (mm Hg):

Property/Value	Units	Temp	Data Type	Reference
WS 4.13E+004	mg/L	25	EXP	YALKOWSKY,SH & DANNENFELSER,RM (1992)
logP -3.64			EXP	CHMELIK,J ET AL. (1991)
VP 1.90E-008	mm Hg	25	EST	NEELY,WB & BLAU,GE (1985)
DC	pKa			
HL 3.01E-016	atm m3/mol	25	EST	MEYLAN,WM & HOWARD,PH (1991)
OH 4.55E-011	cm3/molc sec	25	EST	MEYLAN,WM & HOWARD,PH (1993)

GLUTAMIC ACID

CAS #: 000056-86-0

Formula: $C_5H_9NO_4$

Mol Weight: 147.13

MP (deg C): 224 dec

FP (deg C):

BP (deg C):

BP pressure (mm Hg):

Property/Value	Units	Temp	Data Type	Reference
WS 8.57E+003	mg/L	25	EXP	BULL,HB ET AL. (1978)
logP -3.69			EXP	HANSCH,C & LEO,AJ (1985)
VP 2.68E-011	mm Hg	25	EXT	DAUBERT,TE & DANNER,RP (1989)
DC 2.23	pKa	0	EXP	KORTUM,G ET AL (1961)
HL 1.47E-014	atm m3/mol	25	EST	MEYLAN,WM & HOWARD,PH (1991)
OH 4.10E-011	cm3/molc sec	25	EST	MEYLAN,WM & HOWARD,PH (1993)

LYSINE

CAS #: 000056-87-1

Formula: $C_6H_{14}N_2O_2$

Mol Weight: 146.19

MP (deg C): -130.8

FP (deg C):

BP (deg C): -9.7

BP pressure (mm Hg):

Property/Value	Units	Temp	Data Type	Reference
WS 1.00E+006	mg/L	20	EXP	YALKOWSKY,SH & DANNENFELSER,RM (1992)
logP -3.05			EXP	HANSCH,C & LEO,AJ (1985)
VP 5.28E-009	mm Hg	25	EXT	DAUBERT,TE & DANNER,RP (1989)
DC 3.12	pKa	0	EXP	KORTUM,G ET AL (1961)
HL 3.59E-013	atm m3/mol	25	EST	MEYLAN,WM & HOWARD,PH (1991)
OH 7.41E-011	cm3/molc sec	25	EST	MEYLAN,WM & HOWARD,PH (1993)

CYSTINE

CAS #: 000056-89-3

Formula: $C_6H_{12}N_2O_4S_2$

Mol Weight: 240.30

MP (deg C): -181

FP (deg C):

BP (deg C): -81.4

BP pressure (mm Hg):

Property/Value	Units	Temp	Data Type	Reference
WS 1.09E+002	mg/L	25	EXP	YALKOWSKY,SH & DANNENFELSER,RM (1992)
logP -5.08			EXP	CHMELIK,J ET AL. (1991)
VP 2.12E-010	mm Hg	25	EST	NEELY,WB & BLAU,GE (1985)
DC	pKa			
HL 4.98E-019	atm m3/mol	25	EST	MEYLAN,WM & HOWARD,PH (1991)
OH 3.19E-010	cm3/molc sec	25	EST	MEYLAN,WM & HOWARD,PH (1993)

4-AMINOMETHYLBENZOIC ACID

CAS #: 000056-91-7

Formula: $C_8H_9NO_2$

Mol Weight: 151.17

MP (deg C): -27.5

FP (deg C):

BP (deg C): 110

BP pressure (mm Hg):

Property/Value	Units	Temp	Data Type	Reference
WS 9.89E+003	mg/L	25	EST	MEYLAN,WM ET AL. (1996)
logP 1.03			EXP	DA,YZ ET AL. (1992)
VP 2.64E-004	mm Hg	25	EST	NEELY,WB & BLAU,GE (1985)
DC	pKa			
HL 1.23E-011	atm m3/mol	25	EST	MEYLAN,WM & HOWARD,PH (1991)
OH 3.20E-011	cm3/molc sec	25	EST	MEYLAN,WM & HOWARD,PH (1993)

CAS #: 000056-93-9	BENZYL TRIMETHYL AMMONIUM CHLORIDE

Formula: $C_{10}H_{16}ClN$

Mol Weight: 185.70

MP (deg C): -99 FP (deg C):

BP (deg C): 26.6

BP pressure (mm Hg):

Property/ Value	Units	Temp	Data Type	Reference
WS 1.00E+006	mg/L	25	EST	MEYLAN,WM ET AL. (1996)
logP -2.17			EXP	HANSCH,C & LEO,AJ (1985)
VP 2.18E-007	mm Hg	25	EST	NEELY,WB & BLAU,GE (1985)
DC	pKa			
HL 3.37E-013	atm m3/mol	25	EST	MEYLAN,WM & HOWARD,PH (1991)
OH 1.73E-011	cm3/molc sec	25	EST	MEYLAN,WM & HOWARD,PH (1993)

CAS #: 000057-00-1	CREATINE

Formula: $C_4H_9N_3O_2$

Mol Weight: 131.14

MP (deg C): -40 FP (deg C):

BP (deg C): 60

BP pressure (mm Hg):

Property/ Value	Units	Temp	Data Type	Reference
WS 1.33E+004	mg/L	18	EXP	YALKOWSKY,SH & DANNENFELSER,RM (1992)
logP -3.72			EST	MEYLAN,WM & HOWARD,PH (1995)
VP 7.90E-004	mm Hg	25	EST	NEELY,WB & BLAU,GE (1985)
DC	pKa			
HL 1.67E-014	atm m3/mol	25	EST	MEYLAN,WM & HOWARD,PH (1991)
OH 9.53E-011	cm3/molc sec	25	EST	MEYLAN,WM & HOWARD,PH (1993)

CAS #: 000057-06-7	3-ISOTHIOCYANATO-1-PROPENE

Formula: C_4H_5NS

Mol Weight: 99.16

MP (deg C): -16 FP (deg C):

BP (deg C): 70.3

BP pressure (mm Hg):

Property/ Value	Units	Temp	Data Type	Reference
WS 2.00E+003	mg/L	20	EXP	YALKOWSKY,SH & DANNENFELSER,RM (1992)
logP 2.15			EST	MEYLAN,WM & HOWARD,PH (1995)
VP 3.70E+000	mm Hg	30	EXP	BOUBLIK,T ET AL. (1984)
DC	pKa			
HL 2.41E-004	atm m3/mol	25	EST	VP/WSOL
OH 2.72E-011	cm3/molc sec	25	EST	MEYLAN,WM & HOWARD,PH (1993)

CAS #: 000057-10-3	PALMITIC ACID

Formula: $C_{16}H_{32}O_2$

Mol Weight: 256.43

MP (deg C): -99 FP (deg C):

BP (deg C): 49.7

BP pressure (mm Hg): 1.50E+001

Property/ Value	Units	Temp	Data Type	Reference
WS 8.21E-001	mg/L	25	EXP	YALKOWSKY,SH & DANNENFELSER,RM (1992)
logP 7.17			EXP	SANGSTER,J (1993)
VP 1.99E-009	mm Hg	25	EXT	DAUBERT,TE & DANNER,RP (1989)
DC	pKa			
HL 2.89E-005	atm m3/mol	25	EST	MEYLAN,WM & HOWARD,PH (1991)
OH 1.97E-011	cm3/molc sec	25	EST	MEYLAN,WM & HOWARD,PH (1993)

CAS #: 000057-11-4	STEARIC ACID

Formula: $C_{18}H_{36}O_2$

Mol Weight: 284.49

MP (deg C): -105.5 FP (deg C):

BP (deg C): 6.1

BP pressure (mm Hg):

Property/ Value	Units	Temp	Data Type	Reference
WS 3.40E+002	mg/L	25	EXP	VERSCHUEREN,K (1977)
logP 8.23			EXP	SANGSTER,J (1993)
VP 7.22E-007	mm Hg	25	EXT	DAUBERT,TE & DANNER,RP (1989)
DC 4.53	pKa		EXP	SERJEANT,EP & DEMPSEY,B (1979)
HL 5.10E-005	atm m3/mol	25	EST	MEYLAN,WM & HOWARD,PH (1991)
OH 2.25E-011	cm3/molc sec	25	EST	MEYLAN,WM & HOWARD,PH (1993)

CAS #: 000057-13-6	UREA

Formula: CH_4N_2O

Mol Weight: 60.06

MP (deg C): -95 FP (deg C):

BP (deg C): 91.5

BP pressure (mm Hg):

Property/ Value	Units	Temp	Data Type	Reference
WS 5.45E+005	mg/L	25	EXP	YALKOWSKY,SH & DANNENFELSER,RM (1992)
logP -2.11			EXP	HANSCH,C & LEO,AJ (1985)
VP 1.20E-005	mm Hg	25	EST	JONES,AH (1960)
DC 0.10	pKa	21	EXP	PERRIN,DD (1965)
HL 1.74E-012	atm m3/mol	25	EST	VP/WSOL
OH 4.00E-011	cm3/molc sec	25	EST	ATKINSON,R (1988)

CAS #: 000057-14-7	1,1-DIMETHYLHYDRAZINE

Formula: $C_2H_8N_2$

Mol Weight: 60.10

MP (deg C): -57 FP (deg C):

BP (deg C): 63.9

BP pressure (mm Hg):

Property/ Value	Units	Temp	Data Type	Reference
WS 1.00E+006	mg/L		EXP	BUDAVARI,S (1996)
logP -1.19			EST	MEYLAN,WM & HOWARD,PH (1995)
VP 1.57E+002	mm Hg	25	EXP	SCHIESSL,HW (1980)
DC 7.21	pKa	25	EXP	BRAUN,BA & ZIRROLI,JA (1983)
HL 1.24E-005	atm m3/mol	25	EST	VP/WSOL
OH 2.53E-012	cm3/molc sec	25	EST	MEYLAN,WM & HOWARD,PH (1993)

CAS #: 000057-15-8	B,B,B-TRICHLORO-T-BUTANOL

Formula: $C_4H_7Cl_3O$

Mol Weight: 177.46

MP (deg C): -100.7 FP (deg C):

BP (deg C): -78.1

BP pressure (mm Hg):

Property/ Value	Units	Temp	Data Type	Reference
WS 8.00E+003	mg/L	20	EXP	YALKOWSKY,SH & DANNENFELSER,RM (1992)
logP 2.03			EXP	HANSCH,C & LEO,AJ (1985)
VP 8.33E-001	mm Hg	25	EST	NEELY,WB & BLAU,GE (1985)
DC	pKa			
HL 2.75E-007	atm m3/mol	25	EST	MEYLAN,WM & HOWARD,PH (1991)
OH 1.17E-012	cm3/molc sec	25	EST	MEYLAN,WM & HOWARD,PH (1993)

STRYCHNINE

CAS #:	000057-24-9				
Formula:	$C_{21}H_{22}N_2O_2$				
Mol Weight:	334.42				
MP (deg C): -147.6			FP (deg C):		
BP (deg C): -36.6					
BP pressure (mm Hg):					

Property/Value	Units	Temp	Data Type	Reference
WS 1.60E+002	mg/L	25	EXP	SEIDELL,A (1941)
logP 1.93			EXP	HANSCH,C ET AL. (1995)
VP 2.93E-009	mm Hg	25	EST	NEELY,WB & BLAU,GE (1985)
DC 6.00	pKa	20	EXP	MERCK INDEX (1983)
HL 7.56E-014	atm m3/mol	25	EST	MEYLAN,WM & HOWARD,PH (1991)
OH 2.54E-010	cm3/molc sec	25	EST	MEYLAN,WM & HOWARD,PH (1993)

MORPHINE

CAS #:	000057-27-2				
Formula:	$C_{17}H_{19}NO_3$				
Mol Weight:	285.35				
MP (deg C): 89.5			FP (deg C):		
BP (deg C): 94					
BP pressure (mm Hg): 1.00E+001					

Property/Value	Units	Temp	Data Type	Reference
WS 1.49E+002	mg/L	20	EXP	YALKOWSKY,SH & DANNENFELSER,RM (1992)
logP 0.76			EXP	HANSCH,C & LEO,AJ (1985)
VP 1.69E-009	mm Hg	25	EST	NEELY,WB & BLAU,GE (1985)
DC	pKa			
HL 1.33E-016	atm m3/mol	25	EST	MEYLAN,WM & HOWARD,PH (1991)
OH 2.57E-010	cm3/molc sec	25	EST	MEYLAN,WM & HOWARD,PH (1993)

PHENYTOIN

CAS #:	000057-41-0				
Formula:	$C_{15}H_{12}N_2O_2$				
Mol Weight:	252.28				
MP (deg C): 295-298			FP (deg C):		
BP (deg C):					
BP pressure (mm Hg):					

Property/Value	Units	Temp	Data Type	Reference
WS 3.20E+001	mg/L	22	EXP	YALKOWSKY,SH & DANNENFELSER,RM (1992)
logP 2.47			EXP	HANSCH,C & LEO,AJ (1985)
VP 1.20E-010	mm Hg	25	EST	NEELY,WB & BLAU,GE (1985)
DC	pKa			
HL 1.02E-011	atm m3/mol	25	EST	MEYLAN,WM & HOWARD,PH (1991)
OH 1.06E-011	cm3/molc sec	25	EST	MEYLAN,WM & HOWARD,PH (1993)

DEMEROL

CAS #:	000057-42-1				
Formula:	$C_{15}H_{21}NO_2$				
Mol Weight:	247.34				
MP (deg C): 270			FP (deg C):		
BP (deg C):					
BP pressure (mm Hg):					

Property/Value	Units	Temp	Data Type	Reference
WS 3.22E+003	mg/L	30	EXP	YALKOWSKY,SH & DANNENFELSER,RM (1992)
logP 2.72			EXP	SANGSTER,J (1994)
VP 6.89E-005	mm Hg	25	EST	NEELY,WB & BLAU,GE (1985)
DC	pKa			
HL 9.61E-009	atm m3/mol	25	EST	MEYLAN,WM & HOWARD,PH (1991)
OH 9.74E-011	cm3/molc sec	25	EST	MEYLAN,WM & HOWARD,PH (1993)

AMOBARBITAL

CAS #:	000057-43-2				
Formula:	$C_{11}H_{18}N_2O_3$				
Mol Weight:	226.28				
MP (deg C): 29			FP (deg C):		
BP (deg C): 221					
BP pressure (mm Hg):					

Property/Value	Units	Temp	Data Type	Reference
WS 4.10E+003	mg/L	25	EXP	YALKOWSKY,SH & DANNENFELSER,RM (1992)
logP 2.07			EXP	HANSCH,C & LEO,AJ (1985)
VP 2.17E-011	mm Hg	25	EST	NEELY,WB & BLAU,GE (1985)
DC	pKa			
HL 8.44E-013	atm m3/mol	25	EST	MEYLAN,WM & HOWARD,PH (1991)
OH 1.44E-011	cm3/molc sec	25	EST	MEYLAN,WM & HOWARD,PH (1993)

BARBITAL

CAS #:	000057-44-3				
Formula:	$C_8H_{12}N_2O_3$				
Mol Weight:	184.20				
MP (deg C): 157.5			FP (deg C):		
BP (deg C): 250					
BP pressure (mm Hg): 2.20E+001					

Property/Value	Units	Temp	Data Type	Reference
WS 7.46E+003	mg/L	25	EXP	YALKOWSKY,SH & DANNENFELSER,RM (1992)
logP 0.65			EXP	HANSCH,C & LEO,AJ (1985)
VP 1.59E-010	mm Hg	25	EST	NEELY,WB & BLAU,GE (1985)
DC 8.14	pKa	15	EXP	KORTUM,G ET AL (1961)
HL 3.61E-013	atm m3/mol	25	EST	MEYLAN,WM & HOWARD,PH (1991)
OH 9.62E-012	cm3/molc sec	25	EST	MEYLAN,WM & HOWARD,PH (1993)

PHYSOSTIGMINE

CAS #:	000057-47-6				
Formula:	$C_{15}H_{21}N_3O_2$				
Mol Weight:	275.35				
MP (deg C): 105-106			FP (deg C):		
BP (deg C):					
BP pressure (mm Hg):					

Property/Value	Units	Temp	Data Type	Reference
WS 7.76E+003	mg/L	25	EST	MEYLAN,WM ET AL. (1996)
logP 1.58			EXP	HANSCH,C ET AL. (1995)
VP 3.66E-006	mm Hg	25	EST	NEELY,WB & BLAU,GE (1985)
DC	pKa			
HL 3.03E-011	atm m3/mol	25	EST	MEYLAN,WM & HOWARD,PH (1991)
OH 3.44E-010	cm3/molc sec	25	EST	MEYLAN,WM & HOWARD,PH (1993)

SUCROSE

CAS #:	000057-50-1				
Formula:	$C_{12}H_{22}O_{11}$				
Mol Weight:	342.30				
MP (deg C): 185.5			FP (deg C):		
BP (deg C):					
BP pressure (mm Hg):					

Property/Value	Units	Temp	Data Type	Reference
WS 2.12E+006	mg/L	25	EXP	YALKOWSKY,SH & DANNENFELSER,RM (1992)
logP -3.70			EXP	HANSCH,C & LEO,AJ (1985)
VP 5.15E-017	mm Hg	25	EST	NEELY,WB & BLAU,GE (1985)
DC	pKa			
HL 4.47E-022	atm m3/mol	25	EST	MEYLAN,WM & HOWARD,PH (1991)
OH 1.12E-010	cm3/molc sec	25	EST	MEYLAN,WM & HOWARD,PH (1993)

CAS #: 000057-53-4 — MEPROBAMATE

Formula: $C_9H_{18}N_2O_4$

Mol Weight: 218.25

MP (deg C): 104-106

FP (deg C):

BP (deg C):

BP pressure (mm Hg):

Property/Value	Units	Temp	Data Type	Reference
WS 3.30E+003	mg/L	25	EXP	YALKOWSKY,SH & DANNENFELSER,RM (1992)
logP 0.70			EXP	HANSCH,C & LEO,AJ (1985)
VP 3.05E-003	mm Hg	25	EST	NEELY,WB & BLAU,GE (1985)
DC	pKa			
HL 1.85E-010	atm m3/mol	25	EST	MEYLAN,WM & HOWARD,PH (1991)
OH 1.06E-011	cm3/molc sec	25	EST	MEYLAN,WM & HOWARD,PH (1993)

CAS #: 000057-55-6 — 1,2-PROPANEDIOL

Formula: $C_3H_8O_2$

Mol Weight: 76.10

MP (deg C): 113.5

FP (deg C):

BP (deg C): 310

BP pressure (mm Hg):

Property/Value	Units	Temp	Data Type	Reference
WS 1.00E+006	mg/L	25	EXP	MERCK INDEX (1976)
logP -0.92			EXP	HANSCH,C & LEO,AJ (1985)
VP 1.29E-001	mm Hg	25	EXP	DAUBERT,TE & DANNER,RP (1989)
DC 14.90	pKa		EXP	SERJEANT,EP & DEMPSEY,B (1979)
HL 1.29E-008	atm m3/mol	25	EST	VP/WSOL
OH 1.20E-011	cm3/molc sec	25	EXP	ATKINSON,R (1989)

CAS #: 000057-57-8 — BETA-PROPIOLACTONE

Formula: $C_3H_4O_2$

Mol Weight: 72.06

MP (deg C): -33.4

FP (deg C):

BP (deg C): 162

BP pressure (mm Hg):

Property/Value	Units	Temp	Data Type	Reference
WS 3.70E+005	mg/L	25	EXP	DEAN,JA (1985)
logP -0.80			EST	MEYLAN,WM & HOWARD,PH (1995)
VP 3.40E+000	mm Hg	25	EST	NEELY,WB & BLAU,GE (1985)
DC	pKa			
HL 1.28E-005	atm m3/mol	25	EST	MEYLAN,WM & HOWARD,PH (1991)
OH 3.57E-013	cm3/molc sec	25	EST	MEYLAN,WM & HOWARD,PH (1993)

CAS #: 000057-62-5 — AUREOMYCIN

Formula: $C_{22}H_{23}ClN_2O_8$

Mol Weight: 478.89

MP (deg C): 168-169

FP (deg C):

BP (deg C):

BP pressure (mm Hg):

Property/Value	Units	Temp	Data Type	Reference
WS 6.30E+002	mg/L	25	EXP	YALKOWSKY,SH & DANNENFELSER,RM (1992)
logP -0.62			EXP	SANGSTER,J (1993)
VP 5.22E-024	mm Hg	25	EST	NEELY,WB & BLAU,GE (1985)
DC 3.30	pKa	25	EXP	KORTUM,G ET AL (1961)
HL 3.45E-024	atm m3/mol	25	EST	MEYLAN,WM & HOWARD,PH (1991)
OH 2.09E-010	cm3/molc sec	25	EST	MEYLAN,WM & HOWARD,PH (1993)

CAS #: 000057-63-6 — 19-NORPREGNA-1,3,5(10)-TRIEN-20-YNE-3,17-DIOL,

Formula: $C_{20}H_{24}O_2$

Mol Weight: 296.41

MP (deg C): 141-146

FP (deg C):

BP (deg C):

BP pressure (mm Hg):

Property/Value	Units	Temp	Data Type	Reference
WS 1.16E+002	mg/L	25	EST	MEYLAN,WM ET AL. (1996)
logP 3.67			EXP	HANSCH,C ET AL. (1995)
VP 2.67E-009	mm Hg	25	EST	NEELY,WB & BLAU,GE (1985)
DC	pKa			
HL 7.94E-012	atm m3/mol	25	EST	MEYLAN,WM & HOWARD,PH (1991)
OH 1.24E-010	cm3/molc sec	25	EST	MEYLAN,WM & HOWARD,PH (1993)

CAS #: 000057-64-7 — PHYSOSTIGMINE, MONOSALICYLATE

Formula: $C_{22}H_{27}N_3O_5$

Mol Weight: 413.48

MP (deg C): 150.1

FP (deg C):

BP (deg C):

BP pressure (mm Hg):

Property/Value	Units	Temp	Data Type	Reference
WS 1.38E+004	mg/L	25	EXP	SEIDELL,A (1941)
logP 2.38			EST	MEYLAN,WM & HOWARD,PH (1995)
VP 3.10E-016	mm Hg	25	EST	NEELY,WB & BLAU,GE (1985)
DC	pKa			
HL 3.36E-022	atm m3/mol	25	EST	MEYLAN,WM & HOWARD,PH (1991)
OH 3.06E-010	cm3/molc sec	25	EST	MEYLAN,WM & HOWARD,PH (1993)

CAS #: 000057-66-9 — PROBENECID

Formula: $C_{13}H_{19}NO_4S$

Mol Weight: 285.36

MP (deg C): 194-196

FP (deg C):

BP (deg C):

BP pressure (mm Hg):

Property/Value	Units	Temp	Data Type	Reference
WS 2.71E+001	mg/L	25	EST	MEYLAN,WM ET AL. (1996)
logP 3.21			EXP	HANSCH,C & LEO,AJ (1985)
VP 3.34E-008	mm Hg	25	EST	NEELY,WB & BLAU,GE (1985)
DC	pKa			
HL 1.27E-010	atm m3/mol	25	EST	MEYLAN,WM & HOWARD,PH (1991)
OH 2.47E-011	cm3/molc sec	25	EST	MEYLAN,WM & HOWARD,PH (1993)

CAS #: 000057-67-0 — SULFANILYL GUANADINE

Formula: $C_7H_{10}N_4O_2S$

Mol Weight: 214.25

MP (deg C): 190-193

FP (deg C):

BP (deg C):

BP pressure (mm Hg):

Property/Value	Units	Temp	Data Type	Reference
WS 2.20E+003	mg/L	25	EXP	YALKOWSKY,SH & DANNENFELSER,RM (1992)
logP -1.22			EXP	HANSCH,C & LEO,AJ (1985)
VP 1.47E-007	mm Hg	25	EST	NEELY,WB & BLAU,GE (1985)
DC 11.25	pKa		EXP	PERRIN,DD (1965)
HL 1.01E-015	atm m3/mol	25	EST	MEYLAN,WM & HOWARD,PH (1991)
OH 4.40E-011	cm3/molc sec	25	EST	MEYLAN,WM & HOWARD,PH (1993)

000057-68-1 — SULFAMETHAZINE

Formula: $C_{12}H_{14}N_4O_2S$

Mol Weight: 278.33

MP (deg C): 176 FP (deg C):

BP (deg C):

BP pressure (mm Hg):

Property/Value	Units	Temp	Data Type	Reference
WS 4.30E+002	mg/L	25	EXP	YALKOWSKY,SH & DANNENFELSER,RM (1992)
logP 0.28			EXP	HANSCH,C & LEO,AJ (1985)
VP 8.62E-009	mm Hg	25	EST	NEELY,WB & BLAU,GE (1985)
DC	pKa			
HL 3.05E-013	atm m3/mol	25	EST	MEYLAN,WM & HOWARD,PH (1991)
OH 1.40E-010	cm3/molc sec	25	EST	MEYLAN,WM & HOWARD,PH (1993)

000057-74-9 — CHLORDANE

Formula: $C_{10}H_6Cl_8$

Mol Weight: 409.78

MP (deg C): 104-107 FP (deg C):

BP (deg C): 175

BP pressure (mm Hg): 2.00E+000

Property/Value	Units	Temp	Data Type	Reference
WS 5.60E-002	mg/L	25	EXP	SANBORN,JR ET AL. (1976)
logP 6.00			EXP	VEITH,GD ET AL. (1979)
VP 9.80E-006	mm Hg	25	EXP	WORTHING,CR & WALKER,SB (1983)
DC	pKa			
HL 4.86E-005	atm m3/mol	25	EXP	WARNER,HP ET AL. (1987)
OH 8.14E-012	cm3/molc sec	25	EST	MEYLAN,WM & HOWARD,PH (1993)

000057-83-0 — PROGESTERONE

Formula: $C_{21}H_{30}O_2$

Mol Weight: 314.47

MP (deg C): 121 FP (deg C):

BP (deg C):

BP pressure (mm Hg):

Property/Value	Units	Temp	Data Type	Reference
WS 8.84E+000	mg/L		EXP	YALKOWSKY,SH & DANNENFELSER,RM (1992)
logP 3.87			EXP	HANSCH,C & LEO,AJ (1985)
VP 1.30E-006	mm Hg	25	EST	NEELY,WB & BLAU,GE (1985)
DC	pKa			
HL 6.49E-008	atm m3/mol	25	EST	MEYLAN,WM & HOWARD,PH (1991)
OH 1.02E-010	cm3/molc sec	25	EST	MEYLAN,WM & HOWARD,PH (1993)

000057-88-5 — CHOLESTEROL

Formula: $C_{27}H_{46}O$

Mol Weight: 386.67

MP (deg C): 148.5 FP (deg C):

BP (deg C): 360

BP pressure (mm Hg):

Property/Value	Units	Temp	Data Type	Reference
WS 9.50E-002	mg/L	30	EXP	YALKOWSKY,SH & DANNENFELSER,RM (1992)
logP 8.74			EST	MEYLAN,WM & HOWARD,PH (1995)
VP 7.79E-010	mm Hg	25	EST	NEELY,WB & BLAU,GE (1985)
DC	pKa			
HL 1.67E-004	atm m3/mol	25	EST	MEYLAN,WM & HOWARD,PH (1991)
OH 1.26E-010	cm3/molc sec	25	EST	MEYLAN,WM & HOWARD,PH (1993)

000057-92-1 — STREPTOMYCIN

Formula: $C_{21}H_{39}N_7O_{12}$

Mol Weight: 581.58

MP (deg C): FP (deg C):

BP (deg C):

BP pressure (mm Hg):

Property/Value	Units	Temp	Data Type	Reference
WS 1.00E+006	mg/L	25	EST	MEYLAN,WM ET AL. (1996)
logP -10.88			EST	MEYLAN,WM & HOWARD,PH (1995)
VP 5.82E-028	mm Hg	25	EST	NEELY,WB & BLAU,GE (1985)
DC	pKa			
HL 8.41E-044	atm m3/mol	25	EST	MEYLAN,WM & HOWARD,PH (1991)
OH 5.33E-010	cm3/molc sec	25	EST	MEYLAN,WM & HOWARD,PH (1993)

000057-96-5 — SULFOXYPHENYL PYRAZOLIDINE

Formula: $C_{23}H_{20}N_2O_3S$

Mol Weight: 404.49

MP (deg C): 136-137 FP (deg C):

BP (deg C):

BP pressure (mm Hg):

Property/Value	Units	Temp	Data Type	Reference
WS 3.12E+001	mg/L	25	EST	MEYLAN,WM ET AL. (1996)
logP 2.30			EXP	HANSCH,C & LEO,AJ (1985)
VP 1.65E-014	mm Hg	25	EST	NEELY,WB & BLAU,GE (1985)
DC	pKa			
HL 5.01E-016	atm m3/mol	25	EST	MEYLAN,WM & HOWARD,PH (1991)
OH 9.98E-011	cm3/molc sec	25	EST	MEYLAN,WM & HOWARD,PH (1993)

000057-97-6 — 7,12-DIME-BENZ(A)ANTHRACENE

Formula: $C_{20}H_{16}$

Mol Weight: 256.35

MP (deg C): 122-123 FP (deg C):

BP (deg C): 422

BP pressure (mm Hg):

Property/Value	Units	Temp	Data Type	Reference
WS 6.10E-002	mg/L	25	EXP	MACKAY,D & SHIU,WY (1977)
logP 5.80			EXP	HANSCH,C & LEO,AJ (1985)
VP 5.63E-009	mm Hg	25	EXT	MURRAY,JJ ET AL. (1974)
DC	pKa			
HL 3.11E-008	atm m3/mol	25	EST	VP/WSOL
OH 2.00E-010	cm3/molc sec	25	EST	MEYLAN,WM & HOWARD,PH (1993)

000058-00-4 — 6A-BETA-APORPHINE-10,11-DIOL

Formula: $C_{17}H_{17}NO_2$

Mol Weight: 267.33

MP (deg C): 195 dec FP (deg C):

BP (deg C):

BP pressure (mm Hg):

Property/Value	Units	Temp	Data Type	Reference
WS 1.66E+004	mg/L	16	EXP	STEHPHEN,H & STEPHEN,T (1963)
logP 2.78			EST	MEYLAN,WM & HOWARD,PH (1995)
VP 1.27E-009	mm Hg	25	EST	NEELY,WB & BLAU,GE (1985)
DC	pKa			
HL 5.80E-016	atm m3/mol	25	EST	MEYLAN,WM & HOWARD,PH (1991)
OH 1.61E-010	cm3/molc sec	25	EST	MEYLAN,WM & HOWARD,PH (1993)

CAS #: 000058-08-2	CAFFEINE

Formula: $C_8H_{10}N_4O_2$

Mol Weight: 194.19

MP (deg C): 238 FP (deg C):

BP (deg C):

BP pressure (mm Hg):

Property/ Value	Units	Temp	Data Type	Reference
WS 2.10E+004	mg/L	25	EXP	STEPHEN,H & STEPHEN,T (1963)
logP -0.07			EXP	HANSCH,C & LEO,AJ (1985)
VP 1.50E+001	mm Hg	89	EXP	DAUBERT,TE & DANNER,RP (1989)
DC 14.00	pKa	25	EXP	WINDHOLZ,M ET AL. (1983); LESS THAN 14
HL 1.90E-019	atm m3/mol	25	EST	MEYLAN,WM & HOWARD,PH (1991)
OH 1.52E-010	cm3/molc sec	25	EST	ATKINSON,R (1988)

CAS #: 000058-14-0	PYRIMETHAMINE

Formula: $C_{12}H_{13}ClN_4$

Mol Weight: 248.72

MP (deg C): 233-234 FP (deg C):

BP (deg C):

BP pressure (mm Hg):

Property/ Value	Units	Temp	Data Type	Reference
WS 1.21E+002	mg/L	25	EST	MEYLAN,WM ET AL. (1996)
logP 2.69			EXP	HANSCH,C & LEO,AJ (1985)
VP 6.23E-008	mm Hg	25	EST	NEELY,WB & BLAU,GE (1985)
DC 7.34	pKa	20	EXP	PERRIN,DD (1972)
HL 1.08E-010	atm m3/mol	25	EST	MEYLAN,WM & HOWARD,PH (1991)
OH 2.01E-010	cm3/molc sec	25	EST	MEYLAN,WM & HOWARD,PH (1993)

CAS #: 000058-15-1	AMINOPYRINE

Formula: $C_{13}H_{17}N_3O$

Mol Weight: 231.30

MP (deg C): 107-109 FP (deg C):

BP (deg C):

BP pressure (mm Hg):

Property/ Value	Units	Temp	Data Type	Reference
WS 5.30E+004	mg/L	20	EXP	YALKOWSKY,SH & DANNENFELSER,RM (1992)
logP 1.00			EXP	HANSCH,C ET AL. (1995)
VP 2.77E-006	mm Hg	25	EST	NEELY,WB & BLAU,GE (1985)
DC	pKa			
HL 1.38E-011	atm m3/mol	25	EST	MEYLAN,WM & HOWARD,PH (1991)
OH 9.62E-011	cm3/molc sec	25	EST	MEYLAN,WM & HOWARD,PH (1993)

CAS #: 000058-18-4	17-A-METHYL TESTOSTERONE

Formula: $C_{20}H_{30}O_2$

Mol Weight: 302.46

MP (deg C): 161-166 FP (deg C):

BP (deg C):

BP pressure (mm Hg):

Property/ Value	Units	Temp	Data Type	Reference
WS 3.39E+001	mg/L	25	EXP	YALKOWSKY,SH & DANNENFELSER,RM (1992)
logP 3.36			EXP	HANSCH,C ET AL. (1995)
VP 1.85E-008	mm Hg	25	EST	NEELY,WB & BLAU,GE (1985)
DC	pKa			
HL 4.68E-009	atm m3/mol	25	EST	MEYLAN,WM & HOWARD,PH (1991)
OH 1.01E-010	cm3/molc sec	25	EST	MEYLAN,WM & HOWARD,PH (1993)

CAS #: 000058-19-5	DROMOSTANOLONE

Formula: $C_{20}H_{32}O_2$

Mol Weight: 304.48

MP (deg C): FP (deg C):

BP (deg C):

BP pressure (mm Hg):

Property/ Value	Units	Temp	Data Type	Reference
WS 1.46E+001	mg/L	25	EST	MEYLAN,WM ET AL. (1996)
logP 3.99			EXP	HANSCH,C & LEO,AJ (1985)
VP 1.67E-008	mm Hg	25	EST	NEELY,WB & BLAU,GE (1985)
DC	pKa			
HL 8.46E-009	atm m3/mol	25	EST	MEYLAN,WM & HOWARD,PH (1991)
OH 3.95E-011	cm3/molc sec	25	EST	MEYLAN,WM & HOWARD,PH (1993)

CAS #: 000058-22-0	TESTOSTERONE

Formula: $C_{19}H_{28}O_2$

Mol Weight: 288.43

MP (deg C): 155 FP (deg C):

BP (deg C):

BP pressure (mm Hg):

Property/ Value	Units	Temp	Data Type	Reference
WS 3.00E+001	mg/L		EXP	YALKOWSKY,SH & DANNENFELSER,RM (1992)
logP 3.32			EXP	HANSCH,C & LEO,AJ (1985)
VP 2.23E-008	mm Hg	25	EST	NEELY,WB & BLAU,GE (1985)
DC	pKa			
HL 3.53E-009	atm m3/mol	25	EST	MEYLAN,WM & HOWARD,PH (1991)
OH 1.06E-010	cm3/molc sec	25	EST	MEYLAN,WM & HOWARD,PH (1993)

CAS #: 000058-25-3	LIBRIUM

Formula: $C_{16}H_{14}ClN_3O$

Mol Weight: 299.76

MP (deg C): 236-236.5 FP (deg C):

BP (deg C):

BP pressure (mm Hg):

Property/ Value	Units	Temp	Data Type	Reference
WS 2.00E+003	mg/L		EXP	YALKOWSKY,SH & DANNENFELSER,RM (1992)
logP 2.44			EXP	HANSCH,C & LEO,AJ (1985)
VP 1.03E-011	mm Hg	25	EST	NEELY,WB & BLAU,GE (1985)
DC	pKa			
HL 4.24E-017	atm m3/mol	25	EST	MEYLAN,WM & HOWARD,PH (1991)
OH 8.26E-011	cm3/molc sec	25	EST	MEYLAN,WM & HOWARD,PH (1993)

CAS #: 000058-27-5	2-METHYL-1,4-NAPHTHOQUINONE

Formula: $C_{11}H_8O_2$

Mol Weight: 172.19

MP (deg C): 105-107 FP (deg C):

BP (deg C):

BP pressure (mm Hg):

Property/ Value	Units	Temp	Data Type	Reference
WS 1.60E+002	mg/L	30	EXP	YALKOWSKY,SH & DANNENFELSER,RM (1992)
logP 2.20			EXP	HANSCH,C & LEO,AJ (1985)
VP 1.88E-004	mm Hg	25	EST	NEELY,WB & BLAU,GE (1985)
DC	pKa			
HL 3.09E-009	atm m3/mol	25	EST	MEYLAN,WM & HOWARD,PH (1991)
OH 1.15E-011	cm3/molc sec	25	EST	MEYLAN,WM & HOWARD,PH (1993)

CAS #: 000058-37-7				AMINOPROMAZINE

Formula: $C_{19}H_{25}N_3S$

Mol Weight: 327.50

MP (deg C): 166-170 FP (deg C):

BP (deg C):

BP pressure (mm Hg):

Property/ Value	Units	Temp	Data Type	Reference
WS 5.76E-001	mg/L	25	EST	MEYLAN,WM ET AL. (1996)
logP 4.76			EXP	HANSCH,C & LEO,AJ (1985)
VP 9.82E-008	mm Hg	25	EST	NEELY,WB & BLAU,GE (1985)
DC	pKa			
HL 2.47E-013	atm m3/mol	25	EST	MEYLAN,WM & HOWARD,PH (1991)
OH 3.66E-010	cm3/molc sec	25	EST	MEYLAN,WM & HOWARD,PH (1993)

CAS #: 000058-39-9				PERPHENAZINE

Formula: $C_{21}H_{26}CIN_3OS$

Mol Weight: 403.98

MP (deg C): 94-100 FP (deg C):

BP (deg C): 278-281

BP pressure (mm Hg): 1.00E+000

Property/ Value	Units	Temp	Data Type	Reference
WS 2.83E+001	mg/L	24	EXP	YALKOWSKY,SH & DANNENFELSER,RM (1992)
logP 4.20			EXP	HANSCH,C ET AL. (1995)
VP 2.84E-013	mm Hg	25	EST	NEELY,WB & BLAU,GE (1985)
DC	pKa			
HL 5.19E-018	atm m3/mol	25	EST	MEYLAN,WM & HOWARD,PH (1991)
OH 3.66E-010	cm3/molc sec	25	EST	MEYLAN,WM & HOWARD,PH (1993)

CAS #: 000058-40-2				PROMAZINE

Formula: $C_{17}H_{20}N_2S$

Mol Weight: 284.43

MP (deg C): FP (deg C):

BP (deg C): 203-210

BP pressure (mm Hg): 3.00E-001

Property/ Value	Units	Temp	Data Type	Reference
WS 1.42E+001	mg/L	24	EXP	YALKOWSKY,SH & DANNENFELSER,RM (1992)
logP 4.55			EXP	HANSCH,C ET AL. (1995)
VP 7.89E-007	mm Hg	25	EST	NEELY,WB & BLAU,GE (1985)
DC	pKa			
HL 4.98E-010	atm m3/mol	25	EST	MEYLAN,WM & HOWARD,PH (1991)
OH 2.71E-010	cm3/molc sec	25	EST	MEYLAN,WM & HOWARD,PH (1993)

CAS #: 000058-55-9				THEOPHYLLINE

Formula: $C_7H_8N_4O_2$

Mol Weight: 180.17

MP (deg C): 270-274 FP (deg C):

BP (deg C):

BP pressure (mm Hg):

Property/ Value	Units	Temp	Data Type	Reference
WS 7.36E+003	mg/L	25	EXP	YALKOWSKY,SH & DANNENFELSER,RM (1992)
logP -0.02			EXP	HANSCH,C & LEO,AJ (1985)
VP 5.12E-009	mm Hg	25	EST	NEELY,WB & BLAU,GE (1985)
DC 8.81	pKa		EXP	KORTUM,G ET AL (1961)
HL 8.62E-020	atm m3/mol	25	EST	MEYLAN,WM & HOWARD,PH (1991)
OH 5.38E-011	cm3/molc sec	25	EST	MEYLAN,WM & HOWARD,PH (1993)

CAS #: 000058-56-0				PYRIDOXINE HYDROCHLORIDE

Formula: $C_8H_{12}CINO_3$

Mol Weight: 205.64

MP (deg C): 204-206 FP (deg C):

BP (deg C):

BP pressure (mm Hg):

Property/ Value	Units	Temp	Data Type	Reference
WS 2.20E+005	mg/L		EXP	COFFEN,DL (1978)
logP -4.32			EST	MEYLAN,WM & HOWARD,PH (1995)
VP 6.41E-008	mm Hg	25	EST	NEELY,WB & BLAU,GE (1985)
DC	pKa			
HL 1.32E-021	atm m3/mol	25	EST	MEYLAN,WM & HOWARD,PH (1991)
OH 4.06E-011	cm3/molc sec	25	EST	MEYLAN,WM & HOWARD,PH (1993)

CAS #: 000058-61-7				ADENOSINE

Formula: $C_{10}H_{13}N_5O_4$

Mol Weight: 267.25

MP (deg C): 234-235 FP (deg C):

BP (deg C):

BP pressure (mm Hg):

Property/ Value	Units	Temp	Data Type	Reference
WS 8.23E+003	mg/L	25	EST	MEYLAN,WM ET AL. (1996)
logP -1.05			EXP	HANSCH,C ET AL. (1995)
VP 7.00E-015	mm Hg	25	EST	NEELY,WB & BLAU,GE (1985)
DC	pKa			
HL 1.11E-022	atm m3/mol	25	EST	MEYLAN,WM & HOWARD,PH (1991)
OH 2.47E-010	cm3/molc sec	25	EST	MEYLAN,WM & HOWARD,PH (1993)

CAS #: 000058-63-9				INOSINE

Formula: $C_{10}H_{12}N_4O_5$

Mol Weight: 268.23

MP (deg C): 90 FP (deg C):

BP (deg C):

BP pressure (mm Hg):

Property/ Value	Units	Temp	Data Type	Reference
WS 1.58E+004	mg/L	20	EXP	YALKOWSKY,SH & DANNENFELSER,RM (1992)
logP -2.10			EXP	HANSCH,C ET AL. (1995)
VP 1.40E-014	mm Hg	25	EST	NEELY,WB & BLAU,GE (1985)
DC	pKa			
HL 3.28E-023	atm m3/mol	25	EST	MEYLAN,WM & HOWARD,PH (1991)
OH 2.47E-010	cm3/molc sec	25	EST	MEYLAN,WM & HOWARD,PH (1993)

CAS #: 000058-71-9				CEPHALOTHIN

Formula: $C_{16}H_{18}N_2O_6S_2$

Mol Weight: 398.46

MP (deg C): FP (deg C):

BP (deg C):

BP pressure (mm Hg):

Property/ Value	Units	Temp	Data Type	Reference
WS 1.58E+002	mg/L	25	EST	MEYLAN,WM ET AL. (1996)
logP 0.00			EXP	HANSCH,C & LEO,AJ (1985)
VP 1.67E-014	mm Hg	25	EST	NEELY,WB & BLAU,GE (1985)
DC	pKa			
HL 1.74E-017	atm m3/mol	25	EST	MEYLAN,WM & HOWARD,PH (1991)
OH 1.35E-010	cm3/molc sec	25	EST	MEYLAN,WM & HOWARD,PH (1993)

DIPHENHYDRAMINE

CAS #: 000058-73-1

Formula: $C_{17}H_{21}NO$
Mol Weight: 255.36
MP (deg C): 166-170
FP (deg C):
BP (deg C): 150-165
BP pressure (mm Hg): 2.00E+000

Property/Value	Units	Temp	Data Type	Reference
WS 3.63E+002	mg/L	25	EST	MEYLAN,WM ET AL. (1996)
logP 3.27			EXP	HANSCH,C ET AL. (1995)
VP 3.70E-005	mm Hg	25	EST	NEELY,WB & BLAU,GE (1985)
DC	pKa			
HL 3.70E-009	atm m3/mol	25	EST	MEYLAN,WM & HOWARD,PH (1991)
OH 1.27E-010	cm3/molc sec	25	EST	MEYLAN,WM & HOWARD,PH (1993)

PAPAVERINE

CAS #: 000058-74-2

Formula: $C_{20}H_{21}NO_4$
Mol Weight: 339.39
MP (deg C): 147
FP (deg C):
BP (deg C):
BP pressure (mm Hg):

Property/Value	Units	Temp	Data Type	Reference
WS 5.09E+000	mg/L	25	EST	MEYLAN,WM ET AL. (1996)
logP 3.69			EST	MEYLAN,WM & HOWARD,PH (1995)
VP 7.49E-009	mm Hg	25	EST	NEELY,WB & BLAU,GE (1985)
DC	pKa			
HL 7.51E-013	atm m3/mol	25	EST	MEYLAN,WM & HOWARD,PH (1991)
OH 2.50E-010	cm3/molc sec	25	EST	MEYLAN,WM & HOWARD,PH (1993)

GAMMA-HEXACHLOROCYCLOHEXANE

CAS #: 000058-89-9

Formula: $C_6H_6Cl_6$
Mol Weight: 290.83
MP (deg C): 112-113
FP (deg C):
BP (deg C): 323.4
BP pressure (mm Hg):

Property/Value	Units	Temp	Data Type	Reference
WS 7.30E+000	mg/L	25	EXP	RICHARDSON,LT & MILLER,DM (1960)
logP 3.72			EXP	HANSCH,C & LEO,AJ (1985)
VP 4.10E-004	mm Hg	25	EXP	BIDLEMAN,TF (1984)
DC	pKa			
HL 2.00E-006	atm m3/mol	23	EXP	FENDINGER,NJ & GLOTFELTY,DE (1988)
OH 1.36E-012	cm3/molc sec	25	EST	MEYLAN,WM & HOWARD,PH (1993)

2,3,4,6-TETRACHLOROPHENOL

CAS #: 000058-90-2

Formula: $C_6H_2Cl_4O$
Mol Weight: 231.89
MP (deg C): 69-70
FP (deg C):
BP (deg C): 150
BP pressure (mm Hg): 1.50E+001

Property/Value	Units	Temp	Data Type	Reference
WS 1.00E+003	mg/L	25	EXP	FREITER,ER (1978A)
logP 4.45			EXP	HANSCH,C & LEO,AJ (1985)
VP 4.23E-003	mm Hg	25	EXP	BIDLEMAN,TF & RENBERG,L (1985)
DC	pKa			
HL 1.29E-004	atm m3/mol	25	EST	VP/WSOL
OH 1.80E-013	cm3/molc sec	25	EST	MEYLAN,WM & HOWARD,PH (1993)

HYDROCHLOROTHIAZIDE

CAS #: 000058-93-5

Formula: $C_7H_8ClN_3O_4S_2$
Mol Weight: 297.74
MP (deg C): 273-275
FP (deg C):
BP (deg C):
BP pressure (mm Hg):

Property/Value	Units	Temp	Data Type	Reference
WS 7.22E+002	mg/L	25	EXP	YALKOWSKY,SH & DANNENFELSER,RM (1992)
logP -0.07			EXP	HANSCH,C & LEO,AJ (1985)
VP 1.32E-009	mm Hg	25	EST	NEELY,WB & BLAU,GE (1985)
DC	pKa			
HL 4.39E-012	atm m3/mol	25	EST	MEYLAN,WM & HOWARD,PH (1991)
OH 8.22E-011	cm3/molc sec	25	EST	MEYLAN,WM & HOWARD,PH (1993)

CHLOROTHIAZIDE

CAS #: 000058-94-6

Formula: $C_7H_6ClN_3O_4S_2$
Mol Weight: 295.72
MP (deg C): 350 dec
FP (deg C):
BP (deg C):
BP pressure (mm Hg):

Property/Value	Units	Temp	Data Type	Reference
WS 2.83E+002	mg/L	25	EXP	YALKOWSKY,SH & DANNENFELSER,RM (1992)
logP -0.24			EXP	HANSCH,C & LEO,AJ (1985)
VP 2.09E-011	mm Hg	25	EST	NEELY,WB & BLAU,GE (1985)
DC	pKa			
HL 4.05E-012	atm m3/mol	25	EST	MEYLAN,WM & HOWARD,PH (1991)
OH 1.81E-011	cm3/molc sec	25	EST	MEYLAN,WM & HOWARD,PH (1993)

URIDINE

CAS #: 000058-96-8

Formula: $C_9H_{12}N_2O_6$
Mol Weight: 244.21
MP (deg C): 165
FP (deg C):
BP (deg C):
BP pressure (mm Hg):

Property/Value	Units	Temp	Data Type	Reference
WS 6.12E+004	mg/L	25	EST	MEYLAN,WM ET AL. (1996)
logP -1.98			EXP	HANSCH,C & LEO,AJ (1985)
VP 1.34E-015	mm Hg	25	EST	NEELY,WB & BLAU,GE (1985)
DC	pKa			
HL 2.82E-019	atm m3/mol	25	EST	MEYLAN,WM & HOWARD,PH (1991)
OH 9.17E-011	cm3/molc sec	25	EST	MEYLAN,WM & HOWARD,PH (1993)

METHOTREXATE

CAS #: 000059-05-2

Formula: $C_{20}H_{22}N_8O_5$
Mol Weight: 454.45
MP (deg C):
FP (deg C):
BP (deg C):
BP pressure (mm Hg):

Property/Value	Units	Temp	Data Type	Reference
WS 2.60E+003	mg/L	25	EST	MEYLAN,WM ET AL. (1996)
logP -1.85			EXP	HANSCH,C ET AL. (1995)
VP 2.09E-019	mm Hg	25	EST	NEELY,WB & BLAU,GE (1985)
DC	pKa			
HL 1.54E-031	atm m3/mol	25	EST	MEYLAN,WM & HOWARD,PH (1991)
OH 3.17E-010	cm3/molc sec	25	EST	MEYLAN,WM & HOWARD,PH (1993)

CAS #: 000059-07-4	2-ETHOXY-4-AMINOBENZOIC ACID

Formula: $C_9H_{11}NO_3$
Mol Weight: 181.19
MP (deg C):
FP (deg C):
BP (deg C):
BP pressure (mm Hg):

Property/Value	Units	Temp	Data Type	Reference
WS 7.76E+003	mg/L	25	EST	MEYLAN,WM ET AL. (1996)
logP 0.99			EXP	HANSCH,C & LEO,AJ (1985)
VP 1.57E-005	mm Hg	25	EST	NEELY,WB & BLAU,GE (1985)
DC	pKa			
HL 3.01E-012	atm m3/mol	25	EST	MEYLAN,WM & HOWARD,PH (1991)
OH 2.07E-010	cm3/molc sec	25	EST	MEYLAN,WM & HOWARD,PH (1993)

CAS #: 000059-14-3	5-BROMODEOXYURIDINE

Formula: $C_9H_{11}BrN_2O_5$
Mol Weight: 307.11
MP (deg C): 191-194
FP (deg C):
BP (deg C):
BP pressure (mm Hg):

Property/Value	Units	Temp	Data Type	Reference
WS 9.67E+002	mg/L	25	EST	MEYLAN,WM ET AL. (1996)
logP -0.29			EXP	HANSCH,C & LEO,AJ (1985)
VP 4.73E-015	mm Hg	25	EST	NEELY,WB & BLAU,GE (1985)
DC	pKa			
HL 1.54E-018	atm m3/mol	25	EST	MEYLAN,WM & HOWARD,PH (1991)
OH 8.25E-011	cm3/molc sec	25	EST	MEYLAN,WM & HOWARD,PH (1993)

CAS #: 000059-23-4	GALACTOSE

Formula: $C_6H_{12}O_6$
Mol Weight: 180.16
MP (deg C): 167
FP (deg C):
BP (deg C):
BP pressure (mm Hg):

Property/Value	Units	Temp	Data Type	Reference
WS 6.83E+005	mg/L		EXP	YALKOWSKY,SH & DANNENFELSER,RM (1992)
logP -2.43			EST	MEYLAN,WM & HOWARD,PH (1995)
VP 1.82E-008	mm Hg	25	EST	NEELY,WB & BLAU,GE (1985)
DC 12.92	pKa	0	EXP	KORTUM,G ET AL (1961)
HL 5.88E-011	atm m3/mol	25	EST	MEYLAN,WM & HOWARD,PH (1991)
OH 8.49E-011	cm3/molc sec	25	EST	MEYLAN,WM & HOWARD,PH (1993)

CAS #: 000059-26-7	NIKETHAMIDE

Formula: $C_{10}H_{14}N_2O$
Mol Weight: 178.24
MP (deg C): 24-26
FP (deg C):
BP (deg C): 296-300
BP pressure (mm Hg):

Property/Value	Units	Temp	Data Type	Reference
WS 2.94E+004	mg/L	25	EST	MEYLAN,WM ET AL. (1996)
logP 0.33			EXP	HANSCH,C & LEO,AJ (1985)
VP 4.68E-004	mm Hg	25	EST	NEELY,WB & BLAU,GE (1985)
DC	pKa			
HL 2.46E-011	atm m3/mol	25	EST	MEYLAN,WM & HOWARD,PH (1991)
OH 2.33E-011	cm3/molc sec	25	EST	MEYLAN,WM & HOWARD,PH (1993)

CAS #: 000059-30-3	FOLIC ACID

Formula: $C_{19}H_{19}N_7O_6$
Mol Weight: 441.41
MP (deg C): 250 dec
FP (deg C):
BP (deg C):
BP pressure (mm Hg):

Property/Value	Units	Temp	Data Type	Reference
WS 1.60E+000	mg/L	25	EXP	MERCK INDEX (1983)
logP -2.00			EST	MEYLAN,WM & HOWARD,PH (1995)
VP 3.37E-021	mm Hg	25	EST	NEELY,WB & BLAU,GE (1985)
DC	pKa			
HL 3.55E-033	atm m3/mol	25	EST	MEYLAN,WM & HOWARD,PH (1991)
OH 2.60E-010	cm3/molc sec	25	EST	ATKINSON,R (1988)

CAS #: 000059-31-4	2-(1H)-QUINOLINONE

Formula: C_9H_7NO
Mol Weight: 145.16
MP (deg C): 178
FP (deg C):
BP (deg C):
BP pressure (mm Hg):

Property/Value	Units	Temp	Data Type	Reference
WS 1.05E+003	mg/L	20	EXP	YALKOWSKY,SH & DANNENFELSER,RM (1992)
logP 1.26			EXP	HANSCH,C & LEO,AJ (1985)
VP 4.47E-005	mm Hg	25	EST	NEELY,WB & BLAU,GE (1985)
DC	pKa			
HL 6.62E-010	atm m3/mol	25	EST	MEYLAN,WM & HOWARD,PH (1991)
OH 5.14E-011	cm3/molc sec	25	EST	MEYLAN,WM & HOWARD,PH (1993)

CAS #: 000059-40-5	BENZENESULFONAMIDE, 4-AMINO-N-2-QUINOXALINYL-

Formula: $C_{14}H_{12}N_4O_2S$
Mol Weight: 300.34
MP (deg C): 247-248
FP (deg C):
BP (deg C):
BP pressure (mm Hg):

Property/Value	Units	Temp	Data Type	Reference
WS 4.49E+002	mg/L	25	EST	MEYLAN,WM ET AL. (1996)
logP 1.68			EXP	HANSCH,C ET AL. (1995)
VP 3.00E-010	mm Hg	25	EST	NEELY,WB & BLAU,GE (1985)
DC	pKa			
HL 4.35E-015	atm m3/mol	25	EST	MEYLAN,WM & HOWARD,PH (1991)
OH 6.68E-011	cm3/molc sec	25	EST	MEYLAN,WM & HOWARD,PH (1993)

CAS #: 000059-46-1	PROCAINE

Formula: $C_{13}H_{20}N_2O_2$
Mol Weight: 236.32
MP (deg C): 61
FP (deg C):
BP (deg C):
BP pressure (mm Hg):

Property/Value	Units	Temp	Data Type	Reference
WS 9.45E+003	mg/L	30	EXP	YALKOWSKY,SH & DANNENFELSER,RM (1992)
logP 1.92			EXP	HANSCH,C ET AL. (1995)
VP 2.03E-005	mm Hg	25	EST	NEELY,WB & BLAU,GE (1985)
DC 8.05	pKa	15	EXP	PERRIN,DD (1965)
HL 1.42E-011	atm m3/mol	25	EST	MEYLAN,WM & HOWARD,PH (1991)
OH 1.31E-010	cm3/molc sec	25	EST	MEYLAN,WM & HOWARD,PH (1993)

000059-47-2 — 1,2-PROPANEDIOL-3-(2-TOLYLOXY)

Formula:	$C_{10}H_{14}O_3$
Mol Weight:	182.22
MP (deg C):	70-72
FP (deg C):	
BP (deg C):	
BP pressure (mm Hg):	

Property/Value	Units	Temp	Data Type	Reference
WS 3.36E+003	mg/L	25	EST	MEYLAN,WM ET AL. (1996)
logP 1.41			EXP	HANSCH,C & LEO,AJ (1985)
VP 5.43E-006	mm Hg	25	EST	NEELY,WB & BLAU,GE (1985)
DC	pKa			
HL 8.29E-010	atm m3/mol	25	EST	MEYLAN,WM & HOWARD,PH (1991)
OH 4.81E-011	cm3/molc sec	25	EST	MEYLAN,WM & HOWARD,PH (1993)

000059-48-3 — OXINDOLE

Formula:	C_8H_7NO
Mol Weight:	133.15
MP (deg C):	128
FP (deg C):	
BP (deg C):	227
BP pressure (mm Hg):	2.30E+001

Property/Value	Units	Temp	Data Type	Reference
WS 9.13E+003	mg/L	25	EST	MEYLAN,WM ET AL. (1996)
logP 1.16			EXP	HANSCH,C & LEO,AJ (1985)
VP 1.32E-004	mm Hg	25	EST	NEELY,WB & BLAU,GE (1985)
DC	pKa			
HL 2.26E-009	atm m3/mol	25	EST	MEYLAN,WM & HOWARD,PH (1991)
OH 1.57E-011	cm3/molc sec	25	EST	MEYLAN,WM & HOWARD,PH (1993)

000059-49-4 — O-PHENYLENE CARBAMATE

Formula:	$C_7H_5NO_2$
Mol Weight:	135.12
MP (deg C):	138
FP (deg C):	
BP (deg C):	335
BP pressure (mm Hg):	

Property/Value	Units	Temp	Data Type	Reference
WS 8.96E+003	mg/L	25	EST	MEYLAN,WM ET AL. (1996)
logP 1.16			EXP	HANSCH,C & LEO,AJ (1985)
VP 7.55E-004	mm Hg	25	EST	NEELY,WB & BLAU,GE (1985)
DC	pKa			
HL 3.66E-008	atm m3/mol	25	EST	MEYLAN,WM & HOWARD,PH (1991)
OH 3.13E-011	cm3/molc sec	25	EST	MEYLAN,WM & HOWARD,PH (1993)

000059-50-7 — 3-METHYL-4-CHLOROPHENOL

Formula:	C_7H_7ClO
Mol Weight:	142.59
MP (deg C):	55.5
FP (deg C):	
BP (deg C):	235
BP pressure (mm Hg):	

Property/Value	Units	Temp	Data Type	Reference
WS 3.83E+003	mg/L	25	EXP	KUHNE,R ET AL. (1995)
logP 3.10			EXP	HANSCH,C & LEO,AJ (1985)
VP 5.00E-002	mm Hg	20	EXP	MABEY,WR ET AL. (1981)
DC 9.55	pKa	25	EXP	SERJEANT,EP & DEMPSEY,B (1979)
HL 2.45E-006	atm m3/mol	25	EST	VP/WSOL
OH 2.57E-011	cm3/molc sec	25	EST	MEYLAN,WM & HOWARD,PH (1993)

000059-61-0 — DICHLORISOPROTERENOL

Formula:	$C_{11}H_{15}Cl_2NO$
Mol Weight:	248.15
MP (deg C):	
FP (deg C):	
BP (deg C):	125-135
BP pressure (mm Hg):	1.00E-002

Property/Value	Units	Temp	Data Type	Reference
WS 3.61E+002	mg/L	25	EST	MEYLAN,WM ET AL. (1996)
logP 3.32			EXP	HANSCH,C ET AL. (1995)
VP 4.28E-006	mm Hg	25	EST	NEELY,WB & BLAU,GE (1985)
DC	pKa			
HL 6.31E-011	atm m3/mol	25	EST	MEYLAN,WM & HOWARD,PH (1991)
OH 1.02E-010	cm3/molc sec	25	EST	MEYLAN,WM & HOWARD,PH (1993)

000059-63-2 — ISOCARBOXAZID

Formula:	$C_{12}H_{14}N_3O_2$
Mol Weight:	232.26
MP (deg C):	105-106
FP (deg C):	
BP (deg C):	
BP pressure (mm Hg):	

Property/Value	Units	Temp	Data Type	Reference
WS 1.60E+003	mg/L	25	EST	MEYLAN,WM ET AL. (1996)
logP 1.49			EXP	HANSCH,C & LEO,AJ (1985)
VP 1.63E-007	mm Hg	25	EST	NEELY,WB & BLAU,GE (1985)
DC	pKa			
HL 1.90E-014	atm m3/mol	25	EST	MEYLAN,WM & HOWARD,PH (1991)
OH 9.22E-011	cm3/molc sec	25	EST	MEYLAN,WM & HOWARD,PH (1993)

000059-66-5 — ACETAZOLEAMIDE

Formula:	$C_4H_6N_4O_3S_2$
Mol Weight:	222.25
MP (deg C):	258-259
FP (deg C):	
BP (deg C):	
BP pressure (mm Hg):	

Property/Value	Units	Temp	Data Type	Reference
WS 9.80E+002	mg/L	30	EXP	YALKOWSKY,SH & DANNENFELSER,RM (1992)
logP -0.26			EXP	HANSCH,C & LEO,AJ (1985)
VP 4.09E-009	mm Hg	25	EST	NEELY,WB & BLAU,GE (1985)
DC	pKa			
HL 1.40E-016	atm m3/mol	25	EST	MEYLAN,WM & HOWARD,PH (1991)
OH 8.02E-013	cm3/molc sec	25	EST	MEYLAN,WM & HOWARD,PH (1993)

000059-67-6 — NICOTINIC ACID

Formula:	$C_6H_5NO_2$
Mol Weight:	123.11
MP (deg C):	236.6
FP (deg C):	
BP (deg C):	SUBLIMES
BP pressure (mm Hg):	

Property/Value	Units	Temp	Data Type	Reference
WS 1.80E+004	mg/L	25	EXP	YALKOWSKY,SH & DANNENFELSER,RM (1992)
logP 0.36			EXP	SANGSTER,J (1993)
VP 5.70E-006	mm Hg	25	EST	WSOL X HL
DC 4.75	pKa	25	EXP	DEAN,JA (1985)
HL 5.11E-011	atm m3/mol	25	EST	MEYLAN,WM & HOWARD,PH (1991)
OH 1.72E-013	cm3/molc sec	25	EST	ATKINSON,R (1988)

28

CAS #: 000059-85-8 — P-(CHLOROMERCURI)BENZOIC ACID

Formula: $C_7H_5ClHgO_2$

Mol Weight: 357.16

MP (deg C):
FP (deg C):
BP (deg C):
BP pressure (mm Hg):

Property/Value	Units	Temp	Data Type	Reference
WS 3.03E+002	mg/L	25	EST	MEYLAN,WM ET AL. (1996)
logP 1.48			EST	MEYLAN,WM & HOWARD,PH (1995)
VP 1.48E-005	mm Hg	25	EST	NEELY,WB & BLAU,GE (1985)
DC	pKa			
HL	atm m3/mol			
OH 1.24E-012	cm3/molc sec	25	EST	MEYLAN,WM & HOWARD,PH (1993)

CAS #: 000059-87-0 — NITROFURAZONE

Formula: $C_6H_6N_4O_4$

Mol Weight: 198.14

MP (deg C): 236-240 de
FP (deg C):
BP (deg C):
BP pressure (mm Hg):

Property/Value	Units	Temp	Data Type	Reference
WS 5.68E+002	mg/L	25	EST	MEYLAN,WM ET AL. (1996)
logP 0.23			EXP	HANSCH,C & LEO,AJ (1985)
VP 4.31E-006	mm Hg	25	EST	NEELY,WB & BLAU,GE (1985)
DC	pKa			
HL 3.10E-013	atm m3/mol	25	EST	MEYLAN,WM & HOWARD,PH (1991)
OH 1.59E-011	cm3/molc sec	25	EST	MEYLAN,WM & HOWARD,PH (1993)

CAS #: 000059-89-2 — N-NITROSOMORPHOLINE

Formula: $C_4H_8N_2O_2$

Mol Weight: 116.12

MP (deg C): 29
FP (deg C):
BP (deg C): 224-225
BP pressure (mm Hg): 7.47E+002

Property/Value	Units	Temp	Data Type	Reference
WS 2.46E+005	mg/L	25	EST	MEYLAN,WM ET AL. (1996)
logP -0.44			EXP	HANSCH,C & LEO,AJ (1985)
VP 3.60E-002	mm Hg	20	EXT	KLEIN,RG (1982)
DC	pKa			
HL 2.45E-008	atm m3/mol	37	EXP	MIRVISH,SS ET AL. (1976)
OH 8.86E-011	cm3/molc sec	25	EST	MEYLAN,WM & HOWARD,PH (1993)

CAS #: 000059-92-7 — DOPA

Formula: $C_9H_{11}NO_4$

Mol Weight: 197.19

MP (deg C): 276-278
FP (deg C):
BP (deg C):
BP pressure (mm Hg):

Property/Value	Units	Temp	Data Type	Reference
WS 5.00E+003	mg/L	20	EXP	YALKOWSKY,SH & DANNENFELSER,RM (1992)
logP -2.39			EXP	SANGSTER,J (1993)
VP 1.24E-010	mm Hg	25	EST	NEELY,WB & BLAU,GE (1985)
DC 2.32	pKa	25	EXP	KORTUM,G ET AL (1961)
HL 1.30E-018	atm m3/mol	25	EST	MEYLAN,WM & HOWARD,PH (1991)
OH 9.66E-011	cm3/molc sec	25	EST	MEYLAN,WM & HOWARD,PH (1993)

CAS #: 000059-98-3 — 1H-IMIDAZOLE, 4,5-DIHYDRO-2-(PHENYLMETHYL)-

Formula: $C_{10}H_{12}N_2$

Mol Weight: 160.22

MP (deg C): 174
FP (deg C):
BP (deg C):
BP pressure (mm Hg):

Property/Value	Units	Temp	Data Type	Reference
WS 3.73E+002	mg/L	25	EST	MEYLAN,WM ET AL. (1996)
logP 2.65			EXP	SANGSTER,J (1993)
VP 1.02E-005	mm Hg	25	EST	NEELY,WB & BLAU,GE (1985)
DC	pKa			
HL 5.18E-008	atm m3/mol	25	EST	MEYLAN,WM & HOWARD,PH (1991)
OH 7.85E-011	cm3/molc sec	25	EST	MEYLAN,WM & HOWARD,PH (1993)

CAS #: 000060-00-4 — ETHYLENEDIAMINETETRAACETIC ACID

Formula: $C_{10}H_{16}N_2O_8$

Mol Weight: 292.25

MP (deg C): 245 dec
FP (deg C):
BP (deg C):
BP pressure (mm Hg):

Property/Value	Units	Temp	Data Type	Reference
WS 1.00E+003	mg/L	25	EXP	WOLF,K & GILBERT,PA (1992)
logP -3.86			EST	MEYLAN,WM & HOWARD,PH (1995)
VP 4.98E-013	mm Hg	25	EST	NEELY,WB & BLAU,GE (1985)
DC 0.26	pKa		EXP	SERJEANT,EP & DEMPSEY,B (1979)
HL 1.17E-023	atm m3/mol	25	EST	MEYLAN,WM & HOWARD,PH (1991)
OH 1.82E-010	cm3/molc sec	25	EST	MEYLAN,WM & HOWARD,PH (1993)

CAS #: 000060-01-5 — TRIBUTYRIN

Formula: $C_{15}H_{26}O_6$

Mol Weight: 302.37

MP (deg C): -75
FP (deg C):
BP (deg C): 305-310
BP pressure (mm Hg):

Property/Value	Units	Temp	Data Type	Reference
WS 1.33E+002	mg/L	37	EXP	FUNASAKI,N ET AL. (1976)
logP 3.31			EST	MEYLAN,WM & HOWARD,PH (1995)
VP 3.82E-004	mm Hg	25	EST	NEELY,WB & BLAU,GE (1985)
DC	pKa			
HL 9.59E-009	atm m3/mol	25	EST	MEYLAN,WM & HOWARD,PH (1991)
OH 1.49E-011	cm3/molc sec	25	EST	MEYLAN,WM & HOWARD,PH (1993)

CAS #: 000060-09-3 — P-PHENYLAZOANILINE

Formula: $C_{12}H_{11}N_3$

Mol Weight: 197.24

MP (deg C): 128
FP (deg C):
BP (deg C): 366
BP pressure (mm Hg):

Property/Value	Units	Temp	Data Type	Reference
WS 3.20E+001	mg/L	25	EXP	BAUGHMAN,GL & PERENICH,TA (1988)
logP 3.41			EXP	HANSCH,C ET AL. (1995)
VP 1.40E-006	mm Hg	25	EXT	SHIMIZU,T ET AL. (1987)
DC 2.82	pKa	25	EXP	PERRIN,DD (1965)
HL 8.70E-011	atm m3/mol	25	EST	MEYLAN,WM & HOWARD,PH (1991)
OH 6.65E-011	cm3/molc sec	25	EST	MEYLAN,WM & HOWARD,PH (1993)

CAS #: 000060-11-7 — 4-(N,N-DIMETHYLAMINO)AZOBENZENE

Formula: $C_{14}H_{15}N_3$
Mol Weight: 225.30
MP (deg C): 114-117
FP (deg C):
BP (deg C):
BP pressure (mm Hg):

Property/Value	Units	Temp	Data Type	Reference
WS 2.30E-001	mg/L	25	EXP	BAUGHMAN,GL & PERENICH,TA (1988)
logP 4.58			EXP	HANSCH,C & LEO,AJ (1985)
VP 7.00E-008	mm Hg	25	EXT	CAMPANELLI,AR ET AL. (1985)
DC 2.96	pKa	25	EXP	PERRIN,DD (1972)
HL 4.00E-010	atm m3/mol	25	EST	MEYLAN,WM & HOWARD,PH (1991)
OH 2.22E-010	cm3/molc sec	25	EST	MEYLAN,WM & HOWARD,PH (1993)

CAS #: 000060-12-8 — 2-PHENYLETHANOL

Formula: $C_8H_{10}O$
Mol Weight: 122.17
MP (deg C): -27
FP (deg C):
BP (deg C): 218.2
BP pressure (mm Hg): 7.50E+002

Property/Value	Units	Temp	Data Type	Reference
WS 1.60E+004	mg/L	20	EXP	VALVANI,SC ET AL. (1981)
logP 1.36			EXP	HANSCH,C & LEO,AJ (1985)
VP 8.68E-002	mm Hg	25	EXP	DAUBERT,TE & DANNER,RP (1989)
DC	pKa			
HL 8.72E-007	atm m3/mol	25	EST	VP/WSOL
OH 1.02E-011	cm3/molc sec	25	EST	MEYLAN,WM & HOWARD,PH (1993)

CAS #: 000060-15-1 — 3-AMINO-1-PROPYLBENZENE

Formula: $C_9H_{13}N$
Mol Weight: 135.21
MP (deg C):
FP (deg C):
BP (deg C):
BP pressure (mm Hg):

Property/Value	Units	Temp	Data Type	Reference
WS 5.22E+002	mg/L	25	EST	MEYLAN,WM ET AL. (1996)
logP 2.61			EST	MEYLAN,WM & HOWARD,PH (1995)
VP 3.37E-002	mm Hg	25	EST	NEELY,WB & BLAU,GE (1985)
DC	pKa			
HL 3.70E-006	atm m3/mol	25	EST	MEYLAN,WM & HOWARD,PH (1991)
OH 2.02E-010	cm3/molc sec	25	EST	MEYLAN,WM & HOWARD,PH (1993)

CAS #: 000060-17-3 — DL-P-FLUOROPHENYLALANINE

Formula: $C_9H_{10}FNO_2$
Mol Weight: 183.18
MP (deg C):
FP (deg C):
BP (deg C):
BP pressure (mm Hg):

Property/Value	Units	Temp	Data Type	Reference
WS 1.86E+004	mg/L	25	EST	MEYLAN,WM ET AL. (1996)
logP -1.89			EXP	HANSCH,C & LEO,AJ (1985)
VP 4.13E-008	mm Hg	25	EST	NEELY,WB & BLAU,GE (1985)
DC 2.13	pKa	24	EXP	KORTUM,G ET AL (1961)
HL 1.41E-010	atm m3/mol	25	EST	MEYLAN,WM & HOWARD,PH (1991)
OH 4.23E-011	cm3/molc sec	25	EST	MEYLAN,WM & HOWARD,PH (1993)

CAS #: 000060-18-4 — TYROSINE

Formula: $C_9H_{11}NO_3$
Mol Weight: 181.19
MP (deg C): 343 dec
FP (deg C):
BP (deg C):
BP pressure (mm Hg):

Property/Value	Units	Temp	Data Type	Reference
WS 4.79E+002	mg/L	25	EXP	SEIDELL,A (1941)
logP -2.26			EXP	HANSCH,C & LEO,AJ (1985)
VP 1.26E-010	mm Hg	25	EST	HL X WSOL
DC 2.20	pKa	25	EXP	LEHNINGER,A (1975)
HL 6.28E-014	atm m3/mol	25	EST	MEYLAN,WM & HOWARD,PH (1991)
OH 8.20E-011	cm3/molc sec	25	EST	ATKINSON,R (1987)

CAS #: 000060-27-5 — CREATININE

Formula: $C_4H_7N_3O$
Mol Weight: 113.12
MP (deg C): 300 dec
FP (deg C):
BP (deg C):
BP pressure (mm Hg):

Property/Value	Units	Temp	Data Type	Reference
WS 8.01E+004	mg/L	16	EXP	YALKOWSKY,SH & DANNENFELSER,RM (1992)
logP -1.76			EXP	HANSCH,C & LEO,AJ (1985)
VP 2.17E-004	mm Hg	25	EST	NEELY,WB & BLAU,GE (1985)
DC 4.80	pKa	25	EXP	PERRIN,DD (1965)
HL 2.42E-012	atm m3/mol	25	EST	MEYLAN,WM & HOWARD,PH (1991)
OH 7.80E-011	cm3/molc sec	25	EST	MEYLAN,WM & HOWARD,PH (1993)

CAS #: 000060-29-7 — DIETHYL ETHER

Formula: $C_4H_{10}O$
Mol Weight: 74.12
MP (deg C): -116.3
FP (deg C):
BP (deg C): 34.5
BP pressure (mm Hg):

Property/Value	Units	Temp	Data Type	Reference
WS 6.04E+004	mg/L	25	EXP	RIDDICK,JA ET AL. (1986)
logP 0.89			EXP	HANSCH,C & LEO,AJ (1985)
VP 5.37E+002	mm Hg	25	EXP	BOUBLIK,T ET AL. (1984)
DC -3.59	pKa	25	EXP	KULEVSKY,N ET AL. (1969)
HL 1.23E-003	atm m3/mol	25	EXP	BOCEK,K (1976)
OH 1.33E-011	cm3/molc sec	25	EXP	ATKINSON,R (1989)

CAS #: 000060-32-2 — 6-AMINOHEXANOIC ACID

Formula: $C_6H_{13}NO_2$
Mol Weight: 131.18
MP (deg C): 204-206
FP (deg C):
BP (deg C):
BP pressure (mm Hg):

Property/Value	Units	Temp	Data Type	Reference
WS 5.05E+005	mg/L	25	EXP	YALKOWSKY,SH & DANNENFELSER,RM (1992)
logP -2.95			EXP	HANSCH,C ET AL. (1995)
VP 3.01E-009	mm Hg	25	EST	NEELY,WB & BLAU,GE (1985)
DC 4.43	pKa	25	EXP	KORTUM,G ET AL (1961)
HL 1.75E-010	atm m3/mol	25	EST	MEYLAN,WM & HOWARD,PH (1991)
OH 3.73E-011	cm3/molc sec	25	EST	MEYLAN,WM & HOWARD,PH (1993)

CAS #: 000060-33-3				LINOLEIC ACID
Formula: $C_{18}H_{32}O_2$				
Mol Weight: 280.45				
MP (deg C): -12		FP (deg C):		
BP (deg C): 230				
BP pressure (mm Hg): 1.60E+001				

Property/Value	Units	Temp	Data Type	Reference
WS 3.77E-002	mg/L	25	EST	MEYLAN,WM ET AL. (1996)
logP 7.05			EXP	SANGSTER,J (1993)
VP 8.68E-007	mm Hg	25	EXP	DAUBERT,TE & DANNER,RP (1989)
DC 4.77	pKa	25	EXP	SERJEANT,EP & DEMPSEY,B (1979)
HL 3.94E-005	atm m3/mol	25	EST	MEYLAN,WM & HOWARD,PH (1991)
OH 1.29E-010	cm3/molc sec	25	EST	MEYLAN,WM & HOWARD,PH (1993)

CAS #: 000060-34-4				METHYL HYDRAZINE
Formula: CH_6N_2				
Mol Weight: 46.07				
MP (deg C): -52.4		FP (deg C):		
BP (deg C): 87.5				
BP pressure (mm Hg):				

Property/Value	Units	Temp	Data Type	Reference
WS 1.00E+006	mg/L		EXP	MERCK INDEX (1983)
logP -1.05			EXP	HANSCH,C ET AL. (1995)
VP 5.00E+001	mm Hg	25	EXP	BOUBLIK,T ET AL. (1984)
DC 7.87	pKa	30	EXP	PERRIN,DD (1965)
HL 3.03E-006	atm m3/mol	25	EST	VP/WSOL
OH 6.50E-011	cm3/molc sec	25	EXP	ATKINSON,R (1989)

CAS #: 000060-35-5				ACETAMIDE
Formula: C_2H_5NO				
Mol Weight: 59.07				
MP (deg C): 69.5		FP (deg C):		
BP (deg C): 221.15				
BP pressure (mm Hg):				

Property/Value	Units	Temp	Data Type	Reference
WS 2.25E+006	mg/L	25	EXP	YALKOWSKY,SH & DANNENFELSER,RM (1992)
logP -1.26			EXP	HANSCH,C & LEO,AJ (1985)
VP 1.09E+000	mm Hg	25	EST	NEELY,WB & BLAU,GE (1985)
DC 0.63	pKa		EXP	WEAST,RC (1972)
HL 1.12E-008	atm m3/mol	25	EST	MEYLAN,WM & HOWARD,PH (1991)
OH 1.19E-011	cm3/molc sec	25	EST	MEYLAN,WM & HOWARD,PH (1993)

CAS #: 000060-51-5				DIMETHOATE
Formula: $C_5H_{12}NO_3PS_2$				
Mol Weight: 229.26				
MP (deg C): 52-52.5		FP (deg C):		
BP (deg C): 107				
BP pressure (mm Hg): 5.00E-002				

Property/Value	Units	Temp	Data Type	Reference
WS 2.38E+004	mg/L	20	EXP	TOMLIN,C (1994)
logP 0.78			EXP	HANSCH,C ET AL. (1995)
VP 8.25E-006	mm Hg	25	EXP	TOMLIN,C (1994)
DC	pKa			
HL 1.05E-010	atm m3/mol	25	EST	VP/WSOL
OH 7.90E-011	cm3/molc sec	25	EST	MEYLAN,WM & HOWARD,PH (1993)

CAS #: 000060-54-8				TETRACYCLINE
Formula: $C_{22}H_{24}N_2O_8$				
Mol Weight: 444.45				
MP (deg C): 170-175 de		FP (deg C):		
BP (deg C):				
BP pressure (mm Hg):				

Property/Value	Units	Temp	Data Type	Reference
WS 2.31E+002	mg/L	25	EXP	YALKOWSKY,SH & DANNENFELSER,RM (1992)
logP -1.30			EXP	HANSCH,C & LEO,AJ (1985)
VP 3.09E-023	mm Hg	25	EST	NEELY,WB & BLAU,GE (1985)
DC 3.30	pKa	25	EXP	KORTUM,G ET AL (1961)
HL 4.66E-024	atm m3/mol	25	EST	MEYLAN,WM & HOWARD,PH (1991)
OH 2.56E-010	cm3/molc sec	25	EST	MEYLAN,WM & HOWARD,PH (1993)

CAS #: 000060-56-0				METHIMAZOLE
Formula: $C_4H_6N_2S$				
Mol Weight: 114.17				
MP (deg C): 146-148		FP (deg C):		
BP (deg C): 280				
BP pressure (mm Hg):				

Property/Value	Units	Temp	Data Type	Reference
WS 1.03E+005	mg/L	25	EST	MEYLAN,WM ET AL. (1996)
logP -0.34			EXP	HANSCH,C ET AL. (1995)
VP 7.81E-003	mm Hg	25	EST	NEELY,WB & BLAU,GE (1985)
DC	pKa			
HL 2.03E-006	atm m3/mol	25	EST	MEYLAN,WM & HOWARD,PH (1991)
OH 1.51E-010	cm3/molc sec	25	EST	MEYLAN,WM & HOWARD,PH (1993)

CAS #: 000060-57-1				DIELDRIN
Formula: $C_{12}H_8Cl_6O$				
Mol Weight: 380.91				
MP (deg C): 175-176		FP (deg C):		
BP (deg C): 330				
BP pressure (mm Hg):				

Property/Value	Units	Temp	Data Type	Reference
WS 1.95E-001	mg/L	25	EXP	BIGGAR,JW & RIGGS,RI (1974)
logP 5.40			EXP	DEBRUIJN,J ET AL. (1989)
VP 5.89E-006	mm Hg	25	EXP	GRAYSON,BT & FOSBRAEY,LA (1982)
DC	pKa			
HL 5.80E-005	atm m3/mol	25	EXP	WARNER,HP ET AL. (1987)
OH 4.87E-012	cm3/molc sec	25	EST	MEYLAN,WM & HOWARD,PH (1993)

CAS #: 000060-80-0				ANTIPYRINE
Formula: $C_{11}H_{14}N_2O$				
Mol Weight: 190.25				
MP (deg C): 111-113		FP (deg C):		
BP (deg C):				
BP pressure (mm Hg):				

Property/Value	Units	Temp	Data Type	Reference
WS 5.19E+004	mg/L	25	EXP	YALKOWSKY,SH & DANNENFELSER,RM (1992)
logP 0.38			EXP	HANSCH,C & LEO,AJ (1985)
VP 3.06E-005	mm Hg	25	EST	NEELY,WB & BLAU,GE (1985)
DC	pKa			
HL 6.65E-010	atm m3/mol	25	EST	MEYLAN,WM & HOWARD,PH (1991)
OH 3.21E-011	cm3/molc sec	25	EST	MEYLAN,WM & HOWARD,PH (1993)

CAS #: 000060-87-7				PROMETHAZINE

Formula: $C_{17}H_{20}N_2S$

Mol Weight: 284.43

MP (deg C): 60   FP (deg C):

BP (deg C): 190-192

BP pressure (mm Hg): 3.00E+000

Property/Value	Units	Temp	Data Type	Reference
WS 1.56E+001	mg/L	24	EXP	YALKOWSKY,SH & DANNENFELSER,RM (1992)
logP 4.81			EXP	HANSCH,C ET AL. (1995)
VP 1.37E-006	mm Hg	25	EST	NEELY,WB & BLAU,GE (1985)
DC	pKa			
HL 4.98E-010	atm m3/mol	25	EST	MEYLAN,WM & HOWARD,PH (1991)
OH 2.82E-010	cm3/molc sec	25	EST	MEYLAN,WM & HOWARD,PH (1993)

CAS #: 000060-91-3				10H-PHENOTHIAZINE-10-ETHANAMINE, N,N-DIETHYL-

Formula: $C_{18}H_{21}ClN_2S$

Mol Weight: 332.90

MP (deg C):   FP (deg C):

BP (deg C): 167

BP pressure (mm Hg): 5.00E-001

Property/Value	Units	Temp	Data Type	Reference
WS 5.25E-002	mg/L	25	EST	MEYLAN,WM ET AL. (1996)
logP 5.94			EXP	SANGSTER,J (1994)
VP 7.53E-008	mm Hg	25	EST	NEELY,WB & BLAU,GE (1985)
DC	pKa			
HL 4.89E-010	atm m3/mol	25	EST	MEYLAN,WM & HOWARD,PH (1991)
OH 2.58E-010	cm3/molc sec	25	EST	MEYLAN,WM & HOWARD,PH (1993)

CAS #: 000060-92-4				ADENOSINE, CYCLIC 3',5'-(HYDROGEN PHOSPHATE)

Formula: $C_{10}H_{12}N_5O_6P$

Mol Weight: 329.21

MP (deg C): 219-220   FP (deg C):

BP (deg C):

BP pressure (mm Hg):

Property/Value	Units	Temp	Data Type	Reference
WS 1.36E+005	mg/L	25	EST	MEYLAN,WM ET AL. (1996)
logP -2.96			EXP	HANSCH,C ET AL. (1995)
VP 6.06E-011	mm Hg	25	EST	NEELY,WB & BLAU,GE (1985)
DC	pKa			
HL 2.92E-028	atm m3/mol	25	EST	MEYLAN,WM & HOWARD,PH (1991)
OH 3.01E-010	cm3/molc sec	25	EST	MEYLAN,WM & HOWARD,PH (1993)

CAS #: 000060-99-1				METHOTRIMEPRAZINE

Formula: $C_{19}H_{24}N_2OS$

Mol Weight: 328.48

MP (deg C):   FP (deg C):

BP (deg C):

BP pressure (mm Hg):

Property/Value	Units	Temp	Data Type	Reference
WS 2.00E+001	mg/L	25	EXP	YALKOWSKY,SH & DANNENFELSER,RM (1992)
logP 4.68			EXP	HANSCH,C & LEO,AJ (1985)
VP 9.20E-008	mm Hg	25	EST	NEELY,WB & BLAU,GE (1985)
DC	pKa			
HL 3.91E-011	atm m3/mol	25	EST	MEYLAN,WM & HOWARD,PH (1991)
OH 2.94E-010	cm3/molc sec	25	EST	MEYLAN,WM & HOWARD,PH (1993)

CAS #: 000061-01-8				METHOXYPROMAZINE

Formula: $C_{18}H_{22}N_2OS$

Mol Weight: 314.45

MP (deg C): 44-48   FP (deg C):

BP (deg C):

BP pressure (mm Hg):

Property/Value	Units	Temp	Data Type	Reference
WS 5.22E-001	mg/L	25	EST	MEYLAN,WM ET AL. (1996)
logP 4.90			EXP	HANSCH,C & LEO,AJ (1985)
VP 1.24E-007	mm Hg	25	EST	NEELY,WB & BLAU,GE (1985)
DC	pKa			
HL 2.94E-011	atm m3/mol	25	EST	MEYLAN,WM & HOWARD,PH (1991)
OH 2.92E-010	cm3/molc sec	25	EST	MEYLAN,WM & HOWARD,PH (1993)

CAS #: 000061-32-5				METHICILLIN

Formula: $C_{17}H_{20}N_2O_6S$

Mol Weight: 380.42

MP (deg C):   FP (deg C):

BP (deg C):

BP pressure (mm Hg):

Property/Value	Units	Temp	Data Type	Reference
WS 3.67E+002	mg/L	25	EST	MEYLAN,WM ET AL. (1996)
logP 1.22			EXP	HANSCH,C ET AL. (1995)
VP 1.88E-013	mm Hg	25	EST	NEELY,WB & BLAU,GE (1985)
DC	pKa			
HL 9.99E-017	atm m3/mol	25	EST	MEYLAN,WM & HOWARD,PH (1991)
OH 2.88E-010	cm3/molc sec	25	EST	MEYLAN,WM & HOWARD,PH (1993)

CAS #: 000061-33-6				BENZYLPENICILLIN

Formula: $C_{15}H_{16}N_2O_4S$

Mol Weight: 320.37

MP (deg C):   FP (deg C):

BP (deg C):

BP pressure (mm Hg):

Property/Value	Units	Temp	Data Type	Reference
WS 2.10E+002	mg/L	25	EST	MEYLAN,WM ET AL. (1996)
logP 1.83			EXP	HANSCH,C & LEO,AJ (1985)
VP 2.55E-012	mm Hg	25	EST	NEELY,WB & BLAU,GE (1985)
DC	pKa			
HL 1.16E-014	atm m3/mol	25	EST	MEYLAN,WM & HOWARD,PH (1991)
OH 9.15E-011	cm3/molc sec	25	EST	MEYLAN,WM & HOWARD,PH (1993)

CAS #: 000061-50-7				N,N-DIMETHYLTRYPTAMINE

Formula: $C_{12}H_{16}N_2$

Mol Weight: 188.27

MP (deg C): 44.6-46.8   FP (deg C):

BP (deg C):

BP pressure (mm Hg):

Property/Value	Units	Temp	Data Type	Reference
WS 1.12E+004	mg/L	25	EST	MEYLAN,WM ET AL. (1996)
logP 1.94			EST	MEYLAN,WM & HOWARD,PH (1995)
VP 1.04E-004	mm Hg	25	EST	NEELY,WB & BLAU,GE (1985)
DC	pKa			
HL 6.44E-010	atm m3/mol	25	EST	MEYLAN,WM & HOWARD,PH (1991)
OH 2.80E-010	cm3/molc sec	25	EST	MEYLAN,WM & HOWARD,PH (1993)

CAS #: 000061-54-1				TRYPTAMINE

Formula: $C_{10}H_{12}N_2$

Mol Weight: 160.22

MP (deg C): 118   FP (deg C):

BP (deg C):

BP pressure (mm Hg):

Property/Value	Units	Temp	Data Type	Reference
WS 3.30E+004	mg/L	25	EST	MEYLAN,WM ET AL. (1996)
logP 1.55			EXP	HANSCH,C ET AL. (1995)
VP 1.28E-004	mm Hg	25	EST	NEELY,WB & BLAU,GE (1985)
DC	pKa			
HL 1.33E-010	atm m3/mol	25	EST	MEYLAN,WM & HOWARD,PH (1991)
OH 2.33E-010	cm3/molc sec	25	EST	MEYLAN,WM & HOWARD,PH (1993)

CAS #: 000061-57-4				NIRIDAZOLE

Formula: $C_6H_6N_4O_3S$

Mol Weight: 214.20

MP (deg C): 260-262   FP (deg C):

BP (deg C):

BP pressure (mm Hg):

Property/Value	Units	Temp	Data Type	Reference
WS 1.30E+002	mg/L	25	EXP	YALKOWSKY,SH & DANNENFELSER,RM (1992)
logP 0.95			EXP	HANSCH,C & LEO,AJ (1985)
VP 3.45E-007	mm Hg	25	EST	NEELY,WB & BLAU,GE (1985)
DC	pKa			
HL 1.12E-014	atm m3/mol	25	EST	MEYLAN,WM & HOWARD,PH (1991)
OH 8.60E-012	cm3/molc sec	25	EST	MEYLAN,WM & HOWARD,PH (1993)

CAS #: 000061-68-7				N-PH-ANTHRANILIC ACID,2',3'-DIMETHYL

Formula: $C_{15}H_{15}NO_2$

Mol Weight: 241.29

MP (deg C): 230-231   FP (deg C):

BP (deg C):

BP pressure (mm Hg):

Property/Value	Units	Temp	Data Type	Reference
WS 2.00E+001	mg/L	30	EXP	YALKOWSKY,SH & DANNENFELSER,RM (1992)
logP 5.12			EXP	HANSCH,C ET AL. (1995)
VP 4.63E-007	mm Hg	25	EST	NEELY,WB & BLAU,GE (1985)
DC	pKa			
HL 2.57E-011	atm m3/mol	25	EST	MEYLAN,WM & HOWARD,PH (1991)
OH 2.01E-010	cm3/molc sec	25	EST	MEYLAN,WM & HOWARD,PH (1993)

CAS #: 000061-72-3				CLOXACILLIN

Formula: $C_{19}H_{19}ClN_3O_5S$

Mol Weight: 436.90

MP (deg C):   FP (deg C):

BP (deg C):

BP pressure (mm Hg):

Property/Value	Units	Temp	Data Type	Reference
WS 1.39E+001	mg/L	25	EST	MEYLAN,WM ET AL. (1996)
logP 2.48			EXP	SANGSTER,J (1994)
VP 1.37E-015	mm Hg	25	EST	NEELY,WB & BLAU,GE (1985)
DC	pKa			
HL 1.89E-017	atm m3/mol	25	EST	MEYLAN,WM & HOWARD,PH (1991)
OH 9.61E-011	cm3/molc sec	25	EST	MEYLAN,WM & HOWARD,PH (1993)

CAS #: 000061-73-4				METHYLENE BLUE

Formula: $C_{16}H_{18}ClN_3S$

Mol Weight: 319.86

MP (deg C): 100-110 de   FP (deg C):

BP (deg C):

BP pressure (mm Hg):

Property/Value	Units	Temp	Data Type	Reference
WS 4.36E+004	mg/L	25	EXP	BAUGHMAN,GL ET AL. (1993)
logP 5.85			EST	MEYLAN,WM & HOWARD,PH (1995)
VP 1.30E-007	mm Hg	25	EST	NEELY,WB & BLAU,GE (1985)
DC	pKa			
HL 1.25E-012	atm m3/mol	25	EST	VP/WSOL
OH 2.05E-010	cm3/molc sec	25	EST	MEYLAN,WM & HOWARD,PH (1993)

CAS #: 000061-76-7				PHENYLEPHRINE

Formula: $C_9H_{14}ClNO_2$

Mol Weight: 203.67

MP (deg C): 140-145   FP (deg C):

BP (deg C):

BP pressure (mm Hg):

Property/Value	Units	Temp	Data Type	Reference
WS 1.00E+006	mg/L	25	EST	MEYLAN,WM ET AL. (1996)
logP -0.31			EXP	HANSCH,C & LEO,AJ (1985)
VP 2.22E-005	mm Hg	25	EST	NEELY,WB & BLAU,GE (1985)
DC 8.86	pKa	25	EXP	PERRIN,DD (1965)
HL 6.78E-015	atm m3/mol	25	EST	MEYLAN,WM & HOWARD,PH (1991)
OH 1.63E-010	cm3/molc sec	25	EST	MEYLAN,WM & HOWARD,PH (1993)

CAS #: 000061-78-9				GLYCINE, N-(4-AMINOBENZOYL)-

Formula: $C_9H_{10}N_2O_3$

Mol Weight: 194.19

MP (deg C): 198-199   FP (deg C):

BP (deg C):

BP pressure (mm Hg):

Property/Value	Units	Temp	Data Type	Reference
WS 1.32E+004	mg/L	25	EST	MEYLAN,WM ET AL. (1996)
logP -0.89			EXP	SANGSTER,J (1993)
VP 3.51E-008	mm Hg	25	EST	NEELY,WB & BLAU,GE (1985)
DC	pKa			
HL 2.55E-016	atm m3/mol	25	EST	MEYLAN,WM & HOWARD,PH (1991)
OH 1.07E-010	cm3/molc sec	25	EST	MEYLAN,WM & HOWARD,PH (1993)

CAS #: 000061-80-3				ZOXAZOLAMINE

Formula: $C_7H_5ClN_2O$

Mol Weight: 168.58

MP (deg C): 184-185   FP (deg C):

BP (deg C):

BP pressure (mm Hg):

Property/Value	Units	Temp	Data Type	Reference
WS 4.95E+002	mg/L	25	EST	MEYLAN,WM ET AL. (1996)
logP 2.46			EXP	HANSCH,C & LEO,AJ (1985)
VP 4.62E-004	mm Hg	25	EST	NEELY,WB & BLAU,GE (1985)
DC	pKa			
HL 1.80E-010	atm m3/mol	25	EST	MEYLAN,WM & HOWARD,PH (1991)
OH 3.88E-011	cm3/molc sec	25	EST	MEYLAN,WM & HOWARD,PH (1993)

CAS #: 000061-82-5	AMITROLE

Formula: $C_2H_4N_4$

Mol Weight: 84.08

MP (deg C): 159

FP (deg C):

BP (deg C):

BP pressure (mm Hg):

Property/ Value	Units	Temp	Data Type	Reference
WS 2.80E+005	mg/L	25	EXP	MARTIN,H & WORTHING,CR (1977)
logP -0.86			EXP	SANGSTER,J (1994)
VP 4.40E-007	mm Hg	25	EXP	WEBER,JB (1994)
DC	pKa			
HL 2.21E-013	atm m3/mol	25	EST	VP/WSOL
OH 5.52E-012	cm3/molc sec	25	EST	MEYLAN,WM & HOWARD,PH (1993)

CAS #: 000061-90-5	LEUCINE

Formula: $C_6H_{13}NO_2$

Mol Weight: 131.18

MP (deg C): 293

FP (deg C):

BP (deg C):

BP pressure (mm Hg):

Property/ Value	Units	Temp	Data Type	Reference
WS 2.15E+004	mg/L	25	EXP	YALKOWSKY,SH & DANNENFELSER,RM (1992)
logP -1.52			EXP	HANSCH,C & LEO,AJ (1985)
VP 1.34E-008	mm Hg	25	EST	NEELY,WB & BLAU,GE (1985)
DC 2.35	pKa	13	EXP	KORTUM,G ET AL (1961)
HL 3.49E-009	atm m3/mol	25	EST	MEYLAN,WM & HOWARD,PH (1991)
OH 4.23E-011	cm3/molc sec	25	EST	MEYLAN,WM & HOWARD,PH (1993)

CAS #: 000062-23-7	P-NITROBENZOIC ACID

Formula: $C_7H_5NO_4$

Mol Weight: 167.12

MP (deg C): 242

FP (deg C):

BP (deg C):

BP pressure (mm Hg):

Property/ Value	Units	Temp	Data Type	Reference
WS 2.00E+002	mg/L	15	EXP	YALKOWSKY,SH & DANNENFELSER,RM (1992)
logP 1.89			EXP	HANSCH,C & LEO,AJ (1985)
VP 2.53E-006	mm Hg	25	EST	NEELY,WB & BLAU,GE (1985)
DC 3.44	pKa		EXP	DEAN,JA (1985)
HL 3.79E-010	atm m3/mol	25	EST	MEYLAN,WM & HOWARD,PH (1991)
OH 5.86E-013	cm3/molc sec	25	EST	MEYLAN,WM & HOWARD,PH (1993)

CAS #: 000062-37-3	CHLORMERODRIN

Formula: $C_5H_{11}ClHgN_2O_2$

Mol Weight: 367.20

MP (deg C): 152-153

FP (deg C):

BP (deg C):

BP pressure (mm Hg):

Property/ Value	Units	Temp	Data Type	Reference
WS 1.15E+003	mg/L	25	EST	MEYLAN,WM ET AL. (1996)
logP -0.80			EXP	HALBACH,S (1985)
VP 1.66E-005	mm Hg	25	EST	NEELY,WB & BLAU,GE (1985)
DC	pKa			
HL	atm m3/mol			
OH 2.69E-011	cm3/molc sec	25	EST	MEYLAN,WM & HOWARD,PH (1993)

CAS #: 000062-38-4	PHENYLMERCURIC ACETATE

Formula: $C_8H_8HgO_2$

Mol Weight: 336.74

MP (deg C): 147-150

FP (deg C):

BP (deg C): DEC

BP pressure (mm Hg):

Property/ Value	Units	Temp	Data Type	Reference
WS 4.70E+003	mg/L	25	EXP	SHIU,WY ET AL. (1990)
logP 0.71			EXP	HANSCH,C & LEO,AJ (1985)
VP 6.00E-006	mm Hg	20	EXP	SUNTIO,LR ET AL. (1988)
DC	pKa			
HL 5.66E-010	atm m3/mol	20	EST	VP/WSOL
OH 2.05E-012	cm3/molc sec	25	EST	MEYLAN,WM & HOWARD,PH (1993)

CAS #: 000062-44-2	P-PHENACETIN

Formula: $C_{10}H_{13}NO_2$

Mol Weight: 179.22

MP (deg C): 134-135

FP (deg C):

BP (deg C): 252

BP pressure (mm Hg):

Property/ Value	Units	Temp	Data Type	Reference
WS 7.66E+002	mg/L	25	EXP	SEIDELL,A (1941)
logP 1.58			EXP	NAKAGAWA,Y ET AL. (1992)
VP 6.92E-007	mm Hg	25	EXP	WIEDEMANN,HG (1972)
DC	pKa			
HL 2.13E-010	atm m3/mol	25	EST	VP/WSOL
OH 2.96E-011	cm3/molc sec	25	EST	MEYLAN,WM & HOWARD,PH (1993)

CAS #: 000062-46-4	1,2-DITHIOLANE-3-PENTANOIC ACID

Formula: $C_8H_{14}O_2S_2$

Mol Weight: 206.33

MP (deg C): 60-61

FP (deg C):

BP (deg C): 160-165

BP pressure (mm Hg):

Property/ Value	Units	Temp	Data Type	Reference
WS 1.27E+002	mg/L	25	EST	MEYLAN,WM ET AL. (1996)
logP 3.40			EST	MEYLAN,WM & HOWARD,PH (1995)
VP 3.75E-005	mm Hg	25	EST	NEELY,WB & BLAU,GE (1985)
DC	pKa			
HL 1.64E-008	atm m3/mol	25	EST	MEYLAN,WM & HOWARD,PH (1991)
OH 2.57E-010	cm3/molc sec	25	EST	MEYLAN,WM & HOWARD,PH (1993)

CAS #: 000062-50-0	ETHYL METHANESULFONATE

Formula: $C_3H_8O_3S$

Mol Weight: 124.16

MP (deg C): -0.41 EST

FP (deg C):

BP (deg C): 213-213.5

BP pressure (mm Hg): 7.61E+002

Property/ Value	Units	Temp	Data Type	Reference
WS 1.35E+005	mg/L	25	EST	MEYLAN,WM ET AL. (1996)
logP -0.17			EST	MEYLAN,WM & HOWARD,PH (1995)
VP 2.06E-001	mm Hg	25	EXP	JABER,HM ET AL. (1984B)
DC	pKa			
HL 2.59E-007	atm m3/mol	25	EST	VP/WSOL
OH 1.13E-012	cm3/molc sec	25	EST	MEYLAN,WM & HOWARD,PH (1993)

000062-53-3 — ANILINE

Formula: C_6H_7N
Mol Weight: 93.13
MP (deg C): -5.98
BP (deg C): 184.40
BP pressure (mm Hg):

Property/Value	Units	Temp	Data Type	Reference
WS 3.60E+004	mg/L	25	EXP	YALKOWSKY,SH & DANNENFELSER,RM (1992)
logP 0.90			EXP	HANSCH,C & LEO,AJ (1985)
VP 4.90E-001	mm Hg	25	EXP	DAUBERT,TE & DANNER,RP (1985)
DC 4.60	pKa	25	EXP	PERRIN,DD (1972)
HL 1.90E-006	atm m3/mol	25	EST	VP/WSOL
OH 1.11E-010	cm3/molc sec	25	EXP	ATKINSON,R (1989)

000062-55-5 — THIOACETAMIDE

Formula: C_2H_5NS
Mol Weight: 75.13
MP (deg C): 113-114
BP (deg C):
BP pressure (mm Hg):

Property/Value	Units	Temp	Data Type	Reference
WS 1.63E+005	mg/L	25	EXP	MERCK INDEX (1983)
logP -0.26			EXP	HANSCH,C ET AL. (1995)
VP 1.52E+001	mm Hg	25	EST	NEELY,WB & BLAU,GE (1985)
DC 13.40	pKa	25	EXP	SERJEANT,EP & DEMPSEY,B (1979)
HL 6.44E-006	atm m3/mol	25	EST	MEYLAN,WM & HOWARD,PH (1991)
OH 2.12E-011	cm3/molc sec	25	EST	MEYLAN,WM & HOWARD,PH (1993)

000062-56-6 — THIOUREA

Formula: CH_4N_2S
Mol Weight: 76.12
MP (deg C): 176-178
BP (deg C):
BP pressure (mm Hg):

Property/Value	Units	Temp	Data Type	Reference
WS 1.42E+005	mg/L	25	EXP	SEIDELL,A (1941)
logP -1.08			EXP	GOVERS,H ET AL. (1986)
VP 2.49E+000	mm Hg	25	EST	NEELY,WB & BLAU,GE (1985)
DC	pKa			
HL 1.29E-010	atm m3/mol	25	EST	MEYLAN,WM & HOWARD,PH (1991)
OH 4.00E-011	cm3/molc sec	25	EST	MEYLAN,WM & HOWARD,PH (1993)

000062-59-9 — CEVADINE

Formula: $C_{32}H_{49}NO_9$
Mol Weight: 591.75
MP (deg C): 213-214 de
BP (deg C):
BP pressure (mm Hg):

Property/Value	Units	Temp	Data Type	Reference
WS 5.70E+002	mg/L	25	EXP	GUNTHER,FA ET AL. (1968)
logP 0.89			EST	MEYLAN,WM & HOWARD,PH (1995)
VP 4.57E-021	mm Hg	25	EST	NEELY,WB & BLAU,GE (1985)
DC	pKa			
HL 8.69E-022	atm m3/mol	25	EST	MEYLAN,WM & HOWARD,PH (1991)
OH 2.02E-010	cm3/molc sec	25	EST	MEYLAN,WM & HOWARD,PH (1993)

000062-67-9 — NALORPHINE

Formula: $C_{19}H_{21}NO_3$
Mol Weight: 311.38
MP (deg C): 28-209
BP (deg C):
BP pressure (mm Hg):

Property/Value	Units	Temp	Data Type	Reference
WS 2.77E+003	mg/L	25	EST	MEYLAN,WM ET AL. (1996)
logP 1.86			EXP	HANSCH,C & LEO,AJ (1985)
VP 2.78E-010	mm Hg	25	EST	NEELY,WB & BLAU,GE (1985)
DC	pKa			
HL 1.75E-016	atm m3/mol	25	EST	MEYLAN,WM & HOWARD,PH (1991)
OH 2.91E-010	cm3/molc sec	25	EST	MEYLAN,WM & HOWARD,PH (1993)

000062-73-7 — DICHLORVOS

Formula: $C_4H_7Cl_2O_4P$
Mol Weight: 220.98
MP (deg C):
BP (deg C): 140
BP pressure (mm Hg): 2.00E+001

Property/Value	Units	Temp	Data Type	Reference
WS 1.00E+004	mg/L	20	EXP	WORTHING,CR & WALKER,SB (1983)
logP 1.16			EXP	KAWAMOTO,K & URANO,K (1989)
VP 5.30E-002	mm Hg	25	EXP	KIM,YH ET AL. (1984)
DC	pKa			
HL 1.39E-005	atm m3/mol	25	EST	VP/WSOL
OH 9.24E-012	cm3/molc sec	25	EST	MEYLAN,WM & HOWARD,PH (1993)

000062-74-8 — FLUOROACETIC ACID, SODIUM SALT

Formula: $C_2H_2FNaO_2$
Mol Weight: 100.03
MP (deg C):
BP (deg C):
BP pressure (mm Hg):

Property/Value	Units	Temp	Data Type	Reference
WS 1.00E+006	mg/L	25	EST	MEYLAN,WM ET AL. (1996)
logP -3.78			EST	MEYLAN,WM & HOWARD,PH (1995)
VP 6.54E-007	mm Hg	25	EST	NEELY,WB & BLAU,GE (1985)
DC	pKa			
HL	atm m3/mol			
OH 6.50E-014	cm3/molc sec	25	EST	MEYLAN,WM & HOWARD,PH (1993)

000062-75-9 — N-NITROSODIMETHYLAMINE

Formula: $C_2H_6N_2O$
Mol Weight: 74.08
MP (deg C):
BP (deg C): 154
BP pressure (mm Hg):

Property/Value	Units	Temp	Data Type	Reference
WS 1.00E+006	mg/L		EXP	CALLAHAN,MA ET AL. (1979A)
logP -0.57			EXP	HANSCH,C & LEO,AJ (1985)
VP 2.70E+000	mm Hg	20	EXT	KLEIN,RG (1982)
DC	pKa			
HL 1.82E-006	atm m3/mol	37	EXP	MIRVISH,SS ET AL. (1976)
OH 2.53E-012	cm3/molc sec	25	EXP	ATKINSON,R (1989)

CAS #:	000063-05-8			ANDROSTENEDIONE

Formula: $C_{19}H_{26}O_2$

Mol Weight: 286.42

MP (deg C): 142-144   FP (deg C):

BP (deg C):

BP pressure (mm Hg):

Property/ Value	Units	Temp	Data Type	Reference
WS 5.73E+001	mg/L	25	EXP	YALKOWSKY,SH & DANNENFELSER,RM (1992)
logP 2.75			EXP	HANSCH,C & LEO,AJ (1985)
VP 2.46E-006	mm Hg	25	EST	NEELY,WB & BLAU,GE (1985)
DC	pKa			
HL	atm m3/mol			
OH 1.09E-010	cm3/molc sec	25	EST	MEYLAN,WM & HOWARD,PH (1993)

CAS #:	000063-25-2			1-NAPHTHYL-N-METHYLCARBAMATE

Formula: $C_{12}H_{11}NO_2$

Mol Weight: 201.23

MP (deg C): 142   FP (deg C):

BP (deg C): 315

BP pressure (mm Hg):

Property/ Value	Units	Temp	Data Type	Reference
WS 8.26E+001	mg/L	25	EXP	SWANN,RL ET AL. (1983)
logP 2.36			EXP	HANSCH,C & LEO,AJ (1985)
VP 1.36E-006	mm Hg	25	EXP	FERREIRA,GAL & SEIBER,JN (1981)
DC	pKa			
HL 1.31E-003	atm m3/mol	25	EST	VP/WSOL
OH 2.84E-011	cm3/molc sec	25	EST	MEYLAN,WM & HOWARD,PH (1993)

CAS #:	000063-42-3			LACTOSE

Formula: $C_{12}H_{22}O_{11}$

Mol Weight: 342.30

MP (deg C): 201-202   FP (deg C):

BP (deg C):

BP pressure (mm Hg):

Property/ Value	Units	Temp	Data Type	Reference
WS 1.95E+005	mg/L	20	EXP	YALKOWSKY,SH & DANNENFELSER,RM (1992)
logP -5.43			EST	MEYLAN,WM & HOWARD,PH (1995)
VP 8.09E-017	mm Hg	25	EST	NEELY,WB & BLAU,GE (1985)
DC	pKa			
HL	atm m3/mol			
OH 1.54E-010	cm3/molc sec	25	EST	MEYLAN,WM & HOWARD,PH (1993)

CAS #:	000063-68-3			METHIONINE

Formula: $C_5H_{11}NO_2S$

Mol Weight: 149.21

MP (deg C): 280-282   FP (deg C):

BP (deg C):

BP pressure (mm Hg):

Property/ Value	Units	Temp	Data Type	Reference
WS 5.70E+004	mg/L	25	EXP	YALKOWSKY,SH & DANNENFELSER,RM (1992)
logP -1.87			EXP	HANSCH,C & LEO,AJ (1985)
VP 5.23E-007	mm Hg	25	EST	NEELY,WB & BLAU,GE (1985)
DC 2.28	pKa	25	EXP	ULLMANN VA2:63 (1985)
HL	atm m3/mol			
OH 5.13E-011	cm3/molc sec	25	EST	MEYLAN,WM & HOWARD,PH (1993)

CAS #:	000063-74-1			SULFANILAMIDE

Formula: $C_6H_8N_2O_2S$

Mol Weight: 172.21

MP (deg C): 165.5   FP (deg C):

BP (deg C):

BP pressure (mm Hg):

Property/ Value	Units	Temp	Data Type	Reference
WS 8.36E+003	mg/L	25	EXP	YALKOWSKY,SH & DANNENFELSER,RM (1992)
logP -0.62			EXP	HANSCH,C ET AL. (1995)
VP 7.30E-006	mm Hg	25	EST	NEELY,WB & BLAU,GE (1985)
DC 10.58	pKa	20	EXP	SERJEANT,EP & DEMPSEY,B (1979)
HL 1.50E-010	atm m3/mol	25	EST	MEYLAN,WM & HOWARD,PH (1991)
OH 2.30E-011	cm3/molc sec	25	EST	MEYLAN,WM & HOWARD,PH (1993)

CAS #:	000063-75-2			ARECOLINE

Formula: $C_8H_{13}NO_2$

Mol Weight: 155.20

MP (deg C):   FP (deg C):

BP (deg C): 209

BP pressure (mm Hg):

Property/ Value	Units	Temp	Data Type	Reference
WS 3.68E+005	mg/L	25	EST	MEYLAN,WM ET AL. (1996)
logP 0.35			EXP	HANSCH,C ET AL. (1995)
VP 1.94E-001	mm Hg	25	EST	NEELY,WB & BLAU,GE (1985)
DC	pKa			
HL	atm m3/mol			
OH 1.18E-010	cm3/molc sec	25	EST	MEYLAN,WM & HOWARD,PH (1993)

CAS #:	000063-84-3			DOPA

Formula: $C_9H_{11}NO_4$

Mol Weight: 197.19

MP (deg C):   FP (deg C):

BP (deg C):

BP pressure (mm Hg):

Property/ Value	Units	Temp	Data Type	Reference
WS 3.20E+005	mg/L	25	EST	MEYLAN,WM ET AL. (1996)
logP -2.74			EXP	HANSCH,C ET AL. (1995)
VP 1.24E-010	mm Hg	25	EST	NEELY,WB & BLAU,GE (1985)
DC 2.32	pKa	25	EXP	KORTUM,G ET AL (1961)
HL 1.30E-018	atm m3/mol	25	EST	MEYLAN,WM & HOWARD,PH (1991)
OH 9.66E-011	cm3/molc sec	25	EST	MEYLAN,WM & HOWARD,PH (1993)

CAS #:	000063-91-2			PHENYLALANINE

Formula: $C_9H_{11}NO_2$

Mol Weight: 165.19

MP (deg C): 283 dec   FP (deg C):

BP (deg C):

BP pressure (mm Hg):

Property/ Value	Units	Temp	Data Type	Reference
WS 1.98E+004	mg/L	25	EXP	YALKOWSKY,SH & DANNENFELSER,RM (1992)
logP -1.52			EXP	HANSCH,C & LEO,AJ (1985)
VP 4.65E-008	mm Hg	25	EST	NEELY,WB & BLAU,GE (1985)
DC 1.24	pKa	25	EXP	KORTUM,G ET AL (1961)
HL 1.20E-010	atm m3/mol	25	EST	MEYLAN,WM & HOWARD,PH (1991)
OH 4.42E-011	cm3/molc sec	25	EST	MEYLAN,WM & HOWARD,PH (1993)

CAS #: **000063-98-9**				PHENACEMIDE

Formula: $C_9H_{10}N_2O_2$

Mol Weight: 178.19

MP (deg C): 212-216   FP (deg C):

BP (deg C):

BP pressure (mm Hg):

Property/Value	Units	Temp	Data Type	Reference
WS 1.02E+004	mg/L	25	EST	MEYLAN,WM ET AL. (1996)
logP 0.87			EXP	HANSCH,C & LEO,AJ (1985)
VP 8.94E-008	mm Hg	25	EST	NEELY,WB & BLAU,GE (1985)
DC	pKa			
HL 8.71E-012	atm m3/mol	25	EST	MEYLAN,WM & HOWARD,PH (1991)
OH 7.55E-012	cm3/molc sec	25	EST	MEYLAN,WM & HOWARD,PH (1993)

CAS #: **000063-99-0**				M-TOLYLUREA

Formula: $C_8H_{10}N_2O$

Mol Weight: 150.18

MP (deg C):   FP (deg C):

BP (deg C):

BP pressure (mm Hg):

Property/Value	Units	Temp	Data Type	Reference
WS 5.99E+003	mg/L	25	EST	MEYLAN,WM ET AL. (1996)
logP 1.29			EXP	HANSCH,C & LEO,AJ (1985)
VP 3.19E-004	mm Hg	25	EST	NEELY,WB & BLAU,GE (1985)
DC	pKa			
HL 2.22E-010	atm m3/mol	25	EST	MEYLAN,WM & HOWARD,PH (1991)
OH 1.12E-010	cm3/molc sec	25	EST	MEYLAN,WM & HOWARD,PH (1993)

CAS #: **000064-00-6**				N-METHYL O-(3-IPRPHENYL)CARBAMATE

Formula: $C_{11}H_{15}NO_2$

Mol Weight: 193.25

MP (deg C):   FP (deg C):

BP (deg C):

BP pressure (mm Hg):

Property/Value	Units	Temp	Data Type	Reference
WS 2.69E+002	mg/L	25	EST	MEYLAN,WM ET AL. (1996)
logP 2.63			EXP	HANSCH,C & LEO,AJ (1985)
VP 4.39E-003	mm Hg	25	EST	NEELY,WB & BLAU,GE (1985)
DC	pKa			
HL 6.25E-008	atm m3/mol	25	EST	MEYLAN,WM & HOWARD,PH (1991)
OH 1.31E-011	cm3/molc sec	25	EST	MEYLAN,WM & HOWARD,PH (1993)

CAS #: **000064-02-8**				EDTA, SODIUM SALT

Formula: $C_{10}H_{12}N_2Na_4O_8$

Mol Weight: 380.17

MP (deg C): >300   FP (deg C):

BP (deg C):

BP pressure (mm Hg):

Property/Value	Units	Temp	Data Type	Reference
WS 1.00E+006	mg/L	25	EXP	BROWN,SL ET AL. (1975B)
logP -13.17			EST	MEYLAN,WM & HOWARD,PH (1995)
VP 1.49E-012	mm Hg	25	EST	NEELY,WB & BLAU,GE (1985)
DC	pKa			
HL	atm m3/mol			
OH 1.79E-010	cm3/molc sec	25	EST	MEYLAN,WM & HOWARD,PH (1993)

CAS #: **000064-04-0**				2-PHENYLETHYLAMINE

Formula: $C_8H_{11}N$

Mol Weight: 121.18

MP (deg C):   FP (deg C):

BP (deg C): 197.5

BP pressure (mm Hg):

Property/Value	Units	Temp	Data Type	Reference
WS 6.33E+004	mg/L	25	EST	MEYLAN,WM ET AL. (1996)
logP 1.41			EXP	HANSCH,C & LEO,AJ (1985)
VP 2.98E-001	mm Hg	25	EST	NEELY,WB & BLAU,GE (1985)
DC 9.96	pKa	19	EXP	PERRIN,DD (1965)
HL 8.12E-007	atm m3/mol	25	EST	MEYLAN,WM & HOWARD,PH (1991)
OH 3.77E-011	cm3/molc sec	25	EST	MEYLAN,WM & HOWARD,PH (1993)

CAS #: **000064-10-8**				PHENYLUREA

Formula: $C_7H_8N_2O$

Mol Weight: 136.15

MP (deg C): 147   FP (deg C):

BP (deg C): 238

BP pressure (mm Hg):

Property/Value	Units	Temp	Data Type	Reference
WS 1.70E+004	mg/L	25	EST	MEYLAN,WM ET AL. (1996)
logP 0.83			EXP	HANSCH,C & LEO,AJ (1985)
VP 9.39E-004	mm Hg	25	EST	NEELY,WB & BLAU,GE (1985)
DC	pKa			
HL 2.01E-010	atm m3/mol	25	EST	MEYLAN,WM & HOWARD,PH (1991)
OH 4.37E-011	cm3/molc sec	25	EST	MEYLAN,WM & HOWARD,PH (1993)

CAS #: **000064-13-1**				P-METHOXYAMPHETAMINE

Formula: $C_{10}H_{15}NO$

Mol Weight: 165.24

MP (deg C):   FP (deg C):

BP (deg C):

BP pressure (mm Hg):

Property/Value	Units	Temp	Data Type	Reference
WS 2.03E+004	mg/L	25	EST	MEYLAN,WM ET AL. (1996)
logP 1.77			EXP	HANSCH,C & LEO,AJ (1985)
VP 1.81E-002	mm Hg	25	EST	NEELY,WB & BLAU,GE (1985)
DC 9.53	pKa	25	EXP	PERRIN,DD (1965)
HL 6.38E-008	atm m3/mol	25	EST	MEYLAN,WM & HOWARD,PH (1991)
OH 7.15E-011	cm3/molc sec	25	EST	MEYLAN,WM & HOWARD,PH (1993)

CAS #: **000064-17-5**				ETHANOL

Formula: C_2H_6O

Mol Weight: 46.07

MP (deg C): -114.49   FP (deg C):

BP (deg C): 78.293

BP pressure (mm Hg):

Property/Value	Units	Temp	Data Type	Reference
WS 1.00E+006	mg/L		EXP	RIDDICK,JA ET AL. (1986)
logP -0.31			EXP	HANSCH,C & LEO,AJ (1985)
VP 5.93E+001	mm Hg	25	EXP	DAUBERT,TE & DANNER,RP (1985)
DC 15.90	pKa	25	EXP	RIDDICK,JA ET AL. (1986)
HL 5.00E-006	atm m3/mol	25	EXP	GAFFNEY,JS ET AL. (1987)
OH 3.27E-012	cm3/molc sec	25	EXP	ATKINSON,R (1989)

CAS #: 000064-18-6	FORMIC ACID

Formula: CH_2O_2

Mol Weight: 46.03

MP (deg C): 8.27   FP (deg C):

BP (deg C): 100.56

BP pressure (mm Hg):

Property/Value	Units	Temp	Data Type	Reference
WS 1.00E+006	mg/L		EXP	RIDDICK,JA ET AL. (1986)
logP -0.54			EXP	HANSCH,C & LEO,AJ (1985)
VP 4.26E+001	mm Hg	25	EXP	DAUBERT,TE & DANNER,RP (1985)
DC 3.75	pKa	25	EXP	RIDDICK,JA ET AL. (1986)
HL 1.67E-007	atm m3/mol	25	EXP	GAFFNEY,JS ET AL. (1987)
OH 4.50E-013	cm3/molc sec	25	EXP	ATKINSON,R (1989)

CAS #: 000064-19-7	ACETIC ACID

Formula: $C_2H_4O_2$

Mol Weight: 60.05

MP (deg C): 16.6   FP (deg C):

BP (deg C): 117.9

BP pressure (mm Hg):

Property/Value	Units	Temp	Data Type	Reference
WS 1.00E+006	mg/L		EXP	US EPA (1981)
logP -0.17			EXP	HANSCH,C & LEO,AJ (1985)
VP 1.57E+001	mm Hg	25	EXP	DAUBERT,TE & DANNER,RP (1985)
DC 4.76	pKa	25	EXP	SERJEANT,EP & DEMPSEY,B (1979)
HL 1.00E-007	atm m3/mol	25	EXP	GAFFNEY,JS ET AL. (1987)
OH 7.40E-013	cm3/molc sec	25	EXP	ATKINSON,R (1989)

CAS #: 000064-39-1	PROMEDOL

Formula: $C_{17}H_{25}NO_2$

Mol Weight: 275.39

MP (deg C):   FP (deg C):

BP (deg C):

BP pressure (mm Hg):

Property/Value	Units	Temp	Data Type	Reference
WS 3.68E+002	mg/L	25	EST	MEYLAN,WM ET AL. (1996)
logP 3.13			EXP	HANSCH,C ET AL. (1995)
VP 2.41E-005	mm Hg	25	EST	NEELY,WB & BLAU,GE (1985)
DC	pKa			
HL 1.69E-008	atm m3/mol	25	EST	MEYLAN,WM & HOWARD,PH (1991)
OH 1.10E-010	cm3/molc sec	25	EST	MEYLAN,WM & HOWARD,PH (1993)

CAS #: 000064-67-5	DIETHYL SULFATE

Formula: $C_4H_{10}O_4S$

Mol Weight: 154.19

MP (deg C): -24.4   FP (deg C): 113

BP (deg C): 208

BP pressure (mm Hg):

Property/Value	Units	Temp	Data Type	Reference
WS 7.00E+003	mg/L	20	EXP	MCCORMACK,WB & LAWES,BC (1983)
logP 1.14			EXP	HANSCH,C & LEO,AJ (1985)
VP 2.90E-001	mm Hg	25	EXT	DAUBERT,TE & DANNER,RP (1991)
DC	pKa			
HL 8.40E-006	atm m3/mol	20	EST	VP/WSOL
OH 1.80E-012	cm3/molc sec	25	EXP	JAPAR,SM ET AL. (1990A)

CAS #: 000064-77-7	TOLBUTAMIDE

Formula: $C_{12}H_{18}N_2O_3S$

Mol Weight: 270.35

MP (deg C): 128.5-9.5   FP (deg C):

BP (deg C):

BP pressure (mm Hg):

Property/Value	Units	Temp	Data Type	Reference
WS 1.09E+002	mg/L	37	EXP	YALKOWSKY,SH & DANNENFELSER,RM (1992)
logP 2.34			EXP	HANSCH,C & LEO,AJ (1985)
VP 2.74E-008	mm Hg	25	EST	NEELY,WB & BLAU,GE (1985)
DC	pKa			
HL 2.53E-010	atm m3/mol	25	EST	MEYLAN,WM & HOWARD,PH (1991)
OH 1.07E-011	cm3/molc sec	25	EST	MEYLAN,WM & HOWARD,PH (1993)

CAS #: 000064-85-7	DEOXYCORTICOSTERONE

Formula: $C_{21}H_{30}O_3$

Mol Weight: 330.47

MP (deg C): 141-142   FP (deg C):

BP (deg C):

BP pressure (mm Hg):

Property/Value	Units	Temp	Data Type	Reference
WS 5.95E+001	mg/L	37	EXP	YALKOWSKY,SH & DANNENFELSER,RM (1992)
logP 2.88			EXP	HANSCH,C & LEO,AJ (1985)
VP 2.85E-010	mm Hg	25	EST	NEELY,WB & BLAU,GE (1985)
DC	pKa			
HL 1.01E-008	atm m3/mol	25	EST	MEYLAN,WM & HOWARD,PH (1991)
OH 1.05E-010	cm3/molc sec	25	EST	MEYLAN,WM & HOWARD,PH (1993)

CAS #: 000064-86-8	COLCHICINE

Formula: $C_{22}H_{25}NO_6$

Mol Weight: 399.45

MP (deg C): 142-150   FP (deg C):

BP (deg C):

BP pressure (mm Hg):

Property/Value	Units	Temp	Data Type	Reference
WS 4.50E+004	mg/L	25	EXP	SEIDELL,A (1941)
logP 1.03			EXP	HANSCH,C & LEO,AJ (1985)
VP 3.13E-012	mm Hg	25	EST	NEELY,WB & BLAU,GE (1985)
DC	pKa			
HL 1.79E-017	atm m3/mol	25	EST	MEYLAN,WM & HOWARD,PH (1991)
OH 3.32E-010	cm3/molc sec	25	EST	MEYLAN,WM & HOWARD,PH (1993)

CAS #: 000065-19-0	YOHIMBINE HYDROCHLORIDE

Formula: $C_{21}H_{27}ClN_2O_3$

Mol Weight: 390.91

MP (deg C): 288-290 de   FP (deg C):

BP (deg C):

BP pressure (mm Hg):

Property/Value	Units	Temp	Data Type	Reference
WS 7.30E+003	mg/L	20	EXP	STEPHEN,H & STEPHEN,T (1963)
logP 0.75			EST	MEYLAN,WM & HOWARD,PH (1995)
VP 1.39E-018	mm Hg	25	EST	NEELY,WB & BLAU,GE (1985)
DC	pKa			
HL 1.83E-024	atm m3/mol	25	EST	MEYLAN,WM & HOWARD,PH (1991)
OH 2.68E-010	cm3/molc sec	25	EST	MEYLAN,WM & HOWARD,PH (1993)

PYRIDOXAL HYDROCHLORIDE

CAS #:	000065-22-5			
Formula:	$C_8H_{10}ClNO_3$			
Mol Weight:	203.63			
MP (deg C):	165	FP (deg C):		
BP (deg C):				
BP pressure (mm Hg):				

Property/Value	Units	Temp	Data Type	Reference
WS 5.00E+005	mg/L		EXP	COFFEN,DL (1978)
logP -3.32			EST	MEYLAN,WM & HOWARD,PH (1995)
VP 4.58E-007	mm Hg	25	EST	NEELY,WB & BLAU,GE (1985)
DC	pKa			
HL 1.03E-019	atm m3/mol	25	EST	MEYLAN,WM & HOWARD,PH (1991)
OH 3.69E-011	cm3/molc sec	25	EST	MEYLAN,WM & HOWARD,PH (1993)

PYRIDOXINE

CAS #:	000065-23-6			
Formula:	$C_8H_{11}NO_3$			
Mol Weight:	169.18			
MP (deg C):	159-162	FP (deg C):		
BP (deg C):				
BP pressure (mm Hg):				

Property/Value	Units	Temp	Data Type	Reference
WS 2.82E+005	mg/L	25	EST	MEYLAN,WM ET AL. (1996)
logP -0.77			EXP	SANGSTER,J (1993)
VP 2.84E-007	mm Hg	25	EST	NEELY,WB & BLAU,GE (1985)
DC	pKa			
HL 1.32E-015	atm m3/mol	25	EST	MEYLAN,WM & HOWARD,PH (1991)
OH 3.06E-011	cm3/molc sec	25	EST	MEYLAN,WM & HOWARD,PH (1993)

O-HYDROXYBENZAMIDE

CAS #:	000065-45-2			
Formula:	$C_7H_7NO_2$			
Mol Weight:	137.14			
MP (deg C):	140	FP (deg C):		
BP (deg C):	181.5			
BP pressure (mm Hg):				

Property/Value	Units	Temp	Data Type	Reference
WS 2.06E+003	mg/L	25	EXP	YALKOWSKY,SH & DANNENFELSER,RM (1992)
logP 1.28			EXP	HANSCH,C & LEO,AJ (1985)
VP 6.90E-006	mm Hg	25	EST	NEELY,WB & BLAU,GE (1985)
DC 8.37	pKa	20	EXP	KORTUM,G ET AL (1961)
HL 2.90E-010	atm m3/mol	25	EST	MEYLAN,WM & HOWARD,PH (1991)
OH 3.25E-011	cm3/molc sec	25	EST	MEYLAN,WM & HOWARD,PH (1993)

CYTIDINE

CAS #:	000065-46-3			
Formula:	$C_9H_{13}N_3O_5$			
Mol Weight:	243.22			
MP (deg C):	230-231 de	FP (deg C):		
BP (deg C):				
BP pressure (mm Hg):				

Property/Value	Units	Temp	Data Type	Reference
WS 1.76E+005	mg/L	25	EST	MEYLAN,WM ET AL. (1996)
logP -2.51			EXP	HANSCH,C ET AL. (1995)
VP 1.65E-011	mm Hg	25	EST	NEELY,WB & BLAU,GE (1985)
DC	pKa			
HL 1.57E-019	atm m3/mol	25	EST	MEYLAN,WM & HOWARD,PH (1991)
OH 1.39E-010	cm3/molc sec	25	EST	MEYLAN,WM & HOWARD,PH (1993)

P-AMINOSALICYLIC ACID

CAS #:	000065-49-6			
Formula:	$C_7H_7NO_3$			
Mol Weight:	153.14			
MP (deg C):	150-151	FP (deg C):		
BP (deg C):				
BP pressure (mm Hg):				

Property/Value	Units	Temp	Data Type	Reference
WS 1.69E+003	mg/L	23	EXP	YALKOWSKY,SH & DANNENFELSER,RM (1992)
logP 0.89			EXP	SANGSTER,J (1994)
VP 4.00E-006	mm Hg	25	EST	NEELY,WB & BLAU,GE (1985)
DC 2.05	pKa	25	EXP	KORTUM,G ET AL (1961)
HL 5.02E-012	atm m3/mol	25	EST	MEYLAN,WM & HOWARD,PH (1991)
OH 2.01E-010	cm3/molc sec	25	EST	MEYLAN,WM & HOWARD,PH (1993)

5-METHYLURACIL

CAS #:	000065-71-4			
Formula:	$C_5H_6N_2O_2$			
Mol Weight:	126.12			
MP (deg C):	316	FP (deg C):		
BP (deg C):				
BP pressure (mm Hg):				

Property/Value	Units	Temp	Data Type	Reference
WS 3.82E+003	mg/L	25	EXP	YALKOWSKY,SH & DANNENFELSER,RM (1992)
logP -0.62			EXP	HANSCH,C & LEO,AJ (1985)
VP 1.33E-006	mm Hg	25	EST	NEELY,WB & BLAU,GE (1985)
DC	pKa			
HL 1.36E-010	atm m3/mol	25	EST	MEYLAN,WM & HOWARD,PH (1991)
OH 2.04E-011	cm3/molc sec	25	EST	MEYLAN,WM & HOWARD,PH (1993)

BENZOIC ACID

CAS #:	000065-85-0			
Formula:	$C_7H_6O_2$			
Mol Weight:	122.12			
MP (deg C):	122.4	FP (deg C):		
BP (deg C):	249.2			
BP pressure (mm Hg):				

Property/Value	Units	Temp	Data Type	Reference
WS 3.40E+003	mg/L	25	EXP	YALKOWSKY,SH & DANNENFELSER,RM (1992)
logP 1.87			EXP	HANSCH,C & LEO,AJ (1985)
VP 1.20E-003	mm Hg	25	EXT	MALASPINA,L ET AL. (1973)
DC 4.21	pKa	25	EXP	SERJEANT,EP & DEMPSEY,B (1979)
HL 1.08E-007	atm m3/mol	25	EST	MEYLAN,WM & HOWARD,PH (1991)
OH 1.15E-012	cm3/molc sec	25	EST	MEYLAN,WM & HOWARD,PH (1993)

OROTIC ACID

CAS #:	000065-86-1			
Formula:	$C_5H_4N_2O_4$			
Mol Weight:	156.10			
MP (deg C):	345-346	FP (deg C):		
BP (deg C):				
BP pressure (mm Hg):				

Property/Value	Units	Temp	Data Type	Reference
WS 1.82E+003	mg/L	18	EXP	YALKOWSKY,SH & DANNENFELSER,RM (1992)
logP -0.83			EXP	SANGSTER,J (1994)
VP 1.03E-008	mm Hg	25	EST	NEELY,WB & BLAU,GE (1985)
DC	pKa			
HL 8.10E-015	atm m3/mol	25	EST	MEYLAN,WM & HOWARD,PH (1991)
OH 8.91E-012	cm3/molc sec	25	EST	MEYLAN,WM & HOWARD,PH (1993)

CAS #:	000066-02-4				3,5-DIIODOTYROSINE

Formula: $C_9H_9I_2NO_3$

Mol Weight: 432.99

MP (deg C): 213 dec

FP (deg C):

BP (deg C):

BP pressure (mm Hg):

Property/ Value	Units	Temp	Data Type	Reference
WS 6.00E+002	mg/L	25	EXP	YALKOWSKY,SH & DANNENFELSER,RM (1992)
logP 0.57			EST	MEYLAN,WM & HOWARD,PH (1995)
VP 8.22E-012	mm Hg	25	EST	NEELY,WB & BLAU,GE (1985)
DC 2.21	pKa	0	EXP	KORTUM,G ET AL (1961)
HL 6.73E-016	atm m3/mol	25	EST	MEYLAN,WM & HOWARD,PH (1991)
OH 4.09E-011	cm3/molc sec	25	EST	MEYLAN,WM & HOWARD,PH (1993)

CAS #:	000066-22-8				URACIL

Formula: $C_4H_4N_2O_2$

Mol Weight: 112.09

MP (deg C): 335

FP (deg C):

BP (deg C):

BP pressure (mm Hg):

Property/ Value	Units	Temp	Data Type	Reference
WS 3.60E+003	mg/L	25	EXP	YALKOWSKY,SH & DANNENFELSER,RM (1992)
logP -1.07			EXP	HANSCH,C & LEO,AJ (1985)
VP 2.77E-006	mm Hg	25	EST	NEELY,WB & BLAU,GE (1985)
DC	pKa			
HL 8.67E-011	atm m3/mol	25	EST	MEYLAN,WM & HOWARD,PH (1991)
OH 8.77E-012	cm3/molc sec	25	EST	MEYLAN,WM & HOWARD,PH (1993)

CAS #:	000066-23-9				ACETYLCHOLINE BROMIDE

Formula: $C_7H_{16}BrNO_2$

Mol Weight: 226.12

MP (deg C): 146

FP (deg C):

BP (deg C):

BP pressure (mm Hg):

Property/ Value	Units	Temp	Data Type	Reference
WS 1.00E+006	mg/L	25	EST	MEYLAN,WM ET AL. (1996)
logP -3.61			EXP	SANGSTER,J (1994)
VP 2.51E-007	mm Hg	25	EST	NEELY,WB & BLAU,GE (1985)
DC	pKa			
HL 1.05E-014	atm m3/mol	25	EST	MEYLAN,WM & HOWARD,PH (1991)
OH 1.64E-011	cm3/molc sec	25	EST	MEYLAN,WM & HOWARD,PH (1993)

CAS #:	000066-25-1				HEXANAL

Formula: $C_6H_{12}O$

Mol Weight: 100.16

MP (deg C): -56

FP (deg C): -56

BP (deg C): 131

BP pressure (mm Hg):

Property/ Value	Units	Temp	Data Type	Reference
WS 6.00E+003	mg/L	20	EXP	FALBE,J & LAPPE,P (1985)
logP 1.78			EXP	HANSCH,C & ET AL. (1995)
VP 1.13E+001	mm Hg	25	EXP	DAUBERT,TE & DANNER,RP (1989)
DC	pKa			
HL 2.13E-004	atm m3/mol	25	EXP	BUTTERY,RG ET AL. (1969)
OH 2.90E-011	cm3/molc sec	25	EST	MEYLAN,WM & HOWARD,PH (1993)

CAS #:	000066-27-3				METHYL METHANE SULFONATE

Formula: $C_2H_6O_3S$

Mol Weight: 110.13

MP (deg C): 20

FP (deg C):

BP (deg C): 203

BP pressure (mm Hg): 7.53E+002

Property/ Value	Units	Temp	Data Type	Reference
WS 1.00E+006	mg/L	25	EST	MEYLAN,WM ET AL (1996)
logP -0.66			EST	MEYLAN,WM & HOWARD,PH (1995)
VP 3.10E-001	mm Hg	25	EST	NEELY,WB & BLAU,GE (1985)
DC	pKa			
HL 4.03E-006	atm m3/mol	25	EST	MEYLAN,WM & HOWARD,PH (1991)
OH 2.30E-013	cm3/molc sec	25	EST	MEYLAN,WM & HOWARD,PH (1993)

CAS #:	000066-28-4				STROPHANTHIDIN

Formula: $C_{23}H_{32}O_6$

Mol Weight: 404.51

MP (deg C): 171-175

FP (deg C):

BP (deg C):

BP pressure (mm Hg):

Property/ Value	Units	Temp	Data Type	Reference
WS 8.16E+002	mg/L	25	EST	MEYLAN,WM ET AL. (1996)
logP 0.64			EXP	HANSCH,C ET AL. (1995)
VP 9.87E-016	mm Hg	25	EST	NEELY,WB & BLAU,GE (1985)
DC	pKa			
HL 1.59E-014	atm m3/mol	25	EST	MEYLAN,WM & HOWARD,PH (1991)
OH 9.14E-011	cm3/molc sec	25	EST	MEYLAN,WM & HOWARD,PH (1993)

CAS #:	000066-32-0				STRYCHNINE NITRATE

Formula: $C_{21}H_{23}N_3O_5$

Mol Weight: 397.43

MP (deg C): 295

FP (deg C):

BP (deg C):

BP pressure (mm Hg):

Property/ Value	Units	Temp	Data Type	Reference
WS 8.46E+001	mg/L	25	EST	MEYLAN,WM ET AL. (1996)
logP 0.31			EST	MEYLAN,WM & HOWARD,PH (1995)
VP 1.47E-015	mm Hg	25	EST	NEELY,WB & BLAU,GE (1985)
DC	pKa			
HL 1.19E-022	atm m3/mol	25	EST	MEYLAN,WM & HOWARD,PH (1991)
OH 1.86E-010	cm3/molc sec	25	EST	MEYLAN,WM & HOWARD,PH (1993)

CAS #:	000066-56-8				2,3-DINITROPHENOL

Formula: $C_6H_4N_2O_5$

Mol Weight: 184.11

MP (deg C): 144.5

FP (deg C):

BP (deg C):

BP pressure (mm Hg):

Property/ Value	Units	Temp	Data Type	Reference
WS 1.77E+003	mg/L	25	EST	MEYLAN,WM ET AL. (1996)
logP 1.73			EST	MEYLAN,WM & HOWARD,PH (1995)
VP 1.20E-005	mm Hg	25	EST	NEELY,WB & BLAU,GE (1985)
DC 4.96	pKa	25	EXP	SERJEANT,EP & DEMPSEY,B (1979)
HL 2.76E-008	atm m3/mol	25	EST	MEYLAN,WM & HOWARD,PH (1991)
OH 5.04E-013	cm3/molc sec	25	EST	MEYLAN,WM & HOWARD,PH (1993)

CAS #: 000066-71-7	O-PHENANTHROLINE

Formula: $C_{12}H_8N_2$

Mol Weight: 180.21

MP (deg C): 93-94

FP (deg C):

BP (deg C): > 300

BP pressure (mm Hg):

Property/Value	Units	Temp	Data Type	Reference
WS 2.69E+003	mg/L	25	EXP	YALKOWSKY,SH & DANNENFELSER,RM (1992)
logP 1.78			EXP	HANSCH,C & LEO,AJ (1985)
VP 5.23E-005	mm Hg	25	EST	NEELY,WB & BLAU,GE (1985)
DC 4.27	pKa	20	EXP	ALBERT,A ET AL. (1948)
HL 4.45E-010	atm m3/mol	25	EST	MEYLAN,WM & HOWARD,PH (1991)
OH 2.00E-012	cm3/molc sec	25	EST	MEYLAN,WM & HOWARD,PH (1993)

CAS #: 000066-72-8	PYRIDOXAL

Formula: $C_8H_9NO_3$

Mol Weight: 167.17

MP (deg C): 165

FP (deg C):

BP (deg C):

BP pressure (mm Hg):

Property/Value	Units	Temp	Data Type	Reference
WS 5.00E+005	mg/L		EXP	COFFEN,DL (1978)
logP 0.45			EST	MEYLAN,WM & HOWARD,PH (1995)
VP 1.62E-006	mm Hg	25	EST	NEELY,WB & BLAU,GE (1985)
DC	pKa			
HL 1.03E-013	atm m3/mol	25	EST	MEYLAN,WM & HOWARD,PH (1991)
OH 2.69E-011	cm3/molc sec	25	EST	MEYLAN,WM & HOWARD,PH (1993)

CAS #: 000066-75-1	URACIL MUSTARD

Formula: $C_8H_{11}Cl_2N_3O_2$

Mol Weight: 252.10

MP (deg C): 206

FP (deg C):

BP (deg C):

BP pressure (mm Hg):

Property/Value	Units	Temp	Data Type	Reference
WS 1.07E+003	mg/L	25	EST	MEYLAN,WM ET AL. (1996)
logP 0.64			EST	MEYLAN,WM & HOWARD,PH (1995)
VP 7.50E-010	mm Hg	25	EST	NEELY,WB & BLAU,GE (1985)
DC	pKa			
HL 3.90E-013	atm m3/mol	25	EST	MEYLAN,WM & HOWARD,PH (1991)
OH 2.01E-011	cm3/molc sec	25	EST	MEYLAN,WM & HOWARD,PH (1993)

CAS #: 000066-76-2	DICOUMAROL

Formula: $C_{19}H_{12}O_6$

Mol Weight: 336.30

MP (deg C): 287-293

FP (deg C):

BP (deg C):

BP pressure (mm Hg):

Property/Value	Units	Temp	Data Type	Reference
WS 1.28E+002	mg/L	25	EST	MEYLAN,WM ET AL. (1996)
logP 2.07			EXP	HANSCH,C & LEO,AJ (1985)
VP 6.02E-017	mm Hg	25	EST	NEELY,WB & BLAU,GE (1985)
DC	pKa			
HL 1.36E-013	atm m3/mol	25	EST	MEYLAN,WM & HOWARD,PH (1991)
OH 8.20E-011	cm3/molc sec	25	EST	MEYLAN,WM & HOWARD,PH (1993)

CAS #: 000066-79-5	OXACILLIN

Formula: $C_{19}H_{20}N_3O_5S$

Mol Weight: 402.45

MP (deg C): 188

FP (deg C):

BP (deg C):

BP pressure (mm Hg):

Property/Value	Units	Temp	Data Type	Reference
WS 2.78E+001	mg/L	25	EST	MEYLAN,WM ET AL. (1996)
logP 2.38			EXP	HANSCH,C & LEO,AJ (1985)
VP 5.51E-015	mm Hg	25	EST	NEELY,WB & BLAU,GE (1985)
DC	pKa			
HL 2.55E-017	atm m3/mol	25	EST	MEYLAN,WM & HOWARD,PH (1991)
OH 9.85E-011	cm3/molc sec	25	EST	MEYLAN,WM & HOWARD,PH (1993)

CAS #: 000066-81-9	CYCLOHEXIMIDE

Formula: $C_{15}H_{23}NO_4$

Mol Weight: 281.35

MP (deg C): 119.5-121

FP (deg C):

BP (deg C):

BP pressure (mm Hg):

Property/Value	Units	Temp	Data Type	Reference
WS 2.10E+004	mg/L	2	EXP	YALKOWSKY,SH & DANNENFELSER,RM (1992)
logP 0.55			EXP	HANSCH,C & LEO,AJ (1985)
VP 1.89E-014	mm Hg	25	EST	NEELY,WB & BLAU,GE (1985)
DC	pKa			
HL 3.52E-015	atm m3/mol	25	EST	MEYLAN,WM & HOWARD,PH (1991)
OH 6.55E-011	cm3/molc sec	25	EST	MEYLAN,WM & HOWARD,PH (1993)

CAS #: 000066-97-7	PSORALEN

Formula: $C_{11}H_6O_3$

Mol Weight: 186.17

MP (deg C): 163-164

FP (deg C):

BP (deg C):

BP pressure (mm Hg):

Property/Value	Units	Temp	Data Type	Reference
WS 1.93E+003	mg/L	25	EST	MEYLAN,WM ET AL. (1996)
logP 1.67			EXP	(in press)
VP 2.24E-005	mm Hg	25	EST	NEELY,WB & BLAU,GE (1985)
DC	pKa			
HL 6.77E-007	atm m3/mol	25	EST	MEYLAN,WM & HOWARD,PH (1991)
OH 4.18E-011	cm3/molc sec	25	EST	MEYLAN,WM & HOWARD,PH (1993)

CAS #: 000067-20-9	NITROFURANTOIN

Formula: $C_8H_6N_4O_5$

Mol Weight: 238.16

MP (deg C): 263

FP (deg C):

BP (deg C):

BP pressure (mm Hg):

Property/Value	Units	Temp	Data Type	Reference
WS 7.95E+001	mg/L	24	EXP	YALKOWSKY,SH & DANNENFELSER,RM (1992)
logP -0.47			EXP	HANSCH,C & LEO,AJ (1985)
VP 2.78E-010	mm Hg	25	EST	NEELY,WB & BLAU,GE (1985)
DC	pKa			
HL 1.33E-012	atm m3/mol	25	EST	MEYLAN,WM & HOWARD,PH (1991)
OH 3.22E-011	cm3/molc sec	25	EST	MEYLAN,WM & HOWARD,PH (1993)

CAS #: 000067-45-8				2-OXAZOLIDINONE, 3- (5-NITRO-2-FURANYL)METHYLEN
Formula: $C_8H_7N_3O_5$				
Mol Weight: 225.16				
MP (deg C): 255			FP (deg C):	
BP (deg C):				
BP pressure (mm Hg):				

Property/Value	Units	Temp	Data Type	Reference
WS 6.97E+002	mg/L	25	EST	MEYLAN,WM ET AL. (1996)
logP -0.04			EXP	DEBNATH,AK ET AL. (1991)
VP 2.60E-006	mm Hg	25	EST	NEELY,WB & BLAU,GE (1985)
DC	pKa			
HL 3.26E-011	atm m3/mol	25	EST	MEYLAN,WM & HOWARD,PH (1991)
OH 3.42E-011	cm3/molc sec	25	EST	MEYLAN,WM & HOWARD,PH (1993)

CAS #: 000067-51-6				1H-PYRAZOLE, 3,5-DIMETHYL-
Formula: $C_5H_8N_2$				
Mol Weight: 96.13				
MP (deg C): 107.5			FP (deg C):	
BP (deg C): 218				
BP pressure (mm Hg):				

Property/Value	Units	Temp	Data Type	Reference
WS 1.65E+004	mg/L	25	EST	MEYLAN,WM ET AL. (1996)
logP 1.01			EXP	HANSCH,C ET AL. (1995)
VP 1.43E-001	mm Hg	25	EST	NEELY,WB & BLAU,GE (1985)
DC 4.38	pKa	25	EXP	PERRIN,DD (1965)
HL 4.50E-006	atm m3/mol	25	EST	MEYLAN,WM & HOWARD,PH (1991)
OH 2.00E-010	cm3/molc sec	25	EST	MEYLAN,WM & HOWARD,PH (1993)

CAS #: 000067-52-7				BARBITURIC ACID
Formula: $C_4H_4N_2O_3$				
Mol Weight: 128.09				
MP (deg C): 248			FP (deg C):	
BP (deg C):				
BP pressure (mm Hg):				

Property/Value	Units	Temp	Data Type	Reference
WS 1.90E-002	mg/L	37	EXP	YALKOWSKY,SH & DANNENFELSER,RM (1992)
logP -1.47			EXP	HANSCH,C & LEO,AJ (1985)
VP 2.17E-009	mm Hg	25	EST	NEELY,WB & BLAU,GE (1985)
DC 4.04	pKa	25	EXP	KORTUM,G ET AL (1961)
HL 1.16E-013	atm m3/mol	25	EST	MEYLAN,WM & HOWARD,PH (1991)
OH 2.53E-012	cm3/molc sec	25	EST	MEYLAN,WM & HOWARD,PH (1993)

CAS #: 000067-56-1				METHANOL
Formula: CH_4O				
Mol Weight: 32.04				
MP (deg C): -97.68			FP (deg C):	
BP (deg C): 64.55				
BP pressure (mm Hg):				

Property/Value	Units	Temp	Data Type	Reference
WS 1.00E+006	mg/L		EXP	RIDDICK,JA ET AL. (1986)
logP -0.77			EXP	HANSCH,C & LEO,AJ (1985)
VP 1.27E+002	mm Hg	25	EXP	BOUBLIK,T ET AL. (1984)
DC 15.30	pKa		EXP	SERJEANT,EP & DEMPSEY,B (1979)
HL 4.55E-006	atm m3/mol	25	EXP	GAFFNEY,JS ET AL. (1987)
OH 9.32E-013	cm3/molc sec	25	EXP	ATKINSON,R (1989)

CAS #: 000067-62-9				O-METHYLHYDROXYLAMINE
Formula: CH_5NO				
Mol Weight: 47.06				
MP (deg C):			FP (deg C):	
BP (deg C): 49-50				
BP pressure (mm Hg):				

Property/Value	Units	Temp	Data Type	Reference
WS 2.95E+005	mg/L	25	EST	MEYLAN,WM ET AL. (1996)
logP -0.40			EST	MEYLAN,WM & HOWARD,PH (1995)
VP 2.91E+002	mm Hg	25	EXP	BOUBLIK,T ET AL. (1984)
DC	pKa			
HL 2.21E-006	atm m3/mol	25	EST	MEYLAN,WM & HOWARD,PH (1991)
OH 2.10E-011	cm3/molc sec	25	EST	MEYLAN,WM & HOWARD,PH (1993)

CAS #: 000067-63-0				ISOPROPANOL
Formula: C_3H_8O				
Mol Weight: 60.10				
MP (deg C): -88.0			FP (deg C):	
BP (deg C): 82.24				
BP pressure (mm Hg):				

Property/Value	Units	Temp	Data Type	Reference
WS 1.00E+006	mg/L		EXP	RIDDICK,JA ET AL. (1986)
logP 0.05			EXP	HANSCH,C & LEO,AJ (1985)
VP 4.54E+001	mm Hg	25	EXP	DAUBERT,TE & DANNER,RP (1985)
DC 17.10	pKa		EXP	SERJEANT,EP & DEMPSEY,B (1979)
HL 8.10E-006	atm m3/mol	25	EXP	TAFT,RW ET AL. (1985)
OH 5.21E-012	cm3/molc sec	25	EXP	ATKINSON,R (1989)

CAS #: 000067-64-1				ACETONE
Formula: C_3H_6O				
Mol Weight: 58.08				
MP (deg C): -94.7			FP (deg C):	
BP (deg C): 56.07				
BP pressure (mm Hg):				

Property/Value	Units	Temp	Data Type	Reference
WS 1.00E+006	mg/L		EXP	RIDDICK,JA ET AL. (1986)
logP -0.24			EXP	HANSCH,C & LEO,AJ (1985)
VP 2.31E+002	mm Hg	25	EXP	BOUBLIK,T ET AL. (1984)
DC 20.00	pKa		EXP	SERJEANT,EP & DEMPSEY,B (1979)
HL 3.97E-005	atm m3/mol	25	EXP	TAFT,RW ET AL. (1985)
OH 2.26E-013	cm3/molc sec	25	EXP	ATKINSON,R (1989)

CAS #: 000067-66-3				CHLOROFORM
Formula: $CHCl_3$				
Mol Weight: 119.38				
MP (deg C): -63.52			FP (deg C):	
BP (deg C): 61.18				
BP pressure (mm Hg):				

Property/Value	Units	Temp	Data Type	Reference
WS 7.95E+003	mg/L	25	EXP	MACKAY,D ET AL. (1980)
logP 1.97			EXP	HANSCH,C & LEO,AJ (1985)
VP 1.97E+002	mm Hg	25	EXP	BOUBLIK,T ET AL. (1984)
DC	pKa			
HL 3.67E-003	atm m3/mol	24	EXP	GOSSETT,JM (1987)
OH 1.03E-013	cm3/molc sec	25	EXP	ATKINSON,R (1989)

DIMETHYL SULFOXIDE

CAS #: 000067-68-5

Formula: C_2H_6OS

Mol Weight: 78.13

MP (deg C): 18.45

FP (deg C):

BP (deg C): 189

BP pressure (mm Hg):

Property/ Value	Units	Temp	Data Type	Reference
WS 1.00E+006	mg/L		EXP	DORIGAN,J ET AL. (1976A)
logP -1.35			EXP	HANSCH,C & LEO,AJ (1985)
VP 6.10E-001	mm Hg	25	EXP	DAUBET,TE & DANNER,RP (1989)
DC	pKa			
HL 1.51E-009	atm m3/mol	25	EXP	TAFT,RW ET AL. (1985)
OH 6.20E-011	cm3/molc sec	25	EXP	ATKINSON,R (1989)

DIMETHYL SULFONE

CAS #: 000067-71-0

Formula: $C_2H_6O_2S$

Mol Weight: 94.13

MP (deg C): 109

FP (deg C):

BP (deg C): 238

BP pressure (mm Hg):

Property/ Value	Units	Temp	Data Type	Reference
WS 1.00E+006	mg/L	25	EST	MEYLAN,WM ET AL. (1996)
logP -1.41			EXP	HANSCH,C & LEO,AJ (1985)
VP 5.15E+000	mm Hg	25	EST	NEELY,WB & BLAU,GE (1985)
DC	pKa			
HL 6.24E-006	atm m3/mol	25	EST	MEYLAN,WM & HOWARD,PH (1991)
OH 2.12E-012	cm3/molc sec	25	EST	MEYLAN,WM & HOWARD,PH (1993)

HEXACHLOROETHANE

CAS #: 000067-72-1

Formula: C_2Cl_6

Mol Weight: 236.74

MP (deg C): 185

FP (deg C):

BP (deg C): 186.8

BP pressure (mm Hg):

Property/ Value	Units	Temp	Data Type	Reference
WS 7.70E+000	mg/L	20	EXP	YALKOWSKY,SH & DANNENFELSER,RM (1992)
logP 3.91			EXP	HANSCH,C ET AL. (1995)
VP 2.10E-001	mm Hg	20	EXP	ARCHER,WL (1979)
DC	pKa			
HL 3.89E-003	atm m3/mol	25	EXP	MUNZ,C & ROBERTS,PV (1987)
OH 0.00E+000	cm3/molc sec	25	EST	MEYLAN,WM & HOWARD,PH (1993)

FLUOCINOLONE ACETONIDE

CAS #: 000067-73-2

Formula: $C_{24}H_{30}F_2O_6$

Mol Weight: 452.50

MP (deg C): 265-266

FP (deg C):

BP (deg C):

BP pressure (mm Hg):

Property/ Value	Units	Temp	Data Type	Reference
WS 1.10E+001	mg/L	25	EST	MEYLAN,WM ET AL. (1996)
logP 2.48			EXP	HANSCH,C ET AL. (1995)
VP 1.00E-013	mm Hg	25	EST	NEELY,WB & BLAU,GE (1985)
DC	pKa			
HL 1.96E-012	atm m3/mol	25	EST	MEYLAN,WM & HOWARD,PH (1991)
OH 8.64E-011	cm3/molc sec	25	EST	MEYLAN,WM & HOWARD,PH (1993)

TRIAMCINOLONE DIACETATE

CAS #: 000067-78-7

Formula: $C_{25}H_{31}FO_8$

Mol Weight: 478.52

MP (deg C):

FP (deg C):

BP (deg C):

BP pressure (mm Hg):

Property/ Value	Units	Temp	Data Type	Reference
WS 2.25E+001	mg/L	25	EST	MEYLAN,WM ET AL. (1996)
logP 1.92			EXP	HANSCH,C & LEO,AJ (1985)
VP 2.74E-014	mm Hg	25	EST	NEELY,WB & BLAU,GE (1985)
DC	pKa			
HL 1.39E-014	atm m3/mol	25	EST	MEYLAN,WM & HOWARD,PH (1991)
OH 7.18E-011	cm3/molc sec	25	EST	MEYLAN,WM & HOWARD,PH (1993)

GLIOTOXIN

CAS #: 000067-99-2

Formula: $C_{13}H_{14}N_2O_4S_2$

Mol Weight: 326.40

MP (deg C):

FP (deg C):

BP (deg C):

BP pressure (mm Hg):

Property/ Value	Units	Temp	Data Type	Reference
WS 7.00E+001	mg/L	30	EXP	SHIU,WY ET AL. (1990)
logP 2.90			EST	MEYLAN,WM & HOWARD,PH (1995)
VP 1.32E-014	mm Hg	25	EST	NEELY,WB & BLAU,GE (1985)
DC	pKa			
HL 1.39E-011	atm m3/mol	25	EST	MEYLAN,WM & HOWARD,PH (1991)
OH 4.58E-010	cm3/molc sec	25	EST	MEYLAN,WM & HOWARD,PH (1993)

TETRAETHYL AMMONIUM IODIDE

CAS #: 000068-05-3

Formula: $C_8H_{20}IN$

Mol Weight: 257.16

MP (deg C):

FP (deg C):

BP (deg C):

BP pressure (mm Hg):

Property/ Value	Units	Temp	Data Type	Reference
WS 4.50E+005	mg/L	25	EXP	PEDDLE,CJ & TURNER,WES (1913)
logP -2.82			EXP	HANSCH,C & LEO,AJ (1985)
VP 2.52E-007	mm Hg	25	EST	NEELY,WB & BLAU,GE (1985)
DC	pKa			
HL 8.18E-012	atm m3/mol	25	EST	MEYLAN,WM & HOWARD,PH (1991)
OH 3.54E-011	cm3/molc sec	25	EST	MEYLAN,WM & HOWARD,PH (1993)

MERCAPTOACETIC ACID

CAS #: 000068-11-1

Formula: $C_2H_4O_2S$

Mol Weight: 92.12

MP (deg C): -16.5

FP (deg C):

BP (deg C): 123

BP pressure (mm Hg): 2.90E+001

Property/ Value	Units	Temp	Data Type	Reference
WS 1.00E+006	mg/L		EXP	MERCK INDEX (1976) @2ND
logP 0.09			EXP	HANSCH,C & LEO,AJ (1985)
VP 8.68E-002	mm Hg	25	EXT	PERRY,RH & GREEN,D (1984)
DC 3.55	pKa		EXP	SERJEANT,EP & DEMPSEY,B (1979)
HL 1.94E-008	atm m3/mol	25	EST	MEYLAN,WM & HOWARD,PH (1991)
OH 3.85E-011	cm3/molc sec	25	EST	MEYLAN,WM & HOWARD,PH (1993)

CAS #: 000068-12-2 — N,N'-DIMETHYLFORMAMIDE

Formula: C_3H_7NO

Mol Weight: 73.10

MP (deg C): -61

FP (deg C):

BP (deg C): 153

BP pressure (mm Hg):

Property/Value	Units	Temp	Data Type	Reference
WS 1.00E+006	mg/L		EXP	RIDDICK,JA ET AL. (1986)
logP -1.01			EXP	HANSCH,C & LEO,AJ (1985)
VP 3.87E+000	mm Hg	25	EXP	DAUBERT,TE & DANNER,RP (1985)
DC -0.30	pKa		EXP	PERRIN,DD (1972)
HL 7.39E-008	atm m3/mol	25	EXP	TAFT,RW ET AL. (1985)
OH 1.76E-010	cm3/molc sec	25	EST	MEYLAN,WM & HOWARD,PH (1993)

CAS #: 000068-35-9 — SULFADIAZINE

Formula: $C_{10}H_{10}N_4O_2S$

Mol Weight: 250.28

MP (deg C): 252-256

FP (deg C):

BP (deg C):

BP pressure (mm Hg):

Property/Value	Units	Temp	Data Type	Reference
WS 7.70E+001	mg/L	25	EXP	YALKOWSKY,SH & DANNENFELSER,RM (1992)
logP -0.09			EXP	HANSCH,C & LEO,AJ (1985)
VP 4.31E-008	mm Hg	25	EST	NEELY,WB & BLAU,GE (1985)
DC	pKa			
HL 1.58E-010	atm m3/mol	25	EST	MEYLAN,WM & HOWARD,PH (1991)
OH 2.80E-011	cm3/molc sec	25	EST	MEYLAN,WM & HOWARD,PH (1993)

CAS #: 000068-36-0 — P-DI(TRICHLOROMETHYL)BENZENE

Formula: $C_8H_4Cl_6$

Mol Weight: 312.84

MP (deg C): 101-110

FP (deg C):

BP (deg C):

BP pressure (mm Hg):

Property/Value	Units	Temp	Data Type	Reference
WS 1.17E+000	mg/L	25	EST	MEYLAN,WM ET AL. (1996)
logP 4.62			EXP	HANSCH,C & LEO,AJ (1985)
VP 1.02E-003	mm Hg	25	EST	NEELY,WB & BLAU,GE (1985)
DC	pKa			
HL 1.25E-005	atm m3/mol	25	EST	MEYLAN,WM & HOWARD,PH (1991)
OH 4.81E-014	cm3/molc sec	25	EST	MEYLAN,WM & HOWARD,PH (1993)

CAS #: 000068-76-8 — TRIAZIQUONE

Formula: $C_{12}H_{13}N_3O_2$

Mol Weight: 231.26

MP (deg C): 162.5-163

FP (deg C):

BP (deg C): 357.5

BP pressure (mm Hg):

Property/Value	Units	Temp	Data Type	Reference
WS 3.94E+005	mg/L	25	EST	MEYLAN,WM ET AL. (1996)
logP -0.13			EXP	HANSCH,C ET AL. (1995)
VP 1.92E-006	mm Hg	25	EST	NEELY,WB & BLAU,GE (1985)
DC	pKa			
HL 9.25E-016	atm m3/mol	25	EST	MEYLAN,WM & HOWARD,PH (1991)
OH 3.05E-011	cm3/molc sec	25	EST	MEYLAN,WM & HOWARD,PH (1993)

CAS #: 000068-90-6 — BENZIODARONE

Formula: $C_{17}H_{12}I_2O_3$

Mol Weight: 518.09

MP (deg C): 167

FP (deg C):

BP (deg C):

BP pressure (mm Hg):

Property/Value	Units	Temp	Data Type	Reference
WS 4.95E-003	mg/L	25	EST	MEYLAN,WM ET AL. (1996)
logP 6.59			EST	MEYLAN,WM & HOWARD,PH (1995)
VP 1.49E-011	mm Hg	25	EST	NEELY,WB & BLAU,GE (1985)
DC	pKa			
HL 1.55E-012	atm m3/mol	25	EST	MEYLAN,WM & HOWARD,PH (1991)
OH 4.17E-011	cm3/molc sec	25	EST	MEYLAN,WM & HOWARD,PH (1993)

CAS #: 000068-94-0 — HYPOXANTHINE

Formula: $C_5H_4N_4O$

Mol Weight: 136.11

MP (deg C): 150 dec

FP (deg C):

BP (deg C):

BP pressure (mm Hg):

Property/Value	Units	Temp	Data Type	Reference
WS 7.00E+002	mg/L	23	EXP	YALKOWSKY,SH & DANNENFELSER,RM (1992)
logP -1.11			EXP	HANSCH,C & LEO,AJ (1985)
VP 7.96E-010	mm Hg	25	EST	NEELY,WB & BLAU,GE (1985)
DC	pKa			
HL 2.55E-013	atm m3/mol	25	EST	MEYLAN,WM & HOWARD,PH (1991)
OH 4.95E-011	cm3/molc sec	25	EST	MEYLAN,WM & HOWARD,PH (1993)

CAS #: 000068-96-2 — 17-ALPHA-HYDROXYPROGESTERONE

Formula: $C_{21}H_{30}O_3$

Mol Weight: 330.47

MP (deg C): 222-223

FP (deg C):

BP (deg C):

BP pressure (mm Hg):

Property/Value	Units	Temp	Data Type	Reference
WS 6.48E+000	mg/L		EXP	YALKOWSKY,SH & DANNENFELSER,RM (1992)
logP 3.17			EXP	HANSCH,C ET AL. (1995)
VP 9.28E-010	mm Hg	25	EST	NEELY,WB & BLAU,GE (1985)
DC	pKa			
HL 1.01E-008	atm m3/mol	25	EST	MEYLAN,WM & HOWARD,PH (1991)
OH 1.01E-010	cm3/molc sec	25	EST	MEYLAN,WM & HOWARD,PH (1993)

CAS #: 000069-23-8 — FLUPHENAZINE

Formula: $C_{22}H_{26}F_3N_3OS$

Mol Weight: 437.53

MP (deg C):

FP (deg C):

BP (deg C): 268-274

BP pressure (mm Hg): 5.00E-001

Property/Value	Units	Temp	Data Type	Reference
WS 3.11E+001	mg/L	37	EXP	YALKOWSKY,SH & DANNENFELSER,RM (1992)
logP 4.36			EXP	HANSCH,C & LEO,AJ (1985)
VP 1.02E-012	mm Hg	25	EST	NEELY,WB & BLAU,GE (1985)
DC	pKa			
HL 6.09E-017	atm m3/mol	25	EST	MEYLAN,WM & HOWARD,PH (1991)
OH 1.84E-010	cm3/molc sec	25	EST	MEYLAN,WM & HOWARD,PH (1993)

CAS #: 000069-33-0	TUBERCIDIN

Formula: $C_{11}H_{16}N_4O_4$

Mol Weight: 268.27

MP (deg C):

FP (deg C):

BP (deg C):

BP pressure (mm Hg):

Property/Value	Units	Temp	Data Type	Reference
WS 9.25E+004	mg/L	25	EST	MEYLAN,WM ET AL. (1996)
logP -0.80			EXP	HANSCH,C & LEO,AJ (1985)
VP 4.27E-013	mm Hg	25	EST	NEELY,WB & BLAU,GE (1985)
DC	pKa			
HL 2.69E-022	atm m3/mol	25	EST	MEYLAN,WM & HOWARD,PH (1991)
OH 2.47E-010	cm3/molc sec	25	EST	MEYLAN,WM & HOWARD,PH (1993)

CAS #: 000069-53-4	AMPICILLIN

Formula: $C_{16}H_{19}N_3O_4S$

Mol Weight: 349.41

MP (deg C): 208 dec

FP (deg C):

BP (deg C):

BP pressure (mm Hg):

Property/Value	Units	Temp	Data Type	Reference
WS 1.01E+004	mg/L	21	EXP	YALKOWSKY,SH & DANNENFELSER,RM (1992)
logP 1.35			EXP	SANGSTER,J (1994)
VP 7.75E-015	mm Hg	25	EST	NEELY,WB & BLAU,GE (1985)
DC	pKa			
HL 2.39E-017	atm m3/mol	25	EST	MEYLAN,WM & HOWARD,PH (1991)
OH 1.25E-010	cm3/molc sec	25	EST	MEYLAN,WM & HOWARD,PH (1993)

CAS #: 000069-65-8	MANNITOL

Formula: $C_6H_{14}O_6$

Mol Weight: 182.17

MP (deg C): 166-168

FP (deg C):

BP (deg C): 290-295

BP pressure (mm Hg): 3.50E+000

Property/Value	Units	Temp	Data Type	Reference
WS 2.16E+005	mg/L	25	EXP	YALKOWSKY,SH & DANNENFELSER,RM (1992)
logP -3.10			EXP	HANSCH,C ET AL. (1995)
VP 4.92E-009	mm Hg	25	EST	NEELY,WB & BLAU,GE (1985)
DC	pKa			
HL 7.26E-013	atm m3/mol	25	EST	MEYLAN,WM & HOWARD,PH (1991)
OH 5.00E-011	cm3/molc sec	25	EST	MEYLAN,WM & HOWARD,PH (1993)

CAS #: 000069-72-7	SALICYLIC ACID

Formula: $C_7H_6O_3$

Mol Weight: 138.12

MP (deg C): 157-159

FP (deg C):

BP (deg C): 211

BP pressure (mm Hg): 2.00E+001

Property/Value	Units	Temp	Data Type	Reference
WS 2.06E+003	mg/L		EXP	YALKOWSKY,SH & DANNENFELSER,RM (1992)
logP 2.26			EXP	HANSCH,C & LEO,AJ (1985)
VP 8.20E-005	mm Hg	25	EXP	DAUBERT,TE & DANNER,RP (1989)
DC 2.97	pKa		EXP	SERJEANT,EP & DEMPSEY,B (1979)
HL 7.34E-009	atm m3/mol	25	EST	VP/WSOL
OH 1.32E-011	cm3/molc sec	25	EST	ATKINSON,R (1987)

CAS #: 000069-79-4	MALTOSE

Formula: $C_{12}H_{22}O_{11}$

Mol Weight: 342.30

MP (deg C): 102-103

FP (deg C):

BP (deg C):

BP pressure (mm Hg):

Property/Value	Units	Temp	Data Type	Reference
WS 7.80E+005	mg/L	20	EXP	YALKOWSKY,SH & DANNENFELSER,RM (1992)
logP -5.43			EST	MEYLAN,WM & HOWARD,PH (1995)
VP 8.09E-017	mm Hg	25	EST	NEELY,WB & BLAU,GE (1985)
DC	pKa			
HL 3.40E-019	atm m3/mol	25	EST	MEYLAN,WM & HOWARD,PH (1991)
OH 1.54E-010	cm3/molc sec	25	EST	MEYLAN,WM & HOWARD,PH (1993)

CAS #: 000069-89-6	XANTHINE

Formula: $C_5H_4N_4O_2$

Mol Weight: 152.11

MP (deg C):

FP (deg C):

BP (deg C):

BP pressure (mm Hg):

Property/Value	Units	Temp	Data Type	Reference
WS 1.53E+004	mg/L	25	EST	MEYLAN,WM ET AL. (1996)
logP -0.73			EXP	HANSCH,C & LEO,AJ (1985)
VP 2.64E-010	mm Hg	25	EST	NEELY,WB & BLAU,GE (1985)
DC 7.53	pKa		EXP	KORTUM,G ET AL (1961)
HL 3.73E-014	atm m3/mol	25	EST	MEYLAN,WM & HOWARD,PH (1991)
OH 5.48E-011	cm3/molc sec	25	EST	MEYLAN,WM & HOWARD,PH (1993)

CAS #: 000069-91-0	ALPHA-PHENYLGLYCINE

Formula: $C_8H_9NO_2$

Mol Weight: 151.17

MP (deg C): 292 dec

FP (deg C):

BP (deg C):

BP pressure (mm Hg):

Property/Value	Units	Temp	Data Type	Reference
WS 1.81E+004	mg/L	25	EST	MEYLAN,WM ET AL. (1996)
logP -1.70			EXP	HANSCH,C ET AL. (1995)
VP 1.18E-007	mm Hg	25	EST	NEELY,WB & BLAU,GE (1985)
DC	pKa			
HL 9.08E-011	atm m3/mol	25	EST	MEYLAN,WM & HOWARD,PH (1991)
OH 3.97E-011	cm3/molc sec	25	EST	MEYLAN,WM & HOWARD,PH (1993)

CAS #: 000069-93-2	URIC ACID

Formula: $C_5H_6N_4O_3$

Mol Weight: 170.13

MP (deg C):

FP (deg C):

BP (deg C):

BP pressure (mm Hg):

Property/Value	Units	Temp	Data Type	Reference
WS 9.00E+001	mg/L	30	EXP	YALKOWSKY,SH & DANNENFELSER,RM (1992)
logP -2.17			EXP	NAHUM,A & HORVATH,C (1980)
VP 7.89E-010	mm Hg	25	EST	NEELY,WB & BLAU,GE (1985)
DC 5.40	pKa		EXP	KORTUM,G ET AL (1961)
HL 1.20E-017	atm m3/mol	25	EST	MEYLAN,WM & HOWARD,PH (1991)
OH 1.23E-011	cm3/molc sec	25	EST	MEYLAN,WM & HOWARD,PH (1993)

CAS #: 000070-00-8 — TRIFLURIDINE

Formula: $C_{10}H_{11}F_3N_2O_5$

Mol Weight: 296.20

MP (deg C): 186-189

FP (deg C):

BP (deg C):

BP pressure (mm Hg):

Property/Value	Units	Temp	Data Type	Reference
WS 1.56E+003	mg/L	25	EST	MEYLAN,WM ET AL. (1996)
logP -0.46			EXP	HANSCH,C & LEO,AJ (1985)
VP 9.27E-014	mm Hg	25	EST	NEELY,WB & BLAU,GE (1985)
DC	pKa			
HL 9.53E-017	atm m3/mol	25	EST	MEYLAN,WM & HOWARD,PH (1991)
OH 8.96E-011	cm3/molc sec	25	EST	MEYLAN,WM & HOWARD,PH (1993)

CAS #: 000070-11-1 — ETHANONE, 2-BROMO-1-PHENYL-

Formula: C_8H_7BrO

Mol Weight: 199.05

MP (deg C): 50.5

FP (deg C):

BP (deg C): 135

BP pressure (mm Hg): 1.80E+001

Property/Value	Units	Temp	Data Type	Reference
WS 5.97E+002	mg/L	25	EST	MEYLAN,WM ET AL. (1996)
logP 2.19			EXP	SANGSTER,J (1993)
VP 1.92E-002	mm Hg	25	EST	NEELY,WB & BLAU,GE (1985)
DC	pKa			
HL 1.13E-006	atm m3/mol	25	EST	MEYLAN,WM & HOWARD,PH (1991)
OH 1.97E-012	cm3/molc sec	25	EST	MEYLAN,WM & HOWARD,PH (1993)

CAS #: 000070-25-7 — N-NITROSO-N-METHYL-N'-NITROGUANIDINE

Formula: $C_2H_5N_5O_3$

Mol Weight: 147.09

MP (deg C): 118

FP (deg C):

BP (deg C):

BP pressure (mm Hg):

Property/Value	Units	Temp	Data Type	Reference
WS 4.78E+005	mg/L	25	EST	MEYLAN,WM ET AL. (1996)
logP -0.92			EST	MEYLAN,WM & HOWARD,PH (1995)
VP 1.48E-004	mm Hg	25	EST	NEELY,WB & BLAU,GE (1985)
DC	pKa			
HL 1.22E-012	atm m3/mol	25	EST	MEYLAN,WM & HOWARD,PH (1991)
OH 1.26E-012	cm3/molc sec	25	EST	MEYLAN,WM & HOWARD,PH (1993)

CAS #: 000070-26-8 — ORNITHINE

Formula: $C_5H_{12}N_2O_2$

Mol Weight: 132.16

MP (deg C): 140

FP (deg C):

BP (deg C):

BP pressure (mm Hg):

Property/Value	Units	Temp	Data Type	Reference
WS 1.00E+006	mg/L	25	EST	MEYLAN,WM ET AL. (1996)
logP -4.22			EXP	SANGSTER,J (1994)
VP 1.11E-009	mm Hg	25	EST	NEELY,WB & BLAU,GE (1985)
DC 1.94	pKa	25	EXP	KORTUM,G ET AL (1961)
HL 2.71E-013	atm m3/mol	25	EST	MEYLAN,WM & HOWARD,PH (1991)
OH 7.27E-011	cm3/molc sec	25	EST	MEYLAN,WM & HOWARD,PH (1993)

CAS #: 000070-30-4 — PHENOL,2,2'-METHYLENEBIS 3,4,6-CL

Formula: $C_{13}H_6Cl_6O_2$

Mol Weight: 406.91

MP (deg C): 164-165

FP (deg C):

BP (deg C): 479

BP pressure (mm Hg):

Property/Value	Units	Temp	Data Type	Reference
WS 1.40E+002	mg/L	25	EXP	YALKOWSKY,SH & DANNENFELSER,RM (1992)
logP 7.54			EXP	HANSCH,C ET AL. (1995)
VP 1.03E-010	mm Hg	25	EST	NEELY,WB & BLAU,GE (1985)
DC 4.95	pKa		EXP	HANSCH,C & LEO,AJ (1985)
HL 5.48E-013	atm m3/mol	25	EST	MEYLAN,WM & HOWARD,PH (1991)
OH 1.91E-012	cm3/molc sec	25	EST	MEYLAN,WM & HOWARD,PH (1993)

CAS #: 000070-34-8 — 1-FLUORO-2,4-DINITROBENZENE

Formula: $C_6H_3FN_2O_4$

Mol Weight: 186.10

MP (deg C): 26

FP (deg C):

BP (deg C): 137

BP pressure (mm Hg): 2.00E+000

Property/Value	Units	Temp	Data Type	Reference
WS 4.00E+002	mg/L	25	EXP	YALKOWSKY,SH & DANNENFELSER,RM (1992)
logP 1.79				HANSCH,C & LEO,AJ (1985)
VP 1.35E-003	mm Hg	25	EST	NEELY,WB & BLAU,GE (1985)
DC	pKa			
HL 9.79E-008	atm m3/mol	25	EST	MEYLAN,WM & HOWARD,PH (1991)
OH 3.82E-014	cm3/molc sec	25	EST	MEYLAN,WM & HOWARD,PH (1993)

CAS #: 000070-38-2 — DIMETHRIN

Formula: $C_{19}H_{26}O_2$

Mol Weight: 286.42

MP (deg C):

FP (deg C):

BP (deg C):

BP pressure (mm Hg):

Property/Value	Units	Temp	Data Type	Reference
WS 3.57E-002	mg/L	25	EST	MEYLAN,WM ET AL. (1996)
logP 6.57			EST	MEYLAN,WM & HOWARD,PH (1995)
VP 1.72E-005	mm Hg	25	EST	NEELY,WB & BLAU,GE (1985)
DC	pKa			
HL 7.61E-005	atm m3/mol	25	EST	MEYLAN,WM & HOWARD,PH (1991)
OH 1.06E-010	cm3/molc sec	25	EST	MEYLAN,WM & HOWARD,PH (1993)

CAS #: 000070-47-3 — ASPARAGINE

Formula: $C_4H_8N_2O_3$

Mol Weight: 132.12

MP (deg C): 234-235

FP (deg C):

BP (deg C):

BP pressure (mm Hg):

Property/Value	Units	Temp	Data Type	Reference
WS 2.94E+004	mg/L	25	EXP	YALKOWSKY,SH & DANNENFELSER,RM (1992)
logP -3.82			EXP	CHMELIK,J ET AL. (1991)
VP 4.84E-008	mm Hg	25	EST	NEELY,WB & BLAU,GE (1985)
DC 8.82	pKa	18	EXP	KORTUM,G ET AL (1961)
HL 2.27E-016	atm m3/mol	25	EST	MEYLAN,WM & HOWARD,PH (1991)
OH 7.72E-011	cm3/molc sec	25	EST	MEYLAN,WM & HOWARD,PH (1993)

CAS #: 000070-55-3	4-METHYLBENZENESULFONAMIDE

Formula: $C_7H_9NO_2S$

Mol Weight: 171.22

MP (deg C): 138.5   FP (deg C):

BP (deg C): 214

BP pressure (mm Hg): 9.98E+000

Property/Value	Units	Temp	Data Type	Reference
WS 3.16E+003	mg/L	25	EXP	YALKOWSKY,SH & DANNENFELSER,RM (1992)
logP 0.82			EXP	HANSCH,C ET AL. (1995)
VP 9.60E-005	mm Hg	25	EST	NEELY,WB & BLAU,GE (1985)
DC 10.17	pKa	20	EXP	SERJEANT,EP & DEMPSEY,B (1979)
HL 4.70E-007	atm m3/mol	25	EST	MEYLAN,WM & HOWARD,PH (1991)
OH 1.22E-012	cm3/molc sec	25	EST	MEYLAN,WM & HOWARD,PH (1993)

CAS #: 000070-70-2	P-HYDROXYPROPIOPHENONE

Formula: $C_9H_{10}O_2$

Mol Weight: 150.18

MP (deg C): 149   FP (deg C):

BP (deg C):

BP pressure (mm Hg):

Property/Value	Units	Temp	Data Type	Reference
WS 5.32E+003	mg/L	25	EST	MEYLAN,WM ET AL. (1996)
logP 2.03			EXP	HANSCH,C & LEO,AJ (1985)
VP 2.26E-003	mm Hg	25	EST	NEELY,WB & BLAU,GE (1985)
DC	pKa			
HL 1.36E-009	atm m3/mol	25	EST	MEYLAN,WM & HOWARD,PH (1991)
OH 3.18E-011	cm3/molc sec	25	EST	MEYLAN,WM & HOWARD,PH (1993)

CAS #: 000071-00-1	HISTIDINE

Formula: $C_6H_9N_3O_2$

Mol Weight: 155.16

MP (deg C): 287 dec   FP (deg C):

BP (deg C):

BP pressure (mm Hg):

Property/Value	Units	Temp	Data Type	Reference
WS 4.55E+004	mg/L	25	EXP	YALKOWSKY,SH & DANNENFELSER,RM (1992)
logP -3.32			EXP	CHMELIK,J ET AL. (1991)
VP 5.99E-009	mm Hg	25	EST	NEELY,WB & BLAU,GE (1985)
DC 2.76	pKa	0	EXP	KORTUM,G ET AL (1961)
HL 8.40E-014	atm m3/mol	25	EST	MEYLAN,WM & HOWARD,PH (1991)
OH 1.29E-010	cm3/molc sec	25	EST	MEYLAN,WM & HOWARD,PH (1993)

CAS #: 000071-23-8	1-PROPANOL

Formula: C_3H_8O

Mol Weight: 60.10

MP (deg C): -127   FP (deg C):

BP (deg C): 97.2

BP pressure (mm Hg):

Property/Value	Units	Temp	Data Type	Reference
WS 1.00E+006	mg/L	25	EXP	RIDDICK,JA ET AL. (1986)
logP 0.25			EXP	HANSCH,C & LEO,AJ (1985)
VP 2.10E+001	mm Hg	25	EXP	RIDDICK,JA ET AL. (1986)
DC 16.10	pKa		EXP	SERJEANT,EP & DEMPSEY,B (1979)
HL 7.41E-006	atm m3/mol	25	EXP	SNIDER,JR & DAWSON,GA (1985)
OH 5.34E-012	cm3/molc sec	25	EXP	ATKINSON,R (1989)

CAS #: 000071-30-7	CYTOSINE

Formula: $C_4H_5N_3O$

Mol Weight: 111.10

MP (deg C): 320-325 de   FP (deg C):

BP (deg C):

BP pressure (mm Hg):

Property/Value	Units	Temp	Data Type	Reference
WS 8.00E+003	mg/L	25	EXP	YALKOWSKY,SH & DANNENFELSER,RM (1992)
logP -1.73			EXP	HANSCH,C & LEO,AJ (1985)
VP 1.09E-003	mm Hg	25	EST	NEELY,WB & BLAU,GE (1985)
DC	pKa			
HL 4.81E-011	atm m3/mol	25	EST	MEYLAN,WM & HOWARD,PH (1991)
OH 5.58E-011	cm3/molc sec	25	EST	MEYLAN,WM & HOWARD,PH (1993)

CAS #: 000071-33-0	5-AZAURACIL

Formula: $C_3H_3N_3O_2$

Mol Weight: 113.08

MP (deg C):   FP (deg C):

BP (deg C):

BP pressure (mm Hg):

Property/Value	Units	Temp	Data Type	Reference
WS 2.06E+005	mg/L	25	EST	MEYLAN,WM ET AL. (1996)
logP -1.87			EXP	HANSCH,C ET AL. (1995)
VP 7.05E-008	mm Hg	25	EST	NEELY,WB & BLAU,GE (1985)
DC	pKa			
HL 9.46E-011	atm m3/mol	25	EST	MEYLAN,WM & HOWARD,PH (1991)
OH 9.95E-012	cm3/molc sec	25	EST	MEYLAN,WM & HOWARD,PH (1993)

CAS #: 000071-36-3	1-BUTANOL

Formula: $C_4H_{10}O$

Mol Weight: 74.12

MP (deg C): -88.62   FP (deg C):

BP (deg C): 117.73

BP pressure (mm Hg):

Property/Value	Units	Temp	Data Type	Reference
WS 6.32E+004	mg/L	25	EXP	TEWARI,YB ET AL. (1982)
logP 0.88			EXP	HANSCH,C & LEO,AJ (1985)
VP 6.70E+000	mm Hg	25	EXP	BOUBLIK,T ET AL. (1984)
DC 16.10	pKa	25	EXP	SERJEANT,EP & DEMPSEY,B (1979)
HL 8.81E-006	atm m3/mol	25	EXP	BUTTERY,RG ET AL. (1969)
OH 8.30E-012	cm3/molc sec	25	EXP	ATKINSON,R (1989)

CAS #: 000071-41-0	1-PENTANOL

Formula: $C_5H_{12}O$

Mol Weight: 88.15

MP (deg C): -79   FP (deg C):

BP (deg C): 137.5

BP pressure (mm Hg):

Property/Value	Units	Temp	Data Type	Reference
WS 2.20E+004	mg/L	25	EXP	YALKOWSKY,SH & DANNENFELSER,RM (1992)
logP 1.51			EXP	SANGSTER,J (1994)
VP 2.20E+000	mm Hg	25	EXP	RIDDICK,JA ET AL. (1986)
DC	pKa			
HL 1.30E-005	atm m3/mol	25	EXP	BUTLER,JAV ET AL. (1935)
OH 1.08E-011	cm3/molc sec	23	EXP	WALLINGTON,TJ & KURYLO,MJ (1987)

000071-43-2 — BENZENE

CAS #: 000071-43-2

Formula:	C_6H_6
Mol Weight:	78.11
MP (deg C):	5.53
FP (deg C):	
BP (deg C):	80.09
BP pressure (mm Hg):	

Property/Value	Units	Temp	Data Type	Reference
WS 1.79E+003	mg/L	25	EXP	MAY,WE ET AL. (1983)
logP 2.13			EXP	HANSCH,C & LEO,AJ (1985)
VP 9.52E+001	mm Hg	25	EXP	BOUBLIK,T ET AL. (1984)
DC	pKa			
HL 5.55E-003	atm m3/mol	25	EXP	MACKAY,D ET AL. (1979)
OH 1.23E-012	cm3/molc sec	25	EXP	ATKINSON,R (1989)

000071-55-6 — 1,1,1-TRICHLOROETHANE

CAS #: 000071-55-6

Formula:	$C_2H_3Cl_3$
Mol Weight:	133.41
MP (deg C):	-30.4
FP (deg C):	
BP (deg C):	74.08
BP pressure (mm Hg):	

Property/Value	Units	Temp	Data Type	Reference
WS 1.50E+003	mg/L	25	EXP	HORVATH,AL (1982)
logP 2.49			EXP	HANSCH,C & LEO,AJ (1985)
VP 1.24E+002	mm Hg	25	EXP	BOUBLIK,T ET AL. (1984)
DC	pKa			
HL 1.72E-002	atm m3/mol	24	EXP	GOSSETT,JM (1987)
OH 1.19E-014	cm3/molc sec	25	EXP	ATKINSON,R (1989)

000071-63-6 — DIGITOXIN

CAS #: 000071-63-6

Formula:	$C_{41}H_{64}O_{13}$
Mol Weight:	764.96
MP (deg C):	256-257
FP (deg C):	
BP (deg C):	
BP pressure (mm Hg):	

Property/Value	Units	Temp	Data Type	Reference
WS 3.90E+000	mg/L	25	EXP	YALKOWSKY,SH & DANNENFELSER,RM (1992)
logP 1.85			EXP	SANGSTER,J (1993)
VP 6.93E-029	mm Hg	25	EST	NEELY,WB & BLAU,GE (1985)
DC	pKa			
HL 1.27E-025	atm m3/mol	25	EST	MEYLAN,WM & HOWARD,PH (1991)
OH 1.65E-010	cm3/molc sec	25	EST	MEYLAN,WM & HOWARD,PH (1993)

000071-91-0 — TETRAETHYL AMMONIUM BROMIDE

CAS #: 000071-91-0

Formula:	$C_8H_{20}BrN$
Mol Weight:	210.16
MP (deg C):	285-287 de
FP (deg C):	
BP (deg C):	
BP pressure (mm Hg):	

Property/Value	Units	Temp	Data Type	Reference
WS 1.00E+006	mg/L	25	EXP	PEDDLE,CJ & TURNER,WES (1913)
logP -2.82			EXP	HANSCH,C & LEO,AJ (1985)
VP 7.92E-007	mm Hg	25	EST	NEELY,WB & BLAU,GE (1985)
DC	pKa			
HL 1.03E-011	atm m3/mol	25	EST	MEYLAN,WM & HOWARD,PH (1991)
OH 3.54E-011	cm3/molc sec	25	EST	MEYLAN,WM & HOWARD,PH (1993)

000072-14-0 — SULFATHIOZOLE

CAS #: 000072-14-0

Formula:	$C_9H_9N_3O_2S_2$
Mol Weight:	255.32
MP (deg C):	200-204
FP (deg C):	
BP (deg C):	
BP pressure (mm Hg):	

Property/Value	Units	Temp	Data Type	Reference
WS 3.73E+002	mg/L	25	EXP	YALKOWSKY,SH & DANNENFELSER,RM (1992)
logP 0.05			EXP	HANSCH,C & LEO,AJ (1985)
VP 4.22E-008	mm Hg	25	EST	NEELY,WB & BLAU,GE (1985)
DC	pKa			
HL 5.85E-014	atm m3/mol	25	EST	MEYLAN,WM & HOWARD,PH (1991)
OH 5.36E-011	cm3/molc sec	25	EST	MEYLAN,WM & HOWARD,PH (1993)

000072-18-4 — VALINE

CAS #: 000072-18-4

Formula:	$C_5H_{11}NO_2$
Mol Weight:	117.15
MP (deg C):	315
FP (deg C):	
BP (deg C):	
BP pressure (mm Hg):	

Property/Value	Units	Temp	Data Type	Reference
WS 5.85E+004	mg/L	25	EXP	YALKOWSKY,SH & DANNENFELSER,RM (1992)
logP -2.26			EXP	HANSCH,C & LEO,AJ (1985)
VP 3.41E-008	mm Hg	25	EST	NEELY,WB & BLAU,GE (1985)
DC 2.30	pKa	13	EXP	KORTUM,G ET AL (1961)
HL 2.63E-009	atm m3/mol	25	EST	MEYLAN,WM & HOWARD,PH (1991)
OH 4.09E-011	cm3/molc sec	25	EST	MEYLAN,WM & HOWARD,PH (1993)

000072-19-5 — THREONINE(L)

CAS #: 000072-19-5

Formula:	$C_4H_9NO_3$
Mol Weight:	119.12
MP (deg C):	256 dec
FP (deg C):	
BP (deg C):	
BP pressure (mm Hg):	

Property/Value	Units	Temp	Data Type	Reference
WS 9.70E+004	mg/L	25	EXP	YALKOWSKY,SH & DANNENFELSER,RM (1992)
logP -2.94			EXP	HANSCH,C & LEO,AJ (1985)
VP 4.09E-008	mm Hg	25	EST	NEELY,WB & BLAU,GE (1985)
DC	pKa			
HL 7.24E-014	atm m3/mol	25	EST	MEYLAN,WM & HOWARD,PH (1991)
OH 4.68E-011	cm3/molc sec	25	EST	MEYLAN,WM & HOWARD,PH (1993)

000072-20-8 — ENDRIN

CAS #: 000072-20-8

Formula:	$C_{12}H_8Cl_6O$
Mol Weight:	380.91
MP (deg C):	226-230
FP (deg C):	
BP (deg C):	
BP pressure (mm Hg):	

Property/Value	Units	Temp	Data Type	Reference
WS 2.50E-001	mg/L	25	EXP	BIGGAR,JW & RIGGS,RI (1974)
logP 5.20			EXP	DEBRUIJN,J ET AL. (1989)
VP 3.00E-006	mm Hg	20	EXP	NASH,RG (1983A)
DC	pKa			
HL 7.52E-006	atm m3/mol	20	EST	VP/WSOL
OH 4.87E-012	cm3/molc sec	25	EST	MEYLAN,WM & HOWARD,PH (1993)

CAS #: 000072-43-5 — METHOXYCHLOR

Formula: $C_{16}H_{15}Cl_3O_2$

Mol Weight: 345.66

MP (deg C): 89

FP (deg C):

BP (deg C): 346

BP pressure (mm Hg):

Property/ Value	Units	Temp	Data Type	Reference
WS 1.00E-001	mg/L	25	EXP	RICHARDSON,LT & MILLER,DM (1960)
logP 5.08			EXP	HANSCH,C & LEO,AJ (1985)
VP 2.58E-006	mm Hg	25	EST	NEELY,WB & BLAU,GE (1985)
DC	pKa			
HL 1.58E-005	atm m3/mol	25	EST	MEYLAN,WM & HOWARD,PH (1991)
OH 5.72E-011	cm3/molc sec	25	EST	MEYLAN,WM & HOWARD,PH (1993)

CAS #: 000072-44-6 — METHAQUALONE

Formula: $C_{16}H_{14}N_2O$

Mol Weight: 250.30

MP (deg C): 120

FP (deg C):

BP (deg C):

BP pressure (mm Hg):

Property/ Value	Units	Temp	Data Type	Reference
WS 4.73E+000	mg/L	25	EST	MEYLAN,WM ET AL. (1996)
logP 3.43			EST	MEYLAN,WM & HOWARD,PH (1995)
VP 1.16E-007	mm Hg	25	EST	NEELY,WB & BLAU,GE (1985)
DC	pKa			
HL 2.76E-009	atm m3/mol	25	EST	MEYLAN,WM & HOWARD,PH (1991)
OH 1.66E-011	cm3/molc sec	25	EST	MEYLAN,WM & HOWARD,PH (1993)

CAS #: 000072-48-0 — ALIZARINE

Formula: $C_{14}H_8O_4$

Mol Weight: 240.22

MP (deg C): 290

FP (deg C):

BP (deg C): 430

BP pressure (mm Hg):

Property/ Value	Units	Temp	Data Type	Reference
WS 4.00E+002	mg/L		EXP	YALKOWSKY,SH & DANNENFELSER,RM (1992)
logP 3.16			EST	MEYLAN,WM & HOWARD,PH (1995)
VP 8.76E-010	mm Hg	25	EXT	MALASPINA,L ET AL. (1973A)
DC	pKa			
HL 4.33E-014	atm m3/mol	25	EST	MEYLAN,WM & HOWARD,PH (1991)
OH 9.84E-012	cm3/molc sec	25	EST	MEYLAN,WM & HOWARD,PH (1993)

CAS #: 000072-54-8 — P,P'-DDD

Formula: $C_{14}H_{10}Cl_4$

Mol Weight: 320.05

MP (deg C): 109-110

FP (deg C):

BP (deg C): 350

BP pressure (mm Hg):

Property/ Value	Units	Temp	Data Type	Reference
WS 9.00E-002	mg/L	25	EXP	BIGGAR,JW & RIGGS,RI (1974)
logP 6.02			EXP	SANGSTER,J (1994)
VP 6.70E-007	mm Hg	20	EXP	BIDLEMAN,TF (1984)
DC	pKa			
HL 4.00E-006	atm m3/mol	20	EST	VP/WSOL
OH 4.84E-012	cm3/molc sec	25	EST	MEYLAN,WM & HOWARD,PH (1993)

CAS #: 000072-55-9 — P,P'-DDE

Formula: $C_{14}H_8Cl_4$

Mol Weight: 318.03

MP (deg C): 89

FP (deg C):

BP (deg C): 336

BP pressure (mm Hg):

Property/ Value	Units	Temp	Data Type	Reference
WS 1.20E-001	mg/L	25	EXP	BIGGAR,JW & RIGGS,RI (1974)
logP 6.51			EXP	SANGSTER,J (1993)
VP 6.00E-006	mm Hg	25	EXT	BIDLEMAN,TF (1984)
DC	pKa			
HL 2.10E-005	atm m3/mol	25	EST	VP/WSOL
OH 7.09E-012	cm3/molc sec	25	EST	MEYLAN,WM & HOWARD,PH (1993)

CAS #: 000072-56-0 — 1,1-DICHLORO-2,2-BIS(ETHYLPHENYL)ETHANE

Formula: $C_{18}H_{20}Cl_2$

Mol Weight: 307.27

MP (deg C): 60

FP (deg C):

BP (deg C):

BP pressure (mm Hg):

Property/ Value	Units	Temp	Data Type	Reference
WS 1.00E-001	mg/L	24	EXP	HOLLIFIELD,HC (1979)
logP 6.66			EST	MEYLAN,WM & HOWARD,PH (1995)
VP 3.99E-006	mm Hg	25	EST	NEELY,WB & BLAU,GE (1985)
DC	pKa			
HL 1.70E-004	atm m3/mol	25	EST	MEYLAN,WM & HOWARD,PH (1991)
OH 1.56E-011	cm3/molc sec	25	EST	MEYLAN,WM & HOWARD,PH (1993)

CAS #: 000073-22-3 — TRYPTOPHAN

Formula: $C_{11}H_{12}N_2O_2$

Mol Weight: 204.23

MP (deg C): 290-292 de

FP (deg C):

BP (deg C):

BP pressure (mm Hg):

Property/ Value	Units	Temp	Data Type	Reference
WS 1.34E+004	mg/L	25	EXP	YALKOWSKY,SH & DANNENFELSER,RM (1992)
logP -1.06			EXP	HANSCH,C & LEO,AJ (1985)
VP 2.12E-009	mm Hg	25	EST	NEELY,WB & BLAU,GE (1985)
DC 7.38	pKa	25	EXP	KORTUM,G ET AL (1961)
HL 1.98E-014	atm m3/mol	25	EST	MEYLAN,WM & HOWARD,PH (1991)
OH 2.39E-010	cm3/molc sec	25	EST	MEYLAN,WM & HOWARD,PH (1993)

CAS #: 000073-24-5 — ADENINE

Formula: $C_5H_5N_5$

Mol Weight: 135.13

MP (deg C): 360 dec

FP (deg C):

BP (deg C):

BP pressure (mm Hg):

Property/ Value	Units	Temp	Data Type	Reference
WS 1.03E+003	mg/L	25	EXP	YALKOWSKY,SH & DANNENFELSER,RM (1992)
logP -0.09			EXP	HANSCH,C & LEO,AJ (1985)
VP 8.38E-007	mm Hg	25	EST	NEELY,WB & BLAU,GE (1985)
DC 4.15	pKa	25	EXP	KORTUM,G ET AL (1961)
HL 7.02E-014	atm m3/mol	25	EST	MEYLAN,WM & HOWARD,PH (1991)
OH 2.00E-010	cm3/molc sec	25	EST	MEYLAN,WM & HOWARD,PH (1993)

CAS #: 000073-32-5 — ISO-LEUCINE

Formula:	$C_6H_{13}NO_2$
Mol Weight:	131.18
MP (deg C):	285-286 de
FP (deg C):	
BP (deg C):	
BP pressure (mm Hg):	

Property/Value	Units	Temp	Data Type	Reference
WS 3.43E+004	mg/L	25	EXP	YALKOWSKY,SH & DANNENFELSER,RM (1992)
logP -1.70			EXP	HANSCH,C & LEO,AJ (1985)
VP 1.34E-008	mm Hg	25	EST	NEELY,WB & BLAU,GE (1985)
DC 2.37	pKa	0	EXP	KORTUM,G ET AL (1961)
HL 3.49E-009	atm m3/mol	25	EST	MEYLAN,WM & HOWARD,PH (1991)
OH 4.26E-011	cm3/molc sec	25	EST	MEYLAN,WM & HOWARD,PH (1993)

CAS #: 000073-40-5 — GUANINE

Formula:	$C_5H_5N_5O$
Mol Weight:	151.13
MP (deg C):	360 dec
FP (deg C):	
BP (deg C):	
BP pressure (mm Hg):	

Property/Value	Units	Temp	Data Type	Reference
WS 2.08E+003	mg/L	37	EXP	YALKOWSKY,SH & DANNENFELSER,RM (1992)
logP -0.91			EXP	HANSCH,C & LEO,AJ (1985)
VP 9.60E-011	mm Hg	25	EST	NEELY,WB & BLAU,GE (1985)
DC	pKa			
HL 2.62E-017	atm m3/mol	25	EST	MEYLAN,WM & HOWARD,PH (1991)
OH 6.25E-011	cm3/molc sec	25	EST	MEYLAN,WM & HOWARD,PH (1993)

CAS #: 000073-48-3 — BENDROFLUMETHIAZIDE

Formula:	$C_{15}H_{14}F_3N_3O_4S_2$
Mol Weight:	421.42
MP (deg C):	221-223
FP (deg C):	
BP (deg C):	
BP pressure (mm Hg):	

Property/Value	Units	Temp	Data Type	Reference
WS 1.08E+002	mg/L	25	EXP	YALKOWSKY,SH & DANNENFELSER,RM (1992)
logP 1.19			EXP	HANSCH,C ET AL. (1995)
VP 5.36E-012	mm Hg	25	EST	NEELY,WB & BLAU,GE (1985)
DC	pKa			
HL 5.51E-012	atm m3/mol	25	EST	MEYLAN,WM & HOWARD,PH (1991)
OH 2.13E-010	cm3/molc sec	25	EST	MEYLAN,WM & HOWARD,PH (1993)

CAS #: 000074-11-3 — 4-CHLOROBENZOIC ACID

Formula:	$C_7H_5ClO_2$
Mol Weight:	156.57
MP (deg C):	243
FP (deg C):	
BP (deg C):	
BP pressure (mm Hg):	

Property/Value	Units	Temp	Data Type	Reference
WS 7.20E+001	mg/L	25	EXP	STEPHEN,H & STEPHEN,T (1963)
logP 2.65			EXP	HANSCH,C & LEO,AJ (1985)
VP 2.33E-003	mm Hg	25	EST	NEELY,WB & BLAU,GE (1985)
DC 3.98	pKa	25	EXP	PEARCE,PJ & SIMKINS,RJJ (1968)
HL 8.03E-008	atm m3/mol	25	EST	MEYLAN,WM & HOWARD,PH (1991)
OH 5.63E-001	cm3/molc sec	25	EST	ATKINSON,R (1988)

CAS #: 000074-31-7 — N,N'-DIPHENYL-P-BENZENEDIAMINE

Formula:	$C_{18}H_{16}N_2$
Mol Weight:	260.34
MP (deg C):	150-151
FP (deg C):	
BP (deg C):	220-225
BP pressure (mm Hg):	5.00E-001

Property/Value	Units	Temp	Data Type	Reference
WS 7.35E+000	mg/L	25	EST	MEYLAN,WM ET AL. (1996)
logP 4.04			EST	MEYLAN,WM & HOWARD,PH (1995)
VP 6.35E-009	mm Hg	25	EXT	DAUBERT,TE & DANNER,RP (1989)
DC	pKa			
HL 2.05E-010	atm m3/mol	25	EST	MEYLAN,WM & HOWARD,PH (1991)
OH 2.00E-010	cm3/molc sec	25	EST	MEYLAN,WM & HOWARD,PH (1993)

CAS #: 000074-79-3 — ARGININE

Formula:	$C_6H_{14}N_4O_2$
Mol Weight:	174.20
MP (deg C):	244 dec
FP (deg C):	
BP (deg C):	
BP pressure (mm Hg):	

Property/Value	Units	Temp	Data Type	Reference
WS 1.82E+005	mg/L	25	EXP	YALKOWSKY,SH & DANNENFELSER,RM (1992)
logP -4.20			EXP	HANSCH,C & LEO,AJ (1985)
VP 1.39E-009	mm Hg	25	EST	NEELY,WB & BLAU,GE (1985)
DC 2.24	pKa	0	EXP	KORTUM,G ET AL (1961)
HL 1.83E-018	atm m3/mol	25	EST	MEYLAN,WM & HOWARD,PH (1991)
OH 1.36E-010	cm3/molc sec	25	EST	MEYLAN,WM & HOWARD,PH (1993)

CAS #: 000074-82-8 — METHANE

Formula:	CH_4
Mol Weight:	16.04
MP (deg C):	-182.6
FP (deg C):	
BP (deg C):	-161.4
BP pressure (mm Hg):	

Property/Value	Units	Temp	Data Type	Reference
WS 2.44E+001	mg/L	25	EXP	YALKOWSKY,SH & DANNENFELSER,RM (1992)
logP 1.09			EXP	HANSCH,C & LEO,AJ (1985)
VP 4.66E+005	mm Hg	25	EXP	DAUBERT,TE & DANNER,RP (1994)
DC	pKa			
HL 6.58E-001	atm m3/mol	25	EST	VP/WSOL
OH 8.41E-015	cm3/molc sec	25	EXP	ATKINSON,R (1989)

CAS #: 000074-83-9 — METHYL BROMIDE

Formula:	CH_3Br
Mol Weight:	94.94
MP (deg C):	-93.66
FP (deg C):	
BP (deg C):	3.56
BP pressure (mm Hg):	

Property/Value	Units	Temp	Data Type	Reference
WS 1.52E+004	mg/L	25	EXP	HORVATH,AL (1982)
logP 1.19			EXP	HANSCH,C & LEO,AJ (1985)
VP 1.62E+003	mm Hg	25	EXP	DAUBERT,TE & DANNER,RP (1985)
DC	pKa			
HL 6.24E-003	atm m3/mol	25	EST	VP/WSOL
OH 4.02E-014	cm3/molc sec	25	EXP	ATKINSON,R (1989)

CAS #: 000074-84-0 — ETHANE

Formula:	C_2H_6
Mol Weight:	30.07
MP (deg C):	-172
FP (deg C):	
BP (deg C):	-88
BP pressure (mm Hg):	

Property/Value	Units	Temp	Data Type	Reference
WS 6.02E+001	mg/L	25	EXP	MCAULIFFE,C (1966)
logP 1.81			EXP	HANSCH,C & LEO,AJ (1985)
VP 3.15E+004	mm Hg	25	EXP	DAUBERT,TE & DANNER,RP (1994)
DC	pKa			
HL 5.00E-001	atm m3/mol	25	EST	VP/WSOL
OH 2.68E-013	cm3/molc sec	25	EXP	ATKINSON,R (1989)

CAS #: 000074-85-1 — ETHYLENE

Formula:	C_2H_4
Mol Weight:	28.05
MP (deg C):	-169
FP (deg C):	
BP (deg C):	103.7
BP pressure (mm Hg):	

Property/Value	Units	Temp	Data Type	Reference
WS 1.31E+002	mg/L	25	EXP	MCAULIFFE,C (1966)
logP 1.13			EXP	HANSCH,C & LEO,AJ (1985)
VP 5.21E+004	mm Hg	25	EXT	DAUBERT,TE & DANNER,RP (1985)
DC	pKa			
HL 2.28E-001	atm m3/mol	25	EXP	WASIK,SP & TSANG,W (1970)
OH 8.52E-012	cm3/molc sec	25	EXP	ATKINSON,R (1989)

CAS #: 000074-86-2 — ACETYLENE

Formula:	C_2H_2
Mol Weight:	26.04
MP (deg C):	-81
FP (deg C):	
BP (deg C):	
BP pressure (mm Hg):	

Property/Value	Units	Temp	Data Type	Reference
WS 1.20E+003	mg/L	20	EXP	YALKOWSKY,SH & DANNENFELSER,RM (1992)
logP 0.37			EXP	HANSCH,C & LEO,AJ (1985)
VP 5.24E+003	mm Hg	25	EXP	DAUBERT,TE & DANNER,RP (1989)
DC	pKa			
HL 2.17E-002	atm m3/mol	25	EST	VP/WSOL
OH 8.15E-013	cm3/molc sec	25	EXP	ATKINSON,R (1989)

CAS #: 000074-87-3 — METHYL CHLORIDE

Formula:	CH_3Cl
Mol Weight:	50.49
MP (deg C):	-97.70
FP (deg C):	
BP (deg C):	-24.20
BP pressure (mm Hg):	

Property/Value	Units	Temp	Data Type	Reference
WS 5.32E+003	mg/L	25	EXP	HORVATH,AL (1982)
logP 0.91			EXP	HANSCH,C & LEO,AJ (1985)
VP 4.30E+003	mm Hg	25	EXP	DAUBERT,TE & DANNER,RP (1985)
DC	pKa			
HL 8.82E-003	atm m3/mol	24	EXP	GOSSETT,JM (1987)
OH 4.36E-014	cm3/molc sec	25	EXP	ATKINSON,R (1989)

CAS #: 000074-88-4 — METHYL IODIDE

Formula:	CH_3I
Mol Weight:	141.94
MP (deg C):	-66.45
FP (deg C):	
BP (deg C):	42.43
BP pressure (mm Hg):	

Property/Value	Units	Temp	Data Type	Reference
WS 1.39E+004	mg/L	20	EXP	HORVATH,AL (1982)
logP 1.51			EXP	HANSCH,C & LEO,AJ (1985)
VP 4.05E+002	mm Hg	25	EXP	BOUBLIK,T ET AL. (1984)
DC	pKa			
HL 5.26E-003	atm m3/mol	25	EXP	HUNTER-SMITH,RJ ET AL. (1983)
OH 7.20E-014	cm3/molc sec	25	EXP	ATKINSON,R (1989)

CAS #: 000074-89-5 — METHYLAMINE

Formula:	CH_5N
Mol Weight:	31.06
MP (deg C):	-93.5
FP (deg C):	
BP (deg C):	-6.3
BP pressure (mm Hg):	

Property/Value	Units	Temp	Data Type	Reference
WS 1.08E+006	mg/L	25	EXP	SCHWEIZER,AE ET AL. (1978)
logP -0.57			EXP	HANSCH,C & LEO,AJ (1985)
VP 2.65E+003	mm Hg	25	EXP	DAUBERT,TE & DANNER,RP (1985)
DC 10.66	pKa		EXP	PERRIN,DD (1965)
HL 1.11E-005	atm m3/mol	25	EXP	CHRISTIE,AO & CRISP,DJ (1967)
OH 2.20E-011	cm3/molc sec	25	EXP	ATKINSON,R (1989)

CAS #: 000074-93-1 — METHANETHIOL

Formula:	CH_4S
Mol Weight:	48.11
MP (deg C):	-123
FP (deg C):	
BP (deg C):	5.95
BP pressure (mm Hg):	

Property/Value	Units	Temp	Data Type	Reference
WS 1.54E+004	mg/L	25	EXP	HINE,J & MOOKERJEE,PK (1975)
logP 0.78			EST	MEYLAN,WM & HOWARD,PH (1995)
VP 1.51E+003	mm Hg	25	EXP	DAUBERT,TE & DANNER,RP (1985)
DC 10.33	pKa	25	EXP	SERJEANT,EP & DEMPSEY,B (1979)
HL 3.12E-003	atm m3/mol	25	EST	VP/WSOL
OH 3.29E-011	cm3/molc sec	25	EXP	ATKINSON,R (1989)

CAS #: 000074-95-3 — DIBROMOMETHANE

Formula:	CH_2Br_2
Mol Weight:	173.85
MP (deg C):	-52.7
FP (deg C):	
BP (deg C):	96.95
BP pressure (mm Hg):	

Property/Value	Units	Temp	Data Type	Reference
WS 1.19E+004	mg/L	30	EXP	GROSS,PM & SAYLOR,JH (1931)
logP 1.70			EXP	MARTISKA,A & BEKAREK,V (1990)
VP 4.44E+001	mm Hg	25	EXP	KUDCHADKER,AP ET AL. (1979)
DC	pKa			
HL 8.22E-004	atm m3/mol	20	EXP	MOORE,RM ET AL. (1995)
OH 7.50E-014	cm3/molc sec	25	EST	MEYLAN,WM & HOWARD,PH (1993)

BROMOETHANE

CAS #: **000074-96-4**

Formula:	C_2H_5Br
Mol Weight:	108.97
MP (deg C):	-118.6
FP (deg C):	
BP (deg C):	38.35
BP pressure (mm Hg):	

Property/Value	Units	Temp	Data Type	Reference
WS 8.94E+003	mg/L	25	EXP	HORVATH,AL (1982)
logP 1.61			EXP	HANSCH,C & LEO,AJ (1985)
VP 4.67E+002	mm Hg	25	EXP	BOUBLIK,T ET AL. (1989)
DC	pKa			
HL 7.49E-003	atm m3/mol	25	EST	VP/WSOL
OH 3.50E-013	cm3/molc sec	25	EXP	ATKINSON,R (1989)

CHLOROBROMOMETHANE

CAS #: **000074-97-5**

Formula:	CH_2BrCl
Mol Weight:	129.39
MP (deg C):	-87.9
FP (deg C):	
BP (deg C):	68
BP pressure (mm Hg):	

Property/Value	Units	Temp	Data Type	Reference
WS 1.67E+004	mg/L	25	EXP	YALKOWSKY,SH & DANNENFELSER,RM (1992)
logP 1.41			EXP	HANSCH,C & LEO,AJ (1985)
VP 1.43E+002	mm Hg	25	EXP	DAUBERT,TE & DANNER,RP (1989)
DC	pKa			
HL 1.46E-003	atm m3/mol	25	EST	VP/WSOL
OH 9.94E-014	cm3/molc sec	25	EST	MEYLAN,WM & HOWARD,PH (1993)

PROPANE

CAS #: **000074-98-6**

Formula:	C_3H_8
Mol Weight:	44.10
MP (deg C):	-187.6
FP (deg C):	
BP (deg C):	-42.1
BP pressure (mm Hg):	

Property/Value	Units	Temp	Data Type	Reference
WS 6.24E+001	mg/L	25	EXP	YALKOWSKY,SH & DANNENFELSER,RM (1992)
logP 2.36			EXP	HANSCH,C & LEO,AJ (1985)
VP 7.15E+003	mm Hg	25	EXP	DAUBERT,TE & DANNER,RP (1994)
DC	pKa			
HL 7.07E-001	atm m3/mol	25	EST	VP/WSOL
OH 1.15E-012	cm3/molc sec	25	EXP	ATKINSON,R (1989)

PROPYNE

CAS #: **000074-99-7**

Formula:	C_3H_4
Mol Weight:	40.07
MP (deg C):	-101.5
FP (deg C):	
BP (deg C):	-23.1
BP pressure (mm Hg):	

Property/Value	Units	Temp	Data Type	Reference
WS 3.64E+003	mg/L	25	EXP	MCAULIFFE,C (1966)
logP 0.94			EXP	HANSCH,C & LEO,AJ (1985)
VP 4.31E+003	mm Hg	25	EXP	DAUBERT,TE & DANNER,RP (1989)
DC	pKa			
HL 1.10E-002	atm m3/mol	25	EST	VP/WSOL
OH 5.90E-012	cm3/molc sec	25	EXP	ATKINSON,R (1989)

CHLOROETHANE

CAS #: **000075-00-3**

Formula:	C_2H_5Cl
Mol Weight:	64.52
MP (deg C):	-136.4
FP (deg C):	
BP (deg C):	12.27
BP pressure (mm Hg):	

Property/Value	Units	Temp	Data Type	Reference
WS 5.68E+003	mg/L	20	EXP	HORVATH,AL (1982)
logP 1.43			EXP	HANSCH,C & LEO,AJ (1985)
VP 1.01E+003	mm Hg	20	EXP	DAUBERT,TE & DANNER,RP (1985)
DC	pKa			
HL 1.11E-002	atm m3/mol	24	EXP	GOSSETT,JM (1987)
OH 3.90E-013	cm3/molc sec	25	EXP	ATKINSON,R (1989)

VINYLCHLORIDE

CAS #: **000075-01-4**

Formula:	C_2H_3Cl
Mol Weight:	62.50
MP (deg C):	-153.8
FP (deg C):	
BP (deg C):	-13.37
BP pressure (mm Hg):	

Property/Value	Units	Temp	Data Type	Reference
WS 8.80E+003	mg/L	25	EXP	DELASSUS,PT & SCHMIDT,DD (1981)
logP 1.62			EST	MEYLAN,WM & HOWARD,PH (1995)
VP 2.98E+003	mm Hg	25	EXT	DAUBERT,TE & DANNER,RP (1985)
DC	pKa			
HL 2.78E-002	atm m3/mol	24	EXP	GOSSETT,JM (1987)
OH 6.60E-012	cm3/molc sec	25	EXP	ATKINSON,R (1989)

VINYLFLUORIDE

CAS #: **000075-02-5**

Formula:	C_2H_3F
Mol Weight:	46.04
MP (deg C):	-160.5
FP (deg C):	
BP (deg C):	-72
BP pressure (mm Hg):	

Property/Value	Units	Temp	Data Type	Reference
WS 1.29E+004	mg/L	25	EST	MEYLAN,WM ET AL. (1996)
logP 1.19			EST	MEYLAN,WM & HOWARD,PH (1995)
VP 1.98E+004	mm Hg	25	EXP	DAUBERT,TE & DANNER,RP (1989)
DC	pKa			
HL 1.18E-001	atm m3/mol	25	EST	MEYLAN,WM & HOWARD,PH (1991)
OH 5.56E-012	cm3/molc sec	25	EXP	ATKINSON,R (1989)

IODOETHANE

CAS #: **000075-03-6**

Formula:	C_2H_5I
Mol Weight:	155.97
MP (deg C):	-108
FP (deg C):	
BP (deg C):	72
BP pressure (mm Hg):	

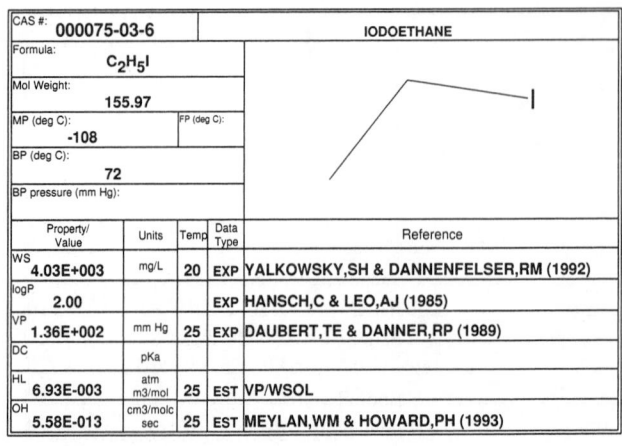

Property/Value	Units	Temp	Data Type	Reference
WS 4.03E+003	mg/L	20	EXP	YALKOWSKY,SH & DANNENFELSER,RM (1992)
logP 2.00			EXP	HANSCH,C & LEO,AJ (1985)
VP 1.36E+002	mm Hg	25	EXP	DAUBERT,TE & DANNER,RP (1989)
DC	pKa			
HL 6.93E-003	atm m3/mol	25	EST	VP/WSOL
OH 5.58E-013	cm3/molc sec	25	EST	MEYLAN,WM & HOWARD,PH (1993)

ETHYLAMINE

CAS #:	000075-04-7

Formula: C_2H_7N

Mol Weight: 45.08

MP (deg C): -81 | FP (deg C):

BP (deg C): 16.6

BP pressure (mm Hg):

Property/Value	Units	Temp	Data Type	Reference
WS 1.00E+006	mg/L		EXP	MERCK INDEX (1983)
logP -0.13			EXP	HANSCH,C & LEO,AJ (1985)
VP 1.00E+000	mm Hg	25	EXP	DAUBERT,TE & DANNER,RP (1985)
DC 10.87	pKa	20	EXP	PERRIN,DD (1965)
HL 1.23E-005	atm m3/mol	25	EXP	CHRISTIE,AO & CRISP,DJ (1967)
OH 2.77E-011	cm3/molc sec	25	EXP	ATKINSON,R (1989)

ACETONITRILE

CAS #:	000075-05-8

Formula: C_2H_3N

Mol Weight: 41.05

MP (deg C): -45 | FP (deg C):

BP (deg C): 81.6

BP pressure (mm Hg):

Property/Value	Units	Temp	Data Type	Reference
WS 1.00E+006	mg/L		EXP	RIDDICK,JA ET AL. (1986)
logP -0.34			EXP	HANSCH,C & LEO,AJ (1985)
VP 8.88E+001	mm Hg	25	EXP	BOUBLIK,T ET AL. (1984)
DC -4.30	pKa		EXP	PERRIN,DD (1965)
HL 3.45E-005	atm m3/mol	25	EXP	GAFFNEY,JS ET AL. (1987)
OH 2.10E-014	cm3/molc sec	25	EXP	ATKINSON,R (1989)

ACETALDEHYDE

CAS #:	000075-07-0

Formula: C_2H_4O

Mol Weight: 44.05

MP (deg C): -123 | FP (deg C):

BP (deg C): 20.4

BP pressure (mm Hg):

Property/Value	Units	Temp	Data Type	Reference
WS 1.00E+006	mg/L		EXP	RIDDICK,JA ET AL. (1986)
logP -0.34			EXP	TSCATS
VP 9.02E+002	mm Hg	25	EXP	BOUBLIK,T ET AL. (1984)
DC 13.57	pKa	25	EXP	SERJEANT,EP & DEMPSEY,B (1979)
HL 6.67E-005	atm m3/mol	25	EXP	GAFFNEY,JS ET AL. (1987)
OH 1.58E-011	cm3/molc sec	25	EXP	ATKINSON,R (1989)

ETHYL MERCAPTAN

CAS #:	000075-08-1

Formula: C_2H_6S

Mol Weight: 62.13

MP (deg C): -144.4 | FP (deg C):

BP (deg C): 35

BP pressure (mm Hg):

Property/Value	Units	Temp	Data Type	Reference
WS 6.80E+003	mg/L	25	EST	INST NATL DE RECHERCHE ET DE SECURITE (1983)
logP 1.27			EST	MEYLAN,WM & HOWARD,PH (1995)
VP 5.29E+002	mm Hg	25	EXP	DAUBERT,TE & DANNER,RP (1989)
DC 10.61	pKa	25	EXP	SERJEANT,EP & DEMPSEY,B (1979)
HL 4.53E-003	atm m3/mol	20	EXP	VITENBERG,AG ET AL. (1975)
OH 4.68E-011	cm3/molc sec	25	EXP	ATKINSON,R (1989)

DICHLOROMETHANE

CAS #:	000075-09-2

Formula: CH_2Cl_2

Mol Weight: 84.93

MP (deg C): -94.92 | FP (deg C):

BP (deg C): 39.64

BP pressure (mm Hg):

Property/Value	Units	Temp	Data Type	Reference
WS 1.30E+004	mg/L	25	EXP	HORVATH,AL (1982)
logP 1.25			EXP	HANSCH,C & LEO,AJ (1985)
VP 4.35E+002	mm Hg	25	EXP	BOUBLIK,T ET AL. (1984)
DC	pKa			
HL 3.25E-003	atm m3/mol	25	EXP	LEIGHTON,DTJR & CALO,JM (1981)
OH 1.42E-013	cm3/molc sec	25	EXP	ATKINSON,R (1989)

DIFLUOROMETHANE

CAS #:	000075-10-5

Formula: CH_2F_2

Mol Weight: 52.02

MP (deg C): -136 | FP (deg C):

BP (deg C): -51.6

BP pressure (mm Hg):

Property/Value	Units	Temp	Data Type	Reference
WS 1.66E+004	mg/L	25	EST	MEYLAN,WM ET AL. (1996)
logP 0.20			EXP	HANSCH,C & LEO,AJ (1985)
VP 1.26E+004	mm Hg	25	EXP	DAUBERT,TE & DANNER,RP (1989)
DC	pKa			
HL 2.92E-001	atm m3/mol	25	EST	MEYLAN,WM & HOWARD,PH (1991)
OH 1.09E-014	cm3/molc sec	25	EXP	ATKINSON,R (1989)

METHYLENE IODIDE

CAS #:	000075-11-6

Formula: CH_2I_2

Mol Weight: 267.84

MP (deg C): 6 | FP (deg C):

BP (deg C): 181

BP pressure (mm Hg):

Property/Value	Units	Temp	Data Type	Reference
WS 1.40E+004	mg/L	20	EXP	YALKOWSKY,SH & DANNENFELSER,RM (1992)
logP 2.30			EXP	HANSCH,C ET AL. (1995)
VP 1.20E+000	mm Hg	25	EXP	DAUBERT,TE & DANNER,RP (1989)
DC	pKa			
HL 3.23E-004	atm m3/mol	20	EXP	MOORE,RM ET AL. (1995)
OH 2.62E-013	cm3/molc sec	25	EST	MEYLAN,WM & HOWARD,PH (1993)

FORMAMIDE

CAS #:	000075-12-7

Formula: CH_3NO

Mol Weight: 45.04

MP (deg C): 2.55 | FP (deg C):

BP (deg C): 210

BP pressure (mm Hg):

Property/Value	Units	Temp	Data Type	Reference
WS 1.00E+006	mg/L		EXP	EBERLING,CL (1980)
logP -1.51			EXP	HANSCH,C & LEO,AJ (1985)
VP 6.10E-002	mm Hg	25	EXP	DAUBERT,TE & DANNER,RP (1989)
DC -0.48	pKa	25	EXP	RIDDICK,JA ET AL. (1986)
HL 1.39E-009	atm m3/mol	25	EST	VP/WSOL
OH 1.73E-010	cm3/molc sec	25	EST	MEYLAN,WM & HOWARD,PH (1993)

CAS #: 000075-15-0 — CARBON DISULFIDE

Formula: CS_2
Mol Weight: 76.14
MP (deg C): -111.6
FP (deg C):
BP (deg C): 46.5
BP pressure (mm Hg):

Property/Value	Units	Temp	Data Type	Reference
WS 1.19E+003	mg/L	25	EXP	SEIDELL,A (1941)
logP 2.14			EXP	MARTISKA,A & BEKAREK,V (1990)
VP 3.58E+002	mm Hg	25	EXP	BOUBLIK,T ET AL. (1984)
DC	pKa			
HL 1.44E-002	atm m3/mol	24	EXP	ELLIOT,S (1989)
OH 2.10E-012	cm3/molc sec	25	EXP	ATKINSON,RA (1992C)

CAS #: 000075-17-2 — FORMAMIDE OXIME

Formula: CH_3NO
Mol Weight: 45.04
MP (deg C): 2.5
FP (deg C):
BP (deg C): 109
BP pressure (mm Hg): 1.50E+001

Property/Value	Units	Temp	Data Type	Reference
WS 1.70E+005	mg/L	20	EXP	YALKOWSKY,SH & DANNENFELSER,RM (1992)
logP 0.68			EST	MEYLAN,WM & HOWARD,PH (1995)
VP 4.31E+001	mm Hg	25	EST	NEELY,WB & BLAU,GE (1985)
DC	pKa			
HL 4.43E-006	atm m3/mol	25	EST	MEYLAN,WM & HOWARD,PH (1991)
OH 6.30E-013	cm3/molc sec	25	EXP	ATKINSON,R (1989)

CAS #: 000075-18-3 — DIMETHYLSULFIDE

Formula: C_2H_6S
Mol Weight: 62.13
MP (deg C): -83
FP (deg C):
BP (deg C): 36.2
BP pressure (mm Hg):

Property/Value	Units	Temp	Data Type	Reference
WS 2.20E+004	mg/L	25	EXP	SUZUKI,T (1991)
logP 0.92			EST	MEYLAN,WM & HOWARD,PH (1995)
VP 5.02E+002	mm Hg	25	EXP	DAUBERT,TE & DANNER,RP (1989)
DC	pKa			
HL 1.61E-003	atm m3/mol	25	EXP	GAFFNEY,JS ET AL. (1987)
OH 4.56E-012	cm3/molc sec	25	EXP	ATKINSON,R (1989)

CAS #: 000075-19-4 — CYCLOPROPANE

Formula: C_3H_6
Mol Weight: 42.08
MP (deg C): -127
FP (deg C):
BP (deg C): -33
BP pressure (mm Hg):

Property/Value	Units	Temp	Data Type	Reference
WS 3.81E+002	mg/L	35	EXP	YALKOWSKY,SH & DANNENFELSER,RM (1992)
logP 1.72			EXP	HANSCH,C & LEO,AJ (1985)
VP 5.41E+003	mm Hg	25	EXP	DAUBERT,TE & DANNER,RP (1989)
DC	pKa			
HL 1.10E-001	atm m3/mol	25	EST	VP/WSOL
OH 7.00E-014	cm3/molc sec	25	EXP	ATKINSON,R (1989)

CAS #: 000075-21-8 — ETHYLENE OXIDE

Formula: C_2H_4O
Mol Weight: 44.05
MP (deg C): -111
FP (deg C):
BP (deg C): 10.7
BP pressure (mm Hg):

Property/Value	Units	Temp	Data Type	Reference
WS 1.00E+006	mg/L		EXP	SCHULTZE,HC (1965)
logP -0.30			EXP	HANSCH,C & LEO,AJ (1985)
VP 1.31E+003	mm Hg	25	EXP	DAUBERT,TE & DANNER,RP (1985)
DC	pKa			
HL 1.48E-004	atm m3/mol	25	EXP	CONWAY,RA ET AL. (1983)
OH 7.60E-014	cm3/molc sec	25	EXP	ATKINSON,R (1989)

CAS #: 000075-25-2 — BROMOFORM

Formula: $CHBr_3$
Mol Weight: 252.75
MP (deg C): 8.05
FP (deg C):
BP (deg C): 149.21
BP pressure (mm Hg):

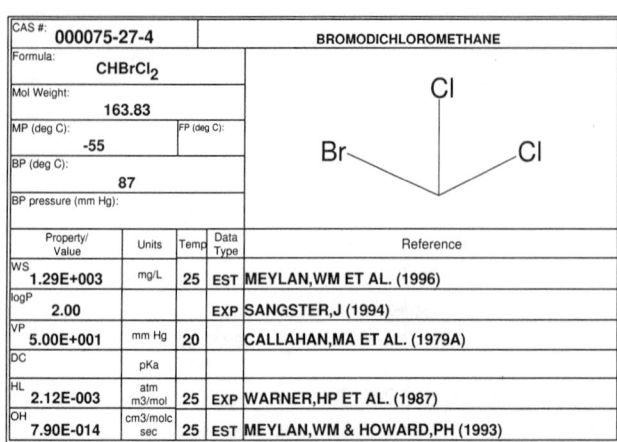

Property/Value	Units	Temp	Data Type	Reference
WS 3.10E+003	mg/L	25	EXP	HORVATH,AL (1982)
logP 2.40			EXP	CHEM INSPECT TEST INST (1992)
VP 5.40E+000	mm Hg	25	EXT	BOUBLIK,T ET AL. (1984)
DC	pKa			
HL 5.35E-004	atm m3/mol	25	EXP	MUNZ,C & ROBERTS,PV (1987)
OH 4.94E-014	cm3/molc sec	25	EST	MEYLAN,WM & HOWARD,PH (1993)

CAS #: 000075-26-3 — 2-BROMOPROPANE

Formula: C_3H_7Br
Mol Weight: 123.00
MP (deg C): -89
FP (deg C):
BP (deg C): 59-60
BP pressure (mm Hg):

Property/Value	Units	Temp	Data Type	Reference
WS 3.18E+003	mg/L	20	EXP	YALKOWSKY,SH & DANNENFELSER,RM (1992)
logP 2.14			EXP	HANSCH,C ET AL. (1995)
VP 2.16E+002	mm Hg	25	EXT	DAUBERT,TE & DANNER,RP (1989)
DC	pKa			
HL 1.10E-002	atm m3/mol	25	EST	VP/WSOL
OH 6.68E-013	cm3/molc sec	25	EST	MEYLAN,WM & HOWARD,PH (1993)

CAS #: 000075-27-4 — BROMODICHLOROMETHANE

Formula: $CHBrCl_2$
Mol Weight: 163.83
MP (deg C): -55
FP (deg C):
BP (deg C): 87
BP pressure (mm Hg):

Property/Value	Units	Temp	Data Type	Reference
WS 1.29E+003	mg/L	25	EST	MEYLAN,WM ET AL. (1996)
logP 2.00			EXP	SANGSTER,J (1994)
VP 5.00E+001	mm Hg	20		CALLAHAN,MA ET AL. (1979A)
DC	pKa			
HL 2.12E-003	atm m3/mol	25	EXP	WARNER,HP ET AL. (1987)
OH 7.90E-014	cm3/molc sec	25	EST	MEYLAN,WM & HOWARD,PH (1993)

000075-28-5 — 2-METHYLPROPANE

Formula: C_4H_{10}

Mol Weight: 58.12

MP (deg C): -138.3

FP (deg C):

BP (deg C): -11.7

BP pressure (mm Hg):

Property/Value	Units	Temp	Data Type	Reference
WS 4.89E+001	mg/L	25	EXP	MCAULIFFE,C (1966)
logP 2.76			EXP	HANSCH,C & LEO,AJ (1985)
VP 2.61E+003	mm Hg	25	EXP	RIDDICK,JA ET AL. (1986)
DC	pKa			
HL 1.19E+000	atm m3/mol	25	EST	VP/WSOL
OH 2.34E-012	cm3/molc sec	25	EXP	ATKINSON,R (1989)

000075-29-6 — 2-CHLOROPROPANE

Formula: C_3H_7Cl

Mol Weight: 78.54

MP (deg C): -117.2

FP (deg C):

BP (deg C): 35.7

BP pressure (mm Hg):

Property/Value	Units	Temp	Data Type	Reference
WS 3.05E+003	mg/L	20	EXP	YALKOWSKY,SH & DANNENFELSER,RM (1992)
logP 1.90			EXP	HANSCH,C & LEO,AJ (1985)
VP 5.15E+002	mm Hg	25	EXP	RIDDICK,JA ET AL. (1986)
DC	pKa			
HL 1.75E-002	atm m3/mol	25	EST	VP/WSOL
OH 8.35E-013	cm3/molc sec	25	EST	MEYLAN,WM & HOWARD,PH (1993)

000075-30-9 — 2-IODOPROPANE

Formula: C_3H_7I

Mol Weight: 169.99

MP (deg C): -90

FP (deg C):

BP (deg C): 89-90

BP pressure (mm Hg):

Property/Value	Units	Temp	Data Type	Reference
WS 1.40E+003	mg/L	20	EXP	YALKOWSKY,SH & DANNENFELSER,RM (1992)
logP 2.89			EXP	HANSCH,C ET AL. (1995)
VP 4.31E+001	mm Hg	25	EXP	DAUBERT,TE & DANNER,RP (1989)
DC	pKa			
HL 6.89E-003	atm m3/mol	25	EST	VP/WSOL
OH 1.17E-012	cm3/molc sec	25	EST	MEYLAN,WM & HOWARD,PH (1993)

000075-31-0 — ISOPROPYLAMINE

Formula: C_3H_9N

Mol Weight: 59.11

MP (deg C): -101

FP (deg C):

BP (deg C): 33-34

BP pressure (mm Hg):

Property/Value	Units	Temp	Data Type	Reference
WS 1.00E+006	mg/L		EXP	DORIGAN,J ET AL. (1976) @2ND
logP 0.26			EXP	HANSCH,C & LEO,AJ (1985)
VP 5.80E+002	mm Hg	25	EXP	DAUBERT,TE & DANNER,RP (1989)
DC 10.60	pKa		EXP	PERRIN,DD (1965)
HL 4.51E-005	atm m3/mol	25	EST	VP/WSOL
OH 3.87E-011	cm3/molc sec	25	EST	MEYLAN,WM & HOWARD,PH (1993)

000075-33-2 — 2-PROPANETHIOL

Formula: C_3H_8S

Mol Weight: 76.16

MP (deg C): -130.5

FP (deg C):

BP (deg C): 52.6

BP pressure (mm Hg):

Property/Value	Units	Temp	Data Type	Reference
WS 4.84E+003	mg/L	25	EST	MEYLAN,WM ET AL. (1996)
logP 1.68			EST	MEYLAN,WM & HOWARD,PH (1995)
VP 2.77E+002	mm Hg	25	EXP	BOUBLIK,T ET AL. (1984)
DC	pKa			
HL 4.61E-003	atm m3/mol	25	EST	MEYLAN,WM & HOWARD,PH (1991)
OH 4.20E-011	cm3/molc sec	25	EXP	ATKINSON,R (1989)

000075-34-3 — 1,1-DICHLOROETHANE

Formula: $C_2H_4Cl_2$

Mol Weight: 98.96

MP (deg C): -96.96

FP (deg C):

BP (deg C): 57.30

BP pressure (mm Hg):

Property/Value	Units	Temp	Data Type	Reference
WS 5.06E+003	mg/L	25	EXP	GROSS,PM (1929)
logP 1.79			EXP	HANSCH,C & LEO,AJ (1985)
VP 2.27E+002	mm Hg	25	EXP	DAUBERT,TE & DANNER,RP (1985)
DC	pKa			
HL 5.62E-003	atm m3/mol	24	EXP	GOSSETT,JM (1987)
OH 2.60E-013	cm3/molc sec	25	EXP	ATKINSON,R (1989)

000075-35-4 — 1,1-DICHLOROETHENE

Formula: $C_2H_2Cl_2$

Mol Weight: 96.94

MP (deg C): -122.56

FP (deg C):

BP (deg C): 31.56

BP pressure (mm Hg):

Property/Value	Units	Temp	Data Type	Reference
WS 2.25E+003	mg/L	25	EXP	DELASSUS,PT & SCHMIDT,DD (1981)
logP 2.13			EXP	HANSCH,C & LEO,AJ (1985)
VP 6.00E+002	mm Hg	25	EXP	BOUBLIK,T ET AL. (1984)
DC	pKa			
HL 2.61E-002	atm m3/mol	24	EXP	GOSSETT,JM (1987)
OH 8.10E-012	cm3/molc sec	25	EXP	ATKINSON,R (1989)

000075-36-5 — ACETYLCHLORIDE

Formula: C_2H_3ClO

Mol Weight: 78.50

MP (deg C): -112

FP (deg C):

BP (deg C): 52

BP pressure (mm Hg):

Property/Value	Units	Temp	Data Type	Reference
WS 3.33E+005	mg/L	25	EST	MEYLAN,WM ET AL. (1996)
logP -0.47			EST	MEYLAN,WM & HOWARD,PH (1995)
VP 2.87E+002	mm Hg	25	EXP	DAUBERT,TE & DANNER,RP (1989)
DC	pKa			
HL 6.65E-004	atm m3/mol	25	EST	MEYLAN,WM & HOWARD,PH (1991)
OH 6.80E-014	cm3/molc sec	25	EXP	ATKINSON,R (1989)

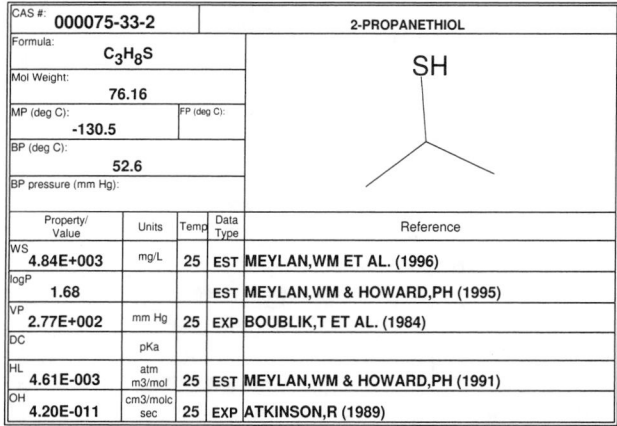

CAS #: 000075-37-6 — 1,1-DIFLUOROETHANE

Formula: $C_2H_4F_2$
Mol Weight: 66.05
MP (deg C): -117
FP (deg C):
BP (deg C): -24.9
BP pressure (mm Hg):

Property/Value	Units	Temp	Data Type	Reference
WS 3.24E+003	mg/L	25	EXP	HINE,J & MOOKERJEE,PK (1975)
logP 0.75			EXP	HANSCH,C & LEO,AJ (1985)
VP 4.55E+003	mm Hg	25	EXP	DAUBERT,TE & DANNER,RP (1989)
DC	pKa			
HL 2.03E-002	atm m3/mol	25	EXP	HINE,J & MOOKERJEE,PK (1975)
OH 3.40E-014	cm3/molc sec	25	EXP	ATKINSON,R (1989)

CAS #: 000075-38-7 — 1,1-DIFLUOROETHENE

Formula: $C_2H_2F_2$
Mol Weight: 64.04
MP (deg C): -144
FP (deg C):
BP (deg C): -83
BP pressure (mm Hg):

Property/Value	Units	Temp	Data Type	Reference
WS 1.80E+002	mg/L	25	EXP	IRAC (1986)
logP 1.24			EXP	HANSCH,C & LEO,AJ (1985)
VP 3.00E+004	mm Hg	25	EXP	DAUBERT,TP & DANNER,RP (1989)
DC	pKa			
HL 3.56E-001	atm m3/mol	25	EST	VP/WSOL
OH 2.10E-012	cm3/molc sec	23	EXP	HOWARD,CJ (1976)

CAS #: 000075-43-4 — DICHLOROFLUOROMETHANE

Formula: $CHCl_2F$
Mol Weight: 102.92
MP (deg C): -158
FP (deg C):
BP (deg C):
BP pressure (mm Hg):

Property/Value	Units	Temp	Data Type	Reference
WS 9.50E+003	mg/L	25	EXP	RIDDICK,JA ET AL. (1986)
logP 1.55			EXP	HANSCH,C & LEO,AJ (1985)
VP 1.36E+003	mm Hg	25	EXP	DAUBERT,TE & DANNER,RP (1989)
DC	pKa			
HL 1.08E-002	atm m3/mol	25	EST	VP/WSOL
OH 3.03E-014	cm3/molc sec	25	EXP	ATKINSON,R (1989)

CAS #: 000075-44-5 — PHOSGENE

Formula: CCl_2O
Mol Weight: 98.92
MP (deg C): -118
FP (deg C):
BP (deg C): 8.2
BP pressure (mm Hg):

Property/Value	Units	Temp	Data Type	Reference
WS 4.75E+005	mg/L	25	EST	MEYLAN,WM ET AL. (1996)
logP -0.71			EST	MEYLAN,WM & HOWARD,PH (1995)
VP 1.42E+003	mm Hg	25	EXP	DAUBERT,TE & DANNER,RP (1985)
DC	pKa			
HL 8.92E-003	atm m3/mol	25	EST	MEYLAN,WM & HOWARD,PH (1991)
OH 0.00E+000	cm3/molc sec	25	EST	MEYLAN,WM & HOWARD,PH (1993)

CAS #: 000075-45-6 — CHLORODIFLUOROMETHANE

Formula: $CHClF_2$
Mol Weight: 86.47
MP (deg C): -157.4
FP (deg C):
BP (deg C): -40.7
BP pressure (mm Hg):

Property/Value	Units	Temp	Data Type	Reference
WS 2.68E+004	mg/L	21	EXP	YALKOWSKY,SH & DANNENFELSER,RM (1992)
logP 1.08			EXP	HANSCH,C & LEO,AJ (1985)
VP 7.25E+003	mm Hg	25	EXP	PERRY,RH & GREEN,D (1984)
DC	pKa			
HL 3.23E-003	atm m3/mol	25	EST	VP/WSOL
OH 4.68E-015	cm3/molc sec	25	EXP	ATKINSON,R (1989)

CAS #: 000075-46-7 — TRIFLUOROMETHANE

Formula: CHF_3
Mol Weight: 70.01
MP (deg C): -160
FP (deg C):
BP (deg C): -84.4
BP pressure (mm Hg):

Property/Value	Units	Temp	Data Type	Reference
WS 7.33E+002	mg/L	25	EXP	HINE,J & MOOKERJEE,PK (1975)
logP 0.64			EXP	HANSCH,C & LEO,AJ (1985)
VP 3.64E+004	mm Hg	25	EXP	DAUBERT,TE & DANNER,RP (1989)
DC	pKa			
HL 9.52E-002	atm m3/mol	25	EXP	HINE,J & MOOKERJEE,PK (1975)
OH 2.40E-016	cm3/molc sec	25	EXP	ATKINSON,R (1989)

CAS #: 000075-47-8 — IODOFORM

Formula: CHI_3
Mol Weight: 393.73
MP (deg C): 120
FP (deg C):
BP (deg C):
BP pressure (mm Hg):

Property/Value	Units	Temp	Data Type	Reference
WS 1.00E+002	mg/L		EXP	YALKOWSKY,SH & DANNENFELSER,RM (1992)
logP 3.03			EST	MEYLAN,WM & HOWARD,PH (1995)
VP 3.96E-002	mm Hg	25	EST	NEELY,WB & BLAU,GE (1985)
DC	pKa			
HL 3.06E-005	atm m3/mol	25	EST	MEYLAN,WM & HOWARD,PH (1991)
OH 2.89E-013	cm3/molc sec	25	EST	MEYLAN,WM & HOWARD,PH (1993)

CAS #: 000075-50-3 — TRIMETHYLAMINE

Formula: C_3H_9N
Mol Weight: 59.11
MP (deg C): -124
FP (deg C):
BP (deg C): 3.2-3.8
BP pressure (mm Hg): 7.47E+002

Property/Value	Units	Temp	Data Type	Reference
WS 8.90E+005	mg/L	30	EXP	SCHWEIZER,AE ET AL. (1978)
logP 0.16			EXP	HANSCH,C & LEO,AJ (1985)
VP 1.61E+003	mm Hg	25	EXP	DAUBERT,TE & DANNER,RP (1985)
DC 9.80	pKa		EXP	PERRIN,DD (1972)
HL 1.04E-004	atm m3/mol	25	EXP	CHRISTIE,AO & CRISP,DJ (1967)
OH 6.09E-011	cm3/molc sec	25	EXP	ATKINSON,R (1989)

000075-52-5 — NITROMETHANE

CAS #:	000075-52-5		NITROMETHANE
Formula:	CH_3NO_2		
Mol Weight:	61.04		
MP (deg C):	-29	FP (deg C):	
BP (deg C):	101.2		
BP pressure (mm Hg):			

Property/Value	Units	Temp	Data Type	Reference
WS 1.11E+005	mg/L	25	EXP	RIDDICK,JA ET AL. (1986)
logP -0.35			EXP	HANSCH,C & LEO,AJ (1985)
VP 3.58E+001	mm Hg	25	EXP	DAUBERT,TE & DANNER,RP (1989)
DC 10.21	pKa	25		RIDDICK,JA ET AL. (1986)
HL 2.86E-005	atm m3/mol	25	EXP	GAFFNEY,JS ET AL. (1987)
OH 1.60E-013	cm3/molc sec	22	EXP	NIELSON,OJ ET AL. (1989)

000075-54-7 — DICHLOROMETHYLSILANE

CAS #:	000075-54-7		DICHLOROMETHYLSILANE
Formula:	CH_4Cl_2Si		
Mol Weight:	115.04		
MP (deg C):	-93	FP (deg C):	
BP (deg C):	41		
BP pressure (mm Hg):			

Property/Value	Units	Temp	Data Type	Reference
WS 3.72E+003	mg/L	25	EST	MEYLAN,WM ET AL. (1996)
logP 1.70			EST	MEYLAN,WM & HOWARD,PH (1995)
VP 4.29E+002	mm Hg	25	EXP	BOUBLIK,T ET AL. (1984)
DC	pKa			
HL 1.34E-002	atm m3/mol	25	EST	MEYLAN,WM & HOWARD,PH (1991)
OH 1.50E-013	cm3/molc sec	25	EST	MEYLAN,WM & HOWARD,PH (1993)

000075-55-8 — 2-METHYLAZIRIDINE

CAS #:	000075-55-8		2-METHYLAZIRIDINE
Formula:	C_3H_7N		
Mol Weight:	57.10		
MP (deg C):	-65.00	FP (deg C):	
BP (deg C):	66.0		
BP pressure (mm Hg):			

Property/Value	Units	Temp	Data Type	Reference
WS 1.00E+006	mg/L		EXP	HAM,GE (1981)
logP 0.13			EST	MEYLAN,WM & HOWARD,PH (1995)
VP 1.12E+002	mm Hg	20	EXP	HAM,GE (1981)
DC	pKa			
HL 8.42E-006	atm m3/mol	25	EST	VP/WSOL
OH 6.08E-011	cm3/molc sec	25	EST	MEYLAN,WM & HOWARD,PH (1993)

000075-56-9 — PROPYLENE OXIDE

CAS #:	000075-56-9		PROPYLENE OXIDE
Formula:	C_3H_6O		
Mol Weight:	58.08		
MP (deg C):	-112.13	FP (deg C):	
BP (deg C):	34.23		
BP pressure (mm Hg):			

Property/Value	Units	Temp	Data Type	Reference
WS 5.90E+005	mg/L	25	EXP	BOGYO,DA ET AL. (1980)
logP 0.03			EXP	HANSCH,C & LEO,AJ (1985)
VP 5.38E+002	mm Hg	25	EXP	BOUBLIK,T ET AL. (1984)
DC	pKa			
HL 6.96E-005	atm m3/mol	25	EST	VP/WSOL
OH 5.20E-013	cm3/molc sec	25	EXP	ATKINSON,R (1989)

000075-60-5 — CACODYLIC ACID

CAS #:	000075-60-5		CACODYLIC ACID
Formula:	$C_2H_7AsO_2$		
Mol Weight:	138.00		
MP (deg C):	195-196	FP (deg C):	
BP (deg C):			
BP pressure (mm Hg):			

Property/Value	Units	Temp	Data Type	Reference
WS 2.00E+006	mg/L	25	EXP	YALKOWSKY,SH & DANNENFELSER,RM (1992)
logP 0.36			EST	MEYLAN,WM & HOWARD,PH (1995)
VP	mm Hg			
DC 6.19	pKa	25	EXP	WEAST,RC ET AL. (1985)
HL	atm m3/mol			
OH	cm3/molc sec			

000075-61-6 — DIBROMODIFLUOROMETHANE

CAS #:	000075-61-6		DIBROMODIFLUOROMETHANE
Formula:	CBr_2F_2		
Mol Weight:	209.83		
MP (deg C):	-110.1	FP (deg C):	
BP (deg C):	25		
BP pressure (mm Hg):			

Property/Value	Units	Temp	Data Type	Reference
WS 1.40E+002	mg/L	25	EST	MEYLAN,WM ET AL. (1996)
logP 1.99			EST	MEYLAN,WM & HOWARD,PH (1995)
VP 8.20E+002	mm Hg	25	EXP	DAUBERT,TE & DANNER,RP (1989)
DC	pKa			
HL 3.08E-002	atm m3/mol	25	EST	MEYLAN,WM & HOWARD,PH (1991)
OH 0.00E+000	cm3/molc sec	25	EST	MEYLAN,WM & HOWARD,PH (1993)

000075-62-7 — BROMOTRICHLOROMETHANE

CAS #:	000075-62-7		BROMOTRICHLOROMETHANE
Formula:	$CBrCl_3$		
Mol Weight:	198.28		
MP (deg C):	-5.7	FP (deg C):	
BP (deg C):	105		
BP pressure (mm Hg):			

Property/Value	Units	Temp	Data Type	Reference
WS 3.08E+002	mg/L	25	EST	MEYLAN,WM ET AL. (1996)
logP 2.53			EST	MEYLAN,WM & HOWARD,PH (1995)
VP 3.90E+001	mm Hg	25	EXP	DAUBERT,TE & DANNER,RP (1989)
DC	pKa			
HL 3.71E-004	atm m3/mol	25	EST	MEYLAN,WM & HOWARD,PH (1991)
OH 0.00E+000	cm3/molc sec	25	EST	MEYLAN,WM & HOWARD,PH (1993)

000075-63-8 — BROMOTRIFLUOROMETHANE

CAS #:	000075-63-8		BROMOTRIFLUOROMETHANE
Formula:	$CBrF_3$		
Mol Weight:	148.92		
MP (deg C):	-172	FP (deg C):	
BP (deg C):	-57.8		
BP pressure (mm Hg):			

Property/Value	Units	Temp	Data Type	Reference
WS 2.97E+002	mg/L	25	EXP	HINE,J & MOOKERJEE,PK (1975)
logP 1.86			EXP	HANSCH,C & LEO,AJ (1985)
VP 1.22E+004	mm Hg	25	EXP	DAUBERT,TE & DANNER,RP (1989)
DC	pKa			
HL 4.99E-001	atm m3/mol	25	EXP	HINE,J & MOOKERJEE,PK (1975)
OH 1.00E-015	cm3/molc sec	25	EXP	ATKINSON,R (1989)

CAS #: 000075-64-9				T-BUTYLAMINE

Formula: $C_4H_{11}N$

Mol Weight: 73.14

MP (deg C): 44-46 | FP (deg C):

BP (deg C): -72.65

BP pressure (mm Hg):

Property/ Value	Units	Temp	Data Type	Reference
WS 1.00E+006	mg/L	20	EXP	YALKOWSKY,SH & DANNENFELSER,RM (1992)
logP 0.40			EXP	HANSCH,C & LEO,AJ (1985)
VP 3.72E+002	mm Hg	25	EXP	DAUBERT,TE & DANNER,RP (1989)
DC 10.68	pKa		EXP	PERRIN,DD (1965)
HL 3.58E-005	atm m3/mol	25	EST	VP/WSOL
OH 2.15E-011	cm3/molc sec	25	EST	MEYLAN,WM & HOWARD,PH (1993)

CAS #: 000075-65-0				T-BUTANOL

Formula: $C_4H_{10}O$

Mol Weight: 74.12

MP (deg C): 25.62 | FP (deg C):

BP (deg C): 82.35

BP pressure (mm Hg):

Property/ Value	Units	Temp	Data Type	Reference
WS 1.00E+006	mg/L		EXP	RIDDICK,JA ET AL. (1986)
logP 0.35			EXP	HANSCH,C & LEO,AJ (1985)
VP 4.17E+001	mm Hg	25	EXT	BOUBLIK,T ET AL. (1984)
DC 19.20	pKa	25	EXP	SERJEANT,EP & DEMPSEY,B (1979)
HL 1.44E-005	atm m3/mol	25	EXP	SNIDER,JR & DAWSON,GA (1985)
OH 1.12E-012	cm3/molc sec	25	EXP	ATKINSON,R (1989)

CAS #: 000075-66-1				2-METHYL-2-PROPANETHIOL

Formula: $C_4H_{10}S$

Mol Weight: 90.19

MP (deg C): -0.5 | FP (deg C):

BP (deg C): 63.7-64.2

BP pressure (mm Hg):

Property/ Value	Units	Temp	Data Type	Reference
WS 2.05E+003	mg/L	25	EST	MEYLAN,WM ET AL (1996)
logP 2.14			EST	MEYLAN,WM & HOWARD,PH (1995)
VP 1.81E+002	mm Hg	25	EXP	DAUBERT,TE & DANNER,RP (1989)
DC	pKa			
HL 6.11E-003	atm m3/mol	25	EST	MEYLAN,WM & HOWARD,PH (1991)
OH 3.31E-011	cm3/molc sec	25	EXP	ATKINSON,R (1989)

CAS #: 000075-68-3				1-CHLORO-1,1-DIFLUOROETHANE

Formula: $C_2H_3ClF_2$

Mol Weight: 100.50

MP (deg C): -130.8 | FP (deg C):

BP (deg C): -9.7

BP pressure (mm Hg):

Property/ Value	Units	Temp	Data Type	Reference
WS 6.31E+003	mg/L	25	EXP	SUZUKI,T (1991)
logP 2.05			EST	MEYLAN,WM & HOWARD,PH (1995)
VP 2.54E+003	mm Hg	25	EXP	DAUBERT,TE & DANNER,RP (1989)
DC	pKa			
HL 1.59E-002	atm m3/mol	25	EST	VP/WSOL
OH 3.58E-015	cm3/molc sec	25	EXP	ATKINSON,R (1989)

CAS #: 000075-69-4				TRICHLOROFLUOROMETHANE

Formula: CCl_3F

Mol Weight: 137.37

MP (deg C): -110.48 | FP (deg C):

BP (deg C): 23.63

BP pressure (mm Hg):

Property/ Value	Units	Temp	Data Type	Reference
WS 1.10E+003	mg/L	25	EXP	DUPONT DE NEMOURS CO (1980)
logP 2.53			EXP	HANSCH,C & LEO,AJ (1985)
VP 8.03E+002	mm Hg	25	EXP	DAUBERT,TE & DANNER,RP (1985)
DC	pKa			
HL 9.70E-002	atm m3/mol	25	EXP	WARNER,MJ & WEISS,RF (1985)
OH 5.00E-016	cm3/molc sec	25	EXP	ATKINSON,R (1989)

CAS #: 000075-71-8				DICHLORODIFLUOROMETHANE

Formula: CCl_2F_2

Mol Weight: 120.91

MP (deg C): -158 | FP (deg C):

BP (deg C): -29.79

BP pressure (mm Hg):

Property/ Value	Units	Temp	Data Type	Reference
WS 2.80E+002	mg/L	25	EXP	SMART,BE (1980)
logP 2.16			EXP	HANSCH,C & LEO,AJ (1985)
VP 4.85E+003	mm Hg	25	EXP	DAUBERT,TE & DANNER,RP (1985)
DC	pKa			
HL 3.43E-001	atm m3/mol	25	EXP	WARNER,MJ & WEISS,RF (1985)
OH 4.00E-016	cm3/molc sec	25	EXP	ATKINSON,R (1989)

CAS #: 000075-72-9				CHLOROTRIFLUOROMETHANE

Formula: $CClF_3$

Mol Weight: 104.46

MP (deg C): -181 | FP (deg C):

BP (deg C): -81.4

BP pressure (mm Hg):

Property/ Value	Units	Temp	Data Type	Reference
WS 6.01E+001	mg/L	25	EXP	HINE,J & MOOKERJEE,PK (1975)
logP 1.65			EXP	HANSCH,C & LEO,AJ (1985)
VP 2.14E+004	mm Hg	25	EXP	PERRY,RH & GREEN,D (1984)
DC	pKa			
HL 1.38E+000	atm m3/mol	25	EXP	PARK,T ET AL. (1982)
OH 7.00E-016	cm3/molc sec	25	EXP	ATKINSON,R (1989)

CAS #: 000075-73-0				TETRAFLUOROMETHANE

Formula: CF_4

Mol Weight: 88.00

MP (deg C): -183.6 | FP (deg C):

BP (deg C): -127.8

BP pressure (mm Hg):

Property/ Value	Units	Temp	Data Type	Reference
WS 1.86E+001	mg/L	25	EXP	YALKOWSKY,SH & DANNENFELSER,RM (1992)
logP 1.18			EXP	HANSCH,C & LEO,AJ (1985)
VP 6.21E+005	mm Hg	25	EXT	DAUBERT,TE & DANNER,RP (1989)
DC	pKa			
HL 5.15E+000	atm m3/mol	25	EXP	PARK,T ET AL. (1982)
OH 4.00E-016	cm3/molc sec	25	EXP	ATKINSON,R (1989)

000075-75-2 — METHANESULFONIC ACID

Formula: CH_4O_3S
Mol Weight: 96.11
MP (deg C): 20
BP (deg C): 167
BP pressure (mm Hg): 1.00E+001

Property/Value	Units	Temp	Data Type	Reference
WS 1.00E+006	mg/L	20	EXP	SEIDELL,A (1941)
logP -2.38			EST	MEYLAN,WM & HOWARD,PH (1995)
VP 4.28E-004	mm Hg	25	EXT	DAUBERT,TE & DANNER,RP (1991)
DC -1.86	pKa	25	EXP	SERJEANT,EP & DEMPSEY,B (1979)
HL 1.26E-008	atm m3/mol	25	EST	MEYLAN,WM & HOWARD,PH (1991)
OH 2.76E-013	cm3/molc sec	25	EST	MEYLAN,WM & HOWARD,PH (1993)

000075-76-3 — TETRAMETHYLSILANE

Formula: $C_4H_{12}Si$
Mol Weight: 88.23
MP (deg C): -99
BP (deg C): 26.6
BP pressure (mm Hg):

Property/Value	Units	Temp	Data Type	Reference
WS 1.96E+001	mg/L	25	EXP	RIDDICK,JA ET AL. (1986)
logP 2.72			EST	MEYLAN,WM & HOWARD,PH (1995)
VP 7.18E+002	mm Hg	25	EXP	BOUBLIK,T ET AL. (1984)
DC	pKa			
HL 4.25E+000	atm m3/mol	25	EST	VP/WSOL
OH 5.98E-013	cm3/molc sec	25	EST	MEYLAN,WM & HOWARD,PH (1993)

000075-77-4 — CHLOROTRIMETHYLSILANE

Formula: C_3H_9ClSi
Mol Weight: 108.64
MP (deg C): -40
BP (deg C): 60
BP pressure (mm Hg):

Property/Value	Units	Temp	Data Type	Reference
WS 8.35E+002	mg/L	25	EST	MEYLAN,WM ET AL. (1996)
logP 2.48			EST	MEYLAN,WM & HOWARD,PH (1995)
VP 2.34E+002	mm Hg	25	EXP	BOUBLIK,T ET AL. (1984)
DC	pKa			
HL 6.39E-002	atm m3/mol	25	EST	MEYLAN,WM & HOWARD,PH (1991)
OH 4.49E-013	cm3/molc sec	25	EST	MEYLAN,WM & HOWARD,PH (1993)

000075-78-5 — DICHLORODIMETHYLSILANE

Formula: $C_2H_6Cl_2Si$
Mol Weight: 129.06
MP (deg C): -16
BP (deg C): 70.3
BP pressure (mm Hg):

Property/Value	Units	Temp	Data Type	Reference
WS 1.12E+003	mg/L	25	EST	MEYLAN,WM ET AL. (1996)
logP 2.24			EST	MEYLAN,WM & HOWARD,PH (1995)
VP 1.44E+002	mm Hg	25	EXP	DAUBERT,TE & DANNER,RP (1989)
DC	pKa			
HL 1.85E-002	atm m3/mol	25	EST	MEYLAN,WM & HOWARD,PH (1991)
OH 2.99E-013	cm3/molc sec	25	EST	MEYLAN,WM & HOWARD,PH (1993)

000075-79-6 — METHYLTRICHLOROSILANE

Formula: CH_3Cl_3Si
Mol Weight: 149.48
MP (deg C): -77.8
BP (deg C): 66.4
BP pressure (mm Hg):

Property/Value	Units	Temp	Data Type	Reference
WS 1.47E+003	mg/L	25	EST	MEYLAN,WM ET AL. (1996)
logP 2.01			EST	MEYLAN,WM & HOWARD,PH (1995)
VP 1.67E+002	mm Hg	25	EXP	RIDDICK,JA ET AL (1986)
DC	pKa			
HL 5.34E-003	atm m3/mol	25	EST	MEYLAN,WM & HOWARD,PH (1991)
OH 1.50E-013	cm3/molc sec	25	EST	MEYLAN,WM & HOWARD,PH (1993)

000075-80-9 — ETHANOL, 2,2,2-TRIBROMO-

Formula: $C_2H_3Br_3O$
Mol Weight: 282.77
MP (deg C): 79-82
BP (deg C): 92-93
BP pressure (mm Hg): 1.00E+001

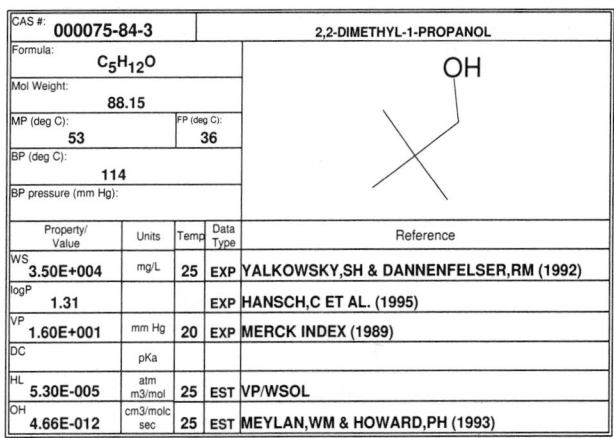

Property/Value	Units	Temp	Data Type	Reference
WS 3.40E+004	mg/L	40	EXP	YALKOWSKY,SH & DANNENFELSER,RM (1992)
logP 2.10			EXP	HANSCH,C ET AL. (1995)
VP 1.72E-003	mm Hg	25	EST	NEELY,WB & BLAU,GE (1985)
DC	pKa			
HL 5.47E-009	atm m3/mol	25	EST	MEYLAN,WM & HOWARD,PH (1991)
OH 3.69E-013	cm3/molc sec	25	EST	MEYLAN,WM & HOWARD,PH (1993)

000075-83-2 — 2,2-DIMETHYLBUTANE

Formula: C_6H_{14}
Mol Weight: 86.18
MP (deg C): -99
BP (deg C): 49.7
BP pressure (mm Hg):

Property/Value	Units	Temp	Data Type	Reference
WS 2.38E+001	mg/L	25	EXP	YALKOWSKY,SH & DANNENFELSER,RM (1992)
logP 3.82			EXP	HANSCH,C & LEO,AJ (1985)
VP 3.19E+002	mm Hg	25	EXP	BOUBLIK,T ET AL. (1984)
DC	pKa			
HL 1.52E+000	atm m3/mol	25	EST	VP/WSOL
OH 2.32E-012	cm3/molc sec	25	EXP	ATKINSON,R (1989)

000075-84-3 — 2,2-DIMETHYL-1-PROPANOL

Formula: $C_5H_{12}O$
Mol Weight: 88.15
MP (deg C): 53   FP (deg C): 36
BP (deg C): 114
BP pressure (mm Hg):

Property/Value	Units	Temp	Data Type	Reference
WS 3.50E+004	mg/L	25	EXP	YALKOWSKY,SH & DANNENFELSER,RM (1992)
logP 1.31			EXP	HANSCH,C ET AL. (1995)
VP 1.60E+001	mm Hg	20	EXP	MERCK INDEX (1989)
DC	pKa			
HL 5.30E-005	atm m3/mol	25	EST	VP/WSOL
OH 4.66E-012	cm3/molc sec	25	EST	MEYLAN,WM & HOWARD,PH (1993)

CAS #: 000075-85-4				2-METHYL-2-BUTANOL

Formula: $C_5H_{12}O$

Mol Weight: 88.15

MP (deg C): -8.8 FP (deg C):

BP (deg C): 102.4

BP pressure (mm Hg): 7.65E+002

Property/ Value	Units	Temp	Data Type	Reference
WS 1.10E+005	mg/L	25	EXP	BARTON,AFM (1984)
logP 0.89			EXP	HANSCH,C & LEO,AJ (1985)
VP 1.68E+001	mm Hg	25	EXP	DAUBERT,TE & DANNER,RP (1985)
DC	pKa			
HL 7.30E-004	atm m3/mol	25	EXP	BUTLER,JAV ET AL. (1935)
OH 4.89E-012	cm3/molc sec	25	EST	MEYLAN,WM & HOWARD,PH (1993)

CAS #: 000075-86-5				ACETONE CYANOHYDRIN

Formula: C_4H_7NO

Mol Weight: 85.11

MP (deg C): -19 FP (deg C):

BP (deg C): 95

BP pressure (mm Hg):

Property/ Value	Units	Temp	Data Type	Reference
WS 1.00E+006	mg/L		EXP	SMILEY,RA (1981)
logP -0.03			EST	MEYLAN,WM & HOWARD,PH (1995)
VP 7.50E-001	mm Hg	20	EXP	WEBER,RC ET AL. (1981)
DC	pKa			
HL 1.97E-009	atm m3/mol	25	EST	MEYLAN,WM & HOWARD,PH (1991)
OH 4.08E-013	cm3/molc sec	25	EST	MEYLAN,WM & HOWARD,PH (1993)

CAS #: 000075-87-6				TRICHLOROACETALDEHYDE

Formula: C_2HCl_3O

Mol Weight: 147.39

MP (deg C): -57.5 FP (deg C):

BP (deg C): 97.8

BP pressure (mm Hg):

Property/ Value	Units	Temp	Data Type	Reference
WS 8.30E+006	mg/L	25	EXP	MERCK INDEX (1983)
logP 0.99			EXP	HANSCH,C & LEO,AJ (1985)
VP 5.00E+001	mm Hg	25	EXP	PERRY,RH & GREEN,D (1984)
DC 9.66	pKa	30	EXP	SERJEANT,EP & DEMPSEY,B (1979)
HL 2.91E-009	atm m3/mol	25	EXP	BETTERTON,EA & HOFFMAN,MR (1988)
OH 1.73E-012	cm3/molc sec	25	EXP	ATKINSON,R (1989)

CAS #: 000075-88-7				1,1,1-TRIFLUORO-2-CHLOROETHANE

Formula: $C_2H_2ClF_3$

Mol Weight: 118.49

MP (deg C): -105.5 FP (deg C):

BP (deg C): 6.1

BP pressure (mm Hg):

Property/ Value	Units	Temp	Data Type	Reference
WS 9.20E+003	mg/L	25	EXP	SUZUKI,T (1991)
logP 1.99			EST	MEYLAN,WM & HOWARD,PH (1995)
VP 9.53E+002	mm Hg	25	EST	NEELY,WB & BLAU,GE (1985)
DC	pKa			
HL 2.71E-001	atm m3/mol	25	EST	MEYLAN,WM & HOWARD,PH (1991)
OH 1.62E-014	cm3/molc sec	25	EXP	ATKINSON,R (1989)

CAS #: 000075-89-8				2,2,2-TRIFLUOROETHANOL

Formula: $C_2H_3F_3O$

Mol Weight: 100.04

MP (deg C): -43.5 FP (deg C):

BP (deg C): 74

BP pressure (mm Hg):

Property/ Value	Units	Temp	Data Type	Reference
WS 1.69E+005	mg/L	25	EST	MEYLAN,WM ET AL. (1996)
logP 0.41			EXP	HANSCH,C & LEO,AJ (1985)
VP 7.13E+001	mm Hg	25	EXP	MEEKS,AC & GOLDFARB,IJ (1967)
DC 12.37	pKa	25	EXP	SERJEANT,EP & DEMPSEY,B (1979)
HL 1.73E-005	atm m3/mol	25	EXP	ROCHESTER,CH & SYMONDS,JR (1973)
OH 9.55E-014	cm3/molc sec	25	EXP	ATKINSON,R (1989)

CAS #: 000075-91-2				TERT-BUTYLHYDROPEROXIDE

Formula: $C_4H_{10}O_2$

Mol Weight: 90.12

MP (deg C): -8 FP (deg C):

BP (deg C): 35

BP pressure (mm Hg): 2.00E+001

Property/ Value	Units	Temp	Data Type	Reference
WS 1.97E+004	mg/L	25	EST	MEYLAN,WM ET AL. (1996)
logP 0.94			EST	MEYLAN,WM & HOWARD,PH (1995)
VP 5.47E+000	mm Hg	25	EXP	DAUBERT,TE & DANNER,RP (1989)
DC 12.80	pKa	20	EXP	SERJEANT,EP & DEMPSEY,B (1979)
HL 1.60E-005	atm m3/mol	25	EST	MEYLAN,WM & HOWARD,PH (1991)
OH 3.00E-012	cm3/molc sec	25	EXP	ATKINSON,R (1989)

CAS #: 000075-94-5				VINYLTRICHLOROSILANE

Formula: $C_2H_3Cl_3Si$

Mol Weight: 161.49

MP (deg C): -95 FP (deg C):

BP (deg C): 91.5

BP pressure (mm Hg):

Property/ Value	Units	Temp	Data Type	Reference
WS 6.48E+002	mg/L	25	EST	MEYLAN,WM ET AL. (1996)
logP 2.36			EST	MEYLAN,WM & HOWARD,PH (1995)
VP 6.59E+001	mm Hg	25	EXP	BOUBLIK,T ET AL. (1984)
DC	pKa			
HL 5.37E-003	atm m3/mol	25	EST	MEYLAN,WM & HOWARD,PH (1991)
OH 2.63E-011	cm3/molc sec	25	EST	MEYLAN,WM & HOWARD,PH (1993)

CAS #: 000075-97-8				3,3-DIMETHYL-2-BUTANONE

Formula: $C_6H_{12}O$

Mol Weight: 100.16

MP (deg C): -52.5 FP (deg C):

BP (deg C): 106.1

BP pressure (mm Hg):

Property/ Value	Units	Temp	Data Type	Reference
WS 1.90E+004	mg/L	25	EXP	YALKOWSKY,SH & DANNENFELSER,RM (1992)
logP 1.20			EXP	HANSCH,C ET AL. (1995)
VP 3.15E+001	mm Hg	25	EXP	DAUBERT,TE & DANNER,RP (1993)
DC	pKa			
HL 2.18E-004	atm m3/mol	25	EST	VP/WSOL
OH 1.21E-012	cm3/molc sec	25	EXP	ATKINSON,R (1989)

CAS #: 000075-98-9 — TRIMETHYLACETIC ACID

Formula: $C_5H_{10}O_2$
Mol Weight: 102.13
MP (deg C): 35.5 FP (deg C):
BP (deg C): 163.8
BP pressure (mm Hg):

Property/Value	Units	Temp	Data Type	Reference
WS 2.17E+004	mg/L	20		YALKOWSKY,SH & DANNENFELSER,RM (1992)
logP 1.48			EXP	CHEM INSPECT TEST INST (1992)
VP 1.82E+000	mm Hg		EST	NEELY,WB & BLAU,GE (1985)
DC 5.03	pKa	25	EXP	DEAN,JA (1985)
HL 1.28E-006	atm m3/mol	25	EST	MEYLAN,WM & HOWARD,PH (1991)
OH 5.90E-013	cm3/molc sec	25	EST	ATKINSON,R (1988)

CAS #: 000075-99-0 — DALAPON

Formula: $C_3H_4Cl_2O_2$
Mol Weight: 142.97
MP (deg C): -4.4 EST FP (deg C):
BP (deg C): 187.5
BP pressure (mm Hg):

Property/Value	Units	Temp	Data Type	Reference
WS 7.43E+003	mg/L	25	EST	MEYLAN,WM ET AL. (1996)
logP 0.76			EXP	RAO,PSC & DAVIDSON,JM (1982)
VP 1.90E-001	mm Hg	25	EST	CLAUSIUS-CLAPEYRON EQN
DC	pKa			
HL 6.43E-008	atm m3/mol	25	EST	MEYLAN,WM & HOWARD,PH (1991)
OH 5.86E-001	cm3/molc sec	25	EST	MEYLAN,WM & HOWARD,PH (1993)

CAS #: 000076-01-7 — PENTACHLOROETHANE

Formula: C_2HCl_5
Mol Weight: 202.30
MP (deg C): -29.0 FP (deg C):
BP (deg C): 159.88
BP pressure (mm Hg):

Property/Value	Units	Temp	Data Type	Reference
WS 4.80E+002	mg/L	25	EXP	DILLING,WL (1977)
logP 3.22			EXP	HANSCH,C ET AL. (1995)
VP 3.50E+000	mm Hg	25	EXP	ENGINEERING SCIENCES DATA UNIT (1976)
DC	pKa			
HL 1.94E-003	atm m3/mol	25	EST	VP/WSOL
OH 2.38E-014	cm3/molc sec	25	EST	MEYLAN,WM & HOWARD,PH (1993)

CAS #: 000076-02-8 — TRICHLORACETYLCHLORIDE

Formula: C_2Cl_4O
Mol Weight: 181.83
MP (deg C): -31.8 FP (deg C):
BP (deg C): 118
BP pressure (mm Hg):

Property/Value	Units	Temp	Data Type	Reference
WS 9.49E+003	mg/L	25	EST	MEYLAN,WM ET AL. (1996)
logP 0.88			EST	MEYLAN,WM & HOWARD,PH (1995)
VP 2.13E+001	mm Hg	25	EXP	DAUBERT,TE & DANNER,RP (1993)
DC	pKa			
HL 2.91E-005	atm m3/mol	25	EST	MEYLAN,WM & HOWARD,PH (1991)
OH 0.00E+000	cm3/molc sec	25	EST	MEYLAN,WM & HOWARD,PH (1993)

CAS #: 000076-03-9 — TRICHLOROACETIC ACID

Formula: $C_2HCl_3O_2$
Mol Weight: 163.39
MP (deg C): 57-58 FP (deg C):
BP (deg C): 196-197
BP pressure (mm Hg):

Property/Value	Units	Temp	Data Type	Reference
WS 1.30E+006	mg/L	25	EXP	WILLIS,GH & MCDOWELL,LL (1982)
logP 1.33			EXP	HANSCH,C & LEO,AJ (1985)
VP 4.54E-009	mm Hg	25	EXT	BOUBLIK,T ET AL. (1984)
DC 0.51	pKa		EXP	SERJEANT,EP & DEMPSEY,B (1979)
HL 2.39E-008	atm m3/mol	25	EST	MEYLAN,WM & HOWARD,PH (1991)
OH 5.20E-013	cm3/molc sec	25	EST	MEYLAN,WM & HOWARD,PH (1993)

CAS #: 000076-05-1 — TRIFLUOROACETATE

Formula: $C_2HF_3O_2$
Mol Weight: 114.02
MP (deg C): -15.4 FP (deg C):
BP (deg C): 72.4
BP pressure (mm Hg):

Property/Value	Units	Temp	Data Type	Reference
WS 1.00E+006	mg/L	20	EXP	RIDDICK,JA ET AL. (1986)
logP 0.50			EST	MEYLAN,WM & HOWARD,PH (1995)
VP 1.10E+002	mm Hg	25	EXP	DAUBERT,TE & DANNER,RP (1989)
DC 0.52	pKa	25	EXP	KORTUM,G ET AL (1961)
HL 1.65E-005	atm m3/mol	25	EST	VP/WSOL
OH 5.20E-013	cm3/molc sec	25	EST	MEYLAN,WM & HOWARD,PH (1993)

CAS #: 000076-06-2 — TRICHLORONITROMETHANE

Formula: CCl_3NO_2
Mol Weight: 164.38
MP (deg C): -64 FP (deg C):
BP (deg C): 112
BP pressure (mm Hg): 7.57E+002

Property/Value	Units	Temp	Data Type	Reference
WS 1.62E+003	mg/L	25	EXP	BUDAVARI,S ET AL. (1989)
logP 2.09			EXP	HANSCH,C & LEO,AJ (1985)
VP 2.38E+001	mm Hg	25	EXP	MARTIN,H & WORTHING,CR (1977)
DC	pKa			
HL 2.05E-003	atm m3/mol	25	EXP	KAWAMOTO,K & URANO,K (1989)
OH 1.30E-013	cm3/molc sec	25	EST	MEYLAN,WM & HOWARD,PH (1993)

CAS #: 000076-09-5 — PINACOL

Formula: $C_6H_{14}O_2$
Mol Weight: 118.18
MP (deg C): 45.4 FP (deg C):
BP (deg C):
BP pressure (mm Hg):

Property/Value	Units	Temp	Data Type	Reference
WS 3.51E+004	mg/L	25	EST	MEYLAN,WM ET AL. (1996)
logP 0.54			EST	MEYLAN,WM & HOWARD,PH (1995)
VP 5.01E-001	mm Hg	25	EST	NEELY,WB & BLAU,GE (1985)
DC	pKa			
HL 4.06E-007	atm m3/mol	25	EST	MEYLAN,WM & HOWARD,PH (1991)
OH 2.35E-012	cm3/molc sec	25	EST	MEYLAN,WM & HOWARD,PH (1993)

CAS #: 000076-11-9 — 1,1,1-TETRACHLORO-2,2-DIFLUOROETHANE

Formula: $C_2Cl_4F_2$

Mol Weight: 203.83

MP (deg C): 40.6   FP (deg C):

BP (deg C): 91.5

BP pressure (mm Hg):

Property/Value	Units	Temp	Data Type	Reference
WS 1.00E+002	mg/L	25	EXP	ROY,WR & GRIFFIN,RA (1985)
logP 3.41			EST	MEYLAN,WM & HOWARD,PH (1995)
VP 5.49E+001	mm Hg	25	EXP	DAUBERT,TE & DANNER,RP (1989)
DC	pKa			
HL 1.47E-001	atm m3/mol	25	EST	VP/WSOL
OH 0.00E+000	cm3/molc sec	25	EST	MEYLAN,WM & HOWARD,PH (1993)

CAS #: 000076-12-0 — 1,1,2,2,-TETRACHLORODIFLUOROETHANE

Formula: $C_2Cl_4F_2$

Mol Weight: 203.83

MP (deg C): 26   FP (deg C):

BP (deg C): 93

BP pressure (mm Hg):

Property/Value	Units	Temp	Data Type	Reference
WS 1.20E+002	mg/L	25	EXP	RIDDICK,JA ET AL. (1986)
logP 3.41			EST	MEYLAN,WM & HOWARD,PH (1995)
VP 2.20E+004	mm Hg	25	EXP	DAUBERT,TE & DANNER,RP (1989)
DC	pKa			
HL 1.70E+000	atm m3/mol	25	EST	VP/WSOL
OH 0.00E+000	cm3/molc sec	25	EST	MEYLAN,WM & HOWARD,PH (1993)

CAS #: 000076-13-1 — 1,1,2-TRICHLOROTRIFLUOROETHANE

Formula: $C_2Cl_3F_3$

Mol Weight: 187.38

MP (deg C): -36.4   FP (deg C):

BP (deg C): 47.633

BP pressure (mm Hg):

Property/Value	Units	Temp	Data Type	Reference
WS 1.70E+002	mg/L	25	EXP	RIDDICK,JA ET AL. (1986)
logP 3.16			EXP	HANSCH,C & LEO,AJ (1985)
VP 3.63E+002	mm Hg	25	EXP	BOUBLIK,T ET AL. (1984)
DC	pKa			
HL 5.26E-001	atm m3/mol	25	EST	VP/WSOL
OH 0.00E+000	cm3/molc sec	25	EST	MEYLAN,WM & HOWARD,PH (1993)

CAS #: 000076-14-2 — 1,2-DICHLOROTETRAFLUOROETHANE

Formula: $C_2Cl_2F_4$

Mol Weight: 170.92

MP (deg C): -94   FP (deg C):

BP (deg C): 4.1

BP pressure (mm Hg):

Property/Value	Units	Temp	Data Type	Reference
WS 1.30E+002	mg/L	25	EXP	RIDDICK,JA ET AL. (1986)
logP 2.82			EXP	HANSCH,C & LEO,AJ (1985)
VP 2.01E+003	mm Hg	25	EXP	RIDDICK,JA ET AL. (1986)
DC	pKa			
HL 2.80E+000	atm m3/mol	25	EST	VP/WSOL
OH 0.00E+000	cm3/molc sec	25	EST	MEYLAN,WM & HOWARD,PH (1993)

CAS #: 000076-15-3 — CHLOROPENTAFLUOROETHANE

Formula: C_2ClF_5

Mol Weight: 154.47

MP (deg C): -99.3   FP (deg C):

BP (deg C): -39

BP pressure (mm Hg):

Property/Value	Units	Temp	Data Type	Reference
WS 5.80E+001	mg/L	25	EXP	HORVATH,AL (1982)
logP 2.47			EST	MEYLAN,WM & HOWARD,PH (1995)
VP 6.86E+003	mm Hg	25	EXP	DAUBERT,TE & DANNER,RP (1989)
DC	pKa			
HL 2.66E+000	atm m3/mol	25	EST	VP/WSOL
OH 0.00E+000	cm3/molc sec	25	EST	MEYLAN,WM & HOWARD,PH (1993)

CAS #: 000076-16-4 — HEXAFLUOROETHANE

Formula: C_2F_6

Mol Weight: 138.01

MP (deg C): -100.7   FP (deg C):

BP (deg C): -78.1

BP pressure (mm Hg):

Property/Value	Units	Temp	Data Type	Reference
WS 7.78E+000	mg/L	25	EXP	YALKOWSKY,SH & DANNENFELSER,RM (1992)
logP 2.00			EXP	HANSCH,C & LEO,AJ (1985)
VP 2.51E+004	mm Hg	25	EXT	DAUBERT,TE & DANNER,RP (1989)
DC	pKa			
HL 2.03E+001	atm m3/mol	25	EXP	PARK,T ET AL. (1982)
OH 0.00E+000	cm3/molc sec	25	EST	MEYLAN,WM & HOWARD,PH (1993)

CAS #: 000076-19-7 — OCTAFLUOROPROPANE

Formula: C_3F_8

Mol Weight: 188.02

MP (deg C): -147.6   FP (deg C):

BP (deg C): -36.6

BP pressure (mm Hg):

Property/Value	Units	Temp	Data Type	Reference
WS 5.70E+000	mg/L	15	EXP	YALKOWSKY,SH & DANNENFELSER,RM (1992)
logP 3.12			EST	MEYLAN,WM & HOWARD,PH (1995)
VP 6.63E+003	mm Hg	25	EXP	DAUBERT,TE & DANNER,RP (1989)
DC	pKa			
HL 3.30E+001	atm m3/mol	25	EST	VP/WSOL
OH 0.00E+000	cm3/molc sec	25	EST	MEYLAN,WM & HOWARD,PH (1993)

CAS #: 000076-20-0 — TRIONAL

Formula: $C_8H_{18}O_4S_2$

Mol Weight: 242.36

MP (deg C): 74-76   FP (deg C):

BP (deg C):

BP pressure (mm Hg):

Property/Value	Units	Temp	Data Type	Reference
WS 5.00E+003	mg/L	16	EXP	YALKOWSKY,SH & DANNENFELSER,RM (1992)
logP 1.91			EST	MEYLAN,WM & HOWARD,PH (1995)
VP 1.71E-005	mm Hg	25	EST	NEELY,WB & BLAU,GE (1985)
DC	pKa			
HL 2.18E-009	atm m3/mol	25	EST	MEYLAN,WM & HOWARD,PH (1991)
OH 1.64E-011	cm3/molc sec	25	EST	MEYLAN,WM & HOWARD,PH (1993)

CAS #: 000076-22-2				CAMPHOR

Formula:	$C_{10}H_{16}O$			
Mol Weight:	152.24			
MP (deg C):	180	FP (deg C):		
BP (deg C):	204			
BP pressure (mm Hg):				

Property/Value	Units	Temp	Data Type	Reference
WS 1.60E+003	mg/L	25	EXP	YALKOWSKY,SH & DANNENFELSER,RM (1992)
logP 2.34			EST	MEYLAN,WM & HOWARD,PH (1995)
VP 6.50E-001	mm Hg	25	EXP	JONES,AH (1960)
DC	pKa			
HL 8.10E-005	atm m3/mol	25	EST	VP/WSOL
OH 1.40E-011	cm3/molc sec	25	EST	MEYLAN,WM & HOWARD,PH (1993)

CAS #: 000076-25-5				TRIAMCINOLONE ACETONIDE

Formula:	$C_{24}H_{31}FO_6$			
Mol Weight:	434.51			
MP (deg C):	292-294	FP (deg C):		
BP (deg C):				
BP pressure (mm Hg):				

Property/Value	Units	Temp	Data Type	Reference
WS 2.10E+001	mg/L	28	EXP	YALKOWSKY,SH & DANNENFELSER,RM (1992)
logP 2.53			EXP	HANSCH,C & LEO,AJ (1985)
VP 8.25E-014	mm Hg	25	EST	NEELY,WB & BLAU,GE (1985)
DC	pKa			
HL 9.83E-013	atm m3/mol	25	EST	MEYLAN,WM & HOWARD,PH (1991)
OH 9.50E-011	cm3/molc sec	25	EST	MEYLAN,WM & HOWARD,PH (1993)

CAS #: 000076-37-9				2,2,3,3-TETRAFLUOROPROPANOL

Formula:	$C_3H_4F_4O$			
Mol Weight:	132.06			
MP (deg C):	-15	FP (deg C):		
BP (deg C):	109-110			
BP pressure (mm Hg):				

Property/Value	Units	Temp	Data Type	Reference
WS 8.41E+004	mg/L	25	EST	MEYLAN,WM ET AL. (1996)
logP 0.63			EST	MEYLAN,WM & HOWARD,PH (1995)
VP 1.47E+001	mm Hg	25	EXP	BOUBLIK,T ET AL. (1984)
DC 12.74	pKa	25	EXP	SERJEANT,EP & DEMPSEY,B (1979)
HL 6.29E-006	atm m3/mol	25	EXP	ROCHESTER,CH & SYMONDS,JR (1973)
OH 3.73E-012	cm3/molc sec	25	EST	MEYLAN,WM & HOWARD,PH (1993)

CAS #: 000076-38-0				METHOXYFLURANE

Formula:	$C_3H_4Cl_2F_2O$			
Mol Weight:	164.97			
MP (deg C):	-35	FP (deg C):		
BP (deg C):	51			
BP pressure (mm Hg):	1.00E+002			

Property/Value	Units	Temp	Data Type	Reference
WS 2.83E+004	mg/L	37	EXP	YALKOWSKY,SH & DANNENFELSER,RM (1992)
logP 2.21			EXP	HANSCH,C & LEO,AJ (1985)
VP 4.91E+001	mm Hg	25	EST	NEELY,WB & BLAU,GE (1985)
DC	pKa			
HL 5.62E-004	atm m3/mol	25	EST	MEYLAN,WM & HOWARD,PH (1991)
OH 3.22E-014	cm3/molc sec	25	EST	MEYLAN,WM & HOWARD,PH (1993)

CAS #: 000076-39-1				2-METHYL-2-NITRO-1-PROPANOL

Formula:	$C_4H_9NO_3$			
Mol Weight:	119.12			
MP (deg C):	89.5	FP (deg C):		
BP (deg C):	94			
BP pressure (mm Hg):	1.00E+001			

Property/Value	Units	Temp	Data Type	Reference
WS 3.50E+006	mg/L	20	EXP	DEWEY,RH & BOLLMEIER,AFJR (1978)
logP -0.14			EST	MEYLAN,WM & HOWARD,PH (1995)
VP 9.01E-002	mm Hg	25	EST	NEELY,WB & BLAU,GE (1985)
DC	pKa			
HL 3.60E-009	atm m3/mol	25	EST	MEYLAN,WM & HOWARD,PH (1991)
OH 7.66E-013	cm3/molc sec	25	EST	MEYLAN,WM & HOWARD,PH (1993)

CAS #: 000076-40-4				2,2,3-TRICHLORO-1,1-BUTANEDIOL

Formula:	$C_4H_7Cl_3O_2$			
Mol Weight:	193.46			
MP (deg C):		FP (deg C):		
BP (deg C):				
BP pressure (mm Hg):				

Property/Value	Units	Temp	Data Type	Reference
WS 2.63E+004	mg/L	20	EXP	STEPHEN,H & STEPHEN,T (1963)
logP 1.89			EST	MEYLAN,WM & HOWARD,PH (1995)
VP 2.52E-003	mm Hg	25	EST	NEELY,WB & BLAU,GE (1985)
DC	pKa			
HL 1.01E-008	atm m3/mol	25	EST	MEYLAN,WM & HOWARD,PH (1991)
OH 6.45E-012	cm3/molc sec	25	EST	MEYLAN,WM & HOWARD,PH (1993)

CAS #: 000076-41-5				OXYMORPHONE

Formula:	$C_{17}H_{19}NO_4$			
Mol Weight:	301.35			
MP (deg C):	248-249	FP (deg C):		
BP (deg C):				
BP pressure (mm Hg):				

Property/Value	Units	Temp	Data Type	Reference
WS 2.40E+004	mg/L	25	EST	MEYLAN,WM ET AL. (1996)
logP 0.83			EXP	HANSCH,C & LEO,AJ (1985)
VP 2.63E-010	mm Hg	25	EST	NEELY,WB & BLAU,GE (1985)
DC	pKa			
HL 4.09E-019	atm m3/mol	25	EST	MEYLAN,WM & HOWARD,PH (1991)
OH 2.43E-010	cm3/molc sec	25	EST	MEYLAN,WM & HOWARD,PH (1993)

CAS #: 000076-43-7				ANDROST-4-EN-3-ONE, 9-FLUORO-11,17-DIHYDROXY-17-

Formula:	$C_{20}H_{31}FO_3$			
Mol Weight:	338.47			
MP (deg C):	270	FP (deg C):		
BP (deg C):				
BP pressure (mm Hg):				

Property/Value	Units	Temp	Data Type	Reference
WS 6.75E+001	mg/L	25	EST	MEYLAN,WM ET AL. (1996)
logP 2.38			EXP	HANSCH,C ET AL. (1995)
VP 6.46E-010	mm Hg	25	EST	NEELY,WB & BLAU,GE (1985)
DC	pKa			
HL 6.15E-010	atm m3/mol	25	EST	MEYLAN,WM & HOWARD,PH (1991)
OH 3.52E-011	cm3/molc sec	25	EST	MEYLAN,WM & HOWARD,PH (1993)

CAS #: 000076-44-8 — HEPTACHLOR

Formula: $C_{10}H_5Cl_7$

Mol Weight: 373.32

MP (deg C): 95-96

BP (deg C): 310

BP pressure (mm Hg):

Property/Value	Units	Temp	Data Type	Reference
WS 1.80E-001	mg/L	25	EXP	BIGGAR,JW & RIGGS,RI (1974)
logP 5.50			EXP	SANGSTER,J (1993)
VP 4.00E-004	mm Hg	25	EXP	WORTHING,CR & WALKER,SB (1983)
DC	pKa			
HL 1.48E-003	atm m3/mol	25	EXP	WARNER,HP ET AL. (1987)
OH 6.34E-011	cm3/molc sec	25	EST	MEYLAN,WM & HOWARD,PH (1993)

CAS #: 000076-49-3 — BORNYL ACETATE

Formula: $C_{12}H_{20}O_2$

Mol Weight: 196.29

MP (deg C): 29

BP (deg C): 221

BP pressure (mm Hg):

Property/Value	Units	Temp	Data Type	Reference
WS 2.32E+001	mg/L	25	EST	MEYLAN,WM ET AL. (1996)
logP 3.86			EST	MEYLAN,WM & HOWARD,PH (1995)
VP 2.28E-001	mm Hg	25	EXP	PERRY,RH & GREEN,D (1984)
DC	pKa			
HL 4.37E-004	atm m3/mol	25	EST	MEYLAN,WM & HOWARD,PH (1991)
OH 8.63E-012	cm3/molc sec	25	EST	MEYLAN,WM & HOWARD,PH (1993)

CAS #: 000076-57-3 — CODEINE

Formula: $C_{18}H_{21}NO_3$

Mol Weight: 299.37

MP (deg C): 157.5

BP (deg C): 250

BP pressure (mm Hg): 2.20E+001

Property/Value	Units	Temp	Data Type	Reference
WS 1.00E+004	mg/L	30	EXP	YALKOWSKY,SH & DANNENFELSER,RM (1992)
logP 1.14			EXP	HANSCH,C & LEO,AJ (1985)
VP 4.15E-009	mm Hg	25	EST	NEELY,WB & BLAU,GE (1985)
DC	pKa			
HL 7.58E-014	atm m3/mol	25	EST	MEYLAN,WM & HOWARD,PH (1991)
OH 2.36E-010	cm3/molc sec	25	EST	MEYLAN,WM & HOWARD,PH (1993)

CAS #: 000076-58-4 — DIONINE

Formula: $C_{19}H_{23}NO_3$

Mol Weight: 313.40

MP (deg C): 199-201

BP (deg C):

BP pressure (mm Hg):

Property/Value	Units	Temp	Data Type	Reference
WS 2.61E+003	mg/L	20	EXP	SEIDELL,A (1941)
logP 1.77			EST	MEYLAN,WM & HOWARD,PH (1995)
VP 1.64E-009	mm Hg	25	EST	NEELY,WB & BLAU,GE (1985)
DC	pKa			
HL 1.01E-013	atm m3/mol	25	EST	MEYLAN,WM & HOWARD,PH (1991)
OH 2.41E-010	cm3/molc sec	25	EST	MEYLAN,WM & HOWARD,PH (1993)

CAS #: 000076-68-6 — BARBITURIC ACID,5-ALLYL,5-(2CYPENTEN1YL)

Formula: $C_{12}H_{14}N_2O_3$

Mol Weight: 234.26

MP (deg C): 139-140

BP (deg C):

BP pressure (mm Hg):

Property/Value	Units	Temp	Data Type	Reference
WS 1.48E+003	mg/L	25	EST	MEYLAN,WM ET AL. (1996)
logP 1.51			EXP	HANSCH,C & LEO,AJ (1985)
VP 4.02E-012	mm Hg	25	EST	NEELY,WB & BLAU,GE (1985)
DC	pKa			
HL 3.24E-013	atm m3/mol	25	EST	MEYLAN,WM & HOWARD,PH (1991)
OH 9.78E-011	cm3/molc sec	25	EST	MEYLAN,WM & HOWARD,PH (1993)

CAS #: 000076-73-3 — SECOBARBITAL

Formula: $C_{12}H_{18}N_2O_3$

Mol Weight: 238.29

MP (deg C):

BP (deg C):

BP pressure (mm Hg):

Property/Value	Units	Temp	Data Type	Reference
WS 5.70E+002	mg/L	25	EST	MEYLAN,WM ET AL. (1996)
logP 1.97			EXP	HANSCH,C & LEO,AJ (1985)
VP 1.00E-011	mm Hg	25	EST	NEELY,WB & BLAU,GE (1985)
DC	pKa			
HL 8.35E-013	atm m3/mol	25	EST	MEYLAN,WM & HOWARD,PH (1991)
OH 4.41E-011	cm3/molc sec	25	EST	MEYLAN,WM & HOWARD,PH (1993)

CAS #: 000076-74-4 — PENTOBARBITAL

Formula: $C_{11}H_{18}N_2O_3$

Mol Weight: 226.28

MP (deg C):

BP (deg C):

BP pressure (mm Hg):

Property/Value	Units	Temp	Data Type	Reference
WS 6.79E+002	mg/L	25	EXP	YALKOWSKY,SH & DANNENFELSER,RM (1992)
logP 2.10			EXP	HANSCH,C & LEO,AJ (1985)
VP 2.17E-011	mm Hg	25	EST	NEELY,WB & BLAU,GE (1985)
DC 7.96	pKa	25	EXP	KORTUM,G ET AL (1961)
HL 8.44E-013	atm m3/mol	25	EST	MEYLAN,WM & HOWARD,PH (1991)
OH 1.80E-011	cm3/molc sec	25	EST	MEYLAN,WM & HOWARD,PH (1993)

CAS #: 000076-75-5 — THIOPENTAL

Formula: $C_{11}H_{18}N_2O_2S$

Mol Weight: 242.34

MP (deg C):

BP (deg C):

BP pressure (mm Hg):

Property/Value	Units	Temp	Data Type	Reference
WS 9.60E+001	mg/L	25	EST	MEYLAN,WM ET AL. (1996)
logP 2.85			EXP	HANSCH,C & LEO,AJ (1985)
VP 2.15E-010	mm Hg	25	EST	NEELY,WB & BLAU,GE (1985)
DC	pKa			
HL 6.35E-009	atm m3/mol	25	EST	MEYLAN,WM & HOWARD,PH (1991)
OH 2.70E-011	cm3/molc sec	25	EST	MEYLAN,WM & HOWARD,PH (1993)

CAS #: 000076-76-6 — PROBARBITAL

Formula: $C_9H_{14}N_2O_3$
Mol Weight: 198.22
MP (deg C): 197-198
FP (deg C):
BP (deg C):
BP pressure (mm Hg):

Property/Value	Units	Temp	Data Type	Reference
WS 1.21E+003	mg/L	25	EXP	YALKOWSKY,SH & DANNENFELSER,RM (1992)
logP 0.97			EXP	HANSCH,C & LEO,AJ (1985)
VP 1.14E-010	mm Hg	25	EST	NEELY,WB & BLAU,GE (1985)
DC	pKa			
HL 4.79E-013	atm m3/mol	25	EST	MEYLAN,WM & HOWARD,PH (1991)
OH 1.37E-011	cm3/molc sec	25	EST	MEYLAN,WM & HOWARD,PH (1993)

CAS #: 000076-83-5 — BENZENE, 1,1',1"-(CHLOROMETHYLIDYNE)TRIS-

Formula: $C_{19}H_{15}Cl$
Mol Weight: 278.78
MP (deg C): 113.5
FP (deg C):
BP (deg C): 310
BP pressure (mm Hg):

Property/Value	Units	Temp	Data Type	Reference
WS 5.35E-001	mg/L	25	EST	MEYLAN,WM ET AL. (1996)
logP 5.25			EXP	CHEM INSPECT TEST INST (1992)
VP 3.87E-006	mm Hg	25	EST	NEELY,WB & BLAU,GE (1985)
DC	pKa			
HL 1.36E-005	atm m3/mol	25	EST	MEYLAN,WM & HOWARD,PH (1991)
OH 1.29E-011	cm3/molc sec	25	EST	MEYLAN,WM & HOWARD,PH (1993)

CAS #: 000076-84-6 — BENZENEMETHANOL, .ALPHA.,.ALPHA.-DIPHENYL-

Formula: $C_{19}H_{16}O$
Mol Weight: 260.34
MP (deg C): 164.2
FP (deg C):
BP (deg C):
BP pressure (mm Hg):

Property/Value	Units	Temp	Data Type	Reference
WS 1.43E+003	mg/L	25	EXP	YALKOWSKY,SH & DANNENFELSER,RM (1992)
logP 3.68			EXP	HANSCH,C ET AL. (1995)
VP 3.15E-008	mm Hg	25	EST	NEELY,WB & BLAU,GE (1985)
DC	pKa			
HL 1.42E-009	atm m3/mol	25	EST	MEYLAN,WM & HOWARD,PH (1991)
OH 1.30E-011	cm3/molc sec	25	EST	MEYLAN,WM & HOWARD,PH (1993)

CAS #: 000076-87-9 — TRIPHENYLTIN HYDROXIDE

Formula: $C_{18}H_{16}OSn$
Mol Weight: 367.02
MP (deg C): 117-119
FP (deg C):
BP (deg C):
BP pressure (mm Hg):

Property/Value	Units	Temp	Data Type	Reference
WS 4.00E-001	mg/L	25	EXP	CHEM INSPECT TEST INST (1992)
logP 2.37			EST	MEYLAN,WM & HOWARD,PH (1995)
VP 3.53E-007	mm Hg	25	EXP	WAUCHOPE,RD ET AL. (1991A)
DC	pKa			
HL 4.26E-007	atm m3/mol	25	EST	VP/WSOL
OH	cm3/molc sec			

CAS #: 000076-93-7 — BENZILIC ACID

Formula: $C_{14}H_{12}O_3$
Mol Weight: 228.25
MP (deg C): 74-75
FP (deg C):
BP (deg C): 187
BP pressure (mm Hg): 1.30E+001

Property/Value	Units	Temp	Data Type	Reference
WS 1.41E+003	mg/L	25	EXP	YALKOWSKY,SH & DANNENFELSER,RM (1992)
logP 2.30			EXP	HANSCH,C & LEO,AJ (1985)
VP 3.00E-001	mm Hg	20	EXP	WEBER,RC ET AL. (1981)
DC 3.05	pKa	18	EXP	KORTUM,G ET AL (1961)
HL 2.28E+002	atm m3/mol	25	EST	VP/WSOL
OH 9.25E-012	cm3/molc sec	25	EST	MEYLAN,WM & HOWARD,PH (1993)

CAS #: 000076-94-8 — BARBITURIC ACID,5-ME-5-PHENYL-

Formula: $C_{11}H_{10}N_2O_3$
Mol Weight: 218.21
MP (deg C): 226
FP (deg C):
BP (deg C):
BP pressure (mm Hg):

Property/Value	Units	Temp	Data Type	Reference
WS 5.88E+003	mg/L	25	EST	MEYLAN,WM ET AL. (1996)
logP 0.91			EXP	PRANKERD,RJ & MCKEOWN,RH (1992)
VP 4.59E-012	mm Hg	25	EST	NEELY,WB & BLAU,GE (1985)
DC 7.73	pKa	25	EXP	KORTUM,G ET AL (1961)
HL 1.24E-014	atm m3/mol	25	EST	MEYLAN,WM & HOWARD,PH (1991)
OH 6.83E-012	cm3/molc sec	25	EST	MEYLAN,WM & HOWARD,PH (1993)

CAS #: 000076-99-3 — METHADONE

Formula: $C_{21}H_{27}NO$
Mol Weight: 309.46
MP (deg C):
FP (deg C):
BP (deg C):
BP pressure (mm Hg):

Property/Value	Units	Temp	Data Type	Reference
WS 4.85E+001	mg/L	25	EST	MEYLAN,WM ET AL. (1996)
logP 3.93			EXP	HANSCH,C & LEO,AJ (1985)
VP 1.12E-006	mm Hg	25	EST	NEELY,WB & BLAU,GE (1985)
DC 8.94	pKa	25	EXP	PERRIN,DD (1965)
HL 4.97E-010	atm m3/mol	25	EST	MEYLAN,WM & HOWARD,PH (1991)
OH 1.05E-010	cm3/molc sec	25	EST	MEYLAN,WM & HOWARD,PH (1993)

CAS #: 000077-02-1 — APROBARBITAL

Formula: $C_{10}H_{14}N_2O_3$
Mol Weight: 210.23
MP (deg C): 140-141.5
FP (deg C):
BP (deg C):
BP pressure (mm Hg):

Property/Value	Units	Temp	Data Type	Reference
WS 4.08E+003	mg/L	25	EXP	YALKOWSKY,SH & DANNENFELSER,RM (1992)
logP 1.15			EXP	HANSCH,C ET AL. (1995)
VP 5.31E-011	mm Hg	25	EST	NEELY,WB & BLAU,GE (1985)
DC 7.99	pKa	25	EXP	KORTUM,G ET AL (1961)
HL 4.74E-013	atm m3/mol	25	EST	MEYLAN,WM & HOWARD,PH (1991)
OH 3.98E-011	cm3/molc sec	25	EST	MEYLAN,WM & HOWARD,PH (1993)

CAS #: 000077-06-5				GIBBERELLIC ACID

Formula: $C_{19}H_{22}O_6$

Mol Weight: 346.38

MP (deg C): 233-235   FP (deg C):

BP (deg C):

BP pressure (mm Hg):

Property/ Value	Units	Temp	Data Type	Reference
WS 5.00E+003	mg/L		EXP	SHIU,WY ET AL. (1990)
logP 0.24			EXP	HANSCH,C & LEO,AJ (1985)
VP 2.06E-013	mm Hg	25	EST	NEELY,WB & BLAU,GE (1985)
DC	pKa			
HL 1.58E-015	atm m3/mol	25	EST	MEYLAN,WM & HOWARD,PH (1991)
OH 1.24E-010	cm3/molc sec	25	EST	MEYLAN,WM & HOWARD,PH (1993)

CAS #: 000077-07-6				LEVORPHANOL

Formula: $C_{17}H_{23}NO$

Mol Weight: 257.38

MP (deg C): 198-199   FP (deg C):

BP (deg C):

BP pressure (mm Hg):

Property/ Value	Units	Temp	Data Type	Reference
WS 1.84E+003	mg/L	25	EST	MEYLAN,WM ET AL. (1996)
logP 3.11			EXP	HANSCH,C & LEO,AJ (1985)
VP 1.04E-006	mm Hg	25	EST	NEELY,WB & BLAU,GE (1985)
DC	pKa			
HL 2.11E-010	atm m3/mol	25	EST	MEYLAN,WM & HOWARD,PH (1991)
OH 2.07E-010	cm3/molc sec	25	EST	MEYLAN,WM & HOWARD,PH (1993)

CAS #: 000077-09-8				PHENOLPHTHALEIN

Formula: $C_{20}H_{14}O_4$

Mol Weight: 318.33

MP (deg C): 258-262   FP (deg C):

BP (deg C):

BP pressure (mm Hg):

Property/ Value	Units	Temp	Data Type	Reference
WS 4.00E+002	mg/L		EXP	YALKOWSKY,SH & DANNENFELSER,RM (1992)
logP 3.06			EST	MEYLAN,WM & HOWARD,PH (1995)
VP 2.16E-012	mm Hg	25	EST	NEELY,WB & BLAU,GE (1985)
DC	pKa			
HL 8.98E-016	atm m3/mol	25	EST	MEYLAN,WM & HOWARD,PH (1991)
OH 8.42E-011	cm3/molc sec	25	EST	MEYLAN,WM & HOWARD,PH (1993)

CAS #: 000077-20-3				ALPHAPRODINE

Formula: $C_{16}H_{23}NO_2$

Mol Weight: 261.37

MP (deg C):   FP (deg C):

BP (deg C):

BP pressure (mm Hg):

Property/ Value	Units	Temp	Data Type	Reference
WS 7.97E+002	mg/L	25	EST	MEYLAN,WM ET AL. (1996)
logP 2.83			EXP	HANSCH,C ET AL. (1995)
VP 4.37E-005	mm Hg	25	EST	NEELY,WB & BLAU,GE (1985)
DC	pKa			
HL 1.28E-008	atm m3/mol	25	EST	MEYLAN,WM & HOWARD,PH (1991)
OH 9.83E-011	cm3/molc sec	25	EST	MEYLAN,WM & HOWARD,PH (1993)

CAS #: 000077-21-4				2-ETHYL-2-PHENYLGLUTERIMIDE

Formula: $C_{13}H_{15}NO_2$

Mol Weight: 217.27

MP (deg C): 84   FP (deg C):

BP (deg C):

BP pressure (mm Hg):

Property/ Value	Units	Temp	Data Type	Reference
WS 9.50E+000	mg/L	27	EXP	YALKOWSKY,SH & DANNENFELSER,RM (1992)
logP 1.90			EXP	HANSCH,C & LEO,AJ (1985)
VP 1.49E-009	mm Hg	25	EST	NEELY,WB & BLAU,GE (1985)
DC	pKa			
HL 4.78E-009	atm m3/mol	25	EST	MEYLAN,WM & HOWARD,PH (1991)
OH 2.87E-011	cm3/molc sec	25	EST	MEYLAN,WM & HOWARD,PH (1993)

CAS #: 000077-27-0				THIAMYLAL

Formula: $C_{12}H_{18}N_2O_2S$

Mol Weight: 254.35

MP (deg C): 132-133   FP (deg C):

BP (deg C):

BP pressure (mm Hg):

Property/ Value	Units	Temp	Data Type	Reference
WS 1.91E+003	mg/L	30	EXP	YALKOWSKY,SH & DANNENFELSER,RM (1992)
logP 3.23			EXP	HANSCH,C & LEO,AJ (1985)
VP 9.99E-011	mm Hg	25	EST	NEELY,WB & BLAU,GE (1985)
DC	pKa			
HL 6.28E-009	atm m3/mol	25	EST	MEYLAN,WM & HOWARD,PH (1991)
OH 5.31E-011	cm3/molc sec	25	EST	MEYLAN,WM & HOWARD,PH (1993)

CAS #: 000077-28-1				5-BUTYL-5-ETHYLBARBITURIC ACID

Formula: $C_{10}H_{16}N_2O_3$

Mol Weight: 212.25

MP (deg C): 124-127   FP (deg C):

BP (deg C):

BP pressure (mm Hg):

Property/ Value	Units	Temp	Data Type	Reference
WS 4.88E+003	mg/L	25	EXP	YALKOWSKY,SH & DANNENFELSER,RM (1992)
logP 1.73			EXP	HANSCH,C ET AL. (1995)
VP 3.02E-011	mm Hg	25	EST	NEELY,WB & BLAU,GE (1985)
DC	pKa			
HL 6.36E-013	atm m3/mol	25	EST	MEYLAN,WM & HOWARD,PH (1991)
OH 1.30E-011	cm3/molc sec	25	EST	MEYLAN,WM & HOWARD,PH (1993)

CAS #: 000077-30-5				5-ETHYL-5-HEXYLBARBITURIC ACID

Formula: $C_{12}H_{20}N_2O_3$

Mol Weight: 240.30

MP (deg C):   FP (deg C):

BP (deg C):

BP pressure (mm Hg):

Property/ Value	Units	Temp	Data Type	Reference
WS 2.12E+002	mg/L	25	EST	MEYLAN,WM ET AL. (1996)
logP 2.46			EXP	HANSCH,C & LEO,AJ (1985)
VP 5.68E-012	mm Hg	25	EST	NEELY,WB & BLAU,GE (1985)
DC	pKa			
HL 1.12E-012	atm m3/mol	25	EST	MEYLAN,WM & HOWARD,PH (1991)
OH 1.58E-011	cm3/molc sec	25	EST	MEYLAN,WM & HOWARD,PH (1993)

CAS #: 000077-32-7				THIOBARBITAL

Formula: $C_8H_{12}N_2O_2S$

Mol Weight: 200.26

MP (deg C): 180 FP (deg C):

BP (deg C):

BP pressure (mm Hg):

Property/ Value	Units	Temp	Data Type	Reference
WS 2.28E+003	mg/L	25	EST	MEYLAN,WM ET AL. (1996)
logP 1.50			EXP	WONG,O & MCKEOWN,RH (1988)
VP 1.54E-009	mm Hg	25	EST	NEELY,WB & BLAU,GE (1985)
DC	pKa			
HL 2.71E-009	atm m3/mol	25	EST	MEYLAN,WM & HOWARD,PH (1991)
OH 1.86E-011	cm3/molc sec	25	EST	MEYLAN,WM & HOWARD,PH (1993)

CAS #: 000077-41-8				METHSUXIMIDE

Formula: $C_{12}H_{13}NO_2$

Mol Weight: 203.24

MP (deg C): 52-53 FP (deg C):

BP (deg C): 121-122

BP pressure (mm Hg): 1.00E-001

Property/ Value	Units	Temp	Data Type	Reference
WS 2.44E+003	mg/L	25	EST	MEYLAN,WM ET AL. (1996)
logP 1.45			EST	MEYLAN,WM & HOWARD,PH (1995)
VP 3.17E-007	mm Hg	25	EST	NEELY,WB & BLAU,GE (1985)
DC	pKa			
HL 5.96E-009	atm m3/mol	25	EST	MEYLAN,WM & HOWARD,PH (1991)
OH 2.26E-011	cm3/molc sec	25	EST	MEYLAN,WM & HOWARD,PH (1993)

CAS #: 000077-47-4				HEXACHLOROCYCLOPENTADIENE

Formula: C_5Cl_6

Mol Weight: 272.77

MP (deg C): 11.34 FP (deg C):

BP (deg C): 239

BP pressure (mm Hg):

Property/ Value	Units	Temp	Data Type	Reference
WS 1.80E+000	mg/L	25	EXP	CALLAHAN,MA ET AL. (1979)
logP 5.04			EXP	HANSCH,C & LEO,AJ (1985)
VP 6.00E-002	mm Hg	25	EXP	DAUBERT,TE & DANNER,RP (1985)
DC	pKa			
HL 2.70E-002	atm m3/mol	25	EXP	WOLFE,NL ET AL. (1982)
OH 5.60E-013	cm3/molc sec	25	EST	MEYLAN,WM & HOWARD,PH (1993)

CAS #: 000077-49-6				2-METHYL-2-NITROPROPANE-1,3-DIOL

Formula: $C_4H_9NO_4$

Mol Weight: 135.12

MP (deg C): 150.1 FP (deg C):

BP (deg C):

BP pressure (mm Hg):

Property/ Value	Units	Temp	Data Type	Reference
WS 8.00E+005	mg/L	20	EXP	DEWEY,RH & BOLLMEIER,AFJR (1978)
logP -1.20			EST	MEYLAN,WM & HOWARD,PH (1995)
VP 3.14E-004	mm Hg	25	EST	NEELY,WB & BLAU,GE (1985)
DC	pKa			
HL 1.32E-010	atm m3/mol	25	EST	MEYLAN,WM & HOWARD,PH (1991)
OH 1.34E-012	cm3/molc sec	25	EST	MEYLAN,WM & HOWARD,PH (1993)

CAS #: 000077-63-4				2,4,6,8-TETRAMETHYL-2,4,6,8-TETRAPHENYLCYCLOTE*

Formula: $C_{28}H_{32}O_4Si_4$

Mol Weight: 544.91

MP (deg C): 99 FP (deg C):

BP (deg C): 237

BP pressure (mm Hg): 1.15E+002

Property/ Value	Units	Temp	Data Type	Reference
WS 1.12E-007	mg/L	25	EST	MEYLAN,WM ET AL. (1996)
logP 9.95			EST	MEYLAN,WM & HOWARD,PH (1995)
VP 1.84E-010	mm Hg	25	EST	NEELY,WB & BLAU,GE (1985)
DC	pKa			
HL 2.88E-013	atm m3/mol	25	EST	MEYLAN,WM & HOWARD,PH (1991)
OH 8.40E-012	cm3/molc sec	25	EST	MEYLAN,WM & HOWARD,PH (1993)

CAS #: 000077-65-6				CARBROMAL

Formula: $C_7H_{13}BrN_2O_2$

Mol Weight: 237.10

MP (deg C): 116-119 FP (deg C):

BP (deg C):

BP pressure (mm Hg):

Property/ Value	Units	Temp	Data Type	Reference
WS 5.00E+002	mg/L	20	EXP	YALKOWSKY,SH & DANNENFELSER,RM (1992)
logP 1.54			EXP	HANSCH,C & LEO,AJ (1985)
VP 2.00E-007	mm Hg	25	EST	NEELY,WB & BLAU,GE (1985)
DC	pKa			
HL 3.86E-011	atm m3/mol	25	EST	MEYLAN,WM & HOWARD,PH (1991)
OH 4.63E-012	cm3/molc sec	25	EST	MEYLAN,WM & HOWARD,PH (1993)

CAS #: 000077-67-8				ETHOSUXIMIDE

Formula: $C_7H_{11}NO_2$

Mol Weight: 141.17

MP (deg C): 64-65 FP (deg C):

BP (deg C):

BP pressure (mm Hg):

Property/ Value	Units	Temp	Data Type	Reference
WS 3.92E+004	mg/L	25	EST	MEYLAN,WM ET AL. (1996)
logP 0.38			EXP	SANGSTER,J (1993)
VP 5.43E-007	mm Hg	25	EST	NEELY,WB & BLAU,GE (1985)
DC	pKa			
HL 5.93E-008	atm m3/mol	25	EST	MEYLAN,WM & HOWARD,PH (1991)
OH 1.20E-011	cm3/molc sec	25	EST	MEYLAN,WM & HOWARD,PH (1993)

CAS #: 000077-71-4				2,4-IMIDAZOLIDINEDIONE, 5,5-DIMETHYL-

Formula: $C_5H_8N_2O_2$

Mol Weight: 128.13

MP (deg C): 178 FP (deg C):

BP (deg C):

BP pressure (mm Hg):

Property/ Value	Units	Temp	Data Type	Reference
WS 1.18E+004	mg/L	25	EST	MEYLAN,WM ET AL. (1996)
logP -0.48			EXP	HANSCH,C & LEO,AJ (1985)
VP 2.80E-006	mm Hg	25	EST	NEELY,WB & BLAU,GE (1985)
DC 9.19	pKa	24	EXP	KORTUM,G ET AL. (1961)
HL 2.77E-009	atm m3/mol	25	EST	MEYLAN,WM & HOWARD,PH (1991)
OH 3.06E-012	cm3/molc sec	25	EST	MEYLAN,WM & HOWARD,PH (1993)

67

CAS #: 000077-73-6				DICYCLOPENTADIENE
Formula: $C_{10}H_{12}$				
Mol Weight: 132.21				
MP (deg C): -1		FP (deg C):		
BP (deg C): 170				
BP pressure (mm Hg):				

Property/Value	Units	Temp	Data Type	Reference
WS 2.65E+001	mg/L	25	EST	MEYLAN,WM ET AL. (1996)
logP 3.51			EST	MEYLAN,WM & HOWARD,PH (1995)
VP 2.29E+000	mm Hg	25	EXT	DAUBERT,TE & DANNER,RP (1989)
DC	pKa			
HL 6.25E-002	atm m3/mol	25	EST	MEYLAN,WM & HOWARD,PH (1991)
OH 1.20E-010	cm3/molc sec	25	EST	MEYLAN,WM & HOWARD,PH (1993)

CAS #: 000077-74-7				3-METHYL-3-PENTANOL
Formula: $C_6H_{14}O$				
Mol Weight: 102.18				
MP (deg C): -23.6		FP (deg C):		
BP (deg C): 122.4				
BP pressure (mm Hg):				

Property/Value	Units	Temp	Data Type	Reference
WS 4.26E+004	mg/L	25	EXP	YALKOWSKY,SH & DANNENFELSER,RM (1992)
logP 1.71			EST	MEYLAN,WM & HOWARD,PH (1995)
VP 5.56E-005	mm Hg	25	EXP	JORDAN,TE (1954)
DC	pKa			
HL 1.76E-010	atm m3/mol	25	EST	VP/WSOL
OH 8.09E-012	cm3/molc sec	25	EST	MEYLAN,WM & HOWARD,PH (1993)

CAS #: 000077-75-8				METHYL PENTYNOL
Formula: $C_6H_{10}O$				
Mol Weight: 98.15				
MP (deg C): 30.5		FP (deg C):		
BP (deg C): 120.5				
BP pressure (mm Hg):				

Property/Value	Units	Temp	Data Type	Reference
WS 9.90E+004	mg/L	20	EXP	HORT,EV (1978)
logP 0.94			EST	MEYLAN,WM & HOWARD,PH (1995)
VP 5.25E+000	mm Hg	20	EXP	WEBER,RC ET AL. (1981)
DC	pKa			
HL 6.85E-006	atm m3/mol	20	EST	VP/WSOL
OH 1.14E-011	cm3/molc sec	25	EST	MEYLAN,WM & HOWARD,PH (1993)

CAS #: 000077-76-9				2,2-DIMETHOXYPROPANE
Formula: $C_5H_{12}O_2$				
Mol Weight: 104.15				
MP (deg C): -47		FP (deg C):		
BP (deg C): 83				
BP pressure (mm Hg):				

Property/Value	Units	Temp	Data Type	Reference
WS 7.52E+003	mg/L	25	EST	MEYLAN,WM ET AL. (1996)
logP 1.38			EST	MEYLAN,WM & HOWARD,PH (1995)
VP 1.02E+002	mm Hg	25	EXT	BAGLAY,AK ET AL. (1988)
DC	pKa			
HL 8.91E-005	atm m3/mol	25	EST	MEYLAN,WM & HOWARD,PH (1991)
OH 3.92E-012	cm3/molc sec	25	EXP	ATKINSON,R (1989)

CAS #: 000077-77-0				1,1'-SULFONYLBISETHENE
Formula: $C_4H_6O_2S$				
Mol Weight: 118.16				
MP (deg C): -26		FP (deg C):		
BP (deg C): 234.5				
BP pressure (mm Hg):				

Property/Value	Units	Temp	Data Type	Reference
WS 2.22E+005	mg/L	25	EST	MEYLAN,WM ET AL. (1996)
logP -0.40			EST	MEYLAN,WM & HOWARD,PH (1995)
VP 7.84E-001	mm Hg	25	EST	NEELY,WB & BLAU,GE (1985)
DC	pKa			
HL 4.91E-005	atm m3/mol	25	EST	MEYLAN,WM & HOWARD,PH (1991)
OH 5.26E-011	cm3/molc sec	25	EST	MEYLAN,WM & HOWARD,PH (1993)

CAS #: 000077-78-1				DIMETHYLSULFATE
Formula: $C_2H_6O_4S$				
Mol Weight: 126.13				
MP (deg C): -27		FP (deg C):		
BP (deg C): 188				
BP pressure (mm Hg):				

Property/Value	Units	Temp	Data Type	Reference
WS 2.80E+004	mg/L	18	EXP	BOULIN,CH & SIMON,LJ (1920)
logP 1.16			EXP	NEELY,WB & BLAU,GE (1985)
VP 6.80E-001	mm Hg	20	EXP	WEBER,RC ET AL. (1981)
DC	pKa			
HL 4.00E-006	atm m3/mol	20	EST	VP/WSOL
OH 5.00E-013	cm3/molc sec	25	EXP	ATKINSON,R (1989)

CAS #: 000077-79-2				2,5-DIHYDROTHIOPHENE 1,1-DIOXIDE
Formula: $C_4H_6O_2S$				
Mol Weight: 118.16				
MP (deg C): 64-66		FP (deg C):		
BP (deg C):				
BP pressure (mm Hg):				

Property/Value	Units	Temp	Data Type	Reference
WS 2.47E+005	mg/L	25	EST	MEYLAN,WM ET AL. (1996)
logP -0.45			EST	MEYLAN,WM & HOWARD,PH (1995)
VP 3.41E-001	mm Hg	25	EST	NEELY,WB & BLAU,GE (1985)
DC	pKa			
HL 4.30E-006	atm m3/mol	25	EST	MEYLAN,WM & HOWARD,PH (1991)
OH 5.74E-011	cm3/molc sec	25	EST	ATKINSON,R (1988)

CAS #: 000077-81-6				TABUN
Formula: $C_5H_{11}N_2O_2P$				
Mol Weight: 162.13				
MP (deg C): -50		FP (deg C):		
BP (deg C): 240				
BP pressure (mm Hg):				

Property/Value	Units	Temp	Data Type	Reference
WS 1.00E+006	mg/L	25	EXP	MERCK INDEX (1983)
logP -1.44			EST	MEYLAN,WM & HOWARD,PH (1995)
VP 7.00E-002	mm Hg	25	EXP	WATKINS,TF ET AL. (1968)
DC	pKa			
HL 1.49E-008	atm m3/mol	25	EST	VP/WSOL
OH 8.03E-011	cm3/molc sec	25	EST	ATKINSON,R (1987)

CAS #: 000077-86-1 — AMINOTRIS(HYDROXYMETHYL)METHANE

Formula: $C_4H_{11}NO_3$

Mol Weight: 121.14

MP (deg C): 171-172   FP (deg C):

BP (deg C): 219-220

BP pressure (mm Hg): $1.00E+001$

Property/Value	Units	Temp	Data Type	Reference
WS $5.50E+005$	mg/L	25	EXP	MERCK INDEX (1996)
logP -1.56			EST	MEYLAN,WM & HOWARD,PH (1995)
VP $2.20E-005$	mm Hg	25	EST	NEELY,WB & BLAU,GE (1985)
DC 8.07	pKa		EXP	PERRIN,DD (1972)
HL $8.67E-013$	atm m3/mol	25	EST	MEYLAN,WM & HOWARD,PH (1991)
OH $3.35E-011$	cm3/molc sec	25	EST	MEYLAN,WM & HOWARD,PH (1993)

CAS #: 000077-92-9 — CITRIC ACID

Formula: $C_6H_8O_7$

Mol Weight: 192.13

MP (deg C): 153   FP (deg C):

BP (deg C):

BP pressure (mm Hg):

Property/Value	Units	Temp	Data Type	Reference
WS $5.92E+005$	mg/L	20	EXP	TOXICOLOGY DATA BASE
logP -1.72			EXP	HANSCH,C & LEO,AJ (1985)
VP $3.70E-009$	mm Hg	25	EST	NEELY,WB & BLAU,GE (1985)
DC 2.79	pKa		EXP	SERJEANT,EP & DEMPSEY,B (1979)
HL $8.33E-018$	atm m3/mol	25	EST	MEYLAN,WM & HOWARD,PH (1991)
OH $7.02E-012$	cm3/molc sec	25	EST	MEYLAN,WM & HOWARD,PH (1993)

CAS #: 000077-93-0 — TRIETHYL CITRATE

Formula: $C_{12}H_{20}O_7$

Mol Weight: 276.29

MP (deg C):   FP (deg C):

BP (deg C): 294

BP pressure (mm Hg):

Property/Value	Units	Temp	Data Type	Reference
WS $2.82E+004$	mg/L	25	EST	MEYLAN,WM ET AL. (1996)
logP 0.33			EST	MEYLAN,WM & HOWARD,PH (1995)
VP $1.89E-003$	mm Hg	25	EXP	PERRY,RH & GREEN,D (1984)
DC	pKa			
HL $6.39E-010$	atm m3/mol	25	EST	MEYLAN,WM & HOWARD,PH (1991)
OH $1.04E-011$	cm3/molc sec	25	EST	MEYLAN,WM & HOWARD,PH (1993)

CAS #: 000077-99-6 — 1,1,1-TRIS(HYDROXYMETHYL)PROPANE

Formula: $C_6H_{14}O_3$

Mol Weight: 134.18

MP (deg C): 58   FP (deg C):

BP (deg C): 160

BP pressure (mm Hg): $5.00E+000$

Property/Value	Units	Temp	Data Type	Reference
WS $1.00E+006$	mg/L		EXP	WEAST,RC (1972)
logP -1.48			EXP	TSCATS
VP $4.49E-005$	mm Hg	25	EXP	DAUBERT,TE & DANNER,RP (1989)
DC	pKa			
HL $7.93E-012$	atm m3/mol	25	EST	VP/WSOL
OH $1.38E-011$	cm3/molc sec	25	EST	MEYLAN,WM & HOWARD,PH (1993)

CAS #: 000078-00-2 — TETRAETHYL LEAD

Formula: $C_8H_{20}Pb$

Mol Weight: 323.44

MP (deg C): -136   FP (deg C):

BP (deg C): 200-227.2

BP pressure (mm Hg):

Property/Value	Units	Temp	Data Type	Reference
WS $2.10E-001$	mg/L	26	EXP	FELDHAKE,CJ & STEVENS,CD (1963)
logP 4.88			EST	MEYLAN,WM & HOWARD,PH (1995)
VP $3.90E-001$	mm Hg	25	EXP	FELDHAKE,CJ & STEVENS,CD (1963)
DC	pKa			
HL $5.68E-001$	atm m3/mol	24	EXP	FELDHAKE,CJ & STEVENS,CD (1963)
OH $4.92E-011$	cm3/molc sec	25	EST	MEYLAN,WM & HOWARD,PH (1993)

CAS #: 000078-08-0 — TRIETHOXYVINYLSILANE

Formula: $C_8H_{18}O_3Si$

Mol Weight: 190.32

MP (deg C):   FP (deg C):

BP (deg C): 160-161

BP pressure (mm Hg): $2.00E+001$

Property/Value	Units	Temp	Data Type	Reference
WS $5.04E+003$	mg/L	25	EST	MEYLAN,WM ET AL. (1996)
logP 1.16			EST	MEYLAN,WM & HOWARD,PH (1995)
VP $2.28E+000$	mm Hg	25	EXP	BOUBLIK,T ET AL. (1984)
DC	pKa			
HL $6.60E-007$	atm m3/mol	25	EST	MEYLAN,WM & HOWARD,PH (1991)
OH $4.48E-011$	cm3/molc sec	25	EST	MEYLAN,WM & HOWARD,PH (1993)

CAS #: 000078-10-4 — ETHYL SILICATE

Formula: $C_8H_{20}O_4Si$

Mol Weight: 208.33

MP (deg C): -77   FP (deg C):

BP (deg C): 165-166

BP pressure (mm Hg):

Property/Value	Units	Temp	Data Type	Reference
WS $3.69E+004$	mg/L	25	EST	MEYLAN,WM ET AL. (1996)
logP 0.04			EST	MEYLAN,WM & HOWARD,PH (1995)
VP $1.88E+000$	mm Hg	25	EXP	PERRY,RH & GREEN,D (1984)
DC	pKa			
HL $9.44E-009$	atm m3/mol	25	EST	MEYLAN,WM & HOWARD,PH (1991)
OH $2.47E-011$	cm3/molc sec	25	EST	MEYLAN,WM & HOWARD,PH (1993)

CAS #: 000078-11-5 — PENTAERYTHRITOL TETRANITRATE

Formula: $C_5H_8N_4O_{12}$

Mol Weight: 316.14

MP (deg C): 140   FP (deg C):

BP (deg C):

BP pressure (mm Hg):

Property/Value	Units	Temp	Data Type	Reference
WS $4.30E+001$	mg/L	25	EXP	RINKENBACK,WH (1965)
logP 2.38			EST	MEYLAN,WM & HOWARD,PH (1995)
VP $5.45E-009$	mm Hg	25	EXT	JONES,AH (1960)
DC	pKa			
HL $1.20E-011$	atm m3/mol	25	EST	MEYLAN,WM & HOWARD,PH (1991)
OH $1.62E-012$	cm3/molc sec	25	EST	MEYLAN,WM & HOWARD,PH (1993)

000078-27-3 — 1-ETHYNYL CYCLOHEXANOL

Formula: $C_8H_{12}O$
Mol Weight: 124.18
MP (deg C): 31.5
FP (deg C):
BP (deg C): 174
BP pressure (mm Hg):

Property/Value	Units	Temp	Data Type	Reference
WS 1.04E+004	mg/L	25	EST	MEYLAN,WM ET AL. (1996)
logP 1.73			EXP	HANSCH,C & LEO,AJ (1985)
VP 1.33E-001	mm Hg		EST	NEELY,WB & BLAU,GE (1985)
DC	pKa			
HL 1.07E-006	atm m3/mol		EST	MEYLAN,WM & HOWARD,PH (1991)
OH 2.01E-011	cm3/molc sec	25	EST	MEYLAN,WM & HOWARD,PH (1993)

000078-32-0 — TRI-P-CRESYL PHOSPHATE

Formula: $C_{21}H_{21}O_4P$
Mol Weight: 368.37
MP (deg C): 77.5
FP (deg C):
BP (deg C): 224
BP pressure (mm Hg): 3.50E+001

Property/Value	Units	Temp	Data Type	Reference
WS 3.00E-001	mg/L	25	EXP	ISHOW (NA--)
logP 6.21			EST	ISHOW (NA--)
VP 2.61E-008	mm Hg	25	EST	NEELY,WB & BLAU,GE (1985)
DC	pKa			
HL 5.35E-008	atm m3/mol	25	EST	MEYLAN,WM & HOWARD,PH (1991)
OH 1.37E-011	cm3/molc sec	25	EST	MEYLAN,WM & HOWARD,PH (1993)

000078-34-2 — DIOXATHION

Formula: $C_{12}H_{26}O_6P_2S_4$
Mol Weight: 456.54
MP (deg C): -20
FP (deg C):
BP (deg C):
BP pressure (mm Hg):

Property/Value	Units	Temp	Data Type	Reference
WS 1.55E+000	mg/L	25	EST	MEYLAN,WM ET AL. (1996)
logP 3.45			EST	MEYLAN,WM & HOWARD,PH (1995)
VP 8.76E-008	mm Hg	25	EST	NEELY,WB & BLAU,GE (1985)
DC	pKa			
HL 1.35E-011	atm m3/mol	25	EST	MEYLAN,WM & HOWARD,PH (1991)
OH 7.94E-010	cm3/molc sec	25	EST	MEYLAN,WM & HOWARD,PH (1993)

000078-38-6 — Phosphonic acid, ethyl-, diethyl ester

Formula: $C_6H_{15}O_3P$
Mol Weight: 166.16
MP (deg C):
FP (deg C):
BP (deg C): 198
BP pressure (mm Hg):

Property/Value	Units	Temp	Data Type	Reference
WS 1.75E+004	mg/L	25	EST	MEYLAN,WM ET AL. (1996)
logP 0.66			EXP	KRIKORIAN,SE ET AL. (1987)
VP 3.16E-001	mm Hg	25	EXP	KOSOLAPOFF,GM (1955)
DC	pKa			
HL 2.92E-006	atm m3/mol	25	EST	MEYLAN,WM & HOWARD,PH (1991)
OH 3.97E-011	cm3/molc sec	25	EST	MEYLAN,WM & HOWARD,PH (1993)

000078-40-0 — TRIETHYL PHOSPHATE

Formula: $C_6H_{15}O_4P$
Mol Weight: 182.16
MP (deg C): -56.4
FP (deg C):
BP (deg C): 215-216
BP pressure (mm Hg):

Property/Value	Units	Temp	Data Type	Reference
WS 5.00E+005	mg/L	25	EXP	YALKOWSKY,SH & DANNENFELSER,RM (1992)
logP 0.80			EXP	HANSCH,C & LEO,AJ (1985)
VP 3.93E-001	mm Hg	25	EXP	DAUBERT,TE & DANNER,RP (1991)
DC	pKa			
HL 1.50E-006	atm m3/mol	20	EXP	WOLFENDEN,R & WILLIAMS,R (1983)
OH 5.50E-011	cm3/molc sec	25	EXP	ATKINSON,R (1989)

000078-42-2 — TRIS(2-ETHYLHEXYL) PHOSPHATE

Formula: $C_{24}H_{51}O_4P$
Mol Weight: 434.65
MP (deg C):
FP (deg C):
BP (deg C): 215
BP pressure (mm Hg): 4.00E+000

Property/Value	Units	Temp	Data Type	Reference
WS 6.00E-001	mg/L	24	EXP	YALKOWSKY,SH & DANNENFELSER,RM (1992)
logP 4.23			EXP	SAEGER,VW ET AL. (1979)
VP 8.25E-008	mm Hg	25	EXP	HINCKLEY,DA ET AL. (1990)
DC	pKa			
HL 7.86E-008	atm m3/mol	25	EST	VP/WSOL
OH 9.79E-011	cm3/molc sec	25	EST	MEYLAN,WM & HOWARD,PH (1993)

000078-48-8 — DEF

Formula: $C_{12}H_{27}OPS_3$
Mol Weight: 314.51
MP (deg C): <-25
FP (deg C):
BP (deg C): 150
BP pressure (mm Hg): 3.00E-001

Property/Value	Units	Temp	Data Type	Reference
WS 7.30E-001	mg/L		EXP	HERMANN,BW (1980)
logP 5.70			EXP	ENVIRONMENTAL RESEARCH LABORATORY (1981)
VP 1.38E-005	mm Hg		EXP	HERMANN,BW (1980)
DC	pKa			
HL 2.94E-007	atm m3/mol	25	EXP	FENDINGER,NJ & GLOFFELTY,DE (1990)
OH 7.88E-011	cm3/molc sec	25	EST	MEYLAN,WM & HOWARD,PH (1993)

000078-51-3 — TRI-2-BUTOXYETHYL PHOSPHATE

Formula: $C_{18}H_{39}O_7P$
Mol Weight: 398.48
MP (deg C): -70
FP (deg C):
BP (deg C):
BP pressure (mm Hg):

Property/Value	Units	Temp	Data Type	Reference
WS 1.10E+003	mg/L	25		YALKOWSKY,SH & DANNENFELSER,RM (1992)
logP 3.75			EXP	CHEM INSPECT TEST INST (1992)
VP 2.50E-008	mm Hg	25	EST	HL X WSOL
DC	pKa			
HL 1.20E-011	atm m3/mol	25	EST	MEYLAN,WM & HOWARD,PH (1991)
OH 1.23E-010	cm3/molc sec	25	EST	ATKINSON,R (1988)

MENAZON

CAS #:	000078-57-9			
Formula:	$C_6H_{12}N_5O_2PS_2$			
Mol Weight:	281.30			
MP (deg C):	164-166	**FP (deg C):**		
BP (deg C):				
BP pressure (mm Hg):				

Property/ Value	Units	Temp	Data Type	Reference
WS 2.40E+002	mg/L	20	EXP	YALKOWSKY,SH & DANNENFELSER,RM (1992)
logP 1.20			EST	MEYLAN,WM & HOWARD,PH (1995)
VP 9.75E-007	mm Hg	25	EXP	METCALF,RA (1989)
DC 3.80	pKa			WEBER,JB (1972)
HL 1.59E-009	atm m3/mol	25	EST	VP/WSOL
OH 7.82E-011	cm3/molc sec	25	EST	MEYLAN,WM & HOWARD,PH (1993)

ISOPHORONE

CAS #:	000078-59-1			
Formula:	$C_9H_{14}O$			
Mol Weight:	138.21			
MP (deg C):	-8.1	**FP (deg C):**		
BP (deg C):	215.3			
BP pressure (mm Hg):				

Property/ Value	Units	Temp	Data Type	Reference
WS 1.20E+004	mg/L	25	EXP	PARRISH,CF (1983)
logP 1.70			EXP	VEITH,GD ET AL. (1980)
VP 4.00E-001	mm Hg	25	EXT	PERRY,RH & GREEN,D (1984)
DC	pKa			
HL 6.06E-006	atm m3/mol	25	EST	VP/WSOL
OH 8.15E-011	cm3/molc sec	25	EST	MEYLAN,WM & HOWARD,PH (1993)

DIMETHYL DIETHOXYSILANE

CAS #:	000078-62-6			
Formula:	$C_6H_{16}O_2Si$			
Mol Weight:	148.28			
MP (deg C):	-87	**FP (deg C):**		
BP (deg C):	114			
BP pressure (mm Hg):				

Property/ Value	Units	Temp	Data Type	Reference
WS 2.33E+004	mg/L	25	EST	MEYLAN,WM ET AL. (1996)
logP 0.61			EXP	HANSCH,C ET AL. (1995)
VP 2.00E+001	mm Hg	25	EXP	JORDAN,TE (1954)
DC	pKa			
HL 4.57E-005	atm m3/mol	25	EST	MEYLAN,WM & HOWARD,PH (1991)
OH 1.26E-011	cm3/molc sec	25	EST	MEYLAN,WM & HOWARD,PH (1993)

LINALOOL

CAS #:	000078-70-6			
Formula:	$C_{10}H_{18}O$			
Mol Weight:	154.25			
MP (deg C):		**FP (deg C):**	76	
BP (deg C):	195-199			
BP pressure (mm Hg):				

Property/ Value	Units	Temp	Data Type	Reference
WS 1.59E+003	mg/L	25	EXP	YALKOWSKY,SH & DANNENFELSER,RM (1992)
logP 2.97			EXP	LI,NY & PERDUE,EM (1995)
VP 1.60E-001	mm Hg	25	EXP	LI,NY & PERDUE,EM (1995)
DC	pKa			
HL 2.04E-005	atm m3/mol	25	EST	VP/WSOL
OH 1.20E-010	cm3/molc sec	25	EST	MEYLAN,WM & HOWARD,PH (1993)

1,2-DIBROMOPROPANE

CAS #:	000078-75-1			
Formula:	$C_3H_6Br_2$			
Mol Weight:	201.90			
MP (deg C):	-55	**FP (deg C):**		
BP (deg C):	140-142			
BP pressure (mm Hg):				

Property/ Value	Units	Temp	Data Type	Reference
WS 1.43E+003	mg/L	25	EXP	YAWS,CL ET AL. (1990A)
logP 2.43			EST	MEYLAN,WM & HOWARD,PH (1995)
VP 7.84E+000	mm Hg	25	EXP	JORDAN,TE (1954)
DC	pKa			
HL 1.46E-003	atm m3/mol	25	EST	VP/WSOL
OH 4.33E-013	cm3/molc sec	25	EST	MEYLAN,WM & HOWARD,PH (1993)

2-BROMOBUTANE

CAS #:	000078-76-2			
Formula:	C_4H_9Br			
Mol Weight:	137.03			
MP (deg C):	-112	**FP (deg C):**		
BP (deg C):	91.2			
BP pressure (mm Hg):				

Property/ Value	Units	Temp	Data Type	Reference
WS 8.45E+002	mg/L	25	EST	MEYLAN,WM ET AL 1996
logP 2.58			EST	MEYLAN,WM & HOWARD,PH (1995)
VP 5.70E+001	mm Hg	25	EXP	DAUBERT,TE & DANNER,RP (1989)
DC	pKa			
HL 1.58E-002	atm m3/mol	25	EST	MEYLAN,WM & HOWARD,PH (1991)
OH 1.33E-012	cm3/molc sec	25	EST	MEYLAN,WM & HOWARD,PH (1993)

ISOBUTYL BROMIDE

CAS #:	000078-77-3			
Formula:	C_4H_9Br			
Mol Weight:	137.03			
MP (deg C):	-119	**FP (deg C):**		
BP (deg C):	91.5			
BP pressure (mm Hg):				

Property/ Value	Units	Temp	Data Type	Reference
WS 5.09E+002	mg/L	25	EXP	SUZUKI,T (1991)
logP 2.58			EST	MEYLAN,WM & HOWARD,PH (1995)
VP 7.20E+001	mm Hg	25	EST	NEELY,WB & BLAU,GE (1985)
DC	pKa			
HL 1.99E-002	atm m3/mol	25	EST	MEYLAN,WM & HOWARD,PH (1991)
OH 1.55E-012	cm3/molc sec	25	EST	MEYLAN,WM & HOWARD,PH (1993)

ISOPENTANE

CAS #:	000078-78-4			
Formula:	C_5H_{12}			
Mol Weight:	72.15			
MP (deg C):	-159.9	**FP (deg C):**		
BP (deg C):	27.8			
BP pressure (mm Hg):				

Property/ Value	Units	Temp	Data Type	Reference
WS 4.80E+001	mg/L	25	EXP	RIDDICK,JA ET AL. (1986)
logP 2.30			EXP	HANSCH,C ET AL. (1968)
VP 6.89E+002	mm Hg	25	EXP	DAUBERT,TE & DANNER,RP (1989)
DC	pKa			
HL 1.40E+000	atm m3/mol	25	EST	VP/WSOL
OH 3.90E-012	cm3/molc sec	25	EXP	ATKINSON,R (1989)

CAS #: 000078-79-5 — 2-METHYL-1,3-BUTADIENE

Formula: C_5H_8
Mol Weight: 68.12
MP (deg C): -145.95
FP (deg C):
BP (deg C): 34.067
BP pressure (mm Hg):

Property/Value	Units	Temp	Data Type	Reference
WS 6.42E+002	mg/L	25	EXP	MCAULIFFE,C (1966)
logP 2.42			EXP	CHEM INSPECT TEST INST (1992)
VP 5.51E+002	mm Hg	25	EXP	DAUBERT,TE & DANNER,RP (1989)
DC	pKa			
HL 7.67E-002	atm m3/mol	25	EST	VP/WSOL
OH 1.01E-010	cm3/molc sec	25	EXP	ATKINSON,R (1989)

CAS #: 000078-81-9 — ISOBUTYLAMINE

Formula: $C_4H_{11}N$
Mol Weight: 73.14
MP (deg C): -85
FP (deg C):
BP (deg C): 68-69
BP pressure (mm Hg):

Property/Value	Units	Temp	Data Type	Reference
WS 1.00E+006	mg/L		EXP	RIDDICK,JA ET AL. (1986)
logP 0.73			EXP	HANSCH,C & LEO,AJ (1985)
VP 1.41E+002	mm Hg	25	EXP	RIDDICK,JA ET AL. (1986)
DC 10.68	pKa	25	EXP	PERRIN,DD (1972)
HL 1.36E-005	atm m3/mol	25	EST	VP/WSOL
OH 3.35E-011	cm3/molc sec	25	EST	MEYLAN,WM & HOWARD,PH (1993)

CAS #: 000078-82-0 — 2-METHYLPROPANENITRILE

Formula: C_4H_7N
Mol Weight: 69.11
MP (deg C): -71.5
FP (deg C):
BP (deg C): 103.85
BP pressure (mm Hg):

Property/Value	Units	Temp	Data Type	Reference
WS 5.50E+004	mg/L	25	EST	VP/H L
logP 0.46			EXP	TANII,H & HASHIMOTO,K (1984)
VP 3.27E+001	mm Hg	25	EXP	DAUBERT,TE & DANNER,RP (1989)
DC	pKa			
HL 5.39E-005	atm m3/mol	25	EST	MEYLAN,WM & HOWARD,PH (1991)
OH 6.28E-013	cm3/molc sec	25	EST	ATKINSON,R (1988)

CAS #: 000078-83-1 — ISOBUTYL ALCOHOL

Formula: $C_4H_{10}O$
Mol Weight: 74.12
MP (deg C): -108
FP (deg C):
BP (deg C): 107.886
BP pressure (mm Hg):

Property/Value	Units	Temp	Data Type	Reference
WS 8.50E+004	mg/L	25	EXP	VALVANI,SC ET AL. (1981)
logP 0.76			EXP	HANSCH,C & LEO,AJ (1985)
VP 1.05E+001	mm Hg	25	EXP	DAUBERT,TE & DANNER,RP (1985)
DC	pKa			
HL 9.78E-006	atm m3/mol	25	EXP	SNIDER,JR & DAWSON,GA (1985)
OH 6.44E+000	cm3/molc sec	25	EST	MEYLAN,WM & HOWARD,PH (1993)

CAS #: 000078-84-2 — ISOBUTYRALDEHYDE

Formula: C_4H_8O
Mol Weight: 72.11
MP (deg C): -65.9
FP (deg C):
BP (deg C): 64
BP pressure (mm Hg):

Property/Value	Units	Temp	Data Type	Reference
WS 8.90E+004	mg/L	25	EXP	YALKOWSKY,SH & DANNENFELSER,RM (1992)
logP 0.74			EST	MEYLAN,WM & HOWARD,PH (1995)
VP 1.73E+002	mm Hg	25	EXP	BOUBLIK,T ET AL. (1984)
DC	pKa			
HL 1.80E-004	atm m3/mol	25	EST	VP/WSOL
OH 2.63E-011	cm3/molc sec	25	EXP	ATKINSON,R (1989)

CAS #: 000078-85-3 — METHACROLEIN

Formula: C_4H_6O
Mol Weight: 70.09
MP (deg C): -81
FP (deg C):
BP (deg C): 68.4
BP pressure (mm Hg):

Property/Value	Units	Temp	Data Type	Reference
WS 5.00E+004	mg/L	20	EXP	COLLIN,G & HOKE,H (1985)
logP 0.74			EST	MEYLAN,WM & HOWARD,PH (1995)
VP 1.55E+002	mm Hg	25	EXP	YAWS,CL (1994)
DC	pKa			
HL 2.86E-004	atm m3/mol	25	EST	VP/WSOL
OH 3.35E-011	cm3/molc sec	25	EXP	ATKINSON,R (1989)

CAS #: 000078-86-4 — 2-CHLOROBUTANE

Formula: C_4H_9Cl
Mol Weight: 92.57
MP (deg C): -140
FP (deg C):
BP (deg C): 68
BP pressure (mm Hg):

Property/Value	Units	Temp	Data Type	Reference
WS 1.00E+003	mg/L	25	EXP	YALKOWSKY,SH & DANNENFELSER,RM (1992)
logP 2.33			EXP	HANSCH,C ET AL. (1995)
VP 1.57E+002	mm Hg	25	EXP	DAUBERT,TE & DANNER,RP (1989)
DC	pKa			
HL 2.41E-002	atm m3/mol	25	EXP	LEIGHTON,DTJR & CALO,JM (1981)
OH 2.30E-012	cm3/molc sec	25	EXP	ATKINSON,R (1989)

CAS #: 000078-87-5 — 1,2-DICHLOROPROPANE

Formula: $C_3H_6Cl_2$
Mol Weight: 112.99
MP (deg C): -100
FP (deg C):
BP (deg C): 95-96
BP pressure (mm Hg):

Property/Value	Units	Temp	Data Type	Reference
WS 2.80E+003	mg/L	25	EXP	HORVATH,AL (1982)
logP 2.25			EST	MEYLAN,WM & HOWARD,PH (1995)
VP 5.33E+001	mm Hg	25	EXP	BOUBLIK,T ET AL. (1984)
DC	pKa			
HL 2.82E-003	atm m3/mol	25	EXP	WARNER,HP ET AL. (1987)
OH 6.60E-013	cm3/molc sec	25	EST	MEYLAN,WM & HOWARD,PH (1993)

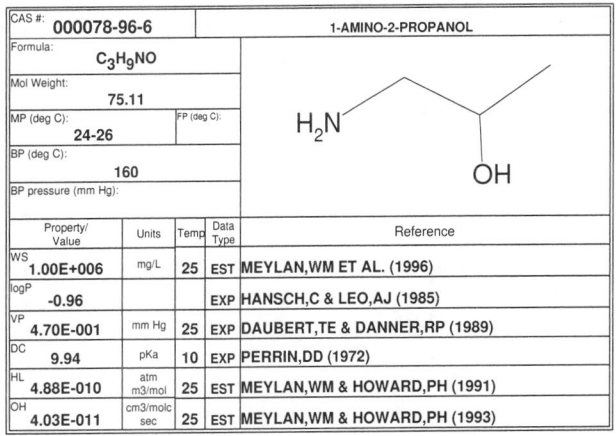

CAS #: 000078-88-6 — 2,3-DICHLOROPROPENE

Formula: $C_3H_4Cl_2$
Mol Weight: 110.97
MP (deg C): 10
FP (deg C):
BP (deg C): 94
BP pressure (mm Hg):

Property/Value	Units	Temp	Data Type	Reference
WS 2.15E+003	mg/L	25	EXP	MACKAY,D & SHIU,WY (1981)
logP 2.42			EST	MEYLAN,WM & HOWARD,PH (1995)
VP 6.12E+001	mm Hg	25	EXP	DAUBERT,TE & DANNER,RP (1989)
DC	pKa			
HL 4.16E-003	atm m3/mol	25	EST	VP/WSOL
OH 8.56E-012	cm3/molc sec	25	EST	MEYLAN,WM & HOWARD,PH (1993)

CAS #: 000078-89-7 — 2-CHLORO-1-PROPANOL

Formula: C_3H_7ClO
Mol Weight: 94.54
MP (deg C):
FP (deg C):
BP (deg C): 133.5
BP pressure (mm Hg):

Property/Value	Units	Temp	Data Type	Reference
WS 1.38E+005	mg/L	25	EST	MEYLAN,WM ET AL. (1996)
logP 0.53			EST	MEYLAN,WM & HOWARD,PH (1995)
VP 5.25E+000	mm Hg	20	EXP	WEBER,RC ET AL. (1981)
DC	pKa			
HL 1.67E-006	atm m3/mol	25	EST	MEYLAN,WM & HOWARD,PH (1991)
OH 2.27E-012	cm3/molc sec	25	EST	MEYLAN,WM & HOWARD,PH (1993)

CAS #: 000078-90-0 — 1,2-DIAMINOPROPANE

Formula: $C_3H_{10}N_2$
Mol Weight: 74.13
MP (deg C):
FP (deg C):
BP (deg C): 119.5
BP pressure (mm Hg):

Property/Value	Units	Temp	Data Type	Reference
WS 1.00E+006	mg/L	25	EST	MEYLAN,WM ET AL. (1996)
logP -1.20			EST	MEYLAN,WM & HOWARD,PH (1995)
VP 1.06E+001	mm Hg	25	EXP	DAUBERT,TE & DANNER,RP (1989)
DC 9.97	pKa	25	EXP	PERRIN,DD (1965)
HL 1.37E-009	atm m3/mol	25	EST	MEYLAN,WM & HOWARD,PH (1991)
OH 7.50E-011	cm3/molc sec	25	EST	MEYLAN,WM & HOWARD,PH (1993)

CAS #: 000078-92-2 — 2-BUTANOL

Formula: $C_4H_{10}O$
Mol Weight: 74.12
MP (deg C): -114.7
FP (deg C):
BP (deg C): 99.512
BP pressure (mm Hg):

Property/Value	Units	Temp	Data Type	Reference
WS 1.81E+005	mg/L	25	EXP	HEFTER,GT (1984A)
logP 0.61			EXP	HANSCH,C & LEO,AJ (1985)
VP 1.83E+001	mm Hg	25	EXP	DAUBERT,TE & DANNER,RP (1985)
DC 17.60	pKa	25	EXP	SERJEANT,EP & DEMPSEY,B (1979)
HL 9.06E-006	atm m3/mol	25	EXP	SNIDER,JR & DAWSON,GA (1985)
OH 9.60E-012	cm3/molc sec	25	EXP	EDNEY,EO & CORSE,EW (1987)

CAS #: 000078-93-3 — 2-BUTANONE

Formula: C_4H_8O
Mol Weight: 72.11
MP (deg C): -86
FP (deg C):
BP (deg C): 79.6
BP pressure (mm Hg):

Property/Value	Units	Temp	Data Type	Reference
WS 2.23E+005	mg/L	25	EXP	TAFT,RW ET AL. (1985)
logP 0.29			EXP	HANSCH,C & LEO,AJ (1985)
VP 9.53E+001	mm Hg	25	EXP	DAUBERT,TE & DANNER,RP (1985)
DC 14.70	pKa	25	EXP	RIDDICK,JA ET AL. (1986)
HL 5.69E-005	atm m3/mol	25	EXP	SNIDER,JR & DAWSON,GA (1985)
OH 1.15E-012	cm3/molc sec	25	EXP	ATKINSON,R (1989)

CAS #: 000078-94-4 — METHYLVINYLKETONE

Formula: C_4H_6O
Mol Weight: 70.09
MP (deg C):
FP (deg C):
BP (deg C): 81.4
BP pressure (mm Hg):

Property/Value	Units	Temp	Data Type	Reference
WS 6.06E+004	mg/L	25	EST	MEYLAN,WM ET AL. (1996)
logP 0.41			EST	MEYLAN,WM & HOWARD,PH (1995)
VP 1.52E+002	mm Hg	25	EST	NEELY,WB & BLAU,GE (1985)
DC	pKa			
HL 2.61E-005	atm m3/mol	25	EST	MEYLAN,WM & HOWARD,PH (1991)
OH 1.88E-011	cm3/molc sec	25	EXP	ATKINSON,R (1989)

CAS #: 000078-95-5 — CHLOROACETONE

Formula: C_3H_5ClO
Mol Weight: 92.53
MP (deg C): -44.5
FP (deg C):
BP (deg C): 119.7
BP pressure (mm Hg):

Property/Value	Units	Temp	Data Type	Reference
WS 9.00E+004	mg/L	20	EXP	YALKOWSKY,SH & DANNENFELSER,RM (1992)
logP 0.28			EST	HANSCH,C & LEO,AJ (1981)
VP 1.20E-001	mm Hg	25	EXP	HANN,RWJR & JENSEN,PA (1977)
DC	pKa			
HL 1.62E-007	atm m3/mol	25	EST	VP/WSOL
OH 3.68E-013	cm3/molc sec	25	EST	MEYLAN,WM & HOWARD,PH (1993)

CAS #: 000078-96-6 — 1-AMINO-2-PROPANOL

Formula: C_3H_9NO
Mol Weight: 75.11
MP (deg C): 24-26
FP (deg C):
BP (deg C): 160
BP pressure (mm Hg):

Property/Value	Units	Temp	Data Type	Reference
WS 1.00E+006	mg/L	25	EST	MEYLAN,WM ET AL. (1996)
logP -0.96			EXP	HANSCH,C & LEO,AJ (1985)
VP 4.70E-001	mm Hg	25	EXP	DAUBERT,TE & DANNER,RP (1989)
DC 9.94	pKa	10	EXP	PERRIN,DD (1972)
HL 4.88E-010	atm m3/mol	25	EST	MEYLAN,WM & HOWARD,PH (1991)
OH 4.03E-011	cm3/molc sec	25	EST	MEYLAN,WM & HOWARD,PH (1993)

000078-97-7 — 2-HYDROXYPROPANENITRILE

Formula:	C₃H₅NO	
Mol Weight:	71.08	
MP (deg C): -40	FP (deg C):	
BP (deg C): 183		
BP pressure (mm Hg):		

Formula: C_3H_5NO
Mol Weight: 71.08
MP (deg C): -40
BP (deg C): 183

Property/ Value	Units	Temp	Data Type	Reference
WS 4.66E+005	mg/L	25	EST	MEYLAN,WM ET AL. (1996)
logP -0.94			EXP	TANII,H & HASHIMOTO,K (1985)
VP 1.19E-001	mm Hg	25	EXP	DAUBERT,TE & DANNER,RP (1989)
DC	pKa			
HL 9.80E-006	atm m3/mol	25	EST	MEYLAN,WM & HOWARD,PH (1991)
OH 1.60E-012	cm3/molc sec	25	EST	MEYLAN,WM & HOWARD,PH (1993)

000078-98-8 — METHYL GLYOXAL

Formula: $C_3H_4O_2$
Mol Weight: 72.06
BP (deg C): 72

Property/ Value	Units	Temp	Data Type	Reference
WS 1.00E+006	mg/L	25	EST	MEYLAN,WM ET AL. (1996)
logP -1.50			EST	MEYLAN,WM & HOWARD,PH (1995)
VP 2.67E+001	mm Hg	25	EST	NEELY,WB & BLAU,GE (1985)
DC	pKa			
HL 2.70E-007	atm m3/mol	25	EXP	BETTERTON,EA & HOFFMAN,MR (1988)
OH 1.72E-011	cm3/molc sec	25	EXP	ATKINSON,R (1989)

000079-00-5 — 1,1,2-TRICHLOROETHANE

Formula: $C_2H_3Cl_3$
Mol Weight: 133.41
MP (deg C): -35
BP (deg C): 113-114

Property/ Value	Units	Temp	Data Type	Reference
WS 4.42E+003	mg/L	25	EXP	DILLING,WL (1977)
logP 1.89			EXP	HANSCH,C ET AL. (1995)
VP 2.30E+001	mm Hg	25	EXP	ENGINEERING SCIENCES DATA UNIT (1976)
DC	pKa			
HL 8.24E-004	atm m3/mol	25	EXP	LEIGHTON,DTJR & CALO,JM (1981)
OH 3.28E-013	cm3/molc sec	25	EXP	ATKINSON,R (1989)

000079-01-6 — TRICHLOROETHENE

Formula: C_2HCl_3
Mol Weight: 131.39
MP (deg C): -84.8
BP (deg C): 86.7

Property/ Value	Units	Temp	Data Type	Reference
WS 1.10E+003	mg/L	25	EXP	HORVATH,AL (1982)
logP 2.42			EXP	HANSCH,C & LEO,AJ (1985)
VP 6.90E+001	mm Hg	25	EXP	BOUBLIK,T ET AL. (1984)
DC	pKa			
HL 9.85E-003	atm m3/mol	25	EXP	LEIGHTON,DTJR & CALO,JM (1981)
OH 2.36E-012	cm3/molc sec	25	EXP	ATKINSON,R (1989)

000079-04-9 — CHLOROACETYLCHLORIDE

Formula: $C_2H_2Cl_2O$
Mol Weight: 112.94
MP (deg C): -21.77
BP (deg C): 106

Property/ Value	Units	Temp	Data Type	Reference
WS 1.63E+005	mg/L	25	EST	MEYLAN,WM ET AL. (1996)
logP -0.22			EST	MEYLAN,WM & HOWARD,PH (1995)
VP 5.55E+003	mm Hg	25	EXP	DAUBERT,TE & DANNER,RP (1993)
DC	pKa			
HL 2.34E-004	atm m3/mol	25	EST	MEYLAN,WM & HOWARD,PH (1991)
OH 2.38E-014	cm3/molc sec	25	EST	MEYLAN,WM & HOWARD,PH (1993)

000079-05-0 — PROPRIONAMIDE

Formula: C_3H_7NO
Mol Weight: 73.10
MP (deg C): 79
BP (deg C): 222.2

Property/ Value	Units	Temp	Data Type	Reference
WS 4.91E+005	mg/L	25	EST	MEYLAN,WM ET AL. (1996)
logP -0.66			EXP	HANSCH,C ET AL. (1995)
VP 3.78E-002	mm Hg	25	EXP	PERRY,RH & GREEN,D (1984)
DC -0.49	pKa	20	EXP	PERRIN,DD (1965)
HL 1.48E-008	atm m3/mol	25	EST	MEYLAN,WM & HOWARD,PH (1991)
OH 3.23E-012	cm3/molc sec	25	EST	MEYLAN,WM & HOWARD,PH (1993)

000079-06-1 — ACRYLAMIDE

Formula: C_3H_5NO
Mol Weight: 71.08
MP (deg C): 84.5
BP (deg C): 125
BP pressure (mm Hg): 2.50E+001

Property/ Value	Units	Temp	Data Type	Reference
WS 6.40E+005	mg/L	25	EXP	YALKOWSKY,SH & DANNENFELSER,RM (1992)
logP -0.67			EXP	HANSCH,C & LEO,AJ (1985)
VP 7.00E-003	mm Hg	25	EXP	BIKALES,NM & KOLODNY,ER (1963)
DC	pKa			
HL 1.00E-009	atm m3/mol	25	EST	VP/WSOL
OH 3.57E-011	cm3/molc sec	25	EST	MEYLAN,WM & HOWARD,PH (1993)

000079-07-2 — CHLOROACETAMIDE

Formula: C_2H_4ClNO
Mol Weight: 93.51
MP (deg C): 119-120
BP (deg C): 225

Property/ Value	Units	Temp	Data Type	Reference
WS 9.00E+004	mg/L	24	EXP	YALKOWSKY,SH & DANNENFELSER,RM (1992)
logP -0.53			EXP	HANSCH,C & LEO,AJ (1985)
VP 5.17E-002	mm Hg	25	EST	NEELY,WB & BLAU,GE (1985)
DC	pKa			
HL 3.94E-009	atm m3/mol	25	EST	MEYLAN,WM & HOWARD,PH (1991)
OH 2.27E-012	cm3/molc sec	25	EST	MEYLAN,WM & HOWARD,PH (1993)

CAS #: 000079-08-3				BROMOACETIC ACID

Formula: $C_2H_3BrO_2$

Mol Weight: 138.95

MP (deg C): 50 FP (deg C):

BP (deg C): 208

BP pressure (mm Hg):

Property/ Value	Units	Temp	Data Type	Reference
WS 9.38E+004	mg/L	25	EST	MEYLAN,WM ET AL. (1996)
logP 0.41			EXP	HANSCH,C & LEO,AJ (1985)
VP 1.19E-001	mm Hg	25	EXP	PERRY,RH & GREEN,D (1984)
DC 2.89	pKa	20	EXP	KORTUM,G ET AL (1961)
HL 6.31E-008	atm m3/mol	25	EST	MEYLAN,WM & HOWARD,PH (1991)
OH 7.16E-013	cm3/molc sec	25	EST	MEYLAN,WM & HOWARD,PH (1993)

CAS #: 000079-09-4				PROPIONIC ACID

Formula: $C_3H_6O_2$

Mol Weight: 74.08

MP (deg C): -20.8 FP (deg C):

BP (deg C): 140.7

BP pressure (mm Hg):

Property/ Value	Units	Temp	Data Type	Reference
WS 1.00E+006	mg/L		EXP	US EPA (1981)
logP 0.33			EXP	HANSCH,C & LEO,AJ (1985)
VP 3.53E+000	mm Hg	25	EXP	DAUBERT,TE & DANNER,RP (1985)
DC 4.88	pKa		EXP	SERJEANT,EP & DEMPSEY,B (1979)
HL 4.45E-007	atm m3/mol	25	EXP	BUTLER,JAV AND RAMCHANDANI,CN (1935)
OH 1.22E-012	cm3/molc sec	25	EXP	ATKINSON,R (1989)

CAS #: 000079-10-7				ACRYLIC ACID

Formula: $C_3H_4O_2$

Mol Weight: 72.06

MP (deg C): 13.5 FP (deg C):

BP (deg C): 141.2

BP pressure (mm Hg):

Property/ Value	Units	Temp	Data Type	Reference
WS 1.00E+006	mg/L		EXP	RIDDICK,JA ET AL. (1986)
logP 0.35			EXP	HANSCH,C ET AL. (1995)
VP 4.00E+000	mm Hg	25	EXP	DAUBERT,TE & DANNER,RP (1985)
DC 4.26	pKa	25	EXP	RIDDICK,JA ET AL. (1986)
HL 1.17E-007	atm m3/mol	25	EST	MEYLAN,WM & HOWARD,PH (1991)
OH 2.43E-011	cm3/molc sec	25	EST	MEYLAN,WM & HOWARD,PH (1993)

CAS #: 000079-11-8				CHLOROACETIC ACID

Formula: $C_2H_3ClO_2$

Mol Weight: 94.50

MP (deg C): 61-63 FP (deg C):

BP (deg C): 189

BP pressure (mm Hg):

Property/ Value	Units	Temp	Data Type	Reference
WS 6.14E+006	mg/L	25	EXP	FREITER,ER (1978)
logP 0.22			EXP	HANSCH,C & LEO,AJ (1985)
VP 6.50E-002	mm Hg	25	EXP	FREITER,ER (1978)
DC 2.87	pKa	25	EXP	SERJEANT,EP & DEMPSEY,B (1979)
HL 1.32E-009	atm m3/mol	25	EST	VP/WSOL
OH 7.86E-013	cm3/molc sec	25	EST	MEYLAN,WM & HOWARD,PH (1993)

CAS #: 000079-14-1				HYDROXYACETIC ACID

Formula: $C_2H_4O_3$

Mol Weight: 76.05

MP (deg C): 80 FP (deg C):

BP (deg C): 100

BP pressure (mm Hg):

Property/ Value	Units	Temp	Data Type	Reference
WS 1.00E+006	mg/L	25	EST	MEYLAN,WM ET AL. (1996)
logP -1.11			EXP	HANSCH,C & LEO,AJ (1985)
VP 2.00E-002	mm Hg	25	EXT	DAUBERT,TE & DANNER,RP (1991)
DC 3.83	pKa		EXP	SERJEANT,EP & DEMPSEY,B (1979)
HL 8.54E-008	atm m3/mol	25	EST	MEYLAN,WM & HOWARD,PH (1991)
OH 3.11E-012	cm3/molc sec	25	EST	MEYLAN,WM & HOWARD,PH (1993)

CAS #: 000079-16-3				N-METHYLACETAMIDE

Formula: C_3H_7NO

Mol Weight: 73.10

MP (deg C): 28 FP (deg C):

BP (deg C): 205

BP pressure (mm Hg):

Property/ Value	Units	Temp	Data Type	Reference
WS 1.00E+006	mg/L	25	EST	MEYLAN,WM ET AL. (1996)
logP -1.05			EXP	HANSCH,C ET AL. (1995)
VP 4.40E-001	mm Hg	23	EXP	YAWS,CL (1994)
DC	pKa			
HL 2.50E-008	atm m3/mol	25	EST	MEYLAN,WM & HOWARD,PH (1991)
OH 6.16E-012	cm3/molc sec	25	EST	MEYLAN,WM & HOWARD,PH (1993)

CAS #: 000079-19-6				THIOSEMICARBAZIDE

Formula: CH_5N_3S

Mol Weight: 91.14

MP (deg C): 180-181 FP (deg C):

BP (deg C):

BP pressure (mm Hg):

Property/ Value	Units	Temp	Data Type	Reference
WS 1.00E+006	mg/L	25	EST	MEYLAN,WM ET AL. (1996)
logP -1.67			EST	MEYLAN,WM & HOWARD,PH (1995)
VP 3.30E-001	mm Hg	25	EST	NEELY,WB & BLAU,GE (1985)
DC	pKa			
HL 6.60E-010	atm m3/mol	25	EST	MEYLAN,WM & HOWARD,PH (1991)
OH 8.40E-011	cm3/molc sec	25	EST	MEYLAN,WM & HOWARD,PH (1993)

CAS #: 000079-20-9				METHYL ACETATE

Formula: $C_3H_6O_2$

Mol Weight: 74.08

MP (deg C): -98.1 FP (deg C):

BP (deg C): 57

BP pressure (mm Hg):

Property/ Value	Units	Temp	Data Type	Reference
WS 2.44E+005	mg/L	20	EXP	STEPHEN,H & STEPHEN,T (1963)
logP 0.18			EXP	HANSCH,C & LEO,AJ (1985)
VP 2.16E+002	mm Hg	25	EXP	AMBROSE,D ET AL. (1981)
DC	pKa			
HL 1.15E-004	atm m3/mol	25	EXP	BUTTERY,RG ET AL. (1969)
OH 3.40E-013	cm3/molc sec	25	EXP	ATKINSON,R (1989)

CAS #: **000079-21-0**				PEROXYACETIC ACID

Formula: $C_2H_4O_3$

Mol Weight: 76.05

MP (deg C): 0.1 FP (deg C):

BP (deg C): 105

BP pressure (mm Hg):

Property/Value	Units	Temp	Data Type	Reference
WS 1.00E+006	mg/L	25	EST	MEYLAN,WM ET AL. (1996)
logP -1.07			EST	MEYLAN,WM & HOWARD,PH (1995)
VP 1.45E+001	mm Hg	25	EXP	DAUBERT,TE & DANNER,RP (1989)
DC 8.20	pKa	25	EXP	DEAN,JA (1985)
HL 2.14E-006	atm m3/mol	25	EXP	LIND,JA & KOK,GL (1986)
OH 4.04E-012	cm3/molc sec	25	EST	MEYLAN,WM & HOWARD,PH (1993)

CAS #: **000079-22-1**				METHYL CHLOROFORMATE

Formula: $C_2H_3ClO_2$

Mol Weight: 94.50

MP (deg C): -72 EST FP (deg C):

BP (deg C): 71

BP pressure (mm Hg):

Property/Value	Units	Temp	Data Type	Reference
WS 9.28E+004	mg/L	25	EST	MEYLAN,WM ET AL. (1996)
logP 0.14			EST	MEYLAN,WM & HOWARD,PH (1995)
VP 1.59E+002	mm Hg	25	EST	NEELY,WB & BLAU,GE (1985)
DC	pKa			
HL 2.35E-003	atm m3/mol	25	EST	MEYLAN,WM & HOWARD,PH (1991)
OH 2.18E-013	cm3/molc sec	25	EST	MEYLAN,WM & HOWARD,PH (1993)

CAS #: **000079-24-3**				NITROETHANE

Formula: $C_2H_5NO_2$

Mol Weight: 75.07

MP (deg C): -50 FP (deg C):

BP (deg C): 114-115

BP pressure (mm Hg):

Property/Value	Units	Temp	Data Type	Reference
WS 4.50E+004	mg/L	20	EXP	YALKOWSKY,SH & DANNENFELSER,RM (1992)
logP 0.18			EXP	HANSCH,C & LEO,AJ (1985)
VP 2.08E+001	mm Hg	25	EXP	DAUBERT,TE & DANNER,RP (1989)
DC 8.60	pKa	25	EXP	SERJEANT,EP & DEMPSEY,B (1979)
HL 4.76E-005	atm m3/mol	25	EXP	GAFFNEY,JS ET AL. (1987)
OH 1.50E-013	cm3/molc sec	25	EXP	ATKINSON,R (1989)

CAS #: **000079-27-6**				1,1,2,2-TETRABROMOETHANE

Formula: $C_2H_2Br_4$

Mol Weight: 345.67

MP (deg C): 0 FP (deg C):

BP (deg C): 151

BP pressure (mm Hg): 5.40E+001

Property/Value	Units	Temp	Data Type	Reference
WS 6.51E+002	mg/L	30	EXP	RIDDICK,JA ET AL (1986)
logP 2.55			EST	MEYLAN,WM & HOWARD,PH (1995)
VP 2.00E-002	mm Hg	25	EXP	DAUBERT,TE & DANNER,RP (1989)
DC	pKa			
HL 1.40E-005	atm m3/mol	25	EST	VP/WSOL
OH 1.40E-013	cm3/molc sec	25	EST	MEYLAN,WM & HOWARD,PH (1993)

CAS #: **000079-29-8**				2,3-DIMETHYLBUTANE

Formula: C_6H_{14}

Mol Weight: 86.18

MP (deg C): -128.8 FP (deg C):

BP (deg C): 57.9

BP pressure (mm Hg):

Property/Value	Units	Temp	Data Type	Reference
WS 2.25E+001	mg/L	25	EXP	YALKOWSKY,SH & DANNENFELSER,RM (1992)
logP 3.42			EXP	HANSCH,C ET AL. (1995)
VP 2.35E+002	mm Hg	25	EXP	DAUBERT,TE & DANNER,RP (1989)
DC	pKa			
HL 1.18E+000	atm m3/mol	25	EST	VP/WSOL
OH 6.20E-012	cm3/molc sec	25	EXP	ATKINSON,R (1989)

CAS #: **000079-31-2**				ISOBUTYRICACID

Formula: $C_4H_8O_2$

Mol Weight: 88.11

MP (deg C): -47 FP (deg C):

BP (deg C): 152-155

BP pressure (mm Hg):

Property/Value	Units	Temp	Data Type	Reference
WS 1.67E+005	mg/L	20	EXP	YALKOWSKY,SH & DANNENFELSER,RM (1992)
logP 0.94			EXP	SANGSTER,J (1993)
VP 1.81E+000	mm Hg	25	EXP	DAUBERT,TE & DANNER,RP (1989)
DC 4.84	pKa	20	EXP	KORTUM,G ET AL (1961)
HL 1.26E-006	atm m3/mol	25	EST	VP/WSOL
OH 2.00E-012	cm3/molc sec	25	EXP	ATKINSON,R (1989)

CAS #: **000079-34-5**				1,1,2,2-TETRACHLOROETHANE

Formula: $C_2H_2Cl_4$

Mol Weight: 167.85

MP (deg C): -44 FP (deg C):

BP (deg C): 146.5

BP pressure (mm Hg):

Property/Value	Units	Temp	Data Type	Reference
WS 2.96E+003	mg/L	25	EXP	HORVATH,AL (1982)
logP 2.39			EXP	HANSCH,C & LEO,AJ (1985)
VP 4.62E+000	mm Hg	25	EXP	DAUBERT,TE & DANNER,RP (1985)
DC	pKa			
HL 3.67E-004	atm m3/mol	25	EXP	LEIGHTON,DTJR & CALO,JM (1981)
OH 2.50E-013	cm3/molc sec	25	EXP	ATKINSON,R (1989)

CAS #: **000079-35-6**				1,1-DICHLORO-2,2-DIFLUOROETHYLENE

Formula: $C_2Cl_2F_2$

Mol Weight: 132.93

MP (deg C): -116 FP (deg C):

BP (deg C): 19

BP pressure (mm Hg):

Property/Value	Units	Temp	Data Type	Reference
WS 1.47E+003	mg/L	25	EST	MEYLAN,WM ET AL. (1996)
logP 2.09			EST	MEYLAN,WM & HOWARD,PH (1995)
VP 3.70E+002	mm Hg	25	EST	NEELY,WB & BLAU,GE (1985)
DC	pKa			
HL 1.17E-001	atm m3/mol	25	EST	MEYLAN,WM & HOWARD,PH (1991)
OH 7.35E-012	cm3/molc sec	25	EXP	ATKINSON,R (1989)

000079-36-7 — DICHLOROACETYL CHLORIDE

Formula: C_2HCl_3O
Mol Weight: 147.39
MP (deg C): — FP (deg C): —
BP (deg C): 108
BP pressure (mm Hg): —

Property/Value	Units	Temp	Data Type	Reference
WS 8.39E+004	mg/L	25	EST	MEYLAN,WM ET AL. (1996)
logP -0.04			EST	MEYLAN,WM & HOWARD,PH (1995)
VP 2.30E+001	mm Hg	25	EXP	DAUBERT,TE & DANNER,RP (1989)
DC	pKa			
HL 8.25E-005	atm m3/mol	25	EST	MEYLAN,WM & HOWARD,PH (1991)
OH 1.88E-014	cm3/molc sec	25	EST	MEYLAN,WM & HOWARD,PH (1993)

000079-38-9 — CHLOROTRIFLUOROETHYLENE

Formula: C_2ClF_3
Mol Weight: 116.47
MP (deg C): -158.2 FP (deg C): -15
BP (deg C): -27.90
BP pressure (mm Hg): —

Property/Value	Units	Temp	Data Type	Reference
WS 4.01E+003	mg/L	25	EST	MEYLAN,WM ET AL. (1996)
logP 1.65			EST	MEYLAN,WM & HOWARD,PH (1995)
VP 4.59E+003	mm Hg	25	EXP	DAUBERT,TE & DANNER,RP (1989)
DC	pKa			
HL 3.11E-001	atm m3/mol	25	EST	MEYLAN,WM & HOWARD,PH (1991)
OH 7.00E-012	cm3/molc sec	25	EXP	DILLING,WL (1982)

000079-40-3 — RUBEANIC ACID

Formula: $C_2H_4N_2S_2$
Mol Weight: 120.20
MP (deg C): 170 dec FP (deg C): —
BP (deg C): —
BP pressure (mm Hg): —

Property/Value	Units	Temp	Data Type	Reference
WS 1.00E+006	mg/L	25	EST	MEYLAN,WM ET AL. (1996)
logP -2.98			EST	MEYLAN,WM & HOWARD,PH (1995)
VP 1.64E-006	mm Hg	25	EXT	JONES,AH (1960)
DC 10.90	pKa		EXP	SERJEANT,EP & DEMPSEY,B (1979)
HL 4.25E-010	atm m3/mol	25	EST	MEYLAN,WM & HOWARD,PH (1991)
OH 4.20E-011	cm3/molc sec	25	EST	MEYLAN,WM & HOWARD,PH (1993)

000079-41-4 — METHACRYLIC ACID

Formula: $C_4H_6O_2$
Mol Weight: 86.09
MP (deg C): 16 FP (deg C): —
BP (deg C): 163
BP pressure (mm Hg): —

Property/Value	Units	Temp	Data Type	Reference
WS 8.90E+004	mg/L	20	EXP	RIDDICK,JA ET AL. (1986)
logP 0.93			EXP	HANSCH,C & LEO,AJ (1985)
VP 9.75E-001	mm Hg	25	EXP	DAUBET,TE & DANNER,RP (1989)
DC 4.65	pKa		EXP	SERJEANT,EP & DEMPSEY,B (1979)
HL 1.24E-006	atm m3/mol	25	EST	VP/WSOL
OH 1.86E-011	cm3/molc sec	25	EST	MEYLAN,WM & HOWARD,PH (1993)

000079-43-6 — DICHLOROACETIC ACID

Formula: $C_2H_2Cl_2O_2$
Mol Weight: 128.94
MP (deg C): 13.5 FP (deg C): —
BP (deg C): 193-194
BP pressure (mm Hg): —

Property/Value	Units	Temp	Data Type	Reference
WS 1.00E+006	mg/L	20	EXP	YALKOWSKY,SH & DANNENFELSER,RM (1992)
logP 0.92			EXP	HANSCH,C & LEO,AJ (1985)
VP 1.79E-001	mm Hg	25	EXP	DAUBERT,TE & DANNER,RP (1989)
DC 1.26	pKa		EXP	MARUTHAMUTHU,P & HUIE,RE (1995)
HL 3.52E-007	atm m3/mol	25	EST	VP/WSOL
OH 7.30E-013	cm3/molc sec	25	EST	MEYLAN,WM & HOWARD,PH (1993)

000079-44-7 — DIMETHYLCARBAMYL CHLORIDE

Formula: C_3H_6ClNO
Mol Weight: 107.54
MP (deg C): -33 FP (deg C): —
BP (deg C): 167-168
BP pressure (mm Hg): 7.75E+002

Property/Value	Units	Temp	Data Type	Reference
WS 4.59E+005	mg/L	25	EST	MEYLAN,WM ET AL. (1996)
logP -0.72			EST	MEYLAN,WM & HOWARD,PH (1995)
VP 1.95E+000	mm Hg	25	EXP	JABER,HM ET AL. (1984B)
DC	pKa			
HL 7.23E-007	atm m3/mol	25	EST	MEYLAN,WM & HOWARD,PH (1991)
OH 1.61E-011	cm3/molc sec	25	EST	MEYLAN,WM & HOWARD,PH (1993)

000079-46-9 — 2-NITROPROPANE

Formula: $C_3H_7NO_2$
Mol Weight: 89.09
MP (deg C): -91.32 FP (deg C): —
BP (deg C): 120.25
BP pressure (mm Hg): —

Property/Value	Units	Temp	Data Type	Reference
WS 1.70E+004	mg/L	25	EXP	BAKER,PJJR & BOLLMEIER,AFJR (1981)
logP 0.93			EXP	CHEM INSPECT TEST INST (1992)
VP 1.72E+001	mm Hg	25	EXP	DAUBERT,TE & DANNER,RP (1987)
DC 7.68	pKa	25	EXP	SERJEANT,EP & DEMPSEY,B (1979)
HL 1.19E-004	atm m3/mol	25	EST	VP/WSOL
OH 2.60E-013	cm3/molc sec	25	EXP	ATKINSON,R (1989)

000079-53-8 — CHLOROPENTAFLUOROACETONE

Formula: C_3ClF_5O
Mol Weight: 182.48
MP (deg C): -133 FP (deg C): —
BP (deg C): 8
BP pressure (mm Hg): —

Property/Value	Units	Temp	Data Type	Reference
WS 8.95E+003	mg/L	25	EST	MEYLAN,WM ET AL. (1996)
logP 0.91			EST	MEYLAN,WM & HOWARD,PH (1995)
VP 1.43E+003	mm Hg	25	EXP	BOUBLIK,T ET AL. (1984)
DC	pKa			
HL 5.44E-004	atm m3/mol	25	EST	MEYLAN,WM & HOWARD,PH (1991)
OH 0.00E+000	cm3/molc sec	25	EST	MEYLAN,WM & HOWARD,PH (1993)

000079-57-2 — OXYTETRACYLCINE

Formula:	$C_{22}H_{24}N_2O_9$		
Mol Weight:	460.44		
MP (deg C):	184.5	FP (deg C):	
BP (deg C):			
BP pressure (mm Hg):			

Property/Value	Units	Temp	Data Type	Reference
WS 3.13E+002	mg/L	25	EXP	YALKOWSKY,SH & DANNENFELSER,RM (1992)
logP -0.90			EXP	SANGSTER,J (1994)
VP 9.68E-025	mm Hg	25	EST	NEELY,WB & BLAU,GE (1985)
DC 3.27	pKa	25	EXP	KORTUM,G ET AL (1961)
HL 1.70E-025	atm m3/mol	25	EST	MEYLAN,WM & HOWARD,PH (1991)
OH 2.65E-010	cm3/molc sec	25	EST	MEYLAN,WM & HOWARD,PH (1993)

000079-60-7 — CORTISONE-9A-FLUORO

Formula:	$C_{21}H_{27}FO_5$		
Mol Weight:	378.44		
MP (deg C):		FP (deg C):	
BP (deg C):			
BP pressure (mm Hg):			

Property/Value	Units	Temp	Data Type	Reference
WS 2.81E+002	mg/L	25	EST	MEYLAN,WM ET AL. (1996)
logP 1.37			EXP	HANSCH,C & LEO,AJ (1985)
VP 7.03E-013	mm Hg	25	EST	NEELY,WB & BLAU,GE (1985)
DC	pKa			
HL 1.20E-009	atm m3/mol	25	EST	MEYLAN,WM & HOWARD,PH (1991)
OH 8.64E-011	cm3/molc sec	25	EST	MEYLAN,WM & HOWARD,PH (1993)

000079-83-4 — D-PANTOTHENIC ACID

Formula:	$C_9H_{17}NO_5$		
Mol Weight:	219.24		
MP (deg C):		FP (deg C):	
BP (deg C):			
BP pressure (mm Hg):			

Property/Value	Units	Temp	Data Type	Reference
WS 9.72E+005	mg/L	25	EST	MEYLAN,WM ET AL. (1996)
logP -1.69			EST	MEYLAN,WM & HOWARD,PH (1995)
VP 1.13E-010	mm Hg	25	EST	NEELY,WB & BLAU,GE (1985)
DC	pKa			
HL 4.30E-015	atm m3/mol	25	EST	MEYLAN,WM & HOWARD,PH (1991)
OH 2.25E-011	cm3/molc sec	25	EST	MEYLAN,WM & HOWARD,PH (1993)

000079-92-5 — CAMPHENE

Formula:	$C_{10}H_{16}$		
Mol Weight:	136.24		
MP (deg C):	51.2	FP (deg C):	
BP (deg C):	160		
BP pressure (mm Hg):			

Property/Value	Units	Temp	Data Type	Reference
WS 4.60E+000	mg/L	25	EXP	CEHM INSPECT TEST INST (1992)
logP 4.35			EST	MEYLAN,WM & HOWARD,PH (1995)
VP 2.51E+000	mm Hg	25	EXT	DAUBERT,TE & DANNER,RP (1989)
DC	pKa			
HL 1.61E-001	atm m3/mol	25	EST	MEYLAN,WM & HOWARD,PH (1991)
OH 5.86E-011	cm3/molc sec	25	EST	MEYLAN,WM & HOWARD,PH (1993)

000079-94-7 — 2,2-BIS(4-HYDROXY-3,5-DIBROMOPHENYL)PROPANE

Formula:	$C_{15}H_{12}Br_4O_2$		
Mol Weight:	543.90		
MP (deg C):	181	FP (deg C):	
BP (deg C):	316		
BP pressure (mm Hg):			

Property/Value	Units	Temp	Data Type	Reference
WS 1.00E-003	mg/L	25	EST	MEYLAN,WM ET AL. (1996)
logP 7.20			EST	MEYLAN,WM & HOWARD,PH (1995)
VP 1.76E-011	mm Hg	25	EST	NEELY,WB & BLAU,GE (1985)
DC	pKa			
HL 7.05E-011	atm m3/mol	25	EST	MEYLAN,WM & HOWARD,PH (1991)
OH 2.89E+000	cm3/molc sec	25	EST	MEYLAN,WM & HOWARD,PH (1993)

000080-00-2 — SULPHENONE

Formula:	$C_{12}H_9ClO_2S$		
Mol Weight:	252.72		
MP (deg C):	94	FP (deg C):	
BP (deg C):			
BP pressure (mm Hg):			

Property/Value	Units	Temp	Data Type	Reference
WS 3.82E+001	mg/L	25	EST	MEYLAN,WM ET AL. (1996)
logP 3.25			EST	MEYLAN,WM & HOWARD,PH (1995)
VP 3.90E-006	mm Hg	25	EST	NEELY,WB & BLAU,GE (1985)
DC	pKa			
HL 1.85E-007	atm m3/mol	25	EST	MEYLAN,WM & HOWARD,PH (1991)
OH 7.10E-013	cm3/molc sec	25	EST	MEYLAN,WM & HOWARD,PH (1993)

000080-05-7 — DIPHENYLOLPROPANE

Formula:	$C_{15}H_{16}O_2$		
Mol Weight:	228.29		
MP (deg C):	150-155	FP (deg C):	
BP (deg C):	220		
BP pressure (mm Hg):	4.00E+000		

Property/Value	Units	Temp	Data Type	Reference
WS 1.20E+002	mg/L	25	EXP	DORN,PB ET AL. (1987)
logP 3.32			EXP	HANSCH,C & LEO,AJ (1985)
VP 3.91E-007	mm Hg	25	EST	NEELY,WB & BLAU,GE (1985)
DC	pKa			
HL 1.00E-011	atm m3/mol	25	EST	MEYLAN,WM & HOWARD,PH (1991)
OH 8.61E-011	cm3/molc sec	25	EST	MEYLAN,WM & HOWARD,PH (1993)

000080-06-8 — DIMITE

Formula:	$C_{14}H_{12}Cl_2O$		
Mol Weight:	267.16		
MP (deg C):	69-69.5	FP (deg C):	
BP (deg C):			
BP pressure (mm Hg):			

Property/Value	Units	Temp	Data Type	Reference
WS 9.67E+000	mg/L	25	EST	MEYLAN,WM ET AL. (1996)
logP 4.45			EST	MEYLAN,WM & HOWARD,PH (1995)
VP 5.38E-007	mm Hg	25	EST	NEELY,WB & BLAU,GE (1985)
DC	pKa			
HL 1.28E-008	atm m3/mol	25	EST	MEYLAN,WM & HOWARD,PH (1991)
OH 3.95E-012	cm3/molc sec	25	EST	MEYLAN,WM & HOWARD,PH (1993)

CAS #: 000080-08-0	DI(P-AMINOPHENYL)SULFONE

Formula: $C_{12}H_{12}N_2O_2S$

Mol Weight: 248.31

MP (deg C): 175-176 FP (deg C):

BP (deg C):

BP pressure (mm Hg):

Property/Value	Units	Temp	Data Type	Reference
WS 3.80E+002	mg/L	37	EXP	YALKOWSKY,SH & DANNENFELSER,RM (1992)
logP 0.97			EXP	HANSCH,C & LEO,AJ (1985)
VP 2.20E-008	mm Hg	25	EST	NEELY,WB & BLAU,GE (1985)
DC 2.41	pKa		EXP	PERRIN,DD (1965)
HL 3.11E-014	atm m3/mol	25	EST	MEYLAN,WM & HOWARD,PH (1991)
OH 4.60E-011	cm3/molc sec	25	EST	MEYLAN,WM & HOWARD,PH (1993)

CAS #: 000080-10-4	DICHLORODIPHENYLSILANE

Formula: $C_{12}H_{10}Cl_2Si$

Mol Weight: 253.21

MP (deg C): FP (deg C):

BP (deg C): 305

BP pressure (mm Hg):

Property/Value	Units	Temp	Data Type	Reference
WS 1.09E+000	mg/L	25	EST	MEYLAN,WM ET AL. (1996)
logP 5.06			EST	MEYLAN,WM & HOWARD,PH (1995)
VP 7.96E-004	mm Hg	25	EST	NEELY,WB & BLAU,GE (1985)
DC	pKa			
HL 7.01E-005	atm m3/mol	25	EST	MEYLAN,WM & HOWARD,PH (1991)
OH 3.90E-012	cm3/molc sec	25	EST	MEYLAN,WM & HOWARD,PH (1993)

CAS #: 000080-11-5	BENZENESULFONAMIDE, N,4-DIMETHYL-N-NITROSO-

Formula: $C_8H_{10}N_2O_3S$

Mol Weight: 214.24

MP (deg C): 62 FP (deg C):

BP (deg C):

BP pressure (mm Hg):

Property/Value	Units	Temp	Data Type	Reference
WS 1.18E+004	mg/L	25	EST	MEYLAN,WM ET AL. (1996)
logP 0.58			EXP	HANSCH,C ET AL. (1995)
VP 4.79E-006	mm Hg	25	EST	NEELY,WB & BLAU,GE (1985)
DC	pKa			
HL 1.27E-007	atm m3/mol	25	EST	MEYLAN,WM & HOWARD,PH (1991)
OH 2.49E-012	cm3/molc sec	25	EST	MEYLAN,WM & HOWARD,PH (1993)

CAS #: 000080-15-9	CUMENE HYDROPEROXIDE

Formula: $C_9H_{12}O_2$

Mol Weight: 152.19

MP (deg C): FP (deg C):

BP (deg C): 100-101

BP pressure (mm Hg): 8.00E+000

Property/Value	Units	Temp	Data Type	Reference
WS 1.39E+004	mg/L	25	EXP	YALKOWSKY,SH & DANNENFELSER,RM (1992)
logP 1.55			EST	NEELY,WB & BLAU,GE (1985)
VP 1.50E-002	mm Hg	25	EST	SHEPPARD,CJ & MAGELI,OL (1982)
DC	pKa			
HL 2.16E-007	atm m3/mol	25	EST	VP/WSOL
OH 8.90E-012	cm3/molc sec	25	EST	MEYLAN,WM & HOWARD,PH (1993)

CAS #: 000080-17-1	BENZENESULFONIC ACID HYDRAZIDE

Formula: $C_6H_8N_2O_2S$

Mol Weight: 172.21

MP (deg C): 101-103 FP (deg C):

BP (deg C):

BP pressure (mm Hg):

Property/Value	Units	Temp	Data Type	Reference
WS 7.91E+004	mg/L	25	EST	MEYLAN,WM ET AL. (1996)
logP -0.14			EXP	HANSCH,C & LEO,AJ (1985)
VP 1.19E-004	mm Hg	25	EST	NEELY,WB & BLAU,GE (1985)
DC	pKa			
HL 1.76E-009	atm m3/mol	25	EST	MEYLAN,WM & HOWARD,PH (1991)
OH 4.17E-013	cm3/molc sec	25	EST	MEYLAN,WM & HOWARD,PH (1993)

CAS #: 000080-33-1	CHLORFENSON

Formula: $C_{12}H_8Cl_2O_3S$

Mol Weight: 303.17

MP (deg C): 86.5-86.8 FP (deg C):

BP (deg C):

BP pressure (mm Hg):

Property/Value	Units	Temp	Data Type	Reference
WS 2.96E+000	mg/L	25	EST	MEYLAN,WM ET AL. (1996)
logP 4.21			EST	MEYLAN,WM & HOWARD,PH (1995)
VP 1.88E-006	mm Hg	25	EST	NEELY,WB & BLAU,GE (1985)
DC	pKa			
HL 1.64E-007	atm m3/mol	25	EST	MEYLAN,WM & HOWARD,PH (1991)
OH 6.57E-012	cm3/molc sec	25	EST	MEYLAN,WM & HOWARD,PH (1993)

CAS #: 000080-35-3	SULFAMETHOXYPYRIDAZINE

Formula: $C_{11}H_{12}N_4O_3S$

Mol Weight: 280.31

MP (deg C): 182-183 FP (deg C):

BP (deg C):

BP pressure (mm Hg):

Property/Value	Units	Temp	Data Type	Reference
WS 7.20E+003	mg/L	37	EXP	YALKOWSKY,SH & DANNENFELSER,RM (1992)
logP 0.32			EXP	HANSCH,C & LEO,AJ (1985)
VP 5.65E-010	mm Hg	25	EST	NEELY,WB & BLAU,GE (1985)
DC	pKa			
HL 2.37E-014	atm m3/mol	25	EST	MEYLAN,WM & HOWARD,PH (1991)
OH 2.82E-011	cm3/molc sec	25	EST	MEYLAN,WM & HOWARD,PH (1993)

CAS #: 000080-38-6	4-CHLOROPHENYL BENZENESULFONATE

Formula: $C_{12}H_9ClO_3S$

Mol Weight: 268.72

MP (deg C): FP (deg C):

BP (deg C):

BP pressure (mm Hg):

Property/Value	Units	Temp	Data Type	Reference
WS 1.66E+001	mg/L	25	EST	MEYLAN,WM ET AL. (1996)
logP 3.57			EST	MEYLAN,WM & HOWARD,PH (1995)
VP 1.52E-006	mm Hg	25	EST	NEELY,WB & BLAU,GE (1985)
DC	pKa			
HL 2.21E-007	atm m3/mol	25	EST	MEYLAN,WM & HOWARD,PH (1991)
OH 6.70E-012	cm3/molc sec	25	EST	MEYLAN,WM & HOWARD,PH (1993)

000080-40-0 — ETHYL-P-METHYLBENZENESULFONATE

Formula: $C_9H_{12}O_3S$
Mol Weight: 200.26
MP (deg C): 33
FP (deg C):
BP (deg C): 173
BP pressure (mm Hg): 1.50E+001

Property/Value	Units	Temp	Data Type	Reference
WS 1.24E+003	mg/L	25	EST	MEYLAN,WM ET AL. (1996)
logP 1.81			EXP	HANSCH,C & LEO,AJ (1985)
VP 2.41E-004	mm Hg	25	EST	NEELY,WB & BLAU,GE (1985)
DC	pKa			
HL 1.18E-006	atm m3/mol	25	EST	MEYLAN,WM & HOWARD,PH (1991)
OH 2.05E-012	cm3/molc sec	25	EST	MEYLAN,WM & HOWARD,PH (1993)

000080-43-3 — PEROXIDE, BIS(1-METHYL-1-PHENYLETHYL)

Formula: $C_{18}H_{22}O_2$
Mol Weight: 270.37
MP (deg C): 40.6
FP (deg C):
BP (deg C):
BP pressure (mm Hg):

Property/Value	Units	Temp	Data Type	Reference
WS 4.60E-001	mg/L	25	EXP	CHEM INSPECT TEST INST (1992)
logP 5.50			EXP	CHEM INSPECT TEST INST (1992)
VP 1.09E+001	mm Hg	25	EXT	DAUBERT,TE & DANNER,RP (1989)
DC	pKa			
HL 4.42E-005	atm m3/mol	25	EST	MEYLAN,WM & HOWARD,PH (1991)
OH 9.26E-012	cm3/molc sec	25	EST	MEYLAN,WM & HOWARD,PH (1993)

000080-46-6 — P-TERT-AMYLPHENOL

Formula: $C_{11}H_{16}O$
Mol Weight: 164.25
MP (deg C): 94-95
FP (deg C):
BP (deg C): 262.5
BP pressure (mm Hg):

Property/Value	Units	Temp	Data Type	Reference
WS 1.68E+002	mg/L	25	EXP	YALKOWSKY,SH & DANNENFELSER,RM (1992)
logP 3.91			EST	MEYLAN,WM & HOWARD,PH (1995)
VP 8.05E-003	mm Hg	25	EXT	DAUBERT,TE & DANNER,RP (1989)
DC 10.43	pKa			SCHULTZ,TW (1987)
HL 2.03E-006	atm m3/mol	25	EST	MEYLAN,WM & HOWARD,PH (1991)
OH 4.45E-011	cm3/molc sec	25	EST	MEYLAN,WM & HOWARD,PH (1993)

000080-56-8 — ALPHA-PINENE

Formula: $C_{10}H_{16}$
Mol Weight: 136.24
MP (deg C): -64
FP (deg C):
BP (deg C): 155.9
BP pressure (mm Hg):

Property/Value	Units	Temp	Data Type	Reference
WS 8.20E+000	mg/L	25	EST	VP/HL
logP 4.83			EXP	LI,NY & PERDUE,EM (1995)
VP 4.75E+000	mm Hg	25	EXP	DAUBERT,TE & DANNER,RP (1989)
DC	pKa			
HL 1.07E-001	atm m3/mol	25	EST	MEYLAN,WM & HOWARD,PH (1991)
OH 5.37E-011	cm3/molc sec	25	EXP	ATKINSON,R (1988)

000080-58-0 — A-BROMOBUTYRIC ACID

Formula: $C_4H_7BrO_2$
Mol Weight: 167.01
MP (deg C): -4
FP (deg C):
BP (deg C): 181-182
BP pressure (mm Hg): 2.50E+002

Property/Value	Units	Temp	Data Type	Reference
WS 7.00E+004	mg/L		EXP	YALKOWSKY,SH & DANNENFELSER,RM (1992)
logP 1.42			EXP	HANSCH,C & LEO,AJ (1985)
VP 1.05E-001	mm Hg	25	EST	NEELY,WB & BLAU,GE (1985)
DC	pKa			
HL 1.11E-007	atm m3/mol	25	EST	MEYLAN,WM & HOWARD,PH (1991)
OH 1.62E-012	cm3/molc sec	25	EST	MEYLAN,WM & HOWARD,PH (1993)

000080-59-1 — TIGLIC ACID

Formula: $C_5H_8O_2$
Mol Weight: 100.12
MP (deg C): 63.5-64
FP (deg C):
BP (deg C): 198.5
BP pressure (mm Hg):

Property/Value	Units	Temp	Data Type	Reference
WS 1.85E+004	mg/L	25	EST	MEYLAN,WM ET AL. (1996)
logP 1.40			EST	MEYLAN,WM & HOWARD,PH (1995)
VP 1.33E-001	mm Hg	25	EXP	PERRY,RH & GREEN,D (1984)
DC 4.96	pKa	18	EXP	KORTUM,G ET AL (1961)
HL 7.09E-007	atm m3/mol	25	EST	MEYLAN,WM & HOWARD,PH (1991)
OH 3.12E-011	cm3/molc sec	25	EST	MEYLAN,WM & HOWARD,PH (1993)

000080-60-4 — ALPHA-AMINOBUTYRIC ACID

Formula: $C_4H_9NO_2$
Mol Weight: 103.12
MP (deg C): 304
FP (deg C):
BP (deg C):
BP pressure (mm Hg):

Property/Value	Units	Temp	Data Type	Reference
WS 2.11E+005	mg/L	25	EXP	YALKOWSKY,SH & DANNENFELSER,RM (1992)
logP -2.54			EXP	TSAI,RS ET AL. (1991)
VP 4.16E-008	mm Hg	25	EST	NEELY,WB & BLAU,GE (1985)
DC 2.31	pKa	13	EXP	KORTUM,G ET AL (1961)
HL 1.98E-009	atm m3/mol	25	EST	MEYLAN,WM & HOWARD,PH (1991)
OH 3.95E-011	cm3/molc sec	25	EST	MEYLAN,WM & HOWARD,PH (1993)

000080-62-6 — METHYL METHACRYLATE

Formula: $C_5H_8O_2$
Mol Weight: 100.12
MP (deg C): -48
FP (deg C):
BP (deg C): 100.5
BP pressure (mm Hg):

Property/Value	Units	Temp	Data Type	Reference
WS 1.50E+004	mg/L	25	EXP	NEMEC,JW & KIRSCH,LS (1981)
logP 1.38			EXP	HANSCH,C & LEO,AJ (1985)
VP 3.84E+001	mm Hg	25	EXP	DAUBERT,TE & DANNER,RP (1985)
DC	pKa			
HL 3.37E-004	atm m3/mol	25	EST	VP/WSOL
OH 4.71E-011	cm3/molc sec	25	EST	MEYLAN,WM & HOWARD,PH (1993)

000080-70-6 — GUANIDINE, N,N,N',N'-TETRAMETHYL-

Formula: $C_5H_{13}N_3$
Mol Weight: 115.18
MP (deg C):
FP (deg C):
BP (deg C): 52-54
BP pressure (mm Hg): 1.10E+001

Property/ Value	Units	Temp	Data Type	Reference
WS 4.66E+004	mg/L	25	EST	MEYLAN,WM ET AL. (1996)
logP 0.41			EXP	HANSCH,C ET AL. (1995)
VP 2.13E+000	mm Hg	25	EST	NEELY,WB & BLAU,GE (1985)
DC	pKa			
HL 5.43E-010	atm m3/mol	25	EST	MEYLAN,WM & HOWARD,PH (1991)
OH 1.37E-010	cm3/molc sec	25	EST	MEYLAN,WM & HOWARD,PH (1993)

000081-04-9 — 1,5-NAPHTHALENEDISULFONIC ACID

Formula: $C_{10}H_8O_6S_2$
Mol Weight: 288.30
MP (deg C): 240-245 de
FP (deg C):
BP (deg C):
BP pressure (mm Hg):

Property/ Value	Units	Temp	Data Type	Reference
WS 9.05E+004	mg/L	25	EST	MEYLAN,WM ET AL. (1996)
logP -3.15			EST	MEYLAN,WM & HOWARD,PH (1995)
VP 6.58E-014	mm Hg	25	EST	NEELY,WB & BLAU,GE (1985)
DC				
HL 1.15E-016	atm m3/mol	25	EST	MEYLAN,WM & HOWARD,PH (1991)
OH 1.27E-012	cm3/molc sec	25	EST	MEYLAN,WM & HOWARD,PH (1993)

000081-06-1 — 1-AMINO-2-NAPHTHALENESULFONIC ACID

Formula: $C_{10}H_9NO_3S$
Mol Weight: 223.25
MP (deg C):
FP (deg C):
BP (deg C):
BP pressure (mm Hg):

Property/ Value	Units	Temp	Data Type	Reference
WS 2.23E+003	mg/L	25	EST	MEYLAN,WM ET AL. (1996)
logP -0.97			EXP	HANSCH,C & LEO,AJ (1985)
VP 5.26E-010	mm Hg	25	EST	NEELY,WB & BLAU,GE (1985)
DC	pKa			
HL 8.68E-014	atm m3/mol	25	EST	MEYLAN,WM & HOWARD,PH (1991)
OH 4.41E-011	cm3/molc sec	25	EST	MEYLAN,WM & HOWARD,PH (1993)

000081-07-2 — 1,1-DIOX-1,2-BENZISOTHIAZOL-3-ONE

Formula: $C_7H_5NO_3S$
Mol Weight: 183.19
MP (deg C): 228.8-9.7
FP (deg C):
BP (deg C):
BP pressure (mm Hg):

Property/ Value	Units	Temp	Data Type	Reference
WS 4.00E+003	mg/L	25	EXP	SEIDELL,A (1941)
logP 0.91			EXP	HANSCH,C & LEO,AJ (1985)
VP 6.45E-007	mm Hg	25	EST	NEELY,WB & BLAU,GE (1985)
DC	pKa			
HL 1.23E-009	atm m3/mol	25	EST	MEYLAN,WM & HOWARD,PH (1991)
OH 5.88E-012	cm3/molc sec	25	EST	MEYLAN,WM & HOWARD,PH (1993)

000081-16-3 — 2-AMINO-1-NAPHTHALENESULFONIC ACID

Formula: $C_{10}H_9NO_3S$
Mol Weight: 223.25
MP (deg C): 180
FP (deg C):
BP (deg C):
BP pressure (mm Hg):

Property/ Value	Units	Temp	Data Type	Reference
WS 4.10E+003	mg/L	20	EXP	YALKOWSKY,SH & DANNENFELSER,RM (1992)
logP -1.16			EXP	HANSCH,C & LEO,AJ (1985)
VP 5.26E-010	mm Hg	25	EST	NEELY,WB & BLAU,GE (1985)
DC	pKa			
HL 8.68E-014	atm m3/mol	25	EST	MEYLAN,WM & HOWARD,PH (1991)
OH 4.41E-011	cm3/molc sec	25	EST	MEYLAN,WM & HOWARD,PH (1993)

000081-20-9 — 1,3-DIMETHYL-2-NITROBENZENE

Formula: $C_8H_9NO_2$
Mol Weight: 151.17
MP (deg C): 15
FP (deg C):
BP (deg C): 226
BP pressure (mm Hg):

Property/ Value	Units	Temp	Data Type	Reference
WS 9.24E+001	mg/L	25	EST	MEYLAN,WM ET AL. (1996)
logP 2.95			EXP	HANSCH,C & LEO,AJ (1985)
VP 1.98E-002	mm Hg	25	EST	NEELY,WB & BLAU,GE (1985)
DC	pKa			
HL 2.59E-005	atm m3/mol	25	EST	MEYLAN,WM & HOWARD,PH (1991)
OH 1.93E-012	cm3/molc sec	25	EST	MEYLAN,WM & HOWARD,PH (1993)

000081-25-4 — CHOLIC ACID

Formula: $C_{24}H_{40}O_5$
Mol Weight: 408.58
MP (deg C): 198
FP (deg C):
BP (deg C):
BP pressure (mm Hg):

Property/ Value	Units	Temp	Data Type	Reference
WS 1.75E+002	mg/L	20	EXP	YALKOWSKY,SH & DANNENFELSER,RM (1992)
logP 2.02			EXP	RODA,A ET AL. (1990)
VP 3.65E-015	mm Hg	25	EST	NEELY,WB & BLAU,GE (1985)
DC	pKa			
HL 5.16E-013	atm m3/mol	25	EST	MEYLAN,WM & HOWARD,PH (1991)
OH 5.37E-011	cm3/molc sec	25	EST	MEYLAN,WM & HOWARD,PH (1993)

000081-49-2 — 1-AMINO-2,4-DIBROMOANTHRAQUINONE

Formula: $C_{14}H_7Br_2NO_2$
Mol Weight: 381.04
MP (deg C): 221
FP (deg C):
BP (deg C):
BP pressure (mm Hg):

Property/ Value	Units	Temp	Data Type	Reference
WS 1.50E-002	mg/L	25	EST	MEYLAN,WM ET AL. (1996)
logP 5.31			EST	MEYLAN,WM & HOWARD,PH (1995)
VP 1.44E-009	mm Hg	25	EST	NEELY,WB & BLAU,GE (1985)
DC	pKa			
HL 1.78E-013	atm m3/mol	25	EST	MEYLAN,WM & HOWARD,PH (1991)
OH 1.23E-012	cm3/molc sec	25	EST	MEYLAN,WM & HOWARD,PH (1993)

000081-54-9 — PURPURIN

CAS #: 000081-54-9

Formula: $C_{14}H_8O_5$

Mol Weight: 256.22

MP (deg C): 257

FP (deg C):

BP (deg C):

BP pressure (mm Hg):

Property/Value	Units	Temp	Data Type	Reference
WS 6.40E+000	mg/L	25	EXP	BAUGHMAN,GL & PERENICH,TA (1988)
logP 3.46			EST	MEYLAN,WM & HOWARD,PH (1995)
VP 7.52E-011	mm Hg	25	EST	NEELY,WB & BLAU,GE (1985)
DC	pKa			
HL 5.68E-015	atm m3/mol	25	EST	MEYLAN,WM & HOWARD,PH (1991)
OH 1.52E-010	cm3/molc sec	25	EST	MEYLAN,WM & HOWARD,PH (1993)

000081-61-8 — QUINALIZARIN

CAS #: 000081-61-8

Formula: $C_{14}H_8O_6$

Mol Weight: 272.22

MP (deg C): >275

FP (deg C):

BP (deg C):

BP pressure (mm Hg):

Property/Value	Units	Temp	Data Type	Reference
WS 2.58E+000	mg/L	25	EXP	BAUGHMAN,GL & PERENICH,TA (1988)
logP 3.75			EST	MEYLAN,WM & HOWARD,PH (1995)
VP 4.64E-012	mm Hg	25	EST	NEELY,WB & BLAU,GE (1985)
DC	pKa			
HL 7.43E-016	atm m3/mol	25	EST	MEYLAN,WM & HOWARD,PH (1991)
OH 1.82E-011	cm3/molc sec	25	EST	MEYLAN,WM & HOWARD,PH (1993)

000081-64-1 — 1,4-DIHYDROXY-9,10-ANTHRACENEDIONE

CAS #: 000081-64-1

Formula: $C_{14}H_8O_4$

Mol Weight: 240.22

MP (deg C): 200-203

FP (deg C):

BP (deg C):

BP pressure (mm Hg):

Property/Value	Units	Temp	Data Type	Reference
WS 4.00E-007	mg/L	25	EXP	BAUGHMAN,GL & PERENICH,TA (1988)
logP 3.94			EST	MEYLAN,WM & HOWARD,PH (1995)
VP 2.43E-008	mm Hg	25	EXP	MALASPINA,L ET AL. (1973A)
DC	pKa			
HL 1.92E-002	atm m3/mol	25	EST	VP/WSOL
OH 9.84E-012	cm3/molc sec	25	EST	MEYLAN,WM & HOWARD,PH (1993)

000081-77-6 — C.I. VAT BLUE 4

CAS #: 000081-77-6

Formula: $C_{28}H_{14}N_2O_4$

Mol Weight: 442.43

MP (deg C): 470-500 de

FP (deg C):

BP (deg C):

BP pressure (mm Hg):

Property/Value	Units	Temp	Data Type	Reference
WS 3.24E-005	mg/L	25	EST	MEYLAN,WM ET AL. (1996)
logP 7.73			EST	MEYLAN,WM & HOWARD,PH (1995)
VP 2.00E-018	mm Hg	25	EXP	BAUGHMAN,GL & PERENICH,TA (1988)
DC	pKa			
HL 3.24E-022	atm m3/mol	25	EST	MEYLAN,WM & HOWARD,PH (1991)
OH 1.86E-010	cm3/molc sec	25	EST	MEYLAN,WM & HOWARD,PH (1993)

000081-81-2 — WARFARIN

CAS #: 000081-81-2

Formula: $C_{19}H_{16}O_4$

Mol Weight: 308.34

MP (deg C): 161

FP (deg C):

BP (deg C):

BP pressure (mm Hg):

Property/Value	Units	Temp	Data Type	Reference
WS 1.70E+001	mg/L	20	EXP	HARTLEY,D & KIDD,H (1983)
logP 2.60			EXP	SANGSTER,J (1994)
VP 1.16E-007	mm Hg	21	EXP	HARTLEY,D & KIDD,H (1983)
DC	pKa			
HL 2.77E-009	atm m3/mol	25	EST	VP/WSOL
OH 1.16E-010	cm3/molc sec	25	EST	MEYLAN,WM & HOWARD,PH (1993)

000081-82-3 — COUMACHLOR

CAS #: 000081-82-3

Formula: $C_{19}H_{15}ClO_4$

Mol Weight: 342.78

MP (deg C): 169

FP (deg C):

BP (deg C):

BP pressure (mm Hg):

Property/Value	Units	Temp	Data Type	Reference
WS 5.00E-001	mg/L	20	EXP	SHIU,WY ET AL. (1990)
logP 2.88			EST	MEYLAN,WM & HOWARD,PH (1995)
VP 7.17E-013	mm Hg	25	EST	NEELY,WB & BLAU,GE (1985)
DC	pKa			
HL 1.69E-013	atm m3/mol	25	EST	MEYLAN,WM & HOWARD,PH (1991)
OH 5.07E-011	cm3/molc sec	25	EST	MEYLAN,WM & HOWARD,PH (1993)

000081-88-9 — RHODAMINE B

CAS #: 000081-88-9

Formula: $C_{28}H_{32}ClN_2O_3$

Mol Weight: 480.03

MP (deg C): 165

FP (deg C):

BP (deg C):

BP pressure (mm Hg):

Property/Value	Units	Temp	Data Type	Reference
WS 1.20E+004	mg/L	25	EXP	BAUGHMAN,GL ET AL. (1993)
logP 1.95			EXP	CHEM INSPECT TEST INST (1992)
VP 1.89E-019	mm Hg	25	EST	NEELY,WB & BLAU,GE (1985)
DC	pKa			
HL	atm m3/mol			
OH 2.37E-010	cm3/molc sec	25	EST	MEYLAN,WM & HOWARD,PH (1993)

000081-92-5 — BENZENEMETHANOL, 2-[BIS(4-HYDROXYPHENYL)METHYL]-

CAS #: 000081-92-5

Formula: $C_{20}H_{18}O_3$

Mol Weight: 306.36

MP (deg C): 201-202

FP (deg C):

BP (deg C):

BP pressure (mm Hg):

Property/Value	Units	Temp	Data Type	Reference
WS 2.24E+002	mg/L	25	EST	MEYLAN,WM ET AL. (1996)
logP 3.27			EXP	SANGSTER,J (1993)
VP 7.03E-012	mm Hg	25	EST	NEELY,WB & BLAU,GE (1985)
DC	pKa			
HL 1.69E-017	atm m3/mol	25	EST	MEYLAN,WM & HOWARD,PH (1991)
OH 9.17E-011	cm3/molc sec	25	EST	MEYLAN,WM & HOWARD,PH (1993)

000082-05-3 — BENZANTHRONE

Formula:	$C_{17}H_{10}O$
Mol Weight:	230.27
MP (deg C):	170
FP (deg C):	
BP (deg C):	
BP pressure (mm Hg):	

Property/Value	Units	Temp	Data Type	Reference
WS 2.40E-001	mg/L	25	EST	MEYLAN,WM ET AL. (1996)
logP 4.81			EXP	CHEM INSPECT TEST INST (1992)
VP 2.21E-007	mm Hg	25	EST	NEELY,WB & BLAU,GE (1986)
DC	pKa			
HL 6.61E-008	atm m3/mol	25	EST	MEYALN,WM & HOWARD,PH (1991)
OH 1.80E-011	cm3/molc sec	25	EST	MEYLAN,WM & HOWARD,PH (1993)

000082-07-5 — XANTHENE-9-CARBOXYLIC ACID

Formula:	$C_{14}H_{10}O_3$
Mol Weight:	226.23
MP (deg C):	217 dec
FP (deg C):	
BP (deg C):	
BP pressure (mm Hg):	

Property/Value	Units	Temp	Data Type	Reference
WS 7.66E+001	mg/L	25	EST	MEYLAN,WM ET AL. (1996)
logP 2.23			EXP	HANSCH,C & LEO,AJ (1985)
VP 1.77E-006	mm Hg	25	EST	NEELY,WB & BLAU,GE (1985)
DC	pKa			
HL 3.54E-010	atm m3/mol	25	EST	MEYLAN,WM & HOWARD,PH (1991)
OH 1.38E-011	cm3/molc sec	25	EST	MEYLAN,WM & HOWARD,PH (1993)

000082-24-6 — 1-AMINOANTHRAQUINONE-2-CARBOXYLIC ACID

Formula:	$C_{15}H_9NO_4$
Mol Weight:	267.24
MP (deg C):	295-296
FP (deg C):	
BP (deg C):	
BP pressure (mm Hg):	

Property/Value	Units	Temp	Data Type	Reference
WS 8.14E-001	mg/L	25	EST	MEYLAN,WM ET AL. (1996)
logP 3.81			EST	MEYLAN,WM & HOWARD,PH (1995)
VP 6.23E-010	mm Hg	25	EST	NEELY,WB & BLAU,GE (1985)
DC	pKa			
HL 2.26E-017	atm m3/mol	25	EST	MEYLAN,WM & HOWARD,PH (1991)
OH 1.66E-011	cm3/molc sec	25	EST	MEYLAN,WM & HOWARD,PH (1993)

000082-28-0 — 1-AMINO-2-METHYL-9,10-ANTHRACENEDIONE

Formula:	$C_{15}H_{11}NO_2$
Mol Weight:	237.26
MP (deg C):	205
FP (deg C):	
BP (deg C):	
BP pressure (mm Hg):	

Property/Value	Units	Temp	Data Type	Reference
WS 3.32E-001	mg/L	25	EXP	BAUGHMAN,GL & PERENICH,TA (1988)
logP 4.07			EST	MEYLAN,WM & HOWARD,PH (1995)
VP 3.82E-008	mm Hg	25	EST	NEELY,WB & BLAU,GE (1985)
DC	pKa			
HL 1.24E-012	atm m3/mol	25	EST	MEYLAN,WM & HOWARD,PH (1991)
OH 5.58E-011	cm3/molc sec	25	EST	MEYLAN,WM & HOWARD,PH (1993)

000082-33-7 — C.I. DISPERSE VIOLET 8

Formula:	$C_{14}H_9N_3O_4$
Mol Weight:	283.25
MP (deg C):	
FP (deg C):	
BP (deg C):	
BP pressure (mm Hg):	

Property/Value	Units	Temp	Data Type	Reference
WS 1.39E+000	mg/L	25	EST	MEYLAN,WM ET AL. (1996)
logP 2.98			EST	MEYLAN,WM & HOWARD,PH (1995)
VP 9.00E-006	mm Hg	25	EXP	BAUGHMAN,GL & PERENICH,TA (1988)
DC	pKa			
HL 1.57E-018	atm m3/mol	25	EST	MEYLAN,WM & HOWARD,PH (1991)
OH 6.79E-011	cm3/molc sec	25	EST	MEYLAN,WM & HOWARD,PH (1993)

000082-34-8 — 1-NITROANTHRAQUINONE

Formula:	$C_{14}H_7NO_4$
Mol Weight:	253.22
MP (deg C):	231.5
FP (deg C):	
BP (deg C):	270
BP pressure (mm Hg):	7.00E+000

Property/Value	Units	Temp	Data Type	Reference
WS 1.43E+000	mg/L	25	EST	MEYLAN,WM ET AL. (1996)
logP 3.16			EST	MEYLAN,WM & HOWARD,PH (1995)
VP 4.00E-009	mm Hg	25	EXP	BAUGHMAN,GL & PERENICH,TA (1988)
DC	pKa			
HL 1.25E-011	atm m3/mol	25	EST	MEYLAN,WM & HOWARD,PH (1991)
OH 8.43E-013	cm3/molc sec	25	EST	MEYLAN,WM & HOWARD,PH (1993)

000082-38-2 — DISPERSE RED 9

Formula:	$C_{15}H_{11}NO_2$
Mol Weight:	237.26
MP (deg C):	170-172
FP (deg C):	
BP (deg C):	
BP pressure (mm Hg):	

Property/Value	Units	Temp	Data Type	Reference
WS 7.35E-002	mg/L	25	EXP	BAUGHMAN,GL & PERENICH,TA (1988)
logP 4.10			EXP	BAUGHMAN,GL & WEBER,EJ (1991)
VP 7.00E-009	mm Hg	25	EXP	BAUGHMAN,GL & PERENICH,TA (1988)
DC	pKa			
HL 2.97E-008	atm m3/mol	25	EST	VP/WSOL
OH 1.84E-011	cm3/molc sec	25	EST	MEYLAN,WM & HOWARD,PH (1993)

000082-39-3 — 1-METHOXY-9,10-ANTHRAQUINONE

Formula:	$C_{15}H_{10}O_3$
Mol Weight:	238.25
MP (deg C):	
FP (deg C):	
BP (deg C):	
BP pressure (mm Hg):	

Property/Value	Units	Temp	Data Type	Reference
WS 2.53E+000	mg/L	25	EST	MEYLAN,WM ET AL. (1996)
logP 3.43			EST	MEYLAN,WM & HOWARD,PH (1995)
VP 2.76E-007	mm Hg	25	EXP	SHIMIZU,T ET AL. (1987)
DC	pKa			
HL 1.88E-010	atm m3/mol	25	EST	MEYLAN,WM & HOWARD,PH (1991)
OH 9.84E-012	cm3/molc sec	25	EST	MEYLAN,WM & HOWARD,PH (1993)

CAS #: 000082-44-0 — 1-CHLOROANTHRAQUINONE

Formula: $C_{14}H_7ClO_2$
Mol Weight: 242.66
MP (deg C): 162
FP (deg C):
BP (deg C):
BP pressure (mm Hg):

Property/Value	Units	Temp	Data Type	Reference
WS 1.68E+000	mg/L	25	EST	MEYLAN,WM ET AL (1996)
logP 3.99			EST	MEYLAN,WM & HOWARD,PH (1995)
VP 8.74E-007	mm Hg	25	EST	NEELY,WB & BLAU,GE (1985)
DC	pKa			
HL 2.36E-009	atm m3/mol	25	EST	MEYLAN,WM & HOWARD,PH (1991)
OH 1.28E-012	cm3/molc sec	25	EST	MEYLAN,WM & HOWARD,PH (1993)

CAS #: 000082-45-1 — 1-AMINOANTHRAQUINONE

Formula: $C_{14}H_9NO_2$
Mol Weight: 223.23
MP (deg C): 250
FP (deg C):
BP (deg C):
BP pressure (mm Hg):

Property/Value	Units	Temp	Data Type	Reference
WS 3.00E-001	mg/L	25	EXP	BAUGHMAN,GL & PERENICH,TA (1988)
logP 3.74			EXP	CHEM INSPECT TEST INST (1992)
VP 6.00E-010	mm Hg	25	EXP	BAUGHMAN,GL & PERENICH,TA (1988)
DC	pKa			
HL 5.87E-010	atm m3/mol	25	EST	VP/WSOL
OH 4.21E-011	cm3/molc sec	25	EST	MEYLAN,WM & HOWARD,PH (1993)

CAS #: 000082-48-4 — 1,8-ANTHRAQUINONEDISULFONIC ACID

Formula: $C_{14}H_8O_8S_2$
Mol Weight: 368.34
MP (deg C):
FP (deg C):
BP (deg C):
BP pressure (mm Hg):

Property/Value	Units	Temp	Data Type	Reference
WS 6.66E+006	mg/L	25	EXP	SEIDELL,A (1941)
logP -2.97			EST	MEYLAN,WM & HOWARD,PH (1995)
VP 5.70E-018	mm Hg	25	EST	NEELY,WB & BLAU,GE (1985)
DC	pKa			
HL 6.94E-022	atm m3/mol	25	EST	MEYLAN,WM & HOWARD,PH (1991)
OH 6.00E-013	cm3/molc sec	25	EST	MEYLAN,WM & HOWARD,PH (1993)

CAS #: 000082-62-2 — 3,4,6-TRICHLORO-2-NITROPHENOL

Formula: $C_6H_2Cl_3NO_3$
Mol Weight: 242.45
MP (deg C): 92-93
FP (deg C):
BP (deg C):
BP pressure (mm Hg):

Property/Value	Units	Temp	Data Type	Reference
WS 1.36E+001	mg/L	25	EST	MEYLAN,WM ET AL. (1996)
logP 3.84			EST	MEYLAN,WM & HOWARD,PH (1995)
VP 1.04E-005	mm Hg	25	EST	NEELY,WB & BLAU,GE (1985)
DC	pKa			
HL 2.85E-006	atm m3/mol	25	EST	MEYLAN,WM & HOWARD,PH (1991)
OH 1.57E-013	cm3/molc sec	25	EST	MEYLAN,WM & HOWARD,PH (1993)

CAS #: 000082-66-6 — DIPHACINONE

Formula: $C_{23}H_{16}O_3$
Mol Weight: 340.38
MP (deg C): 145
FP (deg C):
BP (deg C):
BP pressure (mm Hg):

Property/Value	Units	Temp	Data Type	Reference
WS 3.00E-001	mg/L		EXP	SHIU,WY ET AL. (1990)
logP 2.95			EST	MEYLAN,WM & HOWARD,PH (1995)
VP 3.71E-010	mm Hg	25	EST	NEELY,WB & BLAU,GE (1985)
DC	pKa			
HL 2.60E-014	atm m3/mol	25	EST	MEYLAN,WM & HOWARD,PH (1991)
OH 1.21E-011	cm3/molc sec	25	EST	MEYLAN,WM & HOWARD,PH (1993)

CAS #: 000082-68-8 — PENTACHLORONITROBENZENE

Formula: $C_6Cl_5NO_2$
Mol Weight: 295.34
MP (deg C): 144
FP (deg C):
BP (deg C): 328
BP pressure (mm Hg):

Property/Value	Units	Temp	Data Type	Reference
WS 5.50E-001	mg/L	25	EXP	KANAZAWA,J (1980)
logP 4.64			EXP	SANGSTER,J (1994)
VP 5.00E-005	mm Hg	20	EXP	GILE,JD & GILLETT,JW (1979)
DC	pKa			
HL 8.00E-005	atm m3/mol	25	EST	VP/WSOL
OH 7.30E-015	cm3/molc sec	25	EST	MEYLAN,WM & HOWARD,PH (1993)

CAS #: 000082-71-3 — 2,4,6-TRINITRORESORCINOL

Formula: $C_6H_3N_3O_8$
Mol Weight: 245.11
MP (deg C): 175.5
FP (deg C):
BP (deg C):
BP pressure (mm Hg):

Property/Value	Units	Temp	Data Type	Reference
WS 5.34E+003	mg/L	25	EXP	YALKOWSKY,SH & DANNENFELSER,RM (1992)
logP 1.06			EST	MEYLAN,WM & HOWARD,PH (1995)
VP 2.57E-009	mm Hg	25	EST	NEELY,WB & BLAU,GE (1985)
DC	pKa			
HL 3.58E-011	atm m3/mol	25	EST	MEYLAN,WM & HOWARD,PH (1991)
OH 2.81E-013	cm3/molc sec	25	EST	MEYLAN,WM & HOWARD,PH (1993)

CAS #: 000082-86-0 — 1,2-ACENAPHTHYLENEDIONE

Formula: $C_{12}H_8O_2$
Mol Weight: 184.20
MP (deg C): 261
FP (deg C):
BP (deg C):
BP pressure (mm Hg):

Property/Value	Units	Temp	Data Type	Reference
WS 9.01E+001	mg/L	25	EST	MEYLAN,WM ET AL. (1996)
logP 1.95			EXP	HANSCH,C ET AL. (1995)
VP 1.57E-005	mm Hg	25	EST	NEELY,WB & BLAU,GE (1985)
DC	pKa			
HL 3.43E-009	atm m3/mol	25	EST	MEYLAN,WM & HOWARD,PH (1991)
OH 8.30E-012	cm3/molc sec	25	EST	MEYLAN,WM & HOWARD,PH (1993)

000083-08-9 — QUINOPHALONE

Formula: $C_{18}H_{11}NO_2$
Mol Weight: 273.29
MP (deg C):
FP (deg C):
BP (deg C):
BP pressure (mm Hg):

Property/Value	Units	Temp	Data Type	Reference
WS 5.52E+000	mg/L	25	EST	MEYLAN,WM ET AL. (1996)
logP 4.13			EST	MEYLAN,WM & HOWARD,PH (1995)
VP 9.88E-009	mm Hg	25	EST	NEELY,WB & BLAU,GE (1985)
DC	pKa			
HL 6.12E-014	atm m3/mol	25	EST	MEYLAN,WM & HOWARD,PH (1991)
OH 3.58E-011	cm3/molc sec	25	EST	MEYLAN,WM & HOWARD,PH (1993)

000083-12-5 — 2-PHENYL-1H-INDENE-1,3(2H)-DIONE

Formula: $C_{15}H_{10}O_2$
Mol Weight: 222.25
MP (deg C): 149-151
FP (deg C):
BP (deg C):
BP pressure (mm Hg):

Property/Value	Units	Temp	Data Type	Reference
WS 1.12E+002	mg/L	25	EST	MEYLAN,WM ET AL. (1996)
logP 2.90			EXP	HANSCH,C & LEO,AJ (1985)
VP 3.39E-006	mm Hg	25	EST	NEELY,WB & BLAU,GE (1985)
DC	pKa			
HL 4.79E-010	atm m3/mol	25	EST	MEYLAN,WM & HOWARD,PH (1991)
OH 6.25E-012	cm3/molc sec	25	EST	MEYLAN,WM & HOWARD,PH (1993)

000083-25-0 — N-PHENYLSUCCINIMIDE

Formula: $C_{10}H_9NO_2$
Mol Weight: 175.19
MP (deg C): 154.5
FP (deg C):
BP (deg C): 400
BP pressure (mm Hg):

Property/Value	Units	Temp	Data Type	Reference
WS 5.16E+004	mg/L	25	EST	MEYLAN,WM ET AL. (1996)
logP 0.06			EXP	HANSCH,C & LEO,AJ (1985)
VP 1.00E-006	mm Hg	25	EST	NEELY,WB & BLAU,GE (1985)
DC	pKa			
HL 1.30E-007	atm m3/mol	25	EST	MEYLAN,WM & HOWARD,PH (1991)
OH 1.68E-011	cm3/molc sec	25	EST	MEYLAN,WM & HOWARD,PH (1993)

000083-26-1 — PINDONE

Formula: $C_{14}H_{14}O_3$
Mol Weight: 230.27
MP (deg C): 108-110
FP (deg C):
BP (deg C):
BP pressure (mm Hg):

Property/Value	Units	Temp	Data Type	Reference
WS 1.80E+001	mg/L	25	EXP	YALKOWSKY,SH & DANNENFELSER,RM (1992)
logP 0.97			EST	MEYLAN,WM & HOWARD,PH (1995)
VP 8.36E-006	mm Hg	25	EST	NEELY,WB & BLAU,GE (1985)
DC	pKa			
HL 9.34E-012	atm m3/mol	25	EST	MEYLAN,WM & HOWARD,PH (1991)
OH 3.00E-012	cm3/molc sec	25	EST	MEYLAN,WM & HOWARD,PH (1993)

000083-28-3 — VALONE

Formula: $C_{14}H_{14}O_3$
Mol Weight: 230.27
MP (deg C): 68-69
FP (deg C):
BP (deg C):
BP pressure (mm Hg):

Property/Value	Units	Temp	Data Type	Reference
WS 9.93E+001	mg/L	25	EST	MEYLAN,WM ET AL. (1996)
logP 1.01			EST	MEYLAN,WM & HOWARD,PH (1995)
VP 6.88E-006	mm Hg	25	EST	NEELY,WB & BLAU,GE (1985)
DC	pKa			
HL 9.34E-012	atm m3/mol	25	EST	MEYLAN,WM & HOWARD,PH (1991)
OH 1.02E-011	cm3/molc sec	25	EST	MEYLAN,WM & HOWARD,PH (1993)

000083-32-9 — ACENAPHTHENE

Formula: $C_{12}H_{10}$
Mol Weight: 154.21
MP (deg C): 95
FP (deg C):
BP (deg C): 279
BP pressure (mm Hg):

Property/Value	Units	Temp	Data Type	Reference
WS 3.90E+000	mg/L	25	EXP	MILLER,MM ET AL. (1985)
logP 3.92			EXP	HANSCH,C & LEO,AJ (1985)
VP 2.50E-003	mm Hg	25	EXP	BOYD,RH ET AL. (1965)
DC	pKa			
HL 1.55E-004	atm m3/mol	25	EXP	MACKAY,D ET AL. (1982A)
OH 1.03E-010	cm3/molc sec	25	EXP	ATKINSON,R (1989)

000083-34-1 — 3-METHYLINDOLE

Formula: C_9H_9N
Mol Weight: 131.18
MP (deg C): 95
FP (deg C):
BP (deg C): 265-266
BP pressure (mm Hg):

Property/Value	Units	Temp	Data Type	Reference
WS 4.98E+002	mg/L	20	EXP	PEARLMAN,RS ET AL. (1984)
logP 2.60			EXP	HANSCH,C & LEO,AJ (1985)
VP 5.55E-003	mm Hg	25	EXT	PERRY,RH & GREEN,D (1984)
DC	pKa			
HL 9.78E-007	atm m3/mol	25	EST	MEYLAN,WM & HOWARD,PH (1991)
OH 2.00E-010	cm3/molc sec	25	EST	MEYLAN,WM & HOWARD,PH (1993)

000083-40-9 — 3-METHYLSALICYLIC ACID

Formula: $C_8H_8O_3$
Mol Weight: 152.15
MP (deg C): 165-166
FP (deg C):
BP (deg C):
BP pressure (mm Hg):

Property/Value	Units	Temp	Data Type	Reference
WS 1.16E+004	mg/L	100	EXP	YALKOWSKY,SH & DANNENFELSER,RM (1992)
logP 2.86			EXP	HANSCH,C & LEO,AJ (1985)
VP 5.14E-005	mm Hg	25	EST	NEELY,WB & BLAU,GE (1985)
DC 2.95	pKa	25	EXP	SERJEANT,EP & DEMPSEY,B (1979)
HL 1.57E-008	atm m3/mol	25	EST	MEYLAN,WM & HOWARD,PH (1991)
OH 1.59E-011	cm3/molc sec	25	EST	MEYLAN,WM & HOWARD,PH (1993)

CAS #: 000083-41-0	1,2-DIMETHYL-3-NITROBENZENE

Formula: $C_8H_9NO_2$

Mol Weight: 151.17

MP (deg C): 15
FP (deg C):

BP (deg C): 240

BP pressure (mm Hg):

Property/Value	Units	Temp	Data Type	Reference
WS 1.17E+002	mg/L	25	EST	MEYLAN,WM ET AL. (1996)
logP 2.83			EXP	DENEER,JW ET AL. (1987)
VP 3.59E-001	mm Hg	25	EXT	BOUBLIK,T ET AL. (1984)
DC	pKa			
HL 2.59E-005	atm m3/mol	25	EST	MEYLAN,WM & HOWARD,PH (1991)
OH 1.05E-012	cm3/molc sec	25	EST	MEYLAN,WM & HOWARD,PH (1993)

CAS #: 000083-42-1	BENZENE, 1-CHLORO-2-METHYL-3-NITRO-

Formula: $C_7H_6ClNO_2$

Mol Weight: 171.58

MP (deg C): 37.8
FP (deg C):

BP (deg C): 238

BP pressure (mm Hg):

Property/Value	Units	Temp	Data Type	Reference
WS 5.66E+001	mg/L	25	EST	MEYLAN,WM ET AL. (1996)
logP 3.09			EXP	DENEER,JW ET AL. (1987)
VP 1.06E-002	mm Hg	25	EST	NEELY,WB & BLAU,GE (1985)
DC	pKa			
HL 1.74E-005	atm m3/mol	25	EST	MEYLAN,WM & HOWARD,PH (1991)
OH 3.22E-013	cm3/molc sec	25	EST	MEYLAN,WM & HOWARD,PH (1993)

CAS #: 000083-44-3	DEOXYCHOLIC ACID

Formula: $C_{24}H_{40}O_4$

Mol Weight: 392.58

MP (deg C): 176-178
FP (deg C):

BP (deg C):

BP pressure (mm Hg):

Property/Value	Units	Temp	Data Type	Reference
WS 4.36E+001	mg/L	20	EXP	YALKOWSKY,SH & DANNENFELSER,RM (1992)
logP 3.50			EXP	RODA,A ET AL. (1990)
VP 3.00E-013	mm Hg	25	EST	NEELY,WB & BLAU,GE (1985)
DC	pKa			
HL 1.41E-011	atm m3/mol	25	EST	MEYLAN,WM & HOWARD,PH (1991)
OH 4.64E-011	cm3/molc sec	25	EST	MEYLAN,WM & HOWARD,PH (1993)

CAS #: 000083-49-8	HYODEOXYCHOLIC ACID

Formula: $C_{24}H_{40}O_4$

Mol Weight: 392.58

MP (deg C): 196-197
FP (deg C):

BP (deg C):

BP pressure (mm Hg):

Property/Value	Units	Temp	Data Type	Reference
WS 2.41E+000	mg/L	25	EST	MEYLAN,WM ET AL. (1996)
logP 3.08			EXP	RODA,A ET AL. (1990)
VP 3.00E-013	mm Hg	25	EST	NEELY,WB & BLAU,GE (1985)
DC	pKa			
HL 1.41E-011	atm m3/mol	25	EST	MEYLAN,WM & HOWARD,PH (1991)
OH 4.64E-011	cm3/molc sec	25	EST	MEYLAN,WM & HOWARD,PH (1993)

CAS #: 000083-56-7	1,5-DIHYDROXYNAPHTHALENE

Formula: $C_{10}H_8O_2$

Mol Weight: 160.17

MP (deg C): 262 dec
FP (deg C):

BP (deg C):

BP pressure (mm Hg):

Property/Value	Units	Temp	Data Type	Reference
WS 1.65E+002	mg/L	20	EXP	KORENMAN,YI ET AL. (1980)
logP 1.94			EXP	HANSCH,C & LEO,AJ (1985)
VP 7.15E-006	mm Hg	25	EST	NEELY,WB & BLAU,GE (1985)
DC	pKa			
HL 5.69E-012	atm m3/mol	25	EST	MEYLAN,WM & HOWARD,PH (1991)
OH 2.00E-010	cm3/molc sec	25	EST	MEYLAN,WM & HOWARD,PH (1993)

CAS #: 000083-59-0	PROPYL ISOME

Formula: $C_{20}H_{26}O_6$

Mol Weight: 362.43

MP (deg C):
FP (deg C):

BP (deg C):

BP pressure (mm Hg):

Property/Value	Units	Temp	Data Type	Reference
WS 1.45E-001	mg/L	25	EST	MEYLAN,WM ET AL. (1996)
logP 5.34			EST	MEYLAN,WM & HOWARD,PH (1995)
VP 2.68E-007	mm Hg	25	EST	NEELY,WB & BLAU,GE (1985)
DC	pKa			
HL 1.38E-010	atm m3/mol	25	EST	MEYLAN,WM & HOWARD,PH (1991)
OH 7.12E-011	cm3/molc sec	25	EST	MEYLAN,WM & HOWARD,PH (1993)

CAS #: 000083-67-0	THEOBROMINE

Formula: $C_7H_8N_4O_2$

Mol Weight: 180.17

MP (deg C): 357
FP (deg C):

BP (deg C):

BP pressure (mm Hg):

Property/Value	Units	Temp	Data Type	Reference
WS 3.30E+002	mg/L	25	EXP	YALKOWSKY,SH & DANNENFELSER,RM (1992)
logP -0.78			EXP	HANSCH,C & LEO,AJ (1985)
VP 8.95E-010	mm Hg	25	EST	NEELY,WB & BLAU,GE (1985)
DC 9.90	pKa		EXP	KORTUM,G ET AL (1961)
HL 1.63E-011	atm m3/mol	25	EST	MEYLAN,WM & HOWARD,PH (1991)
OH 1.89E-011	cm3/molc sec	25	EST	MEYLAN,WM & HOWARD,PH (1993)

CAS #: 000083-72-7	1,4-NAPHOQUINONE, 2-HYDROXY

Formula: $C_{10}H_6O_3$

Mol Weight: 174.16

MP (deg C): 195-196 de
FP (deg C):

BP (deg C):

BP pressure (mm Hg):

Property/Value	Units	Temp	Data Type	Reference
WS 3.90E+003	mg/L	25	EST	MEYLAN,WM ET AL. (1996)
logP 1.38			EXP	HANSCH,C & LEO,AJ (1985)
VP 3.63E-007	mm Hg	25	EST	NEELY,WB & BLAU,GE (1985)
DC	pKa			
HL 2.44E-009	atm m3/mol	25	EST	MEYLAN,WM & HOWARD,PH (1991)
OH 1.15E-011	cm3/molc sec	25	EST	MEYLAN,WM & HOWARD,PH (1993)

CAS #: 000083-79-4 — ROTENONE

Formula: $C_{23}H_{22}O_6$

Mol Weight: 394.43

MP (deg C): 165-166

FP (deg C):

BP (deg C):

BP pressure (mm Hg):

Property/Value	Units	Temp	Data Type	Reference
WS 1.50E+001	mg/L	100	EXP	YALKOWSKY,SH & DANNENFELSER,RM (1992)
logP 4.10			EXP	HANSCH,C & LEO,AJ (1985)
VP 2.55E-010	mm Hg	25	EST	NEELY,WB & BLAU,GE (1985)
DC	pKa			
HL 1.12E-013	atm m3/mol	25	EST	MEYLAN,WM & HOWARD,PH (1991)
OH 3.18E-010	cm3/molc sec	25	EST	MEYLAN,WM & HOWARD,PH (1993)

CAS #: 000083-88-5 — RIBOFLAVIN

Formula: $C_{17}H_{20}N_4O_6$

Mol Weight: 376.37

MP (deg C): 280 dec

FP (deg C):

BP (deg C):

BP pressure (mm Hg):

Property/Value	Units	Temp	Data Type	Reference
WS 8.47E+001	mg/L	25	EXP	YALKOWSKY,SH & DANNENFELSER,RM (1992)
logP -1.46			EXP	HANSCH,C & LEO,AJ (1985)
VP 8.92E-022	mm Hg	25	EST	NEELY,WB & BLAU,GE (1985)
DC	pKa			
HL 3.59E-019	atm m3/mol	25	EST	MEYLAN,WM & HOWARD,PH (1991)
OH 2.47E-010	cm3/molc sec	25	EST	MEYLAN,WM & HOWARD,PH (1993)

CAS #: 000083-98-7 — ORPHENADRINE

Formula: $C_{18}H_{23}NO$

Mol Weight: 269.39

MP (deg C):

FP (deg C):

BP (deg C): 195

BP pressure (mm Hg): 1.20E+001

Property/Value	Units	Temp	Data Type	Reference
WS 1.13E+002	mg/L	25	EST	MEYLAN,WM ET AL. (1996)
logP 3.77			EXP	SANGSTER,J (1993)
VP 1.36E-005	mm Hg	25	EST	NEELY,WB & BLAU,GE (1985)
DC	pKa			
HL 4.08E-009	atm m3/mol	25	EST	MEYLAN,WM & HOWARD,PH (1991)
OH 1.29E-010	cm3/molc sec	25	EST	MEYLAN,WM & HOWARD,PH (1993)

CAS #: 000084-01-5 — CHLORPROETHAZINE

Formula: $C_{19}H_{23}ClN_2S$

Mol Weight: 346.93

MP (deg C):

FP (deg C):

BP (deg C):

BP pressure (mm Hg):

Property/Value	Units	Temp	Data Type	Reference
WS 1.56E-002	mg/L	25	EST	MEYLAN,WM ET AL. (1996)
logP 6.46			EXP	SANGSTER,J (1994)
VP 3.34E-008	mm Hg	25	EST	NEELY,WB & BLAU,GE (1985)
DC	pKa			
HL 6.50E-010	atm m3/mol	25	EST	MEYLAN,WM & HOWARD,PH (1991)
OH 2.60E-010	cm3/molc sec	25	EST	MEYLAN,WM & HOWARD,PH (1993)

CAS #: 000084-04-8 — PIPAMAZINE

Formula: $C_{21}H_{24}ClN_3OS$

Mol Weight: 401.96

MP (deg C): 139

FP (deg C):

BP (deg C):

BP pressure (mm Hg):

Property/Value	Units	Temp	Data Type	Reference
WS 3.73E-002	mg/L	25	EST	MEYLAN,WM ET AL. (1996)
logP 4.44			EXP	HANSCH,C & LEO,AJ (1985)
VP 3.20E-012	mm Hg	25	EST	NEELY,WB & BLAU,GE (1985)
DC	pKa			
HL 5.78E-017	atm m3/mol	25	EST	MEYLAN,WM & HOWARD,PH (1991)
OH 2.77E-010	cm3/molc sec	25	EST	MEYLAN,WM & HOWARD,PH (1993)

CAS #: 000084-11-7 — 9,10-PHENANTHRENEDIONE

Formula: $C_{14}H_8O_2$

Mol Weight: 208.22

MP (deg C): 206-207

FP (deg C):

BP (deg C): 360

BP pressure (mm Hg):

Property/Value	Units	Temp	Data Type	Reference
WS 2.17E+001	mg/L	25	EST	MEYLAN,WM ET AL. (1996)
logP 2.52			EXP	HANSCH,C ET AL. (1995)
VP 2.33E-006	mm Hg	25	EST	NEELY,WB & BLAU,GE (1985)
DC	pKa			
HL 2.70E-009	atm m3/mol	25	EST	MEYLAN,WM & HOWARD,PH (1991)
OH 6.18E-012	cm3/molc sec	25	EST	MEYLAN,WM & HOWARD,PH (1993)

CAS #: 000084-15-1 — O-TERPHENYL

Formula: $C_{18}H_{14}$

Mol Weight: 230.31

MP (deg C): 56.2

FP (deg C):

BP (deg C): 332

BP pressure (mm Hg):

Property/Value	Units	Temp	Data Type	Reference
WS 1.24E+000	mg/L	25	EXP	YALKOWSKY,SH & DANNENFELSER,RM (1992)
logP 5.52			EST	MEYLAN,WM & HOWARD,PH (1995)
VP 2.50E-004	mm Hg	25	EXT	DAUBERT,TE & DANNER,RP (1989)
DC	pKa			
HL 6.11E-005	atm m3/mol	25	EST	VP/WSOL
OH 9.19E-012	cm3/molc sec	25	EST	MEYLAN,WM & HOWARD,PH (1993)

CAS #: 000084-51-5 — 2-ETHYLANTHRAQUINONE

Formula: $C_{16}H_{12}O_2$

Mol Weight: 236.27

MP (deg C): 108-111

FP (deg C):

BP (deg C):

BP pressure (mm Hg):

Property/Value	Units	Temp	Data Type	Reference
WS 4.05E-001	mg/L	25	EST	MEYLAN,WM ET AL. (1996)
logP 4.37			EXP	DEBRUIJN,J ET AL. (1989)
VP 1.02E-006	mm Hg	25	EST	NEELY,WB & BLAU,GE (1985)
DC	pKa			
HL 4.66E-009	atm m3/mol	25	EST	MEYLAN,WM & HOWARD,PH (1991)
OH 3.71E-012	cm3/molc sec	25	EST	MEYLAN,WM & HOWARD,PH (1993)

	CAS #:	000084-60-6				2,6-DIHYDROXY-ANTHRAQUINONE

Formula: $C_{14}H_8O_4$

Mol Weight: 240.22

MP (deg C): >320 **FP (deg C):**

BP (deg C):

BP pressure (mm Hg):

Property/ Value	Units	Temp	Data Type	Reference
WS 3.10E-001	mg/L	25	EXP	STEPHEN,H & STEPHEN,T (1963)
logP 2.38			EST	MEYLAN,WM & HOWARD,PH (1995)
VP 1.17E-009	mm Hg	25	EST	NEELY,WB & BLAU,GE (1985)
DC	pKa			
HL 3.44E-017	atm m3/mol	25	EST	MEYLAN,WM & HOWARD,PH (1991)
OH 2.59E-011	cm3/molc sec	25	EST	MEYLAN,WM & HOWARD,PH (1993)

	CAS #:	000084-61-7				DICYCLOHEXYL PHTHALATE

Formula: $C_{20}H_{26}O_4$

Mol Weight: 330.43

MP (deg C): 66 **FP (deg C):**

BP (deg C):

BP pressure (mm Hg):

Property/ Value	Units	Temp	Data Type	Reference
WS 4.00E+000	mg/L	24	EXP	YALKOWSKY,SH & DANNENFELSER,RM (1992)
logP 6.20			EST	MEYLAN,WM & HOWARD,PH (1995)
VP 7.00E-004	mm Hg	25	EXP	GIAM,CS ET AL. (1984A)
DC	pKa			
HL 7.61E-005	atm m3/mol	25	EST	VP/WSOL
OH 2.43E-011	cm3/molc sec	25	EST	MEYLAN,WM & HOWARD,PH (1993)

	CAS #:	000084-62-8				DIPHENYL PHTHALATE

Formula: $C_{20}H_{14}O_4$

Mol Weight: 318.33

MP (deg C): 70-73 **FP (deg C):**

BP (deg C): 255

BP pressure (mm Hg): 1.40E+001

Property/ Value	Units	Temp	Data Type	Reference
WS 8.20E-002	mg/L	25	EXP	HOLLIFIELD,HC (1979)
logP 4.10			EST	MEYLAN,WM & HOWARD,PH (1995)
VP 7.53E-007	mm Hg	25	EST	NEELY,WB & BLAU,GE (1985)
DC	pKa			
HL 3.06E-008	atm m3/mol	25	EST	MEYLAN,WM & HOWARD,PH (1991)
OH 4.77E-012	cm3/molc sec	25	EST	MEYLAN,WM & HOWARD,PH (1993)

	CAS #:	000084-64-0				BUTYLCYCLOHEXYL PHTHALATE

Formula: $C_{18}H_{24}O_4$

Mol Weight: 304.39

MP (deg C): **FP (deg C):**

BP (deg C):

BP pressure (mm Hg):

Property/ Value	Units	Temp	Data Type	Reference
WS 2.80E-001	mg/L	25	EST	MEYLAN,WM ET AL. (1996)
logP 5.41			EST	MEYLAN,WM & HOWARD,PH (1995)
VP 4.77E-003	mm Hg	25	EXT	WERNER,AC (1952)
DC	pKa			
HL 9.51E-007	atm m3/mol	25	EST	MEYLAN,WM & HOWARD,PH (1991)
OH 1.67E-011	cm3/molc sec	25	EST	MEYLAN,WM & HOWARD,PH (1993)

	CAS #:	000084-65-1				ANTHRAQUINONE

Formula: $C_{14}H_8O_2$

Mol Weight: 208.22

MP (deg C): 283.5-85 **FP (deg C):**

BP (deg C): 379-81

BP pressure (mm Hg):

Property/ Value	Units	Temp	Data Type	Reference
WS 1.35E+000	mg/L	25	EXP	YALKOWSKY,SH & DANNENFELSER,RM (1992)
logP 3.39			EXP	HANSCH,C & LEO,AJ (1985)
VP 1.16E-007	mm Hg	25	EXP	SHIMIZU,T ET AL. (1987)
DC 7.40	pKa		EXP	CHUNG,RH (1978)
HL 2.35E-008	atm m3/mol	25	EST	VP/WSOL
OH 1.50E-012	cm3/molc sec	25	EST	MEYLAN,WM & HOWARD,PH (1993)

	CAS #:	000084-66-2				DIETHYL PHTHALATE

Formula: $C_{12}H_{14}O_4$

Mol Weight: 222.24

MP (deg C): -40.5 **FP (deg C):**

BP (deg C): 295

BP pressure (mm Hg):

Property/ Value	Units	Temp	Data Type	Reference
WS 1.08E+003	mg/L	25	EXP	HOWARD,PH ET AL. (1985)
logP 2.47			EXP	HANSCH,C & LEO,AJ (1985)
VP 1.65E-003	mm Hg	25	EXP	HOWARD,PH ET AL. (1985)
DC	pKa			
HL 4.50E-007	atm m3/mol	25	EST	VP/WSOL
OH 3.03E-012	cm3/molc sec	25	EST	MEYLAN,WM & HOWARD,PH (1993)

	CAS #:	000084-69-5				DI-ISOBUTYLPHTHALATE

Formula: $C_{16}H_{22}O_4$

Mol Weight: 278.35

MP (deg C): **FP (deg C):**

BP (deg C): 296.5

BP pressure (mm Hg):

Property/ Value	Units	Temp	Data Type	Reference
WS 2.03E+001	mg/L	20	EXP	YALKOWSKY,SH & DANNENFELSER,RM (1992)
logP 4.11			EXP	HANSCH,C & LEO,AJ (1985)
VP 8.82E-005	mm Hg	25	EXT	PERRY,RH & GREEN,D (1984)
DC	pKa			
HL 1.22E-006	atm m3/mol	25	EST	MEYLAN,WM & HOWARD,PH (1991)
OH 9.26E-012	cm3/molc sec	25	EST	MEYLAN,WM & HOWARD,PH (1993)

	CAS #:	000084-72-0				ETHYL CARBETHOXYMETHYL PHTHALATE

Formula: $C_{14}H_{16}O_6$

Mol Weight: 280.28

MP (deg C): **FP (deg C):**

BP (deg C):

BP pressure (mm Hg):

Property/ Value	Units	Temp	Data Type	Reference
WS 2.17E+002	mg/L	25	EST	MEYLAN,WM ET AL. (1996)
logP 2.19			EST	MEYLAN,WM & HOWARD,PH (1995)
VP 2.16E-004	mm Hg	25	EST	NEELY,WB & BLAU,GE (1985)
DC	pKa			
HL 6.64E-009	atm m3/mol	25	EST	MEYLAN,WM & HOWARD,PH (1991)
OH 4.57E-012	cm3/molc sec	25	EST	MEYLAN,WM & HOWARD,PH (1993)

000084-74-2 — DIBUTYL PHTHALATE

CAS #:	000084-74-2
Formula:	$C_{16}H_{22}O_4$
Mol Weight:	278.35
MP (deg C):	-35
FP (deg C):	
BP (deg C):	340
BP pressure (mm Hg):	

Property/Value	Units	Temp	Data Type	Reference
WS 1.12E+001	mg/L	25	EXP	HOWARD,PH ET AL. (1985)
logP 4.72			EXP	HANSCH,C & LEO,AJ (1985)
VP 7.30E-005	mm Hg	25	EXP	HOWARD,PH ET AL. (1985)
DC	pKa			
HL 1.81E-006	atm m3/mol	23	EXP	ATLAS,E ET AL. (1983)
OH 8.71E-012	cm3/molc sec	25	EST	MEYLAN,WM & HOWARD,PH (1993)

000084-75-3 — DIHEXYL PHTHALATE

CAS #:	000084-75-3
Formula:	$C_{20}H_{30}O_4$
Mol Weight:	334.46
MP (deg C):	
FP (deg C):	
BP (deg C):	
BP pressure (mm Hg):	

Property/Value	Units	Temp	Data Type	Reference
WS 2.40E-001	mg/L	20	EXP	HOWARD,PH ET AL. (1985)
logP 6.57			EST	MEYLAN,WM & HOWARD,PH (1995)
VP 3.50E-006	mm Hg	25	EXP	GIAM,CS ET AL. (1984A)
DC	pKa			
HL 6.42E-006	atm m3/mol	25	EST	VP/WSOL
OH 1.49E-011	cm3/molc sec	25	EST	MEYLAN,WM & HOWARD,PH (1993)

000084-77-5 — DIDECYL PHTHALATE

CAS #:	000084-77-5
Formula:	$C_{28}H_{46}O_4$
Mol Weight:	446.68
MP (deg C):	
FP (deg C):	
BP (deg C):	
BP pressure (mm Hg):	

Property/Value	Units	Temp	Data Type	Reference
WS 3.30E-001	mg/L	24	EXP	YALKOWSKY,SH & DANNENFELSER,RM (1992)
logP 10.50			EST	MEYLAN,WM & HOWARD,PH (1995)
VP 7.89E-009	mm Hg	25	EST	NEELY,WB & BLAU,GE (1985)
DC	pKa			
HL 3.67E-005	atm m3/mol	25	EST	MEYLAN,WM & HOWARD,PH (1991)
OH 2.62E-011	cm3/molc sec	25	EST	MEYLAN,WM & HOWARD,PH (1993)

000084-86-6 — 4-AMINO-1-NAPHTHALENESULFONIC ACID

CAS #:	000084-86-6
Formula:	$C_{10}H_9NO_3S$
Mol Weight:	223.25
MP (deg C):	
FP (deg C):	
BP (deg C):	
BP pressure (mm Hg):	

Property/Value	Units	Temp	Data Type	Reference
WS 3.10E+002	mg/L	20	EXP	YALKOWSKY,SH & DANNENFELSER,RM (1992)
logP -0.91			EST	MEYLAN,WM & HOWARD,PH (1995)
VP 5.26E-010	mm Hg	25	EST	NEELY,WB & BLAU,GE (1985)
DC	pKa			
HL 8.68E-014	atm m3/mol	25	EST	MEYLAN,WM & HOWARD,PH (1991)
OH 4.41E-011	cm3/molc sec	25	EST	MEYLAN,WM & HOWARD,PH (1993)

000084-87-7 — 1-NAPHTHOL-4-SULFONIC ACID

CAS #:	000084-87-7
Formula:	$C_{10}H_8O_4S$
Mol Weight:	224.24
MP (deg C):	170 dec
FP (deg C):	
BP (deg C):	
BP pressure (mm Hg):	

Property/Value	Units	Temp	Data Type	Reference
WS 2.95E+005	mg/L	25	EST	MEYLAN,WM ET AL. (1996)
logP -0.44			EXP	HANSCH,C & LEO,AJ (1985)
VP 9.03E-010	mm Hg	25	EST	NEELY,WB & BLAU,GE (1985)
DC	pKa			
HL 2.56E-014	atm m3/mol	25	EST	MEYLAN,WM & HOWARD,PH (1991)
OH 1.61E-011	cm3/molc sec	25	EST	MEYLAN,WM & HOWARD,PH (1993)

000084-96-8 — TRIMEPRAZINE

CAS #:	000084-96-8
Formula:	$C_{18}H_{22}N_2S$
Mol Weight:	298.45
MP (deg C):	68
FP (deg C):	
BP (deg C):	150-175
BP pressure (mm Hg):	3.00E-001

Property/Value	Units	Temp	Data Type	Reference
WS 9.42E-001	mg/L	25	EST	MEYLAN,WM ET AL. (1996)
logP 4.71			EXP	HANSCH,C & LEO,AJ (1985)
VP 5.88E-007	mm Hg	25	EST	NEELY,WB & BLAU,GE (1985)
DC	pKa			
HL 6.60E-010	atm m3/mol	25	EST	MEYLAN,WM & HOWARD,PH (1991)
OH 2.73E-010	cm3/molc sec	25	EST	MEYLAN,WM & HOWARD,PH (1993)

000084-97-9 — PERAZINE

CAS #:	000084-97-9
Formula:	$C_{20}H_{25}N_3S$
Mol Weight:	339.51
MP (deg C):	51-53
FP (deg C):	
BP (deg C):	160-170
BP pressure (mm Hg):	1.00E-003

Property/Value	Units	Temp	Data Type	Reference
WS 2.17E+000	mg/L	25	EST	MEYLAN,WM ET AL. (1996)
logP 4.00			EXP	MANNHOLD,R ET AL. (1990)
VP 1.01E-008	mm Hg	25	EST	NEELY,WB & BLAU,GE (1985)
DC	pKa			
HL 1.44E-013	atm m3/mol	25	EST	MEYLAN,WM & HOWARD,PH (1991)
OH 3.79E-010	cm3/molc sec	25	EST	MEYLAN,WM & HOWARD,PH (1993)

000085-00-7 — DIQUAT DIBROMIDE

CAS #:	000085-00-7
Formula:	$C_{12}H_{12}Br_2N_2$
Mol Weight:	344.06
MP (deg C):	337
FP (deg C):	
BP (deg C):	
BP pressure (mm Hg):	

Property/Value	Units	Temp	Data Type	Reference
WS 7.08E+005	mg/L	20	EXP	SHIU,WY ET AL. (1990)
logP -2.82			EST	MEYLAN,WM & HOWARD,PH (1995)
VP 1.81E-006	mm Hg	25	EST	NEELY,WB & BLAU,GE (1985)
DC	pKa			
HL 1.42E-013	atm m3/mol	25	EST	MEYLAN,WM & HOWARD,PH (1991)
OH 2.32E-011	cm3/molc sec	25	EST	MEYLAN,WM & HOWARD,PH (1993)

000085-01-8 — PHENANTHRENE

Formula: $C_{14}H_{10}$
Mol Weight: 178.24
MP (deg C): 100
FP (deg C):
BP (deg C): 340
BP pressure (mm Hg):

Property/Value	Units	Temp	Data Type	Reference
WS 1.15E+000	mg/L	25	EXP	SCHWARZ,FP (1977)
logP 4.46			EXP	HANSCH,C & LEO,AJ (1985)
VP 1.12E-004	mm Hg	25	EXP	SCALA,AJ & BANERJEE,S (1982)
DC	pKa			
HL 3.60E-005	atm m3/mol	25	EXP	MACKAY,D ET AL. (1982A)
OH 3.10E-011	cm3/molc sec	25	EXP	ATKINSON,R (1989)

000085-02-9 — BENZO(F)QUINOLINE

Formula: $C_{13}H_9N$
Mol Weight: 179.22
MP (deg C): 93
FP (deg C):
BP (deg C): 349-350
BP pressure (mm Hg): 7.21E+002

Property/Value	Units	Temp	Data Type	Reference
WS 7.61E+001	mg/L	25	EXP	PEARLMAN,RS ET AL. (1984)
logP 3.43			EXP	HANSCH,C & LEO,AJ (1985)
VP 5.60E-005	mm Hg	25	EXP	MCEACHERN,DM ET AL. (1975)
DC 4.21	pKa	20	EXP	PERRIN,DD (1965)
HL 1.74E-007	atm m3/mol	25	EST	VP/WSOL
OH 9.00E-012	cm3/molc sec	25	EST	MEYLAN,WM & HOWARD,PH (1993)

000085-18-7 — 8-CHLOROTHEOPHYLLINE

Formula: $C_7H_7ClN_4O_2$
Mol Weight: 214.61
MP (deg C): 290 dec
FP (deg C):
BP (deg C):
BP pressure (mm Hg):

Property/Value	Units	Temp	Data Type	Reference
WS 1.09E+003	mg/L	25	EST	MEYLAN,WM ET AL. (1996)
logP -0.85			EXP	HANSCH,C & LEO,AJ (1985)
VP 1.38E-009	mm Hg	25	EST	NEELY,WB & BLAU,GE (1985)
DC	pKa			
HL 1.24E-012	atm m3/mol	25	EST	MEYLAN,WM & HOWARD,PH (1991)
OH 3.81E-011	cm3/molc sec	25	EST	MEYLAN,WM & HOWARD,PH (1993)

000085-22-3 — PENTABROMOETHYLBENZENE

Formula: $C_8H_5Br_5$
Mol Weight: 500.67
MP (deg C): 137-39
FP (deg C):
BP (deg C):
BP pressure (mm Hg):

Property/Value	Units	Temp	Data Type	Reference
WS 4.67E-002	mg/L	25	EST	MEYLAN,WM ET AL. (1996)
logP 7.48			EST	MEYLAN,WM & HOWARD,PH (1995)
VP 4.69E-006	mm Hg	25	EST	NEELY,WB & BLAU,GE (1985)
DC	pKa			
HL 5.73E-005	atm m3/mol	25	EST	MEYLAN,WM & HOWARD,PH (1991)
OH 1.07E-012	cm3/molc sec	25	EST	MEYLAN,WM & HOWARD,PH (1993)

000085-34-7 — FENAC

Formula: $C_8H_5Cl_3O_2$
Mol Weight: 239.49
MP (deg C): 161
FP (deg C):
BP (deg C):
BP pressure (mm Hg):

Property/Value	Units	Temp	Data Type	Reference
WS 2.00E+002	mg/L	28	EXP	WORTHING,CR & WALKER,SB (1987)
logP 3.20			EXP	HANSCH,C ET AL. (1995)
VP 3.39E-005	mm Hg	25	EST	NEELY,WB & BLAU,GE (1985)
DC 3.70	pKa	25	EXP	NELSON,NH & FAUST,SD (1969)
HL 1.80E-008	atm m3/mol	25	EST	MEYLAN,WM & HOWARD,PH (1991)
OH 1.51E-012	cm3/molc sec	25	EST	MEYLAN,WM & HOWARD,PH (1993)

000085-38-1 — P-NITROSALICYLIC ACID

Formula: $C_7H_5NO_5$
Mol Weight: 183.12
MP (deg C): 123
FP (deg C):
BP (deg C):
BP pressure (mm Hg):

Property/Value	Units	Temp	Data Type	Reference
WS 1.30E+003	mg/L	16	EXP	YALKOWSKY,SH & DANNENFELSER,RM (1992)
logP 2.64			EST	MEYLAN,WM & HOWARD,PH (1995)
VP 1.16E-006	mm Hg	25	EST	NEELY,WB & BLAU,GE (1985)
DC	pKa			
HL 1.77E-007	atm m3/mol	25	EST	MEYLAN,WM & HOWARD,PH (1991)
OH 2.20E-012	cm3/molc sec	25	EST	MEYLAN,WM & HOWARD,PH (1993)

000085-41-6 — PHTHALIMIDE

Formula: $C_8H_5NO_2$
Mol Weight: 147.13
MP (deg C): 238
FP (deg C):
BP (deg C):
BP pressure (mm Hg):

Property/Value	Units	Temp	Data Type	Reference
WS 3.60E+002	mg/L	25	EXP	YALKOWSKY,SH & DANNENFELSER,RM (1992)
logP 1.15			EXP	HANSCH,C ET AL. (1995)
VP 1.00E-008	mm Hg	25	EST	NEELY,WB & BLAU,GE (1985)
DC 8.30	pKa		EXP	MERCK INDEX (1989)
HL 1.00E-008	atm m3/mol	25	EST	MEYLAN,WM & HOWARD,PH (1991)
OH 6.25E-012	cm3/molc sec	25	EST	MEYLAN,WM & HOWARD,PH (1993)

000085-44-9 — PHTHALIC ANHYDRIDE

Formula: $C_8H_4O_3$
Mol Weight: 148.12
MP (deg C): 130.8
FP (deg C):
BP (deg C): 295
BP pressure (mm Hg):

Property/Value	Units	Temp	Data Type	Reference
WS 6.20E+003	mg/L	25	EXP	TOWLE,PH ET AL. (1968)
logP 1.60			EXP	HANSCH,C ET AL. (1995)
VP 5.17E-004	mm Hg	25	EXT	JONES,AH (1960)
DC	pKa			
HL 1.63E-008	atm m3/mol	25	EST	VP/WSOL
OH 5.04E-013	cm3/molc sec	25	EST	MEYLAN,WM & HOWARD,PH (1993)

CAS #: 000085-45-0				BENZENAMINE, 2-METHOXY-3-NITRO-

Formula: $C_7H_8N_2O_3$

Mol Weight: 168.15

MP (deg C): | FP (deg C):

BP (deg C):

BP pressure (mm Hg):

Property/ Value	Units	Temp	Data Type	Reference
WS 1.89E+003	mg/L	25	EST	MEYLAN,WM ET AL. (1996)
logP 1.78			EXP	CHEM INSPECT TEST INST (1992)
VP 3.19E-004	mm Hg	25	EST	NEELY,WB & BLAU,GE (1985)
DC	pKa			
HL 4.44E-010	atm m3/mol	25	EST	MEYLAN,WM & HOWARD,PH (1991)
OH 8.97E-012	cm3/molc sec	25	EST	MEYLAN,WM & HOWARD,PH (1993)

CAS #: 000085-68-7				BUTYL BENZYL PHTHALATE

Formula: $C_{19}H_{20}O_4$

Mol Weight: 312.37

MP (deg C): | FP (deg C):

BP (deg C): 370

BP pressure (mm Hg):

Property/ Value	Units	Temp	Data Type	Reference
WS 2.69E+000	mg/L	25	EXP	HOWARD,PH ET AL. (1985)
logP 4.91			EXP	HANSCH,C & LEO,AJ (1985)
VP 8.25E-006	mm Hg	25	EXP	HOWARD,PH ET AL. (1985)
DC	pKa			
HL 1.26E-006	atm m3/mol	25	EST	VP/WSOL
OH 1.08E-011	cm3/molc sec	25	EST	MEYLAN,WM & HOWARD,PH (1993)

CAS #: 000085-69-8				BUTYL (2-ETHYLHEXYL) PHTHALATE

Formula: $C_{20}H_{30}O_4$

Mol Weight: 334.46

MP (deg C): | FP (deg C):

BP (deg C):

BP pressure (mm Hg):

Property/ Value	Units	Temp	Data Type	Reference
WS 4.36E-003	mg/L	25	EST	MEYLAN,WM ET AL. (1996)
logP 6.50			EST	MEYLAN,WM & HOWARD,PH (1995)
VP 2.36E-005	mm Hg	25	EST	NEELY,WB & BLAU,GE (1985)
DC	pKa			
HL 7.39E-006	atm m3/mol	25	EST	MEYLAN,WM & HOWARD,PH (1991)
OH 1.53E-011	cm3/molc sec	25	EST	MEYLAN,WM & HOWARD,PH (1993)

CAS #: 000085-70-1				BUTYLGLYCOLYL BUTYL PHTHALATE

Formula: $C_{18}H_{24}O_6$

Mol Weight: 336.39

MP (deg C): | FP (deg C):

BP (deg C): 219

BP pressure (mm Hg): 1.50E+001

Property/ Value	Units	Temp	Data Type	Reference
WS 8.80E+000	mg/L	25	EST	MEYLAN,WM ET AL. (1996)
logP 4.15			EST	MEYLAN,WM & HOWARD,PH (1995)
VP 7.07E-006	mm Hg	25	EST	NEELY,WB & BLAU,GE (1985)
DC	pKa			
HL 8.32E-008	atm m3/mol	25	EST	MEYLAN,WM & HOWARD,PH (1991)
OH 8.71E+000	cm3/molc sec	25	EST	MEYLAN,WM & HOWARD,PH (1993)

CAS #: 000085-79-0				DIBUCAINE

Formula: $C_{20}H_{29}N_3O_2$

Mol Weight: 343.47

MP (deg C): | FP (deg C):

BP (deg C):

BP pressure (mm Hg):

Property/ Value	Units	Temp	Data Type	Reference
WS 1.19E+000	mg/L	25	EST	MEYLAN,WM ET AL. (1996)
logP 4.40			EXP	HANSCH,C & LEO,AJ (1985)
VP 1.36E-010	mm Hg	25	EST	NEELY,WB & BLAU,GE (1985)
DC	pKa			
HL 9.10E-016	atm m3/mol	25	EST	MEYLAN,WM & HOWARD,PH (1991)
OH 1.45E-010	cm3/molc sec	25	EST	MEYLAN,WM & HOWARD,PH (1993)

CAS #: 000085-81-4				QUINOLINE, 6-METHOXY-8-NITRO-

Formula: $C_{10}H_8N_2O_3$

Mol Weight: 204.19

MP (deg C): 158-160 | FP (deg C):

BP (deg C):

BP pressure (mm Hg):

Property/ Value	Units	Temp	Data Type	Reference
WS 4.29E+002	mg/L	25	EST	MEYLAN,WM ET AL. (1996)
logP 1.87			EXP	DEBNATH,AK ET AL. (1991)
VP 2.38E-005	mm Hg	25	EST	NEELY,WB & BLAU,GE (1985)
DC	pKa			
HL 1.61E-010	atm m3/mol	25	EST	MEYLAN,WM & HOWARD,PH (1991)
OH 5.48E-012	cm3/molc sec	25	EST	MEYLAN,WM & HOWARD,PH (1993)

CAS #: 000085-87-0				PYRIDOXAMINE

Formula: $C_8H_{12}N_2O_2$

Mol Weight: 168.20

MP (deg C): | FP (deg C):

BP (deg C):

BP pressure (mm Hg):

Property/ Value	Units	Temp	Data Type	Reference
WS 1.00E+006	mg/L	25	EST	MEYLAN,WM ET AL. (1996)
logP -1.33			EST	MEYLAN,WM & HOWARD,PH (1995)
VP 1.44E-006	mm Hg	25	EST	NEELY,WB & BLAU,GE (1985)
DC	pKa			
HL 3.71E-018	atm m3/mol	25	EST	MEYLAN,WM & HOWARD,PH (1991)
OH 5.68E-011	cm3/molc sec	25	EST	MEYLAN,WM & HOWARD,PH (1993)

CAS #: 000085-98-3				DIETHYL DIPHENYL UREA

Formula: $C_{17}H_{20}N_2O$

Mol Weight: 268.36

MP (deg C): 79 | FP (deg C):

BP (deg C):

BP pressure (mm Hg):

Property/ Value	Units	Temp	Data Type	Reference
WS 4.79E+000	mg/L	25	EST	MEYLAN,WM ET AL. (1996)
logP 4.20			EST	MEYLAN,WM & HOWARD,PH (1995)
VP 2.05E-006	mm Hg	25	EST	NEELY,WB & BLAU,GE (1985)
DC	pKa			
HL 8.22E-008	atm m3/mol	25	EST	MEYLAN,WM & HOWARD,PH (1991)
OH 3.28E-011	cm3/molc sec	25	EST	MEYLAN,WM & HOWARD,PH (1993)

CAS #: 000086-28-2				N-ETHYLCARBAZOL

Fórmula: $C_{14}H_{13}N$

Mol Weight: 195.27

MP (deg C): 68 | FP (deg C):

BP (deg C): 190

BP pressure (mm Hg): 1.00E+001

Property/Value	Units	Temp	Data Type	Reference
WS 7.20E-001	mg/L	25	EST	MEYLAN,WM ET AL. (1996)
logP 4.33			EST	MEYLAN,WM & HOWARD,PH (1995)
VP 2.20E-004	mm Hg	25	EXT	CHAO,J ET AL. (1983)
DC	pKa			
HL 9.96E-006	atm m3/mol	25	EST	MEYLAN,WM & HOWARD,PH (1991)
OH 2.09E-010	cm3/molc sec	25	EST	MEYLAN,WM & HOWARD,PH (1993)

CAS #: 000086-29-3				DIPHENYLACETONITRILE

Formula: $C_{14}H_{11}N$

Mol Weight: 193.25

MP (deg C): 193.2 | FP (deg C):

BP (deg C):

BP pressure (mm Hg):

Property/Value	Units	Temp	Data Type	Reference
WS 2.20E+002	mg/L		EXP	SHIU,WY ET AL. (1990)
logP 3.20			EST	MEYLAN,WM & HOWARD,PH (1995)
VP 1.02E-004	mm Hg	25	EST	NEELY,WB & BLAU,GE (1985)
DC	pKa			
HL 1.99E-007	atm m3/mol	25	EST	MEYLAN,WM & HOWARD,PH (1991)
OH 9.62E-012	cm3/molc sec	25	EST	MEYLAN,WM & HOWARD,PH (1993)

CAS #: 000086-30-6				DIPHENYLNITROSAMINE

Formula: $C_{12}H_{10}N_2O$

Mol Weight: 198.23

MP (deg C): 66.5 | FP (deg C):

BP (deg C):

BP pressure (mm Hg):

Property/Value	Units	Temp	Data Type	Reference
WS 3.50E+001	mg/L	25	EXP	BANERJEE,S ET AL. (1980)
logP 3.13			EXP	HANSCH,C & LEO,AJ (1985)
VP 2.97E-005	mm Hg	25	EST	NEELY,WB & BLAU,GE (1985)
DC	pKa			
HL 1.21E-006	atm m3/mol	25	EST	MEYLAN,WM & HOWARD,PH (1991)
OH 4.48E-012	cm3/molc sec	25	EST	MEYLAN,WM & HOWARD,PH (1993)

CAS #: 000086-34-0				N-METHYL-2-PHENYL SUCCINIMIDE

Formula: $C_{11}H_{11}NO_2$

Mol Weight: 189.22

MP (deg C): 71-73 | FP (deg C):

BP (deg C):

BP pressure (mm Hg):

Property/Value	Units	Temp	Data Type	Reference
WS 7.02E+003	mg/L	25	EST	MEYLAN,WM ET AL. (1996)
logP 0.99			EST	MEYLAN,WM & HOWARD,PH (1995)
VP 5.52E-007	mm Hg	25	EST	NEELY,WB & BLAU,GE (1985)
DC	pKa			
HL 4.49E-009	atm m3/mol	25	EST	MEYLAN,WM & HOWARD,PH (1991)
OH 2.69E-011	cm3/molc sec	25	EST	MEYLAN,WM & HOWARD,PH (1993)

CAS #: 000086-35-1				2,4-IMIDAZOLIDINEDIONE, 3-ETHYL-5-PHENYL-

Formula: $C_{11}H_{12}N_2O_2$

Mol Weight: 204.23

MP (deg C): 94 | FP (deg C):

BP (deg C):

BP pressure (mm Hg):

Property/Value	Units	Temp	Data Type	Reference
WS 5.28E+003	mg/L	25	EST	MEYLAN,WM ET AL. (1996)
logP 1.05			EXP	SANGSTER,J (1994)
VP 5.54E-008	mm Hg	25	EST	NEELY,WB & BLAU,GE (1985)
DC	pKa			
HL 3.69E-010	atm m3/mol	25	EST	MEYLAN,WM & HOWARD,PH (1991)
OH 1.54E-011	cm3/molc sec	25	EST	MEYLAN,WM & HOWARD,PH (1993)

CAS #: 000086-40-8				3,6-DIAMINO-10-METHYL-ACRIDINIUM CHLORIDE

Formula: $C_{14}H_{14}ClN_3$

Mol Weight: 259.74

MP (deg C): | FP (deg C):

BP (deg C):

BP pressure (mm Hg):

Property/Value	Units	Temp	Data Type	Reference
WS 5.37E+004	mg/L	25	EST	MEYLAN,WM ET AL. (1996)
logP -1.78			EXP	HANSCH,C ET AL. (1995)
VP 1.22E-008	mm Hg	25	EST	NEELY,WB & BLAU,GE (1985)
DC	pKa			
HL 1.79E-019	atm m3/mol	25	EST	MEYLAN,WM & HOWARD,PH (1991)
OH 2.10E-010	cm3/molc sec	25	EST	MEYLAN,WM & HOWARD,PH (1993)

CAS #: 000086-50-0				METHYL AZINPHOS

Formula: $C_{10}H_{12}N_3O_3PS_2$

Mol Weight: 317.33

MP (deg C): 73-74 | FP (deg C):

BP (deg C):

BP pressure (mm Hg):

Property/Value	Units	Temp	Data Type	Reference
WS 2.09E+001	mg/L	20	EXP	BOWMAN,BT & SANS,WW (1983)
logP 2.75			EXP	HANSCH,C & LEO,AJ (1985)
VP 7.50E-009	mm Hg	20	EXP	WORTHING,CR & WALKER,SB (1983)
DC	pKa			
HL 1.50E-010	atm m3/mol	20	EST	VP/WSOL
OH 1.45E-010	cm3/molc sec	25	EST	MEYLAN,WM & HOWARD,PH (1993)

CAS #: 000086-54-4				HYDRALAZINE

Formula: $C_8H_8N_4$

Mol Weight: 160.18

MP (deg C): 172-173 | FP (deg C):

BP (deg C):

BP pressure (mm Hg):

Property/Value	Units	Temp	Data Type	Reference
WS 9.56E+003	mg/L	25	EST	MEYLAN,WM ET AL. (1996)
logP 1.00			EXP	HANSCH,C ET AL. (1995)
VP 7.47E-006	mm Hg	25	EST	NEELY,WB & BLAU,GE (1985)
DC	pKa			
HL 4.14E-013	atm m3/mol	25	EST	MEYLAN,WM & HOWARD,PH (1991)
OH 4.38E-011	cm3/molc sec	25	EST	MEYLAN,WM & HOWARD,PH (1993)

CAS #: 000086-55-5 — 1-NAPTHOIC ACID

Formula: $C_{11}H_8O_2$

Mol Weight: 172.19

MP (deg C): 160.5-162 FP (deg C):

BP (deg C): 300

BP pressure (mm Hg):

Property/ Value	Units	Temp	Data Type	Reference
WS 8.60E+001	mg/L	25	EXP	CHEM INSPECT TEST INST (1992)
logP 3.10			EXP	HANSCH,C & LEO,AJ (1985)
VP 8.84E-006	mm Hg	25	EXT	PERRY,RH & GREEN,D (1984)
DC 3.60	pKa	25	EXP	KORTUM,G ET AL (1961)
HL 1.06E-008	atm m3/mol	25	EST	MEYLAN,WM & HOWARD,PH (1991)
OH 8.52E-012	cm3/molc sec	25	EST	MEYLAN,WM & HOWARD,PH (1993)

CAS #: 000086-56-6 — 1-NAPHTHALENAMINE, N,N-DIMETHYL-

Formula: $C_{12}H_{13}N$

Mol Weight: 171.24

MP (deg C): FP (deg C):

BP (deg C): 274.5

BP pressure (mm Hg): 7.11E+002

Property/ Value	Units	Temp	Data Type	Reference
WS 6.73E+001	mg/L	25	EST	MEYLAN,WM ET AL. (1996)
logP 3.46			EXP	HANSCH,C ET AL. (1995)
VP 2.70E-003	mm Hg	25	EST	NEELY,WB & BLAU,GE (1985)
DC 4.83	pKa	28	EXP	PERRIN,DD (1965)
HL 8.36E-006	atm m3/mol	25	EST	MEYLAN,WM & HOWARD,PH (1991)
OH 2.03E-010	cm3/molc sec	25	EST	MEYLAN,WM & HOWARD,PH (1993)

CAS #: 000086-57-7 — 1-NITRONAPHTHALENE

Formula: $C_{10}H_7NO_2$

Mol Weight: 173.17

MP (deg C): 59-61 FP (deg C):

BP (deg C): 304

BP pressure (mm Hg):

Property/ Value	Units	Temp	Data Type	Reference
WS 1.75E+001	mg/L	25	EST	MEYLAN,WM ET AL. (1996)
logP 3.19			EXP	HANSCH,C & LEO,AJ (1985)
VP 4.80E-004	mm Hg	25	EXT	ALDRICH (1988)
DC	pKa			
HL 3.08E-006	atm m3/mol	25	EST	MEYLAN,WM & HOWARD,PH (1991)
OH 5.40E-012	cm3/molc sec	25	EXP	ATKINSON,R (1989)

CAS #: 000086-65-7 — 2-NAPHTHYLAMINE-6,8-DISULFONIC ACID

Formula: $C_{10}H_9NO_6S_2$

Mol Weight: 303.31

MP (deg C): 274 FP (deg C):

BP (deg C):

BP pressure (mm Hg):

Property/ Value	Units	Temp	Data Type	Reference
WS 9.21E+004	mg/L	15	EXP	STEPHEN,H & STEPHEN,T (1963)
logP -4.06			EST	MEYLAN,WM & HOWARD,PH (1995)
VP 1.61E-015	mm Hg	25	EST	NEELY,WB & BLAU,GE (1985)
DC	pKa			
HL 4.06E-020	atm m3/mol	25	EST	MEYLAN,WM & HOWARD,PH (1991)
OH 9.67E-012	cm3/molc sec	25	EST	MEYLAN,WM & HOWARD,PH (1993)

CAS #: 000086-73-7 — 9H-FLUORENE

Formula: $C_{13}H_{10}$

Mol Weight: 166.22

MP (deg C): 116-117 FP (deg C):

BP (deg C): 295

BP pressure (mm Hg):

Property/ Value	Units	Temp	Data Type	Reference
WS 1.89E+000	mg/L	25	EXP	WAUCHOPE,RD & GETZEN,FW (1972)
logP 4.18			EXP	HANSCH,C & LEO,AJ (1985)
VP 8.42E-003	mm Hg	25	EXT	BOUBLIK,T ET AL. (1984)
DC	pKa			
HL 1.00E-004	atm m3/mol	25	EXP	MACKAY,D ET AL. (1982A)
OH 1.30E-011	cm3/molc sec	25	EXP	ATKINSON,R (1989)

CAS #: 000086-74-8 — CARBAZOLE

Formula: $C_{12}H_9N$

Mol Weight: 167.21

MP (deg C): 245 FP (deg C):

BP (deg C): 355

BP pressure (mm Hg):

Property/ Value	Units	Temp	Data Type	Reference
WS 1.80E+000	mg/L	25	EXP	AINSWORTH,CC ET AL. (1989)
logP 3.72			EXP	HANSCH,C & LEO,AJ (1985)
VP 1.10E-004	mm Hg	25	EXT	BOUBLIK,T ET AL. (1984)
DC -3.00	pKa		EXP	SMITH,JH ET AL. (1978)
HL 8.65E-008	atm m3/mol	25	EST	MEYLAN,WM & HOWARD,PH (1991)
OH 1.30E-010	cm3/molc sec	25	EST	ATKINSON,R (1987)

CAS #: 000086-87-3 — NAPHTHALENEACETIC ACID

Formula: $C_{12}H_{10}O_2$

Mol Weight: 186.21

MP (deg C): 129-131.5 FP (deg C):

BP (deg C):

BP pressure (mm Hg):

Property/ Value	Units	Temp	Data Type	Reference
WS 4.20E+002	mg/L	20	EXP	YALKOWSKY,SH & DANNENFELSER,RM (1992)
logP 2.24			EXP	SANGSTER,J (1993)
VP 1.59E-005	mm Hg	25	EST	NEELY,WB & BLAU,GE (1985)
DC 4.23	pKa	25	EXP	KORTUM,G ET AL (1961)
HL 1.17E-009	atm m3/mol	25	EST	MEYLAN,WM & HOWARD,PH (1991)
OH 3.70E-011	cm3/molc sec	25	EST	MEYLAN,WM & HOWARD,PH (1993)

CAS #: 000086-88-4 — -NAPTHYLTHIOUREA

Formula: $C_{11}H_{10}N_2S$

Mol Weight: 202.28

MP (deg C): 198 FP (deg C):

BP (deg C):

BP pressure (mm Hg):

Property/ Value	Units	Temp	Data Type	Reference
WS 6.00E+002	mg/L	25	EXP	YALKOWSKY,SH & DANNENFELSER,RM (1992)
logP 1.65			EXP	GOVERS,H ET AL. (1986)
VP 6.61E-006	mm Hg	25	EST	NEELY,WB & BLAU,GE (1985)
DC	pKa			
HL 8.51E-009	atm m3/mol	25	EST	MEYLAN,WM & HOWARD,PH (1991)
OH 2.21E-010	cm3/molc sec	25	EST	MEYLAN,WM & HOWARD,PH (1993)

000086-96-4 — 7-METHYL QUINAZOLINE-2,4-DIONE

Formula:	$C_9H_8N_2O_2$		
Mol Weight:	176.18		
MP (deg C):	300	FP (deg C):	
BP (deg C):			
BP pressure (mm Hg):			

Property/Value	Units	Temp	Data Type	Reference
WS 1.47E+004	mg/L	25	EST	MEYLAN,WM ET AL. (1996)
logP 0.77			EXP	HANSCH,C ET AL. (1995)
VP 5.49E-008	mm Hg	25	EST	NEELY,WB & BLAU,GE (1985)
DC	pKa			
HL 5.35E-011	atm m3/mol	25	EST	MEYLAN,WM & HOWARD,PH (1991)
OH 3.99E-011	cm3/molc sec	25	EST	MEYLAN,WM & HOWARD,PH (1993)

000086-98-6 — QUINOLINE, 4,7-DICHLORO-

Formula:	$C_9H_5Cl_2N$		
Mol Weight:	198.05		
MP (deg C):	93	FP (deg C):	
BP (deg C):	148		
BP pressure (mm Hg):	1.00E+001		

Property/Value	Units	Temp	Data Type	Reference
WS 4.00E+001	mg/L	25	EST	MEYLAN,WM ET AL. (1996)
logP 3.57			EXP	GO,ML & NGIAM,TL (1988)
VP 7.70E-004	mm Hg	25	EST	NEELY,WB & BLAU,GE (1985)
DC 2.80	pKa	25	EXP	PERRIN,DD (1965)
HL 3.78E-007	atm m3/mol	25	EST	MEYLAN,WM & HOWARD,PH (1991)
OH 2.96E-012	cm3/molc sec	25	EST	MEYLAN,WM & HOWARD,PH (1993)

000087-00-3 — HOMATROPIN

Formula:	$C_{16}H_{21}NO_3$		
Mol Weight:	275.35		
MP (deg C):	99-100	FP (deg C):	
BP (deg C):			
BP pressure (mm Hg):			

Property/Value	Units	Temp	Data Type	Reference
WS 1.07E+004	mg/L	25	EST	MEYLAN,WM ET AL. (1996)
logP 1.42			EST	MEYLAN,WM & HOWARD,PH (1995)
VP 6.98E-008	mm Hg	25	EST	NEELY,WB & BLAU,GE (1985)
DC	pKa			
HL 8.78E-010	atm m3/mol	25	EST	MEYLAN,WM & HOWARD,PH (1991)
OH 6.90E-011	cm3/molc sec	25	EST	MEYLAN,WM & HOWARD,PH (1993)

000087-08-1 — PHENOXYMETHYLPENICILLIN

Formula:	$C_{16}H_{18}N_2O_5S$		
Mol Weight:	350.40		
MP (deg C):		FP (deg C):	
BP (deg C):			
BP pressure (mm Hg):			

Property/Value	Units	Temp	Data Type	Reference
WS 1.01E+002	mg/L	25	EST	MEYLAN,WM ET AL. (1996)
logP 2.09			EXP	HANSCH,C & LEO,AJ (1985)
VP 1.06E-012	mm Hg	25	EST	NEELY,WB & BLAU,GE (1985)
DC	pKa			
HL 4.42E-015	atm m3/mol	25	EST	MEYLAN,WM & HOWARD,PH (1991)
OH 1.12E-010	cm3/molc sec	25	EST	MEYLAN,WM & HOWARD,PH (1993)

000087-17-2 — SALICYLANILIDE

Formula:	$C_{13}H_{11}NO_2$		
Mol Weight:	213.24		
MP (deg C):	135.8-6.2	FP (deg C):	
BP (deg C):			
BP pressure (mm Hg):			

Property/Value	Units	Temp	Data Type	Reference
WS 5.50E+001	mg/L	25	EXP	YALKOWSKY,SH & DANNENFELSER,RM (1992)
logP 3.27			EXP	HANSCH,C & LEO,AJ (1985)
VP 2.45E-008	mm Hg	25	EST	NEELY,WB & BLAU,GE (1985)
DC	pKa			
HL 1.60E-010	atm m3/mol	25	EST	MEYLAN,WM & HOWARD,PH (1991)
OH 4.29E-011	cm3/molc sec	25	EST	MEYLAN,WM & HOWARD,PH (1993)

000087-25-2 — O-AMINOBENZOIC ACID, ETHYL ESTER

Formula:	$C_9H_{11}NO_2$		
Mol Weight:	165.19		
MP (deg C):	13	FP (deg C):	
BP (deg C):	268		
BP pressure (mm Hg):			

Property/Value	Units	Temp	Data Type	Reference
WS 4.14E+002	mg/L	25	EST	MEYLAN,WM ET AL. (1996)
logP 2.57			EXP	HANSCH,C & LEO,AJ (1985)
VP 2.32E-003	mm Hg	25	EST	NEELY,WB & BLAU,GE (1985)
DC 2.18	pKa	25	EXP	PERRIN,DD (1965)
HL 1.63E-008	atm m3/mol	25	EST	MEYLAN,WM & HOWARD,PH (1991)
OH 3.65E-011	cm3/molc sec	25	EST	MEYLAN,WM & HOWARD,PH (1993)

000087-33-2 — D-GLUCITOL, 1,4:3,6-DIANHYDRO-, DINITRATE

Formula:	$C_6H_8N_2O_8$		
Mol Weight:	236.14		
MP (deg C):	70	FP (deg C):	
BP (deg C):			
BP pressure (mm Hg):			

Property/Value	Units	Temp	Data Type	Reference
WS 5.50E+002	mg/L	25	EXP	YALKOWSKY,SH & DANNENFELSER,RM (1992)
logP 1.31			EXP	HANSCH,C ET AL. (1995)
VP 4.55E-004	mm Hg	25	EST	NEELY,WB & BLAU,GE (1985)
DC	pKa			
HL 4.19E-012	atm m3/mol	25	EST	MEYLAN,WM & HOWARD,PH (1991)
OH 6.33E-012	cm3/molc sec	25	EST	MEYLAN,WM & HOWARD,PH (1993)

000087-40-1 — 2,4,6-TRICHLOROANISOLE

Formula:	$C_7H_5Cl_3O$		
Mol Weight:	211.48		
MP (deg C):	61.5	FP (deg C):	
BP (deg C):	241		
BP pressure (mm Hg):			

Property/Value	Units	Temp	Data Type	Reference
WS 1.18E+001	mg/L	25	EST	MEYLAN,WM ET AL. (1996)
logP 4.11			EXP	OPPERHUIZEN,A & VOORS,PI (1987)
VP 2.28E-002	mm Hg	25	EST	NEELY,WB & BLAU,GE (1985)
DC	pKa			
HL 1.30E-004	atm m3/mol	25	EST	MEYLAN,WM & HOWARD,PH (1991)
OH 1.42E-012	cm3/molc sec	25	EST	MEYLAN,WM & HOWARD,PH (1993)

CAS #: 000087-41-2 — PHTHALIDE

Formula: $C_8H_6O_2$
Mol Weight: 134.14
MP (deg C): 75
FP (deg C):
BP (deg C): 290
BP pressure (mm Hg):

Property/Value	Units	Temp	Data Type	Reference
WS 1.84E+004	mg/L	25	EST	MEYLAN,WM ET AL. (1996)
logP 0.80			EXP	HANSCH,C & LEO,AJ (1985)
VP 7.07E-003	mm Hg	25	EXT	PERRY,RH & GREEN,D (1984)
DC	pKa			
HL 1.27E-005	atm m3/mol	25	EST	MEYLAN,WM & HOWARD,PH (1991)
OH 2.76E-012	cm3/molc sec	25	EST	MEYLAN,WM & HOWARD,PH (1993)

CAS #: 000087-47-8 — PYROLAN

Formula: $C_{13}H_{15}N_3O_2$
Mol Weight: 245.28
MP (deg C): 50
FP (deg C):
BP (deg C): 160-162
BP pressure (mm Hg): 2.00E-001

Property/Value	Units	Temp	Data Type	Reference
WS 2.00E+003	mg/L		EXP	YALKOWSKY,SH & DANNENFELSER,RM (1992)
logP 1.96			EST	MEYLAN,WM & HOWARD,PH (1995)
VP 7.77E-006	mm Hg	25	EST	NEELY,WB & BLAU,GE (1985)
DC	pKa			
HL 1.94E-013	atm m3/mol	25	EST	MEYLAN,WM & HOWARD,PH (1991)
OH 1.17E-010	cm3/molc sec	25	EST	MEYLAN,WM & HOWARD,PH (1993)

CAS #: 000087-51-4 — INDOLE-3-ACETIC ACID

Formula: $C_{10}H_9NO_2$
Mol Weight: 175.19
MP (deg C): 168-170
FP (deg C):
BP (deg C):
BP pressure (mm Hg):

Property/Value	Units	Temp	Data Type	Reference
WS 1.50E+003	mg/L	25	EXP	SHIU,WY ET AL. (1990)
logP 1.41			EXP	HANSCH,C & LEO,AJ (1985)
VP 5.26E-006	mm Hg	25	EST	NEELY,WB & BLAU,GE (1985)
DC	pKa			
HL 7.27E-012	atm m3/mol	25	EST	MEYLAN,WM & HOWARD,PH (1991)
OH 1.84E-010	cm3/molc sec	25	EST	MEYLAN,WM & HOWARD,PH (1993)

CAS #: 000087-59-2 — 2,3-DIMETHYLANILINE

Formula: $C_8H_{11}N$
Mol Weight: 121.18
MP (deg C): 2.5
FP (deg C):
BP (deg C): 221.5
BP pressure (mm Hg):

Property/Value	Units	Temp	Data Type	Reference
WS 1.39E+003	mg/L	25	EST	MEYLAN,WM ET AL. (1996)
logP 2.17			EST	MEYLAN,WM & HOWARD,PH (1995)
VP 7.50E-002	mm Hg	25	EXP	WEBER,RC ET AL. (1981)
DC 4.70	pKa	25	EXP	PERRIN,DD (1965)
HL 2.32E-006	atm m3/mol	25	EST	MEYLAN,WM & HOWARD,PH (1991)
OH 2.00E-010	cm3/molc sec	25	EST	MEYLAN,WM & HOWARD,PH (1993)

CAS #: 000087-60-5 — 2-METHYL-3-CHLOROANILINE

Formula: C_7H_8ClN
Mol Weight: 141.60
MP (deg C): 1
FP (deg C):
BP (deg C): 245
BP pressure (mm Hg):

Property/Value	Units	Temp	Data Type	Reference
WS 9.54E+002	mg/L	25	EST	MEYLAN,WM ET AL. (1996)
logP 2.27			EST	MEYLAN,WM & HOWARD,PH (1995)
VP 4.08E-002	mm Hg	25	EST	NEELY,WB & BLAU,GE (1985)
DC	pKa			
HL 1.56E-006	atm m3/mol	25	EST	MEYLAN,WM & HOWARD,PH (1991)
OH 1.20E-010	cm3/molc sec	25	EST	ATKINSON,R (1987)

CAS #: 000087-61-6 — 1,2,3-TRICHLOROBENZENE

Formula: $C_6H_3Cl_3$
Mol Weight: 181.45
MP (deg C): 52.6
FP (deg C):
BP (deg C): 221
BP pressure (mm Hg):

Property/Value	Units	Temp	Data Type	Reference
WS 1.80E+001	mg/L	25	EXP	CHIOU,CT ET AL. (1986)
logP 4.05			EXP	SANGSTER,J (1994)
VP 2.10E-001	mm Hg	25	EXP	MACKAY,D ET AL. (1982)
DC	pKa			
HL 1.25E-003	atm m3/mol	25	EXP	MACKAY,D ET AL. (1982A)
OH 2.91E-013	cm3/molc sec	25	EST	MEYLAN,WM & HOWARD,PH (1993)

CAS #: 000087-62-7 — 2,6-DIMETHYLANILINE

Formula: $C_8H_{11}N$
Mol Weight: 121.18
MP (deg C): 11.2
FP (deg C):
BP (deg C): 215
BP pressure (mm Hg):

Property/Value	Units	Temp	Data Type	Reference
WS 8.24E+003	mg/L	25	EXP	HUYSKENS,P ET AL. (1975)
logP 2.17			EST	MEYLAN,WM & HOWARD,PH (1995)
VP 1.30E-001	mm Hg	25	EXP	CHAO,J ET AL. (1983)
DC 3.95	pKa	25	EXP	HUYSKENS,P ET AL. (1975)
HL 2.52E-006	atm m3/mol	25	EST	VP/WSOL
OH 1.95E-010	cm3/molc sec	25	EST	MEYLAN,WM & HOWARD,PH (1993)

CAS #: 000087-64-9 — 2-METHYL-6-CHLOROPHENOL

Formula: C_7H_7ClO
Mol Weight: 142.59
MP (deg C):
FP (deg C):
BP (deg C): 189
BP pressure (mm Hg):

Property/Value	Units	Temp	Data Type	Reference
WS 1.26E+003	mg/L	25	EST	MEYLAN,WM ET AL. (1996)
logP 2.80			EXP	SOTOMATSU,T ET AL. (1993)
VP 4.05E-002	mm Hg	25	EST	NEELY,WB & BLAU,GE (1985)
DC 8.69	pKa	20	EXP	SERJEANT,EP & DEMPSEY,B (1979)
HL 4.58E-007	atm m3/mol	25	EST	MEYLAN,WM & HOWARD,PH (1991)
OH 1.22E-011	cm3/molc sec	25	EST	MEYLAN,WM & HOWARD,PH (1993)

000087-65-0 — 2,6-DICHLOROPHENOL

Formula: $C_6H_4Cl_2O$

Mol Weight: 163.00
MP (deg C): 64.5-5.5
FP (deg C):
BP (deg C): 219
BP pressure (mm Hg):

Property/Value	Units	Temp	Data Type	Reference
WS 2.65E+003	mg/L	25	EXP	SHIU,WY ET AL. (1994)
logP 2.75			EXP	HANSCH,C ET AL. (1995)
VP 9.53E-002	mm Hg	25	EXP	BIDLEMAN,TF & RENBERG,L (1985)
DC 6.79	pKa		EXP	SERJEANT,EP & DEMPSEY,B (1979)
HL 7.71E-006	atm m3/mol	25	EST	VP/WSOL
OH 2.98E-012	cm3/molc sec	25	EST	MEYLAN,WM & HOWARD,PH (1993)

000087-66-1 — 1,2,3-TRIHYDROXYBENZENE

Formula: $C_6H_6O_3$

Mol Weight: 126.11
MP (deg C): 131-133
FP (deg C):
BP (deg C): 309
BP pressure (mm Hg):

Property/Value	Units	Temp	Data Type	Reference
WS 5.07E+005	mg/L	25	EXP	YALKOWSKY,SH & DANNENFELSER,RM (1992)
logP 0.97			EST	MEYLAN,WM & HOWARD,PH (1995)
VP 1.17E-003	mm Hg	25	EXT	PERRY,RH & GREEN,D (1984)
DC 9.01	pKa		EXP	KORTUM,G ET AL (1961)
HL 6.07E-015	atm m3/mol	25	EST	MEYLAN,WM & HOWARD,PH (1991)
OH 2.00E-010	cm3/molc sec	25	EST	MEYLAN,WM & HOWARD,PH (1993)

000087-68-3 — HEXACHLOROBUTADIENE

Formula: C_4Cl_6

Mol Weight: 260.76
MP (deg C): -21
FP (deg C):
BP (deg C): 215
BP pressure (mm Hg):

Property/Value	Units	Temp	Data Type	Reference
WS 3.20E+000	mg/L	25	EXP	BANERJEE,S ET AL. (1980)
logP 4.78			EXP	HANSCH,C & LEO,AJ (1985)
VP 2.20E-001	mm Hg	25	EXP	DAUBERT,TE & DANNER,RP (1989)
DC	pKa			
HL 1.03E-002	atm m3/mol	20	EXP	WARNER,HP ET AL. (1987)
OH 2.24E-014	cm3/molc sec	25	EST	MEYLAN,WM & HOWARD,PH (1993)

000087-69-4 — TARTARIC ACID

Formula: $C_4H_6O_6$

Mol Weight: 150.09
MP (deg C): 169
FP (deg C):
BP (deg C):
BP pressure (mm Hg):

Property/Value	Units	Temp	Data Type	Reference
WS 5.82E+005	mg/L	20	EXP	YALKOWSKY,SH & DANNENFELSER,RM (1992)
logP -1.00			EST	MEYLAN,WM & HOWARD,PH (1995)
VP 1.48E-007	mm Hg	25	EST	NEELY,WB & BLAU,GE (1985)
DC 3.06	pKa	15	EXP	KORTUM,G ET AL (1961)
HL 1.32E-010	atm m3/mol	25	EST	MEYLAN,WM & HOWARD,PH (1991)
OH 1.38E-011	cm3/molc sec	25	EST	MEYLAN,WM & HOWARD,PH (1993)

000087-82-1 — HEXABROMOBENZENE

Formula: C_6Br_6

Mol Weight: 551.52
MP (deg C): 326
FP (deg C):
BP (deg C):
BP pressure (mm Hg):

Property/Value	Units	Temp	Data Type	Reference
WS 1.60E-004	mg/L	25	EXP	OPPERHUIZEN,A (1986)
logP 6.07			EXP	HANSCH,C ET AL. (1995)
VP 1.63E-008	mm Hg	25	EST	NEELY,WB & BLAU,GE (1985)
DC	pKa			
HL 2.81E-005	atm m3/mol	25	EST	MEYLAN,WM & HOWARD,PH (1991)
OH 1.15E-014	cm3/molc sec	25	EST	MEYLAN,WM & HOWARD,PH (1993)

000087-83-2 — PENTABROMOTOLUENE

Formula: $C_7H_3Br_5$

Mol Weight: 486.65
MP (deg C): 288
FP (deg C):
BP (deg C):
BP pressure (mm Hg):

Property/Value	Units	Temp	Data Type	Reference
WS 9.35E-004	mg/L	25	EST	MEYLAN,WM ET AL. (1996)
logP 6.99			EST	MEYLAN,WM & HOWARD,PH (1995)
VP 1.08E-005	mm Hg	25	EST	NEELY,WB & BLAU,GE (1985)
DC	pKa			
HL 5.97E-005	atm m3/mol	25	EST	MEYLAN,WM & HOWARD,PH (1991)
OH 1.85E-013	cm3/molc sec	25	EST	MEYLAN,WM & HOWARD,PH (1993)

000087-84-3 — CYCLOHEXANE, 1,2,3,4,5-PENTABROMO-6-CHLORO-

Formula: $C_6H_6Br_5Cl$

Mol Weight: 513.11
MP (deg C):
FP (deg C):
BP (deg C):
BP pressure (mm Hg):

Property/Value	Units	Temp	Data Type	Reference
WS 5.50E-002	mg/L	25	EST	MEYLAN,WM ET AL. (1996)
logP 4.72			EXP	CHEM INSPECT TEST INST (1992)
VP 3.46E-006	mm Hg	25	EST	NEELY,WB & BLAU,GE (1985)
DC	pKa			
HL 9.59E-007	atm m3/mol	25	EST	MEYLAN,WM & HOWARD,PH (1991)
OH 6.81E-013	cm3/molc sec	25	EST	MEYLAN,WM & HOWARD,PH (1993)

000087-85-4 — HEXAMETHYLBENZENE

Formula: $C_{12}H_{18}$

Mol Weight: 162.28
MP (deg C): 166.5
FP (deg C):
BP (deg C): 263.4
BP pressure (mm Hg):

Property/Value	Units	Temp	Data Type	Reference
WS 2.35E-001	mg/L	25	EXP	WASIK,SP ET AL. (1981)
logP 5.11			EXP	SANGSTER,J (1994)
VP 8.60E-004	mm Hg	25	EXP	JONES,AH (1960)
DC	pKa			
HL 7.81E-004	atm m3/mol	25	EST	VP/WSOL
OH 6.47E-011	cm3/molc sec	25	EST	MEYLAN,WM & HOWARD,PH (1993)

000087-86-5 — PENTACHLOROPHENOL

Formula: C_6HCl_5O
Mol Weight: 266.34
MP (deg C): 174
BP (deg C): 309-310
BP pressure (mm Hg):

Property/Value	Units	Temp	Data Type	Reference
WS 1.95E+003	mg/L	25	EXP	SCOW,K ET AL. (1980)
logP 5.12			EXP	HANSCH,C & LEO,AJ (1985)
VP 1.10E-004	mm Hg	25	EXP	SCOW,K ET AL. (1980)
DC 4.70	pKa	25	EXP	CESSNA,AJ & GROVER,R (1978)
HL 2.45E-008	atm m3/mol	22	EXP	HELLMANN,H (1987)
OH 4.61E-013	cm3/molc sec	25	EST	MEYLAN,WM & HOWARD,PH (1993)

000087-89-8 — INOSITOL

Formula: $C_6H_{12}O_6$
Mol Weight: 180.16
MP (deg C): 225-227
BP (deg C):
BP pressure (mm Hg):

Property/Value	Units	Temp	Data Type	Reference
WS 1.43E+005	mg/L	19	EXP	YALKOWSKY,SH & DANNENFELSER,RM (1992)
logP -2.08			EST	MEYLAN,WM & HOWARD,PH (1995)
VP 2.05E-009	mm Hg	25	EST	NEELY,WB & BLAU,GE (1985)
DC	pKa			
HL 3.20E-013	atm m3/mol	25	EST	MEYLAN,WM & HOWARD,PH (1991)
OH 6.25E-011	cm3/molc sec	25	EST	MEYLAN,WM & HOWARD,PH (1993)

000087-90-1 — 1,3,5-TRICHLOROISOCYANURIC ACID

Formula: $C_3Cl_3N_3O_3$
Mol Weight: 232.41
MP (deg C): 246-247
BP (deg C):
BP pressure (mm Hg):

Property/Value	Units	Temp	Data Type	Reference
WS 1.20E+004	mg/L	25	EXP	BURAKEVICH,JV (1979)
logP 0.94			EST	MEYLAN,WM & HOWARD,PH (1995)
VP 1.25E-007	mm Hg	25	EST	NEELY,WB & BLAU,GE (1985)
DC	pKa			
HL 6.19E-011	atm m3/mol	25	EST	MEYLAN,WM & HOWARD,PH (1991)
OH 3.00E-012	cm3/molc sec	25	EST	MEYLAN,WM & HOWARD,PH (1993)

000087-91-2 — TARTARIC ACID, DIETHYL ESTER

Formula: $C_8H_{14}O_6$
Mol Weight: 206.20
MP (deg C): 17
BP (deg C): 280
BP pressure (mm Hg):

Property/Value	Units	Temp	Data Type	Reference
WS 7.19E+004	mg/L	25	EST	MEYLAN,WM ET AL. (1996)
logP -0.29			EXP	HANSCH,C & LEO,AJ (1985)
VP 1.62E-004	mm Hg	25	EST	NEELY,WB & BLAU,GE (1985)
DC	pKa			
HL 2.38E-005	atm m3/mol	25	EST	MEYLAN,WM & HOWARD,PH (1991)
OH 1.60E-011	cm3/molc sec	25	EST	MEYLAN,WM & HOWARD,PH (1993)

000088-04-0 — 4-CHLORO-3,5-DIMETHYL PHENOL

Formula: C_8H_9ClO
Mol Weight: 156.61
MP (deg C): 115.5
BP (deg C): 246
BP pressure (mm Hg):

Property/Value	Units	Temp	Data Type	Reference
WS 2.50E+002	mg/L	20	EXP	YALKOWSKY,SH & DANNENFELSER,RM (1992)
logP 3.25			EST	MEYLAN,WM & HOWARD,PH (1995)
VP 1.07E-002	mm Hg	25	EST	NEELY,WB & BLAU,GE (1985)
DC 9.70	pKa	25	EXP	KORTUM,G ET AL (1961)
HL 5.06E-007	atm m3/mol	25	EST	MEYLAN,WM & HOWARD,PH (1991)
OH 6.67E-011	cm3/molc sec	25	EST	MEYLAN,WM & HOWARD,PH (1993)

000088-06-2 — 2,4,6-TRICHLOROPHENOL

Formula: $C_6H_3Cl_3O$
Mol Weight: 197.45
MP (deg C): 69
BP (deg C): 246
BP pressure (mm Hg):

Property/Value	Units	Temp	Data Type	Reference
WS 8.00E+002	mg/L	25	EXP	NEELY,WB (1984)
logP 3.69			EXP	HANSCH,C & LEO,AJ (1985)
VP 2.40E-002	mm Hg	25	EXP	BIDLEMAN,TF & RENBERG,L (1985)
DC 5.99	pKa	25	EXP	DRAHONOVSKY,J & VACEK,Z (1971)
HL 7.79E-006	atm m3/mol	25	EST	VP/WSOL
OH 5.20E-013	cm3/molc sec	25	EST	MEYLAN,WM & HOWARD,PH (1993)

000088-09-5 — 2-ETHYLBUTYRIC ACID

Formula: $C_6H_{12}O_2$
Mol Weight: 116.16
MP (deg C): -15
BP (deg C): 194-195
BP pressure (mm Hg):

Property/Value	Units	Temp	Data Type	Reference
WS 1.80E+004	mg/L	20	EXP	RIEMENSCHNEIDER,W (1986)
logP 1.68			EXP	SANGSTER,J (1993)
VP 1.88E-001	mm Hg	25	EXP	DAUBERT,TE & DANNER,RP (1989)
DC 4.71	pKa	20	EXP	KORTUM,G ET AL (1961)
HL 1.60E-006	atm m3/mol	25	EST	VP/WSOL
OH 5.35E-012	cm3/molc sec	25	EST	MEYLAN,WM & HOWARD,PH (1993)

000088-12-0 — N-VINYL-2-PYRROLIDINONE

Formula: C_6H_9NO
Mol Weight: 111.14
MP (deg C):
BP (deg C): 93
BP pressure (mm Hg): 1.10E+001

Property/Value	Units	Temp	Data Type	Reference
WS 5.21E+004	mg/L	25	EST	MEYLAN,WM ET AL. (1996)
logP 0.37			EXP	HANSCH,C & LEO,AJ (1985)
VP 1.14E-001	mm Hg	25	EST	NEELY,WB & BLAU,GE (1985)
DC	pKa			
HL 5.53E-008	atm m3/mol	25	EST	MEYLAN,WM & HOWARD,PH (1991)
OH 3.88E-011	cm3/molc sec	25	EST	MEYLAN,WM & HOWARD,PH (1993)

000088-13-1 — THIOPHENE-3-CARBOXYLIC ACID

CAS #: 000088-13-1

Formula: $C_5H_4O_2S$

Mol Weight: 128.15

MP (deg C): 137-138

FP (deg C):

BP (deg C):

BP pressure (mm Hg):

Property/Value	Units	Temp	Data Type	Reference
WS 4.30E+003	mg/L	25	EXP	YALKOWSKY,SH & DANNENFELSER,RM (1992)
logP 1.50			EXP	HANSCH,C & LEO,AJ (1985)
VP 7.80E-003	mm Hg	25	EST	NEELY,WB & BLAU,GE (1985)
DC	pKa			
HL 5.89E-008	atm m3/mol	25	EST	MEYLAN,WM & HOWARD,PH (1991)
OH 4.05E-012	cm3/molc sec	25	EST	MEYLAN,WM & HOWARD,PH (1993)

000088-14-2 — 2-FURANCARBOXYLIC ACID

CAS #: 000088-14-2

Formula: $C_5H_4O_3$

Mol Weight: 112.09

MP (deg C): 133.5

FP (deg C):

BP (deg C): 231

BP pressure (mm Hg):

Property/Value	Units	Temp	Data Type	Reference
WS 3.71E+004	mg/L	15	EXP	YALKOWSKY,SH & DANNENFELSER,RM (1992)
logP 0.64			EXP	HANSCH,C ET AL. (1995)
VP 1.03E-001	mm Hg	25	EST	NEELY,WB & BLAU,GE (1985)
DC	pKa			
HL 1.08E-007	atm m3/mol	25	EST	MEYLAN,WM & HOWARD,PH (1991)
OH 1.55E-011	cm3/molc sec	25	EST	MEYLAN,WM & HOWARD,PH (1993)

000088-15-3 — 2-ACETYLTHIOPHENE

CAS #: 000088-15-3

Formula: C_6H_6OS

Mol Weight: 126.18

MP (deg C): 10.5

FP (deg C):

BP (deg C): 213.5

BP pressure (mm Hg):

Property/Value	Units	Temp	Data Type	Reference
WS 8.15E+003	mg/L	25	EST	MEYLAN,WM ET AL. (1996)
logP 1.25			EXP	HANSCH,C & LEO,AJ (1985)
VP 3.65E-001	mm Hg	25	EST	NEELY,WB & BLAU,GE (1985)
DC	pKa			
HL 5.33E-006	atm m3/mol	25	EST	MEYLAN,WM & HOWARD,PH (1991)
OH 8.79E-012	cm3/molc sec	25	EST	MEYLAN,WM & HOWARD,PH (1993)

000088-16-4 — CHLORO(TRIFLUOROMETHYL)BENZENE

CAS #: 000088-16-4

Formula: $C_7H_4ClF_3$

Mol Weight: 180.56

MP (deg C): -6

FP (deg C):

BP (deg C): 152.2

BP pressure (mm Hg):

Property/Value	Units	Temp	Data Type	Reference
WS 5.29E+001	mg/L	25	EST	MEYLAN,WM ET AL. (1996)
logP 3.53			EXP	HANSCH,C ET AL. (1995)
VP 5.08E+000	mm Hg	25	EXP	PERRY,RH & GREEN,D (1984)
DC	pKa			
HL 3.47E-002	atm m3/mol	25	EST	MEYLAN,WM & HOWARD,PH (1991)
OH 2.76E-013	cm3/molc sec	25	EST	MEYLAN,WM & HOWARD,PH (1993)

000088-17-5 — BENZENAMINE, 2-(TRIFLUOROMETHYL)-

CAS #: 000088-17-5

Formula: $C_7H_6F_3N$

Mol Weight: 161.13

MP (deg C): 5.5

FP (deg C):

BP (deg C): 187

BP pressure (mm Hg):

Property/Value	Units	Temp	Data Type	Reference
WS 5.92E+002	mg/L	25	EST	MEYLAN,WM ET AL. (1996)
logP 2.41			EXP	HANSCH,C ET AL. (1995)
VP 5.92E-001	mm Hg	25	EST	NEELY,WB & BLAU,GE (1985)
DC	pKa			
HL 1.65E-005	atm m3/mol	25	EST	MEYLAN,WM & HOWARD,PH (1991)
OH 2.16E-011	cm3/molc sec	25	EST	MEYLAN,WM & HOWARD,PH (1993)

000088-18-6 — O-T-BUTYLPHENOL

CAS #: 000088-18-6

Formula: $C_{10}H_{14}O$

Mol Weight: 150.22

MP (deg C): -6.8

FP (deg C): -7.

BP (deg C): 223

BP pressure (mm Hg):

Property/Value	Units	Temp	Data Type	Reference
WS 1.52E+003	mg/L	25	EST	MEYLAN,WM ET AL. (1996)
logP 3.31			EXP	HANSCH,C ET AL. (1995)
VP 1.50E-004	mm Hg	25	EXP	SHIU,WY ET AL. (1994)
DC 10.28	pKa		EXP	SCHUEUERMAN,G (1991)
HL 1.44E-006	atm m3/mol	25	EST	MEYLAN, WM & HOWARD, PH (1991)
OH 4.06E-011	cm3/molc sec	25	EST	MEYLAN,WM & HOWARD,PH (1993)

000088-19-7 — O-METHYLBENZENESULFONAMIDE

CAS #: 000088-19-7

Formula: $C_7H_9NO_2S$

Mol Weight: 171.22

MP (deg C): 156.3

FP (deg C):

BP (deg C): 214

BP pressure (mm Hg): 9.98E+000

Property/Value	Units	Temp	Data Type	Reference
WS 1.62E+003	mg/L	25	EXP	YALKOWSKY,SH & DANNENFELSER,RM (1992)
logP 0.84			EXP	HANSCH,C ET AL. (1995)
VP 6.00E-005	mm Hg	25	EST	NEELY,WB & BLAU,GE (1985)
DC	pKa			
HL 4.70E-007	atm m3/mol	25	EST	MEYLAN,WM & HOWARD,PH (1991)
OH 1.22E-012	cm3/molc sec	25	EST	MEYLAN,WM & HOWARD,PH (1993)

000088-20-0 — O-TOLUENE SULFONIC ACID

CAS #: 000088-20-0

Formula: $C_6H_7NO_3S$

Mol Weight: 173.19

MP (deg C): 65.7

FP (deg C):

BP (deg C): 128.8

BP pressure (mm Hg): 2.50E+001

Property/Value	Units	Temp	Data Type	Reference
WS 2.02E+005	mg/L	25	EST	MEYLAN,WM ET AL. (1996)
logP -0.62			EST	MEYLAN,WM & HOWARD,PH (1995)
VP 6.75E+000	mm Hg	20	EXP	WEBER,RC ET AL. (1981)
DC	pKa			
HL 2.78E-009	atm m3/mol	25	EST	MEYLAN,WM & HOWARD,PH (1991)
OH 1.36E-012	cm3/molc sec	25	EST	MEYLAN,WM & HOWARD,PH (1993)

CAS #: 000088-21-1				O-ANILINESULFONIC ACID

Formula: $C_6H_7NO_3S$

Mol Weight: 173.19

MP (deg C): >320 FP (deg C):

BP (deg C):

BP pressure (mm Hg):

Property/Value	Units	Temp	Data Type	Reference
WS 1.06E+004	mg/L	7	EXP	STEPHEN,H & STEPHEN,T (1963)
logP -2.08			EST	MEYLAN,WM & HOWARD,PH (1995)
VP 2.02E-007	mm Hg	25	EST	NEELY,WB & BLAU,GE (1985)
DC	pKa			
HL 8.89E-013	atm m3/mol	25	EST	MEYLAN,WM & HOWARD,PH (1991)
OH 2.32E-011	cm3/molc sec	25	EST	MEYLAN,WM & HOWARD,PH (1993)

CAS #: 000088-30-2				3-TRIFLUOROMETHYL-4-NITROPHENOL

Formula: $C_7H_4F_3NO_3$

Mol Weight: 207.11

MP (deg C): 76 FP (deg C):

BP (deg C):

BP pressure (mm Hg):

Property/Value	Units	Temp	Data Type	Reference
WS 5.00E+003	mg/L	25	EXP	THINGVOLD,DA & LEE,GF (1981)
logP 2.87			EST	MEYLAN,WM & HOWARD,PH (1995)
VP 1.33E-003	mm Hg	25	EST	NEELY,WB & BLAU,GE (1985)
DC 6.07	pKa		EXP	SMITH,MA ET AL. (1961)
HL 1.92E-008	atm m3/mol	25	EST	MEYLAN,WM & HOWARD,PH (1991)
OH 7.70E-013	cm3/molc sec	25	EST	MEYLAN,WM & HOWARD,PH (1993)

CAS #: 000088-43-7				4-CHLOROANILINE-3-SULFONIC ACID

Formula: $C_6H_6ClNO_3S$

Mol Weight: 207.64

MP (deg C): FP (deg C):

BP (deg C):

BP pressure (mm Hg):

Property/Value	Units	Temp	Data Type	Reference
WS 1.13E+004	mg/L	0	EXP	STEPHEN,H & STEPHEN,T (1963)
logP -1.44			EST	MEYLAN,WM & HOWARD,PH (1995)
VP 4.37E-008	mm Hg	25	EST	NEELY,WB & BLAU,GE (1985)
DC	pKa			
HL 6.59E-013	atm m3/mol	25	EST	MEYLAN,WM & HOWARD,PH (1991)
OH 6.86E-012	cm3/molc sec	25	EST	MEYLAN,WM & HOWARD,PH (1993)

CAS #: 000088-65-3				O-BROMOBENZOIC ACID

Formula: $C_7H_5BrO_2$

Mol Weight: 201.03

MP (deg C): 150 FP (deg C):

BP (deg C):

BP pressure (mm Hg):

Property/Value	Units	Temp	Data Type	Reference
WS 1.86E+003	mg/L	25	EXP	STEPHEN,H & STEPHEN,T (1963)
logP 2.20			EXP	HANSCH,C & LEO,AJ (1985)
VP 8.27E-004	mm Hg	25	EST	NEELY,WB & BLAU,GE (1985)
DC 2.88	pKa	20	EXP	KORTUM,G ET AL (1961)
HL 4.32E-008	atm m3/mol	25	EST	MEYLAN,WM & HOWARD,PH (1991)
OH 9.74E-013	cm3/molc sec	25	EST	MEYLAN,WM & HOWARD,PH (1993)

CAS #: 000088-67-5				2-IODOBENZOIC ACID

Formula: $C_7H_5IO_2$

Mol Weight: 248.02

MP (deg C): 162 FP (deg C):

BP (deg C):

BP pressure (mm Hg):

Property/Value	Units	Temp	Data Type	Reference
WS 4.61E+002	mg/L	15	EXP	YALKOWSKY,SH & DANNENFELSER,RM (1992)
logP 2.40			EXP	HANSCH,C & LEO,AJ (1985)
VP 1.98E-004	mm Hg	25	EST	NEELY,WB & BLAU,GE (1985)
DC 2.93	pKa	20	EXP	KORTUM,G ET AL (1961)
HL 2.51E-008	atm m3/mol	25	EST	MEYLAN,WM & HOWARD,PH (1991)
OH 9.96E-013	cm3/molc sec	25	EST	MEYLAN,WM & HOWARD,PH (1993)

CAS #: 000088-68-6				BENZAMIDE, 2-AMINO-

Formula: $C_7H_8N_2O$

Mol Weight: 136.15

MP (deg C): 110.5 dec FP (deg C):

BP (deg C): 300

BP pressure (mm Hg):

Property/Value	Units	Temp	Data Type	Reference
WS 2.14E+003	mg/L	25	EST	MEYLAN,WM ET AL. (1996)
logP 0.35			EXP	HANSCH,C ET AL. (1995)
VP 2.63E-005	mm Hg	25	EST	NEELY,WB & BLAU,GE (1985)
DC	pKa			
HL 7.83E-013	atm m3/mol	25	EST	MEYLAN,WM & HOWARD,PH (1991)
OH 1.00E-010	cm3/molc sec	25	EST	MEYLAN,WM & HOWARD,PH (1993)

CAS #: 000088-69-7				O-ISOPROPYLPHENOL

Formula: $C_9H_{12}O$

Mol Weight: 136.20

MP (deg C): 15.5 FP (deg C):

BP (deg C): 213.5

BP pressure (mm Hg):

Property/Value	Units	Temp	Data Type	Reference
WS 1.15E+003	mg/L	25	EST	MEYLAN,WM ET AL. (1996)
logP 2.88			EXP	HANSCH,C & LEO,AJ (1985)
VP 9.89E-002	mm Hg	25	EXT	PERRY,RH & GREEN,D (1984)
DC 10.47	pKa	20	EXP	SERJEANT,EP & DEMPSEY,B (1979)
HL 1.09E-006	atm m3/mol	25	EST	MEYLAN,WM & HOWARD,PH (1991)
OH 4.25E-011	cm3/molc sec	25	EST	MEYLAN,WM & HOWARD,PH (1993)

CAS #: 000088-72-2				2-NITROTOLUENE

Formula: $C_7H_7NO_2$

Mol Weight: 137.14

MP (deg C): -10 FP (deg C):

BP (deg C): 222

BP pressure (mm Hg):

Property/Value	Units	Temp	Data Type	Reference
WS 6.50E+002	mg/L	30	EXP	YALKOWSKY,SH & DANNENFELSER,RM (1992)
logP 2.30			EXP	HANSCH,C & LEO,AJ (1985)
VP 1.88E-001	mm Hg	25	EXT	PERRY,RH & GREEN,D (1984)
DC	pKa			
HL 2.35E-005	atm m3/mol	25	EST	MEYLAN,WM & HOWARD,PH (1991)
OH 7.00E-013	cm3/molc sec	25	EXP	ATKINSON,R (1989)

CAS #: 000088-73-3				2-CHLORO-1-NITROBENZENE

Formula: $C_6H_4ClNO_2$
Mol Weight: 157.56
MP (deg C): 32.5 FP (deg C):
BP (deg C): 245.5
BP pressure (mm Hg):

Property/Value	Units	Temp	Data Type	Reference
WS 4.41E+002	mg/L	20	EXP	YALKOWSKY,SH & DANNENFELSER,RM (1992)
logP 2.24			EXP	HANSCH,C & LEO,AJ (1985)
VP 1.82E-002	mm Hg	25	EXT	DAUBERT,TE & DANNER,RP (1989)
DC	pKa			
HL 4.45E-005	atm m3/mol	22	EXP	HELLMANN,H (1987)
OH 1.76E-013	cm3/molc sec	25	EST	MEYLAN,WM & HOWARD,PH (1993)

CAS #: 000088-74-4				2-NITROANILINE

Formula: $C_6H_6N_2O_2$
Mol Weight: 138.13
MP (deg C): 69-71 FP (deg C):
BP (deg C):
BP pressure (mm Hg):

Property/Value	Units	Temp	Data Type	Reference
WS 1.47E+003	mg/L	30	EXP	SUZUKI,T (1991)
logP 1.85			EXP	HANSCH,C & LEO,AJ (1985)
VP 2.77E-003	mm Hg	25	EXP	DAUBERT,TE & DANNER,RP (1989)
DC -0.28	pKa	25	EXP	DEAN,JA (1985)
HL 1.81E-008	atm m3/mol	25	EST	MEYLAN,WM & HOWARD,PH (1991)
OH 1.35E-011	cm3/molc sec	25	EST	MEYLAN,WM & HOWARD,PH (1993)

CAS #: 000088-75-5				2-NITROPHENOL

Formula: $C_6H_5NO_3$
Mol Weight: 139.11
MP (deg C): 44-45 FP (deg C):
BP (deg C): 214-216
BP pressure (mm Hg):

Property/Value	Units	Temp	Data Type	Reference
WS 2.19E+003	mg/L	25	EXP	YALKOWSKY,SH & DANNENFELSER,RM (1992)
logP 1.79			EXP	HANSCH,C & LEO,AJ (1985)
VP 1.13E-001	mm Hg	25	EXP	SCALA,AJ & BANERJEE,S (1982)
DC 7.23	pKa	25	EXP	SERJEANT,EP & DEMPSEY,B (1979)
HL 9.47E-004	atm m3/mol	25	EST	VP/WSOL
OH 9.00E-013	cm3/molc sec	25	EXP	ATKINSON,R (1989)

CAS #: 000088-82-4				2,3,5-TRIIODOBENZOIC ACID

Formula: $C_7H_3I_3O_2$
Mol Weight: 499.81
MP (deg C): FP (deg C):
BP (deg C):
BP pressure (mm Hg):

Property/Value	Units	Temp	Data Type	Reference
WS 3.61E-002	mg/L	25	EST	MEYLAN,WM ET AL. (1996)
logP 5.03			EST	MEYLAN,WM & HOWARD,PH (1995)
VP 1.34E-007	mm Hg	25	EST	NEELY,WB & BLAU,GE (1985)
DC	pKa			
HL 1.35E-009	atm m3/mol	25	EST	MEYLAN,WM & HOWARD,PH (1991)
OH 5.96E-013	cm3/molc sec	25	EST	MEYLAN,WM & HOWARD,PH (1993)

CAS #: 000088-85-7				DINOSEB

Formula: $C_{10}H_{12}N_2O_5$
Mol Weight: 240.22
MP (deg C): 40 FP (deg C):
BP (deg C):
BP pressure (mm Hg):

Property/Value	Units	Temp	Data Type	Reference
WS 5.20E+001	mg/L	25	EXP	WSSA (1983)
logP 3.56			EXP	HANSCH,C ET AL. (1995)
VP 7.50E-005	mm Hg	24	NR	HARTLEY,D & KIDD,H (1983)
DC 4.62	pKa		EXP	MERCK INDEX (1983)
HL 4.56E-007	atm m3/mol	24	EST	VP/WSOL
OH 3.87E-012	cm3/molc sec	25	EST	MEYLAN,WM & HOWARD,PH (1993)

CAS #: 000088-86-8				DINOBEN

Formula: $C_7H_3Cl_2NO_4$
Mol Weight: 236.01
MP (deg C): FP (deg C):
BP (deg C):
BP pressure (mm Hg):

Property/Value	Units	Temp	Data Type	Reference
WS 6.42E+001	mg/L	25	EST	MEYLAN,WM ET AL. (1996)
logP 2.64			EST	MEYLAN,WM & HOWARD,PH (1995)
VP 4.39E-006	mm Hg	25	EST	NEELY,WB & BLAU,GE (1985)
DC	pKa			
HL 2.35E-010	atm m3/mol	25	EST	MEYLAN,WM & HOWARD,PH (1991)
OH 5.30E-013	cm3/molc sec	25	EST	MEYLAN,WM & HOWARD,PH (1993)

CAS #: 000088-87-9				4-CHLORO-2,6-DINITROPHENOL

Formula: $C_6H_3ClN_2O_5$
Mol Weight: 218.55
MP (deg C): 81 FP (deg C):
BP (deg C):
BP pressure (mm Hg):

Property/Value	Units	Temp	Data Type	Reference
WS 3.31E+002	mg/L	25	EST	MEYLAN,WM ET AL. (1996)
logP 2.37			EST	MEYLAN,WM & HOWARD,PH (1995)
VP 3.01E-006	mm Hg	25	EST	NEELY,WB & BLAU,GE (1985)
DC 2.96	pKa	25	EXP	KORTUM,G ET AL (1961)
HL 2.05E-008	atm m3/mol	25	EST	MEYLAN,WM & HOWARD,PH (1991)
OH 1.47E-013	cm3/molc sec	25	EST	MEYLAN,WM & HOWARD,PH (1993)

CAS #: 000088-88-0				2,4,6-TRINITROCHLOROBENZENE

Formula: $C_6H_2ClN_3O_6$
Mol Weight: 247.55
MP (deg C): 83 FP (deg C):
BP (deg C):
BP pressure (mm Hg):

Property/Value	Units	Temp	Data Type	Reference
WS 5.30E+002	mg/L	16	EXP	YALKOWSKY,SH & DANNENFELSER,RM (1992)
logP 2.09			EST	MEYLAN,WM & HOWARD,PH (1995)
VP 3.83E-006	mm Hg	25	EST	NEELY,WB & BLAU,GE (1985)
DC	pKa			
HL 2.45E-010	atm m3/mol	25	EST	MEYLAN,WM & HOWARD,PH (1991)
OH 3.80E-016	cm3/molc sec	25	EST	MEYLAN,WM & HOWARD,PH (1993)

CAS #: 000088-89-1 — 2,4,6-TRINITROPHENOL

Formula:	$C_6H_3N_3O_7$
Mol Weight:	229.11
MP (deg C):	122-123
FP (deg C):	
BP (deg C):	
BP pressure (mm Hg):	

Property/Value	Units	Temp	Data Type	Reference
WS 1.27E+004	mg/L	25	EXP	YALKOWSKY,SH & DANNENFELSER,RM (1992)
logP 1.33			EXP	SANGSTER,J (1994)
VP 2.27E-007	mm Hg	25	EST	NEELY,WB & BLAU,GE (1985)
DC 0.38	pKa	25	EXP	WEAST,RC ET AL. (1985)
HL 3.79E-013	atm m3/mol	25	EST	MEYLAN,WM & HOWARD,PH (1991)
OH 3.69E-014	cm3/molc sec	25	EST	MEYLAN,WM & HOWARD,PH (1993)

CAS #: 000088-95-9 — PHTHALOYL CHLORIDE

Formula:	$C_8H_4Cl_2O_2$
Mol Weight:	203.03
MP (deg C):	15-16
FP (deg C):	
BP (deg C):	280-282
BP pressure (mm Hg):	

Property/Value	Units	Temp	Data Type	Reference
WS 7.48E+003	mg/L	25	EST	MEYLAN,WM ET AL. (1996)
logP 0.88			EST	MEYLAN,WM & HOWARD,PH (1995)
VP 1.62E-002	mm Hg	25	EXT	PERRY,RH & GREEN,D (1984)
DC	pKa			
HL 3.22E-006	atm m3/mol	25	EST	MEYLAN,WM & HOWARD,PH (1991)
OH 7.49E-013	cm3/molc sec	25	EST	MEYLAN,WM & HOWARD,PH (1993)

CAS #: 000088-96-0 — 1,2-BENZENEDICARBOXAMIDE

Formula:	$C_8H_8N_2O_2$
Mol Weight:	164.17
MP (deg C):	228
FP (deg C):	
BP (deg C):	
BP pressure (mm Hg):	

Property/Value	Units	Temp	Data Type	Reference
WS 2.00E+002	mg/L	20	EXP	YALKOWSKY,SH & DANNENFELSER,RM (1992)
logP -1.73			EXP	HANSCH,C ET AL. (1995)
VP 3.25E-008	mm Hg	25	EST	NEELY,WB & BLAU,GE (1985)
DC	pKa			
HL 1.40E-012	atm m3/mol	25	EST	MEYLAN,WM & HOWARD,PH (1991)
OH 4.75E-012	cm3/molc sec	25	EST	MEYLAN,WM & HOWARD,PH (1993)

CAS #: 000088-99-3 — O-PHTHALIC ACID

Formula:	$C_8H_6O_4$
Mol Weight:	166.13
MP (deg C):	230
FP (deg C):	
BP (deg C):	
BP pressure (mm Hg):	

Property/Value	Units	Temp	Data Type	Reference
WS 1.42E+004	mg/L	25	EXP	YALKOWSKY,SH & DANNENFELSER,RM (1992)
logP 0.73			EXP	HANSCH,C & LEO,AJ (1985)
VP 6.36E-007	mm Hg	25	EXT	DAUBERT,TE & DANNER,RP (1991)
DC 2.76	pKa	25	EXP	SERJEANT,EP & DEMPSEY,B (1979)
HL 2.18E-012	atm m3/mol	25	EST	MEYLAN,WM & HOWARD,PH (1991)
OH 1.24E-012	cm3/molc sec	25	EST	MEYLAN,WM & HOWARD,PH (1993)

CAS #: 000089-00-9 — QUINOLINIC ACID

Formula:	$C_7H_5NO_4$
Mol Weight:	167.12
MP (deg C):	190
FP (deg C):	
BP (deg C):	
BP pressure (mm Hg):	

Property/Value	Units	Temp	Data Type	Reference
WS 1.10E+004	mg/L	25	EXP	YALKOWSKY,SH & DANNENFELSER,RM (1992)
logP -0.12			EST	MEYLAN,WM & HOWARD,PH (1995)
VP 6.10E-006	mm Hg	25	EST	NEELY,WB & BLAU,GE (1985)
DC 2.43	pKa		EXP	PERRIN,DD (1972)
HL 2.86E-015	atm m3/mol	25	EST	MEYLAN,WM & HOWARD,PH (1991)
OH 1.08E-012	cm3/molc sec	25	EST	MEYLAN,WM & HOWARD,PH (1993)

CAS #: 000089-05-4 — PYROMELLITIC ACID

Formula:	$C_{10}H_6O_8$
Mol Weight:	254.15
MP (deg C):	276
FP (deg C):	
BP (deg C):	
BP pressure (mm Hg):	

Property/Value	Units	Temp	Data Type	Reference
WS 1.50E+004	mg/L	25	EXP	BEMIS,AG ET AL. (1982)
logP 0.15			EST	MEYLAN,WM & HOWARD,PH (1995)
VP 8.74E-011	mm Hg	25	EST	NEELY,WB & BLAU,GE (1985)
DC 1.87	pKa	25	EXP	KORTUM,G ET AL (1961)
HL 8.84E-022	atm m3/mol	25	EST	MEYLAN,WM & HOWARD,PH (1991)
OH 2.10E-012	cm3/molc sec	25	EST	MEYLAN,WM & HOWARD,PH (1993)

CAS #: 000089-24-7 — 5-PHENYL-2,4-IMIDAZOLIDIONE

Formula:	$C_9H_8N_2O_2$
Mol Weight:	176.18
MP (deg C):	
FP (deg C):	
BP (deg C):	
BP pressure (mm Hg):	

Property/Value	Units	Temp	Data Type	Reference
WS 2.33E+004	mg/L	25	EST	MEYLAN,WM ET AL. (1996)
logP 0.46			EXP	LIPINSKI,CA ET AL. (1991)
VP 2.89E-008	mm Hg	25	EST	NEELY,WB & BLAU,GE (1985)
DC	pKa			
HL 1.27E-010	atm m3/mol	25	EST	MEYLAN,WM & HOWARD,PH (1991)
OH 1.14E-011	cm3/molc sec	25	EST	MEYLAN,WM & HOWARD,PH (1993)

CAS #: 000089-32-7 — PYROMELLITIC DIANHYDRIDE

Formula:	$C_{10}H_2O_6$
Mol Weight:	218.12
MP (deg C):	152.5
FP (deg C):	
BP (deg C):	
BP pressure (mm Hg):	

Property/Value	Units	Temp	Data Type	Reference
WS 5.97E+001	mg/L	25	EST	MEYLAN,WM ET AL. (1996)
logP 2.14			EST	MEYLAN,WM & HOWARD,PH (1995)
VP 1.47E-005	mm Hg	25	EST	NEELY,WB & BLAU,GE (1985)
DC	pKa			
HL 7.47E-009	atm m3/mol	25	EST	MEYLAN,WM & HOWARD,PH (1991)
OH 2.88E-013	cm3/molc sec	25	EST	MEYLAN,WM & HOWARD,PH (1993)

CAS #: 000089-40-7 — 4-NITROPHTHALIMIDE

Formula: $C_8H_4N_2O_4$

Mol Weight: 192.13

MP (deg C): 200-202 FP (deg C):

BP (deg C):

BP pressure (mm Hg):

Property/Value	Units	Temp	Data Type	Reference
WS 2.18E+003	mg/L	25	EST	MEYLAN,WM ET AL. (1996)
logP 1.12			EST	MEYLAN,WM & HOWARD,PH (1995)
VP 1.13E-009	mm Hg	25	EST	NEELY,WB & BLAU,GE (1985)
DC	pKa			
HL 4.03E-011	atm m3/mol	25	EST	MEYLAN,WM & HOWARD,PH (1991)
OH 5.59E-012	cm3/molc sec	25	EST	MEYLAN,WM & HOWARD,PH (1993)

CAS #: 000089-52-1 — N-ACETYL O-AMINOBENZOIC ACID

Formula: $C_9H_9NO_3$

Mol Weight: 179.18

MP (deg C): 184-187 FP (deg C):

BP (deg C):

BP pressure (mm Hg):

Property/Value	Units	Temp	Data Type	Reference
WS 1.38E+003	mg/L	25	EST	MEYLAN,WM ET AL. (1996)
logP 1.88			EXP	HANSCH,C & LEO,AJ (1985)
VP 7.14E-007	mm Hg	25	EST	NEELY,WB & BLAU,GE (1985)
DC	pKa			
HL 1.24E-013	atm m3/mol	25	EST	MEYLAN,WM & HOWARD,PH (1991)
OH 5.22E-012	cm3/molc sec	25	EST	MEYLAN,WM & HOWARD,PH (1993)

CAS #: 000089-55-4 — 5-BROMOSALICYLIC ACID

Formula: $C_7H_5BrO_3$

Mol Weight: 217.03

MP (deg C): 169.8 FP (deg C):

BP (deg C):

BP pressure (mm Hg):

Property/Value	Units	Temp	Data Type	Reference
WS 2.37E+002	mg/L	25	EST	MEYLAN,WM ET AL. (1996)
logP 3.23			EXP	HANSCH,C & LEO,AJ (1985)
VP 1.07E-005	mm Hg	25	EST	NEELY,WB & BLAU,GE (1985)
DC 2.66	pKa	25	EXP	SERJEANT,EP & DEMPSEY,B (1979)
HL 5.66E-009	atm m3/mol	25	EST	MEYLAN,WM & HOWARD,PH (1991)
OH 4.20E-012	cm3/molc sec	25	EST	MEYLAN,WM & HOWARD,PH (1993)

CAS #: 000089-56-5 — 5-METHYLSALICYLIC ACID

Formula: $C_8H_8O_3$

Mol Weight: 152.15

MP (deg C): 151 FP (deg C):

BP (deg C):

BP pressure (mm Hg):

Property/Value	Units	Temp	Data Type	Reference
WS 2.19E+004	mg/L	100	EXP	YALKOWSKY,SH & DANNENFELSER,RM (1992)
logP 2.78			EXP	HANSCH,C & LEO,AJ (1985)
VP 5.14E-005	mm Hg	25	EST	NEELY,WB & BLAU,GE (1985)
DC 3.15	pKa	25	EXP	SERJEANT,EP & DEMPSEY,B (1979)
HL 1.57E-008	atm m3/mol	25	EST	MEYLAN,WM & HOWARD,PH (1991)
OH 1.59E-011	cm3/molc sec	25	EST	MEYLAN,WM & HOWARD,PH (1993)

CAS #: 000089-59-8 — BENZENE, 4-CHLORO-1-METHYL-2-NITRO-

Formula: $C_7H_6ClNO_2$

Mol Weight: 171.58

MP (deg C): 38 FP (deg C):

BP (deg C): 242

BP pressure (mm Hg):

Property/Value	Units	Temp	Data Type	Reference
WS 6.12E+001	mg/L	25	EST	MEYLAN,WM ET AL. (1996)
logP 3.05			EXP	DENEER,JW ET AL. (1987)
VP 1.06E-002	mm Hg	25	EST	NEELY,WB & BLAU,GE (1985)
DC	pKa			
HL 1.74E-005	atm m3/mol	25	EST	MEYLAN,WM & HOWARD,PH (1991)
OH 3.22E-013	cm3/molc sec	25	EST	MEYLAN,WM & HOWARD,PH (1993)

CAS #: 000089-61-2 — 2,5-DICHLORONITROBENZENE

Formula: $C_6H_3Cl_2NO_2$

Mol Weight: 192.00

MP (deg C): 56 FP (deg C):

BP (deg C): 267

BP pressure (mm Hg):

Property/Value	Units	Temp	Data Type	Reference
WS 9.22E+001	mg/L	20	EXP	YALKOWSKY,SH & DANNENFELSER,RM (1992)
logP 3.09			EXP	HANSCH,C & LEO,AJ (1985)
VP 5.05E-003	mm Hg	25	EST	NEELY,WB & BLAU,GE (1985)
DC	pKa			
HL 1.17E-005	atm m3/mol	25	EST	MEYLAN,WM & HOWARD,PH (1991)
OH 5.01E-014	cm3/molc sec	25	EST	MEYLAN,WM & HOWARD,PH (1993)

CAS #: 000089-62-3 — 2-NITRO-P-TOLUIDINE

Formula: $C_7H_8N_2O_2$

Mol Weight: 152.15

MP (deg C): 116.3 FP (deg C):

BP (deg C):

BP pressure (mm Hg):

Property/Value	Units	Temp	Data Type	Reference
WS 4.78E+002	mg/L	25	EST	MEYLAN,WM ET AL. (1996)
logP 2.57			EST	MEYLAN,WM & HOWARD,PH (1995)
VP 9.72E-004	mm Hg	25	EST	NEELY,WB & BLAU,GE (1985)
DC 0.40	pKa	25	EXP	PERRIN,DD (1972)
HL 1.81E-007	atm m3/mol	25	EST	MEYLAN,WM & HOWARD,PH (1991)
OH 1.66E-011	cm3/molc sec	25	EST	MEYLAN,WM & HOWARD,PH (1993)

CAS #: 000089-63-4 — 4-CHLORO-2-NITROANILINE

Formula: $C_6H_5ClN_2O_2$

Mol Weight: 172.57

MP (deg C): 117-119 FP (deg C):

BP (deg C):

BP pressure (mm Hg):

Property/Value	Units	Temp	Data Type	Reference
WS 2.84E+002	mg/L	25	EST	MEYLAN,WM ET AL. (1996)
logP 2.72			EXP	DEBNATH,AK ET AL. (1992)
VP 4.85E-004	mm Hg	25	EST	NEELY,WB & BLAU,GE (1985)
DC -1.02	pKa	25	EXP	PERRIN,DD (1965)
HL 1.22E-007	atm m3/mol	25	EST	MEYLAN,WM & HOWARD,PH (1991)
OH 3.93E-012	cm3/molc sec	25	EST	MEYLAN,WM & HOWARD,PH (1993)

CAS #: 000089-64-5				4-CHLORO-2-NITROPHENOL

Formula: $C_6H_4ClNO_3$

Mol Weight: 173.56

MP (deg C): 85-87 FP (deg C):

BP (deg C):

BP pressure (mm Hg):

Property/ Value	Units	Temp	Data Type	Reference
WS 1.41E+002	mg/L	20	EXP	SCHWARZENBACH,RP ET AL. (1988)
logP 2.46			EXP	SCHWARZENBACH,RP ET AL. (1988)
VP 7.78E-003	mm Hg	20	EXP	SCHWARZENBACH,RP ET AL. (1988)
DC 6.46	pKa	25	EXP	SERJEANT,EP & DEMPSEY,B (1979)
HL 1.26E-005	atm m3/mol	25	EST	VP/WSOL
OH 1.36E-012	cm3/molc sec	25	EST	MEYLAN,WM & HOWARD,PH (1993)

CAS #: 000089-68-9				CHLOROTHYMOL

Formula: $C_{10}H_{13}ClO$

Mol Weight: 184.67

MP (deg C): 62-64 FP (deg C):

BP (deg C):

BP pressure (mm Hg):

Property/ Value	Units	Temp	Data Type	Reference
WS 8.92E+001	mg/L	25	EST	MEYLAN,WM ET AL. (1996)
logP 3.92			EXP	HANSCH,C & LEO,AJ (1985)
VP 2.11E-003	mm Hg	25	EST	NEELY,WB & BLAU,GE (1985)
DC 9.98	pKa	20	EXP	SERJEANT,EP & DEMPSEY,B (1979)
HL 8.92E-007	atm m3/mol	25	EST	MEYLAN,WM & HOWARD,PH (1991)
OH 3.31E-011	cm3/molc sec	25	EST	MEYLAN,WM & HOWARD,PH (1993)

CAS #: 000089-69-0				2,4,5-TRICHLORONITROBENZENE

Formula: $C_6H_2Cl_3NO_2$

Mol Weight: 226.45

MP (deg C): 57.5 FP (deg C):

BP (deg C): 288

BP pressure (mm Hg):

Property/ Value	Units	Temp	Data Type	Reference
WS 2.94E+001	mg/L	20	EXP	YALKOWSKY,SH & DANNENFELSER,RM (1992)
logP 3.48			EXP	HANSCH,C & LEO,AJ (1985)
VP 7.92E-004	mm Hg	25	EST	NEELY,WB & BLAU,GE (1985)
DC	pKa			
HL 8.65E-006	atm m3/mol	25	EST	MEYLAN,WM & HOWARD,PH (1991)
OH 3.52E-014	cm3/molc sec	25	EST	MEYLAN,WM & HOWARD,PH (1993)

CAS #: 000089-71-4				O-TOLUIC ACID, METHYL ESTER

Formula: $C_9H_{10}O_2$

Mol Weight: 150.18

MP (deg C): <-50 FP (deg C):

BP (deg C): 215

BP pressure (mm Hg):

Property/ Value	Units	Temp	Data Type	Reference
WS 3.40E+002	mg/L	25	EST	MEYLAN,WM ET AL. (1996)
logP 2.75			EXP	HANSCH,C & LEO,AJ (1985)
VP 1.64E-001	mm Hg	25	EST	NEELY,WB & BLAU,GE (1985)
DC	pKa			
HL 3.83E-005	atm m3/mol	25	EST	MEYLAN,WM & HOWARD,PH (1991)
OH 1.99E-012	cm3/molc sec	25	EST	MEYLAN,WM & HOWARD,PH (1993)

CAS #: 000089-72-5				O-SEC-BUTYLPHENOL

Formula: $C_{10}H_{14}O$

Mol Weight: 150.22

MP (deg C): 14 FP (deg C):

BP (deg C): 224-237

BP pressure (mm Hg):

Property/ Value	Units	Temp	Data Type	Reference
WS 1.66E+003	mg/L	25	EST	MEYLAN,WM ET AL. (1996)
logP 3.27			EXP	HANSCH,C & LEO,AJ (1985)
VP 5.00E-002	mm Hg	25	EST	NEELY,WB & BLAU,GE (1985)
DC	pKa			
HL 2.13E-006	atm m3/mol	25	EST	MEYLAN,WM & HOWARD,PH (1991)
OH 4.40E-011	cm3/molc sec	25	EST	MEYLAN,WM & HOWARD,PH (1993)

CAS #: 000089-78-1				MENTHOL

Formula: $C_{10}H_{20}O$

Mol Weight: 156.27

MP (deg C): 41-43 FP (deg C):

BP (deg C): 212

BP pressure (mm Hg):

Property/ Value	Units	Temp	Data Type	Reference
WS 4.56E+002	mg/L	25	EXP	YALKOWSKY,SH & DANNENFELSER,RM (1992)
logP 3.38			EST	MEYLAN,WM & HOWARD,PH (1995)
VP 1.10E-001	mm Hg	25	EXT	JORDAN,TE (1954)
DC	pKa			
HL 1.52E-005	atm m3/mol	25	EST	MEYLAN,WM & HOWARD,PH (1991)
OH 2.41E-011	cm3/molc sec	25	EST	MEYLAN,WM & HOWARD,PH (1993)

CAS #: 000089-80-5				5-METHYL-2-(1-METHYLETHYL)-CYCLOHEXANONE (TR*)

Formula: $C_{10}H_{18}O$

Mol Weight: 154.25

MP (deg C): FP (deg C):

BP (deg C):

BP pressure (mm Hg):

Property/ Value	Units	Temp	Data Type	Reference
WS 6.88E+002	mg/L	25	EXP	SUZUKI,T (1991)
logP 2.87			EST	MEYLAN,WM & HOWARD,PH (1995)
VP 2.78E-001	mm Hg	25	EST	NEELY,WB & BLAU,GE (1985)
DC	pKa			
HL 1.59E-004	atm m3/mol	25	EST	MEYLAN,WM & HOWARD,PH (1991)
OH 2.63E-011	cm3/molc sec	25	EST	MEYLAN,WM & HOWARD,PH (1993)

CAS #: 000089-83-8				THYMOL

Formula: $C_{10}H_{14}O$

Mol Weight: 150.22

MP (deg C): 51.5 FP (deg C):

BP (deg C): 233

BP pressure (mm Hg):

Property/ Value	Units	Temp	Data Type	Reference
WS 9.00E+002	mg/L	20	EXP	YALKOWSKY,SH & DANNENFELSER,RM (1992)
logP 3.30			EXP	HANSCH,C & LEO,AJ (1985)
VP 6.01E-002	mm Hg	25	EXT	PERRY,RH & GREEN,D (1984)
DC 10.62	pKa	20	EXP	SERJEANT,EP & DEMPSEY,B (1979)
HL 1.20E-006	atm m3/mol	25	EST	MEYLAN,WM & HOWARD,PH (1991)
OH 1.07E-010	cm3/molc sec	25	EST	MEYLAN,WM & HOWARD,PH (1993)

CAS #: 000089-86-1 — 2,4-DIHYDROXYBENZOIC ACID

Formula: $C_7H_6O_4$

Mol Weight: 154.12

MP (deg C): 213

FP (deg C):

BP (deg C):

BP pressure (mm Hg):

Property/Value	Units	Temp	Data Type	Reference
WS 1.12E+004	mg/L	25	EST	MEYLAN,WM ET AL. (1996)
logP 1.63			EXP	HANSCH,C & LEO,AJ (1985)
VP 2.10E-006	mm Hg	25	EST	NEELY,WB & BLAU,GE (1985)
DC 3.11	pKa	25	EXP	KORTUM,G ET AL (1961)
HL 1.48E-012	atm m3/mol	25	EST	MEYLAN,WM & HOWARD,PH (1991)
OH 2.01E-010	cm3/molc sec	25	EST	MEYLAN,WM & HOWARD,PH (1993)

CAS #: 000089-87-2 — 4-NITRO-M-XYLENE

Formula: $C_8H_9NO_2$

Mol Weight: 151.17

MP (deg C): 9

FP (deg C):

BP (deg C): 247

BP pressure (mm Hg):

Property/Value	Units	Temp	Data Type	Reference
WS 1.01E+002	mg/L	25	EST	MEYLAN,WM ET AL. (1996)
logP 2.91			EST	MEYLAN,WM & HOWARD,PH (1995)
VP 5.73E-002	mm Hg	25	EXP	PERRY,RH & GREEN,D (1984)
DC	pKa			
HL 2.59E-005	atm m3/mol	25	EST	MEYLAN,WM & HOWARD,PH (1991)
OH 1.93E-012	cm3/molc sec	25	EST	MEYLAN,WM & HOWARD,PH (1993)

CAS #: 000089-98-5 — 2-CHLOROBENZALDEHYDE

Formula: C_7H_5ClO

Mol Weight: 140.57

MP (deg C): 17.5

FP (deg C):

BP (deg C): 213.5

BP pressure (mm Hg):

Property/Value	Units	Temp	Data Type	Reference
WS 8.53E+002	mg/L	25	EST	MEYLAN,WM ET AL. (1996)
logP 2.33			EXP	HANSCH,C & LEO,AJ (1985)
VP 2.30E-001	mm Hg	20	EXP	WEBER,RC ET AL. (1981)
DC	pKa			
HL 9.95E-006	atm m3/mol	25	EST	MEYLAN,WM & HOWARD,PH (1991)
OH 1.73E-011	cm3/molc sec	25	EST	MEYLAN,WM & HOWARD,PH (1993)

CAS #: 000090-00-6 — O-ETHYLPHENOL

Formula: $C_8H_{10}O$

Mol Weight: 122.17

MP (deg C): 18

FP (deg C):

BP (deg C): 204.52

BP pressure (mm Hg):

Property/Value	Units	Temp	Data Type	Reference
WS 5.34E+003	mg/L	25	EXP	MUELLER,M & KLEIN,W (1992)
logP 2.47			EXP	HANSCH,C & LEO,AJ (1985)
VP 1.53E-001	mm Hg	25	EXP	BIDDISCOMBE,DP ET AL. (1963)
DC 10.20	pKa	25	EXP	PEARCE,PJ & SIMKINS,RJJ (1968)
HL 4.61E-006	atm m3/mol	25	EST	VP/WSOL
OH 4.18E-011	cm3/molc sec	25	EST	MEYLAN,WM & HOWARD,PH (1993)

CAS #: 000090-01-7 — O-HYDROXYBENZYL ALCOHOL

Formula: $C_7H_8O_2$

Mol Weight: 124.14

MP (deg C): 86-87

FP (deg C):

BP (deg C):

BP pressure (mm Hg):

Property/Value	Units	Temp	Data Type	Reference
WS 6.30E+004	mg/L	22	EXP	YALKOWSKY,SH & DANNENFELSER,RM (1992)
logP 0.73			EXP	HANSCH,C & LEO,AJ (1985)
VP 8.25E-004	mm Hg	25	EST	NEELY,WB & BLAU,GE (1985)
DC 9.84	pKa	25	EXP	KORTUM,G ET AL (1961)
HL 2.26E-011	atm m3/mol	25	EST	MEYLAN,WM & HOWARD,PH (1991)
OH 4.42E-011	cm3/molc sec	25	EST	MEYLAN,WM & HOWARD,PH (1993)

CAS #: 000090-02-8 — O-HYDROXYBENZALDEHYDE

Formula: $C_7H_6O_2$

Mol Weight: 122.12

MP (deg C): -7

FP (deg C):

BP (deg C): 196-197

BP pressure (mm Hg):

Property/Value	Units	Temp	Data Type	Reference
WS 1.70E+004	mg/L	86	EXP	YALKOWSKY,SH & DANNENFELSER,RM (1992)
logP 1.81			EXP	HANSCH,C & LEO,AJ (1985)
VP 5.93E-001	mm Hg	25	EXP	DAUBERT,TE & DANNER,RP (1989)
DC	pKa			
HL 1.76E-006	atm m3/mol	25	EST	MEYLAN,WM & HOWARD,PH (1991)
OH 2.80E-011	cm3/molc sec	25	EST	MEYLAN,WM & HOWARD,PH (1993)

CAS #: 000090-03-9 — CHLOROMERCURIPHENOL

Formula: C_6H_5ClHgO

Mol Weight: 329.15

MP (deg C): 150-152

FP (deg C):

BP (deg C):

BP pressure (mm Hg):

Property/Value	Units	Temp	Data Type	Reference
WS 4.24E+002	mg/L	25	EST	MEYLAN,WM ET AL. (1996)
logP 1.51			EXP	HANSCH,C & LEO,AJ (1985)
VP 7.67E-003	mm Hg	25	EST	NEELY,WB & BLAU,GE (1985)
DC	pKa			
HL	atm m3/mol			
OH 2.15E-011	cm3/molc sec	25	EST	MEYLAN,WM & HOWARD,PH (1993)

CAS #: 000090-04-0 — 2-METHOXYANILINE

Formula: C_7H_9NO

Mol Weight: 123.16

MP (deg C): 5

FP (deg C):

BP (deg C): 225

BP pressure (mm Hg):

Property/Value	Units	Temp	Data Type	Reference
WS 9.60E+003	mg/L	25	EST	MEYLAN,WM ET AL. (1996)
logP 1.18			EXP	HANSCH,C & LEO,AJ (1985)
VP 8.00E-002	mm Hg	25	EXT	PERRY,RH & GREEN,D (1984)
DC 4.53	pKa	25	EXP	PERRIN,DD (1972)
HL 1.77E-006	atm m3/mol	25	EST	MEYLAN,WM & HOWARD,PH (1991)
OH 1.21E-010	cm3/molc sec	25	EST	ATKINSON,R (1988)

2-METHOXYPHENOL

CAS #: 000090-05-1

Formula: $C_7H_8O_2$

Mol Weight: 124.14

MP (deg C): 28.2-32
FP (deg C):

BP (deg C): 205

BP pressure (mm Hg):

Property/Value	Units	Temp	Data Type	Reference
WS 2.60E+004	mg/L	25	EXP	CHEM INSPECT TEST INST (1992)
logP 1.32			EXP	HANSCH,C & LEO,AJ (1985)
VP 1.03E-001	mm Hg	25	EXP	VERSCHUEREN,K (1983)
DC 9.98	pKa		EXP	PEARCE,PJ & SIMKINS,RJJ (1968)
HL 6.47E-007	atm m3/mol	25	EST	VP/WSOL
OH 3.17E-011	cm3/molc sec	25	EST	ATKINSON,R (1988)

1-BROMONAPHTHALENE

CAS #: 000090-11-9

Formula: $C_{10}H_7Br$

Mol Weight: 207.08

MP (deg C): -1.8
FP (deg C):

BP (deg C): 281.1

BP pressure (mm Hg):

Property/Value	Units	Temp	Data Type	Reference
WS 9.32E+000	mg/L	21	EXP	YALKOWSKY,SH & DANNENFELSER,RM (1992)
logP 4.06			EST	MEYLAN,WM & HOWARD,PH (1995)
VP 9.73E-003	mm Hg	25	EXP	DAUBERT,TE & DANNER,RP (1989)
DC	pKa			
HL 2.85E-004	atm m3/mol	25	EST	VP/WSOL
OH 1.36E-011	cm3/molc sec	25	EST	MEYLAN,WM & HOWARD,PH (1993)

1-METHYLNAPHTHALENE

CAS #: 000090-12-0

Formula: $C_{11}H_{10}$

Mol Weight: 142.20

MP (deg C): -30.5
FP (deg C):

BP (deg C): 244.7

BP pressure (mm Hg):

Property/Value	Units	Temp	Data Type	Reference
WS 2.58E+001	mg/L	25	EXP	YALKOWSKY,SH & DANNENFELSER,RM (1992)
logP 3.87			EXP	HANSCH,C & LEO,AJ (1985)
VP 6.70E-002	mm Hg	25	EXP	MACKNICK,AB & PRAUSNITZ,JM (1979)
DC	pKa			
HL 2.60E-004	atm m3/mol	25	EXP	MACKAY,D ET AL. (1982A)
OH 5.30E-011	cm3/molc sec	25	EXP	ATKINSON,R (1989)

1-CHLORONAPHTHALENE

CAS #: 000090-13-1

Formula: $C_{10}H_7Cl$

Mol Weight: 162.62

MP (deg C): -2.5
FP (deg C):

BP (deg C): 259.3

BP pressure (mm Hg):

Property/Value	Units	Temp	Data Type	Reference
WS 1.74E+001	mg/L	25	EXP	WASIK,SP ET AL. (1981)
logP 3.90			EXP	SANGSTER,J (1994)
VP 2.90E-002	mm Hg	25	EXP	SCHOENE,K ET AL. (1984)
DC	pKa			
HL 3.55E-004	atm m3/mol	25	EXP	MACKAY,D ET AL. (1982A)
OH 1.52E-011	cm3/molc sec	25	EST	ATKINSON,R (1988)

1-IODONAPHTHALENE

CAS #: 000090-14-2

Formula: $C_{10}H_{17}I$

Mol Weight: 264.15

MP (deg C): 4.2
FP (deg C):

BP (deg C): 302

BP pressure (mm Hg):

Property/Value	Units	Temp	Data Type	Reference
WS 7.44E+000	mg/L	25	EXP	YALKOWSKY,SH ET AL. (1983)
logP 4.34			EST	MEYLAN,WM & HOWARD,PH (1995)
VP 8.49E-004	mm Hg	25	EST	NEELY,WB & BLAU,GE (1985)
DC	pKa			
HL 1.22E-004	atm m3/mol	25	EST	MEYLAN,WM & HOWARD,PH (1991)
OH 1.42E-011	cm3/molc sec	25	EST	MEYLAN,WM & HOWARD,PH (1993)

1-NAPHTHOL

CAS #: 000090-15-3

Formula: $C_{10}H_8O$

Mol Weight: 144.17

MP (deg C): 96
FP (deg C):

BP (deg C): 288

BP pressure (mm Hg):

Property/Value	Units	Temp	Data Type	Reference
WS 8.66E+002	mg/L	25	EXP	HASSETT,JJ ET AL. (1980)
logP 2.85			EXP	HANSCH,C & LEO,AJ (1985)
VP 2.74E-004	mm Hg	25	EXT	JORDAN,TE (1954)
DC 9.34	pKa	25	EXP	SERJEANT,EP & DEMPSEY,B (1979)
HL 4.55E-008	atm m3/mol	25	EST	MEYLEN,WM & HOWARD,PH (1991)
OH 2.00E-010	cm3/molc sec	25	EST	MEYLAN,WM & HOWARD,PH (1993)

1,2,3-BENZOTRIAZIN-4(1H)-ONE

CAS #: 000090-16-4

Formula: $C_7H_5N_3O$

Mol Weight: 147.14

MP (deg C): 216-218
FP (deg C):

BP (deg C):

BP pressure (mm Hg):

Property/Value	Units	Temp	Data Type	Reference
WS 1.53E+004	mg/L	25	EST	MEYLAN,WM ET AL. (1996)
logP 0.83			EXP	HANSCH,C ET AL. (1995)
VP 1.14E-006	mm Hg	25	EST	NEELY,WB & BLAU,GE (1985)
DC	pKa			
HL 3.24E-010	atm m3/mol	25	EST	MEYLAN,WM & HOWARD,PH (1991)
OH 5.83E-012	cm3/molc sec	25	EST	MEYLAN,WM & HOWARD,PH (1993)

1-NAPHTHALENAMINE, N-PHENYL-

CAS #: 000090-30-2

Formula: $C_{16}H_{13}N$

Mol Weight: 219.29

MP (deg C): 62
FP (deg C):

BP (deg C): 335

BP pressure (mm Hg): 5.28E+002

Property/Value	Units	Temp	Data Type	Reference
WS 6.00E+001	mg/L	20	EXP	CHEM INSPECT TEST INST (1992)
logP 4.20			EXP	HANSCH,C & LEO,AJ (1985)
VP 8.29E-006	mm Hg	25	EST	NEELY,WB & BLAU,GE (1985)
DC	pKa			
HL 1.03E-007	atm m3/mol	25	EST	MEYLAN,WM & HOWARD,PH (1991)
OH 3.47E-010	cm3/molc sec	25	EST	MEYLAN,WM & HOWARD,PH (1993)

000090-33-5 — 7-HYDROXY-4-METHYLCOUMARIN

CAS #: 000090-33-5

Formula:	$C_{10}H_8O_3$
Mol Weight:	176.17
MP (deg C):	194-195
FP (deg C):	
BP (deg C):	
BP pressure (mm Hg):	

Property/Value	Units	Temp	Data Type	Reference
WS 5.21E+003	mg/L	25	EST	MEYLAN,WM ET AL. (1996)
logP 1.58			EXP	HANSCH,C & LEO,AJ (1985)
VP 7.24E-006	mm Hg	25	EST	NEELY,WB & BLAU,GE (1985)
DC	pKa			
HL 1.13E-009	atm m3/mol	25	EST	MEYLAN,WM & HOWARD,PH (1991)
OH 6.30E-011	cm3/molc sec	25	EST	MEYLAN,WM & HOWARD,PH (1993)

000090-39-1 — L-SPARTEINE

CAS #: 000090-39-1

Formula:	$C_{15}H_{26}N_2$
Mol Weight:	234.39
MP (deg C):	30.5
FP (deg C):	
BP (deg C):	325
BP pressure (mm Hg):	

Property/Value	Units	Temp	Data Type	Reference
WS 3.04E+003	mg/L	22	EXP	YALKOWSKY,SH & DANNENFELSER,RM (1992)
logP 3.41			EST	MEYLAN,WM & HOWARD,PH (1995)
VP 1.79E-004	mm Hg	25	EST	NEELY,WB & BLAU,GE (1985)
DC	pKa			
HL 1.16E-008	atm m3/mol	25	EST	MEYLAN,WM & HOWARD,PH (1991)
OH 1.26E-010	cm3/molc sec	25	EST	MEYLAN,WM & HOWARD,PH (1993)

000090-41-5 — 2-AMINOBIPHENYL

CAS #: 000090-41-5

Formula:	$C_{12}H_{11}N$
Mol Weight:	169.23
MP (deg C):	51
FP (deg C):	
BP (deg C):	299
BP pressure (mm Hg):	

Property/Value	Units	Temp	Data Type	Reference
WS 2.33E+002	mg/L	25	EST	MEYLAN,WM ET AL. (1996)
logP 2.84			EXP	HANSCH,C & LEO,AJ (1985)
VP 1.17E-004	mm Hg	25	EST	NEELY,WB & BLAU,GE (1985)
DC 3.83	pKa	18	EXP	PERRIN,DD (1965)
HL 1.46E-007	atm m3/mol	25	EST	MEYLAN,WM & HOWARD,PH (1991)
OH 8.03E-011	cm3/molc sec	25	EST	MEYLAN,WM & HOWARD,PH (1993)

000090-43-7 — 2-PHENYLPHENOL

CAS #: 000090-43-7

Formula:	$C_{12}H_{10}O$
Mol Weight:	170.21
MP (deg C):	57
FP (deg C):	
BP (deg C):	286
BP pressure (mm Hg):	

Property/Value	Units	Temp	Data Type	Reference
WS 7.00E+002	mg/L	25	EXP	IARC (1983)
logP 3.09			EXP	HANSCH,C ET AL. (1995)
VP 2.00E-003	mm Hg	25	EXT	KUNDEL,H ET AL. (1975)
DC 9.92	pKa	25	EST	LIPNICK,RL ET AL. (1986)
HL 1.05E-006	atm m3/mol	25	EST	VP/WSOL
OH 2.90E-011	cm3/molc sec	25	EST	MEYLAN,WM & HOWARD,PH (1993)

000090-44-8 — 9(10H)-ANTHRACENONE

CAS #: 000090-44-8

Formula:	$C_{14}H_{10}O$
Mol Weight:	194.24
MP (deg C):	155
FP (deg C):	
BP (deg C):	
BP pressure (mm Hg):	

Property/Value	Units	Temp	Data Type	Reference
WS 4.60E+000	mg/L	25	EST	MEYLAN,WM ET AL. (1996)
logP 3.66			EXP	HANSCH,C ET AL. (1995)
VP 1.83E-005	mm Hg		EST	NEELY,WB & BLAU,GE (1985)
DC	pKa			
HL 7.86E-007	atm m3/mol	25	EST	MEYLAN,WM & HOWARD,PH (1991)
OH 9.77E-012	cm3/molc sec	25	EST	MEYLAN,WM & HOWARD,PH (1993)

000090-45-9 — 9-AMINOACRIDINE

CAS #: 000090-45-9

Formula:	$C_{13}H_{10}N_2$
Mol Weight:	194.24
MP (deg C):	241
FP (deg C):	
BP (deg C):	
BP pressure (mm Hg):	

Property/Value	Units	Temp	Data Type	Reference
WS 1.66E+001	mg/L	25	EST	MEYLAN,WM ET AL. (1996)
logP 2.74			EXP	HANSCH,C & LEO,AJ (1985)
VP 2.24E-006	mm Hg	25	EST	NEELY,WB & BLAU,GE (1985)
DC 9.95	pKa	20	EXP	PERRIN,DD (1965)
HL 2.37E-011	atm m3/mol	25	EST	MEYLAN,WM & HOWARD,PH (1991)
OH 2.00E-010	cm3/molc sec	25	EST	MEYLAN,WM & HOWARD,PH (1993)

000090-47-1 — 9H-XANTHEN-9-ONE

CAS #: 000090-47-1

Formula:	$C_{13}H_9O_2$
Mol Weight:	197.22
MP (deg C):	174
FP (deg C):	
BP (deg C):	351
BP pressure (mm Hg):	7.30E+002

Property/Value	Units	Temp	Data Type	Reference
WS 4.52E+000	mg/L	25	EST	MEYLAN,WM ET AL. (1996)
logP 3.39			EXP	HANSCH,C ET AL. (1995)
VP 5.85E-005	mm Hg	25	EST	NEELY,WB & BLAU,GE (1985)
DC	pKa			
HL 1.93E-007	atm m3/mol	25	EST	MEYLAN,WM & HOWARD,PH (1991)
OH 8.97E-012	cm3/molc sec	25	EST	MEYLAN,WM & HOWARD,PH (1993)

000090-64-2 — A-HYDROXYPHENYLACETIC ACID

CAS #: 000090-64-2

Formula:	$C_8H_8O_3$
Mol Weight:	152.15
MP (deg C):	119
FP (deg C):	
BP (deg C):	
BP pressure (mm Hg):	

Property/Value	Units	Temp	Data Type	Reference
WS 1.81E+005	mg/L	25	EXP	YALKOWSKY,SH & DANNENFELSER,RM (1992)
logP 0.62			EXP	HANSCH,C & LEO,AJ (1985)
VP 1.64E-005	mm Hg	25	EST	NEELY,WB & BLAU,GE (1985)
DC 3.41	pKa	25	EXP	KORTUM,G ET AL (1961)
HL 6.90E-009	atm m3/mol	25	EST	MEYLAN,WM & HOWARD,PH (1991)
OH 1.04E-011	cm3/molc sec	25	EST	MEYLAN,WM & HOWARD,PH (1993)

CAS #: 000090-69-7 — LOBELINE

Formula: $C_{22}H_{27}NO_2$

Mol Weight: 337.47

MP (deg C): 130-131 **FP (deg C):**

BP (deg C):

BP pressure (mm Hg):

Property/Value	Units	Temp	Data Type	Reference
WS 4.84E+001	mg/L	25	EST	MEYLAN,WM ET AL. (1996)
logP 3.74			EST	MEYLAN,WM & HOWARD,PH (1995)
VP 8.78E-011	mm Hg	25	EST	NEELY,WB & BLAU,GE (1985)
DC	pKa			
HL 3.47E-014	atm m3/mol	25	EST	MEYLAN,WM & HOWARD,PH (1991)
OH 2.03E-010	cm3/molc sec	25	EST	MEYLAN,WM & HOWARD,PH (1993)

CAS #: 000090-82-4 — BENZENEMETHANOL, .ALPHA.- 1-(METHYLAMINO)ETHYL -

Formula: $C_{10}H_{15}NO$

Mol Weight: 165.24

MP (deg C): 119 **FP (deg C):**

3P (deg C):

BP pressure (mm Hg):

Property/Value	Units	Temp	Data Type	Reference
WS 1.06E+005	mg/L	25	EST	MEYLAN,WM ET AL. (1996)
logP 1.40			EXP	HANSCH,C ET AL. (1995)
VP 8.30E-004	mm Hg	25	EST	NEELY,WB & BLAU,GE (1985)
DC 10.25	pKa	0	EXP	PERRIN,DD (1965)
HL 8.65E-011	atm m3/mol	25	EST	MEYLAN,WM & HOWARD,PH (1991)
OH 9.97E-011	cm3/molc sec	25	EST	MEYLAN,WM & HOWARD,PH (1993)

CAS #: 000090-94-8 — 4,4-BIS(DIMETHYLAMINO)BENZOPHENONE

Formula: $C_{17}H_{20}N_2O$

Mol Weight: 268.36

MP (deg C): 172 **FP (deg C):**

BP (deg C):

BP pressure (mm Hg):

Property/Value	Units	Temp	Data Type	Reference
WS 4.00E+002	mg/L	20	EXP	DEHN,WM (1917)
logP 3.87			EXP	HANSCH,C ET AL. (1995)
VP 1.07E-006	mm Hg	25	EST	NEELY,WB & BLAU,GE (1985)
DC	pKa			
HL 4.91E-010	atm m3/mol	25	EST	MEYLAN,WM & HOWARD,PH (1991)
OH 2.05E-010	cm3/molc sec	25	EST	MEYLAN,WM & HOWARD,PH (1993)

CAS #: 000091-01-0 — BENZHYDROL

Formula: $C_{13}H_{12}O$

Mol Weight: 184.24

MP (deg C): 69 **FP (deg C):**

BP (deg C): 298

BP pressure (mm Hg): 7.48E+002

Property/Value	Units	Temp	Data Type	Reference
WS 5.00E+002	mg/L	20	EXP	MERCK INDEX (1976)
logP 2.67			EXP	HANSCH,C & LEO,AJ (1985)
VP 6.12E-003	mm Hg	25	EXP	OHE,S (1976)
DC	pKa			
HL 2.97E-006	atm m3/mol	25	EST	VP/WSOL
OH 1.62E-011	cm3/molc sec	25	EST	MEYLAN,WM & HOWARD,PH (1993)

CAS #: 000091-02-1 — PHENYL-A-PYRIDYLKETONE

Formula: $C_{12}H_9NO$

Mol Weight: 183.21

MP (deg C): 42 **FP (deg C):**

BP (deg C): 317

BP pressure (mm Hg):

Property/Value	Units	Temp	Data Type	Reference
WS 1.32E+003	mg/L	25	EST	MEYLAN,WM ET AL. (1996)
logP 1.88			EXP	HANSCH,C & LEO,AJ (1985)
VP 3.67E-004	mm Hg	25	EST	NEELY,WB & BLAU,GE (1985)
DC	pKa			
HL 2.54E-009	atm m3/mol	25	EST	MEYLAN,WM & HOWARD,PH (1991)
OH 2.11E-012	cm3/molc sec	25	EST	MEYLAN,WM & HOWARD,PH (1993)

CAS #: 000091-08-7 — TOLUENE-2,6-DIISOCYANATE

Formula: $C_9H_6N_2O_2$

Mol Weight: 174.16

MP (deg C): 18.3 **FP (deg C):**

BP (deg C):

BP pressure (mm Hg):

Property/Value	Units	Temp	Data Type	Reference
WS 3.76E+001	mg/L	25	EST	MEYLAN,WM ET AL. (1996)
logP 3.74			EST	MEYLAN,WM & HOWARD,PH (1995)
VP 9.11E-002	mm Hg	25	EST	NEELY,WB & BLAU,GE (1985)
DC	pKa			
HL 1.11E-005	atm m3/mol	25	EST	MEYLAN,WM & HOWARD,PH (1991)
OH 6.26E-012	cm3/molc sec	25	EST	MEYLAN,WM & HOWARD,PH (1993)

CAS #: 000091-10-1 — 2,6-DIMETHOXYPHENOL

Formula: $C_8H_{10}O_3$

Mol Weight: 154.17

MP (deg C): 56.5 **FP (deg C):**

BP (deg C): 261

BP pressure (mm Hg):

Property/Value	Units	Temp	Data Type	Reference
WS 1.72E+004	mg/L	13	EXP	YALKOWSKY,SH & DANNENFELSER,RM (1992)
logP 1.15			EXP	SOTOMATSU,T ET AL. (1993)
VP 6.17E-003	mm Hg	25	EST	NEELY,WB & BLAU,GE (1985)
DC	pKa			
HL 1.96E-009	atm m3/mol	25	EST	MEYLAN,WM & HOWARD,PH (1991)
OH 1.65E-010	cm3/molc sec	25	EST	MEYLAN,WM & HOWARD,PH (1993)

CAS #: 000091-15-6 — 1,2-DICYANOBENZENE

Formula: $C_8H_4N_2$

Mol Weight: 128.13

MP (deg C): 139-141 **FP (deg C):**

BP (deg C):

BP pressure (mm Hg):

Property/Value	Units	Temp	Data Type	Reference
WS 3.95E+002	mg/L	25	EXP	CHEM INSPECT TEST INST (1992)
logP 0.99			EXP	NAKAGAWA,Y ET AL. (1992)
VP 5.69E-003	mm Hg	25	EST	NEELY,WB & BLAU,GE (1985)
DC	pKa			
HL 5.03E-007	atm m3/mol	25	EST	MEYLAN,WM & HOWARD,PH (1991)
OH 4.51E-014	cm3/molc sec	25	EST	MEYLAN,WM & HOWARD,PH (1993)

CAS #: 000091-16-7				1,2-DIMETHOXYBENZENE

Formula: $C_8H_{10}O_2$

Mol Weight: 138.17

MP (deg C): 22-23   **FP (deg C):**

BP (deg C): 206-207

BP pressure (mm Hg):

Property/Value	Units	Temp	Data Type	Reference
WS 3.67E+003	mg/L	25	EST	MEYLAN,WM ET AL. (1996)
logP 1.60			EXP	HANSCH,C & LEO,AJ (1985)
VP 4.70E-001	mm Hg	25	EXP	RIDDICK,JA ET AL. (1986)
DC	pKa			
HL 1.89E-005	atm m3/mol	25	EST	MEYLAN,WM & HOWARD,PH (1991)
OH 2.03E-011	cm3/molc sec	25	EST	MEYLAN,WM & HOWARD,PH (1993)

CAS #: 000091-17-8				DECAHYDRONAPHTHALENE

Formula: $C_{10}H_{18}$

Mol Weight: 138.25

MP (deg C): -30.4   **FP (deg C):**

BP (deg C): 187.25

BP pressure (mm Hg):

Property/Value	Units	Temp	Data Type	Reference
WS 8.89E-001	mg/L	25	EXP	YALKOWSKY,SH & DANNENFELSER,RM (1992)
logP 4.20			EST	MEYLAN,WM & HOWARD,PH (1995)
VP 2.30E+000	mm Hg	25	EXP	BOUBLIK,T ET AL. (1984)
DC	pKa			
HL 4.70E-001	atm m3/mol	25	EST	VP/WSOL
OH 2.02E-011	cm3/molc sec	25	EXP	ATKINSON,R (1989)

CAS #: 000091-18-9				PTERIDINE

Formula: $C_6H_4N_4$

Mol Weight: 132.13

MP (deg C): 138-138.5   **FP (deg C):**

BP (deg C):

BP pressure (mm Hg):

Property/Value	Units	Temp	Data Type	Reference
WS 1.43E+005	mg/L	20	EXP	YALKOWSKY,SH & DANNENFELSER,RM (1992)
logP -0.58			EXP	HANSCH,C & LEO,AJ (1985)
VP 3.11E-003	mm Hg	25	EST	NEELY,WB & BLAU,GE (1985)
DC	pKa			
HL 1.54E-010	atm m3/mol	25	EST	MEYLAN,WM & HOWARD,PH (1991)
OH 1.00E-012	cm3/molc sec	25	EST	MEYLAN,WM & HOWARD,PH (1993)

CAS #: 000091-19-0				QUINOXALINE

Formula: $C_8H_6N_2$

Mol Weight: 130.15

MP (deg C): 29-30   **FP (deg C):**

BP (deg C): 229.5

BP pressure (mm Hg):

Property/Value	Units	Temp	Data Type	Reference
WS 1.00E+006	mg/L	20	EXP	YALKOWSKY,SH & DANNENFELSER,RM (1992)
logP 1.32			EXP	HANSCH,C & LEO,AJ (1985)
VP 1.44E-002	mm Hg	25	EST	NEELY,WB & BLAU,GE (1985)
DC	pKa			
HL 2.85E-007	atm m3/mol	25	EST	MEYLAN,WM & HOWARD,PH (1991)
OH 2.00E-012	cm3/molc sec	25	EST	MEYLAN,WM & HOWARD,PH (1993)

CAS #: 000091-20-3				NAPHTHALENE

Formula: $C_{10}H_8$

Mol Weight: 128.18

MP (deg C): 80.29   **FP (deg C):**

BP (deg C): 217.942

BP pressure (mm Hg):

Property/Value	Units	Temp	Data Type	Reference
WS 3.10E+001	mg/L	25	EXP	PEARLMAN,RS ET AL. (1984)
logP 3.30			EXP	HANSCH,C & LEO,AJ (1985)
VP 8.50E-002	mm Hg	25	EXP	AMBROSE,D ET AL. (1975C)
DC	pKa			
HL 4.83E-004	atm m3/mol	25	EXP	MACKAY,D ET AL. (1979)
OH 2.16E-011	cm3/molc sec	25	EXP	ATKINSON,R (1989)

CAS #: 000091-22-5				QUINOLINE

Formula: C_9H_7N

Mol Weight: 129.16

MP (deg C): -14.85   **FP (deg C):**

BP (deg C): 237.10

BP pressure (mm Hg):

Property/Value	Units	Temp	Data Type	Reference
WS 6.11E+003	mg/L	25	EXP	SMITH,JH ET AL. (1978)
logP 2.03			EXP	HANSCH,C & LEO,AJ (1985)
VP 6.00E-002	mm Hg	25	EXP	DAUBERT,TE & DANNER,RP (1989)
DC 4.90	pKa	20	EXP	WEAST,RC ET AL. (1985)
HL 1.67E-006	atm m3/mol	25	EST	VP/WSOL
OH 6.50E-012	cm3/molc sec	25	EST	MEYLAN,WM & HOWARD,PH (1993)

CAS #: 000091-23-6				2-NITROANISOLE

Formula: $C_7H_7NO_3$

Mol Weight: 153.14

MP (deg C): 9.4   **FP (deg C):**

BP (deg C): 277

BP pressure (mm Hg):

Property/Value	Units	Temp	Data Type	Reference
WS 1.69E+003	mg/L	30	EXP	RIDDICK,JA ET AL. (1986)
logP 1.73			EXP	HANSCH,C ET AL. (1995)
VP 3.60E-003	mm Hg	25	EXP	YAWS,CL (1994A)
DC	pKa			
HL 4.29E-007	atm m3/mol	25	EST	VP/WSOL
OH 3.52E-012	cm3/molc sec	25	EST	MEYLAN,WM & HOWARD,PH (1993)

CAS #: 000091-40-7				N-PHENYL O-AMINOBENZOIC ACID

Formula: $C_{13}H_{11}NO_2$

Mol Weight: 213.24

MP (deg C): 183.5   **FP (deg C):**

BP (deg C):

BP pressure (mm Hg):

Property/Value	Units	Temp	Data Type	Reference
WS 7.06E+000	mg/L	25	EST	MEYLAN,WM ET AL. (1996)
logP 4.36			EXP	HANSCH,C & LEO,AJ (1985)
VP 2.38E-006	mm Hg	25	EST	NEELY,WB & BLAU,GE (1985)
DC 3.99	pKa		EXP	SERJEANT,EP & DEMPSEY,B (1979)
HL 2.11E-011	atm m3/mol	25	EST	MEYLAN,WM & HOWARD,PH (1991)
OH 2.01E-010	cm3/molc sec	25	EST	MEYLAN,WM & HOWARD,PH (1993)

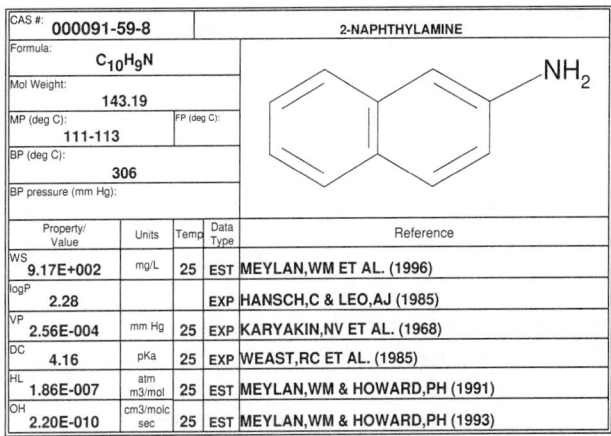

CAS #: 000091-46-3				ETHANAMINE, N,N-DIMETHYL-2-[5-METHYL-2-(1-METHYL
Formula: $C_{14}H_{23}NO$				
Mol Weight: 221.35				
MP (deg C):		FP (deg C):		
BP (deg C):				
BP pressure (mm Hg):				

Property/ Value	Units	Temp	Data Type	Reference
WS 2.00E+002	mg/L	25	EST	MEYLAN,WM ET AL. (1996)
logP 3.79			EXP	DALLET ET AL. (1985)
VP 1.80E-003	mm Hg	25	EST	NEELY,WB & BLAU,GE (1985)
DC	pKa			
HL 4.50E-007	atm m3/mol	25	EST	MEYLAN,WM & HOWARD,PH (1991)
OH 1.76E-010	cm3/molc sec	25	EST	MEYLAN,WM & HOWARD,PH (1993)

CAS #: 000091-52-1				2,4-DIMETHOXYBENZOIC ACID
Formula: $C_9H_{10}O_4$				
Mol Weight: 182.18				
MP (deg C): 108-110		FP (deg C):		
BP (deg C):				
BP pressure (mm Hg):				

Property/ Value	Units	Temp	Data Type	Reference
WS 1.92E+003	mg/L	25	EST	MEYLAN,WM ET AL. (1996)
logP 1.69			EST	MEYLAN,WM & HOWARD,PH (1995)
VP 3.52E-003	mm Hg	25	EXT	COLOMINA,M ET AL. (1985)
DC 4.36	pKa	25	EXP	SERJEANT,EP & DEMPSEY,B (1979)
HL 3.79E-010	atm m3/mol	25	EST	MEYLAN,WM & HOWARD,PH (1991)
OH 9.00E-011	cm3/molc sec	25	EST	MEYLAN,WM & HOWARD,PH (1993)

CAS #: 000091-53-2				ETHOXYQUIN
Formula: $C_{14}H_{19}NO$				
Mol Weight: 217.31				
MP (deg C):		FP (deg C):		
BP (deg C): 123-125				
BP pressure (mm Hg): 2.00E+000				

Property/ Value	Units	Temp	Data Type	Reference
WS 1.75E+001	mg/L	25	EST	MEYLAN,WM ET AL. (1996)
logP 3.87			EST	MEYLAN,WM & HOWARD,PH (1995)
VP 1.32E-004	mm Hg	25	EST	NEELY,WB & BLAU,GE (1985)
DC	pKa			
HL 2.42E-007	atm m3/mol	25	EST	MEYLAN,WM & HOWARD,PH (1991)
OH 1.28E-010	cm3/molc sec	25	EST	MEYLAN,WM & HOWARD,PH (1993)

CAS #: 000091-55-4				1H-INDOLE, 2,3-DIMETHYL-
Formula: $C_{10}H_{11}N$				
Mol Weight: 145.21				
MP (deg C): 105-107		FP (deg C):		
BP (deg C): 285				
BP pressure (mm Hg):				

Property/ Value	Units	Temp	Data Type	Reference
WS 1.98E+002	mg/L	25	EST	MEYLAN,WM ET AL. (1996)
logP 3.05			EXP	SANGSTER,J (1993)
VP 1.79E-003	mm Hg	25	EST	NEELY,WB & BLAU,GE (1985)
DC	pKa			
HL 1.08E-006	atm m3/mol	25	EST	MEYLAN,WM & HOWARD,PH (1991)
OH 2.00E-010	cm3/molc sec	25	EST	MEYLAN,WM & HOWARD,PH (1993)

CAS #: 000091-56-5				1H-INDOLE-2,3-DIONE
Formula: $C_8H_5NO_2$				
Mol Weight: 147.13				
MP (deg C): 203.5		FP (deg C):		
BP (deg C):				
BP pressure (mm Hg):				

Property/ Value	Units	Temp	Data Type	Reference
WS 1.53E+004	mg/L	25	EST	MEYLAN,WM ET AL. (1996)
logP 0.83			EXP	HANSCH,C ET AL. (1995)
VP 1.11E-005	mm Hg	25	EST	NEELY,WB & BLAU,GE (1985)
DC	pKa			
HL 2.21E-011	atm m3/mol	25	EST	MEYLAN,WM & HOWARD,PH (1991)
OH 3.89E-011	cm3/molc sec	25	EST	MEYLAN,WM & HOWARD,PH (1993)

CAS #: 000091-57-6				2-METHYLNAPHTHALENE
Formula: $C_{11}H_{10}$				
Mol Weight: 142.20				
MP (deg C): 34.6		FP (deg C):		
BP (deg C): 241.4				
BP pressure (mm Hg):				

Property/ Value	Units	Temp	Data Type	Reference
WS 2.46E+001	mg/L	25	EXP	YALKOWSKY,SH & DANNENFELSER,RM (1992)
logP 3.86			EXP	HANSCH,C & LEO,AJ (1985)
VP 5.50E-002	mm Hg	25	EXP	KARYAKIN,NV ET AL. (1968)
DC	pKa			
HL 5.18E-004	atm m3/mol	25	EST	VP/WSOL
OH 5.23E-011	cm3/molc sec	25	EXP	ATKINSON,R (1989)

CAS #: 000091-58-7				2-CHLORONAPHTHALENE
Formula: $C_{10}H_7Cl$				
Mol Weight: 162.62				
MP (deg C): 59.5		FP (deg C):		
BP (deg C): 256				
BP pressure (mm Hg):				

Property/ Value	Units	Temp	Data Type	Reference
WS 1.17E+001	mg/L	25	EXP	MACKAY,D & SHIN,WY (1981)
logP 3.98			EXP	SANGSTER,J (1994)
VP 7.98E-003	mm Hg	20	EST	MERCK INDEX (1983)
DC	pKa			
HL 3.20E-004	atm m3/mol	25	EXP	MACKAY,D ET AL. (1982A)
OH 1.51E-011	cm3/molc sec	25	EST	MEYLAN,WM & HOWARD,PH (1993)

CAS #: 000091-59-8				2-NAPHTHYLAMINE
Formula: $C_{10}H_9N$				
Mol Weight: 143.19				
MP (deg C): 111-113		FP (deg C):		
BP (deg C): 306				
BP pressure (mm Hg):				

Property/ Value	Units	Temp	Data Type	Reference
WS 9.17E+002	mg/L	25	EST	MEYLAN,WM ET AL. (1996)
logP 2.28			EXP	HANSCH,C & LEO,AJ (1985)
VP 2.56E-004	mm Hg	25	EXP	KARYAKIN,NV ET AL. (1968)
DC 4.16	pKa	25	EXP	WEAST,RC ET AL. (1985)
HL 1.86E-007	atm m3/mol	25	EST	MEYLAN,WM & HOWARD,PH (1991)
OH 2.20E-010	cm3/molc sec	25	EST	MEYLAN,WM & HOWARD,PH (1993)

CAS #: 000091-62-3				6-METHYLQUINOLINE

Formula: $C_{10}H_9N$

Mol Weight: 143.19

MP (deg C): -22　　FP (deg C):

BP (deg C): 258.6

BP pressure (mm Hg):

Property/Value	Units	Temp	Data Type	Reference
WS 6.31E+002	mg/L	25	EST	MEYLAN,WM ET AL. (1996)
logP 2.57			EXP	HANSCH,C & LEO,AJ (1985)
VP 6.41E-003	mm Hg	25	EXT	BOUBLIK,T ET AL. (1984)
DC 5.34	pKa	20	EXP	PERRIN,DD (1965)
HL 7.60E-007	atm m3/mol	25	EST	MEYLAN,WM & HOWARD,PH (1991)
OH 3.77E-011	cm3/molc sec	25	EST	MEYLAN,WM & HOWARD,PH (1993)

CAS #: 000091-63-4				2-METHYLQUINOLINE

Formula: $C_{10}H_9N$

Mol Weight: 143.19

MP (deg C): 246-247　　FP (deg C):

BP (deg C):

BP pressure (mm Hg):

Property/Value	Units	Temp	Data Type	Reference
WS 4.99E+002	mg/L	25	EST	MEYLAN,WM ET AL. (1996)
logP 2.59			EXP	HANSCH,C & LEO,AJ (1985)
VP 9.46E-003	mm Hg	25	EXP	DAUBERT,TE & DANNER,RP (1989)
DC 5.71	pKa	22	EXP	PERRIN,DD (1965)
HL 7.60E-007	atm m3/mol	25	EST	MEYLAN,WM & HOWARD,PH (1991)
OH 3.77E-011	cm3/molc sec	25	EST	MEYLAN,WM & HOWARD,PH (1993)

CAS #: 000091-64-5				COUMARIN

Formula: $C_9H_6O_2$

Mol Weight: 146.15

MP (deg C): 68-70　　FP (deg C):

BP (deg C): 297-299

BP pressure (mm Hg):

Property/Value	Units	Temp	Data Type	Reference
WS 2.50E+003	mg/L	25	EXP	SEIDELL,A (1941)
logP 1.39			EXP	HANSCH,C & LEO,AJ (1985)
VP 3.39E-003	mm Hg	25	EXT	PERRY,RH & GREEN,D (1984)
DC	pKa			
HL 6.95E-006	atm m3/mol	25	EST	MEYLAN,WM & HOWARD,PH (1991)
OH 1.32E-011	cm3/molc sec	25	EST	MEYLAN,WM & HOWARD,PH (1993)

CAS #: 000091-66-7				N,N-DIETHYLANILINE

Formula: $C_{10}H_{15}N$

Mol Weight: 149.24

MP (deg C): -38　　FP (deg C):

BP (deg C): 215-216

BP pressure (mm Hg):

Property/Value	Units	Temp	Data Type	Reference
WS 1.39E+002	mg/L	25	EXP	KUHNE,R ET AL. (1995)
logP 3.31			EXP	HANSCH,C & LEO,AJ (1985)
VP 1.36E-001	mm Hg	25	EXP	DAUBERT,TE & DANNER,RP (1991)
DC 6.57	pKa		EXP	PERRIN,DD (1972)
HL 1.92E-004	atm m3/mol	25	EST	VP/WSOL
OH 1.64E-010	cm3/molc sec	25	EST	MEYLAN,WM & HOWARD,PH (1993)

CAS #: 000091-76-9				BENZOGUANAMINE

Formula: $C_9H_9N_5$

Mol Weight: 187.21

MP (deg C): 227-228　　FP (deg C):

BP (deg C):

BP pressure (mm Hg):

Property/Value	Units	Temp	Data Type	Reference
WS 3.50E+003	mg/L	25	EST	MEYLAN,WM ET AL. (1996)
logP 1.36			EXP	HANSCH,C & LEO,AJ (1985)
VP 7.52E-007	mm Hg	25	EST	NEELY,WB & BLAU,GE (1985)
DC	pKa			
HL 4.11E-011	atm m3/mol	25	EST	MEYLAN,WM & HOWARD,PH (1991)
OH 3.67E-012	cm3/molc sec	25	EST	MEYLAN,WM & HOWARD,PH (1993)

CAS #: 000091-80-5				METHAPYRILENE

Formula: $C_{14}H_{19}N_3S$

Mol Weight: 261.39

MP (deg C):　　FP (deg C):

BP (deg C): 173-175

BP pressure (mm Hg): 3.00E+000

Property/Value	Units	Temp	Data Type	Reference
WS 6.01E+002	mg/L	30	EXP	YALKOWSKY,SH & DANNENFELSER,RM (1992)
logP 2.87			EXP	SANGSTER,J (1994)
VP 5.33E-006	mm Hg	25	EST	NEELY,WB & BLAU,GE (1985)
DC 8.85	pKa	20	EXP	PERRIN,DD (1965)
HL 3.24E-012	atm m3/mol	25	EST	MEYLAN,WM & HOWARD,PH (1991)
OH 1.50E-010	cm3/molc sec	25	EST	MEYLAN,WM & HOWARD,PH (1993)

CAS #: 000091-84-9				PYRILAMINE

Formula: $C_{17}H_{23}N_3O$

Mol Weight: 285.39

MP (deg C):　　FP (deg C):

BP (deg C): 201

BP pressure (mm Hg): 5.00E+000

Property/Value	Units	Temp	Data Type	Reference
WS 2.45E+002	mg/L	25	EST	MEYLAN,WM ET AL. (1996)
logP 3.27			EXP	HANSCH,C ET AL. (1995)
VP 1.75E-006	mm Hg	25	EST	NEELY,WB & BLAU,GE (1985)
DC 8.85	pKa	20	EXP	PERRIN,DD (1965)
HL 3.52E-013	atm m3/mol	25	EST	MEYLAN,WM & HOWARD,PH (1991)
OH 1.53E-010	cm3/molc sec	25	EST	MEYLAN,WM & HOWARD,PH (1993)

CAS #: 000091-92-9				NAPHTHOL AS-BR

Formula: $C_{36}H_{28}N_2O_6$

Mol Weight: 584.63

MP (deg C):　　FP (deg C):

BP (deg C):

BP pressure (mm Hg):

Property/Value	Units	Temp	Data Type	Reference
WS 1.44E-005	mg/L	25	EST	MEYLAN,WM ET AL. (1996)
logP 8.88			EST	MEYLAN,WM & HOWARD,PH (1995)
VP 5.77E-027	mm Hg	25	EST	NEELY,WB & BLAU,GE (1985)
DC	pKa			
HL 1.93E-020	atm m3/mol	25	EST	MEYLAN,WM & HOWARD,PH (1991)
OH 1.33E-010	cm3/molc sec	25	EST	MEYLAN,WM & HOWARD,PH (1993)

000091-94-1 — 3,3'-DICHLOROBENZIDINE

Formula: $C_{12}H_{10}Cl_2N_2$
Mol Weight: 253.13
MP (deg C): 132-133
FP (deg C):
BP (deg C): 368
BP pressure (mm Hg):

Property/Value	Units	Temp	Data Type	Reference
WS 3.10E+000	mg/L	25	EXP	BANERJEE,S ET AL. (1980)
logP 3.51			EXP	HANSCH,C & LEO,AJ (1985)
VP 2.56E-007	mm Hg	25	EST	NEELY,WB & BLAU,GE (1985)
DC	pKa			
HL 2.84E-011	atm m3/mol	25	EST	MEYLAN,WM & HOWARD,PH (1991)
OH 7.09E-011	cm3/molc sec	25	EST	MEYLAN,WM & HOWARD,PH (1993)

000091-96-3 — C.I. AZOIC COUPLING COMPONENT 5

Formula: $C_{22}H_{24}N_2O_4$
Mol Weight: 380.45
MP (deg C): 212
FP (deg C):
BP (deg C): 353.4
BP pressure (mm Hg):

Property/Value	Units	Temp	Data Type	Reference
WS 1.30E+002	mg/L	25	EST	MEYLAN,WM ET AL. (1996)
logP 1.75			EXP	CHEM INSPECT TEST INST (1992)
VP 1.18E-014	mm Hg	25	EST	NEELY,WB & BLAU,GE (1985)
DC	pKa			
HL 4.74E-019	atm m3/mol	25	EST	MEYLAN,WM & HOWARD,PH (1991)
OH 4.81E+001	cm3/molc sec	25	EST	MEYLAN,WM & HOWARD,PH (1993)

000092-06-8 — m-TERPHENYL

Formula: $C_{18}H_{14}$
Mol Weight: 230.31
MP (deg C): 87
FP (deg C):
BP (deg C): 363
BP pressure (mm Hg):

Property/Value	Units	Temp	Data Type	Reference
WS 1.51E+000	mg/L	25	EXP	YALKOWSKY,SH & DANNENFELSER,RM (1992)
logP 5.52			EST	MEYLAN,WM & HOWARD,PH (1995)
VP 1.75E-005	mm Hg	25	EXT	DAUBERT,TE & DANNER,RP (1992)
DC	pKa			
HL 3.51E-006	atm m3/mol	25	EST	VP/WSOL
OH 1.26E-011	cm3/molc sec	25	EST	MEYLAN,WM & HOWARD,PH (1993)

000092-13-7 — PILOCARPOL

Formula: $C_{11}H_{16}N_2O_2$
Mol Weight: 208.26
MP (deg C): 34
FP (deg C):
BP (deg C): 260
BP pressure (mm Hg): 5.00E+000

Property/Value	Units	Temp	Data Type	Reference
WS 3.13E+004	mg/L	25	EST	MEYLAN,WM ET AL. (1996)
logP 0.12			EXP	SANGSTER,J (1993)
VP 1.83E-006	mm Hg	25	EST	NEELY,WB & BLAU,GE (1985)
DC	pKa			
HL 2.88E-007	atm m3/mol	25	EST	MEYLAN,WM & HOWARD,PH (1991)
OH 9.82E-011	cm3/molc sec	25	EST	MEYLAN,WM & HOWARD,PH (1993)

000092-24-0 — TETRACENE

Formula: $C_{18}H_{12}$
Mol Weight: 228.30
MP (deg C): 341
FP (deg C):
BP (deg C):
BP pressure (mm Hg):

Property/Value	Units	Temp	Data Type	Reference
WS 1.51E-003	mg/L	25	EXP	YALKOWSKY,SH & DANNENFELSER,RM (1992)
logP 5.76			EXP	SANGSTER,J (1993)
VP 6.66E-007	mm Hg	25	EST	NEELY,WB & BLAU,GE (1985)
DC	pKa			
HL 5.01E-006	atm m3/mol	25	EST	MEYLAN,WM & HOWARD,PH (1991)
OH 5.00E-011	cm3/molc sec	25	EST	MEYLAN,WM & HOWARD,PH (1993)

000092-43-3 — 1-PHENYL-3-PYRAZOLIDINONE

Formula: $C_9H_{10}N_2O$
Mol Weight: 162.19
MP (deg C): 121
FP (deg C):
BP (deg C):
BP pressure (mm Hg):

Property/Value	Units	Temp	Data Type	Reference
WS 1.16E+004	mg/L	25	EST	MEYLAN,WM ET AL. (1996)
logP 0.89			EXP	HANSCH,C & LEO,AJ (1985)
VP 1.63E-005	mm Hg	25	EST	NEELY,WB & BLAU,GE (1985)
DC	pKa			
HL 2.33E-010	atm m3/mol	25	EST	MEYLAN,WM & HOWARD,PH (1991)
OH 4.57E-011	cm3/molc sec	25	EST	MEYLAN,WM & HOWARD,PH (1993)

000092-44-4 — 2,3-NAPHTHALENEDIOL

Formula: $C_{10}H_8O_2$
Mol Weight: 160.17
MP (deg C): 162
FP (deg C):
BP (deg C):
BP pressure (mm Hg):

Property/Value	Units	Temp	Data Type	Reference
WS 2.93E+002	mg/L	20	EXP	YALKOWSKY,SH & DANNENFELSER,RM (1992)
logP 2.24			EXP	SANGSTER,J (1993)
VP 7.15E-006	mm Hg	25	EST	NEELY,WB & BLAU,GE (1985)
DC	pKa			
HL 5.69E-012	atm m3/mol	25	EST	MEYLAN,WM & HOWARD,PH (1991)
OH 2.00E-010	cm3/molc sec	25	EST	MEYLAN,WM & HOWARD,PH (1993)

000092-50-2 — ETHANOL, 2-(ETHYLPHENYLAMINO)-

Formula: $C_{10}H_{15}NO$
Mol Weight: 165.24
MP (deg C): 36-38
FP (deg C):
BP (deg C): 268
BP pressure (mm Hg):

Property/Value	Units	Temp	Data Type	Reference
WS 1.04E+003	mg/L	25	EST	MEYLAN,WM ET AL. (1996)
logP 2.10			EXP	HANSCH,C ET AL. (1995)
VP 3.66E-004	mm Hg	25	EST	NEELY,WB & BLAU,GE (1985)
DC	pKa			
HL 5.52E-009	atm m3/mol	25	EST	MEYLAN,WM & HOWARD,PH (1991)
OH 1.70E-010	cm3/molc sec	25	EST	MEYLAN,WM & HOWARD,PH (1993)

000092-51-3 — DICYCLOHEXYL

Formula: $C_{12}H_{22}$
Mol Weight: 166.31
MP (deg C): 4
FP (deg C):
BP (deg C): 238
BP pressure (mm Hg):

Property/Value	Units	Temp	Data Type	Reference
WS 1.83E-001	mg/L	25	EST	MEYLAN,WM ET AL. (1996)
logP 5.86			EST	MEYLAN,WM & HOWARD,PH (1995)
VP 1.08E-001	mm Hg	25	EXP	DAUBERT,TE & DANNER,RP (1989)
DC	pKa			
HL 6.16E-001	atm m3/mol	25	EST	MEYLAN,WM & HOWARD,PH (1991)
OH 2.14E-011	cm3/molc sec	25	EST	MEYLAN,WM & HOWARD,PH (1993)

000092-52-4 — BIPHENYL

Formula: $C_{12}H_{10}$
Mol Weight: 154.21
MP (deg C): 69-71
FP (deg C):
BP (deg C): 254-255
BP pressure (mm Hg):

Property/Value	Units	Temp	Data Type	Reference
WS 6.94E+000	mg/L	25	EXP	PEARLMAN,RS ET AL. (1984)
logP 3.98			EXP	SANGSTER,J (1993)
VP 8.93E-003	mm Hg	25	EXP	BURKHARD,LP ET AL. (1984)
DC	pKa			
HL 4.08E-004	atm m3/mol	25	EXP	MACKAY,D ET AL. (1979)
OH 7.20E-012	cm3/molc sec	25	EXP	ATKINSON,R (1989)

000092-53-5 — N-PHENYLMORPHOLINE

Formula: $C_{10}H_{19}NO$
Mol Weight: 169.27
MP (deg C): 51-54
FP (deg C):
BP (deg C): 165-170
BP pressure (mm Hg): 4.50E+001

Property/Value	Units	Temp	Data Type	Reference
WS 4.56E+003	mg/L	25	EST	MEYLAN,WM ET AL. (1996)
logP 1.36			EXP	HANSCH,C & LEO,AJ (1985)
VP 1.53E-002	mm Hg	25	EST	NEELY,WB & BLAU,GE (1985)
DC	pKa			
HL 5.87E-007	atm m3/mol	25	EST	MEYLAN,WM & HOWARD,PH (1991)
OH 2.21E-010	cm3/molc sec	25	EST	MEYLAN,WM & HOWARD,PH (1993)

000092-54-6 — 1-PHENYLPIPERAZINE

Formula: $C_{10}H_{14}N_2$
Mol Weight: 162.24
MP (deg C):
FP (deg C):
BP (deg C): 286.5
BP pressure (mm Hg):

Property/Value	Units	Temp	Data Type	Reference
WS 7.83E+004	mg/L	25	EST	MEYLAN,WM ET AL. (1996)
logP 1.10			EXP	CACCIA,S ET AL. (1985)
VP 5.45E-003	mm Hg	25	EXT	CHAO,J ET AL. (1983)
DC	pKa			
HL 1.13E-008	atm m3/mol	25	EST	MEYLAN,WM & HOWARD,PH (1991)
OH 2.52E-010	cm3/molc sec	25	EST	MEYLAN,WM & HOWARD,PH (1993)

000092-55-7 — 2-NO2-5-DI-(ACETOXYMETHYL)FURANE

Formula: $C_9H_9NO_7$
Mol Weight: 243.17
MP (deg C): 90-92
FP (deg C):
BP (deg C):
BP pressure (mm Hg):

Property/Value	Units	Temp	Data Type	Reference
WS 1.31E+003	mg/L	25	EST	MEYLAN,WM ET AL. (1996)
logP 1.06			EXP	HANSCH,C & LEO,AJ (1985)
VP 6.35E-004	mm Hg	25	EST	NEELY,WB & BLAU,GE (1985)
DC	pKa			
HL 1.33E-010	atm m3/mol	25	EST	MEYLAN,WM & HOWARD,PH (1991)
OH 1.72E-011	cm3/molc sec	25	EST	MEYLAN,WM & HOWARD,PH (1993)

000092-62-6 — 3,6-DIAMINOACRIDINE

Formula: $C_{13}H_{11}N_3$
Mol Weight: 209.25
MP (deg C): 281
FP (deg C):
BP (deg C):
BP pressure (mm Hg):

Property/Value	Units	Temp	Data Type	Reference
WS 5.00E+005	mg/L	20	EXP	YALKOWSKY,SH & DANNENFELSER,RM (1992)
logP 1.83			EXP	HANSCH,C ET AL. (1995)
VP 8.94E-008	mm Hg	25	EST	NEELY,WB & BLAU,GE (1985)
DC 8.06	pKa	20	EXP	PERRIN,DD (1972)
HL 8.39E-015	atm m3/mol	25	EST	MEYLAN,WM & HOWARD,PH (1991)
OH 2.00E-010	cm3/molc sec	25	EST	MEYLAN,WM & HOWARD,PH (1993)

000092-66-0 — 4-BROMOBIPHENYL

Formula: $C_{12}H_9Br$
Mol Weight: 233.11
MP (deg C): 91.5
FP (deg C):
BP (deg C): 310
BP pressure (mm Hg):

Property/Value	Units	Temp	Data Type	Reference
WS 6.53E-001	mg/L	25	EXP	YALKOWSKY,SH & DANNENFELSER,RM (1992)
logP 4.96			EXP	DOUCETTE,WJ & ANDREN,AW (1987)
VP 1.07E-002	mm Hg	25	EXT	PERRY,RH & GREEN,D (1984)
DC	pKa			
HL 1.65E-004	atm m3/mol	25	EST	MEYLAN,WM & HOWARD,PH (1991)
OH 3.47E-012	cm3/molc sec	25	EST	MEYLAN,WM & HOWARD,PH (1993)

000092-67-1 — 4-AMINOBIPHENYL

Formula: $C_{12}H_{11}N$
Mol Weight: 169.23
MP (deg C): 53.5
FP (deg C):
BP (deg C): 302
BP pressure (mm Hg):

Property/Value	Units	Temp	Data Type	Reference
WS 2.24E+002	mg/L	25	EST	MEYLAN,WM ET AL. (1996)
logP 2.86			EXP	MARTIN-VILLODRE,A ET AL. (1986)
VP 1.17E-004	mm Hg	25	EST	NEELY,WB & BLAU,GE (1985)
DC 4.35	pKa	18	EXP	PERRIN,DD (1965)
HL 1.46E-007	atm m3/mol	25	EST	MEYLAN,WM & HOWARD,PH (1991)
OH 1.06E-010	cm3/molc sec	25	EST	MEYLAN,WM & HOWARD,PH (1993)

CAS #: 000092-69-3				P-PHENYLPHENOL

Formula: $C_{12}H_{10}O$

Mol Weight: 170.21

MP (deg C): 164-165 **FP (deg C):**

BP (deg C): 305-308

BP pressure (mm Hg):

Property/ Value	Units	Temp	Data Type	Reference
WS 5.62E+001	mg/L	25	EXP	YALKOWSKY,SH & DANNENFELSER,RM (1992)
logP 3.20			EXP	HANSCH,C & LEO,AJ (1985)
VP 1.86E-005	mm Hg	25	EST	NEELY,WB & BLAU,GE (1985)
DC 9.55	pKa		EXP	PEARCE,PJ & SIMKINS,RJJ (1968)
HL 5.23E-008	atm m3/mol	25	EST	MEYLAN,WN HOWARD,PH (1993)
OH 2.73E-011	cm3/molc sec	25	EST	MEYLAN,WM & HOWARD,PH (1993)

CAS #: 000092-70-6				2-NAPHTHALENECARBOXYLIC ACID, 3-HYDROXY-

Formula: $C_{11}H_8O_3$

Mol Weight: 188.18

MP (deg C): 222.5 **FP (deg C):**

BP (deg C):

BP pressure (mm Hg):

Property/ Value	Units	Temp	Data Type	Reference
WS 4.74E+002	mg/L	25	EST	MEYLAN,WM ET AL. (1996)
logP 3.05			EXP	SANGSTER,J (1993)
VP 5.45E-007	mm Hg	25	EST	NEELY,WB & BLAU,GE (1985)
DC 2.79	pKa	25	EXP	SERJEANT,EP & DEMPSEY,B (1979)
HL 1.39E-009	atm m3/mol	25	EST	MEYLAN,WM & HOWARD,PH (1991)
OH 2.42E-011	cm3/molc sec	25	EST	MEYLAN,WM & HOWARD,PH (1993)

CAS #: 000092-71-7				OXAZOLE, 2,5-DIPHENYL-

Formula: $C_{15}H_{12}NO$

Mol Weight: 222.27

MP (deg C): 74 **FP (deg C):**

BP (deg C): 360

BP pressure (mm Hg):

Property/ Value	Units	Temp	Data Type	Reference
WS 3.48E+000	mg/L	25	EST	MEYLAN,WM ET AL. (1996)
logP 4.67			EXP	HANSCH,C ET AL. (1995)
VP 4.65E-006	mm Hg	25	EST	NEELY,WB & BLAU,GE (1985)
DC	pKa			
HL 4.15E-008	atm m3/mol	25	EST	MEYLAN,WM & HOWARD,PH (1991)
OH 1.81E-011	cm3/molc sec	25	EST	MEYLAN,WM & HOWARD,PH (1993)

CAS #: 000092-82-0				PHENAZINE

Formula: $C_{12}H_8N_2$

Mol Weight: 180.21

MP (deg C): 171 **FP (deg C):**

BP (deg C): >360

BP pressure (mm Hg):

Property/ Value	Units	Temp	Data Type	Reference
WS 1.60E+001	mg/L	25	EST	MEYLAN,WM ET AL. (1996)
logP 2.84			EXP	HANSCH,C & LEO,AJ (1985)
VP 3.26E-005	mm Hg	25	EST	NEELY,WB & BLAU,GE (1985)
DC 1.19	pKa	20	EXP	PERRIN,DD (1965)
HL 2.78E-008	atm m3/mol	25	EST	MEYLAN,WM & HOWARD,PH (1991)
OH 7.50E-012	cm3/molc sec	25	EST	MEYLAN,WM & HOWARD,PH (1993)

CAS #: 000092-83-1				9H-XANTHENE

Formula: $C_{13}H_{10}O$

Mol Weight: 182.22

MP (deg C): 100.5 **FP (deg C):**

BP (deg C): 311

BP pressure (mm Hg):

Property/ Value	Units	Temp	Data Type	Reference
WS 1.02E+000	mg/L	25	EST	MEYLAN,WM ET AL. (1996)
logP 4.23			EXP	DE VOOGT,P ET AL. (1990)
VP 8.47E-004	mm Hg	25	EST	NEELY,WB & BLAU,GE (1985)
DC	pKa			
HL 4.77E-005	atm m3/mol	25	EST	MEYLAN,WM & HOWARD,PH (1991)
OH 1.29E-011	cm3/molc sec	25	EST	MEYLAN,WM & HOWARD,PH (1993)

CAS #: 000092-84-2				PHENOTHIAZINE

Formula: $C_{12}H_9NS$

Mol Weight: 199.28

MP (deg C): 185.1 **FP (deg C):**

BP (deg C): 371

BP pressure (mm Hg):

Property/ Value	Units	Temp	Data Type	Reference
WS 1.59E+000	mg/L	25	EXP	YALKOWSKY,SH & DANNENFELSER,RM (1992)
logP 4.15			EXP	HANSCH,C ET AL. (1985)
VP 8.90E-007	mm Hg	25	EST	NEELY,WB & BLAU,GE (1985)
DC 2.52	pKa		EXP	SANGSTER,J (1994)
HL 2.80E-008	atm m3/mol	25	EST	MEYLAN,WM & HOWARD,PH (1991)
OH 1.80E-010	cm3/molc sec	25	EST	MEYLAN,WM & HOWARD,PH (1993)

CAS #: 000092-85-3				THIANTHRENE

Formula: $C_{12}H_8S_2$

Mol Weight: 216.33

MP (deg C): 159.3 **FP (deg C):**

BP (deg C): 365

BP pressure (mm Hg):

Property/ Value	Units	Temp	Data Type	Reference
WS 4.26E-001	mg/L	25	EST	MEYLAN,WM ET AL. (1996)
logP 4.47			EXP	DE VOOGT,P ET AL. (1990)
VP 3.10E-005	mm Hg	25	EST	NEELY,WB & BLAU,GE (1985)
DC	pKa			
HL 8.30E-007	atm m3/mol	25	EST	MEYLAN,WM & HOWARD,PH (1991)
OH 1.54E-011	cm3/molc sec	25	EST	MEYLAN,WM & HOWARD,PH (1993)

CAS #: 000092-86-4				4,4'-DIBROMOBIPHENYL

Formula: $C_{12}H_8Br_2$

Mol Weight: 312.02

MP (deg C): 164 **FP (deg C):**

BP (deg C): 357.5

BP pressure (mm Hg):

Property/ Value	Units	Temp	Data Type	Reference
WS 1.36E-001	mg/L	25	EST	MEYLAN,WM ET AL. (1996)
logP 5.72			EXP	HANSCH,C ET AL. (1995)
VP 2.46E-005	mm Hg	25	EST	NEELY,WB & BLAU,GE (1985)
DC	pKa			
HL 6.57E-005	atm m3/mol	25	EST	MEYLAN,WM & HOWARD,PH (1991)
OH 1.55E-012	cm3/molc sec	25	EST	MEYLAN,WM & HOWARD,PH (1993)

CAS #: 000092-87-5				BENZIDINE

Formula: $C_{12}H_{12}N_2$

Mol Weight: 184.24

MP (deg C): 120

FP (deg C):

BP (deg C): 400

BP pressure (mm Hg):

Property/Value	Units	Temp	Data Type	Reference
WS 3.22E+002	mg/L	25	EXP	YALKOWSKY,SH & DANNENFELSER,RM (1992)
logP 1.34			EXP	HANSCH,C & LEO,AJ (1985)
VP 7.50E-008	mm Hg	20	EST	SCHMIDT-BLEEK,F ET AL. (1982)
DC 4.66	pKa	20	EXP	PERRIN,DD (1965)
HL 3.88E-011	atm m3/mol	25	EST	MEYLAN,WM & HOWARD,PH (1991)
OH 2.05E-010	cm3/molc sec	25	EST	MEYLAN,WM & HOWARD,PH (1993)

CAS #: 000092-88-6				4,4'-BIPHENOL

Formula: $C_{12}H_{10}O_2$

Mol Weight: 186.21

MP (deg C): 282-284

FP (deg C):

BP (deg C):

BP pressure (mm Hg):

Property/Value	Units	Temp	Data Type	Reference
WS 7.98E+002	mg/L	25	EST	MEYLAN,WM ET AL. (1996)
logP 2.80			EST	MEYLAN,WM & HOWARD,PH (1995)
VP 7.89E-007	mm Hg	25	EST	NEELY,WB & BLAU,GE (1985)
DC	pKa			
HL 4.48E-012	atm m3/mol	25	EST	MEYLAN,WM & HOWARD,PH (1991)
OH 4.79E-011	cm3/molc sec	25	EST	MEYLAN,WM & HOWARD,PH (1993)

CAS #: 000092-91-1				4-ACETYLBIPHENYL

Formula: $C_{14}H_{12}O$

Mol Weight: 196.25

MP (deg C): 121

FP (deg C):

BP (deg C): 326

BP pressure (mm Hg):

Property/Value	Units	Temp	Data Type	Reference
WS 9.60E+001	mg/L	25	EXP	SOUTHWORTH,GR & KELLER,JL (1986)
logP 3.44			EST	MEYLAN,WM & HOWARD,PH (1995)
VP 1.07E-004	mm Hg	25	EST	NEELY,WB & BLAU,GE (1985)
DC	pKa			
HL 7.53E-007	atm m3/mol	25	EST	MEYLAN,WM & HOWARD,PH (1991)
OH 6.58E-012	cm3/molc sec	25	EST	MEYLAN,WM & HOWARD,PH (1993)

CAS #: 000092-92-2				1,1'-BIPHENYL -4-CARBOXYLIC ACID

Formula: $C_{13}H_{10}O_2$

Mol Weight: 198.22

MP (deg C): 228

FP (deg C):

BP (deg C):

BP pressure (mm Hg):

Property/Value	Units	Temp	Data Type	Reference
WS 2.80E+001	mg/L	25	EST	MEYLAN,WM ET AL. (1996)
logP 3.75			EXP	MIYAKE,KN ET AL. (1986)
VP 5.48E-006	mm Hg	25	EST	NEELY,WB & BLAU,GE (1985)
DC	pKa			
HL 8.33E-009	atm m3/mol	25	EST	MEYLAN,WM & HOWARD,PH (1991)
OH 5.16E-012	cm3/molc sec	25	EST	MEYLAN,WM & HOWARD,PH (1993)

CAS #: 000092-93-3				P-NITROBIPHENYL

Formula: $C_{12}H_9NO_2$

Mol Weight: 199.21

MP (deg C): 113.7

FP (deg C):

BP (deg C): 340

BP pressure (mm Hg):

Property/Value	Units	Temp	Data Type	Reference
WS 9.84E+000	mg/L	25	EST	MEYLAN,WM ET AL. (1996)
logP 3.82			EXP	HANSCH,C ET AL. (1995)
VP 3.25E-005	mm Hg	25	EST	NEELY,WB & BLAU,GE (1985)
DC	pKa			
HL 3.54E-006	atm m3/mol	25	EST	MEYLAN,WM & HOWARD,PH (1991)
OH 2.36E-012	cm3/molc sec	25	EST	MEYLAN,WM & HOWARD,PH (1993)

CAS #: 000092-94-4				1,1':4',1''-TERPHENYL

Formula: $C_{18}H_{14}$

Mol Weight: 230.31

MP (deg C): 210.1

FP (deg C):

BP (deg C): 376

BP pressure (mm Hg):

Property/Value	Units	Temp	Data Type	Reference
WS 1.80E-002	mg/L	25	EXP	YALKOWSKY,SH & DANNENFELSER,RM (1992)
logP 6.03			EXP	HANSCH,C ET AL. (1995)
VP 3.42E-007	mm Hg	25	EST	NEELY,WB & BLAU,GE (1985)
DC	pKa			
HL 3.38E-005	atm m3/mol	25	EST	MEYLAN,WM & HOWARD,PH (1991)
OH 9.19E-012	cm3/molc sec	25	EST	MEYLAN,WM & HOWARD,PH (1993)

CAS #: 000093-04-9				NAPHTHALENE, 2-METHOXY-

Formula: $C_{11}H_{10}O$

Mol Weight: 158.20

MP (deg C): 72

FP (deg C):

BP (deg C): 272

BP pressure (mm Hg):

Property/Value	Units	Temp	Data Type	Reference
WS 7.59E+001	mg/L	25	EST	MEYLAN,WM ET AL. (1996)
logP 3.47			EXP	HANSCH,C ET AL. (1995)
VP 8.23E-003	mm Hg	25	EST	NEELY,WB & BLAU,GE (1985)
DC	pKa			
HL 3.11E-005	atm m3/mol	25	EST	MEYLAN,WM & HOWARD,PH (1991)
OH 2.01E-010	cm3/molc sec	25	EST	MEYLAN,WM & HOWARD,PH (1993)

CAS #: 000093-05-0				N,N-DIETHYL-1,4-BENZENEDIAMINE

Formula: $C_{10}H_{16}N_2$

Mol Weight: 164.25

MP (deg C):

FP (deg C):

BP (deg C): 261

BP pressure (mm Hg):

Property/Value	Units	Temp	Data Type	Reference
WS 8.06E+002	mg/L	25	EST	MEYLAN,WM ET AL. (1996)
logP 2.24			EST	MEYLAN,WM & HOWARD,PH (1995)
VP 2.98E-003	mm Hg	25	EST	NEELY,WB & BLAU,GE (1985)
DC 7.96	pKa		EXP	PERRIN,DD (1972)
HL 5.33E-008	atm m3/mol	25	EST	MEYLAN,WM & HOWARD,PH (1991)
OH 2.18E-010	cm3/molc sec	25	EST	MEYLAN,WM & HOWARD,PH (1993)

CAS #: 000093-07-2	3,4-DIMETHOXYBENZOIC ACID

Formula: $C_9H_{10}O_4$

Mol Weight: 182.18

MP (deg C): 180-181   FP (deg C):

BP (deg C):

BP pressure (mm Hg):

Property/Value	Units	Temp	Data Type	Reference
WS 2.27E+003	mg/L	25	EST	MEYLAN,WM ET AL. (1996)
logP 1.61			EXP	HANSCH,C & LEO,AJ (1985)
VP 2.47E-004	mm Hg	25	EXT	COLOMINA,M ET AL. (1985)
DC 4.36	pKa	25	EXP	SERJEANT,EP & DEMPSEY,B (1979)
HL 3.79E-010	atm m3/mol	25	EST	MEYLAN,WM & HOWARD,PH (1991)
OH 9.07E-012	cm3/molc sec	25	EST	MEYLAN,WM & HOWARD,PH (1993)

CAS #: 000093-08-3	1-(2-NAPHTHALENYL)ETHANONE

Formula: $C_{12}H_{10}O$

Mol Weight: 170.21

MP (deg C): 56   FP (deg C):

BP (deg C): 302

BP pressure (mm Hg):

Property/Value	Units	Temp	Data Type	Reference
WS 2.72E+002	mg/L	25	EXP	SOUTHWORTH,GR & KELLER,JL (1986)
logP 2.85			EST	MEYLAN,WM & HOWARD,PH (1995)
VP 6.11E-004	mm Hg	25	EXT	PERRY,RH & GREEN,D (1984)
DC	pKa			
HL 9.58E-007	atm m3/mol	25	EST	MEYLAN,WM & HOWARD,PH (1991)
OH 1.98E-011	cm3/molc sec	25	EST	MEYLAN,WM & HOWARD,PH (1993)

CAS #: 000093-09-4	2-NAPHTHOIC ACID

Formula: $C_{11}H_8O_2$

Mol Weight: 172.19

MP (deg C): 184-185   FP (deg C):

BP (deg C): >300

BP pressure (mm Hg):

Property/Value	Units	Temp	Data Type	Reference
WS 2.24E+001	mg/L	25	EXP	YALKOWSKY,SH & DANNENFELSER,RM (1992)
logP 3.28			EXP	HANSCH,C ET AL. (1995)
VP 6.63E-006	mm Hg	25	EXT	PERRY,RH & GREEN,D (1984)
DC 4.17	pKa	20	EXP	KORTUM,G ET AL (1961)
HL 1.06E-008	atm m3/mol	25	EST	MEYLAN,WM & HOWARD,PH (1991)
OH 8.52E-012	cm3/molc sec	25	EST	MEYLAN,WM & HOWARD,PH (1993)

CAS #: 000093-15-2	METHYLEUGENOL

Formula: $C_{11}H_{14}O_2$

Mol Weight: 178.23

MP (deg C): -4   FP (deg C):

BP (deg C): -254.7

BP pressure (mm Hg):

Property/Value	Units	Temp	Data Type	Reference
WS 5.00E+002	mg/L	25	EXP	CHEM INSPECT TEST INST (1992)
logP 3.03			EST	MEYLAN,WM & HOWARD,PH (1995)
VP 1.20E-002	mm Hg	25	EXT	PERRY,RH & GREEN,D (1984)
DC	pKa			
HL 5.60E-006	atm m3/mol	25	EST	VP/WSOL
OH 7.50E-011	cm3/molc sec	25	EST	MEYLAN,WM & HOWARD,PH (1993)

CAS #: 000093-26-5	O-METHOXYACETANILIDE

Formula: $C_9H_{11}NO_2$

Mol Weight: 165.19

MP (deg C): 87.5   FP (deg C):

BP (deg C): 304

BP pressure (mm Hg):

Property/Value	Units	Temp	Data Type	Reference
WS 1.74E+004	mg/L	25	EST	MEYLAN,WM ET AL. (1996)
logP 0.67			EXP	HANSCH,C & LEO,AJ (1985)
VP 6.59E-005	mm Hg	25	EST	NEELY,WB & BLAU,GE (1985)
DC	pKa			
HL 3.65E-010	atm m3/mol	25	EST	MEYLAN,WM & HOWARD,PH (1991)
OH 1.22E-011	cm3/molc sec	25	EST	MEYLAN,WM & HOWARD,PH (1993)

CAS #: 000093-28-7	EUGENYL ACETATE

Formula: $C_{12}H_{14}O_3$

Mol Weight: 206.24

MP (deg C): 30.5   FP (deg C):

BP (deg C): 281

BP pressure (mm Hg):

Property/Value	Units	Temp	Data Type	Reference
WS 9.83E+001	mg/L	25	EST	MEYLAN,WM ET AL. (1996)
logP 3.06			EST	MEYLAN,WM & HOWARD,PH (1995)
VP 3.19E-003	mm Hg	25	EXT	JORDAN,TE (1954)
DC	pKa			
HL 5.56E-006	atm m3/mol	25	EST	MEYLAN,WM & HOWARD,PH (1991)
OH 6.74E-011	cm3/molc sec	25	EST	MEYLAN,WM & HOWARD,PH (1993)

CAS #: 000093-45-8	4-(2-NAPHTHYLAMINO)PHENOL

Formula: $C_{16}H_{13}NO$

Mol Weight: 235.29

MP (deg C): 139-142   FP (deg C):

BP (deg C):

BP pressure (mm Hg):

Property/Value	Units	Temp	Data Type	Reference
WS 2.24E+001	mg/L	25	EST	MEYLAN,WM ET AL. (1996)
logP 3.64			EST	MEYLAN,WM & HOWARD,PH (1995)
VP 1.14E-007	mm Hg	25	EST	NEELY,WB & BLAU,GE (1985)
DC	pKa			
HL 1.07E-011	atm m3/mol	25	EST	MEYLAN,WM & HOWARD,PH (1991)
OH 3.01E-010	cm3/molc sec	25	EST	MEYLAN,WM & HOWARD,PH (1993)

CAS #: 000093-51-6	2-METHOXY-4-METHYLPHENOL

Formula: $C_8H_{10}O_2$

Mol Weight: 138.17

MP (deg C): 5.5   FP (deg C):

BP (deg C): 220

BP pressure (mm Hg):

Property/Value	Units	Temp	Data Type	Reference
WS 2.09E+003	mg/L	25	EST	MEYLAN,WM ET AL. (1996)
logP 1.88			EST	MEYLAN,WM & HOWARD,PH (1995)
VP 2.39E-002	mm Hg	25	EST	NEELY,WB & BLAU,GE (1985)
DC 10.28	pKa	25	EXP	SERJEANT,EP & DEMPSEY,B (1979)
HL 3.66E-008	atm m3/mol	25	EST	MEYLAN,WM & HOWARD,PH (1991)
OH 3.98E-011	cm3/molc sec	25	EST	MEYLAN,WM & HOWARD,PH (1993)

CAS #: 000093-52-7 — (1,2-DIBROMOETHYL)BENZENE

Formula: $C_8H_8Br_2$
Mol Weight: 263.97
MP (deg C): 75
FP (deg C):
BP (deg C): 133
BP pressure (mm Hg): 1.90E+001

Property/Value	Units	Temp	Data Type	Reference
WS 1.53E+001	mg/L	25	EST	MEYLAN,WM ET AL. (1996)
logP 3.64			EST	MEYLAN,WM & HOWARD,PH (1995)
VP 1.06E-002	mm Hg	25	EXT	OHE,S (1976)
DC	pKa			
HL 1.05E-004	atm m3/mol	25	EST	MEYLAN,WM & HOWARD,PH (1991)
OH 2.38E-012	cm3/molc sec	25	EST	MEYLAN,WM & HOWARD,PH (1993)

CAS #: 000093-55-0 — PROPIOPHENONE

Formula: $C_9H_{10}O$
Mol Weight: 134.18
MP (deg C): 21
FP (deg C): 18.
BP (deg C): 218
BP pressure (mm Hg):

Property/Value	Units	Temp	Data Type	Reference
WS 2.00E+003	mg/L	20	EXP	PAPA,AJ & SHERMAN,PDJR (1981)
logP 2.19			EXP	HANSCH,C ET AL. (1995)
VP 1.50E+000	mm Hg	20	EXP	PAPA,AJ & SHERMAN,PDJR (1981)
DC	pKa			
HL 1.30E-004	atm m3/mol	20	EST	VP/WSOL
OH 3.01E-012	cm3/molc sec	25	EST	MEYLAN,WM & HOWARD,PH (1993)

CAS #: 000093-58-3 — BENZOIC ACID, METHYL ESTER

Formula: $C_8H_8O_2$
Mol Weight: 136.15
MP (deg C): -15
FP (deg C):
BP (deg C): 199
BP pressure (mm Hg):

Property/Value	Units	Temp	Data Type	Reference
WS 2.10E+003	mg/L	20	EXP	RIDDICK,JA ET AL. (1986)
logP 2.12			EXP	HANSCH,C & LEO,AJ (1985)
VP 3.80E-001	mm Hg	25	EXP	DAUBERT,TE & DANNER,RP (1986)
DC	pKa			
HL 3.24E-005	atm m3/mol	25	EST	VP/WSOL
OH 8.67E-013	cm3/molc sec	25	EST	ATKINSON,R (1987)

CAS #: 000093-60-7 — NICOTINIC ACID, METHYL ESTER

Formula: $C_7H_7NO_2$
Mol Weight: 137.14
MP (deg C): 39
FP (deg C):
BP (deg C): 209
BP pressure (mm Hg):

Property/Value	Units	Temp	Data Type	Reference
WS 4.76E+004	mg/L	20	EXP	YALKOWSKY,SH & DANNENFELSER,RM (1992)
logP 0.83			EXP	HANSCH,C & LEO,AJ (1985)
VP 2.77E-001	mm Hg	25	EST	NEELY,WB & BLAU,GE (1985)
DC 3.13	pKa	22	EXP	PERRIN,DD (1965)
HL 4.54E-008	atm m3/mol	25	EST	MEYLAN,WM & HOWARD,PH (1991)
OH 3.36E-013	cm3/molc sec	25	EST	MEYLAN,WM & HOWARD,PH (1993)

CAS #: 000093-61-8 — N-METHYLFORMANILIDE

Formula: C_8H_9NO
Mol Weight: 135.17
MP (deg C): 14.5
FP (deg C):
BP (deg C): 243
BP pressure (mm Hg):

Property/Value	Units	Temp	Data Type	Reference
WS 1.03E+004	mg/L	25	EST	MEYLAN,WM ET AL. (1996)
logP 1.09			EXP	HANSCH,C & LEO,AJ (1985)
VP 2.08E-002	mm Hg	25	EST	NEELY,WB & BLAU,GE (1985)
DC	pKa			
HL 1.73E-007	atm m3/mol	25	EST	MEYLAN,WM & HOWARD,PH (1991)
OH 1.37E-011	cm3/molc sec	25	EST	MEYLAN,WM & HOWARD,PH (1993)

CAS #: 000093-65-2 — MECOPROP

Formula: $C_{10}H_{11}ClO_3$
Mol Weight: 214.65
MP (deg C): 94-95
FP (deg C):
BP (deg C): 298
BP pressure (mm Hg):

Property/Value	Units	Temp	Data Type	Reference
WS 6.20E+002	mg/L	20	EXP	MARTIN,H & WORTHING,CR (1977)
logP 3.13			EXP	ILCHMANN,A ET AL. (1993)
VP 7.50E-007	mm Hg	20	NR	HARTLEY,D & KIDD,H (1983); <7.5E-7
DC 3.10	pKa	24	EXP	CESSNA,AJ & GROVER,R (1978)
HL 1.82E-008	atm m3/mol	25	EST	MEYLAN,WM & HOWARD,PH (1991)
OH 1.74E-011	cm3/molc sec	25	EST	MEYLAN,WM & HOWARD,PH (1993)

CAS #: 000093-71-0 — 2-CHLORO-N,N-DIALLYLACETAMIDE

Formula: $C_8H_{12}ClNO$
Mol Weight: 173.64
MP (deg C):
FP (deg C):
BP (deg C): 92
BP pressure (mm Hg): 2.00E+000

Property/Value	Units	Temp	Data Type	Reference
WS 2.00E+004	mg/L	25	EXP	SANBORN,JR ET AL. (1977)
logP 1.79			EST	MEYLAN,WM & HOWARD,PH (1995)
VP 9.40E-003	mm Hg	20	EXP	SANBORN,JR ET AL. (1977)
DC	pKa			
HL 1.07E-007	atm m3/mol	25	EST	VP/WSOL
OH 7.55E-011	cm3/molc sec	25	EST	MEYLAN,WM & HOWARD,PH (1993)

CAS #: 000093-72-1 — SILVEX

Formula: $C_9H_7Cl_3O_3$
Mol Weight: 269.51
MP (deg C): 181.6
FP (deg C):
BP (deg C):
BP pressure (mm Hg):

Property/Value	Units	Temp	Data Type	Reference
WS 1.40E+002	mg/L	25	EXP	YALKOWSKY,SH & DANNENFELSER,RM (1992)
logP 3.80			EXP	HANSCH,C ET AL. (1995)
VP 9.97E-006	mm Hg	25	EST	NEELY,WB & BLAU,GE (1985)
DC	pKa			
HL 9.06E-009	atm m3/mol	25	EST	MEYLAN,WM & HOWARD,PH (1991)
OH 1.09E-011	cm3/molc sec	25	EST	MEYLAN,WM & HOWARD,PH (1993)

000093-75-4 — BAYER 30686

Formula: $C_9H_4N_2S_3$
Mol Weight: 236.34
MP (deg C): 180
FP (deg C):
BP (deg C):
BP pressure (mm Hg):

Property/Value	Units	Temp	Data Type	Reference
WS 2.16E+003	mg/L	25	EST	MEYLAN,WM ET AL. (1996)
logP 1.31			EST	MEYLAN,WM & HOWARD,PH (1995)
VP 6.33E-008	mm Hg	25	EST	NEELY,WB & BLAU,GE (1985)
DC	pKa			
HL 7.80E-008	atm m3/mol	25	EST	MEYLAN,WM & HOWARD,PH (1991)
OH 7.92E-012	cm3/molc sec	25	EST	MEYLAN,WM & HOWARD,PH (1993)

000093-76-5 — 2,4,5-TRICHLOROPHENOXYACETIC ACID

Formula: $C_8H_5Cl_3O_3$
Mol Weight: 255.49
MP (deg C): 154-155
FP (deg C):
BP (deg C):
BP pressure (mm Hg):

Property/Value	Units	Temp	Data Type	Reference
WS 2.78E+002	mg/L	25	EXP	HARTLEY,D & KIDD,H (1983)
logP 3.31			EXP	HANSCH,C & LEO,AJ (1985)
VP 3.75E-005	mm Hg	20	EXT	RIEDERER,M (1990)
DC 2.83	pKa	25	EXP	SERJEANT,EP & DEMPSEY,B (1979)
HL 8.68E-009	atm m3/mol	25	EST	MEYLAN,WM & HOWARD,PH (1991)
OH 5.63E-012	cm3/molc sec	25	EST	MEYLAN,WM & HOWARD,PH (1993)

000093-79-8 — 2,4,5-T, N-BUTYL ESTER

Formula: $C_{12}H_{13}Cl_3O_3$
Mol Weight: 311.59
MP (deg C): 28.5
FP (deg C):
BP (deg C):
BP pressure (mm Hg):

Property/Value	Units	Temp	Data Type	Reference
WS 8.19E-001	mg/L	25	EST	MEYLAN,WM ET AL. (1996)
logP 4.81			EXP	SANGSTER,J (1993)
VP 1.50E-005	mm Hg	25	EXT	HAMILTON,DJ (1980)
DC	pKa			
HL 5.11E-006	atm m3/mol	25	EST	MEYLAN,WM & HOWARD,PH (1991)
OH 1.01E-011	cm3/molc sec	25	EST	MEYLAN,WM & HOWARD,PH (1993)

000093-80-1 — 2,4,5-TB

Formula: $C_{10}H_9Cl_3O_3$
Mol Weight: 283.54
MP (deg C):
FP (deg C):
BP (deg C):
BP pressure (mm Hg):

Property/Value	Units	Temp	Data Type	Reference
WS 4.20E+001	mg/L	25	EXP	BAILEY,GW & WHITE,JL (1965)
logP 4.24			EST	MEYLAN,WM & HOWARD,PH (1995)
VP 3.04E-006	mm Hg	25	EST	NEELY,WB & BLAU,GE (1985)
DC	pKa			
HL 1.70E-009	atm m3/mol	25	EST	MEYLAN,WM & HOWARD,PH (1991)
OH 1.37E-011	cm3/molc sec	25	EST	MEYLAN,WM & HOWARD,PH (1993)

000093-88-9 — BENZENEETHANAMINE, N,á-DIMETHYL-

Formula: $C_{10}H_{15}N$
Mol Weight: 149.24
MP (deg C):
FP (deg C):
BP (deg C): 207.5
BP pressure (mm Hg):

Property/Value	Units	Temp	Data Type	Reference
WS 1.33E+004	mg/L	25	EST	MEYLAN,WM ET AL. (1996)
logP 2.28			EXP	DUNCAN,JD ET AL. (1983)
VP 1.63E-001	mm Hg	25	EST	NEELY,WB & BLAU,GE (1985)
DC 9.87	pKa	25	EXP	PERRIN,DD (1965)
HL 2.37E-006	atm m3/mol	25	EST	MEYLAN,WM & HOWARD,PH (1991)
OH 9.26E-011	cm3/molc sec	25	EST	MEYLAN,WM & HOWARD,PH (1993)

000093-89-0 — ETHYL BENZOATE

Formula: $C_9H_{10}O_2$
Mol Weight: 150.18
MP (deg C): -334
FP (deg C):
BP (deg C): 211-213
BP pressure (mm Hg):

Property/Value	Units	Temp	Data Type	Reference
WS 7.20E+002	mg/L	25	EXP	YALKOWSKY,SH & DANNENFELSER,RM (1992)
logP 2.64			EXP	HANSCH,C & LEO,AJ (1985)
VP 2.67E-001	mm Hg	25	EXP	DAUBERT,TE & DANNER,RP (1989)
DC	pKa			
HL 7.33E-005	atm m3/mol	25	EST	VP/WSOL
OH 2.29E-012	cm3/molc sec	25	EST	MEYLAN,WM & HOWARD,PH (1993)

000093-97-0 — BENZOIC ACID, ANHYDRIDE

Formula: $C_{14}H_{10}O_3$
Mol Weight: 226.23
MP (deg C): 42
FP (deg C):
BP (deg C): 360
BP pressure (mm Hg):

Property/Value	Units	Temp	Data Type	Reference
WS 3.75E+002	mg/L	25	EST	MEYLAN,WM ET AL. (1996)
logP 2.26			EXP	HANSCH,C & LEO,AJ (1985)
VP 1.18E-004	mm Hg	25	EXT	PERRY,RH & GREEN,D (1984)
DC	pKa			
HL 1.40E-006	atm m3/mol	25	EST	MEYLAN,WM & HOWARD,PH (1991)
OH 3.55E-012	cm3/molc sec	25	EST	MEYLAN,WM & HOWARD,PH (1993)

000093-98-1 — BENZANILIDE

Formula: $C_{13}H_{11}NO$
Mol Weight: 197.24
MP (deg C): 163
FP (deg C):
BP (deg C): 117-119
BP pressure (mm Hg): 1.00E+001

Property/Value	Units	Temp	Data Type	Reference
WS 2.62E+002	mg/L	25	EST	MEYLAN,WM ET AL. (1996)
logP 2.62			EXP	HANSCH,C & LEO,AJ (1985)
VP 2.01E-006	mm Hg	25	EST	NEELY,WB & BLAU,GE (1985)
DC 1.26	pKa	55	EXP	PERRIN,DD (1965)
HL 1.22E-009	atm m3/mol	25	EST	MEYLAN,WM & HOWARD,PH (1991)
OH 1.42E-011	cm3/molc sec	25	EST	MEYLAN,WM & HOWARD,PH (1993)

CAS #: 000093-99-2				PHENYL BENZOATE

Formula: $C_{13}H_{10}O_2$				
Mol Weight: 198.22				
MP (deg C): 70		FP (deg C):		
BP (deg C): 314				
BP pressure (mm Hg):				

Property/Value	Units	Temp	Data Type	Reference
WS 3.84E+001	mg/L	25	EST	MEYLAN,WM ET AL. (1996)
logP 3.59			EXP	HANSCH,C & LEO,AJ (1985)
VP 4.02E-003	mm Hg	25	EXT	PERRY,RH & GREEN,D (1984)
DC	pKa			
HL 1.28E-005	atm m3/mol	25	EST	MEYLAN,WM & HOWARD,PH (1991)
OH 3.79E-012	cm3/molc sec	25	EST	MEYLAN,WM & HOWARD,PH (1993)

CAS #: 000094-02-0				ETHYL BENZOYLACETATE

Formula: $C_{11}H_{12}O_3$				
Mol Weight: 192.22				
MP (deg C): <0		FP (deg C):		
BP (deg C): 167				
BP pressure (mm Hg): 2.00E+001				

Property/Value	Units	Temp	Data Type	Reference
WS 1.21E+003	mg/L	25	EST	MEYLAN,WM ET AL. (1996)
logP 1.87			EXP	HANSCH,C ET AL. (1995)
VP 4.08E-004	mm Hg	25	EXT	PERRY,RH & GREEN,D (1984)
DC	pKa			
HL 3.10E-008	atm m3/mol	25	EST	MEYLAN,WM & HOWARD,PH (1991)
OH 3.96E-012	cm3/molc sec	25	EST	MEYLAN,WM & HOWARD,PH (1993)

CAS #: 000094-07-5				SYNEPHRINE

Formula: $C_9H_{13}NO_2$				
Mol Weight: 167.21				
MP (deg C): 184-185		FP (deg C):		
BP (deg C):				
BP pressure (mm Hg):				

Property/Value	Units	Temp	Data Type	Reference
WS 1.00E+006	mg/L	25	EST	MEYLAN,WM ET AL. (1996)
logP -0.45			EXP	HANSCH,C & LEO,AJ (1985)
VP 2.22E-005	mm Hg	25	EST	NEELY,WB & BLAU,GE (1985)
DC 8.90	pKa	25	EXP	PERRIN,DD (1965)
HL 6.78E-015	atm m3/mol	25	EST	MEYLAN,WM & HOWARD,PH (1991)
OH 1.24E-010	cm3/molc sec	25	EST	MEYLAN,WM & HOWARD,PH (1993)

CAS #: 000094-09-7				P-AMINOBENZOIC ACID, ETHYL ESTER

Formula: $C_9H_{11}NO_2$				
Mol Weight: 165.19				
MP (deg C): 88-90		FP (deg C):		
BP (deg C):				
BP pressure (mm Hg):				

Property/Value	Units	Temp	Data Type	Reference
WS 1.69E+003	mg/L	25	EXP	SUZUKI,T (1991)
logP 1.86			EXP	HANSCH,C & LEO,AJ (1985)
VP 6.59E-002	mm Hg	25	EXT	OHE,S (1976)
DC 2.51	pKa	25	EXP	PERRIN,DD (1965)
HL 1.63E-008	atm m3/mol	25	EST	MEYLAN,WM & HOWARD,PH (1991)
OH 3.65E-011	cm3/molc sec	25	EST	MEYLAN,WM & HOWARD,PH (1993)

CAS #: 000094-11-1				2,4-D, ISOPROPYL ESTER

Formula: $C_{11}H_{12}Cl_2O_3$				
Mol Weight: 263.12				
MP (deg C): 5		FP (deg C):		
BP (deg C): 140				
BP pressure (mm Hg): 1.00E+000				

Property/Value	Units	Temp	Data Type	Reference
WS 3.73E+001	mg/L		EXP	YALKOWSKY,SH & DANNENFELSER,RM (1992)
logP 3.81			EST	MEYLAN,WM & HOWARD,PH (1995)
VP 2.32E-004	mm Hg	25	EXP	BOSCH,SJ (1983)
DC	pKa			
HL 5.20E-006	atm m3/mol	25	EST	MEYLAN,WM & HOWARD,PH (1991)
OH 9.49E-012	cm3/molc sec	25	EST	MEYLAN,WM & HOWARD,PH (1993)

CAS #: 000094-12-2				PROPYL-P-AMINOBENZOATE

Formula: $C_{10}H_{13}NO_2$				
Mol Weight: 179.22				
MP (deg C): 75-76		FP (deg C):		
BP (deg C):				
BP pressure (mm Hg):				

Property/Value	Units	Temp	Data Type	Reference
WS 8.37E+002	mg/L	25	EXP	SUZUKI,T (1991)
logP 2.43			EXP	HANSCH,C ET AL. (1995)
VP 8.15E-004	mm Hg	25	EST	NEELY,WB & BLAU,GE (1985)
DC 2.49	pKa	25	EXP	PERRIN,DD (1965)
HL 2.16E-008	atm m3/mol	25	EST	MEYLAN,WM & HOWARD,PH (1991)
OH 3.80E-011	cm3/molc sec	25	EST	MEYLAN,WM & HOWARD,PH (1993)

CAS #: 000094-13-3				P-HYDROXY PROPYL BENZOATE

Formula: $C_{10}H_{12}O_3$				
Mol Weight: 180.21				
MP (deg C): 96-97		FP (deg C):		
BP (deg C):				
BP pressure (mm Hg):				

Property/Value	Units	Temp	Data Type	Reference
WS 4.63E+002	mg/L	25	EXP	SUZUKI,T (1991)
logP 3.04			EXP	HANSCH,C & LEO,AJ (1985)
VP 5.55E-004	mm Hg	25	EST	NEELY,WB & BLAU,GE (1985)
DC	pKa			
HL 6.37E-009	atm m3/mol	25	EST	MEYLAN,WM & HOWARD,PH (1991)
OH 1.41E-011	cm3/molc sec	25	EST	MEYLAN,WM & HOWARD,PH (1993)

CAS #: 000094-18-8				BENZYL-4-AMINOBENZOATE

Formula: $C_{14}H_{13}NO_2$				
Mol Weight: 227.27				
MP (deg C): 110-112		FP (deg C):		
BP (deg C):				
BP pressure (mm Hg):				

Property/Value	Units	Temp	Data Type	Reference
WS 1.08E+002	mg/L	25	EST	MEYLAN,WM ET AL. (1996)
logP 3.56			EXP	LEHNER,SJ ET AL. (1993)
VP 3.37E-006	mm Hg	25	EST	NEELY,WB & BLAU,GE (1985)
DC	pKa			
HL 2.92E-010	atm m3/mol	25	EST	MEYLAN,WM & HOWARD,PH (1991)
OH 1.73E-011	cm3/molc sec	25	EST	MEYLAN,WM & HOWARD,PH (1993)

CAS #: 000094-19-9 — ETHAZOLE

Formula: $C_{10}H_{12}N_4O_2S_2$

Mol Weight: 284.36

MP (deg C):
FP (deg C):
BP (deg C): 185.5-186
BP pressure (mm Hg):

Property/Value	Units	Temp	Data Type	Reference
WS 3.25E+003	mg/L	37	EXP	YALKOWSKY,SH & DANNENFELSER,RM (1992)
logP 1.01			EXP	HANSCH,C & LEO,AJ (1985)
VP 9.08E-010	mm Hg	25	EST	NEELY,WB & BLAU,GE (1985)
DC	pKa			
HL 3.49E-014	atm m3/mol	25	EST	MEYLAN,WM & HOWARD,PH (1991)
OH 2.48E-011	cm3/molc sec	25	EST	MEYLAN,WM & HOWARD,PH (1993)

CAS #: 000094-20-2 — N2-BUTYL-N1-P-CHLOROBENZENESULFONYLUREA

Formula: $C_{11}H_{15}ClN_2O_3S$

Mol Weight: 290.77

MP (deg C): 127-129
FP (deg C):
BP (deg C):
BP pressure (mm Hg):

Property/Value	Units	Temp	Data Type	Reference
WS 2.58E+002	mg/L	37	EXP	YALKOWSKY,SH & DANNENFELSER,RM (1992)
logP 2.27			EXP	HANSCH,C ET AL. (1995)
VP 3.74E-008	mm Hg	25	EST	NEELY,WB & BLAU,GE (1985)
DC	pKa			
HL 1.28E-010	atm m3/mol	25	EST	MEYLAN,WM & HOWARD,PH (1991)
OH 8.32E-012	cm3/molc sec	25	EST	MEYLAN,WM & HOWARD,PH (1993)

CAS #: 000094-24-6 — TETRACAINE

Formula: $C_{15}H_{24}N_2O_2$

Mol Weight: 264.37

MP (deg C):
FP (deg C):
BP (deg C):
BP pressure (mm Hg):

Property/Value	Units	Temp	Data Type	Reference
WS 1.31E+002	mg/L	25	EST	MEYLAN,WM ET AL. (1996)
logP 3.73			EXP	HANSCH,C & LEO,AJ (1985)
VP 1.87E-005	mm Hg	25	EST	NEELY,WB & BLAU,GE (1985)
DC 8.20	pKa	25	EXP	PERRIN,DD (1965)
HL 4.15E-011	atm m3/mol	25	EST	MEYLAN,WM & HOWARD,PH (1991)
OH 1.08E-010	cm3/molc sec	25	EST	MEYLAN,WM & HOWARD,PH (1993)

CAS #: 000094-25-7 — BUTYL-P-AMINOBENZOATE

Formula: $C_{11}H_{15}NO_2$

Mol Weight: 193.25

MP (deg C): 57-59
FP (deg C):
BP (deg C): 174
BP pressure (mm Hg): 8.00E+000

Property/Value	Units	Temp	Data Type	Reference
WS 3.35E+002	mg/L	25	EXP	SUZUKI,T (1991)
logP 2.87			EXP	HANSCH,C ET AL. (1995)
VP 2.99E-004	mm Hg	25	EST	NEELY,WB & BLAU,GE (1985)
DC 2.47	pKa	25	EXP	PERRIN,DD (1965)
HL 2.87E-008	atm m3/mol	25	EST	MEYLAN,WM & HOWARD,PH (1991)
OH 3.94E-011	cm3/molc sec	25	EST	MEYLAN,WM & HOWARD,PH (1993)

CAS #: 000094-26-8 — P-HYDROXY BUTYL BENZOATE

Formula: $C_{11}H_{14}O_3$

Mol Weight: 194.23

MP (deg C): 68-69
FP (deg C):
BP (deg C):
BP pressure (mm Hg):

Property/Value	Units	Temp	Data Type	Reference
WS 2.50E+002	mg/L	25	EXP	SUZUKI,T (1991)
logP 3.57			EXP	HANSCH,C & LEO,AJ (1985)
VP 1.86E-004	mm Hg	25	EST	NEELY,WB & BLAU,GE (1985)
DC	pKa			
HL 8.45E-009	atm m3/mol	25	EST	MEYLAN,WM & HOWARD,PH (1991)
OH 1.55E-011	cm3/molc sec	25	EST	MEYLAN,WM & HOWARD,PH (1993)

CAS #: 000094-33-7 — 1,2-ETHANEDIOL, MONOBENZOATE

Formula: $C_9H_{10}O_3$

Mol Weight: 166.18

MP (deg C): 45
FP (deg C):
BP (deg C): 150
BP pressure (mm Hg): 1.00E+001

Property/Value	Units	Temp	Data Type	Reference
WS 2.29E+004	mg/L	25	EST	MEYLAN,WM ET AL. (1996)
logP 1.12			EXP	LEAHY,DE ET AL. (1989)
VP 2.71E-004	mm Hg	25	EST	NEELY,WB & BLAU,GE (1985)
DC	pKa			
HL 1.68E-009	atm m3/mol	25	EST	MEYLAN,WM & HOWARD,PH (1991)
OH 6.63E-012	cm3/molc sec	25	EST	MEYLAN,WM & HOWARD,PH (1993)

CAS #: 000094-36-0 — BENZOYL PEROXIDE

Formula: $C_{14}H_{10}O_4$

Mol Weight: 242.23

MP (deg C): 104-106
FP (deg C):
BP (deg C):
BP pressure (mm Hg):

Property/Value	Units	Temp	Data Type	Reference
WS 9.10E+000	mg/L	25	EXP	CHEM INSPECT TEST INST (1992)
logP 3.46			EXP	SANGSTER,J (1993)
VP 4.23E-004	mm Hg	25	EST	NEELY,WB & BLAU,GE (1985)
DC	pKa			
HL 1.31E-006	atm m3/mol	25	EST	MEYLAN,WM & HOWARD,PH (1991)
OH 2.99E-012	cm3/molc sec	25	EST	MEYLAN,WM & HOWARD,PH (1993)

CAS #: 000094-41-7 — BENZALACETOPHENONE

Formula: $C_{15}H_{12}O$

Mol Weight: 208.26

MP (deg C): 57-58
FP (deg C):
BP (deg C): 345-348
BP pressure (mm Hg):

Property/Value	Units	Temp	Data Type	Reference
WS 9.29E+001	mg/L	25	EST	MEYLAN,WM ET AL. (1996)
logP 3.08			EXP	HANSCH,C & LEO,AJ (1985)
VP 1.07E-004	mm Hg	25	EST	NEELY,WB & BLAU,GE (1985)
DC	pKa			
HL 2.32E-007	atm m3/mol	25	EST	MEYLAN,WM & HOWARD,PH (1991)
OH 2.34E-011	cm3/molc sec	25	EST	MEYLAN,WM & HOWARD,PH (1993)

CAS #: 000094-44-0 — 3-PYRIDINECARBOXYLIC ACID, PHENYLMETHYL ESTER

Formula: $C_{13}H_{11}NO_2$
Mol Weight: 213.24
MP (deg C): | FP (deg C):
BP (deg C): 170
BP pressure (mm Hg): 3.00E+000

Property/Value	Units	Temp	Data Type	Reference
WS 3.33E+002	mg/L	25	EST	MEYLAN,WM ET AL. (1996)
logP 2.40			EXP	GUY,RH ET AL. (1986)
VP 1.16E-004	mm Hg	25	EST	NEELY,WB & BLAU,GE (1985)
DC	pKa			
HL 3.67E-009	atm m3/mol	25	EST	MEYLAN,WM & HOWARD,PH (1991)
OH 6.46E-012	cm3/molc sec	25	EST	MEYLAN,WM & HOWARD,PH (1993)

CAS #: 000094-46-2 — 1-BUTANOL, 3-METHYL-, BENZOATE

Formula: $C_{12}H_{16}O_2$
Mol Weight: 192.26
MP (deg C): | FP (deg C):
BP (deg C): 261
BP pressure (mm Hg):

Property/Value	Units	Temp	Data Type	Reference
WS 1.37E+001	mg/L	25	EST	MEYLAN,WM ET AL. (1996)
logP 4.15			EXP	SANGSTER,J (1993)
VP 1.42E-001	mm Hg	25	EXT	JORDAN,TE (1954)
DC	pKa			
HL 1.08E-004	atm m3/mol	25	EST	MEYLAN,WM & HOWARD,PH (1991)
OH 6.60E-012	cm3/molc sec	25	EST	MEYLAN,WM & HOWARD,PH (1993)

CAS #: 000094-47-3 — BENZOIC ACID, 2-PHENYLETHYL ESTER

Formula: $C_{15}H_{14}O_2$
Mol Weight: 226.28
MP (deg C): | FP (deg C):
BP (deg C):
BP pressure (mm Hg):

Property/Value	Units	Temp	Data Type	Reference
WS 1.20E+001	mg/L	25	EST	MEYLAN,WM ET AL. (1996)
logP 4.01			EXP	HANSCH,C ET AL. (1995)
VP 1.10E-004	mm Hg	25	EST	NEELY,WB & BLAU,GE (1985)
DC	pKa			
HL 3.72E-006	atm m3/mol	25	EST	MEYLAN,WM & HOWARD,PH (1991)
OH 8.46E-012	cm3/molc sec	25	EST	MEYLAN,WM & HOWARD,PH (1993)

CAS #: 000094-58-6 — DIHYDROSAFROLE

Formula: $C_{10}H_{12}O_2$
Mol Weight: 164.21
MP (deg C): 94.05 EST | FP (deg C):
BP (deg C): 225.52
BP pressure (mm Hg):

Property/Value	Units	Temp	Data Type	Reference
WS 5.69E+001	mg/L	25	EST	MEYLAN,WM ET AL. (1996)
logP 3.58			EST	MEYLAN,WM & HOWARD,PH (1995)
VP 5.60E-002	mm Hg	25	EST	NEELY,WB & BLAU,GE (1985)
DC	pKa			
HL 1.22E-005	atm m3/mol	25	EST	MEYLAN,WM & HOWARD,PH (1991)
OH 5.14E-011	cm3/molc sec	25	EST	MEYLAN,WM & HOWARD,PH (1993)

CAS #: 000094-59-7 — SAFROLE

Formula: $C_{10}H_{10}O_2$
Mol Weight: 162.19
MP (deg C): 11.2 | FP (deg C):
BP (deg C): 234.5
BP pressure (mm Hg):

Property/Value	Units	Temp	Data Type	Reference
WS 1.21E+002	mg/L	25	EST	MEYLAN,WM ET AL. (1996)
logP 3.45			EST	MEYLAN,WM & HOWARD,PH (1995)
VP 6.18E-002	mm Hg	25	EST	NEELY,WB & BLAU,GE (1985)
DC	pKa			
HL 9.07E-006	atm m3/mol	25	EST	MEYLAN,WM & HOWARD,PH (1991)
OH 7.60E-011	cm3/molc sec	25	EST	MEYLAN,WM & HOWARD,PH (1993)

CAS #: 000094-62-2 — 1-PIPEROYL-(E,E)-PIPERIDINE

Formula: $C_{17}H_{19}NO_3$
Mol Weight: 285.35
MP (deg C): 130 | FP (deg C):
BP (deg C):
BP pressure (mm Hg):

Property/Value	Units	Temp	Data Type	Reference
WS 4.00E+001	mg/L	18	EXP	YALKOWSKY,SH & DANNENFELSER,RM (1992)
logP 3.69			EST	MEYLAN,WM & HOWARD,PH (1995)
VP 1.30E-007	mm Hg	25	EST	NEELY,WB & BLAU,GE (1985)
DC	pKa			
HL 1.93E-012	atm m3/mol	25	EST	MEYLAN,WM & HOWARD,PH (1991)
OH 9.59E-011	cm3/molc sec	25	EST	MEYLAN,WM & HOWARD,PH (1993)

CAS #: 000094-70-2 — 2-ETHOXYANILINE

Formula: $C_8H_{11}NO$
Mol Weight: 137.18
MP (deg C): <-20 | FP (deg C):
BP (deg C): 228-230
BP pressure (mm Hg):

Property/Value	Units	Temp	Data Type	Reference
WS 3.37E+003	mg/L	25	EST	MEYLAN,WM ET AL. (1996)
logP 1.65			EST	MEYLAN,WM & HOWARD,PH (1995)
VP 2.89E-002	mm Hg	25	EST	NEELY,WB & BLAU,GE (1985)
DC 4.43	pKa	28	EXP	PERRIN,DD (1965)
HL 1.50E-007	atm m3/mol	25	EST	MEYLAN,WM & HOWARD,PH (1991)
OH 9.93E-011	cm3/molc sec	25	EST	MEYLAN,WM & HOWARD,PH (1993)

CAS #: 000094-71-3 — O-ETHOXYPHENOL

Formula: $C_8H_{10}O_2$
Mol Weight: 138.17
MP (deg C): 29 | FP (deg C):
BP (deg C): 217
BP pressure (mm Hg):

Property/Value	Units	Temp	Data Type	Reference
WS 3.13E+003	mg/L	25	EST	MEYLAN,WM ET AL. (1996)
logP 1.68			EXP	HANSCH,C & LEO,AJ (1985)
VP 2.61E-002	mm Hg	25	EST	NEELY,WB & BLAU,GE (1985)
DC 10.11	pKa	25	EXP	SERJEANT,EP & DEMPSEY,B (1979)
HL 4.40E-008	atm m3/mol	25	EST	MEYLAN,WM & HOWARD,PH (1991)
OH 3.51E-011	cm3/molc sec	25	EST	MEYLAN,WM & HOWARD,PH (1993)

000094-74-6 — 2-METHYL-4-CHLOROPHENOXYACETIC ACID

Formula:	$C_9H_9ClO_3$
Mol Weight:	200.62
MP (deg C):	120
FP (deg C):	
BP (deg C):	286.74
BP pressure (mm Hg):	

Property/Value	Units	Temp	Data Type	Reference
WS 1.17E+003	mg/L	25	EXP	YALKOWSKY,SH & DANNENFELSER,RM (1992)
logP 3.25			EXP	ILCHMANN,A ET AL. (1993)
VP 5.90E-006	mm Hg	25	EXP	WOODROW,JE ET AL. (1990)
DC 3.13	pKa	25	EXP	CESSNA,AJ & GROVER,R (1978)
HL 1.33E-009	atm m3/mol	25	EST	VP/WSOL
OH 1.26E-011	cm3/molc sec	25	EST	MEYLAN,WM & HOWARD,PH (1993)

000094-75-7 — 2,4-DICHLOROPHENOXYACETIC ACID

Formula:	$C_8H_6Cl_2O_3$
Mol Weight:	221.04
MP (deg C):	140.5
FP (deg C):	
BP (deg C):	160
BP pressure (mm Hg):	4.00E-001

Property/Value	Units	Temp	Data Type	Reference
WS 6.77E+002	mg/L	25	EXP	YALKOWSKY,SH & DANNENFELSER,RM (1992)
logP 2.81			EXP	HANSCH,C & LEO,AJ (1985)
VP 6.00E-007	mm Hg	25	NR	NEELY,WB (1982)
DC 2.87	pKa	25	EXP	CESSNA,AJ & GROVER,R (1978)
HL 1.02E-008	atm m3/mol	25	EST	MEYLAN,WM & HOWARD,PH (1991)
OH 6.20E-012	cm3/molc sec	25	EST	MEYLAN,WM & HOWARD,PH (1993)

000094-78-0 — PHENAZOPYRIDINE

Formula:	$C_{11}H_{11}N_5$
Mol Weight:	213.24
MP (deg C):	139
FP (deg C):	
BP (deg C):	
BP pressure (mm Hg):	

Property/Value	Units	Temp	Data Type	Reference
WS 3.88E+002	mg/L	25	EST	MEYLAN,WM ET AL. (1996)
logP 2.77			EST	MEYLAN,WM & HOWARD,PH (1995)
VP 4.57E-012	mm Hg	25	EST	WSOL X HL
DC	pKa			
HL 3.30E-015	atm m3/mol	25	EST	MEYLAN,WM & HOWARD,PH (1991)
OH 2.40E-010	cm3/molc sec	25	EST	MEYLAN,WM & HOWARD,PH (1993)

000094-80-4 — 2,4-D, BUTYL ESTER

Formula:	$C_{12}H_{14}Cl_2O_3$
Mol Weight:	277.15
MP (deg C):	
FP (deg C):	
BP (deg C):	
BP pressure (mm Hg):	

Property/Value	Units	Temp	Data Type	Reference
WS 4.60E+001	mg/L	25	EXP	BOSCH,SJ (1983)
logP 4.38			EST	MEYLAN,WM & HOWARD,PH (1995)
VP 6.16E-005	mm Hg	25	EXP	BOSCH,SJ (1983)
DC	pKa			
HL 4.88E-007	atm m3/mol	25	EST	VP/WSOL
OH 1.06E-011	cm3/molc sec	25	EST	MEYLAN,WM & HOWARD,PH (1993)

000094-81-5 — MCPB

Formula:	$C_{11}H_{13}ClO_3$
Mol Weight:	228.68
MP (deg C):	100
FP (deg C):	
BP (deg C):	
BP pressure (mm Hg):	

Property/Value	Units	Temp	Data Type	Reference
WS 4.80E+001	mg/L	25	EXP	YALKOWSKY,SH & DANNENFELSER,RM (1992)
logP 3.50			EST	MEYLAN,WM & HOWARD,PH (1995)
VP 5.00E-007	mm Hg	25	EST	HL X WSOL
DC 6.20	pKa	25	EXP	SMITH,AE (1989A)
HL 3.42E-009	atm m3/mol	25	EST	MEYLAN,WM & HOWARD,PH (1991)
OH 2.10E-011	cm3/molc sec	25	EST	MEYLAN,WM & HOWARD,PH (1993)

000094-82-6 — 2,4-DB

Formula:	$C_{10}H_{10}Cl_2O_3$
Mol Weight:	249.10
MP (deg C):	117-119
FP (deg C):	
BP (deg C):	324.35
BP pressure (mm Hg):	

Property/Value	Units	Temp	Data Type	Reference
WS 5.30E+001	mg/L	25	EXP	BAILEY,GW & WHITE,JL (1965)
logP 3.53			EXP	HANSCH,C ET AL. (1995)
VP 1.11E-005	mm Hg	25	EST	NEELY,WB & BLAU,GE (1985)
DC 4.95	pKa	25	EXP	JAFVERT,CT ET AL. (1990)
HL 2.29E-009	atm m3/mol	25	EST	MEYLAN,WM & HOWARD,PH (1991)
OH 1.42E-011	cm3/molc sec	25	EST	ATKINSON,R (1987)

000094-83-7 — 2,4-DEB

Formula:	$C_{15}H_{12}Cl_2O_3$
Mol Weight:	311.17
MP (deg C):	66
FP (deg C):	
BP (deg C):	185
BP pressure (mm Hg):	1.50E+000

Property/Value	Units	Temp	Data Type	Reference
WS 4.80E+001	mg/L		EXP	SHIU,WY ET AL. (1990)
logP 4.85			EST	MEYLAN,WM & HOWARD,PH (1995)
VP 1.86E-006	mm Hg	25	EST	NEELY,WB & BLAU,GE (1985)
DC	pKa			
HL 1.09E-007	atm m3/mol	25	EST	MEYLAN,WM & HOWARD,PH (1991)
OH 1.13E-011	cm3/molc sec	25	EST	MEYLAN,WM & HOWARD,PH (1993)

000094-84-8 — 2,4-DEP

Formula:	$C_{24}H_{21}Cl_6O_6P$
Mol Weight:	649.12
MP (deg C):	
FP (deg C):	
BP (deg C):	
BP pressure (mm Hg):	

Property/Value	Units	Temp	Data Type	Reference
WS 1.03E-007	mg/L	25	EST	MEYLAN,WM ET AL. (1996)
logP 10.39			EST	MEYLAN,WM & HOWARD,PH (1995)
VP 2.43E-014	mm Hg	25	EST	NEELY,WB & BLAU,GE (1985)
DC	pKa			
HL 3.36E-013	atm m3/mol	25	EST	MEYLAN,WM & HOWARD,PH (1991)
OH 9.72E-011	cm3/molc sec	25	EST	MEYLAN,WM & HOWARD,PH (1993)

000094-96-2 — 2-ETHYL-1,3-HEXANDIOL

CAS #: 000094-96-2

Formula: $C_8H_{18}O_2$

Mol Weight: 146.23
MP (deg C): -40
FP (deg C):
BP (deg C): 244.2
BP pressure (mm Hg):

Property/Value	Units	Temp	Data Type	Reference
WS 4.20E+004	mg/L	25	EXP	YALKOWSKY,SH & DANNENFELSER,RM (1992)
logP 1.60			EST	MEYLAN,WM & HOWARD,PH (1995)
VP 1.00E+001	mm Hg	25	EXP	RIDDICK,JA ET AL. (1986)
DC	pKa			
HL 4.58E-005	atm m3/mol	25	EST	VP/WSOL
OH 2.22E-011	cm3/molc sec	25	EST	MEYLAN,WM & HOWARD,PH (1993)

000094-99-5 — 2,4-DICHLOROBENZYLCHLORIDE

CAS #: 000094-99-5

Formula: $C_7H_5Cl_3$

Mol Weight: 195.48
MP (deg C): -2.6
FP (deg C):
BP (deg C): 248
BP pressure (mm Hg):

Property/Value	Units	Temp	Data Type	Reference
WS 2.52E+001	mg/L	25	EST	MEYLAN,WM ET AL. (1996)
logP 3.82			EXP	HANSCH,C & LEO,AJ (1985)
VP 3.27E-002	mm Hg	25	EXP	DYKYJ,J & VANKO,A (1970)
DC	pKa			
HL 1.15E-003	atm m3/mol	25	EST	MEYLAN,WM & HOWARD,PH (1991)
OH 9.81E-013	cm3/molc sec	25	EST	MEYLAN,WM & HOWARD,PH (1993)

000095-06-7 — SULFALLATE

CAS #: 000095-06-7

Formula: $C_8H_{14}ClNS_2$

Mol Weight: 223.79
MP (deg C):
FP (deg C):
BP (deg C): 129
BP pressure (mm Hg): 1.00E+000

Property/Value	Units	Temp	Data Type	Reference
WS 1.00E+002	mg/L	25	EXP	YALKOWSKY,SH & DANNENFELSER,RM (1992)
logP 3.15			EST	MEYLAN,WM & HOWARD,PH (1995)
VP 2.20E-003	mm Hg	20	EXP	IARC (1983)
DC	pKa			
HL 6.48E-006	atm m3/mol	25	EST	VP/WSOL
OH 1.03E-010	cm3/molc sec	25	EST	MEYLAN,WM & HOWARD,PH (1993)

000095-08-9 — TRIETHYLENE GLYCOL BIS(2-ETHYLBUTYRATE)

CAS #: 000095-08-9

Formula: $C_{18}H_{34}O_6$

Mol Weight: 346.47
MP (deg C): -10
FP (deg C):
BP (deg C): 196
BP pressure (mm Hg): 5.00E+000

Property/Value	Units	Temp	Data Type	Reference
WS 2.00E+002	mg/L	20	EXP	LEWIS,RJSR (1993)
logP 3.64			EST	MEYLAN,WM & HOWARD,PH (1995)
VP 9.05E-005	mm Hg	25	EST	NEELY,WB & BLAU,GE (1985)
DC	pKa			
HL 1.00E-011	atm m3/mol	25	EST	MEYLAN,WM & HOWARD,PH (1991)
OH 3.90E-011	cm3/molc sec	25	EST	MEYLAN,WM & HOWARD,PH (1993)

000095-13-6 — INDENE

CAS #: 000095-13-6

Formula: C_9H_8

Mol Weight: 116.16
MP (deg C): -1.8
FP (deg C):
BP (deg C): 181.6
BP pressure (mm Hg):

Property/Value	Units	Temp	Data Type	Reference
WS 3.32E+002	mg/L	25	EST	MEYLAN,WM ET AL. (1996)
logP 2.92			EXP	HANSCH,C & LEO,AJ (1985)
VP 1.10E+000	mm Hg	25	EXP	DAUBERT,TE & DANNER,RP (1989)
DC	pKa			
HL 1.59E-003	atm m3/mol	25	EST	MEYLAN,WM & HOWARD,PH (1991)
OH 6.16E-011	cm3/molc sec	25	EST	MEYLAN,WM & HOWARD,PH (1993)

000095-14-7 — 1H-BENZOTRIAZOLE

CAS #: 000095-14-7

Formula: $C_6H_5N_3$

Mol Weight: 119.13
MP (deg C): 98.5
FP (deg C):
BP (deg C): 204
BP pressure (mm Hg): 1.30E+001

Property/Value	Units	Temp	Data Type	Reference
WS 1.98E+004	mg/L	25	EXP	DAVIS,LN ET AL. (1977)
logP 1.44			EXP	HANSCH,C ET AL. (1995)
VP 4.00E-002	mm Hg	20	EXP	DAVIS,LN ET AL. (1977)
DC 1.60	pKa	20	EXP	PERRIN,DD (1965)
HL 3.17E-007	atm m3/mol	25	EST	VP/WSOL
OH 1.00E-012	cm3/molc sec	25	EST	MEYLAN,WM & HOWARD,PH (1993)

000095-15-8 — BENZO(B)THIOPHENE

CAS #: 000095-15-8

Formula: C_8H_6S

Mol Weight: 134.20
MP (deg C): 32
FP (deg C):
BP (deg C): 221
BP pressure (mm Hg):

Property/Value	Units	Temp	Data Type	Reference
WS 1.30E+002	mg/L	20	EXP	PEARLMAN,RS ET AL. (1984)
logP 3.12			EXP	HANSCH,C & LEO,AJ (1985)
VP 2.38E-001	mm Hg	25	EXT	DAUBERT,TE & DANNER,RP (1991)
DC	pKa			
HL 2.86E-004	atm m3/mol	25	EST	MEYLAN,WM & HOWARD,PH (1991)
OH 3.00E-011	cm3/molc sec	25	EST	MEYLAN,WM & HOWARD,PH (1993)

000095-16-9 — BENZOTHIAZOLE

CAS #: 000095-16-9

Formula: C_7H_5NS

Mol Weight: 135.19
MP (deg C): 2
FP (deg C):
BP (deg C): 231
BP pressure (mm Hg):

Property/Value	Units	Temp	Data Type	Reference
WS 4.30E+003	mg/L	25	EXP	CHEM INSPECT TEST INST (1992)
logP 2.01			EXP	HANSCH,C ET AL. (1995)
VP 1.43E-002	mm Hg	25	EST	NEELY,WB & BLAU,GE (1985)
DC	pKa			
HL 3.74E-007	atm m3/mol	25	EST	MEYLAN,WM & HOWARD,PH (1991)
OH 7.00E-012	cm3/molc sec	25	EST	MEYLAN,WM & HOWARD,PH (1993)

000095-19-2 — 2-HEPTADECYL-4,5-DIHYDRO-2-IMIDAZOLINE-1-ETHANOL

CAS #:	000095-19-2		
Formula:	$C_{22}H_{46}N_2O$		
Mol Weight:	354.62		
MP (deg C):		FP (deg C):	
BP (deg C):			
BP pressure (mm Hg):			

Property/Value	Units	Temp	Data Type	Reference
WS 1.00E+002	mg/L	75	EXP	SHIU,WY ET AL. (1990)
logP 8.33			EST	MEYLAN,WM & HOWARD,PH (1995)
VP 3.02E-010	mm Hg	25	EST	NEELY,WB & BLAU,GE (1985)
DC	pKa			
HL 6.36E-009	atm m3/mol	25	EST	MEYLAN,WM & HOWARD,PH (1991)
OH 1.13E-010	cm3/molc sec	25	EST	MEYLAN,WM & HOWARD,PH (1993)

000095-20-5 — 2-METHYLINDOLE

CAS #:	000095-20-5		
Formula:	C_9H_9N		
Mol Weight:	131.18		
MP (deg C):	61	FP (deg C):	
BP (deg C):	272		
BP pressure (mm Hg):			

Property/Value	Units	Temp	Data Type	Reference
WS 6.28E+002	mg/L	25	EST	MEYLAN,WM ET AL. (1996)
logP 2.53			EXP	HANSCH,C & LEO,AJ (1985)
VP 6.03E-003	mm Hg	25	EST	NEELY,WB & BLAU,GE (1985)
DC	pKa			
HL 9.78E-007	atm m3/mol	25	EST	MEYLAN,WM & HOWARD,PH (1991)
OH 2.00E-010	cm3/molc sec	25	EST	MEYLAN,WM & HOWARD,PH (1993)

000095-45-4 — DIMETHYL GLYOXIME

CAS #:	000095-45-4		
Formula:	$C_4H_8N_2O_2$		
Mol Weight:	116.12		
MP (deg C):	238-240	FP (deg C):	
BP (deg C):			
BP pressure (mm Hg):			

Property/Value	Units	Temp	Data Type	Reference
WS 6.00E+002	mg/L	20	EXP	YALKOWSKY,SH & DANNENFELSER,RM (1992)
logP 1.08			EST	MEYLAN,WM & HOWARD,PH (1995)
VP 6.90E-004	mm Hg	25	EST	NEELY,WB & BLAU,GE (1985)
DC	pKa			
HL 6.22E-010	atm m3/mol	25	EST	MEYLAN,WM & HOWARD,PH (1991)
OH 3.35E-013	cm3/molc sec	25	EST	MEYLAN,WM & HOWARD,PH (1993)

000095-46-5 — O-BROMOTOLUENE

CAS #:	000095-46-5		
Formula:	C_7H_7Br		
Mol Weight:	171.04		
MP (deg C):	-27.8	FP (deg C):	
BP (deg C):	181.7		
BP pressure (mm Hg):			

Property/Value	Units	Temp	Data Type	Reference
WS 1.07E+002	mg/L	25	EST	MEYLAN,WM ET AL. (1996)
logP 3.50			EXP	ABRAHAM,MH ET AL. (1994)
VP 1.04E+000	mm Hg	25	EXP	PERRY,RH & GREEN,D (1984)
DC	pKa			
HL 2.39E-003	atm m3/mol	25	EST	MEYLAN,WM & HOWARD,PH (1991)
OH 1.64E-012	cm3/molc sec	25	EST	MEYLAN,WM & HOWARD,PH (1993)

000095-47-6 — O-XYLENE

CAS #:	000095-47-6		
Formula:	C_8H_{10}		
Mol Weight:	106.17		
MP (deg C):	-25.182	FP (deg C):	
BP (deg C):	144.429		
BP pressure (mm Hg):			

Property/Value	Units	Temp	Data Type	Reference
WS 1.78E+002	mg/L	25	EXP	SANEMASA,I ET AL. (1982)
logP 3.12			EXP	HANSCH,C & LEO,AJ (1985)
VP 6.61E+000	mm Hg	25	EXP	DAUBERT,TE & DANNER,RP (1985)
DC	pKa			
HL 5.18E-003	atm m3/mol	25	EXP	SANEMASA,I ET AL. (1982)
OH 1.37E-011	cm3/molc sec	25	EXP	ATKINSON,R (1989)

000095-48-7 — O-CRESOL

CAS #:	000095-48-7		
Formula:	C_7H_8O		
Mol Weight:	108.14		
MP (deg C):	30.944	FP (deg C):	
BP (deg C):	191.004		
BP pressure (mm Hg):			

Property/Value	Units	Temp	Data Type	Reference
WS 2.60E+004	mg/L	25	EXP	YALKOWSKY,SH & DANNENFELSER,RM (1992)
logP 1.95			EXP	HANSCH,C & LEO,AJ (1985)
VP 2.99E-001	mm Hg	25	EXT	DAUBERT,TE & DANNER,RP (1985)
DC 10.28	pKa	25	EXP	PEARCE,PJ & SIMKINS,RJJ (1968)
HL 1.20E-006	atm m3/mol	25	EXP	GAFFNEY,JS ET AL. (1987)
OH 4.20E-011	cm3/molc sec	25	EXP	ATKINSON,R (1989)

000095-49-8 — 2-CHLOROTOLUENE

CAS #:	000095-49-8		
Formula:	C_7H_7Cl		
Mol Weight:	126.59		
MP (deg C):	-35.59	FP (deg C):	
BP (deg C):	158.97		
BP pressure (mm Hg):			

Property/Value	Units	Temp	Data Type	Reference
WS 3.74E+002	mg/L	25	EXP	VALVANI,SC ET AL. (1981)
logP 3.42			EXP	HANSCH,C & LEO,AJ (1985)
VP 3.43E+000	mm Hg	25	EXP	DAUBERT,TE & DANNER,RP (1985)
DC	pKa			
HL 3.57E-003	atm m3/mol	25	EXP	LEIGHTON,DTJR & CALO,JM (1981)
OH 1.90E-012	cm3/molc sec	25	EST	MEYLAN,WM & HOWARD,PH (1993)

000095-50-1 — 1,2-DICHLOROBENZENE

CAS #:	000095-50-1		
Formula:	$C_6H_4Cl_2$		
Mol Weight:	147.00		
MP (deg C):	-17.01	FP (deg C):	
BP (deg C):	180.48		
BP pressure (mm Hg):			

Property/Value	Units	Temp	Data Type	Reference
WS 1.56E+002	mg/L	25	EXP	BANERJEE,S ET AL. (1980)
logP 3.43			EXP	HANSCH,C ET AL. (1995)
VP 1.36E+000	mm Hg	25	EXP	DAUBERT,TE & DANNER,RP (1989)
DC	pKa			
HL 1.90E-003	atm m3/mol	25	EXP	MACKAY,D ET AL. (1982A)
OH 4.20E-013	cm3/molc sec	25	EXP	ATKINSON,R (1989)

000095-51-2 — 2-CHLOROANILINE

CAS #: 000095-51-2

Formula: C_6H_6ClN

Mol Weight: 127.57

MP (deg C): -1.94

FP (deg C):

BP (deg C): 208.84

BP pressure (mm Hg):

Property/Value	Units	Temp	Data Type	Reference
WS 8.17E+003	mg/L	25	EXP	HUYSKENS,P ET AL. (1975)
logP 1.90			EXP	HANSCH,C & LEO,AJ (1985)
VP 2.04E-001	mm Hg	25	EXP	DAUBERT,TE & DANNER,RP (1989)
DC 2.66	pKa	25	EXP	PERRIN,DD (1972)
HL 5.39E-006	atm m3/mol	25	EST	VP/WSOL
OH 5.36E-011	cm3/molc sec	25	EST	MEYLAN,WM & HOWARD,PH (1993)

000095-52-3 — O-FLUOROTOLUENE

CAS #: 000095-52-3

Formula: C_7H_7F

Mol Weight: 110.13

MP (deg C): -62

FP (deg C):

BP (deg C): 115

BP pressure (mm Hg):

Property/Value	Units	Temp	Data Type	Reference
WS 4.96E+002	mg/L	25	EST	MEYLAN,WM ET AL. (1996)
logP 2.74			EST	MEYLAN,WM & HOWARD,PH (1995)
VP 2.44E+001	mm Hg	25	EXP	PERRY,RH & GREEN,D (1984)
DC	pKa			
HL 6.94E-003	atm m3/mol	25	EST	MEYLAN,WM & HOWARD,PH (1991)
OH 3.13E-012	cm3/molc sec	25	EST	MEYLAN,WM & HOWARD,PH (1993)

000095-53-4 — O-TOLUIDINE

CAS #: 000095-53-4

Formula: C_7H_9N

Mol Weight: 107.16

MP (deg C): -16.10

FP (deg C):

BP (deg C): 200.40

BP pressure (mm Hg):

Property/Value	Units	Temp	Data Type	Reference
WS 1.66E+004	mg/L	25	EXP	HUYSKENS,P ET AL. (1975)
logP 1.32			EXP	HANSCH,C & LEO,AJ (1985)
VP 2.60E-001	mm Hg	25	EXP	DAUBERT,TE & DANNER,RP (1989)
DC 4.44	pKa	25	EXP	HUYSKENS,P ET AL. (1975)
HL 2.72E+000	atm m3/mol	25	EST	VP/WSOL
OH 1.63E-010	cm3/molc sec	25	EST	MEYLAN,WM & HOWARD,PH (1993)

000095-54-5 — 1,2-BENZENEDIAMINE

CAS #: 000095-54-5

Formula: $C_6H_8N_2$

Mol Weight: 108.14

MP (deg C): 103.5

FP (deg C):

BP (deg C): 257

BP pressure (mm Hg):

Property/Value	Units	Temp	Data Type	Reference
WS 8.22E+004	mg/L	25	EST	MEYLAN,WM ET AL. (1996)
logP 0.15			EXP	HANSCH,C & LEO,AJ (1985)
VP 1.23E-002	mm Hg	25	EST	NEELY,WB & BLAU,GE (1985)
DC 4.47	pKa	25	EST	DEAN,JA (1985)
HL 6.73E-010	atm m3/mol	25	EST	MEYLAN,WM & HOWARD,PH (1991)
OH 2.31E-010	cm3/molc sec	25	EST	MEYLAN,WM & HOWARD,PH (1993)

000095-55-6 — O-AMINOPHENOL

CAS #: 000095-55-6

Formula: C_6H_7NO

Mol Weight: 109.13

MP (deg C): 170-174

FP (deg C):

BP (deg C):

BP pressure (mm Hg):

Property/Value	Units	Temp	Data Type	Reference
WS 2.00E+004	mg/L	20	EXP	YALKOWSKY,SH & DANNENFELSER,RM (1992)
logP 0.62			EXP	HANSCH,C & LEO,AJ (1985)
VP 9.55E-003	mm Hg	25	EST	NEELY,WB & BLAU,GE (1985)
DC 4.84	pKa	20	EXP	PERRIN,DD (1965)
HL 1.98E-010	atm m3/mol	25	EST	MEYLAN,WM & HOWARD,PH (1991)
OH 7.42E-011	cm3/molc sec	25	EST	MEYLAN,WM & HOWARD,PH (1993)

000095-56-7 — O-BROMOPHENOL

CAS #: 000095-56-7

Formula: C_6H_5BrO

Mol Weight: 173.02

MP (deg C): 6

FP (deg C):

BP (deg C): 194

BP pressure (mm Hg):

Property/Value	Units	Temp	Data Type	Reference
WS 2.23E+003	mg/L	25	EST	MEYLAN,WM ET AL. (1996)
logP 2.35			EXP	HANSCH,C & LEO,AJ (1985)
VP 3.73E-002	mm Hg	25	EST	NEELY,WB & BLAU,GE (1985)
DC 8.45	pKa	25	EXP	KORTUM,G ET AL (1961)
HL 2.23E-007	atm m3/mol	25	EST	MEYLAN,WM & HOWARD,PH (1991)
OH 9.69E-012	cm3/molc sec	25	EST	MEYLAN,WM & HOWARD,PH (1993)

000095-57-8 — 2-CHLOROPHENOL

CAS #: 000095-57-8

Formula: C_6H_5ClO

Mol Weight: 128.56

MP (deg C): 9.8

FP (deg C):

BP (deg C): 174.9

BP pressure (mm Hg):

Property/Value	Units	Temp	Data Type	Reference
WS 1.14E+004	mg/L	25	EXP	BANERJEE,S ET AL. (1980)
logP 2.15			EXP	HANSCH,C & LEO,AJ (1985)
VP 1.42E+000	mm Hg	25	EXP	SHIU,WY ET AL. (1994)
DC 8.56	pKa		EXP	SERJEANT,EP & DEMPSEY,B (1979)
HL 3.91E-004	atm m3/mol		EST	CHIOU,CT ET AL. (1980)
OH 1.03E-011	cm3/molc sec	25	EST	MEYLAN,WM & HOWARD,PH (1993)

000095-62-5 — CYCLOPROPANAMINE, 2-PHENYL-, TRANS

CAS #: 000095-62-5

Formula: $C_9H_{11}N$

Mol Weight: 133.19

MP (deg C):

FP (deg C):

BP (deg C):

BP pressure (mm Hg):

Property/Value	Units	Temp	Data Type	Reference
WS 4.86E+004	mg/L	25	EST	MEYLAN,WM ET AL. (1996)
logP 1.58			EXP	SANGSTER,J (1993)
VP 1.09E-001	mm Hg	25	EST	NEELY,WB & BLAU,GE (1985)
DC	pKa			
HL 4.75E-007	atm m3/mol	25	EST	MEYLAN,WM & HOWARD,PH (1991)
OH 2.63E-011	cm3/molc sec	25	EST	MEYLAN,WM & HOWARD,PH (1993)

1,2,4-TRIMETHYLBENZENE

CAS #: 000095-63-6

Formula: C_9H_{12}

Mol Weight: 120.20

MP (deg C): -43.80

FP (deg C):

BP (deg C): 168.89

BP pressure (mm Hg):

Property/Value	Units	Temp	Data Type	Reference
WS 5.70E+001	mg/L	25	EXP	MCAULIFFE,C (1966)
logP 3.78			EXP	HANSCH,C ET AL. (1995)
VP 2.10E+000	mm Hg	25	EXP	CHAO,J ET AL. (1983)
DC	pKa			
HL 6.16E-003	atm m3/mol	25	EXP	SANEMASA,I ET AL. (1982)
OH 3.25E-011	cm3/molc sec	25	EXP	ATKINSON,R (1989)

3,4-DIMETHYLPHENOL

CAS #: 000095-65-8

Formula: $C_8H_{10}O$

Mol Weight: 122.17

MP (deg C): 60.8

FP (deg C):

BP (deg C): 227

BP pressure (mm Hg):

Property/Value	Units	Temp	Data Type	Reference
WS 4.76E+003	mg/L	25	EXP	YALKOWSKY,SH & DANNENFELSER,RM (1992)
logP 2.23			EXP	HANSCH,C & LEO,AJ (1985)
VP 1.17E-002	mm Hg	25	EXT	DAUBERT,TE & DANNER,RP (1989)
DC 10.36	pKa	25	EXP	KORTUM,G ET AL (1961)
HL 6.83E-007	atm m3/mol	25	EST	MEYLAN,WM & HOWARD,PH (1991)
OH 8.14E-011	cm3/molc sec	25	EXP	ATKINSON,R (1989)

2,4-DIMETHYLANILINE

CAS #: 000095-68-1

Formula: $C_8H_{11}N$

Mol Weight: 121.18

MP (deg C): -14.3

FP (deg C):

BP (deg C): 214

BP pressure (mm Hg):

Property/Value	Units	Temp	Data Type	Reference
WS 1.39E+003	mg/L	25	EST	MEYLAN,WM ET AL. (1996)
logP 2.17			EST	MEYLAN,WM & HOWARD,PH (1995)
VP 1.33E-001	mm Hg	25	EXP	CHAO,J ET AL. (1983)
DC 4.89	pKa	25	EXP	PERRIN,DD (1965)
HL 2.32E-006	atm m3/mol	25	EST	MEYLAN,WM & HOWARD,PH (1991)
OH 1.62E-010	cm3/molc sec	25	EST	MEYLAN,WM & HOWARD,PH (1993)

2-METHYL-4-CHLOROANILINE

CAS #: 000095-69-2

Formula: C_7H_8ClN

Mol Weight: 141.60

MP (deg C): 30.3

FP (deg C):

BP (deg C): 244

BP pressure (mm Hg):

Property/Value	Units	Temp	Data Type	Reference
WS 9.54E+002	mg/L	25	EST	MEYLAN,WM ET AL. (1996)
logP 2.27			EST	MEYLAN,WM & HOWARD,PH (1995)
VP 4.08E-002	mm Hg	25	EST	NEELY,WB & BLAU,GE (1985)
DC 3.85	pKa	25	EXP	ROBINSON,RA (1967)
HL 1.99E-006	atm m3/mol	25	EST	MEYLAN,WM & HOWARD,PH (1991)
OH 4.14E-011	cm3/molc sec	25	EST	MEYLAN,WM & HOWARD,PH (1993)

2,5-DIAMINOTOLUENE

CAS #: 000095-70-5

Formula: $C_7H_{10}N_2$

Mol Weight: 122.17

MP (deg C): 64

FP (deg C):

BP (deg C): 273-274

BP pressure (mm Hg):

Property/Value	Units	Temp	Data Type	Reference
WS 7.25E+004	mg/L	25	EST	MEYLAN,WM ET AL. (1996)
logP 0.16			EST	MEYLAN,WM & HOWARD,PH (1995)
VP 3.70E-003	mm Hg	25	EST	NEELY,WB & BLAU,GE (1985)
DC	pKa			
HL 7.49E-009	atm m3/mol	25	EST	MEYLAN,WM & HOWARD,PH (1991)
OH 2.40E-010	cm3/molc sec	25	EST	MEYLAN,WM & HOWARD,PH (1993)

1,4-BENZENEDIOL, 2-METHYL-

CAS #: 000095-71-6

Formula: $C_7H_8O_2$

Mol Weight: 124.14

MP (deg C): 72

FP (deg C):

BP (deg C): 272

BP pressure (mm Hg):

Property/Value	Units	Temp	Data Type	Reference
WS 6.15E+004	mg/L	25	EST	MEYLAN,WM ET AL. (1996)
logP 0.91			EXP	SANGSTER,J (1994)
VP 2.80E-003	mm Hg	25	EST	NEELY,WB & BLAU,GE (1985)
DC 10.20	pKa	13	EXP	KORTUM,G ET AL (1961)
HL 6.44E-011	atm m3/mol	25	EST	MEYLAN,WM & HOWARD,PH (1991)
OH 6.03E-011	cm3/molc sec	25	EST	MEYLAN,WM & HOWARD,PH (1993)

2,4-DICHLOROTOLUENE

CAS #: 000095-73-8

Formula: $C_7H_6Cl_2$

Mol Weight: 161.03

MP (deg C): -13.5

FP (deg C):

BP (deg C): 201

BP pressure (mm Hg):

Property/Value	Units	Temp	Data Type	Reference
WS 1.62E+001	mg/L	25	EST	MEYLAN,WM ET AL. (1996)
logP 4.24			EXP	HANSCH,C & LEO,AJ (1985)
VP 4.58E-001	mm Hg	25	EXP	DAUBERT,TE & DANNER,RP (1989)
DC	pKa			
HL 4.15E-003	atm m3/mol	25	EST	MEYLAN,WM & HOWARD,PH (1991)
OH 1.32E-012	cm3/molc sec	25	EST	MEYLAN,WM & HOWARD,PH (1993)

3-CHLORO-P-TOLUIDINE

CAS #: 000095-74-9

Formula: C_7H_8ClN

Mol Weight: 141.60

MP (deg C): 26

FP (deg C):

BP (deg C): 242-244

BP pressure (mm Hg):

Property/Value	Units	Temp	Data Type	Reference
WS 8.32E+002	mg/L	25	EST	MEYLAN,WM ET AL. (1996)
logP 2.27			EST	MEYLAN,WM & HOWARD,PH (1995)
VP 4.08E-002	mm Hg	25	EST	NEELY,WB & BLAU,GE (1985)
DC 4.05	pKa	25	EXP	PERRIN,DD (1965)
HL 1.56E-006	atm m3/mol	25	EST	MEYLAN,WM & HOWARD,PH (1991)
OH 1.20E-010	cm3/molc sec	25	EST	ATKINSON,R (1988)

125

CAS #: 000095-75-0 — 3,4-DICHLOROTOLUENE

Formula: $C_7H_6Cl_2$

Mol Weight: 161.03

MP (deg C): FP (deg C):

BP (deg C): 200.5

BP pressure (mm Hg): 7.41E+002

Property/Value	Units	Temp	Data Type	Reference
WS 2.60E+001	mg/L	30	EXP	RIDDICK,JA ET AL. (1986)
logP 3.83			EST	MEYLAN,WM & HOWARD,PH (1995)
VP 3.15E-001	mm Hg	25	EXP	RIDDICK,JA ET AL. (1986)
DC	pKa			
HL 2.57E-003	atm m3/mol	25	EST	VP/WSOL
OH 1.18E-012	cm3/molc sec	25	EST	MEYLAN,WM & HOWARD,PH (1993)

CAS #: 000095-76-1 — 3,4-DICHLOROANILINE

Formula: $C_6H_5Cl_2N$

Mol Weight: 162.02

MP (deg C): 71-72 FP (deg C):

BP (deg C): 272

BP pressure (mm Hg):

Property/Value	Units	Temp	Data Type	Reference
WS 9.20E+001	mg/L	20	EXP	CROSSLAND,NO (1986)
logP 2.69			EXP	HANSCH,C & LEO,AJ (1985)
VP 9.75E-003	mm Hg	20	EXP	CROSSLAND,NO (1986)
DC 2.97	pKa	25	EXP	PERRIN,DD (1972)
HL 2.26E-005	atm m3/mol	20	EST	VP/WSOL
OH 4.36E-011	cm3/molc sec	25	EST	MEYLAN,WM & HOWARD,PH (1993)

CAS #: 000095-77-2 — 3,4-DICHLOROPHENOL

Formula: $C_6H_4Cl_2O$

Mol Weight: 163.00

MP (deg C): 66-68 FP (deg C):

BP (deg C): 218-220

BP pressure (mm Hg):

Property/Value	Units	Temp	Data Type	Reference
WS 9.26E+003	mg/L	25	EXP	SHIU,WY ET AL. (1994)
logP 3.33			EXP	HANSCH,C & LEO,AJ (1985)
VP 1.73E-002	mm Hg	25	EST	NEELY,WB & BLAU,GE (1985)
DC 8.63	pKa		EXP	SERJEANT,EP & DEMPSEY,B (1979)
HL 3.08E-007	atm m3/mol	25	EST	MEYLAN,WM & HOWARD,PH (1991)
OH 6.99E-012	cm3/molc sec	25	EST	MEYLAN,WM & HOWARD,PH (1993)

CAS #: 000095-78-3 — 2,5-DIMETHYLANILINE

Formula: $C_8H_{11}N$

Mol Weight: 121.18

MP (deg C): 15.5 FP (deg C):

BP (deg C): 214

BP pressure (mm Hg):

Property/Value	Units	Temp	Data Type	Reference
WS 5.60E+003	mg/L	12	EXP	CHEM INSPECT TEST INST (1992)
logP 1.83			EXP	DEBNATH,AK ET AL. (1992)
VP 1.50E-001	mm Hg	20	EXP	WEBER,RC ET AL. (1981)
DC 4.53	pKa	25	EXP	PERRIN,DD (1965)
HL 2.32E-006	atm m3/mol	25	EST	MEYLAN,WM & HOWARD,PH (1991)
OH 2.00E-010	cm3/molc sec	25	EST	MEYLAN,WM & HOWARD,PH (1993)

CAS #: 000095-79-4 — 5-CHLORO-O-TOLUIDINE

Formula: C_7H_8ClN

Mol Weight: 141.60

MP (deg C): 26 FP (deg C):

BP (deg C): 237

BP pressure (mm Hg): 7.22E+002

Property/Value	Units	Temp	Data Type	Reference
WS 8.32E+002	mg/L	25	EST	MEYLAN,WM ET AL. (1996)
logP 2.27			EST	MEYLAN,WM & HOWARD,PH (1995)
VP 4.08E-002	mm Hg	25	EST	NEELY,WB & BLAU,GE (1985)
DC	pKa			
HL 1.56E-006	atm m3/mol	25	EST	MEYLAN,WM & HOWARD,PH (1991)
OH 1.20E-010	cm3/molc sec	25	EST	ATKINSON,R (1988)

CAS #: 000095-80-7 — 2,4-TOLUENEDIAMINE

Formula: $C_7H_{10}N_2$

Mol Weight: 122.17

MP (deg C): 97-99 FP (deg C):

BP (deg C): 283-285

BP pressure (mm Hg):

Property/Value	Units	Temp	Data Type	Reference
WS 7.48E+004	mg/L	25	EST	MEYLAN,WM ET AL. (1996)
logP 0.14			EXP	DEBNATH,AK ET AL. (1992)
VP 1.70E-004	mm Hg	25	EXP	DAUBERT,TE ET AL. (1987)
DC	pKa			
HL 7.92E-010	atm m3/mol	25	EST	MEYLAN,WM & HOWARD,PH (1991)
OH 1.92E-010	cm3/molc sec	25	EXP	ATKINSON,R (1989)

CAS #: 000095-82-9 — 2,5-DICHLOROANILINE

Formula: $C_6H_5Cl_2N$

Mol Weight: 162.02

MP (deg C): 50 FP (deg C):

BP (deg C): 251

BP pressure (mm Hg):

Property/Value	Units	Temp	Data Type	Reference
WS 3.00E+002	mg/L	25	EST	MEYLAN,WM ET AL. (1996)
logP 2.75			EXP	SANGSTER,J (1994)
VP 2.12E-002	mm Hg	25	EST	NEELY,WB & BLAU,GE (1985)
DC 2.05	pKa	25	EST	WEAST,RC ET AL. (1985); FOR 2,4-ISOMER
HL 1.38E-006	atm m3/mol	25	EST	MEYLAN,WM & HOWARD,PH (1991)
OH 4.36E-011	cm3/molc sec	25	EST	MEYLAN,WM & HOWARD,PH (1993)

CAS #: 000095-83-0 — 4-CHLORO-1,2-BENZENEDIAMINE

Formula: $C_6H_7ClN_2$

Mol Weight: 142.59

MP (deg C): 67-70 FP (deg C):

BP (deg C): 229.3

BP pressure (mm Hg):

Property/Value	Units	Temp	Data Type	Reference
WS 6.59E+003	mg/L	25	EST	MEYLAN,WM ET AL. (1996)
logP 1.28			EXP	HANSCH,C ET AL. (1995)
VP 2.06E-003	mm Hg	25	EST	NEELY,WB & BLAU,GE (1985)
DC 3.83	pKa	25	EXP	PERRIN,DD (1972)
HL 4.99E-010	atm m3/mol	25	EST	MEYLAN,WM & HOWARD,PH (1991)
OH 1.74E+002	cm3/molc sec	25	EST	MEYLAN,WM & HOWARD,PH (1993)

CAS #: 000095-84-1 — PHENOL, 2-AMINO-4-METHYL-

Formula:	C_7H_9NO			
Mol Weight:	123.16			
MP (deg C): 136		FP (deg C):		
BP (deg C):				
BP pressure (mm Hg):				

Property/Value	Units	Temp	Data Type	Reference
WS 9.98E+003	mg/L	25	EST	MEYLAN,WM ET AL. (1996)
logP 1.16			EXP	DEBNATH,AK ET AL. (1992)
VP 3.08E-003	mm Hg	25	EST	NEELY,WB & BLAU,GE (1985)
DC	pKa			
HL 2.19E-010	atm m3/mol	25	EST	MEYLAN,WM & HOWARD,PH (1991)
OH 1.94E-010	cm3/molc sec	25	EST	MEYLAN,WM & HOWARD,PH (1993)

CAS #: 000095-85-2 — PHENOL, 2-AMINO-4-CHLORO-

Formula:	C_6H_6ClNO			
Mol Weight:	143.57			
MP (deg C): 140		FP (deg C):		
BP (deg C):				
BP pressure (mm Hg):				

Property/Value	Units	Temp	Data Type	Reference
WS 2.30E+003	mg/L	25	EST	MEYLAN,WM ET AL. (1996)
logP 1.81			EXP	DEBNATH,AK ET AL. (1992)
VP 1.41E-003	mm Hg	25	EST	NEELY,WB & BLAU,GE (1985)
DC	pKa			
HL 1.47E-010	atm m3/mol	25	EST	MEYLAN,WM & HOWARD,PH (1991)
OH 5.23E-011	cm3/molc sec	25	EST	MEYLAN,WM & HOWARD,PH (1993)

CAS #: 000095-87-4 — 2,5-DIMETHYLPHENOL

Formula:	$C_8H_{10}O$			
Mol Weight:	122.17			
MP (deg C): 78.4		FP (deg C):		
BP (deg C): 211.1				
BP pressure (mm Hg):				

Property/Value	Units	Temp	Data Type	Reference
WS 3.54E+003	mg/L	25	EXP	YALKOWSKY,SH & DANNENFELSER,RM (1992)
logP 2.33			EXP	HANSCH,C & LEO,AJ (1985)
VP 1.05E-001	mm Hg	25	EXT	CHAO,J ET AL. (1983)
DC 10.41	pKa	25	EXP	KORTUM,G ET AL (1961)
HL 6.83E-007	atm m3/mol	25	EST	MEYLAN,WM & HOWARD,PH (1991)
OH 8.00E-011	cm3/molc sec	25	EXP	ATKINSON,R (1989)

CAS #: 000095-88-5 — 4-CHLORORESORCINOL

Formula:	$C_6H_5ClO_2$			
Mol Weight:	144.56			
MP (deg C): 106--108		FP (deg C):		
BP (deg C): 147				
BP pressure (mm Hg): 1.80E+001				

Property/Value	Units	Temp	Data Type	Reference
WS 8.84E+003	mg/L	25	EST	MEYLAN,WM ET AL. (1996)
logP 1.80			EXP	HANSCH,C & LEO,AJ (1985)
VP 1.20E-003	mm Hg	25	EST	NEELY,WB & BLAU,GE (1985)
DC	pKa			
HL 4.32E-011	atm m3/mol	25	EST	MEYLAN,WM & HOWARD,PH (1991)
OH 1.67E-010	cm3/molc sec	25	EST	MEYLAN,WM & HOWARD,PH (1993)

CAS #: 000095-89-6 — PYRAZINE, 3-CHLORO-2,5-DIMETHYL-

Formula:	$C_6H_7ClN_2$			
Mol Weight:	142.59			
MP (deg C):		FP (deg C):		
BP (deg C): 66-69				
BP pressure (mm Hg): 9.00E+000				

Property/Value	Units	Temp	Data Type	Reference
WS 4.28E+003	mg/L	25	EST	MEYLAN,WM ET AL. (1996) ,
logP 1.50			EXP	YAMAGAMI,C ET AL. (1991)
VP 2.72E-001	mm Hg	25	EST	NEELY,WB & BLAU,GE (1985)
DC	pKa			
HL 1.66E-004	atm m3/mol	25	EST	MEYLAN,WM & HOWARD,PH (1991)
OH 7.90E-013	cm3/molc sec	25	EST	MEYLAN,WM & HOWARD,PH (1993)

CAS #: 000095-92-1 — ETHYL OXALATE

Formula:	$C_6H_{10}O_4$			
Mol Weight:	146.14			
MP (deg C): -40.6		FP (deg C):		
BP (deg C): 185.7				
BP pressure (mm Hg):				

Property/Value	Units	Temp	Data Type	Reference
WS 2.62E+004	mg/L	25	EST	MEYLAN,WM ET AL. (1996)
logP 0.56			EXP	HANSCH,C ET AL. (1995)
VP 4.14E-001	mm Hg	25	EXP	DAUBERT,TE & DANNER,RP (1989)
DC	pKa			
HL 4.35E-006	atm m3/mol	25	EST	MEYLAN,WM & HOWARD,PH (1991)
OH 3.32E-012	cm3/molc sec	25	EST	MEYLAN,WM & HOWARD,PH (1993)

CAS #: 000095-93-2 — 1,2,4,5-TETRAMETHYLBENZENE

Formula:	$C_{10}H_{14}$			
Mol Weight:	134.22			
MP (deg C): 80		FP (deg C):		
BP (deg C): 191-193				
BP pressure (mm Hg):				

Property/Value	Units	Temp	Data Type	Reference
WS 3.48E+000	mg/L	25	EXP	YALKOWSKY,SH & DANNENFELSER,RM (1992)
logP 4.00			EXP	HANSCH,C & LEO,AJ (1985)
VP 5.28E-001	mm Hg	25	EXT	CHAO,J ET AL. (1983)
DC	pKa			
HL 7.99E-003	atm m3/mol	25	EST	MEYLAN,WM & HOWARD,PH (1991)
OH 2.05E-011	cm3/molc sec	25	EST	MEYLAN,WM & HOWARD,PH (1993)

CAS #: 000095-94-3 — 1,2,4,5-TETRACHLOROBENZENE

Formula:	$C_6H_2Cl_4$			
Mol Weight:	215.89			
MP (deg C): 139.5		FP (deg C):		
BP (deg C): 248				
BP pressure (mm Hg):				

Property/Value	Units	Temp	Data Type	Reference
WS 5.95E-001	mg/L	25	EXP	YALKOWSKY,SH & DANNENFELSER,RM (1992)
logP 4.64			EXP	HANSCH,C & LEO,AJ (1985)
VP 5.40E-003	mm Hg	25	EXT	MACKAY,D ET AL. (1982)
DC	pKa			
HL 1.00E-003	atm m3/mol	20	EXP	OLIVER,BG (1985)
OH 8.40E-014	cm3/molc sec	25	EST	MEYLAN,WM & HOWARD,PH (1993)

CAS #: 000095-95-4 — 2,4,5-TRICHLOROPHENOL

Formula: $C_6H_3Cl_3O$

Mol Weight: 197.45
MP (deg C): 68
FP (deg C):
BP (deg C): 245-246
BP pressure (mm Hg):

Property/Value	Units	Temp	Data Type	Reference
WS 1.20E+003	mg/L	25	EXP	LEUENBERGER,C ET AL. (1985A)
logP 3.72			EXP	HANSCH,C & LEO,AJ (1985)
VP 2.00E-002	mm Hg	25	EXP	BIDLEMAN,TF & RENBERG,L (1985)
DC 6.94	pKa	20	EST	SCHELLENBERG,K ET AL. (1984)
HL 4.33E-006	atm m3/mol	25		VP/WSOL
OH 2.13E-012	cm3/molc sec	25	EST	MEYLAN,WM & HOWARD,PH (1993)

CAS #: 000096-09-3 — STYRENE OXIDE

Formula: C_8H_8O

Mol Weight: 120.15
MP (deg C): -37
FP (deg C):
BP (deg C): 194
BP pressure (mm Hg):

Property/Value	Units	Temp	Data Type	Reference
WS 3.00E+003	mg/L	20	EXP	LAPKIN,M (1965)
logP 1.61			EXP	HANSCH,C & LEO,AJ (1985)
VP 3.00E-001	mm Hg	20	EXP	LAPKIN,M (1965)
DC	pKa			
HL 1.58E-005	atm m3/mol	20	EST	VP/WSOL
OH 5.23E-012	cm3/molc sec	25	EST	MEYLAN,WM & HOWARD,PH (1993)

CAS #: 000096-12-8 — 1,2-DIBROMO-3-CHLOROPROPANE

Formula: $C_3H_5Br_2Cl$

Mol Weight: 236.34
MP (deg C): 6
FP (deg C):
BP (deg C): 196
BP pressure (mm Hg):

Property/Value	Units	Temp	Data Type	Reference
WS 1.23E+003	mg/L	20	EXP	MUNNECKE,DE & VANGUNDY,SD (1979)
logP 2.96			EXP	CHEM INSPECT TEST INST (1992)
VP 5.80E-001	mm Hg	20	EXP	MUNNECKE,DE & VANGUNDY,SD (1979)
DC	pKa			
HL 1.47E-004	atm m3/mol	20	EST	VP/WSOL
OH 4.40E-013	cm3/molc sec	25	EXP	ATKINSON,R (1989)

CAS #: 000096-13-9 — 2,3-DIBROMO-1-PROPANOL

Formula: $C_3H_6Br_2O$

Mol Weight: 217.90
MP (deg C):
FP (deg C):
BP (deg C): 95-97
BP pressure (mm Hg): 1.00E+001

Property/Value	Units	Temp	Data Type	Reference
WS 5.20E+004	mg/L	25	EXP	LANDE,SS ET AL. (1980)
logP 0.96			EST	MEYLAN,WM & HOWARD,PH (1995)
VP 9.18E-002	mm Hg	25	EXT	PERRY,RH & GREEN,D (1984)
DC	pKa			
HL 6.30E-008	atm m3/mol	25	EST	MEYLAN,WM & HOWARD,PH (1991)
OH 2.07E-012	cm3/molc sec	25	EST	MEYLAN,WM & HOWARD,PH (1993)

CAS #: 000096-14-0 — 3-METHYLPENTANE

Formula: C_6H_{14}

Mol Weight: 86.18
MP (deg C): -162.9
FP (deg C):
BP (deg C): 63.2
BP pressure (mm Hg):

Property/Value	Units	Temp	Data Type	Reference
WS 1.28E+001	mg/L	25	EXP	MCAULIFFE,C (1966)
logP 3.60			EXP	SANGSTER,J (1993)
VP 1.90E+002	mm Hg	25	EXP	RIDDICK,JA ET AL. (1986)
DC	pKa			
HL 1.68E+000	atm m3/mol	25	EST	VP/WSOL
OH 5.70E-012	cm3/molc sec	25	EXP	ATKINSON,R (1989)

CAS #: 000096-17-3 — 2-METHYLBUTANAL

Formula: $C_5H_{10}O$

Mol Weight: 86.13
MP (deg C): 90-92
FP (deg C):
BP (deg C):
BP pressure (mm Hg):

Property/Value	Units	Temp	Data Type	Reference
WS 1.12E+004	mg/L	25	EST	MEYLAN,WM ET AL. (1996)
logP 1.23			EST	MEYLAN,WM & HOWARD,PH (1995)
VP 4.74E+001	mm Hg	25	EST	NEELY,WB & BLAU,GE (1985)
DC	pKa			
HL 1.59E-004	atm m3/mol	25	EST	MEYLAN,WM & HOWARD,PH (1991)
OH 2.69E-011	cm3/molc sec	25	EST	MEYLAN,WM & HOWARD,PH (1993)

CAS #: 000096-18-4 — 1,2,3-TRICHLOROPROPANE

Formula: $C_3H_5Cl_3$

Mol Weight: 147.43
MP (deg C): -14.7
FP (deg C):
BP (deg C): 156.85
BP pressure (mm Hg):

Property/Value	Units	Temp	Data Type	Reference
WS 1.75E+003	mg/L	25	EXP	ALBANESE,V ET AL. (1987)
logP 2.27			EXP	CHEM INSPECT TEST INST (1992)
VP 3.69E+000	mm Hg	25	EXP	DAUBERT,TE & DANNER,RP (1989)
DC	pKa			
HL 3.43E-004	atm m3/mol	25	EXP	LEIGHTON,DTJR & CALO,JM (1981)
OH 5.89E-013	cm3/molc sec	25	EST	MEYLAN,WM & HOWARD,PH (1993)

CAS #: 000096-19-5 — 1,2,3-TRICHLOROPROPENE

Formula: $C_3H_3Cl_3$

Mol Weight: 145.42
MP (deg C):
FP (deg C): -78
BP (deg C): 142
BP pressure (mm Hg):

Property/Value	Units	Temp	Data Type	Reference
WS 3.34E+002	mg/L	25	EST	MEYLAN,WM ET AL. (1996)
logP 2.78			EST	MEYLAN,WM & HOWARD,PH (1995)
VP 4.40E+000	mm Hg	25	EXP	HUTCHINSON,TC ET AL. (1979)
DC	pKa			
HL 1.76E-002	atm m3/mol	25	EST	MEYLAN,WM & HOWARD,PH (1991)
OH 3.27E-012	cm3/molc sec	25	EST	MEYLAN,WM & HOWARD,PH (1993)

000096-22-0 — 3-PENTANONE

Formula: $C_5H_{10}O$

Mol Weight: 86.13

MP (deg C): -42

BP (deg C): 101.5

BP pressure (mm Hg):

Property/Value	Units	Temp	Data Type	Reference
WS 4.81E+004	mg/L	25	EXP	YALKOWSKY,SH & DANNENFELSER,RM (1992)
logP 0.99			EXP	HANSCH,C & LEO,AJ (1985)
VP 3.77E+001	mm Hg	25	EXP	DAUBERT,TE & DANNER,RP (1989)
DC	pKa			
HL 8.88E-005	atm m3/mol	25	EST	VP/WSOL
OH 2.00E-012	cm3/molc sec	25	EXP	ATKINSON,R (1989)

000096-23-1 — 1,3-DICHLORO-2-PROPANOL

Formula: $C_3H_6Cl_2O$

Mol Weight: 128.99

MP (deg C): -4

BP (deg C): 174.3

BP pressure (mm Hg):

Property/Value	Units	Temp	Data Type	Reference
WS 9.90E+004	mg/L	19	EXP	YALKOWSKY,SH & DANNENFELSER,RM (1992)
logP 0.78			EST	MEYLAN,WM & HOWARD,PH (1995)
VP 7.50E-001	mm Hg	20	EXP	WEBER,RC ET AL. (1981)
DC	pKa			
HL 1.29E-006	atm m3/mol	20	EST	VP/WSOL
OH 2.88E-012	cm3/molc sec	25	EST	MEYLAN,WM & HOWARD,PH (1993)

000096-24-2 — 3-CHLORO-1,2-PROPANEDIOL

Formula: $C_3H_7ClO_2$

Mol Weight: 110.54

MP (deg C):

BP (deg C): 116

BP pressure (mm Hg): 1.10E+001

Property/Value	Units	Temp	Data Type	Reference
WS 1.00E+006	mg/L	20	EXP	LANDE,SS ET AL. (1980)
logP -0.53			EST	MEYLAN,WM & HOWARD,PH (1995)
VP 2.00E-001	mm Hg	20	EXP	LANDE,SS ET AL. (1980)
DC	pKa			
HL 2.91E-008	atm m3/mol	20	EST	VP/WSOL
OH 7.74E-012	cm3/molc sec	25	EST	MEYLAN,WM & HOWARD,PH (1993)

000096-26-4 — DIHYDROXY ACETONE

Formula: $C_3H_6O_3$

Mol Weight: 90.08

MP (deg C): 90

BP (deg C):

BP pressure (mm Hg):

Property/Value	Units	Temp	Data Type	Reference
WS 5.03E+005	mg/L	25	EST	MEYLAN,WM ET AL. (1996)
logP -0.71			EST	MEYLAN,WM & HOWARD,PH (1995)
VP 2.08E-002	mm Hg	25	EST	NEELY,WB & BLAU,GE (1985)
DC	pKa			
HL 1.21E-003	atm m3/mol	25	EST	MEYLAN,WM & HOWARD,PH (1991)
OH 5.18E-012	cm3/molc sec	25	EST	MEYLAN,WM & HOWARD,PH (1993)

000096-29-7 — 2-BUTANONE OXIME

Formula: C_4H_9NO

Mol Weight: 87.12

MP (deg C): -29.5

BP (deg C): 152.5

BP pressure (mm Hg):

Property/Value	Units	Temp	Data Type	Reference
WS 3.66E+004	mg/L	25	EST	MEYLAN,WM ET AL. (1996)
logP 0.63			EXP	CHEM INSPECT TEST INST (1992)
VP 9.04E-001	mm Hg	25	EST	NEELY,WB & BLAU,GE (1985)
DC 12.45	pKa		EXP	SERJEANT,EP & DEMPSEY,B (1979)
HL 1.04E-005	atm m3/mol	25	EST	MEYLAN,WM & HOWARD,PH (1991)
OH 1.48E-012	cm3/molc sec	25	EST	MEYLAN,WM & HOWARD,PH (1993)

000096-31-1 — DIMETHYLUREA, SYM

Formula: $C_3H_8N_2O$

Mol Weight: 88.11

MP (deg C): 108

BP (deg C): 269

BP pressure (mm Hg):

Property/Value	Units	Temp	Data Type	Reference
WS 1.62E+004	mg/L	25	EST	MEYLAN,WM ET AL. (1996)
logP -0.49			EXP	HANSCH,C & LEO,AJ (1985)
VP 5.47E-001	mm Hg	25	EST	NEELY,WB & BLAU,GE (1985)
DC	pKa			
HL 1.76E-009	atm m3/mol	25	EST	MEYLAN,WM & HOWARD,PH (1991)
OH 3.12E-012	cm3/molc sec	25	EST	MEYLAN,WM & HOWARD,PH (1993)

000096-33-3 — METHYL ACRYLATE

Formula: $C_4H_6O_2$

Mol Weight: 86.09

MP (deg C): -76.5

BP (deg C): 80.2

BP pressure (mm Hg):

Property/Value	Units	Temp	Data Type	Reference
WS 4.94E+004	mg/L	25	EXP	RIDDICK,JA ET AL. (1986)
logP 0.80			EXP	HANSCH,C & LEO,AJ (1985)
VP 8.60E+001	mm Hg	25	EXP	DAUBERT,TE & DANNER,RP (1985)
DC	pKa			
HL 1.97E-004	atm m3/mol	25	EST	VP/WSOL
OH 2.41E-011	cm3/molc sec	25	EST	MEYLAN,WM & HOWARD,PH (1993)

000096-34-4 — METHYL CHLOROACETATE

Formula: $C_3H_5ClO_2$

Mol Weight: 108.53

MP (deg C): -33

BP (deg C): 131

BP pressure (mm Hg):

Property/Value	Units	Temp	Data Type	Reference
WS 1.70E+004	mg/L	25	EST	VP/HL
logP 0.63			EST	MEYLAN,WM & HOWARD,PH (1995)
VP 7.63E+000	mm Hg	25	EXP	DAUBERT,TE & DANNER,RP (1989)
DC	pKa			
HL 6.18E-005	atm m3/mol	25	EST	MEYLAN,WM & HOWARD,PH (1991)
OH 2.16E-013	cm3/molc sec	25	EST	ATKINSON,R (1988)

METHYLCYCLOPENTANE

CAS #: 000096-37-7				

Formula: C_6H_{12}

Mol Weight: 84.16

MP (deg C): -142.2   **FP (deg C):**

BP (deg C): 71.8

BP pressure (mm Hg):

Property/Value	Units	Temp	Data Type	Reference
WS 4.20E+001	mg/L	25	EXP	MCAULIFFE,C (1966)
logP 3.37			EXP	HANSCH,C ET AL. (1995)
VP 1.38E+002	mm Hg	25	EXP	BOUBLIK,T ET AL. (1984)
DC	pKa			
HL 3.63E-001	atm m3/mol	25	EST	VP/WSOL
OH 7.08E+000	cm3/molc sec	25	EST	MEYLAN,WM & HOWARD,PH (1993)

CYCLOPENTANOL

CAS #: 000096-41-3				

Formula: $C_5H_{10}O$

Mol Weight: 86.13

MP (deg C): -19   **FP (deg C):**

BP (deg C): 140.85

BP pressure (mm Hg):

Property/Value	Units	Temp	Data Type	Reference
WS 1.30E+005	mg/L	25	EST	MEYLAN,WM ET AL. (1996)
logP 0.71			EXP	ABRAHAM,MH ET AL. (1994)
VP 2.49E+000	mm Hg	25	EXP	PATTE,F ET AL. (1982)
DC	pKa			
HL 2.62E-006	atm m3/mol	25	EST	MEYLAN,WM & HOWARD,PH (1991)
OH 1.07E-011	cm3/molc sec	25	EXP	ATKINSON,R (1989)

2-CHLOROTHIOPHENE

CAS #: 000096-43-5				

Formula: C_4H_3ClS

Mol Weight: 118.59

MP (deg C): -71.9   **FP (deg C):**

BP (deg C): 128.3

BP pressure (mm Hg):

Property/Value	Units	Temp	Data Type	Reference
WS 6.88E+002	mg/L	25	EST	MEYLAN,WM ET AL. (1996)
logP 2.54			EXP	HANSCH,C & LEO,AJ (1985)
VP 2.38E+001	mm Hg	25	EXT	BOUBLIK,T ET AL. (1984)
DC	pKa			
HL 2.17E-003	atm m3/mol	25	EST	MEYLAN,WM & HOWARD,PH (1991)
OH 6.70E-012	cm3/molc sec	25	EST	MEYLAN,WM & HOWARD,PH (1993)

ETHYLENETHIOUREA

CAS #: 000096-45-7				

Formula: $C_3H_6N_2S$

Mol Weight: 102.16

MP (deg C): 203-204   **FP (deg C):**

BP (deg C): 347.18

BP pressure (mm Hg):

Property/Value	Units	Temp	Data Type	Reference
WS 2.00E+004	mg/L	30	EXP	MERCK INDEX (1976)
logP -0.66			EXP	GOVERS,H ET AL. (1986)
VP 5.01E-003	mm Hg	25	EST	NEELY,WB & BLAU,GE (1985)
DC	pKa			
HL 3.08E-010	atm m3/mol	25	EST	MEYLAN,WM & HOWARD,PH (1991)
OH 1.37E-010	cm3/molc sec	25	EST	MEYLAN,WM & HOWARD,PH (1993)

2-METHYLTETRAHYDROFURAN

CAS #: 000096-47-9				

Formula: $C_5H_{10}O$

Mol Weight: 86.13

MP (deg C):   **FP (deg C):**

BP (deg C): 78

BP pressure (mm Hg):

Property/Value	Units	Temp	Data Type	Reference
WS 1.39E+005	mg/L	25	EXP	RIDDICK,JA ET AL. (1986)
logP 1.35			EST	MEYLAN,WM & HOWARD,PH (1995)
VP 9.73E+001	mm Hg	25	EXP	CABANI,S ET AL. (1971A)
DC	pKa			
HL 9.30E-005	atm m3/mol	25	EXP	CABANI,S ET AL. (1971A)
OH 2.20E-011	cm3/molc sec	25	EXP	ATKINSON,R (1989)

GAMMA-BUTYROLACTONE

CAS #: 000096-48-0				

Formula: $C_4H_6O_2$

Mol Weight: 86.09

MP (deg C): -44   **FP (deg C):**

BP (deg C): 204

BP pressure (mm Hg):

Property/Value	Units	Temp	Data Type	Reference
WS 1.00E+006	mg/L		EXP	WEAST,RC (1972)
logP -0.64			EXP	HANSCH,C & LEO,AJ (1985)
VP 4.50E-001	mm Hg	25	EXP	DAUBERT,TE & DANNER,RP (1989)
DC	pKa			
HL 5.27E-008	atm m3/mol	25	EST	VP/WSOL
OH 2.41E-012	cm3/molc sec	25	EST	ATKINSON,R (1987)

1,3-DIOXOLAN-2-ONE

CAS #: 000096-49-1				

Formula: $C_3H_4O_3$

Mol Weight: 88.06

MP (deg C): 39.5   **FP (deg C):**

BP (deg C): 248

BP pressure (mm Hg):

Property/Value	Units	Temp	Data Type	Reference
WS 2.45E+005	mg/L	25	EST	MEYLAN,WM ET AL. (1996)
logP 0.12			EST	MEYLAN,WM & HOWARD,PH (1995)
VP 9.80E-003	mm Hg	25	EXT	DAUBERT,TE & DANNER,RP (1989)
DC	pKa			
HL 2.74E-004	atm m3/mol	25	EST	MEYLAN,WM & HOWARD,PH (1991)
OH 2.94E-012	cm3/molc sec	25	EST	MEYLAN,WM & HOWARD,PH (1993)

2-AMINOTHIAZOLE

CAS #: 000096-50-4				

Formula: $C_3H_4N_2S$

Mol Weight: 100.14

MP (deg C): 93   **FP (deg C):**

BP (deg C):

BP pressure (mm Hg):

Property/Value	Units	Temp	Data Type	Reference
WS 5.53E+004	mg/L	25	EST	MEYLAN,WM ET AL. (1996)
logP 0.38			EXP	HANSCH,C & LEO,AJ (1985)
VP 2.43E-001	mm Hg	25	EST	NEELY,WB & BLAU,GE (1985)
DC	pKa			
HL 1.35E-009	atm m3/mol	25	EST	MEYLAN,WM & HOWARD,PH (1991)
OH 7.73E-011	cm3/molc sec	25	EST	MEYLAN,WM & HOWARD,PH (1993)

CAS #: 000096-54-8 — N-METHYLPYRROLE

Formula: C_5H_7N
Mol Weight: 81.12
MP (deg C):
FP (deg C):
BP (deg C): 115
BP pressure (mm Hg):

Property/Value	Units	Temp	Data Type	Reference
WS 1.21E+004	mg/L	25	EST	MEYLAN,WM ET AL. (1996)
logP 1.21			EXP	HANSCH,C & LEO,AJ (1985)
VP 2.14E+001	mm Hg	25	EXP	CHAO,J ET AL. (1983)
DC	pKa			
HL 1.94E-004	atm m3/mol	25	EST	MEYLAN,WM & HOWARD,PH (1991)
OH 1.10E-010	cm3/molc sec	25	EST	MEYLAN,WM & HOWARD,PH (1993)

CAS #: 000096-64-0 — SOMAN

Formula: $C_7H_{16}FO_2P$
Mol Weight: 182.18
MP (deg C): -42
FP (deg C):
BP (deg C): 198
BP pressure (mm Hg):

Property/Value	Units	Temp	Data Type	Reference
WS 2.10E+004	mg/L	25	EXP	BENNETT,SR ET AL. (1984)
logP 1.78			EXP	BENSCHOP,HP & WESSELMAN,HC (1989)
VP 4.00E-001	mm Hg	25	EXP	BENNETT,SR ET AL. (1984)
DC	pKa			
HL 4.57E-006	atm m3/mol	25	EST	VP/WSOL
OH 4.97E-011	cm3/molc sec	25	EST	MEYLAN,WM & HOWARD,PH (1993)

CAS #: 000096-70-8 — 2-(T-BUTYL)-4-ETHYLPHENOL

Formula: $C_{12}H_{18}O$
Mol Weight: 178.28
MP (deg C):
FP (deg C):
BP (deg C):
BP pressure (mm Hg):

Property/Value	Units	Temp	Data Type	Reference
WS 3.31E+001	mg/L	25	EST	MEYLAN,WM ET AL. (1996)
logP 4.46			EST	MEYLAN,WM & HOWARD,PH (1995)
VP 2.63E-002	mm Hg	25	EXT	CHAO,J ET AL. (1983)
DC	pKa			
HL 2.12E-006	atm m3/mol	25	EST	MEYLAN,WM & HOWARD,PH (1991)
OH 5.05E-011	cm3/molc sec	25	EST	MEYLAN,WM & HOWARD,PH (1993)

CAS #: 000096-76-4 — 2,4-DI-T-BUTYLPHENOL

Formula: $C_{14}H_{22}O$
Mol Weight: 206.33
MP (deg C): 53
FP (deg C):
BP (deg C):
BP pressure (mm Hg):

Property/Value	Units	Temp	Data Type	Reference
WS 3.50E+001	mg/L	25	EXP	CHEM INSPECT TEST INST (1992)
logP 5.19			EXP	CHEM INSPECT TEST INST (1992)
VP 1.69E-002	mm Hg	25	EXT	PERRY,RH & GREEN,D (1984)
DC 11.72	pKa	20	EXP	SERJEANT,EP & DEMPSEY,B (1979)
HL 3.74E-006	atm m3/mol	25	EST	MEYLAN,WM & HOWARD,PH (1991)
OH 4.91E-011	cm3/molc sec	25	EST	MEYLAN,WM & HOWARD,PH (1993)

CAS #: 000096-80-0 — 2-(BIS(1-METHYLETHYL)AMINO)ETHANOL

Formula: $C_8H_{19}NO$
Mol Weight: 145.25
MP (deg C): -39
FP (deg C):
BP (deg C): 192
BP pressure (mm Hg):

Property/Value	Units	Temp	Data Type	Reference
WS 1.20E+004	mg/L	25	EXP	BOLLMEIER,AF (1991)
logP 0.88			EST	MEYLAN,WM & HOWARD,PH (1995)
VP 8.00E-002	mm Hg	20	EXP	HAWLEY,GG (1981)
DC 10.08	pKa	20	EXP	PERRIN,DD (1965)
HL 5.51E-009	atm m3/mol	25	EST	MEYLAN,WM & HOWARD,PH (1991)
OH 1.12E-010	cm3/molc sec	25	EST	MEYLAN,WM & HOWARD,PH (1993)

CAS #: 000096-88-8 — 2-PIPERIDINECARBOXAMIDE, N-(2,6-DIMETHYLPHENYL)-

Formula: $C_{15}H_{22}N_2O$
Mol Weight: 246.36
MP (deg C): 150-151
FP (deg C):
BP (deg C):
BP pressure (mm Hg):

Property/Value	Units	Temp	Data Type	Reference
WS 7.00E+003	mg/L	23	EXP	YALKOWSKY,SH & DANNENFELSER,RM (1992)
logP 1.95			EXP	HANSCH,C ET AL. (1995)
VP 2.18E-007	mm Hg	25	EST	NEELY,WB & BLAU,GE (1985)
DC	pKa			
HL 7.67E-011	atm m3/mol	25	EST	MEYLAN,WM & HOWARD,PH (1991)
OH 1.21E-010	cm3/molc sec	25	EST	MEYLAN,WM & HOWARD,PH (1993)

CAS #: 000096-91-3 — 2-AMINO-4,6-DINITROPHENOL

Formula: $C_6H_5N_3O_5$
Mol Weight: 199.12
MP (deg C): 169-170
FP (deg C):
BP (deg C):
BP pressure (mm Hg):

Property/Value	Units	Temp	Data Type	Reference
WS 1.40E+003	mg/L	22	EXP	YALKOWSKY,SH & DANNENFELSER,RM (1992)
logP 0.93			EXP	HANSCH,C & LEO,AJ (1985)
VP 4.16E-007	mm Hg	25	EST	NEELY,WB & BLAU,GE (1985)
DC 1.00	pKa		EXP	SERJEANT,EP & DEMPSEY,B (1979)
HL 9.75E-012	atm m3/mol	25	EST	MEYLAN,WM & HOWARD,PH (1991)
OH 7.06E-013	cm3/molc sec	25	EST	MEYLAN,WM & HOWARD,PH (1993)

CAS #: 000096-96-8 — BENZENAMINE, 4-METHOXY-2-NITRO-

Formula: $C_7H_8N_2O_3$
Mol Weight: 168.15
MP (deg C): 123-125
FP (deg C):
BP (deg C):
BP pressure (mm Hg):

Property/Value	Units	Temp	Data Type	Reference
WS 1.20E+002	mg/L	25	EXP	CHEM INSPECT TEST INST (1992)
logP 1.94			EXP	CHEM INSPECT TEST INST (1992)
VP 3.19E-004	mm Hg	25	EST	NEELY,WB & BLAU,GE (1985)
DC 0.77	pKa	25	EXP	PERRIN,DD (1972)
HL 9.72E-009	atm m3/mol	25	EST	MEYLAN,WM & HOWARD,PH (1991)
OH 1.25E-011	cm3/molc sec	25	EST	MEYLAN,WM & HOWARD,PH (1993)

CAS #: 000096-97-9 — 5-NITROSALICYLIC ACID

Formula: $C_7H_5NO_5$

Mol Weight: 183.12

MP (deg C): 228-230

FP (deg C):

BP (deg C):

BP pressure (mm Hg):

Property/Value	Units	Temp	Data Type	Reference
WS 2.00E+003	mg/L	45	EXP	YALKOWSKY,SH & DANNENFELSER,RM (1992)
logP 2.34			EXP	HANSCH,C & LEO,AJ (1985)
VP 1.16E-006	mm Hg	25	EST	NEELY,WB & BLAU,GE (1985)
DC 2.12	pKa	25	EXP	SERJEANT,EP & DEMPSEY,B (1979)
HL 5.60E-011	atm m3/mol	25	EST	MEYLAN,WM & HOWARD,PH (1991)
OH 2.20E-012	cm3/molc sec	25	EST	MEYLAN,WM & HOWARD,PH (1993)

CAS #: 000097-00-7 — 2,4-DINITROCHLOROBENZENE

Formula: $C_6H_3ClN_2O_4$

Mol Weight: 202.55

MP (deg C): 54

FP (deg C):

BP (deg C): 315

BP pressure (mm Hg):

Property/Value	Units	Temp	Data Type	Reference
WS 8.00E+000	mg/L	15	EXP	YALKOWSKY,SH & DANNENFELSER,RM (1992)
logP 2.17			EXP	DEBNATH,AK ET AL. (1991)
VP 8.49E-005	mm Hg	25	EXT	DAUBERT,TE & DANNER,RP (1991)
DC	pKa			
HL 3.15E-007	atm m3/mol	25	EST	MEYLAN,WM & HOWARD,PH (1991)
OH 2.17E-014	cm3/molc sec	25	EST	ATKINSON,R (1987)

CAS #: 000097-02-9 — 2,4-DINITROANILINE

Formula: $C_6H_5N_3O_4$

Mol Weight: 183.12

MP (deg C): 187.5-188

FP (deg C):

BP (deg C): 333.6

BP pressure (mm Hg):

Property/Value	Units	Temp	Data Type	Reference
WS 1.44E+003	mg/L	25	EST	MEYLAN,WM ET AL. (1996)
logP 1.84			EST	MEYLAN,WM & HOWARD,PH (1995)
VP 2.68E-005	mm Hg	25	EST	NEELY,WB & BLAU,GE (1985)
DC -4.25	pKa	25	EXP	DEAN,JA (1985)
HL 1.51E-010	atm m3/mol	25	EST	MEYLAN,WM & HOWARD,PH (1991)
OH 2.18E-011	cm3/molc sec	25	EST	MEYLAN,WM & HOWARD,PH (1993)

CAS #: 000097-09-6 — BENZENESULFONAMIDE, 4-CHLORO-3-NITRO-

Formula: $C_6H_5ClN_2O_4S$

Mol Weight: 236.63

MP (deg C):

FP (deg C):

BP (deg C):

BP pressure (mm Hg):

Property/Value	Units	Temp	Data Type	Reference
WS 3.12E+003	mg/L	25	EST	MEYLAN,WM ET AL. (1996)
logP 0.66			EXP	SANGSTER,J (1994)
VP 1.80E-006	mm Hg	25	EST	NEELY,WB & BLAU,GE (1985)
DC	pKa			
HL 1.23E-009	atm m3/mol	25	EST	MEYLAN,WM & HOWARD,PH (1991)
OH 3.67E-014	cm3/molc sec	25	EST	MEYLAN,WM & HOWARD,PH (1993)

CAS #: 000097-16-5 — GENITE

Formula: $C_{12}H_8Cl_2O_3S$

Mol Weight: 303.17

MP (deg C): 45.5

FP (deg C):

BP (deg C):

BP pressure (mm Hg):

Property/Value	Units	Temp	Data Type	Reference
WS 2.96E+000	mg/L	25	EST	MEYLAN,WM ET AL. (1996)
logP 4.21			EST	MEYLAN,WM & HOWARD,PH (1995)
VP 3.97E-007	mm Hg	25	EST	NEELY,WB & BLAU,GE (1985)
DC	pKa			
HL 1.64E-007	atm m3/mol	25	EST	MEYLAN,WM & HOWARD,PH (1991)
OH 2.25E-012	cm3/molc sec	25	EST	MEYLAN,WM & HOWARD,PH (1993)

CAS #: 000097-17-6 — DICHLOFENTHION

Formula: $C_{10}H_{13}Cl_2O_3PS$

Mol Weight: 315.16

MP (deg C):

FP (deg C):

BP (deg C): 164-169

BP pressure (mm Hg): 1.00E-001

Property/Value	Units	Temp	Data Type	Reference
WS 2.45E-001	mg/L	25	EXP	CHIOU,CT ET AL. (1977)
logP 5.14			EXP	HANSCH,C & LEO,AJ (1985)
VP 5.60E-004	mm Hg	25	EXP	FREED,VH ET AL. (1979A)
DC	pKa			
HL 9.48E-004	atm m3/mol	25	EST	VP/WSOL
OH 9.23E-011	cm3/molc sec	25	EST	MEYLAN,WM & HOWARD,PH (1993)

CAS #: 000097-23-4 — PHENOL,2,2'-METHYLENEBIS 4-CHLORO-

Formula: $C_{13}H_{10}Cl_2O_2$

Mol Weight: 269.13

MP (deg C): 177-178

FP (deg C):

BP (deg C):

BP pressure (mm Hg):

Property/Value	Units	Temp	Data Type	Reference
WS 3.00E+001	mg/L	25	EXP	TOMLIN,C (1994)
logP 4.26			EXP	HANSCH,C & LEO,AJ (1985)
VP 9.75E-011	mm Hg	25	EXP	TOMLIN,C (1994)
DC 7.60	pKa		EXP	TOMLIN,C (1994)
HL 1.15E-012	atm m3/mol	25	EST	MEYLAN,WM & HOWARD,PH (1991)
OH 2.49E-011	cm3/molc sec	25	EST	MEYLAN,WM & HOWARD,PH (1993)

CAS #: 000097-31-4 — NORMETANEPHRINE

Formula: $C_9H_{13}NO_3$

Mol Weight: 183.21

MP (deg C):

FP (deg C):

BP (deg C):

BP pressure (mm Hg):

Property/Value	Units	Temp	Data Type	Reference
WS 1.00E+006	mg/L	25	EST	MEYLAN,WM ET AL. (1996)
logP -1.05			EXP	HANSCH,C & LEO,AJ (1985)
VP 2.36E-006	mm Hg	25	EST	NEELY,WB & BLAU,GE (1985)
DC	pKa			
HL 1.83E-016	atm m3/mol	25	EST	MEYLAN,WM & HOWARD,PH (1991)
OH 7.63E-011	cm3/molc sec	25	EST	MEYLAN,WM & HOWARD,PH (1993)

CAS #: 000097-53-0				EUGENOL	

Formula: $C_{10}H_{12}O_2$

Mol Weight: 164.21

MP (deg C): -9.2--9.1 FP (deg C):

BP (deg C): 255

BP pressure (mm Hg):

Property/Value	Units	Temp	Data Type	Reference
WS 2.46E+003	mg/L	25	EXP	YALKOWSKY,SH & DANNENFELSER,RM (1992)
logP 2.27			EXP	SANGSTER,J (1993)
VP 2.26E-002	mm Hg	25	EXT	PERRY,RH & GREEN,D (1984)
DC 10.19	pKa	25	EXP	KORTUM,G ET AL (1961)
HL 4.81E-008	atm m3/mol	25	EST	MEYLAN,WM & HOWARD,PH (1991)
OH 6.50E-011	cm3/molc sec	25	EST	MEYLAN,WM & HOWARD,PH (1993)

CAS #: 000097-54-1				2-METHOXY-4-(1-PROPENYL)PHENOL	

Formula: $C_{10}H_{12}O_2$

Mol Weight: 164.21

MP (deg C): -10 FP (deg C):

BP (deg C): 266

BP pressure (mm Hg):

Property/Value	Units	Temp	Data Type	Reference
WS 3.56E+002	mg/L	25	EST	MEYLAN,WM ET AL. (1996)
logP 2.65			EST	MEYLAN,WM & HOWARD,PH (1995)
VP 1.20E-002	mm Hg	25	EXT	PERRY,RH & GREEN,D (1984)
DC 9.88	pKa	25	EXP	SERJEANT,EP & DEMPSEY,B (1979)
HL 2.67E-008	atm m3/mol	25	EST	MEYLAN,WM & HOWARD,PH (1991)
OH 8.46E-011	cm3/molc sec	25	EST	MEYLAN,WM & HOWARD,PH (1993)

CAS #: 000097-56-3				2-METHYL-4-((2-METHYLPHENYL)AZO)BENZENAMINE	

Formula: $C_{14}H_{15}N_3$

Mol Weight: 225.30

MP (deg C): 101-102 FP (deg C):

BP (deg C):

BP pressure (mm Hg):

Property/Value	Units	Temp	Data Type	Reference
WS 7.00E+000	mg/L	25	EXP	YALKOWSKY,SH & DANNENFELSER,RM (1992)
logP 4.92			EST	MEYLAN,WM & HOWARD,PH (1995)
VP 7.50E-007	mm Hg	25	EXT	SHIMIZU,T ET AL. (1987)
DC	pKa			
HL 3.18E-008	atm m3/mol	25	EST	VP/WSOL
OH 5.45E-011	cm3/molc sec	25	EST	MEYLAN,WM & HOWARD,PH (1993)

CAS #: 000097-59-6				ALLANTOIN	

Formula: $C_4H_6N_4O_3$

Mol Weight: 158.12

MP (deg C): 238 FP (deg C):

BP (deg C):

BP pressure (mm Hg):

Property/Value	Units	Temp	Data Type	Reference
WS 5.26E+003	mg/L	25	EXP	YALKOWSKY,SH & DANNENFELSER,RM (1992)
logP -3.14			EST	MEYLAN,WM & HOWARD,PH (1995)
VP 4.32E-009	mm Hg	25	EST	NEELY,WB & BLAU,GE (1985)
DC	pKa			
HL 3.41E-018	atm m3/mol	25	EST	MEYLAN,WM & HOWARD,PH (1991)
OH 2.36E-011	cm3/molc sec	25	EST	MEYLAN,WM & HOWARD,PH (1993)

CAS #: 000097-61-0				PENTANOIC ACID, 2-METHYL-	

Formula: $C_6H_{12}O_2$

Mol Weight: 116.16

MP (deg C): FP (deg C):

BP (deg C): 196-197

BP pressure (mm Hg):

Property/Value	Units	Temp	Data Type	Reference
WS 1.50E+004	mg/L	20	EXP	RIEMENSCHNEIDER,W (1986)
logP 1.80			EXP	CATZ,P & FRIEND,DR (1989)
VP 4.45E-001	mm Hg	25	EST	NEELY,WB & BLAU,GE (1985)
DC 4.79	pKa	18	EXP	KORTUM,G ET AL (1961)
HL 1.70E-006	atm m3/mol	25	EST	MEYLAN,WM & HOWARD,PH (1991)
OH 5.21E-012	cm3/molc sec	25	EST	MEYLAN,WM & HOWARD,PH (1993)

CAS #: 000097-62-1				ETHYL ISOBUTYRATE	

Formula: $C_6H_{12}O_2$

Mol Weight: 116.16

MP (deg C): -88 FP (deg C):

BP (deg C): 110-111

BP pressure (mm Hg):

Property/Value	Units	Temp	Data Type	Reference
WS 3.17E+003	mg/L	25	EST	MEYLAN,WM ET AL. (1996)
logP 1.77			EST	MEYLAN,WM & HOWARD,PH (1995)
VP 2.54E+001	mm Hg	25	EXP	DAUBERT,TE & DANNER,RP (1989)
DC	pKa			
HL 4.10E-004	atm m3/mol	25	EST	MEYLAN,WM & HOWARD,PH (1991)
OH 3.43E-012	cm3/molc sec	25	EST	MEYLAN,WM & HOWARD,PH (1993)

CAS #: 000097-63-2				ETHYL METHACRYLATE	

Formula: $C_6H_{10}O_2$

Mol Weight: 114.15

MP (deg C): -44.7 EST FP (deg C):

BP (deg C): 118-119

BP pressure (mm Hg):

Property/Value	Units	Temp	Data Type	Reference
WS 2.32E+003	mg/L	25	EST	MEYLAN,WM ET AL. (1996)
logP 1.94			EXP	HANSCH,C & LEO,AJ (1985)
VP 2.06E+001	mm Hg	25	EXP	DAUBERT,TE & DANNER,RP (1985)
DC	pKa			
HL 1.92E-004	atm m3/mol	25	EST	MEYLAN,WM & HOWARD,PH (1991)
OH 4.84E-011	cm3/molc sec	25	EST	MEYLAN,WM & HOWARD,PH (1993)

CAS #: 000097-64-3				ETHYL LACTATE	

Formula: $C_5H_{10}O_3$

Mol Weight: 118.13

MP (deg C): FP (deg C):

BP (deg C): 154

BP pressure (mm Hg):

Property/Value	Units	Temp	Data Type	Reference
WS 1.00E+006	mg/L	20	EXP	YALKOWSKY,SH & DANNENFELSER,RM (1992)
logP -0.18			EST	MEYLAN,WM & HOWARD,PH (1995)
VP 3.75E+000	mm Hg	25	EXP	DAUBERT,TE & DANNER,RP (1989)
DC	pKa			
HL 5.83E-007	atm m3/mol	25	EST	VP/WSOL
OH 6.99E-012	cm3/molc sec	25	EST	MEYLAN,WM & HOWARD,PH (1993)

CAS #: 000097-74-5 — TETRAMETHYLTHIURAM

Formula: $C_6H_{12}N_2S_3$

Mol Weight: 208.37

MP (deg C): 109.5

FP (deg C):

BP (deg C):

BP pressure (mm Hg):

Property/Value	Units	Temp	Data Type	Reference
WS 5.00E+004	mg/L	25	EST	MEYLAN,WM ET AL. (1996)
logP 0.75			EST	MEYLAN,WM & HOWARD,PH (1995)
VP 2.70E-004	mm Hg	25	EST	NEELY,WB & BLAU,GE (1985)
DC	pKa			
HL 1.74E-005	atm m3/mol	25	EST	MEYLAN,WM & HOWARD,PH (1991)
OH 1.39E-010	cm3/molc sec	25	EST	MEYLAN,WM & HOWARD,PH (1993)

CAS #: 000097-77-8 — DISULFIRAM

Formula: $C_{10}H_{20}N_2S_4$

Mol Weight: 296.54

MP (deg C): 70

FP (deg C):

BP (deg C):

BP pressure (mm Hg):

Property/Value	Units	Temp	Data Type	Reference
WS 4.09E+000	mg/L	25	EXP	YALKOWSKY,SH & DANNENFELSER,RM (1992)
logP 3.88			EXP	HANSCH,C & LEO,AJ (1985)
VP 8.70E-004	mm Hg	25	EST	WSOL X HL
DC	pKa			
HL 8.32E-005	atm m3/mol	25	EST	MEYLAN,WM & HOWARD,PH (1991)
OH 3.54E-010	cm3/molc sec	25	EST	MEYLAN,WM & HOWARD,PH (1993)

CAS #: 000097-85-8 — ISOBUTYL ISOBUTYRATE

Formula: $C_8H_{16}O_2$

Mol Weight: 144.22

MP (deg C): -80.7

FP (deg C):

BP (deg C): 147.51

BP pressure (mm Hg):

Property/Value	Units	Temp	Data Type	Reference
WS 1.00E+003	mg/L	20	EXP	FLICK,EW (1991)
logP 2.68			EST	MEYLAN,WM & HOWARD,PH (1995)
VP 4.33E+000	mm Hg	25	EXP	DAUBERT,TE & DANNER,RP (1989)
DC	pKa			
HL 8.22E-004	atm m3/mol	25	EST	VP/WSOL
OH 4.72E-012	cm3/molc sec	25	EST	ATKINSON,R (1988)

CAS #: 000097-86-9 — METHACRYLIC ACID, I-BUTYL ESTER

Formula: $C_8H_{14}O_2$

Mol Weight: 142.20

MP (deg C):

FP (deg C):

BP (deg C): 155

BP pressure (mm Hg):

Property/Value	Units	Temp	Data Type	Reference
WS 4.39E+002	mg/L	25	EST	MEYLAN,WM ET AL. (1996)
logP 2.66			EXP	HANSCH,C & LEO,AJ (1985)
VP 4.11E+000	mm Hg	25	EST	NEELY,WB & BLAU,GE (1985)
DC	pKa			
HL 3.39E-004	atm m3/mol	25	EST	MEYLAN,WM & HOWARD,PH (1991)
OH 2.27E-011	cm3/molc sec	25	EST	MEYLAN,WM & HOWARD,PH (1993)

CAS #: 000097-88-1 — METHACRYLIC ACID, N-BUTYL ESTER

Formula: $C_8H_{14}O_2$

Mol Weight: 142.20

MP (deg C):

FP (deg C):

BP (deg C): 160

BP pressure (mm Hg):

Property/Value	Units	Temp	Data Type	Reference
WS 2.85E+002	mg/L	25	EST	MEYLAN,WM ET AL. (1996)
logP 2.88			EXP	HANSCH,C & LEO,AJ (1985)
VP 2.12E+000	mm Hg	25	EXP	DAUBERT,TE & DANNER,RP (1989)
DC	pKa			
HL 3.39E-004	atm m3/mol	25	EST	MEYLAN,WM & HOWARD,PH (1991)
OH 2.27E-011	cm3/molc sec	25	EST	MEYLAN,WM & HOWARD,PH (1993)

CAS #: 000097-90-5 — ETHYLENE DIMETHYLACRYLATE

Formula: $C_{10}H_{14}O_4$

Mol Weight: 198.22

MP (deg C): < -75

FP (deg C):

BP (deg C): 80.7

BP pressure (mm Hg):

Property/Value	Units	Temp	Data Type	Reference
WS 5.81E+002	mg/L	25	EST	MEYLAN,WM ET AL. (1996)
logP 2.21			EST	MEYLAN,WM & HOWARD,PH (1995)
VP 1.88E-001	mm Hg	25	EST	NEELY,WB & BLAU,GE (1985)
DC	pKa			
HL 3.78E-007	atm m3/mol	25	EST	MEYLAN,WM & HOWARD,PH (1991)
OH 3.99E-011	cm3/molc sec	25	EST	MEYLAN,WM & HOWARD,PH (1993)

CAS #: 000097-95-0 — 2-ETHYL-1-BUTANOL

Formula: $C_6H_{14}O$

Mol Weight: 102.18

MP (deg C): < -15

FP (deg C):

BP (deg C): 147

BP pressure (mm Hg):

Property/Value	Units	Temp	Data Type	Reference
WS 4.00E+003	mg/L	25	EXP	YALKOWSKY,SH & DANNENFELSER,RM (1992)
logP 1.75			EST	MEYLAN,WM & HOWARD,PH (1995)
VP 1.53E+000	mm Hg	25	EXP	DAUBERT,TE & DANNER,RP (1989)
DC	pKa			
HL 5.14E-005	atm m3/mol	25	EST	VP/WSOL
OH 1.04E-011	cm3/molc sec	25	EST	MEYLAN,WM & HOWARD,PH (1993)

CAS #: 000097-96-1 — 2-ETHYLBUTYRALDEHYDE

Formula: $C_6H_{12}O$

Mol Weight: 100.16

MP (deg C):

FP (deg C):

BP (deg C): 117-119

BP pressure (mm Hg): 1.60E+002

Property/Value	Units	Temp	Data Type	Reference
WS 3.93E+003	mg/L	25	EST	MEYLAN,WM ET AL. (1996)
logP 1.73			EST	MEYLAN,WM & HOWARD,PH (1995)
VP 1.69E+001	mm Hg	25	EST	NEELY,WB & BLAU,GE (1985)
DC	pKa			
HL 2.11E-004	atm m3/mol	25	EST	MEYLAN,WM & HOWARD,PH (1991)
OH 3.06E-011	cm3/molc sec	25	EST	MEYLAN,WM & HOWARD,PH (1993)

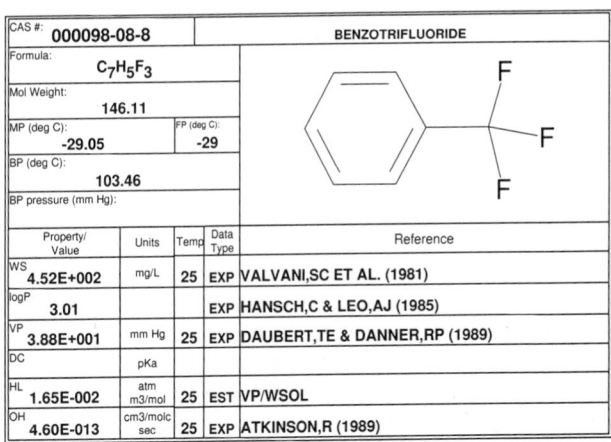

CAS #:	000097-99-4			TETRAHYDROFURFURYL ALCOHOL
Formula:	$C_5H_{10}O_2$			
Mol Weight:	102.13			
MP (deg C):	<80	FP (deg C):		
BP (deg C):	178			
BP pressure (mm Hg):				

Property/Value	Units	Temp	Data Type	Reference
WS 1.00E+006	mg/L	25	EXP	RIDDICK,JA ET AL. (1986)
logP -0.11			EST	MEYLAN,WM & HOWARD,PH (1995)
VP 8.01E-001	mm Hg	25	EXP	DAUBERT,TE & DANNER,RP (1989)
DC	pKa			
HL 4.09E-009	atm m3/mol	25	EST	MEYLAN,WM & HOWARD,PH (1991)
OH 2.88E-011	cm3/molc sec	25	EST	ATKINSON,R (1988)

CAS #:	000098-00-0			2-HYDROXYMETHYLFURAN
Formula:	$C_5H_6O_2$			
Mol Weight:	98.10			
MP (deg C):	-31	FP (deg C):		
BP (deg C):	171			
BP pressure (mm Hg):				

Property/Value	Units	Temp	Data Type	Reference
WS 1.00E+006	mg/L		EXP	MCKILLIP,WJ & SHERMAN,E (1978)
logP 0.28			EXP	HANSCH,C & LEO,AJ (1985)
VP 6.09E-001	mm Hg	25	EXP	DAUBET,TE & DANNER,RP (1989)
DC	pKa			
HL 7.86E-008	atm m3/mol	25	EST	VP/WSOL
OH 1.04E-010	cm3/molc sec	25	EST	MEYLAN,WM & HOWARD,PH (1993)

CAS #:	000098-01-1			FUFURAL
Formula:	$C_5H_4O_2$			
Mol Weight:	96.09			
MP (deg C):	-36.5	FP (deg C):		
BP (deg C):	161.8			
BP pressure (mm Hg):				

Property/Value	Units	Temp	Data Type	Reference
WS 8.30E+004	mg/L	25	EXP	CHEM INSPECT TEST INST (1992)
logP 0.41			EXP	HANSCH,C & LEO,AJ (1985)
VP 2.21E+000	mm Hg	25	EXP	DAUBERT,TE & DANNER,RP (1989)
DC	pKa			
HL 3.37E-006	atm m3/mol	25	EST	VP/WSOL
OH 3.74E-011	cm3/molc sec	25	EST	MEYLAN,WM & HOWARD,PH (1993)

CAS #:	000098-03-3			THIOPHENE-2-CARBOXALDEHYDE
Formula:	C_5H_4OS			
Mol Weight:	112.15			
MP (deg C):		FP (deg C):		
BP (deg C):	198			
BP pressure (mm Hg):				

Property/Value	Units	Temp	Data Type	Reference
WS 1.44E+004	mg/L	25	EST	MEYLAN,WM ET AL. (1996)
logP 1.02			EXP	HANSCH,C & LEO,AJ (1985)
VP 5.55E-001	mm Hg	25	EST	NEELY,WB & BLAU,GE (1985)
DC	pKa			
HL 7.30E-006	atm m3/mol	25	EST	MEYLAN,WM & HOWARD,PH (1991)
OH 2.17E-011	cm3/molc sec	25	EST	MEYLAN,WM & HOWARD,PH (1993)

CAS #:	000098-04-4			PHENYL TRIMETHYL AMMONIUM IODIDE
Formula:	$C_9H_{14}IN$			
Mol Weight:	263.12			
MP (deg C):	175	FP (deg C):		
BP (deg C):				
BP pressure (mm Hg):				

Property/Value	Units	Temp	Data Type	Reference
WS 1.00E+006	mg/L	25	EST	MEYLAN,WM ET AL. (1996)
logP -2.74			EXP	HANSCH,C & LEO,AJ (1985)
VP 3.44E-008	mm Hg	25	EST	NEELY,WB & BLAU,GE (1985)
DC	pKa			
HL 6.62E-013	atm m3/mol	25	EST	MEYLAN,WM & HOWARD,PH (1991)
OH 4.21E-012	cm3/molc sec	25	EST	MEYLAN,WM & HOWARD,PH (1993)

CAS #:	000098-06-6			T-BUTYLBENZENE
Formula:	$C_{10}H_{14}$			
Mol Weight:	134.22			
MP (deg C):	-57.8	FP (deg C):		
BP (deg C):	169.1			
BP pressure (mm Hg):				

Property/Value	Units	Temp	Data Type	Reference
WS 2.95E+001	mg/L	25	EXP	YALKOWSKY,SH & DANNENFELSER,RM (1992)
logP 4.11			EXP	HANSCH,C & LEO,AJ (1985)
VP 2.20E+000	mm Hg	25	EXP	CHAO,J ET AL. (1983)
DC	pKa			
HL 1.32E-002	atm m3/mol	25	EST	VP/WSOL
OH 4.60E-012	cm3/molc sec	25	EXP	ATKINSON,R (1989)

CAS #:	000098-07-7			BENZOTRICHLORIDE
Formula:	$C_7H_5Cl_3$			
Mol Weight:	195.48			
MP (deg C):	-5	FP (deg C):		
BP (deg C):	220.8			
BP pressure (mm Hg):				

Property/Value	Units	Temp	Data Type	Reference
WS 5.30E+001	mg/L	5	EXP	OHNISHI,R & TANABE,K (1971)
logP 3.90			EST	MEYLAN,WM & HOWARD,PH (1995)
VP 4.14E-001	mm Hg	25	EXT	YAWS,CL (1994A)
DC	pKa			
HL 2.60E-004	atm m3/mol	25	EST	MEYLAN,WM & HOWARD,PH (1991)
OH 3.57E-013	cm3/molc sec	25	EST	MEYLAN,WM & HOWARD,PH (1993)

CAS #:	000098-08-8			BENZOTRIFLUORIDE
Formula:	$C_7H_5F_3$			
Mol Weight:	146.11			
MP (deg C):	-29.05	FP (deg C):	-29	
BP (deg C):	103.46			
BP pressure (mm Hg):				

Property/Value	Units	Temp	Data Type	Reference
WS 4.52E+002	mg/L	25	EXP	VALVANI,SC ET AL. (1981)
logP 3.01			EXP	HANSCH,C & LEO,AJ (1985)
VP 3.88E+001	mm Hg	25	EXP	DAUBERT,TE & DANNER,RP (1989)
DC	pKa			
HL 1.65E-002	atm m3/mol	25	EST	VP/WSOL
OH 4.60E-013	cm3/molc sec	25	EXP	ATKINSON,R (1989)

135

000098-09-9 — BENZENESULFONYL CHLORIDE

CAS #: 000098-09-9

Formula: $C_6H_5ClO_2S$

Mol Weight: 176.62

MP (deg C): 14.5

FP (deg C):

BP (deg C): 251-252

BP pressure (mm Hg):

Property/Value	Units	Temp	Data Type	Reference
WS 1.76E+002	mg/L	25	EST	MEYLAN,WM ET AL. (1996)
logP 2.94			EST	MEYLAN,WM & HOWARD,PH (1995)
VP 6.80E-001	mm Hg	25	EST	PERRY,RH & GREEN,D (1984)
DC	pKa			
HL 8.76E-006	atm m3/mol	25	EST	MEYLAN,WM & HOWARD,PH (1991)
OH 4.17E-013	cm3/molc sec	25	EST	MEYLAN,WM & HOWARD,PH (1993)

000098-10-2 — BENZENESULFONAMIDE

CAS #: 000098-10-2

Formula: $C_6H_7NO_2S$

Mol Weight: 157.19

MP (deg C): 150-152

FP (deg C):

BP (deg C):

BP pressure (mm Hg):

Property/Value	Units	Temp	Data Type	Reference
WS 4.30E+003	mg/L	16	EXP	YALKOWSKY,SH & DANNENFELSER,RM (1992)
logP 0.31			EXP	HANSCH,C & LEO,AJ (1985)
VP 8.94E-004	mm Hg	25	EST	NEELY,WB & BLAU,GE (1985)
DC	pKa			
HL 4.22E-007	atm m3/mol	25	EST	MEYLAN,WM & HOWARD,PH (1991)
OH 4.17E-013	cm3/molc sec	25	EST	MEYLAN,WM & HOWARD,PH (1993)

000098-11-3 — BENZENESULFONIC ACID

CAS #: 000098-11-3

Formula: $C_6H_6O_3S$

Mol Weight: 158.18

MP (deg C): 50-51

FP (deg C):

BP (deg C):

BP pressure (mm Hg):

Property/Value	Units	Temp	Data Type	Reference
WS 1.00E+006	mg/L	25	EST	MEYLAN,WM ET AL (1996)
logP -2.25			EST	HANSCH,C & LEO,AJ (1981)
VP 2.36E-005	mm Hg	25	EST	NEELY,WB & BLAU,GE (1985)
DC 2.55	pKa		EXP	DEAN,JA (1987)
HL 2.52E-009	atm m3/mol	25	EST	MEYLAN,WM & HOWARD,PH (1991)
OH 5.57E-013	cm3/molc sec	25	EST	MEYLAN,WM & HOWARD,PH (1993)

000098-13-5 — PHENYLTRICHLOROSILANE

CAS #: 000098-13-5

Formula: $C_6H_5Cl_3Si$

Mol Weight: 211.55

MP (deg C):

FP (deg C):

BP (deg C): 201

BP pressure (mm Hg):

Property/Value	Units	Temp	Data Type	Reference
WS 3.19E+001	mg/L	25	EST	MEYLAN,WM ET AL. (1996)
logP 3.60			EST	MEYLAN,WM & HOWARD,PH (1995)
VP 4.26E-001	mm Hg	25	EXT	BOUBLIK,T ET AL. (1984)
DC	pKa			
HL 3.29E-004	atm m3/mol	25	EST	MEYLAN,WM & HOWARD,PH (1991)
OH 1.95E-012	cm3/molc sec	25	EST	MEYLAN,WM & HOWARD,PH (1993)

000098-16-8 — 3-TRIFLUOROMETHYLANILINE

CAS #: 000098-16-8

Formula: $C_7H_6F_3N$

Mol Weight: 161.13

MP (deg C): 5.5

FP (deg C):

BP (deg C): 187

BP pressure (mm Hg):

Property/Value	Units	Temp	Data Type	Reference
WS 5.46E+003	mg/L	25	EXP	KUHNE,R ET AL. (1995)
logP 2.29			EXP	DEBNATH,AK ET AL. (1992)
VP 5.92E-001	mm Hg	25	EST	NEELY,WB & BLAU,GE (1985)
DC 3.49	pKa	25	EXP	PERRIN,DD (1965)
HL 1.65E-005	atm m3/mol	25	EST	MEYLAN,WM & HOWARD,PH (1991)
OH 1.63E-011	cm3/molc sec	25	EST	MEYLAN,WM & HOWARD,PH (1993)

000098-17-9 — M-TRIFLUOROMETHYLPHENOL

CAS #: 000098-17-9

Formula: $C_7H_5F_3O$

Mol Weight: 162.11

MP (deg C): -1.8

FP (deg C):

BP (deg C): 187

BP pressure (mm Hg):

Property/Value	Units	Temp	Data Type	Reference
WS 7.70E+002	mg/L	25	EST	MEYLAN,WM ET AL. (1996)
logP 2.95			EXP	HANSCH,C & LEO,AJ (1985)
VP 6.68E-001	mm Hg	25	EST	NEELY,WB & BLAU,GE (1985)
DC 8.95	pKa	25	EXP	SERJEANT,EP & DEMPSEY,B (1979)
HL 4.87E-006	atm m3/mol	25	EST	MEYLAN,WM & HOWARD,PH (1991)
OH 5.18E-012	cm3/molc sec	25	EST	MEYLAN,WM & HOWARD,PH (1993)

000098-18-0 — M-AMINOBENZENESULFONAMIDE

CAS #: 000098-18-0

Formula: $C_6H_8N_2O_2S$

Mol Weight: 172.21

MP (deg C):

FP (deg C):

BP (deg C):

BP pressure (mm Hg):

Property/Value	Units	Temp	Data Type	Reference
WS 1.27E+005	mg/L	25	EST	MEYLAN,WM ET AL. (1996)
logP -0.38			EXP	HANSCH,C ET AL. (1995)
VP 2.09E-005	mm Hg	25	EST	NEELY,WB & BLAU,GE (1985)
DC 2.90	pKa	25	EXP	PERRIN,DD (1965)
HL 1.49E-010	atm m3/mol	25	EST	MEYLAN,WM & HOWARD,PH (1991)
OH 2.30E-011	cm3/molc sec	25	EST	MEYLAN,WM & HOWARD,PH (1993)

000098-27-1 — 4-(T-BUTYL)-2-CRESOL

CAS #: 000098-27-1

Formula: $C_{11}H_{16}O$

Mol Weight: 164.25

MP (deg C): 27.5

FP (deg C):

BP (deg C): 237

BP pressure (mm Hg):

Property/Value	Units	Temp	Data Type	Reference
WS 1.01E+002	mg/L	25	EST	MEYLAN,WM ET AL. (1996)
logP 3.97			EST	MEYLAN,WM & HOWARD,PH (1995)
VP 2.82E-002	mm Hg	25	EXT	CHAO,J ET AL. (1983)
DC 10.59	pKa	20	EXP	SERJEANT,EP & DEMPSEY,B (1979)
HL 1.60E-006	atm m3/mol	25	EST	MEYLAN,WM & HOWARD,PH (1991)
OH 4.98E-011	cm3/molc sec	25	EST	MEYLAN,WM & HOWARD,PH (1993)

CAS #: 000098-28-2				4-(TERT-BUTYL)-2-CHLOROPHENOL

Formula: $C_{10}H_{13}ClO$

Mol Weight: 184.67

MP (deg C):   **FP (deg C):**

BP (deg C):

BP pressure (mm Hg):

Property/Value	Units	Temp	Data Type	Reference
WS 6.68E+001	mg/L	25	EST	MEYLAN,WM ET AL. (1996)
logP 4.07			EST	MEYLAN,WM & HOWARD,PH (1995)
VP 3.14E-003	mm Hg	25	EST	NEELY,WB & BLAU,GE (1985)
DC 8.58	pKa	25	EXP	SERJEANT,EP & DEMPSEY,B (1979)
HL 1.07E-006	atm m3/mol	25	EST	MEYLAN,WM & HOWARD,PH (1991)
OH 1.23E-011	cm3/molc sec	25	EST	MEYLAN,WM & HOWARD,PH (1993)

CAS #: 000098-46-4				M-TRIFLUOROMETHYLNITROBENZENE

Formula: $C_7H_4F_3NO_2$

Mol Weight: 191.11

MP (deg C): -2.4   **FP (deg C):**

BP (deg C): 202.8

BP pressure (mm Hg):

Property/Value	Units	Temp	Data Type	Reference
WS 1.14E+002	mg/L	25	EST	MEYLAN,WM ET AL. (1996)
logP 2.62			EXP	HANSCH,C & LEO,AJ (1985)
VP 2.59E-001	mm Hg	25	EXP	DAUBERT,TE & DANNER,RP (1991)
DC	pKa			
HL 1.85E-004	atm m3/mol	25	EST	MEYLAN,WM & HOWARD,PH (1991)
OH 4.90E-014	cm3/molc sec	25	EST	MEYLAN,WM & HOWARD,PH (1993)

CAS #: 000098-51-1				P-(T-BUTYL)TOLUENE

Formula: $C_{11}H_{16}$

Mol Weight: 148.25

MP (deg C): -53   **FP (deg C):**

BP (deg C): 192.7

BP pressure (mm Hg):

Property/Value	Units	Temp	Data Type	Reference
WS 5.50E+000	mg/L	25	EST	AMOORE,JE & HAUTALA,E (1983)
logP 5.17			EXP	CHEM INSPECT TEST INST (1992)
VP 6.71E-001	mm Hg	25	EST	DREISBACH,RR (1955)
DC	pKa			
HL 1.54E-002	atm m3/mol	25	EST	MEYLAN,WM & HOWARD,PH (1991)
OH 7.15E-012	cm3/molc sec	25	EST	ATKINSON,R (1988)

CAS #: 000098-54-4				P-T-BUTYLPHENOL

Formula: $C_{10}H_{14}O$

Mol Weight: 150.22

MP (deg C): 98   **FP (deg C):**

BP (deg C): 237

BP pressure (mm Hg):

Property/Value	Units	Temp	Data Type	Reference
WS 5.80E+002	mg/L	25	EXP	YALKOWSKY,SH & DANNENFELSER,RM (1992)
logP 3.31	~		EXP	HANSCH,C & LEO,AJ (1985)
VP 3.81E-002	mm Hg	25	EXP	CHAO,J ET AL. (1983)
DC 10.39	pKa	25	EXP	SERJEANT,EP & DEMPSEY,B (1979)
HL 1.19E-006	atm m3/mol	25	EXP	PARSONS,GH ET AL. (1972)
OH 4.34E+001	cm3/molc sec	25	EST	MEYLAN,WM & HOWARD,PH (1993)

CAS #: 000098-55-5				ALPHA-TERPINEOL

Formula: $C_{10}H_{18}O$

Mol Weight: 154.25

MP (deg C):   **FP (deg C):**

BP (deg C): 220

BP pressure (mm Hg):

Property/Value	Units	Temp	Data Type	Reference
WS 1.98E+003	mg/L	20	EXP	SEIDELL,A (1941)
logP 2.98			EXP	LI,NY & PERDUE,EM (1995)
VP 2.30E-002	mm Hg	20	EXP	JONES,AH (1960)
DC	pKa			
HL 2.36E-006	atm m3/mol	20	EST	VP/WSOL
OH 9.49E-011	cm3/molc sec	25	EST	MEYLAN,WM & HOWARD,PH (1993)

CAS #: 000098-56-6				4-(TRIFLUOROMETHYL)CHLOROBENZENE

Formula: $C_7H_4ClF_3$

Mol Weight: 180.56

MP (deg C): -33.2   **FP (deg C):**

BP (deg C): 138.7

BP pressure (mm Hg):

Property/Value	Units	Temp	Data Type	Reference
WS 4.61E+001	mg/L	25	EST	MEYLAN,WM ET AL. (1996)
logP 3.60			EST	MEYLAN,WM & HOWARD,PH (1995)
VP 7.63E+000	mm Hg	25	EXP	DAUBERT,TE & DANNER,RP (1989)
DC	pKa			
HL 3.47E-002	atm m3/mol	25	EST	MEYLAN,WM & HOWARD,PH (1991)
OH 2.40E-013	cm3/molc sec	25	EXP	ATKINSON,R (1989)

CAS #: 000098-57-7				4-METHYLSULFONYLCHLOROBENZENE

Formula: $C_7H_7ClO_2S$

Mol Weight: 190.65

MP (deg C):   **FP (deg C):**

BP (deg C):

BP pressure (mm Hg):

Property/Value	Units	Temp	Data Type	Reference
WS 6.07E+003	mg/L	25	EST	MEYLAN,WM ET AL. (1996)
logP 1.06			EXP	HANSCH,C & LEO,AJ (1985)
VP 9.87E-004	mm Hg	25	EST	NEELY,WB & BLAU,GE (1985)
DC	pKa			
HL 9.24E-007	atm m3/mol	25	EST	MEYLAN,WM & HOWARD,PH (1991)
OH 1.35E-012	cm3/molc sec	25	EST	MEYLAN,WM & HOWARD,PH (1993)

CAS #: 000098-59-9				P-TOLUENESULFONYL CHLORIDE

Formula: $C_7H_7ClO_2S$

Mol Weight: 190.65

MP (deg C): 69-71   **FP (deg C):**

BP (deg C): 146

BP pressure (mm Hg): 1.50E+001

Property/Value	Units	Temp	Data Type	Reference
WS 5.12E+001	mg/L	25	EST	MEYLAN,WM ET AL. (1996)
logP 3.49			EST	MEYLAN,WM & HOWARD,PH (1995)
VP 5.00E-001	mm Hg	75	EXP	MONSANTO MATERIAL SAFETY DATA SHEET
DC	pKa			
HL 9.67E-006	atm m3/mol	25	EST	MEYLAN,WM & HOWARD,PH (1991)
OH 1.22E-012	cm3/molc sec	25	EST	MEYLAN,WM & HOWARD,PH (1993)

4-CHLOROBENZENESULFONAMIDE

CAS #:	000098-64-6
Formula:	$C_6H_6ClNO_2S$
Mol Weight:	191.64
MP (deg C):	145-146.5
FP (deg C):	
BP (deg C):	
BP pressure (mm Hg):	

Property/Value	Units	Temp	Data Type	Reference
WS 5.03E+003	mg/L	25	EST	MEYLAN,WM ET AL. (1996)
logP 1.15			EXP	SANGSTER,J (1994)
VP 1.59E-004	mm Hg	25	EST	NEELY,WB & BLAU,GE (1985)
DC	pKa			
HL 3.13E-007	atm m3/mol	25	EST	MEYLAN,WM & HOWARD,PH (1991)
OH 2.93E-013	cm3/molc sec	25	EST	MEYLAN,WM & HOWARD,PH (1993)

4-HYDROXYPHENYLSULFONIC ACID

CAS #:	000098-67-9
Formula:	$C_6H_6O_4S$
Mol Weight:	174.18
MP (deg C):	
FP (deg C):	
BP (deg C):	
BP pressure (mm Hg):	

Property/Value	Units	Temp	Data Type	Reference
WS 1.00E+006	mg/L	25	EST	MEYLAN,WM ET AL (1996)
logP -1.65			EST	MEYLAN,WM & HOWARD,PH (1995)
VP 3.33E-007	mm Hg	25	EST	NEELY,WB & BLAU,GE (1985)
DC	pKa			
HL 2.62E-013	atm m3/mol	25	EST	MEYLAN,WM & HOWARD,PH (1991)
OH 7.41E-012	cm3/molc sec	25	EST	MEYLAN,WM & HOWARD,PH (1993)

BENZOIC ACID, 4-(TERT-BUTYL)-

CAS #:	000098-73-7
Formula:	$C_{11}H_{14}O_2$
Mol Weight:	178.23
MP (deg C):	166.9
FP (deg C):	
BP (deg C):	
BP pressure (mm Hg):	

Property/Value	Units	Temp	Data Type	Reference
WS 2.80E+001	mg/L	25	EXP	CHEM INSPECT TEST INST (1992)
logP 3.85			EXP	HANSCH,C ET AL. (1995)
VP 6.36E-004	mm Hg	25	EST	NEELY,WB & BLAU,GE (1985)
DC 4.40	pKa	25	EXP	KORTUM,G ET AL (1961)
HL 2.80E-007	atm m3/mol	25	EST	MEYLAN,WM & HOWARD,PH (1991)
OH 2.61E-012	cm3/molc sec	25	EST	MEYLAN,WM & HOWARD,PH (1993)

CUMENE

CAS #:	000098-82-8
Formula:	C_9H_{12}
Mol Weight:	120.20
MP (deg C):	-96.033
FP (deg C):	
BP (deg C):	152.411
BP pressure (mm Hg):	

Property/Value	Units	Temp	Data Type	Reference
WS 6.13E+001	mg/L	25	EXP	SANEMASA,I ET AL. (1982)
logP 3.66			EXP	HANSCH,C & LEO,AJ (1985)
VP 4.50E+000	mm Hg	25	EXP	DAUBERT,TE & DANNER,RP (1985)
DC	pKa			
HL 1.15E-002	atm m3/mol	25	EXP	SANEMASA,I ET AL. (1982)
OH 6.50E-012	cm3/molc sec	25	EXP	ATKINSON,R (1989)

ALPHA-METHYLSTYRENE

CAS #:	000098-83-9
Formula:	C_9H_{10}
Mol Weight:	118.18
MP (deg C):	-23.2
FP (deg C):	
BP (deg C):	165.4
BP pressure (mm Hg):	

Property/Value	Units	Temp	Data Type	Reference
WS 1.16E+002	mg/L		EXP	YALKOWSKY,SH & DANNENFELSER,RM (1992)
logP 3.48			EXP	HANSCH,C ET AL. (1995)
VP 2.90E-001	mm Hg	32	EXP	BOUBLIK,T ET AL. (1973)
DC	pKa			
HL 4.33E-010	atm m3/mol	25	EST	MEYLAN,WM & HOWARD,PH (1991)
OH 5.20E-011	cm3/molc sec	25	EXP	ATKINSON,R (1989)

ALPHA-METHYLBENZYL ALCOHOL

CAS #:	000098-85-1
Formula:	$C_8H_{10}O$
Mol Weight:	122.17
MP (deg C):	20.7
FP (deg C):	
BP (deg C):	204
BP pressure (mm Hg):	

Property/Value	Units	Temp	Data Type	Reference
WS 1.95E+003	mg/L	25	EXP	SOUTHWORTH,GR & KELLER,JL (1986)
logP 1.49			EST	MEYLAN,WM & HOWARD,PH (1995)
VP 5.80E-002	mm Hg	25	EST	NEELY,WB, & BLAU,BE (1985)
DC	pKa			
HL 2.89E-007	atm m3/mol	25	EST	MEYLAN,WH & HOWARD,PH (1991)
OH 1.17E-011	cm3/molc sec	25	EST	MEYLAN,WM & HOWARD,PH (1993)

ACETOPHENONE

CAS #:	000098-86-2
Formula:	C_8H_8O
Mol Weight:	120.15
MP (deg C):	19.62
FP (deg C):	
BP (deg C):	202.0
BP pressure (mm Hg):	

Property/Value	Units	Temp	Data Type	Reference
WS 6.13E+003	mg/L	25	EXP	SOUTHWORTH,GR & KELLER,JL (1986)
logP 1.58			EXP	HANSCH,C & LEO,AJ (1985)
VP 3.97E-001	mm Hg	25	EXP	DAUBERT,TE & DANNER,RP (1985)
DC 21.55	pKa	25	EXP	RIDDICK,JA ET AL. (1986)
HL 1.07E-005	atm m3/mol	25	EXP	MACKAY,D ET AL. (1982A)
OH 2.74E-012	cm3/molc sec	25	EXP	ATKINSON,R (1989)

(DICHLOROMETHYL)BENZENE

CAS #:	000098-87-3
Formula:	$C_7H_6Cl_2$
Mol Weight:	161.03
MP (deg C):	-16.4
FP (deg C):	
BP (deg C):	205.2
BP pressure (mm Hg):	

Property/Value	Units	Temp	Data Type	Reference
WS 2.50E+002	mg/L	30	EXP	OHNISHI,R & TANABE,K (1971)
logP 2.97			EST	MEYLAN,WM & HOWARD,PH (1995)
VP 4.70E-001	mm Hg	25	EXP	PERRY,RH & GREEN,D (1984)
DC	pKa			
HL 7.38E-004	atm m3/mol	25	EST	MEYLAN,WM & HOWARD,PH (1991)
OH 2.29E-012	cm3/molc sec	25	EST	MEYLAN,WM & HOWARD,PH (1993)

CAS #: 000098-88-4			BENZOYL CHLORIDE	

Formula: C_7H_5ClO

Mol Weight: 140.57

MP (deg C): -1.0   FP (deg C):

BP (deg C): 197.2

BP pressure (mm Hg):

Property/Value	Units	Temp	Data Type	Reference
WS 4.94E+003	mg/L	25	EST	MEYLAN,WM ET AL. (1996)
logP 1.44			EST	MEYLAN,WM & HOWARD,PH (1995)
VP 7.00E-001	mm Hg	25	EXT	PERRY,RH & GREEN,D (1984)
DC	pKa			
HL 1.23E-004	atm m3/mol	25	EST	MEYLAN,WM & HOWARD,PH (1991)
OH 1.50E-012	cm3/molc sec	25	EST	MEYLAN,WM & HOWARD,PH (1993)

CAS #: 000098-89-5			CYCLOHEXANECARBOXYLIC ACID	

Formula: $C_7H_{12}O_2$

Mol Weight: 128.17

MP (deg C): 29   FP (deg C):

BP (deg C): 232.5

BP pressure (mm Hg):

Property/Value	Units	Temp	Data Type	Reference
WS 2.00E+003	mg/L	20	EXP	YALKOWSKY,SH & DANNENFELSER,RM (1992)
logP 1.96			EXP	HANSCH,C & LEO,AJ (1985)
VP 3.87E-002	mm Hg	25	EST	NEELY,WB & BLAU,GE (1985)
DC 4.90	pKa		EXP	SERJEANT,EP & DEMPSEY,B (1979)
HL 9.96E-007	atm m3/mol	25	EST	MEYLAN,WM & HOWARD,PH (1991)
OH 9.79E-012	cm3/molc sec	25	EST	MEYLAN,WM & HOWARD,PH (1993)

CAS #: 000098-92-0			NICOTINAMIDE	

Formula: $C_6H_6N_2O$

Mol Weight: 122.13

MP (deg C): 130   FP (deg C):

BP (deg C): 157

BP pressure (mm Hg): 5.00E-004

Property/Value	Units	Temp	Data Type	Reference
WS 2.04E+005	mg/L	25	EST	MEYLAN,WM ET AL. (1996)
logP -0.37			EXP	HANSCH,C & LEO,AJ (1985)
VP 4.20E-004	mm Hg	25	EST	NEELY,WB & BLAU,GE (1985)
DC 3.35	pKa	20	EXP	PERRIN,DD (1965)
HL 2.90E-012	atm m3/mol	25	EST	MEYLAN,WM & HOWARD,PH (1991)
OH 2.34E-012	cm3/molc sec	25	EST	MEYLAN,WM & HOWARD,PH (1993)

CAS #: 000098-95-3			NITROBENZENE	

Formula: $C_6H_5NO_2$

Mol Weight: 123.11

MP (deg C): 5.76   FP (deg C):

BP (deg C): 210.80

BP pressure (mm Hg):

Property/Value	Units	Temp	Data Type	Reference
WS 2.09E+003	mg/L	25	EXP	BANERJEE,S ET AL. (1980)
logP 1.85			EXP	HANSCH,C & LEO,AJ (1985)
VP 2.45E-001	mm Hg	25	EXP	DAUBERT,TE & DANNER,RP (1985)
DC	pKa			
HL 2.40E-005	atm m3/mol	25	EXP	WARNER,HP ET AL. (1987)
OH 1.40E-013	cm3/molc sec	25	EXP	ATKINSON,R (1989)

CAS #: 000098-96-4			2-PYRAZINECARBOXAMIDE	

Formula: $C_5H_5N_3O$

Mol Weight: 123.12

MP (deg C): 192   FP (deg C):

BP (deg C):

BP pressure (mm Hg):

Property/Value	Units	Temp	Data Type	Reference
WS 1.56E+004	mg/L	25	EST	MEYLAN,WM ET AL. (1996)
logP -0.60			EXP	HANSCH,C ET AL (1995)
VP 2.44E-004	mm Hg	25	EST	NEELY,WB & BLAU,GE (1985)
DC -0.50	pKa		EXP	PERRIN,DD (1965)
HL 1.20E-012	atm m3/mol	25	EST	MEYLAN,WM & HOWARD,PH (1991)
OH 2.21E-012	cm3/molc sec	25	EST	MEYLAN,WM & HOWARD,PH (1993)

CAS #: 000098-98-6			PICOLINIC ACID	

Formula: $C_6H_5NO_2$

Mol Weight: 123.11

MP (deg C): 134-136   FP (deg C):

BP (deg C):

BP pressure (mm Hg):

Property/Value	Units	Temp	Data Type	Reference
WS 2.37E+004	mg/L	25	EST	MEYLAN,WM ET AL. (1996)
logP 0.72			EXP	HANSCH,C ET AL. (1995)
VP 7.89E-003	mm Hg	25	EST	NEELY,WB & BLAU,GE (1985)
DC 5.39	pKa	25	EXP	KORTUM,G ET AL (1961)
HL 1.42E-010	atm m3/mol	25	EST	MEYLAN,WM & HOWARD,PH (1991)
OH 6.57E-013	cm3/molc sec	25	EST	MEYLAN,WM & HOWARD,PH (1993)

CAS #: 000099-02-5			M-CHLOROACETOPHENONE	

Formula: C_8H_7ClO

Mol Weight: 154.60

MP (deg C):   FP (deg C):

BP (deg C): 244

BP pressure (mm Hg):

Property/Value	Units	Temp	Data Type	Reference
WS 5.20E+002	mg/L	25	EST	MEYLAN,WM ET AL. (1996)
logP 2.51			EXP	HANSCH,C & LEO,AJ (1985)
VP 1.22E-001	mm Hg	25	EST	NEELY,WB & BLAU,GE (1985)
DC	pKa			
HL 7.27E-006	atm m3/mol	25	EST	MEYLAN,WM & HOWARD,PH (1991)
OH 6.80E-013	cm3/molc sec	25	EST	MEYLAN,WM & HOWARD,PH (1993)

CAS #: 000099-03-6			M-AMINOACETOPHENONE	

Formula: C_8H_9NO

Mol Weight: 135.17

MP (deg C): -80   FP (deg C):

BP (deg C): 194

BP pressure (mm Hg):

Property/Value	Units	Temp	Data Type	Reference
WS 1.71E+004	mg/L	25	EST	MEYLAN,WM ET AL. (1996)
logP 0.83			EXP	DUNN,WJ ET AL. (1983)
VP 9.89E-003	mm Hg	25	EST	NEELY,WB & BLAU,GE (1985)
DC 3.56	pKa	25	EXP	PERRIN,DD (1965)
HL 3.47E-009	atm m3/mol	25	EST	MEYLAN,WM & HOWARD,PH (1991)
OH 4.55E-011	cm3/molc sec	25	EST	MEYLAN,WM & HOWARD,PH (1993)

CAS #: 000099-04-7				M-TOLUIC ACID

Formula: $C_8H_8O_2$

Mol Weight: 136.15

MP (deg C): 108.7   FP (deg C):

BP (deg C):

BP pressure (mm Hg):

Property/Value	Units	Temp	Data Type	Reference
WS 9.80E+002	mg/L	25	EXP	YALKOWSKY,SH & DANNENFELSER,RM (1992)
logP 2.37			EXP	HANSCH,C & LEO,AJ (1985)
VP 2.36E-004	mm Hg	25	EXT	COLOMINA,M ET AL. (1986)
DC 4.27	pKa	25	EXP	KORTUM,G ET AL (1961)
HL 1.20E-007	atm m3/mol	25	EST	MEYLAN,WM & HOWARD,PH (1991)
OH 2.04E-012	cm3/molc sec	25	EST	MEYLAN,WM & HOWARD,PH (1993)

CAS #: 000099-05-8				3-AMINOBENZOIC ACID

Formula: $C_7H_7NO_2$

Mol Weight: 137.14

MP (deg C): 174   FP (deg C):

BP (deg C):

BP pressure (mm Hg):

Property/Value	Units	Temp	Data Type	Reference
WS 5.90E+003	mg/L		EXP	YALKOWSKY,SH & DANNENFELSER,RM (1992)
logP 0.65			EXP	SANGSTER,J (1994)
VP 2.78E-004	mm Hg	25	EST	NEELY,WB & BLAU,GE (1985)
DC 3.07	pKa	25	EXP	KORTUM,G ET AL (1961)
HL 3.83E-011	atm m3/mol	25	EST	MEYLAN,WM & HOWARD,PH (1991)
OH 2.99E-011	cm3/molc sec	25	EST	MEYLAN,WM & HOWARD,PH (1993)

CAS #: 000099-06-9				M-HYDROXYBENZOIC ACID

Formula: $C_7H_6O_3$

Mol Weight: 138.12

MP (deg C): 201-203   FP (deg C):

BP (deg C):

BP pressure (mm Hg):

Property/Value	Units	Temp	Data Type	Reference
WS 7.25E+003	mg/L		EXP	YALKOWSKY,SH & DANNENFELSER,RM (1992)
logP 1.50			EXP	HANSCH,C & LEO,AJ (1985)
VP 1.58E-004	mm Hg	25	EST	NEELY,WB & BLAU,GE (1985)
DC 4.30	pKa	258	EXP	KORTUM,G ET AL (1961)
HL 1.13E-011	atm m3/mol	25	EST	MEYLAN,WM & HOWARD,PH (1991)
OH 9.75E-012	cm3/molc sec	25	EST	MEYLAN,WM & HOWARD,PH (1993)

CAS #: 000099-07-0				PHENOL, 3-(DIMETHYLAMINO)-

Formula: $C_8H_{11}NO$

Mol Weight: 137.18

MP (deg C): 86   FP (deg C):

BP (deg C): 266.5

BP pressure (mm Hg):

Property/Value	Units	Temp	Data Type	Reference
WS 4.00E+003	mg/L	25	EST	MEYLAN,WM ET AL. (1996)
logP 1.56			EXP	HANSCH,C & LEO,AJ (1985)
VP 2.63E-002	mm Hg	25	EST	NEELY,WB & BLAU,GE (1985)
DC	pKa			
HL 8.91E-009	atm m3/mol	25	EST	MEYLAN,WM & HOWARD,PH (1991)
OH 2.03E-010	cm3/molc sec	25	EST	MEYLAN,WM & HOWARD,PH (1993)

CAS #: 000099-08-1				3-NITROTOLUENE

Formula: $C_7H_7NO_2$

Mol Weight: 137.14

MP (deg C): 15.5   FP (deg C):

BP (deg C): 232

BP pressure (mm Hg):

Property/Value	Units	Temp	Data Type	Reference
WS 5.00E+002	mg/L	30	EXP	YALKOWSKY,SH & DANNENFELSER,RM (1992)
logP 2.45			EXP	HANSCH,C & LEO,AJ (1985)
VP 2.05E-001	mm Hg	25	EXT	PERRY,RH & GREEN,D (1984)
DC	pKa			
HL 2.35E-005	atm m3/mol	25	EST	MEYLAN,WM & HOWARD,PH (1991)
OH 9.50E-013	cm3/molc sec	25	EXP	ATKINSON,R (1989)

CAS #: 000099-09-2				3-NITROANILINE

Formula: $C_6H_6N_2O_2$

Mol Weight: 138.13

MP (deg C): 114   FP (deg C):

BP (deg C): 305-307

BP pressure (mm Hg):

Property/Value	Units	Temp	Data Type	Reference
WS 1.21E+003	mg/L	24	EXP	SEIDELL,A (1941)
logP 1.37			EXP	HANSCH,C & LEO,AJ (1985)
VP 9.56E-005	mm Hg	25	EXP	MALASPINA,L ET AL. (1973B)
DC 2.47	pKa	25	EXP	WEAST,RC ET AL. (1985)
HL 1.44E-007	atm m3/mol	25	EST	VP/WSOL
OH 3.00E-011	cm3/molc sec	25	EST	MEYLAN,WM & HOWARD,PH (1993)

CAS #: 000099-10-5				3,5-DIHYDROXYBENZOIC ACID

Formula: $C_7H_6O_4$

Mol Weight: 154.12

MP (deg C): 236-238 de   FP (deg C):

BP (deg C):

BP pressure (mm Hg):

Property/Value	Units	Temp	Data Type	Reference
WS 5.10E+004	mg/L	25	EST	MEYLAN,WM ET AL. (1996)
logP 0.86			EXP	HANSCH,C & LEO,AJ (1985)
VP 2.10E-006	mm Hg	25	EST	NEELY,WB & BLAU,GE (1985)
DC 4.04	pKa	25	EXP	SERJEANT,EP & DEMPSEY,B (1979)
HL 1.17E-015	atm m3/mol	25	EST	MEYLAN,WM & HOWARD,PH (1991)
OH 1.56E-010	cm3/molc sec	25	EST	MEYLAN,WM & HOWARD,PH (1993)

CAS #: 000099-24-1				METHYL GALLATE

Formula: $C_8H_8O_5$

Mol Weight: 184.15

MP (deg C): 258-265   FP (deg C):

BP (deg C):

BP pressure (mm Hg):

Property/Value	Units	Temp	Data Type	Reference
WS 1.06E+004	mg/L		EXP	YALKOWSKY,SH & DANNENFELSER,RM (1992)
logP 0.86			EXP	HANSCH,C & LEO,AJ (1985)
VP 1.78E-006	mm Hg	25	EST	NEELY,WB & BLAU,GE (1985)
DC	pKa			
HL 3.91E-017	atm m3/mol	25	EST	MEYLAN,WM & HOWARD,PH (1991)
OH 8.74E-011	cm3/molc sec	25	EST	MEYLAN,WM & HOWARD,PH (1993)

000099-28-5 — 2,6-DIBROMO-4-NITROPHENOL

CAS #: 000099-28-5

Formula: $C_6H_3Br_2NO_3$

Mol Weight: 296.91

MP (deg C): 145 dec

FP (deg C):

BP (deg C):

BP pressure (mm Hg):

Property/Value	Units	Temp	Data Type	Reference
WS 4.35E+001	mg/L	25	EST	MEYLAN,WM ET AL. (1996)
logP 3.57			EXP	HANSCH,C ET AL. (1995)
VP 6.41E-006	mm Hg	25	EST	NEELY,WB & BLAU,GE (1985)
DC 3.39	pKa	25	EXP	SERJEANT,EP & DEMPSEY,B (1979)
HL 3.51E-010	atm m3/mol	25	EST	MEYLAN,WM & HOWARD,PH (1991)
OH 1.86E-013	cm3/molc sec	25	EST	MEYLAN,WM & HOWARD,PH (1993)

000099-30-9 — 2,6-DICHLORO-4-NITROANILINE

CAS #: 000099-30-9

Formula: $C_6H_4Cl_2N_2O_2$

Mol Weight: 207.02

MP (deg C): 191

FP (deg C):

BP (deg C):

BP pressure (mm Hg):

Property/Value	Units	Temp	Data Type	Reference
WS 7.00E+000	mg/L	20	EXP	WORTHING,CR & WALKER,SB (1987)
logP 2.76			EST	MEYLAN,WM & HOWARD,PH (1995)
VP 1.20E-006	mm Hg	20	EXP	WORTHING,CR & WALKER,SB (1987)
DC -2.55	pKa	20	EXP	PERRIN,DD (1965)
HL 4.67E-008	atm m3/mol	20	EST	VP/WSOL
OH 2.01E-011	cm3/molc sec	25	EST	MEYLAN,WM & HOWARD,PH (1993)

000099-32-1 — 4-OXO-4H-PYRAN-2,6-DICARBOXYLIC ACID

CAS #: 000099-32-1

Formula: $C_7H_4O_6$

Mol Weight: 184.11

MP (deg C): 262

FP (deg C):

BP (deg C):

BP pressure (mm Hg):

Property/Value	Units	Temp	Data Type	Reference
WS 1.43E+004	mg/L	25	EXP	YALKOWSKY,SH & DANNENFELSER,RM (1992)
logP 0.57			EST	MEYLAN,WM & HOWARD,PH (1995)
VP 9.80E-007	mm Hg	25	EST	NEELY,WB & BLAU,GE (1985)
DC	pKa			
HL 3.70E-015	atm m3/mol	25	EST	MEYLAN,WM & HOWARD,PH (1991)
OH 2.23E-011	cm3/molc sec	25	EST	MEYLAN,WM & HOWARD,PH (1993)

000099-34-3 — 3,5-DINITROBENZOIC ACID

CAS #: 000099-34-3

Formula: $C_7H_4N_2O_6$

Mol Weight: 212.12

MP (deg C): 205-207

FP (deg C):

BP (deg C):

BP pressure (mm Hg):

Property/Value	Units	Temp	Data Type	Reference
WS 1.35E+003	mg/L	25	EXP	YALKOWSKY,SH & DANNENFELSER,RM (1992)
logP 1.55			EXP	HANSCH,C & LEO,AJ (1985)
VP 1.41E-006	mm Hg	25	EST	NEELY,WB & BLAU,GE (1985)
DC 2.82	pKa	25	EXP	KORTUM,G ET AL (1961)
HL 1.69E-012	atm m3/mol	25	EST	MEYLAN,WM & HOWARD,PH (1991)
OH 5.24E-013	cm3/molc sec	25	EST	MEYLAN,WM & HOWARD,PH (1993)

000099-35-4 — 1,3,5-TRINITROBENZENE

CAS #: 000099-35-4

Formula: $C_6H_3N_3O_6$

Mol Weight: 213.11

MP (deg C): 122.5

FP (deg C):

BP (deg C):

BP pressure (mm Hg):

Property/Value	Units	Temp	Data Type	Reference
WS 2.78E+002	mg/L	15	EXP	YALKOWSKY,SH & DANNENFELSER,RM (1992)
logP 1.18			EXP	HANSCH,C & LEO,AJ (1985)
VP 6.44E-006	mm Hg	25	EXT	OHE,S (1976)
DC	pKa			
HL 3.31E-010	atm m3/mol	25	EST	MEYLAN,WM & HOWARD,PH (1991)
OH 1.30E-015	cm3/molc sec	25	EST	MEYLAN,WM & HOWARD,PH (1993)

000099-36-5 — BENZOIC ACID, 3-METHYL-, METHYL ESTER

CAS #: 000099-36-5

Formula: $C_9H_{10}O_2$

Mol Weight: 150.18

MP (deg C):

FP (deg C):

BP (deg C): 221

BP pressure (mm Hg):

Property/Value	Units	Temp	Data Type	Reference
WS 3.26E+002	mg/L	25	EST	MEYLAN,WM ET AL. (1996)
logP 2.77			EXP	SOTOMATSU,T ET AL. (1993)
VP 1.64E-001	mm Hg	25	EST	NEELY,WB & BLAU,GE (1985)
DC	pKa			
HL 3.83E-005	atm m3/mol	25	EST	MEYLAN,WM & HOWARD,PH (1991)
OH 1.48E-012	cm3/molc sec	25	EST	MEYLAN,WM & HOWARD,PH (1993)

000099-47-8 — ALPHA-CHLORO-M-NITROACETOPHENONE

CAS #: 000099-47-8

Formula: $C_8H_6ClNO_3$

Mol Weight: 199.59

MP (deg C):

FP (deg C):

BP (deg C):

BP pressure (mm Hg):

Property/Value	Units	Temp	Data Type	Reference
WS 5.79E+002	mg/L	25	EST	MEYLAN,WM ET AL. (1996)
logP 1.75			EST	MEYLAN,WM & HOWARD,PH (1995)
VP 8.70E-006	mm Hg	25	EXP	HAMAKER,JW & KERLINGER,HO (1969)
DC	pKa			
HL 1.36E-008	atm m3/mol	25	EST	MEYLAN,WM & HOWARD,PH (1991)
OH 4.88E-013	cm3/molc sec	25	EST	MEYLAN,WM & HOWARD,PH (1993)

000099-49-0 — CARVONE

CAS #: 000099-49-0

Formula: $C_{10}H_{14}O$

Mol Weight: 150.22

MP (deg C):

FP (deg C): 88

BP (deg C): 231

BP pressure (mm Hg): 1.18E+003

Property/Value	Units	Temp	Data Type	Reference
WS 1.30E+003	mg/L	25	EXP	YALKOWSKY,SH & DANNENFELSER,RM (1992)
logP 3.07			EST	MEYLAN,WM & HOWARD,PH (1995)
VP 1.60E-001	mm Hg	25	EST	MEYLAN,WM & BLAN,GE (1985)
DC	pKa			
HL 7.73E-005	atm m3/mol	25	EST	MEYLAN,WM & HOWARD,PH (1991)
OH 1.41E-010	cm3/molc sec	25	EST	MEYLAN,WM & HOWARD,PH (1993)

CAS #:	000099-50-3				3,4-DIHYDROXYBENZOIC ACID
Formula:	$C_7H_6O_4$				
Mol Weight:	154.12				
MP (deg C): 200		FP (deg C):			
BP (deg C):					
BP pressure (mm Hg):					

Property/ Value	Units	Temp	Data Type	Reference
WS 1.82E+004	mg/L	14	EXP	YALKOWSKY,SH & DANNENFELSER,RM (1992)
logP 0.86			EXP	HANSCH,C & LEO,AJ (1985)
VP 2.10E-006	mm Hg	25	EST	NEELY,WB & BLAU,GE (1985)
DC 4.26	pKa		EXP	SERJEANT,EP & DEMPSEY,B (1979)
HL 1.17E-015	atm m3/mol	25	EST	MEYLAN,WM & HOWARD,PH (1991)
OH 9.30E-012	cm3/molc sec	25	EST	MEYLAN,WM & HOWARD,PH (1993)

CAS #:	000099-51-4				1,2-DIMETHYL-4-NITROBENZENE
Formula:	$C_8H_9NO_2$				
Mol Weight:	151.17				
MP (deg C): 30.5		FP (deg C):			
BP (deg C): 251					
BP pressure (mm Hg):					

Property/ Value	Units	Temp	Data Type	Reference
WS 1.00E+002	mg/L	25	EST	MEYLAN,WM ET AL. (1996)
logP 2.91			EXP	DENEER,JW ET AL. (1987)
VP 3.59E-003	mm Hg	25	EXT	BOUBLIK,T ET AL. (1984)
DC	pKa			
HL 2.59E-005	atm m3/mol	25	EST	MEYLAN,WM & HOWARD,PH (1991)
OH 1.05E-012	cm3/molc sec	25	EST	MEYLAN,WM & HOWARD,PH (1993)

CAS #:	000099-52-5				4-NITRO-2-TOLUIDINE
Formula:	$C_7H_8N_2O_2$				
Mol Weight:	152.15				
MP (deg C): 133.5		FP (deg C):			
BP (deg C):					
BP pressure (mm Hg):					

Property/ Value	Units	Temp	Data Type	Reference
WS 1.40E+003	mg/L	25	EST	MEYLAN,WM ET AL. (1996)
logP 2.02			EST	MEYLAN,WM & HOWARD,PH (1995)
VP 9.72E-004	mm Hg	25	EST	NEELY,WB & BLAU,GE (1985)
DC 1.04	pKa	25	EXP	PERRIN,DD (1965)
HL 8.29E-009	atm m3/mol	25	EST	MEYLAN,WM & HOWARD,PH (1991)
OH 1.66E-011	cm3/molc sec	25	EST	MEYLAN,WM & HOWARD,PH (1993)

CAS #:	000099-54-7				3,4-DICHLORONITROBENZENE
Formula:	$C_6H_3Cl_2NO_2$				
Mol Weight:	192.00				
MP (deg C): 43		FP (deg C):			
BP (deg C): 255.5					
BP pressure (mm Hg):					

Property/ Value	Units	Temp	Data Type	Reference
WS 1.21E+002	mg/L	20	EXP	YALKOWSKY,SH & DANNENFELSER,RM (1992)
logP 3.12			EXP	HANSCH,C & LEO,AJ (1985)
VP 1.03E-002	mm Hg	25	EXT	DAUBERT,TE & DANNER,RP (1991)
DC	pKa			
HL 1.17E-005	atm m3/mol	25	EST	MEYLAN,WM & HOWARD,PH (1991)
OH 5.01E-014	cm3/molc sec	25	EST	MEYLAN,WM & HOWARD,PH (1993)

CAS #:	000099-55-8				5-NITRO-2-TOLUIDINE
Formula:	$C_7H_8N_2O_2$				
Mol Weight:	152.15				
MP (deg C): 104-107		FP (deg C):			
BP (deg C):					
BP pressure (mm Hg):					

Property/ Value	Units	Temp	Data Type	Reference
WS 1.88E+003	mg/L	25	EST	MEYLAN,WM ET AL. (1996)
logP 1.87			EXP	HANSCH,C ET AL. (1995)
VP 9.75E-004	mm Hg	25	EST	NEELY,WB & BLAU,GE (1985)
DC 2.35	pKa	25	EXP	PERRIN,DD (1972)
HL 1.77E-008	atm m3/mol	25	EST	MEYLAN,WM & HOWARD,PH (1991)
OH 3.24E-011	cm3/molc sec	25	EST	MEYLAN,WM & HOWARD,PH (1993)

CAS #:	000099-56-9				2-AMINO-4-NITROANILINE
Formula:	$C_6H_7N_3O_2$				
Mol Weight:	153.14				
MP (deg C): 199-201		FP (deg C):			
BP (deg C):					
BP pressure (mm Hg):					

Property/ Value	Units	Temp	Data Type	Reference
WS 1.30E+004	mg/L	25	EST	MEYLAN,WM ET AL. (1996)
logP 0.88			EXP	HANSCH,C ET AL. (1995)
VP 1.09E-004	mm Hg	25	EST	NEELY,WB & BLAU,GE (1985)
DC 2.61	pKa	25	EXP	PERRIN,DD (1972)
HL 7.39E-012	atm m3/mol	25	EST	MEYLAN,WM & HOWARD,PH (1991)
OH 6.35E-011	cm3/molc sec	25	EST	MEYLAN,WM & HOWARD,PH (1993)

CAS #:	000099-57-0				2-AMINO-4-NITROPHENOL
Formula:	$C_6H_6N_2O_3$				
Mol Weight:	154.13				
MP (deg C): 143-145		FP (deg C):			
BP (deg C):					
BP pressure (mm Hg):					

Property/ Value	Units	Temp	Data Type	Reference
WS 6.11E+003	mg/L	25	EST	MEYLAN,WM ET AL. (1996)
logP 1.26			EXP	SANGSTER,J (1994)
VP 3.52E-005	mm Hg	25	EST	NEELY,WB & BLAU,GE (1985)
DC 3.10	pKa	25	EXP	STONE,AT ET AL. (1993); PKA1; PKA2=7.6
HL 2.23E-012	atm m3/mol	25	EST	MEYLAN,WM & HOWARD,PH (1991)
OH 7.24E-012	cm3/molc sec	25	EST	MEYLAN,WM & HOWARD,PH (1993)

CAS #:	000099-59-2				2-METHOXY-5-NITROANILINE
Formula:	$C_7H_8N_2O_3$				
Mol Weight:	168.15				
MP (deg C): 117-119		FP (deg C):			
BP (deg C):					
BP pressure (mm Hg):					

Property/ Value	Units	Temp	Data Type	Reference
WS 3.49E+003	mg/L	25	EST	MEYLAN,WM ET AL. (1996)
logP 1.47			EXP	HANSCH,C ET AL. (1995)
VP 3.19E-004	mm Hg	25	EST	NEELY,WB & BLAU,GE (1985)
DC 2.49	pKa	25	EXP	PERRIN,DD (1972)
HL 1.25E-008	atm m3/mol	25	EST	HINE,J & MOOKERJEE,PK (1975)
OH 3.32E-011	cm3/molc sec	25	EST	ATKINSON,R (1988)

CAS #: 000099-60-5	BENZOIC ACID, 2-CHLORO-4-NITRO-

Formula: $C_7H_4ClNO_4$
Mol Weight: 201.57
MP (deg C): 139-141 FP (deg C):
BP (deg C):
BP pressure (mm Hg):

Property/Value	Units	Temp	Data Type	Reference
WS 3.23E+002	mg/L	25	EST	MEYLAN,WM ET AL. (1996)
logP 2.03			EXP	HANSCH,C & LEO,AJ (1985)
VP 1.89E-005	mm Hg	25	EST	NEELY,WB & BLAU,GE (1985)
DC 2.14	pKa		EXP	KORTUM,G ET AL (1961)
HL 3.17E-010	atm m3/mol	25	EST	MEYLAN,WM & HOWARD,PH (1991)
OH 5.64E-013	cm3/molc sec	25	EST	MEYLAN,WM & HOWARD,PH (1993)

CAS #: 000099-61-6	M-NITROBENZALDEHYDE

Formula: $C_7H_5NO_3$
Mol Weight: 151.12
MP (deg C): 58.5 FP (deg C):
BP (deg C): 164
BP pressure (mm Hg): 2.30E+001

Property/Value	Units	Temp	Data Type	Reference
WS 1.00E+001	mg/L	25	EXP	YALKOWSKY,SH & DANNENFELSER,RM (1992)
logP 1.47			EXP	HANSCH,C & LEO,AJ (1985)
VP 7.25E-003	mm Hg	25	EXT	PERRY,RH & GREEN,D (1984)
DC	pKa			
HL 5.30E-008	atm m3/mol	25	EST	MEYLAN,WM & HOWARD,PH (1991)
OH 1.70E-011	cm3/molc sec	25	EST	MEYLAN,WM & HOWARD,PH (1993)

CAS #: 000099-62-7	1,3-DIISOPROPYLBENZENE

Formula: $C_{12}H_{18}$
Mol Weight: 162.28
MP (deg C): -63.1 FP (deg C):
BP (deg C): 203.2
BP pressure (mm Hg):

Property/Value	Units	Temp	Data Type	Reference
WS 4.33E+000	mg/L	25	EST	MEYLAN,WM ET AL. (1996)
logP 4.90			EST	MEYLAN,WM & HOWARD,PH (1995)
VP 3.93E-001	mm Hg	25	EXP	DAUBERT,TE & DANNER,RP (1989)
DC	pKa			
HL 2.04E-002	atm m3/mol	25	EST	MEYLAN,WM & HOWARD,PH (1991)
OH 1.55E-011	cm3/molc sec	25	EST	MEYLAN,WM & HOWARD,PH (1993)

CAS #: 000099-65-0	1,3-DINITROBENZENE

Formula: $C_6H_4N_2O_4$
Mol Weight: 168.11
MP (deg C): 89-90 FP (deg C):
BP (deg C): 300-303
BP pressure (mm Hg):

Property/Value	Units	Temp	Data Type	Reference
WS 5.33E+002	mg/L	25	EXP	SPANGGORD,RJ ET AL. (1980)
logP 1.49			EXP	HANSCH,C & LEO,AJ (1985)
VP 9.00E-004	mm Hg	25	EXT	MAKSIMOV,YY (1968)
DC	pKa			
HL 3.74E-007	atm m3/mol	25	EST	VP/WSOL
OH 3.09E-014	cm3/molc sec	25	EST	MEYLAN,WM & HOWARD,PH (1993)

CAS #: 000099-66-1	VALPROIC ACID

Formula: $C_8H_{16}O_2$
Mol Weight: 144.22
MP (deg C): FP (deg C):
BP (deg C): 128-130
BP pressure (mm Hg): 2.00E+001

Property/Value	Units	Temp	Data Type	Reference
WS 2.00E+003	mg/L	20	EXP	RIEMENSCHNEIDER,W (1986)
logP 2.75			EXP	SANGSTER,J (1993)
VP 4.58E-002	mm Hg	25	EST	NEELY,WB & BLAU,GE (1985)
DC	pKa			
HL 3.00E-006	atm m3/mol	25	EST	MEYLAN,WM & HOWARD,PH (1991)
OH 8.18E-012	cm3/molc sec	25	EST	MEYLAN,WM & HOWARD,PH (1993)

CAS #: 000099-71-8	P-(SEC-BUTYL)PHENOL

Formula: $C_{10}H_{14}O$
Mol Weight: 150.22
MP (deg C): 61.5 FP (deg C):
BP (deg C): 241
BP pressure (mm Hg):

Property/Value	Units	Temp	Data Type	Reference
WS 9.60E+002	mg/L	25	EXP	YALKOWSKY,SH & DANNENFELSER,RM (1992)
logP 3.08			EXP	HANSCH,C ET AL. (1995)
VP 3.72E-002	mm Hg	25	EXT	PERRY,RH & GREEN,D (1984)
DC	pKa			
HL 1.45E-006	atm m3/mol	25	EST	MEYLAN,WM & HOWARD,PH (1991)
OH 4.41E-011	cm3/molc sec	25	EST	MEYLAN,WM & HOWARD,PH (1993)

CAS #: 000099-75-2	BENZOIC ACID, 4-METHYL-, METHYL ESTER

Formula: $C_9H_{10}O_2$
Mol Weight: 150.18
MP (deg C): 33.2 FP (deg C):
BP (deg C): 220
BP pressure (mm Hg):

Property/Value	Units	Temp	Data Type	Reference
WS 3.75E+002	mg/L	25	EST	MEYLAN,WM ET AL. (1996)
logP 2.70			EXP	SOTOMATSU,T ET AL. (1993)
VP 1.64E-001	mm Hg	25	EST	NEELY,WB & BLAU,GE (1985)
DC	pKa			
HL 3.83E-005	atm m3/mol	25	EST	MEYLAN,WM & HOWARD,PH (1991)
OH 1.99E-012	cm3/molc sec	25	EST	MEYLAN,WM & HOWARD,PH (1993)

CAS #: 000099-76-3	4-HYDROXY METHYL BENZOATE

Formula: $C_8H_8O_3$
Mol Weight: 152.15
MP (deg C): 131 FP (deg C):
BP (deg C): 270-280
BP pressure (mm Hg):

Property/Value	Units	Temp	Data Type	Reference
WS 2.20E+003	mg/L	25	EXP	SUZUKI,T (1991)
logP 1.96			EXP	HANSCH,C & LEO,AJ (1985)
VP 7.00E-002	mm Hg	25	EXP	US EPA (1989A)
DC	pKa			
HL 6.37E-006	atm m3/mol	25	EST	VP/WSOL
OH 1.11E-011	cm3/molc sec	25	EST	MEYLAN,WM & HOWARD,PH (1993)

CAS #:	000099-77-4				ETHYL-P-NITROBENZOATE

Formula: $C_9H_9NO_4$

Mol Weight: 195.18

MP (deg C): 57 **FP (deg C):**

BP (deg C): 186.3

BP pressure (mm Hg):

Property/ Value	Units	Temp	Data Type	Reference
WS 1.93E+002	mg/L	25	EST	MEYLAN,WM ET AL. (1996)
logP 2.33			EXP	HANSCH,C & LEO,AJ (1985)
VP 6.65E-004	mm Hg	25	EST	NEELY,WB & BLAU,GE (1985)
DC	pKa			
HL 1.82E-007	atm m3/mol	25	EST	MEYLAN,WM & HOWARD,PH (1991)
OH 1.72E-012	cm3/molc sec	25	EST	MEYLAN,WM & HOWARD,PH (1993)

CAS #:	000099-82-1				P-MENTHANE

Formula: $C_{10}H_{20}$

Mol Weight: 140.27

MP (deg C): <-30 **FP (deg C):**

BP (deg C): 167-168.5

BP pressure (mm Hg):

Property/ Value	Units	Temp	Data Type	Reference
WS 2.80E-001	mg/L	25	EXP	CHEM INSPECT TEST INST (1992)
logP 5.56			EXP	CHEM INSPECT TEST INST (1992)
VP 2.69E+000	mm Hg	25	EXP	PERRY,RH & GREEN,D (1984)
DC	pKa			
HL 1.76E+000	atm m3/mol	25	EST	VP/WSOL
OH 1.51E-011	cm3/molc sec	25	EST	MEYLAN,WM & HOWARD,PH (1993)

CAS #:	000099-83-2				à-PHELLANDRENE

Formula: $C_{10}H_{16}$

Mol Weight: 136.24

MP (deg C): **FP (deg C):**

BP (deg C):

BP pressure (mm Hg):

Property/ Value	Units	Temp	Data Type	Reference
WS 2.86E+000	mg/L	25	EST	MEYLAN,WM ET AL. (1996)
logP 4.62			EST	MEYLAN,WM & HOWARD,PH (1995)
VP 1.40E+000	mm Hg	25	EXP	PERRY,RH & GREEN,D (1984)
DC	pKa			
HL 3.09E-001	atm m3/mol	25	EST	MEYLAN,WM & HOWARD,PH (1991)
OH 3.13E-010	cm3/molc sec	25	EXP	ATKINSON,R (1989)

CAS #:	000099-85-4				GAMMA-TERPINENE

Formula: $C_{10}H_{16}$

Mol Weight: 136.24

MP (deg C): -10 **FP (deg C):**

BP (deg C): 183

BP pressure (mm Hg):

Property/ Value	Units	Temp	Data Type	Reference
WS 7.64E+002	mg/L	20	EXP	LI,J & PERDUE,EM (1995)
logP 4.50			EXP	LI,J & PERDUE,EM (1995)
VP 1.09E+000	mm Hg	25	EXP	DAUBERT,TE & DANNER,RP (1989)
DC	pKa			
HL 2.56E-004	atm m3/mol	25	EST	VP/WSOL
OH 1.77E-010	cm3/molc sec	25	EXP	ATKINSON,R (1989)

CAS #:	000099-87-6				P-CYMENE

Formula: $C_{10}H_{14}$

Mol Weight: 134.22

MP (deg C): -67.935 **FP (deg C):**

BP (deg C): 177.10

BP pressure (mm Hg):

Property/ Value	Units	Temp	Data Type	Reference
WS 2.34E+001	mg/L	25	EXP	BANERJEE,S ET AL (1980)
logP 4.10			EXP	HANSCH,C & LEO,AJ (1985)
VP 1.46E+000	mm Hg	25	EXP	DAUBERT,TE & DANNER,RP (1989)
DC	pKa			
HL 1.10E-002	atm m3/mol	25	EST	VP/WSOL
OH 1.51E-011	cm3/molc sec	22	EXP	CORCHNOY,SB & ATKINSON,R (1990)

CAS #:	000099-88-7				P-ISOPROPYLANILINE

Formula: $C_9H_{13}N$

Mol Weight: 135.21

MP (deg C): **FP (deg C):**

BP (deg C): 225

BP pressure (mm Hg):

Property/ Value	Units	Temp	Data Type	Reference
WS 6.55E+002	mg/L	25	EST	MEYLAN,WM ET AL. (1996)
logP 2.49			EXP	HANSCH,C & LEO,AJ (1985)
VP 8.35E-002	mm Hg	25	EXT	CHAO,J ET AL. (1983)
DC 4.85	pKa	20	EXP	PERRIN,DD (1972)
HL 3.70E-006	atm m3/mol	25	EST	MEYLAN,WM & HOWARD,PH (1991)
OH 1.32E-010	cm3/molc sec	25	EST	MEYLAN,WM & HOWARD,PH (1993)

CAS #:	000099-89-8				P-ISOPROPYLPHENOL

Formula: $C_9H_{12}O$

Mol Weight: 136.20

MP (deg C): 62.3 **FP (deg C):**

BP (deg C): 230

BP pressure (mm Hg):

Property/ Value	Units	Temp	Data Type	Reference
WS 1.10E+003	mg/L	25	EST	MEYLAN,WM ET AL. (1996)
logP 2.90			EXP	HANSCH,C ET AL. (1995)
VP 4.52E-002	mm Hg	25	EXT	CHAO,J ET AL. (1983)
DC 10.24	pKa	25	EXP	SERJEANT,EP & DEMPSEY,B (1979)
HL 1.09E-006	atm m3/mol	25	EST	MEYLAN,WM & HOWARD,PH (1991)
OH 4.25E-011	cm3/molc sec	25	EST	MEYLAN,WM & HOWARD,PH (1993)

CAS #:	000099-90-1				P-BROMOACETOPHENONE

Formula: C_8H_7BrO

Mol Weight: 199.05

MP (deg C): 54 **FP (deg C):**

BP (deg C): 255.5

BP pressure (mm Hg): 7.36E+002

Property/ Value	Units	Temp	Data Type	Reference
WS 3.72E+002	mg/L	25	EST	MEYLAN,WM ET AL. (1996)
logP 2.43			EXP	HANSCH,C & LEO,AJ (1985)
VP 3.08E-002	mm Hg	25	EST	NEELY,WB & BLAU,GE (1985)
DC	pKa			
HL 3.91E-006	atm m3/mol	25	EST	MEYLAN,WM & HOWARD,PH (1991)
OH 1.22E-012	cm3/molc sec	25	EST	MEYLAN,WM & HOWARD,PH (1993)

144

000099-91-2 — 1-(4-CHLOROPHENYL)ETHANONE

Formula: C_8H_7ClO
Mol Weight: 154.60
MP (deg C): 20-21
BP (deg C): 237
BP pressure (mm Hg):

Property/Value	Units	Temp	Data Type	Reference
WS 1.11E+002	mg/L	25	EST	MEYLAN,WM ET AL. (1996)
logP 2.32			EXP	HANSCH,C & LEO,AJ (1985)
VP 1.22E-001	mm Hg	25	EST	NEELY,WB & BLAU,GE (1985)
DC	pKa			
HL 9.30E-004	atm m3/mol	25	EST	MEYLAN,WM & HOWARD,PH (1991)
OH 1.16E-012	cm3/molc sec	25	EST	ATKINSON,R (1988)

000099-92-3 — 4-AMINOACETOPHENONE

Formula: C_8H_9NO
Mol Weight: 135.17
MP (deg C): 106
BP (deg C): 294
BP pressure (mm Hg):

Property/Value	Units	Temp	Data Type	Reference
WS 1.71E+004	mg/L	25	EST	MEYLAN,WM ET AL. (1996)
logP 0.83			EXP	HANSCH,C & LEO,AJ (1985)
VP 9.89E-003	mm Hg	25	EST	NEELY,WB & BLAU,GE (1985)
DC 2.75	pKa	25	EXP	PERRIN,DD (1965)
HL 3.47E-009	atm m3/mol	25	EST	MEYLAN,WM & HOWARD,PH (1991)
OH 1.05E-010	cm3/molc sec	25	EST	ATKINSON,R (1988)

000099-93-4 — P-HYDROXYACETOPHENONE

Formula: $C_8H_8O_2$
Mol Weight: 136.15
MP (deg C): 109.5
BP (deg C): 147-148
BP pressure (mm Hg): 3.00E+000

Property/Value	Units	Temp	Data Type	Reference
WS 9.90E+003	mg/L	22	EXP	YALKOWSKY,SH & DANNENFELSER,RM (1992)
logP 1.35			EXP	HANSCH,C & LEO,AJ (1985)
VP 7.48E-003	mm Hg	25	EST	NEELY,WB & BLAU,GE (1985)
DC 8.05	pKa	25	EXP	KORTUM,G ET AL (1961)
HL 1.02E-009	atm m3/mol	25	EST	MEYLAN,WM & HOWARD,PH (1991)
OH 3.06E-011	cm3/molc sec	25	EST	MEYLAN,WM & HOWARD,PH (1993)

000099-94-5 — P-TOLUIC ACID

Formula: $C_8H_8O_2$
Mol Weight: 136.15
MP (deg C): 179.6
BP (deg C):
BP pressure (mm Hg):

Property/Value	Units	Temp	Data Type	Reference
WS 4.00E+002	mg/L	25	EXP	YALKOWSKY,SH & DANNENFELSER,RM (1992)
logP 2.27			EXP	HANSCH,C & LEO,AJ (1985)
VP 5.08E-005	mm Hg	25	EXT	COLOMINA,M ET AL. (1986)
DC 4.37	pKa	25	EXP	KORTUM,G ET AL (1961)
HL 1.20E-007	atm m3/mol	25	EST	MEYLAN,WM & HOWARD,PH (1991)
OH 2.54E-012	cm3/molc sec	25	EST	MEYLAN,WM & HOWARD,PH (1993)

000099-96-7 — P-HYDROXYBENZOIC ACID

Formula: $C_7H_6O_3$
Mol Weight: 138.12
MP (deg C): 213-214
BP (deg C):
BP pressure (mm Hg):

Property/Value	Units	Temp	Data Type	Reference
WS 5.35E+003	mg/L		EXP	YALKOWSKY,SH & DANNENFELSER,RM (1992)
logP 1.58			EXP	HANSCH,C & LEO,AJ (1985)
VP 1.98E-007	mm Hg		EXT	JONES,AH (1960)
DC 4.54	pKa		EXP	BYKOVA,LN ET AL. (1970)
HL 1.13E-011	atm m3/mol	25	EST	MEYLAN,WM & HOWARD,PH (1991)
OH 1.30E-011	cm3/molc sec	25	EST	MEYLAN,WM & HOWARD,PH (1993)

000099-97-8 — N,N,4-TRIMETHYLANILINE

Formula: $C_9H_{13}N$
Mol Weight: 135.21
MP (deg C):
BP (deg C): 211
BP pressure (mm Hg):

Property/Value	Units	Temp	Data Type	Reference
WS 3.49E+002	mg/L	25	EST	MEYLAN,WM ET AL. (1996)
logP 2.81			EXP	SANGSTER,J (1993)
VP 1.78E-001	mm Hg	25	EXP	CHAO,J ET AL. (1983)
DC 5.63	pKa	25	EXP	PERRIN,DD (1965)
HL 9.45E-005	atm m3/mol	25	EST	MEYLAN,WM & HOWARD,PH (1991)
OH 2.03E-010	cm3/molc sec	25	EST	MEYLAN,WM & HOWARD,PH (1993)

000099-98-9 — N,N-DIMETHYL-P-PHENYLENEDIAMINE

Formula: $C_8H_{12}N_2$
Mol Weight: 136.20
MP (deg C): 53
BP (deg C): 262
BP pressure (mm Hg):

Property/Value	Units	Temp	Data Type	Reference
WS 7.38E+003	mg/L	25	EST	MEYLAN,WM ET AL. (1996)
logP 1.25			EST	MEYLAN,WM & HOWARD,PH (1995)
VP 2.77E-002	mm Hg	25	EST	NEELY,WB & BLAU,GE (1985)
DC 6.59	pKa	25	EXP	PERRIN,DD (1965)
HL 3.03E-008	atm m3/mol	25	EST	MEYLAN,WM & HOWARD,PH (1991)
OH 2.03E-010	cm3/molc sec	25	EST	MEYLAN,WM & HOWARD,PH (1993)

000099-99-0 — P-NITROTOLUENE

Formula: $C_7H_7NO_2$
Mol Weight: 137.14
MP (deg C): 51.6
BP (deg C): 238.3
BP pressure (mm Hg):

Property/Value	Units	Temp	Data Type	Reference
WS 4.00E+001	mg/L	15	EXP	YALKOWSKY,SH & DANNENFELSER,RM (1992)
logP 2.37			EXP	HANSCH,C & LEO,AJ (1985)
VP 1.64E-001	mm Hg	25	EXP	PERRY,RH & GREEN,D (1984)
DC	pKa			
HL 4.77E-005	atm m3/mol	25	EST	MEYLAN,WM & HOWARD,PH (1991)
OH 7.72E-013	cm3/molc sec	25	EST	MEYLAN,WM & HOWARD,PH (1993)

CAS #: 000100-00-5 — P-CHLORONITROBENZENE

Formula: $C_6H_4ClNO_2$

Mol Weight: 157.56
MP (deg C): 83.5
FP (deg C):
BP (deg C): 242
BP pressure (mm Hg):

Property/Value	Units	Temp	Data Type	Reference
WS 4.53E+002	mg/L	20	EXP	HOWARD,PH ET AL. (1976)
logP 2.39			EXP	HANSCH,C & LEO,AJ (1985)
VP 2.19E-002	mm Hg	25	EXT	DAUBERT,TE & DANNER,RP (1989)
DC	pKa			
HL 5.44E-005	atm m3/mol	22	EXP	HELLMANN,H (1987)
OH 1.76E-013	cm3/molc sec	25	EST	MEYLAN,WM & HOWARD,PH (1993)

CAS #: 000100-01-6 — 4-NITROANILINE

Formula: $C_6H_6N_2O_2$

Mol Weight: 138.13
MP (deg C): 148-149
FP (deg C):
BP (deg C): 331.7
BP pressure (mm Hg):

Property/Value	Units	Temp	Data Type	Reference
WS 7.28E+002	mg/L	30	EXP	GROSS,PM & SAYLOR,JH (1931)
logP 1.39			EXP	HANSCH,C & LEO,AJ (1985)
VP 3.20E-006	mm Hg	25	EXT	FERRO,D & PIACENTE,V (1985)
DC 1.00	pKa	25	EXP	WEAST,RC ET AL. (1985)
HL 1.81E-008	atm m3/mol	25	EST	MEYLAN,WM & HOWARD,PH (1991)
OH 1.35E-011	cm3/molc sec	25	EST	MEYLAN,WM & HOWARD,PH (1993)

CAS #: 000100-02-7 — 4-NITROPHENOL

Formula: $C_6H_5NO_3$

Mol Weight: 139.11
MP (deg C): 113-115
FP (deg C):
BP (deg C): 279
BP pressure (mm Hg):

Property/Value	Units	Temp	Data Type	Reference
WS 1.16E+004	mg/L	20	EXP	SCHWARZENBACH,RP ET AL. (1988)
logP 1.91			EXP	HANSCH,C & LEO,AJ (1985)
VP 4.10E-005	mm Hg	25	EXT	SCHMIDT-BLEEK,F ET AL. (1982)
DC 7.08	pKa	22	EXP	SCHWARZENBACH,RP ET AL. (1988)
HL 4.15E-001	atm m3/mol	25	EXP	PARSONS,GH ET AL. (1971)
OH 4.42E-012	cm3/molc sec	25	EST	ATKINSON,R (1988)

CAS #: 000100-06-1 — P-METHOXYACETOPHENONE

Formula: $C_9H_{10}O_2$

Mol Weight: 150.18
MP (deg C): 38.5
FP (deg C):
BP (deg C): 258
BP pressure (mm Hg):

Property/Value	Units	Temp	Data Type	Reference
WS 2.47E+003	mg/L	25	EST	MEYLAN,WM ET AL. (1996)
logP 1.74			EXP	HANSCH,C & LEO,AJ (1985)
VP 6.44E-003	mm Hg	25	EXP	JONES,AH (1960)
DC	pKa			
HL 5.80E-007	atm m3/mol	25	EST	MEYLAN,WM & HOWARD,PH (1991)
OH 2.05E-011	cm3/molc sec	25	EST	MEYLAN,WM & HOWARD,PH (1993)

CAS #: 000100-09-4 — P-METHOXYBENZOIC ACID

Formula: $C_8H_8O_3$

Mol Weight: 152.15
MP (deg C): 184
FP (deg C):
BP (deg C): 275-280
BP pressure (mm Hg):

Property/Value	Units	Temp	Data Type	Reference
WS 5.30E+002	mg/L	37	EXP	YALKOWSKY,SH & DANNENFELSER,RM (1992)
logP 1.96			EXP	HANSCH,C & LEO,AJ (1985)
VP 1.50E-003	mm Hg	25	EST	NEELY,WB & BLAU,GE (1985)
DC 4.47	pKa	25	EXP	KORTUM,G ET AL (1961)
HL 6.41E-009	atm m3/mol	25	EST	MEYLAN,WM & HOWARD,PH (1991)
OH 9.31E-012	cm3/molc sec	25	EST	MEYLAN,WM & HOWARD,PH (1993)

CAS #: 000100-10-7 — P-FORMYL-N,N-DIMETHYLANILINE

Formula: $C_9H_{11}NO$

Mol Weight: 149.19
MP (deg C): 74
FP (deg C):
BP (deg C): 176-177
BP pressure (mm Hg): 1.70E+001

Property/Value	Units	Temp	Data Type	Reference
WS 2.18E+003	mg/L	25	EST	MEYLAN,WM ET AL. (1996)
logP 1.81			EXP	HANSCH,C & LEO,AJ (1985)
VP 3.68E-003	mm Hg	25	EXT	DAUBERT,TE & DANNER,RP (1989)
DC	pKa			
HL 2.14E-007	atm m3/mol	25	EST	MEYLAN,WM & HOWARD,PH (1991)
OH 1.41E-010	cm3/molc sec	25	EST	MEYLAN,WM & HOWARD,PH (1993)

CAS #: 000100-12-9 — 4-ETHYLNITROBENZENE

Formula: $C_8H_9NO_2$

Mol Weight: 151.17
MP (deg C): -12.3
FP (deg C):
BP (deg C): 245.5
BP pressure (mm Hg):

Property/Value	Units	Temp	Data Type	Reference
WS 7.90E+001	mg/L	25	EST	MEYLAN,WM ET AL. (1996)
logP 3.03			EXP	HANSCH,C ET AL. (1995)
VP 2.17E-002	mm Hg	25	EST	NEELY,WB & BLAU,GE (1985)
DC	pKa			
HL 3.11E-005	atm m3/mol	25	EST	MEYLAN,WM & HOWARD,PH (1991)
OH 1.71E-012	cm3/molc sec	25	EST	MEYLAN,WM & HOWARD,PH (1993)

CAS #: 000100-15-2 — P-NITRO-N-METHYLANILINE

Formula: $C_7H_8N_2O_2$

Mol Weight: 152.15
MP (deg C): 152
FP (deg C):
BP (deg C):
BP pressure (mm Hg):

Property/Value	Units	Temp	Data Type	Reference
WS 1.34E+003	mg/L	25	EST	MEYLAN,WM ET AL. (1996)
logP 2.04			EXP	HANSCH,C & LEO,AJ (1985)
VP 7.64E-003	mm Hg	25	EST	NEELY,WB & BLAU,GE (1985)
DC	pKa			
HL 1.65E-008	atm m3/mol	25	EST	MEYLAN,WM & HOWARD,PH (1991)
OH 6.60E-012	cm3/molc sec	25	EST	MEYLAN,WM & HOWARD,PH (1993)

CAS #: 000100-16-3 — HYDRAZINE, (4-NITROPHENYL)-

Formula: $C_6H_7N_3O_2$
Mol Weight: 153.14
MP (deg C): 158 dec
FP (deg C):
BP (deg C):
BP pressure (mm Hg):

Property/Value	Units	Temp	Data Type	Reference
WS 4.59E+003	mg/L	25	EST	MEYLAN,WM ET AL. (1996)
logP 1.41			EXP	HANSCH,C ET AL. (1995)
VP 9.46E-004	mm Hg	25	EST	NEELY,WB & BLAU,GE (1985)
DC 3.70	pKa	25	EXP	PERRIN,DD (1972)
HL 3.14E-011	atm m3/mol	25	EST	MEYLAN,WM & HOWARD,PH (1991)
OH 5.33E-012	cm3/molc sec	25	EST	MEYLAN,WM & HOWARD,PH (1993)

CAS #: 000100-17-4 — P-NITROANISOLE

Formula: $C_7H_7NO_3$
Mol Weight: 153.14
MP (deg C): 54
FP (deg C):
BP (deg C): 274
BP pressure (mm Hg):

Property/Value	Units	Temp	Data Type	Reference
WS 5.90E+002	mg/L	30	EXP	YALKOWSKY,SH & DANNENFELSER,RM (1992)
logP 2.03			EXP	HANSCH,C & LEO,AJ (1985)
VP 1.92E-002	mm Hg	25	EST	NEELY,WB & BLAU,GE (1985)
DC -10.01	pKa	25	EXP	PERRIN,DD (1972)
HL 1.26E-006	atm m3/mol	25	EST	MEYLAN,WM & HOWARD,PH (1991)
OH 3.52E-012	cm3/molc sec	25	EST	MEYLAN,WM & HOWARD,PH (1993)

CAS #: 000100-18-5 — 1,4-DIISOPROPYLBENZENE

Formula: $C_{12}H_{18}$
Mol Weight: 162.28
MP (deg C): -17
FP (deg C):
BP (deg C): 210.3
BP pressure (mm Hg):

Property/Value	Units	Temp	Data Type	Reference
WS 4.33E+000	mg/L	25	EST	MEYLAN,WM ET AL. (1996)
logP 4.90			EST	MEYLAN,WM & HOWARD,PH (1995)
VP 2.46E-001	mm Hg	25	EXP	DAUBERT,TE & DANNER,RP (1989)
DC	pKa			
HL 2.04E-002	atm m3/mol	25	EST	MEYLAN,WM & HOWARD,PH (1991)
OH 1.01E-011	cm3/molc sec	25	EST	MEYLAN,WM & HOWARD,PH (1993)

CAS #: 000100-19-6 — P-NITROACETOPHENONE

Formula: $C_8H_7NO_3$
Mol Weight: 165.15
MP (deg C): 81.8
FP (deg C):
BP (deg C): 165
BP pressure (mm Hg): 5.00E+000

Property/Value	Units	Temp	Data Type	Reference
WS 1.30E+003	mg/L	25	EST	MEYLAN,WM ET AL. (1996)
logP 1.53			EXP	HANSCH,C & LEO,AJ (1985)
VP 2.40E-003	mm Hg	25	EST	NEELY,WB & BLAU,GE (1985)
DC	pKa			
HL 3.87E-008	atm m3/mol	25	EST	MEYLAN,WM & HOWARD,PH (1991)
OH 2.57E-013	cm3/molc sec	25	EST	MEYLAN,WM & HOWARD,PH (1993)

CAS #: 000100-21-0 — TEREPHTHALIC ACID

Formula: $C_8H_6O_4$
Mol Weight: 166.13
MP (deg C): >300
FP (deg C):
BP (deg C):
BP pressure (mm Hg):

Property/Value	Units	Temp	Data Type	Reference
WS 1.50E+001	mg/L	20	EXP	YALKOWSKY,SH & DANNENFELSER,RM (1992)
logP 2.00			EXP	HANSCH,C ET AL (1995)
VP 9.20E-006	mm Hg	25	EXT	TOWLE,PH ET AL. (1968)
DC 3.51	pKa	25	EXP	TOWLE,PH ET AL. (1968)
HL 3.88E-013	atm m3/mol	25	EST	MEYLAN,WM & HOWARD,PH (1991)
OH 2.75E-013	cm3/molc sec	25	EST	ATKINSON,R (1987)

CAS #: 000100-22-1 — N,N,N',N'-TETRAMETHYL-P-PHENYLENEDIAMINE

Formula: $C_{10}H_{16}N_2$
Mol Weight: 164.25
MP (deg C): 51-52
FP (deg C):
BP (deg C): 260
BP pressure (mm Hg):

Property/Value	Units	Temp	Data Type	Reference
WS 6.46E+002	mg/L	25	EST	MEYLAN,WM ET AL. (1996)
logP 2.35			EST	MEYLAN,WM & HOWARD,PH (1995)
VP 7.69E-002	mm Hg	25	EST	NEELY,WB & BLAU,GE (1985)
DC 6.35	pKa	20	EXP	PERRIN,DD (1965)
HL 1.36E-006	atm m3/mol	25	EST	MEYLAN,WM & HOWARD,PH (1991)
OH 2.05E-010	cm3/molc sec	25	EST	MEYLAN,WM & HOWARD,PH (1993)

CAS #: 000100-23-2 — P-NITRO-N,N-DIMETHYLANILINE

Formula: $C_8H_{10}N_2O_2$
Mol Weight: 166.18
MP (deg C):
FP (deg C):
BP (deg C):
BP pressure (mm Hg):

Property/Value	Units	Temp	Data Type	Reference
WS 7.38E+002	mg/L	25	EST	MEYLAN,WM ET AL. (1996)
logP 2.27			EXP	HANSCH,C & LEO,AJ (1985)
VP 6.07E-006	mm Hg	25	EXT	MAJURY,TG (1956)
DC 0.61	pKa	25	EXP	PERRIN,DD (1965)
HL 3.38E-007	atm m3/mol	25	EST	MEYLAN,WM & HOWARD,PH (1991)
OH 4.88E-011	cm3/molc sec	25	EST	MEYLAN,WM & HOWARD,PH (1993)

CAS #: 000100-25-4 — P-DINITROBENZENE

Formula: $C_6H_4N_2O_4$
Mol Weight: 168.11
MP (deg C): 172-173
FP (deg C):
BP (deg C): 299
BP pressure (mm Hg):

Property/Value	Units	Temp	Data Type	Reference
WS 6.90E+001	mg/L	25	EXP	YALKOWSKY,SH & DANNENFELSER,RM (1992)
logP 1.46			EXP	HANSCH,C & LEO,AJ (1985)
VP 2.61E-005	mm Hg	25	EST	HLX WSOL
DC	pKa			
HL 8.39E-008	atm m3/mol	25	EST	MEYLAN,WM & HOWARD,PH (1991)
OH 2.16E-014	cm3/molc sec	25	EST	MEYLAN,WM & HOWARD,PH (1993)

CAS #: 000100-26-5 — PYRIDINE-2,5-DICARBOXYLIC ACID

Formula: $C_7H_5NO_4$

Mol Weight: 167.12

MP (deg C): 268-269

FP (deg C):

BP (deg C):

BP pressure (mm Hg):

Property/Value	Units	Temp	Data Type	Reference
WS 1.24E+003	mg/L	25	EXP	YALKOWSKY,SH & DANNENFELSER,RM (1992)
logP 0.57			EST	MEYLAN,WM & HOWARD,PH (1995)
VP 6.10E-006	mm Hg	25	EST	NEELY,WB & BLAU,GE (1985)
DC	pKa			
HL 2.86E-015	atm m3/mol	25	EST	MEYLAN,WM & HOWARD,PH (1991)
OH 1.08E-012	cm3/molc sec	25	EST	MEYLAN,WM & HOWARD,PH (1993)

CAS #: 000100-27-6 — 2-(P-NITROPHENYL)ETHANOL

Formula: $C_8H_9NO_3$

Mol Weight: 167.17

MP (deg C): 62-64

FP (deg C):

BP (deg C): 177

BP pressure (mm Hg): 1.60E+001

Property/Value	Units	Temp	Data Type	Reference
WS 2.86E+003	mg/L	25	EST	MEYLAN,WM ET AL. (1996)
logP 1.12			EXP	HANSCH,C & LEO,AJ (1985)
VP 2.67E-005	mm Hg	25	EST	NEELY,WB & BLAU,GE (1985)
DC	pKa			
HL 1.14E-009	atm m3/mol	25	EST	MEYLAN,WM & HOWARD,PH (1991)
OH 5.92E-012	cm3/molc sec	25	EST	MEYLAN,WM & HOWARD,PH (1993)

CAS #: 000100-29-8 — P-NITROPHENETOLE

Formula: $C_8H_9NO_3$

Mol Weight: 167.17

MP (deg C): 60

FP (deg C):

BP (deg C): 283

BP pressure (mm Hg): 3.00E+000

Property/Value	Units	Temp	Data Type	Reference
WS 1.79E+002	mg/L	25	EST	MEYLAN,WM ET AL. (1996)
logP 2.53			EXP	HANSCH,C & LEO,AJ (1985)
VP 6.23E-003	mm Hg	25	EST	NEELY,WB & BLAU,GE (1985)
DC	pKa			
HL 1.67E-006	atm m3/mol	25	EST	MEYLAN,WM & HOWARD,PH (1991)
OH 8.86E-012	cm3/molc sec	25	EST	MEYLAN,WM & HOWARD,PH (1993)

CAS #: 000100-36-7 — 1,2-ETHANEDIAMINE, N,N-DIETHYL-

Formula: $C_6H_{16}N_2$

Mol Weight: 116.21

MP (deg C):

FP (deg C):

BP (deg C): 144

BP pressure (mm Hg):

Property/Value	Units	Temp	Data Type	Reference
WS 6.98E+005	mg/L	25	EST	MEYLAN,WM ET AL. (1996)
logP 0.21			EXP	HANSCH,C ET AL. (1995)
VP 2.97E+000	mm Hg	25	EST	NEELY,WB & BLAU,GE (1985)
DC 10.02	pKa	25	EXP	PERRIN,DD (1965)
HL 8.79E-009	atm m3/mol	25	EST	MEYLAN,WM & HOWARD,PH (1991)
OH 1.26E-010	cm3/molc sec	25	EST	MEYLAN,WM & HOWARD,PH (1993)

CAS #: 000100-37-8 — 2-(DIETHYLAMINO)-ETHANOL

Formula: $C_6H_{15}NO$

Mol Weight: 117.19

MP (deg C):

FP (deg C):

BP (deg C): 163

BP pressure (mm Hg):

Property/Value	Units	Temp	Data Type	Reference
WS 9.54E+005	mg/L	25	EST	MEYLAN,WM ET AL. (1996)
logP 0.05			EST	MEYLAN,WM & HOWARD,PH (1995)
VP 1.40E+000	mm Hg	25	EXP	US EPA (1989A)
DC 9.87	pKa	20	EXP	PERRIN,DD (1965)
HL 3.12E-009	atm m3/mol	25	EST	MEYLAN,WM & HOWARD,PH (1991)
OH 9.86E-011	cm3/molc sec	25	EST	MEYLAN,WM & HOWARD,PH (1993)

CAS #: 000100-39-0 — ALPHA-BROMOTOLUENE

Formula: C_7H_7Br

Mol Weight: 171.04

MP (deg C): -4.0

FP (deg C):

BP (deg C): 198-199

BP pressure (mm Hg):

Property/Value	Units	Temp	Data Type	Reference
WS 3.85E+002	mg/L	25	EST	MEYLAN,WM ET AL. (1996)
logP 2.92			EXP	HANSCH,C & LEO,AJ (1985)
VP 3.90E-001	mm Hg		EST	NEELY,WB & BLAU,GE (1985)
DC	pKa			
HL 6.90E-004	atm m3/mol	25	EST	MEYLAN,WM & HOWARD,PH (1991)
OH 2.30E-012	cm3/molc sec	25	EST	MEYLAN,WM & HOWARD,PH (1993)

CAS #: 000100-40-3 — 4-VINYLCYCLOHEXENE

Formula: C_8H_{12}

Mol Weight: 108.18

MP (deg C): -108.9

FP (deg C):

BP (deg C): 128

BP pressure (mm Hg):

Property/Value	Units	Temp	Data Type	Reference
WS 5.00E+001	mg/L	25	EXP	YALKOWSKY,SH & DANNENFELSER,RM (1992)
logP 3.93			EXP	CHEM INSPECT TEST INST (1992)
VP 1.57E+001	mm Hg	25	EXP	DAUBERT,TE & DANNER,RP (1989)
DC	pKa			
HL 4.48E-002	atm m3/mol	25	EST	VP/WSOL
OH 8.93E-011	cm3/molc sec	25	EST	MEYLAN,WM & HOWARD,PH (1993)

CAS #: 000100-41-4 — ETHYLBENZENE

Formula: C_8H_{10}

Mol Weight: 106.17

MP (deg C): -94.975

FP (deg C):

BP (deg C): 136.193

BP pressure (mm Hg):

Property/Value	Units	Temp	Data Type	Reference
WS 1.69E+002	mg/L	25	EXP	SANEMASA,I ET AL. (1982)
logP 3.15			EXP	HANSCH,C & LEO,AJ (1985)
VP 9.60E+000	mm Hg	25	EXP	DAUBERT,TE & DANNER,RP (1985)
DC	pKa			
HL 7.88E-003	atm m3/mol	25	EXP	SANEMASA,I ET AL. (1982)
OH 7.10E-012	cm3/molc sec	25	EXP	ATKINSON,R (1989)

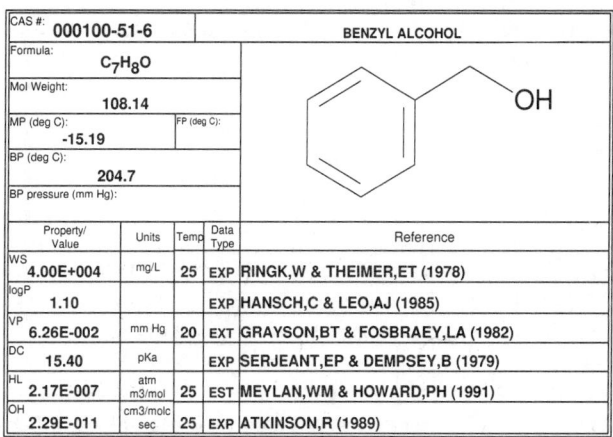

000100-42-5 — STYRENE

CAS #:	000100-42-5
Formula:	C_8H_8
Mol Weight:	104.15
MP (deg C):	-30.628
FP (deg C):	
BP (deg C):	145.14
BP pressure (mm Hg):	

Property/Value	Units	Temp	Data Type	Reference
WS 3.10E+002	mg/L	25	EXP	YALKOWSKY,SH & DANNENFELSER,RM (1992)
logP 2.95			EXP	HANSCH,C & LEO,AJ (1985)
VP 6.40E+000	mm Hg	25	EXP	CHAO,J ET AL. (1983)
DC	pKa			
HL 2.75E-003	atm m3/mol	25	EXP	BOCEK,K (1976)
OH 5.80E-011	cm3/molc sec	25	EXP	ATKINSON,R (1989)

000100-43-6 — 4-VINYLPYRIDINE

CAS #:	000100-43-6
Formula:	C_7H_7N
Mol Weight:	105.14
MP (deg C):	
FP (deg C):	
BP (deg C):	121
BP pressure (mm Hg):	1.50E+002

Property/Value	Units	Temp	Data Type	Reference
WS 2.91E+004	mg/L	20	EXP	GOE,GL (1982)
logP 1.71			EST	MEYLAN,WM & HOWARD,PH (1995)
VP 1.70E+000	mm Hg	25	EXT	WEAST,RC ET AL. (1983) & DEAN,JA (1985)
DC 5.62	pKa	25	EXP	PERRIN,DD (1965)
HL 3.15E-006	atm m3/mol	25	EST	MEYLAN,WM & HOWARD,PH (1991)
OH 2.66E-011	cm3/molc sec	25	EST	MEYLAN,WM & HOWARD,PH (1993)

000100-44-7 — ALPHA-CHLOROTOLUENE

CAS #:	000100-44-7
Formula:	C_7H_7Cl
Mol Weight:	126.59
MP (deg C):	-39.2
FP (deg C):	
BP (deg C):	179.4
BP pressure (mm Hg):	

Property/Value	Units	Temp	Data Type	Reference
WS 5.25E+002	mg/L	25	EXP	OHNISHI,R & TANABE,K (1971)
logP 2.30			EXP	HANSCH,C & LEO,AJ (1985)
VP 1.30E+000	mm Hg	25	EXP	MACKAY,D ET AL. (1982)
DC	pKa			
HL 4.12E-004	atm m3/mol	25	EST	VP/WSOL
OH 2.90E-012	cm3/molc sec	25	EXP	ATKINSON,R (1989)

000100-46-9 — BENZYLAMINE

CAS #:	000100-46-9
Formula:	C_7H_9N
Mol Weight:	107.16
MP (deg C):	10
FP (deg C):	
BP (deg C):	184.5
BP pressure (mm Hg):	

Property/Value	Units	Temp	Data Type	Reference
WS 1.00E+006	mg/L	20	EXP	YALKOWSKY,SH & DANNENFELSER,RM (1992)
logP 1.09			EXP	HANSCH,C & LEO,AJ (1985)
VP 6.53E-001	mm Hg	25	EXP	DAUBERT,TE & DANNER,RP (1989)
DC 9.33	pKa	25	EXP	WEAST,RC (1985)
HL 6.12E-011	atm m3/mol	25	EST	MEYLAN,WM & HOWARD,PH (1991)
OH 3.35E-011	cm3/molc sec	25	EST	ATKINSON,R (1988)

000100-47-0 — BENZONITRILE

CAS #:	000100-47-0
Formula:	C_7H_5N
Mol Weight:	103.12
MP (deg C):	-12.75
FP (deg C):	
BP (deg C):	191.10
BP pressure (mm Hg):	

Property/Value	Units	Temp	Data Type	Reference
WS 2.00E+003	mg/L	25	EXP	RIDDICK,JA ET AL. (1986)
logP 1.56			EXP	HANSCH,C & LEO,AJ (1985)
VP 7.68E-001	mm Hg	25	EXP	DAUBERT,TE & DANNER,RP (1989)
DC	pKa			
HL 5.21E-005	atm m3/mol	25	EST	VP/WSOL
OH 3.30E-013	cm3/molc sec	25	EXP	ATKINSON,R (1989)

000100-48-1 — 4-CYANOPYRIDINE

CAS #:	000100-48-1
Formula:	$C_6H_4N_2$
Mol Weight:	104.11
MP (deg C):	78-80
FP (deg C):	
BP (deg C):	212-215
BP pressure (mm Hg):	

Property/Value	Units	Temp	Data Type	Reference
WS 2.50E+004	mg/L	25	EST	MEYLAN,WM ET AL. (1996)
logP 0.46			EXP	HANSCH,C & LEO,AJ (1985)
VP 3.14E-001	mm Hg	25	EST	NEELY,WB & BLAU,GE (1985)
DC 1.90	pKa		EXP	PERRIN,DD (1965)
HL 6.81E-008	atm m3/mol	25	EST	MEYLAN,WM & HOWARD,PH (1991)
OH 6.53E-014	cm3/molc sec	25	EST	MEYLAN,WM & HOWARD,PH (1993)

000100-50-5 — 3-CYCLOHEXENE-1-CARBOXALDEHYDE

CAS #:	000100-50-5
Formula:	$C_7H_{10}O$
Mol Weight:	110.16
MP (deg C):	2
FP (deg C):	
BP (deg C):	105
BP pressure (mm Hg):	

Property/Value	Units	Temp	Data Type	Reference
WS 2.65E+003	mg/L	25	EST	MEYLAN,WM ET AL. (1996)
logP 1.89			EST	MEYLAN,WM & HOWARD,PH (1995)
VP 1.80E+000	mm Hg	25	EST	NEELY,WB & BLAU,GE (1985)
DC	pKa			
HL 1.08E-004	atm m3/mol	25	EST	MEYLAN,WM & HOWARD,PH (1991)
OH 8.86E-011	cm3/molc sec	25	EST	MEYLAN,WM & HOWARD,PH (1993)

000100-51-6 — BENZYL ALCOHOL

CAS #:	000100-51-6
Formula:	C_7H_8O
Mol Weight:	108.14
MP (deg C):	-15.19
FP (deg C):	
BP (deg C):	204.7
BP pressure (mm Hg):	

Property/Value	Units	Temp	Data Type	Reference
WS 4.00E+004	mg/L	25	EXP	RINGK,W & THEIMER,ET (1978)
logP 1.10			EXP	HANSCH,C & LEO,AJ (1985)
VP 6.26E-002	mm Hg	20	EXT	GRAYSON,BT & FOSBRAEY,LA (1982)
DC 15.40	pKa		EXP	SERJEANT,EP & DEMPSEY,B (1979)
HL 2.17E-007	atm m3/mol	25	EST	MEYLAN,WM & HOWARD,PH (1991)
OH 2.29E-011	cm3/molc sec	25	EXP	ATKINSON,R (1989)

CAS #: 000100-52-7				BENZALDEHYDE

Formula: C_7H_6O

Mol Weight: 106.13

MP (deg C): -56.5   FP (deg C):

BP (deg C): 179

BP pressure (mm Hg):

Property/Value	Units	Temp	Data Type	Reference
WS 3.00E+003	mg/L	25	EXP	SHERMAN,PD (1978)
logP 1.48			EXP	HANSCH,C & LEO,AJ (1985)
VP 1.27E-001	mm Hg	25	EXP	AMBROSE,D ET AL. (1975A)
DC	pKa			
HL 2.67E-005	atm m3/mol	25	EXP	BETTERTON,EA & HOFFMAN,MR (1988)
OH 1.29E-011	cm3/molc sec	25	EXP	ATKINSON,R (1989)

CAS #: 000100-54-9				3-CYANOPYRIDINE

Formula: $C_6H_4N_2$

Mol Weight: 104.11

MP (deg C): 52   FP (deg C):

BP (deg C): 240-5

BP pressure (mm Hg):

Property/Value	Units	Temp	Data Type	Reference
WS 3.00E+004	mg/L	25	EST	MEYLAN,WM ET AL. (1996)
logP 0.36			EXP	SANGSTER,J (1994)
VP 4.00E-001	mm Hg	25	EST	WSOL X HL
DC 1.39	pKa	24	EXP	PERRIN,DD (1965)
HL 6.89E-008	atm m3/mol	25	EST	MEYLAN,WM & HOWARD,PH (1991)
OH 6.45E-014	cm3/molc sec	25	EST	MEYLAN,WM & HOWARD,PH (1993)

CAS #: 000100-55-0				3-PYRIDINEMETHANOL

Formula: C_6H_7NO

Mol Weight: 109.13

MP (deg C): -6.5   FP (deg C):

BP (deg C): 266

BP pressure (mm Hg):

Property/Value	Units	Temp	Data Type	Reference
WS 1.00E+006	mg/L	25	EST	MEYLAN,WM ET AL. (1996)
logP -0.02			EXP	HANSCH,C & LEO,AJ (1985)
VP 2.60E-002	mm Hg	25	EST	NEELY,WB & BLAU,GE (1985)
DC 4.90	pKa	25	EXP	PERRIN,DD (1972)
HL 2.85E-010	atm m3/mol	25	EST	MEYLAN,WM & HOWARD,PH (1991)
OH 4.33E-012	cm3/molc sec	25	EST	MEYLAN,WM & HOWARD,PH (1993)

CAS #: 000100-56-1				PHENYL MERCURIC CHLORIDE

Formula: C_6H_5ClHg

Mol Weight: 313.15

MP (deg C): 250-252   FP (deg C):

BP (deg C):

BP pressure (mm Hg):

Property/Value	Units	Temp	Data Type	Reference
WS 3.10E+002	mg/L	25	EST	MEYLAN,WM ET AL. (1996)
logP 1.78			EXP	HANSCH,C ET AL. (1995)
VP 2.75E-002	mm Hg	25	EST	NEELY,WB & BLAU,GE (1985)
DC	pKa			
HL	atm m3/mol			
OH 1.95E-012	cm3/molc sec	25	EST	MEYLAN,WM & HOWARD,PH (1993)

CAS #: 000100-57-2				PHENYLMERCURIC HYDROXIDE

Formula: C_6H_6HgO

Mol Weight: 294.70

MP (deg C): 190 dec   FP (deg C):

BP (deg C):

BP pressure (mm Hg):

Property/Value	Units	Temp	Data Type	Reference
WS 1.40E+004	mg/L	18	EXP	ULFVARSON,U (1969)
logP -0.12			EST	MEYLAN,WM & HOWARD,PH (1995)
VP 8.32E-005	mm Hg	25	EST	NEELY,WB & BLAU,GE (1985)
DC	pKa			
HL	atm m3/mol			
OH 2.09E-012	cm3/molc sec	25	EST	MEYLAN,WM & HOWARD,PH (1993)

CAS #: 000100-61-8				N-METHYLANILINE

Formula: C_7H_9N

Mol Weight: 107.16

MP (deg C): -57   FP (deg C): -57

BP (deg C): 194-196

BP pressure (mm Hg):

Property/Value	Units	Temp	Data Type	Reference
WS 5.62E+003	mg/L	25	EXP	YALKOWSKY,SH & DANNENFELSER,RM (1992)
logP 1.66			EXP	HANSCH,C & LEO,AJ (1985)
VP 4.53E-001	mm Hg	25	EXP	DAUBERT,TE & DANNER,RP (1989)
DC 4.85	pKa	25	EXP	RIDDICK,JA ET AL. (1986)
HL 1.14E-005	atm m3/mol	25	EST	VP/WSOL
OH 4.39E-011	cm3/molc sec	25	EST	MEYLAN,WM & HOWARD,PH (1993)

CAS #: 000100-63-0				PHENYL HYDRAZINE

Formula: $C_6H_8N_2$

Mol Weight: 108.14

MP (deg C): 19.5   FP (deg C):

BP (deg C): 243.5 dec

BP pressure (mm Hg):

Property/Value	Units	Temp	Data Type	Reference
WS 8.37E+005	mg/L	24	EXP	SEIDELL,A (1941)
logP 1.25			EXP	HANSCH,C ET AL. (1995)
VP 2.60E-002	mm Hg	25	EXP	YAWS,CL (1994A)
DC 8.79	pKa	25	EXP	BUDAVARI,S (1996)
HL 4.42E-009	atm m3/mol	25	EST	VP/WSOL
OH 4.27E-011	cm3/molc sec	25	EST	MEYLAN,WM & HOWARD,PH (1993)

CAS #: 000100-64-1				CYCLOHEXANONE OXIME

Formula: $C_6H_{11}NO$

Mol Weight: 113.16

MP (deg C): 90   FP (deg C):

BP (deg C): 206

BP pressure (mm Hg):

Property/Value	Units	Temp	Data Type	Reference
WS 2.04E+004	mg/L	25	EST	MEYLAN,WM ET AL. (1996)
logP 0.84			EXP	TSCATS
VP 1.20E-001	mm Hg	25	EXT	DAUBERT,TE & DANNER,RP (1989)
DC	pKa			
HL 8.05E-006	atm m3/mol	25	EST	MEYLAN,WM & HOWARD,PH (1991)
OH 7.07E-012	cm3/molc sec	25	EST	MEYLAN,WM & HOWARD,PH (1993)

CAS #: 000100-65-2				PHENYLHYDROXYLAMINE

Formula: C_6H_7NO
Mol Weight: 109.13
MP (deg C): 82	FP (deg C):
BP (deg C):
BP pressure (mm Hg):

Property/Value	Units	Temp	Data Type	Reference
WS 2.00E+004	mg/L	5	EXP	YALKOWSKY,SH & DANNENFELSER,RM (1992)
logP 0.79			EXP	HANSCH,C & LEO,AJ (1985)
VP 7.95E-003	mm Hg	25	EST	NEELY,WB & BLAU,GE (1985)
DC	pKa			
HL 3.81E-009	atm m3/mol	25	EST	MEYLAN,WM & HOWARD,PH (1991)
OH 4.27E-011	cm3/molc sec	25	EST	MEYLAN,WM & HOWARD,PH (1993)

CAS #: 000100-66-3				ANISOLE

Formula: C_7H_8O
Mol Weight: 108.14
MP (deg C): -37.3	FP (deg C):
BP (deg C): 155.5
BP pressure (mm Hg):

Property/Value	Units	Temp	Data Type	Reference
WS 1.53E+003	mg/L	25	EXP	CHIOU,CT ET AL. (1983)
logP 2.11			EXP	HANSCH,C & LEO,AJ (1985)
VP 3.54E+000	mm Hg	25	EXP	AMBROSE,D ET AL. (1976)
DC -6.51	pKa	0	EXP	PERRIN,DD (1965)
HL 4.35E-003	atm m3/mol	25	EST	MEYLAN,WM & HOWARD,PH (1991)
OH 1.73E-011	cm3/molc sec	25	EXP	ATKINSON,R (1989)

CAS #: 000100-68-5				THIOANISOLE

Formula: C_7H_8S
Mol Weight: 124.21
MP (deg C):	FP (deg C):
BP (deg C): 193
BP pressure (mm Hg):

Property/Value	Units	Temp	Data Type	Reference
WS 5.06E+002	mg/L	25	EXP	SUZUKI,T (1991)
logP 2.74			EXP	HANSCH,C & LEO,AJ (1985)
VP 4.85E-001	mm Hg	25	EXT	BOUBLIK,T ET AL. (1984)
DC	pKa			
HL 1.57E-004	atm m3/mol	25	EST	MEYLAN,WM & HOWARD,PH (1991)
OH 1.36E-011	cm3/molc sec	25	EST	MEYLAN,WM & HOWARD,PH (1993)

CAS #: 000100-69-6				2-VINYLPYRIDINE

Formula: C_7H_7N
Mol Weight: 105.14
MP (deg C):	FP (deg C):
BP (deg C): 110
BP pressure (mm Hg): 1.50E+002

Property/Value	Units	Temp	Data Type	Reference
WS 2.75E+004	mg/L	20	EXP	GOE,GL (1982)
logP 1.54			EXP	CHEM INSPECT TEST INST (1992)
VP 1.34E+000	mm Hg	20	EST	WSOL X HL
DC 4.98	pKa	25	EXP	GOE,GL (1982)
HL 6.74E-006	atm m3/mol	25	EST	MEYLAN,WM & HOWARD,PH (1991)
OH 2.66E-011	cm3/molc sec	25	EST	MEYLAN,WM & HOWARD,PH (1993)

CAS #: 000100-70-9				2-CYANOPYRIDINE

Formula: $C_6H_4N_2$
Mol Weight: 104.11
MP (deg C): 28	FP (deg C):
BP (deg C): 215
BP pressure (mm Hg):

Property/Value	Units	Temp	Data Type	Reference
WS 2.55E+004	mg/L	25	EST	MEYLAN,WM ET AL. (1996)
logP 0.45			EXP	SANGSTER,J (1994)
VP 5.00E-001	mm Hg	25	EST	WSOL X HL
DC -0.26	pKa		EXP	PERRIN,DD (1965)
HL 6.89E-008	atm m3/mol	25	EST	MEYLAN,WM & HOWARD,PH (1991)
OH 6.45E-014	cm3/molc sec	25	EST	MEYLAN,WM & HOWARD,PH (1993)

CAS #: 000100-71-0				2-ETHYLPYRIDINE

Formula: C_7H_9N
Mol Weight: 107.16
MP (deg C): -63.1	FP (deg C):
BP (deg C): 148.6
BP pressure (mm Hg):

Property/Value	Units	Temp	Data Type	Reference
WS 3.40E+005	mg/L	-5	EXP	YALKOWSKY,SH & DANNENFELSER,RM (1992)
logP 1.69			EXP	HANSCH,C & LEO,AJ (1985)
VP 4.89E+000	mm Hg	25	EXP	CHAO,J ET AL. (1983)
DC 5.89	pKa	25	EXP	PERRIN,DD (1965)
HL 1.65E-005	atm m3/mol	25	EXP	ANDON,RJL ET AL. (1954)
OH 1.95E-012	cm3/molc sec	25	EST	MEYLAN,WM & HOWARD,PH (1993)

CAS #: 000100-72-1				TETRAHYDROPYRAN-2-METHANOL

Formula: $C_6H_{12}O_2$
Mol Weight: 116.16
MP (deg C):	FP (deg C):
BP (deg C): 187.2
BP pressure (mm Hg):

Property/Value	Units	Temp	Data Type	Reference
WS 1.00E+006	mg/L	25	EXP	SAX,NI & LEWIS,RJ (1987)
logP 0.38			EST	MEYLAN,WM & HOWARD,PH (1995)
VP 1.00E-001	mm Hg	20	EXP	FLICK,EW (1991)
DC	pKa			
HL 1.53E-008	atm m3/mol	25	EST	VP/WSOL
OH 3.59E-011	cm3/molc sec	25	EST	MEYLAN,WM & HOWARD,PH (1993)

CAS #: 000100-74-3				4-ETHYLMORPHOLINE

Formula: $C_6H_{13}NO$
Mol Weight: 115.18
MP (deg C): -64	FP (deg C):
BP (deg C): 138-139
BP pressure (mm Hg):

Property/Value	Units	Temp	Data Type	Reference
WS 1.00E+006	mg/L	25	EXP	YALKOWSKY,SH & DANNENFELSER,RM (1992)
logP 0.14			EST	MEYLAN,WM & HOWARD,PH (1995)
VP 5.03E+000	mm Hg	20	EXP	BROWN,ES ET AL. (1980)
DC 8.08	pKa	15	EXP	PERRIN,DD (1965)
HL 2.74E-008	atm m3/mol	25	EST	MEYLAN,WM & HOWARD,PH (1991)
OH 1.49E-010	cm3/molc sec	25	EST	MEYLAN,WM & HOWARD,PH (1993)

000100-75-4 — N-NITROSOPIPERIDINE

CAS #: 000100-75-4				
Formula: $C_5H_{10}N_2O$				
Mol Weight: 114.15				
MP (deg C):		FP (deg C):		
BP (deg C): 219				
BP pressure (mm Hg):				

Property/Value	Units	Temp	Data Type	Reference
WS 7.65E+004	mg/L	24	EXP	YALKOWSKY,SH & DANNENFELSER,RM (1992)
logP 0.36			EXP	HANSCH,C & LEO,AJ (1985)
VP 1.40E-001	mm Hg	25	EXT	KLEIN,RG (1982)
DC	pKa			
HL 8.44E-007	atm m3/mol	37	EXP	MIRVISH,SS ET AL. (1076)
OH 2.58E+001	cm3/molc sec	25	EST	MEYLAN,WM & HOWARD,PH (1993)

000100-76-5 — QUINUCLIDINE

CAS #: 000100-76-5				
Formula: $C_7H_{13}N$				
Mol Weight: 111.19				
MP (deg C): 156		FP (deg C):		
BP (deg C):				
BP pressure (mm Hg):				

Property/Value	Units	Temp	Data Type	Reference
WS 7.28E+004	mg/L	25	EST	MEYLAN,WM ET AL. (1996)
logP 1.38			EXP	HANSCH,C & LEO,AJ (1985)
VP 1.91E+000	mm Hg	25	EXP	JONES,AH (1960)
DC	pKa			
HL 2.21E-005	atm m3/mol	25	EST	MEYLAN,WM & HOWARD,PH (1991)
OH 4.59E-011	cm3/molc sec	25	EST	MEYLAN,WM & HOWARD,PH (1993)

000100-79-8 — DIOXOLANE

CAS #: 000100-79-8				
Formula: $C_6H_{12}O_3$				
Mol Weight: 132.16				
MP (deg C):		FP (deg C):		
BP (deg C): 82				
BP pressure (mm Hg): 1.00E+001				

Property/Value	Units	Temp	Data Type	Reference
WS 3.55E+004	mg/L	25	EST	MEYLAN,WM ET AL. (1996)
logP 0.22			EST	MEYLAN,WM & HOWARD,PH (1995)
VP 1.07E+002	mm Hg	25	EXP	BOUBLIK,T ET AL. (1984)
DC	pKa			
HL 1.91E-009	atm m3/mol	25	EST	MEYLAN,WM & HOWARD,PH (1991)
OH 2.50E-011	cm3/molc sec	25	EST	MEYLAN,WM & HOWARD,PH (1993)

000100-80-1 — M-METHYLSTYRENE

CAS #: 000100-80-1				
Formula: C_9H_{10}				
Mol Weight: 118.18				
MP (deg C): -86.3		FP (deg C):		
BP (deg C): 164				
BP pressure (mm Hg):				

Property/Value	Units	Temp	Data Type	Reference
WS 1.17E+002	mg/L	25	EST	MEYLAN,WM ET AL. (1996)
logP 3.44			EST	MEYLAN,WM & HOWARD,PH (1995)
VP 1.70E+000	mm Hg	25	EXT	BOUBLIK,T ET AL. (1984)
DC	pKa			
HL 3.05E-003	atm m3/mol	25	EST	MEYLAN,WM & HOWARD,PH (1991)
OH 3.12E-011	cm3/molc sec	25	EST	MEYLAN,WM & HOWARD,PH (1993)

000100-83-4 — 3-HYDROXYBENZALDEHYDE

CAS #: 000100-83-4				
Formula: $C_7H_6O_2$				
Mol Weight: 122.12				
MP (deg C): 108		FP (deg C):		
BP (deg C): 240				
BP pressure (mm Hg):				

Property/Value	Units	Temp	Data Type	Reference
WS 2.75E+004	mg/L	43	EXP	YALKOWSKY,SH & DANNENFELSER,RM (1992)
logP 1.38			EXP	HANSCH,C & LEO,AJ (1985)
VP 1.38E-002	mm Hg	25	EST	NEELY,WB & BLAU,GE (1985)
DC 8.98	pKa	25	EXP	SERJEANT,EP & DEMPSEY,B (1979)
HL 2.50E-009	atm m3/mol	25	EXP	GAFFNEY,JS ET AL. (1987)
OH 3.41E-011	cm3/molc sec	25	EST	MEYLAN,WM & HOWARD,PH (1993)

000100-84-5 — M-METHYLANISOLE

CAS #: 000100-84-5				
Formula: $C_8H_{10}O$				
Mol Weight: 122.17				
MP (deg C): -47		FP (deg C):		
BP (deg C): 175.5				
BP pressure (mm Hg):				

Property/Value	Units	Temp	Data Type	Reference
WS 5.27E+002	mg/L	25	EST	MEYLAN,WM ET AL. (1996)
logP 2.66			EXP	HANSCH,C & LEO,AJ (1985)
VP 1.50E+000	mm Hg	25	EST	NEELY,WB & BLAU,GE (1985)
DC	pKa			
HL 3.52E-004	atm m3/mol	25	EST	MEYLAN,WM & HOWARD,PH (1991)
OH 5.71E-011	cm3/molc sec	25	EST	MEYLAN,WM & HOWARD,PH (1993)

000100-97-0 — HEXAMETHYLENETETRAMINE

CAS #: 000100-97-0				
Formula: $C_6H_{12}N_4$				
Mol Weight: 140.19				
MP (deg C): > 250		FP (deg C):		
BP (deg C):				
BP pressure (mm Hg):				

Property/Value	Units	Temp	Data Type	Reference
WS 4.49E+005	mg/L	12	EXP	YALKOWSKY,SH & DANNENFELSER,RM (1992)
logP -4.15			EST	MEYLAN,WM & HOWARD,PH (1995)
VP 4.00E-003	mm Hg	25	EXP	STRANSKI,IN ET AL. (1957)
DC	pKa			
HL 1.63E-001	atm m3/mol	25	EST	MEYLAN,WM & HOWARD,PH (1991)
OH 5.09E-010	cm3/molc sec	25	EST	MEYLAN,WM & HOWARD,PH (1993)

000101-05-3 — ANILAZINE

CAS #: 000101-05-3				
Formula: $C_9H_5Cl_3N_4$				
Mol Weight: 275.53				
MP (deg C): 159-160		FP (deg C):		
BP (deg C):				
BP pressure (mm Hg):				

Property/Value	Units	Temp	Data Type	Reference
WS 8.00E+000	mg/L	20	EXP	TOMLIN,C (1994)
logP 3.88			EXP	SAITO,H ET AL. (1993)
VP 6.20E-009	mm Hg	20	EXP	TOMLIN,C (1994)
DC	pKa			
HL 2.81E-010	atm m3/mol	20	EST	MEYLAN,WM & HOWARD,PH (1993)
OH 4.28E-011	cm3/molc sec	25	EST	MEYLAN,WM & HOWARD,PH (1993)

CAS #: 000101-14-4				4,4'-METHYLENEBIS(2-CHLOROANILINE)

Formula: $C_{13}H_{12}Cl_2N_2$

Mol Weight: 267.16

MP (deg C): 110

FP (deg C):

BP (deg C): 378.9

BP pressure (mm Hg):

Property/Value	Units	Temp	Data Type	Reference
WS 1.39E+001	mg/L	24	EXP	VOORMAN,R & PENNER,D (1986A)
logP 3.91			EXP	CHEM INSPECT TEST INST (1992)
VP 2.86E-007	mm Hg	25	EST	NEELY,WB & BLAU,GE (1985)
DC	pKa			
HL 4.06E-011	atm m3/mol	25	EST	MEYLAN,WM & HOWARD,PH (1991)
OH 1.23E-010	cm3/molc sec	25	EST	MEYLAN,WM & HOWARD,PH (1993)

CAS #: 000101-16-6				3-METHOXY-N-PHENYLBENZENAMINE

Formula: $C_{13}H_{13}NO$

Mol Weight: 199.25

MP (deg C): 72-74

FP (deg C):

BP (deg C):

BP pressure (mm Hg):

Property/Value	Units	Temp	Data Type	Reference
WS 5.82E+001	mg/L	25	EST	MEYLAN,WM ET AL. (1996)
logP 3.37			EST	MEYLAN,WM & HOWARD,PH (1995)
VP 2.50E-004	mm Hg	25	EST	NEELY,WB & BLAU,GE (1985)
DC	pKa			
HL 6.21E-008	atm m3/mol	25	EST	MEYLAN,WM & HOWARD,PH (1991)
OH 2.01E-010	cm3/molc sec	25	EST	MEYLAN,WM & HOWARD,PH (1993)

CAS #: 000101-17-7				3-CHLORODIPHENYL ETHER

Formula: $C_{12}H_{10}ClN$

Mol Weight: 203.67

MP (deg C):

FP (deg C):

BP (deg C): 338

BP pressure (mm Hg):

Property/Value	Units	Temp	Data Type	Reference
WS 1.82E+001	mg/L	25	EST	MEYLAN,WM ET AL. (1996)
logP 3.94			EST	MEYLAN,WM & HOWARD,PH (1995)
VP 3.75E-004	mm Hg	25	EST	NEELY,WB & BLAU,GE (1985)
DC	pKa			
HL 7.78E-007	atm m3/mol	25	EST	MEYLAN,WM & HOWARD,PH (1991)
OH 2.00E-010	cm3/molc sec	25	EST	MEYLAN,WM & HOWARD,PH (1993)

CAS #: 000101-21-3				CHLORPROPHAM

Formula: $C_{10}H_{12}ClNO_2$

Mol Weight: 213.67

MP (deg C): 41.4

FP (deg C):

BP (deg C):

BP pressure (mm Hg):

Property/Value	Units	Temp	Data Type	Reference
WS 8.90E+001	mg/L	25	EXP	WAUCHOPE,RD ET AL. (1992)
logP 3.51			EXP	SANGSTER,J (1993)
VP 7.50E-006	mm Hg	25	EST	SUNTIO,LR ET AL. (1988)
DC	pKa			
HL 2.40E-008	atm m3/mol	25	EST	VP/WSOL
OH 4.70E-011	cm3/molc sec	25	EST	ATKINSON,R (1988)

CAS #: 000101-23-5				N-PHENYL-3-(TRIFLUOROMETHYL)BENZENAMINE

Formula: $C_{13}H_{10}F_3N$

Mol Weight: 237.23

MP (deg C):

FP (deg C):

BP (deg C):

BP pressure (mm Hg):

Property/Value	Units	Temp	Data Type	Reference
WS 6.47E+000	mg/L	25	EST	MEYLAN,WM ET AL. (1996)
logP 4.25			EST	MEYLAN,WM & HOWARD,PH (1995)
VP 1.42E-003	mm Hg	25	EST	NEELY,WB & BLAU,GE (1985)
DC	pKa			
HL 9.13E-006	atm m3/mol	25	EST	MEYLAN,WM & HOWARD,PH (1991)
OH 1.69E-010	cm3/molc sec	25	EST	MEYLAN,WM & HOWARD,PH (1993)

CAS #: 000101-27-9				BARBAN

Formula: $C_{11}H_9Cl_2NO_2$

Mol Weight: 258.11

MP (deg C): 75

FP (deg C):

BP (deg C):

BP pressure (mm Hg):

Property/Value	Units	Temp	Data Type	Reference
WS 1.10E+001	mg/L	25	EXP	YALKOWSKY,SH & DANNENFELSER,RM (1992)
logP 3.41			EST	MEYLAN,WM & HOWARD,PH (1995)
VP 3.80E-007	mm Hg	25	EXP	AUGUSTIJN-BECKERS,PWM ET AL. (1994)
DC	pKa			
HL 1.17E-008	atm m3/mol	25	EST	VP/WSOL
OH 5.24E-011	cm3/molc sec	25	EST	MEYLAN,WM & HOWARD,PH (1993)

CAS #: 000101-41-7				PHENYLACETIC ACID, METHYL ESTER

Formula: $C_9H_{10}O_2$

Mol Weight: 150.18

MP (deg C):

FP (deg C):

BP (deg C): 216.5

BP pressure (mm Hg):

Property/Value	Units	Temp	Data Type	Reference
WS 2.07E+003	mg/L	25	EST	MEYLAN,WM ET AL. (1996)
logP 1.83			EXP	HANSCH,C & LEO,AJ (1985)
VP 1.64E-001	mm Hg	25	EST	NEELY,WB & BLAU,GE (1985)
DC	pKa			
HL 1.42E-005	atm m3/mol	25	EST	MEYLAN,WM & HOWARD,PH (1991)
OH 4.14E-012	cm3/molc sec	25	EST	MEYLAN,WM & HOWARD,PH (1993)

CAS #: 000101-42-8				FENURON

Formula: $C_9H_{12}N_2O$

Mol Weight: 164.21

MP (deg C): 133-4

FP (deg C):

BP (deg C):

BP pressure (mm Hg):

Property/Value	Units	Temp	Data Type	Reference
WS 4.03E+003	mg/L	25	EXP	YALKOWSKY,SH & DANNENFELSER,RM (1992)
logP 0.98			EXP	HANSCH,C & LEO,AJ (1985)
VP 1.60E-004	mm Hg	60		WORTHING,CR & WALKER,SB (1987)
DC	pKa			
HL 9.71E-010	atm m3/mol	25	EST	MEYLAN,WM & HOWARD,PH (1991)
OH 1.68E-010	cm3/molc sec	25	EST	ATKINSON,R (1988)

000101-43-9 — 2-METHYLCYCLOHEXYL ACRYLATE

Formula: $C_{10}H_{16}O_2$

Mol Weight: 168.24

MP (deg C): | FP (deg C):

BP (deg C): 210

BP pressure (mm Hg):

Property/Value	Units	Temp	Data Type	Reference
WS 1.33E+002	mg/L	25	EST	MEYLAN,WM ET AL. (1996)
logP 3.13			EXP	SANGSTER,J (1993)
VP 1.91E-001	mm Hg		EST	NEELY,WB & BLAU,GE (1985)
DC	pKa			
HL 2.63E-004	atm m3/mol	25	EST	MEYLAN,WM & HOWARD,PH (1991)
OH 2.99E-011	cm3/molc sec	25	EST	MEYLAN,WM & HOWARD,PH (1993)

000101-53-1 — 4-HYDROXYDIPHENYLMETHANE

Formula: $C_{13}H_{12}O$

Mol Weight: 184.24

MP (deg C): 84 | FP (deg C):

BP (deg C): 320-322

BP pressure (mm Hg):

Property/Value	Units	Temp	Data Type	Reference
WS 7.32E+001	mg/L	25	EXP	CHEM INSPECT TEST INST (1992)
logP 3.47			EXP	CHEM INSPECT TEST INST (1992)
VP 6.85E-005	mm Hg	25	EST	NEELY,WB & BLAU,GE (1985)
DC	pKa			
HL 4.99E-008	atm m3/mol	25	EST	MEYLAN,WM & HOWARD,PH (1991)
OH 4.65E-011	cm3/molc sec	25	EST	MEYLAN,WM & HOWARD,PH (1993)

000101-54-2 — 4-AMINODIPHENYLAMINE

Formula: $C_{12}H_{12}N_2$

Mol Weight: 184.24

MP (deg C): 66 | FP (deg C):

BP (deg C): 354

BP pressure (mm Hg):

Property/Value	Units	Temp	Data Type	Reference
WS 1.45E+003	mg/L	25	EST	MEYLAN,WM ET AL. (1996)
logP 1.82			EST	MEYLAN,WM & HOWARD,PH (1995)
VP 1.21E-005	mm Hg	25	EXT	DAUBERT,TE & DANNER,RP (1989)
DC 5.20	pKa	25	EXP	PERRIN,DD (1972)
HL 3.71E-010	atm m3/mol	25	EST	MEYLAN,WM & HOWARD,PH (1991)
OH 2.00E-010	cm3/molc sec	25	EST	MEYLAN,WM & HOWARD,PH (1993)

000101-55-3 — 4-BROMOPHENYL PHENYL ETHER

Formula: $C_{12}H_9BrO$

Mol Weight: 249.11

MP (deg C): 18.7 | FP (deg C):

BP (deg C): 310.1

BP pressure (mm Hg):

Property/Value	Units	Temp	Data Type	Reference
WS 1.45E+000	mg/L	25	EST	MEYLAN,WM ET AL. (1996)
logP 4.94			EST	MEYLAN,WM & HOWARD,PH (1995)
VP 1.50E-003	mm Hg	20	EXT	CALLAHAN,MA ET AL. (1979)
DC	pKa			
HL 1.17E-004	atm m3/mol	25	EST	MEYLAN,WM & HOWARD,PH (1991)
OH 1.24E-011	cm3/molc sec	25	EST	MEYLAN,WM & HOWARD,PH (1993)

000101-61-1 — BIS(4-DIMETHYLAMINOPHENYL)METHANE

Formula: $C_{17}H_{22}N_2$

Mol Weight: 254.38

MP (deg C): 90-91 | FP (deg C):

BP (deg C): 390

BP pressure (mm Hg):

Property/Value	Units	Temp	Data Type	Reference
WS 4.14E+000	mg/L	25	EST	MEYLAN,WM ET AL. (1996)
logP 4.37			EST	MEYLAN,WM & HOWARD,PH (1995)
VP 1.75E-005	mm Hg	25	EST	NEELY,WB & BLAU,GE (1985)
DC	pKa			
HL 1.07E-009	atm m3/mol	25	EST	MEYLAN,WM & HOWARD,PH (1991)
OH 3.27E-010	cm3/molc sec	25	EST	MEYLAN,WM & HOWARD,PH (1993)

000101-68-8 — DIPHENYL METHANE DIISOCYANATE

Formula: $C_{15}H_{10}N_2O_2$

Mol Weight: 250.26

MP (deg C): 38 | FP (deg C):

BP (deg C): 196

BP pressure (mm Hg): 7.00E-001

Property/Value	Units	Temp	Data Type	Reference
WS 8.29E-001	mg/L	25	EST	MEYLAN,WM ET AL. (1996)
logP 5.22			EST	MEYLAN,WM & HOWARD,PH (1995)
VP 1.17E-005	mm Hg	25	EST	DABERT,TE & DANNER,RP (1989); SUPERCOOLED
DC	pKa			
HL 8.95E-007	atm m3/mol	25	EST	MEYLAN,WM & HOWARD,PH (1991)
OH 1.20E-011	cm3/molc sec	25	EST	ATKINSON,R (1988)

000101-77-9 — DI-(P-AMINOPHENYL)METHANE

Formula: $C_{13}H_{14}N_2$

Mol Weight: 198.27

MP (deg C): 89.0 | FP (deg C):

BP (deg C): 398-399

BP pressure (mm Hg): 7.68E+002

Property/Value	Units	Temp	Data Type	Reference
WS 1.00E+003	mg/L	25	EXP	MOORE,WM (1978)
logP 1.59			EXP	HANSCH,C & LEO,AJ (1985)
VP 2.97E+000	mm Hg	25	EST	CLAUSIUS-CLAPEYRON EQN
DC	pKa			
HL 5.60E-011	atm m3/mol	25	EST	MEYLAN,WM & HOWARD,PH (1991)
OH 3.00E-011	cm3/molc sec	25	EXP	ATKINSON,R (1989)

000101-80-4 — 4,4'-DIAMINODIPHENYL ETHER

Formula: $C_{12}H_{12}N_2O$

Mol Weight: 200.24

MP (deg C): 186-187 | FP (deg C):

BP (deg C): 350

BP pressure (mm Hg):

Property/Value	Units	Temp	Data Type	Reference
WS 5.60E+002	mg/L	25	EST	MEYLAN,WM ET AL. (1996)
logP 2.22			EST	MEYLAN,WM & HOWARD,PH (1995)
VP 4.36E-006	mm Hg	25	EST	NEELY,WB & BLAU,GE (1985)
DC	pKa			
HL 1.50E-011	atm m3/mol	25	EST	MEYLAN,WM & HOWARD,PH (1991)
OH 2.10E-010	cm3/molc sec	25	EST	ATKINSON,R (1988)

000101-81-5 — BENZENE, 1,1'-METHYLENEBIS-

Formula: $C_{13}H_{12}$
Mol Weight: 168.24
MP (deg C): 25.9
FP (deg C):
BP (deg C): 264.5
BP pressure (mm Hg):

Property/Value	Units	Temp	Data Type	Reference
WS 1.41E+001	mg/L	25	EXP	YALKOWSKY,SH & DANNENFELSER,RM (1992)
logP 4.14			EXP	HANSCH,C & LEO,AJ (1985)
VP 8.21E-003	mm Hg	25	EXP	JONES,AH (1960)
DC	pKa			
HL 1.29E-004	atm m3/mol	25	EST	VP/WSOL
OH 1.06E-011	cm3/molc sec	25	EST	MEYLAN,WM & HOWARD,PH (1993)

000101-82-6 — 2-BENZYLPYRIDINE

Formula: $C_{12}H_{11}N$
Mol Weight: 169.23
MP (deg C): 12.5
FP (deg C):
BP (deg C): 277
BP pressure (mm Hg):

Property/Value	Units	Temp	Data Type	Reference
WS 3.02E+003	mg/L	25	EST	MEYLAN,WM ET AL. (1996)
logP 3.06			EST	MEYLAN,WM & HOWARD,PH (1995)
VP 3.02E-003	mm Hg	25	EST	NEELY,WB & BLAU,GE (1985)
DC 5.13	pKa	25	EXP	PERRIN,DD (1965)
HL 6.28E-007	atm m3/mol	25	EST	MEYLAN,WM & HOWARD,PH (1991)
OH 6.70E-012	cm3/molc sec	25	EST	MEYLAN,WM & HOWARD,PH (1993)

000101-84-8 — DIPHENYL ETHER

Formula: $C_{12}H_{10}O$
Mol Weight: 170.21
MP (deg C): 28
FP (deg C):
BP (deg C): 259
BP pressure (mm Hg):

Property/Value	Units	Temp	Data Type	Reference
WS 1.80E+001	mg/L	25	EXP	BANERJEE,S ET AL. (1980)
logP 4.21			EXP	HANSCH,C & LEO,AJ (1985)
VP 2.25E-002	mm Hg	25	EXP	AMBROSE,D ET AL. (1976)
DC	pKa			
HL 2.79E-004	atm m3/mol	25	EST	VP/WSOL
OH 1.93E-011	cm3/molc sec	25	EST	MEYLAN,WM & HOWARD,PH (1993)

000101-97-3 — BENZENEACETIC ACID, ETHYL ESTER

Formula: $C_{10}H_{12}O_2$
Mol Weight: 164.21
MP (deg C): -29.4
FP (deg C):
BP (deg C): 226
BP pressure (mm Hg):

Property/Value	Units	Temp	Data Type	Reference
WS 7.39E+002	mg/L	25	EST	MEYLAN,WM ET AL. (1996)
logP 2.28			EXP	HANSCH,C ET AL. (1995)
VP 6.24E-002	mm Hg	25	EST	NEELY,WB & BLAU,GE (1985)
DC	pKa			
HL 1.88E-005	atm m3/mol	25	EST	MEYLAN,WM & HOWARD,PH (1991)
OH 5.59E-012	cm3/molc sec	25	EST	MEYLAN,WM & HOWARD,PH (1993)

000101-99-5 — N-PHENYL ETHYLCARBAMATE

Formula: $C_9H_{11}NO_2$
Mol Weight: 165.19
MP (deg C): 53
FP (deg C):
BP (deg C): 238
BP pressure (mm Hg):

Property/Value	Units	Temp	Data Type	Reference
WS 7.03E+002	mg/L	25	EST	MEYLAN,WM ET AL. (1996)
logP 2.30			EXP	HANSCH,C & LEO,AJ (1985)
VP 2.14E-002	mm Hg	25	EST	NEELY,WB & BLAU,GE (1985)
DC	pKa			
HL 2.90E-008	atm m3/mol	25	EST	MEYLAN,WM & HOWARD,PH (1991)
OH 4.43E-011	cm3/molc sec	25	EST	MEYLAN,WM & HOWARD,PH (1993)

000102-01-2 — ACETOACETANILIDE

Formula: $C_{10}H_{11}NO_2$
Mol Weight: 177.20
MP (deg C): 85
FP (deg C):
BP (deg C):
BP pressure (mm Hg):

Property/Value	Units	Temp	Data Type	Reference
WS 7.83E+003	mg/L	25	EST	MEYLAN,WM ET AL. (1996)
logP 1.01			EST	MEYLAN,WM & HOWARD,PH (1995)
VP 4.29E-005	mm Hg	25	EXT	DAUBERT,TE & DANNER,RP (1993)
DC 10.68	pKa	20	EXP	SERJEANT,EP & DEMPSEY,B (1979)
HL 4.15E-012	atm m3/mol	25	EST	MEYLAN,WM & HOWARD,PH (1991)
OH 1.30E-011	cm3/molc sec	25	EST	MEYLAN,WM & HOWARD,PH (1993)

000102-03-4 — ACETAMIDE, N-[(PHENYLAMINO)CARBONYL]-

Formula: $C_9H_{10}N_2O_2$
Mol Weight: 178.19
MP (deg C):
FP (deg C):
BP (deg C):
BP pressure (mm Hg):

Property/Value	Units	Temp	Data Type	Reference
WS 2.42E+003	mg/L	25	EST	MEYLAN,WM ET AL. (1996)
logP 1.60			EXP	SOTOMATSU,T ET AL. (1987)
VP 1.15E-007	mm Hg	25	EST	NEELY,WB & BLAU,GE (1985)
DC	pKa			
HL 5.95E-011	atm m3/mol	25	EST	MEYLAN,WM & HOWARD,PH (1991)
OH 4.38E-011	cm3/molc sec	25	EST	MEYLAN,WM & HOWARD,PH (1993)

000102-04-5 — DIBENZYL KETONE

Formula: $C_{15}H_{14}O$
Mol Weight: 210.28
MP (deg C): 35
FP (deg C):
BP (deg C): 331
BP pressure (mm Hg):

Property/Value	Units	Temp	Data Type	Reference
WS 7.44E+001	mg/L	25	EST	MEYLAN,WM ET AL. (1996)
logP 3.18			EST	MEYLAN,WM & HOWARD,PH (1995)
VP 7.39E-004	mm Hg	25	EXT	PERRY,RH & GREEN,D (1984)
DC	pKa			
HL 3.23E-007	atm m3/mol	25	EST	MEYLAN,WM & HOWARD,PH (1991)
OH 1.11E-011	cm3/molc sec	25	EST	MEYLAN,WM & HOWARD,PH (1993)

CAS #: 000102-06-7 — N,N'-DIPHENYLGUANIDINE

Formula: $C_{13}H_{13}N_3$
Mol Weight: 211.27
MP (deg C): 150
FP (deg C):
BP (deg C):
BP pressure (mm Hg):

Property/Value	Units	Temp	Data Type	Reference
WS 1.00E+003	mg/L	25	EXP	CHEM INSPECT TEST INST (1992)
logP 2.89			EST	MEYLAN,WM & HOWARD,PH (1995)
VP 8.45E-006	mm Hg	25	EST	NEELY,WB & BLAU,GE (1985)
DC 10.12	pKa		EXP	PERRIN,DD (1965)
HL 7.12E-012	atm m3/mol	25	EST	MEYLAN,WM & HOWARD,PH (1991)
OH 8.53E-011	cm3/molc sec	25	EST	MEYLAN,WM & HOWARD,PH (1993)

CAS #: 000102-07-8 — DIPHENYLUREA, SYM

Formula: $C_{13}H_{12}N_2O$
Mol Weight: 212.25
MP (deg C): 238
FP (deg C):
BP (deg C): 260
BP pressure (mm Hg):

Property/Value	Units	Temp	Data Type	Reference
WS 1.50E+002	mg/L		EXP	YALKOWSKY,SH & DANNENFELSER,RM (1992)
logP 3.00			EXP	HANSCH,C ET AL. (1995)
VP 2.01E-006	mm Hg	25	EST	NEELY,WB & BLAU,GE (1985)
DC	pKa			
HL 1.11E-010	atm m3/mol	25	EST	MEYLAN,WM & HOWARD,PH (1991)
OH 8.53E-011	cm3/molc sec	25	EST	MEYLAN,WM & HOWARD,PH (1993)

CAS #: 000102-08-9 — THIOCARBANILIDE

Formula: $C_{13}H_{12}N_2S$
Mol Weight: 228.32
MP (deg C): 153-154
FP (deg C):
BP (deg C):
BP pressure (mm Hg):

Property/Value	Units	Temp	Data Type	Reference
WS 5.60E+001	mg/L	25	EST	MEYLAN,WM ET AL. (1996)
logP 3.21			EST	MEYLAN,WM & HOWARD,PH (1995)
VP 5.94E-006	mm Hg	25	EST	NEELY,WB & BLAU,GE (1985)
DC	pKa			
HL 4.81E-008	atm m3/mol	25	EST	MEYLAN,WM & HOWARD,PH (1991)
OH 8.53E-011	cm3/molc sec	25	EST	MEYLAN,WM & HOWARD,PH (1993)

CAS #: 000102-09-0 — CARBONIC ACID, DIPHENYL ESTER

Formula: $C_{13}H_{10}O_3$
Mol Weight: 214.22
MP (deg C): 80-81
FP (deg C):
BP (deg C): 302-306
BP pressure (mm Hg):

Property/Value	Units	Temp	Data Type	Reference
WS 5.84E+001	mg/L	25	EST	MEYLAN,WM ET AL. (1996)
logP 3.28			EXP	HANSCH,C ET AL. (1995)
VP 4.23E-004	mm Hg	25	EST	NEELY,WB & BLAU,GE (1985)
DC	pKa			
HL 8.48E-005	atm m3/mol	25	EST	MEYLAN,WM & HOWARD,PH (1991)
OH 4.02E-012	cm3/molc sec	25	EST	MEYLAN,WM & HOWARD,PH (1993)

CAS #: 000102-25-0 — 1,3,5-TRIETHYLBENZENE

Formula: $C_{12}H_{18}$
Mol Weight: 162.28
MP (deg C): -66.5
FP (deg C):
BP (deg C): 215.9
BP pressure (mm Hg):

Property/Value	Units	Temp	Data Type	Reference
WS 1.40E-002	mg/L	20	EXP	MACINTYRE,WG & DEFUR,PO (1985)
logP 5.11			EST	MEYLAN,WM & HOWARD,PH (1995)
VP 7.98E-002	mm Hg	25	EST	NEELY,WB & BLAU,GE (1985)
DC	pKa			
HL 1.69E-002	atm m3/mol	25	EST	MEYLAN,WM & HOWARD,PH (1991)
OH 3.32E-011	cm3/molc sec	25	EST	MEYLAN,WM & HOWARD,PH (1993)

CAS #: 000102-29-4 — M-HYDROXYPHENYL ACETATE

Formula: $C_8H_8O_3$
Mol Weight: 152.15
MP (deg C):
FP (deg C):
BP (deg C): 283
BP pressure (mm Hg):

Property/Value	Units	Temp	Data Type	Reference
WS 2.51E+004	mg/L	25	EST	MEYLAN,WM ET AL. (1996)
logP 1.23			EXP	HANSCH,C & LEO,AJ (1985)
VP 5.63E-003	mm Hg	25	EST	NEELY,WB & BLAU,GE (1985)
DC	pKa			
HL 6.74E-009	atm m3/mol	25	EST	MEYLAN,WM & HOWARD,PH (1991)
OH 3.46E-011	cm3/molc sec	25	EST	MEYLAN,WM & HOWARD,PH (1993)

CAS #: 000102-32-9 — 3,4-DIHYDROXYPHENYLACETIC ACID

Formula: $C_8H_8O_4$
Mol Weight: 168.15
MP (deg C): 127-130
FP (deg C):
BP (deg C):
BP pressure (mm Hg):

Property/Value	Units	Temp	Data Type	Reference
WS 8.62E+004	mg/L	25	EST	MEYLAN,WM ET AL. (1996)
logP 0.98			EXP	SANGSTER,J (1994)
VP 8.18E-007	mm Hg	25	EST	NEELY,WB & BLAU,GE (1985)
DC 4.25	pKa	30	EXP	SERJEANT,EP & DEMPSEY,B (1979)
HL 4.79E-016	atm m3/mol	25	EST	MEYLAN,WM & HOWARD,PH (1991)
OH 3.96E-011	cm3/molc sec	25	EST	MEYLAN,WM & HOWARD,PH (1993)

CAS #: 000102-36-3 — 1,2-DICHLORO-4-ISOCYANATOBENZENE

Formula: $C_7H_3Cl_2NO$
Mol Weight: 188.01
MP (deg C): 43
FP (deg C):
BP (deg C): 113
BP pressure (mm Hg): 9.75E+000

Property/Value	Units	Temp	Data Type	Reference
WS 2.43E+001	mg/L	25	EST	MEYLAN,WM ET AL. (1996)
logP 3.88			EST	MEYLAN,WM & HOWARD,PH (1995)
VP 5.60E-001	mm Hg	43	EXP	DAUBERT,TE & DANNER,RP (1989)
DC	pKa			
HL 1.28E-004	atm m3/mol	25	EST	MEYLAN,WM & HOWARD,PH (1991)
OH 4.30E-011	cm3/molc sec	25	EST	MEYLAN,WM & HOWARD,PH (1993)

CAS #:	000102-37-4		ETHYL CINNAMATE,3,4-DIHYDROXY

Formula: $C_{11}H_{12}O_4$

Mol Weight: 208.22

MP (deg C):

FP (deg C):

BP (deg C):

BP pressure (mm Hg):

Property/Value	Units	Temp	Data Type	Reference
WS 9.83E+002	mg/L	25	EST	MEYLAN,WM ET AL. (1996)
logP 2.56			EXP	HANSCH,C & LEO,AJ (1985)
VP 1.73E-006	mm Hg	25	EST	NEELY,WB & BLAU,GE (1985)
DC	pKa			
HL 5.95E-014	atm m3/mol	25	EST	MEYLAN,WM & HOWARD,PH (1991)
OH 4.33E-011	cm3/molc sec	25	EST	MEYLAN,WM & HOWARD,PH (1993)

CAS #:	000102-38-5		3-NITROFORMANILIDE

Formula: $C_7H_6N_2O_3$

Mol Weight: 166.14

MP (deg C):

FP (deg C):

BP (deg C):

BP pressure (mm Hg):

Property/Value	Units	Temp	Data Type	Reference
WS 4.09E+003	mg/L	25	EST	MEYLAN,WM ET AL. (1996)
logP 1.40			EXP	HANSCH,C & LEO,AJ (1985)
VP 9.83E-006	mm Hg	25	EST	NEELY,WB & BLAU,GE (1985)
DC	pKa			
HL 3.33E-011	atm m3/mol	25	EST	MEYLAN,WM & HOWARD,PH (1991)
OH 3.73E-012	cm3/molc sec	25	EST	MEYLAN,WM & HOWARD,PH (1993)

CAS #:	000102-50-1		BENZENAMINE, 4-METHOXY-2-METHYL-

Formula: $C_8H_{11}NO$

Mol Weight: 137.18

MP (deg C): 29.5

FP (deg C):

BP (deg C): 248.5

BP pressure (mm Hg):

Property/Value	Units	Temp	Data Type	Reference
WS 7.66E+003	mg/L	25	EST	MEYLAN,WM ET AL. (1996)
logP 1.23			EXP	DEBNATH,AK ET AL. (1992)
VP 2.52E-002	mm Hg	25	EST	NEELY,WB & BLAU,GE (1985)
DC	pKa			
HL 1.24E-007	atm m3/mol	25	EST	MEYLAN,WM & HOWARD,PH (1991)
OH 1.15E-010	cm3/molc sec	25	EST	MEYLAN,WM & HOWARD,PH (1993)

CAS #:	000102-56-7		2,5-DIMETHOXYANILINE

Formula: $C_8H_{11}NO_2$

Mol Weight: 153.18

MP (deg C): 82.5

FP (deg C):

BP (deg C): 270

BP pressure (mm Hg):

Property/Value	Units	Temp	Data Type	Reference
WS 6.94E+003	mg/L	25	EST	MEYLAN,WM ET AL. (1996)
logP 1.20			EXP	HANSCH,C & LEO,AJ (1985)
VP 8.28E-003	mm Hg	25	EST	NEELY,WB & BLAU,GE (1985)
DC 3.93	pKa		EXP	PERRIN,DD (1972)
HL 6.66E-009	atm m3/mol	25	EST	MEYLAN,WM & HOWARD,PH (1991)
OH 2.02E-010	cm3/molc sec	25	EST	MEYLAN,WM & HOWARD,PH (1993)

CAS #:	000102-69-2		TRIPROPYLAMINE

Formula: $C_9H_{21}N$

Mol Weight: 143.27

MP (deg C): -93.5

FP (deg C):

BP (deg C): 156

BP pressure (mm Hg):

Property/Value	Units	Temp	Data Type	Reference
WS 7.48E+002	mg/L	25	EXP	YALKOWSKY,SH & DANNENFELSER,RM (1992)
logP 2.79			EXP	HANSCH,C & LEO,AJ (1985)
VP 1.51E+000	mm Hg	25	EXP	DAUBERT,TE & DANNER,RP (1989)
DC 10.65	pKa		EXP	PERRIN,DD (1965)
HL 3.81E-004	atm m3/mol	25	EST	VP/WSOL
OH 1.02E-010	cm3/molc sec	25	EST	MEYLAN,WM & HOWARD,PH (1993)

CAS #:	000102-70-5		2-PROPEN-1-AMINE, N,N-DI-2-PROPENYL-

Formula: $C_9H_{15}N$

Mol Weight: 137.23

MP (deg C): 94

FP (deg C):

BP (deg C): 155.5

BP pressure (mm Hg):

Property/Value	Units	Temp	Data Type	Reference
WS 5.38E+003	mg/L	25	EST	MEYLAN,WM ET AL. (1996)
logP 2.59			EXP	HANSCH,C ET AL. (1995)
VP 2.61E+000	mm Hg	25	EST	NEELY,WB & BLAU,GE (1985)
DC 8.31	pKa	25	EXP	PERRIN,DD (1965)
HL 8.27E-005	atm m3/mol	25	EST	MEYLAN,WM & HOWARD,PH (1991)
OH 1.71E-010	cm3/molc sec	25	EST	MEYLAN,WM & HOWARD,PH (1993)

CAS #:	000102-71-6		TRIETHANOLAMINE

Formula: $C_6H_{15}NO_3$

Mol Weight: 149.19

MP (deg C): 21.57

FP (deg C):

BP (deg C): 335.4

BP pressure (mm Hg):

Property/Value	Units	Temp	Data Type	Reference
WS 1.00E+006	mg/L		EXP	RIDDICK,JA ET AL. (1986)
logP -1.00			EXP	HANSCH,C ET AL. (1995)
VP 3.59E-006	mm Hg	25	EXP	DAUBERT,TE & DANNER,RP (1989)
DC 7.76	pKa	25	EXP	PERRIN,DD (1965)
HL 7.05E-013	atm m3/mol	25	EST	VP/WSOL
OH 1.04E-010	cm3/molc sec	25	EST	MEYLAN,WM & HOWARD,PH (1993)

CAS #:	000102-76-1		TRIACETIN

Formula: $C_9H_{14}O_6$

Mol Weight: 218.21

MP (deg C): -78

FP (deg C):

BP (deg C): 258

BP pressure (mm Hg):

Property/Value	Units	Temp	Data Type	Reference
WS 5.80E+004	mg/L	25	EXP	RIDDICK,JA ET AL. (1986)
logP 0.25			EXP	HANSCH,C ET AL. (1995)
VP 2.48E-003	mm Hg	25	EXP	DAUBERT,TE & DANNER,RP (1989)
DC	pKa			
HL 1.23E-008	atm m3/mol	25	EST	VP/WSOL
OH 7.81E-012	cm3/molc sec	25	EST	ATKINSON,R (1988)

157

CAS #: 000102-82-9 — TRI N-BUTYLAMINE

Formula:	$C_{12}H_{27}N$
Mol Weight:	185.36
MP (deg C): -70	FP (deg C):
BP (deg C): 216	
BP pressure (mm Hg):	

Property/Value	Units	Temp	Data Type	Reference
WS 1.42E+002	mg/L	25	EXP	YALKOWSKY,SH & DANNENFELSER,RM (1992)
logP 4.46			EST	MEYLAN,WM & HOWARD,PH (1995)
VP 1.05E-001	mm Hg	25	EST	NEELY,WB & BLAU,GE (1985)
DC 10.89	pKa		EXP	RIDDICK,JA ET AL. (1986)
HL 1.28E-004	atm m3/mol	25	EST	MEYLAN,WM & HOWARD,PH (1991)
OH 1.00E-010	cm3/molc sec	25	EST	MEYLAN,WM & HOWARD,PH (1993)

CAS #: 000102-93-2 — PROPIONAMIDE, G-PHENYL

Formula:	$C_9H_{11}NO$
Mol Weight:	149.19
MP (deg C):	FP (deg C):
BP (deg C):	
BP pressure (mm Hg):	

Property/Value	Units	Temp	Data Type	Reference
WS 1.28E+004	mg/L	25	EST	MEYLAN,WM ET AL. (1996)
logP 0.91			EXP	HANSCH,C & LEO,AJ (1985)
VP 1.32E-004	mm Hg	25	EST	NEELY,WB & BLAU,GE (1985)
DC	pKa			
HL 1.20E-009	atm m3/mol	25	EST	MEYLAN,WM & HOWARD,PH (1991)
OH 1.13E-011	cm3/molc sec	25	EST	MEYLAN,WM & HOWARD,PH (1993)

CAS #: 000102-96-5 — BETA-NITROSTYRENE

Formula:	$C_8H_7NO_2$
Mol Weight:	149.15
MP (deg C):	FP (deg C):
BP (deg C):	
BP pressure (mm Hg):	

Property/Value	Units	Temp	Data Type	Reference
WS 4.92E+002	mg/L	25	EST	MEYLAN,WM ET AL. (1996)
logP 2.11			EXP	HANSCH,C & LEO,AJ (1985)
VP 1.54E-002	mm Hg	25	EST	NEELY,WB & BLAU,GE (1985)
DC	pKa			
HL 3.48E-006	atm m3/mol	25	EST	MEYLAN,WM & HOWARD,PH (1991)
OH 1.32E-011	cm3/molc sec	25	EST	MEYLAN,WM & HOWARD,PH (1993)

CAS #: 000102-97-6 — BENZENEMETHANAMINE, N-(1-METHYLETHYL)-

Formula:	$C_{10}H_{15}N$
Mol Weight:	149.24
MP (deg C):	FP (deg C):
BP (deg C): 200	
BP pressure (mm Hg):	

Property/Value	Units	Temp	Data Type	Reference
WS 2.13E+004	mg/L	25	EST	MEYLAN,WM ET AL. (1996)
logP 1.83			EXP	KRIL,MB & FUNG,HL (1990)
VP 1.63E-001	mm Hg	25	EST	NEELY,WB & BLAU,GE (1985)
DC	pKa			
HL 2.37E-006	atm m3/mol	25	EST	MEYLAN,WM & HOWARD,PH (1991)
OH 9.49E-011	cm3/molc sec	25	EST	MEYLAN,WM & HOWARD,PH (1993)

CAS #: 000103-00-4 — 1-CYCLOHEXYLAMINO-2-PROPANOL

Formula:	$C_9H_{19}NO$
Mol Weight:	157.26
MP (deg C):	FP (deg C):
BP (deg C):	
BP pressure (mm Hg):	

Property/Value	Units	Temp	Data Type	Reference
WS 3.47E+004	mg/L	25	EST	MEYLAN,WM ET AL. (1996)
logP 1.54			EST	MEYLAN,WM & HOWARD,PH (1995)
VP 1.58E-002	mm Hg	25	EXT	BOUBLIK,T ET AL. (1984)
DC	pKa			
HL 1.95E-009	atm m3/mol	25	EST	MEYLAN,WM & HOWARD,PH (1991)
OH 1.17E-010	cm3/molc sec	25	EST	MEYLAN,WM & HOWARD,PH (1993)

CAS #: 000103-01-5 — N-PHENYLGLYCINE

Formula:	$C_8H_9NO_2$
Mol Weight:	151.17
MP (deg C): 127-128	FP (deg C):
BP (deg C):	
BP pressure (mm Hg):	

Property/Value	Units	Temp	Data Type	Reference
WS 5.50E+004	mg/L	25	EST	MEYLAN,WM ET AL. (1996)
logP 0.62			EXP	HANSCH,C & LEO,AJ (1985)
VP 7.03E-004	mm Hg	25	EST	NEELY,WB & BLAU,GE (1985)
DC	pKa			
HL 6.20E-010	atm m3/mol	25	EST	MEYLAN,WM & HOWARD,PH (1991)
OH 4.97E-011	cm3/molc sec	25	EST	MEYLAN,WM & HOWARD,PH (1993)

CAS #: 000103-03-7 — HYDRAZINECARBOXAMIDE, 2-PHENYL-

Formula:	$C_7H_9N_3O$
Mol Weight:	151.17
MP (deg C): 172	FP (deg C):
BP (deg C):	
BP pressure (mm Hg):	

Property/Value	Units	Temp	Data Type	Reference
WS 2.84E+003	mg/L	25	EST	MEYLAN,WM ET AL. (1996)
logP 0.13			EXP	HANSCH,C ET AL. (1995)
VP 1.23E-004	mm Hg	25	EST	NEELY,WB & BLAU,GE (1985)
DC	pKa			
HL 8.41E-013	atm m3/mol	25	EST	MEYLAN,WM & HOWARD,PH (1991)
OH 4.47E-011	cm3/molc sec	25	EST	MEYLAN,WM & HOWARD,PH (1993)

CAS #: 000103-04-8 — S-PHENYLMERCAPTOACETIC ACID

Formula:	$C_8H_8O_2S$
Mol Weight:	168.22
MP (deg C): 64-66	FP (deg C):
BP (deg C):	
BP pressure (mm Hg):	

Property/Value	Units	Temp	Data Type	Reference
WS 3.86E+003	mg/L	25	EST	MEYLAN,WM ET AL. (1996)
logP 1.88			EXP	HANSCH,C & LEO,AJ (1985)
VP 2.78E-004	mm Hg	25	EST	NEELY,WB & BLAU,GE (1985)
DC	pKa			
HL 1.17E-009	atm m3/mol	25	EST	MEYLAN,WM & HOWARD,PH (1991)
OH 1.86E-011	cm3/molc sec	25	EST	MEYLAN,WM & HOWARD,PH (1993)

CAS #: 000103-09-3	2-ETHYLHEXYLACETATE
Formula: $C_{10}H_{20}O_2$	
Mol Weight: 172.27	
MP (deg C): -93	FP (deg C):
BP (deg C): 199	
BP pressure (mm Hg):	

Property/Value	Units	Temp	Data Type	Reference
WS 9.84E+001	mg/L	25	EST	MEYLAN,WM ET AL. (1996)
logP 3.74			EST	MEYLAN,WM & HOWARD,PH (1995)
VP 2.30E-001	mm Hg	25	EXP	DAUBERT,TE & DANNER,RP (1989)
DC	pKa			
HL 2.81E-003	atm m3/mol	25	EST	MEYLAN,WM & HOWARD,PH (1991)
OH 1.09E+001	cm3/molc sec	25	EST	MEYLAN,WM & HOWARD,PH (1993)

CAS #: 000103-11-7	2-ETHYLHEXYL ACRYLATE
Formula: $C_{11}H_{20}O_2$	
Mol Weight: 184.28	
MP (deg C): -90	FP (deg C):
BP (deg C): 213.5	
BP pressure (mm Hg):	

Property/Value	Units	Temp	Data Type	Reference
WS 1.00E+002	mg/L	25	EXP	CHEM INSPEC TEST INST
logP 4.09			EST	MEYLAN,WM & HOWARD,PH (1995)
VP 1.78E-001	mm Hg	25	EXP	DAUBERT,TE & DANNER,RP (1989)
DC	pKa			
HL 4.32E-004	atm m3/mol	25	EST	VP/WSOL
OH 2.01E-011	cm3/molc sec	25	EST	MEYLAN,WM & HOWARD,PH (1993)

CAS #: 000103-16-2	P-(BENZYLOXY) PHENOL
Formula: $C_{13}H_{12}O_2$	
Mol Weight: 200.24	
MP (deg C): 122.5	FP (deg C):
BP (deg C):	
BP pressure (mm Hg):	

Property/Value	Units	Temp	Data Type	Reference
WS 2.51E+002	mg/L	25	EST	MEYLAN,WM ET AL. (1996)
logP 3.30			EST	MEYLAN,WM & HOWARD,PH (1995)
VP 2.25E-005	mm Hg	25	EST	NEELY,WB & BLAU,GE (1985)
DC	pKa			
HL 2.68E-009	atm m3/mol	25	EST	MEYLAN,WM & HOWARD,PH (1991)
OH 3.95E-011	cm3/molc sec	25	EST	MEYLAN,WM & HOWARD,PH (1993)

CAS #: 000103-17-3	CHLORBENSIDE
Formula: $C_{13}H_{10}Cl_2S$	
Mol Weight: 269.19	
MP (deg C): 75-76	FP (deg C):
BP (deg C):	
BP pressure (mm Hg):	

Property/Value	Units	Temp	Data Type	Reference
WS 3.10E-001	mg/L	25	EST	MEYLAN,WM ET AL. (1996)
logP 5.59			EST	MEYLAN,WM & HOWARD,PH (1995)
VP 1.78E-005	mm Hg	25	EST	NEELY,WB & BLAU,GE (1985)
DC	pKa			
HL 6.96E-006	atm m3/mol	25	EST	MEYLAN,WM & HOWARD,PH (1991)
OH 1.26E-011	cm3/molc sec	25	EST	MEYLAN,WM & HOWARD,PH (1993)

CAS #: 000103-18-4	4-HYDROXY-4'-AMINOAZOBENZENE
Formula: $C_{12}H_{11}N_3O$	
Mol Weight: 213.24	
MP (deg C):	FP (deg C):
BP (deg C):	
BP pressure (mm Hg):	

Property/Value	Units	Temp	Data Type	Reference
WS 2.53E+002	mg/L	25	EST	MEYLAN,WM ET AL. (1996) .
logP 2.71			EST	MEYLAN,WM & HOWARD,PH (1995)
VP 5.39E-007	mm Hg	25	EST	NEELY,WB & BLAU,GE (1985)
DC	pKa			
HL 5.41E-013	atm m3/mol	25	EST	MEYLAN,WM & HOWARD,PH (1991)
OH 5.60E-011	cm3/molc sec	25	EST	MEYLAN,WM & HOWARD,PH (1993)

CAS #: 000103-23-1	DI-2-ETHYLHEXYL ADIPATE
Formula: $C_{22}H_{42}O_4$	
Mol Weight: 370.58	
MP (deg C): -67.8	FP (deg C):
BP (deg C): 417	
BP pressure (mm Hg):	

Property/Value	Units	Temp	Data Type	Reference
WS 7.80E-001	mg/L	22	EXP	FELDER,JD ET AL. (1986)
logP 6.11			EST	FELDER,JD ET AL. (1986); > VALUE
VP 8.50E-007	mm Hg	20	EXP	FELDER,JD ET AL. (1986)
DC	pKa			
HL 4.34E-007	atm m3/mol	20	EXP	FELDER,JD ET AL. (1986)
OH 2.45E-011	cm3/molc sec	25	EST	ATKINSON,R (1988)

CAS #: 000103-25-3	METHYL B-PHENYLPROPIONATE
Formula: $C_{10}H_{12}O_2$	
Mol Weight: 164.21	
MP (deg C):	FP (deg C):
BP (deg C): 238.5	
BP pressure (mm Hg):	

Property/Value	Units	Temp	Data Type	Reference
WS 6.83E+002	mg/L	25	EST	MEYLAN,WM ET AL. (1996)
logP 2.32			EXP	HANSCH,C & LEO,AJ (1985)
VP 6.24E-002	mm Hg	25	EST	NEELY,WB & BLAU,GE (1985)
DC	pKa			
HL 1.88E-005	atm m3/mol	25	EST	MEYLAN,WM & HOWARD,PH (1991)
OH 7.06E-012	cm3/molc sec	25	EST	MEYLAN,WM & HOWARD,PH (1993)

CAS #: 000103-26-4	CINNAMIC ACID, METHYL ESTER
Formula: $C_{10}H_{10}O_2$	
Mol Weight: 162.19	
MP (deg C):	FP (deg C):
BP (deg C):	
BP pressure (mm Hg):	

Property/Value	Units	Temp	Data Type	Reference
WS 3.87E+002	mg/L	25	EST	MEYLAN,WM ET AL. (1996)
logP 2.62			EXP	HANSCH,C & LEO,AJ (1985)
VP 4.65E-002	mm Hg	25	EST	NEELY,WB & BLAU,GE (1985)
DC	pKa			
HL 4.14E-006	atm m3/mol	25	EST	MEYLAN,WM & HOWARD,PH (1991)
OH 2.18E-011	cm3/molc sec	25	EST	MEYLAN,WM & HOWARD,PH (1993)

CAS #: 000103-29-7				BIBENZYL

Formula: $C_{14}H_{14}$
Mol Weight: 182.27
MP (deg C): 52-52.5
FP (deg C):
BP (deg C):
BP pressure (mm Hg):

Property/Value	Units	Temp	Data Type	Reference
WS 4.30E+000	mg/L	25	EXP	YALKOWSKY,SH & DANNENFELSER,RM (1992)
logP 4.79			EXP	HANSCH,C & LEO,AJ (1985)
VP 7.28E-003	mm Hg	25	EXT	CHAO,J ET AL. 91983)
DC	pKa			
HL 6.37E-004	atm m3/mol	25	EST	MEYLAN,WM & HOWARD,PH (1991)
OH 1.20E-011	cm3/molc sec	25	EST	MEYLAN,WM & HOWARD,PH (1993)

CAS #: 000103-30-0				STILBENE

Formula: $C_{14}H_{12}$
Mol Weight: 180.25
MP (deg C): 123
FP (deg C):
BP (deg C): 307
BP pressure (mm Hg):

Property/Value	Units	Temp	Data Type	Reference
WS 2.90E-001	mg/L	25	EXP	YALKOWSKY,SH & DANNENFELSER,RM (1992)
logP 4.81			EXP	HANSCH,C & LEO,AJ (1985)
VP 2.71E-003	mm Hg	25	EXT	JORDAN,TE (1954)
DC	pKa			
HL 1.24E-004	atm m3/mol	25	EST	MEYLAN,WM & HOWARD,PH (1991)
OH 6.01E-011	cm3/molc sec	25	EST	MEYLAN,WM & HOWARD,PH (1993)

CAS #: 000103-32-2				N-BENZYLANILINE

Formula: $C_{13}H_{13}N$
Mol Weight: 183.26
MP (deg C): 37-38
FP (deg C):
BP (deg C): 306-307
BP pressure (mm Hg):

Property/Value	Units	Temp	Data Type	Reference
WS 1.13E+002	mg/L	25	EST	MEYLAN,WM ET AL. (1996)
logP 3.13			EXP	HANSCH,C & LEO,AJ (1985)
VP 9.13E-004	mm Hg	25	EST	NEELY,WB & BLAU,GE (1985)
DC	pKa			
HL 3.37E-007	atm m3/mol	25	EST	MEYLAN,WM & HOWARD,PH (1991)
OH 5.62E-011	cm3/molc sec	25	EST	MEYLAN,WM & HOWARD,PH (1993)

CAS #: 000103-33-3				AZOBENZENE

Formula: $C_{12}H_{10}N_2$
Mol Weight: 182.23
MP (deg C): 68
FP (deg C):
BP (deg C): 293
BP pressure (mm Hg):

Property/Value	Units	Temp	Data Type	Reference
WS 6.40E+000	mg/L	25	EXP	TAKAGISHI,T ET AL. (1969)
logP 3.82			EXP	HANSCH,C & LEO,AJ (1985)
VP 3.61E-004	mm Hg	25	EXP	JONES,AH (1960)
DC -2.95	pKa		EXP	HOEFNAGEL,MA ET AL. (1969)
HL 1.35E-005	atm m3/mol	25	EST	VP/WSOL
OH 1.60E-012	cm3/molc sec	25	EST	MEYLAN,WM & HOWARD,PH (1993)

CAS #: 000103-36-6				CINNAMIC ACID, ETHYL ESTER

Formula: $C_{11}H_{12}O_2$
Mol Weight: 176.22
MP (deg C): 6.5-7.5
FP (deg C):
BP (deg C): 271
BP pressure (mm Hg):

Property/Value	Units	Temp	Data Type	Reference
WS 1.78E+002	mg/L	25	EXP	YALKOWSKY,SH & DANNENFELSER,RM (1992)
logP 2.99			EXP	HANSCH,C & LEO,AJ (1985)
VP 1.25E-002	mm Hg	25	EXT	OHE,S (1976)
DC	pKa			
HL 5.50E-006	atm m3/mol	25	EST	MEYLAN,WM & HOWARD,PH (1991)
OH 2.32E-011	cm3/molc sec	25	EST	MEYLAN,WM & HOWARD,PH (1993)

CAS #: 000103-38-8				BUTANOIC ACID, 3-METHYL-, PHENYLMETHYL ESTER

Formula: $C_{12}H_{16}O_2$
Mol Weight: 192.26
MP (deg C):
FP (deg C):
BP (deg C): 245
BP pressure (mm Hg):

Property/Value	Units	Temp	Data Type	Reference
WS 7.87E+001	mg/L	25	EST	MEYLAN,WM ET AL. (1996)
logP 3.26			EXP	YANG,HZ ET AL. (1987)
VP 1.68E-002	mm Hg	25	EST	NEELY,WB & BLAU,GE (1985)
DC	pKa			
HL 3.31E-005	atm m3/mol	25	EST	MEYLAN,WM & HOWARD,PH (1991)
OH 9.91E-012	cm3/molc sec	25	EST	MEYLAN,WM & HOWARD,PH (1993)

CAS #: 000103-41-3				BENZYL CINNAMATE

Formula: $C_{16}H_{14}O_2$
Mol Weight: 238.29
MP (deg C): 39
FP (deg C):
BP (deg C):
BP pressure (mm Hg):

Property/Value	Units	Temp	Data Type	Reference
WS 9.27E+000	mg/L	25	EST	MEYLAN,WM ET AL. (1996)
logP 4.06			EST	MEYLAN,WM & HOWARD,PH (1995)
VP 4.82E-007	mm Hg	25	EXT	PERRY,RH & GREEN,D (1984)
DC	pKa			
HL 3.34E-007	atm m3/mol	25	EST	MEYLAN,WM & HOWARD,PH (1991)
OH 2.79E-011	cm3/molc sec	25	EST	MEYLAN,WM & HOWARD,PH (1993)

CAS #: 000103-45-7				B-PHENYLETHYL ACETATE

Formula: $C_{10}H_{12}O_2$
Mol Weight: 164.21
MP (deg C): -31.1
FP (deg C):
BP (deg C): 232.6
BP pressure (mm Hg):

Property/Value	Units	Temp	Data Type	Reference
WS 7.11E+002	mg/L	25	EST	MEYLAN,WM ET AL. (1996)
logP 2.30			EXP	HANSCH,C & LEO,AJ (1985)
VP 3.14E-002	mm Hg	25	EXT	OHE,S (1976)
DC	pKa			
HL 1.88E-005	atm m3/mol	25	EST	MEYLAN,WM & HOWARD,PH (1991)
OH 7.93E-012	cm3/molc sec	25	EST	MEYLAN,WM & HOWARD,PH (1993)

CAS #: 000103-46-8				S-BENZYLMERCAPTOACETIC ACID

Formula: $C_9H_{10}O_2S$

Mol Weight: 182.24

MP (deg C): 59-63　　FP (deg C):

BP (deg C):

BP pressure (mm Hg):

Property/ Value	Units	Temp	Data Type	Reference
WS 2.15E+003	mg/L	25	EST	MEYLAN,WM ET AL. (1996)
logP 2.10			EXP	HANSCH,C & LEO,AJ (1985)
VP 1.08E-004	mm Hg	25	EST	NEELY,WB & BLAU,GE (1985)
DC	pKa			
HL 4.72E-010	atm m3/mol	25	EST	MEYLAN,WM & HOWARD,PH (1991)
OH 1.98E-011	cm3/molc sec	25	EST	MEYLAN,WM & HOWARD,PH (1993)

CAS #: 000103-49-1				DIBENZYL AMINE

Formula: $C_{14}H_{15}N$

Mol Weight: 197.28

MP (deg C): -26　　FP (deg C):

BP (deg C): 270

BP pressure (mm Hg): 2.50E+002

Property/ Value	Units	Temp	Data Type	Reference
WS 7.82E+002	mg/L	25	EST	MEYLAN,WM ET AL. (1996)
logP 3.24			EST	MEYLAN,WM & HOWARD,PH (1995)
VP 7.91E-004	mm Hg	25	EXT	PERRY,RH & GREEN,D (1984)
DC	pKa			
HL 1.08E-007	atm m3/mol	25	EST	MEYLAN,WM & HOWARD,PH (1991)
OH 9.01E-011	cm3/molc sec	25	EST	MEYLAN,WM & HOWARD,PH (1993)

CAS #: 000103-50-4				DIBENZYL ETHER

Formula: $C_{14}H_{14}O$

Mol Weight: 198.27

MP (deg C): 3.6　　FP (deg C):

BP (deg C): 295-298

BP pressure (mm Hg):

Property/ Value	Units	Temp	Data Type	Reference
WS 4.00E+001	mg/L	35	EXP	RIDDICK,JA ET AL. (1986)
logP 3.31			EXP	HANSCH,C ET AL. (1995)
VP 1.03E-003	mm Hg	25	EXP	DAUBERT,TE & DANNER,RP (1989)
DC	pKa			
HL 8.30E-008	atm m3/mol	25	EST	MEYLAN,WM & HOWARD,PH (1991)
OH 2.11E-011	cm3/molc sec	25	EST	MEYLAN,WM & HOWARD,PH (1993)

CAS #: 000103-62-8				N-BUTYL-P-AMINOPHENOL

Formula: $C_{10}H_{15}NO$

Mol Weight: 165.24

MP (deg C):　　FP (deg C):

BP (deg C):

BP pressure (mm Hg):

Property/ Value	Units	Temp	Data Type	Reference
WS 7.53E+002	mg/L	25	EST	MEYLAN,WM ET AL. (1996)
logP 2.27			EST	MEYLAN,WM & HOWARD,PH (1995)
VP 7.65E-004	mm Hg	25	EST	NEELY,WB & BLAU,GE (1985)
DC	pKa			
HL 1.02E-009	atm m3/mol	25	EST	MEYLAN,WM & HOWARD,PH (1991)
OH 6.82E-011	cm3/molc sec	25	EST	MEYLAN,WM & HOWARD,PH (1993)

CAS #: 000103-63-9				(2-BROMOETHYL)BENZENE

Formula: C_8H_9Br

Mol Weight: 185.07

MP (deg C): -65.7　　FP (deg C):

BP (deg C): 219

BP pressure (mm Hg):

Property/ Value	Units	Temp	Data Type	Reference
WS 3.91E+001	mg/L	25	EXP	MACKAY,D & SHIU,WY (1981)
logP 3.09			EXP	HANSCH,C & LEO,AJ (1985)
VP 2.45E-001	mm Hg	25	EXP	MACKAY,D & SHIU,WY (1981)
DC	pKa			
HL 1.52E-003	atm m3/mol	25	EST	VP/WSOL
OH 5.60E-012	cm3/molc sec	25	EST	MEYLAN,WM & HOWARD,PH (1993)

CAS #: 000103-64-0				BETA-BROMOSTYRENE

Formula: C_8H_7Br

Mol Weight: 183.05

MP (deg C): 7　　FP (deg C):

BP (deg C): 112

BP pressure (mm Hg): 2.00E+001

Property/ Value	Units	Temp	Data Type	Reference
WS 1.08E+002	mg/L	25	EST	MEYLAN,WM ET AL. (1996)
logP 3.15			EST	MEYLAN,WM & HOWARD,PH (1995)
VP 1.23E-001	mm Hg	25	EST	NEELY,WB & BLAU,GE (1985)
DC	pKa			
HL 5.51E-004	atm m3/mol	25	EST	MEYLAN,WM & HOWARD,PH (1991)
OH 1.65E-011	cm3/molc sec	25	EST	MEYLAN,WM & HOWARD,PH (1993)

CAS #: 000103-65-1				N-PROPYLBENZENE

Formula: C_9H_{12}

Mol Weight: 120.20

MP (deg C): -99.5　　FP (deg C):

BP (deg C): 159.2

BP pressure (mm Hg):

Property/ Value	Units	Temp	Data Type	Reference
WS 5.22E+001	mg/L	25	EXP	TEWARI,YB ET AL (1982A)
logP 3.69			EXP	SANGSTER,J (1994)
VP 3.42E+000	mm Hg	25	EXP	DAUBERT,TE & DANNER,RP (1989)
DC	pKa			
HL 1.05E-002	atm m3/mol	25	EXP	SANEMASA,I ET AL. (1982)
OH 6.00E-012	cm3/molc sec	25	EXP	ATKINSON,R (1989)

CAS #: 000103-67-3				N-METHYLBENZYLAMINE

Formula: $C_8H_{11}N$

Mol Weight: 121.18

MP (deg C):　　FP (deg C):

BP (deg C): 180.5

BP pressure (mm Hg):

Property/ Value	Units	Temp	Data Type	Reference
WS 1.00E+006	mg/L	25	EXP	PEDDLE,CJ & TURNER,WES (1913)
logP 1.52			EXP	HANSCH,C & LEO,AJ (1985)
VP 6.50E-001	mm Hg	25	EST	NEELY,WB & BLAU,GE (1985)
DC 9.54	pKa	25	EXP	PERRIN,DD (1965)
HL 1.34E-006	atm m3/mol	25	EST	MEYLAN,WM & HOWARD,PH (1991)
OH 7.78E-011	cm3/molc sec	25	EST	MEYLAN,WM & HOWARD,PH (1993)

CAS #:	000103-69-5			N-ETHYLANILINE

Formula: $C_8H_{11}N$

Mol Weight: 121.18

MP (deg C): -63.5

FP (deg C):

BP (deg C): 204.5

BP pressure (mm Hg):

Property/Value	Units	Temp	Data Type	Reference
WS 2.41E+003	mg/L	25	EXP	KUHNE,R ET AL. (1995)
logP 2.16			EXP	HANSCH,C & LEO,AJ (1985)
VP 2.46E-001	mm Hg	25	EXP	CHAO,J ET AL. (1983)
DC 5.04	pKa	24	EXP	PERRIN,DD (1965)
HL 1.63E-005	atm m3/mol	25	EST	VP/WSOL
OH 5.15E-011	cm3/molc sec	25	EST	MEYLAN,WM & HOWARD,PH (1993)

CAS #:	000103-70-8			FORMANILIDE

Formula: C_7H_7NO

Mol Weight: 121.14

MP (deg C): 46.6-7.5

FP (deg C):

BP (deg C): 271

BP pressure (mm Hg):

Property/Value	Units	Temp	Data Type	Reference
WS 1.04E+004	mg/L	25	EST	MEYLAN,WM ET AL. (1996)
logP 1.15			EXP	HANSCH,C & LEO,AJ (1985)
VP 3.57E-003	mm Hg	25	EXT	DAUBERT,TE & DANNER,RP (1989)
DC	pKa			
HL 8.45E-009	atm m3/mol	25	EST	MEYLAN,WM & HOWARD,PH (1991)
OH 4.27E-011	cm3/molc sec	25	EST	MEYLAN,WM & HOWARD,PH (1993)

CAS #:	000103-71-9			PHENYL ISOCYANATE

Formula: C_7H_5NO

Mol Weight: 119.12

MP (deg C): -30

FP (deg C): -30

BP (deg C): 163

BP pressure (mm Hg):

Property/Value	Units	Temp	Data Type	Reference
WS 6.17E+002	mg/L	25	EST	MEYLAN,WM ET AL. (1996)
logP 2.59			EST	MEYLAN,WM & HOWARD,PH (1995)
VP 2.57E+000	mm Hg	25	EXP	DAUBERT,TE & DANNER,RP (1989)
DC	pKa			
HL 2.33E-004	atm m3/mol	25	EST	MEYLAN,WM & HOWARD,PH (1991)
OH 2.14E-012	cm3/molc sec	25	EST	MEYLAN,WM & HOWARD,PH (1993)

CAS #:	000103-72-0			PHENYLISOTHIOCYANATE

Formula: C_7H_5NS

Mol Weight: 135.19

MP (deg C): -21

FP (deg C):

BP (deg C): 221

BP pressure (mm Hg):

Property/Value	Units	Temp	Data Type	Reference
WS 8.99E+001	mg/L	25	EXP	YALKOWSKY,SH & DANNENFELSER,RM (1992)
logP 3.28			EXP	HANSCH,C & LEO,AJ (1985)
VP 1.50E+000	mm Hg	25	EXP	BOUBLIK,T ET AL. (1984)
DC	pKa			
HL 2.97E-003	atm m3/mol	25	EST	VP/WSOL
OH 2.14E-012	cm3/molc sec	25	EST	MEYLAN,WM & HOWARD,PH (1993)

CAS #:	000103-73-1			ETHOXYBENZENE

Formula: $C_8H_{10}O$

Mol Weight: 122.17

MP (deg C): -30

FP (deg C):

BP (deg C): 171-173

BP pressure (mm Hg):

Property/Value	Units	Temp	Data Type	Reference
WS 5.69E+002	mg/L	25	EXP	YALKOWSKY,SH & DANNENFELSER,RM (1992)
logP 2.51			EXP	HANSCH,C ET AL. (1995)
VP 1.56E+000	mm Hg	25	EXP	DAUBERT,TE & DANNER,RP (1989)
DC	pKa			
HL 4.41E-004	atm m3/mol	25	EST	VP/WSOL
OH 2.77E-011	cm3/molc sec	25	EST	MEYLAN,WM & HOWARD,PH (1993)

CAS #:	000103-74-2			2-PYRIDINEETHANOL

Formula: C_7H_9NO

Mol Weight: 123.16

MP (deg C): -7.8

FP (deg C):

BP (deg C): 116

BP pressure (mm Hg): 9.00E+000

Property/Value	Units	Temp	Data Type	Reference
WS 1.00E+006	mg/L	25	EST	MEYLAN,WM ET AL. (1996)
logP 0.12			EXP	HANSCH,C & LEO,AJ (1985)
VP 7.03E-005	mm Hg	25	EST	WSOL X HL
DC 5.31	pKa	25	EXP	PERRIN,DD (1972)
HL 1.47E-010	atm m3/mol	25	EST	MEYLAN,WM & HOWARD,PH (1991)
OH 5.72E-012	cm3/molc sec	25	EST	MEYLAN,WM & HOWARD,PH (1993)

CAS #:	000103-79-7			1-PHENYL-2-PROPANONE

Formula: $C_9H_{10}O$

Mol Weight: 134.18

MP (deg C): -15

FP (deg C):

BP (deg C): 216.5

BP pressure (mm Hg):

Property/Value	Units	Temp	Data Type	Reference
WS 5.21E+003	mg/L	25	EST	MEYLAN,WM ET AL. (1996)
logP 1.44			EXP	HANSCH,C & LEO,AJ (1985)
VP 2.21E-001	mm Hg	25	EST	NEELY,WB & BLAU,GE (1985)
DC	pKa			
HL 4.00E-006	atm m3/mol	25	EST	MEYLAN,WM & HOWARD,PH (1991)
OH 5.65E-012	cm3/molc sec	25	EST	MEYLAN,WM & HOWARD,PH (1993)

CAS #:	000103-80-0			PHENYLACETYL CHLORIDE

Formula: C_8H_7ClO

Mol Weight: 154.60

MP (deg C):

FP (deg C):

BP (deg C): 170

BP pressure (mm Hg): 2.50E+002

Property/Value	Units	Temp	Data Type	Reference
WS 6.37E+003	mg/L	25	EST	MEYLAN,WM ET AL. (1996)
logP 1.24			EST	MEYLAN,WM & HOWARD,PH (1995)
VP 1.99E-001	mm Hg	25	EXT	PERRY,RH & GREEN,D (1984)
DC	pKa			
HL 5.37E-005	atm m3/mol	25	EST	MEYLAN,WM & HOWARD,PH (1991)
OH 4.91E-012	cm3/molc sec	25	EST	MEYLAN,WM & HOWARD,PH (1993)

CAS #: 000103-81-1		PHENYLACETAMIDE

Formula: C_8H_9NO
Mol Weight: 135.17
MP (deg C): 155
FP (deg C):
BP (deg C):
BP pressure (mm Hg):

Property/ Value	Units	Temp	Data Type	Reference
WS 3.62E+004	mg/L	25	EST	MEYLAN,WM ET AL. (1996)
logP 0.45			EXP	HANSCH,C & LEO,AJ (1985)
VP 3.44E-004	mm Hg	25	EST	NEELY,WB & BLAU,GE (1985)
DC	pKa			
HL 9.03E-010	atm m3/mol	25	EST	MEYLAN,WM & HOWARD,PH (1991)
OH 7.55E-012	cm3/molc sec	25	EST	MEYLAN,WM & HOWARD,PH (1993)

CAS #: 000103-82-2		PHENYLACETIC ACID

Formula: $C_8H_8O_2$
Mol Weight: 136.15
MP (deg C): 76.5
FP (deg C):
BP (deg C): 265.5
BP pressure (mm Hg):

Property/ Value	Units	Temp	Data Type	Reference
WS 1.66E+004	mg/L	20	EXP	CHIOU,CT ET AL. (1977)
logP 1.41			EXP	HANSCH,C & LEO,AJ (1985)
VP 3.80E-003	mm Hg	25	EXT	PERRY,RH & GREEN,D (1984)
DC 4.31	pKa		EXP	DEAN,JA (1985)
HL 4.10E-008	atm m3/mol	25	EST	VP/WSOL
OH 4.45E-012	cm3/molc sec	25	EST	MEYLAN,WM & HOWARD,PH (1993)

CAS #: 000103-83-3		N,N-DIMETHYLBENZYLAMINE

Formula: $C_9H_{13}N$
Mol Weight: 135.21
MP (deg C): <-10
FP (deg C):
BP (deg C): 181
BP pressure (mm Hg):

Property/ Value	Units	Temp	Data Type	Reference
WS 1.20E+004	mg/L	25	EXP	CHEM INSPECT TEST INST (1992)
logP 1.98			EXP	HANSCH,C & LEO,AJ (1985)
VP 5.87E-001	mm Hg	25	EST	NEELY,WB & BLAU,GE (1985)
DC 8.91	pKa	25	EXP	PERRIN,DD (1965)
HL 2.95E-006	atm m3/mol	25	EST	MEYLAN,WM & HOWARD,PH (1991)
OH 8.21E-011	cm3/molc sec	25	EST	MEYLAN,WM & HOWARD,PH (1993)

CAS #: 000103-84-4		ACETANILIDE

Formula: C_8H_9NO
Mol Weight: 135.17
MP (deg C): 113-115
FP (deg C):
BP (deg C): 304-305
BP pressure (mm Hg):

Property/ Value	Units	Temp	Data Type	Reference
WS 6.93E+003	mg/L	25	EXP	YALKOWSKY,SH & DANNENFELSER,RM (1992)
logP 1.16			EXP	HANSCH,C & LEO,AJ (1985)
VP 1.22E-003	mm Hg	25	EXT	PERRY,RH & GREEN,D (1984)
DC 0.50	pKa	25	EXP	PERRIN,DD (1965)
HL 6.17E-009	atm m3/mol	25	EST	MEYLAN,WM & HOWARD,PH (1991)
OH 1.25E-011	cm3/molc sec	25	EST	MEYLAN,WM & HOWARD,PH (1993)

CAS #: 000103-85-5		N-PHENYLTHIOUREA

Formula: $C_7H_8N_2S$
Mol Weight: 152.22
MP (deg C): 154
FP (deg C):
BP (deg C):
BP pressure (mm Hg):

Property/ Value	Units	Temp	Data Type	Reference
WS 2.47E+003	mg/L	25	EXP	SEIDELL,A (1941)
logP 0.71			EXP	GOVERS,H ET AL. (1986)
VP 2.09E-003	mm Hg	25	EST	NEELY,WB & BLAU,GE (1985)
DC	pKa			
HL 1.04E-010	atm m3/mol	25	EST	MEYLAN,WM & HOWARD,PH (1991)
OH 1.26E-010	cm3/molc sec	25	EST	MEYLAN,WM & HOWARD,PH (1993)

CAS #: 000103-88-8		4-BROMOACETANILIDE

Formula: C_8H_8BrNO
Mol Weight: 214.07
MP (deg C): 168
FP (deg C):
BP (deg C):
BP pressure (mm Hg):

Property/ Value	Units	Temp	Data Type	Reference
WS 4.10E+002	mg/L	25	EST	MEYLAN,WM ET AL. (1996)
logP 2.29			EXP	HANSCH,C & LEO,AJ (1985)
VP 3.42E-005	mm Hg	25	EST	NEELY,WB & BLAU,GE (1985)
DC	pKa			
HL 2.46E-009	atm m3/mol	25	EST	MEYLAN,WM & HOWARD,PH (1991)
OH 3.66E-012	cm3/molc sec	25	EST	MEYLAN,WM & HOWARD,PH (1993)

CAS #: 000103-89-9		4-METHYLACETANILIDE

Formula: $C_9H_{11}NO$
Mol Weight: 149.19
MP (deg C): 152
FP (deg C):
BP (deg C): 307
BP pressure (mm Hg):

Property/ Value	Units	Temp	Data Type	Reference
WS 2.70E+003	mg/L	25	EST	MEYLAN,WM ET AL. (1996)
logP 1.70			EXP	NAKAGAWA,Y ET AL. (1992)
VP 1.83E-004	mm Hg	25	EST	NEELY,WB & BLAU,GE (1985)
DC	pKa			
HL 6.81E-009	atm m3/mol	25	EST	MEYLAN,WM & HOWARD,PH (1991)
OH 1.55E-011	cm3/molc sec	25	EST	MEYLAN,WM & HOWARD,PH (1993)

CAS #: 000103-90-2		ACETAMIDE, N-(4-HYDROXYPHENYL)

Formula: $C_8H_9NO_2$
Mol Weight: 151.17
MP (deg C): 187
FP (deg C):
BP (deg C):
BP pressure (mm Hg):

Property/ Value	Units	Temp	Data Type	Reference
WS 1.40E+004	mg/L	25	EXP	YALKOWSKY,SH & DANNENFELSER,RM (1992)
logP 0.46			EXP	SANGSTER,J (1994)
VP 7.00E-006	mm Hg	25	EST	NEELY,WB & BLAU,GE (1985)
DC 9.38	pKa		EXP	DASTMALCHI,S ET AL. (1995)
HL 6.42E-013	atm m3/mol	25	EST	MEYLAN,WM & HOWARD,PH (1991)
OH 1.77E-011	cm3/molc sec	25	EST	MEYLAN,WM & HOWARD,PH (1993)

163

CAS #: 000104-15-4				P-TOLUENESULFONIC ACID

Formula: $C_7H_8O_3S$

Mol Weight: 172.20

MP (deg C): 106-107   **FP (deg C):**

BP (deg C): 140

BP pressure (mm Hg): 2.00E+001

Property/Value	Units	Temp	Data Type	Reference
WS 6.20E+005	mg/L		EXP	BUDAVARI,S (1989)
logP -0.62			EST	MEYLAN,WM & HOWARD,PH (1995)
VP 2.70E-006	mm Hg	25	EST	MPBPVP
DC -1.34	pKa		EXP	SERJEANT,EP & DEMPSEY,B (1979)
HL 2.78E-009	atm m3/mol	25	EST	MEYLAN,WM & HOWARD,PH (1991)
OH 1.36E-012	cm3/molc sec	25	EST	MEYLAN,WM & HOWARD,PH (1993)

CAS #: 000104-31-4				BENZONATATE

Formula: $C_{30}H_{53}NO_{11}$

Mol Weight: 603.76

MP (deg C):   **FP (deg C):**

BP (deg C):

BP pressure (mm Hg):

Property/Value	Units	Temp	Data Type	Reference
WS 1.23E+000	mg/L	25	EST	MEYLAN,WM ET AL. (1996)
logP 2.45			EXP	HANSCH,C ET AL. (1995)
VP 4.22E-014	mm Hg	25	EST	NEELY,WB & BLAU,GE (1985)
DC	pKa			
HL 3.33E-024	atm m3/mol	25	EST	MEYLAN,WM & HOWARD,PH (1991)
OH 1.49E-010	cm3/molc sec	25	EST	MEYLAN,WM & HOWARD,PH (1993)

CAS #: 000104-40-5				P-NONYLPHENOL

Formula: $C_{15}H_{24}O$

Mol Weight: 220.36

MP (deg C): 42   **FP (deg C):**

BP (deg C):

BP pressure (mm Hg):

Property/Value	Units	Temp	Data Type	Reference
WS 7.00E+000	mg/L	25	EXP	YALKOWSKY,SH & DANNENFELSER,RM (1992)
logP 5.76			EXP	ITOKAWA,H ET AL. (1989)
VP 8.18E-004	mm Hg	25	EXP	BIDLEMAN,TF & RENBERG,L (1985)
DC	pKa			
HL 3.40E-005	atm m3/mol	25	EST	VP/WSOL
OH 5.17E-011	cm3/molc sec	25	EST	MEYLAN,WM & HOWARD,PH (1993)

CAS #: 000104-42-7				P-DODECYLANILINE

Formula: $C_{18}H_{31}N$

Mol Weight: 261.45

MP (deg C): 40-41   **FP (deg C):**

BP (deg C): 220-221

BP pressure (mm Hg): 1.50E+001

Property/Value	Units	Temp	Data Type	Reference
WS 6.00E+002	mg/L	25	EXP	MONSANTO MATERIAL SAFETY DATA SHEET
logP 7.03			EST	MEYLAN,WM & HOWARD,PH (1995)
VP 6.27E-006	mm Hg	25	EST	NEELY,WB & BLAU,GE (1985)
DC	pKa			
HL 4.74E-005	atm m3/mol	25	EST	MEYLAN,WM & HOWARD,PH (1991)
OH 1.46E-010	cm3/molc sec	25	EST	MEYLAN,WM & HOWARD,PH (1993)

CAS #: 000104-43-8				PHENOL, 4-DODECYL-

Formula: $C_{18}H_{30}O$

Mol Weight: 262.44

MP (deg C):   **FP (deg C):**

BP (deg C):

BP pressure (mm Hg):

Property/Value	Units	Temp	Data Type	Reference
WS 1.35E-002	mg/L	25	EST	MEYLAN,WM ET AL. (1996)
logP 7.91			EXP	ITOKAWA,H ET AL. (1989)
VP 2.30E-006	mm Hg	25	EST	NEELY,WB & BLAU,GE (1985)
DC	pKa			
HL 1.40E-005	atm m3/mol	25	EST	MEYLAN,WM & HOWARD,PH (1991)
OH 5.59E-011	cm3/molc sec	25	EST	MEYLAN,WM & HOWARD,PH (1993)

CAS #: 000104-46-1				ANETHOLE

Formula: $C_{10}H_{12}O$

Mol Weight: 148.21

MP (deg C): 21.3   **FP (deg C):**

BP (deg C): 234

BP pressure (mm Hg):

Property/Value	Units	Temp	Data Type	Reference
WS 1.11E+002	mg/L	25	EXP	YALKOWSKY,SH & DANNENFELSER,RM (1992)
logP 3.39			EST	MEYLAN,WM & HOWARD,PH (1992)
VP 5.00E-003	mm Hg	25	EXT	DAUBERT,TE & DANNER,RP (1993)
DC	pKa			
HL 7.18E-005	atm m3/mol	25	EST	MEYLAN,WM & HOWARD,PH (1991)
OH 8.63E-011	cm3/molc sec	25	EST	MEYLAN,WM & HOWARD,PH (1993)

CAS #: 000104-51-8				N-BUTYLBENZENE

Formula: $C_{10}H_{14}$

Mol Weight: 134.22

MP (deg C): -88.5   **FP (deg C):**

BP (deg C): 183.1

BP pressure (mm Hg):

Property/Value	Units	Temp	Data Type	Reference
WS 1.18E+001	mg/L	25	EXP	YALKOWSKY,SH & DANNENFELSER,RM (1992)
logP 4.38			EXP	DEBRUIJN,J ET AL. (1989)
VP 1.06E+000	mm Hg	25	EXP	CHAO,J ET AL. (1983)
DC	pKa			
HL 1.59E-002	atm m3/mol	25	EST	VP/WSOL
OH 8.72E-012	cm3/molc sec	25	EST	MEYLAN,WM & HOWARD,PH (1993)

CAS #: 000104-52-9				3-PHENYL-1-CHLOROPROPANE

Formula: $C_9H_{11}Cl$

Mol Weight: 154.64

MP (deg C):   **FP (deg C):**

BP (deg C): 219.5

BP pressure (mm Hg):

Property/Value	Units	Temp	Data Type	Reference
WS 6.73E+001	mg/L	25	EST	MEYLAN,WM ET AL. (1996)
logP 3.55			EXP	HANSCH,C & LEO,AJ (1985)
VP 1.09E-001	mm Hg	25	EST	NEELY,WB & BLAU,GE (1985)
DC	pKa			
HL 3.69E-003	atm m3/mol	25	EST	MEYLAN,WM & HOWARD,PH (1991)
OH 6.84E-012	cm3/molc sec	25	EST	MEYLAN,WM & HOWARD,PH (1993)

CAS #: 000104-54-1				CINNAMYL ALCOHOL

Formula: $C_9H_{10}O$

Mol Weight: 134.18

MP (deg C): 33   FP (deg C):

BP (deg C): 250

BP pressure (mm Hg):

Property/ Value	Units	Temp	Data Type	Reference
WS 6.19E+003	mg/L	25	EST	MEYLAN,WM ET AL. (1996)
logP 1.95			EXP	HANSCH,C & LEO,AJ (1985)
VP 3.62E-002	mm Hg	25	EXT	PERRY,RH & GREEN,D (1984)
DC	pKa			
HL 1.58E-007	atm m3/mol	25	EST	MEYLAN,WM & HOWARD,PH (1991)
OH 6.16E-011	cm3/molc sec	25	EST	MEYLAN,WM & HOWARD,PH (1993)

CAS #: 000104-55-2				3-PHENYL-2-PROPENAL

Formula: C_9H_8O

Mol Weight: 132.16

MP (deg C): -7.5   FP (deg C):

BP (deg C): 246

BP pressure (mm Hg):

Property/ Value	Units	Temp	Data Type	Reference
WS 1.42E+003	mg/L	25	EXP	VALVANI,SC ET AL. (1981)
logP 1.90			EXP	HANSCH,C & LEO,AJ (1985)
VP 2.89E-002	mm Hg	25	EXT	PERRY,RH & GREEN,D (1984)
DC	pKa			
HL 1.60E-006	atm m3/mol	25	EST	MEYLAN,WM & HOWARD,PH (1991)
OH 3.79E-011	cm3/molc sec	25	EST	MEYLAN,WM & HOWARD,PH (1993)

CAS #: 000104-63-2				ETHANOL, 2- (PHENYLMETHYL)AMINO -

Formula: $C_9H_{13}NO$

Mol Weight: 151.21

MP (deg C):   FP (deg C):

BP (deg C): 153-156

BP pressure (mm Hg): 1.20E+001

Property/ Value	Units	Temp	Data Type	Reference
WS 1.62E+005	mg/L	25	EST	MEYLAN,WM ET AL. (1996)
logP 0.79			EXP	HANSCH,C ET AL. (1995)
VP 3.72E-004	mm Hg	25	EST	NEELY,WB & BLAU,GE (1985)
DC	pKa			
HL 6.52E-011	atm m3/mol	25	EST	MEYLAN,WM & HOWARD,PH (1991)
OH 9.14E-011	cm3/molc sec	25	EST	MEYLAN,WM & HOWARD,PH (1993)

CAS #: 000104-66-5				1,2-DIPHENOXYETHANE

Formula: $C_{14}H_{14}O_2$

Mol Weight: 214.27

MP (deg C): 98   FP (deg C):

BP (deg C): 182

BP pressure (mm Hg): 1.20E+001

Property/ Value	Units	Temp	Data Type	Reference
WS 2.06E+001	mg/L	25	EST	MEYLAN,WM ET AL. (1996)
logP 3.81			EXP	HANSCH,C & LEO,AJ (1985)
VP 3.14E-004	mm Hg	25	EST	NEELY,WB & BLAU,GE (1985)
DC	pKa			
HL 1.83E-006	atm m3/mol	25	EST	MEYLAN,WM & HOWARD,PH (1991)
OH 5.70E-011	cm3/molc sec	25	EST	MEYLAN,WM & HOWARD,PH (1993)

CAS #: 000104-72-3				1-PHENYLDECANE

Formula: $C_{16}H_{26}$

Mol Weight: 218.39

MP (deg C): -14.4   FP (deg C):

BP (deg C): 298

BP pressure (mm Hg):

Property/ Value	Units	Temp	Data Type	Reference
WS 1.86E-002	mg/L	25	EST	MEYLAN,WM ET AL. (1996)
logP 7.35			EXP	SANGSTER,J (1993)
VP 1.28E-003	mm Hg	25	EXP	DAUBERT,TE & DANNER,RP (1989)
DC	pKa			
HL 7.61E-002	atm m3/mol	25	EST	MEYLAN,WM & HOWARD,PH (1991)
OH 1.72E-011	cm3/mol sec	25	EST	MEYLAN,WM & HOWARD,PH (1993)

CAS #: 000104-73-4				DODECYLPYRIDINIUM BROMIDE

Formula: $C_{17}H_{30}BrN$

Mol Weight: 328.34

MP (deg C):   FP (deg C):

BP (deg C):

BP pressure (mm Hg):

Property/ Value	Units	Temp	Data Type	Reference
WS 3.52E+003	mg/L	25	EST	MEYLAN,WM ET AL. (1996)
logP 0.44			EXP	HANSCH,C & LEO,AJ (1985)
VP 1.17E-005	mm Hg	25	EST	NEELY,WB & BLAU,GE (1985)
DC	pKa			
HL 3.39E-009	atm m3/mol	25	EST	MEYLAN,WM & HOWARD,PH (1991)
OH 2.56E-011	cm3/molc sec	25	EST	MEYLAN,WM & HOWARD,PH (1993)

CAS #: 000104-75-6				2-ETHYLHEXYLAMINE

Formula: $C_8H_{19}N$

Mol Weight: 129.25

MP (deg C): -76   FP (deg C):

BP (deg C): 169

BP pressure (mm Hg):

Property/ Value	Units	Temp	Data Type	Reference
WS 2.50E+003	mg/L	20	EXP	RIDDICK,JA ET AL. (1986)
logP 2.82			EXP	HANSCH,C & LEO,AJ (1985)
VP 1.00E+001	mm Hg	54	EXP	RIDDICK,JA ET AL. (1986)
DC	pKa			
HL 8.29E-005	atm m3/mol	25	EST	MEYLAN,WM & HOWARD,PH (1991)
OH 4.08E-011	cm3/molc sec	25	EST	MEYLAN,WM & HOWARD,PH (1993)

CAS #: 000104-76-7				2-ETHYL-1-HEXANOL

Formula: $C_8H_{18}O$

Mol Weight: 130.23

MP (deg C): -76   FP (deg C):

BP (deg C): 184.34

BP pressure (mm Hg):

Property/ Value	Units	Temp	Data Type	Reference
WS 8.80E+002	mg/L	25	EXP	AMIDON,GL ET AL. (1974)
logP 2.73			EST	MEYLAN,WM & HOWARD,PH (1995)
VP 1.36E-001	mm Hg	25	EXP	DAUBERT,TE & DANNER,RP (1985)
DC	pKa			
HL 2.65E-005	atm m3/mol	25	EST	VP/WSOL
OH 1.30E-011	cm3/molc sec	25	EST	MEYLAN,WM & HOWARD,PH (1993)

CAS #: 000104-82-5 — P-(CHLOROMETHYL)TOLUENE

Formula: C_8H_9Cl
Mol Weight: 140.61
MP (deg C):
FP (deg C):
BP (deg C): 201
BP pressure (mm Hg):

Property/Value	Units	Temp	Data Type	Reference
WS 1.17E+002	mg/L	25	EST	MEYLAN,WM ET AL. (1996)
logP 3.34			EST	MEYLAN,WM & HOWARD,PH (1995)
VP 2.97E+000	mm Hg	25	EXP	JORDAN,TE (1954)
DC	pKa			
HL 2.31E-003	atm m3/mol	25	EST	MEYLAN,WM & HOWARD,PH (1991)
OH 3.80E-012	cm3/molc sec	25	EST	MEYLAN,WM & HOWARD,PH (1993)

CAS #: 000104-83-6 — BENZENE, 1-CHLORO-4-(CHLOROMETHYL)-

Formula: $C_7H_6Cl_2$
Mol Weight: 161.03
MP (deg C): 31
FP (deg C):
BP (deg C): 223
BP pressure (mm Hg):

Property/Value	Units	Temp	Data Type	Reference
WS 1.30E+002	mg/L	25	EST	MEYLAN,WM ET AL. (1996)
logP 3.18			EXP	WANG,W ET AL. (1987)
VP 1.59E-001	mm Hg	25	EST	NEELY,WB & BLAU,GE (1985)
DC	pKa			
HL 1.55E-003	atm m3/mol	25	EST	MEYLAN,WM & HOWARD,PH (1991)
OH 1.25E-012	cm3/molc sec	25	EST	MEYLAN,WM & HOWARD,PH (1993)

CAS #: 000104-84-7 — BENZENEMETHANAMINE, 4-METHYL-

Formula: $C_8H_{11}N$
Mol Weight: 121.18
MP (deg C): 12.5
FP (deg C):
BP (deg C): 195
BP pressure (mm Hg):

Property/Value	Units	Temp	Data Type	Reference
WS 5.73E+004	mg/L	25	EST	MEYLAN,WM ET AL. (1996)
logP 1.46			EXP	KRIL,MB & FUNG,HL (1990)
VP 2.98E-001	mm Hg	25	EST	NEELY,WB & BLAU,GE (1985)
DC 9.36	pKa	25	EXP	PERRIN,DD (1965)
HL 6.75E-007	atm m3/mol	25	EST	MEYLAN,WM & HOWARD,PH (1991)
OH 3.60E-011	cm3/molc sec	25	EST	MEYLAN,WM & HOWARD,PH (1993)

CAS #: 000104-85-8 — 4-METHYLBENZONITRILE

Formula: C_8H_7N
Mol Weight: 117.15
MP (deg C): 29.5
FP (deg C):
BP (deg C): 217.6
BP pressure (mm Hg):

Property/Value	Units	Temp	Data Type	Reference
WS 9.21E+002	mg/L	25	EST	MEYLAN,WM ET AL. (1996)
logP 2.09			EST	MEYLAN,WM & HOWARD,PH (1995)
VP 3.13E-001	mm Hg	25	EXT	PERRY,RH & GREEN,D (1984)
DC	pKa			
HL 5.74E-005	atm m3/mol	25	EST	MEYLAN,WM & HOWARD,PH (1991)
OH 1.03E-012	cm3/molc sec	25	EST	MEYLAN,WM & HOWARD,PH (1993)

CAS #: 000104-87-0 — P-TOLUALDEHYDE

Formula: C_8H_8O
Mol Weight: 120.15
MP (deg C): -6
FP (deg C):
BP (deg C): 204
BP pressure (mm Hg):

Property/Value	Units	Temp	Data Type	Reference
WS 2.27E+003	mg/L	25	EXP	YALKOWSKY,SH & DANNENFELSER,RM (1992)
logP 2.26			EST	MEYLAN,WM & HOWARD,PH (1995)
VP 2.50E-001	mm Hg	25	EXP	DAUBERT,TE & DANNER,RP (1989)
DC	pKa			
HL 1.74E-005	atm m3/mol	25	EST	VP/WSOL
OH 1.87E-011	cm3/molc sec	25	EST	MEYLAN,WM & HOWARD,PH (1993)

CAS #: 000104-88-1 — 4-CHLOROBENZALDEHYDE

Formula: C_7H_5ClO
Mol Weight: 140.57
MP (deg C): 47.5
FP (deg C):
BP (deg C): 213.5
BP pressure (mm Hg):

Property/Value	Units	Temp	Data Type	Reference
WS 1.34E+003	mg/L	25	EST	MEYLAN,WM ET AL. (1996)
logP 2.10			EXP	HANSCH,C & LEO,AJ (1985)
VP 1.83E-001	mm Hg	25	EST	NEELY,WB & BLAU,GE (1985)
DC	pKa			
HL 9.95E-006	atm m3/mol	25	EST	MEYLAN,WM & HOWARD,PH (1991)
OH 1.73E-011	cm3/molc sec	25	EST	MEYLAN,WM & HOWARD,PH (1993)

CAS #: 000104-89-2 — 2-METHYL-5-ETHYLPIPERIDINE

Formula: $C_8H_{17}N$
Mol Weight: 127.23
MP (deg C):
FP (deg C):
BP (deg C):
BP pressure (mm Hg):

Property/Value	Units	Temp	Data Type	Reference
WS 6.84E+003	mg/L	25	EST	MEYLAN,WM ET AL. (1996)
logP 2.51			EST	MEYLAN,WM & HOWARD,PH (1995)
VP 2.87E+000	mm Hg	25	EXP	CHAO,J ET AL. (1983)
DC	pKa			
HL 4.01E-005	atm m3/mol	25	EST	MEYLAN,WM & HOWARD,PH (1991)
OH 1.04E-010	cm3/molc sec	25	EST	MEYLAN,WM & HOWARD,PH (1993)

CAS #: 000104-90-5 — 5-ETHYL-2-METHYLPYRIDINE

Formula: $C_8H_{11}N$
Mol Weight: 121.18
MP (deg C): -70.9
FP (deg C):
BP (deg C): 178.3
BP pressure (mm Hg):

Property/Value	Units	Temp	Data Type	Reference
WS 1.20E+004	mg/L	25	EXP	GOE,GL (1982)
logP 2.39			EST	MEYLAN,WM & HOWARD,PH (1995)
VP 1.43E+000	mm Hg	25	EXP	CHAO,J ET AL. (1983)
DC 6.51	pKa	20	EXP	PERRIN,DD (1965)
HL 1.90E-005	atm m3/mol	25	EST	VP/WSOL
OH 2.35E-012	cm3/molc sec	25	EST	MEYLAN,WM & HOWARD,PH (1993)

CAS #: 000104-91-6 — P-NITROSOPHENOL

Formula: $C_6H_5NO_2$

Mol Weight: 123.11

MP (deg C): 144 dec

FP (deg C):

BP (deg C):

BP pressure (mm Hg):

Property/ Value	Units	Temp	Data Type	Reference
WS 2.94E+004	mg/L	25	EST	MEYLAN,WM ET AL. (1996)
logP 1.29			EXP	HANSCH,C & LEO,AJ (1985)
VP 1.53E-001	mm Hg	25	EST	NEELY,WB & BLAU,GE (1985)
DC	pKa			
HL 5.87E-009	atm m3/mol	25	EST	MEYLAN,WM & HOWARD,PH (1991)
OH 1.76E-011	cm3/molc sec	25	EST	MEYLAN,WM & HOWARD,PH (1993)

CAS #: 000104-92-7 — 4-BROMOANISOLE

Formula: C_7H_7BrO

Mol Weight: 187.04

MP (deg C): 13.5

FP (deg C):

BP (deg C): 215

BP pressure (mm Hg):

Property/ Value	Units	Temp	Data Type	Reference
WS 1.26E+002	mg/L	25	EST	MEYLAN,WM ET AL. (1996)
logP 2.96			EST	MEYLAN,WM & HOWARD,PH (1995)
VP 2.03E-001	mm Hg	25	EXT	PERRY,RH & GREEN,D (1984)
DC	pKa			
HL 1.27E-004	atm m3/mol	25	EST	MEYLAN,WM & HOWARD,PH (1991)
OH 6.99E-012	cm3/molc sec	25	EST	MEYLAN,WM & HOWARD,PH (1993)

CAS #: 000104-93-8 — P-METHYLANISOLE

Formula: $C_8H_{10}O$

Mol Weight: 122.17

MP (deg C): -32

FP (deg C):

BP (deg C): 176.5

BP pressure (mm Hg):

Property/ Value	Units	Temp	Data Type	Reference
WS 6.70E+002	mg/L	25	EST	LYMAN,WJ ET AL. (1990)
logP 2.66			EXP	HANSCH,C ET AL. (1995)
VP 1.14E+000	mm Hg	25	EST	NEELY,WB & BLAU,GE (1985)
DC	pKa			
HL 4.66E-003	atm m3/mol	25	EST	MEYLAN,WM & HOWARD,PH (1991)
OH 2.73E-011	cm3/molc sec	25	EST	MEYLAN,WM & HOWARD,PH (1993)

CAS #: 000104-94-9 — 4-METHOXYANILINE

Formula: C_7H_9NO

Mol Weight: 123.16

MP (deg C): 57.2

FP (deg C):

BP (deg C): 243

BP pressure (mm Hg):

Property/ Value	Units	Temp	Data Type	Reference
WS 1.54E+004	mg/L	25	EST	MEYLAN,WM ET AL. (1996)
logP 0.95			EXP	HANSCH,C & LEO,AJ (1985)
VP 1.90E-002	mm Hg	25	EST	MPBPVP
DC 5.34	pKa	25	EXP	WEAST,RC ET AL. (1985)
HL 1.77E-006	atm m3/mol	25	EST	MEYLAN,WM & HOWARD,PH (1991)
OH 1.21E-010	cm3/molc sec	25	EST	MEYLAN,WM & HOWARD,PH (1993)

CAS #: 000105-05-5 — P-DIETHYLBENZENE

Formula: $C_{10}H_{14}$

Mol Weight: 134.22

MP (deg C): -42.83

FP (deg C):

BP (deg C): 183.7

BP pressure (mm Hg):

Property/ Value	Units	Temp	Data Type	Reference
WS 2.48E+001	mg/L	20	EXP	YALKOWSKY,SH & DANNENFELSER,RM (1992)
logP 4.58			EXP	SHERBLOM,PM & EGANHOUSE,RP (1988)
VP 1.06E+000	mm Hg	25	EXP	DAUBERT,TE & DANNER,RP (1989)
DC	pKa			
HL 7.55E-003	atm m3/mol	25	EST	VP/WSOL
OH 8.11E-012	cm3/molc sec	25	EST	MEYLAN,WM & HOWARD,PH (1993)

CAS #: 000105-11-3 — 2,5-CYCLOHEXADIENE-1,4-DIONE, DIOXIME

Formula: $C_6H_6N_2O_2$

Mol Weight: 138.13

MP (deg C): 243 dec

FP (deg C):

BP (deg C):

BP pressure (mm Hg):

Property/ Value	Units	Temp	Data Type	Reference
WS 4.55E+003	mg/L	25	EST	MEYLAN,WM ET AL. (1996)
logP 1.49			EXP	HANSCH,C ET AL. (1995)
VP 1.41E-005	mm Hg	25	EST	NEELY,WB & BLAU,GE (1985)
DC	pKa			
HL 3.74E-010	atm m3/mol	25	EST	MEYLAN,WM & HOWARD,PH (1991)
OH 1.13E-010	cm3/molc sec	25	EST	MEYLAN,WM & HOWARD,PH (1993)

CAS #: 000105-13-5 — P-METHOXYBENZYL ALCOHOL

Formula: $C_8H_{10}O_2$

Mol Weight: 138.17

MP (deg C): 24-25

FP (deg C):

BP (deg C): 259

BP pressure (mm Hg):

Property/ Value	Units	Temp	Data Type	Reference
WS 3.17E+004	mg/L	25	EST	MEYLAN,WM ET AL. (1996)
logP 1.10			EXP	HANSCH,C & LEO,AJ (1985)
VP 3.80E-003	mm Hg	20	EST	ARTHURDLITTLE,INC (1982)
DC	pKa			
HL 1.29E-008	atm m3/mol	25	EST	MEYLAN,WM & HOWARD,PH (1991)
OH 3.04E-011	cm3/molc sec	25	EST	MEYLAN,WM & HOWARD,PH (1993)

CAS #: 000105-30-6 — 2-METHYL-1-PENTANOL

Formula: $C_6H_{14}O$

Mol Weight: 102.18

MP (deg C):

FP (deg C):

BP (deg C): 148

BP pressure (mm Hg):

Property/ Value	Units	Temp	Data Type	Reference
WS 6.00E+003	mg/L	25	EXP	YALKOWSKY,SH & DANNENFELSER,RM (1992)
logP 1.75			EST	MEYLAN,WM & HOWARD,PH (1995)
VP 1.92E+000	mm Hg	25	EXP	DAUBERT,TE & DANNER,RP (1989)
DC	pKa			
HL 4.30E-005	atm m3/mol	25	EST	VP/WSOL
OH 9.99E-012	cm3/molc sec	25	EST	MEYLAN,WM & HOWARD,PH (1993)

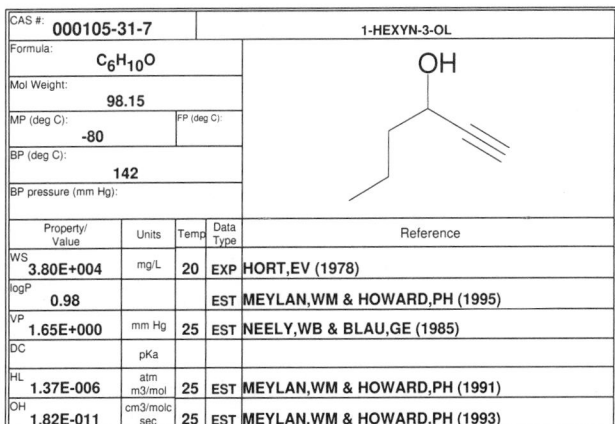

000105-31-7 — 1-HEXYN-3-OL

CAS #: 000105-31-7

Formula:	$C_6H_{10}O$
Mol Weight:	98.15
MP (deg C):	-80
FP (deg C):	
BP (deg C):	142
BP pressure (mm Hg):	

Property/Value	Units	Temp	Data Type	Reference
WS 3.80E+004	mg/L	20	EXP	HORT,EV (1978)
logP 0.98			EST	MEYLAN,WM & HOWARD,PH (1995)
VP 1.65E+000	mm Hg	25	EST	NEELY,WB & BLAU,GE (1985)
DC	pKa			
HL 1.37E-006	atm m3/mol	25	EST	MEYLAN,WM & HOWARD,PH (1991)
OH 1.82E-011	cm3/molc sec	25	EST	MEYLAN,WM & HOWARD,PH (1993)

000105-34-0 — METHYL CYANOACETATE

CAS #: 000105-34-0

Formula:	$C_4H_5NO_2$
Mol Weight:	99.09
MP (deg C):	-22.5
FP (deg C):	
BP (deg C):	200.5
BP pressure (mm Hg):	

Property/Value	Units	Temp	Data Type	Reference
WS 1.61E+005	mg/L	25	EST	MEYLAN,WM ET AL. (1996)
logP -0.47			EXP	HANSCH,C ET AL. (1995)
VP 1.38E-001	mm Hg	25	EXP	DAUBERT,TE & DANNER,RP (1989)
DC	pKa			
HL 7.27E-008	atm m3/mol	25	EST	MEYLAN,WM & HOWARD,PH (1991)
OH 3.49E-013	cm3/molc sec	25	EST	MEYLAN,WM & HOWARD,PH (1993)

000105-36-2 — BROMOACETIC ACID, ETHYL ESTER

CAS #: 000105-36-2

Formula:	$C_4H_7BrO_2$
Mol Weight:	167.01
MP (deg C):	
FP (deg C):	
BP (deg C):	168.5
BP pressure (mm Hg):	

Property/Value	Units	Temp	Data Type	Reference
WS 7.02E+003	mg/L	25	EST	MEYLAN,WM ET AL. (1996)
logP 1.12			EXP	HANSCH,C & LEO,AJ (1985)
VP 3.37E+000	mm Hg	25	EST	NEELY,WB & BLAU,GE (1985)
DC	pKa			
HL 2.68E-005	atm m3/mol	25	EST	MEYLAN,WM & HOWARD,PH (1991)
OH 1.29E-012	cm3/molc sec	25	EST	MEYLAN,WM & HOWARD,PH (1993)

000105-37-3 — ETHYL PROPIONATE

CAS #: 000105-37-3

Formula:	$C_5H_{10}O_2$
Mol Weight:	102.13
MP (deg C):	-73
FP (deg C):	
BP (deg C):	99
BP pressure (mm Hg):	

Property/Value	Units	Temp	Data Type	Reference
WS 1.92E+004	mg/L	20	EXP	YALKOWSKY,SH & DANNENFELSER,RM (1992)
logP 1.21			EXP	HANSCH,C & LEO,AJ (1985)
VP 3.58E+001	mm Hg	25	EXP	PERRY,RH & GREEN,D (1984)
DC	pKa			
HL 2.51E-004	atm m3/mol	25	EST	VP/WSOL
OH 2.10E-012	cm3/molc sec	25	EXP	ATKINSON,R (1989)

000105-38-4 — VINYL PROPIONATE

CAS #: 000105-38-4

Formula:	$C_5H_8O_2$
Mol Weight:	100.12
MP (deg C):	-80
FP (deg C):	
BP (deg C):	91.2
BP pressure (mm Hg):	

Property/Value	Units	Temp	Data Type	Reference
WS 1.06E+004	mg/L	25	EST	MEYLAN,WM ET AL. (1996)
logP 1.22			EST	MEYLAN,WM & HOWARD,PH (1995)
VP 3.66E+001	mm Hg	25	EST	NEELY,WB & BLAU,GE (1985)
DC	pKa			
HL 1.55E-003	atm m3/mol	25	EST	MEYLAN,WM & HOWARD,PH (1991)
OH 2.72E-011	cm3/molc sec	25	EST	MEYLAN,WM & HOWARD,PH (1993)

000105-39-5 — ETHYL CHLOROACETATE

CAS #: 000105-39-5

Formula:	$C_4H_7ClO_2$
Mol Weight:	122.55
MP (deg C):	-26
FP (deg C):	
BP (deg C):	144-146
BP pressure (mm Hg):	

Property/Value	Units	Temp	Data Type	Reference
WS 1.09E+004	mg/L	25	EST	MEYLAN,WM ET AL. (1996)
logP 1.12			EST	MEYLAN,WM & HOWARD,PH (1995)
VP 4.87E+000	mm Hg	25	EXP	PERRY,RH & GREEN,D (1984)
DC	pKa			
HL 8.20E-005	atm m3/mol	25	EST	MEYLAN,WM & HOWARD,PH (1991)
OH 1.36E-012	cm3/molc sec	25	EST	MEYLAN,WM & HOWARD,PH (1993)

000105-40-8 — N-METHYLCARBAMIC ACID, ET ESTER

CAS #: 000105-40-8

Formula:	$C_4H_9NO_2$
Mol Weight:	103.12
MP (deg C):	
FP (deg C):	
BP (deg C):	170
BP pressure (mm Hg):	

Property/Value	Units	Temp	Data Type	Reference
WS 6.90E+005	mg/L	16	EXP	STEPHEN,H & STEPHEN,T (1963)
logP 0.34			EXP	HANSCH,C & LEO,AJ (1985)
VP 9.05E-001	mm Hg	25	EXT	OHE,S (1976)
DC	pKa			
HL 1.15E-007	atm m3/mol	25	EST	MEYLAN,WM & HOWARD,PH (1991)
OH 7.72E-012	cm3/molc sec	25	EST	MEYLAN,WM & HOWARD,PH (1993)

000105-45-3 — METHYL ACETOACETATE

CAS #: 000105-45-3

Formula:	$C_5H_8O_3$
Mol Weight:	116.12
MP (deg C):	-80
FP (deg C):	
BP (deg C):	171.7
BP pressure (mm Hg):	

Property/Value	Units	Temp	Data Type	Reference
WS 3.33E+005	mg/L		EST	WINDHOLZ,M ET AL. (1983)
logP -0.69			EST	MEYLAN,WM & HOWARD,PH (1995)
VP 8.92E-001	mm Hg	25	EXP	DAUBERT,TE & DANNER,RP (1989)
DC	pKa			
HL 1.18E-007	atm m3/mol	25	EST	MEYLAN,WM & HOWARD,PH (1991)
OH 3.25E-013	cm3/molc sec	25	EST	ATKINSON,R (1988)

		CAS #: 000105-46-4		S-BUTYLACETATE

CAS #: 000105-46-4 — S-BUTYLACETATE

Formula: $C_6H_{12}O_2$
Mol Weight: 116.16
MP (deg C): -98.9
FP (deg C):
BP (deg C): 112.2
BP pressure (mm Hg):

Property/Value	Units	Temp	Data Type	Reference
WS 6.20E+003	mg/L	20	EXP	YALKOWSKY,SH & DANNENFELSER,RM (1992)
logP 1.72			EXP	HANSCH,C ET AL. (1995)
VP 1.70E+001	mm Hg	20	EXP	DAUBERT,TE & DANNER,RP (1989)
DC	pKa			
HL 4.19E-004	atm m3/mol	20	EST	VP/WSOL
OH 5.50E-012	cm3/molc sec	25	EXP	ATKINSON,R (1989)

CAS #: 000105-48-6 — ISOPROPYL CHLOROACETATE

Formula: $C_5H_9ClO_2$
Mol Weight: 136.58
MP (deg C):
FP (deg C):
BP (deg C): 150.5
BP pressure (mm Hg):

Property/Value	Units	Temp	Data Type	Reference
WS 4.22E+003	mg/L	25	EST	MEYLAN,WM ET AL. (1996)
logP 1.54			EST	MEYLAN,WM & HOWARD,PH (1995)
VP 4.10E+000	mm Hg	25	EXT	OHE,S (1976)
DC	pKa			
HL 1.09E-004	atm m3/mol	25	EST	MEYLAN,WM & HOWARD,PH (1991)
OH 2.54E-012	cm3/molc sec	25	EST	MEYLAN,WM & HOWARD,PH (1993)

CAS #: 000105-53-3 — DIETHYL MALONATE

Formula: $C_7H_{12}O_4$
Mol Weight: 160.17
MP (deg C): -50
FP (deg C):
BP (deg C): 198-199
BP pressure (mm Hg):

Property/Value	Units	Temp	Data Type	Reference
WS 2.70E+004	mg/L	20	EXP	RIDDICK,JA ET AL. (1986)
logP 0.96			EXP	HANSCH,C ET AL. (1995)
VP 2.69E-001	mm Hg	25	EXP	DAUBERT,TE & DANNER,RP (1989)
DC	pKa			
HL 2.10E-006	atm m3/mol	25	EST	VP/WSOL
OH 3.83E-012	cm3/molc sec	25	EST	MEYLAN,WM & HOWARD,PH (1993)

CAS #: 000105-54-4 — ETHYLBUTYRATE

Formula: $C_6H_{12}O_2$
Mol Weight: 116.16
MP (deg C): -93
FP (deg C):
BP (deg C): 120
BP pressure (mm Hg):

Property/Value	Units	Temp	Data Type	Reference
WS 4.90E+003	mg/L	20	EXP	RIDDICK,JA ET AL. (1986)
logP 1.85			EST	MEYLAN,WM & HOWARD,PH (1995)
VP 1.28E+001	mm Hg	20	EXP	DAUBERT,TE & DANNER,RP (1989)
DC	pKa			
HL 3.99E-004	atm m3/mol	20	EST	VP/WSOL
OH 4.90E-012	cm3/molc sec	25	EXP	ATKINSON,R (1989)

CAS #: 000105-55-5 — N,N-DIETHYLTHIOUREA

Formula: $C_5H_{12}N_2S$
Mol Weight: 132.23
MP (deg C): 68-71
FP (deg C):
BP (deg C):
BP pressure (mm Hg):

Property/Value	Units	Temp	Data Type	Reference
WS 4.56E+003	mg/L	25	EST	MEYLAN,WM ET AL. (1996)
logP 0.57			EXP	GOVERS,H ET AL. (1986)
VP 2.40E-001	mm Hg	25	EST	NEELY,WB & BLAU,GE (1985)
DC	pKa			
HL 6.85E-008	atm m3/mol	25	EST	MEYLAN,WM & HOWARD,PH (1991)
OH 1.01E-010	cm3/molc sec	25	EST	ATKINSON,R (1988)

CAS #: 000105-56-6 — ETHYL CYANOACETATE

Formula: $C_5H_7NO_2$
Mol Weight: 113.12
MP (deg C): -22
FP (deg C):
BP (deg C): 206
BP pressure (mm Hg):

Property/Value	Units	Temp	Data Type	Reference
WS 2.00E+004	mg/L	25	EXP	YALKOWSKY,SH & DANNENFELSER,RM (1992)
logP 1.01			EST	MEYLAN,WM & HOWARD,PH (1995)
VP 3.88E-002	mm Hg	25	EXP	DAUBERT,TE & DANNER,RP (1989)
DC	pKa			
HL 2.89E-007	atm m3/mol	25	EST	VP/WSOL
OH 1.79E-012	cm3/molc sec	25	EST	MEYLAN,WM & HOWARD,PH (1993)

CAS #: 000105-57-7 — 1,1-DIETHOXYETHANE

Formula: $C_6H_{14}O_2$
Mol Weight: 118.18
MP (deg C): -100
FP (deg C):
BP (deg C): 102.7
BP pressure (mm Hg):

Property/Value	Units	Temp	Data Type	Reference
WS 4.40E+004	mg/L	25	EXP	YALKOWSKY,SH & DANNENFELSER,RM (1992)
logP 0.84			EXP	HANSCH,C & LEO,AJ (1985)
VP 2.76E+001	mm Hg	25	EXP	DAUBERT,TE & DANNER,RP (1989)
DC	pKa			
HL 9.75E-005	atm m3/mol	25	EST	VP/WSOL
OH 1.96E-011	cm3/molc sec	25	EST	MEYLAN,WM & HOWARD,PH (1993)

CAS #: 000105-58-8 — DIETHYL CARBONATE

Formula: $C_5H_{10}O_3$
Mol Weight: 118.13
MP (deg C): -43
FP (deg C):
BP (deg C): 126
BP pressure (mm Hg):

Property/Value	Units	Temp	Data Type	Reference
WS 1.88E+004	mg/L	20	EXP	YALKOWSKY,SH & DANNENFELSER,RM (1992)
logP 1.21			EXP	HANSCH,C & LEO,AJ (1985)
VP 1.08E+001	mm Hg	25	EXP	DAUBERT,TE & DANNER,RP (1989)
DC	pKa			
HL 8.93E-005	atm m3/mol	25	EST	VP/WSOL
OH 3.32E-012	cm3/molc sec	25	EST	MEYLAN,WM & HOWARD,PH (1993)

CAS #: 000105-59-9				N-METHYLDIETHANOLAMINE

Formula: $C_5H_{13}NO_2$

Mol Weight: 119.16

MP (deg C): -21   FP (deg C):

BP (deg C): 247

BP pressure (mm Hg):

Property/Value	Units	Temp	Data Type	Reference
WS 1.00E+006	mg/L		EXP	MULLINS,RM (1978)
logP -1.50			EST	MEYLAN,WM & HOWARD,PH (1995)
VP 2.00E-004	mm Hg	25	EXP	DAUBERT,TE & DANNER,RP (1989)
DC 8.52	pKa	25	EXP	PERRIN,DD (1965)
HL 3.14E-011	atm m3/mol	25	EST	VP/WSOL
OH 9.70E-011	cm3/molc sec	25	EST	MEYLAN,WM & HOWARD,PH (1993)

CAS #: 000105-60-2				CAPROLACTAM

Formula: $C_6H_{11}NO$

Mol Weight: 113.16

MP (deg C): 70   FP (deg C):

BP (deg C): 180

BP pressure (mm Hg): 5.00E+001

Property/Value	Units	Temp	Data Type	Reference
WS 7.72E+005	mg/L		EXP	YALKOWSKY,SH & DANNENFELSER,RM (1992)
logP -0.19			EXP	HANSCH,C & LEO,AJ (1981)
VP 1.60E-003	mm Hg	25	EXT	RITZ,J & FUCHS,H (1986)
DC	pKa			
HL 2.53E-008	atm m3/mol	25	EST	MEYLAN,WM & HOWARD,PH (1991)
OH 1.84E-011	cm3/molc sec	25	EST	MEYLAN,WM & HOWARD,PH (1993)

CAS #: 000105-66-8				N-PROPYLBUTYRATE

Formula: $C_7H_{14}O_2$

Mol Weight: 130.19

MP (deg C): -95   FP (deg C):

BP (deg C): 143

BP pressure (mm Hg):

Property/Value	Units	Temp	Data Type	Reference
WS 1.60E+003	mg/L	25	EXP	SUZUKI,T (1991)
logP 2.34			EST	MEYLAN,WM & HOWARD,PH (1995)
VP 5.95E+000	mm Hg	25	EXP	PERRY,RH & GREEN,D (1984)
DC	pKa			
HL 6.36E-004	atm m3/mol	25	EST	VP/WSOL
OH 7.40E-012	cm3/molc sec	25	EXP	ATKINSON,R (1989)

CAS #: 000105-67-9				2,4-DIMETHYLPHENOL

Formula: $C_8H_{10}O$

Mol Weight: 122.17

MP (deg C): 24.54   FP (deg C):

BP (deg C): 210.931

BP pressure (mm Hg):

Property/Value	Units	Temp	Data Type	Reference
WS 7.87E+003	mg/L	25	EXP	BANERJEE,S ET AL. (1980)
logP 2.30			EXP	HANSCH,C & LEO,AJ (1985)
VP 9.80E-002	mm Hg	25	EXP	CHAO,J ET AL. (1983)
DC 10.61	pKa	25	EXP	PEARCE,PJ & SIMKINS,RJJ (1968)
HL 2.00E-006	atm m3/mol	25	EST	VP/WSOL
OH 7.15E-011	cm3/molc sec	25	EXP	ATKINSON,R (1989)

CAS #: 000105-76-0				DIBUTYL MALEATE

Formula: $C_{12}H_{20}O_4$

Mol Weight: 228.29

MP (deg C):   FP (deg C):

BP (deg C): 281

BP pressure (mm Hg):

Property/Value	Units	Temp	Data Type	Reference
WS 8.71E+000	mg/L	25	EST	MEYLAN,WM ET AL. (1996)
logP 4.16			EST	MEYLAN,WM & HOWARD,PH (1995)
VP 6.23E-004	mm Hg	25	EXP	DAUBERT,TE & DANNER,RP (1989)
DC	pKa			
HL 7.58E-007	atm m3/mol	25	EST	MEYLAN,WM & HOWARD,PH (1991)
OH 1.60E-011	cm3/molc sec	25	EST	MEYLAN,WM & HOWARD,PH (1993)

CAS #: 000105-87-3				GERANIOL ACETATE

Formula: $C_{12}H_{20}O_2$

Mol Weight: 196.29

MP (deg C):   FP (deg C):

BP (deg C): 138

BP pressure (mm Hg): 2.50E+001

Property/Value	Units	Temp	Data Type	Reference
WS 6.89E+000	mg/L	25	EST	MEYLAN,WM ET AL. (1996)
logP 4.48			EST	MEYLAN,WM & HOWARD,PH (1995)
VP 3.30E-002	mm Hg	25	EXT	OHE,S (1976)
DC	pKa			
HL 2.42E-003	atm m3/mol	25	EST	MEYLAN,WM & HOWARD,PH (1991)
OH 1.78E-010	cm3/molc sec	25	EST	MEYLAN,WM & HOWARD,PH (1993)

CAS #: 000106-22-9				3,7-DIMETHYL-6-OCTEN-1-OL

Formula: $C_{10}H_{20}O$

Mol Weight: 156.27

MP (deg C):   FP (deg C):

BP (deg C): 224.5

BP pressure (mm Hg):

Property/Value	Units	Temp	Data Type	Reference
WS 2.12E+002	mg/L	25	EST	MEYLAN,WM ET AL. (1996)
logP 3.56			EST	MEYLAN,WM & HOWARD,PH (1995)
VP 4.41E-002	mm Hg	25	EXT	PERRY,RH & GREEN,D (1984)
DC	pKa			
HL 5.68E-005	atm m3/mol	25	EST	MEYLAN,WM & HOWARD,PH (1991)
OH 9.84E-011	cm3/molc sec	25	EST	MEYLAN,WM & HOWARD,PH (1993)

CAS #: 000106-23-0				CITRONELLAL

Formula: $C_{10}H_{18}O$

Mol Weight: 154.25

MP (deg C):   FP (deg C):

BP (deg C): 47

BP pressure (mm Hg): 1.00E+000

Property/Value	Units	Temp	Data Type	Reference
WS 7.02E+001	mg/L	25	EST	MEYLAN,WM ET AL. (1996)
logP 3.53			EST	MEYLAN,WM & HOWARD,PH (1995)
VP 2.80E-001	mm Hg	25	EST	NEELY,WB & BLAU,GE (1990)
DC	pKa			
HL 2.62E-004	atm m3/mol	25	EST	MEYLAN,WM & HOWARD,PH (1991)
OH 1.21E-010	cm3/molc sec	25	EST	MEYLAN,WM & HOWARD,PH (1993)

000106-24-1 — GERANIOL

Formula: $C_{10}H_{18}O$

Mol Weight: 154.25

MP (deg C): < -15
FP (deg C):
BP (deg C): 229-230
BP pressure (mm Hg): 7.57E+002

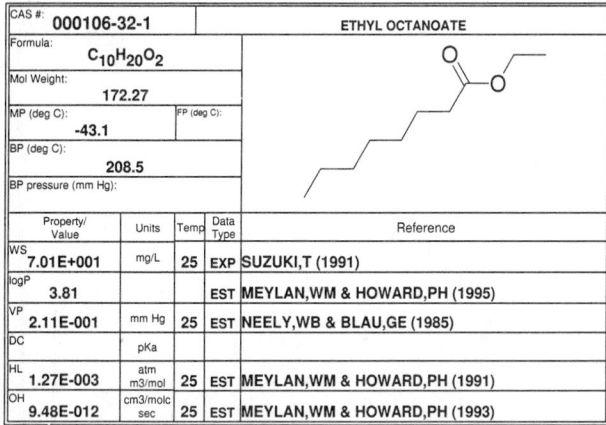

Property/Value	Units	Temp	Data Type	Reference
WS 1.00E+002	mg/L	25	EXP	CHEM INSPECT TEST INST (1992)
logP 3.47			EST	MEYLAN,WM & HOWARD,PH (1995)
VP 3.00E-002	mm Hg	25	EXT	PERRY,RH & GREEN,D (1984)
DC	pKa			
HL 5.89E-005	atm m3/mol	25	EST	MEYLAN,WM & HOWARD,PH (1991)
OH 1.80E-010	cm3/molc sec	25	EST	MEYLAN,WM & HOWARD,PH (1993)

000106-27-4 — ISOPENTYL BUTANOATE

Formula: $C_9H_{18}O_2$

Mol Weight: 158.24

MP (deg C):
FP (deg C):
BP (deg C): 179
BP pressure (mm Hg):

Property/Value	Units	Temp	Data Type	Reference
WS 1.18E+002	mg/L	25	EST	MEYLAN,WM ET AL. (1996)
logP 3.25			EST	MEYLAN,WM & HOWARD,PH (1995)
VP 9.50E-001	mm Hg	20	EXP	VOITKEVICH,SA (1963)
DC	pKa			
HL 9.60E-004	atm m3/mol	25	EST	MEYLAN,WM & HOWARD,PH (1991)
OH 8.14E-012	cm3/molc sec	25	EST	MEYLAN,WM & HOWARD,PH (1993)

000106-30-9 — ETHYL HEPTANOATE

Formula: $C_9H_{18}O_2$

Mol Weight: 158.24

MP (deg C): -66.1
FP (deg C):
BP (deg C): 189
BP pressure (mm Hg):

Property/Value	Units	Temp	Data Type	Reference
WS 2.90E+002	mg/L	25	EXP	SUZUKI,T (1991)
logP 3.32			EST	MEYLAN,WM & HOWARD,PH (1995)
VP 6.80E-001	mm Hg	25	EXP	HINE,J & MOOKERJEE,PK (1975)
DC	pKa			
HL 4.99E-004	atm m3/mol	25	EXP	HINE,J & MOOKERJEE,PK (1975)
OH 8.07E-012	cm3/molc sec	25	EST	MEYLAN,WM & HOWARD,PH (1993)

000106-31-0 — BUTYRIC ANHYDRIDE

Formula: $C_8H_{14}O_3$

Mol Weight: 158.20

MP (deg C): -75
FP (deg C):
BP (deg C): 199.4-1.4
BP pressure (mm Hg):

Property/Value	Units	Temp	Data Type	Reference
WS 4.56E+003	mg/L	25	EST	MEYLAN,WM ET AL. (1996)
logP 1.39			EST	MEYLAN,WM & HOWARD,PH (1995)
VP 2.83E-001	mm Hg	25	EXP	DAUBERT,TE & DANNER,RP (1989)
DC	pKa			
HL 1.11E-004	atm m3/mol	25	EST	MEYLAN,WM & HOWARD,PH (1991)
OH 4.33E-012	cm3/molc sec	25	EST	MEYLAN,WM & HOWARD,PH (1993)

000106-32-1 — ETHYL OCTANOATE

Formula: $C_{10}H_{20}O_2$

Mol Weight: 172.27

MP (deg C): -43.1
FP (deg C):
BP (deg C): 208.5
BP pressure (mm Hg):

Property/Value	Units	Temp	Data Type	Reference
WS 7.01E+001	mg/L	25	EXP	SUZUKI,T (1991)
logP 3.81			EST	MEYLAN,WM & HOWARD,PH (1995)
VP 2.11E-001	mm Hg	25	EST	NEELY,WB & BLAU,GE (1985)
DC	pKa			
HL 1.27E-003	atm m3/mol	25	EST	MEYLAN,WM & HOWARD,PH (1991)
OH 9.48E-012	cm3/molc sec	25	EST	MEYLAN,WM & HOWARD,PH (1993)

000106-33-2 — ETHYL LAURATE

Formula: $C_{14}H_{28}O_2$

Mol Weight: 228.38

MP (deg C): -10
FP (deg C):
BP (deg C): 269
BP pressure (mm Hg):

Property/Value	Units	Temp	Data Type	Reference
WS 3.63E-001	mg/L	25	EST	MEYLAN,WM ET AL. (1996)
logP 5.78			EST	MEYLAN,WM & HOWARD,PH (1995)
VP 5.26E-003	mm Hg	25	EST	NEELY,WB & BLAU,GE (1985)
DC	pKa			
HL 3.96E-003	atm m3/mol	25	EST	MEYLAN,WM & HOWARD,PH (1991)
OH 1.51E-011	cm3/molc sec	25	EST	MEYLAN,WM & HOWARD,PH (1993)

000106-34-3 — 2,5-CYCLOHEXADIENE-1,4-DIONE, W/1,4-BENZENEDIOL

Formula: $C_{12}H_{10}O_4$

Mol Weight: 218.21

MP (deg C): 171
FP (deg C):
BP (deg C):
BP pressure (mm Hg):

Property/Value	Units	Temp	Data Type	Reference
WS 4.06E+003	mg/L	25	EXP	YALKOWSKY,SH & DANNENFELSER,RM (1992)
logP 0.16			EST	MEYLAN,WM & HOWARD,PH (1995)
VP 9.84E-009	mm Hg	25	EST	NEELY,WB & BLAU,GE (1985)
DC	pKa			
HL 3.59E-019	atm m3/mol	25	EST	MEYLAN,WM & HOWARD,PH (1991)
OH 7.35E-011	cm3/molc sec	25	EST	MEYLAN,WM & HOWARD,PH (1993)

000106-35-4 — 3-HEPTANONE

Formula: $C_7H_{14}O$

Mol Weight: 114.19

MP (deg C): -39
FP (deg C): -39
BP (deg C): 147.8
BP pressure (mm Hg):

Property/Value	Units	Temp	Data Type	Reference
WS 4.30E+003	mg/L	20	EXP	FLICK,EW (1991)
logP 1.73			EST	MEYLAN,WM & HOWARD,PH (1995)
VP 2.60E+000	mm Hg	20	EXP	DAUBERT,TE & DANNER,RP (1993)
DC	pKa			
HL 9.08E-005	atm m3/mol	20	EST	VP/WSOL
OH 8.12E-012	cm3/molc sec	25	EST	MEYLAN,WM & HOWARD,PH (1993)

CAS #: 000106-36-5				N-PROPYLPROPIONATE

Formula: $C_6H_{12}O_2$

Mol Weight: 116.16

MP (deg C): -76 **FP (deg C):**

BP (deg C): 122-124

BP pressure (mm Hg):

Property/ Value	Units	Temp	Data Type	Reference
WS 5.30E+003	mg/L	25	EXP	SUZUKI,T (1991)
logP 1.85			EST	MEYLAN,WM & HOWARD,PH (1995)
VP 1.39E+001	mm Hg	25	EXP	PERRY,RH & GREEN,D (1984)
DC	pKa			
HL 4.01E-004	atm m3/mol	25	EST	VP/WSOL
OH 4.00E-012	cm3/molc sec	25	EXP	ATKINSON,R (1989)

CAS #: 000106-37-6				1,4-DIBROMOBENZENE

Formula: $C_6H_4Br_2$

Mol Weight: 235.92

MP (deg C): 87.31 **FP (deg C):**

BP (deg C): 220.40

BP pressure (mm Hg):

Property/ Value	Units	Temp	Data Type	Reference
WS 2.00E+001	mg/L	25	EXP	YALKOWSKY,SH & DANNENFELSER,RM (1992)
logP 3.79			EXP	HANSCH,C & LEO,AJ (1985)
VP 5.75E-002	mm Hg	25	EXP	WALSH,PN & SMITH,NO (1961)
DC	pKa			
HL 8.93E-004	atm m3/mol	25	EST	VP/WSOL
OH 3.52E-013	cm3/molc sec	25	EST	MEYLAN,WM & HOWARD,PH (1993)

CAS #: 000106-38-7				P-BROMOTOLUENE

Formula: C_7H_7Br

Mol Weight: 171.04

MP (deg C): 28.5 **FP (deg C):**

BP (deg C): 184.3

BP pressure (mm Hg):

Property/ Value	Units	Temp	Data Type	Reference
WS 1.10E+002	mg/L	25	EXP	HINE,J & MOOKERJEE,PK (1975)
logP 3.42			EXP	DUNN,WJ ET AL. (1983)
VP 1.15E+000	mm Hg	25	EXP	HINE,J & MOOKERJEE,PK (1975)
DC	pKa			
HL 2.33E-003	atm m3/mol	25	EXP	HINE,J & MOOKERJEE,PK (1975)
OH 1.64E-012	cm3/molc sec	25	EST	MEYLAN,WM & HOWARD,PH (1993)

CAS #: 000106-39-8				P-BROMOCHLOROBENZENE

Formula: C_6H_4BrCl

Mol Weight: 191.46

MP (deg C): 68 **FP (deg C):**

BP (deg C): 196

BP pressure (mm Hg):

Property/ Value	Units	Temp	Data Type	Reference
WS 4.49E+001	mg/L	25	EXP	SUZUKI,T (1991)
logP 3.54			EXP	DUNN,WJ ET AL. (1983)
VP 2.58E-001	mm Hg	25	EXP	MACKAY,D & SHIU,WY (1981)
DC	pKa			
HL 1.45E-003	atm m3/mol	25	EST	VP/WSOL
OH 3.93E-013	cm3/molc sec	25	EST	MEYLAN,WM & HOWARD,PH (1993)

CAS #: 000106-40-1				P-BROMOANILINE

Formula: C_6H_6BrN

Mol Weight: 172.03

MP (deg C): 66.4 **FP (deg C):**

BP (deg C):

BP pressure (mm Hg):

Property/ Value	Units	Temp	Data Type	Reference
WS 7.07E+002	mg/L	25	EST	MEYLAN,WM ET AL. (1996)
logP 2.26			EXP	HANSCH,C & LEO,AJ (1985)
VP 4.29E-002	mm Hg	25	EST	NEELY,WB & BLAU,GE (1985)
DC 3.86	pKa	25	EXP	PERRIN,DD (1965)
HL 7.59E-007	atm m3/mol	25	EST	MEYLAN,WM & HOWARD,PH (1991)
OH 3.09E-011	cm3/molc sec	25	EST	MEYLAN,WM & HOWARD,PH (1993)

CAS #: 000106-41-2				P-BROMOPHENOL

Formula: C_6H_5BrO

Mol Weight: 173.02

MP (deg C): 64 **FP (deg C):**

BP (deg C): 238

BP pressure (mm Hg):

Property/ Value	Units	Temp	Data Type	Reference
WS 1.40E+004	mg/L	25	EXP	CHEM INSPECT TEST INST
logP 2.59			EXP	HANSCH,C & LEO,AJ (1985)
VP 1.17E-002	mm Hg	25	EXT	PARSONS,GH ET AL. (1971)
DC 9.17	pKa	25	EXP	KORTUM,G ET AL (1961)
HL 1.51E-007	atm m3/mol	25	EXP	PARSONS,GH ET AL. (1971)
OH 1.02E-011	cm3/molc sec	25	EST	MEYLAN,WM & HOWARD,PH (1993)

CAS #: 000106-42-3				P-XYLENE

Formula: C_8H_{10}

Mol Weight: 106.17

MP (deg C): 13.263 **FP (deg C):**

BP (deg C): 138.359

BP pressure (mm Hg):

Property/ Value	Units	Temp	Data Type	Reference
WS 1.62E+002	mg/L	25	EXP	SANEMASA,I ET AL. (1982)
logP 3.15			EXP	HANSCH,C & LEO,AJ (1985)
VP 8.84E+000	mm Hg	25	EXP	CHAO,J ET AL. (1983)
DC	pKa			
HL 7.53E-003	atm m3/mol	25	EXP	SANEMASA,I ET AL. (1982)
OH 1.43E-011	cm3/molc sec	25	EXP	ATKINSON,R (1989)

CAS #: 000106-43-4				P-CHLOROTOLUENE

Formula: C_7H_7Cl

Mol Weight: 126.59

MP (deg C): 7.5 **FP (deg C):**

BP (deg C): 162.4

BP pressure (mm Hg):

Property/ Value	Units	Temp	Data Type	Reference
WS 1.06E+000	mg/L	20	EXP	YALKOWSKY,SH & DANNENFELSER,RM (1992)
logP 3.33			EXP	HANSCH,C & LEO,AJ (1985)
VP 2.69E+000	mm Hg	25	EXP	DAUBERT,TE & DANNER,RP (1989)
DC	pKa			
HL 4.38E-003	atm m3/mol	25	EST	VP/WSOL
OH 1.82E-012	cm3/molc sec	25	EST	MEYLAN,WM & HOWARD,PH (1993)

CAS #: 000106-44-5				P-CRESOL

Formula: C_7H_8O

Mol Weight: 108.14

MP (deg C): 34.739 FP (deg C):

BP (deg C): 201.94

BP pressure (mm Hg):

Property/Value	Units	Temp	Data Type	Reference
WS 2.15E+004	mg/L	25	EXP	YALKOWSKY,SH & DANNENFELSER,RM (1992)
logP 1.94			EXP	HANSCH,C & LEO,AJ (1985)
VP 1.10E-001	mm Hg	25	EXP	CHAO,J ET AL. (1983)
DC 10.26	pKa	25	EXP	PEARCE,PJ & SIMKINS,RJJ (1968)
HL 1.00E-006	atm m3/mol	25	EXP	GAFFNEY,JS ET AL. (1987)
OH 4.70E-011	cm3/molc sec	25	EXP	ATKINSON,R (1989)

CAS #: 000106-45-6				4-METHYLBENZENETHIOL

Formula: C_7H_8S

Mol Weight: 124.21

MP (deg C): 43 FP (deg C):

BP (deg C): 195

BP pressure (mm Hg):

Property/Value	Units	Temp	Data Type	Reference
WS 1.68E+002	mg/L	25	EST	MEYLAN,WM ET AL. (1996)
logP 3.23			EST	MEYLAN,WM & HOWARD,PH (1995)
VP 6.45E-001	mm Hg	25	EXT	CHAO,J ET AL. (1983)
DC	pKa			
HL 5.76E-004	atm m3/mol	25	EST	MEYLAN,WM & HOWARD,PH (1991)
OH 1.40E-011	cm3/molc sec	25	EST	MEYLAN,WM & HOWARD,PH (1993)

CAS #: 000106-46-7				1,4-DICHLOROBENZENE

Formula: $C_6H_4Cl_2$

Mol Weight: 147.00

MP (deg C): 53.13 FP (deg C):

BP (deg C): 174.12

BP pressure (mm Hg):

Property/Value	Units	Temp	Data Type	Reference
WS 7.60E+001	mg/L	25	EXP	YALKOWSKY,SH & DANNENFELSER,RM (1992)
logP 3.44			EXP	HANSCH,C ET AL. (1995)
VP 1.00E+000	mm Hg	25	EXP	WALSH,PN & SMITH,NO (1961)
DC	pKa			
HL 2.40E-003	atm m3/mol	25	EXP	MACKAY,D ET AL. (1982A)
OH 3.20E-013	cm3/molc sec	25	EXP	ATKINSON,R (1989)

CAS #: 000106-47-8				4-CHLOROANILINE

Formula: C_6H_6ClN

Mol Weight: 127.57

MP (deg C): 72.5 FP (deg C):

BP (deg C): 232

BP pressure (mm Hg):

Property/Value	Units	Temp	Data Type	Reference
WS 3.90E+003	mg/L	25	EXP	KILZER,L ET AL. (1979)
logP 1.83			EXP	SANGSTER,J (1994)
VP 2.70E-002	mm Hg	26	EXP	PIACENTE,V ET AL. (1985)
DC 3.98	pKa	25	EXP	PERRIN,DD (1972)
HL 1.16E-006	atm m3/mol	25	EST	VP/WSOL
OH 8.30E-011	cm3/molc sec	25	EXP	ATKINSON,R (1989)

CAS #: 000106-48-9				4-CHLOROPHENOL

Formula: C_6H_5ClO

Mol Weight: 128.56

MP (deg C): 42.7 FP (deg C):

BP (deg C): 220

BP pressure (mm Hg):

Property/Value	Units	Temp	Data Type	Reference
WS 2.40E+004	mg/L	25	EXP	ROBERTS,MS ET AL. (1977)
logP 2.39			EXP	HANSCH,C & LEO,AJ (1985)
VP 8.90E-002	mm Hg	25	EXP	SHIU,WY ET AL. (1994)
DC 9.41	pKa	26	EXP	SERJEANT,EP & DEMPSEY,B (1979)
HL 6.27E-007	atm m3/mol	25	EST	VP/WSOL
OH 1.03E+001	cm3/molc sec	25	EST	MEYLAN,WM & HOWARD,PH (1993)

CAS #: 000106-49-0				P-TOLUIDINE

Formula: C_7H_9N

Mol Weight: 107.16

MP (deg C): 44-45 FP (deg C):

BP (deg C): 200-201

BP pressure (mm Hg):

Property/Value	Units	Temp	Data Type	Reference
WS 7.35E+004	mg/L	20	EXP	RIDDICK,JA ET AL. (1986)
logP 1.39			EXP	HANSCH,C & LEO,AJ (1985)
VP 2.86E-001	mm Hg	25	EXP	CHAO,J ET AL. (1983)
DC 5.10	pKa	25	EXP	PERRIN,DD (1965)
HL 5.49E-007	atm m3/mol	25	EST	VP/WSOL
OH 1.32E-010	cm3/molc sec	25	EST	MEYLAN,WM & HOWARD,PH (1993)

CAS #: 000106-50-3				1,4-BENZENEDIAMINE

Formula: $C_6H_8N_2$

Mol Weight: 108.14

MP (deg C): 145-147 FP (deg C):

BP (deg C): 267

BP pressure (mm Hg):

Property/Value	Units	Temp	Data Type	Reference
WS 3.70E+004	mg/L	23	EXP	SEIDELL,A (1941)
logP -0.30			EXP	HANSCH,C ET AL. (1995)
VP 1.23E-002	mm Hg	25	EST	NEELY,WB & BLAU,GE (1985)
DC 6.16	pKa	25	EXP	PERRIN,DD (1965)
HL 6.73E-010	atm m3/mol	25	EST	MEYLAN,WM & HOWARD,PH (1991)
OH 2.31E-010	cm3/molc sec	25	EST	MEYLAN,WM & HOWARD,PH (1993)

CAS #: 000106-51-4				2,5-CYCLOHEXADIENE-1,4-DIONE

Formula: $C_6H_4O_2$

Mol Weight: 108.10

MP (deg C): 115.7 FP (deg C):

BP (deg C):

BP pressure (mm Hg):

Property/Value	Units	Temp	Data Type	Reference
WS 1.11E+004	mg/L	18	EXP	YALKOWSKY,SH & DANNENFELSER,RM (1992)
logP 0.20			EXP	HANSCH,C ET AL. (1995)
VP 9.00E-001	mm Hg	20	EXP	FINLEY,KT (1985)
DC	pKa			
HL 4.79E-004	atm m3/mol	20	EST	VP/WSOL
OH 4.49E-012	cm3/molc sec	25	EST	MEYLAN,WM & HOWARD,PH (1993)

CAS #: 000106-54-7	4-CHLOROBENZENETHIOL

Formula: C_6H_5ClS

Mol Weight: 144.62

MP (deg C): 61 FP (deg C):

BP (deg C): 206

BP pressure (mm Hg):

Property/Value	Units	Temp	Data Type	Reference
WS 1.15E+002	mg/L	25	EST	MEYLAN,WM ET AL. (1996)
logP 3.33				MEYLAN,WM & HOWARD,PH (1995)
VP 2.32E-001	mm Hg	25	EST	NEELY,WB & BLAU,GE (1985)
DC	pKa			
HL 3.87E-004	atm m3/mol	25	EST	MEYLAN,WM & HOWARD,PH (1991)
OH 3.30E-012	cm3/molc sec	25	EST	MEYLAN,WM & HOWARD,PH (1993)

CAS #: 000106-55-8	2,5-DIMETHYLPIPERAZINE

Formula: $C_6H_{14}N_2$

Mol Weight: 114.19

MP (deg C): FP (deg C):

BP (deg C):

BP pressure (mm Hg):

Property/Value	Units	Temp	Data Type	Reference
WS 9.99E+005	mg/L	25	EST	MEYLAN,WM ET AL. (1996)
logP 0.04				MEYLAN,WM & HOWARD,PH (1995)
VP 2.65E+000	mm Hg	25	EXT	CHAO,J ET AL. (1983)
DC 9.66	pKa	25	EXP	PERRIN,DD (1965)
HL 3.88E-009	atm m3/mol	25	EST	MEYLAN,WM & HOWARD,PH (1991)
OH 1.92E-010	cm3/molc sec	25	EST	MEYLAN,WM & HOWARD,PH (1993)

CAS #: 000106-58-1	PIPERAZINE, 1,4-DIMETHYL-

Formula: $C_6H_{14}N_2$

Mol Weight: 114.19

MP (deg C): FP (deg C):

BP (deg C): 131

BP pressure (mm Hg):

Property/Value	Units	Temp	Data Type	Reference
WS 1.00E+006	mg/L	25	EST	MEYLAN,WM ET AL. (1996)
logP -0.40			EXP	HANSCH,C ET AL. (1995)
VP 5.22E+000	mm Hg	25	EST	NEELY,WB & BLAU,GE (1985)
DC	pKa			
HL 1.06E-008	atm m3/mol	25	EST	MEYLAN,WM & HOWARD,PH (1991)
OH 1.77E-010	cm3/molc sec	25	EST	MEYLAN,WM & HOWARD,PH (1993)

CAS #: 000106-62-7	1-PROPANOL, 2-(2-HYDROXYPROPOXY)-

Formula: $C_6H_{14}O_3$

Mol Weight: 134.18

MP (deg C): FP (deg C):

BP (deg C):

BP pressure (mm Hg):

Property/Value	Units	Temp	Data Type	Reference
WS 3.30E+005	mg/L	25	EST	MEYLAN,WM ET AL. (1996)
logP -0.67			EXP	HANSCH,C ET AL. (1995)
VP 1.05E-002	mm Hg	25	EST	NEELY,WB & BLAU,GE (1985)
DC	pKa			
HL 3.58E-009	atm m3/mol	25	EST	MEYLAN,WM & HOWARD,PH (1991)
OH 3.46E-011	cm3/molc sec	25	EST	MEYLAN,WM & HOWARD,PH (1993)

CAS #: 000106-63-8	ISOBUTYL ACRYLATE

Formula: $C_7H_{12}O_2$

Mol Weight: 128.17

MP (deg C): -61 FP (deg C):

BP (deg C): 132

BP pressure (mm Hg):

Property/Value	Units	Temp	Data Type	Reference
WS 1.19E+003	mg/L	25	EST	MEYLAN,WM ET AL. (1996)
logP 2.22			EXP	HANSCH,C & LEO,AJ (1985)
VP 8.07E+000	mm Hg	25	EXP	DAUBERT,TE & DANNER,RP (1989)
DC	pKa			
HL 2.16E-004	atm m3/mol	25	EST	MEYLAN,WM & HOWARD,PH (1991)
OH 1.38E-011	cm3/molc sec	25	EST	MEYLAN,WM & HOWARD,PH (1993)

CAS #: 000106-65-0	DIMETHYL SUCCINATE

Formula: $C_6H_{10}O_4$

Mol Weight: 146.14

MP (deg C): 19 FP (deg C):

BP (deg C): 196.4

BP pressure (mm Hg):

Property/Value	Units	Temp	Data Type	Reference
WS 8.30E+003	mg/L	25	EXP	MERCK INDEX (1976)
logP 0.35			EXP	HANSCH,C ET AL. (1995)
VP 2.50E-001	mm Hg	103	EXP	MERCK INDEX (1976)
DC	pKa			
HL 6.43E-008	atm m3/mol	25	EST	MEYLAN,WM & HOWARD,PH (1991)
OH 2.14E-012	cm3/molc sec	25	EST	MEYLAN,WM & HOWARD,PH (1993)

CAS #: 000106-67-2	4-METHYL-2-ETHYL-1-PENTANOL

Formula: $C_8H_{18}O$

Mol Weight: 130.23

MP (deg C): FP (deg C):

BP (deg C):

BP pressure (mm Hg):

Property/Value	Units	Temp	Data Type	Reference
WS 1.59E+003	mg/L	25	EST	MEYLAN,WM ET AL. (1996)
logP 2.66				MEYLAN,WM & HOWARD,PH (1995)
VP 7.55E-003	mm Hg	25	EXT	DYKYJ,J ET AL. (1961)
DC	pKa			
HL 3.10E-005	atm m3/mol	25	EST	MEYLAN,WM & HOWARD,PH (1991)
OH 1.32E-011	cm3/molc sec	25	EST	MEYLAN,WM & HOWARD,PH (1993)

CAS #: 000106-68-3	3-OCTANONE

Formula: $C_8H_{16}O$

Mol Weight: 128.22

MP (deg C): FP (deg C): 46

BP (deg C): 167.5

BP pressure (mm Hg):

Property/Value	Units	Temp	Data Type	Reference
WS 2.60E+003	mg/L	20	EXP	LANDE,SS ET AL. (1976)
logP 2.22				MEYLAN,WM & HOWARD,PH (1995)
VP 2.00E+000	mm Hg	20	EXP	LANDE,SS ET AL. (1976)
DC	pKa			
HL 1.30E-004	atm m3/mol	20	EST	VP/WSOL
OH 9.30E-012	cm3/molc sec	25	EST	MEYLAN,WM & HOWARD,PH (1993)

000106-69-4 — 1,2,6-HEXANETRIOL

Formula: $C_6H_{14}O_3$
Mol Weight: 134.18
MP (deg C):
FP (deg C):
BP (deg C): 170
BP pressure (mm Hg): 3.00E+000

Property/Value	Units	Temp	Data Type	Reference
WS 4.05E+005	mg/L	25	EST	MEYLAN,WM ET AL. (1996)
logP -0.77			EST	MEYLAN,WM & HOWARD,PH (1995)
VP 7.91E-005	mm Hg	25	EST	NEELY,WB & BLAU,GE (1985)
DC	pKa			
HL 1.48E-008	atm m3/mol	25	EST	MEYLAN,WM & HOWARD,PH (1991)
OH 2.30E-011	cm3/molc sec	25	EST	MEYLAN,WM & HOWARD,PH (1993)

000106-70-7 — METHYL HEXANOATE

Formula: $C_7H_{14}O_2$
Mol Weight: 130.19
MP (deg C): -71
FP (deg C):
BP (deg C): 149.5
BP pressure (mm Hg):

Property/Value	Units	Temp	Data Type	Reference
WS 1.33E+003	mg/L	20	EXP	YALKOWSKY,SH & DANNENFELSER,RM (1992)
logP 2.34			EST	MEYLAN,WM & HOWARD,PH (1995)
VP 3.72E+000	mm Hg	25	EXP	PERRY,RH & GREEN,D (1984)
DC	pKa			
HL 3.67E-004	atm m3/mol	25	EXP	BUTTERY,RG ET AL. (1969)
OH 4.27E-012	cm3/molc sec	25	EST	MEYLAN,WM & HOWARD,PH (1993)

000106-73-0 — METHYL HEPTANOATE

Formula: $C_8H_{16}O_2$
Mol Weight: 144.22
MP (deg C): -56
FP (deg C):
BP (deg C): 174
BP pressure (mm Hg):

Property/Value	Units	Temp	Data Type	Reference
WS 3.09E+002	mg/L	25	EST	MEYLAN,WM ET AL. (1996)
logP 2.83			EST	MEYLAN,WM & HOWARD,PH (1995)
VP 1.56E+000	mm Hg	25	EST	NEELY,WB & BLAU,GE (1985)
DC	pKa			
HL 7.23E-004	atm m3/mol	25	EST	MEYLAN,WM & HOWARD,PH (1991)
OH 6.62E-012	cm3/molc sec	25	EST	MEYLAN,WM & HOWARD,PH (1993)

000106-86-5 — 7-OXABICYCLO 4.1.0 HEPTANE, 3-ETHENYL-

Formula: $C_8H_{12}O$
Mol Weight: 124.18
MP (deg C): < -100
FP (deg C):
BP (deg C): 169
BP pressure (mm Hg):

Property/Value	Units	Temp	Data Type	Reference
WS 1.62E+003	mg/L	25	EST	MEYLAN,WM ET AL. (1996)
logP 2.08			EXP	SANGSTER,J (1993)
VP 5.09E+000	mm Hg	25	EST	NEELY,WB & BLAU,GE (1985)
DC	pKa			
HL 2.16E-004	atm m3/mol	25	EST	MEYLAN,WM & HOWARD,PH (1991)
OH 3.06E-011	cm3/molc sec	25	EST	MEYLAN,WM & HOWARD,PH (1993)

000106-87-6 — 7-OXABICYCLO 4.1.0 HEPTANE, 3-OXIRANYL-

Formula: $C_8H_{12}O_2$
Mol Weight: 140.18
MP (deg C): < -55
FP (deg C):
BP (deg C): 227
BP pressure (mm Hg):

Property/Value	Units	Temp	Data Type	Reference
WS 3.52E+004	mg/L	25	EST	MEYLAN,WM ET AL. (1996)
logP 0.44			EXP	SANGSTER,J (1993)
VP 1.17E+000	mm Hg	25	EST	NEELY,WB & BLAU,GE (1985)
DC	pKa			
HL 1.42E-007	atm m3/mol	25	EST	MEYLAN,WM & HOWARD,PH (1991)
OH 5.49E-012	cm3/molc sec	25	EST	MEYLAN,WM & HOWARD,PH (1993)

000106-88-7 — 1,2-BUTYLENEOXIDE

Formula: C_4H_8O
Mol Weight: 72.11
MP (deg C): -150
FP (deg C):
BP (deg C): 63
BP pressure (mm Hg):

Property/Value	Units	Temp	Data Type	Reference
WS 9.50E+004	mg/L	25	EXP	BOGYO,DA ET AL. (1980)
logP 0.86			EST	MEYLAN,WM & HOWARD,PH (1995)
VP 1.80E+002	mm Hg	25	EXP	OSBORN,AG & SCOTT,DW (1980)
DC	pKa			
HL 1.80E-004	atm m3/mol	25	EST	VP/WSOL
OH 2.10E-012	cm3/molc sec	25	EXP	ATKINSON,R (1989)

000106-89-8 — 1-CHLORO-2,3-EPOXYPROPANE

Formula: C_3H_5ClO
Mol Weight: 92.53
MP (deg C): -57.2
FP (deg C):
BP (deg C): 116.11
BP pressure (mm Hg):

Property/Value	Units	Temp	Data Type	Reference
WS 6.59E+004	mg/L	25	EXP	YALKOWSKY,SH & DANNENFELSER,RM (1992)
logP 0.45			EXP	DENEER,JW ET AL. (1988)
VP 1.64E+001	mm Hg	25	EXP	DAUBERT,TE & DANNER,RP (1985)
DC	pKa			
HL 3.04E-005	atm m3/mol	25	EST	VP/WSOL
OH 4.40E-013	cm3/molc sec	25	EXP	ATKINSON,R (1989)

000106-93-4 — 1,2-DIBROMOETHANE

Formula: $C_2H_4Br_2$
Mol Weight: 187.87
MP (deg C): 9.79
FP (deg C):
BP (deg C): 131.36
BP pressure (mm Hg):

Property/Value	Units	Temp	Data Type	Reference
WS 4.15E+003	mg/L	25	EXP	HORVATH,AL (1982)
logP 1.96			EXP	HANSCH,C ET AL. (1995)
VP 1.12E+001	mm Hg	25	EXP	CALL,F (1957)
DC	pKa			
HL 6.67E-004	atm m3/mol	25	EST	VP/WSOL
OH 2.50E-013	cm3/molc sec	25	EXP	ATKINSON,R (1989)

CAS #: 000106-94-5 — 1-BROMOPROPANE

Formula: C_3H_7Br

Mol Weight: 123.00

MP (deg C): -110

FP (deg C):

BP (deg C): 71

BP pressure (mm Hg):

Property/Value	Units	Temp	Data Type	Reference
WS 2.45E+003	mg/L	20	EXP	YALKOWSKY,SH & DANNENFELSER,RM (1992)
logP 2.10			EXP	HANSCH,C & LEO,AJ (1985)
VP 1.11E+002	mm Hg	20	EXP	BOUBLIK,T ET AL (1984)
DC	pKa			
HL 7.32E-003	atm m3/mol	20	EST	VP/WSOL
OH 1.18E-012	cm3/molc sec	25	EXP	DONAGHY,T ET AL (1993)

CAS #: 000106-95-6 — 3-BROMO-1-PROPENE

Formula: C_3H_5Br

Mol Weight: 120.98

MP (deg C): -119

FP (deg C):

BP (deg C): 71.3

BP pressure (mm Hg):

Property/Value	Units	Temp	Data Type	Reference
WS 3.84E+003	mg/L	25	EXP	YALKOWSKY,SH & DANNENFELSER,RM (1992)
logP 1.79			EXP	HANSCH,C & LEO,AJ (1985)
VP 1.36E+002	mm Hg	25	EST	NEELY,WB & BLAU,GE (1985)
DC	pKa			
HL 1.11E-002	atm m3/mol	25	EST	MEYLAN,WM & HOWARD,PH (1991)
OH 2.02E-011	cm3/molc sec	25	EST	MEYLAN,WM & HOWARD,PH (1993)

CAS #: 000106-97-8 — N-BUTANE

Formula: C_4H_{10}

Mol Weight: 58.12

MP (deg C): -138.2

FP (deg C):

BP (deg C): -0.5

BP pressure (mm Hg):

Property/Value	Units	Temp	Data Type	Reference
WS 6.12E+001	mg/L	25	EXP	MCAULIFFE,C (1966)
logP 2.89			EXP	HANSCH,C ET AL. (1995)
VP 1.82E+003	mm Hg	25	EXP	RIDDICK,JA ET AL. (1986)
DC	pKa			
HL 9.50E-001	atm m3/mol	25	EST	VP/WSOL
OH 2.54E-012	cm3/molc sec	25	EXP	ATKINSON,R (1989)

CAS #: 000106-98-9 — 1-BUTENE

Formula: C_4H_8

Mol Weight: 56.11

MP (deg C): -185

FP (deg C):

BP (deg C): -6.3

BP pressure (mm Hg):

Property/Value	Units	Temp	Data Type	Reference
WS 2.21E+002	mg/L	25	EXP	MCAULIFFE,C (1966)
logP 2.40			EXP	HANSCH,C & LEO,AJ (1985)
VP 1.90E+003	mm Hg	20	EXP	OBENAUS,F ET AL. (1985)
DC	pKa			
HL 2.33E-001	atm m3/mol	25	EXP	WASIK,SP & TSANG,W (1970)
OH 3.13E-011	cm3/molc sec	25	EXP	ATKINSON,R (1987)

CAS #: 000106-99-0 — 1,3-BUTADIENE

Formula: C_4H_6

Mol Weight: 54.09

MP (deg C): -108.966

FP (deg C):

BP (deg C): -4.5

BP pressure (mm Hg):

Property/Value	Units	Temp	Data Type	Reference
WS 7.35E+002	mg/L	25	EXP	MCAULIFFE,C (1966)
logP 1.99			EXP	HANSCH,C & LEO,AJ (1985)
VP 2.11E+003	mm Hg	25	EXP	DAUBERT,TE & DANNER,RP (1985)
DC	pKa			
HL 7.36E-002	atm m3/mol	25	EST	VP/WSOL
OH 6.66E-011	cm3/molc sec	25	EXP	ATKINSON,R (1989)

CAS #: 000107-00-6 — 1-BUTYNE

Formula: C_4H_6

Mol Weight: 54.09

MP (deg C): -125.7

FP (deg C):

BP (deg C): 8.0

BP pressure (mm Hg):

Property/Value	Units	Temp	Data Type	Reference
WS 2.87E+003	mg/L	25	EXP	YALKOWSKY,SH & DANNENFELSER,RM (1992)
logP 1.53			EST	MEYLAN,WM & HOWARD,PH (1995)
VP 1.41E+003	mm Hg	25	EXT	BOUBLIK,T ET AL. (1984)
DC	pKa			
HL 2.13E-002	atm m3/mol	25	EST	MEYLAN,WM & HOWARD,PH (1991)
OH 8.00E-012	cm3/molc sec	25	EXP	ATKINSON,R (1989)

CAS #: 000107-02-8 — ACROLEIN

Formula: C_3H_4O

Mol Weight: 56.06

MP (deg C): -86.95

FP (deg C):

BP (deg C): 52.69

BP pressure (mm Hg):

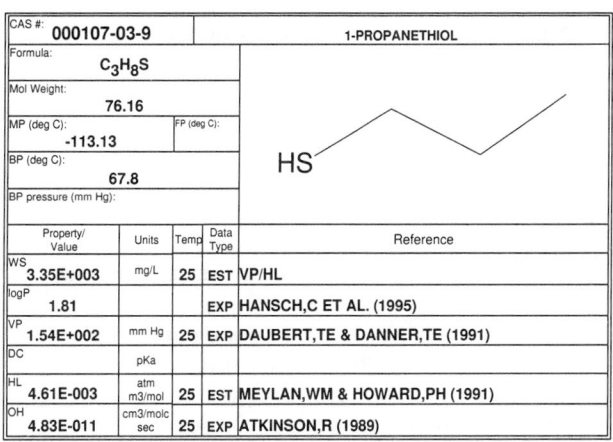

Property/Value	Units	Temp	Data Type	Reference
WS 2.13E+005	mg/L	25	EXP	SEIDELL,A (1941)
logP -0.01			EXP	HANSCH,C & LEO,AJ (1985)
VP 2.74E+002	mm Hg	25	EXP	DAUBERT,TE & DANNER,RP (1985)
DC	pKa			
HL 1.22E-004	atm m3/mol	25	EXP	GAFFNEY,JS ET AL. (1987)
OH 1.99E-011	cm3/molc sec	25	EXP	ATKINSON,R (1989)

CAS #: 000107-03-9 — 1-PROPANETHIOL

Formula: C_3H_8S

Mol Weight: 76.16

MP (deg C): -113.13

FP (deg C):

BP (deg C): 67.8

BP pressure (mm Hg):

Property/Value	Units	Temp	Data Type	Reference
WS 3.35E+003	mg/L	25	EST	VP/HL
logP 1.81			EXP	HANSCH,C ET AL. (1995)
VP 1.54E+002	mm Hg	25	EXP	DAUBERT,TE & DANNER,TE (1991)
DC	pKa			
HL 4.61E-003	atm m3/mol	25	EST	MEYLAN,WM & HOWARD,PH (1991)
OH 4.83E-011	cm3/molc sec	25	EXP	ATKINSON,R (1989)

CAS #: 000107-04-0				1-BROMO-2-CHLOROETHANE
Formula: C_2H_4BrCl				
Mol Weight: 143.42				
MP (deg C): -16.7		FP (deg C):		
BP (deg C): 107				
BP pressure (mm Hg):				

Property/Value	Units	Temp	Data Type	Reference
WS 6.86E+003	mg/L	25	EXP	HINE,J & MOOKERJEE,PK (1975)
logP 1.92			EST	MEYLAN,WM & HOWARD,PH (1995)
VP 3.31E+001	mm Hg	25	EXP	HINE,J & MOOKERJEE,PK (1975)
DC	pKa			
HL 9.09E-004	atm m3/mol	25	EXP	HINE,J & MOOKERJEE,PK (1975)
OH 2.57E-013	cm3/molc sec	25	EST	MEYLAN,WM & HOWARD,PH (1993)

CAS #: 000107-05-1				3-CHLOROPROPYLENE
Formula: C_3H_5Cl				
Mol Weight: 76.53				
MP (deg C): -134.5		FP (deg C):		
BP (deg C): 45.1				
BP pressure (mm Hg):				

Property/Value	Units	Temp	Data Type	Reference
WS 3.37E+003	mg/L	25	EXP	DILLING,WL (1977)
logP 1.93			EST	MEYLAN,WM & HOWARD,PH (1995)
VP 3.68E+002	mm Hg	25	EXP	BOUBLIK,T ET AL. (1984)
DC	pKa			
HL 1.10E-002	atm m3/mol	25	EST	VP/WSOL
OH 1.70E-011	cm3/molc sec	25	EXP	ATKINSON,R (1989)

CAS #: 000107-06-2				1,2-DICHLOROETHANE
Formula: $C_2H_4Cl_2$				
Mol Weight: 98.96				
MP (deg C): -35.66		FP (deg C):		
BP (deg C): 83.483				
BP pressure (mm Hg):				

Property/Value	Units	Temp	Data Type	Reference
WS 8.52E+003	mg/L	20	EXP	HORVATH,AL (1982)
logP 1.48			EXP	HANSCH,C & LEO,AJ (1985)
VP 7.89E+001	mm Hg	25	EXP	DAUBERT,TE & DANNER,RP (1985)
DC	pKa			
HL 1.18E-003	atm m3/mol	25	EXP	LEIGHTON,DTJR & CALO,JM (1981)
OH 2.20E-013	cm3/molc sec	25	EXP	ATKINSON,R (1989)

CAS #: 000107-07-3				2-CHLOROETHANOL
Formula: C_2H_5ClO				
Mol Weight: 80.51				
MP (deg C): -67		FP (deg C): -68		
BP (deg C): 128-130				
BP pressure (mm Hg):				

Property/Value	Units	Temp	Data Type	Reference
WS 1.00E+006	mg/L		EXP	RIDDICK,JA ET AL. (1986)
logP 0.03			EXP	HANSCH,C & LEO,AJ (1985)
VP 7.18E+000	mm Hg	25	EXP	DAUBERT,TE & DANNER,RP (1989)
DC 14.31	pKa		EXP	RIDDICK,JA ET AL. (1986)
HL 7.61E-007	atm m3/mol	25	EST	VP/WSOL
OH 1.40E-012	cm3/molc sec	25	EXP	ATKINSON,R (1989)

CAS #: 000107-08-4				1-IODOPROPANE
Formula: C_3H_7I				
Mol Weight: 169.99				
MP (deg C): -98		FP (deg C):		
BP (deg C): 102-103				
BP pressure (mm Hg):				

Property/Value	Units	Temp	Data Type	Reference
WS 1.07E+003	mg/L	20	EXP	YALKOWSKY,SH & DANNENFELSER,RM (1992)
logP 2.57			EST	MEYLAN,WM & HOWARD,PH (1995)
VP 4.31E+001	mm Hg	25	EXP	RIDDICK,JA ET AL. (1986)
DC	pKa			
HL 9.01E-003	atm m3/mol	25	EXP	VP/WSOL
OH 1.21E-012	cm3/molc sec	25	EST	MEYLAN,WM & HOWARD,PH (1993)

CAS #: 000107-10-8				PROPYLAMINE
Formula: C_3H_9N				
Mol Weight: 59.11				
MP (deg C): -83		FP (deg C):		
BP (deg C): 48-49				
BP pressure (mm Hg):				

Property/Value	Units	Temp	Data Type	Reference
WS 1.00E+006	mg/L	25	EXP	YALKOWSKY,SH & DANNENFELSER,RM (1992)
logP 0.48			EXP	HANSCH,C & LEO,AJ (1985)
VP 3.08E+002	mm Hg	25	EXP	RIDDICK,JA ET AL. (1986)
DC 10.71	pKa	20	EXP	PERRIN,DD (1965)
HL 1.48E-005	atm m3/mol	25	EXP	CHRISTIE,AO & CRISP,DJ (1967)
OH 3.21E-011	cm3/molc sec	25	EST	MEYLAN,WM & HOWARD,PH (1993)

CAS #: 000107-11-9				ALLYLAMINE
Formula: C_3H_7N				
Mol Weight: 57.10				
MP (deg C): -88		FP (deg C):		
BP (deg C): 55-58				
BP pressure (mm Hg):				

Property/Value	Units	Temp	Data Type	Reference
WS 1.00E+006	mg/L	25	EXP	YALKOWSKY,SH & DANNENFELSER,RM (1990)
logP 0.03			EXP	HANSCH,C & LEO,AJ (1985)
VP 2.42E+002	mm Hg	25	EXP	DAUBERT,TE & DANNER,RP (1989)
DC 9.70	pKa		EXP	LEUNG,HW & PAUSTENBACH,DJ (1990)
HL 9.95E-006	atm m3/mol	25	EST	MEYLAN,WM & HOWARD,PH (1991)
OH 5.59E-011	cm3/molc sec	25	EST	MEYLAN,WM & HOWARD,PH (1993)

CAS #: 000107-12-0				PROPIONITRILE
Formula: C_3H_5N				
Mol Weight: 55.08				
MP (deg C): -91.8		FP (deg C):		
BP (deg C): 97.2				
BP pressure (mm Hg):				

Property/Value	Units	Temp	Data Type	Reference
WS 1.03E+005	mg/L	25	EXP	RIDDICK,JA ET AL. (1986)
logP 0.16			EXP	HANSCH,C & LEO,AJ (1985)
VP 4.74E+001	mm Hg	25	EXP	DAUBERT,TE & DANNER,RP (1985)
DC	pKa			
HL 3.70E-005	atm m3/mol	25	EXP	BUTLER,JAV AND RAMCHANDANI,CN (1935)
OH 1.90E-013	cm3/molc sec	25	EXP	ATKINSON,R (1989)

CAS #: 000107-13-1				ACRYLONITRILE

Formula: C_3H_3N

Mol Weight: 53.06

MP (deg C): -83.55

FP (deg C):

BP (deg C): 77.3

BP pressure (mm Hg):

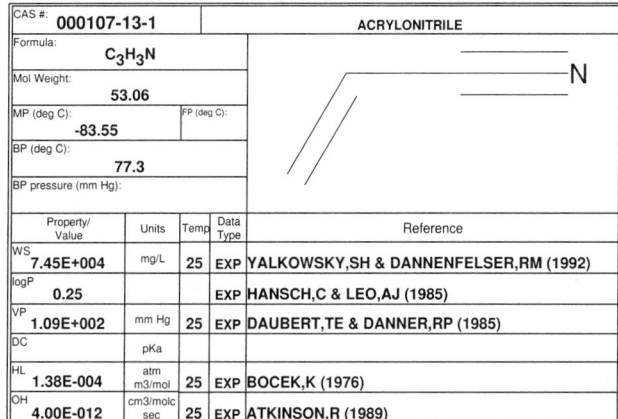

Property/Value	Units	Temp	Data Type	Reference
WS 7.45E+004	mg/L	25	EXP	YALKOWSKY,SH & DANNENFELSER,RM (1992)
logP 0.25			EXP	HANSCH,C & LEO,AJ (1985)
VP 1.09E+002	mm Hg	25	EXP	DAUBERT,TE & DANNER,RP (1985)
DC	pKa			
HL 1.38E-004	atm m3/mol	25	EXP	BOCEK,K (1976)
OH 4.00E-012	cm3/molc sec	25	EXP	ATKINSON,R (1989)

CAS #: 000107-14-2				CHLOROACETONITRILE

Formula: C_2H_2ClN

Mol Weight: 75.50

MP (deg C):

FP (deg C):

BP (deg C): 126

BP pressure (mm Hg):

Property/Value	Units	Temp	Data Type	Reference
WS 2.99E+004	mg/L	25	EST	MEYLAN,WM ET AL. (1996)
logP 0.45			EXP	HANSCH,C & LEO,AJ (1985)
VP 1.50E+001	mm Hg	31	EXP	WEAST,RC (1985)
DC	pKa			
HL 1.08E-005	atm m3/mol	25	EST	MEYLAN,WM & HOWARD,PH (1991)
OH 4.46E-014	cm3/molc sec	25	EST	ATKINSON,R (1988)

CAS #: 000107-15-3				1,2-DIAMINOETHANE

Formula: $C_2H_8N_2$

Mol Weight: 60.10

MP (deg C): 8.5

FP (deg C):

BP (deg C): 118

BP pressure (mm Hg):

Property/Value	Units	Temp	Data Type	Reference
WS 1.00E+006	mg/L		EXP	US EPA (1981)
logP -2.04			EXP	HANSCH,C ET AL. (1995)
VP 1.20E+001	mm Hg	25	EXP	BOUBLIK,T ET AL. (1984)
DC 9.92	pKa		EXP	PERRIN,DD (1972)
HL 1.73E-009	atm m3/mol	25	EXP	HINE,J & MOOKERJEE,PK (1975)
OH 6.16E-011	cm3/molc sec	25	EST	MEYLAN,WM & HOWARD,PH (1993)

CAS #: 000107-18-6				ALLYL ALCOHOL

Formula: C_3H_6O

Mol Weight: 58.08

MP (deg C): -129

FP (deg C):

BP (deg C): 96-98

BP pressure (mm Hg):

Property/Value	Units	Temp	Data Type	Reference
WS 3.17E+005	mg/L	25	EST	MEYLAN,WM ET AL. (1996)
logP 0.17			EXP	HANSCH,C & LEO,AJ (1985)
VP 2.61E+001	mm Hg	25	EXP	DAUBERT,TE & DANNER,RP (1989)
DC 15.50	pKa	25	EXP	SERJEANT,EP & DEMPSEY,B (1979)
HL 4.99E-006	atm m3/mol	25	EXP	HINE,J & MOOKERJEE,PK (1975)
OH 2.59E-011	cm3/molc sec	25	EXP	ATKINSON,R (1989)

CAS #: 000107-19-7				PROPARGYL ALCOHOL

Formula: C_3H_4O

Mol Weight: 56.06

MP (deg C): -51.8

FP (deg C):

BP (deg C): 113.6

BP pressure (mm Hg):

Property/Value	Units	Temp	Data Type	Reference
WS 1.00E+006	mg/L	20	EXP	YALKOWSKY,SH & DANNENFELSER,RM (1992)
logP -0.38			EXP	HANSCH,C ET AL. (1995)
VP 1.56E+001	mm Hg	25	EXP	DAUBERT,TE & DANNER,RP (1989)
DC	pKa			
HL 1.15E-006	atm m3/mol	25	EST	VP/WSOL
OH 1.04E-011	cm3/molc sec	25	EST	MEYLAN,WM & HOWARD,PH (1993)

CAS #: 000107-20-0				CHLOROACETALDEHYDE

Formula: C_2H_3ClO

Mol Weight: 78.50

MP (deg C): -16.3

FP (deg C):

BP (deg C): 85.5

BP pressure (mm Hg):

Property/Value	Units	Temp	Data Type	Reference
WS 1.11E+005	mg/L	25	EST	MEYLAN,WM ET AL. (1996)
logP 0.09			EST	MEYLAN,WM & HOWARD,PH (1995)
VP 6.43E+001	mm Hg	25	EXP	DAUBERT,TE & DANNER,RP (1994)
DC	pKa			
HL 2.39E-005	atm m3/mol	25	EST	MEYLAN,WM & HOWARD,PH (1991)
OH 6.34E-012	cm3/molc sec	25	EST	MEYLAN,WM & HOWARD,PH (1993)

CAS #: 000107-21-1				ETHYLENE GLYCOL

Formula: $C_2H_6O_2$

Mol Weight: 62.07

MP (deg C): -12.6

FP (deg C):

BP (deg C): 197.54

BP pressure (mm Hg):

Property/Value	Units	Temp	Data Type	Reference
WS 1.00E+006	mg/L		EXP	RIDDICK,JA ET AL. (1986)
logP -1.36			EXP	HANSCH,C & LEO,AJ (1985)
VP 9.20E-002	mm Hg	25	EXP	DAUBERT,TE & DANNER,RP (1985)
DC 15.10	pKa		EXP	SERJEANT,EP & DEMPSEY,B (1979)
HL 6.00E-008	atm m3/mol	25	EXP	BUTLER,JAV & RAMCHANDANI,CN (1935)
OH 7.70E-012	cm3/molc sec	25	EXP	ATKINSON,R (1989)

CAS #: 000107-22-2				GLYOXAL

Formula: $C_2H_2O_2$

Mol Weight: 58.04

MP (deg C): 15

FP (deg C):

BP (deg C): 51

BP pressure (mm Hg): 7.76E+002

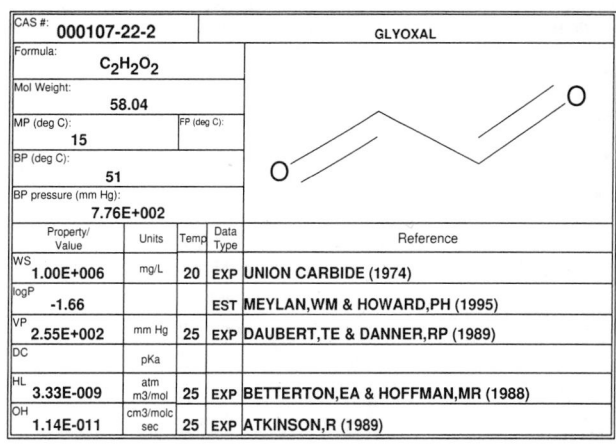

Property/Value	Units	Temp	Data Type	Reference
WS 1.00E+006	mg/L	20	EXP	UNION CARBIDE (1974)
logP -1.66			EST	MEYLAN,WM & HOWARD,PH (1995)
VP 2.55E+002	mm Hg	25	EXP	DAUBERT,TE & DANNER,RP (1989)
DC	pKa			
HL 3.33E-009	atm m3/mol	25	EXP	BETTERTON,EA & HOFFMAN,MR (1988)
OH 1.14E-011	cm3/molc sec	25	EXP	ATKINSON,R (1989)

000107-25-5 — METHYLVINYLETHER

Formula: C_3H_6O

Mol Weight: 58.08

MP (deg C): -122.8

BP (deg C): 5.5

Property/Value	Units	Temp	Data Type	Reference
WS 1.50E+004	mg/L	20	EXP	HORT,EV (1978)
logP 0.42			EST	MEYLAN,WM & HOWARD,PH (1995)
VP 1.32E+003	mm Hg	20	EXP	DAUBERT,TE & DANNER,RP (1989)
DC	pKa			
HL 6.71E-003	atm m3/mol	20	EST	VP/WSOL
OH 3.35E-011	cm3/molc sec	25	EXP	ATKINSON,R (1989)

000107-27-7 — ETHYLMERCURY CHLORIDE

Formula: C_2H_5ClHg

Mol Weight: 265.11

MP (deg C): 192

BP (deg C):

Property/Value	Units	Temp	Data Type	Reference
WS 1.50E+000	mg/L		EXP	SHIU,WY ET AL. (1990)
logP 0.88			EST	MEYLAN,WM & HOWARD,PH (1995)
VP 6.44E+000	mm Hg	25	EST	NEELY,WB & BLAU,GE (1985)
DC	pKa			
HL	atm m3/mol			
OH 1.10E-012	cm3/molc sec	25	EST	MEYLAN,WM & HOWARD,PH (1993)

000107-29-9 — ACETALDEHYDE OXIME

Formula: C_2H_5NO

Mol Weight: 59.07

MP (deg C): 46.5

BP (deg C): 114.5

Property/Value	Units	Temp	Data Type	Reference
WS 1.77E+005	mg/L	25	EST	MEYLAN,WM ET AL. (1996)
logP -0.13			EXP	HANSCH,C & LEO,AJ (1985)
VP 7.71E+000	mm Hg	25	EST	NEELY,WB & BLAU,GE (1985)
DC	pKa			
HL 5.88E-006	atm m3/mol	25	EST	MEYLAN,WM & HOWARD,PH (1991)
OH 2.20E-012	cm3/molc sec	25	EXP	ATKINSON,R (1989)

000107-30-2 — CHLOROMETHOXYMETHANE

Formula: C_2H_5ClO

Mol Weight: 80.51

MP (deg C): 103.5

BP (deg C): 59

Property/Value	Units	Temp	Data Type	Reference
WS 6.94E+004	mg/L	25	EST	MEYLAN,WM ET AL. (1996)
logP 0.32			EST	MEYLAN,WM & HOWARD,PH (1995)
VP 1.65E+002	mm Hg	25	EST	NEELY,WB & BLAU,GE (1985)
DC	pKa			
HL 3.04E-004	atm m3/mol	25	EST	MEYLAN,WM & HOWARD,PH (1991)
OH 2.30E-012	cm3/molc sec	25	EST	MEYLAN,WM & HOWARD,PH (1993)

000107-31-3 — METHYLFORMATE

Formula: $C_2H_4O_2$

Mol Weight: 60.05

MP (deg C): -99

BP (deg C): 31.75

Property/Value	Units	Temp	Data Type	Reference
WS 2.30E+005	mg/L	25	EXP	RIDDICK,JA ET AL. (1986)
logP 0.03			EXP	HANSCH,C ET AL. (1995)
VP 5.86E+002	mm Hg	25	EXP	DAUBERT,TE & DANNER,RP (1989)
DC	pKa			
HL 2.23E-004	atm m3/mol	25	EXP	HINE,J & MOOKERJEE,PK (1975)
OH 2.30E-013	cm3/molc sec	25	EXP	ATKINSON,R (1989)

000107-35-7 — TAURINE

Formula: $C_2H_7NO_3S$

Mol Weight: 125.15

MP (deg C): 328

BP (deg C):

Property/Value	Units	Temp	Data Type	Reference
WS 8.07E+004	mg/L	20	EXP	YALKOWSKY,SH & DANNENFELSER,RM (1992)
logP -3.36			EST	MEYLAN,WM & HOWARD,PH (1995)
VP 1.72E-004	mm Hg	25	EST	NEELY,WB & BLAU,GE (1985)
DC	pKa			
HL 1.72E-012	atm m3/mol	25	EST	MEYLAN,WM & HOWARD,PH (1991)
OH 3.30E-011	cm3/molc sec	25	EST	MEYLAN,WM & HOWARD,PH (1993)

000107-37-9 — ALLYTRICHLOROSILANE

Formula: $C_3H_5Cl_3Si$

Mol Weight: 175.52

MP (deg C): 35

BP (deg C): 117.5

Property/Value	Units	Temp	Data Type	Reference
WS 2.12E+002	mg/L	25	EST	MEYLAN,WM ET AL. (1996)
logP 2.85			EST	MEYLAN,WM & HOWARD,PH (1995)
VP 1.76E+001	mm Hg	25	EXT	BOUBLIK,T ET AL. (1984)
DC	pKa			
HL 7.01E-003	atm m3/mol	25	EST	MEYLAN,WM & HOWARD,PH (1991)
OH 2.73E-011	cm3/molc sec	25	EST	MEYLAN,WM & HOWARD,PH (1993)

000107-39-1 — 2,4,4-TRIMETHYL-1-PENTENE

Formula: C_8H_{16}

Mol Weight: 112.22

MP (deg C): -93.5

BP (deg C): 101.4

Property/Value	Units	Temp	Data Type	Reference
WS 4.04E+000	mg/L	25	EST	MEYLAN,WM ET AL. (1996)
logP 4.55			EXP	CHEM INSPECT TEST INST (1992)
VP 4.47E+001	mm Hg	25	EXP	DAUBERT,TE & DANNER,RP (1989)
DC	pKa			
HL 7.46E-001	atm m3/mol	25	EST	MEYLAN,WM & HOWARD,PH (1991)
OH 5.32E-011	cm3/molc sec	25	EST	MEYLAN,WM & HOWARD,PH (1993)

000107-40-4 — 2,4,4-TRIMETHYL-2-PENTENE

Formula: C_8H_{16}
Mol Weight: 112.22
MP (deg C): -106.3
FP (deg C):
BP (deg C): 104.9
BP pressure (mm Hg):

Property/Value	Units	Temp	Data Type	Reference
WS 1.19E+001	mg/L	25	EST	MEYLAN,WM ET AL. (1996)
logP 4.00			EST	MEYLAN,WM & HOWARD,PH (1995)
VP 3.59E+001	mm Hg	25	EXP	DAUBERT,TE & DANNER,RP (1989)
DC	pKa			
HL 8.81E-001	atm m3/mol	25	EST	MEYLAN,WM & HOWARD,PH (1991)
OH 8.77E-011	cm3/molc sec	25	EST	MEYLAN,WM & HOWARD,PH (1993)

000107-41-5 — 2-METHYL-2,4-PENTANEDIOL

Formula: $C_6H_{14}O_2$
Mol Weight: 118.18
MP (deg C): -50
FP (deg C):
BP (deg C): -197.5
BP pressure (mm Hg):

Property/Value	Units	Temp	Data Type	Reference
WS 1.00E+006	mg/L		EXP	RIDDICK,JA ET AL. (1986)
logP 0.58			EST	MEYLAN,WM & HOWARD,PH (1995)
VP 1.30E-002	mm Hg	25	EXP	DAUBERT,TE & DANNER,RP (1989)
DC	pKa			
HL 4.06E-007	atm m3/mol	25	EST	MEYLAN,WM & HOWARD,PH (1991)
OH 1.00E-011	cm3/molc sec	25	EST	MEYLAN,WM & HOWARD,PH (1993)

000107-43-7 — BETAINE

Formula: $C_5H_{11}NO_2$
Mol Weight: 117.15
MP (deg C): 293 dec
FP (deg C):
BP (deg C):
BP pressure (mm Hg):

Property/Value	Units	Temp	Data Type	Reference
WS 6.11E+005	mg/L	19	EXP	YALKOWSKY,SH & DANENFELSER,RM (1992)
logP -4.93			EST	MEYLAN,WM & HOWARD,PH (1995)
VP 5.34E-007	mm Hg	25	EST	NEELY,WB & BLAU,GE (1985)
DC	pKa			
HL 6.19E-016	atm m3/mol	25	EST	MEYLAN,WM & HOWARD,PH (1991)
OH 1.08E-011	cm3/molc sec	25	EST	MEYLAN,WM & HOWARD,PH (1993)

000107-44-8 — SARIN

Formula: $C_4H_{10}FO_2P$
Mol Weight: 140.10
MP (deg C): -57
FP (deg C): -38
BP (deg C): 147
BP pressure (mm Hg):

Property/Value	Units	Temp	Data Type	Reference
WS 1.00E+006	mg/L	25	EXP	BRITTON,KB (1986); MISCIBLE
logP 0.72			EST	SMALL,MJ (1983)
VP 2.86E+000	mm Hg	25	EXP	EPSTEIN,J (1974)
DC	pKa			
HL 5.72E-007	atm m3/mol	25	EST	VP/WSOL
OH 3.71E-011	cm3/molc sec	25	EST	ATKINSON,R (1987)

000107-46-0 — HEXAMETHYLDISILOXANE

Formula: $C_6H_{18}OSi_2$
Mol Weight: 162.38
MP (deg C): -67
FP (deg C):
BP (deg C): 99.5
BP pressure (mm Hg):

Property/Value	Units	Temp	Data Type	Reference
WS 2.00E+000	mg/L	25	EST	MEYLAN,WM ET AL. (1996)
logP 4.20			EXP	SANGSTER,J (1994)
VP 4.21E+001	mm Hg	25	EXT	FLANINGAM,OL (1986)
DC	pKa			
HL 4.50E+000	atm m3/mol	25	EST	VP/WSOL
OH 1.38E-012	cm3/molc sec	25	EXP	ATKINSON,R (1991)

000107-49-3 — TETRAETHYL PYROPHOSPHATE

Formula: $C_8H_{20}O_7P_2$
Mol Weight: 290.19
MP (deg C): 170 dec
FP (deg C):
BP (deg C): 138
BP pressure (mm Hg): 2.30E+000

Property/Value	Units	Temp	Data Type	Reference
WS 1.00E+006	mg/L		EXP	MERCK INDEX (1976)
logP -0.55			EST	MEYLAN,WM & HOWARD,PH (1995)
VP 4.70E-001	mm Hg	30	EXP	MERCK INDEX (1976)
DC	pKa			
HL 1.80E-007	atm m3/mol	30	EST	VP/WSOL
OH 7.73E-011	cm3/molc sec	25	EST	MEYLAN,WM & HOWARD,PH (1993)

000107-50-6 — TETRADECAMETHYLCYCLOHEPTASILOXANE

Formula: $C_{14}H_{42}O_7Si_7$
Mol Weight: 519.09
MP (deg C): -26
FP (deg C):
BP (deg C): 154
BP pressure (mm Hg): 2.00E+001

Property/Value	Units	Temp	Data Type	Reference
WS 1.43E-003	mg/L	25	EST	MEYLAN,WM ET AL. (1996)
logP 6.95			EST	MEYLAN,WM & HOWARD,PH (1995)
VP 1.12E-002	mm Hg	25	EXT	JORDAN,TE (1954)
DC	pKa			
HL	atm m3/mol			
OH 2.09E-012	cm3/molc sec	25	EST	MEYLAN,WM & HOWARD,PH (1993)

000107-51-7 — OCTAMETHYLTRISILOXANE

Formula: $C_8H_{24}O_2Si_3$
Mol Weight: 236.54
MP (deg C): -80
FP (deg C):
BP (deg C): 153
BP pressure (mm Hg):

Property/Value	Units	Temp	Data Type	Reference
WS 1.50E-001	mg/L	25	EST	MEYLAN,WM ET AL. (1996)
logP 4.80			EXP	SANGSTER,J (1994)
VP 3.34E+000	mm Hg	25	EXP	JORDAN,TE (1954)
DC	pKa			
HL 2.00E-004	atm m3/mol	25	EST	MEYLAN,WM & HOWARD,PH (1991)
OH 1.20E-012	cm3/molc sec	25	EST	MEYLAN,WM & HOWARD,PH (1993)

000107-52-8 — TETRADECAMETHYLHEXASILOXANE

Formula: $C_{14}H_{42}O_5Si_6$
Mol Weight: 459.00
MP (deg C): -59
FP (deg C):
BP (deg C): 245.5
BP pressure (mm Hg):

Property/Value	Units	Temp	Data Type	Reference
WS 1.31E-005	mg/L	25	EST	MEYLAN,WM ET AL. (1996)
logP 6.60			EXP	SANGSTER,J (1994)
VP 2.94E-002	mm Hg	25	EXP	JORDAN,TE (1954)
DC	pKa			
HL 5.43E-009	atm m3/mol	25	EST	MEYLAN,WM & HOWARD,PH (1991)
OH 2.09E-012	cm3/molc sec	25	EST	MEYLAN,WM & HOWARD,PH (1993)

000107-66-4 — DI-N-BUTYL PHOSPHATE

Formula: $C_8H_{19}O_4P$
Mol Weight: 210.21
MP (deg C):
FP (deg C):
BP (deg C):
BP pressure (mm Hg):

Property/Value	Units	Temp	Data Type	Reference
WS 4.30E+002	mg/L	25	EST	MEYLAN,WM ET AL. (1996)
logP 2.29			EST	MEYLAN,WM & HOWARD,PH (1995)
VP 9.60E-005	mm Hg	25	EST	HL X WSOL
DC	pKa			
HL 4.26E-009	atm m3/mol	25	EST	MEYLAN,WM & HOWARD,PH (1991)
OH 5.55E-011	cm3/molc sec	25	EST	MEYLAN,WM & HOWARD, PH (1993)

000107-70-0 — 4-METHOXY-4-METHYL-2-PENTANONE

Formula: $C_7H_{14}O_2$
Mol Weight: 130.19
MP (deg C): -20 (EST)
FP (deg C):
BP (deg C): 160
BP pressure (mm Hg):

Property/Value	Units	Temp	Data Type	Reference
WS 2.80E+005	mg/L	25	EXP	NORTON,LC & SCHERZINGER,RA (1961)
logP 0.36			EST	MEYLAN,WM & HOWARD,PH (1995)
VP 2.20E+000	mm Hg	20	EXP	LANDE,SS ET AL. (1976)
DC	pKa			
HL 1.93E-006	atm m3/mol	25	EST	VP/WSOL
OH 5.15E-012	cm3/molc sec	25	EST	MEYLAN,WM & HOWARD,PH (1993)

000107-72-2 — TRICHLOROPENTYLSILANE

Formula: $C_5H_{11}Cl_3Si$
Mol Weight: 205.59
MP (deg C):
FP (deg C):
BP (deg C): 172
BP pressure (mm Hg):

Property/Value	Units	Temp	Data Type	Reference
WS 1.66E+001	mg/L	25	EST	MEYLAN,WM ET AL. (1996)
logP 3.97			EST	MEYLAN,WM & HOWARD,PH (1995)
VP 3.45E+000	mm Hg	25	EST	NEELY,WB & BLAU,GE (1985)
DC	pKa			
HL 1.66E-002	atm m3/mol	25	EST	MEYLAN,WM & HOWARD,PH (1991)
OH 5.41E-012	cm3/molc sec	25	EST	MEYLAN,WM & HOWARD,PH (1993)

000107-81-3 — 2-BROMOPENTANE

Formula: $C_5H_{11}Br$
Mol Weight: 151.05
MP (deg C): -95.5
FP (deg C):
BP (deg C): 117.4
BP pressure (mm Hg):

Property/Value	Units	Temp	Data Type	Reference
WS 1.80E+002	mg/L	25	EST	MEYLAN,WM ET AL. (1996)
logP 3.07			EST	MEYLAN,WM & HOWARD,PH (1995)
VP 2.13E+001	mm Hg	25	EXP	LEVANOVA,SV ET AL. (1967)
DC	pKa			
HL 2.64E-002	atm m3/mol	25	EST	MEYLAN,WM & HOWARD,PH (1991)
OH 2.58E-012	cm3/molc sec	25	EST	MEYLAN,WM & HOWARD,PH (1993)

000107-82-4 — 1-BROMO-3-METHYLBUTANE

Formula: $C_5H_{11}Br$
Mol Weight: 151.05
MP (deg C): -112
FP (deg C):
BP (deg C): 120-121
BP pressure (mm Hg):

Property/Value	Units	Temp	Data Type	Reference
WS 1.99E+002	mg/L	25	EXP	HINE,J & MOOKERJEE,PJ (1975)
logP 3.07			EST	MEYLAN,WM & HOWARD,PH (1995)
VP 3.46E+001	mm Hg	25	EXP	HINE,J & MOOKERJEE,PJ (1975)
DC	pKa			
HL 3.46E-002	atm m3/mol	25	EXP	HINE,J & MOOKERJEE,PJ (1975)
OH 3.57E-012	cm3/molc sec	25	EST	MEYLAN,WM & HOWARD,PH (1993)

000107-83-5 — 2-METHYLPENTANE

Formula: C_6H_{14}
Mol Weight: 86.18
MP (deg C): -153.7
FP (deg C):
BP (deg C): 60.27
BP pressure (mm Hg):

Property/Value	Units	Temp	Data Type	Reference
WS 1.40E+001	mg/L	25	EXP	RIDDICK,JA ET AL. (1986)
logP 3.21			EST	MEYLAN,WM & HOWARD,PH (1995)
VP 2.11E+002	mm Hg	25	EXP	DAUBERT,TE & DANNER,RP (1989)
DC	pKa			
HL 1.71E+000	atm m3/mol	25	EST	VP/WSOL
OH 5.57E-012	cm3/molc sec	24	EXP	ATKINSON,R (1990)

000107-87-9 — 2-PENTANONE

Formula: $C_5H_{10}O$
Mol Weight: 86.13
MP (deg C): -78
FP (deg C):
BP (deg C): 102
BP pressure (mm Hg):

Property/Value	Units	Temp	Data Type	Reference
WS 4.30E+004	mg/L	25	EXP	YALKOWSKY,SH & DANNENFELSER,RM (1992)
logP 0.91			EXP	HANSCH,C & LEO,AJ (1985)
VP 3.54E+001	mm Hg	25	EXP	RIDDICK,JA ET AL. (1986)
DC	pKa			
HL 8.80E-005	atm m3/mol	25	EXP	MACKAY,D ET AL. (1982A)
OH 4.90E-012	cm3/molc sec	25	EXP	ATKINSON,R (1989)

000107-88-0 — 1,3-BUTANEDIOL

Formula: $C_4H_{10}O_2$
Mol Weight: 90.12
MP (deg C): -77 FP (deg C):
BP (deg C): 207.5
BP pressure (mm Hg):

Property/Value	Units	Temp	Data Type	Reference
WS 1.00E+006	mg/L	25	EXP	RIDDICK,JA ET AL (1986)
logP -0.29			EST	MEYLAN,WM & HOWARD,PH (1995)
VP 2.01E-002	mm Hg	25	EXP	DAUBERT,TE & DANNER,RP (1989)
DC 15.10	pKa		EXP	RIDDICK,JA ET AL (1986)
HL 2.38E-009	atm m3/mol	25	EST	VP/WSOL
OH 1.34E-011	cm3/molc sec	25	EST	ATKINSON,R (1988)

000107-92-6 — BUTYRIC ACID

Formula: $C_4H_8O_2$
Mol Weight: 88.11
MP (deg C): -5.7 FP (deg C):
BP (deg C): 163.7
BP pressure (mm Hg):

Property/Value	Units	Temp	Data Type	Reference
WS 5.64E+004	mg/L	25	EXP	HEMOHILL,L & SWANSON,WS (1964)
logP 0.79			EXP	HANSCH,C & LEO,AJ (1985)
VP 1.00E-003	mm Hg	25	EXP	RIDDICK,JA ET AL. (1986)
DC 4.82	pKa	25	EXP	RIDDICK,JA ET AL. (1986)
HL 5.35E-007	atm m3/mol	25	EXP	BUTLER,JAV AND RAMCHANDANI,CN (1935)
OH 2.40E-012	cm3/molc sec	25	EXP	ATKINSON,R (1985)

000107-93-7 — CROTONIC ACID

Formula: $C_4H_6O_2$
Mol Weight: 86.09
MP (deg C): 72 FP (deg C):
BP (deg C): 184.7
BP pressure (mm Hg):

Property/Value	Units	Temp	Data Type	Reference
WS 7.66E+004	mg/L	25	EST	MEYLAN,WM ET AL. (1996)
logP 0.85			EST	MEYLAN,WM & HOWARD,PH (1995)
VP 2.54E-001	mm Hg	25	EXT	YAWS,CL (1994)
DC 4.17	pKa	18	EXP	KORTUM,G ET AL (1961)
HL 4.53E-007	atm m3/mol	25	EST	MEYLAN,WM & HOWARD,PH (1991)
OH 2.04E-011	cm3/molc sec	25	EST	MEYLAN,WM & HOWARD,PH (1993)

000107-94-8 — 3-CHLOROPROPIONIC ACID

Formula: $C_3H_5ClO_2$
Mol Weight: 108.53
MP (deg C): 41 FP (deg C):
BP (deg C): 155-157
BP pressure (mm Hg):

Property/Value	Units	Temp	Data Type	Reference
WS 1.22E+005	mg/L	25	EST	MEYLAN,WM ET AL. (1996)
logP 0.41			EXP	HANSCH,C ET AL. (1995)
VP 3.36E-001	mm Hg	25	EST	NEELY,WB & BLAU,GE (1985)
DC 3.99	pKa	25	EXP	KORTUM,G ET AL (1961)
HL 2.56E-007	atm m3/mol	25	EST	MEYLAN,WM & HOWARD,PH (1991)
OH 1.21E-012	cm3/molc sec	25	EST	MEYLAN,WM & HOWARD,PH (1993)

000107-95-9 — BETA-ALANINE

Formula: $C_3H_7NO_2$
Mol Weight: 89.09
MP (deg C): 200 dec FP (deg C):
BP (deg C):
BP pressure (mm Hg):

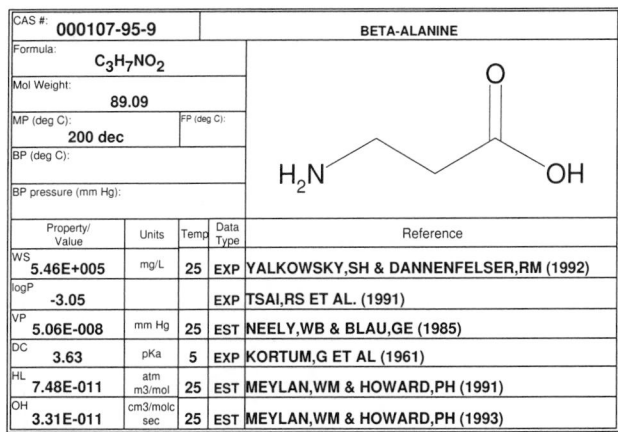

Property/Value	Units	Temp	Data Type	Reference
WS 5.46E+005	mg/L	25	EXP	YALKOWSKY,SH & DANNENFELSER,RM (1992)
logP -3.05			EXP	TSAI,RS ET AL. (1991)
VP 5.06E-008	mm Hg	25	EST	NEELY,WB & BLAU,GE (1985)
DC 3.63	pKa	5	EXP	KORTUM,G ET AL (1961)
HL 7.48E-011	atm m3/mol	25	EST	MEYLAN,WM & HOWARD,PH (1991)
OH 3.31E-011	cm3/molc sec	25	EST	MEYLAN,WM & HOWARD,PH (1993)

000107-96-0 — 3-MERCAPTOPROPIONIC ACID

Formula: $C_3H_6O_2S$
Mol Weight: 106.14
MP (deg C): 18 FP (deg C):
BP (deg C): 111
BP pressure (mm Hg): 1.50E+001

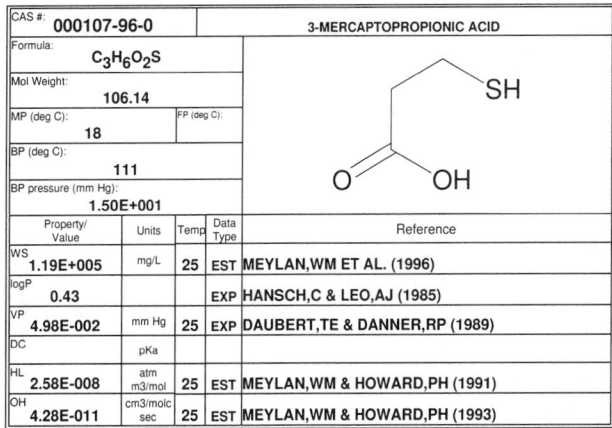

Property/Value	Units	Temp	Data Type	Reference
WS 1.19E+005	mg/L	25	EST	MEYLAN,WM ET AL. (1996)
logP 0.43			EXP	HANSCH,C & LEO,AJ (1985)
VP 4.98E-002	mm Hg	25	EXP	DAUBERT,TE & DANNER,RP (1989)
DC	pKa			
HL 2.58E-008	atm m3/mol	25	EST	MEYLAN,WM & HOWARD,PH (1991)
OH 4.28E-011	cm3/molc sec	25	EST	MEYLAN,WM & HOWARD,PH (1993)

000107-97-1 — N-METHYL GLYCINE

Formula: $C_3H_7NO_2$
Mol Weight: 89.09
MP (deg C): 212-213 de FP (deg C):
BP (deg C):
BP pressure (mm Hg):

Property/Value	Units	Temp	Data Type	Reference
WS 3.00E+005	mg/L	20	EXP	YALKOWSKY,SH & DANNENFELSER,RM (1992)
logP -2.78			EXP	HANSCH,C ET AL. (1995)
VP 1.65E-007	mm Hg	25	EST	NEELY,WB & BLAU,GE (1985)
DC 2.35	pKa	25	EXP	KORTUM,G ET AL (1961)
HL 2.47E-009	atm m3/mol	25	EST	MEYLAN,WM & HOWARD,PH (1991)
OH 7.13E-011	cm3/molc sec	25	EST	MEYLAN,WM & HOWARD,PH (1993)

000107-98-2 — 1-METHOXY-2-PROPANOL

Formula: $C_4H_{10}O_2$
Mol Weight: 90.12
MP (deg C): -142 FP (deg C):
BP (deg C): 118-119
BP pressure (mm Hg):

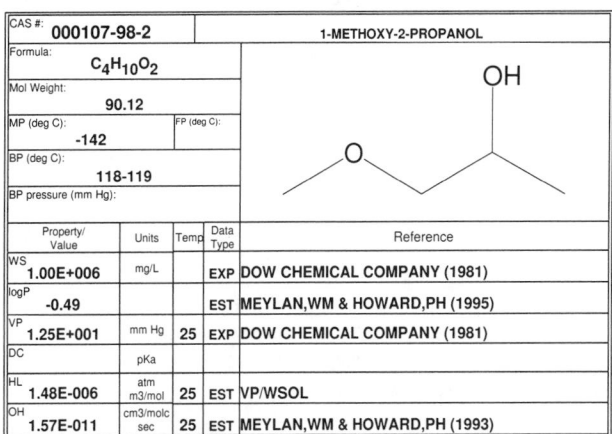

Property/Value	Units	Temp	Data Type	Reference
WS 1.00E+006	mg/L		EXP	DOW CHEMICAL COMPANY (1981)
logP -0.49			EST	MEYLAN,WM & HOWARD,PH (1995)
VP 1.25E+001	mm Hg	25	EXP	DOW CHEMICAL COMPANY (1981)
DC	pKa			
HL 1.48E-006	atm m3/mol	25	EST	VP/WSOL
OH 1.57E-011	cm3/molc sec	25	EST	MEYLAN,WM & HOWARD,PH (1993)

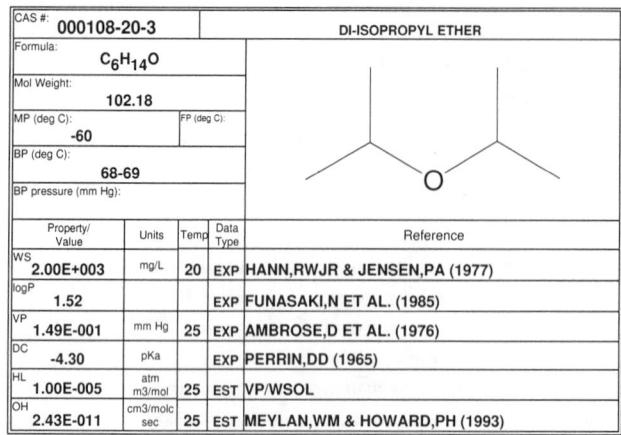

000108-01-0 — 2-DIMETHYLAMINOETHANOL

Formula: $C_4H_{11}NO$
Mol Weight: 89.14
MP (deg C): -59
FP (deg C):
BP (deg C): 135
BP pressure (mm Hg): 7.58E+002

Property/Value	Units	Temp	Data Type	Reference
WS 1.00E+006	mg/L		EXP	HANN,RWJR & JENSEN,PA (1977)
logP -0.94			EST	MEYLAN,WM & HOWARD,PH (1995)
VP 4.20E-001	mm Hg	20	EXP	HANN,RWJR & JENSEN,PA (1977)
DC 9.31	pKa	20	EXP	PERRIN,DD (1965)
HL 4.93E-008	atm m3/mol	20	EST	VP/WSOL
OH 9.00E-011	cm3/molc sec	25	EXP	ATKINSON,R (1989)

000108-03-2 — 1-NITROPROPANE

Formula: $C_3H_7NO_2$
Mol Weight: 89.09
MP (deg C): -108
FP (deg C):
BP (deg C): 131.6
BP pressure (mm Hg):

Property/Value	Units	Temp	Data Type	Reference
WS 1.50E+004	mg/L	25	EXP	RIDDICK,JA ET AL. (1986)
logP 0.87			EXP	HANSCH,C & LEO,AJ (1985)
VP 1.01E+001	mm Hg	25	EXP	DAUBERT,TE & DANNER,RP (1989)
DC 8.98	pKa		EXP	RIDDICK,JA ET AL. (1986)
HL 8.70E-005	atm m3/mol	25	EXP	HINE,J & MOOKERJEE,PK (1975)
OH 2.40E-013	cm3/molc sec	25	EXP	ATKINSON,R (1989)

000108-05-4 — VINYL ACETATE

Formula: $C_4H_6O_2$
Mol Weight: 86.09
MP (deg C): -92.8
FP (deg C):
BP (deg C): 72.5
BP pressure (mm Hg):

Property/Value	Units	Temp	Data Type	Reference
WS 2.00E+004	mg/L	20	EXP	RIDDICK,JA ET AL. (1986)
logP 0.73			EXP	HANSCH,C & LEO,AJ (1985)
VP 9.02E+001	mm Hg	20	EXP	DAUBERT,TE & DANNER,RP (1985)
DC	pKa			
HL 5.11E-004	atm m3/mol	20	EST	VP/WSOL
OH 2.63E-011	cm3/molc sec	25	EST	MEYLAN,WM & HOWARD,PH (1993)

000108-08-7 — 2,4-DIMETHYLPENTANE

Formula: C_7H_{16}
Mol Weight: 100.21
MP (deg C): -119.9
FP (deg C):
BP (deg C): 80.4
BP pressure (mm Hg):

Property/Value	Units	Temp	Data Type	Reference
WS 5.50E+000	mg/L	25	EXP	YALKOWSKY,SH & DANNENFELSER,RM (1992)
logP 3.63			EST	MEYLAN,WM & HOWARD,PH (1995)
VP 7.94E+001	mm Hg	25	EXP	DAUBERT,TE & DANNER,RP (1989)
DC	pKa			
HL 1.90E+000	atm m3/mol	25	EST	VP/WSOL
OH 5.16E-012	cm3/molc sec	25	EXP	ATKINSON,R (1989)

000108-10-1 — 4-METHYL-2-PENTANONE

Formula: $C_6H_{12}O$
Mol Weight: 100.16
MP (deg C): -84
FP (deg C):
BP (deg C): 117.4
BP pressure (mm Hg):

Property/Value	Units	Temp	Data Type	Reference
WS 1.90E+004	mg/L	25	EXP	YALKOWSKY,SH & DANNENFELSER,RM (1992)
logP 1.31			EXP	TANII,H ET AL. (1986)
VP 1.99E+001	mm Hg	25	EXP	DAUBERT,TE & DANNER,RP (1985)
DC	pKa			
HL 1.38E-004	atm m3/mol	25	EST	VP/WSOL
OH 1.41E-011	cm3/molc sec	25	EXP	ATKINSON,R (1989)

000108-11-2 — 4-METHYL-2-PENTANOL

Formula: $C_6H_{14}O$
Mol Weight: 102.18
MP (deg C): -90
FP (deg C):
BP (deg C): 131.6
BP pressure (mm Hg):

Property/Value	Units	Temp	Data Type	Reference
WS 1.64E+004	mg/L	25	EXP	RIDDICK,JA ET AL (9186)
logP 1.43			EST	HANSCH,C & LEO,AJ (1981)
VP 5.30E+000	mm Hg	25	EXP	DAUBERT,TE & DANNER,RP (1989)
DC	pKa			
HL 4.45E-005	atm m3/mol	25	EXP	HINE,J & MOOKERJEE,PK (1975)
OH 1.24E-011	cm3/molc sec	25	EST	MEYLAN,WM & HOWARD,PH (1993)

000108-18-9 — DIISOPROPYLAMINE

Formula: $C_6H_{15}N$
Mol Weight: 101.19
MP (deg C): -61
FP (deg C):
BP (deg C): 83.57
BP pressure (mm Hg):

Property/Value	Units	Temp	Data Type	Reference
WS 1.10E+005	mg/L	25	EXP	PARRISH,CF (1983)
logP 1.40			EXP	HANSCH,C ET AL. (1995)
VP 7.94E+001	mm Hg	25	EXP	DAUBERT,TE & DANNER,RP (1989)
DC 11.07	pKa	25	EXP	RIDDICK,JA ET AL. (1986)
HL 9.60E-005	atm m3/mol	25	EST	VP/WSOL
OH 9.70E-011	cm3/molc sec	25	EST	MEYLAN,WM & HOWARD,PH (1993)

000108-20-3 — DI-ISOPROPYL ETHER

Formula: $C_6H_{14}O$
Mol Weight: 102.18
MP (deg C): -60
FP (deg C):
BP (deg C): 68-69
BP pressure (mm Hg):

Property/Value	Units	Temp	Data Type	Reference
WS 2.00E+003	mg/L	20	EXP	HANN,RWJR & JENSEN,PA (1977)
logP 1.52			EXP	FUNASAKI,N ET AL. (1985)
VP 1.49E-001	mm Hg	25	EXP	AMBROSE,D ET AL. (1976)
DC -4.30	pKa		EXP	PERRIN,DD (1965)
HL 1.00E-005	atm m3/mol	25	EST	VP/WSOL
OH 2.43E-011	cm3/molc sec	25	EST	MEYLAN,WM & HOWARD,PH (1993)

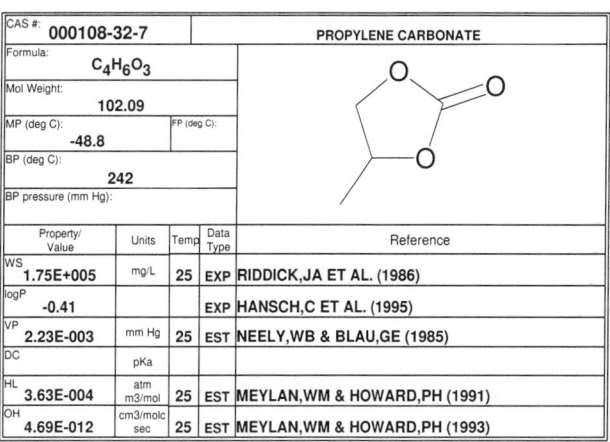

CAS #: 000108-21-4 — ISO-PROPYLACETATE

Formula: $C_5H_{10}O_2$
Mol Weight: 102.13
MP (deg C): -73.4
FP (deg C):
BP (deg C): 90
BP pressure (mm Hg):

Property/Value	Units	Temp	Data Type	Reference
WS 3.09E+004	mg/L	20	EXP	STEPHEN,H & STEPHEN,T (1963)
logP 1.03			EST	LYMAN,WJ ET AL. (1982)
VP 5.92E+001	mm Hg	25	EXP	AMBROSE,D ET AL. (1981)
DC	pKa			
HL 2.81E-004	atm m3/mol	25	EST	NIRMALAKHANDRAN,NN & SPEECE,RE (1988A)
OH 3.40E-012	cm3/molc sec	25	EXP	ATKINSON,R (1989)

CAS #: 000108-23-6 — ISOPROPYL CHLOROFORMATE

Formula: $C_4H_7ClO_2$
Mol Weight: 122.55
MP (deg C):
FP (deg C):
BP (deg C): 105
BP pressure (mm Hg):

Property/Value	Units	Temp	Data Type	Reference
WS 1.26E+004	mg/L	25	EST	MEYLAN,WM ET AL. (1996)
logP 1.04			EST	MEYLAN,WM & HOWARD,PH (1995)
VP 1.00E+002	mm Hg	47	EXP	DAMLE,SB (1993)
DC	pKa			
HL 4.15E-003	atm m3/mol	25	EST	MEYLAN,WM & HOWARD,PH (1991)
OH 3.12E-012	cm3/molc sec	25	EST	MEYLAN,WM & HOWARD,PH (1993)

CAS #: 000108-24-7 — ACETIC ANHYDRIDE

Formula: $C_4H_6O_3$
Mol Weight: 102.09
MP (deg C): -73
FP (deg C):
BP (deg C): 139
BP pressure (mm Hg):

Property/Value	Units	Temp	Data Type	Reference
WS 1.20E+005	mg/L	20	EXP	YALKOWSKY,SH & DANNENFELSER,RM (1992)
logP -0.58			EST	MEYLAN,WM & HOWARD,PH (1995)
VP 4.98E-001	mm Hg	25	EST	WEAST,RC (1972)
DC	pKa			
HL 3.57E-005	atm m3/mol	25	EST	MEYLAN,WM & HOWARD,PH (1991)
OH 2.01E-013	cm3/molc sec	25	EST	MEYLAN,WM & HOWARD,PH (1993)

CAS #: 000108-25-8 — O-ISOPROPYL DITHIOCARBONATE

Formula: $C_4H_8OS_2$
Mol Weight: 136.24
MP (deg C):
FP (deg C):
BP (deg C):
BP pressure (mm Hg):

Property/Value	Units	Temp	Data Type	Reference
WS 6.94E+003	mg/L	25	EST	MEYLAN,WM ET AL. (1996)
logP 1.28			EST	MEYLAN,WM & HOWARD,PH (1995)
VP 5.51E-001	mm Hg	25	EST	NEELY,WB & BLAU,GE (1985)
DC	pKa			
HL 3.45E-005	atm m3/mol	25	EST	MEYLAN,WM & HOWARD,PH (1991)
OH 4.47E-011	cm3/molc sec	25	EST	MEYLAN,WM & HOWARD,PH (1993)

CAS #: 000108-27-0 — 2-PYRROLIDINONE, 5-METHYL-

Formula: C_5H_9NO
Mol Weight: 99.13
MP (deg C): 43
FP (deg C):
BP (deg C): 248
BP pressure (mm Hg):

Property/Value	Units	Temp	Data Type	Reference
WS 1.43E+005	mg/L	25	EST	MEYLAN,WM ET AL. (1996)
logP -0.10			EXP	KIM,KH ET AL. (1993)
VP 1.51E-002	mm Hg	25	EST	NEELY,WB & BLAU,GE (1985)
DC	pKa			
HL 1.91E-008	atm m3/mol	25	EST	MEYLAN,WM & HOWARD,PH (1991)
OH 1.78E-011	cm3/molc sec	25	EST	MEYLAN,WM & HOWARD,PH (1993)

CAS #: 000108-30-5 — SUCCINIC ANHYDRIDE

Formula: $C_4H_4O_3$
Mol Weight: 100.07
MP (deg C): 119.6
FP (deg C):
BP (deg C): 261
BP pressure (mm Hg):

Property/Value	Units	Temp	Data Type	Reference
WS 2.38E+004	mg/L	25	EST	MEYLAN,WM ET AL. (1996)
logP 0.81			EST	MEYLAN,WM & HOWARD,PH (1995)
VP 2.86E-006	mm Hg	25	EXT	PERRY,RH & GREEN,D (1984)
DC	pKa			
HL 1.57E-005	atm m3/mol	25	EST	MEYLAN,WM & HOWARD,PH (1991)
OH 1.36E-012	cm3/molc sec	25	EST	MEYLAN,WM & HOWARD,PH (1993)

CAS #: 000108-31-6 — MALEIC ANHYDRIDE

Formula: $C_4H_2O_3$
Mol Weight: 98.06
MP (deg C): 52.8
FP (deg C):
BP (deg C): 202.0
BP pressure (mm Hg):

Property/Value	Units	Temp	Data Type	Reference
WS 4.91E+003	mg/L	25	EST	MEYLAN,WM ET AL. (1996)
logP 1.62			EST	MEYLAN,WM & HOWARD,PH (1995)
VP 2.50E-001	mm Hg	25	EXT	DAUBERT,TE & DANNER,RP (1985)
DC	pKa			
HL 3.93E-006	atm m3/mol	25	EST	MEYLAN,WM & HOWARD,PH (1991)
OH 2.26E-012	cm3/molc sec	25	EST	MEYLAN,WM & HOWARD,PH (1993)

CAS #: 000108-32-7 — PROPYLENE CARBONATE

Formula: $C_4H_6O_3$
Mol Weight: 102.09
MP (deg C): -48.8
FP (deg C):
BP (deg C): 242
BP pressure (mm Hg):

Property/Value	Units	Temp	Data Type	Reference
WS 1.75E+005	mg/L	25	EXP	RIDDICK,JA ET AL. (1986)
logP -0.41			EXP	HANSCH,C ET AL. (1995)
VP 2.23E-003	mm Hg	25	EST	NEELY,WB & BLAU,GE (1985)
DC	pKa			
HL 3.63E-004	atm m3/mol	25	EST	MEYLAN,WM & HOWARD,PH (1991)
OH 4.69E-012	cm3/molc sec	25	EST	MEYLAN,WM & HOWARD,PH (1993)

000108-34-9 — DIETHYL 3-METHYL-5-PYRAZOLYL PHOSPHATE

CAS #: 000108-34-9
Formula: $C_8H_{15}N_2O_4P$
Mol Weight: 234.19
MP (deg C):
FP (deg C):
BP (deg C):
BP pressure (mm Hg):

Property/Value	Units	Temp	Data Type	Reference
WS 1.00E+003	mg/L		EXP	GUNTHER,FA ET AL. (1968)
logP 0.76			EST	MEYLAN,WM & HOWARD,PH (1995)
VP 1.15E-005	mm Hg	25	EST	NEELY,WB & BLAU,GE (1985)
DC	pKa			
HL 1.23E-010	atm m3/mol	25	EST	MEYLAN,WM & HOWARD,PH (1991)
OH 2.13E-010	cm3/molc sec	25	EST	MEYLAN,WM & HOWARD,PH (1993)

000108-35-0 — PYRAZOTHION

CAS #: 000108-35-0
Formula: $C_8H_{15}N_2O_3PS$
Mol Weight: 250.26
MP (deg C):
FP (deg C):
BP (deg C):
BP pressure (mm Hg):

Property/Value	Units	Temp	Data Type	Reference
WS 1.64E+002	mg/L	25	EST	MEYLAN,WM ET AL. (1996)
logP 2.53			EST	MEYLAN,WM & HOWARD,PH (1995)
VP 1.47E-005	mm Hg	25	EST	NEELY,WB & BLAU,GE (1985)
DC	pKa			
HL 5.68E-008	atm m3/mol	25	EST	MEYLAN,WM & HOWARD,PH (1991)
OH 2.66E-010	cm3/molc sec	25	EST	MEYLAN,WM & HOWARD,PH (1993)

000108-36-1 — M-DIBROMOBENZENE

CAS #: 000108-36-1
Formula: $C_6H_4Br_2$
Mol Weight: 235.92
MP (deg C): -7
FP (deg C):
BP (deg C): 218
BP pressure (mm Hg):

Property/Value	Units	Temp	Data Type	Reference
WS 6.75E+001	mg/L	35	EXP	YALKOWSKY,SH & DANNENFELSER,RM (1992)
logP 3.75			EXP	HANSCH,C & LEO,AJ (1985)
VP 2.69E-001	mm Hg	25	EXP	DAUBERT,TE & DANNER,RP (1989)
DC	pKa			
HL 1.24E-003	atm m3/mol	25	EST	VP/WSOL
OH 7.73E-013	cm3/molc sec	25	EST	MEYLAN,WM & HOWARD,PH (1993)

000108-37-2 — M-BROMOCHLOROBENZENE

CAS #: 000108-37-2
Formula: C_6H_4BrCl
Mol Weight: 191.46
MP (deg C): -21.5
FP (deg C):
BP (deg C): 196
BP pressure (mm Hg):

Property/Value	Units	Temp	Data Type	Reference
WS 1.18E+002	mg/L	25	EXP	SUZUKI,T (1991)
logP 3.70			EXP	DUNN,WJ ET AL. (1983)
VP 4.41E-001	mm Hg	25	EST	NEELY,WB & BLAU,GE (1985)
DC	pKa			
HL 1.59E-003	atm m3/mol	25	EST	MEYLAN,WM & HOWARD,PH (1991)
OH 8.63E-013	cm3/molc sec	25	EST	MEYLAN,WM & HOWARD,PH (1993)

000108-38-3 — M-XYLENE

CAS #: 000108-38-3
Formula: C_8H_{10}
Mol Weight: 106.17
MP (deg C): -47.872
FP (deg C):
BP (deg C): 139.12
BP pressure (mm Hg):

Property/Value	Units	Temp	Data Type	Reference
WS 1.61E+002	mg/L	25	EXP	SANEMASA,I ET AL. (1982)
logP 3.20			EXP	HANSCH,C & LEO,AJ (1985)
VP 8.29E+000	mm Hg	25	EXP	CHAO,J ET AL. (1983)
DC	pKa			
HL 7.18E-003	atm m3/mol	25	EXP	SANEMASA,I ET AL. (1982)
OH 2.36E-011	cm3/molc sec	25	EXP	ATKINSON,R (1989)

000108-39-4 — M-CRESOL

CAS #: 000108-39-4
Formula: C_7H_8O
Mol Weight: 108.14
MP (deg C): 12.22
FP (deg C):
BP (deg C): 202.232
BP pressure (mm Hg):

Property/Value	Units	Temp	Data Type	Reference
WS 2.27E+004	mg/L	25	EXP	YALKOWSKY,SH & DANNENFELSER,RM (1992)
logP 1.96			EXP	HANSCH,C & LEO,AJ (1985)
VP 1.38E-001	mm Hg	25	EXP	CHAO,J ET AL. (1983)
DC 10.09	pKa	25	EXP	PEARCE,PJ & SIMKINS,RJJ (1968)
HL 8.65E-007	atm m3/mol	25	EST	VP/WSOL
OH 6.40E-011	cm3/molc sec	25	EXP	ATKINSON,R (1989)

000108-40-7 — 3-METHYLBENZENETHIOL

CAS #: 000108-40-7
Formula: C_7H_8S
Mol Weight: 124.21
MP (deg C): -20
FP (deg C):
BP (deg C): 195
BP pressure (mm Hg):

Property/Value	Units	Temp	Data Type	Reference
WS 1.68E+002	mg/L	25	EST	MEYLAN,WM ET AL. (1996)
logP 3.23			EST	MEYLAN,WM & HOWARD,PH (1995)
VP 4.21E+000	mm Hg	25	EXT	CHAO,J ET AL. (1983)
DC	pKa			
HL 5.76E-004	atm m3/mol	25	EST	MEYLAN,WM & HOWARD,PH (1991)
OH 2.97E-011	cm3/molc sec	25	EST	MEYLAN,WM & HOWARD,PH (1993)

000108-41-8 — M-CHLOROTOLUENE

CAS #: 000108-41-8
Formula: C_7H_7Cl
Mol Weight: 126.59
MP (deg C): -47.8
FP (deg C):
BP (deg C): 161.75
BP pressure (mm Hg):

Property/Value	Units	Temp	Data Type	Reference
WS 3.80E+001	mg/L		EXP	YALKOWSKY,SH & DANNENFELSER,RM (1992)
logP 3.28			EXP	HANSCH,C & LEO,AJ (1985)
VP 3.68E+000	mm Hg	25	EXP	PERRY,RH & GREEN,D (1984)
DC	pKa			
HL 1.61E-002	atm m3/mol	25	EST	VP/WSOL
OH 3.72E-012	cm3/molc sec	25	EST	MEYLAN,WM & HOWARD,PH (1993)

000108-42-9 — 3-CHLOROANILINE

Formula: C_6H_6ClN
Mol Weight: 127.57
MP (deg C): -10.3
FP (deg C):
BP (deg C): 230.5
BP pressure (mm Hg):

Property/Value	Units	Temp	Data Type	Reference
WS 5.40E+003	mg/L	20	EXP	CHIOU,CT ET AL. (1982)
logP 1.88			EXP	HANSCH,C & LEO,AJ (1985)
VP 5.40E-002	mm Hg	20	EXP	PECANTE,V ET AL. (1985)
DC 3.52	pKa		EXP	PERRIN,DD (1972)
HL 1.70E-006	atm m3/mol	20	EST	VP/WSOL
OH 7.57E-011	cm3/molc sec	25	EST	MEYLAN,WM & HOWARD,PH (1993)

000108-43-0 — 3-CHLOROPHENOL

Formula: C_6H_5ClO
Mol Weight: 128.56
MP (deg C): 32.6
FP (deg C):
BP (deg C): 214
BP pressure (mm Hg):

Property/Value	Units	Temp	Data Type	Reference
WS 2.60E+004	mg/L	20	EXP	VERSCHUEREN,K (1977)
logP 2.50			EXP	HANSCH,C ET AL. (1995)
VP 1.25E-001	mm Hg	25	EXP	SHIU,WY ET AL. (1994)
DC 9.12	pKa		EXP	SERJEANT,EP & DEMPSEY,B (1979)
HL 8.13E-007	atm m3/mol	25	EST	VP/WSOL
OH 2.36E-011	cm3/molc sec	25	EST	MEYLAN,WM & HOWARD,PH (1993)

000108-44-1 — M-TOLUIDINE

Formula: C_7H_9N
Mol Weight: 107.16
MP (deg C): -50
FP (deg C):
BP (deg C): 203-204
BP pressure (mm Hg):

Property/Value	Units	Temp	Data Type	Reference
WS 1.50E+004	mg/L	20	EXP	YALKOWSKY,SH & DANNENFELSER,RM (1992)
logP 1.40			EXP	HANSCH,C & LEO,AJ (1985)
VP 3.03E-001	mm Hg	25	EXP	DAUBERT,TE & DANNER,RP (1989)
DC 4.69	pKa	25	EXP	PERRIN,DD (1965)
HL 2.85E-006	atm m3/mol	25	EST	VP/WSOL
OH 2.00E-010	cm3/molc sec	25	EST	MEYLAN,WM & HOWARD,PH (1993)

000108-45-2 — 1,3-BENZENEDIAMINE

Formula: $C_6H_8N_2$
Mol Weight: 108.14
MP (deg C): 62.8
FP (deg C):
BP (deg C): 282-284
BP pressure (mm Hg):

Property/Value	Units	Temp	Data Type	Reference
WS 2.38E+005	mg/L	20	EXP	STEPHEN,H & STEPHEN,T (1963)
logP -0.33			EXP	HANSCH,C ET AL. (1995)
VP 1.38E-003	mm Hg	25	EXT	CHAO,J ET AL. (1983)
DC 4.98	pKa	25	EXP	PERRIN,DD (1965)
HL 9.53E-011	atm m3/mol	25	EST	VP/WSOL
OH 2.40E-010	cm3/molc sec	25	EST	MEYLAN,WM & HOWARD,PH (1993)

000108-46-3 — RESORCINOL

Formula: $C_6H_6O_2$
Mol Weight: 110.11
MP (deg C): 109-111
FP (deg C):
BP (deg C): 280
BP pressure (mm Hg):

Property/Value	Units	Temp	Data Type	Reference
WS 2.29E+006	mg/L	30	EXP	VERSCHUEREN,K (1977)
logP 0.80			EXP	HANSCH,C & LEO,AJ (1985)
VP 4.89E-004	mm Hg	25	EXT	YAWS,CL (1994A)
DC 9.32	pKa		EXP	SERJEANT,EP & DEMPSEY,B (1979)
HL 8.10E-011	atm m3/mol	25	EST	MEYLAN,WM & HOWARD,PH (1991)
OH 2.00E-010	cm3/molc sec	25	EST	MEYLAN,WM & HOWARD,PH (1993)

000108-47-4 — 2,4-DIMETHYLPYRIDINE

Formula: C_7H_9N
Mol Weight: 107.16
MP (deg C): -64
FP (deg C):
BP (deg C): 158.5
BP pressure (mm Hg):

Property/Value	Units	Temp	Data Type	Reference
WS 3.50E+005	mg/L	25	EXP	YALKOWSKY,SH & DANNENFELSER,RM (1992)
logP 1.90			EST	MEYLAN,WM & HOWARD,PH (1995)
VP 3.17E+000	mm Hg	25	EXP	CHAO,J ET AL. (1983)
DC 6.99	pKa	25	EXP	PERRIN,DD (1965)
HL 6.74E-006	atm m3/mol	25	EXP	ANDON,RJL ET AL. (1954)
OH 2.85E-012	cm3/molc sec	25	EST	MEYLAN,WM & HOWARD,PH (1993)

000108-48-5 — 2,6-LUTIDINE

Formula: C_7H_9N
Mol Weight: 107.16
MP (deg C): -5.8
FP (deg C):
BP (deg C): 144
BP pressure (mm Hg):

Property/Value	Units	Temp	Data Type	Reference
WS 3.00E+005	mg/L	34	EXP	YALKOWSKY,SH & DANNENFELSER,RM (1992)
logP 1.68			EXP	HANSCH,C & LEO,AJ (1985)
VP 5.65E+000	mm Hg	25	EXP	CHAO,J ET AL. (1983)
DC 6.60	pKa	25	EXP	PERRIN,DD (1965)
HL 1.04E-005	atm m3/mol	25	EXP	ANDON,RJL ET AL. (1954)
OH 2.85E-012	cm3/molc sec	25	EST	MEYLAN,WM & HOWARD,PH (1993)

000108-50-9 — PYRAZINE, 2,6-DIMETHYL-

Formula: $C_6H_8N_2$
Mol Weight: 108.14
MP (deg C): 47.5
FP (deg C):
BP (deg C): 155.6
BP pressure (mm Hg):

Property/Value	Units	Temp	Data Type	Reference
WS 3.82E+004	mg/L	25	EST	MEYLAN,WM ET AL. (1996)
logP 0.54			EXP	YAMAGAMI,C ET AL. (1991)
VP 1.50E+000	mm Hg	25	EST	NEELY,WB & BLAU,GE (1985)
DC 1.90	pKa	27	EXP	PERRIN,DD (1965)
HL 3.55E-006	atm m3/mol	25	EST	MEYLAN,WM & HOWARD,PH (1991)
OH 1.84E-012	cm3/molc sec	25	EST	MEYLAN,WM & HOWARD,PH (1993)

2-AMINO-4-METHYLPYRIMIDINE

CAS #: 000108-52-1

Formula: $C_5H_7N_3$

Mol Weight: 109.13

MP (deg C): 160.3

FP (deg C):

BP (deg C):

BP pressure (mm Hg):

Property/Value	Units	Temp	Data Type	Reference
WS 4.73E+004	mg/L	25	EST	MEYLAN,WM ET AL. (1996)
logP 0.43			EST	MEYLAN,WM & HOWARD,PH (1995)
VP 6.92E-002	mm Hg	25	EST	NEELY,WB & BLAU,GE (1985)
DC 4.11	pKa	20	EXP	PERRIN,DD (1965)
HL 4.04E-006	atm m3/mol	25	EST	MEYLAN,WM & HOWARD,PH (1991)
OH 3.33E-011	cm3/molc sec	25	EST	MEYLAN,WM & HOWARD,PH (1993)

2-AMINO-4-PYRIMIDONE

CAS #: 000108-53-2

Formula: $C_4H_5N_3O$

Mol Weight: 111.10

MP (deg C):

FP (deg C):

BP (deg C):

BP pressure (mm Hg):

Property/Value	Units	Temp	Data Type	Reference
WS 3.70E+004	mg/L	25	EST	MEYLAN,WM ET AL. (1996)
logP -0.99			EXP	HANSCH,C & LEO,AJ (1985)
VP 1.51E-006	mm Hg	25	EST	NEELY,WB & BLAU,GE (1985)
DC	pKa			
HL 6.09E-014	atm m3/mol	25	EST	MEYLAN,WM & HOWARD,PH (1991)
OH 3.33E-011	cm3/molc sec	25	EST	MEYLAN,WM & HOWARD,PH (1993)

GLUTARIC ANHYDRIDE

CAS #: 000108-55-4

Formula: $C_5H_6O_3$

Mol Weight: 114.10

MP (deg C): 56.3

FP (deg C):

BP (deg C): 158

BP pressure (mm Hg): 1.50E+001

Property/Value	Units	Temp	Data Type	Reference
WS 8.15E+003	mg/L	25	EST	MEYLAN,WM ET AL. (1996)
logP 1.30			EST	MEYLAN,WM & HOWARD,PH (1995)
VP 5.98E-003	mm Hg	25	EXT	PERRY,RH & GREEN,D (1984)
DC	pKa			
HL 2.09E-005	atm m3/mol	25	EST	MEYLAN,WM & HOWARD,PH (1991)
OH 3.11E-012	cm3/molc sec	25	EST	MEYLAN,WM & HOWARD,PH (1993)

M-DIETHENYLBENZENE

CAS #: 000108-57-6

Formula: $C_{10}H_{10}$

Mol Weight: 130.19

MP (deg C): -52.3

FP (deg C):

BP (deg C): 121

BP pressure (mm Hg): 7.60E+001

Property/Value	Units	Temp	Data Type	Reference
WS 5.25E+001	mg/L	25	EST	MEYLAN,WM ET AL. (1996)
logP 3.80			EST	MEYLAN,WM & HOWARD,PH (1995)
VP 5.79E-001	mm Hg	25	EXP	DAUBERT,TE & DANNER,RP (1989)
DC	pKa			
HL 1.42E-003	atm m3/mol	25	EST	MEYLAN,WM & HOWARD,PH (1991)
OH 5.43E-011	cm3/molc sec	25	EST	MEYLAN,WM & HOWARD,PH (1993)

PROPANEDIOIC ACID, DIMETHYL ESTER

CAS #: 000108-59-8

Formula: $C_5H_8O_4$

Mol Weight: 132.12

MP (deg C): -62

FP (deg C):

BP (deg C): 180-181

BP pressure (mm Hg):

Property/Value	Units	Temp	Data Type	Reference
WS 9.95E+004	mg/L	25	EST	MEYLAN,WM ET AL. (1996)
logP -0.05			EXP	HANSCH,C ET AL. (1995)
VP 1.47E+001	mm Hg	25	EST	NEELY,WB & BLAU,GE (1985)
DC	pKa			
HL 4.17E-007	atm m3/mol	25	EST	MEYLAN,WM & HOWARD,PH (1991)
OH 9.47E-013	cm3/molc sec	25	EST	MEYLAN,WM & HOWARD,PH (1993)

DCIP (2,2'-OXYBIS-1-CHLOROPROPANE)

CAS #: 000108-60-1

Formula: $C_6H_{12}Cl_2O$

Mol Weight: 171.07

MP (deg C): -97

FP (deg C):

BP (deg C): 189

BP pressure (mm Hg):

Property/Value	Units	Temp	Data Type	Reference
WS 1.70E+003	mg/L	20	EXP	YALKOWSKY,SH & DANNENFELSER,RM (1992)
logP 2.48			EXP	KAWAMOTO,K & URANO,K (1989)
VP 8.50E-001	mm Hg	20	NR	CALLAHAN,MA ET AL. (1979A)
DC	pKa			
HL 1.13E-004	atm m3/mol	20	EST	VP/WSOL
OH 3.28E+000	cm3/molc sec	25	EST	MEYLAN,WM & HOWARD,PH (1993)

ETHYL ISOVALERATE

CAS #: 000108-64-5

Formula: $C_7H_{14}O_2$

Mol Weight: 130.19

MP (deg C): -99

FP (deg C):

BP (deg C): 135

BP pressure (mm Hg):

Property/Value	Units	Temp	Data Type	Reference
WS 2.00E+003	mg/L	20	EXP	RIDDICK,JA ET AL. (1986)
logP 2.26			EST	MEYLAN,WM & HOWARD,PH (1995)
VP 8.30E+000	mm Hg	25	EXP	DAUBERT,TE & DANNER,RP (1989)
DC	pKa			
HL 7.11E-004	atm m3/mol	25	EST	VP/WSOL
OH 5.23E-012	cm3/molc sec	25	EST	MEYLAN,WM & HOWARD,PH (1993)

PROPYLENE GLYCOL ME ETHER ACETATE

CAS #: 000108-65-6

Formula: $C_6H_{12}O_3$

Mol Weight: 132.16

MP (deg C):

FP (deg C):

BP (deg C): 145-146

BP pressure (mm Hg):

Property/Value	Units	Temp	Data Type	Reference
WS 1.98E+005	mg/L	25	EXP	DANIELS,SL ET AL. (1985)
logP 0.56			EXP	DANIELS,SL ET AL. (1985)
VP 3.92E+000	mm Hg	25	EXP	DAUBERT,TE & DANNER,RP (1994)
DC	pKa			
HL 3.44E-006	atm m3/mol	25	EST	VP/WSOL
OH 1.19E-011	cm3/molc sec	25	EST	MEYLAN,WM & HOWARD,PH (1993)

000108-67-8 — 1,3,5-TRIMETHYLBENZENE

Formula:	C_9H_{12}	
Mol Weight:	120.20	
MP (deg C):	-44.8	FP (deg C):
BP (deg C):	164.7	
BP pressure (mm Hg):		

Property/Value	Units	Temp	Data Type	Reference
WS 4.82E+001	mg/L	25	EXP	YALKOWSKY,SH & DANNENFELSER,RM (1992)
logP 3.42			EXP	HANSCH,C & LEO,AJ (1985)
VP 2.48E+000	mm Hg	25	EXP	DAUBERT,TE & DANNER,RP (1989)
DC	pKa			
HL 8.77E-003	atm m3/mol	25	EXP	SANEMASA,I ET AL. (1982)
OH 5.75E-011	cm3/molc sec	25	EXP	ATKINSON,R (1989)

000108-68-9 — 3,5-DIMETHYLPHENOL

Formula:	$C_8H_{10}O$	
Mol Weight:	122.17	
MP (deg C):	63.6	FP (deg C):
BP (deg C):	221.7	
BP pressure (mm Hg):		

Property/Value	Units	Temp	Data Type	Reference
WS 4.88E+003	mg/L	25	EXP	YALKOWSKY,SH & DANNENFELSER,RM (1992)
logP 2.35			EXP	HANSCH,C & LEO,AJ (1985)
VP 3.58E-002	mm Hg	25	EXT	CHAO,J ET AL. (1983)
DC 10.19	pKa	25	EXP	KORTUM,G ET AL (1961)
HL 6.83E-007	atm m3/mol	25	EST	MEYLAN,WM & HOWARD,PH (1991)
OH 1.13E-010	cm3/molc sec	25	EXP	ATKINSON,R (1989)

000108-69-0 — 3,5-DIMETHYLANILINE

Formula:	$C_8H_{11}N$	
Mol Weight:	121.18	
MP (deg C):	9.8	FP (deg C):
BP (deg C):	220.5	
BP pressure (mm Hg):		

Property/Value	Units	Temp	Data Type	Reference
WS 1.39E+003	mg/L	25	EST	MEYLAN,WM ET AL. (1996)
logP 2.17			EST	MEYLAN,WM & HOWARD,PH (1995)
VP 9.44E-002	mm Hg	25	EST	NEELY,WB & BLAU,GE (1985)
DC 4.79	pKa	19	EXP	PERRIN,DD (1965)
HL 2.32E-006	atm m3/mol	25	EST	MEYLAN,WM & HOWARD,PH (1991)
OH 2.00E-010	cm3/molc sec	25	EST	MEYLAN,WM & HOWARD,PH (1993)

000108-70-3 — 1,3,5-TRICHLOROBENZENE

Formula:	$C_6H_3Cl_3$	
Mol Weight:	181.45	
MP (deg C):	63.4	FP (deg C):
BP (deg C):	208.4	
BP pressure (mm Hg):		

Property/Value	Units	Temp	Data Type	Reference
WS 6.01E+000	mg/L	25	EXP	YALKOWSKY,SH & DANNENFELSER,RM (1992)
logP 4.19			EXP	HANSCH,C & LEO,AJ (1985)
VP 5.28E-001	mm Hg	25	EXT	PERRY,RH & GREEN,D (1984)
DC	pKa			
HL 2.19E-003	atm m3/mol	25	EST	MEYLAN,WM & HOWARD,PH (1991)
OH 6.79E-013	cm3/molc sec	25	EST	MEYLAN,WM & HOWARD,PH (1993)

000108-71-4 — 5-METHYL-1,3-BENZENEDIAMINE

Formula:	$C_7H_{10}N_2$	
Mol Weight:	122.17	
MP (deg C):		FP (deg C):
BP (deg C):		
BP pressure (mm Hg):		

Property/Value	Units	Temp	Data Type	Reference
WS 7.25E+004	mg/L	25	EST	MEYLAN,WM ET AL. (1996)
logP 0.16			EST	MEYLAN,WM & HOWARD,PH (1995)
VP 3.70E-003	mm Hg	25	EST	NEELY,WB & BLAU,GE (1985)
DC	pKa			
HL 7.43E-010	atm m3/mol	25	EST	MEYLAN,WM & HOWARD,PH (1991)
OH 2.00E-010	cm3/molc sec	25	EST	MEYLAN,WM & HOWARD,PH (1993)

000108-73-6 — 1,3,5-TRIHYDROXYBENZENE

Formula:	$C_6H_6O_3$	
Mol Weight:	126.11	
MP (deg C):	218	FP (deg C):
BP (deg C):		
BP pressure (mm Hg):		

Property/Value	Units	Temp	Data Type	Reference
WS 1.06E+004	mg/L	20	EXP	YALKOWSKY,SH & DANNENFELSER,RM (1992)
logP 0.16			EXP	HANSCH,C & LEO,AJ (1985)
VP 1.60E-004	mm Hg	25	EST	NEELY,WB & BLAU,GE (1985)
DC 8.45	pKa		EXP	KORTUM,G ET AL (1961)
HL 6.07E-015	atm m3/mol	25	EST	MEYLAN,WM & HOWARD,PH (1991)
OH 2.00E-010	cm3/molc sec	25	EST	MEYLAN,WM & HOWARD,PH (1993)

000108-75-8 — 2,4,6-COLLIDINE

Formula:	$C_8H_{11}N$	
Mol Weight:	121.18	
MP (deg C):	-46	FP (deg C):
BP (deg C):	171	
BP pressure (mm Hg):		

Property/Value	Units	Temp	Data Type	Reference
WS 3.60E+004	mg/L	20	EXP	GOE,GL (1982)
logP 1.88			EXP	HANSCH,C & LEO,AJ (1985)
VP 1.99E+000	mm Hg	25	EXP	CHAO,J ET AL. (1983)
DC 7.43	pKa	25	EXP	PERRIN,DD (1965)
HL 8.81E-006	atm m3/mol	25	EST	VP/WSOL
OH 6.99E-012	cm3/molc sec	25	EST	MEYLAN,WM & HOWARD,PH (1993)

000108-78-1 — MELAMINE

Formula:	$C_3H_6N_6$	
Mol Weight:	126.12	
MP (deg C):	345	FP (deg C):
BP (deg C):		
BP pressure (mm Hg):		

Property/Value	Units	Temp	Data Type	Reference
WS 3.24E+003	mg/L	20	EXP	YALKOWSKY,SH & DANNENFELSER,RM (1992)
logP -1.37			EXP	HANSCH,C ET AL. (1995)
VP 3.59E-010	mm Hg	20	EXT	HIRT,RC ET AL. (1960)
DC 5.16	pKa	20	EXP	ALBERT,A ET AL. (1948)
HL 1.84E-014	atm m3/mol	20	EST	VP/WSOL
OH 6.07E-011	cm3/molc sec	25	EST	MEYLAN,WM & HOWARD,PH (1993)

CAS #: 000108-80-5				CYANURIC ACID

Formula: $C_3H_3N_3O_3$

Mol Weight: 129.08

MP (deg C): 360 FP (deg C):

BP (deg C):

BP pressure (mm Hg):

Property/ Value	Units	Temp	Data Type	Reference
WS 2.00E+003	mg/L	25	EXP	BURAKEVICH,JV (1979)
logP 0.61			EST	MEYLAN,WM & HOWARD,PH (1995)
VP 3.05E-008	mm Hg	25	EST	MPBPVP
DC 7.20	pKa		EXP	BUDAVARI,S (1989)
HL 1.36E-018	atm m3/mol	25	EST	MEYLAN,WM & HOWARD,PH (1991)
OH 1.57E-013	cm3/molc sec	25	EST	MEYLAN,WM & HOWARD,PH (1993)

CAS #: 000108-82-7				2,6-DIMETHYL-4-HEPTANOL

Formula: $C_9H_{20}O$

Mol Weight: 144.26

MP (deg C): FP (deg C):

BP (deg C): 174.5

BP pressure (mm Hg):

Property/ Value	Units	Temp	Data Type	Reference
WS 4.45E+002	mg/L	25	EXP	SUZUKI,T (1991)
logP 3.08			EST	MEYLAN,WM & HOWARD,PH (1995)
VP 3.03E-001	mm Hg	25	EXP	DAUBERT,TE & DANNER,RP (1989)
DC	pKa			
HL 1.29E-004	atm m3/mol	25	EST	VP/WSOL
OH 1.87E-011	cm3/molc sec	25	EST	MEYLAN,WM & HOWARD,PH (1993)

CAS #: 000108-83-8				2,6-DIMETHYL-4-HEPTANONE

Formula: $C_9H_{18}O$

Mol Weight: 142.24

MP (deg C): -41.5 FP (deg C):

BP (deg C): 169.4

BP pressure (mm Hg):

Property/ Value	Units	Temp	Data Type	Reference
WS 2.64E+003	mg/L	24	EXP	YALKOWSKY,SH & DANNENFELSER,RM (1992)
logP 2.56			EST	MEYLAN,WM & HOWARD,PH (1995)
VP 1.65E+000	mm Hg	25	EXP	RIDDICK,JA ET AL. (1986)
DC	pKa			
HL 1.17E-004	atm m3/mol	25	EST	VP/WSOL
OH 2.75E-011	cm3/molc sec	25	EXP	ATKINSON,R (1989)

CAS #: 000108-84-9				4-METHYL-2-PENTYL ACETATE

Formula: $C_8H_{16}O_2$

Mol Weight: 144.22

MP (deg C): FP (deg C):

BP (deg C): 147.5

BP pressure (mm Hg):

Property/ Value	Units	Temp	Data Type	Reference
WS 1.30E+003	mg/L	20	EXP	RIDDICK,JA ET AL. (1986)
logP 2.68			EST	MEYLAN,WM & HOWARD,PH (1995)
VP 4.00E+000	mm Hg	20	EXP	RIDDICK,JA ET AL. (1986)
DC	pKa			
HL 5.84E-004	atm m3/mol	20	EST	VP/WSOL
OH 8.22E-012	cm3/molc sec	25	EST	MEYLAN,WM & HOWARD,PH (1993)

CAS #: 000108-85-0				BROMOCYCLOHEXANE

Formula: $C_6H_{11}Br$

Mol Weight: 163.06

MP (deg C): -56.5 FP (deg C):

BP (deg C): 166.2

BP pressure (mm Hg):

Property/ Value	Units	Temp	Data Type	Reference
WS 1.23E+002	mg/L	25	EST	MEYLAN,WM ET AL. (1996)
logP 3.20			EXP	HANSCH,C & LEO,AJ (1985)
VP 3.62E+000	mm Hg	25	EST	NEELY,WB & BLAU,GE (1985)
DC	pKa			
HL 1.54E-002	atm m3/mol	25	EST	MEYLAN,WM & HOWARD,PH (1991)
OH 6.12E-012	cm3/molc sec	25	EST	MEYLAN,WM & HOWARD,PH (1993)

CAS #: 000108-86-1				BROMOBENZENE

Formula: C_6H_5Br

Mol Weight: 157.02

MP (deg C): -30.6 FP (deg C): -30

BP (deg C): 156.2

BP pressure (mm Hg):

Property/ Value	Units	Temp	Data Type	Reference
WS 4.10E+002	mg/L	25	EXP	YALKOWSKY,SH & DANNENFELSER,RM (1992)
logP 2.99			EXP	HANSCH,C & LEO,AJ (1985)
VP 4.18E+000	mm Hg	25	EXP	RIDDICK,JA ET AL. (1986)
DC	pKa			
HL 2.08E-003	atm m3/mol	25	EXP	MACKAY,D ET AL. (1982A)
OH 7.70E-013	cm3/molc sec	25	EXP	ATKINSON,R (1989)

CAS #: 000108-87-2				1-METHYLCYCLOHEXANE

Formula: C_7H_{14}

Mol Weight: 98.19

MP (deg C): -126.6 FP (deg C):

BP (deg C): 100.8

BP pressure (mm Hg):

Property/ Value	Units	Temp	Data Type	Reference
WS 1.40E+001	mg/L	25	EXP	YALKOWSKY,SH & DANNENFELSER,RM (1992)
logP 3.61			EXP	HANSCH,C ET AL. (1995)
VP 4.60E+001	mm Hg	25	EXP	DAUBERT,TE & DANNER,RP (1989)
DC	pKa			
HL 4.30E-001	atm m3/mol	25	EXP	HINE,J & MOOKERJEE,PK (1975)
OH 1.04E-011	cm3/molc sec	25	EXP	ATKINSON,R (1989)

CAS #: 000108-88-3				TOLUENE

Formula: C_7H_8

Mol Weight: 92.14

MP (deg C): -94.991 FP (deg C):

BP (deg C): 110.63

BP pressure (mm Hg):

Property/ Value	Units	Temp	Data Type	Reference
WS 5.26E+002	mg/L	25	EXP	SANEMASA,I ET AL. (1982)
logP 2.73			EXP	HANSCH,C & LEO,AJ (1985)
VP 2.84E+001	mm Hg	25	EXP	DAUBERT,TE & DANNER,RP (1985)
DC	pKa			
HL 6.64E-003	atm m3/mol	25	EXP	MACKAY,D ET AL. (1979)
OH 5.96E-012	cm3/molc sec	25	EXP	ATKINSON,R (1989)

CAS #: 000108-89-4				4-METHYLPYRIDINE

Formula: C_6H_7N

Mol Weight: 93.13

MP (deg C): 3.6 FP (deg C):

BP (deg C): 145

BP pressure (mm Hg):

Property/Value	Units	Temp	Data Type	Reference
WS 1.00E+006	mg/L	25	EXP	GOE,GL (1982)
logP 1.22			EXP	HANSCH,C & LEO,AJ (1985)
VP 5.77E+000	mm Hg	25	EXP	CHAO,J ET AL. (1983)
DC 5.98	pKa		EXP	GOE,GL (1982)
HL 6.00E-006	atm m3/mol	25	EXP	ANDON,RJL ET AL. (1954)
OH 1.10E-012	cm3/molc sec	25	EST	MEYLAN,WM & HOWARD,PH (1993)

CAS #: 000108-90-7				CHLOROBENZENE

Formula: C_6H_5Cl

Mol Weight: 112.56

MP (deg C): -45.58 FP (deg C):

BP (deg C): 131.687

BP pressure (mm Hg):

Property/Value	Units	Temp	Data Type	Reference
WS 4.98E+002	mg/L	25	EXP	HORVATH,AL (1982)
logP 2.84			EXP	SANGSTER,J (1994)
VP 1.20E+001	mm Hg	25	EXP	DAUBERT,TE & DANNER,RP (1985)
DC	pKa			
HL 3.77E-003	atm m3/mol	25	EXP	MACKAY,D ET AL. (1979)
OH 7.70E-013	cm3/molc sec	25	EXP	ATKINSON,R (1989)

CAS #: 000108-91-8				CYCLOHEXANAMINE

Formula: $C_6H_{13}N$

Mol Weight: 99.18

MP (deg C): -17.7 FP (deg C):

BP (deg C): 134.5

BP pressure (mm Hg):

Property/Value	Units	Temp	Data Type	Reference
WS 1.00E+006	mg/L	25	EXP	RIDDICK,JA ET AL. (1986)
logP 1.49			EXP	HANSCH,C & LEO,AJ (1985)
VP 1.01E+001	mm Hg	25	EXP	DAUBERT,TE & DANNER,RP (1989)
DC 10.63	pKa	23	EXP	PERRIN,DD (1972)
HL 1.32E-006	atm m3/mol	25	EST	VP/WSOL
OH 5.54E-011	cm3/molc sec	25	EST	MEYLAN,WM & HOWARD,PH (1993)

CAS #: 000108-93-0				CYCLOHEXANOL

Formula: $C_6H_{12}O$

Mol Weight: 100.16

MP (deg C): 23-25 FP (deg C):

BP (deg C): 161

BP pressure (mm Hg):

Property/Value	Units	Temp	Data Type	Reference
WS 3.75E+004	mg/L	25	EXP	RIDDICK,JA ET AL. (1986)
logP 1.23			EXP	HANSCH,C & LEO,AJ (1985)
VP 8.00E-001	mm Hg	25	EXP	DAUBERT,TE & DANNER,RP (1989)
DC	pKa			
HL 2.81E-006	atm m3/mol	25	EST	VP/WSOL
OH 1.74E-011	cm3/molc sec	25	EST	ATKINSON,R (1988)

CAS #: 000108-94-1				CYCLOHEXANONE

Formula: $C_6H_{10}O$

Mol Weight: 98.15

MP (deg C): -47 FP (deg C):

BP (deg C): 156.4

BP pressure (mm Hg):

Property/Value	Units	Temp	Data Type	Reference
WS 2.30E+004	mg/L	20	EXP	RIDDICK,JA ET AL. (1986)
logP 0.81			EXP	HANSCH,C & LEO,AJ (1985)
VP 4.33E+000	mm Hg	25	EXP	DAUBERT,TE & DANNER,RP (1985)
DC 11.30	pKa		EXP	SERJEANT,EP & DEMPSEY,B (1979)
HL 9.00E-006	atm m3/mol	25	EXP	HAWTHORNE,SB ET AL. (1985)
OH 6.39E-012	cm3/molc sec	25	EXP	ATKINSON,R (1989)

CAS #: 000108-95-2				PHENOL

Formula: C_6H_6O

Mol Weight: 94.11

MP (deg C): 40.90 FP (deg C):

BP (deg C): 181.839

BP pressure (mm Hg):

Property/Value	Units	Temp	Data Type	Reference
WS 8.28E+004	mg/L	25	EXP	SOUTHWORTH,GR & KELLER,JL (1986)
logP 1.46			EXP	HANSCH,C & LEO,AJ (1985)
VP 3.50E-001	mm Hg	25	EXP	JONES,AH (1960)
DC 9.99	pKa	25	EXP	SERJEANT,EP & DEMPSEY,B (1979)
HL 3.33E-007	atm m3/mol	25	EXP	GAFFNEY,JS ET AL. (1987)
OH 2.63E-011	cm3/molc sec	25	EXP	ATKINSON,R (1989)

CAS #: 000108-98-5				THIOPHENOL

Formula: C_6H_6S

Mol Weight: 110.18

MP (deg C): -14.9 FP (deg C):

BP (deg C): 169.1

BP pressure (mm Hg):

Property/Value	Units	Temp	Data Type	Reference
WS 8.34E+002	mg/L	25	EXP	SUZUKI,T (1991)
logP 2.52			EXP	HANSCH,C & LEO,AJ (1985)
VP 1.93E+000	mm Hg	25	EXP	CHAO,J ET AL. (1983)
DC 6.61	pKa		EXP	SERJEANT,EP & DEMPSEY,B (1979)
HL 3.35E-004	atm m3/mol	25	EST	VP/WSOL
OH 1.12E-011	cm3/molc sec	25	EXP	ATKINSON,R (1989)

CAS #: 000108-99-6				3-METHYLPYRIDINE

Formula: C_6H_7N

Mol Weight: 93.13

MP (deg C): -18.1 FP (deg C): -18

BP (deg C): 144

BP pressure (mm Hg):

Property/Value	Units	Temp	Data Type	Reference
WS 1.00E+006	mg/L	25	EXP	GOE,GL (1982)
logP 1.20			EXP	HANSCH,C & LEO,AJ (1985)
VP 6.05E+000	mm Hg	25	EXP	CHAO,J ET AL. (1983)
DC 5.63	pKa	25	EXP	GOE,GL (1982)
HL 7.73E-006	atm m3/mol	25	EXP	ANDON,RJL ET AL. (1954)
OH 1.10E-012	cm3/molc sec	25	EST	MEYLAN,WM & HOWARD,PH (1993)

CAS #: 000109-00-2				3-HYDROXYPYRIDINE
Formula: C_5H_5NO				
Mol Weight: 95.10				
MP (deg C): 126-129		FP (deg C):		
BP (deg C):				
BP pressure (mm Hg):				

Property/Value	Units	Temp	Data Type	Reference
WS 3.33E+004	mg/L	20	EXP	YALKOWSKY,SH & DANNENFELSER,RM (1992)
logP 0.48			EXP	HANSCH,C & LEO,AJ (1985)
VP 5.52E-001	mm Hg	25	EST	NEELY,WB & BLAU,GE (1985)
DC 8.72	pKa	20	EXP	KORTUM,G ET AL (1961)
HL 7.34E-010	atm m3/mol	25	EST	MEYLAN,WM & HOWARD,PH (1991)
OH 6.46E-012	cm3/molc sec	25	EST	MEYLAN,WM & HOWARD,PH (1993)

CAS #: 000109-02-4				N-METHYLMORPHOLINE
Formula: $C_5H_{11}NO$				
Mol Weight: 101.15				
MP (deg C):		FP (deg C):		
BP (deg C): 116				
BP pressure (mm Hg):				

Property/Value	Units	Temp	Data Type	Reference
WS 1.00E+006	mg/L	25	EST	MEYLAN,WM ET AL. (1996)
logP -0.33			EXP	HANSCH,C & LEO,AJ (1985)
VP 1.32E+001	mm Hg	25	EST	NEELY,WB & BLAU,GE (1985)
DC 7.38	pKa	25	EXP	PERRIN,DD (1965)
HL 2.50E-007	atm m3/mol	25	EST	MEYLAN,WM & HOWARD,PH (1991)
OH 1.42E-010	cm3/molc sec	25	EST	MEYLAN,WM & HOWARD,PH (1993)

CAS #: 000109-04-6				2-BROMOPYRIDINE
Formula: C_5H_4BrN				
Mol Weight: 158.00				
MP (deg C):		FP (deg C):		
BP (deg C): 193				
BP pressure (mm Hg):				

Property/Value	Units	Temp	Data Type	Reference
WS 4.28E+003	mg/L	25	EST	MEYLAN,WM ET AL. (1996)
logP 1.42			EXP	HANSCH,C & LEO,AJ (1985)
VP 1.53E+003	mm Hg			
DC 0.90	pKa	25	EXP	PERRIN,DD (1965)
HL 2.81E-006	atm m3/mol	25	EST	MEYLAN,WM & HOWARD,PH (1991)
OH 2.33E-013	cm3/molc sec	25	EST	MEYLAN,WM & HOWARD,PH (1993)

CAS #: 000109-05-7				2-METHYLPIPERIDINE
Formula: $C_6H_{13}N$				
Mol Weight: 99.18				
MP (deg C):		FP (deg C):		
BP (deg C):				
BP pressure (mm Hg):				

Property/Value	Units	Temp	Data Type	Reference
WS 5.09E+004	mg/L	25	EST	MEYLAN,WM ET AL. (1996)
logP 1.61			EST	MEYLAN,WM & HOWARD,PH (1995)
VP 1.93E+001	mm Hg	25	EXP	CHAO,J ET AL. (1983)
DC 11.08	pKa	25	EXP	PERRIN,DD (1965)
HL 2.28E-005	atm m3/mol	25	EST	MEYLAN,WM & HOWARD,PH (1991)
OH 1.00E-010	cm3/molc sec	25	EST	MEYLAN,WM & HOWARD,PH (1993)

CAS #: 000109-06-8				2-METHYLPYRIDINE
Formula: C_6H_7N				
Mol Weight: 93.13				
MP (deg C): -66.7		FP (deg C): 26		
BP (deg C): 129.3				
BP pressure (mm Hg):				

Property/Value	Units	Temp	Data Type	Reference
WS 1.00E+006	mg/L		EXP	RIDDICK,JA ET AL. (1986)
logP 1.11			EXP	HANSCH,C ET AL. (1995)
VP 1.12E+001	mm Hg	25	EXP	CHAO,J ET AL. (1983)
DC 6.00	pKa		EXP	RIDDICK,JA ET AL. (1986)
HL 9.96E-006	atm m3/mol	25	EXP	ANDON,RJL ET AL. (1954)
OH 1.12E-012	cm3/molc sec	25	EST	MEYLAN,WM & HOWARD,PH (1993)

CAS #: 000109-08-0				2-METHYLPYRAZINE
Formula: $C_5H_6N_2$				
Mol Weight: 94.12				
MP (deg C): -29		FP (deg C):		
BP (deg C): 137				
BP pressure (mm Hg):				

Property/Value	Units	Temp	Data Type	Reference
WS 8.04E+004	mg/L	25	EST	MEYLAN,WM ET AL. (1996)
logP 0.21			EXP	HANSCH,C & LEO,AJ (1985)
VP 8.06E+000	mm Hg	25	EXT	CHAO,J ET AL. (1983)
DC 1.45	pKa	27	EXP	PERRIN,DD (1965)
HL 2.20E-006	atm m3/mol	25	EXP	BUTTERY,RG ET AL. (1971)
OH 7.49E-013	cm3/molc sec	25	EST	MEYLAN,WM & HOWARD,PH (1993)

CAS #: 000109-09-1				2-CHLOROPYRIDINE
Formula: C_5H_4ClN				
Mol Weight: 113.55				
MP (deg C):		FP (deg C):		
BP (deg C): 170				
BP pressure (mm Hg):				

Property/Value	Units	Temp	Data Type	Reference
WS 2.00E+004	mg/L	25	EXP	CHEM INSPEC TEST INST
logP 1.22			EXP	HANSCH,C ET AL. (1995)
VP 2.18E+000	mm Hg	25	EXP	OHE,S (1976)
DC 0.49	pKa	25	EXP	PERRIN,DD (1965)
HL 1.63E-005	atm m3/mol	25	EST	VP/WSOL
OH 2.60E-013	cm3/molc sec	25	EST	MEYLAN,WM & HOWARD,PH (1993)

CAS #: 000109-10-4				A-PYRIDONE
Formula: C_5H_5NO				
Mol Weight: 95.10				
MP (deg C):		FP (deg C):		
BP (deg C):				
BP pressure (mm Hg):				

Property/Value	Units	Temp	Data Type	Reference
WS 3.78E+005	mg/L	25	EST	MEYLAN,WM ET AL. (1996)
logP -0.58			EXP	HANSCH,C & LEO,AJ (1985)
VP 7.57E-003	mm Hg	25	EST	NEELY,WB & BLAU,GE (1985)
DC	pKa			
HL 1.14E-008	atm m3/mol	25	EST	MEYLAN,WM & HOWARD,PH (1991)
OH 3.53E-011	cm3/molc sec	25	EST	MEYLAN,WM & HOWARD,PH (1993)

CAS #: 000109-12-6				2-AMINOPYRIMIDINE

Formula: $C_4H_5N_3$

Mol Weight: 95.10

MP (deg C): 127.5   FP (deg C):

BP (deg C):

BP pressure (mm Hg):

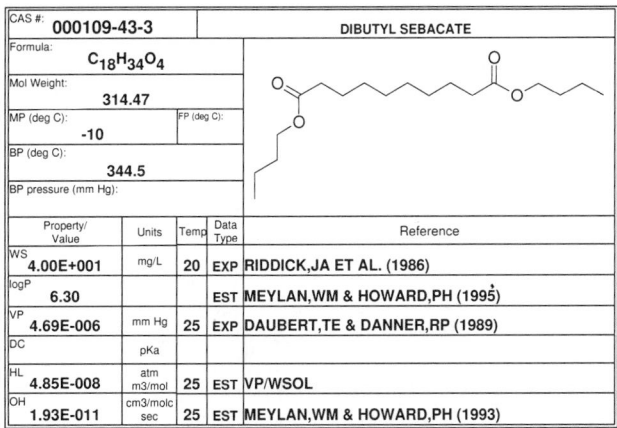

Property/Value	Units	Temp	Data Type	Reference
WS 1.86E+005	mg/L	25	EST	MEYLAN,WM ET AL. (1996)
logP -0.12			EST	MEYLAN,WM & HOWARD,PH (1995)
VP 2.46E-001	mm Hg	25	EST	NEELY,WB & BLAU,GE (1985)
DC 3.45	pKa	20	EXP	PERRIN,DD (1965)
HL 3.66E-006	atm m3/mol	25	EST	MEYLAN,WM & HOWARD,PH (1991)
OH 1.27E-011	cm3/molc sec	25	EST	MEYLAN,WM & HOWARD,PH (1993)

CAS #: 000109-16-0				TRIETHYLENEGLYCOL DIMETHACRYLATE

Formula: $C_{14}H_{22}O_6$

Mol Weight: 286.33

MP (deg C):   FP (deg C):

BP (deg C): 170-172

BP pressure (mm Hg): 5.00E+000

Property/Value	Units	Temp	Data Type	Reference
WS 3.66E+002	mg/L	25	EST	MEYLAN,WM ET AL. (1996)
logP 1.88			EXP	SANGSTER,J (1993)
VP 9.40E-004	mm Hg	25	EST	NEELY,WB & BLAU,GE (1985)
DC	pKa			
HL 1.70E-012	atm m3/mol	25	EST	MEYLAN,WM & HOWARD,PH (1991)
OH 6.80E-011	cm3/molc sec	25	EST	MEYLAN,WM & HOWARD,PH (1993)

CAS #: 000109-17-1				TETRAETHYLENEGLYCOL DIMETHACRYLATE

Formula: $C_{16}H_{26}O_7$

Mol Weight: 330.38

MP (deg C):   FP (deg C):

BP (deg C): 220

BP pressure (mm Hg):

Property/Value	Units	Temp	Data Type	Reference
WS 5.33E+002	mg/L	25	EST	MEYLAN,WM ET AL. (1996)
logP 1.39			EST	MEYLAN,WM & HOWARD,PH (1995)
VP 1.20E-004	mm Hg	25	EST	NEELY,WB & BLAU,GE (1985)
DC	pKa			
HL 1.42E-012	atm m3/mol	25	EST	MEYLAN,WM & HOWARD,PH (1991)
OH 8.20E-011	cm3/molc sec	25	EST	MEYLAN,WM & HOWARD,PH (1993)

CAS #: 000109-21-7				N-BUTYLBUTYRATE

Formula: $C_8H_{16}O_2$

Mol Weight: 144.22

MP (deg C): -91.5   FP (deg C):

BP (deg C): 165

BP pressure (mm Hg):

Property/Value	Units	Temp	Data Type	Reference
WS 5.00E+002	mg/L	20	EXP	YALKOWSKY,SH & DANNENFELSER,RM (1992)
logP 2.83			EST	MEYLAN,WM & HOWARD,PH (1995)
VP 1.81E+000	mm Hg	25	EXP	DAUBERT,TE & DANNER,RP (1989)
DC	pKa			
HL 6.87E-004	atm m3/mol	25	EST	VP/WSOL
OH 1.06E-011	cm3/molc sec	25	EXP	ATKINSON,R (1989)

CAS #: 000109-43-3				DIBUTYL SEBACATE

Formula: $C_{18}H_{34}O_4$

Mol Weight: 314.47

MP (deg C): -10   FP (deg C):

BP (deg C): 344.5

BP pressure (mm Hg):

Property/Value	Units	Temp	Data Type	Reference
WS 4.00E+001	mg/L	20	EXP	RIDDICK,JA ET AL. (1986)
logP 6.30			EST	MEYLAN,WM & HOWARD,PH (1995)
VP 4.69E-006	mm Hg	25	EXP	DAUBERT,TE & DANNER,RP (1989)
DC	pKa			
HL 4.85E-008	atm m3/mol	25	EST	VP/WSOL
OH 1.93E-011	cm3/molc sec	25	EST	MEYLAN,WM & HOWARD,PH (1993)

CAS #: 000109-46-6				N,N-DIBUTYLTHIOUREA

Formula: $C_9H_{20}N_2S$

Mol Weight: 188.34

MP (deg C): 63-65   FP (deg C):

BP (deg C):

BP pressure (mm Hg):

Property/Value	Units	Temp	Data Type	Reference
WS 2.29E+003	mg/L	25	EST	MEYLAN,WM ET AL. (1996)
logP 2.75			EXP	GOVERS,H ET AL. (1986)
VP 2.38E-003	mm Hg	25	EST	NEELY,WB & BLAU,GE (1985)
DC	pKa			
HL 4.17E-006	atm m3/mol	25	EST	MEYLAN,WM & HOWARD,PH (1991)
OH 1.53E-010	cm3/molc sec	25	EST	MEYLAN,WM & HOWARD,PH (1993)

CAS #: 000109-49-9				1-HEXEN-5-ONE

Formula: $C_6H_{10}O$

Mol Weight: 98.15

MP (deg C):   FP (deg C):

BP (deg C): 129.5

BP pressure (mm Hg):

Property/Value	Units	Temp	Data Type	Reference
WS 1.59E+004	mg/L	25	EST	MEYLAN,WM ET AL. (1996)
logP 1.02			EXP	HANSCH,C & LEO,AJ (1985)
VP 2.13E+001	mm Hg	25	EST	NEELY,WB & BLAU,GE (1985)
DC	pKa			
HL 8.64E-005	atm m3/mol	25	EST	MEYLAN,WM & HOWARD,PH (1991)
OH 3.09E-011	cm3/molc sec	25	EST	MEYLAN,WM & HOWARD,PH (1993)

CAS #: 000109-50-2				3-HEXYN-2-OL

Formula: $C_6H_{10}O$

Mol Weight: 98.15

MP (deg C):   FP (deg C):

BP (deg C):

BP pressure (mm Hg):

Property/Value	Units	Temp	Data Type	Reference
WS 3.80E+004	mg/L	20	EXP	HORT,EV (1978)
logP 1.03			EST	MEYLAN,WM & HOWARD,PH (1995)
VP 1.04E+000	mm Hg	25	EST	NEELY,WB & BLAU,GE (1985)
DC	pKa			
HL 6.93E-007	atm m3/mol	25	EST	MEYLAN,WM & HOWARD,PH (1991)
OH 3.52E-011	cm3/molc sec	25	EST	MEYLAN,WM & HOWARD,PH (1993)

CAS #: 000109-52-4				PENTANOIC ACID

Formula: $C_5H_{10}O_2$

Mol Weight: 102.13

MP (deg C): -34.5 | FP (deg C):

BP (deg C): 186-187

BP pressure (mm Hg):

Property/Value	Units	Temp	Data Type	Reference
WS 3.92E+004	mg/L	25	EXP	YALKOWSKY,SH & DANNENFELSER,RM (1992)
logP 1.39			EXP	HANSCH,C & LEO,AJ (1985)
VP 2.46E-001	mm Hg	25	EXP	DAUBERT,TE & DANNER,RP (1989)
DC 4.83	pKa	20	EXP	KORTUM,G ET AL (1961)
HL 8.43E-007	atm m3/mol		EST	VP/WSOL
OH 4.11E-012	cm3/molc sec	25	EST	MEYLAN,WM & HOWARD,PH (1993)

CAS #: 000109-53-5				ISOBUTYL VINYL ETHER

Formula: $C_6H_{12}O$

Mol Weight: 100.16

MP (deg C): -112 | FP (deg C):

BP (deg C): 83

BP pressure (mm Hg):

Property/Value	Units	Temp	Data Type	Reference
WS 3.00E+003	mg/L	20	EXP	HORT,EV (1978)
logP 1.82			EST	MEYLAN,WM & HOWARD,PH (1995)
VP 5.95E+001	mm Hg	20	EXP	FLICK,EW (1991)
DC	pKa			
HL 2.61E-003	atm m3/mol	20	EST	VP/WSOL
OH 4.83E-011	cm3/molc sec	25	EST	MEYLAN,WM & HOWARD,PH (1993)

CAS #: 000109-57-9				ALLYLTHIOUREA

Formula: $C_4H_8N_2S$

Mol Weight: 116.19

MP (deg C): 78 | FP (deg C):

BP (deg C):

BP pressure (mm Hg):

Property/Value	Units	Temp	Data Type	Reference
WS 5.90E+004	mg/L	15	EXP	STEPHEN,H & STEPHEN,T (1963)
logP 0.00			EST	MEYLAN,WM & HOWARD,PH (1995)
VP 2.77E-001	mm Hg	25	EST	NEELY,WB & BLAU,GE (1985)
DC	pKa			
HL 4.55E-007	atm m3/mol	25	EST	MEYLAN,WM & HOWARD,PH (1991)
OH 1.19E-010	cm3/molc sec	25	EST	MEYLAN,WM & HOWARD,PH (1993)

CAS #: 000109-59-1				ISOPROPOXYETHANOL

Formula: $C_5H_{12}O_2$

Mol Weight: 104.15

MP (deg C): | FP (deg C):

BP (deg C): 145

BP pressure (mm Hg):

Property/Value	Units	Temp	Data Type	Reference
WS 3.33E+005	mg/L	25	EST	MEYLAN,WM ET AL. (1996)
logP 0.05			EXP	HANSCH,C & LEO,AJ (1985)
VP 1.62E+000	mm Hg	25	EST	NEELY,WB & BLAU,GE (1985)
DC	pKa			
HL 7.38E-008	atm m3/mol	25	EST	MEYLAN,WM & HOWARD,PH (1991)
OH 2.33E-011	cm3/molc sec	25	EST	MEYLAN,WM & HOWARD,PH (1993)

CAS #: 000109-60-4				N-PROPYL ACETATE

Formula: $C_5H_{10}O_2$

Mol Weight: 102.13

MP (deg C): -95 | FP (deg C):

BP (deg C): 101.6

BP pressure (mm Hg):

Property/Value	Units	Temp	Data Type	Reference
WS 1.89E+004	mg/L	20	EXP	STEPHEN,H & STEPHEN,T (1963)
logP 1.24			EXP	HANSCH,C & LEO,AJ (1985)
VP 3.37E+001	mm Hg	25	EXP	AMBROSE,D ET AL. (1981)
DC	pKa			
HL 2.18E-004	atm m3/mol	25	EXP	KIECKBUSCH,TG & KING,CJ (1979)
OH 3.40E-012	cm3/molc sec	25	EXP	ATKINSON,R (1989)

CAS #: 000109-64-8				1,3-DIBROMOPROPANE

Formula: $C_3H_6Br_2$

Mol Weight: 201.90

MP (deg C): -36 | FP (deg C):

BP (deg C): 167

BP pressure (mm Hg):

Property/Value	Units	Temp	Data Type	Reference
WS 1.70E+003	mg/L	20	EXP	YALKOWSKY,SH & DANNENFELSER,RM (1992)
logP 2.37			EXP	HANSCH,C ET AL. (1995)
VP 1.36E+000	mm Hg	25	EXP	HINE,J & MOOKERJEE,PK (1975)
DC	pKa			
HL 8.88E-004	atm m3/mol	25	EXP	HINE,J & MOOKERJEE,PK (1975)
OH 8.41E-013	cm3/molc sec	25	EST	MEYLAN,WM & HOWARD,PH (1993)

CAS #: 000109-65-9				1-BROMOBUTANE

Formula: C_4H_9Br

Mol Weight: 137.03

MP (deg C): -112 | FP (deg C):

BP (deg C): 101.3

BP pressure (mm Hg):

Property/Value	Units	Temp	Data Type	Reference
WS 8.69E+002	mg/L	25	EXP	YALKOWSKY,SH & DANNENFELSER,RM (1992)
logP 2.75			EXP	HANSCH,C & LEO,AJ (1985)
VP 4.20E+001	mm Hg	25	EXP	DAUBERT,TE & DANNER,RP (1989)
DC	pKa			
HL 8.71E-003	atm m3/mol	25	EST	VP/WSOL
OH 2.17E-012	cm3/molc sec	25	EST	MEYLAN,WM & HOWARD,PH (1993)

CAS #: 000109-66-0				N-PENTANE

Formula: C_5H_{12}

Mol Weight: 72.15

MP (deg C): -129.7 | FP (deg C):

BP (deg C): 36.1

BP pressure (mm Hg):

Property/Value	Units	Temp	Data Type	Reference
WS 3.80E+001	mg/L	25	EXP	RIDDICK,JA ET AL. (1986)
logP 3.39			EXP	HANSCH,C & LEO,AJ (1985)
VP 5.14E+002	mm Hg	25	EXP	DAUBERT,TE & DANNER,RP (1989)
DC	pKa			
HL 1.25E+000	atm m3/mol	25	EXP	HINE,J & MOOKERJEE,PK (1975)
OH 3.94E-012	cm3/molc sec	25	EXP	ATKINSON,R (1989)

CAS #: 000109-67-1				1-PENTENE
Formula: C_5H_{10}				
Mol Weight: 70.14				
MP (deg C): -165.2	FP (deg C):			
BP (deg C): 30.1				
BP pressure (mm Hg):				

Property/Value	Units	Temp	Data Type	Reference
WS 1.48E+002	mg/L	25	EXP	YALKOWSKY,SH & DANNENFELSER,RM (1992)
logP 2.66			EST	MEYLAN,WM & HOWARD,PH (1995)
VP 6.35E+002	mm Hg	25	EXP	PERRY,RH & GREEN,D (1984)
DC	pKa			
HL 3.96E-001	atm m3/mol	25	EST	VP/WSOL
OH 3.14E-011	cm3/molc sec	25	EXP	ATKINSON,R (1989)

CAS #: 000109-68-2				2-PENTENE
Formula: C_5H_{10}				
Mol Weight: 70.14				
MP (deg C):	FP (deg C):			
BP (deg C): 37				
BP pressure (mm Hg):				

Property/Value	Units	Temp	Data Type	Reference
WS 2.03E+002	mg/L	25	EXP	YALKOWSKY,SH & DANNENFELSER,RM (1992)
logP 2.58			EST	MEYLAN,WM & HOWARD,PH (1995)
VP 5.28E+002	mm Hg	25	EXP	RIDDICK,JA ET AL. (1986)
DC	pKa			
HL 2.40E-001	atm m3/mol	25	EST	VP/WSOL
OH 5.76E-011	cm3/molc sec	25	EST	MEYLAN,WM & HOWARD,PH (1993)

CAS #: 000109-69-3				1-CHLOROBUTANE
Formula: C_4H_9Cl				
Mol Weight: 92.57				
MP (deg C): -123.1	FP (deg C):			
BP (deg C): 78.5				
BP pressure (mm Hg):				

Property/Value	Units	Temp	Data Type	Reference
WS 1.10E+003	mg/L	25	EXP	RIDDICK,JA ET AL. (1986)
logP 2.64			EXP	HANSCH,C & LEO,AJ (1985)
VP 1.01E+002	mm Hg	25	EXP	DAUBERT,TE & DANNER,RP (1989)
DC	pKa			
HL 1.67E-002	atm m3/mol	25	EXP	LEIGHTON,DTJR & CALO,JM (1981)
OH 1.50E-012	cm3/molc sec	25	EXP	ATKINSON,R (1989)

CAS #: 000109-70-6				1-CHLORO-3-BROMOPROPANE
Formula: C_3H_6BrCl				
Mol Weight: 157.44				
MP (deg C): -58.9	FP (deg C):			
BP (deg C): 143.3				
BP pressure (mm Hg):				

Property/Value	Units	Temp	Data Type	Reference
WS 2.24E+003	mg/L	25	EXP	YALKOWSKY,SH & DANNENFELSER,RM (1992)
logP 2.18			EXP	TEWARI,YB ET AL. (1982)
VP 6.40E+000	mm Hg	25	EXT	DYKYJ,J (1970)
DC	pKa			
HL 2.45E-004	atm m3/mol	25	EST	MEYLAN,WM & HOWARD,PH (1991)
OH 9.13E-013	cm3/molc sec	25	EST	MEYLAN,WM & HOWARD,PH (1993)

CAS #: 000109-73-9				N-BUTYLAMINE
Formula: $C_4H_{11}N$				
Mol Weight: 73.14				
MP (deg C): -50	FP (deg C):			
BP (deg C): 78				
BP pressure (mm Hg):				

Property/Value	Units	Temp	Data Type	Reference
WS 1.00E+006	mg/L	25	EXP	YALKOWSKY,SH & DANNENFELSER,RM (1992)
logP 0.97			EXP	HANSCH,C & LEO,AJ (1985)
VP 9.15E+001	mm Hg	25	EXP	YAWS,CL (1994)
DC 10.78	pKa	20	EXP	PERRIN,DD (1965)
HL 1.74E-005	atm m3/mol	25	EXP	CHRISTIE,AO & CRISP,DJ (1967)
OH 3.35E-011	cm3/molc sec	25	EST	MEYLAN,WM & HOWARD,PH (1993)

CAS #: 000109-74-0				BUTYRONITRILE
Formula: C_4H_7N				
Mol Weight: 69.11				
MP (deg C): -111.9	FP (deg C):			
BP (deg C): 117.62				
BP pressure (mm Hg):				

Property/Value	Units	Temp	Data Type	Reference
WS 3.30E+004	mg/L	25	EXP	RIDDICK,JA ET AL. (1986)
logP 0.53			EXP	TANII,H & HASHIMOTO,K (1984)
VP 1.95E+001	mm Hg	25	EXP	DAUBERT,TE & DANNER,RP (1989)
DC	pKa			
HL 5.23E-005	atm m3/mol	25	EXP	BUTLER,JAV AND RAMCHANDANI,CN (1935)
OH 7.56E-013	cm3/molc sec	25	EST	ATKINSON,R (1988)

CAS #: 000109-75-1				3-BUTENENITRILE
Formula: C_4H_5N				
Mol Weight: 67.09				
MP (deg C): -87	FP (deg C):			
BP (deg C): 119				
BP pressure (mm Hg):				

Property/Value	Units	Temp	Data Type	Reference
WS 3.37E+004	mg/L	25	EST	MEYLAN,WM ET AL. (1996)
logP 0.40			EXP	TANII,H & HASHIMOTO,K (1984)
VP 1.85E+001	mm Hg	25	EXP	YAWS,CL (1994)
DC	pKa			
HL 4.01E-005	atm m3/mol	25	EST	MEYLAN,WM & HOWARD,PH (1991)
OH 1.33E-011	cm3/molc sec	25	EST	MEYLAN,WM & HOWARD,PH (1993)

CAS #: 000109-76-2				1,3-PROPANEDIAMINE
Formula: $C_3H_{10}N_2$				
Mol Weight: 74.13				
MP (deg C):	FP (deg C):			
BP (deg C): 139.8				
BP pressure (mm Hg):				

Property/Value	Units	Temp	Data Type	Reference
WS 1.00E+006	mg/L	25	EST	MEYLAN,WM ET AL. (1996)
logP -1.43			EXP	HANSCH,C ET AL. (1995)
VP 1.15E+001	mm Hg	25	EST	NEELY,WB & BLAU,GE (1985)
DC 10.62	pKa	20	EXP	PERRIN,DD (1965)
HL 1.37E-009	atm m3/mol	25	EST	MEYLAN,WM & HOWARD,PH (1991)
OH 6.48E-011	cm3/molc sec	25	EST	MEYLAN,WM & HOWARD,PH (1993)

CAS #: 000109-77-3				MALONONITRILE
Formula: $C_3H_2N_2$				
Mol Weight: 66.06				
MP (deg C): 30-1		FP (deg C):		
BP (deg C): 218-9				
BP pressure (mm Hg):				

Property/Value	Units	Temp	Data Type	Reference
WS 1.33E+005	mg/L	25	EXP	HUGHES,DW (1981)
logP -0.60			EXP	TANII,H & HASHIMOTO,K (1985)
VP 2.00E-001	mm Hg	25	EXT	DAUBERT,TE & DANNER,RP (1989)
DC 11.40	pKa		EXP	HUGHS,DW (1981)
HL 1.27E-008	atm m3/mol	25	EST	MEYLAN,WM & HOWARD,PH (1991)
OH 1.64E-014	cm3/molc sec	25	EST	MEYLAN,WM & HOWARD,PH (1993)

CAS #: 000109-78-4				ETHYLENE CYANOHYDRIN
Formula: C_3H_5NO				
Mol Weight: 71.08				
MP (deg C): -46		FP (deg C):		
BP (deg C): 228				
BP pressure (mm Hg):				

Property/Value	Units	Temp	Data Type	Reference
WS 1.00E+006	mg/L	20	EXP	YALKOWSKY,SH & DANNENFELSER,RM (1992)
logP -1.12			EST	MEYLAN,WM & HOWARD,PH (1995)
VP 8.02E-002	mm Hg	25	EXP	DAUBERT,TE & DANNER,RP (1989)
DC	pKa			
HL 7.50E-009	atm m3/mol	25	EST	VP/WSOL
OH 7.51E-013	cm3/molc sec	25	EST	MEYLAN,WM & HOWARD,PH (1993)

CAS #: 000109-79-5				BUTANETHIOL
Formula: $C_4H_{10}S$				
Mol Weight: 90.19				
MP (deg C): -115.9		FP (deg C):		
BP (deg C): 98.4				
BP pressure (mm Hg):				

Property/Value	Units	Temp	Data Type	Reference
WS 5.95E+002	mg/L	25	EXP	KUHNE,R ET AL. (1995)
logP 2.28			EXP	HANSCH,C & LEO,AJ (1985)
VP 4.55E+001	mm Hg	25	EXP	YAWS,CL (1994)
DC	pKa			
HL 9.08E-003	atm m3/mol	25	EST	VP/WSOL
OH 5.10E-011	cm3/molc sec	25	EXP	ATKINSON,R (1989)

CAS #: 000109-83-1				2-(METHYLAMINO)ETHANOL
Formula: C_3H_9NO				
Mol Weight: 75.11				
MP (deg C): 102-104		FP (deg C):		
BP (deg C): 253				
BP pressure (mm Hg): 7.12E+002				

Property/Value	Units	Temp	Data Type	Reference
WS 1.00E+006	mg/L	25	EST	MEYLAN,WM ET AL. (1996)
logP -0.94			EXP	HANSCH,C ET AL. (1995)
VP 1.08E+000	mm Hg	25	EXP	YAWS,CL (1994)
DC	pKa			
HL 8.07E-010	atm m3/mol	25	EST	MEYLAN,WM & HOWARD,PH (1991)
OH 7.91E-011	cm3/molc sec	25	EST	MEYLAN,WM & HOWARD,PH (1993)

CAS #: 000109-86-4				2-METHOXYETHANOL
Formula: $C_3H_8O_2$				
Mol Weight: 76.10				
MP (deg C): -85.1		FP (deg C):		
BP (deg C): 124.6				
BP pressure (mm Hg):				

Property/Value	Units	Temp	Data Type	Reference
WS 1.00E+006	mg/L		EXP	DOW CHEMICAL COMPANY (1981)
logP -0.77			EXP	HANSCH,C & LEO,AJ (1985)
VP 9.50E+000	mm Hg	25	EXP	DOW CHEMICAL COMPANY (1981)
DC 14.80	pKa	25	EXP	SERJEANT,EP & DEMPSEY,B (1979)
HL 9.51E-007	atm m3/mol	25	EST	VP/WSOL
OH 1.25E-011	cm3/molc sec	25	EXP	ATKINSON,R (1989)

CAS #: 000109-87-5				DIMETHOXYMETHANE
Formula: $C_3H_8O_2$				
Mol Weight: 76.10				
MP (deg C): -105		FP (deg C):		
BP (deg C): 41.6				
BP pressure (mm Hg):				

Property/Value	Units	Temp	Data Type	Reference
WS 2.30E+005	mg/L	20	EXP	YALKOWSKY,SH & DANNENFELSER,RM (1992)
logP 0.00			EXP	HANSCH,C & LEO,AJ (1985)
VP 3.98E+002	mm Hg	25	EXP	DAUBERT,TE & DANNER,RP (1989)
DC	pKa			
HL 1.73E-004	atm m3/mol	25	EXP	HINE,J & MOOKERJEE,PK (1975)
OH 6.87E-012	cm3/molc sec	25	EST	MEYLAN,WM & HOWARD,PH (1993)

CAS #: 000109-89-7				DIETHYLAMINE
Formula: $C_4H_{11}N$				
Mol Weight: 73.14				
MP (deg C): -50		FP (deg C):		
BP (deg C): 55.5				
BP pressure (mm Hg):				

Property/Value	Units	Temp	Data Type	Reference
WS 1.00E+006	mg/L	25	EXP	RIDDICK,JA ET AL. (1986)
logP 0.58			EXP	HANSCH,C & LEO,AJ (1985)
VP 2.37E+002	mm Hg	25	EXP	YAWS,CL (1994)
DC 11.09	pKa	20	EXP	PERRIN,DD (1965)
HL 2.55E-005	atm m3/mol	25	EXP	CHRISTIE,AO & CRISP,DJ (1967)
OH 7.71E-011	cm3/molc sec	25	EST	MEYLAN,WM & HOWARD,PH (1993)

CAS #: 000109-92-2				ETHYL VINYL ETHER
Formula: C_4H_8O				
Mol Weight: 72.11				
MP (deg C): -115.8		FP (deg C):		
BP (deg C): 35.5				
BP pressure (mm Hg):				

Property/Value	Units	Temp	Data Type	Reference
WS 9.00E+003	mg/L	20	EXP	RIDDICK,JA ET AL. (1986)
logP 1.04			EXP	HANSCH,C & LEO,AJ (1985)
VP 5.11E+002	mm Hg	25	EXP	DAUBERT,TE & DANNER,RP (1989)
DC	pKa			
HL 5.39E-003	atm m3/mol	25	EST	VP/WSOL
OH 4.04E-011	cm3/molc sec	25	EST	MEYLAN,WM & HOWARD,PH (1993)

DIVINYL ETHER

CAS #:	000109-93-3		DIVINYL ETHER
Formula:	C_4H_6O		
Mol Weight:	70.09		
MP (deg C):	-101	FP (deg C):	
BP (deg C):	28.4		
BP pressure (mm Hg):			

Property/Value	Units	Temp	Data Type	Reference
WS 7.69E+003	mg/L	25	EXP	SUZUKI,T (1991)
logP 0.78			EST	MEYLAN,WM & HOWARD,PH (1995)
VP 6.70E+002	mm Hg	25	EXP	DAUBERT,TE & DANNER,RP (1989)
DC	pKa			
HL 8.04E-003	atm m3/mol	25	EST	VP/WSOL
OH 5.26E-011	cm3/molc sec	25	EST	MEYLAN,WM & HOWARD,PH (1993)

ETHYL FORMATE

CAS #:	000109-94-4		ETHYL FORMATE
Formula:	$C_3H_6O_2$		
Mol Weight:	74.08		
MP (deg C):	-80	FP (deg C):	
BP (deg C):	53-54		
BP pressure (mm Hg):			

Property/Value	Units	Temp	Data Type	Reference
WS 8.83E+004	mg/L	25	EXP	HANSCH,C ET AL. (1968)
logP 0.23			EXP	HANSCH,C & LEO,AJ (1985)
VP 2.45E+002	mm Hg	25	EXP	DAUBERT,TE & DANNER,RP (1985)
DC	pKa			
HL 3.85E-004	atm m3/mol	25	EXP	BOCEK,K (1976)
OH 1.02E-012	cm3/molc sec	25	EXP	ATKINSON,R (1989)

ETHYLNITRITE

CAS #:	000109-95-5		ETHYLNITRITE
Formula:	$C_2H_5NO_2$		
Mol Weight:	75.07		
MP (deg C):		FP (deg C):	
BP (deg C):	17		
BP pressure (mm Hg):			

Property/Value	Units	Temp	Data Type	Reference
WS 8.97E+003	mg/L	25	EST	MEYLAN,WM ET AL. (1996)
logP 1.37			EST	MEYLAN,WM & HOWARD,PH (1995)
VP 6.20E+002	mm Hg	25	EST	NEELY,WB & BLAU,GE (1985)
DC	pKa			
HL 8.70E-005	atm m3/mol	25	EST	MEYLAN,WM & HOWARD,PH (1991)
OH 1.75E-012	cm3/molc sec	25	EXP	ATKINSON,R (1989)

PYRROLE

CAS #:	000109-97-7		PYRROLE
Formula:	C_4H_5N		
Mol Weight:	67.09		
MP (deg C):	-23.4	FP (deg C):	
BP (deg C):	129.8		
BP pressure (mm Hg):			

Property/Value	Units	Temp	Data Type	Reference
WS 4.70E+004	mg/L		EXP	GHERINI,SA ET AL. (1989)
logP 0.75			EXP	HANSCH,C & LEO,AJ (1985)
VP 8.36E+000	mm Hg	25	EXP	YAWS,CL (1994)
DC -3.80	pKa		EXP	PERRIN,DD (1965)
HL 1.80E-005	atm m3/mol	25	EXP	HAWTHORNE,SB ET AL. (1985)
OH 1.10E-010	cm3/molc sec	25	EXP	ATKINSON,R (1989)

TETRAHYDROFURAN

CAS #:	000109-99-9		TETRAHYDROFURAN
Formula:	C_4H_8O		
Mol Weight:	72.11		
MP (deg C):	-108.5	FP (deg C):	
BP (deg C):	66		
BP pressure (mm Hg):			

Property/Value	Units	Temp	Data Type	Reference
WS 1.00E+006	mg/L		EXP	RIDDICK,JA ET AL. (1986)
logP 0.46			EXP	HANSCH,C & LEO,AJ (1985)
VP 1.62E+002	mm Hg	25	EXP	DAUBERT,TE & DANNER,RP (1989)
DC -2.08	pKa		EXP	RIDDICK,JA ET AL. (1986)
HL 7.05E-005	atm m3/mol	25	EXP	CABANI,S ET AL. (1971A)
OH 1.61E-011	cm3/molc sec	25	EXP	ATKINSON,R (1989)

FURAN

CAS #:	000110-00-9		FURAN
Formula:	C_4H_4O		
Mol Weight:	68.08		
MP (deg C):	-85.61	FP (deg C):	
BP (deg C):	31.36		
BP pressure (mm Hg):			

Property/Value	Units	Temp	Data Type	Reference
WS 1.00E+004	mg/L	25	EXP	VALVANI,SC ET AL. (1981)
logP 1.34			EXP	HANSCH,C ET AL. (1995)
VP 6.00E+002	mm Hg	25	EXP	DAUBERT,TE & DANNER,RP (1989)
DC	pKa			
HL 5.40E-003	atm m3/mol	25	EST	VP/WSOL
OH 4.05E-011	cm3/molc sec	25	EXP	ATKINSON,R (1989)

TETRAHYDROTHIOPHENE

CAS #:	000110-01-0		TETRAHYDROTHIOPHENE
Formula:	C_4H_8S		
Mol Weight:	88.17		
MP (deg C):	-96.1	FP (deg C):	
BP (deg C):	121		
BP pressure (mm Hg):			

Property/Value	Units	Temp	Data Type	Reference
WS 3.73E+003	mg/L	25	EST	MEYLAN,WM ET AL. (1996)
logP 1.79			EST	MEYLAN,WM & HOWARD,PH (1995)
VP 1.81E-001	mm Hg	25	EXP	DREISBACH,RR (1955)
DC	pKa			
HL 6.11E-004	atm m3/mol	25	EST	MEYLAN,WM & HOWARD,PH (1991)
OH 1.97E-011	cm3/molc sec	25	EXP	ATKINSON,R (1989)

THIOPHENE

CAS #:	000110-02-1		THIOPHENE
Formula:	C_4H_4S		
Mol Weight:	84.14		
MP (deg C):	-38.3	FP (deg C):	
BP (deg C):	84.4		
BP pressure (mm Hg):			

Property/Value	Units	Temp	Data Type	Reference
WS 3.02E+003	mg/L	25	EXP	YALKOWSKY,SH & DANNENFELSER,RM (1992)
logP 1.81			EXP	HANSCH,C & LEO,AJ (1985)
VP 7.97E+001	mm Hg	25	EXP	BOUBLIK,T ET AL. (1984)
DC	pKa			
HL 2.92E-003	atm m3/mol	25	EST	VP/WSOL
OH 9.53E-012	cm3/molc sec	25	EXP	ATKINSON,R (1989)

CAS #: 000110-12-3				5-METHYL-2-HEXANONE
Formula: $C_7H_{14}O$				
Mol Weight: 114.19				
MP (deg C):		FP (deg C):		
BP (deg C): 144				
BP pressure (mm Hg):				

Property/Value	Units	Temp	Data Type	Reference
WS 5.40E+003	mg/L	20	EXP	LANDE,SS ET AL. (1976)
logP 1.88			EXP	TANII,H & HASHIMOTO,K (1986)
VP 5.20E+000	mm Hg	25	EXP	DAUBERT,TE & DANNER,RP (1989)
DC	pKa			
HL 1.45E-004	atm m3/mol	25	EST	VP/WSOL
OH 8.16E-012	cm3/molc sec	25	EST	MEYLAN,WM & HOWARD,PH (1993)

CAS #: 000110-13-4				2,5-HEXANEDIONE
Formula: $C_6H_{10}O_2$				
Mol Weight: 114.15				
MP (deg C): -9		FP (deg C):		
BP (deg C): 188				
BP pressure (mm Hg):				

Property/Value	Units	Temp	Data Type	Reference
WS 1.79E+005	mg/L	25	EST	MEYLAN,WM ET AL. (1996)
logP -0.27			EXP	HANSCH,C ET AL. (1995)
VP 2.93E+000	mm Hg	25	EST	NEELY,WB & BLAU,GE (1985)
DC	pKa			
HL 4.43E-008	atm m3/mol	25	EST	MEYLAN,WM & HOWARD,PH (1991)
OH 7.13E-012	cm3/molc sec	25	EXP	ATKINSON,R (1989)

CAS #: 000110-14-5				SUCCINAMIDE
Formula: $C_4H_8N_2O_2$				
Mol Weight: 116.12				
MP (deg C): 260		FP (deg C):		
BP (deg C):				
BP pressure (mm Hg):				

Property/Value	Units	Temp	Data Type	Reference
WS 3.30E+003	mg/L	15	EXP	YALKOWSKY,SH & DANNENFELSER,RM (1992)
logP -2.67			EST	MEYLAN,WM & HOWARD,PH (1995)
VP 1.04E-005	mm Hg	25	EST	NEELY,WB & BLAU,GE (1985)
DC	pKa			
HL 3.57E-012	atm m3/mol	25	EST	MEYLAN,WM & HOWARD,PH (1991)
OH 9.46E-012	cm3/molc sec	25	EST	MEYLAN,WM & HOWARD,PH (1993)

CAS #: 000110-15-6				BUTANEDIOC ACID
Formula: $C_4H_6O_4$				
Mol Weight: 118.09				
MP (deg C): 185-187		FP (deg C):		
BP (deg C): 235				
BP pressure (mm Hg):				

Property/Value	Units	Temp	Data Type	Reference
WS 8.32E+004	mg/L	25	EXP	YALKOWSKY,SH & DANNENFELSER,RM (1992)
logP -0.59			EXP	HANSCH,C & LEO,AJ (1985)
VP 7.86E-006	mm Hg	25	EXT	YAWS,CL (1994)
DC 4.28	pKa	20	EXP	KORTUM,G ET AL (1961)
HL 5.41E-012	atm m3/mol	25	EST	MEYLAN,WM & HOWARD,PH (1991)
OH 2.76E-012	cm3/molc sec	25	EST	MEYLAN,WM & HOWARD,PH (1993)

CAS #: 000110-16-7				MALEIC ACID
Formula: $C_4H_4O_4$				
Mol Weight: 116.07				
MP (deg C): 138-139		FP (deg C):		
BP (deg C):				
BP pressure (mm Hg):				

Property/Value	Units	Temp	Data Type	Reference
WS 4.41E+005	mg/L	25	EXP	YALKOWSKY,SH & DANNENFELSER,RM (1992)
logP -0.48			EXP	SANGSTER,J (1994)
VP 3.59E-005	mm Hg	25	EXT	YAWS,CL (1994)
DC 1.83	pKa		EXP	WEAST,RC (1972)
HL 1.35E-012	atm m3/mol	25	EST	MEYLAN,WM & HOWARD,PH (1991)
OH 7.95E-012	cm3/molc sec	25	EST	MEYLAN,WM & HOWARD,PH (1993)

CAS #: 000110-17-8				FUMARIC ACID
Formula: $C_4H_4O_4$				
Mol Weight: 116.07				
MP (deg C): 287		FP (deg C):		
BP (deg C): 522				
BP pressure (mm Hg):				

Property/Value	Units	Temp	Data Type	Reference
WS 7.00E+003	mg/L	25	EXP	US EPA (1981)
logP 0.46			EXP	HANSCH,C ET AL. (1995)
VP 1.54E-004	mm Hg	25	EXT	YAWS,CL (1994)
DC 3.03	pKa	18	EXP	WEAST,RC (1972)
HL 8.50E-014	atm m3/mol	25	EST	MEYLAN,WM & HOWARD,PH (1991)
OH 5.04E-011	cm3/molc sec	25	EST	MEYLAN,WM & HOWARD,PH (1993)

CAS #: 000110-18-9				N,N,N',N'-TETRAME ETHYLENEDIAMINE
Formula: $C_6H_{16}N_2$				
Mol Weight: 116.21				
MP (deg C): -55		FP (deg C):		
BP (deg C): 121				
BP pressure (mm Hg):				

Property/Value	Units	Temp	Data Type	Reference
WS 5.85E+005	mg/L	25	EST	MEYLAN,WM ET AL. (1996)
logP 0.30			EXP	HANSCH,C & LEO,AJ (1985)
VP 1.67E+001	mm Hg	25	EST	NEELY,WB & BLAU,GE (1985)
DC 9.10	pKa	22	EXP	PERRIN,DD (1965)
HL 2.40E-008	atm m3/mol	25	EST	MEYLAN,WM & HOWARD,PH (1991)
OH 1.58E-010	cm3/molc sec	25	EST	MEYLAN,WM & HOWARD,PH (1993)

CAS #: 000110-19-0				SEC-BUTYLACETATE
Formula: $C_6H_{12}O_2$				
Mol Weight: 116.16				
MP (deg C): -99		FP (deg C):		
BP (deg C): 118				
BP pressure (mm Hg):				

Property/Value	Units	Temp	Data Type	Reference
WS 6.30E+003	mg/L	25	EXP	YALKOWSKY,SH & DANNENFELSER,RM (1992)
logP 1.78			EXP	HANSCH,C ET AL. (1995)
VP 1.78E+001	mm Hg	25	EXP	YAWS,CL (1994)
DC	pKa			
HL 4.32E-004	atm m3/mol	25	EST	VP/WSOL
OH 5.50E-012	cm3/molc sec	25	EXP	ATKINSON,R (1989)

CAS #: 000110-27-0				ISOPROPYL MYRISTATE

Formula: $C_{17}H_{34}O_2$

Mol Weight: 270.46

MP (deg C): 3   FP (deg C):

BP (deg C): 192.6

BP pressure (mm Hg):

Property/ Value	Units	Temp	Data Type	Reference
WS 1.35E-002	mg/L	25	EST	MEYLAN,WM ET AL. (1996)
logP 7.17			EST	MEYLAN,WM & HOWARD,PH (1995)
VP 2.00E-001	mm Hg	140	EXP	BONOHORST,CW ET AL. (1948)
DC	pKa			
HL 9.26E-003	atm m3/mol	25	EST	MEYLAN,WM & HOWARD,PH (1991)
OH 1.97E-011	cm3/molc sec	25	EST	MEYLAN,WM & HOWARD,PH (1993)

CAS #: 000110-33-8				DIHEXYL ADIPATE

Formula: $C_{18}H_{34}O_4$

Mol Weight: 314.47

MP (deg C):   FP (deg C):

BP (deg C):

BP pressure (mm Hg):

Property/ Value	Units	Temp	Data Type	Reference
WS 4.22E-002	mg/L	25	EST	MEYLAN,WM ET AL. (1996)
logP 6.30			EST	MEYLAN,WM & HOWARD,PH (1995)
VP 2.88E-006	mm Hg	25	EXP	DAUBERT,TE & DANNER,RP (1989)
DC	pKa			
HL 1.66E-005	atm m3/mol	25	EST	MEYLAN,WM & HOWARD,PH (1991)
OH 1.93E-011	cm3/molc sec	25	EST	MEYLAN,WM & HOWARD,PH (1993)

CAS #: 000110-38-3				ETHYL DECANOATE

Formula: $C_{12}H_{24}O_2$

Mol Weight: 200.32

MP (deg C): -20   FP (deg C):

BP (deg C): 241.5

BP pressure (mm Hg):

Property/ Value	Units	Temp	Data Type	Reference
WS 1.59E+001	mg/L	25	EXP	SUZUKI,T (1991)
logP 4.79			EST	MEYLAN,WM & HOWARD,PH (1995)
VP 3.07E-002	mm Hg	25	EST	NEELY,WB & BLAU,GE (1985)
DC	pKa			
HL 2.25E-003	atm m3/mol	25	EST	MEYLAN,WM & HOWARD,PH (1991)
OH 1.23E-011	cm3/molc sec	25	EST	MEYLAN,WM & HOWARD,PH (1993)

CAS #: 000110-40-7				DIETHYL SEBACATE

Formula: $C_{14}H_{26}O_4$

Mol Weight: 258.36

MP (deg C): 5   FP (deg C):

BP (deg C): 305

BP pressure (mm Hg):

Property/ Value	Units	Temp	Data Type	Reference
WS 8.00E+001	mg/L	20	EXP	STEPHEN,H & STEPHEN,T (1963)
logP 4.33			EST	MEYLAN,WM & HOWARD,PH (1995)
VP 5.08E-004	mm Hg	25	EXT	OHE,S (1976)
DC	pKa			
HL 5.35E-006	atm m3/mol	25	EST	MEYLAN,WM & HOWARD,PH (1991)
OH 1.35E-011	cm3/molc sec	25	EST	MEYLAN,WM & HOWARD,PH (1993)

CAS #: 000110-42-9				METHYL DECANOATE

Formula: $C_{11}H_{22}O_2$

Mol Weight: 186.30

MP (deg C): -18   FP (deg C):

BP (deg C): 224

BP pressure (mm Hg):

Property/ Value	Units	Temp	Data Type	Reference
WS 4.40E+000	mg/L	20	EXP	YALKOWSKY,SH & DANNENFELSER,RM (1992)
logP 4.41			EXP	HANSCH,C & LEO,AJ (1985)
VP 4.87E-002	mm Hg	25	EXT	OHE,S (1976)
DC	pKa			
HL 1.69E-003	atm m3/mol	25	EST	MEYLAN,WM & HOWARD,PH (1991)
OH 1.09E-011	cm3/molc sec	25	EST	MEYLAN,WM & HOWARD,PH (1993)

CAS #: 000110-43-0				2-HEPTANONE

Formula: $C_7H_{14}O$

Mol Weight: 114.19

MP (deg C): -35   FP (deg C):

BP (deg C): 151.1

BP pressure (mm Hg):

Property/ Value	Units	Temp	Data Type	Reference
WS 4.30E+003	mg/L	25	EXP	RIDDICK,JA ET AL. (1986)
logP 1.98			EXP	HANSCH,C & LEO,AJ (1985)
VP 3.86E+000	mm Hg	25	EXP	AMBROSE,D ET AL. (1975)
DC	pKa			
HL 1.77E-004	atm m3/mol	25	EXP	MACKAY,D ET AL. (1982A)
OH 8.67E-011	cm3/molc sec	23	EXP	WALLINGTON,TJ & KURYCO,MJ (1987)

CAS #: 000110-44-1				SORBIC ACID

Formula: $C_6H_8O_2$

Mol Weight: 112.13

MP (deg C): 134.5   FP (deg C):

BP (deg C):

BP pressure (mm Hg):

Property/ Value	Units	Temp	Data Type	Reference
WS 1.91E+003	mg/L	30	EXP	YALKOWSKY,SH & DANNENFELSER,RM (1992)
logP 1.33			EXP	HANSCH,C & LEO,AJ (1985)
VP 1.34E-001	mm Hg	25	EST	NEELY,WB & BLAU,GE (1985)
DC	pKa			
HL 5.72E-007	atm m3/mol	25	EST	MEYLAN,WM & HOWARD,PH (1991)
OH 5.04E-011	cm3/molc sec	25	EST	MEYLAN,WM & HOWARD,PH (1993)

CAS #: 000110-45-2				ISOAMYL FORMATE

Formula: $C_6H_{12}O_2$

Mol Weight: 116.16

MP (deg C): -93.5   FP (deg C):

BP (deg C): 123.5

BP pressure (mm Hg):

Property/ Value	Units	Temp	Data Type	Reference
WS 3.50E+003	mg/L	25	EXP	SUZUKI,T (1991)
logP 1.72			EST	MEYLAN,WM & HOWARD,PH (1995)
VP 1.53E+001	mm Hg	25	EXP	HINE,J & MOOKERJEE,PK (1975)
DC	pKa			
HL 6.74E-004	atm m3/mol	25	EXP	HINE,J & MOOKERJEE,PK (1975)
OH 5.97E-012	cm3/molc sec	25	EST	MEYLAN,WM & HOWARD,PH (1993)

CAS #: 000110-49-6				2-METHOXYETHYL ACETATE

Formula: $C_5H_{10}O_3$
Mol Weight: 118.13
MP (deg C): -65.1
FP (deg C):
BP (deg C): 145
BP pressure (mm Hg):

Property/ Value	Units	Temp	Data Type	Reference
WS 8.40E+004	mg/L	25	EST	MEYLAN,WM ET AL. (1996)
logP 0.10			EST	MEYLAN,WM & HOWARD,PH (1995)
VP 4.01E+000	mm Hg	25	EXT	OHE,S (1976)
DC	pKa			
HL 2.72E-006	atm m3/mol	25	EST	MEYLAN,WM & HOWARD,PH (1991)
OH 9.78E-012	cm3/molc sec	25	EST	MEYLAN,WM & HOWARD,PH (1993)

CAS #: 000110-50-9				BUTYL XANTHATE

Formula: $C_5H_{10}OS_2$
Mol Weight: 150.26
MP (deg C):
FP (deg C):
BP (deg C):
BP pressure (mm Hg):

Property/ Value	Units	Temp	Data Type	Reference
WS 1.99E+003	mg/L	25	EST	MEYLAN,WM ET AL. (1996)
logP 1.85			EST	MEYLAN,WM & HOWARD,PH (1995)
VP 1.13E-001	mm Hg	25	EST	NEELY,WB & BLAU,GE (1985)
DC	pKa			
HL 4.59E-005	atm m3/mol	25	EST	MEYLAN,WM & HOWARD,PH (1991)
OH 4.48E-011	cm3/molc sec	25	EST	MEYLAN,WM & HOWARD,PH (1993)

CAS #: 000110-52-1				1,4-DIBROMOBUTANE

Formula: $C_4H_8Br_2$
Mol Weight: 215.93
MP (deg C): -16.5
FP (deg C):
BP (deg C): 197
BP pressure (mm Hg):

Property/ Value	Units	Temp	Data Type	Reference
WS 3.50E+002	mg/L	25	EXP	ALBANESE,V ET AL. (1987)
logP 2.99			EST	MEYLAN,WM & HOWARD,PH (1995)
VP 6.16E-001	mm Hg	25	EXT	PERRY,RH & GREEN,D (1984)
DC	pKa			
HL 2.29E-003	atm m3/mol	25	EST	MEYLAN,WM & HOWARD,PH (1991)
OH 1.70E-012	cm3/molc sec	25	EST	MEYLAN,WM & HOWARD,PH (1993)

CAS #: 000110-53-2				1-BROMOPENTANE

Formula: $C_5H_{11}Br$
Mol Weight: 151.05
MP (deg C): -95
FP (deg C):
BP (deg C): 129.7
BP pressure (mm Hg): 7.40E+002

Property/ Value	Units	Temp	Data Type	Reference
WS 1.27E+002	mg/L	25	EXP	YALKOWSKY,SH & DANNENFELSER,RM (1992)
logP 3.37			EXP	HANSCH,C & LEO,AJ (1985)
VP 1.26E+001	mm Hg	25	EXP	RIDDICK,JA ET AL. (1986)
DC	pKa			
HL 1.97E-002	atm m3/mol	25	EST	VP/WSOL
OH 3.58E-012	cm3/molc sec	25	EST	MEYLAN,WM & HOWARD,PH (1993)

CAS #: 000110-54-3				N-HEXANE

Formula: C_6H_{14}
Mol Weight: 86.18
MP (deg C): -95.3
FP (deg C):
BP (deg C): 69
BP pressure (mm Hg):

Property/ Value	Units	Temp	Data Type	Reference
WS 9.50E+000	mg/L	25	EXP	MCAULIFFE,C (1966)
logP 3.90			EXP	HANSCH,C & LEO,AJ (1985)
VP 1.51E+002	mm Hg	25	EXP	BOUBLIK,T ET AL. (1984)
DC	pKa			
HL 1.80E+000	atm m3/mol	25	EST	VP/WSOL
OH 5.61E-012	cm3/molc sec	25	EXP	ATKINSON,R (1989)

CAS #: 000110-56-5				1,4-DICHLOROBUTANE

Formula: $C_4H_8Cl_2$
Mol Weight: 127.01
MP (deg C): -37.3
FP (deg C):
BP (deg C): 161
BP pressure (mm Hg):

Property/ Value	Units	Temp	Data Type	Reference
WS 1.53E+004	mg/L		EXP	YALKOWSKY,SH & DANNENFELSER,RM (1992)
logP 2.81			EST	MEYLAN,WM & HOWARD,PH (1995)
VP 4.13E+000	mm Hg	25	EXP	DAUBERT,TE & DANNER,RP (1989)
DC	pKa			
HL 4.86E-004	atm m3/mol	25	EXP	LEIGHTON,DTJR & CALO,JM (1981)
OH 2.05E-012	cm3/molc sec	25	EST	MEYLAN,WM & HOWARD,PH (1993)

CAS #: 000110-57-6				1,4-DICHLORO-2-BUTENE(TRANS)

Formula: $C_4H_6Cl_2$
Mol Weight: 125.00
MP (deg C): 2.0
FP (deg C):
BP (deg C): 155.4
BP pressure (mm Hg):

Property/ Value	Units	Temp	Data Type	Reference
WS 8.50E+002	mg/L	25	EXP	ALBANESE,V ET AL. (1987)
logP 2.60			EST	MEYLAN,WM & HOWARD,PH (1995)
VP 3.43E+000	mm Hg	25	EXP	DAUBERT,TE & DANNER,RP (1989)
DC	pKa			
HL 6.64E-004	atm m3/mol	25	EST	VP/WSOL
OH 3.33E-011	cm3/molc sec	25	EST	MEYLAN,WM & HOWARD,PH (1993)

CAS #: 000110-58-7				N-PENTYLAMINE

Formula: $C_5H_{13}N$
Mol Weight: 87.17
MP (deg C): -55
FP (deg C):
BP (deg C): 104
BP pressure (mm Hg):

Property/ Value	Units	Temp	Data Type	Reference
WS 1.00E+006	mg/L		EXP	RIDDICK,JA ET AL. (1986)
logP 1.49			EXP	HANSCH,C & LEO,AJ (1985)
VP 3.00E+001	mm Hg	25	EXP	YAWS,CL (1994)
DC 10.63	pKa	25	EXP	PERRIN,DD (1965)
HL 2.43E-005	atm m3/mol	25	EXP	CHRISTIE,AO & CRISP,DJ (1967)
OH 3.49E-011	cm3/molc sec	25	EST	MEYLAN,WM & HOWARD,PH (1993)

CAS #: 000110-59-8 — VALERONITRILE

Formula: C_5H_9N

Mol Weight: 83.13

MP (deg C): -96.2

FP (deg C):

BP (deg C): 141.3

BP pressure (mm Hg):

Property/Value	Units	Temp	Data Type	Reference
WS 7.75E+003	mg/L	25	EST	MEYLAN,WM ET AL. (1996)
logP 1.12			EXP	HANSCH,C ET AL. (1995)
VP 7.30E+000	mm Hg	25	EXP	DAUBERT,TE & DANNER,RP (1989)
DC	pKa			
HL 7.15E-005	atm m3/mol	25	EST	MEYLAN,WM & HOWARD,PH (1991)
OH 1.67E-012	cm3/molc sec	25	EST	MEYLAN,WM & HOWARD,PH (1993)

CAS #: 000110-60-1 — 1,4-BUTANEDIAMINE

Formula: $C_4H_{12}N_2$

Mol Weight: 88.15

MP (deg C): 23-24

FP (deg C):

BP (deg C): 158-160

BP pressure (mm Hg):

Property/Value	Units	Temp	Data Type	Reference
WS 1.00E+006	mg/L	25	EST	MEYLAN,WM ET AL. (1996)
logP -0.70			EXP	SANGSTER,J (1994)
VP 4.12E+000	mm Hg	25	EST	NEELY,WB & BLAU,GE (1985)
DC 10.80	pKa	20	EXP	PERRIN,DD (1965)
HL 1.82E-009	atm m3/mol	25	EST	MEYLAN,WM & HOWARD,PH (1991)
OH 6.62E-011	cm3/molc sec	25	EST	MEYLAN,WM & HOWARD,PH (1993)

CAS #: 000110-61-2 — SUCCINONITRILE

Formula: $C_4H_4N_2$

Mol Weight: 80.09

MP (deg C): 57.15

FP (deg C):

BP (deg C): 265-267

BP pressure (mm Hg):

Property/Value	Units	Temp	Data Type	Reference
WS 1.28E+005	mg/L	25	EXP	VERSCHUEREN,K (1977)
logP -0.99			EXP	TANII,H & HASHIMOTO,K (1985)
VP 5.39E-003	mm Hg		EXT	YAWS,CL (1994)
DC	pKa			
HL 1.68E-008	atm m3/mol	25	EST	MEYLAN,WM & HOWARD,PH (1991)
OH 4.26E-014	cm3/molc sec	25	EST	MEYLAN,WM & HOWARD,PH (1993)

CAS #: 000110-62-3 — 1-PENTANAL

Formula: $C_5H_{10}O$

Mol Weight: 86.13

MP (deg C): -91.5

FP (deg C):

BP (deg C): 103

BP pressure (mm Hg):

Property/Value	Units	Temp	Data Type	Reference
WS 1.40E+001	mg/L	20	EXP	FABLE,J ET AL. (1985)
logP 1.31			EST	MEYLAN,WM & HOWARD,PH (1995)
VP 2.60E+001	mm Hg	20	EXP	FLICK,EW (1991)
DC	pKa			
HL 1.47E-004	atm m3/mol	25	EXP	BUTTERY,RG ET AL. (1969)
OH 2.85E-011	cm3/molc sec	25	EXP	ATKINSON,R (1989)

CAS #: 000110-63-4 — 1,4-BUTANEDIOL

Formula: $C_4H_{10}O_2$

Mol Weight: 90.12

MP (deg C): 20

FP (deg C):

BP (deg C): 235

BP pressure (mm Hg):

Property/Value	Units	Temp	Data Type	Reference
WS 1.00E+006	mg/L	20	EXP	YALKOWSKY,SH & DANNENFELSER,RM (1992)
logP -0.83			EXP	HANSCH,C ET AL. (1995)
VP 1.05E-002	mm Hg	25	EXP	DAUBERT,TE & DANNER,RP (1989)
DC 14.50	pKa	25	EXP	RIDDICK,JA ET AL. (1986)
HL 1.30E-009	atm m3/mol	25	EST	VP/WSOL
OH 1.02E-011	cm3/molc sec	25	EST	MEYLAN,WM & HOWARD,PH (1993)

CAS #: 000110-64-5 — 2-BUTENE-1,4-DIOL

Formula: $C_4H_8O_2$

Mol Weight: 88.11

MP (deg C):

FP (deg C):

BP (deg C):

BP pressure (mm Hg):

Property/Value	Units	Temp	Data Type	Reference
WS 6.19E+005	mg/L	25	EST	MEYLAN,WM ET AL. (1996)
logP -0.81			EXP	HANSCH,C ET AL. (1995)
VP 4.62E-002	mm Hg	25	EST	NEELY,WB & BLAU,GE (1985)
DC	pKa			
HL 2.03E-007	atm m3/mol	25	EST	MEYLAN,WM & HOWARD,PH (1991)
OH 6.32E-011	cm3/molc sec	25	EST	MEYLAN,WM & HOWARD,PH (1993)

CAS #: 000110-65-6 — 2-BUTYNE-1,4-DIOL

Formula: $C_4H_6O_2$

Mol Weight: 86.09

MP (deg C): 56-58

FP (deg C):

BP (deg C): 238

BP pressure (mm Hg):

Property/Value	Units	Temp	Data Type	Reference
WS 3.74E+005	mg/L	25	EXP	HORT,EV (1978)
logP -0.93			EST	MEYLAN,WM & HOWARD,PH (1995)
VP 5.56E-004	mm Hg	25	EXT	YAWS,CL (1994)
DC	pKa			
HL 1.44E-008	atm m3/mol	25	EST	MEYLAN,WM & HOWARD,PH (1991)
OH 3.38E-011	cm3/molc sec	25	EST	MEYLAN,WM & HOWARD,PH (1993)

CAS #: 000110-66-7 — 1-PENTANETHIOL

Formula: $C_5H_{12}S$

Mol Weight: 104.22

MP (deg C): -75.7

FP (deg C):

BP (deg C): 126.6

BP pressure (mm Hg):

Property/Value	Units	Temp	Data Type	Reference
WS 5.19E+002	mg/L	25	EST	MEYLAN,WM ET AL. (1996)
logP 2.74			EST	MEYLAN,WM & HOWARD,PH (1995)
VP 1.38E+001	mm Hg	25	EXP	DAUBERT,TE & DANNER,RP (1989)
DC	pKa			
HL 8.12E-003	atm m3/mol	25	EST	MEYLAN,WM & HOWARD,PH (1991)
OH 4.56E-011	cm3/molc sec	25	EST	MEYLAN,WM & HOWARD,PH (1993)

CAS #:	000110-68-9		METHYLBUTYLAMINE

Formula: $C_5H_{13}N$

Mol Weight: 87.17

MP (deg C): | FP (deg C):

BP (deg C): 91

BP pressure (mm Hg):

Property/ Value	Units	Temp	Data Type	Reference
WS 9.42E+004	mg/L	25	EST	MEYLAN,WM ET AL. (1996)
logP 1.33			EXP	HANSCH,C & LEO,AJ (1985)
VP 5.05E+001	mm Hg	25	EST	NEELY,WB & BLAU,GE (1985)
DC	pKa			
HL 3.89E-005	atm m3/mol	25	EST	MEYLAN,WM & HOWARD,PH (1991)
OH 7.77E-011	cm3/molc sec	25	EST	MEYLAN,WM & HOWARD,PH (1993)

CAS #:	000110-70-3		1,2-ETHANEDIAMINE, N,N'-DIMETHYL-

Formula: $C_4H_{12}N_2$

Mol Weight: 88.15

MP (deg C): | FP (deg C):

BP (deg C): 120

BP pressure (mm Hg):

Property/ Value	Units	Temp	Data Type	Reference
WS 1.00E+006	mg/L	25	EST	MEYLAN,WM ET AL. (1996)
logP -0.62			EXP	HANSCH,C ET AL. (1995)
VP 2.06E+001	mm Hg	25	EST	NEELY,WB & BLAU,GE (1985)
DC 10.16	pKa	25	EXP	PERRIN,DD (1965)
HL 4.99E-009	atm m3/mol	25	EST	MEYLAN,WM & HOWARD,PH (1991)
OH 1.50E-010	cm3/molc sec	25	EST	MEYLAN,WM & HOWARD,PH (1993)

CAS #:	000110-71-4		ETHYLENE GLYCOL DIMETHYL ETHER

Formula: $C_4H_{10}O_2$

Mol Weight: 90.12

MP (deg C): -58 | FP (deg C):

BP (deg C): 82-83

BP pressure (mm Hg):

Property/ Value	Units	Temp	Data Type	Reference
WS 1.00E+006	mg/L	25	EXP	RIDDICK,JA ET AL. (1986)
logP -0.21			EXP	HANSCH,C ET AL. (1995)
VP 4.80E+001	mm Hg	20	EXP	RIDDICK,JA ET AL. (1986)
DC	pKa			
HL 1.10E-006	atm m3/mol	25	EST	MEYLAN,WM & HOWARD,PH (1991)
OH 1.57E-011	cm3/molc sec	25	EST	MEYLAN,WM & HOWARD,PH (1993)

CAS #:	000110-74-7		PROPYL FORMATE

Formula: $C_4H_8O_2$

Mol Weight: 88.11

MP (deg C): -93 | FP (deg C):

BP (deg C): 81-82

BP pressure (mm Hg):

Property/ Value	Units	Temp	Data Type	Reference
WS 2.20E+004	mg/L	22	EXP	YALKOWSKY,SH & DANNENFELSER,RM (1992)
logP 0.83			EXP	HANSCH,C & LEO,AJ (1985)
VP 8.26E+001	mm Hg	25	EXP	YAWS,CL (1994)
DC	pKa			
HL 4.35E-004	atm m3/mol	25	EST	VP/WSOL
OH 2.40E-012	cm3/molc sec	25	EXP	ATKINSON,R (1989)

CAS #:	000110-75-8		2-CHLOROETHYL VINYL ETHER

Formula: C_4H_7ClO

Mol Weight: 106.55

MP (deg C): -70 | FP (deg C):

BP (deg C): 108

BP pressure (mm Hg):

Property/ Value	Units	Temp	Data Type	Reference
WS 4.29E+002	mg/L	25	EXP	ENFIELD,CG ET AL. (1986)
logP 1.17			EST	MEYLAN,WM & HOWARD,PH (1995)
VP 2.68E+001	mm Hg	20	EXP	CALLAHAN,MA ET AL. (1979A)
DC	pKa			
HL 8.76E-003	atm m3/mol	25	EST	VP/WSOL
OH 3.75E-011	cm3/molc sec	25	EST	MEYLAN,WM & HOWARD,PH (1993)

CAS #:	000110-80-5		2-ETHOXYETHANOL

Formula: $C_4H_{10}O_2$

Mol Weight: 90.12

MP (deg C): -70 | FP (deg C): -90

BP (deg C): 135

BP pressure (mm Hg):

Property/ Value	Units	Temp	Data Type	Reference
WS 1.00E+006	mg/L		EXP	DOW CHEMICAL COMPANY (1981)
logP -0.32			EXP	HANSCH,C ET AL. (1995)
VP 5.31E+000	mm Hg	25	EXP	DAUBERT,TE & DANNER,RP (1989)
DC 14.80	pKa		EXP	RIDDICK,JA ET AL. (1986)
HL 1.00E-008	atm m3/mol	25	EST	MEYLAN,WM & HOWARD,PH (1991)
OH 1.20E-011	cm3/molc sec	25	EXP	ATKINSON,R (1989)

CAS #:	000110-81-6		DIETHYL DISULFIDE

Formula: $C_4H_{10}S_2$

Mol Weight: 122.25

MP (deg C): -101.5 | FP (deg C):

BP (deg C): 154.1

BP pressure (mm Hg):

Property/ Value	Units	Temp	Data Type	Reference
WS 3.59E+002	mg/L	25	EST	MEYLAN,WM ET AL. (1996)
logP 2.86			EST	MEYLAN,WM & HOWARD,PH (1995)
VP 4.28E+000	mm Hg	25	EXP	YAWS,CL (1994)
DC	pKa			
HL 2.15E-003	atm m3/mol	20	EXP	VITENBERG,AG ET AL. (1975)
OH 2.15E+002	cm3/molc sec	25	EST	MEYLAN,WM & HOWARD,PH (1993)

CAS #:	000110-82-7		CYCLOHEXANE

Formula: C_6H_{12}

Mol Weight: 84.16

MP (deg C): 6.554 | FP (deg C):

BP (deg C): 80.738

BP pressure (mm Hg):

Property/ Value	Units	Temp	Data Type	Reference
WS 5.50E+001	mg/L	25	EXP	MCAULIFFE,C (1966)
logP 3.44			EXP	HANSCH,C & LEO,AJ (1985)
VP 9.69E+001	mm Hg	25	EXP	CHAO,J ET AL. (1983)
DC	pKa			
HL 1.50E-003	atm m3/mol	25	EXP	BOCEK,K (1976)
OH 7.49E-012	cm3/molc sec	25	EXP	ATKINSON,R (1989)

CYCLOHEXENE

CAS #:	000110-83-8			
Formula:	C_6H_{10}			
Mol Weight:	82.15			
MP (deg C):	-103.5	FP (deg C):		
BP (deg C):	83			
BP pressure (mm Hg):				

Property/Value	Units	Temp	Data Type	Reference
WS 2.13E+002	mg/L	25	EXP	YALKOWSKY,SH & DANNENFELSER,RM (1992)
logP 2.86			EXP	HANSCH,C & LEO,AJ (1985)
VP 8.90E+001	mm Hg	25	EXP	DAUBERT,TE & DANNER,RP (1989)
DC	pKa			
HL 4.55E-002	atm m3/mol	25	EXP	HINE,J & MOOKERJEE,PK (1975)
OH 6.77E-011	cm3/molc sec	25	EXP	ATKINSON,R (1989)

PIPERAZINE

CAS #:	000110-85-0			
Formula:	$C_4H_{10}N_2$			
Mol Weight:	86.14			
MP (deg C):	106	FP (deg C):		
BP (deg C):	146			
BP pressure (mm Hg):				

Property/Value	Units	Temp	Data Type	Reference
WS 1.00E+006	mg/L	25	EST	MEYLAN,WM ET AL. (1996)
logP -1.50			EXP	HANSCH,C ET AL. (1995)
VP 1.60E-001	mm Hg	20	EXP	MJOS,K (1978)
DC 9.73	pKa		EXP	PERRIN,DD (1972)
HL 2.20E-009	atm m3/mol	25	EST	MEYLAN,WM & HOWARD,PH (1991)
OH 1.69E-010	cm3/molc sec	25	EST	MEYLAN,WM & HOWARD,PH (1993)

PYRIDINE

CAS #:	000110-86-1			
Formula:	C_5H_5N			
Mol Weight:	79.10			
MP (deg C):	-41.6	FP (deg C):		
BP (deg C):	115-116			
BP pressure (mm Hg):				

Property/Value	Units	Temp	Data Type	Reference
WS 1.00E+006	mg/L	25	EXP	YALKOWSKY,SH & DANNENFELSER,RM (1992)
logP 0.65			EXP	HANSCH,C & LEO,AJ (1985)
VP 2.08E+001	mm Hg	25	EXP	DAUBERT,TE & DANNER,RP (1989)
DC 5.23	pKa		EXP	BINTEIN,S & DEVILLERS,J (1994)
HL 1.10E-005	atm m3/mol	25	EXP	HAWTHORNE,SB ET AL. (1985)
OH 3.70E-013	cm3/molc sec	25	EXP	ATKINSON,R (1989)

2H-PYRAN, 3,4-DIHYDRO-

CAS #:	000110-87-2			
Formula:	C_5H_8O			
Mol Weight:	84.12			
MP (deg C):		FP (deg C):		
BP (deg C):	86			
BP pressure (mm Hg):				

Property/Value	Units	Temp	Data Type	Reference
WS 3.31E+004	mg/L	25	EST	MEYLAN,WM ET AL. (1996)
logP 0.69			EXP	HANSCH,C ET AL. (1995)
VP 7.49E+002	mm Hg		EXP	FLICK,EW (1991)
DC	pKa			
HL 5.26E-003	atm m3/mol	25	EST	MEYLAN,WM & HOWARD,PH (1991)
OH 8.55E-011	cm3/molc sec	25	EST	MEYLAN,WM & HOWARD,PH (1993)

1,3,5-TRIOXANE

CAS #:	000110-88-3			
Formula:	$C_3H_6O_3$			
Mol Weight:	90.08			
MP (deg C):	64	FP (deg C):		
BP (deg C):	114.5			
BP pressure (mm Hg):	7.59E+002			

Property/Value	Units	Temp	Data Type	Reference
WS 1.75E+005	mg/L	25	EXP	YALKOWSKY,SH & DANNENFELSER,RM (1992)
logP -0.43			EXP	HANSCH,C ET AL. (1995)
VP 1.75E+001	mm Hg	25	EXT	YAWS,CL (1994)
DC	pKa			
HL 1.97E-007	atm m3/mol	25	EST	MEYLAN,WM & HOWARD,PH (1991)
OH 6.20E-012	cm3/molc sec	25	EXP	ATKINSON,R (1989)

PIPERADINE

CAS #:	000110-89-4			
Formula:	$C_5H_{11}N$			
Mol Weight:	85.15			
MP (deg C):	-7	FP (deg C):	-11	
BP (deg C):	106.3			
BP pressure (mm Hg):				

Property/Value	Units	Temp	Data Type	Reference
WS 1.00E+006	mg/L	20	EXP	GOE,GL (1982)
logP 0.84			EXP	HANSCH,C & LEO,AJ (1985)
VP 3.21E+001	mm Hg	25	EXP	YAWS,CL (1994)
DC 11.28	pKa	20	EXP	PERRIN,DD (1965)
HL 4.45E-006	atm m3/mol	25	EXP	CABANI,S ET AL. (1971)
OH 8.58E-011	cm3/molc sec	25	EST	MEYLAN,WM & HOWARD,PH (1993)

MORPHOLINE

CAS #:	000110-91-8			
Formula:	C_4H_9NO			
Mol Weight:	87.12			
MP (deg C):	-4.9	FP (deg C):		
BP (deg C):	128.9			
BP pressure (mm Hg):				

Property/Value	Units	Temp	Data Type	Reference
WS 1.00E+006	mg/L		EXP	RIDDICK,JA ET AL. (1986)
logP -0.86			EXP	HANSCH,C & LEO,AJ (1985)
VP 1.01E+001	mm Hg	25	EXP	RIDDICK,JA ET AL. (1986)
DC 8.49	pKa	25	EXP	PERRIN,DD (1972)
HL 1.16E-006	atm m3/mol	25	EST	VP/WSOL
OH 1.49E-010	cm3/molc sec	25	EST	MEYLAN,WM & HOWARD,PH (1993)

PENTANEDIOC ACID

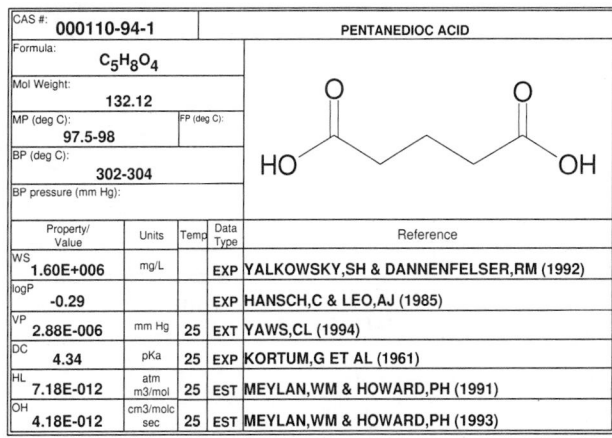

CAS #:	000110-94-1			
Formula:	$C_5H_8O_4$			
Mol Weight:	132.12			
MP (deg C):	97.5-98	FP (deg C):		
BP (deg C):	302-304			
BP pressure (mm Hg):				

Property/Value	Units	Temp	Data Type	Reference
WS 1.60E+006	mg/L		EXP	YALKOWSKY,SH & DANNENFELSER,RM (1992)
logP -0.29			EXP	HANSCH,C & LEO,AJ (1985)
VP 2.88E-006	mm Hg	25	EXT	YAWS,CL (1994)
DC 4.34	pKa	25	EXP	KORTUM,G ET AL (1961)
HL 7.18E-012	atm m3/mol	25	EST	MEYLAN,WM & HOWARD,PH (1991)
OH 4.18E-012	cm3/molc sec	25	EST	MEYLAN,WM & HOWARD,PH (1993)

CAS #: 000110-96-3				DIISOBUTYLAMINE

Formula: $C_8H_{19}N$

Mol Weight: 129.25

MP (deg C): -77   FP (deg C):

BP (deg C): 139-140

BP pressure (mm Hg):

Property/ Value	Units	Temp	Data Type	Reference
WS 2.20E+003	mg/L	25	EXP	CHEM INSPECT TEST INST (1992)
logP 2.63			EST	MEYLAN,WM & HOWARD,PH (1995)
VP 7.27E+000	mm Hg	25	EXP	YAWS,CL (1994)
DC 10.91	pKa	21	EXP	PERRIN,DD (1965)
HL 5.62E-004	atm m3/mol	25	EST	VP/WSOL
OH 8.98E-011	cm3/molc sec	25	EST	MEYLAN,WM & HOWARD,PH (1993)

CAS #: 000110-97-4				DIISOPROPANOLAMINE

Formula: $C_6H_{15}NO_2$

Mol Weight: 133.19

MP (deg C): 44.5   FP (deg C):

BP (deg C): 250

BP pressure (mm Hg):

Property/ Value	Units	Temp	Data Type	Reference
WS 8.60E+005	mg/L	25	EXP	KUHNE,R ET AL. (1995)
logP -0.82			EXP	HANSCH,C & LEO,AJ (1985)
VP 1.25E-004	mm Hg	25	EXT	YAWS,CL (1994)
DC	pKa			
HL 6.91E-011	atm m3/mol	25	EST	MEYLAN,WM & HOWARD,PH (1991)
OH 1.02E-010	cm3/molc sec	25	EST	MEYLAN,WM & HOWARD,PH (1993)

CAS #: 000111-06-8				BUTYL PALMITATE

Formula: $C_{20}H_{40}O_2$

Mol Weight: 312.54

MP (deg C): 16.9   FP (deg C):

BP (deg C):

BP pressure (mm Hg):

Property/ Value	Units	Temp	Data Type	Reference
WS 3.69E-004	mg/L	25	EST	MEYLAN,WM ET AL. (1996)
logP 8.72			EST	MEYLAN,WM & HOWARD,PH (1995)
VP 1.72E-005	mm Hg	25	EST	NEELY,WB & BLAU,GE (1985)
DC	pKa			
HL 2.17E-002	atm m3/mol	25	EST	MEYLAN,WM & HOWARD,PH (1991)
OH 2.37E-011	cm3/molc sec	25	EST	MEYLAN,WM & HOWARD,PH (1993)

CAS #: 000111-11-5				METHYL OCTANOATE

Formula: $C_9H_{18}O_2$

Mol Weight: 158.24

MP (deg C): -40   FP (deg C):

BP (deg C): 192.9

BP pressure (mm Hg):

Property/ Value	Units	Temp	Data Type	Reference
WS 6.44E+001	mg/L	20	EXP	YALKOWSKY,SH & DANNENFELSER,RM (1992)
logP 3.32			EST	MEYLAN,WM & HOWARD,PH (1995)
VP 5.40E-001	mm Hg	25	EXT	PERRY,RH & GREEN,D (1984)
DC	pKa			
HL 7.83E-004	atm m3/mol	25	EXP	BUTTERY,RG ET AL. (1969)
OH 7.06E-012	cm3/molc sec	25	EST	MEYLAN,WM & HOWARD,PH (1993)

CAS #: 000111-13-7				2-OCTANONE

Formula: $C_8H_{16}O$

Mol Weight: 128.22

MP (deg C): -16   FP (deg C): -20

BP (deg C): 172-173

BP pressure (mm Hg):

Property/ Value	Units	Temp	Data Type	Reference
WS 9.00E+002	mg/L	20	EXP	YALKOWSKY,SH & DANNENFELSER,RM (1992)
logP 2.37			EXP	HANSCH,C ET AL. (1995)
VP 1.35E+000	mm Hg	25	EXP	RIDDICK,JA ET AL. (1986)
DC	pKa			
HL 1.88E-004	atm m3/mol	25	EXP	BUTTERY,RG ET AL. (1969)
OH 1.00E-011	cm3/molc sec	25	EXP	ATKINSON,R (1989)

CAS #: 000111-14-8				HEPTANOIC ACID

Formula: $C_7H_{14}O_2$

Mol Weight: 130.19

MP (deg C): -55.8   FP (deg C):

BP (deg C): 173.8

BP pressure (mm Hg):

Property/ Value	Units	Temp	Data Type	Reference
WS 2.82E+003	mg/L	25	EXP	YALKOWSKY,SH & DANNENFELSER,RM (1992)
logP 2.42			EXP	SANGSTER,J (1994)
VP 5.55E-003	mm Hg	25	EXP	DAUBERT,TE & DANNER,RP (1989)
DC 4.89	pKa	25	EXP	KORTUM,G ET AL (1961)
HL 3.37E-007	atm m3/mol	25	EST	VP/WSOL
OH 6.94E-012	cm3/molc sec	25	EST	MEYLAN,WM & HOWARD,PH (1993)

CAS #: 000111-15-9				ETHOXYETHYLACETATE

Formula: $C_6H_{12}O_3$

Mol Weight: 132.16

MP (deg C): -61.7   FP (deg C): -62

BP (deg C): 156.4

BP pressure (mm Hg):

Property/ Value	Units	Temp	Data Type	Reference
WS 2.29E+005	mg/L	20	EXP	RIDDICK,JA ET AL. (1986)
logP 0.59			EST	MEYLAN,WM & HOWARD,PH (1995)
VP 2.34E+000	mm Hg	25	EXP	DAUBERT,TE & DANNER,RP (1989)
DC	pKa			
HL 1.78E-006	atm m3/mol	25	EST	VP/WSOL
OH 1.30E-011	cm3/molc sec	25	EXP	ATKINSON,R (1989)

CAS #: 000111-16-0				PIMELIC ACID

Formula: $C_7H_{12}O_4$

Mol Weight: 160.17

MP (deg C): 106   FP (deg C):

BP (deg C): 272

BP pressure (mm Hg): 1.00E+002

Property/ Value	Units	Temp	Data Type	Reference
WS 5.34E+005	mg/L	21	EXP	YALKOWSKY,SH & DANNENFELSER,RM (1992)
logP 0.61			EXP	HANSCH,C & LEO,AJ (1985)
VP 2.08E-005	mm Hg	25	EXT	PERRY,RH & GREEN,D (1984)
DC 4.51	pKa	25	EXP	KORTUM,G ET AL (1961)
HL 1.27E-011	atm m3/mol	25	EST	MEYLAN,WM & HOWARD,PH (1991)
OH 7.00E-012	cm3/molc sec	25	EST	MEYLAN,WM & HOWARD,PH (1993)

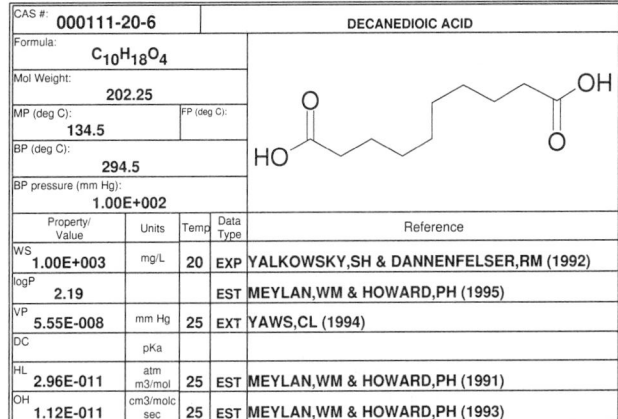

CAS #: 000111-20-6		DECANEDIOIC ACID			
Formula: $C_{10}H_{18}O_4$					
Mol Weight: 202.25					
MP (deg C): 134.5		FP (deg C):			
BP (deg C): 294.5					
BP pressure (mm Hg): 1.00E+002					

Property/Value	Units	Temp	Data Type	Reference
WS 1.00E+003	mg/L	20	EXP	YALKOWSKY,SH & DANNENFELSER,RM (1992)
logP 2.19			EST	MEYLAN,WM & HOWARD,PH (1995)
VP 5.55E-008	mm Hg	25	EXT	YAWS,CL (1994)
DC	pKa			
HL 2.96E-011	atm m3/mol	25	EST	MEYLAN,WM & HOWARD,PH (1991)
OH 1.12E-011	cm3/molc sec	25	EST	MEYLAN,WM & HOWARD,PH (1993)

CAS #: 000111-21-7		ETHYLENEBIS(OXYETHYLENE)DIACETATE			
Formula: $C_{10}H_{18}O_6$					
Mol Weight: 234.25					
MP (deg C): -50		FP (deg C):			
BP (deg C): 286					
BP pressure (mm Hg):					

Property/Value	Units	Temp	Data Type	Reference
WS 3.83E+004	mg/L	25	EST	MEYLAN,WM ET AL. (1996)
logP -0.79			EST	HANSCH,C & LEO,AJ (1981)
VP 1.35E-002	mm Hg	25	EST	NEELY,WB & BLAU,GE (1985)
DC	pKa			
HL 1.34E-010	atm m3/mol	25	EST	MEYLAN,WM & HOWARD,PH (1991)
OH 3.19E-011	cm3/molc sec	25	EST	MEYLAN,WM & HOWARD,PH (1993)

CAS #: 000111-25-1		1-BROMOHEXANE			
Formula: $C_6H_{13}Br$					
Mol Weight: 165.08					
MP (deg C): -84.7		FP (deg C):			
BP (deg C): 155.3					
BP pressure (mm Hg):					

Property/Value	Units	Temp	Data Type	Reference
WS 2.58E+001	mg/L	25	EXP	YALKOWSKY,SH & DANNENFELSER,RM (1992)
logP 3.80			EXP	HANSCH,C & LEO,AJ (1985)
VP 3.90E+000	mm Hg	25	EXP	LI,JCM & ROSSINI,FD (1953)
DC	pKa			
HL 3.28E-002	atm m3/mol	25	EST	VP/WSOL
OH 4.99E-012	cm3/molc sec	25	EST	MEYLAN,WM & HOWARD,PH (1993)

CAS #: 000111-26-2		N-HEXYLAMINE			
Formula: $C_6H_{15}N$					
Mol Weight: 101.19					
MP (deg C): -22.9		FP (deg C):			
BP (deg C): 132.8					
BP pressure (mm Hg):					

Property/Value	Units	Temp	Data Type	Reference
WS 2.06E+004	mg/L	25	EST	MEYLAN,WM ET AL. (1996)
logP 2.06			EXP	HANSCH,C & LEO,AJ (1985)
VP 8.99E+000	mm Hg	25	EXP	DAUBERT,TE & DANNER,RP (1989)
DC 10.64	pKa	25	EXP	PERRIN,DD (1965)
HL 2.68E-005	atm m3/mol	25	EXP	CHRISTIE,AO & CRISP,DJ (1967)
OH 3.63E-011	cm3/molc sec	25	EST	MEYLAN,WM & HOWARD,PH (1993)

CAS #: 000111-27-3		1-HEXANOL			
Formula: $C_6H_{14}O$					
Mol Weight: 102.18					
MP (deg C): -44.6		FP (deg C):			
BP (deg C): 157.1					
BP pressure (mm Hg):					

Property/Value	Units	Temp	Data Type	Reference
WS 5.90E+003	mg/L	25	EXP	YALKOWSKY,SH & DANNENFELSER,RM (1992)
logP 2.03			EXP	HANSCH,C & LEO,AJ (1985)
VP 9.28E-001	mm Hg	25	EXP	DAUBERT,TE & DANNER,RP (1989)
DC	pKa			
HL 1.71E-005	atm m3/mol	25	EXP	BUTTERY,RG ET AL. (1969)
OH 1.24E-011	cm3/molc sec	25	EXP	ATKINSON,R (1989)

CAS #: 000111-29-5		1,5-PENTANEDIOL			
Formula: $C_5H_{12}O_2$					
Mol Weight: 104.15					
MP (deg C): -18		FP (deg C):			
BP (deg C): 239					
BP pressure (mm Hg):					

Property/Value	Units	Temp	Data Type	Reference
WS 6.63E+004	mg/L	25	EST	MEYLAN,WM ET AL. (1996)
logP 0.27			EST	MEYLAN,WM & HOWARD,PH (1995)
VP 3.90E-003	mm Hg	25	EXP	DAUBERT,TE & DANNER,RP (1989)
DC	pKa			
HL 3.06E-007	atm m3/mol	25	EST	MEYLAN,WM & HOWARD,PH (1991)
OH 1.26E-011	cm3/molc sec	25	EST	MEYLAN,WM & HOWARD,PH (1993)

CAS #: 000111-30-8		PENTANE-1,5-DIAL			
Formula: $C_5H_8O_2$					
Mol Weight: 100.12					
MP (deg C):		FP (deg C):			
BP (deg C): 188 dec					
BP pressure (mm Hg):					

Property/Value	Units	Temp	Data Type	Reference
WS 1.67E+005	mg/L	25	EST	MEYLAN,WM ET AL. (1996)
logP -0.18			EST	MEYLAN,WM & HOWARD,PH (1995)
VP 6.00E-001	mm Hg	30	EXP	HEISLER,SL & FRIEDLANDER,SK (1977)
DC	pKa			
HL 1.10E-007	atm m3/mol	25	EST	MEYLAN,WM & HOWARD,PH (1991)
OH 2.38E-011	cm3/molc sec	25	EXP	ATKINSON,R (1989)

CAS #: 000111-31-9		1-HEXANETHIOL			
Formula: $C_6H_{14}S$					
Mol Weight: 118.24					
MP (deg C): -81		FP (deg C):			
BP (deg C): 151					
BP pressure (mm Hg):					

Property/Value	Units	Temp	Data Type	Reference
WS 1.77E+002	mg/L	25	EST	MEYLAN,WM ET AL. (1996)
logP 3.23			EST	MEYLAN,WM & HOWARD,PH (1995)
VP 4.20E+000	mm Hg	25	EXT	OHE,S (1976)
DC	pKa			
HL 1.08E-002	atm m3/mol	25	EST	MEYLAN,WM & HOWARD,PH (1991)
OH 4.70E-011	cm3/molc sec	25	EST	MEYLAN,WM & HOWARD,PH (1993)

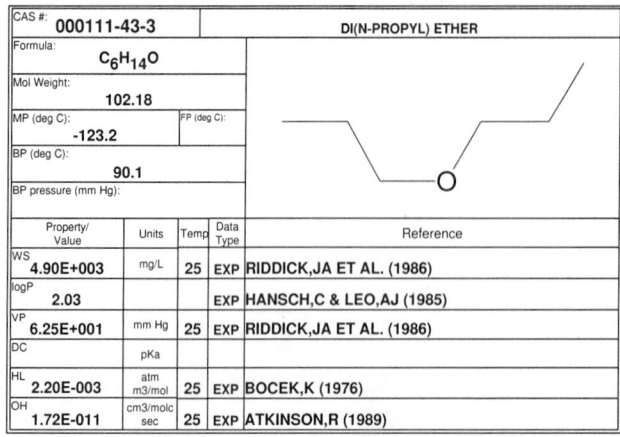

CAS #: 000111-32-0 — 1,3-BUTYLENE GLYCOL METHYL ETHER

Formula: $C_5H_{12}O_2$
Mol Weight: 104.15
MP (deg C):
FP (deg C):
BP (deg C):
BP pressure (mm Hg):

Property/Value	Units	Temp	Data Type	Reference
WS 3.17E+005	mg/L	25	EST	MEYLAN,WM ET AL. (1996)
logP 0.08			EST	MEYLAN,WM & HOWARD,PH (1995)
VP 7.71E-001	mm Hg	25	EST	NEELY,WB & BLAU,GE (1985)
DC	pKa			
HL 7.38E-008	atm m3/mol	25	EST	MEYLAN,WM & HOWARD,PH (1991)
OH 1.74E-011	cm3/molc sec	25	EST	MEYLAN,WM & HOWARD,PH (1993)

CAS #: 000111-34-2 — BUTYL VINYL ETHER

Formula: $C_6H_{12}O$
Mol Weight: 100.16
MP (deg C): -92
FP (deg C):
BP (deg C): 94
BP pressure (mm Hg):

Property/Value	Units	Temp	Data Type	Reference
WS 3.00E+003	mg/L	20	EXP	RIDDICK,JA ET AL. (1986)
logP 1.89			EST	MEYLAN,WM & HOWARD,PH (1995)
VP 4.91E+001	mm Hg	25	EXP	DAUBERT,TE & DANNER,RP (1989)
DC	pKa			
HL 2.16E-003	atm m3/mol	25	EST	VP/WSOL
OH 4.65E-011	cm3/molc sec	25	EST	MEYLAN,WM & HOWARD,PH (1993)

CAS #: 000111-35-3 — 3-ETHOXY-1-PROPANOL

Formula: $C_5H_{12}O_2$
Mol Weight: 104.15
MP (deg C): 160-161
FP (deg C):
BP (deg C):
BP pressure (mm Hg):

Property/Value	Units	Temp	Data Type	Reference
WS 3.17E+005	mg/L	25	EST	MEYLAN,WM ET AL. (1996)
logP 0.08			EST	MEYLAN,WM & HOWARD,PH (1995)
VP 7.71E-001	mm Hg	25	EST	NEELY,WB & BLAU,GE (1985)
DC	pKa			
HL 7.38E-008	atm m3/mol	25	EST	MEYLAN,WM & HOWARD,PH (1991)
OH 2.20E-011	cm3/molc sec	25	EXP	ATKINSON,R (1989)

CAS #: 000111-36-4 — BUTYL ISOCYANATE

Formula: C_5H_9NO
Mol Weight: 99.13
MP (deg C): <-70
FP (deg C):
BP (deg C): 115
BP pressure (mm Hg):

Property/Value	Units	Temp	Data Type	Reference
WS 1.38E+003	mg/L	25	EST	MEYLAN,WM ET AL. (1996)
logP 2.26			EST	MEYLAN,WM & HOWARD,PH (1995)
VP 1.76E+001	mm Hg	25	EXP	DAUBERT,TE & DANNER,RP (1989)
DC	pKa			
HL 2.17E-003	atm m3/mol	25	EST	MEYLAN,WM & HOWARD,PH (1991)
OH 3.74E-012	cm3/molc sec	25	EST	MEYLAN,WM & HOWARD,PH (1993)

CAS #: 000111-40-0 — DIETHYLENETRIAMINE

Formula: $C_4H_{13}N_3$
Mol Weight: 103.17
MP (deg C): -39
FP (deg C):
BP (deg C): 206.9
BP pressure (mm Hg):

Property/Value	Units	Temp	Data Type	Reference
WS 1.00E+006	mg/L		EXP	RIDDICK,JA ET AL. (1986)
logP -2.13			EST	MEYLAN,WM & HOWARD,PH (1995)
VP 2.32E-001	mm Hg	25	EXP	DAUBERT,TE & DANNER,RP (1989)
DC 10.45	pKa	20	EXP	RIDDICK,JA ET AL. (1986)
HL 3.15E-007	atm m3/mol	25	EST	VP/WSOL
OH 1.43E-010	cm3/molc sec	25	EST	MEYLAN,WM & HOWARD,PH (1993)

CAS #: 000111-41-1 — 2-(2-AMINOETHYLAMINO)ETHANOL

Formula: $C_4H_{12}N_2O$
Mol Weight: 104.15
MP (deg C):
FP (deg C):
BP (deg C): 243.8
BP pressure (mm Hg):

Property/Value	Units	Temp	Data Type	Reference
WS 1.00E+006	mg/L	20	EXP	YALKOWSKY,SH & DANNENFELSER,RM (1992)
logP -2.13			EST	MEYLAN,WM & HOWARD,PH (1995)
VP 8.19E-004	mm Hg	25	EXP	DAUBERT,TE & DANNER,RP (1989)
DC 7.21	pKa	25	EXP	DEAN,J (1985)
HL 1.10E-013	atm m3/mol	25	EST	MEYLAN,WM & HOWARD,PH (1991)
OH 1.16E-010	cm3/molc sec	25	EST	ATKINSON,R (1988)

CAS #: 000111-42-2 — DIETHANOLAMINE

Formula: $C_4H_{11}NO_2$
Mol Weight: 105.14
MP (deg C): 27.95
FP (deg C):
BP (deg C): 268.39
BP pressure (mm Hg):

Property/Value	Units	Temp	Data Type	Reference
WS 1.00E+006	mg/L		EXP	DOW CHEMICAL COMPANY (1980)
logP -1.43			EXP	HANSCH,C & LEO,AJ (1985)
VP 2.80E-004	mm Hg	25	EXP	DOW CHEMICAL COMPANY (1980)
DC 8.96	pKa	25	EXP	CHREMOS,G & ZIMMERMAN,HKJR (1959)
HL 3.87E-011	atm m3/mol	25	EST	VP/WSOL
OH 8.90E-011	cm3/molc sec	25	EST	MEYLAN,WM & HOWARD,PH (1993)

CAS #: 000111-43-3 — DI(N-PROPYL) ETHER

Formula: $C_6H_{14}O$
Mol Weight: 102.18
MP (deg C): -123.2
FP (deg C):
BP (deg C): 90.1
BP pressure (mm Hg):

Property/Value	Units	Temp	Data Type	Reference
WS 4.90E+003	mg/L	25	EXP	RIDDICK,JA ET AL. (1986)
logP 2.03			EXP	HANSCH,C & LEO,AJ (1985)
VP 6.25E+001	mm Hg	25	EXP	RIDDICK,JA ET AL. (1986)
DC	pKa			
HL 2.20E-003	atm m3/mol	25	EXP	BOCEK,K (1976)
OH 1.72E-011	cm3/molc sec	25	EXP	ATKINSON,R (1989)

CAS #: 000111-44-4				BIS(2-CHLOROETHYL) ETHER
Formula: $C_4H_8Cl_2O$				
Mol Weight: 143.01				
MP (deg C): -46.8		FP (deg C):		
BP (deg C): 178.75				
BP pressure (mm Hg):				

Property/Value	Units	Temp	Data Type	Reference
WS 1.72E+004	mg/L	20	EXP	VEITH,GD ET AL. (1980)
logP 1.29			EXP	HANSCH,C & LEO,AJ (1985)
VP 1.55E+000	mm Hg	25	EXP	RIDDICK,JA ET AL. (1986)
DC	pKa			
HL 1.70E-005	atm m3/mol	25	EST	VP/WSOL
OH 8.69E-012	cm3/molc sec	25	EST	ATKINSON,R (1988)

CAS #: 000111-45-5				ETHANOL, 2-(2-PROPENYLOXY)-
Formula: $C_5H_{10}O_2$				
Mol Weight: 102.13				
MP (deg C):		FP (deg C):		
BP (deg C): 158.5				
BP pressure (mm Hg):				

Property/Value	Units	Temp	Data Type	Reference
WS 4.54E+005	mg/L	25	EST	MEYLAN,WM ET AL. (1996)
logP -0.10			EXP	TANII,H ET AL. (1992)
VP 8.46E-001	mm Hg	25	EST	NEELY,WB & BLAU,GE (1985)
DC	pKa			
HL 5.50E-008	atm m3/mol	25	EST	MEYLAN,WM & HOWARD,PH (1991)
OH 4.32E-011	cm3/molc sec	25	EST	MEYLAN,WM & HOWARD,PH (1993)

CAS #: 000111-46-6				DIETHYLENE GLYCOL
Formula: $C_4H_{10}O_3$				
Mol Weight: 106.12				
MP (deg C): -6.5		FP (deg C):		
BP (deg C): 244-245				
BP pressure (mm Hg):				

Property/Value	Units	Temp	Data Type	Reference
WS 1.00E+006	mg/L	25	EXP	RIDDICK,JA ET AL. (1986)
logP -1.47			EST	MEYLAN,WM & HOWARD,PH (1995)
VP 5.70E-003	mm Hg	25	EXP	DAUBERT,TE & DANNER,RP (1991)
DC	pKa			
HL 2.00E-009	atm m3/mol	25	EST	MEYLAN,WM & HOWARD,PH (1991)
OH 2.23E-011	cm3/molc sec	25	EST	MEYLAN,WM & HOWARD,PH (1993)

CAS #: 000111-47-7				DI-N-PROPYLSULFIDE
Formula: $C_7H_{16}S$				
Mol Weight: 132.27				
MP (deg C): -102		FP (deg C):		
BP (deg C): 142				
BP pressure (mm Hg):				

Property/Value	Units	Temp	Data Type	Reference
WS 3.51E+002	mg/L	25	EST	MEYLAN,WM ET AL. (1996)
logP 2.88			EST	MEYLAN,WM & HOWARD,PH (1995)
VP 6.45E+000	mm Hg	25	EXT	ZWOLINSKI,BJ & WILHOIT,RC (1971)
DC	pKa			
HL 2.44E-003	atm m3/mol	25	EST	MEYLAN,WM & HOWARD,PH (1991)
OH 2.00E-011	cm3/molc sec	25	EXP	ATKINSON,R (1989)

CAS #: 000111-48-8				2,2'-THIOBISETHANOL
Formula: $C_4H_{10}O_2S$				
Mol Weight: 122.19				
MP (deg C): -16		FP (deg C):		
BP (deg C): 168				
BP pressure (mm Hg): 1.40E+001				

Property/Value	Units	Temp	Data Type	Reference
WS 3.40E+005	mg/L	25	EST	MEYLAN,WM ET AL. (1996)
logP -0.63			EXP	HANSCH,C ET AL. (1995)
VP 5.42E-001	mm Hg	25	EXT	OHE,S (1976)
DC	pKa			
HL 1.85E-009	atm m3/mol	25	EST	MEYLAN,WM & HOWARD,PH (1991)
OH 2.79E-011	cm3/molc sec	25	EST	MEYLAN,WM & HOWARD,PH (1993)

CAS #: 000111-49-9				HEXAMETHYLENEIMINE
Formula: $C_6H_{13}N$				
Mol Weight: 99.18				
MP (deg C):		FP (deg C):		
BP (deg C): 138				
BP pressure (mm Hg):				

Property/Value	Units	Temp	Data Type	Reference
WS 3.19E+004	mg/L	25	EXP	YALKOWSKY,SH & DANNENFELSER,RM (1992)
logP 1.68			EST	MEYLAN,WM & HOWARD,PH (1995)
VP 8.09E+000	mm Hg	25	EXP	DAUBERT,TE & DANNER,RP (1989)
DC 11.07	pKa	25	EXP	PERRIN,DD (1965)
HL 6.14E-006	atm m3/mol	25	EXP	CABANI,S ET AL. (1971)
OH 8.72E+001	cm3/molc sec	25	EST	MEYLAN,WM & HOWARD,PH (1993)

CAS #: 000111-54-6				1,2-ETHANEDIYLBISCARBAMODITHIOIC ACID
Formula: $C_4H_8N_2S_4$				
Mol Weight: 212.38				
MP (deg C):		FP (deg C):		
BP (deg C):				
BP pressure (mm Hg):				

Property/Value	Units	Temp	Data Type	Reference
WS 1.14E+005	mg/L	25	EST	MEYLAN,WM ET AL. (1996)
logP 0.62			EST	MEYLAN,WM & HOWARD,PH (1995)
VP 1.49E-005	mm Hg	25	EST	NEELY,WB & BLAU,GE (1985)
DC	pKa			
HL 5.64E-007	atm m3/mol	25	EST	MEYLAN,WM & HOWARD,PH (1991)
OH 2.12E-010	cm3/molc sec	25	EST	MEYLAN,WM & HOWARD,PH (1993)

CAS #: 000111-55-7				ETHYLENE GLYCOL DIACETATE
Formula: $C_6H_{10}O_4$				
Mol Weight: 146.14				
MP (deg C): -31		FP (deg C): -31		
BP (deg C): 190-191				
BP pressure (mm Hg):				

Property/Value	Units	Temp	Data Type	Reference
WS 1.78E+005	mg/L	25	EXP	YALKOWSKY,SH & DANNENFELSER,RM (1992)
logP 0.40			EST	MEYLAN,WM & HOWARD,PH (1995)
VP 7.74E-002	mm Hg	25	EXP	DAUBERT,TE & DANNER,RP (1989)
DC	pKa			
HL 8.40E-008	atm m3/mol	25	EST	VP/WSOL
OH 3.88E-012	cm3/molc sec	25	EST	MEYLAN,WM & HOWARD,PH (1993)

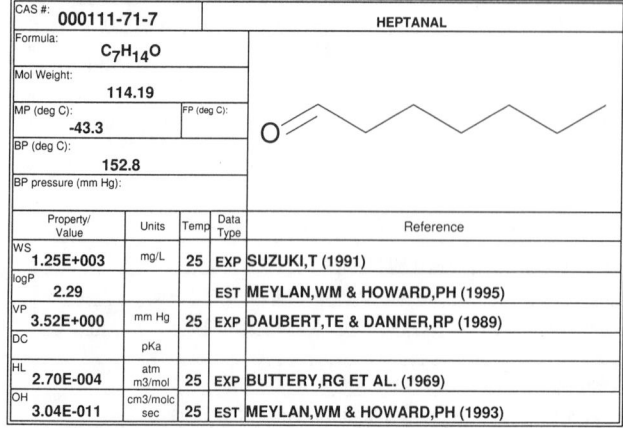

CAS #:	000111-61-5				ETHYL STEARATE

Formula: $C_{20}H_{40}O_2$

Mol Weight: 312.54

MP (deg C): 33 **FP (deg C):**

BP (deg C): 199

BP pressure (mm Hg): 1.00E+001

Property/Value	Units	Temp	Data Type	Reference
WS 3.69E-004	mg/L	25	EST	MEYLAN,WM ET AL. (1996)
logP 8.72			EST	MEYLAN,WM & HOWARD,PH (1995)
VP 3.27E-007	mm Hg	25	EXT	OMAR,MM (1967)
DC	pKa			
HL 2.17E-002	atm m3/mol	25	EST	MEYLAN,WM & HOWARD,PH (1991)
OH 2.36E-011	cm3/molc sec	25	EST	MEYLAN,WM & HOWARD,PH (1993)

CAS #:	000111-65-9				OCTANE

Formula: C_8H_{18}

Mol Weight: 114.23

MP (deg C): -56.8 **FP (deg C):**

BP (deg C): 125.6

BP pressure (mm Hg):

Property/Value	Units	Temp	Data Type	Reference
WS 6.60E-001	mg/L	25	EXP	YALKOWSKY,SH & DANNENFELSER,RM (1992)
logP 5.18			EXP	MILLER,MM ET AL. (1985)
VP 1.41E+001	mm Hg	25	EXP	YAWS,CL (1994)
DC	pKa			
HL 3.21E+000	atm m3/mol	25	EST	VP/WSOL
OH 8.68E-012	cm3/molc sec	25	EXP	ATKINSON,R (1989)

CAS #:	000111-66-0				1-OCTENE

Formula: C_8H_{16}

Mol Weight: 112.22

MP (deg C): -102 **FP (deg C):**

BP (deg C): 121

BP pressure (mm Hg):

Property/Value	Units	Temp	Data Type	Reference
WS 4.10E+000	mg/L	25	EXP	YALKOWSKY,SH & DANNENFELSER,RM (1992)
logP 4.57			EXP	HANSCH,C & LEO,AJ (1985)
VP 1.74E+001	mm Hg	25	EXP	YAWS,CL (1994)
DC	pKa			
HL 6.27E-001	atm m3/mol	25	EST	VP/WSOL
OH 3.30E-011	cm3/molc sec	25	EST	MEYLAN,WM & HOWARD,PH (1993)

CAS #:	000111-67-1				2-OCTENE

Formula: C_8H_{16}

Mol Weight: 112.22

MP (deg C): **FP (deg C):**

BP (deg C):

BP pressure (mm Hg):

Property/Value	Units	Temp	Data Type	Reference
WS 1.07E+001	mg/L	25	EST	MEYLAN,WM ET AL. (1996)
logP 4.06			EST	MEYLAN,WM & HOWARD,PH (1995)
VP 1.63E+001	mm Hg	25	EXT	ZWOLINSKI,BJ & WILHOIT,RC (1971)
DC	pKa			
HL 7.46E-001	atm m3/mol	25	EST	MEYLAN,WM & HOWARD,PH (1991)
OH 6.18E-011	cm3/molc sec	25	EST	MEYLAN,WM & HOWARD,PH (1993)

CAS #:	000111-68-2				HEPTYLAMINE

Formula: $C_7H_{17}N$

Mol Weight: 115.22

MP (deg C): -18 **FP (deg C):**

BP (deg C): 156

BP pressure (mm Hg):

Property/Value	Units	Temp	Data Type	Reference
WS 6.79E+003	mg/L	25	EST	MEYLAN,WM ET AL. (1996)
logP 2.57			EXP	HANSCH,C & LEO,AJ (1985)
VP 2.74E+000	mm Hg	25	EXP	DAUBERT,TE & DANNER,RP (1989)
DC 10.66	pKa	25	EXP	PERRIN,DD (1965)
HL 4.15E-005	atm m3/mol	25	EST	MEYLAN,WM & HOWARD,PH (1991)
OH 3.87E-011	cm3/molc sec	25	EST	MEYLAN,WM & HOWARD,PH (1993)

CAS #:	000111-69-3				ADIPONITRILE

Formula: $C_6H_8N_2$

Mol Weight: 108.14

MP (deg C): 1 **FP (deg C):**

BP (deg C): 295

BP pressure (mm Hg):

Property/Value	Units	Temp	Data Type	Reference
WS 8.00E+004	mg/L	20	EXP	SMILEY,RA (1981)
logP -0.32			EXP	TANII,H & HASHIMOTO,K (1985)
VP 6.79E-004	mm Hg	25	EXP	DAUBERT,TE & DANNER,RP (1989)
DC	pKa			
HL 1.21E-009	atm m3/mol	25	EST	VP/WSOL
OH 1.38E-012	cm3/molc sec	25	EST	MEYLAN,WM & HOWARD,PH (1993)

CAS #:	000111-70-6				1-HEPTANOL

Formula: $C_7H_{16}O$

Mol Weight: 116.20

MP (deg C): -34.6 **FP (deg C):**

BP (deg C): 175.8

BP pressure (mm Hg):

Property/Value	Units	Temp	Data Type	Reference
WS 1.31E+003	mg/L		EXP	TEWARI,YB (1982)
logP 2.62			EXP	SANGSTER,J (1994)
VP 2.16E-001	mm Hg	25	EXP	YAWS,CL (1994A)
DC	pKa			
HL 1.88E-005	atm m3/mol	25	EXP	BUTLER,JAV ET AL. (1935)
OH 1.36E-011	cm3/molc sec	25	EXP	ATKINSON,R (1989)

CAS #:	000111-71-7				HEPTANAL

Formula: $C_7H_{14}O$

Mol Weight: 114.19

MP (deg C): -43.3 **FP (deg C):**

BP (deg C): 152.8

BP pressure (mm Hg):

Property/Value	Units	Temp	Data Type	Reference
WS 1.25E+003	mg/L	25	EXP	SUZUKI,T (1991)
logP 2.29			EST	MEYLAN,WM & HOWARD,PH (1995)
VP 3.52E+000	mm Hg	25	EXP	DAUBERT,TE & DANNER,RP (1989)
DC	pKa			
HL 2.70E-004	atm m3/mol	25	EXP	BUTTERY,RG ET AL. (1969)
OH 3.04E-011	cm3/molc sec	25	EST	MEYLAN,WM & HOWARD,PH (1993)

000111-74-0 — 1,2-ETHANEDIAMINE, N,N'-DIETHYL-

Formula:	$C_6H_{16}N_2$
Mol Weight:	116.21
MP (deg C):	FP (deg C):
BP (deg C):	146
BP pressure (mm Hg):	

Property/Value	Units	Temp	Data Type	Reference
WS 6.98E+005	mg/L	25	EST	MEYLAN,WM ET AL. (1996)
logP 0.21			EXP	HANSCH,C ET AL. (1995)
VP 2.64E+000	mm Hg	25	EST	NEELY,WB & BLAU,GE (1985)
DC 11.06	pKa	0	EXP	PERRIN,DD (1965)
HL 8.79E-009	atm m3/mol	25	EST	MEYLAN,WM & HOWARD,PH (1991)
OH 1.65E-010	cm3/molc sec	25	EST	MEYLAN,WM & HOWARD,PH (1993)

000111-76-2 — ETHYLENE GLYCOL N-BUTYL ETHER

Formula:	$C_6H_{14}O_2$
Mol Weight:	118.18
MP (deg C):	-70	FP (deg C):
BP (deg C):	171-172
BP pressure (mm Hg):	

Property/Value	Units	Temp	Data Type	Reference
WS 1.00E+006	mg/L	25	EXP	RIDDICK,JA ET AL. (1986)
logP 0.83			EXP	HANSCH,C ET AL. (1995)
VP 8.80E-001	mm Hg	25	EXP	DOW CHEMICAL COMPANY (1981)
DC	pKa			
HL 2.08E-008	atm m3/mol	25	EST	MEYLAN,WM & HOWARD,PH (1991)
OH 1.86E-011	cm3/molc sec	25	EXP	ATKINSON,R (1989)

000111-77-3 — DIETHYLENE GLYCOL MONOMETHYL ETHER

Formula:	$C_5H_{12}O_3$
Mol Weight:	120.15
MP (deg C):	<-84	FP (deg C):
BP (deg C):	193
BP pressure (mm Hg):	

Property/Value	Units	Temp	Data Type	Reference
WS 1.00E+006	mg/L	25	EXP	RIDDICK,JA ET AL (1986)
logP -1.18			EST	MEYLAN,WM & HOWARD,PH (1995)
VP 2.50E-001	mm Hg	25	EXP	DOW CHEMICAL COMPANY (1990)
DC	pKa			
HL 1.65E-011	atm m3/mol	25	EST	MEYLAN,WM & HOWARD,PH (1991)
OH 2.44E-011	cm3/molc sec	25	EST	MEYLAN,WM & HOWARD,PH (1993)

000111-78-4 — CYCLOOCTA-1,5-DIENE

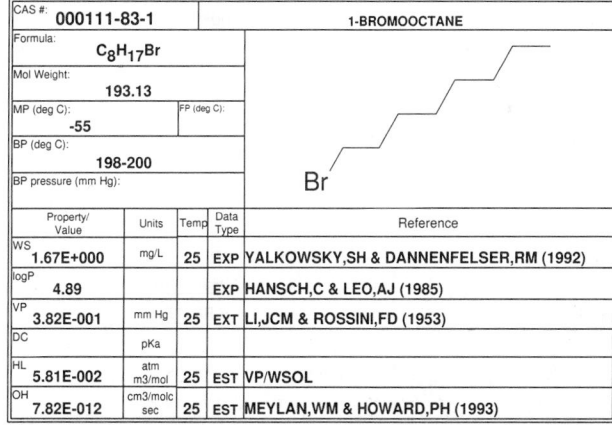

Formula:	C_8H_{12}
Mol Weight:	108.18
MP (deg C):	-56.4	FP (deg C):
BP (deg C):	150.8
BP pressure (mm Hg):	

Property/Value	Units	Temp	Data Type	Reference
WS 6.41E+001	mg/L	25	EST	MEYLAN,WM ET AL. (1996)
logP 3.16			EXP	HANSCH,C & LEO,AJ (1985)
VP 4.95E+000	mm Hg	25	EXP	YAWS,CL (1994)
DC	pKa			
HL 1.83E-001	atm m3/mol	25	EST	MEYLAN,WM & HOWARD,PH (1991)
OH 1.17E-010	cm3/molc sec	25	EST	MEYLAN,WM & HOWARD,PH (1993)

000111-82-0 — METHYL LAURATE

Formula:	$C_{13}H_{26}O_2$
Mol Weight:	214.35
MP (deg C):	5.2	FP (deg C):
BP (deg C):	267
BP pressure (mm Hg):	

Property/Value	Units	Temp	Data Type	Reference
WS 1.13E+000	mg/L	25	EST	MEYLAN,WM ET AL. (1996)
logP 5.28			EST	MEYLAN,WM & HOWARD,PH (1995)
VP 4.11E-003	mm Hg	25	EXP	DAUBERT,TE & DANNER,RP (1989)
DC	pKa			
HL 2.98E-003	atm m3/mol	25	EST	MEYLAN,WM & HOWARD,PH (1991)
OH 1.37E-011	cm3/molc sec	25	EST	MEYLAN,WM & HOWARD,PH (1993)

000111-83-1 — 1-BROMOOCTANE

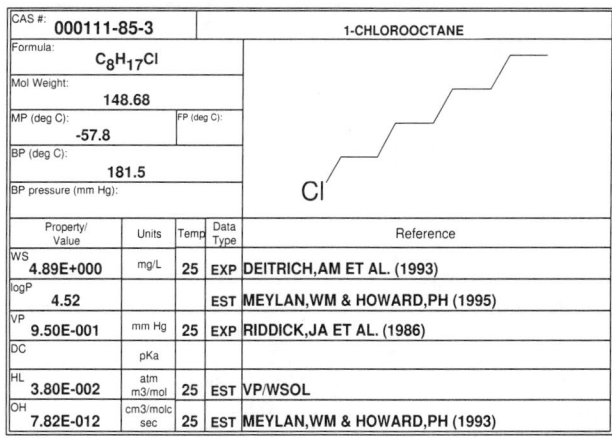

Formula:	$C_8H_{17}Br$
Mol Weight:	193.13
MP (deg C):	-55	FP (deg C):
BP (deg C):	198-200
BP pressure (mm Hg):	

Property/Value	Units	Temp	Data Type	Reference
WS 1.67E+000	mg/L	25	EXP	YALKOWSKY,SH & DANNENFELSER,RM (1992)
logP 4.89			EXP	HANSCH,C & LEO,AJ (1985)
VP 3.82E-001	mm Hg	25	EXT	LI,JCM & ROSSINI,FD (1953)
DC	pKa			
HL 5.81E-002	atm m3/mol	25	EST	VP/WSOL
OH 7.82E-012	cm3/molc sec	25	EST	MEYLAN,WM & HOWARD,PH (1993)

000111-84-2 — N-NONANE

Formula:	C_9H_{20}
Mol Weight:	128.26
MP (deg C):	-53.5	FP (deg C): -54
BP (deg C):	150.7
BP pressure (mm Hg):	

Property/Value	Units	Temp	Data Type	Reference
WS 2.20E+002	mg/L	25	EXP	RIDDICK,JA ET AL. (1986)
logP 4.76			EST	MEYLAN,WM & HOWARD,PH (1995)
VP 4.45E+000	mm Hg	25	EXP	DAUBERT,TE & DANNER,RP (1989)
DC	pKa			
HL 3.40E+000	atm m3/mol	25	EST	VP/WSOL
OH 1.02E-011	cm3/molc sec	25	EXP	ATKINSON,R (1989)

000111-85-3 — 1-CHLOROOCTANE

Formula:	$C_8H_{17}Cl$
Mol Weight:	148.68
MP (deg C):	-57.8	FP (deg C):
BP (deg C):	181.5
BP pressure (mm Hg):	

Property/Value	Units	Temp	Data Type	Reference
WS 4.89E+000	mg/L	25	EXP	DEITRICH,AM ET AL. (1993)
logP 4.52			EST	MEYLAN,WM & HOWARD,PH (1995)
VP 9.50E-001	mm Hg	25	EXP	RIDDICK,JA ET AL. (1986)
DC	pKa			
HL 3.80E-002	atm m3/mol	25	EST	VP/WSOL
OH 7.82E-012	cm3/molc sec	25	EST	MEYLAN,WM & HOWARD,PH (1993)

CAS #: 000111-86-4 — N-OCTYLAMINE

Formula: $C_8H_{19}N$
Mol Weight: 129.25
MP (deg C): 0	FP (deg C):
BP (deg C): 179.6
BP pressure (mm Hg):

Property/Value	Units	Temp	Data Type	Reference
WS 2.00E+002	mg/L	25	EXP	YALKOWSKY,SH & DANNENFELSER,RM (1992)
logP 2.90			EXP	SANGSTER,J (1993)
VP 9.69E-001	mm Hg	25	EXP	DAUBERT,TE & DANNER,RP (1989)
DC 10.65	pKa	25	EXP	PERRIN,DD (1965)
HL 8.24E-004	atm m3/mol	25	EST	VP/WSOL
OH 4.01E-011	cm3/molc sec	25	EST	MEYLAN,WM & HOWARD,PH (1993)

CAS #: 000111-87-5 — 1-OCTANOL

Formula: $C_8H_{18}O$
Mol Weight: 130.23
MP (deg C): -15.5	FP (deg C):
BP (deg C): 195.1
BP pressure (mm Hg):

Property/Value	Units	Temp	Data Type	Reference
WS 5.40E+002	mg/L	25	EXP	BARTON,AFM (1984)
logP 3.00			EXP	HANSCH,C ET AL. (1995)
VP 7.94E-002	mm Hg	25	EXP	DAUBERT,TE & DANNER,RP (1989)
DC	pKa			
HL 2.45E-005	atm m3/mol	25	EXP	BUTTERY,RG ET AL. (1969)
OH 1.19E-011	cm3/molc sec	25	EST	MEYLAN,WM & HOWARD,PH (1993)

CAS #: 000111-88-6 — 1-OCTANETHIOL

Formula: $C_8H_{18}S$
Mol Weight: 146.30
MP (deg C): -49.2	FP (deg C):
BP (deg C): 199.1
BP pressure (mm Hg):

Property/Value	Units	Temp	Data Type	Reference
WS 2.26E+001	mg/L	25	EST	MEYLAN,WM ET AL (1996)
logP 4.21			EST	MEYLAN,WM & HOWARD,PH (1995)
VP 4.25E-001	mm Hg	25	EXP	DAUBERT,TE & DANNER,RP (1989)
DC	pKa			
HL 2.34E-002	atm m3/mol	25	EST	MEYLAN,WM & HOWARD,PH (1991)
OH 4.98E-011	cm3/molc sec	25	EST	MEYLAN,WM & HOWARD,PH (1993)

CAS #: 000111-90-0 — DIETHYLENE GLYCOL MONOETHYL ETHER

Formula: $C_6H_{14}O_3$
Mol Weight: 134.18
MP (deg C): -76	FP (deg C):
BP (deg C): 196
BP pressure (mm Hg):

Property/Value	Units	Temp	Data Type	Reference
WS 1.00E+006	mg/L	20	EXP	RIDDICK,JA ET AL. (1986)
logP -0.54			EXP	HANSCH,C ET AL. (1995)
VP 1.26E-001	mm Hg	25	EXP	DAUBERT,TE & DANNER,RP (1989)
DC	pKa			
HL 2.23E-008	atm m3/mol	25	EST	VP/WSOL
OH 3.13E-011	cm3/molc sec	25	EST	MEYLAN,WM & HOWARD,PH (1993)

CAS #: 000111-91-1 — BIS(2-CHLOROETHOXY)METHANE

Formula: $C_5H_{10}Cl_2O_2$
Mol Weight: 173.04
MP (deg C): -32	FP (deg C):
BP (deg C): 218
BP pressure (mm Hg):

Property/Value	Units	Temp	Data Type	Reference
WS 4.66E+003	mg/L	25	EST	MEYLAN,WM ET AL. (1996)
logP 1.30			EST	MEYLAN,WM & HOWARD,PH (1995)
VP 1.40E-004	mm Hg	25	EXT	PERRY,RH ET AL. (1984)
DC	pKa			
HL 1.70E-007	atm m3/mol	25	EST	MEYLAN,WM & HOWARD,PH (1991)
OH 9.53E-012	cm3/molc sec	25	EST	MEYLAN,WM & HOWARD,PH (1993)

CAS #: 000111-92-2 — DIBUTYLAMINE

Formula: $C_8H_{19}N$
Mol Weight: 129.25
MP (deg C): -62	FP (deg C):
BP (deg C): 159.6
BP pressure (mm Hg):

Property/Value	Units	Temp	Data Type	Reference
WS 3.50E+003	mg/L	25	EXP	CHEM INSPECT TEST INST (1992)
logP 2.83			EXP	HANSCH,C & LEO,AJ (1985)
VP 2.59E+000	mm Hg	25	EXP	YAWS,CL (1994)
DC 11.39	pKa	21	EXP	PERRIN,DD (1965)
HL 8.90E-005	atm m3/mol	25	EXP	CHRISTIE,AO & CRISP,DJ (1967)
OH 8.69E+001	cm3/molc sec	25	EST	MEYLAN,WM & HOWARD,PH (1993)

CAS #: 000111-94-4 — PROPANENITRILE, 3,3'-IMINOBIS-

Formula: $C_6H_9N_3$
Mol Weight: 123.16
MP (deg C):	FP (deg C):
BP (deg C): 205
BP pressure (mm Hg): 2.50E+001

Property/Value	Units	Temp	Data Type	Reference
WS 1.00E+006	mg/L	25	EST	MEYLAN,WM ET AL. (1996)
logP -1.34			EXP	CHEM INSPECT TEST INST (1992)
VP 3.61E-003	mm Hg	25	EST	NEELY,WB & BLAU,GE (1985)
DC 5.26	pKa	25	EXP	PERRIN,DD (1965)
HL 5.05E-012	atm m3/mol	25	EST	MEYLAN,WM & HOWARD,PH (1991)
OH 2.52E-012	cm3/molc sec	25	EST	MEYLAN,WM & HOWARD,PH (1993)

CAS #: 000111-96-6 — 2-METHOXYETHYLETHER

Formula: $C_6H_{14}O_3$
Mol Weight: 134.18
MP (deg C): -68	FP (deg C):
BP (deg C): 162
BP pressure (mm Hg):

Property/Value	Units	Temp	Data Type	Reference
WS 1.00E+006	mg/L		EXP	RIDDICK,JA ET AL. (1986)
logP -0.36			EXP	HANSCH,C ET AL. (1995)
VP 2.96E+000	mm Hg	25	EXP	DAUBERT,TE & DANNER,RP (1989)
DC	pKa			
HL 5.23E-007	atm m3/mol	25	EST	VP/WSOL
OH 1.75E-011	cm3/molc sec	25	EXP	ATKINSON,R (1989)

000112-02-7 — CETRIMONIUM CHLORIDE

Formula: $C_{19}H_{42}ClN$

Mol Weight: 320.01

MP (deg C): | FP (deg C):

BP (deg C):

BP pressure (mm Hg):

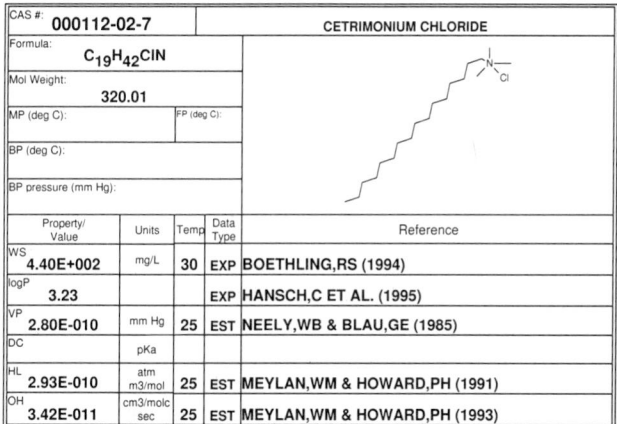

Property/Value	Units	Temp	Data Type	Reference
WS 4.40E+002	mg/L	30	EXP	BOETHLING,RS (1994)
logP 3.23			EXP	HANSCH,C ET AL. (1995)
VP 2.80E-010	mm Hg	25	EST	NEELY,WB & BLAU,GE (1985)
DC	pKa			
HL 2.93E-010	atm m3/mol	25	EST	MEYLAN,WM & HOWARD,PH (1991)
OH 3.42E-011	cm3/molc sec	25	EST	MEYLAN,WM & HOWARD,PH (1993)

000112-04-9 — OCTADECYLTRICHLOROSILANE

Formula: $C_{18}H_{37}Cl_3Si$

Mol Weight: 387.94

MP (deg C): | FP (deg C):

BP (deg C): 223

BP pressure (mm Hg): 1.00E+001

Property/Value	Units	Temp	Data Type	Reference
WS 5.21E-006	mg/L	25	EST	MEYLAN,WM ET AL. (1996)
logP 10.36			EST	MEYLAN,WM & HOWARD,PH (1995)
VP 9.92E-006	mm Hg	25	EST	NEELY,WB & BLAU,GE (1985)
DC	pKa			
HL 6.60E-001	atm m3/mol	25	EST	MEYLAN,WM & HOWARD,PH (1991)
OH 2.38E-011	cm3/molc sec	25	EST	MEYLAN,WM & HOWARD,PH (1993)

000112-05-0 — NONANOIC ACID

Formula: $C_9H_{18}O_2$

Mol Weight: 158.24

MP (deg C): 12.5 | FP (deg C):

BP (deg C): 252-253

BP pressure (mm Hg): 7.52E+002

Property/Value	Units	Temp	Data Type	Reference
WS 2.12E+002	mg/L	30	EXP	YALKOWSKY,SH & DANNENFELSER,RM (1992)
logP 3.42			EXP	SANGSTER,J (1993)
VP 1.17E-003	mm Hg	25	EXP	DAUBERT,TE & DANNER,RP (1989)
DC 4.95	pKa	25	EXP	KORTUM,G ET AL (1961)
HL 1.15E-006	atm m3/mol	25	EST	VP/WSOL
OH 9.76E-012	cm3/molc sec	25	EST	MEYLAN,WM & HOWARD,PH (1993)

000112-07-2 — 2-BUTOXYETHANOL ACETATE

Formula: $C_8H_{16}O_3$

Mol Weight: 160.21

MP (deg C): | FP (deg C):

BP (deg C): 192

BP pressure (mm Hg):

Property/Value	Units	Temp	Data Type	Reference
WS 1.50E+004	mg/L	20	EXP	ASHFORD,RD (1994)
logP 1.57			EST	MEYLAN,WM & HOWARD,PH (1995)
VP 3.75E-001	mm Hg	20	EXP	WEBER,RC ET AL. (1981)
DC	pKa			
HL 5.27E-006	atm m3/mol	20	EST	VP/WSOL
OH 2.13E-011	cm3/molc sec	25	EST	MEYLAN,WM & HOWARD,PH (1993)

000112-10-7 — ISOPROPYL STEARATE

Formula: $C_{21}H_{42}O_2$

Mol Weight: 326.57

MP (deg C): 28 | FP (deg C):

BP (deg C): 207

BP pressure (mm Hg): 6.00E+000

Property/Value	Units	Temp	Data Type	Reference
WS 1.34E-004	mg/L	25	EST	MEYLAN,WM ET AL. (1996)
logP 9.14			EST	MEYLAN,WM & HOWARD,PH (1995)
VP 3.97E-005	mm Hg	25	EXT	BOUBLIK,T ET AL. (1984)
DC	pKa			
HL 2.88E-002	atm m3/mol	25	EST	MEYLAN,WM & HOWARD,PH (1991)
OH 2.54E-011	cm3/molc sec	25	EST	MEYLAN,WM & HOWARD,PH (1993)

000112-12-9 — UNDECANONE

Formula: $C_{11}H_{22}O$

Mol Weight: 170.30

MP (deg C): 15 | FP (deg C):

BP (deg C): 231.5

BP pressure (mm Hg):

Property/Value	Units	Temp	Data Type	Reference
WS 1.97E+001	mg/L	25	EST	MEYLAN,WM ET AL. (1996)
logP 4.09			EXP	TANII,H & HASHIMOTO,K (1986)
VP 4.14E-002	mm Hg	25	EXT	PERRY,RH & GREEN,D (1984)
DC	pKa			
HL 6.36E-005	atm m3/mol	25	EXP	BUTTERY,RG ET AL. (1969)
OH 1.39E-011	cm3/molc sec	25	EST	MEYLAN,WM & HOWARD,PH (1993)

000112-15-2 — CARBITOL ACETATE

Formula: $C_8H_{16}O_4$

Mol Weight: 176.21

MP (deg C): -25 | FP (deg C):

BP (deg C): 218

BP pressure (mm Hg):

Property/Value	Units	Temp	Data Type	Reference
WS 1.00E+006	mg/L	20	EXP	YALKOWSKY,SH & DANNENFELSER,RM (1992)
logP 0.32			EST	MEYLAN,WM & HOWARD,PH (1995)
VP 1.48E-001	mm Hg	25	EXP	YAWS,CL (1994)
DC	pKa			
HL 3.43E-008	atm m3/mol	25	EST	VP/WSOL
OH 2.91E-011	cm3/molc sec	25	EST	MEYLAN,WM & HOWARD,PH (1993)

000112-17-4 — DECYL ACETATE

Formula: $C_{12}H_{24}O_2$

Mol Weight: 200.32

MP (deg C): -15 | FP (deg C):

BP (deg C): 244

BP pressure (mm Hg):

Property/Value	Units	Temp	Data Type	Reference
WS 3.52E+000	mg/L	25	EST	MEYLAN,WM ET AL. (1996)
logP 4.79			EST	MEYLAN,WM & HOWARD,PH (1995)
VP 3.47E-003	mm Hg		EXP	HEATH,RR & TUMLINSON,JH (1986)
DC	pKa			
HL 2.25E-003	atm m3/mol	25	EST	MEYLAN,WM & HOWARD,PH (1991)
OH 1.31E-011	cm3/molc sec	25	EST	MEYLAN,WM & HOWARD,PH (1993)

000112-20-9 — N-NONYLAMINE

Formula: $C_9H_{21}N$		
Mol Weight: 143.27		
MP (deg C): -1	FP (deg C):	
BP (deg C): 202.2		
BP pressure (mm Hg):		

Property/Value	Units	Temp	Data Type	Reference
WS 1.28E+003	mg/L	25	EST	MEYLAN,WM ET AL. (1996)
logP 3.29			EST	MEYLAN,WM & HOWARD,PH (1995)
VP 2.78E-001	mm Hg	25	EXP	DAUBERT,TE & DANNER,RP (1989)
DC 10.64	pKa	25	EXP	PERRIN,DD (1965)
HL 7.31E-005	atm m3/mol	25	EST	MEYLAN,WM & HOWARD,PH (1991)
OH 4.15E-011	cm3/molc sec	25	EST	MEYLAN,WM & HOWARD,PH (1993)

000112-24-3 — TRIETHYLENETETRAAMINE

Formula: $C_6H_{18}N_4$		
Mol Weight: 146.24		
MP (deg C): 12	FP (deg C):	
BP (deg C): 266-267		
BP pressure (mm Hg):		

Property/Value	Units	Temp	Data Type	Reference
WS 4.77E+006	mg/L		EXP	YALKOWSKY,SH & DANNENFELSER,RM (1992)
logP -2.65			EST	MEYLAN,WM & HOWARD,PH (1995)
VP 4.12E-004	mm Hg	25	EXP	YAWS,CL (1994)
DC 9.92	pKa	20	EXP	PERRIN,DD (1965)
HL 1.66E-011	atm m3/mol	25		VP/WSOL
OH 2.32E-010	cm3/molc sec	25	EST	MEYLAN,WM & HOWARD,PH (1993)

000112-25-4 — ETHANOL, 2-(HEXYLOXY)-

Formula: $C_8H_{18}O_2$		
Mol Weight: 146.23		
MP (deg C): -45.1	FP (deg C): -50	
BP (deg C): 208		
BP pressure (mm Hg):		

Property/Value	Units	Temp	Data Type	Reference
WS 6.58E+003	mg/L	25	EST	MEYLAN,WM ET AL. (1996)
logP 1.86			EXP	FUNASAKI,N ET AL. (1984)
VP 5.10E-002	mm Hg		EXP	FLICK,EW (1991)
DC	pKa			
HL 1.73E-007	atm m3/mol	25	EST	MEYLAN,WM & HOWARD,PH (1991)
OH 2.63E-011	cm3/molc sec	25	EST	MEYLAN,WM & HOWARD,PH (1993)

000112-26-5 — 1,2-BIS(2-CHLOROETHOXY)ETHANE

Formula: $C_6H_{12}Cl_2O_2$		
Mol Weight: 187.07		
MP (deg C): -31.5	FP (deg C):	
BP (deg C): 241.3		
BP pressure (mm Hg):		

Property/Value	Units	Temp	Data Type	Reference
WS 1.89E+004	mg/L	20	EXP	FLICK,EW (1991)
logP 1.28			EST	MEYLAN,WM & HOWARD,PH (1995)
VP 6.00E-002	mm Hg	20	EXP	FLICK,EW (1991)
DC	pKa			
HL 7.81E-007	atm m3/mol	20	EST	VP/WSOL
OH 1.09E-011	cm3/molc sec	25	EST	MEYLAN,WM & HOWARD,PH (1993)

000112-27-6 — TRIETHYLENE GLYCOL

Formula: $C_6H_{14}O_4$		
Mol Weight: 150.18		
MP (deg C): -7.2	FP (deg C):	
BP (deg C): 287.4		
BP pressure (mm Hg):		

Property/Value	Units	Temp	Data Type	Reference
WS 1.00E+006	mg/L		EXP	RIDDICK,JA ET AL. (1986)
logP -1.98			EST	HANSCH,C & LEO,AJ (1981)
VP 1.32E-003	mm Hg	25	EXP	DAUBERT,TE & DANNER,RP (1989)
DC	pKa			
HL 2.61E-010	atm m3/mol	25	EST	VP/WSOL
OH 3.38E-011	cm3/molc sec	25	EST	MEYLAN,WM & HOWARD,PH (1993)

000112-29-8 — 1-BROMODECANE

Formula: $C_{10}H_{21}Br$		
Mol Weight: 221.19		
MP (deg C): -29.2	FP (deg C):	
BP (deg C): 240.6		
BP pressure (mm Hg):		

Property/Value	Units	Temp	Data Type	Reference
WS 1.97E-001	mg/L	25	EST	MUELLER,M & KLEIN,W (1992)
logP 5.60			EST	MEYLAN,WM & HOWARD,PH (1995)
VP 4.00E-002	mm Hg	25	EXP	RIDDICK,JA ET AL. (1986)
DC	pKa			
HL 1.09E-001	atm m3/mol	25	EST	MEYLAN,WM & HOWARD,PH (1991)
OH 1.06E-011	cm3/molc sec	25	EST	MEYLAN,WM & HOWARD,PH (1993)

000112-30-1 — 1-DECANOL

Formula: $C_{10}H_{22}O$		
Mol Weight: 158.29		
MP (deg C): 7	FP (deg C):	
BP (deg C): 230		
BP pressure (mm Hg):		

Property/Value	Units	Temp	Data Type	Reference
WS 3.70E+001	mg/L	25	EXP	BARTON,AFM (1984)
logP 4.57			EXP	HANSCH,C ET AL. (1995)
VP 8.51E-003	mm Hg	25	EXP	DAUBERT,TE & DANNER,RP (1989)
DC	pKa			
HL 4.79E-005	atm m3/mol	25	EST	VP/WSOL
OH 1.47E-011	cm3/molc sec	25	EST	MEYLAN,WM & HOWARD,PH (1993)

000112-31-2 — N-DECANAL

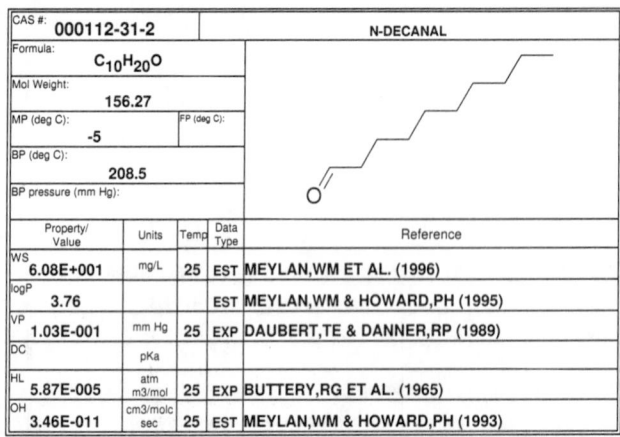

Formula: $C_{10}H_{20}O$		
Mol Weight: 156.27		
MP (deg C): -5	FP (deg C):	
BP (deg C): 208.5		
BP pressure (mm Hg):		

Property/Value	Units	Temp	Data Type	Reference
WS 6.08E+001	mg/L	25	EST	MEYLAN,WM ET AL. (1996)
logP 3.76			EST	MEYLAN,WM & HOWARD,PH (1995)
VP 1.03E-001	mm Hg	25	EXP	DAUBERT,TE & DANNER,RP (1989)
DC	pKa			
HL 5.87E-005	atm m3/mol	25	EXP	BUTTERY,RG ET AL. (1965)
OH 3.46E-011	cm3/molc sec	25	EST	MEYLAN,WM & HOWARD,PH (1993)

000112-34-5 — DIETHYLENE GLYCOL MONO-N-BUTYL ETHER

Formula: $C_8H_{18}O_3$
Mol Weight: 162.23
MP (deg C): -68.1
FP (deg C):
BP (deg C): 230.6
BP pressure (mm Hg):

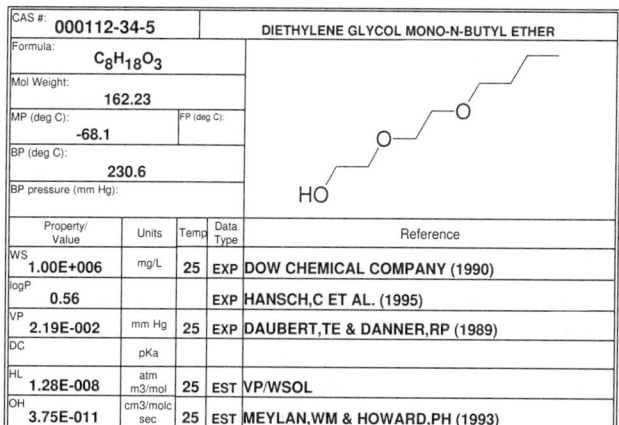

Property/Value	Units	Temp	Data Type	Reference
WS 1.00E+006	mg/L	25	EXP	DOW CHEMICAL COMPANY (1990)
logP 0.56			EXP	HANSCH,C ET AL. (1995)
VP 2.19E-002	mm Hg	25	EXP	DAUBERT,TE & DANNER,RP (1989)
DC	pKa			
HL 1.28E-008	atm m3/mol	25	EST	VP/WSOL
OH 3.75E-011	cm3/molc sec	25	EST	MEYLAN,WM & HOWARD,PH (1993)

000112-35-6 — METHOXY TRIETHYLENE GLYCOL

Formula: $C_7H_{16}O_4$
Mol Weight: 164.20
MP (deg C):
FP (deg C): -44
BP (deg C): 249
BP pressure (mm Hg):

Property/Value	Units	Temp	Data Type	Reference
WS 1.00E+006	mg/L	25	EXP	DOW CHEMICAL COMPANY (1990)
logP -1.46			EST	MEYLAN,WM & HOWARD,PH (1995)
VP 3.50E-003	mm Hg	25	EST	NEELY,WB & BLAU,GE (1985)
DC	pKa			
HL 3.50E-014	atm m3/mol	25	EST	MEYLAN,WM & HOWARD,PH (1991)
OH 4.00E-011	cm3/molc sec	25	EST	MEYLAN,WM & HOWARD,PH (1993)

000112-36-7 — BIS (2-ETHOXY ETHYL) ETHER

Formula: $C_8H_{18}O_3$
Mol Weight: 162.23
MP (deg C): -45
FP (deg C): -44
BP (deg C): 188
BP pressure (mm Hg):

Property/Value	Units	Temp	Data Type	Reference
WS 1.00E+006	mg/L	20	EXP	YALKOWSKY,SH & DANNENFELSER,RM (1992)
logP 0.39			EXP	FUNASAKI,N ET AL. (1984)
VP 5.21E-001	mm Hg	25	EXP	YAWS,CL (1994)
DC	pKa			
HL 1.11E-007	atm m3/mol	25	EST	VP/WSOL
OH 2.68E-011	cm3/molc sec	25	EXP	ATKINSON,R (1989)

000112-37-8 — UNDECANOIC ACID

Formula: $C_{11}H_{22}O_2$
Mol Weight: 186.30
MP (deg C): 28.6
FP (deg C):
BP (deg C): 280
BP pressure (mm Hg):

Property/Value	Units	Temp	Data Type	Reference
WS 5.22E+001	mg/L	30	EXP	YALKOWSKY,SH & DANNENFELSER,RM (1992)
logP 4.42			EXP	SANGSTER,J (1993)
VP 3.81E-003	mm Hg	25	EXT	PERRY,RH & GREEN,D (1984)
DC	pKa			
HL 7.01E-006	atm m3/mol	25	EST	MEYLAN,WM & HOWARD,PH (1991)
OH 1.26E-011	cm3/molc sec	25	EST	MEYLAN,WM & HOWARD,PH (1993)

000112-38-9 — UNDECYLENIC ACID

Formula: $C_{11}H_{20}O_2$
Mol Weight: 184.28
MP (deg C): 115-116
FP (deg C):
BP (deg C):
BP pressure (mm Hg):

Property/Value	Units	Temp	Data Type	Reference
WS 1.07E+002	mg/L	20	EXP	YALKOWSKY,SH & DANNENFELSER,RM (1992)
logP 3.86			EXP	HANSCH,C & LEO,AJ (1985)
VP 4.27E-004	mm Hg	25	EXT	PERRY,RH & GREEN,D (1984)
DC	pKa			
HL 5.23E-006	atm m3/mol	25	EST	MEYLAN,WM & HOWARD,PH (1991)
OH 3.73E-011	cm3/molc sec	25	EST	MEYLAN,WM & HOWARD,PH (1993)

000112-39-0 — METHYL HEXADECANOATE

Formula: $C_{17}H_{34}O_2$
Mol Weight: 270.46
MP (deg C): 20
FP (deg C):
BP (deg C): 417
BP pressure (mm Hg):

Property/Value	Units	Temp	Data Type	Reference
WS 1.17E-002	mg/L	25	EST	MEYLAN,WM ET AL. (1996)
logP 7.25			EST	MEYLAN,WM & HOWARD,PH (1995)
VP 6.04E-005	mm Hg	25	EXT	PERRY,RH & GREEN,D (1984)
DC	pKa			
HL 9.26E-003	atm m3/mol	25	EST	MEYLAN,WM & HOWARD,PH (1991)
OH 1.93E-011	cm3/molc sec	25	EST	MEYLAN,WM & HOWARD,PH (1993)

000112-40-3 — DODECANE

Formula: $C_{12}H_{26}$
Mol Weight: 170.34
MP (deg C): -9.6
FP (deg C):
BP (deg C): 216.3
BP pressure (mm Hg):

Property/Value	Units	Temp	Data Type	Reference
WS 3.70E-003	mg/L	25	EXP	YALKOWSKY,SH & DANNENFELSER,RM (1992)
logP 6.10			EXP	COATES,M ET AL. (1985)
VP 1.36E-001	mm Hg	25	EXP	DAUBERT,TE & DANNER,RP (1989)
DC	pKa			
HL 8.24E+000	atm m3/mol	25	EST	VP/WSOL
OH 1.42E-011	cm3/molc sec	25	EXP	ATKINSON,R (1989)

000112-41-4 — N-DODECENE

Formula: $C_{12}H_{24}$
Mol Weight: 168.33
MP (deg C): -35
FP (deg C):
BP (deg C): 213
BP pressure (mm Hg):

Property/Value	Units	Temp	Data Type	Reference
WS 1.13E-001	mg/L	25	EST	MEYLAN,WM ET AL. (1996)
logP 6.10			EST	MEYLAN,WM & HOWARD,PH (1995)
VP 1.59E-001	mm Hg	25	EXP	DAUBERT,TE & DANNER,RP (1989)
DC	pKa			
HL 4.25E+000	atm m3/mol	25	EST	MEYLAN,WM & HOWARD,PH (1991)
OH 3.84E-011	cm3/molc sec	25	EST	ATKINSON,R (1985)

CAS #: 000112-42-5				1-UNDECANOL

Formula: $C_{11}H_{24}O$

Mol Weight: 172.31

MP (deg C): 19 FP (deg C):

BP (deg C): 243

BP pressure (mm Hg):

Property/Value	Units	Temp	Data Type	Reference
WS 1.91E+001	mg/L	25	EST	MEYLAN,WM ET AL. (1996)
logP 4.72			EXP	ABRAHAM,MH ET AL. (1994)
VP 2.97E-003	mm Hg	25	EXP	DAUBERT,TE & DANNER,RP (1989)
DC	pKa			
HL 7.26E-005	atm m3/mol	25	EST	MEYLAN,WM & HOWARD,PH (1991)
OH 1.68E-011	cm3/molc sec	25	EST	MEYLAN,WM & HOWARD,PH (1993)

CAS #: 000112-48-1				1,2-DIBUTYOXYETHANE

Formula: $C_{10}H_{22}O_2$

Mol Weight: 174.29

MP (deg C): -69.1 FP (deg C):

BP (deg C): 203.3

BP pressure (mm Hg):

Property/Value	Units	Temp	Data Type	Reference
WS 2.00E+003	mg/L	20	EXP	RIDDICK,JA ET AL. (1986)
logP 2.48			EXP	FUNASAKI,N ET AL. (1984)
VP 9.00E-002	mm Hg	20	EXP	FLICK,EW (1991)
DC	pKa			
HL 1.03E-005	atm m3/mol	20	EST	VP/WSOL
OH 3.87E-011	cm3/molc sec	25	EST	MEYLAN,WM & HOWARD,PH (1993)

CAS #: 000112-49-2				2,5,8,11-TETRAOXADODECANE

Formula: $C_8H_{18}O_4$

Mol Weight: 178.23

MP (deg C): -45 FP (deg C):

BP (deg C): 216

BP pressure (mm Hg):

Property/Value	Units	Temp	Data Type	Reference
WS 2.49E+005	mg/L	25	EST	MEYLAN,WM ET AL. (1996)
logP -0.76			EST	MEYLAN,WM & HOWARD,PH (1995)
VP 4.03E-002	mm Hg	25	EXP	YAWS,CL (1994)
DC	pKa			
HL 3.24E-009	atm m3/mol	25	EST	MEYLAN,WM & HOWARD,PH (1991)
OH 4.37E-011	cm3/molc sec	25	EST	MEYLAN,WM & HOWARD,PH (1993)

CAS #: 000112-50-5				ETHOXYTRIETHYLENE GLYCOL

Formula: $C_8H_{18}O_4$

Mol Weight: 178.23

MP (deg C): -18.7 FP (deg C):

BP (deg C): 255.8

BP pressure (mm Hg):

Property/Value	Units	Temp	Data Type	Reference
WS 1.00E+006	mg/L	25	EXP	LOVELL,RJ ET AL. (1980)
logP -0.96			EST	MEYLAN,WM & HOWARD,PH (1995)
VP 2.24E-003	mm Hg	25	EST	NEELY,WB & BLAU,GE (1985)
DC	pKa			
HL 4.77E-014	atm m3/mol	25	EST	MEYLAN,WM & HOWARD,PH (1991)
OH 4.54E-011	cm3/molc sec	25	EST	MEYLAN,WM & HOWARD,PH (1993)

CAS #: 000112-51-6				6,7-DITHIADODECANE

Formula: $C_{10}H_{22}S_2$

Mol Weight: 206.41

MP (deg C): FP (deg C):

BP (deg C): 119

BP pressure (mm Hg): 7.00E+000

Property/Value	Units	Temp	Data Type	Reference
WS 4.50E-001	mg/L	25	EST	MEYLAN,WM ET AL. (1996)
logP 5.80			EST	MEYLAN,WM & HOWARD,PH (1995)
VP 6.35E-003	mm Hg	25	EXT	ZWOLINSKI,BJ & WILHOIT,RC (1971)
DC	pKa			
HL 1.17E-002	atm m3/mol	25	EST	MEYLAN,WM & HOWARD,PH (1991)
OH 2.51E-010	cm3/molc sec	25	EST	MEYLAN,WM & HOWARD,PH (1993)

CAS #: 000112-52-7				1-CHLORODODECANE

Formula: $C_{12}H_{25}Cl$

Mol Weight: 204.79

MP (deg C): -9.3 FP (deg C):

BP (deg C): 260

BP pressure (mm Hg):

Property/Value	Units	Temp	Data Type	Reference
WS 1.26E-002	mg/L	25	EST	MUELLER,M & KLEIN,W (1992)
logP 6.49			EST	MEYLAN,WM & HOWARD,PH (1995)
VP 6.30E-003	mm Hg	25	EXT	OHE,S (1976)
DC	pKa			
HL 1.85E-001	atm m3/mol	25	EST	MEYLAN,WM & HOWARD,PH (1991)
OH 1.35E-011	cm3/molc sec	25	EST	MEYLAN,WM & HOWARD,PH (1993)

CAS #: 000112-53-8				DODECANOL

Formula: $C_{12}H_{26}O$

Mol Weight: 186.34

MP (deg C): 24 FP (deg C):

BP (deg C): 259

BP pressure (mm Hg):

Property/Value	Units	Temp	Data Type	Reference
WS 4.00E+000	mg/L	25	EXP	BARTON,AFM (1984)
logP 5.13			EXP	HANSCH,C & LEO,AJ (1985)
VP 8.48E-004	mm Hg	25	EXP	DAUBERT,TE & DANNER,RP (1989)
DC	pKa			
HL 5.20E-005	atm m3/mol	25	EST	VP/WSOL
OH 1.75E-011	cm3/molc sec	25	EST	MEYLAN,WM & HOWARD,PH (1993)

CAS #: 000112-54-9				DODECANAL

Formula: $C_{12}H_{24}O$

Mol Weight: 184.32

MP (deg C): 44.5 FP (deg C):

BP (deg C): 185

BP pressure (mm Hg): 1.00E+002

Property/Value	Units	Temp	Data Type	Reference
WS 4.65E+000	mg/L	25	EST	MEYLAN,WM ET AL. (1996)
logP 4.75			EST	MEYLAN,WM & HOWARD,PH (1995)
VP 1.53E-002	mm Hg	25	EXP	DAUBERT,TE & DANNER,RP (1989)
DC	pKa			
HL 1.15E-003	atm m3/mol	25	EST	MEYLAN,WM & HOWARD,PH (1991)
OH 3.73E-011	cm3/molc sec	25	EST	MEYLAN,WM & HOWARD,PH (1993)

214

CAS #: 000112-55-0			DODECYL MERCAPTAN	
Formula: $C_{12}H_{26}S$				
Mol Weight: 202.41				
MP (deg C):		FP (deg C):		
BP (deg C): 142-145				
BP pressure (mm Hg): 1.50E+001				

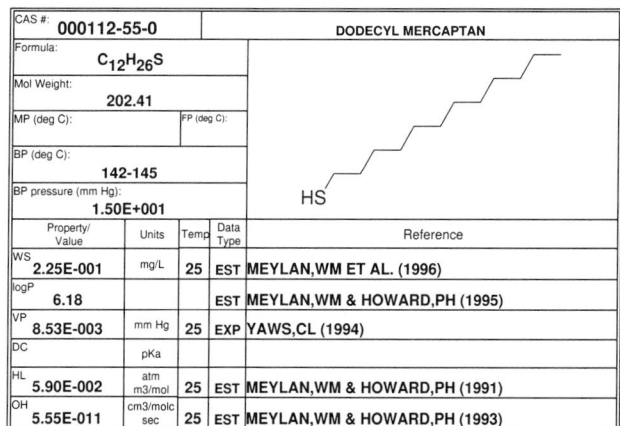

Property/Value	Units	Temp	Data Type	Reference
WS 2.25E-001	mg/L	25	EST	MEYLAN,WM ET AL. (1996)
logP 6.18			EST	MEYLAN,WM & HOWARD,PH (1995)
VP 8.53E-003	mm Hg	25	EXP	YAWS,CL (1994)
DC	pKa			
HL 5.90E-002	atm m3/mol	25	EST	MEYLAN,WM & HOWARD,PH (1991)
OH 5.55E-011	cm3/molc sec	25	EST	MEYLAN,WM & HOWARD,PH (1993)

CAS #: 000112-56-1			2(2-BUTOXYETHOXY)ETHYLTHIOCYANATE	
Formula: $C_9H_{17}NO_2S$				
Mol Weight: 203.31				
MP (deg C):		FP (deg C):		
BP (deg C): 120-125				
BP pressure (mm Hg): 2.50E-001				

Property/Value	Units	Temp	Data Type	Reference
WS 1.55E+003	mg/L	25	EST	MEYLAN,WM ET AL. (1996)
logP 1.68			EXP	HANSCH,C & LEO,AJ (1985)
VP 5.94E-004	mm Hg	25	EST	NEELY,WB & BLAU,GE (1985)
DC	pKa			
HL 2.48E-008	atm m3/mol	25	EST	MEYLAN,WM & HOWARD,PH (1991)
OH 5.89E-011	cm3/molc sec	25	EST	MEYLAN,WM & HOWARD,PH (1993)

CAS #: 000112-57-2			TETRAETHYLENEPENTAMINE	
Formula: $C_8H_{23}N_5$				
Mol Weight: 189.31				
MP (deg C):		FP (deg C):		
BP (deg C): 333				
BP pressure (mm Hg):				

Property/Value	Units	Temp	Data Type	Reference
WS 6.54E+006	mg/L		EXP	YALKOWSKY,SH & DANNENFELSER,RM (1992)
logP -3.16			EST	MEYLAN,WM & HOWARD,PH (1995)
VP 8.00E-007	mm Hg	25	EXP	DAUBERT,TE & DANNER,RP (1991)
DC 9.68	pKa		EXP	PERRIN,DD (1965); pKa1
HL 3.00E-020	atm m3/mol	25	EST	MEYLAN,WM & HOWARD,PH (1991)
OH 3.06E-010	cm3/molc sec	25	EST	ATKINSON,R (1988)

CAS #: 000112-58-3			DI-N-HEXYL ETHER	
Formula: $C_{12}H_{26}O$				
Mol Weight: 186.34				
MP (deg C):		FP (deg C):		
BP (deg C): 223				
BP pressure (mm Hg):				

Property/Value	Units	Temp	Data Type	Reference
WS 2.87E+000	mg/L	25	EST	MEYLAN,WM ET AL. (1996)
logP 4.98			EST	MEYLAN,WM & HOWARD,PH (1995)
VP 4.61E-002	mm Hg	25	EXP	DAUBERT,TE & DANNER,RP (1989)
DC	pKa			
HL 1.47E-002	atm m3/mol	25	EST	MEYLAN,WM & HOWARD,PH (1991)
OH 3.03E-011	cm3/molc sec	25	EST	MEYLAN,WM & HOWARD,PH (1993)

CAS #: 000112-59-4			ETHANOL, 2- 2-(HEXYLOXY)ETHOXY -	
Formula: $C_{10}H_{22}O_3$				
Mol Weight: 190.29				
MP (deg C): -40		FP (deg C): -40		
BP (deg C): 260				
BP pressure (mm Hg):				

Property/Value	Units	Temp	Data Type	Reference
WS 1.70E+004	mg/L	20	EXP	FLICK,EW (1991)
logP 1.70			EXP	FUNASAKI,N ET AL. (1984)
VP 5.03E-004	mm Hg	25	EST	NEELY,WB & BLAU,GE (1985)
DC	pKa			
HL 8.90E-011	atm m3/mol	25	EST	MEYLAN,WM & HOWARD,PH (1991)
OH 4.04E-011	cm3/molc sec	25	EST	MEYLAN,WM & HOWARD,PH (1993)

CAS #: 000112-60-7			TETRAETHYLENE GLYCOL	
Formula: $C_8H_{18}O_5$				
Mol Weight: 194.23				
MP (deg C): -6.2		FP (deg C):		
BP (deg C): 327.3				
BP pressure (mm Hg):				

Property/Value	Units	Temp	Data Type	Reference
WS 1.00E+006	mg/L	20	EXP	YALKOWSKY,SH & DANNENFELSER,RM (1992)
logP -2.02			EST	MEYLAN,WM & HOWARD,PH (1995)
VP 4.65E-005	mm Hg	26	EXP	YAWS,CL (1994B)
DC	pKa			
HL 5.50E-019	atm m3/mol	25	EST	MEYLAN,WM & HOWARD,PH (1991)
OH 4.70E-011	cm3/molc sec	25	EST	MEYLAN,WM & HOWARD,PH (1993)

CAS #: 000112-61-8			METHYL STEARATE	
Formula: $C_{19}H_{38}O_2$				
Mol Weight: 298.51				
MP (deg C): 39.1		FP (deg C):		
BP (deg C): 443				
BP pressure (mm Hg):				

Property/Value	Units	Temp	Data Type	Reference
WS 1.17E-003	mg/L	25	EST	MEYLAN,WM ET AL. (1996)
logP 8.23			EST	MEYLAN,WM & HOWARD,PH (1995)
VP 1.36E-005	mm Hg	25	EXT	OHE,S (1976)
DC	pKa			
HL 1.63E-002	atm m3/mol	25	EST	MEYLAN,WM & HOWARD,PH (1991)
OH 2.22E-011	cm3/molc sec	25	EST	MEYLAN,WM & HOWARD,PH (1993)

CAS #: 000112-62-9			METHYL OLEATE	
Formula: $C_{19}H_{36}O_2$				
Mol Weight: 296.50				
MP (deg C): -19.9		FP (deg C):		
BP (deg C): 218.5				
BP pressure (mm Hg): 2.00E+001				

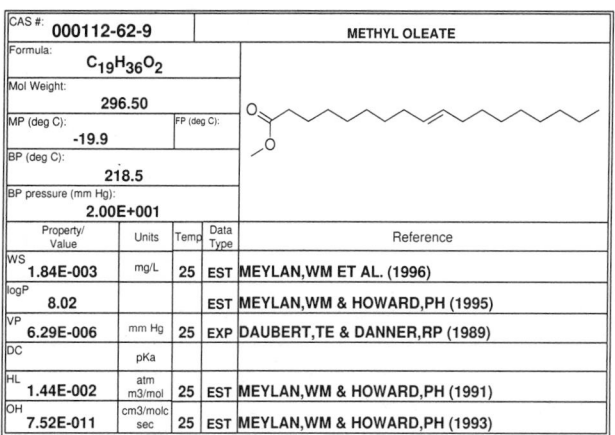

Property/Value	Units	Temp	Data Type	Reference
WS 1.84E-003	mg/L	25	EST	MEYLAN,WM ET AL. (1996)
logP 8.02			EST	MEYLAN,WM & HOWARD,PH (1995)
VP 6.29E-006	mm Hg	25	EXP	DAUBERT,TE & DANNER,RP (1989)
DC	pKa			
HL 1.44E-002	atm m3/mol	25	EST	MEYLAN,WM & HOWARD,PH (1991)
OH 7.52E-011	cm3/molc sec	25	EST	MEYLAN,WM & HOWARD,PH (1993)

CAS #: 000112-63-0			METHYL LINOLEATE	

Formula: $C_{19}H_{34}O_2$

Mol Weight: 294.48

MP (deg C): -35 | FP (deg C):

BP (deg C): 215

BP pressure (mm Hg): 2.00E+001

Property/Value	Units	Temp	Data Type	Reference
WS 2.88E-003	mg/L	25	EST	MEYLAN,WM ET AL. (1996)
logP 7.80			EST	MEYLAN,WM & HOWARD,PH (1995)
VP 3.67E-006	mm Hg	25	EXT	BOUBLIK,T ET AL. (1984)
DC	pKa			
HL 1.26E-002	atm m3/mol	25	EST	MEYLAN,WM & HOWARD,PH (1991)
OH 1.28E-010	cm3/molc sec	25	EST	MEYLAN,WM & HOWARD,PH (1993)

CAS #: 000112-66-3			DODECYL ACETATE	

Formula: $C_{14}H_{28}O_2$

Mol Weight: 228.38

MP (deg C): 1.3 | FP (deg C):

BP (deg C): 265

BP pressure (mm Hg):

Property/Value	Units	Temp	Data Type	Reference
WS 3.63E-001	mg/L	25	EST	MEYLAN,WM ET AL. (1996)
logP 5.78			EST	MEYLAN,WM & HOWARD,PH (1995)
VP 4.68E-004	mm Hg		EXP	HEATH,R & TUMLINSON,JH (1986)
DC	pKa			
HL 3.96E-003	atm m3/mol	25	EST	MEYLAN,WM & HOWARD,PH (1991)
OH 1.60E-011	cm3/molc sec	25	EST	MEYLAN,WM & HOWARD,PH (1993)

CAS #: 000112-70-9			1-TRIDECANOL	

Formula: $C_{13}H_{28}O$

Mol Weight: 200.37

MP (deg C): 32.5 | FP (deg C):

BP (deg C): 152

BP pressure (mm Hg): 1.40E+001

Property/Value	Units	Temp	Data Type	Reference
WS 1.36E+000	mg/L	25	EST	MEYLAN,WM ET AL. (1996)
logP 5.82			EXP	ABRAHAM,MH (1994)
VP 4.36E-004	mm Hg	25	EXP	DAUBERT,TE & DANNER,RP (1989)
DC	pKa			
HL 1.28E-004	atm m3/mol	25	EST	MEYLAN,WM & HOWARD,PH (1991)
OH 1.96E-011	cm3/molc sec	25	EST	MEYLAN,WM & HOWARD,PH (1993)

CAS #: 000112-71-0			1-BROMO-N-TETRADECANE	

Formula: $C_{14}H_{29}Br$

Mol Weight: 277.30

MP (deg C): 5.6 | FP (deg C):

BP (deg C): 307

BP pressure (mm Hg):

Property/Value	Units	Temp	Data Type	Reference
WS 5.80E-003	mg/L	25	EST	MEYLAN,WM ET AL. (1996)
logP 7.56			EST	MEYLAN,WM & HOWARD,PH (1995)
VP 4.49E-004	mm Hg	25	EXT	LI,JCM & ROSSINI,FD (1953)
DC	pKa			
HL 3.38E-001	atm m3/mol	25	EST	MEYLAN,WM & HOWARD,PH (1991)
OH 1.63E-011	cm3/molc sec	25	EST	MEYLAN,WM & HOWARD,PH (1993)

CAS #: 000112-72-1			1-TETRADECANOL	

Formula: $C_{14}H_{30}O$

Mol Weight: 214.39

MP (deg C): 39.5 | FP (deg C):

BP (deg C): 289

BP pressure (mm Hg): 1.50E+001

Property/Value	Units	Temp	Data Type	Reference
WS 1.91E-001	mg/L	25	EXP	YALKOWSKY,SH & DANNENFELSER,RM (1992)
logP 6.03			EXP	BURKHARD,LP ET AL. (1985B)
VP 1.10E-004	mm Hg	25	EST	DAUBERT,TE & DANNER,RP (1989)
DC	pKa			
HL 1.60E-004	atm m3/mol	25	EST	VP/WSOL
OH 2.10E-011	cm3/molc sec	25	EST	MEYLAN,WM & HOWARD,PH (1993)

CAS #: 000112-73-2			DIETHYLENE GLYCOL DIBUTYL ETHER	

Formula: $C_{12}H_{26}O_3$

Mol Weight: 218.34

MP (deg C): -60 | FP (deg C):

BP (deg C): 256

BP pressure (mm Hg):

Property/Value	Units	Temp	Data Type	Reference
WS 3.00E+003	mg/L	20	EXP	RIDDICK,JA ET AL. (1986)
logP 1.92			EXP	FUNASAKI,N ET AL. (1984)
VP 2.61E-002	mm Hg	25	EXP	YAWS,CL (1994)
DC	pKa			
HL 2.50E-006	atm m3/mol	25	EST	VP/WSOL
OH 5.27E-011	cm3/molc sec	25	EST	MEYLAN,WM & HOWARD,PH (1993)

CAS #: 000112-80-1			OLEIC ACID	

Formula: $C_{18}H_{34}O_2$

Mol Weight: 282.47

MP (deg C): 13.4 | FP (deg C):

BP (deg C): 286

BP pressure (mm Hg): 1.00E+002

Property/Value	Units	Temp	Data Type	Reference
WS 1.15E-002	mg/L	25	EST	MEYLAN,WM ET AL. (1996)
logP 7.73			EST	MEYLAN,WM & HOWARD,PH (1995)
VP 5.46E-007	mm Hg	25	EXP	DAUBERT,TE & DANNER,RP (1989)
DC	pKa			
HL 4.48E-005	atm m3/mol	25	EST	MEYLAN,WM & HOWARD,PH (1991)
OH 7.55E-011	cm3/molc sec	25	EST	MEYLAN,WM & HOWARD,PH (1993)

CAS #: 000112-82-3			1-BROMOHEXADECANE	

Formula: $C_{16}H_{33}Br$

Mol Weight: 305.35

MP (deg C): 18 | FP (deg C):

BP (deg C): 336

BP pressure (mm Hg):

Property/Value	Units	Temp	Data Type	Reference
WS 5.79E-004	mg/L	25	EST	MEYLAN,WM ET AL. (1996)
logP 8.54			EST	MEYLAN,WM & HOWARD,PH (1995)
VP 4.85E-005	mm Hg	25	EXT	LI,JCM & ROSSINI,FD (1953)
DC	pKa			
HL 5.95E-001	atm m3/mol	25	EST	MEYLAN,WM & HOWARD,PH (1991)
OH 1.91E-011	cm3/molc sec	25	EST	MEYLAN,WM & HOWARD,PH (1993)

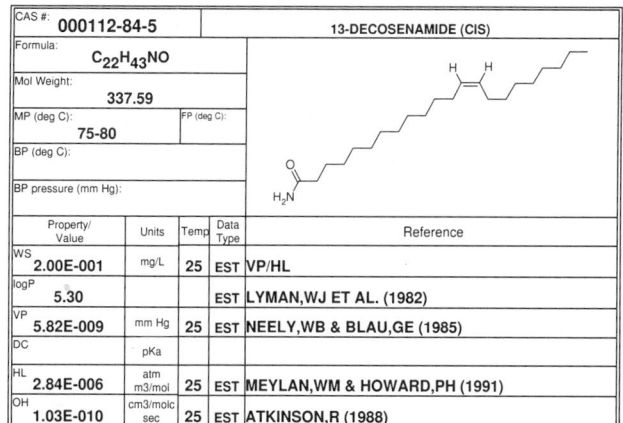

CAS #: 000112-84-5 — 13-DECOSENAMIDE (CIS)

Formula: $C_{22}H_{43}NO$
Mol Weight: 337.59
MP (deg C): 75-80
FP (deg C):
BP (deg C):
BP pressure (mm Hg):

Property/Value	Units	Temp	Data Type	Reference
WS 2.00E-001	mg/L	25	EST	VP/HL
logP 5.30			EST	LYMAN,WJ ET AL. (1982)
VP 5.82E-009	mm Hg	25	EST	NEELY,WB & BLAU,GE (1985)
DC	pKa			
HL 2.84E-006	atm m3/mol	25	EST	MEYLAN,WM & HOWARD,PH (1991)
OH 1.03E-010	cm3/molc sec	25	EST	ATKINSON,R (1988)

CAS #: 000112-85-6 — DOCOSANOIC ACID

Formula: $C_{22}H_{44}O_2$
Mol Weight: 340.59
MP (deg C): 79.95
FP (deg C):
BP (deg C): 306
BP pressure (mm Hg): 6.00E+001

Property/Value	Units	Temp	Data Type	Reference
WS 6.05E-005	mg/L	25	EST	MEYLAN,WM ET AL. (1996)
logP 9.91			EST	MEYLAN,WM & HOWARD,PH (1995)
VP 7.15E-008	mm Hg	25	EST	NEELY,WB & BLAU,GE (1985)
DC	pKa			
HL 1.58E-004	atm m3/mol	25	EST	MEYLAN,WM & HOWARD,PH (1991)
OH 2.81E-011	cm3/molc sec	25	EST	MEYLAN,WM & HOWARD,PH (1993)

CAS #: 000112-88-9 — 1-OCTADECENE

Formula: $C_{18}H_{36}$
Mol Weight: 252.49
MP (deg C): 17.5
FP (deg C):
BP (deg C): 179
BP pressure (mm Hg): 1.50E+001

Property/Value	Units	Temp	Data Type	Reference
WS 1.26E-004	mg/L	25	EST	MEYLAN,WM ET AL. (1996)
logP 9.04			EST	MEYLAN,WM & HOWARD,PH (1995)
VP 6.75E-005	mm Hg	25	EXP	DAUBERT,TE & DANNER,RP (1989)
DC	pKa			
HL 1.07E+001	atm m3/mol	25	EST	MEYLAN,WM & HOWARD,PH (1991)
OH 4.71E-011	cm3/molc sec	25	EST	MEYLAN,WM & HOWARD,PH (1993)

CAS #: 000112-89-0 — 1-BROMOOCTADECANE

Formula: $C_{18}H_{37}Br$
Mol Weight: 333.40
MP (deg C): 28.2
FP (deg C):
BP (deg C): 210
BP pressure (mm Hg): 1.00E+001

Property/Value	Units	Temp	Data Type	Reference
WS 5.72E-005	mg/L	25	EST	MEYLAN,WM ET AL. (1996)
logP 9.52			EST	MEYLAN,WM & HOWARD,PH (1995)
VP 5.94E-006	mm Hg	25	EXP	LI,JCM & ROSSINI,FD (1953
DC	pKa			
HL 1.05E+000	atm m3/mol	25	EST	MEYLAN,WM & HOWARD,PH (1991)
OH 2.19E-011	cm3/molc sec	25	EST	MEYLAN,WM & HOWARD,PH (1993)

CAS #: 000112-92-5 — 1-OCTADECANOL

Formula: $C_{18}H_{38}O$
Mol Weight: 270.50
MP (deg C): 59.4
FP (deg C):
BP (deg C): 210
BP pressure (mm Hg): 1.50E+001

Property/Value	Units	Temp	Data Type	Reference
WS 1.10E-003	mg/L	25	EXP	YALKOWSKY,SH & VALVANI,SC (1980)
logP 7.70			EST	MEYLAN,WM & HOWARD,PH (1995)
VP 2.70E-006	mm Hg	25	EXP	DAUBERT,TE & DANNER,RP (1989)
DC	pKa			
HL 8.41E-004	atm m3/mol		EST	VP/WSOL
OH 2.67E-011	cm3/molc sec	25	EST	MEYLAN,WM & HOWARD,PH (1993)

CAS #: 000112-95-8 — EICOSANE

Formula: $C_{20}H_{42}$
Mol Weight: 282.56
MP (deg C): 36.8
FP (deg C):
BP (deg C): 343
BP pressure (mm Hg):

Property/Value	Units	Temp	Data Type	Reference
WS 1.90E-003	mg/L	25	EXP	MACKAY,D & SHIU,WY (1981)
logP 10.16			EST	MEYLAN,WM & HOWARD,PH (1995)
VP 4.62E-006	mm Hg	25	EXT	ZWOLINSKI,BJ & WILHOIT,RC (1971)
DC	pKa			
HL 9.02E+001	atm m3/mol	25	EST	MEYLAN,WM & HOWARD,PH (1991)
OH 2.52E-011	cm3/molc sec	25	EST	MEYLAN,WM & HOWARD,PH (1993)

CAS #: 000113-00-8 — GUANIDINE

Formula: CH_5N_3
Mol Weight: 59.07
MP (deg C): 50
FP (deg C):
BP (deg C):
BP pressure (mm Hg):

Property/Value	Units	Temp	Data Type	Reference
WS 1.84E+003	mg/L	20	EXP	GREENWALD,I (1926)
logP -1.63			EST	MEYLAN,WM & HOWARD,PH (1995)
VP 4.10E+000	mm Hg	25	EST	NEELY,WB & BLAU,GE (1985)
DC	pKa			
HL 2.34E-011	atm m3/mol	25	EST	MEYLAN,WM & HOWARD,PH (1991)
OH 4.20E-011	cm3/molc sec	25	EST	MEYLAN,WM & HOWARD,PH (1993)

CAS #: 000113-48-4 — MGK 264

Formula: $C_{17}H_{25}NO_2$
Mol Weight: 275.39
MP (deg C): < -20
FP (deg C):
BP (deg C): 157
BP pressure (mm Hg):

Property/Value	Units	Temp	Data Type	Reference
WS 1.37E+001	mg/L	25	EST	MEYLAN,WM ET AL. (1996)
logP 3.70			EXP	TOMLIN,C (1994)
VP 1.53E-008	mm Hg	25	EST	MPBPVP
DC	pKa			
HL 2.85E-007	atm m3/mol	25	EST	MEYLAN,WM & HOWARD,PH (1991)
OH 1.01E-010	cm3/molc sec	25	EST	MEYLAN,WM & HOWARD,PH (1993)

CAS #: 000113-53-1				DOLSULEPINE

Formula: $C_{19}H_{21}NS$

Mol Weight: 295.45

MP (deg C): 55-57 FP (deg C):

BP (deg C): 171-172

BP pressure (mm Hg): 5.00E-002

Property/Value	Units	Temp	Data Type	Reference
WS 1.51E+000	mg/L	25	EST	MEYLAN,WM ET AL. (1996)
logP 4.49			EXP	HANSCH,C & LEO,AJ (1985)
VP 4.28E-007	mm Hg	25	EST	NEELY,WB & BLAU,GE (1985)
DC	pKa			
HL 1.36E-009	atm m3/mol	25	EST	MEYLAN,WM & HOWARD,PH (1991)
OH 1.91E-010	cm3/molc sec	25	EST	MEYLAN,WM & HOWARD,PH (1993)

CAS #: 000113-59-7				CHLORPROTHIXENE

Formula: $C_{18}H_{18}ClNS$

Mol Weight: 315.87

MP (deg C): 97-98 FP (deg C):

BP (deg C):

BP pressure (mm Hg):

Property/Value	Units	Temp	Data Type	Reference
WS 2.95E-001	mg/L	25	EST	MEYLAN,WM ET AL. (1996)
logP 5.18			EXP	HANSCH,C & LEO,AJ (1985)
VP 2.37E-007	mm Hg	25	EST	NEELY,WB & BLAU,GE (1985)
DC	pKa			
HL 2.50E-009	atm m3/mol	25	EST	MEYLAN,WM & HOWARD,PH (1991)
OH 1.83E-010	cm3/molc sec	25	EST	MEYLAN,WM & HOWARD,PH (1993)

CAS #: 000114-07-8				ERYTHROMYCIN

Formula: $C_{37}H_{67}NO_{13}$

Mol Weight: 733.95

MP (deg C): 135-140 FP (deg C):

BP (deg C):

BP pressure (mm Hg):

Property/Value	Units	Temp	Data Type	Reference
WS 1.44E+000	mg/L	25	EST	MEYLAN,WM ET AL. (1996)
logP 2.54			EXP	HANSCH,C ET AL. (1995)
VP 2.28E-027	mm Hg	25	EST	NEELY,WB & BLAU,GE (1985)
DC	pKa			
HL 5.42E-029	atm m3/mol	25	EST	MEYLAN,WM & HOWARD,PH (1991)
OH 4.04E-010	cm3/molc sec	25	EST	MEYLAN,WM & HOWARD,PH (1993)

CAS #: 000114-26-1				PROPOXUR

Formula: $C_{11}H_{15}NO_3$

Mol Weight: 209.25

MP (deg C): 91.5 FP (deg C):

BP (deg C):

BP pressure (mm Hg):

Property/Value	Units	Temp	Data Type	Reference
WS 1.86E+003	mg/L	30	EXP	BOWMAN,BT & SANS,WW (1983)
logP 1.52			EXP	HANSCH,C ET AL. (1995)
VP 3.00E-006	mm Hg	20	EXP	WRIGHT,CG ET AL. (1981)
DC	pKa			
HL 4.00E-007	atm m3/mol	25	EST	VP/WSOL
OH 4.07E-011	cm3/molc sec	25	EST	MEYLAN,WM & HOWARD,PH (1993)

CAS #: 000114-33-0				N-METHYLNICOTINAMIDE

Formula: $C_7H_8N_2O$

Mol Weight: 136.15

MP (deg C): 102-105 FP (deg C):

BP (deg C):

BP pressure (mm Hg):

Property/Value	Units	Temp	Data Type	Reference
WS 8.69E+004	mg/L	25	EST	MEYLAN,WM ET AL. (1996)
logP 0.00			EXP	HANSCH,C & LEO,AJ (1985)
VP 2.36E-004	mm Hg	25	EST	NEELY,WB & BLAU,GE (1985)
DC	pKa			
HL 6.36E-012	atm m3/mol	25	EST	MEYLAN,WM & HOWARD,PH (1991)
OH 6.39E-012	cm3/molc sec	25	EST	MEYLAN,WM & HOWARD,PH (1993)

CAS #: 000114-38-5				O-CHLOROPHENYLUREA

Formula: $C_7H_7ClN_2O$

Mol Weight: 170.60

MP (deg C): FP (deg C):

BP (deg C):

BP pressure (mm Hg):

Property/Value	Units	Temp	Data Type	Reference
WS 5.03E+003	mg/L	25	EST	MEYLAN,WM ET AL. (1996)
logP 1.27			EXP	HANSCH,C & LEO,AJ (1985)
VP 1.65E-004	mm Hg	25	EST	NEELY,WB & BLAU,GE (1985)
DC	pKa			
HL 1.49E-010	atm m3/mol	25	EST	MEYLAN,WM & HOWARD,PH (1991)
OH 1.35E-011	cm3/molc sec	25	EST	MEYLAN,WM & HOWARD,PH (1993)

CAS #: 000114-49-8				SCOPOLAMINE BROMIDE

Formula: $C_{17}H_{22}BrNO_4$

Mol Weight: 384.28

MP (deg C): 196-197 FP (deg C):

BP (deg C):

BP pressure (mm Hg):

Property/Value	Units	Temp	Data Type	Reference
WS 6.66E+005	mg/L	25	EXP	SEIDELL,A (1941)
logP -1.13			EST	MEYLAN,WM & HOWARD,PH (1995)
VP 1.13E-015	mm Hg	25	EST	NEELY,WB & BLAU,GE (1985)
DC	pKa			
HL 1.68E-023	atm m3/mol	25	EST	MEYLAN,WM & HOWARD,PH (1991)
OH 5.27E-011	cm3/molc sec	25	EST	MEYLAN,WM & HOWARD,PH (1993)

CAS #: 000114-86-3				PHENFORMIN

Formula: $C_{10}H_{15}N_5$

Mol Weight: 205.26

MP (deg C): 175-178 FP (deg C):

BP (deg C):

BP pressure (mm Hg):

Property/Value	Units	Temp	Data Type	Reference
WS 2.10E+005	mg/L	25	EST	MEYLAN,WM ET AL. (1996)
logP -0.83			EXP	HANSCH,C & LEO,AJ (1985)
VP 2.01E-006	mm Hg	25	EST	NEELY,WB & BLAU,GE (1985)
DC	pKa			
HL 3.73E-017	atm m3/mol	25	EST	MEYLAN,WM & HOWARD,PH (1991)
OH 1.64E-010	cm3/molc sec	25	EST	MEYLAN,WM & HOWARD,PH (1993)

CAS #: 000115-02-6 — AZASERINE

Formula: $C_5H_7N_3O_4$

Mol Weight: 173.13

MP (deg C): 146-162

FP (deg C):

BP (deg C):

BP pressure (mm Hg):

Property/Value	Units	Temp	Data Type	Reference
WS 2.58E+004	mg/L	25	EST	MEYLAN,WM ET AL. (1996)
logP -2.36			EST	MEYLAN,WM & HOWARD,PH (1995)
VP 4.70E-013	mm Hg	25	EST	NEELY,WB & BLAU,GE (1985)
DC 8.55	pKa		EXP	MERCK INDEX (1983)
HL 3.25E-017	atm m3/mol	25	EST	MEYLAN,WM & HOWARD,PH (1991)
OH 4.00E-011	cm3/molc sec	25	EST	MEYLAN,WM & HOWARD,PH (1993)

CAS #: 000115-07-1 — PROPENE

Formula: C_3H_6

Mol Weight: 42.08

MP (deg C): -185.2

FP (deg C):

BP (deg C): -48

BP pressure (mm Hg):

Property/Value	Units	Temp	Data Type	Reference
WS 2.00E+002	mg/L	25	EXP	MCAULIFFE,C (1966)
logP 1.77			EXP	HANSCH,C & LEO,AJ (1985)
VP 8.69E+003	mm Hg	25	EXP	DAUBERT,TE & DANNER,RP (1985)
DC	pKa			
HL 1.96E-001	atm m3/mol	25	EXP	WASIK,SP & TSANG,W (1970)
OH 2.63E+001	cm3/molc sec	25	EXP	ATKINSON,R (1989)

CAS #: 000115-10-6 — DIMETHYL ETHER

Formula: C_2H_6O

Mol Weight: 46.07

MP (deg C): -141.5

FP (deg C):

BP (deg C): -24.8

BP pressure (mm Hg):

Property/Value	Units	Temp	Data Type	Reference
WS 3.53E+004	mg/L	25	EXP	RIDDICK,JA ET AL. (1986)
logP 0.10			EXP	HANSCH,C & LEO,AJ (1985)
VP 4.45E+003	mm Hg	25	EXP	RIDDICK,JA ET AL. (1986)
DC	pKa			
HL 1.31E-003	atm m3/mol	25	EST	VP/WSOL
OH 2.98E-012	cm3/molc sec	25	EXP	ATKINSON,R (1989)

CAS #: 000115-11-7 — ISOBUTENE

Formula: C_4H_8

Mol Weight: 56.11

MP (deg C): -140.4

FP (deg C):

BP (deg C): -6.9

BP pressure (mm Hg):

Property/Value	Units	Temp	Data Type	Reference
WS 2.63E+002	mg/L	25	EXP	YALKOWSKY,SH & DANNENFELSER,RM (1992)
logP 2.34			EXP	HANSCH,C & LEO,AJ (1985)
VP 7.80E-001	mm Hg	12	EST	HOFF,MC ET AL. (1978)
DC	pKa			
HL 2.18E-001	atm m3/mol	25	EXP	WASIK,SP & TSANG,W (1970)
OH 5.14E-011	cm3/molc sec	25	EXP	ATKINSON,R (1989)

CAS #: 000115-17-3 — TRIBROMOACETALDEHYDE

Formula: C_2HBr_3O

Mol Weight: 280.76

MP (deg C): -57.5

FP (deg C):

BP (deg C): 174

BP pressure (mm Hg):

Property/Value	Units	Temp	Data Type	Reference
WS 9.02E+002	mg/L	25	EST	MEYLAN,WM ET AL. (1996)
logP 1.46			EST	MEYLAN,WM & HOWARD,PH (1995)
VP 1.52E+000	mm Hg	25	EXP	PERRY,RH & GREEN,D (1984)
DC	pKa			
HL 1.04E-007	atm m3/mol	25	EST	MEYLAN,WM & HOWARD,PH (1991)
OH 1.18E-012	cm3/molc sec	25	EST	MEYLAN,WM & HOWARD,PH (1993)

CAS #: 000115-19-5 — 2-METHYL-3-BUTYN-2-OL

Formula: C_5H_8O

Mol Weight: 84.12

MP (deg C): 3

FP (deg C):

BP (deg C): 104

BP pressure (mm Hg):

Property/Value	Units	Temp	Data Type	Reference
WS 2.40E+005	mg/L	25	EST	MEYLAN,WM ET AL. (1996)
logP 0.28			EXP	HANSCH,C ET AL. (1995)
VP 1.60E+001	mm Hg	25	EXP	BOUBLIK,T ET AL. (1984)
DC	pKa			
HL 1.04E-006	atm m3/mol	25	EST	MEYLAN,WM & HOWARD,PH (1991)
OH 8.17E-012	cm3/molc sec	25	EST	MEYLAN,WM & HOWARD,PH (1993)

CAS #: 000115-20-8 — 2,2,2-TRICHLOROETHANOL

Formula: $C_2H_3Cl_3O$

Mol Weight: 149.40

MP (deg C): 18

FP (deg C):

BP (deg C): 151-153

BP pressure (mm Hg):

Property/Value	Units	Temp	Data Type	Reference
WS 1.51E+004	mg/L	25	EST	MEYLAN,WM ET AL. (1996)
logP 1.42			EXP	HANSCH,C & LEO,AJ (1985)
VP 1.08E+000	mm Hg	25	EST	NEELY,WB & BLAU,GE (1985)
DC 12.24	pKa	25	EXP	SERJEANT,EP & DEMPSEY,B (1979)
HL 1.56E-007	atm m3/mol	25	EST	MEYLAN,WM & HOWARD,PH (1991)
OH 2.45E-013	cm3/molc sec	25	EXP	ATKINSON,R (1989)

CAS #: 000115-21-9 — ETHYLTRICHLOROSILANE

Formula: $C_2H_5Cl_3Si$

Mol Weight: 163.51

MP (deg C): -105.6

FP (deg C):

BP (deg C): 100.5

BP pressure (mm Hg):

Property/Value	Units	Temp	Data Type	Reference
WS 4.85E+002	mg/L	25	EST	MEYLAN,WM ET AL. (1996)
logP 2.50			EST	MEYLAN,WM & HOWARD,PH (1995)
VP 4.72E+001	mm Hg	25	EXP	BOUBLIK,T ET AL. (1984)
DC	pKa			
HL 7.09E-003	atm m3/mol	25	EST	MEYLAN,WM & HOWARD,PH (1991)
OH 1.19E-012	cm3/molc sec	25	EST	MEYLAN,WM & HOWARD,PH (1993)

000115-22-0 — 3-HYDROXY-3-METHYL-2-BUTANONE

Formula: $C_5H_{10}O_2$
Mol Weight: 102.13
MP (deg C): 140-141
FP (deg C):
BP (deg C):
BP pressure (mm Hg):

Property/Value	Units	Temp	Data Type	Reference
WS 3.13E+005	mg/L	25	EST	MEYLAN,WM ET AL. (1996)
logP 0.09			EST	MEYLAN,WM & HOWARD,PH (1995)
VP 6.19E+000	mm Hg	25	EXT	BOUBLIK,T ET AL. (1984)
DC	pKa			
HL 1.36E-005	atm m3/mol	25	EST	MEYLAN,WM & HOWARD,PH (1991)
OH 1.30E-012	cm3/molc sec	25	EST	MEYLAN,WM & HOWARD,PH (1993)

000115-24-2 — 2,2-BIS(ETHYLSULFONYL)PROPANE

Formula: $C_7H_{16}O_4S_2$
Mol Weight: 228.33
MP (deg C): 124-126
FP (deg C):
BP (deg C): 300
BP pressure (mm Hg):

Property/Value	Units	Temp	Data Type	Reference
WS 1.38E+004	mg/L	16	EXP	YALKOWSKY,SH & DANNENFELSER,RM (1992)
logP 1.41			EST	MEYLAN,WM & HOWARD,PH (1995)
VP 3.97E-005	mm Hg	25	EST	NEELY,WB & BLAU,GE (1985)
DC	pKa			
HL 1.64E-009	atm m3/mol	25	EST	MEYLAN,WM & HOWARD,PH (1991)
OH 1.52E-011	cm3/molc sec	25	EST	MEYLAN,WM & HOWARD,PH (1993)

000115-25-3 — PERFLUOROCYCLOBUTANE

Formula: C_4F_8
Mol Weight: 200.03
MP (deg C): -41.4
FP (deg C):
BP (deg C): -6.04
BP pressure (mm Hg):

Property/Value	Units	Temp	Data Type	Reference
WS 2.36E+001	mg/L	25	EXP	YALKOWSKY,SH & DANNENFELSER,RM (1992)
logP 1.29			EST	MEYLAN,WM & HOWARD,PH (1995)
VP 2.35E+003	mm Hg	25	EXP	DAUBERT,TE & DANNER,RP (1989)
DC	pKa			
HL 8.48E+000	atm m3/mol	25	EST	VP/WSOL
OH 0.00E+000	cm3/molc sec	25	EST	MEYLAN,WM & HOWARD,PH (1993)

000115-26-4 — DIMEFOX

Formula: $C_4H_{12}FN_2OP$
Mol Weight: 154.13
MP (deg C):
FP (deg C):
BP (deg C): 86
BP pressure (mm Hg): 1.50E+001

Property/Value	Units	Temp	Data Type	Reference
WS 1.00E+006	mg/L	20	EXP	GUNTHER,FA ET AL. (1968)
logP -0.43			EST	MEYLAN,WM & HOWARD,PH (1995)
VP 1.38E+000	mm Hg	25	EST	NEELY,WB & BLAU,GE (1985)
DC	pKa			
HL 9.71E-010	atm m3/mol	25	EST	MEYLAN,WM & HOWARD,PH (1991)
OH 6.01E-011	cm3/molc sec	25	EST	MEYLAN,WM & HOWARD,PH (1993)

000115-28-6 — CHLORENDIC ACID

Formula: $C_9H_4Cl_6O_4$
Mol Weight: 388.85
MP (deg C): 208-210
FP (deg C):
BP (deg C):
BP pressure (mm Hg):

Property/Value	Units	Temp	Data Type	Reference
WS 3.50E+003	mg/L	25	EXP	CHEM INSPECT TEST INST (1992)
logP 3.14			EST	MEYLAN,WM & HOWARD,PH (1995)
VP 3.04E-008	mm Hg	25	EST	NEELY,WB & BLAU,GE (1985)
DC	pKa			
HL 1.12E-013	atm m3/mol	25	EST	MEYLAN,WM & HOWARD,PH (1991)
OH 8.16E+000	cm3/molc sec	25	EST	MEYLAN,WM & HOWARD,PH (1993)

000115-29-7 — ENDOSULFAN

Formula: $C_9H_6Cl_6O_3S$
Mol Weight: 406.93
MP (deg C): 106
FP (deg C):
BP (deg C):
BP pressure (mm Hg):

Property/Value	Units	Temp	Data Type	Reference
WS 4.50E-001	mg/L	20	EST	BOWMAN,BT & SANS,WW (1983)
logP 3.83			EST	FROM ALPHA; HANSCH,C & LEO,AJ (1985)
VP 1.00E-005	mm Hg	25	EXP	SUNTIO,LR ET AL. (1988)
DC	pKa			
HL 1.12E-005	atm m3/mol	20	EST	VP/WSOL
OH 8.25E-011	cm3/molc sec	25	EST	MEYLAN,WM & HOWARD,PH (1993)

000115-31-1 — THANITE

Formula: $C_{13}H_{19}NO_2S$
Mol Weight: 253.37
MP (deg C):
FP (deg C):
BP (deg C): 95
BP pressure (mm Hg): 6.00E-002

Property/Value	Units	Temp	Data Type	Reference
WS 1.42E+001	mg/L	25	EST	MEYLAN,WM ET AL. (1996)
logP 3.75			EST	MEYLAN,WM & HOWARD,PH (1995)
VP 2.50E-005	mm Hg	25	EST	NEELY,WB & BLAU,GE (1985)
DC	pKa			
HL 2.60E-007	atm m3/mol	25	EST	MEYLAN,WM & HOWARD,PH (1991)
OH 1.39E-011	cm3/molc sec	25	EST	MEYLAN,WM & HOWARD,PH (1993)

000115-32-2 — DICOFOL

Formula: $C_{14}H_9Cl_5O$
Mol Weight: 370.49
MP (deg C): 77-78
FP (deg C):
BP (deg C):
BP pressure (mm Hg):

Property/Value	Units	Temp	Data Type	Reference
WS 1.32E+000	mg/L	25	EXP	WEIL,L ET AL. (1974)
logP 5.02			EXP	SAITO,H ET AL. (1993)
VP 1.21E-008	mm Hg	25	EST	NEELY,WB & BLAU,GE (1985)
DC	pKa			
HL 5.59E-010	atm m3/mol	25	EST	MEYLAN,WM & HOWARD,PH (1991)
OH 3.43E-012	cm3/molc sec	25	EST	MEYLAN,WM & HOWARD,PH (1993)

CAS #: 000115-37-7	PARAMORPHINE

Formula: $C_{19}H_{21}NO_3$

Mol Weight: 311.38

MP (deg C): 193 FP (deg C):

BP (deg C): 170-180

BP pressure (mm Hg):

Property/Value	Units	Temp	Data Type	Reference
WS 2.02E+003	mg/L	25	EST	MEYLAN,WM ET AL. (1996)
logP 2.02			EST	MEYLAN,WM & HOWARD,PH (1995)
VP 3.88E-007	mm Hg	25	EST	NEELY,WB & BLAU,GE (1985)
DC	pKa			
HL 1.63E-010	atm m3/mol	25	EST	MEYLAN,WM & HOWARD,PH (1991)
OH 5.02E-010	cm3/molc sec	25	EST	MEYLAN,WM & HOWARD,PH (1993)

CAS #: 000115-38-8	MEPHOBARBITAL

Formula: $C_{13}H_{14}N_2O_3$

Mol Weight: 246.27

MP (deg C): 176 FP (deg C):

BP (deg C):

BP pressure (mm Hg):

Property/Value	Units	Temp	Data Type	Reference
WS 6.66E+002	mg/L	25	EST	MEYLAN,WM ET AL. (1996)
logP 1.84			EXP	HANSCH,C & LEO,AJ (1985)
VP 9.09E-012	mm Hg	25	EST	NEELY,WB & BLAU,GE (1985)
DC	pKa			
HL 3.63E-014	atm m3/mol	25	EST	MEYLAN,WM & HOWARD,PH (1991)
OH 1.07E-011	cm3/molc sec	25	EST	MEYLAN,WM & HOWARD,PH (1993)

CAS #: 000115-43-5	PHENALLYMAL

Formula: $C_{13}H_{12}N_2O_3$

Mol Weight: 244.25

MP (deg C): 156-157.5 FP (deg C):

BP (deg C):

BP pressure (mm Hg):

Property/Value	Units	Temp	Data Type	Reference
WS 9.17E+002	mg/L	25	EST	MEYLAN,WM ET AL. (1996)
logP 1.69			EXP	SANGSTER,J (1993)
VP 9.06E-013	mm Hg	25	EST	NEELY,WB & BLAU,GE (1985)
DC	pKa			
HL 1.63E-014	atm m3/mol	25	EST	MEYLAN,WM & HOWARD,PH (1991)
OH 3.62E-011	cm3/molc sec	25	EST	MEYLAN,WM & HOWARD,PH (1993)

CAS #: 000115-44-6	BARBITURIC ACID, ALLYL,SEC-BUTYL

Formula: $C_{11}H_{16}N_2O_3$

Mol Weight: 224.26

MP (deg C): 108-110 FP (deg C):

BP (deg C):

BP pressure (mm Hg):

Property/Value	Units	Temp	Data Type	Reference
WS 1.81E+003	mg/L	25	EST	MEYLAN,WM ET AL. (1996)
logP 1.47			EXP	HANSCH,C ET AL. (1995)
VP 2.31E-011	mm Hg	25	EST	NEELY,WB & BLAU,GE (1985)
DC 7.79	pKa	25	EXP	KORTUM,G ET AL (1961)
HL 6.29E-013	atm m3/mol	25	EST	MEYLAN,WM & HOWARD,PH (1991)
OH 4.27E-011	cm3/molc sec	25	EST	MEYLAN,WM & HOWARD,PH (1993)

CAS #: 000115-56-0	BARBITURIC,2-THIO-5-ET-5-MEALLYL

Formula: $C_{10}H_{14}N_2O_2S$

Mol Weight: 226.30

MP (deg C): 160-161 FP (deg C):

BP (deg C):

BP pressure (mm Hg):

Property/Value	Units	Temp	Data Type	Reference
WS 4.30E+002	mg/L	25	EST	MEYLAN,WM ET AL. (1996)
logP 2.19			EXP	HANSCH,C & LEO,AJ (1985)
VP 3.96E-010	mm Hg	25	EST	NEELY,WB & BLAU,GE (1985)
DC	pKa			
HL 4.20E-009	atm m3/mol	25	EST	MEYLAN,WM & HOWARD,PH (1991)
OH 7.00E-011	cm3/molc sec	25	EST	MEYLAN,WM & HOWARD,PH (1993)

CAS #: 000115-58-2	BARBITURIC ACID, 5-ET-5-AMYL

Formula: $C_{11}H_{18}N_2O_3$

Mol Weight: 226.28

MP (deg C): FP (deg C):

BP (deg C):

BP pressure (mm Hg):

Property/Value	Units	Temp	Data Type	Reference
WS 1.51E+003	mg/L	25	EXP	YALKOWSKY,SH & DANNENFELSER,RM (1992)
logP 2.24			EXP	HANSCH,C & LEO,AJ (1985)
VP 1.31E-011	mm Hg	25	EST	NEELY,WB & BLAU,GE (1985)
DC	pKa			
HL 8.44E-013	atm m3/mol	25	EST	MEYLAN,WM & HOWARD,PH (1991)
OH 1.44E-011	cm3/molc sec	25	EST	MEYLAN,WM & HOWARD,PH (1993)

CAS #: 000115-69-5	2-AMINO-2-METHYLPROPANEDIOL

Formula: $C_4H_{11}NO_2$

Mol Weight: 105.14

MP (deg C): 109-111 FP (deg C):

BP (deg C): 151-152

BP pressure (mm Hg): 1.00E+001

Property/Value	Units	Temp	Data Type	Reference
WS 1.00E+006	mg/L	25	EST	MEYLAN,WM ET AL. (1996)
logP -1.10			EST	MEYLAN,WM & HOWARD,PH (1995)
VP 6.54E-003	mm Hg	25	EST	NEELY,WB & BLAU,GE (1985)
DC 8.80	pKa	25	EXP	PERRIN,DD (1965)
HL 2.37E-011	atm m3/mol	25	EST	MEYLAN,WM & HOWARD,PH (1991)
OH 2.95E-011	cm3/molc sec	25	EST	MEYLAN,WM & HOWARD,PH (1993)

CAS #: 000115-76-4	2,2-DIETHYL-1,3-PROPANEDIOL

Formula: $C_7H_{16}O_2$

Mol Weight: 132.20

MP (deg C): 61.5 FP (deg C):

BP (deg C): 240.5

BP pressure (mm Hg):

Property/Value	Units	Temp	Data Type	Reference
WS 9.49E+003	mg/L	25	EST	MEYLAN,WM ET AL. (1996)
logP 1.14			EST	MEYLAN,WM & HOWARD,PH (1995)
VP 5.81E-003	mm Hg	25	EST	NEELY,WB & BLAU,GE (1985)
DC	pKa			
HL 5.39E-007	atm m3/mol	25	EST	MEYLAN,WM & HOWARD,PH (1991)
OH 1.10E-011	cm3/molc sec	25	EST	MEYLAN,WM & HOWARD,PH (1993)

CAS #: 000115-77-5 — PENTAERYTHRITOL

Formula:	$C_5H_{12}O_4$	
Mol Weight:	136.15	
MP (deg C): 261-262	FP (deg C):	
BP (deg C): 276		
BP pressure (mm Hg): 3.00E+001		

Property/Value	Units	Temp	Data Type	Reference
WS 7.23E+004	mg/L	25	EXP	WEBER,J & DALEY,J (1978)
logP -1.69			EXP	TSCATS
VP 1.51E+001	mm Hg	261	EXP	DAUBERT,TE & DANNER,RP (1992)
DC 14.10	pKa		EXP	SERJEANT,EP & DEMPSEY,B (1979)
HL 4.10E-010	atm m3/mol	25	EST	MEYLAN,WM & HOWARD,PH (1991)
OH 1.48E-011	cm3/molc sec	25	EST	ATKINSON,R (1988)

CAS #: 000115-86-6 — TRIPHENYLPHOSPHATE

Formula:	$C_{18}H_{15}O_4P$	
Mol Weight:	326.29	
MP (deg C): 49-50	FP (deg C):	
BP (deg C): 245		
BP pressure (mm Hg):		

Property/Value	Units	Temp	Data Type	Reference
WS 1.90E+000	mg/L	25	EXP	SAEGER,VW ET AL. (1979)
logP 4.59			EXP	HANSCH,C & LEO,AJ (1985)
VP 6.28E-006	mm Hg	25	EXT	DOBRY,A & KELLER,R (1957)
DC	pKa			
HL 3.31E-006	atm m3/mol	25	EST	VP/WSOL
OH 1.14E-011	cm3/molc sec	25	EST	ATKINSON,R (1988)

CAS #: 000115-88-8 — OCTYLDIPHENYL PHOSPHATE

Formula:	$C_{20}H_{27}O_4P$	
Mol Weight:	362.41	
MP (deg C):	FP (deg C):	
BP (deg C):		
BP pressure (mm Hg):		

Property/Value	Units	Temp	Data Type	Reference
WS 1.40E-001	mg/L	25	EXP	HOLLIFIELD,HC (1979)
logP 6.37			EST	MEYLAN,WM & HOWARD,PH (1995)
VP 1.24E-007	mm Hg	25	EST	NEELY,WB & BLAU,GE (1985)
DC	pKa			
HL 2.48E-007	atm m3/mol	25	EST	MEYLAN,WM & HOWARD,PH (1991)
OH 3.74E-011	cm3/molc sec	25	EST	MEYLAN,WM & HOWARD,PH (1993)

CAS #: 000115-89-9 — DIPHENYLMETHYL PHOSPHATE

Formula:	$C_{13}H_{13}O_4P$	
Mol Weight:	264.22	
MP (deg C):	FP (deg C):	
BP (deg C): 125		
BP pressure (mm Hg): 3.00E-002		

Property/Value	Units	Temp	Data Type	Reference
WS 2.00E+003	mg/L	25	EXP	YALKOWSKY,SH & DANNENFELSER,RM (1992)
logP 2.93			EST	MEYLAN,WM & HOWARD,PH (1995)
VP 1.16E-005	mm Hg	25	EST	NEELY,WB & BLAU,GE (1985)
DC	pKa			
HL 3.41E-008	atm m3/mol	25	EST	MEYLAN,WM & HOWARD,PH (1991)
OH 1.00E-011	cm3/molc sec	25	EST	MEYLAN,WM & HOWARD,PH (1993)

CAS #: 000115-90-2 — FENSULFOTHION

Formula:	$C_{11}H_{17}O_4PS_2$	
Mol Weight:	308.36	
MP (deg C): 214-217	FP (deg C):	
BP (deg C): 140		
BP pressure (mm Hg): 1.00E-002		

Property/Value	Units	Temp	Data Type	Reference
WS 2.00E+003	mg/L	20	EXP	YALKOWSKY,SH & DANNENFELSER,RM (1992)
logP 2.23			EXP	HANSCH,C & LEO,AJ (1985)
VP 5.00E-005	mm Hg	25	EXP	AUGUSTIJN-BECKERS,PWM ET AL, (1994)
DC	pKa			
HL 1.01E-008	atm m3/mol	25	EST	VP/WSOL
OH 1.53E-010	cm3/molc sec	25	EST	MEYLAN,WM & HOWARD,PH (1993)

CAS #: 000115-93-5 — CYTHIOATE

Formula:	$C_8H_{12}NO_5PS_2$	
Mol Weight:	297.29	
MP (deg C):	FP (deg C):	
BP (deg C):		
BP pressure (mm Hg):		

Property/Value	Units	Temp	Data Type	Reference
WS 9.72E+002	mg/L	25	EST	MEYLAN,WM ET AL. (1996)
logP 1.31			EST	MEYLAN,WM & HOWARD,PH (1995)
VP 1.53E-006	mm Hg	25	EST	NEELY,WB & BLAU,GE (1985)
DC	pKa			
HL 3.34E-009	atm m3/mol	25	EST	MEYLAN,WM & HOWARD,PH (1991)
OH 5.93E-011	cm3/molc sec	25	EST	MEYLAN,WM & HOWARD,PH (1993)

CAS #: 000115-95-7 — LINALYL ACETATE

Formula:	$C_{12}H_{20}O_2$	
Mol Weight:	196.29	
MP (deg C):	FP (deg C):	
BP (deg C): 220		
BP pressure (mm Hg):		

Property/Value	Units	Temp	Data Type	Reference
WS 8.20E+000	mg/L	25	EST	MEYLAN,WM ET AL. (1996)
logP 4.39			EST	MEYLAN,WM & HOWARD,PH (1995)
VP 9.72E-002	mm Hg	25	EXT	PERRY,RH & GREEN,D (1984)
DC	pKa			
HL 1.74E-003	atm m3/mol	25	EST	MEYLAN,WM & HOWARD,PH (1991)
OH 1.16E-010	cm3/molc sec	25	EST	MEYLAN,WM & HOWARD,PH (1993)

CAS #: 000115-96-8 — TRI-2-CHLOROETHYL PHOSPHATE

Formula:	$C_6H_{12}Cl_3O_4P$	
Mol Weight:	285.49	
MP (deg C): -35	FP (deg C):	
BP (deg C): 214		
BP pressure (mm Hg): 2.50E+001		

Property/Value	Units	Temp	Data Type	Reference
WS 7.00E+003	mg/L		EXP	MUIR,DCG (1984)
logP 1.44			EXP	CHEM INSPECT TEST INST (1992)
VP 3.36E-005	mm Hg	25	EST	NEELY,WB & BLAU,GE (1985)
DC	pKa			
HL 2.55E-008	atm m3/mol	25	EST	MEYLAN,WM & HOWARD,PH (1991)
OH 2.20E-011	cm3/molc sec	25	EST	MEYLAN,WM & HOWARD,PH (1993)

000116-01-8 — ETHOATE METHYL

Formula: $C_6H_{14}NO_3PS_2$
Mol Weight: 243.29
MP (deg C):
FP (deg C):
BP (deg C):
BP pressure (mm Hg):

Property/Value	Units	Temp	Data Type	Reference
WS 8.50E+003	mg/L	25	EXP	YALKOWSKY,SH & DANNENFELSER,RM (1992)
logP 0.77			EST	MEYLAN,WM & HOWARD,PH (1995)
VP 9.25E-006	mm Hg	25	EST	NEELY,WB & BLAU,GE (1985)
DC	pKa			
HL 2.80E-011	atm m3/mol	25	EST	MEYLAN,WM & HOWARD,PH (1991)
OH 8.24E-011	cm3/molc sec	25	EST	MEYLAN,WM & HOWARD,PH (1993)

000116-06-3 — ALDICARB

Formula: $C_7H_{14}N_2O_2S$
Mol Weight: 190.27
MP (deg C): 98-100
FP (deg C):
BP (deg C):
BP pressure (mm Hg):

Property/Value	Units	Temp	Data Type	Reference
WS 6.03E+003	mg/L	25	EXP	BOWMAN,BT & SANS,WW (1983A)
logP 1.13			EXP	HANSCH,C & LEO,AJ (1985)
VP 3.47E-005	mm Hg	25	EXP	FERREIRA,GAL & SEIBER,JN (1981)
DC	pKa			
HL 1.44E-009	atm m3/mol	25	EST	VP/WSOL
OH 1.84E-011	cm3/molc sec	25	EST	MEYLAN,WM & HOWARD,PH (1993)

000116-09-6 — HYDROXYACETONE

Formula: $C_3H_6O_2$
Mol Weight: 74.08
MP (deg C): -17
FP (deg C):
BP (deg C): 145.5
BP pressure (mm Hg):

Property/Value	Units	Temp	Data Type	Reference
WS 1.00E+006	mg/L	20	EXP	YALKOWSKY,SH & DANNENFELSER,RM (1992)
logP -0.78			EST	MEYLAN,WM & HOWARD,PH (1995)
VP 2.95E+000	mm Hg	25	EST	NEELY,WB & BLAU,GE (1985)
DC	pKa			
HL 7.73E-006	atm m3/mol	25	EST	MEYLAN,WM & HOWARD,PH (1991)
OH 3.02E-012	cm3/molc sec	25	EXP	ATKINSON,R (1989)

000116-11-0 — ISOPROPENYL METHYL ETHER

Formula: C_4H_8O
Mol Weight: 72.11
MP (deg C):
FP (deg C):
BP (deg C): 34-36
BP pressure (mm Hg):

Property/Value	Units	Temp	Data Type	Reference
WS 2.00E+004	mg/L	25	EST	MEYLAN,WM ET AL. (1996)
logP 0.97			EST	MEYLAN,WM & HOWARD,PH (1995)
VP 5.17E+002	mm Hg	25	EXP	BAGLAY,AK ET AL. (1988)
DC	pKa			
HL 8.98E-003	atm m3/mol	25	EST	MEYLAN,WM & HOWARD,PH (1991)
OH 6.78E-011	cm3/molc sec	25	EST	MEYLAN,WM & HOWARD,PH (1993)

000116-14-3 — TETRAFLUOROETHYLENE

Formula: C_2F_4
Mol Weight: 100.02
MP (deg C): -142.5
FP (deg C):
BP (deg C): -75.9
BP pressure (mm Hg):

Property/Value	Units	Temp	Data Type	Reference
WS 2.51E+003	mg/L	25	EXP	SUZUKI,T (1991)
logP 1.21			EST	MEYLAN,WM & HOWARD,PH (1995)
VP 2.45E+004	mm Hg	25	EXP	DAUBERT,TE & DANNER,RP (1989)
DC	pKa			
HL 3.99E-002	atm m3/mol	25	EST	VP/WSOL
OH 2.14E-013	cm3/molc sec	25	EST	MEYLAN,WM & HOWARD,PH (1993)

000116-15-4 — HEXAFLUOROPROPENE

Formula: C_3F_6
Mol Weight: 150.02
MP (deg C): -156.5
FP (deg C):
BP (deg C): -29.6
BP pressure (mm Hg):

Property/Value	Units	Temp	Data Type	Reference
WS 1.17E+003	mg/L	25	EST	MEYLAN,WM ET AL. (1996)
logP 2.12			EST	MEYLAN,WM & HOWARD,PH (1995)
VP 4.90E+003	mm Hg	25	EXP	DAUBERT,TE & DANNER,RP (1989)
DC	pKa			
HL 5.34E+000	atm m3/mol	25	EST	MEYLAN,WM & HOWARD,PH (1991)
OH 7.74E-013	cm3/molc sec	25	EST	MEYLAN,WM & HOWARD,PH (1993)

000116-16-5 — HEXACHLOROACETONE

Formula: C_3Cl_6O
Mol Weight: 264.75
MP (deg C): -2
FP (deg C):
BP (deg C): 203
BP pressure (mm Hg):

Property/Value	Units	Temp	Data Type	Reference
WS 1.50E+002	mg/L	25	EST	MEYLAN,WM ET AL. (1996)
logP 2.48			EST	MEYLAN,WM & HOWARD,PH (1995)
VP 1.23E-001	mm Hg	25	EST	NEELY,WB & BLAU,GE (1985)
DC	pKa			
HL 9.47E-008	atm m3/mol	25	EST	MEYLAN,WM & HOWARD,PH (1991)
OH 0.00E+000	cm3/molc sec	25	EST	MEYLAN,WM & HOWARD,PH (1993)

000116-29-0 — TETRADIFON

Formula: $C_{12}H_6Cl_4O_2S$
Mol Weight: 356.06
MP (deg C): 147
FP (deg C):
BP (deg C):
BP pressure (mm Hg):

Property/Value	Units	Temp	Data Type	Reference
WS 5.00E-002	mg/L	10	EXP	WORTHING,CR & WALKER,SB (1987)
logP 4.61			EXP	TOMLIN,C (1994)
VP 2.40E-007	mm Hg	20	EXP	WORTHING,CR & WALKER,SB (1987)
DC	pKa			
HL 7.51E-008	atm m3/mol	25	EST	MEYLAN,WM & HOWARD,PH (1991)
OH 3.64E-013	cm3/molc sec	25	EST	ATKINSON,R (1988)

223

CAS #: 000116-52-9 — DICHLORAL UREA

Formula: $C_5H_6Cl_6N_2O_3$

Mol Weight: 354.83

MP (deg C): FP (deg C):

BP (deg C):

BP pressure (mm Hg):

Property/Value	Units	Temp	Data Type	Reference
WS 2.64E+003	mg/L	25	EST	MEYLAN,WM ET AL. (1996)
logP 0.40			EST	MEYLAN,WM & HOWARD,PH (1995)
VP 1.27E-009	mm Hg	25	EST	NEELY,WB & BLAU,GE (1985)
DC	pKa			
HL 3.15E-015	atm m3/mol	25	EST	MEYLAN,WM & HOWARD,PH (1991)
OH 4.12E-012	cm3/molc sec	25	EST	MEYLAN,WM & HOWARD,PH (1993)

CAS #: 000116-53-0 — BUTANOIC ACID, 2-METHYL-

Formula: $C_5H_{10}O_2$

Mol Weight: 102.13

MP (deg C): FP (deg C):

BP (deg C):

BP pressure (mm Hg):

Property/Value	Units	Temp	Data Type	Reference
WS 4.50E+004	mg/L	20	EXP	RIEMENSCHNEIDER,W (1986)
logP 1.18			EXP	HANSCH,C ET AL. (1995)
VP 4.91E-001	mm Hg	25	EXP	YAWS,CL (1994A)
DC 4.81	pKa	18	EXP	KORTUM,G ET AL (1961)
HL 1.47E-006	atm m3/mol		EST	VP/WSOL
OH 3.79E-012	cm3/molc sec	25	EST	MEYLAN,WM & HOWARD,PH (1993)

CAS #: 000116-54-1 — METHYL DICHLOROACETATE

Formula: $C_3H_4Cl_2O_2$

Mol Weight: 142.97

MP (deg C): -51.9 FP (deg C):

BP (deg C): 142.9

BP pressure (mm Hg):

Property/Value	Units	Temp	Data Type	Reference
WS 1.66E+004	mg/L	25	EST	MEYLAN,WM ET AL. (1996)
logP 0.81			EST	MEYLAN,WM & HOWARD,PH (1995)
VP 4.50E+000	mm Hg	25	EXP	PERRY,RH & GREEN,D (1984)
DC	pKa			
HL 2.18E-005	atm m3/mol	25	EST	MEYLAN,WM & HOWARD,PH (1991)
OH 3.16E-013	cm3/molc sec	25	EST	MEYLAN,WM & HOWARD,PH (1993)

CAS #: 000116-82-5 — 1-AMINO-2-BROMO-4-HYDROXY-9,10-ANTHRACENEDIONE

Formula: $C_{14}H_8BrNO_3$

Mol Weight: 318.13

MP (deg C): FP (deg C):

BP (deg C):

BP pressure (mm Hg):

Property/Value	Units	Temp	Data Type	Reference
WS 1.40E-001	mg/L	25	EST	MEYLAN,WM ET AL. (1996)
logP 4.36			EST	MEYLAN,WM & HOWARD,PH (1995)
VP 3.06E-010	mm Hg	25	EST	NEELY,WB & BLAU,GE (1985)
DC	pKa			
HL 5.86E-014	atm m3/mol	25	EST	MEYLAN,WM & HOWARD,PH (1991)
OH 1.41E-011	cm3/molc sec	25	EST	MEYLAN,WM & HOWARD,PH (1993)

CAS #: 000116-85-8 — 1-AMINO-4-HYDROXYANTHRAQUINONE

Formula: $C_{14}H_9NO_3$

Mol Weight: 239.23

MP (deg C): FP (deg C):

BP (deg C):

BP pressure (mm Hg):

Property/Value	Units	Temp	Data Type	Reference
WS 3.59E+001	mg/L		EXP	YALKOWSKY,SH & DANNENFELSER,RM (1992)
logP 3.47			EST	MEYLAN,WM & HOWARD,PH (1995)
VP 3.00E-009	mm Hg	25	EXP	BAUGHMAN,GL & PERENICH,TA (1988)
DC	pKa			
HL 2.63E-011	atm m3/mol	25	EST	VP/WSOL
OH 2.94E-011	cm3/molc sec	25	EST	MEYLAN,WM & HOWARD,PH (1993)

CAS #: 000117-12-4 — 1,5-DIHYDROXYANTHRAQUINONE

Formula: $C_{14}H_8O_4$

Mol Weight: 240.22

MP (deg C): 280 FP (deg C):

BP (deg C):

BP pressure (mm Hg):

Property/Value	Units	Temp	Data Type	Reference
WS 3.43E+000	mg/L	25	EST	MEYLAN,WM ET AL. (1996)
logP 3.94			EST	MEYLAN,WM & HOWARD,PH (1995)
VP 1.17E-009	mm Hg	25	EST	NEELY,WB & BLAU,GE (1985)
DC	pKa			
HL 5.45E-011	atm m3/mol	25	EST	MEYLAN,WM & HOWARD,PH (1991)
OH 2.59E-011	cm3/molc sec	25	EST	MEYLAN,WM & HOWARD,PH (1993)

CAS #: 000117-14-6 — 1,5-ANTHRAQUINONEDISULFONIC ACID

Formula: $C_{14}H_8O_8S_2$

Mol Weight: 368.34

MP (deg C): FP (deg C):

BP (deg C):

BP pressure (mm Hg):

Property/Value	Units	Temp	Data Type	Reference
WS 6.66E+005	mg/L	25	EXP	SEIDELL,A (1941)
logP -2.97			EST	MEYLAN,WM & HOWARD,PH (1995)
VP 5.70E-018	mm Hg	25	EST	NEELY,WB & BLAU,GE (1985)
DC	pKa			
HL 6.94E-022	atm m3/mol	25	EST	MEYLAN,WM & HOWARD,PH (1991)
OH 6.00E-013	cm3/molc sec	25	EST	MEYLAN,WM & HOWARD,PH (1993)

CAS #: 000117-18-0 — 2,3,5,6-TETRACHLORONITROBENZENE

Formula: $C_6HCl_4NO_2$

Mol Weight: 260.89

MP (deg C): 99 FP (deg C):

BP (deg C): 304 dec

BP pressure (mm Hg):

Property/Value	Units	Temp	Data Type	Reference
WS 2.09E+000	mg/L	20	EXP	YALKOWSKY,SH & DANNENFELSER,RM (1992)
logP 3.89			EXP	HANSCH,C & LEO,AJ (1985)
VP 1.80E-003	mm Hg	15	EXP	HARTLEY,D & KIDD,H (1983)
DC	pKa			
HL 2.34E-005	atm m3/mol	25	EST	MEYLAN,WM & HOWARD,PH (1991)
OH 7.20E-015	cm3/molc sec	25	EST	ATKINSON,R (1988)

CAS #: 000117-26-0 — BULAN

Formula: $C_{16}H_{15}Cl_2NO_2$
Mol Weight: 324.21
MP (deg C):
FP (deg C):
BP (deg C):
BP pressure (mm Hg):

Property/Value	Units	Temp	Data Type	Reference
WS 7.55E-002	mg/L	25	EST	MEYLAN,WM ET AL. (1996)
logP 5.48			EST	MEYLAN,WM & HOWARD,PH (1995)
VP 5.47E-007	mm Hg	25	EST	NEELY,WB & BLAU,GE (1985)
DC	pKa			
HL 3.53E-007	atm m3/mol	25	EST	MEYLAN,WM & HOWARD,PH (1991)
OH 4.00E-012	cm3/molc sec	25	EST	MEYLAN,WM & HOWARD,PH (1993)

CAS #: 000117-27-1 — PROLAN

Formula: $C_{15}H_{13}Cl_2NO_2$
Mol Weight: 310.18
MP (deg C):
FP (deg C):
BP (deg C): 180
BP pressure (mm Hg): 1.60E-001

Property/Value	Units	Temp	Data Type	Reference
WS 2.40E-001	mg/L	25	EST	MEYLAN,WM ET AL. (1996)
logP 4.99			EST	MEYLAN,WM & HOWARD,PH (1995)
VP 1.28E-006	mm Hg	25	EST	NEELY,WB & BLAU,GE (1985)
DC	pKa			
HL 2.66E-007	atm m3/mol	25	EST	MEYLAN,WM & HOWARD,PH (1991)
OH 3.72E-012	cm3/molc sec	25	EST	MEYLAN,WM & HOWARD,PH (1993)

CAS #: 000117-34-0 — DIPHENYLACETIC ACID

Formula: $C_{14}H_{12}O_2$
Mol Weight: 212.25
MP (deg C): 148
FP (deg C):
BP (deg C):
BP pressure (mm Hg):

Property/Value	Units	Temp	Data Type	Reference
WS 1.27E+002	mg/L	25	EXP	YALKOWSKY,SH & DANNENFELSER,RM (1992)
logP 3.09			EXP	HANSCH,C & LEO,AJ (1985)
VP 9.63E-006	mm Hg	25	EST	NEELY,WB & BLAU,GE (1985)
DC 3.94	pKa	25	EXP	KORTUM,G ET AL (1961)
HL 3.57E-009	atm m3/mol	25	EST	MEYLAN,WM & HOWARD,PH (1991)
OH 1.12E-011	cm3/molc sec	25	EST	MEYLAN,WM & HOWARD,PH (1993)

CAS #: 000117-37-3 — 1H-INDENE-1,3(2H)-DIONE, 2-(4-METHOXYPHENYL)-

Formula: $C_{16}H_{12}O_3$
Mol Weight: 252.27
MP (deg C): 156-157
FP (deg C):
BP (deg C):
BP pressure (mm Hg):

Property/Value	Units	Temp	Data Type	Reference
WS 7.98E+001	mg/L	25	EST	MEYLAN,WM ET AL. (1996)
logP 2.88			EXP	SANGSTER,J (1993)
VP 6.20E-007	mm Hg	25	EST	NEELY,WB & BLAU,GE (1985)
DC	pKa			
HL 2.83E-011	atm m3/mol	25	EST	MEYLAN,WM & HOWARD,PH (1991)
OH 2.83E-011	cm3/molc sec	25	EST	MEYLAN,WM & HOWARD,PH (1993)

CAS #: 000117-39-5 — QUERCETIN

Formula: $C_{15}H_{10}O_7$
Mol Weight: 302.24
MP (deg C): 310-311
FP (deg C):
BP (deg C):
BP pressure (mm Hg):

Property/Value	Units	Temp	Data Type	Reference
WS 6.00E+001	mg/L	16	EXP	SEIDELL,A (1941)
logP 1.48			EST	MEYLAN,WM & HOWARD,PH (1995)
VP 2.81E-014	mm Hg	25	EST	NEELY,WB & BLAU,GE (1985)
DC	pKa			
HL 6.60E-021	atm m3/mol	25	EST	MEYLAN,WM & HOWARD,PH (1991)
OH 2.39E-010	cm3/molc sec	25	EST	MEYLAN,WM & HOWARD,PH (1993)

CAS #: 000117-52-2 — COUMAFURYL

Formula: $C_{17}H_{14}O_5$
Mol Weight: 298.30
MP (deg C): 124
FP (deg C):
BP (deg C):
BP pressure (mm Hg):

Property/Value	Units	Temp	Data Type	Reference
WS 5.38E+002	mg/L	25	EST	MEYLAN,WM ET AL. (1996)
logP 1.60			EST	MEYLAN,WM & HOWARD,PH (1995)
VP 2.17E-011	mm Hg	25	EST	NEELY,WB & BLAU,GE (1985)
DC	pKa			
HL 2.27E-013	atm m3/mol	25	EST	MEYLAN,WM & HOWARD,PH (1991)
OH 1.45E-010	cm3/molc sec	25	EST	MEYLAN,WM & HOWARD,PH (1993)

CAS #: 000117-59-9 — 1-NAPHTHALENESULFONIC ACID, 5-HYDROXY-

Formula: $C_{10}H_8O_4S$
Mol Weight: 224.24
MP (deg C):
FP (deg C):
BP (deg C):
BP pressure (mm Hg):

Property/Value	Units	Temp	Data Type	Reference
WS 1.74E+005	mg/L	25	EST	MEYLAN,WM ET AL. (1996)
logP -0.17			EXP	SANGSTER,J (1993)
VP 9.03E-010	mm Hg	25	EST	NEELY,WB & BLAU,GE (1985)
DC	pKa			
HL 2.56E-014	atm m3/mol	25	EST	MEYLAN,WM & HOWARD,PH (1991)
OH 1.61E-011	cm3/molc sec	25	EST	MEYLAN,WM & HOWARD,PH (1993)

CAS #: 000117-79-3 — 2-AMINOANTHRAQUINONE

Formula: $C_{14}H_9NO_2$
Mol Weight: 223.23
MP (deg C): 292-295
FP (deg C):
BP (deg C):
BP pressure (mm Hg):

Property/Value	Units	Temp	Data Type	Reference
WS 1.63E-001	mg/L	25	EXP	KUROIWA,S & OGASAWARA,S (1973)
logP 3.31			EXP	CHEMICAL INSP TESTING INSTITUTE (1992)
VP 5.10E-011	mm Hg	25	EXT	KUROIWA,S & OGASAWARA,S (1973)
DC	pKa			
HL 9.19E-011	atm m3/mol	25	EST	VP/WSOL
OH 4.92E-011	cm3/molc sec	25	EST	MEYLAN,WM & HOWARD,PH (1993)

DICHLONE — CAS #: 000117-80-6

Formula: $C_{10}H_4Cl_2O_2$
Mol Weight: 227.05
MP (deg C): 193
FP (deg C):
BP (deg C):
BP pressure (mm Hg):

Property/Value	Units	Temp	Data Type	Reference
WS 1.00E-001	mg/L	25	EXP	YALKOWSKY,SH & DANNENFELSER,RM (1992)
logP 2.65			EST	MEYLAN,WM & HOWARD,PH (1995)
VP 8.20E+001	mm Hg	25	EST	AUGUSTIJN-BECKERS,PWM ET AL. (1994)
DC	pKa			
HL 1.02E-009	atm m3/mol	25	EST	MEYLAN,WM & HOWARD,PH (1991)
OH 1.34E-012	cm3/molc sec	25	EST	MEYLAN,WM & HOWARD,PH (1993)

BIS(2-ETHYLHEXYL)PHTHALATE — CAS #: 000117-81-7

Formula: $C_{24}H_{38}O_4$
Mol Weight: 390.57
MP (deg C): -50
FP (deg C):
BP (deg C): 384
BP pressure (mm Hg):

Property/Value	Units	Temp	Data Type	Reference
WS 3.40E-001	mg/L	25	EXP	HOWARD,PH ET AL. (1985)
logP 7.60			EXP	DEBRUIJN,J ET AL. (1989)
VP 9.75E-006	mm Hg	25	EXP	HOWARD,PH ET AL. (1985)
DC	pKa			
HL 1.47E-005	atm m3/mol	25	EST	VP/WSOL
OH 2.19E-011	cm3/molc sec	25	EST	MEYLAN,WM & HOWARD,PH (1993)

BIS(METHOXYETHYL)PHTHALATE — CAS #: 000117-82-8

Formula: $C_{14}H_{18}O_6$
Mol Weight: 282.30
MP (deg C): -45
FP (deg C):
BP (deg C): 340
BP pressure (mm Hg):

Property/Value	Units	Temp	Data Type	Reference
WS 8.50E+003	mg/L	25	EXP	US EPA (1991S)
logP 1.11			EST	MEYLAN,WM & HOWARD,PH (1995)
VP 2.43E-004	mm Hg	25	EST	NEELY,WB & BLAU,GE (1985)
DC	pKa			
HL 5.40E-011	atm m3/mol	25	EST	MEYLAN,WM & HOWARD,PH (1991)
OH 1.95E-011	cm3/molc sec	25	EST	MEYLAN,WM & HOWARD,PH (1993)

DIOCTYL PHTHALATE — CAS #: 000117-84-0

Formula: $C_{24}H_{38}O_4$
Mol Weight: 390.57
MP (deg C): -25
FP (deg C):
BP (deg C): 220-248
BP pressure (mm Hg): 4.00E+000

Property/Value	Units	Temp	Data Type	Reference
WS 3.00E+000	mg/L	25	EXP	WOLFE,NL ET AL. (1980B)
logP 8.06			EXP	HANSCH,C & LEO,AJ (1985)
VP 2.60E-006	mm Hg	25	EXP	PERWAK,J ET AL. (1981)
DC	pKa			
HL 4.45E-007	atm m3/mol	25	EST	VP/WSOL
OH 1.99E-011	cm3/molc sec	25	EST	MEYLAN,WM & HOWARD,PH (1993)

TRIFLUORPERAZINE — CAS #: 000117-89-5

Formula: $C_{21}H_{24}F_3N_3S$
Mol Weight: 407.50
MP (deg C):
FP (deg C):
BP (deg C): 202-210
BP pressure (mm Hg): 6.00E-001

Property/Value	Units	Temp	Data Type	Reference
WS 1.22E+001	mg/L	24	EXP	YALKOWSKY,SH & DANNENFELSER,RM (1992)
logP 5.03			EXP	HANSCH,C & LEO,AJ (1985)
VP 7.94E-009	mm Hg	25	EST	NEELY,WB & BLAU,GE (1985)
DC	pKa			
HL 1.26E-012	atm m3/mol	25	EST	MEYLAN,WM & HOWARD,PH (1991)
OH 1.70E-010	cm3/molc sec	25	EST	MEYLAN,WM & HOWARD,PH (1993)

O-HYDROXYBENZOPHENONE — CAS #: 000117-99-7

Formula: $C_{13}H_{10}O_2$
Mol Weight: 198.22
MP (deg C): 40
FP (deg C):
BP (deg C): 250
BP pressure (mm Hg): 5.60E+002

Property/Value	Units	Temp	Data Type	Reference
WS 1.68E+002	mg/L	25	EST	MEYLAN,WM ET AL. (1996)
logP 3.52			EXP	HANSCH,C & LEO,AJ (1985)
VP 1.00E-005	mm Hg	25	EST	NEELY,WB & BLAU,GE (1985)
DC	pKa			
HL 2.54E-007	atm m3/mol	25	EST	MEYLAN,WM & HOWARD,PH (1991)
OH 3.23E-011	cm3/molc sec	25	EST	MEYLAN,WM & HOWARD,PH (1993)

GUANOSINE — CAS #: 000118-00-3

Formula: $C_{10}H_{13}N_5O_5$
Mol Weight: 283.25
MP (deg C): 239 dec
FP (deg C):
BP (deg C):
BP pressure (mm Hg):

Property/Value	Units	Temp	Data Type	Reference
WS 7.00E+002	mg/L	18	EXP	YALKOWSKY,SH & DANNENFELSER,RM (1992)
logP -1.90			EXP	SANGSTER,J (1993)
VP 5.84E-020	mm Hg	25	EST	NEELY,WB & BLAU,GE (1985)
DC	pKa			
HL 4.15E-026	atm m3/mol	25	EST	MEYLAN,WM & HOWARD,PH (1991)
OH 1.09E-010	cm3/molc sec	25	EST	MEYLAN,WM & HOWARD,PH (1993)

HYDRASTINE — CAS #: 000118-08-1

Formula: $C_{21}H_{21}NO_6$
Mol Weight: 383.40
MP (deg C): 132
FP (deg C):
BP (deg C):
BP pressure (mm Hg):

Property/Value	Units	Temp	Data Type	Reference
WS 3.00E+001	mg/L	20	EXP	YALKOWSKY,SH & DANNENFELSER,RM (1992)
logP 1.89			EST	MEYLAN,WM & HOWARD,PH (1995)
VP 6.01E-011	mm Hg	25	EST	NEELY,WB & BLAU,GE (1985)
DC	pKa			
HL 1.34E-015	atm m3/mol	25	EST	MEYLAN,WM & HOWARD,PH (1991)
OH 1.91E-010	cm3/molc sec	25	EST	MEYLAN,WM & HOWARD,PH (1993)

000118-10-5 — CINCHONINE

CAS #	000118-10-5
Formula	$C_{19}H_{22}N_2O$
Mol Weight	294.40
MP (deg C)	265
FP (deg C)	
BP (deg C)	
BP pressure (mm Hg)	

Property/Value	Units	Temp	Data Type	Reference
WS 2.40E+002	mg/L		EXP	YALKOWSKY,SH & DANNENFELSER,RM (1992)
logP 2.68			EXP	SANGSTER,J (1994)
VP 9.04E-010	mm Hg	25	EST	NEELY,WB & BLAU,GE (1985)
DC	pKa			
HL 1.45E-014	atm m3/mol	25	EST	MEYLAN,WM & HOWARD,PH (1991)
OH 1.33E-010	cm3/molc sec	25	EST	MEYLAN,WM & HOWARD,PH (1993)

000118-33-2 — 2-NAPHTHYLAMINE-5,7-DISULFONIC ACID

CAS #	000118-33-2
Formula	$C_{10}H_9NO_6S_2$
Mol Weight	303.31
MP (deg C)	
FP (deg C)	
BP (deg C)	
BP pressure (mm Hg)	

Property/Value	Units	Temp	Data Type	Reference
WS 2.29E+005	mg/L	15	EXP	STEPHEN,H & STEPHEN,T (1963)
logP -4.06			EST	MEYLAN,WM & HOWARD,PH (1995)
VP 1.61E-015	mm Hg	25	EST	NEELY,WB & BLAU,GE (1985)
DC	pKa			
HL 4.06E-020	atm m3/mol	25	EST	MEYLAN,WM & HOWARD,PH (1991)
OH 9.67E-012	cm3/molc sec	25	EST	MEYLAN,WM & HOWARD,PH (1993)

000118-44-5 — 1-NAPHTHALENAMINE, N-ETHYL-

CAS #	000118-44-5
Formula	$C_{12}H_{13}N$
Mol Weight	171.24
MP (deg C)	305
FP (deg C)	
BP (deg C)	175-176
BP pressure (mm Hg)	1.50E+001

Property/Value	Units	Temp	Data Type	Reference
WS 1.10E+002	mg/L	25	EXP	CHEM INSPECT TEST INST (1992)
logP 3.45			EXP	CHEM INSPECT TEST INST (1992)
VP 9.51E-004	mm Hg	25	EST	NEELY,WB & BLAU,GE (1985)
DC 4.19	pKa	29	EXP	PERRIN,DD (1965)
HL 5.42E-007	atm m3/mol	25	EST	MEYLAN,WM & HOWARD,PH (1991)
OH 2.09E-010	cm3/molc sec	25	EST	MEYLAN,WM & HOWARD,PH (1993)

000118-55-8 — PHENYL 2-HYDROXYBENZOATE

CAS #	000118-55-8
Formula	$C_{13}H_{10}O_3$
Mol Weight	214.22
MP (deg C)	41-43
FP (deg C)	
BP (deg C)	173
BP pressure (mm Hg)	1.20E+001

Property/Value	Units	Temp	Data Type	Reference
WS 1.50E+002	mg/L	25	EXP	YALKOWSKY,SH & DANNENFELSER,RM (1992)
logP 3.82			EST	MEYLAN,WM & HOWARD,PH (1995)
VP 8.17E-006	mm Hg	25	EST	NEELY,WB & BLAU,GE (1985)
DC	pKa			
HL 1.68E-006	atm m3/mol	25	EST	MEYLAN,WM & HOWARD,PH (1991)
OH 3.25E-011	cm3/molc sec	25	EST	MEYLAN,WM & HOWARD,PH (1993)

000118-58-1 — BENZYL SALICYLATE

CAS #	000118-58-1
Formula	$C_{15}H_{14}O_2$
Mol Weight	226.28
MP (deg C)	130.5
FP (deg C)	
BP (deg C)	173
BP pressure (mm Hg)	1.20E+001

Property/Value	Units	Temp	Data Type	Reference
WS 2.46E+001	mg/L	25	EST	MEYLAN,WM ET AL. (1996)
logP 4.31			EST	MEYLAN,WM & HOWARD,PH (1995)
VP 3.37E-006	mm Hg	25	EST	NEELY,WB & BLAU,GE (1985)
DC	pKa			
HL 3.67E-007	atm m3/mol	25	EST	MEYLAN,WM & HOWARD,PH (1991)
OH 1.73E-011	cm3/molc sec	25	EST	MEYLAN,WM & HOWARD,PH (1993)

000118-61-6 — ETHYL 2-HYDROXYBENZOATE

CAS #	000118-61-6
Formula	$C_9H_{10}O_3$
Mol Weight	166.18
MP (deg C)	1
FP (deg C)	
BP (deg C)	231-234
BP pressure (mm Hg)	

Property/Value	Units	Temp	Data Type	Reference
WS 7.37E+002	mg/L	25	EST	MEYLAN,WM ET AL. (1996)
logP 2.95			EXP	KORENMAN,YI & DANILOV,VN (1990)
VP 8.46E-002	mm Hg	25	EXT	PERRY,RH & GREEN,D (1984)
DC	pKa			
HL 6.04E-006	atm m3/mol	25	EST	MEYLAN,WM & HOWARD,PH (1991)
OH 1.26E-011	cm3/molc sec	25	EST	MEYLAN,WM & HOWARD,PH (1993)

000118-69-4 — 2,6-DICHLOROTOLUENE

CAS #	000118-69-4
Formula	$C_7H_6Cl_2$
Mol Weight	161.03
MP (deg C)	25.8
FP (deg C)	
BP (deg C)	198
BP pressure (mm Hg)	

Property/Value	Units	Temp	Data Type	Reference
WS 1.85E+001	mg/L	25	EST	MEYLAN,WM ET AL. (1996)
logP 4.29			EXP	HANSCH,C & LEO,AJ (1985)
VP 3.58E-001	mm Hg	25	EST	NEELY,WB & BLAU,GE (1985)
DC	pKa			
HL 4.15E-003	atm m3/mol	25	EST	MEYLAN,WM & HOWARD,PH (1991)
OH 1.32E-012	cm3/molc sec	25	EST	MEYLAN,WM & HOWARD,PH (1993)

000118-71-8 — 4H-PYRAN-4-ONE, 3-HYDROXY-2-METHYL-

CAS #	000118-71-8
Formula	$C_6H_6O_3$
Mol Weight	126.11
MP (deg C)	161-162
FP (deg C)	
BP (deg C)	
BP pressure (mm Hg)	

Property/Value	Units	Temp	Data Type	Reference
WS 1.09E+004	mg/L	15	EXP	YALKOWSKY,SH & DANNENFELSER,RM (1992)
logP 0.09			EXP	SANGSTER,J (1993)
VP 5.07E-004	mm Hg	25	EST	NEELY,WB & BLAU,GE (1985)
DC	pKa			
HL 6.54E-006	atm m3/mol	25	EST	MEYLAN,WM & HOWARD,PH (1991)
OH 5.01E-011	cm3/molc sec	25	EST	MEYLAN,WM & HOWARD,PH (1993)

CAS #: 000118-74-1				HEXACHLOROBENZENE

Formula: C_6Cl_6
Mol Weight: 284.78
MP (deg C): 228.7　　FP (deg C):
BP (deg C): 319.3
BP pressure (mm Hg):

Property/Value	Units	Temp	Data Type	Reference
WS 6.20E-003	mg/L	25	EXP	FARMER,WJ ET AL. (1976)
logP 5.73			EXP	DEBRUIJN,J ET AL. (1989)
VP 1.80E-005	mm Hg	25	EXP	SCHOENE,K ET AL. (1984)
DC	pKa			
HL 1.70E-003	atm m3/mol	25	EXP	WARNER,HP ET AL. (1987)
OH 1.71E-014	cm3/molc sec	25	EST	MEYLAN,WM & HOWARD,PH (1993)

CAS #: 000118-75-2				CHLORANIL

Formula: $C_6Cl_4O_2$
Mol Weight: 245.88
MP (deg C): 290　　FP (deg C):
BP (deg C):
BP pressure (mm Hg):

Property/Value	Units	Temp	Data Type	Reference
WS 2.50E+002	mg/L	25	EXP	YALKOWSKY,SH & DANNENFELSER,RM (1992)
logP 2.22			EST	MEYLAN,WM & HOWARD,PH (1995)
VP 7.61E-003	mm Hg	25	EXT	PERRY,RH & GREEN,D (1984)
DC	pKa			
HL 3.27E-010	atm m3/mol	25	EST	MEYLAN,WM & HOWARD,PH (1991)
OH 1.19E-012	cm3/molc sec	25	EST	MEYLAN,WM & HOWARD,PH (1993)

CAS #: 000118-79-6				2,4,6-TRIBROMOPHENOL

Formula: $C_6H_3Br_3O$
Mol Weight: 330.82
MP (deg C): 94-96　　FP (deg C):
BP (deg C): 244
BP pressure (mm Hg):

Property/Value	Units	Temp	Data Type	Reference
WS 7.00E+001	mg/L	15	EXP	YALKOWSKY,SH & DANNENFELSER,RM (1992)
logP 4.13			EXP	HANSCH,C & LEO,AJ (1985)
VP 5.72E-005	mm Hg	25	EST	NEELY,WB & BLAU,GE (1985)
DC	pKa			
HL 3.55E-008	atm m3/mol	25	EST	MEYLAN,WM & HOWARD,PH (1991)
OH 4.75E-013	cm3/molc sec	25	EST	MEYLAN,WM & HOWARD,PH (1993)

CAS #: 000118-83-2				4-CHLORO-1-NITRO-2(TRIFLUOROMETHYL)BENZENE

Formula: $C_7H_3ClF_3NO_2$
Mol Weight: 225.56
MP (deg C):　　FP (deg C):
BP (deg C): 1.526
BP pressure (mm Hg): 2.50E+001

Property/Value	Units	Temp	Data Type	Reference
WS 1.00E+002	mg/L	25	EXP	CHEM INSPECT TEST INST (1992)
logP 3.20			EXP	CHEM INSPECT TEST INST (1992)
VP 2.21E-002	mm Hg	25	EST	NEELY,WB & BLAU,GE (1985)
DC	pKa			
HL 1.37E-004	atm m3/mol	25	EST	MEYLAN,WM & HOWARD,PH (1991)
OH 2.59E-014	cm3/molc sec	25	EST	MEYLAN,WM & HOWARD,PH (1993)

CAS #: 000118-90-1				O-TOLUIC ACID

Formula: $C_8H_8O_2$
Mol Weight: 136.15
MP (deg C): 103.7　　FP (deg C):
BP (deg C): 259
BP pressure (mm Hg):

Property/Value	Units	Temp	Data Type	Reference
WS 1.19E+003	mg/L	25	EXP	YALKOWSKY,SH & DANNENFELSER,RM (1992)
logP 2.46			EXP	HANSCH,C ET AL. (1995)
VP 6.57E-004	mm Hg	25	EXT	YAWS,CL (1994B)
DC 3.98	pKa	20	EXP	KORTUM,G ET AL (1961)
HL 1.20E-007	atm m3/mol	25	EST	MEYLAN,WM & HOWARD,PH (1991)
OH 2.54E-012	cm3/molc sec	25	EST	MEYLAN,WM & HOWARD,PH (1993)

CAS #: 000118-91-2				2-CHLOROBENZOIC ACID

Formula: $C_7H_5ClO_2$
Mol Weight: 156.57
MP (deg C): 142　　FP (deg C):
BP (deg C): 287
BP pressure (mm Hg):

Property/Value	Units	Temp	Data Type	Reference
WS 2.09E+003	mg/L	25	EXP	SEIDELL,A (1941)
logP 2.05			EXP	HANSCH,C & LEO,AJ (1985)
VP 6.60E-004	mm Hg	25	EXT	DAUBERT,TE & DANNER,RP (1989)
DC 2.89	pKa	25	EXP	BYKOVA,LN ET AL. (1970)
HL 6.50E-008	atm m3/mol	25	EST	VP/WSOL
OH 5.62E-013	cm3/molc sec	25	EST	ATKINSON,R (1987)

CAS #: 000118-92-3				2-AMINOBENZOIC ACID

Formula: $C_7H_7NO_2$
Mol Weight: 137.14
MP (deg C): 144-146　　FP (deg C):
BP (deg C):
BP pressure (mm Hg):

Property/Value	Units	Temp	Data Type	Reference
WS 3.50E+003	mg/L	20	EXP	YALKOWSKY,SH & DANNENFELSER,RM (1992)
logP 1.21			EXP	HANSCH,C & LEO,AJ (1985)
VP 2.78E-004	mm Hg	25	EST	NEELY,WB & BLAU,GE (1985)
DC 2.14	pKa		EXP	KORTUM,G ET AL (1961)
HL 3.83E-011	atm m3/mol	25	EST	MEYLAN,WM & HOWARD,PH (1991)
OH 4.04E-011	cm3/molc sec	25	EST	MEYLAN,WM & HOWARD,PH (1993)

CAS #: 000118-93-4				O-HYDROXYACETOPHENONE

Formula: $C_8H_8O_2$
Mol Weight: 136.15
MP (deg C): 5　　FP (deg C):
BP (deg C): 218
BP pressure (mm Hg):

Property/Value	Units	Temp	Data Type	Reference
WS 7.57E+003	mg/L	25	EST	MEYLAN,WM ET AL. (1996)
logP 1.92			EXP	HANSCH,C & LEO,AJ (1985)
VP 7.48E-003	mm Hg	25	EST	NEELY,WB & BLAU,GE (1985)
DC	pKa			
HL 1.29E-006	atm m3/mol	25	EST	MEYLAN,WM & HOWARD,PH (1991)
OH 3.06E-011	cm3/molc sec	25	EST	MEYLAN,WM & HOWARD,PH (1993)

CAS #: 000118-96-7				2,4,6-TRINITROTOLUENE
Formula: $C_7H_5N_3O_6$				
Mol Weight: 227.13				
MP (deg C): 80.1		FP (deg C):		
BP (deg C): 186.6				
BP pressure (mm Hg): 7.50E+000				

Property/ Value	Units	Temp	Data Type	Reference
WS 1.30E+002	mg/L	20	EXP	SEIDELL,A (1941)
logP 1.60			EXP	HANSCH,C & LEO,AJ (1985)
VP 8.02E-006	mm Hg	25	EXP	SPANGGORD,RJ ET AL. (1980)
DC	pKa			
HL 4.57E-007	atm m3/mol	20	EST	VP/WSOL
OH 1.46E-013	cm3/molc sec	25	EST	MEYLAN,WM & HOWARD,PH (1993)

CAS #: 000118-97-8				4-CHLORO-3,5-DINITROBENZOIC ACID
Formula: $C_7H_3ClN_2O_6$				
Mol Weight: 246.56				
MP (deg C):		FP (deg C)		
BP (deg C): 159-162				
BP pressure (mm Hg):				

Property/ Value	Units	Temp	Data Type	Reference
WS 7.38E+001	mg/L	25	EST	MEYLAN,WM ET AL. (1996)
logP 2.50			EXP	HANSCH,C & LEO,AJ (1985)
VP 3.18E-007	mm Hg	25	EST	NEELY,WB & BLAU,GE (1985)
DC	pKa			
HL 1.25E-012	atm m3/mol	25	EST	MEYLAN,WM & HOWARD,PH (1991)
OH 5.21E-013	cm3/molc sec	25	EST	MEYLAN,WM & HOWARD,PH (1993)

CAS #: 000119-06-2				DITRIDECYL PHTHALATE
Formula: $C_{34}H_{58}O_4$				
Mol Weight: 530.84				
MP (deg C):		FP (deg C):		
BP (deg C):				
BP pressure (mm Hg):				

Property/ Value	Units	Temp	Data Type	Reference
WS 1.48E-009	mg/L	25	EST	MEYLAN,WM ET AL. (1996)
logP 13.45			EST	MEYLAN,WM & HOWARD,PH (1995)
VP 2.51E-011	mm Hg	25	EST	NEELY,WB & BLAU,GE (1985)
DC	pKa			
HL 2.01E-004	atm m3/mol	25	EST	MEYLAN,WM & HOWARD,PH (1991)
OH 3.47E-011	cm3/molc sec	25	EST	MEYLAN,WM & HOWARD,PH (1993)

CAS #: 000119-07-3				N-OCTYL N-DECYL PHTHALATE
Formula: $C_{26}H_{42}O_4$				
Mol Weight: 418.62				
MP (deg C):		FP (deg C):		
BP (deg C):				
BP pressure (mm Hg):				

Property/ Value	Units	Temp	Data Type	Reference
WS 1.74E-005	mg/L	25	EST	MEYLAN,WM ET AL. (1996)
logP 9.52			EST	MEYLAN,WM & HOWARD,PH (1995)
VP 1.00E-013	mm Hg	25	EXP	GIAM,CS ET AL. (1984A)
DC	pKa			
HL 2.08E-005	atm m3/mol	25	EST	MEYLAN,WM & HOWARD,PH (1991)
OH 2.34E-011	cm3/molc sec	25	EST	MEYLAN,WM & HOWARD,PH (1993)

CAS #: 000119-12-0				PYRIDAPHENTHION
Formula: $C_{14}H_{17}N_2O_4PS$				
Mol Weight: 340.34				
MP (deg C):		FP (deg C):		
BP (deg C):				
BP pressure (mm Hg):				

Property/ Value	Units	Temp	Data Type	Reference
WS 1.31E+001	mg/L	25	EST	MEYLAN,WM ET AL. (1996)
logP 3.20			EXP	TOMLIN,C (1994)
VP 5.54E-008	mm Hg	25	EST	NEELY,WB & BLAU,GE (1985)
DC	pKa			
HL 1.17E-010	atm m3/mol	25	EST	MEYLAN,WM & HOWARD,PH (1991)
OH 1.15E-010	cm3/molc sec	25	EST	MEYLAN,WM & HOWARD,PH (1993)

CAS #: 000119-15-3				P-(2,4-DINITROANILINE)PHENOL
Formula: $C_{12}H_9N_3O_5$				
Mol Weight: 275.22				
MP (deg C): 194		FP (deg C):		
BP (deg C):				
BP pressure (mm Hg):				

Property/ Value	Units	Temp	Data Type	Reference
WS 9.90E-001	mg/L	25	EXP	BAUGHMAN,GL & PERENICH,TA (1988)
logP 2.67			EST	MEYLAN,WM & HOWARD,PH (1995)
VP 4.00E-012	mm Hg	25	EXP	BAUGHMAN,GL & PERENICH,TA (1988)
DC	pKa			
HL 1.46E-012	atm m3/mol	25	EST	VP/WSOL
OH 1.03E-010	cm3/molc sec	25	EST	MEYLAN,WM & HOWARD,PH (1993)

CAS #: 000119-26-6				HYDRAZINE, (2,4-DINITROPHENYL)-
Formula: $C_6H_6N_4O_4$				
Mol Weight: 198.14				
MP (deg C): 200		FP (deg C):		
BP (deg C):				
BP pressure (mm Hg):				

Property/ Value	Units	Temp	Data Type	Reference
WS 2.48E+003	mg/L	25	EST	MEYLAN,WM ET AL. (1996)
logP 1.47			EXP	HANSCH,C ET AL. (1995)
VP 1.25E-005	mm Hg	25	EST	NEELY,WB & BLAU,GE (1985)
DC	pKa			
HL 2.71E-012	atm m3/mol	25	EST	MEYLAN,WM & HOWARD,PH (1991)
OH 6.66E-013	cm3/molc sec	25	EST	MEYLAN,WM & HOWARD,PH (1993)

CAS #: 000119-27-7				2,4-DINITROANISOLE
Formula: $C_7H_6N_2O_5$				
Mol Weight: 198.14				
MP (deg C): 94.5		FP (deg C):		
BP (deg C): 206				
BP pressure (mm Hg): 1.20E+001				

Property/ Value	Units	Temp	Data Type	Reference
WS 1.55E+002	mg/L	15	EXP	YALKOWSKY,SH & DANNENFELSER,RM (1992)
logP 1.71			EST	MEYLAN,WM & HOWARD,PH (1995)
VP 1.38E-004	mm Hg	25	EST	NEELY,WB & BLAU,GE (1985)
DC	pKa			
HL 4.96E-009	atm m3/mol	25	EST	MEYLAN,WM & HOWARD,PH (1991)
OH 1.17E-012	cm3/molc sec	25	EST	MEYLAN,WM & HOWARD,PH (1993)

229

000119-30-2 — BENZOIC ACID, 2-HYDROXY-5-IODO-

Formula: $C_7H_5IO_3$

Mol Weight: 264.02

MP (deg C): 189-191

FP (deg C):

BP (deg C):

BP pressure (mm Hg):

Property/Value	Units	Temp	Data Type	Reference
WS 1.06E+002	mg/L	25	EST	MEYLAN,WM ET AL. (1996)
logP 3.34			EXP	SANGSTER,J (1994)
VP 2.89E-006	mm Hg	25	EST	NEELY,WB & BLAU,GE (1985)
DC 2.62	pKa	25	EXP	SERJEANT,EP & DEMPSEY,B (1979)
HL 3.29E-009	atm m3/mol	25	EST	MEYLAN,WM & HOWARD,PH (1991)
OH 4.74E-012	cm3/molc sec	25	EST	MEYLAN,WM & HOWARD,PH (1993)

000119-32-4 — 3-NITRO-4-TOLUIDINE

Formula: $C_7H_8N_2O_2$

Mol Weight: 152.15

MP (deg C): 44-47

FP (deg C):

BP (deg C):

BP pressure (mm Hg):

Property/Value	Units	Temp	Data Type	Reference
WS 1.40E+003	mg/L	25	EST	MEYLAN,WM ET AL. (1996)
logP 2.02			EST	MEYLAN,WM & HOWARD,PH (1995)
VP 9.72E-004	mm Hg	25	EST	NEELY,WB & BLAU,GE (1985)
DC 3.03	pKa	17	EXP	PERRIN,DD (1965)
HL 8.29E-009	atm m3/mol	25	EST	MEYLAN,WM & HOWARD,PH (1991)
OH 1.17E-011	cm3/molc sec	25	EST	MEYLAN,WM & HOWARD,PH (1993)

000119-33-5 — 4-METHYL-2-NITROPHENOL

Formula: $C_7H_7NO_3$

Mol Weight: 153.14

MP (deg C): 36.5

FP (deg C):

BP (deg C): 125

BP pressure (mm Hg): 2.20E+001

Property/Value	Units	Temp	Data Type	Reference
WS 6.33E+002	mg/L	25	EXP	CHEM INSPECT TEST INST (1992)
logP 2.37			EXP	HANSCH,C ET AL. (1995)
VP 6.32E-004	mm Hg	25	EST	NEELY,WB & BLAU,GE (1985)
DC 7.60	pKa	25	EXP	SERJEANT,EP & DEMPSEY,B (1979)
HL 7.72E-006	atm m3/mol	25	EST	MEYLAN,WM & HOWARD,PH (1991)
OH 5.38E-012	cm3/molc sec	25	EST	MEYLAN,WM & HOWARD,PH (1993)

000119-34-6 — PHENOL, 4-AMINO-2-NITRO-

Formula: $C_6H_6N_2O_3$

Mol Weight: 154.13

MP (deg C): 125-127

FP (deg C):

BP (deg C):

BP pressure (mm Hg):

Property/Value	Units	Temp	Data Type	Reference
WS 1.10E+004	mg/L	25	EST	MEYLAN,WM ET AL. (1996)
logP 0.96			EXP	BRONAUGH,RL & CONGDON,ER (1984)
VP 3.52E-005	mm Hg	25	EST	NEELY,WB & BLAU,GE (1985)
DC 7.81	pKa	25	EXP	SERJEANT,EP & DEMPSEY,B (1979)
HL 2.23E-012	atm m3/mol	25	EST	MEYLAN,WM & HOWARD,PH (1991)
OH 7.25E-012	cm3/molc sec	25	EST	MEYLAN,WM & HOWARD,PH (1993)

000119-36-8 — METHYL SALICYLATE

Formula: $C_8H_8O_3$

Mol Weight: 152.15

MP (deg C): -8

FP (deg C):

BP (deg C): 222.9

BP pressure (mm Hg):

Property/Value	Units	Temp	Data Type	Reference
WS 7.00E+002	mg/L	30	EXP	YALKOWSKY,SH & DANNENFELSER,RM (1992)
logP 2.55			EXP	SANGSTER,J (1994)
VP 3.43E-002	mm Hg	25	EXP	DAUBERT,TE & DANNER,RP (1989)
DC 9.87	pKa	20	EXP	SERJEANT,EP & DEMPSEY,B (1979)
HL 9.81E-005	atm m3/mol	25	EST	VP/WSOL
OH 1.11E-011	cm3/molc sec	25	EST	MEYLAN,WM & HOWARD,PH (1993)

000119-38-0 — ISOLAN

Formula: $C_{10}H_{17}N_3O_2$

Mol Weight: 211.27

MP (deg C):

FP (deg C):

BP (deg C): 117.5-18

BP pressure (mm Hg): 2.50E+000

Property/Value	Units	Temp	Data Type	Reference
WS 1.00E+006	mg/L	20	EXP	YALKOWSKY,SH & DANNENFELSER,RM (1992)
logP 1.65			EST	MEYLAN,WM & HOWARD,PH (1995)
VP 1.27E-003	mm Hg	25	EST	NEELY,WB & BLAU,GE (1985)
DC	pKa			
HL 2.01E-009	atm m3/mol	25	EST	MEYLAN,WM & HOWARD,PH (1991)
OH 1.15E-010	cm3/molc sec	25	EST	MEYLAN,WM & HOWARD,PH (1993)

000119-39-1 — 1,2(H)-PHTHALAZINONE

Formula: $C_8H_6N_2O$

Mol Weight: 146.15

MP (deg C): 187-189

FP (deg C):

BP (deg C):

BP pressure (mm Hg):

Property/Value	Units	Temp	Data Type	Reference
WS 1.80E+004	mg/L	25	EST	MEYLAN,WM ET AL. (1996)
logP 0.75			EXP	HANSCH,C & LEO,AJ (1985)
VP 2.57E-006	mm Hg	25	EST	NEELY,WB & BLAU,GE (1985)
DC	pKa			
HL 2.17E-009	atm m3/mol	25	EST	MEYLAN,WM & HOWARD,PH (1991)
OH 1.17E-011	cm3/molc sec	25	EST	MEYLAN,WM & HOWARD,PH (1993)

000119-47-1 — BIS (2-HYDROXY-3TERT-BUTYL-5-METHYLPHENYL) METHA

Formula: $C_{23}H_{32}O_2$

Mol Weight: 340.51

MP (deg C): 118-128

FP (deg C):

BP (deg C):

BP pressure (mm Hg):

Property/Value	Units	Temp	Data Type	Reference
WS 2.00E-002	mg/L	25	EXP	CHEM INSPECT TEST INST (1992)
logP 6.25			EXP	CHEM INSPECT TEST INST (1992)
VP 2.50E-009	mm Hg	25	EST	NEELY,WB & BLAU,GE (1985)
DC	pKa			
HL 7.90E-012	atm m3/mol	25	EST	MEYLAN,WM & HOWARD,PH (1991)
OH 4.10E-011	cm3/molc sec	25	EST	MEYLAN,WM & HOWARD,PH (1993)

230

CAS #: 000119-53-9 — BENZOIN

Formula: $C_{14}H_{12}O_2$
Mol Weight: 212.25
MP (deg C): 137
FP (deg C):
BP (deg C): 344
BP pressure (mm Hg): 7.68E+002

Property/Value	Units	Temp	Data Type	Reference
WS 3.00E+002	mg/L	25	EXP	YALKOWSKY,SH & DANNENFELSER,RM (1992)
logP 1.84			EST	MEYLAN,WM & HOWARD,PH (1995)
VP 2.91E-005	mm Hg	25	EXT	OHE,S (1976)
DC	pKa			
HL 1.24E-007	atm m3/mol	25	EST	MEYLAN,WM & HOWARD,PH (1991)
OH 1.16E-011	cm3/molc sec	25	EST	MEYLAN,WM & HOWARD,PH (1993)

CAS #: 000119-56-2 — BENZENEMETHANOL, 4-CHLORO-.ALPHA.-PHENYL-

Formula: $C_{13}H_{11}ClO$
Mol Weight: 218.69
MP (deg C): 60-61.5
FP (deg C):
BP (deg C):
BP pressure (mm Hg):

Property/Value	Units	Temp	Data Type	Reference
WS 7.10E+001	mg/L	25	EXP	CHEM INSPECT TEST INST (1992)
logP 3.61			EXP	CHEM INSPECT TEST INST (1992)
VP 3.51E-006	mm Hg	25	EST	NEELY,WB & BLAU,GE (1985)
DC	pKa			
HL 1.30E-008	atm m3/mol	25	EST	MEYLAN,WM & HOWARD,PH (1991)
OH 1.32E-011	cm3/molc sec	25	EST	MEYLAN,WM & HOWARD,PH (1993)

CAS #: 000119-61-9 — BENZOPHENONE

Formula: $C_{13}H_{10}O$
Mol Weight: 182.22
MP (deg C): 48.5
FP (deg C):
BP (deg C): 305.4
BP pressure (mm Hg):

Property/Value	Units	Temp	Data Type	Reference
WS 1.37E+002	mg/L	25	EXP	YALKOWSKY,SH & DANNENFELSER,RM (1992)
logP 3.18			EXP	HANSCH,C & LEO,AJ (1985)
VP 1.93E-003	mm Hg	25	EXT	YAWS,CL (1994B)
DC	pKa			
HL 1.94E-006	atm m3/mol	25	EST	MEYLAN,WM & HOWARD,PH (1991)
OH 3.55E-012	cm3/molc sec	25	EST	MEYLAN,WM & HOWARD,PH (1993)

CAS #: 000119-64-2 — TETRALIN

Formula: $C_{10}H_{12}$
Mol Weight: 132.21
MP (deg C): -31.0
FP (deg C):
BP (deg C): 207.2
BP pressure (mm Hg):

Property/Value	Units	Temp	Data Type	Reference
WS 4.70E+001	mg/L	28	EXP	YALKOWSKY,SH & DANNENFELSER,RM (1992)
logP 3.49			EXP	HANSCH,C ET AL. (1995)
VP 3.68E-001	mm Hg	25	EXP	DAUBERT,TE & DANNER,RP (1989)
DC	pKa			
HL 5.10E-003	atm m3/mol	25	EST	MEYLAN,WM & HOWARD,PH (1991)
OH 3.43E-011	cm3/molc sec	25	EXP	ATKINSON,R (1989)

CAS #: 000119-65-3 — ISOQUINOLINE

Formula: C_9H_7N
Mol Weight: 129.16
MP (deg C): 26.48
FP (deg C):
BP (deg C): 243.25
BP pressure (mm Hg):

Property/Value	Units	Temp	Data Type	Reference
WS 4.52E+003	mg/L	20	EXP	PEARLMAN,RS ET AL. (1984)
logP 2.08			EXP	HANSCH,C & LEO,AJ (1985)
VP 7.00E-002	mm Hg	25	EXT	YAWS,CL (1994B)
DC 5.42	pKa	20	EXP	PERRIN,DD (1965)
HL 6.88E-007	atm m3/mol	25	EST	MEYLAN,WM & HOWARD,PH (1991)
OH 1.44E-011	cm3/molc sec	25	EST	MEYLAN,WM & HOWARD,PH (1993)

CAS #: 000119-75-5 — 2-NITRO-N-PHENYLBENZENAMINE

Formula: $C_{12}H_{10}N_2O_2$
Mol Weight: 214.23
MP (deg C): 74-76
FP (deg C):
BP (deg C):
BP pressure (mm Hg):

Property/Value	Units	Temp	Data Type	Reference
WS 2.77E+001	mg/L	25	EST	MEYLAN,WM ET AL. (1996)
logP 3.69			EST	MEYLAN,WM & HOWARD,PH (1995)
VP 1.00E-005	mm Hg	25	EXP	BAUGHMAN,GL & PERENICH,TA (1988)
DC	pKa			
HL 9.07E-008	atm m3/mol	25	EST	MEYLAN,WM & HOWARD,PH (1991)
OH 1.65E-010	cm3/molc sec	25	EST	MEYLAN,WM & HOWARD,PH (1993)

CAS #: 000119-79-9 — 1-NAPHTHYLAMINE-6-SULFONIC ACID

Formula: $C_{10}H_9NO_3S$
Mol Weight: 223.25
MP (deg C):
FP (deg C):
BP (deg C):
BP pressure (mm Hg):

Property/Value	Units	Temp	Data Type	Reference
WS 1.00E+003	mg/L	25	EXP	MERCK INDEX (1976)
logP -0.91			EST	MEYLAN,WM & HOWARD,PH (1995)
VP 5.26E-010	mm Hg	25	EST	NEELY,WB & BLAU,GE (1985)
DC	pKa			
HL 8.68E-014	atm m3/mol	25	EST	MEYLAN,WM & HOWARD,PH (1991)
OH 4.41E-011	cm3/molc sec	25	EST	MEYLAN,WM & HOWARD,PH (1993)

CAS #: 000119-81-3 — 2-NITRO-4-METHOXYACETANILIDE

Formula: $C_9H_{10}N_2O_4$
Mol Weight: 210.19
MP (deg C):
FP (deg C):
BP (deg C):
BP pressure (mm Hg):

Property/Value	Units	Temp	Data Type	Reference
WS 1.85E+003	mg/L	25	EST	MEYLAN,WM ET AL. (1996)
logP 1.09			EXP	NAKAGAWA,Y ET AL. (1992)
VP 8.04E-007	mm Hg	25	EST	NEELY,WB & BLAU,GE (1985)
DC	pKa			
HL 3.15E-011	atm m3/mol	25	EST	MEYLAN,WM & HOWARD,PH (1991)
OH 2.27E-012	cm3/molc sec	25	EST	MEYLAN,WM & HOWARD,PH (1993)

231

CAS #: 000119-89-1 — PIPERONYL CYCLONENE

Formula: $C_{19}H_{24}O_3$

Mol Weight: 300.40

MP (deg C):

FP (deg C):

BP (deg C):

BP pressure (mm Hg):

Property/Value	Units	Temp	Data Type	Reference
WS 1.12E-001	mg/L	25	EST	MEYLAN,WM ET AL. (1996)
logP 5.90			EST	MEYLAN,WM & HOWARD,PH (1995)
VP 2.10E-007	mm Hg	25	EST	NEELY,WB & BLAU,GE (1985)
DC	pKa			
HL 1.45E-008	atm m3/mol	25	EST	MEYLAN,WM & HOWARD,PH (1991)
OH 1.41E-010	cm3/molc sec	25	EST	MEYLAN,WM & HOWARD,PH (1993)

CAS #: 000119-90-4 — 3,3'-DIMETHOXYBENZIDINE

Formula: $C_{14}H_{16}N_2O_2$

Mol Weight: 244.30

MP (deg C): 137-138

FP (deg C):

BP (deg C): 356

BP pressure (mm Hg):

Property/Value	Units	Temp	Data Type	Reference
WS 6.00E+001	mg/L	25	EXP	YALKOWSKY,SH & DANNENFELSER,RM (1992)
logP 1.81			EXP	DEBNATH,AK ET AL. (1992)
VP 1.25E-007	mm Hg	25	EST	NEELY,WB & BLAU,GE (1985)
DC	pKa			
HL 4.70E-011	atm m3/mol	25	EST	MEYLAN,WM & HOWARD,PH (1991)
OH 1.85E-010	cm3/molc sec	25	EST	ATKINSON,R (1988)

CAS #: 000119-91-5 — 2,2'-BIQUINOLINE

Formula: $C_{18}H_{12}N_2$

Mol Weight: 256.31

MP (deg C): 193-196

FP (deg C):

BP (deg C):

BP pressure (mm Hg):

Property/Value	Units	Temp	Data Type	Reference
WS 1.02E+000	mg/L	25	EXP	BANWART,WL ET AL. (1982)
logP 4.31			EXP	HANSCH,C & LEO,AJ (1985)
VP 2.84E-008	mm Hg	25	EST	NEELY,WB & BLAU,GE (1985)
DC	pKa			
HL 6.75E-012	atm m3/mol	25	EST	MEYLAN,WM & HOWARD,PH (1991)
OH 5.00E-011	cm3/molc sec	25	EST	MEYLAN,WM & HOWARD,PH (1993)

CAS #: 000119-93-7 — BIANISIDINE

Formula: $C_{14}H_{16}N_2$

Mol Weight: 212.30

MP (deg C): 129-131

FP (deg C):

BP (deg C): 339

BP pressure (mm Hg):

Property/Value	Units	Temp	Data Type	Reference
WS 1.30E+003	mg/L	25	EXP	YALKOWSKY,SH & DANNENFELSER,RM (1992)
logP 2.34			EXP	HANSCH,C & LEO,AJ (1985)
VP 6.92E-007	mm Hg	25	EST	NEELY,WB & BLAU,GE (1985)
DC 4.50	pKa	25	EXP	PERRIN,DD (1965)
HL 6.29E-011	atm m3/mol	25	EST	MEYLAN,WM & HOWARD,PH (1991)
OH 2.40E-010	cm3/molc sec	25	EST	MEYLAN,WM & HOWARD,PH (1993)

CAS #: 000120-07-0 — PHENYLDIETHANOLAMINE

Formula: $C_{10}H_{15}NO_2$

Mol Weight: 181.24

MP (deg C): 57

FP (deg C):

BP (deg C): 200

BP pressure (mm Hg): 1.00E+001

Property/Value	Units	Temp	Data Type	Reference
WS 3.34E+004	mg/L	20	EXP	YALKOWSKY,SH & DANNENFELSER,RM (1992)
logP 0.63			EST	MEYLAN,WM & HOWARD,PH (1995)
VP 2.02E-006	mm Hg	25	EST	NEELY,WB & BLAU,GE (1985)
DC	pKa			
HL 2.02E-010	atm m3/mol	25	EST	MEYLAN,WM & HOWARD,PH (1991)
OH 1.76E-010	cm3/molc sec	25	EST	MEYLAN,WM & HOWARD,PH (1993)

CAS #: 000120-12-7 — ANTHRACENE

Formula: $C_{14}H_{10}$

Mol Weight: 178.24

MP (deg C): 215

FP (deg C):

BP (deg C): 340

BP pressure (mm Hg):

Property/Value	Units	Temp	Data Type	Reference
WS 4.34E-002	mg/L	24	EXP	MAY,WE ET AL. (1983)
logP 4.45			EXP	HANSCH,C & LEO,AJ (1985)
VP 2.67E-006	mm Hg	25	EXT	SCALA,AJ & BANERJEE,S (1982)
DC	pKa			
HL 7.20E-004	atm m3/mol	25	EXP	MACKAY,D ET AL. (1982A)
OH 1.10E-010	cm3/molc sec	25	EXP	ATKINSON,R (1989)

CAS #: 000120-14-9 — BENZALDEHYDE, 3,4-DIMETHOXY-

Formula: $C_9H_{10}O_3$

Mol Weight: 166.18

MP (deg C): 42-43

FP (deg C):

BP (deg C): 281

BP pressure (mm Hg):

Property/Value	Units	Temp	Data Type	Reference
WS 5.82E+003	mg/L	25	EST	MEYLAN,WM ET AL. (1996)
logP 1.22			EXP	BAZACO,JF & COCA,CM (1989)
VP 1.10E-002	mm Hg	25	EST	NEELY,WB & BLAU,GE (1985)
DC	pKa			
HL 4.70E-008	atm m3/mol	25	EST	MEYLAN,WM & HOWARD,PH (1991)
OH 2.80E-011	cm3/molc sec	25	EST	MEYLAN,WM & HOWARD,PH (1993)

CAS #: 000120-18-3 — NAPTHALENE-2-SULFONIC ACID

Formula: $C_{10}H_8O_3S$

Mol Weight: 208.24

MP (deg C): 91

FP (deg C):

BP (deg C): 124-125

BP pressure (mm Hg):

Property/Value	Units	Temp	Data Type	Reference
WS 6.79E+005	mg/L	30	EXP	BAUGHMAN,GL & PERENICH,TA (1988A)
logP 0.63			EXP	HANSCH,C & LEO,AJ (1985)
VP 2.30E-008	mm Hg	25	EST	NEELY,WB & BLAU,GE (1985)
DC	pKa			
HL 2.46E-010	atm m3/mol	25	EST	MEYLAN,WM & HOWARD,PH (1991)
OH 4.76E-012	cm3/molc sec	25	EST	MEYLAN,WM & HOWARD,PH (1993)

CAS #: 000120-20-7	3,4-DIMETHOXYPHENETHYLAMINE

Formula: $C_{10}H_{15}NO_2$
Mol Weight: 181.24
MP (deg C): FP (deg C):
BP (deg C): 163-165
BP pressure (mm Hg): 1.40E+001

Property/Value	Units	Temp	Data Type	Reference
WS 1.22E+005	mg/L	25	EST	MEYLAN,WM ET AL. (1996)
logP 0.77			EXP	HANSCH,C & LEO,AJ (1985)
VP 2.61E-003	mm Hg	25	EST	NEELY,WB & BLAU,GE (1985)
DC	pKa			
HL 2.84E-009	atm m3/mol	25	EST	MEYLAN,WM & HOWARD,PH (1991)
OH 8.07E-011	cm3/molc sec	25	EST	MEYLAN,WM & HOWARD,PH (1993)

CAS #: 000120-23-0	ACETIC ACID, 2-NAPHTHYLOXY-

Formula: $C_{12}H_{10}O_3$
Mol Weight: 202.21
MP (deg C): 156 FP (deg C):
BP (deg C):
BP pressure (mm Hg):

Property/Value	Units	Temp	Data Type	Reference
WS 7.31E+002	mg/L	25	EST	MEYLAN,WM ET AL. (1996)
logP 2.53			EXP	HANSCH,C & LEO,AJ (1985)
VP 6.27E-006	mm Hg	25	EST	NEELY,WB & BLAU,GE (1985)
DC	pKa			
HL 1.64E-009	atm m3/mol	25	EST	MEYLAN,WM & HOWARD,PH (1991)
OH 2.05E-010	cm3/molc sec	25	EST	MEYLAN,WM & HOWARD,PH (1993)

CAS #: 000120-26-3	BENZENEETHANAMINE, 3,4-DIMETHOXY-A-METHYL

Formula: $C_{11}H_{17}NO_2$
Mol Weight: 195.26
MP (deg C): FP (deg C):
BP (deg C):
BP pressure (mm Hg):

Property/Value	Units	Temp	Data Type	Reference
WS 4.45E+004	mg/L	25	EST	MEYLAN,WM ET AL. (1996)
logP 1.20			EXP	SANGSTER,J (1993)
VP 1.83E-003	mm Hg	25	EST	NEELY,WB & BLAU,GE (1985)
DC	pKa			
HL 3.77E-009	atm m3/mol	25	EST	MEYLAN,WM & HOWARD,PH (1991)
OH 9.24E-011	cm3/molc sec	25	EST	MEYLAN,WM & HOWARD,PH (1993)

CAS #: 000120-29-6	ENDO-8-METHYL-8-AZABICYCLO(3.2.1)OCTAN-3-OL

Formula: $C_8H_{15}NO$
Mol Weight: 141.21
MP (deg C): 63 FP (deg C):
BP (deg C): 233
BP pressure (mm Hg):

Property/Value	Units	Temp	Data Type	Reference
WS 5.24E+005	mg/L	25	EST	MEYLAN,WM ET AL. (1996)
logP 0.24			EST	MEYLAN,WM & HOWARD,PH (1995)
VP 1.57E-002	mm Hg	25	EST	NEELY,WB & BLAU,GE (1985)
DC	pKa			
HL 1.07E-009	atm m3/mol	25	EST	MEYLAN,WM & HOWARD,PH (1991)
OH 6.38E-011	cm3/molc sec	25	EST	MEYLAN,WM & HOWARD,PH (1993)

CAS #: 000120-32-1	5-CHLORO-2-HYDROXYDIPHENYLMETHANE

Formula: $C_{13}H_{11}ClO$
Mol Weight: 218.69
MP (deg C): 48.5 FP (deg C):
BP (deg C): 160-162
BP pressure (mm Hg): 3.50E+000

Property/Value	Units	Temp	Data Type	Reference
WS 1.49E+002	mg/L	25	EXP	WERNER,FA ET AL. (1983)
logP 4.18			EST	MEYLAN,WM & HOWARD,PH (1995)
VP 5.06E-005	mm Hg	25	EST	NEELY,WB & BLAU,GE (1985)
DC	pKa			
HL 9.96E-009	atm m3/mol	25	EST	MEYLAN,WM & HOWARD,PH (1991)
OH 1.85E-011	cm3/molc sec	25	EST	MEYLAN,WM & HOWARD,PH (1993)

CAS #: 000120-36-5	DICHLORPROP

Formula: $C_9H_8Cl_2O_3$
Mol Weight: 235.07
MP (deg C): 117.5-118 FP (deg C):
BP (deg C):
BP pressure (mm Hg):

Property/Value	Units	Temp	Data Type	Reference
WS 3.50E+002	mg/L	20	EXP	WORTHING,CR & WALKER,SB (1987)
logP 3.43			EXP	ILCHMANN,A ET AL. (1993)
VP 1.40E-005	mm Hg		EST	HL X WSOL
DC 3.10	pKa		EXP	KEARNEY,PC & KAUFMAN,DD (1975)
HL 1.22E-008	atm m3/mol	25	EST	MEYLAN,WM & HOWARD,PH (1991)
OH 1.10E-011	cm3/molc sec	25	EST	MEYLAN,WM & HOWARD,PH (1993)

CAS #: 000120-40-1	N,N-DI(2-HYDROXYETHYL)LAURAMIDE

Formula: $C_{15}H_{31}NO_4$
Mol Weight: 289.42
MP (deg C): 38.7 FP (deg C):
BP (deg C):
BP pressure (mm Hg):

Property/Value	Units	Temp	Data Type	Reference
WS 2.26E+002	mg/L	25	EST	MEYLAN,WM ET AL. (1996)
logP 2.89			EST	MEYLAN,WM & HOWARD,PH (1992)
VP 6.70E-009	mm Hg	25	EST	NEELY,WB & BLAU,GE (1985)
DC	pKa			
HL 2.16E-012	atm m3/mol	25	EST	MEYLAN,WM & HOWARD,PH (1991)
OH 4.81E-011	cm3/molc sec	25	EST	MEYLAN,WM & HOWARD,PH (1993)

CAS #: 000120-47-8	P-HYDROXYBENZOIC ACID,ETHYL ESTER

Formula: $C_9H_{10}O_3$
Mol Weight: 166.18
MP (deg C): 116 FP (deg C):
BP (deg C): 297-298
BP pressure (mm Hg):

Property/Value	Units	Temp	Data Type	Reference
WS 8.85E+002	mg/L	25	EXP	YALKOWSKY,SH & DANNENFELSER,RM (1992)
logP 2.47			EXP	HANSCH,C & LEO,AJ (1985)
VP 1.73E-003	mm Hg	25	EST	NEELY,WB & BLAU,GE (1985)
DC	pKa			
HL 4.79E-009	atm m3/mol	25	EST	MEYLAN,WM & HOWARD,PH (1991)
OH 1.26E-011	cm3/molc sec	25	EST	MEYLAN,WM & HOWARD,PH (1993)

CAS #: 000120-51-4				BENZYLBENZOATE
Formula: $C_{14}H_{12}O_2$				
Mol Weight: 212.25				
MP (deg C): 21		FP (deg C):		
BP (deg C): 323-324				
BP pressure (mm Hg):				

Property/Value	Units	Temp	Data Type	Reference
WS 1.54E+001	mg/L	25	EST	MEYLAN,WM ET AL. (1996)
logP 3.97			EXP	HANSCH,C & LEO,AJ (1985)
VP 2.24E-004	mm Hg	25	EXP	DAUBERT,TE & DANNER,RP (1989)
DC	pKa			
HL 2.80E-006	atm m3/mol	25	EST	MEYLAN,WM & HOWARD,PH (1991)
OH 6.97E-012	cm3/molc sec	25	EST	MEYLAN,WM & HOWARD,PH (1993)

CAS #: 000120-55-8				DIETHYLENE GLYCOL DIBENZOATE
Formula: $C_{18}H_{18}O_5$				
Mol Weight: 314.34				
MP (deg C): -28		FP (deg C):		
BP (deg C): 225-225				
BP pressure (mm Hg):				

Property/Value	Units	Temp	Data Type	Reference
WS 1.93E+002	mg/L	25	EST	MEYLAN,WM ET AL. (1996)
logP 3.04			EST	MEYLAN,WM & HOWARD,PH (1995)
VP 9.60E-002	mm Hg	25	EST	NEELY,WB & BLAU,GE (1985)
DC	pKa			
HL 3.00E-012	atm m3/mol	25	EST	MEYLAN,WM & HOWARD,PH (1991)
OH 1.90E-011	cm3/molc sec	25	EST	MEYLAN,WM & HOWARD,PH (1993)

CAS #: 000120-57-0				PIPERONAL
Formula: $C_8H_6O_3$				
Mol Weight: 150.14				
MP (deg C): 37		FP (deg C):		
BP (deg C): 263				
BP pressure (mm Hg):				

Property/Value	Units	Temp	Data Type	Reference
WS 3.50E+003	mg/L	20	EXP	YALKOWSKY,SH & DANNENFELSER,RM (1992)
logP 1.05			EXP	HANSCH,C ET AL. (1995)
VP 1.00E-002	mm Hg	25	EXT	PERRY,RH & GREEN,D (1984)
DC	pKa			
HL 5.60E-007	atm m3/mol	25	EST	VP/WSOL
OH 2.90E-011	cm3/molc sec	25	EST	MEYLAN,WM & HOWARD,PH (1993)

CAS #: 000120-58-1				1,2-METHYLENEDIOXY-4-PROPENYL BENZENE
Formula: $C_{10}H_{10}O_2$				
Mol Weight: 162.19				
MP (deg C): 6.7-6.8		FP (deg C):		
BP (deg C): 252				
BP pressure (mm Hg):				

Property/Value	Units	Temp	Data Type	Reference
WS 1.44E+002	mg/L	25	EST	MEYLAN,WM ET AL. (1996)
logP 3.37			EST	MEYLAN,WM & HOWARD,PH (1995)
VP 2.45E-002	mm Hg	25	EST	NEELY,WB & BLAU,GE (1985)
DC	pKa			
HL 3.60E-002	atm m3/mol	25	EST	MEYLAN,WM & HOWARD,PH (1991)
OH 8.27E-011	cm3/molc sec	25	EST	MEYLAN,WM & HOWARD,PH (1993)

CAS #: 000120-61-6				DIMETHYLTEREPHTHALATE
Formula: $C_{10}H_{10}O_4$				
Mol Weight: 194.19				
MP (deg C): 140-142		FP (deg C): 140		
BP (deg C): 288				
BP pressure (mm Hg):				

Property/Value	Units	Temp	Data Type	Reference
WS 1.90E+001	mg/L	25	EXP	BEMIS,AG ET AL. (1982)
logP 2.25			EXP	HANSCH,C ET AL (1995)
VP 1.00E-002	mm Hg	25	EXP	DAUBERT,TE & DANNER,RP (1991)
DC	pKa			
HL 1.34E-004	atm m3/mol	25	EST	VP/WSOL
OH 5.74E-013	cm3/molc sec	25	EST	MEYLAN,WM & HOWARD,PH (1993)

CAS #: 000120-62-7				SULFOXIDE
Formula: $C_{18}H_{28}O_3S$				
Mol Weight: 324.49				
MP (deg C):		FP (deg C):		
BP (deg C):				
BP pressure (mm Hg):				

Property/Value	Units	Temp	Data Type	Reference
WS 5.83E-001	mg/L	25	EST	MEYLAN,WM ET AL. (1996)
logP 4.89			EST	MEYLAN,WM & HOWARD,PH (1995)
VP 7.30E-008	mm Hg	25	EST	NEELY,WB & BLAU,GE (1985)
DC	pKa			
HL 5.96E-011	atm m3/mol	25	EST	MEYLAN,WM & HOWARD,PH (1991)
OH 1.46E-010	cm3/molc sec	25	EST	MEYLAN,WM & HOWARD,PH (1993)

CAS #: 000120-66-1				ACETAMIDE, N-(2-METHYLPHENYL)-
Formula: $C_9H_{11}NO$				
Mol Weight: 149.19				
MP (deg C): 110		FP (deg C):		
BP (deg C): 296				
BP pressure (mm Hg):				

Property/Value	Units	Temp	Data Type	Reference
WS 1.44E+004	mg/L	25	EST	MEYLAN,WM ET AL. (1996)
logP 0.85			EXP	HANSCH,C & LEO,AJ (1985)
VP 1.83E-004	mm Hg	25	EST	NEELY,WB & BLAU,GE (1985)
DC	pKa			
HL 6.81E-009	atm m3/mol	25	EST	MEYLAN,WM & HOWARD,PH (1991)
OH 1.55E-011	cm3/molc sec	25	EST	MEYLAN,WM & HOWARD,PH (1993)

CAS #: 000120-67-2				2-(2,4-DICHLOROPHENOXY)ETHANOL
Formula: $C_8H_8Cl_2O_2$				
Mol Weight: 207.06				
MP (deg C):		FP (deg C):		
BP (deg C):				
BP pressure (mm Hg):				

Property/Value	Units	Temp	Data Type	Reference
WS 1.19E+003	mg/L	25	EST	MEYLAN,WM ET AL. (1996)
logP 2.39			EST	MEYLAN,WM & HOWARD,PH (1995)
VP 1.99E-004	mm Hg	25	EXT	OHE,S (1976)
DC	pKa			
HL 8.49E-009	atm m3/mol	25	EST	MEYLAN,WM & HOWARD,PH (1991)
OH 1.30E-011	cm3/molc sec	25	EST	MEYLAN,WM & HOWARD,PH (1993)

CAS #: 000120-71-8				2-METHOXY-5-METHYLBENZENAMINE

Formula: $C_8H_{11}NO$

Mol Weight: 137.18

MP (deg C): 52-54   **FP (deg C):**

BP (deg C): 235

BP pressure (mm Hg):

Property/ Value	Units	Temp	Data Type	Reference
WS 2.81E+003	mg/L	25	EST	MEYLAN,WM ET AL. (1996)
logP 1.74			EXP	HANSCH,C ET AL. (1995)
VP 2.52E-002	mm Hg	25	EST	NEELY,WB & BLAU,GE (1985)
DC	pKa			
HL 1.24E-007	atm m3/mol	25	EST	MEYLAN,WM & HOWARD,PH (1991)
OH 2.21E-010	cm3/molc sec	25	EST	ATKINSON,R (1988)

CAS #: 000120-72-9				INDOLE

Formula: C_8H_7N

Mol Weight: 117.15

MP (deg C): 52   **FP (deg C):**

BP (deg C): 253

BP pressure (mm Hg): 7.62E+002

Property/ Value	Units	Temp	Data Type	Reference
WS 3.56E+003	mg/L	25	EXP	YALKOWSKY,SH & DANNENFELSER,RM (1992)
logP 2.14			EXP	HANSCH,C ET AL. (1995)
VP 1.22E-002	mm Hg	25	EXP	YAWS,CL (1994B)
DC -2.40	pKa	25	EXP	ADLER,TK & ALBERT,A (1963)
HL 5.28E-007	atm m3/mol	25	EST	VP/WSOL
OH 1.54E-010	cm3/molc sec	25	EXP	ATKINSON,R ET AL. (1995B)

CAS #: 000120-73-0				1H-PURINE

Formula: $C_5H_4N_4$

Mol Weight: 120.11

MP (deg C): 216-217   **FP (deg C):**

BP (deg C):

BP pressure (mm Hg):

Property/ Value	Units	Temp	Data Type	Reference
WS 5.00E+005	mg/L	20	EXP	YALKOWSKY,SH & DANNENFELSER,RM (1992)
logP -0.37			EXP	HANSCH,C & LEO,AJ (1985)
VP 1.99E-005	mm Hg	25	EST	NEELY,WB & BLAU,GE (1985)
DC 2.52	pKa	25	EXP	KORTUM,G ET AL (1961)
HL 1.99E-010	atm m3/mol	25	EST	MEYLAN,WM & HOWARD,PH (1991)
OH 3.60E-011	cm3/molc sec	25	EST	MEYLAN,WM & HOWARD,PH (1993)

CAS #: 000120-75-2				2-METHYBENZOTHIAZOLE

Formula: C_8H_7NS

Mol Weight: 149.22

MP (deg C): 14   **FP (deg C):**

BP (deg C): 238

BP pressure (mm Hg):

Property/ Value	Units	Temp	Data Type	Reference
WS 3.66E+002	mg/L	25	EST	MEYLAN,WM ET AL. (1996)
logP 2.72			EST	MEYLAN,WM & HOWARD,PH (1995)
VP 3.61E-001	mm Hg	25	EXT	PERRY,RH & GREEN,D (1984)
DC	pKa			
HL 4.13E-007	atm m3/mol	25	EST	MEYLAN,WM & HOWARD,PH (1991)
OH 1.84E-011	cm3/molc sec	25	EST	MEYLAN,WM & HOWARD,PH (1993)

CAS #: 000120-78-5				2,2'-DITHIOBISBENZOTHIAZOLE

Formula: $C_{14}H_8N_2S_4$

Mol Weight: 332.49

MP (deg C): 180   **FP (deg C):**

BP (deg C):

BP pressure (mm Hg):

Property/ Value	Units	Temp	Data Type	Reference
WS 1.00E+001	mg/L	25	EXP	CHEMICAL INSPECTION TESTING INST (1992);<
logP 4.66			EST	MEYLAN,WM & HOWARD,PH (1995)
VP 2.54E-010	mm Hg	25	EST	NEELY,WB & BLAU,GE (1985)
DC	pKa			
HL 2.34E-013	atm m3/mol	25	EST	MEYLAN,WM & HOWARD,PH (1991)
OH 2.91E-010	cm3/molc sec	25	EST	MEYLAN,WM & HOWARD,PH (1993)

CAS #: 000120-80-9				CATECHOL

Formula: $C_6H_6O_2$

Mol Weight: 110.11

MP (deg C): 105   **FP (deg C):**

BP (deg C): 245.5

BP pressure (mm Hg):

Property/ Value	Units	Temp	Data Type	Reference
WS 4.61E+005	mg/L	25	EXP	YALKOWSKY,SH & DANNENFELSER,RM (1992)
logP 0.88			EXP	HANSCH,C & LEO,AJ (1985)
VP 1.00E-002	mm Hg	25	EXT	BOUBLIK,T ET AL. (1984)
DC 9.45	pKa	25	EXP	SERJEANT,EP & DEMPSEY,B (1979)
HL 3.14E-009	atm m3/mol	25	EST	VP/WSOL
OH 2.45E-011	cm3/molc sec	25	EST	MEYLAN,WM & HOWARD,PH (1993)

CAS #: 000120-82-1				1,2,4-TRICHLOROBENZENE

Formula: $C_6H_3Cl_3$

Mol Weight: 181.45

MP (deg C): 17   **FP (deg C):**

BP (deg C): 213

BP pressure (mm Hg):

Property/ Value	Units	Temp	Data Type	Reference
WS 4.90E+001	mg/L	25	EXP	SOUTHWORTH,GR & KELLER,JL (1986)
logP 4.02			EXP	HANSCH,C ET AL. (1995)
VP 2.90E-001	mm Hg	25	EXP	SEARS,GW & HOPKE,ER (1949A)
DC	pKa			
HL 1.42E-003	atm m3/mol	25	EXP	WARNER,HP ET AL. (1987)
OH 5.32E-013	cm3/molc sec	25	EXP	ATKINSON,R (1989)

CAS #: 000120-83-2				2,4-DICHLOROPHENOL

Formula: $C_6H_4Cl_2O$

Mol Weight: 163.00

MP (deg C): 42-43   **FP (deg C):**

BP (deg C): 209-210

BP pressure (mm Hg):

Property/ Value	Units	Temp	Data Type	Reference
WS 4.50E+003	mg/L	20	EXP	YALKOWSKY,SH & DANNENFELSER,RM (1992)
logP 3.06			EXP	HANSCH,C ET AL. (1995)
VP 6.70E-002	mm Hg	25	EXP	BIDLEMAN,TF & RENBERG,L (1985)
DC 7.89	pKa		EXP	SERJEANT,EP & DEMPSEY,B (1979)
HL 3.16E-006	atm m3/mol	25	EST	VP/WSOL
OH 1.06E-012	cm3/molc sec	25	EXP	ATKINSON,R (1989)

CAS #: 000120-92-3	CYCLOPENTANONE

Formula: C_5H_8O

Mol Weight: 84.12

MP (deg C): -58.2 **FP (deg C):**

BP (deg C): 130.6

BP pressure (mm Hg):

Property/Value	Units	Temp	Data Type	Reference
WS 9.18E+003	mg/L	25	EST	MEYLAN,WM ET AL. (1996)
logP 0.63			EST	MEYLAN,WM & HOWARD,PH (1995)
VP 1.14E+001	mm Hg	25	EXP	DAUBERT,TE & DANNER,RP (1989)
DC	pKa			
HL 1.00E-005	atm m3/mol	25	EXP	HAWTHORNE,SB ET AL. (1985)
OH 1.00E-005	cm3/molc sec	25	EXP	HAWTHORNE,SB ET AL. (1985)

CAS #: 000120-94-5	1-METHYL-PYRROLIDINE

Formula: $C_5H_{11}N$

Mol Weight: 85.15

MP (deg C): **FP (deg C):**

BP (deg C): 81

BP pressure (mm Hg):

Property/Value	Units	Temp	Data Type	Reference
WS 2.13E+005	mg/L	25	EST	MEYLAN,WM ET AL. (1996)
logP 0.92			EXP	HANSCH,C ET AL. (1995)
VP 1.00E+002	mm Hg	25	EXP	DAUBERT,TE & DANNER,RP (1989)
DC 10.32	pKa		EXP	PERRIN,DD (1965)
HL 3.01E-005	atm m3/mol	25	EXP	CABANI,S ET AL. (1971)
OH 8.10E-011	cm3/molc sec	25	EST	MEYLAN,WM & HOWARD,PH (1993)

CAS #: 000121-14-2	2,4-DINITROTOLUENE

Formula: $C_7H_6N_2O_4$

Mol Weight: 182.14

MP (deg C): 67-70 **FP (deg C):**

BP (deg C): 300

BP pressure (mm Hg):

Property/Value	Units	Temp	Data Type	Reference
WS 2.70E+002	mg/L	22	EXP	SPANGGORD,RJ ET AL. (1980)
logP 1.98			EXP	HANSCH,C & LEO,AJ (1985)
VP 1.47E-004	mm Hg	22	EXP	PELLA,PA (1977)
DC	pKa			
HL 1.30E-007	atm m3/mol	22	EST	VP/WSOL
OH 2.25E-013	cm3/molc sec	25	EST	MEYLAN,WM & HOWARD,PH (1993)

CAS #: 000121-21-1	PYRETHRIN I

Formula: $C_{21}H_{28}O_3$

Mol Weight: 328.46

MP (deg C): 146 **FP (deg C):**

BP (deg C):

BP pressure (mm Hg): 0.00E+000

Property/Value	Units	Temp	Data Type	Reference
WS 3.60E-002	mg/L	25	EST	MEYLAN,WM ET AL. (1996)
logP 6.28			EST	MEYLAN,WM & HOWARD,PH (1995)
VP 5.18E-007	mm Hg	25	EST	NEELY,WB & BLAU,GE (1985)
DC	pKa			
HL 7.73E-007	atm m3/mol	25	EST	MEYLAN,WM & HOWARD,PH (1991)
OH 3.04E-010	cm3/molc sec	25	EST	ATKINSON,R (1988)

CAS #: 000121-29-9	PYRETHRIN II

Formula: $C_{22}H_{32}O_5$

Mol Weight: 376.50

MP (deg C): **FP (deg C):**

BP (deg C): 200

BP pressure (mm Hg): 1.00E-001

Property/Value	Units	Temp	Data Type	Reference
WS 1.27E-001	mg/L	25	EST	MEYLAN,WM ET AL. (1996)
logP 5.33			EST	MEYLAN,WM & HOWARD,PH (1995)
VP 2.06E-007	mm Hg	25	EST	NEELY,WB & BLAU,GE (1985)
DC	pKa			
HL 7.39E-010	atm m3/mol	25	EST	MEYLAN,WM & HOWARD,PH (1991)
OH 1.34E-011	cm3/molc sec	25	EST	HINE,J & MOOKERJEE,PK (1975)

CAS #: 000121-32-4	BENZALDEHYDE, 3-ETHOXY-4-HYDROXY-

Formula: $C_9H_{10}O_3$

Mol Weight: 166.18

MP (deg C): 77-78 **FP (deg C):**

BP (deg C):

BP pressure (mm Hg):

Property/Value	Units	Temp	Data Type	Reference
WS 2.87E+003	mg/L	25	EST	MEYLAN,WM ET AL. (1996)
logP 1.58			EXP	BAZACO,JF & COCA,CM (1989)
VP 1.04E-005	mm Hg	25	EXT	YAWS,CL (1994B)
DC	pKa			
HL 1.10E-010	atm m3/mol	25	EST	MEYLAN,WM & HOWARD,PH (1991)
OH 3.27E-011	cm3/molc sec	25	EST	MEYLAN,WM & HOWARD,PH (1993)

CAS #: 000121-33-5	VANILLIN

Formula: $C_8H_8O_3$

Mol Weight: 152.15

MP (deg C): 80-81 **FP (deg C):**

BP (deg C): 285

BP pressure (mm Hg):

Property/Value	Units	Temp	Data Type	Reference
WS 1.10E+004	mg/L	25	EXP	YALKOWSKY,SH & DANNENFELSER,RM (1992)
logP 1.21			EXP	HANSCH,C & LEO,AJ (1985)
VP 1.18E-004	mm Hg	25	EXT	YAWS,CL (1994B)
DC 7.40	pKa	25	EXP	KORTUM,G ET AL (1961)
HL 8.27E-011	atm m3/mol	25	EST	MEYLAN,WM & HOWARD,PH (1991)
OH 2.73E-011	cm3/molc sec	25	EST	MEYLAN,WM & HOWARD,PH (1993)

CAS #: 000121-34-6	4-HYDROXY-3-METHOXYBENZOIC ACID

Formula: $C_8H_8O_4$

Mol Weight: 168.15

MP (deg C): 211.5 **FP (deg C):**

BP (deg C):

BP pressure (mm Hg):

Property/Value	Units	Temp	Data Type	Reference
WS 1.50E+003	mg/L	14	EXP	YALKOWSKY,SH & DANNENFELSER,RM (1992)
logP 1.43			EXP	HANSCH,C & LEO,AJ (1985)
VP 1.71E-005	mm Hg	25	EST	NEELY,WB & BLAU,GE (1985)
DC 4.51	pKa	25	EXP	SERJEANT,EP & DEMPSEY,B (1979)
HL 6.67E-013	atm m3/mol	25	EST	MEYLAN,WM & HOWARD,PH (1991)
OH 1.22E-011	cm3/molc sec	25	EST	MEYLAN,WM & HOWARD,PH (1993)

CAS #: 000121-44-8				TRIETHYLAMINE

Formula: $C_6H_{15}N$

Mol Weight: 101.19

MP (deg C): -115   FP (deg C):

BP (deg C): 89-90

BP pressure (mm Hg):

Property/Value	Units	Temp	Data Type	Reference
WS 7.37E+004	mg/L		EXP	YALKOWSKY,SH & DANNENFELSER,RM (1992)
logP 1.45			EXP	HANSCH,C & LEO,AJ (1985)
VP 5.71E+001	mm Hg	25	EXP	RIDDICK,JA ET AL. (1986)
DC 10.78	pKa	25	EXP	RIDDICK,JA ET AL. (1986)
HL 1.49E-004	atm m3/mol	25	EXP	CHRISTIE,AO & CRISP,DJ (1967)
OH 8.57E-011	cm3/molc sec	25	EST	MEYLAN,WM & HOWARD,PH (1993)

CAS #: 000121-45-9				TRIMETHYL PHOSPHITE

Formula: $C_3H_9O_3P$

Mol Weight: 124.08

MP (deg C):   FP (deg C):

BP (deg C): 111.5

BP pressure (mm Hg):

Property/Value	Units	Temp	Data Type	Reference
WS 7.20E+003	mg/L	25	EST	MEYLAN,WM ET AL. (1996)
logP 1.32			EST	MEYLAN,WM & HOWARD,PH (1995)
VP 1.87E+001	mm Hg	25	EST	NEELY,WB & BLAU,GE (1985)
DC	pKa			
HL 1.07E-005	atm m3/mol	25	EST	MEYLAN,WM & HOWARD,PH (1991)
OH 8.36E-012	cm3/molc sec	25	EST	MEYLAN,WM & HOWARD,PH (1993)

CAS #: 000121-46-0				BICYCLO(2.2.1)HEPTA-2,5-DIENE

Formula: C_7H_8

Mol Weight: 92.14

MP (deg C): -19.1   FP (deg C):

BP (deg C): 89.5

BP pressure (mm Hg):

Property/Value	Units	Temp	Data Type	Reference
WS 2.01E+002	mg/L	25	EST	MEYLAN,WM ET AL. (1996)
logP 2.64			EST	MEYLAN,WM & HOWARD,PH (1995)
VP 2.55E+001	mm Hg	25	EST	NEELY,WB & BLAU,GE (1985)
DC	pKa			
HL 6.06E-002	atm m3/mol	25	EST	MEYLAN,WM & HOWARD,PH (1991)
OH 1.20E-010	cm3/molc sec	25	EXP	ATKINSON,R (1989)

CAS #: 000121-47-1				3-AMINOBENZENESULFONIC ACID

Formula: $C_6H_7NO_3S$

Mol Weight: 173.19

MP (deg C): dec   FP (deg C):

BP (deg C):

BP pressure (mm Hg):

Property/Value	Units	Temp	Data Type	Reference
WS 1.28E+004	mg/L	7	EXP	STEPHEN,H & STEPHEN,T (1963)
logP -2.08			EST	MEYLAN,WM & HOWARD,PH (1995)
VP 2.02E-007	mm Hg	25	EST	NEELY,WB & BLAU,GE (1985)
DC 3.75	pKa		EXP	SERJEANT,EP & DEMPSEY,B (1979)
HL 8.89E-013	atm m3/mol	25	EST	MEYLAN,WM & HOWARD,PH (1991)
OH 2.32E-011	cm3/molc sec	25	EST	MEYLAN,WM & HOWARD,PH (1993)

CAS #: 000121-50-6				BENZENAMINE, 2-CHLORO-5-(TRIFLUOROMETHYL)-

Formula: $C_7H_5ClF_3N$

Mol Weight: 195.57

MP (deg C):   FP (deg C):

BP (deg C):

BP pressure (mm Hg):

Property/Value	Units	Temp	Data Type	Reference
WS 1.12E+002	mg/L	25	EST	MEYLAN,WM ET AL. (1996)
logP 3.06			EXP	HANSCH,C ET AL. (1995)
VP 9.74E-002	mm Hg	25	EST	NEELY,WB & BLAU,GE (1985)
DC	pKa			
HL 1.23E-005	atm m3/mol	25	EST	MEYLAN,WM & HOWARD,PH (1991)
OH 4.76E-012	cm3/molc sec	25	EST	MEYLAN,WM & HOWARD,PH (1993)

CAS #: 000121-52-8				M-NITROBENZENESULFONAMIDE

Formula: $C_6H_6N_2O_4S$

Mol Weight: 202.19

MP (deg C): 166-168   FP (deg C):

BP (deg C):

BP pressure (mm Hg):

Property/Value	Units	Temp	Data Type	Reference
WS 5.89E+003	mg/L	25	EST	MEYLAN,WM ET AL. (1996)
logP 0.55			EXP	HANSCH,C & LEO,AJ (1985)
VP 7.82E-006	mm Hg	25	EST	NEELY,WB & BLAU,GE (1985)
DC	pKa			
HL 1.67E-009	atm m3/mol	25	EST	MEYLAN,WM & HOWARD,PH (1991)
OH 5.21E-014	cm3/molc sec	25	EST	MEYLAN,WM & HOWARD,PH (1993)

CAS #: 000121-57-3				BENZENESULFONIC ACID, 4-AMINO-

Formula: $C_6H_7NO_3S$

Mol Weight: 173.19

MP (deg C): 288   FP (deg C):

BP (deg C):

BP pressure (mm Hg):

Property/Value	Units	Temp	Data Type	Reference
WS 4.50E+003	mg/L	0	EXP	YALKOWSKY,SH & DANNENFELSER,RM (1992)
logP -2.16			EXP	OKAMOTO,H ET AL. (1991)
VP 2.02E-007	mm Hg	25	EST	NEELY,WB & BLAU,GE (1985)
DC 3.34	pKa	15	EXP	KORTUM,G ET AL (1961)
HL 8.89E-013	atm m3/mol	25	EST	MEYLAN,WM & HOWARD,PH (1991)
OH 2.32E-011	cm3/molc sec	25	EST	MEYLAN,WM & HOWARD,PH (1993)

CAS #: 000121-61-9				ACETAMIDE, N-[4-(AMINOSULFONYL)PHENYL]

Formula: $C_8H_{10}N_2O_3S$

Mol Weight: 214.24

MP (deg C): 216   FP (deg C):

BP (deg C):

BP pressure (mm Hg):

Property/Value	Units	Temp	Data Type	Reference
WS 5.30E+003	mg/L	37	EXP	YALKOWSKY,SH & DANNENFELSER,RM (1992)
logP -0.04			EXP	CAROTTI,A ET AL. (1989)
VP 8.06E-008	mm Hg	25	EST	NEELY,WB & BLAU,GE (1985)
DC 5.32	pKa		EXP	PERRIN,DD (1965)
HL 4.84E-013	atm m3/mol	25	EST	MEYLAN,WM & HOWARD,PH (1991)
OH 2.76E-012	cm3/molc sec	25	EST	MEYLAN,WM & HOWARD,PH (1993)

000121-66-4 — 2-THIAZOLAMINE, 5-NITRO-

Formula:	$C_3H_3N_3O_2S$			
Mol Weight:	145.14			
MP (deg C):	202 dec	FP (deg C):		
BP (deg C):				
BP pressure (mm Hg):				

Property/Value	Units	Temp	Data Type	Reference
WS 6.34E+003	mg/L	25	EST	MEYLAN,WM ET AL. (1996)
logP 0.83			EXP	HANSCH,C ET AL. (1995)
VP 7.43E-004	mm Hg	25	EST	NEELY,WB & BLAU,GE (1985)
DC	pKa			
HL 5.34E-012	atm m3/mol	25	EST	MEYLAN,WM & HOWARD,PH (1991)
OH 2.00E-013	cm3/molc sec	25	EST	MEYLAN,WM & HOWARD,PH (1993)

000121-69-7 — N,N-DIMETHYLANILINE

Formula:	$C_8H_{11}N$			
Mol Weight:	121.18			
MP (deg C):	2.45	FP (deg C):		
BP (deg C):	194.05			
BP pressure (mm Hg):				

Property/Value	Units	Temp	Data Type	Reference
WS 1.45E+003	mg/L	25	EXP	HUYSKENS,P ET AL. (1975)
logP 2.31			EXP	HANSCH,C & LEO,AJ (1985)
VP 7.00E-001	mm Hg	25	EXP	DAUBERT,TE & DANNER,RP (1989)
DC 5.15	pKa	25	EXP	WEAST,RC ET AL. (1985)
HL 5.68E-005	atm m3/mol	25	EST	VP/WSOL
OH 1.48E-010	cm3/molc sec	25	EXP	ATKINSON,R (1989)

000121-71-1 — M-HYDROXYACETOPHENONE

Formula:	$C_8H_8O_2$			
Mol Weight:	136.15			
MP (deg C):	96	FP (deg C):		
BP (deg C):	296			
BP pressure (mm Hg):				

Property/Value	Units	Temp	Data Type	Reference
WS 2.15E+004	mg/L	25	EST	MEYLAN,WM ET AL. (1996)
logP 1.39			EXP	HANSCH,C & LEO,AJ (1985)
VP 7.48E-003	mm Hg	25	EST	NEELY,WB & BLAU,GE (1985)
DC 9.25	pKa	25	EXP	KORTUM,G ET AL (1961)
HL 1.02E-009	atm m3/mol	25	EST	MEYLAN,WM & HOWARD,PH (1991)
OH 1.43E-011	cm3/molc sec	25	EST	MEYLAN,WM & HOWARD,PH (1993)

000121-72-2 — M-TOLUIDINE-N,N-DIMETHYL

Formula:	$C_9H_{13}N$			
Mol Weight:	135.21			
MP (deg C):		FP (deg C):		
BP (deg C):	212			
BP pressure (mm Hg):				

Property/Value	Units	Temp	Data Type	Reference
WS 3.56E+002	mg/L	25	EST	MEYLAN,WM ET AL. (1996)
logP 2.80			EXP	HANSCH,C & LEO,AJ (1985)
VP 5.87E-001	mm Hg	25	EST	NEELY,WB & BLAU,GE (1985)
DC 5.34	pKa	25	EXP	PERRIN,DD (1965)
HL 9.45E-005	atm m3/mol	25	EST	MEYLAN,WM & HOWARD,PH (1991)
OH 2.03E-010	cm3/molc sec	25	EST	MEYLAN,WM & HOWARD,PH (1993)

000121-73-3 — 3-CHLORO-NITROBENZENE

Formula:	$C_6H_4ClNO_2$			
Mol Weight:	157.56			
MP (deg C):	46	FP (deg C):		
BP (deg C):	236			
BP pressure (mm Hg):				

Property/Value	Units	Temp	Data Type	Reference
WS 2.73E+002	mg/L	25	EXP	YALKOWSKY,SH & DANNENFELSER,RM (1992)
logP 2.46			EXP	HANSCH,C ET AL. (1995)
VP 9.70E-002	mm Hg	25	EXP	DAUBERT,TE & DANNER,RP (1994)
DC	pKa			
HL 3.79E-005	atm m3/mol	25	EST	MEYLAN,WM & HOWARD,PH (1991)
OH 1.20E-013	cm3/molc sec	25	EST	MEYLAN,WM & HOWARD,PH (1993)

000121-75-5 — MALATHION

Formula:	$C_{10}H_{19}O_6PS_2$			
Mol Weight:	330.36			
MP (deg C):	2.85	FP (deg C):		
BP (deg C):	156-157			
BP pressure (mm Hg):	7.00E-001			

Property/Value	Units	Temp	Data Type	Reference
WS 1.43E+002	mg/L	20	EXP	BOWMAN,BT & SANS,WW (1985)
logP 2.36			EXP	HANSCH,C & LEO,AJ (1985)
VP 7.90E-006	mm Hg	25	EXT	KIM,YH ET AL. (1984)
DC	pKa			
HL 2.40E-008	atm m3/mol	25	EST	VP/WSOL
OH 6.56E-011	cm3/molc sec	25	EST	MEYLAN,WM & HOWARD,PH (1993)

000121-79-9 — PROPYL GALLATE

Formula:	$C_{10}H_{12}O_5$			
Mol Weight:	212.20			
MP (deg C):	150	FP (deg C):		
BP (deg C):				
BP pressure (mm Hg):				

Property/Value	Units	Temp	Data Type	Reference
WS 3.50E+003	mg/L	25	EXP	MERCK INDEX (1989)
logP 1.80			EXP	HANSCH,C & LEO,AJ (1985)
VP 3.30E-007	mm Hg	25	EST	PCGEMS (1987)
DC	pKa			
HL 6.30E-012	atm m3/mol	25	EST	MEYLAN,WM & HOWARD,PH (1991)
OH 9.85E-011	cm3/molc sec	25	EST	ATKINSON,R (1987)

000121-81-3 — 3,5-DINITROBENZAMIDE

Formula:	$C_7H_5N_3O_5$			
Mol Weight:	211.14			
MP (deg C):	183	FP (deg C):		
BP (deg C):				
BP pressure (mm Hg):				

Property/Value	Units	Temp	Data Type	Reference
WS 3.05E+003	mg/L	25	EST	MEYLAN,WM ET AL. (1996)
logP 0.83			EXP	HANSCH,C & LEO,AJ (1985)
VP 1.56E-007	mm Hg	25	EST	NEELY,WB & BLAU,GE (1985)
DC	pKa			
HL 3.45E-014	atm m3/mol	25	EST	MEYLAN,WM & HOWARD,PH (1991)
OH 2.01E-012	cm3/molc sec	25	EST	MEYLAN,WM & HOWARD,PH (1993)

000121-82-4 — 1,3,5,-TRINITROHEXAHYDRO-1,3,5-TRIAZINE

Formula: $C_3H_6N_6O_6$
Mol Weight: 222.12
MP (deg C): 205-206
FP (deg C):
BP (deg C):
BP pressure (mm Hg):

Property/Value	Units	Temp	Data Type	Reference
WS 5.98E+001	mg/L	25	EXP	YALKOWSKY,SH & DANNENFELSER,RM (1992)
logP 0.87			EXP	SANGSTER,J (1993)
VP 4.10E-009	mm Hg	20	EXT	SPANGGORD,RJ ET AL. (1980)
DC	pKa			
HL 6.32E-008	atm m3/mol	25	EST	MEYLAN,WM & HOWARD,PH (1991)
OH 2.46E-010	cm3/molc sec	25	EST	MEYLAN,WM & HOWARD,PH (1993)

000121-86-8 — 2-CHLORO-4-NITROTOLUENE

Formula: $C_7H_6ClNO_2$
Mol Weight: 171.58
MP (deg C): 68
FP (deg C):
BP (deg C): 260
BP pressure (mm Hg):

Property/Value	Units	Temp	Data Type	Reference
WS 6.72E+001	mg/L	25	EST	MEYLAN,WM ET AL. (1996)
logP 3.00			EST	MEYLAN,WM & HOWARD,PH (1995)
VP 1.06E-002	mm Hg	25	EST	NEELY,WB & BLAU,GE (1985)
DC	pKa			
HL 1.74E-005	atm m3/mol	25	EST	MEYLAN,WM & HOWARD,PH (1991)
OH 4.06E-005	cm3/molc sec	25	EST	HINE,J & MOOKERJEE,PK (1975)

000121-87-9 — 2-CHLORO-4-NITROANILINE

Formula: $C_6H_5ClN_2O_2$
Mol Weight: 172.57
MP (deg C): 107-109
FP (deg C):
BP (deg C):
BP pressure (mm Hg):

Property/Value	Units	Temp	Data Type	Reference
WS 9.33E+002	mg/L	25	EST	MEYLAN,WM ET AL. (1996)
logP 2.12			EST	MEYLAN,WM & HOWARD,PH (1995)
VP 4.85E-004	mm Hg	25	EST	NEELY,WB & BLAU,GE (1985)
DC -0.94	pKa	25	EXP	PERRIN,DD (1972)
HL 5.57E-009	atm m3/mol	25	EST	MEYLAN,WM & HOWARD,PH (1991)
OH 3.93E-012	cm3/molc sec	25	EST	MEYLAN,WM & HOWARD,PH (1993)

000121-89-1 — 3-NITROACETOPHENONE

Formula: $C_8H_7NO_3$
Mol Weight: 165.15
MP (deg C): 114
FP (deg C):
BP (deg C): 305-307
BP pressure (mm Hg):

Property/Value	Units	Temp	Data Type	Reference
WS 7.50E+002	mg/L	25	EST	MEYLAN,WM ET AL. (1996)
logP 1.42			EXP	HANSCH,C & LEO,AJ (1985)
VP 2.40E-003	mm Hg	25	EST	NEELY,WB & BLAU,GE (1985)
DC	pKa			
HL 9.10E-008	atm m3/mol	25	EST	MEYLAN,WM & HOWARD,PH (1991)
OH 2.94E-013	cm3/molc sec	25	EST	ATKINSON,R (1988)

000121-91-5 — ISOPHTHALIC ACID

Formula: $C_8H_6O_4$
Mol Weight: 166.13
MP (deg C): 345-348
FP (deg C):
BP (deg C):
BP pressure (mm Hg):

Property/Value	Units	Temp	Data Type	Reference
WS 1.30E+002	mg/L	25	EXP	BEMIS,AG ET AL. (1982)
logP 1.66			EXP	HANSCH,C ET AL. (1995)
VP 6.80E-002	mm Hg	100	EXP	BEMIS,AG ET AL. (1982)
DC 3.70	pKa		EXP	SERJEANT,EP & DEMPSEY,B (1979)
HL 3.90E-013	atm m3/mol	25	EST	MEYLAN,WM & HOWARD,PH (1991)
OH 1.31E-012	cm3/molc sec	25	EST	MEYLAN,WM & HOWARD,PH (1993)

000121-92-6 — M-NITROBENZOIC ACID

Formula: $C_7H_5NO_4$
Mol Weight: 167.12
MP (deg C): 142
FP (deg C):
BP (deg C):
BP pressure (mm Hg):

Property/Value	Units	Temp	Data Type	Reference
WS 3.58E+003	mg/L	25	EXP	YALKOWSKY,SH & DANNENFELSER,RM (1992)
logP 1.83			EXP	HANSCH,C & LEO,AJ (1985)
VP 3.71E-005	mm Hg	25	EST	NEELY,WB & BLAU,GE (1985)
DC 3.46	pKa	25	EXP	BUDAVARI,S ET AL (1989)
HL 3.79E-010	atm m3/mol	25	EST	MEYLAN,WM & HOWARD,PH (1991)
OH 6.10E-013	cm3/molc sec	25	EST	MEYLAN,WM & HOWARD,PH (1993)

000121-98-2 — P-ANISIC ACID, METHYL ESTER

Formula: $C_9H_{10}O_3$
Mol Weight: 166.18
MP (deg C): 49
FP (deg C):
BP (deg C): 244
BP pressure (mm Hg):

Property/Value	Units	Temp	Data Type	Reference
WS 6.43E+002	mg/L	20	EXP	YALKOWSKY,SH & DANNENFELSER,RM (1992)
logP 2.27			EXP	HANSCH,C & LEO,AJ (1985)
VP 4.63E-004	mm Hg	25	EXT	OHE,S (1989)
DC	pKa			
HL 2.05E-006	atm m3/mol	25	EST	MEYLAN,WM & HOWARD,PH (1991)
OH 7.96E-012	cm3/molc sec	25	EST	MEYLAN,WM & HOWARD,PH (1993)

000122-00-9 — P-METHYLACETOPHENONE

Formula: $C_9H_{10}O$
Mol Weight: 134.18
MP (deg C): 28
FP (deg C):
BP (deg C): 226
BP pressure (mm Hg):

Property/Value	Units	Temp	Data Type	Reference
WS 1.42E+003	mg/L	25	EST	MEYLAN,WM ET AL. (1996)
logP 2.10			EXP	HANSCH,C & LEO,AJ (1985)
VP 2.21E-001	mm Hg	25	EST	NEELY,WB & BLAU,GE (1985)
DC	pKa			
HL 1.08E-005	atm m3/mol	25	EST	MEYLAN,WM & HOWARD,PH (1991)
OH 4.88E-012	cm3/molc sec	25	EST	MEYLAN,WM & HOWARD,PH (1993)

CAS #: 000122-09-8				PHENTERMINE

Formula: $C_{10}H_{15}N$

Mol Weight: 149.24

MP (deg C): | FP (deg C):

BP (deg C): 205

BP pressure (mm Hg): 7.50E+002

Property/Value	Units	Temp	Data Type	Reference
WS 1.86E+004	mg/L	25	EST	MEYLAN,WM ET AL. (1996)
logP 1.90			EXP	HANSCH,C & LEO,AJ (1985)
VP 9.61E-002	mm Hg	25	EST	NEELY,WB & BLAU,GE (1985)
DC	pKa			
HL 1.43E-006	atm m3/mol	25	EST	MEYLAN,WM & HOWARD,PH (1991)
OH 2.73E-011	cm3/molc sec	25	EST	MEYLAN,WM & HOWARD,PH (1993)

CAS #: 000122-11-2				SULFADIMETHOXINE

Formula: $C_{12}H_{14}N_4O_4S$

Mol Weight: 310.33

MP (deg C): 201-203 | FP (deg C):

BP (deg C):

BP pressure (mm Hg):

Property/Value	Units	Temp	Data Type	Reference
WS 3.43E+002	mg/L		EXP	YALKOWSKY,SH & DANNENFELSER,RM (1992)
logP 1.63			EXP	HANSCH,C ET AL. (1995)
VP 1.59E-009	mm Hg	25	EST	NEELY,WB & BLAU,GE (1985)
DC	pKa			
HL 1.30E-014	atm m3/mol	25	EST	MEYLAN,WM & HOWARD,PH (1991)
OH 2.02E-010	cm3/molc sec	25	EST	MEYLAN,WM & HOWARD,PH (1993)

CAS #: 000122-14-5				FENITROTHION

Formula: $C_9H_{12}NO_5PS$

Mol Weight: 277.24

MP (deg C): | FP (deg C):

BP (deg C): 118

BP pressure (mm Hg): 5.00E-002

Property/Value	Units	Temp	Data Type	Reference
WS 3.00E+001	mg/L	20	EXP	SHIU,WY ET AL. (1990)
logP 3.30			EXP	HANSCH,C ET AL. (1995)
VP 5.40E-005	mm Hg	20	EXP	MACKAY,D & SHIU,WY (1981)
DC	pKa			
HL 9.30E-007	atm m3/mol	25	EXP	METCALFE,CD ET AL. (1980)
OH 6.21E-011	cm3/molc sec		EST	ATKINSON,R (1988)

CAS #: 000122-15-6				DIMETAN

Formula: $C_{11}H_{17}NO_3$

Mol Weight: 211.26

MP (deg C): 45-46 | FP (deg C):

BP (deg C): 170-180

BP pressure (mm Hg): 1.10E+001

Property/Value	Units	Temp	Data Type	Reference
WS 3.00E+004	mg/L		EXP	YALKOWSKY,SH & DANNENFELSER,RM (1992)
logP 1.66			EST	MEYLAN,WM & HOWARD,PH (1995)
VP 8.35E-004	mm Hg	25	EST	NEELY,WB & BLAU,GE (1985)
DC	pKa			
HL 8.68E-010	atm m3/mol	25	EST	MEYLAN,WM & HOWARD,PH (1991)
OH 9.67E-011	cm3/molc sec	25	EST	MEYLAN,WM & HOWARD,PH (1993)

CAS #: 000122-28-1				M-NITROACETANILIDE

Formula: $C_8H_8N_2O_3$

Mol Weight: 180.16

MP (deg C): 155 | FP (deg C):

BP (deg C): 100

BP pressure (mm Hg): 7.00E-003

Property/Value	Units	Temp	Data Type	Reference
WS 9.17E+005	mg/L	124	EXP	SEIDELL,A (1941)
logP 1.47			EXP	NAKAGAWA,Y ET AL. (1992)
VP 5.00E-006	mm Hg	25	EST	NEELY,WB & BLAU,GE (1985)
DC	pKa			
HL 2.44E-011	atm m3/mol	25	EST	MEYLAN,WM & HOWARD,PH (1991)
OH 1.19E-012	cm3/molc sec	25	EST	MEYLAN,WM & HOWARD,PH (1993)

CAS #: 000122-34-9				SIMAZINE

Formula: $C_7H_{12}ClN_5$

Mol Weight: 201.66

MP (deg C): 226-227 | FP (deg C):

BP (deg C):

BP pressure (mm Hg):

Property/Value	Units	Temp	Data Type	Reference
WS 5.70E+000	mg/L	20	EXP	YALKOWSKY,SH & DANNENFELSER,RM (1992)
logP 2.18			EXP	HANSCH,C & LEO,AJ (1985)
VP 2.21E-008	mm Hg	25	EXT	WAUCHOPE,RD ET AL. (1991A)
DC	pKa			
HL 3.37E-009	atm m3/mol	25	EST	MEYLAN,WM & HOWARD,PH (1991)
OH 1.78E-011	cm3/molc sec	25	EST	MEYLAN,WM & HOWARD,PH (1993)

CAS #: 000122-37-2				PHENOL, 4-(PHENYLAMINO)-

Formula: $C_{12}H_{11}NO$

Mol Weight: 185.23

MP (deg C): 73 | FP (deg C):

BP (deg C): 330

BP pressure (mm Hg):

Property/Value	Units	Temp	Data Type	Reference
WS 2.03E+002	mg/L	25	EST	MEYLAN,WM ET AL. (1996)
logP 2.82			EXP	HANSCH,C ET AL. (1995)
VP 2.70E-005	mm Hg	25	EST	NEELY,WB & BLAU,GE (1985)
DC	pKa			
HL 1.09E-010	atm m3/mol	25	EST	MEYLAN,WM & HOWARD,PH (1991)
OH 2.00E-010	cm3/molc sec	25	EST	MEYLAN,WM & HOWARD,PH (1993)

CAS #: 000122-39-4				DIPHENYLAMINE

Formula: $C_{12}H_{11}N$

Mol Weight: 169.23

MP (deg C): 53-54 | FP (deg C):

BP (deg C): 302

BP pressure (mm Hg):

Property/Value	Units	Temp	Data Type	Reference
WS 5.30E+001	mg/L	20	EXP	YALKOWSKY,SH & DANNENFELSER,RM (1992)
logP 3.50			EXP	HANSCH,C & LEO,AJ (1985)
VP 8.06E-004	mm Hg	20	EXT	CHAO,J ET AL. (1983)
DC 0.78	pKa	24	EXP	PERRIN,DD (1972)
HL 3.39E-006	atm m3/mol	20	EST	VP/WSOL
OH 2.60E-010	cm3/molc sec	25	EST	MEYLAN,WM & HOWARD,PH (1993)

CAS #: 000122-42-9	ISOPROPYL PHENYL CARBAMATE

Formula: $C_{10}H_{13}NO_2$

Mol Weight: 179.22

MP (deg C): 90   FP (deg C):

BP (deg C):

BP pressure (mm Hg):

Property/ Value	Units	Temp	Data Type	Reference
WS 3.20E+001	mg/L	25	EXP	BAILEY,GW ET AL. (1968)
logP 2.60			EXP	HANSCH,C & LEO,AJ (1985)
VP 1.50E-002	mm Hg	25	EST	NEELY,WB & BLAU,GE (1985)
DC	pKa			
HL 3.85E-008	atm m3/mol	25	EST	MEYLAN,WM & HOWARD,PH (1991)
OH 4.61E-011	cm3/molc sec	25	EST	MEYLAN,WM & HOWARD,PH (1993)

CAS #: 000122-46-3	M-TOLYLACETATE

Formula: $C_9H_{10}O_2$

Mol Weight: 150.18

MP (deg C): 12   FP (deg C):

BP (deg C): 212

BP pressure (mm Hg):

Property/ Value	Units	Temp	Data Type	Reference
WS 1.24E+003	mg/L	25	EST	MEYLAN,WM ET AL. (1996)
logP 2.09			EXP	HANSCH,C & LEO,AJ (1985)
VP 1.64E-001	mm Hg	25	EST	NEELY,WB & BLAU,GE (1985)
DC	pKa			
HL 7.15E-005	atm m3/mol	25	EST	MEYLAN,WM & HOWARD,PH (1991)
OH 5.49E-012	cm3/molc sec	25	EST	MEYLAN,WM & HOWARD,PH (1993)

CAS #: 000122-51-0	ETHYL ORTHOFORMATE

Formula: $C_7H_{16}O_3$

Mol Weight: 148.20

MP (deg C):   FP (deg C):

BP (deg C): 143

BP pressure (mm Hg):

Property/ Value	Units	Temp	Data Type	Reference
WS 7.30E+003	mg/L	25	EST	MEYLAN,WM ET AL. (1996)
logP 1.20			EXP	HANSCH,C ET AL. (1995)
VP 3.86E+000	mm Hg	25	EXP	PERRY,RH & GREEN,D (1984)
DC	pKa			
HL 1.39E-006	atm m3/mol	25	EST	MEYLAN,WM & HOWARD,PH (1991)
OH 6.18E-011	cm3/molc sec	25	EST	MEYLAN,WM & HOWARD,PH (1993)

CAS #: 000122-57-6	METHYL STYRYL KETONE

Formula: $C_{10}H_{10}O$

Mol Weight: 146.19

MP (deg C): 41.5   FP (deg C):

BP (deg C): 261

BP pressure (mm Hg):

Property/ Value	Units	Temp	Data Type	Reference
WS 1.35E+003	mg/L	25	EST	MEYLAN,WM ET AL. (1996)
logP 2.07			EXP	HANSCH,C & LEO,AJ (1985)
VP 1.24E-002	mm Hg	25	EXT	PERRY,RH & GREEN,D (1984)
DC	pKa			
HL 1.17E-006	atm m3/mol	25	EST	MEYLAN,WM & HOWARD,PH (1991)
OH 5.27E-011	cm3/molc sec	25	EST	MEYLAN,WM & HOWARD,PH (1993)

CAS #: 000122-59-8	PHENOXYACETIC ACID

Formula: $C_8H_8O_3$

Mol Weight: 152.15

MP (deg C): 98   FP (deg C):

BP (deg C): 285

BP pressure (mm Hg):

Property/ Value	Units	Temp	Data Type	Reference
WS 1.67E+001	mg/L	25	EXP	YALKOWSKY,SH & DANNENFELSER,RM (1992)
logP 1.34			EXP	HANSCH,C & LEO,AJ (1985)
VP 1.65E-003	mm Hg	25	EST	NEELY,WB & BLAU,GE (1985)
DC 3.17	pKa	25	EXP	KORTUM,G ET AL (1961)
HL 1.68E-008	atm m3/mol	25	EST	MEYLAN,WM & HOWARD,PH (1991)
OH 2.63E-011	cm3/molc sec	25	EST	MEYLAN,WM & HOWARD,PH (1993)

CAS #: 000122-60-1	PHENYL GLYDIDYL ETHER

Formula: $C_9H_{10}O_2$

Mol Weight: 150.18

MP (deg C): 3.5   FP (deg C):

BP (deg C): 245

BP pressure (mm Hg):

Property/ Value	Units	Temp	Data Type	Reference
WS 2.40E+003	mg/L	25	EXP	CLAYTON,GD & CLAYTON,FE (1981)
logP 1.61			EST	MEYLAN,WM & HOWARD,PH (1995)
VP 1.00E-002	mm Hg	25	EXP	CLAYTON,GD & CLAYTON,FE (1981)
DC	pKa			
HL 8.23E-007	atm m3/mol	25	EST	VP/WSOL
OH 2.99E-011	cm3/molc sec	25	EST	MEYLAN,WM & HOWARD,PH (1993)

CAS #: 000122-66-7	HYDRAZOBENZENE

Formula: $C_{12}H_{12}N_2$

Mol Weight: 184.24

MP (deg C): 123-126   FP (deg C):

BP (deg C): 309

BP pressure (mm Hg):

Property/ Value	Units	Temp	Data Type	Reference
WS 1.62E+002	mg/L	25	EST	MEYLAN,WM ET AL. (1996)
logP 2.94			EXP	HANSCH,C & LEO,AJ (1985)
VP 3.15E-004	mm Hg	25	EST	NEELY,WB & BLAU,GE (1985)
DC	pKa			
HL 4.39E-009	atm m3/mol	25	EST	MEYLAN,WM & HOWARD,PH (1991)
OH 2.11E-010	cm3/molc sec	25	EST	MEYLAN,WM & HOWARD,PH (1993)

CAS #: 000122-69-0	3-PHENYLPROP-2-ENYL CINNAMATE

Formula: $C_{18}H_{16}O_2$

Mol Weight: 264.33

MP (deg C): 44   FP (deg C):

BP (deg C):

BP pressure (mm Hg):

Property/ Value	Units	Temp	Data Type	Reference
WS 3.12E+000	mg/L	25	EST	MEYLAN,WM ET AL. (1996)
logP 4.45			EXP	HANSCH,C & LEO,AJ (1985)
VP 6.03E-006	mm Hg	25	EST	NEELY,WB & BLAU,GE (1985)
DC	pKa			
HL 2.43E-007	atm m3/mol	25	EST	MEYLAN,WM & HOWARD,PH (1991)
OH 8.13E-011	cm3/molc sec	25	EST	MEYLAN,WM & HOWARD,PH (1993)

CAS #: 000122-78-1				PHENYLACETALDEHYDE

Formula: C_8H_8O

Mol Weight: 120.15

MP (deg C): 33.5   FP (deg C):

BP (deg C): 195

BP pressure (mm Hg):

Property/Value	Units	Temp	Data Type	Reference
WS 3.03E+003	mg/L	25	EST	MEYLAN,WM ET AL. (1996)
logP 1.78			EXP	HANSCH,C & LEO,AJ (1985)
VP 3.92E-001	mm Hg	25	EXP	JONES,AH (1960)
DC	pKa			
HL 5.48E-006	atm m3/mol	25	EST	MEYLAN,WM & HOWARD,PH (1991)
OH 2.63E-011	cm3/molc sec	25	EST	MEYLAN,WM & HOWARD,PH (1993)

CAS #: 000122-79-2				PHENYL ACETATE

Formula: $C_8H_8O_2$

Mol Weight: 136.15

MP (deg C):   FP (deg C):

BP (deg C): 195-196

BP pressure (mm Hg):

Property/Value	Units	Temp	Data Type	Reference
WS 4.64E+003	mg/L	25	EST	MEYLAN,WM ET AL. (1996)
logP 1.49			EXP	HANSCH,C & LEO,AJ (1985)
VP 3.98E-001	mm Hg	25	EXT	PERRY,RH & GREEN,D (1984)
DC	pKa			
HL 6.48E-005	atm m3/mol	25	EST	MEYLAN,WM & HOWARD,PH (1991)
OH 2.11E-012	cm3/molc sec	25	EST	MEYLAN,WM & HOWARD,PH (1993)

CAS #: 000122-80-5				P-AMINOACETANILIDE

Formula: $C_8H_{10}N_2O$

Mol Weight: 150.18

MP (deg C): 166.5   FP (deg C):

BP (deg C): 267

BP pressure (mm Hg):

Property/Value	Units	Temp	Data Type	Reference
WS 3.17E+003	mg/L	25	EST	MEYLAN,WM ET AL. (1996)
logP 0.08			EXP	HANSCH,C & LEO,AJ (1985)
VP 1.34E-005	mm Hg	25	EST	NEELY,WB & BLAU,GE (1985)
DC	pKa			
HL 2.18E-012	atm m3/mol	25	EST	MEYLAN,WM & HOWARD,PH (1991)
OH 5.64E-011	cm3/molc sec	25	EST	MEYLAN,WM & HOWARD,PH (1993)

CAS #: 000122-85-0				P-FORMYLACETANILIDE

Formula: $C_9H_9NO_2$

Mol Weight: 163.18

MP (deg C): 156-158   FP (deg C):

BP (deg C):

BP pressure (mm Hg):

Property/Value	Units	Temp	Data Type	Reference
WS 5.67E+003	mg/L	25	EST	MEYLAN,WM ET AL. (1996)
logP 1.25			EXP	HANSCH,C & LEO,AJ (1985)
VP 1.79E-005	mm Hg	25	EST	NEELY,WB & BLAU,GE (1985)
DC	pKa			
HL 1.54E-011	atm m3/mol	25	EST	MEYLAN,WM & HOWARD,PH (1991)
OH 2.11E-011	cm3/molc sec	25	EST	MEYLAN,WM & HOWARD,PH (1993)

CAS #: 000122-88-3				P-CHLOROPHENOXYACETIC ACID

Formula: $C_8H_7ClO_3$

Mol Weight: 186.60

MP (deg C): 156.5   FP (deg C):

BP (deg C):

BP pressure (mm Hg):

Property/Value	Units	Temp	Data Type	Reference
WS 3.81E+002	mg/L	25	EXP	BAILEY,GW & WHITE,JL (1965)
logP 2.25			EXP	HANSCH,C ET AL. (1995)
VP 3.18E-004	mm Hg	25	EST	NEELY,WB & BLAU,GE (1985)
DC 3.56	pKa	25	EXP	BAILEY,GW & WHITE,JL (1965)
HL 6.43E-008	atm m3/mol	25	EST	MEYLAN,WM & HOWARD,PH (1991)
OH 1.09E-011	cm3/molc sec	25	EST	ATKINSON,R (1985)

CAS #: 000122-94-1				4-BUTOXYPHENOL

Formula: $C_{10}H_{14}O_2$

Mol Weight: 166.22

MP (deg C): 65-66   FP (deg C):

BP (deg C):

BP pressure (mm Hg):

Property/Value	Units	Temp	Data Type	Reference
WS 8.13E+002	mg/L	25	EST	MEYLAN,WM ET AL. (1996)
logP 2.90			EXP	HANSCH,C ET AL. (1995)
VP 2.33E-003	mm Hg	25	EST	NEELY,WB & BLAU,GE (1985)
DC	pKa			
HL 7.76E-008	atm m3/mol	25	EST	MEYLAN,WM & HOWARD,PH (1991)
OH 4.13E-011	cm3/molc sec	25	EST	MEYLAN,WM & HOWARD,PH (1993)

CAS #: 000122-97-4				3-PHENYLPROPANOL

Formula: $C_9H_{12}O$

Mol Weight: 136.20

MP (deg C): < -18   FP (deg C):

BP (deg C): 235

BP pressure (mm Hg):

Property/Value	Units	Temp	Data Type	Reference
WS 5.63E+003	mg/L	25	EXP	ISNARD,P & LAMBERT,S (1989)
logP 1.88			EXP	HANSCH,C & LEO,AJ (1985)
VP 1.99E-002	mm Hg	25	EXT	PERRY,RH & GREEN,D (1984)
DC	pKa			
HL 3.83E-007	atm m3/mol	25	EST	MEYLAN,WM & HOWARD,PH (1991)
OH 1.16E-011	cm3/molc sec	25	EST	MEYLAN,WM & HOWARD,PH (1993)

CAS #: 000122-98-5				2-ANILINOETHANOL

Formula: $C_8H_{11}NO$

Mol Weight: 137.18

MP (deg C):   FP (deg C):

BP (deg C): 279.5

BP pressure (mm Hg):

Property/Value	Units	Temp	Data Type	Reference
WS 1.72E+004	mg/L	25	EST	MEYLAN,WM ET AL. (1996)
logP 0.82			EXP	HANSCH,C & LEO,AJ (1985)
VP 2.21E-003	mm Hg	25	EXT	PERRY,RH & GREEN,D (1984)
DC 4.06	pKa		EXP	PERRIN,DD (1965)
HL 2.03E-010	atm m3/mol	25	EST	MEYLAN,WM & HOWARD,PH (1991)
OH 5.75E-011	cm3/molc sec	25	EST	MEYLAN,WM & HOWARD,PH (1993)

CAS #: 000122-99-6				2-PHENOXYETHANOL
Formula: $C_8H_{10}O_2$				
Mol Weight: 138.17				
MP (deg C): 14		FP (deg C):		
BP (deg C): 245.2				
BP pressure (mm Hg):				

Property/Value	Units	Temp	Data Type	Reference
WS 2.69E+004	mg/L	25	EXP	VALVANI,SC ET AL. (1981)
logP 1.16			EXP	HANSCH,C & LEO,AJ (1985)
VP 7.00E-003	mm Hg	25	EXP	DOW CHEMICAL COMPANY (1990)
DC 15.10	pKa		EXP	SERJEANT,EP & DEMPSEY,B (1979)
HL 4.72E-008	atm m3/mol	25	EST	VP/WSOL
OH 3.32E-011	cm3/molc sec	25	EST	MEYLAN,WM & HOWARD,PH (1993)

CAS #: 000123-01-3				DODECYLBENZENE
Formula: $C_{18}H_{30}$				
Mol Weight: 246.44				
MP (deg C): 3		FP (deg C):		
BP (deg C): 328				
BP pressure (mm Hg):				

Property/Value	Units	Temp	Data Type	Reference
WS 1.02E-003	mg/L	25	EST	MEYLAN,WM ET AL. (1996)
logP 8.65			EXP	SHERBLOM,PM ET AL. (1992)
VP 5.11E-005	mm Hg	25	EXP	DAUBERT,TE & DANNER,RP (1989)
DC	pKa			
HL 1.34E-001	atm m3/mol	25	EST	MEYLAN,WM & HOWARD,PH (1991)
OH 2.00E-011	cm3/molc sec	25	EST	MEYLAN,WM & HOWARD,PH (1993)

CAS #: 000123-02-4				1-PHENYLTRIDECANE
Formula: $C_{19}H_{32}$				
Mol Weight: 260.47				
MP (deg C): 10		FP (deg C):		
BP (deg C): 346				
BP pressure (mm Hg):				

Property/Value	Units	Temp	Data Type	Reference
WS 2.10E-004	mg/L	25	EST	MEYLAN,WM ET AL. (1996)
logP 9.36			EXP	SHERBLOM,PM ET AL. (1992)
VP 1.27E-005	mm Hg	25	EXP	DAUBERT,TE & DANNER,RP (1989)
DC	pKa			
HL 1.78E-001	atm m3/mol	25	EST	MEYLAN,WM & HOWARD,PH (1991)
OH 2.14E-011	cm3/molc sec	25	EST	MEYLAN,WM & HOWARD,PH (1993)

CAS #: 000123-04-6				3-(CHLOROMETHYL)HEPTANE
Formula: $C_8H_{17}Cl$				
Mol Weight: 148.68				
MP (deg C):		FP (deg C):		
BP (deg C): 172				
BP pressure (mm Hg):				

Property/Value	Units	Temp	Data Type	Reference
WS 1.00E+002	mg/L	20	EXP	RIDDICK,JA ET AL. (1986)
logP 4.45			EST	MEYLAN,WM & HOWARD,PH (1995)
VP 1.20E+000	mm Hg	20	EXP	RIDDICK,JA ET AL. (1986)
DC	pKa			
HL 2.35E-003	atm m3/mol	20	EST	VP/WSOL
OH 6.95E-012	cm3/molc sec	25	EST	MEYLAN,WM & HOWARD,PH (1993)

CAS #: 000123-05-7				2-ETHYLHEXANAL
Formula: $C_8H_{16}O$				
Mol Weight: 128.22				
MP (deg C): <-100		FP (deg C):		
BP (deg C): 163				
BP pressure (mm Hg):				

Property/Value	Units	Temp	Data Type	Reference
WS 4.00E+002	mg/L	25	EXP	HANN,RWJR & JENSEN,PA (1977)
logP 2.71			EST	MEYLAN,WM & HOWARD,PH (1995)
VP 1.80E+000	mm Hg	20	EXP	LEWIS,RJ (1993)
DC	pKa			
HL 7.59E-004	atm m3/mol	25	EST	VP/WSOL
OH 3.40E-011	cm3/molc sec	25	EST	MEYLAN,WM & HOWARD,PH (1993)

CAS #: 000123-07-9				P-ETHYLPHENOL
Formula: $C_8H_{10}O$				
Mol Weight: 122.17				
MP (deg C): 46		FP (deg C):		
BP (deg C): 217.9				
BP pressure (mm Hg):				

Property/Value	Units	Temp	Data Type	Reference
WS 4.90E+003	mg/L	25	EXP	YALKOWSKY,SH & DANNENFELSER,RM (1992)
logP 2.58			EXP	HANSCH,C & LEO,AJ (1985)
VP 3.72E-002	mm Hg	25	EXP	BIDDISCOMBE,DP ET AL. (1963)
DC 10.00	pKa	25	EXP	SCHULTZ,TW (1987A)
HL 1.22E-006	atm m3/mol	25	EST	VP/WSOL
OH 4.18E-011	cm3/molc sec	25	EST	MEYLAN,WM & HOWARD,PH (1993)

CAS #: 000123-08-0				4-HYDROXYBENZALDEHYDE
Formula: $C_7H_6O_2$				
Mol Weight: 122.12				
MP (deg C): 117		FP (deg C):		
BP (deg C):				
BP pressure (mm Hg):				

Property/Value	Units	Temp	Data Type	Reference
WS 1.29E+004	mg/L	30	EXP	YALKOWSKY,SH & DANNENFELSER,RM (1992)
logP 1.35			EXP	HANSCH,C & LEO,AJ (1985)
VP 1.13E-004	mm Hg	25	EXT	YAWS,CL (1994A)
DC 7.61	pKa	25	EXP	SERJEANT,EP & DEMPSEY,B (1979)
HL 5.11E-001	atm m3/mol	25	EXP	PARSONS,GH ET AL. (1971)
OH 2.78E-011	cm3/molc sec	25	EST	MEYLAN,WM & HOWARD,PH (1993)

CAS #: 000123-11-5				P-ANISALDEHYDE
Formula: $C_8H_8O_2$				
Mol Weight: 136.15				
MP (deg C): 0		FP (deg C):		
BP (deg C): 248				
BP pressure (mm Hg):				

Property/Value	Units	Temp	Data Type	Reference
WS 4.29E+003	mg/L	25	EXP	YALKOWSKY,SH & DANNENFELSER,RM (1992)
logP 1.76			EXP	HANSCH,C & LEO,AJ (1985)
VP 3.29E-002	mm Hg	25	EXT	OHE,S (1989)
DC	pKa			
HL 7.94E-007	atm m3/mol	25	EST	MEYLAN,WM & HOWARD,PH (1991)
OH 2.48E-011	cm3/molc sec	25	EST	MEYLAN,WM & HOWARD,PH (1993)

243

CAS #: 000123-15-9				2-METHYLPENTALDEHYDE

Formula: $C_6H_{12}O$

Mol Weight: 100.16

MP (deg C): | FP (deg C): -10

BP (deg C): 117

BP pressure (mm Hg):

Property/Value	Units	Temp	Data Type	Reference
WS 4.20E+003	mg/L	25	EXP	LEWIS,RJ ET AL. (1993)
logP 1.73			EST	MEYLAN,WM & HOWARD,PH (1995)
VP 1.79E+001	mm Hg	25	EST	NEELY,WB & BLAU,GE (1985)
DC	pKa			
HL 3.70E-004	atm m3/mol	25	EST	MEYLAN,WM & HOWARD,PH (1991)
OH 2.90E-011	cm3/molc sec	25	EST	MEYLAN,WM & HOWARD,PH (1993)

CAS #: 000123-19-3				DI(N-PROPYL) KETONE

Formula: $C_7H_{14}O$

Mol Weight: 114.19

MP (deg C): -32.6 | FP (deg C):

BP (deg C): 144

BP pressure (mm Hg):

Property/Value	Units	Temp	Data Type	Reference
WS 3.20E+003	mg/L	26	EXP	YALKOWSKY,SH & DANNENFELSER,RM (1992)
logP 1.73			EST	MEYLAN,WM & HOWARD,PH (1995)
VP 1.17E+000	mm Hg	25	EXP	OHE,S (1976)
DC	pKa			
HL 5.49E-008	atm m3/mol	25	EST	VP/WSOL
OH 9.34E-012	cm3/molc sec	25	EST	MEYLAN,WM & HOWARD,PH (1993)

CAS #: 000123-25-1				DIETHYL SUCCINATE

Formula: $C_8H_{14}O_4$

Mol Weight: 174.20

MP (deg C): -21 | FP (deg C):

BP (deg C): 217.7

BP pressure (mm Hg):

Property/Value	Units	Temp	Data Type	Reference
WS 1.91E+004	mg/L	25	EXP	SUZUKI,T (1990)
logP 1.20			EXP	CATZ,P & FRIEND,DR (1989)
VP 4.39E-002	mm Hg	25	EXP	DAUBERT,TE & DANNER,RP (1989)
DC	pKa			
HL 5.27E-007	atm m3/mol	25	EST	VP/WSOL
OH 5.02E-012	cm3/molc sec	25	EST	MEYLAN,WM & HOWARD,PH (1993)

CAS #: 000123-29-5				ETHYL NONANOATE

Formula: $C_{11}H_{22}O_2$

Mol Weight: 186.30

MP (deg C): -44 | FP (deg C):

BP (deg C): 220

BP pressure (mm Hg):

Property/Value	Units	Temp	Data Type	Reference
WS 2.95E+001	mg/L	25	EXP	SUZUKI,T (1991)
logP 4.30			EST	MEYLAN,WM & HOWARD,PH (1995)
VP 7.95E-002	mm Hg	25	EST	NEELY,WB & BLAU,GE (1985)
DC	pKa			
HL 1.69E-003	atm m3/mol	25	EST	MEYLAN,WM & HOWARD,PH (1991)
OH 1.09E-011	cm3/molc sec	25	EST	MEYLAN,WM & HOWARD,PH (1993)

CAS #: 000123-30-8				PHENOL, 4-AMINO-

Formula: C_6H_7NO

Mol Weight: 109.13

MP (deg C): 186 | FP (deg C):

BP (deg C): 284

BP pressure (mm Hg):

Property/Value	Units	Temp	Data Type	Reference
WS 1.60E+004	mg/L	20	EXP	YALKOWSKY,SH & DANNENFELSER,RM (1992)
logP 0.04			EXP	HANSCH,C & LEO,AJ (1985)
VP 7.50E-002	mm Hg	20	EXP	WEBER,RC ET AL. (1981)
DC 10.46	pKa		EXP	SERJEANT,EP & DEMPSEY,B (1979)
HL 6.73E-007	atm m3/mol	20	EST	VP/WSOL
OH 7.42E-011	cm3/molc sec	25	EST	MEYLAN,WM & HOWARD,PH (1993)

CAS #: 000123-31-9				HYDROQUINONE

Formula: $C_6H_6O_2$

Mol Weight: 110.11

MP (deg C): 172 | FP (deg C):

BP (deg C): 287

BP pressure (mm Hg):

Property/Value	Units	Temp	Data Type	Reference
WS 7.33E+004	mg/L	25	EXP	YALKOWSKY,SH & DANNENFELSER,RM (1992)
logP 0.59			EXP	HANSCH,C ET AL (1995)
VP 6.70E-004	mm Hg	25	EXT	DAUBERT,TE & DANNER,RP (1991)
DC 10.85	pKa	25	EXP	PEARCE,PJ & SIMKINS,RJJ (1968)
HL 1.32E-009	atm m3/mol	25	EST	VP/WSOL
OH 2.30E-011	cm3/molc sec	25	EST	MEYLAN,WM & HOWARD,PH (1993)

CAS #: 000123-32-0				2,5-DIMETHYLPYRAZINE

Formula: $C_6H_8N_2$

Mol Weight: 108.14

MP (deg C): 15 | FP (deg C):

BP (deg C): 155

BP pressure (mm Hg):

Property/Value	Units	Temp	Data Type	Reference
WS 3.20E+004	mg/L	25	EST	MEYLAN,WM ET AL. (1996)
logP 0.63			EXP	HANSCH,C & LEO,AJ (1985)
VP 1.50E+000	mm Hg	25	EST	NEELY,WB & BLAU,GE (1985)
DC 1.85	pKa	27	EXP	PERRIN,DD (1965)
HL 3.55E-006	atm m3/mol	25	EST	MEYLAN,WM & HOWARD,PH (1991)
OH 1.01E-012	cm3/molc sec	25	EST	MEYLAN,WM & HOWARD,PH (1993)

CAS #: 000123-33-1				MALEIC HYDRAZIDE

Formula: $C_4H_4N_2O_2$

Mol Weight: 112.09

MP (deg C): 306-308 | FP (deg C):

BP (deg C):

BP pressure (mm Hg):

Property/Value	Units	Temp	Data Type	Reference
WS 6.00E+003	mg/L	25	EXP	MARTIN,H & WORTHING,CR (1977)
logP -0.84			EXP	HANSCH,C & LEO,AJ (1985)
VP 2.77E-006	mm Hg	25	EST	NEELY,WB & BLAU,GE (1985)
DC 5.67	pKa	25	EXP	SERJEANT,EP & DEMPSEY,B (1979)
HL 2.65E-011	atm m3/mol	25	EST	MEYLAN,WM & HOWARD,PH (1991)
OH 1.33E-011	cm3/molc sec	25	EST	MEYLAN,WM & HOWARD,PH (1993)

CAS #:	000123-35-3			3-METHYLENE-7-METHYL-1,6-OCTADIENE (MYRCENE)
Formula:	$C_{10}H_{16}$			
Mol Weight:	136.24			
MP (deg C): <-10		FP (deg C):		
BP (deg C): 93				
BP pressure (mm Hg): 3.50E+001				

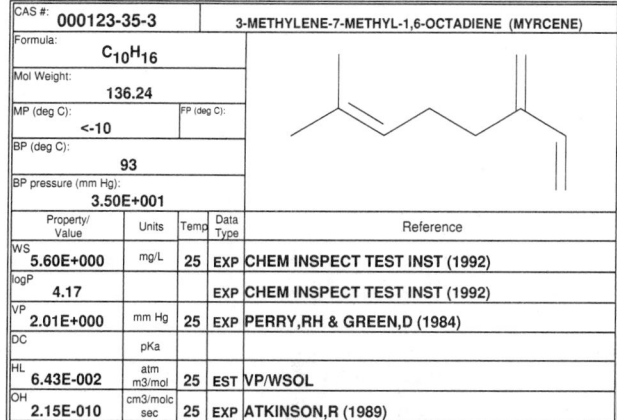

Property/Value	Units	Temp	Data Type	Reference
WS 5.60E+000	mg/L	25	EXP	CHEM INSPECT TEST INST (1992)
logP 4.17			EXP	CHEM INSPECT TEST INST (1992)
VP 2.01E+000	mm Hg	25	EXP	PERRY,RH & GREEN,D (1984)
DC	pKa			
HL 6.43E-002	atm m3/mol	25	EST	VP/WSOL
OH 2.15E-010	cm3/molc sec	25	EXP	ATKINSON,R (1989)

CAS #:	000123-38-6			PROPIONALDEHYDE
Formula:	C_3H_6O			
Mol Weight:	58.08			
MP (deg C): -80		FP (deg C):		
BP (deg C): 48.0				
BP pressure (mm Hg):				

Property/Value	Units	Temp	Data Type	Reference
WS 3.06E+005	mg/L	25	EXP	RIDDICK,JA ET AL. (1986)
logP 0.59			EXP	HANSCH,C & LEO,AJ (1985)
VP 3.17E+002	mm Hg	25	EXP	DAUBERT,TE & DANNER,RP (1985)
DC	pKa			
HL 7.34E-005	atm m3/mol	25	EXP	BUTTERY,RG ET AL. (1969)
OH 1.96E-011	cm3/molc sec	25	EXP	ATKINSON,R (1989)

CAS #:	000123-39-7			METHYLFORMAMIDE
Formula:	C_2H_5NO			
Mol Weight:	59.07			
MP (deg C): -3.8		FP (deg C):		
BP (deg C): 199.5				
BP pressure (mm Hg):				

Property/Value	Units	Temp	Data Type	Reference
WS 1.00E+006	mg/L	25	EXP	RIDDICK,JA ET AL. (1986)
logP -0.97			EXP	HANSCH,C ET AL. (1995)
VP 2.53E-001	mm Hg	25	EXP	DAUBERT,TE & DANNER,RP (1989)
DC	pKa			
HL 1.97E-008	atm m3/mol	25	EST	VP/WSOL
OH 6.76E-012	cm3/molc sec	25	EST	MEYLAN,WM & HOWARD,PH (1993)

CAS #:	000123-42-2			4-HYDROXY-4-METHYL-2-PENTANONE
Formula:	$C_6H_{12}O_2$			
Mol Weight:	116.16			
MP (deg C): -44		FP (deg C):		
BP (deg C): 167.9				
BP pressure (mm Hg):				

Property/Value	Units	Temp	Data Type	Reference
WS 1.00E+006	mg/L	25	EXP	LANDE,SS ET AL. (1976)
logP -0.34			EST	MEYLAN,WM & HOWARD,PH (1995)
VP 1.71E+000	mm Hg	25	EXP	DAUBERT,TE & DANNER,RP (1989)
DC	pKa			
HL 2.61E-007	atm m3/mol	25	EST	VP/WSOL
OH 3.94E-012	cm3/molc sec	25	EST	MEYLAN,WM & HOWARD,PH (1993)

CAS #:	000123-44-4			2,2,4-TRIMETHYL-1-PENTANOL
Formula:	$C_8H_{18}O$			
Mol Weight:	130.23			
MP (deg C):		FP (deg C):		
BP (deg C):				
BP pressure (mm Hg):				

Property/Value	Units	Temp	Data Type	Reference
WS 1.72E+003	mg/L	25	EST	MEYLAN,WM ET AL. (1996)
logP 2.62			EST	MEYLAN,WM & HOWARD,PH (1995)
VP 4.96E-001	mm Hg	25	EST	NEELY,WB & BLAU,GE (1985)
DC	pKa			
HL 3.10E-005	atm m3/mol	25	EST	MEYLAN,WM & HOWARD,PH (1991)
OH 8.63E-012	cm3/molc sec	25	EST	MEYLAN,WM & HOWARD,PH (1993)

CAS #:	000123-51-3			ISOPENTANOL
Formula:	$C_5H_{12}O$			
Mol Weight:	88.15			
MP (deg C): -117.2		FP (deg C):		
BP (deg C):				
BP pressure (mm Hg):				

Property/Value	Units	Temp	Data Type	Reference
WS 2.67E+001	mg/L	25	EXP	RIDDICK,JA ET AL. (1986)
logP 1.16			EXP	HANSCH,C ET AL. (1995)
VP 2.37E+000	mm Hg	25	EXP	RIDDICK,JA ET AL. (1986)
DC	pKa			
HL 1.41E-005	atm m3/mol	25	EXP	BUTLER,JAV ET AL. (1935)
OH 7.84E-012	cm3/molc sec	25	EST	MEYLAN,WM & HOWARD,PH (1993)

CAS #:	000123-54-6			2,4-PENTANEDIONE
Formula:	$C_5H_8O_2$			
Mol Weight:	100.12			
MP (deg C): 23		FP (deg C):		
BP (deg C): 141				
BP pressure (mm Hg):				

Property/Value	Units	Temp	Data Type	Reference
WS 1.66E+005	mg/L	20	EXP	RIDDICK,JA ET AL. (1986)
logP 0.40			EXP	HANSCH,C ET AL. (1995)
VP 2.96E+000	mm Hg	20	EXP	DAUBERT,TE & DANNER,RP (1989)
DC 8.93	pKa	25	EXP	RIDDICK,JA ET AL. (1986)
HL 2.35E-006	atm m3/mol	20	EST	VP/WSOL
OH 1.15E-012	cm3/molc sec	25	EXP	ATKINSON,R (1989)

CAS #:	000123-56-8			SUCCINIMIDE
Formula:	$C_4H_5NO_2$			
Mol Weight:	99.09			
MP (deg C): 125-127		FP (deg C):		
BP (deg C): 287-289				
BP pressure (mm Hg):				

Property/Value	Units	Temp	Data Type	Reference
WS 1.96E+005	mg/L	21	EXP	YALKOWSKY,SH & DANNENFELSER,RM (1992)
logP -0.85			EST	MEYLAN,WM & HOWARD,PH (1995)
VP 2.80E+000	mm Hg	25	EXT	PERRY,RH & GREEN,D (1984)
DC	pKa			
HL 2.53E-008	atm m3/mol	25	EST	MEYLAN,WM & HOWARD,PH (1991)
OH 9.87E-012	cm3/molc sec	25	EST	MEYLAN,WM & HOWARD,PH (1993)

CAS #: 000123-62-6				PROPIONIC ANHYDRIDE

Formula: $C_6H_{10}O_3$

Mol Weight: 130.14

MP (deg C): -45   FP (deg C):

BP (deg C): 167

BP pressure (mm Hg):

Property/Value	Units	Temp	Data Type	Reference
WS 4.14E+004	mg/L	25	EST	MEYLAN,WM ET AL. (1996)
logP 0.40			EST	MEYLAN,WM & HOWARD,PH (1995)
VP 1.36E+000	mm Hg	25	EXP	DAUBERT,TE & DANNER,RP (1989)
DC	pKa			
HL 6.28E-005	atm m3/mol	25	EST	MEYLAN,WM & HOWARD,PH (1991)
OH 1.72E-012	cm3/molc sec	25	EST	MEYLAN,WM & HOWARD,PH (1993)

CAS #: 000123-63-7				PARALDEHYDE

Formula: $C_6H_{12}O_3$

Mol Weight: 132.16

MP (deg C): 12   FP (deg C):

BP (deg C): 124

BP pressure (mm Hg):

Property/Value	Units	Temp	Data Type	Reference
WS 1.12E+005	mg/L	30	EXP	YALKOWSKY,SH & DANNENFELSER,RM (1992)
logP 0.67			EXP	HANSCH,C & LEO,AJ (1985)
VP 1.10E+001	mm Hg	25	EXP	DAUBERT,TE & DANNER,RP (1989)
DC	pKa			
HL 1.71E-005	atm m3/mol	25	EST	VP/WSOL
OH 2.18E-011	cm3/molc sec	25	EST	MEYLAN,WM & HOWARD,PH (1993)

CAS #: 000123-66-0				ETHYL HEXANOATE

Formula: $C_8H_{16}O_2$

Mol Weight: 144.22

MP (deg C): -67   FP (deg C):

BP (deg C): 167

BP pressure (mm Hg):

Property/Value	Units	Temp	Data Type	Reference
WS 6.29E+002	mg/L	25	EXP	SUZUKI,T (1991)
logP 2.83			EST	MEYLAN,WM & HOWARD,PH (1995)
VP 1.56E+000	mm Hg	25	EST	NEELY,WB & BLAU,GE (1985)
DC	pKa			
HL 7.23E-004	atm m3/mol	25	EST	MEYLAN,WM & HOWARD,PH (1991)
OH 6.65E-012	cm3/molc sec	25	EST	MEYLAN,WM & HOWARD,PH (1993)

CAS #: 000123-72-8				BUTYRALDEHYDE

Formula: C_4H_8O

Mol Weight: 72.11

MP (deg C): -96.4   FP (deg C):

BP (deg C): 74.8

BP pressure (mm Hg):

Property/Value	Units	Temp	Data Type	Reference
WS 7.10E+004	mg/L	25	EXP	UNION CARBIDE (1974)
logP 0.88			EXP	HANSCH,C & LEO,AJ (1985)
VP 1.11E+002	mm Hg	25	EXP	DAUBERT,TE & DANNER,RP (1985)
DC	pKa			
HL 1.15E-004	atm m3/mol	25	EXP	BUTTERY,RG ET AL. (1969)
OH 2.35E-011	cm3/molc sec	25	EXP	ATKINSON,R (1985)

CAS #: 000123-73-9				TRANS-CROTONALDEHYDE

Formula: C_4H_6O

Mol Weight: 70.09

MP (deg C): -74   FP (deg C): -69

BP (deg C): 104

BP pressure (mm Hg):

Property/Value	Units	Temp	Data Type	Reference
WS 1.50E+005	mg/L	20	EXP	YALKOWSKY,SH & DANNENFELSER,RM (1992)
logP 0.60			EST	MEYLAN,WM & HOWARD,PH (1995)
VP 3.00E+001	mm Hg	25	EXP	LEWIS,RJ (1993)
DC	pKa			
HL 1.94E-005	atm m3/mol	25	EXP	HINE,J & MOOKERJEE,PK (1975)
OH 3.60E-011	cm3/molc sec	25	EXP	ATKINSON,R (1989)

CAS #: 000123-75-1				PYRROLIDINE

Formula: C_4H_9N

Mol Weight: 71.12

MP (deg C): -57.8   FP (deg C):

BP (deg C): 86.5

BP pressure (mm Hg):

Property/Value	Units	Temp	Data Type	Reference
WS 1.00E+006	mg/L	20	EXP	YALKOWSKY,SH & DANNENFELSER,RM (1992)
logP 0.46			EXP	HANSCH,C & LEO,AJ (1985)
VP 6.27E+001	mm Hg	25	EXP	YAWS,CL (1994A)
DC 11.27	pKa	25	EXP	PERRIN,DD (1965)
HL 2.39E-006	atm m3/mol	25	EXP	CABANI,S ET AL. (1971)
OH 7.95E+001	cm3/molc sec	25	EST	MEYLAN,WM & HOWARD,PH (1993)

CAS #: 000123-76-2				LEVULINIC ACID

Formula: $C_5H_8O_3$

Mol Weight: 116.12

MP (deg C): 33-35   FP (deg C):

BP (deg C): 245-246

BP pressure (mm Hg):

Property/Value	Units	Temp	Data Type	Reference
WS 6.75E+005	mg/L	25	EST	MEYLAN,WM ET AL. (1996)
logP -0.49			EXP	HANSCH,C & LEO,AJ (1985)
VP 1.12E-003	mm Hg	25	EXT	YAWS,CL (1994A)
DC 4.64	pKa	18	EXP	KORTUM,G ET AL (1961)
HL 4.89E-010	atm m3/mol	25	EST	MEYLAN,WM & HOWARD,PH (1991)
OH 4.22E-012	cm3/molc sec	25	EST	MEYLAN,WM & HOWARD,PH (1993)

CAS #: 000123-79-5				DIOCTYL ADIPATE

Formula: $C_{22}H_{42}O_4$

Mol Weight: 370.58

MP (deg C): -70   FP (deg C):

BP (deg C): 214

BP pressure (mm Hg): 5.00E+000

Property/Value	Units	Temp	Data Type	Reference
WS 7.80E-001	mg/L	22	EST	FELDER,JD ET AL. (1986)
logP 6.11			EST	FELDER,JD ET AL. (1986); > VALUE
VP 8.50E-007	mm Hg	20	EST	FELDER,JD ET AL. (1986)
DC	pKa			
HL 4.34E-007	atm m3/mol	20	EST	FELDER,JD ET AL. (1986)
OH 2.25E-011	cm3/molc sec	25	EST	ATKINSON,R (1988)

CAS #: 000123-86-4 — N-BUTYL ACETATE

Formula:	$C_6H_{12}O_2$
Mol Weight:	116.16
MP (deg C):	-77
FP (deg C):	
BP (deg C):	125-126
BP pressure (mm Hg):	

Property/Value	Units	Temp	Data Type	Reference
WS 6.29E+003	mg/L	25	EXP	SYRACUSE RESEARCH CORP (1979)
logP 1.78			EXP	HANSCH,C ET AL. (1995)
VP 1.15E+001	mm Hg	25	EXP	DAUBERT,TE & DANNER,RP (1985)
DC	pKa			
HL 2.81E-004	atm m3/mol	25	EXP	KIECKBUSCH,TG & KING,CJ (1979)
OH 4.20E-012	cm3/molc sec	25	EXP	ATKINSON,R (1989)

CAS #: 000123-88-6 — ARETAN

Formula:	C_3H_7ClHgO
Mol Weight:	295.13
MP (deg C):	
FP (deg C):	
BP (deg C):	
BP pressure (mm Hg):	

Property/Value	Units	Temp	Data Type	Reference
WS 5.00E+004	mg/L		EXP	GUNTHER,FA ET AL. (1968)
logP 0.11			EST	MEYLAN,WM & HOWARD,PH (1995)
VP 8.08E-001	mm Hg	25	EST	NEELY,WB & BLAU,GE (1985)
DC	pKa			
HL	atm m3/mol			
OH 1.11E-011	cm3/molc sec	25	EST	MEYLAN,WM & HOWARD,PH (1993)

CAS #: 000123-91-1 — 1,4-DIOXANE

Formula:	$C_4H_8O_2$
Mol Weight:	88.11
MP (deg C):	11.80
FP (deg C):	
BP (deg C):	101.32
BP pressure (mm Hg):	

Property/Value	Units	Temp	Data Type	Reference
WS 1.00E+006	mg/L		EXP	RIDDICK,JA ET AL. (1986)
logP -0.27			EXP	HANSCH,C & LEO,AJ (1985)
VP 3.81E+001	mm Hg	25	EXP	DAUBERT,TE & DANNER,RP (1985)
DC -2.92	pKa		EXP	PERRIN,DD (1965)
HL 4.80E-006	atm m3/mol	25	EXP	PARK,JH ET AL. (1987)
OH 1.09E-011	cm3/molc sec	25	EXP	ATKINSON,R (1989)

CAS #: 000123-92-2 — ISOAMYL ACETATE

Formula:	$C_7H_{14}O_2$
Mol Weight:	130.19
MP (deg C):	-78.5
FP (deg C):	-99
BP (deg C):	142.5
BP pressure (mm Hg):	

Property/Value	Units	Temp	Data Type	Reference
WS 2.00E+003	mg/L	25	EXP	YALKOWSKY,SH & DANNENFELSER,RM (1992)
logP 2.26			EST	MEYLAN,WM & HOWARD,PH (1995)
VP 5.60E+000	mm Hg	25	EXP	DAUBERT,TE & DANNER,RP (1989)
DC	pKa			
HL 5.87E-004	atm m3/mol	25	EXP	HINE,J & MOOKERJEE,PK (1975)
OH 5.75E-012	cm3/molc sec	25	EST	MEYLAN,WM & HOWARD,PH (1993)

CAS #: 000123-95-5 — N-BUTYL STEARATE

Formula:	$C_{22}H_{44}O_2$
Mol Weight:	340.59
MP (deg C):	26.3
FP (deg C):	
BP (deg C):	343
BP pressure (mm Hg):	

Property/Value	Units	Temp	Data Type	Reference
WS 6.81E-005	mg/L	25	EST	VP/HL
logP 9.70			EST	MEYLAN,WM & HOWARD,PH (1995)
VP 5.80E-006	mm Hg	25	EST	DAUBERT,TE & DANNER,RP (1989)
DC	pKa			
HL 3.82E-002	atm m3/mol	25	EST	MEYLAN,WM & HOWARD,PH (1991)
OH 2.51E-011	cm3/molc sec	25	EST	ATKINSON,R (1988)

CAS #: 000123-96-6 — 2-OCTANOL

Formula:	$C_8H_{18}O$
Mol Weight:	130.23
MP (deg C):	38.6
FP (deg C):	
BP (deg C):	178.5
BP pressure (mm Hg):	

Property/Value	Units	Temp	Data Type	Reference
WS 1.12E+003	mg/L	25	EXP	AMIDON,GL ET AL (1974)
logP 2.90			EXP	FUNASAKI,N ET AL. (1986)
VP 2.42E-001	mm Hg	25	EXP	DAUBERT,TE & DANNER,RP (1989)
DC	pKa			
HL 1.23E-004	atm m3/mol	25	EXP	BOCEK,K (1976)
OH 1.51E-011	cm3/molc sec	25	EST	MEYLAN,WM & HOWARD,PH (1993)

CAS #: 000123-99-9 — NONANEDIOC ACID

Formula:	$C_9H_{16}O_4$
Mol Weight:	188.23
MP (deg C):	106.5
FP (deg C):	
BP (deg C):	286.5
BP pressure (mm Hg):	1.00E+002

Property/Value	Units	Temp	Data Type	Reference
WS 2.40E+003	mg/L	20	EXP	YALKOWSKY,SH & DANNENFELSER,RM (1992)
logP 1.57			EXP	HANSCH,C & LEO,AJ (1985)
VP 1.07E-008	mm Hg	25	EXT	YAWS,CL (1994B)
DC 4.55	pKa	25	EXP	KORTUM,G ET AL (1961)
HL 2.23E-011	atm m3/mol	25	EST	MEYLAN,WM & HOWARD,PH (1991)
OH 9.83E-012	cm3/molc sec	25	EST	MEYLAN,WM & HOWARD,PH (1993)

CAS #: 000124-02-7 — DIALLYLAMINE

Formula:	$C_6H_{11}N$
Mol Weight:	97.16
MP (deg C):	
FP (deg C):	
BP (deg C):	112
BP pressure (mm Hg):	

Property/Value	Units	Temp	Data Type	Reference
WS 1.37E+005	mg/L	25	EST	MEYLAN,WM ET AL. (1996)
logP 1.11			EXP	HANSCH,C & LEO,AJ (1985)
VP 2.10E+001	mm Hg	25	EST	NEELY,WB & BLAU,GE (1985)
DC 9.29	pKa	25	EXP	PERRIN,DD (1965)
HL 2.87E-005	atm m3/mol	25	EST	MEYLAN,WM & HOWARD,PH (1991)
OH 1.33E-010	cm3/molc sec	25	EST	MEYLAN,WM & HOWARD,PH (1993)

CAS #: 000124-04-9 — HEXANEDIOIC ACID

Formula: $C_6H_{10}O_4$
Mol Weight: 146.14
MP (deg C): 152
FP (deg C):
BP (deg C): 337.5
BP pressure (mm Hg):

Property/Value	Units	Temp	Data Type	Reference
WS 3.24E+002	mg/L	25	EXP	YALKOWSKY,SH & DANNENFELSER,RM (1992)
logP 0.08			EXP	HANSCH,C & LEO,AJ (1985)
VP 7.30E-002	mm Hg	18	EXP	DANLY,DE & CAMPBELL,CR (1978)
DC 4.44	pKa		EXP	SERJEANT,EP & DEMPSEY,B (1979)
HL 8.10E-013	atm m3/mol	25	EST	MEYLAN,WM & HOWARD,PH (1991)
OH 5.23E-012	cm3/molc sec	25	EST	MEYLAN,WM & HOWARD,PH (1993)

CAS #: 000124-06-1 — ETHYLMYRISTATE

Formula: $C_{16}H_{32}O_2$
Mol Weight: 256.43
MP (deg C): 12.3
FP (deg C):
BP (deg C): 265
BP pressure (mm Hg):

Property/Value	Units	Temp	Data Type	Reference
WS 3.69E-002	mg/L	25	EST	MEYLAN,WM ET AL. (1996)
logP 6.76			EST	MEYLAN,WM & HOWARD,PH (1995)
VP 5.94E-004	mm Hg	25	EST	NEELY,WB & BLAU,GE (1985)
DC	pKa			
HL 6.97E-003	atm m3/mol	25	EST	MEYLAN,WM & HOWARD,PH (1991)
OH 1.80E-011	cm3/molc sec	25	EST	MEYLAN,WM & HOWARD,PH (1993)

CAS #: 000124-07-2 — OCTANOIC ACID

Formula: $C_8H_{16}O_2$
Mol Weight: 144.22
MP (deg C): 16.7
FP (deg C):
BP (deg C): 239.7
BP pressure (mm Hg):

Property/Value	Units	Temp	Data Type	Reference
WS 7.89E+002	mg/L	30	EXP	YALKOWSKY,SH & DANNENFELSER,RM (1992)
logP 3.05			EXP	HANSCH,C & LEO,AJ (1985)
VP 3.44E-003	mm Hg	25	EXP	DAUBERT,TE & DANNER,RP (1989)
DC 4.89	pKa	25	EXP	KORTUM,G ET AL (1961)
HL 8.27E-007	atm m3/mol	25	EST	VP/WSOL
OH 8.35E-012	cm3/molc sec	25	EST	MEYLAN,WM & HOWARD,PH (1993)

CAS #: 000124-09-4 — HEXAMETHYLENEDIAMINE

Formula: $C_6H_{16}N_2$
Mol Weight: 116.21
MP (deg C): 42
FP (deg C):
BP (deg C): 205
BP pressure (mm Hg):

Property/Value	Units	Temp	Data Type	Reference
WS 2.46E+006	mg/L		EXP	YALKOWSKY,SH & DANNENFELSER,RM (1992)
logP 0.35			EST	MEYLAN,WM & HOWARD,PH (1995)
VP 1.18E-001	mm Hg	25	EXT	YAWS,CL (1994A)
DC 11.02	pKa		EXP	PERRIN,DD (1972)
HL 3.21E-009	atm m3/mol	25	EST	MEYLAN,WM & HOWARD,PH (1991)
OH 6.90E-011	cm3/molc sec	25	EST	MEYLAN,WM & HOWARD,PH (1993)

CAS #: 000124-10-7 — METHYL MYRISTATE

Formula: $C_{15}H_{30}O_2$
Mol Weight: 242.41
MP (deg C): 19
FP (deg C):
BP (deg C): 295
BP pressure (mm Hg):

Property/Value	Units	Temp	Data Type	Reference
WS 1.16E-001	mg/L	25	EST	MEYLAN,WM ET AL. (1996)
logP 6.27			EST	MEYLAN,WM & HOWARD,PH (1995)
VP 4.90E-004	mm Hg	25	EXT	PERRY,RH & GREEN,D (1984)
DC	pKa			
HL 5.25E-003	atm m3/mol	25	EST	MEYLAN,WM & HOWARD,PH (1991)
OH 1.65E-011	cm3/molc sec	25	EST	MEYLAN,WM & HOWARD,PH (1993)

CAS #: 000124-11-8 — ISONONENE

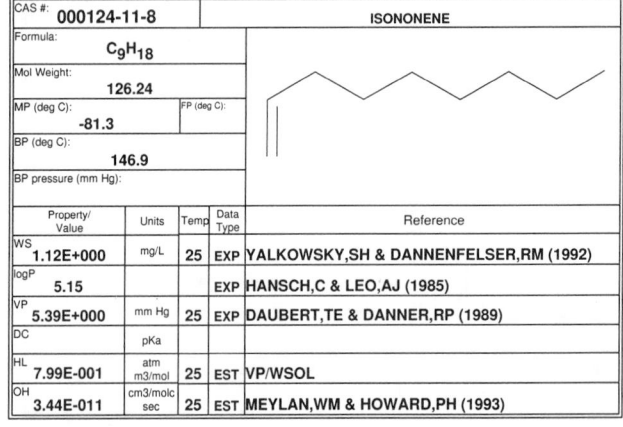

Formula: C_9H_{18}
Mol Weight: 126.24
MP (deg C): -81.3
FP (deg C):
BP (deg C): 146.9
BP pressure (mm Hg):

Property/Value	Units	Temp	Data Type	Reference
WS 1.12E+000	mg/L	25	EXP	YALKOWSKY,SH & DANNENFELSER,RM (1992)
logP 5.15			EXP	HANSCH,C & LEO,AJ (1985)
VP 5.39E+000	mm Hg	25	EXP	DAUBERT,TE & DANNER,RP (1989)
DC	pKa			
HL 7.99E-001	atm m3/mol	25	EST	VP/WSOL
OH 3.44E-011	cm3/molc sec	25	EST	MEYLAN,WM & HOWARD,PH (1993)

CAS #: 000124-12-9 — OCTANONITRILE

Formula: $C_8H_{15}N$
Mol Weight: 125.22
MP (deg C): -45.6
FP (deg C):
BP (deg C): 205.2
BP pressure (mm Hg):

Property/Value	Units	Temp	Data Type	Reference
WS 2.34E+002	mg/L	25	EST	MEYLAN,WM ET AL. (1996)
logP 2.75			EXP	TANII,H & HASHIMOTO,K (1984)
VP 3.90E-001	mm Hg	25	EXP	RIDDICK,JA ET AL. (1986)
DC	pKa			
HL 1.67E-004	atm m3/mol	25	EST	MEYLAN,WM & HOWARD,PH (1991)
OH 5.91E-012	cm3/molc sec	25	EST	MEYLAN,WM & HOWARD,PH (1993)

CAS #: 000124-13-0 — OCTANAL

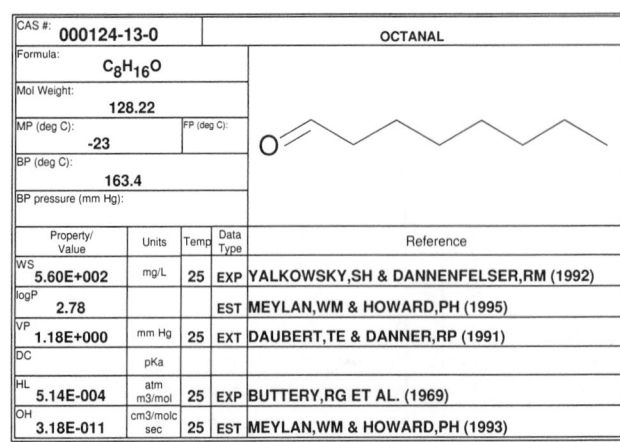

Formula: $C_8H_{16}O$
Mol Weight: 128.22
MP (deg C): -23
FP (deg C):
BP (deg C): 163.4
BP pressure (mm Hg):

Property/Value	Units	Temp	Data Type	Reference
WS 5.60E+002	mg/L	25	EXP	YALKOWSKY,SH & DANNENFELSER,RM (1992)
logP 2.78			EST	MEYLAN,WM & HOWARD,PH (1995)
VP 1.18E+000	mm Hg	25	EXT	DAUBERT,TE & DANNER,RP (1991)
DC	pKa			
HL 5.14E-004	atm m3/mol	25	EXP	BUTTERY,RG ET AL. (1969)
OH 3.18E-011	cm3/molc sec	25	EST	MEYLAN,WM & HOWARD,PH (1993)

CAS #: 000124-17-4				DIETHYLENE GLYCOL MONOBUTYL ETHER ACETATE
Formula: $C_{10}H_{20}O_4$				
Mol Weight: 204.27				
MP (deg C): -32		FP (deg C): -32		
BP (deg C): 245				
BP pressure (mm Hg):				

Property/ Value	Units	Temp	Data Type	Reference
WS 6.50E+004	mg/L	20	EXP	FLICK,EW (1991)
logP 1.30			EST	MEYLAN,WM & HOWARD,PH (1995)
VP 4.00E-002	mm Hg	20	EXP	FLICK,EW (1991)
DC	pKa			
HL 1.65E-007	atm m3/mol	25	EST	VP/WSOL
OH 3.53E-011	cm3/molc sec	25	EST	MEYLAN,WM & HOWARD,PH (1993)

CAS #: 000124-18-5				N-DECANE
Formula: $C_{10}H_{22}$				
Mol Weight: 142.29				
MP (deg C): -29.7		FP (deg C):		
BP (deg C): 174.1				
BP pressure (mm Hg):				

Property/ Value	Units	Temp	Data Type	Reference
WS 5.20E-002	mg/L	25	EXP	YALKOWSKY,SH & DANNENFELSER,RM (1992)
logP 5.01			EXP	COATES,M ET AL. (1985)
VP 1.43E+000	mm Hg	25	EXP	DAUBERT,TE & DANNER,RP (1989)
DC	pKa			
HL 5.15E+000	atm m3/mol	25	EST	VP/WSOL
OH 1.16E-011	cm3/molc sec	25	EXP	ATKINSON,R (1989)

CAS #: 000124-19-6				NONANAL
Formula: $C_9H_{18}O$				
Mol Weight: 142.24				
MP (deg C):		FP (deg C):		
BP (deg C): 191				
BP pressure (mm Hg):				

Property/ Value	Units	Temp	Data Type	Reference
WS 9.60E+001	mg/L	25	EXP	YALKOWSKY,SH & DANNENFELSER,RM (1992)
logP 3.27			EST	MEYLAN,WM & HOWARD,PH (1995)
VP 3.70E-001	mm Hg	25	EXP	DAUBERT,TE & DANNER,RP (1989)
DC	pKa			
HL 7.34E-004	atm m3/mol	25	EXP	BUTTERY,RG ET AL. (1969)
OH 3.32E-011	cm3/molc sec	25	EST	MEYLAN,WM & HOWARD,PH (1993)

CAS #: 000124-38-9				CARBON DIOXIDE
Formula: CO_2				
Mol Weight: 44.01				
MP (deg C): -56.5 5atm		FP (deg C):		
BP (deg C):				
BP pressure (mm Hg):				

Property/ Value	Units	Temp	Data Type	Reference
WS 2.90E+003	mg/L	25	EXP	YALKOWSKY,SH & DANNENFELSER,RM (1992)
logP -1.33			EST	MEYLAN,WM & HOWARD,PH (1995)
VP 4.83E+004	mm Hg	25	EXP	DAUBERT,TE & DANNER,RP (1989)
DC	pKa			
HL 1.52E-002	atm m3/mol	25	EST	VP/WSOL
OH 0.00E+000	cm3/molc sec	25	EST	MEYLAN,WM & HOWARD,PH (1993)

CAS #: 000124-40-3				DIMETHYLAMINE
Formula: C_2H_7N				
Mol Weight: 45.08				
MP (deg C): -92.2		FP (deg C):		
BP (deg C): 6.89				
BP pressure (mm Hg):				

Property/ Value	Units	Temp	Data Type	Reference
WS 1.63E+006	mg/L	40	EXP	SCHWEIZER,AE ET AL. (1978)
logP -0.38			EXP	HANSCH,C & LEO,AJ (1985)
VP 1.52E+003	mm Hg	25	EXT	DAUBERT,TE & DANNER,RP (1985)
DC 10.73	pKa		EXP	PERRIN,DD (1965)
HL 1.77E-005	atm m3/mol	25	EXP	CHRISTIE,AO & CRISP,DJ (1967)
OH 6.54E-011	cm3/molc sec	25	EXP	ATKINSON,R (1989)

CAS #: 000124-48-1				DIBROMOCHLOROMETHANE
Formula: $CHBr_2Cl$				
Mol Weight: 208.29				
MP (deg C): -22		FP (deg C):		
BP (deg C): 134				
BP pressure (mm Hg):				

Property/ Value	Units	Temp	Data Type	Reference
WS 2.70E+003	mg/L	20	EXP	HEIKES,DL (1987)
logP 2.16			EXP	SANGSTER,J (1994)
VP 5.54E+000	mm Hg	25	EST	NEELY,WB & BLAU,GE (1985)
DC	pKa			
HL 7.83E-004	atm m3/mol	20	EXP	WARNER,HP ET AL. (1987)
OH 6.26E-014	cm3/molc sec	25	EST	MEYLAN,WM & HOWARD,PH (1993)

CAS #: 000124-63-0				METHYLSULFONYL CHLORIDE
Formula: CH_3ClO_2S				
Mol Weight: 114.55				
MP (deg C):		FP (deg C): -32		
BP (deg C): 161				
BP pressure (mm Hg): 7.30E+002				

Property/ Value	Units	Temp	Data Type	Reference
WS 8.59E+003	mg/L	25	EST	MEYLAN,WM ET AL. (1996)
logP 1.27			EST	MEYLAN,WM & HOWARD,PH (1995)
VP 3.09E+000	mm Hg	25	EST	NEELY,WB & BLAU,GE (1985)
DC	pKa			
HL 4.38E-005	atm m3/mol	25	EST	MEYLAN,WM & HOWARD,PH (1991)
OH 1.30E-012	cm3/molc sec	25	EST	MEYLAN,WM & HOWARD,PH (1993)

CAS #: 000124-64-1				TETRAKIS(HYDROXYMETHYL)PHOSPHONIUM CHLORIDE
Formula: $C_4H_{12}ClO_4P$				
Mol Weight: 190.56				
MP (deg C):		FP (deg C):		
BP (deg C):				
BP pressure (mm Hg):				

Property/ Value	Units	Temp	Data Type	Reference
WS 4.00E+005	mg/L		EXP	BROWN,SL ET AL. (1975A)
logP -9.77			EST	MEYLAN,WM & HOWARD,PH (1995)
VP 1.13E-007	mm Hg	25	EST	NEELY,WB & BLAU,GE (1985)
DC	pKa			
HL 2.30E-013	atm m3/mol	25	EST	MEYLAN,WM & HOWARD,PH (1991)
OH 1.36E-011	cm3/molc sec	25	EST	MEYLAN,WM & HOWARD,PH (1993)

CAS #: 000124-68-5 — 2-AMINO-2-METHYL-1-PROPANOL

Formula: $C_4H_{11}NO$

Mol Weight: 89.14

MP (deg C): 30-31 FP (deg C):

BP (deg C): 165

BP pressure (mm Hg):

Property/Value	Units	Temp	Data Type	Reference
WS 1.00E+006	mg/L	25	EST	MEYLAN,WM ET AL. (1996)
logP -0.74			EST	MEYLAN,WM & HOWARD,PH (1995)
VP 1.00E+000	mm Hg	20	EXP	FLICK,EW (1991)
DC 10.19	pKa	10	EXP	PERRIN,DD (1972)
HL 6.48E-010	atm m3/mol	25	EST	MEYLAN,WM & HOWARD,PH (1991)
OH 2.80E-011	cm3/molc sec	25	EXP	ATKINSON,R (1989)

CAS #: 000124-73-2 — 1,2-DIBROMOTETRAFLUOROETHANE

Formula: $C_2Br_2F_4$

Mol Weight: 259.83

MP (deg C): -110.4 FP (deg C):

BP (deg C): 47.3

BP pressure (mm Hg):

Property/Value	Units	Temp	Data Type	Reference
WS 1.12E+001	mg/L	25	EST	MEYLAN,WM ET AL. (1996)
logP 2.96			EST	MEYLAN,WM & HOWARD,PH (1995)
VP 3.25E+002	mm Hg	25	EXP	DAUBERT,TE & DANNER,RP (1989)
DC	pKa			
HL 1.62E-001	atm m3/mol	25	EST	MEYLAN,WM & HOWARD,PH (1991)
OH 0.00E+000	cm3/molc sec	25	EST	MEYLAN,WM & HOWARD,PH (1993)

CAS #: 000124-76-5 — BICYCLO 2.2.1 HEPTAN-2-OL, 1,7,7-TRIMETHYL-, EXO

Formula: $C_{10}H_{18}O$

Mol Weight: 154.25

MP (deg C): FP (deg C):

BP (deg C):

BP pressure (mm Hg):

Property/Value	Units	Temp	Data Type	Reference
WS 1.19E+003	mg/L	25	EST	MEYLAN,WM ET AL. (1996)
logP 2.32			EXP	HANSCH,C ET AL. (1995)
VP 3.53E-002	mm Hg	25	EST	NEELY,WB & BLAU,GE (1985)
DC	pKa			
HL 6.70E-006	atm m3/mol	25	EST	MEYLAN,WM & HOWARD,PH (1991)
OH 1.22E-011	cm3/molc sec	25	EST	MEYLAN,WM & HOWARD,PH (1993)

CAS #: 000124-83-4 — CAMPHORIC ACID

Formula: $C_{10}H_{16}O_4$

Mol Weight: 200.24

MP (deg C): 189.1 FP (deg C):

BP (deg C):

BP pressure (mm Hg):

Property/Value	Units	Temp	Data Type	Reference
WS 7.60E+003	mg/L	25	EXP	YALKOWSKY,SH & DANNENFELSER,RM (1992)
logP 1.78			EST	MEYLAN,WM & HOWARD,PH (1995)
VP 1.51E-005	mm Hg	25	EST	NEELY,WB & BLAU,GE (1985)
DC	pKa			
HL 1.31E-011	atm m3/mol	25	EST	MEYLAN,WM & HOWARD,PH (1991)
OH 5.56E-012	cm3/molc sec	25	EST	MEYLAN,WM & HOWARD,PH (1993)

CAS #: 000124-94-7 — TRIAMCINOLONE

Formula: $C_{21}H_{27}FO_6$

Mol Weight: 394.44

MP (deg C): 269-271 FP (deg C):

BP (deg C):

BP pressure (mm Hg):

Property/Value	Units	Temp	Data Type	Reference
WS 8.00E+001	mg/L	25	EXP	YALKOWSKY,SH & DANNENFELSER,RM (1992)
logP 1.16			EXP	HANSCH,C & LEO,AJ (1985)
VP 7.20E-015	mm Hg	25	EST	NEELY,WB & BLAU,GE (1985)
DC	pKa			
HL 1.97E-009	atm m3/mol	25	EST	MEYLAN,WM & HOWARD,PH (1991)
OH 8.73E-011	cm3/molc sec	25	EST	MEYLAN,WM & HOWARD,PH (1993)

CAS #: 000125-29-1 — DIHYDROCODEINONE

Formula: $C_{18}H_{21}NO_3$

Mol Weight: 299.37

MP (deg C): 198 FP (deg C):

BP (deg C):

BP pressure (mm Hg):

Property/Value	Units	Temp	Data Type	Reference
WS 6.87E+003	mg/L	25	EST	MEYLAN,WM ET AL. (1996)
logP 1.48			EST	MEYLAN,WM & HOWARD,PH (1995)
VP 2.83E-007	mm Hg	25	EST	NEELY,WB & BLAU,GE (1985)
DC	pKa			
HL 6.37E-012	atm m3/mol	25	EST	MEYLAN,WM & HOWARD,PH (1991)
OH 1.72E-010	cm3/molc sec	25	EST	MEYLAN,WM & HOWARD,PH (1993)

CAS #: 000125-30-4 — CODETHYLINE HYDROCHLORIDE

Formula: $C_{19}H_{26}ClNO_3$

Mol Weight: 351.88

MP (deg C): FP (deg C):

BP (deg C):

BP pressure (mm Hg):

Property/Value	Units	Temp	Data Type	Reference
WS 2.08E+003	mg/L	25	EXP	STEPHEN,H & STEPHEN,T (1963)
logP 0.69			EST	MEYLAN,WM & HOWARD,PH (1995)
VP 4.20E-016	mm Hg	25	EST	NEELY,WB & BLAU,GE (1985)
DC	pKa			
HL 5.73E-021	atm m3/mol	25	EST	MEYLAN,WM & HOWARD,PH (1991)
OH 1.23E-010	cm3/molc sec	25	EST	MEYLAN,WM & HOWARD,PH (1993)

CAS #: 000125-33-7 — 5-ETHYLDIHYDRO-5-PHENYL-4,6(1H,5H)-PYRINIDINEDI*

Formula: $C_{12}H_{14}N_2O_2$

Mol Weight: 218.26

MP (deg C): 281-282 FP (deg C):

BP (deg C):

BP pressure (mm Hg):

Property/Value	Units	Temp	Data Type	Reference
WS 5.00E+002	mg/L	25	EXP	YALKOWSKY,SH & DANNENFELSER,RM (1992)
logP 0.91			EXP	HANSCH,C ET AL. (1995)
VP 3.64E-009	mm Hg	25	EST	NEELY,WB & BLAU,GE (1985)
DC	pKa			
HL 1.94E-010	atm m3/mol	25	EST	MEYLAN,WM & HOWARD,PH (1991)
OH 3.48E-011	cm3/molc sec	25	EST	MEYLAN,WM & HOWARD,PH (1993)

CAS #: 000125-40-6 — BUTABARBITAL

Formula: $C_{10}H_{16}N_2O_3$

Mol Weight: 212.25

MP (deg C): 165-168

FP (deg C):

BP (deg C):

BP pressure (mm Hg):

Property/Value	Units	Temp	Data Type	Reference
WS 1.36E+003	mg/L	25	EST	MEYLAN,WM ET AL. (1996)
logP 1.65			EXP	WONG,O & MCKEOWN,RH (1988)
VP 4.99E-011	mm Hg	25	EST	NEELY,WB & BLAU,GE (1985)
DC	pKa			
HL 6.36E-013	atm m3/mol	25	EST	MEYLAN,WM & HOWARD,PH (1991)
OH 1.66E-011	cm3/molc sec	25	EST	MEYLAN,WM & HOWARD,PH (1993)

CAS #: 000125-46-2 — 1,3(2H,9BH)-DIBENZOFURANDIONE, 2,6-DIACETYL-7,9-

Formula: $C_{18}H_{16}O_7$

Mol Weight: 344.32

MP (deg C): 204

FP (deg C):

BP (deg C):

BP pressure (mm Hg):

Property/Value	Units	Temp	Data Type	Reference
WS 8.85E+001	mg/L	25	EST	MEYLAN,WM ET AL. (1996)
logP 2.88			EXP	SANGSTER,J (1993)
VP 1.14E-012	mm Hg	25	EST	NEELY,WB & BLAU,GE (1985)
DC	pKa			
HL 2.82E-021	atm m3/mol	25	EST	MEYLAN,WM & HOWARD,PH (1991)
OH 2.80E-010	cm3/molc sec	25	EST	MEYLAN,WM & HOWARD,PH (1993)

CAS #: 000125-55-3 — BARBITURIC ACID,5-IPR,5-BR-ALLYL,N-ME

Formula: $C_{11}H_{15}BrN_2O_3$

Mol Weight: 303.16

MP (deg C): 115

FP (deg C):

BP (deg C):

BP pressure (mm Hg):

Property/Value	Units	Temp	Data Type	Reference
WS 1.86E+002	mg/L	25	EST	MEYLAN,WM ET AL. (1996)
logP 2.11			EXP	HANSCH,C & LEO,AJ (1985)
VP 2.03E-011	mm Hg	25	EST	NEELY,WB & BLAU,GE (1985)
DC	pKa			
HL 2.07E-013	atm m3/mol	25	EST	MEYLAN,WM & HOWARD,PH (1991)
OH 2.75E-011	cm3/molc sec	25	EST	MEYLAN,WM & HOWARD,PH (1993)

CAS #: 000125-64-4 — DIMERIN

Formula: $C_{10}H_{17}NO_2$

Mol Weight: 183.25

MP (deg C):

FP (deg C):

BP (deg C):

BP pressure (mm Hg):

Property/Value	Units	Temp	Data Type	Reference
WS 1.15E+004	mg/L	25	EST	MEYLAN,WM ET AL. (1996)
logP 0.78			EXP	SANGSTER,J (1994)
VP 1.26E-005	mm Hg	25	EST	NEELY,WB & BLAU,GE (1985)
DC	pKa			
HL 3.00E-011	atm m3/mol	25	EST	MEYLAN,WM & HOWARD,PH (1991)
OH 3.04E-011	cm3/molc sec	25	EST	MEYLAN,WM & HOWARD,PH (1993)

CAS #: 000125-73-5 — MORPHINAN-3-OL, 17-METHYL-, (9_,13_,14_)-

Formula: $C_{17}H_{23}NO$

Mol Weight: 257.38

MP (deg C):

FP (deg C):

BP (deg C):

BP pressure (mm Hg):

Property/Value	Units	Temp	Data Type	Reference
WS 1.84E+003	mg/L	25	EST	MEYLAN,WM ET AL. (1996)
logP 3.11			EXP	SANGSTER,J (1994)
VP 1.04E-006	mm Hg	25	EST	NEELY,WB & BLAU,GE (1985)
DC	pKa			
HL 2.11E-010	atm m3/mol	25	EST	MEYLAN,WM & HOWARD,PH (1991)
OH 2.07E-010	cm3/molc sec	25	EST	MEYLAN,WM & HOWARD,PH (1993)

CAS #: 000126-04-5 — PROPOXYPHENE CARBINOL

Formula: $C_{19}H_{25}NO$

Mol Weight: 283.42

MP (deg C):

FP (deg C):

BP (deg C):

BP pressure (mm Hg):

Property/Value	Units	Temp	Data Type	Reference
WS 8.36E+001	mg/L	25	EST	MEYLAN,WM ET AL. (1996)
logP 3.83			EXP	HANSCH,C & LEO,AJ (1985)
VP 7.06E-008	mm Hg	25	EST	NEELY,WB & BLAU,GE (1985)
DC	pKa			
HL 2.70E-011	atm m3/mol	25	EST	MEYLAN,WM & HOWARD,PH (1991)
OH 1.03E-010	cm3/molc sec	25	EST	MEYLAN,WM & HOWARD,PH (1993)

CAS #: 000126-07-8 — GRISEOFULVIN

Formula: $C_{17}H_{17}ClO_6$

Mol Weight: 352.77

MP (deg C): 220

FP (deg C):

BP (deg C):

BP pressure (mm Hg):

Property/Value	Units	Temp	Data Type	Reference
WS 8.64E+000	mg/L	25	EXP	YALKOWSKY,SH & DANNENFELSER,RM (1992)
logP 2.18			EXP	HANSCH,C & LEO,AJ (1985)
VP 9.59E-009	mm Hg	25	EST	NEELY,WB & BLAU,GE (1985)
DC	pKa			
HL 1.42E-013	atm m3/mol	25	EST	MEYLAN,WM & HOWARD,PH (1991)
OH 3.29E-010	cm3/molc sec	25	EST	MEYLAN,WM & HOWARD,PH (1993)

CAS #: 000126-11-4 — 2-HYDROXYMETHYL-2-NITRO-1,3-PROPANEDIOL

Formula: $C_4H_9NO_5$

Mol Weight: 151.12

MP (deg C): 214

FP (deg C):

BP (deg C):

BP pressure (mm Hg):

Property/Value	Units	Temp	Data Type	Reference
WS 2.20E+006	mg/L	20	EXP	DEWEY,RH & BOLLMEIER,AFJR (1978)
logP -1.66			EST	MEYLAN,WM & HOWARD,PH (1995)
VP 1.54E-006	mm Hg	25	EST	NEELY,WB & BLAU,GE (1985)
DC	pKa			
HL 4.82E-012	atm m3/mol	25	EST	MEYLAN,WM & HOWARD,PH (1991)
OH 1.92E-012	cm3/molc sec	25	EST	MEYLAN,WM & HOWARD,PH (1993)

CAS #: 000126-22-7				BUTONATE

Formula: $C_8H_{14}Cl_3O_5P$

Mol Weight: 327.53

MP (deg C):

FP (deg C):

BP (deg C): 129

BP pressure (mm Hg): 5.00E-001

Property/Value	Units	Temp	Data Type	Reference
WS 2.92E+002	mg/L	25	EST	MEYLAN,WM ET AL. (1996)
logP 1.71			EST	MEYLAN,WM & HOWARD,PH (1995)
VP 1.12E-004	mm Hg	25	EST	NEELY,WB & BLAU,GE (1985)
DC	pKa			
HL 3.04E-010	atm m3/mol	25	EST	MEYLAN,WM & HOWARD,PH (1991)
OH 7.96E-012	cm3/molc sec	25	EST	MEYLAN,WM & HOWARD,PH (1993)

CAS #: 000126-33-0				TETRAHYDROTHIOPHENE-1,1-DIOXIDE

Formula: $C_4H_8O_2S$

Mol Weight: 120.17

MP (deg C): 28.45

FP (deg C):

BP (deg C): 287.3

BP pressure (mm Hg):

Property/Value	Units	Temp	Data Type	Reference
WS 1.00E+006	mg/L			RIDDICK,JA ET AL. (1986)
logP -0.77			EXP	HANSCH,C & LEO,AJ (1985)
VP 6.20E-003	mm Hg		EXP	DAUBERT,TE & DANNER,RP (1989)
DC 15.30	pKa	25	EXP	RIDDICK,JA ET AL. (1986)
HL 4.85E-006	atm m3/mol	25	EST	MEYLAN,WM & HOWARD,PH (1991)
OH 4.46E-012	cm3/molc sec		EST	ATKINSON,R (1988)

CAS #: 000126-72-7				TRIS(2,3-DIBROMOPROPYL) PHOSPHATE

Formula: $C_9H_{15}Br_6O_4P$

Mol Weight: 697.65

MP (deg C): 5.5

FP (deg C):

BP (deg C):

BP pressure (mm Hg):

Property/Value	Units	Temp	Data Type	Reference
WS 8.00E+000	mg/L	24	EXP	HOLLIFIELD,HC (1979)
logP 4.29			EXP	SANGSTER,J (1994) (AVG)
VP 1.90E-004	mm Hg	25	EXP	IARC (1979)
DC	pKa			
HL 2.18E-005	atm m3/mol	25	EST	VP/WSOL
OH 3.33E-011	cm3/molc sec	25	EST	MEYLAN,WM & HOWARD,PH (1993)

CAS #: 000126-73-8				TRIBUTYLPHOSPHATE

Formula: $C_{12}H_{27}O_4P$

Mol Weight: 266.32

MP (deg C): -79

FP (deg C):

BP (deg C): 289

BP pressure (mm Hg):

Property/Value	Units	Temp	Data Type	Reference
WS 2.80E+002	mg/L	25	EXP	SAEGER,VW ET AL (1979)
logP 4.00			EXP	HANSCH,C & LEO,AJ (1985)
VP 1.20E-004	mm Hg	25	EXT	RIDDICK,JA ET AL. (1986)
DC	pKa			
HL 1.50E-007	atm m3/mol	25	EST	VP/WSOL
OH 8.33E+001	cm3/molc sec	25	EST	MEYLAN,WM & HOWARD,PH (1993)

CAS #: 000126-81-8				5,5-DIMETHYL-1,3-CYCLOHEXANEDIONE

Formula: $C_8H_{12}O_2$

Mol Weight: 140.18

MP (deg C): 149-151 de

FP (deg C):

BP (deg C):

BP pressure (mm Hg):

Property/Value	Units	Temp	Data Type	Reference
WS 6.46E+003	mg/L	25	EST	MEYLAN,WM ET AL. (1996)
logP 1.30			EST	MEYLAN,WM & HOWARD,PH (1995)
VP 6.39E-002	mm Hg	25	EST	NEELY,WB & BLAU,GE (1985)
DC	pKa			
HL 3.44E-008	atm m3/mol	25	EST	MEYLAN,WM & HOWARD,PH (1991)
OH 2.58E-012	cm3/molc sec	25	EST	MEYLAN,WM & HOWARD,PH (1993)

CAS #: 000126-84-1				2,2-DIETHOXYPROPANE

Formula: $C_7H_{16}O_2$

Mol Weight: 132.20

MP (deg C):

FP (deg C):

BP (deg C): 114

BP pressure (mm Hg):

Property/Value	Units	Temp	Data Type	Reference
WS 8.64E+002	mg/L	25	EST	MEYLAN,WM ET AL. (1996)
logP 2.36			EST	MEYLAN,WM & HOWARD,PH (1995)
VP 1.75E+001	mm Hg	25	EST	NEELY,WB & BLAU,GE (1985)
DC	pKa			
HL 1.57E-004	atm m3/mol	25	EST	MEYLAN,WM & HOWARD,PH (1991)
OH 1.17E-011	cm3/molc sec	25	EXP	ATKINSON,R (1989)

CAS #: 000126-98-7				METHACRYLONITRILE

Formula: C_4H_5N

Mol Weight: 67.09

MP (deg C): -35.8

FP (deg C):

BP (deg C): 90.3

BP pressure (mm Hg):

Property/Value	Units	Temp	Data Type	Reference
WS 2.54E+004	mg/L	25	EXP	YALKOWSKY,SH & DANNENFELSER,RM (1992)
logP 0.68			EXP	TANII,H & HASHIMOTO,K (1984)
VP 7.12E+001	mm Hg	25	EXP	DAUBERT,TE & DANNER,RP (1989)
DC	pKa			
HL 2.47E-004	atm m3/mol	25	EST	VP/WSOL
OH 7.85E-012	cm3/molc sec	25	EST	MEYLAN,WM & HOWARD,PH (1993)

CAS #: 000126-99-8				CHLOROPRENE

Formula: C_4H_5Cl

Mol Weight: 88.54

MP (deg C): -130

FP (deg C):

BP (deg C): 59.4

BP pressure (mm Hg):

Property/Value	Units	Temp	Data Type	Reference
WS 8.75E+002	mg/L	25	EST	MEYLAN,WM ET AL. (1996)
logP 2.53			EST	MEYLAN,WM & HOWARD,PH (1995)
VP 2.16E+002	mm Hg	25	EXP	BOUBLIK,T ET AL. (1984)
DC	pKa			
HL 5.61E-002	atm m3/mol	25	EST	MEYLAN,WM & HOWARD,PH (1991)
OH 2.10E-011	cm3/molc sec	25	EST	MEYLAN,WM & HOWARD,PH (1993)

CAS #: 000127-00-4				1-CHLORO-2-PROPANOL

Formula: C_3H_7ClO

Mol Weight: 94.54

MP (deg C): FP (deg C):

BP (deg C): 126-127

BP pressure (mm Hg):

Property/Value	Units	Temp	Data Type	Reference
WS 1.38E+005	mg/L	25	EST	MEYLAN,WM ET AL. (1996)
logP 0.53			EST	MEYLAN,WM & HOWARD,PH (1995)
VP 4.90E+000	mm Hg	20	EXP	YANG,RSH (1987)
DC	pKa			
HL 1.67E-006	atm m3/mol	25	EST	MEYLAN,WM & HOWARD,PH (1991)
OH 3.19E-012	cm3/molc sec	25	EST	MEYLAN,WM & HOWARD,PH (1993)

CAS #: 000127-06-0				2-PROPANONE, OXIME

Formula: C_3H_7NO

Mol Weight: 73.10

MP (deg C): 61 FP (deg C):

BP (deg C): 136

BP pressure (mm Hg):

Property/Value	Units	Temp	Data Type	Reference
WS 1.06E+005	mg/L	25	EST	MEYLAN,WM ET AL. (1996)
logP 0.12			EXP	HANSCH,C ET AL. (1995)
VP 3.12E+000	mm Hg	25	EST	NEELY,WB & BLAU,GE (1985)
DC 12.42	pKa	25	EXP	KORTUM,G ET AL (1961)
HL 7.80E-006	atm m3/mol	25	EST	MEYLAN,WM & HOWARD,PH (1991)
OH 3.35E-013	cm3/molc sec	25	EST	MEYLAN,WM & HOWARD,PH (1993)

CAS #: 000127-07-1				HYDROXYUREA

Formula: $CH_4N_2O_2$

Mol Weight: 76.06

MP (deg C): 133-136 FP (deg C):

BP (deg C):

BP pressure (mm Hg):

Property/Value	Units	Temp	Data Type	Reference
WS 2.24E+005	mg/L	25	EST	MEYLAN,WM ET AL. (1996)
logP -1.80			EXP	HANSCH,C & LEO,AJ (1985)
VP 2.43E-003	mm Hg	25	EST	NEELY,WB & BLAU,GE (1985)
DC	pKa			
HL 5.42E-011	atm m3/mol	25	EST	MEYLAN,WM & HOWARD,PH (1991)
OH 2.00E-012	cm3/molc sec	25	EST	MEYLAN,WM & HOWARD,PH (1993)

CAS #: 000127-09-3				ACETIC ACID, SODIUM SALT

Formula: $C_2H_3NaO_2$

Mol Weight: 82.03

MP (deg C): 58 FP (deg C):

BP (deg C):

BP pressure (mm Hg):

Property/Value	Units	Temp	Data Type	Reference
WS 1.00E+006	mg/L	25	EST	MEYLAN,WM ET AL. (1996)
logP -3.72			EST	MEYLAN,WM & HOWARD,PH (1995)
VP 7.08E-007	mm Hg	25	EST	NEELY,WB & BLAU,GE (1985)
DC	pKa			
HL	atm m3/mol			
OH 1.01E-013	cm3/molc sec	25	EST	MEYLAN,WM & HOWARD,PH (1993)

CAS #: 000127-17-3				PYRUVIC ACID

Formula: $C_3H_4O_3$

Mol Weight: 88.06

MP (deg C): 13.8 FP (deg C):

BP (deg C): 54

BP pressure (mm Hg): 1.00E+001

Property/Value	Units	Temp	Data Type	Reference
WS 1.00E+006	mg/L	20	EXP	YALKOWSKY,SH & DANNENFELSER,RM (1992)
logP -1.24			EST	MEYLAN,WM & HOWARD,PH (1995)
VP 1.29E+000	mm Hg	25	EXP	PERRY,RH & GREEN,D (1984)
DC 2.45	pKa	25	EXP	KORTUM,G ET AL (1961)
HL 1.50E-007	atm m3/mol	25	EST	VP/WSOL
OH 6.22E-013	cm3/molc sec	25	EST	MEYLAN,WM & HOWARD,PH (1993)

CAS #: 000127-18-4				TETRACHLOROETHENE

Formula: C_2Cl_4

Mol Weight: 165.83

MP (deg C): -22.35 FP (deg C):

BP (deg C): 121.07

BP pressure (mm Hg):

Property/Value	Units	Temp	Data Type	Reference
WS 2.00E+002	mg/L	25	EXP	COCA,J & DIAZ,R (1980)
logP 3.40			EXP	HANSCH,C & LEO,AJ (1985)
VP 1.86E+001	mm Hg	25	EXP	DAUBERT,TE & DANNER,RP (1985)
DC	pKa			
HL 1.77E-002	atm m3/mol	24	EXP	GOSSETT,JM (1987)
OH 1.67E-013	cm3/molc sec	25	EXP	ATKINSON,R (1989)

CAS #: 000127-19-5				N,N'-DIMETHYLACETAMIDE

Formula: C_4H_9NO

Mol Weight: 87.12

MP (deg C): -20 FP (deg C):

BP (deg C): 166.1

BP pressure (mm Hg):

Property/Value	Units	Temp	Data Type	Reference
WS 1.00E+006	mg/L	25	EXP	RIDDICK,JA ET AL. (1986)
logP -0.77			EXP	HANSCH,C & LEO,AJ (1985)
VP 2.00E+000	mm Hg	25	EXP	DAUBERT,TE & DANNER,RP (1989)
DC 10.78	pKa	25	EXP	RIDDICK,JA ET AL. (1986)
HL 1.31E-008	atm m3/mol	25	EXP	TAFT,RW ET AL. (1985)
OH 6.20E-011	cm3/molc sec	25	EST	ATKINSON,R (1988)

CAS #: 000127-20-8				DALAPON (SODIUM SALT)

Formula: $C_3H_3Cl_2NaO_2$

Mol Weight: 164.95

MP (deg C): 166.5 FP (deg C):

BP (deg C):

BP pressure (mm Hg):

Property/Value	Units	Temp	Data Type	Reference
WS 9.00E+005	mg/L	25	EXP	SHIU,WY ET AL. (1990)
logP -2.13			EST	MEYLAN,WM & HOWARD,PH (1995)
VP 4.60E-008	mm Hg	25	EST	NEELY,WB & BLAU,GE (1985)
DC	pKa			
HL	atm m3/mol			
OH 3.40E-014	cm3/molc sec	25	EST	MEYLAN,WM & HOWARD,PH (1993)

CAS #: 000127-31-1				HYDROCORTISONE-9A-FLUORO

Formula: $C_{21}H_{29}FO_5$
Mol Weight: 380.46
MP (deg C): FP (deg C):
BP (deg C):
BP pressure (mm Hg):

Property/ Value	Units	Temp	Data Type	Reference
WS 1.52E+002	mg/L	25	EST	MEYLAN,WM ET AL. (1996)
logP 1.67			EXP	HANSCH,C & LEO,AJ (1985)
VP 1.65E-013	mm Hg	25	EST	NEELY,WB & BLAU,GE (1985)
DC	pKa			
HL 1.15E-007	atm m3/mol	25	EST	MEYLAN,WM & HOWARD,PH (1991)
OH 9.00E-011	cm3/molc sec	25	EST	MEYLAN,WM & HOWARD,PH (1993)

CAS #: 000127-41-3				ALPHA-IONONE

Formula: $C_{13}H_{20}O$
Mol Weight: 192.30
MP (deg C): FP (deg C):
BP (deg C): 131
BP pressure (mm Hg): 1.30E+001

Property/ Value	Units	Temp	Data Type	Reference
WS 1.00E+006	mg/L	25	EXP	ETZWEILER,F ET AL. (1995)
logP 4.29			EST	MEYLAN,WM & HOWARD,PH (1995)
VP 1.73E-002	mm Hg	25	EXT	PERRY,RH & GREEN,D (1984)
DC	pKa			
HL 1.81E-004	atm m3/mol	25	EST	MEYLAN,WM & HOWARD,PH (1991)
OH 1.43E-010	cm3/molc sec	25	EST	MEYLAN,WM & HOWARD,PH (1993)

CAS #: 000127-63-9				DIPHENYL SULFONE

Formula: $C_{12}H_{10}O_2S$
Mol Weight: 218.28
MP (deg C): 128-129 FP (deg C):
BP (deg C): 378-379
BP pressure (mm Hg):

Property/ Value	Units	Temp	Data Type	Reference
WS 3.14E+002	mg/L	25	EXP	MEYLAN,WM ET AL. (1996)
logP 2.40			EXP	HANSCH,C & LEO,AJ (1985)
VP 1.53E-005	mm Hg	25	EST	NEELY,WB & BLAU,GE (1985)
DC	pKa			
HL 2.49E-007	atm m3/mol	25	EST	MEYLAN,WM & HOWARD,PH (1991)
OH 8.34E-013	cm3/molc sec	25	EST	MEYLAN,WM & HOWARD,PH (1993)

CAS #: 000127-68-4				M-NITROBENZENESULFONIC ACID, SODIUM SALT

Formula: $C_6H_4NNaO_5S$
Mol Weight: 225.16
MP (deg C): 52.3 FP (deg C):
BP (deg C): 217.5
BP pressure (mm Hg):

Property/ Value	Units	Temp	Data Type	Reference
WS 2.77E+005	mg/L	20	EXP	CHEM INSPECT TEST INST (1992)
logP -2.61			EXP	SANGSTER,J (1994)
VP 1.08E-012	mm Hg	25	EST	NEELY,WB & BLAU,GE (1985)
DC	pKa			
HL	atm m3/mol			
OH 5.21E-014	cm3/molc sec	25	EST	MEYLAN,WM & HOWARD,PH (1993)

CAS #: 000127-69-5				SULFISOXAZOLE

Formula: $C_{11}H_{13}N_3O_3S$
Mol Weight: 267.31
MP (deg C): 194 FP (deg C):
BP (deg C):
BP pressure (mm Hg):

Property/ Value	Units	Temp	Data Type	Reference
WS 3.00E+002	mg/L	37	EXP	YALKOWSKY,SH & DANNENFELSER,RM (1992)
logP 1.01			EXP	HANSCH,C & LEO,AJ (1985)
VP 5.08E-008	mm Hg	25	EST	NEELY,WB & BLAU,GE (1985)
DC	pKa			
HL 1.06E-012	atm m3/mol	25	EST	MEYLAN,WM & HOWARD,PH (1991)
OH 3.24E-011	cm3/molc sec	25	EST	MEYLAN,WM & HOWARD,PH (1993)

CAS #: 000127-77-5				N1-PHENYLSULFANILIDE

Formula: $C_{12}H_{12}N_2O_2S$
Mol Weight: 248.31
MP (deg C): 200 FP (deg C):
BP (deg C):
BP pressure (mm Hg):

Property/ Value	Units	Temp	Data Type	Reference
WS 7.00E+003	mg/L	100	EXP	YALKOWSKY,SH & DANNENFELSER,RM (1992)
logP 1.55			EXP	HANSCH,C & LEO,AJ (1985)
VP 1.17E-007	mm Hg	25	EST	NEELY,WB & BLAU,GE (1985)
DC	pKa			
HL 8.23E-011	atm m3/mol	25	EST	MEYLAN,WM & HOWARD,PH (1991)
OH 6.57E-011	cm3/molc sec	25	EST	MEYLAN,WM & HOWARD,PH (1993)

CAS #: 000127-79-7				SULFAMERAZINE

Formula: $C_{11}H_{12}N_4O_2S$
Mol Weight: 264.31
MP (deg C): 236 FP (deg C):
BP (deg C):
BP pressure (mm Hg):

Property/ Value	Units	Temp	Data Type	Reference
WS 2.02E+002	mg/L	20	EXP	YALKOWSKY,SH & DANNENFELSER,RM (1992)
logP 0.14			EXP	HANSCH,C & LEO,AJ (1985)
VP 1.93E-008	mm Hg	25	EST	NEELY,WB & BLAU,GE (1985)
DC	pKa			
HL 1.75E-010	atm m3/mol	25	EST	MEYLAN,WM & HOWARD,PH (1991)
OH 3.63E-011	cm3/molc sec	25	EST	MEYLAN,WM & HOWARD,PH (1993)

CAS #: 000127-91-3				á-PINENE

Formula: $C_{10}H_{16}$
Mol Weight: 136.24
MP (deg C): -61.5 FP (deg C):
BP (deg C): 166
BP pressure (mm Hg):

Property/ Value	Units	Temp	Data Type	Reference
WS 4.89E+000	mg/L	25	EST	MEYLAN,WM ET AL. (1996)
logP 4.35			EST	MEYLAN,WM & HOWARD,PH (1995)
VP 2.93E+000	mm Hg	25	EXP	DAUBERT,TE & DANNER,RP (1989)
DC	pKa			
HL 1.61E-001	atm m3/mol	25	EST	MEYLAN,WM & HOWARD,PH (1991)
OH 7.89E-011	cm3/molc sec	25	EXP	ATKINSON,R (1989)

000128-04-1 — DIMETHYLDITHIOCARBAMIC ACID, SODIUM SALT

Formula: $C_3H_6NNaS_2$
Mol Weight: 143.21
MP (deg C): — FP (deg C): —
BP (deg C): —
BP pressure (mm Hg): —

Property/Value	Units	Temp	Data Type	Reference
WS 1.00E+006	mg/L	25	EST	MEYLAN,WM ET AL. (1996)
logP -0.71			EST	MEYLAN,WM & HOWARD,PH (1995)
VP 4.17E-009	mm Hg	25	EST	NEELY,WB & BLAU,GE (1985)
DC	pKa			
HL	atm m3/mol			
OH 6.85E-011	cm3/molc sec	25	EST	MEYLAN,WM & HOWARD,PH (1993)

000128-13-2 — URSODEOXYCHOLIC ACID

Formula: $C_{24}H_{40}O_4$
Mol Weight: 392.58
MP (deg C): 203 FP (deg C): —
BP (deg C): —
BP pressure (mm Hg): —

Property/Value	Units	Temp	Data Type	Reference
WS 2.00E+001	mg/L	20	EXP	YALKOWSKY,SH & DANNENFELSER,RM (1992)
logP 3.00			EXP	RODA,A ET AL. (1990)
VP 3.00E-013	mm Hg	25	EST	NEELY,WB & BLAU,GE (1985)
DC	pKa			
HL 1.41E-011	atm m3/mol	25	EST	MEYLAN,WM & HOWARD,PH (1991)
OH 4.64E-011	cm3/molc sec	25	EST	MEYLAN,WM & HOWARD,PH (1993)

000128-37-0 — 2,6-DI-T-BUTYL-4-METHYLPHENOL (BHT)

Formula: $C_{15}H_{24}O$
Mol Weight: 220.36
MP (deg C): 70 FP (deg C): —
BP (deg C): 265
BP pressure (mm Hg): —

Property/Value	Units	Temp	Data Type	Reference
WS 6.00E-001	mg/L	25	EXP	INUI,H ET AL. (1979A)
logP 5.10			EXP	TSCATS
VP 5.16E-003	mm Hg	25	EXT	PERRY,RH & GREEN,D (1984)
DC 12.23	pKa		EXP	SERJEANT,EP & DEMPSEY,B (1979)
HL 4.12E-006	atm m3/mol	25	EST	MEYLAN,WM & HOWARD,PH (1991)
OH 1.83E-011	cm3/molc sec	25	EST	MEYLAN,WM & HOWARD,PH (1993)

000128-39-2 — 2,6-DI-T-BUTYLPHENOL

Formula: $C_{14}H_{22}O$
Mol Weight: 206.33
MP (deg C): 37 FP (deg C): —
BP (deg C): 253
BP pressure (mm Hg): —

Property/Value	Units	Temp	Data Type	Reference
WS 2.50E+000	mg/L	25	EXP	GEYER,H ET AL. (1981)
logP 4.92			EXP	HANSCH,C ET AL. (1995)
VP 7.30E-003	mm Hg	25	EST	NEELY,WB & BLAU,GE (1985)
DC 11.70	pKa	25	EXP	SERJEANT,EP & DEMPSEY,B (1979)
HL 3.15E-006	atm m3/mol	25	EST	MEYLAN,WM & HOWARD,PH (1991)
OH 4.90E-011	cm3/molc sec	25	EST	MEYLAN,WM & HOWARD,PH (1993)

000128-66-5 — DIBENZO(B,DEF)CHRYSENE-7,14-DIONE

Formula: $C_{24}H_{12}O_2$
Mol Weight: 332.36
MP (deg C): 158 EST FP (deg C): —
BP (deg C): —
BP pressure (mm Hg): —

Property/Value	Units	Temp	Data Type	Reference
WS 2.64E-003	mg/L	25	EST	MEYLAN,WM ET AL. (1996)
logP 6.28			EST	MEYLAN,WM & HOWARD,PH (1995)
VP 8.34E-012	mm Hg	25	EST	NEELY,WB & BLAU,GE (1985)
DC	pKa			
HL 8.31E-012	atm m3/mol	25	EST	MEYLAN,WM & HOWARD,PH (1991)
OH 1.65E-011	cm3/molc sec	25	EST	MEYLAN,WM & HOWARD,PH (1993)

000128-95-0 — DISPERSE VIOLET 1

Formula: $C_{14}H_{10}N_2O_2$
Mol Weight: 238.25
MP (deg C): 265-269 FP (deg C): —
BP (deg C): —
BP pressure (mm Hg): —

Property/Value	Units	Temp	Data Type	Reference
WS 1.31E-001	mg/L	25	EXP	BAUGHMAN,GL & PERENICH,TA (1988)
logP 3.00			EXP	BAUGHMAN,GL & WEBER,EJ (1991)
VP 2.00E-010	mm Hg	25	EXP	BAUGHMAN,GL & PERENICH,TA (1988)
DC	pKa			
HL 4.79E-010	atm m3/mol	25	EST	VP/WSOL
OH 6.85E-011	cm3/molc sec	25	EST	MEYLAN,WM & HOWARD,PH (1993)

000129-00-0 — PYRENE

Formula: $C_{16}H_{10}$
Mol Weight: 202.26
MP (deg C): 149-151 FP (deg C): —
BP (deg C): 393
BP pressure (mm Hg): —

Property/Value	Units	Temp	Data Type	Reference
WS 1.35E-007	mg/L	25	EXP	MILLER,MM ET AL. (1985)
logP 4.88			EXP	HANSCH,C & LEO,AJ (1985)
VP 2.45E-006	mm Hg	25	EXT	HOYER,H & PEPERLE,W (1958)
DC	pKa			
HL 1.10E-005	atm m3/mol	25	EXP	MACKAY,D ET AL. (1982A)
OH 5.00E-011	cm3/molc sec	25	EXP	ATKISON,R (1989)

000129-03-3 — CYPROHEPTADINE

Formula: $C_{21}H_{21}N$
Mol Weight: 287.41
MP (deg C): 112.3-3.3 FP (deg C): —
BP (deg C): —
BP pressure (mm Hg): —

Property/Value	Units	Temp	Data Type	Reference
WS 1.14E+000	mg/L	25	EST	MEYLAN,WM ET AL. (1996)
logP 4.69			EXP	SANGSTER,J (1993)
VP 3.61E-007	mm Hg	25	EST	NEELY,WB & BLAU,GE (1985)
DC	pKa			
HL 9.20E-009	atm m3/mol	25	EST	MEYLAN,WM & HOWARD,PH (1991)
OH 2.61E-010	cm3/molc sec	25	EST	MEYLAN,WM & HOWARD,PH (1993)

CAS #: 000129-20-4				OXYPHENBUTAZONE

Formula: $C_{19}H_{20}N_2O_3$

Mol Weight: 324.38

MP (deg C): 96 | FP (deg C):

BP (deg C):

BP pressure (mm Hg):

Property/Value	Units	Temp	Data Type	Reference
WS 6.00E+001	mg/L	30	EXP	YALKOWSKY,SH & DANNENFELSER,RM (1992)
logP 2.72			EXP	HANSCH,C ET AL. (1995)
VP 2.70E-012	mm Hg	25	EST	NEELY,WB & BLAU,GE (1985)
DC	pKa			
HL 6.83E-013	atm m3/mol	25	EST	MEYLAN,WM & HOWARD,PH (1991)
OH 3.83E-011	cm3/molc sec	25	EST	MEYLAN,WM & HOWARD,PH (1993)

CAS #: 000129-35-1				1-CHLORO-2-METHYLANTHRAQUINONE

Formula: $C_{15}H_9ClO_2$

Mol Weight: 256.69

MP (deg C): 170-171 | FP (deg C):

BP (deg C):

BP pressure (mm Hg):

Property/Value	Units	Temp	Data Type	Reference
WS 2.25E-001	mg/L	25	EST	MEYLAN,WM ET AL. (1996)
logP 4.54			EST	MEYLAN,WM & HOWARD,PH (1995)
VP 6.39E-007	mm Hg	25	EST	NEELY,WB & BLAU,GE (1985)
DC	pKa			
HL 2.60E-009	atm m3/mol	25	EST	MEYLAN,WM & HOWARD,PH (1991)
OH 1.53E-012	cm3/molc sec	25	EST	MEYLAN,WM & HOWARD,PH (1993)

CAS #: 000129-43-1				1-HYDROXYANTHRAQUINONE

Formula: $C_{14}H_8O_3$

Mol Weight: 224.22

MP (deg C): 193.8 | FP (deg C):

BP (deg C):

BP pressure (mm Hg):

Property/Value	Units	Temp	Data Type	Reference
WS 8.50E+000	mg/L	25	EXP	BAUGHMAN,GL ET AL. (1993)
logP 3.52			EXP	BROOKE,D ET AL. (1990)
VP 2.09E-007	mm Hg	25	EXP	SHIMIZU,T ET AL. (1987)
DC	pKa			
HL 7.25E-009	atm m3/mol	25	EST	VP/WSOL
OH 1.37E-011	cm3/molc sec	25	EST	MEYLAN,WM & HOWARD,PH (1993)

CAS #: 000129-66-8				2,4,6-TRINITROBENZOIC ACID

Formula: $C_7H_3N_3O_8$

Mol Weight: 257.12

MP (deg C): 228.7 | FP (deg C):

BP (deg C):

BP pressure (mm Hg):

Property/Value	Units	Temp	Data Type	Reference
WS 2.05E+004	mg/L	25	EXP	YALKOWSKY,SH & DANNENFELSER,RM (1992)
logP 0.23			EST	MEYLAN,WM & HOWARD,PH (1995)
VP 7.09E-009	mm Hg	25	EST	NEELY,WB & BLAU,GE (1985)
DC 0.65	pKa	25	EXP	KORTUM,G ET AL (1961)
HL 2.62E-014	atm m3/mol	25	EST	MEYLAN,WM & HOWARD,PH (1991)
OH 5.20E-013	cm3/molc sec	25	EST	MEYLAN,WM & HOWARD,PH (1993)

CAS #: 000129-79-3				9H-FLUOREN-9-ONE, 2,4,7-TRINITRO-

Formula: $C_{13}H_5N_3O_7$

Mol Weight: 315.20

MP (deg C): 175.2-76 | FP (deg C):

BP (deg C):

BP pressure (mm Hg):

Property/Value	Units	Temp	Data Type	Reference
WS 2.71E+000	mg/L	25	EST	MEYLAN,WM ET AL. (1996)
logP 2.42			EXP	DEBNATH,AK & HANSCH,C (1992)
VP 2.92E-010	mm Hg	25	EST	NEELY,WB & BLAU,GE (1985)
DC	pKa			
HL 4.16E-014	atm m3/mol	25	EST	MEYLAN,WM & HOWARD,PH (1991)
OH 2.06E-013	cm3/molc sec	25	EST	MEYLAN,WM & HOWARD,PH (1993)

CAS #: 000129-81-7				4-IODOANTIPYRINE

Formula: $C_{11}H_{11}IN_2O$

Mol Weight: 314.13

MP (deg C): | FP (deg C):

BP (deg C):

BP pressure (mm Hg):

Property/Value	Units	Temp	Data Type	Reference
WS 8.35E+002	mg/L	25	EST	MEYLAN,WM ET AL. (1996)
logP 1.27			EXP	HANSCH,C ET AL. (1995)
VP 7.03E-007	mm Hg	25	EST	NEELY,WB & BLAU,GE (1985)
DC	pKa			
HL 8.37E-011	atm m3/mol	25	EST	MEYLAN,WM & HOWARD,PH (1991)
OH 1.98E-011	cm3/molc sec	25	EST	MEYLAN,WM & HOWARD,PH (1993)

CAS #: 000130-13-2				1-NAPHTHALENESULFONIC ACID, 4-AMINO-, SODIUM SAL

Formula: $C_{10}H_8NNaO_3S$

Mol Weight: 245.23

MP (deg C): | FP (deg C):

BP (deg.C):

BP pressure (mm Hg):

Property/Value	Units	Temp	Data Type	Reference
WS 1.00E+006	mg/L	25	EST	MEYLAN,WM ET AL. (1996)
logP -2.34			EXP	HANSCH,C ET AL. (1995)
VP 1.55E-014	mm Hg	25	EST	NEELY,WB & BLAU,GE (1985)
DC	pKa			
HL	atm m3/mol			
OH 4.39E-011	cm3/molc sec	25	EST	MEYLAN,WM & HOWARD,PH (1993)

CAS #: 000130-15-4				1,4-NAPHTHOQUINONE

Formula: $C_{10}H_6O_2$

Mol Weight: 158.16

MP (deg C): 126 | FP (deg C):

BP (deg C):

BP pressure (mm Hg):

Property/Value	Units	Temp	Data Type	Reference
WS 6.68E+002	mg/L	25	EST	MEYLAN,WM ET AL (1996)
logP 1.71			EXP	HANSCH,C ET AL. (1995)
VP 1.80E-004	mm Hg	25	EST	NEELY,WB & BLAU,GE (1985)
DC	pKa			
HL 1.97E-009	atm m3/mol	25	EST	MEYLAN,WM & HOWARD,PH (1991)
OH 3.10E-012	cm3/molc sec	25	EXP	LEIFER,A (1993)

CAS #: 000131-18-0 — DI-N-AMYL PHTHALATE

Formula: $C_{18}H_{26}O_4$

Mol Weight: 306.41

MP (deg C):   FP (deg C):

BP (deg C):

BP pressure (mm Hg):

Property/Value	Units	Temp	Data Type	Reference
WS 1.89E-001	mg/L	25	EST	MEYLAN,WM ET AL. (1996)
logP 5.59			EST	MEYLAN,WM & HOWARD,PH (1995)
VP 6.80E-005	mm Hg	25	EST	NEELY,WB & BLAU,GE (1985)
DC	pKa			
HL 2.16E-006	atm m3/mol	25	EST	MEYLAN,WM & HOWARD,PH (1991)
OH 1.21E-011	cm3/molc sec	25	EST	MEYLAN,WM & HOWARD,PH (1993)

CAS #: 000131-22-6 — 4-(PHENYLAZO)-1-NAPHTHALENAMINE

Formula: $C_{16}H_{13}N_3$

Mol Weight: 247.30

MP (deg C):   FP (deg C):

BP (deg C):

BP pressure (mm Hg):

Property/Value	Units	Temp	Data Type	Reference
WS 1.68E+000	mg/L	25	EST	MEYLAN,WM ET AL. (1996)
logP 4.37			EST	MEYLAN,WM & HOWARD,PH (1995)
VP 2.06E-007	mm Hg	25	EST	NEELY,WB & BLAU,GE (1985)
DC	pKa			
HL 5.07E-010	atm m3/mol	25	EST	MEYLAN,WM & HOWARD,PH (1991)
OH 4.90E-011	cm3/molc sec	25	EST	MEYLAN,WM & HOWARD,PH (1993)

CAS #: 000131-27-1 — 3-AMINO-1,5-NAPHTHALENE DISULFONIC ACID

Formula: $C_{10}H_9NO_6S_2$

Mol Weight: 303.31

MP (deg C):   FP (deg C):

BP (deg C):

BP pressure (mm Hg):

Property/Value	Units	Temp	Data Type	Reference
WS 4.49E+003	mg/L	25	EST	MEYLAN,WM ET AL. (1996)
logP -4.06			EST	MEYLAN,WM & HOWARD,PH (1995)
VP 1.61E-015	mm Hg	25	EST	NEELY,WB & BLAU,GE (1985)
DC	pKa			
HL 4.06E-020	atm m3/mol	25	EST	MEYLAN,WM & HOWARD,PH (1991)
OH 9.67E-012	cm3/molc sec	25	EST	MEYLAN,WM & HOWARD,PH (1993)

CAS #: 000131-28-2 — NARCEINE

Formula: $C_{23}H_{27}NO_8$

Mol Weight: 445.47

MP (deg C): 138   FP (deg C):

BP (deg C):

BP pressure (mm Hg):

Property/Value	Units	Temp	Data Type	Reference
WS 7.80E+002	mg/L	13	EXP	STEPHEN,H & STEPHEN,T (1963)
logP 1.97			EST	MEYLAN,WM & HOWARD,PH (1995)
VP 1.35E-012	mm Hg	25	EST	NEELY,WB & BLAU,GE (1985)
DC	pKa			
HL 2.79E-021	atm m3/mol	25	EST	MEYLAN,WM & HOWARD,PH (1991)
OH 2.87E-010	cm3/molc sec	25	EST	MEYLAN,WM & HOWARD,PH (1993)

CAS #: 000131-52-2 — PENTACHLOROPHENOL, NA SALT

Formula: C_6Cl_5NaO

Mol Weight: 288.32

MP (deg C):   FP (deg C):

BP (deg C):

BP pressure (mm Hg):

Property/Value	Units	Temp	Data Type	Reference
WS 3.30E+005	mg/L	25	EXP	SHIU,WY ET AL. (1990)
logP 2.05			EST	MEYLAN,WM & HOWARD,PH (1995)
VP 2.70E-011	mm Hg	25	EST	NEELY,WB & BLAU,GE (1985)
DC	pKa			
HL	atm m3/mol			
OH 2.65E-013	cm3/molc sec	25	EST	MEYLAN,WM & HOWARD,PH (1993)

CAS #: 000131-57-7 — METHANONE, (2-HYDROXY-4-METHOXYPHENYL)PHENYL-

Formula: $C_{14}H_{12}O_3$

Mol Weight: 228.25

MP (deg C): 66   FP (deg C):

BP (deg C):

BP pressure (mm Hg):

Property/Value	Units	Temp	Data Type	Reference
WS 6.86E+001	mg/L	25	EST	MEYLAN,WM ET AL. (1996)
logP 3.79			EXP	CHEM INSPECT TEST INST (1992)
VP 1.42E-006	mm Hg	25	EST	NEELY,WB & BLAU,GE (1985)
DC	pKa			
HL 1.50E-008	atm m3/mol	25	EST	MEYLAN,WM & HOWARD,PH (1991)
OH 2.01E-010	cm3/molc sec	25	EST	MEYLAN,WM & HOWARD,PH (1993)

CAS #: 000131-72-6 — KARATHANE

Formula: $C_{18}H_{24}N_2O_6$

Mol Weight: 364.40

MP (deg C):   FP (deg C):

BP (deg C):

BP pressure (mm Hg):

Property/Value	Units	Temp	Data Type	Reference
WS 1.62E-002	mg/L	25	EST	MEYLAN,WM ET AL. (1996)
logP 5.98			EST	MEYLAN,WM & HOWARD,PH (1995)
VP 3.41E-009	mm Hg	25	EST	NEELY,WB & BLAU,GE (1985)
DC	pKa			
HL 6.69E-009	atm m3/mol	25	EST	MEYLAN,WM & HOWARD,PH (1991)
OH 2.94E-011	cm3/molc sec	25	EST	MEYLAN,WM & HOWARD,PH (1993)

CAS #: 000131-73-7 — 2,2',4,4',6,6'-HEXANITRODIPHENYLAMINE

Formula: $C_{12}H_5N_7O_{12}$

Mol Weight: 439.21

MP (deg C): 238   FP (deg C):

BP (deg C):

BP pressure (mm Hg):

Property/Value	Units	Temp	Data Type	Reference
WS 6.00E+001	mg/L	17	EXP	SEIDELL,A (1941)
logP 3.35			EST	MEYLAN,WM & HOWARD,PH (1995)
VP 2.89E-014	mm Hg	25	EST	NEELY,WB & BLAU,GE (1985)
DC	pKa			
HL 2.33E-017	atm m3/mol	25	EST	MEYLAN,WM & HOWARD,PH (1991)
OH 4.26E-015	cm3/molc sec	25	EST	MEYLAN,WM & HOWARD,PH (1993)

000131-74-8 — AMMONIUM PICRATE

CAS #: 000131-74-8

Formula: $C_6H_6N_4O_7$

Mol Weight: 246.14

MP (deg C):

FP (deg C):

BP (deg C):

BP pressure (mm Hg):

Property/Value	Units	Temp	Data Type	Reference
WS 1.60E+005	mg/L	25	EST	MEYLAN,WM ET AL. (1996)
logP -1.40			EST	MEYLAN,WM & HOWARD,PH (1995)
VP 3.37E-011	mm Hg	25	EST	NEELY,WB & BLAU,GE (1985)
DC	pKa			
HL 2.94E-022	atm m3/mol	25	EST	MEYLAN,WM & HOWARD,PH (1991)
OH 1.13E-015	cm3/molc sec	25	EST	MEYLAN,WM & HOWARD,PH (1993)

000131-89-5 — 2-CYCLOHEXYL-4,6-DINITROPHENOL

CAS #: 000131-89-5

Formula: $C_{12}H_{14}N_2O_5$

Mol Weight: 266.26

MP (deg C): 106.5-7.5

FP (deg C):

BP (deg C):

BP pressure (mm Hg):

Property/Value	Units	Temp	Data Type	Reference
WS 1.50E+001	mg/L	25	EXP	YALKOWSKY,SH & DANNENFELSER,RM (1992)
logP 4.54			EST	MEYLAN,WM & HOWARD,PH (1995)
VP 4.19E-008	mm Hg	25	EST	NEELY,WB & BLAU,GE (1985)
DC	pKa			
HL 5.54E-008	atm m3/mol	25	EST	MEYLAN,WM & HOWARD,PH (1991)
OH 1.02E-011	cm3/molc sec	25	EST	MEYLAN,WM & HOWARD,PH (1993)

000131-91-9 — 1-NITROSO-2-NAPHTHOL

CAS #: 000131-91-9

Formula: $C_{10}H_7NO_2$

Mol Weight: 173.17

MP (deg C): 109-110

FP (deg C):

BP (deg C):

BP pressure (mm Hg):

Property/Value	Units	Temp	Data Type	Reference
WS 2.55E+003	mg/L	25	EST	MEYLAN,WM ET AL. (1996)
logP 2.28			EXP	HANSCH,C & LEO,AJ (1985)
VP 9.11E-005	mm Hg	25	EST	NEELY,WB & BLAU,GE (1985)
DC	pKa			
HL 5.73E-010	atm m3/mol	25	EST	MEYLAN,WM & HOWARD,PH (1991)
OH 1.35E-010	cm3/molc sec	25	EST	MEYLAN,WM & HOWARD,PH (1993)

000132-22-9 — CHLORPHENIRAMINE

CAS #: 000132-22-9

Formula: $C_{16}H_{19}ClN_2$

Mol Weight: 274.80

MP (deg C):

FP (deg C):

BP (deg C): 142

BP pressure (mm Hg): 1.00E+000

Property/Value	Units	Temp	Data Type	Reference
WS 4.53E+003	mg/L	25	EST	MEYLAN,WM ET AL. (1996)
logP 3.38			EXP	HANSCH,C & LEO,AJ (1985)
VP 1.12E-005	mm Hg	25	EST	NEELY,WB & BLAU,GE (1985)
DC 9.13	pKa	25	EXP	PERRIN,DD (1965)
HL 4.07E-010	atm m3/mol	25	EST	MEYLAN,WM & HOWARD,PH (1991)
OH 8.55E-011	cm3/molc sec	25	EST	MEYLAN,WM & HOWARD,PH (1993)

000132-27-4 — 2-PHENYLPHENOL, SODIUM SALT

CAS #: 000132-27-4

Formula: $C_{12}H_9NaO$

Mol Weight: 192.19

MP (deg C):

FP (deg C):

BP (deg C):

BP pressure (mm Hg):

Property/Value	Units	Temp	Data Type	Reference
WS 1.10E+006	mg/L	25	EXP	KENAGA,EE (1980)
logP 0.59			EST	MEYLAN,WM & HOWARD,PH (1995)
VP 1.91E-011	mm Hg	25	EST	NEELY,WB & BLAU,GE (1985)
DC	pKa			
HL	atm m3/mol			
OH 1.87E-011	cm3/molc sec	25	EST	MEYLAN,WM & HOWARD,PH (1993)

000132-53-6 — 2-NITROSO-1-NAPHTHOL

CAS #: 000132-53-6

Formula: $C_{10}H_7NO_2$

Mol Weight: 173.17

MP (deg C): 158 dec

FP (deg C):

BP (deg C):

BP pressure (mm Hg):

Property/Value	Units	Temp	Data Type	Reference
WS 1.79E+003	mg/L	25	EST	MEYLAN,WM ET AL. (1996)
logP 2.46			EXP	HANSCH,C ET AL. (1995)
VP 9.11E-005	mm Hg	25	EST	NEELY,WB & BLAU,GE (1985)
DC	pKa -			
HL 5.73E-010	atm m3/mol	25	EST	MEYLAN,WM & HOWARD,PH (1991)
OH 1.35E-010	cm3/molc sec	25	EST	MEYLAN,WM & HOWARD,PH (1993)

000132-60-5 — 2-PHENYLCINCHONINIC ACID

CAS #: 000132-60-5

Formula: $C_{16}H_{11}NO_2$

Mol Weight: 249.27

MP (deg C): 213-216

FP (deg C):

BP (deg C):

BP pressure (mm Hg):

Property/Value	Units	Temp	Data Type	Reference
WS 1.60E+002	mg/L	25	EXP	YALKOWSKY,SH & DANNENFELSER,RM (1992)
logP 3.79			EST	MEYLAN,WM & HOWARD,PH (1995)
VP 1.55E-008	mm Hg	25	EST	NEELY,WB & BLAU,GE (1985)
DC	pKa			
HL 1.06E-012	atm m3/mol	25	EST	MEYLAN,WM & HOWARD,PH (1991)
OH 9.01E-012	cm3/molc sec	25	EST	MEYLAN,WM & HOWARD,PH (1993)

000132-64-9 — DIBENZOFURAN

CAS #: 000132-64-9

Formula: $C_{12}H_8O$

Mol Weight: 168.20

MP (deg C): 86.5

FP (deg C):

BP (deg C): 287

BP pressure (mm Hg):

Property/Value	Units	Temp	Data Type	Reference
WS 4.22E+000	mg/L	25	EXP	DOUCETTE,WJ & ANDREN,AW (1988)
logP 4.12			EXP	HANSCH,C & LEO,AJ (1985)
VP 2.48E-003	mm Hg	25	EXP	CHIRICO,RD ET AL. (1990)
DC	pKa			
HL 2.13E-004	atm m3/mol	25	EST	VP/WSOL
OH 3.91E-012	cm3/molc sec	25	EXP	KWOK,ESC ET AL. (1994A)

DIBENZOTHIOPHENE

CAS #: 000132-65-0

Formula: $C_{12}H_8S$

Mol Weight: 184.26

MP (deg C): 99.5

FP (deg C):

BP (deg C): 332.5

BP pressure (mm Hg):

Property/Value	Units	Temp	Data Type	Reference
WS 1.47E+000	mg/L	25	EXP	HASSETT,JJ ET AL. (1980A)
logP 4.38			EXP	HANSCH,C & LEO,AJ (1985)
VP 2.04E-006	mm Hg	25	EXP	AUBRY,M ET AL. (1975)
DC	pKa			
HL 3.37E-007	atm m3/mol	25	EST	VP/WSOL
OH 3.00E-011	cm3/molc sec	25	EST	MEYLAN,WM & HOWARD,PH (1993)

NPA

CAS #: 000132-66-1

Formula: $C_{18}H_{13}NO_3$

Mol Weight: 291.31

MP (deg C): 203

FP (deg C):

BP (deg C):

BP pressure (mm Hg):

Property/Value	Units	Temp	Data Type	Reference
WS 2.00E+002	mg/L	25	EXP	YALKOWSKY,SH & DANNENFELSER,RM (1992)
logP 3.71			EST	MEYLAN,WM & HOWARD,PH (1995)
VP 2.86E-011	mm Hg	25	EST	NEELY,WB & BLAU,GE (1985)
DC	pKa			
HL 2.40E-015	atm m3/mol	25	EST	MEYLAN,WM & HOWARD,PH (1991)
OH 1.39E-010	cm3/molc sec	25	EST	MEYLAN,WM & HOWARD,PH (1993)

N-1-NAPHTHYLPHTHALMIC ACID

CAS #: 000132-67-2

Formula: $C_{18}H_{13}NNaO_3$

Mol Weight: 314.30

MP (deg C):

FP (deg C):

BP (deg C):

BP pressure (mm Hg):

Property/Value	Units	Temp	Data Type	Reference
WS 2.31E+005	mg/L	25	EXP	SHIU,WY ET AL. (1990)
logP -0.39			EST	MEYLAN,WM & HOWARD,PH (1995)
VP 1.46E-017	mm Hg	25	EST	NEELY,WB & BLAU,GE (1985)
DC	pKa			
HL	atm m3/mol			
OH 1.38E-010	cm3/molc sec	25	EST	MEYLAN,WM & HOWARD,PH (1993)

NAPHTHALENE-1,3-DIOL

CAS #: 000132-86-5

Formula: $C_{10}H_8O_2$

Mol Weight: 160.17

MP (deg C): 124-125

FP (deg C):

BP (deg C):

BP pressure (mm Hg):

Property/Value	Units	Temp	Data Type	Reference
WS 1.82E+007	mg/L	20	EXP	KORENMAN,YI ET AL. (1980)
logP 1.97			EXP	HANSCH,C & LEO,AJ (1985)
VP 7.15E-006	mm Hg	25	EST	NEELY,WB & BLAU,GE (1985)
DC	pKa			
HL 5.69E-012	atm m3/mol	25	EST	MEYLAN,WM & HOWARD,PH (1991)
OH 2.00E-010	cm3/molc sec	25	EST	MEYLAN,WM & HOWARD,PH (1993)

CAPTAN

CAS #: 000133-06-2

Formula: $C_9H_8Cl_3NO_2S$

Mol Weight: 300.59

MP (deg C): 178

FP (deg C):

BP (deg C):

BP pressure (mm Hg):

Property/Value	Units	Temp	Data Type	Reference
WS 3.30E+000	mg/L	25	EXP	MIDWEST RESEARCH INSTITUTE (1975B)
logP 2.35			EXP	HANSCH,C & LEO,AJ (1985)
VP 6.00E-005	mm Hg	25	EXP	GILE,JD & GILLETT,JW (1979)
DC	pKa			
HL 7.19E-006	atm m3/mol	25	EST	VP/WSOL
OH 8.84E-011	cm3/molc sec	25	EST	MEYLAN,WM & HOWARD,PH (1993)

FOLPET

CAS #: 000133-07-3

Formula: $C_9H_4Cl_3NO_2S$

Mol Weight: 296.56

MP (deg C): 177

FP (deg C):

BP (deg C):

BP pressure (mm Hg):

Property/Value	Units	Temp	Data Type	Reference
WS 1.00E+000	mg/L	20	EXP	FURER,R & GEIGER,M (1977)
logP 2.85			EXP	HANSCH,C & LEO,AJ (1985)
VP 9.75E-006	mm Hg	20	EST	WORTHING,CR & WALKER,SB (1987)
DC	pKa			
HL 3.81E-006	atm m3/mol	20	EST	VP/WSOL
OH 1.43E-011	cm3/molc sec	25	EST	MEYLAN,WM & HOWARD,PH (1993)

PHENYL-4-AMINOSALICYLATE

CAS #: 000133-11-9

Formula: $C_{13}H_{11}NO_3$

Mol Weight: 229.24

MP (deg C): 153

FP (deg C):

BP (deg C):

BP pressure (mm Hg):

Property/Value	Units	Temp	Data Type	Reference
WS 6.27E+001	mg/L	25	EST	MEYLAN,WM ET AL. (1996)
logP 3.15			EXP	HANSCH,C & LEO,AJ (1985)
VP 2.68E-007	mm Hg	25	EST	NEELY,WB & BLAU,GE (1985)
DC	pKa			
HL 5.94E-010	atm m3/mol	25	EST	MEYLAN,WM & HOWARD,PH (1991)
OH 2.00E-010	cm3/molc sec	25	EST	MEYLAN,WM & HOWARD,PH (1993)

CHLOROPROCAINE

CAS #: 000133-16-4

Formula: $C_{13}H_{19}ClN_2O_2$

Mol Weight: 270.76

MP (deg C):

FP (deg C):

BP (deg C):

BP pressure (mm Hg):

Property/Value	Units	Temp	Data Type	Reference
WS 6.65E+002	mg/L	25	EST	MEYLAN,WM ET AL. (1996)
logP 2.86			EXP	HANSCH,C ET AL. (1995)
VP 5.29E-006	mm Hg	25	EST	NEELY,WB & BLAU,GE (1985)
DC	pKa			
HL 1.05E-011	atm m3/mol	25	EST	MEYLAN,WM & HOWARD,PH (1991)
OH 1.21E-010	cm3/molc sec	25	EST	MEYLAN,WM & HOWARD,PH (1993)

CAS #: 000133-32-4	3-INDOLEBUTYRIC ACID

Formula: $C_{12}H_{13}NO_2$

Mol Weight: 203.24

MP (deg C): 123-125 FP (deg C):

BP (deg C):

BP pressure (mm Hg):

Property/Value	Units	Temp	Data Type	Reference
WS 1.14E+003	mg/L	25	EST	MEYLAN,WM ET AL. (1996)
logP 2.30			EXP	HANSCH,C & LEO,AJ (1985)
VP 1.17E-006	mm Hg	25	EST	NEELY,WB & BLAU,GE (1985)
DC	pKa			
HL 1.28E-011	atm m3/mol	25	EST	MEYLAN,WM & HOWARD,PH (1991)
OH 2.04E-010	cm3/molc sec	25	EST	MEYLAN,WM & HOWARD,PH (1993)

CAS #: 000133-43-7	DL-MANNITOL

Formula: $C_6H_{14}O_6$

Mol Weight: 182.17

MP (deg C): FP (deg C):

BP (deg C):

BP pressure (mm Hg):

Property/Value	Units	Temp	Data Type	Reference
WS 1.00E+006	mg/L	25	EST	MEYLAN,WM ET AL. (1996)
logP -2.47			EXP	SANGSTER,J (1994)
VP 4.92E-009	mm Hg	25	EST	NEELY,WB & BLAU,GE (1985)
DC	pKa			
HL 7.26E-013	atm m3/mol	25	EST	MEYLAN,WM & HOWARD,PH (1991)
OH 5.00E-011	cm3/molc sec	25	EST	MEYLAN,WM & HOWARD,PH (1993)

CAS #: 000133-59-5	O-TOLUENESULFONYL CHLORIDE

Formula: $C_7H_7ClO_2S$

Mol Weight: 190.65

MP (deg C): 10.2 FP (deg C):

BP (deg C): 154

BP pressure (mm Hg): 3.60E+001

Property/Value	Units	Temp	Data Type	Reference
WS 5.12E+001	mg/L	25	EST	MEYLAN,WM ET AL. (1996)
logP 3.49			EST	MEYLAN,WM & HOWARD,PH (1995)
VP 3.75E+000	mm Hg	20	EXP	WEBER,RC ET AL. (1981)
DC	pKa			
HL 9.67E-006	atm m3/mol	25	EST	MEYLAN,WM & HOWARD,PH (1991)
OH 1.22E-012	cm3/molc sec	25	EST	MEYLAN,WM & HOWARD,PH (1993)

CAS #: 000133-67-5	TRICHLORMETHIAZIDE

Formula: $C_8H_8Cl_3N_3O_4S_2$

Mol Weight: 380.66

MP (deg C): 270 dec FP (deg C):

BP (deg C):

BP pressure (mm Hg):

Property/Value	Units	Temp	Data Type	Reference
WS 1.06E+002	mg/L	25	EST	MEYLAN,WM ET AL. (1996)
logP 0.62			EXP	SANGSTER,J (1994)
VP 3.59E-011	mm Hg	25	EST	NEELY,WB & BLAU,GE (1985)
DC	pKa			
HL 7.23E-013	atm m3/mol	25	EST	MEYLAN,WM & HOWARD,PH (1991)
OH 6.21E-011	cm3/molc sec	25	EST	MEYLAN,WM & HOWARD,PH (1993)

CAS #: 000133-74-4	4-CHLOROANILINE-2-SULFONIC ACID

Formula: $C_6H_6ClNO_3S$

Mol Weight: 207.64

MP (deg C): FP (deg C):

BP (deg C):

BP pressure (mm Hg):

Property/Value	Units	Temp	Data Type	Reference
WS 3.13E+003	mg/L	0	EXP	STEPHEN,H & STEPHEN,T (1963)
logP -1.44			EST	MEYLAN,WM & HOWARD,PH (1995)
VP 4.37E-008	mm Hg	25	EST	NEELY,WB & BLAU,GE (1985)
DC	pKa			
HL 6.59E-013	atm m3/mol	25	EST	MEYLAN,WM & HOWARD,PH (1991)
OH 6.86E-012	cm3/molc sec	25	EST	MEYLAN,WM & HOWARD,PH (1993)

CAS #: 000133-90-4	3-AMINO-2,5-DICHLOROBENZOIC ACID

Formula: $C_7H_5Cl_2NO_2$

Mol Weight: 206.03

MP (deg C): 200-201 FP (deg C):

BP (deg C): 312

BP pressure (mm Hg):

Property/Value	Units	Temp	Data Type	Reference
WS 7.00E+002	mg/L	25	EXP	MARTIN,H & WORTHING,CR (1977)
logP 1.11			EXP	HANSCH,C & LEO,AJ (1985)
VP 1.00E-007	mm Hg	25	EST	CLAUSIUS-CLAPEYRON EQN
DC 3.40	pKa		EST	WEBER,JB (1972)
HL 3.87E-007	atm m3/mol	25	EST	VP/WSOL
OH 2.68E-011	cm3/molc sec	25	EST	MEYLAN,WM & HOWARD,PH (1993)

CAS #: 000133-91-5	3,5-DIIODOSALICYLIC ACID

Formula: $C_7H_4I_2O_3$

Mol Weight: 389.92

MP (deg C): 235-236 FP (deg C):

BP (deg C):

BP pressure (mm Hg):

Property/Value	Units	Temp	Data Type	Reference
WS 1.92E+002	mg/L	25	EXP	YALKOWSKY,SH & DANNENFELSER,RM (1992)
logP 4.56			EXP	HANSCH,C & LEO,AJ (1985)
VP 5.93E-008	mm Hg	25	EST	NEELY,WB & BLAU,GE (1985)
DC 2.30	pKa	25	EXP	SERJEANT,EP & DEMPSEY,B (1979)
HL 7.63E-010	atm m3/mol	25	EST	MEYLAN,WM & HOWARD,PH (1991)
OH 8.19E-013	cm3/molc sec	25	EST	MEYLAN,WM & HOWARD,PH (1993)

CAS #: 000134-20-3	METHYL ANTHRANILATE

Formula: $C_8H_9NO_2$

Mol Weight: 151.17

MP (deg C): 24-25 FP (deg C):

BP (deg C): 135.5

BP pressure (mm Hg): 1.50E+001

Property/Value	Units	Temp	Data Type	Reference
WS 1.86E+003	mg/L	25	EST	MEYLAN,WM ET AL. (1996)
logP 1.88			EXP	HANSCH,C ET AL. (1995)
VP 2.71E-002	mm Hg	25	EXT	PERRY,RH & GREEN,D (1984)
DC 2.23	pKa	25	EXP	PERRIN,DD (1965)
HL 1.23E-008	atm m3/mol	25	EST	MEYLAN,WM & HOWARD,PH (1991)
OH 3.48E-011	cm3/molc sec	25	EST	MEYLAN,WM & HOWARD,PH (1993)

000134-31-6 — OXYQUINOLINESULPHATE

Formula: $C_{18}H_{16}N_2O_6S$
Mol Weight: 388.40
MP (deg C): 175-178
FP (deg C):
BP (deg C):
BP pressure (mm Hg):

Property/Value	Units	Temp	Data Type	Reference
WS 1.00E+006	mg/L		EXP	SHIU,WY ET AL. (1990)
logP 1.73			EST	MEYLAN,WM & HOWARD,PH (1995)
VP 2.10E-015	mm Hg	25	EST	NEELY,WB & BLAU,GE (1985)
DC	pKa			
HL	atm m3/mol			
OH 2.20E-010	cm3/molc sec	25	EST	MEYLAN,WM & HOWARD,PH (1993)

000134-32-7 — 1-NAPHTHYLAMINE

Formula: $C_{10}H_9N$
Mol Weight: 143.19
MP (deg C): 50
FP (deg C):
BP (deg C): 301
BP pressure (mm Hg):

Property/Value	Units	Temp	Data Type	Reference
WS 1.70E+003	mg/L	20	EXP	DRAGUN,J & HELLING,CS (1985)
logP 2.25			EXP	HANSCH,C & LEO,AJ (1985)
VP 4.18E-003	mm Hg	25	EXT	CHAO,J ET AL. (1983)
DC 3.92	pKa	25	EXP	WEAST,RC (1985)
HL 4.64E-007	atm m3/mol	25	EST	VP/WSOL
OH 2.20E-010	cm3/molc sec	25	EST	MEYLAN,WM & HOWARD,PH (1993)

000134-55-4 — BENZOIC ACID, 2-(ACETYLOXY)-, PHENYL ESTER

Formula: $C_{15}H_{12}O_4$
Mol Weight: 256.26
MP (deg C): 97
FP (deg C):
BP (deg C): 198
BP pressure (mm Hg): 1.10E+001

Property/Value	Units	Temp	Data Type	Reference
WS 7.58E+001	mg/L	25	EST	MEYLAN,WM ET AL. (1996)
logP 2.88			EXP	NIELSEN,LS & BUNDGAARD,H (1989)
VP 2.16E-004	mm Hg	25	EST	NEELY,WB & BLAU,GE (1985)
DC	pKa			
HL 1.54E-007	atm m3/mol	25	EST	MEYLAN,WM & HOWARD,PH (1991)
OH 3.94E-012	cm3/molc sec	25	EST	MEYLAN,WM & HOWARD,PH (1993)

000134-58-7 — 8-AZAGUANINE

Formula: $C_4H_4N_6O$
Mol Weight: 152.12
MP (deg C): 300
FP (deg C):
BP (deg C):
BP pressure (mm Hg):

Property/Value	Units	Temp	Data Type	Reference
WS 1.47E+004	mg/L	25	EST	MEYLAN,WM ET AL. (1996)
logP -0.71			EXP	HANSCH,C & LEO,AJ (1985)
VP 8.37E-010	mm Hg	25	EST	NEELY,WB & BLAU,GE (1985)
DC	pKa			
HL 1.05E-017	atm m3/mol	25	EST	MEYLAN,WM & HOWARD,PH (1991)
OH 2.66E-011	cm3/molc sec	25	EST	MEYLAN,WM & HOWARD,PH (1993)

000134-62-3 — DEET [N,N,-DIET-3-ME BENZAMIDE]

Formula: $C_{12}H_{17}NO$
Mol Weight: 191.28
MP (deg C):
FP (deg C):
BP (deg C): 288-292
BP pressure (mm Hg):

Property/Value	Units	Temp	Data Type	Reference
WS 9.12E+002	mg/L	25	EST	MEYLAN,WM ET AL. (1996)
logP 2.02			EXP	HANSCH,C ET AL. (1995)
VP 5.60E-003	mm Hg	20	EXP	LITTLE,AD (1982)
DC	pKa			
HL 2.08E-008	atm m3/mol	25	EST	MEYLAN,WM & HOWARD,PH (1991)
OH 2.53E-011	cm3/molc sec	25	EST	MEYLAN,WM & HOWARD,PH (1993)

000134-81-6 — BENZIL

Formula: $C_{14}H_{10}O_2$
Mol Weight: 210.23
MP (deg C): 95
FP (deg C):
BP (deg C): 346-348
BP pressure (mm Hg):

Property/Value	Units	Temp	Data Type	Reference
WS 5.03E+001	mg/L	25	EST	MEYLAN,WM ET AL. (1996)
logP 3.38			EXP	HANSCH,C & LEO,AJ (1985)
VP 2.38E-004	mm Hg	25	EXT	PERRY,RH & GREEN,D (1984)
DC	pKa			
HL 7.73E-009	atm m3/mol	25	EST	MEYLAN,WM & HOWARD,PH (1991)
OH 3.55E-012	cm3/molc sec	25	EST	MEYLAN,WM & HOWARD,PH (1993)

000134-98-5 — 2,3-DIMETHYLACETANILIDE

Formula: $C_{10}H_{13}NO$
Mol Weight: 163.22
MP (deg C):
FP (deg C):
BP (deg C):
BP pressure (mm Hg):

Property/Value	Units	Temp	Data Type	Reference
WS 4.65E+003	mg/L	25	EST	MEYLAN,WM ET AL. (1996)
logP 1.35			EXP	NAKAGAWA,Y ET AL. (1992)
VP 6.74E-005	mm Hg	25	EST	NEELY,WB & BLAU,GE (1985)
DC	pKa			
HL 7.52E-009	atm m3/mol	25	EST	MEYLAN,WM & HOWARD,PH (1991)
OH 4.01E-011	cm3/molc sec	25	EST	MEYLAN,WM & HOWARD,PH (1993)

000135-01-3 — O-DIETHYLBENZENE

Formula: $C_{10}H_{14}$
Mol Weight: 134.22
MP (deg C): -31.2
FP (deg C):
BP (deg C): 184
BP pressure (mm Hg):

Property/Value	Units	Temp	Data Type	Reference
WS 7.11E+001	mg/L	20	EXP	YALKOWSKY,SH & DANNENFELSER,RM (1992)
logP 4.42			EXP	SHERBLOM,PM & EGANHOUSE,RP (1988)
VP 1.05E+000	mm Hg	25	EXP	DAUBERT,TE & DANNER,RP (1989)
DC	pKa			
HL 2.61E-003	atm m3/mol	25	EST	VP/WSOL
OH 8.11E-012	cm3/molc sec	25	EST	MEYLAN,WM & HOWARD,PH (1993)

262

CAS #: 000135-02-4 — BENZALDEHYDE, 2-METHOXY-

Formula: $C_8H_8O_2$

Mol Weight: 136.15

MP (deg C): 37.5

BP (deg C): 243.5

BP pressure (mm Hg):

Property/ Value	Units	Temp	Data Type	Reference
WS 2.95E+003	mg/L	25	EST	MEYLAN,WM ET AL. (1996)
logP 1.72			EXP	BAZACO,JF & COCA,CM (1989)
VP 1.21E-001	mm Hg		EST	NEELY,WB & BLAU,GE (1985)
DC	pKa			
HL 7.94E-007	atm m3/mol	25	EST	MEYLAN,WM & HOWARD,PH (1991)
OH 2.48E-011	cm3/molc sec	25	EST	MEYLAN,WM & HOWARD,PH (1993)

CAS #: 000135-07-9 — 2-METHYL-3-CHLOROMETHYLHYDROCHLOROTHIAZIDE

Formula: $C_9H_{11}Cl_2N_3O_4S_2$

Mol Weight: 360.24

MP (deg C): 225

BP (deg C):

BP pressure (mm Hg):

Property/ Value	Units	Temp	Data Type	Reference
WS 2.93E+001	mg/L	25	EST	MEYLAN,WM ET AL. (1996)
logP 1.42			EXP	SANGSTER,J (1993)
VP 1.17E-010	mm Hg	25	EST	NEELY,WB & BLAU,GE (1985)
DC	pKa			
HL 4.50E-012	atm m3/mol	25	EST	MEYLAN,WM & HOWARD,PH (1991)
OH 6.35E-011	cm3/molc sec	25	EST	MEYLAN,WM & HOWARD,PH (1993)

CAS #: 000135-09-1 — 3,4-DIHYDRO-6-(TRIFLUOROMETHYL)-2H(1,2,4)-BENZO*

Formula: $C_8H_8F_3N_3O_4S_2$

Mol Weight: 331.29

MP (deg C): 272-273

BP (deg C):

BP pressure (mm Hg):

Property/ Value	Units	Temp	Data Type	Reference
WS 3.52E+002	mg/L	25	EST	MEYLAN,WM ET AL. (1996)
logP 0.36			EXP	HANSCH,C ET AL. (1995)
VP 3.87E-009	mm Hg	25	EST	NEELY,WB & BLAU,GE (1985)
DC	pKa			
HL 5.15E-011	atm m3/mol	25	EST	MEYLAN,WM & HOWARD,PH (1991)
OH 8.11E-011	cm3/molc sec	25	EST	MEYLAN,WM & HOWARD,PH (1993)

CAS #: 000135-19-3 — 2-NAPHTHOL

Formula: $C_{10}H_8O$

Mol Weight: 144.17

MP (deg C): 121.6

BP (deg C): 285-286

BP pressure (mm Hg):

Property/ Value	Units	Temp	Data Type	Reference
WS 7.56E+002	mg/L	25	EXP	YALKOWSKY,SH & DANNENFELSER,RM (1992)
logP 2.70			EXP	HANSCH,C ET AL. (1995)
VP 3.20E-004	mm Hg	25	EXT	BOUBLIK,T ET AL. (1984)
DC 9.51	pKa	25	EXP	BHAITACHARYYA,D ET AL. (1986)
HL 4.60E-008	atm m3/mol	25	EST	MEYLAN,WH & HOWARD,PH (1991)
OH 2.00E-010	cm3/molc sec	25	EST	MEYLAN,WM & HOWARD,PH (1993)

CAS #: 000135-20-6 — N-HYDROXY-N-NITROSOBENZENAMINE AMMONIUM SALT

Formula: $C_6H_9N_3O_2$

Mol Weight: 155.16

MP (deg C): 163-164

BP (deg C):

BP pressure (mm Hg):

Property/ Value	Units	Temp	Data Type	Reference
WS 6.08E+005	mg/L	25	EST	MEYLAN,WM ET AL. (1996)
logP -1.73			EST	MEYLAN,WM & HOWARD,PH (1995)
VP 6.29E-005	mm Hg	25	EST	NEELY,WB & BLAU,GE (1985)
DC	pKa			
HL 3.62E-009	atm m3/mol	25	EST	MEYLAN,WM & HOWARD,PH (1991)
OH 3.34E-011	cm3/molc sec	25	EST	MEYLAN,WM & HOWARD,PH (1993)

CAS #: 000135-48-8 — PENTACENE

Formula: $C_{22}H_{14}$

Mol Weight: 278.36

MP (deg C): 257

BP (deg C):

BP pressure (mm Hg):

Property/ Value	Units	Temp	Data Type	Reference
WS 1.08E-003	mg/L	25	EST	MEYLAN,WM ET AL. (1996)
logP 7.11			EXP	DE VOOGT,P ET AL. (1990)
VP 3.73E-009	mm Hg	25	EST	NEELY,WB & BLAU,GE (1985)
DC	pKa			
HL 4.89E-007	atm m3/mol	25	EST	MEYLAN,WM & HOWARD,PH (1991)
OH 5.00E-011	cm3/molc sec	25	EST	MEYLAN,WM & HOWARD,PH (1993)

CAS #: 000135-67-1 — PHENOXAZINE

Formula: $C_{12}H_9NO$

Mol Weight: 183.21

MP (deg C): 156

BP (deg C):

BP pressure (mm Hg):

Property/ Value	Units	Temp	Data Type	Reference
WS 2.13E+000	mg/L	25	EST	MEYLAN,WM ET AL. (1996)
logP 3.85			EXP	HANSCH,C & LEO,AJ (1985)
VP 1.38E-004	mm Hg	25	EST	NEELY,WB & BLAU,GE (1985)
DC	pKa			
HL 1.04E-007	atm m3/mol	25	EST	MEYLAN,WM & HOWARD,PH (1991)
OH 2.00E-010	cm3/molc sec	25	EST	MEYLAN,WM & HOWARD,PH (1993)

CAS #: 000135-88-6 — N-PHENYL-2-NAPHTHYLAMINE

Formula: $C_{16}H_{13}N$

Mol Weight: 219.29

MP (deg C): 108

BP (deg C): 395.5

BP pressure (mm Hg):

Property/ Value	Units	Temp	Data Type	Reference
WS 6.31E+000	mg/L	25	EST	MEYLAN,WM ET AL. (1996)
logP 4.38			EXP	HANSCH,C & LEO,AJ (1985)
VP 8.29E-006	mm Hg	25	EST	NEELY,WB & BLAU,GE (1985)
DC	pKa			
HL 1.03E-007	atm m3/mol	25	EST	MEYLAN,WM & HOWARD,PH (1991)
OH 4.18E-010	cm3/molc sec	25	EST	MEYLAN,WM & HOWARD,PH (1993)

SEC-BUTYLBENZENE

CAS #:	000135-98-8		SEC-BUTYLBENZENE
Formula:	$C_{10}H_{14}$		
Mol Weight:	134.22		
MP (deg C): -82.7		FP (deg C):	
BP (deg C): 173.5			
BP pressure (mm Hg):			

Property/Value	Units	Temp	Data Type	Reference
WS 1.76E+001	mg/L	25	EXP	YALKOWSKY,SH & DANNENFELSER,RM (1992)
logP 4.57			EXP	SHERBLOM,PM & EGANHOUSE,RP (1988)
VP 1.75E+000	mm Hg	25	EXP	YAWS,CL (1994B)
DC	pKa			
HL 1.76E-002	atm m3/mol	25	EST	VP/WSOL
OH 8.50E-012	cm3/molc sec	25	EST	MEYLAN,WM & HOWARD,PH (1993)

ERBON

CAS #:	000136-25-4		ERBON
Formula:	$C_{11}H_9Cl_5O_3$		
Mol Weight:	366.46		
MP (deg C): 49-50		FP (deg C):	
BP (deg C): 161-164			
BP pressure (mm Hg): 5.00E-001			

Property/Value	Units	Temp	Data Type	Reference
WS 7.63E-002	mg/L	25	EST	MEYLAN,WM ET AL. (1996)
logP 5.63			EST	MEYLAN,WM & HOWARD,PH (1995)
VP 5.49E-006	mm Hg	25	EST	NEELY,WB & BLAU,GE (1985)
DC	pKa			
HL 6.75E-008	atm m3/mol	25	EST	MEYLAN,WM & HOWARD,PH (1991)
OH 9.25E-012	cm3/molc sec	25	EST	MEYLAN,WM & HOWARD,PH (1993)

1,3-DIPHENYL-1-TRIAZENE

CAS #:	000136-35-6		1,3-DIPHENYL-1-TRIAZENE
Formula:	$C_{12}H_{11}N_3$		
Mol Weight:	197.24		
MP (deg C): 98		FP (deg C):	
BP (deg C):			
BP pressure (mm Hg):			

Property/Value	Units	Temp	Data Type	Reference
WS 5.00E+002	mg/L		EXP	YALKOWSKY,SH & DANNENFELSER,RM (1992)
logP 3.99			EST	MEYLAN,WM & HOWARD,PH (1995)
VP 1.91E-005	mm Hg	25	EXT	YAWS,CL (1994B)
DC	pKa			
HL 6.14E-008	atm m3/mol	25	EST	MEYLAN,WM & HOWARD,PH (1991)
OH 4.34E-011	cm3/molc sec	25	EST	MEYLAN,WM & HOWARD,PH (1993)

DIPROPYL PYRIDINE-2,5-DICARBOXYLATE

CAS #:	000136-45-8		DIPROPYL PYRIDINE-2,5-DICARBOXYLATE
Formula:	$C_{13}H_{17}NO_4$		
Mol Weight:	251.28		
MP (deg C):		FP (deg C):	
BP (deg C):			
BP pressure (mm Hg):			

Property/Value	Units	Temp	Data Type	Reference
WS 2.08E+001	mg/L	25	EST	MEYLAN,WM ET AL. (1996)
logP 3.57			EXP	TOMLIN,C (1994)
VP 5.48E-004	mm Hg	25	EST	NEELY,WB & BLAU,GE (1985)
DC	pKa			
HL 9.09E-010	atm m3/mol	25	EST	MEYLAN,WM & HOWARD,PH (1991)
OH 6.34E-012	cm3/molc sec	25	EST	MEYLAN,WM & HOWARD,PH (1993)

BUTYL BENZOATE

CAS #:	000136-60-7		BUTYL BENZOATE
Formula:	$C_{11}H_{14}O_2$		
Mol Weight:	178.23		
MP (deg C): -22		FP (deg C):	
BP (deg C): 250			
BP pressure (mm Hg):			

Property/Value	Units	Temp	Data Type	Reference
WS 5.90E+001	mg/L	25	EXP	HAAS,JM ET AL. (1975)
logP 3.84			EXP	HANSCH,C ET AL. (1995)
VP 1.00E-002	mm Hg	25	EXT	KATAYAMA,H (1988)
DC	pKa			
HL 3.97E-005	atm m3/mol	25	EST	VP/WSOL
OH 4.94E-012	cm3/molc sec	25	EST	ATKINSON,R (1988)

4-HEXYLRESORCINOL

CAS #:	000136-77-6		4-HEXYLRESORCINOL
Formula:	$C_{12}H_{18}O_2$		
Mol Weight:	194.28		
MP (deg C): 67.5-69		FP (deg C):	
BP (deg C): 333-335			
BP pressure (mm Hg):			

Property/Value	Units	Temp	Data Type	Reference
WS 5.00E+002	mg/L	18	EXP	YALKOWSKY,SH & DANNENFELSER,RM (1992)
logP 3.45			EXP	HANSCH,C & LEO,AJ (1985)
VP 8.45E-006	mm Hg	25	EST	NEELY,WB & BLAU,GE (1985)
DC	pKa			
HL 2.65E-010	atm m3/mol	25	EST	MEYLAN,WM & HOWARD,PH (1991)
OH 2.07E-010	cm3/molc sec	25	EST	MEYLAN,WM & HOWARD,PH (1993)

SESONE

CAS #:	000136-78-7		SESONE
Formula:	$C_8H_8Cl_2NaO_5S$		
Mol Weight:	310.11		
MP (deg C): 245		FP (deg C):	
BP (deg C):			
BP pressure (mm Hg):			

Property/Value	Units	Temp	Data Type	Reference
WS 2.50E+005	mg/L	25	EXP	GUNTHER,FA ET AL. (1968)
logP -0.69			EST	MEYLAN,WM & HOWARD,PH (1995)
VP 1.30E-013	mm Hg	25	EST	NEELY,WB & BLAU,GE (1985)
DC	pKa			
HL	atm m3/mol			
OH 9.65E-012	cm3/molc sec	25	EST	MEYLAN,WM & HOWARD,PH (1993)

PHENOL, 2-NONYL-

CAS #:	000136-83-4		PHENOL, 2-NONYL-
Formula:	$C_{15}H_{24}O$		
Mol Weight:	220.36		
MP (deg C):		FP (deg C):	
BP (deg C):			
BP pressure (mm Hg):			

Property/Value	Units	Temp	Data Type	Reference
WS 1.57E+000	mg/L	25	EST	MEYLAN,WM ET AL. (1996)
logP 5.76			EXP	ITOKAWA,H ET AL. (1989)
VP 3.58E-005	mm Hg	25	EST	NEELY,WB & BLAU,GE (1985)
DC	pKa			
HL 5.97E-006	atm m3/mol	25	EST	MEYLAN,WM & HOWARD,PH (1991)
OH 5.17E-011	cm3/molc sec	25	EST	MEYLAN,WM & HOWARD,PH (1993)

CAS #: 000137-05-3			METHYL 2-CYANO-2-PROPENOATE
Formula: $C_5H_5NO_2$			
Mol Weight: 111.10			
MP (deg C):	FP (deg C):		
BP (deg C): 47-49			
BP pressure (mm Hg): 1.80E+000			

Property/Value	Units	Temp	Data Type	Reference
WS 9.48E+003	mg/L	25	EST	MEYLAN,WM ET AL. (1996)
logP 0.93			EST	MEYLAN,WM & HOWARD,PH (1995)
VP 7.96E-001	mm Hg	25	EST	NEELY,WB & BLAU,GE (1985)
DC	pKa			
HL 2.06E-007	atm m3/mol	25	EST	MEYLAN,WM & HOWARD,PH (1991)
OH 3.10E-012	cm3/molc sec	25	EST	MEYLAN,WM & HOWARD,PH (1993)

CAS #: 000137-06-4			2-METHYLBENZENETHIOL
Formula: C_7H_8S			
Mol Weight: 124.21			
MP (deg C): 15	FP (deg C):		
BP (deg C): 195			
BP pressure (mm Hg):			

Property/Value	Units	Temp	Data Type	Reference
WS 1.68E+002	mg/L	25	EST	MEYLAN,WM ET AL. (1996)
logP 3.23			EST	MEYLAN,WM & HOWARD,PH (1995)
VP 6.42E-001	mm Hg	25	EXT	CHAO,J ET AL. (1983)
DC	pKa			
HL 5.76E-004	atm m3/mol	25	EST	MEYLAN,WM & HOWARD,PH (1991)
OH 1.40E-011	cm3/molc sec	25	EST	MEYLAN,WM & HOWARD,PH (1993)

CAS #: 000137-07-5			O-AMINOTHIOPHENOL
Formula: C_6H_7NS			
Mol Weight: 125.19			
MP (deg C): 26	FP (deg C):		
BP (deg C): 234			
BP pressure (mm Hg):			

Property/Value	Units	Temp	Data Type	Reference
WS 5.77E+003	mg/L	25	EST	MEYLAN,WM ET AL. (1996)
logP 1.43			EXP	HANSCH,C & LEO,AJ (1985)
VP 2.44E-002	mm Hg	25	EST	NEELY,WB & BLAU,GE (1985)
DC 3.00	pKa	20	EXP	PERRIN,DD (1965)
HL 1.85E-007	atm m3/mol	25	EST	MEYLAN,WM & HOWARD,PH (1991)
OH 5.81E-011	cm3/molc sec	25	EST	MEYLAN,WM & HOWARD,PH (1993)

CAS #: 000137-17-7			2,4,5-TRIMETHYLANILINE
Formula: $C_9H_{13}N$			
Mol Weight: 135.21			
MP (deg C): 68	FP (deg C):		
BP (deg C): 234.5			
BP pressure (mm Hg):			

Property/Value	Units	Temp	Data Type	Reference
WS 1.50E+003	mg/L		EXP	YALKOWSKY,SH & DANNENFELSER,RM (1992)
logP 2.72			EST	MEYLAN,WM & HOWARD,PH (1995)
VP 2.09E-002	mm Hg	25	EXT	CHAO,J ET AL. (1983)
DC	pKa			
HL 2.56E-006	atm m3/mol	25	EST	MEYLAN,WM & HOWARD,PH (1991)
OH 2.00E-010	cm3/molc sec	25	EST	MEYLAN,WM & HOWARD,PH (1993)

CAS #: 000137-18-8			2,5-CYCLOHEXADIENE-1,4-DIONE, 2,5-DIMETHYL-
Formula: $C_8H_8O_2$			
Mol Weight: 136.15			
MP (deg C):	FP (deg C):		
BP (deg C):			
BP pressure (mm Hg):			

Property/Value	Units	Temp	Data Type	Reference
WS 7.01E+003	mg/L	25	EST	MEYLAN,WM ET AL. (1996)
logP 1.28			EXP	HANSCH,C ET AL. (1995)
VP 2.41E-002	mm Hg	25	EST	NEELY,WB & BLAU,GE (1985)
DC	pKa			
HL 3.00E-009	atm m3/mol	25	EST	MEYLAN,WM & HOWARD,PH (1991)
OH 2.16E-011	cm3/molc sec	25	EST	MEYLAN,WM & HOWARD,PH (1993)

CAS #: 000137-26-8			THIRAM
Formula: $C_6H_{12}N_2S_4$			
Mol Weight: 240.43			
MP (deg C): 155-156	FP (deg C):		
BP (deg C): 129			
BP pressure (mm Hg): 2.00E+001			

Property/Value	Units	Temp	Data Type	Reference
WS 3.00E+001	mg/L	25	EXP	YALKOWSKY,SH & DANNENFELSER,RM (1992)
logP 1.70			EST	MEYLAN,WM & HOWARD,PH (1995)
VP 1.73E-005	mm Hg	25	EXP	TOMLIN,C (1994)
DC	pKa			
HL 1.82E-007	atm m3/mol	25	EST	VP/WSOL
OH 3.62E-010	cm3/molc sec	25	EST	MEYLAN,WM & HOWARD,PH (1993)

CAS #: 000137-30-4			ZIRAM
Formula: $C_6H_{12}N_2S_4Zn$			
Mol Weight: 305.80			
MP (deg C): 250	FP (deg C):		
BP (deg C):			
BP pressure (mm Hg):			

Property/Value	Units	Temp	Data Type	Reference
WS 6.50E+001	mg/L	25	EXP	GUNTHER,FA ET AL. (1968)
logP 1.09			EXP	TOMLIN,C (1994)
VP 1.00E-007	mm Hg	25	EXP	AUGUSTIJN-BECKERS,PWM (1994)
DC	pKa			
HL 6.19E-010	atm m3/mol	25	EST	VP/WSOL
OH	cm3/molc sec			

CAS #: 000137-32-6			2-METHYL-1-BUTANOL
Formula: $C_5H_{12}O$			
Mol Weight: 88.15			
MP (deg C):	FP (deg C):		
BP (deg C): 128			
BP pressure (mm Hg):			

Property/Value	Units	Temp	Data Type	Reference
WS 2.97E+004	mg/L	25	EXP	YALKOWSKY,SH & DANNENFELSER,RM (1992)
logP 1.29			EXP	VALVANI,SC ET AL. (1981)
VP 3.12E+000	mm Hg	25	EXP	RIDDICK,JA ET AL. (1986)
DC	pKa			
HL 1.41E-005	atm m3/mol	25	EXP	HINE,J & MOOKERJEE,PK (1975)
OH 8.58E-012	cm3/molc sec	25	EST	MEYLAN,WM & HOWARD,PH (1993)

CAS #: 000137-42-8	METHYLDITHIOCARBAMIC ACID, NA SALT
Formula: C₂H₄NNaS₂	

Formula: $C_2H_4NNaS_2$

Mol Weight: 129.18

MP (deg C):

FP (deg C):

BP (deg C):

BP pressure (mm Hg):

Property/Value	Units	Temp	Data Type	Reference
WS 7.22E+005	mg/L	20	EXP	SHIU,WY ET AL. (1990)
logP -0.92			EST	MEYLAN,WM & HOWARD,PH (1995)
VP 4.53E-009	mm Hg	25	EST	NEELY,WB & BLAU,GE (1985)
DC	pKa			
HL	atm m3/mol			
OH 6.43E-011	cm3/molc sec	25	EST	MEYLAN,WM & HOWARD,PH (1993)

CAS #: 000137-58-6 — LIDOCAINE

Formula: $C_{14}H_{22}N_2O$

Mol Weight: 234.34

MP (deg C): 68-69

FP (deg C):

BP (deg C): 159-160

BP pressure (mm Hg):

Property/Value	Units	Temp	Data Type	Reference
WS 4.10E+003	mg/L	30	EXP	YALKOWSKY,SH & DANNENFELSER,RM (1992)
logP 2.26			EXP	HANSCH,C & LEO,AJ (1985)
VP 1.08E-006	mm Hg	25	EST	NEELY,WB & BLAU,GE (1985)
DC	pKa			
HL 1.31E-010	atm m3/mol	25	EST	MEYLAN,WM & HOWARD,PH (1991)
OH 1.09E-010	cm3/molc sec	25	EST	MEYLAN,WM & HOWARD,PH (1993)

CAS #: 000138-15-8 — GLUTAMIC ACID HYDROCHLORIDE

Formula: $C_5H_{10}ClNO_4$

Mol Weight: 183.59

MP (deg C): 214 dec

FP (deg C):

BP (deg C):

BP pressure (mm Hg):

Property/Value	Units	Temp	Data Type	Reference
WS 3.80E+005	mg/L	20	EXP	SEIDELL,A (1941)
logP -4.99			EST	MEYLAN,WM & HOWARD,PH (1995)
VP 1.59E-009	mm Hg	25	EST	NEELY,WB & BLAU,GE (1985)
DC	pKa			
HL 1.37E-021	atm m3/mol	25	EST	MEYLAN,WM & HOWARD,PH (1991)
OH 2.00E-011	cm3/molc sec	25	EST	MEYLAN,WM & HOWARD,PH (1993)

CAS #: 000138-22-7 — BUTYL LACTATE

Formula: $C_7H_{14}O_3$

Mol Weight: 146.19

MP (deg C): -28

FP (deg C):

BP (deg C): 185-187

BP pressure (mm Hg):

Property/Value	Units	Temp	Data Type	Reference
WS 4.00E+004	mg/L	20	EXP	YALKOWSKY,SH & DANNENFELSER,RM (1992)
logP 0.80			EST	MEYLAN,WM & HOWARD,PH (1995)
VP 4.00E-001	mm Hg	20	EXP	FLICK,EW (1991)
DC	pKa			
HL 1.92E-006	atm m3/mol	20	EST	VP/WSOL
OH 9.90E-012	cm3/molc sec	25	EST	MEYLAN,WM & HOWARD,PH (1993)

CAS #: 000138-38-5 — 4-ETHYLBENZENESULFONAMIDE

Formula: $C_8H_{11}NO_2S$

Mol Weight: 185.25

MP (deg C):

FP (deg C):

BP (deg C):

BP pressure (mm Hg):

Property/Value	Units	Temp	Data Type	Reference
WS 3.95E+003	mg/L	25	EST	MEYLAN,WM ET AL. (1996)
logP 1.31			EXP	HANSCH,C & LEO,AJ (1985)
VP 1.19E-004	mm Hg	25	EST	NEELY,WB & BLAU,GE (1985)
DC	pKa			
HL 6.19E-007	atm m3/mol	25	EST	MEYLAN,WM & HOWARD,PH (1991)
OH 2.14E-012	cm3/molc sec	25	EST	MEYLAN,WM & HOWARD,PH (1993)

CAS #: 000138-41-0 — 4-AMINOSULFONYLBENZOIC ACID

Formula: $C_7H_7NO_4S$

Mol Weight: 201.20

MP (deg C): 290-292 de

FP (deg C):

BP (deg C):

BP pressure (mm Hg):

Property/Value	Units	Temp	Data Type	Reference
WS 1.61E+004	mg/L	25	EST	MEYLAN,WM ET AL. (1996)
logP 0.50			EXP	HANSCH,C ET AL. (1995)
VP 1.12E-006	mm Hg	25	EST	NEELY,WB & BLAU,GE (1985)
DC	pKa			
HL 8.50E-012	atm m3/mol	25	EST	MEYLAN,WM & HOWARD,PH (1991)
OH 6.74E-013	cm3/molc sec	25	EST	MEYLAN,WM & HOWARD,PH (1993)

CAS #: 000138-52-3 — SALICIN

Formula: $C_{13}H_{18}O_7$

Mol Weight: 286.28

MP (deg C): 199-202

FP (deg C):

BP (deg C):

BP pressure (mm Hg):

Property/Value	Units	Temp	Data Type	Reference
WS 4.00E+004	mg/L	25	EXP	YALKOWSKY,SH & DANNENFELSER,RM (1992)
logP -1.22			EXP	HANSCH,C & LEO,AJ (1985)
VP 3.98E-013	mm Hg	25	EST	NEELY,WB & BLAU,GE (1985)
DC	pKa			
HL 4.64E-017	atm m3/mol	25	EST	MEYLAN,WM & HOWARD,PH (1991)
OH 9.16E-011	cm3/molc sec	25	EST	MEYLAN,WM & HOWARD,PH (1993)

CAS #: 000138-56-7 — TRIMETHOBENZAMIDE (TIGAN)

Formula: $C_{21}H_{28}N_2O_5$

Mol Weight: 388.47

MP (deg C): 187.5-190

FP (deg C):

BP (deg C):

BP pressure (mm Hg):

Property/Value	Units	Temp	Data Type	Reference
WS 4.00E+001	mg/L	25	EST	MEYLAN,WM ET AL. (1996)
logP 2.29			EXP	EL TAYER,N ET AL. (1985)
VP 3.08E-011	mm Hg	25	EST	NEELY,WB & BLAU,GE (1985)
DC	pKa			
HL 3.16E-018	atm m3/mol	25	EST	MEYLAN,WM & HOWARD,PH (1991)
OH 2.30E-010	cm3/molc sec	25	EST	MEYLAN,WM & HOWARD,PH (1993)

000138-86-3 — LIMONENE

CAS #:	000138-86-3		LIMONENE

Formula: $C_{10}H_{16}$
Mol Weight: 136.24
MP (deg C): -95.5
FP (deg C):
BP (deg C): 176
BP pressure (mm Hg):

Property/Value	Units	Temp	Data Type	Reference
WS 1.38E+001	mg/L	25	EXP	MASSALDI,HA & KING,CJ (1973)
logP 4.57			EXP	LI,NY & PERDUE,EM (1995)
VP 1.55E+000	mm Hg	25	EXP	BOUBLIK,T ET AL. (1984)
DC	pKa			
HL 2.01E-002	atm m3/mol	25	EST	VP/WSOL
OH 1.49E-010	cm3/molc sec	25	EXP	WINER,AM ET AL. (1976)

000138-87-4 — BETA-TERPINEOL

CAS #:	000138-87-4		BETA-TERPINEOL

Formula: $C_{10}H_{18}O$
Mol Weight: 154.25
MP (deg C): 32.5
FP (deg C):
BP (deg C): 210
BP pressure (mm Hg):

Property/Value	Units	Temp	Data Type	Reference
WS 2.20E+003	mg/L	15	EXP	STEPHEN,H & STEPHEN,T (1963)
logP 3.41			EST	MEYLAN,WM & HOWARD,PH (1995)
VP 3.95E-002	mm Hg	25	EST	NEELY,WB & BLAU,GE (1985)
DC	pKa			
HL 1.34E-005	atm m3/mol	25	EST	MEYLAN,WM & HOWARD,PH (1991)
OH 6.67E-011	cm3/molc sec	25	EST	MEYLAN,WM & HOWARD,PH (1993)

000138-89-6 — P-NITROSO-N,N-DIMETHYLANILINE

CAS #:	000138-89-6		P-NITROSO-N,N-DIMETHYLANILINE

Formula: $C_8H_{10}N_2O$
Mol Weight: 150.18
MP (deg C): 92.5-93.5
FP (deg C):
BP (deg C):
BP pressure (mm Hg):

Property/Value	Units	Temp	Data Type	Reference
WS 1.37E+003	mg/L	25	EST	MEYLAN,WM ET AL. (1996)
logP 2.04			EST	MEYLAN,WM & HOWARD,PH (1995)
VP 4.06E-001	mm Hg	25	EST	NEELY,WB & BLAU,GE (1985)
DC 4.54	pKa	25	EXP	PERRIN,DD (1965)
HL 8.96E-007	atm m3/mol	25	EST	MEYLAN,WM & HOWARD,PH (1991)
OH 1.96E-010	cm3/molc sec	25	EST	MEYLAN,WM & HOWARD,PH (1993)

000139-02-6 — PHENOL, SODIUM SALT

CAS #:	000139-02-6		PHENOL, SODIUM SALT

Formula: C_6H_5NaO
Mol Weight: 116.10
MP (deg C):
FP (deg C):
BP (deg C):
BP pressure (mm Hg):

Property/Value	Units	Temp	Data Type	Reference
WS 1.00E+006	mg/L	25	EST	MEYLAN,WM ET AL. (1996)
logP -1.17			EST	MEYLAN,WM & HOWARD,PH (1995)
VP 2.60E-008	mm Hg	25	EST	NEELY,WB & BLAU,GE (1985)
DC	pKa			
HL	atm m3/mol			
OH 2.15E-011	cm3/molc sec	25	EST	MEYLAN,WM & HOWARD,PH (1993)

000139-13-9 — NITRILOTRIACETIC ACID

CAS #:	000139-13-9		NITRILOTRIACETIC ACID

Formula: $C_6H_9NO_6$
Mol Weight: 191.14
MP (deg C): 241.5
FP (deg C):
BP (deg C):
BP pressure (mm Hg):

Property/Value	Units	Temp	Data Type	Reference
WS 5.91E+004	mg/L	25	EXP	YALKOWSKY,SH & DANNENFELSER,RM (1992)
logP -3.81			EST	MEYLAN,WM & HOWARD,PH (1995)
VP 3.00E-005	mm Hg	25	EST	BANERJEE,S ET AL. (1990)
DC 3.03	pKa	20	EXP	MERCK INDEX (1983)
HL 1.30E-010	atm m3/mol	25	EST	VP/WSOL
OH 7.90E-011	cm3/molc sec	25	EST	ATKINSON,R (1988)

000139-40-2 — PROPAZINE

CAS #:	000139-40-2		PROPAZINE

Formula: $C_9H_{16}ClN_5$
Mol Weight: 229.71
MP (deg C): 213
FP (deg C):
BP (deg C):
BP pressure (mm Hg):

Property/Value	Units	Temp	Data Type	Reference
WS 8.60E+000	mg/L	22	EXP	YALKOWSKY,SH & DANNENFELSER,RM (1992)
logP 2.93			EXP	HANSCH,C & LEO,AJ (1985)
VP 1.31E-007	mm Hg	25	EXP	WAUCHOPE,RD ET AL. (1991A)
DC	pKa			
HL 4.60E-009	atm m3/mol	25	EST	VP/WSOL
OH 3.69E-011	cm3/molc sec	25	EST	MEYLAN,WM & HOWARD,PH (1993)

000139-45-7 — TRIPROPIONIN

CAS #:	000139-45-7		TRIPROPIONIN

Formula: $C_{12}H_{20}O_6$
Mol Weight: 260.29
MP (deg C):
FP (deg C):
BP (deg C): 175-176
BP pressure (mm Hg): 2.00E+001

Property/Value	Units	Temp	Data Type	Reference
WS 3.07E+003	mg/L	37	EXP	FUNASAKI,N ET AL. (1976)
logP 1.84			EST	MEYLAN,WM & HOWARD,PH (1995)
VP 3.61E-003	mm Hg	25	EST	NEELY,WB & BLAU,GE (1985)
DC	pKa			
HL 4.10E-009	atm m3/mol	25	EST	MEYLAN,WM & HOWARD,PH (1991)
OH 1.09E-011	cm3/molc sec	25	EST	MEYLAN,WM & HOWARD,PH (1993)

000139-59-3 — P-PHENOXYANILINE

CAS #:	000139-59-3		P-PHENOXYANILINE

Formula: $C_{12}H_{11}NO$
Mol Weight: 185.23
MP (deg C): 82-84
FP (deg C):
BP (deg C):
BP pressure (mm Hg):

Property/Value	Units	Temp	Data Type	Reference
WS 1.54E+002	mg/L	25	EST	MEYLAN,WM ET AL. (1996)
logP 2.96			EXP	DEBNATH,AK ET AL. (1992)
VP 1.25E-004	mm Hg	25	EST	NEELY,WB & BLAU,GE (1985)
DC	pKa			
HL 4.16E-008	atm m3/mol	25	EST	MEYLAN,WM & HOWARD,PH (1991)
OH 1.13E-010	cm3/molc sec	25	EST	MEYLAN,WM & HOWARD,PH (1993)

CAS #: 000139-65-1	4,4'-THIODIANILINE

Formula: $C_{12}H_{12}N_2S$

Mol Weight: 216.31

MP (deg C): 108-109   FP (deg C):

BP (deg C): 361

BP pressure (mm Hg):

Property/Value	Units	Temp	Data Type	Reference
WS 3.10E+002	mg/L	25	EST	MEYLAN,WM ET AL. (1996)
logP 2.18			EXP	HANSCH,C ET AL. (1995)
VP 1.11E-005	mm Hg	25	EST	NEELY,WB & BLAU,GE (1985)
DC	pKa			
HL 3.92E-012	atm m3/mol	25	EST	MEYLAN,WM & HOWARD,PH (1991)
OH 1.32E-010	cm3/molc sec	25	EST	MEYLAN,WM & HOWARD,PH (1993)

CAS #: 000139-66-2	DIPHENYLSULFIDE

Formula: $C_{12}H_{10}S$

Mol Weight: 186.28

MP (deg C): -40   FP (deg C):

BP (deg C): 295-297

BP pressure (mm Hg):

Property/Value	Units	Temp	Data Type	Reference
WS 8.12E+000	mg/L	25	EST	MEYLAN,WM ET AL. (1996)
logP 4.45			EXP	HANSCH,C & LEO,AJ (1985)
VP 7.54E-003	mm Hg	25	EXT	OHE,S (1976)
DC	pKa			
HL 3.14E-005	atm m3/mol	25	EST	MEYLAN,WM & HOWARD,PH (1991)
OH 2.51E-011	cm3/molc sec	25	EST	MEYLAN,WM & HOWARD,PH (1993)

CAS #: 000139-71-9	FORMAMIDE, N-(3-CHLOROPHENYL)-

Formula: C_7H_6ClNO

Mol Weight: 155.58

MP (deg C):   FP (deg C):

BP (deg C):

BP pressure (mm Hg):

Property/Value	Units	Temp	Data Type	Reference
WS 1.71E+003	mg/L	25	EST	MEYLAN,WM ET AL. (1996)
logP 1.90			EXP	HANSCH,C & LEO,AJ (1985)
VP 2.06E-004	mm Hg	25	EST	NEELY,WB & BLAU,GE (1985)
DC	pKa			
HL 6.26E-009	atm m3/mol	25	EST	MEYLAN,WM & HOWARD,PH (1991)
OH 3.00E-011	cm3/molc sec	25	EST	MEYLAN,WM & HOWARD,PH (1993)

CAS #: 000139-84-4	PHENOL, 3-NONYL-

Formula: $C_{15}H_{24}O$

Mol Weight: 220.36

MP (deg C):   FP (deg C):

BP (deg C):

BP pressure (mm Hg):

Property/Value	Units	Temp	Data Type	Reference
WS 2.11E+000	mg/L	25	EST	MEYLAN,WM ET AL. (1996)
logP 5.61			EXP	ITOKAWA,H ET AL. (1989)
VP 3.58E-005	mm Hg	25	EST	NEELY,WB & BLAU,GE (1985)
DC	pKa			
HL 5.97E-006	atm m3/mol	25	EST	MEYLAN,WM & HOWARD,PH (1991)
OH 9.39E-011	cm3/molc sec	25	EST	MEYLAN,WM & HOWARD,PH (1993)

CAS #: 000140-01-2	DIETHYLENETRIAMINEPENTAACETIC ACID, PENTA SODIU*

Formula: $C_{14}H_{18}N_3Na_5O_{10}$

Mol Weight: 503.26

MP (deg C):   FP (deg C):

BP (deg C):

BP pressure (mm Hg):

Property/Value	Units	Temp	Data Type	Reference
WS 1.00E+006	mg/L	25	EST	MEYLAN,WM ET AL. (1996)
logP -16.25			EST	MEYLAN,WM & HOWARD,PH (1995)
VP 2.85E-015	mm Hg	25	EST	NEELY,WB & BLAU,GE (1985)
DC	pKa			
HL	atm m3/mol			
OH 2.54E-010	cm3/molc sec	25	EST	MEYLAN,WM & HOWARD,PH (1993)

CAS #: 000140-10-3	2-PROPENOIC ACID, 3-PHENYL-, (E)-

Formula: $C_9H_8O_2$

Mol Weight: 148.16

MP (deg C): 133   FP (deg C):

BP (deg C): 300

BP pressure (mm Hg):

Property/Value	Units	Temp	Data Type	Reference
WS 5.46E+002	mg/L	25	EXP	YALKOWSKY,SH & DANNENFELSER,RM (1992)
logP 2.13			EXP	HANSCH,C ET AL. (1995)
VP 2.06E-005	mm Hg	25	EXT	OHE,S (1976)
DC 4.44	pKa	25	EXP	KORTUM,G ET AL (1961)
HL 1.29E-008	atm m3/mol	25	EST	MEYLAN,WM & HOWARD,PH (1991)
OH 2.21E-011	cm3/molc sec	25	EST	MEYLAN,WM & HOWARD,PH (1993)

CAS #: 000140-11-4	BENZYL ACETATE

Formula: $C_9H_{10}O_2$

Mol Weight: 150.18

MP (deg C): -51   FP (deg C):

BP (deg C): 213

BP pressure (mm Hg):

Property/Value	Units	Temp	Data Type	Reference
WS 3.10E+003	mg/L	25	EXP	CHEM INSPECT TEST INST (1992)
logP 1.96			EXP	HANSCH,C & LEO,AJ (1985)
VP 1.77E-001	mm Hg	25	EXP	DAUBERT,TE & DANNER,RP (1989)
DC	pKa			
HL 1.13E-005	atm m3/mol	25	EST	VP/WSOL
OH 6.44E-012	cm3/molc sec	25	EST	MEYLAN,WM & HOWARD,PH (1993)

CAS #: 000140-29-4	PHENYLACETONITRILE

Formula: C_8H_7N

Mol Weight: 117.15

MP (deg C): -23.8   FP (deg C):

BP (deg C): 233.5

BP pressure (mm Hg):

Property/Value	Units	Temp	Data Type	Reference
WS 1.00E+002	mg/L	25	EXP	CHEM INSPECT TEST INST (1992)
logP 1.56			EXP	HANSCH,C & LEO,AJ (1985)
VP 9.00E-001	mm Hg	25	EXT	OHE,S (1976)
DC	pKa			
HL 2.47E-006	atm m3/mol	25	EST	MEYLAN,WM & HOWARD,PH (1991)
OH 2.07E-012	cm3/molc sec	25	EST	MEYLAN,WM & HOWARD,PH (1993)

CAS #: 000140-38-5 — P-CHLOROPHENYLUREA

Formula: $C_7H_7ClN_2O$
Mol Weight: 170.60
MP (deg C):
FP (deg C):
BP (deg C):
BP pressure (mm Hg):

Property/ Value	Units	Temp	Data Type	Reference
WS 1.77E+003	mg/L	25	EST	MEYLAN,WM ET AL. (1996)
logP 1.80			EXP	HANSCH,C & LEO,AJ (1985)
VP 1.65E-004	mm Hg	25	EST	NEELY,WB & BLAU,GE (1985)
DC	pKa			
HL 1.49E-010	atm m3/mol	25	EST	MEYLAN,WM & HOWARD,PH (1991)
OH 1.35E-011	cm3/molc sec	25	EST	MEYLAN,WM & HOWARD,PH (1993)

CAS #: 000140-39-6 — P-TOLYLACETATE

Formula: $C_9H_{10}O_2$
Mol Weight: 150.18
MP (deg C):
FP (deg C):
BP (deg C): 212.5
BP pressure (mm Hg):

Property/ Value	Units	Temp	Data Type	Reference
WS 1.20E+003	mg/L	25	EST	MEYLAN,WM ET AL. (1996)
logP 2.11			EXP	HANSCH,C & LEO,AJ (1985)
VP 1.64E-001	mm Hg	25	EST	NEELY,WB & BLAU,GE (1985)
DC	pKa			
HL 7.15E-005	atm m3/mol	25	EST	MEYLAN,WM & HOWARD,PH (1991)
OH 3.98E-012	cm3/molc sec	25	EST	MEYLAN,WM & HOWARD,PH (1993)

CAS #: 000140-40-9 — NITHIAMIDE

Formula: $C_5H_7N_3O_3S$
Mol Weight: 189.19
MP (deg C): 264-265
FP (deg C):
BP (deg C):
BP pressure (mm Hg):

Property/ Value	Units	Temp	Data Type	Reference
WS 1.61E+003	mg/L	25	EST	MEYLAN,WM ET AL. (1996)
logP 1.30			EXP	HANSCH,C ET AL. (1995)
VP 1.86E-006	mm Hg	25	EST	NEELY,WB & BLAU,GE (1985)
DC	pKa			
HL 1.73E-014	atm m3/mol	25	EST	MEYLAN,WM & HOWARD,PH (1991)
OH 1.66E-013	cm3/molc sec	25	EST	MEYLAN,WM & HOWARD,PH (1993)

CAS #: 000140-56-7 — FENAMINOSULF

Formula: $C_8H_{10}N_3NaO_3S$
Mol Weight: 251.24
MP (deg C):
FP (deg C):
BP (deg C):
BP pressure (mm Hg):

Property/ Value	Units	Temp	Data Type	Reference
WS 2.00E+004	mg/L	25	EXP	AUGUSTIJN-BECKERS,PWM ET AL. (1994)
logP -1.66			EST	MEYLAN,WM & HOWARD,PH (1995)
VP 5.70E-013	mm Hg	25	EST	NEELY,WB & BLAU,GE (1985)
DC	pKa			
HL	atm m3/mol			
OH 1.49E-010	cm3/molc sec	25	EST	MEYLAN,WM & HOWARD,PH (1993)

CAS #: 000140-57-8 — ARAMITE

Formula: $C_{15}H_{23}ClO_4S$
Mol Weight: 334.87
MP (deg C): -37.3
FP (deg C):
BP (deg C): 195
BP pressure (mm Hg): 2.00E+000

Property/ Value	Units	Temp	Data Type	Reference
WS 5.90E-001	mg/L	25	EST	MEYLAN,WM ET AL. (1996)
logP 4.82			EST	MEYLAN,WM & HOWARD,PH (1995)
VP 2.18E-007	mm Hg	25	EST	NEELY,WB & BLAU,GE (1985)
DC	pKa			
HL 1.90E-007	atm m3/mol	25	EST	MEYLAN,WM & HOWARD,PH (1991)
OH 3.58E-011	cm3/molc sec	25	EST	MEYLAN,WM & HOWARD,PH (1993)

CAS #: 000140-66-9 — P-(1,1,3,3-TETRAMETHYLBUTYL)PHENOL

Formula: $C_{14}H_{22}O$
Mol Weight: 206.33
MP (deg C): 84-85
FP (deg C):
BP (deg C): 158
BP pressure (mm Hg): 1.50E+001

Property/ Value	Units	Temp	Data Type	Reference
WS 5.00E+000	mg/L	25	EST	MEYLAN,WM ET AL. (1996)
logP 5.28			EST	MEYLAN,WM & HOWARD,PH (1995)
VP 4.78E-004	mm Hg	25	EXT	BOUBLIK,T ET AL. (1984)
DC	pKa			
HL 6.89E-006	atm m3/mol	25	EST	MEYLAN,WM & HOWARD,PH (1991)
OH 4.24E-011	cm3/molc sec	25	EST	MEYLAN,WM & HOWARD,PH (1993)

CAS #: 000140-76-1 — 2-METHYL-5-VINYLPYRIDINE

Formula: C_8H_9N
Mol Weight: 119.17
MP (deg C):
FP (deg C):
BP (deg C): 181
BP pressure (mm Hg):

Property/ Value	Units	Temp	Data Type	Reference
WS 4.25E+004	mg/L	25	EST	VP/HENL
logP 2.25			EST	MEYLAN,WM & HOWARD,PH (1995)
VP 1.18E+000	mm Hg	25	EXP	CHAO,J ET AL. (1983)
DC 5.67	pKa	25	EXP	PERRIN,DD (1965)
HL 4.35E-006	atm m3/mol	25	EST	MEYLAN,WM & HOWARD,PH (1991)
OH 2.74E-011	cm3/molc sec	25	EST	MEYLAN,WM & HOWARD,PH (1993)

CAS #: 000140-77-2 — CYCLOPENTYLPROPRIONIC ACID

Formula: $C_8H_{14}O_2$
Mol Weight: 142.20
MP (deg C):
FP (deg C):
BP (deg C): 130-132
BP pressure (mm Hg): 1.20E+001

Property/ Value	Units	Temp	Data Type	Reference
WS 2.42E+003	mg/L	25	EXP	NIYAZOV,AN ET AL. (1975)
logP 2.85			EST	MEYLAN,WM & HOWARD,PH (1995)
VP 1.53E-002	mm Hg	25	EST	NEELY,WB & BLAU,GE (1985)
DC	pKa			
HL 1.32E-006	atm m3/mol	25	EST	MEYLAN,WM & HOWARD,PH (1991)
OH 1.02E-011	cm3/molc sec	25	EST	MEYLAN,WM & HOWARD,PH (1993)

000140-79-4 — N,N'-DINITROSOPIPERAZINE

Formula: $C_4H_8N_4O_2$

Mol Weight: 144.13

MP (deg C):
FP (deg C):
BP (deg C):
BP pressure (mm Hg):

Property/Value	Units	Temp	Data Type	Reference
WS 4.28E+005	mg/L	25	EST	MEYLAN,WM ET AL. (1996)
logP -0.85			EXP	HANSCH,C & LEO,AJ (1985)
VP 1.90E-004	mm Hg	25	EST	NEELY,WB & BLAU,GE (1985)
DC	pKa			
HL 3.36E-011	atm m3/mol	25	EST	MEYLAN,WM & HOWARD,PH (1991)
OH 4.27E-011	cm3/molc sec	25	EST	MEYLAN,WM & HOWARD,PH (1993)

000140-88-5 — ETHYL ACRYLATE

Formula: $C_5H_8O_2$

Mol Weight: 100.12

MP (deg C): -71.2
FP (deg C):
BP (deg C): 99.5
BP pressure (mm Hg):

Property/Value	Units	Temp	Data Type	Reference
WS 1.50E+004	mg/L	25	EXP	RIDDICK,JA ET AL. (1986)
logP 1.32			EXP	HANSCH,C & LEO,AJ (1985)
VP 3.86E+001	mm Hg	25	EXP	DAUBERT,TE & DANNER,RP (1985)
DC	pKa			
HL 3.93E-004	atm m3/mol	25	EST	VP/WSOL
OH 1.46E-011	cm3/molc sec	25	EST	MEYLAN,WM & HOWARD,PH (1993)

000140-89-6 — POTASSIUM ETHYLXANTHATE

Formula: $C_3H_5KOS_2$

Mol Weight: 160.30

MP (deg C): 210 dec
FP (deg C):
BP (deg C):
BP pressure (mm Hg):

Property/Value	Units	Temp	Data Type	Reference
WS 1.96E+005	mg/L	25	EST	MEYLAN,WM ET AL. (1996)
logP -0.54			EST	MEYLAN,WM & HOWARD,PH (1995)
VP 3.95E-009	mm Hg	25	EST	NEELY,WB & BLAU,GE (1985)
DC	pKa			
HL	atm m3/mol			
OH 6.17E-012	cm3/molc sec	25	EST	MEYLAN,WM & HOWARD,PH (1993)

000140-93-2 — ISOPROPYLXANTHIC ACID, SODIUM SALT

Formula: $C_4H_7NaOS_2$

Mol Weight: 158.22

MP (deg C):
FP (deg C):
BP (deg C):
BP pressure (mm Hg):

Property/Value	Units	Temp	Data Type	Reference
WS 4.60E+005	mg/L	24	EXP	SHIU,WY ET AL. (1990)
logP -0.12			EST	MEYLAN,WM & HOWARD,PH (1995)
VP 2.88E-009	mm Hg	25	EST	NEELY,WB & BLAU,GE (1985)
DC	pKa			
HL	atm m3/mol			
OH 1.22E-011	cm3/molc sec	25	EST	MEYLAN,WM & HOWARD,PH (1993)

000140-95-4 — DIMETHYLOL UREA

Formula: $C_3H_8N_2O_3$

Mol Weight: 120.11

MP (deg C): 137-139
FP (deg C):
BP (deg C):
BP pressure (mm Hg):

Property/Value	Units	Temp	Data Type	Reference
WS 4.00E+001	mg/L	25	EXP	GUNTHER,FA ET AL. (1968)
logP -3.15			EST	MEYLAN,WM & HOWARD,PH (1995)
VP 3.99E-006	mm Hg	25	EST	NEELY,WB & BLAU,GE (1985)
DC	pKa			
HL 9.36E-013	atm m3/mol	25	EST	MEYLAN,WM & HOWARD,PH (1991)
OH 2.91E-011	cm3/molc sec	25	EST	MEYLAN,WM & HOWARD,PH (1993)

000141-05-9 — DIETHYL MALEATE

Formula: $C_8H_{12}O_4$

Mol Weight: 172.18

MP (deg C): -8.8
FP (deg C):
BP (deg C): 223
BP pressure (mm Hg):

Property/Value	Units	Temp	Data Type	Reference
WS 8.01E+002	mg/L	25	EST	MEYLAN,WM ET AL. (1996)
logP 2.20			EST	MEYLAN,WM & HOWARD,PH (1995)
VP 1.05E-001	mm Hg	25	EXT	OHE,S (1976)
DC	pKa			
HL 2.44E-007	atm m3/mol	25	EST	MEYLAN,WM & HOWARD,PH (1991)
OH 1.02E-011	cm3/molc sec	25	EST	MEYLAN,WM & HOWARD,PH (1993)

000141-22-0 — 12-HYDROXY-9-OCTADECENOIC ACID (CIS)

Formula: $C_{18}H_{34}O_3$

Mol Weight: 298.47

MP (deg C): 5.5
FP (deg C):
BP (deg C): 245
BP pressure (mm Hg):

Property/Value	Units	Temp	Data Type	Reference
WS 3.46E+003	mg/L	25	EXP	SEIDELL,A (1941)
logP 6.19			EST	MEYLAN,WM & HOWARD,PH (1995)
VP 2.63E-009	mm Hg	25	EST	NEELY,WB & BLAU,GE (1985)
DC	pKa			
HL 1.64E-009	atm m3/mol	25	EST	MEYLAN,WM & HOWARD,PH (1991)
OH 8.45E-011	cm3/molc sec	25	EST	MEYLAN,WM & HOWARD,PH (1993)

000141-28-6 — DIETHYLHEXANEDIOATE

Formula: $C_{10}H_{18}O_4$

Mol Weight: 202.25

MP (deg C): -19.8
FP (deg C):
BP (deg C): 245
BP pressure (mm Hg):

Property/Value	Units	Temp	Data Type	Reference
WS 4.23E+003	mg/L	20	EXP	STEPHEN,H & STEPHEN,T (1963)
logP 2.37			EST	MEYLAN,WM & HOWARD,PH (1995)
VP 5.77E-002	mm Hg	25	EXT	OHE,S (1976)
DC	pKa			
HL 1.72E-006	atm m3/mol	25	EST	MEYLAN,WM & HOWARD,PH (1991)
OH 7.85E-012	cm3/molc sec	25	EST	MEYLAN,WM & HOWARD,PH (1993)

CAS #: 000141-32-2				BUTYL ACRYLATE
Formula: $C_7H_{12}O_2$				
Mol Weight: 128.17				
MP (deg C): -64.6		FP (deg C):		
BP (deg C): 146-148				
BP pressure (mm Hg):				

Property/ Value	Units	Temp	Data Type	Reference
WS 2.00E+003	mg/L	23	EXP	NEMEC,JW & BAUER,WJR (1978)
logP 2.36			EXP	HANSCH,C & LEO,AJ (1985)
VP 5.45E+000	mm Hg	25	EXP	DAUBERT,TE & DANNER,RP (1985)
DC	pKa			
HL 4.60E-004	atm m3/mol	25	EST	VP/WSOL
OH 2.82E-011	cm3/molc sec	25	EST	MEYLAN,WM & HOWARD,PH (1993)

CAS #: 000141-43-5				ETHANOLAMINE
Formula: C_2H_7NO				
Mol Weight: 61.08				
MP (deg C): 10.3		FP (deg C):		
BP (deg C): 170.8				
BP pressure (mm Hg):				

Property/ Value	Units	Temp	Data Type	Reference
WS 1.00E+006	mg/L		EXP	RIDDICK,JA ET AL. (1986)
logP -1.31			EXP	HANSCH,C & LEO,AJ (1985)
VP 4.04E-001	mm Hg	25	EXP	DOW CHEMICAL COMPANY (1980)
DC 9.50	pKa		EXP	PERRIN,DD (1972)
HL 3.25E-008	atm m3/mol	25	EST	VP/WSOL
OH 3.45E-011	cm3/molc sec	25	EST	MEYLAN,WM & HOWARD,PH (1993)

CAS #: 000141-46-8				GLYCOLALDEHYDE
Formula: $C_2H_4O_2$				
Mol Weight: 60.05				
MP (deg C): 97		FP (deg C):		
BP (deg C):				
BP pressure (mm Hg):				

Property/ Value	Units	Temp	Data Type	Reference
WS 1.00E+006	mg/L	25	EST	MEYLAN,WM ET AL. (1996)
logP -1.63			EST	MEYLAN,WM & HOWARD,PH (1995)
VP 4.97E+000	mm Hg	25	EST	NEELY,WB & BLAU,GE (1985)
DC	pKa			
HL 2.42E-008	atm m3/mol	25	EXP	BETTERTON,EA & HOFFMAN,MR (1988)
OH 9.90E-012	cm3/molc sec	25	EXP	ATKINSON,R (1989)

CAS #: 000141-57-1				PROPYLTRICHLOROSILANE
Formula: $C_3H_7Cl_3Si$				
Mol Weight: 177.53				
MP (deg C):		FP (deg C):		
BP (deg C): 123.5				
BP pressure (mm Hg):				

Property/ Value	Units	Temp	Data Type	Reference
WS 1.58E+002	mg/L	25	EST	MEYLAN,WM ET AL. (1996)
logP 2.99			EST	MEYLAN,WM & HOWARD,PH (1995)
VP 7.49E+002	mm Hg	25	EXP	DYKYJ,J (1970)
DC	pKa			
HL 9.41E-003	atm m3/mol	25	EST	MEYLAN,WM & HOWARD,PH (1991)
OH 2.58E-012	cm3/molc sec	25	EST	MEYLAN,WM & HOWARD,PH (1993)

CAS #: 000141-59-3				T-OCTYL MERCAPTAN
Formula: $C_8H_{18}S$				
Mol Weight: 146.30				
MP (deg C):		FP (deg C):		
BP (deg C): 154-166				
BP pressure (mm Hg):				

Property/ Value	Units	Temp	Data Type	Reference
WS 3.07E+001	mg/L	25	EST	MEYLAN,WM ET AL. (1996)
logP 3.99			EST	MEYLAN,WM & HOWARD,PH (1995)
VP 4.93E+000	mm Hg	25	EXP	DAUBERT,TE & DANNER,RP (1989)
DC	pKa			
HL 1.90E-002	atm m3/mol	25	EST	MEYLAN,WM & HOWARD,PH (1991)
OH 3.47E-011	cm3/molc sec	25	EST	MEYLAN,WM & HOWARD,PH (1993)

CAS #: 000141-62-8				DECAMETHYLTETRASILOXANE
Formula: $C_{10}H_{30}O_3Si_4$				
Mol Weight: 310.69				
MP (deg C): -70		FP (deg C):		
BP (deg C): 194				
BP pressure (mm Hg):				

Property/ Value	Units	Temp	Data Type	Reference
WS 7.00E-003	mg/L	25	EST	MEYLAN,WM ET AL. (1996)
logP 5.40			EXP	SANGSTER,J (1994)
VP 4.89E-001	mm Hg	25	EXT	JORDAN,TE (1954)
DC	pKa			
HL 6.01E-006	atm m3/mol	25	EST	MEYLAN,WM & HOWARD,PH (1991)
OH 1.50E-012	cm3/molc sec	25	EST	MEYLAN,WM & HOWARD,PH (1993)

CAS #: 000141-63-9				DODECAMETHYLPENTASILOXANE
Formula: $C_{12}H_{36}O_4Si_5$				
Mol Weight: 384.85				
MP (deg C): -80		FP (deg C):		
BP (deg C): 229				
BP pressure (mm Hg): 7.10E+002				

Property/ Value	Units	Temp	Data Type	Reference
WS 3.09E-004	mg/L	25	EST	MEYLAN,WM ET AL. (1996)
logP 6.00			EXP	SANGSTER,J (1994)
VP 1.02E-001	mm Hg	25	EXT	JORDAN,TE (1954)
DC	pKa			
HL 1.81E-007	atm m3/mol	25	EST	MEYLAN,WM & HOWARD,PH (1991)
OH 1.80E-012	cm3/molc sec	25	EST	MEYLAN,WM & HOWARD,PH (1993)

CAS #: 000141-66-2				DICROTOPHOS
Formula: $C_8H_{16}NO_5P$				
Mol Weight: 237.19				
MP (deg C):		FP (deg C):		
BP (deg C): 400				
BP pressure (mm Hg):				

Property/ Value	Units	Temp	Data Type	Reference
WS 1.00E+006	mg/L	25	EXP	SHIU,WY ET AL. (1990)
logP -0.49			EXP	HANSCH,C ET AL. (1995)
VP 8.63E-005	mm Hg	20	EXP	GRAYSON,BT & FOSBRAEY,LA (1982)
DC	pKa			
HL 1.20E-012	atm m3/mol	25	EST	MEYLAN,WM & HOWARD,PH (1991)
OH 9.80E-011	cm3/molc sec	25	EST	MEYLAN,WM & HOWARD,PH (1993)

271

000141-76-4 — 3-IODOPROPIONIC ACID

Formula: $C_3H_5IO_2$

Mol Weight: 199.98

MP (deg C): 80-83

FP (deg C):

BP (deg C):

BP pressure (mm Hg):

Property/Value	Units	Temp	Data Type	Reference
WS 7.43E+004	mg/L	25	EXP	YALKOWSKY,SH & DANNENFELSER,RM (1992)
logP 1.34			EST	MEYLAN,WM & HOWARD,PH (1995)
VP 3.25E-002	mm Hg	25	EST	NEELY,WB & BLAU,GE (1985)
DC 4.09	pKa	18	EXP	KORTUM,G ET AL (1961)
HL 5.43E-008	atm m3/mol	25	EST	MEYLAN,WM & HOWARD,PH (1991)
OH 1.45E-012	cm3/molc sec	25	EST	MEYLAN,WM & HOWARD,PH (1993)

000141-78-6 — ETHYL ACETATE

Formula: $C_4H_8O_2$

Mol Weight: 88.11

MP (deg C): -83.6

FP (deg C):

BP (deg C): 77.06

BP pressure (mm Hg):

Property/Value	Units	Temp	Data Type	Reference
WS 8.00E+004	mg/L	25	EXP	BANERJEE,S (1984)
logP 0.73			EXP	HANSCH,C & LEO,AJ (1985)
VP 9.37E+001	mm Hg	25	EXP	DAUBERT,TE & DANNER,RP (1985)
DC	pKa			
HL 1.34E-004	atm m3/mol	25	EXP	BUTLER,JAV AND RAMCHANDANI,CN (1935)
OH 1.60E-012	cm3/molc sec	25	EXP	ATKINSON,R (1989)

000141-79-7 — MESITYL OXIDE

Formula: $C_6H_{10}O$

Mol Weight: 98.15

MP (deg C): -57

FP (deg C):

BP (deg C): 130

BP pressure (mm Hg):

Property/Value	Units	Temp	Data Type	Reference
WS 2.90E+004	mg/L	20		YALKOWSKY,SH & DANNENFELSER,RM (1992)
logP 1.37			EST	MEYLAN,WM & HOWARD,PH (1995)
VP 8.21E+000	mm Hg	25	EXP	DAUBERT,TE & DANNER,RP (1989)
DC	pKa			
HL 3.67E-005	atm m3/mol	20	EST	VP/WSOL
OH 7.95E-011	cm3/molc sec	25	EST	MEYLAN,WM & HOWARD,PH (1993)

000141-82-2 — MALONIC ACID

Formula: $C_3H_4O_4$

Mol Weight: 104.06

MP (deg C): 135 dec

FP (deg C):

BP (deg C):

BP pressure (mm Hg):

Property/Value	Units	Temp	Data Type	Reference
WS 7.63E+005	mg/L	25	EXP	YALKOWSKY,SH & DANNENFELSER,RM (1992)
logP -0.81			EXP	HANSCH,C ET AL. (1995)
VP 1.50E-003	mm Hg	25	EXP	JORDAN,TE (1954)
DC 2.85	pKa	25	EXP	KORTUM,G ET AL (1961)
HL 2.69E-010	atm m3/mol	25	EST	VP/WSOL
OH 1.57E-012	cm3/molc sec	25	EST	MEYLAN,WM & HOWARD,PH (1993)

000141-93-5 — M-DIETHYLBENZENE

Formula: $C_{10}H_{14}$

Mol Weight: 134.22

MP (deg C): <-10

FP (deg C):

BP (deg C): 181-182

BP pressure (mm Hg):

Property/Value	Units	Temp	Data Type	Reference
WS 2.40E+001	mg/L	25	EXP	CHEM INSPECT TEST INST (1992)
logP 4.57			EXP	SHERBLOM,PM & EGANHOUSE,RP (1988)
VP 1.13E+000	mm Hg	25	EXP	DAUBERT,TE & DANNER,RP (1989)
DC	pKa			
HL 8.31E-003	atm m3/mol	25	EST	VP/WSOL
OH 1.42E-011	cm3/molc sec	25	EST	MEYLAN,WM & HOWARD,PH (1993)

000141-97-9 — ACETOACETIC ESTER

Formula: $C_6H_{10}O_3$

Mol Weight: 130.14

MP (deg C): -45

FP (deg C):

BP (deg C): 180.8

BP pressure (mm Hg):

Property/Value	Units	Temp	Data Type	Reference
WS 1.11E+005	mg/L	17	EXP	YALKOWSKY,SH & DANNENFELSER,RM (1992)
logP 0.25			EXP	CATZ,P & FRIEND,DR (1989)
VP 7.80E-001	mm Hg	25	EXP	DAUBERT,TE & DANNER,RP (1989)
DC	pKa			
HL 1.20E-006	atm m3/mol	25	EST	VP/WSOL
OH 2.28E-012	cm3/molc sec	25	EST	MEYLAN,WM & HOWARD,PH (1993)

000142-04-1 — ANILINE HYDROCHLORIDE

Formula: C_6H_8ClN

Mol Weight: 129.59

MP (deg C): 198

FP (deg C):

BP (deg C):

BP pressure (mm Hg):

Property/Value	Units	Temp	Data Type	Reference
WS 1.78E+005	mg/L	15	EXP	STEPHEN,H & STEPHEN,T (1963)
logP 1.08			EST	MEYLAN,WM & HOWARD,PH (1995)
VP 3.00E+001	mm Hg	20	EXP	WEBER,RC ET AL. (1981)
DC	pKa			
HL 2.87E-005	atm m3/mol	20	EST	VP/WSOL
OH 4.27E-011	cm3/molc sec	25	EST	MEYLAN,WM & HOWARD,PH (1993)

000142-08-5 — 2-HYDROXYPYRIDINE

Formula: C_5H_5NO

Mol Weight: 95.10

MP (deg C): 107.8

FP (deg C):

BP (deg C): 280

BP pressure (mm Hg):

Property/Value	Units	Temp	Data Type	Reference
WS 1.00E+006	mg/L	20	EXP	YALKOWSKY,SH & DANNENFELSER,RM (1992)
logP 1.07			EST	MEYLAN,WM & HOWARD,PH (1995)
VP 5.52E-001	mm Hg	25	EST	NEELY,WB & BLAU,GE (1985)
DC 11.62	pKa	20	EXP	KORTUM,G ET AL (1961)
HL 7.34E-010	atm m3/mol	25	EST	MEYLAN,WM & HOWARD,PH (1991)
OH 6.46E-012	cm3/molc sec	25	EST	MEYLAN,WM & HOWARD,PH (1993)

CAS #: 000142-14-3			ZINEB (ALSO SEE CAS NO. 12122-67-7)

Formula: $C_4H_6N_2S_4Zn$
Mol Weight: 275.73
MP (deg C): 146-60
FP (deg C):
BP (deg C):
BP pressure (mm Hg):

Property/ Value	Units	Temp	Data Type	Reference
WS 1.00E+001	mg/L	25	EXP	SHIU,WY ET AL. (1990)
logP				TOMLIN,C (1994); < 1.3
VP	mm Hg	20		TOMLIN,C (1994) ; < 7.5E-8
DC	pKa			
HL	atm m3/mol			
OH	cm3/molc sec			

CAS #: 000142-25-6			1,2-ETHANEDIAMINE, N,N,N'-TRIMETHYL-

Formula: $C_5H_{14}N_2$
Mol Weight: 102.18
MP (deg C):
FP (deg C):
BP (deg C): 116-118
BP pressure (mm Hg):

Property/ Value	Units	Temp	Data Type	Reference
WS 1.00E+006	mg/L	25	EST	MEYLAN,WM ET AL. (1996)
logP -0.20			EXP	HANSCH,C ET AL. (1995)
VP 1.85E+001	mm Hg	25	EST	NEELY,WB & BLAU,GE (1985)
DC	pKa			
HL 1.10E-008	atm m3/mol	25	EST	MEYLAN,WM & HOWARD,PH (1991)
OH 1.54E-010	cm3/molc sec	25	EST	MEYLAN,WM & HOWARD,PH (1993)

CAS #: 000142-28-9			1,3-DICHLOROPROPANE

Formula: $C_3H_6Cl_2$
Mol Weight: 112.99
MP (deg C): -99.5
FP (deg C):
BP (deg C): 120.9
BP pressure (mm Hg):

Property/ Value	Units	Temp	Data Type	Reference
WS 2.75E+003	mg/L	25	EXP	YALKOWSKY,SH & DANNENFELSER,RM (1992)
logP 2.00			EXP	HANSCH,C & LEO,AJ (1985)
VP 1.82E+001	mm Hg	25	EXP	YAWS,CL (1994)
DC	pKa			
HL 9.76E-004	atm m3/mol	25	EXP	LEIGHTON,DTJR & CALO,JM 1981)
OH 1.09E-012	cm3/molc sec	25	EST	MEYLAN,WM & HOWARD,PH (1993)

CAS #: 000142-29-0			CYCLOPENTENE

Formula: C_5H_8
Mol Weight: 68.12
MP (deg C): -135.1
FP (deg C):
BP (deg C): 44.2
BP pressure (mm Hg):

Property/ Value	Units	Temp	Data Type	Reference
WS 5.35E+002	mg/L	25	EXP	YALKOWSKY,SH & DANNENFELSER,RM (1992)
logP 2.47			EST	MEYLAN,WM & HOWARD,PH (1995)
VP 3.80E+002	mm Hg	25	EXP	YAWS,CL (1994A)
DC	pKa			
HL 6.37E-003	atm m3/mol	25	EST	VP/WSOL
OH 6.73E-011	cm3/molc sec	25	EXP	ATKINSON,R (1989)

CAS #: 000142-30-3			2,5-DIMETHYL-3-HEXYNE-2,5-DIOL

Formula: $C_8H_{14}O_2$
Mol Weight: 142.20
MP (deg C): 95
FP (deg C):
BP (deg C): 205
BP pressure (mm Hg):

Property/ Value	Units	Temp	Data Type	Reference
WS 1.67E+004	mg/L	25	EST	MEYLAN,WM ET AL. (1996)
logP 0.81			EST	MEYLAN,WM & HOWARD,PH (1995)
VP 9.81E-003	mm Hg	25	EST	NEELY,WB & BLAU,GE (1985)
DC	pKa			
HL 4.47E-008	atm m3/mol	25	EST	MEYLAN,WM & HOWARD,PH (1991)
OH 2.93E-011	cm3/molc sec	25	EST	MEYLAN,WM & HOWARD,PH (1993)

CAS #: 000142-47-2			MONOSODIUM L-GLUTAMATE

Formula: $C_5H_8NNaO_4$
Mol Weight: 169.11
MP (deg C):
FP (deg C):
BP (deg C):
BP pressure (mm Hg):

Property/ Value	Units	Temp	Data Type	Reference
WS 3.85E+005	mg/L	25	EXP	YOSHIDA,T (1978)
logP -7.06			EST	MEYLAN,WM & HOWARD,PH (1995)
VP 3.70E-014	mm Hg	25	EST	NEELY,WB & BLAU,GE (1985)
DC	pKa			
HL	atm m3/mol			
OH 4.04E-011	cm3/molc sec	25	EST	MEYLAN,WM & HOWARD,PH (1993)

CAS #: 000142-59-6			NABAM

Formula: $C_4H_6N_2Na_2S_4$
Mol Weight: 256.34
MP (deg C):
FP (deg C):
BP (deg C):
BP pressure (mm Hg):

Property/ Value	Units	Temp	Data Type	Reference
WS 2.00E+005	mg/L		EST	WORTHING,CR & WALKER,SB (1987)
logP -2.19			EST	MEYLAN,WM & HOWARD,PH (1995)
VP 9.48E-013	mm Hg	25	EST	NEELY,WB & BLAU,GE (1985)
DC	pKa			
HL	atm m3/mol			
OH 1.47E-010	cm3/molc sec	25	EST	MEYLAN,WM & HOWARD,PH (1993)

CAS #: 000142-62-1			HEXANOIC ACID

Formula: $C_6H_{12}O_2$
Mol Weight: 116.16
MP (deg C): -3.44
FP (deg C):
BP (deg C): 205.02
BP pressure (mm Hg):

Property/ Value	Units	Temp	Data Type	Reference
WS 1.03E+004	mg/L	25	EXP	YALKOWSKY,SH & DANNENFELSER,RM (1992)
logP 1.92			EXP	HANSCH,C & LEO,AJ (1985)
VP 4.35E-002	mm Hg	25	EXP	DAUBERT,TE & DANNER,RP (1989)
DC 4.88	pKa	25	EXP	RIDDICK,JA ET AL. (1986)
HL 6.47E-007	atm m3/mol	25	EST	VP/WSOL
OH 4.91E-012	cm3/molc sec	25	EST	ATKINSON,R (1987)

273

CAS #:	000142-68-7		TETRAHYDROPYRAN

Formula: $C_5H_{10}O$

Mol Weight: 86.13

MP (deg C): -49.2

FP (deg C):

BP (deg C): 88

BP pressure (mm Hg):

Property/ Value	Units	Temp	Data Type	Reference
WS 8.02E+004	mg/L	25	EXP	YALKOWSKY,SH & DANNENFELSER,RM (1992)
logP 0.95			EXP	HANSCH,C ET AL. (1995)
VP 7.15E+001	mm Hg	25	EXP	CABANI,S ET AL. (1971A)
DC	pKa			
HL 1.25E-004	atm m3/mol	25	EXP	CABANI,S ET AL. (1971A)
OH 1.38E-011	cm3/molc sec	25	EXP	ATKINSON,R (1989)

CAS #:	000142-77-8		BUTYL OLEATE

Formula: $C_{22}H_{42}O_2$

Mol Weight: 338.58

MP (deg C): -26.4

FP (deg C):

BP (deg C): 227-228

BP pressure (mm Hg): 1.50E+001

Property/ Value	Units	Temp	Data Type	Reference
WS 5.72E-005	mg/L	25	EST	MEYLAN,WM ET AL. (1996)
logP 9.49			EST	MEYLAN,WM & HOWARD,PH (1995)
VP 2.63E-006	mm Hg	25	EST	NEELY,WB & BLAU,GE (1985)
DC	pKa			
HL 3.36E-002	atm m3/mol	25	EST	MEYLAN,WM & HOWARD,PH (1991)
OH 7.96E-011	cm3/molc sec	25	EST	MEYLAN,WM & HOWARD,PH (1993)

CAS #:	000142-78-9		2-AMINOETHANE DODECANOATE

Formula: $C_{14}H_{29}NO_2$

Mol Weight: 243.39

MP (deg C):

FP (deg C):

BP (deg C):

BP pressure (mm Hg):

Property/ Value	Units	Temp	Data Type	Reference
WS 4.39E+001	mg/L	25	EST	MEYLAN,WM ET AL. (1996)
logP 3.24			EST	MEYLAN,WM & HOWARD,PH (1995)
VP 6.57E-009	mm Hg	25	EST	NEELY,WB & BLAU,GE (1985)
DC	pKa			
HL 2.03E-011	atm m3/mol	25	EST	MEYLAN,WM & HOWARD,PH (1991)
OH 3.09E-011	cm3/molc sec	25	EST	MEYLAN,WM & HOWARD,PH (1993)

CAS #:	000142-82-5		N-HEPTANE

Formula: C_7H_{16}

Mol Weight: 100.21

MP (deg C): -90.6

FP (deg C):

BP (deg C): 98.4

BP pressure (mm Hg):

Property/ Value	Units	Temp	Data Type	Reference
WS 3.40E+000	mg/L	25	EXP	YALKOWSKY,SH & DANNENFELSER,RM (1992)
logP 4.66			EXP	MILLER,MM ET AL. (1985)
VP 4.60E+001	mm Hg	25	EXP	DAUBERT,TE & DANNER,RP (1989)
DC	pKa			
HL 2.00E+000	atm m3/mol	25	EXP	HINE,J & MOOKERJEE,PK (1975)
OH 7.15E-012	cm3/molc sec	25	EXP	ATKINSON,R (1989)

CAS #:	000142-83-6		2,4-HEXADIENAL

Formula: C_6H_8O

Mol Weight: 96.13

MP (deg C):

FP (deg C):

BP (deg C): 174

BP pressure (mm Hg):

Property/ Value	Units	Temp	Data Type	Reference
WS 8.14E+003	mg/L	25	EST	MEYLAN,WM ET AL. (1996)
logP 1.37			EST	MEYLAN,WM & HOWARD,PH (1995)
VP 4.81E+000	mm Hg	25	EST	NEELY,WB & BLAU,GE (1985)
DC	pKa			
HL 9.78E-006	atm m3/mol	25	EXP	BUTTERY,RG ET AL. (1971)
OH 5.13E-011	cm3/molc sec	25	EST	MEYLAN,WM & HOWARD,PH (1993)

CAS #:	000142-84-7		DIPROPYLAMINE

Formula: $C_6H_{15}N$

Mol Weight: 101.19

MP (deg C): -63

FP (deg C):

BP (deg C): 110

BP pressure (mm Hg):

Property/ Value	Units	Temp	Data Type	Reference
WS 5.86E+004	mg/L		EXP	YALKOWSKY,SH & DANNENFELSER,RM (1992)
logP 1.67			EXP	HANSCH,C & LEO,AJ (1985)
VP 2.01E+001	mm Hg	25	EXP	YAWS,CL (1994A)
DC 11.00	pKa	25	EXP	PERRIN,DD (1965)
HL 5.10E-005	atm m3/mol	25	EXP	CHRISTIE,AO & CRISP,DJ (1967)
OH 8.41E-011	cm3/molc sec	25	EST	MEYLAN,WM & HOWARD,PH (1993)

CAS #:	000142-90-5		DODECYL 2-METHYL-2-PROPENOATE

Formula: $C_{16}H_{30}O_2$

Mol Weight: 254.42

MP (deg C): -7

FP (deg C):

BP (deg C): 142

BP pressure (mm Hg): 4.00E+000

Property/ Value	Units	Temp	Data Type	Reference
WS 4.44E-002	mg/L	25	EST	MEYLAN,WM ET AL. (1996)
logP 6.68			EST	MEYLAN,WM & HOWARD,PH (1995)
VP 1.18E-003	mm Hg	25	EST	NEELY,WB & BLAU,GE (1985)
DC	pKa			
HL 3.27E-003	atm m3/mol	25	EST	MEYLAN,WM & HOWARD,PH (1991)
OH 3.40E-011	cm3/molc sec	25	EST	MEYLAN,WM & HOWARD,PH (1993)

CAS #:	000142-91-6		ISOPROPYL PALMITATE

Formula: $C_{19}H_{38}O_2$

Mol Weight: 298.51

MP (deg C): 13.5

FP (deg C):

BP (deg C): 160

BP pressure (mm Hg): 2.00E+000

Property/ Value	Units	Temp	Data Type	Reference
WS 1.35E-003	mg/L	25	EST	MEYLAN,WM ET AL. (1996)
logP 8.16			EST	MEYLAN,WM & HOWARD,PH (1995)
VP 5.59E-005	mm Hg	25	EXT	BOUBLIK,T ET AL. (1984)
DC	pKa			
HL 1.63E-002	atm m3/mol	25	EST	MEYLAN,WM & HOWARD,PH (1991)
OH 2.26E-011	cm3/molc sec	25	EST	MEYLAN,WM & HOWARD,PH (1993)

274

N-HEXYL ACETATE

CAS #:	000142-92-7	N-HEXYL ACETATE

Formula: $C_8H_{16}O_2$
Mol Weight: 144.22
MP (deg C): -80.9
FP (deg C):
BP (deg C): 171.5
BP pressure (mm Hg):

Property/Value	Units	Temp	Data Type	Reference
WS 5.11E+002	mg/L	25	EXP	YALKOWSKY,SH & DANNENFELSER,RM (1992)
logP 2.83			EST	MEYLAN,WM & HOWARD,PH (1995)
VP 1.32E+000	mm Hg	25	EXP	YAWS,CL (1994B)
DC	pKa			
HL 4.90E-004	atm m3/mol	25	EST	VP/WSOL
OH 7.49E-012	cm3/molc sec	25	EST	MEYLAN,WM & HOWARD,PH (1993)

DI-N-BUTYL ETHER

CAS #:	000142-96-1	DI-N-BUTYL ETHER

Formula: $C_8H_{18}O$
Mol Weight: 130.23
MP (deg C): -95.2
FP (deg C):
BP (deg C): 140.3
BP pressure (mm Hg):

Property/Value	Units	Temp	Data Type	Reference
WS 3.00E+002	mg/L	25	EXP	RIDDICK,JA ET AL. (1986)
logP 3.21			EXP	HANSCH,C ET AL. (1995)
VP 6.01E+000	mm Hg	25	EXP	DAUBERT,TE & DANNER,RP (1989)
DC	pKa			
HL 6.00E-003	atm m3/mol	25	EXP	HINE,J & MOOKERJEE,PK (1975)
OH 2.24E-011	cm3/molc sec	25	EXP	ATKINSON,R (1989)

DODECANOIC ACID

CAS #:	000143-07-7	DODECANOIC ACID

Formula: $C_{12}H_{24}O_2$
Mol Weight: 200.32
MP (deg C): 44
FP (deg C):
BP (deg C): 225
BP pressure (mm Hg): 1.00E+002

Property/Value	Units	Temp	Data Type	Reference
WS 4.81E+000	mg/L	25	EXP	YALKOWSKY,SH & DANNENFELSER,RM (1992)
logP 4.60			EXP	SANGSTER,J (1993)
VP 1.60E-005	mm Hg	25	EXT	YAWS,CL (1994B)
DC 5.30	pKa	20	EXP	SERJEANT,EP & DEMPSEY,B (1979)
HL 9.31E-006	atm m3/mol	25	EST	MEYLAN,WM & HOWARD,PH (1991)
OH 1.40E-011	cm3/molc sec	25	EST	MEYLAN,WM & HOWARD,PH (1993)

1-NONANOL

CAS #:	000143-08-8	1-NONANOL

Formula: $C_9H_{20}O$
Mol Weight: 144.26
MP (deg C): -5
FP (deg C):
BP (deg C): 213.4
BP pressure (mm Hg):

Property/Value	Units	Temp	Data Type	Reference
WS 1.40E+002	mg/L	25	EXP	BARTON,AFM (1984)
logP 3.77			EXP	TEWARI,YB ET AL. (1982)
VP 2.27E-002	mm Hg	25	EXP	DAUBERT,TE & DANNER,RP (1989)
DC	pKa			
HL 3.08E-005	atm m3/mol	25	EST	VP/WSOL
OH 1.40E-011	cm3/molc sec	25	EST	MEYLAN,WM & HOWARD,PH (1993)

1-DECANETHIOL

CAS #:	000143-10-2	1-DECANETHIOL

Formula: $C_{10}H_{22}S$
Mol Weight: 174.35
MP (deg C): -26
FP (deg C):
BP (deg C): 240.6
BP pressure (mm Hg):

Property/Value	Units	Temp	Data Type	Reference
WS 2.14E+000	mg/L	25	EST	MEYLAN,WM ET AL. (1996)
logP 5.20			EST	MEYLAN,WM & HOWARD,PH (1995)
VP 3.67E-002	mm Hg	25	EXP	YAWS,CL (1994B)
DC	pKa			
HL 3.35E-002	atm m3/mol	25	EST	MEYLAN,WM & HOWARD,PH (1991)
OH 5.27E-011	cm3/molc sec	25	EST	MEYLAN,WM & HOWARD,PH (1993)

1-BROMO-N-DODECANE

CAS #:	000143-15-7	1-BROMO-N-DODECANE

Formula: $C_{12}H_{25}Br$
Mol Weight: 249.24
MP (deg C): -9.5
FP (deg C):
BP (deg C): 276
BP pressure (mm Hg):

Property/Value	Units	Temp	Data Type	Reference
WS 5.76E-002	mg/L	25	EST	MEYLAN,WM ET AL. (1996)
logP 6.58			EST	MEYLAN,WM & HOWARD,PH (1995)
VP 3.93E-003	mm Hg	25	EXT	LI,JCM & ROSSINI,FD (1953)
DC	pKa			
HL 1.92E-001	atm m3/mol	25	EST	MEYLAN,WM & HOWARD,PH (1991)
OH 1.35E-011	cm3/molc sec	25	EST	MEYLAN,WM & HOWARD,PH (1993)

OLEIC ACID, SODIUM SALT

CAS #:	000143-19-1	OLEIC ACID, SODIUM SALT

Formula: $C_{18}H_{34}NaO_2$
Mol Weight: 305.46
MP (deg C): 232-235
FP (deg C):
BP (deg C):
BP pressure (mm Hg):

Property/Value	Units	Temp	Data Type	Reference
WS 5.21E+000	mg/L	25	EST	MEYLAN,WM ET AL. (1996)
logP 3.92			EST	MEYLAN,WM & HOWARD,PH (1995)
VP 7.80E-013	mm Hg	25	EST	NEELY,WB & BLAU,GE (1985)
DC	pKa			
HL	atm m3/mol			
OH 7.45E-011	cm3/molc sec	25	EST	MEYLAN,WM & HOWARD,PH (1993)

TRIETHYLENE GLYCOL BUTYL ETHER

CAS #:	000143-22-6	TRIETHYLENE GLYCOL BUTYL ETHER

Formula: $C_{10}H_{22}O_4$
Mol Weight: 206.28
MP (deg C):
FP (deg C): -35
BP (deg C): 278
BP pressure (mm Hg):

Property/Value	Units	Temp	Data Type	Reference
WS 1.00E+006	mg/L	25	EXP	DOW CHEMICAL COMPANY (1990)
logP 0.02			EST	MEYLAN,WM & HOWARD,PH (1995)
VP 2.50E-003	mm Hg	25	EXP	CLAYTON,GD & CLAYTON,FE (1982)
DC	pKa			
HL 9.50E-014	atm m3/mol	25	EST	MEYLAN,WM & HOWARD,PH (1991)
OH 5.15E-011	cm3/molc sec	25	EST	MEYLAN,WM & HOWARD,PH (1993)

000143-27-1 — N-HEXADECYLAMINE

Formula: $C_{16}H_{35}N$
Mol Weight: 241.46
MP (deg C): 46.8
FP (deg C):
BP (deg C): 322.5
BP pressure (mm Hg):

Property/Value	Units	Temp	Data Type	Reference
WS 4.82E-001	mg/L	25	EST	MEYLAN,WM ET AL. (1996)
logP 6.73			EST	MEYLAN,WM & HOWARD,PH (1995)
VP 1.33E-004	mm Hg	25	EXT	PERRY,RH & GREEN,D (1984)
DC 10.61	pKa	25	EXP	PERRIN,DD (1965)
HL 5.31E-004	atm m3/mol	25	EST	MEYLAN,WM & HOWARD,PH (1991)
OH 5.14E-011	cm3/molc sec	25	EST	MEYLAN,WM & HOWARD,PH (1993)

000143-28-2 — OLEYL ALCOHOL (CIS)

Formula: $C_{18}H_{36}$
Mol Weight: 252.49
MP (deg C): 5.5-7.5
FP (deg C):
BP (deg C): 333-335
BP pressure (mm Hg):

Property/Value	Units	Temp	Data Type	Reference
WS 7.00E-002	mg/L	25	EST	VP/HL
logP 7.50			EST	MEYLAN,WM & HOWARD,PH (1995)
VP 9.30E-005	mm Hg	25	EST	ANTOINE EQUATION
DC	pKa			
HL 4.60E-004	atm m3/mol	25	EST	MEYLAN,WM & HOWARD,PH (1991)
OH 7.85E-011	cm3/molc sec	25	EST	ATKINSON,R (1988)

000143-50-0 — KEPONE

Formula: $C_{10}Cl_{10}O$
Mol Weight: 490.64
MP (deg C): 350 dec
FP (deg C):
BP (deg C):
BP pressure (mm Hg):

Property/Value	Units	Temp	Data Type	Reference
WS 7.60E+000	mg/L	24	EXP	HOLLIFIELD,HC (1979)
logP 5.41			EXP	HANSCH,C ET AL. (1995)
VP 3.00E-007	mm Hg	25	EXP	IARC (1979)
DC	pKa			
HL 2.50E-008	atm m3/mol		EST	VP/WSOL
OH 0.00E+000	cm3/molc sec	25	EST	MEYLAN,WM & HOWARD,PH (1993)

000143-62-4 — DIGITOXIGENIN

Formula: $C_{23}H_{34}O_4$
Mol Weight: 374.53
MP (deg C): 253
FP (deg C):
BP (deg C):
BP pressure (mm Hg):

Property/Value	Units	Temp	Data Type	Reference
WS 2.45E+001	mg/L	25	EST	MEYLAN,WM ET AL. (1996)
logP 2.64			EXP	HANSCH,C ET AL. (1995)
VP 5.34E-013	mm Hg	25	EST	NEELY,WB & BLAU,GE (1985)
DC	pKa			
HL 6.25E-010	atm m3/mol	25	EST	MEYLAN,WM & HOWARD,PH (1991)
OH 6.22E-011	cm3/molc sec	25	EST	MEYLAN,WM & HOWARD,PH (1993)

000143-74-8 — PHENOLSULPHONEPHTHALEIN

Formula: $C_{19}H_{14}O_5S$
Mol Weight: 354.38
MP (deg C): > 300
FP (deg C):
BP (deg C):
BP pressure (mm Hg):

Property/Value	Units	Temp	Data Type	Reference
WS 3.10E+002	mg/L	100	EXP	YALKOWSKY,SH & DANNENFELSER,RM (1992)
logP 3.02			EXP	HANSCH,C & LEO,AJ (1985)
VP 6.22E-013	mm Hg	25	EST	NEELY,WB & BLAU,GE (1985)
DC 8.08	pKa	20	EXP	KORTUM,G ET AL (1961)
HL 2.08E-017	atm m3/mol	25	EST	MEYLAN,WM & HOWARD,PH (1991)
OH 8.12E-011	cm3/molc sec	25	EST	MEYLAN,WM & HOWARD,PH (1993)

000144-11-6 — TRIHEXYLPHENEDYL

Formula: $C_{20}H_{31}NO$
Mol Weight: 301.48
MP (deg C):
FP (deg C):
BP (deg C):
BP pressure (mm Hg):

Property/Value	Units	Temp	Data Type	Reference
WS 1.80E+001	mg/L	25	EST	MEYLAN,WM ET AL. (1996)
logP 4.49			EXP	SANGSTER,J (1993)
VP 1.21E-008	mm Hg	25	EST	NEELY,WB & BLAU,GE (1985)
DC	pKa			
HL 4.73E-010	atm m3/mol	25	EST	MEYLAN,WM & HOWARD,PH (1991)
OH 1.29E-010	cm3/molc sec	25	EST	MEYLAN,WM & HOWARD,PH (1993)

000144-19-4 — 2,2,4-TRIMETHYL-1,3-PENTANEDIOL

Formula: $C_8H_{18}O_2$
Mol Weight: 146.23
MP (deg C): 51.5
FP (deg C):
BP (deg C): 235
BP pressure (mm Hg):

Property/Value	Units	Temp	Data Type	Reference
WS 1.90E+004	mg/L	25	EXP	FLICK,EW (1991)
logP 1.24			EXP	CHEM INSPECT TEST INST (1992)
VP 1.10E-002	mm Hg	25	EST	NEELY,WB & BLAU,GE (1985)
DC	pKa			
HL 7.16E-007	atm m3/mol	25	EST	MEYLAN,WM & HOWARD,PH (1991)
OH 1.72E-011	cm3/molc sec	25	EST	ATKINSON,R (1988)

000144-21-8 — DSMA

Formula: $CH_3AsNa_2O_3$
Mol Weight: 183.93
MP (deg C): 132-139
FP (deg C):
BP (deg C):
BP pressure (mm Hg):

Property/Value	Units	Temp	Data Type	Reference
WS 4.32E+005	mg/L	25	EXP	SHIU,WY ET AL. (1990)
logP -5.34			EST	MEYLAN,WM & HOWARD,PH (1995)
VP	mm Hg			
DC	pKa			
HL	atm m3/mol			
OH	cm3/molc sec			

CAS #:	000144-48-9	IODOACETAMIDE

Formula: C_2H_4INO

Mol Weight: 184.96

MP (deg C): 93-96

FP (deg C):

BP (deg C):

BP pressure (mm Hg):

Property/Value	Units	Temp	Data Type	Reference
WS 7.57E+004	mg/L	25	EST	MEYLAN,WM ET AL. (1996)
logP -0.19			EXP	HANSCH,C & LEO,AJ (1985)
VP 5.06E-003	mm Hg	25	EST	NEELY,WB & BLAU,GE (1985)
DC	pKa			
HL 8.35E-010	atm m3/mol	25	EST	MEYLAN,WM & HOWARD,PH (1991)
OH 2.37E-012	cm3/molc sec	25	EST	MEYLAN,WM & HOWARD,PH (1993)

CAS #:	000144-49-0	FLUOROACETIC ACID

Formula: $C_2H_3FO_2$

Mol Weight: 78.04

MP (deg C): 35.2

FP (deg C):

BP (deg C): 165

BP pressure (mm Hg):

Property/Value	Units	Temp	Data Type	Reference
WS 3.10E+005	mg/L	25	EST	MEYLAN,WM ET AL. (1996)
logP 0.03			EST	MEYLAN,WM & HOWARD,PH (1995)
VP 1.28E+001	mm Hg	25	EST	NEELY,WB & BLAU,GE (1985)
DC 2.59	pKa	25	EXP	DEAN,JA (1985)
HL 1.09E-006	atm m3/mol	25	EST	MEYLAN,WM & HOWARD,PH (1991)
OH 4.63E-013	cm3/molc sec	25	EST	MEYLAN,WM & HOWARD,PH (1993)

CAS #:	000144-62-7	OXALIC ACID

Formula: $C_2H_2O_4$

Mol Weight: 90.04

MP (deg C): 189.5 dec

FP (deg C):

BP (deg C):

BP pressure (mm Hg):

Property/Value	Units	Temp	Data Type	Reference
WS 2.20E+005	mg/L	25	EXP	YALKOWSKY,SH & DANNENFELSER,RM (1992)
logP -2.22			EST	JABER,HM ET AL. (1984)
VP 2.34E-004	mm Hg	25	EXP	BARNALES,CA ET AL. (1981)
DC 1.25	pKa			BUXTON,GV ET AL. (1988)
HL 1.43E-001	atm m3/mol	25	EXP	GAFFNEY,JS ET AL. (1987)
OH 7.20E-014	cm3/molc sec	25	EST	ATKINSON,R ET AL. (1988)

CAS #:	000144-80-9	SULFANILACETAMIDE

Formula: $C_8H_{10}N_2O_3S$

Mol Weight: 214.24

MP (deg C): 182-184

FP (deg C):

BP (deg C):

BP pressure (mm Hg):

Property/Value	Units	Temp	Data Type	Reference
WS 1.25E+004	mg/L	37	EXP	YALKOWSKY,SH & DANNENFELSER,RM (1992)
logP -0.96			EXP	HANSCH,C & LEO,AJ (1985)
VP 3.57E-008	mm Hg	25	EST	NEELY,WB & BLAU,GE (1985)
DC	pKa			
HL 4.84E-013	atm m3/mol	25	EST	MEYLAN,WM & HOWARD,PH (1991)
OH 2.86E-011	cm3/molc sec	25	EST	MEYLAN,WM & HOWARD,PH (1993)

CAS #:	000144-82-1	SULFAMETHIAZOLE

Formula: $C_9H_{10}N_4O_2S_2$

Mol Weight: 270.33

MP (deg C): 208

FP (deg C):

BP (deg C):

BP pressure (mm Hg):

Property/Value	Units	Temp	Data Type	Reference
WS 1.05E+003	mg/L	37	EXP	YALKOWSKY,SH & DANNENFELSER,RM (1992)
logP 0.54			EXP	HANSCH,C & LEO,AJ (1985)
VP 2.06E-009	mm Hg	25	EST	NEELY,WB & BLAU,GE (1985)
DC	pKa			
HL 2.63E-014	atm m3/mol	25	EST	MEYLAN,WM & HOWARD,PH (1991)
OH 2.39E-011	cm3/molc sec	25	EST	MEYLAN,WM & HOWARD,PH (1993)

CAS #:	000144-83-2	SULFAPYRIDINE

Formula: $C_{11}H_{11}N_3O_2S$

Mol Weight: 249.29

MP (deg C): 191-193

FP (deg C):

BP (deg C):

BP pressure (mm Hg):

Property/Value	Units	Temp	Data Type	Reference
WS 2.68E+002	mg/L	25	EXP	YALKOWSKY,SH & DANNENFELSER,RM (1992)
logP 0.35			EXP	HANSCH,C ET AL. (1995)
VP 6.27E-008	mm Hg	25	EST	NEELY,WB & BLAU,GE (1985)
DC 8.43	pKa		EXP	PERRIN,DD (1965)
HL 1.08E-013	atm m3/mol	25	EST	MEYLAN,WM & HOWARD,PH (1991)
OH 3.11E-011	cm3/molc sec	25	EST	MEYLAN,WM & HOWARD,PH (1993)

CAS #:	000145-13-1	PREGNENOLONE

Formula: $C_{21}H_{32}O_2$

Mol Weight: 316.49

MP (deg C): 193

FP (deg C):

BP (deg C):

BP pressure (mm Hg):

Property/Value	Units	Temp	Data Type	Reference
WS 7.06E+000	mg/L	37	EXP	YALKOWSKY,SH & DANNENFELSER,RM (1992)
logP 4.22			EXP	HANSCH,C ET AL. (1995)
VP 1.03E-008	mm Hg	25	EST	NEELY,WB & BLAU,GE (1985)
DC	pKa			
HL 1.17E-008	atm m3/mol	25	EST	MEYLAN,WM & HOWARD,PH (1991)
OH 1.17E-010	cm3/molc sec	25	EST	MEYLAN,WM & HOWARD,PH (1993)

CAS #:	000145-73-3	ENDOTHAL

Formula: $C_8H_{10}O_5$

Mol Weight: 186.17

MP (deg C): 144

FP (deg C):

BP (deg C):

BP pressure (mm Hg):

Property/Value	Units	Temp	Data Type	Reference
WS 1.00E+005	mg/L	20	EXP	WSSA (1983)
logP 1.91			EXP	REINERT,KH & ROGERS,JHJR (1984)
VP 1.24E-005	mm Hg	25	EST	NEELY,WB & BLAU,GE (1985)
DC 3.40	pKa			WSSA (1983)
HL 3.62E-015	atm m3/mol	25	EST	MEYLAN,WM & HOWARD,PH (1991)
OH 2.86E-011	cm3/molc sec	25	EST	MEYLAN,WM & HOWARD,PH (1993)

000146-22-5 — NITRAZEPAM

CAS #:	000146-22-5
Formula:	$C_{15}H_{11}N_3O_3$
Mol Weight:	281.27
MP (deg C):	224-226
FP (deg C):	
BP (deg C):	
BP pressure (mm Hg):	

Property/Value	Units	Temp	Data Type	Reference
WS 7.69E+001	mg/L	25	EST	MEYLAN,WM ET AL. (1996)
logP 2.25			EXP	HANSCH,C & LEO,AJ (1985)
VP 4.78E-010	mm Hg	25	EST	NEELY,WB & BLAU,GE (1985)
DC	pKa			
HL 9.47E-013	atm m3/mol	25	EST	MEYLAN,WM & HOWARD,PH (1991)
OH 6.86E-012	cm3/molc sec	25	EST	MEYLAN,WM & HOWARD,PH (1993)

000146-48-5 — APHRODINE

CAS #:	000146-48-5
Formula:	$C_{21}H_{26}N_2O_3$
Mol Weight:	354.45
MP (deg C):	235-237
FP (deg C):	
BP (deg C):	
BP pressure (mm Hg):	

Property/Value	Units	Temp	Data Type	Reference
WS 2.77E+002	mg/L	25	EST	MEYLAN,WM ET AL. (1996)
logP 2.73			EXP	HANSCH,C ET AL. (1995)
VP 1.61E-012	mm Hg	25	EST	NEELY,WB & BLAU,GE (1985)
DC	pKa			
HL 2.90E-017	atm m3/mol	25	EST	MEYLAN,WM & HOWARD,PH (1991)
OH 2.75E-010	cm3/molc sec	25	EST	MEYLAN,WM & HOWARD,PH (1993)

000146-50-9 — DIISOHEXYL PHTHALATE

CAS #:	000146-50-9
Formula:	$C_{20}H_{30}O_4$
Mol Weight:	334.46
MP (deg C):	
FP (deg C):	
BP (deg C):	
BP pressure (mm Hg):	

Property/Value	Units	Temp	Data Type	Reference
WS 2.49E-002	mg/L	25	EST	MEYLAN,WM ET AL. (1996)
logP 6.43			EST	MEYLAN,WM & HOWARD,PH (1995)
VP 4.28E-005	mm Hg	25	EST	NEELY,WB & BLAU,GE (1985)
DC	pKa			
HL 3.80E-006	atm m3/mol	25	EST	MEYLAN,WM & HOWARD,PH (1991)
OH 1.49E-011	cm3/molc sec	25	EST	MEYLAN,WM & HOWARD,PH (1993)

000146-54-3 — TRIFLUPROMAZINE

CAS #:	000146-54-3
Formula:	$C_{18}H_{19}F_3N_2S$
Mol Weight:	352.42
MP (deg C):	
FP (deg C):	
BP (deg C):	176
BP pressure (mm Hg):	7.00E-001

Property/Value	Units	Temp	Data Type	Reference
WS 1.75E-001	mg/L	25	EST	MEYLAN,WM ET AL. (1996)
logP 5.19			EXP	HANSCH,C & LEO,AJ (1985)
VP 5.42E-007	mm Hg	25	EST	NEELY,WB & BLAU,GE (1985)
DC	pKa			
HL 4.32E-009	atm m3/mol	25	EST	MEYLAN,WM & HOWARD,PH (1991)
OH 1.29E-010	cm3/molc sec	25	EST	MEYLAN,WM & HOWARD,PH (1993)

000147-55-7 — PHENETHICILLIN

CAS #:	000147-55-7
Formula:	$C_{17}H_{20}N_2O_5S$
Mol Weight:	364.42
MP (deg C):	
FP (deg C):	
BP (deg C):	
BP pressure (mm Hg):	

Property/Value	Units	Temp	Data Type	Reference
WS 5.72E+001	mg/L	25	EST	MEYLAN,WM ET AL. (1996)
logP 2.20			EXP	HANSCH,C & LEO,AJ (1985)
VP 7.59E-013	mm Hg	25	EST	NEELY,WB & BLAU,GE (1985)
DC	pKa			
HL 5.87E-015	atm m3/mol	25	EST	MEYLAN,WM & HOWARD,PH (1991)
OH 1.17E-010	cm3/molc sec	25	EST	MEYLAN,WM & HOWARD,PH (1993)

000147-58-0 — GLYCINE, N-(2-IODOBENZOYL)-

CAS #:	000147-58-0
Formula:	$C_9H_8INO_3$
Mol Weight:	305.07
MP (deg C):	
FP (deg C):	
BP (deg C):	
BP pressure (mm Hg):	

Property/Value	Units	Temp	Data Type	Reference
WS 1.70E+003	mg/L	25	EST	MEYLAN,WM ET AL. (1996)
logP 0.97			EXP	LAZNICEK,M & KVETINA,J (1988)
VP 2.26E-008	mm Hg	25	EST	NEELY,WB & BLAU,GE (1985)
DC	pKa			
HL 1.67E-013	atm m3/mol	25	EST	MEYLAN,WM & HOWARD,PH (1991)
OH 1.01E-011	cm3/molc sec	25	EST	MEYLAN,WM & HOWARD,PH (1993)

000147-81-9 — ARABINOSE

CAS #:	000147-81-9
Formula:	$C_5H_{10}O_5$
Mol Weight:	150.13
MP (deg C):	157-160
FP (deg C):	
BP (deg C):	
BP pressure (mm Hg):	

Property/Value	Units	Temp	Data Type	Reference
WS 1.00E+006	mg/L	25	EST	MEYLAN,WM ET AL. (1996)
logP -3.02			EXP	HANSCH,C & LEO,AJ (1985)
VP 1.11E-006	mm Hg	25	EST	NEELY,WB & BLAU,GE (1985)
DC	pKa			
HL 2.00E-013	atm m3/mol	25	EST	MEYLAN,WM & HOWARD,PH (1991)
OH 8.93E-011	cm3/molc sec	25	EST	MEYLAN,WM & HOWARD,PH (1993)

000147-85-3 — PROLINE

CAS #:	000147-85-3
Formula:	$C_5H_9NO_2$
Mol Weight:	115.13
MP (deg C):	220-222 de
FP (deg C):	
BP (deg C):	
BP pressure (mm Hg):	

Property/Value	Units	Temp	Data Type	Reference
WS 1.31E+005	mg/L	25	EST	MEYLAN,WM ET AL. (1996)
logP -2.54			EXP	HANSCH,C & LEO,AJ (1985)
VP 3.77E-009	mm Hg	25	EST	NEELY,WB & BLAU,GE (1985)
DC	pKa			
HL 1.92E-009	atm m3/mol	25	EST	MEYLAN,WM & HOWARD,PH (1991)
OH 8.76E-011	cm3/molc sec	25	EST	MEYLAN,WM & HOWARD,PH (1993)

CAS #: 000147-93-3				2-MERCAPTOBENZOIC ACID

Formula: $C_7H_6O_2S$

Mol Weight: 154.19

MP (deg C): 164-165   FP (deg C):

BP (deg C):

BP pressure (mm Hg):

Property/Value	Units	Temp	Data Type	Reference
WS 6.62E+002	mg/L	25	EST	MEYLAN,WM ET AL. (1996)
logP 2.39			EXP	HANSCH,C & LEO,AJ (1985)
VP 4.52E-004	mm Hg	25	EST	NEELY,WB & BLAU,GE (1985)
DC	pKa			
HL 1.05E-008	atm m3/mol	25	EST	MEYLAN,WM & HOWARD,PH (1991)
OH 4.71E-012	cm3/molc sec	25	EST	MEYLAN,WM & HOWARD,PH (1993)

CAS #: 000148-18-5				DIETHYLCARBAMODITHIOIC ACID, SODIUM SALT

Formula: $C_5H_{11}NNaS_2$

Mol Weight: 172.27

MP (deg C): 94-102   FP (deg C):

BP (deg C):

BP pressure (mm Hg):

Property/Value	Units	Temp	Data Type	Reference
WS 3.64E+005	mg/L	25	EST	MEYLAN,WM ET AL. (1996)
logP 0.27			EST	MEYLAN,WM & HOWARD,PH (1995)
VP 8.15E-010	mm Hg	25	EST	NEELY,WB & BLAU,GE (1985)
DC	pKa			
HL	atm m3/mol			
OH 8.37E-011	cm3/molc sec	25	EST	MEYLAN,WM & HOWARD,PH (1993)

CAS #: 000148-24-3				8-QUINOLINOL

Formula: C_9H_7NO

Mol Weight: 145.16

MP (deg C): 76   FP (deg C):

BP (deg C): 267

BP pressure (mm Hg):

Property/Value	Units	Temp	Data Type	Reference
WS 5.56E+002	mg/L	20	EXP	YALKOWSKY,SH & DANNENFELSER,RM (1992)
logP 2.02			EXP	HANSCH,C & LEO,AJ (1985)
VP 1.69E-003	mm Hg	25	EXT	YAWS,CL (1994B)
DC 5.02	pKa	20	EXP	KORTUM,G ET AL (1961)
HL 7.16E-011	atm m3/mol	25	EST	MEYLAN,WM & HOWARD,PH (1991)
OH 2.00E-010	cm3/molc sec	25	EST	MEYLAN,WM & HOWARD,PH (1993)

CAS #: 000148-53-8				O-VANILLIN

Formula: $C_8H_8O_3$

Mol Weight: 152.15

MP (deg C): 44.5   FP (deg C):

BP (deg C): 265.5

BP pressure (mm Hg):

Property/Value	Units	Temp	Data Type	Reference
WS 5.02E+003	mg/L	25	EST	MEYLAN,WM ET AL. (1996)
logP 1.37			EXP	HANSCH,C & LEO,AJ (1985)
VP 1.03E-003	mm Hg	25	EST	NEELY,WB & BLAU,GE (1985)
DC 7.91	pKa	25	EXP	KORTUM,G ET AL (1961)
HL 1.04E-007	atm m3/mol	25	EST	MEYLAN,WM & HOWARD,PH (1991)
OH 2.73E-011	cm3/molc sec	25	EST	MEYLAN,WM & HOWARD,PH (1993)

CAS #: 000148-56-1				FLUMETHAZIDE

Formula: $C_8H_6F_3N_3O_4S_2$

Mol Weight: 329.28

MP (deg C):   FP (deg C):

BP (deg C):

BP pressure (mm Hg):

Property/Value	Units	Temp	Data Type	Reference
WS 3.91E+002	mg/L	25	EST	MEYLAN,WM ET AL. (1996)
logP 0.32			EXP	SANGSTER,J (1994)
VP 6.25E-011	mm Hg	25	EST	NEELY,WB & BLAU,GE (1985)
DC	pKa			
HL 4.75E-011	atm m3/mol	25	EST	MEYLAN,WM & HOWARD,PH (1991)
OH 1.81E-011	cm3/molc sec	25	EST	MEYLAN,WM & HOWARD,PH (1993)

CAS #: 000148-79-8				THIABENDAZOLE

Formula: $C_{10}H_7N_3S$

Mol Weight: 201.25

MP (deg C): 304-305   FP (deg C):

BP (deg C):

BP pressure (mm Hg):

Property/Value	Units	Temp	Data Type	Reference
WS 5.00E+001	mg/L	25	EXP	WAUCHOPE,RD ET AL. (1991A)
logP 2.47			EXP	NIELSEN,LS ET AL. (1992)
VP 4.00E-009	mm Hg	25	EXP	WAUCHOPE,RD ET AL. (1991A)
DC 4.70	pKa		EXP	WAUCHOPE,RD ET AL. (1991A)
HL 2.12E-011	atm m3/mol	25	EST	MEYLAN,WM & HOWARD,PH (1991)
OH 6.52E-011	cm3/molc sec	25	EST	MEYLAN,WM & HOWARD,PH (1993)

CAS #: 000148-82-3				4-(BIS(2-CHLOROETHYL)AMINO)-L-PHENYLALANINE

Formula: $C_{13}H_{18}Cl_2N_2O_2$

Mol Weight: 305.21

MP (deg C): 182-183   FP (deg C):

BP (deg C):

BP pressure (mm Hg):

Property/Value	Units	Temp	Data Type	Reference
WS 4.57E+001	mg/L	25	EST	MEYLAN,WM ET AL. (1996)
logP -0.52			EXP	SANGSTER,J (1994) @pH=7.4
VP 3.00E-010	mm Hg	25	EST	NEELY,WB & BLAU,GE (1985)
DC	pKa			
HL 4.20E-013	atm m3/mol	25	EST	MEYLAN,WM & HOWARD,PH (1991)
OH 2.25E-010	cm3/molc sec	25	EST	MEYLAN,WM & HOWARD,PH (1993)

CAS #: 000149-30-4				2-MERCAPTOBENZOTHIAZOLE

Formula: $C_7H_5NS_2$

Mol Weight: 167.25

MP (deg C): 180.2-81.7   FP (deg C):

BP (deg C):

BP pressure (mm Hg):

Property/Value	Units	Temp	Data Type	Reference
WS 2.50E+003	mg/L	25	EXP	SANTODONATO,J ET AL. (1976)
logP 2.42			EXP	TSCATS
VP 4.64E-004	mm Hg	25	EST	NEELY,WB & BLAU,GE (1985)
DC 6.93	pKa	20	EXP	SERJEANT,EP & DEMPSEY,B (1979)
HL 3.63E-008	atm m3/mol	25	EST	MEYLAN,WM & HOWARD,PH (1991)
OH 4.06E-011	cm3/molc sec	25	EST	MEYLAN,WM & HOWARD,PH (1993)

ERYTHRITOL

CAS #: 000149-32-6

Formula: $C_4H_{10}O_4$

Mol Weight: 122.12

MP (deg C): 121.5

BP (deg C): 330.5

BP pressure (mm Hg):

Property/ Value	Units	Temp	Data Type	Reference
WS 6.10E+005	mg/L		EXP	YALKOWSKY,SH & DANNENFELSER,RM (1992)
logP -2.29			EXP	HANSCH,C & LEO,AJ (1985)
VP 2.37E-005	mm Hg	25	EST	NEELY,WB & BLAU,GE (1985)
DC	pKa			
HL 3.08E-010	atm m3/mol	25	EST	MEYLAN,WM & HOWARD,PH (1991)
OH 2.91E-011	cm3/molc sec	25	EST	MEYLAN,WM & HOWARD,PH (1993)

HEXANOIC ACID, 2-ETHYL-

CAS #: 000149-57-5

Formula: $C_8H_{16}O_2$

Mol Weight: 144.22

MP (deg C):

BP (deg C): 228

BP pressure (mm Hg):

Property/ Value	Units	Temp	Data Type	Reference
WS 2.00E+003	mg/L	20	EXP	RIEMENSCHNEIDER,W (1986)
logP 2.64			EXP	SANGSTER,J (1993)
VP 3.00E-002	mm Hg	20	EXP	FLICK,EW (1991)
DC	pKa			
HL 2.85E-006	atm m3/mol	20	EST	VP/WSOL
OH 8.18E-012	cm3/molc sec	25	EST	MEYLAN,WM & HOWARD,PH (1993)

TRIMETHOXYMETHANE

CAS #: 000149-73-5

Formula: $C_4H_{10}O_3$

Mol Weight: 106.12

MP (deg C): 15

BP (deg C): 104

BP pressure (mm Hg):

Property/ Value	Units	Temp	Data Type	Reference
WS 6.85E+004	mg/L	25	EST	MEYLAN,WM ET AL. (1996)
logP 0.25			EXP	HANSCH,C & LEO,AJ (1985)
VP 8.14E+001	mm Hg	25	EST	NEELY,WB & BLAU,GE (1985)
DC	pKa			
HL 5.92E-007	atm m3/mol	25	EST	MEYLAN,WM & HOWARD,PH (1991)
OH 4.58E-011	cm3/molc sec	25	EST	MEYLAN,WM & HOWARD,PH (1993)

DICHLOROMETHYLPHENYLSILANE

CAS #: 000149-74-6

Formula: $C_7H_8Cl_2Si$

Mol Weight: 191.13

MP (deg C):

BP (deg C): 206.5

BP pressure (mm Hg):

Property/ Value	Units	Temp	Data Type	Reference
WS 2.55E+001	mg/L	25	EST	MEYLAN,WM ET AL. (1996)
logP 3.84			EST	MEYLAN,WM & HOWARD,PH (1995)
VP 4.88E-001	mm Hg	25	EXT	OHE,S (1976)
DC	pKa			
HL 1.14E-003	atm m3/mol	25	EST	MEYLAN,WM & HOWARD,PH (1991)
OH 2.10E-012	cm3/molc sec	25	EST	MEYLAN,WM & HOWARD,PH (1993)

3,4,5-TRIHYDROXYBENZOIC ACID

CAS #: 000149-91-7

Formula: $C_7H_6O_5$

Mol Weight: 170.12

MP (deg C): 253 dec

BP (deg C):

BP pressure (mm Hg):

Property/ Value	Units	Temp	Data Type	Reference
WS 1.19E+004	mg/L	20	EXP	YALKOWSKY,SH & DANNENFELSER,RM (1992)
logP 0.70			EXP	HANSCH,C & LEO,AJ (1985)
VP 1.19E-007	mm Hg	25	EST	NEELY,WB & BLAU,GE (1985)
DC 3.13	pKa	20	EXP	KORTUM,G ET AL (1961)
HL 1.22E-019	atm m3/mol	25	EST	MEYLAN,WM & HOWARD,PH (1991)
OH 1.08E-010	cm3/molc sec	25	EST	MEYLAN,WM & HOWARD,PH (1993)

4-AMINOBENZOIC ACID

CAS #: 000150-13-0

Formula: $C_7H_7NO_2$

Mol Weight: 137.14

MP (deg C): 187-187.5

BP (deg C):

BP pressure (mm Hg):

Property/ Value	Units	Temp	Data Type	Reference
WS 6.11E+004	mg/L	30	EXP	YALKOWSKY,SH & DANNENFELSER,RM (1992)
logP 0.83			EXP	HANSCH,C & LEO,AJ (1985)
VP 2.78E-004	mm Hg	25	EST	NEELY,WB & BLAU,GE (1985)
DC 2.38	pKa	25	EXP	KORTUM,G ET AL (1961)
HL 3.83E-011	atm m3/mol	25	EST	MEYLAN,WM & HOWARD,PH (1991)
OH 4.04E-011	cm3/molc sec	25	EST	MEYLAN,WM & HOWARD,PH (1993)

M-METHOXYPHENOL

CAS #: 000150-19-6

Formula: $C_7H_8O_2$

Mol Weight: 124.14

MP (deg C): < -17

BP (deg C): 114

BP pressure (mm Hg): 5.00E+000

Property/ Value	Units	Temp	Data Type	Reference
WS 4.97E+002	mg/L	37	EXP	YALKOWSKY,SH & DANNENFELSER,RM (1992)
logP 1.34			EXP	HANSCH,C & LEO,AJ (1985)
VP 1.96E-003	mm Hg	25	EXT	DYKYJ,J & REPAS,M (1973)
DC 9.65	pKa	25	EXP	KORTUM,G ET AL (1961)
HL 3.32E-008	atm m3/mol	25	EST	MEYLAN,WM & HOWARD,PH (1991)
OH 2.01E-010	cm3/molc sec	25	EST	MEYLAN,WM & HOWARD,PH (1993)

DL-PHENYLALANINE

CAS #: 000150-30-1

Formula: $C_9H_{11}NO_2$

Mol Weight: 165.19

MP (deg C): 284-288 de

BP (deg C):

BP pressure (mm Hg):

Property/ Value	Units	Temp	Data Type	Reference
WS 1.42E+004	mg/L	25	EXP	YALKOWSKY,SH & DANNENFELSER,RM (1992)
logP -1.44			EXP	HANSCH,C ET AL. (1995)
VP 4.65E-008	mm Hg	25	EST	NEELY,WB & BLAU,GE (1985)
DC 1.24	pKa	25	EXP	KORTUM,G ET AL (1961)
HL 1.20E-010	atm m3/mol	25	EST	MEYLAN,WM & HOWARD,PH (1991)
OH 4.42E-011	cm3/molc sec	25	EST	MEYLAN,WM & HOWARD,PH (1993)

000150-50-5 — MERPHOS

CAS #	000150-50-5
Formula	$C_{12}H_{27}PS_3$
Mol Weight	298.51
MP (deg C)	83.0
FP (deg C)	
BP (deg C)	150-152
BP pressure (mm Hg)	2.00E+000

Property/Value	Units	Temp	Data Type	Reference
WS 3.50E-003	mg/L	25	EST	MEYLAN,WM ET AL. (1996)
logP 7.67			EST	MEYLAN,WM & HOWARD,PH (1995)
VP 2.00E-005	mm Hg	25	EST	NEELY,WB & BLAU,GE (1985)
DC	pKa			
HL 2.27E-005	atm m3/mol	25	EST	MEYLAN,WM & HOWARD,PH (1991)
OH 7.88E-011	cm3/molc sec	25	EST	MEYLAN,WM & HOWARD,PH (1993)

000150-68-5 — MONURON

CAS #	000150-68-5
Formula	$C_9H_{11}ClN_2O$
Mol Weight	198.65
MP (deg C)	176-177
FP (deg C)	
BP (deg C)	
BP pressure (mm Hg)	

Property/Value	Units	Temp	Data Type	Reference
WS 2.30E+002	mg/L	25	EXP	BAILEY,GW ET AL. (1968)
logP 1.94			EXP	HANSCH,C & LEO,AJ (1985)
VP 5.03E-007	mm Hg	25	EXP	WORTHING,CR & WALKER,SB (1987)
DC	pKa			
HL 5.72E-010	atm m3/mol	25	EST	VP/WSOL
OH 3.80E-011	cm3/molc sec	25	EST	ATKINSON,R (1988)

000150-76-5 — P-METHOXYPHENOL

CAS #	000150-76-5
Formula	$C_7H_8O_2$
Mol Weight	124.14
MP (deg C)	57
FP (deg C)	
BP (deg C)	243
BP pressure (mm Hg)	

Property/Value	Units	Temp	Data Type	Reference
WS 4.00E+004	mg/L		EXP	CHEM INSPECT TEST INST (1992)
logP 1.34			EXP	HANSCH,C ET AL. (1995)
VP 8.30E-003	mm Hg	25	EST	NEELY,WB & BLAU,GE (1985)
DC 10.10	pKa	25	EST	RICCO,G ET AL. (1994)
HL 5.53E-007	atm m3/mol	25	EST	MEYLAN,WM & HOWARD,PH (1991)
OH 2.78E-011	cm3/molc sec	25	EST	MEYLAN,WM & HOWARD,PH (1993)

000150-78-7 — P-DIMETHOXYBENZENE

CAS #	000150-78-7
Formula	$C_8H_{10}O_2$
Mol Weight	138.17
MP (deg C)	58-60
FP (deg C)	
BP (deg C)	212.6
BP pressure (mm Hg)	

Property/Value	Units	Temp	Data Type	Reference
WS 1.37E+003	mg/L	25	EST	MEYLAN,WM ET AL (1996)
logP 2.15			EST	MEYLAN,WM & HOWARD,PH (1995)
VP 8.73E-002	mm Hg	25	EST	NEELY,WB & BLAU,GE (1985)
DC	pKa			
HL 3.54E-003	atm m3/mol	25	EST	MEYLAN,WM & HOWARD,PH (1991)
OH 2.03E-011	cm3/molc sec	25	EST	MEYLAN,WM & HOWARD,PH (1993)

000151-10-0 — M-DIMETHOXYBENZENE

CAS #	000151-10-0
Formula	$C_8H_{10}O_2$
Mol Weight	138.17
MP (deg C)	-52
FP (deg C)	
BP (deg C)	217.5
BP pressure (mm Hg)	

Property/Value	Units	Temp	Data Type	Reference
WS 1.11E+003	mg/L	25	EST	MEYLAN,WM ET AL. (1996)
logP 2.21			EXP	NAKAGAWA,Y ET AL. (1992)
VP 5.28E-001	mm Hg	25	EST	NEELY,WB & BLAU,GE (1985)
DC	pKa			
HL 1.89E-005	atm m3/mol	25	EST	MEYLAN,WM & HOWARD,PH (1991)
OH 2.02E-010	cm3/molc sec	25	EST	MEYLAN,WM & HOWARD,PH (1993)

000151-21-3 — DODECYL SULFATE, SODIUM SALT

CAS #	000151-21-3
Formula	$C_{12}H_{25}NaO_4S$
Mol Weight	288.38
MP (deg C)	204-207
FP (deg C)	
BP (deg C)	
BP pressure (mm Hg)	

Property/Value	Units	Temp	Data Type	Reference
WS 1.00E+005	mg/L		EXP	SINGER,MM & TJEERDEMA,RS (1993)
logP 1.60			EXP	HANSCH,C & LEO,AJ (1985)
VP 4.70E-013	mm Hg	25	EST	NEELY,WB & BLAU,GE (1985)
DC	pKa			
HL	atm m3/mol			
OH 1.48E-011	cm3/molc sec	25	EST	MEYLAN,WM & HOWARD,PH (1993)

000151-33-7 — SODIUM HEXANOATE

CAS #	000151-33-7
Formula	$C_6H_{11}NaO_2$
Mol Weight	138.14
MP (deg C)	
FP (deg C)	
BP (deg C)	
BP pressure (mm Hg)	

Property/Value	Units	Temp	Data Type	Reference
WS 1.00E+006	mg/L	25	EST	MEYLAN,WM ET AL. (1996)
logP -2.17			EXP	HANSCH,C & LEO,AJ (1985)
VP 2.77E-008	mm Hg	25	EST	NEELY,WB & BLAU,GE (1985)
DC	pKa			
HL	atm m3/mol			
OH 4.99E-012	cm3/molc sec	25	EST	MEYLAN,WM & HOWARD,PH (1993)

000151-38-2 — (METHOXYETHYL)MERCURIC ACETATE

CAS #	000151-38-2
Formula	$C_5H_{10}HgO_3$
Mol Weight	318.72
MP (deg C)	
FP (deg C)	
BP (deg C)	
BP pressure (mm Hg)	

Property/Value	Units	Temp	Data Type	Reference
WS 3.11E+004	mg/L	25	EST	MEYLAN,WM ET AL. (1996)
logP -0.60			EST	MEYLAN,WM & HOWARD,PH (1995)
VP 4.53E-002	mm Hg	25	EST	NEELY,WB & BLAU,GE (1985)
DC	pKa			
HL	atm m3/mol			
OH 2.04E-010	cm3/molc sec	25	EST	MEYLAN,WM & HOWARD,PH (1993)

AZIRIDINE

CAS #:	000151-56-4			
Formula:	C_2H_5N			
Mol Weight:	43.07			
MP (deg C):	-78.0	FP (deg C):		
BP (deg C):	57.0			
BP pressure (mm Hg):				

Property/Value	Units	Temp	Data Type	Reference
WS 1.00E+006	mg/L		EXP	RIDDICK,JA ET AL. (1986)
logP -0.28			EST	MEYLAN,WM & HOWARD,PH (1995)
VP 2.13E+002	mm Hg	25	EXP	CHAO,J ET AL. (1983)
DC 8.04	pKa	25	EXP	RIDDICK,JA ET AL. (1986)
HL 1.21E-005	atm m3/mol	25	EST	VP/WSOL
OH 6.10E-012	cm3/molc sec	25	EXP	ATKINSON,R (1989)

HALOTHANE

CAS #:	000151-67-7			
Formula:	$C_2HBrClF_3$			
Mol Weight:	197.39			
MP (deg C):	-118	FP (deg C):		
BP (deg C):	50.2			
BP pressure (mm Hg):				

Property/Value	Units	Temp	Data Type	Reference
WS 3.90E+003	mg/L	25	EXP	SUZUKI,T (1991)
logP 2.30			EXP	HANSCH,C ET AL (1995)
VP 3.02E+002	mm Hg	25	EXP	DAUBERT,TE & DANNER,RP (1989)
DC	pKa			
HL 3.13E-002	atm m3/mol	25	EST	MEYLAN,WM & HOWARD,PH (1991)
OH 6.00E-014	cm3/molc sec	25	EXP	BROWN,AC ET AL. (1990A)

LEVALLORPHAN

CAS #:	000152-02-3			
Formula:	$C_{19}H_{25}NO$			
Mol Weight:	283.42			
MP (deg C):	180-182	FP (deg C):		
BP (deg C):				
BP pressure (mm Hg):				

Property/Value	Units	Temp	Data Type	Reference
WS 6.33E+002	mg/L	25	EST	MEYLAN,WM ET AL. (1996)
logP 3.48			EXP	HANSCH,C & LEO,AJ (1985)
VP 1.87E-007	mm Hg	25	EST	NEELY,WB & BLAU,GE (1985)
DC	pKa			
HL 2.77E-010	atm m3/mol	25	EST	MEYLAN,WM & HOWARD,PH (1991)
OH 2.40E-010	cm3/molc sec	25	EST	MEYLAN,WM & HOWARD,PH (1993)

OCTAMETHYLPYROPHOSPHORAMIDE

CAS #:	000152-16-9			
Formula:	$C_8H_{24}N_4O_3P_2$			
Mol Weight:	286.25			
MP (deg C):	14-20	FP (deg C):		
BP (deg C):	154			
BP pressure (mm Hg):	2.00E+000			

Property/Value	Units	Temp	Data Type	Reference
WS 1.00E+006	mg/L	20	EXP	YALKOWSKY,SH & DANNENFELSER,RM (1992)
logP -2.00			EST	MEYLAN,WM & HOWARD,PH (1995)
VP 1.00E-003	mm Hg	25	EXP	MARTIN,H & WORTHING,CR (1977)
DC	pKa			
HL 3.77E-010	atm m3/mol	25	EST	MEYLAN,WM & HOWARD,PH (1991)
OH 1.20E-010	cm3/molc sec	25	EST	MEYLAN,WM & HOWARD,PH (1993)

O,O,O-TRIMETHYL PHOSPHOROTHIOATE

CAS #:	000152-18-1			
Formula:	$C_3H_9O_3PS$			
Mol Weight:	156.14			
MP (deg C):		FP (deg C):		
BP (deg C):				
BP pressure (mm Hg):				

Property/Value	Units	Temp	Data Type	Reference
WS 7.23E+003	mg/L	25	EST	MEYLAN,WM ET AL. (1996)
logP 1.16			EST	MEYLAN,WM & HOWARD,PH (1995)
VP 1.73E+000	mm Hg	25	EST	NEELY,WB & BLAU,GE (1985)
DC	pKa			
HL 1.15E-004	atm m3/mol	25	EST	MEYLAN,WM & HOWARD,PH (1991)
OH 6.97E-011	cm3/molc sec	25	EXP	ATKINSON,R (1989)

PHOSPHOROTHIOIC ACID, O,O,S-TRIMETHYL ESTER

CAS #:	000152-20-5			
Formula:	$C_3H_9O_3PS$			
Mol Weight:	156.14			
MP (deg C):		FP (deg C):		
BP (deg C):				
BP pressure (mm Hg):				

Property/Value	Units	Temp	Data Type	Reference
WS 6.57E+004	mg/L	25	EST	MEYLAN,WM ET AL. (1996)
logP 0.04			EST	MEYLAN,WM & HOWARD,PH (1995)
VP 2.06E-001	mm Hg	25	EST	NEELY,WB & BLAU,GE (1985)
DC	pKa			
HL 1.37E-007	atm m3/mol	25	EST	MEYLAN,WM & HOWARD,PH (1991)
OH 9.30E-012	cm3/molc sec	25	EXP	ATKINSON,R (1989)

SULFALENE

CAS #:	000152-47-6			
Formula:	$C_{11}H_{12}N_4O_3S$			
Mol Weight:	280.31			
MP (deg C):	176	FP (deg C):		
BP (deg C):				
BP pressure (mm Hg):				

Property/Value	Units	Temp	Data Type	Reference
WS 4.03E+003	mg/L	25	EST	MEYLAN,WM ET AL. (1996)
logP 0.70			EXP	HANSCH,C & LEO,AJ (1985)
VP 8.35E-009	mm Hg	25	EST	NEELY,WB & BLAU,GE (1985)
DC	pKa			
HL 2.41E-014	atm m3/mol	25	EST	MEYLAN,WM & HOWARD,PH (1991)
OH 2.82E-011	cm3/molc sec	25	EST	MEYLAN,WM & HOWARD,PH (1993)

CORTEXOLONE

CAS #:	000152-58-9			
Formula:	$C_{21}H_{30}O_4$			
Mol Weight:	346.47			
MP (deg C):	212.8-6.8	FP (deg C):		
BP (deg C):				
BP pressure (mm Hg):				

Property/Value	Units	Temp	Data Type	Reference
WS 1.52E+001	mg/L	25	EST	MEYLAN,WM ET AL. (1996)
logP 3.08			EXP	SANGSTER,J (1994)
VP 9.49E-012	mm Hg	25	EST	NEELY,WB & BLAU,GE (1985)
DC	pKa			
HL 1.58E-006	atm m3/mol	25	EST	MEYLAN,WM & HOWARD,PH (1991)
OH 1.03E-010	cm3/molc sec	25	EST	MEYLAN,WM & HOWARD,PH (1993)

CAS #: 000152-72-7	ACENOCOUMARIN

Formula: $C_{19}H_{15}NO_6$

Mol Weight: 353.33

MP (deg C): 197　FP (deg C):

BP (deg C):

BP pressure (mm Hg):

Property/Value	Units	Temp	Data Type	Reference
WS 4.91E+001	mg/L	25	EST	MEYLAN,WM ET AL. (1996)
logP 1.98			EXP	SANGSTER,J (1994)
VP 2.85E-014	mm Hg	25	EST	NEELY,WB & BLAU,GE (1985)
DC	pKa			
HL 8.99E-016	atm m3/mol	25	EST	MEYLAN,WM & HOWARD,PH (1991)
OH 4.96E-011	cm3/molc sec	25	EST	MEYLAN,WM & HOWARD,PH (1993)

CAS #: 000153-61-7	CEPHALOTHIN

Formula: $C_{16}H_{16}N_2O_6S_2$

Mol Weight: 396.44

MP (deg C): 160-160.5　FP (deg C):

BP (deg C):

BP pressure (mm Hg):

Property/Value	Units	Temp	Data Type	Reference
WS 1.58E+002	mg/L	25	EST	MEYLAN,WM ET AL. (1996)
logP 0.00			EXP	HANSCH,C ET AL. (1995)
VP 1.67E-014	mm Hg	25	EST	NEELY,WB & BLAU,GE (1985)
DC	pKa			
HL 1.74E-017	atm m3/mol	25	EST	MEYLAN,WM & HOWARD,PH (1991)
OH 1.35E-010	cm3/molc sec	25	EST	MEYLAN,WM & HOWARD,PH (1993)

CAS #: 000153-78-6	2-AMINOFLUORENE

Formula: $C_{13}H_{11}N$

Mol Weight: 181.24

MP (deg C): 127-131　FP (deg C):

BP (deg C):

BP pressure (mm Hg):

Property/Value	Units	Temp	Data Type	Reference
WS 8.78E+000	mg/L	25	EST	MEYLAN,WM ET AL. (1996)
logP 3.14			EXP	DEBNATH,AK ET AL. (1992)
VP 2.79E-005	mm Hg	25	EST	NEELY,WB & BLAU,GE (1985)
DC 4.64	pKa	25	EXP	PERRIN,DD (1965)
HL 4.54E-008	atm m3/mol	25	EST	MEYLAN,WM & HOWARD,PH (1991)
OH 1.96E-010	cm3/molc sec	25	EST	MEYLAN,WM & HOWARD,PH (1993)

CAS #: 000154-21-2	LINCOMYCIN

Formula: $C_{18}H_{34}N_2O_6S$

Mol Weight: 406.55

MP (deg C):　FP (deg C):

BP (deg C):

BP pressure (mm Hg):

Property/Value	Units	Temp	Data Type	Reference
WS 9.27E+002	mg/L	25	EST	MEYLAN,WM ET AL. (1996)
logP 0.56			EXP	HANSCH,C ET AL. (1995)
VP 1.34E-017	mm Hg	25	EST	NEELY,WB & BLAU,GE (1985)
DC	pKa			
HL 3.00E-023	atm m3/mol	25	EST	MEYLAN,WM & HOWARD,PH (1991)
OH 2.58E-010	cm3/molc sec	25	EST	MEYLAN,WM & HOWARD,PH (1993)

CAS #: 000154-23-4	CIANIDANOL

Formula: $C_{15}H_{14}O_6$

Mol Weight: 290.28

MP (deg C): 212-216　FP (deg C):

BP (deg C):

BP pressure (mm Hg):

Property/Value	Units	Temp	Data Type	Reference
WS 6.31E+004	mg/L	25	EST	MEYLAN,WM ET AL. (1996)
logP 0.51			EXP	PERRISSOUD,D & TESTA,B (1986)
VP 2.07E-012	mm Hg	25	EST	NEELY,WB & BLAU,GE (1985)
DC	pKa			
HL 7.12E-026	atm m3/mol	25	EST	MEYLAN,WM & HOWARD,PH (1991)
OH 2.27E-010	cm3/molc sec	25	EST	MEYLAN,WM & HOWARD,PH (1993)

CAS #: 000154-42-7	THIOGUANINE

Formula: $C_5H_5N_5S$

Mol Weight: 167.19

MP (deg C): 161.5　FP (deg C):

BP (deg C):

BP pressure (mm Hg):

Property/Value	Units	Temp	Data Type	Reference
WS 3.63E+004	mg/L	25	EST	MEYLAN,WM ET AL. (1996)
logP -0.07			EXP	HANSCH,C & LEO,AJ (1985)
VP 1.01E-010	mm Hg	25	EST	NEELY,WB & BLAU,GE (1985)
DC	pKa			
HL 4.64E-015	atm m3/mol	25	EST	MEYLAN,WM & HOWARD,PH (1991)
OH 1.31E-010	cm3/molc sec	25	EST	MEYLAN,WM & HOWARD,PH (1993)

CAS #: 000154-93-8	1,3-BIS(2-CHLOROET)-1-NITROSOUREA

Formula: $C_5H_9Cl_2N_3O_2$

Mol Weight: 214.05

MP (deg C): 30-32　FP (deg C):

BP (deg C):

BP pressure (mm Hg):

Property/Value	Units	Temp	Data Type	Reference
WS 1.83E+003	mg/L	25	EST	MEYLAN,WM ET AL. (1996)
logP 1.53			EXP	HANSCH,C & LEO,AJ (1985)
VP 3.69E-005	mm Hg	25	EST	NEELY,WB & BLAU,GE (1985)
DC	pKa			
HL 4.76E-011	atm m3/mol	25	EST	MEYLAN,WM & HOWARD,PH (1991)
OH 3.63E-012	cm3/molc sec	25	EST	MEYLAN,WM & HOWARD,PH (1993)

CAS #: 000154-97-2	PYRIDINIUM, 2-[(HYDROXYIMINO)METHYL]-1-METHYL-,

Formula: $C_8H_{12}N_2O_4S$

Mol Weight: 232.26

MP (deg C): 235-238　FP (deg C):

BP (deg C):

BP pressure (mm Hg):

Property/Value	Units	Temp	Data Type	Reference
WS 1.00E+006	mg/L	25	EST	MEYLAN,WM ET AL. (1996)
logP -3.24			EXP	SANGSTER,J (1994)
VP 3.33E-008	mm Hg	25	EST	NEELY,WB & BLAU,GE (1985)
DC	pKa			
HL 9.17E-018	atm m3/mol	25	EST	MEYLAN,WM & HOWARD,PH (1991)
OH 1.30E-011	cm3/molc sec	25	EST	MEYLAN,WM & HOWARD,PH (1993)

CAS #: 000156-10-5 — 4-NITROSO-N-PHENYLBENZENAMINE

Formula: $C_{12}H_{10}N_2O$

Mol Weight: 198.23

MP (deg C): 144-145

FP (deg C):

BP (deg C): 292

BP pressure (mm Hg):

Property/Value	Units	Temp	Data Type	Reference
WS 1.73E+001	mg/L	25	EST	MEYLAN,WM ET AL. (1996)
logP 3.16			EST	MEYLAN,WM & HOWARD,PH (1995)
VP 1.85E-004	mm Hg	25	EST	NEELY,WB & BLAU,GE (1985)
DC	pKa			
HL 1.10E-008	atm m3/mol	25	EST	MEYLAN,WM & HOWARD,PH (1991)
OH 2.00E-010	cm3/molc sec	25	EST	MEYLAN,WM & HOWARD,PH (1993)

CAS #: 000156-34-3 — BENZENEETHANAMINE, A-METHYL-, (R)-

Formula: $C_9H_{13}N$

Mol Weight: 135.21

MP (deg C):

FP (deg C):

BP (deg C):

BP pressure (mm Hg):

Property/Value	Units	Temp	Data Type	Reference
WS 2.80E+004	mg/L	25	EST	MEYLAN,WM ET AL. (1996)
logP 1.76			EXP	SANGSTER,J (1994)
VP 2.01E-001	mm Hg	25	EST	NEELY,WB & BLAU,GE (1985)
DC 10.13	pKa	20	EXP	PERRIN,DD (1965)
HL 1.08E-006	atm m3/mol	25	EST	MEYLAN,WM & HOWARD,PH (1991)
OH 4.94E-011	cm3/molc sec	25	EST	MEYLAN,WM & HOWARD,PH (1993)

CAS #: 000156-38-7 — P-HYDROXYPHENYLACETIC ACID

Formula: $C_8H_8O_3$

Mol Weight: 152.15

MP (deg C): 152

FP (deg C):

BP (deg C):

BP pressure (mm Hg):

Property/Value	Units	Temp	Data Type	Reference
WS 1.60E+005	mg/L	25	EST	MEYLAN,WM ET AL. (1996)
logP 0.75			EXP	HANSCH,C & LEO,AJ (1985)
VP 5.55E-005	mm Hg	25	EST	NEELY,WB & BLAU,GE (1985)
DC	pKa			
HL 4.60E-012	atm m3/mol	25	EST	MEYLAN,WM & HOWARD,PH (1991)
OH 3.57E-011	cm3/molc sec	25	EST	MEYLAN,WM & HOWARD,PH (1993)

CAS #: 000156-43-4 — P-ETHOXYANILINE

Formula: $C_8H_{11}NO$

Mol Weight: 137.18

MP (deg C): 186-188

FP (deg C):

BP (deg C):

BP pressure (mm Hg):

Property/Value	Units	Temp	Data Type	Reference
WS 7.51E+003	mg/L	25	EST	MEYLAN,WM ET AL. (1996)
logP 1.24			EXP	HANSCH,C & LEO,AJ (1985)
VP 1.05E-002	mm Hg	25	EXP	YAWS,CL (1994B)
DC 5.20	pKa	28	EXP	PERRIN,DD (1965)
HL 1.50E-007	atm m3/mol	25	EST	MEYLAN,WM & HOWARD,PH (1991)
OH 9.93E-011	cm3/molc sec	25	EST	MEYLAN,WM & HOWARD,PH (1993)

CAS #: 000156-54-7 — BUTANOIC ACID, SODIUM SALT

Formula: $C_4H_7NaO_2$

Mol Weight: 110.09

MP (deg C): 250-253

FP (deg C):

BP (deg C):

BP pressure (mm Hg):

Property/Value	Units	Temp	Data Type	Reference
WS 1.00E+006	mg/L	25	EST	MEYLAN,WM ET AL. (1996)
logP -3.20			EXP	HANSCH,C ET AL. (1995)
VP 1.54E-007	mm Hg	25	EST	NEELY,WB & BLAU,GE (1985)
DC	pKa			
HL	atm m3/mol			
OH 2.17E-012	cm3/molc sec	25	EST	MEYLAN,WM & HOWARD,PH (1993)

CAS #: 000156-59-2 — 1,2-DICHLOROETHENE (CIS)

Formula: $C_2H_2Cl_2$

Mol Weight: 96.94

MP (deg C): -80.1

FP (deg C):

BP (deg C): 60.2

BP pressure (mm Hg):

Property/Value	Units	Temp	Data Type	Reference
WS 3.50E+003	mg/L	25	EXP	YALKOWSKY,SH & DANNENFELSER,RM (1992)
logP 1.86			EXP	HANSCH,C & LEO,AJ (1985)
VP 2.01E+002	mm Hg	25	EXT	BOUBLIK,T ET AL. (1984)
DC	pKa			
HL 4.08E-003	atm m3/mol	24	EXP	GOSSETT,JM (1987)
OH 2.38E-012	cm3/molc sec	25	EXP	ATKINSON,R (1989)

CAS #: 000156-60-5 — 1,2-DICHLOROETHENE (TRANS)

Formula: $C_2H_2Cl_2$

Mol Weight: 96.94

MP (deg C): -49.8

FP (deg C):

BP (deg C): 47.7

BP pressure (mm Hg):

Property/Value	Units	Temp	Data Type	Reference
WS 6.30E+003	mg/L	25	EXP	YALKOWSKY,SH & DANNENFELSER,RM (1992)
logP 2.09			EXP	HANSCH,C & LEO,AJ (1985)
VP 3.31E+002	mm Hg	25	EXT	BOUBLIK,T ET AL. (1984)
DC	pKa			
HL 9.38E-003	atm m3/mol	24	EXP	GOSSETT,JM (1987)
OH 1.80E-012	cm3/molc sec	25	EXP	ATKINSON,R (1989)

CAS #: 000156-82-1 — 2-THIOURACIL

Formula: $C_4H_4N_2OS$

Mol Weight: 128.15

MP (deg C):

FP (deg C):

BP (deg C):

BP pressure (mm Hg):

Property/Value	Units	Temp	Data Type	Reference
WS 7.94E+003	mg/L	25	EST	MEYLAN,WM ET AL. (1996)
logP -0.28			EXP	HANSCH,C & LEO,AJ (1985)
VP 3.71E-006	mm Hg	25	EST	NEELY,WB & BLAU,GE (1985)
DC	pKa			
HL 4.11E-010	atm m3/mol	25	EST	MEYLAN,WM & HOWARD,PH (1991)
OH 7.53E-011	cm3/molc sec	25	EST	MEYLAN,WM & HOWARD,PH (1993)

CAS #: 000156-87-6				PROPANOLAMINE

Formula: C_3H_9NO

Mol Weight: 75.11

MP (deg C): 11	FP (deg C):

BP (deg C): 187.5

BP pressure (mm Hg):

Property/Value	Units	Temp	Data Type	Reference
WS 1.00E+006	mg/L	25	EST	MEYLAN,WM ET AL. (1996)
logP -1.12			EXP	HANSCH,C ET AL. (1995)
VP 7.71E-002	mm Hg	25	EXP	DAUBERT,TE & DANNER,RP (1989)
DC 9.96	pKa	25	EXP	PERRIN,DD (1965)
HL 4.88E-010	atm m3/mol	25	EST	MEYLAN,WM & HOWARD,PH (1991)
OH 3.73E-011	cm3/molc sec	25	EST	MEYLAN,WM & HOWARD,PH (1993)

CAS #: 000157-40-4				SPIROPENTANE

Formula: C_5H_8

Mol Weight: 68.12

MP (deg C): -134.6	FP (deg C):

BP (deg C): 39

BP pressure (mm Hg):

Property/Value	Units	Temp	Data Type	Reference
WS 3.12E+002	mg/L	25	EST	MEYLAN,WM ET AL. (1996)
logP 2.46			EST	MEYLAN,WM & HOWARD,PH (1995)
VP 4.58E+002	mm Hg	25	EXP	OHE,S (1976)
DC	pKa			
HL 8.47E-002	atm m3/mol	25	EST	MEYLAN,WM & HOWARD,PH (1991)
OH 1.13E-013	cm3/molc sec	25	EST	MEYLAN,WM & HOWARD,PH (1993)

CAS #: 000177-49-1				TETRACYCLOTETRAMETHYLENECYCLOTETRASILOXANE

Formula: $C_{16}H_{32}O_4Si_4$

Mol Weight: 400.78

MP (deg C):	FP (deg C):

BP (deg C):

BP pressure (mm Hg):

Property/Value	Units	Temp	Data Type	Reference
WS 1.41E-005	mg/L	25	EST	MEYLAN,WM ET AL. (1996)
logP 8.57			EST	MEYLAN,WM & HOWARD,PH (1995)
VP 2.47E-001	mm Hg	25	EXP	DITSENT,VE ET AL. (1974A)
DC	pKa			
HL	atm m3/mol			
OH 8.77E-012	cm3/molc sec	25	EST	MEYLAN,WM & HOWARD,PH (1993)

CAS #: 000185-94-4				BICYCLO(2.1.0)PENTANE

Formula: C_5H_8

Mol Weight: 68.12

MP (deg C):	FP (deg C):

BP (deg C):

BP pressure (mm Hg):

Property/Value	Units	Temp	Data Type	Reference
WS 3.35E+002	mg/L	25	EST	MEYLAN,WM ET AL. (1996)
logP 2.43			EST	MEYLAN,WM & HOWARD,PH (1995)
VP 4.69E-001	mm Hg	25	EXP	VARUSHCHENKO,RM ET AL. (1974)
DC	pKa			
HL 8.47E-002	atm m3/mol	25	EST	MEYLAN,WM & HOWARD,PH (1991)
OH 9.17E-013	cm3/molc sec	25	EST	MEYLAN,WM & HOWARD,PH (1993)

CAS #: 000187-26-8				TRICYCLO(4.1.0.2,4)HEPTANE

Formula: C_7H_{10}

Mol Weight: 94.16

MP (deg C):	FP (deg C):

BP (deg C):

BP pressure (mm Hg):

Property/Value	Units	Temp	Data Type	Reference
WS 7.23E+001	mg/L	25	EST	MEYLAN,WM ET AL. (1996)
logP 3.15			EST	MEYLAN,WM & HOWARD,PH (1995)
VP 5.00E-002	mm Hg	25	EXP	VARUSHCHENKO,RM ET AL. (1974)
DC	pKa			
HL 6.58E-002	atm m3/mol	25	EST	MEYLAN,WM & HOWARD,PH (1991)
OH 1.12E-012	cm3/molc sec	25	EST	MEYLAN,WM & HOWARD,PH (1993)

CAS #: 000187-78-0				CYCLOPENT[FG]ACENAPHTHYLENE

Formula: $C_{14}H_{16}$

Mol Weight: 184.28

MP (deg C):	FP (deg C):

BP (deg C):

BP pressure (mm Hg):

Property/Value	Units	Temp	Data Type	Reference
WS 2.09E-001	mg/L	25	EST	MEYLAN,WM ET AL. (1996)
logP 5.07			EXP	DE VOOGT,P ET AL. (1990)
VP 3.03E-005	mm Hg	25	EST	NEELY,WB & BLAU,GE (1985)
DC	pKa			
HL 1.37E-005	atm m3/mol	25	EST	MEYLAN,WM & HOWARD,PH (1991)
OH 3.33E-011	cm3/molc sec	25	EST	MEYLAN,WM & HOWARD,PH (1993)

CAS #: 000189-55-9				DIBENZO(A,I)PYRENE

Formula: $C_{24}H_{14}$

Mol Weight: 302.38

MP (deg C): 281.5-282	FP (deg C):

BP (deg C): 275

BP pressure (mm Hg): 5.00E-002

Property/Value	Units	Temp	Data Type	Reference
WS 5.54E-004	mg/L	25	EST	MEYLAN,WM ET AL. (1996)
logP 7.28			EST	MEYLAN,WM & HOWARD,PH (1995)
VP 2.00E-012	mm Hg	25	EST	H X WSOL
DC	pKa			
HL 1.41E-008	atm m3/mol	25	EST	MEYLAN,WM & HOWARD,PH (1991)
OH 2.00E-010	cm3/molc sec	25	EST	MEYLAN,WM & HOWARD,PH (1993)

CAS #: 000189-64-0				DIBENZO[A,H]PYRENE

Formula: $C_{24}H_{14}$

Mol Weight: 302.38

MP (deg C):	FP (deg C):

BP (deg C):

BP pressure (mm Hg):

Property/Value	Units	Temp	Data Type	Reference
WS 4.06E-005	mg/L	25	EST	MEYLAN,WM ET AL. (1996)
logP 7.28			EST	MEYLAN,WM & HOWARD,PH (1995)
VP 7.87E-012	mm Hg	25	EST	NEELY,WB & BLAU,GE (1985)
DC	pKa			
HL 1.41E-008	atm m3/mol	25	EST	MEYLAN,WM & HOWARD,PH (1991)
OH 5.00E-011	cm3/molc sec	25	EST	MEYLAN,WM & HOWARD,PH (1993)

000191-07-1 — CORONENE

CAS #:	000191-07-1
Formula:	$C_{24}H_{12}$
Mol Weight:	300.36
MP (deg C):	437.3
FP (deg C):	
BP (deg C):	525
BP pressure (mm Hg):	

Property/Value	Units	Temp	Data Type	Reference
WS 1.40E-004	mg/L	25	EXP	YALKOWSKY,SH & DANNENFELSER,RM (1992)
logP 7.64			EXP	DE VOOGT,P ET AL. (1990)
VP 3.29E-011	mm Hg	25	EST	NEELY,WB & BLAU,GE (1985)
DC	pKa			
HL 2.12E-008	atm m3/mol	25	EST	MEYLAN,WM & HOWARD,PH (1991)
OH 5.00E-011	cm3/molc sec	25	EST	MEYLAN,WM & HOWARD,PH (1993)

000191-24-2 — BENZO(GHI)PERYLENE

CAS #:	000191-24-2
Formula:	$C_{22}H_{12}$
Mol Weight:	276.34
MP (deg C):	277-279
FP (deg C):	
BP (deg C):	>500
BP pressure (mm Hg):	

Property/Value	Units	Temp	Data Type	Reference
WS 2.60E-004	mg/L	25	EXP	MACKAY,D & SHIU,WY (1977)
logP 6.63			EXP	HANSCH,C ET AL. (1995)
VP 1.01E-010	mm Hg	25	EXT	MURRAY,JJ ET AL. (1974)
DC	pKa			
HL 1.60E-006	atm m3/mol	25	EST	VP/WSOL
OH 2.00E-010	cm3/molc sec	25	EST	MEYLAN,WM & HOWARD,PH (1993)

000191-26-4 — DIBENZO(DEF,MNO)CHRYSENE

CAS #:	000191-26-4
Formula:	$C_{22}H_{12}$
Mol Weight:	276.34
MP (deg C):	
FP (deg C):	
BP (deg C):	
BP pressure (mm Hg):	

Property/Value	Units	Temp	Data Type	Reference
WS 1.27E-003	mg/L	25	EST	MEYLAN,WM ET AL. (1996)
logP 7.04			EXP	DEVOOGT,P ET AL. (1990)
VP 8.75E-010	mm Hg	25	EST	NEELY,WB & BLAU,GE (1985)
DC	pKa			
HL 1.31E-007	atm m3/mol	25	EST	MEYLAN,WM & HOWARD,PH (1991)
OH 5.00E-011	cm3/molc sec	25	EST	MEYLAN,WM & HOWARD,PH (1993)

000191-30-0 — DIBENZO[a,l]PYRENE

CAS #:	000191-30-0
Formula:	$C_{24}H_{14}$
Mol Weight:	302.38
MP (deg C):	162.4
FP (deg C):	
BP (deg C):	
BP pressure (mm Hg):	

Property/Value	Units	Temp	Data Type	Reference
WS 3.60E-004	mg/L	25	EST	MEYLAN,WM ET AL. (1996)
logP 7.71			EXP	DEVOOGT,P ET AL. (1990)
VP 4.80E-010	mm Hg	25	EST	NEELY,WB & BLAU,GE (1985)
DC	pKa			
HL 1.41E-008	atm m3/mol	25	EST	MEYLAN,WM & HOWARD,PH (1991)
OH 5.00E-011	cm3/molc sec	25	EST	MEYLAN,WM & HOWARD,PH (1993)

000191-68-4 — DIBENZO[G,P]CHRYSENE

CAS #:	000191-68-4
Formula:	$C_{26}H_{16}$
Mol Weight:	328.42
MP (deg C):	
FP (deg C):	
BP (deg C):	
BP pressure (mm Hg):	

Property/Value	Units	Temp	Data Type	Reference
WS 4.43E-005	mg/L	25	EST	MEYLAN,WM ET AL. (1996)
logP 8.39			EXP	DEVOOGT,P ET AL. (1990)
VP 1.48E-011	mm Hg	25	EST	NEELY,WB & BLAU,GE (1985)
DC	pKa			
HL 4.78E-008	atm m3/mol	25	EST	MEYLAN,WM & HOWARD,PH (1991)
OH 5.00E-011	cm3/molc sec	25	EST	MEYLAN,WM & HOWARD,PH (1993)

000192-65-4 — DIBENZO[a,e]PYRENE

CAS #:	000192-65-4
Formula:	$C_{24}H_{14}$
Mol Weight:	302.38
MP (deg C):	233
FP (deg C):	
BP (deg C):	
BP pressure (mm Hg):	

Property/Value	Units	Temp	Data Type	Reference
WS 8.02E-005	mg/L	25	EST	MEYLAN,WM ET AL. (1996)
logP 7.28			EST	MEYLAM,WM & HOWARD,PH (1995)
VP 7.03E-011	mm Hg	25	EST	NEELY,WB & BLAU,GE (1985)
DC	pKa			
HL 1.41E-008	atm m3/mol	25	EST	MEYLAN,WM & HOWARD,PH (1991)
OH 5.00E-011	cm3/molc sec	25	EST	MEYLAN,WM & HOWARD,PH (1993)

000192-97-2 — BENZO(E)PYRENE

CAS #:	000192-97-2
Formula:	$C_{20}H_{12}$
Mol Weight:	252.32
MP (deg C):	178.7
FP (deg C):	
BP (deg C):	310-312
BP pressure (mm Hg):	1.00E+001

Property/Value	Units	Temp	Data Type	Reference
WS 6.30E-003	mg/L	25	EXP	PEARLMAN,RS ET AL. (1984)
logP 6.44			EXP	DEVOOGT,P ET AL. (1990)
VP 5.70E-009	mm Hg	25	EXT	MURRAY,JJ ET AL. (1974)
DC	pKa			
HL 1.07E-006	atm m3/mol	25	EST	MEYLAN,WM & HOWARD,PH (1991)
OH 1.50E-010	cm3/molc sec	25	EST	ATKINSON,R (1988)

000193-39-5 — INDENO(1,2,3-CD)PYRENE

CAS #:	000193-39-5
Formula:	$C_{22}H_{12}$
Mol Weight:	276.34
MP (deg C):	163.6
FP (deg C):	
BP (deg C):	497
BP pressure (mm Hg):	

Property/Value	Units	Temp	Data Type	Reference
WS 2.20E-005	mg/L	20	EXP	COOVER,MP & SIMS,RC (1987)
logP 6.70			EST	MEYLAN,WM & HOWARD,PH (1995)
VP 1.00E-010	mm Hg	20	EXP	COOVER,MP & SIMS,RC (1987)
DC	pKa			
HL 1.60E-006	atm m3/mol	20	EST	VP/WSOL
OH 1.02E-010	cm3/molc sec	25	EST	MEYLAN,WM & HOWARD,PH (1993)

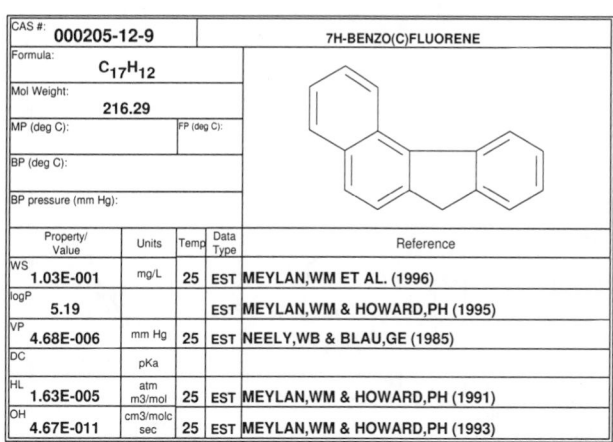
287

CAS #: 000205-82-3				BENZO(J)FLUORANTHENE

Formula: $C_{20}H_{12}$

Mol Weight: 252.32

MP (deg C): 166   FP (deg C):

BP (deg C):

BP pressure (mm Hg):

Property/Value	Units	Temp	Data Type	Reference
WS 2.50E-003	mg/L		EXP	YALKOWSKY,SH & DANNENFELSER,RM (1992)
logP 6.11			EST	MEYLAN,WM & HOWARD,PH (1995)
VP 2.40E-008	mm Hg	25	EST	NEELY,WB & BLAU,GE (1985)
DC	pKa			
HL 8.10E-007	atm m3/mol	25	EST	MEYLAN,WM & HOWARD,PH (1991)
OH 5.36E-011	cm3/molc sec	25	EST	MEYLAN,WM & HOWARD,PH (1993)

CAS #: 000205-99-2				BENZO(B)FLUORANTHENE

Formula: $C_{20}H_{12}$

Mol Weight: 252.32

MP (deg C): 167-168   FP (deg C):

BP (deg C):

BP pressure (mm Hg):

Property/Value	Units	Temp	Data Type	Reference
WS 1.50E-003	mg/L		EXP	YALKOWSKY,SH & DANNENFELSER,RM (1992)
logP 5.78			EXP	WANG,L ET AL. (1986)
VP 5.00E-007	mm Hg	25	NR	PERWAK,J ET AL. (1982B)
DC	pKa			
HL 1.11E-004	atm m3/mol	25	EST	VP/WSOL
OH 4.11E+001	cm3/molc sec	25	EST	MEYLAN,WM & HOWARD,PH (1993)

CAS #: 000206-44-0				FLUORANTHENE

Formula: $C_{16}H_{10}$

Mol Weight: 202.26

MP (deg C): 111   FP (deg C):

BP (deg C): 384

BP pressure (mm Hg):

Property/Value	Units	Temp	Data Type	Reference
WS 2.60E-001	mg/L	25	EXP	MACKAY,D & SHIU,WY (1977)
logP 5.16			EXP	HANSCH,C & LEO,AJ (1985)
VP 1.23E-008	mm Hg	25	EXT	BOYD,RH ET AL. (1965)
DC	pKa			
HL 1.36E-004	atm m3/mol	22	EXP	HELLMANN,H (1987)
OH 5.00E-011	cm3/molc sec	25	EXP	ATKINSON,R (1989)

CAS #: 000207-08-9				BENZO(K)FLUORANTHENE

Formula: $C_{20}H_{12}$

Mol Weight: 252.32

MP (deg C): 217   FP (deg C):

BP (deg C): 480

BP pressure (mm Hg):

Property/Value	Units	Temp	Data Type	Reference
WS 8.00E-004	mg/L	25	EXP	PEARLMAN,RS ET AL. (1984)
logP 6.11			EST	MEYLAN,WM & HOWARD,PH (1995)
VP 9.65E-010	mm Hg	25	EXT	MURRAY,JJ ET AL. (1974)
DC	pKa			
HL 4.00E-007	atm m3/mol	25	EST	VP/WSOL
OH 5.37E-011	cm3/molc sec	25	EST	MEYLAN,WM & HOWARD,PH (1993)

CAS #: 000208-96-8				ACENAPHTHYLENE

Formula: $C_{12}H_{8}$

Mol Weight: 152.20

MP (deg C): 88-91   FP (deg C):

BP (deg C): 280

BP pressure (mm Hg):

Property/Value	Units	Temp	Data Type	Reference
WS 1.61E+001	mg/L	25	EXP	WALTERS,RW & LUTHY,RG (1984A)
logP 3.94			EXP	HANSCH,C & LEO,AJ (1985)
VP 9.12E-004	mm Hg	25	EXP	BOYD,RH ET AL. (1965)
DC	pKa			
HL 1.14E-004	atm m3/mol	25	EXP	WARNER,HP ET AL. (1987)
OH 1.10E-010	cm3/molc sec	25	EXP	ATKINSON,R (1989)

CAS #: 000213-46-7				PICENE

Formula: $C_{22}H_{14}$

Mol Weight: 278.36

MP (deg C): 366-367   FP (deg C):

BP (deg C): 518-520

BP pressure (mm Hg):

Property/Value	Units	Temp	Data Type	Reference
WS 4.31E-003	mg/L	20	EXP	YALKOWSKY,SH & DANNENFELSER,RM (1992)
logP 7.11			EXP	DEVOOGT,P ET AL. (1990)
VP 3.73E-009	mm Hg	25	EST	NEELY,WB & BLAU,GE (1985)
DC	pKa			
HL 4.89E-007	atm m3/mol	25	EST	MEYLAN,WM & HOWARD,PH (1991)
OH 5.00E-011	cm3/molc sec	25	EST	MEYLAN,WM & HOWARD,PH (1993)

CAS #: 000214-17-5				BENZO(B)CHRYSENE

Formula: $C_{22}H_{14}$

Mol Weight: 278.36

MP (deg C):   FP (deg C):

BP (deg C):

BP pressure (mm Hg):

Property/Value	Units	Temp	Data Type	Reference
WS 1.08E-003	mg/L	25	EST	MEYLAN,WM ET AL. (1996)
logP 7.11			EXP	DEVOOGT,P ET AL. (1990)
VP 3.73E-009	mm Hg	25	EST	NEELY,WB & BLAU,GE (1985)
DC	pKa			
HL 4.89E-007	atm m3/mol	25	EST	MEYLAN,WM & HOWARD,PH (1991)
OH 5.00E-011	cm3/molc sec	25	EST	MEYLAN,WM & HOWARD,PH (1993)

CAS #: 000215-58-7				1,2,3,4-DIBENZANTHRACENE

Formula: $C_{22}H_{14}$

Mol Weight: 278.36

MP (deg C): 205-207   FP (deg C):

BP (deg C): 524

BP pressure (mm Hg):

Property/Value	Units	Temp	Data Type	Reference
WS 1.60E-003	mg/L	25	EXP	YALKOWSKY,SH & DANNENFELSER,RM (1992)
logP 7.11			EXP	DEVOOGT,P ET AL. (1990)
VP 3.73E-009	mm Hg	25	EST	NEELY,WB & BLAU,GE (1985)
DC	pKa			
HL 4.89E-007	atm m3/mol	25	EST	MEYLAN,WM & HOWARD,PH (1991)
OH 5.00E-011	cm3/molc sec	25	EST	MEYLAN,WM & HOWARD,PH (1993)

CAS #: 000217-59-4				TRIPHENYLENE

Formula: $C_{18}H_{12}$

Mol Weight: 228.30

MP (deg C): 199 FP (deg C):

BP (deg C): 425

BP pressure (mm Hg):

Property/Value	Units	Temp	Data Type	Reference
WS 4.11E-002	mg/L	25	EXP	YALKOWSKY,SH & DANNENFELSER,RM (1992)
logP 5.49			EXP	WANG,L ET AL. (1986)
VP 1.70E-006	mm Hg	25	EXP	ATKINSON,R (1990A)
DC	pKa			
HL 1.24E-005	atm m3/mol	25	EST	VP/WSOL
OH 5.00E-011	cm3/molc sec	25	EST	MEYLAN,WM & HOWARD,PH (1993)

CAS #: 000218-01-9				CHRYSENE

Formula: $C_{18}H_{12}$

Mol Weight: 228.30

MP (deg C): 254-255 FP (deg C):

BP (deg C): 448

BP pressure (mm Hg):

Property/Value	Units	Temp	Data Type	Reference
WS 2.00E-003	mg/L	25	EXP	MILLER,MM ET AL. (1985)
logP 5.50			EXP	ALCORN,CJ ET AL. (1993)
VP 6.23E-009	mm Hg	25	EXP	HOYER,H & PEPERLE,W (1958)
DC	pKa			
HL 9.46E-005	atm m3/mol	25	EST	VP/WSOL
OH 8.00E-011	cm3/molc sec	25	EST	ATKINSON,R (1988)

CAS #: 000224-41-9				DIBENZ(A,J)ANTHRACENE

Formula: $C_{22}H_{14}$

Mol Weight: 278.36

MP (deg C): FP (deg C):

BP (deg C):

BP pressure (mm Hg):

Property/Value	Units	Temp	Data Type	Reference
WS 1.20E-002	mg/L	27	EXP	YALKOWSKY,SH & DANNENFELSER,RM (1992)
logP 7.11			EXP	DEVOOGT,P ET AL. (1990)
VP 3.73E-009	mm Hg	25	EST	NEELY,WB & BLAU,GE (1985)
DC	pKa			
HL 4.89E-007	atm m3/mol	25	EST	MEYLAN,WM & HOWARD,PH (1991)
OH 5.00E-011	cm3/molc sec	25	EST	MEYLAN,WM & HOWARD,PH (1993)

CAS #: 000224-42-0				DIBENZ[a,j]ACRIDINE

Formula: $C_{21}H_{13}N$

Mol Weight: 279.34

MP (deg C): 216 FP (deg C):

BP (deg C):

BP pressure (mm Hg):

Property/Value	Units	Temp	Data Type	Reference
WS 1.80E-002	mg/L	25	EST	MEYLAN,WM ET AL. (1996)
logP 5.67			EST	MEYLAN,WM & HOWARD,PH (1995)
VP 1.85E-009	mm Hg	25	EST	NEELY,WB & BLAU,GE (1985)
DC	pKa			
HL 1.90E-009	atm m3/mol	25	EST	MEYLAN,WM & HWOARD,PH (1991)
OH 2.80E-011	cm3/molc sec	25	EST	MEYLAN,WM & HOWARD,PH (1993)

CAS #: 000225-11-6				BENZ(A)ACRIDINE

Formula: $C_{17}H_{11}N$

Mol Weight: 229.28

MP (deg C): FP (deg C):

BP (deg C):

BP pressure (mm Hg):

Property/Value	Units	Temp	Data Type	Reference
WS 3.46E-001	mg/L	25	EST	MEYLAN,WM ET AL. (1996)
logP 4.49			EST	MEYLAN,WM & HOWARD,PH (1995)
VP 3.49E-007	mm Hg	25	EST	NEELY,WB & BLAU,GE (1985)
DC 4.70	pKa	20	EXP	PERRIN,DD (1965)
HL 6.56E-009	atm m3/mol	25	EST	MEYLAN,WM & HOWARD,PH (1991)
OH 2.80E-011	cm3/molc sec	25	EST	MEYLAN,WM & HOWARD,PH (1993)

CAS #: 000225-51-4				BENZ(C)ACRIDINE

Formula: $C_{17}H_{11}N$

Mol Weight: 229.28

MP (deg C): 132 FP (deg C):

BP (deg C):

BP pressure (mm Hg):

Property/Value	Units	Temp	Data Type	Reference
WS 3.46E-001	mg/L	25	EST	MEYLAN,WM ET AL. (1996)
logP 4.49			EST	MEYLAN,WM & HOWARD,PH (1995)
VP 3.49E-007	mm Hg	25	EST	NEELY,WB & BLAU,GE (1985)
DC 4.70	pKa	20	EXP	ALBERT,A ET AL. (1948)
HL 1.41E-009	atm m3/mol	25	EST	MEYLAN,WM & HOWARD,PH (1991)
OH 2.80E-011	cm3/molc sec	25	EST	MEYLAN,WM & HOWARD,PH (1993)

CAS #: 000226-36-8				DIBENZ(A,H)ACRIDINE

Formula: $C_{21}H_{13}N$

Mol Weight: 279.34

MP (deg C): 226 FP (deg C):

BP (deg C):

BP pressure (mm Hg):

Property/Value	Units	Temp	Data Type	Reference
WS 1.59E-001	mg/L	25	EXP	PEARLMAN,RS ET AL. (1984)
logP 5.67			EST	MEYLAN,WM & HOWARD,PH (1995)
VP 1.85E-009	mm Hg	25	EST	NEELY,WB & BLAU,GE (1985)
DC	pKa			
HL 1.90E-009	atm m3/mol	25	EST	MEYLAN,WM & HOWARD,PH (1991)
OH 5.00E-011	cm3/molc sec	25	EST	MEYLAN,WM & HOWARD,PH (1993)

CAS #: 000226-88-0				BENZO[A]NAPHTHACENE

Formula: $C_{22}H_{14}$

Mol Weight: 278.36

MP (deg C): FP (deg C):

BP (deg C):

BP pressure (mm Hg):

Property/Value	Units	Temp	Data Type	Reference
WS 1.94E-003	mg/L	25	EST	MEYLAN,WM ET AL. (1996)
logP 6.81			EXP	WANG,L ET AL. (1986)
VP 3.73E-009	mm Hg	25	EST	NEELY,WB & BLAU,GE (1985)
DC	pKa			
HL 4.89E-007	atm m3/mol	25	EST	MEYLAN,WM & HOWARD,PH (1991)
OH 5.00E-011	cm3/molc sec	25	EST	MEYLAN,WM & HOWARD,PH (1993)

CAS #: 000229-87-8				PHENANTHRIDINE

Formula: $C_{13}H_9N$
Mol Weight: 179.22
MP (deg C): 107.4   **FP (deg C):**
BP (deg C): 349
BP pressure (mm Hg):

Property/ Value	Units	Temp	Data Type	Reference
WS 3.00E+002	mg/L	20	EXP	YALKOWSKY,SH & DANNENFELSER,RM (1992)
logP 3.48			EXP	HANSCH,C ET AL. (1995)
VP 2.08E-005	mm Hg	25	EXP	MCEACHERN,DM ET AL. (1975)
DC 4.61	pKa	20	EXP	PERRIN,DD (1965)
HL 1.64E-008	atm m3/mol	25	EST	VP/WSOL
OH 9.00E-012	cm3/molc sec	25	EST	MEYLAN,WM & HOWARD,PH (1993)

CAS #: 000230-07-9				4,7-PHENANTHROLINE

Formula: $C_{12}H_8N_2$
Mol Weight: 180.21
MP (deg C): 177   **FP (deg C):**
BP (deg C):
BP pressure (mm Hg):

Property/ Value	Units	Temp	Data Type	Reference
WS 7.57E+001	mg/L	25	EST	MEYLAN,WM ET AL. (1996)
logP 2.05			EXP	HANSCH,C & LEO,AJ (1985)
VP 3.26E-005	mm Hg	25	EST	NEELY,WB & BLAU,GE (1985)
DC	pKa			
HL 8.80E-011	atm m3/mol	25	EST	MEYLAN,WM & HOWARD,PH (1991)
OH 2.00E-012	cm3/molc sec	25	EST	MEYLAN,WM & HOWARD,PH (1993)

CAS #: 000230-17-1				5,6-DIAZAPHENANTHRENE

Formula: $C_{12}H_8N_2$
Mol Weight: 180.21
MP (deg C): 157-159   **FP (deg C):**
BP (deg C):
BP pressure (mm Hg):

Property/ Value	Units	Temp	Data Type	Reference
WS 3.80E+001	mg/L	25	EST	MEYLAN,WM ET AL. (1996)
logP 2.40			EXP	HANSCH,C ET AL. (1995)
VP 2.60E-006	mm Hg	25	EST	NEELY,WB & BLAU,GE (1985)
DC	pKa			
HL 2.73E-008	atm m3/mol	25	EST	MEYLAN,WM & HOWARD,PH (1991)
OH 2.00E-012	cm3/molc sec	25	EST	MEYLAN,WM & HOWARD,PH (1993)

CAS #: 000230-27-3				BENZO(H)QUINOLINE

Formula: $C_{13}H_9N$
Mol Weight: 179.22
MP (deg C): 52   **FP (deg C):**
BP (deg C): 339
BP pressure (mm Hg):

Property/ Value	Units	Temp	Data Type	Reference
WS 5.08E+000	mg/L	25	EST	MEYLAN,WM ET AL. (1996)
logP 3.43			EXP	SANGSTER,J (1993)
VP 2.18E-004	mm Hg	25	EXP	MCEACHERN,DM ET AL. (1975)
DC 4.21	pKa	20	EXP	PERRIN,DD (1965)
HL 6.72E-008	atm m3/mol	25	EST	MEYLAN,WM & HOWARD,PH (1991)
OH 9.00E-012	cm3/molc sec	25	EST	MEYLAN,WM & HOWARD,PH (1993)

CAS #: 000230-46-6				1,7-PHENANTHROLINE

Formula: $C_{12}H_8N_2$
Mol Weight: 180.21
MP (deg C): 177   **FP (deg C):**
BP (deg C):
BP pressure (mm Hg):

Property/ Value	Units	Temp	Data Type	Reference
WS 3.06E+001	mg/L	25	EST	MEYLAN,WM ET AL. (1996)
logP 2.51			EXP	HANSCH,C & LEO,AJ (1985)
VP 3.26E-005	mm Hg	25	EST	NEELY,WB & BLAU,GE (1985)
DC	pKa			
HL 2.78E-008	atm m3/mol	25	EST	MEYLAN,WM & HOWARD,PH (1991)
OH 2.00E-012	cm3/molc sec	25	EST	MEYLAN,WM & HOWARD,PH (1993)

CAS #: 000238-84-6				11H-BENZO(A)FLUORENE

Formula: $C_{17}H_{12}$
Mol Weight: 216.29
MP (deg C): 189.5   **FP (deg C):**
BP (deg C): 405
BP pressure (mm Hg):

Property/ Value	Units	Temp	Data Type	Reference
WS 4.54E-002	mg/L	25	EXP	YALKOWSKY,SH & DANNENFELSER,RM (1992)
logP 5.40			EXP	SANGSTER,J (1993)
VP 4.68E-006	mm Hg	25	EST	NEELY,WB & BLAU,GE (1985)
DC	pKa			
HL 1.63E-005	atm m3/mol	25	EST	MEYLAN,WM & HOWARD,PH (1991)
OH 4.67E-011	cm3/molc sec	25	EST	MEYLAN,WM & HOWARD,PH (1993)

CAS #: 000239-35-0				BENZO(B)NAPHTHO(2,1-D)THIOPHENE

Formula: $C_{16}H_{10}S$
Mol Weight: 234.32
MP (deg C): 188-190   **FP (deg C):**
BP (deg C):
BP pressure (mm Hg):

Property/ Value	Units	Temp	Data Type	Reference
WS 6.15E-002	mg/L	25	EST	MEYLAN,WM ET AL. (1996)
logP 5.34			EST	MEYLAN,WM & HOWARD,PH (1995)
VP 3.49E-007	mm Hg	25	EST	NEELY,WB & BLAU,GE (1985)
DC	pKa			
HL 2.72E-006	atm m3/mol	25	EST	MEYLAN,WM & HOWARD,PH (1991)
OH 5.00E-011	cm3/molc sec	25	EST	MEYLAN,WM & HOWARD,PH (1993)

CAS #: 000239-64-5				13H-DIBENZO(A,I)CARBAZOLE

Formula: $C_{20}H_{21}N$
Mol Weight: 275.40
MP (deg C):   **FP (deg C):**
BP (deg C):
BP pressure (mm Hg):

Property/ Value	Units	Temp	Data Type	Reference
WS 1.04E-002	mg/L	25	EXP	BANWART,WL ET AL. (1982)
logP 6.40			EXP	HANSCH,C & LEO,AJ (1985)
VP 1.22E-009	mm Hg	25	EST	NEELY,WB & BLAU,GE (1985)
DC	pKa			
HL 8.24E-010	atm m3/mol	25	EST	MEYLAN,WM & HOWARD,PH (1991)
OH 2.80E-011	cm3/molc sec	25	EST	MEYLAN,WM & HOWARD,PH (1993)

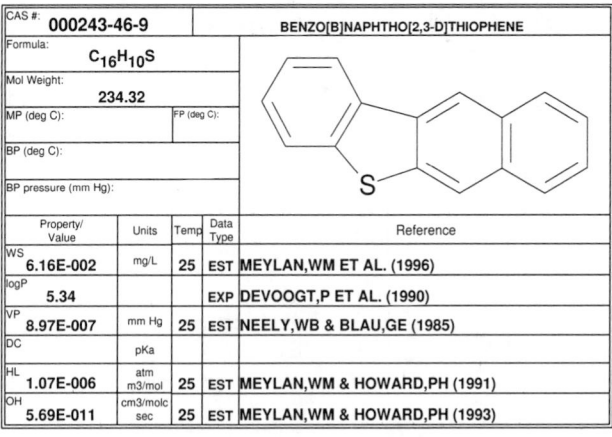

CAS #:	000243-17-4			11H-BENZO(B)FLUORENE
Formula:	$C_{17}H_{12}$			
Mol Weight:	216.29			
MP (deg C): 211-213		FP (deg C):		
BP (deg C):				
BP pressure (mm Hg):				

Property/ Value	Units	Temp	Data Type	Reference
WS 2.00E-003	mg/L	25	EXP	YALKOWSKY,SH & DANNENFELSER,RM (1992)
logP 5.77			EXP	WANG,L ET AL. (1986)
VP 4.68E-006	mm Hg	25	EST	NEELY,WB & BLAU,GE (1985)
DC	pKa			
HL 1.63E-005	atm m3/mol	25	EST	MEYLAN,WM & HOWARD,PH (1991)
OH 4.67E-011	cm3/molc sec	25	EST	MEYLAN,WM & HOWARD,PH (1993)

CAS #:	000243-42-5			BENZO[B]NAPHTHO[2,3-D]FURAN
Formula:	$C_{16}H_{10}O$			
Mol Weight:	218.26			
MP (deg C):		FP (deg C):		
BP (deg C):				
BP pressure (mm Hg):				

Property/ Value	Units	Temp	Data Type	Reference
WS 1.33E-001	mg/L	25	EST	MEYLAN,WM ET AL. (1996)
logP 5.05			EXP	DEVOOGT,P ET AL. (1990)
VP 3.36E-006	mm Hg	25	EST	NEELY,WB & BLAU,GE (1985)
DC	pKa			
HL 4.01E-006	atm m3/mol	25	EST	MEYLAN,WM & HOWARD,PH (1991)
OH 4.17E-011	cm3/molc sec	25	EST	MEYLAN,WM & HOWARD,PH (1993)

CAS #:	000243-46-9			BENZO[B]NAPHTHO[2,3-D]THIOPHENE
Formula:	$C_{16}H_{10}S$			
Mol Weight:	234.32			
MP (deg C):		FP (deg C):		
BP (deg C):				
BP pressure (mm Hg):				

Property/ Value	Units	Temp	Data Type	Reference
WS 6.16E-002	mg/L	25	EST	MEYLAN,WM ET AL. (1996)
logP 5.34			EXP	DEVOOGT,P ET AL. (1990)
VP 8.97E-007	mm Hg	25	EST	NEELY,WB & BLAU,GE (1985)
DC	pKa			
HL 1.07E-006	atm m3/mol	25	EST	MEYLAN,WM & HOWARD,PH (1991)
OH 5.69E-011	cm3/molc sec	25	EST	MEYLAN,WM & HOWARD,PH (1993)

CAS #:	000244-36-0			1H-FLUORENE
Formula:	$C_{13}H_{10}$			
Mol Weight:	166.22			
MP (deg C):		FP (deg C):		
BP (deg C):				
BP pressure (mm Hg):				

Property/ Value	Units	Temp	Data Type	Reference
WS 1.34E+000	mg/L	25	EST	MEYLAN,WM ET AL. (1996)
logP 4.02			EST	MEYLAN,WM & HOWARD,PH (1995)
VP 1.26E-003	mm Hg	25	EST	NEELY,WB & BLAU,GE (1985)
DC	pKa			
HL 1.67E-004	atm m3/mol	25	EST	MEYLAN,WM & HOWARD,PH (1991)
OH 9.00E-012	cm3/molc sec	25	EST	MEYLAN,WM & HOWARD,PH (1993)

CAS #:	000244-63-3			9H-PYRIDO[3,4-B]INDOLE
Formula:	$C_{11}H_8N_2$			
Mol Weight:	168.20			
MP (deg C): 198-200		FP (deg C):		
BP (deg C):				
BP pressure (mm Hg):				

Property/ Value	Units	Temp	Data Type	Reference
WS 9.55E+000	mg/L	25	EST	MEYLAN,WM ET AL. (1996)
logP 3.17			EXP	BIAGI,GL ET AL. (1989)
VP 2.23E-005	mm Hg	25	EST	NEELY,WB & BLAU,GE (1985)
DC	pKa			
HL 1.13E-010	atm m3/mol	25	EST	MEYLAN,WM & HOWARD,PH (1991)
OH 1.30E-010	cm3/molc sec	25	EST	MEYLAN,WM & HOWARD,PH (1993)

CAS #:	000253-52-1			PHTHALAZINE
Formula:	$C_8H_6N_2$			
Mol Weight:	130.15			
MP (deg C): 90-91		FP (deg C):		
BP (deg C): 315-317				
BP pressure (mm Hg):				

Property/ Value	Units	Temp	Data Type	Reference
WS 2.99E+004	mg/L	25	EST	MEYLAN,WM ET AL. (1996)
logP 0.57			EXP	HANSCH,C & LEO,AJ (1985)
VP 4.20E-001	mm Hg	25	EXP	LANDE,SS ET AL. (1987)
DC	pKa			
HL 2.80E-007	atm m3/mol	25	EST	MEYLAN,WM & HOWARD,PH (1991)
OH 2.00E-012	cm3/molc sec	25	EST	MEYLAN,WM & HOWARD,PH (1993)

CAS #:	000253-66-7			CINNOLINE
Formula:	$C_8H_6N_2$			
Mol Weight:	130.15			
MP (deg C): 38		FP (deg C):		
BP (deg C): 114				
BP pressure (mm Hg): 3.50E-001				

Property/ Value	Units	Temp	Data Type	Reference
WS 1.48E+004	mg/L	25	EST	MEYLAN,WM ET AL. (1996)
logP 0.93			EXP	HANSCH,C & LEO,AJ (1985)
VP 5.62E-004	mm Hg	25	EST	NEELY,WB & BLAU,GE (1985)
DC	pKa			
HL 2.80E-007	atm m3/mol	25	EST	MEYLAN,WM & HOWARD,PH (1991)
OH 2.00E-012	cm3/molc sec	25	EST	MEYLAN,WM & HOWARD,PH (1993)

CAS #:	000253-82-7			QUINAZOLINE
Formula:	$C_8H_6N_2$			
Mol Weight:	130.15			
MP (deg C): 48-48.5		FP (deg C):		
BP (deg C): 241.5				
BP pressure (mm Hg): 7.64E+002				

Property/ Value	Units	Temp	Data Type	Reference
WS 1.29E+004	mg/L	25	EST	MEYLAN,WM ET AL. (1996)
logP 1.00			EXP	HANSCH,C & LEO,AJ (1985)
VP 1.44E-002	mm Hg	25	EST	NEELY,WB & BLAU,GE (1985)
DC 3.43	pKa	20	EXP	PERRIN,DD (1965)
HL 2.85E-007	atm m3/mol	25	EST	MEYLAN,WM & HOWARD,PH (1991)
OH 2.00E-012	cm3/molc sec	25	EST	MEYLAN,WM & HOWARD,PH (1993)

CAS #: 000256-96-2	5-AZADIBENZO(A,E)CYCLOHEPTATRIENE

Formula: $C_{14}H_{11}N$

Mol Weight: 193.25

MP (deg C): 197-199   FP (deg C):

BP (deg C):

BP pressure (mm Hg):

Property/Value	Units	Temp	Data Type	Reference
WS 1.26E+000	mg/L	25	EST	MEYLAN,WM ET AL. (1996)
logP 4.06			EST	MEYLAN,WM & HOWARD,PH (1995)
VP 6.43E-005	mm Hg	25	EST	NEELY,WB & BLAU,GE (1985)
DC	pKa			
HL 1.10E-007	atm m3/mol	25	EST	MEYLAN,WM & HOWARD,PH (1991)
OH 2.56E-010	cm3/molc sec	25	EST	MEYLAN,WM & HOWARD,PH (1993)

CAS #: 000260-94-6	ACRIDINE

Formula: $C_{13}H_9N$

Mol Weight: 179.22

MP (deg C): 106-110   FP (deg C):

BP (deg C): 346

BP pressure (mm Hg):

Property/Value	Units	Temp	Data Type	Reference
WS 3.84E+001	mg/L	24	EXP	BANWART,WL ET AL. (1982)
logP 3.40			EXP	HANSCH,C ET AL. (1995)
VP 1.35E-004	mm Hg	25	EXT	PERRY,RH & GREEN,D (1984)
DC 5.45	pKa	15	EXP	PERRIN,DD (1965)
HL 3.97E-007	atm m3/mol	25	EST	MEYLAN,WM & HOWARD,PH (1991)
OH 2.75E-011	cm3/molc sec	25	EST	MEYLAN,WM & HOWARD,PH (1993)

CAS #: 000262-12-4	DIBENZO-P-DIOXIN

Formula: $C_{12}H_8O_2$

Mol Weight: 184.20

MP (deg C):   FP (deg C):

BP (deg C):

BP pressure (mm Hg):

Property/Value	Units	Temp	Data Type	Reference
WS 9.01E-001	mg/L	25	EXP	YALKOWSKY,SH & DANNENFELSER,RM (1992)
logP 4.30			EXP	SHIU,WY ET AL. (1988)
VP 4.12E-004	mm Hg	25	EXP	SHIU,WY ET AL. (1988)
DC	pKa			
HL 1.11E-004	atm m3/mol	25	EST	VP/WSOL
OH 2.52E-011	cm3/molc sec	25	EXP	ATKINSON,R (1989)

CAS #: 000262-20-4	PHENOTHIOXIN

Formula: $C_{12}H_8OS$

Mol Weight: 200.26

MP (deg C):   FP (deg C):

BP (deg C):

BP pressure (mm Hg):

Property/Value	Units	Temp	Data Type	Reference
WS 4.49E-001	mg/L	25	EST	MEYLAN,WM ET AL. (1996)
logP 4.54			EXP	HANSCH,C ET AL. (1995)
VP 1.44E-004	mm Hg	25	EST	NEELY,WB & BLAU,GE (1985)
DC	pKa			
HL 3.12E-006	atm m3/mol	25	EST	MEYLAN,WM & HOWARD,PH (1991)
OH 2.51E-011	cm3/molc sec	25	EST	MEYLAN,WM & HOWARD,PH (1993)

CAS #: 000271-06-7	6-AZATHIANAPHTHENE

Formula: C_7H_5NS

Mol Weight: 135.19

MP (deg C):   FP (deg C):

BP (deg C):

BP pressure (mm Hg):

Property/Value	Units	Temp	Data Type	Reference
WS 2.86E+003	mg/L	25	EST	MEYLAN,WM ET AL. (1996)
logP 1.74			EXP	HANSCH,C & LEO,AJ (1985)
VP 1.43E-002	mm Hg	25	EST	NEELY,WB & BLAU,GE (1985)
DC	pKa			
HL 3.74E-007	atm m3/mol	25	EST	MEYLAN,WM & HOWARD,PH (1991)
OH 3.00E-011	cm3/molc sec	25	EST	MEYLAN,WM & HOWARD,PH (1993)

CAS #: 000271-29-4	6-AZAINDOLE

Formula: $C_7H_6N_2$

Mol Weight: 118.14

MP (deg C):   FP (deg C):

BP (deg C):

BP pressure (mm Hg):

Property/Value	Units	Temp	Data Type	Reference
WS 1.35E+004	mg/L	25	EST	MEYLAN,WM ET AL. (1996)
logP 1.03			EST	MEYLAN,WM & HOWARD,PH (1995)
VP 8.58E-003	mm Hg	25	EST	NEELY,WB & BLAU,GE (1985)
DC	pKa			
HL 1.16E-009	atm m3/mol	25	EST	MEYLAN,WM & HOWARD,PH (1991)
OH 1.10E-010	cm3/molc sec	25	EST	MEYLAN,WM & HOWARD,PH (1993)

CAS #: 000271-34-1	5-AZAINDOLE

Formula: $C_7H_6N_2$

Mol Weight: 118.14

MP (deg C):   FP (deg C):

BP (deg C):

BP pressure (mm Hg):

Property/Value	Units	Temp	Data Type	Reference
WS 1.35E+004	mg/L	25	EST	MEYLAN,WM ET AL. (1996)
logP 1.03			EST	MEYLAN,WM & HOWARD,PH (1995)
VP 8.58E-003	mm Hg	25	EST	NEELY,WB & BLAU,GE (1985)
DC	pKa			
HL 1.16E-009	atm m3/mol	25	EST	MEYLAN,WM & HOWARD,PH (1991)
OH 1.10E-010	cm3/molc sec	25	EST	MEYLAN,WM & HOWARD,PH (1993)

CAS #: 000271-44-3	INDAZOLE

Formula: $C_7H_6N_2$

Mol Weight: 118.14

MP (deg C): 146.5   FP (deg C):

BP (deg C): 267-270

BP pressure (mm Hg): 7.43E+002

Property/Value	Units	Temp	Data Type	Reference
WS 8.26E+002	mg/L	20	EXP	PEARLMAN,RS ET AL. (1984)
logP 1.77			EXP	HANSCH,C & LEO,AJ (1985)
VP 1.48E-003	mm Hg	25	EST	NEELY,WB & BLAU,GE (1985)
DC	pKa			
HL 3.60E-007	atm m3/mol	25	EST	MEYLAN,WM & HOWARD,PH (1991)
OH 3.60E-011	cm3/molc sec	25	EST	MEYLAN,WM & HOWARD,PH (1993)

CAS #: 000271-58-9				BENZISOXAZOLE

Formula: C_7H_5NO

Mol Weight: 119.12

MP (deg C): < -18   **FP (deg C):**

BP (deg C): 215

BP pressure (mm Hg):

Property/Value	Units	Temp	Data Type	Reference
WS 4.10E+003	mg/L	25	EST	MEYLAN,WM ET AL. (1996)
logP 1.52			EXP	HANSCH,C & LEO,AJ (1985)
VP 2.47E-001	mm Hg	25	EST	NEELY,WB & BLAU,GE (1985)
DC	pKa			
HL 5.54E-006	atm m3/mol	25	EST	MEYLAN,WM & HOWARD,PH (1991)
OH 9.00E-012	cm3/molc sec	25	EST	MEYLAN,WM & HOWARD,PH (1993)

CAS #: 000271-89-6				BENZOFURAN

Formula: C_8H_6O

Mol Weight: 118.14

MP (deg C): < -18   **FP (deg C):**

BP (deg C): 173-5

BP pressure (mm Hg):

Property/Value	Units	Temp	Data Type	Reference
WS 6.78E+002	mg/L	25	EST	MEYLAN,WM ET AL. (1996)
logP 2.67			EXP	HANSCH,C & LEO,AJ (1985)
VP 4.40E-001	mm Hg	25	EXP	CHAO,J ET AL. (1983)
DC	pKa			
HL 1.01E-004	atm m3/mol	25	EST	MEYLAN,WM & HOWARD,PH (1991)
OH 3.73E-011	cm3/molc sec	25	EXP	ATKINSON,R (1989)

CAS #: 000271-95-4				1,2-BENZISOXAZOLE

Formula: C_7H_5NO

Mol Weight: 119.12

MP (deg C):   **FP (deg C):**

BP (deg C): 100

BP pressure (mm Hg): 2.60E+001

Property/Value	Units	Temp	Data Type	Reference
WS 4.10E+003	mg/L	25	EST	MEYLAN,WM ET AL. (1996)
logP 1.63			EXP	HANSCH,C & LEO,AJ (1985)
VP 2.47E-001	mm Hg	25	EST	NEELY,WB & BLAU,GE (1985)
DC	pKa			
HL 5.54E-006	atm m3/mol	25	EST	MEYLAN,WM & HOWARD,PH (1991)
OH 9.00E-012	cm3/molc sec	25	EST	MEYLAN,WM & HOWARD,PH (1993)

CAS #: 000272-30-0				SELANONAPHTHENE

Formula: C_8H_6Se

Mol Weight: 181.10

MP (deg C):   **FP (deg C):**

BP (deg C):

BP pressure (mm Hg):

Property/Value	Units	Temp	Data Type	Reference
WS 1.03E+002	mg/L	25	EST	MEYLAN,WM ET AL. (1996)
logP 3.19			EXP	HANSCH,C & LEO,AJ (1985)
VP 1.03E-002	mm Hg	25	EST	NEELY,WB & BLAU,GE (1985)
DC	pKa			
HL 2.91E-004	atm m3/mol	25	EST	MEYLAN,WM & HOWARD,PH (1991)
OH 3.00E-011	cm3/molc sec	25	EST	MEYLAN,WM & HOWARD,PH (1993)

CAS #: 000272-49-1				4-AZAINDOLE

Formula: $C_7H_6N_2$

Mol Weight: 118.14

MP (deg C):   **FP (deg C):**

BP (deg C):

BP pressure (mm Hg):

Property/Value	Units	Temp	Data Type	Reference
WS 2.85E+003	mg/L	25	EST	MEYLAN,WM ET AL. (1996)
logP 1.03			EST	MEYLAN,WM & HOWARD,PH (1995)
VP 8.58E-003	mm Hg	25	EST	NEELY,WB & BLAU,GE (1985)
DC	pKa			
HL 1.16E-009	atm m3/mol	25	EST	MEYLAN,WM & HOWARD,PH (1991)
OH 1.10E-010	cm3/molc sec	25	EST	MEYLAN,WM & HOWARD,PH (1993)

CAS #: 000272-97-9				3,5-DIAZAINDOLE

Formula: $C_6H_5N_3$

Mol Weight: 119.13

MP (deg C):   **FP (deg C):**

BP (deg C):

BP pressure (mm Hg):

Property/Value	Units	Temp	Data Type	Reference
WS 5.61E+004	mg/L	25	EST	MEYLAN,WM ET AL. (1996)
logP 0.18			EXP	HANSCH,C ET AL. (1995)
VP 3.54E-005	mm Hg	25	EST	NEELY,WB & BLAU,GE (1985)
DC	pKa			
HL 4.80E-010	atm m3/mol	25	EST	MEYLAN,WM & HOWARD,PH (1991)
OH 3.60E-011	cm3/molc sec	25	EST	MEYLAN,WM & HOWARD,PH (1993)

CAS #: 000273-02-9				2H-BENZOTRIAZOLE

Formula: $C_6H_5N_3$

Mol Weight: 119.13

MP (deg C):   **FP (deg C):**

BP (deg C):

BP pressure (mm Hg):

Property/Value	Units	Temp	Data Type	Reference
WS 5.96E+003	mg/L	25	EST	MEYLAN,WM ET AL. (1996)
logP 1.44			EXP	SANGSTER,J (1993)
VP 5.98E-004	mm Hg	25	EST	NEELY,WB & BLAU,GE (1985)
DC	pKa			
HL 1.47E-007	atm m3/mol	25	EST	MEYLAN,WM & HOWARD,PH (1991)
OH 1.00E-012	cm3/molc sec	25	EST	MEYLAN,WM & HOWARD,PH (1993)

CAS #: 000273-09-6				BENZOFURAZAN

Formula: $C_6H_4N_2O$

Mol Weight: 120.11

MP (deg C):   **FP (deg C):**

BP (deg C):

BP pressure (mm Hg):

Property/Value	Units	Temp	Data Type	Reference
WS 3.61E+003	mg/L	25	EST	MEYLAN,WM ET AL. (1996)
logP 1.69			EXP	CALVINO,R ET AL. (1992)
VP 1.09E-001	mm Hg	25	EST	NEELY,WB & BLAU,GE (1985)
DC	pKa			
HL 1.85E-005	atm m3/mol	25	EST	MEYLAN,WM & HOWARD,PH (1991)
OH 9.00E-012	cm3/molc sec	25	EST	MEYLAN,WM & HOWARD,PH (1993)

000273-13-2 — 2,1,3-BENZOTHIADIAZOLE

CAS #: 000273-13-2

Formula: $C_6H_6N_2S$

Mol Weight: 138.19

MP (deg C): 42-44

FP (deg C):

BP (deg C): 203

BP pressure (mm Hg):

Property/Value	Units	Temp	Data Type	Reference
WS 1.77E+003	mg/L	25	EST	MEYLAN,WM ET AL. (1996)
logP 1.98			EXP	CALVINO,R ET AL. (1992)
VP 7.13E-003	mm Hg	25	EST	NEELY,WB & BLAU,GE (1985)
DC	pKa			
HL 1.93E-005	atm m3/mol	25	EST	MEYLAN,WM & HOWARD,PH (1991)
OH 7.00E-012	cm3/molc sec	25	EST	MEYLAN,WM & HOWARD,PH (1993)

000273-15-4 — 2,1,3-BENZOSELENADIAZOLE

CAS #: 000273-15-4

Formula: $C_6H_4N_2Se$

Mol Weight: 183.07

MP (deg C):

FP (deg C):

BP (deg C):

BP pressure (mm Hg):

Property/Value	Units	Temp	Data Type	Reference
WS 2.12E+003	mg/L	25	EST	MEYLAN,WM ET AL. (1996)
logP 1.64			EXP	CALVINO,R ET AL. (1992)
VP 6.77E-006	mm Hg	25	EST	NEELY,WB & BLAU,GE (1985)
DC	pKa			
HL 1.93E-005	atm m3/mol	25	EST	MEYLAN,WM & HOWARD,PH (1991)
OH 7.00E-012	cm3/molc sec	25	EST	MEYLAN,WM & HOWARD,PH (1993)

000273-21-2 — 1H-IMIDAZO[4,5-B]PYRIDINE

CAS #: 000273-21-2

Formula: $C_6H_5N_3$

Mol Weight: 119.13

MP (deg C): 148-151

FP (deg C):

BP (deg C):

BP pressure (mm Hg):

Property/Value	Units	Temp	Data Type	Reference
WS 5.61E+004	mg/L	25	EST	MEYLAN,WM ET AL. (1996)
logP 0.30			EXP	HANSCH,C ET AL. (1995)
VP 3.54E-005	mm Hg	25	EST	NEELY,WB & BLAU,GE (1985)
DC	pKa			
HL 4.80E-010	atm m3/mol	25	EST	MEYLAN,WM & HOWARD,PH (1991)
OH 3.60E-011	cm3/molc sec	25	EST	MEYLAN,WM & HOWARD,PH (1993)

000273-53-0 — BENZOXAZOLE

CAS #: 000273-53-0

Formula: C_7H_5NO

Mol Weight: 119.12

MP (deg C): 31

FP (deg C):

BP (deg C): 182.5

BP pressure (mm Hg):

Property/Value	Units	Temp	Data Type	Reference
WS 8.33E+003	mg/L	20	EXP	PEARLMAN,RS ET AL. (1984)
logP 1.59			EXP	HANSCH,C & LEO,AJ (1985)
VP 2.47E-001	mm Hg	25	EST	NEELY,WB & BLAU,GE (1985)
DC	pKa			
HL 6.87E-007	atm m3/mol	25	EST	MEYLAN,WM & HOWARD,PH (1991)
OH 9.00E-012	cm3/molc sec	25	EST	MEYLAN,WM & HOWARD,PH (1993)

000274-09-9 — 1,3-BENZODIOXOLE

CAS #: 000274-09-9

Formula: $C_7H_6O_2$

Mol Weight: 122.12

MP (deg C):

FP (deg C):

BP (deg C): 172.5

BP pressure (mm Hg):

Property/Value	Units	Temp	Data Type	Reference
WS 1.65E+003	mg/L	25	EST	MEYLAN,WM ET AL. (1996)
logP 2.08			EXP	HANSCH,C & LEO,AJ (1985)
VP 5.71E-001	mm Hg	25	EST	NEELY,WB & BLAU,GE (1985)
DC	pKa			
HL 6.26E-006	atm m3/mol	25	EST	MEYLAN,WM & HOWARD,PH (1991)
OH 2.13E-011	cm3/molc sec	25	EST	MEYLAN,WM & HOWARD,PH (1993)

000274-40-8 — INDOLIZINE

CAS #: 000274-40-8

Formula: C_8H_7N

Mol Weight: 117.15

MP (deg C): 75

FP (deg C):

BP (deg C): 205

BP pressure (mm Hg):

Property/Value	Units	Temp	Data Type	Reference
WS 7.68E+002	mg/L	25	EST	MEYLAN,WM ET AL. (1996)
logP 2.49			EXP	HANSCH,C & LEO,AJ (1985)
VP 2.84E-001	mm Hg	25	EST	NEELY,WB & BLAU,GE (1985)
DC	pKa			
HL 5.46E-008	atm m3/mol	25	EST	MEYLAN,WM & HOWARD,PH (1991)
OH 1.10E-010	cm3/molc sec	25	EST	MEYLAN,WM & HOWARD,PH (1993)

000274-55-5 — 7-AZAINDOLE

CAS #: 000274-55-5

Formula: $C_7H_6N_2$

Mol Weight: 118.14

MP (deg C):

FP (deg C):

BP (deg C):

BP pressure (mm Hg):

Property/Value	Units	Temp	Data Type	Reference
WS 2.85E+003	mg/L	25	EST	MEYLAN,WM ET AL. (1996)
logP 1.82			EXP	HANSCH,C & LEO,AJ (1985)
VP 8.58E-003	mm Hg	25	EST	NEELY,WB & BLAU,GE (1985)
DC	pKa			
HL 1.16E-009	atm m3/mol	25	EST	MEYLAN,WM & HOWARD,PH (1991)
OH 1.10E-010	cm3/molc sec	25	EST	MEYLAN,WM & HOWARD,PH (1993)

000274-87-3 — TETRAZOLO[1,5-A]PYRIDINE

CAS #: 000274-87-3

Formula: $C_5H_6N_4$

Mol Weight: 122.13

MP (deg C):

FP (deg C):

BP (deg C):

BP pressure (mm Hg):

Property/Value	Units	Temp	Data Type	Reference
WS 1.20E+005	mg/L	25	EST	MEYLAN,WM ET AL. (1996)
logP -0.09			EXP	HANSCH,C ET AL. (1995)
VP 2.92E-002	mm Hg	25	EST	NEELY,WB & BLAU,GE (1985)
DC	pKa			
HL 3.68E-009	atm m3/mol	25	EST	MEYLAN,WM & HOWARD,PH (1991)
OH 1.00E-012	cm3/molc sec	25	EST	MEYLAN,WM & HOWARD,PH (1993)

CAS #: 000275-51-4				AZULENE

Formula: $C_{10}H_8$

Mol Weight: 128.18

MP (deg C): 98.5-99   FP (deg C):

BP (deg C):

BP pressure (mm Hg):

Property/Value	Units	Temp	Data Type	Reference
WS 5.02E+001	mg/L	25	EST	MEYLAN,WM ET AL. (1996)
logP 3.20			EXP	HANSCH,C & LEO,AJ (1985)
VP 4.89E-001	mm Hg		EST	NEELY,WB & BLAU,GE (1985)
DC	pKa			
HL 3.93E-002	atm m3/mol	25	EST	MEYLAN,WM & HOWARD,PH (1991)
OH 2.55E-010	cm3/molc sec	25	EST	MEYLAN,WM & HOWARD,PH (1993)

CAS #: 000279-23-2				BICYCLO(2.2.1.)HEPTANE

Formula: C_7H_{12}

Mol Weight: 96.17

MP (deg C): 87.5   FP (deg C):

BP (deg C): 105.3

BP pressure (mm Hg):

Property/Value	Units	Temp	Data Type	Reference
WS 8.40E+001	mg/L	25	EST	MEYLAN,WM ET AL. (1996)
logP 3.07			EST	MEYLAN,WM & HOWARD,PH (1995)
VP 3.30E+001	mm Hg	25	EST	NEELY,WB & BLAU,GE (1985)
DC	pKa			
HL 1.49E-001	atm m3/mol	25	EST	MEYLAN,WM & HOWARD,PH (1991)
OH 5.49E-012	cm3/molc sec	25	EXP	ATKINSON,R (1989)

CAS #: 000280-33-1				BICYCLO[2.2.2]OCTANE

Formula: C_8H_{14}

Mol Weight: 110.20

MP (deg C):   FP (deg C):

BP (deg C):

BP pressure (mm Hg):

Property/Value	Units	Temp	Data Type	Reference
WS 2.90E+001	mg/L	25	EST	MEYLAN,WM ET AL. (1996)
logP 3.56			EST	MEYLAN,WM & HOWARD,PH (1995)
VP 3.34E+000	mm Hg	25	EXP	BOYD,RH ET AL. (1971)
DC	pKa			
HL 1.98E-001	atm m3/mol	25	EST	MEYLAN,WM & HOWARD,PH (1991)
OH 1.47E-011	cm3/molc sec	25	EXP	ATKINSON,R (1989)

CAS #: 000280-57-9				1,4-DIAZABICYCLO(2,2,2)OCTANE (DABCO)

Formula: $C_6H_{12}N_2$

Mol Weight: 112.18

MP (deg C): 158   FP (deg C):

BP (deg C): 174

BP pressure (mm Hg):

Property/Value	Units	Temp	Data Type	Reference
WS 1.00E+006	mg/L	25	EST	MEYLAN,WM ET AL. (1996)
logP -0.83			EST	MEYLAN,WM & HOWARD,PH (1995)
VP 7.42E-001	mm Hg	25	EXT	YAWS,CL (1994A)
DC 8.82	pKa		EXP	PERRIN,DD (1972)
HL 4.67E-009	atm m3/mol	25	EST	MEYLAN,WM & HOWARD,PH (1991)
OH 7.61E-011	cm3/molc sec	25	EST	MEYLAN,WM & HOWARD,PH (1993)

CAS #: 000281-23-2				TRICYCLO[3.3.1.1]DECANE

Formula: $C_{10}H_{16}$

Mol Weight: 136.24

MP (deg C): 269-270   FP (deg C):

BP (deg C):

BP pressure (mm Hg):

Property/Value	Units	Temp	Data Type	Reference
WS 6.03E+000	mg/L	25	EST	MEYLAN,WM ET AL. (1996)
logP 4.24			EXP	SANGSTER,J (1994)
VP 1.30E-001	mm Hg	25	EXT	BOYD,RH ET AL. (1971)
DC	pKa			
HL 1.54E-001	atm m3/mol	25	EST	MEYLAN,WM & HOWARD,PH (1991)
OH 2.26E-011	cm3/molc sec	25	EXP	ATKINSON,R (1989)

CAS #: 000286-08-8				BICYCLO(4.1.0)HEPTANE

Formula: C_7H_{12}

Mol Weight: 96.17

MP (deg C): 116-117   FP (deg C):

BP (deg C):

BP pressure (mm Hg):

Property/Value	Units	Temp	Data Type	Reference
WS 4.29E+001	mg/L	25	EST	MEYLAN,WM ET AL. (1996)
logP 3.41			EST	MEYLAN,WM & HOWARD,PH (1995)
VP 3.07E-002	mm Hg	25	EXP	VARUSHCHENKO,RM ET AL. (1974)
DC	pKa			
HL 1.49E-001	atm m3/mol	25	EST	MEYLAN,WM & HOWARD,PH (1991)
OH 1.71E-012	cm3/molc sec	25	EST	MEYLAN,WM & HOWARD,PH (1993)

CAS #: 000286-16-8				2-OXABICYCLO[4.1.0]HEPTANE

Formula: $C_7H_{12}O$

Mol Weight: 112.17

MP (deg C):   FP (deg C):

BP (deg C):

BP pressure (mm Hg):

Property/Value	Units	Temp	Data Type	Reference
WS 8.98E+003	mg/L	25	EST	MEYLAN,WM ET AL. (1996)
logP 1.26			EXP	SANGSTER,J (1994)
VP 8.67E+000	mm Hg	25	EST	NEELY,WB & BLAU,GE (1985)
DC	pKa			
HL 2.19E-004	atm m3/mol	25	EST	MEYLAN,WM & HOWARD,PH (1991)
OH 5.14E-012	cm3/molc sec	25	EST	MEYLAN,WM & HOWARD,PH (1993)

CAS #: 000286-99-7				CYCLODODECANE EPOXIDE

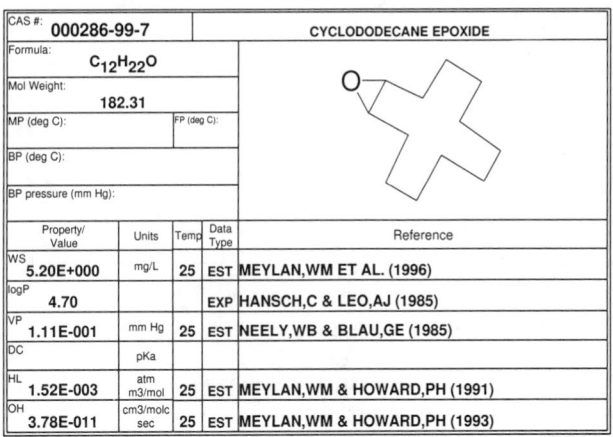

Formula: $C_{12}H_{22}O$

Mol Weight: 182.31

MP (deg C):   FP (deg C):

BP (deg C):

BP pressure (mm Hg):

Property/Value	Units	Temp	Data Type	Reference
WS 5.20E+000	mg/L	25	EST	MEYLAN,WM ET AL. (1996)
logP 4.70			EXP	HANSCH,C & LEO,AJ (1985)
VP 1.11E-001	mm Hg	25	EST	NEELY,WB & BLAU,GE (1985)
DC	pKa			
HL 1.52E-003	atm m3/mol	25	EST	MEYLAN,WM & HOWARD,PH (1991)
OH 3.78E-011	cm3/molc sec	25	EST	MEYLAN,WM & HOWARD,PH (1993)

CAS #:	000287-13-8			TRICYCLO(4.1.0.02,7)HEPTANE

Formula: C_7H_{10}

Mol Weight: 94.16

MP (deg C):   FP (deg C):

BP (deg C):

BP pressure (mm Hg):

Property/Value	Units	Temp	Data Type	Reference
WS 7.23E+001	mg/L	25	EST	MEYLAN,WM ET AL. (1996)
logP 3.15			EST	MEYLAN,WM & HOWARD,PH (1995)
VP 4.96E-002	mm Hg	25	EXP	VARUSHCHENKO,RM ET AL. (1974)
DC	pKa			
HL 6.58E-002	atm m3/mol	25	EST	MEYLAN,WM & HOWARD,PH (1991)
OH 1.12E-012	cm3/molc sec	25	EST	MEYLAN,WM & HOWARD,PH (1993)

CAS #:	000287-23-0			CYCLOBUTANE

Formula: C_4H_8

Mol Weight: 56.11

MP (deg C): -80   FP (deg C):

BP (deg C): 13.08

BP pressure (mm Hg): 7.41E+002

Property/Value	Units	Temp	Data Type	Reference
WS 5.33E+002	mg/L	25	EST	MEYLAN,WM ET AL. (1996)
logP 2.19			EST	MEYLAN,WM & HOWARD,PH (1995)
VP 1.18E+003	mm Hg	25	EXP	DAUBERT,TE & DANNER,RP (1989)
DC	pKa			
HL 1.45E-001	atm m3/mol	25	EST	MEYLAN,WM & HOWARD,PH (1991)
OH 1.20E-012	cm3/molc sec	25	EXP	ATKINSON,R (1989)

CAS #:	000287-27-4			THIACYCLOBUTANE

Formula: C_3H_6S

Mol Weight: 74.15

MP (deg C): -73.2   FP (deg C):

BP (deg C): 95

BP pressure (mm Hg):

Property/Value	Units	Temp	Data Type	Reference
WS 1.04E+004	mg/L	25	EST	MEYLAN,WM ET AL. (1996)
logP 1.30			EST	MEYLAN,WM & HOWARD,PH (1995)
VP 5.27E+001	mm Hg	25	EXP	CHAO,J ET AL. (1983)
DC	pKa			
HL 4.60E-004	atm m3/mol	25	EST	MEYLAN,WM & HOWARD,PH (1991)
OH 7.11E-012	cm3/molc sec	25	EST	MEYLAN,WM & HOWARD,PH (1993)

CAS #:	000287-92-3			CYCLOPENTANE

Formula: C_5H_{10}

Mol Weight: 70.14

MP (deg C): -94.4   FP (deg C):

BP (deg C): 49.3

BP pressure (mm Hg):

Property/Value	Units	Temp	Data Type	Reference
WS 1.56E+002	mg/L	25	EXP	YALKOWSKY,SH & DANNENFELSER,RM (1992)
logP 3.00			EXP	HANSCH,C & LEO,AJ (1985)
VP 3.18E+002	mm Hg	25	EXP	DAUBERT,TE & DANNER,RP (1989)
DC	pKa			
HL 1.88E-001	atm m3/mol	25	EST	VP/WSOL
OH 5.16E-012	cm3/molc sec	25	EXP	ATKINSON,R (1989)

CAS #:	000288-13-1			PYRAZOLE

Formula: $C_3H_4N_2$

Mol Weight: 68.08

MP (deg C): 69.5-70   FP (deg C):

BP (deg C): 186-188

BP pressure (mm Hg): 7.58E+002

Property/Value	Units	Temp	Data Type	Reference
WS 8.15E+004	mg/L	25	EST	MEYLAN,WM ET AL. (1996)
logP 0.26			EXP	HANSCH,C & LEO,AJ (1985)
VP 1.19E+000	mm Hg	25	EST	NEELY,WB & BLAU,GE (1985)
DC 2.48	pKa	25	EXP	PERRIN,DD (1965)
HL 3.69E-006	atm m3/mol	25	EST	MEYLAN,WM & HOWARD,PH (1991)
OH 3.60E-011	cm3/molc sec	25	EST	MEYLAN,WM & HOWARD,PH (1993)

CAS #:	000288-14-2			ISO-OXAZOLE

Formula: C_3H_3NO

Mol Weight: 69.06

MP (deg C):   FP (deg C):

BP (deg C): 95

BP pressure (mm Hg):

Property/Value	Units	Temp	Data Type	Reference
WS 1.16E+005	mg/L	25	EST	MEYLAN,WM ET AL. (1996)
logP 0.08			EXP	HANSCH,C & LEO,AJ (1985)
VP 4.55E+001	mm Hg	25	EXP	CHAO,J ET AL. (1983)
DC 1.30	pKa	25	EXP	PERRIN,DD (1965)
HL 5.67E-005	atm m3/mol	25	EST	MEYLAN,WM & HOWARD,PH (1991)
OH 9.10E-012	cm3/molc sec	25	EST	MEYLAN,WM & HOWARD,PH (1993)

CAS #:	000288-32-4			IMIDAZOLE

Formula: $C_3H_4N_2$

Mol Weight: 68.08

MP (deg C): 90-91   FP (deg C):

BP (deg C): 257

BP pressure (mm Hg):

Property/Value	Units	Temp	Data Type	Reference
WS 1.59E+005	mg/L	25	EST	MEYLAN,WM ET AL. (1996)
logP -0.08			EXP	HANSCH,C & LEO,AJ (1985)
VP 4.62E-002	mm Hg	25	EST	NEELY,WB & BLAU,GE (1985)
DC 6.95	pKa	25	EXP	PERRIN,DD (1965)
HL 3.76E-006	atm m3/mol	25	EST	MEYLAN,WM & HOWARD,PH (1991)
OH 3.60E-011	cm3/molc sec	25	EXP	ATKINSON,R (1989)

CAS #:	000288-36-8			2H-1,2,3-TRIAZOLE

Formula: $C_2H_3N_3$

Mol Weight: 69.07

MP (deg C): 23   FP (deg C):

BP (deg C): 204

BP pressure (mm Hg):

Property/Value	Units	Temp	Data Type	Reference
WS 2.40E+005	mg/L	25	EST	MEYLAN,WM ET AL. (1996)
logP -0.29			EXP	HANSCH,C & LEO,AJ (1985)
VP 6.03E-001	mm Hg	25	EST	NEELY,WB & BLAU,GE (1985)
DC	pKa			
HL 1.50E-006	atm m3/mol	25	EST	MEYLAN,WM & HOWARD,PH (1991)
OH 1.00E-013	cm3/molc sec	25	EST	MEYLAN,WM & HOWARD,PH (1993)

FURAZAN

CAS #:	000288-37-9	FURAZAN
Formula:	$C_2H_2N_2O$	
Mol Weight:	70.05	
MP (deg C):		FP (deg C):
BP (deg C):		
BP pressure (mm Hg):		

Property/ Value	Units	Temp	Data Type	Reference
WS 9.88E+004	mg/L	25	EST	MEYLAN,WM ET AL. (1996)
logP 0.16			EXP	CALVINO,R ET AL. (1992)
VP 7.76E+001	mm Hg	25	EST	NEELY,WB & BLAU,GE (1985)
DC	pKa			
HL 1.89E-004	atm m3/mol	25	EST	MEYLAN,WM & HOWARD,PH (1991)
OH 4.00E-012	cm3/molc sec	25	EST	MEYLAN,WM & HOWARD,PH (1993)

OXAZOLE

CAS #:	000288-42-6	OXAZOLE
Formula:	C_3H_3NO	
Mol Weight:	69.06	
MP (deg C):		FP (deg C):
BP (deg C):	69.5	
BP pressure (mm Hg):		

Property/ Value	Units	Temp	Data Type	Reference
WS 1.07E+005	mg/L	25	EST	MEYLAN,WM ET AL. (1996)
logP 0.12			EXP	HANSCH,C & LEO,AJ (1985)
VP 1.32E+002	mm Hg	25	EXP	DAUBERT,TE & DANNER,RP (1989)
DC	pKa			
HL 7.04E-006	atm m3/mol	25	EST	MEYLAN,WM & HOWARD,PH (1991)
OH 9.10E-012	cm3/molc sec	25	EXP	ATKINSON,R (1989)

THIAZOLE

CAS #:	000288-47-1	THIAZOLE
Formula:	C_3H_3NS	
Mol Weight:	85.13	
MP (deg C):		FP (deg C):
BP (deg C):	118	
BP pressure (mm Hg):		

Property/ Value	Units	Temp	Data Type	Reference
WS 5.38E+004	mg/L	25	EST	MEYLAN,WM ET AL. (1996)
logP 0.44			EXP	HANSCH,C & LEO,AJ (1985)
VP 1.72E+001	mm Hg	25	EXT	OHE,S (1976)
DC	pKa			
HL 3.83E-006	atm m3/mol	25	EST	MEYLAN,WM & HOWARD,PH (1991)
OH 1.40E-012	cm3/molc sec	25	EXP	ATKINSON,R (1989)

1H-1,2,4-TRIAZOLE

CAS #:	000288-88-0	1H-1,2,4-TRIAZOLE
Formula:	$C_2H_3N_3$	
Mol Weight:	69.07	
MP (deg C):		FP (deg C):
BP (deg C):	270-280	
BP pressure (mm Hg):	1.20E+000	

Property/ Value	Units	Temp	Data Type	Reference
WS 4.24E+005	mg/L	25	EST	MEYLAN,WM ET AL. (1996)
logP -0.58			EXP	HANSCH,C ET AL. (1995)
VP 6.03E-001	mm Hg	25	EST	NEELY,WB & BLAU,GE (1985)
DC	pKa			
HL 1.53E-006	atm m3/mol	25	EST	MEYLAN,WM & HOWARD,PH (1991)
OH 1.00E-013	cm3/molc sec	25	EST	MEYLAN,WM & HOWARD,PH (1993)

1H-TETRAZOLE

CAS #:	000288-94-8	1H-TETRAZOLE
Formula:	CH_2N_4	
Mol Weight:	70.05	
MP (deg C):	157	FP (deg C):
BP (deg C):		
BP pressure (mm Hg):		

Property/ Value	Units	Temp	Data Type	Reference
WS 2.16E+004	mg/L	25	EST	MEYLAN,WM ET AL. (1996)
logP -0.60			EXP	HANSCH,C ET AL. (1995)
VP 2.83E-001	mm Hg	25	EST	NEELY,WB & BLAU,GE (1985)
DC 4.89	pKa		EXP	KORTUM,G ET AL (1961)
HL 6.11E-007	atm m3/mol	25	EST	MEYLAN,WM & HOWARD,PH (1991)
OH 1.00E-013	cm3/molc sec	25	EST	MEYLAN,WM & HOWARD,PH (1993)

1,2,4-TRITHIOLANE

CAS #:	000289-16-7	1,2,4-TRITHIOLANE
Formula:	$C_2H_4S_3$	
Mol Weight:	124.25	
MP (deg C):		FP (deg C):
BP (deg C):		
BP pressure (mm Hg):		

Property/ Value	Units	Temp	Data Type	Reference
WS 6.72E+003	mg/L	25	EST	MEYLAN,WM ET AL. (1996)
logP 1.36			EST	MEYLAN,WM & HOWARD,PH (1995)
VP 8.00E-001	mm Hg	25	EST	NEELY,WB & BLAU,GE (1985)
DC	pKa			
HL 4.30E-006	atm m3/mol	25	EST	MEYLAN,WM & HOWARD,PH (1991)
OH 3.18E-010	cm3/molc sec	25	EST	MEYLAN,WM & HOWARD,PH (1993)

PYRIDAZINE

CAS #:	000289-80-5	PYRIDAZINE
Formula:	$C_4H_4N_2$	
Mol Weight:	80.09	
MP (deg C):	-8	FP (deg C):
BP (deg C):	208	
BP pressure (mm Hg):		

Property/ Value	Units	Temp	Data Type	Reference
WS 5.39E+005	mg/L	25	EST	MEYLAN,WM ET AL. (1996)
logP -0.72			EXP	HANSCH,C & LEO,AJ (1985)
VP 4.48E-001	mm Hg	25	EST	NEELY,WB & BLAU,GE (1985)
DC 2.24	pKa	20	EXP	PERRIN,DD (1965)
HL 2.87E-006	atm m3/mol	25	EST	MEYLAN,WM & HOWARD,PH (1991)
OH 2.30E-013	cm3/molc sec	25	EST	MEYLAN,WM & HOWARD,PH (1993)

PYRIMIDINE

CAS #:	000289-95-2	PYRIMIDINE
Formula:	$C_4H_4N_2$	
Mol Weight:	80.09	
MP (deg C):	20-22	FP (deg C):
BP (deg C):	123-124	
BP pressure (mm Hg):	7.62E+002	

Property/ Value	Units	Temp	Data Type	Reference
WS 2.87E+005	mg/L	25	EST	MEYLAN,WM ET AL. (1996)
logP -0.40			EXP	HANSCH,C & LEO,AJ (1985)
VP 1.29E+001	mm Hg	25	EST	NEELY,WB & BLAU,GE (1985)
DC 1.23	pKa	20	EXP	PERRIN,DD (1965)
HL 2.92E-006	atm m3/mol	25	EST	MEYLAN,WM & HOWARD,PH (1991)
OH 2.30E-013	cm3/molc sec	25	EST	MEYLAN,WM & HOWARD,PH (1993)

CAS #: 000290-37-9				PYRAZINE

Formula: $C_4H_4N_2$

Mol Weight: 80.09

MP (deg C): 53

FP (deg C):

BP (deg C): 115-118

BP pressure (mm Hg):

Property/ Value	Units	Temp	Data Type	Reference
WS 2.18E+005	mg/L	25	EST	MEYLAN,WM ET AL. (1996)
logP -0.26			EXP	HANSCH,C & LEO,AJ (1985)
VP 1.08E+001	mm Hg	25	EXT	CHAO,J ET AL. (1983)
DC 0.51	pKa	20	EXP	PERRIN,DD (1965)
HL 2.92E-006	atm m3/mol	25	EST	MEYLAN,WM & HOWARD,PH (1991)
OH 2.30E-013	cm3/molc sec	25	EST	MEYLAN,WM & HOWARD,PH (1993)

CAS #: 000290-87-9				SYM-TRIAZINE

Formula: $C_3H_3N_3$

Mol Weight: 81.08

MP (deg C): 86

FP (deg C):

BP (deg C): 114

BP pressure (mm Hg):

Property/ Value	Units	Temp	Data Type	Reference
WS 5.47E+005	mg/L	25	EST	MEYLAN,WM ET AL. (1996)
logP -0.73			EXP	HANSCH,C & LEO,AJ (1985)
VP 7.98E+000	mm Hg	25	EST	NEELY,WB & BLAU,GE (1985)
DC	pKa			
HL 1.21E-006	atm m3/mol	25	EST	MEYLAN,WM & HOWARD,PH (1991)
OH 1.50E-013	cm3/molc sec	25	EXP	ATKINSON,R (1989)

CAS #: 000291-64-5				CYCLOHEPTANE

Formula: C_7H_{14}

Mol Weight: 98.19

MP (deg C): -8

FP (deg C):

BP (deg C): 118.4

BP pressure (mm Hg):

Property/ Value	Units	Temp	Data Type	Reference
WS 3.00E+001	mg/L	25	EXP	YALKOWSKY,SH & DANNENFELSER,RM (1992)
logP 4.00			EXP	HANSCH,C ET AL. (1995)
VP 2.16E+001	mm Hg	25	EXP	DAUBERT,TE & DANNER,RP (1989)
DC	pKa			
HL 9.30E-002	atm m3/mol	25	EST	VP/WSOL
OH 1.25E-011	cm3/molc sec	25	EXP	ATKINSON,R (1989)

CAS #: 000292-64-8				CYCLOOCTANE

Formula: C_8H_{16}

Mol Weight: 112.22

MP (deg C): 14.8

FP (deg C):

BP (deg C): 149

BP pressure (mm Hg):

Property/ Value	Units	Temp	Data Type	Reference
WS 1.82E+002	mg/L	20	EXP	YALKOWSKY,SH & DANNENFELSER,RM (1992)
logP 4.45			EXP	SANGSTER,J (1993)
VP 5.51E+000	mm Hg	25	EXT	OHE,S (1976)
DC	pKa			
HL 4.50E-001	atm m3/mol	25	EST	MEYLAN,WM & HOWARD,PH (1991)
OH 1.37E-011	cm3/molc sec	25	EXP	ATKINSON,R (1989)

CAS #: 000293-96-9				CYCLODECANE

Formula: $C_{10}H_{20}$

Mol Weight: 140.27

MP (deg C): 10

FP (deg C):

BP (deg C): 202

BP pressure (mm Hg):

Property/ Value	Units	Temp	Data Type	Reference
WS 9.89E-001	mg/L	25	EST	MEYLAN,WM ET AL. (1996)
logP 5.14			EST	MEYLAN,WM & HOWARD,PH (1995)
VP 5.60E-001	mm Hg	25	EXT	BOUBLIK,T ET AL. (1984)
DC	pKa			
HL 7.92E-001	atm m3/mol	25	EST	MEYLAN,WM & HOWARD,PH (1991)
OH 1.41E-011	cm3/molc sec	25	EST	MEYLAN,WM & HOWARD,PH (1993)

CAS #: 000294-62-2				CYCLODODECANE

Formula: $C_{12}H_{24}$

Mol Weight: 168.33

MP (deg C): 60

FP (deg C):

BP (deg C): 244

BP pressure (mm Hg):

Property/ Value	Units	Temp	Data Type	Reference
WS 4.70E-003	mg/L	25	EST	VP/HL
logP 6.12			EST	MEYLAN,WM & HOWARD,PH (1995)
VP 2.95E-002	mm Hg	25	EST	BOUBLIK,T ET AL. (1984)
DC	pKa			
HL 1.40E+000	atm m3/mol	25	EST	MEYLAN,WM & HOWARD,PH (1991)
OH 1.67E-011	cm3/molc sec	25	EST	ATKINSON,R (1988)

CAS #: 000294-93-9				1,4,7,10-TETRAOXACYCLODODECANE

Formula: $C_8H_{16}O_4$

Mol Weight: 176.21

MP (deg C): 16

FP (deg C):

BP (deg C): 61-70

BP pressure (mm Hg): 5.00E-001

Property/ Value	Units	Temp	Data Type	Reference
WS 6.21E+004	mg/L	25	EST	MEYLAN,WM ET AL. (1996)
logP -0.04			EXP	HANSCH,C ET AL. (1995)
VP 4.49E-002	mm Hg	25	EST	NEELY,WB & BLAU,GE (1985)
DC	pKa			
HL 1.43E-009	atm m3/mol	25	EST	MEYLAN,WM & HOWARD,PH (1991)
OH 5.61E-011	cm3/molc sec	25	EST	MEYLAN,WM & HOWARD,PH (1993)

CAS #: 000297-78-9				ISOBENZAN

Formula: $C_9H_4Cl_8O$

Mol Weight: 411.76

MP (deg C): 121

FP (deg C):

BP (deg C):

BP pressure (mm Hg):

Property/ Value	Units	Temp	Data Type	Reference
WS 9.45E-002	mg/L	25	EST	MEYLAN,WM ET AL. (1996)
logP 5.20			EST	MEYLAN,WM & HOWARD,PH (1995)
VP 2.92E-006	mm Hg	25	EXP	ULLMANN'S ENCYC INDUST CHEM (1989)
DC	pKa			
HL 5.88E-008	atm m3/mol	25	EST	MEYLAN,WM & HOWARD,PH (1991)
OH 4.60E-013	cm3/molc sec	25	EST	MEYLAN,WM & HOWARD,PH (1993)

CAS #: 000297-97-2		THIONAZIN

Formula: $C_8H_{13}N_2O_3PS$

Mol Weight: 248.24

MP (deg C): -1.7

FP (deg C):

BP (deg C): 80

BP pressure (mm Hg):

Property/Value	Units	Temp	Data Type	Reference
WS 1.14E+003	mg/L	25	EXP	YALKOWSKY,SH & DANNENFELSER,RM (1992)
logP 1.86			EST	MEYLAN,WM & HOWARD,PH (1995)
VP 3.00E-003	mm Hg	25	EXP	BACCI,E ET AL. (1990)
DC	pKa			
HL 8.60E-007	atm m3/mol	25	EST	VP/WSOL
OH 9.21E-011	cm3/molc sec	25	EST	MEYLAN,WM & HOWARD,PH (1993)

CAS #: 000297-99-4		PHOSPHAMIDON

Formula: $C_{10}H_{19}ClNO_5P$

Mol Weight: 299.69

MP (deg C): -45

FP (deg C):

BP (deg C):

BP pressure (mm Hg):

Property/Value	Units	Temp	Data Type	Reference
WS 1.00E+006	mg/L	20	EXP	SHIU,WY ET AL. (1990)
logP 0.38			EST	MEYLAN,WM & HOWARD,PH (1995)
VP 1.65E-005	mm Hg	25	EXP	WAUCHOPE,RD ET AL. (1991A)
DC	pKa			
HL 6.51E-012	atm m3/mol	25	EST	VP/WSOL
OH 3.68E-011	cm3/molc sec	25	EST	MEYLAN,WM & HOWARD,PH (1993)

CAS #: 000298-00-0		PARATHION METHYL

Formula: $C_8H_{10}NO_5PS$

Mol Weight: 263.21

MP (deg C): 35-36

FP (deg C):

BP (deg C): 109

BP pressure (mm Hg): 5.00E-002

Property/Value	Units	Temp	Data Type	Reference
WS 3.77E+001	mg/L	20	EXP	BOWMAN,BT & SANS,WW (1983)
logP 2.86			EXP	HANSCH,C & LEO,AJ (1985)
VP 1.72E-005	mm Hg	25	EXT	SPENCER,WF ET AL. (1979)
DC 7.15	pKa		EXP	WOLFE,NL (1980)
HL 1.00E-007	atm m3/mol	25	EXP	METCALFE,CD ET AL. (1980)
OH 6.12E-011	cm3/molc sec	25	EST	MEYLAN,WM & HOWARD,PH (1993)

CAS #: 000298-01-1		MEVINPHOS (TRANS)

Formula: $C_7H_{13}O_6P$

Mol Weight: 224.15

MP (deg C):

FP (deg C):

BP (deg C):

BP pressure (mm Hg):

Property/Value	Units	Temp	Data Type	Reference
WS 1.00E+006	mg/L		EXP	SHAROM,MS ET AL. (1980A) @2ND
logP -0.24			EST	MEYLAN,WM & HOWARD,PH (1995)
VP 3.33E-003	mm Hg	25	EST	NEELY,WB & BLAU,GE (1985)
DC	pKa			
HL 3.89E-009	atm m3/mol	25	EST	MEYLAN,WM & HOWARD,PH (1991)
OH 3.63E-011	cm3/molc sec	25	EST	MEYLAN,WM & HOWARD,PH (1993)

CAS #: 000298-02-2		PHORATE

Formula: $C_7H_{17}O_2PS_3$

Mol Weight: 260.38

MP (deg C): < -15

FP (deg C):

BP (deg C): 125-127

BP pressure (mm Hg): 2.00E+000

Property/Value	Units	Temp	Data Type	Reference
WS 5.00E+001	mg/L	25	EXP	MARTIN,H & WORTHING,CR (1977)
logP 3.56			EXP	HANSCH,C ET AL. (1995)
VP 8.40E-004	mm Hg	20	EXP	MARTIN,H & WORTHING,CR (1977)
DC	pKa			
HL 5.76E-006	atm m3/mol	25	EST	VP/WSOL
OH 2.50E-010	cm3/molc sec	25	EST	MEYLAN,WM & HOWARD,PH (1993)

CAS #: 000298-04-4		DISULFOTON

Formula: $C_8H_{19}O_2PS_3$

Mol Weight: 274.41

MP (deg C): -25

FP (deg C):

BP (deg C): 132-133

BP pressure (mm Hg): 1.50E+000

Property/Value	Units	Temp	Data Type	Reference
WS 1.63E+001	mg/L	20	EXP	BOWMAN,BT & SANS,WW (1983A)
logP 4.02			EXP	HANSCH,C & LEO,AJ (1985)
VP 1.80E-004	mm Hg	20	EXP	WORTHING,CR & WALKER,SB (1983)
DC	pKa			
HL 3.99E-006	atm m3/mol	25	EST	VP/WSOL
OH 1.33E-010	cm3/molc sec	25	EST	MEYLAN,WM & HOWARD,PH (1993)

CAS #: 000298-07-7		BIS(2-ETHYLHEXYL)PHOSPHATE

Formula: $C_{16}H_{35}O_4P$

Mol Weight: 322.43

MP (deg C): -60

FP (deg C):

BP (deg C):

BP pressure (mm Hg):

Property/Value	Units	Temp	Data Type	Reference
WS 1.00E+002	mg/L	25	EXP	YALKOWSKY,SH & DANNENFELSER,RM (1992)
logP 4.37			EST	LYMAN,WJ ET AL. (1990)
VP 9.69E-006	mm Hg	25	EST	LYMAN,WJ ET AL. (1990)
DC	pKa			
HL 4.11E-008	atm m3/mol	25	EST	MEYLAN,WM & HOWARD,PH (1991)
OH 6.18E-011	cm3/molc sec	25	EST	ATKINSON,R (1988)

CAS #: 000298-12-4		GLYOXYLIC ACID

Formula: $C_2H_2O_3$

Mol Weight: 74.04

MP (deg C): 98

FP (deg C):

BP (deg C):

BP pressure (mm Hg):

Property/Value	Units	Temp	Data Type	Reference
WS 1.00E+006	mg/L	25	EST	MEYLAN,WM ET AL. (1996)
logP -1.40			EST	MEYLAN,WM & HOWARD,PH (1995)
VP 1.06E+000	mm Hg	25	EST	NEELY,WB & BLAU,GE (1985)
DC 3.30	pKa	25	EXP	DEAN,JA (1985)
HL 2.99E-009	atm m3/mol	25	EST	MEYLAN,WM & HOWARD,PH (1991)
OH 1.23E-011	cm3/molc sec	25	EST	ATKINSON,R (1988)

1,2:3,4-DIEPOXYBUTANE DL

CAS #:	000298-18-0
Formula:	$C_4H_6O_2$
Mol Weight:	86.09
MP (deg C):	2-4
FP (deg C):	
BP (deg C):	144
BP pressure (mm Hg):	

Property/Value	Units	Temp	Data Type	Reference
WS 1.00E+006	mg/L		EXP	DEAN,JA (1985)
logP -0.58			EST	MEYLAN,WM & HOWARD,PH (1995)
VP 6.11E+000	mm Hg	25	EST	CLAUSIUS-CLAPEYRON EQN; ALDRICH (1988)
DC	pKa			
HL 3.54E-008	atm m3/mol	25	EST	MEYLAN,WM & HOWARD,PH (1991)
OH 8.56E-013	cm3/molc sec	25	EST	MEYLAN,WM & HOWARD,PH (1993)

CARBAMAZEPINE

CAS #:	000298-46-4
Formula:	$C_{15}H_{12}N_2O$
Mol Weight:	236.28
MP (deg C):	190-193
FP (deg C):	
BP (deg C):	
BP pressure (mm Hg):	

Property/Value	Units	Temp	Data Type	Reference
WS 1.77E+001	mg/L	25	EST	MEYLAN,WM ET AL. (1996)
logP 2.45			EXP	DAL POZZO,A ET AL. (1989)
VP 1.84E-007	mm Hg	25	EST	NEELY,WB & BLAU,GE (1985)
DC	pKa			
HL 1.08E-010	atm m3/mol	25	EST	MEYLAN,WM & HOWARD,PH (1991)
OH 8.07E-011	cm3/molc sec	25	EST	MEYLAN,WM & HOWARD,PH (1993)

EPHEDRINE

CAS #:	000299-42-3
Formula:	$C_{10}H_{15}NO$
Mol Weight:	165.24
MP (deg C):	34
FP (deg C):	
BP (deg C):	255
BP pressure (mm Hg):	

Property/Value	Units	Temp	Data Type	Reference
WS 6.36E+004	mg/L	30	EXP	YALKOWSKY,SH & DANNENFELSER,RM (1992)
logP 0.93			EXP	HANSCH,C ET AL. (1995)
VP 8.30E-004	mm Hg	25	EST	NEELY,WB & BLAU,GE (1985)
DC 10.25	pKa	0	EXP	PERRIN,DD (1965)
HL 8.65E-011	atm m3/mol	25	EST	MEYLAN,WM & HOWARD,PH (1991)
OH 9.97E-011	cm3/molc sec	25	EST	MEYLAN,WM & HOWARD,PH (1993)

E-838

CAS #:	000299-45-6
Formula:	$C_{14}H_{17}O_5PS$
Mol Weight:	328.33
MP (deg C):	38
FP (deg C):	
BP (deg C):	210
BP pressure (mm Hg):	1.00E+000

Property/Value	Units	Temp	Data Type	Reference
WS 3.33E+000	mg/L	25	EST	MEYLAN,WM ET AL. (1996)
logP 3.98			EST	MEYLAN,WM & HOWARD,PH (1995)
VP 2.50E-007	mm Hg	25	EST	NEELY,WB & BLAU,GE (1985)
DC	pKa			
HL 1.52E-007	atm m3/mol	25	EST	MEYLAN,WM & HOWARD,PH (1991)
OH 1.26E-010	cm3/molc sec	25	EST	MEYLAN,WM & HOWARD,PH (1993)

RONNEL

CAS #:	000299-84-3
Formula:	$C_8H_8Cl_3O_3PS$
Mol Weight:	321.55
MP (deg C):	41
FP (deg C):	
BP (deg C):	151-154
BP pressure (mm Hg):	4.00E-001

Property/Value	Units	Temp	Data Type	Reference
WS 1.00E+000	mg/L	20	EXP	SUNTIO,LR ET AL. (1988)
logP 4.88			EXP	SANGSTER,J (1994)
VP 7.50E-005	mm Hg	20	EXP	SUNTIO,LR ET AL. (1988)
DC	pKa			
HL 3.20E-005	atm m3/mol	20	EST	SUNTIO,LR ET AL. (1988)
OH 6.10E-011	cm3/molc sec	25	EST	MEYLAN,WM & HOWARD,PH (1993)

DMPA

CAS #:	000299-85-4
Formula:	$C_{10}H_{14}Cl_2NO_2PS$
Mol Weight:	314.17
MP (deg C):	51.4
FP (deg C):	
BP (deg C):	
BP pressure (mm Hg):	

Property/Value	Units	Temp	Data Type	Reference
WS 5.00E+000	mg/L	25	EXP	YALKOWSKY,SH & DANNENFELSER,RM (1992)
logP 4.30			EXP	HANSCH,C & LEO,AJ (1985)
VP 6.14E-005	mm Hg	25	EST	NEELY,WB & BLAU,GE (1985)
DC	pKa			
HL 5.74E-007	atm m3/mol	25	EST	MEYLAN,WM & HOWARD,PH (1991)
OH 2.53E-010	cm3/molc sec	25	EST	MEYLAN,WM & HOWARD,PH (1993)

CRUFOMATE

CAS #:	000299-86-5
Formula:	$C_{12}H_{19}ClNO_3P$
Mol Weight:	291.72
MP (deg C):	60-60.5
FP (deg C):	
BP (deg C):	117-118
BP pressure (mm Hg):	1.00E-002

Property/Value	Units	Temp	Data Type	Reference
WS 5.00E+003	mg/L		EXP	YALKOWSKY,SH & DANNENFELSER,RM (1992)
logP 3.42			EXP	HANSCH,C & LEO,AJ (1985)
VP 2.08E-005	mm Hg	25	EST	NEELY,WB & BLAU,GE (1985)
DC	pKa			
HL 2.46E-009	atm m3/mol	25	EST	MEYLAN,WM & HOWARD,PH (1991)
OH 3.43E-011	cm3/molc sec	25	EST	MEYLAN,WM & HOWARD,PH (1993)

TYROSINE, 3-METHOXY-

CAS #:	000300-48-1
Formula:	$C_{10}H_{13}NO_4$
Mol Weight:	211.22
MP (deg C):	
FP (deg C):	
BP (deg C):	
BP pressure (mm Hg):	

Property/Value	Units	Temp	Data Type	Reference
WS 4.81E+004	mg/L	25	EST	MEYLAN,WM ET AL. (1996)
logP -2.54			EXP	HANSCH,C & LEO,AJ (1985)
VP 1.35E-009	mm Hg	25	EST	NEELY,WB & BLAU,GE (1985)
DC	pKa			
HL 7.42E-016	atm m3/mol	25	EST	MEYLAN,WM & HOWARD,PH (1991)
OH 7.71E-011	cm3/molc sec	25	EST	MEYLAN,WM & HOWARD,PH (1993)

000300-57-2 — ALLYLBENZENE

Formula: C_9H_{10}

Mol Weight: 118.18

MP (deg C): -40

FP (deg C):

BP (deg C): 153

BP pressure (mm Hg):

Property/Value	Units	Temp	Data Type	Reference
WS 1.78E+002	mg/L	25	EST	MEYLAN,WM ET AL. (1996)
logP 3.23			EXP	HANSCH,C & LEO,AJ (1985)
VP 1.69E+000	mm Hg	25	EST	NEELY,WB & BLAU,GE (1985)
DC	pKa			
HL 7.81E-003	atm m3/mol	25	EST	MEYLAN,WM & HOWARD,PH (1991)
OH 3.21E-011	cm3/molc sec	25	EST	MEYLAN,WM & HOWARD,PH (1993)

000300-62-9 — ALPHA-METHYLPHENETHYLAMINE

Formula: $C_9H_{13}N$

Mol Weight: 135.21

MP (deg C):

FP (deg C):

BP (deg C): 203

BP pressure (mm Hg):

Property/Value	Units	Temp	Data Type	Reference
WS 2.80E+004	mg/L	25	EST	MEYLAN,WM ET AL. (1996)
logP 1.76			EXP	HANSCH,C & LEO,AJ (1985)
VP 2.40E-001	mm Hg	20	EXP	LAWRENCE,AH ET AL. (1984)
DC 10.13	pKa	20	EXP	PERRIN,DD (1965)
HL 1.08E-006	atm m3/mol	25	EST	MEYLAN,WM & HOWARD,PH (1991)
OH 4.94E-011	cm3/molc sec	25	EST	MEYLAN,WM & HOWARD,PH (1993)

000300-76-5 — NALED

Formula: $C_4H_7Br_2Cl_2O_4P$

Mol Weight: 380.79

MP (deg C): 26.5-27.5

FP (deg C):

BP (deg C): 120

BP pressure (mm Hg): 5.00E-001

Property/Value	Units	Temp	Data Type	Reference
WS 2.00E+003	mg/L	25	EXP	WAUCHOPE,RD ET AL. (1991A)
logP 1.38			EXP	HANSCH,C & LEO,AJ (1985)
VP 2.00E-003	mm Hg	20	EXP	MARTIN,H & WORTHING,CR (1977)
DC	pKa			
HL 5.00E-007	atm m3/mol	25	EST	MEYLAN,WM & HOWARD,PH (1991)
OH 6.70E-012	cm3/molc sec	25	EST	MEYLAN,WM & HOWARD,PH (1993)

000301-11-1 — LETHANE 60

Formula: $C_{15}H_{27}NO_2S$

Mol Weight: 285.45

MP (deg C):

FP (deg C):

BP (deg C): 160-190

BP pressure (mm Hg): 1.00E-001

Property/Value	Units	Temp	Data Type	Reference
WS 2.15E-001	mg/L	25	EST	MEYLAN,WM ET AL. (1996)
logP 5.67			EST	MEYLAN,WM & HOWARD,PH (1995)
VP 2.63E-006	mm Hg	25	EST	NEELY,WB & BLAU,GE (1985)
DC	pKa			
HL 2.36E-006	atm m3/mol	25	EST	MEYLAN,WM & HOWARD,PH (1991)
OH 2.43E-011	cm3/molc sec	25	EST	MEYLAN,WM & HOWARD,PH (1993)

000301-12-2 — OXYDEMETON METHYL

Formula: $C_6H_{15}O_4PS_2$

Mol Weight: 246.29

MP (deg C): < -20

FP (deg C):

BP (deg C): 106

BP pressure (mm Hg): 1.00E-002

Property/Value	Units	Temp	Data Type	Reference
WS 1.00E+006	mg/L		EXP	WORTHING,R (1987)
logP -0.75			EXP	TOMLIN,C (1994)
VP 2.85E-005	mm Hg	20	EXP	WORTHING,R (1987)
DC	pKa			
HL 1.62E-013	atm m3/mol	25	EST	MEYLAN,WM & HOWARD,PH (1991)
OH 1.07E-010	cm3/molc sec	25	EST	MEYLAN,WM & HOWARD,PH (1993)

000302-17-0 — CHORAL HYDRATE

Formula: $C_2H_3Cl_3O_2$

Mol Weight: 165.40

MP (deg C): 57

FP (deg C):

BP (deg C): 98

BP pressure (mm Hg):

Property/Value	Units	Temp	Data Type	Reference
WS 9.31E+006	mg/L	38	EXP	YALKOWSKY,SH & DANNENFELSER,RM (1992)
logP 0.99			EXP	HANSCH,C & LEO,AJ (1985)
VP 1.50E+001	mm Hg	25	EXP	PERRY,RH & GREEN,D (1984)
DC	pKa			
HL 5.71E-009	atm m3/mol	25	EST	MEYLAN,WM & HOWARD,PH (1991)
OH 1.92E-012	cm3/molc sec	25	EST	MEYLAN,WM & HOWARD,PH (1993)

000302-27-2 — ACONITINE

Formula: $C_{34}H_{47}NO_{11}$

Mol Weight: 645.75

MP (deg C): 204

FP (deg C):

BP (deg C):

BP pressure (mm Hg):

Property/Value	Units	Temp	Data Type	Reference
WS 3.10E+002	mg/L	25	EXP	SEIDELL,A (1941)
logP 0.13			EST	MEYLAN,WM & HOWARD,PH (1995)
VP 1.38E-020	mm Hg	25	EST	NEELY,WB & BLAU,GE (1985)
DC	pKa			
HL 4.53E-026	atm m3/mol	25	EST	MEYLAN,WM & HOWARD,PH (1991)
OH 2.05E-010	cm3/molc sec	25	EST	MEYLAN,WM & HOWARD,PH (1993)

000302-41-0 — PIRITRAMIDE

Formula: $C_{27}H_{34}N_4O$

Mol Weight: 430.60

MP (deg C): 149-150

FP (deg C):

BP (deg C):

BP pressure (mm Hg):

Property/Value	Units	Temp	Data Type	Reference
WS 4.91E-001	mg/L	25	EST	MEYLAN,WM ET AL. (1996)
logP 3.91			EXP	SANGSTER,J (1993)
VP 7.13E-014	mm Hg	25	EST	NEELY,WB & BLAU,GE (1985)
DC	pKa			
HL 2.78E-019	atm m3/mol	25	EST	MEYLAN,WM & HOWARD,PH (1991)
OH 2.11E-010	cm3/molc sec	25	EST	MEYLAN,WM & HOWARD,PH (1993)

CAS #: 000302-57-8 — TRIETHYL METHYL AMMONIUM IODIDE

Formula: $C_7H_{18}IN$

Mol Weight: 243.13

MP (deg C):
FP (deg C):
BP (deg C):
BP pressure (mm Hg):

Property/Value	Units	Temp	Data Type	Reference
WS 1.00E+006	mg/L	25	EST	MEYLAN,WM ET AL. (1996)
logP -2.13			EXP	NEEF,C & MEIJER,DKF (1984)
VP 5.01E-007	mm Hg	25	EST	NEELY,WB & BLAU,GE (1985)
DC	pKa			
HL 6.16E-012	atm m3/mol	25	EST	MEYLAN,WM & HOWARD,PH (1991)
OH 2.78E-011	cm3/molc sec	25	EST	MEYLAN,WM & HOWARD,PH (1993)

CAS #: 000302-72-7 — ALANINE

Formula: $C_3H_7NO_2$

Mol Weight: 89.09

MP (deg C): 300 dec
FP (deg C):
BP (deg C):
BP pressure (mm Hg):

Property/Value	Units	Temp	Data Type	Reference
WS 1.66E+005	mg/L	25	EXP	YALKOWSKY,SH & DANNENFELSER,RM (1992)
logP -2.96			EXP	HANSCH,C & LEO,AJ (1985)
VP 1.05E-007	mm Hg	25	EST	NEELY,WB & BLAU,GE (1985)
DC 2.34	pKa	25	EXP	KORTUM,G ET AL (1961)
HL 1.49E-009	atm m3/mol	25	EST	MEYLAN,WM & HOWARD,PH (1991)
OH 3.52E-011	cm3/molc sec	25	EST	MEYLAN,WM & HOWARD,PH (1993)

CAS #: 000302-79-4 — RETINOIC ACID

Formula: $C_{20}H_{28}O_2$

Mol Weight: 300.44

MP (deg C): 180-182
FP (deg C):
BP (deg C):
BP pressure (mm Hg):

Property/Value	Units	Temp	Data Type	Reference
WS 1.26E-001	mg/L	25	EST	MEYLAN,WM ET AL. (1996)
logP 6.30			EXP	HANSCH,C ET AL. (1995)
VP 2.42E-007	mm Hg	25	EST	NEELY,WB & BLAU,GE (1985)
DC	pKa			
HL 9.50E-006	atm m3/mol	25	EST	MEYLAN,WM & HOWARD,PH (1991)
OH 2.93E-010	cm3/molc sec	25	EST	MEYLAN,WM & HOWARD,PH (1993)

CAS #: 000303-07-1 — 2,6-DIHYDROXYBENZOIC ACID

Formula: $C_7H_6O_4$

Mol Weight: 154.12

MP (deg C): 165 dec
FP (deg C):
BP (deg C):
BP pressure (mm Hg):

Property/Value	Units	Temp	Data Type	Reference
WS 9.56E+003	mg/L	25	EXP	YALKOWSKY,SH & DANNENFELSER,RM (1992)
logP 2.20			EXP	HANSCH,C & LEO,AJ (1985)
VP 2.10E-006	mm Hg	25	EST	NEELY,WB & BLAU,GE (1985)
DC 1.05	pKa	25	EXP	KORTUM,G ET AL (1961)
HL 1.86E-009	atm m3/mol	25	EST	MEYLAN,WM & HOWARD,PH (1991)
OH 2.01E-010	cm3/molc sec	25	EST	MEYLAN,WM & HOWARD,PH (1993)

CAS #: 000303-34-4 — LASIOCARPINE

Formula: $C_{21}H_{33}NO_7$

Mol Weight: 411.50

MP (deg C): 94-95.5
FP (deg C):
BP (deg C):
BP pressure (mm Hg):

Property/Value	Units	Temp	Data Type	Reference
WS 2.14E+003	mg/L	25	EST	MEYLAN,WM ET AL. (1996)
logP 2.43			EST	MEYLAN,WM & HOWARD,PH (1995)
VP 5.58E-011	mm Hg	25	EST	NEELY,WB & BLAU,GE (1985)
DC	pKa			
HL 6.14E-014	atm m3/mol	25	EST	MEYLAN,WM & HOWARD,PH (1991)
OH 2.03E-010	cm3/molc sec	25	EST	MEYLAN,WM & HOWARD,PH (1993)

CAS #: 000303-38-8 — 2,3-DIHYDROXYBENZOIC ACID

Formula: $C_7H_6O_4$

Mol Weight: 154.12

MP (deg C): 204
FP (deg C):
BP (deg C):
BP pressure (mm Hg):

Property/Value	Units	Temp	Data Type	Reference
WS 2.61E+004	mg/L	25	EST	MEYLAN,WM ET AL. (1996)
logP 1.20			EXP	HANSCH,C & LEO,AJ (1985)
VP 2.10E-006	mm Hg	25	EST	NEELY,WB & BLAU,GE (1985)
DC 2.91	pKa	25	EXP	SERJEANT,EP & DEMPSEY,B (1979)
HL 1.48E-012	atm m3/mol	25	EST	MEYLAN,WM & HOWARD,PH (1991)
OH 9.30E-012	cm3/molc sec	25	EST	MEYLAN,WM & HOWARD,PH (1993)

CAS #: 000303-47-9 — OCHRATOXIN A

Formula: $C_{20}H_{18}ClNO_6$

Mol Weight: 403.82

MP (deg C): 169
FP (deg C):
BP (deg C):
BP pressure (mm Hg):

Property/Value	Units	Temp	Data Type	Reference
WS 9.87E-001	mg/L	25	EST	MEYLAN,WM ET AL. (1996)
logP 4.74			EXP	SANGSTER,J (1993)
VP 3.31E-016	mm Hg	25	EST	NEELY,WB & BLAU,GE (1985)
DC	pKa			
HL 3.12E-017	atm m3/mol	25	EST	MEYLAN,WM & HOWARD,PH (1991)
OH 2.46E-012	cm3/molc sec	25	EST	MEYLAN,WM & HOWARD,PH (1993)

CAS #: 000303-49-1 — CHLORIMIPRAMINE

Formula: $C_{19}H_{23}ClN_2$

Mol Weight: 314.86

MP (deg C): 189-190
FP (deg C):
BP (deg C): 160-170
BP pressure (mm Hg): 3.00E-001

Property/Value	Units	Temp	Data Type	Reference
WS 2.94E-001	mg/L	25	EST	MEYLAN,WM ET AL. (1996)
logP 5.19			EXP	HANSCH,C & LEO,AJ (1985)
VP 4.07E-007	mm Hg	25	EST	NEELY,WB & BLAU,GE (1985)
DC	pKa			
HL 7.48E-009	atm m3/mol	25	EST	MEYLAN,WM & HOWARD,PH (1991)
OH 2.94E-010	cm3/molc sec	25	EST	MEYLAN,WM & HOWARD,PH (1993)

CAS #: 000305-03-3				4-(BIS(2-CHLOROETHYL)AMINO)BENZENEBUTANOIC AC*
Formula: $C_{14}H_{19}Cl_2NO_2$				
Mol Weight: 304.22				
MP (deg C): 64-66		FP (deg C):		
BP (deg C):				
BP pressure (mm Hg):				

Property/Value	Units	Temp	Data Type	Reference
WS 1.24E+004	mg/L	25	EST	MEYLAN,WM ET AL. (1996)
logP 1.47			EXP	HANSCH,C ET AL. (1995)
VP 9.30E-007	mm Hg	25	EST	NEELY,WB & BLAU,GE (1985)
DC 5.75	pKa		EXP	HANSCH,C & LEO,AJ (1987)
HL 2.71E-010	atm m3/mol	25	EST	MEYLAN,WM & HOWARD,PH (1991)
OH 1.89E-010	cm3/molc sec	25	EST	MEYLAN,WM & HOWARD,PH (1993)

CAS #: 000305-33-9				4-PYRIDINECARBOXYLIC ACID, 2-(1-METHYLETHYL)HYDR
Formula: $C_9H_{13}N_3O$				
Mol Weight: 179.22				
MP (deg C): 180-182		FP (deg C):		
BP (deg C):				
BP pressure (mm Hg):				

Property/Value	Units	Temp	Data Type	Reference
WS 2.68E+004	mg/L	25	EST	MEYLAN,WM ET AL. (1996)
logP 0.37			EXP	HANSCH,C ET AL. (1995)
VP 1.36E-005	mm Hg	25	EST	NEELY,WB & BLAU,GE (1985)
DC	pKa			
HL 4.68E-014	atm m3/mol	25	EST	MEYLAN,WM & HOWARD,PH (1991)
OH 8.72E-011	cm3/molc sec	25	EST	MEYLAN,WM & HOWARD,PH (1993)

CAS #: 000305-62-4				2,4-DIAMINOBUTYRIC ACID
Formula: $C_4H_{10}N_2O_2$				
Mol Weight: 118.14				
MP (deg C):		FP (deg C):		
BP (deg C):				
BP pressure (mm Hg):				

Property/Value	Units	Temp	Data Type	Reference
WS 1.00E+006	mg/L	25	EST	MEYLAN,WM ET AL. (1996)
logP -4.64			EXP	HANSCH,C ET AL. (1995)
VP 2.25E-009	mm Hg	25	EST	NEELY,WB & BLAU,GE (1985)
DC	pKa			
HL 2.04E-013	atm m3/mol	25	EST	MEYLAN,WM & HOWARD,PH (1991)
OH 7.13E-011	cm3/molc sec	25	EST	MEYLAN,WM & HOWARD,PH (1993)

CAS #: 000306-08-1				BENZENEACETIC ACID, 4-HYDROXY-3-METHOXY-
Formula: $C_9H_{10}O_4$				
Mol Weight: 182.18				
MP (deg C): 143		FP (deg C):		
BP (deg C):				
BP pressure (mm Hg):				

Property/Value	Units	Temp	Data Type	Reference
WS 6.97E+004	mg/L	25	EST	MEYLAN,WM ET AL. (1996)
logP 0.33			EXP	LAHANN,TR ET AL. (1989)
VP 8.23E-006	mm Hg	25	EST	NEELY,WB & BLAU,GE (1985)
DC 4.41	pKa	25	EXP	SERJEANT,EP & DEMPSEY,B (1979)
HL 2.72E-013	atm m3/mol	25	EST	MEYLAN,WM & HOWARD,PH (1991)
OH 3.19E-011	cm3/molc sec	25	EST	MEYLAN,WM & HOWARD,PH (1993)

CAS #: 000306-18-3				CARBONYL CYANIDE,PHENYLHYDRAZONE
Formula: $C_9H_6N_4$				
Mol Weight: 170.17				
MP (deg C):		FP (deg C):		
BP (deg C):				
BP pressure (mm Hg):				

Property/Value	Units	Temp	Data Type	Reference
WS 3.22E+002	mg/L	25	EST	MEYLAN,WM ET AL. (1996)
logP 2.67			EXP	HANSCH,C & LEO,AJ (1985)
VP 2.46E-005	mm Hg	25	EST	NEELY,WB & BLAU,GE (1985)
DC	pKa			
HL 5.79E-009	atm m3/mol	25	EST	MEYLAN,WM & HOWARD,PH (1991)
OH 4.27E-011	cm3/molc sec	25	EST	MEYLAN,WM & HOWARD,PH (1993)

CAS #: 000306-44-5				HYDROXYIMINOACETONE
Formula: $C_3H_5NO_2$				
Mol Weight: 87.08				
MP (deg C): 69		FP (deg C):		
BP (deg C):				
BP pressure (mm Hg):				

Property/Value	Units	Temp	Data Type	Reference
WS 6.48E+004	mg/L	25	EST	MEYLAN,WM ET AL. (1996)
logP 0.34			EXP	HANSCH,C & LEO,AJ (1985)
VP 2.01E-001	mm Hg	25	EST	NEELY,WB & BLAU,GE (1985)
DC	pKa			
HL 5.94E-008	atm m3/mol	25	EST	MEYLAN,WM & HOWARD,PH (1991)
OH 1.56E-012	cm3/molc sec	25	EST	MEYLAN,WM & HOWARD,PH (1993)

CAS #: 000306-52-5				TRICHLOROETHYL PHOSPHATE
Formula: $C_2H_4Cl_3O_4P$				
Mol Weight: 229.38				
MP (deg C):		FP (deg C):		
BP (deg C):				
BP pressure (mm Hg):				

Property/Value	Units	Temp	Data Type	Reference
WS 3.31E+003	mg/L	25	EST	MEYLAN,WM ET AL. (1996)
logP 1.13			EST	MEYLAN,WM & HOWARD,PH (1995)
VP 2.13E-006	mm Hg	25	EST	NEELY,WB & BLAU,GE (1985)
DC	pKa			
HL 1.41E-013	atm m3/mol	25	EST	MEYLAN,WM & HOWARD,PH (1991)
OH 1.58E-012	cm3/molc sec	25	EST	MEYLAN,WM & HOWARD,PH (1993)

CAS #: 000306-83-2				1,1,1-TRIFLUORO-2,2-DICHLOROETHANE
Formula: $C_2HCl_2F_3$				
Mol Weight: 152.93				
MP (deg C):		FP (deg C):		
BP (deg C):				
BP pressure (mm Hg):				

Property/Value	Units	Temp	Data Type	Reference
WS 1.49E+003	mg/L	25	EST	VP/HL
logP 2.17			EST	MEYLAN,WM & HOWARD,PH (1995)
VP 3.58E+002	mm Hg	25	EST	NEELY,WB & BLAU,GE (1985)
DC	pKa			
HL 9.55E-002	atm m3/mol	25	EST	MEYLAN,WM & HOWARD,PH (1991)
OH 3.35E-012	cm3/molc sec	25	EXP	ATKINSON,R (1989)

303

000309-00-2 — ALDRIN

Formula: $C_{12}H_8Cl_6$				
Mol Weight: 364.92				
MP (deg C): 104-105			**FP (deg C):**	
BP (deg C): 145				
BP pressure (mm Hg): 2.00E+000				

Property/Value	Units	Temp	Data Type	Reference
WS 1.80E-001	mg/L	25	EXP	YALKOWSKY,SH & DANNENFELSER,RM (1992)
logP 6.50			EXP	DEBRUIJN,J ET AL. (1989)
VP 1.20E-004	mm Hg	25	EXP	GRAYSON,BT & FOSBRAEY,LA (1982)
DC	pKa			
HL 4.93E-004	atm m3/mol	20	EXP	WARNER,HP ET AL. (1987)
OH 6.84E-011	cm3/molc sec	25	EST	MEYLAN,WM & HOWARD,PH (1993)

000309-11-5 — PENTAFLUOROETHYLBENZENE

Formula: $C_8H_5F_5$				
Mol Weight: 196.12				
MP (deg C):			**FP (deg C):**	
BP (deg C):				
BP pressure (mm Hg):				

Property/Value	Units	Temp	Data Type	Reference
WS 6.19E+001	mg/L	25	EST	MEYLAN,WM ET AL. (1996)
logP 3.36			EXP	HANSCH,C & LEO,AJ (1985)
VP 1.42E+001	mm Hg	25	EST	NEELY,WB & BLAU,GE (1985)
DC	pKa			
HL 1.71E-002	atm m3/mol	25	EST	MEYLAN,WM & HOWARD,PH (1991)
OH 1.63E-012	cm3/molc sec	25	EST	MEYLAN,WM & HOWARD,PH (1993)

000311-45-5 — PARAOXON

Formula: $C_{10}H_{14}NO_6P$				
Mol Weight: 275.20				
MP (deg C): 300			**FP (deg C):**	
BP (deg C):				
BP pressure (mm Hg):				

Property/Value	Units	Temp	Data Type	Reference
WS 3.64E+003	mg/L	20	EXP	BOWMAN,BT & SANS,WW (1983A)
logP 1.98			EXP	HANSCH,C & LEO,AJ (1985)
VP 1.10E-006	mm Hg	25	EXP	BENNETT,SR ET AL. (1984)
DC	pKa			
HL 9.95E-011	atm m3/mol	25	EST	VP/WSOL
OH 3.91E-011	cm3/molc sec	25	EST	MEYLAN,WM & HOWARD,PH (1993)

000311-47-7 — 2-CHLOROVINYL DIETHYL PHOSPHATE

Formula: $C_6H_{12}ClO_4P$				
Mol Weight: 214.59				
MP (deg C):			**FP (deg C):**	
BP (deg C):				
BP pressure (mm Hg):				

Property/Value	Units	Temp	Data Type	Reference
WS 1.00E+004	mg/L	25	EXP	GUNTHER,FA ET AL. (1968)
logP 1.09			EST	MEYLAN,WM & HOWARD,PH (1995)
VP 1.22E-002	mm Hg	25	EST	NEELY,WB & BLAU,GE (1985)
DC	pKa			
HL 2.10E-006	atm m3/mol	25	EST	MEYLAN,WM & HOWARD,PH (1991)
OH 5.05E-011	cm3/molc sec	25	EST	MEYLAN,WM & HOWARD,PH (1993)

000311-89-7 — HEPTACOSAFLUOROTRIBUTYLAMINE

Formula: $C_{12}F_{27}N$				
Mol Weight: 671.10				
MP (deg C):			**FP (deg C):**	
BP (deg C): 178				
BP pressure (mm Hg):				

Property/Value	Units	Temp	Data Type	Reference
WS 1.67E-006	mg/L	25	EST	MEYLAN,WM ET AL. (1996)
logP 9.98			EST	MEYLAN,WM & HOWARD,PH (1995)
VP 5.52E-001	mm Hg	25	EXP	RIDDICK,JA ET AL. (1986)
DC	pKa			
HL 5.45E+004	atm m3/mol	25	EST	MEYLAN,WM & HOWARD,PH (1991)
OH 0.00E+000	cm3/molc sec	25	EST	MEYLAN,WM & HOWARD,PH (1993)

000312-35-6 — BENZENAMINE, 4-[(4-FLUOROPHENYL)SULFONYL]-

Formula: $C_{12}H_{10}FNO_2S$				
Mol Weight: 251.28				
MP (deg C):			**FP (deg C):**	
BP (deg C):				
BP pressure (mm Hg):				

Property/Value	Units	Temp	Data Type	Reference
WS 3.26E+002	mg/L	25	EST	MEYLAN,WM ET AL. (1996)
logP 2.17			EXP	HANSCH,C ET AL. (1995)
VP 7.53E-007	mm Hg	25	EST	NEELY,WB & BLAU,GE (1985)
DC	pKa			
HL 1.03E-010	atm m3/mol	25	EST	MEYLAN,WM & HOWARD,PH (1991)
OH 2.35E-011	cm3/molc sec	25	EST	MEYLAN,WM & HOWARD,PH (1993)

000312-73-2 — 2-CF3 BENZIMIDAZOLE

Formula: $C_8H_5F_3N_2$				
Mol Weight: 186.14				
MP (deg C): 208-211			**FP (deg C):**	
BP (deg C):				
BP pressure (mm Hg):				

Property/Value	Units	Temp	Data Type	Reference
WS 2.70E+002	mg/L	25	EST	MEYLAN,WM ET AL. (1996)
logP 2.67			EXP	HANSCH,C & LEO,AJ (1985)
VP 5.26E-005	mm Hg	25	EST	NEELY,WB & BLAU,GE (1985)
DC	pKa			
HL 3.19E-006	atm m3/mol	25	EST	MEYLAN,WM & HOWARD,PH (1991)
OH 7.24E-012	cm3/molc sec	25	EST	MEYLAN,WM & HOWARD,PH (1993)

000312-90-3 — 11-A-HYDROXYPROGESTERONE

Formula: $C_{21}H_{30}O_3$				
Mol Weight: 330.47				
MP (deg C):			**FP (deg C):**	
BP (deg C):				
BP pressure (mm Hg):				

Property/Value	Units	Temp	Data Type	Reference
WS 2.53E+002	mg/L	25	EST	MEYLAN,WM ET AL. (1996)
logP 2.36			EXP	HANSCH,C & LEO,AJ (1985)
VP 5.19E-010	mm Hg	25	EST	NEELY,WB & BLAU,GE (1985)
DC	pKa			
HL 2.37E-012	atm m3/mol	25	EST	MEYLAN,WM & HOWARD,PH (1991)
OH 1.10E-010	cm3/molc sec	25	EST	MEYLAN,WM & HOWARD,PH (1993)

CAS #: 000314-19-2				6,ALPHA,BETA-APORPHINE-10,11-DIOL, HYDROCHLOR*
Formula: $C_{17}H_{18}ClNO_2$				
Mol Weight: 303.79				
MP (deg C):			FP (deg C):	
BP (deg C):				
BP pressure (mm Hg):				

Property/ Value	Units	Temp	Data Type	Reference
WS 2.00E+004	mg/L	25	EXP	SEIDELL,A (1941)
logP 0.24			EST	MEYLAN,WM & HOWARD,PH (1995)
VP 2.36E-015	mm Hg	25	EST	NEELY,WB & BLAU,GE (1985)
DC	pKa			
HL 3.66E-023	atm m3/mol	25	EST	MEYLAN,WM & HOWARD,PH (1991)
OH 9.54E-011	cm3/molc sec	25	EST	MEYLAN,WM & HOWARD,PH (1993)

CAS #: 000314-40-9				BROMACIL
Formula: $C_9H_{13}BrN_2O_2$				
Mol Weight: 261.13				
MP (deg C): 158.3			FP (deg C):	
BP (deg C):				
BP pressure (mm Hg):				

Property/ Value	Units	Temp	Data Type	Reference
WS 8.15E+002	mg/L	25	EXP	MARTIN,H & WORTHING,CR (1977)
logP 2.11			EXP	HANSCH,C & LEO,AJ (1985)
VP 2.50E-007	mm Hg	25	EXP	WORTHING,CR & WALKER,SB (1987)
DC 9.30	pKa	25	EXP	WEBER,JB (1972)
HL 1.05E-010	atm m3/mol	25	EST	VP/WSOL
OH 5.07E-011	cm3/molc sec	25	EST	MEYLAN,WM & HOWARD,PH (1993)

CAS #: 000314-42-1				ISOCIL
Formula: $C_8H_{11}BrN_2O_2$				
Mol Weight: 247.10				
MP (deg C): 158-159			FP (deg C):	
BP (deg C):				
BP pressure (mm Hg):				

Property/ Value	Units	Temp	Data Type	Reference
WS 2.15E+003	mg/L		EXP	SHIU,WY ET AL. (1990)
logP 1.19			EST	MEYLAN,WM & HOWARD,PH (1995)
VP 1.53E-007	mm Hg	25	EST	NEELY,WB & BLAU,GE (1985)
DC	pKa			
HL 1.05E-010	atm m3/mol	25	EST	MEYLAN,WM & HOWARD,PH (1991)
OH 1.64E-011	cm3/molc sec	25	EST	MEYLAN,WM & HOWARD,PH (1993)

CAS #: 000315-18-4				MEXACARBATE
Formula: $C_{12}H_{18}N_2O_2$				
Mol Weight: 222.29				
MP (deg C): 85			FP (deg C):	
BP (deg C):				
BP pressure (mm Hg):				

Property/ Value	Units	Temp	Data Type	Reference
WS 1.00E+002	mg/L	25	EXP	SHIU,WY ET AL. (1990)
logP 2.56			EXP	SANGSTER,J (1994)
VP 1.00E-001	mm Hg	25	EXP	AUGUSTIJN-BECKERS,PWM ET AL. (1994)
DC	pKa			
HL 2.93E-004	atm m3/mol	25	EST	VP/WSOL
OH 3.13E-011	cm3/molc sec	25	EST	MEYLAN,WM & HOWARD,PH (1993)

CAS #: 000315-30-0				ALLOPURINOL
Formula: $C_5H_4N_4O$				
Mol Weight: 136.11				
MP (deg C): >350			FP (deg C):	
BP (deg C):				
BP pressure (mm Hg):				

Property/ Value	Units	Temp	Data Type	Reference
WS 5.69E+002	mg/L	25	EXP	YALKOWSKY,SH & DANNENFELSER,RM (1992)
logP -0.55			EXP	BUNDGAARD,H & FALCH,E (1985)
VP 4.73E-006	mm Hg	25	EST	NEELY,WB & BLAU,GE (1985)
DC	pKa			
HL 2.03E-014	atm m3/mol	25	EST	MEYLAN,WM & HOWARD,PH (1991)
OH 2.00E-010	cm3/molc sec	25	EST	MEYLAN,WM & HOWARD,PH (1993)

CAS #: 000315-37-7				TESTOSTERONE ENANTYHATE
Formula: $C_{26}H_{40}O_3$				
Mol Weight: 400.61				
MP (deg C): 36-37.5			FP (deg C):	
BP (deg C):				
BP pressure (mm Hg):				

Property/ Value	Units	Temp	Data Type	Reference
WS 5.43E-003	mg/L	25	EST	MEYLAN,WM ET AL. (1996)
logP 6.73			EST	MEYLAN,WM & HOWARD,PH (1995)
VP 6.18E-009	mm Hg	25	EST	NEELY,WB & BLAU,GE (1985)
DC	pKa			
HL 9.47E-007	atm m3/mol	25	EST	MEYLAN,WM & HOWARD,PH (1991)
OH 1.08E-010	cm3/molc sec	25	EST	MEYLAN,WM & HOWARD,PH (1993)

CAS #: 000316-42-7				EMETINE DIHYDROCHLORIDE
Formula: $C_{29}H_{42}Cl_2N_2O_4$				
Mol Weight: 553.58				
MP (deg C):			FP (deg C):	
BP (deg C):				
BP pressure (mm Hg):				

Property/ Value	Units	Temp	Data Type	Reference
WS 1.65E+005	mg/L	25	EXP	STEPHEN,H & STEPHEN,T (1963)
logP 5.20			EST	MEYLAN,WM & HOWARD,PH (1995)
VP 2.47E-012	mm Hg	25	EST	NEELY,WB & BLAU,GE (1985)
DC	pKa			
HL 4.99E-016	atm m3/mol	25	EST	MEYLAN,WM & HOWARD,PH (1991)
OH 2.74E-010	cm3/molc sec	25	EST	MEYLAN,WM & HOWARD,PH (1993)

CAS #: 000317-57-7				8-CF3 QUINOLINE
Formula: $C_{10}H_6F_3N$				
Mol Weight: 197.16				
MP (deg C):			FP (deg C):	
BP (deg C):				
BP pressure (mm Hg):				

Property/ Value	Units	Temp	Data Type	Reference
WS 3.32E+002	mg/L	25	EST	MEYLAN,WM ET AL. (1996)
logP 2.50			EXP	HANSCH,C ET AL. (1995)
VP 2.20E-002	mm Hg	25	EST	NEELY,WB & BLAU,GE (1985)
DC	pKa			
HL 5.98E-006	atm m3/mol	25	EST	MEYLAN,WM & HOWARD,PH (1991)
OH 2.89E-012	cm3/molc sec	25	EST	MEYLAN,WM & HOWARD,PH (1993)

CAS #: 000319-78-8				ISOLEUCINE

Formula: $C_6H_{13}NO_2$

Mol Weight: 131.18

MP (deg C): FP (deg C):

BP (deg C):

BP pressure (mm Hg):

Property/Value	Units	Temp	Data Type	Reference
WS 4.12E+004	mg/L	25	EXP	SEIDELL,A (1941)
logP -1.59			EST	MEYLAN,WM & HOWARD,PH (1995)
VP 1.34E-008	mm Hg	25	EST	NEELY,WB & BLAU,GE (1985)
DC 2.37	pKa	0	EXP	KORTUM,G ET AL (1961)
HL 3.49E-009	atm m3/mol	25	EST	MEYLAN,WM & HOWARD,PH (1991)
OH 4.26E-011	cm3/molc sec	25	EST	MEYLAN,WM & HOWARD,PH (1993)

CAS #: 000319-81-3				TETRACHLOROCYCLOHEXENE

Formula: $C_6H_6Cl_4$

Mol Weight: 219.93

MP (deg C): FP (deg C):

BP (deg C):

BP pressure (mm Hg):

Property/Value	Units	Temp	Data Type	Reference
WS 4.30E+001	mg/L	25	EST	MEYLAN,WM ET AL. (1996)
logP 3.08			EXP	HANSCH,C & LEO,AJ (1985)
VP 3.13E-002	mm Hg	25	EST	NEELY,WB & BLAU,GE (1985)
DC	pKa			
HL 1.81E-003	atm m3/mol	25	EST	MEYLAN,WM & HOWARD,PH (1991)
OH 4.36E-011	cm3/molc sec	25	EST	MEYLAN,WM & HOWARD,PH (1993)

CAS #: 000319-84-6				ALPHA-HEXACHLOROCYCLOHEXANE

Formula: $C_6H_6Cl_6$

Mol Weight: 290.83

MP (deg C): 159-160 FP (deg C):

BP (deg C): 288

BP pressure (mm Hg):

Property/Value	Units	Temp	Data Type	Reference
WS 2.00E+000	mg/L	25	EXP	WEIL,L ET AL. (1974)
logP 3.80			EXP	HANSCH,C & LEO,AJ (1985)
VP 4.50E-005	mm Hg	25	EXT	SCHWABE,K & LEGLER,C (1960)
DC	pKa			
HL 1.06E-005	atm m3/mol	25	EST	VP/WSOL
OH 1.36E-012	cm3/molc sec	25	EST	MEYLAN,WM & HOWARD,PH (1993)

CAS #: 000319-85-7				BETA-HEXACHLOROCYCLOHEXANE

Formula: $C_6H_6Cl_6$

Mol Weight: 290.83

MP (deg C): 314-315 FP (deg C):

BP (deg C): 60

BP pressure (mm Hg): 5.80E-001

Property/Value	Units	Temp	Data Type	Reference
WS 7.30E+000	mg/L	25	EXP	SHIU,WY ET AL. (1990)
logP 3.78			EXP	HANSCH,C & LEO,AJ (1985)
VP 4.66E-007	mm Hg	25	EXT	SCHWABE,K & LEGLER,C (1960)
DC	pKa			
HL 7.43E-007	atm m3/mol	25	EST	VP/WSOL
OH 1.36E-012	cm3/molc sec	25	EST	MEYLAN,WM & HOWARD,PH (1993)

CAS #: 000319-86-8				DELTA-HEXACHLOROCYCLOHEXANE

Formula: $C_6H_6Cl_6$

Mol Weight: 290.83

MP (deg C): 141.5-142 FP (deg C):

BP (deg C): 60

BP pressure (mm Hg): 3.40E-001

Property/Value	Units	Temp	Data Type	Reference
WS 3.14E+001	mg/L	25	EXP	SHIU,WY ET AL. (1990)
logP 4.14			EXP	HANSCH,C & LEO,AJ (1985)
VP 3.52E-005	mm Hg	25	EXP	SCHWABE,K & LEGLER,C (1960)
DC	pKa			
HL 4.29E-007	atm m3/mol	25	EST	VP/WSOL
OH 1.36E-012	cm3/molc sec	25	EST	MEYLAN,WM & HOWARD,PH (1993)

CAS #: 000319-94-8				PENTACHLOROCYCLOHEXENE

Formula: $C_6H_5Cl_5$

Mol Weight: 254.37

MP (deg C): FP (deg C):

BP (deg C):

BP pressure (mm Hg):

Property/Value	Units	Temp	Data Type	Reference
WS 1.15E+001	mg/L	25	EST	MEYLAN,WM ET AL. (1996)
logP 3.95			EXP	HANSCH,C & LEO,AJ (1985)
VP 5.99E-003	mm Hg	25	EST	NEELY,WB & BLAU,GE (1985)
DC	pKa			
HL 1.30E-003	atm m3/mol	25	EST	MEYLAN,WM & HOWARD,PH (1991)
OH 1.46E-011	cm3/molc sec	25	EST	MEYLAN,WM & HOWARD,PH (1993)

CAS #: 000320-60-5				2,4-DICHLORO(TRIFLUOROMETHYL)BENZENE

Formula: $C_7H_3Cl_2F_3$

Mol Weight: 215.00

MP (deg C): FP (deg C):

BP (deg C): 117-118

BP pressure (mm Hg):

Property/Value	Units	Temp	Data Type	Reference
WS 8.67E+000	mg/L	25	EST	MEYLAN,WM ET AL. (1996)
logP 4.24			EST	MEYLAN,WM & HOWARD,PH (1995)
VP 1.11E+000	mm Hg	25	EXP	YAWS,CL (1994A)
DC	pKa			
HL 2.57E-002	atm m3/mol	25	EST	MEYLAN,WM & HOWARD,PH (1991)
OH 1.94E-013	cm3/molc sec	25	EST	MEYLAN,WM & HOWARD,PH (1993)

CAS #: 000321-14-2				5-CHLOROSALICYLIC ACID

Formula: $C_7H_5ClO_3$

Mol Weight: 172.57

MP (deg C): 171-172 FP (deg C):

BP (deg C):

BP pressure (mm Hg):

Property/Value	Units	Temp	Data Type	Reference
WS 5.22E+002	mg/L	25	EST	MEYLAN,WM ET AL. (1996)
logP 3.09			EXP	HANSCH,C & LEO,AJ (1985)
VP 2.58E-005	mm Hg	25	EST	NEELY,WB & BLAU,GE (1985)
DC 2.65	pKa	25	EXP	SERJEANT,EP & DEMPSEY,B (1979)
HL 1.05E-008	atm m3/mol	25	EST	MEYLAN,WM & HOWARD,PH (1991)
OH 4.26E-012	cm3/molc sec	25	EST	MEYLAN,WM & HOWARD,PH (1993)

CAS #: 000321-28-8				2-FLUOROANISOLE

Formula: C_7H_7FO
Mol Weight: 126.13
MP (deg C): -39
FP (deg C):
BP (deg C): 154.5
BP pressure (mm Hg):

Property/ Value	Units	Temp	Data Type	Reference
WS 1.42E+003	mg/L	25	EST	MEYLAN,WM ET AL. (1996)
logP 2.14			EXP	NAKAGAWA,Y ET AL. (1992)
VP 5.28E+000	mm Hg	25	EST	NEELY,WB & BLAU,GE (1985)
DC	pKa			
HL 3.72E-004	atm m3/mol	25	EST	MEYLAN,WM & HOWARD,PH (1991)
OH 8.09E-012	cm3/molc sec	25	EST	MEYLAN,WM & HOWARD,PH (1993)

CAS #: 000321-38-0				1-FLUORONAPHTHALENE

Formula: $C_{10}H_7F$
Mol Weight: 146.17
MP (deg C): -9
FP (deg C):
BP (deg C): 215
BP pressure (mm Hg):

Property/ Value	Units	Temp	Data Type	Reference
WS 5.15E+001	mg/L	25	EXP	WASIK,SP ET AL. (1981)
logP 3.37			EST	MEYLAN,WM & HOWARD,PH (1995)
VP 8.96E-002	mm Hg	25	EST	NEELY,WB & BLAU,GE (1985)
DC	pKa			
HL 6.14E-004	atm m3/mol	25	EST	MEYLAN,WM & HOWARD,PH (1991)
OH 2.71E-011	cm3/molc sec	25	EST	MEYLAN,WM & HOWARD,PH (1993)

CAS #: 000321-64-2				TACRINE

Formula: $C_{13}H_{14}N_2$
Mol Weight: 198.27
MP (deg C): 183-184
FP (deg C):
BP (deg C):
BP pressure (mm Hg):

Property/ Value	Units	Temp	Data Type	Reference
WS 2.17E+002	mg/L	25	EST	MEYLAN,WM ET AL. (1996)
logP 2.71			EXP	HANSCH,C ET AL. (1995)
VP 7.50E-006	mm Hg	25	EST	NEELY,WB & BLAU,GE (1985)
DC 9.95	pKa	20	EXP	PERRIN,DD (1972)
HL 2.30E-010	atm m3/mol	25	EST	MEYLAN,WM & HOWARD,PH (1991)
OH 2.05E-010	cm3/molc sec	25	EST	MEYLAN,WM & HOWARD,PH (1993)

CAS #: 000322-46-3				PYRIDO(2,3)PYRAZINE

Formula: $C_7H_5N_3$
Mol Weight: 131.14
MP (deg C):
FP (deg C):
BP (deg C):
BP pressure (mm Hg):

Property/ Value	Units	Temp	Data Type	Reference
WS 9.85E+004	mg/L	25	EST	MEYLAN,WM ET AL. (1996)
logP -0.04			EXP	HANSCH,C & LEO,AJ (1985)
VP 7.21E-003	mm Hg	25	EST	NEELY,WB & BLAU,GE (1985)
DC 1.20	pKa	20	EXP	PERRIN,DD (1965)
HL 3.73E-010	atm m3/mol	25	EST	MEYLAN,WM & HOWARD,PH (1991)
OH 1.00E-012	cm3/molc sec	25	EST	MEYLAN,WM & HOWARD,PH (1993)

CAS #: 000322-97-4				4-HYDROXY-7-TRIFLUOROMEQUINOLINE

Formula: $C_{10}H_6F_3NO$
Mol Weight: 213.16
MP (deg C): 266-269
FP (deg C):
BP (deg C):
BP pressure (mm Hg):

Property/ Value	Units	Temp	Data Type	Reference
WS 6.64E+002	mg/L	25	EST	MEYLAN,WM ET AL. (1996)
logP 2.05			EXP	HANSCH,C & LEO,AJ (1985)
VP 2.07E-004	mm Hg	25	EST	NEELY,WB & BLAU,GE (1985)
DC	pKa			
HL 6.23E-010	atm m3/mol	25	EST	MEYLAN,WM & HOWARD,PH (1991)
OH 8.75E-012	cm3/molc sec	25	EST	MEYLAN,WM & HOWARD,PH (1993)

CAS #: 000324-74-3				4-FLUOROBIPHENYL

Formula: $C_{12}H_9F$
Mol Weight: 172.20
MP (deg C): 74.2
FP (deg C):
BP (deg C): 253
BP pressure (mm Hg):

Property/ Value	Units	Temp	Data Type	Reference
WS 2.51E+001	mg/L	25	EST	MEYLAN,WM ET AL. (1996)
logP 3.96			EST	MEYLAN,WM & HOWARD,PH (1995)
VP 9.01E-003	mm Hg	25	EST	NEELY,WB & BLAU,GE (1985)
DC	pKa			
HL 4.83E-004	atm m3/mol	25	EST	MEYLAN,WM & HOWARD,PH (1991)
OH 4.24E-012	cm3/molc sec	25	EST	MEYLAN,WM & HOWARD,PH (1993)

CAS #: 000325-14-4				7-CF3 QUINOLINE

Formula: $C_{10}H_6F_3N$
Mol Weight: 197.16
MP (deg C):
FP (deg C):
BP (deg C):
BP pressure (mm Hg):

Property/ Value	Units	Temp	Data Type	Reference
WS 1.19E+002	mg/L	25	EST	MEYLAN,WM ET AL. (1996)
logP 3.02			EXP	HANSCH,C & LEO,AJ (1985)
VP 2.20E-002	mm Hg	25	EST	NEELY,WB & BLAU,GE (1985)
DC	pKa			
HL 5.98E-006	atm m3/mol	25	EST	MEYLAN,WM & HOWARD,PH (1991)
OH 2.89E-012	cm3/molc sec	25	EST	MEYLAN,WM & HOWARD,PH (1993)

CAS #: 000326-91-0				THENOYLTRIFLUOROMETHYLACETONE

Formula: $C_8H_5F_3O_2S$
Mol Weight: 222.19
MP (deg C): 40-44
FP (deg C):
BP (deg C): 96-98
BP pressure (mm Hg): 8.00E+000

Property/ Value	Units	Temp	Data Type	Reference
WS 1.90E+003	mg/L	25	EST	MEYLAN,WM ET AL. (1996)
logP 1.46			EXP	HANSCH,C & LEO,AJ (1985)
VP 6.90E-003	mm Hg	25	EST	NEELY,WB & BLAU,GE (1985)
DC	pKa			
HL 2.82E-008	atm m3/mol	25	EST	MEYLAN,WM & HOWARD,PH (1991)
OH 8.76E-012	cm3/molc sec	25	EST	MEYLAN,WM & HOWARD,PH (1993)

CAS #: 000327-19-5 — 2-CF3-5-NO2 BENZIMADAZOLE

Formula: $C_8H_4F_3N_3O_2$

Mol Weight: 231.14

MP (deg C): FP (deg C):

BP (deg C):

BP pressure (mm Hg):

Property/Value	Units	Temp	Data Type	Reference
WS 6.29E+001	mg/L	25	EST	MEYLAN,WM ET AL. (1996)
logP 2.68			EXP	HANSCH,C & LEO,AJ (1985)
VP 6.75E-007	mm Hg	25	EST	NEELY,WB & BLAU,GE (1985)
DC	pKa			
HL 1.26E-008	atm m3/mol	25	EST	MEYLAN,WM & HOWARD,PH (1991)
OH 6.56E-013	cm3/molc sec	25	EST	MEYLAN,WM & HOWARD,PH (1993)

CAS #: 000327-54-8 — 1,2,4,5-TETRAFLUOROBENZENE

Formula: $C_6H_2F_4$

Mol Weight: 150.08

MP (deg C): FP (deg C):

BP (deg C): 132

BP pressure (mm Hg): 1.02E+002

Property/Value	Units	Temp	Data Type	Reference
WS 6.32E+002	mg/L	25	EXP	YALKOWSKY,SH & DANNENFELSER,RM (1992)
logP 2.79			EST	MEYLAN,WM & HOWARD,PH (1995)
VP 5.67E+001	mm Hg	25	EXP	BOUBLIK,T ET AL. (1984)
DC	pKa			
HL 1.77E-002	atm m3/mol	25	EST	VP/WSOL
OH 3.49E-013	cm3/molc sec	25	EST	MEYLAN,WM & HOWARD,PH (1993)

CAS #: 000327-57-1 — A-AMINOCAPROIC ACID

Formula: $C_6H_{13}NO_2$

Mol Weight: 131.18

MP (deg C): 298.5 FP (deg C):

BP (deg C):

BP pressure (mm Hg):

Property/Value	Units	Temp	Data Type	Reference
WS 1.64E+004	mg/L	25	EXP	YALKOWSKY,SH & DANNENFELSER,RM (1992)
logP -1.53			EXP	HANSCH,C & LEO,AJ (1985)
VP 6.35E-009	mm Hg	25	EST	NEELY,WB & BLAU,GE (1985)
DC 2.36	pKa	13	EXP	KORTUM,G ET AL (1961)
HL 3.49E-009	atm m3/mol	25	EST	MEYLAN,WM & HOWARD,PH (1991)
OH 4.23E-011	cm3/molc sec	25	EST	MEYLAN,WM & HOWARD,PH (1993)

CAS #: 000327-98-0 — TRICHLORONATE

Formula: $C_{10}H_{12}Cl_3O_2PS$

Mol Weight: 333.60

MP (deg C): FP (deg C):

BP (deg C): 108

BP pressure (mm Hg): 1.00E-002

Property/Value	Units	Temp	Data Type	Reference
WS 5.90E-001	mg/L	20	EXP	BOWMAN,BT & SANS,WW (1983)
logP 5.23			EXP	HANSCH,C ET AL. (1995)
VP 1.50E-005	mm Hg	25	EXP	AUGUSTIJN-BECKERS,PWM ET AL. (1994)
DC	pKa			
HL 1.12E-005	atm m3/mol	25	EST	VP/WSOL
OH 7.36E-011	cm3/molc sec	25	EST	MEYLAN,WM & HOWARD,PH (1993)

CAS #: 000328-39-2 — LEUCINE

Formula: $C_6H_{13}NO_2$

Mol Weight: 131.18

MP (deg C): 293 FP (deg C):

BP (deg C):

BP pressure (mm Hg):

Property/Value	Units	Temp	Data Type	Reference
WS 1.18E+004	mg/L	25	EXP	YALKOWSKY,SH & DANNENFELSER,RM (1992)
logP -1.59			EST	MEYLAN,WM & HOWARD,PH (1995)
VP 1.34E-008	mm Hg	25	EST	NEELY,WB & BLAU,GE (1985)
DC 2.35	pKa	13	EXP	KORTUM,G ET AL (1961)
HL 3.49E-009	atm m3/mol	25	EST	MEYLAN,WM & HOWARD,PH (1991)
OH 4.23E-011	cm3/molc sec	25	EST	MEYLAN,WM & HOWARD,PH (1993)

CAS #: 000328-57-4 — DIFLUOROMETHYLPHENYLSILANE

Formula: $C_7H_8F_2Si$

Mol Weight: 158.22

MP (deg C): FP (deg C):

BP (deg C):

BP pressure (mm Hg):

Property/Value	Units	Temp	Data Type	Reference
WS 1.26E+002	mg/L	25	EST	MEYLAN,WM ET AL. (1996)
logP 3.21			EST	MEYLAN,WM & HOWARD,PH (1995)
VP 7.60E+000	mm Hg	25	EXP	SOKOLOV,VB ET AL. (1975)
DC	pKa			
HL	atm m3/mol			
OH 2.10E-012	cm3/molc sec	25	EST	MEYLAN,WM & HOWARD,PH (1993)

CAS #: 000328-74-5 — 3,5-BIS(TRIFLUOROME)ANILINE

Formula: $C_8H_5F_6N$

Mol Weight: 229.13

MP (deg C): FP (deg C):

BP (deg C): 85

BP pressure (mm Hg): 1.50E+001

Property/Value	Units	Temp	Data Type	Reference
WS 2.75E+001	mg/L	25	EST	MEYLAN,WM ET AL. (1996)
logP 3.57			EXP	HANSCH,C & LEO,AJ (1985)
VP 4.00E-001	mm Hg	25	EST	NEELY,WB & BLAU,GE (1985)
DC	pKa			
HL 1.44E-004	atm m3/mol	25	EST	MEYLAN,WM & HOWARD,PH (1991)
OH 2.47E-012	cm3/molc sec	25	EST	MEYLAN,WM & HOWARD,PH (1993)

CAS #: 000328-84-7 — 3,4-DICHLOROBENZOTRIFLUORIDE

Formula: $C_7H_3Cl_2F_3$

Mol Weight: 215.00

MP (deg C): -12.5 FP (deg C):

BP (deg C): 173-174

BP pressure (mm Hg):

Property/Value	Units	Temp	Data Type	Reference
WS 8.67E+000	mg/L	25	EST	MEYLAN,WM ET AL. (1996)
logP 4.24			EST	MEYLAN,WM & HOWARD,PH (1995)
VP 2.36E+000	mm Hg	25	EXP	OHE,S (1976)
DC	pKa			
HL 2.57E-002	atm m3/mol	25	EST	MEYLAN,WM & HOWARD,PH (1991)
OH 8.05E-014	cm3/molc sec	25	EST	MEYLAN,WM & HOWARD,PH (1993)

308

CAS #: 000329-20-4 — P-ACETAMIDO-BENZENESO2-FLUORIDE

Formula: $C_8H_8FNO_3S$

Mol Weight: 217.22

MP (deg C): FP (deg C):

BP (deg C):

BP pressure (mm Hg):

Property/Value	Units	Temp	Data Type	Reference
WS 4.99E+002	mg/L	25	EST	MEYLAN,WM ET AL. (1996)
logP 2.17			EXP	HANSCH,C & LEO,AJ (1985)
VP 7.50E-007	mm Hg		EST	NEELY,WB & BLAU,GE (1985)
DC	pKa			
HL 1.78E-011	atm m3/mol	25	EST	MEYLAN,WM & HOWARD,PH (1991)
OH 2.76E-012	cm3/molc sec	25	EST	MEYLAN,WM & HOWARD,PH (1993)

CAS #: 000329-63-5 — EPINEPHRINE SALT (HCL)

Formula: $C_9H_{14}ClNO_3$

Mol Weight: 219.67

MP (deg C): FP (deg C):

BP (deg C):

BP pressure (mm Hg):

Property/Value	Units	Temp	Data Type	Reference
WS 1.00E+006	mg/L	25	EST	MEYLAN,WM ET AL. (1996)
logP -3.58			EST	MEYLAN,WM & HOWARD,PH (1995)
VP 1.48E-012	mm Hg		EST	NEELY,WB & BLAU,GE (1985)
DC	pKa			
HL 8.55E-025	atm m3/mol	25	EST	MEYLAN,WM & HOWARD,PH (1991)
OH 7.52E-011	cm3/molc sec	25	EST	MEYLAN,WM & HOWARD,PH (1993)

CAS #: 000329-71-5 — 2,5-DINITROPHENOL

Formula: $C_6H_4N_2O_5$

Mol Weight: 184.11

MP (deg C): 108 FP (deg C):

BP (deg C):

BP pressure (mm Hg):

Property/Value	Units	Temp	Data Type	Reference
WS 1.68E+003	mg/L	25	EST	MEYLAN,WM ET AL. (1996)
logP 1.75			EXP	HANSCH,C & LEO,AJ (1985)
VP 1.05E-003	mm Hg	20	EXP	SCHWARZENBACH,RP ET AL. (1988)
DC 5.21	pKa	25	EXP	KORTUM,G ET AL (1961)
HL 2.76E-008	atm m3/mol	25	EST	MEYLAN,WM & HOWARD,PH (1991)
OH 5.04E-013	cm3/molc sec	25	EST	MEYLAN,WM & HOWARD,PH (1993)

CAS #: 000330-39-2 — 1,1-DIMETHYL-3-M-FLUOROPHENYLUREA

Formula: $C_9H_{11}FN_2O$

Mol Weight: 182.20

MP (deg C): FP (deg C):

BP (deg C):

BP pressure (mm Hg):

Property/Value	Units	Temp	Data Type	Reference
WS 3.63E+003	mg/L	25	EST	MEYLAN,WM ET AL. (1996)
logP 1.37			EXP	HANSCH,C & LEO,AJ (1985)
VP 2.67E-004	mm Hg	25	EST	NEELY,WB & BLAU,GE (1985)
DC	pKa			
HL 1.13E-009	atm m3/mol	25	EST	MEYLAN,WM & HOWARD,PH (1991)
OH 5.55E-011	cm3/molc sec	25	EST	MEYLAN,WM & HOWARD,PH (1993)

CAS #: 000330-54-1 — DIURON

Formula: $C_9H_{10}Cl_2N_2O$

Mol Weight: 233.10

MP (deg C): 158-159 FP (deg C):

BP (deg C):

BP pressure (mm Hg):

Property/Value	Units	Temp	Data Type	Reference
WS 4.20E+001	mg/L	25	EXP	BAILEY,GW ET AL. (1968)
logP 2.68			EXP	HANSCH,C ET AL. (1995)
VP 2.70E-006	mm Hg	30	EXP	JURY,WA (1983)
DC	pKa			
HL 2.70E-006	atm m3/mol	25	EST	VP/WSOL
OH 3.41E-011	cm3/molc sec	25	EST	MEYLAN,WM & HOWARD,PH (1993)

CAS #: 000330-55-2 — LINURON

Formula: $C_9H_{10}Cl_2N_2O_2$

Mol Weight: 249.10

MP (deg C): 93-94 FP (deg C):

BP (deg C):

BP pressure (mm Hg):

Property/Value	Units	Temp	Data Type	Reference
WS 7.50E+001	mg/L	25	EXP	BAILEY,GW & WHITE,JL (1965)
logP 3.20			EXP	HANSCH,C & LEO,AJ (1985)
VP 1.50E-005	mm Hg	24	EXP	WORTHING,CR & WALKER,SB (1983)
DC	pKa			
HL 6.60E-008	atm m3/mol	24	EST	VP/WSOL
OH 3.35E-011	cm3/molc sec	25	EST	MEYLAN,WM & HOWARD,PH (1993)

CAS #: 000330-64-3 — 3,5-DI(IPR)-N-ME-PHENYLCARBAMATE

Formula: $C_{14}H_{21}NO_2$

Mol Weight: 235.33

MP (deg C): FP (deg C):

BP (deg C):

BP pressure (mm Hg):

Property/Value	Units	Temp	Data Type	Reference
WS 1.65E+001	mg/L	25	EST	MEYLAN,WM ET AL. (1996)
logP 3.79			EXP	HANSCH,C & LEO,AJ (1985)
VP 3.55E-004	mm Hg	25	EST	NEELY,WB & BLAU,GE (1985)
DC	pKa			
HL 1.22E-007	atm m3/mol	25	EST	MEYLAN,WM & HOWARD,PH (1991)
OH 2.19E-011	cm3/molc sec	25	EST	MEYLAN,WM & HOWARD,PH (1993)

CAS #: 000331-25-9 — M-FLUOROPHENYLACETIC ACID

Formula: $C_8H_7FO_2$

Mol Weight: 154.14

MP (deg C): 42-44 FP (deg C):

BP (deg C):

BP pressure (mm Hg):

Property/Value	Units	Temp	Data Type	Reference
WS 7.04E+003	mg/L	25	EST	MEYLAN,WM ET AL. (1996)
logP 1.65			EXP	HANSCH,C & LEO,AJ (1985)
VP 6.22E-003	mm Hg	25	EST	NEELY,WB & BLAU,GE (1985)
DC 4.13	pKa	25	EXP	SERJEANT,EP & DEMPSEY,B (1979)
HL 5.16E-008	atm m3/mol	25	EST	MEYLAN,WM & HOWARD,PH (1991)
OH 5.27E-012	cm3/molc sec	25	EST	MEYLAN,WM & HOWARD,PH (1993)

CAS #:	000331-39-5		2-PROPENOIC ACID, 3-(3,4-DIHYDROXYPHENYL)-, (E)-

Formula: $C_9H_8O_4$

Mol Weight: 180.16

MP (deg C): 225 dec
FP (deg C):

BP (deg C):

BP pressure (mm Hg):

Property/Value	Units	Temp	Data Type	Reference
WS 5.41E+004	mg/L	25	EST	MEYLAN,WM ET AL. (1996)
logP 1.15			EXP	SANGSTER,J (1993)
VP 2.52E-007	mm Hg	25	EST	NEELY,WB & BLAU,GE (1985)
DC 4.62	pKa	25	EXP	SERJEANT,EP & DEMPSEY,B (1979)
HL 1.40E-016	atm m3/mol	25	EST	MEYLAN,WM & HOWARD,PH (1991)
OH 4.21E-011	cm3/molc sec	25	EST	MEYLAN,WM & HOWARD,PH (1993)

CAS #:	000332-33-2		1,1-DIMETHYL-3-P-FLUOROPHENYLUREA

Formula: $C_9H_{11}FN_2O$

Mol Weight: 182.20

MP (deg C):
FP (deg C):

BP (deg C):

BP pressure (mm Hg):

Property/Value	Units	Temp	Data Type	Reference
WS 5.82E+003	mg/L	25	EST	MEYLAN,WM ET AL. (1996)
logP 1.13			EXP	HANSCH,C & LEO,AJ (1985)
VP 2.67E-004	mm Hg	25	EST	NEELY,WB & BLAU,GE (1985)
DC	pKa			
HL 1.13E-009	atm m3/mol	25	EST	MEYLAN,WM & HOWARD,PH (1991)
OH 1.65E-011	cm3/molc sec	25	EST	MEYLAN,WM & HOWARD,PH (1993)

CAS #:	000333-41-5		DIAZINON

Formula: $C_{12}H_{21}N_2O_3PS$

Mol Weight: 304.35

MP (deg C):
FP (deg C):

BP (deg C): 83-84

BP pressure (mm Hg): 2.00E-003

Property/Value	Units	Temp	Data Type	Reference
WS 4.00E+001	mg/L	25	EXP	SHAROM,MS ET AL. (1980A)
logP 3.81			EXP	HANSCH,C & LEO,AJ (1985)
VP 1.12E-002	mm Hg	25	EXP	SUNTIO,LR (1988)
DC	pKa			
HL 1.13E-007	atm m3/mol	23	EXP	FENDINGER,NJ & GLOTFELTY,DE (1988)
OH 9.48E-011	cm3/molc sec	25	EST	MEYLAN,WM & HOWARD,PH (1993)

CAS #:	000334-48-5		DECANOIC ACID

Formula: $C_{10}H_{20}O_2$

Mol Weight: 172.27

MP (deg C): 31.4
FP (deg C):

BP (deg C): 270

BP pressure (mm Hg):

Property/Value	Units	Temp	Data Type	Reference
WS 6.18E+001	mg/L	25	EXP	YALKOWSKY,SH & DANNENFELSER,RM (1992)
logP 4.09			EXP	HANSCH,C & LEO,AJ (1985)
VP 3.00E-005	mm Hg	25	EXT	YAWS,CL (1994B)
DC	pKa			
HL 5.28E-006	atm m3/mol	25	EST	MEYLAN,WM & HOWARD,PH (1991)
OH 1.12E-011	cm3/molc sec	25	EST	MEYLAN,WM & HOWARD,PH (1993)

CAS #:	000334-56-5		1-FLUORODECANE

Formula: $C_{10}H_{21}F$

Mol Weight: 160.28

MP (deg C): -35
FP (deg C):

BP (deg C): 186.2

BP pressure (mm Hg):

Property/Value	Units	Temp	Data Type	Reference
WS 2.51E+000	mg/L	25	EST	MEYLAN,WM ET AL. (1996)
logP 5.19			EST	MEYLAN,WM & HOWARD,PH (1995)
VP 6.26E-001	mm Hg	25	EXT	LI,JCM & ROSSINI,FD (1953)
DC	pKa			
HL 2.11E-001	atm m3/mol	25	EST	MEYLAN,WM & HOWARD,PH (1991)
OH 1.06E-011	cm3/molc sec	25	EST	MEYLAN,WM & HOWARD,PH (1993)

CAS #:	000334-68-9		1-FLUORODODECANE

Formula: $C_{12}H_{25}F$

Mol Weight: 188.33

MP (deg C):
FP (deg C):

BP (deg C):

BP pressure (mm Hg):

Property/Value	Units	Temp	Data Type	Reference
WS 2.67E-001	mg/L	25	EST	MEYLAN,WM ET AL. (1996)
logP 6.18			EST	MEYLAN,WM & HOWARD,PH (1995)
VP 2.08E-001	mm Hg	25	EXT	LI,JCM & ROSSINI,FD (1953)
DC	pKa			
HL 3.71E-001	atm m3/mol	25	EST	MEYLAN,WM & HOWARD,PH (1991)
OH 1.34E-011	cm3/molc sec	25	EST	MEYLAN,WM & HOWARD,PH (1993)

CAS #:	000334-88-3		DIAZOMETHANE

Formula: CH_2N_2

Mol Weight: 42.04

MP (deg C): -145
FP (deg C):

BP (deg C): -23

BP pressure (mm Hg):

Property/Value	Units	Temp	Data Type	Reference
WS 2.55E+003	mg/L	25	EST	MEYLAN,WM ET AL. (1996)
logP 2.00			EST	MEYLAN,WM & HOWARD,PH (1995)
VP 7.60E+002	mm Hg	-23	EXP	MERCK INDEX (1983)
DC	pKa			
HL 3.37E-008	atm m3/mol	25	EST	MEYLAN,WM & HOWARD,PH (1991)
OH 9.34E-013	cm3/molc sec	25	EST	MEYLAN,WM & HOWARD,PH (1993)

CAS #:	000335-02-4		TRIFLUORONITROMETHANE

Formula: CF_3NO_2

Mol Weight: 115.01

MP (deg C):
FP (deg C):

BP (deg C):

BP pressure (mm Hg):

Property/Value	Units	Temp	Data Type	Reference
WS 2.02E+004	mg/L	25	EST	MEYLAN,WM ET AL. (1996)
logP 0.38			EST	MEYLAN,WM & HOWARD,PH (1995)
VP 7.05E+003	mm Hg	25	EXP	FAZEKAS,GB & TAKACS,GA (1983)
DC	pKa			
HL 3.32E-004	atm m3/mol	25	EST	MEYLAN,WM & HOWARD,PH (1991)
OH 1.30E-013	cm3/molc sec	25	EST	MEYLAN,WM & HOWARD,PH (1993)

CAS #: 000335-57-9				PERFLUOROHEPTANE

Formula: C_7F_{16}

Mol Weight: 388.05

MP (deg C): -78

FP (deg C):

BP (deg C): 82.5

BP pressure (mm Hg):

Property/Value	Units	Temp	Data Type	Reference
WS 7.12E-004	mg/L	25	EST	MEYLAN,WM ET AL. (1996)
logP 6.99			EST	MEYLAN,WM & HOWARD,PH (1995)
VP 7.64E+001	mm Hg	25	EXP	DYKYJ,J & REPAS,M (1973)
DC	pKa			
HL 9.67E+004	atm m3/mol	25	EST	MEYLAN,WM & HOWARD,PH (1991)
OH 0.00E+000	cm3/molc sec	25	EST	MEYLAN,WM & HOWARD,PH (1993)

CAS #: 000338-95-4				PREDNISOLONE-9A-FLUORO

Formula: $C_{21}H_{27}FO_5$

Mol Weight: 378.44

MP (deg C): 263-266

FP (deg C):

BP (deg C):

BP pressure (mm Hg):

Property/Value	Units	Temp	Data Type	Reference
WS 2.14E+002	mg/L	25	EST	MEYLAN,WM ET AL. (1996)
logP 1.51			EXP	HANSCH,C & LEO,AJ (1985)
VP 1.44E-013	mm Hg	25	EST	NEELY,WB & BLAU,GE (1985)
DC	pKa			
HL 5.39E-008	atm m3/mol	25	EST	MEYLAN,WM & HOWARD,PH (1991)
OH 6.07E-011	cm3/molc sec	25	EST	MEYLAN,WM & HOWARD,PH (1993)

CAS #: 000339-43-5				BENZENESULFONAMIDE, 4-AMINO-N- (BUTYLAMINO)CARBO

Formula: $C_{11}H_{17}N_3O_3S$

Mol Weight: 271.34

MP (deg C): 144-145

FP (deg C):

BP (deg C):

BP pressure (mm Hg):

Property/Value	Units	Temp	Data Type	Reference
WS 5.29E+002	mg/L	37	EXP	YALKOWSKY,SH & DANNENFELSER,RM (1992)
logP 1.01			EXP	SANGSTER,J (1993)
VP 3.16E-009	mm Hg	25	EST	NEELY,WB & BLAU,GE (1985)
DC	pKa			
HL 8.11E-014	atm m3/mol	25	EST	MEYLAN,WM & HOWARD,PH (1991)
OH 3.25E-011	cm3/molc sec	25	EST	MEYLAN,WM & HOWARD,PH (1993)

CAS #: 000342-69-8				9H-PURINE, 6-(METHYLTHIO)-9-(beta-D-RIBOFURANOSY

Formula: $C_{11}H_{14}N_4O_4S$

Mol Weight: 298.32

MP (deg C): 163-165

FP (deg C):

BP (deg C):

BP pressure (mm Hg):

Property/Value	Units	Temp	Data Type	Reference
WS 5.15E+002	mg/L	25	EST	MEYLAN,WM ET AL. (1996)
logP 0.09			EXP	SANGSTER,J (1994)
VP 5.70E-015	mm Hg	25	EST	NEELY,WB & BLAU,GE (1985)
DC	pKa			
HL 3.57E-019	atm m3/mol	25	EST	MEYLAN,WM & HOWARD,PH (1991)
OH 2.48E-010	cm3/molc sec	25	EST	MEYLAN,WM & HOWARD,PH (1993)

CAS #: 000344-04-7				BROMOPENTAFLUOROBENZENE

Formula: C_6BrF_5

Mol Weight: 246.97

MP (deg C): -31

FP (deg C):

BP (deg C): 137

BP pressure (mm Hg):

Property/Value	Units	Temp	Data Type	Reference
WS 1.18E+001	mg/L	25	EST	MEYLAN,WM ET AL. (1996)
logP 3.88			EST	MEYLAN,WM & HOWARD,PH (1995)
VP 8.00E+000	mm Hg	25	EXT	OHE,S (1976)
DC	pKa			
HL 4.65E-003	atm m3/mol	25	EST	MEYLAN,WM & HOWARD,PH (1991)
OH 1.25E-013	cm3/molc sec	25	EST	MEYLAN,WM & HOWARD,PH (1993)

CAS #: 000344-07-0				CHLOROPENTAFLUOROBENZENE

Formula: C_6ClF_5

Mol Weight: 202.51

MP (deg C):

FP (deg C):

BP (deg C):

BP pressure (mm Hg):

Property/Value	Units	Temp	Data Type	Reference
WS 3.31E+001	mg/L	25	EST	MEYLAN,WM ET AL. (1996)
logP 3.64			EST	MEYLAN,WM & HOWARD,PH (1995)
VP 1.78E+001	mm Hg	25	EXT	BOUBLIK,T ET AL. (1984)
DC	pKa			
HL 8.64E-003	atm m3/mol	25	EST	MEYLAN,WM & HOWARD,PH (1991)
OH 1.27E-013	cm3/molc sec	25	EST	MEYLAN,WM & HOWARD,PH (1993)

CAS #: 000344-53-6				2-CF3-4-CHLOROACETANILIDE

Formula: $C_9H_7ClF_3NO$

Mol Weight: 237.61

MP (deg C):

FP (deg C):

BP (deg C):

BP pressure (mm Hg):

Property/Value	Units	Temp	Data Type	Reference
WS 5.64E+002	mg/L	25	EST	MEYLAN,WM ET AL. (1996)
logP 1.98			EXP	NAKAGAWA,Y ET AL. (1992)
VP 6.34E-005	mm Hg	25	EST	NEELY,WB & BLAU,GE (1985)
DC	pKa			
HL 3.97E-008	atm m3/mol	25	EST	MEYLAN,WM & HOWARD,PH (1991)
OH 8.31E-013	cm3/molc sec	25	EST	MEYLAN,WM & HOWARD,PH (1993)

CAS #: 000344-62-7				O-CF3 ACETANILIDE

Formula: $C_9H_8F_3NO$

Mol Weight: 203.17

MP (deg C):

FP (deg C):

BP (deg C):

BP pressure (mm Hg):

Property/Value	Units	Temp	Data Type	Reference
WS 4.22E+003	mg/L	25	EST	MEYLAN,WM ET AL. (1996)
logP 1.17			EXP	NAKAGAWA,Y ET AL. (1992)
VP 3.36E-004	mm Hg	25	EST	NEELY,WB & BLAU,GE (1985)
DC	pKa			
HL 5.36E-008	atm m3/mol	25	EST	MEYLAN,WM & HOWARD,PH (1991)
OH 2.60E-012	cm3/molc sec	25	EST	MEYLAN,WM & HOWARD,PH (1993)

CAS #: 000345-35-7				O-FLUOROBENZYL CHLORIDE
Formula: C_7H_6ClF				
Mol Weight: 144.58				
MP (deg C):		FP (deg C):		
BP (deg C): 86				
BP pressure (mm Hg): 4.00E+001				

Property/Value	Units	Temp	Data Type	Reference
WS 4.16E+002	mg/L	25	EXP	YALKOWSKY,SH & DANNENFELSER,RM (1992)
logP 2.67			EXP	HANSCH,C & LEO,AJ (1985)
VP 9.94E-001	mm Hg	25	EST	NEELY,WB & BLAU,GE (1985)
DC	pKa			
HL 2.44E-003	atm m3/mol	25	EST	MEYLAN,WM & HOWARD,PH (1991)
OH 1.94E-012	cm3/molc sec	25	EST	MEYLAN,WM & HOWARD,PH (1993)

CAS #: 000346-18-9				POLYTHIAZIDE
Formula: $C_{11}H_{13}ClF_3N_3O_4S_3$				
Mol Weight: 439.88				
MP (deg C): 202.5		FP (deg C):		
BP (deg C):				
BP pressure (mm Hg):				

Property/Value	Units	Temp	Data Type	Reference
WS 3.66E+000	mg/L	25	EST	MEYLAN,WM ET AL. (1996)
logP 1.90			EXP	SANGSTER,J (1994)
VP 3.43E-011	mm Hg	25	EST	NEELY,WB & BLAU,GE (1985)
DC	pKa			
HL 1.43E-012	atm m3/mol	25	EST	MEYLAN,WM & HOWARD,PH (1991)
OH 2.20E-010	cm3/molc sec	25	EST	MEYLAN,WM & HOWARD,PH (1993)

CAS #: 000348-10-7				O-FLUOROPHENOXYACETIC ACID
Formula: $C_8H_7FO_3$				
Mol Weight: 170.14				
MP (deg C):		FP (deg C):		
BP (deg C):				
BP pressure (mm Hg):				

Property/Value	Units	Temp	Data Type	Reference
WS 9.91E+003	mg/L	25	EST	MEYLAN,WM ET AL. (1996)
logP 1.39			EXP	HANSCH,C & LEO,AJ (1985)
VP 1.86E-003	mm Hg	25	EST	NEELY,WB & BLAU,GE (1985)
DC 3.08	pKa	25	EXP	KORTUM,G ET AL (1961)
HL 1.96E-008	atm m3/mol	25	EST	MEYLAN,WM & HOWARD,PH (1991)
OH 1.21E-011	cm3/molc sec	25	EST	MEYLAN,WM & HOWARD,PH (1993)

CAS #: 000348-51-6				1-CHLORO-2-FLUOROBENZENE
Formula: C_6H_4ClF				
Mol Weight: 130.55				
MP (deg C): -43		FP (deg C):		
BP (deg C): 137.6				
BP pressure (mm Hg):				

Property/Value	Units	Temp	Data Type	Reference
WS 5.01E+002	mg/L	25	EXP	YALKOWSKY,SH & DANNENFELSER,RM (1992)
logP 2.84			EST	MEYLAN,WM & HOWARD,PH (1995)
VP 8.19E+000	mm Hg	25	EST	NEELY,WB & BLAU,GE (1985)
DC	pKa			
HL 4.66E-003	atm m3/mol	25	EST	MEYLAN,WM & HOWARD,PH (1991)
OH 7.13E-013	cm3/molc sec	25	EST	MEYLAN,WM & HOWARD,PH (1993)

CAS #: 000348-54-9				2-FLUOROANILINE
Formula: C_6H_6FN				
Mol Weight: 111.12				
MP (deg C): -34.6		FP (deg C):		
BP (deg C): 175				
BP pressure (mm Hg):				

Property/Value	Units	Temp	Data Type	Reference
WS 9.05E+003	mg/L	25	EST	MEYLAN,WM ET AL. (1996)
logP 1.26			EXP	HANSCH,C & LEO,AJ (1985)
VP 1.00E+000	mm Hg	25	EST	NEELY,WB & BLAU,GE (1985)
DC 3.20	pKa	25	EXP	PERRIN,DD (1965)
HL 2.22E-006	atm m3/mol	25	EST	MEYLAN,WM & HOWARD,PH (1991)
OH 3.63E-011	cm3/molc sec	25	EST	MEYLAN,WM & HOWARD,PH (1993)

CAS #: 000348-90-3				3-CF3-4-CHLOROACETANILIDE
Formula: $C_9H_7ClF_3NO$				
Mol Weight: 237.61				
MP (deg C):		FP (deg C):		
BP (deg C):				
BP pressure (mm Hg):				

Property/Value	Units	Temp	Data Type	Reference
WS 3.97E+001	mg/L	25	EST	MEYLAN,WM ET AL. (1996)
logP 3.33			EXP	NAKAGAWA,Y ET AL. (1992)
VP 6.34E-005	mm Hg	25	EST	NEELY,WB & BLAU,GE (1985)
DC	pKa			
HL 3.97E-008	atm m3/mol	25	EST	MEYLAN,WM & HOWARD,PH (1991)
OH 6.51E-013	cm3/molc sec	25	EST	MEYLAN,WM & HOWARD,PH (1993)

CAS #: 000349-27-9				2-FLUORO-5-CF3-ACETANILIDE
Formula: $C_9H_7F_4NO$				
Mol Weight: 221.16				
MP (deg C):		FP (deg C):		
BP (deg C):				
BP pressure (mm Hg):				

Property/Value	Units	Temp	Data Type	Reference
WS 4.15E+002	mg/L	25	EST	MEYLAN,WM ET AL. (1996)
logP 2.24			EXP	NAKAGAWA,Y ET AL. (1992)
VP 3.84E-004	mm Hg	25	EST	NEELY,WB & BLAU,GE (1985)
DC	pKa			
HL 6.26E-008	atm m3/mol	25	EST	MEYLAN,WM & HOWARD,PH (1991)
OH 7.36E-013	cm3/molc sec	25	EST	MEYLAN,WM & HOWARD,PH (1993)

CAS #: 000349-75-7				M-(TRIFLUOROME)BENZYL ALCOHOL
Formula: $C_8H_7F_3O$				
Mol Weight: 176.14				
MP (deg C):		FP (deg C):		
BP (deg C): 68				
BP pressure (mm Hg): 2.00E+000				

Property/Value	Units	Temp	Data Type	Reference
WS 2.27E+003	mg/L	25	EST	MEYLAN,WM ET AL. (1996)
logP 2.24			EXP	HANSCH,C & LEO,AJ (1985)
VP 3.69E-002	mm Hg	25	EST	NEELY,WB & BLAU,GE (1985)
DC	pKa			
HL 1.89E-006	atm m3/mol	25	EST	MEYLAN,WM & HOWARD,PH (1991)
OH 4.14E-012	cm3/molc sec	25	EST	MEYLAN,WM & HOWARD,PH (1993)

CAS #: 000349-82-6 — M-(TRIFLUOROME)PHENOXYACETIC ACID

Formula: $C_9H_7F_3O_3$

Mol Weight: 220.15

MP (deg C):

FP (deg C):

BP (deg C):

BP pressure (mm Hg):

Property/Value	Units	Temp	Data Type	Reference
WS 8.23E+002	mg/L	25	EST	MEYLAN,WM ET AL. (1996)
logP 2.36			EXP	HANSCH,C & LEO,AJ (1985)
VP 1.19E-003	mm Hg	25	EST	NEELY,WB & BLAU,GE (1985)
DC	pKa			
HL 1.46E-007	atm m3/mol	25	EST	MEYLAN,WM & HOWARD,PH (1991)
OH 8.05E-012	cm3/molc sec	25	EST	MEYLAN,WM & HOWARD,PH (1993)

CAS #: 000349-97-3 — ACETAMIDE, N-[4-(TRIFLUOROMETHYL)PHENYL]-

Formula: $C_9H_8F_3NO$

Mol Weight: 203.17

MP (deg C):

FP (deg C):

BP (deg C):

BP pressure (mm Hg):

Property/Value	Units	Temp	Data Type	Reference
WS 2.97E+002	mg/L	25	EST	MEYLAN,WM ET AL. (1996)
logP 2.52			EXP	HANSCH,C ET AL. (1995)
VP 3.36E-004	mm Hg	25	EST	NEELY,WB & BLAU,GE (1985)
DC	pKa			
HL 5.36E-008	atm m3/mol	25	EST	MEYLAN,WM & HOWARD,PH (1991)
OH 2.60E-012	cm3/molc sec	25	EST	MEYLAN,WM & HOWARD,PH (1993)

CAS #: 000350-03-8 — 3-ACETYLPYRIDINE

Formula: C_7H_7NO

Mol Weight: 121.14

MP (deg C): 13.5

FP (deg C):

BP (deg C): 220

BP pressure (mm Hg):

Property/Value	Units	Temp	Data Type	Reference
WS 4.27E+004	mg/L	25	EST	MEYLAN,WM ET AL. (1996)
logP 0.43			EXP	HANSCH,C & LEO,AJ (1985)
VP 3.74E-001	mm Hg	25	EST	NEELY,WB & BLAU,GE (1985)
DC 3.18	pKa	25	EXP	PERRIN,DD (1965)
HL 1.28E-008	atm m3/mol	25	EST	MEYLAN,WM & HOWARD,PH (1991)
OH 4.39E-013	cm3/molc sec	25	EST	MEYLAN,WM & HOWARD,PH (1993)

CAS #: 000350-46-9 — 4-FLUORONITROBENZENE

Formula: $C_6H_4FNO_2$

Mol Weight: 141.10

MP (deg C): 21

FP (deg C):

BP (deg C): 205

BP pressure (mm Hg):

Property/Value	Units	Temp	Data Type	Reference
WS 9.80E+002	mg/L	25	EST	MEYLAN,WM ET AL. (1996)
logP 1.80			EXP	HANSCH,C & LEO,AJ (1985)
VP 2.94E-001	mm Hg	25	EST	NEELY,WB & BLAU,GE (1985)
DC	pKa			
HL 2.48E-005	atm m3/mol	25	EST	MEYLAN,WM & HOWARD,PH (1991)
OH 3.05E-013	cm3/molc sec	25	EST	MEYLAN,WM & HOWARD,PH (1993)

CAS #: 000350-50-5 — BENZYL FLUORIDE

Formula: C_7H_7F

Mol Weight: 110.13

MP (deg C): -35

FP (deg C):

BP (deg C): 140

BP pressure (mm Hg):

Property/Value	Units	Temp	Data Type	Reference
WS 8.27E+002	mg/L	25	EST	MEYLAN,WM ET AL. (1996)
logP 2.48			EST	MEYLAN,WM & HOWARD,PH (1995)
VP 6.79E+000	mm Hg	25	EXP	ASHCROFT,SJ (1976)
DC	pKa			
HL 1.18E-002	atm m3/mol	25	EST	MEYLAN,WM & HOWARD,PH (1991)
OH 2.36E-012	cm3/molc sec	25	EST	MEYLAN,WM & HOWARD,PH (1993)

CAS #: 000351-28-0 — 3-FLUOROACETANILIDE

Formula: C_8H_8FNO

Mol Weight: 153.16

MP (deg C): 82-84

FP (deg C):

BP (deg C):

BP pressure (mm Hg):

Property/Value	Units	Temp	Data Type	Reference
WS 2.87E+003	mg/L	25	EST	MEYLAN,WM ET AL. (1996)
logP 1.65			EXP	NAKAGAWA,Y ET AL. (1992)
VP 5.97E-004	mm Hg	25	EST	NEELY,WB & BLAU,GE (1985)
DC	pKa			
HL 7.20E-009	atm m3/mol	25	EST	MEYLAN,WM & HOWARD,PH (1991)
OH 1.57E-011	cm3/molc sec	25	EST	MEYLAN,WM & HOWARD,PH (1993)

CAS #: 000351-35-9 — M-(TRIFLUOROME)PHENYLACETIC ACID

Formula: $C_9H_7F_3O_2$

Mol Weight: 204.15

MP (deg C): 76-79

FP (deg C):

BP (deg C):

BP pressure (mm Hg):

Property/Value	Units	Temp	Data Type	Reference
WS 5.99E+002	mg/L	25	EST	MEYLAN,WM ET AL. (1996)
logP 2.62			EXP	HANSCH,C & LEO,AJ (1985)
VP 3.27E-003	mm Hg	25	EST	NEELY,WB & BLAU,GE (1985)
DC	pKa			
HL 3.84E-007	atm m3/mol	25	EST	MEYLAN,WM & HOWARD,PH (1991)
OH 1.71E-012	cm3/molc sec	25	EST	MEYLAN,WM & HOWARD,PH (1993)

CAS #: 000351-36-0 — M-(TRIFLUOROME)ACETANILIDE

Formula: $C_9H_8F_3NO$

Mol Weight: 203.17

MP (deg C): 103-106

FP (deg C):

BP (deg C):

BP pressure (mm Hg):

Property/Value	Units	Temp	Data Type	Reference
WS 2.75E+002	mg/L	25	EST	MEYLAN,WM ET AL. (1996)
logP 2.56			EXP	HANSCH,C & LEO,AJ (1985)
VP 3.36E-004	mm Hg	25	EST	NEELY,WB & BLAU,GE (1985)
DC	pKa			
HL 5.36E-008	atm m3/mol	25	EST	MEYLAN,WM & HOWARD,PH (1991)
OH 1.98E-012	cm3/molc sec	25	EST	MEYLAN,WM & HOWARD,PH (1993)

CAS #: 000351-83-7				4-FLUOROACETANILIDE

Formula: C_8H_8FNO
Mol Weight: 153.16
MP (deg C): 153-155
FP (deg C):
BP (deg C):
BP pressure (mm Hg):

Property/Value	Units	Temp	Data Type	Reference
WS 4.08E+003	mg/L	25	EST	MEYLAN,WM ET AL. (1996)
logP 1.47			EXP	HANSCH,C & LEO,AJ (1985)
VP 5.97E-004	mm Hg	25	EST	NEELY,WB & BLAU,GE (1985)
DC	pKa			
HL 7.20E-009	atm m3/mol	25	EST	MEYLAN,WM & HOWARD,PH (1991)
OH 4.29E-012	cm3/molc sec	25	EST	MEYLAN,WM & HOWARD,PH (1993)

CAS #: 000352-11-4				P-FLUOROBENZYL CHLORIDE

Formula: C_7H_6ClF
Mol Weight: 144.58
MP (deg C):
FP (deg C):
BP (deg C): 82
BP pressure (mm Hg): 2.60E+001

Property/Value	Units	Temp	Data Type	Reference
WS 4.13E+002	mg/L	25	EXP	WASIK,SP ET AL. (1981)
logP 2.73			EXP	HANSCH,C & LEO,AJ (1985)
VP 9.94E-001	mm Hg	25	EST	NEELY,WB & BLAU,GE (1985)
DC	pKa			
HL 2.44E-003	atm m3/mol	25	EST	MEYLAN,WM & HOWARD,PH (1991)
OH 1.94E-012	cm3/molc sec	25	EST	MEYLAN,WM & HOWARD,PH (1993)

CAS #: 000352-32-9				P-FLUOROTOLUENE

Formula: C_7H_7F
Mol Weight: 110.13
MP (deg C): -56
FP (deg C):
BP (deg C): 116.6
BP pressure (mm Hg):

Property/Value	Units	Temp	Data Type	Reference
WS 6.81E+002	mg/L	25	EST	MEYLAN,WM ET AL. (1996)
logP 2.58			EXP	DUNN,WJ ET AL. (1983)
VP 2.10E+001	mm Hg	25	EXP	PERRY,RH & GREEN,D (1984)
DC	pKa			
HL 6.94E-003	atm m3/mol	25	EST	MEYLAN,WM & HOWARD,PH (1991)
OH 3.13E-012	cm3/molc sec	25	EST	MEYLAN,WM & HOWARD,PH (1993)

CAS #: 000352-33-0				P-CHLOROFLUOROBENZENE

Formula: C_6H_4ClF
Mol Weight: 130.55
MP (deg C): -26.8
FP (deg C):
BP (deg C): 130
BP pressure (mm Hg):

Property/Value	Units	Temp	Data Type	Reference
WS 3.87E+002	mg/L	25	EST	MEYLAN,WM ET AL. (1996)
logP 2.78			EXP	DUNN,WJ ET AL. (1983)
VP 8.19E+000	mm Hg	25	EST	NEELY,WB & BLAU,GE (1985)
DC	pKa			
HL 4.66E-003	atm m3/mol	25	EST	MEYLAN,WM & HOWARD,PH (1991)
OH 7.13E-013	cm3/molc sec	25	EST	MEYLAN,WM & HOWARD,PH (1993)

CAS #: 000352-61-4				5,5,5-TRIFLUORO-1-PENTANOL

Formula: $C_5H_9F_3O$
Mol Weight: 142.12
MP (deg C):
FP (deg C):
BP (deg C):
BP pressure (mm Hg):

Property/Value	Units	Temp	Data Type	Reference
WS 2.77E+004	mg/L	25	EST	MEYLAN,WM ET AL. (1996)
logP 1.15			EXP	MULLER,N (1986)
VP 6.70E+000	mm Hg	25	EST	NEELY,WB & BLAU,GE (1985)
DC	pKa			
HL 6.59E-005	atm m3/mol	25	EST	MEYLAN,WM & HOWARD,PH (1991)
OH 7.07E-012	cm3/molc sec	25	EST	MEYLAN,WM & HOWARD,PH (1993)

CAS #: 000352-70-5				M-FLUOROTOLUENE

Formula: C_7H_7F
Mol Weight: 110.13
MP (deg C): -87
FP (deg C):
BP (deg C): 115
BP pressure (mm Hg):

Property/Value	Units	Temp	Data Type	Reference
WS 4.87E+002	mg/L	25	EST	MEYLAN,WM ET AL. (1996)
logP 2.75			EXP	DUNN,WJ ET AL. (1983)
VP 2.18E+001	mm Hg	25	EXP	PERRY,RH & GREEN,D (1984)
DC	pKa			
HL 6.94E-003	atm m3/mol	25	EST	MEYLAN,WM & HOWARD,PH (1991)
OH 6.51E-012	cm3/molc sec	25	EST	MEYLAN,WM & HOWARD,PH (1993)

CAS #: 000352-87-4				2,2,2-TRIFLUOROETHYL METHACRYLATE

Formula: $C_6H_7F_3O_2$
Mol Weight: 168.12
MP (deg C):
FP (deg C):
BP (deg C): 59
BP pressure (mm Hg): 1.00E+002

Property/Value	Units	Temp	Data Type	Reference
WS 2.92E+003	mg/L	25	EST	MEYLAN,WM ET AL. (1996)
logP 1.56			EXP	SANGSTER,J (1993)
VP 3.58E+001	mm Hg	25	EST	NEELY,WB & BLAU,GE (1985)
DC	pKa			
HL 1.51E-003	atm m3/mol	25	EST	MEYLAN,WM & HOWARD,PH (1991)
OH 1.82E-011	cm3/molc sec	25	EST	MEYLAN,WM & HOWARD,PH (1993)

CAS #: 000352-93-2				DIETHYLSULFIDE

Formula: $C_4H_{10}S$
Mol Weight: 90.19
MP (deg C): -103.9
FP (deg C):
BP (deg C): 92.1
BP pressure (mm Hg):

Property/Value	Units	Temp	Data Type	Reference
WS 8.02E+003	mg/L	25	EXP	HINE,J & MOOKERJEE,PK (1975)
logP 1.95			EXP	HANSCH,C & LEO,AJ (1985)
VP 5.84E+001	mm Hg	25	EXP	RIDDICK,JA ET AL. (1986)
DC	pKa			
HL 8.64E-004	atm m3/mol	25	EST	VP/WSOL
OH 1.50E-011	cm3/molc sec	25	EXP	ATKINSON,R (1989)

000353-36-6 — FLUOROETHANE

Formula: C_2H_5F

Mol Weight: 48.06

MP (deg C): -143.2

FP (deg C):

BP (deg C): -37.6

BP pressure (mm Hg):

Property/Value	Units	Temp	Data Type	Reference
WS 2.16E+003	mg/L	25	EXP	HORVATH,AL (1975) @ 1 atm
logP 1.26			EST	MEYLAN,WM & HOWARD,PH (1995)
VP 6.84E+003	mm Hg	25	EXP	DAUBERT,TE & DANNER,RP (1989)
DC	pKa			
HL 2.18E-002	atm m3/mol	25	EST	MEYLAN,WM & HOWARD,PH (1991)
OH 2.32E-013	cm3/molc sec	25	EXP	ATKINSON,R (1989)

000353-50-4 — CARBONIC DIFLUORIDE

Formula: CF_2O

Mol Weight: 66.01

MP (deg C): -83.1

FP (deg C):

BP (deg C): -114

BP pressure (mm Hg):

Property/Value	Units	Temp	Data Type	Reference
WS 1.00E+006	mg/L	25	EST	MEYLAN,WM ET AL. (1996)
logP -1.34			EST	MEYLAN,WM & HOWARD,PH (1995)
VP 4.45E+004	mm Hg	25	EXP	DAUBERT,TE & DANNER,RP (1989)
DC	pKa			
HL 4.88E-002	atm m3/mol	25	EST	MEYLAN,WM & HOWARD,PH (1991)
OH 0.00E+000	cm3/molc sec	25	EST	MEYLAN,WM & HOWARD,PH (1993)

000353-59-3 — BROMOCHLORODIFLUOROMETHANE

Formula: $CBrClF_2$

Mol Weight: 165.37

MP (deg C): -159.5

FP (deg C):

BP (deg C): -3.7

BP pressure (mm Hg):

Property/Value	Units	Temp	Data Type	Reference
WS 2.77E+002	mg/L	25	EST	MEYLAN,WM ET AL. (1996)
logP 1.90			EST	MEYLAN,WM & HOWARD,PH (1995)
VP 2.07E+003	mm Hg	25	EXP	DAUBERT,TE & DANNER,RP (1989)
DC	pKa			
HL 9.40E-002	atm m3/mol	25	EST	MEYLAN,WM & HOWARD,PH (1991)
OH 1.00E-015	cm3/molc sec	25	EXP	ATKINSON,R (1989)

000354-33-6 — 1,1,1,2,2-PENTAFLUOROETHANE

Formula: C_2HF_5

Mol Weight: 120.02

MP (deg C): -103

FP (deg C):

BP (deg C): -48.5

BP pressure (mm Hg):

Property/Value	Units	Temp	Data Type	Reference
WS 3.90E+001	mg/L	25	EST	VP/HL
logP 1.55			EST	MEYLAN,WM & HOWARD,PH (1995)
VP 8.94E+003	mm Hg	25	EST	NEELY,WB & BLAU,GE (1985)
DC	pKa			
HL 3.05E+000	atm m3/mol	25	EST	MEYLAN,WM & HOWARD,PH (1991)
OH 2.50E-015	cm3/molc sec	25	EXP	ATKINSON,R (1989)

000354-38-1 — TRIFLUOROACETAMIDE

Formula: $C_2H_2F_3NO$

Mol Weight: 113.04

MP (deg C): 70-75

FP (deg C):

BP (deg C): 162.5

BP pressure (mm Hg):

Property/Value	Units	Temp	Data Type	Reference
WS 8.39E+004	mg/L	25	EST	MEYLAN,WM ET AL. (1996)
logP 0.12			EXP	HANSCH,C & LEO,AJ (1985)
VP 1.85E+000	mm Hg	25	EST	NEELY,WB & BLAU,GE (1985)
DC	pKa			
HL 8.81E-008	atm m3/mol	25	EST	MEYLAN,WM & HOWARD,PH (1991)
OH 0.00E+000	cm3/molc sec	25	EST	MEYLAN,WM & HOWARD,PH (1993)

000354-58-5 — 1,1,1-TRICHLORO-2,2,2-TRIFLUOROETHANE

Formula: $C_2Cl_3F_3$

Mol Weight: 187.38

MP (deg C): 13-14

FP (deg C):

BP (deg C): 46

BP pressure (mm Hg):

Property/Value	Units	Temp	Data Type	Reference
WS 2.09E+001	mg/L	25	EST	MEYLAN,WM ET AL. (1996)
logP 3.09			EST	MEYLAN,WM & HOWARD,PH (1995)
VP 3.60E+002	mm Hg	25	EXP	BOUBLIK,T ET AL. (1984)
DC	pKa			
HL 3.88E-002	atm m3/mol	25	EST	MEYLAN,WM & HOWARD,PH (1991)
OH 0.00E+000	cm3/molc sec	25	EST	MEYLAN,WM & HOWARD,PH (1993)

000355-02-2 — PERFLUOROMETHYLCYCLOHEXANE

Formula: C_7F_{14}

Mol Weight: 350.06

MP (deg C): -44.7

FP (deg C):

BP (deg C): 76.3

BP pressure (mm Hg):

Property/Value	Units	Temp	Data Type	Reference
WS 4.67E+000	mg/L	25	EST	MEYLAN,WM ET AL. (1996)
logP 2.79			EST	MEYLAN,WM & HOWARD,PH (1995)
VP 1.06E+002	mm Hg	25	EXT	BOUBLIK,T ET AL. (1984)
DC	pKa			
HL 1.08E+004	atm m3/mol	25	EST	MEYLAN,WM & HOWARD,PH (1991)
OH 0.00E+000	cm3/molc sec	25	EST	MEYLAN,WM & HOWARD,PH (1993)

000355-25-9 — DECAFLUOROBUTANE

Formula: C_4F_{10}

Mol Weight: 238.03

MP (deg C): -128.2

FP (deg C):

BP (deg C): -1.9

BP pressure (mm Hg):

Property/Value	Units	Temp	Data Type	Reference
WS 1.61E+000	mg/L	25	EST	MEYLAN,WM ET AL. (1996)
logP 4.09			EST	MEYLAN,WM & HOWARD,PH (1995)
VP 2.01E+003	mm Hg	25	EXP	DAUBERT,TE & DANNER,RP (1989)
DC	pKa			
HL 6.67E+002	atm m3/mol	25	EST	MEYLAN,WM & HOWARD,PH (1991)
OH 0.00E+000	cm3/molc sec	25	EST	MEYLAN,WM & HOWARD,PH (1993)

000355-42-0 — PERFLUORO-N-HEXANE

Formula: C_6F_{14}
Mol Weight: 338.04
MP (deg C): -87.1
FP (deg C):
BP (deg C): 56.6
BP pressure (mm Hg):

Property/Value	Units	Temp	Data Type	Reference
WS 9.58E-003	mg/L	25	EST	MEYLAN,WM ET AL. (1996)
logP 6.02			EST	MEYLAN,WM & HOWARD,PH (1995)
VP 2.18E+002	mm Hg	25	EXP	BOUBLIK,T ET AL. (1984)
DC	pKa			
HL 1.84E+004	atm m3/mol	25	EST	MEYLAN,WM & HOWARD,PH (1991)
OH 0.00E+000	cm3/molc sec	25	EST	MEYLAN,WM & HOWARD,PH (1993)

000355-68-0 — PERFLUOROCYCLOHEXANE

Formula: C_6F_{12}
Mol Weight: 300.05
MP (deg C): 48.5
FP (deg C):
BP (deg C): 50.6
BP pressure (mm Hg):

Property/Value	Units	Temp	Data Type	Reference
WS 7.27E+000	mg/L	25	EST	MEYLAN,WM ET AL. (1996)
logP 2.91			EXP	HANSCH,C ET AL. (1995)
VP 1.71E+002	mm Hg	25	EXP	BOUBLIK,T ET AL. (1984)
DC	pKa			
HL 2.05E+003	atm m3/mol	25	EST	MEYLAN,WM & HOWARD,PH (1991)
OH 0.00E+000	cm3/molc sec	25	EST	MEYLAN,WM & HOWARD,PH (1993)

000356-12-7 — FLUOCINONIDE

Formula: $C_{26}H_{32}F_2O_7$
Mol Weight: 494.54
MP (deg C): 308-311
FP (deg C):
BP (deg C):
BP pressure (mm Hg):

Property/Value	Units	Temp	Data Type	Reference
WS 4.74E+000	mg/L	25	EST	MEYLAN,WM ET AL. (1996)
logP 3.19			EXP	HANSCH,C & LEO,AJ (1985)
VP 2.00E-013	mm Hg	25	EST	NEELY,WB & BLAU,GE (1985)
DC	pKa			
HL 2.11E-016	atm m3/mol	25	EST	MEYLAN,WM & HOWARD,PH (1991)
OH 8.50E-011	cm3/molc sec	25	EST	MEYLAN,WM & HOWARD,PH (1993)

000357-56-2 — DEXTROMORAMIDE

Formula: $C_{25}H_{32}N_2O_2$
Mol Weight: 392.55
MP (deg C): 180-184
FP (deg C):
BP (deg C):
BP pressure (mm Hg):

Property/Value	Units	Temp	Data Type	Reference
WS 2.81E+000	mg/L	25	EST	MEYLAN,WM ET AL. (1996)
logP 3.61			EXP	SANGSTER,J (1994)
VP 2.43E-010	mm Hg	25	EST	NEELY,WB & BLAU,GE (1985)
DC	pKa			
HL 2.17E-015	atm m3/mol	25	EST	MEYLAN,WM & HOWARD,PH (1991)
OH 1.79E-010	cm3/molc sec	25	EST	MEYLAN,WM & HOWARD,PH (1993)

000357-57-3 — BRUCINE

Formula: $C_{23}H_{26}N_2O_4$
Mol Weight: 394.47
MP (deg C): 178
FP (deg C):
BP (deg C):
BP pressure (mm Hg):

Property/Value	Units	Temp	Data Type	Reference
WS 3.20E+003	mg/L	15	EXP	YALKOWSKY,SH & DANNENFELSER,RM (1992)
logP 0.98			EXP	HANSCH,C & LEO,AJ (1985)
VP 1.03E-010	mm Hg	25	EST	NEELY,WB & BLAU,GE (1985)
DC	pKa			
HL 2.09E-016	atm m3/mol	25	EST	MEYLAN,WM & HOWARD,PH (1991)
OH 3.21E-010	cm3/molc sec	25	EST	MEYLAN,WM & HOWARD,PH (1993)

000358-43-0 — TRIETHYLFLUOROSILANE

Formula: $C_6H_{15}FSi$
Mol Weight: 134.27
MP (deg C):
FP (deg C):
BP (deg C): 110
BP pressure (mm Hg):

Property/Value	Units	Temp	Data Type	Reference
WS 6.87E+001	mg/L	25	EST	MEYLAN,WM ET AL. (1996)
logP 3.64			EST	MEYLAN,WM & HOWARD,PH (1995)
VP 3.68E+001	mm Hg	25	EXP	SOKOLOV,VB ET AL. (1975)
DC	pKa			
HL	atm m3/mol			
OH 3.58E-012	cm3/molc sec	25	EST	MEYLAN,WM & HOWARD,PH (1993)

000359-10-4 — CHLORO-1,1-DIFLUOROETHYLENE

Formula: C_2HClF_2
Mol Weight: 98.48
MP (deg C): -138.5
FP (deg C):
BP (deg C): -18.5
BP pressure (mm Hg):

Property/Value	Units	Temp	Data Type	Reference
WS 5.11E+003	mg/L	25	EST	MEYLAN,WM ET AL. (1996)
logP 1.60			EST	MEYLAN,WM & HOWARD,PH (1995)
VP 3.55E+003	mm Hg	25	EXP	DAUBERT,TE & DANNER,RP (1989)
DC	pKa			
HL 1.63E-001	atm m3/mol	25	EST	MEYLAN,WM & HOWARD,PH (1991)
OH 8.05E-013	cm3/molc sec	25	EST	MEYLAN,WM & HOWARD,PH (1993)

000359-29-5 — CHLOROTRIFLUOROETHENE

Formula: C_2ClF_3
Mol Weight: 116.47
MP (deg C): -108.9
FP (deg C):
BP (deg C): 71
BP pressure (mm Hg):

Property/Value	Units	Temp	Data Type	Reference
WS 4.01E+003	mg/L	25	EST	MEYLAN,WM ET AL. (1996)
logP 1.65			EST	MEYLAN,WM & HOWARD,PH (1995)
VP 1.39E+003	mm Hg	25	EST	NEELY,WB & BLAU,GE (1985)
DC	pKa			
HL 3.11E-001	atm m3/mol	25	EST	MEYLAN,WM & HOWARD,PH (1991)
OH 7.00E-012	cm3/molc sec	25	EXP	ATKINSON,R (1989)

CAS #: 000359-85-3			124-BENZOTHIADIAZINE-1,1-O2

Formula: $C_7H_6N_2O_2S$
Mol Weight: 182.20
MP (deg C): | FP (deg C):
BP (deg C):
BP pressure (mm Hg):

Property/ Value	Units	Temp	Data Type	Reference
WS 3.92E+004	mg/L	25	EST	MEYLAN,WM ET AL. (1996)
logP 0.16			EXP	HANSCH,C & LEO,AJ (1985)
VP 3.04E-007	mm Hg	25	EST	NEELY,WB & BLAU,GE (1985)
DC	pKa			
HL 6.97E-008	atm m3/mol	25	EST	MEYLAN,WM & HOWARD,PH (1991)
OH 1.85E-011	cm3/molc sec	25	EST	MEYLAN,WM & HOWARD,PH (1993)

CAS #: 000360-64-5			2-TRIFLUOROMETHYLBENZAMIDE

Formula: $C_8H_6F_3NO$
Mol Weight: 189.14
MP (deg C): | FP (deg C):
BP (deg C):
BP pressure (mm Hg):

Property/ Value	Units	Temp	Data Type	Reference
WS 1.30E+004	mg/L	25	EST	MEYLAN,WM ET AL. (1996)
logP 0.68			EXP	NAKAGAWA,Y ET AL. (1992)
VP 6.01E-004	mm Hg	25	EST	NEELY,WB & BLAU,GE (1985)
DC	pKa			
HL 1.92E-008	atm m3/mol	25	EST	MEYLAN,WM & HOWARD,PH (1991)
OH 2.69E-013	cm3/molc sec	25	EST	MEYLAN,WM & HOWARD,PH (1993)

CAS #: 000360-65-6			GLYCODEOXYCHOLIC ACID

Formula: $C_{26}H_{43}NO_5$
Mol Weight: 449.64
MP (deg C): | FP (deg C):
BP (deg C):
BP pressure (mm Hg):

Property/ Value	Units	Temp	Data Type	Reference
WS 1.80E+001	mg/L	25	EST	MEYLAN,WM ET AL. (1996)
logP 2.25			EXP	RODA,A ET AL. (1990)
VP 8.36E-018	mm Hg	25	EST	NEELY,WB & BLAU,GE (1985)
DC	pKa			
HL 9.38E-017	atm m3/mol	25	EST	MEYLAN,WM & HOWARD,PH (1991)
OH 5.79E-011	cm3/molc sec	25	EST	MEYLAN,WM & HOWARD,PH (1993)

CAS #: 000360-81-6			124-BENZOTHIADIAZINE-1,1-O2-3-ME

Formula: $C_8H_8N_2O_2S$
Mol Weight: 196.23
MP (deg C): | FP (deg C):
BP (deg C):
BP pressure (mm Hg):

Property/ Value	Units	Temp	Data Type	Reference
WS 2.59E+004	mg/L	25	EST	MEYLAN,WM ET AL. (1996)
logP 0.29			EXP	HANSCH,C & LEO,AJ (1985)
VP 1.30E-007	mm Hg	25	EST	NEELY,WB & BLAU,GE (1985)
DC	pKa			
HL 9.25E-008	atm m3/mol	25	EST	MEYLAN,WM & HOWARD,PH (1991)
OH 6.25E-013	cm3/molc sec	25	EST	MEYLAN,WM & HOWARD,PH (1993)

CAS #: 000360-89-4			OCTAFLUORO-2-BUTENE

Formula: C_4F_8
Mol Weight: 200.03
MP (deg C): -129 | FP (deg C):
BP (deg C): 1.5
BP pressure (mm Hg):

Property/ Value	Units	Temp	Data Type	Reference
WS 1.13E+002	mg/L	25	EST	MEYLAN,WM ET AL. (1996)
logP 3.03			EST	MEYLAN,WM & HOWARD,PH (1995)
VP 2.00E+003	mm Hg	25	EXP	DAUBERT,TE & DANNER,RP (1989)
DC	pKa			
HL 3.44E+001	atm m3/mol	25	EST	MEYLAN,WM & HOWARD,PH (1991)
OH 2.80E-012	cm3/molc sec	25	EST	MEYLAN,WM & HOWARD,PH (1993)

CAS #: 000363-03-1			2,5-CYCLOHEXADIENE-1,4-DIONE, 2-PHENYL-

Formula: $C_{12}H_8O_2$
Mol Weight: 184.20
MP (deg C): 113-115 | FP (deg C):
BP (deg C):
BP pressure (mm Hg):

Property/ Value	Units	Temp	Data Type	Reference
WS 1.14E+003	mg/L	25	EST	MEYLAN,WM ET AL. (1996)
logP 1.95			EXP	SANGSTER,J (1993)
VP 6.69E-005	mm Hg	25	EST	NEELY,WB & BLAU,GE (1985)
DC	pKa			
HL 5.46E-011	atm m3/mol	25	EST	MEYLAN,WM & HOWARD,PH (1991)
OH 1.47E-011	cm3/molc sec	25	EST	MEYLAN,WM & HOWARD,PH (1993)

CAS #: 000363-24-6			PROSTAGLANDIN E2

Formula: $C_{20}H_{32}O_5$
Mol Weight: 352.48
MP (deg C): 115-116 | FP (deg C):
BP (deg C):
BP pressure (mm Hg):

Property/ Value	Units	Temp	Data Type	Reference
WS 5.81E+001	mg/L	25	EST	MEYLAN,WM ET AL. (1996)
logP 2.82			EXP	HANSCH,C ET AL. (1995)
VP 1.68E-013	mm Hg	25	EST	NEELY,WB & BLAU,GE (1985)
DC	pKa			
HL 1.56E-014	atm m3/mol	25	EST	MEYLAN,WM & HOWARD,PH (1991)
OH 1.71E-010	cm3/molc sec	25	EST	MEYLAN,WM & HOWARD,PH (1993)

CAS #: 000363-72-4			PENTAFLUOROBENZENE

Formula: C_6HF_5
Mol Weight: 168.07
MP (deg C): -47.3 | FP (deg C):
BP (deg C): 85.7
BP pressure (mm Hg):

Property/ Value	Units	Temp	Data Type	Reference
WS 4.34E+002	mg/L	25	EST	MEYLAN,WM ET AL. (1996)
logP 2.53			EXP	HANSCH,C ET AL. (1995)
VP 6.76E+001	mm Hg	25	EXP	BOUBLIK,T ET AL. (1984)
DC	pKa			
HL 1.17E-002	atm m3/mol	25	EST	MEYLAN,WM & HOWARD,PH (1991)
OH 4.37E-013	cm3/molc sec	25	EST	MEYLAN,WM & HOWARD,PH (1993)

CAS #: 000364-62-5 — METOCLOPRAMIDE

Formula: $C_{14}H_{22}ClN_3O_2$

Mol Weight: 299.80

MP (deg C): 146.5-148

FP (deg C):

BP (deg C):

BP pressure (mm Hg):

Property/Value	Units	Temp	Data Type	Reference
WS 7.13E+001	mg/L	25	EST	MEYLAN,WM ET AL. (1996)
logP 2.62			EXP	HANSCH,C & LEO,AJ (1985)
VP 4.61E-009	mm Hg	25	EST	NEELY,WB & BLAU,GE (1985)
DC	pKa			
HL 8.74E-017	atm m3/mol	25	EST	MEYLAN,WM & HOWARD,PH (1991)
OH 3.05E-010	cm3/molc sec	25	EST	MEYLAN,WM & HOWARD,PH (1993)

CAS #: 000364-74-9 — BENZENE, 1,4-DIFLUORO-2-NITRO-

Formula: $C_6H_3F_2NO_2$

Mol Weight: 159.09

MP (deg C): -11.7

FP (deg C):

BP (deg C): 206.5

BP pressure (mm Hg):

Property/Value	Units	Temp	Data Type	Reference
WS 7.26E+002	mg/L	25	EST	MEYLAN,WM ET AL. (1996)
logP 1.86			EXP	SANGSTER,J (1993)
VP 3.50E-001	mm Hg	25	EST	NEELY,WB & BLAU,GE (1985)
DC	pKa			
HL 2.90E-005	atm m3/mol	25	EST	MEYLAN,WM & HOWARD,PH (1991)
OH 1.03E-013	cm3/molc sec	25	EST	MEYLAN,WM & HOWARD,PH (1993)

CAS #: 000364-98-7 — 124-BENZTHIDIAZN-1,1-O2-3-ME-7-CL

Formula: $C_8H_7ClN_2O_2S$

Mol Weight: 230.67

MP (deg C): 330-331

FP (deg C):

BP (deg C):

BP pressure (mm Hg):

Property/Value	Units	Temp	Data Type	Reference
WS 2.85E+003	mg/L	25	EST	MEYLAN,WM ET AL. (1996)
logP 1.20			EXP	HANSCH,C & LEO,AJ (1985)
VP 2.88E-008	mm Hg	25	EST	NEELY,WB & BLAU,GE (1985)
DC	pKa			
HL 6.85E-008	atm m3/mol	25	EST	MEYLAN,WM & HOWARD,PH (1991)
OH 4.89E-013	cm3/molc sec	25	EST	MEYLAN,WM & HOWARD,PH (1993)

CAS #: 000366-18-7 — 2,2-BIPYRIDINE

Formula: $C_{10}H_8N_2$

Mol Weight: 156.19

MP (deg C): 69.7

FP (deg C):

BP (deg C): 273

BP pressure (mm Hg):

Property/Value	Units	Temp	Data Type	Reference
WS 5.00E+003	mg/L	25	EXP	DEAN,JA (1985)
logP 1.50			EXP	DE VOOGT,P ET AL. (1990)
VP 1.30E-005	mm Hg	25	EST	WSOL/HL
DC 4.33	pKa	25	EXP	PERRIN,DD (1964)
HL 5.35E-010	atm m3/mol	25	EST	HINE,J & MOOKERJEE,PR (1975)
OH 1.29E-012	cm3/molc sec	25	EST	MEYLAN,WM & HOWARD,PH (1993)

CAS #: 000366-29-0 — N,N,N',N'-TETRAMETHYLBENZIDINE

Formula: $C_{16}H_{20}N_2$

Mol Weight: 240.35

MP (deg C): 193-195

FP (deg C):

BP (deg C):

BP pressure (mm Hg):

Property/Value	Units	Temp	Data Type	Reference
WS 8.23E+000	mg/L	25	EST	MEYLAN,WM ET AL. (1996)
logP 4.11			EST	MEYLAN,WM & HOWARD,PH (1995)
VP 1.63E-005	mm Hg	25	EST	NEELY,WB & BLAU,GE (1985)
DC	pKa			
HL 1.05E-007	atm m3/mol	25	EST	MEYLAN,WM & HOWARD,PH (1991)
OH 2.05E-010	cm3/molc sec	25	EST	MEYLAN,WM & HOWARD,PH (1993)

CAS #: 000367-11-3 — 1,2-DIFLUOROBENZENE

Formula: $C_6H_4F_2$

Mol Weight: 114.10

MP (deg C): -34

FP (deg C):

BP (deg C): 94

BP pressure (mm Hg):

Property/Value	Units	Temp	Data Type	Reference
WS 1.14E+004	mg/L	25	EXP	MUELLER,M & KLEIN,W (1992)
logP 2.37			EXP	HANSCH,C ET AL. (1995)
VP 5.34E+001	mm Hg	25	EXT	BOUBLIK,T ET AL. (1984)
DC	pKa			
HL 7.34E-003	atm m3/mol	25	EST	MEYLAN,WM & HOWARD,PH (1991)
OH 8.24E-013	cm3/molc sec	25	EST	MEYLAN,WM & HOWARD,PH (1993)

CAS #: 000367-12-4 — O-FLUOROPHENOL

Formula: C_6H_5FO

Mol Weight: 112.10

MP (deg C): 16.1

FP (deg C):

BP (deg C): 151.5

BP pressure (mm Hg):

Property/Value	Units	Temp	Data Type	Reference
WS 1.41E+004	mg/L	25	EST	MEYLAN,WM ET AL. (1996)
logP 1.71			EXP	HANSCH,C & LEO,AJ (1985)
VP 1.19E+000	mm Hg	25	EST	NEELY,WB & BLAU,GE (1985)
DC 8.70	pKa		EXP	KORTUM,G ET AL (1961)
HL 6.54E-007	atm m3/mol	25	EST	MEYLAN,WM & HOWARD,PH (1991)
OH 1.14E-011	cm3/molc sec	25	EST	MEYLAN,WM & HOWARD,PH (1993)

CAS #: 000367-21-5 — 3-CHLORO-4-FLUOROANILINE

Formula: C_6H_5ClFN

Mol Weight: 145.56

MP (deg C): 44-47

FP (deg C):

BP (deg C): 227-228

BP pressure (mm Hg):

Property/Value	Units	Temp	Data Type	Reference
WS 1.38E+003	mg/L	25	EST	MEYLAN,WM ET AL. (1996)
logP 2.06			EXP	HANSCH,C & LEO,AJ (1985)
VP 1.69E-001	mm Hg	25	EST	NEELY,WB & BLAU,GE (1985)
DC 3.66	pKa	25	EXP	PERRIN,DD (1972)
HL 1.65E-006	atm m3/mol	25	EST	MEYLAN,WM & HOWARD,PH (1991)
OH 2.56E-011	cm3/molc sec	25	EST	MEYLAN,WM & HOWARD,PH (1993)

CAS #: 000367-23-7 — BENZENE, 1,2,4-TRIFLUORO-

Formula:	$C_6H_3F_3$			
Mol Weight:	132.09			
MP (deg C):		FP (deg C):		
BP (deg C):	90			
BP pressure (mm Hg):				

Property/ Value	Units	Temp	Data Type	Reference
WS 6.36E+002	mg/L	25	EST	MEYLAN,WM ET AL. (1996)
logP 2.52			EXP	SANGSTER,J (1993)
VP 6.94E+001	mm Hg	25	EST	NEELY,WB & BLAU,GE (1985)
DC	pKa			
HL 8.56E-003	atm m3/mol	25	EST	MEYLAN,WM & HOWARD,PH (1991)
OH 1.03E-012	cm3/molc sec	25	EST	MEYLAN,WM & HOWARD,PH (1993)

CAS #: 000367-25-9 — 2,4-DIFLUOROANILINE

Formula:	$C_6H_5F_2N$			
Mol Weight:	129.11			
MP (deg C):	-7.5	FP (deg C):		
BP (deg C):	170			
BP pressure (mm Hg):				

Property/ Value	Units	Temp	Data Type	Reference
WS 4.49E+003	mg/L	25	EST	MEYLAN,WM ET AL. (1996)
logP 1.54			EXP	HANSCH,C & LEO,AJ (1985)
VP 1.27E+000	mm Hg	25	EST	NEELY,WB & BLAU,GE (1985)
DC	pKa			
HL 2.59E-006	atm m3/mol	25	EST	MEYLAN,WM & HOWARD,PH (1991)
OH 1.23E-011	cm3/molc sec	25	EST	MEYLAN,WM & HOWARD,PH (1993)

CAS #: 000367-49-7 — A-FLUORO-N-METHYLACETAMIDE

Formula:	C_3H_6FNO			
Mol Weight:	91.09			
MP (deg C):		FP (deg C):		
BP (deg C):				
BP pressure (mm Hg):				

Property/ Value	Units	Temp	Data Type	Reference
WS 6.33E+005	mg/L	25	EST	MEYLAN,WM ET AL. (1996)
logP -0.83			EXP	HANSCH,C ET AL. (1995)
VP 4.96E-001	mm Hg	25	EST	NEELY,WB & BLAU,GE (1985)
DC	pKa			
HL 4.89E-008	atm m3/mol	25	EST	MEYLAN,WM & HOWARD,PH (1991)
OH 6.12E-012	cm3/molc sec	25	EST	MEYLAN,WM & HOWARD,PH (1993)

CAS #: 000367-93-1 — 1-ISOPROPYLTHIO-B-GALACTOPYRANSIDE

Formula:	$C_9H_{18}O_5S$			
Mol Weight:	238.30			
MP (deg C):	120-122	FP (deg C):		
BP (deg C):				
BP pressure (mm Hg):				

Property/ Value	Units	Temp	Data Type	Reference
WS 3.27E+005	mg/L	25	EST	MEYLAN,WM ET AL. (1996)
logP -1.26			EXP	HANSCH,C & LEO,AJ (1985)
VP 1.65E-009	mm Hg	25	EST	NEELY,WB & BLAU,GE (1985)
DC	pKa			
HL 5.00E-015	atm m3/mol	25	EST	MEYLAN,WM & HOWARD,PH (1991)
OH 1.84E-010	cm3/molc sec	25	EST	MEYLAN,WM & HOWARD,PH (1993)

CAS #: 000369-24-4 — 2-FLUOROISONIAZID

Formula:	$C_6H_6FN_3O$			
Mol Weight:	155.13			
MP (deg C):		FP (deg C):		
BP (deg C):				
BP pressure (mm Hg):				

Property/ Value	Units	Temp	Data Type	Reference
WS 4.38E+003	mg/L	25	EST	MEYLAN,WM ET AL. (1996)
logP -0.11			EXP	HANSCH,C & LEO,AJ (1985)
VP 5.27E-005	mm Hg	25	EST	NEELY,WB & BLAU,GE (1985)
DC	pKa			
HL 1.41E-014	atm m3/mol	25	EST	MEYLAN,WM & HOWARD,PH (1991)
OH 5.70E-012	cm3/molc sec	25	EST	MEYLAN,WM & HOWARD,PH (1993)

CAS #: 000369-90-4 — N-(4-NITROPHENYL)-3-(TRIFLUOROMETHYL)BENZENAMINE

Formula:	$C_{13}H_9F_3N_2O_2$			
Mol Weight:	282.22			
MP (deg C):		FP (deg C):		
BP (deg C):				
BP pressure (mm Hg):				

Property/ Value	Units	Temp	Data Type	Reference
WS 1.66E+000	mg/L	25	EST	MEYLAN,WM ET AL. (1996)
logP 4.65			EST	MEYLAN,WM & HOWARD,PH (1995)
VP 1.11E-005	mm Hg	25	EST	NEELY,WB & BLAU,GE (1985)
DC	pKa			
HL 3.60E-008	atm m3/mol	25	EST	MEYLAN,WM & HOWARD,PH (1991)
OH 4.05E-011	cm3/molc sec	25	EST	MEYLAN,WM & HOWARD,PH (1993)

CAS #: 000370-09-2 — 3-PYRIDYLMETHYL-N-PYRROLIDINE

Formula:	$C_{10}H_{14}N_2$			
Mol Weight:	162.24			
MP (deg C):		FP (deg C):		
BP (deg C):				
BP pressure (mm Hg):				

Property/ Value	Units	Temp	Data Type	Reference
WS 1.00E+006	mg/L	25	EST	MEYLAN,WM ET AL. (1996)
logP 1.10			EXP	HANSCH,C & LEO,AJ (1985)
VP 1.26E-002	mm Hg	25	EST	NEELY,WB & BLAU,GE (1985)
DC	pKa			
HL 3.00E-009	atm m3/mol	25	EST	MEYLAN,WM & HOWARD,PH (1991)
OH 9.50E-011	cm3/molc sec	25	EST	MEYLAN,WM & HOWARD,PH (1993)

CAS #: 000370-86-5 — PROPANEDINITRILE, [[4-(TRIFLUOROMETHOXY)PHENYL]H

Formula:	$C_{10}H_5F_3N_4O$			
Mol Weight:	254.17			
MP (deg C):	174-175 de	FP (deg C):		
BP (deg C):				
BP pressure (mm Hg):				

Property/ Value	Units	Temp	Data Type	Reference
WS 1.61E+001	mg/L	25	EST	MEYLAN,WM ET AL. (1996)
logP 3.68			EXP	STURDIK,E ET AL. (1985)
VP 6.55E-006	mm Hg	25	EST	NEELY,WB & BLAU,GE (1985)
DC	pKa			
HL 2.70E-009	atm m3/mol	25	EST	MEYLAN,WM & HOWARD,PH (1991)
OH 3.69E-011	cm3/molc sec	25	EST	MEYLAN,WM & HOWARD,PH (1993)

319

CAS #: 000371-40-4				4-FLUOROANILINE

Formula: C_6H_6FN

Mol Weight: 111.12

MP (deg C): -1.9 | FP (deg C):

BP (deg C): 188

BP pressure (mm Hg):

Property/Value	Units	Temp	Data Type	Reference
WS 1.12E+004	mg/L	25	EST	MEYLAN,WM ET AL. (1996)
logP 1.15			EXP	HANSCH,C & LEO,AJ (1985)
VP 1.00E+000	mm Hg	25	EST	NEELY,WB & BLAU,GE (1985)
DC 4.65	pKa	25	EXP	PERRIN,DD (1965)
HL 2.22E-006	atm m3/mol	25	EST	MEYLAN,WM & HOWARD,PH (1991)
OH 3.63E-011	cm3/molc sec	25	EST	MEYLAN,WM & HOWARD,PH (1993)

CAS #: 000371-41-5				P-FLUOROPHENOL

Formula: C_6H_5FO

Mol Weight: 112.10

MP (deg C): 48 | FP (deg C):

BP (deg C): 185.5

BP pressure (mm Hg):

Property/Value	Units	Temp	Data Type	Reference
WS 1.25E+004	mg/L	25	EST	MEYLAN,WM ET AL. (1996)
logP 1.77			EXP	HANSCH,C & LEO,AJ (1985)
VP 1.19E+000	mm Hg	25	EST	NEELY,WB & BLAU,GE (1985)
DC 9.91	pKa	25	EXP	KORTUM,G ET AL (1961)
HL 6.54E-007	atm m3/mol	25	EST	MEYLAN,WM & HOWARD,PH (1991)
OH 1.14E-011	cm3/molc sec	25	EST	MEYLAN,WM & HOWARD,PH (1993)

CAS #: 000371-62-0				2-FLUOROETHANOL

Formula: C_2H_5FO

Mol Weight: 64.06

MP (deg C): -26.4 | FP (deg C):

BP (deg C): 103.5

BP pressure (mm Hg):

Property/Value	Units	Temp	Data Type	Reference
WS 1.00E+006	mg/L	25	EST	MEYLAN,WM ET AL. (1996)
logP -0.67			EXP	HANSCH,C & LEO,AJ (1985)
VP 1.03E+002	mm Hg	25	EST	NEELY,WB & BLAU,GE (1985)
DC	pKa			
HL 7.11E-006	atm m3/mol	25	EST	MEYLAN,WM & HOWARD,PH (1991)
OH 2.24E-012	cm3/molc sec	25	EST	MEYLAN,WM & HOWARD,PH (1993)

CAS #: 000371-86-8				ISOPESTOX

Formula: $C_6H_{16}FN_2OP$

Mol Weight: 182.18

MP (deg C): 65 | FP (deg C):

BP (deg C):

BP pressure (mm Hg):

Property/Value	Units	Temp	Data Type	Reference
WS 8.00E+004	mg/L	25	EXP	SHIU,WY ET AL. (1990)
logP 0.29			EST	MEYLAN,WM & HOWARD,PH (1995)
VP 1.05E-001	mm Hg	25	EST	NEELY,WB & BLAU,GE (1985)
DC	pKa			
HL 6.26E-010	atm m3/mol	25	EST	MEYLAN,WM & HOWARD,PH (1991)
OH 9.18E-011	cm3/molc sec	25	EST	MEYLAN,WM & HOWARD,PH (1993)

CAS #: 000372-09-8				CYANOACETIC ACID

Formula: $C_3H_3NO_2$

Mol Weight: 85.06

MP (deg C): 66 | FP (deg C):

BP (deg C): 108

BP pressure (mm Hg): 1.50E+001

Property/Value	Units	Temp	Data Type	Reference
WS 7.68E+005	mg/L	25	EST	MEYLAN,WM ET AL. (1996)
logP -0.76			EXP	HANSCH,C ET AL. (1995)
VP 3.59E+001	mm Hg	25	EXP	DYKYJ,J (1970)
DC 2.45	pKa	25	EXP	KORTUM,G ET AL (1961)
HL 2.27E-010	atm m3/mol	25	EST	MEYLAN,WM & HOWARD,PH (1991)
OH 6.53E-013	cm3/molc sec	25	EST	MEYLAN,WM & HOWARD,PH (1993)

CAS #: 000372-18-9				M-DIFLUOROBENZENE

Formula: $C_6H_4F_2$

Mol Weight: 114.10

MP (deg C): -69 | FP (deg C):

BP (deg C): 82.6

BP pressure (mm Hg):

Property/Value	Units	Temp	Data Type	Reference
WS 1.14E+004	mg/L	25	EXP	SUZUKI,T (1991)
logP 2.21			EXP	DUNN,WJ ET AL. (1983)
VP 8.28E+001	mm Hg	25	EXT	OSBORN,AG & SCOTT,DW (1980)
DC	pKa			
HL 7.34E-003	atm m3/mol	25	EST	MEYLAN,WM & HOWARD,PH (1991)
OH 3.06E-012	cm3/molc sec	25	EST	MEYLAN,WM & HOWARD,PH (1993)

CAS #: 000372-19-0				3-FLUOROANILINE

Formula: C_6H_6FN

Mol Weight: 111.12

MP (deg C): | FP (deg C):

BP (deg C): 188

BP pressure (mm Hg):

Property/Value	Units	Temp	Data Type	Reference
WS 8.37E+003	mg/L	25	EST	MEYLAN,WM ET AL. (1996)
logP 1.30			EXP	HANSCH,C & LEO,AJ (1985)
VP 1.00E+000	mm Hg	25	EST	NEELY,WB & BLAU,GE (1985)
DC 3.50	pKa	25	EXP	PERRIN,DD (1965)
HL 2.22E-006	atm m3/mol	25	EST	MEYLAN,WM & HOWARD,PH (1991)
OH 1.35E-010	cm3/molc sec	25	EST	MEYLAN,WM & HOWARD,PH (1993)

CAS #: 000372-20-3				M-FLUOROPHENOL

Formula: C_6H_5FO

Mol Weight: 112.10

MP (deg C): 13.7 | FP (deg C):

BP (deg C): 178

BP pressure (mm Hg):

Property/Value	Units	Temp	Data Type	Reference
WS 9.15E+003	mg/L	25	EST	MEYLAN,WM ET AL. (1996)
logP 1.93			EXP	HANSCH,C ET AL. (1995)
VP 1.19E+000	mm Hg	25	EST	NEELY,WB & BLAU,GE (1985)
DC 9.21	pKa	25	EXP	KORTUM,G ET AL (1961)
HL 6.54E-007	atm m3/mol	25	EST	MEYLAN,WM & HOWARD,PH (1991)
OH 4.19E-011	cm3/molc sec	25	EST	MEYLAN,WM & HOWARD,PH (1993)

CAS #: 000372-38-3		1,3,5-TRIFLUOROBENZENE

Formula: $C_6H_3F_3$
Mol Weight: 132.09
MP (deg C): -5.5
FP (deg C):
BP (deg C): 75.5
BP pressure (mm Hg):

Property/ Value	Units	Temp	Data Type	Reference
WS 5.49E+002	mg/L	25	EST	MEYLAN,WM ET AL. (1996)
logP 2.59			EST	MEYLAN,WM & HOWARD,PH (1995)
VP 1.04E+002	mm Hg	25	EXP	BOUBLIK,T ET AL. (1984)
DC	pKa			
HL 8.56E-003	atm m3/mol	25	EST	MEYLAN,WM & HOWARD,PH (1991)
OH 3.83E-012	cm3/molc sec	25	EST	MEYLAN,WM & HOWARD,PH (1993)

CAS #: 000372-47-4		3-FLUOROPYRIDINE

Formula: C_5H_4FN
Mol Weight: 97.09
MP (deg C):
FP (deg C):
BP (deg C): 107-108
BP pressure (mm Hg):

Property/ Value	Units	Temp	Data Type	Reference
WS 2.62E+004	mg/L	25	EST	MEYLAN,WM ET AL. (1996)
logP 0.77			EXP	HANSCH,C ET AL. (1995)
VP 2.66E+001	mm Hg	25	EST	NEELY,WB & BLAU,GE (1985)
DC 2.97	pKa	25	EXP	PERRIN,DD (1965)
HL 8.23E-006	atm m3/mol	25	EST	MEYLAN,WM & HOWARD,PH (1991)
OH 4.63E-013	cm3/molc sec	25	EST	MEYLAN,WM & HOWARD,PH (1993)

CAS #: 000372-48-5		2-FLUOROPYRIDINE

Formula: C_5H_4FN
Mol Weight: 97.09
MP (deg C):
FP (deg C):
BP (deg C): 126
BP pressure (mm Hg): 7.53E+002

Property/ Value	Units	Temp	Data Type	Reference
WS 2.29E+004	mg/L	25	EST	MEYLAN,WM ET AL. (1996)
logP 0.84			EXP	YAMAGAMI,C ET AL. (1990A)
VP 2.66E+001	mm Hg	25	EST	NEELY,WB & BLAU,GE (1985)
DC -0.44	pKa	25	EXP	PERRIN,DD (1965)
HL 8.23E-006	atm m3/mol	25	EST	MEYLAN,WM & HOWARD,PH (1991)
OH 4.63E-013	cm3/molc sec	25	EST	MEYLAN,WM & HOWARD,PH (1993)

CAS #: 000372-75-8		L-ORNITHINE, N5-(AMINOCARBONYL)-

Formula: $C_6H_{13}N_3O_3$
Mol Weight: 175.19
MP (deg C): 222
FP (deg C):
BP (deg C):
BP pressure (mm Hg):

Property/ Value	Units	Temp	Data Type	Reference
WS 1.28E+004	mg/L	25	EST	MEYLAN,WM ET AL. (1996)
logP -3.19			EXP	SANGSTER,J (1994)
VP 8.87E-009	mm Hg	25	EST	NEELY,WB & BLAU,GE (1985)
DC 2.43	pKa	25	EXP	KORTUM,G ET AL (1961)
HL 2.86E-017	atm m3/mol	25	EST	MEYLAN,WM & HOWARD,PH (1991)
OH 4.77E-011	cm3/molc sec	25	EST	MEYLAN,WM & HOWARD,PH (1993)

CAS #: 000373-14-8		1-FLUOROHEXANE

Formula: $C_6H_{13}F$
Mol Weight: 104.17
MP (deg C): -103
FP (deg C):
BP (deg C): 91.5
BP pressure (mm Hg):

Property/ Value	Units	Temp	Data Type	Reference
WS 1.99E+002	mg/L	25	EST	MEYLAN,WM ET AL. (1996)
logP 3.23			EST	MEYLAN,WM & HOWARD,PH (1995)
VP 5.78E+001	mm Hg	25	EXP	LI,JCM & ROSSINI,FD (1953)
DC	pKa			
HL 6.78E-002	atm m3/mol	25	EST	MEYLAN,WM & HOWARD,PH (1991)
OH 4.95E-012	cm3/molc sec	25	EST	MEYLAN,WM & HOWARD,PH (1993)

CAS #: 000374-01-6		1,1,1-TRIFLUORO-2-PROPANOL

Formula: $C_3H_5F_3O$
Mol Weight: 114.07
MP (deg C):
FP (deg C):
BP (deg C):
BP pressure (mm Hg):

Property/ Value	Units	Temp	Data Type	Reference
WS 8.44E+004	mg/L	25	EST	MEYLAN,WM ET AL. (1996)
logP 0.71			EXP	HANSCH,C & LEO,AJ (1985)
VP 6.09E+001	mm Hg	25	EXP	ROCHESTER,CH & SYMONDS,JR (1973)
DC	pKa			
HL 2.18E-005	atm m3/mol	25	EXP	ROCHESTER,CH & SYMONDS,JR (1973)
OH 6.88E-013	cm3/molc sec	25	EST	MEYLAN,WM & HOWARD,PH (1993)

CAS #: 000374-07-2		1,1-DICHLORO-1,2,2,2-TETRAFLUOROETHANE

Formula: $C_2Cl_2F_4$
Mol Weight: 170.92
MP (deg C): -56.6
FP (deg C):
BP (deg C): 4
BP pressure (mm Hg):

Property/ Value	Units	Temp	Data Type	Reference
WS 2.93E+000	mg/L	25	EXP	SUZUKI,T (1991)
logP 2.78			EST	MEYLAN,WM & HOWARD,PH (1995)
VP 7.60E+002	mm Hg	25	EXP	HINE,J & MOOKERJEE,PK (1975)
DC	pKa			
HL 5.83E+001	atm m3/mol	25	EST	VP/WSOL
OH 0.00E+000	cm3/molc sec	25	EST	MEYLAN,WM & HOWARD,PH (1993)

CAS #: 000375-01-9		2,2,3,3,4,4,-HEPTAFLUOROBUTANOL

Formula: $C_4H_3F_7O$
Mol Weight: 200.06
MP (deg C):
FP (deg C):
BP (deg C): 95
BP pressure (mm Hg):

Property/ Value	Units	Temp	Data Type	Reference
WS 3.12E+003	mg/L	25	EST	MEYLAN,WM ET AL. (1996)
logP 1.94			EXP	HANSCH,C ET AL. (1995)
VP 3.31E+001	mm Hg	25	EXP	BOUBLIK,T ET AL. (1984)
DC	pKa			
HL 7.78E-004	atm m3/mol	25	EST	MEYLAN,WM & HOWARD,PH (1991)
OH 1.99E-013	cm3/molc sec	25	EST	MEYLAN,WM & HOWARD,PH (1993)

321

BETAMETHASONE

CAS #: 000378-44-9

Formula:	$C_{22}H_{29}FO_5$	
Mol Weight:	392.47	
MP (deg C): 231-234	FP (deg C):	
BP (deg C):		
BP pressure (mm Hg):		

Property/Value	Units	Temp	Data Type	Reference
WS 7.51E+001	mg/L	25	EST	MEYLAN,WM ET AL. (1996)
logP 1.94			EXP	HANSCH,C ET AL. (1995)
VP 8.86E-014	mm Hg	25	EST	NEELY,WB & BLAU,GE (1985)
DC	pKa			
HL 7.15E-008	atm m3/mol	25	EST	MEYLAN,WM & HOWARD,PH (1991)
OH 6.87E-011	cm3/molc sec	25	EST	MEYLAN,WM & HOWARD,PH (1993)

BETAMETHASONE-21-DESOXY

CAS #: 000382-67-2

Formula:	$C_{22}H_{29}FO_4$	
Mol Weight:	376.47	
MP (deg C): 217	FP (deg C):	
BP (deg C):		
BP pressure (mm Hg):		

Property/Value	Units	Temp	Data Type	Reference
WS 4.21E+001	mg/L	25	EST	MEYLAN,WM ET AL. (1996)
logP 2.35			EXP	HANSCH,C & LEO,AJ (1985)
VP 3.15E-012	mm Hg	25	EST	NEELY,WB & BLAU,GE (1985)
DC	pKa			
HL 4.59E-010	atm m3/mol	25	EST	MEYLAN,WM & HOWARD,PH (1991)
OH 7.04E-011	cm3/molc sec	25	EST	MEYLAN,WM & HOWARD,PH (1993)

TRIFLUOROETHYL ACETATE

CAS #: 000383-63-1

Formula:	$C_4H_5F_3O_2$	
Mol Weight:	142.08	
MP (deg C):	FP (deg C):	
BP (deg C): 61		
BP pressure (mm Hg):		

Property/Value	Units	Temp	Data Type	Reference
WS 8.06E+003	mg/L	25	EST	MEYLAN,WM ET AL. (1996)
logP 1.18			EXP	HANSCH,C & LEO,AJ (1985)
VP 1.96E+002	mm Hg	25	EST	NEELY,WB & BLAU,GE (1985)
DC	pKa			
HL 1.83E-003	atm m3/mol	25	EST	MEYLAN,WM & HOWARD,PH (1991)
OH 5.04E-013	cm3/molc sec	25	EST	MEYLAN,WM & HOWARD,PH (1993)

1-NO2 2-CF3 BENZENE

CAS #: 000384-22-5

Formula:	$C_7H_4F_3NO_2$	
Mol Weight:	191.11	
MP (deg C): 32.5	FP (deg C):	
BP (deg C): 217		
BP pressure (mm Hg):		

Property/Value	Units	Temp	Data Type	Reference
WS 1.24E+002	mg/L	25	EST	MEYLAN,WM ET AL. (1996)
logP 2.58			EXP	HANSCH,C & LEO,AJ (1985)
VP 1.47E-001	mm Hg	25	EST	NEELY,WB & BLAU,GE (1985)
DC	pKa			
HL 1.85E-004	atm m3/mol	25	EST	MEYLAN,WM & HOWARD,PH (1991)
OH 3.69E-014	cm3/molc sec	25	EST	MEYLAN,WM & HOWARD,PH (1993)

111F3-22-DI(P-MEOPHENYL)ETHANE

CAS #: 000384-97-4

Formula:	$C_{16}H_{15}F_3O_2$	
Mol Weight:	296.29	
MP (deg C):	FP (deg C):	
BP (deg C):		
BP pressure (mm Hg):		

Property/Value	Units	Temp	Data Type	Reference
WS 1.11E+000	mg/L	25	EST	MEYLAN,WM ET AL. (1996)
logP 4.76			EXP	HANSCH,C & LEO,AJ (1985)
VP 1.17E-004	mm Hg	25	EST	NEELY,WB & BLAU,GE (1985)
DC	pKa			
HL 1.76E-005	atm m3/mol	25	EST	MEYLAN,WM & HOWARD,PH (1991)
OH 5.36E-011	cm3/molc sec	25	EST	MEYLAN,WM & HOWARD,PH (1993)

BENZOIC ACID, 2,6-DIFLUORO-

CAS #: 000385-00-2

Formula:	$C_7H_4F_2O_2$	
Mol Weight:	158.11	
MP (deg C): 158-160	FP (deg C):	
BP (deg C):		
BP pressure (mm Hg):		

Property/Value	Units	Temp	Data Type	Reference
WS 3.06E+003	mg/L	25	EST	MEYLAN,WM ET AL. (1996)
logP 1.59			EXP	SOTOMATSU,T ET AL. (1993)
VP 2.57E-002	mm Hg	25	EST	NEELY,WB & BLAU,GE (1985)
DC	pKa			
HL 1.48E-007	atm m3/mol	25	EST	MEYLAN,WM & HOWARD,PH (1991)
OH 1.65E-012	cm3/molc sec	25	EST	MEYLAN,WM & HOWARD,PH (1993)

NALIDIXIC ACID

CAS #: 000389-08-2

Formula:	$C_{12}H_{12}N_2O_3$	
Mol Weight:	232.24	
MP (deg C): 229.5	FP (deg C):	
BP (deg C):		
BP pressure (mm Hg):		

Property/Value	Units	Temp	Data Type	Reference
WS 4.59E+003	mg/L	25	EST	MEYLAN,WM ET AL. (1996)
logP 1.41			EXP	HANSCH,C & LEO,AJ (1985)
VP 3.56E-007	mm Hg	25	EST	NEELY,WB & BLAU,GE (1985)
DC	pKa			
HL 5.12E-016	atm m3/mol	25	EST	MEYLAN,WM & HOWARD,PH (1991)
OH 2.15E-011	cm3/molc sec	25	EST	MEYLAN,WM & HOWARD,PH (1993)

HEXAFLUOROBENZENE

CAS #: 000392-56-3

Formula:	C_6F_6	
Mol Weight:	186.06	
MP (deg C): 5.3	FP (deg C):	
BP (deg C): 80.2		
BP pressure (mm Hg):		

Property/Value	Units	Temp	Data Type	Reference
WS 3.42E+002	mg/L	25	EST	MEYLAN,WM ET AL. (1996)
logP 2.55			EXP	HANSCH,C ET AL. (1995)
VP 8.45E+001	mm Hg	25	EXP	DAUBERT,TE & DANNER,RP (1989)
DC	pKa			
HL 1.36E-002	atm m3/mol	25	EST	MEYLAN,WM & HOWARD,PH (1991)
OH 1.72E-013	cm3/molc sec	25	EXP	ATKINSON,R (1989)

CAS #: 000392-85-8	1-FLUORO-2-TRIFLUOROMETHYLBENZENE

Formula: $C_7H_4F_4$

Mol Weight: 164.10

MP (deg C):

FP (deg C):

BP (deg C): 114.5

BP pressure (mm Hg):

Property/Value	Units	Temp	Data Type	Reference
WS 1.32E+002	mg/L	25	EST	MEYLAN,WM ET AL. (1996)
logP 3.16			EST	MEYLAN,WM & HOWARD,PH (1995)
VP 3.84E+001	mm Hg	25	EXT	BOUBLIK,T ET AL. (1984)
DC	pKa			
HL 5.46E-002	atm m3/mol	25	EST	MEYLAN,WM & HOWARD,PH (1991)
OH 4.91E-013	cm3/molc sec	25	EST	MEYLAN,WM & HOWARD,PH (1993)

CAS #: 000393-75-9	1,3-DINO2 2-CL 5-CF3 BENZENE

Formula: $C_7H_2ClF_3N_2O_4$

Mol Weight: 270.55

MP (deg C): 56-58

FP (deg C):

BP (deg C):

BP pressure (mm Hg):

Property/Value	Units	Temp	Data Type	Reference
WS 5.42E+001	mg/L	25	EST	MEYLAN,WM ET AL. (1996)
logP 2.50			EXP	NAKAGAWA,Y ET AL. (1992)
VP 1.34E-004	mm Hg	25	EST	NEELY,WB & BLAU,GE (1985)
DC	pKa			
HL 5.40E-007	atm m3/mol	25	EST	MEYLAN,WM & HOWARD,PH (1991)
OH 6.60E-016	cm3/molc sec	25	EST	MEYLAN,WM & HOWARD,PH (1993)

CAS #: 000394-35-4	BENZOIC ACID, 2-FLUORO-, METHYL ESTER

Formula: $C_8H_7FO_2$

Mol Weight: 154.14

MP (deg C):

FP (deg C):

BP (deg C):

BP pressure (mm Hg):

Property/Value	Units	Temp	Data Type	Reference
WS 2.03E+003	mg/L	25	EST	MEYLAN,WM ET AL. (1996)
logP 1.82			EXP	SOTOMATSU,T ET AL. (1993)
VP 5.56E-001	mm Hg	25	EST	NEELY,WB & BLAU,GE (1985)
DC	pKa			
HL 4.05E-005	atm m3/mol	25	EST	MEYLAN,WM & HOWARD,PH (1991)
OH 1.00E-012	cm3/molc sec	25	EST	MEYLAN,WM & HOWARD,PH (1993)

CAS #: 000394-68-3	8-FLUOROQUINOLINE

Formula: C_9H_6FN

Mol Weight: 147.15

MP (deg C):

FP (deg C):

BP (deg C):

BP pressure (mm Hg):

Property/Value	Units	Temp	Data Type	Reference
WS 1.05E+003	mg/L	25	EST	MEYLAN,WM ET AL. (1996)
logP 2.19			EXP	HANSCH,C ET AL. (1995)
VP 3.73E-002	mm Hg	25	EST	NEELY,WB & BLAU,GE (1985)
DC 3.34	pKa	25	EXP	PERRIN,DD (1965)
HL 8.03E-007	atm m3/mol	25	EST	MEYLAN,WM & HOWARD,PH (1991)
OH 1.80E-011	cm3/molc sec	25	EST	MEYLAN,WM & HOWARD,PH (1993)

CAS #: 000396-01-0	2,4,7-TRIAMINO-6-PH PTERIDINE

Formula: $C_{12}H_{11}N_7$

Mol Weight: 253.27

MP (deg C): 316

FP (deg C):

BP (deg C):

BP pressure (mm Hg):

Property/Value	Units	Temp	Data Type	Reference
WS 3.30E+003	mg/L	25	EST	MEYLAN,WM ET AL. (1996)
logP 0.98			EXP	HANSCH,C & LEO,AJ (1985)
VP 1.93E-010	mm Hg	25	EST	NEELY,WB & BLAU,GE (1985)
DC	pKa			
HL 1.86E-018	atm m3/mol	25	EST	MEYLAN,WM & HOWARD,PH (1991)
OH 2.03E-010	cm3/molc sec	25	EST	MEYLAN,WM & HOWARD,PH (1993)

CAS #: 000399-31-5	2-FLUOROACETANILIDE

Formula: C_8H_8FNO

Mol Weight: 153.16

MP (deg C): 77-79

FP (deg C):

BP (deg C): 140-142

BP pressure (mm Hg): 1.40E+001

Property/Value	Units	Temp	Data Type	Reference
WS 9.70E+003	mg/L	25	EST	MEYLAN,WM ET AL. (1996)
logP 1.03			EXP	NAKAGAWA,Y ET AL. (1992)
VP 5.97E-004	mm Hg	25	EST	NEELY,WB & BLAU,GE (1985)
DC	pKa			
HL 7.20E-009	atm m3/mol	25	EST	MEYLAN,WM & HOWARD,PH (1991)
OH 4.29E-012	cm3/molc sec	25	EST	MEYLAN,WM & HOWARD,PH (1993)

CAS #: 000400-99-7	4-TRIFLUOROMETHYL-2-NITROPHENOL

Formula: $C_7H_4F_3NO_3$

Mol Weight: 207.11

MP (deg C):

FP (deg C):

BP (deg C): 92-94

BP pressure (mm Hg): 1.20E+001

Property/Value	Units	Temp	Data Type	Reference
WS 4.04E+002	mg/L	25	EST	MEYLAN,WM ET AL. (1996)
logP 2.34			EXP	SCHWARZENBACH,RP ET AL. (1988)
VP 1.33E-003	mm Hg	25	EST	NEELY,WB & BLAU,GE (1985)
DC	pKa			
HL 6.08E-005	atm m3/mol	25	EST	MEYLAN,WM & HOWARD,PH (1991)
OH 9.77E-013	cm3/molc sec	25	EST	MEYLAN,WM & HOWARD,PH (1993)

CAS #: 000402-31-3	BENZENE, 1,3-BIS(TRIFLUOROMETHYL)-

Formula: $C_8H_4F_6$

Mol Weight: 214.11

MP (deg C):

FP (deg C):

BP (deg C): 116

BP pressure (mm Hg):

Property/Value	Units	Temp	Data Type	Reference
WS 1.98E+001	mg/L	25	EST	MEYLAN,WM ET AL. (1996)
logP 3.83			EXP	HANSCH,C ET AL. (1995)
VP 1.85E+001	mm Hg	25	EST	NEELY,WB & BLAU,GE (1985)
DC	pKa			
HL 4.07E-001	atm m3/mol	25	EST	MEYLAN,WM & HOWARD,PH (1991)
OH 7.88E-014	cm3/molc sec	25	EST	MEYLAN,WM & HOWARD,PH (1993)

| CAS #: 000402-38-0 | | | | 3,3-DIME-1-(3-CF3 PH)TRIAZENE |

Formula: $C_9H_{10}F_3N_3$
Mol Weight: 217.20
MP (deg C):
FP (deg C):
BP (deg C):
BP pressure (mm Hg):

Property/Value	Units	Temp	Data Type	Reference
WS 2.11E+001	mg/L	25	EST	MEYLAN,WM ET AL. (1996)
logP 3.78			EXP	HANSCH,C & LEO,AJ (1985)
VP 9.20E-002	mm Hg	25	EST	NEELY,WB & BLAU,GE (1985)
DC	pKa			
HL 4.66E-006	atm m3/mol	25	EST	MEYLAN,WM & HOWARD,PH (1991)
OH 2.68E-012	cm3/molc sec	25	EST	MEYLAN,WM & HOWARD,PH (1993)

| CAS #: 000402-45-9 | | | | P-TRIFLUOROMETHYLPHENOL |

Formula: $C_7H_5F_3O$
Mol Weight: 162.11
MP (deg C): 45-47
FP (deg C):
BP (deg C):
BP pressure (mm Hg):

Property/Value	Units	Temp	Data Type	Reference
WS 9.94E+002	mg/L	25	EST	MEYLAN,WM ET AL. (1996)
logP 2.82			EXP	HANSCH,C ET AL. (1995)
VP 6.68E-001	mm Hg	25	EST	NEELY,WB & BLAU,GE (1985)
DC 8.68	pKa	25	EXP	SERJEANT,EP & DEMPSEY,B (1979)
HL 4.87E-006	atm m3/mol	25	EST	MEYLAN,WM & HOWARD,PH (1991)
OH 6.84E-012	cm3/molc sec	25	EST	MEYLAN,WM & HOWARD,PH (1993)

| CAS #: 000402-54-0 | | | | BENZENE, 1-NITRO-4-(TRIFLUOROMETHYL)- |

Formula: $C_7H_4F_3NO_2$
Mol Weight: 191.11
MP (deg C): 38-40
FP (deg C):
BP (deg C): 81-83
BP pressure (mm Hg): 1.00E+001

Property/Value	Units	Temp	Data Type	Reference
WS 1.31E+002	mg/L	25	EST	MEYLAN,WM ET AL. (1996)
logP 2.55			EXP	HANSCH,C ET AL. (1995)
VP 1.47E-001	mm Hg	25	EST	NEELY,WB & BLAU,GE (1985)
DC	pKa			
HL 1.85E-004	atm m3/mol	25	EST	MEYLAN,WM & HOWARD,PH (1991)
OH 3.69E-014	cm3/molc sec	25	EST	MEYLAN,WM & HOWARD,PH (1993)

| CAS #: 000402-67-5 | | | | 3-FLUORONITROBENZENE |

Formula: $C_6H_4FNO_2$
Mol Weight: 141.10
MP (deg C): 41
FP (deg C):
BP (deg C): 199
BP pressure (mm Hg):

Property/Value	Units	Temp	Data Type	Reference
WS 8.05E+002	mg/L	25	EST	MEYLAN,WM ET AL. (1996)
logP 1.90			EXP	HANSCH,C & LEO,AJ (1985)
VP 2.94E-001	mm Hg	25	EST	NEELY,WB & BLAU,GE (1985)
DC	pKa			
HL 2.48E-005	atm m3/mol	25	EST	MEYLAN,WM & HOWARD,PH (1991)
OH 2.13E-013	cm3/molc sec	25	EST	MEYLAN,WM & HOWARD,PH (1993)

| CAS #: 000403-19-0 | | | | 2-FLUORO-4-NITROPHENOL |

Formula: $C_6H_4FNO_3$
Mol Weight: 157.10
MP (deg C): 120-122
FP (deg C):
BP (deg C):
BP pressure (mm Hg):

Property/Value	Units	Temp	Data Type	Reference
WS 4.24E+003	mg/L	25	EST	MEYLAN,WM ET AL. (1996)
logP 2.11			EST	MEYLAN,WM & HOWARD,PH (1995)
VP 2.60E-003	mm Hg	25	EST	NEELY,WB & BLAU,GE (1985)
DC	pKa			
HL 2.58E-009	atm m3/mol	25	EST	MEYLAN,WM & HOWARD,PH (1991)
OH 1.55E-012	cm3/molc sec	25	EST	MEYLAN,WM & HOWARD,PH (1993)

| CAS #: 000403-33-8 | | | | BENZOIC ACID, 4-FLUORO-, METHYL ESTER |

Formula: $C_8H_7FO_2$
Mol Weight: 154.14
MP (deg C):
FP (deg C):
BP (deg C):
BP pressure (mm Hg):

Property/Value	Units	Temp	Data Type	Reference
WS 8.22E+002	mg/L	25	EST	MEYLAN,WM ET AL. (1996)
logP 2.28			EXP	SOTOMATSU,T ET AL. (1993)
VP 5.56E-001	mm Hg	25	EST	NEELY,WB & BLAU,GE (1985)
DC	pKa			
HL 4.05E-005	atm m3/mol	25	EST	MEYLAN,WM & HOWARD,PH (1991)
OH 1.00E-012	cm3/molc sec	25	EST	MEYLAN,WM & HOWARD,PH (1993)

| CAS #: 000403-42-9 | | | | 4-FLUOROACETOPHENONE |

Formula: C_8H_7FO
Mol Weight: 138.14
MP (deg C): -45
FP (deg C):
BP (deg C): 196
BP pressure (mm Hg):

Property/Value	Units	Temp	Data Type	Reference
WS 2.90E+003	mg/L	25	EST	MEYLAN,WM ET AL. (1996)
logP 1.72			EXP	HANSCH,C & LEO,AJ (1985)
VP 7.53E-001	mm Hg	25	EST	NEELY,WB & BLAU,GE (1985)
DC	pKa			
HL 1.14E-005	atm m3/mol	25	EST	MEYLAN,WM & HOWARD,PH (1991)
OH 2.33E-012	cm3/molc sec	25	EST	MEYLAN,WM & HOWARD,PH (1993)

| CAS #: 000403-54-3 | | | | BENZONITRILE, 3-FLUORO- |

Formula: C_7H_4FN
Mol Weight: 121.12
MP (deg C): -16
FP (deg C):
BP (deg C): 182-183
BP pressure (mm Hg): 7.53E+002

Property/Value	Units	Temp	Data Type	Reference
WS 3.00E+003	mg/L	25	EST	MEYLAN,WM ET AL. (1996)
logP 1.47			EXP	HANSCH,C ET AL. (1995)
VP 6.96E-001	mm Hg	25	EST	NEELY,WB & BLAU,GE (1985)
DC	pKa			
HL 6.07E-005	atm m3/mol	25	EST	MEYLAN,WM & HOWARD,PH (1991)
OH 3.20E-013	cm3/molc sec	25	EST	MEYLAN,WM & HOWARD,PH (1993)

CAS #: 000404-24-0				TRIFLUOROACETANILIDE

Formula: $C_8H_6F_3NO$
Mol Weight: 189.14
MP (deg C): | FP (deg C):
BP (deg C):
BP pressure (mm Hg):

Property/ Value	Units	Temp	Data Type	Reference
WS 6.43E+002	mg/L	25	EST	MEYLAN,WM ET AL. (1996)
logP 2.21			EXP	HANSCH,C & LEO,AJ (1985)
VP 9.88E-004	mm Hg	25	EST	NEELY,WB & BLAU,GE (1985)
DC	pKa			
HL 4.86E-008	atm m3/mol	25	EST	MEYLAN,WM & HOWARD,PH (1991)
OH 1.24E-011	cm3/molc sec	25	EST	MEYLAN,WM & HOWARD,PH (1993)

CAS #: 000404-98-8				M-FLUOROPHENOXYACETIC ACID

Formula: $C_8H_7FO_3$
Mol Weight: 170.14
MP (deg C): | FP (deg C):
BP (deg C):
BP pressure (mm Hg):

Property/ Value	Units	Temp	Data Type	Reference
WS 6.06E+003	mg/L	25	EST	MEYLAN,WM ET AL. (1996)
logP 1.48			EXP	HANSCH,C & LEO,AJ (1985)
VP 1.86E-003	mm Hg	25	EST	NEELY,WB & BLAU,GE (1985)
DC 3.13	pKa	25	EXP	KORTUM,G ET AL (1961)
HL 1.96E-008	atm m3/mol	25	EST	MEYLAN,WM & HOWARD,PH (1991)
OH 1.21E-011	cm3/molc sec	25	EST	MEYLAN,WM & HOWARD,PH (1993)

CAS #: 000405-30-1				FLUORBENSIDE

Formula: $C_{13}H_{10}ClFS$
Mol Weight: 252.74
MP (deg C): | FP (deg C):
BP (deg C):
BP pressure (mm Hg):

Property/ Value	Units	Temp	Data Type	Reference
WS 9.20E-001	mg/L	25	EST	MEYLAN,WM ET AL. (1996)
logP 5.15			EST	MEYLAN,WM & HOWARD,PH (1995)
VP 8.28E-005	mm Hg	25	EST	NEELY,WB & BLAU,GE (1985)
DC	pKa			
HL 1.10E-005	atm m3/mol	25	EST	MEYLAN,WM & HOWARD,PH (1991)
OH 1.32E-011	cm3/molc sec	25	EST	MEYLAN,WM & HOWARD,PH (1993)

CAS #: 000405-50-5				P-FLUOROPHENYLACETIC ACID

Formula: $C_8H_7FO_2$
Mol Weight: 154.14
MP (deg C): 86 | FP (deg C):
BP (deg C): 164
BP pressure (mm Hg): 2.00E+000

Property/ Value	Units	Temp	Data Type	Reference
WS 8.57E+003	mg/L	25	EST	MEYLAN,WM ET AL. (1996)
logP 1.55			EXP	HANSCH,C & LEO,AJ (1985)
VP 6.22E-003	mm Hg	25	EST	NEELY,WB & BLAU,GE (1985)
DC 4.24	pKa	25	EXP	KORTUM,G ET AL (1961)
HL 5.16E-008	atm m3/mol	25	EST	MEYLAN,WM & HOWARD,PH (1991)
OH 3.74E-012	cm3/molc sec	25	EST	MEYLAN,WM & HOWARD,PH (1993)

CAS #: 000405-79-8				P-FLUOROPHENOXYACETIC ACID

Formula: $C_8H_7FO_3$
Mol Weight: 170.14
MP (deg C): 104-104.5 | FP (deg C):
BP (deg C):
BP pressure (mm Hg):

Property/ Value	Units	Temp	Data Type	Reference
WS 6.06E+003	mg/L	25	EST	MEYLAN,WM ET AL. (1996)
logP 1.64			EXP	HANSCH,C ET AL. (1995)
VP 1.86E-003	mm Hg	25	EST	NEELY,WB & BLAU,GE (1985)
DC 3.13	pKa	25	EXP	KORTUM,G ET AL (1961)
HL 1.96E-008	atm m3/mol	25	EST	MEYLAN,WM & HOWARD,PH (1991)
OH 1.21E-011	cm3/molc sec	25	EST	MEYLAN,WM & HOWARD,PH (1993)

CAS #: 000406-90-6				FLUROXENE

Formula: $C_4H_5F_3O$
Mol Weight: 126.08
MP (deg C): | FP (deg C):
BP (deg C): 42.5
BP pressure (mm Hg): 7.51E+002

Property/ Value	Units	Temp	Data Type	Reference
WS 4.19E+003	mg/L	37	EXP	YALKOWSKY,SH & DANNENFELSER,RM (1992)
logP 1.33			EST	MEYLAN,WM & HOWARD,PH (1995)
VP 7.21E+002	mm Hg	25	EST	NEELY,WB & BLAU,GE (1985)
DC	pKa			
HL 5.99E-002	atm m3/mol	25	EST	MEYLAN,WM & HOWARD,PH (1991)
OH 3.46E-011	cm3/molc sec	25	EST	MEYLAN,WM & HOWARD,PH (1993)

CAS #: 000407-25-0				TRIFLUOROACETIC ANHYDRIDE

Formula: $C_4F_6O_3$
Mol Weight: 210.03
MP (deg C): -65 | FP (deg C):
BP (deg C): 39.5
BP pressure (mm Hg):

Property/ Value	Units	Temp	Data Type	Reference
WS 2.36E+004	mg/L	25	EST	MEYLAN,WM ET AL. (1996)
logP 0.25			EST	MEYLAN,WM & HOWARD,PH (1995)
VP 4.12E+001	mm Hg	25	EXP	BOUBLIK,T ET AL. (1984)
DC	pKa			
HL 2.21E-003	atm m3/mol	25	EST	MEYLAN,WM & HOWARD,PH (1991)
OH 0.00E+000	cm3/molc sec	25	EST	MEYLAN,WM & HOWARD,PH (1993)

CAS #: 000408-35-5				HEXADECANOIC ACID, SODIUM SALT

Formula: $C_{16}H_{32}NaO_2$
Mol Weight: 279.42
MP (deg C): 283-290 | FP (deg C):
BP (deg C):
BP pressure (mm Hg):

Property/ Value	Units	Temp	Data Type	Reference
WS 3.33E+001	mg/L	25	EST	MEYLAN,WM ET AL. (1996)
logP 3.15			EST	MEYLAN,WM & HOWARD,PH (1995)
VP 5.46E-012	mm Hg	25	EST	NEELY,WB & BLAU,GE (1985)
DC	pKa			
HL	atm m3/mol			
OH 1.86E-011	cm3/molc sec	25	EST	MEYLAN,WM & HOWARD,PH (1993)

CAS #: 000420-12-2				ETHYLENE SULFIDE

Formula: C_2H_4S

Mol Weight: 60.12

MP (deg C):   FP (deg C):

BP (deg C): 55-56

BP pressure (mm Hg):

Property/ Value	Units	Temp	Data Type	Reference
WS 2.81E+004	mg/L	25	EST	MEYLAN,WM ET AL. (1996)
logP 0.81			EST	MEYLAN,WM & HOWARD,PH (1995)
VP 2.50E+002	mm Hg	25	EXP	ZWOLINSKI,BJ & WILHOIT,RC (1971)
DC	pKa			
HL 3.46E-004	atm m3/mol	25	EST	MEYLAN,WM & HOWARD,PH (1991)
OH 2.06E-012	cm3/molc sec	25	EST	MEYLAN,WM & HOWARD,PH (1993)

CAS #: 000420-46-2				1,1,1-TRIFLUOROETHANE

Formula: $C_2H_3F_3$

Mol Weight: 84.04

MP (deg C): -111.3   FP (deg C):

BP (deg C): -47.5

BP pressure (mm Hg):

Property/ Value	Units	Temp	Data Type	Reference
WS 7.61E+002	mg/L	25	EST	MEYLAN,WM ET AL. (1996)
logP 1.74			EST	MEYLAN,WM & HOWARD,PH (1995)
VP 9.54E+003	mm Hg	25	EXP	DAUBERT,TE & DANNER,RP (1989)
DC	pKa			
HL 7.70E-001	atm m3/mol	25	EST	MEYLAN,WM & HOWARD,PH (1991)
OH 1.70E-015	cm3/molc sec	25	EXP	ATKINSON,R (1989)

CAS #: 000421-50-1				1,1,1-TRIFLUOROACETONE

Formula: $C_3H_3F_3O$

Mol Weight: 112.05

MP (deg C):   FP (deg C):

BP (deg C): 22

BP pressure (mm Hg):

Property/ Value	Units	Temp	Data Type	Reference
WS 7.23E+004	mg/L	25	EST	MEYLAN,WM ET AL. (1996)
logP 0.20			EXP	HANSCH,C & LEO,AJ (1985)
VP 7.28E+002	mm Hg	25	EST	NEELY,WB & BLAU,GE (1985)
DC	pKa			
HL 3.90E-004	atm m3/mol	25	EST	MEYLAN,WM & HOWARD,PH (1991)
OH 1.51E-014	cm3/molc sec	25	EXP	ATKINSON,R (1989)

CAS #: 000422-05-9				PENTAFLUORO-1-PROPANOL

Formula: $C_3H_3F_5O$

Mol Weight: 150.05

MP (deg C):   FP (deg C):

BP (deg C): 80

BP pressure (mm Hg): 7.48E+002

Property/ Value	Units	Temp	Data Type	Reference
WS 2.19E+004	mg/L	25	EST	MEYLAN,WM ET AL. (1996)
logP 1.23			EXP	HANSCH,C & LEO,AJ (1985)
VP 4.61E+001	mm Hg	25	EXT	BOUBLIK,T ET AL. (1984)
DC	pKa			
HL 2.23E-005	atm m3/mol	25	EXP	ROCHESTER,CH & SYMONDS,JR (1973)
OH 3.71E-012	cm3/molc sec	25	EST	MEYLAN,WM & HOWARD,PH (1993)

CAS #: 000425-32-1				BENZENE, 1,1'-(1,1,2,2-TETRAFLUORO-1,2-ETHANEDIY

Formula: $C_{14}H_{10}F_4$

Mol Weight: 254.23

MP (deg C):   FP (deg C):

BP (deg C):

BP pressure (mm Hg):

Property/ Value	Units	Temp	Data Type	Reference
WS 1.20E+000	mg/L	25	EST	MEYLAN,WM ET AL. (1996)
logP 5.00			EXP	SANGSTER,J (1993)
VP 1.34E-002	mm Hg	25	EST	NEELY,WB & BLAU,GE (1985)
DC	pKa			
HL 9.98E-003	atm m3/mol	25	EST	MEYLAN,WM & HOWARD,PH (1991)
OH 8.59E-012	cm3/molc sec	25	EST	MEYLAN,WM & HOWARD,PH (1993)

CAS #: 000426-13-1				FLUOROMETHALONE

Formula: $C_{22}H_{29}FO_4$

Mol Weight: 376.47

MP (deg C): 292-303   FP (deg C):

BP (deg C):

BP pressure (mm Hg):

Property/ Value	Units	Temp	Data Type	Reference
WS 3.00E+001	mg/L	25	EXP	YALKOWSKY,SH & DANNENFELSER,RM (1992)
logP 2.00			EXP	HANSCH,C & LEO,AJ (1985)
VP 1.16E-011	mm Hg	25	EST	NEELY,WB & BLAU,GE (1985)
DC	pKa			
HL 4.59E-010	atm m3/mol	25	EST	MEYLAN,WM & HOWARD,PH (1991)
OH 6.63E-011	cm3/molc sec	25	EST	MEYLAN,WM & HOWARD,PH (1993)

CAS #: 000426-58-4				SULFONE, PHENYL, METHYL-

Formula: $C_7H_8O_2S$

Mol Weight: 156.20

MP (deg C):   FP (deg C):

BP (deg C):

BP pressure (mm Hg):

Property/ Value	Units	Temp	Data Type	Reference
WS 2.66E+004	mg/L	25	EST	MEYLAN,WM ET AL. (1996)
logP 0.50			EXP	HANSCH,C & LEO,AJ (1985)
VP 5.83E-003	mm Hg	25	EST	NEELY,WB & BLAU,GE (1985)
DC	pKa			
HL 1.25E-006	atm m3/mol	25	EST	MEYLAN,WM & HOWARD,PH (1991)
OH 1.48E-012	cm3/molc sec	25	EST	MEYLAN,WM & HOWARD,PH (1993)

CAS #: 000426-65-3				ETHYL PENTAFLUOROPROPIONATE

Formula: $C_5H_5F_5O_2$

Mol Weight: 192.09

MP (deg C): 75-76   FP (deg C):

BP (deg C):

BP pressure (mm Hg):

Property/ Value	Units	Temp	Data Type	Reference
WS 7.42E+002	mg/L	25	EST	MEYLAN,WM ET AL. (1996)
logP 2.12			EXP	HANSCH,C & LEO,AJ (1985)
VP 1.60E+002	mm Hg	25	EST	NEELY,WB & BLAU,GE (1985)
DC	pKa			
HL 9.63E-003	atm m3/mol	25	EST	MEYLAN,WM & HOWARD,PH (1991)
OH 9.14E-013	cm3/molc sec	25	EST	MEYLAN,WM & HOWARD,PH (1993)

CAS #: 000430-51-3				FLUOROACETONE

Formula: C_3H_5FO

Mol Weight: 76.07

MP (deg C): | FP (deg C):

BP (deg C): 77

BP pressure (mm Hg):

Property/ Value	Units	Temp	Data Type	Reference
WS 2.86E+005	mg/L	25	EST	MEYLAN,WM ET AL. (1996)
logP -0.39			EXP	HANSCH,C & LEO,AJ (1985)
VP 6.34E+001	mm Hg	25	EXP	DYKYJ,J (1970)
DC	pKa			
HL 9.86E-005	atm m3/mol	25	EST	MEYLAN,WM & HOWARD,PH (1991)
OH 1.68E-013	cm3/molc sec	25	EST	MEYLAN,WM & HOWARD,PH (1993)

CAS #: 000430-66-0				1,1,2-TRIFLUOROETHANE

Formula: $C_2H_3F_3$

Mol Weight: 84.04

MP (deg C): -84 | FP (deg C):

BP (deg C): 5

BP pressure (mm Hg):

Property/ Value	Units	Temp	Data Type	Reference
WS 2.83E+003	mg/L	25	EST	MEYLAN,WM ET AL. (1996)
logP 1.07			EST	MEYLAN,WM & HOWARD,PH (1995)
VP 6.57E+003	mm Hg	25	EST	NEELY,WB & BLAU,GE (1985)
DC	pKa			
HL 7.70E-001	atm m3/mol	25	EST	MEYLAN,WM & HOWARD,PH (1991)
OH 1.80E-014	cm3/molc sec	25	EXP	ATKINSON,R (1989)

CAS #: 000431-03-8				2,3-BUTANEDIONE

Formula: $C_4H_6O_2$

Mol Weight: 86.09

MP (deg C): 173 | FP (deg C):

BP (deg C): 272-274

BP pressure (mm Hg): 1.20E+001

Property/ Value	Units	Temp	Data Type	Reference
WS 2.00E+005	mg/L	15	EXP	YALKOWSKY,SH & DANNENFELSER,RM (1992)
logP -1.34			EXP	HANSCH,C & LEO,AJ (1985)
VP 5.68E+001	mm Hg	25	EXP	BOUBLIK,T ET AL. (1984)
DC	pKa			
HL 1.75E-005	atm m3/mol	25	EXP	SNIDER,JR & DAWSON,GA (1985)
OH 2.38E-013	cm3/molc sec	25	EXP	ATKINSON,R (1989)

CAS #: 000431-47-0				METHYLTRIFLUOROACETATE

Formula: $C_3H_3F_3O_2$

Mol Weight: 128.05

MP (deg C): | FP (deg C):

BP (deg C): 43-43.5

BP pressure (mm Hg):

Property/ Value	Units	Temp	Data Type	Reference
WS 2.88E+004	mg/L	25	EST	MEYLAN,WM ET AL. (1996)
logP 0.60			EXP	HANSCH,C ET AL. (1995)
VP 5.38E+002	mm Hg	25	EST	NEELY,WB & BLAU,GE (1985)
DC	pKa			
HL 1.38E-003	atm m3/mol	25	EST	MEYLAN,WM & HOWARD,PH (1991)
OH 5.00E-014	cm3/molc sec	25	EXP	ATKINSON,R (1989)

CAS #: 000433-17-0				BENZOTHIAZOLE-2-SULFONAMIDE

Formula: $C_7H_6N_2O_2S_2$

Mol Weight: 214.27

MP (deg C): | FP (deg C):

BP (deg C):

BP pressure (mm Hg):

Property/ Value	Units	Temp	Data Type	Reference
WS 2.70E+003	mg/L	25	EST	MEYLAN,WM ET AL. (1996)
logP 1.33			EXP	HANSCH,C & LEO,AJ (1985)
VP 1.03E-006	mm Hg	25	EST	NEELY,WB & BLAU,GE (1985)
DC	pKa			
HL 2.93E-011	atm m3/mol	25	EST	MEYLAN,WM & HOWARD,PH (1991)
OH 1.50E-012	cm3/molc sec	25	EST	MEYLAN,WM & HOWARD,PH (1993)

CAS #: 000433-19-2				BENZENE, 1,4-BIS(TRIFLUOROMETHYL)-

Formula: $C_8H_4F_6$

Mol Weight: 214.11

MP (deg C): | FP (deg C):

BP (deg C): 116

BP pressure (mm Hg):

Property/ Value	Units	Temp	Data Type	Reference
WS 1.98E+001	mg/L	25	EST	MEYLAN,WM ET AL. (1996)
logP 3.83			EXP	HANSCH,C ET AL. (1995)
VP 1.85E+001	mm Hg	25	EST	NEELY,WB & BLAU,GE (1985)
DC	pKa			
HL 4.07E-001	atm m3/mol	25	EST	MEYLAN,WM & HOWARD,PH (1991)
OH 5.93E-014	cm3/molc sec	25	EST	MEYLAN,WM & HOWARD,PH (1993)

CAS #: 000434-22-0				NANDROLONE

Formula: $C_{18}H_{26}O_2$

Mol Weight: 274.41

MP (deg C): | FP (deg C):

BP (deg C):

BP pressure (mm Hg):

Property/ Value	Units	Temp	Data Type	Reference
WS 3.23E+002	mg/L	25	EST	MEYLAN,WM ET AL. (1996)
logP 2.62			EXP	HANSCH,C & LEO,AJ (1985)
VP 3.53E-008	mm Hg	25	EST	NEELY,WB & BLAU,GE (1985)
DC	pKa			
HL 2.66E-009	atm m3/mol	25	EST	MEYLAN,WM & HOWARD,PH (1991)
OH 1.09E-010	cm3/molc sec	25	EST	MEYLAN,WM & HOWARD,PH (1993)

CAS #: 000434-45-7				TRIFLUOROACETOPHENONE

Formula: $C_8H_5F_3O$

Mol Weight: 174.12

MP (deg C): -40 | FP (deg C):

BP (deg C): 153

BP pressure (mm Hg):

Property/ Value	Units	Temp	Data Type	Reference
WS 8.57E+002	mg/L	25	EST	MEYLAN,WM ET AL. (1996)
logP 2.15			EXP	HANSCH,C & LEO,AJ (1985)
VP 1.22E+000	mm Hg	25	EST	NEELY,WB & BLAU,GE (1985)
DC	pKa			
HL 7.73E-005	atm m3/mol	25	EST	MEYLAN,WM & HOWARD,PH (1991)
OH 1.78E-012	cm3/molc sec	25	EST	MEYLAN,WM & HOWARD,PH (1993)

CAS #: 000434-64-0				OCTAFLUOROTOLUENE

Formula: C_7F_8

Mol Weight: 236.07

MP (deg C): -65.6 | FP (deg C):

BP (deg C): 104.5

BP pressure (mm Hg):

Property/Value	Units	Temp	Data Type	Reference
WS 1.18E+001	mg/L	25	EST	MEYLAN,WM ET AL. (1996)
logP 3.96			EST	MEYLAN,WM & HOWARD,PH (1995)
VP 2.90E+001	mm Hg	25	EXP	BOUBLIK,T ET AL. (1984)
DC	pKa			
HL 1.01E-001	atm m3/mol	25	EST	MEYLAN,WM & HOWARD,PH (1991)
OH 8.78E-014	cm3/molc sec	25	EST	MEYLAN,WM & HOWARD,PH (1993)

CAS #: 000434-75-3				BENZOIC ACID, 2-CHLORO-6-FLUORO-

Formula: $C_7H_4ClFO_2$

Mol Weight: 174.56

MP (deg C): 159-161 | FP (deg C):

BP (deg C):

BP pressure (mm Hg):

Property/Value	Units	Temp	Data Type	Reference
WS 9.23E+002	mg/L	25	EST	MEYLAN,WM ET AL. (1996)
logP 2.11			EXP	HANSCH,C ET AL. (1995)
VP 3.16E-003	mm Hg	25	EST	NEELY,WB & BLAU,GE (1985)
DC	pKa			
HL 9.38E-008	atm m3/mol	25	EST	MEYLAN,WM & HOWARD,PH (1991)
OH 1.16E-012	cm3/molc sec	25	EST	MEYLAN,WM & HOWARD,PH (1993)

CAS #: 000435-97-2				PHENPROCOUMON

Formula: $C_{18}H_{16}O_3$

Mol Weight: 280.33

MP (deg C): 179-180 | FP (deg C):

BP (deg C):

BP pressure (mm Hg):

Property/Value	Units	Temp	Data Type	Reference
WS 1.29E+001	mg/L	25	EST	MEYLAN,WM ET AL. (1996)
logP 3.62			EXP	HANSCH,C & LEO,AJ (1985)
VP 7.75E-011	mm Hg	25	EST	NEELY,WB & BLAU,GE (1985)
DC	pKa			
HL 4.50E-010	atm m3/mol	25	EST	MEYLAN,WM & HOWARD,PH (1991)
OH 4.89E-011	cm3/molc sec	25	EST	MEYLAN,WM & HOWARD,PH (1993)

CAS #: 000437-38-7				FENTANYL

Formula: $C_{22}H_{28}N_2O$

Mol Weight: 336.48

MP (deg C): 83-84 | FP (deg C):

BP (deg C):

BP pressure (mm Hg):

Property/Value	Units	Temp	Data Type	Reference
WS 2.00E+002	mg/L	25	EXP	ROY,SD & FLYNN,GL (1988)
logP 4.05			EXP	SANGSTER,J (1993)
VP 5.29E-009	mm Hg	25	EST	NEELY,WB & BLAU,GE (1985)
DC	pKa			
HL 9.20E-012	atm m3/mol	25	EST	MEYLAN,WM & HOWARD,PH (1991)
OH 1.33E-010	cm3/molc sec	25	EST	MEYLAN,WM & HOWARD,PH (1993)

CAS #: 000439-14-5				DIAZEPAM

Formula: $C_{16}H_{13}ClN_2O$

Mol Weight: 284.75

MP (deg C): 125-126 | FP (deg C):

BP (deg C):

BP pressure (mm Hg):

Property/Value	Units	Temp	Data Type	Reference
WS 5.00E+001	mg/L	25	EXP	YALKOWSKY,SH & DANNENFELSER,RM (1992)
logP 2.82			EXP	SANGSTER,J (1994)
VP 2.78E-008	mm Hg	25	EST	NEELY,WB & BLAU,GE (1985)
DC	pKa			
HL 3.64E-009	atm m3/mol	25	EST	MEYLAN,WM & HOWARD,PH (1991)
OH 9.90E-012	cm3/molc sec	25	EST	MEYLAN,WM & HOWARD,PH (1993)

CAS #: 000442-51-3				9H-PYRIDO[3,4-B]INDOLE-, 7-METHOXY-1-METHYL-

Formula: $C_{13}H_{12}N_2O$

Mol Weight: 212.25

MP (deg C): 261 | FP (deg C):

BP (deg C):

BP pressure (mm Hg):

Property/Value	Units	Temp	Data Type	Reference
WS 2.68E+000	mg/L	25	EST	MEYLAN,WM ET AL. (1996)
logP 3.56			EXP	SANGSTER,J (1994)
VP 6.26E-006	mm Hg	25	EST	NEELY,WB & BLAU,GE (1985)
DC	pKa			
HL 3.13E-011	atm m3/mol	25	EST	MEYLAN,WM & HOWARD,PH (1991)
OH 2.01E-010	cm3/molc sec	25	EST	MEYLAN,WM & HOWARD,PH (1993)

CAS #: 000443-48-1				METRONIDAZOLE

Formula: $C_6H_9N_3O_3$

Mol Weight: 171.16

MP (deg C): 158-160 | FP (deg C):

BP (deg C):

BP pressure (mm Hg):

Property/Value	Units	Temp	Data Type	Reference
WS 9.50E+003	mg/L	25	EXP	YALKOWSKY,SH & DANNENFELSER,RM (1992)
logP -0.02			EXP	HANSCH,C & LEO,AJ (1985)
VP 3.05E-007	mm Hg	25	EST	NEELY,WB & BLAU,GE (1985)
DC	pKa			
HL 1.69E-011	atm m3/mol	25	EST	MEYLAN,WM & HOWARD,PH (1991)
OH 9.30E-012	cm3/molc sec	25	EST	MEYLAN,WM & HOWARD,PH (1993)

CAS #: 000444-30-4				O-TRIFLUOROMETHYLPHENOL

Formula: $C_7H_5F_3O$

Mol Weight: 162.11

MP (deg C): 45-46 | FP (deg C):

BP (deg C): 147-148

BP pressure (mm Hg):

Property/Value	Units	Temp	Data Type	Reference
WS 1.03E+003	mg/L	25	EST	MEYLAN,WM ET AL. (1996)
logP 2.80			EXP	HANSCH,C & LEO,AJ (1985)
VP 6.68E-001	mm Hg	25	EST	NEELY,WB & BLAU,GE (1985)
DC	pKa			
HL 4.87E-006	atm m3/mol	25	EST	MEYLAN,WM & HOWARD,PH (1991)
OH 6.84E-012	cm3/molc sec	25	EST	MEYLAN,WM & HOWARD,PH (1993)

CAS #: 000445-28-3 — 2-FLUOROBENZAMIDE

Formula: C_7H_6FNO
Mol Weight: 139.13
MP (deg C): 117-119
FP (deg C):
BP (deg C):
BP pressure (mm Hg):

Property/Value	Units	Temp	Data Type	Reference
WS 2.65E+004	mg/L	25	EST	MEYLAN,WM ET AL. (1996)
logP 0.59			EXP	NAKAGAWA,Y ET AL. (1992)
VP 1.09E-003	mm Hg	25	EST	NEELY,WB & BLAU,GE (1985)
DC	pKa			
HL 2.58E-009	atm m3/mol	25	EST	MEYLAN,WM & HOWARD,PH (1991)
OH 4.23E-012	cm3/molc sec	25	EST	MEYLAN,WM & HOWARD,PH (1993)

CAS #: 000445-29-4 — 2-FLUOROBENZOIC ACID

Formula: $C_7H_5FO_2$
Mol Weight: 140.12
MP (deg C): 126.5
FP (deg C):
BP (deg C):
BP pressure (mm Hg):

Property/Value	Units	Temp	Data Type	Reference
WS 7.20E+003	mg/L	25	EXP	YALKOWSKY,SH & DANNENFELSER,RM (1992)
logP 1.77			EXP	HANSCH,C & LEO,AJ (1985)
VP 5.85E-002	mm Hg	25	EXT	DOLFING,J & HARRISON,BK (1992)
DC 3.27	pKa		EXP	KORTUM,G ET AL (1961)
HL 1.27E-007	atm m3/mol	25	EST	MEYLAN,WM & HOWARD,PH (1991)
OH 1.42E-012	cm3/molc sec	25	EST	MEYLAN,WM & HOWARD,PH (1993)

CAS #: 000445-66-9 — 2,6-NITRO-4-TRIFLUORO-ANILINE

Formula: $C_7H_4F_3N_3O_4$
Mol Weight: 251.12
MP (deg C):
FP (deg C):
BP (deg C):
BP pressure (mm Hg):

Property/Value	Units	Temp	Data Type	Reference
WS 2.58E+002	mg/L	25	EST	MEYLAN,WM ET AL. (1996)
logP 2.29			EXP	HANSCH,C & LEO,AJ (1985)
VP 1.80E-005	mm Hg	25	EST	NEELY,WB & BLAU,GE (1985)
DC	pKa			
HL 1.52E-006	atm m3/mol	25	EST	MEYLAN,WM & HOWARD,PH (1991)
OH 3.69E-015	cm3/molc sec	25	EST	MEYLAN,WM & HOWARD,PH (1993)

CAS #: 000446-31-1 — 2-FLUORO-4-AMINOBENZOIC ACID

Formula: $C_7H_6FNO_2$
Mol Weight: 155.13
MP (deg C):
FP (deg C):
BP (deg C):
BP pressure (mm Hg):

Property/Value	Units	Temp	Data Type	Reference
WS 5.70E+003	mg/L	25	EST	MEYLAN,WM ET AL. (1996)
logP 1.29			EXP	HANSCH,C & LEO,AJ (1985)
VP 3.17E-004	mm Hg	25	EST	NEELY,WB & BLAU,GE (1985)
DC	pKa			
HL 4.47E-011	atm m3/mol	25	EST	MEYLAN,WM & HOWARD,PH (1991)
OH 5.04E-011	cm3/molc sec	25	EST	MEYLAN,WM & HOWARD,PH (1993)

CAS #: 000446-36-6 — 5-FLUORO-2-NITROPHENOL

Formula: $C_6H_4FNO_3$
Mol Weight: 157.10
MP (deg C): 35-37
FP (deg C):
BP (deg C):
BP pressure (mm Hg):

Property/Value	Units	Temp	Data Type	Reference
WS 1.40E+003	mg/L	20	EXP	SCHWARZENBACH,RP ET AL. (1988)
logP 1.91			EXP	SCHWARZENBACH,RP ET AL. (1988)
VP 2.29E-002	mm Hg	25	EXP	SCHWARZENBACH,RP ET AL. (1988)
DC 6.07	pKa	25	EXP	SERJEANT,EP & DEMPSEY,B (1979)
HL 3.38E-006	atm m3/mol	25	EST	VP/WSOL
OH 5.36E-012	cm3/molc sec	25	EST	MEYLAN,WM & HOWARD,PH (1993)

CAS #: 000446-51-5 — BENZENEMETHANOL, 2-FLUORO-

Formula: C_7H_7FO
Mol Weight: 126.13
MP (deg C):
FP (deg C):
BP (deg C):
BP pressure (mm Hg):

Property/Value	Units	Temp	Data Type	Reference
WS 2.34E+004	mg/L	25	EST	MEYLAN,WM ET AL. (1996)
logP 1.31			EXP	EL TAYAR,N ET AL. (1991)
VP 6.90E-002	mm Hg	25	EST	NEELY,WB & BLAU,GE (1985)
DC	pKa			
HL 2.54E-007	atm m3/mol	25	EST	MEYLAN,WM & HOWARD,PH (1991)
OH 6.38E-012	cm3/molc sec	25	EST	MEYLAN,WM & HOWARD,PH (1993)

CAS #: 000446-86-6 — AZATHIOPRINE

Formula: $C_9H_7N_7O_2S$
Mol Weight: 277.27
MP (deg C): 243-244
FP (deg C):
BP (deg C):
BP pressure (mm Hg):

Property/Value	Units	Temp	Data Type	Reference
WS 2.72E+002	mg/L	25	EST	MEYLAN,WM ET AL. (1996)
logP 0.10			EXP	HANSCH,C & LEO,AJ (1985)
VP 2.41E-012	mm Hg	25	EST	NEELY,WB & BLAU,GE (1985)
DC	pKa			
HL 2.64E-015	atm m3/mol	25	EST	MEYLAN,WM & HOWARD,PH (1991)
OH 2.03E-010	cm3/molc sec	25	EST	MEYLAN,WM & HOWARD,PH (1993)

CAS #: 000447-05-2 — PYRIDOXINE 5'-PHOSPHATE

Formula: $C_8H_{12}NO_6P$
Mol Weight: 249.16
MP (deg C):
FP (deg C):
BP (deg C):
BP pressure (mm Hg):

Property/Value	Units	Temp	Data Type	Reference
WS 1.87E+005	mg/L	25	EST	MEYLAN,WM ET AL. (1996)
logP -1.04			EST	MEYLAN,WM & HOWARD,PH (1995)
VP 4.02E-011	mm Hg	25	EST	NEELY,WB & BLAU,GE (1985)
DC	pKa			
HL 1.19E-024	atm m3/mol	25	EST	MEYLAN,WM & HOWARD,PH (1991)
OH 4.66E-011	cm3/molc sec	25	EST	MEYLAN,WM & HOWARD,PH (1993)

CAS #: 000447-53-0 — 1,2-DIHYDRONAPHTHALENE

Formula: $C_{10}H_{10}$

Mol Weight: 130.19

MP (deg C): -8

FP (deg C):

BP (deg C): 206.5

BP pressure (mm Hg):

Property/Value	Units	Temp	Data Type	Reference
WS 5.85E+001	mg/L	25	EST	MEYLAN,WM ET AL. (1996)
logP 3.74			EST	MEYLAN,WM & HOWARD,PH (1995)
VP 2.89E-001	mm Hg	25	EST	NEELY,WB & BLAU,GE (1985)
DC	pKa			
HL 2.11E-003	atm m3/mol	25	EST	MEYLAN,WM & HOWARD,PH (1991)
OH 6.33E-011	cm3/molc sec	25	EST	MEYLAN,WM & HOWARD,PH (1993)

CAS #: 000451-13-8 — BENZENEACETIC ACID, 2,5-DIHYDROXY-

Formula: $C_8H_8O_4$

Mol Weight: 168.15

MP (deg C): 152

FP (deg C):

BP (deg C):

BP pressure (mm Hg):

Property/Value	Units	Temp	Data Type	Reference
WS 1.09E+005	mg/L	25	EST	MEYLAN,WM ET AL. (1996)
logP 0.86			EXP	SANGSTER,J (1994)
VP 8.18E-007	mm Hg	25	EST	NEELY,WB & BLAU,GE (1985)
DC	pKa			
HL 4.79E-016	atm m3/mol	25	EST	MEYLAN,WM & HOWARD,PH (1991)
OH 3.96E-011	cm3/molc sec	25	EST	MEYLAN,WM & HOWARD,PH (1993)

CAS #: 000451-40-1 — 2-PHENYLACETOPHENONE

Formula: $C_{14}H_{12}O$

Mol Weight: 196.25

MP (deg C): 60

FP (deg C):

BP (deg C): 320

BP pressure (mm Hg):

Property/Value	Units	Temp	Data Type	Reference
WS 8.80E+001	mg/L	25	EST	MEYLAN,WM ET AL. (1996)
logP 3.18			EXP	HANSCH,C ET AL. (1995)
VP 1.84E-004	mm Hg	25	EXT	PERRY,RH & GREEN,D (1984)
DC	pKa			
HL 7.92E-007	atm m3/mol	25	EST	MEYLAN,WM & HOWARD,PH (1991)
OH 7.32E-012	cm3/molc sec	25	EST	MEYLAN,WM & HOWARD,PH (1993)

CAS #: 000451-82-1 — O-FLUOROPHENYLACETIC ACID

Formula: $C_8H_7FO_2$

Mol Weight: 154.14

MP (deg C): 62-64

FP (deg C):

BP (deg C):

BP pressure (mm Hg):

Property/Value	Units	Temp	Data Type	Reference
WS 9.46E+003	mg/L	25	EST	MEYLAN,WM ET AL. (1996)
logP 1.50			EXP	HANSCH,C & LEO,AJ (1985)
VP 6.22E-003	mm Hg	25	EST	NEELY,WB & BLAU,GE (1985)
DC	pKa			
HL 5.16E-008	atm m3/mol	25	EST	MEYLAN,WM & HOWARD,PH (1991)
OH 3.74E-012	cm3/molc sec	25	EST	MEYLAN,WM & HOWARD,PH (1993)

CAS #: 000452-35-7 — ETHOXYZOLAMIDE

Formula: $C_9H_{10}N_2O_3S_2$

Mol Weight: 258.32

MP (deg C): 188-190.5

FP (deg C):

BP (deg C):

BP pressure (mm Hg):

Property/Value	Units	Temp	Data Type	Reference
WS 4.00E+001	mg/L		EXP	YALKOWSKY,SH & DANNENFELSER,RM (1992)
logP 2.01			EXP	HANSCH,C & LEO,AJ (1985)
VP 9.33E-008	mm Hg	25	EST	NEELY,WB & BLAU,GE (1985)
DC	pKa			
HL 2.30E-012	atm m3/mol	25	EST	MEYLAN,WM & HOWARD,PH (1991)
OH 1.08E-011	cm3/molc sec	25	EST	MEYLAN,WM & HOWARD,PH (1993)

CAS #: 000452-86-8 — 1,2-BENZENEDIOL, 4-METHYL-

Formula: $C_7H_8O_2$

Mol Weight: 124.14

MP (deg C): 65

FP (deg C):

BP (deg C): 251

BP pressure (mm Hg):

Property/Value	Units	Temp	Data Type	Reference
WS 2.49E+004	mg/L	25	EST	MEYLAN,WM ET AL. (1996)
logP 1.37			EXP	HANSCH,C ET AL. (1995)
VP 2.80E-003	mm Hg	25	EST	NEELY,WB & BLAU,GE (1985)
DC	pKa			
HL 6.44E-011	atm m3/mol	25	EST	MEYLAN,WM & HOWARD,PH (1991)
OH 6.03E-011	cm3/molc sec	25	EST	MEYLAN,WM & HOWARD,PH (1993)

CAS #: 000453-13-4 — 1,3-DIFLUORO-2-PROPANOL

Formula: $C_3H_6F_2O$

Mol Weight: 96.08

MP (deg C):

FP (deg C):

BP (deg C): 54-55

BP pressure (mm Hg): 3.40E+001

Property/Value	Units	Temp	Data Type	Reference
WS 7.88E+005	mg/L	25	EST	MEYLAN,WM ET AL. (1996)
logP -0.36			EXP	HANSCH,C & LEO,AJ (1985)
VP 8.33E+001	mm Hg	25	EST	NEELY,WB & BLAU,GE (1985)
DC	pKa			
HL 1.88E-005	atm m3/mol	25	EST	MEYLAN,WM & HOWARD,PH (1991)
OH 2.88E-012	cm3/molc sec	25	EST	MEYLAN,WM & HOWARD,PH (1993)

CAS #: 000454-31-9 — ACETIC ACID, DIFLUORO-, ETHYL ESTER

Formula: $C_4H_6F_2O_2$

Mol Weight: 124.09

MP (deg C):

FP (deg C):

BP (deg C): 100

BP pressure (mm Hg):

Property/Value	Units	Temp	Data Type	Reference
WS 1.75E+004	mg/L	25	EST	MEYLAN,WM ET AL. (1996)
logP 0.87			EXP	HANSCH,C ET AL. (1995)
VP 1.44E+002	mm Hg	25	EST	NEELY,WB & BLAU,GE (1985)
DC	pKa			
HL 9.21E-004	atm m3/mol	25	EST	MEYLAN,WM & HOWARD,PH (1991)
OH 9.27E-013	cm3/molc sec	25	EST	MEYLAN,WM & HOWARD,PH (1993)

000454-57-9 — FLUORODIMETHYLPHENYLSILANE

Formula:	$C_8H_{11}FSi$
Mol Weight:	154.26
MP (deg C):	FP (deg C):
BP (deg C):	
BP pressure (mm Hg):	

Property/Value	Units	Temp	Data Type	Reference
WS 4.43E+001	mg/L	25	EST	MEYLAN,WM ET AL. (1996)
logP 3.76			EST	MEYLAN,WM & HOWARD,PH (1995)
VP 2.90E+000	mm Hg	25	EXP	SOKOLOV,VB ET AL. (1975)
DC	pKa			
HL	atm m3/mol			
OH 2.25E-012	cm3/molc sec	25	EST	MEYLAN,WM & HOWARD,PH (1993)

000454-89-7 — M-(TRIFLUOROMETHYL)BENZALDEHYDE

Formula:	$C_8H_5F_3O$
Mol Weight:	174.12
MP (deg C):	FP (deg C):
BP (deg C):	83-86
BP pressure (mm Hg):	3.00E+001

Property/Value	Units	Temp	Data Type	Reference
WS 4.57E+002	mg/L	25	EST	MEYLAN,WM ET AL. (1996)
logP 2.47			EXP	HANSCH,C & LEO,AJ (1985)
VP 6.77E-001	mm Hg	25	EST	NEELY,WB & BLAU,GE (1985)
DC	pKa			
HL 1.17E-004	atm m3/mol	25	EST	MEYLAN,WM & HOWARD,PH (1991)
OH 1.70E-011	cm3/molc sec	25	EST	MEYLAN,WM & HOWARD,PH (1993)

000454-92-2 — M-TRIFLUOROMETHYLBENZOIC ACID

Formula:	$C_8H_5F_3O_2$
Mol Weight:	190.12
MP (deg C):	105-106 FP (deg C):
BP (deg C):	238.5
BP pressure (mm Hg):	7.75E+002

Property/Value	Units	Temp	Data Type	Reference
WS 1.49E+002	mg/L	25	EST	MEYLAN,WM ET AL. (1996)
logP 2.95			EXP	HANSCH,C & LEO,AJ (1985)
VP 9.87E-003	mm Hg	25	EST	NEELY,WB & BLAU,GE (1985)
DC	pKa			
HL 9.42E-007	atm m3/mol	25	EST	MEYLAN,WM & HOWARD,PH (1991)
OH 6.65E-013	cm3/molc sec	25	EST	MEYLAN,WM & HOWARD,PH (1993)

000455-14-1 — P-TRIFLUOROMETHYLANILINE

Formula:	$C_7H_6F_3N$
Mol Weight:	161.13
MP (deg C):	38 FP (deg C):
BP (deg C):	117.5
BP pressure (mm Hg):	6.00E+001

Property/Value	Units	Temp	Data Type	Reference
WS 1.46E+003	mg/L	25	EST	MEYLAN,WM ET AL. (1996)
logP 1.95			EXP	HANSCH,C & LEO,AJ (1985)
VP 5.92E-001	mm Hg	25	EST	NEELY,WB & BLAU,GE (1985)
DC 2.45	pKa	25	EXP	PERRIN,DD (1965)
HL 1.65E-005	atm m3/mol	25	EST	MEYLAN,WM & HOWARD,PH (1991)
OH 2.16E-011	cm3/molc sec	25	EST	MEYLAN,WM & HOWARD,PH (1993)

000455-16-3 — P-FLUOROSULFONYLTOLUENE

Formula:	$C_7H_7FO_2S$
Mol Weight:	174.20
MP (deg C):	41-42 FP (deg C):
BP (deg C):	112
BP pressure (mm Hg):	1.60E+001

Property/Value	Units	Temp	Data Type	Reference
WS 2.69E+002	mg/L	25	EST	MEYLAN,WM ET AL. (1996)
logP 2.74			EXP	HANSCH,C & LEO,AJ (1985)
VP 5.36E-003	mm Hg	25	EST	NEELY,WB & BLAU,GE (1985)
DC	pKa			
HL 1.72E-005	atm m3/mol	25	EST	MEYLAN,WM & HOWARD,PH (1991)
OH 1.22E-012	cm3/molc sec	25	EST	MEYLAN,WM & HOWARD,PH (1993)

000455-24-3 — P-CF3 BENZOIC ACID

Formula:	$C_8H_5F_3O_2$
Mol Weight:	190.12
MP (deg C):	219-220 FP (deg C):
BP (deg C):	
BP pressure (mm Hg):	

Property/Value	Units	Temp	Data Type	Reference
WS 1.11E+002	mg/L	25	EST	MEYLAN,WM ET AL. (1996)
logP 3.10			EXP	HANSCH,C ET AL. (1995)
VP 9.87E-003	mm Hg	25	EST	NEELY,WB & BLAU,GE (1985)
DC	pKa			
HL 9.42E-007	atm m3/mol	25	EST	MEYLAN,WM & HOWARD,PH (1991)
OH 6.29E-013	cm3/molc sec	25	EST	MEYLAN,WM & HOWARD,PH (1993)

000455-36-7 — M-FLUOROACETOPHENONE

Formula:	C_8H_7FO
Mol Weight:	138.14
MP (deg C):	FP (deg C):
BP (deg C):	81
BP pressure (mm Hg):	9.00E+000

Property/Value	Units	Temp	Data Type	Reference
WS 2.63E+003	mg/L	25	EST	MEYLAN,WM ET AL. (1996)
logP 1.77			EXP	HANSCH,C & LEO,AJ (1985)
VP 7.53E-001	mm Hg	25	EST	NEELY,WB & BLAU,GE (1985)
DC	pKa			
HL 1.14E-005	atm m3/mol	25	EST	MEYLAN,WM & HOWARD,PH (1991)
OH 1.13E-012	cm3/molc sec	25	EST	MEYLAN,WM & HOWARD,PH (1993)

000455-37-8 — 3-FLUOROBENZAMIDE

Formula:	C_7H_6FNO
Mol Weight:	139.13
MP (deg C):	FP (deg C):
BP (deg C):	
BP pressure (mm Hg):	

Property/Value	Units	Temp	Data Type	Reference
WS 1.41E+004	mg/L	25	EST	MEYLAN,WM ET AL. (1996)
logP 0.91			EXP	HANSCH,C & LEO,AJ (1985)
VP 1.09E-003	mm Hg	25	EST	NEELY,WB & BLAU,GE (1985)
DC	pKa			
HL 2.58E-009	atm m3/mol	25	EST	MEYLAN,WM & HOWARD,PH (1991)
OH 4.23E-012	cm3/molc sec	25	EST	MEYLAN,WM & HOWARD,PH (1993)

000455-38-9 — M-FLUOROBENZOIC ACID

Formula: $C_7H_5FO_2$
Mol Weight: 140.12
MP (deg C): 124
FP (deg C):
BP (deg C):
BP pressure (mm Hg):

Property/Value	Units	Temp	Data Type	Reference
WS 1.50E+003	mg/L	25	EXP	YALKOWSKY,SH & DANNENFELSER,RM (1992)
logP 2.15			EXP	HANSCH,C & LEO,AJ (1985)
VP 5.85E-002	mm Hg	25	EXT	DOLFING,J & HARRISON,BK (1992)
DC 3.86	pKa	25	EXP	KORTUM,G ET AL (1961)
HL 1.27E-007	atm m3/mol	25	EST	MEYLAN,WM & HOWARD,PH (1991)
OH 1.19E-012	cm3/molc sec	25	EST	MEYLAN,WM & HOWARD,PH (1993)

000455-87-8 — BENZOIC ACID, 4-AMINO-3-FLUORO-

Formula: $C_7H_6FNO_2$
Mol Weight: 155.13
MP (deg C):
FP (deg C):
BP (deg C):
BP pressure (mm Hg):

Property/Value	Units	Temp	Data Type	Reference
WS 4.50E+003	mg/L	25	EST	MEYLAN,WM ET AL. (1996)
logP 1.41			EXP	SANGSTER,J (1994)
VP 3.17E-004	mm Hg	25	EST	NEELY,WB & BLAU,GE (1985)
DC	pKa			
HL 4.47E-011	atm m3/mol	25	EST	MEYLAN,WM & HOWARD,PH (1991)
OH 1.40E-011	cm3/molc sec	25	EST	MEYLAN,WM & HOWARD,PH (1993)

000456-22-4 — P-FLUOROBENZOIC ACID

Formula: $C_7H_5FO_2$
Mol Weight: 140.12
MP (deg C): 182.6
FP (deg C):
BP (deg C):
BP pressure (mm Hg):

Property/Value	Units	Temp	Data Type	Reference
WS 1.20E+003	mg/L	25	EXP	YALKOWSKY,SH & DANNENFELSER,RM (1992)
logP 2.07			EXP	HANSCH,C & LEO,AJ (1985)
VP 1.91E-002	mm Hg	25	EST	NEELY,WB & BLAU,GE (1985)
DC 4.14	pKa	25	EXP	KORTUM,G ET AL (1961)
HL 1.27E-007	atm m3/mol	25	EST	MEYLAN,WM & HOWARD,PH (1991)
OH 1.42E-012	cm3/molc sec	25	EST	MEYLAN,WM & HOWARD,PH (1993)

000456-42-8 — M-FLUOROBENZYL CHLORIDE

Formula: C_7H_6ClF
Mol Weight: 144.58
MP (deg C):
FP (deg C):
BP (deg C):
BP pressure (mm Hg):

Property/Value	Units	Temp	Data Type	Reference
WS 4.14E+002	mg/L	25	EXP	YALKOWSKY,SH & DANNENFELSER,RM (1992)
logP 2.77			EXP	HANSCH,C & LEO,AJ (1985)
VP 9.94E-001	mm Hg	25	EST	NEELY,WB & BLAU,GE (1985)
DC	pKa			
HL 2.44E-003	atm m3/mol	25	EST	MEYLAN,WM & HOWARD,PH (1991)
OH 2.87E-012	cm3/molc sec	25	EST	MEYLAN,WM & HOWARD,PH (1993)

000456-49-5 — M-FLUOROANISOLE

Formula: C_7H_7FO
Mol Weight: 126.13
MP (deg C): -35
FP (deg C):
BP (deg C): 159
BP pressure (mm Hg):

Property/Value	Units	Temp	Data Type	Reference
WS 9.19E+002	mg/L	25	EST	MEYLAN,WM ET AL. (1996)
logP 2.36			EXP	DUNN,WJ ET AL. (1983)
VP 5.28E+000	mm Hg	25	EST	NEELY,WB & BLAU,GE (1985)
DC	pKa			
HL 3.72E-004	atm m3/mol	25	EST	MEYLAN,WM & HOWARD,PH (1991)
OH 2.78E-011	cm3/molc sec	25	EST	MEYLAN,WM & HOWARD,PH (1993)

000456-55-3 — TRIFLUOROMETHOXYBENZENE

Formula: $C_7H_5F_3O$
Mol Weight: 162.11
MP (deg C):
FP (deg C):
BP (deg C): 102
BP pressure (mm Hg):

Property/Value	Units	Temp	Data Type	Reference
WS 1.31E+002	mg/L	25	EST	MEYLAN,WM ET AL. (1996)
logP 3.17			EXP	HANSCH,C & LEO,AJ (1985)
VP 8.61E+000	mm Hg	25	EST	NEELY,WB & BLAU,GE (1985)
DC	pKa			
HL 2.51E-003	atm m3/mol	25	EST	MEYLAN,WM & HOWARD,PH (1991)
OH 2.15E-011	cm3/molc sec	25	EST	MEYLAN,WM & HOWARD,PH (1993)

000456-56-4 — TRIFLUOROMETHYLTHIOBENZENE

Formula: $C_7H_5F_3S$
Mol Weight: 178.18
MP (deg C):
FP (deg C):
BP (deg C):
BP pressure (mm Hg):

Property/Value	Units	Temp	Data Type	Reference
WS 5.02E+001	mg/L	25	EST	MEYLAN,WM ET AL. (1996)
logP 3.57			EXP	HANSCH,C & LEO,AJ (1985)
VP 1.33E+000	mm Hg	25	EST	NEELY,WB & BLAU,GE (1985)
DC	pKa			
HL 1.24E-003	atm m3/mol	25	EST	MEYLAN,WM & HOWARD,PH (1991)
OH 1.26E-011	cm3/molc sec	25	EST	MEYLAN,WM & HOWARD,PH (1993)

000456-64-4 — CF3-METHANESULFONANILIDE

Formula: $C_7H_6F_3NO_2S$
Mol Weight: 225.19
MP (deg C):
FP (deg C):
BP (deg C):
BP pressure (mm Hg):

Property/Value	Units	Temp	Data Type	Reference
WS 8.03E+001	mg/L	25	EST	MEYLAN,WM ET AL. (1996)
logP 3.05			EXP	HANSCH,C & LEO,AJ (1985)
VP 1.35E-003	mm Hg	25	EST	NEELY,WB & BLAU,GE (1985)
DC	pKa			
HL 9.18E-006	atm m3/mol	25	EST	MEYLAN,WM & HOWARD,PH (1991)
OH 4.27E-011	cm3/molc sec	25	EST	MEYLAN,WM & HOWARD,PH (1993)

000458-24-2 — FENFLURAMINE

CAS #:	000458-24-2
Formula:	$C_{12}H_{16}F_3N$
Mol Weight:	231.26
MP (deg C):	
FP (deg C):	
BP (deg C):	108-112
BP pressure (mm Hg):	1.20E+001

Property/ Value	Units	Temp	Data Type	Reference
WS 4.12E+002	mg/L	25	EST	MEYLAN,WM ET AL. (1996)
logP 3.36			EXP	SANGSTER,J (1993)
VP 4.08E-002	mm Hg	25	EST	NEELY,WB & BLAU,GE (1985)
DC	pKa			
HL 2.73E-005	atm m3/mol	25	EST	MEYLAN,WM & HOWARD,PH (1991)
OH 3.31E-011	cm3/molc sec	25	EST	MEYLAN,WM & HOWARD,PH (1993)

000458-88-8 — 2-PROPYLPIPERIDINE

CAS #:	000458-88-8
Formula:	$C_8H_{17}N$
Mol Weight:	127.23
MP (deg C):	-2
FP (deg C):	
BP (deg C):	166-166.5
BP pressure (mm Hg):	

Property/ Value	Units	Temp	Data Type	Reference
WS 1.80E+004	mg/L		EXP	YALKOWSKY,SH & DANNENFELSER,RM (1992)
logP 2.59			EST	MEYLAN,WM & HOWARD,PH (1995)
VP 6.93E-001	mm Hg	25	EST	NEELY,WB & BLAU,GE (1985)
DC 11.00	pKa		EXP	PERRIN,DD (1965)
HL 4.01E-005	atm m3/mol	25	EST	MEYLAN,WM & HOWARD,PH (1991)
OH 1.08E-010	cm3/molc sec	25	EST	MEYLAN,WM & HOWARD,PH (1993)

000458-92-4 — DIFLUOROMETHOXYBENZENE

CAS #:	000458-92-4
Formula:	$C_7H_6F_2O$
Mol Weight:	144.12
MP (deg C):	
FP (deg C):	
BP (deg C):	
BP pressure (mm Hg):	

Property/ Value	Units	Temp	Data Type	Reference
WS 6.63E+002	mg/L	25	EST	MEYLAN,WM ET AL. (1996)
logP 2.44			EXP	HANSCH,C & LEO,AJ (1985)
VP 6.33E+000	mm Hg	25	EST	NEELY,WB & BLAU,GE (1985)
DC	pKa			
HL 1.26E-003	atm m3/mol	25	EST	MEYLAN,WM & HOWARD,PH (1991)
OH 2.16E-011	cm3/molc sec	25	EST	MEYLAN,WM & HOWARD,PH (1993)

000459-23-4 — P-FLUOROBENZALDOXIME

CAS #:	000459-23-4
Formula:	C_7H_6FNO
Mol Weight:	139.13
MP (deg C):	82-85
FP (deg C):	
BP (deg C):	
BP pressure (mm Hg):	

Property/ Value	Units	Temp	Data Type	Reference
WS 1.56E+003	mg/L	25	EST	MEYLAN,WM ET AL. (1996)
logP 2.03			EXP	HANSCH,C ET AL. (1995)
VP 5.34E-003	mm Hg	25	EST	NEELY,WB & BLAU,GE (1985)
DC	pKa			
HL 4.17E-007	atm m3/mol	25	EST	MEYLAN,WM & HOWARD,PH (1991)
OH 4.88E-012	cm3/molc sec	25	EST	MEYLAN,WM & HOWARD,PH (1993)

000459-25-6 — P-FLUOROFORMANILIDE

CAS #:	000459-25-6
Formula:	C_7H_6FNO
Mol Weight:	139.13
MP (deg C):	
FP (deg C):	
BP (deg C):	
BP pressure (mm Hg):	

Property/ Value	Units	Temp	Data Type	Reference
WS 6.95E+003	mg/L	25	EST	MEYLAN,WM ET AL. (1996)
logP 1.27			EXP	HANSCH,C & LEO,AJ (1985)
VP 1.39E-003	mm Hg	25	EST	NEELY,WB & BLAU,GE (1985)
DC	pKa			
HL 9.86E-009	atm m3/mol	25	EST	MEYLAN,WM & HOWARD,PH (1991)
OH 1.44E-011	cm3/molc sec	25	EST	MEYLAN,WM & HOWARD,PH (1993)

000459-56-3 — BENZENEMETHANOL, 4-FLUORO-

CAS #:	000459-56-3
Formula:	C_7H_7FO
Mol Weight:	126.13
MP (deg C):	23
FP (deg C):	
BP (deg C):	210
BP pressure (mm Hg):	

Property/ Value	Units	Temp	Data Type	Reference
WS 2.12E+004	mg/L	25	EST	MEYLAN,WM ET AL. (1996)
logP 1.36			EXP	EL TAYAR,N ET AL. (1991)
VP 6.90E-002	mm Hg	25	EST	NEELY,WB & BLAU,GE (1985)
DC	pKa			
HL 2.54E-007	atm m3/mol	25	EST	MEYLAN,WM & HOWARD,PH (1991)
OH 6.38E-012	cm3/molc sec	25	EST	MEYLAN,WM & HOWARD,PH (1993)

000459-57-4 — BENZALDEHYDE, 4-FLUORO-

CAS #:	000459-57-4
Formula:	C_7H_5FO
Mol Weight:	124.12
MP (deg C):	-10
FP (deg C):	
BP (deg C):	181.5
BP pressure (mm Hg):	

Property/ Value	Units	Temp	Data Type	Reference
WS 4.69E+003	mg/L	25	EST	MEYLAN,WM ET AL. (1996)
logP 1.54			EXP	BAZACO,JF & COCA,CM (1989)
VP 1.15E+000	mm Hg	25	EST	NEELY,WB & BLAU,GE (1985)
DC	pKa			
HL 1.57E-005	atm m3/mol	25	EST	MEYLAN,WM & HOWARD,PH (1991)
OH 1.77E-011	cm3/molc sec	25	EST	MEYLAN,WM & HOWARD,PH (1993)

000459-60-9 — P-FLUOROANISOLE

CAS #:	000459-60-9
Formula:	C_7H_7FO
Mol Weight:	126.13
MP (deg C):	-45
FP (deg C):	
BP (deg C):	157
BP pressure (mm Hg):	

Property/ Value	Units	Temp	Data Type	Reference
WS 1.12E+003	mg/L	25	EST	MEYLAN,WM ET AL. (1996)
logP 2.26			EXP	DUNN,WJ ET AL. (1983)
VP 5.28E+000	mm Hg	25	EST	NEELY,WB & BLAU,GE (1985)
DC	pKa			
HL 3.72E-004	atm m3/mol	25	EST	MEYLAN,WM & HOWARD,PH (1991)
OH 8.09E-012	cm3/molc sec	25	EST	MEYLAN,WM & HOWARD,PH (1993)

P-BROMOFLUOROBENZENE

CAS #: 000460-00-4

Formula: C_6H_4BrF

Mol Weight: 175.01

MP (deg C): -17.4

BP (deg C): 151.5

BP pressure (mm Hg):

Property/Value	Units	Temp	Data Type	Reference
WS 1.36E+002	mg/L	25	EST	MEYLAN,WM ET AL. (1996)
logP 3.08			EXP	DUNN,WJ ET AL. (1983)
VP 2.84E+000	mm Hg	25	EST	NEELY,WB & BLAU,GE (1985)
DC	pKa			
HL 2.50E-003	atm m3/mol	25	EST	MEYLAN,WM & HOWARD,PH (1991)
OH 7.00E-013	cm3/molc sec	25	EST	MEYLAN,WM & HOWARD,PH (1993)

DIACETYLENE

CAS #: 000460-12-8

Formula: C_4H_2

Mol Weight: 50.06

MP (deg C): -36.4

BP (deg C): 10.3

BP pressure (mm Hg):

Property/Value	Units	Temp	Data Type	Reference
WS 1.00E+002	mg/L	25	EXP	YALKOWSKY,SH & DANNENFELSER,RM (1992)
logP 1.30			EST	MEYLAN,WM & HOWARD,PH (1995)
VP 1.32E+003	mm Hg	25	EXT	PERRY,RH & GREEN,D (1984)
DC	pKa			
HL 1.91E-002	atm m3/mol	25	EST	MEYLAN,WM & HOWARD,PH (1991)
OH 1.89E-011	cm3/molc sec	25	EXP	ATKINSON,R (1989)

1-FLUOROPROPANE

CAS #: 000460-13-9

Formula: C_3H_7F

Mol Weight: 62.09

MP (deg C): -159

BP (deg C): 2.5

BP pressure (mm Hg):

Property/Value	Units	Temp	Data Type	Reference
WS 4.34E+003	mg/L	25	EST	MEYLAN,WM ET AL. (1996)
logP 1.76			EST	MEYLAN,WM & HOWARD,PH (1995)
VP 2.01E+003	mm Hg	25	EXT	DYKYJ,J (1970)
DC	pKa			
HL 2.90E-002	atm m3/mol	25	EST	MEYLAN,WM & HOWARD,PH (1991)
OH 8.45E-013	cm3/molc sec	25	EST	MEYLAN,WM & HOWARD,PH (1993)

CYANOGEN

CAS #: 000460-19-5

Formula: C_2N_2

Mol Weight: 52.04

MP (deg C): -27.9

BP (deg C): 21.17

BP pressure (mm Hg):

Property/Value	Units	Temp	Data Type	Reference
WS 1.19E+005	mg/L	25	EST	MEYLAN,WM ET AL. (1996)
logP 0.07			EXP	HANSCH,C ET AL. (1995)
VP 4.30E+003	mm Hg	25	EXP	DAUBERT,TE & DANNER,RP (1989)
DC	pKa			
HL	atm m3/mol			
OH 0.00E+000	cm3/molc sec	25	EST	MEYLAN,WM & HOWARD,PH (1993)

4,4,4-TRIFLUORO-1-BUTANOL

CAS #: 000461-18-7

Formula: $C_4H_7F_3O$

Mol Weight: 128.09

MP (deg C):

BP (deg C):

BP pressure (mm Hg):

Property/Value	Units	Temp	Data Type	Reference
WS 5.16E+004	mg/L	25	EST	MEYLAN,WM ET AL. (1996)
logP 0.90			EXP	MULLER,N (1986)
VP 2.33E+001	mm Hg	25	EST	NEELY,WB & BLAU,GE (1985)
DC	pKa			
HL 4.96E-005	atm m3/mol	25	EST	MEYLAN,WM & HOWARD,PH (1991)
OH 5.66E-012	cm3/molc sec	25	EST	MEYLAN,WM & HOWARD,PH (1993)

2-CYANOGUANIDINE

CAS #: 000461-58-5

Formula: $C_2H_4N_4$

Mol Weight: 84.08

MP (deg C): 209.5

BP (deg C):

BP pressure (mm Hg):

Property/Value	Units	Temp	Data Type	Reference
WS 4.13E+004	mg/L	25	EXP	YALKOWSKY,SH & DANNENFELSER,RM (1992)
logP -1.15			EXP	HANSCH,C & LEO,AJ (1985)
VP 1.52E-001	mm Hg	25	EST	NEELY,WB & BLAU,GE (1985)
DC	pKa			
HL 2.25E-010	atm m3/mol	25	EST	MEYLAN,WM & HOWARD,PH (1991)
OH 4.20E-011	cm3/molc sec	25	EST	MEYLAN,WM & HOWARD,PH (1993)

HYDANTOIN

CAS #: 000461-72-3

Formula: $C_3H_4N_2O_2$

Mol Weight: 100.08

MP (deg C): 220

BP (deg C):

BP pressure (mm Hg):

Property/Value	Units	Temp	Data Type	Reference
WS 2.95E+005	mg/L	100	EXP	YALKOWSKY,SH & DANNENFELSER,RM (1992)
logP -1.69			EXP	HANSCH,C & LEO,AJ (1985)
VP 7.47E-006	mm Hg	25	EST	NEELY,WB & BLAU,GE (1985)
DC 9.12	pKa	24	EXP	KORTUM,G ET AL (1961)
HL 1.57E-009	atm m3/mol	25	EST	MEYLAN,WM & HOWARD,PH (1991)
OH 4.30E-012	cm3/molc sec	25	EST	MEYLAN,WM & HOWARD,PH (1993)

CHLORPHENTERMINE

CAS #: 000461-78-9

Formula: $C_{10}H_{14}ClN$

Mol Weight: 183.68

MP (deg C):

BP (deg C): 100-102

BP pressure (mm Hg): 2.00E+000

Property/Value	Units	Temp	Data Type	Reference
WS 3.24E+003	mg/L	25	EST	MEYLAN,WM ET AL. (1996)
logP 2.60			EXP	HANSCH,C & LEO,AJ (1985)
VP 1.38E-002	mm Hg	25	EST	NEELY,WB & BLAU,GE (1985)
DC	pKa			
HL 1.06E-006	atm m3/mol	25	EST	MEYLAN,WM & HOWARD,PH (1991)
OH 2.42E-011	cm3/molc sec	25	EST	MEYLAN,WM & HOWARD,PH (1993)

CAS #: 000461-89-2 — 1,2,4-TRIAZINE-3,5(2H,4H)-DIONE

Formula: $C_3H_3N_3O_2$
Mol Weight: 113.08
MP (deg C): 274-275

Property/Value	Units	Temp	Data Type	Reference
WS 1.66E+004	mg/L	25	EST	MEYLAN,WM ET AL. (1996)
logP -0.59			EXP	HANSCH,C ET AL. (1995)
VP 1.74E-007	mm Hg	25	EST	NEELY,WB & BLAU,GE (1985)
DC	pKa			
HL 1.91E-009	atm m3/mol	25	EST	MEYLAN,WM & HOWARD,PH (1991)
OH 3.46E-012	cm3/molc sec	25	EST	MEYLAN,WM & HOWARD,PH (1993)

CAS #: 000461-98-3 — 2,6-DIMETHYL-4-PYRIMIDINAMINE

Formula: $C_6H_9N_3$
Mol Weight: 123.16
MP (deg C): 182-183

Property/Value	Units	Temp	Data Type	Reference
WS 6.40E+003	mg/L	18	EXP	YALKOWSKY,SH & DANNENFELSER,RM (1992)
logP 0.39			EXP	HANSCH,C & LEO,AJ (1985)
VP 2.00E-002	mm Hg	25	EST	NEELY,WB & BLAU,GE (1985)
DC 6.97	pKa	20	EXP	PERRIN,DD (1972)
HL 1.26E-009	atm m3/mol	25	EST	MEYLAN,WM & HOWARD,PH (1991)
OH 8.68E-011	cm3/molc sec	25	EST	MEYLAN,WM & HOWARD,PH (1993)

CAS #: 000462-06-6 — FLUOROBENZENE

Formula: C_6H_5F
Mol Weight: 96.11
MP (deg C): -40
BP (deg C): 84.73

Property/Value	Units	Temp	Data Type	Reference
WS 1.55E+003	mg/L	25	EXP	YALKOWSKY,SH & DANNENFELSER,RM (1992)
logP 2.27			EXP	HANSCH,C & LEO,AJ (1985)
VP 7.72E+001	mm Hg	25	EXP	DAUBERT,TE & DANNER,RP (1989)
DC	pKa			
HL 6.30E-003	atm m3/mol	25	EST	VP/WSOL
OH 6.90E-013	cm3/molc sec	25	EXP	ATKINSON,R (1989)

CAS #: 000462-08-8 — 3-AMINOPYRIDINE

Formula: $C_5H_6N_2$
Mol Weight: 94.12
MP (deg C): 64
BP (deg C): 250-252

Property/Value	Units	Temp	Data Type	Reference
WS 9.79E+004	mg/L	25	EST	MEYLAN,WM ET AL. (1996)
logP 0.11			EXP	HANSCH,C & LEO,AJ (1985)
VP 4.37E-001	mm Hg	25	EST	NEELY,WB & BLAU,GE (1985)
DC 6.03	pKa	24	EXP	PERRIN,DD (1965)
HL 2.49E-009	atm m3/mol	25	EST	MEYLAN,WM & HOWARD,PH (1991)
OH 2.04E-011	cm3/molc sec	25	EST	MEYLAN,WM & HOWARD,PH (1993)

CAS #: 000462-18-0 — TRIDECAN-7-ONE

Formula: $C_{13}H_{26}O$
Mol Weight: 198.35
MP (deg C): 33
BP (deg C): 261

Property/Value	Units	Temp	Data Type	Reference
WS 4.53E+000	mg/L	25	EST	MEYLAN,WM ET AL. (1996)
logP 4.68			EST	MEYLAN,WM & HOWARD,PH (1995)
VP 6.68E-003	mm Hg	25	EXT	ENGINEERING SCIENCE DATA UNIT (1975)
DC	pKa			
HL 8.43E-004	atm m3/mol	25	EST	MEYLAN,WM & HOWARD,PH (1991)
OH 1.90E-011	cm3/molc sec	25	EST	MEYLAN,WM & HOWARD,PH (1993)

CAS #: 000462-43-1 — 3-FLUOROPROPANOL

Formula: C_3H_7FO
Mol Weight: 78.09
BP (deg C): 127.8

Property/Value	Units	Temp	Data Type	Reference
WS 7.40E+005	mg/L	25	EST	MEYLAN,WM ET AL. (1996)
logP -0.28			EXP	HANSCH,C ET AL. (1995)
VP 2.98E+001	mm Hg	25	EST	NEELY,WB & BLAU,GE (1985)
DC	pKa			
HL 9.44E-006	atm m3/mol	25	EST	MEYLAN,WM & HOWARD,PH (1991)
OH 4.97E-012	cm3/molc sec	25	EST	MEYLAN,WM & HOWARD,PH (1993)

CAS #: 000462-60-2 — N-(AMINOCARBONYL)GLYCINE

Formula: $C_3H_6N_2O_3$
Mol Weight: 118.09

Property/Value	Units	Temp	Data Type	Reference
WS 8.08E+004	mg/L	25	EST	MEYLAN,WM ET AL. (1996)
logP -1.42			EST	MEYLAN,WM & HOWARD,PH (1995)
VP 2.47E-004	mm Hg	25	EST	NEELY,WB & BLAU,GE (1985)
DC 3.89	pKa	15	EXP	KORTUM,G ET AL (1961)
HL 1.19E-013	atm m3/mol	25	EST	MEYLAN,WM & HOWARD,PH (1991)
OH 5.39E-012	cm3/molc sec	25	EST	MEYLAN,WM & HOWARD,PH (1993)

CAS #: 000462-95-3 — DIETHOXYMETHANE

Formula: $C_5H_{12}O_2$
Mol Weight: 104.15
MP (deg C): -66.5
BP (deg C): 88

Property/Value	Units	Temp	Data Type	Reference
WS 7.00E+004	mg/L	18	EXP	YALKOWSKY,SH & DANNENFELSER,RM (1992)
logP 0.84			EXP	HANSCH,C & LEO,AJ (1985)
VP 3.41E+001	mm Hg	25	EXP	BOUBLIK,T ET AL. (1984)
DC	pKa			
HL 6.68E-005	atm m3/mol	25	EST	VP/WSOL
OH 1.68E-011	cm3/molc sec	25	EXP	ATKINSON,R (1989)

000463-04-7 — AMYL NITRITE

Formula: $C_5H_{11}NO_2$

Mol Weight: 117.15

MP (deg C):

FP (deg C):

BP (deg C): 104.5

BP pressure (mm Hg):

Property/Value	Units	Temp	Data Type	Reference
WS 3.81E+002	mg/L	25	EST	MEYLAN,WM ET AL. (1996)
logP 2.85			EST	MEYLAN,WM & HOWARD,PH (1995)
VP 3.97E+000	mm Hg	25	EXP	PERRY,RH & GREEN,D (1984)
DC	pKa			
HL 2.04E-004	atm m3/mol	25	EST	MEYLAN,WM & HOWARD,PH (1991)
OH 4.48E-012	cm3/molc sec	25	EST	MEYLAN,WM & HOWARD,PH (1993)

000463-11-6 — 1-FLUOROOCTANE

Formula: $C_8H_{17}F$

Mol Weight: 132.22

MP (deg C):

FP (deg C):

BP (deg C): 142.4

BP pressure (mm Hg):

Property/Value	Units	Temp	Data Type	Reference
WS 2.28E+001	mg/L	25	EST	MEYLAN,WM ET AL. (1996)
logP 4.21			EST	MEYLAN,WM & HOWARD,PH (1995)
VP 5.92E+000	mm Hg	25	EXP	LI,JCM & ROSSINI,FD (1953)
DC	pKa			
HL 1.19E-001	atm m3/mol	25	EST	MEYLAN,WM & HOWARD,PH (1991)
OH 7.78E-012	cm3/molc sec	25	EST	MEYLAN,WM & HOWARD,PH (1993)

000463-18-3 — 1-FLUORONONANE

Formula: $C_9H_{19}F$

Mol Weight: 146.25

MP (deg C): 1

FP (deg C):

BP (deg C): 166-169

BP pressure (mm Hg):

Property/Value	Units	Temp	Data Type	Reference
WS 7.60E+000	mg/L	25	EST	MEYLAN,WM ET AL. (1996)
logP 4.70			EST	MEYLAN,WM & HOWARD,PH (1995)
VP 6.44E+000	mm Hg	25	EST	NEELY,WB & BLAU,GE (1985)
DC	pKa			
HL 1.59E-001	atm m3/mol	25	EST	MEYLAN,WM & HOWARD,PH (1991)
OH 9.19E-012	cm3/molc sec	25	EST	MEYLAN,WM & HOWARD,PH (1993)

000463-40-1 — 9,12,15-OCTADECATRIENOIC ACID, (Z,Z,Z)-

Formula: $C_{18}H_{30}O_2$

Mol Weight: 278.44

MP (deg C): -16.5

FP (deg C):

BP (deg C): 231

BP pressure (mm Hg): 1.70E+001

Property/Value	Units	Temp	Data Type	Reference
WS 1.24E-001	mg/L	25	EST	MEYLAN,WM ET AL. (1996)
logP 6.46			EXP	SANGSTER,J (1993)
VP 5.40E-007	mm Hg	25	EXP	DAUBERT,TE & DANNER,RP (1993)
DC	pKa			
HL 3.47E-005	atm m3/mol	25	EST	MEYLAN,WM & HOWARD,PH (1991)
OH 1.82E-010	cm3/molc sec	25	EST	MEYLAN,WM & HOWARD,PH (1993)

000463-49-0 — 1,2-PROPADIENE

Formula: C_3H_4

Mol Weight: 40.07

MP (deg C): -136.2

FP (deg C):

BP (deg C): -34.4

BP pressure (mm Hg):

Property/Value	Units	Temp	Data Type	Reference
WS 2.15E+003	mg/L	25	EST	MEYLAN,WM ET AL. (1996)
logP 1.45			EXP	HANSCH,C & LEO,AJ (1985)
VP 5.43E+003	mm Hg	25	EXP	DAUBERT,TE & DANNER,RP (1989)
DC	pKa			
HL 6.17E-002	atm m3/mol	25	EST	MEYLAN,WM & HOWARD,PH (1991)
OH 9.82E-012	cm3/molc sec	25	EXP	ATKINSON,R (1989)

000463-51-4 — KETENE

Formula: C_2H_2O

Mol Weight: 42.04

MP (deg C): -150

FP (deg C):

BP (deg C): -56

BP pressure (mm Hg):

Property/Value	Units	Temp	Data Type	Reference
WS 3.63E+005	mg/L	25	EST	MEYLAN,WM ET AL. (1996)
logP -0.52			EST	MEYLAN,WM & HOWARD,PH (1995)
VP 1.04E+004	mm Hg	25	EXP	DAUBERT,TE & DANNER,RP (1989)
DC	pKa			
HL	atm m3/mol			
OH 1.73E-011	cm3/molc sec	25	EXP	ATKINSON,R (1989)

000463-58-1 — CARBONYL SULFIDE

Formula: COS

Mol Weight: 60.07

MP (deg C): -138.8

FP (deg C):

BP (deg C): -50.2

BP pressure (mm Hg):

Property/Value	Units	Temp	Data Type	Reference
WS 1.22E+003	mg/L	25	EXP	MACALUSO,P (1969)
logP -1.33			EST	MEYLAN,WM & HOWARD,PH (1995)
VP 9.41E+003	mm Hg	25	EXP	DAUBERT,TE & DANNER,RP (1985)
DC	pKa			
HL 6.10E-001	atm m3/mol	25	EST	VP/WSOL
OH 2.00E-015	cm3/molc sec	25	EXP	ATKINSON,R ET AL. (1992C)

000463-82-1 — 2,2-DIMETHYLPROPANE

Formula: C_5H_{12}

Mol Weight: 72.15

MP (deg C): -19.8

FP (deg C):

BP (deg C): 9.5

BP pressure (mm Hg):

Property/Value	Units	Temp	Data Type	Reference
WS 3.32E+001	mg/L	25	EXP	YALKOWSKY,SH & DANNENFELSER,RM (1992)
logP 3.11			EXP	HANSCH,C & LEO,AJ (1985)
VP 1.29E+003	mm Hg	25	EXP	RIDDICK,JA ET AL. (1986)
DC	pKa			
HL 2.17E+000	atm m3/mol	25	EST	VP/WSOL
OH 8.49E-013	cm3/molc sec	25	EXP	ATKINSON,R (1989)

CAS #: 000464-06-2				2,2,3-TRIMETHYLBUTANE

Formula: C_7H_{16}
Mol Weight: 100.21
MP (deg C): -25   FP (deg C):
BP (deg C): 80.8
BP pressure (mm Hg):

Property/Value	Units	Temp	Data Type	Reference
WS 2.89E+001	mg/L	25	EST	MEYLAN,WM ET AL. (1996)
logP 3.59			EST	MEYLAN,WM & HOWARD,PH (1995)
VP 1.02E+002	mm Hg	25	EXP	DAUBERT,TE & DANNER,RP (1989)
DC	pKa			
HL 2.27E+000	atm m3/mol	25	EST	MEYLAN,WM & HOWARD,PH (1991)
OH 4.23E-012	cm3/molc sec	25	EXP	ATKINSON,R (1989)

CAS #: 000464-07-3				3,3-DIMETHYL-2-BUTANOL

Formula: $C_6H_{14}O$
Mol Weight: 102.18
MP (deg C): 4.8   FP (deg C):
BP (deg C): 119-121
BP pressure (mm Hg):

Property/Value	Units	Temp	Data Type	Reference
WS 2.43E+004	mg/L	25	EXP	YALKOWSKY,SH & DANNENFELSER,RM (1992)
logP 1.48			EXP	HANSCH,C & LEO,AJ (1985)
VP 8.81E+000	mm Hg	25	EST	NEELY,WB & BLAU,GE (1985)
DC	pKa			
HL 1.76E-005	atm m3/mol	25	EST	MEYLAN,WM & HOWARD,PH (1991)
OH 9.16E-012	cm3/molc sec	25	EST	MEYLAN,WM & HOWARD,PH (1993)

CAS #: 000464-49-3				(1R,4R)-(+)-CAMPHOR

Formula: $C_{10}H_{16}O$
Mol Weight: 152.24
MP (deg C): 178.8   FP (deg C):
BP (deg C): 207.4
BP pressure (mm Hg):

Property/Value	Units	Temp	Data Type	Reference
WS 7.44E+002	mg/L	25	EST	MEYLAN,WM ET AL. (1996)
logP 2.34			EST	MEYLAN,WM & HOWARD,PH (1995)
VP 7.20E-002	mm Hg	25	EXT	OHE,S (1976)
DC	pKa			
HL 7.00E-005	atm m3/mol	25	EST	MEYLAN,WM & HOWARD,PH (1991)
OH 1.25E-011	cm3/molc sec	25	EST	MEYLAN,WM & HOWARD,PH (1993)

CAS #: 000465-15-6				CARD-20(22)-ENOLIDE, 16-(ACETYLOXY)-3,14-DIHYDRO

Formula: $C_{25}H_{36}O_6$
Mol Weight: 432.56
MP (deg C):   FP (deg C):
BP (deg C):
BP pressure (mm Hg):

Property/Value	Units	Temp	Data Type	Reference
WS 4.31E+001	mg/L	25	EST	MEYLAN,WM ET AL. (1996)
logP 1.93			EXP	SANGSTER,J (1993)
VP 9.93E-015	mm Hg	25	EST	NEELY,WB & BLAU,GE (1985)
DC	pKa			
HL 1.49E-012	atm m3/mol	25	EST	MEYLAN,WM & HOWARD,PH (1991)
OH 6.44E-011	cm3/molc sec	25	EST	MEYLAN,WM & HOWARD,PH (1993)

CAS #: 000465-16-7				OLEANDRIN

Formula: $C_{32}H_{48}O_9$
Mol Weight: 576.73
MP (deg C):   FP (deg C):
BP (deg C):
BP pressure (mm Hg):

Property/Value	Units	Temp	Data Type	Reference
WS 2.68E+000	mg/L	25	EST	MEYLAN,WM ET AL. (1996)
logP 2.53			EXP	HANSCH,C & LEO,AJ (1985)
VP 3.61E-019	mm Hg	25	EST	NEELY,WB & BLAU,GE (1985)
DC	pKa			
HL 2.06E-018	atm m3/mol	25	EST	MEYLAN,WM & HOWARD,PH (1991)
OH 1.04E-010	cm3/molc sec	25	EST	MEYLAN,WM & HOWARD,PH (1993)

CAS #: 000465-29-2				CAMPHORQUINONE

Formula: $C_{10}H_{14}O_2$
Mol Weight: 166.22
MP (deg C):   FP (deg C):
BP (deg C):
BP pressure (mm Hg):

Property/Value	Units	Temp	Data Type	Reference
WS 3.23E+003	mg/L	25	EST	MEYLAN,WM ET AL. (1996)
logP 1.52			EXP	HANSCH,C & LEO,AJ (1985)
VP 1.45E-002	mm Hg	25	EST	NEELY,WB & BLAU,GE (1985)
DC	pKa			
HL 2.67E-008	atm m3/mol	25	EST	MEYLAN,WM & HOWARD,PH (1991)
OH 8.83E-012	cm3/molc sec	25	EST	MEYLAN,WM & HOWARD,PH (1993)

CAS #: 000465-62-3				PSEUDO-STRYCHNINE

Formula: $C_{21}H_{22}N_2O_3$
Mol Weight: 350.42
MP (deg C):   FP (deg C):
BP (deg C):
BP pressure (mm Hg):

Property/Value	Units	Temp	Data Type	Reference
WS 1.72E+002	mg/L	25	EST	MEYLAN,WM ET AL. (1996)
logP 1.82			EXP	HANSCH,C & LEO,AJ (1985)
VP 1.48E-012	mm Hg	25	EST	NEELY,WB & BLAU,GE (1985)
DC	pKa			
HL 2.18E-018	atm m3/mol	25	EST	MEYLAN,WM & HOWARD,PH (1991)
OH 1.98E-010	cm3/molc sec	25	EST	MEYLAN,WM & HOWARD,PH (1993)

CAS #: 000465-65-6				NALOXONE

Formula: $C_{19}H_{21}NO_4$
Mol Weight: 327.38
MP (deg C):   FP (deg C):
BP (deg C):
BP pressure (mm Hg):

Property/Value	Units	Temp	Data Type	Reference
WS 1.42E+003	mg/L	25	EST	MEYLAN,WM ET AL. (1996)
logP 2.09			EXP	HANSCH,C & LEO,AJ (1985)
VP 4.25E-011	mm Hg	25	EST	NEELY,WB & BLAU,GE (1985)
DC	pKa			
HL 5.38E-019	atm m3/mol	25	EST	MEYLAN,WM & HOWARD,PH (1991)
OH 2.17E-010	cm3/molc sec	25	EST	MEYLAN,WM & HOWARD,PH (1993)

CAS #: 000465-73-6	ISODRIN

Formula: $C_{12}H_8Cl_6$

Mol Weight: 364.92

MP (deg C): 239-241

FP (deg C):

BP (deg C):

BP pressure (mm Hg):

Property/Value	Units	Temp	Data Type	Reference
WS 1.42E-002	mg/L	25	EST	MEYLAN,WM ET AL. (1996)
logP 6.75			EST	MEYLAN,WM & HOWARD,PH (1995)
VP 4.40E-005	mm Hg	25	EST	NEELY,WB & BLAU,GE (1985)
DC	pKa			
HL 3.87E-004	atm m3/mol	25	EST	MEYLAN,WM & HOWARD,PH (1991)
OH 6.66E-011	cm3/molc sec	25	EST	MEYLAN,WM & HOWARD,PH (1993)

CAS #: 000465-84-9	CYMAROL

Formula: $C_{30}H_{46}O_9$

Mol Weight: 550.70

MP (deg C):

FP (deg C):

BP (deg C):

BP pressure (mm Hg):

Property/Value	Units	Temp	Data Type	Reference
WS 1.12E+002	mg/L	25	EST	MEYLAN,WM ET AL. (1996)
logP 0.56			EXP	HANSCH,C & LEO,AJ (1985)
VP 1.09E-020	mm Hg	25	EST	NEELY,WB & BLAU,GE (1985)
DC	pKa			
HL 9.81E-019	atm m3/mol	25	EST	MEYLAN,WM & HOWARD,PH (1991)
OH 1.04E-010	cm3/molc sec	25	EST	MEYLAN,WM & HOWARD,PH (1993)

CAS #: 000466-06-8	PROSCILLARIDIN

Formula: $C_{30}H_{42}O_8$

Mol Weight: 530.66

MP (deg C): 219-222

FP (deg C):

BP (deg C):

BP pressure (mm Hg):

Property/Value	Units	Temp	Data Type	Reference
WS 3.47E+000	mg/L	25	EST	MEYLAN,WM ET AL. (1996)
logP 2.48			EXP	SANGSTER,J (1994)
VP 7.49E-021	mm Hg	25	EST	NEELY,WB & BLAU,GE (1985)
DC	pKa			
HL 6.57E-016	atm m3/mol	25	EST	MEYLAN,WM & HOWARD,PH (1991)
OH 2.35E-010	cm3/molc sec	25	EST	MEYLAN,WM & HOWARD,PH (1993)

CAS #: 000466-07-9	NERIIFOLIN

Formula: $C_{30}H_{46}O_8$

Mol Weight: 534.70

MP (deg C): 218-225

FP (deg C):

BP (deg C):

BP pressure (mm Hg):

Property/Value	Units	Temp	Data Type	Reference
WS 6.89E+000	mg/L	25	EST	MEYLAN,WM ET AL. (1996)
logP 2.10			EXP	HANSCH,C & LEO,AJ (1985)
VP 1.81E-019	mm Hg	25	EST	NEELY,WB & BLAU,GE (1985)
DC	pKa			
HL 3.17E-017	atm m3/mol	25	EST	MEYLAN,WM & HOWARD,PH (1991)
OH 1.08E-010	cm3/molc sec	25	EST	MEYLAN,WM & HOWARD,PH (1993)

CAS #: 000468-50-8	4-PIPERIDINOL, 3-ETHYL-1-METHYL-4-PHENYL-, PROPI

Formula: $C_{17}H_{25}NO_2$

Mol Weight: 275.39

MP (deg C):

FP (deg C):

BP (deg C):

BP pressure (mm Hg):

Property/Value	Units	Temp	Data Type	Reference
WS 1.43E+002	mg/L	25	EST	MEYLAN,WM ET AL. (1996)
logP 3.61			EXP	CRAIG (1990)
VP 1.89E-005	mm Hg	25	EST	NEELY,WB & BLAU,GE (1985)
DC	pKa			
HL 1.69E-008	atm m3/mol	25	EST	MEYLAN,WM & HOWARD,PH (1991)
OH 1.00E-010	cm3/molc sec	25	EST	MEYLAN,WM & HOWARD,PH (1993)

CAS #: 000469-01-2	6H-[1,3]DIOXOLO[5,6]BENZOFURO[3,2-C][1]BENZOPYRA

Formula: $C_{17}H_{14}O_6$

Mol Weight: 314.30

MP (deg C):

FP (deg C):

BP (deg C):

BP pressure (mm Hg):

Property/Value	Units	Temp	Data Type	Reference
WS 2.05E+002	mg/L	25	EST	MEYLAN,WM ET AL. (1996)
logP 2.58			EXP	ARNOLDI,A & MERLINI,L (1990)
VP 3.61E-010	mm Hg	25	EST	NEELY,WB & BLAU,GE (1985)
DC	pKa			
HL 8.21E-016	atm m3/mol	25	EST	MEYLAN,WM & HOWARD,PH (1991)
OH 2.46E-010	cm3/molc sec	25	EST	MEYLAN,WM & HOWARD,PH (1993)

CAS #: 000469-62-5	PROPOXYPHENE

Formula: $C_{22}H_{29}NO_2$

Mol Weight: 339.48

MP (deg C): 75-76

FP (deg C):

BP (deg C):

BP pressure (mm Hg):

Property/Value	Units	Temp	Data Type	Reference
WS 1.97E+001	mg/L	25	EST	MEYLAN,WM ET AL. (1996)
logP 4.18			EXP	HANSCH,C & LEO,AJ (1985)
VP 5.54E-007	mm Hg	25	EST	NEELY,WB & BLAU,GE (1985)
DC	pKa			
HL 2.34E-009	atm m3/mol	25	EST	MEYLAN,WM & HOWARD,PH (1991)
OH 9.92E-011	cm3/molc sec	25	EST	MEYLAN,WM & HOWARD,PH (1993)

CAS #: 000470-17-7	NAPHTHO[2,3-B]FURAN-2(3H)-ONE, DECAHYDRO-8A-METH

Formula: $C_{15}H_{20}O_2$

Mol Weight: 232.33

MP (deg C):

FP (deg C):

BP (deg C):

BP pressure (mm Hg):

Property/Value	Units	Temp	Data Type	Reference
WS 3.55E+001	mg/L	25	EST	MEYLAN,WM ET AL. (1996)
logP 3.42			EXP	SANGSTER,J (1993)
VP 3.89E-005	mm Hg	25	EST	NEELY,WB & BLAU,GE (1985)
DC	pKa			
HL 1.86E-004	atm m3/mol	25	EST	MEYLAN,WM & HOWARD,PH (1991)
OH 8.35E-011	cm3/molc sec	25	EST	MEYLAN,WM & HOWARD,PH (1993)

000470-67-7 — 1,4-CINEOL

Formula: $C_{10}H_{18}O$
Mol Weight: 154.25
MP (deg C): 1
FP (deg C):
BP (deg C): 173.5
BP pressure (mm Hg):

Property/Value	Units	Temp	Data Type	Reference
WS 1.53E+002	mg/L	25	EST	MEYLAN,WM ET AL. (1996)
logP 3.13			EST	MEYLAN,WM & HOWARD,PH (1995)
VP 1.93E+000	mm Hg	25	EXP	WAYAKU,M (1982)
DC	pKa			
HL 2.04E-004	atm m3/mol	25	EST	MEYLAN,WM & HOWARD,PH (1991)
OH 1.79E-011	cm3/molc sec	25	EST	MEYLAN,WM & HOWARD,PH (1993)

000470-82-6 — 1,8-CINEOLE

Formula: $C_{10}H_{18}O$
Mol Weight: 154.25
MP (deg C): 1.5
FP (deg C):
BP (deg C): 176-177
BP pressure (mm Hg):

Property/Value	Units	Temp	Data Type	Reference
WS 3.50E+003	mg/L	21	EXP	YALKOWSKY,SH & DANNENFELSER,RM (1992)
logP 2.50			EXP	HANSCH,C ET AL. (1995)
VP 1.90E+000	mm Hg	25	EXP	RIDDICK,JA ET AL. (1986)
DC	pKa			
HL 1.10E-004	atm m3/mol	25	EST	VP/WSOL
OH 1.11E-011	cm3/molc sec	25	EXP	ATKINSON,R (1989)

000470-90-6 — CHLORFENVINPHOS

Formula: $C_{12}H_{14}Cl_3O_4P$
Mol Weight: 359.58
MP (deg C): -20
FP (deg C):
BP (deg C): 167-170
BP pressure (mm Hg): 5.00E-001

Property/Value	Units	Temp	Data Type	Reference
WS 1.24E+002	mg/L	20	EXP	BOWMAN,BT & SANS,WW (1983)
logP 3.81			EXP	SANGSTER,J (1994)
VP 4.00E-006	mm Hg	20	EST	WORTHING,CR & WALKER,SB (1987)
DC	pKa			
HL 1.53E-008	atm m3/mol	20	EST	VP/WSOL
OH 5.31E-011	cm3/molc sec	25	EST	MEYLAN,WM & HOWARD,PH (1993)

000471-03-4 — 1,1'-SULFONYLBIS(2-CHLOROETHANE)

Formula: $C_4H_8Cl_2O_2S$
Mol Weight: 191.08
MP (deg C):
FP (deg C):
BP (deg C):
BP pressure (mm Hg):

Property/Value	Units	Temp	Data Type	Reference
WS 2.28E+004	mg/L	25	EST	MEYLAN,WM ET AL. (1996)
logP 0.38			EST	MEYLAN,WM & HOWARD,PH (1995)
VP 2.05E-003	mm Hg	25	EST	NEELY,WB & BLAU,GE (1985)
DC	pKa			
HL 1.36E-006	atm m3/mol	25	EST	MEYLAN,WM & HOWARD,PH (1991)
OH 6.12E-012	cm3/molc sec	25	EST	MEYLAN,WM & HOWARD,PH (1993)

000471-29-4 — METHYLGUANIDINE

Formula: $C_2H_7N_3$
Mol Weight: 73.10
MP (deg C):
FP (deg C):
BP (deg C):
BP pressure (mm Hg):

Property/Value	Units	Temp	Data Type	Reference
WS 1.78E+003	mg/L	20	EXP	GREENWALD,I (1926)
logP -1.16			EST	MEYLAN,WM & HOWARD,PH (1995)
VP 3.28E+000	mm Hg	25	EST	NEELY,WB & BLAU,GE (1985)
DC	pKa			
HL 5.13E-011	atm m3/mol	25	EST	MEYLAN,WM & HOWARD,PH (1991)
OH 8.53E-011	cm3/molc sec	25	EST	MEYLAN,WM & HOWARD,PH (1993)

000471-46-5 — OXAMIDE

Formula: $C_2H_4N_2O_2$
Mol Weight: 88.07
MP (deg C): 350 dec
FP (deg C):
BP (deg C):
BP pressure (mm Hg):

Property/Value	Units	Temp	Data Type	Reference
WS 3.70E+002	mg/L	7	EXP	YALKOWSKY,SH & DANNENFELSER,RM (1992)
logP -2.45			EST	MEYLAN,WM & HOWARD,PH (1995)
VP 6.35E-008	mm Hg	25	EXP	JONES,AH (1960)
DC	pKa			
HL 1.59E-011	atm m3/mol	25	EST	MEYLAN,WM & HOWARD,PH (1991)
OH 4.00E-012	cm3/molc sec	25	EST	MEYLAN,WM & HOWARD,PH (1993)

000471-47-6 — OXAMIC ACID

Formula: $C_2H_3NO_3$
Mol Weight: 89.05
MP (deg C): 210 dec
FP (deg C):
BP (deg C):
BP pressure (mm Hg):

Property/Value	Units	Temp	Data Type	Reference
WS 1.00E+006	mg/L	25	EST	MEYLAN,WM ET AL. (1996)
logP -2.40			EST	MEYLAN,WM & HOWARD,PH (1995)
VP 4.64E-007	mm Hg	25	EXP	JONES,AH (1960)
DC	pKa			
HL 4.92E-013	atm m3/mol	25	EST	MEYLAN,WM & HOWARD,PH (1991)
OH 2.52E-012	cm3/molc sec	25	EST	MEYLAN,WM & HOWARD,PH (1993)

000473-03-0 — AMBREIN

Formula: $C_{30}H_{52}O$
Mol Weight: 428.75
MP (deg C):
FP (deg C):
BP (deg C):
BP pressure (mm Hg):

Property/Value	Units	Temp	Data Type	Reference
WS 6.90E-007	mg/L	25	EST	MEYLAN,WM ET AL. (1996)
logP 11.68			EST	MEYLAN,WM & HOWARD,PH (1995)
VP 7.35E-011	mm Hg	25	EST	NEELY,WB & BLAU,GE (1985)
DC	pKa			
HL 7.80E-004	atm m3/mol	25	EST	MEYLAN,WM & HOWARD,PH (1991)
OH 1.77E-010	cm3/molc sec	25	EST	MEYLAN,WM & HOWARD,PH (1993)

000474-25-9 — CHENODEOXYCHOLIC ACID

Formula: $C_{24}H_{40}O_4$
Mol Weight: 392.58
MP (deg C): 119

Property/Value	Units	Temp	Data Type	Reference
WS 8.99E+001	mg/L	20	EXP	YALKOWSKY,SH & DANNENFELSER,RM (1992)
logP 4.15			EXP	SANGSTER,J (1993)
VP 3.00E-013	mm Hg	25	EST	NEELY,WB & BLAU,GE (1985)
DC	pKa			
HL 1.41E-011	atm m3/mol	25	EST	MEYLAN,WM & HOWARD,PH (1991)
OH 4.64E-011	cm3/molc sec	25	EST	MEYLAN,WM & HOWARD,PH (1993)

000475-26-3 — DFDT

Formula: $C_{14}H_9Cl_3F_2$
Mol Weight: 321.58
MP (deg C): 44-45

Property/Value	Units	Temp	Data Type	Reference
WS 8.28E-002	mg/L	25	EST	MEYLAN,WM ET AL. (1996)
logP 5.91			EST	MEYLAN,WM & HOWARD,PH (1995)
VP 1.18E-004	mm Hg	25	EST	NEELY,WB & BLAU,GE (1985)
DC	pKa			
HL 3.79E-005	atm m3/mol	25	EST	MEYLAN,WM & HOWARD,PH (1991)
OH 6.01E-012	cm3/molc sec	25	EST	MEYLAN,WM & HOWARD,PH (1993)

000475-31-0 — GLYCOCHOLIC ACID

Formula: $C_{26}H_{43}NO_6$
Mol Weight: 465.64
MP (deg C): 130

Property/Value	Units	Temp	Data Type	Reference
WS 3.30E+000	mg/L	20	EXP	YALKOWSKY,SH & DANNENFELSER,RM (1992)
logP 1.65			EXP	RODA,A ET AL. (1990)
VP 6.48E-020	mm Hg	25	EST	NEELY,WB & BLAU,GE (1985)
DC	pKa			
HL 3.43E-018	atm m3/mol	25	EST	MEYLAN,WM & HOWARD,PH (1991)
OH 6.51E-011	cm3/molc sec	25	EST	MEYLAN,WM & HOWARD,PH (1993)

000475-38-7 — 1,4-NAPHTHALENEDIONE, 5,8-DIHYDROXY-

Formula: $C_{10}H_6O_4$
Mol Weight: 190.16
MP (deg C): 232

Property/Value	Units	Temp	Data Type	Reference
WS 5.52E+003	mg/L	25	EST	MEYLAN,WM ET AL. (1996)
logP 1.79			EXP	SANGSTER,J (1994)
VP 9.95E-008	mm Hg	25	EST	NEELY,WB & BLAU,GE (1985)
DC	pKa			
HL 3.38E-011	atm m3/mol	25	EST	MEYLAN,WM & HOWARD,PH (1991)
OH 1.14E-011	cm3/molc sec	25	EST	MEYLAN,WM & HOWARD,PH (1993)

000476-73-3 — MELLOPHANIC ACID

Formula: $C_{10}H_6O_8$
Mol Weight: 254.15

Property/Value	Units	Temp	Data Type	Reference
WS 1.68E+004	mg/L	25	EST	MEYLAN,WM ET AL. (1996)
logP 0.15			EST	MEYLAN,WM & HOWARD,PH (1995)
VP 8.74E-011	mm Hg	25	EST	NEELY,WB & BLAU,GE (1985)
DC 2.05	pKa		EXP	KORTUM,G ET AL (1961)
HL 8.84E-022	atm m3/mol	25	EST	MEYLAN,WM & HOWARD,PH (1991)
OH 2.10E-012	cm3/molc sec	25	EST	MEYLAN,WM & HOWARD,PH (1993)

000477-27-0 — COLCHICEINE

Formula: $C_{21}H_{23}NO_6$
Mol Weight: 385.42
MP (deg C): 178-179

Property/Value	Units	Temp	Data Type	Reference
WS 3.56E+002	mg/L	25	EST	MEYLAN,WM ET AL. (1996)
logP 1.20			EXP	HANSCH,C ET AL. (1995)
VP 8.88E-016	mm Hg	25	EST	NEELY,WB & BLAU,GE (1985)
DC	pKa			
HL 3.37E-017	atm m3/mol	25	EST	MEYLAN,WM & HOWARD,PH (1991)
OH 3.24E-010	cm3/molc sec	25	EST	MEYLAN,WM & HOWARD,PH (1993)

000479-23-2 — CHOLANTHRENE

Formula: $C_{20}H_{14}$
Mol Weight: 254.33
MP (deg C): 173

Property/Value	Units	Temp	Data Type	Reference
WS 3.50E-003	mg/L	27	EXP	YALKOWSKY,SH & DANNENFELSER,RM (1992)
logP 6.50			EST	MEYLAN,WM & HOWARD,PH (1995)
VP 7.23E-008	mm Hg	25	EST	NEELY,WB & BLAU,GE (1985)
DC	pKa			
HL 2.69E-006	atm m3/mol	25	EST	MEYLAN,WM & HOWARD,PH (1991)
OH 1.53E-010	cm3/molc sec	25	EST	MEYLAN,WM & HOWARD,PH (1993)

000479-27-6 — 1,8-NAPHTHALENEDIAMINE

Formula: $C_{10}H_{10}N_2$
Mol Weight: 158.20
MP (deg C): 66.5
BP (deg C): 205
BP pressure (mm Hg): 1.20E+001

Property/Value	Units	Temp	Data Type	Reference
WS 8.50E+002	mg/L	25	EXP	CHEM INSPECT TEST INST (1992)
logP 1.78			EXP	SCHULTZ,TW & APPLEHANS,FM (1985)
VP 2.93E-005	mm Hg	25	EST	NEELY,WB & BLAU,GE (1985)
DC 4.44	pKa	25	EXP	PERRIN,DD (1972)
HL 6.57E-011	atm m3/mol	25	EST	MEYLAN,WM & HOWARD,PH (1991)
OH 2.00E-010	cm3/molc sec	25	EST	MEYLAN,WM & HOWARD,PH (1993)

CAS #: 000479-45-8 — TETRYL

Formula: $C_7H_5N_5O_8$
Mol Weight: 287.15
MP (deg C): 130-132
FP (deg C):
BP (deg C):
BP pressure (mm Hg):

Property/Value	Units	Temp	Data Type	Reference
WS 7.40E+001	mg/L		EXP	YALKOWSKY,SH & DANNENFELSER,RM (1992)
logP 1.64			EST	MEYLAN,WM & HOWARD,PH (1995)
VP 5.66E-008	mm Hg	25	EST	NEELY,WB & BLAU,GE (1985)
DC	pKa			
HL 2.71E-009	atm m3/mol	25	EST	MEYLAN,WM & HOWARD,PH (1991)
OH 1.27E-012	cm3/molc sec	25	EST	MEYLAN,WM & HOWARD,PH (1993)

CAS #: 000479-47-0 — 1,2,3,5-BENZENETETRACARBOXYLIC ACID

Formula: $C_{10}H_6O_8$
Mol Weight: 254.15
MP (deg C):
FP (deg C):
BP (deg C):
BP pressure (mm Hg):

Property/Value	Units	Temp	Data Type	Reference
WS 8.54E+003	mg/L	25	EST	MEYLAN,WM ET AL. (1996)
logP 0.49			EST	MEYLAN,WM & HOWARD,PH (1995)
VP 8.74E-011	mm Hg	25	EST	NEELY,WB & BLAU,GE (1985)
DC 2.38	pKa		EXP	KORTUM,G ET AL (1961)
HL 8.84E-022	atm m3/mol	25	EST	MEYLAN,WM & HOWARD,PH (1991)
OH 2.09E-012	cm3/molc sec	25	EST	MEYLAN,WM & HOWARD,PH (1993)

CAS #: 000479-92-5 — 3H-PYRAZOL-3-ONE, 1,2-DIHYDRO-1,5-DIMETHYL-4-(1-

Formula: $C_{14}H_{18}N_2O$
Mol Weight: 230.31
MP (deg C): 103
FP (deg C):
BP (deg C):
BP pressure (mm Hg):

Property/Value	Units	Temp	Data Type	Reference
WS 3.00E+006	mg/L	15	EXP	YALKOWSKY,SH & DANNENFELSER,RM (1992)
logP 1.94			EXP	SANGSTER,J (1994)
VP 5.20E-006	mm Hg	25	EST	NEELY,WB & BLAU,GE (1985)
DC	pKa			
HL 1.84E-009	atm m3/mol	25	EST	MEYLAN,WM & HOWARD,PH (1991)
OH 3.92E-011	cm3/molc sec	25	EST	MEYLAN,WM & HOWARD,PH (1993)

CAS #: 000480-16-0 — MORIN

Formula: $C_{15}H_{10}O_7$
Mol Weight: 302.24
MP (deg C): 303.5
FP (deg C):
BP (deg C):
BP pressure (mm Hg):

Property/Value	Units	Temp	Data Type	Reference
WS 2.50E+002	mg/L	20	EXP	YALKOWSKY,SH & DANNENFELSER,RM (1992)
logP 1.54			EXP	HANSCH,C ET AL. (1995)
VP 2.81E-014	mm Hg	25	EST	NEELY,WB & BLAU,GE (1985)
DC	pKa			
HL 6.60E-021	atm m3/mol	25	EST	MEYLAN,WM & HOWARD,PH (1991)
OH 2.39E-010	cm3/molc sec	25	EST	MEYLAN,WM & HOWARD,PH (1993)

CAS #: 000480-18-2 — 3,5,7,3',4'-PENTAHYDROXYFLAVANONE

Formula: $C_{15}H_{12}O_7$
Mol Weight: 304.26
MP (deg C):
FP (deg C):
BP (deg C):
BP pressure (mm Hg):

Property/Value	Units	Temp	Data Type	Reference
WS 2.20E+004	mg/L	25	EST	MEYLAN,WM ET AL. (1996)
logP 0.95			EXP	PERRISSOUD,D & TESTA,B (1986)
VP 1.34E-013	mm Hg	25	EST	NEELY,WB & BLAU,GE (1985)
DC	pKa			
HL 4.76E-022	atm m3/mol	25	EST	MEYLAN,WM & HOWARD,PH (1991)
OH 2.53E-010	cm3/molc sec	25	EST	MEYLAN,WM & HOWARD,PH (1993)

CAS #: 000480-40-0 — 5,7-HYDROXYFLAVONE

Formula: $C_{15}H_{10}O_4$
Mol Weight: 254.24
MP (deg C): 285
FP (deg C):
BP (deg C):
BP pressure (mm Hg):

Property/Value	Units	Temp	Data Type	Reference
WS 8.40E+001	mg/L	25	EST	MEYLAN,WM ET AL. (1996)
logP 3.52			EXP	PERRISSOUD,D & TESTA,B (1986)
VP 1.57E-009	mm Hg	25	EST	NEELY,WB & BLAU,GE (1985)
DC	pKa			
HL 4.92E-013	atm m3/mol	25	EST	MEYLAN,WM & HOWARD,PH (1991)
OH 2.31E-010	cm3/molc sec	25	EST	MEYLAN,WM & HOWARD,PH (1993)

CAS #: 000480-41-1 — NARINGENINE

Formula: $C_{15}H_{12}O_5$
Mol Weight: 272.26
MP (deg C): 251
FP (deg C):
BP (deg C):
BP pressure (mm Hg):

Property/Value	Units	Temp	Data Type	Reference
WS 4.75E+002	mg/L	25	EST	MEYLAN,WM ET AL. (1996)
logP 2.52			EXP	PERRISSOUD,D & TESTA,B (1986)
VP 1.38E-010	mm Hg	25	EST	NEELY,WB & BLAU,GE (1985)
DC	pKa			
HL 2.93E-017	atm m3/mol	25	EST	MEYLAN,WM & HOWARD,PH (1991)
OH 2.47E-010	cm3/molc sec	25	EST	MEYLAN,WM & HOWARD,PH (1993)

CAS #: 000480-96-6 — BENZOFURAZAN, 1-OXIDE

Formula: $C_6H_4N_2O_2$
Mol Weight: 136.11
MP (deg C): 69-71
FP (deg C):
BP (deg C):
BP pressure (mm Hg):

Property/Value	Units	Temp	Data Type	Reference
WS 5.22E+003	mg/L	25	EST	MEYLAN,WM ET AL. (1996)
logP 1.43			EXP	CALVINO,R ET AL. (1992)
VP 5.23E-004	mm Hg	25	EST	NEELY,WB & BLAU,GE (1985)
DC	pKa			
HL 1.85E-010	atm m3/mol	25	EST	MEYLAN,WM & HOWARD,PH (1991)
OH 9.00E-012	cm3/molc sec	25	EST	MEYLAN,WM & HOWARD,PH (1993)

CAS #:	000481-06-1			SANTONIN

Formula: $C_{15}H_{18}O_3$

Mol Weight: 246.31

MP (deg C): 175   FP (deg C):

BP (deg C):

BP pressure (mm Hg):

Property/ Value	Units	Temp	Data Type	Reference
WS 2.00E+002	mg/L		EXP	YALKOWSKY,SH & DANNENFELSER,RM (1992)
logP 1.78			EST	MEYLAN,WM & HOWARD,PH (1995)
VP 2.38E-006	mm Hg	25	EST	NEELY,WB & BLAU,GE (1985)
DC	pKa			
HL 5.27E-008	atm m3/mol	25	EST	MEYLAN,WM & HOWARD,PH (1991)
OH 5.99E-011	cm3/molc sec	25	EST	MEYLAN,WM & HOWARD,PH (1993)

CAS #:	000481-39-0			5-HYDROXY-1,4-NAPHTHOQUINONE

Formula: $C_{10}H_6O_3$

Mol Weight: 174.16

MP (deg C): 155   FP (deg C):

BP (deg C):

BP pressure (mm Hg):

Property/ Value	Units	Temp	Data Type	Reference
WS 5.12E+003	mg/L	25	EST	MEYLAN,WM ET AL. (1996)
logP 1.92			EXP	HANSCH,C & LEO,AJ (1985)
VP 6.34E-006	mm Hg	25	EST	NEELY,WB & BLAU,GE (1985)
DC	pKa			
HL 2.58E-010	atm m3/mol	25	EST	MEYLAN,WM & HOWARD,PH (1991)
OH 1.52E-011	cm3/molc sec	25	EST	MEYLAN,WM & HOWARD,PH (1993)

CAS #:	000482-05-3			1,1'-BIPHENYL -2,2'-DICARBOXYLIC ACID

Formula: $C_{14}H_{10}O_4$

Mol Weight: 242.23

MP (deg C): 228-229   FP (deg C):

BP (deg C):

BP pressure (mm Hg):

Property/ Value	Units	Temp	Data Type	Reference
WS 1.00E+002	mg/L	25	EXP	CHEM INSPECT TEST INST (1992)
logP 2.07			EXP	HANSCH,C ET AL. (1995)
VP 9.63E-009	mm Hg	25	EST	NEELY,WB & BLAU,GE (1985)
DC	pKa			
HL 1.68E-013	atm m3/mol	25	EST	MEYLAN,WM & HOWARD,PH (1991)
OH 3.55E-012	cm3/molc sec	25	EST	MEYLAN,WM & HOWARD,PH (1993)

CAS #:	000482-89-3			2-(1,3-DIHYDRO-3-OXO-2H-INDOL-2-YLIDENE)-1,2-DI*

Formula: $C_{16}H_{10}N_2O_2$

Mol Weight: 262.27

MP (deg C): 390 dec   FP (deg C):

BP (deg C):

BP pressure (mm Hg):

Property/ Value	Units	Temp	Data Type	Reference
WS 1.34E+001	mg/L	25	EST	MEYLAN,WM ET AL. (1996)
logP 3.72			EXP	HANSCH,C ET AL. (1995)
VP 9.00E-013	mm Hg	25	EXP	BAUGHMAN,GL & PERENICH,TA (1988)
DC	pKa			
HL 5.01E-014	atm m3/mol	25	EST	MEYLAN,WM & HOWARD,PH (1991)
OH 8.26E-011	cm3/molc sec	25	EST	MEYLAN,WM & HOWARD,PH (1993)

CAS #:	000483-18-1			6',7',10,11-TETRAMETHOXYEMETAN

Formula: $C_{29}H_{40}N_2O_4$

Mol Weight: 480.65

MP (deg C): 74   FP (deg C):

BP (deg C):

BP pressure (mm Hg):

Property/ Value	Units	Temp	Data Type	Reference
WS 9.84E+002	mg/L	15	EXP	STEPHEN,H & STEPHEN,T (1963)
logP 5.20			EST	MEYLAN,WM & HOWARD,PH (1995)
VP 2.47E-012	mm Hg	25	EST	NEELY,WB & BLAU,GE (1985)
DC	pKa			
HL 4.99E-016	atm m3/mol	25	EST	MEYLAN,WM & HOWARD,PH (1991)
OH 2.74E-010	cm3/molc sec	25	EST	MEYLAN,WM & HOWARD,PH (1993)

CAS #:	000483-54-5			2,3-DIMETHOXY-5,6-DIMETHYLBENZOQUINONE

Formula: $C_{10}H_{12}O_4$

Mol Weight: 196.20

MP (deg C): 62.5   FP (deg C):

BP (deg C):

BP pressure (mm Hg):

Property/ Value	Units	Temp	Data Type	Reference
WS 3.84E+003	mg/L	25	EST	MEYLAN,WM ET AL. (1996)
logP 1.26			EXP	HANSCH,C ET AL. (1995)
VP 3.97E-004	mm Hg	25	EST	NEELY,WB & BLAU,GE (1985)
DC	pKa			
HL 1.30E-009	atm m3/mol	25	EST	MEYLAN,WM & HOWARD,PH (1991)
OH 3.82E-011	cm3/molc sec	25	EST	MEYLAN,WM & HOWARD,PH (1993)

CAS #:	000483-55-6			2-ME-3-OH-1,4-NAPHTHOQUINONE

Formula: $C_{11}H_8O_3$

Mol Weight: 188.18

MP (deg C): 173.5   FP (deg C):

BP (deg C):

BP pressure (mm Hg):

Property/ Value	Units	Temp	Data Type	Reference
WS 4.74E+003	mg/L	25	EST	MEYLAN,WM ET AL. (1996)
logP 1.20			EXP	HANSCH,C & LEO,AJ (1985)
VP 1.38E-007	mm Hg	25	EST	NEELY,WB & BLAU,GE (1985)
DC	pKa			
HL 3.82E-009	atm m3/mol	25	EST	MEYLAN,WM & HOWARD,PH (1991)
OH 1.45E-011	cm3/molc sec	25	EST	MEYLAN,WM & HOWARD,PH (1993)

CAS #:	000485-31-4			BINAPACRYL

Formula: $C_{15}H_{18}N_2O_6$

Mol Weight: 322.32

MP (deg C): 66-67   FP (deg C):

BP (deg C):

BP pressure (mm Hg):

Property/ Value	Units	Temp	Data Type	Reference
WS 1.00E+000	mg/L	20	EXP	WORTHING,CR & WALKER,SR (1987)
logP 4.49			EST	MEYLAN,WM & HOWARD,PH (1995)
VP 3.16E-007	mm Hg	20	EXP	WORTHING,CR & WALKER,SR (1987)
DC	pKa			
HL 1.34E-007	atm m3/mol	20	EST	VP/WSOL
OH 8.32E-011	cm3/molc sec	25	EST	ATKINSON,R (1987)

000485-35-8 — CYTISINE

Formula: $C_{11}H_{14}N_2O$
Mol Weight: 190.25
MP (deg C): 152-153
FP (deg C):
BP (deg C):
BP pressure (mm Hg):

Property/Value	Units	Temp	Data Type	Reference
WS 4.39E+005	mg/L	16	EXP	YALKOWSKY,SH & DANNENFELSER,RM (1992)
logP 0.26			EST	MEYLAN,WM & HOWARD,PH (1995)
VP 1.57E-005	mm Hg	25	EST	NEELY,WB & BLAU,GE (1985)
DC	pKa			
HL 4.01E-012	atm m3/mol	25	EST	MEYLAN,WM & HOWARD,PH (1991)
OH 8.59E-011	cm3/molc sec	25	EST	MEYLAN,WM & HOWARD,PH (1993)

000485-47-2 — 1H-INDENE-1,3(2H)-DIONE, 2,2-DIHYDROXY-

Formula: $C_9H_6O_4$
Mol Weight: 178.15
MP (deg C): 241-243 de
FP (deg C):
BP (deg C):
BP pressure (mm Hg):

Property/Value	Units	Temp	Data Type	Reference
WS 1.51E+004	mg/L	25	EST	MEYLAN,WM ET AL. (1996)
logP 0.67			EXP	HANSCH,C ET AL. (1995)
VP 2.36E-007	mm Hg	25	EST	NEELY,WB & BLAU,GE (1985)
DC	pKa			
HL 3.38E-011	atm m3/mol	25	EST	MEYLAN,WM & HOWARD,PH (1991)
OH 1.03E-012	cm3/molc sec	25	EST	MEYLAN,WM & HOWARD,PH (1993)

000485-71-2 — CINCHONIDINE

Formula: $C_{19}H_{22}N_2O$
Mol Weight: 294.40
MP (deg C): 210
FP (deg C):
BP (deg C):
BP pressure (mm Hg):

Property/Value	Units	Temp	Data Type	Reference
WS 2.00E+002	mg/L	25	EXP	YALKOWSKY,SH & DANNENFELSER,RM (1992)
logP 2.82			EXP	SANGSTER,J (1994)
VP 9.04E-010	mm Hg	25	EST	NEELY,WB & BLAU,GE (1985)
DC	pKa			
HL 1.45E-014	atm m3/mol	25	EST	MEYLAN,WM & HOWARD,PH (1991)
OH 1.33E-010	cm3/molc sec	25	EST	MEYLAN,WM & HOWARD,PH (1993)

000486-12-4 — TRIPOLIDINE

Formula: $C_{19}H_{22}N_2$
Mol Weight: 278.40
MP (deg C): 59-61
FP (deg C):
BP (deg C):
BP pressure (mm Hg):

Property/Value	Units	Temp	Data Type	Reference
WS 7.49E+001	mg/L	25	EST	MEYLAN,WM ET AL. (1996)
logP 3.92			EXP	HANSCH,C ET AL. (1995)
VP 8.74E-007	mm Hg	25	EST	NEELY,WB & BLAU,GE (1985)
DC	pKa			
HL 1.08E-010	atm m3/mol	25	EST	MEYLAN,WM & HOWARD,PH (1991)
OH 1.86E-010	cm3/molc sec	25	EST	MEYLAN,WM & HOWARD,PH (1993)

000486-25-9 — FLUORENONE

Formula: $C_{13}H_8O$
Mol Weight: 180.21
MP (deg C): 84
FP (deg C):
BP (deg C): 341.5
BP pressure (mm Hg):

Property/Value	Units	Temp	Data Type	Reference
WS 2.53E+001	mg/L	25	EST	MEYLAN,WM ET AL. (1996)
logP 3.58			EXP	HANSCH,C & LEO,AJ (1985)
VP 5.72E-005	mm Hg	25	EST	NEELY,WB & BLAU,GE (1985)
DC	pKa			
HL 6.77E-007	atm m3/mol	25	EST	MEYLAN,WM & HOWARD,PH (1993)
OH 6.18E-012	cm3/molc sec	25	EST	MEYLAN,WM & HOWARD,PH (1993)

000486-56-6 — 2-PYRROLIDINONE, 1-METHYL-5-(3-PYRIDINYL)-, (S)-

Formula: $C_{10}H_{12}N_2O$
Mol Weight: 176.22
MP (deg C): 40-42
FP (deg C):
BP (deg C): 250
BP pressure (mm Hg): 1.50E+002

Property/Value	Units	Temp	Data Type	Reference
WS 9.99E+005	mg/L	25	EST	MEYLAN,WM ET AL. (1996)
logP 0.07			EXP	LI,NY & GORROD,JW (1992)
VP 8.55E-005	mm Hg	25	EST	NEELY,WB & BLAU,GE (1985)
DC	pKa			
HL 3.33E-012	atm m3/mol	25	EST	MEYLAN,WM & HOWARD,PH (1991)
OH 2.85E-011	cm3/molc sec	25	EST	MEYLAN,WM & HOWARD,PH (1993)

000487-26-3 — FLAVANONE

Formula: $C_{15}H_{12}O_2$
Mol Weight: 224.26
MP (deg C): 76-78
FP (deg C):
BP (deg C):
BP pressure (mm Hg):

Property/Value	Units	Temp	Data Type	Reference
WS 6.80E+001	mg/L	25	EST	MEYLAN,WM ET AL. (1996)
logP 3.14			EXP	BEYELER,S ET AL. (1988)
VP 1.20E-005	mm Hg	25	EST	NEELY,WB & BLAU,GE (1985)
DC	pKa			
HL 2.07E-008	atm m3/mol	25	EST	MEYLAN,WM & HOWARD,PH (1991)
OH 7.12E-011	cm3/molc sec	25	EST	MEYLAN,WM & HOWARD,PH (1993)

000487-54-7 — GLYCINE, N-(2-HYDROXYBENZOYL)-

Formula: $C_9H_9NO_4$
Mol Weight: 195.18
MP (deg C): 167-169
FP (deg C):
BP (deg C):
BP pressure (mm Hg):

Property/Value	Units	Temp	Data Type	Reference
WS 2.72E+004	mg/L	25	EST	MEYLAN,WM ET AL. (1996)
logP 0.95			EXP	HANSCH,C ET AL. (1995)
VP 1.13E-008	mm Hg	25	EST	NEELY,WB & BLAU,GE (1985)
DC	pKa			
HL 9.45E-014	atm m3/mol	25	EST	MEYLAN,WM & HOWARD,PH (1991)
OH 3.94E-011	cm3/molc sec	25	EST	MEYLAN,WM & HOWARD,PH (1993)

CAS #: 000487-89-8				1H-INDOLE-3-CARBOXALDEHYDE

Formula: C_9H_7NO

Mol Weight: 145.16

MP (deg C): 195-198 **FP (deg C):**

BP (deg C):

BP pressure (mm Hg):

Property/Value	Units	Temp	Data Type	Reference
WS 2.93E+003	mg/L	25	EST	MEYLAN,WM ET AL. (1996)
logP 1.68			EXP	HANSCH,C ET AL. (1995)
VP 4.30E-004	mm Hg	25	EST	NEELY,WB & BLAU,GE (1985)
DC	pKa			
HL 2.21E-009	atm m3/mol	25	EST	MEYLAN,WM & HOWARD,PH (1991)
OH 7.27E-011	cm3/molc sec	25	EST	MEYLAN,WM & HOWARD,PH (1993)

CAS #: 000488-23-3				1,2,3,4-TETRAMETHYLBENZENE

Formula: $C_{10}H_{14}$

Mol Weight: 134.22

MP (deg C): -6.2 **FP (deg C):**

BP (deg C): 205

BP pressure (mm Hg):

Property/Value	Units	Temp	Data Type	Reference
WS 3.39E+001	mg/L	25	EST	MEYLAN,WM ET AL. (1996)
logP 4.00			EXP	SANGSTER,J (1994)
VP 3.61E-001	mm Hg	25	EXT	CHAO,J ET AL. (1983)
DC				
HL 7.99E-003	atm m3/mol	25	EST	MEYLAN,WM & HOWARD,PH (1991)
OH 2.05E-011	cm3/molc sec	25	EST	MEYLAN,WM & HOWARD,PH (1993)

CAS #: 000488-41-5				1,6-DIBR-1,6-DIDEOXYMANNITOL

Formula: $C_6H_{12}Br_2O_4$

Mol Weight: 307.98

MP (deg C): **FP (deg C):**

BP (deg C):

BP pressure (mm Hg):

Property/Value	Units	Temp	Data Type	Reference
WS 1.95E+004	mg/L	25	EST	MEYLAN,WM ET AL. (1996)
logP -0.24			EXP	HANSCH,C & LEO,AJ (1985)
VP 1.54E-008	mm Hg	25	EST	NEELY,WB & BLAU,GE (1985)
DC	pKa			
HL 7.21E-012	atm m3/mol	25	EST	MEYLAN,WM & HOWARD,PH (1991)
OH 2.94E-011	cm3/molc sec	25	EST	MEYLAN,WM & HOWARD,PH (1993)

CAS #: 000488-93-7				FURAN-3-CARBOXYLIC ACID

Formula: $C_5H_4O_3$

Mol Weight: 112.09

MP (deg C): 122.5 **FP (deg C):**

BP (deg C):

BP pressure (mm Hg):

Property/Value	Units	Temp	Data Type	Reference
WS 1.41E+004	mg/L	25	EST	MEYLAN,WM ET AL. (1996)
logP 1.03			EXP	HANSCH,C & LEO,AJ (1985)
VP 1.03E-001	mm Hg	25	EST	NEELY,WB & BLAU,GE (1985)
DC	pKa			
HL 1.08E-007	atm m3/mol	25	EST	MEYLAN,WM & HOWARD,PH (1991)
OH 1.55E-011	cm3/molc sec	25	EST	MEYLAN,WM & HOWARD,PH (1993)

CAS #: 000490-11-9				CINCHOMERONIC ACID

Formula: $C_7H_5NO_4$

Mol Weight: 167.12

MP (deg C): 256 **FP (deg C):**

BP (deg C):

BP pressure (mm Hg):

Property/Value	Units	Temp	Data Type	Reference
WS 2.34E+003	mg/L	25	EXP	YALKOWSKY,SH & DANNENFELSER,RM (1992)
logP -0.12			EST	MEYLAN,WM & HOWARD,PH (1995)
VP 6.10E-006	mm Hg	25	EST	NEELY,WB & BLAU,GE (1985)
DC 2.63	pKa		EXP	PERRIN,DD (1972)
HL 2.86E-015	atm m3/mol	25	EST	MEYLAN,WM & HOWARD,PH (1991)
OH 1.08E-012	cm3/molc sec	25	EST	MEYLAN,WM & HOWARD,PH (1993)

CAS #: 000490-67-5				MONOTROPITOSIDE

Formula: $C_{19}H_{26}O_{12}$

Mol Weight: 446.41

MP (deg C): 180 **FP (deg C):**

BP (deg C):

BP pressure (mm Hg):

Property/Value	Units	Temp	Data Type	Reference
WS 8.09E+004	mg/L	20	EXP	SEIDELL,A (1941)
logP -2.22			EST	MEYLAN,WM & HOWARD,PH (1995)
VP 4.96E-019	mm Hg	25	EST	NEELY,WB & BLAU,GE (1985)
DC	pKa			
HL 7.02E-024	atm m3/mol	25	EST	MEYLAN,WM & HOWARD,PH (1991)
OH 1.19E-010	cm3/molc sec	25	EST	MEYLAN,WM & HOWARD,PH (1993)

CAS #: 000490-79-9				2,5-DIHYDROXYBENZOIC ACID

Formula: $C_7H_6O_4$

Mol Weight: 154.12

MP (deg C): 199-200 **FP (deg C):**

BP (deg C):

BP pressure (mm Hg):

Property/Value	Units	Temp	Data Type	Reference
WS 9.03E+003	mg/L	25	EST	MEYLAN,WM ET AL. (1996)
logP 1.74			EXP	HANSCH,C & LEO,AJ (1985)
VP 2.10E-006	mm Hg	25	EST	NEELY,WB & BLAU,GE (1985)
DC 2.95	pKa	25	EXP	SERJEANT,EP & DEMPSEY,B (1979)
HL 1.48E-012	atm m3/mol	25	EST	MEYLAN,WM & HOWARD,PH (1991)
OH 9.30E-012	cm3/molc sec	25	EST	MEYLAN,WM & HOWARD,PH (1993)

CAS #: 000490-91-5				2,5-CYCLOHEXADIENE-1,4-DIONE, 2-METHYL-5-(1-METH

Formula: $C_{10}H_{12}O_2$

Mol Weight: 164.21

MP (deg C): 45.5 **FP (deg C):**

BP (deg C): 232

BP pressure (mm Hg):

Property/Value	Units	Temp	Data Type	Reference
WS 8.65E+002	mg/L	25	EST	MEYLAN,WM ET AL. (1996)
logP 2.20			EXP	SANGSTER,J (1993)
VP 5.85E-003	mm Hg	25	EST	NEELY,WB & BLAU,GE (1985)
DC	pKa			
HL 5.28E-009	atm m3/mol	25	EST	MEYLAN,WM & HOWARD,PH (1991)
OH 2.37E-011	cm3/molc sec	25	EST	MEYLAN,WM & HOWARD,PH (1993)

CAS #: 000491-11-2				3-CHLORO-4-NITROPHENOL

Formula: $C_6H_4ClNO_3$
Mol Weight: 173.56
MP (deg C):   **FP (deg C):**
BP (deg C):
BP pressure (mm Hg):

Property/Value	Units	Temp	Data Type	Reference
WS 1.49E+003	mg/L	25	EST	MEYLAN,WM ET AL. (1996)
logP 2.55			EST	MEYLAN,WM & HOWARD,PH (1995)
VP 2.96E-004	mm Hg	25	EST	NEELY,WB & BLAU,GE (1985)
DC	pKa			
HL 1.64E-009	atm m3/mol	25	EST	MEYLAN,WM & HOWARD,PH (1991)
OH 3.07E-012	cm3/molc sec	25	EST	MEYLAN,WM & HOWARD,PH (1993)

CAS #: 000491-30-5				1(2H)-ISOQUINOLINONE

Formula: C_9H_7NO
Mol Weight: 145.16
MP (deg C): 211-214   **FP (deg C):**
BP (deg C):
BP pressure (mm Hg):

Property/Value	Units	Temp	Data Type	Reference
WS 4.84E+003	mg/L	25	EST	MEYLAN,WM ET AL. (1996)
logP 1.42			EST	MEYLAN,WM & HOWARD,PH (1995)
VP 4.47E-005	mm Hg	25	EST	NEELY,WB & BLAU,GE (1985)
DC	pKa			
HL 1.73E-009	atm m3/mol	25	EST	MEYLAN,WM & HOWARD,PH (1991)
OH 4.10E-011	cm3/molc sec	25	EST	MEYLAN,WM & HOWARD,PH (1993)

CAS #: 000491-35-0				4-METHYLQUINOLINE

Formula: $C_{10}H_9N$
Mol Weight: 143.19
MP (deg C): 9.5   **FP (deg C):**
BP (deg C): 262
BP pressure (mm Hg):

Property/Value	Units	Temp	Data Type	Reference
WS 4.79E+002	mg/L	25	EST	MEYLAN,WM ET AL. (1996)
logP 2.61			EXP	HANSCH,C ET AL. (1995)
VP 6.37E-003	mm Hg	25	EXT	WALTON,J (1977)
DC 5.67	pKa	20	EXP	PERRIN,DD (1965)
HL 7.60E-007	atm m3/mol	25	EST	MEYLAN,WM & HOWARD,PH (1991)
OH 3.77E-011	cm3/molc sec	25	EST	MEYLAN,WM & HOWARD,PH (1993)

CAS #: 000491-36-1				QUINAZOLIDIN-4-ONE

Formula: $C_8H_6N_2O$
Mol Weight: 146.15
MP (deg C): 216-219   **FP (deg C):**
BP (deg C):
BP pressure (mm Hg):

Property/Value	Units	Temp	Data Type	Reference
WS 1.74E+004	mg/L	25	EST	MEYLAN,WM ET AL. (1996)
logP 0.77			EXP	HANSCH,C & LEO,AJ (1985)
VP 2.40E-007	mm Hg	25	EST	NEELY,WB & BLAU,GE (1985)
DC	pKa			
HL 3.65E-010	atm m3/mol	25	EST	MEYLAN,WM & HOWARD,PH (1991)
OH 1.46E-011	cm3/molc sec	25	EST	MEYLAN,WM & HOWARD,PH (1993)

CAS #: 000491-38-3				4H-1-BENZOPYRAN-4-ONE

Formula: $C_9H_6O_2$
Mol Weight: 146.15
MP (deg C): 59   **FP (deg C):**
BP (deg C):
BP pressure (mm Hg):

Property/Value	Units	Temp	Data Type	Reference
WS 5.23E+003	mg/L	25	EST	MEYLAN,WM ET AL. (1996)
logP 1.38			EXP	HANSCH,C ET AL. (1995)
VP 7.96E-003	mm Hg	25	EST	NEELY,WB & BLAU,GE (1985)
DC	pKa			
HL 8.07E-007	atm m3/mol	25	EST	MEYLAN,WM & HOWARD,PH (1991)
OH 3.09E-011	cm3/molc sec	25	EST	MEYLAN,WM & HOWARD,PH (1993)

CAS #: 000491-59-8				CHRYSAROBIN

Formula: $C_{15}H_{12}O_3$
Mol Weight: 240.26
MP (deg C): 203.4-204   **FP (deg C):**
BP (deg C):
BP pressure (mm Hg):

Property/Value	Units	Temp	Data Type	Reference
WS 2.10E+002	mg/L	25	EXP	SEIDELL,A (1941)
logP 3.45			EST	MEYLAN,WM & HOWARD,PH (1995)
VP 4.89E-010	mm Hg	25	EST	NEELY,WB & BLAU,GE (1985)
DC	pKa			
HL 6.38E-017	atm m3/mol	25	EST	MEYLAN,WM & HOWARD,PH (1991)
OH 2.01E-010	cm3/molc sec	25	EST	MEYLAN,WM & HOWARD,PH (1993)

CAS #: 000491-70-3				LUTEOLIN

Formula: $C_{15}H_{10}O_6$
Mol Weight: 286.24
MP (deg C): 229-230 de   **FP (deg C):**
BP (deg C):
BP pressure (mm Hg):

Property/Value	Units	Temp	Data Type	Reference
WS 3.88E+002	mg/L	25	EST	MEYLAN,WM ET AL. (1996)
logP 2.53			EXP	PERRISSOUD,D & TESTA,B (1986)
VP 6.27E-012	mm Hg	25	EST	NEELY,WB & BLAU,GE (1985)
DC	pKa			
HL 5.33E-021	atm m3/mol	25	EST	MEYLAN,WM & HOWARD,PH (1991)
OH 2.31E-010	cm3/molc sec	25	EST	MEYLAN,WM & HOWARD,PH (1993)

CAS #: 000491-78-1				5-HYDROXYFLAVONE

Formula: $C_{15}H_{10}O_3$
Mol Weight: 238.25
MP (deg C):   **FP (deg C):**
BP (deg C):
BP pressure (mm Hg):

Property/Value	Units	Temp	Data Type	Reference
WS 2.22E+001	mg/L	25	EST	MEYLAN,WM ET AL. (1996)
logP 4.30			EXP	PERRISSOUD,D & TESTA,B (1986)
VP 1.25E-007	mm Hg	25	EST	NEELY,WB & BLAU,GE (1985)
DC	pKa			
HL 4.73E-009	atm m3/mol	25	EST	MEYLAN,WM & HOWARD,PH (1991)
OH 2.31E-010	cm3/molc sec	25	EST	MEYLAN,WM & HOWARD,PH (1993)

CAS #:	000492-22-8			9H-THIOXANTHEN-9-ONE

Formula: $C_{13}H_8OS$
Mol Weight: 212.27
MP (deg C): 211   FP (deg C):
BP (deg C): 273
BP pressure (mm Hg): 7.15E+002

Property/ Value	Units	Temp	Data Type	Reference
WS 1.15E+000	mg/L	25	EST	MEYLAN,WM ET AL. (1996)
logP 3.99			EXP	HANSCH,C & LEO,AJ (1985)
VP 1.35E-005	mm Hg	25	EST	NEELY,WB & BLAU,GE (1985)
DC	pKa			
HL 5.14E-008	atm m3/mol	25	EST	MEYLAN,WM & HOWARD,PH (1991)
OH 2.29E-011	cm3/molc sec	25	EST	MEYLAN,WM & HOWARD,PH (1993)

CAS #:	000492-37-5			ALPHA-PHENYLPROPIONIC ACID

Formula: $C_9H_{10}O_2$
Mol Weight: 150.18
MP (deg C):   FP (deg C):
BP (deg C): 260-262
BP pressure (mm Hg):

Property/ Value	Units	Temp	Data Type	Reference
WS 4.99E+003	mg/L	25	EST	MEYLAN,WM ET AL. (1996)
logP 1.85			EST	MEYLAN,WM & HOWARD,PH (1995)
VP 3.75E-003	mm Hg	25	EST	NEELY,WB & BLAU,GE (1985)
DC	pKa			
HL 5.87E-008	atm m3/mol	25	EST	MEYLAN,WM & HOWARD,PH (1991)
OH 6.77E-012	cm3/molc sec	25	EST	MEYLAN,WM & HOWARD,PH (1993)

CAS #:	000492-80-8			AURAMINE

Formula: $C_{17}H_{21}N_3$
Mol Weight: 267.38
MP (deg C): 136   FP (deg C):
BP (deg C): 330.5
BP pressure (mm Hg):

Property/ Value	Units	Temp	Data Type	Reference
WS 5.35E+001	mg/L	25	EST	MEYLAN,WM ET AL. (1996)
logP 2.98			EST	MEYLAN,WM & HOWARD,PH (1995)
VP 1.29E-006	mm Hg	25	EST	NEELY,WB & BLAU,GE (1985)
DC	pKa			
HL 3.64E-009	atm m3/mol	25	EST	MEYLAN,WM & HOWARD,PH (1991)
OH 3.26E-010	cm3/molc sec	25	EST	MEYLAN,WM & HOWARD,PH (1993)

CAS #:	000492-94-4			FURAN, 2-COCO-FURYL

Formula: $C_{10}H_8O_4$
Mol Weight: 192.17
MP (deg C): 163-165   FP (deg C):
BP (deg C):
BP pressure (mm Hg):

Property/ Value	Units	Temp	Data Type	Reference
WS 4.28E+003	mg/L	25	EST	MEYLAN,WM ET AL. (1996)
logP 1.24			EXP	HANSCH,C & LEO,AJ (1985)
VP 9.55E-004	mm Hg	25	EST	NEELY,WB & BLAU,GE (1985)
DC	pKa			
HL 7.70E-009	atm m3/mol	25	EST	MEYLAN,WM & HOWARD,PH (1991)
OH 7.38E-011	cm3/molc sec	25	EST	MEYLAN,WM & HOWARD,PH (1993)

CAS #:	000492-97-7			2,2'-BITHIOPHENE

Formula: $C_8H_6S_2$
Mol Weight: 166.27
MP (deg C): 33   FP (deg C):
BP (deg C): 260
BP pressure (mm Hg):

Property/ Value	Units	Temp	Data Type	Reference
WS 4.02E+001	mg/L	25	EST	MEYLAN,WM ET AL. (1996)
logP 3.75			EXP	DEVOOGT,P ET AL. (1990)
VP 1.19E-003	mm Hg	25	EST	NEELY,WB & BLAU,GE (1985)
DC	pKa			
HL 1.22E-004	atm m3/mol	25	EST	MEYLAN,WM & HOWARD,PH (1991)
OH 3.31E-011	cm3/molc sec	25	EST	MEYLAN,WM & HOWARD,PH (1993)

CAS #:	000493-01-6			CIS-BICYCLO[4.4.0]DECANE

Formula: $C_{10}H_{18}$
Mol Weight: 138.25
MP (deg C): -42.9   FP (deg C):
BP (deg C): 195.8
BP pressure (mm Hg):

Property/ Value	Units	Temp	Data Type	Reference
WS 6.45E+000	mg/L	25	EST	MEYLAN,WM ET AL. (1996)
logP 4.20			EST	MEYLAN,WM & HOWARD,PH (1995)
VP 7.86E-001	mm Hg	25	EXP	DAUBERT,TE & DANNER,RP (1989)
DC	pKa			
HL 3.49E-001	atm m3/mol	25	EST	MEYLAN,WM & HOWARD,PH (1991)
OH 1.99E-011	cm3/molc sec	25	EXP	ATKINSON,R (1989)

CAS #:	000493-02-7			TRANS-BICYCLO[4.4.0]DECANE

Formula: $C_{10}H_{18}$
Mol Weight: 138.25
MP (deg C): -30.3   FP (deg C):
BP (deg C): 187.3
BP pressure (mm Hg):

Property/ Value	Units	Temp	Data Type	Reference
WS 6.45E+000	mg/L	25	EST	MEYLAN,WM ET AL. (1996)
logP 4.20			EST	MEYLAN,WM & HOWARD,PH (1995)
VP 1.22E+000	mm Hg	25	EXP	DAUBERT,TE & DANNER,RP (1989)
DC	pKa			
HL 3.49E-001	atm m3/mol	25	EST	MEYLAN,WM & HOWARD,PH (1991)
OH 2.04E-011	cm3/molc sec	25	EXP	ATKINSON,R (1989)

CAS #:	000493-09-4			1,4-BENZODIOXAN

Formula: $C_8H_8O_2$
Mol Weight: 136.15
MP (deg C):   FP (deg C):
BP (deg C): 103
BP pressure (mm Hg): 6.00E+000

Property/ Value	Units	Temp	Data Type	Reference
WS 1.67E+003	mg/L	25	EST	MEYLAN,WM ET AL. (1996)
logP 2.01			EXP	HANSCH,C & LEO,AJ (1985)
VP 1.93E-001	mm Hg	25	EST	NEELY,WB & BLAU,GE (1985)
DC	pKa			
HL 8.31E-006	atm m3/mol	25	EST	MEYLAN,WM & HOWARD,PH (1991)
OH 3.26E-011	cm3/molc sec	25	EST	MEYLAN,WM & HOWARD,PH (1993)

346

000493-52-7 — 2-((4-(DIMETHYLAMINO)PHENYL)AZO)BENZOIC ACID

Formula: $C_{15}H_{15}N_3O_2$
Mol Weight: 269.31
MP (deg C): 181-182
FP (deg C):
BP (deg C):
BP pressure (mm Hg):

Property/ Value	Units	Temp	Data Type	Reference
WS 3.67E+000	mg/L	25	EST	MEYLAN,WM ET AL. (1996)
logP 3.83			EST	MEYLAN,WM & HOWARD,PH (1995)
VP 1.44E-007	mm Hg	25	EST	NEELY,WB & BLAU,GE (1985)
DC 2.63	pKa	28	EXP	KORTUM,G ET AL (1961)
HL 4.71E-012	atm m3/mol	25	EST	MEYLAN,WM & HOWARD,PH (1991)
OH 1.50E-010	cm3/molc sec	25	EST	MEYLAN,WM & HOWARD,PH (1993)

000494-03-1 — CHLORONAPHAZINE

Formula: $C_{14}H_{15}Cl_2N$
Mol Weight: 268.19
MP (deg C): 54-56
FP (deg C):
BP (deg C): 210
BP pressure (mm Hg): 5.00E+000

Property/ Value	Units	Temp	Data Type	Reference
WS 1.39E+000	mg/L	25	EST	MEYLAN,WM ET AL. (1996)
logP 4.84			EST	MEYLAN,WM & HOWARD,PH (1995)
VP 3.17E-006	mm Hg	25	EST	NEELY,WB & BLAU,GE (1985)
DC	pKa			
HL 1.83E-006	atm m3/mol	25	EST	MEYLAN,WM & HOWARD,PH (1991)
OH 2.07E-010	cm3/molc sec	25	EST	MEYLAN,WM & HOWARD,PH (1993)

000494-52-0 — ANABASINE

Formula: $C_{10}H_{14}N_2$
Mol Weight: 162.24
MP (deg C): 9
FP (deg C):
BP (deg C): 276
BP pressure (mm Hg):

Property/ Value	Units	Temp	Data Type	Reference
WS 1.00E+006	mg/L	25	EXP	SHIU,WY ET AL. (1990)
logP 0.97			EXP	HANSCH,C & LEO,AJ (1985)
VP 3.01E-003	mm Hg	25	EST	NEELY,WB & BLAU,GE (1985)
DC 11.00	pKa	30	EXP	PERRIN,DD (1965)
HL 1.81E-009	atm m3/mol	25	EST	MEYLAN,WM & HOWARD,PH (1991)
OH 1.01E-010	cm3/molc sec	25	EST	MEYLAN,WM & HOWARD,PH (1993)

000494-97-3 — NORNICOTINE

Formula: $C_9H_{12}N_2$
Mol Weight: 148.21
MP (deg C):
FP (deg C):
BP (deg C): 270
BP pressure (mm Hg):

Property/ Value	Units	Temp	Data Type	Reference
WS 1.00E+006	mg/L	25	EST	MEYLAN,WM ET AL. (1996)
logP 0.17			EXP	HANSCH,C & LEO,AJ (1985)
VP 9.70E-003	mm Hg	25	EST	NEELY,WB & BLAU,GE (1985)
DC	pKa			
HL 1.37E-009	atm m3/mol	25	EST	MEYLAN,WM & HOWARD,FH (1991)
OH 9.24E-011	cm3/molc sec	25	EST	MEYLAN,WM & HOWARD,PH (1993)

000495-18-1 — BENZOHYDROXAMIC ACID

Formula: $C_7H_7NO_2$
Mol Weight: 137.14
MP (deg C): 126-130
FP (deg C):
BP (deg C):
BP pressure (mm Hg):

Property/ Value	Units	Temp	Data Type	Reference
WS 5.16E+004	mg/L	25	EST	MEYLAN,WM ET AL. (1996)
logP 0.26			EXP	HANSCH,C & LEO,AJ (1985)
VP 8.48E-007	mm Hg	25	EST	NEELY,WB & BLAU,GE (1985)
DC	pKa			
HL 4.44E-012	atm m3/mol	25	EST	MEYLAN,WM & HOWARD,PH (1991)
OH 7.28E-012	cm3/molc sec	25	EST	MEYLAN,WM & HOWARD,PH (1993)

000495-40-9 — 1-BUTANONE, 1-PHENYL-

Formula: $C_{10}H_{12}O$
Mol Weight: 148.21
MP (deg C): 12
FP (deg C):
BP (deg C): 228.5
BP pressure (mm Hg):

Property/ Value	Units	Temp	Data Type	Reference
WS 3.33E+002	mg/L	25	EST	MEYLAN,WM ET AL. (1996)
logP 2.77			EXP	SANGSTER,J (1993)
VP 8.35E-002	mm Hg	25	EST	NEELY,WB & BLAU,GE (1985)
DC	pKa			
HL 1.73E-005	atm m3/mol	25	EST	MEYLAN,WM & HOWARD,PH (1991)
OH 6.45E-012	cm3/molc sec	25	EST	MEYLAN,WM & HOWARD,PH (1993)

000495-54-5 — 2,4-DIAMINOAZOBENZENE

Formula: $C_{12}H_{12}N_4$
Mol Weight: 212.26
MP (deg C):
FP (deg C):
BP (deg C):
BP pressure (mm Hg):

Property/ Value	Units	Temp	Data Type	Reference
WS 2.13E+002	mg/L	25	EST	MEYLAN,WM ET AL. (1996)
logP 2.13			EST	MEYLAN,WM & HOWARD,PH (1995)
VP 1.52E-006	mm Hg	25	EST	NEELY,WB & BLAU,GE (1985)
DC	pKa			
HL 1.84E-012	atm m3/mol	25	EST	MEYLAN,WM & HOWARD,PH (1991)
OH 2.00E-010	cm3/molc sec	25	EST	MEYLAN,WM & HOWARD,PH (1993)

000495-69-2 — HIPPURIC ACID

Formula: $C_9H_9NO_3$
Mol Weight: 179.18
MP (deg C): 118-118.5
FP (deg C):
BP (deg C):
BP pressure (mm Hg):

Property/ Value	Units	Temp	Data Type	Reference
WS 3.75E+003	mg/L	25	EXP	YALKOWSKY,SH & DANNENFELSER,RM (1992)
logP 0.31			EXP	HANSCH,C & LEO,AJ (1985)
VP 7.62E-007	mm Hg	25	EST	NEELY,WB & BLAU,GE (1985)
DC	pKa			
HL 7.21E-013	atm m3/mol	25	EST	MEYLAN,WM & HOWARD,PH (1991)
OH 1.07E-011	cm3/molc sec	25	EST	MEYLAN,WM & HOWARD,PH (1993)

CAS #:	000495-73-8			BAYER 15080
Formula: $C_{13}H_{11}N_3O_2$				
Mol Weight: 241.25				
MP (deg C):		FP (deg C):		
BP (deg C):				
BP pressure (mm Hg):				

Property/Value	Units	Temp	Data Type	Reference
WS 5.00E+000	mg/L		EXP	GUNTHER,FA ET AL. (1968)
logP 1.39			EST	MEYLAN,WM & HOWARD,PH (1995)
VP 8.37E-011	mm Hg	25	EST	NEELY,WB & BLAU,GE (1985)
DC	pKa			
HL 5.01E-013	atm m3/mol	25	EST	MEYLAN,WM & HOWARD,PH (1991)
OH 1.20E-010	cm3/molc sec	25	EST	MEYLAN,WM & HOWARD,PH (1993)

CAS #:	000495-76-1			1,3-BENZODIOXOLE-5-METHANOL
Formula: $C_8H_8O_3$				
Mol Weight: 152.15				
MP (deg C): 58		FP (deg C):		
BP (deg C): 157				
BP pressure (mm Hg): 1.60E+001				

Property/Value	Units	Temp	Data Type	Reference
WS 3.05E+004	mg/L	25	EST	MEYLAN,WM ET AL. (1996)
logP 1.05			EXP	HANSCH,C & LEO,AJ (1985)
VP 2.54E-004	mm Hg		EST	NEELY,WB & BLAU,GE (1985)
DC	pKa			
HL 2.53E-010	atm m3/mol	25	EST	MEYLAN,WM & HOWARD,PH (1991)
OH 5.24E-011	cm3/molc sec	25	EST	MEYLAN,WM & HOWARD,PH (1993)

CAS #:	000496-11-7			INDANE
Formula: C_9H_{10}				
Mol Weight: 118.18				
MP (deg C): -51.4		FP (deg C):		
BP (deg C): 176.5				
BP pressure (mm Hg): 7.62E+002				

Property/Value	Units	Temp	Data Type	Reference
WS 1.09E+002	mg/L	25	EXP	YALKOWSKY,SH & DANNENFELSER,RM (1992)
logP 3.18			EXP	HANSCH,C & LEO,AJ (1985)
VP 1.47E+000	mm Hg	25	EXT	BOUBLIK,T ET AL. (1984)
DC	pKa			
HL 2.01E-004	atm m3/mol	25	EXP	LION,LW & GARBARINI,D (1983)
OH 9.20E-012	cm3/molc sec	25	EXP	ATKINSON,R (1989)

CAS #:	000496-14-0			PHTHALAN
Formula: C_8H_8O				
Mol Weight: 120.15				
MP (deg C):		FP (deg C):		
BP (deg C): 192				
BP pressure (mm Hg):				

Property/Value	Units	Temp	Data Type	Reference
WS 3.15E+003	mg/L	25	EST	MEYLAN,WM ET AL. (1996)
logP 1.76			EXP	HANSCH,C & LEO,AJ (1985)
VP 7.48E-001	mm Hg	25	EST	NEELY,WB & BLAU,GE (1985)
DC	pKa			
HL 2.55E-005	atm m3/mol	25	EST	MEYLAN,WM & HOWARD,PH (1991)
OH 1.50E-011	cm3/molc sec	25	EST	MEYLAN,WM & HOWARD,PH (1993)

CAS #:	000496-16-2			2,3-DIHYDROBENZOFURAN
Formula: C_8H_8O				
Mol Weight: 120.15				
MP (deg C): -21.5		FP (deg C):		
BP (deg C): 188.5				
BP pressure (mm Hg):				

Property/Value	Units	Temp	Data Type	Reference
WS 1.49E+003	mg/L	25	EST	MEYLAN,WM ET AL. (1996)
logP 2.14			EXP	HANSCH,C & LEO,AJ (1985)
VP 7.48E-001	mm Hg	25	EST	NEELY,WB & BLAU,GE (1985)
DC	pKa			
HL 1.55E-004	atm m3/mol	25	EST	MEYLAN,WM & HOWARD,PH (1991)
OH 3.66E-011	cm3/molc sec	25	EXP	ATKINSON,R (1989)

CAS #:	000496-41-3			2-BENZOFURANCARBOXYLIC ACID
Formula: $C_9H_6O_3$				
Mol Weight: 162.15				
MP (deg C): 192-193		FP (deg C):		
BP (deg C): 310-315				
BP pressure (mm Hg):				

Property/Value	Units	Temp	Data Type	Reference
WS 5.85E+002	mg/L	25	EST	MEYLAN,WM ET AL. (1996)
logP 2.41			EXP	HANSCH,C ET AL. (1995)
VP 1.48E-004	mm Hg	25	EST	NEELY,WB & BLAU,GE (1985)
DC	pKa			
HL 1.06E-008	atm m3/mol	25	EST	MEYLAN,WM & HOWARD,PH (1991)
OH 1.43E-011	cm3/molc sec	25	EST	MEYLAN,WM & HOWARD,PH (1993)

CAS #:	000496-67-3			A-BROMO-I-VALERYLUREA
Formula: $C_6H_{11}BrN_2O_2$				
Mol Weight: 223.07				
MP (deg C): 147-149		FP (deg C):		
BP (deg C):				
BP pressure (mm Hg):				

Property/Value	Units	Temp	Data Type	Reference
WS 3.52E+003	mg/L	25	EST	MEYLAN,WM ET AL. (1996)
logP 1.14			EXP	HANSCH,C & LEO,AJ (1985)
VP 5.18E-007	mm Hg	25	EST	NEELY,WB & BLAU,GE (1985)
DC	pKa			
HL 2.91E-011	atm m3/mol	25	EST	MEYLAN,WM & HOWARD,PH (1991)
OH 8.85E-012	cm3/molc sec	25	EST	MEYLAN,WM & HOWARD,PH (1993)

CAS #:	000496-72-0			3,4-DIAMINOTOLUENE
Formula: $C_7H_{10}N_2$				
Mol Weight: 122.17				
MP (deg C): 88.5		FP (deg C):		
BP (deg C): 265				
BP pressure (mm Hg):				

Property/Value	Units	Temp	Data Type	Reference
WS 2.69E+004	mg/L	25	EST	MEYLAN,WM ET AL. (1996)
logP 0.66			EXP	DEBNATH,AK ET AL. (1992)
VP 6.29E-004	mm Hg	25	EXT	MILLIGAN,B & GILBERT,KE (1978)
DC 4.68	pKa	25	EXP	PERRIN,DD (1972)
HL 7.49E-009	atm m3/mol	25	EST	MEYLAN,WM & HOWARD,PH (1991)
OH 2.40E-010	cm3/molc sec	25	EST	MEYLAN,WM & HOWARD,PH (1993)

CAS #: 000496-78-6		2,4,5-TRIMETHYLPHENOL

Formula: $C_9H_{12}O$

Mol Weight: 136.20

MP (deg C): 72   FP (deg C):

BP (deg C): 232

BP pressure (mm Hg):

Property/ Value	Units	Temp	Data Type	Reference
WS 6.68E+002	mg/L	25	EST	MEYLAN,WM ET AL. (1996)
logP 3.15			EST	MEYLAN,WM & HOWARD,PH (1995)
VP 6.79E-003	mm Hg	25	EXT	BOUBLIK,T ET AL. (1984)
DC 10.57	pKa	25	EXP	SERJEANT,EP & DEMPSEY,B (1979)
HL 7.53E-007	atm m3/mol	25	EST	MEYLAN,WM & HOWARD,PH (1991)
OH 1.31E-010	cm3/molc sec	25	EST	MEYLAN,WM & HOWARD,PH (1993)

CAS #: 000497-06-3		3-BUTENE-1,2-DIOL

Formula: $C_4H_8O_2$

Mol Weight: 88.11

MP (deg C):   FP (deg C):

BP (deg C): 196.5

BP pressure (mm Hg):

Property/ Value	Units	Temp	Data Type	Reference
WS 2.91E+005	mg/L	25	EST	MEYLAN,WM ET AL. (1996)
logP -0.43			EST	MEYLAN,WM & HOWARD,PH (1995)
VP 2.40E-001	mm Hg	25	EST	NEELY,WB & BLAU,GE (1985)
DC	pKa			
HL 1.72E-007	atm m3/mol	25	EST	MEYLAN,WM & HOWARD,PH (1991)
OH 3.90E-011	cm3/molc sec	25	EST	MEYLAN,WM & HOWARD,PH (1993)

CAS #: 000497-23-4		2(5H)-FURANONE

Formula: $C_4H_4O_2$

Mol Weight: 84.08

MP (deg C): 4-5   FP (deg C):

BP (deg C): 86-87

BP pressure (mm Hg): 1.20E+001

Property/ Value	Units	Temp	Data Type	Reference
WS 4.18E+005	mg/L	25	EST	MEYLAN,WM ET AL. (1996)
logP -0.60			EXP	HANSCH,C ET AL. (1995)
VP 9.81E-001	mm Hg	25	EST	NEELY,WB & BLAU,GE (1985)
DC	pKa			
HL 4.81E-005	atm m3/mol	25	EST	MEYLAN,WM & HOWARD,PH (1991)
OH 1.25E-011	cm3/molc sec	25	EST	MEYLAN,WM & HOWARD,PH (1993)

CAS #: 000497-38-1		NORCAMPHOR

Formula: $C_7H_{10}O$

Mol Weight: 110.16

MP (deg C):   FP (deg C):

BP (deg C):

BP pressure (mm Hg):

Property/ Value	Units	Temp	Data Type	Reference
WS 1.93E+003	mg/L	20	EXP	SEIDALL, A (1941)
logP 1.02			EST	MEYLAN,WM & HOWARD,PH (1995)
VP 2.09E+000	mm Hg	25	EST	NEELY,WB & BLAU,GE (1985)
DC	pKa			
HL 2.99E-005	atm m3/mol	25	EST	MEYLAN,WM & HOWARD,PH (1991)
OH 1.59E-011	cm3/molc sec	25	EST	MEYLAN,WM & HOWARD,PH (1993)

CAS #: 000497-59-6		3-HYDROXY-4-OXO-4H-PYRAN-2,6-DICARBOXYLIC ACID

Formula: $C_7H_4O_7$

Mol Weight: 200.11

MP (deg C): 120 dec   FP (deg C):

BP (deg C):

BP pressure (mm Hg):

Property/ Value	Units	Temp	Data Type	Reference
WS 8.40E+003	mg/L	25	EXP	YALKOWSKY,SH & DANNENFELSER,RM (1992)
logP 0.61			EST	MEYLAN,WM & HOWARD,PH (1995)
VP 4.51E-010	mm Hg	25	EST	NEELY,WB & BLAU,GE (1985)
DC	pKa			
HL 4.58E-015	atm m3/mol	25	EST	MEYLAN,WM & HOWARD,PH (1991)
OH 2.53E-011	cm3/molc sec	25	EST	MEYLAN,WM & HOWARD,PH (1993)

CAS #: 000497-76-7		ARBUTIN

Formula: $C_{12}H_{16}O_7$

Mol Weight: 272.26

MP (deg C): 165   FP (deg C):

BP (deg C):

BP pressure (mm Hg):

Property/ Value	Units	Temp	Data Type	Reference
WS 9.59E+005	mg/L	25	EST	MEYLAN,WM ET AL. (1996)
logP -1.35			EXP	HANSCH,C & LEO,AJ (1985)
VP 2.30E-012	mm Hg	25	EST	NEELY,WB & BLAU,GE (1985)
DC	pKa			
HL 1.20E-019	atm m3/mol	25	EST	MEYLAN,WM & HOWARD,PH (1991)
OH 9.10E-011	cm3/molc sec	25	EST	MEYLAN,WM & HOWARD,PH (1993)

CAS #: 000498-21-5		METHYLSUCCINIC ACID

Formula: $C_5H_8O_4$

Mol Weight: 132.12

MP (deg C): 116-119   FP (deg C):

BP (deg C):

BP pressure (mm Hg):

Property/ Value	Units	Temp	Data Type	Reference
WS 6.66E+005	mg/L	15	EXP	YALKOWSKY,SH & DANNENFELSER,RM (1992)
logP -0.34			EST	MEYLAN,WM & HOWARD,PH (1995)
VP 1.05E-003	mm Hg	25	EST	NEELY,WB & BLAU,GE (1985)
DC	pKa			
HL 7.18E-012	atm m3/mol	25	EST	MEYLAN,WM & HOWARD,PH (1991)
OH 3.86E-012	cm3/molc sec	25	EST	MEYLAN,WM & HOWARD,PH (1993)

CAS #: 000498-60-2		FURAN-3-CARBOXALDEHYDE

Formula: $C_5H_4O_2$

Mol Weight: 96.09

MP (deg C):   FP (deg C):

BP (deg C): 145

BP pressure (mm Hg):

Property/ Value	Units	Temp	Data Type	Reference
WS 4.40E+004	mg/L	25	EST	MEYLAN,WM ET AL. (1996)
logP 0.51			EXP	HANSCH,C & LEO,AJ (1985)
VP 5.33E+000	mm Hg	25	EST	NEELY,WB & BLAU,GE (1985)
DC	pKa			
HL 1.34E-005	atm m3/mol	25	EST	MEYLAN,WM & HOWARD,PH (1991)
OH 3.74E-011	cm3/molc sec	25	EST	MEYLAN,WM & HOWARD,PH (1993)

CAS #: 000498-62-4				3-THIOPHENECARBOXALDEHYDE

Formula: C_5H_4OS

Mol Weight: 112.15

MP (deg C):
FP (deg C):

BP (deg C): 86.7

BP pressure (mm Hg): 2.00E+001

Property/Value	Units	Temp	Data Type	Reference
WS 1.47E+004	mg/L	25	EST	MEYLAN,WM ET AL. (1996)
logP 1.01			EXP	HANSCH,C & LEO,AJ (1985)
VP 5.55E-001	mm Hg	25	EST	NEELY,WB & BLAU,GE (1985)
DC	pKa			
HL 7.30E-006	atm m3/mol	25	EST	MEYLAN,WM & HOWARD,PH (1991)
OH 2.17E-011	cm3/molc sec	25	EST	MEYLAN,WM & HOWARD,PH (1993)

CAS #: 000498-66-8				BICYCLO[2.2.1]-2-HEPTENE

Formula: C_7H_{10}

Mol Weight: 94.16

MP (deg C): 44-46
FP (deg C):

BP (deg C): 96

BP pressure (mm Hg):

Property/Value	Units	Temp	Data Type	Reference
WS 1.30E+002	mg/L	25	EST	MEYLAN,WM ET AL. (1996)
logP 2.85			EST	MEYLAN,WM & HOWARD,PH (1995)
VP 3.92E+001	mm Hg	25	EXT	YAWS,CL (1994A)
DC	pKa			
HL 3.88E-002	atm m3/mol	25	EST	MEYLAN,WM & HOWARD,PH (1991)
OH 4.93E-011	cm3/molc sec	25	EXP	ATKINSON,R (1989)

CAS #: 000498-94-2				ISONIPECOTIC ACID

Formula: $C_6H_{11}NO_2$

Mol Weight: 129.16

MP (deg C): 336
FP (deg C):

BP (deg C):

BP pressure (mm Hg):

Property/Value	Units	Temp	Data Type	Reference
WS 3.17E+005	mg/L	25	EST	MEYLAN,WM ET AL. (1996)
logP -3.05			EXP	TSAI,RS ET AL. (1991)
VP 1.75E-009	mm Hg	25	EST	NEELY,WB & BLAU,GE (1985)
DC	pKa			
HL 1.28E-010	atm m3/mol	25	EST	MEYLAN,WM & HOWARD,PH (1991)
OH 8.99E-011	cm3/molc sec	25	EST	MEYLAN,WM & HOWARD,PH (1993)

CAS #: 000498-95-3				NIPECOTIC ACID

Formula: $C_6H_{11}NO_2$

Mol Weight: 129.16

MP (deg C): 261
FP (deg C):

BP (deg C):

BP pressure (mm Hg):

Property/Value	Units	Temp	Data Type	Reference
WS 2.32E+005	mg/L	25	EST	MEYLAN,WM ET AL. (1996)
logP -2.89			EXP	TSAI,RS ET AL. (1991)
VP 1.75E-009	mm Hg	25	EST	NEELY,WB & BLAU,GE (1985)
DC	pKa			
HL 1.28E-010	atm m3/mol	25	EST	MEYLAN,WM & HOWARD,PH (1991)
OH 8.99E-011	cm3/molc sec	25	EST	MEYLAN,WM & HOWARD,PH (1993)

CAS #: 000499-74-1				6-METHYL-3-ISOPROPYL-2-CYCLOHEXEN-1-ONE

Formula: $C_{10}H_{16}O$

Mol Weight: 152.24

MP (deg C):
FP (deg C):

BP (deg C):

BP pressure (mm Hg):

Property/Value	Units	Temp	Data Type	Reference
WS 2.18E+003	mg/L	15	EXP	STEPHEN,H & STEPHEN,T (1963)
logP 3.07			EST	MEYLAN,WM & HOWARD,PH (1995)
VP 2.03E-001	mm Hg	25	EST	NEELY,WB & BLAU,GE (1985)
DC	pKa			
HL 8.79E-005	atm m3/mol	25	EST	MEYLAN,WM & HOWARD,PH (1991)
OH 8.84E-011	cm3/molc sec	25	EST	MEYLAN,WM & HOWARD,PH (1993)

CAS #: 000499-80-9				2,4-PYRIDINEDICARBOXYLIC ACID

Formula: $C_7H_5NO_4$

Mol Weight: 167.12

MP (deg C): 249
FP (deg C):

BP (deg C):

BP pressure (mm Hg):

Property/Value	Units	Temp	Data Type	Reference
WS 2.49E+003	mg/L	25	EXP	YALKOWSKY,SH & DANNENFELSER,RM (1992)
logP 0.57			EST	MEYLAN,WM & HOWARD,PH (1995)
VP 6.10E-006	mm Hg	25	EST	NEELY,WB & BLAU,GE (1985)
DC	pKa			
HL 2.86E-015	atm m3/mol	25	EST	MEYLAN,WM & HOWARD,PH (1991)
OH 1.09E-012	cm3/molc sec	25	EST	MEYLAN,WM & HOWARD,PH (1993)

CAS #: 000499-81-0				DINICOTINIC ACID

Formula: $C_7H_5NO_4$

Mol Weight: 167.12

MP (deg C): 324
FP (deg C):

BP (deg C):

BP pressure (mm Hg):

Property/Value	Units	Temp	Data Type	Reference
WS 1.07E+003	mg/L	25	EXP	YALKOWSKY,SH & DANNENFELSER,RM (1992)
logP 0.57			EST	MEYLAN,WM & HOWARD,PH (1995)
VP 6.10E-006	mm Hg	25	EST	NEELY,WB & BLAU,GE (1985)
DC 1.10	pKa	25	EXP	PERRIN,DD (1972)
HL 2.86E-015	atm m3/mol	25	EST	MEYLAN,WM & HOWARD,PH (1991)
OH 1.07E-012	cm3/molc sec	25	EST	MEYLAN,WM & HOWARD,PH (1993)

CAS #: 000499-83-2				PYRIDINE-2,6-DICARBOXYLIC ACID

Formula: $C_7H_5NO_4$

Mol Weight: 167.12

MP (deg C): 248-250 de
FP (deg C):

BP (deg C):

BP pressure (mm Hg):

Property/Value	Units	Temp	Data Type	Reference
WS 5.00E+003	mg/L	20	EXP	GOE,GL (1982)
logP 0.57			EST	MEYLAN,WM & HOWARD,PH (1995)
VP 6.10E-006	mm Hg	25	EST	NEELY,WB & BLAU,GE (1985)
DC 2.16	pKa	30	EXP	PERRIN,DD (1972)
HL 2.86E-015	atm m3/mol	25	EST	MEYLAN,WM & HOWARD,PH (1991)
OH 1.09E-012	cm3/molc sec	25	EST	MEYLAN,WM & HOWARD,PH (1993)

CAS #: 000499-94-5 — CARVOMETHENE

Formula	$C_{10}H_{18}$
Mol Weight	138.25
MP (deg C)	
FP (deg C)	
BP (deg C)	
BP pressure (mm Hg)	

Property/Value		Units	Temp	Data Type	Reference
WS	1.84E+000	mg/L	25	EST	MEYLAN,WM ET AL. (1996)
logP	4.83			EST	MEYLAN,WM & HOWARD,PH (1995)
VP	2.98E+000	mm Hg	25	EST	NEELY,WB & BLAU,GE (1985)
DC		pKa			
HL	2.43E-001	atm m3/mol	25	EST	MEYLAN,WM & HOWARD,PH (1991)
OH	1.27E-010	cm3/molc sec	25	EXP	ATKINSON,R (1989)

CAS #: 000500-22-1 — 3-FORMYLPYRIDINE

Formula	C_6H_5NO
Mol Weight	107.11
MP (deg C)	
FP (deg C)	
BP (deg C)	89.5
BP pressure (mm Hg)	1.40E+001

Property/Value		Units	Temp	Data Type	Reference
WS	6.29E+004	mg/L	25	EST	MEYLAN,WM ET AL. (1996)
logP	0.29			EXP	HANSCH,C ET AL. (1995)
VP	5.68E-001	mm Hg	25	EST	NEELY,WB & BLAU,GE (1985)
DC	3.80	pKa	20	EXP	PERRIN,DD (1965)
HL	1.76E-008	atm m3/mol	25	EST	MEYLAN,WM & HOWARD,PH (1991)
OH	1.71E-011	cm3/molc sec	25	EST	MEYLAN,WM & HOWARD,PH (1993)

CAS #: 000500-28-7 — 3-CHLORO-DIMETHYL PARATHION

Formula	$C_8H_9ClNO_5PS$
Mol Weight	297.66
MP (deg C)	21
FP (deg C)	
BP (deg C)	
BP pressure (mm Hg)	

Property/Value		Units	Temp	Data Type	Reference
WS	4.00E+001	mg/L	20	EXP	YALKOWSKY,SH & DANNENFELSER,RM (1992)
logP	3.45			EXP	HANSCH,C & LEO,AJ (1985)
VP	1.29E-005	mm Hg	25	EST	NEELY,WB & BLAU,GE (1985)
DC		pKa			
HL	1.25E-007	atm m3/mol	25	EST	MEYLAN,WM & HOWARD,PH (1991)
OH	5.89E-011	cm3/molc sec	25	EST	MEYLAN,WM & HOWARD,PH (1993)

CAS #: 000500-34-5 — BETA-EUCAINE

Formula	$C_{15}H_{21}NO_2$
Mol Weight	247.34
MP (deg C)	70-71
FP (deg C)	
BP (deg C)	
BP pressure (mm Hg)	

Property/Value		Units	Temp	Data Type	Reference
WS	2.96E+003	mg/L	20	EXP	STEPHEN,H & STEPHEN,T (1963)
logP	2.95			EST	MEYLAN,WM & HOWARD,PH (1995)
VP	3.48E-005	mm Hg	25	EST	NEELY,WB & BLAU,GE (1985)
DC		pKa			
HL	1.89E-008	atm m3/mol	25	EST	MEYLAN,WM & HOWARD,PH (1991)
OH	9.50E-011	cm3/molc sec	25	EST	MEYLAN,WM & HOWARD,PH (1993)

CAS #: 000500-42-5 — 2-AMINO-4-(P-CHLOROANILINO)-S-TRIAZINE

Formula	$C_9H_8ClN_5$
Mol Weight	221.65
MP (deg C)	233-234
FP (deg C)	
BP (deg C)	
BP pressure (mm Hg)	

Property/Value		Units	Temp	Data Type	Reference
WS	2.29E+002	mg/L	25	EST	MEYLAN,WM ET AL. (1996)
logP	2.54			EST	MEYLAN,WM & HOWARD,PH (1995)
VP	2.25E-006	mm Hg	25	EST	NEELY,WB & BLAU,GE (1985)
DC		pKa			
HL	2.19E-010	atm m3/mol	25	EST	MEYLAN,WM & HOWARD,PH (1991)
OH	4.32E-011	cm3/molc sec	25	EST	MEYLAN,WM & HOWARD,PH (1993)

CAS #: 000500-92-5 — IMIDODICARBONIMIDIC DIAMIDE, N-(4-CHLOROPHENYL)-

Formula	$C_{11}H_{16}ClN_5$
Mol Weight	253.74
MP (deg C)	129
FP (deg C)	
BP (deg C)	
BP pressure (mm Hg)	

Property/Value		Units	Temp	Data Type	Reference
WS	1.56E+002	mg/L	25	EST	MEYLAN,WM ET AL. (1996)
logP	2.53			EXP	SANGSTER,J (1994)
VP	7.40E-007	mm Hg	25	EST	NEELY,WB & BLAU,GE (1985)
DC		pKa			
HL	2.51E-016	atm m3/mol	25	EST	MEYLAN,WM & HOWARD,PH (1991)
OH	1.57E-010	cm3/molc sec	25	EST	MEYLAN,WM & HOWARD,PH (1993)

CAS #: 000500-99-2 — 3,5-DIMETHOXYPHENOL

Formula	$C_8H_{10}O_3$
Mol Weight	154.17
MP (deg C)	37
FP (deg C)	
BP (deg C)	199
BP pressure (mm Hg)	3.50E+001

Property/Value		Units	Temp	Data Type	Reference
WS	1.10E+004	mg/L	25	EST	MEYLAN,WM ET AL. (1996)
logP	1.64			EXP	HANSCH,C & LEO,AJ (1985)
VP	6.17E-003	mm Hg	25	EST	NEELY,WB & BLAU,GE (1985)
DC	9.34	pKa		EXP	SERJEANT,EP & DEMPSEY,B (1979)
HL	1.96E-009	atm m3/mol	25	EST	MEYLAN,WM & HOWARD,PH (1991)
OH	2.02E-010	cm3/molc sec	25	EST	MEYLAN,WM & HOWARD,PH (1993)

CAS #: 000501-30-4 — 4H-PYRAN-4-ONE, 5-HYDROXY-2-(HYDROXYMETHYL)-

Formula	$C_6H_6O_4$
Mol Weight	142.11
MP (deg C)	153-154
FP (deg C)	
BP (deg C)	
BP pressure (mm Hg)	

Property/Value		Units	Temp	Data Type	Reference
WS	9.35E+005	mg/L	25	EST	MEYLAN,WM ET AL. (1996)
logP	-0.64			EXP	KONTOGHIORGHES,GJ (1988)
VP	3.21E-006	mm Hg	25	EST	NEELY,WB & BLAU,GE (1985)
DC		pKa			
HL	2.39E-007	atm m3/mol	25	EST	MEYLAN,WM & HOWARD,PH (1991)
OH	6.44E-011	cm3/molc sec	25	EST	MEYLAN,WM & HOWARD,PH (1993)

CAS #: 000501-52-0 — BETA-PHENYLPROPIONIC ACID

Formula: $C_9H_{10}O_2$

Mol Weight: 150.18

MP (deg C): 47-48 FP (deg C):

BP (deg C): 280

BP pressure (mm Hg):

Property/Value	Units	Temp	Data Type	Reference
WS 5.90E+003	mg/L	20	EXP	YALKOWSKY,SH & DANNENFELSER,RM (1992)
logP 1.84			EXP	HANSCH,C & LEO,AJ (1985)
VP 2.67E-003	mm Hg	25	EXT	PERRY,RH & GREEN,D (1984)
DC 4.66	pKa	18	EXP	KORTUM,G ET AL (1961)
HL 5.87E-008	atm m3/mol	25	EST	MEYLAN,WM & HOWARD,PH (1991)
OH 7.38E-012	cm3/molc sec	25	EST	MEYLAN,WM & HOWARD,PH (1993)

CAS #: 000501-53-1 — BENZYL CHLOROCARBONATE

Formula: $C_8H_7ClO_2$

Mol Weight: 170.60

MP (deg C): FP (deg C):

BP (deg C): 103

BP pressure (mm Hg): 2.00E+001

Property/Value	Units	Temp	Data Type	Reference
WS 1.63E+003	mg/L	25	EST	MEYLAN,WM ET AL. (1996)
logP 1.84			EST	MEYLAN,WM & HOWARD,PH (1995)
VP 1.01E-001	mm Hg	25	EST	NEELY,WB & BLAU,GE (1985)
DC	pKa			
HL 1.90E-004	atm m3/mol	25	EST	MEYLAN,WM & HOWARD,PH (1991)
OH 6.34E-012	cm3/molc sec	25	EST	MEYLAN,WM & HOWARD,PH (1993)

CAS #: 000501-58-6 — BIS(4-METHOXYPHENOL)DIAZENE

Formula: $C_{14}H_{14}N_2O_2$

Mol Weight: 242.28

MP (deg C): FP (deg C):

BP (deg C):

BP pressure (mm Hg):

Property/Value	Units	Temp	Data Type	Reference
WS 2.17E+000	mg/L	25	EST	MEYLAN,WM ET AL. (1996)
logP 4.27			EST	MEYLAN,WM & HOWARD,PH (1995)
VP 3.24E-005	mm Hg	25	EST	NEELY,WB & BLAU,GE (1985)
DC	pKa			
HL 5.15E-008	atm m3/mol	25	EST	MEYLAN,WM & HOWARD,PH (1991)
OH 1.87E-011	cm3/molc sec	25	EST	MEYLAN,WM & HOWARD,PH (1993)

CAS #: 000501-60-0 — BIS(4-METHYLPHENYL)DIAZENE

Formula: $C_{14}H_{14}N_2$

Mol Weight: 210.28

MP (deg C): FP (deg C):

BP (deg C):

BP pressure (mm Hg):

Property/Value	Units	Temp	Data Type	Reference
WS 5.13E-001	mg/L	25	EST	MEYLAN,WM ET AL. (1996)
logP 5.21			EST	MEYLAN,WM & HOWARD,PH (1995)
VP 2.59E-004	mm Hg	25	EST	NEELY,WB & BLAU,GE (1985)
DC	pKa			
HL 1.79E-005	atm m3/mol	25	EST	MEYLAN,WM & HOWARD,PH (1991)
OH 4.31E-012	cm3/molc sec	25	EST	MEYLAN,WM & HOWARD,PH (1993)

CAS #: 000501-65-5 — DIPHENYLACETYLENE

Formula: $C_{14}H_{10}$

Mol Weight: 178.24

MP (deg C): 60-61 FP (deg C):

BP (deg C): 300

BP pressure (mm Hg):

Property/Value	Units	Temp	Data Type	Reference
WS 4.65E+000	mg/L	25	EST	MEYLAN,WM ET AL. (1996)
logP 4.78			EXP	HANSCH,C & LEO,AJ (1985)
VP 2.28E-003	mm Hg	25	EXT	YAWS,CL (1994B)
DC	pKa			
HL 5.83E-005	atm m3/mol	25	EST	MEYLAN,WM & HOWARD,PH (1991)
OH 2.90E-011	cm3/molc sec	25	EST	MEYLAN,WM & HOWARD,PH (1993)

CAS #: 000501-89-3 — BENZENEACETIC ACID, 4-CARBOXY-

Formula: $C_9H_8O_4$

Mol Weight: 180.16

MP (deg C): FP (deg C):

BP (deg C):

BP pressure (mm Hg):

Property/Value	Units	Temp	Data Type	Reference
WS 1.19E+004	mg/L	25	EST	MEYLAN,WM ET AL. (1996)
logP 1.24			EXP	SANGSTER,J (1994)
VP 5.18E-006	mm Hg	25	EST	NEELY,WB & BLAU,GE (1985)
DC	pKa			
HL 8.90E-013	atm m3/mol	25	EST	MEYLAN,WM & HOWARD,PH (1991)
OH 2.94E-012	cm3/molc sec	25	EST	MEYLAN,WM & HOWARD,PH (1993)

CAS #: 000501-94-0 — 2-(P-HYDROXYPHENYL)ETHANOL

Formula: $C_8H_{10}O_2$

Mol Weight: 138.17

MP (deg C): 89-92 FP (deg C):

BP (deg C):

BP pressure (mm Hg):

Property/Value	Units	Temp	Data Type	Reference
WS 1.24E+005	mg/L	25	EST	MEYLAN,WM ET AL. (1996)
logP 1.09			EST	MEYLAN,WM & HOWARD,PH (1995)
VP 2.33E-004	mm Hg	25	EST	NEELY,WB & BLAU,GE (1985)
DC	pKa			
HL 3.00E-011	atm m3/mol	25	EST	MEYLAN,WM & HOWARD,PH (1991)
OH 4.61E-011	cm3/molc sec	25	EST	MEYLAN,WM & HOWARD,PH (1993)

CAS #: 000501-97-3 — P-HYDROXY-B-PROPIONIC ACID

Formula: $C_9H_{10}O_3$

Mol Weight: 166.18

MP (deg C): 130.8 FP (deg C):

BP (deg C): 208-210

BP pressure (mm Hg): 1.40E+001

Property/Value	Units	Temp	Data Type	Reference
WS 6.18E+004	mg/L	25	EST	MEYLAN,WM ET AL. (1996)
logP 1.83			EXP	HANSCH,C & LEO,AJ (1985)
VP 2.05E-005	mm Hg	25	EST	NEELY,WB & BLAU,GE (1985)
DC	pKa			
HL 6.11E-012	atm m3/mol	25	EST	MEYLAN,WM & HOWARD,PH (1991)
OH 4.33E-011	cm3/molc sec	25	EST	MEYLAN,WM & HOWARD,PH (1993)

CAS #: 000501-98-4 — 2-PROPENOIC ACID, 3-(4-HYDROXYPHENYL)-, (E)-

Formula: $C_9H_8O_3$
Mol Weight: 164.16
MP (deg C): 214 dec
FP (deg C):
BP (deg C):
BP pressure (mm Hg):

Property/Value	Units	Temp	Data Type	Reference
WS 1.83E+004	mg/L	25	EST	MEYLAN,WM ET AL. (1996)
logP 1.46			EXP	SANGSTER,J (1994)
VP 1.61E-005	mm Hg	25	EST	NEELY,WB & BLAU,GE (1985)
DC 4.64	pKa	25	EXP	SERJEANT,EP & DEMPSEY,B (1979)
HL 1.35E-012	atm m3/mol	25	EST	MEYLAN,WM & HOWARD,PH (1991)
OH 5.17E-011	cm3/molc sec	25	EST	MEYLAN,WM & HOWARD,PH (1993)

CAS #: 000502-41-0 — CYCLOHEPTANOL

Formula: $C_7H_{14}O$
Mol Weight: 114.19
MP (deg C): 2
FP (deg C):
BP (deg C): 185
BP pressure (mm Hg):

Property/Value	Units	Temp	Data Type	Reference
WS 5.18E+003	mg/L	25	EST	MEYLAN,WM ET AL. (1996)
logP 2.13			EST	MEYLAN,WM & HOWARD,PH (1995)
VP 1.83E-001	mm Hg	25	EST	NEELY,WB & BLAU,GE (1985)
DC	pKa			
HL 6.50E-006	atm m3/mol	25	EST	MEYLAN,WM & HOWARD,PH (1991)
OH 1.89E-011	cm3/molc sec	25	EST	MEYLAN,WM & HOWARD,PH (1993)

CAS #: 000502-42-1 — CYCLOHEPTANONE

Formula: $C_7H_{12}O$
Mol Weight: 112.17
MP (deg C): 179-181
FP (deg C):
BP (deg C):
BP pressure (mm Hg):

Property/Value	Units	Temp	Data Type	Reference
WS 4.45E+003	mg/L	25	EST	MEYLAN,WM ET AL. (1996)
logP 1.62			EST	MEYLAN,WM & HOWARD,PH (1995)
VP 3.67E-002	mm Hg	25	EXT	BOUBLIK,T ET AL. (1984)
DC	pKa			
HL 6.79E-005	atm m3/mol	25	EST	MEYLAN,WM & HOWARD,PH (1991)
OH 1.35E-011	cm3/molc sec	25	EST	MEYLAN,WM & HOWARD,PH (1993)

CAS #: 000502-44-3 — E-CAPROLACTONE

Formula: $C_6H_{10}O_2$
Mol Weight: 114.15
MP (deg C): -18
FP (deg C):
BP (deg C): 215
BP pressure (mm Hg):

Property/Value	Units	Temp	Data Type	Reference
WS 1.12E+002	mg/L	25	EST	VP/HL
logP 0.68			EST	MEYLAN,WM & HOWARD,PH (1995)
VP 1.35E-001	mm Hg	25	EXP	DAUBERT,TE & DANNER,RP (1989)
DC	pKa			
HL 1.81E-004	atm m3/mol	25	EST	MEYLAN,WM & HOWARD,PH (1991)
OH 5.81E-012	cm3/molc sec		EST	ATKINSON,R (1988)

CAS #: 000502-49-8 — CYCLOOCTANONE

Formula: $C_8H_{14}O$
Mol Weight: 126.20
MP (deg C): 29
FP (deg C):
BP (deg C): 196
BP pressure (mm Hg):

Property/Value	Units	Temp	Data Type	Reference
WS 1.51E+003	mg/L	25	EST	MEYLAN,WM ET AL. (1996)
logP 2.11			EST	MEYLAN,WM & HOWARD,PH (1995)
VP 4.08E-001	mm Hg	25	EXT	BOUBLIK,T ET AL. (1984)
DC	pKa			
HL 9.01E-005	atm m3/mol	25	EST	MEYLAN,WM & HOWARD,PH (1991)
OH 1.49E-011	cm3/molc sec	25	EST	MEYLAN,WM & HOWARD,PH (1993)

CAS #: 000502-55-6 — BIS(ETHYLXANTHOGEN)

Formula: $C_6H_{10}O_2S_4$
Mol Weight: 242.40
MP (deg C): 28-32
FP (deg C):
BP (deg C):
BP pressure (mm Hg):

Property/Value	Units	Temp	Data Type	Reference
WS 2.76E+000	mg/L	25	EXP	YALKOWSKY,SH & DANNENFELSER,RM (1992)
logP 2.06			EST	MEYLAN,WM & HOWARD,PH (1995)
VP 3.25E-005	mm Hg	25	EST	NEELY,WB & BLAU,GE (1985)
DC	pKa			
HL 1.20E-007	atm m3/mol	25	EST	MEYLAN,WM & HOWARD,PH (1991)
OH 2.37E-010	cm3/molc sec	25	EST	MEYLAN,WM & HOWARD,PH (1993)

CAS #: 000502-56-7 — 5-NONANONE

Formula: $C_9H_{18}O$
Mol Weight: 142.24
MP (deg C): -5.9
FP (deg C):
BP (deg C): 188.4
BP pressure (mm Hg):

Property/Value	Units	Temp	Data Type	Reference
WS 3.63E+002	mg/L	30	EXP	YALKOWSKY,SH & DANNENFELSER,RM (1992)
logP 2.71			EST	MEYLAN,WM & HOWARD,PH (1995)
VP 5.44E-001	mm Hg	25	EXT	BOUBLIK,T ET AL. (1984)
DC	pKa			
HL 2.71E-004	atm m3/mol	25	EST	MEYLAN,WM & HOWARD,PH (1991)
OH 1.33E-011	cm3/molc sec	25	EST	MEYLAN,WM & HOWARD,PH (1993)

CAS #: 000502-72-7 — CYCLOPENTADECANONE

Formula: $C_{15}H_{28}O$
Mol Weight: 224.39
MP (deg C): 63
FP (deg C):
BP (deg C): 120
BP pressure (mm Hg): 3.00E-001

Property/Value	Units	Temp	Data Type	Reference
WS 5.99E-001	mg/L	25	EST	MEYLAN,WM ET AL. (1996)
logP 5.55			EST	MEYLAN,WM & HOWARD,PH (1995)
VP 6.28E-004	mm Hg	25	EST	NEELY,WB & BLAU,GE (1985)
DC	pKa			
HL 6.55E-004	atm m3/mol	25	EST	MEYLAN,WM & HOWARD,PH (1991)
OH 2.48E-011	cm3/molc sec	25	EST	MEYLAN,WM & HOWARD,PH (1993)

EPIFLUOROHYDRIN

CAS #: 000503-09-3				
Formula: C_3H_5FO				
Mol Weight: 76.07				
MP (deg C):		FP (deg C):		
BP (deg C): 85.5				
BP pressure (mm Hg):				

Property/Value	Units	Temp	Data Type	Reference
WS 1.82E+005	mg/L	25	EST	MEYLAN,WM ET AL. (1996)
logP -0.16			EXP	HANSCH,C ET AL. (1995)
VP 5.53E+002	mm Hg	25	EST	NEELY,WB & BLAU,GE (1985)
DC	pKa			
HL 3.17E-004	atm m3/mol	25	EST	MEYLAN,WM & HOWARD,PH (1991)
OH 3.03E-013	cm3/molc sec	25	EST	MEYLAN,WM & HOWARD,PH (1993)

2-BUTYNE

CAS #: 000503-17-3				
Formula: C_4H_6				
Mol Weight: 54.09				
MP (deg C): -32.3		FP (deg C):		
BP (deg C): 26.9				
BP pressure (mm Hg):				

Property/Value	Units	Temp	Data Type	Reference
WS 2.25E+003	mg/L	25	EST	MEYLAN,WM ET AL. (1996)
logP 1.46			EXP	HANSCH,C & LEO,AJ (1985)
VP 7.06E+002	mm Hg	25	EXP	DAUBERT,TE & DANNER,RP (1989)
DC	pKa			
HL 1.08E-002	atm m3/mol	25	EST	MEYLAN,WM & HOWARD,PH (1991)
OH 2.74E-011	cm3/molc sec	25	EXP	ATKINSON,R (1989)

AZETIDINE

CAS #: 000503-29-7				
Formula: C_3H_7N				
Mol Weight: 57.10				
MP (deg C):		FP (deg C):		
BP (deg C): 63				
BP pressure (mm Hg):				

Property/Value	Units	Temp	Data Type	Reference
WS 9.31E+005	mg/L	25	EST	MEYLAN,WM ET AL. (1996)
logP 0.21			EST	MEYLAN,WM & HOWARD,PH (1995)
VP 1.78E+002	mm Hg	25	EXP	CHAO,J ET AL. (1983)
DC 11.29	pKa	25	EXP	PERRIN,DD (1965)
HL 9.74E-006	atm m3/mol	25	EST	MEYLAN,WM & HOWARD,PH (1991)
OH 2.14E-011	cm3/molc sec	25	EST	MEYLAN,WM & HOWARD,PH (1993)

TRIMETHYLENEOXIDE

CAS #: 000503-30-0				
Formula: C_3H_6O				
Mol Weight: 58.08				
MP (deg C): -97		FP (deg C):		
BP (deg C): 47.6				
BP pressure (mm Hg):				

Property/Value	Units	Temp	Data Type	Reference
WS 1.00E+006	mg/L	20	EXP	YALKOWSKY,SH & DANNENFELSER,RM (1992)
logP -0.14			EXP	HANSCH,C ET AL. (1995)
VP 3.24E+002	mm Hg	25	EXP	DAUBERT,TE & DANNER,RP (1989)
DC	pKa			
HL 2.48E-005	atm m3/mol	25	EST	VP/WSOL
OH 1.03E-011	cm3/molc sec	25	EXP	ATKINSON,R (1989)

METHANEDISULFONIC ACID

CAS #: 000503-40-2				
Formula: $CH_4O_6S_2$				
Mol Weight: 176.17				
MP (deg C): 98		FP (deg C):		
BP (deg C):				
BP pressure (mm Hg):				

Property/Value	Units	Temp	Data Type	Reference
WS 2.46E+006	mg/L	25	EXP	YALKOWSKY,SH & DANNENFELSER,RM (1992)
logP -3.70			EST	MEYLAN,WM & HOWARD,PH (1995)
VP 1.65E-008	mm Hg	25	EST	NEELY,WB & BLAU,GE (1985)
DC	pKa			
HL 2.15E-015	atm m3/mol	25	EST	MEYLAN,WM & HOWARD,PH (1991)
OH 1.21E-012	cm3/molc sec	25	EST	MEYLAN,WM & HOWARD,PH (1993)

(Z)-2-BUTENOIC ACID (ISOCROTONIC ACID)

CAS #: 000503-64-0				
Formula: $C_4H_6O_2$				
Mol Weight: 86.09				
MP (deg C): 13.5-14.5		FP (deg C):		
BP (deg C): 168-169				
BP pressure (mm Hg):				

Property/Value	Units	Temp	Data Type	Reference
WS 1.00E+006	mg/L	20	EXP	YALKOWSKY,SH & DANNENFELSER,RM (1992)
logP 0.85			EST	MEYLAN,WM & HOWARD,PH (1995)
VP 5.88E-001	mm Hg	25	EXP	DAUBERT,TE & DANNER,RP (1989)
DC 4.17	pKa	18	EXP	KORTUM,G ET AL (1961)
HL 6.66E-008	atm m3/mol	25	EST	VP/WSOL
OH 2.04E-011	cm3/molc sec	25	EST	MEYLAN,WM & HOWARD,PH (1993)

ISOVALERIC ACID

CAS #: 000503-74-2				
Formula: $C_5H_{10}O_2$				
Mol Weight: 102.13				
MP (deg C): -37		FP (deg C):		
BP (deg C): 175-177				
BP pressure (mm Hg):				

Property/Value	Units	Temp	Data Type	Reference
WS 4.07E+004	mg/L	20	EXP	YALKOWSKY,SH & DANNENFELSER,RM (1992)
logP 1.16			EXP	HANSCH,C ET AL. (1995)
VP 4.40E-001	mm Hg	25	EXP	DAUBERT,TE & DANNER,RP (1989)
DC 4.77	pKa	20	EXP	KORTUM,G ET AL (1961)
HL 1.45E-006	atm m3/mol	25	EST	VP/WSOL
OH 4.10E-012	cm3/molc sec	25	EST	MEYLAN,WM & HOWARD,PH (1993)

5-METHYLRESORCINOL

CAS #: 000504-15-4				
Formula: $C_7H_8O_2$				
Mol Weight: 124.14				
MP (deg C): 58		FP (deg C):		
BP (deg C): 290				
BP pressure (mm Hg):				

Property/Value	Units	Temp	Data Type	Reference
WS 1.65E+004	mg/L	25	EST	MEYLAN,WM ET AL. (1996)
logP 1.58			EST	MEYLAN,WM & HOWARD,PH (1995)
VP 2.80E-003	mm Hg	25	EST	NEELY,WB & BLAU,GE (1985)
DC	pKa			
HL 6.44E-011	atm m3/mol	25	EST	MEYLAN,WM & HOWARD,PH (1991)
OH 2.00E-010	cm3/molc sec	25	EST	MEYLAN,WM & HOWARD,PH (1993)

000504-20-1 — PHORONE

Formula: $C_9H_{14}O$
Mol Weight: 138.21
MP (deg C): 28
FP (deg C):
BP (deg C): 198-199
BP pressure (mm Hg):

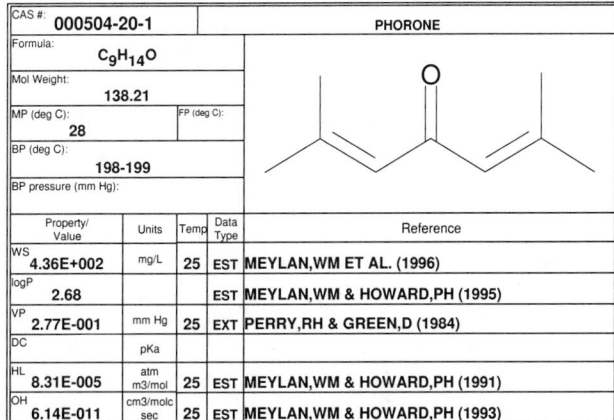

Property/Value	Units	Temp	Data Type	Reference
WS 4.36E+002	mg/L	25	EST	MEYLAN,WM ET AL. (1996)
logP 2.68			EST	MEYLAN,WM & HOWARD,PH (1995)
VP 2.77E-001	mm Hg	25	EXT	PERRY,RH & GREEN,D (1984)
DC	pKa			
HL 8.31E-005	atm m3/mol	25	EST	MEYLAN,WM & HOWARD,PH (1991)
OH 6.14E-011	cm3/molc sec	25	EST	MEYLAN,WM & HOWARD,PH (1993)

000504-24-5 — 4-AMINOPYRIDINE

Formula: $C_5H_6N_2$
Mol Weight: 94.12
MP (deg C): 158.5
FP (deg C):
BP (deg C): 273
BP pressure (mm Hg):

Property/Value	Units	Temp	Data Type	Reference
WS 8.80E+003	mg/L	25	EST	MEYLAN,WM ET AL. (1996)
logP 0.26			EXP	HANSCH,C & LEO,AJ (1985)
VP 3.70E-004	mm Hg	25	EXT	WEAST,RC ET AL. (1983) & DEAN,JA (1985)
DC 9.17	pKa	20	EXP	ALBERT,A ET AL. (1948)
HL 2.81E-009	atm m3/mol	25	EST	MEYLAN,WM & HOWARD,PH (1991)
OH 4.10E+001	cm3/molc sec	25	EST	MEYLAN,WM & HOWARD,PH (1993)

000504-29-0 — 2-AMINOPYRIDINE

Formula: $C_5H_6N_2$
Mol Weight: 94.12
MP (deg C): 57-58
FP (deg C):
BP (deg C): 210.6
BP pressure (mm Hg):

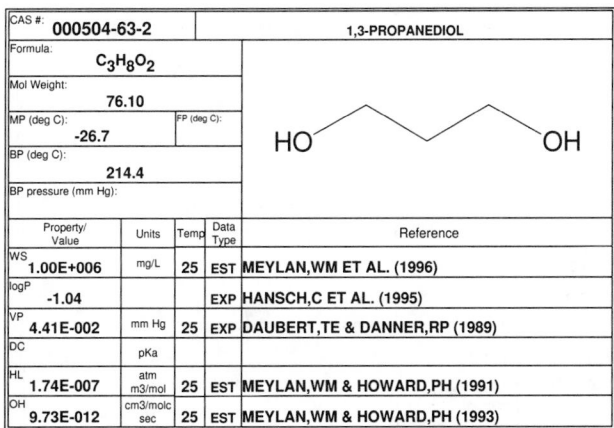

Property/Value	Units	Temp	Data Type	Reference
WS 5.40E+003	mg/L	25	EST	MEYLAN,WM ET AL. (1996)
logP 0.48			EXP	HANSCH,C & LEO,AJ (1985)
VP 8.00E-001	mm Hg	25	EXT	WEAST,RC ET AL. (1983) & DEAN,JA (1985)
DC 6.86	pKa	20	EXP	ALBERT,A ET AL. (1948)
HL 2.81E+000	atm m3/mol	25	EST	MEYLAN,WM & HOWARD,PH (1991)
OH 4.10E-011	cm3/molc sec	25	EST	MEYLAN,WM & HOWARD,PH (1993)

000504-60-9 — 1,3-PENTADIENE

Formula: C_5H_8
Mol Weight: 68.12
MP (deg C):
FP (deg C):
BP (deg C): 42
BP pressure (mm Hg):

Property/Value	Units	Temp	Data Type	Reference
WS 3.26E+002	mg/L	25	EST	MEYLAN,WM ET AL. (1996)
logP 2.45			EST	MEYLAN,WM & HOWARD,PH (1995)
VP 4.05E+002	mm Hg	25	EXP	PERRY,RH & GREEN,D (1984)
DC	pKa			
HL 1.22E-001	atm m3/mol	25	EST	MEYLAN,WM & HOWARD,PH (1991)
OH 1.05E-010	cm3/molc sec	25	EST	MEYLAN,WM & HOWARD,PH (1993)

000504-63-2 — 1,3-PROPANEDIOL

Formula: $C_3H_8O_2$
Mol Weight: 76.10
MP (deg C): -26.7
FP (deg C):
BP (deg C): 214.4
BP pressure (mm Hg):

Property/Value	Units	Temp	Data Type	Reference
WS 1.00E+006	mg/L	25	EST	MEYLAN,WM ET AL. (1996)
logP -1.04			EXP	HANSCH,C ET AL. (1995)
VP 4.41E-002	mm Hg	25	EXP	DAUBERT,TE & DANNER,RP (1989)
DC	pKa			
HL 1.74E-007	atm m3/mol	25	EST	MEYLAN,WM & HOWARD,PH (1991)
OH 9.73E-012	cm3/molc sec	25	EST	MEYLAN,WM & HOWARD,PH (1993)

000505-02-2 — 2,8-DECADIENE-4,6-DIYNOIC ACID, METHYL ESTER

Formula: $C_{11}H_{10}O_2$
Mol Weight: 174.20
MP (deg C):
FP (deg C):
BP (deg C):
BP pressure (mm Hg):

Property/Value	Units	Temp	Data Type	Reference
WS 4.75E+002	mg/L	25	EST	MEYLAN,WM ET AL. (1996)
logP 2.45			EXP	MCLACHLAN,D ET AL. (1986)
VP 3.70E-003	mm Hg	25	EST	NEELY,WB & BLAU,GE (1985)
DC	pKa			
HL 2.90E-005	atm m3/mol	25	EST	MEYLAN,WM & HOWARD,PH (1991)
OH 1.59E-010	cm3/molc sec	25	EST	MEYLAN,WM & HOWARD,PH (1993)

000505-22-6 — 1,3-DIOXANE

Formula: $C_4H_8O_2$
Mol Weight: 88.11
MP (deg C): -45
FP (deg C):
BP (deg C): 106.1
BP pressure (mm Hg):

Property/Value	Units	Temp	Data Type	Reference
WS 8.76E+004	mg/L	25	EST	MEYLAN,WM ET AL. (1996)
logP 0.18			EST	MEYLAN,WM & HOWARD,PH (1995)
VP 3.85E+001	mm Hg	25	EST	NEELY,WB & BLAU,GE (1985)
DC	pKa			
HL 2.96E-005	atm m3/mol	25	EST	MEYLAN,WM & HOWARD,PH (1991)
OH 9.15E-012	cm3/molc sec	25	EXP	ATKINSON,R (1989)

000505-29-3 — 1,4-DITHIANE

Formula: $C_4H_8S_2$
Mol Weight: 120.24
MP (deg C): 112.3
FP (deg C):
BP (deg C): 199.5
BP pressure (mm Hg):

Property/Value	Units	Temp	Data Type	Reference
WS 6.58E+003	mg/L	25	EST	MEYLAN,WM ET AL. (1996)
logP 1.38			EST	MEYLAN,WM & HOWARD,PH (1995)
VP 1.57E+000	mm Hg	25	EST	NEELY,WB & BLAU,GE (1985)
DC	pKa			
HL 4.91E-006	atm m3/mol	25	EST	MEYLAN,WM & HOWARD,PH (1991)
OH 3.92E-011	cm3/molc sec	25	EST	MEYLAN,WM & HOWARD,PH (1993)

CAS #: 000505-48-6				SUBERIC ACID

Formula: $C_8H_{14}O_4$

Mol Weight: 174.20

MP (deg C): 140-144 FP (deg C):

BP (deg C): 279

BP pressure (mm Hg): 1.00E+002

Property/Value	Units	Temp	Data Type	Reference
WS 1.19E+004	mg/L	25	EXP	YALKOWSKY,SH & DANNENFELSER,RM (1992)
logP 1.21			EST	MEYLAN,WM & HOWARD,PH (1995)
VP 1.69E-008	mm Hg	25	EXT	PERRY,RH & GREEN,D (1984)
DC 4.52	pKa	25	EXP	KORTUM,G ET AL (1961)
HL 1.68E-011	atm m3/mol	25	EST	MEYLAN,WM & HOWARD,PH (1991)
OH 8.42E-012	cm3/molc sec	25	EST	MEYLAN,WM & HOWARD,PH (1993)

CAS #: 000505-57-7				2-HEXENAL(TRANS)

Formula: $C_6H_{10}O$

Mol Weight: 98.15

MP (deg C): FP (deg C):

BP (deg C):

BP pressure (mm Hg):

Property/Value	Units	Temp	Data Type	Reference
WS 5.26E+003	mg/L	25	EST	MEYLAN,WM ET AL. (1996)
logP 1.58			EST	MEYLAN,WM & HOWARD,PH (1995)
VP 6.60E+000	mm Hg	25	EST	NEELY,WB & BLAU,GE (1985)
DC	pKa			
HL 4.89E-005	atm m3/mol	25	EXP	BUTTERY,RG ET AL. (1971)
OH 3.40E-011	cm3/molc sec	25	EST	MEYLAN,WM & HOWARD,PH (1993)

CAS #: 000505-60-2				DI-2-CHLOROETHYL SULFIDE

Formula: $C_4H_8Cl_2S$

Mol Weight: 159.08

MP (deg C): 13-14 FP (deg C):

BP (deg C): 215-217

BP pressure (mm Hg):

Property/Value	Units	Temp	Data Type	Reference
WS 6.84E+002	mg/L	25	EXP	SEIDELL,A (1941)
logP 2.41			EST	MEYLAN,WM & HOWARD,PH (1995)
VP 1.10E-001	mm Hg	25	EXP	BOUBLIK,T ET AL. (1984)
DC	pKa			
HL 3.37E-005	atm m3/mol	25	EST	VP/WSOL
OH 1.14E-011	cm3/molc sec	25	EST	MEYLAN,WM & HOWARD,PH (1993)

CAS #: 000506-05-8				1-FLUOROUNDECANE

Formula: $C_{11}H_{23}F$

Mol Weight: 174.30

MP (deg C): FP (deg C):

BP (deg C):

BP pressure (mm Hg):

Property/Value	Units	Temp	Data Type	Reference
WS 8.21E-001	mg/L	25	EST	MEYLAN,WM ET AL. (1996)
logP 5.68			EST	MEYLAN,WM & HOWARD,PH (1995)
VP 2.08E-001	mm Hg	25	EXT	LI,JCM & ROSSINI,FD (1953)
DC	pKa			
HL 2.80E-001	atm m3/mol	25	EST	MEYLAN,WM & HOWARD,PH (1991)
OH 1.20E-011	cm3/molc sec	25	EST	MEYLAN,WM & HOWARD,PH (1993)

CAS #: 000506-12-7				HEPTADECANOIC ACID

Formula: $C_{17}H_{34}O_2$

Mol Weight: 270.46

MP (deg C): 61 FP (deg C):

BP (deg C): 227

BP pressure (mm Hg): 1.00E+002

Property/Value	Units	Temp	Data Type	Reference
WS 4.20E+000	mg/L	20	EXP	YALKOWSKY,SH & DANNENFELSER,RM (1992)
logP 7.45			EST	MEYLAN,WM & HOWARD,PH (1995)
VP 7.02E-009	mm Hg	25	EXT	DAUBERT,TE & DANNER,RP (1993)
DC	pKa			
HL 3.84E-005	atm m3/mol	25	EST	MEYLAN,WM & HOWARD,PH (1991)
OH 2.11E-011	cm3/molc sec	25	EST	MEYLAN,WM & HOWARD,PH (1993)

CAS #: 000506-30-9				EICOSANOIC ACID

Formula: $C_{20}H_{40}O_2$

Mol Weight: 312.54

MP (deg C): 75.5 FP (deg C):

BP (deg C): 328

BP pressure (mm Hg):

Property/Value	Units	Temp	Data Type	Reference
WS 3.00E-004	mg/L	25	EST	MEYLAN,WM ET AL. (1996)
logP 9.29			EXP	SANGSTER,J (1993)
VP 1.81E-009	mm Hg	25	EXT	DAUBERT,TE & DANNER,RP (1993)
DC	pKa			
HL 8.98E-005	atm m3/mol	25	EST	MEYLAN,WM & HOWARD,PH (1991)
OH 2.53E-011	cm3/molc sec	25	EST	MEYLAN,WM & HOWARD,PH (1993)

CAS #: 000506-32-1				5,8,11,14-EICOSATETRAENOIC ACID

Formula: $C_{20}H_{32}O_2$

Mol Weight: 304.48

MP (deg C): -49.5 FP (deg C):

BP (deg C):

BP pressure (mm Hg):

Property/Value	Units	Temp	Data Type	Reference
WS 3.14E-002	mg/L	25	EST	MEYLAN,WM ET AL. (1996)
logP 6.98			EXP	SANGSTER,J (1993)
VP 1.55E-007	mm Hg	25	EST	NEELY,WB & BLAU,GE (1985)
DC	pKa			
HL 5.38E-005	atm m3/mol	25	EST	MEYLAN,WM & HOWARD,PH (1991)
OH 2.38E-010	cm3/molc sec	25	EST	MEYLAN,WM & HOWARD,PH (1993)

CAS #: 000506-59-2				N-METHYLMETHANAMINE, HYDROCHLORIDE

Formula: C_2H_8ClN

Mol Weight: 81.55

MP (deg C): 170-173 FP (deg C):

BP (deg C):

BP pressure (mm Hg):

Property/Value	Units	Temp	Data Type	Reference
WS 1.00E+006	mg/L	25	EXP	SEIDELL,A (1941)
logP -3.28			EST	MEYLAN,WM & HOWARD,PH (1995)
VP 4.35E-004	mm Hg	25	EST	NEELY,WB & BLAU,GE (1985)
DC	pKa			
HL 1.27E-012	atm m3/mol	25	EST	MEYLAN,WM & HOWARD,PH (1991)
OH 2.53E-012	cm3/molc sec	25	EST	MEYLAN,WM & HOWARD,PH (1993)

ACETYL BROMIDE

CAS #:	000506-96-7				
Formula:	C_2H_3BrO				
Mol Weight:	122.95				
MP (deg C):	-96	FP (deg C):			
BP (deg C):	76				
BP pressure (mm Hg):					

Property/ Value	Units	Temp	Data Type	Reference
WS 2.08E+005	mg/L	25	EST	MEYLAN,WM ET AL. (1996)
logP -0.38			EST	MEYLAN,WM & HOWARD,PH (1995)
VP 1.22E+002	mm Hg	25	EXP	OHE,S (1976)
DC	pKa			
HL 3.91E-004	atm m3/mol	25	EST	MEYLAN,WM & HOWARD,PH (1991)
OH 9.11E-015	cm3/molc sec	25	EST	MEYLAN,WM & HOWARD,PH (1993)

ACETYL IODIDE

CAS #:	000507-02-8				
Formula:	C_2H_3IO				
Mol Weight:	169.95				
MP (deg C):		FP (deg C):			
BP (deg C):	108				
BP pressure (mm Hg):					

Property/ Value	Units	Temp	Data Type	Reference
WS 5.77E+004	mg/L	25	EST	MEYLAN,WM ET AL. (1996)
logP 0.03			EST	MEYLAN,WM & HOWARD,PH (1995)
VP 3.24E+001	mm Hg	25	EXP	OHE,S (1976)
DC	pKa			
HL 1.96E-004	atm m3/mol	25	EST	MEYLAN,WM & HOWARD,PH (1991)
OH 9.11E-015	cm3/molc sec	25	EST	MEYLAN,WM & HOWARD,PH (1993)

2-BROMO-2-METHYLPROPANE

CAS #:	000507-19-7				
Formula:	C_4H_9Br				
Mol Weight:	137.03				
MP (deg C):	-16.3	FP (deg C):			
BP (deg C):	73.3				
BP pressure (mm Hg):					

Property/ Value	Units	Temp	Data Type	Reference
WS 6.00E+002	mg/L	25	EXP	YAWS,CL & YANG,HC (1990)
logP 2.54			EST	MEYLAN,WM & HOWARD,PH (1995)
VP 1.35E+002	mm Hg	25	EXP	BOUBLIK,T ET AL (1984)
DC	pKa			
HL 4.07E-002	atm m3/mol	25	EST	VP/WSOL
OH 4.08E-013	cm3/molc sec	25	EST	MEYLAN,WM & HOWARD,PH (1993)

T-BUTYL CHLORIDE

CAS #:	000507-20-0				
Formula:	C_4H_9Cl				
Mol Weight:	92.57				
MP (deg C):	-26.5	FP (deg C):			
BP (deg C):	51				
BP pressure (mm Hg):					

Property/ Value	Units	Temp	Data Type	Reference
WS 2.88E+003	mg/L	15	EXP	YALKOWSKY,SH & DANNENFELSER,RM (1992)
logP 2.45			EST	MEYLAN,WM & HOWARD,PH (1995)
VP 3.02E+002	mm Hg	25	EXP	DAUBERT,TE & DANNER,RP (1989)
DC	pKa			
HL 1.28E-002	atm m3/mol	25	EST	MEYLAN,WM & HOWARD,PH (1993)
OH 4.08E-013	cm3/molc sec	25	EST	MEYLAN,WM & HOWARD,PH (1993)

2-(TRIFLUORMETHYL)-2-PROPANOL

CAS #:	000507-52-8				
Formula:	$C_4H_7F_3O$				
Mol Weight:	128.09				
MP (deg C):		FP (deg C):			
BP (deg C):					
BP pressure (mm Hg):					

Property/ Value	Units	Temp	Data Type	Reference
WS 3.92E+004	mg/L	25	EST	MEYLAN,WM ET AL. (1996)
logP 1.04			EXP	MULLER,N (1986)
VP 2.14E+002	mm Hg	25	EST	NEELY,WB & BLAU,GE (1985)
DC	pKa			
HL 4.96E-005	atm m3/mol	25	EST	MEYLAN,WM & HOWARD,PH (1991)
OH 1.17E-012	cm3/molc sec	25	EST	MEYLAN,WM & HOWARD,PH (1993)

BORNEOL

CAS #:	000507-70-0				
Formula:	$C_{10}H_{18}O$				
Mol Weight:	154.25				
MP (deg C):		FP (deg C):			
BP (deg C):					
BP pressure (mm Hg):					

Property/ Value	Units	Temp	Data Type	Reference
WS 7.38E+002	mg/L	25	EXP	YALKOWSKY,SH & DANNENFELSER,RM (1992)
logP 2.69			EXP	CHEM INSPECT TEST INST (1992)
VP 3.53E-002	mm Hg	25	EST	NEELY,WB & BLAU,GE (1985)
DC	pKa			
HL 6.70E-006	atm m3/mol	25	EST	MEYLAN,WM & HOWARD,PH (1991)
OH 1.22E-011	cm3/molc sec	25	EST	MEYLAN,WM & HOWARD,PH (1993)

4'-EPIOLEANDRIN

CAS #:	000508-22-5				
Formula:	$C_{32}H_{48}O_9$				
Mol Weight:	576.73				
MP (deg C):		FP (deg C):			
BP (deg C):					
BP pressure (mm Hg):					

Property/ Value	Units	Temp	Data Type	Reference
WS 2.68E+000	mg/L	25	EST	MEYLAN,WM ET AL. (1996)
logP 2.26			EXP	HANSCH,C & LEO,AJ (1985)
VP 3.61E-019	mm Hg	25	EST	NEELY,WB & BLAU,GE (1985)
DC	pKa			
HL 2.06E-018	atm m3/mol	25	EST	MEYLAN,WM & HOWARD,PH (1991)
OH 1.04E-010	cm3/molc sec	25	EST	MEYLAN,WM & HOWARD,PH (1993)

OUABAGENIN

CAS #:	000508-52-1				
Formula:	$C_{23}H_{34}O_8$				
Mol Weight:	438.52				
MP (deg C):	235-238	FP (deg C):			
BP (deg C):					
BP pressure (mm Hg):					

Property/ Value	Units	Temp	Data Type	Reference
WS 1.83E+003	mg/L	25	EST	MEYLAN,WM ET AL. (1996)
logP -0.02			EXP	HANSCH,C ET AL. (1995)
VP 3.52E-019	mm Hg	25	EST	NEELY,WB & BLAU,GE (1985)
DC	pKa			
HL 1.12E-015	atm m3/mol	25	EST	MEYLAN,WM & HOWARD,PH (1991)
OH 7.94E-011	cm3/molc sec	25	EST	MEYLAN,WM & HOWARD,PH (1993)

CAS #: 000508-59-8	PARTHENIN

Formula: $C_{15}H_{18}O_4$

Mol Weight: 262.31

MP (deg C): 163-166   FP (deg C):

BP (deg C):

BP pressure (mm Hg):

Property/Value	Units	Temp	Data Type	Reference
WS 1.44E+004	mg/L	25	EST	MEYLAN,WM ET AL. (1996)
logP 0.77			EXP	HANSCH,C & LEO,AJ (1985)
VP 8.90E-009	mm Hg	25	EST	NEELY,WB & BLAU,GE (1985)
DC	pKa			
HL 1.38E-012	atm m3/mol	25	EST	MEYLAN,WM & HOWARD,PH (1991)
OH 8.86E-011	cm3/molc sec	25	EST	MEYLAN,WM & HOWARD,PH (1993)

CAS #: 000508-75-8	CONVALLATOXIN

Formula: $C_{29}H_{42}O_{10}$

Mol Weight: 550.65

MP (deg C): 235-242   FP (deg C):

BP (deg C):

BP pressure (mm Hg):

Property/Value	Units	Temp	Data Type	Reference
WS 8.67E+002	mg/L	25	EST	MEYLAN,WM ET AL. (1996)
logP -0.48			EXP	SANGSTER,J (1993)
VP 1.82E-022	mm Hg	25	EST	NEELY,WB & BLAU,GE (1985)
DC	pKa			
HL 2.51E-021	atm m3/mol	25	EST	MEYLAN,WM & HOWARD,PH (1991)
OH 1.31E-010	cm3/molc sec	25	EST	MEYLAN,WM & HOWARD,PH (1993)

CAS #: 000508-77-0	CYMARIN

Formula: $C_{30}H_{44}O_9$

Mol Weight: 548.68

MP (deg C): 148   FP (deg C):

BP (deg C):

BP pressure (mm Hg):

Property/Value	Units	Temp	Data Type	Reference
WS 9.87E+001	mg/L	25	EST	MEYLAN,WM ET AL. (1996)
logP 0.64			EXP	SANGSTER,J (1993)
VP 2.50E-020	mm Hg	25	EST	NEELY,WB & BLAU,GE (1985)
DC	pKa			
HL 2.20E-020	atm m3/mol	25	EST	MEYLAN,WM & HOWARD,PH (1991)
OH 1.31E-010	cm3/molc sec	25	EST	MEYLAN,WM & HOWARD,PH (1993)

CAS #: 000508-93-0	EVOMONOSIDE

Formula: $C_{29}H_{44}O_8$

Mol Weight: 520.67

MP (deg C):   FP (deg C):

BP (deg C):

BP pressure (mm Hg):

Property/Value	Units	Temp	Data Type	Reference
WS 8.49E+000	mg/L	25	EST	MEYLAN,WM ET AL. (1996)
logP 2.10			EXP	HANSCH,C & LEO,AJ (1985)
VP 6.54E-020	mm Hg	25	EST	NEELY,WB & BLAU,GE (1985)
DC	pKa			
HL 8.38E-017	atm m3/mol	25	EST	MEYLAN,WM & HOWARD,PH (1991)
OH 9.87E-011	cm3/molc sec	25	EST	MEYLAN,WM & HOWARD,PH (1993)

CAS #: 000509-14-8	TETRANITROMETHANE

Formula: CN_4O_8

Mol Weight: 196.03

MP (deg C): 13.8   FP (deg C):

BP (deg C): 126

BP pressure (mm Hg):

Property/Value	Units	Temp	Data Type	Reference
WS 8.50E+004	mg/L	25	EST	MEYLAN,WM ET AL. (1996)
logP -2.05			EST	MEYLAN,WM & HOWARD,PH (1995)
VP 8.42E+000	mm Hg	25	EXP	BOUBLIK,T ET AL. (1984)
DC	pKa			
HL 2.56E-005	atm m3/mol	25	EST	VP/WSOL
OH 0.00E+000	cm3/molc sec	25	EST	MEYLAN,WM & HOWARD,PH (1993)

CAS #: 000509-36-4	2-METHOXYSTRYCHNINE

Formula: $C_{22}H_{24}N_2O_3$

Mol Weight: 364.45

MP (deg C): 222   FP (deg C):

BP (deg C):

BP pressure (mm Hg):

Property/Value	Units	Temp	Data Type	Reference
WS 1.33E+002	mg/L	25	EST	MEYLAN,WM ET AL. (1996)
logP 1.85			EXP	HANSCH,C & LEO,AJ (1985)
VP 5.53E-010	mm Hg	25	EST	NEELY,WB & BLAU,GE (1985)
DC	pKa			
HL 3.53E-015	atm m3/mol	25	EST	MEYLAN,WM & HOWARD,PH (1991)
OH 2.03E-010	cm3/molc sec	25	EST	MEYLAN,WM & HOWARD,PH (1993)

CAS #: 000509-74-0	ACETYLMETHADOL

Formula: $C_{23}H_{31}NO_2$

Mol Weight: 353.51

MP (deg C):   FP (deg C):

BP (deg C):

BP pressure (mm Hg):

Property/Value	Units	Temp	Data Type	Reference
WS 1.36E+001	mg/L	25	EST	MEYLAN,WM ET AL. (1996)
logP 4.27			EXP	HANSCH,C & LEO,AJ (1985)
VP 3.05E-007	mm Hg	25	EST	NEELY,WB & BLAU,GE (1985)
DC	pKa			
HL 3.10E-009	atm m3/mol	25	EST	MEYLAN,WM & HOWARD,PH (1991)
OH 1.07E-010	cm3/molc sec	25	EST	MEYLAN,WM & HOWARD,PH (1993)

CAS #: 000509-86-4	HEPTABARBITAL

Formula: $C_{13}H_{18}N_2O_3$

Mol Weight: 250.30

MP (deg C): 174   FP (deg C):

BP (deg C):

BP pressure (mm Hg):

Property/Value	Units	Temp	Data Type	Reference
WS 2.50E+002	mg/L	25	EXP	YALKOWSKY,SH & DANNENFELSER,RM (1992)
logP 2.03			EXP	HANSCH,C ET AL. (1995)
VP 1.17E-012	mm Hg	25	EST	NEELY,WB & BLAU,GE (1985)
DC	pKa			
HL 6.81E-013	atm m3/mol	25	EST	MEYLAN,WM & HOWARD,PH (1991)
OH 9.92E-011	cm3/molc sec	25	EST	MEYLAN,WM & HOWARD,PH (1993)

AMBROSIN

CAS #:	000509-93-3				AMBROSIN

Formula: $C_{15}H_{18}O_3$

Mol Weight: 246.31

MP (deg C): 146   FP (deg C):

BP (deg C):

BP pressure (mm Hg):

Property/ Value	Units	Temp	Data Type	Reference
WS 3.27E+003	mg/L	25	EST	MEYLAN,WM ET AL. (1996)
logP 1.03			EXP	HANSCH,C & LEO,AJ (1985)
VP 3.35E-006	mm Hg	25	EST	NEELY,WB & BLAU,GE (1985)
DC	pKa			
HL 3.78E-008	atm m3/mol	25	EST	MEYLAN,WM & HOWARD,PH (1991)
OH 9.05E-011	cm3/molc sec	25	EST	MEYLAN,WM & HOWARD,PH (1993)

CHLOROBENZILATE

CAS #:	000510-15-6				CHLOROBENZILATE

Formula: $C_{16}H_{14}Cl_2O_3$

Mol Weight: 325.19

MP (deg C): 35-37   FP (deg C):

BP (deg C): 415

BP pressure (mm Hg):

Property/ Value	Units	Temp	Data Type	Reference
WS 1.30E+001	mg/L	20	EXP	FURER,R & GEIGER,M (1977)
logP 4.74			EXP	CHEM INSPECT TEST INST (1992)
VP 2.20E-006	mm Hg	20	EXP	MARTIN,H & WORTHING,CR (1977)
DC	pKa			
HL 7.24E-008	atm m3/mol	20	EST	VP/WSOL
OH 4.92E-012	cm3/molc sec	25	EST	MEYLAN,WM & HOWARD,PH (1993)

TRIMETHYL PHOSPHATE

CAS #:	000512-56-1				TRIMETHYL PHOSPHATE

Formula: $C_3H_9O_4P$

Mol Weight: 140.08

MP (deg C): -46   FP (deg C):

BP (deg C): 192.85

BP pressure (mm Hg):

Property/ Value	Units	Temp	Data Type	Reference
WS 5.00E+005	mg/L	25	EXP	YALKOWSKY,SH & DANNENFELSER,RM (1992)
logP -0.65			EXP	HANSCH,C & LEO,AJ (1985)
VP 8.50E-001	mm Hg	25	EXT	DAUBERT,TE & DANNER,RP (1991)
DC	pKa			
HL 7.20E-009	atm m3/mol	20	EXP	WOLFENDEN,R & WILLIAMS,R (1983)
OH 7.40E-012	cm3/molc sec	25	EXP	ATKINSON,R (1989)

RAFFINOSE

CAS #:	000512-69-6				RAFFINOSE

Formula: $C_{18}H_{32}O_{16}$

Mol Weight: 504.45

MP (deg C): 80   FP (deg C):

BP (deg C):

BP pressure (mm Hg):

Property/ Value	Units	Temp	Data Type	Reference
WS 2.03E+005	mg/L		EXP	YALKOWSKY,SH & DANNENFELSER,RM (1992)
logP -7.61			EST	MEYLAN,WM & HOWARD,PH (1995)
VP 2.00E-025	mm Hg	25	EST	NEELY,WB & BLAU,GE (1985)
DC	pKa			
HL 4.10E-030	atm m3/mol	25	EST	MEYLAN,WM & HOWARD,PH (1991)
OH 1.66E-010	cm3/molc sec	25	EST	MEYLAN,WM & HOWARD,PH (1993)

TRIPROPYL PHOSPHATE

CAS #:	000513-08-6				TRIPROPYL PHOSPHATE

Formula: $C_9H_{21}O_4P$

Mol Weight: 224.24

MP (deg C):   FP (deg C):

BP (deg C): 252

BP pressure (mm Hg):

Property/ Value	Units	Temp	Data Type	Reference
WS 8.27E+002	mg/L	25	EST	MEYLAN,WM ET AL. (1996)
logP 1.87			EXP	HANSCH,C & LEO,AJ (1985)
VP 4.33E-003	mm Hg	25	EST	NEELY,WB & BLAU,GE (1985)
DC	pKa			
HL 1.36E-006	atm m3/mol	25	EST	MEYLAN,WM & HOWARD,PH (1991)
OH 7.46E-011	cm3/molc sec	25	EST	MEYLAN,WM & HOWARD,PH (1993)

2,3-DIBROMO-1-PROPENE

CAS #:	000513-31-5				2,3-DIBROMO-1-PROPENE

Formula: $C_3H_4Br_2$

Mol Weight: 199.88

MP (deg C):   FP (deg C):

BP (deg C): 141

BP pressure (mm Hg):

Property/ Value	Units	Temp	Data Type	Reference
WS 3.80E+002	mg/L	25	EST	MEYLAN,WM ET AL. (1996)
logP 2.42			EST	MEYLAN,WM & HOWARD,PH (1995)
VP 7.58E+000	mm Hg	25	EXP	PERRY,RH & GREEN,D (1984)
DC	pKa			
HL 2.22E-003	atm m3/mol	25	EST	MEYLAN,WM & HOWARD,PH (1991)
OH 1.04E-011	cm3/molc sec	25	EST	MEYLAN,WM & HOWARD,PH (1993)

2-METHYL-2-BUTENE

CAS #:	000513-35-9				2-METHYL-2-BUTENE

Formula: C_5H_{10}

Mol Weight: 70.14

MP (deg C): -134   FP (deg C):

BP (deg C): 35-38

BP pressure (mm Hg):

Property/ Value	Units	Temp	Data Type	Reference
WS 1.93E+002	mg/L	25	EXP	HINE,J & MOOKERJEE,PK (1975)
logP 2.64			EST	MEYLAN,WM & HOWARD,PH (1995)
VP 4.60E+002	mm Hg	25	EXP	PERRY,RH & GREEN,D (1984)
DC	pKa			
HL 2.20E-001	atm m3/mol	25	EST	VP/WSOL
OH 8.69E-011	cm3/molc sec	25	EXP	ATKINSON,R (1989)

ISOBUTYL CHLORIDE

CAS #:	000513-36-0				ISOBUTYL CHLORIDE

Formula: C_4H_9Cl

Mol Weight: 92.57

MP (deg C): -131   FP (deg C):

BP (deg C): 68-69

BP pressure (mm Hg):

Property/ Value	Units	Temp	Data Type	Reference
WS 9.26E+002	mg/L	13	EXP	YALKOWSKY,SH & DANNENFELSER,RM (1992)
logP 2.49			EST	MEYLAN,WM & HOWARD,PH (1995)
VP 1.50E+002	mm Hg	25	EXP	PERRY,RH & GREEN,D (1984)
DC	pKa			
HL 2.18E-002	atm m3/mol	25	EST	MEYLAN,WM & HOWARD,PH (1991)
OH 1.47E-012	cm3/molc sec	25	EST	MEYLAN,WM & HOWARD,PH (1993)

CAS #: 000513-37-1				1-CHLORO-2-METHYLPROPENE

Formula: C_4H_7Cl

Mol Weight: 90.55

MP (deg C): | **FP (deg C):**

BP (deg C): 68.1

BP pressure (mm Hg):

Property/ Value	Units	Temp	Data Type	Reference
WS 1.00E+003	mg/L	25	EXP	SHIU,WY ET AL. (1990)
logP 2.58			EST	MEYLAN,WM & HOWARD,PH (1995)
VP 2.12E+002	mm Hg	25	EST	NEELY,WB & BLAU,GE (1985)
DC	pKa			
HL 1.09E-001	atm m3/mol	25	EST	MEYLAN,WM & HOWARD,PH (1991)
OH 1.85E-011	cm3/molc sec	25	EST	MEYLAN,WM & HOWARD,PH (1993)

CAS #: 000513-38-2				1-IODO-2-METHYLPROPANE

Formula: C_4H_9I

Mol Weight: 184.02

MP (deg C): -93 | **FP (deg C):**

BP (deg C): 120

BP pressure (mm Hg):

Property/ Value	Units	Temp	Data Type	Reference
WS 1.47E+002	mg/L	25	EST	MEYLAN,WM ET AL. (1996)
logP 2.99			EST	MEYLAN,WM & HOWARD,PH (1995)
VP 1.57E+001	mm Hg	25	EXP	OHE,S (1976)
DC	pKa			
HL 1.29E-002	atm m3/mol	25	EST	MEYLAN,WM & HOWARD,PH (1991)
OH 1.84E-012	cm3/molc sec	25	EST	MEYLAN,WM & HOWARD,PH (1993)

CAS #: 000513-44-0				2-METHYL-1-PROPANETHIOL

Formula: $C_4H_{10}S$

Mol Weight: 90.19

MP (deg C): -79 | **FP (deg C):**

BP (deg C): 88

BP pressure (mm Hg):

Property/ Value	Units	Temp	Data Type	Reference
WS 1.72E+003	mg/L	25	EST	MEYLAN,WM ET AL. (1996)
logP 2.18			EST	MEYLAN,WM & HOWARD,PH (1995)
VP 6.98E+001	mm Hg	25	EXT	BOUBLIK,T ET AL. (1984)
DC	pKa			
HL 6.11E-003	atm m3/mol	25	EST	MEYLAN,WM & HOWARD,PH (1991)
OH 4.50E-011	cm3/molc sec	25	EXP	ATKINSON,R (1989)

CAS #: 000513-48-4				2-IODOBUTANE

Formula: C_4H_9I

Mol Weight: 184.02

MP (deg C): -104 | **FP (deg C):**

BP (deg C): 120

BP pressure (mm Hg):

Property/ Value	Units	Temp	Data Type	Reference
WS 2.10E+002	mg/L	18	EXP	STEPHEN,H & STEPHEN,T (1963)
logP 2.99			EST	MEYLAN,WM & HOWARD,PH (1995)
VP 1.68E+000	mm Hg	25	EXP	PATTE,F ET AL. (1982)
DC	pKa			
HL 1.94E-003	atm m3/mol	25	EST	VP/WSOL
OH 2.00E-012	cm3/molc sec	25	EST	MEYLAN,WM & HOWARD,PH (1993)

CAS #: 000513-49-5				SEC-BUTYLAMINE

Formula: $C_4H_{11}N$

Mol Weight: 73.14

MP (deg C): | **FP (deg C):**

BP (deg C): 62.5

BP pressure (mm Hg):

Property/ Value	Units	Temp	Data Type	Reference
WS 3.19E+005	mg/L	25	EST	MEYLAN,WM ET AL. (1996)
logP 0.74			EXP	HANSCH,C & LEO,AJ (1985)
VP 1.18E+002	mm Hg	25	EST	NEELY,WB & BLAU,GE (1985)
DC 10.56	pKa		EXP	PERRIN,DD (1965)
HL 1.77E-005	atm m3/mol	25	EST	MEYLAN,WM & HOWARD,PH (1991)
OH 4.47E-011	cm3/molc sec	25	EST	MEYLAN,WM & HOWARD,PH (1993)

CAS #: 000513-53-1				2-BUTANETHIOL

Formula: $C_4H_{10}S$

Mol Weight: 90.19

MP (deg C): -165 | **FP (deg C):** -14

BP (deg C): 84-85

BP pressure (mm Hg):

Property/ Value	Units	Temp	Data Type	Reference
WS 1.89E+003	mg/L	25	EST	MEYLAN,WM ET AL (1996)
logP 2.18			EST	MEYLAN,WM & HOWARD,PH (1995)
VP 8.07E+001	mm Hg	25	EXP	DAUBERT,TE & DANNER,RP (1989)
DC	pKa			
HL 6.11E-003	atm m3/mol	25	EST	MEYLAN,WM & HOWARD,PH (1991)
OH 4.00E-011	cm3/molc sec	25	EXP	ATKINSON,R (1989)

CAS #: 000513-81-5				2,3-DIMETHYL-1,3-BUTADIENE

Formula: C_6H_{10}

Mol Weight: 82.15

MP (deg C): -76 | **FP (deg C):**

BP (deg C): 68.8

BP pressure (mm Hg):

Property/ Value	Units	Temp	Data Type	Reference
WS 3.26E+002	mg/L	25	EXP	HINE,J & MOOKERJEE,PK (1975)
logP 3.13			EST	MEYLAN,WM & HOWARD,PH (1995)
VP 1.52E+002	mm Hg	25	EXP	YAWS,CL (1994A)
DC	pKa			
HL 5.04E-002	atm m3/mol	25	EST	VP/WSOL
OH 1.22E-010	cm3/molc sec	25	EXP	ATKINSON,R (1989)

CAS #: 000513-85-9				2,3-BUTANEDIOL

Formula: $C_4H_{10}O_2$

Mol Weight: 90.12

MP (deg C): 25 | **FP (deg C):**

BP (deg C): 182

BP pressure (mm Hg):

Property/ Value	Units	Temp	Data Type	Reference
WS 1.00E+006	mg/L	20	EXP	RIDDICK,JA ET AL. (1986)
logP -0.92			EXP	HANSCH,C & LEO,AJ (1985)
VP 1.70E+001	mm Hg	20	EXP	FLICK,EW (1991)
DC 14.90	pKa	25	EXP	RIDDICK,JA ET AL. (1986)
HL 2.02E-006	atm m3/mol	20	EST	VP/WSOL
OH 1.65E-011	cm3/molc sec	25	EST	MEYLAN,WM & HOWARD,PH (1993)

ACETOIN

CAS #: 000513-86-0				
Formula: $C_4H_8O_2$				
Mol Weight: 88.11				
MP (deg C): 15		**FP (deg C):**		
BP (deg C): 148				
BP pressure (mm Hg):				

Property/Value	Units	Temp	Data Type	Reference
WS 1.00E+006	mg/L	20	EXP	YALKOWSKY,SH & DANNENFELSER,RM (1992)
logP -0.36			EST	MEYLAN,WM & HOWARD,PH (1995)
VP 2.69E+000	mm Hg	25	EST	NEELY,WB & BLAU,GE (1985)
DC	pKa			
HL 1.03E-005	atm m3/mol	25	EST	MEYLAN,WM & HOWARD,PH (1991)
OH 5.86E-012	cm3/molc sec	25	EST	MEYLAN,WM & HOWARD,PH (1993)

1,1-DICHLOROPROPANONE

CAS #: 000513-88-2				
Formula: $C_3H_4Cl_2O$				
Mol Weight: 126.97				
MP (deg C):		**FP (deg C):**		
BP (deg C): 120				
BP pressure (mm Hg):				

Property/Value	Units	Temp	Data Type	Reference
WS 6.38E+004	mg/L	25	EST	MEYLAN,WM ET AL. (1996)
logP 0.20			EST	MEYLAN,WM & HOWARD,PH (1995)
VP 2.70E+001	mm Hg	25	EXP	OHE,S (1976)
DC	pKa			
HL 6.15E-006	atm m3/mol	25	EST	MEYLAN,WM & HOWARD,PH (1991)
OH 2.65E-013	cm3/molc sec	25	EST	MEYLAN,WM & HOWARD,PH (1993)

BIPERIDEN

CAS #: 000514-65-8				
Formula: $C_{21}H_{29}NO$				
Mol Weight: 311.47				
MP (deg C): 112-116		**FP (deg C):**		
BP (deg C):				
BP pressure (mm Hg):				

Property/Value	Units	Temp	Data Type	Reference
WS 2.51E+001	mg/L	25	EST	MEYLAN,WM ET AL. (1996)
logP 4.25			EXP	SANGSTER,J (1993)
VP 5.42E-009	mm Hg	25	EST	NEELY,WB & BLAU,GE (1985)
DC	pKa			
HL 2.44E-010	atm m3/mol	25	EST	MEYLAN,WM & HOWARD,PH (1991)
OH 1.71E-010	cm3/molc sec	25	EST	MEYLAN,WM & HOWARD,PH (1993)

BENZENEACETIC ACID, .ALPHA.-HYDROXY-.ALPHA.-METH

CAS #: 000515-30-0				
Formula: $C_9H_{10}O_3$				
Mol Weight: 166.18				
MP (deg C): 116.5		**FP (deg C):**		
BP (deg C):				
BP pressure (mm Hg):				

Property/Value	Units	Temp	Data Type	Reference
WS 1.07E+005	mg/L	25	EST	MEYLAN,WM ET AL. (1996)
logP 0.80			EXP	KUCHAR,M ET AL. (1985)
VP 1.39E-005	mm Hg	25	EST	NEELY,WB & BLAU,GE (1985)
DC 3.53	pKa	18	EXP	KORTUM,G ET AL (1961)
HL 9.16E-009	atm m3/mol	25	EST	MEYLAN,WM & HOWARD,PH (1991)
OH 5.47E-012	cm3/molc sec	25	EST	MEYLAN,WM & HOWARD,PH (1993)

ETHYL BENZENESULFONATE

CAS #: 000515-46-8				
Formula: $C_8H_{10}O_3S$				
Mol Weight: 186.23				
MP (deg C):		**FP (deg C):**		
BP (deg C): 156				
BP pressure (mm Hg): 1.50E+001				

Property/Value	Units	Temp	Data Type	Reference
WS 7.39E+000	mg/L	25	EXP	STEPHEN,H & STEPHEN,T (1963)
logP 1.41			EST	MEYLAN,WM & HOWARD,PH (1995)
VP 7.11E-004	mm Hg	25	EST	NEELY,WB & BLAU,GE (1985)
DC	pKa			
HL 1.07E-006	atm m3/mol	25	EST	MEYLAN,WM & HOWARD,PH (1991)
OH 1.24E-012	cm3/molc sec	25	EST	MEYLAN,WM & HOWARD,PH (1993)

SULFISOMIDINE

CAS #: 000515-64-0				
Formula: $C_{12}H_{14}N_4O_2S$				
Mol Weight: 278.33				
MP (deg C): 243		**FP (deg C):**		
BP (deg C):				
BP pressure (mm Hg):				

Property/Value	Units	Temp	Data Type	Reference
WS 1.62E+003	mg/L		EXP	YALKOWSKY,SH & DANNENFELSER,RM (1992)
logP -0.33			EXP	HANSCH,C ET AL. (1995)
VP 8.62E-009	mm Hg	25	EST	NEELY,WB & BLAU,GE (1985)
DC	pKa			
HL 5.43E-014	atm m3/mol	25	EST	MEYLAN,WM & HOWARD,PH (1991)
OH 5.76E-011	cm3/molc sec	25	EST	MEYLAN,WM & HOWARD,PH (1993)

ETHYL TRICHLOROACETATE

CAS #: 000515-84-4				
Formula: $C_4H_5Cl_3O_2$				
Mol Weight: 191.44				
MP (deg C):		**FP (deg C):**		
BP (deg C): 167.5				
BP pressure (mm Hg):				

Property/Value	Units	Temp	Data Type	Reference
WS 4.40E+002	mg/L	25	EST	MEYLAN,WM ET AL. (1996)
logP 2.39			EXP	HANSCH,C ET AL. (1995)
VP 3.81E+000	mm Hg	25	EXP	PERRY,RH & GREEN,D (1984)
DC	pKa			
HL 1.02E-005	atm m3/mol	25	EST	MEYLAN,WM & HOWARD,PH (1991)
OH 5.04E-013	cm3/molc sec	25	EST	MEYLAN,WM & HOWARD,PH (1993)

METHYL MALONIC ACID

CAS #: 000516-05-2				
Formula: $C_4H_6O_4$				
Mol Weight: 118.09				
MP (deg C): 135 dec		**FP (deg C):**		
BP (deg C):				
BP pressure (mm Hg):				

Property/Value	Units	Temp	Data Type	Reference
WS 6.79E+005	mg/L	25	EXP	YALKOWSKY,SH & DANNENFELSER,RM (1992)
logP -0.83			EST	MEYLAN,WM & HOWARD,PH (1995)
VP 3.04E-003	mm Hg	25	EST	NEELY,WB & BLAU,GE (1985)
DC 3.12	pKa	20	EXP	KORTUM,G ET AL (1961)
HL 5.41E-012	atm m3/mol	25	EST	MEYLAN,WM & HOWARD,PH (1991)
OH 2.30E-012	cm3/molc sec	25	EST	MEYLAN,WM & HOWARD,PH (1993)

CAS #: 000517-16-8				CERESAN M
Formula: $C_{15}H_{16}HgNO_2S$				
Mol Weight: 474.95				
MP (deg C):			FP (deg C):	
BP (deg C):				
BP pressure (mm Hg):				

Property/Value	Units	Temp	Data Type	Reference
WS 1.15E+000	mg/L	25	EST	MEYLAN,WM ET AL. (1996)
logP 3.46			EST	MEYLAN,WM & HOWARD,PH (1995)
VP	mm Hg			
DC	pKa			
HL	atm m3/mol			
OH 9.33E-011	cm3/molc sec	25	EST	MEYLAN,WM & HOWARD,PH (1993)

CAS #: 000517-60-2				MELLITIC ACID
Formula: $C_{12}H_6O_{12}$				
Mol Weight: 342.17				
MP (deg C): 286-288			FP (deg C):	
BP (deg C):				
BP pressure (mm Hg):				

Property/Value	Units	Temp	Data Type	Reference
WS 3.62E+002	mg/L	25	EST	MEYLAN,WM ET AL. (1996)
logP 1.50			EXP	AVDEEF,A (1993)
VP 4.94E-016	mm Hg	25	EST	NEELY,WB & BLAU,GE (1985)
DC 0.80	pKa	25	EXP	KORTUM,G ET AL (1961)
HL 3.58E-031	atm m3/mol	25	EST	MEYLAN,WM & HOWARD,PH (1991)
OH 3.12E-012	cm3/molc sec	25	EST	MEYLAN,WM & HOWARD,PH (1993)

CAS #: 000518-28-5				PODOPHYLLOTOXIN
Formula: $C_{22}H_{22}O_8$				
Mol Weight: 414.42				
MP (deg C): 114-118			FP (deg C):	
BP (deg C):				
BP pressure (mm Hg):				

Property/Value	Units	Temp	Data Type	Reference
WS 1.55E+002	mg/L	25	EST	MEYLAN,WM ET AL. (1996)
logP 2.01			EXP	HANSCH,C & LEO,AJ (1985)
VP 8.34E-015	mm Hg	25	EST	NEELY,WB & BLAU,GE (1985)
DC	pKa			
HL 3.80E-018	atm m3/mol	25	EST	MEYLAN,WM & HOWARD,PH (1991)
OH 2.17E-010	cm3/molc sec	25	EST	MEYLAN,WM & HOWARD,PH (1993)

CAS #: 000518-45-6				FLUORESCEIN
Formula: $C_{20}H_{12}O_5$				
Mol Weight: 332.32				
MP (deg C):			FP (deg C):	
BP (deg C):				
BP pressure (mm Hg):				

Property/Value	Units	Temp	Data Type	Reference
WS 5.00E+001	mg/L	20	EXP	DEHN,WM (1917)
logP 3.35			EST	MEYLAN,WM & HOWARD,PH (1995)
VP 5.05E-013	mm Hg	25	EST	NEELY,WB & BLAU,GE (1985)
DC	pKa			
HL 8.91E-017	atm m3/mol	25	EST	MEYLAN,WM & HOWARD,PH (1991)
OH 2.00E-010	cm3/molc sec	25	EST	MEYLAN,WM & HOWARD,PH (1993)

CAS #: 000519-23-3				ELLIPTICINE
Formula: $C_{17}H_{14}N_2$				
Mol Weight: 246.31				
MP (deg C): 311-315			FP (deg C):	
BP (deg C):				
BP pressure (mm Hg):				

Property/Value	Units	Temp	Data Type	Reference
WS 1.53E-001	mg/L	25	EST	MEYLAN,WM ET AL. (1996)
logP 4.80			EXP	HANSCH,C & LEO,AJ (1985)
VP 2.14E-008	mm Hg	25	EST	NEELY,WB & BLAU,GE (1985)
DC	pKa			
HL 1.35E-011	atm m3/mol	25	EST	MEYLAN,WM & HOWARD,PH (1991)
OH 8.99E-011	cm3/molc sec	25	EST	MEYLAN,WM & HOWARD,PH (1993)

CAS #: 000519-44-8				2,4-DINITRORESORCINOL
Formula: $C_6H_4N_2O_6$				
Mol Weight: 200.11				
MP (deg C): 146-148			FP (deg C):	
BP (deg C):				
BP pressure (mm Hg):				

Property/Value	Units	Temp	Data Type	Reference
WS 7.03E+002	mg/L	25	EST	MEYLAN,WM ET AL. (1996)
logP 2.10			EXP	HANSCH,C & LEO,AJ (1985)
VP 1.92E-007	mm Hg	25	EST	NEELY,WB & BLAU,GE (1985)
DC	pKa			
HL 9.08E-009	atm m3/mol	25	EST	MEYLAN,WM & HOWARD,PH (1991)
OH 9.18E-012	cm3/molc sec	25	EST	MEYLAN,WM & HOWARD,PH (1993)

CAS #: 000519-73-3				TRIPHENYLMETHANE
Formula: $C_{19}H_{16}$				
Mol Weight: 244.34				
MP (deg C): 78.2			FP (deg C):	
BP (deg C): 360				
BP pressure (mm Hg):				

Property/Value	Units	Temp	Data Type	Reference
WS 6.64E-001	mg/L	25	EST	MEYLAN,WM ET AL. (1996)
logP 5.37			EST	MEYLAN,WM & HOWARD,PH (1995)
VP 1.26E-009	mm Hg	25	EXT	OHE,S (1976)
DC	pKa			
HL 3.87E-005	atm m3/mol	25	EST	MEYLAN,WM & HOWARD,PH (1991)
OH 1.58E-011	cm3/molc sec	25	EST	MEYLAN,WM & HOWARD,PH (1993)

CAS #: 000519-87-9				N,N-DIPHENYLACETAMIDE
Formula: $C_{14}H_{13}NO$				
Mol Weight: 211.27				
MP (deg C): 103			FP (deg C):	
BP (deg C):				
BP pressure (mm Hg):				

Property/Value	Units	Temp	Data Type	Reference
WS 7.21E+002	mg/L	25	EST	MEYLAN,WM ET AL. (1996)
logP 2.02			EXP	HANSCH,C & LEO,AJ (1985)
VP 2.10E-005	mm Hg	25	EST	NEELY,WB & BLAU,GE (1985)
DC	pKa			
HL 3.18E-008	atm m3/mol	25	EST	MEYLAN,WM & HOWARD,PH (1991)
OH 2.49E-011	cm3/molc sec	25	EST	MEYLAN,WM & HOWARD,PH (1993)

CAS #: 000520-03-6				1H-ISOINDOLE-1,3(2H)-DIONE, 2-PHENYL-
Formula: $C_{14}H_9NO_2$				
Mol Weight: 223.23				
MP (deg C): 210			FP (deg C):	
BP (deg C):				
BP pressure (mm Hg):				

Property/ Value	Units	Temp	Data Type	Reference
WS 2.95E+002	mg/L	25	EST	MEYLAN,WM ET AL. (1996)
logP 2.40			EXP	HANSCH,C ET AL. (1995)
VP 1.38E-008	mm Hg	25	EST	NEELY,WB & BLAU,GE (1985)
DC	pKa			
HL 5.26E-008	atm m3/mol	25	EST	MEYLAN,WM & HOWARD,PH (1991)
OH 1.32E-011	cm3/molc sec	25	EST	MEYLAN,WM & HOWARD,PH (1993)

CAS #: 000520-27-4				DIOSMINE
Formula: $C_{28}H_{32}O_{15}$				
Mol Weight: 608.56				
MP (deg C): 274 dec			FP (deg C):	
BP (deg C):				
BP pressure (mm Hg):				

Property/ Value	Units	Temp	Data Type	Reference
WS 4.08E+002	mg/L	25	EST	MEYLAN,WM ET AL. (1996)
logP 0.14			EXP	PERRISSOUD,D & TESTA,B (1986)
VP 1.35E-027	mm Hg	25	EST	NEELY,WB & BLAU,GE (1985)
DC	pKa			
HL 7.82E-033	atm m3/mol	25	EST	MEYLAN,WM & HOWARD,PH (1991)
OH 3.71E-010	cm3/molc sec	25	EST	MEYLAN,WM & HOWARD,PH (1993)

CAS #: 000520-33-2				HESPERETINE
Formula: $C_{16}H_{14}O_6$				
Mol Weight: 302.29				
MP (deg C): 226-228			FP (deg C):	
BP (deg C):				
BP pressure (mm Hg):				

Property/ Value	Units	Temp	Data Type	Reference
WS 2.73E+002	mg/L	25	EST	MEYLAN,WM ET AL. (1996)
logP 2.60			EXP	PERRISSOUD,D & TESTA,B (1986)
VP 2.10E-011	mm Hg	25	EST	NEELY,WB & BLAU,GE (1985)
DC	pKa			
HL 1.73E-018	atm m3/mol	25	EST	MEYLAN,WM & HOWARD,PH (1991)
OH 2.48E-010	cm3/molc sec	25	EST	MEYLAN,WM & HOWARD,PH (1993)

CAS #: 000520-34-3				DIOSMETINE
Formula: $C_{16}H_{12}O_6$				
Mol Weight: 300.27				
MP (deg C): 253-254			FP (deg C):	
BP (deg C):				
BP pressure (mm Hg):				

Property/ Value	Units	Temp	Data Type	Reference
WS 1.05E+002	mg/L	25	EST	MEYLAN,WM ET AL. (1996)
logP 3.10			EXP	PERRISSOUD,D & TESTA,B (1986)
VP 1.54E-011	mm Hg	25	EST	NEELY,WB & BLAU,GE (1985)
DC	pKa			
HL 3.03E-018	atm m3/mol	25	EST	MEYLAN,WM & HOWARD,PH (1991)
OH 2.32E-010	cm3/molc sec	25	EST	MEYLAN,WM & HOWARD,PH (1993)

CAS #: 000520-36-5				APIGENIN
Formula: $C_{15}H_{10}O_5$				
Mol Weight: 270.24				
MP (deg C): 347.5			FP (deg C):	
BP (deg C):				
BP pressure (mm Hg):				

Property/ Value	Units	Temp	Data Type	Reference
WS 1.83E+002	mg/L	25	EST	MEYLAN,WM ET AL. (1996)
logP 3.02			EXP	HANSCH,C & LEO,AJ (1985)
VP 1.01E-010	mm Hg	25	EST	NEELY,WB & BLAU,GE (1985)
DC	pKa			
HL 5.12E-017	atm m3/mol	25	EST	MEYLAN,WM & HOWARD,PH (1991)
OH 2.31E-010	cm3/molc sec	25	EST	MEYLAN,WM & HOWARD,PH (1993)

CAS #: 000520-45-6				DEHYDROACETIC ACID
Formula: $C_8H_8O_4$				
Mol Weight: 168.15				
MP (deg C): 109-111			FP (deg C):	
BP (deg C): 269.9				
BP pressure (mm Hg):				

Property/ Value	Units	Temp	Data Type	Reference
WS 6.90E+002	mg/L	25	EXP	CHEM INSPECT TEST INST (1992)
logP 0.78			EST	MEYLAN,WM & HOWARD,PH (1995)
VP 1.25E-003	mm Hg	25	EXT	PERRY,RH & GREEN,D (1984)
DC	pKa			
HL 1.73E-010	atm m3/mol	25	EST	MEYLAN,WM & HOWARD,PH (1991)
OH 7.93E-011	cm3/molc sec	25	EST	MEYLAN,WM & HOWARD,PH (1993)

CAS #: 000521-18-6				ANDROSTAN-3-ONE, 17-HYDROXY-, (5ALPHA,17BETA)-
Formula: $C_{19}H_{30}O_2$				
Mol Weight: 290.45				
MP (deg C): 181			FP (deg C):	
BP (deg C):				
BP pressure (mm Hg):				

Property/ Value	Units	Temp	Data Type	Reference
WS 4.20E+001	mg/L	25	EST	MEYLAN,WM ET AL. (1996)
logP 3.55			EXP	HANSCH,C ET AL. (1995)
VP 2.79E-008	mm Hg	25	EST	NEELY,WB & BLAU,GE (1985)
DC	pKa			
HL 6.37E-009	atm m3/mol	25	EST	MEYLAN,WM & HOWARD,PH (1991)
OH 3.83E-011	cm3/molc sec	25	EST	MEYLAN,WM & HOWARD,PH (1993)

CAS #: 000522-00-9				ETHOPROPAZINE
Formula: $C_{19}H_{24}N_2S$				
Mol Weight: 312.48				
MP (deg C): 64-65			FP (deg C):	
BP (deg C):				
BP pressure (mm Hg):				

Property/ Value	Units	Temp	Data Type	Reference
WS 6.93E-001	mg/L	25	EST	MEYLAN,WM ET AL. (1996)
logP 4.77			EXP	HANSCH,C ET AL. (1995)
VP 2.52E-007	mm Hg	25	EST	NEELY,WB & BLAU,GE (1985)
DC	pKa			
HL 8.77E-010	atm m3/mol	25	EST	MEYLAN,WM & HOWARD,PH (1991)
OH 2.97E-010	cm3/molc sec	25	EST	MEYLAN,WM & HOWARD,PH (1993)

QUERCITRIN

CAS #:	000522-12-3		

Formula: $C_{21}H_{20}O_{11}$

Mol Weight: 448.39

MP (deg C): 176-179 **FP (deg C):**

BP (deg C):

BP pressure (mm Hg):

Property/Value	Units	Temp	Data Type	Reference
WS 6.40E+001	mg/L	16	EXP	SEIDELL,A (1941)
logP 0.43			EST	MEYLAN,WM & HOWARD,PH (1995)
VP 1.34E-021	mm Hg	25	EST	NEELY,WB & BLAU,GE (1985)
DC	pKa			
HL 1.73E-030	atm m3/mol	25	EST	MEYLAN,WM & HOWARD,PH (1991)
OH 3.05E-010	cm3/molc sec	25	EST	MEYLAN,WM & HOWARD,PH (1993)

FENETHAZINE

CAS #:	000522-24-7		

Formula: $C_{16}H_{18}N_2S$

Mol Weight: 270.40

MP (deg C): **FP (deg C):**

BP (deg C): 183-187

BP pressure (mm Hg): 1.00E+000

Property/Value	Units	Temp	Data Type	Reference
WS 3.72E+000	mg/L	25	EST	MEYLAN,WM ET AL. (1996)
logP 4.20			EXP	HANSCH,C & LEO,AJ (1985)
VP 1.83E-006	mm Hg	25	EST	NEELY,WB & BLAU,GE (1985)
DC	pKa			
HL 3.75E-010	atm m3/mol	25	EST	MEYLAN,WM & HOWARD,PH (1991)
OH 2.70E-010	cm3/molc sec	25	EST	MEYLAN,WM & HOWARD,PH (1993)

HYDROQUININE

CAS #:	000522-66-7		

Formula: $C_{20}H_{26}N_2O_2$

Mol Weight: 326.44

MP (deg C): 172 **FP (deg C):**

BP (deg C):

BP pressure (mm Hg):

Property/Value	Units	Temp	Data Type	Reference
WS 2.90E+002	mg/L	25	EXP	SEIDELL,A (1941)
logP 3.43			EST	MEYLAN,WM & HOWARD,PH (1995)
VP 1.03E-010	mm Hg	25	EST	NEELY,WB & BLAU,GE (1985)
DC	pKa			
HL 1.15E-015	atm m3/mol	25	EST	MEYLAN,WM & HOWARD,PH (1991)
OH 1.50E-010	cm3/molc sec	25	EST	MEYLAN,WM & HOWARD,PH (1993)

FURO(2,3-H)COUMARIN

CAS #:	000523-50-2		

Formula: $C_{11}H_6O_3$

Mol Weight: 186.17

MP (deg C): **FP (deg C):**

BP (deg C):

BP pressure (mm Hg):

Property/Value	Units	Temp	Data Type	Reference
WS 8.60E+002	mg/L	25	EST	MEYLAN,WM ET AL. (1996)
logP 2.08			EXP	SANGSTER,J (1993)
VP 2.24E-005	mm Hg	25	EST	NEELY,WB & BLAU,GE (1985)
DC	pKa			
HL 6.77E-007	atm m3/mol	25	EST	MEYLAN,WM & HOWARD,PH (1991)
OH 4.18E-011	cm3/molc sec	25	EST	MEYLAN,WM & HOWARD,PH (1993)

PYRIDOXAMINE

CAS #:	000524-36-7		

Formula: $C_8H_{12}N_2O_2$

Mol Weight: 168.20

MP (deg C): 226-227 **FP (deg C):**

BP (deg C):

BP pressure (mm Hg):

Property/Value	Units	Temp	Data Type	Reference
WS 5.00E+005	mg/L		EXP	COFFEN,DL (1978)
logP -1.30			EXP	HANSCH,C & LEO,AJ (1985)
VP 1.44E-006	mm Hg	25	EST	NEELY,WB & BLAU,GE (1985)
DC	pKa			
HL 3.71E-018	atm m3/mol	25	EST	MEYLAN,WM & HOWARD,PH (1991)
OH 5.68E-011	cm3/molc sec	25	EST	MEYLAN,WM & HOWARD,PH (1993)

N-HYDROXYPHTHALIMIDE

CAS #:	000524-38-9		

Formula: $C_8H_5NO_3$

Mol Weight: 163.13

MP (deg C): 233 dec **FP (deg C):**

BP (deg C):

BP pressure (mm Hg):

Property/Value	Units	Temp	Data Type	Reference
WS 5.05E+004	mg/L	25	EST	MEYLAN,WM ET AL. (1996)
logP 0.14			EST	MEYLAN,WM & HOWARD,PH (1995)
VP 2.46E-009	mm Hg	25	EST	NEELY,WB & BLAU,GE (1985)
DC	pKa			
HL 2.05E-011	atm m3/mol	25	EST	MEYLAN,WM & HOWARD,PH (1991)
OH 1.57E-011	cm3/molc sec	25	EST	MEYLAN,WM & HOWARD,PH (1993)

RICIDINE

CAS #:	000524-40-3		

Formula: $C_8H_8N_2O_2$

Mol Weight: 164.17

MP (deg C): 201.5 **FP (deg C):**

BP (deg C):

BP pressure (mm Hg):

Property/Value	Units	Temp	Data Type	Reference
WS 2.70E+003	mg/L	10	EXP	YALKOWSKY,SH & DANNENFELSER,RM (1992)
logP -0.45			EST	MEYLAN,WM & HOWARD,PH (1995)
VP 4.57E-005	mm Hg	25	EST	NEELY,WB & BLAU,GE (1985)
DC	pKa			
HL 5.16E-012	atm m3/mol	25	EST	MEYLAN,WM & HOWARD,PH (1991)
OH 2.77E-011	cm3/molc sec	25	EST	MEYLAN,WM & HOWARD,PH (1993)

HARMALOL

CAS #:	000525-57-5		

Formula: $C_{12}H_{10}N_2O$

Mol Weight: 198.23

MP (deg C): **FP (deg C):**

BP (deg C):

BP pressure (mm Hg):

Property/Value	Units	Temp	Data Type	Reference
WS 1.78E+002	mg/L	25	EST	MEYLAN,WM ET AL. (1996)
logP 2.19			EXP	HANSCH,C & LEO,AJ (1985)
VP 1.25E-007	mm Hg	25	EST	NEELY,WB & BLAU,GE (1985)
DC	pKa			
HL 1.30E-014	atm m3/mol	25	EST	MEYLAN,WM & HOWARD,PH (1991)
OH 2.00E-010	cm3/molc sec	25	EST	MEYLAN,WM & HOWARD,PH (1993)

CAS #: 000525-66-6	PROPRANOL

Formula: $C_{16}H_{19}NO_2$
Mol Weight: 257.34
MP (deg C): 96
FP (deg C):
BP (deg C):
BP pressure (mm Hg):

Property/ Value	Units	Temp	Data Type	Reference
WS 6.09E+002	mg/L	25	EST	MEYLAN,WM ET AL. (1996)
logP 2.98			EXP	HANSCH,C ET AL. (1995)
VP 3.82E-008	mm Hg	25	EST	NEELY,WB & BLAU,GE (1985)
DC	pKa			
HL 7.98E-013	atm m3/mol	25	EST	MEYLAN,WM & HOWARD,PH (1991)
OH 3.09E-010	cm3/molc sec	25	EST	MEYLAN,WM & HOWARD,PH (1993)

CAS #: 000525-74-6	4,5-DIHYDRONICOTYRINE

Formula: $C_{10}H_{12}N_2$
Mol Weight: 160.22
MP (deg C):
FP (deg C):
BP (deg C): 60
BP pressure (mm Hg): 1.00E-001

Property/ Value	Units	Temp	Data Type	Reference
WS 7.54E+004	mg/L	25	EST	MEYLAN,WM ET AL. (1996)
logP 1.13			EXP	HANSCH,C & LEO,AJ (1985)
VP 1.11E-002	mm Hg	25	EST	NEELY,WB & BLAU,GE (1985)
DC	pKa			
HL 2.59E-009	atm m3/mol	25	EST	MEYLAN,WM & HOWARD,PH (1991)
OH 1.29E-010	cm3/molc sec	25	EST	MEYLAN,WM & HOWARD,PH (1993)

CAS #: 000525-82-6	FLAVONE

Formula: $C_{15}H_{10}O_2$
Mol Weight: 222.25
MP (deg C): 99-100
FP (deg C):
BP (deg C):
BP pressure (mm Hg):

Property/ Value	Units	Temp	Data Type	Reference
WS 3.05E+001	mg/L	25	EST	MEYLAN,WM ET AL. (1996)
logP 3.56			EXP	HANSCH,C ET AL. (1995)
VP 9.52E-006	mm Hg	25	EST	NEELY,WB & BLAU,GE (1985)
DC	pKa			
HL 3.61E-008	atm m3/mol	25	EST	MEYLAN,WM & HOWARD,PH (1991)
OH 5.19E-011	cm3/molc sec	25	EST	MEYLAN,WM & HOWARD,PH (1993)

CAS #: 000525-97-3	6-(OCTANOYLAMINO)PENICILLANIC ACID

Formula: $C_{16}H_{26}N_2O_4S$
Mol Weight: 342.46
MP (deg C):
FP (deg C):
BP (deg C):
BP pressure (mm Hg):

Property/ Value	Units	Temp	Data Type	Reference
WS 1.01E+001	mg/L	25	EST	MEYLAN,WM ET AL. (1996)
logP 3.32			EXP	SANGSTER,J (1994)
VP 7.31E-012	mm Hg	25	EST	NEELY,WB & BLAU,GE (1985)
DC	pKa			
HL 7.90E-013	atm m3/mol	25	EST	MEYLAN,WM & HOWARD,PH (1991)
OH 9.68E-011	cm3/molc sec	25	EST	MEYLAN,WM & HOWARD,PH (1993)

CAS #: 000526-08-9	SULFAPHENAZOLE

Formula: $C_{15}H_{14}N_4O_2S$
Mol Weight: 314.37
MP (deg C): 179-183
FP (deg C):
BP (deg C):
BP pressure (mm Hg):

Property/ Value	Units	Temp	Data Type	Reference
WS 5.09E+002	mg/L	25	EST	MEYLAN,WM ET AL. (1996)
logP 1.52			EXP	HANSCH,C & LEO,AJ (1985)
VP 2.02E-010	mm Hg	25	EST	NEELY,WB & BLAU,GE (1985)
DC	pKa			
HL 2.05E-016	atm m3/mol	25	EST	MEYLAN,WM & HOWARD,PH (1991)
OH 2.00E-010	cm3/molc sec	25	EST	MEYLAN,WM & HOWARD,PH (1993)

CAS #: 000526-73-8	1,2,3-TRIMETHYLBENZENE

Formula: C_9H_{12}
Mol Weight: 120.20
MP (deg C): -25.4
FP (deg C):
BP (deg C): 176.1
BP pressure (mm Hg):

Property/ Value	Units	Temp	Data Type	Reference
WS 7.52E+001	mg/L	25	EXP	YALKOWSKY,SH & DANNENFELSER,RM (1992)
logP 3.66			EXP	HANSCH,C & LEO,AJ (1985)
VP 1.69E+000	mm Hg	25	EXP	YAWS,CL (1994B)
DC	pKa			
HL 4.36E-003	atm m3/mol	25	EXP	SANEMASA,I ET AL. (1982)
OH 3.27E-011	cm3/molc sec	25	EXP	ATKINSON,R (1989)

CAS #: 000526-75-0	2,3-DIMETHYLPHENOL

Formula: $C_8H_{10}O$
Mol Weight: 122.17
MP (deg C): 72.8
FP (deg C):
BP (deg C): 216.9
BP pressure (mm Hg):

Property/ Value	Units	Temp	Data Type	Reference
WS 4.57E+003	mg/L	25	EXP	YALKOWSKY,SH & DANNENFELSER,RM (1992)
logP 2.61			EST	MEYLAN,WM & HOWARD,PH (1995)
VP 2.69E-002	mm Hg	25	EXT	YAWS,CL (1994B)
DC 10.54	pKa	25	EXP	KORTUM,G ET AL (1961)
HL 6.83E-007	atm m3/mol	25	EST	MEYLAN,WM & HOWARD,PH (1991)
OH 8.02E-011	cm3/molc sec	25	EXP	ATKINSON,R (1989)

CAS #: 000526-86-3	2,5-CYCLOHEXADIENE-1,4-DIONE, 2,3-DIMETHYL-

Formula: $C_8H_8O_2$
Mol Weight: 136.15
MP (deg C): 55
FP (deg C):
BP (deg C):
BP pressure (mm Hg):

Property/ Value	Units	Temp	Data Type	Reference
WS 7.73E+003	mg/L	25	EST	MEYLAN,WM ET AL. (1996)
logP 1.23			EXP	RICH,PR (1990)
VP 2.41E-002	mm Hg	25	EST	NEELY,WB & BLAU,GE (1985)
DC	pKa			
HL 3.00E-009	atm m3/mol	25	EST	MEYLAN,WM & HOWARD,PH (1991)
OH 1.60E-011	cm3/molc sec	25	EST	MEYLAN,WM & HOWARD,PH (1993)

D-GLUCONIC ACID

CAS #: 000526-95-4

Formula: $C_6H_{12}O_7$

Mol Weight: 196.16

MP (deg C): 131

FP (deg C):

BP (deg C):

BP pressure (mm Hg):

Property/ Value	Units	Temp	Data Type	Reference
WS 1.00E+006	mg/L	25	EST	MEYLAN,WM ET AL. (1996)
logP -1.87			EST	MEYLAN,WM & HOWARD,PH (1995)
VP 3.72E-010	mm Hg	25	EST	NEELY,WB & BLAU,GE (1985)
DC	pKa			
HL 4.74E-013	atm m3/mol	25	EST	MEYLAN,WM & HOWARD,PH (1991)
OH 4.23E-011	cm3/molc sec	25	EST	MEYLAN,WM & HOWARD,PH (1993)

DUROQUINONE

CAS #: 000527-17-3

Formula: $C_{10}H_{12}O_2$

Mol Weight: 164.21

MP (deg C): 111-112

FP (deg C):

BP (deg C):

BP pressure (mm Hg):

Property/ Value	Units	Temp	Data Type	Reference
WS 4.43E+002	mg/L	25	EST	MEYLAN,WM ET AL. (1996)
logP 2.54			EXP	RICH,PR (1990)
VP 2.88E-003	mm Hg	25	EST	NEELY,WB & BLAU,GE (1985)
DC	pKa			
HL 7.36E-009	atm m3/mol	25	EST	MEYLAN,WM & HOWARD,PH (1991)
OH 2.75E-011	cm3/molc sec	25	EST	MEYLAN,WM & HOWARD,PH (1993)

PENTACHLOROANILINE

CAS #: 000527-20-8

Formula: $C_6H_2Cl_5N$

Mol Weight: 265.35

MP (deg C):

FP (deg C):

BP (deg C):

BP pressure (mm Hg):

Property/ Value	Units	Temp	Data Type	Reference
WS 1.49E+000	mg/L	25	EST	MEYLAN,WM ET AL. (1996)
logP 4.82			EXP	SANGSTER,J (1994)
VP 1.01E-004	mm Hg	25	EST	NEELY,WB & BLAU,GE (1985)
DC	pKa			
HL 4.25E-007	atm m3/mol	25	EST	MEYLAN,WM & HOWARD,PH (1991)
OH 1.33E-012	cm3/molc sec	25	EST	MEYLAN,WM & HOWARD,PH (1993)

1,2,3,5-TETRAMETHYLBENZENE

CAS #: 000527-53-7

Formula: $C_{10}H_{14}$

Mol Weight: 134.22

MP (deg C): -24

FP (deg C):

BP (deg C): 197.9

BP pressure (mm Hg):

Property/ Value	Units	Temp	Data Type	Reference
WS 2.79E+001	mg/L	25	EST	MEYLAN,WM ET AL. (1996)
logP 4.10			EXP	SANGSTER,J (1994)
VP 4.98E-001	mm Hg	25	EXP	YAWS,CL (1994B)
DC	pKa			
HL 7.99E-003	atm m3/mol	25	EST	MEYLAN,WM & HOWARD,PH (1991)
OH 4.31E-011	cm3/molc sec	25	EST	MEYLAN,WM & HOWARD,PH (1993)

2,4,6-TRIMETHYLPHENOL

CAS #: 000527-60-6

Formula: $C_9H_{12}O$

Mol Weight: 136.20

MP (deg C): 73

FP (deg C):

BP (deg C): 220

BP pressure (mm Hg):

Property/ Value	Units	Temp	Data Type	Reference
WS 1.01E+003	mg/L	25	EXP	SHIU,WY ET AL. (1994)
logP 2.73			EXP	SANGSTER,J (1993)
VP 1.76E-002	mm Hg	25	EXT	BOUBLIK,T ET AL. (1984)
DC 10.86	pKa	25	EXP	KORTUM,G ET AL (1961)
HL 7.53E-007	atm m3/mol	25	EST	MEYLAN,WM & HOWARD,PH (1991)
OH 2.44E-011	cm3/molc sec	25	EST	MEYLAN,WM & HOWARD,PH (1993)

2,5-CYCLOHEXADIENE-1,4-DIONE, 2,6-DIMETHYL

CAS #: 000527-61-7

Formula: $C_8H_8O_2$

Mol Weight: 136.15

MP (deg C): 72.5

FP (deg C):

BP (deg C):

BP pressure (mm Hg):

Property/ Value	Units	Temp	Data Type	Reference
WS 7.89E+003	mg/L	25	EST	MEYLAN,WM ET AL. (1996)
logP 1.22			EXP	HANSCH,C ET AL. (1995)
VP 2.41E-002	mm Hg	25	EST	NEELY,WB & BLAU,GE (1985)
DC	pKa			
HL 3.00E-009	atm m3/mol	25	EST	MEYLAN,WM & HOWARD,PH (1991)
OH 2.16E-011	cm3/molc sec	25	EST	MEYLAN,WM & HOWARD,PH (1993)

THIOPHENE-2-CARBOXYLIC ACID

CAS #: 000527-72-0

Formula: $C_5H_4O_2S$

Mol Weight: 128.15

MP (deg C): 128.5

FP (deg C):

BP (deg C):

BP pressure (mm Hg):

Property/ Value	Units	Temp	Data Type	Reference
WS 4.27E+003	mg/L	25	EST	MEYLAN,WM ET AL. (1996)
logP 1.57			EXP	HANSCH,C & LEO,AJ (1985)
VP 7.80E-003	mm Hg	25	EST	NEELY,WB & BLAU,GE (1985)
DC	pKa			
HL 5.89E-008	atm m3/mol	25	EST	MEYLAN,WM & HOWARD,PH (1991)
OH 4.05E-012	cm3/molc sec	25	EST	MEYLAN,WM & HOWARD,PH (1993)

2-NITROIMIDAZOLE

CAS #: 000527-73-1

Formula: $C_3H_3N_3O_2$

Mol Weight: 113.08

MP (deg C): 287 dec

FP (deg C):

BP (deg C):

BP pressure (mm Hg):

Property/ Value	Units	Temp	Data Type	Reference
WS 3.22E+004	mg/L	25	EST	MEYLAN,WM ET AL. (1996)
logP 0.15			EXP	HANSCH,C & LEO,AJ (1985)
VP 1.75E-004	mm Hg	25	EST	NEELY,WB & BLAU,GE (1985)
DC	pKa			
HL 1.48E-008	atm m3/mol	25	EST	MEYLAN,WM & HOWARD,PH (1991)
OH 4.50E-012	cm3/molc sec	25	EST	MEYLAN,WM & HOWARD,PH (1993)

CAS #: 000527-75-3				ERYTHROMYCIN, 12-DEOXY-

Formula: $C_{37}H_{67}NO_{12}$

Mol Weight: 717.95

MP (deg C): | FP (deg C):

BP (deg C):

BP pressure (mm Hg):

Property/Value	Units	Temp	Data Type	Reference
WS 5.87E-001	mg/L	25	EST	MEYLAN,WM ET AL. (1996)
logP 3.12			EXP	HANSCH,C ET AL. (1995)
VP 2.34E-026	mm Hg	25	EST	NEELY,WB & BLAU,GE (1985)
DC	pKa			
HL 1.48E-027	atm m3/mol	25	EST	MEYLAN,WM & HOWARD,PH (1991)
OH 3.28E-010	cm3/molc sec	25	EST	MEYLAN,WM & HOWARD,PH (1993)

CAS #: 000527-84-4				1-METHYL-2-ISOPROPYLBENZENE

Formula: $C_{10}H_{14}$

Mol Weight: 134.22

MP (deg C): -71.5 | FP (deg C):

BP (deg C): 178.1

BP pressure (mm Hg):

Property/Value	Units	Temp	Data Type	Reference
WS 2.33E+001	mg/L	25	EXP	OKOUCHI,S ET AL. (1992)
logP 4.38			EXP	SHERBLOM,PM & EGANHOUSE,RP (1988)
VP 1.50E+000	mm Hg	25	EXP	DAUBERT,TE & DANNER,RP (1989)
DC	pKa			
HL 1.14E-002	atm m3/mol	25	EXP	VP/WSOL
OH 8.54E-012	cm3/molc sec	25	EST	MEYLAN,WM & HOWARD,PH (1993)

CAS #: 000527-85-5				2-METHYLBENZAMIDE

Formula: C_8H_9NO

Mol Weight: 135.17

MP (deg C): 144-145 | FP (deg C):

BP (deg C):

BP pressure (mm Hg):

Property/Value	Units	Temp	Data Type	Reference
WS 1.97E+004	mg/L	25	EST	MEYLAN,WM ET AL. (1996)
logP 0.76			EXP	NAKAGAWA,Y ET AL. (1992)
VP 3.19E-004	mm Hg	25	EST	NEELY,WB & BLAU,GE (1985)
DC	pKa			
HL 2.44E-009	atm m3/mol	25	EST	MEYLAN,WM & HOWARD,PH (1991)
OH 6.78E-012	cm3/molc sec	25	EST	MEYLAN,WM & HOWARD,PH (1993)

CAS #: 000528-29-0				O-DINITROBENZENE

Formula: $C_6H_4N_2O_4$

Mol Weight: 168.11

MP (deg C): 118.5 | FP (deg C):

BP (deg C): 318

BP pressure (mm Hg):

Property/Value	Units	Temp	Data Type	Reference
WS 1.33E+002	mg/L	25	EXP	YALKOWSKY,SH & DANNENFELSER,RM (1992)
logP 1.69			EXP	HANSCH,C ET AL. (1995)
VP 4.55E-005	mm Hg	25	EXT	YAWS,CL (1994A)
DC	pKa			
HL 8.39E-008	atm m3/mol	25	EST	MEYLAN,WM & HOWARD,PH (1991)
OH 2.13E-014	cm3/molc sec	25	EST	MEYLAN,WM & HOWARD,PH (1993)

CAS #: 000528-44-9				TRIMELLITIC ACID

Formula: $C_9H_6O_6$

Mol Weight: 210.14

MP (deg C): 238 | FP (deg C):

BP (deg C):

BP pressure (mm Hg):

Property/Value	Units	Temp	Data Type	Reference
WS 2.10E+004	mg/L	25	EXP	BEMIS,AG ET AL. (1982)
logP 0.95			EST	MEYLAN,WM & HOWARD,PH (1995)
VP 2.88E-008	mm Hg	25	EST	NEELY,WB & BLAU,GE (1985)
DC 2.52	pKa		EXP	KORTUM,G ET AL (1961)
HL 4.39E-017	atm m3/mol	25	EST	MEYLAN,WM & HOWARD,PH (1991)
OH 1.63E-012	cm3/molc sec	25	EST	MEYLAN,WM & HOWARD,PH (1993)

CAS #: 000528-45-0				3,4-DINITROBENZOIC ACID

Formula: $C_7H_4N_2O_6$

Mol Weight: 212.12

MP (deg C): 166 | FP (deg C):

BP (deg C):

BP pressure (mm Hg):

Property/Value	Units	Temp	Data Type	Reference
WS 6.70E+003	mg/L	25	EXP	YALKOWSKY,SH & DANNENFELSER,RM (1992)
logP 1.51			EST	MEYLAN,WM & HOWARD,PH (1995)
VP 1.41E-006	mm Hg	25	EST	NEELY,WB & BLAU,GE (1985)
DC 2.82	pKa	25	EXP	KORTUM,G ET AL (1961)
HL 1.69E-012	atm m3/mol	25	EST	MEYLAN,WM & HOWARD,PH (1991)
OH 5.28E-013	cm3/molc sec	25	EST	MEYLAN,WM & HOWARD,PH (1993)

CAS #: 000528-46-1				PHTHALONIC ACID

Formula: $C_9H_6O_5$

Mol Weight: 194.15

MP (deg C): | FP (deg C):

BP (deg C):

BP pressure (mm Hg):

Property/Value	Units	Temp	Data Type	Reference
WS 6.44E+005	mg/L	15	EXP	STEPHEN,H & STEPHEN,T (1963)
logP 0.74			EST	MEYLAN,WM & HOWARD,PH (1995)
VP 8.74E-007	mm Hg	25	EST	NEELY,WB & BLAU,GE (1985)
DC	pKa			
HL 8.69E-015	atm m3/mol	25	EST	MEYLAN,WM & HOWARD,PH (1991)
OH 1.52E-012	cm3/molc sec	25	EST	MEYLAN,WM & HOWARD,PH (1993)

CAS #: 000528-74-5				DICHLOROMETHOTREXATE

Formula: $C_{20}H_{20}Cl_2N_8O_5$

Mol Weight: 523.34

MP (deg C): | FP (deg C):

BP (deg C):

BP pressure (mm Hg):

Property/Value	Units	Temp	Data Type	Reference
WS 3.02E+001	mg/L	25	EST	MEYLAN,WM ET AL. (1996)
logP -0.10			EXP	HANSCH,C & LEO,AJ (1985)
VP 1.71E-020	mm Hg	25	EST	NEELY,WB & BLAU,GE (1985)
DC	pKa			
HL 8.46E-032	atm m3/mol	25	EST	MEYLAN,WM & HOWARD,PH (1991)
OH 1.84E-010	cm3/molc sec	25	EST	MEYLAN,WM & HOWARD,PH (1993)

CAS #: 000529-19-1				2-METHYLBENZONITRILE

Formula: C_8H_7N

Mol Weight: 117.15

MP (deg C): -13 | FP (deg C):

BP (deg C): 205.2

BP pressure (mm Hg):

Property/Value	Units	Temp	Data Type	Reference
WS 7.24E+002	mg/L	25	EST	MEYLAN,WM ET AL. (1996)
logP 2.21			EXP	HANSCH,C ET AL. (1995)
VP 4.48E-001	mm Hg	25	EXT	PERRY,RH & GREEN,D (1984)
DC	pKa			
HL 5.74E-005	atm m3/mol	25	EST	MEYLAN,WM & HOWARD,PH (1991)
OH 1.03E-012	cm3/molc sec	25	EST	MEYLAN,WM & HOWARD,PH (1993)

CAS #: 000529-20-4				2-METHYLBENZALDEHYDE

Formula: C_8H_8O

Mol Weight: 120.15

MP (deg C): | FP (deg C):

BP (deg C): 200-202

BP pressure (mm Hg):

Property/Value	Units	Temp	Data Type	Reference
WS 1.18E+003	mg/L	25	EST	MEYLAN,WM ET AL. (1996)
logP 2.26			EXP	HANSCH,C & LEO,AJ (1985)
VP 3.35E-001	mm Hg	25	EST	NEELY,WB & BLAU,GE (1985)
DC	pKa			
HL 1.48E-005	atm m3/mol	25	EST	MEYLAN,WM & HOWARD,PH (1991)
OH 1.87E-011	cm3/molc sec	25	EST	MEYLAN,WM & HOWARD,PH (1993)

CAS #: 000529-33-9				1-NAPHTHALENOL, 1,2,3,4-TETRAHYDRO-

Formula: $C_{10}H_{12}O$

Mol Weight: 148.21

MP (deg C): | FP (deg C):

BP (deg C): 102-104

BP pressure (mm Hg): 2.00E+000

Property/Value	Units	Temp	Data Type	Reference
WS 5.09E+003	mg/L	25	EST	MEYLAN,WM ET AL. (1996)
logP 1.98			EXP	SANGSTER,J (1993)
VP 1.26E-003	mm Hg	25	EST	NEELY,WB & BLAU,GE (1985)
DC	pKa			
HL 1.86E-007	atm m3/mol	25	EST	MEYLAN,WM & HOWARD,PH (1991)
OH 1.83E-011	cm3/molc sec	25	EST	MEYLAN,WM & HOWARD,PH (1993)

CAS #: 000529-34-0				3,4-DIHYDRO-1(2H)-NAPHTHALENONE

Formula: $C_{10}H_{10}O$

Mol Weight: 146.19

MP (deg C): 8 | FP (deg C):

BP (deg C): 115

BP pressure (mm Hg): 6.00E+000

Property/Value	Units	Temp	Data Type	Reference
WS 4.75E+002	mg/L	25	EST	MEYLAN,WM ET AL. (1996)
logP 2.60				MEYLAN,WM & HOWARD,PH (1995)
VP 1.41E-002	mm Hg	25	EST	NEELY,WB & BLAU,GE (1985)
DC	pKa			
HL 6.34E-006	atm m3/mol	25	EST	MEYLAN,WM & HOWARD,PH (1991)
OH 1.09E-011	cm3/molc sec	25	EST	MEYLAN,WM & HOWARD,PH (1993)

CAS #: 000529-35-1				1-NAPHTHALENOL, 5,6,7,8-TETRAHYDRO-

Formula: $C_{10}H_{12}O$

Mol Weight: 148.21

MP (deg C): 70 | FP (deg C):

BP (deg C): 266

BP pressure (mm Hg):

Property/Value	Units	Temp	Data Type	Reference
WS 8.88E+002	mg/L	25	EST	MEYLAN,WM ET AL. (1996)
logP 2.95			EXP	SANGSTER,J (1993)
VP 3.72E-003	mm Hg	25	EST	NEELY,WB & BLAU,GE (1985)
DC 10.28	pKa	25	EXP	KORTUM,G ET AL (1961)
HL 5.31E-007	atm m3/mol	25	EST	MEYLAN,WM & HOWARD,PH (1991)
OH 1.06E-010	cm3/molc sec	25	EST	MEYLAN,WM & HOWARD,PH (1993)

CAS #: 000529-37-3				4(1H)-QUINOLINONE

Formula: C_9H_7NO

Mol Weight: 145.16

MP (deg C): | FP (deg C):

BP (deg C):

BP pressure (mm Hg):

Property/Value	Units	Temp	Data Type	Reference
WS 2.55E+004	mg/L	25	EST	MEYLAN,WM ET AL. (1996)
logP 0.58			EXP	HANSCH,C ET AL. (1995)
VP 1.58E-003	mm Hg	25	EST	NEELY,WB & BLAU,GE (1985)
DC	pKa			
HL 2.79E-009	atm m3/mol	25	EST	MEYLAN,WM & HOWARD,PH (1991)
OH 4.57E-011	cm3/molc sec	25	EST	MEYLAN,WM & HOWARD,PH (1993)

CAS #: 000529-64-6				TROPIC ACID

Formula: $C_9H_{10}O_3$

Mol Weight: 166.18

MP (deg C): 118 | FP (deg C):

BP (deg C):

BP pressure (mm Hg):

Property/Value	Units	Temp	Data Type	Reference
WS 1.95E+004	mg/L	20	EXP	YALKOWSKY,SH & DANNENFELSER,RM (1992)
logP 0.77			EXP	HANSCH,C & LEO,AJ (1985)
VP 4.00E-006	mm Hg	25	EST	NEELY,WB & BLAU,GE (1985)
DC	pKa			
HL 2.15E-012	atm m3/mol	25	EST	MEYLAN,WM & HOWARD,PH (1991)
OH 1.11E-011	cm3/molc sec	25	EST	MEYLAN,WM & HOWARD,PH (1993)

CAS #: 000529-96-4				PYRIDOXAMINE-5'-PHOSPHATE

Formula: $C_8H_{13}N_2O_5P$

Mol Weight: 248.18

MP (deg C): | FP (deg C):

BP (deg C):

BP pressure (mm Hg):

Property/Value	Units	Temp	Data Type	Reference
WS 1.00E+006	mg/L	25	EST	MEYLAN,WM ET AL. (1996)
logP -1.41			EST	MEYLAN,WM & HOWARD,PH (1995)
VP 2.97E-010	mm Hg	25	EST	NEELY,WB & BLAU,GE (1985)
DC	pKa			
HL 3.35E-024	atm m3/mol	25	EST	MEYLAN,WM & HOWARD,PH (1991)
OH 7.29E-011	cm3/molc sec	25	EST	MEYLAN,WM & HOWARD,PH (1993)

CAS #: 000530-48-3				1,1-DIPHENYLETHYLENE

Formula: $C_{14}H_{12}$
Mol Weight: 180.25
MP (deg C): 8.2
FP (deg C):
BP (deg C): 277
BP pressure (mm Hg):

Property/Value	Units	Temp	Data Type	Reference
WS 6.60E+000	mg/L	25	EXP	YALKOWSKY,SH & DANNENFELSER,RM (1992)
logP 3.86			EST	MEYLAN,WM & HOWARD,PH (1995)
VP 1.31E-002	mm Hg	25	EXT	OHE,S (1976)
DC	pKa			
HL 1.24E-004	atm m3/mol	25	EST	MEYLAN,WM & HOWARD,PH (1991)
OH 5.51E-011	cm3/molc sec	25	EST	MEYLAN,WM & HOWARD,PH (1993)

CAS #: 000530-50-7				HYDRAZINE, 1,1-DIPHENYL-

Formula: $C_{12}H_{12}N_2$
Mol Weight: 184.24
MP (deg C): 50.5
FP (deg C):
BP (deg C): 220
BP pressure (mm Hg): 4.00E+001

Property/Value	Units	Temp	Data Type	Reference
WS 2.87E+002	mg/L	25	EST	MEYLAN,WM ET AL. (1996)
logP 2.80			EXP	HANSCH,C ET AL. (1995)
VP 4.91E-004	mm Hg	25	EXT	OHE,S (1976)
DC	pKa			
HL 4.10E-008	atm m3/mol	25	EST	MEYLAN,WM & HOWARD,PH (1991)
OH 2.48E-011	cm3/molc sec	25	EST	MEYLAN,WM & HOWARD,PH (1993)

CAS #: 000530-53-0				PYRROLO[2,1-B]QUINAZOLIN-9(1H)-ONE, 2,3-DIHYDRO-

Formula: $C_{11}H_{10}N_2O$
Mol Weight: 186.22
MP (deg C):
FP (deg C):
BP (deg C):
BP pressure (mm Hg):

Property/Value	Units	Temp	Data Type	Reference
WS 6.26E+003	mg/L	25	EST	MEYLAN,WM ET AL. (1996)
logP 1.07			EXP	SANGSTER,J (1993)
VP 1.57E-005	mm Hg	25	EST	NEELY,WB & BLAU,GE (1985)
DC	pKa			
HL 6.23E-010	atm m3/mol	25	EST	MEYLAN,WM & HOWARD,PH (1991)
OH 1.31E-011	cm3/molc sec	25	EST	MEYLAN,WM & HOWARD,PH (1993)

CAS #: 000530-55-2				2,6-DIMETHOXY-P-BENZOQUINONE

Formula: $C_8H_8O_4$
Mol Weight: 168.15
MP (deg C): 256
FP (deg C):
BP (deg C):
BP pressure (mm Hg):

Property/Value	Units	Temp	Data Type	Reference
WS 7.06E+004	mg/L	25	EST	MEYLAN,WM ET AL. (1996)
logP -0.06			EXP	HUANG,JX ET AL. (1985)
VP 2.34E-003	mm Hg	25	EST	NEELY,WB & BLAU,GE (1985)
DC	pKa			
HL 5.27E-010	atm m3/mol	25	EST	MEYLAN,WM & HOWARD,PH (1991)
OH 2.93E-011	cm3/molc sec	25	EST	MEYLAN,WM & HOWARD,PH (1993)

CAS #: 000530-57-4				4-OH-3,5-DIMETHOXYBENZIOC ACID

Formula: $C_9H_{10}O_5$
Mol Weight: 198.18
MP (deg C): 206-209
FP (deg C):
BP (deg C):
BP pressure (mm Hg):

Property/Value	Units	Temp	Data Type	Reference
WS 5.78E+003	mg/L	25	EST	MEYLAN,WM ET AL. (1996)
logP 1.04			EXP	HANSCH,C & LEO,AJ (1985)
VP 2.91E-006	mm Hg	25	EST	NEELY,WB & BLAU,GE (1985)
DC 4.34	pKa	25	EXP	SERJEANT,EP & DEMPSEY,B (1979)
HL 3.95E-014	atm m3/mol	25	EST	MEYLAN,WM & HOWARD,PH (1991)
OH 4.69E-011	cm3/molc sec	25	EST	MEYLAN,WM & HOWARD,PH (1993)

CAS #: 000530-78-9				FLUFENAMIC ACID

Formula: $C_{14}H_{10}F_3NO_2$
Mol Weight: 281.24
MP (deg C): 125
FP (deg C):
BP (deg C):
BP pressure (mm Hg):

Property/Value	Units	Temp	Data Type	Reference
WS 9.09E+000	mg/L	25	EXP	YALKOWSKY,SH & DANNENFELSER,RM (1992)
logP 5.25			EXP	HANSCH,C & LEO,AJ (1985)
VP 1.95E-006	mm Hg	25	EST	NEELY,WB & BLAU,GE (1985)
DC	pKa			
HL 1.84E-010	atm m3/mol	25	EST	MEYLAN,WM & HOWARD,PH (1991)
OH 7.70E-011	cm3/molc sec	25	EST	MEYLAN,WM & HOWARD,PH (1993)

CAS #: 000531-85-1				BENZIDINE DIHYDROCHLORIDE

Formula: $C_{12}H_{14}Cl_2N_2$
Mol Weight: 257.16
MP (deg C):
FP (deg C):
BP (deg C):
BP pressure (mm Hg):

Property/Value	Units	Temp	Data Type	Reference
WS 3.77E+003	mg/L	25	EST	MEYLAN,WM ET AL. (1996)
logP 1.92			EST	MEYLAN,WM & HOWARD,PH (1995)
VP 4.38E-006	mm Hg	25	EST	NEELY,WB & BLAU,GE (1985)
DC 4.95	pKa	25	EXP	PERRIN,DD (1965)
HL 5.17E-011	atm m3/mol	25	EST	MEYLAN,WM & HOWARD,PH (1991)
OH 1.54E-010	cm3/molc sec	25	EST	MEYLAN,WM & HOWARD,PH (1993)

CAS #: 000532-03-6				1,2-PROPANEDIOL, 3-(2-METHOXYPHENOXY)-, 1-CARBAM

Formula: $C_{11}H_{15}NO_5$
Mol Weight: 241.25
MP (deg C): 92-94
FP (deg C):
BP (deg C):
BP pressure (mm Hg):

Property/Value	Units	Temp	Data Type	Reference
WS 7.20E+003	mg/L	25	EXP	CHEM INSPECT TEST INST (1992)
logP 0.61			EXP	CHEM INSPECT TEST INST (1992)
VP 3.41E-007	mm Hg	25	EST	NEELY,WB & BLAU,GE (1985)
DC	pKa			
HL 6.53E-016	atm m3/mol	25	EST	MEYLAN,WM & HOWARD,PH (1991)
OH 4.07E-011	cm3/molc sec	25	EST	MEYLAN,WM & HOWARD,PH (1993)

CAS #: 000532-27-4				2-CHLORO-1-PHENYLETHANONE

Formula: C8H7ClO

Mol Weight: 154.60

MP (deg C): 58-59   FP (deg C):

BP (deg C): 244-245

BP pressure (mm Hg):

Property/ Value	Units	Temp	Data Type	Reference
WS 1.64E+003	mg/L	25	EST	MEYLAN,WM ET AL. (1996)
logP 1.93			EST	MEYLAN,WM & HOWARD,PH (1995)
VP 5.40E-003	mm Hg	20	EXP	MERCK INDEX (1983)
DC	pKa			
HL 3.46E-006	atm m3/mol	25	EST	MEYLAN,WM & HOWARD,PH (1991)
OH 1.74E-012	cm3/molc sec	25	EST	ATKINSON,R (1987A)

CAS #: 000532-32-1				BENZOIC ACID, SODIUM SALT

Formula: C7H5NaO2

Mol Weight: 144.11

MP (deg C): > 300   FP (deg C):

BP (deg C):

BP pressure (mm Hg):

Property/ Value	Units	Temp	Data Type	Reference
WS 1.00E+006	mg/L	25	EST	MEYLAN,WM ET AL. (1996)
logP -2.27			EST	MEYLAN,WM & HOWARD,PH (1995)
VP 3.67E-009	mm Hg	25	EST	NEELY,WB & BLAU,GE (1985)
DC	pKa			
HL	atm m3/mol			
OH 1.78E-012	cm3/molc sec		EST	MEYLAN,WM & HOWARD,PH (1993)

CAS #: 000532-34-3				INDALONE

Formula: C12H18O4

Mol Weight: 226.27

MP (deg C):   FP (deg C):

BP (deg C): 256-270

BP pressure (mm Hg):

Property/ Value	Units	Temp	Data Type	Reference
WS 2.75E+002	mg/L	25	EST	MEYLAN,WM ET AL. (1996)
logP 2.42			EST	MEYLAN,WM & HOWARD,PH (1995)
VP 2.47E-004	mm Hg	25	EST	NEELY,WB & BLAU,GE (1985)
DC	pKa			
HL 4.68E-008	atm m3/mol	25	EST	MEYLAN,WM & HOWARD,PH (1991)
OH 4.36E-011	cm3/molc sec	25	EST	MEYLAN,WM & HOWARD,PH (1993)

CAS #: 000533-17-5				2-CHLOROACETANILIDE

Formula: C8H8ClNO

Mol Weight: 169.61

MP (deg C): 88-90   FP (deg C):

BP (deg C):

BP pressure (mm Hg):

Property/ Value	Units	Temp	Data Type	Reference
WS 3.08E+003	mg/L		EXP	BRIGGS,GG (1981)
logP 1.28			EXP	HANSCH,C & LEO,AJ (1985)
VP 9.63E-005	mm Hg	25	EST	NEELY,WB & BLAU,GE (1985)
DC	pKa			
HL 4.57E-009	atm m3/mol	25	EST	MEYLAN,WM & HOWARD,PH (1991)
OH 3.73E-012	cm3/molc sec	25	EST	MEYLAN,WM & HOWARD,PH (1993)

CAS #: 000533-18-6				O-TOLYLACETATE

Formula: C9H10O2

Mol Weight: 150.18

MP (deg C):   FP (deg C):

BP (deg C): 208

BP pressure (mm Hg):

Property/ Value	Units	Temp	Data Type	Reference
WS 1.70E+003	mg/L	25	EST	MEYLAN,WM ET AL. (1996)
logP 1.93			EXP	HANSCH,C & LEO,AJ (1985)
VP 1.64E-001	mm Hg	25	EST	NEELY,WB & BLAU,GE (1985)
DC	pKa			
HL 7.15E-005	atm m3/mol	25	EST	MEYLAN,WM & HOWARD,PH (1991)
OH 3.98E-012	cm3/molc sec	25	EST	MEYLAN,WM & HOWARD,PH (1993)

CAS #: 000533-23-3				2,4-D, ETHYL ESTER

Formula: C10H10Cl2O3

Mol Weight: 249.10

MP (deg C):   FP (deg C):

BP (deg C): 136

BP pressure (mm Hg): 1.44E+000

Property/ Value	Units	Temp	Data Type	Reference
WS 1.11E+002	mg/L	25	EXP	PARIS,DF ET AL. (1984)
logP 3.39			EST	MEYLAN,WM & HOWARD,PH (1995)
VP 1.10E-003	mm Hg	25	EXP	BOSCH,SJ (1983)
DC	pKa			
HL 2.24E+000	atm m3/mol	25	EST	VP/WSOL
OH 5.26E-012	cm3/molc sec	25	EST	MEYLAN,WM & HOWARD,PH (1993)

CAS #: 000533-45-9				CLOMETHIAZOLE

Formula: C6H10ClNS

Mol Weight: 163.67

MP (deg C):   FP (deg C):

BP (deg C): 92

BP pressure (mm Hg): 7.00E+000

Property/ Value	Units	Temp	Data Type	Reference
WS 1.04E+003	mg/L	25	EST	MEYLAN,WM ET AL. (1996)
logP 2.12			EXP	SANGSTER,J (1993)
VP 2.29E-002	mm Hg	25	EST	NEELY,WB & BLAU,GE (1985)
DC	pKa			
HL 2.18E-006	atm m3/mol	25	EST	MEYLAN,WM & HOWARD,PH (1991)
OH 3.00E-012	cm3/molc sec	25	EST	MEYLAN,WM & HOWARD,PH (1993)

CAS #: 000533-58-4				2-IODOPHENOL

Formula: C6H5IO

Mol Weight: 220.01

MP (deg C): 43   FP (deg C):

BP (deg C): 186-187

BP pressure (mm Hg): 1.60E+002

Property/ Value	Units	Temp	Data Type	Reference
WS 7.14E+002	mg/L	25	EST	MEYLAN,WM ET AL. (1996)
logP 2.65			EXP	HANSCH,C & LEO,AJ (1985)
VP 6.16E-003	mm Hg	25	EST	NEELY,WB & BLAU,GE (1985)
DC 8.51	pKa	25	EXP	KORTUM,G ET AL (1961)
HL 1.30E-007	atm m3/mol	25	EST	MEYLAN,WM & HOWARD,PH (1991)
OH 1.11E-011	cm3/molc sec	25	EST	MEYLAN,WM & HOWARD,PH (1993)

CAS #: 000533-73-3 — 1,2,4-TRIHYDROXYBENZENE (PYROGALLOL)

Formula:	$C_6H_6O_3$
Mol Weight:	126.11
MP (deg C):	141
FP (deg C):	
BP (deg C):	
BP pressure (mm Hg):	

Property/Value	Units	Temp	Data Type	Reference
WS 1.22E+005	mg/L	25	EST	MEYLAN,WM ET AL. (1996)
logP 0.55			EST	MEYLAN,WM & HOWARD,PH (1995)
VP 1.60E-004	mm Hg	25	EST	NEELY,WB & BLAU,GE (1985)
DC	pKa			
HL 6.07E-015	atm m3/mol	25	EST	MEYLAN,WM & HOWARD,PH (1991)
OH 2.00E-010	cm3/molc sec	25	EST	MEYLAN,WM & HOWARD,PH (1993)

CAS #: 000533-74-4 — DMTT

Formula:	$C_5H_{10}N_2S_2$
Mol Weight:	162.28
MP (deg C):	106
FP (deg C):	
BP (deg C):	
BP pressure (mm Hg):	

Property/Value	Units	Temp	Data Type	Reference
WS 1.20E+003	mg/L	25	EXP	YALKOWSKY,SH & DANNENFELSER,RM (1992)
logP 1.40			EXP	TOMLIN,C (1994)
VP 2.80E-006	mm Hg	20	EXP	WORTHING,CR & WALKER,SB (1987)
DC	pKa			
HL 4.98E-010	atm m3/mol	25	EST	VP/WSOL
OH 2.84E-010	cm3/molc sec	25	EST	MEYLAN,WM & HOWARD,PH (1993)

CAS #: 000533-75-5 — TROPOLONE

Formula:	$C_7H_6O_2$
Mol Weight:	122.12
MP (deg C):	50.8
FP (deg C):	
BP (deg C):	
BP pressure (mm Hg):	

Property/Value	Units	Temp	Data Type	Reference
WS 3.48E+004	mg/L	25	EST	MEYLAN,WM ET AL. (1996)
logP 0.53			EXP	HANSCH,C ET AL. (1995)
VP 6.79E-004	mm Hg	25	EST	NEELY,WB & BLAU,GE (1985)
DC	pKa			
HL 1.08E-005	atm m3/mol	25	EST	MEYLAN,WM & HOWARD,PH (1991)
OH 6.82E-011	cm3/molc sec	25	EST	MEYLAN,WM & HOWARD,PH (1993)

CAS #: 000533-98-2 — 1,2-DIBROMOBUTANE

Formula:	$C_4H_8Br_2$
Mol Weight:	215.93
MP (deg C):	-65
FP (deg C):	
BP (deg C):	166
BP pressure (mm Hg):	

Property/Value	Units	Temp	Data Type	Reference
WS 1.16E+002	mg/L	25	EST	MEYLAN,WM ET AL. (1996)
logP 2.92			EST	MEYLAN,WM & HOWARD,PH (1995)
VP 3.10E+000	mm Hg	25	EXP	OHE,S (1976)
DC	pKa			
HL 2.29E-003	atm m3/mol	25	EST	MEYLAN,WM & HOWARD,PH (1991)
OH 1.02E-012	cm3/molc sec	25	EST	MEYLAN,WM & HOWARD,PH (1993)

CAS #: 000534-07-6 — 1,3-DICHLOROACETONE

Formula:	$C_3H_4Cl_2O$
Mol Weight:	126.97
MP (deg C):	45
FP (deg C):	
BP (deg C):	173
BP pressure (mm Hg):	

Property/Value	Units	Temp	Data Type	Reference
WS 5.52E+004	mg/L	25	EST	MEYLAN,WM ET AL. (1996)
logP 0.27			EST	MEYLAN,WM & HOWARD,PH (1995)
VP 6.28E-001	mm Hg	25	EXT	OHE,S (1976)
DC	pKa			
HL 6.15E-006	atm m3/mol	25	EST	MEYLAN,WM & HOWARD,PH (1991)
OH 5.32E-013	cm3/molc sec	25	EST	MEYLAN,WM & HOWARD,PH (1993)

CAS #: 000534-13-4 — N,N-DIMETHYLTHIOUREA

Formula:	$C_3H_8N_2S$
Mol Weight:	104.17
MP (deg C):	60-62
FP (deg C):	
BP (deg C):	
BP pressure (mm Hg):	

Property/Value	Units	Temp	Data Type	Reference
WS 1.00E+006	mg/L	20	EXP	YALKOWSKY,SH & DANNENFELSER,RM (1992)
logP -0.24			EXP	HANSCH,C ET AL. (1995)
VP 1.78E+000	mm Hg	25	EST	NEELY,WB & BLAU,GE (1985)
DC	pKa			
HL 7.61E-007	atm m3/mol	25	EST	MEYLAN,WM & HOWARD,PH (1991)
OH 1.29E-010	cm3/molc sec	25	EST	MEYLAN,WM & HOWARD,PH (1993)

CAS #: 000534-15-6 — 1,1-DIMETHOXYETHANE

Formula:	$C_4H_{10}O_2$
Mol Weight:	90.12
MP (deg C):	-113.2
FP (deg C):	
BP (deg C):	64.5
BP pressure (mm Hg):	

Property/Value	Units	Temp	Data Type	Reference
WS 1.00E+006	mg/L	25	EXP	MERCK INDEX (1989)
logP 0.22			EST	MEYLAN,WM & HOWARD,PH (1995)
VP 1.71E+002	mm Hg	25	EXP	BOUBLIK,T ET AL. (1984)
DC	pKa			
HL 6.71E-005	atm m3/mol	25	EST	MEYLAN,WM & HOWARD,PH (1991)
OH 8.89E-012	cm3/molc sec	25	EXP	ATKINSON,R (1989)

CAS #: 000534-22-5 — 2-METHYLFURAN

Formula:	C_5H_6O
Mol Weight:	82.10
MP (deg C):	-87.5
FP (deg C):	
BP (deg C):	65
BP pressure (mm Hg):	

Property/Value	Units	Temp	Data Type	Reference
WS 3.41E+003	mg/L	25	EST	MEYLAN,WM ET AL. (1996)
logP 1.85			EXP	HANSCH,C & LEO,AJ (1985)
VP 1.56E+002	mm Hg	25	EXT	CHAO,J ET AL. (1983)
DC	pKa			
HL 5.93E-003	atm m3/mol	25	EST	MEYLAN,WM & HOWARD,PH (1991)
OH 1.06E-010	cm3/molc sec	25	EST	MEYLAN,WM & HOWARD,PH (1993)

000534-26-9 — 2-METHYL-2-IMIDAZOLINE

CAS #:	000534-26-9
Formula:	$C_4H_8N_2$
Mol Weight:	84.12
MP (deg C):	105
FP (deg C):	
BP (deg C):	198-200
BP pressure (mm Hg):	

Property/Value	Units	Temp	Data Type	Reference
WS 4.62E+004	mg/L	25	EST	MEYLAN,WM ET AL. (1996)
logP 0.52			EXP	HANSCH,C & LEO,AJ (1985)
VP 1.04E-002	mm Hg	25	EST	NEELY,WB & BLAU,GE (1985)
DC	pKa			
HL 6.41E-007	atm m3/mol	25	EST	MEYLAN,WM & HOWARD,PH (1991)
OH 7.26E-011	cm3/molc sec	25	EST	MEYLAN,WM & HOWARD,PH (1993)

000534-52-1 — 4,6-DINITRO-O-CRESOL

CAS #:	000534-52-1
Formula:	$C_7H_6N_2O_5$
Mol Weight:	198.14
MP (deg C):	84-86
FP (deg C):	
BP (deg C):	378
BP pressure (mm Hg):	

Property/Value	Units	Temp	Data Type	Reference
WS 1.98E+002	mg/L	20	EXP	SCHWARZENBACH,RP ET AL. (1988)
logP 2.12			EXP	HANSCH,C & LEO,AJ (1985)
VP 3.24E-004	mm Hg	20	EXP	SCHWARZENBACH,RP ET AL. (1988)
DC 4.31	pKa	21	EXP	SCHWARZENBACH,RP ET AL. (1988)
HL 1.40E-006	atm m3/mol	25	EXP	WARNER,HP ET AL. (1987)
OH 2.07E-013	cm3/molc sec	25	EST	MEYLAN,WM & HOWARD,PH (1993)

000534-59-8 — BUTYL MALONIC ACID

CAS #:	000534-59-8
Formula:	$C_7H_{12}O_4$
Mol Weight:	160.17
MP (deg C):	104.5
FP (deg C):	
BP (deg C):	235-240
BP pressure (mm Hg):	

Property/Value	Units	Temp	Data Type	Reference
WS 4.38E+005	mg/L	25	EXP	YALKOWSKY,SH & DANNENFELSER,RM (1992)
logP 0.65			EST	MEYLAN,WM & HOWARD,PH (1995)
VP 1.44E-004	mm Hg	25	EST	NEELY,WB & BLAU,GE (1985)
DC	pKa			
HL 1.27E-011	atm m3/mol	25	EST	MEYLAN,WM & HOWARD,PH (1991)
OH 6.52E-012	cm3/molc sec	25	EST	MEYLAN,WM & HOWARD,PH (1993)

000534-82-7 — 3-METHOXY-4-HYDROXYPHENYLGLYCOL

CAS #:	000534-82-7
Formula:	$C_9H_{12}O_4$
Mol Weight:	184.19
MP (deg C):	
FP (deg C):	
BP (deg C):	
BP pressure (mm Hg):	

Property/Value	Units	Temp	Data Type	Reference
WS 1.64E+005	mg/L	25	EST	MEYLAN,WM ET AL. (1996)
logP -0.58			EXP	HANSCH,C & LEO,AJ (1985)
VP 4.79E-007	mm Hg	25	EST	NEELY,WB & BLAU,GE (1985)
DC	pKa			
HL 6.50E-014	atm m3/mol	25	EST	MEYLAN,WM & HOWARD,PH (1991)
OH 4.87E-011	cm3/molc sec	25	EST	MEYLAN,WM & HOWARD,PH (1993)

000535-13-7 — ETHYL ALPHA-CHLOROPROPIONATE

CAS #:	000535-13-7
Formula:	$C_5H_9ClO_2$
Mol Weight:	136.58
MP (deg C):	
FP (deg C):	
BP (deg C):	147-148
BP pressure (mm Hg):	

Property/Value	Units	Temp	Data Type	Reference
WS 4.22E+003	mg/L	25	EST	MEYLAN,WM ET AL. (1996)
logP 1.54			EST	MEYLAN,WM & HOWARD,PH (1995)
VP 3.61E+000	mm Hg	25	EXP	OHE,S (1976)
DC	pKa			
HL 1.09E-004	atm m3/mol	25	EST	MEYLAN,WM & HOWARD,PH (1991)
OH 1.70E-012	cm3/molc sec	25	EST	MEYLAN,WM & HOWARD,PH (1993)

000535-15-9 — ETHYL DICHLOROACETATE

CAS #:	000535-15-9
Formula:	$C_4H_6Cl_2O_2$
Mol Weight:	157.00
MP (deg C):	
FP (deg C):	
BP (deg C):	155
BP pressure (mm Hg):	

Property/Value	Units	Temp	Data Type	Reference
WS 5.49E+003	mg/L	25	EST	MEYLAN,WM ET AL. (1996)
logP 1.30			EST	MEYLAN,WM & HOWARD,PH (1995)
VP 2.86E+000	mm Hg	25	EXP	OHE,S (1976)
DC	pKa			
HL 2.89E-005	atm m3/mol	25	EST	MEYLAN,WM & HOWARD,PH (1991)
OH 1.12E-012	cm3/molc sec	25	EST	MEYLAN,WM & HOWARD,PH (1993)

000535-75-1 — 2-PIPERIDINECARBOXYLIC ACID

CAS #:	000535-75-1
Formula:	$C_6H_{11}NO_2$
Mol Weight:	129.16
MP (deg C):	
FP (deg C):	
BP (deg C):	
BP pressure (mm Hg):	

Property/Value	Units	Temp	Data Type	Reference
WS 7.41E+004	mg/L	25	EST	MEYLAN,WM ET AL. (1996)
logP -2.31			EXP	TSAI,RS ET AL. (1991)
VP 1.75E-009	mm Hg	25	EST	NEELY,WB & BLAU,GE (1985)
DC	pKa			
HL 2.55E-009	atm m3/mol	25	EST	MEYLAN,WM & HOWARD,PH (1991)
OH 9.51E-011	cm3/molc sec	25	EST	MEYLAN,WM & HOWARD,PH (1993)

000535-77-3 — M-CYMENE

CAS #:	000535-77-3
Formula:	$C_{10}H_{14}$
Mol Weight:	134.22
MP (deg C):	-63.7
FP (deg C):	
BP (deg C):	175.1
BP pressure (mm Hg):	

Property/Value	Units	Temp	Data Type	Reference
WS 1.27E+001	mg/L	25	EST	MEYLAN,WM ET AL. (1996)
logP 4.50			EXP	SHERBLOM,PM & EGANHOUSE,RP (1988)
VP 1.72E+000	mm Hg	25	EXP	DAUBERT,TE & DANNER,RP (1989)
DC	pKa			
HL 1.16E-002	atm m3/mol	25	EST	MEYLAN,WM & HOWARD,PH (1991)
OH 1.45E-011	cm3/molc sec	25	EST	MEYLAN,WM & HOWARD,PH (1993)

000535-80-8 — M-CHLOROBENZOIC ACID

CAS #: 000535-80-8

Formula: $C_7H_5ClO_2$

Mol Weight: 156.57

MP (deg C): 158

FP (deg C):

BP (deg C):

BP pressure (mm Hg):

Property/Value	Units	Temp	Data Type	Reference
WS 4.50E+002	mg/L	25	EXP	SEIDELL,A (1941)
logP 2.68			EXP	HANSCH,C & LEO,AJ (1985)
VP 2.33E-003	mm Hg	25	EST	NEELY,WB & BLAU,GE (1985)
DC 3.81	pKa	25	EXP	BYKOVA,N ET AL. (1970)
HL 3.88E-008	atm m3/mol	25	EST	MEYLAN,WM & HOWARD,PH (1991)
OH 4.23E-013	cm3/molc sec	25	EST	ATKINSON,R (1987)

000535-89-7 — 2-CHLORO-4-METHYL-6-(DIMETHYLAMINO)PYRIMIDINE

CAS #: 000535-89-7

Formula: $C_7H_{10}ClN_3$

Mol Weight: 171.63

MP (deg C): 87

FP (deg C):

BP (deg C): 140-147

BP pressure (mm Hg): 4.00E+000

Property/Value	Units	Temp	Data Type	Reference
WS 9.40E+003	mg/L	20	EXP	SHIU,WY ET AL. (1990)
logP 1.31			EST	MEYLAN,WM & HOWARD,PH (1995)
VP 2.32E-002	mm Hg	25	EST	NEELY,WB & BLAU,GE (1985)
DC	pKa			
HL 3.79E-008	atm m3/mol	25	EST	MEYLAN,WM & HOWARD,PH (1991)
OH 8.28E-011	cm3/molc sec	25	EST	MEYLAN,WM & HOWARD,PH (1993)

000536-40-3 — BENZOYLHYDRAZINE, P-CHLORO

CAS #: 000536-40-3

Formula: $C_7H_7ClN_2O$

Mol Weight: 170.60

MP (deg C): 163-165

FP (deg C):

BP (deg C):

BP pressure (mm Hg):

Property/Value	Units	Temp	Data Type	Reference
WS 6.00E+003	mg/L	25	EST	MEYLAN,WM ET AL. (1996)
logP 1.18			EXP	HANSCH,C & LEO,AJ (1985)
VP 2.02E-005	mm Hg	25	EST	NEELY,WB & BLAU,GE (1985)
DC	pKa			
HL 6.85E-012	atm m3/mol	25	EST	MEYLAN,WM & HOWARD,PH (1991)
OH 6.75E-012	cm3/molc sec	25	EST	MEYLAN,WM & HOWARD,PH (1993)

000536-57-2 — P-TOLUENESULFINIC ACID

CAS #: 000536-57-2

Formula: $C_7H_8O_2S$

Mol Weight: 156.20

MP (deg C): 85

FP (deg C):

BP (deg C):

BP pressure (mm Hg):

Property/Value	Units	Temp	Data Type	Reference
WS 2.41E+005	mg/L	25	EST	MEYLAN,WM ET AL. (1996)
logP -0.62			EST	MEYLAN,WM & HOWARD,PH (1995)
VP 3.00E+000	mm Hg	20	EXP	WEBER,RC ET AL. (1981)
DC	pKa			
HL 2.21E-011	atm m3/mol	25	EST	MEYLAN,WM & HOWARD,PH (1991)
OH 1.36E-012	cm3/molc sec	25	EST	MEYLAN,WM & HOWARD,PH (1993)

000536-66-3 — CUMIC ACID

CAS #: 000536-66-3

Formula: $C_{10}H_{12}O_2$

Mol Weight: 164.21

MP (deg C): 115-117

FP (deg C):

BP (deg C):

BP pressure (mm Hg):

Property/Value	Units	Temp	Data Type	Reference
WS 1.52E+002	mg/L	25	EXP	STEPHEN,H & STEPHEN,T (1963)
logP 3.40			EXP	HANSCH,C ET AL. (1995)
VP 1.19E-003	mm Hg	25	EST	NEELY,WB & BLAU,GE (1985)
DC 4.35	pKa	25	EXP	KORTUM,G ET AL. (1961)
HL 2.11E-007	atm m3/mol	25	EST	MEYLAN,WM & HOWARD,PH (1991)
OH 4.51E-012	cm3/molc sec	25	EST	MEYLAN,WM & HOWARD,PH (1993)

000536-74-3 — ETHYNYL BENZENE

CAS #: 000536-74-3

Formula: C_8H_6

Mol Weight: 102.14

MP (deg C): -44.8

FP (deg C):

BP (deg C): 142.4

BP pressure (mm Hg):

Property/Value	Units	Temp	Data Type	Reference
WS 4.56E+002	mg/L		EXP	YALKOWSKY,SH & DANNENFELSER,RM (1992)
logP 2.53			EXP	HANSCH,C & LEO,AJ (1985)
VP 2.06E+000	mm Hg	25	EXP	PATTE,F ET AL. (1982)
DC	pKa			
HL 6.07E-004	atm m3/mol	25	EST	VP/WSOL
OH 8.02E-012	cm3/molc sec	25	EST	MEYLAN,WM & HOWARD,PH (1993)

000536-75-4 — 4-ETHYLPYRIDINE

CAS #: 000536-75-4

Formula: C_7H_9N

Mol Weight: 107.16

MP (deg C): -90.5

FP (deg C):

BP (deg C): 168.3

BP pressure (mm Hg):

Property/Value	Units	Temp	Data Type	Reference
WS 1.07E+005	mg/L	25	EXP	KUHNE,R ET AL. (1995)
logP 1.65			EXP	HANSCH,C ET AL. (1995)
VP 2.27E+000	mm Hg	25	EXP	CHAO,J ET AL. (1983)
DC 5.87	pKa	25	EXP	PERRIN,DD (1965)
HL 8.48E-006	atm m3/mol	25	EXP	ANDON,RJL ET AL. (1954)
OH 1.95E+000	cm3/molc sec	25	EST	MEYLAN,WM & HOWARD,PH (1993)

000536-78-7 — 3-ETHYLPYRIDINE

CAS #: 000536-78-7

Formula: C_7H_9N

Mol Weight: 107.16

MP (deg C): -76.9

FP (deg C):

BP (deg C): 165

BP pressure (mm Hg):

Property/Value	Units	Temp	Data Type	Reference
WS 3.70E+005	mg/L	196	EXP	YALKOWSKY,SH & DANNENFELSER,RM (1992)
logP 1.66			EXP	HANSCH,C ET AL. (1995)
VP 2.53E+000	mm Hg	25	EXP	CHAO,J ET AL. (1983)
DC 5.56	pKa	25	EXP	PERRIN,DD (1965)
HL 1.04E-005	atm m3/mol	25	EXP	ANDON,RJL ET AL. (1954)
OH 1.95E-012	cm3/molc sec	25	EST	MEYLAN,WM & HOWARD,PH (1993)

000536-90-3 — M-METHOXYANILINE

CAS #:	000536-90-3
Formula:	C_7H_9NO
Mol Weight:	123.16
MP (deg C):	-1
FP (deg C):	
BP (deg C):	251
BP pressure (mm Hg):	

Property/Value	Units	Temp	Data Type	Reference
WS 1.57E+004	mg/L	25	EST	MEYLAN,WM ET AL. (1996)
logP 0.93			EXP	HANSCH,C & LEO,AJ (1985)
VP 7.50E-002	mm Hg	25	EXP	WEBER,RC ET AL. (1981)
DC 4.24	pKa	22	EXP	PERRIN,DD (1965)
HL 1.13E-007	atm m3/mol	25	EST	MEYLAN,WM & HOWARD,PH (1991)
OH 2.01E-010	cm3/molc sec	25	EST	MEYLAN,WM & HOWARD,PH (1993)

000537-24-6 — 3,5-DIBROMOTYROSINE

CAS #:	000537-24-6
Formula:	$C_9H_9Br_2NO_3$
Mol Weight:	338.99
MP (deg C):	
FP (deg C):	
BP (deg C):	
BP pressure (mm Hg):	

Property/Value	Units	Temp	Data Type	Reference
WS 2.26E+002	mg/L	25	EST	MEYLAN,WM ET AL. (1996)
logP 0.02			EST	MEYLAN,WM & HOWARD,PH (1995)
VP 9.34E-011	mm Hg	25	EST	NEELY,WB & BLAU,GE (1985)
DC	pKa			
HL 1.99E-015	atm m3/mol	25	EST	MEYLAN,WM & HOWARD,PH (1991)
OH 4.08E-011	cm3/molc sec	25	EST	MEYLAN,WM & HOWARD,PH (1993)

000537-26-8 — TROPACOCAINE

CAS #:	000537-26-8
Formula:	$C_{15}H_{19}NO_2$
Mol Weight:	245.32
MP (deg C):	49
FP (deg C):	
BP (deg C):	
BP pressure (mm Hg):	

Property/Value	Units	Temp	Data Type	Reference
WS 1.70E+003	mg/L	25	EXP	SEIDELL,A (1941)
logP 2.70			EST	MEYLAN,WM & HOWARD,PH (1995)
VP 5.64E-005	mm Hg	25	EST	NEELY,WB & BLAU,GE (1985)
DC	pKa			
HL 1.38E-008	atm m3/mol	25	EST	MEYLAN,WM & HOWARD,PH (1991)
OH 6.10E-011	cm3/molc sec	25	EST	MEYLAN,WM & HOWARD,PH (1993)

000537-45-1 — 2,6-DIBROMOQUINONECHLOROIMIDE

CAS #:	000537-45-1
Formula:	$C_6H_2Br_2ClNO$
Mol Weight:	299.36
MP (deg C):	83
FP (deg C):	
BP (deg C):	
BP pressure (mm Hg):	

Property/Value	Units	Temp	Data Type	Reference
WS 1.48E+002	mg/L	25	EST	MEYLAN,WM ET AL. (1996)
logP 2.25			EST	MEYLAN,WM & HOWARD,PH (1995)
VP 1.27E-004	mm Hg	25	EST	NEELY,WB & BLAU,GE (1985)
DC	pKa			
HL 2.57E-007	atm m3/mol	25	EST	MEYLAN,WM & HOWARD,PH (1991)
OH 1.58E-011	cm3/molc sec	25	EST	MEYLAN,WM & HOWARD,PH (1993)

000537-46-2 — METHAMPHETAMINE

CAS #:	000537-46-2
Formula:	$C_{10}H_{15}N$
Mol Weight:	149.24
MP (deg C):	170-175
FP (deg C):	
BP (deg C):	
BP pressure (mm Hg):	

Property/Value	Units	Temp	Data Type	Reference
WS 1.33E+004	mg/L	25	EST	MEYLAN,WM ET AL. (1996)
logP 2.07			EXP	HANSCH,C & LEO,AJ (1985)
VP 1.63E-001	mm Hg	25	EST	NEELY,WB & BLAU,GE (1985)
DC 9.87	pKa	25	EXP	PERRIN,DD (1965)
HL 2.37E-006	atm m3/mol	25	EST	MEYLAN,WM & HOWARD,PH (1991)
OH 9.26E-011	cm3/molc sec	25	EST	MEYLAN,WM & HOWARD,PH (1993)

000537-47-3 — HYDRAZINECARBOXAMIDE, N-PHENYL-

CAS #:	000537-47-3
Formula:	$C_7H_9N_3O$
Mol Weight:	151.17
MP (deg C):	122
FP (deg C):	
BP (deg C):	
BP pressure (mm Hg):	

Property/Value	Units	Temp	Data Type	Reference
WS 7.00E+002	mg/L	15	EXP	YALKOWSKY,SH & DANNENFELSER,RM (1992)
logP 0.16			EXP	KRAMER,CR & BECK,L (1981)
VP 7.78E-005	mm Hg	25	EST	NEELY,WB & BLAU,GE (1985)
DC	pKa			
HL 8.41E-013	atm m3/mol	25	EST	MEYLAN,WM & HOWARD,PH (1991)
OH 4.37E-011	cm3/molc sec	25	EST	MEYLAN,WM & HOWARD,PH (1993)

000537-55-3 — N-ACETYL-L-TYROSINE

CAS #:	000537-55-3
Formula:	$C_{11}H_{13}NO_4$
Mol Weight:	223.23
MP (deg C):	
FP (deg C):	
BP (deg C):	
BP pressure (mm Hg):	

Property/Value	Units	Temp	Data Type	Reference
WS 9.38E+003	mg/L	25	EST	MEYLAN,WM ET AL. (1996)
logP 1.32			EXP	HUANG,CH ET AL. (1985)
VP 3.41E-009	mm Hg	25	EST	NEELY,WB & BLAU,GE (1985)
DC	pKa			
HL 4.06E-017	atm m3/mol	25	EST	MEYLAN,WM & HOWARD,PH (1991)
OH 5.54E-011	cm3/molc sec	25	EST	MEYLAN,WM & HOWARD,PH (1993)

000537-91-7 — DISULFIDE, BIS(3-NITROPHENYL)

CAS #:	000537-91-7
Formula:	$C_{12}H_8N_2O_4S_2$
Mol Weight:	308.34
MP (deg C):	83
FP (deg C):	
BP (deg C):	
BP pressure (mm Hg):	

Property/Value	Units	Temp	Data Type	Reference
WS 1.53E+000	mg/L	25	EST	MEYLAN,WM ET AL. (1996)
logP 4.06			EXP	HANSCH,C ET AL. (1995)
VP 1.38E-008	mm Hg	25	EST	NEELY,WB & BLAU,GE (1985)
DC	pKa			
HL 7.54E-010	atm m3/mol	25	EST	MEYLAN,WM & HOWARD,PH (1991)
OH 2.27E-010	cm3/molc sec	25	EST	MEYLAN,WM & HOWARD,PH (1993)

CAS #: 000537-92-8				3-METHYLACETANILIDE

Formula: $C_9H_{11}NO$

Mol Weight: 149.19

MP (deg C): 65.5 FP (deg C):

BP (deg C): 303

BP pressure (mm Hg):

Property/ Value	Units	Temp	Data Type	Reference
WS 2.81E+003	mg/L	25	EST	MEYLAN,WM ET AL. (1996)
logP 1.68			EXP	HANSCH,C ET AL. (1995)
VP 1.83E-004	mm Hg	25	EST	NEELY,WB & BLAU,GE (1985)
DC	pKa			
HL 6.81E-009	atm m3/mol	25	EST	MEYLAN,WM & HOWARD,PH (1991)
OH 3.27E-011	cm3/molc sec	25	EST	MEYLAN,WM & HOWARD,PH (1993)

CAS #: 000538-23-8				TRIOCTONOIN

Formula: $C_{27}H_{50}O_6$

Mol Weight: 470.70

MP (deg C): 10 FP (deg C):

BP (deg C): 233

BP pressure (mm Hg):

Property/ Value	Units	Temp	Data Type	Reference
WS 4.00E-001	mg/L	37	EXP	FUNASAKI,N ET AL. (1976)
logP 9.20			EST	MEYLAN,WM & HOWARD,PH (1995)
VP 2.09E-008	mm Hg	25	EST	NEELY,WB & BLAU,GE (1985)
DC	pKa			
HL 2.87E-007	atm m3/mol	25	EST	MEYLAN,WM & HOWARD,PH (1991)
OH 3.18E-011	cm3/molc sec	25	EST	MEYLAN,WM & HOWARD,PH (1993)

CAS #: 000538-32-9				BENZYLUREA

Formula: $C_8H_{10}N_2O$

Mol Weight: 150.18

MP (deg C): 147-148 FP (deg C):

BP (deg C):

BP pressure (mm Hg):

Property/ Value	Units	Temp	Data Type	Reference
WS 1.70E+004	mg/L	45	EXP	YALKOWSKY,SH & DANNENFELSER,RM (1992)
logP 0.73			EXP	HANSCH,C & LEO,AJ (1985)
VP 3.44E-004	mm Hg	25	EST	NEELY,WB & BLAU,GE (1985)
DC	pKa			
HL 6.47E-011	atm m3/mol	25	EST	MEYLAN,WM & HOWARD,PH (1991)
OH 1.07E-011	cm3/molc sec	25	EST	MEYLAN,WM & HOWARD,PH (1993)

CAS #: 000538-41-0				4,4'-DIAMINOAZOBENZENE

Formula: $C_{12}H_{12}N_4$

Mol Weight: 212.26

MP (deg C): 250.5 FP (deg C):

BP (deg C):

BP pressure (mm Hg):

Property/ Value	Units	Temp	Data Type	Reference
WS 1.59E+002	mg/L	25	EST	MEYLAN,WM ET AL. (1996)
logP 2.28			EST	MEYLAN,WM & HOWARD,PH (1995)
VP 1.52E-006	mm Hg	25	EST	NEELY,WB & BLAU,GE (1985)
DC	pKa			
HL 1.84E-012	atm m3/mol	25	EST	MEYLAN,WM & HOWARD,PH (1991)
OH 8.53E-011	cm3/molc sec	25	EST	MEYLAN,WM & HOWARD,PH (1993)

CAS #: 000538-43-2				PHENYL GLYCEROL

Formula: $C_9H_{12}O_3$

Mol Weight: 168.19

MP (deg C): 50-52 FP (deg C):

BP (deg C): 129-142

BP pressure (mm Hg): 6.00E-001

Property/ Value	Units	Temp	Data Type	Reference
WS 1.58E+004	mg/L	25	EST	MEYLAN,WM ET AL. (1996)
logP 0.70			EXP	HANSCH,C & LEO,AJ (1985)
VP 2.00E-005	mm Hg	25	EST	NEELY,WB & BLAU,GE (1985)
DC	pKa			
HL 7.51E-010	atm m3/mol	25	EST	MEYLAN,WM & HOWARD,PH (1991)
OH 4.31E-011	cm3/molc sec	25	EST	MEYLAN,WM & HOWARD,PH (1993)

CAS #: 000538-51-2				BENZALANILINE

Formula: $C_{13}H_{11}N$

Mol Weight: 181.24

MP (deg C): 48 FP (deg C):

BP (deg C): 300

BP pressure (mm Hg):

Property/ Value	Units	Temp	Data Type	Reference
WS 2.12E+002	mg/L	25	EST	MEYLAN,WM ET AL. (1996)
logP 2.82			EXP	HANSCH,C & LEO,AJ (1985)
VP 2.99E-003	mm Hg	25	EST	NEELY,WB & BLAU,GE (1985)
DC	pKa			
HL 9.84E-005	atm m3/mol	25	EST	MEYLAN,WM & HOWARD,PH (1991)
OH 8.70E-012	cm3/molc sec	25	EST	MEYLAN,WM & HOWARD,PH (1993)

CAS #: 000538-68-1				PENTYLBENZENE

Formula: $C_{11}H_{16}$

Mol Weight: 148.25

MP (deg C): -78.25 FP (deg C):

BP (deg C): 202.2

BP pressure (mm Hg):

Property/ Value	Units	Temp	Data Type	Reference
WS 3.37E+000	mg/L	25	EXP	YALKOWSKY,SH & DANNENFELSER,RM (1992)
logP 4.90			EXP	HANSCH,C & LEO,AJ (1985)
VP 4.39E-001	mm Hg	25	EXP	YAWS,CL (1994B)
DC	pKa			
HL 2.54E-002	atm m3/mol	25	EST	VP/WSOL
OH 1.01E-011	cm3/molc sec	25	EST	MEYLAN,WM & HOWARD,PH (1993)

CAS #: 000538-86-3				BENZYL METHYL ETHER

Formula: $C_8H_{10}O$

Mol Weight: 122.17

MP (deg C): -52.6 FP (deg C):

BP (deg C): 174

BP pressure (mm Hg):

Property/ Value	Units	Temp	Data Type	Reference
WS 2.26E+003	mg/L	25	EST	MEYLAN,WM ET AL. (1996)
logP 1.35			EXP	HANSCH,C & LEO,AJ (1985)
VP 1.50E+000	mm Hg	25	EST	NEELY,WB & BLAU,GE (1985)
DC	pKa			
HL 6.96E-005	atm m3/mol	25	EST	MEYLAN,WM & HOWARD,PH (1991)
OH 1.14E-011	cm3/molc sec	25	EST	MEYLAN,WM & HOWARD,PH (1993)

000538-93-2 — ISOBUTYLBENZENE

Formula: $C_{10}H_{14}$
Mol Weight: 134.22
MP (deg C): -51.4
FP (deg C):
BP (deg C): 172.7
BP pressure (mm Hg):

Property/Value	Units	Temp	Data Type	Reference
WS 1.01E+001	mg/L	25	EXP	YALKOWSKY,SH & DANNENFELSER,RM (1992)
logP 4.68			EXP	SHERBLOM,PM & EGANHOUSE,RP (1988)
VP 1.93E+000	mm Hg	25	EXP	DAUBERT,TE & DANNER,RP (1989)
DC	pKa			
HL 3.38E-002	atm m3/mol	25	EST	VP/WSOL
OH 8.71E-012	cm3/molc sec	25	EST	MEYLAN,WM & HOWARD,PH (1993)

000539-03-7 — P-CHLOROACETANILIDE

Formula: C_8H_8ClNO
Mol Weight: 169.61
MP (deg C): 179
FP (deg C):
BP (deg C): 333
BP pressure (mm Hg):

Property/Value	Units	Temp	Data Type	Reference
WS 9.50E+002	mg/L	65	EXP	SEIDELL,A (1941)
logP 2.09			EXP	SANGSTER,J (1994)
VP 9.63E-005	mm Hg	25	EST	NEELY,WB & BLAU,GE (1985)
DC	pKa			
HL 4.57E-009	atm m3/mol	25	EST	MEYLAN,WM & HOWARD,PH (1991)
OH 3.92E-012	cm3/molc sec	25	EST	MEYLAN,WM & HOWARD,PH (1993)

000539-17-3 — C.I. DISPERSE BLACK 3

Formula: $C_{14}H_{16}N_4$
Mol Weight: 240.31
MP (deg C): 190 dec
FP (deg C):
BP (deg C):
BP pressure (mm Hg):

Property/Value	Units	Temp	Data Type	Reference
WS 6.12E+000	mg/L	25	EXP	BAUGHMAN,GL & PERENICH,TA (1988)
logP 3.37			EST	MEYLAN,WM & HOWARD,PH (1995)
VP 2.74E-006	mm Hg	25	EST	NEELY,WB & BLAU,GE (1985)
DC	pKa			
HL 8.26E-011	atm m3/mol	25	EST	MEYLAN,WM & HOWARD,PH (1991)
OH 1.92E-010	cm3/molc sec	25	EST	MEYLAN,WM & HOWARD,PH (1993)

000539-21-9 — AMBAZONE

Formula: $C_8H_{11}N_7S$
Mol Weight: 237.29
MP (deg C):
FP (deg C):
BP (deg C):
BP pressure (mm Hg):

Property/Value	Units	Temp	Data Type	Reference
WS 2.47E+003	mg/L	25	EST	MEYLAN,WM ET AL. (1996)
logP 1.23			EXP	KRAMARCZYK,K (1987)
VP 1.03E-007	mm Hg	25	EST	NEELY,WB & BLAU,GE (1985)
DC	pKa			
HL 5.04E-016	atm m3/mol	25	EST	MEYLAN,WM & HOWARD,PH (1991)
OH 2.81E-010	cm3/molc sec	25	EST	MEYLAN,WM & HOWARD,PH (1993)

000539-30-0 — BENZYL ETHYL ETHER

Formula: $C_9H_{12}O$
Mol Weight: 136.20
MP (deg C):
FP (deg C):
BP (deg C): 186
BP pressure (mm Hg):

Property/Value	Units	Temp	Data Type	Reference
WS 1.24E+003	mg/L	25	EST	MEYLAN,WM ET AL. (1996)
logP 2.16			EXP	SANGSTER,J (1993)
VP 9.25E-001	mm Hg	25	EXP	DAUBERT,TE & DANNER,RP (1989)
DC	pKa			
HL 9.24E-005	atm m3/mol	25	EST	MEYLAN,WM & HOWARD,PH (1991)
OH 1.67E-011	cm3/molc sec	25	EST	MEYLAN,WM & HOWARD,PH (1993)

000539-82-2 — ETHYL VALERATE

Formula: $C_7H_{14}O_2$
Mol Weight: 130.19
MP (deg C): -91.2
FP (deg C):
BP (deg C): 146.1
BP pressure (mm Hg):

Property/Value	Units	Temp	Data Type	Reference
WS 2.21E+003	mg/L	25	EXP	SUZUKI,T (1991)
logP 2.34			EST	MEYLAN,WM & HOWARD,PH (1995)
VP 4.80E+000	mm Hg	25	EXP	HINE,J & MOOKERJEE,PK (1975)
DC	pKa			
HL 3.72E-004	atm m3/mol	25	EST	VP/WSOL
OH 5.24E-012	cm3/molc sec	25	EST	MEYLAN,WM & HOWARD,PH (1993)

000539-88-8 — ETHYL LEVULINATE

Formula: $C_7H_{12}O_3$
Mol Weight: 144.17
MP (deg C): 205-206
FP (deg C):
BP (deg C):
BP pressure (mm Hg):

Property/Value	Units	Temp	Data Type	Reference
WS 4.57E+004	mg/L	25	EST	MEYLAN,WM ET AL. (1996)
logP 0.29			EST	MEYLAN,WM & HOWARD,PH (1995)
VP 3.25E-003	mm Hg	25	EXT	PERRY,RH & GREEN,D (1984)
DC	pKa			
HL 2.08E-007	atm m3/mol	25	EST	MEYLAN,WM & HOWARD,PH (1991)
OH 5.32E-012	cm3/molc sec	25	EST	MEYLAN,WM & HOWARD,PH (1993)

000540-18-1 — PENTYL BUTYRATE

Formula: $C_9H_{18}O_2$
Mol Weight: 158.24
MP (deg C): -73.2
FP (deg C):
BP (deg C): 185
BP pressure (mm Hg):

Property/Value	Units	Temp	Data Type	Reference
WS 6.00E+001	mg/L	20	EXP	STEPHEN,H & STEPHEN,T (1963)
logP 3.32			EST	MEYLAN,WM & HOWARD,PH (1995)
VP 5.69E-001	mm Hg	25	EST	NEELY,WB & BLAU,GE (1985)
DC	pKa			
HL 9.60E-004	atm m3/mol	25	EST	MEYLAN,WM & HOWARD,PH (1991)
OH 8.15E-012	cm3/molc sec	25	EST	MEYLAN,WM & HOWARD,PH (1993)

CAS #: 000540-36-3				P-DIFLUOROBENZENE

Formula: $C_6H_4F_2$

Mol Weight: 114.10

MP (deg C): -23.7   FP (deg C):

BP (deg C): 88.82

BP pressure (mm Hg):

Property/ Value	Units	Temp	Data Type	Reference
WS 1.22E+003	mg/L	25	EXP	SUZUKI,T (1991)
logP 2.13			EXP	DUNN,WJ ET AL. (1983)
VP 6.42E+001	mm Hg	25	EXT	OSBORN,AG & SCOTT,DW (1980)
DC	pKa			
HL 7.34E-003	atm m3/mol	25	EST	MEYLAN,WM & HOWARD,PH (1991)
OH 8.24E-013	cm3/molc sec	25	EST	MEYLAN,WM & HOWARD,PH (1993)

CAS #: 000540-37-4				BENZENAMINE, 4-IODO-

Formula: C_6H_6IN

Mol Weight: 219.03

MP (deg C): 67-68   FP (deg C):

BP (deg C):

BP pressure (mm Hg):

Property/ Value	Units	Temp	Data Type	Reference
WS 3.50E+002	mg/L	25	EST	MEYLAN,WM ET AL. (1996)
logP 2.34			EXP	HANSCH,C ET AL. (1995)
VP 7.67E-003	mm Hg	25	EST	NEELY,WB & BLAU,GE (1985)
DC 3.78	pKa	25	EXP	PERRIN,DD (1965)
HL 4.41E-007	atm m3/mol	25	EST	MEYLAN,WM & HOWARD,PH (1991)
OH 3.56E-011	cm3/molc sec	25	EST	MEYLAN,WM & HOWARD,PH (1993)

CAS #: 000540-38-5				4-IODOPHENOL

Formula: C_6H_5IO

Mol Weight: 220.01

MP (deg C): 93-94   FP (deg C):

BP (deg C):

BP pressure (mm Hg):

Property/ Value	Units	Temp	Data Type	Reference
WS 4.28E+002	mg/L	25	EST	MEYLAN,WM ET AL. (1996)
logP 2.91			EXP	HANSCH,C & LEO,AJ (1985)
VP 6.16E-003	mm Hg	25	EST	NEELY,WB & BLAU,GE (1985)
DC 9.21	a pKa	25	EXP	KORTUM,G ET AL (1961)
HL 1.30E-007	atm m3/mol	25	EST	MEYLAN,WM & HOWARD,PH (1991)
OH 1.11E-011	cm3/molc sec	25	EST	MEYLAN,WM & HOWARD,PH (1993)

CAS #: 000540-42-1				2-METHYLPROPYL PROPANOATE

Formula: $C_7H_{14}O_2$

Mol Weight: 130.19

MP (deg C): -71   FP (deg C):

BP (deg C): 137

BP pressure (mm Hg):

Property/ Value	Units	Temp	Data Type	Reference
WS 1.07E+003	mg/L	25	EST	MEYLAN,WM ET AL. (1996)
logP 2.26			EST	MEYLAN,WM & HOWARD,PH (1995)
VP 6.47E+000	mm Hg	25	EXP	OHE,S (1976)
DC	pKa			
HL 5.45E-004	atm m3/mol	25	EST	MEYLAN,WM & HOWARD,PH (1991)
OH 5.42E-012	cm3/molc sec	25	EST	MEYLAN,WM & HOWARD,PH (1993)

CAS #: 000540-51-2				2-BROMOETHANOL

Formula: C_2H_5BrO

Mol Weight: 124.97

MP (deg C):   FP (deg C):

BP (deg C): 150

BP pressure (mm Hg):

Property/ Value	Units	Temp	Data Type	Reference
WS 1.98E+005	mg/L	25	EST	MEYLAN,WM ET AL. (1996)
logP 0.23			EXP	HANSCH,C & LEO,AJ (1985)
VP 2.06E+000	mm Hg	25	EST	NEELY,WB & BLAU,GE (1985)
DC	pKa			
HL 4.12E-007	atm m3/mol	25	EST	MEYLAN,WM & HOWARD,PH (1991)
OH 1.97E-012	cm3/molc sec	25	EST	MEYLAN,WM & HOWARD,PH (1993)

CAS #: 000540-54-5				1-CHLOROPROPANE

Formula: C_3H_7Cl

Mol Weight: 78.54

MP (deg C): -122.8   FP (deg C):

BP (deg C): 46.6

BP pressure (mm Hg):

Property/ Value	Units	Temp	Data Type	Reference
WS 2.72E+003	mg/L	20	EXP	YALKOWSKY,SH & DANNENFELSER,RM (1992)
logP 2.04			EXP	HANSCH,C ET AL. (1995)
VP 3.45E+002	mm Hg	25	EXP	YAWS,CL (1994)
DC	pKa			
HL 1.31E-002	atm m3/mol	25	EST	VP/WSOL
OH 1.12E-012	cm3/molc sec	25	EXP	DONAGHY,T ET AL. (1993)

CAS #: 000540-59-0				1,2-DICHLOROETHYLENE

Formula: $C_2H_2Cl_2$

Mol Weight: 96.94

MP (deg C): -57   FP (deg C):

BP (deg C): 48-60

BP pressure (mm Hg):

Property/ Value	Units	Temp	Data Type	Reference
WS 3.50E+003	mg/L	25	EXP	YALKOWSKY,SH & DANNENFELSER,RM (1992)
logP 1.86			EXP	HANSCH,C & LEO,AJ (1985)
VP 2.01E+002	mm Hg	25	EXP	BOUBLIK,T ET AL. (1984)
DC	pKa			
HL 4.08E-003	atm m3/mol	25	EXP	GOSSETT,JM (1987)
OH 2.38E-012	cm3/molc sec	25	EXP	ATKINSON,R (1989)

CAS #: 000540-61-4				AMINOACETONITRILE

Formula: $C_2H_4N_2$

Mol Weight: 56.07

MP (deg C): 101   FP (deg C):

BP (deg C):

BP pressure (mm Hg):

Property/ Value	Units	Temp	Data Type	Reference
WS 1.00E+006	mg/L	25	EST	MEYLAN,WM ET AL. (1996)
logP -1.37			EXP	HANSCH,C ET AL. (1995)
VP 7.94E+000	mm Hg	25	EST	NEELY,WB & BLAU,GE (1985)
DC 5.34	pKa	25	EXP	PERRIN,DD (1965)
HL 3.14E-009	atm m3/mol	25	EST	MEYLAN,WM & HOWARD,PH (1991)
OH 1.65E-012	cm3/molc sec	25	EST	MEYLAN,WM & HOWARD,PH (1993)

CAS #: 000540-63-6 — 1,2-ETHANEDITHIOL

Formula: $C_2H_6S_2$
Mol Weight: 94.20
MP (deg C): -41.2
FP (deg C):
BP (deg C): 146
BP pressure (mm Hg):

Property/Value	Units	Temp	Data Type	Reference
WS 1.12E+004	mg/L	25	EST	MEYLAN,WM ET AL. (1996)
logP 1.21			EST	MEYLAN,WM & HOWARD,PH (1995)
VP 5.61E+000	mm Hg	25	EXP	DAUBERT,TE & DANNER,RP (1993)
DC 8.96	pKa	30	EXP	SERJEANT,EP & DEMPSEY,B (1979)
HL 1.23E-004	atm m3/mol	25	EST	MEYLAN,WM & HOWARD,PH (1991)
OH 8.29E-011	cm3/molc sec	25	EST	MEYLAN,WM & HOWARD,PH (1993)

CAS #: 000540-67-0 — ETHYL METHYL ETHER

Formula: C_3H_8O
Mol Weight: 60.10
MP (deg C): -113
FP (deg C):
BP (deg C): 7.35
BP pressure (mm Hg):

Property/Value	Units	Temp	Data Type	Reference
WS 2.05E+005	mg/L	25	EST	MEYLAN,WM ET AL. (1996)
logP 0.56			EST	MEYLAN,WM & HOWARD,PH (1995)
VP 1.51E+003	mm Hg	25	EXP	DAUBERT,TE & DANNER,RP (1989)
DC	pKa			
HL 2.93E-004	atm m3/mol	25	EST	VP/WSOL
OH 7.00E-012	cm3/molc sec	25	EST	MEYLAN,WM & HOWARD,PH (1993)

CAS #: 000540-73-8 — 1,2-DIMETHYLHYDRAZINE

Formula: $C_2H_8N_2$
Mol Weight: 60.10
MP (deg C): -9
FP (deg C):
BP (deg C): 81
BP pressure (mm Hg): 7.53E+002

Property/Value	Units	Temp	Data Type	Reference
WS 1.00E+006	mg/L	25	EXP	BUDAVARI,S (1996)
logP -0.54			EST	MEYLAN,WM & HOWARD,PH (1995)
VP 6.99E+001	mm Hg	25	EXP	BOUBLIK,T ET AL. (1984)
DC	pKa			
HL 6.95E-008	atm m3/mol	25	EST	MEYLAN,WM & HOWARD,PH (1991)
OH 1.28E-010	cm3/molc sec	25	EST	MEYLAN,WM & HOWARD,PH (1993)

CAS #: 000540-84-1 — 2,2,4-TRIMETHYLPENTANE

Formula: C_8H_{18}
Mol Weight: 114.23
MP (deg C): -107.39
FP (deg C):
BP (deg C): 99.24
BP pressure (mm Hg):

Property/Value	Units	Temp	Data Type	Reference
WS 2.44E+000	mg/L	25	EXP	YALKOWSKY,SH & DANNENFELSER,RM (1992)
logP 4.09			EST	MEYLAN,WM & HOWARD,PH (1995)
VP 4.93E+001	mm Hg	25	EXP	DAUBERT,TE & DANNER,RP (1991)
DC	pKa			
HL 3.01E+000	atm m3/mol	25	EXP	NIRMALAKHANDAN,NN & SPEECE,RE (1988)
OH 3.68E-012	cm3/molc sec	25	EXP	ATKINSON,R (1989)

CAS #: 000540-88-5 — T-BUTYL ACETATE

Formula: $C_6H_{12}O_2$
Mol Weight: 116.16
MP (deg C):
FP (deg C):
BP (deg C): 97.8
BP pressure (mm Hg):

Property/Value	Units	Temp	Data Type	Reference
WS 1.75E+004	mg/L	25	EST	VP/HL
logP 1.76			EXP	HANSCH,C ET AL. (1995)
VP 4.70E+001	mm Hg	25	EXP	DAUBERT,TE & DANNER,RP (1991)
DC	pKa			
HL 4.10E-004	atm m3/mol	25	EST	MEYLAN,WM & HOWARD,PH (1991)
OH 1.94E-012	cm3/molc sec	25	EST	MEYLAN,WM & HOWARD (1993)

CAS #: 000540-97-6 — DODECAMETHYLCYCLOHEXASILOXANE

Formula: $C_{12}H_{36}O_6Si_6$
Mol Weight: 444.93
MP (deg C): -3
FP (deg C):
BP (deg C): 245
BP pressure (mm Hg):

Property/Value	Units	Temp	Data Type	Reference
WS 5.01E-003	mg/L	25	EST	MEYLAN,WM ET AL. (1996)
logP 6.33			EST	MEYLAN,WM & HOWARD,PH (1995)
VP 4.94E-002	mm Hg	25	EXT	JORDAN,TE (1954)
DC	pKa			
HL 1.81E-011	atm m3/mol	25	EST	MEYLAN,WM & HOWARD,PH (1991)
OH 1.80E-012	cm3/molc sec	25	EST	MEYLAN,WM & HOWARD,PH (1993)

CAS #: 000541-01-5 — HEXADECAMETHYLHEPTASILOXANE

Formula: $C_{16}H_{48}O_6Si_7$
Mol Weight: 533.16
MP (deg C): -78
FP (deg C):
BP (deg C): 270
BP pressure (mm Hg):

Property/Value	Units	Temp	Data Type	Reference
WS 5.43E-007	mg/L	25	EST	MEYLAN,WM ET AL. (1996)
logP 7.20			EXP	SANGSTER,J (1994)
VP 3.19E-003	mm Hg	25	EXT	JORDAN,TE (1954)
DC	pKa			
HL 1.63E-010	atm m3/mol	25	EST	MEYLAN,WM & HOWARD,PH (1991)
OH 2.39E-012	cm3/molc sec	25	EST	MEYLAN,WM & HOWARD,PH (1993)

CAS #: 000541-02-6 — DECAMETHYLCYCLOPENTASILOXANE

Formula: $C_{10}H_{30}O_5Si_5$
Mol Weight: 370.78
MP (deg C): -44
FP (deg C):
BP (deg C): 210
BP pressure (mm Hg):

Property/Value	Units	Temp	Data Type	Reference
WS 2.40E-001	mg/L	25	EST	MEYLAN,WM ET AL. (1996)
logP 5.20			EXP	BRUGGEMAN,WA ET AL. (1984)
VP 2.00E-001	mm Hg	25	EXT	FLANINGAM,OL (1986)
DC	pKa			
HL 4.00E-001	atm m3/mol	25	EST	VP/WSOL
OH 1.55E-012	cm3/molc sec	25	EXP	ATKINSON,R (1991)

000541-05-9 — HEXAMETHYLCYCLOTRISILOXANE

Formula: $C_6H_{18}O_3Si_3$

Mol Weight: 222.47

MP (deg C): 64.5

FP (deg C):

BP (deg C): 134

BP pressure (mm Hg):

Property/Value	Units	Temp	Data Type	Reference
WS 1.57E+000	mg/L	25	EST	MEYLAN,WM ET AL. (1996)
logP 4.47			EST	MEYLAN,WM & HOWARD,PH (1995)
VP 3.53E+000	mm Hg	25	EXT	YAWS,CL (1994A)
DC	pKa			
HL 6.65E-007	atm m3/mol	25	EST	MEYLAN,WM & HOWARD,PH (1991)
OH 8.98E-013	cm3/molc sec	25	EST	MEYLAN,WM & HOWARD,PH (1993)

000541-23-1 — ISOAMYLAMINE HYDROCHLORIDE

Formula: $C_5H_{14}ClN$

Mol Weight: 123.63

MP (deg C):

FP (deg C):

BP (deg C):

BP pressure (mm Hg):

Property/Value	Units	Temp	Data Type	Reference
WS 1.00E+006	mg/L	25	EXP	SEIDELL,A (1941)
logP -1.93			EST	MEYLAN,WM & HOWARD,PH (1995)
VP 3.99E-005	mm Hg	25	EST	NEELY,WB & BLAU,GE (1985)
DC	pKa			
HL 2.18E-012	atm m3/mol	25	EST	MEYLAN,WM & HOWARD,PH (1991)
OH 1.48E-011	cm3/molc sec	25	EST	MEYLAN,WM & HOWARD,PH (1993)

000541-25-3 — DICHLORO(2-CHLOROVINYL)ARSINE (TRANS)

Formula: $C_2H_2AsCl_3$

Mol Weight: 207.32

MP (deg C): -1.2

FP (deg C):

BP (deg C): 196.6

BP pressure (mm Hg):

Property/Value	Units	Temp	Data Type	Reference
WS 5.00E+002	mg/L		EXP	GOLDMAN,M & DACRE,JC (1989)
logP 2.51			EST	MEYLAN,WM & HOWARD,PH (1995)
VP 4.00E-001	mm Hg	25	EXP	GOLDMAN,M & DACRE,JC (1989)
DC	pKa			
HL 2.18E-004	atm m3/mol	25	EST	VP/WSOL
OH 1.30E-011	cm3/molc sec	25	EXP	MEYLAN,WM & HOWARD,PH (1993)

000541-31-1 — ISOAMYLTHIOL

Formula: $C_5H_{12}S$

Mol Weight: 104.22

MP (deg C):

FP (deg C):

BP (deg C): 120

BP pressure (mm Hg):

Property/Value	Units	Temp	Data Type	Reference
WS 5.99E+002	mg/L	25	EST	MEYLAN,WM ET AL. (1996)
logP 2.67			EST	MEYLAN,WM & HOWARD,PH (1995)
VP 2.03E+001	mm Hg	25	EXT	BOUBLIK,T ET AL. (1984)
DC	pKa			
HL 8.12E-003	atm m3/mol	25	EST	MEYLAN,WM & HOWARD,PH (1991)
OH 4.56E-011	cm3/molc sec	25	EST	MEYLAN,WM & HOWARD,PH (1993)

000541-33-3 — 1,1-DICHLOROBUTANE

Formula: $C_4H_8Cl_2$

Mol Weight: 127.01

MP (deg C):

FP (deg C):

BP (deg C): 113.8

BP pressure (mm Hg):

Property/Value	Units	Temp	Data Type	Reference
WS 5.01E+002	mg/L	25	EXP	SUZUKI,T (1991)
logP 2.74			EST	MEYLAN,WM & HOWARD,PH (1995)
VP 2.28E+001	mm Hg	25	EXP	HINE,J & MOOKERJEE,PK (1975)
DC	pKa			
HL 7.61E-003	atm m3/mol	25	EST	VP/WSOL
OH 2.07E-012	cm3/molc sec	25	EST	MEYLAN,WM & HOWARD,PH (1993)

000541-35-5 — BUTYRAMIDE

Formula: C_4H_9NO

Mol Weight: 87.12

MP (deg C): 115-116

FP (deg C):

BP (deg C): 216

BP pressure (mm Hg):

Property/Value	Units	Temp	Data Type	Reference
WS 1.91E+005	mg/L	25	EST	MEYLAN,WM ET AL. (1996)
logP -0.21			EXP	HANSCH,C & LEO,AJ (1985)
VP 4.86E-004	mm Hg	25	EXT	JONES,AH (1960)
DC -0.43	pKa	20	EXP	PERRIN,DD (1965)
HL 1.97E-008	atm m3/mol	25	EST	MEYLAN,WM & HOWARD,PH (1991)
OH 6.67E-012	cm3/molc sec	25	EST	MEYLAN,WM & HOWARD,PH (1993)

000541-41-3 — ETHYL CHLOROCARBONATE

Formula: $C_3H_5ClO_2$

Mol Weight: 108.53

MP (deg C): -81

FP (deg C):

BP (deg C): 95

BP pressure (mm Hg):

Property/Value	Units	Temp	Data Type	Reference
WS 3.21E+004	mg/L	25	EST	MEYLAN,WM ET AL. (1996)
logP 0.63			EST	MEYLAN,WM & HOWARD,PH (1995)
VP 2.24E+001	mm Hg	25	EXP	DAUBERT,TE & DANNER,RP (1991)
DC	pKa			
HL 3.12E-003	atm m3/mol	25	EST	MEYLAN,WM & HOWARD,PH (1991)
OH 1.44E-012	cm3/molc sec	25	EST	MEYLAN,WM & HOWARD,PH (1993)

000541-50-4 — ACETOACETIC ACID

Formula: $C_4H_6O_3$

Mol Weight: 102.09

MP (deg C): 36-37

FP (deg C):

BP (deg C):

BP pressure (mm Hg):

Property/Value	Units	Temp	Data Type	Reference
WS 1.00E+006	mg/L	20	EXP	YALKOWSKY,SH & DANNENFELSER,RM (1992)
logP -0.98			EST	MEYLAN,WM & HOWARD,PH (1995)
VP 2.18E-001	mm Hg	25	EST	NEELY,WB & BLAU,GE (1985)
DC 3.59	pKa	0	EXP	KORTUM,G ET AL (1961)
HL 3.69E-010	atm m3/mol	25	EST	MEYLAN,WM & HOWARD,PH (1991)
OH 1.15E-012	cm3/molc sec	25	EST	MEYLAN,WM & HOWARD,PH (1993)

CAS #: 000541-58-2				2,4-DIMETHYLTHIAZOLE

Formula: C_5H_7NS

Mol Weight: 113.18

MP (deg C):   FP (deg C):

BP (deg C): 146

BP pressure (mm Hg):

Property/Value	Units	Temp	Data Type	Reference
WS 1.75E+003	mg/L	25	EST	MEYLAN,WM ET AL. (1996)
logP 2.09			EST	MEYLAN,WM & HOWARD,PH (1995)
VP 5.08E+000	mm Hg	25	EXP	SOULIE,MA ET AL. (1975)
DC	pKa			
HL 4.67E-006	atm m3/mol	25	EST	MEYLAN,WM & HOWARD,PH (1991)
OH 4.75E-012	cm3/molc sec	25	EST	MEYLAN,WM & HOWARD,PH (1993)

CAS #: 000541-59-3				1H-PYRROLE-2,5-DIONE

Formula: $C_4H_3NO_2$

Mol Weight: 97.07

MP (deg C): 94   FP (deg C):

BP (deg C):

BP pressure (mm Hg):

Property/Value	Units	Temp	Data Type	Reference
WS 2.11E+005	mg/L	25	EST	MEYLAN,WM ET AL. (1996)
logP -0.29			EXP	HANSCH,C & LEO,AJ (1985)
VP 2.77E-006	mm Hg	25	EST	NEELY,WB & BLAU,GE (1985)
DC	pKa			
HL 6.33E-009	atm m3/mol	25	EST	MEYLAN,WM & HOWARD,PH (1991)
OH 7.76E-012	cm3/molc sec	25	EST	MEYLAN,WM & HOWARD,PH (1993)

CAS #: 000541-64-0				2-FURANMETHANAMINIUM, N,N,N-TRIMETHYL-, IODIDE

Formula: $C_8H_{14}INO$

Mol Weight: 267.11

MP (deg C):   FP (deg C):

BP (deg C):

BP pressure (mm Hg):

Property/Value	Units	Temp	Data Type	Reference
WS 1.00E+006	mg/L	25	EST	MEYLAN,WM ET AL. (1996)
logP -2.40			EXP	SANGSTER,J (1994)
VP 5.65E-008	mm Hg	25	EST	NEELY,WB & BLAU,GE (1985)
DC	pKa			
HL 2.12E-013	atm m3/mol	25	EST	MEYLAN,WM & HOWARD,PH (1991)
OH 1.13E-010	cm3/molc sec	25	EST	MEYLAN,WM & HOWARD,PH (1993)

CAS #: 000541-66-2				1,3-DIOXOLANE-4-METHANAMINIUM, N,N,N-TRIMETHYL-,

Formula: $C_7H_{16}INO_2$

Mol Weight: 273.12

MP (deg C): 158-160   FP (deg C):

BP (deg C):

BP pressure (mm Hg):

Property/Value	Units	Temp	Data Type	Reference
WS 1.00E+006	mg/L	25	EST	MEYLAN,WM ET AL. (1996)
logP -3.03			EXP	SANGSTER,J (1994)
VP 4.24E-008	mm Hg	25	EST	NEELY,WB & BLAU,GE (1985)
DC	pKa			
HL 1.06E-015	atm m3/mol	25	EST	MEYLAN,WM & HOWARD,PH (1991)
OH 3.71E-011	cm3/molc sec	25	EST	MEYLAN,WM & HOWARD,PH (1993)

CAS #: 000541-73-1				3-DICHLOROBENZENE

Formula: $C_6H_4Cl_2$

Mol Weight: 147.00

MP (deg C): -24.76   FP (deg C):

BP (deg C): 173.00

BP pressure (mm Hg):

Property/Value	Units	Temp	Data Type	Reference
WS 1.25E+002	mg/L	25	EXP	MILLER,MM ET AL. (1984)
logP 3.53			EXP	HANSCH,C ET AL. (1995)
VP 2.15E+000	mm Hg	25	EXP	DAUBERT,TE & DANNER,RP (1985)
DC	pKa			
HL 2.63E-003	atm m3/mol	25	EXP	WARNER,HP ET AL. (1987)
OH 7.20E-013	cm3/molc sec	25	EXP	ATKINSON,R (1989)

CAS #: 000541-79-7				CARBAMIC ACID, (2,2,2-TRICHLORO-1-HYDROXYETHYL)-

Formula: $C_5H_8Cl_3NO_3$

Mol Weight: 236.48

MP (deg C):   FP (deg C):

BP (deg C):

BP pressure (mm Hg):

Property/Value	Units	Temp	Data Type	Reference
WS 7.53E+002	mg/L	25	EST	MEYLAN,WM ET AL. (1996)
logP 1.84			EXP	HANSCH,C ET AL. (1995)
VP 3.25E-004	mm Hg	25	EST	NEELY,WB & BLAU,GE (1985)
DC	pKa			
HL 4.88E-012	atm m3/mol	25	EST	MEYLAN,WM & HOWARD,PH (1991)
OH 3.72E-012	cm3/molc sec	25	EST	MEYLAN,WM & HOWARD,PH (1993)

CAS #: 000541-85-5				ETHYL AMYLKETONE

Formula: $C_8H_{16}O$

Mol Weight: 128.22

MP (deg C):   FP (deg C):

BP (deg C): 157-162

BP pressure (mm Hg):

Property/Value	Units	Temp	Data Type	Reference
WS 1.37E+003	mg/L	25	EST	MEYLAN,WM ET AL. (1996)
logP 2.15			EST	MEYLAN,WM & HOWARD,PH (1995)
VP 5.03E+000	mm Hg	25	EST	NEELY,WB & BLAU,GE (1985)
DC	pKa			
HL 2.04E-004	atm m3/mol	25	EST	MEYLAN,WM & HOWARD,PH (1991)
OH 1.29E-011	cm3/molc sec	25	EST	MEYLAN,WM & HOWARD,PH (1993)

CAS #: 000541-88-8				CHLOROACETIC ANHYDRIDE

Formula: $C_4H_4Cl_2O_3$

Mol Weight: 170.98

MP (deg C): 46   FP (deg C):

BP (deg C): 203

BP pressure (mm Hg):

Property/Value	Units	Temp	Data Type	Reference
WS 6.97E+004	mg/L	25	EST	MEYLAN,WM ET AL. (1996)
logP -0.07			EST	MEYLAN,WM & HOWARD,PH (1995)
VP 3.72E-002	mm Hg	25	EXT	PERRY,RH & GREEN,D (1984)
DC	pKa			
HL 4.42E-006	atm m3/mol	25	EST	MEYLAN,WM & HOWARD,PH (1991)
OH 5.25E-013	cm3/molc sec	25	EST	MEYLAN,WM & HOWARD,PH (1993)

000542-16-5 — ANILINE SULFATE

Formula: $C_{12}H_{16}N_2O_4S$

Mol Weight: 284.34

MP (deg C):

FP (deg C):

BP (deg C):

BP pressure (mm Hg):

Property/Value	Units	Temp	Data Type	Reference
WS 6.60E+004	mg/L	15	EXP	SEIDELL,A (1941)
logP -0.59			EST	MEYLAN,WM & HOWARD,PH (1995)
VP 9.70E-017	mm Hg	25	EST	NEELY,WB & BLAU,GE (1985)
DC	pKa			
HL 1.22E-022	atm m3/mol	25	EST	MEYLAN,WM & HOWARD,PH (1991)
OH 8.34E-013	cm3/molc sec	25	EST	MEYLAN,WM & HOWARD,PH (1993)

000542-18-7 — CHLOROCYCLOHEXANE

Formula: $C_6H_{11}Cl$

Mol Weight: 118.61

MP (deg C): -43.9

FP (deg C):

BP (deg C): 143

BP pressure (mm Hg):

Property/Value	Units	Temp	Data Type	Reference
WS 5.00E+002	mg/L	25	EXP	CHEM INSPECT TEST INST (1992)
logP 3.36			EST	MEYLAN,WM & HOWARD,PH (1995)
VP 6.73E+000	mm Hg	25	EXT	OHE,S (1989)
DC	pKa			
HL 1.49E-002	atm m3/mol	25	EST	MEYLAN,WM & HOWARD,PH (1991)
OH 6.18E-012	cm3/molc sec	25	EST	MEYLAN,WM & HOWARD,PH (1993)

000542-28-9 — D-VALEROLACTONE

Formula: $C_5H_4O_2$

Mol Weight: 96.09

MP (deg C): -12.5

FP (deg C):

BP (deg C): 219

BP pressure (mm Hg):

Property/Value	Units	Temp	Data Type	Reference
WS 2.33E+005	mg/L	25	EST	MEYLAN,WM ET AL. (1996)
logP -0.35			EXP	HANSCH,C & LEO,AJ (1985)
VP 3.75E-001	mm Hg	25	EST	NEELY,WB & BLAU,GE (1985)
DC	pKa			
HL 1.36E-004	atm m3/mol	25	EST	MEYLAN,WM & HOWARD,PH (1991)
OH 5.51E-012	cm3/molc sec	25	EST	MEYLAN,WM & HOWARD,PH (1993)

000542-54-1 — 4-METHYLPENTANITRILE

Formula: $C_6H_{11}N$

Mol Weight: 97.16

MP (deg C): -51

FP (deg C):

BP (deg C): 155-156

BP pressure (mm Hg):

Property/Value	Units	Temp	Data Type	Reference
WS 3.13E+003	mg/L	25	EST	MEYLAN,WM ET AL. (1996)
logP 1.54			EXP	TANII,H & HASHIMOTO,K (1984)
VP 3.69E+000	mm Hg	25	EST	NEELY,WB & BLAU,GE (1985)
DC	pKa			
HL 9.49E-005	atm m3/mol	25	EST	MEYLAN,WM & HOWARD,PH (1991)
OH 3.08E-012	cm3/molc sec	25	EST	MEYLAN,WM & HOWARD,PH (1993)

000542-55-2 — ISOBUTYL FORMATE

Formula: $C_5H_{10}O_2$

Mol Weight: 102.13

MP (deg C): -95

FP (deg C):

BP (deg C): 98

BP pressure (mm Hg):

Property/Value	Units	Temp	Data Type	Reference
WS 1.02E+004	mg/L	25	EXP	SUZUKI,T (1991)
logP 1.23			EST	MEYLAN,WM & HOWARD,PH (1995)
VP 4.01E+001	mm Hg	25	EXP	RIDDICK,JA ET AL. (1986)
DC	pKa			
HL 5.28E-004	atm m3/mol	25	EST	VP/WSOL
OH 4.56E-012	cm3/molc sec	25	EST	MEYLAN,WM & HOWARD,PH (1993)

000542-69-8 — N-BUTYL IODIDE

Formula: C_4H_9I

Mol Weight: 184.02

MP (deg C): -103

FP (deg C):

BP (deg C): 130.4

BP pressure (mm Hg):

Property/Value	Units	Temp	Data Type	Reference
WS 2.02E+002	mg/L	20	EXP	YALKOWSKY,SH & DANNENFELSER,RM (1992)
logP 3.06			EST	MEYLAN,WM & HOWARD,PH (1995)
VP 1.39E+001	mm Hg	25	EXP	RIDDICK,JA ET AL. (1986)
DC	pKa			
HL 1.67E-002	atm m3/mol	25	EST	VP/WSOL
OH 2.45E-012	cm3/molc sec	25	EST	MEYLAN,WM & HOWARD,PH (1993)

000542-75-6 — TRANS-1,3-DICHLOROPROPENE

Formula: $C_3H_4Cl_2$

Mol Weight: 110.97

MP (deg C): -48 EST

FP (deg C):

BP (deg C): 112

BP pressure (mm Hg):

Property/Value	Units	Temp	Data Type	Reference
WS 2.80E+003	mg/L	20	EXP	DILLING,WL (1977)
logP 2.29			EST	MEYLAN,WM & HOWARD,PH (1995)
VP 3.40E+001	mm Hg	25	EXP	DILLING,WL (1977)
DC	pKa			
HL 3.55E-003	atm m3/mol	20	EXP	WARNER,HP ET AL. (1987)
OH 1.43E-011	cm3/molc sec	25	EXP	ATKINSON,R (1989)

000542-76-7 — 3-CHLOROPROPIONITRILE

Formula: C_3H_4ClN

Mol Weight: 89.53

MP (deg C): -51

FP (deg C):

BP (deg C): 176

BP pressure (mm Hg):

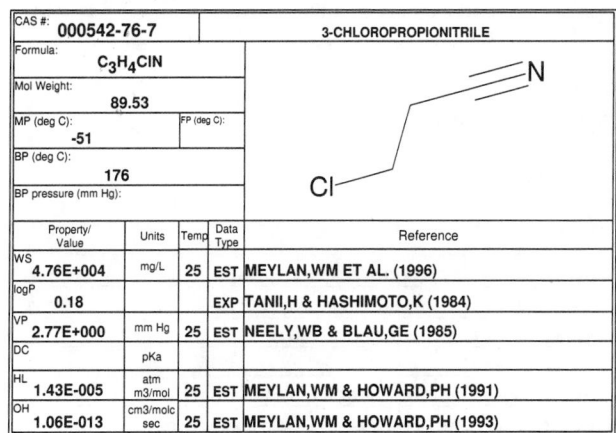

Property/Value	Units	Temp	Data Type	Reference
WS 4.76E+004	mg/L	25	EST	MEYLAN,WM ET AL. (1996)
logP 0.18			EXP	TANII,H & HASHIMOTO,K (1984)
VP 2.77E+000	mm Hg	25	EST	NEELY,WB & BLAU,GE (1985)
DC	pKa			
HL 1.43E-005	atm m3/mol	25	EST	MEYLAN,WM & HOWARD,PH (1991)
OH 1.06E-013	cm3/molc sec	25	EST	MEYLAN,WM & HOWARD,PH (1993)

CAS #: 000542-85-8	ETHYL ISOTHIOCYANATE
Formula: C_3H_7NS	
Mol Weight: 89.16	
MP (deg C): -6	FP (deg C):
BP (deg C): 130-132	
BP pressure (mm Hg):	

Property/Value	Units	Temp	Data Type	Reference
WS 7.02E+003	mg/L	25	EST	MEYLAN,WM ET AL. (1996)
logP 1.47			EXP	HANSCH,C & LEO,AJ (1985)
VP 1.14E+001	mm Hg	25	EXP	BOUBLIK,T ET AL. (1984)
DC	pKa			
HL 4.13E-003	atm m3/mol	25	EST	MEYLAN,WM & HOWARD,PH (1991)
OH 1.10E-012	cm3/molc sec	25	EST	MEYLAN,WM & HOWARD,PH (1993)

CAS #: 000542-88-1	BIS(CHLOROMETHYL) ETHER
Formula: $C_2H_4Cl_2O$	
Mol Weight: 114.96	
MP (deg C): -41.5	FP (deg C):
BP (deg C): 105	
BP pressure (mm Hg):	

Property/Value	Units	Temp	Data Type	Reference
WS 2.20E+004	mg/L	25	EXP	MABEY,WR ET AL. (1981)
logP 0.57			EST	MEYLAN,WM & HOWARD,PH (1995)
VP 3.00E+001	mm Hg	22	EXP	CALLAHAN,MA ET AL. (1979A)
DC	pKa			
HL 2.06E-004	atm m3/mol	25	EST	VP/WSOL
OH 7.10E-013	cm3/molc sec	25	EST	MEYLAN,WM & HOWARD,PH (1993)

CAS #: 000542-92-7	CYCLOPENTADIENE
Formula: C_5H_6	
Mol Weight: 66.10	
MP (deg C): -85	FP (deg C):
BP (deg C): 41.5-42	
BP pressure (mm Hg):	

Property/Value	Units	Temp	Data Type	Reference
WS 4.71E+002	mg/L	25	EST	MEYLAN,WM ET AL. (1996)
logP 2.25			EST	MEYLAN,WM & HOWARD,PH (1995)
VP 4.35E+002	mm Hg	25	EXP	DAUBERT,TE & DANNER,RP (1989)
DC	pKa			
HL 6.36E-002	atm m3/mol	25	EST	MEYLAN,WM & HOWARD,PH (1991)
OH 1.43E-010	cm3/molc sec	25	EST	MEYLAN,WM & HOWARD,PH (1993)

CAS #: 000543-20-4	SUCCINYL CHLORIDE
Formula: $C_4H_4Cl_2O_2$	
Mol Weight: 154.98	
MP (deg C): 17	FP (deg C):
BP (deg C): 192-193	
BP pressure (mm Hg):	

Property/Value	Units	Temp	Data Type	Reference
WS 9.04E+005	mg/L	25	EST	MEYLAN,WM ET AL. (1996)
logP -1.29			EST	MEYLAN,WM & HOWARD,PH (1995)
VP 3.70E-001	mm Hg	25	EXT	PERRY,RH & GREEN,D (1984)
DC	pKa			
HL 7.97E-006	atm m3/mol	25	EST	MEYLAN,WM & HOWARD,PH (1991)
OH 4.88E-013	cm3/molc sec	25	EST	MEYLAN,WM & HOWARD,PH (1993)

CAS #: 000543-29-3	ISOBUTYL NITRATE
Formula: $C_4H_9NO_3$	
Mol Weight: 119.12	
MP (deg C):	FP (deg C):
BP (deg C): 123.4	
BP pressure (mm Hg):	

Property/Value	Units	Temp	Data Type	Reference
WS 1.46E+003	mg/L	25	EST	MEYLAN,WM ET AL. (1996)
logP 2.16			EST	MEYLAN,WM & HOWARD,PH (1995)
VP 8.30E+000	mm Hg	25	EXP	BOUBLIK,T ET AL. (1984)
DC	pKa			
HL 4.64E-004	atm m3/mol	25	EST	MEYLAN,WM & HOWARD,PH (1991)
OH 8.37E-013	cm3/molc sec	25	EST	MEYLAN,WM & HOWARD,PH (1993)

CAS #: 000543-49-7	2-HEPTANOL
Formula: $C_7H_{16}O$	
Mol Weight: 116.20	
MP (deg C):	FP (deg C):
BP (deg C): 158-160	
BP pressure (mm Hg):	

Property/Value	Units	Temp	Data Type	Reference
WS 3.38E+003	mg/L	25	EXP	SUZUKI,T (1991)
logP 2.31			EXP	HANSCH,C ET AL. (1995)
VP 1.23E+000	mm Hg	25	EXP	DAUBERT,TE & DANNER,RP (1989)
DC	pKa			
HL 5.56E-005	atm m3/mol	25	EST	VP/WSOL
OH 1.42E-011	cm3/molc sec	25	EST	MEYLAN,WM & HOWARD,PH (1993)

CAS #: 000543-59-9	1-CHLOROPENTANE
Formula: $C_5H_{11}Cl$	
Mol Weight: 106.60	
MP (deg C): -99	FP (deg C):
BP (deg C): 107.8	
BP pressure (mm Hg):	

Property/Value	Units	Temp	Data Type	Reference
WS 1.97E+002	mg/L	25	EXP	KUHNE,R ET AL. (1995)
logP 2.73			EXP	SUZUKI,T (1991)
VP 3.29E+001	mm Hg	25	EXP	YAWS,CL (1994A)
DC	pKa			
HL 2.38E-002	atm m3/mol	25	EXP	LEIGHTON,DTJR & CALO,JM (1981)
OH 3.36E-012	cm3/molc sec	22	EXP	MARKERT,R & NIELSON,OJ (1992A)

CAS #: 000543-67-9	N-PROPYLNITRITE
Formula: $C_3H_7NO_2$	
Mol Weight: 89.09	
MP (deg C):	FP (deg C):
BP (deg C): 46-48	
BP pressure (mm Hg):	

Property/Value	Units	Temp	Data Type	Reference
WS 3.20E+003	mg/L	25	EST	MEYLAN,WM ET AL. (1996)
logP 1.86			EST	MEYLAN,WM & HOWARD,PH (1995)
VP 2.26E+002	mm Hg	25	EST	NEELY,WB & BLAU,GE (1985)
DC	pKa			
HL 1.15E-004	atm m3/mol	25	EST	MEYLAN,WM & HOWARD,PH (1991)
OH 2.38E-012	cm3/molc sec	25	EXP	ATKINSON,R (1989)

000543-87-3 — 1-BUTANOL, 3-METHYL-, NITRATE

Formula:	$C_5H_{11}NO_3$
Mol Weight:	133.15
MP (deg C):	
FP (deg C):	
BP (deg C):	147-148
BP pressure (mm Hg):	

Property/Value	Units	Temp	Data Type	Reference
WS 3.36E+002	mg/L	25	EST	MEYLAN,WM ET AL. (1996)
logP 2.84			EXP	HANSCH,C ET AL. (1995)
VP 5.44E+000	mm Hg	25	EST	NEELY,WB & BLAU,GE (1985)
DC	pKa			
HL 6.16E-004	atm m3/mol	25	EST	MEYLAN,WM & HOWARD,PH (1991)
OH 3.36E-012	cm3/molc sec	25	EST	MEYLAN,WM & HOWARD,PH (1993)

000544-01-4 — DIISOPENTYL ETHER

Formula:	$C_{10}H_{22}O$
Mol Weight:	158.29
MP (deg C):	
FP (deg C):	
BP (deg C):	172
BP pressure (mm Hg):	

Property/Value	Units	Temp	Data Type	Reference
WS 2.00E+002	mg/L	25	EXP	RIDDICK,JA ET AL. (1986)
logP 4.25			EXP	CHEM INSPECT TEST INST (1992)
VP 1.40E+000	mm Hg	25	EXP	RIDDICK,JA ET AL. (1986)
DC	pKa			
HL 1.46E-003	atm m3/mol	25	EST	VP/WSOL
OH 2.84E-011	cm3/molc sec	25	EST	MEYLAN,WM & HOWARD,PH (1993)

000544-02-5 — DIISOAMYL SULFIDE

Formula:	$C_{10}H_{22}S$
Mol Weight:	174.35
MP (deg C):	-74.6
FP (deg C):	
BP (deg C):	211
BP pressure (mm Hg):	

Property/Value	Units	Temp	Data Type	Reference
WS 5.67E+000	mg/L	25	EST	MEYLAN,WM ET AL. (1996)
logP 4.70			EST	MEYLAN,WM & HOWARD,PH (1995)
VP 3.19E-001	mm Hg	25	EXT	PERRY,RH & GREEN,D (1984)
DC	pKa			
HL 7.58E-003	atm m3/mol	25	EST	MEYLAN,WM & HOWARD,PH (1991)
OH 2.79E-011	cm3/molc sec	25	EST	MEYLAN,WM & HOWARD,PH (1993)

000544-10-5 — 1-CHLOROHEXANE

Formula:	$C_6H_{13}Cl$
Mol Weight:	120.62
MP (deg C):	-94
FP (deg C):	
BP (deg C):	135
BP pressure (mm Hg):	

Property/Value	Units	Temp	Data Type	Reference
WS 9.10E+001	mg/L	25	EXP	KUHNE,R ET AL. (1995)
logP 3.54			EST	MEYLAN,WM & HOWARD,PH (1995)
VP 9.37E+000	mm Hg	25	EXP	OHE,S (1976)
DC	pKa			
HL 2.42E-002	atm m3/mol	25	EXP	LEIGHTON,DTJR & CALO,JM (1981)
OH 3.80E-012	cm3/molc sec	25	EXP	ATKINSON,R (1989)

000544-12-7 — 3-HEXENE-1-OL

Formula:	$C_6H_{12}O$
Mol Weight:	100.16
MP (deg C):	
FP (deg C):	
BP (deg C):	156-157
BP pressure (mm Hg):	

Property/Value	Units	Temp	Data Type	Reference
WS 1.60E+004	mg/L	25	EST	MEYLAN,WM ET AL. (1996)
logP 1.61			EST	MEYLAN,WM & HOWARD,PH (1995)
VP 5.54E-001	mm Hg	25	EST	NEELY,WB & BLAU,GE (1985)
DC	pKa			
HL 1.55E-005	atm m3/mol	25	EST	MEYLAN,WM & HOWARD,PH (1991)
OH 6.28E-011	cm3/molc sec	25	EST	MEYLAN,WM & HOWARD,PH (1993)

000544-13-8 — PENTANEDINITRILE

Formula:	$C_5H_6N_2$
Mol Weight:	94.12
MP (deg C):	-29
FP (deg C):	
BP (deg C):	286
BP pressure (mm Hg):	

Property/Value	Units	Temp	Data Type	Reference
WS 2.72E+005	mg/L	25	EST	MEYLAN,WM ET AL. (1996)
logP -0.72			EXP	TANII,H & HASHIMOTO,K (1985)
VP 8.24E-003	mm Hg	25	EXP	YAWS,CL (1994A)
DC	pKa			
HL 2.23E-008	atm m3/mol	25	EST	MEYLAN,WM & HOWARD,PH (1991)
OH 4.50E-013	cm3/molc sec	25	EST	MEYLAN,WM & HOWARD,PH (1993)

000544-16-1 — 1-BUTYLNITRITE

Formula:	$C_4H_9NO_2$
Mol Weight:	103.12
MP (deg C):	
FP (deg C):	
BP (deg C):	78.2
BP pressure (mm Hg):	

Property/Value	Units	Temp	Data Type	Reference
WS 1.12E+003	mg/L	25	EST	MEYLAN,WM ET AL. (1996)
logP 2.35			EST	MEYLAN,WM & HOWARD,PH (1995)
VP 8.13E+001	mm Hg	25	EST	NEELY,WB & BLAU,GE (1985)
DC	pKa			
HL 1.53E-004	atm m3/mol	25	EST	MEYLAN,WM & HOWARD,PH (1991)
OH 3.80E-012	cm3/molc sec	25	EXP	ATKINSON,R (1989)

000544-25-2 — CYCLOHEPTATRIENE

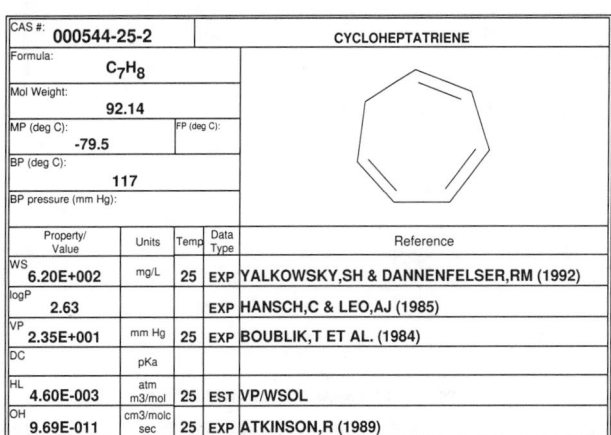

Formula:	C_7H_8
Mol Weight:	92.14
MP (deg C):	-79.5
FP (deg C):	
BP (deg C):	117
BP pressure (mm Hg):	

Property/Value	Units	Temp	Data Type	Reference
WS 6.20E+002	mg/L	25	EXP	YALKOWSKY,SH & DANNENFELSER,RM (1992)
logP 2.63			EXP	HANSCH,C & LEO,AJ (1985)
VP 2.35E+001	mm Hg	25	EXP	BOUBLIK,T ET AL. (1984)
DC	pKa			
HL 4.60E-003	atm m3/mol	25	EST	VP/WSOL
OH 9.69E-011	cm3/molc sec	25	EXP	ATKINSON,R (1989)

CAS #: 000544-40-1				5-THIANONANE
Formula: $C_8H_{18}S$				
Mol Weight: 146.30				
MP (deg C): -79.7		FP (deg C):		
BP (deg C): 182				
BP pressure (mm Hg):				

Property/Value	Units	Temp	Data Type	Reference
WS 3.94E+001	mg/L	25	EST	MEYLAN,WM ET AL. (1996)
logP 3.87			EST	MEYLAN,WM & HOWARD,PH (1995)
VP 6.09E+001	mm Hg	25	EXP	BOUBLIK,T ET AL. (1984)
DC	pKa			
HL 4.30E-003	atm m3/mol	25	EST	MEYLAN,WM & HOWARD,PH (1991)
OH 2.51E-011	cm3/molc sec	25	EST	MEYLAN,WM & HOWARD,PH (1993)

CAS #: 000544-63-8				TETRADECANOIC ACID
Formula: $C_{14}H_{28}O_2$				
Mol Weight: 228.38				
MP (deg C): 58.5		FP (deg C):		
BP (deg C): 250.5				
BP pressure (mm Hg): 1.00E+002				

Property/Value	Units	Temp	Data Type	Reference
WS 2.20E+001	mg/L	30	EXP	YALKOWSKY,SH & DANNENFELSER,RM (1992)
logP 6.11			EXP	SANGSTER,J (1993)
VP 3.18E-006	mm Hg	25	EXT	YAWS,CL (1994B)
DC	pKa			
HL 1.64E-005	atm m3/mol	25	EST	MEYLAN,WM & HOWARD,PH (1991)
OH 1.68E-011	cm3/molc sec	25	EST	MEYLAN,WM & HOWARD,PH (1993)

CAS #: 000544-76-3				N-HEXADECANE
Formula: $C_{16}H_{34}$				
Mol Weight: 226.45				
MP (deg C): 18.14		FP (deg C): 93		
BP (deg C): 286.5				
BP pressure (mm Hg):				

Property/Value	Units	Temp	Data Type	Reference
WS 9.00E-004	mg/L	25	EXP	SUTTON,C & CALDER,JA (1974)
logP 8.25			EST	COATES,M ET AL. (1985)
VP 1.43E-003	mm Hg	25	EXP	DAUBERT,TE & DANNER,RP (1989)
DC	pKa			
HL 2.28E-001	atm m3/mol	25	EST	YAWS,C ET AL. (1991)
OH 2.49E-011	cm3/molc sec	25	EXP	ATKINSON,R (1989)

CAS #: 000544-77-4				1-IODOHEXADECANE
Formula: $C_{16}H_{33}I$				
Mol Weight: 352.35				
MP (deg C): 22		FP (deg C):		
BP (deg C): 212				
BP pressure (mm Hg): 1.50E+001				

Property/Value	Units	Temp	Data Type	Reference
WS 1.34E-004	mg/L	25	EST	MEYLAN,WM ET AL. (1996)
logP 8.96			EST	MEYLAN,WM & HOWARD,PH (1995)
VP 1.65E-005	mm Hg	25	EXT	LI,JCM & ROSSINI,FD (1953)
DC	pKa			
HL 3.85E-001	atm m3/mol	25	EST	MEYLAN,WM & HOWARD,PH (1991)
OH 1.94E-011	cm3/molc sec	25	EST	MEYLAN,WM & HOWARD,PH (1993)

CAS #: 000544-85-4				DOTRIACONTANE
Formula: $C_{32}H_{66}$				
Mol Weight: 450.88				
MP (deg C): 69.7		FP (deg C):		
BP (deg C): 467				
BP pressure (mm Hg):				

Property/Value	Units	Temp	Data Type	Reference
WS 8.29E-012	mg/L	25	EST	MEYLAN,WM ET AL. (1996)
logP 16.06			EST	MEYLAN,WM & HOWARD,PH (1995)
VP 8.49E-018	mm Hg	25	EXT	ZWOLINSKI,BJ & WILHOIT,RC (1971)
DC	pKa			
HL 2.70E+003	atm m3/mol	25	EST	MEYLAN,WM & HOWARD,PH (1991)
OH 4.22E-011	cm3/molc sec	25	EST	MEYLAN,WM & HOWARD,PH (1993)

CAS #: 000545-06-2				TRICHLOROACETONITRILE
Formula: C_2Cl_3N				
Mol Weight: 144.39				
MP (deg C): -42		FP (deg C):		
BP (deg C): 85.7				
BP pressure (mm Hg):				

Property/Value	Units	Temp	Data Type	Reference
WS 7.15E+002	mg/L	25	EST	MEYLAN,WM ET AL. (1996)
logP 2.09			EXP	HANSCH,C & LEO,AJ (1985)
VP 7.41E+001	mm Hg	25	EXP	BOUBLIK,T ET AL. (1984)
DC	pKa			
HL 1.34E-006	atm m3/mol	25	EST	MEYLAN,WM & HOWARD,PH (1991)
OH 0.00E+000	cm3/molc sec	25	EST	MEYLAN,WM & HOWARD,PH (1993)

CAS #: 000545-26-6				GITOXIGENIN
Formula: $C_{23}H_{34}O_5$				
Mol Weight: 390.52				
MP (deg C): 234		FP (deg C):		
BP (deg C):				
BP pressure (mm Hg):				

Property/Value	Units	Temp	Data Type	Reference
WS 1.51E+002	mg/L	25	EST	MEYLAN,WM ET AL. (1996)
logP 1.60			EXP	HANSCH,C & LEO,AJ (1985)
VP 6.69E-015	mm Hg	25	EST	NEELY,WB & BLAU,GE (1985)
DC	pKa			
HL 2.29E-011	atm m3/mol	25	EST	MEYLAN,WM & HOWARD,PH (1991)
OH 6.80E-011	cm3/molc sec	25	EST	MEYLAN,WM & HOWARD,PH (1993)

CAS #: 000545-27-7				CARD-20(22)-ENOLIDE, 3-[(2,6-DIDEOXY-BETA-D-RIBO
Formula: $C_{29}H_{44}O_8$				
Mol Weight: 520.67				
MP (deg C):		FP (deg C):		
BP (deg C):				
BP pressure (mm Hg):				

Property/Value	Units	Temp	Data Type	Reference
WS 1.62E+001	mg/L	25	EST	MEYLAN,WM ET AL. (1996)
logP 1.77			EXP	SANGSTER,J (1993)
VP 4.06E-020	mm Hg	25	EST	NEELY,WB & BLAU,GE (1985)
DC	pKa			
HL 9.89E-017	atm m3/mol	25	EST	MEYLAN,WM & HOWARD,PH (1991)
OH 1.02E-010	cm3/molc sec	25	EST	MEYLAN,WM & HOWARD,PH (1993)

CAS #: 000545-55-1				TEPA

Formula: $C_6H_{12}N_3OP$

Mol Weight: 173.16

MP (deg C): 41   FP (deg C):

BP (deg C): 90-91

BP pressure (mm Hg): 2.30E+001

Property/ Value	Units	Temp	Data Type	Reference
WS 2.01E+005	mg/L	25	EST	MEYLAN,WM ET AL. (1996)
logP -0.62			EXP	SOSNOVSKY,G ET AL. (1985)
VP 6.30E-003	mm Hg	25	EST	NEELY,WB & BLAU,GE (1985)
DC	pKa			
HL 6.11E-013	atm m3/mol	25	EST	MEYLAN,WM & HOWARD,PH (1991)
OH 1.93E-011	cm3/molc sec	25	EST	MEYLAN,WM & HOWARD,PH (1993)

CAS #: 000545-93-7				BARBITURIC ACID,5-IPR,5(2BR ALLYL)

Formula: $C_{10}H_{13}BrN_2O_3$

Mol Weight: 289.14

MP (deg C): 177-179   FP (deg C):

BP (deg C):

BP pressure (mm Hg):

Property/ Value	Units	Temp	Data Type	Reference
WS 7.89E+002	mg/L	25	EST	MEYLAN,WM ET AL. (1996)
logP 1.47			EXP	HANSCH,C & LEO,AJ (1985)
VP 4.44E-012	mm Hg	25	EST	NEELY,WB & BLAU,GE (1985)
DC	pKa			
HL 9.44E-014	atm m3/mol	25	EST	MEYLAN,WM & HOWARD,PH (1991)
OH 2.69E-011	cm3/molc sec	25	EST	MEYLAN,WM & HOWARD,PH (1993)

CAS #: 000546-43-0				ALANTOLACTONE

Formula: $C_{15}H_{20}O_2$

Mol Weight: 232.33

MP (deg C): 78-79   FP (deg C):

BP (deg C): 275

BP pressure (mm Hg):

Property/ Value	Units	Temp	Data Type	Reference
WS 3.84E+001	mg/L	25	EST	MEYLAN,WM ET AL. (1996)
logP 3.38			EXP	SANGSTER,J (1993)
VP 3.16E-005	mm Hg	25	EST	NEELY,WB & BLAU,GE (1985)
DC	pKa			
HL 2.19E-004	atm m3/mol	25	EST	MEYLAN,WM & HOWARD,PH (1991)
OH 1.17E-010	cm3/molc sec	25	EST	MEYLAN,WM & HOWARD,PH (1993)

CAS #: 000546-80-5				THUJONE

Formula: $C_{10}H_{16}O$

Mol Weight: 152.24

MP (deg C):   FP (deg C):

BP (deg C):

BP pressure (mm Hg):

Property/ Value	Units	Temp	Data Type	Reference
WS 4.08E+002	mg/L	25	EST	MEYLAN,WM ET AL. (1996)
logP 2.65			EST	MEYLAN,WM & HOWARD,PH (1995)
VP 4.12E-001	mm Hg	25	EXT	PERRY,RH & GREEN,D (1984)
DC	pKa			
HL 7.00E-005	atm m3/mol	25	EST	MEYLAN,WM & HOWARD,PH (1991)
OH 1.07E-011	cm3/molc sec	25	EST	MEYLAN,WM & HOWARD,PH (1993)

CAS #: 000546-88-3				ACETAMIDE, N-HYDROXY-

Formula: $C_2H_5NO_2$

Mol Weight: 75.07

MP (deg C): 89-92   FP (deg C):

BP (deg C):

BP pressure (mm Hg):

Property/ Value	Units	Temp	Data Type	Reference
WS 1.00E+006	mg/L	25	EST	MEYLAN,WM ET AL. (1996)
logP -1.59			EXP	HANSCH,C ET AL. (1995)
VP 9.39E-004	mm Hg	25	EST	NEELY,WB & BLAU,GE (1985)
DC 8.70	pKa	25	EXP	SERJEANT,EP & DEMPSEY,B (1979)
HL 2.24E-011	atm m3/mol	25	EST	MEYLAN,WM & HOWARD,PH (1991)
OH 5.60E-012	cm3/molc sec	25	EST	MEYLAN,WM & HOWARD,PH (1993)

CAS #: 000547-52-4				N1-(4-SO2NH2-PH)SULFANILAMIDE

Formula: $C_{12}H_{13}N_3O_4S_2$

Mol Weight: 327.38

MP (deg C): 133-134   FP (deg C):

BP (deg C):

BP pressure (mm Hg):

Property/ Value	Units	Temp	Data Type	Reference
WS 4.26E+003	mg/L	25	EST	MEYLAN,WM ET AL. (1996)
logP 0.35			EXP	HANSCH,C & LEO,AJ (1985)
VP 3.24E-011	mm Hg	25	EST	NEELY,WB & BLAU,GE (1985)
DC	pKa			
HL 6.45E-015	atm m3/mol	25	EST	MEYLAN,WM & HOWARD,PH (1991)
OH 3.21E-011	cm3/molc sec	25	EST	MEYLAN,WM & HOWARD,PH (1993)

CAS #: 000547-57-9				ACID ORANGE 6

Formula: $C_{12}H_9N_2NaO_5S$

Mol Weight: 316.27

MP (deg C):   FP (deg C):

BP (deg C):

BP pressure (mm Hg):

Property/ Value	Units	Temp	Data Type	Reference
WS 4.77E+003	mg/L	25	EST	MEYLAN,WM ET AL. (1996)
logP 0.69			EST	MEYLAN,WM & HOWARD,PH (1995)
VP 1.30E-012	mm Hg	25	EST	NEELY,WB & BLAU,GE (1985)
DC	pKa			
HL 7.44E-020	atm m3/mol	25	EST	MEYLAN,WM & HOWARD,PH (1991)
OH 2.00E-010	cm3/molc sec	25	EST	MEYLAN,WM & HOWARD,PH (1993)

CAS #: 000547-58-0				METHYL ORANGE

Formula: $C_{14}H_{14}N_3NaO_3S$

Mol Weight: 327.34

MP (deg C):   FP (deg C):

BP (deg C):

BP pressure (mm Hg):

Property/ Value	Units	Temp	Data Type	Reference
WS 2.00E+002	mg/L	25	EXP	DEHN,WM (1917)
logP -0.66			EST	MEYLAN,WM & HOWARD,PH (1995)
VP 1.13E-015	mm Hg	25	EST	NEELY,WB & BLAU,GE (1985)
DC	pKa			
HL	atm m3/mol			
OH 1.49E-010	cm3/molc sec	25	EST	MEYLAN,WM & HOWARD,PH (1993)

CAS #: 000547-63-7	METHYL ISOBUTYRATE

Formula: $C_5H_{10}O_2$

Mol Weight: 102.13

MP (deg C): -84.7   FP (deg C):

BP (deg C): 92.5

BP pressure (mm Hg):

Property/ Value	Units	Temp	Data Type	Reference
WS 9.27E+003	mg/L	25	EST	MEYLAN,WM ET AL. (1996)
logP 1.28			EST	MEYLAN,WM & HOWARD,PH (1995)
VP 4.93E+001	mm Hg	25	EXP	PERRY,RH & GREEN,D (1984)
DC	pKa			
HL 3.09E-004	atm m3/mol	25	EST	MEYLAN,WM & HOWARD,PH (1991)
OH 1.99E-012	cm3/molc sec	25	EST	MEYLAN,WM & HOWARD,PH (1993)

CAS #: 000547-64-8	2-HYDROXYPROPANOIC ACID, METHYL ESTER

Formula: $C_4H_8O_3$

Mol Weight: 104.11

MP (deg C): -66   FP (deg C):

BP (deg C): 145

BP pressure (mm Hg):

Property/ Value	Units	Temp	Data Type	Reference
WS 1.00E+006	mg/L	25	EXP	DEAN,JA (1985)
logP -0.67			EST	MEYLAN,WM & HOWARD,PH (1995)
VP 1.85E+000	mm Hg	25	EST	NEELY,WB & BLAU,GE (1985)
DC	pKa			
HL 8.51E-009	atm m3/mol	25	EST	MEYLAN,WM & HOWARD,PH (1991)
OH 4.38E-013	cm3/molc sec	25	EST	ATKINSON,R (1988)

CAS #: 000547-75-1	HYOCHOLIC ACID

Formula: $C_{24}H_{40}O_5$

Mol Weight: 408.58

MP (deg C):   FP (deg C):

BP (deg C):

BP pressure (mm Hg):

Property/ Value	Units	Temp	Data Type	Reference
WS 2.73E+001	mg/L	25	EST	MEYLAN,WM ET AL. (1996)
logP 2.80			EXP	RODA,A ET AL. (1990)
VP 3.65E-015	mm Hg	25	EST	NEELY,WB & BLAU,GE (1985)
DC	pKa			
HL 5.16E-013	atm m3/mol	25	EST	MEYLAN,WM & HOWARD,PH (1991)
OH 5.37E-011	cm3/molc sec	25	EST	MEYLAN,WM & HOWARD,PH (1993)

CAS #: 000548-62-9	CRYSTAL VIOLET (BASIC VIOLET 3)

Formula: $C_{25}H_{30}ClN_3$

Mol Weight: 407.99

MP (deg C): 215 dec   FP (deg C):

BP (deg C):

BP pressure (mm Hg):

Property/ Value	Units	Temp	Data Type	Reference
WS 1.00E+003	mg/L	25	EXP	BAUGHMAN,GL ET AL. (1993)
logP 0.51			EXP	TSAI,RS ET AL. (1991)
VP 1.93E-014	mm Hg	25	EST	NEELY,WB & BLAU,GE (1985)
DC	pKa			
HL 3.06E-016	atm m3/mol	25	EST	MEYLAN,WM & HOWARD,PH (1991)
OH 4.43E-010	cm3/molc sec	25	EST	MEYLAN,WM & HOWARD,PH (1993)

CAS #: 000548-73-2	DROPERIDOL

Formula: $C_{22}H_{22}FN_3O_2$

Mol Weight: 379.44

MP (deg C): 145-146.5   FP (deg C):

BP (deg C):

BP pressure (mm Hg):

Property/ Value	Units	Temp	Data Type	Reference
WS 4.21E+000	mg/L	25	EST	MEYLAN,WM ET AL. (1996)
logP 3.50			EXP	SANGSTER,J (1993)
VP 2.02E-011	mm Hg	25	EST	NEELY,WB & BLAU,GE (1985)
DC	pKa			
HL 2.66E-015	atm m3/mol	25	EST	MEYLAN,WM & HOWARD,PH (1991)
OH 1.75E-010	cm3/molc sec	25	EST	MEYLAN,WM & HOWARD,PH (1993)

CAS #: 000549-49-5	QUININE HYDROBROMIDE

Formula: $C_{20}H_{25}BrN_2O_2$

Mol Weight: 405.34

MP (deg C):   FP (deg C):

BP (deg C):

BP pressure (mm Hg):

Property/ Value	Units	Temp	Data Type	Reference
WS 2.33E+004	mg/L	25	EXP	SEIDELL,A (1941)
logP 1.78			EST	MEYLAN,WM & HOWARD,PH (1995)
VP 2.08E-017	mm Hg	25	EST	NEELY,WB & BLAU,GE (1985)
DC	pKa			
HL 4.30E-023	atm m3/mol	25	EST	MEYLAN,WM & HOWARD,PH (1991)
OH 1.69E-010	cm3/molc sec	25	EST	MEYLAN,WM & HOWARD,PH (1993)

CAS #: 000550-44-7	PHTHALIMIDE,N-METHYL

Formula: $C_9H_7NO_2$

Mol Weight: 161.16

MP (deg C): 134   FP (deg C):

BP (deg C): 286

BP pressure (mm Hg):

Property/ Value	Units	Temp	Data Type	Reference
WS 5.14E+003	mg/L	25	EST	MEYLAN,WM ET AL. (1996)
logP 1.31			EXP	HANSCH,C & LEO,AJ (1985)
VP 2.42E-006	mm Hg	25	EST	NEELY,WB & BLAU,GE (1985)
DC	pKa			
HL 2.24E-008	atm m3/mol	25	EST	MEYLAN,WM & HOWARD,PH (1991)
OH 1.63E-011	cm3/molc sec	25	EST	MEYLAN,WM & HOWARD,PH (1993)

CAS #: 000551-06-4	1-ISOTHIOCYANONAPHTHALENE

Formula: $C_{11}H_7NS$

Mol Weight: 185.25

MP (deg C): 58   FP (deg C):

BP (deg C):

BP pressure (mm Hg):

Property/ Value	Units	Temp	Data Type	Reference
WS 4.63E+000	mg/L	25	EXP	YALKOWSKY,SH & DANNENFELSER,RM (1992)
logP 4.34			EXP	HANSCH,C & LEO,AJ (1985)
VP 1.32E-004	mm Hg	25	EST	NEELY,WB & BLAU,GE (1985)
DC	pKa			
HL 7.62E-005	atm m3/mol	25	EST	MEYLAN,WM & HOWARD,PH (1991)
OH 2.37E-011	cm3/molc sec	25	EST	MEYLAN,WM & HOWARD,PH (1993)

000551-11-1 — PROSTAGLANDIN

Formula: $C_{20}H_{34}O_5$
Mol Weight: 354.49
MP (deg C): 25-35
FP (deg C):
BP (deg C):
BP pressure (mm Hg):

Property/Value	Units	Temp	Data Type	Reference
WS 2.58E+000	mg/L	25	EST	MEYLAN,WM ET AL. (1996)
logP 4.39			EXP	BODOR,H & HUANG,M (1992)
VP 3.77E-014	mm Hg	25	EST	NEELY,WB & BLAU,GE (1985)
DC	pKa			
HL 1.50E-012	atm m3/mol	25	EST	MEYLAN,WM & HOWARD,PH (1991)
OH 1.55E-010	cm3/molc sec	25	EST	MEYLAN,WM & HOWARD,PH (1993)

000551-27-9 — PROPICILLIN

Formula: $C_{18}H_{22}N_2O_5S$
Mol Weight: 378.45
MP (deg C):
FP (deg C):
BP (deg C):
BP pressure (mm Hg):

Property/Value	Units	Temp	Data Type	Reference
WS 2.27E+001	mg/L	25	EST	MEYLAN,WM ET AL. (1996)
logP 2.65			EXP	HANSCH,C & LEO,AJ (1985)
VP 3.25E-013	mm Hg	25	EST	NEELY,WB & BLAU,GE (1985)
DC	pKa			
HL 7.79E-015	atm m3/mol	25	EST	MEYLAN,WM & HOWARD,PH (1991)
OH 1.22E-010	cm3/molc sec	25	EST	MEYLAN,WM & HOWARD,PH (1993)

000551-62-2 — 1,2,3,4-TETRAFLUOROBENZENE

Formula: $C_6H_2F_4$
Mol Weight: 150.08
MP (deg C): -42
FP (deg C):
BP (deg C): 94.3
BP pressure (mm Hg):

Property/Value	Units	Temp	Data Type	Reference
WS 3.11E+002	mg/L	25	EST	MEYLAN,WM ET AL. (1996)
logP 2.79			EST	MEYLAN,WM & HOWARD,PH (1995)
VP 4.95E+001	mm Hg	25	EXP	BOUBLIK,T ET AL. (1984)
DC	pKa			
HL 9.99E-003	atm m3/mol	25	EST	MEYLAN,WM & HOWARD,PH (1991)
OH 3.49E-013	cm3/molc sec	25	EST	MEYLAN,WM & HOWARD,PH (1993)

000551-76-8 — 2,4,6-TRICHLORO-3-METHYLPHENOL

Formula: $C_7H_5Cl_3O$
Mol Weight: 211.48
MP (deg C): 45-47
FP (deg C):
BP (deg C): 265
BP pressure (mm Hg):

Property/Value	Units	Temp	Data Type	Reference
WS 5.64E+001	mg/L	25	EST	MEYLAN,WM ET AL. (1996)
logP 3.99			EST	MEYLAN,WM & HOWARD,PH (1995)
VP 7.20E-004	mm Hg	25	EST	NEELY,WB & BLAU,GE (1985)
DC	pKa			
HL 2.52E-007	atm m3/mol	25	EST	MEYLAN,WM & HOWARD,PH (1991)
OH 8.49E-013	cm3/molc sec	25	EST	MEYLAN,WM & HOWARD,PH (1993)

000551-92-8 — 1,2-DIMETHYL-5-NITRO-1H-IMIDAZOLE

Formula: $C_5H_7N_3O_2$
Mol Weight: 141.13
MP (deg C): 138-139
FP (deg C):
BP (deg C):
BP pressure (mm Hg):

Property/Value	Units	Temp	Data Type	Reference
WS 1.83E+004	mg/L	25	EST	MEYLAN,WM ET AL. (1996)
logP 0.31			EXP	KOSANOVIC,D ET AL. (1988)
VP 5.58E-004	mm Hg	25	EST	NEELY,WB & BLAU,GE (1985)
DC	pKa			
HL 3.49E-007	atm m3/mol	25	EST	MEYLAN,WM & HOWARD,PH (1991)
OH 4.13E-012	cm3/molc sec	25	EST	MEYLAN,WM & HOWARD,PH (1993)

000551-93-9 — o-AMINOACETOPHENONE

Formula: C_8H_9NO
Mol Weight: 135.17
MP (deg C):
FP (deg C):
BP (deg C): 85-90
BP pressure (mm Hg): 5.00E-001

Property/Value	Units	Temp	Data Type	Reference
WS 3.56E+003	mg/L	25	EST	MEYLAN,WM ET AL. (1996)
logP 1.63			EXP	HANSCH,C & LEO,AJ (1985)
VP 9.89E-003	mm Hg	25	EST	NEELY,WB & BLAU,GE (1985)
DC 2.22	pKa	25	EXP	PERRIN,DD (1965)
HL 3.47E-009	atm m3/mol	25	EST	MEYLAN,WM & HOWARD,PH (1991)
OH 9.82E-011	cm3/molc sec	25	EST	MEYLAN,WM & HOWARD,PH (1993)

000552-16-9 — 2-NITROBENZOIC ACID

Formula: $C_7H_5NO_4$
Mol Weight: 167.12
MP (deg C): 147.5
FP (deg C):
BP (deg C):
BP pressure (mm Hg):

Property/Value	Units	Temp	Data Type	Reference
WS 7.86E+003	mg/L	25	EXP	YALKOWSKY,SH & DANNENFELSER,RM (1992)
logP 1.46			EXP	HANSCH,C & LEO,AJ (1985)
VP 7.14E-005	mm Hg	25	EST	NEELY,WB & BLAU,GE (1985)
DC 2.47	pKa	20	EXP	KORTUM,G ET AL (1961)
HL 4.28E-010	atm m3/mol	25	EST	MEYLAN,WM & HOWARD,PH (1991)
OH 5.86E-013	cm3/molc sec	25	EST	MEYLAN,WM & HOWARD,PH (1993)

000552-30-7 — TRIMELLITIC ANHYDRIDE

Formula: $C_9H_4O_5$
Mol Weight: 192.13
MP (deg C): 161-163.5
FP (deg C):
BP (deg C): 240-245
BP pressure (mm Hg): 1.40E+001

Property/Value	Units	Temp	Data Type	Reference
WS 1.04E+003	mg/L	25	EST	MEYLAN,WM ET AL. (1996)
logP 1.95			EST	MEYLAN,WM & HOWARD,PH (1995)
VP 1.16E-009	mm Hg	25	EXT	YAWS,CL (1994B)
DC	pKa			
HL 1.28E-010	atm m3/mol	25	EST	MEYLAN,WM & HOWARD,PH (1991)
OH 7.97E-013	cm3/molc sec	25	EST	MEYLAN,WM & HOWARD,PH (1993)

CAS #: 000552-32-9				2-NITROACETANILIDE

Formula: $C_8H_8N_2O_3$

Mol Weight: 180.16

MP (deg C): 94 FP (deg C):

BP (deg C): 100

BP pressure (mm Hg): 1.00E-001

Property/ Value	Units	Temp	Data Type	Reference
WS 2.21E+003	mg/L	151	EXP	SUZUKI,T (1991)
logP 1.00			EXP	HANSCH,C & LEO,AJ (1985)
VP 5.00E-006	mm Hg	25	EST	NEELY,WB & BLAU,GE (1985)
DC	pKa			
HL 5.33E-010	atm m3/mol	25	EST	MEYLAN,WM & HOWARD,PH (1991)
OH 1.65E-012	cm3/molc sec	25	EST	MEYLAN,WM & HOWARD,PH (1993)

CAS #: 000552-58-9				ERIODICTYOL

Formula: $C_{15}H_{12}O_6$

Mol Weight: 288.26

MP (deg C): 267 dec FP (deg C):

BP (deg C):

BP pressure (mm Hg):

Property/ Value	Units	Temp	Data Type	Reference
WS 7.00E+001	mg/L	20	EXP	YALKOWSKY,SH & DANNENFELSER,RM (1992)
logP 2.02			EXP	PERRISSOUD,D & TESTA,B (1986)
VP 8.56E-012	mm Hg	25	EST	NEELY,WB & BLAU,GE (1985)
DC	pKa			
HL 3.05E-021	atm m3/mol	25	EST	MEYLAN,WM & HOWARD,PH (1991)
OH 2.48E-010	cm3/molc sec	25	EST	MEYLAN,WM & HOWARD,PH (1993)

CAS #: 000552-62-5				7-METHYLXANTHINE

Formula: $C_6H_8N_4O_2$

Mol Weight: 168.16

MP (deg C): > 300 FP (deg C):

BP (deg C):

BP pressure (mm Hg):

Property/ Value	Units	Temp	Data Type	Reference
WS 1.81E+004	mg/L	25	EST	MEYLAN,WM ET AL. (1996)
logP -0.89			EXP	GASPARI,F & BONATI,M (1987)
VP 1.88E-009	mm Hg	25	EST	NEELY,WB & BLAU,GE (1985)
DC 8.33	pKa		EXP	KORTUM,G ET AL (1961)
HL 7.95E-013	atm m3/mol	25	EST	MEYLAN,WM & HOWARD,PH (1991)
OH 5.49E-011	cm3/molc sec	25	EST	MEYLAN,WM & HOWARD,PH (1993)

CAS #: 000552-82-9				DIPHENYL METHYLAMINE

Formula: $C_{13}H_{13}N$

Mol Weight: 183.26

MP (deg C): -7.6 FP (deg C):

BP (deg C): 296-297

BP pressure (mm Hg):

Property/ Value	Units	Temp	Data Type	Reference
WS 2.48E+001	mg/L	25	EST	MEYLAN,WM ET AL. (1996)
logP 3.90			EXP	HANSCH,C & LEO,AJ (1985)
VP 3.11E-003	mm Hg	25	EXT	OHE,S (1976)
DC	pKa			
HL 2.15E-005	atm m3/mol	25	EST	MEYLAN,WM & HOWARD,PH (1991)
OH 2.01E-010	cm3/molc sec	25	EST	MEYLAN,WM & HOWARD,PH (1993)

CAS #: 000552-86-3				FURAN, 2-CH(OH)CO-FURYL

Formula: $C_{10}H_8O_4$

Mol Weight: 192.17

MP (deg C): 134-137 FP (deg C):

BP (deg C):

BP pressure (mm Hg):

Property/ Value	Units	Temp	Data Type	Reference
WS 5.36E+004	mg/L	25	EST	MEYLAN,WM ET AL. (1996)
logP 0.54			EXP	HANSCH,C & LEO,AJ (1985)
VP 2.49E-005	mm Hg	25	EST	NEELY,WB & BLAU,GE (1985)
DC	pKa			
HL 1.23E-007	atm m3/mol	25	EST	MEYLAN,WM & HOWARD,PH (1991)
OH 1.38E-010	cm3/molc sec	25	EST	MEYLAN,WM & HOWARD,PH (1993)

CAS #: 000552-89-6				O-NITROBENZALDEHYDE

Formula: $C_7H_5NO_3$

Mol Weight: 151.12

MP (deg C): FP (deg C):

BP (deg C):

BP pressure (mm Hg):

Property/ Value	Units	Temp	Data Type	Reference
WS 2.00E+001	mg/L	25	EXP	YALKOWSKY,SH & DANNENFELSER,RM (1992)
logP 1.74			EXP	HANSCH,C & LEO,AJ (1985)
VP 1.79E-002	mm Hg	25	EXT	PERRY,RH & GREEN,D (1984)
DC	pKa			
HL 5.30E-008	atm m3/mol	25	EST	MEYLAN,WM & HOWARD,PH (1991)
OH 1.70E-011	cm3/molc sec	25	EST	MEYLAN,WM & HOWARD,PH (1993)

CAS #: 000552-94-3				2-CARBOXYPHENYL 2-HYDROXYBENZOATE

Formula: $C_{14}H_{10}O_5$

Mol Weight: 258.23

MP (deg C): 148-149 FP (deg C):

BP (deg C):

BP pressure (mm Hg):

Property/ Value	Units	Temp	Data Type	Reference
WS 1.10E+002	mg/L	25	EST	MEYLAN,WM ET AL. (1996)
logP 3.36			EST	MEYLAN,WM & HOWARD,PH (1995)
VP 9.42E-009	mm Hg	25	EST	NEELY,WB & BLAU,GE (1985)
DC	pKa			
HL 3.38E-011	atm m3/mol	25	EST	MEYLAN,WM & HOWARD,PH (1991)
OH 3.18E-011	cm3/molc sec	25	EST	MEYLAN,WM & HOWARD,PH (1993)

CAS #: 000553-03-7				2(1H)-QUINOLINONE, 3,4-DIHYDRO-

Formula: C_9H_9NO

Mol Weight: 147.18

MP (deg C): 165-166.5 FP (deg C):

BP (deg C):

BP pressure (mm Hg):

Property/ Value	Units	Temp	Data Type	Reference
WS 5.60E+003	mg/L	25	EST	MEYLAN,WM ET AL. (1996)
logP 1.34			EXP	SANGSTER,J (1993)
VP 5.00E-005	mm Hg	25	EST	NEELY,WB & BLAU,GE (1985)
DC	pKa			
HL 3.00E-009	atm m3/mol	25	EST	MEYLAN,WM & HOWARD,PH (1991)
OH 1.96E-011	cm3/molc sec	25	EST	MEYLAN,WM & HOWARD,PH (1993)

000553-20-8 — ACETAMIDE, N-(5-NITRO-2-PROPOXYPHENYL)-

CAS #: 000553-20-8

Formula: $C_{11}H_{14}N_2O_4$

Mol Weight: 238.25

MP (deg C): 102.5-3.5

FP (deg C):

BP (deg C):

BP pressure (mm Hg):

Property/Value	Units	Temp	Data Type	Reference
WS 4.81E+002	mg/L	25	EST	MEYLAN,WM ET AL. (1996)
logP 1.60			EXP	FURST,W & BECHER,M (1990)
VP 1.65E-007	mm Hg	25	EST	NEELY,WB & BLAU,GE (1985)
DC	pKa			
HL 2.54E-012	atm m3/mol	25	EST	MEYLAN,WM & HOWARD,PH (1991)
OH 1.20E-011	cm3/molc sec	25	EST	MEYLAN,WM & HOWARD,PH (1993)

000553-26-4 — 4,4'-DIPYRIDYL

CAS #: 000553-26-4

Formula: $C_{10}H_8N_2$

Mol Weight: 156.19

MP (deg C): 73

FP (deg C):

BP (deg C): 304.8

BP pressure (mm Hg):

Property/Value	Units	Temp	Data Type	Reference
WS 4.53E+003	mg/L	25	EXP	YALKOWSKY,SH & DANNENFELSER,RM (1992)
logP 1.28			EXP	DEVOOGT,P ET AL. (1990)
VP 5.56E-005	mm Hg	20	EXP	KUO,HW ET AL. (1992)
DC 4.82	pKa	20	EXP	PERRIN,DD (1965)
HL 2.52E-009	atm m3/mol	25	EST	VP/WSOL
OH 1.29E-012	cm3/molc sec	25	EST	MEYLAN,WM & HOWARD,PH (1993)

000553-27-5 — N,N-DI-B-CHLOROETHYLANILINE

CAS #: 000553-27-5

Formula: $C_{10}H_{13}Cl_2N$

Mol Weight: 218.13

MP (deg C): 45

FP (deg C):

BP (deg C): 164

BP pressure (mm Hg): 1.40E+001

Property/Value	Units	Temp	Data Type	Reference
WS 1.18E+002	mg/L	25	EST	MEYLAN,WM ET AL. (1996)
logP 2.90			EXP	HANSCH,C & LEO,AJ (1985)
VP 9.59E-004	mm Hg	25	EST	NEELY,WB & BLAU,GE (1985)
DC	pKa			
HL 1.87E-005	atm m3/mol	25	EST	MEYLAN,WM & HOWARD,PH (1991)
OH 1.54E-010	cm3/molc sec	25	EST	MEYLAN,WM & HOWARD,PH (1993)

000553-84-4 — 1-PENTANONE, 1-(3-FURANYL)-4-METHYL-

CAS #: 000553-84-4

Formula: $C_{10}H_{14}O_2$

Mol Weight: 166.22

MP (deg C):

FP (deg C):

BP (deg C): 196

BP pressure (mm Hg):

Property/Value	Units	Temp	Data Type	Reference
WS 2.82E+002	mg/L	25	EST	MEYLAN,WM ET AL. (1996)
logP 2.76			EXP	HANSCH,C ET AL. (1995)
VP 1.16E-001	mm Hg	25	EST	NEELY,WB & BLAU,GE (1985)
DC	pKa			
HL 3.04E-005	atm m3/mol	25	EST	MEYLAN,WM & HOWARD,PH (1991)
OH 4.50E-011	cm3/molc sec	25	EST	MEYLAN,WM & HOWARD,PH (1993)

000553-90-2 — METHYL OXALATE

CAS #: 000553-90-2

Formula: $C_4H_6O_4$

Mol Weight: 118.09

MP (deg C): 54

FP (deg C):

BP (deg C): 163-164

BP pressure (mm Hg):

Property/Value	Units	Temp	Data Type	Reference
WS 6.03E+004	mg/L	25	EXP	YALKOWSKY,SH & DANNENFELSER,RM (1992)
logP -0.17			EXP	HANSCH,C ET AL. (1995)
VP 1.14E+000	mm Hg	25	EXP	OHE,S (1976)
DC	pKa			
HL 2.94E-006	atm m3/mol	25	EST	VP/WSOL
OH 4.35E-013	cm3/molc sec	25	EST	MEYLAN,WM & HOWARD,PH (1993)

000553-97-9 — 2-METHYL-1,4-BENZOQUINONE

CAS #: 000553-97-9

Formula: $C_7H_6O_2$

Mol Weight: 122.12

MP (deg C): 69

FP (deg C):

BP (deg C):

BP pressure (mm Hg):

Property/Value	Units	Temp	Data Type	Reference
WS 1.99E+004	mg/L	25	EST	MEYLAN,WM ET AL. (1996)
logP 0.72			EXP	HANSCH,C ET AL. (1995)
VP 3.43E-002	mm Hg	25	EST	NEELY,WB & BLAU,GE (1985)
DC	pKa			
HL 1.91E-009	atm m3/mol	25	EST	MEYLAN,WM & HOWARD,PH (1991)
OH 1.30E-011	cm3/molc sec	25	EST	MEYLAN,WM & HOWARD,PH (1993)

000554-00-7 — 2,4-DICHLOROANILINE

CAS #: 000554-00-7

Formula: $C_6H_5Cl_2N$

Mol Weight: 162.02

MP (deg C): 63.5

FP (deg C):

BP (deg C): 245

BP pressure (mm Hg):

Property/Value	Units	Temp	Data Type	Reference
WS 6.20E+002	mg/L	60	EXP	CHEM INSPECT TEST INST
logP 2.78			EXP	SANGSTER,J (1994)
VP 2.12E-002	mm Hg	25	EST	NEELY,WB & BLAU,GE (1985)
DC 2.00	pKa	25	EXP	PERRIN,DD (1965)
HL 1.05E-006	atm m3/mol	25	EST	MEYLAN,WM & HOWARD,PH (1991)
OH 9.18E-012	cm3/molc sec	25	EST	MEYLAN,WM & HOWARD,PH (1993)

000554-12-1 — METHYL PROPIONATE

CAS #: 000554-12-1

Formula: $C_4H_8O_2$

Mol Weight: 88.11

MP (deg C): -88

FP (deg C):

BP (deg C): 79.7

BP pressure (mm Hg):

Property/Value	Units	Temp	Data Type	Reference
WS 6.24E+004	mg/L	25	EXP	HINE,J & MOOKERJEE,PK (1975)
logP 0.84			EXP	CATZ,P & FRIEND,DR (1989)
VP 8.40E+001	mm Hg	25	EXP	DAUBERT,TE & DANNER,RP (1989)
DC	pKa			
HL 1.74E-004	atm m3/mol	25	EXP	BUTTERY,RG ET AL. (1969)
OH 1.00E-012	cm3/molc sec	25	EXP	ATKINSON,R (1989)

CAS #:	000554-14-3				2-METHYLTHIOPHENE

Formula:		C_5H_6S			
Mol Weight:	98.17				
MP (deg C): -63.4			FP (deg C):		
BP (deg C): 112.6					
BP pressure (mm Hg):					

Property/ Value	Units	Temp	Data Type	Reference
WS 1.21E+003	mg/L	25	EST	MEYLAN,WM ET AL. (1996)
logP 2.33			EXP	HANSCH,C & LEO,AJ (1985)
VP 2.49E+001	mm Hg	25	EXP	RIDDICK,JA ET AL. (1986)
DC	pKa			
HL 3.23E-003	atm m3/mol	25	EST	MEYLAN,WM & HOWARD,PH (1991)
OH 2.50E-011	cm3/molc sec	25	EST	MEYLAN,WM & HOWARD,PH (1993)

CAS #:	000554-52-9				O-METHYLDOPAMINE

Formula:		$C_9H_{13}NO_2$			
Mol Weight:	167.21				
MP (deg C):			FP (deg C):		
BP (deg C):					
BP pressure (mm Hg):					

Property/ Value	Units	Temp	Data Type	Reference
WS 1.00E+006	mg/L	25	EST	MEYLAN,WM ET AL. (1996)
logP -0.08			EXP	HANSCH,C & LEO,AJ (1985)
VP 1.02E-005	mm Hg	25	EST	NEELY,WB & BLAU,GE (1985)
DC	pKa			
HL 9.70E-015	atm m3/mol	25	EST	MEYLAN,WM & HOWARD,PH (1991)
OH 1.06E-010	cm3/molc sec	25	EST	MEYLAN,WM & HOWARD,PH (1993)

CAS #:	000554-57-4				METHAZOLAMIDE

Formula:		$C_5H_8N_4O_3S_2$			
Mol Weight:	236.27				
MP (deg C): 213-214			FP (deg C):		
BP (deg C):					
BP pressure (mm Hg):					

Property/ Value	Units	Temp	Data Type	Reference
WS 3.50E+003	mg/L		EXP	YALKOWSKY,SH & DANNENFELSER,RM (1992)
logP 0.13			EXP	HANSCH,C ET AL. (1995)
VP 2.48E-007	mm Hg	25	EST	NEELY,WB & BLAU,GE (1985)
DC	pKa			
HL 4.40E-016	atm m3/mol	25	EST	MEYLAN,WM & HOWARD,PH (1991)
OH 3.07E-012	cm3/molc sec	25	EST	MEYLAN,WM & HOWARD,PH (1993)

CAS #:	000554-61-0				2-CARENE

Formula:		$C_{10}H_{16}$			
Mol Weight:	136.24				
MP (deg C):			FP (deg C):		
BP (deg C): 167-168					
BP pressure (mm Hg):					

Property/ Value	Units	Temp	Data Type	Reference
WS 2.91E+000	mg/L	25	EST	MEYLAN,WM ET AL. (1996)
logP 4.61			EST	MEYLAN,WM & HOWARD,PH (1995)
VP 3.72E+000	mm Hg	25	EST	NEELY,WB & BLAU,GE (1985)
DC	pKa			
HL 1.07E-001	atm m3/mol	25	EST	MEYLAN,WM & HOWARD,PH (1991)
OH 7.95E-011	cm3/molc sec	25	EXP	ATKINSON,R (1989)

CAS #:	000554-68-7				TRIETHYLAMINE HYDROCHLORIDE

Formula:		$C_6H_{16}ClN$			
Mol Weight:	137.65				
MP (deg C): 260 dec			FP (deg C):		
BP (deg C):					
BP pressure (mm Hg):					

Property/ Value	Units	Temp	Data Type	Reference
WS 1.00E+006	mg/L	25	EXP	PEDDLE,CJ & TURNER,WES (1913)
logP -1.26			EST	MEYLAN,WM & HOWARD,PH (1995)
VP 1.38E-005	mm Hg	25	EST	NEELY,WB & BLAU,GE (1985)
DC	pKa			
HL 5.39E-012	atm m3/mol	25	EST	MEYLAN,WM & HOWARD,PH (1991)
OH 2.66E-011	cm3/molc sec	25	EST	MEYLAN,WM & HOWARD,PH (1993)

CAS #:	000554-77-8				P-CHLOROMERCURIPHENYLSULFONIC ACID

Formula:		$C_6H_5ClHgO_3S$			
Mol Weight:	393.21				
MP (deg C):			FP (deg C):		
BP (deg C):					
BP pressure (mm Hg):					

Property/ Value	Units	Temp	Data Type	Reference
WS 4.60E+005	mg/L	25	EST	MEYLAN,WM ET AL. (1996)
logP -1.55			EST	MEYLAN,WM & HOWARD,PH (1995)
VP 9.94E-009	mm Hg	25	EST	NEELY,WB & BLAU,GE (1985)
DC	pKa			
HL	atm m3/mol			
OH 5.57E-013	cm3/molc sec	25	EST	MEYLAN,WM & HOWARD,PH (1993)

CAS #:	000554-84-7				3-NITROPHENOL

Formula:		$C_6H_5NO_3$			
Mol Weight:	139.11				
MP (deg C): 96.8			FP (deg C):		
BP (deg C): 194					
BP pressure (mm Hg): 7.00E+001					

Property/ Value	Units	Temp	Data Type	Reference
WS 1.35E+004	mg/L	25	EXP	MATSUGUMA,HJ (1963)
logP 2.00			EXP	HANSCH,C & LEO,AJ (1985)
VP 1.00E-001	mm Hg	25	EXP	DORIGAN,J ET AL. (1976)
DC 8.36	pKa		EXP	SERJEANT,EP & DEMPSEY,B (1979)
HL 2.00E-009	atm m3/mol	25	EXP	GAFFNEY,JS ET AL. (1987)
OH 3.09E+000	cm3/molc sec	25	EST	MEYLAN,WM & HOWARD,PH (1993)

CAS #:	000554-95-0				1,3,5-BENZENETRICARBOXYLIC ACID

Formula:		$C_9H_6O_6$			
Mol Weight:	210.14				
MP (deg C): > 300			FP (deg C):		
BP (deg C):					
BP pressure (mm Hg):					

Property/ Value	Units	Temp	Data Type	Reference
WS 2.63E+004	mg/L	23	EXP	YALKOWSKY,SH & DANNENFELSER,RM (1992)
logP 1.64			EST	MEYLAN,WM & HOWARD,PH (1995)
VP 2.88E-008	mm Hg	25	EST	NEELY,WB & BLAU,GE (1985)
DC 3.12	pKa		EXP	KORTUM,G ET AL (1961)
HL 4.39E-017	atm m3/mol	25	EST	MEYLAN,WM & HOWARD,PH (1991)
OH 1.60E-012	cm3/molc sec	25	EST	MEYLAN,WM & HOWARD,PH (1993)

CAS #: 000555-03-3				M-NITROANISOLE

Formula: $C_7H_7NO_3$
Mol Weight: 153.14
MP (deg C): 38.5
FP (deg C):
BP (deg C): 258
BP pressure (mm Hg):

Property/ Value	Units	Temp	Data Type	Reference
WS 4.28E+002	mg/L	25	EST	MEYLAN,WM ET AL. (1996)
logP 2.16			EXP	HANSCH,C & LEO,AJ (1985)
VP 1.92E-002	mm Hg	25	EST	NEELY,WB & BLAU,GE (1985)
DC	pKa			
HL 1.26E-006	atm m3/mol	25	EST	MEYLAN,WM & HOWARD,PH (1991)
OH 2.71E-012	cm3/molc sec	25	EST	MEYLAN,WM & HOWARD,PH (1993)

CAS #: 000555-10-2				BETA-PHELLANDRENE

Formula: $C_{10}H_{16}$
Mol Weight: 136.24
MP (deg C):
FP (deg C):
BP (deg C): 171.5
BP pressure (mm Hg):

Property/ Value	Units	Temp	Data Type	Reference
WS 2.45E+000	mg/L	25	EST	MEYLAN,WM ET AL. (1996)
logP 4.70			EST	MEYLAN,WM & HOWARD,PH (1995)
VP 1.59E+000	mm Hg	25	EXP	YAWS,CL (1994B)
DC	pKa			
HL 2.62E-001	atm m3/mol	25	EST	MEYLAN,WM & HOWARD,PH (1991)
OH 1.68E-010	cm3/molc sec	25	EXP	ATKINSON,R (1989)

CAS #: 000555-16-8				4-NITROBENZALDEHYDE

Formula: $C_7H_5NO_3$
Mol Weight: 151.12
MP (deg C): 107
FP (deg C):
BP (deg C):
BP pressure (mm Hg):

Property/ Value	Units	Temp	Data Type	Reference
WS 2.32E+003	mg/L	25	EXP	STEPHEN,H & STEPHEN,T (1963)
logP 1.56			EXP	HANSCH,C & LEO,AJ (1985)
VP 3.54E-003	mm Hg	25	EST	NEELY,WB & BLAU,GE (1985)
DC	pKa			
HL 5.30E-008	atm m3/mol	25	EST	MEYLAN,WM & HOWARD,PH (1991)
OH 1.70E-011	cm3/molc sec	25	EST	MEYLAN,WM & HOWARD,PH (1993)

CAS #: 000555-37-3				NEBURON

Formula: $C_{12}H_{16}Cl_2N_2O$
Mol Weight: 275.18
MP (deg C): 101.5-103
FP (deg C):
BP (deg C):
BP pressure (mm Hg):

Property/ Value	Units	Temp	Data Type	Reference
WS 4.80E+000	mg/L	25	EXP	YALKOWSKY,SH & DANNENFELSER,RM (1992)
logP 3.80			EXP	HANSCH,C & LEO,AJ (1985)
VP 1.01E-006	mm Hg	25	EST	NEELY,WB & BLAU,GE (1985)
DC	pKa			
HL 1.25E-009	atm m3/mol	25	EST	MEYLAN,WM & HOWARD,PH (1991)
OH 1.78E-011	cm3/molc sec	25	EST	MEYLAN,WM & HOWARD,PH (1993)

CAS #: 000555-60-2				PROPANEDINITRILE, (3-CHLOROPHENYL)HYDRAZONO -

Formula: $C_9H_5ClN_4$
Mol Weight: 204.62
MP (deg C): 175-177 de
FP (deg C):
BP (deg C):
BP pressure (mm Hg):

Property/ Value	Units	Temp	Data Type	Reference
WS 5.38E+001	mg/L	25	EST	MEYLAN,WM ET AL. (1996)
logP 3.38			EXP	STURDIK,E ET AL. (1985)
VP 5.67E-006	mm Hg	25	EST	NEELY,WB & BLAU,GE (1985)
DC	pKa			
HL 4.29E-009	atm m3/mol	25	EST	MEYLAN,WM & HOWARD,PH (1991)
OH 3.00E-011	cm3/molc sec	25	EST	MEYLAN,WM & HOWARD,PH (1993)

CAS #: 000555-77-1				TRIS(BETA-CHLOROETHYL)AMINE

Formula: $C_6H_{12}Cl_3N$
Mol Weight: 204.53
MP (deg C): -40
FP (deg C):
BP (deg C): 144
BP pressure (mm Hg): 1.50E+001

Property/ Value	Units	Temp	Data Type	Reference
WS 4.83E+003	mg/L	25	EST	MEYLAN,WM ET AL. (1996)
logP 2.27			EST	MEYLAN,WM & HOWARD,PH (1995)
VP 1.53E-002	mm Hg	25	EST	NEELY,WB & BLAU,GE (1985)
DC 4.64	pKa	0	EXP	PERRIN,DD (1965)
HL 2.18E-009	atm m3/mol	25	EST	MEYLAN,WM & HOWARD,PH (1991)
OH 7.55E-011	cm3/molc sec	25	EST	MEYLAN,WM & HOWARD,PH (1993)

CAS #: 000555-89-5				NEOTRAN

Formula: $C_{13}H_{10}Cl_2O_2$
Mol Weight: 269.13
MP (deg C): 69.7-70.2
FP (deg C):
BP (deg C): 189-194
BP pressure (mm Hg): 6.00E+000

Property/ Value	Units	Temp	Data Type	Reference
WS 8.03E-001	mg/L	25	EST	MEYLAN,WM ET AL. (1996)
logP 5.11			EST	MEYLAN,WM & HOWARD,PH (1995)
VP 2.97E-005	mm Hg	25	EST	NEELY,WB & BLAU,GE (1985)
DC	pKa			
HL 7.57E-007	atm m3/mol	25	EST	MEYLAN,WM & HOWARD,PH (1991)
OH 1.60E-011	cm3/molc sec	25	EST	MEYLAN,WM & HOWARD,PH (1993)

CAS #: 000556-03-6				DL-TYROSINE

Formula: $C_9H_{11}NO_3$
Mol Weight: 181.19
MP (deg C): 325 dec
FP (deg C):
BP (deg C):
BP pressure (mm Hg):

Property/ Value	Units	Temp	Data Type	Reference
WS 4.00E+002	mg/L	25	EXP	YALKOWSKY,SH & DANNENFELER,RM (1992)
logP -2.04			EXP	SANGSTER,J (1993)
VP 1.67E-008	mm Hg	25	EST	NEELY,WB & BLAU,GE (1985)
DC	pKa			
HL 1.25E-014	atm m3/mol	25	EST	MEYLAN,WM & HOWARD,PH (1991)
OH 8.01E-011	cm3/molc sec	25	EST	MEYLAN,WM & HOWARD,PH (1993)

000556-08-1 — P-CARBOXYACETANILIDE

CAS #	000556-08-1
Formula	$C_9H_9NO_3$
Mol Weight	179.18
MP (deg C)	259-262 de
FP (deg C)	
BP (deg C)	
BP pressure (mm Hg)	

Property/Value	Units	Temp	Data Type	Reference
WS 4.23E+003	mg/L	25	EST	MEYLAN,WM ET AL. (1996)
logP 1.31			EXP	HANSCH,C & LEO,AJ (1985)
VP 7.14E-007	mm Hg	25	EST	NEELY,WB & BLAU,GE (1985)
DC	pKa			
HL 1.24E-013	atm m3/mol	25	EST	MEYLAN,WM & HOWARD,PH (1991)
OH 5.22E-012	cm3/molc sec	25	EST	MEYLAN,WM & HOWARD,PH (1993)

000556-24-1 — METHYL (3-METHYL)BUTANOATE

CAS #	000556-24-1
Formula	$C_6H_{12}O_2$
Mol Weight	116.16
MP (deg C)	
FP (deg C)	
BP (deg C)	116-117
BP pressure (mm Hg)	

Property/Value	Units	Temp	Data Type	Reference
WS 2.89E+003	mg/L	25	EST	MEYLAN,WM ET AL. (1996)
logP 1.82			EXP	HANSCH,C ET AL. (1995)
VP 1.40E+001	mm Hg	20	EXP	VOITKEVICH,SA (1963)
DC	pKa			
HL 4.10E-004	atm m3/mol	25	EST	MEYLAN,WM & HOWARD,PH (1991)
OH 3.79E-012	cm3/molc sec	25	EST	MEYLAN,WM & HOWARD,PH (1993)

000556-33-2 — TRIGLYCINE

CAS #	000556-33-2
Formula	$C_6H_{11}N_3O_4$
Mol Weight	189.17
MP (deg C)	
FP (deg C)	
BP (deg C)	
BP pressure (mm Hg)	

Property/Value	Units	Temp	Data Type	Reference
WS 1.02E+004	mg/L	25	EXP	SEIDELL,A (1941)
logP -2.68			EXP	HANSCH,C ET AL. (1995)
VP 1.96E-011	mm Hg	25	EST	NEELY,WB & BLAU,GE (1985)
DC	pKa			
HL 7.89E-017	atm m3/mol	25	EST	MEYLAN,WM & HOWARD,PH (1991)
OH 4.48E-011	cm3/molc sec	25	EST	MEYLAN,WM & HOWARD,PH (1993)

000556-50-3 — GYCYLGLYCINE

CAS #	000556-50-3
Formula	$C_4H_8N_2O_3$
Mol Weight	132.12
MP (deg C)	215 dec
FP (deg C)	
BP (deg C)	
BP pressure (mm Hg)	

Property/Value	Units	Temp	Data Type	Reference
WS 1.66E+005	mg/L	21	EXP	YALKOWSKY,SH & DANNENFELSER,RM (1992)
logP -2.92			EXP	HANSCH,C & LEO,AJ (1985)
VP 4.54E-008	mm Hg	25	EST	NEELY,WB & BLAU,GE (1985)
DC	pKa			
HL 7.48E-015	atm m3/mol	25	EST	MEYLAN,WM & HOWARD,PH (1991)
OH 3.64E-011	cm3/molc sec	25	EST	MEYLAN,WM & HOWARD,PH (1993)

000556-52-5 — OXIRANEMETHANOL

CAS #	000556-52-5
Formula	$C_3H_6O_2$
Mol Weight	74.08
MP (deg C)	
FP (deg C)	
BP (deg C)	66
BP pressure (mm Hg)	2.50E+000

Property/Value	Units	Temp	Data Type	Reference
WS 1.00E+006	mg/L	25	EST	MEYLAN,WM ET AL. (1996)
logP -0.95			EXP	DENEER,JW ET AL. (1988)
VP 5.59E+000	mm Hg	25	EST	NEELY,WB & BLAU,GE (1985)
DC	pKa			
HL 5.84E-009	atm m3/mol	25	EST	MEYLAN,WM & HOWARD,PH (1991)
OH 4.66E-012	cm3/molc sec	25	EST	MEYLAN,WM & HOWARD,PH (1993)

000556-53-6 — PROPYLAMINE HYDROCHLORIDE

CAS #	000556-53-6
Formula	$C_3H_{10}ClN$
Mol Weight	95.57
MP (deg C)	160-162
FP (deg C)	
BP (deg C)	
BP pressure (mm Hg)	

Property/Value	Units	Temp	Data Type	Reference
WS 1.00E+006	mg/L	25	EXP	SEIDELL,A (1941)
logP -2.84			EST	MEYLAN,WM & HOWARD,PH (1995)
VP 1.36E-004	mm Hg	25	EST	NEELY,WB & BLAU,GE (1985)
DC	pKa			
HL 1.24E-012	atm m3/mol	25	EST	MEYLAN,WM & HOWARD,PH (1991)
OH 1.20E-011	cm3/molc sec	25	EST	MEYLAN,WM & HOWARD,PH (1993)

000556-61-6 — METHYL ISOTHIOCYANATE

CAS #	000556-61-6
Formula	C_2H_3NS
Mol Weight	73.12
MP (deg C)	35-36
FP (deg C)	
BP (deg C)	119
BP pressure (mm Hg)	

Property/Value	Units	Temp	Data Type	Reference
WS 7.60E+003	mg/L	20	EXP	YALKOWSKY,SH & DANNENFELSER,RM (1992)
logP 0.94			EXP	HANSCH,C ET AL. (1995)
VP 3.54E+000	mm Hg	25	EXP	BOUBLIK,T ET AL. (1984)
DC	pKa			
HL 4.48E-005	atm m3/mol	25	EST	VP/WSOL
OH 1.36E-013	cm3/molc sec	25	EST	MEYLAN,WM & HOWARD,PH (1993)

000556-64-9 — METHYL THIOCYANATE

CAS #	000556-64-9
Formula	C_2H_3NS
Mol Weight	73.12
MP (deg C)	-5
FP (deg C)	
BP (deg C)	132.9
BP pressure (mm Hg)	

Property/Value	Units	Temp	Data Type	Reference
WS 3.22E+004	mg/L	25	EST	MEYLAN,WM ET AL. (1996)
logP 0.73			EST	MEYLAN,WM & HOWARD,PH (1995)
VP 1.21E+001	mm Hg	25	EXP	PERRY,RH & GREEN,D (1984)
DC	pKa			
HL 4.38E-005	atm m3/mol	25	EST	MEYLAN,WM & HOWARD,PH (1991)
OH 1.06E-012	cm3/molc sec	25	EST	MEYLAN,WM & HOWARD,PH (1993)

CAS #: 000556-67-2 — OCTAMETHYLTETRASILOXANE

Formula: $C_8H_{24}O_4Si_4$

Mol Weight: 296.62

MP (deg C): 17.5   FP (deg C):

BP (deg C): 175

BP pressure (mm Hg):

Property/Value	Units	Temp	Data Type	Reference
WS 5.00E-003	mg/L	25	EXP	DOW CORNING (1987)
logP 5.10			EXP	TSCATS
VP 1.05E+000	mm Hg	25	EXT	FLANINGAM,OL (1986)
DC	pKa			
HL 4.20E-001	atm m3/mol	20	EXP	SILICONES HEALTH COUNCIL (1989)
OH 1.01E-012	cm3/molc sec	25	EXP	ATKINSON,R (1991)

CAS #: 000556-68-3 — HEXADECAMETHYLCYCLOOCTASILOXANE

Formula: $C_{16}H_{48}O_8Si_8$

Mol Weight: 593.24

MP (deg C): 31.5   FP (deg C):

BP (deg C): 290

BP pressure (mm Hg):

Property/Value	Units	Temp	Data Type	Reference
WS 3.97E-004	mg/L	25	EST	MEYLAN,WM ET AL. (1996)
logP 7.57			EST	MEYLAN,WM & HOWARD,PH (1995)
VP 2.89E-003	mm Hg	25	EXT	JORDAN,TE (1954)
DC	pKa			
HL	atm m3/mol			
OH 2.39E-012	cm3/molc sec	25	EST	MEYLAN,WM & HOWARD,PH (1993)

CAS #: 000556-69-4 — OCTADECAMETHYLOCTASILOXANE

Formula: $C_{18}H_{54}O_7Si_8$

Mol Weight: 607.31

MP (deg C):   FP (deg C):

BP (deg C): 153

BP pressure (mm Hg): 5.10E+000

Property/Value	Units	Temp	Data Type	Reference
WS 2.68E-008	mg/L	25	EST	MEYLAN,WM ET AL. (1996)
logP 7.70			EXP	SANGSTER,J (1994)
VP 3.19E-003	mm Hg	25	EXP	JORDAN,TE (1954)
DC	pKa			
HL 4.91E-012	atm m3/mol	25	EST	MEYLAN,WM & HOWARD,PH (1991)
OH 2.69E-012	cm3/molc sec	25	EST	MEYLAN,WM & HOWARD,PH (1993)

CAS #: 000556-71-8 — OCTADECAMETHYLCYCLONONASILOXANE

Formula: $C_{18}H_{54}O_9Si_9$

Mol Weight: 667.40

MP (deg C):   FP (deg C):

BP (deg C): 188

BP pressure (mm Hg): 2.00E+001

Property/Value	Units	Temp	Data Type	Reference
WS 1.09E-004	mg/L	25	EST	MEYLAN,WM ET AL. (1996)
logP 8.19			EST	MEYLAN,WM & HOWARD,PH (1995)
VP 1.77E-005	mm Hg	25	EXP	FLANINGAM,OL (1986)
DC	pKa			
HL	atm m3/mol			
OH 2.69E-012	cm3/molc sec	25	EST	MEYLAN,WM & HOWARD,PH (1993)

CAS #: 000556-75-2 — O-ET-S-(ETSET)-ME-PHOSPHONATE

Formula: $C_7H_{17}O_2PS_2$

Mol Weight: 228.31

MP (deg C):   FP (deg C):

BP (deg C):

BP pressure (mm Hg):

Property/Value	Units	Temp	Data Type	Reference
WS 1.69E+003	mg/L	25	EST	MEYLAN,WM ET AL. (1996)
logP 1.48			EXP	HANSCH,C & LEO,AJ (1985)
VP 1.06E-003	mm Hg	25	EST	NEELY,WB & BLAU,GE (1985)
DC	pKa			
HL 1.71E-008	atm m3/mol	25	EST	MEYLAN,WM & HOWARD,PH (1991)
OH 6.11E-011	cm3/molc sec	25	EST	MEYLAN,WM & HOWARD,PH (1993)

CAS #: 000556-88-7 — NITROGUANIDINE

Formula: $CH_4N_4O_2$

Mol Weight: 104.07

MP (deg C): 239 dec   FP (deg C):

BP (deg C):

BP pressure (mm Hg):

Property/Value	Units	Temp	Data Type	Reference
WS 4.40E+003	mg/L	25	EXP	YALKOWSKY,SH & DANNENFELSER,RM (1992)
logP -0.89			EXP	HANSCH,C ET AL. (1995)
VP 3.33E-002	mm Hg	25	EST	NEELY,WB & BLAU,GE (1985)
DC	pKa			
HL 4.49E-012	atm m3/mol	25	EST	MEYLAN,WM & HOWARD,PH (1991)
OH 2.10E-011	cm3/molc sec	25	EST	MEYLAN,WM & HOWARD,PH (1993)

CAS #: 000557-00-6 — PROPYL ISOVALERATE

Formula: $C_8H_{16}O_2$

Mol Weight: 144.22

MP (deg C):   FP (deg C):

BP (deg C): 155.9

BP pressure (mm Hg):

Property/Value	Units	Temp	Data Type	Reference
WS 3.57E+002	mg/L	25	EST	MEYLAN,WM ET AL. (1996)
logP 2.76			EST	MEYLAN,WM & HOWARD,PH (1995)
VP 2.59E+000	mm Hg	25	EXP	JORDAN,TE (1954)
DC	pKa			
HL 7.23E-004	atm m3/mol	25	EST	MEYLAN,WM & HOWARD,PH (1991)
OH 6.73E-012	cm3/molc sec	25	EST	MEYLAN,WM & HOWARD,PH (1993)

CAS #: 000557-01-7 — 2-PYRIMIDONE

Formula: $C_4H_4N_2O$

Mol Weight: 96.09

MP (deg C):   FP (deg C):

BP (deg C):

BP pressure (mm Hg):

Property/Value	Units	Temp	Data Type	Reference
WS 1.42E+005	mg/L	25	EST	MEYLAN,WM ET AL. (1996)
logP -1.62			EXP	HANSCH,C & LEO,AJ (1985)
VP 2.72E-002	mm Hg	25	EST	NEELY,WB & BLAU,GE (1985)
DC	pKa			
HL 4.68E-007	atm m3/mol	25	EST	MEYLAN,WM & HOWARD,PH (1991)
OH 3.68E-011	cm3/molc sec	25	EST	MEYLAN,WM & HOWARD,PH (1993)

METHYL PROPYL ETHER

CAS #:	000557-17-5			

Formula: $C_4H_{10}O$

Mol Weight: 74.12

MP (deg C): | FP (deg C):

BP (deg C): 38.8

BP pressure (mm Hg):

Property/Value	Units	Temp	Data Type	Reference
WS 3.05E+004	mg/L	25	EXP	YALKOWSKY,SH & DANNENFELSER,RM (1992)
logP 1.21			EXP	HANSCH,C ET AL. (1995)
VP 4.65E+002	mm Hg	25	EXP	BOUBLIK,T ET AL. (1984)
DC	pKa			
HL 1.49E-003	atm m3/mol	25	EST	VP/WSOL
OH 1.13E-011	cm3/molc sec	25	EST	MEYLAN,WM & HOWARD,PH (1993)

1-MONOBUTYRIN

CAS #:	000557-25-5			

Formula: $C_7H_{14}O_4$

Mol Weight: 162.19

MP (deg C): | FP (deg C):

BP (deg C): 280

BP pressure (mm Hg):

Property/Value	Units	Temp	Data Type	Reference
WS 9.34E+004	mg/L	25	EST	MEYLAN,WM ET AL. (1996)
logP -0.17			EXP	HANSCH,C & LEO,AJ (1985)
VP 2.64E-004	mm Hg	25	EST	NEELY,WB & BLAU,GE (1985)
DC	pKa			
HL 7.28E-010	atm m3/mol	25	EST	MEYLAN,WM & HOWARD,PH (1991)
OH 1.86E-011	cm3/molc sec	25	EST	MEYLAN,WM & HOWARD,PH (1993)

ALLYL ETHYL ETHER

CAS #:	000557-31-3			

Formula: $C_5H_{10}O$

Mol Weight: 86.13

MP (deg C): | FP (deg C):

BP (deg C): 67.6

BP pressure (mm Hg):

Property/Value	Units	Temp	Data Type	Reference
WS 8.04E+003	mg/L	25	EST	MEYLAN,WM ET AL. (1996)
logP 1.40			EST	MEYLAN,WM & HOWARD,PH (1995)
VP 1.53E+002	mm Hg	20	EXP	AMBROSE,D ET AL. (1976)
DC	pKa			
HL 1.50E-003	atm m3/mol	25	EST	MEYLAN,WM & HOWARD,PH (1991)
OH 3.82E-011	cm3/molc sec	25	EST	MEYLAN,WM & HOWARD,PH (1993)

2-IODOOCTANE

CAS #:	000557-36-8			

Formula: $C_8H_{17}I$

Mol Weight: 240.13

MP (deg C): | FP (deg C):

BP (deg C): 210

BP pressure (mm Hg):

Property/Value	Units	Temp	Data Type	Reference
WS 1.57E+000	mg/L	25	EST	MEYLAN,WM ET AL. (1996)
logP 4.96			EST	MEYLAN,WM & HOWARD,PH (1995)
VP 1.37E+000	mm Hg	25	EXP	LEVANOVA,SV ET AL. (1967)
DC	pKa			
HL 3.99E-002	atm m3/mol	25	EST	MEYLAN,WM & HOWARD,PH (1991)
OH 7.50E-012	cm3/molc sec	25	EST	MEYLAN,WM & HOWARD,PH (1993)

ALLYL ETHER

CAS #:	000557-40-4			

Formula: $C_6H_{10}O$

Mol Weight: 98.15

MP (deg C): -6 | FP (deg C):

BP (deg C): 94

BP pressure (mm Hg):

Property/Value	Units	Temp	Data Type	Reference
WS 9.36E+004	mg/L	25	EXP	SUZUKI,T (1991)
logP 1.76			EST	MEYLAN,WM & HOWARD,PH (1995)
VP 5.63E+001	mm Hg	25	EST	NEELY,WB & BLAU,GE (1985)
DC	pKa			
HL 1.49E-003	atm m3/mol	25	EST	MEYLAN,WM & HOWARD,PH (1991)
OH 6.40E-011	cm3/molc sec	25	EST	MEYLAN,WM & HOWARD,PH (1993)

ETHYLAMINE, HYDROCHLORIDE

CAS #:	000557-66-4			

Formula: C_2H_8ClN

Mol Weight: 81.55

MP (deg C): 109.5 | FP (deg C):

BP (deg C):

BP pressure (mm Hg):

Property/Value	Units	Temp	Data Type	Reference
WS 1.00E+006	mg/L	25	EST	MEYLAN,WM ET AL. (1996)
logP -3.33			EST	MEYLAN,WM & HOWARD,PH (1995)
VP 3.62E-004	mm Hg	25	EST	NEELY,WB & BLAU,GE (1985)
DC	pKa			
HL 9.32E-013	atm m3/mol	25	EST	MEYLAN,WM & HOWARD,PH (1991)
OH 8.85E-012	cm3/molc sec	25	EST	MEYLAN,WM & HOWARD,PH (1993)

1,1-DIBROMOETHANE

CAS #:	000557-91-5			

Formula: $C_2H_4Br_2$

Mol Weight: 187.87

MP (deg C): -63 | FP (deg C):

BP (deg C): 109

BP pressure (mm Hg):

Property/Value	Units	Temp	Data Type	Reference
WS 1.12E+003	mg/L	25	EST	MEYLAN,WM ET AL. (1996)
logP 1.94			EST	MEYLAN,WM & HOWARD,PH (1995)
VP 2.56E+001	mm Hg	25	EXP	YAWS,CL (1994)
DC	pKa			
HL 1.30E-003	atm m3/mol	25	EST	MEYLAN,WM & HOWARD,PH (1991)
OH 2.15E-013	cm3/molc sec	25	EST	MEYLAN,WM & HOWARD,PH (1993)

2-CHLOROPROPENE

CAS #:	000557-98-2			

Formula: C_3H_5Cl

Mol Weight: 76.53

MP (deg C): -137.4 | FP (deg C):

BP (deg C): 22.6

BP pressure (mm Hg):

Property/Value	Units	Temp	Data Type	Reference
WS 2.60E+003	mg/L	25	EST	MEYLAN,WM ET AL. (1996)
logP 2.00			EXP	HANSCH,C ET AL. (1995)
VP 8.19E+002	mm Hg	25	EXP	DAUBERT,TE & DANNER,RP (1989)
DC	pKa			
HL 6.96E-002	atm m3/mol	25	EST	MEYLAN,WM & HOWARD,PH (1991)
OH 1.09E-011	cm3/molc sec	25	EST	MEYLAN,WM & HOWARD,PH (1993)

CARBON TETRABROMIDE

CAS #:	000558-13-4			CARBON TETRABROMIDE

Formula: CBr_4

Mol Weight: 331.65

MP (deg C): 90.1   FP (deg C):

BP (deg C): 189.5

BP pressure (mm Hg):

Property/Value	Units	Temp	Data Type	Reference
WS 2.40E+002	mg/L	30	EXP	GROSS,PM & SAYLOR,JH (1931)
logP 3.42			EXP	HANSCH,C ET AL. (1995)
VP 2.70E-001	mm Hg	25	EXT	KUDCHADKER,AP ET AL. (1979)
DC	pKa			
HL 1.82E+000	atm m3/mol	25	EST	VP/WSOL
OH 0.00E+000	cm3/molc sec	25	EST	MEYLAN,WM & HOWARD,PH (1993)

FLUOROMETHYL SULFONE

CAS #:	000558-25-8			FLUOROMETHYL SULFONE

Formula: CH_3FO_2S

Mol Weight: 98.10

MP (deg C):   FP (deg C):

BP (deg C):

BP pressure (mm Hg):

Property/Value	Units	Temp	Data Type	Reference
WS 5.00E+004	mg/L		EXP	SHIU,WY ET AL. (1990)
logP 0.55			EXP	HANSCH,C ET AL. (1995)
VP 1.09E+001	mm Hg	20	EXP	SNOW,AW & BARGER,WR (1988)
DC	pKa			
HL 2.81E-005	atm m3/mol	20	EST	VP/WSOL
OH 1.06E-012	cm3/molc sec	25	EST	MEYLAN,WM & HOWARD,PH (1993)

2,2-DIMETHYLOXIRANE

CAS #:	000558-30-5			2,2-DIMETHYLOXIRANE

Formula: C_4H_8O

Mol Weight: 72.11

MP (deg C):   FP (deg C):

BP (deg C): 52

BP pressure (mm Hg):

Property/Value	Units	Temp	Data Type	Reference
WS 2.65E+004	mg/L	25	EST	MEYLAN,WM ET AL. (1996)
logP 0.83			EST	MEYLAN,WM & HOWARD,PH (1995)
VP 3.87E+002	mm Hg	25	EST	NEELY,WB & BLAU,GE (1985)
DC	pKa			
HL 2.12E-004	atm m3/mol	25	EST	MEYLAN,WM & HOWARD,PH (1991)
OH 1.35E-012	cm3/molc sec	25	EST	MEYLAN,WM & HOWARD,PH (1993)

3,3-DIMETHYL-1-BUTENE

CAS #:	000558-37-2			3,3-DIMETHYL-1-BUTENE

Formula: C_6H_{12}

Mol Weight: 84.16

MP (deg C): -115.2   FP (deg C):

BP (deg C): 41.2

BP pressure (mm Hg):

Property/Value	Units	Temp	Data Type	Reference
WS 9.44E+001	mg/L	25	EST	MEYLAN,WM ET AL. (1996)
logP 3.04			EST	MEYLAN,WM & HOWARD,PH (1995)
VP 4.31E+002	mm Hg	25	EXP	YAWS,CL (1994A)
DC	pKa			
HL 3.59E-001	atm m3/mol	25	EST	MEYLAN,WM & HOWARD,PH (1991)
OH 2.84E-011	cm3/molc sec	25	EXP	ATKINSON,R (1989)

2,3,3-TRIMETHYLPENTANE

CAS #:	000560-21-4			2,3,3-TRIMETHYLPENTANE

Formula: C_8H_{18}

Mol Weight: 114.23

MP (deg C): -100.9   FP (deg C):

BP (deg C): 114.8

BP pressure (mm Hg):

Property/Value	Units	Temp	Data Type	Reference
WS 9.91E+000	mg/L	25	EST	MEYLAN,WM ET AL. (1996)
logP 4.09			EST	MEYLAN,WM & HOWARD,PH (1995)
VP 2.70E+001	mm Hg	25	EXP	DAUBERT,TE & DANNER,RP (1989)
DC	pKa			
HL 3.01E+000	atm m3/mol	25	EST	MEYLAN,WM & HOWARD,PH (1991)
OH 4.37E-012	cm3/molc sec	25	EST	MEYLAN,WM & HOWARD,PH (1993)

2,3,3-TRIMETHYL-1-PENTENE

CAS #:	000560-23-6			2,3,3-TRIMETHYL-1-PENTENE

Formula: C_8H_{16}

Mol Weight: 112.22

MP (deg C): -69   FP (deg C):

BP (deg C): 108.3

BP pressure (mm Hg):

Property/Value	Units	Temp	Data Type	Reference
WS 1.02E+001	mg/L	25	EST	MEYLAN,WM ET AL. (1996)
logP 4.08			EST	MEYLAN,WM & HOWARD,PH (1995)
VP 3.02E+001	mm Hg	25	EXP	ZWOLINSKI,BJ & WILHOIT,RC (1971)
DC	pKa			
HL 7.46E-001	atm m3/mol	25	EST	MEYLAN,WM & HOWARD,PH (1991)
OH 5.32E-011	cm3/molc sec	25	EST	MEYLAN,WM & HOWARD,PH (1993)

HEROIN

CAS #:	000561-27-3			HEROIN

Formula: $C_{21}H_{23}NO_5$

Mol Weight: 369.42

MP (deg C): 173   FP (deg C):

BP (deg C): 272-274

BP pressure (mm Hg): 1.20E+001

Property/Value	Units	Temp	Data Type	Reference
WS 6.00E+002	mg/L	25	EXP	SEIDELL,A (1941)
logP 0.96			EXP	SANGSTER,J (1994)
VP 7.59E-010	mm Hg	25	EXP	LAWRENCE,AH ET AL. (1984)
DC	pKa			
HL 6.15E-013	atm m3/mol	25	EST	VP/WSOL
OH 2.23E-010	cm3/molc sec	25	EST	MEYLAN,WM & HOWARD,PH (1993)

BARBITURIC ACID,5-ALLYL-5-NEOPENTYL

CAS #:	000561-83-1			BARBITURIC ACID,5-ALLYL-5-NEOPENTYL

Formula: $C_{12}H_{18}N_2O_3$

Mol Weight: 238.29

MP (deg C): 155-157   FP (deg C):

BP (deg C):

BP pressure (mm Hg):

Property/Value	Units	Temp	Data Type	Reference
WS 4.78E+002	mg/L	25	EST	MEYLAN,WM ET AL. (1996)
logP 2.06			EXP	HANSCH,C ET AL. (1995)
VP 1.40E-011	mm Hg	25	EST	NEELY,WB & BLAU,GE (1985)
DC	pKa			
HL 8.35E-013	atm m3/mol	25	EST	MEYLAN,WM & HOWARD,PH (1991)
OH 3.36E-011	cm3/molc sec	25	EST	MEYLAN,WM & HOWARD,PH (1993)

CAS #: 000561-86-4 — BARBITURIC ACID,5-ALLYL-5(2BR ALLYL)

Formula: $C_{10}H_{11}BrN_2O_3$

Mol Weight: 287.12

MP (deg C): 168-169

FP (deg C):

BP (deg C):

BP pressure (mm Hg):

Property/Value	Units	Temp	Data Type	Reference
WS 9.86E+002	mg/L	25	EST	MEYLAN,WM ET AL. (1996)
logP 1.37			EXP	HANSCH,C & LEO,AJ (1985)
VP 2.85E-012	mm Hg	25	EST	NEELY,WB & BLAU,GE (1985)
DC	pKa			
HL 7.03E-014	atm m3/mol	25	EST	MEYLAN,WM & HOWARD,PH (1991)
OH 4.89E-011	cm3/molc sec	25	EST	MEYLAN,WM & HOWARD,PH (1993)

CAS #: 000562-26-5 — 4-PIPERIDINECARBOXYLIC ACID, 1-(3-HYDROXY-3-PHEN

Formula: $C_{23}H_{29}NO_3$

Mol Weight: 367.49

MP (deg C):

FP (deg C):

BP (deg C):

BP pressure (mm Hg):

Property/Value	Units	Temp	Data Type	Reference
WS 1.75E+001	mg/L	25	EST	MEYLAN,WM ET AL. (1996)
logP 4.04			EXP	SANGSTER,J (1994)
VP 2.69E-011	mm Hg	25	EST	NEELY,WB & BLAU,GE (1985)
DC	pKa			
HL 5.00E-014	atm m3/mol	25	EST	MEYLAN,WM & HOWARD,PH (1991)
OH 1.21E-010	cm3/molc sec	25	EST	MEYLAN,WM & HOWARD,PH (1993)

CAS #: 000562-49-2 — 3,3-DIMETHYLPENTANE

Formula: C_7H_{16}

Mol Weight: 100.21

MP (deg C): -134.9

FP (deg C):

BP (deg C): 86.0

BP pressure (mm Hg):

Property/Value	Units	Temp	Data Type	Reference
WS 5.92E+000	mg/L	25	EXP	YALKOWSKY,SH & DANNENFELSER,RM (1992)
logP 3.67			EST	MEYLAN,WM & HOWARD,PH (1995)
VP 8.27E+001	mm Hg	25	EXP	DAUBERT,TE & DANNER,RP (1989)
DC	pKa			
HL 1.84E+000	atm m3/mol	25	EST	VP/WSOL
OH 2.97E-012	cm3/molc sec	25	EST	MEYLAN,WM & HOWARD,PH (1993)

CAS #: 000563-12-2 — ETHION

Formula: $C_9H_{22}O_4P_2S_4$

Mol Weight: 384.48

MP (deg C): -13

FP (deg C):

BP (deg C): 165

BP pressure (mm Hg): 3.00E-001

Property/Value	Units	Temp	Data Type	Reference
WS 6.00E-001	mg/L	25	EXP	SHAROM,MS ET AL. (1980A)
logP 5.07			EXP	HANSCH,C & LEO,AJ (1985)
VP 1.13E-006	mm Hg	25	EXP	SUNTIO,LR ET AL. (1988)
DC	pKa			
HL 9.53E-007	atm m3/mol	25	EST	VP/WSOL
OH 5.76E-010	cm3/molc sec	25	EST	MEYLAN,WM & HOWARD,PH (1993)

CAS #: 000563-16-6 — 3,3-DIMETHYL HEXANE

Formula: C_8H_{18}

Mol Weight: 114.23

MP (deg C): -126.1

FP (deg C):

BP (deg C): 111.9

BP pressure (mm Hg):

Property/Value	Units	Temp	Data Type	Reference
WS 8.58E+000	mg/L	25	EST	MEYLAN,WM ET AL. (1996)
logP 4.16			EST	MEYLAN,WM & HOWARD,PH (1995)
VP 2.86E+001	mm Hg	25	EXP	DAUBERT,TE & DANNER,RP (1989)
DC	pKa			
HL 3.01E+000	atm m3/mol	25	EST	MEYLAN,WM & HOWARD,PH (1991)
OH 4.38E-012	cm3/molc sec	25	EST	MEYLAN,WM & HOWARD,PH (1993)

CAS #: 000563-41-7 — SEMICARBAZIDE

Formula: CH_6ClN_3O

Mol Weight: 111.53

MP (deg C): 96

FP (deg C):

BP (deg C):

BP pressure (mm Hg):

Property/Value	Units	Temp	Data Type	Reference
WS 1.00E+006	mg/L	25	EST	MEYLAN,WM ET AL. (1996)
logP -2.75			EXP	HANSCH,C & LEO,AJ (1985)
VP 1.31E-001	mm Hg	25	EST	NEELY,WB & BLAU,GE (1985)
DC	pKa			
HL 1.52E-012	atm m3/mol	25	EST	MEYLAN,WM & HOWARD,PH (1991)
OH 2.00E-012	cm3/molc sec	25	EST	MEYLAN,WM & HOWARD,PH (1993)

CAS #: 000563-45-1 — 3-METHYL-1-BUTENE

Formula: C_5H_{10}

Mol Weight: 70.14

MP (deg C): -168.5

FP (deg C):

BP (deg C): 20.1

BP pressure (mm Hg):

Property/Value	Units	Temp	Data Type	Reference
WS 1.30E+002	mg/L	25	EXP	YALKOWSKY,SH & DANNENFELSER,RM (1992)
logP 2.59			EST	MEYLAN,WM & HOWARD,PH (1995)
VP 9.03E+002	mm Hg	25	EXP	YAWS,CL (1994A)
DC	pKa			
HL 5.40E-001	atm m3/mol	25	EST	VP/WSOL
OH 3.18E-011	cm3/molc sec	25	EXP	ATKINSON,R (1989)

CAS #: 000563-46-2 — 2-METHYL-1-BUTENE

Formula: C_5H_{10}

Mol Weight: 70.14

MP (deg C): -137.5

FP (deg C):

BP (deg C): 31.2

BP pressure (mm Hg):

Property/Value	Units	Temp	Data Type	Reference
WS 1.30E+002	mg/L	25	EXP	SUZUKI,T (1991)
logP 2.72			EST	MEYLAN,WM & HOWARD,PH (1995)
VP 6.10E+002	mm Hg	25	EXP	DAUBERT,TE & DANNER,RP (1989)
DC	pKa			
HL 4.31E-001	atm m3/mol	25	EST	VP/WSOL
OH 6.07E-011	cm3/molc sec	25	EXP	ATKINSON,R (1989)

CAS #: 000563-47-3				3-CHLORO-2-METHYLPROPENE

Formula: C_4H_7Cl

Mol Weight: 90.55

MP (deg C):   FP (deg C):

BP (deg C): 71.5

BP pressure (mm Hg):

Property/ Value	Units	Temp	Data Type	Reference
WS 1.40E+003	mg/L	25	EXP	CHEM INSPECT TEST INST (1992)
logP 2.48			EST	MEYLAN,WM & HOWARD,PH (1995)
VP 1.02E+002	mm Hg	20	EXP	PARMEGGIANI,L (1983)
DC	pKa			
HL 8.70E-003	atm m3/mol	25	EST	VP/WSOL
OH 3.96E-011	cm3/molc sec	25	EST	MEYLAN,WM & HOWARD,PH (1993)

CAS #: 000563-52-0				3-CHLORO-1-BUTENE

Formula: C_4H_7Cl

Mol Weight: 90.55

MP (deg C):   FP (deg C):

BP (deg C): 63.9-64.2

BP pressure (mm Hg):

Property/ Value	Units	Temp	Data Type	Reference
WS 1.22E+003	mg/L	25	EST	MEYLAN,WM ET AL. (1996)
logP 2.35			EST	MEYLAN,WM & HOWARD,PH (1995)
VP 1.53E+002	mm Hg	20	EXP	WEBER,RC ET AL. (1981)
DC	pKa			
HL 4.52E-002	atm m3/mol	25	EST	MEYLAN,WM & HOWARD,PH (1991)
OH 2.08E-011	cm3/molc sec	25	EST	MEYLAN,WM & HOWARD,PH (1993)

CAS #: 000563-54-2				1,2-DICHLOROPROPENE

Formula: $C_3H_4Cl_2$

Mol Weight: 110.97

MP (deg C):   FP (deg C):

BP (deg C):

BP pressure (mm Hg):

Property/ Value	Units	Temp	Data Type	Reference
WS 2.70E+004	mg/L	25	EXP	GUNTHER,FA ET AL. (1968)
logP 2.53			EST	MEYLAN,WM & HOWARD,PH (1995)
VP 9.08E+001	mm Hg	20	EXP	WEBER,RC ET AL. (1981)
DC	pKa			
HL 4.91E-004	atm m3/mol	20	EST	VP/WSOL
OH 3.97E-012	cm3/molc sec	25	EST	MEYLAN,WM & HOWARD,PH (1993)

CAS #: 000563-57-5				3,3-DICHLOROPROPENE

Formula: $C_3H_4Cl_2$

Mol Weight: 110.97

MP (deg C):   FP (deg C):

BP (deg C): 84.4

BP pressure (mm Hg):

Property/ Value	Units	Temp	Data Type	Reference
WS 1.69E+003	mg/L	25	EST	MEYLAN,WM ET AL. (1996)
logP 2.11			EST	MEYLAN,WM & HOWARD,PH (1995)
VP 7.05E+001	mm Hg	20	EXP	WEBER,RC ET AL. (1981)
DC	pKa			
HL 1.20E-002	atm m3/mol	25	EST	MEYLAN,WM & HOWARD,PH (1991)
OH 2.03E-011	cm3/molc sec	25	EST	MEYLAN,WM & HOWARD,PH (1993)

CAS #: 000563-58-6				1,1-DICHLOROPROPENE

Formula: $C_3H_4Cl_2$

Mol Weight: 110.97

MP (deg C):   FP (deg C):

BP (deg C): 76.5

BP pressure (mm Hg):

Property/ Value	Units	Temp	Data Type	Reference
WS 7.49E+002	mg/L	25	EST	MEYLAN,WM ET AL. (1996)
logP 2.53			EST	MEYLAN,WM & HOWARD,PH (1995)
VP 9.08E+001	mm Hg	20	EXP	WEBER,RC ET AL. (1981)
DC	pKa			
HL 5.00E-002	atm m3/mol	25	EST	MEYLAN,WM & HOWARD,PH (1991)
OH 3.97E-012	cm3/molc sec	25	EST	MEYLAN,WM & HOWARD,PH (1993)

CAS #: 000563-78-0				2,3-DIMETHYL-1-BUTENE

Formula: C_6H_{12}

Mol Weight: 84.16

MP (deg C): -157.3   FP (deg C):

BP (deg C): 55.6

BP pressure (mm Hg):

Property/ Value	Units	Temp	Data Type	Reference
WS 7.85E+001	mg/L	25	EST	MEYLAN,WM ET AL. (1996)
logP 3.13			EST	MEYLAN,WM & HOWARD,PH (1995)
VP 2.52E+002	mm Hg	25	EXP	DAUBERT,TE & DANNER,RP (1989)
DC	pKa			
HL 4.23E-001	atm m3/mol	25	EST	MEYLAN,WM & HOWARD,PH (1991)
OH 5.38E-011	cm3/molc sec	25	EST	MEYLAN,WM & HOWARD,PH (1993)

CAS #: 000563-79-1				2,3-DIMETHYL-2-BUTENE

Formula: C_6H_{12}

Mol Weight: 84.16

MP (deg C): -74.6   FP (deg C):

BP (deg C): 73.3

BP pressure (mm Hg):

Property/ Value	Units	Temp	Data Type	Reference
WS 7.10E+001	mg/L	25	EST	MEYLAN,WM ET AL. (1996)
logP 3.19			EST	MEYLAN,WM & HOWARD,PH (1995)
VP 1.26E+002	mm Hg	25	EXP	DAUBERT,TE & DANNER,RP (1989)
DC	pKa			
HL 5.90E-001	atm m3/mol	25	EST	MEYLAN,WM & HOWARD,PH (1991)
OH 1.10E-010	cm3/molc sec	25	EXP	ATKINSON,R (1989)

CAS #: 000563-80-4				3-METHYL-2-BUTANONE

Formula: $C_5H_{10}O$

Mol Weight: 86.13

MP (deg C): -92   FP (deg C):

BP (deg C): 94.3

BP pressure (mm Hg):

Property/ Value	Units	Temp	Data Type	Reference
WS 6.08E+004	mg/L	25	EXP	YALKOWSKY,SH & DANNENFELSER,RM (1992)
logP 0.84			EXP	HANSCH,C ET AL. (1995)
VP 5.22E+001	mm Hg	25	EXP	DAUBERT,TE & DANNER,RP (1989)
DC	pKa			
HL 9.73E-005	atm m3/mol	25	EST	VP/WSOL
OH 2.62E-012	cm3/molc sec	25	EST	MEYLAN,WM & HOWARD,PH (1993)

CAS #: 000564-00-1				1,2:3,4-DIEPOXYBUTANE

Formula: $C_4H_6O_2$

Mol Weight: 86.09

MP (deg C): -19   FP (deg C):

BP (deg C): 140-142

BP pressure (mm Hg): 7.61E+002

Property/Value	Units	Temp	Data Type	Reference
WS 1.00E+006	mg/L		EXP	MERCK INDEX (1983)
logP -0.52			EXP	HANSCH,C ET AL. (1995)
VP 7.31E+000	mm Hg	25	EST	MERCK INDEX (1983)
DC	pKa			
HL 3.54E-008	atm m3/mol	25	EST	MEYLAN,WM & HOWARD,PH (1991)
OH 8.56E-013	cm3/molc sec	25	EST	MEYLAN,WM & HOWARD,PH (1993)

CAS #: 000564-02-3				2,2,3-TRIMETHYLPENTANE

Formula: C_8H_{18}

Mol Weight: 114.23

MP (deg C): -112.2   FP (deg C):

BP (deg C): 110

BP pressure (mm Hg):

Property/Value	Units	Temp	Data Type	Reference
WS 9.91E+000	mg/L	25	EST	MEYLAN,WM ET AL. (1996)
logP 4.09			EST	MEYLAN,WM & HOWARD,PH (1995)
VP 3.21E+001	mm Hg	25	EXP	DAUBERT,TE & DANNER,RP (1989)
DC	pKa			
HL 3.01E+000	atm m3/mol	25	EST	MEYLAN,WM & HOWARD,PH (1991)
OH 4.92E-012	cm3/molc sec	25	EST	MEYLAN,WM & HOWARD,PH (1993)

CAS #: 000564-25-0				VIBRAMYCIN

Formula: $C_{22}H_{24}N_2O_8$

Mol Weight: 444.45

MP (deg C):   FP (deg C):

BP (deg C):

BP pressure (mm Hg):

Property/Value	Units	Temp	Data Type	Reference
WS 6.30E+002	mg/L	25	EXP	YALKOWSKY,SH & DANNENFELSER,RM (1992)
logP -0.02			EXP	SANGSTER,J (1994)
VP 1.42E-023	mm Hg	25	EST	NEELY,WB & BLAU,GE (1985)
DC	pKa			
HL 4.66E-024	atm m3/mol	25	EST	MEYLAN,WM & HOWARD,PH (1991)
OH 2.66E-010	cm3/molc sec	25	EST	MEYLAN,WM & HOWARD,PH (1993)

CAS #: 000565-33-3				1-CYCLOHEXYL-3-(4-METHYLMETANILYL)UREA

Formula: $C_{14}H_{21}N_3O_3S$

Mol Weight: 311.41

MP (deg C):   FP (deg C):

BP (deg C):

BP pressure (mm Hg):

Property/Value	Units	Temp	Data Type	Reference
WS 1.18E+002	mg/L	25	EST	MEYLAN,WM ET AL. (1996)
logP 2.29			EST	MEYLAN,WM & HOWARD,PH (1995)
VP 1.88E-010	mm Hg	25	EST	NEELY,WB & BLAU,GE (1985)
DC	pKa			
HL 6.95E-014	atm m3/mol	25	EST	MEYLAN,WM & HOWARD,PH (1991)
OH 4.94E-011	cm3/molc sec	25	EST	MEYLAN,WM & HOWARD,PH (1993)

CAS #: 000565-59-3				2,3-DIMETHYLPENTANE

Formula: C_7H_{16}

Mol Weight: 100.21

MP (deg C):   FP (deg C):

BP (deg C): 89.7

BP pressure (mm Hg):

Property/Value	Units	Temp	Data Type	Reference
WS 5.25E+000	mg/L	25	EXP	YALKOWSKY,SH & DANNENFELSER,RM (1992)
logP 3.63			EST	MEYLAN,WM & HOWARD,PH (1995)
VP 6.89E+001	mm Hg	25	EXP	DAUBERT,TE & DANNER,RP (1989)
DC	pKa			
HL 1.73E+000	atm m3/mol	25	EST	VP/WSOL
OH 7.14E-012	cm3/molc sec	25	EST	MEYLAN,WM & HOWARD,PH (1993)

CAS #: 000565-60-6				3-METHYL-2-PENTANOL

Formula: $C_6H_{14}O$

Mol Weight: 102.18

MP (deg C):   FP (deg C):

BP (deg C): 134.3

BP pressure (mm Hg):

Property/Value	Units	Temp	Data Type	Reference
WS 1.94E+004	mg/L	25	EXP	YALKOWSKY,SH & DANNENFELSER,RM (1992)
logP 1.68			EST	MEYLAN,WM & HOWARD,PH (1995)
VP 5.38E+000	mm Hg	25	EST	NEELY,WB & BLAU,GE (1985)
DC	pKa			
HL 1.76E-005	atm m3/mol	25	EST	MEYLAN,WM & HOWARD,PH (1991)
OH 1.31E-011	cm3/molc sec	25	EST	MEYLAN,WM & HOWARD,PH (1993)

CAS #: 000565-61-7				3-METHYLPENTAN-2-ONE

Formula: $C_6H_{12}O$

Mol Weight: 100.16

MP (deg C):   FP (deg C):

BP (deg C): 118

BP pressure (mm Hg): 7.58E+002

Property/Value	Units	Temp	Data Type	Reference
WS 2.09E+004	mg/L	25	EXP	YALKOWSKY,SH & DANNENFELSER,RM (1992)
logP 1.16			EST	MEYLAN,WM & HOWARD,PH (1995)
VP 1.16E+001	mm Hg	25	EXP	AMBROSE,D ET AL. (1975)
DC	pKa			
HL 7.32E-005	atm m3/mol	25	EST	VP/WSOL
OH 6.23E-012	cm3/molc sec	25	EST	MEYLAN,WM & HOWARD,PH (1993)

CAS #: 000565-64-0				2,3-DICHLOROPROPIONIC ACID

Formula: $C_3H_4Cl_2O_2$

Mol Weight: 142.97

MP (deg C): 50   FP (deg C):

BP (deg C): 210

BP pressure (mm Hg):

Property/Value	Units	Temp	Data Type	Reference
WS 2.76E+004	mg/L	25	EST	MEYLAN,WM ET AL. (1996)
logP 1.01			EST	MEYLAN,WM & HOWARD,PH (1995)
VP 5.37E-002	mm Hg	25	EST	NEELY,WB & BLAU,GE (1985)
DC	pKa			
HL 9.02E-008	atm m3/mol	25	EST	MEYLAN,WM & HOWARD,PH (1991)
OH 8.47E-013	cm3/molc sec	25	EST	MEYLAN,WM & HOWARD,PH (1993)

000565-67-3 — 2-METHYL-3-PENTANOL

Formula: $C_6H_{14}O$

Mol Weight: 102.18

MP (deg C): FP (deg C):

BP (deg C): 126.5

BP pressure (mm Hg):

Property/Value	Units	Temp	Data Type	Reference
WS 2.01E+004	mg/L	25	EXP	YALKOWSKY,SH & DANNENFELSER,RM (1992)
logP 1.68			EST	MEYLAN,WM & HOWARD,PH (1995)
VP 5.38E+000	mm Hg	25	EST	NEELY,WB & BLAU,GE (1985)
DC	pKa			
HL 1.76E-005	atm m3/mol	25	EST	MEYLAN,WM & HOWARD,PH (1991)
OH 1.44E-011	cm3/molc sec	25	EST	MEYLAN,WM & HOWARD,PH (1993)

000565-69-5 — 2-METHYL-3-PENTANONE

Formula: $C_6H_{12}O$

Mol Weight: 100.16

MP (deg C): FP (deg C):

BP (deg C): 113.5

BP pressure (mm Hg):

Property/Value	Units	Temp	Data Type	Reference
WS 1.55E+004	mg/L	25	EXP	SUZUKI,T (1991)
logP 1.16			EST	MEYLAN,WM & HOWARD,PH (1995)
VP 1.81E+001	mm Hg	25	EXP	DAUBERT,TE & DANNER,RP (1989)
DC	pKa			
HL 1.54E-004	atm m3/mol	25	EST	VP/WSOL
OH 3.75E-012	cm3/molc sec	25	EST	MEYLAN,WM & HOWARD,PH (1993)

000565-74-2 — 2-BROMO-3-METHYL-BUTANOIC ACID

Formula: $C_5H_9BrO_2$

Mol Weight: 181.04

MP (deg C): 44 FP (deg C):

BP (deg C): 230

BP pressure (mm Hg):

Property/Value	Units	Temp	Data Type	Reference
WS 4.27E+003	mg/L	25	EST	MEYLAN,WM ET AL. (1996)
logP 1.76			EST	MEYLAN,WM & HOWARD,PH (1995)
VP 6.95E-002	mm Hg	25	EST	NEELY,WB & BLAU,GE (1985)
DC	pKa			
HL 1.48E-007	atm m3/mol	25	EST	MEYLAN,WM & HOWARD,PH (1991)
OH 2.25E-012	cm3/molc sec	25	EST	MEYLAN,WM & HOWARD,PH (1993)

000565-75-3 — 2,3,4-TRIMETHYLPENTANE

Formula: C_8H_{18}

Mol Weight: 114.23

MP (deg C): -109.2 FP (deg C):

BP (deg C): 113.5

BP pressure (mm Hg):

Property/Value	Units	Temp	Data Type	Reference
WS 2.30E+000	mg/L	25	EXP	YALKOWSKY,SH & DANNENFELSER,RM (1992)
logP 4.05			EST	MEYLAN,WM & HOWARD,PH (1995)
VP 2.71E+001	mm Hg	25	EXP	DAUBERT,TE & DANNER,RP (1989)
DC	pKa			
HL 1.77E+000	atm m3/mol	25	EST	VP/WSOL
OH 7.00E-012	cm3/molc sec	25	EXP	ATKINSON,R (1989)

000565-77-5 — 2,3,4-TRIMETHYL-2-PENTENE

Formula: C_8H_{16}

Mol Weight: 112.22

MP (deg C): -113.4 FP (deg C):

BP (deg C): 116.5

BP pressure (mm Hg):

Property/Value	Units	Temp	Data Type	Reference
WS 9.90E+000	mg/L	25	EST	MEYLAN,WM ET AL. (1996)
logP 4.09			EST	MEYLAN,WM & HOWARD,PH (1995)
VP 2.25E+001	mm Hg	25	EXP	ZWOLINSKI,BJ & WILHOIT,RC (1971)
DC	pKa			
HL 1.04E+000	atm m3/mol	25	EST	MEYLAN,WM & HOWARD,PH (1991)
OH 1.13E-010	cm3/molc sec	25	EST	MEYLAN,WM & HOWARD,PH (1993)

000565-80-0 — 2,4-DIMETHYL-3-PENTANONE

Formula: $C_7H_{14}O$

Mol Weight: 114.19

MP (deg C): -69 FP (deg C):

BP (deg C): 125.4

BP pressure (mm Hg):

Property/Value	Units	Temp	Data Type	Reference
WS 5.70E+003	mg/L	25	EXP	YALKOWSKY,SH & DANNENFELSER,RM (1992)
logP 1.86			EXP	HANSCH,C ET AL. (1995)
VP 1.34E+001	mm Hg	25	EXP	YAWS,CL (1994A)
DC	pKa			
HL 3.53E-004	atm m3/mol	25	EST	VP/WSOL
OH 5.38E-012	cm3/molc sec	25	EXP	ATKINSON,R (1989)

000569-41-5 — 1,8-DIMETHYLNAPHTHALENE

Formula: $C_{12}H_{12}$

Mol Weight: 156.23

MP (deg C): 65 FP (deg C):

BP (deg C): 270

BP pressure (mm Hg):

Property/Value	Units	Temp	Data Type	Reference
WS 1.29E+001	mg/L	25	EST	MEYLAN,WM ET AL. (1996)
logP 4.26			EXP	HANSCH,C & LEO,AJ (1985)
VP 4.79E-003	mm Hg	25	EXT	OHE,S (1976)
DC	pKa			
HL 6.41E-004	atm m3/mol	25	EST	MEYLAN,WM & HOWARD,PH (1991)
OH 6.94E-011	cm3/molc sec	25	EST	MEYLAN,WM & HOWARD,PH (1993)

000569-51-7 — HEMIMELLITIC ACID

Formula: $C_9H_6O_6$

Mol Weight: 210.14

MP (deg C): 200 FP (deg C):

BP (deg C):

BP pressure (mm Hg):

Property/Value	Units	Temp	Data Type	Reference
WS 3.06E+004	mg/L	19	EXP	YALKOWSKY,SH & DANNENFELSER,RM (1992)
logP 0.61			EST	MEYLAN,WM & HOWARD,PH (1995)
VP 2.88E-008	mm Hg	25	EST	NEELY,WB & BLAU,GE (1985)
DC 2.84	pKa	25	EXP	KORTUM,G ET AL (1961)
HL 4.39E-017	atm m3/mol	25	EST	MEYLAN,WM & HOWARD,PH (1991)
OH 1.63E-012	cm3/molc sec	25	EST	MEYLAN,WM & HOWARD,PH (1993)

CAS #: 000569-61-9				PARAROSANILINE

Formula: $C_{19}H_{17}N_3$
Mol Weight: 287.37
MP (deg C): FP (deg C):
BP (deg C):
BP pressure (mm Hg):

Property/Value	Units	Temp	Data Type	Reference
WS 2.20E+004	mg/L	25	EST	MEYLAN,WM ET AL. (1996)
logP -0.21			EXP	TSAI,RS ET AL. (1991)
VP 9.26E-010	mm Hg	25	EST	NEELY,WB & BLAU,GE (1985)
DC	pKa			
HL 6.03E-013	atm m3/mol	25	EST	MEYLAN,WM & HOWARD,PH (1991)
OH 4.36E-010	cm3/molc sec	25	EST	MEYLAN,WM & HOWARD,PH (1993)

CAS #: 000569-64-2				MALACHITE GREEN

Formula: $C_{23}H_{25}ClN_2$
Mol Weight: 364.92
MP (deg C): FP (deg C):
BP (deg C):
BP pressure (mm Hg):

Property/Value	Units	Temp	Data Type	Reference
WS 4.00E+004	mg/L	25	EXP	BAUGHMAN,GL ET AL. (1993)
logP 0.80			EST	MEYLAN,WM & HOWARD,PH (1995)
VP 2.45E-013	mm Hg	25	EST	NEELY,WB & BLAU,GE (1985)
DC	pKa			
HL 1.93E-014	atm m3/mol	25	EST	MEYLAN,WM & HOWARD,PH (1991)
OH 4.41E-010	cm3/molc sec	25	EST	MEYLAN,WM & HOWARD,PH (1993)

CAS #: 000571-58-4				1,4-DIMETHYLNAPHTHALENE

Formula: $C_{12}H_{12}$
Mol Weight: 156.23
MP (deg C): 7.6 FP (deg C):
BP (deg C): 268
BP pressure (mm Hg):

Property/Value	Units	Temp	Data Type	Reference
WS 1.14E+001	mg/L	25	EXP	YALKOWSKY,SH & DANNENFELSER,RM (1992)
logP 4.37			EXP	HANSCH,C & LEO,AJ (1985)
VP 8.40E-003	mm Hg	25	EST	NEELY,WB & BLAU,GE (1985)
DC	pKa			
HL 6.41E-004	atm m3/mol	25	EST	MEYLAN,WM & HOWARD,PH (1991)
OH 6.94E-011	cm3/molc sec	25	EST	MEYLAN,WM & HOWARD,PH (1993)

CAS #: 000571-60-8				1,4-NAPHTHALENEDIOL

Formula: $C_{10}H_8O_2$
Mol Weight: 160.17
MP (deg C): FP (deg C):
BP (deg C):
BP pressure (mm Hg):

Property/Value	Units	Temp	Data Type	Reference
WS 6.19E+003	mg/L	25	EST	MEYLAN,WM ET AL. (1996)
logP 1.90			EXP	SANGSTER,J (1993)
VP 7.15E-006	mm Hg	25	EST	NEELY,WB & BLAU,GE (1985)
DC 9.58	pKa	14	EXP	KORTUM,G ET AL (1961)
HL 5.69E-012	atm m3/mol	25	EST	MEYLAN,WM & HOWARD,PH (1991)
OH 2.00E-010	cm3/molc sec	25	EST	MEYLAN,WM & HOWARD,PH (1993)

CAS #: 000571-61-9				1,5-DIMETHYLNAPHTHALENE

Formula: $C_{12}H_{12}$
Mol Weight: 156.23
MP (deg C): 82 FP (deg C):
BP (deg C): 265
BP pressure (mm Hg):

Property/Value	Units	Temp	Data Type	Reference
WS 2.74E+000	mg/L	25	EXP	YALKOWSKY,SH & DANNENFELSER,RM (1992)
logP 4.38			EXP	HANSCH,C & LEO,AJ (1985)
VP 8.40E-003	mm Hg	25	EST	NEELY,WB & BLAU,GE (1985)
DC	pKa			
HL 3.50E-004	atm m3/mol	25	EXP	MACKAY,D ET AL. (1982A)
OH 7.00E-011	cm3/molc sec	25	EST	MEYLAN,WM & HOWARD,PH (1993)

CAS #: 000573-11-5				2,3,4-TRIMETHOXYBENZOIC ACID

Formula: $C_{10}H_{12}O_5$
Mol Weight: 212.20
MP (deg C): 99-102 FP (deg C):
BP (deg C):
BP pressure (mm Hg):

Property/Value	Units	Temp	Data Type	Reference
WS 4.87E+003	mg/L	25	EST	MEYLAN,WM ET AL. (1996)
logP 1.04			EST	MEYLAN,WM & HOWARD,PH (1995)
VP 2.72E-005	mm Hg	25	EST	NEELY,WB & BLAU,GE (1985)
DC	pKa			
HL 2.24E-011	atm m3/mol	25	EST	MEYLAN,WM & HOWARD,PH (1991)
OH 7.90E-011	cm3/molc sec	25	EST	MEYLAN,WM & HOWARD,PH (1993)

CAS #: 000573-56-8				2,6-DINITROPHENOL

Formula: $C_6H_4N_2O_5$
Mol Weight: 184.11
MP (deg C): 63.5 FP (deg C):
BP (deg C):
BP pressure (mm Hg):

Property/Value	Units	Temp	Data Type	Reference
WS 3.15E+002	mg/L	15	EXP	YALKOWSKY,SH & DANNENFELSER,RM (1992)
logP 1.37			EXP	HANSCH,C & LEO,AJ (1985)
VP 1.20E-005	mm Hg	25	EST	NEELY,WB & BLAU,GE (1985)
DC 3.97	pKa	25	EXP	KORTUM,G ET AL (1961)
HL 2.76E-008	atm m3/mol	25	EST	MEYLAN,WM & HOWARD,PH (1991)
OH 6.61E-013	cm3/molc sec	25	EST	MEYLAN,WM & HOWARD,PH (1993)

CAS #: 000573-58-0				CONGO RED

Formula: $C_{32}H_{22}N_6Na_2O_6S_2$
Mol Weight: 696.68
MP (deg C): > 360 FP (deg C):
BP (deg C):
BP pressure (mm Hg):

Property/Value	Units	Temp	Data Type	Reference
WS 1.16E+005	mg/L	25	EXP	DEHN,WM (1917)
logP 2.63			EST	MEYLAN,WM & HOWARD,PH (1995)
VP 2.24E-030	mm Hg	25	EST	NEELY,WB & BLAU,GE (1985)
DC	pKa			
HL	atm m3/mol			
OH 2.33E-011	cm3/molc sec	25	EST	MEYLAN,WM & HOWARD,PH (1993)

CAS #: 000573-98-8 — 1,2-DIMETHYLNAPHTHALENE

Formula: $C_{12}H_{12}$
Mol Weight: 156.23
MP (deg C): 1.6
FP (deg C):
BP (deg C): 266.5
BP pressure (mm Hg):

Property/Value	Units	Temp	Data Type	Reference
WS 1.49E+001	mg/L	25	EST	MEYLAN,WM ET AL. (1996)
logP 4.31			EXP	HANSCH,C & LEO,AJ (1985)
VP 1.27E-002	mm Hg	25	EXP	RIDDICK,JA ET AL. (1986)
DC	pKa			
HL 6.41E-004	atm m3/mol	25	EST	MEYLAN,WM & HOWARD,PH (1991)
OH 6.94E-011	cm3/molc sec	25	EST	MEYLAN,WM & HOWARD,PH (1993)

CAS #: 000574-66-3 — METHANONE, DIPHENYL-, OXIME

Formula: $C_{13}H_{11}NO$
Mol Weight: 197.24
MP (deg C):
FP (deg C):
BP (deg C):
BP pressure (mm Hg):

Property/Value	Units	Temp	Data Type	Reference
WS 8.36E+001	mg/L	25	EST	MEYLAN,WM ET AL. (1996)
logP 3.20			EXP	HANSCH,C ET AL. (1995)
VP 2.02E-006	mm Hg	25	EST	NEELY,WB & BLAU,GE (1985)
DC	pKa			
HL 2.88E-008	atm m3/mol	25	EST	MEYLAN,WM & HOWARD,PH (1991)
OH 8.59E-012	cm3/molc sec	25	EST	MEYLAN,WM & HOWARD,PH (1993)

CAS #: 000574-98-1 — N-(2-BROMOETHYL)PHTHALIMIDE

Formula: $C_{10}H_8BrNO_2$
Mol Weight: 254.09
MP (deg C): 81-84
FP (deg C):
BP (deg C):
BP pressure (mm Hg):

Property/Value	Units	Temp	Data Type	Reference
WS 2.24E+002	mg/L	25	EST	MEYLAN,WM ET AL. (1996)
logP 2.34			EST	MEYLAN,WM & HOWARD,PH (1995)
VP 6.94E-008	mm Hg	25	EST	NEELY,WB & BLAU,GE (1985)
DC	pKa			
HL 3.43E-009	atm m3/mol	25	EST	MEYLAN,WM & HOWARD,PH (1991)
OH 2.83E-012	cm3/molc sec	25	EST	MEYLAN,WM & HOWARD,PH (1993)

CAS #: 000575-37-1 — 1,7-DIMETHYLNAPHTHALENE

Formula: $C_{12}H_{12}$
Mol Weight: 156.23
MP (deg C): -13.9
FP (deg C):
BP (deg C): 263
BP pressure (mm Hg):

Property/Value	Units	Temp	Data Type	Reference
WS 1.15E+001	mg/L	25	EST	MEYLAN,WM ET AL. (1996)
logP 4.44			EXP	HANSCH,C & LEO,AJ (1985)
VP 8.40E-003	mm Hg	25	EST	NEELY,WB & BLAU,GE (1985)
DC	pKa			
HL 6.41E-004	atm m3/mol	25	EST	MEYLAN,WM & HOWARD,PH (1991)
OH 6.94E-011	cm3/molc sec	25	EST	MEYLAN,WM & HOWARD,PH (1993)

CAS #: 000575-38-2 — NAPTHALENE-1,7-DIOL

Formula: $C_{10}H_8O_2$
Mol Weight: 160.17
MP (deg C): 180.5
FP (deg C):
BP (deg C):
BP pressure (mm Hg):

Property/Value	Units	Temp	Data Type	Reference
WS 5.73E+003	mg/L	25	EST	MEYLAN,WM ET AL. (1996)
logP 1.94			EXP	HANSCH,C & LEO,AJ (1985)
VP 7.15E-006	mm Hg	25	EST	NEELY,WB & BLAU,GE (1985)
DC	pKa			
HL 5.69E-012	atm m3/mol	25	EST	MEYLAN,WM & HOWARD,PH (1991)
OH 2.00E-010	cm3/molc sec	25	EST	MEYLAN,WM & HOWARD,PH (1993)

CAS #: 000575-41-7 — 1,3-DIMETHYLNAPTHALENE

Formula: $C_{12}H_{12}$
Mol Weight: 156.23
MP (deg C): -6
FP (deg C):
BP (deg C): 263
BP pressure (mm Hg):

Property/Value	Units	Temp	Data Type	Reference
WS 8.00E+000	mg/L	25	EXP	YALKOWSKY,SH & DANNENFELSER,RM (1992)
logP 4.42			EXP	HANSCH,C & LEO,AJ (1985)
VP 8.40E-003	mm Hg	25	EST	NEELY,WB & BLAU,GE (1985)
DC	pKa			
HL 6.41E-004	atm m3/mol	25	EST	MEYLAN,WM & HOWARD,PH (1991)
OH 6.94E-011	cm3/molc sec	25	EST	MEYLAN,WM & HOWARD,PH (1993)

CAS #: 000575-43-9 — 1,6-DIMETHYLNAPHTHALENE

Formula: $C_{12}H_{12}$
Mol Weight: 156.23
MP (deg C): -16.9
FP (deg C):
BP (deg C): 264
BP pressure (mm Hg):

Property/Value	Units	Temp	Data Type	Reference
WS 9.00E-001	mg/L	25	EST	MEYLAN,WM ET AL. (1996)
logP 4.26			EST	MEYLAN,WM & HOWARD,PH (1995)
VP 1.46E-002	mm Hg	25	EXP	RIDDICK,JA ET AL. (1986)
DC	pKa			
HL 4.25E-004	atm m3/mol	25	EST	MEYLAN,WM & HOWARD,PH (1991)
OH 7.00E-011	cm3/molc sec	25	EST	MEYLAN,WM & HOWARD,PH (1993)

CAS #: 000575-89-3 — 2,4,6-TRICHLOROPHENOXYACETIC ACID

Formula: $C_8H_5Cl_3O_3$
Mol Weight: 255.49
MP (deg C):
FP (deg C):
BP (deg C):
BP pressure (mm Hg):

Property/Value	Units	Temp	Data Type	Reference
WS 2.48E+002	mg/L	25	EXP	YALKOWSKY,SH & DANNENFELSER,RM (1992)
logP 3.26			EST	MEYLAN,WM & HOWARD,PH (1995)
VP 1.32E-005	mm Hg	25	EST	NEELY,WB & BLAU,GE (1985)
DC	pKa			
HL 6.83E-009	atm m3/mol	25	EST	MEYLAN,WM & HOWARD,PH (1991)
OH 5.38E-012	cm3/molc sec	25	EST	MEYLAN,WM & HOWARD,PH (1993)

000576-24-9 — 2,3-DICHLOROPHENOL

CAS #:	000576-24-9
Formula:	$C_6H_4Cl_2O$
Mol Weight:	163.00
MP (deg C):	58
FP (deg C):	
BP (deg C):	
BP pressure (mm Hg):	

Property/Value	Units	Temp	Data Type	Reference
WS 8.22E+003	mg/L	59	EXP	SHIU,WY ET AL. (1994)
logP 2.84			EXP	HANSCH,C & LEO,AJ (1985)
VP 5.80E-002	mm Hg	25	EXT	DOLFING,J & HARRISON,BK (1992)
DC 7.70	pKa		EXP	SERJEANT,EP & DEMPSEY,B (1979)
HL 3.08E-007	atm m3/mol	25	EST	MEYLAN,WM & HOWARD,PH (1991)
OH 1.66E-012	cm3/molc sec	25	EXP	ATKINSON,R (1989)

000576-26-1 — 2,6-DIMETHYLPHENOL

CAS #:	000576-26-1
Formula:	$C_8H_{10}O$
Mol Weight:	122.17
MP (deg C):	45.7
FP (deg C):	
BP (deg C):	201.0
BP pressure (mm Hg):	

Property/Value	Units	Temp	Data Type	Reference
WS 6.05E+003	mg/L	25	EXP	YALKOWSKY,SH & DANNENFELSER,RM (1992)
logP 2.36			EXP	HANSCH,C & LEO,AJ (1985)
VP 1.58E-001	mm Hg	25	EXT	CHAO,J ET AL. (1983)
DC 10.22	pKa		EXP	ARTIOLA-FORTUNY,J & FULLER,WH (1982)
HL 6.65E-006	atm m3/mol	25	EXP	HAWTHORNE,SB ET AL. (1985)
OH 6.59E-011	cm3/molc sec	25	EXP	ATKINSON,R (1989)

000577-11-7 — BIS(2-ETHYLHEXYL) SODIUM SULFOSUCCINATE

CAS #:	000577-11-7
Formula:	$C_{20}H_{37}NaO_7S$
Mol Weight:	444.57
MP (deg C):	173-179
FP (deg C):	
BP (deg C):	
BP pressure (mm Hg):	

Property/Value	Units	Temp	Data Type	Reference
WS 7.10E+004	mg/L	25	EXP	CHEMICAL INSPECTION TESTING INST (1992)
logP 6.10			EST	MEYLAN,WM & HOWARD,PH (1992)
VP 2.17E-011	mm Hg	25	EST	NEELY,WB & BLAU,GE (1985)
DC	pKa			
HL 5.00E-012	atm m3/mol	25	EST	MEYLAN,WM & HOWARD,PH (1991)
OH 2.18E-011	cm3/molc sec	25	EST	MEYLAN,WM & HOWARD,PH (1993)

000577-19-5 — O-BROMONITROBENZENE

CAS #:	000577-19-5
Formula:	$C_6H_4BrNO_2$
Mol Weight:	202.01
MP (deg C):	43
FP (deg C):	
BP (deg C):	258
BP pressure (mm Hg):	

Property/Value	Units	Temp	Data Type	Reference
WS 1.23E+002	mg/L	25	EST	MEYLAN,WM ET AL. (1996)
logP 2.52			EXP	HANSCH,C & LEO,AJ (1985)
VP 9.89E-003	mm Hg	25	EST	NEELY,WB & BLAU,GE (1985)
DC	pKa			
HL 8.47E-006	atm m3/mol	25	EST	MEYLAN,WM & HOWARD,PH (1991)
OH 1.53E-013	cm3/molc sec	25	EST	MEYLAN,WM & HOWARD,PH (1993)

000577-55-9 — 1,2-DIISOPROPYLBENZENE

CAS #:	000577-55-9
Formula:	$C_{12}H_{18}$
Mol Weight:	162.28
MP (deg C):	-57
FP (deg C):	
BP (deg C):	204
BP pressure (mm Hg):	

Property/Value	Units	Temp	Data Type	Reference
WS 4.33E+000	mg/L	25	EST	MEYLAN,WM ET AL. (1996)
logP 4.90			EST	MEYLAN,WM & HOWARD,PH (1995)
VP 3.62E-001	mm Hg	25	EXT	CHAO,J ET AL. (1983)
DC	pKa			
HL 2.04E-002	atm m3/mol	25	EST	MEYLAN,WM & HOWARD,PH (1991)
OH 1.01E-011	cm3/molc sec	25	EST	MEYLAN,WM & HOWARD,PH (1993)

000577-56-0 — O-ACETYLBENZOIC ACID

CAS #:	000577-56-0
Formula:	$C_9H_8O_3$
Mol Weight:	164.16
MP (deg C):	114.5
FP (deg C):	
BP (deg C):	110-112
BP pressure (mm Hg):	2.00E+000

Property/Value	Units	Temp	Data Type	Reference
WS 1.33E+004	mg/L	25	EST	MEYLAN,WM ET AL. (1996)
logP 0.81			EXP	HANSCH,C & LEO,AJ (1985)
VP 2.30E-004	mm Hg	25	EST	NEELY,WB & BLAU,GE (1985)
DC 4.13	pKa	25	EXP	SERJEANT,EP & DEMPSEY,B (1979)
HL 1.97E-010	atm m3/mol	25	EST	MEYLAN,WM & HOWARD,PH (1991)
OH 1.11E-012	cm3/molc sec	25	EST	MEYLAN,WM & HOWARD,PH (1993)

000577-59-3 — 2-NITROACETOPHENONE

CAS #:	000577-59-3
Formula:	$C_8H_7NO_3$
Mol Weight:	165.15
MP (deg C):	28.5
FP (deg C):	
BP (deg C):	178
BP pressure (mm Hg):	3.20E+001

Property/Value	Units	Temp	Data Type	Reference
WS 2.13E+003	mg/L	25	EST	MEYLAN,WM ET AL. (1996)
logP 1.28			EXP	HANSCH,C & LEO,AJ (1985)
VP 2.40E-003	mm Hg	25	EST	NEELY,WB & BLAU,GE (1985)
DC	pKa			
HL 3.87E-008	atm m3/mol	25	EST	MEYLAN,WM & HOWARD,PH (1991)
OH 2.57E-013	cm3/molc sec	25	EST	MEYLAN,WM & HOWARD,PH (1993)

000577-71-9 — 3,4-DINITROPHENOL

CAS #:	000577-71-9
Formula:	$C_6H_4N_2O_5$
Mol Weight:	184.11
MP (deg C):	134
FP (deg C):	
BP (deg C):	
BP pressure (mm Hg):	

Property/Value	Units	Temp	Data Type	Reference
WS 6.05E+004	mg/L	82	EXP	SEIDELL,A (1941)
logP 1.73			EST	MEYLAN,WM & HOWARD,PH (1995)
VP 1.20E-005	mm Hg	25	EST	NEELY,WB & BLAU,GE (1985)
DC 5.42	pKa	25	EXP	KORTUM,G ET AL (1961)
HL 8.73E-012	atm m3/mol	25	EST	MEYLAN,WM & HOWARD,PH (1991)
OH 5.04E-013	cm3/molc sec	25	EST	MEYLAN,WM & HOWARD,PH (1993)

CAS #: 000578-06-3				1-AMINOACRIDINE

Formula: $C_{13}H_{10}N_2$
Mol Weight: 194.24
MP (deg C):
FP (deg C):
BP (deg C):
BP pressure (mm Hg):

Property/ Value	Units	Temp	Data Type	Reference
WS 1.17E+001	mg/L	24	EXP	YALKOWSKY,SH & DANNENFELSER,RM (1992)
logP 2.47			EXP	HANSCH,C & LEO,AJ (1985)
VP 2.24E-006	mm Hg	25	EST	NEELY,WB & BLAU,GE (1985)
DC 4.36	pKa	20	EXP	PERRIN,DD (1965)
HL 2.37E-011	atm m3/mol	25	EST	MEYLAN,WM & HOWARD,PH (1991)
OH 2.00E-010	cm3/molc sec	25	EST	MEYLAN,WM & HOWARD,PH (1993)

CAS #: 000578-07-4				4-AMINOACRIDINE

Formula: $C_{13}H_{10}N_2$
Mol Weight: 194.24
MP (deg C): 108
FP (deg C):
BP (deg C): 183.5
BP pressure (mm Hg):

Property/ Value	Units	Temp	Data Type	Reference
WS 5.97E+000	mg/L	25	EST	MEYLAN,WM ET AL. (1996)
logP 3.26			EXP	HANSCH,C & LEO,AJ (1985)
VP 2.24E-006	mm Hg	25	EST	NEELY,WB & BLAU,GE (1985)
DC 4.36	pKa	20	EXP	PERRIN,DD (1965)
HL 2.37E-011	atm m3/mol	25	EST	MEYLAN,WM & HOWARD,PH (1991)
OH 2.00E-010	cm3/molc sec	25	EST	MEYLAN,WM & HOWARD,PH (1993)

CAS #: 000578-54-1				O-ETHYLANILINE

Formula: $C_8H_{11}N$
Mol Weight: 121.18
MP (deg C): -43
FP (deg C):
BP (deg C): 209.5
BP pressure (mm Hg):

Property/ Value	Units	Temp	Data Type	Reference
WS 3.25E+003	mg/L	25	EST	MEYLAN,WM ET AL. (1996)
logP 1.74			EXP	HANSCH,C & LEO,AJ (1985)
VP 1.71E-001	mm Hg	25	EXP	YAWS,CL (1994B)
DC 4.37	pKa	25	EXP	PERRIN,DD (1965)
HL 2.79E-006	atm m3/mol	25	EST	MEYLAN,WM & HOWARD,PH (1991)
OH 1.32E-010	cm3/molc sec	25	EST	MEYLAN,WM & HOWARD,PH (1993)

CAS #: 000578-57-4				1-BROMO-2-METHOXYBENZENE

Formula: C_7H_7BrO
Mol Weight: 187.04
MP (deg C): 2.5
FP (deg C):
BP (deg C): 216
BP pressure (mm Hg):

Property/ Value	Units	Temp	Data Type	Reference
WS 1.84E+002	mg/L	25	EST	MEYLAN,WM ET AL. (1996)
logP 2.86			EXP	NAKAGAWA,Y ET AL. (1992)
VP 2.88E-001	mm Hg	25	EST	NEELY,WB & BLAU,GE (1985)
DC	pKa			
HL 1.27E-004	atm m3/mol	25	EST	MEYLAN,WM & HOWARD,PH (1991)
OH 6.99E-012	cm3/molc sec	25	EST	MEYLAN,WM & HOWARD,PH (1993)

CAS #: 000578-58-5				2-METHYLANISOLE

Formula: $C_8H_{10}O$
Mol Weight: 122.17
MP (deg C): -34.1
FP (deg C):
BP (deg C): 171
BP pressure (mm Hg):

Property/ Value	Units	Temp	Data Type	Reference
WS 4.50E+002	mg/L	25	EST	MEYLAN,WM ET AL. (1996)
logP 2.74			EXP	HANSCH,C & LEO,AJ (1985)
VP 1.50E+000	mm Hg	25	EST	NEELY,WB & BLAU,GE (1985)
DC	pKa			
HL 3.52E-004	atm m3/mol	25	EST	MEYLAN,WM & HOWARD,PH (1991)
OH 2.73E-011	cm3/molc sec	25	EST	MEYLAN,WM & HOWARD,PH (1993)

CAS #: 000578-66-5				8-AMINOQUINOLINE

Formula: $C_9H_8N_2$
Mol Weight: 144.18
MP (deg C): 70
FP (deg C):
BP (deg C): 157
BP pressure (mm Hg): 1.90E+001

Property/ Value	Units	Temp	Data Type	Reference
WS 8.22E+003	mg/L	25	EST	MEYLAN,WM ET AL. (1996)
logP 1.79			EXP	HANSCH,C & LEO,AJ (1985)
VP 5.50E-004	mm Hg	25	EST	NEELY,WB & BLAU,GE (1985)
DC 3.95	pKa	20	EXP	PERRIN,DD (1965)
HL 2.43E-010	atm m3/mol	25	EST	MEYLAN,WM & HOWARD,PH (1991)
OH 2.00E-010	cm3/molc sec	25	EST	MEYLAN,WM & HOWARD,PH (1993)

CAS #: 000578-67-6				5-QUINOLINOL

Formula: C_9H_7NO
Mol Weight: 145.16
MP (deg C): 226 dec
FP (deg C):
BP (deg C):
BP pressure (mm Hg):

Property/ Value	Units	Temp	Data Type	Reference
WS 4.17E+002	mg/L	20	EXP	YALKOWSKY,SH & DANNENFELSER,RM (1992)
logP 2.27			EXP	HANSCH,C ET AL. (1995)
VP 3.30E-004	mm Hg	25	EST	NEELY,WB & BLAU,GE (1985)
DC 5.02	pKa	20	EXP	KORTUM,G ET AL (1961)
HL 7.16E-011	atm m3/mol	25	EST	MEYLAN,WM & HOWARD,PH (1991)
OH 2.00E-010	cm3/molc sec	25	EST	MEYLAN,WM & HOWARD,PH (1993)

CAS #: 000578-68-7				4-AMINOQUINOLINE

Formula: $C_9H_8N_2$
Mol Weight: 144.18
MP (deg C): 154.8
FP (deg C):
BP (deg C): 180
BP pressure (mm Hg): 1.20E+001

Property/ Value	Units	Temp	Data Type	Reference
WS 3.26E+003	mg/L	25	EST	MEYLAN,WM ET AL. (1996)
logP 1.63			EXP	HANSCH,C & LEO,AJ (1985)
VP 5.50E-004	mm Hg	25	EST	NEELY,WB & BLAU,GE (1985)
DC 9.13	pKa	20	EXP	PERRIN,DD (1965)
HL 2.43E-010	atm m3/mol	25	EST	MEYLAN,WM & HOWARD,PH (1991)
OH 2.00E-010	cm3/molc sec	25	EST	MEYLAN,WM & HOWARD,PH (1993)

CAS #: 000578-96-1 — 11H-PYRIDO[2,1-B]QUINAZOLIN-11-ONE

Formula: $C_{12}H_8N_2O$

Mol Weight: 196.21

MP (deg C): FP (deg C):

BP (deg C):

BP pressure (mm Hg):

Property/Value	Units	Temp	Data Type	Reference
WS 7.66E+002	mg/L	25	EST	MEYLAN,WM ET AL. (1996)
logP 2.08			EXP	SANGSTER,J (1994)
VP 5.26E-006	mm Hg	25	EST	NEELY,WB & BLAU,GE (1985)
DC	pKa			
HL 9.23E-010	atm m3/mol	25	EST	MEYLAN,WM & HOWARD,PH (1991)
OH 9.23E-011	cm3/molc sec	25	EST	MEYLAN,WM & HOWARD,PH (1993)

CAS #: 000579-10-2 — N-METHYLACETANILIDE

Formula: $C_9H_{11}NO$

Mol Weight: 149.19

MP (deg C): 102-104 FP (deg C):

BP (deg C): 253

BP pressure (mm Hg): 7.12E+002

Property/Value	Units	Temp	Data Type	Reference
WS 1.67E+004	mg/L	25	EXP	CHEM INSPECT TEST INST (1992)
logP 1.12			EXP	HANSCH,C & LEO,AJ (1985)
VP 8.87E-003	mm Hg	25	EST	NEELY,WB & BLAU,GE (1985)
DC -0.50	pKa	25	EXP	PERRIN,DD (1965)
HL 1.26E-007	atm m3/mol	25	EST	MEYLAN,WM & HOWARD,PH (1991)
OH 1.31E-011	cm3/molc sec	25	EST	MEYLAN,WM & HOWARD,PH (1993)

CAS #: 000579-38-4 — DILOXANIDE

Formula: $C_9H_9Cl_2NO_2$

Mol Weight: 234.08

MP (deg C): 175 FP (deg C):

BP (deg C):

BP pressure (mm Hg):

Property/Value	Units	Temp	Data Type	Reference
WS 8.73E+002	mg/L	25	EST	MEYLAN,WM ET AL. (1996)
logP 1.78			EXP	DUTTA,H ET AL. (1988)
VP 2.82E-006	mm Hg	25	EST	NEELY,WB & BLAU,GE (1985)
DC	pKa			
HL 1.63E-012	atm m3/mol	25	EST	MEYLAN,WM & HOWARD,PH (1991)
OH 1.83E-011	cm3/molc sec	25	EST	MEYLAN,WM & HOWARD,PH (1993)

CAS #: 000579-43-1 — MESO-HYDROBENZOIN

Formula: $C_{14}H_{14}O_2$

Mol Weight: 214.27

MP (deg C): 137-139 FP (deg C):

BP (deg C):

BP pressure (mm Hg):

Property/Value	Units	Temp	Data Type	Reference
WS 8.63E+002	mg/L	25	EST	MEYLAN,WM ET AL. (1996)
logP 1.56			EXP	SANGSTER,J (1994)
VP 2.06E-007	mm Hg	25	EST	NEELY,WB & BLAU,GE (1985)
DC	pKa			
HL 8.52E-010	atm m3/mol	25	EST	MEYLAN,WM & HOWARD,PH (1991)
OH 2.62E-011	cm3/molc sec	25	EST	MEYLAN,WM & HOWARD,PH (1993)

CAS #: 000579-44-2 — DL-BENZOIN

Formula: $C_{13}H_{10}O_2$

Mol Weight: 198.22

MP (deg C): 137 FP (deg C):

BP (deg C): 344

BP pressure (mm Hg):

Property/Value	Units	Temp	Data Type	Reference
WS 3.00E+002	mg/L	25	EXP	YALKOWSKY,SH & DANNENFELSER,RM (1992)
logP 3.04			EST	MEYLAN,WM & HOWARD,PH (1995)
VP 1.26E-005	mm Hg	25	EXT	PERRY,RH & GREEN,D (1984)
DC	pKa			
HL 1.28E-005	atm m3/mol	25	EST	MEYLAN,WM & HOWARD,PH (1991)
OH 3.79E-012	cm3/molc sec	25	EST	MEYLAN,WM & HOWARD,PH (1993)

CAS #: 000579-58-8 — N-METHYL P-HYDROXYACETANILIDE

Formula: $C_9H_{11}NO_2$

Mol Weight: 165.19

MP (deg C): FP (deg C):

BP (deg C):

BP pressure (mm Hg):

Property/Value	Units	Temp	Data Type	Reference
WS 1.95E+004	mg/L	25	EST	MEYLAN,WM ET AL. (1996)
logP 0.61			EXP	HANSCH,C & LEO,AJ (1985)
VP 9.15E-005	mm Hg	25	EST	NEELY,WB & BLAU,GE (1985)
DC	pKa			
HL 1.32E-011	atm m3/mol	25	EST	MEYLAN,WM & HOWARD,PH (1991)
OH 1.82E-011	cm3/molc sec	25	EST	MEYLAN,WM & HOWARD,PH (1993)

CAS #: 000579-66-8 — 2,6-DIETHYLANILINE

Formula: $C_{10}H_{15}N$

Mol Weight: 149.24

MP (deg C): 3 FP (deg C):

BP (deg C): 243

BP pressure (mm Hg):

Property/Value	Units	Temp	Data Type	Reference
WS 6.70E+002	mg/L		EXP	YALKOWSKY,SH & DANNENFELSER,RM (1992)
logP 3.15			EST	MEYLAN,WM & HOWARD,PH (1995)
VP 3.82E-003	mm Hg	25	EXP	YAWS,CL (1994B)
DC	pKa			
HL 1.12E-006	atm m3/mol	25	EST	VP/WSOL
OH 1.62E-010	cm3/molc sec	25	EST	MEYLAN,WM & HOWARD,PH (1993)

CAS #: 000579-75-9 — O-METHOXYBENZOIC ACID

Formula: $C_8H_8O_3$

Mol Weight: 152.15

MP (deg C): 101 FP (deg C):

BP (deg C): 200

BP pressure (mm Hg):

Property/Value	Units	Temp	Data Type	Reference
WS 5.00E+003	mg/L	30	EXP	YALKOWSKY,SH & DANNENFELSER,RM (1992)
logP 1.59			EXP	HANSCH,C & LEO,AJ (1985)
VP 1.59E-005	mm Hg	25	EXT	JONES,AH (1960)
DC 3.90	pKa	25	EXP	KORTUM,G ET AL (1961)
HL 6.41E-009	atm m3/mol	25	EST	MEYLAN,WM & HOWARD,PH (1991)
OH 9.31E-012	cm3/molc sec	25	EST	MEYLAN,WM & HOWARD,PH (1993)

ACETYLSALICYLIC ACID, METHYL ESTER

CAS #: 000580-02-9

Formula: $C_{10}H_{10}O_4$

Mol Weight: 194.19

MP (deg C): 51-52 FP (deg C):

BP (deg C): 134-136

BP pressure (mm Hg): 9.00E+000

Property/Value	Units	Temp	Data Type	Reference
WS 2.65E+003	mg/L	25	EST	MEYLAN,WM ET AL. (1996)
logP 1.46			EXP	HANSCH,C ET AL. (1995)
VP 2.90E-002	mm Hg	25	EST	NEELY,WB & BLAU,GE (1985)
DC	pKa			
HL 4.18E-007	atm m3/mol	25	EST	MEYLAN,WM & HOWARD,PH (1991)
OH 9.64E-013	cm3/molc sec	25	EST	MEYLAN,WM & HOWARD,PH (1993)

2-BROMONAPHTHALENE

CAS #: 000580-13-2

Formula: $C_{10}H_7Br$

Mol Weight: 207.08

MP (deg C): 59 FP (deg C):

BP (deg C): 281-282

BP pressure (mm Hg):

Property/Value	Units	Temp	Data Type	Reference
WS 8.24E+000	mg/L	25	EXP	KUHNE,R ET AL. (1995)
logP 4.06			EST	MEYLAN,WM & HOWARD,PH (1995)
VP 3.47E-003	mm Hg	25	EST	NEELY,WB & BLAU,GE (1985)
DC	pKa			
HL 2.10E-004	atm m3/mol	25	EST	MEYLAN,WM & HOWARD,PH (1991)
OH 1.36E-011	cm3/molc sec	25	EST	MEYLAN,WM & HOWARD,PH (1993)

6-AMINOQUINOLINE

CAS #: 000580-15-4

Formula: $C_9H_8N_2$

Mol Weight: 144.18

MP (deg C): 114 FP (deg C):

BP (deg C): 187

BP pressure (mm Hg): 1.20E+001

Property/Value	Units	Temp	Data Type	Reference
WS 6.49E+003	mg/L	25	EST	MEYLAN,WM ET AL. (1996)
logP 1.28			EXP	HANSCH,C & LEO,AJ (1985)
VP 7.41E-007	mm Hg	25	EXT	RIBEIRODASILVA,MAV ET AL. (1993)
DC 6.61	pKa	20	EXP	PERRIN,DD (1965)
HL 2.43E-010	atm m3/mol	25	EST	MEYLAN,WM & HOWARD,PH (1991)
OH 2.00E-010	cm3/molc sec	25	EST	MEYLAN,WM & HOWARD,PH (1993)

6-HDROXYQUINOLINE

CAS #: 000580-16-5

Formula: C_9H_7NO

Mol Weight: 145.16

MP (deg C): 195 FP (deg C):

BP (deg C): 360

BP pressure (mm Hg):

Property/Value	Units	Temp	Data Type	Reference
WS 1.00E+003	mg/L	20	EXP	YALKOWSKY,SH & DANNENFELSER,RM (1992)
logP 1.80			EXP	HANSCH,C & LEO,AJ (1985)
VP 3.30E-004	mm Hg	25	EST	NEELY,WB & BLAU,GE (1985)
DC 5.46	pKa	20	EXP	PERRIN,DD (1965)
HL 7.16E-011	atm m3/mol	25	EST	MEYLAN,WM & HOWARD,PH (1991)
OH 2.00E-010	cm3/molc sec	25	EST	MEYLAN,WM & HOWARD,PH (1993)

3-AMINOQUINOLINE

CAS #: 000580-17-6

Formula: $C_9H_8N_2$

Mol Weight: 144.18

MP (deg C): 91-92 FP (deg C):

BP (deg C):

BP pressure (mm Hg):

Property/Value	Units	Temp	Data Type	Reference
WS 3.26E+003	mg/L	25	EST	MEYLAN,WM ET AL. (1996)
logP 1.63			EXP	HANSCH,C & LEO,AJ (1985)
VP 7.70E-006	mm Hg	25	EXT	RIBEIRODASILVA,MAV ET AL. (1993)
DC 4.91	pKa	20	EXP	PERRIN,DD (1965)
HL 2.43E-010	atm m3/mol	25	EST	MEYLAN,WM & HOWARD,PH (1991)
OH 2.00E-010	cm3/molc sec	25	EST	MEYLAN,WM & HOWARD,PH (1993)

3-QUINOLINOL

CAS #: 000580-18-7

Formula: C_9H_7NO

Mol Weight: 145.16

MP (deg C): FP (deg C):

BP (deg C):

BP pressure (mm Hg):

Property/Value	Units	Temp	Data Type	Reference
WS 1.33E+003	mg/L	25	EST	MEYLAN,WM ET AL. (1996)
logP 2.08			EXP	HANSCH,C & LEO,AJ (1985)
VP 3.30E-004	mm Hg	25	EST	NEELY,WB & BLAU,GE (1985)
DC 4.28	pKa	20	EXP	PERRIN,DD (1965)
HL 7.16E-011	atm m3/mol	25	EST	MEYLAN,WM & HOWARD,PH (1991)
OH 2.00E-010	cm3/molc sec	25	EST	MEYLAN,WM & HOWARD,PH (1993)

7-QUINOLINOL

CAS #: 000580-20-1

Formula: C_9H_7NO

Mol Weight: 145.16

MP (deg C): 239 FP (deg C):

BP (deg C):

BP pressure (mm Hg):

Property/Value	Units	Temp	Data Type	Reference
WS 4.55E+002	mg/L	20	EXP	YALKOWSKY,SH & DANNENFELSER,RM (1992)
logP 1.98			EXP	HANSCH,C ET AL. (1995)
VP 3.30E-004	mm Hg	25	EST	NEELY,WB & BLAU,GE (1985)
DC 5.46	pKa	20	EXP	PERRIN,DD (1965)
HL 7.16E-011	atm m3/mol	25	EST	MEYLAN,WM & HOWARD,PH (1991)
OH 2.00E-010	cm3/molc sec	25	EST	MEYLAN,WM & HOWARD,PH (1993)

2-AMINOQUINOLINE

CAS #: 000580-22-3

Formula: $C_9H_8N_2$

Mol Weight: 144.18

MP (deg C): 131.5 FP (deg C):

BP (deg C):

BP pressure (mm Hg):

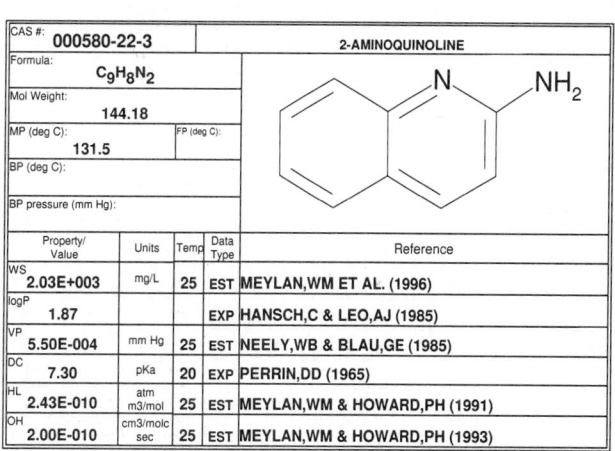

Property/Value	Units	Temp	Data Type	Reference
WS 2.03E+003	mg/L	25	EST	MEYLAN,WM ET AL. (1996)
logP 1.87			EXP	HANSCH,C & LEO,AJ (1985)
VP 5.50E-004	mm Hg	25	EST	NEELY,WB & BLAU,GE (1985)
DC 7.30	pKa	20	EXP	PERRIN,DD (1965)
HL 2.43E-010	atm m3/mol	25	EST	MEYLAN,WM & HOWARD,PH (1991)
OH 2.00E-010	cm3/molc sec	25	EST	MEYLAN,WM & HOWARD,PH (1993)

CAS #: 000580-48-3				CHLORAZINE

Formula: $C_{11}H_{20}ClN_5$

Mol Weight: 257.77

MP (deg C): 27   FP (deg C):

BP (deg C): 154-156

BP pressure (mm Hg): 9.00E+000

Property/Value	Units	Temp	Data Type	Reference
WS 1.00E+001	mg/L	21	EXP	YALKOWSKY,SH & DANNENFELSER,RM (1992)
logP 4.48			EST	MEYLAN,WM & HOWARD,PH (1995)
VP 2.65E-005	mm Hg	25	EST	NEELY,WB & BLAU,GE (1985)
DC	pKa			
HL 2.49E-006	atm m3/mol	25	EST	MEYLAN,WM & HOWARD,PH (1991)
OH 3.55E-011	cm3/molc sec	25	EST	MEYLAN,WM & HOWARD,PH (1993)

CAS #: 000580-51-8				3-PHENYLPHENOL

Formula: $C_{12}H_{10}O$

Mol Weight: 170.21

MP (deg C): 78   FP (deg C):

BP (deg C): > 300

BP pressure (mm Hg):

Property/Value	Units	Temp	Data Type	Reference
WS 4.07E+002	mg/L	25	EST	MEYLAN,WM ET AL. (1996)
logP 3.23			EXP	HANSCH,C & LEO,AJ (1985)
VP 6.14E-005	mm Hg	25	EST	NEELY,WB & BLAU,GE (1985)
DC 9.64	pKa	23	EXP	KORTUM,G ET AL (1961)
HL 4.30E-008	atm m3/mol	25	EST	MEYLAN,WM & HOWARD,PH (1991)
OH 6.14E-011	cm3/molc sec	25	EST	MEYLAN,WM & HOWARD,PH (1993)

CAS #: 000581-08-8				ACETAMIDE, N-(2-ETHOXYPHENYL)-

Formula: $C_{10}H_{13}NO_2$

Mol Weight: 179.22

MP (deg C): 79   FP (deg C):

BP (deg C): > 240

BP pressure (mm Hg):

Property/Value	Units	Temp	Data Type	Reference
WS 2.97E+003	mg/L	25	EST	MEYLAN,WM ET AL. (1996)
logP 1.49			EXP	HANSCH,C ET AL. (1995)
VP 2.76E-005	mm Hg	25	EST	NEELY,WB & BLAU,GE (1985)
DC	pKa			
HL 4.85E-010	atm m3/mol	25	EST	MEYLAN,WM & HOWARD,PH (1991)
OH 1.75E-011	cm3/molc sec	25	EST	MEYLAN,WM & HOWARD,PH (1993)

CAS #: 000581-28-2				2-AMINOACRIDINE

Formula: $C_{13}H_{10}N_2$

Mol Weight: 194.24

MP (deg C):   FP (deg C):

BP (deg C):

BP pressure (mm Hg):

Property/Value	Units	Temp	Data Type	Reference
WS 1.57E+001	mg/L	25	EST	MEYLAN,WM ET AL. (1996)
logP 2.62			EXP	HANSCH,C & LEO,AJ (1985)
VP 2.24E-006	mm Hg	25	EST	NEELY,WB & BLAU,GE (1985)
DC 8.00	pKa	20	EXP	PERRIN,DD (1965)
HL 2.37E-011	atm m3/mol	25	EST	MEYLAN,WM & HOWARD,PH (1991)
OH 2.00E-010	cm3/molc sec	25	EST	MEYLAN,WM & HOWARD,PH (1993)

CAS #: 000581-29-3				3-AMINOACRIDINE

Formula: $C_{13}H_{10}N_2$

Mol Weight: 194.24

MP (deg C):   FP (deg C):

BP (deg C):

BP pressure (mm Hg):

Property/Value	Units	Temp	Data Type	Reference
WS 1.57E+001	mg/L	25	EST	MEYLAN,WM ET AL. (1996)
logP 2.77			EXP	HANSCH,C ET AL. (1995)
VP 2.24E-006	mm Hg	25	EST	NEELY,WB & BLAU,GE (1985)
DC 8.00	pKa	20	EXP	PERRIN,DD (1965)
HL 2.37E-011	atm m3/mol	25	EST	MEYLAN,WM & HOWARD,PH (1991)
OH 2.00E-010	cm3/molc sec	25	EST	MEYLAN,WM & HOWARD,PH (1993)

CAS #: 000581-40-8				2,3-DIMETHYLNAPHTHALENE

Formula: $C_{12}H_{12}$

Mol Weight: 156.23

MP (deg C): 105   FP (deg C):

BP (deg C): 268

BP pressure (mm Hg):

Property/Value	Units	Temp	Data Type	Reference
WS 1.99E+000	mg/L	25	EXP	YALKOWSKY,SH & DANNENFELSER,RM (1992)
logP 4.40			EXP	HANSCH,C & LEO,AJ (1985)
VP 1.10E-003	mm Hg	25	EXT	BOUBLIK,T ET AL. (1984)
DC	pKa			
HL 9.20E-004	atm m3/mol	25	EXP	LION,LW & GARBARINI,D (1983)
OH 7.68E-011	cm3/molc sec	25	EXP	ATKINSON,R (1989)

CAS #: 000581-42-0				2,6-DIMETHYLNAPHTHALENE

Formula: $C_{12}H_{12}$

Mol Weight: 156.23

MP (deg C): 112   FP (deg C):

BP (deg C): 262

BP pressure (mm Hg):

Property/Value	Units	Temp	Data Type	Reference
WS 2.00E+000	mg/L	25	EXP	YALKOWSKY,SH & DANNENFELSER,RM (1992)
logP 4.31			EXP	HANSCH,C & LEO,AJ (1985)
VP 4.30E-003	mm Hg	25	EXT	YAWS,CL (1994B)
DC	pKa			
HL 6.41E-004	atm m3/mol	25	EST	MEYLAN,WM & HOWARD,PH (1991)
OH 6.94E-011	cm3/molc sec	25	EST	MEYLAN,WM & HOWARD,PH (1993)

CAS #: 000581-43-1				2,6-NAPHTHALENEDIOL

Formula: $C_{10}H_8O_2$

Mol Weight: 160.17

MP (deg C): 220   FP (deg C):

BP (deg C):

BP pressure (mm Hg):

Property/Value	Units	Temp	Data Type	Reference
WS 1.00E+003	mg/L	25	EXP	YALKOWSKY,SH & DANNENFELSER,RM (1992)
logP 1.90			EXP	SANGSTER,J (1993)
VP 7.15E-006	mm Hg	25	EST	NEELY,WB & BLAU,GE (1985)
DC	pKa			
HL 5.69E-012	atm m3/mol	25	EST	MEYLAN,WM & HOWARD,PH (1991)
OH 2.00E-010	cm3/molc sec	25	EST	MEYLAN,WM & HOWARD,PH (1993)

000581-47-5 — (4-PYRIDYL)-2-PYRIDINE

Formula: $C_{10}H_8N_2$
Mol Weight: 156.19
MP (deg C): 61.5
FP (deg C):
BP (deg C): 281
BP pressure (mm Hg):

Property/Value	Units	Temp	Data Type	Reference
WS 3.19E+003	mg/L	25	EST	MEYLAN,WM ET AL. (1996)
logP 1.58			EXP	HANSCH,C ET AL. (1995)
VP 1.20E-003	mm Hg	25	EST	NEELY,WB & BLAU,GE (1985)
DC 4.77	pKa	20	EXP	PERRIN,DD (1965)
HL 7.09E-010	atm m3/mol	25	EST	MEYLAN,WM & HOWARD,PH (1991)
OH 1.29E-012	cm3/molc sec	25	EST	MEYLAN,WM & HOWARD,PH (1993)

000581-50-0 — 2,3'-BIPYRIDINE

Formula: $C_{10}H_8N_2$
Mol Weight: 156.19
MP (deg C):
FP (deg C):
BP (deg C): 295.5
BP pressure (mm Hg):

Property/Value	Units	Temp	Data Type	Reference
WS 4.03E+003	mg/L	25	EST	MEYLAN,WM ET AL. (1996)
logP 1.46			EXP	HANSCH,C ET AL. (1995)
VP 1.20E-003	mm Hg	25	EST	NEELY,WB & BLAU,GE (1985)
DC 4.42	pKa	20	EXP	PERRIN,DD (1965)
HL 7.09E-010	atm m3/mol	25	EST	MEYLAN,WM & HOWARD,PH (1991)
OH 1.29E-012	cm3/molc sec	25	EST	MEYLAN,WM & HOWARD,PH (1993)

000581-55-5 — BENZAL DIACETATE

Formula: $C_{11}H_{12}O_4$
Mol Weight: 208.22
MP (deg C): 46
FP (deg C):
BP (deg C): 220
BP pressure (mm Hg):

Property/Value	Units	Temp	Data Type	Reference
WS 1.64E+003	mg/L	25	EST	MEYLAN,WM ET AL. (1996)
logP 1.62			EXP	HANSCH,C & LEO,AJ (1985)
VP 2.00E-002	mm Hg	25	EST	NEELY,WB & BLAU,GE (1985)
DC	pKa			
HL 3.37E-008	atm m3/mol	25	EST	MEYLAN,WM & HOWARD,PH (1991)
OH 9.79E-012	cm3/molc sec	25	EST	MEYLAN,WM & HOWARD,PH (1993)

000581-89-5 — 2-NITRONAPHTHALENE

Formula: $C_{10}H_7NO_2$
Mol Weight: 173.17
MP (deg C): 79
FP (deg C):
BP (deg C): 314
BP pressure (mm Hg):

Property/Value	Units	Temp	Data Type	Reference
WS 9.24E+000	mg/L	25	EXP	GANG,Y & XIABAI,XU (1992)
logP 3.24			EXP	DEBNATH,AK ET AL. (1992)
VP 2.58E-004	mm Hg	25	EST	NEELY,WB & BLAU,GE (1985)
DC	pKa			
HL 2.08E-006	atm m3/mol	25	EST	MEYLAN,WM & HOWARD,PH (1991)
OH 5.60E-012	cm3/molc sec	25	EXP	ATKINSON,R (1989)

000581-96-4 — 2-NAPHTHALENEACETIC ACID

Formula: $C_{12}H_{10}O_2$
Mol Weight: 186.21
MP (deg C): 141-143
FP (deg C):
BP (deg C):
BP pressure (mm Hg):

Property/Value	Units	Temp	Data Type	Reference
WS 5.08E+002	mg/L	25	EST	MEYLAN,WM ET AL. (1996)
logP 2.81			EXP	HANSCH,C ET AL. (1995)
VP 1.59E-005	mm Hg	25	EST	NEELY,WB & BLAU,GE (1985)
DC 4.25	pKa	25	EXP	KORTUM,G ET AL (1961)
HL 4.32E-009	atm m3/mol	25	EST	MEYLAN,WM & HOWARD,PH (1991)
OH 3.70E-011	cm3/molc sec	25	EST	MEYLAN,WM & HOWARD,PH (1993)

000582-16-1 — 2,7-DIMETHYLNAPHTHALENE

Formula: $C_{12}H_{12}$
Mol Weight: 156.23
MP (deg C): 97
FP (deg C):
BP (deg C): 265
BP pressure (mm Hg):

Property/Value	Units	Temp	Data Type	Reference
WS 1.49E+001	mg/L	25	EST	MEYLAN,WM ET AL. (1996)
logP 4.26			EST	MEYLAN,WM & HOWARD,PH (1995)
VP 5.68E-003	mm Hg	25	EXT	YAWS,CL (1994B)
DC	pKa			
HL 6.41E-004	atm m3/mol	25	EST	MEYLAN,WM & HOWARD,PH (1991)
OH 6.94E-011	cm3/molc sec	25	EST	MEYLAN,WM & HOWARD,PH (1993)

000582-60-5 — 5,6-DIMETHYLBENZIMIDAZOLE

Formula: $C_9H_{10}N_2$
Mol Weight: 146.19
MP (deg C): 205-206
FP (deg C):
BP (deg C):
BP pressure (mm Hg):

Property/Value	Units	Temp	Data Type	Reference
WS 7.76E+002	mg/L	25	EST	MEYLAN,WM ET AL. (1996)
logP 2.35			EXP	HANSCH,C & LEO,AJ (1985)
VP 1.24E-005	mm Hg	25	EST	NEELY,WB & BLAU,GE (1985)
DC	pKa			
HL 4.46E-007	atm m3/mol	25	EST	MEYLAN,WM & HOWARD,PH (1991)
OH 1.15E-010	cm3/molc sec	25	EST	MEYLAN,WM & HOWARD,PH (1993)

000583-15-3 — BENZOIC ACID, MERCURIC SALT

Formula: $C_{14}H_{10}HgO_4$
Mol Weight: 442.82
MP (deg C):
FP (deg C):
BP (deg C):
BP pressure (mm Hg):

Property/Value	Units	Temp	Data Type	Reference
WS 6.67E+001	mg/L	25	EST	MEYLAN,WM ET AL. (1996)
logP 1.63			EST	MEYLAN,WM & HOWARD,PH (1995)
VP 3.11E-006	mm Hg	25	EST	NEELY,WB & BLAU,GE (1985)
DC	pKa			
HL	atm m3/mol			
OH 3.55E-012	cm3/molc sec	25	EST	MEYLAN,WM & HOWARD,PH (1993)

CAS #:	000583-39-1	O-PHENYLENETHIOUREA			

Formula: $C_7H_6N_2S$

Mol Weight: 150.20

MP (deg C): 303-304 **FP (deg C):**

BP (deg C):

BP pressure (mm Hg):

Property/Value	Units	Temp	Data Type	Reference
WS 2.90E+003	mg/L	25	EST	MEYLAN,WM ET AL. (1996)
logP 1.66			EXP	HANSCH,C & LEO,AJ (1985)
VP 4.66E-006	mm Hg	25	EST	NEELY,WB & BLAU,GE (1985)
DC	pKa			
HL 3.55E-008	atm m3/mol	25	EST	MEYLAN,WM & HOWARD,PH (1991)
OH 2.00E-010	cm3/molc sec	25	EST	MEYLAN,WM & HOWARD,PH (1993)

CAS #:	000583-48-2	3,4-DIMETHYLHEXANE			

Formula: C_8H_{18}

Mol Weight: 114.23

MP (deg C): **FP (deg C):**

BP (deg C): 117.7

BP pressure (mm Hg):

Property/Value	Units	Temp	Data Type	Reference
WS 9.20E+000	mg/L	25	EST	MEYLAN,WM ET AL. (1996)
logP 4.12			EST	MEYLAN,WM & HOWARD,PH (1995)
VP 2.17E+001	mm Hg	25	EXP	DAUBERT,TE & DANNER,RP (1989)
DC	pKa			
HL 3.01E+000	atm m3/mol	25	EST	MEYLAN,WM & HOWARD,PH (1991)
OH 8.84E-012	cm3/molc sec	25	EST	MEYLAN,WM & HOWARD,PH (1993)

CAS #:	000583-53-9	O-DIBROMOBENZENE			

Formula: $C_6H_4Br_2$

Mol Weight: 235.92

MP (deg C): 7.1 **FP (deg C):**

BP (deg C): 225

BP pressure (mm Hg):

Property/Value	Units	Temp	Data Type	Reference
WS 7.46E+001	mg/L	25	EXP	SUZUKI,T (1991)
logP 3.64			EXP	HANSCH,C & LEO,AJ (1985)
VP 1.39E-001	mm Hg	25	EST	NEELY,WB & BLAU,GE (1985)
DC	pKa			
HL 8.55E-004	atm m3/mol	25	EST	MEYLAN,WM & HOWARD,PH (1991)
OH 3.52E-013	cm3/molc sec	25	EST	MEYLAN,WM & HOWARD,PH (1993)

CAS #:	000583-57-3	1,2-DIMETHYLCYCLOHEXANE			

Formula: C_8H_{16}

Mol Weight: 112.22

MP (deg C): **FP (deg C):**

BP (deg C): 124

BP pressure (mm Hg):

Property/Value	Units	Temp	Data Type	Reference
WS 6.03E+000	mg/L	25	EXP	HINE,J & MOOKERJEE,PK (1975)
logP 4.01			EST	MEYLAN,WM & HOWARD,PH (1995)
VP 1.45E+001	mm Hg	25	EXP	HINE,J & MOOKERJEE,PK (1975)
DC	pKa			
HL 3.55E-001	atm m3/mol	25	EST	VP/WSOL
OH 1.19E-011	cm3/molc sec	25	EST	MEYLAN,WM & HOWARD,PH (1993)

CAS #:	000583-58-4	3,4-DIMETHYLPYRIDINE			

Formula: C_7H_9N

Mol Weight: 107.16

MP (deg C): -11 **FP (deg C):**

BP (deg C): 179.1

BP pressure (mm Hg):

Property/Value	Units	Temp	Data Type	Reference
WS 2.45E+005	mg/L	-4	EXP	YALKOWSKY,SH & DANNENFELSER,RM (1992)
logP 1.90			EST	MEYLAN,WM & HOWARD,PH (1995)
VP 1.24E+000	mm Hg	25	EXP	CHAO,J ET AL. (1983)
DC 6.46	pKa	25	EXP	PERRIN,DD (1965)
HL 3.70E-006	atm m3/mol	25	EXP	ANDON,RJL ET AL. (1954)
OH 1.48E-012	cm3/molc sec	25	EST	MEYLAN,WM & HOWARD,PH (1993)

CAS #:	000583-59-5	2-METHYLCYCLOHEXANOL			

Formula: $C_7H_{14}O$

Mol Weight: 114.19

MP (deg C): **FP (deg C):**

BP (deg C): 163-166

BP pressure (mm Hg):

Property/Value	Units	Temp	Data Type	Reference
WS 9.14E+003	mg/L	25	EST	MEYLAN,WM ET AL. (1996)
logP 2.05			EST	MEYLAN,WM & HOWARD,PH (1995)
VP 2.91E-001	mm Hg	25	EST	NEELY,WB & BLAU,GE (1985)
DC	pKa			
HL 6.50E-006	atm m3/mol	25	EST	MEYLAN,WM & HOWARD,PH (1991)
OH 1.92E-011	cm3/molc sec	25	EST	MEYLAN,WM & HOWARD,PH (1993)

CAS #:	000583-60-8	2-METHYLCYCLOHEXANONE			

Formula: $C_7H_{12}O$

Mol Weight: 112.17

MP (deg C): **FP (deg C):**

BP (deg C): 162-163

BP pressure (mm Hg):

Property/Value	Units	Temp	Data Type	Reference
WS 5.14E+003	mg/L	25	EST	MEYLAN,WM ET AL. (1996)
logP 1.54			EST	MEYLAN,WM & HOWARD,PH (1995)
VP 2.25E+000	mm Hg	20	EXP	WEBER,RC ET AL. (1981)
DC	pKa			
HL 6.79E-005	atm m3/mol	25	EST	MEYLAN,WM & HOWARD,PH (1991)
OH 1.36E-011	cm3/molc sec	25	EST	MEYLAN,WM & HOWARD,PH (1993)

CAS #:	000583-61-9	2,3-DIMETHYLPYRIDINE			

Formula: C_7H_9N

Mol Weight: 107.16

MP (deg C): **FP (deg C):**

BP (deg C): 161.2

BP pressure (mm Hg):

Property/Value	Units	Temp	Data Type	Reference
WS 2.60E+005	mg/L	17	EXP	YALKOWSKY,SH & DANNENFELSER,RM (1992)
logP 1.90			EST	MEYLAN,WM & HOWARD,PH (1995)
VP 2.69E+000	mm Hg	25	EXP	CHAO,J ET AL. (1983)
DC 6.57	pKa	25	EXP	PERRIN,DD (1965)
HL 7.22E-006	atm m3/mol	25	EXP	ANDON,RJL ET AL. (1954)
OH 1.48E+000	cm3/molc sec	25	EST	MEYLAN,WM & HOWARD,PH (1993)

CAS #: 000583-78-8				2,5-DICHLOROPHENOL

Formula: $C_6H_4Cl_2O$
Mol Weight: 163.00
MP (deg C): 59   FP (deg C):
BP (deg C): 211
BP pressure (mm Hg):

Property/Value	Units	Temp	Data Type	Reference
WS 6.14E+002	mg/L	25	EST	MEYLAN,WM ET AL. (1996)
logP 3.06			EXP	HANSCH,C & LEO,AJ (1985)
VP 5.62E-002	mm Hg	25	EXT	DOLFING,J & HARRISON,BK (1992)
DC 7.51	pKa		EXP	SERJEANT,EP & DEMPSEY,B (1979)
HL 3.08E-007	atm m3/mol	25	EST	MEYLAN,WM & HOWARD,PH (1991)
OH 6.99E-012	cm3/molc sec	25	EST	MEYLAN,WM & HOWARD,PH (1993)

CAS #: 000584-02-1				3-PENTANOL

Formula: $C_5H_{12}O$
Mol Weight: 88.15
MP (deg C): -69   FP (deg C):
BP (deg C): 116.2
BP pressure (mm Hg):

Property/Value	Units	Temp	Data Type	Reference
WS 5.15E+004	mg/L	25	EXP	YALKOWSKY,SH & DANNENFELSER,RM (1992)
logP 1.21			EXP	HANSCH,C & LEO,AJ (1985)
VP 8.77E+000	mm Hg	25	EXP	DAUBERT,TE & DANNER,RP (1989)
DC	pKa			
HL 1.98E-005	atm m3/mol	25	EST	VP/WSOL
OH 1.22E-011	cm3/molc sec	25	EXP	ATKINSON,R (1989)

CAS #: 000584-03-2				1,2-BUTANEDIOL

Formula: $C_4H_{10}O_2$
Mol Weight: 90.12
MP (deg C):   FP (deg C):
BP (deg C):
BP pressure (mm Hg):

Property/Value	Units	Temp	Data Type	Reference
WS 1.00E+006	mg/L	25	EXP	BOSCHE,H & SCHNEIDER,K (1985)
logP -0.29			EST	MEYLAN,WM & HOWARD,PH (1995)
VP 5.01E-002	mm Hg	25	EXP	DAUBERT,TE & DANNER,RP (1993)
DC	pKa			
HL 5.94E-009	atm m3/mol	25	EST	VP/WSOL
OH 1.54E-011	cm3/molc sec	25	EST	MEYLAN,WM & HOWARD,PH (1993)

CAS #: 000584-79-2				ALLETHRIN

Formula: $C_{19}H_{26}O_3$
Mol Weight: 302.42
MP (deg C):   FP (deg C): 4
BP (deg C): 140
BP pressure (mm Hg): 1.00E-001

Property/Value	Units	Temp	Data Type	Reference
WS 9.84E-001	mg/L	25	EST	MEYLAN,WM ET AL. (1996)
logP 4.78			EXP	HANSCH,C & LEO,AJ (1985)
VP 1.20E-004	mm Hg	30	EXP	WORTHING,CR & WALKER,SB (1987)
DC	pKa			
HL 6.10E-007	atm m3/mol	25	EST	MEYLAN,WM & HOWARD,PH (1991)
OH 2.25E-010	cm3/molc sec	25	EST	ATKINSON,R (1988)

CAS #: 000584-84-9				2,4-TOLUENE DIISOCYANATE

Formula: $C_9H_6N_2O_2$
Mol Weight: 174.16
MP (deg C): 19.5-21.5   FP (deg C):
BP (deg C): 251
BP pressure (mm Hg):

Property/Value	Units	Temp	Data Type	Reference
WS 3.76E+001	mg/L	25	EST	MEYLAN,WM ET AL. (1996)
logP 3.74			EST	MEYLAN,WM & HOWARD,PH (1995)
VP 8.00E-003	mm Hg	20	EXP	BOUBLIK,T ET AL. (1984)
DC	pKa			
HL 1.11E-005	atm m3/mol	25	EST	MEYLAN,WM & HOWARD,PH (1991)
OH 6.26E-012	cm3/molc sec	25	EST	MEYLAN,WM & HOWARD,PH (1993)

CAS #: 000584-90-7				BIS(2-METHYLPHENYL)DIAZENE

Formula: $C_{14}H_{14}N_2$
Mol Weight: 210.28
MP (deg C): 55.5   FP (deg C):
BP (deg C):
BP pressure (mm Hg):

Property/Value	Units	Temp	Data Type	Reference
WS 5.13E-001	mg/L	25	EST	MEYLAN,WM ET AL. (1996)
logP 5.21			EST	MEYLAN,WM & HOWARD,PH (1995)
VP 2.59E-004	mm Hg	25	EST	NEELY,WB & BLAU,GE (1985)
DC	pKa			
HL 1.79E-005	atm m3/mol	25	EST	MEYLAN,WM & HOWARD,PH (1991)
OH 4.31E-012	cm3/molc sec	25	EST	MEYLAN,WM & HOWARD,PH (1993)

CAS #: 000584-93-0				2-BROMOPENTANOIC ACID

Formula: $C_6H_{11}BrO_2$
Mol Weight: 195.06
MP (deg C):   FP (deg C):
BP (deg C): 132-135
BP pressure (mm Hg): 2.50E+001

Property/Value	Units	Temp	Data Type	Reference
WS 3.70E+003	mg/L	25	EST	MEYLAN,WM ET AL. (1996)
logP 1.83			EST	MEYLAN,WM & HOWARD,PH (1995)
VP 3.25E-002	mm Hg	25	EST	NEELY,WB & BLAU,GE (1985)
DC	pKa			
HL 1.48E-007	atm m3/mol	25	EST	MEYLAN,WM & HOWARD,PH (1991)
OH 2.87E-012	cm3/molc sec	25	EST	MEYLAN,WM & HOWARD,PH (1993)

CAS #: 000584-94-1				2,3-DIMETHYLHEXANE

Formula: C_8H_{18}
Mol Weight: 114.23
MP (deg C):   FP (deg C):
BP (deg C): 115.6
BP pressure (mm Hg):

Property/Value	Units	Temp	Data Type	Reference
WS 9.20E+000	mg/L	25	EST	MEYLAN,WM ET AL. (1996)
logP 4.12			EST	MEYLAN,WM & HOWARD,PH (1995)
VP 2.34E+001	mm Hg	25	EXP	DAUBERT,TE & DANNER,RP (1989)
DC	pKa			
HL 3.01E+000	atm m3/mol	25	EST	MEYLAN,WM & HOWARD,PH (1991)
OH 8.55E-012	cm3/molc sec	25	EST	MEYLAN,WM & HOWARD,PH (1993)

CAS #: 000585-07-9				T-BUTYL METHACRYLATE
Formula: $C_8H_{14}O_2$				
Mol Weight: 142.20				
MP (deg C):		FP (deg C):		
BP (deg C): 67				
BP pressure (mm Hg): 7.00E+001				

Property/Value	Units	Temp	Data Type	Reference
WS 5.00E+001	mg/L	25	EXP	CHEM INSPECT TEST INST (1992)
logP 2.54			EXP	HANSCH,C & LEO,AJ (1985)
VP 6.17E+000	mm Hg	25	EST	NEELY,WB & BLAU,GE (1985)
DC	pKa			
HL 3.39E-004	atm m3/mol	25	EST	MEYLAN,WM & HOWARD,PH (1991)
OH 1.86E-011	cm3/molc sec	25	EST	MEYLAN,WM & HOWARD,PH (1993)

CAS #: 000585-34-2				PHENOL, 3-(1,1-DIMETHYLETHYL)-
Formula: $C_{10}H_{14}O$				
Mol Weight: 150.22				
MP (deg C): 42.3		FP (deg C):		
BP (deg C): 240				
BP pressure (mm Hg):				

Property/Value	Units	Temp	Data Type	Reference
WS 2.07E+003	mg/L	25	EXP	SHIU,WY ET AL. (1994)
logP 3.30			EXP	SANGSTER,J (1993)
VP 1.59E-003	mm Hg	25	EXT	JORDAN,TE (1954)
DC 10.12	pKa	25	EXP	SERJEANT,EP & DEMPSEY,B (1979)
HL 1.45E-006	atm m3/mol	25	EST	MEYLAN,WM & HOWARD,PH (1991)
OH 7.41E-011	cm3/molc sec	25	EST	MEYLAN,WM & HOWARD,PH (1993)

CAS #: 000585-48-8				2,6-DI-T-BUTYLPYRIDINE
Formula: $C_{13}H_{21}N$				
Mol Weight: 191.32				
MP (deg C): 2.2		FP (deg C):		
BP (deg C): 100-101				
BP pressure (mm Hg): 2.30E+001				

Property/Value	Units	Temp	Data Type	Reference
WS 1.41E+001	mg/L	25	EST	MEYLAN,WM ET AL. (1996)
logP 4.14			EXP	HANSCH,C ET AL. (1995)
VP 3.33E-002	mm Hg	25	EST	NEELY,WB & BLAU,GE (1985)
DC 5.02	pKa	24	EXP	PERRIN,DD (1972)
HL 4.70E-005	atm m3/mol	25	EST	MEYLAN,WM & HOWARD,PH (1991)
OH 2.80E-012	cm3/molc sec	25	EST	MEYLAN,WM & HOWARD,PH (1993)

CAS #: 000585-74-0				M-METHYLACETOPHENONE
Formula: $C_9H_{10}O$				
Mol Weight: 134.18				
MP (deg C):		FP (deg C):		
BP (deg C): 220				
BP pressure (mm Hg):				

Property/Value	Units	Temp	Data Type	Reference
WS 1.29E+003	mg/L	25	EST	MEYLAN,WM ET AL. (1996)
logP 2.15			EXP	DUNN,WJ ET AL. (1983)
VP 2.21E-001	mm Hg	25	EST	NEELY,WB & BLAU,GE (1985)
DC	pKa			
HL 1.08E-005	atm m3/mol	25	EST	MEYLAN,WM & HOWARD,PH (1991)
OH 2.42E-012	cm3/molc sec	25	EST	MEYLAN,WM & HOWARD,PH (1993)

CAS #: 000585-76-2				M-BROMOBENZOIC ACID
Formula: $C_7H_5BrO_2$				
Mol Weight: 201.03				
MP (deg C): 155		FP (deg C):		
BP (deg C): > 280				
BP pressure (mm Hg):				

Property/Value	Units	Temp	Data Type	Reference
WS 4.02E+002	mg/L	25	EXP	STEPHEN,H & STEPHEN,T (1963)
logP 2.87			EXP	HANSCH,C & LEO,AJ (1985)
VP 9.06E-005	mm Hg	25	EXT	DOLFING,J & HARRISON,BK (1992)
DC 3.81	pKa	20	EXP	KORTUM,G ET AL (1961)
HL 4.32E-008	atm m3/mol	25	EST	MEYLAN,WM & HOWARD,PH (1991)
OH 8.55E-013	cm3/molc sec	25	EST	MEYLAN,WM & HOWARD,PH (1993)

CAS #: 000585-79-5				3-BROMO-1-NITROBENZENE
Formula: $C_6H_4BrNO_2$				
Mol Weight: 202.01				
MP (deg C): 56		FP (deg C):		
BP (deg C): 265				
BP pressure (mm Hg):				

Property/Value	Units	Temp	Data Type	Reference
WS 1.00E+004	mg/L	25	EXP	NEELY,WB (1976)
logP 2.64			EXP	HANSCH,C & LEO,AJ (1985)
VP 7.00E-002	mm Hg	25	EXP	NEELY,WB (1976)
DC	pKa			
HL 1.86E-006	atm m3/mol	25	EST	VP/WSOL
OH 1.07E-013	cm3/molc sec	25	EST	MEYLAN,WM & HOWARD,PH (1993)

CAS #: 000586-11-8				3,5-DINITROPHENOL
Formula: $C_6H_4N_2O_5$				
Mol Weight: 184.11				
MP (deg C): 125.1		FP (deg C):		
BP (deg C):				
BP pressure (mm Hg):				

Property/Value	Units	Temp	Data Type	Reference
WS 1.35E+004	mg/L	52	EXP	BUIKEMA,ALJR ET AL. (1979)
logP 2.36			EXP	HANSCH,C & LEO,AJ (1985)
VP 1.20E-005	mm Hg	25	EST	NEELY,WB & BLAU,GE (1985)
DC 6.69	pKa	25	EXP	SERJEANT,EP & DEMPSEY,B (1979)
HL 8.73E-012	atm m3/mol	25	EST	MEYLAN,WM & HOWARD,PH (1991)
OH 3.94E-013	cm3/molc sec	25	EST	MEYLAN,WM & HOWARD,PH (1993)

CAS #: 000586-37-8				M-METHOXYACETOPHENONE
Formula: $C_9H_{10}O_2$				
Mol Weight: 150.18				
MP (deg C): 95.5		FP (deg C):		
BP (deg C): 240				
BP pressure (mm Hg):				

Property/Value	Units	Temp	Data Type	Reference
WS 2.03E+003	mg/L	25	EST	MEYLAN,WM ET AL. (1996)
logP 1.84			EXP	HANSCH,C & LEO,AJ (1985)
VP 8.04E-002	mm Hg	25	EST	NEELY,WB & BLAU,GE (1985)
DC	pKa			
HL 5.80E-007	atm m3/mol	25	EST	MEYLAN,WM & HOWARD,PH (1991)
OH 1.00E-011	cm3/molc sec	25	EST	MEYLAN,WM & HOWARD,PH (1993)

CAS #: 000586-38-9				M-METHOXYBENZOIC ACID

Formula: $C_8H_8O_3$

Mol Weight: 152.15

MP (deg C): 107 FP (deg C):

BP (deg C): 170

BP pressure (mm Hg): 1.00E+001

Property/Value	Units	Temp	Data Type	Reference
WS 1.40E+003	mg/L	25	EST	MEYLAN,WM ET AL. (1996)
logP 2.02			EXP	HANSCH,C & LEO,AJ (1985)
VP 1.50E-003	mm Hg	25	EST	NEELY,WB & BLAU,GE (1985)
DC 4.09	pKa	25	EXP	KORTUM,G ET AL (1961)
HL 6.41E-009	atm m3/mol	25	EST	MEYLAN,WM & HOWARD,PH (1991)
OH 7.22E-012	cm3/molc sec	25	EST	MEYLAN,WM & HOWARD,PH (1993)

CAS #: 000586-61-8				1-ISOPROPYL-4-BROMOBENZENE

Formula: $C_9H_{11}Br$

Mol Weight: 199.10

MP (deg C): -22.5 FP (deg C):

BP (deg C): 218.7

BP pressure (mm Hg):

Property/Value	Units	Temp	Data Type	Reference
WS 8.71E+000	mg/L	25	EST	MEYLAN,WM ET AL. (1996)
logP 4.34			EST	MEYLAN,WM & HOWARD,PH (1995)
VP 2.03E-001	mm Hg	25	EST	NEELY,WB & BLAU,GE (1985)
DC	pKa			
HL 4.17E-003	atm m3/mol	25	EST	MEYLAN,WM & HOWARD,PH (1991)
OH 3.75E-012	cm3/molc sec	25	EST	MEYLAN,WM & HOWARD,PH (1993)

CAS #: 000586-62-9				TERPINOLENE

Formula: $C_{10}H_{16}$

Mol Weight: 136.24

MP (deg C): FP (deg C):

BP (deg C): 185

BP pressure (mm Hg):

Property/Value	Units	Temp	Data Type	Reference
WS 1.74E+000	mg/L	25	EST	VP/HL
logP 4.47			EXP	LI,NY & PERDUE,EM (1995)
VP 5.95E-001	mm Hg	25	EXT	PERRY,RH & GREEN,D (1984)
DC	pKa			
HL 6.14E-002	atm m3/mol	25	EST	MEYLAN,WM & HOWARD,PH (1991)
OH 2.25E-010	cm3/molc sec	25	EXP	ATKINSON,R (1989)

CAS #: 000586-76-5				P-BROMOBENZOIC ACID

Formula: $C_7H_5BrO_2$

Mol Weight: 201.03

MP (deg C): 251-253 FP (deg C):

BP (deg C):

BP pressure (mm Hg):

Property/Value	Units	Temp	Data Type	Reference
WS 6.00E+001	mg/L		EXP	YALKOWSKY,SH & DANNENFELSER,RM (1992)
logP 2.86			EXP	HANSCH,C & LEO,AJ (1985)
VP 9.95E-006	mm Hg	25	EXT	DOLFING,J & HARRISON,BK (1992)
DC 4.00	pKa	20	EXP	KORTUM,G ET AL (1961)
HL 4.32E-008	atm m3/mol	25	EST	MEYLAN,WM & HOWARD,PH (1991)
OH 9.74E-013	cm3/molc sec	25	EST	MEYLAN,WM & HOWARD,PH (1993)

CAS #: 000586-78-7				4-BROMO-1-NITROBENZENE

Formula: $C_6H_4BrNO_2$

Mol Weight: 202.01

MP (deg C): 127 FP (deg C):

BP (deg C): 256

BP pressure (mm Hg):

Property/Value	Units	Temp	Data Type	Reference
WS 1.16E+002	mg/L	25	EST	MEYLAN,WM ET AL. (1996)
logP 2.55			EXP	HANSCH,C & LEO,AJ (1985)
VP 9.89E-003	mm Hg	25	EST	NEELY,WB & BLAU,GE (1985)
DC	pKa			
HL 8.47E-006	atm m3/mol	25	EST	MEYLAN,WM & HOWARD,PH (1991)
OH 1.53E-013	cm3/molc sec	25	EST	MEYLAN,WM & HOWARD,PH (1993)

CAS #: 000586-84-5				2-(METHOXYMETHYL)-5-NO2 FURAN

Formula: $C_6H_8NO_4$

Mol Weight: 158.13

MP (deg C): FP (deg C):

BP (deg C): 104-105

BP pressure (mm Hg): 3.00E+000

Property/Value	Units	Temp	Data Type	Reference
WS 4.99E+003	mg/L	25	EST	MEYLAN,WM ET AL. (1996)
logP 0.89			EXP	HANSCH,C & LEO,AJ (1985)
VP 4.43E-002	mm Hg	25	EST	NEELY,WB & BLAU,GE (1985)
DC	pKa			
HL 2.74E-007	atm m3/mol	25	EST	MEYLAN,WM & HOWARD,PH (1991)
OH 1.91E-011	cm3/molc sec	25	EST	MEYLAN,WM & HOWARD,PH (1993)

CAS #: 000586-89-0				P-ACETYLBENZOIC ACID

Formula: $C_9H_8O_3$

Mol Weight: 164.16

MP (deg C): 208 FP (deg C):

BP (deg C):

BP pressure (mm Hg):

Property/Value	Units	Temp	Data Type	Reference
WS 2.76E+003	mg/L	25	EST	MEYLAN,WM ET AL. (1996)
logP 1.61			EXP	HANSCH,C ET AL. (1995)
VP 2.30E-004	mm Hg	25	EST	NEELY,WB & BLAU,GE (1985)
DC 3.70	pKa	25	EXP	SERJEANT,EP & DEMPSEY,B (1979)
HL 1.97E-010	atm m3/mol	25	EST	MEYLAN,WM & HOWARD,PH (1991)
OH 1.11E-012	cm3/molc sec	25	EST	MEYLAN,WM & HOWARD,PH (1993)

CAS #: 000586-95-8				4-PYRIDINEMETHANOL

Formula: C_6H_7NO

Mol Weight: 109.13

MP (deg C): 53 FP (deg C):

BP (deg C): 140-142

BP pressure (mm Hg): 1.20E+001

Property/Value	Units	Temp	Data Type	Reference
WS 1.00E+006	mg/L	25	EST	MEYLAN,WM ET AL. (1996)
logP 0.06			EXP	HANSCH,C & LEO,AJ (1985)
VP 2.60E-002	mm Hg	25	EST	NEELY,WB & BLAU,GE (1985)
DC 5.33	pKa	25	EXP	PERRIN,DD (1972)
HL 2.85E-010	atm m3/mol	25	EST	MEYLAN,WM & HOWARD,PH (1991)
OH 4.33E-012	cm3/molc sec	25	EST	MEYLAN,WM & HOWARD,PH (1993)

CAS #: 000586-96-9				NITROSOBENZENE

Formula: C_6H_5NO

Mol Weight: 107.11

MP (deg C): 68.5　　FP (deg C):

BP (deg C): 58

BP pressure (mm Hg): 1.80E+001

Property/Value	Units	Temp	Data Type	Reference
WS 2.14E+003	mg/L	25	EST	MEYLAN,WM ET AL. (1996)
logP 2.01			EXP	HANSCH,C & LEO,AJ (1985)
VP 8.30E+000	mm Hg	25	EST	NEELY,WB & BLAU,GE (1985)
DC	pKa			
HL 5.64E-005	atm m3/mol	25	EST	MEYLAN,WM & HOWARD,PH (1991)
OH 1.24E-011	cm3/molc sec	25	EST	MEYLAN,WM & HOWARD,PH (1993)

CAS #: 000586-98-1				2-PYRIDINEMETHANOL

Formula: C_6H_7NO

Mol Weight: 109.13

MP (deg C):　　FP (deg C):

BP (deg C): 112

BP pressure (mm Hg): 1.60E+001

Property/Value	Units	Temp	Data Type	Reference
WS 1.00E+006	mg/L	25	EST	MEYLAN,WM ET AL. (1996)
logP 0.06			EXP	HANSCH,C & LEO,AJ (1985)
VP 2.60E-002	mm Hg	25	EST	NEELY,WB & BLAU,GE (1985)
DC 4.86	pKa		EXP	PERRIN,DD (1972)
HL 2.85E-010	atm m3/mol	25	EST	MEYLAN,WM & HOWARD,PH (1991)
OH 4.33E-012	cm3/molc sec	25	EST	MEYLAN,WM & HOWARD,PH (1993)

CAS #: 000587-03-1				M-METHYLBENZYL ALCOHOL

Formula: $C_8H_{10}O$

Mol Weight: 122.17

MP (deg C): < -20　　FP (deg C):

BP (deg C): 215.5

BP pressure (mm Hg):

Property/Value	Units	Temp	Data Type	Reference
WS 1.37E+004	mg/L	25	EST	MEYLAN,WM ET AL. (1996)
logP 1.60			EXP	HANSCH,C & LEO,AJ (1985)
VP 1.61E-002	mm Hg	25	EST	NEELY,WB & BLAU,GE (1985)
DC	pKa			
HL 2.40E-007	atm m3/mol	25	EST	MEYLAN,WM & HOWARD,PH (1991)
OH 1.62E-011	cm3/molc sec	25	EST	MEYLAN,WM & HOWARD,PH (1993)

CAS #: 000587-04-2				M-CHLOROBENZALDEHYDE

Formula: C_7H_5ClO

Mol Weight: 140.57

MP (deg C): 17.5　　FP (deg C):

BP (deg C): 213.5

BP pressure (mm Hg):

Property/Value	Units	Temp	Data Type	Reference
WS 9.78E+002	mg/L	25	EST	MEYLAN,WM ET AL. (1996)
logP 2.26			EXP	SANGSTER,J (1993)
VP 1.83E-001	mm Hg	25	EST	NEELY,WB & BLAU,GE (1985)
DC	pKa			
HL 9.95E-006	atm m3/mol	25	EST	MEYLAN,WM & HOWARD,PH (1991)
OH 1.76E-011	cm3/molc sec	25	EST	MEYLAN,WM & HOWARD,PH (1993)

CAS #: 000587-34-8				1,1-DIMETHYL-3(M-CHLOROPHENYL)

Formula: $C_9H_{11}ClN_2O$

Mol Weight: 198.65

MP (deg C):　　FP (deg C):

BP (deg C):

BP pressure (mm Hg):

Property/Value	Units	Temp	Data Type	Reference
WS 8.71E+002	mg/L	25	EST	MEYLAN,WM ET AL. (1996)
logP 2.00			EXP	HANSCH,C & LEO,AJ (1985)
VP 4.70E-005	mm Hg	25	EST	NEELY,WB & BLAU,GE (1985)
DC	pKa			
HL 7.19E-010	atm m3/mol	25	EST	MEYLAN,WM & HOWARD,PH (1991)
OH 3.21E-011	cm3/molc sec	25	EST	MEYLAN,WM & HOWARD,PH (1993)

CAS #: 000587-48-4				3-ACETYLAMINOBENZOIC ACID

Formula: $C_9H_9NO_3$

Mol Weight: 179.18

MP (deg C):　　FP (deg C):

BP (deg C):

BP pressure (mm Hg):

Property/Value	Units	Temp	Data Type	Reference
WS 4.15E+003	mg/L	25	EST	MEYLAN,WM ET AL. (1996)
logP 1.32			EXP	DA,YZ ET AL. (1992)
VP 7.14E-007	mm Hg	25	EST	NEELY,WB & BLAU,GE (1985)
DC	pKa			
HL 1.24E-013	atm m3/mol	25	EST	MEYLAN,WM & HOWARD,PH (1991)
OH 4.01E-012	cm3/molc sec	25	EST	MEYLAN,WM & HOWARD,PH (1993)

CAS #: 000587-56-4				2-CHLOROETHYL (3-CHLOROPHENYL)CARBAMATE

Formula: $C_9H_9Cl_2NO_2$

Mol Weight: 234.08

MP (deg C):　　FP (deg C):

BP (deg C):

BP pressure (mm Hg):

Property/Value	Units	Temp	Data Type	Reference
WS 6.01E+001	mg/L	25	EST	MEYLAN,WM ET AL. (1996)
logP 3.14			EST	MEYLAN,WM & HOWARD,PH (1995)
VP 2.15E-004	mm Hg	25	EST	NEELY,WB & BLAU,GE (1985)
DC	pKa			
HL 7.57E-009	atm m3/mol	25	EST	MEYLAN,WM & HOWARD,PH (1991)
OH 3.21E-011	cm3/molc sec	25	EST	MEYLAN,WM & HOWARD,PH (1993)

CAS #: 000587-64-4				3,5-DICHLOROPHENOXYACETIC ACID

Formula: $C_8H_6Cl_2O_3$

Mol Weight: 221.04

MP (deg C):　　FP (deg C):

BP (deg C):

BP pressure (mm Hg):

Property/Value	Units	Temp	Data Type	Reference
WS 9.61E+002	mg/L	25	EXP	SHIU,WY ET AL. (1990)
logP 2.84			EXP	SANGSTER,J (1994)
VP 6.11E-005	mm Hg	25	EST	NEELY,WB & BLAU,GE (1985)
DC	pKa			
HL 9.21E-009	atm m3/mol	25	EST	MEYLAN,WM & HOWARD,PH (1991)
OH 1.54E-011	cm3/molc sec	25	EST	MEYLAN,WM & HOWARD,PH (1993)

CAS #: 000587-65-5				A-CHLOROACETANILIDE

Formula: C_8H_8ClNO
Mol Weight: 169.61
MP (deg C):
FP (deg C):
BP (deg C):
BP pressure (mm Hg):

Property/Value	Units	Temp	Data Type	Reference
WS 2.50E+003	mg/L	25	EST	MEYLAN,WM ET AL. (1996)
logP 1.63			EXP	HANSCH,C & LEO,AJ (1985)
VP 4.33E-005	mm Hg	25	EST	NEELY,WB & BLAU,GE (1985)
DC	pKa			
HL 2.17E-009	atm m3/mol	25	EST	MEYLAN,WM & HOWARD,PH (1991)
OH 1.27E-011	cm3/molc sec	25	EST	MEYLAN,WM & HOWARD,PH (1993)

CAS #: 000588-04-5				BIS(3-METHYLPHENYL)DIAZENE

Formula: $C_{14}H_{14}N_2$
Mol Weight: 210.28
MP (deg C):
FP (deg C):
BP (deg C):
BP pressure (mm Hg):

Property/Value	Units	Temp	Data Type	Reference
WS 5.13E-001	mg/L	25	EST	MEYLAN,WM ET AL. (1996)
logP 5.21			EST	MEYLAN,WM & HOWARD,PH (1995)
VP 2.59E-004	mm Hg	25	EST	NEELY,WB & BLAU,GE (1985)
DC	pKa			
HL 1.79E-005	atm m3/mol	25	EST	MEYLAN,WM & HOWARD,PH (1991)
OH 2.17E-012	cm3/molc sec	25	EST	MEYLAN,WM & HOWARD,PH (1993)

CAS #: 000588-07-8				3-CHLOROACETANILIDE

Formula: C_8H_8ClNO
Mol Weight: 169.61
MP (deg C): 79-81
FP (deg C):
BP (deg C):
BP pressure (mm Hg):

Property/Value	Units	Temp	Data Type	Reference
WS 9.01E+002	mg/L	25	EST	MEYLAN,WM ET AL. (1996)
logP 2.15			EXP	NAKAGAWA,Y ET AL. (1992)
VP 9.63E-005	mm Hg	25	EST	NEELY,WB & BLAU,GE (1985)
DC	pKa			
HL 4.57E-009	atm m3/mol	25	EST	MEYLAN,WM & HOWARD,PH (1991)
OH 8.84E-012	cm3/molc sec	25	EST	MEYLAN,WM & HOWARD,PH (1993)

CAS #: 000588-16-9				3-METHOXYACETANILIDE

Formula: $C_9H_{11}NO_2$
Mol Weight: 165.19
MP (deg C):
FP (deg C):
BP (deg C):
BP pressure (mm Hg):

Property/Value	Units	Temp	Data Type	Reference
WS 5.03E+003	mg/L	25	EST	MEYLAN,WM ET AL. (1996)
logP 1.30			EXP	NAKAGAWA,Y ET AL. (1992)
VP 6.59E-005	mm Hg	25	EST	NEELY,WB & BLAU,GE (1985)
DC	pKa			
HL 3.65E-010	atm m3/mol	25	EST	MEYLAN,WM & HOWARD,PH (1991)
OH 1.38E-010	cm3/molc sec	25	EST	MEYLAN,WM & HOWARD,PH (1993)

CAS #: 000588-22-7				3,4-DICHLOROPHENOXYACETIC ACID

Formula: $C_8H_6Cl_2O_3$
Mol Weight: 221.04
MP (deg C): 138-140
FP (deg C):
BP (deg C):
BP pressure (mm Hg):

Property/Value	Units	Temp	Data Type	Reference
WS 4.57E+002	mg/L	25	EXP	YALKOWSKY,SH & DANNENFELSER,RM (1992)
logP 2.81			EXP	HANSCH,C & LEO,AJ (1985)
VP 6.11E-005	mm Hg	25	EST	NEELY,WB & BLAU,GE (1985)
DC 2.92	pKa	20	EXP	SERJEANT,EP & DEMPSEY,B (1979)
HL 9.21E-009	atm m3/mol	25	EST	MEYLAN,WM & HOWARD,PH (1991)
OH 9.21E-012	cm3/molc sec	25	EST	MEYLAN,WM & HOWARD,PH (1993)

CAS #: 000588-32-9				M-CHLOROPHENOXYACETIC ACID

Formula: $C_8H_7ClO_3$
Mol Weight: 186.60
MP (deg C):
FP (deg C):
BP (deg C):
BP pressure (mm Hg):

Property/Value	Units	Temp	Data Type	Reference
WS 2.34E+003	mg/L	25	EST	MEYLAN,WM ET AL. (1996)
logP 2.03			EXP	HANSCH,C & LEO,AJ (1985)
VP 3.18E-004	mm Hg	25	EST	NEELY,WB & BLAU,GE (1985)
DC 3.07	pKa	25	EXP	KORTUM,G ET AL (1961)
HL 1.24E-008	atm m3/mol	25	EST	MEYLAN,WM & HOWARD,PH (1991)
OH 1.99E-011	cm3/molc sec	25	EST	MEYLAN,WM & HOWARD,PH (1993)

CAS #: 000588-46-5				N-BENZYLACETAMIDE

Formula: $C_9H_{11}NO$
Mol Weight: 149.19
MP (deg C): 61
FP (deg C):
BP (deg C): 157
BP pressure (mm Hg): 2.00E+000

Property/Value	Units	Temp	Data Type	Reference
WS 1.18E+004	mg/L	25	EST	MEYLAN,WM ET AL. (1996)
logP 0.95			EXP	HANSCH,C & LEO,AJ (1985)
VP 1.97E-004	mm Hg	25	EST	NEELY,WB & BLAU,GE (1985)
DC	pKa			
HL 1.98E-009	atm m3/mol	25	EST	MEYLAN,WM & HOWARD,PH (1991)
OH 1.43E-011	cm3/molc sec	25	EST	MEYLAN,WM & HOWARD,PH (1993)

CAS #: 000588-59-0				STILBENE

Formula: $C_{14}H_{12}$
Mol Weight: 180.25
MP (deg C): 124
FP (deg C):
BP (deg C): 306-307
BP pressure (mm Hg):

Property/Value	Units	Temp	Data Type	Reference
WS 2.92E-001	mg/L	25	EXP	YALKOWSKY,SH & DANNENFELSER,RM (1992)
logP 4.53			EXP	YALKOWSKY,SH ET AL. (1983)
VP 2.99E-003	mm Hg	25	EST	NEELY,WB & BLAU,GE (1985)
DC	pKa			
HL 1.24E-004	atm m3/mol	25	EST	MEYLAN,WM & HOWARD,PH (1991)
OH 6.01E-011	cm3/molc sec	25	EST	MEYLAN,WM & HOWARD,PH (1993)

CAS #:	000589-08-2			BENZENEETHANAMINE, N-METHYL-

Formula: $C_9H_{13}N$

Mol Weight: 135.21

MP (deg C):

FP (deg C):

BP (deg C): 206

BP pressure (mm Hg):

Property/Value	Units	Temp	Data Type	Reference
WS 2.09E+004	mg/L	25	EST	MEYLAN,WM ET AL. (1996)
logP 1.91			EXP	SANGSTER,J (1993)
VP 2.40E-001	mm Hg	25	EST	NEELY,WB & BLAU,GE (1985)
DC 10.08	pKa	20	EXP	PERRIN,DD (1965)
HL 1.78E-006	atm m3/mol	25	EST	MEYLAN,WM & HOWARD,PH (1991)
OH 8.09E-011	cm3/molc sec	25	EST	MEYLAN,WM & HOWARD,PH (1993)

CAS #:	000589-16-2			P-ETHYLANILINE

Formula: $C_8H_{11}N$

Mol Weight: 121.18

MP (deg C): -4.8

FP (deg C):

BP (deg C): 217.5

BP pressure (mm Hg):

Property/Value	Units	Temp	Data Type	Reference
WS 2.11E+003	mg/L	25	EST	MEYLAN,WM ET AL. (1996)
logP 1.96			EXP	HANSCH,C & LEO,AJ (1985)
VP 1.43E-001	mm Hg	25	EXP	CHAO,J ET AL. (1983)
DC 5.00	pKa		EXP	PERRIN,DD (1972)
HL 2.79E-006	atm m3/mol	25	EST	MEYLAN,WM & HOWARD,PH (1991)
OH 1.32E-010	cm3/molc sec	25	EST	MEYLAN,WM & HOWARD,PH (1993)

CAS #:	000589-18-4			P-METHYL BENZYL ALCOHOL

Formula: $C_8H_{10}O$

Mol Weight: 122.17

MP (deg C): 61.5

FP (deg C):

BP (deg C): 217

BP pressure (mm Hg):

Property/Value	Units	Temp	Data Type	Reference
WS 7.70E+003	mg/L	25	EXP	VALVANI,SC ET AL. (1981)
logP 1.58			EXP	HANSCH,C & LEO,AJ (1985)
VP 1.61E-002	mm Hg	25	EST	NEELY,WB & BLAU,GE (1985)
DC	pKa			
HL 2.40E-007	atm m3/mol	25	EST	MEYLAN,WM & HOWARD,PH (1991)
OH 9.75E-012	cm3/molc sec	25	EST	MEYLAN,WM & HOWARD,PH (1993)

CAS #:	000589-21-9			HYDRAZINE, (4-BROMOPHENYL)-

Formula: $C_6H_7BrN_2$

Mol Weight: 187.05

MP (deg C): 108-109

FP (deg C):

BP (deg C):

BP pressure (mm Hg):

Property/Value	Units	Temp	Data Type	Reference
WS 3.31E+003	mg/L	25	EST	MEYLAN,WM ET AL. (1996)
logP 1.39			EXP	HANSCH,C ET AL. (1995)
VP 1.39E-002	mm Hg	25	EST	NEELY,WB & BLAU,GE (1985)
DC 5.05	pKa	25	EXP	PERRIN,DD (1972)
HL 3.17E-009	atm m3/mol	25	EST	MEYLAN,WM & HOWARD,PH (1991)
OH 1.22E-011	cm3/molc sec	25	EST	MEYLAN,WM & HOWARD,PH (1993)

CAS #:	000589-34-4			3-METHYLHEXANE

Formula: C_7H_{16}

Mol Weight: 100.21

MP (deg C): -119

FP (deg C):

BP (deg C): 91

BP pressure (mm Hg):

Property/Value	Units	Temp	Data Type	Reference
WS 4.95E+000	mg/L	25	EXP	YALKOWSKY,SH & DANNENFELSER,RM (1992)
logP 3.71			EST	MEYLAN,WM & HOWARD,PH (1995)
VP 6.15E+001	mm Hg	25	EXP	DAUBERT,TE & DANNER,RP (1989)
DC	pKa			
HL 1.64E+000	atm m3/mol	25	EST	VP/WSOL
OH 7.15E-012	cm3/molc sec	25	EST	MEYLAN,WM & HOWARD,PH (1993)

CAS #:	000589-38-8			3-HEXANONE

Formula: $C_6H_{12}O$

Mol Weight: 100.16

MP (deg C): -55.5

FP (deg C):

BP (deg C): 123.5

BP pressure (mm Hg):

Property/Value	Units	Temp	Data Type	Reference
WS 1.47E+004	mg/L	25	EXP	YALKOWSKY,SH & DANNENFELSER,RM (1992)
logP 1.24			EST	MEYLAN,WM & HOWARD,PH (1995)
VP 1.39E+001	mm Hg	25	EXP	YAWS,CL (1994A)
DC	pKa			
HL 1.25E-004	atm m3/mol	25	EXP	VP/WSOL
OH 6.90E-012	cm3/molc sec	25	EXP	ATKINSON,R (1989)

CAS #:	000589-43-5			2,4-DIMETHYLHEXANE

Formula: C_8H_{18}

Mol Weight: 114.23

MP (deg C):

FP (deg C):

BP (deg C): 109.5

BP pressure (mm Hg):

Property/Value	Units	Temp	Data Type	Reference
WS 9.80E+001	mg/L	20	EXP	KERTES,AS (1989)
logP 4.12			EST	MEYLAN,WM & HOWARD,PH (1995)
VP 3.04E+001	mm Hg	25	EXP	DAUBERT,TE & DANNER,RP (1989)
DC	pKa			
HL 4.66E-002	atm m3/mol	25	EST	VP/WSOL
OH 8.55E-012	cm3/molc sec	25	EST	MEYLAN,WM & HOWARD,PH (1993)

CAS #:	000589-53-7			4-METHYLHEPTANE

Formula: C_8H_{18}

Mol Weight: 114.23

MP (deg C): -121

FP (deg C):

BP (deg C): 117.7

BP pressure (mm Hg):

Property/Value	Units	Temp	Data Type	Reference
WS 7.97E+000	mg/L	25	EST	MEYLAN,WM ET AL. (1996)
logP 4.20			EST	MEYLAN,WM & HOWARD,PH (1995)
VP 2.05E+001	mm Hg	25	EXP	DAUBERT,TE & DANNER,RP (1989)
DC	pKa			
HL 3.01E+000	atm m3/mol	25	EST	MEYLAN,WM & HOWARD,PH (1991)
OH 8.56E-012	cm3/molc sec	25	EST	MEYLAN,WM & HOWARD,PH (1993)

000589-55-9 — 4-HEPTANOL

CAS #: 000589-55-9				4-HEPTANOL

Formula: $C_7H_{16}O$

Mol Weight: 116.20

MP (deg C): -41.2

FP (deg C):

BP (deg C): 156

BP pressure (mm Hg):

Property/Value	Units	Temp	Data Type	Reference
WS 4.70E+003	mg/L	25	EXP	BARTON,AFM (1984)
logP 2.22			EXP	HANSCH,C ET AL. (1995)
VP 9.89E-001	mm Hg	25	EXT	WILHOIT,RC & ZWOLINSKI,BJ (1973)
DC	pKa			
HL 2.34E-005	atm m3/mol	25	EST	MEYLAN,WM & HOWARD,PH (1991)
OH 1.59E-011	cm3/molc sec	25	EST	MEYLAN,WM & HOWARD,PH (1993)

000589-59-3 — ISOBUTYL ISOVALERATE

Formula: $C_9H_{18}O_2$

Mol Weight: 158.24

MP (deg C):

FP (deg C):

BP (deg C): 170-172

BP pressure (mm Hg):

Property/Value	Units	Temp	Data Type	Reference
WS 1.36E+002	mg/L	25	EST	MEYLAN,WM ET AL. (1996)
logP 3.17			EST	MEYLAN,WM & HOWARD,PH (1995)
VP 1.84E+000	mm Hg	25	EXP	OHE,S (1976)
DC	pKa			
HL 9.60E-004	atm m3/mol	25	EST	MEYLAN,WM & HOWARD,PH (1991)
OH 8.13E-012	cm3/molc sec	25	EST	MEYLAN,WM & HOWARD,PH (1993)

000589-62-8 — 4-OCTANOL

Formula: $C_8H_{18}O$

Mol Weight: 130.23

MP (deg C):

FP (deg C):

BP (deg C):

BP pressure (mm Hg):

Property/Value	Units	Temp	Data Type	Reference
WS 1.53E+003	mg/L	25	EST	MEYLAN,WM ET AL. (1996)
logP 2.68			EXP	HANSCH,C ET AL. (1995)
VP 3.11E-001	mm Hg	25	EXP	BOUBLIK,T ET AL. (1984)
DC	pKa			
HL 3.10E-005	atm m3/mol	25	EST	MEYLAN,WM & HOWARD,PH (1991)
OH 1.73E-011	cm3/molc sec	25	EST	MEYLAN,WM & HOWARD,PH (1993)

000589-75-3 — BUTYL OCTANOATE

Formula: $C_{12}H_{24}O_2$

Mol Weight: 200.32

MP (deg C): -42.9

FP (deg C):

BP (deg C): 240.5

BP pressure (mm Hg):

Property/Value	Units	Temp	Data Type	Reference
WS 3.52E+000	mg/L	25	EST	MEYLAN,WM ET AL. (1996)
logP 4.79			EST	MEYLAN,WM & HOWARD,PH (1995)
VP 3.07E-002	mm Hg	25	EST	NEELY,WB & BLAU,GE (1985)
DC	pKa			
HL 2.25E-003	atm m3/mol	25	EST	MEYLAN,WM & HOWARD,PH (1991)
OH 1.24E-011	cm3/molc sec	25	EST	MEYLAN,WM & HOWARD,PH (1993)

000589-81-1 — 3-METHYLHEPTANE

Formula: C_8H_{18}

Mol Weight: 114.23

MP (deg C):

FP (deg C):

BP (deg C):

BP pressure (mm Hg):

Property/Value	Units	Temp	Data Type	Reference
WS 7.92E-001	mg/L	25	EXP	YALKOWSKY,SH & DANNENFELSER,RM (1992)
logP 4.20			EST	MEYLAN,WM & HOWARD,PH (1995)
VP 1.96E+001	mm Hg	25	EXP	DAUBERT,TE & DANNER,RP (1989)
DC	pKa			
HL 3.72E+000	atm m3/mol	25	EST	VP/WSOL
OH 8.56E-012	cm3/molc sec	25	EST	MEYLAN,WM & HOWARD,PH (1993)

000589-82-2 — 3-HEPTANOL

Formula: $C_7H_{16}O$

Mol Weight: 116.20

MP (deg C):

FP (deg C):

BP (deg C): 66

BP pressure (mm Hg): 2.00E+001

Property/Value	Units	Temp	Data Type	Reference
WS 4.00E+003	mg/L	25	EXP	YALKOWSKY,SH & DANNENFELSER,RM (1992)
logP 2.24			EXP	HANSCH,C ET AL. (1995)
VP 7.40E-001	mm Hg	22	EXP	RIDDICK,JA ET AL. (1986)
DC	pKa			
HL 2.83E-005	atm m3/mol	25	EST	VP/WSOL
OH 1.59E-011	cm3/molc sec	25	EST	MEYLAN,WM & HOWARD,PH (1993)

000589-87-7 — 4-BROMOIODOBENZENE

Formula: C_6H_4BrI

Mol Weight: 282.91

MP (deg C): 92

FP (deg C):

BP (deg C): 252

BP pressure (mm Hg):

Property/Value	Units	Temp	Data Type	Reference
WS 7.79E+000	mg/L	25	EXP	OKOUCHI,S ET AL. (1992)
logP 4.05			EST	MEYLAN,WM & HOWARD,PH (1995)
VP 2.74E-002	mm Hg	25	EST	NEELY,WB & BLAU,GE (1985)
DC	pKa			
HL 4.97E-004	atm m3/mol	25	EST	MEYLAN,WM & HOWARD,PH (1991)
OH 4.05E-013	cm3/molc sec	25	EST	MEYLAN,WM & HOWARD,PH (1993)

000589-90-2 — 1,4-DIMETHYLCYCLOHEXANE

Formula: C_8H_{16}

Mol Weight: 112.22

MP (deg C): -87

FP (deg C):

BP (deg C): 120

BP pressure (mm Hg):

Property/Value	Units	Temp	Data Type	Reference
WS 3.84E+000	mg/L	25	EXP	YALKOWSKY,SH & DANNENFELSER,RM (1992)
logP 4.01			EST	MEYLAN,WM & HOWARD,PH (1995)
VP 1.79E+001	mm Hg	25	EXT	CHAO,J ET AL. (1983) [CIS]
DC	pKa			
HL 4.50E-001	atm m3/mol	25	EST	MEYLAN,WM & HOWARD,PH (1991)
OH 1.19E-011	cm3/molc sec	25	EST	MEYLAN,WM & HOWARD,PH (1993)

CAS #: 000589-91-3				4-METHYLCYCLOHEXANOL

Formula: $C_7H_{14}O$

Mol Weight: 114.19

MP (deg C):   FP (deg C):

BP (deg C): 171-173

BP pressure (mm Hg):

Property/Value	Units	Temp	Data Type	Reference
WS 1.01E+004	mg/L	25	EST	MEYLAN,WM ET AL. (1996)
logP 1.79			EXP	HANSCH,C ET AL. (1995)
VP 2.91E-001	mm Hg	25	EST	NEELY,WB & BLAU,GE (1985)
DC	pKa			
HL 6.50E-006	atm m3/mol	25	EST	MEYLAN,WM & HOWARD,PH (1991)
OH 1.92E-011	cm3/molc sec	25	EST	MEYLAN,WM & HOWARD,PH (1993)

CAS #: 000589-92-4				4-METHYLCYCLOHEXANONE

Formula: $C_7H_{12}O$

Mol Weight: 112.17

MP (deg C): -40.6   FP (deg C):

BP (deg C): 170

BP pressure (mm Hg):

Property/Value	Units	Temp	Data Type	Reference
WS 7.09E+003	mg/L	25	EST	MEYLAN,WM ET AL. (1996)
logP 1.38			EXP	HANSCH,C ET AL. (1995)
VP 1.50E+000	mm Hg	20	EXP	WEBER,RC ET AL. (1981)
DC	pKa			
HL 6.79E-005	atm m3/mol	25	EST	MEYLAN,WM & HOWARD,PH (1991)
OH 1.38E-011	cm3/molc sec	25	EST	MEYLAN,WM & HOWARD,PH (1993)

CAS #: 000589-93-5				2,5-DIMETHYLPYRIDINE

Formula: C_7H_9N

Mol Weight: 107.16

MP (deg C): -16   FP (deg C):

BP (deg C): 157

BP pressure (mm Hg):

Property/Value	Units	Temp	Data Type	Reference
WS 2.69E+005	mg/L	25	EXP	KUHNE,R ET AL. (1995)
logP 1.90			EST	MEYLAN,WM & HOWARD,PH (1995)
VP 3.36E+000	mm Hg	25	EXP	CHAO,J ET AL. (1983)
DC 6.40	pKa	25	EXP	PERRIN,DD (1965)
HL 8.68E-006	atm m3/mol	25	EXP	ANDON,RJL ET AL. (1954)
OH 1.48E-012	cm3/molc sec	25	EST	MEYLAN,WM & HOWARD,PH (1993)

CAS #: 000589-98-0				3-OCTANOL

Formula: $C_8H_{18}O$

Mol Weight: 130.23

MP (deg C): -45   FP (deg C):

BP (deg C): 171

BP pressure (mm Hg):

Property/Value	Units	Temp	Data Type	Reference
WS 1.38E+003	mg/L	25	EST	MEYLAN,WM ET AL. (1996)
logP 2.73			EST	MEYLAN,WM & HOWARD,PH (1995)
VP 2.56E-001	mm Hg	25	EXP	BOUBLIK,T ET AL. (1984)
DC	pKa			
HL 3.10E-005	atm m3/mol	25	EST	MEYLAN,WM & HOWARD,PH (1991)
OH 1.73E-011	cm3/molc sec	25	EST	MEYLAN,WM & HOWARD,PH (1993)

CAS #: 000590-01-2				N-BUTYL PROPIONATE

Formula: $C_7H_{14}O_2$

Mol Weight: 130.19

MP (deg C): -89   FP (deg C):

BP (deg C): 146.8

BP pressure (mm Hg):

Property/Value	Units	Temp	Data Type	Reference
WS 1.50E+003	mg/L	20	EXP	YALKOWSKY,SH & DANNENFELSER,RM (1992)
logP 2.34			EST	MEYLAN,WM & HOWARD,PH (1995)
VP 4.42E+000	mm Hg	25	EXP	YAWS,CL (1994A)
DC	pKa			
HL 5.05E-004	atm m3/mol	25	EST	VP/WSOL
OH 5.43E-012	cm3/molc sec	25	EST	MEYLAN,WM & HOWARD,PH (1993)

CAS #: 000590-18-1				CIS-2-BUTENE

Formula: C_4H_8

Mol Weight: 56.11

MP (deg C): -138.9   FP (deg C):

BP (deg C): 3.7

BP pressure (mm Hg):

Property/Value	Units	Temp	Data Type	Reference
WS 6.58E+002	mg/L	25	EXP	SUZUKI,T (1991)
logP 2.33			EXP	HANSCH,C & LEO,AJ (1985)
VP 1.60E+003	mm Hg	25	EXP	DAUBERT,TE & DANNER,RP (1989)
DC	pKa			
HL 2.31E-001	atm m3/mol	25	EXP	WASIK,SP & TSANG,W (1970)
OH 6.40E-011	cm3/molc sec	25	EXP	ATKINSON,R (1989)

CAS #: 000590-19-2				1,2-BUTADIENE

Formula: C_4H_6

Mol Weight: 54.09

MP (deg C): -136.2   FP (deg C):

BP (deg C): 10.9

BP pressure (mm Hg):

Property/Value	Units	Temp	Data Type	Reference
WS 6.88E+002	mg/L	25	EST	MEYLAN,WM ET AL. (1996)
logP 2.20			EST	MEYLAN,WM & HOWARD,PH (1995)
VP 1.26E+003	mm Hg	25	EXP	DAUBERT,TE & DANNER,RP (1989)
DC	pKa			
HL 9.67E-002	atm m3/mol	25	EST	MEYLAN,WM & HOWARD,PH (1991)
OH 2.61E-011	cm3/molc sec	25	EXP	ATKINSON,R (1989)

CAS #: 000590-21-6				1-CHLOROPROPENE

Formula: C_3H_5Cl

Mol Weight: 76.53

MP (deg C):   FP (deg C):

BP (deg C): 35-36

BP pressure (mm Hg):

Property/Value	Units	Temp	Data Type	Reference
WS 2.42E+003	mg/L	25	EST	MEYLAN,WM ET AL. (1996)
logP 2.04			EST	MEYLAN,WM & HOWARD,PH (1995)
VP 5.07E+002	mm Hg	25	EXP	PERRY,RH & GREEN,D (1984)
DC	pKa			
HL 5.50E-002	atm m3/mol	25	EST	MEYLAN,WM & HOWARD,PH (1991)
OH 1.20E-011	cm3/molc sec	25	EST	MEYLAN,WM & HOWARD,PH (1993)

000590-26-1 — 1,2-DIIODOETHENE (CIS)

Formula:	$C_2H_2I_2$
Mol Weight:	279.85
MP (deg C):	-14
FP (deg C):	
BP (deg C):	72.5
BP pressure (mm Hg):	1.60E+001

Property/Value	Units	Temp	Data Type	Reference
WS 9.20E+001	mg/L	25	EST	MEYLAN,WM ET AL. (1996)
logP 2.63			EST	MEYLAN,WM & HOWARD,PH (1995)
VP 3.07E-001	mm Hg	25	EXT	BOUBLIK,T ET AL. (1984)
DC	pKa			
HL 9.76E-004	atm m3/mol	25	EST	MEYLAN,WM & HOWARD,PH (1991)
OH 3.81E-012	cm3/molc sec	25	EST	MEYLAN,WM & HOWARD,PH (1993)

000590-27-2 — 1,2-DIIODOETHENE (TRANS)

Formula:	$C_2H_2I_2$
Mol Weight:	279.85
MP (deg C):	
FP (deg C):	
BP (deg C):	
BP pressure (mm Hg):	

Property/Value	Units	Temp	Data Type	Reference
WS 9.20E+001	mg/L	25	EST	MEYLAN,WM ET AL. (1996)
logP 2.63			EST	MEYLAN,WM & HOWARD,PH (1995)
VP 1.13E+000	mm Hg	25	EXT	OHE,S (1976)
DC	pKa			
HL 9.76E-004	atm m3/mol	25	EST	MEYLAN,WM & HOWARD,PH (1991)
OH 3.81E-012	cm3/molc sec	25	EST	MEYLAN,WM & HOWARD,PH (1993)

000590-35-2 — 2,2-DIMETHYLPENTANE

Formula:	C_7H_{16}
Mol Weight:	100.21
MP (deg C):	-123.8
FP (deg C):	
BP (deg C):	79.2
BP pressure (mm Hg):	

Property/Value	Units	Temp	Data Type	Reference
WS 4.40E+000	mg/L	25	EXP	YALKOWSKY,SH & DANNENFELSER,RM (1992)
logP 3.67			EST	MEYLAN,WM & HOWARD,PH (1995)
VP 1.05E+002	mm Hg	25	EXP	DAUBERT,TE & DANNER,RP (1989)
DC	pKa			
HL 3.15E+000	atm m3/mol	25	EST	VP/WSOL
OH 3.37E-012	cm3/molc sec	25	EXP	ATKINSON,R (1989)

000590-36-3 — 2-METHYL-2-PENTANOL

Formula:	$C_6H_{14}O$
Mol Weight:	102.18
MP (deg C):	-103
FP (deg C):	
BP (deg C):	121.1
BP pressure (mm Hg):	

Property/Value	Units	Temp	Data Type	Reference
WS 3.24E+004	mg/L	25	EXP	YALKOWSKY,SH & DANNENFELSER,RM (1992)
logP 1.71			EST	MEYLAN,WM & HOWARD,PH (1995)
VP 8.60E+000	mm Hg	25	EXP	RIDDICK,JA ET AL. (1986)
DC	pKa			
HL 3.57E-005	atm m3/mol	25	EST	VP/WSOL
OH 6.86E-012	cm3/molc sec	25	EST	MEYLAN,WM & HOWARD,PH (1993)

000590-46-5 — ACIDINE

Formula:	$C_5H_{12}ClNO_2$
Mol Weight:	153.61
MP (deg C):	241-242
FP (deg C):	
BP (deg C):	
BP pressure (mm Hg):	

Property/Value	Units	Temp	Data Type	Reference
WS 1.00E+006	mg/L	25	EST	MEYLAN,WM ET AL. (1996)
logP -4.93			EST	MEYLAN,WM & HOWARD,PH (1995)
VP 1.67E-007	mm Hg	25	EST	NEELY,WB & BLAU,GE (1985)
DC	pKa			
HL 6.19E-016	atm m3/mol	25	EST	MEYLAN,WM & HOWARD,PH (1991)
OH 1.08E-011	cm3/molc sec	25	EST	MEYLAN,WM & HOWARD,PH (1993)

000590-66-9 — 1,1-DIMETHYL CYCLOHEXANE

Formula:	C_8H_{16}
Mol Weight:	112.22
MP (deg C):	-33.3
FP (deg C):	
BP (deg C):	119.6
BP pressure (mm Hg):	

Property/Value	Units	Temp	Data Type	Reference
WS 1.09E+001	mg/L	25	EST	MEYLAN,WM ET AL. (1996)
logP 4.05			EST	MEYLAN,WM & HOWARD,PH (1995)
VP 2.27E+001	mm Hg	25	EXP	DAUBERT,TE & DANNER,RP (1989)
DC	pKa			
HL 4.50E-001	atm m3/mol	25	EST	MEYLAN,WM & HOWARD,PH (1991)
OH 7.40E-012	cm3/molc sec	25	EST	MEYLAN,WM & HOWARD,PH (1993)

000590-67-0 — 1-METHYLCYCLOHEXANOL

Formula:	$C_7H_{14}O$
Mol Weight:	114.19
MP (deg C):	25
FP (deg C):	
BP (deg C):	155
BP pressure (mm Hg):	

Property/Value	Units	Temp	Data Type	Reference
WS 5.58E+003	mg/L	25	EST	MEYLAN,WM ET AL. (1996)
logP 2.09			EST	MEYLAN,WM & HOWARD,PH (1995)
VP 1.13E+000	mm Hg	26	EXP	DAUBERT,TE & DANNER,RP (1989)
DC	pKa			
HL 6.50E-006	atm m3/mol	25	EST	MEYLAN,WM & HOWARD,PH (1991)
OH 1.36E-011	cm3/molc sec	25	EST	MEYLAN,WM & HOWARD,PH (1993)

000590-73-8 — 2,2-DIMETHYLHEXANE

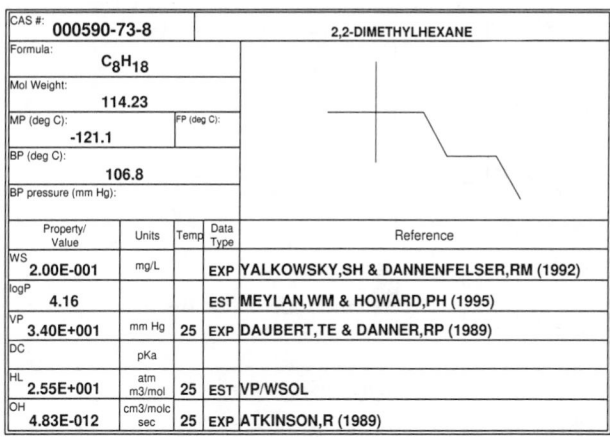

Formula:	C_8H_{18}
Mol Weight:	114.23
MP (deg C):	-121.1
FP (deg C):	
BP (deg C):	106.8
BP pressure (mm Hg):	

Property/Value	Units	Temp	Data Type	Reference
WS 2.00E-001	mg/L	25	EXP	YALKOWSKY,SH & DANNENFELSER,RM (1992)
logP 4.16			EST	MEYLAN,WM & HOWARD,PH (1995)
VP 3.40E+001	mm Hg	25	EXP	DAUBERT,TE & DANNER,RP (1989)
DC	pKa			
HL 2.55E+001	atm m3/mol	25	EST	VP/WSOL
OH 4.83E-012	cm3/molc sec	25	EXP	ATKINSON,R (1989)

CAS #: 000590-86-3 — 3-METHYL-1-BUTANAL

Formula: $C_5H_{10}O$
Mol Weight: 86.13
MP (deg C): -51
FP (deg C):
BP (deg C): 92-93
BP pressure (mm Hg):

Property/Value	Units	Temp	Data Type	Reference
WS 1.40E+004	mg/L	20	EXP	FALBE,J ET AL. (1985)
logP 1.23			EST	MEYLAN,WM & HOWARD,PH (1995)
VP 5.00E+001	mm Hg	25	EXP	CLAYTON,GD & CLAYTON,FE (1985)
DC	pKa			
HL 4.05E-004	atm m3/mol	25	EST	VP/WSOL
OH 2.74E-011	cm3/molc sec	25	EXP	ATKINSON,R (1989)

CAS #: 000590-90-9 — 3-HYDROXY-2-BUTANONE

Formula: $C_4H_8O_2$
Mol Weight: 88.11
MP (deg C):
FP (deg C):
BP (deg C): 109
BP pressure (mm Hg): 3.00E+001

Property/Value	Units	Temp	Data Type	Reference
WS 1.00E+006	mg/L	25	EST	MEYLAN,WM ET AL. (1996)
logP -1.21			EST	MEYLAN,WM & HOWARD,PH (1995)
VP 8.57E-001	mm Hg	25	EST	NEELY,WB & BLAU,GE (1985)
DC	pKa			
HL 2.41E-009	atm m3/mol	25	EST	MEYLAN,WM & HOWARD,PH (1991)
OH 1.39E-011	cm3/molc sec	25	EST	MEYLAN,WM & HOWARD,PH (1993)

CAS #: 000591-08-2 — 1-ACETYL-2-THIOUREA

Formula: $C_3H_6N_2OS$
Mol Weight: 118.16
MP (deg C): 165
FP (deg C):
BP (deg C):
BP pressure (mm Hg):

Property/Value	Units	Temp	Data Type	Reference
WS 8.53E+003	mg/L	25	EST	MEYLAN,WM ET AL. (1996)
logP -0.27			EST	MEYLAN,WM & HOWARD,PH (1995)
VP 3.72E-004	mm Hg	25	EST	NEELY,WB & BLAU,GE (1985)
DC	pKa			
HL 2.61E-011	atm m3/mol	25	EST	MEYLAN,WM & HOWARD,PH (1991)
OH 3.81E-011	cm3/molc sec	25	EST	MEYLAN,WM & HOWARD,PH (1993)

CAS #: 000591-17-3 — M-BROMOTOLUENE

Formula: C_7H_7Br
Mol Weight: 171.04
MP (deg C): -39.8
FP (deg C):
BP (deg C): 183.7
BP pressure (mm Hg):

Property/Value	Units	Temp	Data Type	Reference
WS 5.13E+001	mg/L		EXP	YALKOWSKY,SH & DANNENFELSER,RM (1992)
logP 3.41			EXP	DUNN,WJ ET AL. (1983)
VP 1.53E+000	mm Hg	25	EXP	OHE,S (1976)
DC	pKa			
HL 6.71E-003	atm m3/mol	25	EST	VP/WSOL
OH 3.34E-012	cm3/molc sec	25	EST	MEYLAN,WM & HOWARD,PH (1993)

CAS #: 000591-19-5 — M-BROMOANILINE

Formula: C_6H_6BrN
Mol Weight: 172.03
MP (deg C): 18.5
FP (deg C):
BP (deg C): 251
BP pressure (mm Hg):

Property/Value	Units	Temp	Data Type	Reference
WS 9.68E+002	mg/L	25	EST	MEYLAN,WM ET AL. (1996)
logP 2.10			EXP	HANSCH,C & LEO,AJ (1985)
VP 4.29E-002	mm Hg	25	EST	NEELY,WB & BLAU,GE (1985)
DC 3.58	pKa	25	EXP	PERRIN,DD (1965)
HL 7.59E-007	atm m3/mol	25	EST	MEYLAN,WM & HOWARD,PH (1991)
OH 6.78E-011	cm3/molc sec	25	EST	MEYLAN,WM & HOWARD,PH (1993)

CAS #: 000591-20-8 — M-BROMOPHENOL

Formula: C_6H_5BrO
Mol Weight: 173.02
MP (deg C): 33
FP (deg C):
BP (deg C): 235-236
BP pressure (mm Hg):

Property/Value	Units	Temp	Data Type	Reference
WS 2.30E+004	mg/L	25	EXP	CHEM INSPECT TEST INST (1992)
logP 2.63			EXP	HANSCH,C & LEO,AJ (1985)
VP 3.73E-002	mm Hg	25	EST	NEELY,WB & BLAU,GE (1985)
DC 9.03	pKa	25	EXP	KORTUM,G ET AL (1961)
HL 2.23E-007	atm m3/mol	25	EST	MEYLAN,WM & HOWARD,PH (1991)
OH 2.11E-011	cm3/molc sec	25	EST	MEYLAN,WM & HOWARD,PH (1993)

CAS #: 000591-21-9 — 1,3-DIMETHYLCYCLOHEXANE

Formula: C_8H_{16}
Mol Weight: 112.22
MP (deg C):
FP (deg C):
BP (deg C): 121-124
BP pressure (mm Hg):

Property/Value	Units	Temp	Data Type	Reference
WS 1.17E+001	mg/L	25	EST	MEYLAN,WM ET AL. (1996)
logP 4.01			EST	MEYLAN,WM & HOWARD,PH (1995)
VP 2.14E+001	mm Hg	25	EXP	CHAO,J ET AL. (1983) [CIS]
DC	pKa			
HL 4.50E-001	atm m3/mol	25	EST	MEYLAN,WM & HOWARD,PH (1991)
OH 1.19E-011	cm3/molc sec	25	EST	MEYLAN,WM & HOWARD,PH (1993)

CAS #: 000591-22-0 — 3,5-DIMETHYLPYRIDINE

Formula: C_7H_9N
Mol Weight: 107.16
MP (deg C): -6.6
FP (deg C):
BP (deg C): 172
BP pressure (mm Hg):

Property/Value	Units	Temp	Data Type	Reference
WS 2.57E+005	mg/L	25	EXP	KUHNE,R ET AL. (1995)
logP 1.78			EXP	HANSCH,C ET AL. (1995)
VP 1.74E+000	mm Hg	25	EXP	CHAO,J ET AL. (1983)
DC 6.15	pKa	25	EXP	PERRIN,DD (1965)
HL 6.89E-006	atm m3/mol	25	EXP	ANDON,RJL ET AL. (1954)
OH 2.85E-012	cm3/molc sec	25	EST	MEYLAN,WM & HOWARD,PH (1993)

CAS #: 000591-23-1				3-METHYLCYCLOHEXANOL

Formula: $C_7H_{14}O$
Mol Weight: 114.19
MP (deg C):   FP (deg C):
BP (deg C): 163
BP pressure (mm Hg):

Property/Value	Units	Temp	Data Type	Reference
WS 5.99E+003	mg/L	25	EST	MEYLAN,WM ET AL. (1996)
logP 2.05			EST	MEYLAN,WM & HOWARD,PH (1995)
VP 4.79E-001	mm Hg	25	EXT	JORDAN,TE (1954)
DC	pKa			
HL 6.50E-006	atm m3/mol	25	EST	MEYLAN,WM & HOWARD,PH (1991)
OH 1.92E-011	cm3/molc sec	25	EST	MEYLAN,WM & HOWARD,PH (1993)

CAS #: 000591-24-2				3-METHYLCYCLOHEXANONE

Formula: $C_7H_{12}O$
Mol Weight: 112.17
MP (deg C):   FP (deg C):
BP (deg C): 169-170
BP pressure (mm Hg):

Property/Value	Units	Temp	Data Type	Reference
WS 1.50E+003	mg/L	20	EXP	YALKOWSKY,SH & DANNENFELSER,RM (1992)
logP 1.54			EST	MEYLAN,WM & HOWARD,PH (1995)
VP 1.50E+000	mm Hg	25	EXP	WEBER,RC ET AL. (1981)
DC	pKa			
HL 2.18E-005	atm m3/mol	25	EST	MEYLAN,WM & HOWARD,PH (1991)
OH 1.71E-011	cm3/molc sec	25	EST	MEYLAN,WM & HOWARD,PH (1993)

CAS #: 000591-27-5				PHENOL, 3-AMINO-

Formula: C_6H_7NO
Mol Weight: 109.13
MP (deg C): 122-123   FP (deg C):
BP (deg C):
BP pressure (mm Hg):

Property/Value	Units	Temp	Data Type	Reference
WS 2.70E+004	mg/L	20	EXP	YALKOWSKY,SH & DANNENFELSER,RM (1992)
logP 0.21			EXP	HANSCH,C ET AL. (1995)
VP 9.55E-003	mm Hg	25	EST	NEELY,WB & BLAU,GE (1985)
DC 4.37	pKa	20	EXP	PERRIN,DD (1965)
HL 1.98E-010	atm m3/mol	25	EST	MEYLAN,WM & HOWARD,PH (1991)
OH 2.00E-010	cm3/molc sec	25	EST	MEYLAN,WM & HOWARD,PH (1993)

CAS #: 000591-31-1				BENZALDEHYDE, 3-METHOXY-

Formula: $C_8H_8O_2$
Mol Weight: 136.15
MP (deg C):   FP (deg C):
BP (deg C): 231
BP pressure (mm Hg):

Property/Value	Units	Temp	Data Type	Reference
WS 3.01E+003	mg/L	25	EST	MEYLAN,WM ET AL. (1996)
logP 1.71			EXP	BAZACO,JF & COCA,CM (1989)
VP 1.21E-001	mm Hg	25	EST	NEELY,WB & BLAU,GE (1985)
DC	pKa			
HL 7.94E-007	atm m3/mol	25	EST	MEYLAN,WM & HOWARD,PH (1991)
OH 2.86E-011	cm3/molc sec	25	EST	MEYLAN,WM & HOWARD,PH (1993)

CAS #: 000591-33-3				3-ETHOXYACETANILIDE

Formula: $C_{10}H_{13}NO_2$
Mol Weight: 179.22
MP (deg C):   FP (deg C):
BP (deg C):
BP pressure (mm Hg):

Property/Value	Units	Temp	Data Type	Reference
WS 1.55E+003	mg/L	25	EST	MEYLAN,WM ET AL. (1996)
logP 1.82			EXP	HANSCH,C & LEO,AJ (1985)
VP 2.76E-005	mm Hg	25	EST	NEELY,WB & BLAU,GE (1985)
DC	pKa			
HL 4.85E-010	atm m3/mol	25	EST	MEYLAN,WM & HOWARD,PH (1991)
OH 1.43E-010	cm3/molc sec	25	EST	MEYLAN,WM & HOWARD,PH (1993)

CAS #: 000591-35-5				3,5-DICHLOROPHENOL

Formula: $C_6H_4Cl_2O$
Mol Weight: 163.00
MP (deg C): 68   FP (deg C):
BP (deg C): 233
BP pressure (mm Hg):

Property/Value	Units	Temp	Data Type	Reference
WS 7.39E+003	mg/L	25	EXP	SHIU,WY ET AL. (1994)
logP 3.62			EXP	HANSCH,C & LEO,AJ (1985)
VP 8.42E-003	mm Hg	25	EXP	SHIU,WY ET AL. (1994)
DC 8.18	pKa		EXP	SERJEANT,EP & DEMPSEY,B (1979)
HL 2.44E-010	atm m3/mol	25	EST	VP/WSOL
OH 1.66E-011	cm3/molc sec	25	EST	MEYLAN,WM & HOWARD,PH (1993)

CAS #: 000591-48-0				3-METHYL-1-CYCLOHEXENE

Formula: C_7H_{12}
Mol Weight: 96.17
MP (deg C):   FP (deg C):
BP (deg C):
BP pressure (mm Hg):

Property/Value	Units	Temp	Data Type	Reference
WS 4.54E+001	mg/L	25	EST	MEYLAN,WM ET AL. (1996)
logP 3.38			EST	MEYLAN,WM & HOWARD,PH (1995)
VP 3.47E+001	mm Hg	25	EXT	BOUBLIK,T ET AL. (1984)
DC	pKa			
HL 8.79E-002	atm m3/mol	25	EST	MEYLAN,WM & HOWARD,PH (1991)
OH 6.29E-011	cm3/molc sec	25	EST	MEYLAN,WM & HOWARD,PH (1993)

CAS #: 000591-49-1				1-METHYLCYCLOHEXENE

Formula: C_7H_{12}
Mol Weight: 96.17
MP (deg C): -120.4   FP (deg C):
BP (deg C): 110.3
BP pressure (mm Hg):

Property/Value	Units	Temp	Data Type	Reference
WS 5.20E+001	mg/L	25	EXP	YALKOWSKY,SH & DANNENFELSER,RM (1992)
logP 3.51			EST	MEYLAN,WM & HOWARD,PH (1995)
VP 3.06E+001	mm Hg	25	EXP	HINE,J & MOOKERJEE,PK (1975)
DC	pKa			
HL 7.45E-002	atm m3/mol	25	EST	VP/WSOL
OH 9.44E-011	cm3/molc sec	25	EXP	ATKINSON,R (1989)

419

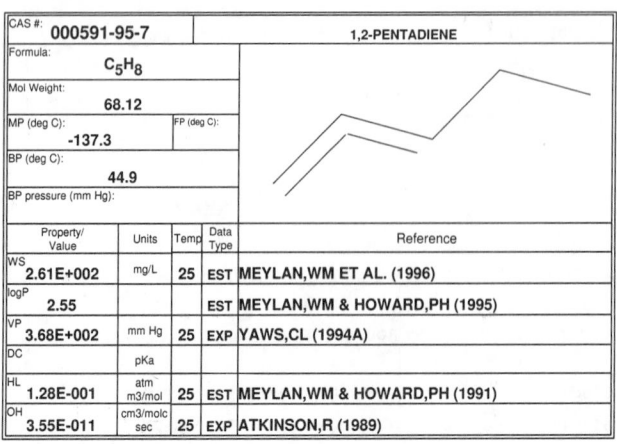

CAS #: 000591-50-4 — IODOBENZENE

Formula: C_6H_5I
Mol Weight: 204.01
MP (deg C): -30
FP (deg C):
BP (deg C): 188-189
BP pressure (mm Hg):

Property/Value	Units	Temp	Data Type	Reference
WS 3.40E+002	mg/L	30	EXP	CHIOU,CT ET AL. (1977)
logP 3.25			EXP	HANSCH,C & LEO,AJ (1985)
VP 1.06E+000	mm Hg	25	EXP	DAUBERT,TE & DANNER,RP (1989)
DC	pKa			
HL 8.37E-004	atm m3/mol	25	EST	VP/WSOL
OH 1.10E-012	cm3/molc sec	25	EXP	ATKINSON,R (1989)

CAS #: 000591-54-8 — 4-PYRIMIDINAMINE

Formula: $C_4H_5N_3$
Mol Weight: 95.10
MP (deg C): 154-156
FP (deg C):
BP (deg C):
BP pressure (mm Hg):

Property/Value	Units	Temp	Data Type	Reference
WS 1.97E+005	mg/L	25	EST	MEYLAN,WM ET AL. (1996)
logP -0.25			EXP	YAMAGAMI,C ET AL. (1990)
VP 2.46E-001	mm Hg	25	EST	NEELY,WB & BLAU,GE (1985)
DC 5.69	pKa	20	EXP	PERRIN,DD (1965)
HL 1.03E-009	atm m3/mol	25	EST	MEYLAN,WM & HOWARD,PH (1991)
OH 1.27E-011	cm3/molc sec	25	EST	MEYLAN,WM & HOWARD,PH (1993)

CAS #: 000591-76-4 — 2-METHYLHEXANE

Formula: C_7H_{16}
Mol Weight: 100.21
MP (deg C): -118.2
FP (deg C):
BP (deg C): 90
BP pressure (mm Hg):

Property/Value	Units	Temp	Data Type	Reference
WS 2.54E+000	mg/L	25	EXP	YALKOWSKY,SH & DANNENFELSER,RM (1992)
logP 3.71			EST	MEYLAN,WM & HOWARD,PH (1995)
VP 6.60E+001	mm Hg	25	EXP	DAUBERT,TE & DANNER,RP (1989)
DC	pKa			
HL 3.43E+000	atm m3/mol	25	EST	VP/WSOL
OH 6.86E-012	cm3/molc sec	25	EST	MEYLAN,WM & HOWARD,PH (1993)

CAS #: 000591-78-6 — 2-HEXANONE

Formula: $C_6H_{12}O$
Mol Weight: 100.16
MP (deg C): -55.5
FP (deg C):
BP (deg C): 127.6
BP pressure (mm Hg):

Property/Value	Units	Temp	Data Type	Reference
WS 1.75E+004	mg/L	20	EXP	PAPA,AJ & SHERMAN,PDJR (1981)
logP 1.38			EXP	HANSCH,C & LEO,AJ (1985)
VP 1.16E-001	mm Hg	25	EXP	AMBROSE,D ET AL. (1975)
DC	pKa			
HL 8.74E-007	atm m3/mol	25	EST	VP/WSOL
OH 9.10E-012	cm3/molc sec	25	EXP	ATKINSON,R (1989)

CAS #: 000591-82-2 — PROPANE, 1-ISOTHIOCYANATO-2-METHYL-

Formula: C_5H_9NS
Mol Weight: 115.20
MP (deg C):
FP (deg C):
BP (deg C): 160
BP pressure (mm Hg):

Property/Value	Units	Temp	Data Type	Reference
WS 4.08E+002	mg/L	25	EST	MEYLAN,WM ET AL. (1996)
logP 2.82			EXP	AUGUSTIN,J ET AL. (1987)
VP 4.73E+000	mm Hg	25	EST	NEELY,WB & BLAU,GE (1985)
DC	pKa			
HL 7.27E-003	atm m3/mol	25	EST	MEYLAN,WM & HOWARD,PH (1991)
OH 3.87E-012	cm3/molc sec	25	EST	MEYLAN,WM & HOWARD,PH (1993)

CAS #: 000591-87-7 — ALLYL ACETATE

Formula: $C_5H_8O_2$
Mol Weight: 100.12
MP (deg C):
FP (deg C):
BP (deg C): 103.5
BP pressure (mm Hg):

Property/Value	Units	Temp	Data Type	Reference
WS 1.74E+004	mg/L	25	EST	MEYLAN,WM ET AL. (1996)
logP 0.97			EXP	HANSCH,C ET AL. (1995)
VP 3.52E+001	mm Hg	25	EXP	DAUBERT,TE & DANNER,RP (1989)
DC	pKa			
HL 2.30E-004	atm m3/mol	25	EST	MEYLAN,WM & HOWARD,PH (1991)
OH 2.79E-011	cm3/molc sec	25	EST	MEYLAN,WM & HOWARD,PH (1993)

CAS #: 000591-93-5 — 1,4-PENTADIENE

Formula: C_5H_8
Mol Weight: 68.12
MP (deg C): -148.8
FP (deg C):
BP (deg C): 26
BP pressure (mm Hg):

Property/Value	Units	Temp	Data Type	Reference
WS 5.58E+002	mg/L	25	EXP	YALKOWSKY,SH & DANNENFELSER,RM (1992)
logP 2.48			EXP	HANSCH,C & LEO,AJ (1985)
VP 7.48E+002	mm Hg	25	EXP	YAWS,CL (1994A)
DC	pKa			
HL 1.20E-001	atm m3/mol	25	EST	VP/WSOL
OH 5.33E-011	cm3/molc sec	25	EXP	ATKINSON,R (1989)

CAS #: 000591-95-7 — 1,2-PENTADIENE

Formula: C_5H_8
Mol Weight: 68.12
MP (deg C): -137.3
FP (deg C):
BP (deg C): 44.9
BP pressure (mm Hg):

Property/Value	Units	Temp	Data Type	Reference
WS 2.61E+002	mg/L	25	EST	MEYLAN,WM ET AL. (1996)
logP 2.55			EST	MEYLAN,WM & HOWARD,PH (1995)
VP 3.68E+002	mm Hg	25	EXP	YAWS,CL (1994A)
DC	pKa			
HL 1.28E-001	atm m3/mol	25	EST	MEYLAN,WM & HOWARD,PH (1991)
OH 3.55E-011	cm3/molc sec	25	EXP	ATKINSON,R (1989)

CAS #: 000591-97-9				1-CHLORO-2-BUTENE

Formula: C_4H_7Cl
Mol Weight: 90.55
MP (deg C): | FP (deg C):
BP (deg C):
BP pressure (mm Hg):

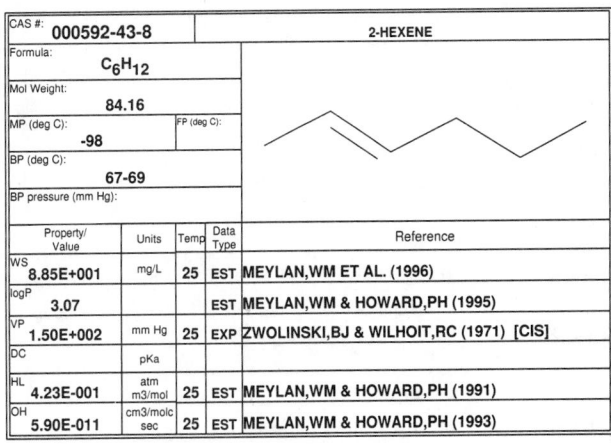

Property/Value	Units	Temp	Data Type	Reference
WS 1.00E+003	mg/L		EXP	YALKOWSKY,SH & DANNENFELSER,RM (1992)
logP 2.34			EST	MEYLAN,WM & HOWARD,PH (1995)
VP 6.90E+001	mm Hg	20	EXP	WEBER,RC ET AL. (1981)
DC	pKa			
HL 8.22E-003	atm m3/mol	20	EST	VP/WSOL
OH 4.34E-011	cm3/molc sec	25	EST	MEYLAN,WM & HOWARD,PH (1993)

CAS #: 000592-13-2				2,5-DIMETHYLHEXANE

Formula: C_8H_{18}
Mol Weight: 114.23
MP (deg C): -91 | FP (deg C):
BP (deg C): 109.1
BP pressure (mm Hg):

Property/Value	Units	Temp	Data Type	Reference
WS 9.20E+000	mg/L	25	EST	MEYLAN,WM ET AL. (1996)
logP 4.12			EST	MEYLAN,WM & HOWARD,PH (1995)
VP 3.03E+001	mm Hg	25	EXP	DAUBERT,TE & DANNER,RP (1989)
DC	pKa			
HL 3.01E+000	atm m3/mol	25	EST	MEYLAN,WM & HOWARD,PH (1991)
OH 8.27E-012	cm3/molc sec	25	EST	MEYLAN,WM & HOWARD,PH (1993)

CAS #: 000592-27-8				2-METHYLHEPTANE

Formula: C_8H_{18}
Mol Weight: 114.23
MP (deg C): -108.9 | FP (deg C):
BP (deg C): 117.6
BP pressure (mm Hg):

Property/Value	Units	Temp	Data Type	Reference
WS 7.97E+000	mg/L	25	EST	MEYLAN,WM ET AL. (1996)
logP 4.20			EST	MEYLAN,WM & HOWARD,PH (1995)
VP 2.07E+001	mm Hg	25	EXP	DAUBERT,TE & DANNER,RP (1989)
DC	pKa			
HL 3.01E+000	atm m3/mol	25	EST	MEYLAN,WM & HOWARD,PH (1991)
OH 8.28E-012	cm3/molc sec	25	EST	MEYLAN,WM & HOWARD,PH (1993)

CAS #: 000592-31-4				BUTYLUREA

Formula: $C_5H_{12}N_2O$
Mol Weight: 116.16
MP (deg C): 96-98 | FP (deg C):
BP (deg C):
BP pressure (mm Hg):

Property/Value	Units	Temp	Data Type	Reference
WS 4.63E+004	mg/L	25	EST	MEYLAN,WM ET AL. (1996)
logP 0.41			EXP	HANSCH,C & LEO,AJ (1985)
VP 3.55E-002	mm Hg	25	EST	NEELY,WB & BLAU,GE (1985)
DC	pKa			
HL 1.87E-009	atm m3/mol	25	EST	MEYLAN,WM & HOWARD,PH (1991)
OH 9.44E-012	cm3/molc sec	25	EST	MEYLAN,WM & HOWARD,PH (1993)

CAS #: 000592-35-8				O-BUTYL CARBAMATE

Formula: $C_5H_{11}NO_2$
Mol Weight: 117.15
MP (deg C): 53 | FP (deg C):
BP (deg C): 108
BP pressure (mm Hg): 1.40E+001

Property/Value	Units	Temp	Data Type	Reference
WS 2.58E+004	mg/L	37	EXP	YALKOWSKY,SH & DANNENFELSER,RM (1992)
logP 0.85			EXP	HANSCH,C & LEO,AJ (1985)
VP 2.22E+000	mm Hg	25	EST	NEELY,WB & BLAU,GE (1985)
DC	pKa			
HL 9.26E-008	atm m3/mol	25	EST	MEYLAN,WM & HOWARD,PH (1991)
OH 6.57E-012	cm3/molc sec	25	EST	MEYLAN,WM & HOWARD,PH (1993)

CAS #: 000592-41-6				1-HEXENE

Formula: C_6H_{12}
Mol Weight: 84.16
MP (deg C): -139.7 | FP (deg C):
BP (deg C): 63.4
BP pressure (mm Hg):

Property/Value	Units	Temp	Data Type	Reference
WS 5.00E+001	mg/L	25	EXP	YALKOWSKY,SH & DANNENFELSER,RM (1992)
logP 3.39			EXP	HANSCH,C & LEO,AJ (1985)
VP 1.86E+002	mm Hg	25	EXP	RIDDICK,JA ET AL. (1986)
DC	pKa			
HL 4.12E-001	atm m3/mol	25	EST	VP/WSOL
OH 3.75E-011	cm3/molc sec	25	EXP	ATKINSON,R (1989)

CAS #: 000592-42-7				1,5-HEXADIENE

Formula: C_6H_{10}
Mol Weight: 82.15
MP (deg C): -140.7 | FP (deg C):
BP (deg C): 59.4
BP pressure (mm Hg):

Property/Value	Units	Temp	Data Type	Reference
WS 1.69E+002	mg/L	25	EXP	YALKOWSKY,SH & DANNENFELSER,RM (1992)
logP 2.87			EXP	HANSCH,C ET AL. (1995)
VP 2.21E+002	mm Hg	25	EXP	YAWS,CL (1994A)
DC	pKa			
HL 1.41E-001	atm m3/mol	25	EST	VP/WSOL
OH 6.21E-011	cm3/molc sec	25	EXP	ATKINSON,R (1989)

CAS #: 000592-43-8				2-HEXENE

Formula: C_6H_{12}
Mol Weight: 84.16
MP (deg C): -98 | FP (deg C):
BP (deg C): 67-69
BP pressure (mm Hg):

Property/Value	Units	Temp	Data Type	Reference
WS 8.85E+001	mg/L	25	EST	MEYLAN,WM ET AL. (1996)
logP 3.07			EST	MEYLAN,WM & HOWARD,PH (1995)
VP 1.50E+002	mm Hg	25	EXP	ZWOLINSKI,BJ & WILHOIT,RC (1971) [CIS]
DC	pKa			
HL 4.23E-001	atm m3/mol	25	EST	MEYLAN,WM & HOWARD,PH (1991)
OH 5.90E-011	cm3/molc sec	25	EST	MEYLAN,WM & HOWARD,PH (1993)

421

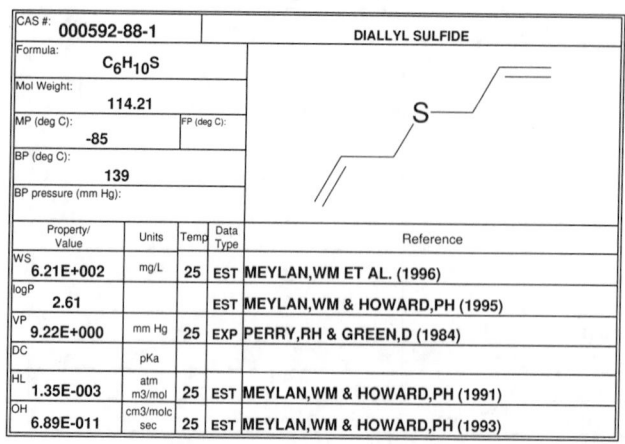

000592-90-5 — OXEPANE

CAS #: 000592-90-5				OXEPANE
Formula: $C_6H_{12}O$				
Mol Weight: 100.16				
MP (deg C):		FP (deg C):		
BP (deg C): 119				
BP pressure (mm Hg):				

Property/Value	Units	Temp	Data Type	Reference
WS 2.68E+003	mg/L	25	EST	MEYLAN,WM ET AL. (1996)
logP 1.92			EST	MEYLAN,WM & HOWARD,PH (1995)
VP 1.72E+001	mm Hg	25	EST	NEELY,WB & BLAU,GE (1985)
DC	pKa			
HL 1.49E-004	atm m3/mol	25	EST	MEYLAN,WM & HOWARD,PH (1991)
OH 1.54E-011	cm3/molc sec	25	EXP	ATKINSON,R (1989)

000592-99-4 — 4-OCTENE

CAS #: 000592-99-4				4-OCTENE
Formula: C_8H_{16}				
Mol Weight: 112.22				
MP (deg C):		FP (deg C):		
BP (deg C):				
BP pressure (mm Hg):				

Property/Value	Units	Temp	Data Type	Reference
WS 1.07E+001	mg/L	25	EST	MEYLAN,WM ET AL. (1996)
logP 4.06			EST	MEYLAN,WM & HOWARD,PH (1995)
VP 1.78E+001	mm Hg	25	EXP	ZWOLINSKI,BJ & WILHOIT,RC (1971) [CIS]
DC	pKa			
HL 7.46E-001	atm m3/mol	25	EST	MEYLAN,WM & HOWARD,PH (1991)
OH 6.13E-011	cm3/molc sec	25	EST	MEYLAN,WM & HOWARD,PH (1993)

000593-08-8 — 2-TRIDECANONE

CAS #: 000593-08-8				2-TRIDECANONE
Formula: $C_{13}H_{26}O$				
Mol Weight: 198.35				
MP (deg C): 30.5		FP (deg C):		
BP (deg C): 263				
BP pressure (mm Hg):				

Property/Value	Units	Temp	Data Type	Reference
WS 4.53E+000	mg/L	25	EST	MEYLAN,WM ET AL. (1996)
logP 4.68			EST	MEYLAN,WM & HOWARD,PH (1995)
VP 1.04E-002	mm Hg	25	EXT	OHE,S (1976)
DC	pKa			
HL 8.43E-004	atm m3/mol	25	EST	MEYLAN,WM & HOWARD,PH (1991)
OH 1.67E-011	cm3/molc sec	25	EST	MEYLAN,WM & HOWARD,PH (1993)

000593-45-3 — OCTADECANE

CAS #: 000593-45-3				OCTADECANE
Formula: $C_{18}H_{38}$				
Mol Weight: 254.50				
MP (deg C): 28.2		FP (deg C):		
BP (deg C): 316.3				
BP pressure (mm Hg):				

Property/Value	Units	Temp	Data Type	Reference
WS 6.00E-003	mg/L	25	EXP	YALKOWSKY,SH & DANNENFELSER,RM (1992)
logP 9.18			EST	MEYLAN,WM & HOWARD,PH (1995)
VP 3.41E-004	mm Hg	25	EXT	PERRY,RH & GREEN,D (1984)
DC	pKa			
HL 5.12E+001	atm m3/mol	25	EST	MEYLAN,WM & HOWARD,PH (1991)
OH 2.24E-011	cm3/molc sec	25	EST	MEYLAN,WM & HOWARD,PH (1993)

000593-49-7 — HEPTACOSANE

CAS #: 000593-49-7				HEPTACOSANE
Formula: $C_{27}H_{56}$				
Mol Weight: 380.75				
MP (deg C): 59.5		FP (deg C):		
BP (deg C): 442				
BP pressure (mm Hg):				

Property/Value	Units	Temp	Data Type	Reference
WS 2.83E-009	mg/L	25	EST	MEYLAN,WM ET AL. (1996)
logP 13.60			EST	MEYLAN,WM & HOWARD,PH (1995)
VP 2.81E-007	mm Hg	25	EXT	PERRY,RH & GREEN,D (1984)
DC	pKa			
HL 6.55E+002	atm m3/mol	25	EST	MEYLAN,WM & HOWARD,PH (1991)
OH 3.51E-011	cm3/molc sec	25	EST	MEYLAN,WM & HOWARD,PH (1993)

000593-51-1 — METHANAMINE, HYDROCHLORIDE

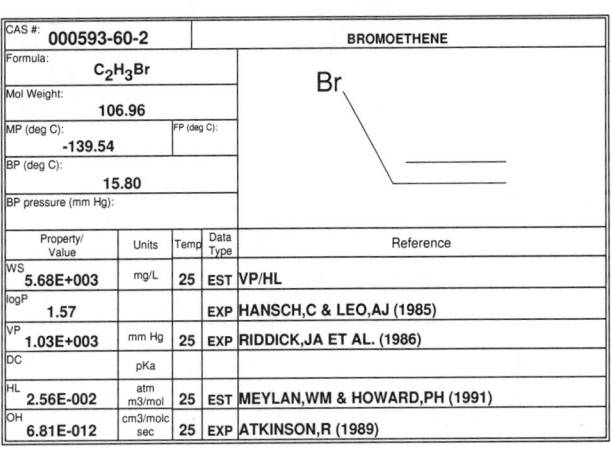

CAS #: 000593-51-1				METHANAMINE, HYDROCHLORIDE
Formula: CH_6ClN				
Mol Weight: 67.52				
MP (deg C): 227.5		FP (deg C):		
BP (deg C): 227				
BP pressure (mm Hg): 1.50E+001				

Property/Value	Units	Temp	Data Type	Reference
WS 1.00E+006	mg/L	25	EST	MEYLAN,WM ET AL. (1996)
logP -3.82			EST	MEYLAN,WM & HOWARD,PH (1995)
VP 1.01E-003	mm Hg	25	EST	NEELY,WB & BLAU,GE (1985)
DC	pKa			
HL 7.02E-013	atm m3/mol	25	EST	MEYLAN,WM & HOWARD,PH (1991)
OH 1.26E-012	cm3/molc sec	25	EST	MEYLAN,WM & HOWARD,PH (1993)

000593-53-3 — FLUOROMETHANE

CAS #: 000593-53-3				FLUOROMETHANE
Formula: CH_3F				
Mol Weight: 34.03				
MP (deg C): -141.8		FP (deg C):		
BP (deg C): -78.2				
BP pressure (mm Hg):				

Property/Value	Units	Temp	Data Type	Reference
WS 2.00E+003	mg/L	25	EXP	HINE,J & MOOKERJEE.PK (1975)
logP 0.51			EXP	HANSCH,C & LEO,AJ (1985)
VP 2.85E+003	mm Hg	25	EXP	YAWS,CL (1994)
DC	pKa			
HL 1.70E-002	atm m3/mol	25	EST	VP/WSOL
OH 1.68E-014	cm3/molc sec	25	EXP	ATKINSON,R (1989)

000593-60-2 — BROMOETHENE

CAS #: 000593-60-2				BROMOETHENE
Formula: C_2H_3Br				
Mol Weight: 106.96				
MP (deg C): -139.54		FP (deg C):		
BP (deg C): 15.80				
BP pressure (mm Hg):				

Property/Value	Units	Temp	Data Type	Reference
WS 5.68E+003	mg/L	25	EST	VP/HL
logP 1.57			EXP	HANSCH,C & LEO,AJ (1985)
VP 1.03E+003	mm Hg	25	EXP	RIDDICK,JA ET AL. (1986)
DC	pKa			
HL 2.56E-002	atm m3/mol	25	EST	MEYLAN,WM & HOWARD,PH (1991)
OH 6.81E-012	cm3/molc sec	25	EXP	ATKINSON,R (1989)

000593-70-4 — CHLOROFLUOROMETHANE

Formula: CH_2ClF
Mol Weight: 68.48
MP (deg C): -133
FP (deg C):
BP (deg C): -9.1
BP pressure (mm Hg):

Property/Value	Units	Temp	Data Type	Reference
WS 1.04E+004	mg/L	25	EXP	HINE,J & MOOKERJEE,PK (1975)
logP 0.51			EXP	HANSCH,C & LEO,AJ (1985)
VP 7.60E+002	mm Hg	25	EXP	HINE,J & MOOKERJEE,PK (1975)
DC	pKa			
HL 6.59E-003	atm m3/mol	25	EST	VP/WSOL
OH 4.41E-014	cm3/molc sec	25	EXP	ATKINSON,R (1989)

000593-71-5 — CHLOROIODOMETHANE

Formula: CH_2ClI
Mol Weight: 176.38
MP (deg C):
FP (deg C):
BP (deg C): 109
BP pressure (mm Hg):

Property/Value	Units	Temp	Data Type	Reference
WS 1.52E+003	mg/L	25	EST	MEYLAN,WM ET AL. (1996)
logP 1.84			EST	MEYLAN,WM & HOWARD,PH (1995)
VP 1.06E+001	mm Hg	25	EST	NEELY,WB & BLAU,GE (1985)
DC	pKa			
HL 8.71E-004	atm m3/mol	20	EXP	MOORE,RM ET AL. (1995)
OH 1.88E-013	cm3/molc sec	25	EST	MEYLAN,WM & HOWARD,PH (1993)

000593-74-8 — DIMETHYL MERCURY

Formula: C_2H_6Hg
Mol Weight: 230.66
MP (deg C):
FP (deg C):
BP (deg C): 92
BP pressure (mm Hg): 7.40E+002

Property/Value	Units	Temp	Data Type	Reference
WS 8.86E+003	mg/L	25	EST	MEYLAN,WM ET AL. (1996)
logP 0.62			EST	MEYLAN,WM & HOWARD,PH (1995)
VP 6.23E+001	mm Hg	25	EXP	LONG,LH & CATTANACH,J (1961)
DC	pKa			
HL	atm m3/mol			
OH 2.72E-013	cm3/molc sec	25	EST	MEYLAN,WM & HOWARD,PH (1993)

000593-75-9 — METHYL ISOCYANIDE

Formula: C_2H_3N
Mol Weight: 41.05
MP (deg C): -45
FP (deg C):
BP (deg C):
BP pressure (mm Hg):

Property/Value	Units	Temp	Data Type	Reference
WS 1.37E+005	mg/L	25	EST	MEYLAN,WM ET AL. (1996)
logP -0.15			EST	MEYLAN,WM & HOWARD,PH (1995)
VP 1.22E+002	mm Hg	25	EST	NEELY,WB & BLAU,GE (1985)
DC -4.30	pKa		EXP	PERRIN,DD (1965)
HL 3.06E-005	atm m3/mol	25	EST	MEYLAN,WM & HOWARD,PH (1991)
OH 2.58E-014	cm3/molc sec	25	EST	MEYLAN,WM & HOWARD,PH (1993)

000593-77-1 — N-METHYLHYDROXYLAMINE

Formula: CH_5NO
Mol Weight: 47.06
MP (deg C): 87.5
FP (deg C):
BP (deg C): 62.5
BP pressure (mm Hg): 1.50E+001

Property/Value	Units	Temp	Data Type	Reference
WS 5.99E+005	mg/L	25	EST	MEYLAN,WM ET AL. (1996)
logP -0.76			EST	MEYLAN,WM & HOWARD,PH (1995)
VP 1.67E+000	mm Hg	25	EXT	BOUBLIK,T ET AL.(1984)
DC 5.96	pKa	25	EXP	PERRIN,DD (1965)
HL 1.52E-008	atm m3/mol	25	EST	MEYLAN,WM & HOWARD,PH (1991)
OH 6.43E-011	cm3/molc sec	25	EST	MEYLAN,WM & HOWARD,PH (1993)

000593-96-4 — 1-BROMO-1-CHLOROETHANE

Formula: C_2H_4BrCl
Mol Weight: 143.42
MP (deg C):
FP (deg C):
BP (deg C): 83
BP pressure (mm Hg):

Property/Value	Units	Temp	Data Type	Reference
WS 2.14E+003	mg/L	25	EST	MEYLAN,WM ET AL. (1996)
logP 1.85			EST	MEYLAN,WM & HOWARD,PH (1995)
VP 8.51E+001	mm Hg	25	EXP	OHE,S (1976)
DC	pKa			
HL 3.97E-003	atm m3/mol	25	EST	MEYLAN,WM & HOWARD,PH (1991)
OH 2.69E-013	cm3/molc sec	25	EST	MEYLAN,WM & HOWARD,PH (1993)

000594-11-6 — METHYLCYCLOPROPANE

Formula: C_4H_8
Mol Weight: 56.11
MP (deg C): -177.3
FP (deg C):
BP (deg C): 0.7
BP pressure (mm Hg):

Property/Value	Units	Temp	Data Type	Reference
WS 6.15E+002	mg/L	25	EST	MEYLAN,WM ET AL. (1996)
logP 2.12			EST	MEYLAN,WM & HOWARD,PH (1995)
VP 1.58E+003	mm Hg	25	EXT	PERRY,RH & GREEN,D (1984)
DC	pKa			
HL 1.45E-001	atm m3/mol	25	EST	MEYLAN,WM & HOWARD,PH (1991)
OH 2.83E-013	cm3/molc sec	25	EST	MEYLAN,WM & HOWARD,PH (1993)

000594-34-3 — 1,2-DIBROMO-2-METHYLPROPANE

Formula: $C_4H_8Br_2$
Mol Weight: 215.93
MP (deg C): 10.5
FP (deg C):
BP (deg C): 150
BP pressure (mm Hg):

Property/Value	Units	Temp	Data Type	Reference
WS 1.25E+002	mg/L	25	EST	MEYLAN,WM ET AL. (1996)
logP 2.88			EST	MEYLAN,WM & HOWARD,PH (1995)
VP 1.95E+001	mm Hg	25	EXP	PERRY,RH & GREEN,D (1984)
DC	pKa			
HL 2.29E-003	atm m3/mol	25	EST	MEYLAN,WM & HOWARD,PH (1991)
OH 5.34E-013	cm3/molc sec	25	EST	MEYLAN,WM & HOWARD,PH (1993)

CAS #: 000594-36-5 — BUTANE, 2-CHLORO-2-METHYL-

Formula: $C_5H_{11}Cl$

Mol Weight: 106.60

MP (deg C): -73.5

FP (deg C):

BP (deg C): 85.6

BP pressure (mm Hg):

Property/Value	Units	Temp	Data Type	Reference
WS 3.29E+002	mg/L	25	EXP	KUHNE,R ET AL. (1995)
logP 2.52			EXP	HANSCH,C ET AL. (1995)
VP 1.88E+002	mm Hg	25	EST	NEELY,WB & BLAU,GE (1985)
DC	pKa			
HL 2.55E-002	atm m3/mol	25	EST	MEYLAN,WM & HOWARD,PH (1991)
OH 1.37E-012	cm3/molc sec	25	EST	MEYLAN,WM & HOWARD,PH (1993)

CAS #: 000594-37-6 — 1,2-DICHLORO-2-METHYLPROPANE

Formula: $C_4H_8Cl_2$

Mol Weight: 127.01

MP (deg C):

FP (deg C):

BP (deg C): 106.5

BP pressure (mm Hg):

Property/Value	Units	Temp	Data Type	Reference
WS 4.65E+002	mg/L	25	EST	MEYLAN,WM ET AL. (1996)
logP 2.70			EST	MEYLAN,WM & HOWARD,PH (1995)
VP 2.81E+001	mm Hg	25	EXP	OHE,S (1976)
DC	pKa			
HL 2.14E-002	atm m3/mol	25	EST	MEYLAN,WM & HOWARD,PH (1991)
OH 6.27E-013	cm3/molc sec	25	EST	MEYLAN,WM & HOWARD,PH (1993)

CAS #: 000594-42-3 — PERCHLOROMETHYL MERCAPTAN

Formula: CCl_4S

Mol Weight: 185.89

MP (deg C):

FP (deg C):

BP (deg C): 149

BP pressure (mm Hg):

Property/Value	Units	Temp	Data Type	Reference
WS 5.63E+001	mg/L	25	EST	MEYLAN,WM ET AL. (1996)
logP 3.47			EST	MEYLAN,WM & HOWARD,PH (1995)
VP 3.00E+000	mm Hg	20	EXP	WEBER,RC ET AL. (1981)
DC	pKa			
HL 2.41E-004	atm m3/mol	25	EST	MEYLAN,WM & HOWARD,PH (1991)
OH 0.00E+000	cm3/molc sec	25	EST	MEYLAN,WM & HOWARD,PH (1993)

CAS #: 000594-56-9 — 2,3,3-TRIMETHYL-1-BUTENE

Formula: C_7H_{14}

Mol Weight: 98.19

MP (deg C): -109.9

FP (deg C):

BP (deg C): 77.9

BP pressure (mm Hg):

Property/Value	Units	Temp	Data Type	Reference
WS 2.97E+001	mg/L	25	EST	MEYLAN,WM ET AL. (1996)
logP 3.59			EST	MEYLAN,WM & HOWARD,PH (1995)
VP 1.12E+002	mm Hg	25	EXP	YAWS,CL (1994A)
DC	pKa			
HL 5.62E-001	atm m3/mol	25	EST	MEYLAN,WM & HOWARD,PH (1991)
OH 5.20E-011	cm3/molc sec	25	EST	MEYLAN,WM & HOWARD,PH (1993)

CAS #: 000594-60-5 — 2,3-DIMETHYL-2-BUTANOL

Formula: $C_6H_{14}O$

Mol Weight: 102.18

MP (deg C): -14

FP (deg C):

BP (deg C): 118.4

BP pressure (mm Hg):

Property/Value	Units	Temp	Data Type	Reference
WS 4.35E+004	mg/L	25	EXP	SUZUKI,T (1991)
logP 1.64			EST	MEYLAN,WM & HOWARD,PH (1995)
VP 3.20E+000	mm Hg	25	EXP	WILHOIT,RC & ZWOLINSKI,BJ (1973)
DC	pKa			
HL 9.98E-006	atm m3/mol	25	EST	VP/WSOL
OH 8.88E-012	cm3/molc sec	25	EST	MEYLAN,WM & HOWARD,PH (1993)

CAS #: 000594-61-6 — A-HYDROXY-I-BUTYRIC ACID

Formula: $C_4H_8O_3$

Mol Weight: 104.11

MP (deg C): 82.5

FP (deg C):

BP (deg C): 212

BP pressure (mm Hg):

Property/Value	Units	Temp	Data Type	Reference
WS 1.00E+006	mg/L	25	EST	MEYLAN,WM ET AL. (1996)
logP -0.36			EXP	HANSCH,C & LEO,AJ (1985)
VP 8.75E-003	mm Hg	25	EXT	PERRY,RH & GREEN,D (1984)
DC 3.61	pKa	25	EXP	KORTUM,G ET AL (1961)
HL 1.51E-007	atm m3/mol	25	EST	MEYLAN,WM & HOWARD,PH (1991)
OH 1.69E-012	cm3/molc sec	25	EST	MEYLAN,WM & HOWARD,PH (1993)

CAS #: 000594-65-0 — TRICHLOROACETAMIDE

Formula: $C_2H_2Cl_3NO$

Mol Weight: 162.40

MP (deg C): 142

FP (deg C):

BP (deg C): 240

BP pressure (mm Hg):

Property/Value	Units	Temp	Data Type	Reference
WS 8.63E+003	mg/L	25	EST	MEYLAN,WM ET AL. (1996)
logP 1.04			EXP	HANSCH,C & LEO,AJ (1985)
VP 1.12E-002	mm Hg	25	EST	NEELY,WB & BLAU,GE (1985)
DC	pKa			
HL 4.89E-010	atm m3/mol	25	EST	MEYLAN,WM & HOWARD,PH (1991)
OH 2.00E-012	cm3/molc sec	25	EST	MEYLAN,WM & HOWARD,PH (1993)

CAS #: 000594-70-7 — 2-METHYL-2-NITROPROPANE

Formula: $C_4H_9NO_2$

Mol Weight: 103.12

MP (deg C): 26.23

FP (deg C):

BP (deg C): 127.16

BP pressure (mm Hg):

Property/Value	Units	Temp	Data Type	Reference
WS 4.67E+003	mg/L	25	EST	MEYLAN,WM ET AL. (1996)
logP 1.17			EXP	HANSCH,C & LEO,AJ (1985)
VP 1.51E+003	mm Hg	25	EXP	PATTE,F ET AL. (1982)
DC	pKa			
HL 9.85E-005	atm m3/mol	25	EST	MEYLAN,WM & HOWARD,PH (1991)
OH 1.87E-013	cm3/molc sec	25	EST	MEYLAN,WM & HOWARD,PH (1993)

000594-72-9 — 1,1-DICHLORO-1-NITROETHANE

Formula: $C_2H_3Cl_2NO_2$
Mol Weight: 143.96
MP (deg C):
FP (deg C):
BP (deg C): 124
BP pressure (mm Hg):

Property/Value	Units	Temp	Data Type	Reference
WS 2.50E+003	mg/L	20	EXP	SHIU,WY ET AL. (1990)
logP 1.56			EST	MEYLAN,WM & HOWARD,PH (1995)
VP 1.69E+001	mm Hg	25	EXP	METCALFE,RL (1978)
DC	pKa			
HL 1.28E-003	atm m3/mol	25	EST	MEYLAN,WM & HOWARD,PH (1991)
OH 1.49E-013	cm3/molc sec	25	EST	MEYLAM,WM & HOWARD,PH (1993)

000594-82-1 — 2,2,3,3-TETRAMETHYLBUTANE

Formula: C_8H_{18}
Mol Weight: 114.23
MP (deg C): 100.7
FP (deg C):
BP (deg C): 106.4
BP pressure (mm Hg):

Property/Value	Units	Temp	Data Type	Reference
WS 1.07E+001	mg/L	25	EST	MEYLAN,WM ET AL. (1996)
logP 4.05			EST	MEYLAN,WM & HOWARD,PH (1995)
VP 2.09E+001	mm Hg	25	EXP	RIDDICK,JA ET AL.(1986)
DC	pKa			
HL 3.01E+000	atm m3/mol	25	EST	MEYLAN,WM & HOWARD,PH (1991)
OH 1.08E-012	cm3/molc sec	25	EXP	ATKINSON,R (1989)

000594-83-2 — 2,3,3,-TRIMETHYL-2-BUTANOL

Formula: $C_7H_{16}O$
Mol Weight: 116.20
MP (deg C): 17
FP (deg C):
BP (deg C): 131
BP pressure (mm Hg):

Property/Value	Units	Temp	Data Type	Reference
WS 2.20E+004	mg/L	40	EXP	YALKOWSKY,SH & DANNENFELSER,RM (1992)
logP 2.09			EST	MEYLAN,WM & HOWARD,PH (1995)
VP 7.43E+000	mm Hg	25	EST	NEELY,WB & BLAU,GE (1985)
DC	pKa			
HL 2.34E-005	atm m3/mol	25	EST	MEYLAN,WM & HOWARD,PH (1991)
OH 1.68E-012	cm3/molc sec	25	EST	MEYLAN,WM & HOWARD,PH (1993)

000594-90-1 — PERCHLOROPROPANE

Formula: C_3Cl_8
Mol Weight: 319.66
MP (deg C):
FP (deg C):
BP (deg C):
BP pressure (mm Hg):

Property/Value	Units	Temp	Data Type	Reference
WS 1.47E-001	mg/L	25	EST	MEYLAN,WM ET AL. (1996)
logP 5.63			EST	MEYLAN,WM & HOWARD,PH (1995)
VP 1.02E-001	mm Hg	25	EST	NEELY,WB & BLAU,GE (1985)
DC	pKa			
HL 6.89E-004	atm m3/mol	25	EST	MEYLAN,WM & HOWARD,PH (1991)
OH 0.00E+000	cm3/molc sec	25	EST	MEYLAN,WM & HOWARD,PH (1993)

000595-21-1 — STROSPESIDE

Formula: $C_{30}H_{46}O_9$
Mol Weight: 550.70
MP (deg C):
FP (deg C):
BP (deg C):
BP pressure (mm Hg):

Property/Value	Units	Temp	Data Type	Reference
WS 1.06E+002	mg/L	25	EST	MEYLAN,WM ET AL. (1996)
logP 0.59			EXP	HANSCH,C & LEO,AJ (1985)
VP 9.84E-021	mm Hg	25	EST	NEELY,WB & BLAU,GE (1985)
DC	pKa			
HL 9.81E-019	atm m3/mol	25	EST	MEYLAN,WM & HOWARD,PH (1991)
OH 1.13E-010	cm3/molc sec	25	EST	MEYLAN,WM & HOWARD,PH (1993)

000595-37-9 — 2,2-DIMETHYL BUTYRIC ACID

Formula: $C_6H_{12}O_2$
Mol Weight: 116.16
MP (deg C): -14
FP (deg C):
BP (deg C): 186
BP pressure (mm Hg):

Property/Value	Units	Temp	Data Type	Reference
WS 6.14E+003	mg/L	25	EST	MEYLAN,WM ET AL. (1996)
logP 1.90			EXP	HANSCH,C ET AL. (1995)
VP 6.61E-001	mm Hg	25	EST	NEELY,WB & BLAU,GE (1985)
DC 5.03	pKa	18	EXP	KORTUM,G ET AL (1961)
HL 1.70E-006	atm m3/mol	25	EST	MEYLAN,WM & HOWARD,PH (1991)
OH 2.17E-012	cm3/molc sec	25	EST	MEYLAN,WM & HOWARD,PH (1993)

000595-41-5 — 2,3-DIMETHYL-3-PENTANOL

Formula: $C_7H_{16}O$
Mol Weight: 116.20
MP (deg C): < -30
FP (deg C):
BP (deg C): 139.7
BP pressure (mm Hg):

Property/Value	Units	Temp	Data Type	Reference
WS 1.64E+004	mg/L	25	EXP	YALKOWSKY,SH & DANNENFELSER,RM (1992)
logP 2.13			EST	MEYLAN,WM & HOWARD,PH (1995)
VP 4.54E+000	mm Hg	25	EST	NEELY,WB & BLAU,GE (1985)
DC	pKa			
HL 2.34E-005	atm m3/mol	25	EST	MEYLAN,WM & HOWARD,PH (1991)
OH 1.21E-011	cm3/molc sec	25	EST	MEYLAN,WM & HOWARD,PH (1993)

000595-46-0 — DIMETHYLMALONIC ACID

Formula: $C_5H_8O_4$
Mol Weight: 132.12
MP (deg C): 192 dec
FP (deg C):
BP (deg C):
BP pressure (mm Hg):

Property/Value	Units	Temp	Data Type	Reference
WS 9.00E+004	mg/L	13	EXP	YALKOWSKY,SH & DANNENFELSER,RM (1992)
logP 0.39			EXP	HANSCH,C ET AL. (1995)
VP 1.40E-003	mm Hg	25	EST	NEELY,WB & BLAU,GE (1985)
DC 3.15	pKa	25	EXP	KORTUM,G ET AL (1961)
HL 7.18E-012	atm m3/mol	25	EST	MEYLAN,WM & HOWARD,PH (1991)
OH 1.37E-012	cm3/molc sec	25	EST	MEYLAN,WM & HOWARD,PH (1993)

CAS #: 000596-15-6				MORPHINE ACETATE
Formula: $C_{19}H_{23}NO_5$				
Mol Weight: 345.40				
MP (deg C):		FP (deg C):		
BP (deg C):				
BP pressure (mm Hg):				

Property/Value	Units	Temp	Data Type	Reference
WS 4.49E+005	mg/L	25	EXP	SEIDELL,A (1941)
logP -0.25			EST	MEYLAN,WM & HOWARD,PH (1995)
VP 1.39E-016	mm Hg	25	EST	NEELY,WB & BLAU,GE (1985)
DC	pKa			
HL 5.71E-026	atm m3/mol	25	EST	MEYLAN,WM & HOWARD,PH (1991)
OH 1.91E-010	cm3/molc sec	25	EST	MEYLAN,WM & HOWARD,PH (1993)

CAS #: 000597-09-1				2-ETHYL-2-NITRO-1,3-PROPANEDIOL
Formula: $C_5H_{11}NO_4$				
Mol Weight: 149.15				
MP (deg C): 57.5		FP (deg C):		
BP (deg C):				
BP pressure (mm Hg):				

Property/Value	Units	Temp	Data Type	Reference
WS 1.25E+005	mg/L	25	EST	MEYLAN,WM ET AL. (1996)
logP -0.71			EST	MEYLAN,WM & HOWARD,PH (1995)
VP 8.65E-005	mm Hg	25	EST	NEELY,WB & BLAU,GE (1985)
DC	pKa			
HL 1.75E-010	atm m3/mol	25	EST	MEYLAN,WM & HOWARD,PH (1991)
OH 1.62E-012	cm3/molc sec	25	EST	MEYLAN,WM & HOWARD,PH (1993)

CAS #: 000597-35-3				DIETHYL SULFONE
Formula: $C_4H_{10}O_2S$				
Mol Weight: 122.19				
MP (deg C): 73.5		FP (deg C):		
BP (deg C): 248				
BP pressure (mm Hg):				

Property/Value	Units	Temp	Data Type	Reference
WS 1.35E+005	mg/L	16	EXP	YALKOWSKY,SH & DANNENFELSER,RM (1992)
logP -0.59			EXP	HANSCH,C ET AL. (1995)
VP 6.75E-001	mm Hg	25	EST	NEELY,WB & BLAU,GE (1985)
DC	pKa			
HL 1.10E-005	atm m3/mol	25	EST	MEYLAN,WM & HOWARD,PH (1991)
OH 1.49E-011	cm3/molc sec	25	EST	MEYLAN,WM & HOWARD,PH (1993)

CAS #: 000597-49-9				3-ETHYL-3-PENTANOL
Formula: $C_7H_{16}O$				
Mol Weight: 116.20				
MP (deg C): -12.5		FP (deg C):		
BP (deg C): 142				
BP pressure (mm Hg):				

Property/Value	Units	Temp	Data Type	Reference
WS 1.68E+004	mg/L	25	EXP	YALKOWSKY,SH & DANNENFELSER,RM (1992)
logP 2.20			EST	MEYLAN,WM & HOWARD,PH (1995)
VP 2.15E+000	mm Hg	25	EST	NEELY,WB & BLAU,GE (1985)
DC	pKa			
HL 2.34E-005	atm m3/mol	25	EST	MEYLAN,WM & HOWARD,PH (1991)
OH 1.13E-011	cm3/molc sec	25	EST	MEYLAN,WM & HOWARD,PH (1993)

CAS #: 000597-96-6				3-METHYL-3-HEXANOL
Formula: $C_7H_{16}O$				
Mol Weight: 116.20				
MP (deg C):		FP (deg C):		
BP (deg C): 143				
BP pressure (mm Hg):				

Property/Value	Units	Temp	Data Type	Reference
WS 1.19E+004	mg/L	25	EXP	BARTON,AFM (1984)
logP 2.20			EST	MEYLAN,WM & HOWARD,PH (1995)
VP 1.96E+000	mm Hg	25	EXT	WILHOIT,RC & ZWOLINSKI,BJ (1973)
DC	pKa			
HL 2.34E-005	atm m3/mol	25	EST	MEYLAN,WM & HOWARD,PH (1991)
OH 1.01E-011	cm3/molc sec	25	EST	MEYLAN,WM & HOWARD,PH (1993)

CAS #: 000598-03-8				PROPYL SULFONE
Formula: $C_6H_{14}O_2S$				
Mol Weight: 150.24				
MP (deg C): 29.5		FP (deg C):		
BP (deg C):				
BP pressure (mm Hg):				

Property/Value	Units	Temp	Data Type	Reference
WS 3.52E+004	mg/L	25	EST	MEYLAN,WM ET AL. (1996)
logP 0.39			EXP	HANSCH,C & LEO,AJ (1985)
VP 4.12E-003	mm Hg	25	EXT	DAUBERT,TE & DANNER,RP (1991)
DC	pKa			
HL 1.94E-005	atm m3/mol	25	EST	MEYLAN,WM & HOWARD,PH (1991)
OH 2.06E-011	cm3/molc sec	25	EST	MEYLAN,WM & HOWARD,PH (1993)

CAS #: 000598-06-1				3-METHYL-3-HEPTANOL
Formula: $C_8H_{18}O$				
Mol Weight: 130.23				
MP (deg C):		FP (deg C):		
BP (deg C):				
BP pressure (mm Hg):				

Property/Value	Units	Temp	Data Type	Reference
WS 3.25E+003	mg/L	30	EXP	BARTON,AFM (1984)
logP 2.69			EST	MEYLAN,WM & HOWARD,PH (1995)
VP 6.24E-001	mm Hg	25	EST	NEELY,WB & BLAU,GE (1985)
DC	pKa			
HL 3.10E-005	atm m3/mol	25	EST	MEYLAN,WM & HOWARD,PH (1991)
OH 1.15E-011	cm3/molc sec	25	EST	MEYLAN,WM & HOWARD,PH (1993)

CAS #: 000598-16-3				ETHENE, TRIBROMO-
Formula: C_2HBr_3				
Mol Weight: 264.76				
MP (deg C):		FP (deg C):		
BP (deg C): 164				
BP pressure (mm Hg):				

Property/Value	Units	Temp	Data Type	Reference
WS 3.62E+001	mg/L	25	EST	MEYLAN,WM ET AL. (1996)
logP 3.20			EXP	CANTON,JH & WEGMAN,RCC (1983)
VP 4.94E-001	mm Hg	25	EST	NEELY,WB & BLAU,GE (1985)
DC	pKa			
HL 4.89E-004	atm m3/mol	25	EST	MEYLAN,WM & HOWARD,PH (1991)
OH 1.53E-012	cm3/molc sec	25	EST	MEYLAN,WM & HOWARD,PH (1993)

CAS #:	000598-25-4			3-METHYL-1,2-BUTADIENE

Formula: C_5H_8

Mol Weight: 68.12

MP (deg C): -113.6   FP (deg C):

BP (deg C): 40.83

BP pressure (mm Hg):

Property/Value	Units	Temp	Data Type	Reference
WS 2.33E+002	mg/L	25	EST	MEYLAN,WM ET AL. (1996)
logP 2.61			EST	MEYLAN,WM & HOWARD,PH (1995)
VP 4.26E+002	mm Hg	25	EXP	YAWS,CL (1994A)
DC	pKa			
HL 1.52E-001	atm m3/mol	25	EST	MEYLAN,WM & HOWARD,PH (1991)
OH 5.69E-011	cm3/molc sec	25	EXP	ATKINSON,R (1989)

CAS #:	000598-26-5			2-METHYL-1-PROPEN-1-ONE(DIMETHYLKETENE)

Formula: C_4H_6O

Mol Weight: 70.09

MP (deg C):   FP (deg C):

BP (deg C):

BP pressure (mm Hg):

Property/Value	Units	Temp	Data Type	Reference
WS 5.73E+004	mg/L	25	EST	MEYLAN,WM ET AL. (1996)
logP 0.44			EST	MEYLAN,WM & HOWARD,PH (1995)
VP 1.34E+003	mm Hg	25	EST	NEELY,WB & BLAU,GE (1985)
DC	pKa			
HL	atm m3/mol			
OH 1.07E-010	cm3/molc sec	25	EXP	ATKINSON,R (1989)

CAS #:	000598-31-2			1-BROMO-2-PROPANONE

Formula: C_3H_5BrO

Mol Weight: 136.98

MP (deg C): -36.5   FP (deg C):

BP (deg C): 137

BP pressure (mm Hg):

Property/Value	Units	Temp	Data Type	Reference
WS 6.96E+004	mg/L	25	EST	MEYLAN,WM ET AL. (1996)
logP 0.11			EST	MEYLAN,WM & HOWARD,PH (1995)
VP 1.56E+001	mm Hg	25	EST	NEELY,WB & BLAU,GE (1985)
DC	pKa			
HL 5.71E-006	atm m3/mol	25	EST	MEYLAN,WM & HOWARD,PH (1991)
OH 2.98E-013	cm3/molc sec	25	EST	MEYLAN,WM & HOWARD,PH (1993)

CAS #:	000598-38-9			2,2-DICHLOROETHANOL

Formula: $C_2H_4Cl_2O$

Mol Weight: 114.96

MP (deg C):   FP (deg C):

BP (deg C): 146

BP pressure (mm Hg):

Property/Value	Units	Temp	Data Type	Reference
WS 1.64E+005	mg/L	25	EST	MEYLAN,WM ET AL. (1996)
logP 0.37			EXP	HANSCH,C & LEO,AJ (1985)
VP 1.23E+000	mm Hg	25	EST	NEELY,WB & BLAU,GE (1985)
DC 12.89	pKa	25	EXP	SERJEANT,EP & DEMPSEY,B (1979)
HL 4.44E-007	atm m3/mol	25	EST	MEYLAN,WM & HOWARD,PH (1991)
OH 1.66E-012	cm3/molc sec	25	EST	MEYLAN,WM & HOWARD,PH (1993)

CAS #:	000598-50-5			UREA, METHYL-

Formula: $C_2H_6N_2O$

Mol Weight: 74.08

MP (deg C): 103   FP (deg C):

BP (deg C):

BP pressure (mm Hg):

Property/Value	Units	Temp	Data Type	Reference
WS 1.03E+005	mg/L	25	EST	MEYLAN,WM ET AL. (1996)
logP -1.40			EXP	SOTOMATSU,T ET AL. (1987)
VP 1.09E+000	mm Hg	25	EST	NEELY,WB & BLAU,GE (1985)
DC	pKa			
HL 8.01E-010	atm m3/mol	25	EST	MEYLAN,WM & HOWARD,PH (1991)
OH 2.56E-012	cm3/molc sec	25	EST	MEYLAN,WM & HOWARD,PH (1993)

CAS #:	000598-52-7			N-METHYLTHIOUREA

Formula: $C_2H_6N_2S$

Mol Weight: 90.15

MP (deg C): 119-121   FP (deg C):

BP (deg C):

BP pressure (mm Hg):

Property/Value	Units	Temp	Data Type	Reference
WS 2.41E+005	mg/L	25	EST	MEYLAN,WM ET AL. (1996)
logP -0.69			EXP	GOVERS,H ET AL. (1986)
VP 2.22E+000	mm Hg	25	EST	NEELY,WB & BLAU,GE (1985)
DC	pKa			
HL 3.47E-007	atm m3/mol	25	EST	MEYLAN,WM & HOWARD,PH (1991)
OH 8.53E-011	cm3/molc sec	25	EST	MEYLAN,WM & HOWARD,PH (1993)

CAS #:	000598-53-8			METHYL ISOPROPYL ETHER

Formula: $C_4H_{10}O$

Mol Weight: 74.12

MP (deg C):   FP (deg C):

BP (deg C): 30.7

BP pressure (mm Hg):

Property/Value	Units	Temp	Data Type	Reference
WS 6.50E+004	mg/L	25	EXP	YALKOWSKY,SH & DANNENFELSER,RM (1992)
logP 0.98			EST	MEYLAN,WM & HOWARD,PH (1995)
VP 6.05E+002	mm Hg	25	EXP	YAWS,CL (1994A)
DC	pKa			
HL 9.08E-005	atm m3/mol	25	EST	VP/WSOL
OH 1.30E-011	cm3/molc sec	25	EST	MEYLAN,WM & HOWARD,PH (1993)

CAS #:	000598-55-0			O-METHYL CARBAMATE

Formula: $C_2H_5NO_2$

Mol Weight: 75.07

MP (deg C): 52-54   FP (deg C):

BP (deg C): 177

BP pressure (mm Hg):

Property/Value	Units	Temp	Data Type	Reference
WS 6.91E+005	mg/L		EXP	YALKOWSKY,SH & DANNENFELSER,RM (1992)
logP -0.66			EXP	HANSCH,C & LEO,AJ (1985)
VP 4.86E+001	mm Hg	25	EST	NEELY,WB & BLAU,GE (1985)
DC	pKa			
HL 3.96E-008	atm m3/mol	25	EST	MEYLAN,WM & HOWARD,PH (1991)
OH 2.22E-012	cm3/molc sec	25	EST	MEYLAN,WM & HOWARD,PH (1993)

CAS #: 000598-56-1				ETHYL DIMETHYLAMINE

Formula: $C_4H_{11}N$

Mol Weight: 73.14

MP (deg C): -140 **FP (deg C):**

BP (deg C): 36.5

BP pressure (mm Hg):

Property/ Value	Units	Temp	Data Type	Reference
WS 3.45E+005	mg/L	25	EST	MEYLAN,WM ET AL. (1996)
logP 0.70			EXP	HANSCH,C & LEO,AJ (1985)
VP 3.52E+002	mm Hg	25	EST	NEELY,WB & BLAU,GE (1985)
DC 10.16	pKa	20	EXP	PERRIN,DD (1965)
HL 4.85E-005	atm m3/mol	25	EST	MEYLAN,WM & HOWARD,PH (1991)
OH 7.74E-011	cm3/molc sec	25	EST	MEYLAN,WM & HOWARD,PH (1993)

CAS #: 000598-58-3				METHYLNITRATE

Formula: CH_3NO_3

Mol Weight: 77.04

MP (deg C): -83 **FP (deg C):**

BP (deg C): 64.6

BP pressure (mm Hg):

Property/ Value	Units	Temp	Data Type	Reference
WS 2.99E+004	mg/L	25	EST	MEYLAN,WM ET AL. (1996)
logP 0.76			EST	MEYLAN,WM & HOWARD,PH (1995)
VP 1.80E+002	mm Hg	25	EST	NEELY,WB & BLAU,GE (1985)
DC	pKa			
HL 1.98E-004	atm m3/mol	25	EST	MEYLAN,WM & HOWARD,PH (1991)
OH 3.40E-014	cm3/molc sec	25	EXP	ATKINSON,R (1989)

CAS #: 000598-61-8				METHYLCYCLOBUTANE

Formula: C_5H_{10}

Mol Weight: 70.14

MP (deg C): -161.5 **FP (deg C):**

BP (deg C): 36.3

BP pressure (mm Hg):

Property/ Value	Units	Temp	Data Type	Reference
WS 2.32E+002	mg/L	25	EST	MEYLAN,WM ET AL. (1996)
logP 2.61			EST	MEYLAN,WM & HOWARD,PH (1995)
VP 1.48E+003	mm Hg	25	EXT	CHAO,J ET AL. (1983)
DC	pKa			
HL 1.92E-001	atm m3/mol	25	EST	MEYLAN,WM & HOWARD,PH (1991)
OH 2.18E-012	cm3/molc sec	25	EST	MEYLAN,WM & HOWARD,PH (1993)

CAS #: 000598-72-1				A-BROMOPROPIONIC ACID

Formula: $C_3H_5BrO_2$

Mol Weight: 152.98

MP (deg C): **FP (deg C):**

BP (deg C): 203

BP pressure (mm Hg):

Property/ Value	Units	Temp	Data Type	Reference
WS 2.99E+004	mg/L	25	EST	MEYLAN,WM ET AL. (1996)
logP 0.92			EXP	HANSCH,C & LEO,AJ (1985)
VP 2.91E-001	mm Hg	25	EST	NEELY,WB & BLAU,GE (1985)
DC 2.97	pKa	18	EXP	KORTUM,G ET AL (1961)
HL 8.38E-008	atm m3/mol	25	EST	MEYLAN,WM & HOWARD,PH (1991)
OH 9.90E-013	cm3/molc sec	25	EST	MEYLAN,WM & HOWARD,PH (1993)

CAS #: 000598-73-2				2-BROMO-1,1,2-TRIFLUOROETHYLENE

Formula: C_2BrF_3

Mol Weight: 160.93

MP (deg C): **FP (deg C):**

BP (deg C):

BP pressure (mm Hg):

Property/ Value	Units	Temp	Data Type	Reference
WS 3.20E+003	mg/L	25	EST	MEYLAN,WM ET AL. (1996)
logP 1.55			EST	MEYLAN,WM & HOWARD,PH (1995)
VP 1.97E+003	mm Hg	25	EXP	DAUBERT,TE & DANNER,RP (1989)
DC	pKa			
HL 8.62E-002	atm m3/mol	25	EST	MEYLAN,WM & HOWARD,PH (1991)
OH 2.65E-013	cm3/molc sec	25	EST	MEYLAN,WM & HOWARD,PH (1993)

CAS #: 000598-75-4				3-METHYL-2-BUTANOL

Formula: $C_5H_{12}O$

Mol Weight: 88.15

MP (deg C): **FP (deg C):**

BP (deg C): 112

BP pressure (mm Hg):

Property/ Value	Units	Temp	Data Type	Reference
WS 6.07E+004	mg/L	20	EXP	YALKOWSKY,SH & DANNENFELSER,RM (1992)
logP 1.28			EXP	HANSCH,C & LEO,AJ (1985)
VP 9.15E+000	mm Hg	25	EXP	DAUBERT,TE & DANNER,RP (1989)
DC	pKa			
HL 1.75E-005	atm m3/mol	25	EST	VP/WSOL
OH 1.24E-011	cm3/molc sec	25	EXP	ATKINSON,R (1989)

CAS #: 000598-77-6				1,1,2-TRICHLOROPROPANE

Formula: $C_3H_5Cl_3$

Mol Weight: 147.43

MP (deg C): **FP (deg C):**

BP (deg C): 132

BP pressure (mm Hg):

Property/ Value	Units	Temp	Data Type	Reference
WS 1.90E+003	mg/L	25	EXP	MACKAY,D & SHIU (1981)
logP 2.43			EST	MEYLAN,WM & HOWARD,PH (1995)
VP 3.10E+000	mm Hg	25	EXP	MACKAY,D & SHIU,WY (1981)
DC	pKa			
HL 3.17E-004	atm m3/mol	25	EST	VP/WSOL
OH 4.15E-013	cm3/molc sec	25	EST	MEYLAN,WM & HOWARD,PH (1993)

CAS #: 000598-78-7				2-CHLOROPROPIONIC ACID

Formula: $C_3H_5ClO_2$

Mol Weight: 108.53

MP (deg C): **FP (deg C):**

BP (deg C): 185

BP pressure (mm Hg):

Property/ Value	Units	Temp	Data Type	Reference
WS 1.00E+006	mg/L	20	EXP	YALKOWSKY,SH & DANNENFELSER,RM (1992)
logP 0.76			EST	MEYLAN,WM & HOWARD,PH (1995)
VP 1.06E+000	mm Hg	25	EST	NEELY,WB & BLAU,GE (1985)
DC 2.80	pKa	25	EXP	FOY,CL (1969)
HL 2.60E-007	atm m3/mol	25	EST	MEYLAN,WM & HOWARD,PH (1991)
OH 1.06E-012	cm3/molc sec	25	EST	ATKINSON,R (1985)

CAS #: 000598-82-3				A-HYDROXYPROPIONIC ACID

Formula: $C_3H_6O_3$

Mol Weight: 90.08

MP (deg C): 16.8

FP (deg C):

BP (deg C): 122

BP pressure (mm Hg):

Property/Value	Units	Temp	Data Type	Reference
WS 1.00E+006	mg/L	25	EST	MEYLAN,WM ET AL. (1996)
logP -0.72			EXP	HANSCH,C & LEO,AJ (1985)
VP 8.13E-002	mm Hg	25	EXP	YAWS,CL (1994)
DC 3.86	pKa	20	EXP	KORTUM,G ET AL (1961)
HL 1.13E-007	atm m3/mol	25	EST	MEYLAN,WM & HOWARD,PH (1991)
OH 5.92E-012	cm3/molc sec	25	EST	MEYLAN,WM & HOWARD,PH (1993)

CAS #: 000598-88-9				1,2-DICHLORO-1,2-DIFLUOROETHYLENE

Formula: $C_2Cl_2F_2$

Mol Weight: 132.93

MP (deg C): -130.5

FP (deg C):

BP (deg C): 21.1

BP pressure (mm Hg):

Property/Value	Units	Temp	Data Type	Reference
WS 1.47E+003	mg/L	25	EST	MEYLAN,WM ET AL. (1996)
logP 2.09			EST	MEYLAN,WM & HOWARD,PH (1995)
VP 8.90E+002	mm Hg	25	EXT	BOUBLIK,T ET AL. (1984)
DC	pKa			
HL 1.17E-001	atm m3/mol	25	EST	MEYLAN,WM & HOWARD,PH (1991)
OH 2.14E-013	cm3/molc sec	25	EST	MEYLAN,WM & HOWARD,PH (1993)

CAS #: 000598-92-5				ETHANE, 1-CHLORO-1-NITRO-

Formula: $C_2H_4ClNO_2$

Mol Weight: 109.51

MP (deg C):

FP (deg C):

BP (deg C): 124.5

BP pressure (mm Hg):

Property/Value	Units	Temp	Data Type	Reference
WS 4.00E+003	mg/L	20	EXP	YALKOWSKY,SH & DANNENFELSER,RM (1992)
logP 1.10			EXP	HANSCH,C ET AL. (1995)
VP 1.12E+001	mm Hg	25	EST	NEELY,WB & BLAU,GE (1985)
DC	pKa			
HL 1.97E-005	atm m3/mol	25	EST	MEYLAN,WM & HOWARD,PH (1991)
OH 1.59E-013	cm3/molc sec	25	EST	MEYLAN,WM & HOWARD,PH (1993)

CAS #: 000598-94-7				1,1-DIMETHYLUREA

Formula: $C_3H_8N_2O$

Mol Weight: 88.11

MP (deg C): 183.5

FP (deg C):

BP (deg C):

BP pressure (mm Hg):

Property/Value	Units	Temp	Data Type	Reference
WS 3.46E+004	mg/L	25	EST	MEYLAN,WM ET AL. (1996)
logP -0.88			EST	MEYLAN,WM & HOWARD,PH (1995)
VP 4.40E-001	mm Hg	25	EST	NEELY,WB & BLAU,GE (1985)
DC	pKa			
HL 1.76E-009	atm m3/mol	25	EST	MEYLAN,WM & HOWARD,PH (1991)
OH 3.12E-012	cm3/molc sec	25	EST	MEYLAN,WM & HOWARD,PH (1993)

CAS #: 000598-98-1				PROPANOIC ACID, 2,2-DIMETHYL-, METHYL ESTER

Formula: $C_6H_{12}O_2$

Mol Weight: 116.16

MP (deg C):

FP (deg C):

BP (deg C): 101.1

BP pressure (mm Hg):

Property/Value	Units	Temp	Data Type	Reference
WS 2.84E+003	mg/L	25	EST	MEYLAN,WM ET AL. (1996)
logP 1.83			EXP	HANSCH,C ET AL. (1995)
VP 3.39E+001	mm Hg	25	EST	NEELY,WB & BLAU,GE (1985)
DC	pKa			
HL 4.10E-004	atm m3/mol	25	EST	MEYLAN,WM & HOWARD,PH (1991)
OH 7.19E-013	cm3/molc sec	25	EST	MEYLAN,WM & HOWARD,PH (1993)

CAS #: 000598-99-2				METHYL TRICHLOROACETATE

Formula: $C_3H_3Cl_3O_2$

Mol Weight: 177.42

MP (deg C): -17.5

FP (deg C):

BP (deg C): 153.8

BP pressure (mm Hg):

Property/Value	Units	Temp	Data Type	Reference
WS 1.05E+003	mg/L	25	EST	MEYLAN,WM ET AL. (1996)
logP 2.03			EXP	HANSCH,C ET AL. (1995)
VP 5.50E+000	mm Hg	25	EST	NEELY,WB & BLAU,GE (1985)
DC	pKa			
HL 7.66E-006	atm m3/mol	25	EST	MEYLAN,WM & HOWARD,PH (1991)
OH 4.90E-014	cm3/molc sec	25	EST	MEYLAN,WM & HOWARD,PH (1993)

CAS #: 000599-71-3				BENZENESULFONAMIDE, N-HYDROXY-

Formula: $C_6H_7NO_3S$

Mol Weight: 173.19

MP (deg C):

FP (deg C):

BP (deg C):

BP pressure (mm Hg):

Property/Value	Units	Temp	Data Type	Reference
WS 2.31E+004	mg/L	25	EST	MEYLAN,WM ET AL. (1996)
logP 0.48			EXP	HANSCH,C ET AL. (1995)
VP 1.32E-006	mm Hg	25	EST	NEELY,WB & BLAU,GE (1985)
DC	pKa			
HL 8.46E-010	atm m3/mol	25	EST	MEYLAN,WM & HOWARD,PH (1991)
OH 4.17E-013	cm3/molc sec	25	EST	MEYLAN,WM & HOWARD,PH (1993)

CAS #: 000599-82-6				2-SULFANILAMIDOPYRIMIDINE

Formula: $C_{10}H_{10}N_4O_2S$

Mol Weight: 250.28

MP (deg C):

FP (deg C):

BP (deg C):

BP pressure (mm Hg):

Property/Value	Units	Temp	Data Type	Reference
WS 3.78E+004	mg/L	25	EST	MEYLAN,WM ET AL. (1996)
logP -0.24			EXP	HANSCH,C & LEO,AJ (1985)
VP 4.31E-008	mm Hg	25	EST	NEELY,WB & BLAU,GE (1985)
DC	pKa			
HL 1.58E-010	atm m3/mol	25	EST	MEYLAN,WM & HOWARD,PH (1991)
OH 2.80E-011	cm3/molc sec	25	EST	MEYLAN,WM & HOWARD,PH (1993)

CAS #: 000599-88-2				SULFAPERINE

Formula: $C_{11}H_{12}N_4O_2S$

Mol Weight: 264.31

MP (deg C): 262-263 FP (deg C):

BP (deg C):

BP pressure (mm Hg):

Property/Value	Units	Temp	Data Type	Reference
WS 1.01E+004	mg/L	25	EST	MEYLAN,WM ET AL. (1996)
logP 0.34			EXP	SANGSTER,J (1993)
VP 1.93E-008	mm Hg	25	EST	NEELY,WB & BLAU,GE (1985)
DC	pKa			
HL 1.75E-010	atm m3/mol	25	EST	MEYLAN,WM & HOWARD,PH (1991)
OH 2.41E-011	cm3/molc sec	25	EST	MEYLAN,WM & HOWARD,PH (1993)

CAS #: 000600-00-0				ETHYL ALPHA-BROMOISOBUTYRATE

Formula: $C_6H_{11}BrO_2$

Mol Weight: 195.06

MP (deg C): FP (deg C):

BP (deg C): 163

BP pressure (mm Hg):

Property/Value	Units	Temp	Data Type	Reference
WS 7.78E+002	mg/L	25	EST	MEYLAN,WM ET AL. (1996)
logP 2.08			EST	MEYLAN,WM & HOWARD,PH (1995)
VP 2.61E+000	mm Hg	25	EXP	PERRY,RH & GREEN,D (1984)
DC	pKa			
HL 4.73E-005	atm m3/mol	25	EST	MEYLAN,WM & HOWARD,PH (1991)
OH 1.37E-012	cm3/molc sec	25	EST	MEYLAN,WM & HOWARD,PH (1993)

CAS #: 000600-24-8				2-NITROBUTANE

Formula: $C_4H_9NO_2$

Mol Weight: 103.12

MP (deg C): FP (deg C):

BP (deg C):

BP pressure (mm Hg):

Property/Value	Units	Temp	Data Type	Reference
WS 4.40E+003	mg/L	25	EST	MEYLAN,WM ET AL. (1996)
logP 1.20			EXP	HANSCH,C ET AL. (1995)
VP 1.29E+001	mm Hg	25	EST	NEELY,WB & BLAU,GE (1985)
DC	pKa			
HL 9.85E-005	atm m3/mol	25	EST	MEYLAN,WM & HOWARD,PH (1991)
OH 4.47E-013	cm3/molc sec	25	EST	MEYLAN,WM & HOWARD,PH (1993)

CAS #: 000600-36-2				2,4-DIMETHYL-3-PENTANOL

Formula: $C_7H_{16}O$

Mol Weight: 116.20

MP (deg C): < -70 FP (deg C):

BP (deg C): 138.7

BP pressure (mm Hg):

Property/Value	Units	Temp	Data Type	Reference
WS 7.00E+003	mg/L	25	EXP	YALKOWSKY,SH & DANNENFELSER,RM (1992)
logP 2.09			EST	MEYLAN,WM & HOWARD,PH (1995)
VP 3.00E+000	mm Hg	25	EXT	WILHOIT,RC & ZWOLINSKI,BJ (1973)
DC	pKa			
HL 2.34E-005	atm m3/mol	25	EST	MEYLAN,WM & HOWARD,PH (1991)
OH 1.59E-011	cm3/molc sec	25	EST	MEYLAN,WM & HOWARD,PH (1993)

CAS #: 000601-75-2				ETHYLMALONIC ACID

Formula: $C_5H_8O_4$

Mol Weight: 132.12

MP (deg C): 114 FP (deg C):

BP (deg C): 180

BP pressure (mm Hg): 5.00E-002

Property/Value	Units	Temp	Data Type	Reference
WS 7.12E+005	mg/L	25	EXP	YALKOWSKY,SH & DANNENFELSER,RM (1992)
logP -0.34			EST	MEYLAN,WM & HOWARD,PH (1995)
VP 1.05E-003	mm Hg	25	EST	NEELY,WB & BLAU,GE (1985)
DC 2.96	pKa	25	EXP	KORTUM,G ET AL (1961)
HL 7.18E-012	atm m3/mol	25	EST	MEYLAN,WM & HOWARD,PH (1991)
OH 3.70E-012	cm3/molc sec	25	EST	MEYLAN,WM & HOWARD,PH (1993)

CAS #: 000601-77-4				DI-I-PROPYLNITROSOAMINE

Formula: $C_6H_{14}N_2O$

Mol Weight: 130.19

MP (deg C): 48 FP (deg C):

BP (deg C): 194.5

BP pressure (mm Hg):

Property/Value	Units	Temp	Data Type	Reference
WS 1.30E+004	mg/L	24	EXP	YALKOWSKY,SH & DANNENFELSER,RM (1992)
logP 1.38			EXP	HANSCH,C ET AL. (1995)
VP 2.66E-001	mm Hg	25	EST	NEELY,WB & BLAU,GE (1985)
DC	pKa			
HL 7.83E-006	atm m3/mol	37	EXP	MIRVISH,SS ET AL. (1976)
OH 3.73E-011	cm3/molc sec	25	EST	MEYLAN,WM & HOWARD,PH (1993)

CAS #: 000601-89-8				2-NITRORESORCINOL

Formula: $C_6H_5NO_4$

Mol Weight: 155.11

MP (deg C): 81-83 FP (deg C):

BP (deg C):

BP pressure (mm Hg):

Property/Value	Units	Temp	Data Type	Reference
WS 1.31E+003	mg/L		EXP	YALKOWSKY,SH & DANNENFELSER,RM (1992)
logP 1.56			EXP	HANSCH,C & LEO,AJ (1985)
VP 2.27E-005	mm Hg	25	EST	NEELY,WB & BLAU,GE (1985)
DC	pKa			
HL 2.30E-006	atm m3/mol	25	EST	MEYLAN,WM & HOWARD,PH (1991)
OH 7.15E-011	cm3/molc sec	25	EST	MEYLAN,WM & HOWARD,PH (1993)

CAS #: 000602-01-7				2,3-DINITROTOLUENE

Formula: $C_7H_6N_2O_4$

Mol Weight: 182.14

MP (deg C): 59-61 FP (deg C):

BP (deg C): 284

BP pressure (mm Hg):

Property/Value	Units	Temp	Data Type	Reference
WS 2.20E+002	mg/L	25	EST	MEYLAN,WM ET AL. (1996)
logP 2.18			EST	MEYLAN,WM & HOWARD,PH (1995)
VP 3.97E-004	mm Hg	25	EST	NEELY,WB & BLAU,GE (1985)
DC	pKa			
HL 9.26E-008	atm m3/mol	25	EST	MEYLAN,WM & HOWARD,PH (1991)
OH 2.01E-013	cm3/molc sec	25	EST	MEYLAN,WM & HOWARD,PH (1993)

CAS #: 000602-38-0				NAPHTHALENE, 1,8-DINITRO-

Formula: $C_{10}H_6N_2O_4$
Mol Weight: 218.17
MP (deg C): 173   FP (deg C):
BP (deg C):
BP pressure (mm Hg):

Property/Value	Units	Temp	Data Type	Reference
WS 3.40E+001	mg/L	15	EXP	YALKOWSKY,SH & DANNENFELSER,RM (1992)
logP 2.52			EXP	DEBNATH,AK ET AL. (1992)
VP 4.28E-006	mm Hg	25	EST	NEELY,WB & BLAU,GE (1985)
DC	pKa			
HL 8.19E-009	atm m3/mol	25	EST	MEYLAN,WM & HOWARD,PH (1991)
OH 2.36E-013	cm3/molc sec	25	EST	MEYLAN,WM & HOWARD,PH (1993)

CAS #: 000602-55-1				9-PHENYLANTHRACENE

Formula: $C_{20}H_{14}$
Mol Weight: 254.33
MP (deg C): 156   FP (deg C):
BP (deg C): 417
BP pressure (mm Hg):

Property/Value	Units	Temp	Data Type	Reference
WS 1.28E-002	mg/L	25	EST	MEYLAN,WM ET AL. (1996)
logP 6.01			EXP	WANG,L ET AL. (1986)
VP 9.48E-008	mm Hg	25	EST	NEELY,WB & BLAU,GE (1985)
DC	pKa			
HL 3.94E-006	atm m3/mol	25	EST	MEYLAN,WM & HOWARD,PH (1991)
OH 7.29E-011	cm3/molc sec	25	EST	MEYLAN,WM & HOWARD,PH (1993)

CAS #: 000602-60-8				ANTHRACENE, 9-NITRO-

Formula: $C_{14}H_9NO_2$
Mol Weight: 223.23
MP (deg C): 146   FP (deg C):
BP (deg C): 275
BP pressure (mm Hg): 1.70E+001

Property/Value	Units	Temp	Data Type	Reference
WS 1.14E-001	mg/L	25	EXP	GANG,Y & XIAOBAI,XU (1992)
logP 4.78			EXP	DEBNATH,AK & HANSCH,C (1992)
VP 1.35E-006	mm Hg	25	EST	NEELY,WB & BLAU,GE (1985)
DC	pKa			
HL 2.03E-007	atm m3/mol	25	EST	MEYLAN,WM & HOWARD,PH (1991)
OH 5.00E-012	cm3/molc sec	25	EST	MEYLAN,WM & HOWARD,PH (1993)

CAS #: 000602-87-9				ACENAPHTHYLENE, 1,2-DIHYDRO-5-NITRO-

Formula: $C_{12}H_9NO_2$
Mol Weight: 199.21
MP (deg C): 100-101   FP (deg C):
BP (deg C):
BP pressure (mm Hg):

Property/Value	Units	Temp	Data Type	Reference
WS 7.20E-001	mg/L	25	EST	MEYLAN,WM ET AL. (1996)
logP 3.85			EXP	DEBNATH,AK ET AL. (1992)
VP 2.65E-005	mm Hg	25	EST	NEELY,WB & BLAU,GE (1985)
DC	pKa			
HL 1.11E-006	atm m3/mol	25	EST	MEYLAN,WM & HOWARD,PH (1991)
OH 8.67E-012	cm3/molc sec	25	EST	MEYLAN,WM & HOWARD,PH (1993)

CAS #: 000603-00-9				PROXYPHYLLINE

Formula: $C_{10}H_{16}N_4O_3$
Mol Weight: 240.26
MP (deg C): 135-136   FP (deg C):
BP (deg C):
BP pressure (mm Hg):

Property/Value	Units	Temp	Data Type	Reference
WS 6.11E+003	mg/L	25	EST	MEYLAN,WM ET AL. (1996)
logP -0.77			EXP	SANGSTER,J (1994)
VP 5.66E-012	mm Hg	25	EST	NEELY,WB & BLAU,GE (1985)
DC	pKa			
HL 2.31E-015	atm m3/mol	25	EST	MEYLAN,WM & HOWARD,PH (1991)
OH 2.91E-011	cm3/molc sec	25	EST	MEYLAN,WM & HOWARD,PH (1993)

CAS #: 000603-11-2				3-NITROPHTHALIC ACID

Formula: $C_8H_5NO_6$
Mol Weight: 211.13
MP (deg C): 213-216 de   FP (deg C):
BP (deg C):
BP pressure (mm Hg):

Property/Value	Units	Temp	Data Type	Reference
WS 2.01E+004	mg/L	25	EXP	YALKOWSKY,SH & DANNENFELSER,RM (1992)
logP 0.75			EXP	HANSCH,C & LEO,AJ (1985)
VP 1.82E-007	mm Hg	25	EST	NEELY,WB & BLAU,GE (1985)
DC	pKa			
HL 8.61E-015	atm m3/mol	25	EST	MEYLAN,WM & HOWARD,PH (1991)
OH 1.06E-012	cm3/molc sec	25	EST	MEYLAN,WM & HOWARD,PH (1993)

CAS #: 000603-34-9				TRIPHENYLAMINE

Formula: $C_{18}H_{15}N$
Mol Weight: 245.33
MP (deg C): 127   FP (deg C):
BP (deg C): 365
BP pressure (mm Hg):

Property/Value	Units	Temp	Data Type	Reference
WS 3.15E-001	mg/L	25	EST	MEYLAN,WM ET AL. (1996)
logP 5.74			EXP	HANSCH,C & LEO,AJ (1985)
VP 3.92E-004	mm Hg	25	EXP	FORWARD,MV ET AL. (1949)
DC	pKa			
HL 5.41E-006	atm m3/mol	25	EST	MEYLAN,WM & HOWARD,PH (1991)
OH 2.00E-010	cm3/molc sec	25	EST	MEYLAN,WM & HOWARD,PH (1993)

CAS #: 000603-35-0				TRIPHENYLPHOSPHINE

Formula: $C_{18}H_{15}P$
Mol Weight: 262.29
MP (deg C): 80.5   FP (deg C):
BP (deg C): >360
BP pressure (mm Hg):

Property/Value	Units	Temp	Data Type	Reference
WS 2.79E-001	mg/L	25	EST	MEYLAN,WM ET AL. (1996)
logP 5.69			EXP	HANSCH,C ET AL. (1995)
VP 7.20E-001	mm Hg	210	EXP	FORWARD,MV ET AL. (1949)
DC	pKa			
HL 2.26E-008	atm m3/mol	25	EST	MEYLAN,WM & HOWARD,PH (1991)
OH 5.85E-012	cm3/molc sec	25	EST	MEYLAN,WM & HOWARD,PH (1993)

CAS #: 000603-45-2				ROSOLIC ACID

Formula: $C_{19}H_{14}O_3$

Mol Weight: 290.32

MP (deg C): 308-310 de   FP (deg C):

BP (deg C):

BP pressure (mm Hg):

Property/Value	Units	Temp	Data Type	Reference
WS 1.20E+003	mg/L		EXP	YALKOWSKY,SH & DANNENFELSER,RM (1992)
logP 3.03			EST	MEYLAN,WM & HOWARD,PH (1995)
VP 1.02E-010	mm Hg	25	EST	NEELY,WB & BLAU,GE (1985)
DC	pKa			
HL 1.89E-016	atm m3/mol	25	EST	MEYLAN,WM & HOWARD,PH (1991)
OH 1.38E-010	cm3/molc sec	25	EST	MEYLAN,WM & HOWARD,PH (1993)

CAS #: 000603-54-3				DIPHENYL UREA, UNSYM

Formula: $C_{13}H_{12}N_2O$

Mol Weight: 212.25

MP (deg C): 189   FP (deg C):

BP (deg C):

BP pressure (mm Hg):

Property/Value	Units	Temp	Data Type	Reference
WS 1.87E+003	mg/L	25	EST	MEYLAN,WM ET AL. (1996)
logP 1.53			EXP	HANSCH,C & LEO,AJ (1985)
VP 1.64E-006	mm Hg	25	EST	NEELY,WB & BLAU,GE (1985)
DC	pKa			
HL 1.04E-009	atm m3/mol	25	EST	MEYLAN,WM & HOWARD,PH (1991)
OH 2.58E-011	cm3/molc sec	25	EST	MEYLAN,WM & HOWARD,PH (1993)

CAS #: 000603-62-3				3-NITROPHTHALIMIDE

Formula: $C_8H_4N_2O_4$

Mol Weight: 192.13

MP (deg C): 213-215   FP (deg C):

BP (deg C):

BP pressure (mm Hg):

Property/Value	Units	Temp	Data Type	Reference
WS 2.18E+003	mg/L	25	EST	MEYLAN,WM ET AL. (1996)
logP 1.12			EST	MEYLAN,WM & HOWARD,PH (1995)
VP 1.13E-009	mm Hg	25	EST	NEELY,WB & BLAU,GE (1985)
DC	pKa			
HL 4.03E-011	atm m3/mol	25	EST	MEYLAN,WM & HOWARD,PH (1991)
OH 5.59E-012	cm3/molc sec	25	EST	MEYLAN,WM & HOWARD,PH (1993)

CAS #: 000603-76-9				1-METHYLINDOLE

Formula: C_9H_9N

Mol Weight: 131.18

MP (deg C):   FP (deg C):

BP (deg C): 237

BP pressure (mm Hg):

Property/Value	Units	Temp	Data Type	Reference
WS 4.33E+002	mg/L	25	EST	MEYLAN,WM ET AL. (1996)
logP 2.72			EXP	HANSCH,C & LEO,AJ (1985)
VP 5.14E-002	mm Hg	25	EST	NEELY,WB & BLAU,GE (1985)
DC	pKa			
HL 1.89E-005	atm m3/mol	25	EST	MEYLAN,WM & HOWARD,PH (1991)
OH 1.10E-010	cm3/molc sec	25	EST	MEYLAN,WM & HOWARD,PH (1993)

CAS #: 000603-86-1				6-CHLORO-2-NITROPHENOL

Formula: $C_6H_4ClNO_3$

Mol Weight: 173.56

MP (deg C):   FP (deg C):

BP (deg C):

BP pressure (mm Hg):

Property/Value	Units	Temp	Data Type	Reference
WS 3.91E+002	mg/L	25	EST	MEYLAN,WM ET AL. (1996)
logP 2.55			EST	MEYLAN,WM & HOWARD,PH (1995)
VP 2.96E-004	mm Hg	25	EST	NEELY,WB & BLAU,GE (1985)
DC 5.48	pKa	25	EXP	SERJEANT,EP & DEMPSEY,B (1979)
HL 5.18E-006	atm m3/mol	25	EST	MEYLAN,WM & HOWARD,PH (1991)
OH 1.36E-012	cm3/molc sec	25	EST	MEYLAN,WM & HOWARD,PH (1993)

CAS #: 000604-44-4				4-CHLORO-1-NAPHTHOL

Formula: $C_{10}H_7ClO$

Mol Weight: 178.62

MP (deg C):   FP (deg C):

BP (deg C):

BP pressure (mm Hg):

Property/Value	Units	Temp	Data Type	Reference
WS 9.18E+001	mg/L	25	EST	MEYLAN,WM ET AL. (1996)
logP 3.94			EXP	HANSCH,C ET AL. (1995)
VP 1.00E-004	mm Hg	25	EST	NEELY,WB & BLAU,GE (1985)
DC 8.86	pKa	25	EXP	SERJEANT,EP & DEMPSEY,B (1979)
HL 4.06E-008	atm m3/mol	25	EST	MEYLAN,WM & HOWARD,PH (1991)
OH 3.37E-011	cm3/molc sec	25	EST	MEYLAN,WM & HOWARD,PH (1993)

CAS #: 000604-68-2				.ALPHA.-D-GLUCOPYRANOSE, PENTAACETATE

Formula: $C_{16}H_{22}O_{11}$

Mol Weight: 390.35

MP (deg C): 113.3   FP (deg C):

BP (deg C):

BP pressure (mm Hg):

Property/Value	Units	Temp	Data Type	Reference
WS 1.02E+003	mg/L	25	EST	MEYLAN,WM ET AL. (1996)
logP 0.63			EXP	HANSCH,C ET AL. (1995)
VP 7.48E-007	mm Hg	25	EST	NEELY,WB & BLAU,GE (1985)
DC	pKa			
HL 1.14E-017	atm m3/mol	25	EST	MEYLAN,WM & HOWARD,PH (1991)
OH 5.76E-011	cm3/molc sec	25	EST	MEYLAN,WM & HOWARD,PH (1993)

CAS #: 000604-75-1				OXAZEPAM

Formula: $C_{15}H_{11}ClN_2O_2$

Mol Weight: 286.72

MP (deg C): 205-206   FP (deg C):

BP (deg C):

BP pressure (mm Hg):

Property/Value	Units	Temp	Data Type	Reference
WS 1.79E+002	mg/L	25	EST	MEYLAN,WM ET AL. (1996)
logP 2.24			EXP	HANSCH,C & LEO,AJ (1985)
VP 3.76E-012	mm Hg	25	EST	NEELY,WB & BLAU,GE (1985)
DC	pKa			
HL 5.53E-010	atm m3/mol	25	EST	MEYLAN,WM & HOWARD,PH (1991)
OH 1.39E-011	cm3/molc sec	25	EST	MEYLAN,WM & HOWARD,PH (1993)

CAS #: 000605-32-3 — 2-HYDROXYANTHRAQUINONE

Formula: $C_{14}H_8O_3$

Mol Weight: 224.22

MP (deg C): 306

FP (deg C):

BP (deg C):

BP pressure (mm Hg):

Property/Value	Units	Temp	Data Type	Reference
WS 1.10E+000	mg/L	25	EXP	STEPHEN,H & STEPHEN,T (1963)
logP 2.86			EST	MEYLAN,WM & HOWARD,PH (1995)
VP 6.85E-011	mm Hg	25	EXP	SHIMIZU,T ET AL. (1987)
DC	pKa			
HL 1.84E-011	atm m3/mol	25	EST	VP/WSOL
OH 1.37E-011	cm3/molc sec	25	EST	MEYLAN,WM & HOWARD,PH (1993)

CAS #: 000605-45-8 — DI-I-PROPYL PHTHALATE

Formula: $C_{14}H_{18}O_4$

Mol Weight: 250.30

MP (deg C):

FP (deg C):

BP (deg C):

BP pressure (mm Hg):

Property/Value	Units	Temp	Data Type	Reference
WS 3.33E+002	mg/L	20	EXP	LEYDER,F & BOULANGER,P (1983)
logP 2.83			EXP	HANSCH,C & LEO,AJ (1985)
VP 2.63E-003	mm Hg	25	EST	NEELY,WB & BLAU,GE (1985)
DC	pKa			
HL 6.95E-007	atm m3/mol	25	EST	MEYLAN,WM & HOWARD,PH (1991)
OH 7.63E-012	cm3/molc sec	25	EST	MEYLAN,WM & HOWARD,PH (1993)

CAS #: 000605-71-0 — NAPHTHALENE, 1,5-DINITRO-

Formula: $C_{10}H_6N_2O_4$

Mol Weight: 218.17

MP (deg C): 219

FP (deg C):

BP (deg C):

BP pressure (mm Hg):

Property/Value	Units	Temp	Data Type	Reference
WS 5.80E+001	mg/L	12	EXP	YALKOWSKY,SH & DANNENFELSER,RM (1992)
logP 2.58			EXP	DEBNATH,AK ET AL. (1992)
VP 4.28E-006	mm Hg	25	EST	NEELY,WB & BLAU,GE (1985)
DC	pKa			
HL 8.19E-009	atm m3/mol	25	EST	MEYLAN,WM & HOWARD,PH (1991)
OH 2.36E-013	cm3/molc sec	25	EST	MEYLAN,WM & HOWARD,PH (1993)

CAS #: 000605-93-6 — 6-METHYL-1,4-NAPHTHOQUINONE

Formula: $C_{11}H_8O_2$

Mol Weight: 172.19

MP (deg C):

FP (deg C):

BP (deg C):

BP pressure (mm Hg):

Property/Value	Units	Temp	Data Type	Reference
WS 9.66E+002	mg/L	25	EST	MEYLAN,WM ET AL. (1996)
logP 2.10			EXP	HANSCH,C & LEO,AJ (1985)
VP 1.75E-004	mm Hg	25	EST	NEELY,WB & BLAU,GE (1985)
DC	pKa			
HL 2.17E-009	atm m3/mol	25	EST	MEYLAN,WM & HOWARD,PH (1991)
OH 4.35E-012	cm3/molc sec	25	EST	MEYLAN,WM & HOWARD,PH (1993)

CAS #: 000605-94-7 — 2,5-CYCLOHEXADIENE-1,4-DIONE, 2,3-DIMETHOXY-5-ME

Formula: $C_9H_{10}O_4$

Mol Weight: 182.18

MP (deg C): 59-60

FP (deg C):

BP (deg C):

BP pressure (mm Hg):

Property/Value	Units	Temp	Data Type	Reference
WS 1.11E+004	mg/L	25	EST	MEYLAN,WM ET AL. (1996)
logP 0.80			EXP	RICH,PR (1990)
VP 1.02E-003	mm Hg	25	EST	NEELY,WB & BLAU,GE (1985)
DC	pKa			
HL 8.27E-010	atm m3/mol	25	EST	MEYLAN,WM & HOWARD,PH (1991)
OH 3.52E-011	cm3/molc sec	25	EST	MEYLAN,WM & HOWARD,PH (1993)

CAS #: 000605-99-2 — 3-METHYL URIC ACID

Formula: $C_6H_6N_4O_3$

Mol Weight: 182.14

MP (deg C): > 350

FP (deg C):

BP (deg C):

BP pressure (mm Hg):

Property/Value	Units	Temp	Data Type	Reference
WS 2.68E+004	mg/L	25	EST	MEYLAN,WM ET AL. (1996)
logP -1.08			EXP	GASPARI,F & BONATI,M (1987)
VP 3.74E-010	mm Hg	25	EST	NEELY,WB & BLAU,GE (1985)
DC	pKa			
HL 2.63E-017	atm m3/mol	25	EST	MEYLAN,WM & HOWARD,PH (1991)
OH 1.29E-011	cm3/molc sec	25	EST	MEYLAN,WM & HOWARD,PH (1993)

CAS #: 000606-20-2 — 1,3-DINO2 2-METHYL BENZENE

Formula: $C_7H_6N_2O_4$

Mol Weight: 182.14

MP (deg C): 71

FP (deg C):

BP (deg C): 300

BP pressure (mm Hg):

Property/Value	Units	Temp	Data Type	Reference
WS 1.82E+002	mg/L	20	EXP	SPANGGORD,RJ ET AL. (1980)
logP 2.10			EXP	NAKAGAWA,Y ET AL. (1992)
VP 5.67E-004	mm Hg	25	EXP	PELLA,PA (1977)
DC 1.80	pKa		EXP	SPANGGORD,RJ ET AL. (1980)
HL 7.47E-007	atm m3/mol	25	EST	VP/WSOL
OH 2.25E-013	cm3/molc sec	25	EST	MEYLAN,WM & HOWARD,PH (1993)

CAS #: 000606-22-4 — 2,6-DINITROANILINE

Formula: $C_6H_5N_3O_4$

Mol Weight: 183.12

MP (deg C): 139-140

FP (deg C):

BP (deg C):

BP pressure (mm Hg):

Property/Value	Units	Temp	Data Type	Reference
WS 1.57E+003	mg/L	25	EST	MEYLAN,WM ET AL. (1996)
logP 1.79			EXP	HANSCH,C ET AL. (1995)
VP 2.68E-005	mm Hg	25	EST	NEELY,WB & BLAU,GE (1985)
DC -5.00	pKa		EXP	PERRIN,DD (1972)
HL 1.75E-007	atm m3/mol	25	EST	MEYLAN,WM & HOWARD,PH (1991)
OH 1.68E-012	cm3/molc sec	25	EST	MEYLAN,WM & HOWARD,PH (1993)

CAS #: 000606-23-5				1,3-INDANDIONE

Formula: $C_9H_6O_2$

Mol Weight: 146.15

MP (deg C): 131-132 de

FP (deg C):

BP (deg C):

BP pressure (mm Hg):

Property/Value	Units	Temp	Data Type	Reference
WS 2.38E+004	mg/L	25	EST	MEYLAN,WM ET AL. (1996)
logP 0.61			EXP	HANSCH,C & LEO,AJ (1985)
VP 1.72E-003	mm Hg	25	EST	NEELY,WB & BLAU,GE (1985)
DC	pKa			
HL 5.93E-009	atm m3/mol	25	EST	MEYLAN,WM & HOWARD,PH (1991)
OH 1.17E-012	cm3/molc sec	25	EST	MEYLAN,WM & HOWARD,PH (1993)

CAS #: 000606-25-7				1-NAPHTHALENESULFONAMIDE

Formula: $C_{10}H_9NO_2S$

Mol Weight: 207.25

MP (deg C):

FP (deg C):

BP (deg C):

BP pressure (mm Hg):

Property/Value	Units	Temp	Data Type	Reference
WS 1.91E+003	mg/L	25	EST	MEYLAN,WM ET AL. (1996)
logP 1.55			EXP	ALCORN,CJ ET AL. (1993)
VP 3.34E-006	mm Hg	25	EST	NEELY,WB & BLAU,GE (1985)
DC	pKa			
HL 4.12E-008	atm m3/mol	25	EST	MEYLAN,WM & HOWARD,PH (1991)
OH 4.62E-012	cm3/molc sec	25	EST	MEYLAN,WM & HOWARD,PH (1993)

CAS #: 000606-26-8				BENZOYLHYDRAZINE, O-NITRO

Formula: $C_7H_7N_3O_3$

Mol Weight: 181.15

MP (deg C): 123-125

FP (deg C):

BP (deg C):

BP pressure (mm Hg):

Property/Value	Units	Temp	Data Type	Reference
WS 3.14E+003	mg/L	25	EST	MEYLAN,WM ET AL. (1996)
logP -0.54			EXP	HANSCH,C & LEO,AJ (1985)
VP 1.08E-006	mm Hg	25	EST	NEELY,WB & BLAU,GE (1985)
DC	pKa			
HL 3.65E-014	atm m3/mol	25	EST	MEYLAN,WM & HOWARD,PH (1991)
OH 5.66E-012	cm3/molc sec	25	EST	MEYLAN,WM & HOWARD,PH (1993)

CAS #: 000606-27-9				BENZOIC ACID, 2-NITRO-, METHYL ESTER

Formula: $C_8H_7NO_4$

Mol Weight: 181.15

MP (deg C): -13

FP (deg C):

BP (deg C): 275

BP pressure (mm Hg):

Property/Value	Units	Temp	Data Type	Reference
WS 8.47E+002	mg/L	25	EST	MEYLAN,WM ET AL. (1996)
logP 1.66			EXP	SOTOMATSU,T ET AL. (1993)
VP 1.86E-003	mm Hg	25	EST	NEELY,WB & BLAU,GE (1985)
DC	pKa			
HL 1.37E-007	atm m3/mol	25	EST	MEYLAN,WM & HOWARD,PH (1991)
OH 2.72E-013	cm3/molc sec	25	EST	MEYLAN,WM & HOWARD,PH (1993)

CAS #: 000606-28-0				BENZOIC ACID, 2-BENZOYL-, METHYL ESTER

Formula: $C_{15}H_{12}O_3$

Mol Weight: 240.26

MP (deg C): 52

FP (deg C):

BP (deg C): 350-352

BP pressure (mm Hg):

Property/Value	Units	Temp	Data Type	Reference
WS 8.00E+001	mg/L	25	EXP	CHEM INSPECT TEST INST (1992)
logP 2.70			EXP	CHEM INSPECT TEST INST (1992)
VP 1.53E-005	mm Hg	25	EST	NEELY,WB & BLAU,GE (1985)
DC	pKa			
HL 1.25E-008	atm m3/mol	25	EST	MEYLAN,WM & HOWARD,PH (1991)
OH 2.39E-012	cm3/molc sec	25	EST	MEYLAN,WM & HOWARD,PH (1993)

CAS #: 000606-35-9				2,4,6-TRINITROANISOLE

Formula: $C_7H_5N_3O_7$

Mol Weight: 243.13

MP (deg C): 69

FP (deg C):

BP (deg C):

BP pressure (mm Hg):

Property/Value	Units	Temp	Data Type	Reference
WS 2.00E+002	mg/L	15	EXP	YALKOWSKY,SH & DANNENFELSER,RM (1992)
logP 1.53			EST	MEYLAN,WM & HOWARD,PH (1995)
VP 2.70E-006	mm Hg	25	EST	NEELY,WB & BLAU,GE (1985)
DC	pKa			
HL 1.96E-011	atm m3/mol	25	EST	MEYLAN,WM & HOWARD,PH (1991)
OH 8.31E-013	cm3/molc sec	25	EST	MEYLAN,WM & HOWARD,PH (1993)

CAS #: 000606-37-1				NAPHTHALENE, 1,3-DINITRO-

Formula: $C_{10}H_6N_2O_4$

Mol Weight: 218.17

MP (deg C): 148

FP (deg C):

BP (deg C):

BP pressure (mm Hg):

Property/Value	Units	Temp	Data Type	Reference
WS 5.49E+001	mg/L	25	EST	MEYLAN,WM ET AL. (1996)
logP 2.83			EXP	DEBNATH,AK ET AL. (1992)
VP 4.28E-006	mm Hg	25	EST	NEELY,WB & BLAU,GE (1985)
DC	pKa			
HL 8.19E-009	atm m3/mol	25	EST	MEYLAN,WM & HOWARD,PH (1991)
OH 2.36E-013	cm3/molc sec	25	EST	MEYLAN,WM & HOWARD,PH (1993)

CAS #: 000606-38-2				2,4-DIME 6-NO2 ACETANILIDE

Formula: $C_{10}H_{12}N_2O_3$

Mol Weight: 208.22

MP (deg C):

FP (deg C):

BP (deg C):

BP pressure (mm Hg):

Property/Value	Units	Temp	Data Type	Reference
WS 1.59E+003	mg/L	25	EST	MEYLAN,WM ET AL. (1996)
logP 1.18			EXP	NAKAGAWA,Y ET AL. (1992)
VP 8.25E-007	mm Hg	25	EST	NEELY,WB & BLAU,GE (1985)
DC	pKa			
HL 6.49E-010	atm m3/mol	25	EST	MEYLAN,WM & HOWARD,PH (1991)
OH 9.81E-013	cm3/molc sec	25	EST	MEYLAN,WM & HOWARD,PH (1993)

CAS #: 000606-43-9				1-METHYL-2(H)-QUINOLINONE

Formula: $C_{10}H_9NO$
Mol Weight: 159.19
MP (deg C): 74
FP (deg C):
BP (deg C): 325
BP pressure (mm Hg):

Property/ Value	Units	Temp	Data Type	Reference
WS 3.99E+003	mg/L	25	EST	MEYLAN,WM ET AL. (1996)
logP 1.45			EXP	HANSCH,C & LEO,AJ (1985)
VP 3.13E-004	mm Hg	25	EST	NEELY,WB & BLAU,GE (1985)
DC	pKa			
HL 1.36E-008	atm m3/mol	25	EST	MEYLAN,WM & HOWARD,PH (1991)
OH 2.35E-011	cm3/molc sec	25	EST	MEYLAN,WM & HOWARD,PH (1993)

CAS #: 000606-45-1				BENZOIC ACID, 2-METHOXY-, METHYL ESTER

Formula: $C_9H_{10}O_3$
Mol Weight: 166.18
MP (deg C):
FP (deg C):
BP (deg C): 246.5
BP pressure (mm Hg):

Property/ Value	Units	Temp	Data Type	Reference
WS 1.16E+003	mg/L	25	EST	MEYLAN,WM ET AL. (1996)
logP 2.04			EXP	SOTOMATSU,T ET AL. (1993)
VP 6.02E-002	mm Hg	25	EST	NEELY,WB & BLAU,GE (1985)
DC	pKa			
HL 2.05E-006	atm m3/mol	25	EST	MEYLAN,WM & HOWARD,PH (1991)
OH 7.96E-012	cm3/molc sec	25	EST	MEYLAN,WM & HOWARD,PH (1993)

CAS #: 000607-12-5				4-CHLORO-2-PHENYLPHENOL

Formula: $C_{12}H_9ClO$
Mol Weight: 204.66
MP (deg C):
FP (deg C):
BP (deg C):
BP pressure (mm Hg):

Property/ Value	Units	Temp	Data Type	Reference
WS 7.05E+001	mg/L	25	EST	MEYLAN,WM ET AL. (1996)
logP 3.92			EST	MEYLAN,WM & HOWARD,PH (1995)
VP 1.07E-005	mm Hg	25	EST	NEELY,WB & BLAU,GE (1985)
DC	pKa			
HL 3.19E-008	atm m3/mol	25	EST	MEYLAN,WM & HOWARD,PH (1991)
OH 9.59E-012	cm3/molc sec	25	EST	MEYLAN,WM & HOWARD,PH (1993)

CAS #: 000607-34-1				5-NITROQUINOLINE

Formula: $C_9H_6N_2O_2$
Mol Weight: 174.16
MP (deg C): 74
FP (deg C):
BP (deg C):
BP pressure (mm Hg):

Property/ Value	Units	Temp	Data Type	Reference
WS 1.53E+003	mg/L	25	EST	MEYLAN,WM ET AL. (1996)
logP 1.86			EXP	HANSCH,C & LEO,AJ (1985)
VP 1.38E-004	mm Hg	25	EST	NEELY,WB & BLAU,GE (1985)
DC	pKa			
HL 2.72E-009	atm m3/mol	25	EST	MEYLAN,WM & HOWARD,PH (1991)
OH 1.80E-012	cm3/molc sec	25	EST	MEYLAN,WM & HOWARD,PH (1993)

CAS #: 000607-35-2				8-NITROQUINOLINE

Formula: $C_9H_6N_2O_2$
Mol Weight: 174.16
MP (deg C): 89-91
FP (deg C):
BP (deg C):
BP pressure (mm Hg):

Property/ Value	Units	Temp	Data Type	Reference
WS 1.53E+003	mg/L	25	EST	MEYLAN,WM ET AL. (1996)
logP 1.40			EXP	HANSCH,C & LEO,AJ (1985)
VP 1.38E-004	mm Hg	25	EST	NEELY,WB & BLAU,GE (1985)
DC	pKa			
HL 2.72E-009	atm m3/mol	25	EST	MEYLAN,WM & HOWARD,PH (1991)
OH 1.80E-012	cm3/molc sec	25	EST	MEYLAN,WM & HOWARD,PH (1993)

CAS #: 000607-57-8				2-NITROFLUORENE

Formula: $C_{13}H_9NO_2$
Mol Weight: 211.22
MP (deg C): 156-158
FP (deg C):
BP (deg C):
BP pressure (mm Hg):

Property/ Value	Units	Temp	Data Type	Reference
WS 2.16E-001	mg/L	25	EXP	GANG,Y & XIABAI,XU (1992)
logP 3.37			EXP	DEBNATH,AK & HANSCH,C (1992)
VP 9.54E-006	mm Hg	25	EST	NEELY,WB & BLAU,GE (1985)
DC	pKa			
HL 6.60E-007	atm m3/mol	25	EST	MEYLAN,WM & HOWARD,PH (1991)
OH 4.31E-012	cm3/molc sec	25	EST	MEYLAN,WM & HOWARD,PH (1993)

CAS #: 000607-81-8				BENZYLMALONIC ACID, DIETHYL ESTER

Formula: $C_{14}H_{18}O_4$
Mol Weight: 250.30
MP (deg C):
FP (deg C):
BP (deg C): 300
BP pressure (mm Hg):

Property/ Value	Units	Temp	Data Type	Reference
WS 1.04E+002	mg/L	25	EST	MEYLAN,WM ET AL. (1996)
logP 2.76			EXP	HANSCH,C & LEO,AJ (1985)
VP 1.64E-003	mm Hg	25	EST	NEELY,WB & BLAU,GE (1985)
DC	pKa			
HL 7.88E-008	atm m3/mol	25	EST	MEYLAN,WM & HOWARD,PH (1991)
OH 1.06E-011	cm3/molc sec	25	EST	MEYLAN,WM & HOWARD,PH (1993)

CAS #: 000607-90-9				BENZOIC ACID, 2-HYDROXY-, PROPYL ESTER

Formula: $C_{10}H_{12}O_3$
Mol Weight: 180.21
MP (deg C): 97
FP (deg C):
BP (deg C): 239
BP pressure (mm Hg):

Property/ Value	Units	Temp	Data Type	Reference
WS 1.19E+002	mg/L	25	EST	MEYLAN,WM ET AL. (1996)
logP 3.80			EXP	KORENMAN,YI & DANILOV,VN (1990)
VP 5.55E-004	mm Hg	25	EST	NEELY,WB & BLAU,GE (1985)
DC	pKa			
HL 8.01E-006	atm m3/mol	25	EST	MEYLAN,WM & HOWARD,PH (1991)
OH 1.41E-011	cm3/molc sec	25	EST	MEYLAN,WM & HOWARD,PH (1993)

CAS #: 000607-91-0 — MYRISTICIN

Formula: $C_{11}H_{12}O_3$
Mol Weight: 192.22
MP (deg C): <-20
FP (deg C):
BP (deg C): 276.5
BP pressure (mm Hg):

Property/Value	Units	Temp	Data Type	Reference
WS 4.64E+001	mg/L	25	EST	MEYLAN,WM ET AL. (1996)
logP 3.53			EST	MEYLAN,WM & HOWARD,PH (1995)
VP 1.37E+004	mm Hg	100	EXP	LIDE,DR (1994)
DC	pKa			
HL 5.37E-007	atm m3/mol	25	EST	MEYLAN,WM & HOWARD,PH (1991)
OH 2.31E-010	cm3/molc sec	25	EST	MEYLAN,WM & HOWARD,PH (1993)

CAS #: 000607-99-8 — 2,4,6-TRIBROMOANISOLE

Formula: $C_7H_5Br_3O$
Mol Weight: 344.84
MP (deg C): 88
FP (deg C):
BP (deg C): 298
BP pressure (mm Hg):

Property/Value	Units	Temp	Data Type	Reference
WS 9.94E-001	mg/L	25	EST	MEYLAN,WM ET AL. (1996)
logP 4.48			EXP	SANGSTER,J (1993)
VP 6.44E-004	mm Hg	25	EST	NEELY,WB & BLAU,GE (1985)
DC	pKa			
HL 2.02E-005	atm m3/mol	25	EST	MEYLAN,WM & HOWARD,PH (1991)
OH 1.25E-012	cm3/molc sec	25	EST	MEYLAN,WM & HOWARD,PH (1993)

CAS #: 000608-08-2 — INDOLE-3-OL,ACETATE ESTER

Formula: $C_{10}H_9NO_2$
Mol Weight: 175.19
MP (deg C): 128-130
FP (deg C):
BP (deg C):
BP pressure (mm Hg):

Property/Value	Units	Temp	Data Type	Reference
WS 1.12E+003	mg/L	25	EST	MEYLAN,WM ET AL. (1996)
logP 2.01			EXP	HANSCH,C ET AL. (1995)
VP 2.09E-004	mm Hg	25	EST	NEELY,WB & BLAU,GE (1985)
DC	pKa			
HL 1.07E-008	atm m3/mol	25	EST	MEYLAN,WM & HOWARD,PH (1991)
OH 1.14E-010	cm3/molc sec	25	EST	MEYLAN,WM & HOWARD,PH (1993)

CAS #: 000608-21-9 — 1,2,3-TRIBROMOBENZENE

Formula: $C_6H_3Br_3$
Mol Weight: 314.82
MP (deg C): 87.5
FP (deg C):
BP (deg C):
BP pressure (mm Hg):

Property/Value	Units	Temp	Data Type	Reference
WS 2.87E+000	mg/L	25	EXP	KUHNE,R ET AL. (1995)
logP 4.66			EST	MEYLAN,WM & HOWARD,PH (1995)
VP 5.48E-003	mm Hg	25	EST	NEELY,WB & BLAU,GE (1985)
DC	pKa			
HL 3.41E-004	atm m3/mol	25	EST	MEYLAN,WM & HOWARD,PH (1991)
OH 2.21E-013	cm3/molc sec	25	EST	MEYLAN,WM & HOWARD,PH (1993)

CAS #: 000608-27-5 — 2,3-DICHLOROANILINE

Formula: $C_6H_5Cl_2N$
Mol Weight: 162.02
MP (deg C): 24
FP (deg C):
BP (deg C): 252
BP pressure (mm Hg):

Property/Value	Units	Temp	Data Type	Reference
WS 2.62E+002	mg/L	25	EST	MEYLAN,WM ET AL. (1996)
logP 2.82			EXP	SANGSTER,J (1994)
VP 2.12E-002	mm Hg	25	EST	NEELY,WB & BLAU,GE (1985)
DC 1.76	pKa	25	EXP	PERRIN,DD (1972)
HL 1.05E-006	atm m3/mol	25	EST	MEYLAN,WM & HOWARD,PH (1991)
OH 2.21E-011	cm3/molc sec	25	EST	MEYLAN,WM & HOWARD,PH (1993)

CAS #: 000608-31-1 — 2,6-DICHLOROANILINE

Formula: $C_6H_5Cl_2N$
Mol Weight: 162.02
MP (deg C): 39
FP (deg C):
BP (deg C):
BP pressure (mm Hg):

Property/Value	Units	Temp	Data Type	Reference
WS 2.95E+002	mg/L	25	EST	MEYLAN,WM ET AL. (1996)
logP 2.76			EXP	SANGSTER,J (1994)
VP 2.12E-002	mm Hg	25	EST	NEELY,WB & BLAU,GE (1985)
DC 0.42	pKa	25	EXP	PERRIN,DD (1972)
HL 1.05E-006	atm m3/mol	25	EST	MEYLAN,WM & HOWARD,PH (1991)
OH 9.18E-012	cm3/molc sec	25	EST	MEYLAN,WM & HOWARD,PH (1993)

CAS #: 000608-33-3 — 2,6-DIBROMOPHENOL

Formula: $C_6H_4Br_2O$
Mol Weight: 251.92
MP (deg C): 56.5
FP (deg C):
BP (deg C): 255
BP pressure (mm Hg):

Property/Value	Units	Temp	Data Type	Reference
WS 1.19E+002	mg/L	25	EST	MEYLAN,WM ET AL. (1996)
logP 3.36			EXP	SOTOMATSU,T ET AL. (1993)
VP 1.34E-003	mm Hg	25	EST	NEELY,WB & BLAU,GE (1985)
DC 6.67	pKa	25	EXP	SERJEANT,EP & DEMPSEY,B (1979)
HL 8.90E-008	atm m3/mol	25	EST	MEYLAN,WM & HOWARD,PH (1991)
OH 2.88E-012	cm3/molc sec	25	EST	MEYLAN,WM & HOWARD,PH (1993)

CAS #: 000608-43-5 — 1,4-BENZENEDIOL, 2,3-DIMETHYL-

Formula: $C_8H_{10}O_2$
Mol Weight: 138.17
MP (deg C): 223-225
FP (deg C):
BP (deg C):
BP pressure (mm Hg):

Property/Value	Units	Temp	Data Type	Reference
WS 2.23E+004	mg/L	25	EST	MEYLAN,WM ET AL. (1996)
logP 1.36			EXP	SANGSTER,J (1994)
VP 7.39E-004	mm Hg	25	EST	NEELY,WB & BLAU,GE (1985)
DC	pKa			
HL 7.10E-011	atm m3/mol	25	EST	MEYLAN,WM & HOWARD,PH (1991)
OH 7.40E-011	cm3/molc sec	25	EST	MEYLAN,WM & HOWARD,PH (1993)

CAS #: 000608-66-2				GALACTITOL

Formula: $C_6H_{14}O_6$

Mol Weight: 182.17

MP (deg C): 188-189 FP (deg C):

BP (deg C):

BP pressure (mm Hg):

Property/Value	Units	Temp	Data Type	Reference
WS 3.10E+004	mg/L	15	EXP	YALKOWSKY,SH & DANNENFELSER,RM (1992)
logP -3.10			EXP	HANSCH,C & LEO,AJ (1985)
VP 4.92E-009	mm Hg	25	EST	NEELY,WB & BLAU,GE (1985)
DC	pKa			
HL 7.26E-013	atm m3/mol	25	EST	MEYLAN,WM & HOWARD,PH (1991)
OH 5.00E-011	cm3/molc sec	25	EST	MEYLAN,WM & HOWARD,PH (1993)

CAS #: 000608-68-4				L-DIMETHYL TARTRATE

Formula: $C_6H_{10}O_6$

Mol Weight: 178.14

MP (deg C): 57-60 dec FP (deg C):

BP (deg C):

BP pressure (mm Hg):

Property/Value	Units	Temp	Data Type	Reference
WS 4.57E+005	mg/L	25	EST	MEYLAN,WM ET AL. (1996)
logP -1.07			EST	MEYLAN,WM & HOWARD,PH (1995)
VP 6.47E-004	mm Hg	25	EXP	JONES,AH (1960)
DC	pKa			
HL 1.35E-005	atm m3/mol	25	EST	MEYLAN,WM & HOWARD,PH (1991)
OH 1.31E-011	cm3/molc sec	25	EST	MEYLAN,WM & HOWARD,PH (1993)

CAS #: 000608-71-9				PENTABROMOPHENOL

Formula: C_6HBr_5O

Mol Weight: 488.62

MP (deg C): 229.5 FP (deg C):

BP (deg C):

BP pressure (mm Hg):

Property/Value	Units	Temp	Data Type	Reference
WS 2.61E-002	mg/L	25	EST	MEYLAN,WM ET AL. (1996)
logP 5.96			EST	MEYLAN,WM & HOWARD,PH (1995)
VP 3.85E-007	mm Hg	25	EST	NEELY,WB & BLAU,GE (1985)
DC	pKa			
HL 5.63E-009	atm m3/mol	25	EST	MEYLAN,WM & HOWARD,PH (1991)
OH 4.51E-013	cm3/molc sec	25	EST	MEYLAN,WM & HOWARD,PH (1993)

CAS #: 000608-73-1				1,2,3,4,5,6-HEXACHLOROCYCLOHEXANE

Formula: $C_6H_6Cl_6$

Mol Weight: 290.83

MP (deg C): 112.5 FP (deg C):

BP (deg C):

BP pressure (mm Hg):

Property/Value	Units	Temp	Data Type	Reference
WS 8.00E+000	mg/L	25	EXP	CHEM INSPECT TEST INST (1992)
logP 4.26			EST	MEYLAN,WM & HOWARD,PH (1995)
VP 7.83E-004	mm Hg	25	EST	NEELY,WB & BLAU,GE (1985)
DC	pKa			
HL 2.56E-004	atm m3/mol	25	EST	MEYLAN,WM & HOWARD,PH (1991)
OH 5.73E-013	cm3/molc sec	25	EST	MEYLAN,WM & HOWARD,PH (1993)

CAS #: 000608-90-2				PENTABROMOBENZENE

Formula: C_6HBr_5

Mol Weight: 472.62

MP (deg C): 160.5 FP (deg C):

BP (deg C):

BP pressure (mm Hg):

Property/Value	Units	Temp	Data Type	Reference
WS 3.37E-003	mg/L	25	EST	MEYLAN,WM ET AL. (1996)
logP 6.44			EST	MEYLAN,WM & HOWARD,PH (1995)
VP 2.63E-005	mm Hg	25	EST	NEELY,WB & BLAU,GE (1985)
DC	pKa			
HL 5.41E-005	atm m3/mol	25	EST	MEYLAN,WM & HOWARD,PH (1991)
OH 4.00E-014	cm3/molc sec	25	EST	MEYLAN,WM & HOWARD,PH (1993)

CAS #: 000608-93-5				PENTACHLOROBENZENE

Formula: C_6HCl_5

Mol Weight: 250.34

MP (deg C): 86 FP (deg C):

BP (deg C): 277

BP pressure (mm Hg):

Property/Value	Units	Temp	Data Type	Reference
WS 1.33E+000	mg/L	25	EXP	YALKOWSKY,SH & DANNENFELSER,RM (1992)
logP 5.17			EXP	HANSCH,C & LEO,AJ (1985)
VP 2.19E-003	mm Hg	25	EXT	MACKAY,D & SHIU,WY (1981)
DC	pKa			
HL 7.10E-004	atm m3/mol	20	EXP	OLIVER,BG (1985)
OH 5.90E-014	cm3/molc sec	25	EST	MEYLAN,WM & HOWARD,PH (1993)

CAS #: 000609-19-8				3,4,5-TRICHLOROPHENOL

Formula: $C_6H_3Cl_3O$

Mol Weight: 197.45

MP (deg C): 101 FP (deg C):

BP (deg C): 275

BP pressure (mm Hg):

Property/Value	Units	Temp	Data Type	Reference
WS 6.45E+001	mg/L	25	EST	MEYLAN,WM ET AL. (1996)
logP 4.01			EXP	HANSCH,C ET AL. (1995)
VP 2.46E-003	mm Hg	25	EST	NEELY,WB & BLAU,GE (1985)
DC 7.84	pKa	25	EXP	SERJEANT,EP & DEMPSEY,B (1979)
HL 2.28E-007	atm m3/mol	25	EST	MEYLAN,WM & HOWARD,PH (1991)
OH 4.96E-012	cm3/molc sec	25	EST	MEYLAN,WM & HOWARD,PH (1993)

CAS #: 000609-23-4				2,4,6-TRIIODOPHENOL

Formula: $C_6H_3I_3O$

Mol Weight: 471.80

MP (deg C): 159.8 FP (deg C):

BP (deg C):

BP pressure (mm Hg):

Property/Value	Units	Temp	Data Type	Reference
WS 2.15E-001	mg/L	25	EST	MEYLAN,WM ET AL. (1996)
logP 5.01			EST	MEYLAN,WM & HOWARD,PH (1995)
VP 1.03E-006	mm Hg	25	EST	NEELY,WB & BLAU,GE (1985)
DC	pKa			
HL 6.98E-009	atm m3/mol	25	EST	MEYLAN,WM & HOWARD,PH (1991)
OH 5.25E-013	cm3/molc sec	25	EST	MEYLAN,WM & HOWARD,PH (1993)

CAS #: 000609-26-7				2-METHYL-3-ETHYLPENTANE
Formula: C$_8$H$_{18}$				
Mol Weight: 114.23				
MP (deg C): -114.9		FP (deg C):		
BP (deg C): 115.6				
BP pressure (mm Hg):				

Property/ Value	Units	Temp	Data Type	Reference
WS 9.20E+000	mg/L	25	EST	MEYLAN,WM ET AL. (1996)
logP 4.12			EST	MEYLAN,WM & HOWARD,PH (1995)
VP 2.40E+001	mm Hg	25	EXP	DAUBERT,TE & DANNER,RP (1989)
DC	pKa			
HL 3.01E+000	atm m3/mol	25	EST	MEYLAN,WM & HOWARD,PH (1991)
OH 8.96E-012	cm3/molc sec	25	EST	MEYLAN,WM & HOWARD,PH (1993)

CAS #: 000609-31-4				2-NITRO-1-BUTANOL
Formula: C$_4$H$_9$NO$_3$				
Mol Weight: 119.12				
MP (deg C): -47		FP (deg C):		
BP (deg C): 105				
BP pressure (mm Hg): 1.00E+001				

Property/ Value	Units	Temp	Data Type	Reference
WS 5.40E+004	mg/L	20	EXP	DEWEY,RH & BOLLMEIER,AFJR (1978)
logP -0.10			EST	MEYLAN,WM & HOWARD,PH (1995)
VP 4.70E-002	mm Hg	25	EST	NEELY,WB & BLAU,GE (1985)
DC	pKa			
HL 3.60E-009	atm m3/mol	25	EST	MEYLAN,WM & HOWARD,PH (1991)
OH 1.03E-012	cm3/molc sec	25	EST	MEYLAN,WM & HOWARD,PH (1993)

CAS #: 000609-35-8				FURAN-3-CARBOXAMIDE
Formula: C$_5$H$_6$NO$_2$				
Mol Weight: 112.11				
MP (deg C):		FP (deg C):		
BP (deg C):				
BP pressure (mm Hg):				

Property/ Value	Units	Temp	Data Type	Reference
WS 9.04E+004	mg/L	25	EST	MEYLAN,WM ET AL. (1996)
logP 0.09			EXP	HANSCH,C & LEO,AJ (1985)
VP 5.14E-003	mm Hg	25	EST	NEELY,WB & BLAU,GE (1985)
DC	pKa			
HL 2.21E-009	atm m3/mol	25	EST	MEYLAN,WM & HOWARD,PH (1991)
OH 3.89E-011	cm3/molc sec	25	EST	MEYLAN,WM & HOWARD,PH (1993)

CAS #: 000609-38-1				FURAN-2-CARBOXAMIDE
Formula: C$_5$H$_5$NO$_2$				
Mol Weight: 111.10				
MP (deg C):		FP (deg C):		
BP (deg C):				
BP pressure (mm Hg):				

Property/ Value	Units	Temp	Data Type	Reference
WS 1.34E+005	mg/L	25	EST	MEYLAN,WM ET AL. (1996)
logP -0.11			EXP	HANSCH,C & LEO,AJ (1985)
VP 5.14E-003	mm Hg	25	EST	NEELY,WB & BLAU,GE (1985)
DC	pKa			
HL 2.21E-009	atm m3/mol	25	EST	MEYLAN,WM & HOWARD,PH (1991)
OH 3.89E-011	cm3/molc sec	25	EST	MEYLAN,WM & HOWARD,PH (1993)

CAS #: 000609-39-2				2-NITROFURAN
Formula: C$_4$H$_4$NO$_3$				
Mol Weight: 114.08				
MP (deg C): 30		FP (deg C):		
BP (deg C): 134				
BP pressure (mm Hg): 1.23E+002				

Property/ Value	Units	Temp	Data Type	Reference
WS 1.18E+004	mg/L	25	EST	MEYLAN,WM ET AL. (1996)
logP 0.66			EXP	HANSCH,C ET AL. (1995)
VP 1.47E+000	mm Hg	25	EST	NEELY,WB & BLAU,GE (1985)
DC	pKa			
HL 2.12E-005	atm m3/mol	25	EST	MEYLAN,WM & HOWARD,PH (1991)
OH 5.06E-012	cm3/molc sec	25	EST	MEYLAN,WM & HOWARD,PH (1993)

CAS #: 000609-40-5				2-NITROTHIOPHENE
Formula: C$_4$H$_3$NO$_2$S				
Mol Weight: 129.14				
MP (deg C): 46.5		FP (deg C):		
BP (deg C): 224.5				
BP pressure (mm Hg):				

Property/ Value	Units	Temp	Data Type	Reference
WS 1.79E+003	mg/L	25	EST	MEYLAN,WM ET AL. (1996)
logP 1.55			EXP	HANSCH,C & LEO,AJ (1985)
VP 1.14E-001	mm Hg	25	EST	NEELY,WB & BLAU,GE (1985)
DC	pKa			
HL 1.16E-005	atm m3/mol	25	EST	MEYLAN,WM & HOWARD,PH (1991)
OH 1.19E-012	cm3/molc sec	25	EST	MEYLAN,WM & HOWARD,PH (1993)

CAS #: 000609-66-5				2-CHLOROBENZAMIDE
Formula: C$_7$H$_6$ClNO				
Mol Weight: 155.58				
MP (deg C): 142-144		FP (deg C):		
BP (deg C):				
BP pressure (mm Hg):				

Property/ Value	Units	Temp	Data Type	Reference
WS 2.04E+004	mg/L	25	EST	MEYLAN,WM ET AL. (1996)
logP 0.64			EXP	NAKAGAWA,Y ET AL. (1992)
VP 1.65E-004	mm Hg	25	EST	NEELY,WB & BLAU,GE (1985)
DC	pKa			
HL 1.64E-009	atm m3/mol	25	EST	MEYLAN,WM & HOWARD,PH (1991)
OH 3.25E-012	cm3/molc sec	25	EST	MEYLAN,WM & HOWARD,PH (1993)

CAS #: 000609-72-3				N,N-DIMETHYL-O-TOLUIDINE
Formula: C$_9$H$_{13}$N				
Mol Weight: 135.21				
MP (deg C): -60		FP (deg C):		
BP (deg C): 194.1				
BP pressure (mm Hg):				

Property/ Value	Units	Temp	Data Type	Reference
WS 3.23E+002	mg/L	25	EST	MEYLAN,WM ET AL. (1996)
logP 2.85			EXP	HANSCH,C & LEO,AJ (1985)
VP 8.62E-001	mm Hg	25	EXT	CHAO,J ET AL. (1983)
DC 5.94	pKa	20	EXP	PERRIN,DD (1965)
HL 9.45E-005	atm m3/mol	25	EST	MEYLAN,WM & HOWARD,PH (1991)
OH 2.03E-010	cm3/molc sec	25	EST	MEYLAN,WM & HOWARD,PH (1993)

CAS #: 000609-89-2				2,4-DICHLORO-6-NITROPHENOL
Formula: $C_6H_3Cl_2NO_3$				
Mol Weight: 208.00				
MP (deg C): 122		FP (deg C):		
BP (deg C):				
BP pressure (mm Hg):				

Property/Value	Units	Temp	Data Type	Reference
WS 1.70E+001	mg/L	25	EXP	CHEM INSPECT TEST INST
logP 3.20			EST	MEYLAN,WM & HOWARD,PH (1995)
VP 5.70E-005	mm Hg	25	EST	NEELY,WB & BLAU,GE (1985)
DC	pKa			
HL 3.84E-006	atm m3/mol	25	EST	MEYLAN,WM & HOWARD,PH (1991)
OH 1.98E-013	cm3/molc sec	25	EST	MEYLAN,WM & HOWARD,PH (1993)

CAS #: 000609-93-8				2,6-DINITRO-P-CRESOL
Formula: $C_7H_6N_2O_5$				
Mol Weight: 198.14				
MP (deg C): 77-79		FP (deg C):		
BP (deg C):				
BP pressure (mm Hg):				

Property/Value	Units	Temp	Data Type	Reference
WS 5.12E+002	mg/L	25	EST	MEYLAN,WM ET AL. (1996)
logP 2.27			EST	MEYLAN,WM & HOWARD,PH (1995)
VP 5.48E-006	mm Hg	25	EST	NEELY,WB & BLAU,GE (1985)
DC 4.23	pKa	25	EXP	SERJEANT,EP & DEMPSEY,B (1979)
HL 3.05E-008	atm m3/mol	25	EST	MEYLAN,WM & HOWARD,PH (1991)
OH 3.03E-013	cm3/molc sec	25	EST	MEYLAN,WM & HOWARD,PH (1993)

CAS #: 000609-99-4				BENZOIC ACID, 2-HYDROXY-3,5-DINITRO-
Formula: $C_7H_4N_2O_7$				
Mol Weight: 228.12				
MP (deg C): 169-172		FP (deg C):		
BP (deg C):				
BP pressure (mm Hg):				

Property/Value	Units	Temp	Data Type	Reference
WS 1.08E+003	mg/L	25	EST	MEYLAN,WM ET AL. (1996)
logP 1.71			EXP	SANGSTER,J (1994)
VP 2.05E-008	mm Hg	25	EST	NEELY,WB & BLAU,GE (1985)
DC 0.70	pKa	25	EXP	SERJEANT,EP & DEMPSEY,B (1979)
HL 6.99E-010	atm m3/mol	25	EST	MEYLAN,WM & HOWARD,PH (1991)
OH 6.63E-013	cm3/molc sec	25	EST	MEYLAN,WM & HOWARD,PH (1993)

CAS #: 000610-02-6				2,3,4-TRIHYDROXYBENZOIC ACID
Formula: $C_7H_6O_5$				
Mol Weight: 170.12				
MP (deg C): 221		FP (deg C):		
BP (deg C):				
BP pressure (mm Hg):				

Property/Value	Units	Temp	Data Type	Reference
WS 1.00E+003	mg/L	25	EXP	YALKOWSKY,SH & DANNENFELSER,RM (1992)
logP 1.05			EXP	HANSCH,C & LEO,AJ (1985)
VP 1.19E-007	mm Hg	25	EST	NEELY,WB & BLAU,GE (1985)
DC	pKa			
HL 1.54E-016	atm m3/mol	25	EST	MEYLAN,WM & HOWARD,PH (1991)
OH 1.46E-010	cm3/molc sec	25	EST	MEYLAN,WM & HOWARD,PH (1993)

CAS #: 000610-15-1				2-NITROBENZAMIDE
Formula: $C_7H_6N_2O_3$				
Mol Weight: 166.14				
MP (deg C): 176.6		FP (deg C):		
BP (deg C): 317				
BP pressure (mm Hg):				

Property/Value	Units	Temp	Data Type	Reference
WS 1.62E+003	mg/L	25	EST	MEYLAN,WM ET AL. (1996)
logP -0.12			EXP	SOTOMATSU,T ET AL. (1993)
VP 8.09E-006	mm Hg	25	EST	NEELY,WB & BLAU,GE (1985)
DC	pKa			
HL 8.74E-012	atm m3/mol	25	EST	MEYLAN,WM & HOWARD,PH (1991)
OH 2.16E-012	cm3/molc sec	25	EST	MEYLAN,WM & HOWARD,PH (1993)

CAS #: 000610-30-0				2,4-DINITROBENZOIC ACID
Formula: $C_7H_4N_2O_6$				
Mol Weight: 212.12				
MP (deg C): 181-183		FP (deg C):		
BP (deg C):				
BP pressure (mm Hg):				

Property/Value	Units	Temp	Data Type	Reference
WS 1.82E+004	mg/L	25	EXP	YALKOWSKY,SH & DANNENFELSER,RM (1992)
logP 1.17			EST	MEYLAN,WM & HOWARD,PH (1995)
VP 1.41E-006	mm Hg	25	EST	NEELY,WB & BLAU,GE (1985)
DC 1.42	pKa	25	EXP	KORTUM,G ET AL (1961)
HL 1.69E-012	atm m3/mol	25	EST	MEYLAN,WM & HOWARD,PH (1991)
OH 5.28E-013	cm3/molc sec	25	EST	MEYLAN,WM & HOWARD,PH (1993)

CAS #: 000610-39-9				1,2-DINO2 4-METHYL BENZENE
Formula: $C_7H_6N_2O_4$				
Mol Weight: 182.14				
MP (deg C): 58.3		FP (deg C):		
BP (deg C): 284				
BP pressure (mm Hg):				

Property/Value	Units	Temp	Data Type	Reference
WS 1.22E+002	mg/L	25	EST	MEYLAN,WM ET AL. (1996)
logP 2.08			EXP	NAKAGAWA,Y ET AL. (1992)
VP 3.97E-004	mm Hg	25	EST	NEELY,WB & BLAU,GE (1985)
DC	pKa			
HL 9.26E-008	atm m3/mol	25	EST	MEYLAN,WM & HOWARD,PH (1991)
OH 2.01E-013	cm3/molc sec	25	EST	MEYLAN,WM & HOWARD,PH (1993)

CAS #: 000610-40-2				1,2-DINO2 4-CL BENZENE
Formula: $C_6H_3ClN_2O_4$				
Mol Weight: 202.55				
MP (deg C):		FP (deg C):		
BP (deg C):				
BP pressure (mm Hg):				

Property/Value	Units	Temp	Data Type	Reference
WS 2.38E+002	mg/L	25	EST	MEYLAN,WM ET AL. (1996)
logP 2.18			EXP	NAKAGAWA,Y ET AL. (1992)
VP 2.05E-004	mm Hg	25	EST	NEELY,WB & BLAU,GE (1985)
DC	pKa			
HL 6.22E-008	atm m3/mol	25	EST	MEYLAN,WM & HOWARD,PH (1991)
OH 1.50E-014	cm3/molc sec	25	EST	MEYLAN,WM & HOWARD,PH (1993)

440

CAS #: 000610-48-0				1-METHYLANTHRACENE

Formula: $C_{15}H_{12}$

Mol Weight: 192.26

MP (deg C): 85.5 FP (deg C):

BP (deg C): 199.5

BP pressure (mm Hg):

Property/Value	Units	Temp	Data Type	Reference
WS 2.47E-001	mg/L	25	EST	MEYLAN,WM ET AL. (1996)
logP 4.89			EST	MEYLAN,WM & HOWARD,PH (1995)
VP 5.01E-005	mm Hg	25	EST	NEELY,WB & BLAU,GE (1985)
DC	pKa			
HL 5.67E-005	atm m3/mol	25	EST	MEYLAN,WM & HOWARD,PH (1991)
OH 1.05E-010	cm3/molc sec	25	EST	MEYLAN,WM & HOWARD,PH (1993)

CAS #: 000610-49-1				1-ANTHRACENAMINE

Formula: $C_{14}H_{11}N$

Mol Weight: 193.25

MP (deg C): 114-118 FP (deg C):

BP (deg C):

BP pressure (mm Hg):

Property/Value	Units	Temp	Data Type	Reference
WS 2.59E+000	mg/L	25	EST	MEYLAN,WM ET AL. (1996)
logP 3.69			EXP	DEBNATH,AK ET AL. (1992)
VP 4.05E-006	mm Hg	25	EST	NEELY,WB & BLAU,GE (1985)
DC 4.10	pKa	25	EXP	PERRIN,DD (1965)
HL 1.81E-008	atm m3/mol	25	EST	MEYLAN,WM & HOWARD,PH (1991)
OH 2.00E-010	cm3/molc sec	25	EST	MEYLAN,WM & HOWARD,PH (1993)

CAS #: 000610-54-8				2,4-DINITROPHENETOLE

Formula: $C_8H_8N_2O_5$

Mol Weight: 212.16

MP (deg C): FP (deg C):

BP (deg C):

BP pressure (mm Hg):

Property/Value	Units	Temp	Data Type	Reference
WS 2.04E+002	mg/L	25	EST	MEYLAN,WM ET AL. (1996)
logP 2.20			EST	MEYLAN,WM & HOWARD,PH (1995)
VP 5.60E-005	mm Hg	25	EST	NEELY,WB & BLAU,GE (1985)
DC	pKa			
HL 6.59E-009	atm m3/mol	25	EST	MEYLAN,WM & HOWARD,PH (1991)
OH 6.51E-012	cm3/molc sec	25	EST	MEYLAN,WM & HOWARD,PH (1993)

CAS #: 000610-69-5				O-NITROPHENYL ACETATE

Formula: $C_8H_7NO_4$

Mol Weight: 181.15

MP (deg C): 40.5 FP (deg C):

BP (deg C): 141

BP pressure (mm Hg): 1.10E+001

Property/Value	Units	Temp	Data Type	Reference
WS 1.05E+003	mg/L	25	EST	MEYLAN,WM ET AL. (1996)
logP 1.55			EXP	HANSCH,C & LEO,AJ (1985)
VP 2.18E-003	mm Hg	25	EXT	PERRY,RH & GREEN,D (1984)
DC	pKa			
HL 2.56E-007	atm m3/mol	25	EST	MEYLAN,WM & HOWARD,PH (1991)
OH 3.52E-013	cm3/molc sec	25	EST	MEYLAN,WM & HOWARD,PH (1993)

CAS #: 000610-94-6				METHYL 2-BROMOBENZOATE

Formula: $C_8H_7BrO_2$

Mol Weight: 215.05

MP (deg C): FP (deg C):

BP (deg C): 252

BP pressure (mm Hg):

Property/Value	Units	Temp	Data Type	Reference
WS 1.74E+002	mg/L	25	EST	MEYLAN,WM ET AL. (1996)
logP 2.72			EST	MEYLAN,WM & HOWARD,PH (1995)
VP 2.46E-002	mm Hg	25	EST	NEELY,WB & BLAU,GE (1985)
DC	pKa			
HL 1.38E-005	atm m3/mol	25	EST	MEYLAN,WM & HOWARD,PH (1991)
OH 6.12E-013	cm3/molc sec	25	EST	MEYLAN,WM & HOWARD,PH (1993)

CAS #: 000610-96-8				METHYL O-CHLOROBENZOATE

Formula: $C_8H_7ClO_2$

Mol Weight: 170.60

MP (deg C): FP (deg C):

BP (deg C):

BP pressure (mm Hg):

Property/Value	Units	Temp	Data Type	Reference
WS 5.67E+002	mg/L	25	EST	MEYLAN,WM ET AL. (1996)
logP 2.38			EXP	HANSCH,C & LEO,AJ (1985)
VP 9.07E-002	mm Hg	25	EST	NEELY,WB & BLAU,GE (1985)
DC	pKa			
HL 2.57E-005	atm m3/mol	25	EST	MEYLAN,WM & HOWARD,PH (1991)
OH 6.58E-013	cm3/molc sec	25	EST	MEYLAN,WM & HOWARD,PH (1993)

CAS #: 000610-97-9				BENZOIC ACID, 2-IODO-, METHYL ESTER

Formula: $C_8H_7IO_2$

Mol Weight: 262.05

MP (deg C): FP (deg C):

BP (deg C): 280

BP pressure (mm Hg):

Property/Value	Units	Temp	Data Type	Reference
WS 9.26E+001	mg/L	25	EST	MEYLAN,WM ET AL. (1996)
logP 2.74			EXP	SOTOMATSU,T ET AL. (1993)
VP 5.32E-003	mm Hg	25	EST	NEELY,WB & BLAU,GE (1985)
DC	pKa			
HL 8.05E-006	atm m3/mol	25	EST	MEYLAN,WM & HOWARD,PH (1991)
OH 6.31E-013	cm3/molc sec	25	EST	MEYLAN,WM & HOWARD,PH (1993)

CAS #: 000610-99-1				O-HYDROXYPROPIOPHENONE

Formula: $C_9H_{10}O_2$

Mol Weight: 150.18

MP (deg C): FP (deg C):

BP (deg C): 150

BP pressure (mm Hg): 8.00E+001

Property/Value	Units	Temp	Data Type	Reference
WS 1.95E+003	mg/L	25	EST	MEYLAN,WM ET AL. (1996)
logP 2.54			EXP	HANSCH,C & LEO,AJ (1985)
VP 2.26E-003	mm Hg	25	EST	NEELY,WB & BLAU,GE (1985)
DC	pKa			
HL 1.71E-006	atm m3/mol	25	EST	MEYLAN,WM & HOWARD,PH (1991)
OH 3.18E-011	cm3/molc sec	25	EST	MEYLAN,WM & HOWARD,PH (1993)

CAS #: 000611-03-0				BENZOIC ACID, 2,4-DIAMINO-

Formula: $C_7H_8N_2O_2$

Mol Weight: 152.15

MP (deg C): 140   FP (deg C):

BP (deg C):

BP pressure (mm Hg):

Property/Value	Units	Temp	Data Type	Reference
WS 1.37E+005	mg/L	25	EST	MEYLAN,WM ET AL. (1996)
logP -0.31			EXP	SANGSTER,J (1994)
VP 9.56E-006	mm Hg		EST	NEELY,WB & BLAU,GE (1985)
DC	pKa			
HL 1.35E-014	atm m3/mol	25	EST	MEYLAN,WM & HOWARD,PH (1991)
OH 2.01E-010	cm3/molc sec	25	EST	MEYLAN,WM & HOWARD,PH (1993)

CAS #: 000611-05-2				3-METHYL-4-NITROANILINE

Formula: $C_7H_8N_2O_2$

Mol Weight: 152.15

MP (deg C):   FP (deg C):

BP (deg C):

BP pressure (mm Hg):

Property/Value	Units	Temp	Data Type	Reference
WS 1.40E+003	mg/L	25	EST	MEYLAN,WM ET AL. (1996)
logP 2.02			EST	MEYLAN,WM & HOWARD,PH (1995)
VP 9.72E-004	mm Hg	25	EST	NEELY,WB & BLAU,GE (1985)
DC 1.64	pKa	20	EXP	PERRIN,DD (1965)
HL 8.29E-009	atm m3/mol	25	EST	MEYLAN,WM & HOWARD,PH (1991)
OH 3.53E-011	cm3/molc sec	25	EST	MEYLAN,WM & HOWARD,PH (1993)

CAS #: 000611-06-3				2,4-DICHLORONITROBENZENE

Formula: $C_6H_3Cl_2NO_2$

Mol Weight: 192.00

MP (deg C): 34   FP (deg C):

BP (deg C): 258.5

BP pressure (mm Hg):

Property/Value	Units	Temp	Data Type	Reference
WS 4.68E+001	mg/L	25	EST	MEYLAN,WM ET AL. (1996)
logP 3.07			EXP	NIIMI,AJ ET AL. (1989)
VP 5.05E-003	mm Hg	25	EST	NEELY,WB & BLAU,GE (1985)
DC	pKa			
HL 1.17E-005	atm m3/mol	25	EST	MEYLAN,WM & HOWARD,PH (1991)
OH 1.21E-013	cm3/molc sec	25	EST	MEYLAN,WM & HOWARD,PH (1993)

CAS #: 000611-07-4				5-CHLORO-2-NITROPHENOL

Formula: $C_6H_4ClNO_3$

Mol Weight: 173.56

MP (deg C): 41   FP (deg C):

BP (deg C):

BP pressure (mm Hg):

Property/Value	Units	Temp	Data Type	Reference
WS 3.91E+002	mg/L	25	EST	MEYLAN,WM ET AL. (1996)
logP 2.55			EST	MEYLAN,WM & HOWARD,PH (1995)
VP 2.96E-004	mm Hg	25	EST	NEELY,WB & BLAU,GE (1985)
DC 6.05	pKa	25	EXP	SERJEANT,EP & DEMPSEY,B (1979)
HL 5.18E-006	atm m3/mol	25	EST	MEYLAN,WM & HOWARD,PH (1991)
OH 3.07E-012	cm3/molc sec	25	EST	MEYLAN,WM & HOWARD,PH (1993)

CAS #: 000611-09-6				1H-INDOLE-2,3-DIONE, 5-NITRO-

Formula: $C_8H_4N_2O_4$

Mol Weight: 192.13

MP (deg C): 251 dec   FP (deg C):

BP (deg C):

BP pressure (mm Hg):

Property/Value	Units	Temp	Data Type	Reference
WS 7.75E+003	mg/L	25	EST	MEYLAN,WM ET AL. (1996)
logP 0.47			EXP	DEBNATH,AK ET AL. (1992)
VP 1.75E-007	mm Hg	25	EST	NEELY,WB & BLAU,GE (1985)
DC	pKa			
HL 8.71E-014	atm m3/mol	25	EST	MEYLAN,WM & HOWARD,PH (1991)
OH 4.86E-012	cm3/molc sec	25	EST	MEYLAN,WM & HOWARD,PH (1993)

CAS #: 000611-13-2				FUROIC ACID, METHYL ESTER

Formula: $C_6H_6O_3$

Mol Weight: 126.11

MP (deg C):   FP (deg C):

BP (deg C): 181.3

BP pressure (mm Hg):

Property/Value	Units	Temp	Data Type	Reference
WS 1.33E+004	mg/L	25	EST	MEYLAN,WM ET AL. (1996)
logP 1.00			EXP	HANSCH,C & LEO,AJ (1985)
VP 2.55E+000	mm Hg	25	EST	NEELY,WB & BLAU,GE (1985)
DC	pKa			
HL 3.46E-005	atm m3/mol	25	EST	MEYLAN,WM & HOWARD,PH (1991)
OH 1.32E-011	cm3/molc sec	25	EST	MEYLAN,WM & HOWARD,PH (1993)

CAS #: 000611-14-3				1-ETHYL-2-METHYLBENZENE

Formula: C_9H_{12}

Mol Weight: 120.20

MP (deg C): -80.8   FP (deg C):

BP (deg C): 165.2

BP pressure (mm Hg):

Property/Value	Units	Temp	Data Type	Reference
WS 7.46E+001	mg/L	25	EXP	YALKOWSKY,SH & DANNENFELSER,RM (1992)
logP 3.53			EXP	HANSCH,C & LEO,AJ (1985)
VP 2.61E+000	mm Hg	25	EXP	DAUBERT,TE & DANNER,RP (1989)
DC	pKa			
HL 4.58E-007	atm m3/mol	25	EST	VP/WSOL
OH 1.23E-011	cm3/molc sec	25	EXP	ATKINSON,R (1989)

CAS #: 000611-15-4				O-METHYLSTYRENE

Formula: C_9H_{10}

Mol Weight: 118.18

MP (deg C): -68.5   FP (deg C):

BP (deg C): 169.8

BP pressure (mm Hg):

Property/Value	Units	Temp	Data Type	Reference
WS 1.17E+002	mg/L	25	EST	MEYLAN,WM ET AL. (1996)
logP 3.44			EST	MEYLAN,WM & HOWARD,PH (1995)
VP 1.85E+000	mm Hg	25	EXP	YAWS,CL (1994B)
DC	pKa			
HL 3.05E-003	atm m3/mol	25	EST	MEYLAN,WM & HOWARD,PH (1991)
OH 3.12E-011	cm3/molc sec	25	EST	MEYLAN,WM & HOWARD,PH (1993)

000611-20-1 — O-CYANOPHENOL

CAS #:	000611-20-1		O-CYANOPHENOL
Formula:	C_7H_5NO		
Mol Weight:	119.12		
MP (deg C):	98	FP (deg C):	
BP (deg C):	149		
BP pressure (mm Hg):	1.40E+001		

Property/Value	Units	Temp	Data Type	Reference
WS 8.81E+003	mg/L	25	EST	MEYLAN,WM ET AL. (1996)
logP 1.61			EXP	HANSCH,C & LEO,AJ (1985)
VP 6.65E-003	mm Hg	25	EST	NEELY,WB & BLAU,GE (1985)
DC 6.86	pKa	25	EXP	SERJEANT,EP & DEMPSEY,B (1979)
HL 5.42E-009	atm m3/mol	25	EST	MEYLAN,WM & HOWARD,PH (1991)
OH 6.02E-012	cm3/molc sec	25	EST	MEYLAN,WM & HOWARD,PH (1993)

000611-21-2 — N-METHYL-O-TOLUIDINE

CAS #:	000611-21-2		N-METHYL-O-TOLUIDINE
Formula:	$C_8H_{11}N$		
Mol Weight:	121.18		
MP (deg C):		FP (deg C):	
BP (deg C):	207.5		
BP pressure (mm Hg):			

Property/Value	Units	Temp	Data Type	Reference
WS 1.42E+003	mg/L	25	EST	MEYLAN,WM ET AL. (1996)
logP 2.16			EXP	HANSCH,C & LEO,AJ (1985)
VP 6.50E-001	mm Hg	25	EST	NEELY,WB & BLAU,GE (1985)
DC 4.62	pKa	23	EXP	PERRIN,DD (1965)
HL 4.61E-006	atm m3/mol	25	EST	MEYLAN,WM & HOWARD,PH (1991)
OH 5.37E-011	cm3/molc sec	25	EST	MEYLAN,WM & HOWARD,PH (1993)

000611-32-5 — 8-METHYLQUINOLINE

CAS #:	000611-32-5		8-METHYLQUINOLINE
Formula:	$C_{10}H_9N$		
Mol Weight:	143.19		
MP (deg C):	-80	FP (deg C):	
BP (deg C):	247.5		
BP pressure (mm Hg):			

Property/Value	Units	Temp	Data Type	Reference
WS 4.89E+002	mg/L	25	EST	MEYLAN,WM ET AL. (1996)
logP 2.60			EXP	HANSCH,C & LEO,AJ (1985)
VP 2.50E-002	mm Hg	25	EXT	CHAO,J ET AL. (1983)
DC 5.05	pKa	20	EXP	PERRIN,DD (1965)
HL 7.60E-007	atm m3/mol	25	EST	MEYLAN,WM & HOWARD,PH (1991)
OH 3.77E-011	cm3/molc sec	25	EST	MEYLAN,WM & HOWARD,PH (1993)

000611-33-6 — 8-CHLOROQUINOLINE

CAS #:	000611-33-6		8-CHLOROQUINOLINE
Formula:	C_9H_6ClN		
Mol Weight:	163.61		
MP (deg C):	-20	FP (deg C):	
BP (deg C):	288.5		
BP pressure (mm Hg):			

Property/Value	Units	Temp	Data Type	Reference
WS 5.43E+002	mg/L	25	EST	MEYLAN,WM ET AL. (1996)
logP 2.44			EXP	HANSCH,C ET AL. (1995)
VP 5.08E-003	mm Hg	25	EST	NEELY,WB & BLAU,GE (1985)
DC 3.12	pKa	25	EXP	PERRIN,DD (1965)
HL 5.10E-007	atm m3/mol	25	EST	MEYLAN,WM & HOWARD,PH (1991)
OH 1.01E-011	cm3/molc sec	25	EST	MEYLAN,WM & HOWARD,PH (1993)

000611-34-7 — 5-AMINOQUINOLINE

CAS #:	000611-34-7		5-AMINOQUINOLINE
Formula:	$C_9H_8N_2$		
Mol Weight:	144.18		
MP (deg C):	110	FP (deg C):	
BP (deg C):	310		
BP pressure (mm Hg):			

Property/Value	Units	Temp	Data Type	Reference
WS 8.22E+003	mg/L	25	EST	MEYLAN,WM ET AL. (1996)
logP 1.16			EXP	HANSCH,C & LEO,AJ (1985)
VP 2.47E-006	mm Hg	25	EXT	RIBEIRODASILVA,MAV ET AL. (1993)
DC 3.95	pKa	20	EXP	PERRIN,DD (1965)
HL 2.43E-010	atm m3/mol	25	EST	MEYLAN,WM & HOWARD,PH (1991)
OH 2.00E-010	cm3/molc sec	25	EST	MEYLAN,WM & HOWARD,PH (1993)

000611-70-1 — 1-PROPANONE, 2-METHYL-1-PHENYL-

CAS #:	000611-70-1		1-PROPANONE, 2-METHYL-1-PHENYL-
Formula:	$C_{10}H_{12}O$		
Mol Weight:	148.21		
MP (deg C):	-1.3	FP (deg C):	
BP (deg C):	220		
BP pressure (mm Hg):			

Property/Value	Units	Temp	Data Type	Reference
WS 3.60E+002	mg/L	25	EST	MEYLAN,WM ET AL. (1996)
logP 2.73			EXP	HANSCH,C ET AL. (1995)
VP 1.50E-001	mm Hg	25	EST	NEELY,WB & BLAU,GE (1985)
DC	pKa			
HL 1.73E-005	atm m3/mol	25	EST	MEYLAN,WM & HOWARD,PH (1991)
OH 4.29E-012	cm3/molc sec	25	EST	MEYLAN,WM & HOWARD,PH (1993)

000611-73-4 — BENZOYLFORMIC ACID

CAS #:	000611-73-4		BENZOYLFORMIC ACID
Formula:	$C_8H_6O_3$		
Mol Weight:	150.14		
MP (deg C):	66	FP (deg C):	
BP (deg C):	163		
BP pressure (mm Hg):	1.50E+001		

Property/Value	Units	Temp	Data Type	Reference
WS 9.20E+005	mg/L	0	EXP	YALKOWSKY,SH & DANNENFELSER,RM (1992)
logP 1.20			EST	MEYLAN,WM & HOWARD,PH (1995)
VP 6.60E-004	mm Hg	25	EST	NEELY,WB & BLAU,GE (1985)
DC	pKa			
HL 4.32E-010	atm m3/mol	25	EST	MEYLAN,WM & HOWARD,PH (1991)
OH 2.30E-012	cm3/molc sec	25	EST	MEYLAN,WM & HOWARD,PH (1993)

000611-74-5 — N,N-DIMETHYLBENZAMIDE

CAS #:	000611-74-5		N,N-DIMETHYLBENZAMIDE
Formula:	$C_9H_{11}NO$		
Mol Weight:	149.19		
MP (deg C):	43-45	FP (deg C):	
BP (deg C):	132-133		
BP pressure (mm Hg):	1.50E+001		

Property/Value	Units	Temp	Data Type	Reference
WS 2.26E+004	mg/L	25	EST	MEYLAN,WM ET AL. (1996)
logP 0.62			EXP	HANSCH,C & LEO,AJ (1985)
VP 8.87E-003	mm Hg	25	EST	NEELY,WB & BLAU,GE (1985)
DC	pKa			
HL 1.07E-008	atm m3/mol	25	EST	MEYLAN,WM & HOWARD,PH (1991)
OH 1.79E-011	cm3/molc sec	25	EST	MEYLAN,WM & HOWARD,PH (1993)

CAS #: 000612-00-0				1,1-DIPHENYLETHANE
Formula: C14H14				
Mol Weight: 182.27				
MP (deg C): -17.9		FP (deg C):		
BP (deg C): 272.6				
BP pressure (mm Hg):				

Property/Value	Units	Temp	Data Type	Reference
WS 1.53E+001	mg/L	25	EST	MEYLAN,WM ET AL. (1996)
logP 4.15			EST	MEYLAN,WM & HOWARD,PH (1995)
VP 1.17E-002	mm Hg	25	EXP	DAUBERT,TE & DANNER,RP (1989)
DC	pKa			
HL 6.37E-004	atm m3/mol	25	EST	MEYLAN,WM & HOWARD,PH (1991)
OH 1.14E-011	cm3/molc sec	25	EST	MEYLAN,WM & HOWARD,PH (1993)

CAS #: 000612-16-8				O-METHOXYBENZYL ALCOHOL
Formula: C8H10O2				
Mol Weight: 138.17				
MP (deg C):		FP (deg C):		
BP (deg C): 249				
BP pressure (mm Hg):				

Property/Value	Units	Temp	Data Type	Reference
WS 2.99E+004	mg/L	25	EST	MEYLAN,WM ET AL. (1996)
logP 1.13			EXP	HANSCH,C & LEO,AJ (1985)
VP 4.29E-003	mm Hg	25	EST	NEELY,WB & BLAU,GE (1985)
DC	pKa			
HL 1.29E-008	atm m3/mol	25	EST	MEYLAN,WM & HOWARD,PH (1991)
OH 3.04E-011	cm3/molc sec	25	EST	MEYLAN,WM & HOWARD,PH (1993)

CAS #: 000612-22-6				2-ETHYLNITROBENZENE
Formula: C8H9NO2				
Mol Weight: 151.17				
MP (deg C): 12.3		FP (deg C):		
BP (deg C): 232.5				
BP pressure (mm Hg):				

Property/Value	Units	Temp	Data Type	Reference
WS 2.40E+002	mg/L	25	EXP	CHEM INSPECT TEST INST (1992)
logP 2.58			EXP	CHEM INSPECT TEST INST (1992)
VP 2.17E-002	mm Hg	25	EST	NEELY,WB & BLAU,GE (1985)
DC	pKa			
HL 3.11E-005	atm m3/mol	25	EST	MEYLAN,WM & HOWARD,PH (1991)
OH 1.71E-012	cm3/molc sec	25	EST	MEYLAN,WM & HOWARD,PH (1993)

CAS #: 000612-24-8				1-NITRO-2-CYANOBENZENE
Formula: C7H4N2O2				
Mol Weight: 148.12				
MP (deg C): 107-109		FP (deg C):		
BP (deg C):				
BP pressure (mm Hg):				

Property/Value	Units	Temp	Data Type	Reference
WS 2.30E+003	mg/L	25	EST	MEYLAN,WM ET AL. (1996)
logP 1.02			EXP	NAKAGAWA,Y ET AL. (1992)
VP 2.17E-003	mm Hg	25	EST	NEELY,WB & BLAU,GE (1985)
DC	pKa			
HL 2.05E-007	atm m3/mol	25	EST	MEYLAN,WM & HOWARD,PH (1991)
OH 3.19E-014	cm3/molc sec	25	EST	MEYLAN,WM & HOWARD,PH (1993)

CAS #: 000612-25-9				2-NITROBENZYL ALCOHOL
Formula: C7H7NO3				
Mol Weight: 153.14				
MP (deg C): 74		FP (deg C):		
BP (deg C): 270				
BP pressure (mm Hg):				

Property/Value	Units	Temp	Data Type	Reference
WS 2.61E+003	mg/L	25	EST	MEYLAN,WM ET AL. (1996)
logP 1.24			EXP	HANSCH,C ET AL. (1995)
VP 8.58E-005	mm Hg	25	EST	NEELY,WB & BLAU,GE (1985)
DC	pKa			
HL 8.58E-010	atm m3/mol	25	EST	MEYLAN,WM & HOWARD,PH (1991)
OH 4.01E-012	cm3/molc sec	25	EST	MEYLAN,WM & HOWARD,PH (1993)

CAS #: 000612-28-2				N-METHYL-O-NITROANILINE
Formula: C7H8N2O2				
Mol Weight: 152.15				
MP (deg C): 38		FP (deg C):		
BP (deg C): 158				
BP pressure (mm Hg): 1.80E+001				

Property/Value	Units	Temp	Data Type	Reference
WS 1.02E+003	mg/L	25	EST	MEYLAN,WM ET AL. (1996)
logP 2.18			EXP	HANSCH,C & LEO,AJ (1985)
VP 7.64E-003	mm Hg	25	EST	NEELY,WB & BLAU,GE (1985)
DC	pKa			
HL 3.61E-007	atm m3/mol	25	EST	MEYLAN,WM & HOWARD,PH (1991)
OH 6.60E-012	cm3/molc sec	25	EST	MEYLAN,WM & HOWARD,PH (1993)

CAS #: 000612-37-3				7-METHYL URIC ACID
Formula: C6H6N4O3				
Mol Weight: 182.14				
MP (deg C): 370-380 de		FP (deg C):		
BP (deg C):				
BP pressure (mm Hg):				

Property/Value	Units	Temp	Data Type	Reference
WS 2.68E+004	mg/L	25	EST	MEYLAN,WM ET AL. (1996)
logP -1.18			EXP	GASPARI,F & BONATI,M (1987)
VP 3.74E-010	mm Hg	25	EST	NEELY,WB & BLAU,GE (1985)
DC	pKa			
HL 2.63E-017	atm m3/mol	25	EST	MEYLAN,WM & HOWARD,PH (1991)
OH 1.29E-011	cm3/molc sec	25	EST	MEYLAN,WM & HOWARD,PH (1993)

CAS #: 000612-45-3				2-NITRO-4-METHYLACETANILIDE
Formula: C9H10N2O3				
Mol Weight: 194.19				
MP (deg C):		FP (deg C):		
BP (deg C):				
BP pressure (mm Hg):				

Property/Value	Units	Temp	Data Type	Reference
WS 1.02E+003	mg/L	25	EST	MEYLAN,WM ET AL. (1996)
logP 1.49			EXP	NAKAGAWA,Y ET AL. (1992)
VP 2.03E-006	mm Hg	25	EST	NEELY,WB & BLAU,GE (1985)
DC	pKa			
HL 5.88E-010	atm m3/mol	25	EST	MEYLAN,WM & HOWARD,PH (1991)
OH 2.14E-012	cm3/molc sec	25	EST	MEYLAN,WM & HOWARD,PH (1993)

CAS #: 000612-57-7				6-CHLOROQUINOLINE

Formula: C_9H_6ClN

Mol Weight: 163.61

MP (deg C): 43.8 FP (deg C):

BP (deg C): 263

BP pressure (mm Hg):

Property/Value	Units	Temp	Data Type	Reference
WS 3.07E+002	mg/L	25	EST	MEYLAN,WM ET AL. (1996)
logP 2.73			EXP	HANSCH,C & LEO,AJ (1985)
VP 5.08E-003	mm Hg	25	EST	NEELY,WB & BLAU,GE (1985)
DC 3.85	pKa	25	EXP	PERRIN,DD (1965)
HL 5.10E-007	atm m3/mol	25	EST	MEYLAN,WM & HOWARD,PH (1991)
OH 1.01E-011	cm3/molc sec	25	EST	MEYLAN,WM & HOWARD,PH (1993)

CAS #: 000612-58-8				3-METHYLQUINOLINE

Formula: $C_{10}H_9N$

Mol Weight: 143.19

MP (deg C): 16.5 FP (deg C):

BP (deg C): 259.8

BP pressure (mm Hg):

Property/Value	Units	Temp	Data Type	Reference
WS 5.61E+002	mg/L	25	EST	MEYLAN,WM ET AL. (1996)
logP 2.53			EXP	HANSCH,C ET AL. (1995)
VP 9.52E-003	mm Hg	25	EST	NEELY,WB & BLAU,GE (1985)
DC 5.17	pKa	20	EXP	PERRIN,DD (1965)
HL 7.60E-007	atm m3/mol	25	EST	MEYLAN,WM & HOWARD,PH (1991)
OH 3.77E-011	cm3/molc sec	25	EST	MEYLAN,WM & HOWARD,PH (1993)

CAS #: 000612-60-2				7-METHYLQUINOLINE

Formula: $C_{10}H_9N$

Mol Weight: 143.19

MP (deg C): 39 FP (deg C):

BP (deg C): 257.6

BP pressure (mm Hg):

Property/Value	Units	Temp	Data Type	Reference
WS 6.31E+002	mg/L	25	EST	MEYLAN,WM ET AL. (1996)
logP 2.47			EXP	HANSCH,C & LEO,AJ (1985)
VP 8.84E-003	mm Hg	25	EXT	CHAO,J ET AL. (1983)
DC 5.34	pKa	20	EXP	PERRIN,DD (1965)
HL 7.60E-007	atm m3/mol	25	EST	MEYLAN,WM & HOWARD,PH (1991)
OH 3.77E-011	cm3/molc sec	25	EST	MEYLAN,WM & HOWARD,PH (1993)

CAS #: 000612-62-4				2-CHLOROQUINOLINE

Formula: C_9H_6ClN

Mol Weight: 163.61

MP (deg C): 38 FP (deg C):

BP (deg C): 266

BP pressure (mm Hg):

Property/Value	Units	Temp	Data Type	Reference
WS 3.19E+002	mg/L	25	EST	MEYLAN,WM ET AL. (1996)
logP 2.71			EXP	HANSCH,C & LEO,AJ (1985)
VP 5.08E-003	mm Hg	25	EST	NEELY,WB & BLAU,GE (1985)
DC	pKa			
HL 3.22E-005	atm m3/mol	25	EST	MEYLAN,WM & HOWARD,PH (1991)
OH 1.01E-011	cm3/molc sec	25	EST	MEYLAN,WM & HOWARD,PH (1993)

CAS #: 000612-71-5				5'-PHENYL-1,1':3',1"-TERPHENYL

Formula: $C_{24}H_{18}$

Mol Weight: 306.41

MP (deg C): 176 FP (deg C):

BP (deg C): 462

BP pressure (mm Hg):

Property/Value	Units	Temp	Data Type	Reference
WS 6.77E-003	mg/L	25	EST	MEYLAN,WM ET AL. (1996)
logP 7.28			EST	MEYLAN,WM & HOWARD,PH (1995)
VP 1.78E-009	mm Hg	25	EST	NEELY,WB & BLAU,GE (1985)
DC	pKa			
HL 2.44E-006	atm m3/mol	25	EST	MEYLAN,WM & HOWARD,PH (1991)
OH 2.04E-011	cm3/molc sec	25	EST	MEYLAN,WM & HOWARD,PH (1993)

CAS #: 000612-75-9				3,3'-DIMETHYLBIPHENYL

Formula: $C_{14}H_{14}$

Mol Weight: 182.27

MP (deg C): 9 FP (deg C):

BP (deg C): 280

BP pressure (mm Hg):

Property/Value	Units	Temp	Data Type	Reference
WS 3.86E+000	mg/L	25	EST	MEYLAN,WM ET AL. (1996)
logP 4.85			EST	MEYLAN,WM & HOWARD,PH (1995)
VP 8.13E-004	mm Hg	25	EST	NEELY,WB & BLAU,GE (1985)
DC	pKa			
HL 5.04E-004	atm m3/mol	25	EST	MEYLAN,WM & HOWARD,PH (1991)
OH 1.80E-011	cm3/molc sec	25	EST	MEYLAN,WM & HOWARD,PH (1993)

CAS #: 000612-96-4				2-PHENYLQUINOLINE

Formula: $C_{15}H_{11}N$

Mol Weight: 205.26

MP (deg C): 86 FP (deg C):

BP (deg C): 363

BP pressure (mm Hg):

Property/Value	Units	Temp	Data Type	Reference
WS 1.92E+001	mg/L	25	EST	MEYLAN,WM ET AL. (1996)
logP 3.90			EXP	HANSCH,C & LEO,AJ (1985)
VP 8.96E-006	mm Hg	25	EST	NEELY,WB & BLAU,GE (1985)
DC	pKa			
HL 5.29E-008	atm m3/mol	25	EST	MEYLAN,WM & HOWARD,PH (1991)
OH 2.84E-011	cm3/molc sec	25	EST	MEYLAN,WM & HOWARD,PH (1993)

CAS #: 000612-98-6				N-BENZYL-N-NITROSOANILINE

Formula: $C_{13}H_{12}N_2O$

Mol Weight: 212.25

MP (deg C): FP (deg C):

BP (deg C):

BP pressure (mm Hg):

Property/Value	Units	Temp	Data Type	Reference
WS 4.91E+001	mg/L	25	EST	MEYLAN,WM ET AL. (1996)
logP 3.38			EXP	HANSCH,C & LEO,AJ (1985)
VP 1.30E-005	mm Hg	25	EST	NEELY,WB & BLAU,GE (1985)
DC	pKa			
HL 3.90E-007	atm m3/mol	25	EST	MEYLAN,WM & HOWARD,PH (1991)
OH 2.59E-011	cm3/molc sec	25	EST	MEYLAN,WM & HOWARD,PH (1993)

CAS #: 000613-12-7				2-METHYLANTHRACENE

Formula: $C_{15}H_{12}$

Mol Weight: 192.26

MP (deg C): 209 FP (deg C):

BP (deg C):

BP pressure (mm Hg):

Property/Value	Units	Temp	Data Type	Reference
WS 2.13E-002	mg/L	25	EXP	YALKOWSKY,SH & DANNENFELSER,RM (1992)
logP 5.00			EXP	ALCORN,CJ ET AL. (1993)
VP 5.01E-005	mm Hg	25	EST	NEELY,WB & BLAU,GE (1985)
DC	pKa			
HL 5.67E-005	atm m3/mol	25	EST	MEYLAN,WM & HOWARD,PH (1991)
OH 1.05E-010	cm3/molc sec	25	EST	MEYLAN,WM & HOWARD,PH (1993)

CAS #: 000613-13-8				2-AMINOANTHRACENE

Formula: $C_{14}H_{11}N$

Mol Weight: 193.25

MP (deg C): 238.8 FP (deg C):

BP (deg C):

BP pressure (mm Hg):

Property/Value	Units	Temp	Data Type	Reference
WS 1.30E-003	mg/L	25	EXP	MEANS,JC ET AL. (1982)
logP 4.13			EXP	MEANS,JC ET AL. (1982)
VP 4.05E-006	mm Hg	25	EST	NEELY,WB & BLAU,GE (1985)
DC	pKa			
HL 1.81E-008	atm m3/mol	25	EST	MEYLAN,WM & HOWARD,PH (1991)
OH 2.00E-010	cm3/molc sec	25	EST	MEYLAN,WM & HOWARD,PH (1993)

CAS #: 000613-26-3				2,6-DIMETHYLANTHRACENE

Formula: $C_{16}H_{14}$

Mol Weight: 206.29

MP (deg C): FP (deg C):

BP (deg C):

BP pressure (mm Hg):

Property/Value	Units	Temp	Data Type	Reference
WS 7.13E-002	mg/L	25	EST	MEYLAN,WM ET AL. (1996)
logP 5.44			EST	MEYLAN,WM & HOWARD,PH (1995)
VP 1.82E-005	mm Hg	25	EST	NEELY,WB & BLAU,GE (1985)
DC	pKa			
HL 6.25E-005	atm m3/mol	25	EST	MEYLAN,WM & HOWARD,PH (1991)
OH 1.28E-010	cm3/molc sec	25	EST	MEYLAN,WM & HOWARD,PH (1993)

CAS #: 000613-31-0				9,10-DIHYDROANTHRACENE

Formula: $C_{14}H_{12}$

Mol Weight: 180.25

MP (deg C): 111 FP (deg C):

BP (deg C): 305

BP pressure (mm Hg):

Property/Value	Units	Temp	Data Type	Reference
WS 1.00E+000	mg/L	25	EST	MEYLAN,WM ET AL. (1996)
logP 4.25			EXP	HANSCH,C & LEO,AJ (1985)
VP 1.29E-003	mm Hg	25	EST	NEELY,WB & BLAU,GE (1985)
DC	pKa			
HL 1.94E-004	atm m3/mol	25	EST	MEYLAN,WM & HOWARD,PH (1991)
OH 1.37E-011	cm3/molc sec	25	EST	MEYLAN,WM & HOWARD,PH (1993)

CAS #: 000613-33-2				P-TOLYLTOLUENE

Formula: $C_{14}H_{14}$

Mol Weight: 182.27

MP (deg C): 125 FP (deg C):

BP (deg C): 295

BP pressure (mm Hg):

Property/Value	Units	Temp	Data Type	Reference
WS 1.75E-001	mg/L	25	EXP	YALKOWSKY,SH & DANNENFELSER,RM (1992)
logP 5.09			EXP	DOUCETTE,WJ & ANDREN,AW (1987)
VP 8.13E-004	mm Hg	25	EST	NEELY,WB & BLAU,GE (1985)
DC	pKa			
HL 5.04E-004	atm m3/mol	25	EST	MEYLAN,WM & HOWARD,PH (1991)
OH 8.58E-012	cm3/molc sec	25	EST	MEYLAN,WM & HOWARD,PH (1993)

CAS #: 000613-37-6				4-METHOXYBIPHENYL

Formula: $C_{13}H_{12}O$

Mol Weight: 184.24

MP (deg C): 90 FP (deg C):

BP (deg C): 157

BP pressure (mm Hg): 1.00E+001

Property/Value	Units	Temp	Data Type	Reference
WS 2.77E+001	mg/L	25	EST	MEYLAN,WM ET AL. (1996)
logP 3.84			EST	MEYLAN,WM & HOWARD,PH (1995)
VP 7.99E-004	mm Hg	25	EST	NEELY,WB & BLAU,GE (1985)
DC	pKa			
HL 2.45E-005	atm m3/mol	25	EST	MEYLAN,WM & HOWARD,PH (1991)
OH 1.96E-011	cm3/molc sec	25	EST	MEYLAN,WM & HOWARD,PH (1993)

CAS #: 000613-50-3				6-NITROQUINOLINE

Formula: $C_9H_6N_2O_2$

Mol Weight: 174.16

MP (deg C): 153.5 FP (deg C):

BP (deg C): 170

BP pressure (mm Hg): 2.00E-001

Property/Value	Units	Temp	Data Type	Reference
WS 6.68E+002	mg/L	25	EST	MEYLAN,WM ET AL. (1996)
logP 1.84			EXP	HANSCH,C & LEO,AJ (1985)
VP 1.38E-004	mm Hg	25	EST	NEELY,WB & BLAU,GE (1985)
DC	pKa			
HL 2.72E-009	atm m3/mol	25	EST	MEYLAN,WM & HOWARD,PH (1991)
OH 1.80E-012	cm3/molc sec	25	EST	MEYLAN,WM & HOWARD,PH (1993)

CAS #: 000613-51-4				7-NITROQUINOLINE

Formula: $C_9H_6N_2O_2$

Mol Weight: 174.16

MP (deg C): 132.5 FP (deg C):

BP (deg C):

BP pressure (mm Hg):

Property/Value	Units	Temp	Data Type	Reference
WS 6.68E+002	mg/L	25	EST	MEYLAN,WM ET AL. (1996)
logP 1.82			EXP	HANSCH,C & LEO,AJ (1985)
VP 1.38E-004	mm Hg	25	EST	NEELY,WB & BLAU,GE (1985)
DC	pKa			
HL 2.72E-009	atm m3/mol	25	EST	MEYLAN,WM & HOWARD,PH (1991)
OH 1.80E-012	cm3/molc sec	25	EST	MEYLAN,WM & HOWARD,PH (1993)

CAS #: 000613-55-8				BIS(2-METHOXYPHENYL)DIAZENE
Formula: $C_{14}H_{14}N_2O_2$				
Mol Weight: 242.28				
MP (deg C):		FP (deg C):		
BP (deg C):				
BP pressure (mm Hg):				

Property/Value	Units	Temp	Data Type	Reference
WS 2.17E+000	mg/L	25	EST	MEYLAN,WM ET AL. (1996)
logP 4.27			EST	MEYLAN,WM & HOWARD,PH (1995)
VP 3.24E-005	mm Hg	25	EST	NEELY,WB & BLAU,GE (1985)
DC	pKa			
HL 5.15E-008	atm m3/mol	25	EST	MEYLAN,WM & HOWARD,PH (1991)
OH 1.87E-011	cm3/molc sec	25	EST	MEYLAN,WM & HOWARD,PH (1993)

CAS #: 000613-70-7				O-METHOXYPHENYL ACETATE
Formula: $C_9H_{10}O_3$				
Mol Weight: 166.18				
MP (deg C): 31.5		FP (deg C):		
BP (deg C): 123-124				
BP pressure (mm Hg): 1.30E+001				

Property/Value	Units	Temp	Data Type	Reference
WS 4.25E+003	mg/L	25	EST	MEYLAN,WM ET AL. (1996)
logP 1.38			EXP	HANSCH,C & LEO,AJ (1985)
VP 6.02E-002	mm Hg	25	EST	NEELY,WB & BLAU,GE (1985)
DC	pKa			
HL 3.83E-006	atm m3/mol	25	EST	MEYLAN,WM & HOWARD,PH (1991)
OH 1.67E-011	cm3/molc sec	25	EST	MEYLAN,WM & HOWARD,PH (1993)

CAS #: 000613-93-4				N-METHYLBENZAMIDE
Formula: C_8H_9NO				
Mol Weight: 135.17				
MP (deg C): 82		FP (deg C):		
BP (deg C): 291				
BP pressure (mm Hg):				

Property/Value	Units	Temp	Data Type	Reference
WS 1.62E+004	mg/L	25	EST	MEYLAN,WM ET AL. (1996)
logP 0.86			EXP	HANSCH,C & LEO,AJ (1985)
VP 5.20E-004	mm Hg	25	EST	NEELY,WB & BLAU,GE (1985)
DC	pKa			
HL 4.86E-009	atm m3/mol	25	EST	MEYLAN,WM & HOWARD,PH (1991)
OH 7.84E-012	cm3/molc sec	25	EST	MEYLAN,WM & HOWARD,PH (1993)

CAS #: 000613-94-5				BENZOYL HYDRAZINE
Formula: $C_7H_8N_2O$				
Mol Weight: 136.15				
MP (deg C): 115		FP (deg C):		
BP (deg C):				
BP pressure (mm Hg):				

Property/Value	Units	Temp	Data Type	Reference
WS 2.93E+003	mg/L	25	EST	MEYLAN,WM ET AL. (1996)
logP 0.19			EXP	HANSCH,C & LEO,AJ (1985)
VP 7.78E-005	mm Hg	25	EST	NEELY,WB & BLAU,GE (1985)
DC 3.03	pKa	20	EXP	PERRIN,DD (1965)
HL 9.25E-012	atm m3/mol	25	EST	MEYLAN,WM & HOWARD,PH (1991)
OH 7.28E-012	cm3/molc sec	25	EST	MEYLAN,WM & HOWARD,PH (1993)

CAS #: 000614-18-6				NICOTINIC ACID, ETHYL ESTER
Formula: $C_8H_9NO_2$				
Mol Weight: 151.17				
MP (deg C): 8.5		FP (deg C):		
BP (deg C): 224				
BP pressure (mm Hg):				

Property/Value	Units	Temp	Data Type	Reference
WS 5.59E+003	mg/L	25	EST	MEYLAN,WM ET AL. (1996)
logP 1.32			EXP	HANSCH,C & LEO,AJ (1985)
VP 8.40E-002	mm Hg	25	EST	NEELY,WB & BLAU,GE (1985)
DC 3.35	pKa	20	EXP	PERRIN,DD (1965)
HL 6.03E-008	atm m3/mol	25	EST	MEYLAN,WM & HOWARD,PH (1991)
OH 1.78E-012	cm3/molc sec	25	EST	MEYLAN,WM & HOWARD,PH (1993)

CAS #: 000614-19-7				3-AMINO-3-PHENYLPROPANOIC ACID
Formula: $C_9H_{11}NO_2$				
Mol Weight: 165.19				
MP (deg C): 222 dec		FP (deg C):		
BP (deg C):				
BP pressure (mm Hg):				

Property/Value	Units	Temp	Data Type	Reference
WS 8.65E+003	mg/L	25	EST	MEYLAN,WM ET AL. (1996)
logP -1.40			EXP	HANSCH,C ET AL. (1995)
VP 4.65E-008	mm Hg	25	EST	NEELY,WB & BLAU,GE (1985)
DC	pKa			
HL 6.04E-012	atm m3/mol	25	EST	MEYLAN,WM & HOWARD,PH (1991)
OH 4.92E-011	cm3/molc sec	25	EST	MEYLAN,WM & HOWARD,PH (1993)

CAS #: 000614-30-2				N-METHYL-N-BENZYLANILINE
Formula: $C_{14}H_{15}N$				
Mol Weight: 197.28				
MP (deg C):		FP (deg C):		
BP (deg C):				
BP pressure (mm Hg):				

Property/Value	Units	Temp	Data Type	Reference
WS 1.13E+001	mg/L	25	EST	MEYLAN,WM ET AL. (1996)
logP 4.22			EXP	HANSCH,C & LEO,AJ (1985)
VP 9.23E-004	mm Hg	25	EST	NEELY,WB & BLAU,GE (1985)
DC	pKa			
HL 6.91E-006	atm m3/mol	25	EST	MEYLAN,WM & HOWARD,PH (1991)
OH 1.61E-010	cm3/molc sec	25	EST	MEYLAN,WM & HOWARD,PH (1993)

CAS #: 000614-61-9				O-CHLOROPHENOXYACETIC ACID
Formula: $C_8H_7ClO_3$				
Mol Weight: 186.60				
MP (deg C):		FP (deg C):		
BP (deg C):				
BP pressure (mm Hg):				

Property/Value	Units	Temp	Data Type	Reference
WS 1.28E+003	mg/L	25	EXP	YALKOWSKY,SH & DANNENFELSER,RM (1992)
logP 1.86			EXP	HANSCH,C & LEO,AJ (1985)
VP 3.18E-004	mm Hg	25	EST	NEELY,WB & BLAU,GE (1985)
DC 3.05	pKa	25	EXP	KORTUM,G ET AL (1961)
HL 1.24E-008	atm m3/mol	25	EST	MEYLAN,WM & HOWARD,PH (1991)
OH 1.11E-011	cm3/molc sec	25	EST	MEYLAN,WM & HOWARD,PH (1993)

447

CAS #: 000614-75-5		O-HYDROXYPHENYLACETIC ACID

Formula: $C_8H_8O_3$

Mol Weight: 152.15

MP (deg C): 148   FP (deg C):

BP (deg C): 240

BP pressure (mm Hg):

	Property/Value	Units	Temp	Data Type	Reference
WS	1.32E+005	mg/L	25	EST	MEYLAN,WM ET AL. (1996)
logP	0.85			EXP	HANSCH,C & LEO,AJ (1985)
VP	5.55E-005	mm Hg	25	EST	NEELY,WB & BLAU,GE (1985)
DC		pKa			
HL	4.60E-012	atm m3/mol	25	EST	MEYLAN,WM & HOWARD,PH (1991)
OH	3.57E-011	cm3/molc sec	25	EST	MEYLAN,WM & HOWARD,PH (1993)

CAS #: 000614-76-6		2-BROMOACETANILIDE

Formula: C_8H_8BrNO

Mol Weight: 214.07

MP (deg C):   FP (deg C):

BP (deg C):

BP pressure (mm Hg):

	Property/Value	Units	Temp	Data Type	Reference
WS	2.65E+003	mg/L	25	EST	MEYLAN,WM ET AL. (1996)
logP	1.34			EXP	NAKAGAWA,Y ET AL. (1992)
VP	3.42E-005	mm Hg	25	EST	NEELY,WB & BLAU,GE (1985)
DC		pKa			
HL	2.46E-009	atm m3/mol	25	EST	MEYLAN,WM & HOWARD,PH (1991)
OH	3.66E-012	cm3/molc sec	25	EST	MEYLAN,WM & HOWARD,PH (1993)

CAS #: 000614-80-2		O-HYDROXYACETANILIDE

Formula: $C_8H_9NO_2$

Mol Weight: 151.17

MP (deg C): 207-209   FP (deg C):

BP (deg C):

BP pressure (mm Hg):

	Property/Value	Units	Temp	Data Type	Reference
WS	8.80E+002	mg/L	25	EXP	CHEM INSPECT TEST INST (1992)
logP	0.72			EXP	HANSCH,C & LEO,AJ (1985)
VP	7.00E-006	mm Hg	25	EST	NEELY,WB & BLAU,GE (1985)
DC		pKa			
HL	6.42E-013	atm m3/mol	25	EST	MEYLAN,WM & HOWARD,PH (1991)
OH	1.77E-011	cm3/molc sec	25	EST	MEYLAN,WM & HOWARD,PH (1993)

CAS #: 000614-95-9		ETHYL N-ETHYLNITROSOCARBAMATE

Formula: $C_5H_{10}N_2O_3$

Mol Weight: 146.15

MP (deg C):   FP (deg C):

BP (deg C):

BP pressure (mm Hg):

	Property/Value	Units	Temp	Data Type	Reference
WS	2.73E+003	mg/L	25	EST	MEYLAN,WM ET AL. (1996)
logP	1.71			EST	MEYLAN,WM & HOWARD,PH (1995)
VP	2.62E-002	mm Hg	25	EST	NEELY,WB & BLAU,GE (1985)
DC		pKa			
HL	1.89E-008	atm m3/mol	25	EST	MEYLAN,WM & HOWARD,PH (1991)
OH	5.66E-012	cm3/molc sec	25	EST	MEYLAN,WM & HOWARD,PH (1993)

CAS #: 000614-96-0		5-METHYLINDOLE

Formula: C_9H_9N

Mol Weight: 131.18

MP (deg C): 60   FP (deg C):

BP (deg C): 267

BP pressure (mm Hg):

	Property/Value	Units	Temp	Data Type	Reference
WS	4.68E+002	mg/L	25	EST	MEYLAN,WM ET AL. (1996)
logP	2.68			EXP	HANSCH,C & LEO,AJ (1985)
VP	6.03E-003	mm Hg	25	EST	NEELY,WB & BLAU,GE (1985)
DC		pKa			
HL	9.78E-007	atm m3/mol	25	EST	MEYLAN,WM & HOWARD,PH (1991)
OH	2.00E-010	cm3/molc sec	25	EST	MEYLAN,WM & HOWARD,PH (1993)

CAS #: 000614-98-2		FURAN-3-CARBOXYLIC AC, ET ESTER

Formula: $C_7H_9O_3$

Mol Weight: 141.15

MP (deg C):   FP (deg C):

BP (deg C): 93-95

BP pressure (mm Hg): 3.50E+001

	Property/Value	Units	Temp	Data Type	Reference
WS	2.53E+003	mg/L	25	EST	MEYLAN,WM ET AL. (1996)
logP	1.78			EXP	HANSCH,C & LEO,AJ (1985)
VP	9.25E-001	mm Hg	25	EST	NEELY,WB & BLAU,GE (1985)
DC		pKa			
HL	4.60E-005	atm m3/mol	25	EST	MEYLAN,WM & HOWARD,PH (1991)
OH	1.48E-011	cm3/molc sec	25	EST	MEYLAN,WM & HOWARD,PH (1993)

CAS #: 000614-99-3		FUROIC ACID, ETHYL ESTER

Formula: $C_7H_8O_3$

Mol Weight: 140.14

MP (deg C): 34.5   FP (deg C):

BP (deg C): 196.8

BP pressure (mm Hg):

	Property/Value	Units	Temp	Data Type	Reference
WS	4.21E+003	mg/L	25	EST	MEYLAN,WM ET AL. (1996)
logP	1.52			EXP	HANSCH,C & LEO,AJ (1985)
VP	3.37E-001	mm Hg	25	EXT	DYKYJ,J & REPAS,M (1973)
DC		pKa			
HL	4.60E-005	atm m3/mol	25	EST	MEYLAN,WM & HOWARD,PH (1991)
OH	1.48E-011	cm3/molc sec	25	EST	MEYLAN,WM & HOWARD,PH (1993)

CAS #: 000615-05-4		2,4 DIAMINOANISOLE

Formula: $C_7H_{10}N_2O$

Mol Weight: 138.17

MP (deg C): 67-68   FP (deg C):

BP (deg C): 221

BP pressure (mm Hg):

	Property/Value	Units	Temp	Data Type	Reference
WS	1.95E+004	mg/L	25	EST	MEYLAN,WM ET AL. (1996)
logP	-0.31			EST	MEYLAN,WM & HOWARD,PH (1995)
VP	4.68E-002	mm Hg	25	EST	NEELY,WB & BLAU,GE (1985)
DC		pKa			
HL	7.22E-010	atm m3/mol	25	EST	MEYLAN,WM & HOWARD,PH (1991)
OH	2.00E-010	cm3/molc sec	25	EST	MEYLAN,WM & HOWARD,PH (1993)

CAS #: 000615-10-1				3-FURANCARBOXYLIC ACID, PROPYL ESTER

Formula: $C_8H_{10}O_3$

Mol Weight: 154.17

MP (deg C): FP (deg C):

BP (deg C): 210.9

BP pressure (mm Hg):

Property/ Value	Units	Temp	Data Type	Reference
WS 7.02E+002	mg/L	25	EST	MEYLAN,WM ET AL. (1996)
logP 2.36			EXP	SANGSTER,J (1994)
VP 3.40E-001	mm Hg	25	EST	NEELY,WB & BLAU,GE (1985)
DC	pKa			
HL 6.10E-005	atm m3/mol	25	EST	MEYLAN,WM & HOWARD,PH (1991)
OH 1.62E-011	cm3/molc sec	25	EST	MEYLAN,WM & HOWARD,PH (1993)

CAS #: 000615-15-6				2-METHYLBENAIMIDAZOLE

Formula: $C_8H_8N_2$

Mol Weight: 132.17

MP (deg C): 176-177 FP (deg C):

BP (deg C):

BP pressure (mm Hg):

Property/ Value	Units	Temp	Data Type	Reference
WS 1.45E+003	mg/L	20	EXP	PEARLMAN,RS ET AL . (1984)
logP 1.43			EXP	HANSCH,C & LEO,AJ (1985)
VP 2.74E-005	mm Hg	25	EST	NEELY,WB & BLAU,GE (1985)
DC	pKa			
HL 4.05E-007	atm m3/mol	25	EST	MEYLAN,WM & HOWARD,PH (1991)
OH 9.41E-011	cm3/molc sec	25	EST	MEYLAN,WM & HOWARD,PH (1993)

CAS #: 000615-16-7				O-PHENYLENE UREA

Formula: $C_7H_6N_2O$

Mol Weight: 134.14

MP (deg C): > 300 FP (deg C):

BP (deg C):

BP pressure (mm Hg):

Property/ Value	Units	Temp	Data Type	Reference
WS 9.79E+003	mg/L	25	EST	MEYLAN,WM ET AL. (1996)
logP 1.12			EXP	HANSCH,C & LEO,AJ (1985)
VP 3.70E-004	mm Hg	25	EST	NEELY,WB & BLAU,GE (1985)
DC	pKa			
HL 5.04E-010	atm m3/mol	25	EST	MEYLAN,WM & HOWARD,PH (1991)
OH 6.99E-011	cm3/molc sec	25	EST	MEYLAN,WM & HOWARD,PH (1993)

CAS #: 000615-22-5				2-METHYLTHIOBENZOTHIAZOLE

Formula: $C_8H_7NS_2$

Mol Weight: 181.28

MP (deg C): 43-46 FP (deg C):

BP (deg C):

BP pressure (mm Hg):

Property/ Value	Units	Temp	Data Type	Reference
WS 1.25E+002	mg/L	24	EXP	BROWNLEE,BG ET AL. (1992)
logP 3.15			EXP	PLATFORD,RF (1983)
VP 2.60E-004	mm Hg	25	EST	NEELY,WB & BLAU,GE (1985)
DC	pKa			
HL 1.09E-008	atm m3/mol	25	EST	MEYLAN,WM & HOWARD,PH (1991)
OH 4.62E-011	cm3/molc sec	25	EST	MEYLAN,WM & HOWARD,PH (1993)

CAS #: 000615-36-1				O-BROMOANILINE

Formula: C_6H_6BrN

Mol Weight: 172.03

MP (deg C): 32 FP (deg C):

BP (deg C): 229

BP pressure (mm Hg):

Property/ Value	Units	Temp	Data Type	Reference
WS 9.49E+002	mg/L	25	EST	MEYLAN,WM ET AL. (1996)
logP 2.11			EXP	HANSCH,C & LEO,AJ (1985)
VP 4.29E-002	mm Hg	25	EST	NEELY,WB & BLAU,GE (1985)
DC 2.53	pKa	25	EXP	PERRIN,DD (1965)
HL 7.59E-007	atm m3/mol	25	EST	MEYLAN,WM & HOWARD,PH (1991)
OH 3.09E-011	cm3/molc sec	25	EST	MEYLAN,WM & HOWARD,PH (1993)

CAS #: 000615-37-2				O-IODOTOLUENE

Formula: C_7H_7I

Mol Weight: 218.04

MP (deg C): FP (deg C):

BP (deg C): 211.5

BP pressure (mm Hg):

Property/ Value	Units	Temp	Data Type	Reference
WS 2.40E+001	mg/L	25	EST	MEYLAN,WM ET AL. (1996)
logP 3.71			EST	MEYLAN,WM & HOWARD,PH (1995)
VP 4.53E-001	mm Hg	25	EXT	PERRY,RH & GREEN,D (1984)
DC	pKa			
HL 1.38E-003	atm m3/mol	25	EST	MEYLAN,WM & HOWARD,PH (1991)
OH 1.82E-012	cm3/molc sec	25	EST	MEYLAN,WM & HOWARD,PH (1993)

CAS #: 000615-41-8				2-CHLOROIODOBENZENE

Formula: C_6H_4ClI

Mol Weight: 238.46

MP (deg C): 0.7 FP (deg C):

BP (deg C): 234.5

BP pressure (mm Hg):

Property/ Value	Units	Temp	Data Type	Reference
WS 6.88E+001	mg/L	25	EXP	SUZUKI,T (1991)
logP 3.80			EST	MEYLAN,WM & HOWARD,PH (1995)
VP 9.41E-002	mm Hg	25	EST	NEELY,WB & BLAU,GE (1985)
DC	pKa			
HL 9.25E-004	atm m3/mol	25	EST	MEYLAN,WM & HOWARD,PH (1991)
OH 4.53E-013	cm3/molc sec	25	EST	MEYLAN,WM & HOWARD,PH (1993)

CAS #: 000615-42-9				1,2-DIIODOBENZENE

Formula: $C_6H_4I_2$

Mol Weight: 329.91

MP (deg C): 27 FP (deg C):

BP (deg C): 287

BP pressure (mm Hg):

Property/ Value	Units	Temp	Data Type	Reference
WS 1.90E+001	mg/L	27	EXP	SUZUKI,T (1991)
logP 4.33			EST	MEYLAN,WM & HOWARD,PH (1995)
VP 5.93E-003	mm Hg	25	EST	NEELY,WB & BLAU,GE (1985)
DC	pKa			
HL 2.89E-004	atm m3/mol	25	EST	MEYLAN,WM & HOWARD,PH (1991)
OH 4.25E-013	cm3/molc sec	25	EST	MEYLAN,WM & HOWARD,PH (1993)

449

000615-43-0 — 2-IODOANILINE

Formula: C_6H_6IN
Mol Weight: 219.03
MP (deg C): 55-58
FP (deg C):
BP (deg C):
BP pressure (mm Hg):

Property/Value	Units	Temp	Data Type	Reference
WS 3.64E+002	mg/L	25	EST	MEYLAN,WM ET AL. (1996)
logP 2.32			EXP	HANSCH,C & LEO,AJ (1985)
VP 7.67E-003	mm Hg	25	EST	NEELY,WB & BLAU,GE (1985)
DC 2.60	pKa		EXP	PERRIN,DD (1965)
HL 4.41E-007	atm m3/mol	25	EST	MEYLAN,WM & HOWARD,PH (1991)
OH 3.56E-011	cm3/molc sec	25	EST	MEYLAN,WM & HOWARD,PH (1993)

000615-53-2 — ETHYL N-METHYLNITROSOCARBAMATE

Formula: $C_4H_8N_2O_3$
Mol Weight: 132.12
MP (deg C):
FP (deg C):
BP (deg C): 170
BP pressure (mm Hg):

Property/Value	Units	Temp	Data Type	Reference
WS 8.21E+003	mg/L	25	EST	MEYLAN,WM ET AL. (1996)
logP 1.22			EST	MEYLAN,WM & HOWARD,PH (1995)
VP 1.19E+000	mm Hg	25	EXT	KLEIN,RG (1982)
DC	pKa			
HL 1.43E-008	atm m3/mol	25	EST	MEYLAN,WM & HOWARD,PH (1991)
OH 2.22E-012	cm3/molc sec	25	EST	MEYLAN,WM & HOWARD,PH (1993)

000615-54-3 — 1,2,4-TRIBROMOBENZENE

Formula: $C_6H_3Br_3$
Mol Weight: 314.82
MP (deg C): 44.5
FP (deg C):
BP (deg C): 275
BP pressure (mm Hg):

Property/Value	Units	Temp	Data Type	Reference
WS 4.90E+000	mg/L	25	EXP	SUZUKI,T (1991)
logP 4.66			EST	MEYLAN,WM & HOWARD,PH (1995)
VP 5.48E-003	mm Hg	25	EST	NEELY,WB & BLAU,GE (1985)
DC	pKa			
HL 3.41E-004	atm m3/mol	25	EST	MEYLAN,WM & HOWARD,PH (1991)
OH 2.21E-013	cm3/molc sec	25	EST	MEYLAN,WM & HOWARD,PH (1993)

000615-56-5 — 3,4-DIBROMOPHENOL

Formula: $C_6H_4Br_2O$
Mol Weight: 251.92
MP (deg C):
FP (deg C):
BP (deg C):
BP pressure (mm Hg):

Property/Value	Units	Temp	Data Type	Reference
WS 1.35E+002	mg/L	25	EST	MEYLAN,WM ET AL. (1996)
logP 3.29			EST	MEYLAN,WM & HOWARD,PH (1995)
VP 1.34E-003	mm Hg	25	EST	NEELY,WB & BLAU,GE (1985)
DC	pKa			
HL 8.90E-008	atm m3/mol	25	EST	MEYLAN,WM & HOWARD,PH (1991)
OH 6.15E-012	cm3/molc sec	25	EST	MEYLAN,WM & HOWARD,PH (1993)

000615-57-6 — 2,4-DIBROMOANILINE

Formula: $C_6H_5Br_2N$
Mol Weight: 250.93
MP (deg C): 79.5
FP (deg C):
BP (deg C): 156
BP pressure (mm Hg): 7.40E+001

Property/Value	Units	Temp	Data Type	Reference
WS 8.51E+001	mg/L	25	EST	MEYLAN,WM ET AL. (1996)
logP 2.86			EST	MEYLAN,WM & HOWARD,PH (1995)
VP 1.96E-003	mm Hg	25	EST	NEELY,WB & BLAU,GE (1985)
DC 2.30	pKa	15	EXP	PERRIN,DD (1965)
HL 3.02E-007	atm m3/mol	25	EST	MEYLAN,WM & HOWARD,PH (1991)
OH 8.84E-012	cm3/molc sec	25	EST	MEYLAN,WM & HOWARD,PH (1993)

000615-58-7 — 2,4-DIBROMOPHENOL

Formula: $C_6H_4Br_2O$
Mol Weight: 251.92
MP (deg C): 38
FP (deg C):
BP (deg C): 238.5
BP pressure (mm Hg):

Property/Value	Units	Temp	Data Type	Reference
WS 1.56E+002	mg/L	25	EST	MEYLAN,WM ET AL. (1996)
logP 3.22			EXP	HANSCH,C & LEO,AJ (1985)
VP 1.34E-003	mm Hg	25	EST	NEELY,WB & BLAU,GE (1985)
DC 7.79	pKa	25	EXP	SERJEANT,EP & DEMPSEY,B (1979)
HL 8.90E-008	atm m3/mol	25	EST	MEYLAN,WM & HOWARD,PH (1991)
OH 2.88E-012	cm3/molc sec	25	EST	MEYLAN,WM & HOWARD,PH (1993)

000615-60-1 — 3,4-DIMETHYLCHLOROBENZENE

Formula: C_8H_9Cl
Mol Weight: 140.61
MP (deg C): -6
FP (deg C):
BP (deg C): 194
BP pressure (mm Hg):

Property/Value	Units	Temp	Data Type	Reference
WS 4.55E+001	mg/L	25	EST	MEYLAN,WM ET AL. (1996)
logP 3.82			EXP	HANSCH,C & LEO,AJ (1985)
VP 8.43E-001	mm Hg	25	EST	NEELY,WB & BLAU,GE (1985)
DC	pKa			
HL 4.86E-003	atm m3/mol	25	EST	MEYLAN,WM & HOWARD,PH (1991)
OH 4.66E-012	cm3/molc sec	25	EST	MEYLAN,WM & HOWARD,PH (1993)

000615-67-8 — CHLOROHYDROQUINONE

Formula: $C_6H_5ClO_2$
Mol Weight: 144.56
MP (deg C): 108
FP (deg C):
BP (deg C): 263
BP pressure (mm Hg):

Property/Value	Units	Temp	Data Type	Reference
WS 1.94E+004	mg/L	25	EST	MEYLAN,WM ET AL. (1996)
logP 1.40			EXP	HANSCH,C & LEO,AJ (1985)
VP 3.14E-006	mm Hg	25	EXT	JONES,AH (1960)
DC	pKa			
HL 4.32E-011	atm m3/mol	25	EST	MEYLAN,WM & HOWARD,PH (1991)
OH 1.64E-011	cm3/molc sec	25	EST	MEYLAN,WM & HOWARD,PH (1993)

CAS #: 000615-74-7				PHENOL, 2-CHLORO-5-METHYL-

Formula: C_7H_7ClO
Mol Weight: 142.59
MP (deg C): 45.5
FP (deg C):
BP (deg C): 196
BP pressure (mm Hg):

Property/ Value	Units	Temp	Data Type	Reference
WS 1.04E+003	mg/L	25	EST	MEYLAN,WM ET AL. (1996)
logP 2.90			EXP	CHEM INSPECT TEST INST (1992)
VP 4.05E-002	mm Hg	25	EST	NEELY,WB & BLAU,GE (1985)
DC	pKa			
HL 4.58E-007	atm m3/mol	25	EST	MEYLAN,WM & HOWARD,PH (1991)
OH 2.57E-011	cm3/molc sec	25	EST	MEYLAN,WM & HOWARD,PH (1993)

CAS #: 000615-77-0				1-METHYLURACIL

Formula: $C_5H_6N_2O_2$
Mol Weight: 126.12
MP (deg C): 236-238
FP (deg C):
BP (deg C):
BP pressure (mm Hg):

Property/ Value	Units	Temp	Data Type	Reference
WS 2.00E+004	mg/L		EXP	YALKOWSKY,SH & DANNENFELSER,RM (1992)
logP -1.20			EXP	HANSCH,C & LEO,AJ (1985)
VP 1.40E-006	mm Hg	25	EST	NEELY,WB & BLAU,GE (1985)
DC	pKa			
HL 1.90E-010	atm m3/mol	25	EST	MEYLAN,WM & HOWARD,PH (1991)
OH 9.33E-012	cm3/molc sec	25	EST	MEYLAN,WM & HOWARD,PH (1993)

CAS #: 000616-03-5				2,4-IMIDAZOLIDINEDIONE, 5-METHYL-

Formula: $C_4H_6N_2O_2$
Mol Weight: 114.10
MP (deg C):
FP (deg C):
BP (deg C):
BP pressure (mm Hg):

Property/ Value	Units	Temp	Data Type	Reference
WS 3.09E+004	mg/L	25	EST	MEYLAN,WM ET AL. (1996)
logP -0.91			EXP	SANGSTER,J (1994)
VP 4.16E-006	mm Hg	25	EST	NEELY,WB & BLAU,GE (1985)
DC	pKa			
HL 2.08E-009	atm m3/mol	25	EST	MEYLAN,WM & HOWARD,PH (1991)
OH 7.30E-012	cm3/molc sec	25	EST	MEYLAN,WM & HOWARD,PH (1993)

CAS #: 000616-20-6				3-CHLOROPENTANE

Formula: $C_5H_{11}Cl$
Mol Weight: 106.60
MP (deg C): 105
FP (deg C):
BP (deg C): 97.5
BP pressure (mm Hg):

Property/ Value	Units	Temp	Data Type	Reference
WS 2.50E+002	mg/L	25	EXP	HINE,J & MOOKERJEE,PK (1975)
logP 2.98			EST	MEYLAN,WM & HOWARD,PH (1995)
VP 4.68E+001	mm Hg	25	EXP	HINE,J & MOOKERJEE,PK (1975)
DC	pKa			
HL 2.62E-002	atm m3/mol	25	EST	VP/WSOL
OH 2.12E-012	cm3/molc sec	25	EST	MEYLAN,WM & HOWARD,PH (1993)

CAS #: 000616-21-7				1,2-DICHLOROBUTANE

Formula: $C_4H_8Cl_2$
Mol Weight: 127.01
MP (deg C):
FP (deg C):
BP (deg C): 124.1
BP pressure (mm Hg):

Property/ Value	Units	Temp	Data Type	Reference
WS 4.32E+002	mg/L	25	EST	MEYLAN,WM ET AL. (1996)
logP 2.74			EST	MEYLAN,WM & HOWARD,PH (1995)
VP 2.09E+001	mm Hg	25	EXP	PERRY,RH & GREEN,D (1984)
DC	pKa			
HL 2.14E-002	atm m3/mol	25	EST	MEYLAN,WM & HOWARD,PH (1991)
OH 9.58E-013	cm3/molc sec	25	EST	MEYLAN,WM & HOWARD,PH (1993)

CAS #: 000616-23-9				2,3-DICHLOROPROPANOL

Formula: $C_3H_6Cl_2O$
Mol Weight: 128.99
MP (deg C):
FP (deg C):
BP (deg C): 184
BP pressure (mm Hg):

Property/ Value	Units	Temp	Data Type	Reference
WS 6.42E+004	mg/L	25	EST	MEYLAN,WM ET AL. (1996)
logP 0.78			EST	MEYLAN,WM & HOWARD,PH (1995)
VP 1.84E-001	mm Hg	25	EST	NEELY,WB & BLAU,GE (1985)
DC	pKa			
HL 5.89E-007	atm m3/mol	25	EST	MEYLAN,WM & HOWARD,PH (1991)
OH 1.77E-012	cm3/molc sec	25	EST	MEYLAN,WM & HOWARD,PH (1993)

CAS #: 000616-25-1				1-PENTEN-3-OL

Formula: $C_5H_{10}O$
Mol Weight: 86.13
MP (deg C):
FP (deg C):
BP (deg C): 115
BP pressure (mm Hg):

Property/ Value	Units	Temp	Data Type	Reference
WS 1.46E+005	mg/L	25	EXP	SUZUKI,T (1991)
logP 1.12			EST	MEYLAN,WM & HOWARD,PH (1995)
VP 9.68E+000	mm Hg	25	EST	NEELY,WB & BLAU,GE (1985)
DC	pKa			
HL 9.88E-006	atm m3/mol	25	EST	MEYLAN,WM & HOWARD,PH (1991)
OH 3.61E-011	cm3/molc sec	25	EST	MEYLAN,WM & HOWARD,PH (1993)

CAS #: 000616-44-4				3-METHYLTHIOPHENE

Formula: C_5H_6S
Mol Weight: 98.17
MP (deg C): 13
FP (deg C):
BP (deg C): 112.5
BP pressure (mm Hg):

Property/ Value	Units	Temp	Data Type	Reference
WS 4.00E+002	mg/L	25	EXP	CHEM INSPECT TEST INST
logP 2.34			EXP	HANSCH,C & LEO,AJ (1985)
VP 2.22E+001	mm Hg	25	EXP	RIDDICK,JA ET AL. (1986)
DC	pKa			
HL 7.15E-003	atm m3/mol	25	EST	VP/WSOL
OH 2.50E-011	cm3/molc sec	25	EST	MEYLAN,WM & HOWARD,PH (1993)

451

CAS #: 000616-45-5 — PYRROLIDONE

Formula:	C_4H_7NO			
Mol Weight:	85.11			
MP (deg C):	25	FP (deg C):		
BP (deg C):	251			
BP pressure (mm Hg):				

Property/Value	Units	Temp	Data Type	Reference
WS 1.00E+006	mg/L	25	EXP	RIDDICK,JA ET AL. (1986)
logP -0.85			EXP	SASAKI,H ET AL. (1991)
VP 2.03E-002	mm Hg	25	EXP	DAUBERT,TE & DANNER,RP (1989)
DC	pKa			
HL 2.27E-009	atm m3/mol	25	EST	VP/WSOL
OH 1.35E-011	cm3/molc sec	25	EST	MEYLAN,WM & HOWARD,PH (1993)

CAS #: 000616-47-7 — 1H-IMIDAZOLE, 1-METHYL-

Formula:	$C_4H_6N_2$			
Mol Weight:	82.11			
MP (deg C):	-6	FP (deg C):		
BP (deg C):	195.5			
BP pressure (mm Hg):				

Property/Value	Units	Temp	Data Type	Reference
WS 1.00E+006	mg/L	20	EXP	YALKOWSKY,SH & DANNENFELSER,RM (1992)
logP -0.06			EXP	HANSCH,C ET AL. (1995)
VP 4.78E-001	mm Hg	25	EST	NEELY,WB & BLAU,GE (1985)
DC 6.95	pKa	25	EXP	PERRIN,DD (1965)
HL 8.01E-005	atm m3/mol	25	EST	MEYLAN,WM & HOWARD,PH (1991)
OH 3.61E-011	cm3/molc sec	25	EST	MEYLAN,WM & HOWARD,PH (1993)

CAS #: 000616-62-6 — PROPYLPROPANEDIOIC ACID

Formula:	$C_6H_{10}O_4$			
Mol Weight:	146.14			
MP (deg C):		FP (deg C):		
BP (deg C):				
BP pressure (mm Hg):				

Property/Value	Units	Temp	Data Type	Reference
WS 7.00E+005	mg/L	25	EXP	YALKOWSKY,SH & DANNENFELSER,RM (1992)
logP 0.15			EST	MEYLAN,WM & HOWARD,PH (1995)
VP 3.80E-004	mm Hg	25	EST	NEELY,WB & BLAU,GE (1985)
DC 2.99	pKa	25	EXP	KORTUM,G ET AL (1961)
HL 9.53E-012	atm m3/mol	25	EST	MEYLAN,WM & HOWARD,PH (1991)
OH 5.11E-012	cm3/molc sec	25	EST	MEYLAN,WM & HOWARD,PH (1993)

CAS #: 000616-68-2 — TRIMECAIN

Formula:	$C_{15}H_{24}N_2O$			
Mol Weight:	248.37			
MP (deg C):	44	FP (deg C):		
BP (deg C):	187			
BP pressure (mm Hg):	6.00E+000			

Property/Value	Units	Temp	Data Type	Reference
WS 2.11E+002	mg/L	25	EST	MEYLAN,WM ET AL. (1996)
logP 2.41			EXP	HANSCH,C & LEO,AJ (1985)
VP 4.34E-007	mm Hg	25	EST	NEELY,WB & BLAU,GE (1985)
DC	pKa			
HL 1.45E-010	atm m3/mol	25	EST	MEYLAN,WM & HOWARD,PH (1991)
OH 1.09E-010	cm3/molc sec	25	EST	MEYLAN,WM & HOWARD,PH (1993)

CAS #: 000616-73-9 — 2,4-DINITRO-5-METHYLPHENOL

Formula:	$C_7H_6N_2O_5$			
Mol Weight:	198.14			
MP (deg C):		FP (deg C):		
BP (deg C):				
BP pressure (mm Hg):				

Property/Value	Units	Temp	Data Type	Reference
WS 5.12E+002	mg/L	25	EST	MEYLAN,WM ET AL. (1996)
logP 2.27			EST	MEYLAN,WM & HOWARD,PH (1995)
VP 5.48E-006	mm Hg	25	EST	NEELY,WB & BLAU,GE (1985)
DC	pKa			
HL 3.05E-008	atm m3/mol	25	EST	MEYLAN,WM & HOWARD,PH (1991)
OH 1.64E-012	cm3/molc sec	25	EST	MEYLAN,WM & HOWARD,PH (1993)

CAS #: 000616-82-0 — 4-HYDROXY-3-NITROBENZOIC ACID

Formula:	$C_7H_5NO_5$			
Mol Weight:	183.12			
MP (deg C):	229-231	FP (deg C):		
BP (deg C):				
BP pressure (mm Hg):				

Property/Value	Units	Temp	Data Type	Reference
WS 1.40E+003	mg/L	25	EST	MEYLAN,WM ET AL. (1996)
logP 1.85			EXP	HANSCH,C & LEO,AJ (1985)
VP 1.16E-006	mm Hg	25	EST	NEELY,WB & BLAU,GE (1985)
DC	pKa			
HL 1.41E-010	atm m3/mol	25	EST	MEYLAN,WM & HOWARD,PH (1991)
OH 2.20E-012	cm3/molc sec	25	EST	MEYLAN,WM & HOWARD,PH (1993)

CAS #: 000616-88-6 — PENTYLMALONIC ACID

Formula:	$C_8H_{14}O_4$			
Mol Weight:	174.20			
MP (deg C):		FP (deg C):		
BP (deg C):				
BP pressure (mm Hg):				

Property/Value	Units	Temp	Data Type	Reference
WS 6.81E+005	mg/L	25	EXP	STEPHEN,H & STEPHEN,T (1963)
logP 1.14			EST	MEYLAN,WM & HOWARD,PH (1995)
VP 5.73E-005	mm Hg	25	EST	NEELY,WB & BLAU,GE (1985)
DC	pKa			
HL 1.68E-011	atm m3/mol	25	EST	MEYLAN,WM & HOWARD,PH (1991)
OH 7.94E-012	cm3/molc sec	25	EST	MEYLAN,WM & HOWARD,PH (1993)

CAS #: 000617-29-8 — 2-METHYL-3-HEXANOL

Formula:	$C_7H_{16}O$			
Mol Weight:	116.20			
MP (deg C):		FP (deg C):		
BP (deg C):	141-143			
BP pressure (mm Hg):	7.65E+002			

Property/Value	Units	Temp	Data Type	Reference
WS 5.70E+003	mg/L	25	EXP	BARTON,AFM (1984)
logP 2.17			EST	MEYLAN,WM & HOWARD,PH (1995)
VP 1.35E+000	mm Hg	25	EXT	WILHOIT,RC & ZWOLINSKI,BJ (1973)
DC	pKa			
HL 2.34E-005	atm m3/mol	25	EST	MEYLAN,WM & HOWARD,PH (1991)
OH 1.59E-011	cm3/molc sec	25	EST	MEYLAN,WM & HOWARD,PH (1993)

CAS #: 000617-31-2				PENTANOIC ACID, 2-HYDROXY-

Formula: $C_5H_{10}O_3$

Mol Weight: 118.13

MP (deg C): 34   FP (deg C):

BP (deg C):

BP pressure (mm Hg):

Property/ Value	Units	Temp	Data Type	Reference
WS 3.38E+005	mg/L	25	EST	MEYLAN,WM ET AL. (1996)
logP 0.45			EXP	SANGSTER,J (1993)
VP 3.46E-003	mm Hg	25	EST	NEELY,WB & BLAU,GE (1985)
DC 3.59	pKa	25	EXP	SERJEANT,EP & DEMPSEY,B (1979)
HL 2.00E-007	atm m3/mol	25	EST	MEYLAN,WM & HOWARD,PH (1991)
OH 9.65E-012	cm3/molc sec	25	EST	MEYLAN,WM & HOWARD,PH (1993)

CAS #: 000617-50-5				ISOPROPYL ISOBUTYRATE

Formula: $C_7H_{14}O_2$

Mol Weight: 130.19

MP (deg C):   FP (deg C):

BP (deg C): 120.7

BP pressure (mm Hg):

Property/ Value	Units	Temp	Data Type	Reference
WS 1.24E+003	mg/L	25	EST	MEYLAN,WM ET AL. (1996)
logP 2.19			EST	MEYLAN,WM & HOWARD,PH (1995)
VP 1.60E+001	mm Hg	25	EXP	OHE,S (1976)
DC	pKa			
HL 5.45E-004	atm m3/mol	25	EST	MEYLAN,WM & HOWARD,PH (1991)
OH 5.21E-012	cm3/molc sec	25	EST	MEYLAN,WM & HOWARD,PH (1993)

CAS #: 000617-65-2				GLUTAMIC ACID (DL)

Formula: $C_5H_9NO_4$

Mol Weight: 147.13

MP (deg C): 199 dec   FP (deg C):

BP (deg C):

BP pressure (mm Hg):

Property/ Value	Units	Temp	Data Type	Reference
WS 2.05E+004	mg/L	25	EXP	HORN,AS (1981)
logP -3.83			EST	MEYLAN,WM & HOWARD,PH (1995)
VP 5.19E-007	mm Hg	25	EST	NEELY,WB & BLAU,GE (1985)
DC 2.23	pKa	0	EXP	KORTUM,G ET AL (1961)
HL 1.47E-014	atm m3/mol	25	EST	MEYLAN,WM & HOWARD,PH (1991)
OH 4.10E-011	cm3/molc sec	25	EST	MEYLAN,WM & HOWARD,PH (1993)

CAS #: 000617-78-7				3-ETHYLPENTANE

Formula: C_7H_{16}

Mol Weight: 100.21

MP (deg C): -118.6   FP (deg C):

BP (deg C): 93.5

BP pressure (mm Hg):

Property/ Value	Units	Temp	Data Type	Reference
WS 2.32E+001	mg/L	25	EST	MEYLAN,WM ET AL. (1996)
logP 3.71			EST	MEYLAN,WM & HOWARD,PH (1995)
VP 5.79E+001	mm Hg	25	EXP	DAUBERT,TE & DANNER,RP (1989)
DC	pKa			
HL 2.27E+000	atm m3/mol	25	EST	MEYLAN,WM & HOWARD,PH (1991)
OH 7.56E-012	cm3/molc sec	25	EST	MEYLAN,WM & HOWARD,PH (1993)

CAS #: 000617-84-5				N,N-DIETHYLFORMAMIDE

Formula: $C_5H_{11}NO$

Mol Weight: 101.15

MP (deg C):   FP (deg C):

BP (deg C): 177.5

BP pressure (mm Hg):

Property/ Value	Units	Temp	Data Type	Reference
WS 1.00E+006	mg/L	20	EXP	YALKOWSKY,SH & DANNENFELSER,RM (1992)
logP 0.05			EST	MEYLAN,WM & HOWARD,PH (1995)
VP 1.21E+000	mm Hg	25	EXT	BOUBLIK,T ET AL. (1984)
DC	pKa			
HL 1.30E-007	atm m3/mol	25	EST	MEYLAN,WM & HOWARD,PH (1991)
OH 3.27E-011	cm3/molc sec	25	EST	MEYLAN,WM & HOWARD,PH (1993)

CAS #: 000617-86-7				TRIETHYLSILANE

Formula: $C_6H_{16}Si$

Mol Weight: 116.28

MP (deg C):   FP (deg C):

BP (deg C): 109

BP pressure (mm Hg):

Property/ Value	Units	Temp	Data Type	Reference
WS 7.99E+001	mg/L	25	EST	MEYLAN,WM ET AL. (1996)
logP 3.64			EST	MEYLAN,WM & HOWARD,PH (1995)
VP 4.29E+001	mm Hg	25	EST	NEELY,WB & BLAU,GE (1985)
DC	pKa			
HL 3.76E-001	atm m3/mol	25	EST	MEYLAN,WM & HOWARD,PH (1991)
OH 3.58E-012	cm3/molc sec	25	EST	MEYLAN,WM & HOWARD,PH (1993)

CAS #: 000617-89-0				2-AMINOMETHYLFURAN

Formula: C_5H_7NO

Mol Weight: 97.12

MP (deg C):   FP (deg C):

BP (deg C): 145.5

BP pressure (mm Hg):

Property/ Value	Units	Temp	Data Type	Reference
WS 5.87E+005	mg/L	25	EST	MEYLAN,WM ET AL. (1996)
logP 0.37			EXP	HANSCH,C & LEO,AJ (1985)
VP 4.73E+000	mm Hg	25	EST	NEELY,WB & BLAU,GE (1985)
DC 8.89	pKa	30	EXP	PERRIN,DD (1965)
HL 6.10E-007	atm m3/mol	25	EST	MEYLAN,WM & HOWARD,PH (1991)
OH 1.30E-010	cm3/molc sec	25	EST	MEYLAN,WM & HOWARD,PH (1993)

CAS #: 000617-90-3				2-CYANOFURAN

Formula: C_5H_4NO

Mol Weight: 94.09

MP (deg C):   FP (deg C):

BP (deg C): 147

BP pressure (mm Hg):

Property/ Value	Units	Temp	Data Type	Reference
WS 1.01E+004	mg/L	25	EST	MEYLAN,WM ET AL. (1996)
logP 0.96			EXP	HANSCH,C & LEO,AJ (1985)
VP 3.20E+000	mm Hg	25	EST	NEELY,WB & BLAU,GE (1985)
DC	pKa			
HL 5.19E-005	atm m3/mol	25	EST	MEYLAN,WM & HOWARD,PH (1991)
OH 7.15E-012	cm3/molc sec	25	EST	MEYLAN,WM & HOWARD,PH (1993)

CAS #: 000617-94-7 — 2-PHENYL ISOPROPANOL

Formula: $C_9H_{12}O$
Mol Weight: 136.20
MP (deg C): 35-37
FP (deg C):
BP (deg C): 202
BP pressure (mm Hg):

Property/Value	Units	Temp	Data Type	Reference
WS 7.14E+003	mg/L	25	EST	MEYLAN,WM ET AL. (1996)
logP 1.95			EST	MEYLAN,WM & HOWARD,PH (1995)
VP 5.20E-001	mm Hg	38	EXP	YAWS,CL (1994B)
DC	pKa			
HL 3.83E-007	atm m3/mol	25	EST	MEYLAN,WM & HOWARD,PH (1991)
OH 5.50E-012	cm3/molc sec	25	EST	MEYLAN,WM & HOWARD,PH (1993)

CAS #: 000618-32-6 — BENZOYL BROMIDE

Formula: C_7H_5BrO
Mol Weight: 185.03
MP (deg C): -24
FP (deg C):
BP (deg C): 218.5
BP pressure (mm Hg):

Property/Value	Units	Temp	Data Type	Reference
WS 2.59E+003	mg/L	25	EST	MEYLAN,WM ET AL. (1996)
logP 1.53			EST	MEYLAN,WM & HOWARD,PH (1995)
VP 2.22E-001	mm Hg	25	EXT	PERRY,RH & GREEN,D (1984)
DC	pKa			
HL 7.73E-005	atm m3/mol	25	EST	MEYLAN,WM & HOWARD,PH (1991)
OH 1.78E-012	cm3/molc sec	25	EST	MEYLAN,WM & HOWARD,PH (1993)

CAS #: 000618-45-1 — M-ISOPROPYLPHENOL

Formula: $C_9H_{12}O$
Mol Weight: 136.20
MP (deg C): 26
FP (deg C):
BP (deg C): 228
BP pressure (mm Hg):

Property/Value	Units	Temp	Data Type	Reference
WS 9.62E+002	mg/L	25	EST	MEYLAN,WM ET AL. (1996)
logP 2.97			EST	MEYLAN,WM & HOWARD,PH (1995)
VP 6.95E-002	mm Hg	25	EXT	CHAO,J ET AL. (1983)
DC 10.16	pKa	20	EXP	SERJEANT,EP & DEMPSEY,B (1979)
HL 1.09E-006	atm m3/mol	25	EST	MEYLAN,WM & HOWARD,PH (1991)
OH 8.15E-011	cm3/molc sec	25	EST	MEYLAN,WM & HOWARD,PH (1993)

CAS #: 000618-46-2 — M-CHLOROBENZOYLCHLORIDE

Formula: $C_7H_4Cl_2O$
Mol Weight: 175.02
MP (deg C):
FP (deg C):
BP (deg C): 225
BP pressure (mm Hg):

Property/Value	Units	Temp	Data Type	Reference
WS 9.72E+002	mg/L	25	EST	MEYLAN,WM ET AL. (1996)
logP 2.08			EST	MEYLAN,WM & HOWARD,PH (1995)
VP 1.40E-001	mm Hg	25	EXP	DAUBERT,TE & DANNER,RP (1989)
DC	pKa			
HL 9.75E-005	atm m3/mol	25	EST	MEYLAN,WM & HOWARD,PH (1991)
OH 5.78E-013	cm3/molc sec	25	EST	MEYLAN,WM & HOWARD,PH (1993)

CAS #: 000618-47-3 — 3-METHYLBENZAMIDE

Formula: C_8H_9NO
Mol Weight: 135.17
MP (deg C): 94-96
FP (deg C):
BP (deg C):
BP pressure (mm Hg):

Property/Value	Units	Temp	Data Type	Reference
WS 8.61E+003	mg/L	25	EST	MEYLAN,WM ET AL. (1996)
logP 1.18			EXP	HANSCH,C & LEO,AJ (1985)
VP 3.19E-004	mm Hg	25	EST	NEELY,WB & BLAU,GE (1985)
DC	pKa			
HL 2.44E-009	atm m3/mol	25	EST	MEYLAN,WM & HOWARD,PH (1991)
OH 4.31E-012	cm3/molc sec	25	EST	MEYLAN,WM & HOWARD,PH (1993)

CAS #: 000618-48-4 — 3-CHLOROBENZAMIDE

Formula: C_7H_6ClNO
Mol Weight: 155.58
MP (deg C):
FP (deg C):
BP (deg C):
BP pressure (mm Hg):

Property/Value	Units	Temp	Data Type	Reference
WS 3.68E+003	mg/L	25	EST	MEYLAN,WM ET AL. (1996)
logP 1.51			EXP	HANSCH,C & LEO,AJ (1985)
VP 1.65E-004	mm Hg	25	EST	NEELY,WB & BLAU,GE (1985)
DC	pKa			
HL 1.64E-009	atm m3/mol	25	EST	MEYLAN,WM & HOWARD,PH (1991)
OH 2.58E-012	cm3/molc sec	25	EST	MEYLAN,WM & HOWARD,PH (1993)

CAS #: 000618-49-5 — 3-HYDROXYBENZAMIDE

Formula: $C_7H_7NO_2$
Mol Weight: 137.14
MP (deg C):
FP (deg C):
BP (deg C):
BP pressure (mm Hg):

Property/Value	Units	Temp	Data Type	Reference
WS 1.52E+005	mg/L	25	EST	MEYLAN,WM ET AL. (1996)
logP 0.39			EXP	HANSCH,C & LEO,AJ (1985)
VP 1.20E-005	mm Hg	25	EST	NEELY,WB & BLAU,GE (1985)
DC	pKa			
HL 2.30E-013	atm m3/mol	25	EST	MEYLAN,WM & HOWARD,PH (1991)
OH 1.62E-011	cm3/molc sec	25	EST	MEYLAN,WM & HOWARD,PH (1993)

CAS #: 000618-51-9 — 3-IODOBENZOIC ACID

Formula: $C_7H_5IO_2$
Mol Weight: 248.02
MP (deg C): 188.3
FP (deg C):
BP (deg C):
BP pressure (mm Hg):

Property/Value	Units	Temp	Data Type	Reference
WS 1.16E+002	mg/L	25	EXP	STEPHEN,H & STEPHEN,T (1963)
logP 3.13			EXP	HANSCH,C & LEO,AJ (1985)
VP 1.98E-004	mm Hg	25	EST	NEELY,WB & BLAU,GE (1985)
DC 3.85	pKa	25	EXP	KORTUM,G ET AL (1961)
HL 2.51E-008	atm m3/mol	25	EST	MEYLAN,WM & HOWARD,PH (1991)
OH 8.71E-013	cm3/molc sec	25	EST	MEYLAN,WM & HOWARD,PH (1993)

CAS #: 000618-62-2 — BENZENE, 1,3-DICHLORO-5-NITRO-

Formula: $C_6H_3Cl_2NO_2$

Mol Weight: 192.00

MP (deg C): 65.4 | FP (deg C):

BP (deg C):

BP pressure (mm Hg):

Property/Value	Units	Temp	Data Type	Reference
WS 4.49E+001	mg/L	25	EST	MEYLAN,WM ET AL. (1996)
logP 3.09			EXP	NIIMI,AJ ET AL. (1989)
VP 5.05E-003	mm Hg	25	EST	NEELY,WB & BLAU,GE (1985)
DC	pKa			
HL 1.17E-005	atm m3/mol	25	EST	MEYLAN,WM & HOWARD,PH (1991)
OH 8.43E-014	cm3/molc sec	25	EST	MEYLAN,WM & HOWARD,PH (1993)

CAS #: 000618-80-4 — 2,6-DICHLORO-4-NITROPHENOL

Formula: $C_6H_3Cl_2NO_3$

Mol Weight: 208.00

MP (deg C): | FP (deg C):

BP (deg C):

BP pressure (mm Hg):

Property/Value	Units	Temp	Data Type	Reference
WS 4.67E+002	mg/L	25	EST	MEYLAN,WM ET AL. (1996)
logP 2.94			EXP	HANSCH,C & LEO,AJ (1985)
VP 5.70E-005	mm Hg	25	EST	NEELY,WB & BLAU,GE (1985)
DC 3.55	pKa		EXP	SERJEANT,EP & DEMPSEY,B (1979)
HL 1.21E-009	atm m3/mol	25	EST	MEYLAN,WM & HOWARD,PH (1991)
OH 1.98E-013	cm3/molc sec	25	EST	MEYLAN,WM & HOWARD,PH (1993)

CAS #: 000618-87-1 — 3,5-DINITROANILINE

Formula: $C_6H_5N_3O_4$

Mol Weight: 183.12

MP (deg C): 163 | FP (deg C):

BP (deg C):

BP pressure (mm Hg):

Property/Value	Units	Temp	Data Type	Reference
WS 1.29E+003	mg/L	25	EST	MEYLAN,WM ET AL. (1996)
logP 1.89			EXP	HANSCH,C ET AL. (1995)
VP 2.68E-005	mm Hg	25	EST	NEELY,WB & BLAU,GE (1985)
DC 0.30	pKa	15	EXP	PERRIN,DD (1972)
HL 2.96E-011	atm m3/mol	25	EST	MEYLAN,WM & HOWARD,PH (1991)
OH 8.22E-013	cm3/molc sec	25	EST	MEYLAN,WM & HOWARD,PH (1993)

CAS #: 000618-89-3 — METHYL 3-BROMOBENZOATE

Formula: $C_8H_7BrO_2$

Mol Weight: 215.05

MP (deg C): 32 | FP (deg C):

BP (deg C): 125

BP pressure (mm Hg): 1.50E+001

Property/Value	Units	Temp	Data Type	Reference
WS 1.74E+002	mg/L	25	EST	MEYLAN,WM ET AL. (1996)
logP 2.72			EST	MEYLAN,WM & HOWARD,PH (1995)
VP 2.46E-002	mm Hg	25	EST	NEELY,WB & BLAU,GE (1985)
DC	pKa			
HL 1.38E-005	atm m3/mol	25	EST	MEYLAN,WM & HOWARD,PH (1991)
OH 4.89E-013	cm3/molc sec	25	EST	MEYLAN,WM & HOWARD,PH (1993)

CAS #: 000618-94-0 — 3-NITROBENZOYLHYDRAZINE

Formula: $C_7H_7N_3O_3$

Mol Weight: 181.15

MP (deg C): | FP (deg C):

BP (deg C):

BP pressure (mm Hg):

Property/Value	Units	Temp	Data Type	Reference
WS 6.90E+002	mg/L	25	EST	MEYLAN,WM ET AL. (1996)
logP 0.23			EXP	HANSCH,C & LEO,AJ (1985)
VP 1.08E-006	mm Hg	25	EST	NEELY,WB & BLAU,GE (1985)
DC	pKa			
HL 3.65E-014	atm m3/mol	25	EST	MEYLAN,WM & HOWARD,PH (1991)
OH 5.72E-012	cm3/molc sec	25	EST	MEYLAN,WM & HOWARD,PH (1993)

CAS #: 000618-95-1 — BENZOIC ACID, 3-NITRO-, METHYL ESTER

Formula: $C_8H_7NO_4$

Mol Weight: 181.15

MP (deg C): 78 | FP (deg C):

BP (deg C): 279

BP pressure (mm Hg): 6.00E+001

Property/Value	Units	Temp	Data Type	Reference
WS 5.39E+002	mg/L	25	EST	MEYLAN,WM ET AL. (1996)
logP 1.89			EXP	SOTOMATSU,T ET AL. (1993)
VP 1.86E-003	mm Hg	25	EST	NEELY,WB & BLAU,GE (1985)
DC	pKa			
HL 1.37E-007	atm m3/mol	25	EST	MEYLAN,WM & HOWARD,PH (1991)
OH 2.96E-013	cm3/molc sec	25	EST	MEYLAN,WM & HOWARD,PH (1993)

CAS #: 000618-98-4 — ETHYL-M-NITROBENZOATE

Formula: $C_9H_9NO_4$

Mol Weight: 195.18

MP (deg C): 47 | FP (deg C):

BP (deg C): 297

BP pressure (mm Hg):

Property/Value	Units	Temp	Data Type	Reference
WS 1.86E+002	mg/L	25	EST	MEYLAN,WM ET AL. (1996)
logP 2.35			EXP	HANSCH,C & LEO,AJ (1985)
VP 6.65E-004	mm Hg	25	EST	NEELY,WB & BLAU,GE (1985)
DC	pKa			
HL 1.82E-007	atm m3/mol	25	EST	MEYLAN,WM & HOWARD,PH (1991)
OH 1.74E-012	cm3/molc sec	25	EST	MEYLAN,WM & HOWARD,PH (1993)

CAS #: 000619-05-6 — BENZOIC ACID, 3,4-DIAMINO-

Formula: $C_7H_8N_2O_2$

Mol Weight: 152.15

MP (deg C): | FP (deg C):

BP (deg C): 215-218 de

BP pressure (mm Hg):

Property/Value	Units	Temp	Data Type	Reference
WS 5.75E+004	mg/L	25	EST	MEYLAN,WM ET AL. (1996)
logP 0.13			EXP	SANGSTER,J (1994)
VP 9.56E-006	mm Hg	25	EST	NEELY,WB & BLAU,GE (1985)
DC 3.49	pKa	25	EXP	PERRIN,DD (1972)
HL 1.35E-014	atm m3/mol	25	EST	MEYLAN,WM & HOWARD,PH (1991)
OH 6.58E-011	cm3/molc sec	25	EST	MEYLAN,WM & HOWARD,PH (1993)

000619-08-9 — 2-CHLORO-4-NITROPHENOL

Formula: $C_6H_4ClNO_3$

Mol Weight: 173.56

MP (deg C):

FP (deg C):

BP (deg C):

BP pressure (mm Hg):

Property/Value	Units	Temp	Data Type	Reference
WS 1.49E+003	mg/L	25	EST	MEYLAN,WM ET AL. (1996)
logP 2.55			EST	MEYLAN,WM & HOWARD,PH (1995)
VP 2.96E-004	mm Hg	25	EST	NEELY,WB & BLAU,GE (1985)
DC 5.45	pKa	25	EXP	SERJEANT,EP & DEMPSEY,B (1979)
HL 1.64E-009	atm m3/mol	25	EST	MEYLAN,WM & HOWARD,PH (1991)
OH 1.36E-012	cm3/molc sec	25	EST	MEYLAN,WM & HOWARD,PH (1993)

000619-14-7 — 3-HYDROXY-4-NITROBENZOIC ACID

Formula: $C_7H_5NO_5$

Mol Weight: 183.12

MP (deg C): 229-231

FP (deg C):

BP (deg C):

BP pressure (mm Hg):

Property/Value	Units	Temp	Data Type	Reference
WS 1.08E+003	mg/L	25	EST	MEYLAN,WM ET AL. (1996)
logP 1.98			EXP	HANSCH,C & LEO,AJ (1985)
VP 1.16E-006	mm Hg	25	EST	NEELY,WB & BLAU,GE (1985)
DC	pKa			
HL 1.41E-010	atm m3/mol	25	EST	MEYLAN,WM & HOWARD,PH (1991)
OH 1.80E-012	cm3/molc sec	25	EST	MEYLAN,WM & HOWARD,PH (1993)

000619-15-8 — 2,5-DINITROTOLUENE

Formula: $C_7H_6N_2O_4$

Mol Weight: 182.14

MP (deg C): 52.5

FP (deg C):

BP (deg C): 284

BP pressure (mm Hg):

Property/Value	Units	Temp	Data Type	Reference
WS 2.20E+002	mg/L	25	EST	MEYLAN,WM ET AL. (1996)
logP 2.18			EST	MEYLAN,WM & HOWARD,PH (1995)
VP 3.97E-004	mm Hg	25	EST	NEELY,WB & BLAU,GE (1985)
DC	pKa			
HL 9.26E-008	atm m3/mol	25	EST	MEYLAN,WM & HOWARD,PH (1991)
OH 2.01E-013	cm3/molc sec	25	EST	MEYLAN,WM & HOWARD,PH (1993)

000619-17-0 — 2-AMINO-4-NITROBENZOIC ACID

Formula: $C_7H_6N_2O_4$

Mol Weight: 182.14

MP (deg C): 257 dec

FP (deg C):

BP (deg C):

BP pressure (mm Hg):

Property/Value	Units	Temp	Data Type	Reference
WS 1.26E+003	mg/L	25	EST	MEYLAN,WM ET AL. (1996)
logP 1.91			EXP	HANSCH,C & LEO,AJ (1985)
VP 2.93E-006	mm Hg	25	EST	NEELY,WB & BLAU,GE (1985)
DC	pKa			
HL 1.51E-013	atm m3/mol	25	EST	MEYLAN,WM & HOWARD,PH (1991)
OH 4.00E-012	cm3/molc sec	25	EST	MEYLAN,WM & HOWARD,PH (1993)

000619-21-6 — 3-FORMYLBENZOIC ACID

Formula: $C_8H_6O_3$

Mol Weight: 150.14

MP (deg C): 173-175

FP (deg C):

BP (deg C):

BP pressure (mm Hg):

Property/Value	Units	Temp	Data Type	Reference
WS 3.31E+003	mg/L	25	EST	MEYLAN,WM ET AL. (1996)
logP 1.59			EST	MEYLAN,WM & HOWARD,PH (1995)
VP 3.75E-004	mm Hg	25	EST	NEELY,WB & BLAU,GE (1985)
DC 3.84	pKa	25	EXP	SERJEANT,EP & DEMPSEY,B (1979)
HL 2.70E-010	atm m3/mol	25	EST	MEYLAN,WM & HOWARD,PH (1991)
OH 1.77E-011	cm3/molc sec	25	EST	MEYLAN,WM & HOWARD,PH (1993)

000619-24-9 — 3-CYANO-1-NITROBENZENE

Formula: $C_7H_4N_2O_2$

Mol Weight: 148.12

MP (deg C): 115-117

FP (deg C):

BP (deg C):

BP pressure (mm Hg):

Property/Value	Units	Temp	Data Type	Reference
WS 1.71E+003	mg/L	25	EST	MEYLAN,WM ET AL. (1996)
logP 1.17			EXP	HANSCH,C & LEO,AJ (1985)
VP 2.17E-003	mm Hg	25	EST	NEELY,WB & BLAU,GE (1985)
DC	pKa			
HL 2.05E-007	atm m3/mol	25	EST	MEYLAN,WM & HOWARD,PH (1991)
OH 4.30E-014	cm3/molc sec	25	EST	MEYLAN,WM & HOWARD,PH (1993)

000619-25-0 — M-NITROBENZYL ALCOHOL

Formula: $C_7H_7NO_3$

Mol Weight: 153.14

MP (deg C): 30.5

FP (deg C):

BP (deg C): 175-180

BP pressure (mm Hg): 3.00E+000

Property/Value	Units	Temp	Data Type	Reference
WS 2.77E+003	mg/L	25	EST	MEYLAN,WM ET AL. (1996)
logP 1.21			EXP	HANSCH,C & LEO,AJ (1985)
VP 8.58E-005	mm Hg	25	EST	NEELY,WB & BLAU,GE (1985)
DC	pKa			
HL 8.58E-010	atm m3/mol	25	EST	MEYLAN,WM & HOWARD,PH (1991)
OH 3.83E-012	cm3/molc sec	25	EST	MEYLAN,WM & HOWARD,PH (1993)

000619-31-8 — BENZENAMINE, N,N-DIMETHYL-3-NITRO-

Formula: $C_8H_{10}N_2O_2$

Mol Weight: 166.18

MP (deg C): 60.5

FP (deg C):

BP (deg C): 282.5

BP pressure (mm Hg):

Property/Value	Units	Temp	Data Type	Reference
WS 9.17E+002	mg/L	25	EST	MEYLAN,WM ET AL. (1996)
logP 2.16			EXP	KRAMER,CR & HENZ,U (1990)
VP 7.22E-003	mm Hg	25	EST	NEELY,WB & BLAU,GE (1985)
DC 2.63	pKa	25	EXP	PERRIN,DD (1965)
HL 3.38E-007	atm m3/mol	25	EST	MEYLAN,WM & HOWARD,PH (1991)
OH 3.48E-011	cm3/molc sec	25	EST	MEYLAN,WM & HOWARD,PH (1993)

CAS #: 000619-33-0 — 1,1-DICHLORO-2,2-DIETHOXYETHANE

Formula: $C_6H_{12}Cl_2O_2$
Mol Weight: 187.07
MP (deg C): FP (deg C):
BP (deg C): 183.5
BP pressure (mm Hg):

Property/Value	Units	Temp	Data Type	Reference
WS 2.02E+003	mg/L	25	EST	MEYLAN,WM ET AL. (1996)
logP 1.64			EST	MEYLAN,WM & HOWARD,PH (1995)
VP 6.40E-001	mm Hg	25	EST	NEELY,WB & BLAU,GE (1985)
DC	pKa			
HL 1.47E-005	atm m3/mol	25	EST	MEYLAN,WM & HOWARD,PH (1991)
OH 5.72E-012	cm3/molc sec	25	EST	MEYLAN,WM & HOWARD,PH (1993)

CAS #: 000619-42-1 — BENZOIC ACID, 4-BROMO-, METHYL ESTER

Formula: $C_8H_7BrO_2$
Mol Weight: 215.05
MP (deg C): 81 FP (deg C):
BP (deg C):
BP pressure (mm Hg):

Property/Value	Units	Temp	Data Type	Reference
WS 1.11E+002	mg/L	25	EST	MEYLAN,WM ET AL. (1996)
logP 2.95			EXP	SOTOMATSU,T ET AL. (1993)
VP 2.46E-002	mm Hg	25	EST	NEELY,WB & BLAU,GE (1985)
DC	pKa			
HL 1.38E-005	atm m3/mol	25	EST	MEYLAN,WM & HOWARD,PH (1991)
OH 6.12E-013	cm3/molc sec	25	EST	MEYLAN,WM & HOWARD,PH (1993)

CAS #: 000619-45-4 — METHYL P-AMINOBENZOATE

Formula: $C_8H_9NO_2$
Mol Weight: 151.17
MP (deg C): 110-111 FP (deg C):
BP (deg C):
BP pressure (mm Hg):

Property/Value	Units	Temp	Data Type	Reference
WS 3.79E+003	mg/L	25	EXP	SUZUKI,T (1991)
logP 1.37			EXP	SANGSTER,J (1993)
VP 6.86E-003	mm Hg	25	EST	NEELY,WB & BLAU,GE (1985)
DC 2.47	pKa	25	EXP	PERRIN,DD (1965)
HL 1.23E-008	atm m3/mol	25	EST	MEYLAN,WM & HOWARD,PH (1991)
OH 3.48E-011	cm3/molc sec	25	EST	MEYLAN,WM & HOWARD,PH (1993)

CAS #: 000619-50-1 — METHYL P-NITROBENZOATE

Formula: $C_8H_7NO_4$
Mol Weight: 181.15
MP (deg C): 94-96 FP (deg C):
BP (deg C):
BP pressure (mm Hg):

Property/Value	Units	Temp	Data Type	Reference
WS 5.39E+002	mg/L	25	EST	MEYLAN,WM ET AL. (1996)
logP 1.89			EXP	SOTOMATSU,T ET AL. (1993)
VP 1.86E-003	mm Hg	25	EST	NEELY,WB & BLAU,GE (1985)
DC	pKa			
HL 1.37E-007	atm m3/mol	25	EST	MEYLAN,WM & HOWARD,PH (1991)
OH 2.72E-013	cm3/molc sec	25	EST	MEYLAN,WM & HOWARD,PH (1993)

CAS #: 000619-55-6 — 4-METHYLBENZAMIDE

Formula: C_8H_9NO
Mol Weight: 135.17
MP (deg C): 161-163 FP (deg C):
BP (deg C):
BP pressure (mm Hg):

Property/Value	Units	Temp	Data Type	Reference
WS 8.61E+003	mg/L	25	EST	MEYLAN,WM ET AL. (1996)
logP 1.18			EXP	HANSCH,C & LEO,AJ (1985)
VP 3.19E-004	mm Hg	25	EST	NEELY,WB & BLAU,GE (1985)
DC	pKa			
HL 2.44E-009	atm m3/mol	25	EST	MEYLAN,WM & HOWARD,PH (1991)
OH 6.78E-012	cm3/molc sec	25	EST	MEYLAN,WM & HOWARD,PH (1993)

CAS #: 000619-56-7 — 4-CHLOROBENZAMIDE

Formula: C_7H_6ClNO
Mol Weight: 155.58
MP (deg C): 172-176 FP (deg C):
BP (deg C):
BP pressure (mm Hg):

Property/Value	Units	Temp	Data Type	Reference
WS 3.40E+003	mg/L	25	EST	MEYLAN,WM ET AL. (1996)
logP 1.55			EXP	HANSCH,C & LEO,AJ (1985)
VP 1.65E-004	mm Hg	25	EST	NEELY,WB & BLAU,GE (1985)
DC	pKa			
HL 1.64E-009	atm m3/mol	25	EST	MEYLAN,WM & HOWARD,PH (1991)
OH 3.25E-012	cm3/molc sec	25	EST	MEYLAN,WM & HOWARD,PH (1993)

CAS #: 000619-57-8 — P-HYDROXYBENZAMIDE

Formula: $C_7H_7NO_2$
Mol Weight: 137.14
MP (deg C): 161-162 FP (deg C):
BP (deg C):
BP pressure (mm Hg):

Property/Value	Units	Temp	Data Type	Reference
WS 1.71E+005	mg/L	25	EST	MEYLAN,WM ET AL. (1996)
logP 0.33			EXP	HANSCH,C & LEO,AJ (1985)
VP 1.20E-005	mm Hg	25	EST	NEELY,WB & BLAU,GE (1985)
DC	pKa			
HL 2.30E-013	atm m3/mol	25	EST	MEYLAN,WM & HOWARD,PH (1991)
OH 3.25E-011	cm3/molc sec	25	EST	MEYLAN,WM & HOWARD,PH (1993)

CAS #: 000619-58-9 — 4-IODOBENZOIC ACID

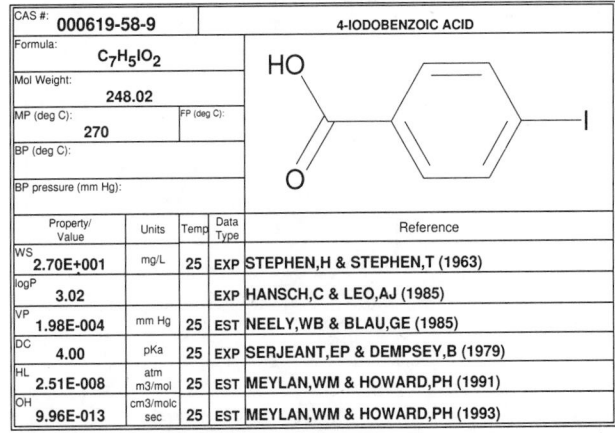

Formula: $C_7H_5IO_2$
Mol Weight: 248.02
MP (deg C): 270 FP (deg C):
BP (deg C):
BP pressure (mm Hg):

Property/Value	Units	Temp	Data Type	Reference
WS 2.70E+001	mg/L	25	EXP	STEPHEN,H & STEPHEN,T (1963)
logP 3.02			EXP	HANSCH,C & LEO,AJ (1985)
VP 1.98E-004	mm Hg	25	EST	NEELY,WB & BLAU,GE (1985)
DC 4.00	pKa	25	EXP	SERJEANT,EP & DEMPSEY,B (1979)
HL 2.51E-008	atm m3/mol	25	EST	MEYLAN,WM & HOWARD,PH (1991)
OH 9.96E-013	cm3/molc sec	25	EST	MEYLAN,WM & HOWARD,PH (1993)

CAS #: 000619-64-7	4-ETHYLBENZOIC ACID

Formula: $C_9H_{10}O_2$

Mol Weight: 150.18

MP (deg C): 112-113   FP (deg C):

BP (deg C):

BP pressure (mm Hg):

Property/ Value	Units	Temp	Data Type	Reference
WS 2.58E+002	mg/L	25	EST	MEYLAN,WM ET AL. (1996)
logP 2.89			EXP	DA,YZ ET AL. (1992)
VP 1.69E-003	mm Hg	25	EST	NEELY,WB & BLAU,GE (1985)
DC 4.35	pKa		EXP	KORTUM,G ET AL (1961)
HL 1.59E-007	atm m3/mol	25	EST	MEYLAN,WM & HOWARD,PH (1991)
OH 3.42E-012	cm3/molc sec	25	EST	MEYLAN,WM & HOWARD,PH (1993)

CAS #: 000619-65-8	P-CYANOBENZOIC ACID

Formula: $C_8H_5NO_2$

Mol Weight: 147.13

MP (deg C): 220-222 de   FP (deg C):

BP (deg C):

BP pressure (mm Hg):

Property/ Value	Units	Temp	Data Type	Reference
WS 1.97E+003	mg/L	25	EST	MEYLAN,WM ET AL. (1996)
logP 1.56			EXP	HANSCH,C & LEO,AJ (1985)
VP 2.09E-004	mm Hg	25	EST	NEELY,WB & BLAU,GE (1985)
DC 3.55	pKa	20	EXP	KORTUM,G ET AL (1961)
HL 1.05E-009	atm m3/mol	25	EST	MEYLAN,WM & HOWARD,PH (1991)
OH 6.15E-013	cm3/molc sec	25	EST	MEYLAN,WM & HOWARD,PH (1993)

CAS #: 000619-66-9	4-FORMYLBENZOIC ACID

Formula: $C_8H_6O_3$

Mol Weight: 150.14

MP (deg C): 247   FP (deg C):

BP (deg C):

BP pressure (mm Hg):

Property/ Value	Units	Temp	Data Type	Reference
WS 2.38E+003	mg/L	25	EST	MEYLAN,WM ET AL. (1996)
logP 1.76			EXP	DA,YZ ET AL. (1992)
VP 3.75E-004	mm Hg	25	EST	NEELY,WB & BLAU,GE (1985)
DC 3.77	pKa	25	EXP	SERJEANT,EP & DEMPSEY,B (1979)
HL 2.70E-010	atm m3/mol	25	EST	MEYLAN,WM & HOWARD,PH (1991)
OH 1.78E-011	cm3/molc sec	25	EST	MEYLAN,WM & HOWARD,PH (1993)

CAS #: 000619-72-7	1-NITRO-4-CYANOBENZENE

Formula: $C_7H_4N_2O_2$

Mol Weight: 148.12

MP (deg C): 146-149   FP (deg C):

BP (deg C):

BP pressure (mm Hg):

Property/ Value	Units	Temp	Data Type	Reference
WS 1.65E+003	mg/L	25	EST	MEYLAN,WM ET AL. (1996)
logP 1.19			EXP	HANSCH,C & LEO,AJ (1985)
VP 2.17E-003	mm Hg	25	EST	NEELY,WB & BLAU,GE (1985)
DC	pKa			
HL 2.05E-007	atm m3/mol	25	EST	MEYLAN,WM & HOWARD,PH (1991)
OH 3.19E-014	cm3/molc sec	25	EST	MEYLAN,WM & HOWARD,PH (1993)

CAS #: 000619-73-8	P-NITROBENZYL ALCOHOL

Formula: $C_7H_7NO_3$

Mol Weight: 153.14

MP (deg C): 96.5   FP (deg C):

BP (deg C): 185

BP pressure (mm Hg): 1.20E+001

Property/ Value	Units	Temp	Data Type	Reference
WS 2.51E+003	mg/L	25	EST	MEYLAN,WM ET AL. (1996)
logP 1.26			EXP	HANSCH,C & LEO,AJ (1985)
VP 8.58E-005	mm Hg	25	EST	NEELY,WB & BLAU,GE (1985)
DC	pKa			
HL 8.58E-010	atm m3/mol	25	EST	MEYLAN,WM & HOWARD,PH (1991)
OH 4.01E-012	cm3/molc sec	25	EST	MEYLAN,WM & HOWARD,PH (1993)

CAS #: 000619-76-1	4-ISOPROPYLBENZAMIDE

Formula: $C_{10}H_{13}NO$

Mol Weight: 163.22

MP (deg C):   FP (deg C):

BP (deg C):

BP pressure (mm Hg):

Property/ Value	Units	Temp	Data Type	Reference
WS 9.84E+002	mg/L	25	EST	MEYLAN,WM ET AL. (1996)
logP 2.14			EXP	HANSCH,C & LEO,AJ (1985)
VP 8.93E-005	mm Hg	25	EST	NEELY,WB & BLAU,GE (1985)
DC	pKa			
HL 4.31E-009	atm m3/mol	25	EST	MEYLAN,WM & HOWARD,PH (1991)
OH 8.49E-012	cm3/molc sec	25	EST	MEYLAN,WM & HOWARD,PH (1993)

CAS #: 000619-80-7	4-NITROBENZAMIDE

Formula: $C_7H_6N_2O_3$

Mol Weight: 166.14

MP (deg C): 199-201   FP (deg C):

BP (deg C):

BP pressure (mm Hg):

Property/ Value	Units	Temp	Data Type	Reference
WS 5.21E+003	mg/L	25	EST	MEYLAN,WM ET AL. (1996)
logP 0.82			EXP	HANSCH,C & LEO,AJ (1985)
VP 8.09E-006	mm Hg	25	EST	NEELY,WB & BLAU,GE (1985)
DC	pKa			
HL 8.74E-012	atm m3/mol	25	EST	MEYLAN,WM & HOWARD,PH (1991)
OH 2.16E-012	cm3/molc sec	25	EST	MEYLAN,WM & HOWARD,PH (1993)

CAS #: 000619-84-1	4-DIMETHYLAMINO BENZOIC ACID

Formula: $C_9H_{11}NO_2$

Mol Weight: 165.19

MP (deg C): 242.5-3.5   FP (deg C):

BP (deg C):

BP pressure (mm Hg):

Property/ Value	Units	Temp	Data Type	Reference
WS 6.61E+001	mg/L		EXP	YALKOWSKY,SH & DANNENFELSER,RM (1992)
logP 1.28			EXP	DA,YZ ET AL. (1992)
VP 6.25E-004	mm Hg	25	EST	NEELY,WB & BLAU,GE (1985)
DC	pKa			
HL 1.72E-009	atm m3/mol	25	EST	MEYLAN,WM & HOWARD,PH (1991)
OH 1.40E-010	cm3/molc sec	25	EST	MEYLAN,WM & HOWARD,PH (1993)

CAS #: 000619-86-3				P-ETHOXYBENZOIC ACID

Formula: $C_9H_{10}O_3$

Mol Weight: 166.18

MP (deg C): 197-199   FP (deg C):

BP (deg C):

BP pressure (mm Hg):

Property/Value	Units	Temp	Data Type	Reference
WS 5.83E+002	mg/L	25	EST	MEYLAN,WM ET AL. (1996)
logP 2.39			EXP	HANSCH,C & LEO,AJ (1985)
VP 5.34E-004	mm Hg	25	EST	NEELY,WB & BLAU,GE (1985)
DC 4.80	pKa	20	EXP	KORTUM,G ET AL (1961)
HL 8.52E-009	atm m3/mol	25	EST	MEYLAN,WM & HOWARD,PH (1991)
OH 1.47E-011	cm3/molc sec	25	EST	MEYLAN,WM & HOWARD,PH (1993)

CAS #: 000619-90-9				ACETIC ACID, NITROBENZYL ESTER

Formula: $C_9H_9NO_4$

Mol Weight: 195.18

MP (deg C):   FP (deg C):

BP (deg C):

BP pressure (mm Hg):

Property/Value	Units	Temp	Data Type	Reference
WS 6.28E+002	mg/L	25	EST	MEYLAN,WM ET AL. (1996)
logP 1.73			EXP	HANSCH,C & LEO,AJ (1985)
VP 6.65E-004	mm Hg	25	EST	NEELY,WB & BLAU,GE (1985)
DC	pKa			
HL 5.59E-008	atm m3/mol	25	EST	MEYLAN,WM & HOWARD,PH (1991)
OH 2.20E-012	cm3/molc sec	25	EST	MEYLAN,WM & HOWARD,PH (1993)

CAS #: 000619-99-8				3-ETHYLHEXANE

Formula: C_8H_{18}

Mol Weight: 114.23

MP (deg C):   FP (deg C):

BP (deg C): 118.6

BP pressure (mm Hg):

Property/Value	Units	Temp	Data Type	Reference
WS 7.97E+000	mg/L	25	EST	MEYLAN,WM ET AL. (1996)
logP 4.20			EST	MEYLAN,WM & HOWARD,PH (1995)
VP 2.00E+001	mm Hg	25	EXP	DAUBERT,TE & DANNER,RP (1989)
DC	pKa			
HL 3.01E+000	atm m3/mol	25	EST	MEYLAN,WM & HOWARD,PH (1991)
OH 8.97E-012	cm3/molc sec	25	EST	MEYLAN,WM & HOWARD,PH (1993)

CAS #: 000620-02-0				2-FURANCARBOXALDEHYDE, 5-METHYL-

Formula: $C_6H_6O_2$

Mol Weight: 110.11

MP (deg C):   FP (deg C):

BP (deg C): 187

BP pressure (mm Hg):

Property/Value	Units	Temp	Data Type	Reference
WS 2.91E+004	mg/L	25	EST	MEYLAN,WM ET AL. (1996)
logP 0.67			EXP	SANGSTER,J (1994)
VP 1.92E+000	mm Hg	25	EST	NEELY,WB & BLAU,GE (1985)
DC	pKa			
HL 1.48E-005	atm m3/mol	25	EST	MEYLAN,WM & HOWARD,PH (1991)
OH 5.18E-011	cm3/molc sec	25	EST	MEYLAN,WM & HOWARD,PH (1993)

CAS #: 000620-05-3				BENZYL IODIDE

Formula: C_7H_7I

Mol Weight: 218.04

MP (deg C): 24.5   FP (deg C):

BP (deg C): 93

BP pressure (mm Hg): 1.00E+001

Property/Value	Units	Temp	Data Type	Reference
WS 5.37E+001	mg/L	25	EST	MEYLAN,WM ET AL. (1996)
logP 3.30			EST	MEYLAN,WM & HOWARD,PH (1995)
VP 3.02E-001	mm Hg	25	EXT	ASHCROFT,SJ (1976)
DC	pKa			
HL 4.44E-004	atm m3/mol	25	EST	MEYLAN,WM & HOWARD,PH (1991)
OH 2.51E-012	cm3/molc sec	25	EST	MEYLAN,WM & HOWARD,PH (1993)

CAS #: 000620-08-6				4-METHOXYPYRIDINE

Formula: C_6H_7NO

Mol Weight: 109.13

MP (deg C):   FP (deg C):

BP (deg C):

BP pressure (mm Hg):

Property/Value	Units	Temp	Data Type	Reference
WS 1.53E+004	mg/L	25	EST	MEYLAN,WM ET AL. (1996)
logP 1.00			EXP	HANSCH,C & LEO,AJ (1985)
VP 2.58E+000	mm Hg	25	EST	NEELY,WB & BLAU,GE (1985)
DC 6.47	pKa	25	EXP	PERRIN,DD (1965)
HL 4.17E-007	atm m3/mol	25	EST	MEYLAN,WM & HOWARD,PH (1991)
OH 4.91E-012	cm3/molc sec	25	EST	MEYLAN,WM & HOWARD,PH (1993)

CAS #: 000620-14-4				M-ETHYLTOLUENE

Formula: C_9H_{12}

Mol Weight: 120.20

MP (deg C): -95.5   FP (deg C):

BP (deg C): 161.3

BP pressure (mm Hg):

Property/Value	Units	Temp	Data Type	Reference
WS 4.00E+001	mg/L	25	EST	MEYLAN,WM ET AL. (1996)
logP 3.98			EXP	SHERBLOM,PM & EGANHOUSE,RP (1988)
VP 3.04E+000	mm Hg	25	EXP	DAUBERT,TE & DANNER,RP (1989)
DC	pKa			
HL 8.71E-003	atm m3/mol	25	EST	MEYLAN,WM & HOWARD,PH (1991)
OH 1.92E-011	cm3/molc sec	25	EXP	ATKINSON,R (1989)

CAS #: 000620-17-7				M-ETHYLPHENOL

Formula: $C_8H_{10}O$

Mol Weight: 122.17

MP (deg C): -4   FP (deg C):

BP (deg C): 218.4

BP pressure (mm Hg):

Property/Value	Units	Temp	Data Type	Reference
WS 1.13E+004	mg/L	25	EST	MEYLAN,WM ET AL. (1996)
logP 2.40			EXP	HANSCH,C & LEO,AJ (1985)
VP 5.00E-002	mm Hg	25	EXP	BIDDISCOMBE,DP ET AL. (1963)
DC 9.90	pKa		EXP	PEARCE,PJ & SIMKINS,RJJ (1968)
HL 1.09E-006	atm m3/mol	25	EST	MEYLAN,WM & HOWARD,PH (1991)
OH 8.41E-011	cm3/molc sec	25	EST	MEYLAN,WM & HOWARD,PH (1993)

CAS #: 000620-24-6				M-HYDROXYBENZYL ALCOHOL

Formula: $C_7H_8O_2$

Mol Weight: 124.14

MP (deg C): 73   FP (deg C):

BP (deg C):

BP pressure (mm Hg):

Property/Value	Units	Temp	Data Type	Reference
WS 4.55E+005	mg/L	25	EST	MEYLAN,WM ET AL. (1996)
logP 0.49			EXP	HANSCH,C & LEO,AJ (1985)
VP 8.25E-004	mm Hg	25	EST	NEELY,WB & BLAU,GE (1985)
DC 9.83	pKa		EXP	KORTUM,G ET AL (1961)
HL 2.26E-011	atm m3/mol	25	EST	MEYLAN,WM & HOWARD,PH (1991)
OH 8.64E-011	cm3/molc sec	25	EST	MEYLAN,WM & HOWARD,PH (1993)

CAS #: 000620-71-3				PROPIONANILIDE

Formula: $C_9H_{11}NO$

Mol Weight: 149.19

MP (deg C): 105.5   FP (deg C):

BP (deg C): 222.2

BP pressure (mm Hg):

Property/Value	Units	Temp	Data Type	Reference
WS 3.23E+003	mg/L	25	EST	MEYLAN,WM ET AL. (1996)
logP 1.61			EXP	HANSCH,C & LEO,AJ (1985)
VP 1.97E-004	mm Hg	25	EST	NEELY,WB & BLAU,GE (1985)
DC	pKa			
HL 8.19E-009	atm m3/mol	25	EST	MEYLAN,WM & HOWARD,PH (1991)
OH 1.36E-011	cm3/molc sec	25	EST	MEYLAN,WM & HOWARD,PH (1993)

CAS #: 000620-79-1				BENZYLACETOACETIC ACID, ETHYL ESTER

Formula: $C_{13}H_{16}O_3$

Mol Weight: 220.27

MP (deg C):   FP (deg C):

BP (deg C): 276

BP pressure (mm Hg):

Property/Value	Units	Temp	Data Type	Reference
WS 2.42E+002	mg/L	25	EST	MEYLAN,WM ET AL. (1996)
logP 2.52			EXP	HANSCH,C & LEO,AJ (1985)
VP 3.09E-004	mm Hg	25	EST	NEELY,WB & BLAU,GE (1985)
DC	pKa			
HL 1.68E-008	atm m3/mol	25	EST	MEYLAN,WM & HOWARD,PH (1991)
OH 1.15E-011	cm3/molc sec	25	EST	MEYLAN,WM & HOWARD,PH (1993)

CAS #: 000620-80-4				ETHYLACETOACETATE, BENZAL

Formula: $C_{13}H_{14}O_3$

Mol Weight: 218.25

MP (deg C): 60.5   FP (deg C):

BP (deg C): 296

BP pressure (mm Hg):

Property/Value	Units	Temp	Data Type	Reference
WS 4.56E+002	mg/L	25	EST	MEYLAN,WM ET AL. (1996)
logP 2.21			EXP	HANSCH,C & LEO,AJ (1985)
VP 3.73E-004	mm Hg	25	EST	NEELY,WB & BLAU,GE (1985)
DC	pKa			
HL 2.33E-009	atm m3/mol	25	EST	MEYLAN,WM & HOWARD,PH (1991)
OH 3.10E-011	cm3/molc sec	25	EST	MEYLAN,WM & HOWARD,PH (1993)

CAS #: 000620-84-8				4-METHYL-N-PHENYLBENZAMINE

Formula: $C_{13}H_{13}N$

Mol Weight: 183.26

MP (deg C):   FP (deg C):

BP (deg C):

BP pressure (mm Hg):

Property/Value	Units	Temp	Data Type	Reference
WS 2.80E+001	mg/L	25	EST	MEYLAN,WM ET AL. (1996)
logP 3.84			EST	MEYLAN,WM & HOWARD,PH (1995)
VP 7.92E-004	mm Hg	25	EST	NEELY,WB & BLAU,GE (1985)
DC	pKa			
HL 1.16E-006	atm m3/mol	25	EST	MEYLAN,WM & HOWARD,PH (1991)
OH 2.00E-010	cm3/molc sec	25	EST	MEYLAN,WM & HOWARD,PH (1993)

CAS #: 000620-88-2				4-NITRO DIPHENYL ETHER

Formula: $C_{12}H_9NO_3$

Mol Weight: 215.21

MP (deg C): 61   FP (deg C):

BP (deg C): 320

BP pressure (mm Hg):

Property/Value	Units	Temp	Data Type	Reference
WS 6.68E+000	mg/L	25	EST	MEYLAN,WM ET AL. (1996)
logP 3.92			EXP	NANDIHALLI,UB ET AL. (1993)
VP 3.80E-005	mm Hg	25	EST	NEELY,WB & BLAU,GE (1985)
DC	pKa			
HL 4.65E-007	atm m3/mol	25	EST	MEYLAN,WM & HOWARD,PH (1991)
OH 5.54E-012	cm3/molc sec	25	EST	MEYLAN,WM & HOWARD,PH (1993)

CAS #: 000620-92-8				4,4'-DIHYDROXYDIPHENYLMETHANE

Formula: $C_{13}H_{12}O_2$

Mol Weight: 200.24

MP (deg C): 162.5   FP (deg C):

BP (deg C):

BP pressure (mm Hg):

Property/Value	Units	Temp	Data Type	Reference
WS 5.43E+002	mg/L	25	EST	MEYLAN,WM ET AL. (1996)
logP 2.91			EXP	HANSCH,C & LEO,AJ (1985)
VP 8.84E-007	mm Hg	25	EST	NEELY,WB & BLAU,GE (1985)
DC 7.55	pKa	25	EXP	SERJEANT,EP & DEMPSEY,B (1979)
HL 5.20E-012	atm m3/mol	25	EST	MEYLAN,WM & HOWARD,PH (1991)
OH 8.24E-011	cm3/molc sec	25	EST	MEYLAN,WM & HOWARD,PH (1993)

CAS #: 000621-04-5				1-PHENYL-3-ETHYLUREA

Formula: $C_9H_{12}N_2O$

Mol Weight: 164.21

MP (deg C):   FP (deg C):

BP (deg C):

BP pressure (mm Hg):

Property/Value	Units	Temp	Data Type	Reference
WS 2.60E+003	mg/L	25	EST	MEYLAN,WM ET AL. (1996)
logP 1.64			EXP	HANSCH,C & LEO,AJ (1985)
VP 1.97E-004	mm Hg	25	EST	NEELY,WB & BLAU,GE (1985)
DC	pKa			
HL 5.87E-010	atm m3/mol	25	EST	MEYLAN,WM & HOWARD,PH (1991)
OH 4.77E-011	cm3/molc sec	25	EST	MEYLAN,WM & HOWARD,PH (1993)

CAS #: 000621-06-7				BENZENEACETAMIDE, N-PHENYL-
Formula: C₁₄H₁₃NO				

CAS #: 000621-06-7			
Formula: $C_{14}H_{13}NO$			
Mol Weight: 211.27			
MP (deg C):		**FP (deg C):**	
BP (deg C):			
BP pressure (mm Hg):			

BENZENEACETAMIDE, N-PHENYL-

Property/Value	Units	Temp	Data Type	Reference
WS 1.89E+002	mg/L	25	EST	MEYLAN,WM ET AL. (1996)
logP 2.70			EXP	YAMAGAMI,C ET AL. (1984)
VP 8.60E-007	mm Hg	25	EST	NEELY,WB & BLAU,GE (1985)
DC	pKa			
HL 4.98E-010	atm m3/mol	25	EST	MEYLAN,WM & HOWARD,PH (1991)
OH 1.80E-011	cm3/molc sec	25	EST	MEYLAN,WM & HOWARD,PH (1993)

CAS #: 000621-09-0			
Formula: $C_{14}H_{14}N_2$			
Mol Weight: 210.28			
MP (deg C):		**FP (deg C):**	
BP (deg C):			
BP pressure (mm Hg):			

N,N'-DIPHENYLACETAMIDINE

Property/Value	Units	Temp	Data Type	Reference
WS 6.22E+000	mg/L	25	EST	MEYLAN,WM ET AL. (1996)
logP 4.44			EST	MEYLAN,WM & HOWARD,PH (1995)
VP 1.59E-004	mm Hg	25	EST	NEELY,WB & BLAU,GE (1985)
DC	pKa			
HL 9.19E-008	atm m3/mol	25	EST	MEYLAN,WM & HOWARD,PH (1991)
OH 4.50E-011	cm3/molc sec	25	EST	MEYLAN,WM & HOWARD,PH (1993)

CAS #: 000621-15-8			
Formula: $C_{12}H_{17}NO$			
Mol Weight: 191.28			
MP (deg C): 95		**FP (deg C):**	
BP (deg C):			
BP pressure (mm Hg):			

CAPRYLANILIDE

Property/Value	Units	Temp	Data Type	Reference
WS 7.66E+001	mg/L	25	EST	MEYLAN,WM ET AL. (1996)
logP 3.28			EXP	HANSCH,C & LEO,AJ (1985)
VP 1.43E-005	mm Hg	25	EST	NEELY,WB & BLAU,GE (1985)
DC	pKa			
HL 1.92E-008	atm m3/mol	25	EST	MEYLAN,WM & HOWARD,PH (1991)
OH 2.05E-011	cm3/molc sec	25	EST	MEYLAN,WM & HOWARD,PH (1993)

CAS #: 000621-34-1			
Formula: $C_8H_{10}O_2$			
Mol Weight: 138.17			
MP (deg C):		**FP (deg C):**	
BP (deg C): 131			
BP pressure (mm Hg): 1.00E+001			

M-ETHOXYPHENOL

Property/Value	Units	Temp	Data Type	Reference
WS 6.60E+003	mg/L	25	EST	MEYLAN,WM ET AL. (1996)
logP 1.98			EXP	HANSCH,C & LEO,AJ (1985)
VP 2.61E-002	mm Hg	25	EST	NEELY,WB & BLAU,GE (1985)
DC 9.65	pKa	25	EXP	SERJEANT,EP & DEMPSEY,B (1979)
HL 4.40E-008	atm m3/mol	25	EST	MEYLAN,WM & HOWARD,PH (1991)
OH 2.06E-010	cm3/molc sec	25	EST	MEYLAN,WM & HOWARD,PH (1993)

CAS #: 000621-36-3			
Formula: $C_9H_{10}O_2$			
Mol Weight: 150.18			
MP (deg C): 62		**FP (deg C):**	
BP (deg C): 120-123			
BP pressure (mm Hg): 2.60E+001			

M-METHYLPHENYLACETIC ACID

Property/Value	Units	Temp	Data Type	Reference
WS 4.07E+003	mg/L	25	EST	MEYLAN,WM ET AL. (1996)
logP 1.95			EXP	HANSCH,C & LEO,AJ (1985)
VP 5.93E-002	mm Hg	25	EXT	BOUBLIK,T ET AL. (1984)
DC	pKa			
HL 4.88E-008	atm m3/mol	25	EST	MEYLAN,WM & HOWARD,PH (1991)
OH 9.80E-012	cm3/molc sec	25	EST	MEYLAN,WM & HOWARD,PH (1993)

CAS #: 000621-37-4			
Formula: $C_8H_8O_3$			
Mol Weight: 152.15			
MP (deg C): 132		**FP (deg C):**	
BP (deg C): 190			
BP pressure (mm Hg): 1.10E+001			

M-HYDROXYPHENYLACETIC ACID

Property/Value	Units	Temp	Data Type	Reference
WS 1.00E+006	mg/L	20	EXP	YALKOWSKY,SH & DANNENFELSER,RM (1992)
logP 0.85			EXP	HANSCH,C & LEO,AJ (1985)
VP 5.55E-005	mm Hg	25	EST	NEELY,WB & BLAU,GE (1985)
DC	pKa			
HL 4.60E-012	atm m3/mol	25	EST	MEYLAN,WM & HOWARD,PH (1991)
OH 5.66E-011	cm3/molc sec	25	EST	MEYLAN,WM & HOWARD,PH (1993)

CAS #: 000621-38-5			
Formula: C_8H_8BrNO			
Mol Weight: 214.07			
MP (deg C):		**FP (deg C):**	
BP (deg C):			
BP pressure (mm Hg):			

3-BROMOACETANILIDE

Property/Value	Units	Temp	Data Type	Reference
WS 3.94E+002	mg/L	25	EST	MEYLAN,WM ET AL. (1996)
logP 2.31			EXP	NAKAGAWA,Y ET AL. (1992)
VP 3.42E-005	mm Hg	25	EST	NEELY,WB & BLAU,GE (1985)
DC	pKa			
HL 2.46E-009	atm m3/mol	25	EST	MEYLAN,WM & HOWARD,PH (1991)
OH 7.92E-012	cm3/molc sec	25	EST	MEYLAN,WM & HOWARD,PH (1993)

CAS #: 000621-42-1			
Formula: $C_8H_9NO_2$			
Mol Weight: 151.17			
MP (deg C): 146-149		**FP (deg C):**	
BP (deg C):			
BP pressure (mm Hg):			

M-HYDROXYACETANILIDE

Property/Value	Units	Temp	Data Type	Reference
WS 1.79E+004	mg/L	25	EST	MEYLAN,WM ET AL. (1996)
logP 0.73			EXP	HANSCH,C & LEO,AJ (1985)
VP 7.00E-006	mm Hg	25	EST	NEELY,WB & BLAU,GE (1985)
DC	pKa			
HL 6.42E-013	atm m3/mol	25	EST	MEYLAN,WM & HOWARD,PH (1991)
OH 2.00E-010	cm3/molc sec	25	EST	MEYLAN,WM & HOWARD,PH (1993)

CAS #: 000621-59-0				ISOVANILLIN

Formula: $C_8H_8O_3$

Mol Weight: 152.15

MP (deg C): 114

FP (deg C):

BP (deg C): 179

BP pressure (mm Hg): 1.50E+001

Property/ Value	Units	Temp	Data Type	Reference
WS 1.10E+004	mg/L	25	EST	MEYLAN,WM ET AL. (1996)
logP 0.97			EXP	HANSCH,C & LEO,AJ (1985)
VP 1.03E-003	mm Hg	25	EST	NEELY,WB & BLAU,GE (1985)
DC 8.89	pKa	25	EXP	KORTUM,G ET AL (1961)
HL 8.27E-011	atm m3/mol	25	EST	MEYLAN,WM & HOWARD,PH (1991)
OH 3.25E-011	cm3/molc sec	25	EST	MEYLAN,WM & HOWARD,PH (1993)

CAS #: 000621-64-7				N-NITROSODIPROPYLAMINE

Formula: $C_6H_{14}N_2O$

Mol Weight: 130.19

MP (deg C): 7 EST

FP (deg C):

BP (deg C): 206

BP pressure (mm Hg):

Property/ Value	Units	Temp	Data Type	Reference
WS 9.89E+003	mg/L	24	EXP	YALKOWSKY,SH & DANNENFELSER,RM (1992)
logP 1.36			EXP	HANSCH,C & LEO,AJ (1985)
VP 1.30E-001	mm Hg	25	EXT	KLEIN,RG (1982)
DC	pKa			
HL 5.38E-006	atm m3/mol	37	EXP	MIRVISH,SS ET AL. (1976)
OH 2.41E-011	cm3/molc sec	25	EST	MEYLAN,WM & HOWARD,PH (1993)

CAS #: 000621-65-8				1,2,3-PROPANETRIOL, 1,2-DINITRATE

Formula: $C_3H_6N_2O_7$

Mol Weight: 182.09

MP (deg C):

FP (deg C):

BP (deg C):

BP pressure (mm Hg):

Property/ Value	Units	Temp	Data Type	Reference
WS 1.05E+004	mg/L	25	EST	MEYLAN,WM ET AL. (1996)
logP 0.83			EXP	KIKKOJI,T ET AL. (1991)
VP 7.66E-004	mm Hg	25	EST	NEELY,WB & BLAU,GE (1985)
DC	pKa			
HL 3.44E-011	atm m3/mol	25	EST	MEYLAN,WM & HOWARD,PH (1991)
OH 1.54E-012	cm3/molc sec	25	EST	MEYLAN,WM & HOWARD,PH (1993)

CAS #: 000621-70-5				TRICAPROIN

Formula: $C_{21}H_{38}O_6$

Mol Weight: 386.53

MP (deg C): -60

FP (deg C):

BP (deg C): > 200

BP pressure (mm Hg):

Property/ Value	Units	Temp	Data Type	Reference
WS 4.50E-001	mg/L	37	EXP	FUNASAKI,N ET AL. (1976)
logP 6.26			EST	MEYLAN,WM & HOWARD,PH (1995)
VP 4.75E-006	mm Hg	25	EST	NEELY,WB & BLAU,GE (1985)
DC	pKa			
HL 5.25E-008	atm m3/mol	25	EST	MEYLAN,WM & HOWARD,PH (1991)
OH 2.33E-011	cm3/molc sec	25	EST	MEYLAN,WM & HOWARD,PH (1993)

CAS #: 000621-79-4				CINNAMAMIDE

Formula: C_9H_9NO

Mol Weight: 147.18

MP (deg C): 148-150

FP (deg C):

BP (deg C):

BP pressure (mm Hg):

Property/ Value	Units	Temp	Data Type	Reference
WS 4.69E+003	mg/L	25	EST	MEYLAN,WM ET AL. (1996)
logP 1.43			EXP	HANSCH,C & LEO,AJ (1985)
VP 1.05E-004	mm Hg	25	EST	NEELY,WB & BLAU,GE (1985)
DC	pKa			
HL 2.64E-010	atm m3/mol	25	EST	MEYLAN,WM & HOWARD,PH (1991)
OH 2.36E-011	cm3/molc sec	25	EST	MEYLAN,WM & HOWARD,PH (1993)

CAS #: 000621-82-9				CINNAMIC ACID

Formula: $C_9H_8O_2$

Mol Weight: 148.16

MP (deg C): 133

FP (deg C):

BP (deg C): 300

BP pressure (mm Hg):

Property/ Value	Units	Temp	Data Type	Reference
WS 5.70E+002	mg/L	25	EXP	YALKOWSKY,SH & DANNENFELSER,RM (1992)
logP 2.13			EXP	HANSCH,C & LEO,AJ (1985)
VP 8.26E-006	mm Hg	25	EXT	PERRY,RH & GREEN,D (1984)
DC 4.44	pKa	25	EXP	KORTUM,G ET AL (1961)
HL 1.29E-008	atm m3/mol	25	EST	MEYLAN,WM & HOWARD,PH (1991)
OH 2.21E-011	cm3/molc sec	25	EST	MEYLAN,WM & HOWARD,PH (1993)

CAS #: 000621-84-1				O-BENZYL CARBAMATE

Formula: $C_8H_9NO_2$

Mol Weight: 151.17

MP (deg C): 87-89

FP (deg C):

BP (deg C):

BP pressure (mm Hg):

Property/ Value	Units	Temp	Data Type	Reference
WS 6.80E+004	mg/L	37	EXP	YALKOWSKY,SH & DANNENFELSER,RM (1992)
logP 1.20			EXP	HANSCH,C & LEO,AJ (1985)
VP 2.17E-002	mm Hg	25	EST	NEELY,WB & BLAU,GE (1985)
DC	pKa			
HL 3.19E-009	atm m3/mol	25	EST	MEYLAN,WM & HOWARD,PH (1991)
OH 8.34E-012	cm3/molc sec	25	EST	MEYLAN,WM & HOWARD,PH (1993)

CAS #: 000621-87-4				2-PROPANONE, 1-PHENOXY-

Formula: $C_9H_{10}O_2$

Mol Weight: 150.18

MP (deg C):

FP (deg C):

BP (deg C): 229.5

BP pressure (mm Hg):

Property/ Value	Units	Temp	Data Type	Reference
WS 9.80E+003	mg/L	25	EST	MEYLAN,WM ET AL. (1996)
logP 1.04			EXP	HANSCH,C ET AL. (1995)
VP 8.04E-002	mm Hg	25	EST	NEELY,WB & BLAU,GE (1985)
DC	pKa			
HL 1.52E-006	atm m3/mol	25	EST	MEYLAN,WM & HOWARD,PH (1991)
OH 2.59E-011	cm3/molc sec	25	EST	MEYLAN,WM & HOWARD,PH (1993)

CAS #: 000621-88-5				PHENOXYACETAMIDE

Formula: $C_8H_9NO_2$
Mol Weight: 151.17
MP (deg C): 101-103
FP (deg C):
BP (deg C):
BP pressure (mm Hg):

Property/Value	Units	Temp	Data Type	Reference
WS 1.68E+004	mg/L	25	EST	MEYLAN,WM ET AL. (1996)
logP 0.76			EXP	HANSCH,C & LEO,AJ (1985)
VP 1.20E-004	mm Hg	25	EST	NEELY,WB & BLAU,GE (1985)
DC	pKa			
HL 3.43E-010	atm m3/mol	25	EST	MEYLAN,WM & HOWARD,PH (1991)
OH 2.78E-011	cm3/molc sec	25	EST	MEYLAN,WM & HOWARD,PH (1993)

CAS #: 000622-08-2				ETHYLENE GLYCOL MONOBENZYL ETHER

Formula: $C_8H_{10}O_2$
Mol Weight: 138.17
MP (deg C): < -75
FP (deg C):
BP (deg C): 256
BP pressure (mm Hg):

Property/Value	Units	Temp	Data Type	Reference
WS 4.97E+004	mg/L	25	EST	MEYLAN,WM ET AL. (1996)
logP 0.80			EST	MEYLAN,WM & HOWARD,PH (1995)
VP 2.00E-002	mm Hg	25	EXP	FLICK,EW (1991)
DC	pKa			
HL 3.38E-009	atm m3/mol	25	EST	MEYLAN,WM & HOWARD,PH (1991)
OH 2.17E-011	cm3/molc sec	25	EST	MEYLAN,WM & HOWARD,PH (1993)

CAS #: 000622-24-2				B-PHENYL ETHYL CHLORIDE

Formula: C_8H_9Cl
Mol Weight: 140.61
MP (deg C):
FP (deg C):
BP (deg C): 197.5
BP pressure (mm Hg):

Property/Value	Units	Temp	Data Type	Reference
WS 2.52E+002	mg/L	25	EST	MEYLAN,WM ET AL. (1996)
logP 2.95			EXP	HANSCH,C & LEO,AJ (1985)
VP 1.69E+000	mm Hg	25	EXP	PERRY,RH & GREEN,D (1984)
DC	pKa			
HL 2.78E-003	atm m3/mol	25	EST	MEYLAN,WM & HOWARD,PH (1991)
OH 5.62E-012	cm3/molc sec	25	EST	MEYLAN,WM & HOWARD,PH (1993)

CAS #: 000622-29-7				METHANAMINE, N-(PHENYLMETHYLENE)-

Formula: C_8H_9N
Mol Weight: 119.17
MP (deg C):
FP (deg C):
BP (deg C): 185
BP pressure (mm Hg):

Property/Value	Units	Temp	Data Type	Reference
WS 4.02E+003	mg/L	25	EST	MEYLAN,WM ET AL. (1996)
logP 1.64			EXP	HANSCH,C & LEO,AJ (1985)
VP 9.94E-001	mm Hg	25	EST	NEELY,WB & BLAU,GE (1985)
DC	pKa			
HL 3.92E-004	atm m3/mol	25	EST	MEYLAN,WM & HOWARD,PH (1991)
OH 6.70E-012	cm3/molc sec	25	EST	MEYLAN,WM & HOWARD,PH (1993)

CAS #: 000622-31-1				BENZALDOXIME,SYN

Formula: C_7H_7NO
Mol Weight: 121.14
MP (deg C): 35
FP (deg C):
BP (deg C): 119
BP pressure (mm Hg): 1.00E+001

Property/Value	Units	Temp	Data Type	Reference
WS 3.18E+003	mg/L	25	EST	MEYLAN,WM ET AL. (1996)
logP 1.85			EXP	HANSCH,C ET AL. (1995)
VP 4.10E-003	mm Hg	25	EST	NEELY,WB & BLAU,GE (1985)
DC	pKa			
HL 3.57E-007	atm m3/mol	25	EST	MEYLAN,WM & HOWARD,PH (1991)
OH 6.57E-012	cm3/molc sec	25	EST	MEYLAN,WM & HOWARD,PH (1993)

CAS #: 000622-38-8				PHENYL ETHYL SULFIDE

Formula: $C_8H_{10}S$
Mol Weight: 138.23
MP (deg C):
FP (deg C):
BP (deg C): 205
BP pressure (mm Hg):

Property/Value	Units	Temp	Data Type	Reference
WS 1.58E+002	mg/L	25	EST	MEYLAN,WM ET AL. (1996)
logP 3.20			EXP	HANSCH,C & LEO,AJ (1985)
VP 2.41E-001	mm Hg	25	EST	NEELY,WB & BLAU,GE (1985)
DC	pKa			
HL 2.08E-004	atm m3/mol	25	EST	MEYLAN,WM & HOWARD,PH (1991)
OH 2.00E-011	cm3/molc sec	25	EST	MEYLAN,WM & HOWARD,PH (1993)

CAS #: 000622-42-4				A-NITROTOLUENE

Formula: $C_7H_7NO_2$
Mol Weight: 137.14
MP (deg C):
FP (deg C):
BP (deg C): 226
BP pressure (mm Hg):

Property/Value	Units	Temp	Data Type	Reference
WS 1.12E+003	mg/L	25	EST	MEYLAN,WM ET AL. (1996)
logP 1.75			EXP	HANSCH,C & LEO,AJ (1985)
VP 7.55E-002	mm Hg	25	EST	NEELY,WB & BLAU,GE (1985)
DC	pKa			
HL 3.40E-006	atm m3/mol	25	EST	MEYLAN,WM & HOWARD,PH (1991)
OH 4.98E-012	cm3/molc sec	25	EST	MEYLAN,WM & HOWARD,PH (1993)

CAS #: 000622-44-6				PHENYLISOCYANIDE DICHLORIDE

Formula: $C_7H_5Cl_2N$
Mol Weight: 174.03
MP (deg C):
FP (deg C):
BP (deg C): 103-106
BP pressure (mm Hg): 3.00E+001

Property/Value	Units	Temp	Data Type	Reference
WS 3.99E+002	mg/L	25	EST	MEYLAN,WM ET AL. (1996)
logP 2.54			EXP	HANSCH,C & LEO,AJ (1985)
VP 1.78E-001	mm Hg	25	EST	NEELY,WB & BLAU,GE (1985)
DC	pKa			
HL 1.51E-004	atm m3/mol	25	EST	MEYLAN,WM & HOWARD,PH (1991)
OH 2.14E-012	cm3/molc sec	25	EST	MEYLAN,WM & HOWARD,PH (1993)

CAS #: 000622-46-8		O-PHENYL CARBAMATE	

Formula: $C_7H_7NO_2$

Mol Weight: 137.14

MP (deg C): 149-152   FP (deg C):

BP (deg C):

BP pressure (mm Hg):

Property/Value	Units	Temp	Data Type	Reference
WS 1.03E+004	mg/L	25	EST	MEYLAN,WM ET AL. (1996)
logP 1.08			EXP	HANSCH,C & LEO,AJ (1985)
VP 6.89E-002	mm Hg	25	EST	NEELY,WB & BLAU,GE (1985)
DC	pKa			
HL 1.46E-008	atm m3/mol	25	EST	MEYLAN,WM & HOWARD,PH (1991)
OH 4.01E-012	cm3/molc sec	25	EST	MEYLAN,WM & HOWARD,PH (1993)

CAS #: 000622-47-9		P-METHYLPHENYLACETIC ACID	

Formula: $C_9H_{10}O_2$

Mol Weight: 150.18

MP (deg C): 93   FP (deg C):

BP (deg C): 265

BP pressure (mm Hg):

Property/Value	Units	Temp	Data Type	Reference
WS 4.85E+003	mg/L	25	EST	MEYLAN,WM ET AL. (1996)
logP 1.86			EXP	HANSCH,C & LEO,AJ (1985)
VP 3.44E-002	mm Hg	25	EXT	SHEEHAN,RJ & LANGER,SH (1969)
DC 4.37	pKa	25	EXP	KORTUM,G ET AL (1961)
HL 4.88E-008	atm m3/mol	25	EST	MEYLAN,WM & HOWARD,PH (1991)
OH 6.61E-012	cm3/molc sec	25	EST	MEYLAN,WM & HOWARD,PH (1993)

CAS #: 000622-50-4		4-IODOACETANILIDE	

Formula: C_8H_8INO

Mol Weight: 261.06

MP (deg C):   FP (deg C):

BP (deg C):

BP pressure (mm Hg):

Property/Value	Units	Temp	Data Type	Reference
WS 1.63E+002	mg/L	25	EST	MEYLAN,WM ET AL. (1996)
logP 2.46			EXP	HANSCH,C & LEO,AJ (1985)
VP 1.03E-005	mm Hg	25	EST	NEELY,WB & BLAU,GE (1985)
DC	pKa			
HL 1.43E-009	atm m3/mol	25	EST	MEYLAN,WM & HOWARD,PH (1991)
OH 4.20E-012	cm3/molc sec	25	EST	MEYLAN,WM & HOWARD,PH (1993)

CAS #: 000622-59-3		4-METHYLPHENYLISOTHIOCYANATE	

Formula: C_8H_7NS

Mol Weight: 149.22

MP (deg C): 25-26   FP (deg C):

BP (deg C): 237

BP pressure (mm Hg):

Property/Value	Units	Temp	Data Type	Reference
WS 2.84E+000	mg/L	25	EXP	YALKOWSKY,SH & DANNENFELSER,RM (1992)
logP 3.92			EXP	HANSCH,C & LEO,AJ (1985)
VP 3.02E-002	mm Hg	25	EST	NEELY,WB & BLAU,GE (1985)
DC	pKa			
HL 8.62E-004	atm m3/mol	25	EST	MEYLAN,WM & HOWARD,PH (1991)
OH 5.72E-012	cm3/molc sec	25	EST	MEYLAN,WM & HOWARD,PH (1993)

CAS #: 000622-62-8		P-ETHOXYPHENOL	

Formula: $C_8H_{10}O_2$

Mol Weight: 138.17

MP (deg C): 66.5   FP (deg C):

BP (deg C): 246.5

BP pressure (mm Hg):

Property/Value	Units	Temp	Data Type	Reference
WS 9.22E+003	mg/L	25	EST	MEYLAN,WM ET AL. (1996)
logP 1.81			EXP	HANSCH,C & LEO,AJ (1985)
VP 2.61E-002	mm Hg	25	EST	NEELY,WB & BLAU,GE (1985)
DC 10.13	pKa	25	EXP	SERJEANT,EP & DEMPSEY,B (1979)
HL 4.40E-008	atm m3/mol	25	EST	MEYLAN,WM & HOWARD,PH (1991)
OH 3.51E-011	cm3/molc sec	25	EST	MEYLAN,WM & HOWARD,PH (1993)

CAS #: 000622-78-6		BENZYLISOTHIOCYANATE	

Formula: C_8H_7NS

Mol Weight: 149.22

MP (deg C):   FP (deg C):

BP (deg C): 243

BP pressure (mm Hg):

Property/Value	Units	Temp	Data Type	Reference
WS 1.09E+002	mg/L	25	EXP	YALKOWSKY,SH & DANNENFELSER,RM (1992)
logP 3.16			EXP	HANSCH,C & LEO,AJ (1985)
VP 1.53E-002	mm Hg	25	EXT	PERRY,RH & GREEN,D (1984)
DC	pKa			
HL 2.51E-004	atm m3/mol	25	EST	MEYLAN,WM & HOWARD,PH (1991)
OH 5.78E-012	cm3/molc sec	25	EST	MEYLAN,WM & HOWARD,PH (1993)

CAS #: 000622-80-0		N-PROPYLANILINE	

Formula: $C_9H_{13}N$

Mol Weight: 135.21

MP (deg C):   FP (deg C):

BP (deg C): 222

BP pressure (mm Hg):

Property/Value	Units	Temp	Data Type	Reference
WS 7.09E+002	mg/L	25	EST	MEYLAN,WM ET AL. (1996)
logP 2.45			EXP	HANSCH,C & LEO,AJ (1985)
VP 2.40E-001	mm Hg	25	EST	NEELY,WB & BLAU,GE (1985)
DC	pKa			
HL 7.37E-006	atm m3/mol	25	EST	MEYLAN,WM & HOWARD,PH (1991)
OH 5.47E-011	cm3/molc sec	25	EST	MEYLAN,WM & HOWARD,PH (1993)

CAS #: 000622-83-3		DIAZENECARBONITRILE, PHENYL-	

Formula: $C_7H_5N_3$

Mol Weight: 131.14

MP (deg C):   FP (deg C):

BP (deg C):

BP pressure (mm Hg):

Property/Value	Units	Temp	Data Type	Reference
WS 3.66E+002	mg/L	25	EST	MEYLAN,WM ET AL. (1996)
logP 2.30			EXP	HANSCH,C ET AL. (1995)
VP 1.77E-002	mm Hg	25	EST	NEELY,WB & BLAU,GE (1985)
DC	pKa			
HL 2.57E-004	atm m3/mol	25	EST	MEYLAN,WM & HOWARD,PH (1991)
OH 7.73E-013	cm3/molc sec	25	EST	MEYLAN,WM & HOWARD,PH (1993)

CAS #: 000622-85-5				PHENYLPROPYL ETHER

Formula: $C_9H_{12}O$

Mol Weight: 136.20

MP (deg C): -27   FP (deg C):

BP (deg C): 189.9

BP pressure (mm Hg):

Property/Value	Units	Temp	Data Type	Reference
WS 1.67E+002	mg/L	25	EST	MEYLAN,WM ET AL. (1996)
logP 3.18			EXP	HANSCH,C & LEO,AJ (1985)
VP 5.49E-001	mm Hg	25	EST	NEELY,WB & BLAU,GE (1985)
DC	pKa			
HL 5.62E-004	atm m3/mol	25	EST	MEYLAN,WM & HOWARD,PH (1991)
OH 3.19E-011	cm3/molc sec	25	EST	MEYLAN,WM & HOWARD,PH (1993)

CAS #: 000622-96-8				P-ETHYLTOLUENE

Formula: C_9H_{12}

Mol Weight: 120.20

MP (deg C): -62.4   FP (deg C):

BP (deg C): 162

BP pressure (mm Hg):

Property/Value	Units	Temp	Data Type	Reference
WS 9.48E+001	mg/L	25	EXP	MACKAY,D & SHIU,WY (1981)
logP 3.63			EXP	SANGSTER,J (1993)
VP 3.00E+000	mm Hg	25	EXP	DAUBERT,TE & DANNER,RP (1989)
DC	pKa			
HL 5.01E-003	atm m3/mol	25	EST	VP/WSOL
OH 1.21E-011	cm3/molc sec	25	EXP	ATKINSON,R (1989)

CAS #: 000622-97-9				P-METHYLSTYRENE

Formula: C_9H_{10}

Mol Weight: 118.18

MP (deg C): -37.8   FP (deg C):

BP (deg C): 169

BP pressure (mm Hg):

Property/Value	Units	Temp	Data Type	Reference
WS 9.35E+001	mg/L	25	EST	VP/HL
logP 3.35			EST	OGATA,M ET AL. (1984)
VP 1.81E+000	mm Hg	25	EXP	BOUBLIK,T ET AL. (1984)
DC	pKa			
HL 3.01E-003	atm m3/mol	25	EST	MEYLAN,WM & HOWARD,PH (1991)
OH 3.15E-011	cm3/molc sec	25	EST	ATKINSON,R (1987)

CAS #: 000623-00-7				4-BROMOBENZONITRILE

Formula: C_7H_4BrN

Mol Weight: 182.03

MP (deg C): 114   FP (deg C):

BP (deg C): 236

BP pressure (mm Hg):

Property/Value	Units	Temp	Data Type	Reference
WS 2.46E+002	mg/L	25	EST	MEYLAN,WM ET AL. (1996)
logP 2.43			EST	MEYLAN,WM & HOWARD,PH (1995)
VP 2.73E-002	mm Hg	25	EST	NEELY,WB & BLAU,GE (1985)
DC	pKa			
HL 2.07E-005	atm m3/mol	25	EST	MEYLAN,WM & HOWARD,PH (1991)
OH 2.17E-013	cm3/molc sec	25	EST	MEYLAN,WM & HOWARD,PH (1993)

CAS #: 000623-03-0				BENZONITRILE, 4-CHLORO-

Formula: C_7H_4ClN

Mol Weight: 137.57

MP (deg C): 95   FP (deg C):

BP (deg C): 223

BP pressure (mm Hg):

Property/Value	Units	Temp	Data Type	Reference
WS 3.62E+002	mg/L	25	EST	MEYLAN,WM ET AL. (1996)
logP 2.47			EXP	SANGSTER,J (1994)
VP 1.13E-001	mm Hg	25	EST	NEELY,WB & BLAU,GE (1985)
DC	pKa			
HL 3.86E-005	atm m3/mol	25	EST	MEYLAN,WM & HOWARD,PH (1991)
OH 2.42E-013	cm3/molc sec	25	EST	MEYLAN,WM & HOWARD,PH (1993)

CAS #: 000623-05-2				P-HYDROXYBENZYL ALCOHOL

Formula: $C_7H_8O_2$

Mol Weight: 124.14

MP (deg C): 124.5   FP (deg C):

BP (deg C): 252

BP pressure (mm Hg):

Property/Value	Units	Temp	Data Type	Reference
WS 7.29E+005	mg/L	25	EST	MEYLAN,WM ET AL. (1996)
logP 0.25			EXP	HANSCH,C & LEO,AJ (1985)
VP 8.25E-004	mm Hg	25	EST	NEELY,WB & BLAU,GE (1985)
DC 9.82	pKa	25	EXP	KORTUM,G ET AL (1961)
HL 2.26E-011	atm m3/mol	25	EST	MEYLAN,WM & HOWARD,PH (1991)
OH 4.42E-011	cm3/molc sec	25	EST	MEYLAN,WM & HOWARD,PH (1993)

CAS #: 000623-08-5				N-METHYL P-TOLUIDINE

Formula: $C_8H_{11}N$

Mol Weight: 121.18

MP (deg C):   FP (deg C):

BP (deg C): 210

BP pressure (mm Hg):

Property/Value	Units	Temp	Data Type	Reference
WS 1.45E+003	mg/L	25	EST	MEYLAN,WM ET AL. (1996)
logP 2.15			EXP	HANSCH,C & LEO,AJ (1985)
VP 6.50E-001	mm Hg	25	EST	NEELY,WB & BLAU,GE (1985)
DC 5.36	pKa	23	EXP	PERRIN,DD (1965)
HL 4.61E-006	atm m3/mol	25	EST	MEYLAN,WM & HOWARD,PH (1991)
OH 5.37E-011	cm3/molc sec	25	EST	MEYLAN,WM & HOWARD,PH (1993)

CAS #: 000623-12-1				4-CHLOROANISOLE

Formula: C_7H_7ClO

Mol Weight: 142.59

MP (deg C): < -18   FP (deg C):

BP (deg C): 197.5

BP pressure (mm Hg):

Property/Value	Units	Temp	Data Type	Reference
WS 3.45E+002	mg/L	25	EST	MEYLAN,WM ET AL. (1996)
logP 2.78			EXP	HANSCH,C & LEO,AJ (1985)
VP 8.11E-001	mm Hg	25	EST	NEELY,WB & BLAU,GE (1985)
DC	pKa			
HL 2.36E-004	atm m3/mol	25	EST	MEYLAN,WM & HOWARD,PH (1991)
OH 7.11E-012	cm3/molc sec	25	EST	MEYLAN,WM & HOWARD,PH (1993)

CAS #: 000623-15-4 — 3-BUTEN-2-ONE, 4-(2-FURANYL)-

Formula: $C_8H_8O_2$

Mol Weight: 136.15

MP (deg C): 39.5

BP (deg C): 113

BP pressure (mm Hg): 1.00E+001

Property/Value	Units	Temp	Data Type	Reference
WS 6.11E+003	mg/L	25	EST	MEYLAN,WM ET AL. (1996)
logP 1.35			EXP	HANSCH,C ET AL. (1995)
VP 3.38E-001	mm Hg	25	EST	NEELY,WB & BLAU,GE (1985)
DC	pKa			
HL 1.17E-006	atm m3/mol	25	EST	MEYLAN,WM & HOWARD,PH (1991)
OH 8.89E-011	cm3/molc sec	25	EST	MEYLAN,WM & HOWARD,PH (1993)

CAS #: 000623-26-7 — 1,4-BENZENEDICARBONITRILE

Formula: $C_8H_4N_2$

Mol Weight: 128.13

MP (deg C): 224

BP (deg C):

BP pressure (mm Hg):

Property/Value	Units	Temp	Data Type	Reference
WS 8.16E+003	mg/L	25	EST	MEYLAN,WM ET AL. (1996)
logP 0.93			EXP	HANSCH,C ET AL. (1995)
VP 5.69E-003	mm Hg	25	EST	NEELY,WB & BLAU,GE (1985)
DC	pKa			
HL 5.03E-007	atm m3/mol	25	EST	MEYLAN,WM & HOWARD,PH (1991)
OH 4.51E-014	cm3/molc sec	25	EST	MEYLAN,WM & HOWARD,PH (1993)

CAS #: 000623-30-3 — 2-PROPENAL, 3-(2-FURANYL)-

Formula: $C_7H_6O_2$

Mol Weight: 122.12

MP (deg C): 54

BP (deg C): 135

BP pressure (mm Hg): 1.40E+001

Property/Value	Units	Temp	Data Type	Reference
WS 1.03E+004	mg/L	25	EST	MEYLAN,WM ET AL. (1996)
logP 1.15			EXP	HANSCH,C ET AL. (1995)
VP 5.13E-001	mm Hg	25	EST	NEELY,WB & BLAU,GE (1985)
DC	pKa			
HL 1.60E-006	atm m3/mol	25	EST	MEYLAN,WM & HOWARD,PH (1991)
OH 7.41E-011	cm3/molc sec	25	EST	MEYLAN,WM & HOWARD,PH (1993)

CAS #: 000623-37-0 — 3-HEXANOL

Formula: $C_6H_{14}O$

Mol Weight: 102.18

MP (deg C):

BP (deg C): 134.5-135.

BP pressure (mm Hg):

Property/Value	Units	Temp	Data Type	Reference
WS 1.61E+004	mg/L	25	EXP	YALKOWSKY,SH & DANNENFELSER,RM (1992)
logP 1.65			EXP	FUNASAKI,N ET AL. (1986)
VP 4.81E+000	mm Hg	25	EXP	PERRY,RH & GREEN,D (1984)
DC	pKa			
HL 4.02E-005	atm m3/mol		EST	VP/WSOL
OH 1.45E-011	cm3/molc sec	25	EST	MEYLAN,WM & HOWARD,PH (1993)

CAS #: 000623-42-7 — METHYLBUTYRATE

Formula: $C_5H_{10}O_2$

Mol Weight: 102.13

MP (deg C): -95

BP (deg C): 102

BP pressure (mm Hg):

Property/Value	Units	Temp	Data Type	Reference
WS 1.50E+004	mg/L	25	EXP	YALKOWSKY,SH & DANNENFELSER,RM (1992)
logP 1.29			EXP	HANSCH,C ET AL. (1995)
VP 3.23E+001	mm Hg	25	EXP	YAWS,CL (1994A)
DC	pKa			
HL 2.05E-004	atm m3/mol	25	EXP	BUTTERY,RG ET AL. (1969)
OH 3.00E-012	cm3/molc sec	25	EXP	ATKINSON,R (1989)

CAS #: 000623-48-3 — ETHYL IODOACETATE

Formula: $C_4H_7IO_2$

Mol Weight: 214.00

MP (deg C):

BP (deg C): 179

BP pressure (mm Hg):

Property/Value	Units	Temp	Data Type	Reference
WS 1.52E+003	mg/L	25	EST	MEYLAN,WM ET AL. (1996)
logP 1.62			EST	MEYLAN,WM & HOWARD,PH (1995)
VP 2.38E-001	mm Hg	25	EXT	BOUBLIK,T ET AL. (1984)
DC	pKa			
HL 1.74E-005	atm m3/mol	25	EST	MEYLAN,WM & HOWARD,PH (1991)
OH 1.47E-012	cm3/molc sec	25	EST	MEYLAN,WM & HOWARD,PH (1993)

CAS #: 000623-59-6 — ACETAMIDE, N- (METHYLAMINO)CARBONYL -

Formula: $C_4H_8N_2O_2$

Mol Weight: 116.12

MP (deg C): 108.5

BP (deg C):

BP pressure (mm Hg):

Property/Value	Units	Temp	Data Type	Reference
WS 2.01E+004	mg/L	25	EST	MEYLAN,WM ET AL. (1996)
logP -0.70			EXP	SOTOMATSU,T ET AL. (1987)
VP 2.09E-005	mm Hg	25	EST	NEELY,WB & BLAU,GE (1985)
DC	pKa			
HL 2.37E-010	atm m3/mol	25	EST	MEYLAN,WM & HOWARD,PH (1991)
OH 2.66E-012	cm3/molc sec	25	EST	MEYLAN,WM & HOWARD,PH (1993)

CAS #: 000623-78-9 — ETHYL ETHYLCARBAMATE

Formula: $C_5H_{11}NO_2$

Mol Weight: 117.15

MP (deg C):

BP (deg C): 176

BP pressure (mm Hg):

Property/Value	Units	Temp	Data Type	Reference
WS 4.20E+005	mg/L	16	EXP	STEPHEN,H & STEPHEN,T (1963)
logP 0.94			EST	MEYLAN,WM & HOWARD,PH (1995)
VP 4.94E+000	mm Hg	25	EST	NEELY,WB & BLAU,GE (1985)
DC	pKa			
HL 1.53E-007	atm m3/mol	25	EST	MEYLAN,WM & HOWARD,PH (1991)
OH 1.12E-011	cm3/molc sec	25	EST	MEYLAN,WM & HOWARD,PH (1993)

CAS #:	000623-80-3	S,S-DIETHYL CARBONODITHIOATE

Formula: $C_5H_{10}OS_2$

Mol Weight: 150.26

MP (deg C): | FP (deg C):

BP (deg C): 197

BP pressure (mm Hg):

Property/Value	Units	Temp	Data Type	Reference
WS 1.47E+003	mg/L	25	EST	MEYLAN,WM ET AL. (1996)
logP 2.01			EST	MEYLAN,WM & HOWARD,PH (1995)
VP 1.38E-001	mm Hg	25	EST	NEELY,WB & BLAU,GE (1985)
DC	pKa			
HL 1.38E-004	atm m3/mol	25	EST	MEYLAN,WM & HOWARD,PH (1991)
OH 9.99E-012	cm3/molc sec	25	EST	MEYLAN,WM & HOWARD,PH (1993)

CAS #:	000623-87-0	1,2,3-PROPANETRIOL, 1,3-DINITRATE

Formula: $C_3H_6N_2O_7$

Mol Weight: 182.09

MP (deg C): 26 | FP (deg C):

BP (deg C): 148

BP pressure (mm Hg): 1.50E+001

Property/Value	Units	Temp	Data Type	Reference
WS 7.84E+004	mg/L	20	EXP	YALKOWLSKY,SH & DANNENFELSER,RM (1992)
logP 0.71			EXP	KIKKOJI,T ET AL. (1991)
VP 1.11E-003	mm Hg	25	EST	NEELY,WB & BLAU,GE (1985)
DC	pKa			
HL 3.44E-011	atm m3/mol	25	EST	MEYLAN,WM & HOWARD,PH (1991)
OH 1.22E-012	cm3/molc sec	25	EST	MEYLAN,WM & HOWARD,PH (1993)

CAS #:	000623-93-8	5-NONANOL

Formula: $C_9H_{20}O$

Mol Weight: 144.26

MP (deg C): | FP (deg C):

BP (deg C): 193

BP pressure (mm Hg):

Property/Value	Units	Temp	Data Type	Reference
WS 4.67E+002	mg/L	25	EXP	SUZUKI,T (1991)
logP 3.22			EST	MEYLAN,WM & HOWARD,PH (1995)
VP 6.48E-002	mm Hg	25	EST	NEELY,WB & BLAU,GE (1985)
DC	pKa			
HL 4.12E-005	atm m3/mol	25	EST	MEYLAN,WM & HOWARD,PH (1991)
OH 1.87E-011	cm3/molc sec	25	EST	MEYLAN,WM & HOWARD,PH (1993)

CAS #:	000624-17-9	DIETHYL AZELATE

Formula: $C_{13}H_{24}O_4$

Mol Weight: 244.33

MP (deg C): -18.5 | FP (deg C):

BP (deg C): 291.5

BP pressure (mm Hg):

Property/Value	Units	Temp	Data Type	Reference
WS 2.50E+002	mg/L	25	EXP	SUZUKI,T (1991)
logP 3.84			EST	MEYLAN,WM & HOWARD,PH (1995)
VP 5.90E-003	mm Hg	25	EST	NEELY,WB & BLAU,GE (1985)
DC	pKa			
HL 4.03E-006	atm m3/mol	25	EST	MEYLAN,WM & HOWARD,PH (1991)
OH 1.21E-011	cm3/molc sec	25	EST	MEYLAN,WM & HOWARD,PH (1993)

CAS #:	000624-24-8	METHYL VALERATE

Formula: $C_6H_{12}O_2$

Mol Weight: 116.16

MP (deg C): | FP (deg C):

BP (deg C): 127.4

BP pressure (mm Hg):

Property/Value	Units	Temp	Data Type	Reference
WS 5.06E+003	mg/L	25	EXP	KUHNE,R ET AL. (1995)
logP 1.96			EXP	HANSCH,C ET AL. (1995)
VP 1.91E+001	mm Hg	25	EXP	JORDAN,TE (1954)
DC	pKa			
HL 3.18E-004	atm m3/mol	25	EXP	BUTTERY,RG ET AL. (1969)
OH 2.88E-012	cm3/molc sec	25	EST	MEYLAN,WM & HOWARD,PH (1993)

CAS #:	000624-29-3	1,4-DIMETHYLCYCLOHEXANE (CIS)

Formula: C_8H_{16}

Mol Weight: 112.22

MP (deg C): -87.4 | FP (deg C):

BP (deg C): 124.4

BP pressure (mm Hg):

Property/Value	Units	Temp	Data Type	Reference
WS 1.17E+001	mg/L	25	EST	MEYLAN,WM ET AL. (1996)
logP 4.01			EST	MEYLAN,WM & HOWARD,PH (1995)
VP 1.79E+001	mm Hg	25	EXP	YAWS,CL (1994B)
DC	pKa			
HL 4.50E-001	atm m3/mol	25	EST	MEYLAN,WM & HOWARD,PH (1991)
OH 1.19E-011	cm3/molc sec	25	EST	MEYLAN,WM & HOWARD,PH (1993)

CAS #:	000624-38-4	1,4-DIIODOBENZENE

Formula: $C_6H_4I_2$

Mol Weight: 329.91

MP (deg C): 131.5 | FP (deg C):

BP (deg C): 285

BP pressure (mm Hg):

Property/Value	Units	Temp	Data Type	Reference
WS 1.40E+000	mg/L	25	EXP	YALKOWSKY,SH & DANNENFELSER,RM (1992)
logP 4.11			EXP	HANSCH,C ET AL. (1995)
VP 5.93E-003	mm Hg	25	EST	NEELY,WB & BLAU,GE (1985)
DC	pKa			
HL 2.89E-004	atm m3/mol	25	EST	MEYLAN,WM & HOWARD,PH (1991)
OH 4.25E-013	cm3/molc sec	25	EST	MEYLAN,WM & HOWARD,PH (1993)

CAS #:	000624-45-3	PENTANOIC ACID, 4-OXO-, METHYL ESTER

Formula: $C_6H_{10}O_3$

Mol Weight: 130.14

MP (deg C): | FP (deg C):

BP (deg C): 196

BP pressure (mm Hg):

Property/Value	Units	Temp	Data Type	Reference
WS 1.19E+005	mg/L	25	EST	MEYLAN,WM ET AL. (1996)
logP -0.13			EXP	SANGSTER,J (1993)
VP 1.63E+000	mm Hg	25	EST	NEELY,WB & BLAU,GE (1985)
DC	pKa			
HL 1.57E-007	atm m3/mol	25	EST	MEYLAN,WM & HOWARD,PH (1991)
OH 3.88E-012	cm3/molc sec	25	EST	MEYLAN,WM & HOWARD,PH (1993)

467

CAS #:	000624-48-6		METHYL MALEATE	
Formula:	$C_6H_8O_4$			
Mol Weight:	144.13			
MP (deg C): -10		FP (deg C):		
BP (deg C): 205-207				
BP pressure (mm Hg):				

Property/Value	Units	Temp	Data Type	Reference
WS 8.00E+004	mg/L	25	EXP	RIDDICK,JA ET AL. (1986)
logP 0.22			EXP	HANSCH,C ET AL. (1995)
VP 3.00E-001	mm Hg	25	EXP	RIDDICK,JA ET AL. (1986)
DC	pKa			
HL 7.11E-007	atm m3/mol	25	EST	VP/WSOL
OH 7.34E-012	cm3/molc sec	25	EST	MEYLAN,WM & HOWARD,PH (1993)

CAS #:	000624-49-7		2-BUTENEDIOIC ACID (E)-, DIMETHYL ESTER	
Formula:	$C_6H_8O_4$			
Mol Weight:	144.13			
MP (deg C): 103.5		FP (deg C):		
BP (deg C): 193				
BP pressure (mm Hg):				

Property/Value	Units	Temp	Data Type	Reference
WS 1.88E+004	mg/L	25	EST	MEYLAN,WM ET AL. (1996)
logP 0.74			EXP	HANSCH,C ET AL. (1995)
VP 3.83E+000	mm Hg	25	EST	NEELY,WB & BLAU,GE (1985)
DC	pKa			
HL 1.39E-007	atm m3/mol	25	EST	MEYLAN,WM & HOWARD,PH (1991)
OH 7.34E-012	cm3/molc sec	25	EST	MEYLAN,WM & HOWARD,PH (1993)

CAS #:	000624-51-1		3-NONANOL	
Formula:	$C_9H_{20}O$			
Mol Weight:	144.26			
MP (deg C):		FP (deg C):		
BP (deg C):				
BP pressure (mm Hg):				

Property/Value	Units	Temp	Data Type	Reference
WS 3.15E+002	mg/L	25	EXP	SUZUKI,T (1991)
logP 3.22			EST	MEYLAN,WM & HOWARD,PH (1995)
VP 6.48E-002	mm Hg	25	EST	NEELY,WB & BLAU,GE (1985)
DC	pKa			
HL 4.12E-005	atm m3/mol	25	EST	MEYLAN,WM & HOWARD,PH (1991)
OH 1.87E-011	cm3/molc sec	25	EST	MEYLAN,WM & HOWARD,PH (1993)

CAS #:	000624-54-4		AMYL PROPIONATE	
Formula:	$C_8H_{16}O_2$			
Mol Weight:	144.22			
MP (deg C): -73.1		FP (deg C):		
BP (deg C): 168.6				
BP pressure (mm Hg):				

Property/Value	Units	Temp	Data Type	Reference
WS 8.10E+002	mg/L	25	EXP	HINE,J & MOOKERJEE,PK (1975)
logP 2.83			EST	MEYLAN,WM & HOWARD,PH (1995)
VP 3.60E+000	mm Hg	25	EXP	HINE,J & MOOKERJEE,PK (1975)
DC	pKa			
HL 8.43E-004	atm m3/mol	25	EST	VP/WSOL
OH 6.84E-012	cm3/molc sec	25	EST	MEYLAN,WM & HOWARD,PH (1993)

CAS #:	000624-64-6		TRANS-2-BUTENE	
Formula:	C_4H_8			
Mol Weight:	56.11			
MP (deg C): -105.8		FP (deg C):		
BP (deg C): 0.88				
BP pressure (mm Hg):				

Property/Value	Units	Temp	Data Type	Reference
WS 5.11E+002	mg/L	25	EXP	SUZUKI,T (1991)
logP 2.31			EXP	HANSCH,C & LEO,AJ (1985)
VP 1.76E+003	mm Hg	25	EXP	DAUBERT,TE & DANNER,RP (1989)
DC	pKa			
HL 2.24E-001	atm m3/mol	25	EXP	WASIK,SP & TSANG,W (1970)
OH 6.39E-011	cm3/molc sec	25	EXP	ATKINSON,R (1985)

CAS #:	000624-72-6		1,2-DIFLUOROETHANE	
Formula:	$C_2H_4F_2$			
Mol Weight:	66.05			
MP (deg C):		FP (deg C):		
BP (deg C): 30.7				
BP pressure (mm Hg):				

Property/Value	Units	Temp	Data Type	Reference
WS 2.31E+003	mg/L	25	EST	MEYLAN,WM ET AL. (1996)
logP 1.21			EST	MEYLAN,WM & HOWARD,PH (1995)
VP 6.16E+002	mm Hg	25	EXP	YAWS,CL (1994)
DC	pKa			
HL 3.87E-001	atm m3/mol	25	EST	MEYLAN,WM & HOWARD,PH (1991)
OH 1.12E-013	cm3/molc sec	25	EXP	ATKINSON,R (1989)

CAS #:	000624-73-7		ETHANE, 1,2-DIIODO-	
Formula:	$C_2H_4I_2$			
Mol Weight:	281.86			
MP (deg C): 83		FP (deg C):		
BP (deg C): 200				
BP pressure (mm Hg):				

Property/Value	Units	Temp	Data Type	Reference
WS 7.58E+001	mg/L	25	EST	MEYLAN,WM ET AL. (1996)
logP 2.71			EXP	HANSCH,C ET AL. (1995)
VP 6.91E-001	mm Hg	25	EST	NEELY,WB & BLAU,GE (1985)
DC	pKa			
HL 5.45E-004	atm m3/mol	25	EST	MEYLAN,WM & HOWARD,PH (1991)
OH 4.55E-013	cm3/molc sec	25	EST	MEYLAN,WM & HOWARD,PH (1993)

CAS #:	000624-78-2		METHYLETHYLAMINE	
Formula:	C_3H_9N			
Mol Weight:	59.11			
MP (deg C):		FP (deg C):		
BP (deg C): 36-37				
BP pressure (mm Hg):				

Property/Value	Units	Temp	Data Type	Reference
WS 1.00E+006	mg/L	25	EST	MEYLAN,WM ET AL. (1996)
logP 0.15			EXP	HANSCH,C & LEO,AJ (1985)
VP 3.90E+002	mm Hg	25	EST	NEELY,WB & BLAU,GE (1985)
DC	pKa			
HL 2.21E-005	atm m3/mol	25	EST	MEYLAN,WM & HOWARD,PH (1991)
OH 7.31E-011	cm3/molc sec	25	EST	MEYLAN,WM & HOWARD,PH (1993)

468

CAS #: 000624-83-9				METHYLISOCYANATE

Formula: C_2H_3NO

Mol Weight: 57.05

MP (deg C): -45 FP (deg C):

BP (deg C): 59.6

BP pressure (mm Hg):

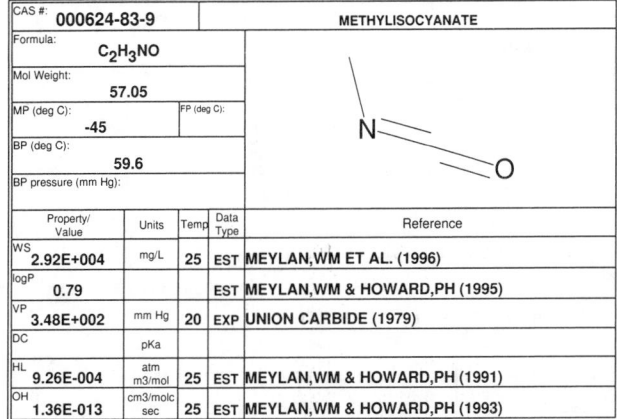

Property/ Value	Units	Temp	Data Type	Reference
WS 2.92E+004	mg/L	25	EST	MEYLAN,WM ET AL. (1996)
logP 0.79			EST	MEYLAN,WM & HOWARD,PH (1995)
VP 3.48E+002	mm Hg	20	EXP	UNION CARBIDE (1979)
DC	pKa			
HL 9.26E-004	atm m3/mol	25	EST	MEYLAN,WM & HOWARD,PH (1991)
OH 1.36E-013	cm3/molc sec	25	EST	MEYLAN,WM & HOWARD,PH (1993)

CAS #: 000624-84-0				N-FORMYLHYDRAZINE

Formula: CH_4N_2O

Mol Weight: 60.06

MP (deg C): 54-56 FP (deg C):

BP (deg C):

BP pressure (mm Hg):

Property/ Value	Units	Temp	Data Type	Reference
WS 1.00E+006	mg/L	25	EST	MEYLAN,WM ET AL. (1996)
logP -2.05			EXP	HANSCH,C & LEO,AJ (1985)
VP 1.74E-001	mm Hg	25	EST	NEELY,WB & BLAU,GE (1985)
DC	pKa			
HL 6.40E-011	atm m3/mol	25	EST	MEYLAN,WM & HOWARD,PH (1991)
OH 5.50E-012	cm3/molc sec	25	EST	MEYLAN,WM & HOWARD,PH (1993)

CAS #: 000624-89-5				ETHYL METHYL SULFIDE

Formula: C_3H_8S

Mol Weight: 76.16

MP (deg C): -105.9 FP (deg C):

BP (deg C): 66.7

BP pressure (mm Hg):

Property/ Value	Units	Temp	Data Type	Reference
WS 6.43E+003	mg/L	25	EST	MEYLAN,WM ET AL. (1996)
logP 1.54			EXP	HANSCH,C ET AL. (1995)
VP 1.60E+002	mm Hg	25	EXP	BOUBLIK,T ET AL. (1984)
DC	pKa			
HL 1.04E-003	atm m3/mol	25	EST	MEYLAN,WM & HOWARD,PH (1991)
OH 8.50E-012	cm3/molc sec	25	EXP	ATKINSON,R (1989)

CAS #: 000624-91-9				METHYLNITRITE

Formula: CH_3NO_2

Mol Weight: 61.04

MP (deg C): -16 FP (deg C):

BP (deg C): -12

BP pressure (mm Hg):

Property/ Value	Units	Temp	Data Type	Reference
WS 2.42E+004	mg/L	25	EST	MEYLAN,WM ET AL. (1996)
logP 0.88			EST	MEYLAN,WM & HOWARD,PH (1995)
VP 1.65E+003	mm Hg	25	EST	NEELY,WB & BLAU,GE (1985)
DC 8.65	pKa		EXP	SERJEANT,EP & DEMPSEY,B (1979)
HL 6.55E-005	atm m3/mol	25	EST	MEYLAN,WM & HOWARD,PH (1991)
OH 5.00E-013	cm3/molc sec	25	EXP	ATKINSON,R (1989)

CAS #: 000624-92-0				DIMETHYLDISULFIDE

Formula: $C_2H_6S_2$

Mol Weight: 94.20

MP (deg C): -84.7 FP (deg C):

BP (deg C): 109.75

BP pressure (mm Hg):

Property/ Value	Units	Temp	Data Type	Reference
WS 3.74E+003	mg/L	25	EST	MEYLAN,WM ET AL. (1996)
logP 1.77			EXP	HANSCH,C & LEO,AJ (1985)
VP 2.87E+001	mm Hg	25	EXP	DAUBERT,TE & DANNER,RP (1989)
DC	pKa			
HL 1.21E-003	atm m3/mol	20	EXP	VITENBERG,AG ET AL. (1975)
OH 2.11E-010	cm3/molc sec	25	EXP	ATKINSON,R (1989)

CAS #: 000624-95-3				3,3-DIMETHYL-1-BUTANOL

Formula: $C_6H_{14}O$

Mol Weight: 102.18

MP (deg C): -60 FP (deg C):

BP (deg C): 143

BP pressure (mm Hg):

Property/ Value	Units	Temp	Data Type	Reference
WS 7.56E+003	mg/L	25	EXP	SUZUKI,T (1991)
logP 1.71			EST	MEYLAN,WM & HOWARD,PH (1995)
VP 2.45E+000	mm Hg	25	EXT	WILHOIT,RC & ZWOLINSKI,BJ (1973)
DC	pKa			
HL 1.76E-005	atm m3/mol	25	EST	MEYLAN,WM & HOWARD,PH (1991)
OH 6.08E-012	cm3/molc sec	25	EST	MEYLAN,WM & HOWARD,PH (1993)

CAS #: 000625-06-9				2,4-DIMETHYL-2-PENTANOL

Formula: $C_7H_{16}O$

Mol Weight: 116.20

MP (deg C): < -20 FP (deg C):

BP (deg C): 133.1

BP pressure (mm Hg):

Property/ Value	Units	Temp	Data Type	Reference
WS 1.36E+004	mg/L	25	EXP	SUZUKI,T (1991)
logP 2.13			EST	MEYLAN,WM & HOWARD,PH (1995)
VP 4.54E+000	mm Hg	25	EST	NEELY,WB & BLAU,GE (1985)
DC	pKa			
HL 2.34E-005	atm m3/mol	25	EST	MEYLAN,WM & HOWARD,PH (1991)
OH 8.26E-012	cm3/molc sec	25	EST	MEYLAN,WM & HOWARD,PH (1993)

CAS #: 000625-16-1				TERT-PENTYLACETATE

Formula: $C_7H_{14}O_2$

Mol Weight: 130.19

MP (deg C): FP (deg C):

BP (deg C):

BP pressure (mm Hg):

Property/ Value	Units	Temp	Data Type	Reference
WS 1.15E+003	mg/L	25	EST	MEYLAN,WM ET AL. (1996)
logP 2.23			EST	MEYLAN,WM & HOWARD,PH (1995)
VP 1.21E+001	mm Hg	25	EST	NEELY,WB & BLAU,GE (1985)
DC	pKa			
HL 5.45E-004	atm m3/mol	25	EST	MEYLAN,WM & HOWARD,PH (1991)
OH 1.75E-012	cm3/molc sec	25	EST	MEYLAN,WM & HOWARD,PH (1993)

000625-23-0 — 2-METHYL-2-HEXANOL

CAS #:	000625-23-0
Formula:	$C_7H_{16}O$
Mol Weight:	116.20
MP (deg C):	
FP (deg C):	
BP (deg C):	143
BP pressure (mm Hg):	

Property/Value	Units	Temp	Data Type	Reference
WS 9.70E+003	mg/L	25	EXP	BARTON,AFM (1984)
logP 2.20			EST	MEYLAN,WM & HOWARD,PH (1995)
VP 2.07E+000	mm Hg	25	EXT	WILHOIT,RC & ZWOLINSKI,BJ (1973)
DC	pKa			
HL 2.34E-005	atm m3/mol	25	EST	MEYLAN,WM & HOWARD,PH (1991)
OH 8.27E-012	cm3/molc sec	25	EST	MEYLAN,WM & HOWARD,PH (1993)

000625-25-2 — 2-METHYL-2-HEPTANOL

CAS #:	000625-25-2
Formula:	$C_8H_{18}O$
Mol Weight:	130.23
MP (deg C):	-50.4
FP (deg C):	
BP (deg C):	156
BP pressure (mm Hg):	

Property/Value	Units	Temp	Data Type	Reference
WS 2.50E+003	mg/L	30	EXP	BARTON,AFM (1984)
logP 2.69			EST	MEYLAN,WM & HOWARD,PH (1995)
VP 6.24E-001	mm Hg	25	EST	NEELY,WB & BLAU,GE (1985)
DC	pKa			
HL 3.10E-005	atm m3/mol	25	EST	MEYLAN,WM & HOWARD,PH (1991)
OH 9.68E-012	cm3/molc sec	25	EST	MEYLAN,WM & HOWARD,PH (1993)

000625-27-4 — 2-METHYL-2-PENTENE

CAS #:	000625-27-4
Formula:	C_6H_{12}
Mol Weight:	84.16
MP (deg C):	-135
FP (deg C):	
BP (deg C):	67.3
BP pressure (mm Hg):	

Property/Value	Units	Temp	Data Type	Reference
WS 7.93E+001	mg/L	25	EST	MEYLAN,WM ET AL. (1996)
logP 3.13			EST	MEYLAN,WM & HOWARD,PH (1995)
VP 1.58E+002	mm Hg	25	EXP	DAUBERT,TE & DANNER,RP (1989)
DC	pKa			
HL 5.00E-001	atm m3/mol	25	EST	MEYLAN,WM & HOWARD,PH (1991)
OH 8.88E-011	cm3/molc sec	25	EXP	ATKINSON,R (1989)

000625-28-5 — BUTANENITRILE, 3-METHYL-

CAS #:	000625-28-5
Formula:	C_5H_9N
Mol Weight:	83.13
MP (deg C):	-101
FP (deg C):	
BP (deg C):	127.5
BP pressure (mm Hg):	

Property/Value	Units	Temp	Data Type	Reference
WS 8.55E+003	mg/L	25	EST	MEYLAN,WM ET AL. (1996)
logP 1.07			EXP	HANSCH,C ET AL. (1995)
VP 1.03E+001	mm Hg	25	EST	NEELY,WB & BLAU,GE (1985)
DC	pKa			
HL 7.15E-005	atm m3/mol	25	EST	MEYLAN,WM & HOWARD,PH (1991)
OH 7.86E-013	cm3/molc sec	25	EST	MEYLAN,WM & HOWARD,PH (1993)

000625-29-6 — 2-CHLOROPENTANE

CAS #:	000625-29-6
Formula:	$C_5H_{11}Cl$
Mol Weight:	106.60
MP (deg C):	
FP (deg C):	
BP (deg C):	
BP pressure (mm Hg):	

Property/Value	Units	Temp	Data Type	Reference
WS 4.65E+002	mg/L	25	EXP	HINE,J & MOOKERJEE,PK (1975)
logP 2.98			EST	MEYLAN,WM & HOWARD,PH (1995)
VP 4.87E+001	mm Hg	25	EXP	HINE,J & MOOKERJEE,PK (1975)
DC	pKa			
HL 1.47E-002	atm m3/mol	25	EST	VP/WSOL
OH 2.69E-012	cm3/molc sec	25	EST	MEYLAN,WM & HOWARD,PH (1993)

000625-33-2 — 3-PENTEN-2-ONE

CAS #:	000625-33-2
Formula:	C_5H_8O
Mol Weight:	84.12
MP (deg C):	
FP (deg C):	
BP (deg C):	121-124
BP pressure (mm Hg):	

Property/Value	Units	Temp	Data Type	Reference
WS 4.62E+004	mg/L	25	EST	MEYLAN,WM ET AL. (1996)
logP 0.52			EXP	HANSCH,C ET AL. (1995)
VP 3.91E+001	mm Hg	25	EST	NEELY,WB & BLAU,GE (1985)
DC	pKa			
HL 4.10E-005	atm m3/mol	25	EST	MEYLAN,WM & HOWARD,PH (1991)
OH 5.10E-011	cm3/molc sec	25	EST	MEYLAN,WM & HOWARD,PH (1993)

000625-44-5 — METHYL ISOBUTYL ETHER

CAS #:	000625-44-5
Formula:	$C_5H_{12}O$
Mol Weight:	88.15
MP (deg C):	
FP (deg C):	
BP (deg C):	58.6
BP pressure (mm Hg):	

Property/Value	Units	Temp	Data Type	Reference
WS 1.11E+004	mg/L	25	EXP	SUZUKI,T (1991)
logP 1.47			EST	MEYLAN,WM & HOWARD,PH (1995)
VP 2.11E+002	mm Hg	25	EXP	YAWS,CL (1994A)
DC	pKa			
HL 2.21E-003	atm m3/mol	25	EST	VP/WSOL
OH 1.50E-011	cm3/molc sec	25	EST	MEYLAN,WM & HOWARD,PH (1993)

000625-45-6 — METHOXYACETIC ACID

CAS #:	000625-45-6
Formula:	$C_3H_6O_3$
Mol Weight:	90.08
MP (deg C):	
FP (deg C):	
BP (deg C):	203.5
BP pressure (mm Hg):	

Property/Value	Units	Temp	Data Type	Reference
WS 1.00E+006	mg/L	20	EXP	YALKOWSKY,SH & DANNENFELSER,RM (1992)
logP -0.68			EST	MEYLAN,WM & HOWARD,PH (1995)
VP 5.42E-002	mm Hg	25	EXP	YAWS,CL (1994)
DC	pKa			
HL 6.42E-009	atm m3/mol	25	EST	VP/WSOL
OH 5.62E-012	cm3/molc sec	25	EST	MEYLAN,WM & HOWARD,PH (1993)

000625-48-9 — 2-NITROETHANOL

Formula: $C_2H_5NO_3$
Mol Weight: 91.07
MP (deg C): -80
FP (deg C):
BP (deg C): 194
BP pressure (mm Hg):

Property/Value	Units	Temp	Data Type	Reference
WS 1.15E+005	mg/L	25	EST	MEYLAN,WM ET AL. (1996)
logP -0.42			EXP	HANSCH,C & LEO,AJ (1985)
VP 3.03E-001	mm Hg	25	EST	NEELY,WB & BLAU,GE (1985)
DC	pKa			
HL 2.04E-009	atm m3/mol	25	EST	MEYLAN,WM & HOWARD,PH (1991)
OH 7.28E-013	cm3/molc sec	25	EST	MEYLAN,WM & HOWARD,PH (1993)

000625-52-5 — ETHYLUREA

Formula: $C_3H_8N_2O$
Mol Weight: 88.11
MP (deg C): 92.5
FP (deg C):
BP (deg C):
BP pressure (mm Hg):

Property/Value	Units	Temp	Data Type	Reference
WS 2.64E+004	mg/L	25	EST	MEYLAN,WM ET AL. (1996)
logP -0.74			EXP	HANSCH,C & LEO,AJ (1985)
VP 4.01E-001	mm Hg	25	EST	NEELY,WB & BLAU,GE (1985)
DC	pKa			
HL 1.06E-009	atm m3/mol	25	EST	MEYLAN,WM & HOWARD,PH (1991)
OH 6.00E-012	cm3/molc sec	25	EST	MEYLAN,WM & HOWARD,PH (1993)

000625-53-6 — N-ETHYLTHIOUREA

Formula: $C_3H_8N_2S$
Mol Weight: 104.17
MP (deg C): 111-113
FP (deg C):
BP (deg C):
BP pressure (mm Hg):

Property/Value	Units	Temp	Data Type	Reference
WS 2.39E+004	mg/L	25	EST	MEYLAN,WM ET AL. (1996)
logP -0.21			EXP	GOVERS,H ET AL. (1986)
VP 6.72E-001	mm Hg	25	EST	NEELY,WB & BLAU,GE (1985)
DC	pKa			
HL 2.35E-008	atm m3/mol	25	EST	MEYLAN,WM & HOWARD,PH (1991)
OH 9.96E-011	cm3/molc sec	25	EST	ATKINSON,R (1988)

000625-54-7 — ETHYL ISOPROPYL ETHER

Formula: $C_5H_{12}O$
Mol Weight: 88.15
MP (deg C):
FP (deg C):
BP (deg C): 54.1
BP pressure (mm Hg):

Property/Value	Units	Temp	Data Type	Reference
WS 2.42E+004	mg/L	25	EXP	SUZUKI,T (1991)
logP 1.47			EST	MEYLAN,WM & HOWARD,PH (1995)
VP 2.33E+002	mm Hg	25	EST	NEELY,WB & BLAU,GE (1985)
DC	pKa			
HL 2.02E-003	atm m3/mol	25	EST	MEYLAN,WM & HOWARD,PH (1991)
OH 1.83E-011	cm3/molc sec	25	EST	MEYLAN,WM & HOWARD,PH (1993)

000625-55-8 — ISOPROPYL FORMATE

Formula: $C_4H_8O_2$
Mol Weight: 88.11
MP (deg C):
FP (deg C):
BP (deg C): 68.2
BP pressure (mm Hg):

Property/Value	Units	Temp	Data Type	Reference
WS 2.07E+004	mg/L	25	EXP	WAKITA,K ET AL. (1986)
logP 0.73			EST	MEYLAN,WM & HOWARD,PH (1995)
VP 1.38E+002	mm Hg	25	EXP	BOUBLIK,T ET AL. (1984)
DC	pKa			
HL 7.75E-004	atm m3/mol	25	EST	VP/WSOL
OH 3.12E-012	cm3/molc sec	25	EST	ATKINSON,R (1988)

000625-58-1 — ETHYLNITRATE

Formula: $C_2H_5NO_3$
Mol Weight: 91.07
MP (deg C): -94.6
FP (deg C):
BP (deg C): 87.2
BP pressure (mm Hg):

Property/Value	Units	Temp	Data Type	Reference
WS 1.06E+004	mg/L	25	EST	MEYLAN,WM ET AL. (1996)
logP 1.25			EST	MEYLAN,WM & HOWARD,PH (1995)
VP 6.40E+001	mm Hg	25	EXP	BOUBLIK,T ET AL. (1984)
DC	pKa			
HL 2.63E-004	atm m3/mol	25	EST	MEYLAN,WM & HOWARD,PH (1991)
OH 4.90E-013	cm3/molc sec	25	EXP	ATKINSON,R (1989)

000625-74-1 — ISONITROBUTANE

Formula: $C_4H_9NO_2$
Mol Weight: 103.12
MP (deg C):
FP (deg C):
BP (deg C): 140.5
BP pressure (mm Hg):

Property/Value	Units	Temp	Data Type	Reference
WS 2.97E+003	mg/L	25	EST	MEYLAN,WM ET AL. (1996)
logP 1.40			EXP	HANSCH,C ET AL. (1995)
VP 1.29E+001	mm Hg	25	EST	NEELY,WB & BLAU,GE (1985)
DC	pKa			
HL 9.85E-005	atm m3/mol	25	EST	MEYLAN,WM & HOWARD,PH (1991)
OH 7.36E-013	cm3/molc sec	25	EST	MEYLAN,WM & HOWARD,PH (1993)

000625-77-4 — DIACETAMIDE

Formula: $C_4H_7NO_2$
Mol Weight: 101.11
MP (deg C): 79
FP (deg C):
BP (deg C): 223.5
BP pressure (mm Hg):

Property/Value	Units	Temp	Data Type	Reference
WS 1.00E+006	mg/L	25	EST	MEYLAN,WM ET AL. (1996)
logP -1.22			EST	MEYLAN,WM & HOWARD,PH (1995)
VP 3.91E-003	mm Hg	25	EXT	PERRY,RH & GREEN,D (1984)
DC	pKa			
HL 5.75E-008	atm m3/mol	25	EST	MEYLAN,WM & HOWARD,PH (1991)
OH 5.70E-012	cm3/molc sec	25	EST	MEYLAN,WM & HOWARD,PH (1993)

CAS #:	000625-80-9				2,4-DIMETHYL-3-THIAPENTANE
Formula:	$C_6H_{14}S$				
Mol Weight:	118.24				
MP (deg C): -78.1		FP (deg C):			
BP (deg C): 120.1					
BP pressure (mm Hg):					

Property/ Value	Units	Temp	Data Type	Reference
WS 3.83E+002	mg/L	25	EST	MEYLAN,WM ET AL. (1996)
logP 2.84			EXP	HANSCH,C ET AL. (1995)
VP 1.92E+001	mm Hg	25	EXT	BOUBLIK,T ET AL. (1984)
DC	pKa			
HL 2.44E-003	atm m3/mol	25	EST	MEYLAN,WM & HOWARD,PH (1991)
OH 3.26E-011	cm3/molc sec	25	EST	MEYLAN,WM & HOWARD,PH (1993)

CAS #:	000625-84-3				2,5-DIMETHYLPYRROLE
Formula:	C_6H_9N				
Mol Weight:	95.15				
MP (deg C):		FP (deg C):			
BP (deg C): 171					
BP pressure (mm Hg):					

Property/ Value	Units	Temp	Data Type	Reference
WS 6.70E+003	mg/L	25	EST	MEYLAN,WM ET AL. (1996)
logP 1.47			EXP	HANSCH,C ET AL. (1995)
VP 3.35E-001	mm Hg	25	EXP	CHAO,J ET AL. (1983)
DC -0.71	pKa		EXP	PERRIN,DD (1965)
HL 1.11E-005	atm m3/mol	25	EST	MEYLAN,WM & HOWARD,PH (1991)
OH 2.00E-010	cm3/molc sec	25	EST	MEYLAN,WM & HOWARD,PH (1993)

CAS #:	000625-86-5				FURAN, 2,5-DIMETHYL-
Formula:	C_6H_8O				
Mol Weight:	96.13				
MP (deg C):		FP (deg C):			
BP (deg C): 179-181					
BP pressure (mm Hg): 1.80E+001					

Property/ Value	Units	Temp	Data Type	Reference
WS 1.47E+003	mg/L	25	EST	MEYLAN,WM ET AL. (1996)
logP 2.24			EXP	HANSCH,C ET AL. (1995)
VP 2.59E+001	mm Hg	25	EST	NEELY,WB & BLAU,GE (1985)
DC	pKa			
HL 6.55E-003	atm m3/mol	25	EST	MEYLAN,WM & HOWARD,PH (1991)
OH 1.30E-010	cm3/molc sec	25	EST	MEYLAN,WM & HOWARD,PH (1993)

CAS #:	000625-89-8				N-NITROSO-BIS(2,2,2TRIFET)AMINE
Formula:	$C_4H_4F_6N_2O$				
Mol Weight:	210.08				
MP (deg C):		FP (deg C):			
BP (deg C):					
BP pressure (mm Hg):					

Property/ Value	Units	Temp	Data Type	Reference
WS 5.66E+002	mg/L	25	EST	MEYLAN,WM ET AL. (1996)
logP 2.15			EXP	HANSCH,C & LEO,AJ (1985)
VP 2.17E+000	mm Hg	25	EST	NEELY,WB & BLAU,GE (1985)
DC	pKa			
HL 2.25E-004	atm m3/mol	25	EST	MEYLAN,WM & HOWARD,PH (1991)
OH 1.23E-012	cm3/molc sec	25	EST	MEYLAN,WM & HOWARD,PH (1993)

CAS #:	000625-98-9				M-CHLOROFLUOROBENZENE
Formula:	C_6H_4ClF				
Mol Weight:	130.55				
MP (deg C):		FP (deg C):			
BP (deg C): 127.6					
BP pressure (mm Hg):					

Property/ Value	Units	Temp	Data Type	Reference
WS 5.89E+002	mg/L	25	EXP	YALKOWSKY,SH & DANNENFELSER,RM (1992)
logP 2.76			EXP	DUNN,WJ ET AL. (1983)
VP 8.19E+000	mm Hg	25	EST	NEELY,WB & BLAU,GE (1985)
DC	pKa			
HL 4.66E-003	atm m3/mol	25	EST	MEYLAN,WM & HOWARD,PH (1991)
OH 1.72E-012	cm3/molc sec	25	EST	MEYLAN,WM & HOWARD,PH (1993)

CAS #:	000625-99-0				3-CHLOROIODOBENZENE
Formula:	C_6H_4ClI				
Mol Weight:	238.46				
MP (deg C):		FP (deg C):			
BP (deg C): 103-104					
BP pressure (mm Hg): 1.50E+001					

Property/ Value	Units	Temp	Data Type	Reference
WS 6.72E+001	mg/L	25	EXP	SUZUKI,T (1991)
logP 3.80			EST	MEYLAN,WM & HOWARD,PH (1995)
VP 9.41E-002	mm Hg	25	EST	NEELY,WB & BLAU,GE (1985)
DC	pKa			
HL 9.25E-004	atm m3/mol	25	EST	MEYLAN,WM & HOWARD,PH (1991)
OH 9.04E-013	cm3/molc sec	25	EST	MEYLAN,WM & HOWARD,PH (1993)

CAS #:	000626-00-6				1,3-DIIODOBENZENE
Formula:	$C_6H_4I_2$				
Mol Weight:	329.91				
MP (deg C): 40.4		FP (deg C):			
BP (deg C): 285					
BP pressure (mm Hg):					

Property/ Value	Units	Temp	Data Type	Reference
WS 8.88E+000	mg/L	25	EXP	SUZUKI,T (1991)
logP 4.33			EST	MEYLAN,WM & HOWARD,PH (1995)
VP 5.93E-003	mm Hg	25	EST	NEELY,WB & BLAU,GE (1985)
DC	pKa			
HL 2.89E-004	atm m3/mol	25	EST	MEYLAN,WM & HOWARD,PH (1991)
OH 8.48E-013	cm3/molc sec	25	EST	MEYLAN,WM & HOWARD,PH (1993)

CAS #:	000626-01-7				BENZENAMINE, 3-IODO-
Formula:	C_6H_6IN				
Mol Weight:	219.03				
MP (deg C): 33		FP (deg C):			
BP (deg C): 145					
BP pressure (mm Hg): 1.50E+001					

Property/ Value	Units	Temp	Data Type	Reference
WS 1.26E+002	mg/L	25	EST	MEYLAN,WM ET AL. (1996)
logP 2.86			EXP	SANGSTER,J (1993)
VP 7.67E-003	mm Hg	25	EST	NEELY,WB & BLAU,GE (1985)
DC 3.61	pKa	25	EXP	PERRIN,DD (1965)
HL 4.41E-007	atm m3/mol	25	EST	MEYLAN,WM & HOWARD,PH (1991)
OH 7.10E-011	cm3/molc sec	25	EST	MEYLAN,WM & HOWARD,PH (1993)

000626-02-8 — 3-IODOPHENOL

Formula:	C_6H_5IO
Mol Weight:	220.01
MP (deg C):	118
FP (deg C):	
BP (deg C):	186
BP pressure (mm Hg):	1.00E+002

Property/Value	Units	Temp	Data Type	Reference
WS 4.12E+002	mg/L	25	EST	MEYLAN,WM ET AL. (1996)
logP 2.93			EXP	HANSCH,C & LEO,AJ (1985)
VP 6.16E-003	mm Hg	25	EST	NEELY,WB & BLAU,GE (1985)
DC 9.03	pKa	25	EXP	KORTUM,G ET AL (1961)
HL 1.30E-007	atm m3/mol	25	EST	MEYLAN,WM & HOWARD,PH (1991)
OH 2.21E-011	cm3/molc sec	25	EST	MEYLAN,WM & HOWARD,PH (1993)

000626-16-4 — 1,3-(BIS-CHLOROMETHYL)BENZENE

Formula:	$C_8H_8Cl_2$
Mol Weight:	175.06
MP (deg C):	34.2
FP (deg C):	
BP (deg C):	251.5
BP pressure (mm Hg):	

Property/Value	Units	Temp	Data Type	Reference
WS 2.77E+002	mg/L	25	EST	MEYLAN,WM ET AL. (1996)
logP 2.72			EXP	HANSCH,C & LEO,AJ (1985)
VP 2.21E-002	mm Hg	25	EST	NEELY,WB & BLAU,GE (1985)
DC	pKa			
HL 8.14E-004	atm m3/mol	25	EST	MEYLAN,WM & HOWARD,PH (1991)
OH 2.78E-012	cm3/molc sec	25	EST	MEYLAN,WM & HOWARD,PH (1993)

000626-17-5 — 1,3-DICYANOBENZENE

Formula:	$C_8H_4N_2$
Mol Weight:	128.13
MP (deg C):	162
FP (deg C):	
BP (deg C):	
BP pressure (mm Hg):	

Property/Value	Units	Temp	Data Type	Reference
WS 1.05E+004	mg/L	25	EST	MEYLAN,WM ET AL. (1996)
logP 0.80			EXP	HANSCH,C ET AL. (1995)
VP 5.69E-003	mm Hg	25	EST	NEELY,WB & BLAU,GE (1985)
DC	pKa			
HL 5.03E-007	atm m3/mol	25	EST	MEYLAN,WM & HOWARD,PH (1991)
OH 6.08E-014	cm3/molc sec	25	EST	MEYLAN,WM & HOWARD,PH (1993)

000626-38-0 — 2-PENTANOL ACETATE

Formula:	$C_7H_{14}O_2$
Mol Weight:	130.19
MP (deg C):	196.5
FP (deg C):	
BP (deg C):	
BP pressure (mm Hg):	

Property/Value	Units	Temp	Data Type	Reference
WS 1.07E+003	mg/L	25	EST	MEYLAN,WM ET AL. (1996)
logP 2.26			EST	MEYLAN,WM & HOWARD,PH (1995)
VP 8.03E+000	mm Hg	25	EST	NEELY,WB & BLAU,GE (1985)
DC	pKa			
HL 5.45E-004	atm m3/mol	25	EST	MEYLAN,WM & HOWARD,PH (1991)
OH 6.81E-012	cm3/molc sec	25	EST	MEYLAN,WM & HOWARD,PH (1993)

000626-39-1 — 1,3,5-TRIBROMOBENZENE

Formula:	$C_6H_3Br_3$
Mol Weight:	314.82
MP (deg C):	122.8
FP (deg C):	
BP (deg C):	271
BP pressure (mm Hg):	

Property/Value	Units	Temp	Data Type	Reference
WS 7.91E-001	mg/L	25	EXP	SUZUKI,T (1991)
logP 4.51			EXP	HANSCH,C & LEO,AJ (1985)
VP 1.50E-003	mm Hg	25	EXP	HUININK,J ET AL. (1988)
DC	pKa			
HL 7.86E-004	atm m3/mol	25	EST	VP/WSOL
OH 4.86E-013	cm3/molc sec	25	EST	MEYLAN,WM & HOWARD,PH (1993)

000626-41-5 — 3,5-DIBROMOPHENOL

Formula:	$C_6H_4Br_2O$
Mol Weight:	251.92
MP (deg C):	81
FP (deg C):	
BP (deg C):	274
BP pressure (mm Hg):	

Property/Value	Units	Temp	Data Type	Reference
WS 1.35E+002	mg/L	25	EST	MEYLAN,WM ET AL. (1996)
logP 3.29			EST	MEYLAN,WM & HOWARD,PH (1995)
VP 1.34E-003	mm Hg	25	EST	NEELY,WB & BLAU,GE (1985)
DC 8.06	pKa	25	EXP	SERJEANT,EP & DEMPSEY,B (1979)
HL 8.90E-008	atm m3/mol	25	EST	MEYLAN,WM & HOWARD,PH (1991)
OH 1.33E-011	cm3/molc sec	25	EST	MEYLAN,WM & HOWARD,PH (1993)

000626-43-7 — 3,5-DICHLOROANILINE

Formula:	$C_6H_5Cl_2N$
Mol Weight:	162.02
MP (deg C):	52
FP (deg C):	
BP (deg C):	261
BP pressure (mm Hg):	

Property/Value	Units	Temp	Data Type	Reference
WS 2.24E+002	mg/L	25	EST	MEYLAN,WM ET AL. (1996)
logP 2.90			EXP	HANSCH,C & LEO,AJ (1985)
VP 2.12E-002	mm Hg	25	EST	NEELY,WB & BLAU,GE (1985)
DC 2.51	pKa	15	EXP	PERRIN,DD (1972)
HL 1.05E-006	atm m3/mol	25	EST	MEYLAN,WM & HOWARD,PH (1991)
OH 5.33E-011	cm3/molc sec	25	EST	MEYLAN,WM & HOWARD,PH (1993)

000626-55-1 — 3-BROMOPYRIDINE

Formula:	C_5H_4BrN
Mol Weight:	158.00
MP (deg C):	-27.3
FP (deg C):	
BP (deg C):	173
BP pressure (mm Hg):	

Property/Value	Units	Temp	Data Type	Reference
WS 3.01E+003	mg/L	25	EST	MEYLAN,WM ET AL. (1996)
logP 1.60			EXP	HANSCH,C & LEO,AJ (1985)
VP 1.75E+000	mm Hg	25	EXP	PERRY,RH & GREEN,D (1984)
DC 2.91	pKa	20	EXP	PERRIN,DD (1965)
HL 2.81E-006	atm m3/mol	25	EST	MEYLAN,WM & HOWARD,PH (1991)
OH 2.33E-013	cm3/molc sec	25	EST	MEYLAN,WM & HOWARD,PH (1993)

CAS #: 000626-60-8 — 3-CHLOROPYRIDINE

Formula: C_5H_4ClN

Mol Weight: 113.55

MP (deg C): FP (deg C):

BP (deg C): 148

BP pressure (mm Hg):

Property/Value	Units	Temp	Data Type	Reference
WS 7.74E+003	mg/L	25	EST	MEYLAN,WM ET AL. (1996)
logP 1.33			EXP	HANSCH,C & LEO,AJ (1985)
VP 4.00E+000	mm Hg	25	EST	NEELY,WB & BLAU,GE (1985)
DC 2.84	pKa	25	EXP	PERRIN,DD (1965)
HL 5.22E-006	atm m3/mol	25	EST	MEYLAN,WM & HOWARD,PH (1991)
OH 2.60E-013	cm3/molc sec	25	EST	MEYLAN,WM & HOWARD,PH (1993)

CAS #: 000626-61-9 — 4-CHLOROPYRIDINE

Formula: C_5H_4ClN

Mol Weight: 113.55

MP (deg C): -43.5 FP (deg C):

BP (deg C): 147.5

BP pressure (mm Hg):

Property/Value	Units	Temp	Data Type	Reference
WS 8.54E+003	mg/L	25	EST	MEYLAN,WM ET AL. (1996)
logP 1.28			EXP	SANGSTER,J (1994)
VP 4.00E+000	mm Hg	25	EST	NEELY,WB & BLAU,GE (1985)
DC 3.84	pKa	20	EXP	PERRIN,DD (1965)
HL 5.22E-006	atm m3/mol	25	EST	MEYLAN,WM & HOWARD,PH (1991)
OH 2.60E-013	cm3/molc sec	25	EST	MEYLAN,WM & HOWARD,PH (1993)

CAS #: 000626-64-2 — 4-HYDROXYPYRIDINE

Formula: C_5H_5NO

Mol Weight: 95.10

MP (deg C): 149.8 FP (deg C):

BP (deg C): 257

BP pressure (mm Hg): 1.00E+001

Property/Value	Units	Temp	Data Type	Reference
WS 1.00E+006	mg/L	20	EXP	YALKOWSKY,SH & DANNENFELSER,RM (1992)
logP 0.32			EST	MEYLAN,WM & HOWARD,PH (1995)
VP 3.96E-001	mm Hg	25	EST	NEELY,WB & BLAU,GE (1985)
DC	pKa			
HL 4.52E-009	atm m3/mol	25	EST	MEYLAN,WM & HOWARD,PH (1991)
OH 7.65E-011	cm3/molc sec	25	EST	MEYLAN,WM & HOWARD,PH (1993)

CAS #: 000626-67-5 — N-METHYLPIPERIDINE

Formula: $C_6H_{13}N$

Mol Weight: 99.18

MP (deg C): FP (deg C):

BP (deg C): 107

BP pressure (mm Hg):

Property/Value	Units	Temp	Data Type	Reference
WS 9.29E+004	mg/L	25	EST	MEYLAN,WM ET AL. (1996)
logP 1.30			EXP	HANSCH,C & LEO,AJ (1985)
VP 1.94E+001	mm Hg	25	EXP	WALTON,J (1977)
DC 10.08	pKa	25	EXP	PERRIN,DD (1965)
HL 3.46E-005	atm m3/mol	25	EXP	CABANI,S ET AL. (1971)
OH 8.72E-011	cm3/molc sec	25	EST	MEYLAN,WM & HOWARD,PH (1993)

CAS #: 000626-70-0 — 2-METHYLHEXANEDIOIC ACID

Formula: $C_7H_{12}O_4$

Mol Weight: 160.17

MP (deg C): 64 FP (deg C):

BP (deg C): 209

BP pressure (mm Hg): 1.30E+001

Property/Value	Units	Temp	Data Type	Reference
WS 4.77E+004	mg/L	25	EST	MEYLAN,WM ET AL. (1996)
logP 0.65			EST	MEYLAN,WM & HOWARD,PH (1995)
VP 1.44E-004	mm Hg	25	EST	NEELY,WB & BLAU,GE (1985)
DC	pKa			
HL 1.27E-011	atm m3/mol	25	EST	MEYLAN,WM & HOWARD,PH (1991)
OH 6.68E-012	cm3/molc sec	25	EST	MEYLAN,WM & HOWARD,PH (1993)

CAS #: 000626-77-7 — PROPYL HEXANOATE

Formula: $C_9H_{18}O_2$

Mol Weight: 158.24

MP (deg C): -68.7 FP (deg C):

BP (deg C): 187

BP pressure (mm Hg):

Property/Value	Units	Temp	Data Type	Reference
WS 1.02E+002	mg/L	25	EST	MEYLAN,WM ET AL. (1996)
logP 3.32			EST	MEYLAN,WM & HOWARD,PH (1995)
VP 2.00E+000	mm Hg	43	EXP	BONHORST,CW ET AL. (1948)
DC	pKa			
HL 9.60E-004	atm m3/mol	25	EST	MEYLAN,WM & HOWARD,PH (1991)
OH 7.65E-012	cm3/molc sec	25	EST	MEYLAN,WM & HOWARD,PH (1993)

CAS #: 000626-82-4 — N-BUTYL HEXANOATE

Formula: $C_{10}H_{20}O_2$

Mol Weight: 172.27

MP (deg C): -64.3 FP (deg C):

BP (deg C): 208

BP pressure (mm Hg):

Property/Value	Units	Temp	Data Type	Reference
WS 3.34E+001	mg/L	25	EST	MEYLAN,WM ET AL. (1996)
logP 3.81			EST	MEYLAN,WM & HOWARD,PH (1995)
VP 2.11E-001	mm Hg	25	EST	NEELY,WB & BLAU,GE (1985)
DC	pKa			
HL 1.27E-003	atm m3/mol	25	EST	MEYLAN,WM & HOWARD,PH (1991)
OH 9.56E-012	cm3/molc sec	25	EST	MEYLAN,WM & HOWARD,PH (1993)

CAS #: 000626-86-8 — HEXANEDIOIC ACID, MONOETHYL ESTER

Formula: $C_8H_{14}O_4$

Mol Weight: 174.20

MP (deg C): 20 FP (deg C):

BP (deg C):

BP pressure (mm Hg):

Property/Value	Units	Temp	Data Type	Reference
WS 7.60E+004	mg/L	25	EXP	CHEM INSPECT TEST INST (1992)
logP 0.91			EXP	CHEM INSPECT TEST INST (1992)
VP 1.79E-003	mm Hg	25	EST	NEELY,WB & BLAU,GE (1985)
DC	pKa			
HL 4.05E-009	atm m3/mol	25	EST	MEYLAN,WM & HOWARD,PH (1991)
OH 6.72E-012	cm3/molc sec	25	EST	MEYLAN,WM & HOWARD,PH (1993)

CAS #: 000626-89-1				4-METHYL-1-PENTANOL

Formula: $C_6H_{14}O$

Mol Weight: 102.18

MP (deg C):   FP (deg C):

BP (deg C): 151.9

BP pressure (mm Hg):

Property/Value	Units	Temp	Data Type	Reference
WS 7.60E+003	mg/L	25	EXP	BARTON,AFM (1984)
logP 1.75			EST	MEYLAN,WM & HOWARD,PH (1995)
VP 1.70E+000	mm Hg	25	EST	NEELY,WB & BLAU,GE (1985)
DC	pKa			
HL 1.76E-005	atm m3/mol	25	EST	MEYLAN,WM & HOWARD,PH (1991)
OH 9.71E-012	cm3/molc sec	25	EST	MEYLAN,WM & HOWARD,PH (1993)

CAS #: 000626-93-7				2-HEXANOL

Formula: $C_6H_{14}O$

Mol Weight: 102.18

MP (deg C):   FP (deg C):

BP (deg C): 136

BP pressure (mm Hg):

Property/Value	Units	Temp	Data Type	Reference
WS 1.37E+004	mg/L	25	EXP	YALKOWSKY,SH & DANNENFELSER,RM (1992)
logP 1.76			EXP	FUNASAKI,N ET AL. (1986)
VP 2.49E+000	mm Hg	25	EXP	YAWS,CL (1994A)
DC	pKa			
HL 2.44E-005	atm m3/mol	25	EST	VP/WSOL
OH 1.21E-011	cm3/molc sec	25	EXP	ATKINSON,R (1989)

CAS #: 000627-05-4				1-NITROBUTANE

Formula: $C_4H_9NO_2$

Mol Weight: 103.12

MP (deg C):   FP (deg C):

BP (deg C): 153

BP pressure (mm Hg):

Property/Value	Units	Temp	Data Type	Reference
WS 4.60E+003	mg/L	25	EXP	SUZUKI,T (1991)
logP 1.47			EXP	HANSCH,C & LEO,AJ (1985)
VP 6.91E+000	mm Hg	25	EST	NEELY,WB & BLAU,GE (1985)
DC	pKa			
HL 9.85E-005	atm m3/mol	25	EST	MEYLAN,WM & HOWARD,PH (1991)
OH 1.45E-012	cm3/molc sec	25	EXP	ATKINSON,R (1989)

CAS #: 000627-08-7				PROPYL ISOPROPYL ETHER

Formula: $C_6H_{14}O$

Mol Weight: 102.18

MP (deg C):   FP (deg C):

BP (deg C): 83

BP pressure (mm Hg):

Property/Value	Units	Temp	Data Type	Reference
WS 4.70E+003	mg/L	25	EXP	YALKOWSKY,SH & DANNENFELSER,RM (1992)
logP 1.96			EST	MEYLAN,WM & HOWARD,PH (1995)
VP 8.79E+001	mm Hg	25	EST	NEELY,WB & BLAU,GE (1985)
DC	pKa			
HL 2.68E-003	atm m3/mol	25	EST	MEYLAN,WM & HOWARD,PH (1991)
OH 2.26E-011	cm3/molc sec	25	EST	MEYLAN,WM & HOWARD,PH (1993)

CAS #: 000627-11-2				CHLOROETHYL CHLOROFORMATE

Formula: $C_3H_4Cl_2O_2$

Mol Weight: 142.97

MP (deg C):   FP (deg C):

BP (deg C): 155

BP pressure (mm Hg):

Property/Value	Units	Temp	Data Type	Reference
WS 1.44E+004	mg/L	25	EST	MEYLAN,WM ET AL. (1996)
logP 0.88			EST	MEYLAN,WM & HOWARD,PH (1995)
VP 1.30E+001	mm Hg	49	EXP	BEILSTEIN (NA--)
DC	pKa			
HL 1.10E-003	atm m3/mol	25	EST	MEYLAN,WM & HOWARD,PH (1991)
OH 1.13E-012	cm3/molc sec	25	EST	MEYLAN,WM & HOWARD,PH (1993)

CAS #: 000627-12-3				O-PROPYLCARBAMATE

Formula: $C_4H_9NO_2$

Mol Weight: 103.12

MP (deg C): 60   FP (deg C):

BP (deg C):

BP pressure (mm Hg):

Property/Value	Units	Temp	Data Type	Reference
WS 7.64E+004	mg/L	25	EXP	SUZUKI,T (1991)
logP 0.36			EXP	HANSCH,C & LEO,AJ (1985)
VP 5.53E-002	mm Hg	25	EXT	JORDAN,TE (1954)
DC	pKa			
HL 6.97E-008	atm m3/mol	25	EST	MEYLAN,WM & HOWARD,PH (1991)
OH 5.15E-012	cm3/molc sec	25	EST	MEYLAN,WM & HOWARD,PH (1993)

CAS #: 000627-13-4				1-PROPYLNITRATE

Formula: $C_3H_7NO_3$

Mol Weight: 105.09

MP (deg C):   FP (deg C):

BP (deg C): 110

BP pressure (mm Hg): 7.62E+002

Property/Value	Units	Temp	Data Type	Reference
WS 3.70E+003	mg/L	25	EST	MEYLAN,WM ET AL. (1996)
logP 1.74			EST	MEYLAN,WM & HOWARD,PH (1995)
VP 2.35E+001	mm Hg	25	EXP	BOUBLIK,T ET AL. (1984)
DC	pKa			
HL 3.50E-004	atm m3/mol	25	EST	MEYLAN,WM & HOWARD,PH (1991)
OH 6.70E-013	cm3/molc sec	25	EXP	ATKINSON,R (1989)

CAS #: 000627-18-9				3-BROMOPROPANOL

Formula: C_3H_7BrO

Mol Weight: 139.00

MP (deg C):   FP (deg C):

BP (deg C): 105

BP pressure (mm Hg): 1.85E+002

Property/Value	Units	Temp	Data Type	Reference
WS 1.42E+005	mg/L	20	EXP	YALKOWSKY,SH & DANNENFELSER,RM (1992)
logP 0.69			EST	MEYLAN,WM & HOWARD,PH (1995)
VP 1.00E-001	mm Hg	20	EXP	NEELY,WB (1976)
DC	pKa			
HL 1.29E-007	atm m3/mol	20	EST	VP/WSOL
OH 5.01E-012	cm3/molc sec	25	EST	MEYLAN,WM & HOWARD,PH (1993)

CAS #: 000627-19-0				1-PENTYNE

Formula: C_5H_8

Mol Weight: 68.12

MP (deg C): -90   FP (deg C):

BP (deg C): 40.1

BP pressure (mm Hg):

Property/ Value	Units	Temp	Data Type	Reference
WS 1.57E+003	mg/L	25	EXP	YALKOWSKY,SH & DANNENFELSER,RM (1992)
logP 1.98			EXP	HANSCH,C & LEO,AJ (1985)
VP 4.36E+002	mm Hg	25	EXP	YAWS,CL (1994A)
DC	pKa			
HL 2.49E-002	atm m3/mol	25	EST	VP/WSOL
OH 1.12E-011	cm3/molc sec	25	EXP	ATKINSON,R (1989)

CAS #: 000627-20-3				CIS-2-PENTENE

Formula: C_5H_{10}

Mol Weight: 70.14

MP (deg C): -151.4   FP (deg C):

BP (deg C): 36.9

BP pressure (mm Hg):

Property/ Value	Units	Temp	Data Type	Reference
WS 2.03E+002	mg/L	25	EXP	MACKAY,D & SHIU,W (1981)
logP 2.58			EST	MEYLAN,WM & HOWARD,PH (1995)
VP 4.95E+002	mm Hg	25	EXP	DAUBERT,TE & DANNER,RP (1989)
DC	pKa			
HL 2.25E-001	atm m3/mol	25	EST	VP/WSOL
OH 6.54E-011	cm3/molc sec	25	EXP	ATKINSON,R (1989)

CAS #: 000627-21-4				2-PENTYNE

Formula: C_5H_8

Mol Weight: 68.12

MP (deg C): -109.3   FP (deg C):

BP (deg C): 56.1

BP pressure (mm Hg):

Property/ Value	Units	Temp	Data Type	Reference
WS 6.59E+002	mg/L	25	EST	MEYLAN,WM ET AL. (1996)
logP 2.08			EST	MEYLAN,WM & HOWARD,PH (1995)
VP 2.36E+002	mm Hg	25	EXP	ZWOLINSKI,BJ & WILHOIT,RC (1971)
DC	pKa			
HL 1.43E-002	atm m3/mol	25	EST	MEYLAN,WM & HOWARD,PH (1991)
OH 2.82E-011	cm3/molc sec	25	EST	MEYLAN,WM & HOWARD,PH (1993)

CAS #: 000627-22-5				1-CHLOROPRENE

Formula: C_4H_5Cl

Mol Weight: 88.54

MP (deg C):   FP (deg C):

BP (deg C): 68

BP pressure (mm Hg):

Property/ Value	Units	Temp	Data Type	Reference
WS 1.14E+003	mg/L	25	EST	MEYLAN,WM ET AL. (1996)
logP 2.39			EST	MEYLAN,WM & HOWARD,PH (1995)
VP 1.75E+002	mm Hg	25	EST	NEELY,WB & BLAU,GE (1985)
DC	pKa			
HL 5.61E-002	atm m3/mol	25	EST	MEYLAN,WM & HOWARD,PH (1991)
OH 2.21E-011	cm3/molc sec	25	EST	MEYLAN,WM & HOWARD,PH (1993)

CAS #: 000627-23-6				1-CHLORO-1,2-BUTADIENE

Formula: C_4H_5Cl

Mol Weight: 88.54

MP (deg C):   FP (deg C):

BP (deg C):

BP pressure (mm Hg):

Property/ Value	Units	Temp	Data Type	Reference
WS 1.08E+003	mg/L	25	EST	MEYLAN,WM ET AL. (1996)
logP 2.42			EST	MEYLAN,WM & HOWARD,PH (1995)
VP 1.50E+002	mm Hg	25	EST	NEELY,WB & BLAU,GE (1985)
DC	pKa			
HL 6.96E-002	atm m3/mol	25	EST	MEYLAN,WM & HOWARD,PH (1991)
OH 1.21E-011	cm3/molc sec	25	EST	MEYLAN,WM & HOWARD,PH (1993)

CAS #: 000627-30-5				3-CHLORO-1-PROPANOL

Formula: C_3H_7ClO

Mol Weight: 94.54

MP (deg C):   FP (deg C):

BP (deg C): 165

BP pressure (mm Hg):

Property/ Value	Units	Temp	Data Type	Reference
WS 2.50E+005	mg/L	20	EXP	YALKOWSKY,SH & DANNENFELSER,RM (1992)
logP 0.50			EXP	HANSCH,C ET AL. (1995)
VP 1.20E+000	mm Hg	25	EST	NEELY,WB & BLAU,GE (1985)
DC	pKa			
HL 1.67E-006	atm m3/mol	25	EST	MEYLAN,WM & HOWARD,PH (1991)
OH 5.01E-012	cm3/molc sec	25	EST	MEYLAN,WM & HOWARD,PH (1993)

CAS #: 000627-31-6				PROPANE, 1,3-DIIODO-

Formula: $C_3H_6I_2$

Mol Weight: 295.89

MP (deg C): -20   FP (deg C):

BP (deg C): 110

BP pressure (mm Hg): 1.90E+001

Property/ Value	Units	Temp	Data Type	Reference
WS 3.42E+001	mg/L	25	EST	MEYLAN,WM ET AL. (1996)
logP 3.02			EXP	HANSCH,C ET AL. (1994)
VP 2.69E-001	mm Hg	25	EST	NEELY,WB & BLAU,GE (1985)
DC	pKa			
HL 7.23E-004	atm m3/mol	25	EST	MEYLAN,WM & HOWARD,PH (1991)
OH 1.42E-012	cm3/molc sec	25	EST	MEYLAN,WM & HOWARD,PH (1993)

CAS #: 000627-35-0				N-METHYLPROPYLAMINE

Formula: $C_4H_{11}N$

Mol Weight: 73.14

MP (deg C):   FP (deg C):

BP (deg C): 61-63

BP pressure (mm Hg):

Property/ Value	Units	Temp	Data Type	Reference
WS 2.62E+005	mg/L	25	EST	MEYLAN,WM ET AL. (1996)
logP 0.84			EXP	HANSCH,C ET AL. (1995)
VP 1.41E+002	mm Hg	25	EST	NEELY,WB & BLAU,GE (1985)
DC	pKa			
HL 2.93E-005	atm m3/mol	25	EST	MEYLAN,WM & HOWARD,PH (1991)
OH 7.63E-011	cm3/molc sec	25	EST	MEYLAN,WM & HOWARD,PH (1993)

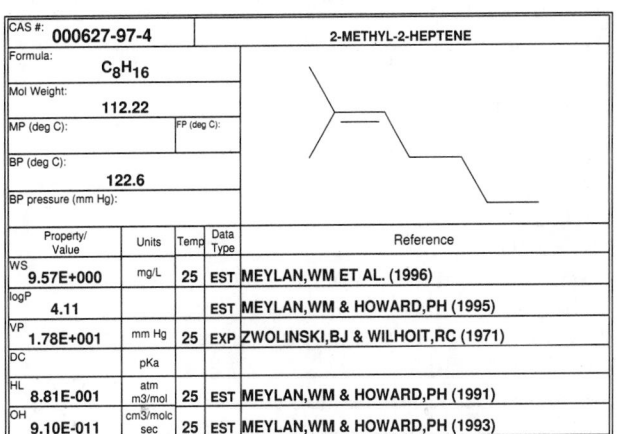

CAS #:	000627-45-2		N-FORMYLETHYLAMINE	
Formula:	C_3H_7NO			
Mol Weight:	73.10			
MP (deg C):		FP (deg C):		
BP (deg C):	198			
BP pressure (mm Hg):				

Property/ Value	Units	Temp	Data Type	Reference
WS 4.85E+005	mg/L	25	EST	MEYLAN,WM ET AL. (1996)
logP -0.65			EST	MEYLAN,WM & HOWARD,PH (1995)
VP 4.99E-001	mm Hg	25	EST	NEELY,WB & BLAU,GE (1985)
DC	pKa			
HL 4.46E-008	atm m3/mol	25	EST	MEYLAN,WM & HOWARD,PH (1991)
OH 1.44E-011	cm3/molc sec	25	EST	MEYLAN,WM & HOWARD,PH (1993)

CAS #:	000627-51-0		1,1'-THIOBISETHENE	
Formula:	C_4H_6S			
Mol Weight:	86.16			
MP (deg C):	20	FP (deg C):		
BP (deg C):	84			
BP pressure (mm Hg):				

Property/ Value	Units	Temp	Data Type	Reference
WS 2.84E+003	mg/L	25	EST	MEYLAN,WM ET AL. (1996)
logP 1.93			EST	MEYLAN,WM & HOWARD,PH (1995)
VP 5.91E+001	mm Hg	25	EST	NEELY,WB & BLAU,GE (1985)
DC	pKa			
HL 6.19E-003	atm m3/mol	25	EST	MEYLAN,WM & HOWARD,PH (1991)
OH 5.43E-011	cm3/molc sec	25	EST	MEYLAN,WM & HOWARD,PH (1993)

CAS #:	000627-58-7		2,5-DIMETHYL-1,5-HEXADIENE	
Formula:	C_8H_{14}			
Mol Weight:	110.20			
MP (deg C):	-75.6	FP (deg C):		
BP (deg C):	114.3			
BP pressure (mm Hg):				

Property/ Value	Units	Temp	Data Type	Reference
WS 9.75E+000	mg/L	25	EST	MEYLAN,WM ET AL. (1996)
logP 4.11			EST	MEYLAN,WM & HOWARD,PH (1995)
VP 3.74E+001	mm Hg	25	EST	NEELY,WB & BLAU,GE (1985)
DC	pKa			
HL 4.14E-001	atm m3/mol	25	EST	MEYLAN,WM & HOWARD,PH (1991)
OH 1.20E-010	cm3/molc sec	25	EXP	ATKINSON,R (1989)

CAS #:	000627-59-8		5-METHYL-2-HEXANOL	
Formula:	$C_7H_{16}O$			
Mol Weight:	116.20			
MP (deg C):		FP (deg C):		
BP (deg C):	151			
BP pressure (mm Hg):				

Property/ Value	Units	Temp	Data Type	Reference
WS 4.90E+003	mg/L	25	EXP	BARTON,AFM (1984)
logP 2.17			EST	MEYLAN,WM & HOWARD,PH (1995)
VP 2.03E+000	mm Hg	25	EXT	WILHOIT,BJ & ZWOLINSKI,BJ (1973)
DC	pKa			
HL 2.34E-005	atm m3/mol	25	EST	MEYLAN,WM & HOWARD,PH (1991)
OH 1.42E-011	cm3/molc sec	25	EST	MEYLAN,WM & HOWARD,PH (1993)

CAS #:	000627-63-4		FUMARYL CHLORIDE	
Formula:	$C_4H_2Cl_2O_2$			
Mol Weight:	152.97			
MP (deg C):		FP (deg C):		
BP (deg C):	159			
BP pressure (mm Hg):				

Property/ Value	Units	Temp	Data Type	Reference
WS 6.05E+004	mg/L	25	EST	MEYLAN,WM ET AL. (1996)
logP 0.10			EST	MEYLAN,WM & HOWARD,PH (1995)
VP 2.07E+000	mm Hg	25	EXP	PERRY,RH & GREEN,D (1984)
DC	pKa			
HL 1.99E-006	atm m3/mol	25	EST	MEYLAN,WM & HOWARD,PH (1991)
OH 6.91E-012	cm3/molc sec	25	EST	MEYLAN,WM & HOWARD,PH (1993)

CAS #:	000627-90-7		ETHYL UNDECANOATE	
Formula:	$C_{13}H_{26}O_2$			
Mol Weight:	214.35			
MP (deg C):	-15	FP (deg C):		
BP (deg C):	131			
BP pressure (mm Hg):	1.40E+001			

Property/ Value	Units	Temp	Data Type	Reference
WS 1.13E+000	mg/L	25	EST	MEYLAN,WM ET AL. (1996)
logP 5.28			EST	MEYLAN,WM & HOWARD,PH (1995)
VP 4.30E-003	mm Hg	20	EXP	VOITKEVICH,SA (1963)
DC	pKa			
HL 2.98E-003	atm m3/mol	25	EST	MEYLAN,WM & HOWARD,PH (1991)
OH 1.37E-011	cm3/molc sec	25	EST	MEYLAN,WM & HOWARD,PH (1993)

CAS #:	000627-93-0		HEXANEDIOIC ACID, DIMETHYL ESTER	
Formula:	$C_8H_{14}O_4$			
Mol Weight:	174.20			
MP (deg C):	10.3	FP (deg C):		
BP (deg C):	115			
BP pressure (mm Hg):	1.30E+001			

Property/ Value	Units	Temp	Data Type	Reference
WS 6.00E+003	mg/L		EXP	BENNETT,SR ET AL. (1984)
logP 1.03			EXP	HANSCH,C ET AL. (1995)
VP 6.04E-002	mm Hg	25	EXT	VLASOV,ON & GRANZHAN,VA (1965)
DC	pKa			
HL 9.77E-007	atm m3/mol	25	EST	MEYLAN,WM & HOWARD,PH (1991)
OH 4.96E-012	cm3/molc sec	25	EST	MEYLAN,WM & HOWARD,PH (1993)

CAS #:	000627-97-4		2-METHYL-2-HEPTENE	
Formula:	C_8H_{16}			
Mol Weight:	112.22			
MP (deg C):		FP (deg C):		
BP (deg C):	122.6			
BP pressure (mm Hg):				

Property/ Value	Units	Temp	Data Type	Reference
WS 9.57E+000	mg/L	25	EST	MEYLAN,WM ET AL. (1996)
logP 4.11			EST	MEYLAN,WM & HOWARD,PH (1995)
VP 1.78E+001	mm Hg	25	EXP	ZWOLINSKI,BJ & WILHOIT,RC (1971)
DC	pKa			
HL 8.81E-001	atm m3/mol	25	EST	MEYLAN,WM & HOWARD,PH (1991)
OH 9.10E-011	cm3/molc sec	25	EST	MEYLAN,WM & HOWARD,PH (1993)

CAS #: 000628-05-7				1-NITROPENTANE

Formula: $C_5H_{11}NO_2$

Mol Weight: 117.15

MP (deg C): FP (deg C):

BP (deg C): 172.5

BP pressure (mm Hg):

Property/Value	Units	Temp	Data Type	Reference
WS 1.30E+003	mg/L	25	EXP	YALKOWSKY,SH & DANNENFELSER,RM (1992)
logP 2.01			EXP	HANSCH,C & LEO,AJ (1985)
VP 2.48E+000	mm Hg	25	EST	NEELY,WB & BLAU,GE (1985)
DC	pKa			
HL 1.31E-004	atm m3/mol	25	EST	MEYLAN,WM & HOWARD,PH (1991)
OH 3.30E-012	cm3/molc sec	25	EXP	ATKINSON,R (1989)

CAS #: 000628-17-1				1-IODOPENTANE

Formula: $C_5H_{11}I$

Mol Weight: 198.05

MP (deg C): -85.6 FP (deg C):

BP (deg C): 155

BP pressure (mm Hg):

Property/Value	Units	Temp	Data Type	Reference
WS 4.12E+001	mg/L	25	EST	MEYLAN,WM ET AL. (1996)
logP 3.56			EST	MEYLAN,WM & HOWARD,PH (1995)
VP 4.39E+000	mm Hg	25	EXP	LI,JCM & ROSSINI,FD (1953)
DC	pKa			
HL 1.71E-002	atm m3/mol	25	EST	MEYLAN,WM & HOWARD,PH (1991)
OH 3.87E-012	cm3/molc sec	25	EST	MEYLAN,WM & HOWARD,PH (1993)

CAS #: 000628-20-6				BUTANENITRILE, 4-CHLORO-

Formula: C_4H_6ClN

Mol Weight: 103.55

MP (deg C): FP (deg C):

BP (deg C): 192

BP pressure (mm Hg):

Property/Value	Units	Temp	Data Type	Reference
WS 2.06E+004	mg/L	25	EST	MEYLAN,WM ET AL. (1996)
logP 0.56			EXP	TANII,H & HASHIMOTO,K (1984)
VP 1.00E+000	mm Hg	25	EST	NEELY,WB & BLAU,GE (1985)
DC	pKa			
HL 1.90E-005	atm m3/mol	25	EST	MEYLAN,WM & HOWARD,PH (1991)
OH 6.95E-013	cm3/molc sec	25	EST	MEYLAN,WM & HOWARD,PH (1993)

CAS #: 000628-28-4				METHYL N-BUTYL ETHER

Formula: $C_5H_{12}O$

Mol Weight: 88.15

MP (deg C): -115.5 FP (deg C):

BP (deg C): 70.1

BP pressure (mm Hg):

Property/Value	Units	Temp	Data Type	Reference
WS 9.01E+003	mg/L	25	EXP	SUZUKI,T (1991)
logP 1.66			EXP	HANSCH,C ET AL. (1995)
VP 1.39E+002	mm Hg	25	EXP	BOUBLIK,T ET AL. (1984)
DC	pKa			
HL 1.79E-003	atm m3/mol	25	EST	VP/WSOL
OH 1.64E-011	cm3/molc sec	25	EXP	ATKINSON,R (1989)

CAS #: 000628-29-5				2-THIAHEXANE

Formula: $C_5H_{12}S$

Mol Weight: 104.22

MP (deg C): -97.8 FP (deg C):

BP (deg C): 123.5

BP pressure (mm Hg):

Property/Value	Units	Temp	Data Type	Reference
WS 1.03E+003	mg/L	25	EST	MEYLAN,WM ET AL. (1996)
logP 2.39			EST	MEYLAN,WM & HOWARD,PH (1995)
VP 1.56E+001	mm Hg	25	EXT	BOUBLIK,T ET AL. (1984)
DC	pKa			
HL 1.84E-003	atm m3/mol	25	EST	MEYLAN,WM & HOWARD,PH (1991)
OH 1.45E-011	cm3/molc sec	25	EST	MEYLAN,WM & HOWARD,PH (1993)

CAS #: 000628-32-0				ETHYL PROPYL ETHER

Formula: $C_5H_{12}O$

Mol Weight: 88.15

MP (deg C): -127.5 FP (deg C):

BP (deg C): 63.2

BP pressure (mm Hg):

Property/Value	Units	Temp	Data Type	Reference
WS 1.84E+004	mg/L	25	EXP	SUZUKI,T (1991)
logP 1.54			EST	MEYLAN,WM & HOWARD,PH (1995)
VP 1.81E+002	mm Hg	25	EXP	YAWS,CL (1994A)
DC	pKa			
HL 1.14E-003	atm m3/mol	25	EST	VP/WSOL
OH 1.66E-011	cm3/molc sec	25	EST	MEYLAN,WM & HOWARD,PH (1993)

CAS #: 000628-34-2				2-CHLOROETHYLETHER

Formula: C_4H_9ClO

Mol Weight: 108.57

MP (deg C): FP (deg C):

BP (deg C): 107.5

BP pressure (mm Hg):

Property/Value	Units	Temp	Data Type	Reference
WS 1.60E+004	mg/L	25	EST	MEYLAN,WM ET AL. (1996)
logP 0.98			EXP	HANSCH,C & LEO,AJ (1985)
VP 2.11E+001	mm Hg	25	EST	NEELY,WB & BLAU,GE (1985)
DC	pKa			
HL 5.35E-004	atm m3/mol	25	EST	MEYLAN,WM & HOWARD,PH (1991)
OH 4.70E-012	cm3/molc sec	25	EST	MEYLAN,WM & HOWARD,PH (1993)

CAS #: 000628-36-4				DIFORMYLHYDRAZINE

Formula: $C_2H_4N_2O_2$

Mol Weight: 88.07

MP (deg C): 155-157 FP (deg C):

BP (deg C):

BP pressure (mm Hg):

Property/Value	Units	Temp	Data Type	Reference
WS 1.00E+006	mg/L	25	EST	MEYLAN,WM ET AL. (1996)
logP -2.48			EST	MEYLAN,WM & HOWARD,PH (1995)
VP 3.79E-008	mm Hg	25	EXT	JONES,AH (1960)
DC	pKa			
HL 4.50E-010	atm m3/mol	25	EST	MEYLAN,WM & HOWARD,PH (1991)
OH 1.10E-011	cm3/molc sec	25	EST	MEYLAN,WM & HOWARD,PH (1993)

CAS #: 000628-41-1				1,4-CYCLOHEXADIENE

Formula: C_6H_8
Mol Weight: 80.13
MP (deg C): -49.2   FP (deg C):
BP (deg C): 85.5
BP pressure (mm Hg):

Property/Value	Units	Temp	Data Type	Reference
WS 7.00E+002	mg/L	25	EXP	YALKOWSKY,SH & DANNENFELSER,RM (1992)
logP 2.30			EXP	HANSCH,C & LEO,AJ (1985)
VP 6.66E+001	mm Hg	25	EXT	CHAO,J ET AL. (1983)
DC	pKa			
HL 1.04E-001	atm m3/mol	25	EST	MEYLAN,WM & HOWARD,PH (1991)
OH 9.94E-011	cm3/molc sec	25	EXP	ATKINSON,R (1989)

CAS #: 000628-55-7				DI-ISOBUTYL ETHER

Formula: $C_8H_{18}O$
Mol Weight: 130.23
MP (deg C):   FP (deg C):
BP (deg C): 122-124
BP pressure (mm Hg):

Property/Value	Units	Temp	Data Type	Reference
WS 3.88E+002	mg/L	25	EST	MEYLAN,WM ET AL. (1996)
logP 2.78			EXP	FUNASAKI,N ET AL. (1985)
VP 1.54E+001	mm Hg	25	EXP	AMBROSE,D ET AL. (1976)
DC	pKa			
HL 4.72E-003	atm m3/mol	25	EST	MEYLAN,WM & HOWARD,PH (1991)
OH 2.60E-011	cm3/molc sec	25	EXP	ATKINSON,R (1989)

CAS #: 000628-61-5				2-CHLOROOCTANE

Formula: $C_8H_{17}Cl$
Mol Weight: 148.68
MP (deg C):   FP (deg C):
BP (deg C): 172
BP pressure (mm Hg):

Property/Value	Units	Temp	Data Type	Reference
WS 1.22E+001	mg/L	25	EST	MEYLAN,WM ET AL. (1996)
logP 4.45			EST	MEYLAN,WM & HOWARD,PH (1995)
VP 1.37E+000	mm Hg	25	EXP	LEVANOVA,SV ET AL. (1967)
DC	pKa			
HL 5.96E-002	atm m3/mol	25	EST	MEYLAN,WM & HOWARD,PH (1991)
OH 6.92E-012	cm3/molc sec	25	EST	MEYLAN,WM & HOWARD,PH (1993)

CAS #: 000628-63-7				N-AMYL ACETATE

Formula: $C_7H_{14}O_2$
Mol Weight: 130.19
MP (deg C): -70.8   FP (deg C):
BP (deg C): 149.25
BP pressure (mm Hg):

Property/Value	Units	Temp	Data Type	Reference
WS 1.70E+003	mg/L	20	EXP	RIDDICK,JA ET AL. (1986)
logP 2.34			EST	MEYLAN,WM & HOWARD,PH (1995)
VP 3.50E+000	mm Hg	25	EXP	DAUBERT,TE & DANNER,RP (1989)
DC	pKa			
HL 3.88E-004	atm m3/mol	25	EXP	TAFT,RW ET AL. (1985)
OH 5.68E-012	cm3/molc sec	25	EST	ATKINSON,R (1988)

CAS #: 000628-71-7				1-HEPTYNE

Formula: C_7H_{12}
Mol Weight: 96.17
MP (deg C): -81   FP (deg C):
BP (deg C): 99.7
BP pressure (mm Hg):

Property/Value	Units	Temp	Data Type	Reference
WS 9.40E+001	mg/L	25	EXP	YALKOWSKY,SH & DANNENFELSER,RM (1992)
logP 3.01			EST	MEYLAN,WM & HOWARD,PH (1995)
VP 5.25E+001	mm Hg	25	EXP	HINE,J & MOOKERJEE,PK (1975)
DC	pKa			
HL 7.07E-002	atm m3/mol	25	EST	VP/WSOL
OH 1.23E-011	cm3/molc sec	25	EST	MEYLAN,WM & HOWARD,PH (1993)

CAS #: 000628-73-9				HEXANENITRILE

Formula: $C_6H_{11}N$
Mol Weight: 97.16
MP (deg C): -80.3   FP (deg C):
BP (deg C): 163.6
BP pressure (mm Hg):

Property/Value	Units	Temp	Data Type	Reference
WS 2.48E+003	mg/L	25	EST	MEYLAN,WM ET AL. (1996)
logP 1.66			EXP	TANII,H & HASHIMOTO,K (1984)
VP 2.85E+000	mm Hg	25	EXP	DAUBERT,TE & DANNER,RP (1989)
DC	pKa			
HL 9.49E-005	atm m3/mol	25	EST	MEYLAN,WM & HOWARD,PH (1991)
OH 3.09E-012	cm3/molc sec	25	EST	MEYLAN,WM & HOWARD,PH (1993)

CAS #: 000628-76-2				1,5-DICHLOROPENTANE

Formula: $C_5H_{10}Cl_2$
Mol Weight: 141.04
MP (deg C): -72.8   FP (deg C):
BP (deg C): 179
BP pressure (mm Hg):

Property/Value	Units	Temp	Data Type	Reference
WS 1.25E+002	mg/L	25	EST	MEYLAN,WM ET AL. (1996)
logP 3.30			EST	MEYLAN,WM & HOWARD,PH (1995)
VP 1.13E+000	mm Hg	25	EXP	DAUBERT,TE & DANNER,RP (1989)
DC	pKa			
HL 5.58E-004	atm m3/mol	25	EXP	LEIGHTON,DTJR & CALO,JM (1981)
OH 3.45E+000	cm3/molc sec	25	EST	MEYLAN,WM & HOWARD,PH (1993)

CAS #: 000628-81-9				BUTYL ETHYL ETHER

Formula: $C_6H_{14}O$
Mol Weight: 102.18
MP (deg C): -124   FP (deg C):
BP (deg C): 92.3
BP pressure (mm Hg):

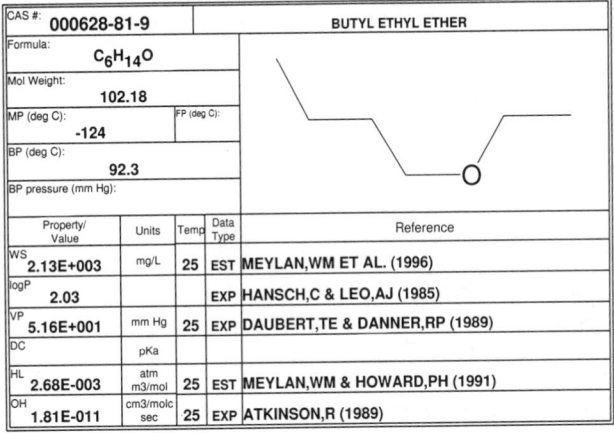

Property/Value	Units	Temp	Data Type	Reference
WS 2.13E+003	mg/L	25	EST	MEYLAN,WM ET AL. (1996)
logP 2.03			EXP	HANSCH,C & LEO,AJ (1985)
VP 5.16E+001	mm Hg	25	EXP	DAUBERT,TE & DANNER,RP (1989)
DC	pKa			
HL 2.68E-003	atm m3/mol	25	EST	MEYLAN,WM & HOWARD,PH (1991)
OH 1.81E-011	cm3/molc sec	25	EXP	ATKINSON,R (1989)

CAS #: 000628-83-1				THIOCYANIC ACID, BUTYL ESTER

Formula: C_5H_9NS

Mol Weight: 115.20

MP (deg C): | FP (deg C):

BP (deg C): 186

BP pressure (mm Hg):

Property/Value	Units	Temp	Data Type	Reference
WS 1.93E+003	mg/L	25	EST	MEYLAN,WM ET AL. (1996)
logP 2.03			EXP	HANSCH,C & LEO,AJ (1985)
VP 3.06E-001	mm Hg	25	EST	NEELY,WB & BLAU,GE (1985)
DC	pKa			
HL 1.03E-004	atm m3/mol	25	EST	MEYLAN,WM & HOWARD,PH (1991)
OH 1.17E-011	cm3/molc sec	25	EST	MEYLAN,WM & HOWARD,PH (1993)

CAS #: 000628-92-2				CYCLOHEPTENE

Formula: C_7H_{12}

Mol Weight: 96.17

MP (deg C): -56 | FP (deg C):

BP (deg C): 115

BP pressure (mm Hg):

Property/Value	Units	Temp	Data Type	Reference
WS 6.60E+001	mg/L	25	EXP	YALKOWSKY,SH & DANNENFELSER,RM (1992)
logP 3.45			EST	MEYLAN,WM & HOWARD,PH (1995)
VP 1.97E+001	mm Hg	25	EST	NEELY,WB & BLAU,GE (1985)
DC	pKa			
HL 8.79E-002	atm m3/mol	25	EST	MEYLAN,WM & HOWARD,PH (1991)
OH 7.44E-011	cm3/molc sec	25	EXP	ATKINSON,R (1989)

CAS #: 000628-96-6				ETHYLENE GLYCOL DINITRATE

Formula: $C_2H_4N_2O_6$

Mol Weight: 152.06

MP (deg C): -22.3 | FP (deg C):

BP (deg C): 198.5

BP pressure (mm Hg):

Property/Value	Units	Temp	Data Type	Reference
WS 7.59E+003	mg/L	25	EXP	CLAYTON,GD & CLAYTON,FE (1977)
logP 1.16			EXP	HANSCH,C ET AL. (1995)
VP 7.20E-002	mm Hg	25	EXP	CLAYTON,GD & CLAYTON,FE (1977)
DC	pKa			
HL 2.77E-006	atm m3/mol	25	EST	VP/WSOL
OH 7.35E-013	cm3/molc sec	25	EST	MEYLAN,WM & HOWARD,PH (1993)

CAS #: 000628-97-7				ETHYL PALMITATE

Formula: $C_{18}H_{36}O_2$

Mol Weight: 284.49

MP (deg C): 24 | FP (deg C):

BP (deg C): 191

BP pressure (mm Hg): 1.00E+001

Property/Value	Units	Temp	Data Type	Reference
WS 3.71E-003	mg/L	25	EST	MEYLAN,WM ET AL. (1996)
logP 7.74			EST	MEYLAN,WM & HOWARD,PH (1995)
VP 2.65E-006	mm Hg	25	EXT	OMAR,MM (1967)
DC	pKa			
HL 1.23E-002	atm m3/mol	25	EST	MEYLAN,WM & HOWARD,PH (1991)
OH 2.08E-011	cm3/molc sec	25	EST	MEYLAN,WM & HOWARD,PH (1993)

CAS #: 000628-99-9				2-NONANOL

Formula: $C_9H_{20}O$

Mol Weight: 144.26

MP (deg C): -36 - -35 | FP (deg C):

BP (deg C): 193-194

BP pressure (mm Hg):

Property/Value	Units	Temp	Data Type	Reference
WS 2.63E+002	mg/L	25	EXP	SUZUKI,T (1991)
logP 3.22			EST	MEYLAN,WM & HOWARD,PH (1995)
VP 6.76E-002	mm Hg	25	EXP	YAWS,CL (1994B)
DC	pKa			
HL 4.88E-005	atm m3/mol	25	EST	VP/WSOL
OH 1.70E-011	cm3/molc sec	25	EST	MEYLAN,WM & HOWARD,PH (1993)

CAS #: 000629-04-9				1-BROMOHEPTANE

Formula: $C_7H_{15}Br$

Mol Weight: 179.11

MP (deg C): -58 | FP (deg C):

BP (deg C): 179

BP pressure (mm Hg):

Property/Value	Units	Temp	Data Type	Reference
WS 6.65E+000	mg/L	25	EXP	YALKOWSKY,SH & DANNENFELSER,RM (1992)
logP 4.36			EXP	HANSCH,C & LEO,AJ (1985)
VP 1.27E+000	mm Hg	25	EXP	YAWS,CL (1994A)
DC	pKa			
HL 4.51E-002	atm m3/mol	25	EST	VP/WSOL
OH 6.41E-012	cm3/molc sec	25	EST	MEYLAN,WM & HOWARD,PH (1993)

CAS #: 000629-05-0				1-OCTYNE

Formula: C_8H_{14}

Mol Weight: 110.20

MP (deg C): -79.3 | FP (deg C):

BP (deg C): 126.3

BP pressure (mm Hg):

Property/Value	Units	Temp	Data Type	Reference
WS 2.40E+001	mg/L	25	EXP	YALKOWSKY,SH & DANNENFELSER,RM (1992)
logP 3.50			EST	MEYLAN,WM & HOWARD,PH (1995)
VP 1.36E+001	mm Hg	25	EXP	HINE,J & MOOKERJEE,PK (1975)
DC	pKa			
HL 8.22E-002	atm m3/mol	25	EST	VP/WSOL
OH 1.37E-011	cm3/molc sec	25	EST	MEYLAN,WM & HOWARD,PH (1993)

CAS #: 000629-06-1				1-CHLOROHEPTANE

Formula: $C_7H_{15}Cl$

Mol Weight: 134.65

MP (deg C): -69.5 | FP (deg C):

BP (deg C): 159

BP pressure (mm Hg):

Property/Value	Units	Temp	Data Type	Reference
WS 1.36E+001	mg/L	25	EXP	YALKOWSKY,SH & DANNENFELSER,RM (1992)
logP 4.15			EXP	HANSCH,C & LEO,AJ (1985)
VP 2.82E+000	mm Hg	25	EXT	BOUBLIK,T ET AL. (1984)
DC	pKa			
HL 4.49E-002	atm m3/mol	25	EST	MEYLAN,WM & HOWARD,PH (1991)
OH 6.41E-012	cm3/molc sec	25	EST	MEYLAN,WM & HOWARD,PH (1993)

CAS #: 000629-08-3				HEPTANENITRILE

Formula: $C_7H_{13}N$

Mol Weight: 111.19

MP (deg C): | FP (deg C):

BP (deg C): 183

BP pressure (mm Hg):

Property/ Value	Units	Temp	Data Type	Reference
WS 6.24E+002	mg/L	25	EST	MEYLAN,WM ET AL. (1996)
logP 2.31			EST	MEYLAN,WM & HOWARD,PH (1995)
VP 7.26E-001	mm Hg	25	EST	NEELY,WB & BLAU,GE (1985)
DC	pKa			
HL 1.26E-004	atm m3/mol	25	EST	MEYLAN,WM & HOWARD,PH (1991)
OH 4.50E-012	cm3/molc sec	25	EST	MEYLAN,WM & HOWARD,PH (1993)

CAS #: 000629-11-8				1,6-HEXANEDIOL

Formula: $C_6H_{14}O_2$

Mol Weight: 118.18

MP (deg C): 42.8 | FP (deg C):

BP (deg C): 250

BP pressure (mm Hg):

Property/ Value	Units	Temp	Data Type	Reference
WS 2.27E+004	mg/L	25	EST	MEYLAN,WM ET AL. (1996)
logP 0.76			EST	MEYLAN,WM & HOWARD,PH (1995)
VP 5.00E-004	mm Hg	25	EXT	DAUBERT,TE & DANNER,RP (1989)
DC	pKa			
HL 2.23E-010	atm m3/mol	25	EST	MEYLAN,WM & HOWARD,PH (1991)
OH 1.30E-011	cm3/molc sec	25	EST	ATKINSON,R (1988)

CAS #: 000629-14-1				1,2-DIETHOXYETHANE

Formula: $C_6H_{14}O_2$

Mol Weight: 118.18

MP (deg C): -74 | FP (deg C):

BP (deg C): 119.4

BP pressure (mm Hg):

Property/ Value	Units	Temp	Data Type	Reference
WS 8.30E+004	mg/L	25	EXP	HINE,J & MOOKERJEE,PK (1975)
logP 0.66			EXP	HANSCH,C ET AL. (1995)
VP 3.37E+001	mm Hg	25	EXP	HINE,J & MOOKERJEE,PK (1975)
DC	pKa			
HL 6.32E-005	atm m3/mol	25	EXP	VP/WSOL
OH 2.64E-011	cm3/molc sec	25	EST	MEYLAN,WM & HOWARD,PH (1993)

CAS #: 000629-19-6				4,5-DITHIAOCTANE

Formula: $C_6H_{14}S_2$

Mol Weight: 150.31

MP (deg C): -85.6 | FP (deg C):

BP (deg C): 193.5

BP pressure (mm Hg):

Property/ Value	Units	Temp	Data Type	Reference
WS 3.99E+001	mg/L	25	EST	MEYLAN,WM ET AL. (1996)
logP 3.84			EST	MEYLAN,WM & HOWARD,PH (1995)
VP 5.12E-001	mm Hg	25	EXT	BOUBLIK,T ET AL. (1984)
DC	pKa			
HL 3.77E-003	atm m3/mol	25	EST	MEYLAN,WM & HOWARD,PH (1991)
OH 2.46E-010	cm3/molc sec	25	EST	MEYLAN,WM & HOWARD,PH (1993)

CAS #: 000629-20-9				1,3,5,7-CYCLOOCTATETRAENE

Formula: C_8H_8

Mol Weight: 104.15

MP (deg C): -4.7 | FP (deg C):

BP (deg C): 140.5

BP pressure (mm Hg):

Property/ Value	Units	Temp	Data Type	Reference
WS 7.73E+001	mg/L	25	EST	MEYLAN,WM ET AL. (1996)
logP 3.08			EXP	HANSCH,C ET AL. (1995)
VP 7.80E+000	mm Hg	25	EXP	BOUBLIK,T ET AL. (1984)
DC	pKa			
HL 6.22E-002	atm m3/mol	25	EST	MEYLAN,WM & HOWARD,PH (1991)
OH 2.05E-010	cm3/molc sec	25	EST	MEYLAN,WM & HOWARD,PH (1993)

CAS #: 000629-27-6				1-IODOOCTANE

Formula: $C_8H_{17}I$

Mol Weight: 240.13

MP (deg C): -45.7 | FP (deg C):

BP (deg C): 225.5

BP pressure (mm Hg):

Property/ Value	Units	Temp	Data Type	Reference
WS 1.36E+000	mg/L	25	EST	MEYLAN,WM ET AL. (1996)
logP 5.03			EST	MEYLAN,WM & HOWARD,PH (1995)
VP 2.50E-001	mm Hg	25	EXT	PERRY,RH & GREEN,D (1984)
DC	pKa			
HL 3.99E-002	atm m3/mol	25	EST	MEYLAN,WM & HOWARD,PH (1991)
OH 8.11E-012	cm3/molc sec	25	EST	MEYLAN,WM & HOWARD,PH (1993)

CAS #: 000629-38-9				DIETHYLENE GLYCOL MONOMETHYL ETHER ACETATE

Formula: $C_7H_{14}O_4$

Mol Weight: 162.19

MP (deg C): | FP (deg C):

BP (deg C):

BP pressure (mm Hg):

Property/ Value	Units	Temp	Data Type	Reference
WS 9.45E+004	mg/L	25	EST	MEYLAN,WM ET AL. (1996)
logP -0.18			EST	MEYLAN,WM & HOWARD,PH (1995)
VP 5.26E-001	mm Hg	25	EST	NEELY,WB & BLAU,GE (1985)
DC	pKa			
HL 4.23E-008	atm m3/mol	25	EST	MEYLAN,WM & HOWARD,PH (1991)
OH 2.94E-011	cm3/molc sec	25	EST	MEYLAN,WM & HOWARD,PH (1993)

CAS #: 000629-40-3				OCTANEDINITRILE

Formula: $C_8H_{12}N_2$

Mol Weight: 136.20

MP (deg C): -3.5 | FP (deg C):

BP (deg C): 185

BP pressure (mm Hg): 1.50E+001

Property/ Value	Units	Temp	Data Type	Reference
WS 1.48E+004	mg/L	25	EST	MEYLAN,WM ET AL. (1996)
logP 0.59			EXP	TANII,H & HASHIMOTO,K (1985)
VP 4.43E-003	mm Hg	25	EST	NEELY,WB & BLAU,GE (1985)
DC	pKa			
HL 5.23E-008	atm m3/mol	25	EST	MEYLAN,WM & HOWARD,PH (1991)
OH 3.54E-012	cm3/molc sec	25	EST	MEYLAN,WM & HOWARD,PH (1993)

CAS #: 000629-45-8				DIBUTYL DISULFIDE

Formula: $C_8H_{18}S_2$

Mol Weight: 178.36

MP (deg C):

FP (deg C):

BP (deg C): 117

BP pressure (mm Hg): 2.00E+001

Property/Value	Units	Temp	Data Type	Reference
WS 4.29E+000	mg/L	25	EST	MEYLAN,WM ET AL. (1996)
logP 4.82			EST	MEYLAN,WM & HOWARD,PH (1995)
VP 6.32E-002	mm Hg	25	EXT	ZWOLINSKI,BJ & WILHOIT,RC (1971)
DC	pKa			
HL 6.64E-003	atm m3/mol	25	EST	MEYLAN,WM & HOWARD,PH (1991)
OH 2.48E-010	cm3/molc sec	25	EST	MEYLAN,WM & HOWARD,PH (1993)

CAS #: 000629-50-5				N-TRIDECANE

Formula: $C_{13}H_{28}$

Mol Weight: 184.37

MP (deg C): -5.3

FP (deg C):

BP (deg C): 235.4

BP pressure (mm Hg):

Property/Value	Units	Temp	Data Type	Reference
WS 6.00E+001	mg/L	25	EXP	YALKOWSKY,SH & DANNENFELSER,RM (1992)
logP 6.73			EST	MEYLAN,WM & HOWARD,PH (1995)
VP 5.58E-002	mm Hg	25	EXP	DAUBERT,TE & DANNER,RP (1989)
DC	pKa			
HL 2.26E-004	atm m3/mol	25	EST	VP/WSOL
OH 1.60E-011	cm3/molc sec	25	EXP	ATKINSON,R (1989)

CAS #: 000629-59-4				TETRADECANE

Formula: $C_{14}H_{30}$

Mol Weight: 198.40

MP (deg C): 6

FP (deg C):

BP (deg C): 254

BP pressure (mm Hg):

Property/Value	Units	Temp	Data Type	Reference
WS 2.20E-003	mg/L	25	EXP	SUTTON,C & CALDER,JA (1974)
logP 7.20			EXP	HANSCH,C & LEO,AJ (1985)
VP 1.16E-002	mm Hg	25	EXP	DAUBERT,TE & DANNER,RP (1989)
DC	pKa			
HL 9.20E+000	atm m3/mol	25	EST	VP/WSOL
OH 1.92E-011	cm3/molc sec	25	EXP	ATKINSON,R (1989)

CAS #: 000629-62-9				N-PENTADECANE

Formula: $C_{15}H_{32}$

Mol Weight: 212.42

MP (deg C): 9.9

FP (deg C):

BP (deg C): 270.6

BP pressure (mm Hg):

Property/Value	Units	Temp	Data Type	Reference
WS 2.87E-003	mg/L	25	EST	MEYLAN,WM ET AL. (1996)
logP 7.71			EST	MEYLAN,WM & HOWARD,PH (1995)
VP 3.43E-003	mm Hg	25	EXP	DAUBERT,TE & DANNER,RP (1989)
DC	pKa			
HL 2.19E+001	atm m3/mol	25	EST	MEYLAN,WM & HOWARD,PH (1991)
OH 2.22E-011	cm3/molc sec	25	EXP	ATKINSON,R (1989)

CAS #: 000629-72-1				1-BROMOPENTADECANE

Formula: $C_{15}H_{31}Br$

Mol Weight: 291.32

MP (deg C): 19

FP (deg C):

BP (deg C): 322

BP pressure (mm Hg):

Property/Value	Units	Temp	Data Type	Reference
WS 1.84E-003	mg/L	25	EST	MEYLAN,WM ET AL. (1996)
logP 8.05			EST	MEYLAN,WM & HOWARD,PH (1995)
VP 1.48E-004	mm Hg	25	EXT	LI,JCM & ROSSINI,FD (1953)
DC	pKa			
HL 4.48E-001	atm m3/mol	25	EST	MEYLAN,WM & HOWARD,PH (1991)
OH 1.77E-011	cm3/molc sec	25	EST	MEYLAN,WM & HOWARD,PH (1993)

CAS #: 000629-73-2				1-HEXADECENE

Formula: $C_{16}H_{32}$

Mol Weight: 224.43

MP (deg C): 4.1

FP (deg C):

BP (deg C): 284.9

BP pressure (mm Hg):

Property/Value	Units	Temp	Data Type	Reference
WS 1.23E-003	mg/L	25	EST	MEYLAN,WM ET AL. (1996)
logP 8.06			EST	MEYLAN,WM & HOWARD,PH (1995)
VP 2.64E-003	mm Hg	25	EXP	DAUBERT,TE & DANNER,RP (1989)
DC	pKa			
HL 6.10E+000	atm m3/mol	25	EST	MEYLAN,WM & HOWARD,PH (1991)
OH 4.43E-011	cm3/molc sec	25	EST	MEYLAN,WM & HOWARD,PH (1993)

CAS #: 000629-76-5				1-PENTADECANOL

Formula: $C_{15}H_{32}O$

Mol Weight: 228.42

MP (deg C): 45-46

FP (deg C):

BP (deg C):

BP pressure (mm Hg):

Property/Value	Units	Temp	Data Type	Reference
WS 1.03E-001	mg/L	25	EXP	BARTON,AFM (1984)
logP 6.24			EST	MEYLAN,WM & HOWARD,PH (1995)
VP 3.85E-005	mm Hg	25	EXT	DAUBERT,TE & DANNER,RP (1993)
DC	pKa			
HL 2.25E-004	atm m3/mol	25	EST	MEYLAN,WM & HOWARD,PH (1991)
OH 2.24E-011	cm3/molc sec	25	EST	MEYLAN,WM & HOWARD,PH (1993)

CAS #: 000629-78-7				HEPTADECANE

Formula: $C_{17}H_{36}$

Mol Weight: 240.48

MP (deg C): 22

FP (deg C):

BP (deg C): 302

BP pressure (mm Hg):

Property/Value	Units	Temp	Data Type	Reference
WS 2.94E-004	mg/L	25	EST	MEYLAN,WM ET AL. (1996)
logP 8.69			EST	MEYLAN,WM & HOWARD,PH (1995)
VP 2.28E-004	mm Hg	25	EXP	DAUBERT,TE & DANNER,RP (1989)
DC	pKa			
HL 3.85E+001	atm m3/mol	25	EST	MEYLAN,WM & HOWARD,PH (1991)
OH 2.10E-011	cm3/molc sec	25	EST	MEYLAN,WM & HOWARD,PH (1993)

NONADECANE

CAS #:	000629-92-5	NONADECANE
Formula:	$C_{19}H_{40}$	
Mol Weight:	268.53	
MP (deg C):	32.1	FP (deg C):
BP (deg C):	329.9	
BP pressure (mm Hg):		

Property/Value	Units	Temp	Data Type	Reference
WS 2.97E-005	mg/L	25	EST	MEYLAN,WM ET AL. (1996)
logP 9.67			EST	MEYLAN,WM & HOWARD,PH (1995)
VP 4.90E-005	mm Hg	25	EXT	YAWS,CL (1994B)
DC	pKa			
HL 6.79E+001	atm m3/mol	25	EST	MEYLAN,WM & HOWARD,PH (1991)
OH 2.38E-011	cm3/molc sec	25	EST	MEYLAN,WM & HOWARD,PH (1993)

1-IODOOCTADECANE

CAS #:	000629-93-6	1-IODOOCTADECANE
Formula:	$C_{18}H_{37}I$	
Mol Weight:	380.40	
MP (deg C):	34	FP (deg C):
BP (deg C):	383	
BP pressure (mm Hg):		

Property/Value	Units	Temp	Data Type	Reference
WS 1.31E-005	mg/L	25	EST	MEYLAN,WM ET AL. (1996)
logP 9.94			EST	MEYLAN,WM & HOWARD,PH (1995)
VP 1.92E-006	mm Hg	25	EXT	LI,JCM & ROSSINI,FD (1953)
DC	pKa			
HL 6.79E-001	atm m3/mol	25	EST	MEYLAN,WM & HOWARD,PH (1991)
OH 2.22E-011	cm3/molc sec	25	EST	MEYLAN,WM & HOWARD,PH (1993)

HENEICOSANE

CAS #:	000629-94-7	HENEICOSANE
Formula:	$C_{21}H_{44}$	
Mol Weight:	296.58	
MP (deg C):	40.5	FP (deg C):
BP (deg C):	356.5	
BP pressure (mm Hg):		

Property/Value	Units	Temp	Data Type	Reference
WS 2.90E-008	mg/L	25	EXT	COATES,M ET AL. (1985)
logP 10.65			EST	MEYLAN,WM & HOWARD,PH (1995)
VP 8.73E-005	mm Hg	25	EXT	PERRY,RH & GREEN,D (1984)
DC	pKa			
HL 1.20E+002	atm m3/mol	25	EST	MEYLAN,WM & HOWARD,PH (1991)
OH 2.67E-011	cm3/molc sec	25	EST	MEYLAN,WM & HOWARD,PH (1993)

1-EICOSANOL

CAS #:	000629-96-9	1-EICOSANOL
Formula:	$C_{20}H_{42}O$	
Mol Weight:	298.56	
MP (deg C):	66.1	FP (deg C):
BP (deg C):	309	
BP pressure (mm Hg):		

Property/Value	Units	Temp	Data Type	Reference
WS 1.51E-003	mg/L	25	EST	MEYLAN,WM ET AL. (1996)
logP 8.70			EST	MEYLAN,WM & HOWARD,PH (1995)
VP 4.52E-008	mm Hg	25	EXT	YAWS,CL (1994B)
DC	pKa			
HL 9.29E-004	atm m3/mol	25	EST	MEYLAN,WM & HOWARD,PH (1991)
OH 2.95E-011	cm3/molc sec	25	EST	MEYLAN,WM & HOWARD,PH (1993)

DOCOSANE

CAS #:	000629-97-0	DOCOSANE
Formula:	$C_{22}H_{46}$	
Mol Weight:	310.61	
MP (deg C):	44.4	FP (deg C):
BP (deg C):	368.6	
BP pressure (mm Hg):		

Property/Value	Units	Temp	Data Type	Reference
WS 3.87E-003	mg/L	25	EXT	COATES,M ET AL. (1985)
logP 11.15			EST	MEYLAN,WM & HOWARD,PH (1995)
VP 5.45E-008	mm Hg	25	EXT	ZWOLINSKI,BJ & WILHOIT,RC (1971)
DC	pKa			
HL 1.59E+002	atm m3/mol	25	EST	MEYLAN,WM & HOWARD,PH (1991)
OH 2.81E-011	cm3/molc sec	25	EST	MEYLAN,WM & HOWARD,PH (1993)

PENTACOSANE

CAS #:	000629-99-2	PENTACOSANE
Formula:	$C_{25}H_{52}$	
Mol Weight:	352.69	
MP (deg C):	53.3	FP (deg C):
BP (deg C):	401.9	
BP pressure (mm Hg):		

Property/Value	Units	Temp	Data Type	Reference
WS 2.90E-008	mg/L	25	EST	MEYLAN,WM ET AL. (1996)
logP 12.62			EST	MEYLAN,WM & HOWARD,PH (1995)
VP 1.51E-006	mm Hg	25	EXT	PERRY,RH & GREEN,D (1984)
DC	pKa			
HL 3.72E+002	atm m3/mol	25	EST	MEYLAN,WM & HOWARD,PH (1991)
OH 3.23E-011	cm3/molc sec	25	EST	MEYLAN,WM & HOWARD,PH (1993)

HEXACOSANE

CAS #:	000630-01-3	HEXACOSANE
Formula:	$C_{26}H_{54}$	
Mol Weight:	366.72	
MP (deg C):	56.4	FP (deg C):
BP (deg C):	412.2	
BP pressure (mm Hg):		

Property/Value	Units	Temp	Data Type	Reference
WS 1.70E-003	mg/L	25	EXP	SHAW,DG (1989)
logP 13.11			EST	MEYLAN,WM & HOWARD,PH (1995)
VP 4.69E-007	mm Hg	25	EXT	PERRY,RH & GREEN,D (1984)
DC	pKa			
HL 4.94E+002	atm m3/mol	25	EST	MEYLAN,WM & HOWARD,PH (1991)
OH 3.37E-011	cm3/molc sec	25	EST	MEYLAN,WM & HOWARD,PH (1993)

OCTACOSANE

CAS #:	000630-02-4	OCTACOSANE
Formula:	$C_{28}H_{58}$	
Mol Weight:	394.77	
MP (deg C):	64.5	FP (deg C):
BP (deg C):	431.6	
BP pressure (mm Hg):		

Property/Value	Units	Temp	Data Type	Reference
WS 8.84E-010	mg/L	25	EST	MEYLAN,WM ET AL. (1996)
logP 14.09			EST	MEYLAN,WM & HOWARD,PH (1995)
VP 9.76E-014	mm Hg	25	EXT	ZWOLINSKI,BJ & WILHOIT,RC (1971)
DC	pKa			
HL 8.70E+002	atm m3/mol	25	EST	MEYLAN,WM & HOWARD,PH (1991)
OH 3.65E-011	cm3/molc sec	25	EST	MEYLAN,WM & HOWARD,PH (1993)

NONACOSANE

CAS #:	000630-03-5

Formula:	$C_{29}H_{60}$
Mol Weight:	408.80
MP (deg C):	63.7
FP (deg C):	
BP (deg C):	440.8
BP pressure (mm Hg):	

Property/Value	Units	Temp	Data Type	Reference
WS 2.76E-010	mg/L	25	EST	MEYLAN,WM ET AL. (1996)
logP 14.58			EST	MEYLAN,WM & HOWARD,PH (1995)
VP 3.24E-014	mm Hg	25	EXT	ZWOLINSKI,BJ & WILHOIT,RC (1971)
DC	pKa			
HL 1.15E+003	atm m3/mol	25	EST	MEYLAN,WM & HOWARD,PH (1991)
OH 3.80E-011	cm3/molc sec	25	EST	MEYLAN,WM & HOWARD,PH (1993)

HENTRIACONTANE

CAS #:	000630-04-6

Formula:	$C_{31}H_{64}$
Mol Weight:	436.86
MP (deg C):	67.9
FP (deg C):	
BP (deg C):	458
BP pressure (mm Hg):	

Property/Value	Units	Temp	Data Type	Reference
WS 2.67E-011	mg/L	25	EST	MEYLAN,WM ET AL. (1996)
logP 15.57			EST	MEYLAN,WM & HOWARD,PH (1995)
VP 1.40E-016	mm Hg	25	EXT	ZWOLINSKI,BJ & WILHOIT,RC (1971)
DC	pKa			
HL 2.04E+003	atm m3/mol	25	EST	MEYLAN,WM & HOWARD,PH (1991)
OH 4.08E-011	cm3/molc sec	25	EST	MEYLAN,WM & HOWARD,PH (1993)

TRITRIACONTANE

CAS #:	000630-05-7

Formula:	$C_{33}H_{68}$
Mol Weight:	464.91
MP (deg C):	71-73
FP (deg C):	
BP (deg C):	
BP pressure (mm Hg):	

Property/Value	Units	Temp	Data Type	Reference
WS 2.57E-012	mg/L	25	EST	MEYLAN,WM ET AL. (1996)
logP 16.55			EST	MEYLAN,WM & HOWARD,PH (1995)
VP 4.02E-019	mm Hg	25	EXT	ZWOLINSKI,BJ & WILHOIT,RC (1971)
DC	pKa			
HL 3.59E+003	atm m3/mol	25	EST	MEYLAN,WM & HOWARD,PH (1991)
OH 4.36E-011	cm3/molc sec	25	EST	MEYLAN,WM & HOWARD,PH (1993)

HEXATRIACONTANE

CAS #:	000630-06-8

Formula:	$C_{36}H_{74}$
Mol Weight:	506.99
MP (deg C):	76.5
FP (deg C):	
BP (deg C):	298.4
BP pressure (mm Hg):	3.00E+000

Property/Value	Units	Temp	Data Type	Reference
WS 1.70E-003	mg/L	25	EXP	YALKOWSKY,SH & DANNENFELSER,RM (1992)
logP 18.02			EST	MEYLAN,WM & HOWARD,PH (1995)
VP 1.34E-023	mm Hg	25	EST	ZWOLINSKI,BJ & WILHOIT,RC (1971)
DC	pKa			
HL 8.39E+003	atm m3/mol	25	EST	MEYLAN,WM & HOWARD,PH (1991)
OH 4.78E-011	cm3/molc sec	25	EST	MEYLAN,WM & HOWARD,PH (1993)

N-PENTATRIACONTANE

CAS #:	000630-07-9

Formula:	$C_{35}H_{72}$
Mol Weight:	492.96
MP (deg C):	75
FP (deg C):	
BP (deg C):	490
BP pressure (mm Hg):	

Property/Value	Units	Temp	Data Type	Reference
WS 2.47E-013	mg/L	25	EST	MEYLAN,WM ET AL. (1996)
logP 17.53			EST	MEYLAN,WM & HOWARD,PH (1995)
VP 5.39E-022	mm Hg	25	EXT	ZWOLINSKI,BJ & WILHOIT,RC (1971)
DC	pKa			
HL 6.32E+003	atm m3/mol	25	EST	MEYLAN,WM & HOWARD,PH (1991)
OH 4.64E-011	cm3/molc sec	25	EST	MEYLAN,WM & HOWARD,PH (1993)

TRIMETHYLACETONITRILE

CAS #:	000630-18-2

Formula:	C_5H_9N
Mol Weight:	83.13
MP (deg C):	15
FP (deg C):	
BP (deg C):	106.1
BP pressure (mm Hg):	

Property/Value	Units	Temp	Data Type	Reference
WS 8.38E+003	mg/L	25	EST	MEYLAN,WM ET AL. (1996)
logP 1.08			EXP	HANSCH,C ET AL. (1995)
VP 3.45E+001	mm Hg	25	EXT	BOUBLIK,T ET AL. (1984)
DC	pKa			
HL 7.15E-005	atm m3/mol	25	EST	MEYLAN,WM & HOWARD,PH (1991)
OH 5.02E-013	cm3/molc sec	25	EST	MEYLAN,WM & HOWARD,PH (1993)

PIVALALDEHYDE

CAS #:	000630-19-3

Formula:	$C_5H_{10}O$
Mol Weight:	86.13
MP (deg C):	6
FP (deg C):	
BP (deg C):	77.5
BP pressure (mm Hg):	

Property/Value	Units	Temp	Data Type	Reference
WS 1.21E+004	mg/L	25	EST	MEYLAN,WM ET AL. (1996)
logP 1.20			EST	MEYLAN,WM & HOWARD,PH (1995)
VP 7.12E+001	mm Hg	25	EST	NEELY,WB & BLAU,GE (1985)
DC	pKa			
HL 1.59E-004	atm m3/mol	25	EST	MEYLAN,WM & HOWARD,PH (1991)
OH 2.65E-011	cm3/molc sec	25	EXP	ATKINSON,R (1989)

1,1,1,2-TETRACHLOROETHANE

CAS #:	000630-20-6

Formula:	$C_2H_2Cl_4$
Mol Weight:	167.85
MP (deg C):	-68.7
FP (deg C):	
BP (deg C):	130.2
BP pressure (mm Hg):	

Property/Value	Units	Temp	Data Type	Reference
WS 1.10E+003	mg/L	25	EXP	HORVATH,AL (1982)
logP 2.93			EST	MEYLAN,WM & HOWARD,PH (1995)
VP 1.20E+001	mm Hg	25	EXP	DAUBERT,TE & DANNER,RP (1989)
DC	pKa			
HL 2.42E-003	atm m3/mol	25	EST	VP/WSOL
OH 2.87E-014	cm3/molc sec	25	EST	ATKINSON,R (1988)

CAS #: 000630-60-4				OUABAIN

Formula: $C_{29}H_{44}O_{12}$

Mol Weight: 584.67

MP (deg C): 200

FP (deg C):

BP (deg C):

BP pressure (mm Hg):

Property/Value	Units	Temp	Data Type	Reference
WS 1.03E+004	mg/L	25	EST	MEYLAN,WM ET AL. (1996)
logP -2.00			EXP	SANGSTER,J (1995)
VP 6.08E-026	mm Hg	25	EST	NEELY,WB & BLAU,GE (1985)
DC	pKa			
HL 1.77E-022	atm m3/mol	25	EST	MEYLAN,WM & HOWARD,PH (1991)
OH 1.19E-010	cm3/molc sec	25	EST	MEYLAN,WM & HOWARD,PH (1993)

CAS #: 000630-64-8				HELVETICOSIDE

Formula: $C_{29}H_{42}O_9$

Mol Weight: 534.65

MP (deg C):

FP (deg C):

BP (deg C):

BP pressure (mm Hg):

Property/Value	Units	Temp	Data Type	Reference
WS 2.38E+002	mg/L	25	EST	MEYLAN,WM ET AL. (1996)
logP 0.30			EXP	HANSCH,C & LEO,AJ (1985)
VP 9.01E-021	mm Hg	25	EST	NEELY,WB & BLAU,GE (1985)
DC	pKa			
HL 5.82E-020	atm m3/mol	25	EST	MEYLAN,WM & HOWARD,PH (1991)
OH 1.23E-010	cm3/molc sec	25	EST	MEYLAN,WM & HOWARD,PH (1993)

CAS #: 000631-07-2				2,4-IMIDAZOLIDINEDIONE, 5-ETHYL-5-PHENYL-

Formula: $C_{11}H_{12}N_2O_2$

Mol Weight: 204.23

MP (deg C):

FP (deg C):

BP (deg C):

BP pressure (mm Hg):

Property/Value	Units	Temp	Data Type	Reference
WS 2.05E+003	mg/L	25	EST	MEYLAN,WM ET AL. (1996)
logP 1.53			EXP	HANSCH,C ET AL. (1995)
VP 8.81E-009	mm Hg	25	EST	NEELY,WB & BLAU,GE (1985)
DC	pKa			
HL 2.23E-010	atm m3/mol	25	EST	MEYLAN,WM & HOWARD,PH (1991)
OH 1.01E-011	cm3/molc sec	25	EST	MEYLAN,WM & HOWARD,PH (1993)

CAS #: 000631-36-7				TETRAETHYLSILANE

Formula: $C_8H_{20}Si$

Mol Weight: 144.33

MP (deg C):

FP (deg C):

BP (deg C): 154.7

BP pressure (mm Hg):

Property/Value	Units	Temp	Data Type	Reference
WS 3.25E-001	mg/L	25	EXP	RIDDICK,JA ET AL. (1986)
logP 4.68			EST	MEYLAN,WM & HOWARD,PH (1995)
VP 5.00E+000	mm Hg	24	EXP	RIDDICK,JA ET AL. (1986)
DC	pKa			
HL 2.92E+000	atm m3/mol	24	EST	VP/WSOL
OH 4.78E-012	cm3/molc sec	25	EST	MEYLAN,WM & HOWARD,PH (1993)

CAS #: 000631-40-3				TETRAPROPYL AMMONIUM IODIDE

Formula: $C_{12}H_{28}IN$

Mol Weight: 313.27

MP (deg C): 280 dec

FP (deg C):

BP (deg C):

BP pressure (mm Hg):

Property/Value	Units	Temp	Data Type	Reference
WS 1.86E+005	mg/L	25	EXP	PEDDLE,CJ & TURNER,WES (1913)
logP -0.25			EST	MEYLAN,WM & HOWARD,PH (1995)
VP 8.10E-009	mm Hg	25	EST	NEELY,WB & BLAU,GE (1985)
DC	pKa			
HL 2.54E-011	atm m3/mol	25	EST	MEYLAN,WM & HOWARD,PH (1991)
OH 4.80E-011	cm3/molc sec	25	EST	MEYLAN,WM & HOWARD,PH (1993)

CAS #: 000632-22-4				TETRAMETHYLUREA

Formula: $C_5H_{12}N_2O$

Mol Weight: 116.16

MP (deg C): -1.2

FP (deg C):

BP (deg C): 176.5

BP pressure (mm Hg):

Property/Value	Units	Temp	Data Type	Reference
WS 1.00E+006	mg/L		EXP	RIDDICK,JA ET AL. (1986)
logP 0.19			EXP	HANSCH,C & LEO,AJ (1985)
VP 1.39E+001	mm Hg	71	EXP	RIDDICK,JA ET AL. (1986)
DC	pKa			
HL 8.48E-009	atm m3/mol	25	EST	MEYLAN,WM & HOWARD,PH (1991)
OH 4.23E-012	cm3/molc sec	25	EST	MEYLAN,WM & HOWARD,PH (1993)

CAS #: 000632-46-2				2,6-DIMETHYLBENZOIC ACID

Formula: $C_9H_{10}O_2$

Mol Weight: 150.18

MP (deg C): 116

FP (deg C):

BP (deg C): 274.5

BP pressure (mm Hg):

Property/Value	Units	Temp	Data Type	Reference
WS 9.82E+002	mg/L	25	EST	MEYLAN,WM ET AL. (1996)
logP 2.21			EXP	SOTOMATSU,T ET AL. (1993)
VP 1.54E-003	mm Hg	25	EST	NEELY,WB & BLAU,GE (1985)
DC 3.35	pKa	25	EXP	KORTUM,G ET AL (1961)
HL 1.32E-007	atm m3/mol	25	EST	MEYLAN,WM & HOWARD,PH (1991)
OH 5.71E-012	cm3/molc sec	25	EST	MEYLAN,WM & HOWARD,PH (1993)

CAS #: 000632-79-1				4,5,6,7-TETRABROMO-1,3-ISOBENZOFURANDIONE

Formula: $C_8Br_4O_3$

Mol Weight: 463.72

MP (deg C): 274-276

FP (deg C):

BP (deg C):

BP pressure (mm Hg):

Property/Value	Units	Temp	Data Type	Reference
WS 1.90E-002	mg/L	25	EST	MEYLAN,WM ET AL. (1996)
logP 5.63			EST	MEYLAN,WM & HOWARD,PH (1995)
VP 5.74E-007	mm Hg	25	EST	NEELY,WB & BLAU,GE (1985)
DC	pKa			
HL 1.60E-007	atm m3/mol	25	EST	MEYLAN,WM & HOWARD,PH (1991)
OH 2.44E-014	cm3/molc sec	25	EST	MEYLAN,WM & HOWARD,PH (1993)

486

000634-91-3

CAS #: 000634-91-3 — 3,4,5-TRICHLOROANILINE

Formula: $C_6H_4Cl_3N$

Mol Weight: 196.46

MP (deg C): 98-100

FP (deg C):

BP (deg C):

BP pressure (mm Hg):

Property/Value	Units	Temp	Data Type	Reference
WS 6.66E+001	mg/L	25	EST	MEYLAN,WM ET AL. (1996)
logP 3.32			EXP	HANSCH,C & LEO,AJ (1985)
VP 2.96E-003	mm Hg	25	EST	NEELY,WB & BLAU,GE (1985)
DC	pKa			
HL 7.75E-007	atm m3/mol	25	EST	MEYLAN,WM & HOWARD,PH (1991)
OH 1.56E-011	cm3/molc sec	25	EST	MEYLAN,WM & HOWARD,PH (1993)

000634-93-5

CAS #: 000634-93-5 — 2,4,6-TRICHLOROANILINE

Formula: $C_6H_4Cl_3N$

Mol Weight: 196.46

MP (deg C): 79

FP (deg C):

BP (deg C):

BP pressure (mm Hg):

Property/Value	Units	Temp	Data Type	Reference
WS 4.00E+001	mg/L	25	EXP	CHEM INSPECT TEST INST (1992)
logP 3.52			EXP	SANGSTER,J (1994)
VP 1.47E-007	mm Hg	25	EXT	PERRY,RH & GREEN,D (1984)
DC -0.03	pKa		EXP	PERRIN,DD (1972)
HL 7.75E-007	atm m3/mol	25	EST	MEYLAN,WM & HOWARD,PH (1991)
OH 1.11E-012	cm3/molc sec	25	EST	MEYLAN,WM & HOWARD,PH (1993)

000634-97-9

CAS #: 000634-97-9 — PYRROLE-2-CARBOXYLIC ACID

Formula: $C_5H_5NO_2$

Mol Weight: 111.10

MP (deg C): 204-208 de

FP (deg C):

BP (deg C):

BP pressure (mm Hg):

Property/Value	Units	Temp	Data Type	Reference
WS 2.03E+004	mg/L	25	EST	MEYLAN,WM ET AL. (1996)
logP 0.85			EXP	HANSCH,C ET AL. (1995)
VP 2.25E-006	mm Hg	25	EXT	JONES,AH (1960)
DC 12.52	pKa		EXP	PERRIN,DD (1965)
HL 1.83E-010	atm m3/mol	25	EST	MEYLAN,WM & HOWARD,PH (1991)
OH 4.12E-011	cm3/molc sec	25	EST	MEYLAN,WM & HOWARD,PH (1993)

000635-21-2

CAS #: 000635-21-2 — 2-AMINO-5-CHLOROBENZOIC ACID

Formula: $C_7H_6ClNO_2$

Mol Weight: 171.58

MP (deg C): 210 dec

FP (deg C):

BP (deg C):

BP pressure (mm Hg):

Property/Value	Units	Temp	Data Type	Reference
WS 2.76E+003	mg/L	25	EST	MEYLAN,WM ET AL. (1996)
logP 1.57			EXP	HANSCH,C & LEO,AJ (1985)
VP 5.32E-005	mm Hg	25	EST	NEELY,WB & BLAU,GE (1985)
DC	pKa			
HL 2.84E-011	atm m3/mol	25	EST	MEYLAN,WM & HOWARD,PH (1991)
OH 1.22E-011	cm3/molc sec	25	EST	MEYLAN,WM & HOWARD,PH (1993)

000635-22-3

CAS #: 000635-22-3 — BENZENAMINE, 4-CHLORO-3-NITRO-

Formula: $C_6H_5ClN_2O_2$

Mol Weight: 172.57

MP (deg C): 99-101

FP (deg C):

BP (deg C):

BP pressure (mm Hg):

Property/Value	Units	Temp	Data Type	Reference
WS 1.04E+003	mg/L	25	EST	MEYLAN,WM ET AL. (1996)
logP 2.06			EXP	HANSCH,C ET AL. (1995)
VP 4.85E-004	mm Hg	25	EST	NEELY,WB & BLAU,GE (1985)
DC 1.90	pKa	25	EXP	PERRIN,DD (1965)
HL 5.57E-009	atm m3/mol	25	EST	MEYLAN,WM & HOWARD,PH (1991)
OH 2.75E-012	cm3/molc sec	25	EST	MEYLAN,WM & HOWARD,PH (1993)

000635-46-1

CAS #: 000635-46-1 — 1,2,3,4-TETRAHYDROQUINOLINE

Formula: $C_9H_{11}N$

Mol Weight: 133.19

MP (deg C): 20

FP (deg C):

BP (deg C): 251

BP pressure (mm Hg):

Property/Value	Units	Temp	Data Type	Reference
WS 9.89E+002	mg/L	25	EST	MEYLAN,WM ET AL. (1996)
logP 2.29			EXP	HANSCH,C & LEO,AJ (1985)
VP 4.56E-002	mm Hg	25	EST	NEELY,WB & BLAU,GE (1985)
DC	pKa			
HL 2.70E-006	atm m3/mol	25	EST	MEYLAN,WM & HOWARD,PH (1991)
OH 6.52E-011	cm3/molc sec	25	EST	MEYLAN,WM & HOWARD,PH (1993)

000635-90-5

CAS #: 000635-90-5 — PYRROLE, 1-PHENYL-

Formula: $C_{10}H_{11}N$

Mol Weight: 145.21

MP (deg C): 62

FP (deg C):

BP (deg C): 234

BP pressure (mm Hg):

Property/Value	Units	Temp	Data Type	Reference
WS 1.90E+002	mg/L	25	EST	MEYLAN,WM ET AL. (1996)
logP 3.08			EXP	HANSCH,C ET AL. (1995)
VP 2.58E-002	mm Hg	25	EST	NEELY,WB & BLAU,GE (1985)
DC	pKa			
HL 3.30E-008	atm m3/mol	25	EST	MEYLAN,WM & HOWARD,PH (1991)
OH 1.13E-010	cm3/molc sec	25	EST	MEYLAN,WM & HOWARD,PH (1993)

000636-04-4

CAS #: 000636-04-4 — BENZENECARBOTHIOAMIDE, N-PHENYL-

Formula: $C_{13}H_{11}NS$

Mol Weight: 213.30

MP (deg C): 102

FP (deg C):

BP (deg C):

BP pressure (mm Hg):

Property/Value	Units	Temp	Data Type	Reference
WS 9.65E+001	mg/L	25	EST	MEYLAN,WM ET AL. (1996)
logP 3.03			EXP	HANSCH,C ET AL. (1995)
VP 3.02E-005	mm Hg	25	EST	NEELY,WB & BLAU,GE (1985)
DC	pKa			
HL 2.16E-007	atm m3/mol	25	EST	MEYLAN,WM & HOWARD,PH (1991)
OH 4.70E-011	cm3/molc sec	25	EST	MEYLAN,WM & HOWARD,PH (1993)

CAS #: 000636-21-5				2-METHYLBENZENAMINE HYDROCHLORIDE
Formula: C$_7$H$_{10}$ClN				
Mol Weight: 143.62				
MP (deg C): 215		FP (deg C):		
BP (deg C): 242.2				
BP pressure (mm Hg):				

Property/Value	Units	Temp	Data Type	Reference
WS 8.29E+003	mg/L	25	EST	MEYLAN,WM ET AL. (1996)
logP 1.62			EST	MEYLAN,WM & HOWARD,PH (1995)
VP 2.93E-001	mm Hg	25	EST	NEELY,WB & BLAU,GE (1985)
DC 4.39	pKa		EXP	PERRIN,DD (1965)
HL 2.10E-006	atm m3/mol	25	EST	MEYLAN,WM & HOWARD,PH (1991)
OH 1.32E-010	cm3/molc sec	25	EST	MEYLAN,WM & HOWARD,PH (1993)

CAS #: 000636-28-2				1,2,4,5-TETRABROMOBENZENE
Formula: C$_6$H$_2$Br$_4$				
Mol Weight: 393.72				
MP (deg C): 182		FP (deg C):		
BP (deg C): 3.1				
BP pressure (mm Hg): 2.00E+001				

Property/Value	Units	Temp	Data Type	Reference
WS 5.57E-002	mg/L	25	EXP	SUZUKI,T (1991)
logP 5.13			EXP	HANSCH,C & LEO,AJ (1985)
VP 3.13E-004	mm Hg	25	EST	NEELY,WB & BLAU,GE (1985)
DC	pKa			
HL 1.36E-004	atm m3/mol	25	EST	MEYLAN,WM & HOWARD,PH (1991)
OH 6.35E-014	cm3/molc sec	25	EST	MEYLAN,WM & HOWARD,PH (1993)

CAS #: 000636-30-6				2,4,5-TRICHLOROANILINE
Formula: C$_6$H$_4$Cl$_3$N				
Mol Weight: 196.46				
MP (deg C): 115		FP (deg C):		
BP (deg C):				
BP pressure (mm Hg):				

Property/Value	Units	Temp	Data Type	Reference
WS 5.16E+001	mg/L	25	EST	MEYLAN,WM ET AL. (1996)
logP 3.45			EXP	SANGSTER,J (1994)
VP 2.96E-003	mm Hg	25	EST	NEELY,WB & BLAU,GE (1985)
DC 1.09	pKa		EXP	PERRIN,DD (1972)
HL 7.75E-007	atm m3/mol	25	EST	MEYLAN,WM & HOWARD,PH (1991)
OH 6.46E-012	cm3/molc sec	25	EST	MEYLAN,WM & HOWARD,PH (1993)

CAS #: 000636-70-4				TRIETHYLAMINE, HYDROBROMIDE
Formula: C$_6$H$_{16}$BrN				
Mol Weight: 182.11				
MP (deg C):		FP (deg C):		
BP (deg C):				
BP pressure (mm Hg):				

Property/Value	Units	Temp	Data Type	Reference
WS 1.00E+006	mg/L	25	EXP	PEDDLE,CJ & TURNER,WES (1913)
logP -1.26			EST	MEYLAN,WM & HOWARD,PH (1995)
VP 3.63E-006	mm Hg	25	EST	NEELY,WB & BLAU,GE (1985)
DC	pKa			
HL 4.28E-012	atm m3/mol	25	EST	MEYLAN,WM & HOWARD,PH (1991)
OH 2.66E-011	cm3/molc sec	25	EST	MEYLAN,WM & HOWARD,PH (1993)

CAS #: 000636-72-6				2-HYDROXYMETHYLTHIOPHENE
Formula: C$_5$H$_8$OS				
Mol Weight: 116.18				
MP (deg C):		FP (deg C):		
BP (deg C): 207				
BP pressure (mm Hg):				

Property/Value	Units	Temp	Data Type	Reference
WS 6.16E+004	mg/L	25	EST	MEYLAN,WM ET AL. (1996)
logP 0.87			EXP	HANSCH,C & LEO,AJ (1985)
VP 2.59E-002	mm Hg	25	EST	NEELY,WB & BLAU,GE (1985)
DC	pKa			
HL 1.18E-007	atm m3/mol	25	EST	MEYLAN,WM & HOWARD,PH (1991)
OH 2.71E-011	cm3/molc sec	25	EST	MEYLAN,WM & HOWARD,PH (1993)

CAS #: 000636-97-5				4-NITROBENZOYLHYDRAZINE
Formula: C$_9$H$_{13}$N$_3$O$_3$				
Mol Weight: 211.22				
MP (deg C): 218 dec		FP (deg C):		
BP (deg C):				
BP pressure (mm Hg):				

Property/Value	Units	Temp	Data Type	Reference
WS 5.45E+002	mg/L	25	EST	MEYLAN,WM ET AL. (1996)
logP 0.35			EXP	HANSCH,C & LEO,AJ (1985)
VP 1.08E-006	mm Hg	25	EST	NEELY,WB & BLAU,GE (1985)
DC	pKa			
HL 3.65E-014	atm m3/mol	25	EST	MEYLAN,WM & HOWARD,PH (1991)
OH 5.66E-012	cm3/molc sec	25	EST	MEYLAN,WM & HOWARD,PH (1993)

CAS #: 000637-50-3				BETA-METHYLSTYRENE
Formula: C$_9$H$_{10}$				
Mol Weight: 118.18				
MP (deg C): -27.1		FP (deg C):		
BP (deg C): 175.5				
BP pressure (mm Hg):				

Property/Value	Units	Temp	Data Type	Reference
WS 5.53E+001	mg/L	25	EST	MEYLAN,WM ET AL. (1996)
logP 3.35			EXP	HANSCH,C & LEO,AJ (1985)
VP 9.70E-001	mm Hg	25	EST	WSOL X HL
DC	pKa			
HL 2.74E-003	atm m3/mol	25	EST	MEYLAN,WM & HOWARD,PH (1991)
OH 5.90E-011	cm3/molc sec	25	EXP	ATKINSON,R (1989)

CAS #: 000637-53-6				THIOACETANILIDE
Formula: C$_8$H$_9$NS				
Mol Weight: 151.23				
MP (deg C): 76-79		FP (deg C):		
BP (deg C):				
BP pressure (mm Hg):				

Property/Value	Units	Temp	Data Type	Reference
WS 2.60E+003	mg/L	25	EST	MEYLAN,WM ET AL. (1996)
logP 1.71			EXP	HANSCH,C & LEO,AJ (1985)
VP 1.44E-002	mm Hg	25	EST	NEELY,WB & BLAU,GE (1985)
DC	pKa			
HL 3.56E-006	atm m3/mol	25	EST	MEYLAN,WM & HOWARD,PH (1991)
OH 4.28E-011	cm3/molc sec	25	EST	MEYLAN,WM & HOWARD,PH (1993)

000637-59-2 — PROPYLBROMIDE, G-PHENYL

Formula: $C_9H_{11}Br$
Mol Weight: 199.10
MP (deg C): FP (deg C):
BP (deg C): 219.5
BP pressure (mm Hg):

Property/Value	Units	Temp	Data Type	Reference
WS 2.94E+001	mg/L	25	EST	MEYLAN,WM ET AL. (1996)
logP 3.72			EXP	HANSCH,C & LEO,AJ (1985)
VP 6.32E-002	mm Hg	25	EST	NEELY,WB & BLAU,GE (1985)
DC	pKa			
HL 1.21E-003	atm m3/mol	25	EST	MEYLAN,WM & HOWARD,PH (1991)
OH 6.84E-012	cm3/molc sec	25	EST	MEYLAN,WM & HOWARD,PH (1993)

000637-61-6 — QUINONE CHLOROIMIDE

Formula: C_6H_4ClNO
Mol Weight: 141.56
MP (deg C): FP (deg C):
BP (deg C):
BP pressure (mm Hg):

Property/Value	Units	Temp	Data Type	Reference
WS 6.92E+003	mg/L	25	EST	MEYLAN,WM ET AL. (1996)
logP 1.26			EXP	HANSCH,C ET AL. (1995)
VP 1.52E-001	mm Hg	25	EST	NEELY,WB & BLAU,GE (1985)
DC	pKa			
HL 6.48E-006	atm m3/mol	25	EST	MEYLAN,WM & HOWARD,PH (1991)
OH 2.26E-011	cm3/molc sec	25	EST	MEYLAN,WM & HOWARD,PH (1993)

000637-62-7 — P-BENZOQUINONE OXIME

Formula: $C_6H_5NO_2$
Mol Weight: 123.11
MP (deg C): FP (deg C):
BP (deg C):
BP pressure (mm Hg):

Property/Value	Units	Temp	Data Type	Reference
WS 1.17E+004	mg/L	25	EST	MEYLAN,WM ET AL. (1996)
logP 1.08			EXP	HANSCH,C ET AL. (1995)
VP 6.35E-004	mm Hg	25	EST	NEELY,WB & BLAU,GE (1985)
DC 6.21	pKa		EXP	KORTUM,G ET AL (1961)
HL 6.76E-010	atm m3/mol	25	EST	MEYLAN,WM & HOWARD,PH (1991)
OH 2.26E-011	cm3/molc sec	25	EST	MEYLAN,WM & HOWARD,PH (1993)

000637-78-5 — ISOPROPYL PROPIONATE

Formula: $C_6H_{12}O_2$
Mol Weight: 116.16
MP (deg C): FP (deg C):
BP (deg C): 109.5
BP pressure (mm Hg):

Property/Value	Units	Temp	Data Type	Reference
WS 5.95E+003	mg/L	25	EXP	SUZUKI,T (1991)
logP 1.77			EST	MEYLAN,WM & HOWARD,PH (1995)
VP 2.25E+001	mm Hg	25	EXP	HINE,J & MOOKERJEE,PK (1975)
DC	pKa			
HL 5.79E-004	atm m3/mol	25	EST	VP/WSOL
OH 4.30E-012	cm3/molc sec	25	EST	MEYLAN,WM & HOWARD,PH (1993)

000637-87-6 — 4-CHLOROIODOBENZENE

Formula: C_6H_4ClI
Mol Weight: 238.46
MP (deg C): 57 FP (deg C):
BP (deg C): 227
BP pressure (mm Hg):

Property/Value	Units	Temp	Data Type	Reference
WS 6.88E+001	mg/L	25	EXP	SUZUKI,T (1991)
logP 3.80			EST	MEYLAN,WM & HOWARD,PH (1995)
VP 5.89E-002	mm Hg	25	EXT	JONES,AH (1960)
DC	pKa			
HL 9.25E-004	atm m3/mol	25	EST	MEYLAN,WM & HOWARD,PH (1991)
OH 4.53E-013	cm3/molc sec	25	EST	MEYLAN,WM & HOWARD,PH (1993)

000637-92-3 — ETHYL T-BUTYL ETHER

Formula: $C_6H_{14}O$
Mol Weight: 102.18
MP (deg C): -94 FP (deg C):
BP (deg C): 73.1
BP pressure (mm Hg):

Property/Value	Units	Temp	Data Type	Reference
WS 1.20E+004	mg/L	20	EXP	EVANS,TW & EDLUND,KR (1936)
logP 1.92			EST	MEYLAN,WM & HOWARD,PH (1995)
VP 1.24E+002	mm Hg	25	EXP	DAUBERT,TE & DANNER,RP (1993)
DC	pKa			
HL 1.39E-003	atm m3/mol	25	EST	VP/WSOL
OH 6.90E-012	cm3/molc sec	25	EXP	ATKINSON,R (1989)

000637-97-8 — 2-IODOPENTANE

Formula: $C_5H_{11}I$
Mol Weight: 198.05
MP (deg C): FP (deg C):
BP (deg C):
BP pressure (mm Hg):

Property/Value	Units	Temp	Data Type	Reference
WS 4.76E+001	mg/L	25	EST	MEYLAN,WM ET AL. (1996)
logP 3.48			EST	MEYLAN,WM & HOWARD,PH (1995)
VP 1.10E+001	mm Hg	25	EXP	LEVANOVA,SV ET AL. (1967)
DC	pKa			
HL 1.71E-002	atm m3/mol	25	EST	MEYLAN,WM & HOWARD,PH (1991)
OH 3.26E-012	cm3/molc sec	25	EST	MEYLAN,WM & HOWARD,PH (1993)

000638-02-8 — 2,5-DIMETHYLTHIOPHENE

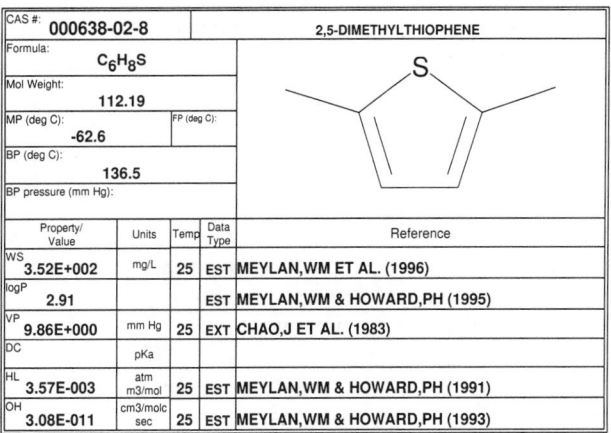

Formula: C_6H_8S
Mol Weight: 112.19
MP (deg C): -62.6 FP (deg C):
BP (deg C): 136.5
BP pressure (mm Hg):

Property/Value	Units	Temp	Data Type	Reference
WS 3.52E+002	mg/L	25	EST	MEYLAN,WM ET AL. (1996)
logP 2.91			EST	MEYLAN,WM & HOWARD,PH (1995)
VP 9.86E+000	mm Hg	25	EXT	CHAO,J ET AL. (1983)
DC	pKa			
HL 3.57E-003	atm m3/mol	25	EST	MEYLAN,WM & HOWARD,PH (1991)
OH 3.08E-011	cm3/molc sec	25	EST	MEYLAN,WM & HOWARD,PH (1993)

CAS #: 000638-04-0				1,3-DIMETHYLCYCLOHEXANE (CIS)
Formula: C_8H_{16}				
Mol Weight: 112.22				
MP (deg C): -75.6		FP (deg C):		
BP (deg C): 120.1				
BP pressure (mm Hg):				

Property/Value	Units	Temp	Data Type	Reference
WS 1.17E+001	mg/L	25	EST	MEYLAN,WM ET AL. (1996)
logP 4.01			EST	MEYLAN,WM & HOWARD,PH (1995)
VP 2.15E+001	mm Hg	25	EXP	DAUBERT,TE & DANNER,RP (1989)
DC	pKa			
HL 4.50E-001	atm m3/mol	25	EST	MEYLAN,WM & HOWARD,PH (1991)
OH 1.19E-011	cm3/molc sec	25	EST	MEYLAN,WM & HOWARD,PH (1993)

CAS #: 000638-20-0				1,3-DIACETYLUREA
Formula: $C_5H_8N_2O_3$				
Mol Weight: 144.13				
MP (deg C): 154.5		FP (deg C):		
BP (deg C):				
BP pressure (mm Hg):				

Property/Value	Units	Temp	Data Type	Reference
WS 3.06E+005	mg/L	25	EST	MEYLAN,WM ET AL. (1996)
logP -0.68			EXP	HANSCH,C & LEO,AJ (1985)
VP 9.99E-009	mm Hg	25	EST	NEELY,WB & BLAU,GE (1985)
DC	pKa			
HL 3.50E-013	atm m3/mol	25	EST	MEYLAN,WM & HOWARD,PH (1991)
OH 2.20E-012	cm3/molc sec	25	EST	MEYLAN,WM & HOWARD,PH (1993)

CAS #: 000638-28-8				2-CHLOROHEXANE
Formula: $C_6H_{13}Cl$				
Mol Weight: 120.62				
MP (deg C):		FP (deg C):		
BP (deg C): 122.5				
BP pressure (mm Hg):				

Property/Value	Units	Temp	Data Type	Reference
WS 1.09E+002	mg/L	25	EST	MEYLAN,WM ET AL. (1996)
logP 3.47			EST	MEYLAN,WM & HOWARD,PH (1995)
VP 1.36E+001	mm Hg	25	EXP	LEVANOVA,SV (1967)
DC	pKa			
HL 3.38E-002	atm m3/mol	25	EST	MEYLAN,WM & HOWARD,PH (1991)
OH 4.10E-012	cm3/molc sec	25	EST	MEYLAN,WM & HOWARD,PH (1993)

CAS #: 000638-42-6				O-PENTYL CARBAMATE
Formula: $C_6H_{13}NO_2$				
Mol Weight: 131.18				
MP (deg C):		FP (deg C):		
BP (deg C):				
BP pressure (mm Hg):				

Property/Value	Units	Temp	Data Type	Reference
WS 6.40E+003	mg/L	25	EST	MEYLAN,WM ET AL. (1996)
logP 1.35			EXP	HANSCH,C & LEO,AJ (1985)
VP 8.07E-001	mm Hg	25	EST	NEELY,WB & BLAU,GE (1985)
DC	pKa			
HL 1.23E-007	atm m3/mol	25	EST	MEYLAN,WM & HOWARD,PH (1991)
OH 7.98E-012	cm3/molc sec	25	EST	MEYLAN,WM & HOWARD,PH (1993)

CAS #: 000638-45-9				1-IODOHEXANE
Formula: $C_6H_{13}I$				
Mol Weight: 212.07				
MP (deg C): -74.2		FP (deg C):		
BP (deg C): 181				
BP pressure (mm Hg):				

Property/Value	Units	Temp	Data Type	Reference
WS 1.33E+001	mg/L	25	EST	MEYLAN,WM ET AL. (1996)
logP 4.05			EST	MEYLAN,WM & HOWARD,PH (1995)
VP 1.40E+000	mm Hg	25	EXP	LI,JCM & ROSSINI,FD (1953)
DC	pKa			
HL 2.27E-002	atm m3/mol	25	EST	MEYLAN,WM & HOWARD,PH (1991)
OH 5.28E-012	cm3/molc sec	25	EST	MEYLAN,WM & HOWARD,PH (1993)

CAS #: 000638-46-0				ETHYL BUTYLSULFIDE
Formula: $C_6H_{14}S$				
Mol Weight: 118.24				
MP (deg C): -95.1		FP (deg C):		
BP (deg C): 144.3				
BP pressure (mm Hg):				

Property/Value	Units	Temp	Data Type	Reference
WS 3.51E+002	mg/L	25	EST	MEYLAN,WM ET AL. (1996)
logP 2.88			EST	MEYLAN,WM & HOWARD,PH (1995)
VP 6.45E+000	mm Hg	25	EST	NEELY,WB & BLAU,GE (1985)
DC	pKa			
HL 2.44E-003	atm m3/mol	25	EST	MEYLAN,WM & HOWARD,PH (1991)
OH 2.08E-011	cm3/molc sec	25	EST	MEYLAN,WM & HOWARD,PH (1993)

CAS #: 000638-53-9				TRIDECANOIC ACID
Formula: $C_{13}H_{26}O_2$				
Mol Weight: 214.35				
MP (deg C): 44.5		FP (deg C):		
BP (deg C): 236				
BP pressure (mm Hg): 1.00E+002				

Property/Value	Units	Temp	Data Type	Reference
WS 3.30E+001	mg/L	20	EXP	YALKOWSKY,SH & DANNENFELSER,RM (1992)
logP 5.49			EST	MEYLAN,WM & HOWARD,PH (1995)
VP 1.26E-005	mm Hg	25	EXT	PERRY,RH & GREEN,D (1984)
DC	pKa			
HL 1.24E-005	atm m3/mol	25	EST	MEYLAN,WM & HOWARD,PH (1991)
OH 1.54E-011	cm3/molc sec	25	EST	MEYLAN,WM & HOWARD,PH (1993)

CAS #: 000638-59-5				1-TETRADECANOL ACETATE
Formula: $C_{16}H_{32}O_2$				
Mol Weight: 256.43				
MP (deg C):		FP (deg C):		
BP (deg C):				
BP pressure (mm Hg):				

Property/Value	Units	Temp	Data Type	Reference
WS 3.69E-002	mg/L	25	EST	MEYLAN,WM ET AL. (1996)
logP 6.76			EST	MEYLAN,WM & HOWARD,PH (1995)
VP 5.94E-004	mm Hg	25	EST	NEELY,WB & BLAU,GE (1985)
DC	pKa			
HL 6.97E-003	atm m3/mol	25	EST	MEYLAN,WM & HOWARD,PH (1991)
OH 1.88E-011	cm3/molc sec	25	EST	MEYLAN,WM & HOWARD,PH (1993)

CAS #: 000638-67-5				TRICOSANE

Formula: $C_{23}H_{48}$

Mol Weight: 324.64

MP (deg C): 47.6   FP (deg C):

BP (deg C): 380

BP pressure (mm Hg):

Property/Value	Units	Temp	Data Type	Reference
WS 2.95E-007	mg/L	25	EST	MEYLAN,WM ET AL. (1996)
logP 11.64			EST	MEYLAN,WM & HOWARD,PH (1995)
VP 1.74E-005	mm Hg	25	EXT	PERRY,RH & GREEN,D (1984)
DC	pKa			
HL 2.11E+002	atm m3/mol	25	EST	MEYLAN,WM & HOWARD,PH (1991)
OH 2.95E-011	cm3/molc sec	25	EST	MEYLAN,WM & HOWARD,PH (1993)

CAS #: 000638-68-6				TRIACONTANE

Formula: $C_{30}H_{62}$

Mol Weight: 422.83

MP (deg C): 65.8   FP (deg C):

BP (deg C): 449.7

BP pressure (mm Hg):

Property/Value	Units	Temp	Data Type	Reference
WS 8.58E-011	mg/L	25	EST	MEYLAN,WM ET AL. (1996)
logP 15.07			EST	MEYLAN,WM & HOWARD,PH (1995)
VP 2.74E-015	mm Hg	25	EXT	ZWOLINSKI,BJ & WILHOIT,RC (1971)
DC	pKa			
HL 1.53E+003	atm m3/mol	25	EST	MEYLAN,WM & HOWARD,PH (1991)
OH 3.94E-011	cm3/molc sec	25	EST	MEYLAN,WM & HOWARD,PH (1993)

CAS #: 000640-15-3				THIOMETON

Formula: $C_6H_{15}O_2PS_3$

Mol Weight: 246.35

MP (deg C):   FP (deg C):

BP (deg C): 110

BP pressure (mm Hg): 1.00E-001

Property/Value	Units	Temp	Data Type	Reference
WS 2.00E+002	mg/L	25	EXP	YALKOWSKY,SH & DANNENFELSER,RM (1992)
logP 2.88			EST	MEYLAN,WM & HOWARD,PH (1995)
VP 1.70E-005	mm Hg	20	EXP	TOMLIN,C (1994)
DC	pKa			
HL 2.76E-008	atm m3/mol	25	EST	MEYLAN,WM & HOWARD,PH (1991)
OH 1.00E-010	cm3/molc sec	25	EST	MEYLAN,WM & HOWARD,PH (1993)

CAS #: 000640-19-7				FLUOROACETAMIDE

Formula: C_2H_4FNO

Mol Weight: 77.06

MP (deg C): 108   FP (deg C):

BP (deg C):

BP pressure (mm Hg):

Property/Value	Units	Temp	Data Type	Reference
WS 1.00E+006	mg/L	25	EST	MEYLAN,WM ET AL. (1996)
logP -1.05			EXP	HANSCH,C & LEO,AJ (1985)
VP 9.91E-001	mm Hg	25	EST	NEELY,WB & BLAU,GE (1985)
DC	pKa			
HL 2.23E-008	atm m3/mol	25	EST	MEYLAN,WM & HOWARD,PH (1991)
OH 2.07E-012	cm3/molc sec	25	EST	MEYLAN,WM & HOWARD,PH (1993)

CAS #: 000640-61-9				BENZENESULFONAMIDE, N,4-DIMETHYL-

Formula: $C_8H_{11}NO_2S$

Mol Weight: 185.25

MP (deg C): 78.5   FP (deg C):

BP (deg C):

BP pressure (mm Hg):

Property/Value	Units	Temp	Data Type	Reference
WS 4.19E+003	mg/L	25	EST	MEYLAN,WM ET AL. (1996)
logP 1.28			EXP	LARSEN,JD ET AL. (1988)
VP 2.82E-004	mm Hg	25	EST	NEELY,WB & BLAU,GE (1985)
DC	pKa			
HL 1.02E-006	atm m3/mol	25	EST	MEYLAN,WM & HOWARD,PH (1991)
OH 2.49E-012	cm3/molc sec	25	EST	MEYLAN,WM & HOWARD,PH (1993)

CAS #: 000640-79-9				GLYCOCHENODEOXYCHOLIC ACID

Formula: $C_{26}H_{43}NO_5$

Mol Weight: 449.64

MP (deg C):   FP (deg C):

BP (deg C):

BP pressure (mm Hg):

Property/Value	Units	Temp	Data Type	Reference
WS 2.82E+001	mg/L	25	EST	MEYLAN,WM ET AL. (1996)
logP 2.12			EXP	RODA,A ET AL. (1990)
VP 8.36E-018	mm Hg	25	EST	NEELY,WB & BLAU,GE (1985)
DC	pKa			
HL 9.38E-017	atm m3/mol	25	EST	MEYLAN,WM & HOWARD,PH (1991)
OH 5.79E-011	cm3/molc sec	25	EST	MEYLAN,WM & HOWARD,PH (1993)

CAS #: 000643-15-2				PENTACHLOROCYCLOHEXENE

Formula: $C_6H_5Cl_5$

Mol Weight: 254.37

MP (deg C):   FP (deg C):

BP (deg C):

BP pressure (mm Hg):

Property/Value	Units	Temp	Data Type	Reference
WS 1.15E+001	mg/L	25	EST	MEYLAN,WM ET AL. (1996)
logP 3.61			EXP	HANSCH,C & LEO,AJ (1985)
VP 5.99E-003	mm Hg	25	EST	NEELY,WB & BLAU,GE (1985)
DC	pKa			
HL 1.30E-003	atm m3/mol	25	EST	MEYLAN,WM & HOWARD,PH (1991)
OH 1.13E-011	cm3/molc sec	25	EST	MEYLAN,WM & HOWARD,PH (1993)

CAS #: 000643-28-7				BENZENAMINE, 2-(1-METHYLETHYL)-

Formula: $C_9H_{13}N$

Mol Weight: 135.21

MP (deg C):   FP (deg C):

BP (deg C): 221

BP pressure (mm Hg):

Property/Value	Units	Temp	Data Type	Reference
WS 1.36E+003	mg/L	25	EST	MEYLAN,WM ET AL. (1996)
logP 2.12			EXP	HANSCH,C ET AL. (1995)
VP 7.49E-002	mm Hg	25	EST	NEELY,WB & BLAU,GE (1985)
DC 4.42	pKa	25	EXP	PERRIN,DD (1965)
HL 3.70E-006	atm m3/mol	25	EST	MEYLAN,WM & HOWARD,PH (1991)
OH 1.32E-010	cm3/molc sec	25	EST	MEYLAN,WM & HOWARD,PH (1993)

491

CAS #: 000643-43-6				DINITROPHENYL ACETATE

Formula: $C_8H_6N_2O_6$

Mol Weight: 226.15

MP (deg C): 169-175 FP (deg C):

BP (deg C):

BP pressure (mm Hg):

Property/Value	Units	Temp	Data Type	Reference
WS 3.99E+003	mg/L	25	EST	MEYLAN,WM ET AL. (1996)
logP 1.06			EST	MEYLAN,WM & HOWARD,PH (1995)
VP 6.06E-007	mm Hg	25	EST	NEELY,WB & BLAU,GE (1985)
DC 3.50	pKa	25	EXP	KORTUM,G ET AL (1961)
HL 6.88E-013	atm m3/mol	25	EST	MEYLAN,WM & HOWARD,PH (1991)
OH 1.27E-012	cm3/molc sec	25	EST	MEYLAN,WM & HOWARD,PH (1993)

CAS #: 000643-58-3				2-METHYLBIPHENYL

Formula: $C_{13}H_{12}$

Mol Weight: 168.24

MP (deg C): -0.2 FP (deg C):

BP (deg C): 255.5

BP pressure (mm Hg):

Property/Value	Units	Temp	Data Type	Reference
WS 1.81E+001	mg/L	25	EST	MEYLAN,WM ET AL. (1996)
logP 4.14			EXP	HANSCH,C & LEO,AJ (1995)
VP 2.03E-002	mm Hg	25	EST	NEELY,WB & BLAU,GE (1985)
DC	pKa			
HL 4.55E-004	atm m3/mol	25	EST	MEYLAN,WM & HOWARD,PH (1991)
OH 7.68E-012	cm3/molc sec	25	EST	MEYLAN,WM & HOWARD,PH (1993)

CAS #: 000643-93-6				3-METHYLBIPHENYL

Formula: $C_{13}H_{12}$

Mol Weight: 168.24

MP (deg C): FP (deg C):

BP (deg C):

BP pressure (mm Hg):

Property/Value	Units	Temp	Data Type	Reference
WS 1.81E+001	mg/L	25	EST	MEYLAN,WM ET AL. (1996)
logP 4.30			EST	MEYLAN,WM & HOWARD,PH (1995)
VP 7.96E-003	mm Hg	25	EST	NEELY,WB & BLAU,GE (1985)
DC	pKa			
HL 4.55E-004	atm m3/mol	25	EST	MEYLAN,WM & HOWARD,PH (1991)
OH 1.24E-005	cm3/molc sec	25	EST	MEYLAN,WM & HOWARD,PH (1993)

CAS #: 000644-08-6				4-METHYL-1,1'-BIPHENYL

Formula: $C_{13}H_{12}$

Mol Weight: 168.24

MP (deg C): 49.5 FP (deg C):

BP (deg C): 267.5

BP pressure (mm Hg):

Property/Value	Units	Temp	Data Type	Reference
WS 4.06E+000	mg/L	25	EXP	YALKOWSKY,SH & DANNENFELSER,RM (1992)
logP 4.63			EXP	DOUCETTE,WJ & ANDREN,AW (1987)
VP 6.42E-003	mm Hg	25	EST	NEELY,WB & BLAU,GE (1985)
DC	pKa			
HL 4.55E-004	atm m3/mol	25	EST	MEYLAN,WM & HOWARD,PH (1991)
OH 7.68E-012	cm3/molc sec	25	EST	MEYLAN,WM & HOWARD,PH (1993)

CAS #: 000644-26-8				1-(DIMETHYLAMINO)-2-METHYL-2-BUTANOL BENZOATE

Formula: $C_{14}H_{21}NO_2$

Mol Weight: 235.33

MP (deg C): FP (deg C):

BP (deg C):

BP pressure (mm Hg):

Property/Value	Units	Temp	Data Type	Reference
WS 9.92E+002	mg/L	25	EST	MEYLAN,WM ET AL. (1996)
logP 2.89			EST	MEYLAN,WM & HOWARD,PH (1995)
VP 7.47E-004	mm Hg	25	EST	NEELY,WB & BLAU,GE (1985)
DC	pKa			
HL 5.35E-008	atm m3/mol	25	EST	MEYLAN,WM & HOWARD,PH (1991)
OH 8.27E-011	cm3/molc sec	25	EST	MEYLAN,WM & HOWARD,PH (1993)

CAS #: 000644-35-9				O-PROPYLPHENOL

Formula: $C_9H_{12}O$

Mol Weight: 136.20

MP (deg C): 7 FP (deg C):

BP (deg C): 220

BP pressure (mm Hg):

Property/Value	Units	Temp	Data Type	Reference
WS 1.04E+003	mg/L	25	EST	MEYLAN,WM ET AL. (1996)
logP 2.93			EXP	HANSCH,C & LEO,AJ (1985)
VP 9.00E-002	mm Hg	25	EXT	BOUBLIK,T ET AL. (1984)
DC 10.47	pKa	25	EXP	SERJEANT,EP & DEMPSEY,B (1979)
HL 1.09E-006	atm m3/mol	25	EST	MEYLAN,WM & HOWARD,PH (1991)
OH 4.32E-011	cm3/molc sec	25	EST	MEYLAN,WM & HOWARD,PH (1993)

CAS #: 000644-49-5				PROPYL ISOBUTYRATE

Formula: $C_7H_{14}O_2$

Mol Weight: 130.19

MP (deg C): FP (deg C):

BP (deg C): 135.4

BP pressure (mm Hg):

Property/Value	Units	Temp	Data Type	Reference
WS 1.07E+003	mg/L	25	EST	MEYLAN,WM ET AL. (1996)
logP 2.26			EST	MEYLAN,WM & HOWARD,PH (1995)
VP 8.40E+000	mm Hg	25	EXP	OHE,S (1976)
DC	pKa			
HL 5.45E-004	atm m3/mol	25	EST	MEYLAN,WM & HOWARD,PH (1991)
OH 4.92E-012	cm3/molc sec	25	EST	MEYLAN,WM & HOWARD,PH (1993)

CAS #: 000644-64-4				DIMETILAN

Formula: $C_{10}H_{16}N_4O_3$

Mol Weight: 240.26

MP (deg C): 68-71 FP (deg C):

BP (deg C):

BP pressure (mm Hg):

Property/Value	Units	Temp	Data Type	Reference
WS 2.99E+003	mg/L	25	EST	MEYLAN,WM ET AL. (1996)
logP -1.07			EST	MEYLAN,WM & HOWARD,PH (1995)
VP 7.36E-006	mm Hg	25	EST	NEELY,WB & BLAU,GE (1985)
DC	pKa			
HL 4.14E-011	atm m3/mol	25	EST	MEYLAN,WM & HOWARD,PH (1991)
OH 1.29E-010	cm3/molc sec	25	EST	MEYLAN,WM & HOWARD,PH (1993)

CAS #: 000645-00-1				3-IODONITROBENZENE

Formula: $C_6H_4INO_2$

Mol Weight: 249.01

MP (deg C): 38.5   FP (deg C):

BP (deg C): 280

BP pressure (mm Hg):

Property/Value	Units	Temp	Data Type	Reference
WS 3.01E+001	mg/L	25	EST	MEYLAN,WM ET AL. (1996)
logP 2.94			EXP	HANSCH,C & LEO,AJ (1985)
VP 1.74E-003	mm Hg	25	EST	NEELY,WB & BLAU,GE (1985)
DC	pKa			
HL 4.93E-006	atm m3/mol	25	EST	MEYLAN,WM & HOWARD,PH (1991)
OH 1.12E-013	cm3/molc sec	25	EST	MEYLAN,WM & HOWARD,PH (1993)

CAS #: 000645-05-6				HEXAMETHYL MELAMINE

Formula: $C_9H_{18}N_6$

Mol Weight: 210.28

MP (deg C): 172-174   FP (deg C):

BP (deg C):

BP pressure (mm Hg):

Property/Value	Units	Temp	Data Type	Reference
WS 9.10E+001	mg/L	25	EXP	YALKOWSKY,SH & DANNENFELSER,RM (1992)
logP 2.73			EXP	HANSCH,C ET AL. (1995)
VP 2.95E-004	mm Hg	25	EST	NEELY,WB & BLAU,GE (1985)
DC	pKa			
HL 1.72E-008	atm m3/mol	25	EST	MEYLAN,WM & HOWARD,PH (1991)
OH 8.25E-012	cm3/molc sec	25	EST	MEYLAN,WM & HOWARD,PH (1993)

CAS #: 000645-09-0				M-NITROBENZAMIDE

Formula: $C_7H_6N_2O_3$

Mol Weight: 166.14

MP (deg C): 142.7   FP (deg C):

BP (deg C): 312.5

BP pressure (mm Hg):

Property/Value	Units	Temp	Data Type	Reference
WS 5.75E+003	mg/L	25	EST	MEYLAN,WM ET AL. (1996)
logP 0.77			EXP	HANSCH,C & LEO,AJ (1985)
VP 8.09E-006	mm Hg	25	EST	NEELY,WB & BLAU,GE (1985)
DC	pKa			
HL 8.74E-012	atm m3/mol	25	EST	MEYLAN,WM & HOWARD,PH (1991)
OH 2.22E-012	cm3/molc sec	25	EST	MEYLAN,WM & HOWARD,PH (1993)

CAS #: 000645-13-6				P-ISOPROPYLACETOPHENONE

Formula: $C_{11}H_{14}O$

Mol Weight: 162.23

MP (deg C):   FP (deg C):

BP (deg C): 254

BP pressure (mm Hg):

Property/Value	Units	Temp	Data Type	Reference
WS 1.91E+002	mg/L	25	EST	MEYLAN,WM ET AL. (1996)
logP 2.98			EXP	HANSCH,C & LEO,AJ (1985)
VP 5.70E-002	mm Hg	25	EST	NEELY,WB & BLAU,GE (1985)
DC	pKa			
HL 1.91E-005	atm m3/mol	25	EST	MEYLAN,WM & HOWARD,PH (1991)
OH 6.59E-012	cm3/molc sec	25	EST	MEYLAN,WM & HOWARD,PH (1993)

CAS #: 000645-49-8				1,1'-(1,2-ETHENEDIYL)BISBENZENE (CIS)

Formula: $C_{14}H_{12}$

Mol Weight: 180.25

MP (deg C): -5   FP (deg C):

BP (deg C): 141

BP pressure (mm Hg): 1.20E+001

Property/Value	Units	Temp	Data Type	Reference
WS 4.29E+000	mg/L	25	EST	MEYLAN,WM ET AL. (1996)
logP 4.52			EST	MEYLAN,WM & HOWARD,PH (1995)
VP 6.86E-003	mm Hg	25	EXP	DAUBERT,TE & DANNER,RP (1989)
DC	pKa			
HL 1.24E-004	atm m3/mol	25	EST	MEYLAN,WM & HOWARD,PH (1991)
OH 6.01E-011	cm3/molc sec	25	EST	MEYLAN,WM & HOWARD,PH (1993)

CAS #: 000645-56-7				4-PROPYLPHENOL

Formula: $C_9H_{12}O$

Mol Weight: 136.20

MP (deg C): 22   FP (deg C):

BP (deg C): 232.6

BP pressure (mm Hg):

Property/Value	Units	Temp	Data Type	Reference
WS 1.28E+003	mg/L	25	EXP	SHIU,WY ET AL. (1994)
logP 3.20			EXP	SANGSTER,J (1993)
VP 3.75E-002	mm Hg	25	EXT	JORDAN,TE (1954)
DC 10.34	pKa	20	EXP	SERJEANT,EP & DEMPSEY,B (1979)
HL 1.09E-006	atm m3/mol	25	EST	MEYLAN,WM & HOWARD,PH (1991)
OH 4.32E-011	cm3/molc sec	25	EST	MEYLAN,WM & HOWARD,PH (1993)

CAS #: 000645-59-0				BENZYLACETONITRILE

Formula: C_9H_9N

Mol Weight: 131.18

MP (deg C): -1   FP (deg C):

BP (deg C): 261

BP pressure (mm Hg):

Property/Value	Units	Temp	Data Type	Reference
WS 1.68E+003	mg/L	25	EST	MEYLAN,WM ET AL. (1996)
logP 1.72			EXP	HANSCH,C & LEO,AJ (1985)
VP 3.05E-002	mm Hg	25	EST	NEELY,WB & BLAU,GE (1985)
DC	pKa			
HL 3.27E-006	atm m3/mol	25	EST	MEYLAN,WM & HOWARD,PH (1991)
OH 5.18E-012	cm3/molc sec	25	EST	MEYLAN,WM & HOWARD,PH (1993)

CAS #: 000645-62-5				2-ETHYL-2-HEXENAL

Formula: $C_8H_{14}O$

Mol Weight: 126.20

MP (deg C):   FP (deg C):

BP (deg C): 195

BP pressure (mm Hg):

Property/Value	Units	Temp	Data Type	Reference
WS 5.86E+002	mg/L	25	EXP	CHEM INSPECT TEST INST (1992)
logP 2.62			EST	MEYLAN,WM & HOWARD,PH (1995)
VP 1.28E+000	mm Hg	25	EXT	BOUBLIK,T ET AL. (1984)
DC	pKa			
HL 2.06E-004	atm m3/mol	25	EST	MEYLAN,WM & HOWARD,PH (1991)
OH 5.00E-011	cm3/molc sec	25	EST	MEYLAN,WM & HOWARD,PH (1993)

000646-04-8 — TRANS-2-PENTENE

Formula	C_5H_{10}
Mol Weight	70.14
MP (deg C)	-140
FP (deg C)	
BP (deg C)	37
BP pressure (mm Hg)	

Property/Value	Units	Temp	Data Type	Reference
WS 2.45E+002	mg/L	25	EST	MEYLAN,WM ET AL. (1996)
logP 2.58			EST	MEYLAN,WM & HOWARD,PH (1995)
VP 5.06E+002	mm Hg	25	EXP	DAUBERT,TE & DANNER,RP (1989)
DC	pKa			
HL 3.19E-001	atm m3/mol	25	EST	MEYLAN,WM & HOWARD,PH (1991)
OH 6.69E-011	cm3/molc sec	25	EXP	ATKINSON,R (1989)

000646-06-0 — 1,3-DIOXALANE

Formula	$C_3H_6O_2$
Mol Weight	74.08
MP (deg C)	-95
FP (deg C)	
BP (deg C)	78
BP pressure (mm Hg)	

Property/Value	Units	Temp	Data Type	Reference
WS 2.77E+005	mg/L	25	EST	MEYLAN,WM ET AL. (1996)
logP -0.37			EXP	HANSCH,C & LEO,AJ (1985)
VP 7.90E+001	mm Hg	20	EXP	RIDDICK,JA ET AL. (1985)
DC -3.80	pKa	25	EXP	RIDDICK,JA ET AL. (1985)
HL 2.45E-005	atm m3/mol	25	EXP	CABANI,S ET AL. (1971A)
OH 1.46E-011	cm3/molc sec	25	EST	ATKINSON,R (1988)

000646-07-1 — 4-METHYLPENTANOIC ACID

Formula	$C_6H_{12}O_2$
Mol Weight	116.16
MP (deg C)	-33
FP (deg C)	
BP (deg C)	200.5
BP pressure (mm Hg)	

Property/Value	Units	Temp	Data Type	Reference
WS 5.27E+003	mg/L	25	EST	MEYLAN,WM ET AL. (1996)
logP 1.98			EST	MEYLAN,WM & HOWARD,PH (1995)
VP 4.45E-001	mm Hg	25	EST	NEELY,WB & BLAU,GE (1985)
DC 4.84	pKa	18	EXP	KORTUM,G ET AL (1961)
HL 1.70E-006	atm m3/mol	25	EST	MEYLAN,WM & HOWARD,PH (1991)
OH 5.52E-012	cm3/molc sec	25	EST	MEYLAN,WM & HOWARD,PH (1993)

000646-14-0 — NITROHEXANE

Formula	$C_6H_{13}NO_2$
Mol Weight	131.18
MP (deg C)	
FP (deg C)	
BP (deg C)	193
BP pressure (mm Hg)	

Property/Value	Units	Temp	Data Type	Reference
WS 1.83E+002	mg/L	25	EST	MEYLAN,WM ET AL. (1996)
logP 2.70			EXP	HANSCH,C ET AL. (1995)
VP 9.00E-001	mm Hg	25	EST	NEELY,WB & BLAU,GE (1985)
DC	pKa			
HL 1.74E-004	atm m3/mol	25	EST	MEYLAN,WM & HOWARD,PH (1991)
OH 4.43E-012	cm3/molc sec	25	EST	MEYLAN,WM & HOWARD,PH (1993)

000646-20-8 — HEPTANEDINITRILE

Formula	$C_7H_{10}N_2$
Mol Weight	122.17
MP (deg C)	-31.4
FP (deg C)	
BP (deg C)	155
BP pressure (mm Hg)	1.40E+001

Property/Value	Units	Temp	Data Type	Reference
WS 4.85E+004	mg/L	25	EST	MEYLAN,WM ET AL. (1996)
logP 0.05			EXP	TANII,H & HASHIMOTO,K (1985)
VP 1.30E-002	mm Hg	25	EST	NEELY,WB & BLAU,GE (1985)
DC	pKa			
HL 3.94E-008	atm m3/mol	25	EST	MEYLAN,WM & HOWARD,PH (1991)
OH 2.13E-012	cm3/molc sec	25	EST	MEYLAN,WM & HOWARD,PH (1993)

000646-31-1 — TETRACOSANE

Formula	$C_{24}H_{50}$
Mol Weight	338.67
MP (deg C)	54
FP (deg C)	
BP (deg C)	391.3
BP pressure (mm Hg)	

Property/Value	Units	Temp	Data Type	Reference
WS 9.25E-008	mg/L	25	EST	MEYLAN,WM ET AL. (1996)
logP 12.13			EST	MEYLAN,WM & HOWARD,PH (1995)
VP 4.07E-006	mm Hg	25	EXT	PERRY,RH & GREEN,D (1984)
DC	pKa			
HL 2.80E+002	atm m3/mol	25	EST	MEYLAN,WM & HOWARD,PH (1991)
OH 3.09E-011	cm3/molc sec	25	EST	MEYLAN,WM & HOWARD,PH (1993)

000650-51-1 — SODIUM TRICHLOROACETATE

Formula	$C_2Cl_3NaO_2$
Mol Weight	185.37
MP (deg C)	> 300
FP (deg C)	
BP (deg C)	
BP pressure (mm Hg)	

Property/Value	Units	Temp	Data Type	Reference
WS 1.00E+006	mg/L	25	EXP	GUNTHER,FA ET AL. (1968)
logP -2.37			EST	MEYLAN,WM & HOWARD,PH (1995)
VP 2.60E-008	mm Hg	25	EST	NEELY,WB & BLAU,GE (1985)
DC	pKa			
HL	atm m3/mol			
OH 0.00E+000	cm3/molc sec	25	EST	MEYLAN,WM & HOWARD,PH (1993)

000651-06-9 — SULFAMETER

Formula	$C_{11}H_{12}N_4O_3S$
Mol Weight	280.31
MP (deg C)	214-216
FP (deg C)	
BP (deg C)	
BP pressure (mm Hg)	

Property/Value	Units	Temp	Data Type	Reference
WS 7.30E+002	mg/L	30	EXP	YALKOWSKY,SH & DANNENFELSER,RM (1992)
logP 0.41			EXP	HANSCH,C & LEO,AJ (1985)
VP 8.35E-009	mm Hg	25	EST	NEELY,WB & BLAU,GE (1985)
DC	pKa			
HL 9.36E-012	atm m3/mol	25	EST	MEYLAN,WM & HOWARD,PH (1991)
OH 2.80E-011	cm3/molc sec	25	EST	MEYLAN,WM & HOWARD,PH (1993)

CAS #: 000651-85-4 — PENTAFLUOROPHENYL METHYL SUFONE

Formula: $C_7H_3F_5O_2S$

Mol Weight: 246.16

MP (deg C): **FP (deg C):**

BP (deg C):

BP pressure (mm Hg):

Property/ Value	Units	Temp	Data Type	Reference
WS 2.80E+003	mg/L	25	EST	MEYLAN,WM ET AL. (1996)
logP 1.11			EXP	HANSCH,C & LEO,AJ (1985)
VP 1.61E-002	mm Hg	25	EST	NEELY,WB & BLAU,GE (1985)
DC	pKa			
HL 2.70E-006	atm m3/mol	25	EST	MEYLAN,WM & HOWARD,PH (1991)
OH 1.15E-012	cm3/molc sec	25	EST	MEYLAN,WM & HOWARD,PH (1993)

CAS #: 000652-32-4 — BENZOIC ACID, 2,3,5,6-TETRAFLUORO-4-METHYL-

Formula: $C_8H_4F_4O_2$

Mol Weight: 208.11

MP (deg C): 173-175 **FP (deg C):**

BP (deg C):

BP pressure (mm Hg):

Property/ Value	Units	Temp	Data Type	Reference
WS 8.11E+001	mg/L	25	EST	MEYLAN,WM ET AL. (1996)
logP 3.15			EXP	SANGSTER,J (1994)
VP 1.09E-002	mm Hg	25	EST	NEELY,WB & BLAU,GE (1985)
DC 2.00	pKa	25	EXP	SERJEANT,EP & DEMPSEY,B (1979)
HL 2.22E-007	atm m3/mol	25	EST	MEYLAN,WM & HOWARD,PH (1991)
OH 9.93E-013	cm3/molc sec	25	EST	MEYLAN,WM & HOWARD,PH (1993)

CAS #: 000653-22-5 — 2,3,4,5,6-PENTAFLUOROACETANILIDE

Formula: $C_8H_4F_5NO$

Mol Weight: 225.12

MP (deg C): **FP (deg C):**

BP (deg C):

BP pressure (mm Hg):

Property/ Value	Units	Temp	Data Type	Reference
WS 1.14E+003	mg/L	25	EST	MEYLAN,WM ET AL. (1996)
logP 1.70			EXP	NAKAGAWA,Y ET AL. (1992)
VP 1.06E-003	mm Hg	25	EST	NEELY,WB & BLAU,GE (1985)
DC	pKa			
HL 1.34E-008	atm m3/mol	25	EST	MEYLAN,WM & HOWARD,PH (1991)
OH 8.51E-013	cm3/molc sec	25	EST	MEYLAN,WM & HOWARD,PH (1993)

CAS #: 000654-13-7 — 1H-BENZOTRIAZOLE-1-PROPANOIC ACID, á-METHYL-

Formula: $C_{10}H_{11}N_3O_2$

Mol Weight: 205.22

MP (deg C): **FP (deg C):**

BP (deg C):

BP pressure (mm Hg):

Property/ Value	Units	Temp	Data Type	Reference
WS 1.51E+005	mg/L	25	EST	MEYLAN,WM ET AL. (1996)
logP -0.20			EXP	SPARATORE,F ET AL. (1988)
VP 2.55E-006	mm Hg	25	EST	NEELY,WB & BLAU,GE (1985)
DC	pKa			
HL 4.10E-011	atm m3/mol	25	EST	MEYLAN,WM & HOWARD,PH (1991)
OH 4.94E-012	cm3/molc sec	25	EST	MEYLAN,WM & HOWARD,PH (1993)

CAS #: 000654-15-9 — 1-(2-CARBOXYET)BENZOTRIAZOLE

Formula: $C_9H_9N_3O_2$

Mol Weight: 191.19

MP (deg C): **FP (deg C):**

BP (deg C):

BP pressure (mm Hg):

Property/ Value	Units	Temp	Data Type	Reference
WS 4.16E+005	mg/L	25	EST	MEYLAN,WM ET AL. (1996)
logP -0.63			EXP	HANSCH,C & LEO,AJ (1985)
VP 3.40E-006	mm Hg	25	EST	NEELY,WB & BLAU,GE (1985)
DC	pKa			
HL 3.09E-011	atm m3/mol	25	EST	MEYLAN,WM & HOWARD,PH (1991)
OH 3.53E-012	cm3/molc sec	25	EST	MEYLAN,WM & HOWARD,PH (1993)

CAS #: 000654-19-3 — 1-(3-CARBOXYPROPYL)BENZOTRIAZOLE

Formula: $C_{10}H_{11}N_3O_2$

Mol Weight: 205.22

MP (deg C): **FP (deg C):**

BP (deg C):

BP pressure (mm Hg):

Property/ Value	Units	Temp	Data Type	Reference
WS 1.00E+005	mg/L	25	EST	MEYLAN,WM ET AL. (1996)
logP 0.01			EXP	HANSCH,C & LEO,AJ (1985)
VP 1.48E-006	mm Hg	25	EST	NEELY,WB & BLAU,GE (1985)
DC	pKa			
HL 4.10E-011	atm m3/mol	25	EST	MEYLAN,WM & HOWARD,PH (1991)
OH 4.94E-012	cm3/molc sec	25	EST	MEYLAN,WM & HOWARD,PH (1993)

CAS #: 000655-48-1 — 1,2-ETHANEDIOL, 1,2-DIPHENYL-, (R*,R*)-(ñ)-

Formula: $C_{14}H_{14}O_2$

Mol Weight: 214.27

MP (deg C): 122.5 **FP (deg C):**

BP (deg C): > 300

BP pressure (mm Hg):

Property/ Value	Units	Temp	Data Type	Reference
WS 8.63E+002	mg/L	25	EST	MEYLAN,WM ET AL. (1996)
logP 1.91			EXP	SANGSTER,J (1994)
VP 2.06E-007	mm Hg	25	EST	NEELY,WB & BLAU,GE (1985)
DC	pKa			
HL 8.52E-010	atm m3/mol	25	EST	MEYLAN,WM & HOWARD,PH (1991)
OH 2.62E-011	cm3/molc sec	25	EST	MEYLAN,WM & HOWARD,PH (1993)

CAS #: 000655-86-7 — 2,3-DIAMINOPHENAZINE

Formula: $C_{12}H_{10}N_4$

Mol Weight: 210.24

MP (deg C): 264 **FP (deg C):**

BP (deg C):

BP pressure (mm Hg):

Property/ Value	Units	Temp	Data Type	Reference
WS 2.53E+002	mg/L	25	EST	MEYLAN,WM ET AL. (1996)
logP 1.26			EST	MEYLAN,WM & HOWARD,PH (1995)
VP 5.55E-008	mm Hg	25	EST	NEELY,WB & BLAU,GE (1985)
DC	pKa			
HL 3.47E-015	atm m3/mol	25	EST	MEYLAN,WM & HOWARD,PH (1991)
OH 2.00E-010	cm3/molc sec	25	EST	MEYLAN,WM & HOWARD,PH (1993)

CAS #: 000656-31-5 — O-FLUOROPHENYLUREA

Formula: $C_7H_7FN_2O$

Mol Weight: 154.15

MP (deg C): FP (deg C):

BP (deg C):

BP pressure (mm Hg):

Property/Value	Units	Temp	Data Type	Reference
WS 1.29E+004	mg/L	25	EST	MEYLAN,WM ET AL. (1996)
logP 0.88			EXP	HANSCH,C & LEO,AJ (1985)
VP 1.09E-003	mm Hg	25	EST	NEELY,WB & BLAU,GE (1985)
DC	pKa			
HL 2.35E-010	atm m3/mol	25	EST	MEYLAN,WM & HOWARD,PH (1991)
OH 1.54E-011	cm3/molc sec	25	EST	MEYLAN,WM & HOWARD,PH (1993)

CAS #: 000656-49-5 — 2-CF3-5-CHLOROBENZIMIDAZOLE

Formula: $C_8H_4ClF_3N_2$

Mol Weight: 220.58

MP (deg C): FP (deg C):

BP (deg C):

BP pressure (mm Hg):

Property/Value	Units	Temp	Data Type	Reference
WS 4.35E+001	mg/L	25	EST	MEYLAN,WM ET AL. (1996)
logP 3.39			EXP	HANSCH,C & LEO,AJ (1985)
VP 1.11E-005	mm Hg	25	EST	NEELY,WB & BLAU,GE (1985)
DC	pKa			
HL 2.36E-006	atm m3/mol	25	EST	MEYLAN,WM & HOWARD,PH (1991)
OH 2.85E-012	cm3/molc sec	25	EST	MEYLAN,WM & HOWARD,PH (1993)

CAS #: 000658-43-5 — DIFLUOROMETHANESULFONANILIDE

Formula: $C_7H_7F_2NO_2S$

Mol Weight: 207.20

MP (deg C): FP (deg C):

BP (deg C):

BP pressure (mm Hg):

Property/Value	Units	Temp	Data Type	Reference
WS 8.68E+002	mg/L	25	EST	MEYLAN,WM ET AL. (1996)
logP 1.95			EXP	HANSCH,C & LEO,AJ (1985)
VP 1.17E-003	mm Hg	25	EST	NEELY,WB & BLAU,GE (1985)
DC	pKa			
HL 4.61E-006	atm m3/mol	25	EST	MEYLAN,WM & HOWARD,PH (1991)
OH 4.28E-011	cm3/molc sec	25	EST	MEYLAN,WM & HOWARD,PH (1993)

CAS #: 000658-48-0 — TYROSINE, A-METHYL-

Formula: $C_{10}H_{13}NO_3$

Mol Weight: 195.22

MP (deg C): > 300 FP (deg C):

BP (deg C):

BP pressure (mm Hg):

Property/Value	Units	Temp	Data Type	Reference
WS 6.16E+004	mg/L	25	EST	MEYLAN,WM ET AL. (1996)
logP -1.89			EXP	HANSCH,C & LEO,AJ (1985)
VP 5.77E-009	mm Hg	25	EST	NEELY,WB & BLAU,GE (1985)
DC	pKa			
HL 1.66E-014	atm m3/mol	25	EST	MEYLAN,WM & HOWARD,PH (1991)
OH 6.36E-011	cm3/molc sec	25	EST	MEYLAN,WM & HOWARD,PH (1993)

CAS #: 000659-30-3 — P-FLUOROPHENYLUREA

Formula: $C_7H_7FN_2O$

Mol Weight: 154.15

MP (deg C): FP (deg C):

BP (deg C):

BP pressure (mm Hg):

Property/Value	Units	Temp	Data Type	Reference
WS 9.41E+003	mg/L	25	EST	MEYLAN,WM ET AL. (1996)
logP 1.04			EXP	HANSCH,C & LEO,AJ (1985)
VP 1.09E-003	mm Hg	25	EST	NEELY,WB & BLAU,GE (1985)
DC	pKa			
HL 2.35E-010	atm m3/mol	25	EST	MEYLAN,WM & HOWARD,PH (1991)
OH 1.54E-011	cm3/molc sec	25	EST	MEYLAN,WM & HOWARD,PH (1993)

CAS #: 000659-70-1 — ISOAMYL ISOVALERATE

Formula: $C_{10}H_{20}O_2$

Mol Weight: 172.27

MP (deg C): FP (deg C):

BP (deg C): 190.4

BP pressure (mm Hg):

Property/Value	Units	Temp	Data Type	Reference
WS 4.46E+001	mg/L	25	EST	MEYLAN,WM ET AL. (1996)
logP 3.66			EST	MEYLAN,WM & HOWARD,PH (1995)
VP 8.86E-001	mm Hg	25	EXP	DAUBERT,TE & DANNER,RP (1989)
DC	pKa			
HL 1.27E-003	atm m3/mol	25	EST	MEYLAN,WM & HOWARD,PH (1991)
OH 9.54E-012	cm3/molc sec	25	EST	MEYLAN,WM & HOWARD,PH (1993)

CAS #: 000660-68-4 — N-ETHYLETHANAMINE, HYDROCHLORIDE

Formula: $C_4H_{12}ClN$

Mol Weight: 109.60

MP (deg C): 228.5 FP (deg C):

BP (deg C):

BP pressure (mm Hg):

Property/Value	Units	Temp	Data Type	Reference
WS 1.00E+006	mg/L	25	EXP	SEIDELL,A (1941)
logP -2.29			EST	MEYLAN,WM & HOWARD,PH (1995)
VP 6.45E-005	mm Hg	25	EST	NEELY,WB & BLAU,GE (1985)
DC	pKa			
HL 2.24E-012	atm m3/mol	25	EST	MEYLAN,WM & HOWARD,PH (1991)
OH 1.77E-011	cm3/molc sec	25	EST	MEYLAN,WM & HOWARD,PH (1993)

CAS #: 000660-88-8 — 5-AMINOPENTANOIC ACID

Formula: $C_5H_{11}NO_2$

Mol Weight: 117.15

MP (deg C): 157-158 de FP (deg C):

BP (deg C):

BP pressure (mm Hg):

Property/Value	Units	Temp	Data Type	Reference
WS 1.00E+006	mg/L	20	EXP	YALKOWSKY,SH & DANNENFELSER,RM (1992)
logP -2.63			EXP	HANSCH,C ET AL. (1995)
VP 7.74E-009	mm Hg	25	EST	NEELY,WB & BLAU,GE (1985)
DC 4.20	pKa	25	EXP	KORTUM,G ET AL (1961)
HL 1.32E-010	atm m3/mol	25	EST	MEYLAN,WM & HOWARD,PH (1991)
OH 3.59E-011	cm3/molc sec	25	EST	MEYLAN,WM & HOWARD,PH (1993)

000661-11-0 — 1-FLUOROHEPTANE

Formula:	$C_7H_{15}F$		
Mol Weight:	118.20		
MP (deg C):	-73	FP (deg C):	
BP (deg C):	117.9		
BP pressure (mm Hg):			

Property/Value	Units	Temp	Data Type	Reference
WS 6.78E+001	mg/L	25	EST	MEYLAN,WM ET AL. (1996)
logP 3.72			EST	MEYLAN,WM & HOWARD,PH (1995)
VP 1.84E+001	mm Hg	25	EXP	LI,JCM & ROSSINI,FD (1953)
DC	pKa			
HL 9.00E-002	atm m3/mol	25	EST	MEYLAN,WM & HOWARD,PH (1991)
OH 6.36E-012	cm3/molc sec	25	EST	MEYLAN,WM & HOWARD,PH (1993)

000664-95-9 — TOLCYCLAMIDE

Formula:	$C_{14}H_{20}N_2O_3S$		
Mol Weight:	296.39		
MP (deg C):	174-176	FP (deg C):	
BP (deg C):			
BP pressure (mm Hg):			

Property/Value	Units	Temp	Data Type	Reference
WS 1.84E+002	mg/L	37	EXP	YALKOWSKY,SH & DANNENFELSER,RM (1992)
logP 2.90			EXP	SANGSTER,J (1993)
VP 3.80E-009	mm Hg	25	EST	NEELY,WB & BLAU,GE (1985)
DC	pKa			
HL 1.97E-010	atm m3/mol	25	EST	MEYLAN,WM & HOWARD,PH (1991)
OH 2.23E-011	cm3/molc sec	25	EST	MEYLAN,WM & HOWARD,PH (1993)

000670-96-2 — 1H-IMIDAZOLE, 2-PHENYL-

Formula:	$C_9H_8N_2$		
Mol Weight:	144.18		
MP (deg C):	149.3	FP (deg C):	
BP (deg C):	340		
BP pressure (mm Hg):			

Property/Value	Units	Temp	Data Type	Reference
WS 1.99E+003	mg/L	25	EST	MEYLAN,WM ET AL. (1996)
logP 1.88			EXP	HANSCH,C & LEO,AJ (1985)
VP 1.13E-005	mm Hg	25	EST	NEELY,WB & BLAU,GE (1985)
DC 6.48	pKa	25	EXP	PERRIN,DD (1965)
HL 2.88E-007	atm m3/mol	25	EST	MEYLAN,WM & HOWARD,PH (1991)
OH 6.59E-011	cm3/molc sec	25	EST	MEYLAN,WM & HOWARD,PH (1993)

000671-03-4 — 2,4,5-TRIMEPH N-ME CARBAMATE

Formula:	$C_{11}H_{15}NO_2$		
Mol Weight:	193.25		
MP (deg C):		FP (deg C):	
BP (deg C):			
BP pressure (mm Hg):			

Property/Value	Units	Temp	Data Type	Reference
WS 3.34E+002	mg/L	25	EST	MEYLAN,WM ET AL. (1996)
logP 2.52			EXP	HANSCH,C & LEO,AJ (1985)
VP 1.90E-003	mm Hg	25	EST	NEELY,WB & BLAU,GE (1985)
DC	pKa			
HL 4.32E-008	atm m3/mol	25	EST	MEYLAN,WM & HOWARD,PH (1991)
OH 1.84E-011	cm3/molc sec	25	EST	MEYLAN,WM & HOWARD,PH (1993)

000671-04-5 — BANOL

Formula:	$C_{10}H_{12}ClNO_2$		
Mol Weight:	213.67		
MP (deg C):		FP (deg C):	
BP (deg C):			
BP pressure (mm Hg):			

Property/Value	Units	Temp	Data Type	Reference
WS 2.02E+002	mg/L	25	EST	MEYLAN,WM ET AL. (1996)
logP 2.65			EST	MEYLAN,WM & HOWARD,PH (1995)
VP 9.22E-004	mm Hg	25	EST	NEELY,WB & BLAU,GE (1985)
DC	pKa			
HL 2.90E-008	atm m3/mol	25	EST	MEYLAN,WM & HOWARD,PH (1991)
OH 9.55E-012	cm3/molc sec	25	EST	MEYLAN,WM & HOWARD,PH (1993)

000671-16-9 — PROCARBAZINE

Formula:	$C_{12}H_{19}N_3O$		
Mol Weight:	221.30		
MP (deg C):		FP (deg C):	
BP (deg C):			
BP pressure (mm Hg):			

Property/Value	Units	Temp	Data Type	Reference
WS 1.42E+003	mg/L	25	EST	MEYLAN,WM ET AL. (1996)
logP 0.08			EXP	HANSCH,C ET AL. (1995)
VP 6.66E+000	mm Hg	25	EXP	HANSCH,C & LEO,AJ (1985)
DC	pKa			
HL 8.92E-015	atm m3/mol	25	EST	MEYLAN,WM & HOWARD,PH (1991)
OH 1.54E-010	cm3/molc sec	25	EST	MEYLAN,WM & HOWARD,PH (1993)

000671-56-7 — 3-HYDROXY-4-CHLORO-1-BUTENE

Formula:	C_4H_7ClO		
Mol Weight:	106.55		
MP (deg C):		FP (deg C):	
BP (deg C):	145.5		
BP pressure (mm Hg):			

Property/Value	Units	Temp	Data Type	Reference
WS 6.34E+004	mg/L	25	EST	MEYLAN,WM ET AL. (1996)
logP 0.89			EST	MEYLAN,WM & HOWARD,PH (1995)
VP 1.20E+000	mm Hg	25	EST	NEELY,WB & BLAU,GE (1985)
DC	pKa			
HL 1.65E-006	atm m3/mol	25	EST	MEYLAN,WM & HOWARD,PH (1991)
OH 2.93E-011	cm3/molc sec	25	EST	MEYLAN,WM & HOWARD,PH (1993)

000672-65-1 — (1-CHLOROETHYL)BENZENE

Formula:	C_8H_9Cl		
Mol Weight:	140.61		
MP (deg C):		FP (deg C):	
BP (deg C):			
BP pressure (mm Hg):			

Property/Value	Units	Temp	Data Type	Reference
WS 1.51E+002	mg/L	25	EST	MEYLAN,WM ET AL. (1996)
logP 3.21			EST	MEYLAN,WM & HOWARD,PH (1995)
VP 9.15E-001	mm Hg	25	EST	NEELY,WB & BLAU,GE (1985)
DC	pKa			
HL 2.78E-003	atm m3/mol	25	EST	MEYLAN,WM & HOWARD,PH (1991)
OH 2.80E-012	cm3/molc sec	25	EST	MEYLAN,WM & HOWARD,PH (1993)

CAS #: 000672-66-2				(DIMETHYL)-PHENYLPHOSPHINE
Formula: $C_8H_{11}P$				
Mol Weight: 138.15				
MP (deg C):		FP (deg C):		
BP (deg C): 74-75				
BP pressure (mm Hg): 1.20E+001				

Property/Value	Units	Temp	Data Type	Reference
WS 5.44E+002	mg/L	25	EST	MEYLAN,WM ET AL. (1996)
logP 2.57			EXP	HANSCH,C & LEO,AJ (1985)
VP 7.60E-001	mm Hg	25	EST	NEELY,WB & BLAU,GE (1985)
DC	pKa			
HL 3.45E-005	atm m3/mol	25	EST	MEYLAN,WM & HOWARD,PH (1991)
OH 2.22E-012	cm3/molc sec	25	EST	MEYLAN,WM & HOWARD,PH (1993)

CAS #: 000672-76-4				5-I-PROPYLTROPALONE
Formula: $C_{10}H_{12}O_2$				
Mol Weight: 164.21				
MP (deg C):		FP (deg C):		
BP (deg C):				
BP pressure (mm Hg):				

Property/Value	Units	Temp	Data Type	Reference
WS 1.59E+003	mg/L	25	EST	MEYLAN,WM ET AL. (1996)
logP 1.89			EXP	HANSCH,C & LEO,AJ (1985)
VP 3.90E-005	mm Hg	25	EST	NEELY,WB & BLAU,GE (1985)
DC	pKa			
HL 2.98E-005	atm m3/mol	25	EST	MEYLAN,WM & HOWARD,PH (1991)
OH 8.48E-011	cm3/molc sec	25	EST	MEYLAN,WM & HOWARD,PH (1993)

CAS #: 000673-04-1				SIMETONE
Formula: $C_8H_{15}N_5O$				
Mol Weight: 197.24				
MP (deg C):		FP (deg C):		
BP (deg C):				
BP pressure (mm Hg):				

Property/Value	Units	Temp	Data Type	Reference
WS 3.20E+003	mg/L	20	EXP	BAILEY,GW ET AL. (1968)
logP 2.73			EST	MEYLAN,WM & HOWARD,PH (1995)
VP 2.40E-006	mm Hg	20	EXP	KEARNEY,PC & KAUFMAN,DD (1975A)
DC 4.17	pKa		EXP	BAILEY,GW ET AL. (1968)
HL 1.95E-010	atm m3/mol	20	EST	VP/WSOL
OH 1.89E-011	cm3/molc sec	25	EST	MEYLAN,WM & HOWARD,PH (1993)

CAS #: 000673-31-4				CARBAMIC ACID, 3-PHENYLPROPYL ESTER
Formula: $C_{10}H_{13}NO_2$				
Mol Weight: 179.22				
MP (deg C): 101-104		FP (deg C):		
BP (deg C):				
BP pressure (mm Hg):				

Property/Value	Units	Temp	Data Type	Reference
WS 1.18E+003	mg/L	25	EST	MEYLAN,WM ET AL. (1996)
logP 1.96			EXP	TANAKA,M ET AL. (1985)
VP 2.39E-003	mm Hg	25	EST	NEELY,WB & BLAU,GE (1985)
DC	pKa			
HL 5.63E-009	atm m3/mol	25	EST	MEYLAN,WM & HOWARD,PH (1991)
OH 1.12E-011	cm3/molc sec	25	EST	MEYLAN,WM & HOWARD,PH (1993)

CAS #: 000673-66-5				2-AZACYCLOOCTANONE
Formula: $C_7H_{13}NO$				
Mol Weight: 127.19				
MP (deg C): 35-38		FP (deg C):		
BP (deg C): 148-150				
BP pressure (mm Hg): 1.00E+001				

Property/Value	Units	Temp	Data Type	Reference
WS 5.88E+004	mg/L	25	EST	MEYLAN,WM ET AL. (1996)
logP 0.24			EXP	HANSCH,C & LEO,AJ (1985)
VP 8.76E-004	mm Hg	25	EST	NEELY,WB & BLAU,GE (1985)
DC	pKa			
HL 3.36E-008	atm m3/mol	25	EST	MEYLAN,WM & HOWARD,PH (1991)
OH 1.98E-011	cm3/molc sec	25	EST	MEYLAN,WM & HOWARD,PH (1993)

CAS #: 000674-76-0				TRANS-4-METHYL-2-PENTENE
Formula: C_6H_{12}				
Mol Weight: 84.16				
MP (deg C): -140.8		FP (deg C):		
BP (deg C): 58.6				
BP pressure (mm Hg):				

Property/Value	Units	Temp	Data Type	Reference
WS 1.02E+002	mg/L	25	EST	MEYLAN,WM ET AL. (1996)
logP 3.00			EST	MEYLAN,WM & HOWARD,PH (1995)
VP 2.23E+002	mm Hg	25	EXP	DAUBERT,TE & DANNER,RP (1989)
DC	pKa			
HL 4.23E-001	atm m3/mol	25	EST	MEYLAN,WM & HOWARD,PH (1991)
OH 6.08E-011	cm3/molc sec	25	EXP	ATKINSON,R (1989)

CAS #: 000674-82-8				DIKETENE
Formula: $C_4H_4O_2$				
Mol Weight: 84.08				
MP (deg C): -6.5		FP (deg C):		
BP (deg C): 127.4				
BP pressure (mm Hg):				

Property/Value	Units	Temp	Data Type	Reference
WS 5.30E+005	mg/L	25	EST	MEYLAN,WM ET AL. (1996)
logP -0.39			EST	MEYLAN,WM & HOWARD,PH (1995)
VP 1.07E+001	mm Hg	25	EXP	DAUBERT,TE & DANNER,RP (1989)
DC	pKa			
HL 6.07E-004	atm m3/mol	25	EST	MEYLAN,WM & HOWARD,PH (1991)
OH 5.16E-011	cm3/molc sec	25	EST	MEYLAN,WM & HOWARD,PH (1993)

CAS #: 000675-20-7				DELTA-VALEROLACTAM
Formula: C_5H_9NO				
Mol Weight: 99.13				
MP (deg C): 39.5		FP (deg C):		
BP (deg C): 256				
BP pressure (mm Hg):				

Property/Value	Units	Temp	Data Type	Reference
WS 2.91E+005	mg/L	25	EST	MEYLAN,WM ET AL. (1996)
logP -0.46			EXP	HANSCH,C & LEO,AJ (1985)
VP 4.38E-003	mm Hg	25	EXP	JONES,AH (1960)
DC	pKa			
HL 1.91E-008	atm m3/mol	25	EST	MEYLAN,WM & HOWARD,PH (1991)
OH 1.70E-011	cm3/molc sec	25	EST	MEYLAN,WM & HOWARD,PH (1993)

CAS #: 000675-21-8 — PYRIMIDINE, 5-FLUORO-

Formula: $C_4H_3FN_2$

Mol Weight: 98.08

MP (deg C): | FP (deg C):
BP (deg C):
BP pressure (mm Hg):

Property/Value	Units	Temp	Data Type	Reference
WS 1.26E+005	mg/L	25	EST	MEYLAN,WM ET AL. (1996)
logP -0.03			EXP	YAMAGAMI,C ET AL. (1990)
VP 1.64E+001	mm Hg	25	EST	NEELY,WB & BLAU,GE (1985)
DC	pKa			
HL 3.41E-006	atm m3/mol	25	EST	MEYLAN,WM & HOWARD,PH (1991)
OH 2.88E-013	cm3/molc sec	25	EST	MEYLAN,WM & HOWARD,PH (1993)

CAS #: 000677-21-4 — 3,3,3-TRIFLUORO-1-PROPENE

Formula: $C_3H_3F_3$

Mol Weight: 96.05

MP (deg C): | FP (deg C):
BP (deg C): -17
BP pressure (mm Hg):

Property/Value	Units	Temp	Data Type	Reference
WS 1.95E+003	mg/L	25	EST	MEYLAN,WM ET AL. (1996)
logP 2.09			EST	MEYLAN,WM & HOWARD,PH (1995)
VP 4.42E+003	mm Hg	25	EST	NEELY,WB & BLAU,GE (1985)
DC	pKa			
HL 7.62E-001	atm m3/mol	25	EST	MEYLAN,WM & HOWARD,PH (1991)
OH 2.00E-011	cm3/molc sec	25	EST	MEYLAN,WM & HOWARD,PH (1993)

CAS #: 000678-26-2 — PENTANE, DODECAFLUORO-

Formula: C_5F_{12}

Mol Weight: 288.04

MP (deg C): -10 | FP (deg C):
BP (deg C): 29.2
BP pressure (mm Hg):

Property/Value	Units	Temp	Data Type	Reference
WS 4.56E-001	mg/L	25	EST	MEYLAN,WM ET AL. (1996)
logP 4.40			EXP	HANSCH,C ET AL. (1995)
VP 8.67E+003	mm Hg	25	EST	NEELY,WB & BLAU,GE (1985)
DC	pKa			
HL 3.50E+003	atm m3/mol	25	EST	MEYLAN,WM & HOWARD,PH (1991)
OH 0.00E+000	cm3/molc sec	25	EST	MEYLAN,WM & HOWARD,PH (1993)

CAS #: 000680-31-9 — HEXAMETHYLPHOSPHORAMIDE

Formula: $C_6H_{18}N_3OP$

Mol Weight: 179.20

MP (deg C): 7.20 | FP (deg C):
BP (deg C): 233
BP pressure (mm Hg):

Property/Value	Units	Temp	Data Type	Reference
WS 1.00E+006	mg/L		EXP	RIDDICK,JA ET AL. (1986)
logP 0.28			EXP	HANSCH,C & LEO,AJ (1985)
VP 4.60E-002	mm Hg	25	EXP	DAUBERT,TE & DANNER,RP (1989)
DC	pKa			
HL 2.00E-008	atm m3/mol	25	EST	VP/WSOL
OH 1.89E-010	cm3/molc sec	25	EST	MEYLAN,WM & HOWARD,PH (1993)

CAS #: 000681-57-2 — GLUTARIC ACID, 2,2-DIMETHYL-

Formula: $C_7H_{12}O_4$

Mol Weight: 160.17

MP (deg C): 83-85 | FP (deg C):
BP (deg C):
BP pressure (mm Hg):

Property/Value	Units	Temp	Data Type	Reference
WS 4.54E+004	mg/L	25	EST	MEYLAN,WM ET AL. (1996)
logP 0.67			EXP	HANSCH,C ET AL. (1995)
VP 1.84E-004	mm Hg	25	EST	NEELY,WB & BLAU,GE (1985)
DC	pKa			
HL 1.27E-011	atm m3/mol	25	EST	MEYLAN,WM & HOWARD,PH (1991)
OH 3.65E-012	cm3/molc sec	25	EST	MEYLAN,WM & HOWARD,PH (1993)

CAS #: 000683-08-9 — DIETHYL METHYLPHOSPHONATE

Formula: $C_5H_{13}O_3P$

Mol Weight: 152.13

MP (deg C): | FP (deg C):
BP (deg C): 194
BP pressure (mm Hg):

Property/Value	Units	Temp	Data Type	Reference
WS 3.40E+004	mg/L	25	EST	MEYLAN,WM ET AL. (1996)
logP 0.40			EST	MEYLAN,WM & HOWARD,PH (1995)
VP 1.77E-003	mm Hg	25	EXT	BOUBLIK,T ET AL. (1984)
DC	pKa			
HL 2.20E-006	atm m3/mol	25	EST	MEYLAN,WM & HOWARD,PH (1991)
OH 3.88E-011	cm3/molc sec	25	EST	MEYLAN,WM & HOWARD,PH (1993)

CAS #: 000683-57-8 — BROMOACETAMIDE

Formula: C_2H_4BrNO

Mol Weight: 137.97

MP (deg C): 88-90 | FP (deg C):
BP (deg C):
BP pressure (mm Hg):

Property/Value	Units	Temp	Data Type	Reference
WS 2.37E+005	mg/L	25	EST	MEYLAN,WM ET AL. (1996)
logP -0.52			EXP	HANSCH,C & LEO,AJ (1985)
VP 2.36E-002	mm Hg	25	EST	NEELY,WB & BLAU,GE (1985)
DC	pKa			
HL 1.29E-009	atm m3/mol	25	EST	MEYLAN,WM & HOWARD,PH (1991)
OH 2.20E-012	cm3/molc sec	25	EST	MEYLAN,WM & HOWARD,PH (1993)

CAS #: 000683-72-7 — ACETAMIDE, 2,2-DICHLORO-

Formula: $C_2H_3Cl_2NO$

Mol Weight: 127.96

MP (deg C): 99.4 | FP (deg C):
BP (deg C): 234
BP pressure (mm Hg):

Property/Value	Units	Temp	Data Type	Reference
WS 6.45E+004	mg/L	25	EST	MEYLAN,WM ET AL. (1996)
logP 0.19			EXP	HANSCH,C ET AL. (1995)
VP 1.63E-002	mm Hg	25	EST	NEELY,WB & BLAU,GE (1985)
DC	pKa			
HL 1.39E-009	atm m3/mol	25	EST	MEYLAN,WM & HOWARD,PH (1991)
OH 2.21E-012	cm3/molc sec	25	EST	MEYLAN,WM & HOWARD,PH (1993)

CAS #: 000684-16-2				HEXAFLUORO-2-PROPANONE
Formula: C_3F_6O				
Mol Weight: 166.02				
MP (deg C): -125		FP (deg C):		
BP (deg C): -27.4				
BP pressure (mm Hg):				

Property/Value	Units	Temp	Data Type	Reference
WS 3.64E+003	mg/L	25	EST	MEYLAN,WM ET AL. (1996)
logP 1.46			EXP	HANSCH,C ET AL. (1995)
VP 5.00E+000	mm Hg	25	EXP	DAUBERT,TE & DANNER,RP (1989)
DC	pKa			
HL 3.07E-003	atm m3/mol	25	EST	MEYLAN,WM & HOWARD,PH (1991)
OH 0.00E+000	cm3/molc sec	25	EST	MEYLAN,WM & HOWARD,PH (1993)

CAS #: 000684-93-5				1-NITROSO-1-METHYLUREA
Formula: $C_2H_5N_3O_2$				
Mol Weight: 103.08				
MP (deg C): 124		FP (deg C):		
BP (deg C):				
BP pressure (mm Hg):				

Property/Value	Units	Temp	Data Type	Reference
WS 1.44E+004	mg/L	23	EXP	MIRVISH,SS ET AL. (1976)
logP -0.03			EXP	HANSCH,C & LEO,AJ (1985)
VP 1.79E-001	mm Hg	25	EST	NEELY,WB & BLAU,GE (1985)
DC	pKa			
HL 9.91E-011	atm m3/mol	25	EST	MEYLAN,WM & HOWARD,PH (1991)
OH 1.56E-012	cm3/molc sec	25	EST	MEYLAN,WM & HOWARD,PH (1993)

CAS #: 000685-91-6				DIETHYLACETAMIDE
Formula: $C_6H_{13}NO$				
Mol Weight: 115.18				
MP (deg C):		FP (deg C):		
BP (deg C): 185.5				
BP pressure (mm Hg):				

Property/Value	Units	Temp	Data Type	Reference
WS 5.35E+004	mg/L	25	EST	MEYLAN,WM ET AL. (1996)
logP 0.34			EXP	HANSCH,C & LEO,AJ (1985)
VP 8.97E-001	mm Hg	25	EST	NEELY,WB & BLAU,GE (1985)
DC	pKa			
HL 9.50E-008	atm m3/mol	25	EST	MEYLAN,WM & HOWARD,PH (1991)
OH 2.31E-011	cm3/molc sec	25	EST	MEYLAN,WM & HOWARD,PH (1993)

CAS #: 000687-48-9				N,N-DIMETHYLETHYLCARBAMATE
Formula: $C_5H_{11}NO_2$				
Mol Weight: 117.15				
MP (deg C):		FP (deg C):		
BP (deg C):				
BP pressure (mm Hg):				

Property/Value	Units	Temp	Data Type	Reference
WS 1.79E+004	mg/L	25	EST	MEYLAN,WM ET AL. (1996)
logP 0.89			EXP	HANSCH,C ET AL. (1995)
VP 1.24E+001	mm Hg	25	EST	NEELY,WB & BLAU,GE (1985)
DC	pKa			
HL 2.53E-007	atm m3/mol	25	EST	MEYLAN,WM & HOWARD,PH (1991)
OH 1.78E-011	cm3/molc sec	25	EST	MEYLAN,WM & HOWARD,PH (1993)

CAS #: 000688-84-6				2-ETHYLHEXYL 2-METHYL-2-PROPENOATE
Formula: $C_{12}H_{22}O_2$				
Mol Weight: 198.31				
MP (deg C):		FP (deg C):		
BP (deg C): 110				
BP pressure (mm Hg): 1.40E+001				

Property/Value	Units	Temp	Data Type	Reference
WS 5.92E+000	mg/L	25	EST	MEYLAN,WM ET AL. (1996)
logP 4.54			EXP	SANGSTER,J (1993)
VP 7.58E-002	mm Hg	25	EST	NEELY,WB & BLAU,GE (1985)
DC	pKa			
HL 1.05E-003	atm m3/mol	25	EST	MEYLAN,WM & HOWARD,PH (1991)
OH 2.90E-011	cm3/molc sec	25	EST	MEYLAN,WM & HOWARD,PH (1993)

CAS #: 000689-11-2				SEC-BUTYLUREA
Formula: $C_5H_{12}N_2O$				
Mol Weight: 116.16				
MP (deg C):		FP (deg C):		
BP (deg C):				
BP pressure (mm Hg):				

Property/Value	Units	Temp	Data Type	Reference
WS 2.76E+003	mg/L	25	EST	MEYLAN,WM ET AL. (1996)
logP 0.31			EST	MEYLAN,WM & HOWARD,PH (1995)
VP 6.56E-002	mm Hg	25	EST	NEELY,WB & BLAU,GE (1985)
DC	pKa			
HL 1.87E-009	atm m3/mol	25	EST	MEYLAN,WM & HOWARD,PH (1991)
OH 1.33E-011	cm3/molc sec	25	EST	MEYLAN,WM & HOWARD,PH (1993)

CAS #: 000689-97-4				1-BUTEN-3-YNE
Formula: C_4H_4				
Mol Weight: 52.08				
MP (deg C):		FP (deg C):		
BP (deg C): 5.1				
BP pressure (mm Hg):				

Property/Value	Units	Temp	Data Type	Reference
WS 1.79E+003	mg/L	30	EXP	YALKOWSKY,SH & DANNENFELSER,RM (1992)
logP 1.40			EST	MEYLAN,WM & HOWARD,PH (1995)
VP 1.35E+003	mm Hg	25	EXP	YAWS,CL (1994)
DC	pKa			
HL 2.91E-002	atm m3/mol	25	EST	VP/WSOL
OH 4.01E-011	cm3/molc sec	25	EST	MEYLAN,WM & HOWARD,PH (1993)

CAS #: 000690-08-4				4,4-DIMETHYL-2-PENTENE (TRANS)
Formula: C_7H_{14}				
Mol Weight: 98.19				
MP (deg C): -115.2		FP (deg C):		
BP (deg C): 76.7				
BP pressure (mm Hg):				

Property/Value	Units	Temp	Data Type	Reference
WS 3.87E+001	mg/L	25	EST	MEYLAN,WM ET AL. (1996)
logP 3.45			EST	MEYLAN,WM & HOWARD,PH (1995)
VP 1.11E+002	mm Hg	25	EXP	DYKYJ,J & REPAS,M (1973)
DC	pKa			
HL 5.62E-001	atm m3/mol	25	EST	MEYLAN,WM & HOWARD,PH (1991)
OH 5.45E-011	cm3/molc sec	25	EXP	ATKINSON,R (1989)

CAS #: 000691-37-2				4-METHYL-1-PENTENE

Formula: C_6H_{12}

Mol Weight: 84.16

MP (deg C): -153.6

FP (deg C):

BP (deg C): 53.9

BP pressure (mm Hg):

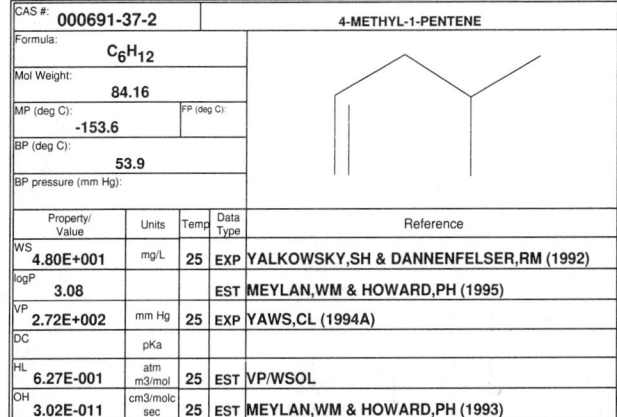

Property/Value	Units	Temp	Data Type	Reference
WS 4.80E+001	mg/L	25	EXP	YALKOWSKY,SH & DANNENFELSER,RM (1992)
logP 3.08			EST	MEYLAN,WM & HOWARD,PH (1995)
VP 2.72E+002	mm Hg	25	EXP	YAWS,CL (1994A)
DC	pKa			
HL 6.27E-001	atm m3/mol	25	EST	VP/WSOL
OH 3.02E-011	cm3/molc sec	25	EST	MEYLAN,WM & HOWARD,PH (1993)

CAS #: 000691-38-3				4-METHYL-2-PENTENE (CIS)

Formula: C_6H_{12}

Mol Weight: 84.16

MP (deg C): -134.8

FP (deg C):

BP (deg C): 56.3

BP pressure (mm Hg):

Property/Value	Units	Temp	Data Type	Reference
WS 1.02E+002	mg/L	25	EST	MEYLAN,WM ET AL. (1996)
logP 3.00			EST	MEYLAN,WM & HOWARD,PH (1995)
VP 2.44E+002	mm Hg	25	EXP	DAUBERT,TE & DANNER,RP (1989)
DC	pKa			
HL 4.23E-001	atm m3/mol	25	EST	MEYLAN,WM & HOWARD,PH (1991)
OH 5.88E-011	cm3/molc sec	25	EST	MEYLAN,WM & HOWARD,PH (1993)

CAS #: 000692-13-7				1-BUTYLBIGUANIDE

Formula: $C_6H_{15}N_5$

Mol Weight: 157.22

MP (deg C):

FP (deg C):

BP (deg C):

BP pressure (mm Hg):

Property/Value	Units	Temp	Data Type	Reference
WS 7.46E+005	mg/L	25	EST	MEYLAN,WM ET AL. (1996)
logP -1.20			EXP	HANSCH,C & LEO,AJ (1985)
VP 1.64E-004	mm Hg	25	EST	NEELY,WB & BLAU,GE (1985)
DC	pKa			
HL 8.14E-016	atm m3/mol	25	EST	MEYLAN,WM & HOWARD,PH (1991)
OH 1.60E-010	cm3/molc sec	25	EST	MEYLAN,WM & HOWARD,PH (1993)

CAS #: 000692-45-5				VINYL FORMATE

Formula: $C_3H_4O_2$

Mol Weight: 72.06

MP (deg C):

FP (deg C):

BP (deg C):

BP pressure (mm Hg):

Property/Value	Units	Temp	Data Type	Reference
WS 9.44E+004	mg/L	25	EST	MEYLAN,WM ET AL. (1996)
logP 0.18			EST	MEYLAN,WM & HOWARD,PH (1995)
VP 3.13E+002	mm Hg	25	EXP	DAUBERT,TE & DANNER,RP (1989)
DC	pKa			
HL 1.60E-003	atm m3/mol	25	EST	MEYLAN,WM & HOWARD,PH (1991)
OH 2.63E-011	cm3/molc sec	25	EST	MEYLAN,WM & HOWARD,PH (1993)

CAS #: 000692-94-4				2-DECENE-4,6,8-TRIYONIC ACID, METHYL ESTER

Formula: $C_{11}H_8O_2$

Mol Weight: 172.19

MP (deg C):

FP (deg C):

BP (deg C):

BP pressure (mm Hg):

Property/Value	Units	Temp	Data Type	Reference
WS 5.15E+002	mg/L	25	EST	MEYLAN,WM ET AL. (1996)
logP 3.85			EXP	MCLACHLAN,D ET AL. (1986)
VP 1.68E-003	mm Hg	25	EST	NEELY,WB & BLAU,GE (1985)
DC	pKa			
HL 1.22E-005	atm m3/mol	25	EST	MEYLAN,WM & HOWARD,PH (1991)
OH 1.37E-010	cm3/molc sec	25	EST	MEYLAN,WM & HOWARD,PH (1993)

CAS #: 000693-02-7				1-HEXYNE

Formula: C_6H_{10}

Mol Weight: 82.15

MP (deg C): -131.9

FP (deg C):

BP (deg C): 71.3

BP pressure (mm Hg):

Property/Value	Units	Temp	Data Type	Reference
WS 3.60E+002	mg/L	25	EXP	YALKOWSKY,SH & DANNENFELSER,RM (1992)
logP 2.73			EXP	HANSCH,C & LEO,AJ (1985)
VP 1.33E+002	mm Hg	25	EXP	YAWS,CL (1994A)
DC	pKa			
HL 3.99E-001	atm m3/mol	25	EST	VP/WSOL
OH 1.19E-011	cm3/molc sec	25	EXP	ATKINSON,R (1989)

CAS #: 000693-13-0				N,N'-METHANETETRAYLBIS-2-PROPANAMINE

Formula: $C_7H_{14}N_2$

Mol Weight: 126.20

MP (deg C):

FP (deg C):

BP (deg C): 147

BP pressure (mm Hg):

Property/Value	Units	Temp	Data Type	Reference
WS 2.93E+001	mg/L	25	EST	MEYLAN,WM ET AL. (1996)
logP 4.11			EST	MEYLAN,WM & HOWARD,PH (1995)
VP 9.79E+000	mm Hg	25	EST	NEELY,WB & BLAU,GE (1985)
DC	pKa			
HL 9.92E-004	atm m3/mol	25	EST	MEYLAN,WM & HOWARD,PH (1991)
OH 4.55E-012	cm3/molc sec	25	EST	MEYLAN,WM & HOWARD,PH (1993)

CAS #: 000693-30-1				2-((2-CHLOROETHYL)THIO)ETHANOL

Formula: C_4H_9ClOS

Mol Weight: 140.63

MP (deg C):

FP (deg C):

BP (deg C): 44.5

BP pressure (mm Hg): 6.00E+000

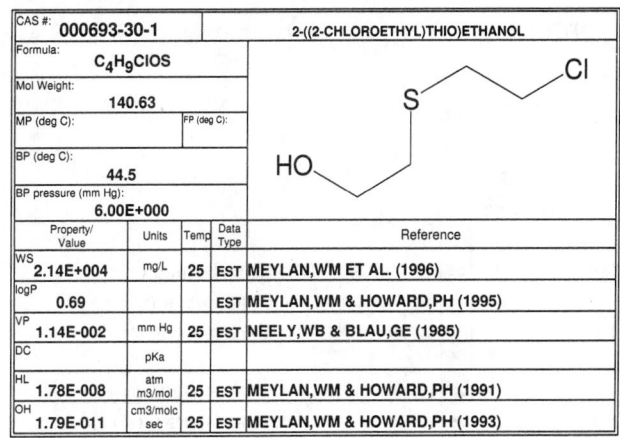

Property/Value	Units	Temp	Data Type	Reference
WS 2.14E+004	mg/L	25	EST	MEYLAN,WM ET AL. (1996)
logP 0.69			EST	MEYLAN,WM & HOWARD,PH (1995)
VP 1.14E-002	mm Hg	25	EST	NEELY,WB & BLAU,GE (1985)
DC	pKa			
HL 1.78E-008	atm m3/mol	25	EST	MEYLAN,WM & HOWARD,PH (1991)
OH 1.79E-011	cm3/molc sec	25	EST	MEYLAN,WM & HOWARD,PH (1993)

CAS #: 000693-54-9				2-DECANONE

Formula: $C_{10}H_{20}O$

Mol Weight: 156.27

MP (deg C): 14   FP (deg C):

BP (deg C): 210

BP pressure (mm Hg):

Property/Value	Units	Temp	Data Type	Reference
WS 7.85E+001	mg/L	25	EXP	WASIK,SP ET AL. (1981)
logP 3.73			EXP	TANII,H ET AL. (1986)
VP 2.69E-001	mm Hg	25	EXT	PERRY,RH & GREEN,D (1984)
DC	pKa			
HL 3.60E-004	atm m3/mol	25	EST	MEYLAN,WM & HOWARD,PH (1991)
OH 1.32E-011	cm3/molc sec	25	EXP	ATKINSON,R (1989)

CAS #: 000693-58-3				1-BROMO-N-NONANE

Formula: $C_9H_{19}Br$

Mol Weight: 207.16

MP (deg C):   FP (deg C):

BP (deg C): 88

BP pressure (mm Hg): 4.00E+000

Property/Value	Units	Temp	Data Type	Reference
WS 7.45E-001	mg/L	25	EST	MUELLER,M & KLEIN,W (1992)
logP 5.10			EST	MEYLAN,WM & HOWARD,PH (1995)
VP 1.20E-001	mm Hg	25	EXT	LI,JCM & ROSSINI,FD (1953)
DC	pKa			
HL 8.19E-002	atm m3/mol	25	EST	MEYLAN,WM & HOWARD,PH (1991)
OH 9.23E-012	cm3/molc sec	25	EST	MEYLAN,WM & HOWARD,PH (1993)

CAS #: 000693-65-2				DI-N-PENTYLETHER

Formula: $C_{10}H_{22}O$

Mol Weight: 158.29

MP (deg C): -69.43   FP (deg C):

BP (deg C): 186.75

BP pressure (mm Hg):

Property/Value	Units	Temp	Data Type	Reference
WS 2.70E+001	mg/L	25	EST	MEYLAN,WM ET AL. (1996)
logP 4.00			EST	MEYLAN,WM & HOWARD,PH (1995)
VP 8.57E-001	mm Hg	25	EXP	YAWS,CL (1994B)
DC	pKa			
HL 8.32E-003	atm m3/mol	25	EST	MEYLAN,WM & HOWARD,PH (1991)
OH 3.47E-011	cm3/molc sec	25	EXP	ATKINSON,R (1989)

CAS #: 000693-67-4				1-BROMOUNDECANE

Formula: $C_{11}H_{23}Br$

Mol Weight: 235.21

MP (deg C): -9   FP (deg C):

BP (deg C): 137-138

BP pressure (mm Hg): 1.80E+001

Property/Value	Units	Temp	Data Type	Reference
WS 1.81E-001	mg/L	25	EST	MEYLAN,WM ET AL. (1996)
logP 6.09			EST	MEYLAN,WM & HOWARD,PH (1995)
VP 1.21E-002	mm Hg	25	EXT	LI,JCM & ROSSINI,FD (1953)
DC	pKa			
HL 1.44E-001	atm m3/mol	25	EST	MEYLAN,WM & HOWARD,PH (1991)
OH 1.21E-011	cm3/molc sec	25	EST	MEYLAN,WM & HOWARD,PH (1993)

CAS #: 000693-95-8				4-METHYLTHIAZOLE

Formula: C_4H_5NS

Mol Weight: 99.16

MP (deg C):   FP (deg C):

BP (deg C): 133.3

BP pressure (mm Hg):

Property/Value	Units	Temp	Data Type	Reference
WS 1.75E+004	mg/L	25	EST	MEYLAN,WM ET AL. (1996)
logP 0.97			EXP	HANSCH,C ET AL. (1995)
VP 8.60E+000	mm Hg	25	EXP	SOULIE,MA ET AL. (1975)
DC	pKa			
HL 4.23E-006	atm m3/mol	25	EST	MEYLAN,WM & HOWARD,PH (1991)
OH 3.79E-012	cm3/molc sec	25	EST	MEYLAN,WM & HOWARD,PH (1993)

CAS #: 000693-98-1				1H-IMIDAZOLE, 2-METHYL-

Formula: $C_4H_6N_2$

Mol Weight: 82.11

MP (deg C): 144   FP (deg C):

BP (deg C): 267

BP pressure (mm Hg):

Property/Value	Units	Temp	Data Type	Reference
WS 8.09E+004	mg/L	25	EST	MEYLAN,WM ET AL. (1996)
logP 0.24			EXP	HANSCH,C ET AL. (1995)
VP 1.45E-002	mm Hg	25	EST	NEELY,WB & BLAU,GE (1985)
DC 7.85	pKa	25	EXP	PERRIN,DD (1965)
HL 4.14E-006	atm m3/mol	25	EST	MEYLAN,WM & HOWARD,PH (1991)
OH 9.41E-011	cm3/molc sec	25	EST	MEYLAN,WM & HOWARD,PH (1993)

CAS #: 000694-59-7				PYRIDINE, 1-OXIDE

Formula: C_5H_5NO

Mol Weight: 95.10

MP (deg C): 66   FP (deg C):

BP (deg C): 100-105

BP pressure (mm Hg): 1.00E+000

Property/Value	Units	Temp	Data Type	Reference
WS 1.00E+006	mg/L	25	EST	MEYLAN,WM ET AL. (1996)
logP -1.20			EXP	HANSCH,C ET AL. (1995)
VP 1.90E-001	mm Hg	25	EST	NEELY,WB & BLAU,GE (1985)
DC	pKa			
HL 7.05E-011	atm m3/mol	25	EST	MEYLAN,WM & HOWARD,PH (1991)
OH 3.70E-013	cm3/molc sec	25	EST	MEYLAN,WM & HOWARD,PH (1993)

CAS #: 000694-80-4				2-BROMOCHLOROBENZENE

Formula: C_6H_4BrCl

Mol Weight: 191.46

MP (deg C): -12.3   FP (deg C):

BP (deg C): 204

BP pressure (mm Hg):

Property/Value	Units	Temp	Data Type	Reference
WS 4.49E+001	mg/L	25	EXP	SUZUKI,T (1991)
logP 3.53			EST	MEYLAN,WM & HOWARD,PH (1995)
VP 4.41E-001	mm Hg	25	EST	NEELY,WB & BLAU,GE (1985)
DC	pKa			
HL 1.59E-003	atm m3/mol	25	EST	MEYLAN,WM & HOWARD,PH (1991)
OH 3.93E-013	cm3/molc sec	25	EST	MEYLAN,WM & HOWARD,PH (1993)

CAS #: 000694-83-7	1,2-CYCLOHEXANEDIAMINE

Formula: $C_6H_{14}N_2$
Mol Weight: 114.19
MP (deg C):   FP (deg C):
BP (deg C): 92-93
BP pressure (mm Hg): 1.80E+001

Property/Value	Units	Temp	Data Type	Reference
WS 9.04E+005	mg/L	25	EST	MEYLAN,WM ET AL. (1996)
logP 0.09			EST	MEYLAN,WM & HOWARD,PH (1995)
VP 4.07E-001	mm Hg	25	EST	NEELY,WB & BLAU,GE (1985)
DC 10.24	pKa	10	EXP	PERRIN,DD (1965)
HL 1.42E-009	atm m3/mol	25	EST	MEYLAN,WM & HOWARD,PH (1991)
OH 1.02E-010	cm3/molc sec	25	EST	MEYLAN,WM & HOWARD,PH (1993)

CAS #: 000694-85-9	2(1H)-PYRIDINONE, 1-METHYL-

Formula: C_6H_7NO
Mol Weight: 109.13
MP (deg C): 31   FP (deg C):
BP (deg C): 250
BP pressure (mm Hg):

Property/Value	Units	Temp	Data Type	Reference
WS 1.72E+005	mg/L	25	EST	MEYLAN,WM ET AL. (1996)
logP -0.23			EXP	HANSCH,C ET AL. (1995)
VP 6.23E-002	mm Hg	25	EST	NEELY,WB & BLAU,GE (1985)
DC 0.32	pKa	20	EXP	PERRIN,DD (1965)
HL 2.49E-008	atm m3/mol	25	EST	MEYLAN,WM & HOWARD,PH (1991)
OH 4.54E-011	cm3/molc sec	25	EST	MEYLAN,WM & HOWARD,PH (1993)

CAS #: 000695-19-2	4(1H)-PYRIDINONE, 1-METHYL-

Formula: C_6H_7NO
Mol Weight: 109.13
MP (deg C):   FP (deg C):
BP (deg C):
BP pressure (mm Hg):

Property/Value	Units	Temp	Data Type	Reference
WS 1.00E+006	mg/L	25	EST	MEYLAN,WM ET AL. (1996)
logP -1.22			EXP	HANSCH,C ET AL. (1995)
VP 7.44E-001	mm Hg	25	EST	NEELY,WB & BLAU,GE (1985)
DC 3.29	pKa	20	EXP	PERRIN,DD (1965)
HL 9.92E-009	atm m3/mol	25	EST	MEYLAN,WM & HOWARD,PH (1991)
OH 8.08E-011	cm3/molc sec	25	EST	MEYLAN,WM & HOWARD,PH (1993)

CAS #: 000695-94-3	1-(2-OHET)-2-ME IMIDAZOLINE

Formula: $C_6H_{12}N_2O$
Mol Weight: 128.18
MP (deg C):   FP (deg C):
BP (deg C):
BP pressure (mm Hg):

Property/Value	Units	Temp	Data Type	Reference
WS 8.64E+004	mg/L	25	EST	MEYLAN,WM ET AL. (1996)
logP 0.04			EXP	HANSCH,C & LEO,AJ (1985)
VP 7.45E-003	mm Hg	25	EST	NEELY,WB & BLAU,GE (1985)
DC	pKa			
HL 6.84E-011	atm m3/mol	25	EST	MEYLAN,WM & HOWARD,PH (1991)
OH 9.05E-011	cm3/molc sec	25	EST	MEYLAN,WM & HOWARD,PH (1993)

CAS #: 000696-07-1	2,4(1H,3H)-PYRIMIDINEDIONE, 5-IODO-

Formula: $C_4H_3IN_2O_2$
Mol Weight: 237.99
MP (deg C): 274-276 de   FP (deg C):
BP (deg C):
BP pressure (mm Hg):

Property/Value	Units	Temp	Data Type	Reference
WS 1.25E+003	mg/L	25	EST	MEYLAN,WM ET AL. (1996)
logP 0.04			EXP	HANSCH,C ET AL. (1995)
VP 7.09E-008	mm Hg	25	EST	NEELY,WB & BLAU,GE (1985)
DC	pKa			
HL 1.09E-011	atm m3/mol	25	EST	MEYLAN,WM & HOWARD,PH (1991)
OH 6.74E-012	cm3/molc sec	25	EST	MEYLAN,WM & HOWARD,PH (1993)

CAS #: 000696-23-1	2-METHYL-4-NITRO-1H-IMIDAZOLE

Formula: $C_4H_5N_3O_2$
Mol Weight: 127.10
MP (deg C):   FP (deg C):
BP (deg C):
BP pressure (mm Hg):

Property/Value	Units	Temp	Data Type	Reference
WS 2.01E+004	mg/L	25	EST	MEYLAN,WM ET AL. (1996)
logP 0.42			EST	MEYLAN,WM & HOWARD,PH (1995)
VP 6.48E-005	mm Hg	25	EST	NEELY,WB & BLAU,GE (1985)
DC	pKa			
HL 1.64E-008	atm m3/mol	25	EST	MEYLAN,WM & HOWARD,PH (1991)
OH 3.99E-012	cm3/molc sec	25	EST	MEYLAN,WM & HOWARD,PH (1993)

CAS #: 000696-29-7	ISOPROPYLCYCLOHEXANE

Formula: C_9H_{18}
Mol Weight: 126.24
MP (deg C): -89.4   FP (deg C):
BP (deg C): 154.8
BP pressure (mm Hg):

Property/Value	Units	Temp	Data Type	Reference
WS 3.95E+000	mg/L	25	EST	MEYLAN,WM ET AL. (1996)
logP 4.50			EST	MEYLAN,WM & HOWARD,PH (1995)
VP 4.80E+000	mm Hg	25	EXP	YAWS,CL (1994B)
DC	pKa			
HL 5.97E-001	atm m3/mol	25	EST	MEYLAN,WM & HOWARD,PH (1991)
OH 1.34E-011	cm3/molc sec	25	EST	MEYLAN,WM & HOWARD,PH (1993)

CAS #: 000696-44-6	BENZENAMINE, N,3-DIMETHYL-

Formula: $C_8H_{11}N$
Mol Weight: 121.18
MP (deg C):   FP (deg C):
BP (deg C): 206-207
BP pressure (mm Hg):

Property/Value	Units	Temp	Data Type	Reference
WS 1.08E+003	mg/L	25	EXP	CHEM INSPECT TEST INST (1992)
logP 2.19			EXP	CHEM INSPECT TEST INST (1992)
VP 6.50E-001	mm Hg	25	EST	NEELY,WB & BLAU,GE (1985)
DC 5.00	pKa	21	EXP	PERRIN,DD (1965)
HL 4.61E-006	atm m3/mol	25	EST	MEYLAN,WM & HOWARD,PH (1991)
OH 1.13E-010	cm3/molc sec	25	EST	MEYLAN,WM & HOWARD,PH (1993)

CAS #: 000696-54-8 — 4-PYRIDINEALDOXIME

Formula: $C_6H_6N_2O$

Mol Weight: 122.13

MP (deg C): 130-133

FP (deg C):

BP (deg C):

BP pressure (mm Hg):

Property/Value	Units	Temp	Data Type	Reference
WS 4.33E+005	mg/L	25	EST	MEYLAN,WM ET AL. (1996)
logP 0.77			EXP	HANSCH,C ET AL. (1995)
VP 1.83E-003	mm Hg	25	EST	NEELY,WB & BLAU,GE (1985)
DC 4.73	pKa	20	EXP	PERRIN,DD (1965)
HL 4.68E-010	atm m3/mol	25	EST	MEYLAN,WM & HOWARD,PH (1991)
OH 2.82E-012	cm3/molc sec	25	EST	MEYLAN,WM & HOWARD,PH (1993)

CAS #: 000697-82-5 — 2,3,5-TRIMETHYLPHENOL

Formula: $C_9H_{12}O$

Mol Weight: 136.20

MP (deg C): 92-95

FP (deg C):

BP (deg C): 230-231

BP pressure (mm Hg):

Property/Value	Units	Temp	Data Type	Reference
WS 7.62E+002	mg/L	25	EXP	SHIU,WY ET AL. (1994)
logP 3.15			EST	MEYLAN,WM & HOWARD,PH (1995)
VP 3.69E-005	mm Hg	25	EXT	BOUBLIK,T ET AL. (1984)
DC 10.67	pKa	25	EXP	SERJEANT,EP & DEMPSEY,B (1979)
HL 7.53E-007	atm m3/mol	25	EST	MEYLAN,WM & HOWARD,PH (1991)
OH 2.01E-010	cm3/molc sec	25	EST	MEYLAN,WM & HOWARD,PH (1993)

CAS #: 000698-63-5 — 5-NITRO-2-FURALDEHYDE

Formula: $C_5H_3NO_4$

Mol Weight: 141.08

MP (deg C):

FP (deg C):

BP (deg C): 228

BP pressure (mm Hg):

Property/Value	Units	Temp	Data Type	Reference
WS 4.63E+003	mg/L	25	EST	MEYLAN,WM ET AL. (1996)
logP 1.01			EXP	HANSCH,C ET AL. (1995)
VP 2.45E-002	mm Hg	25	EST	NEELY,WB & BLAU,GE (1985)
DC	pKa			
HL 5.29E-008	atm m3/mol	25	EST	MEYLAN,WM & HOWARD,PH (1991)
OH 1.94E-011	cm3/molc sec	25	EST	MEYLAN,WM & HOWARD,PH (1993)

CAS #: 000698-67-9 — 4-BROMOBENZAMIDE

Formula: C_7H_6BrNO

Mol Weight: 200.04

MP (deg C): 190-193

FP (deg C):

BP (deg C):

BP pressure (mm Hg):

Property/Value	Units	Temp	Data Type	Reference
WS 1.37E+003	mg/L	25	EST	MEYLAN,WM ET AL. (1996)
logP 1.76			EXP	HANSCH,C & LEO,AJ (1985)
VP 5.64E-005	mm Hg	25	EST	NEELY,WB & BLAU,GE (1985)
DC	pKa			
HL 8.82E-010	atm m3/mol	25	EST	MEYLAN,WM & HOWARD,PH (1991)
OH 3.12E-012	cm3/molc sec	25	EST	MEYLAN,WM & HOWARD,PH (1993)

CAS #: 000698-71-5 — 3-ETHYL-5-METHYLPHENOL

Formula: $C_9H_{12}O$

Mol Weight: 136.20

MP (deg C):

FP (deg C):

BP (deg C):

BP pressure (mm Hg):

Property/Value	Units	Temp	Data Type	Reference
WS 7.46E+002	mg/L	25	EST	MEYLAN,WM ET AL. (1996)
logP 3.10			EST	MEYLAN,WM & HOWARD,PH (1995)
VP 1.17E+000	mm Hg	25	EXP	SHIU,WY ET AL. (1994)
DC 10.10	pKa	28	EXP	KORTUM,G ET AL (1961)
HL 9.06E-007	atm m3/mol	25	EST	MEYLAN,WM & HOWARD,PH (1991)
OH 2.01E-010	cm3/molc sec	25	EST	MEYLAN,WM & HOWARD,PH (1993)

CAS #: 000698-87-3 — BENZENEETHANOL, .ALPHA.-METHYL-

Formula: $C_9H_{12}O$

Mol Weight: 136.20

MP (deg C):

FP (deg C):

BP (deg C): 125

BP pressure (mm Hg): 2.50E+001

Property/Value	Units	Temp	Data Type	Reference
WS 5.84E+003	mg/L	25	EST	MEYLAN,WM ET AL. (1996)
logP 1.97			EXP	HANSCH,C ET AL. (1995)
VP 1.47E-002	mm Hg	25	EST	NEELY,WB & BLAU,GE (1985)
DC	pKa			
HL 3.83E-007	atm m3/mol	25	EST	MEYLAN,WM & HOWARD,PH (1991)
OH 1.47E-011	cm3/molc sec	25	EST	MEYLAN,WM & HOWARD,PH (1993)

CAS #: 000699-18-3 — 2-(B-NITROVINYL)FURAN

Formula: $C_6H_6NO_3$

Mol Weight: 140.12

MP (deg C):

FP (deg C):

BP (deg C):

BP pressure (mm Hg):

Property/Value	Units	Temp	Data Type	Reference
WS 1.60E+003	mg/L	25	EST	MEYLAN,WM ET AL. (1996)
logP 1.56			EXP	HANSCH,C & LEO,AJ (1985)
VP 1.11E-001	mm Hg	25	EST	NEELY,WB & BLAU,GE (1985)
DC	pKa			
HL 3.47E-006	atm m3/mol	25	EST	MEYLAN,WM & HOWARD,PH (1991)
OH 4.95E-011	cm3/molc sec	25	EST	MEYLAN,WM & HOWARD,PH (1993)

CAS #: 000700-00-5 — 9H-PURIN-6-AMINE, 9-METHYL-

Formula: $C_6H_7N_5$

Mol Weight: 149.16

MP (deg C):

FP (deg C):

BP (deg C):

BP pressure (mm Hg):

Property/Value	Units	Temp	Data Type	Reference
WS 3.98E+003	mg/L	25	EST	MEYLAN,WM ET AL. (1996)
logP -0.03			EXP	LAM,SP ET AL. (1989)
VP 5.57E-006	mm Hg	25	EST	NEELY,WB & BLAU,GE (1985)
DC	pKa			
HL 1.50E-012	atm m3/mol	25	EST	MEYLAN,WM & HOWARD,PH (1991)
OH 2.00E-010	cm3/molc sec	25	EST	MEYLAN,WM & HOWARD,PH (1993)

CAS #: 000700-12-9				PENTAMETHYLBENZENE

Formula: $C_{11}H_{16}$

Mol Weight: 148.25

MP (deg C): 54.5 **FP (deg C):**

BP (deg C): 232

BP pressure (mm Hg):

Property/Value	Units	Temp	Data Type	Reference
WS 1.55E+001	mg/L		EXP	YALKOWSKY,SH & DANNENFELSER,RM (1992)
logP 4.56			EXP	HANSCH,C & LEO,AJ (1985)
VP 3.48E-002	mm Hg	20	EXP	RUZICKA,V ET AL. (1994)
DC	pKa			
HL 4.38E-004	atm m3/mol	20	EST	VP/WSOL
OH 5.28E-011	cm3/molc sec	25	EST	MEYLAN,WM & HOWARD,PH (1993)

CAS #: 000700-38-9				5-METHYL-2-NITROPHENOL

Formula: $C_7H_7NO_3$

Mol Weight: 153.14

MP (deg C): 53-56 **FP (deg C):**

BP (deg C):

BP pressure (mm Hg):

Property/Value	Units	Temp	Data Type	Reference
WS 2.72E+002	mg/L	20	EXP	SCHWARZENBACH,RP ET AL. (1988)
logP 2.31			EXP	SCHWARZENBACH,RP ET AL. (1988)
VP 8.86E-005	mm Hg	20	EXP	SCHWARZENBACH,RP ET AL. (1998)
DC 7.41	pKa	25	EXP	SERJEANT,EP & DEMPSEY,B (1979)
HL 6.56E-008	atm m3/mol	20	EST	VP/WSOL
OH 1.11E-011	cm3/molc sec	25	EST	MEYLAN,WM & HOWARD,PH (1993)

CAS #: 000700-75-4				N,N-DIMETHYL O-ANISIDINE

Formula: $C_9H_{13}NO$

Mol Weight: 151.21

MP (deg C): **FP (deg C):**

BP (deg C): 210

BP pressure (mm Hg):

Property/Value	Units	Temp	Data Type	Reference
WS 3.04E+003	mg/L	25	EST	MEYLAN,WM ET AL. (1996)
logP 1.63			EXP	HANSCH,C & LEO,AJ (1985)
VP 2.09E-001	mm Hg	25	EST	NEELY,WB & BLAU,GE (1985)
DC	pKa			
HL 5.07E-006	atm m3/mol	25	EST	MEYLAN,WM & HOWARD,PH (1991)
OH 2.03E-010	cm3/molc sec	25	EST	MEYLAN,WM & HOWARD,PH (1993)

CAS #: 000701-34-8				4-BROMOBENZENESULFONAMIDE

Formula: $C_6H_6BrNO_2S$

Mol Weight: 236.09

MP (deg C): **FP (deg C):**

BP (deg C):

BP pressure (mm Hg):

Property/Value	Units	Temp	Data Type	Reference
WS 1.95E+003	mg/L	25	EST	MEYLAN,WM ET AL. (1996)
logP 1.36			EXP	HANSCH,C & LEO,AJ (1985)
VP 5.48E-005	mm Hg	25	EST	NEELY,WB & BLAU,GE (1985)
DC	pKa			
HL 1.68E-007	atm m3/mol	25	EST	MEYLAN,WM & HOWARD,PH (1991)
OH 2.62E-013	cm3/molc sec	25	EST	MEYLAN,WM & HOWARD,PH (1993)

CAS #: 000701-56-4				N,N-DIMETHYL P-ANISIDINE

Formula: $C_9H_{13}NO$

Mol Weight: 151.21

MP (deg C): **FP (deg C):**

BP (deg C):

BP pressure (mm Hg):

Property/Value	Units	Temp	Data Type	Reference
WS 4.59E+003	mg/L	25	EST	MEYLAN,WM ET AL. (1996)
logP 1.42			EXP	HANSCH,C & LEO,AJ (1985)
VP 2.09E-001	mm Hg	25	EST	NEELY,WB & BLAU,GE (1985)
DC 5.85	pKa	25	EXP	PERRIN,DD (1965)
HL 5.07E-006	atm m3/mol	25	EST	MEYLAN,WM & HOWARD,PH (1991)
OH 2.03E-010	cm3/molc sec	25	EST	MEYLAN,WM & HOWARD,PH (1993)

CAS #: 000701-65-5				TRICHLOROMETHYLTHIOBENZENE

Formula: $C_7H_5Cl_3S$

Mol Weight: 227.54

MP (deg C): **FP (deg C):**

BP (deg C):

BP pressure (mm Hg):

Property/Value	Units	Temp	Data Type	Reference
WS 1.86E+001	mg/L	25	EST	MEYLAN,WM ET AL. (1996)
logP 3.78			EXP	HANSCH,C & LEO,AJ (1985)
VP 1.17E-002	mm Hg	25	EST	NEELY,WB & BLAU,GE (1985)
DC	pKa			
HL 6.86E-006	atm m3/mol	25	EST	MEYLAN,WM & HOWARD,PH (1991)
OH 1.26E-011	cm3/molc sec	25	EST	MEYLAN,WM & HOWARD,PH (1993)

CAS #: 000701-83-7				3-FLUOROPHENYL ACETATE

Formula: $C_8H_7FO_2$

Mol Weight: 154.14

MP (deg C): **FP (deg C):**

BP (deg C):

BP pressure (mm Hg):

Property/Value	Units	Temp	Data Type	Reference
WS 2.38E+003	mg/L	25	EST	MEYLAN,WM ET AL. (1996)
logP 1.74			EXP	HANSCH,C & LEO,AJ (1985)
VP 5.56E-001	mm Hg	25	EST	NEELY,WB & BLAU,GE (1985)
DC	pKa			
HL 7.56E-005	atm m3/mol	25	EST	MEYLAN,WM & HOWARD,PH (1991)
OH 2.62E-012	cm3/molc sec	25	EST	MEYLAN,WM & HOWARD,PH (1993)

CAS #: 000704-73-4				N-METHYL-2-FLUOROPHENYLCARBAMATE

Formula: $C_8H_8FNO_2$

Mol Weight: 169.16

MP (deg C): **FP (deg C):**

BP (deg C):

BP pressure (mm Hg):

Property/Value	Units	Temp	Data Type	Reference
WS 5.31E+003	mg/L	25	EST	MEYLAN,WM ET AL. (1996)
logP 1.25			EXP	HANSCH,C & LEO,AJ (1985)
VP 7.45E-002	mm Hg	25	EST	NEELY,WB & BLAU,GE (1985)
DC	pKa			
HL 3.75E-008	atm m3/mol	25	EST	MEYLAN,WM & HOWARD,PH (1991)
OH 7.85E-012	cm3/molc sec	25	EST	MEYLAN,WM & HOWARD,PH (1993)

CAS #: 000705-19-1				1-(2-OHET)-2-ME-4-NO2 IMIDAZOLE
Formula: $C_6H_{11}N_3O_3$				
Mol Weight: 173.17				
MP (deg C):		FP (deg C):		
BP (deg C):				
BP pressure (mm Hg):				

Property/Value	Units	Temp	Data Type	Reference
WS 7.89E+004	mg/L	25	EST	MEYLAN,WM ET AL. (1996)
logP -0.59			EXP	HANSCH,C & LEO,AJ (1985)
VP 3.05E-007	mm Hg	25	EST	NEELY,WB & BLAU,GE (1985)
DC	pKa			
HL 1.69E-011	atm m3/mol	25	EST	MEYLAN,WM & HOWARD,PH (1991)
OH 9.30E-012	cm3/molc sec	25	EST	MEYLAN,WM & HOWARD,PH (1993)

CAS #: 000705-48-6				N-METHYL-3-FLUOROPHENYLCARBAMATE
Formula: $C_8H_8FNO_2$				
Mol Weight: 169.16				
MP (deg C):		FP (deg C):		
BP (deg C):				
BP pressure (mm Hg):				

Property/Value	Units	Temp	Data Type	Reference
WS 3.38E+003	mg/L	25	EST	MEYLAN,WM ET AL. (1996)
logP 1.48			EXP	HANSCH,C & LEO,AJ (1985)
VP 7.45E-002	mm Hg	25	EST	NEELY,WB & BLAU,GE (1985)
DC	pKa			
HL 3.75E-008	atm m3/mol	25	EST	MEYLAN,WM & HOWARD,PH (1991)
OH 8.58E-012	cm3/molc sec	25	EST	MEYLAN,WM & HOWARD,PH (1993)

CAS #: 000705-60-2				B-METHYL B-NITROSTYRENE
Formula: $C_9H_9NO_2$				
Mol Weight: 163.18				
MP (deg C):		FP (deg C):		
BP (deg C):				
BP pressure (mm Hg):				

Property/Value	Units	Temp	Data Type	Reference
WS 1.90E+002	mg/L	25	EST	MEYLAN,WM ET AL. (1996)
logP 2.52			EXP	HANSCH,C & LEO,AJ (1985)
VP 7.68E-003	mm Hg	25	EST	NEELY,WB & BLAU,GE (1985)
DC	pKa			
HL 5.45E-006	atm m3/mol	25	EST	MEYLAN,WM & HOWARD,PH (1991)
OH 1.95E-011	cm3/molc sec	25	EST	MEYLAN,WM & HOWARD,PH (1993)

CAS #: 000705-62-4				THIOUREA, N,N-DIMETHYL-N'-PHENYL-
Formula: $C_9H_{12}N_2S$				
Mol Weight: 180.27				
MP (deg C):		FP (deg C):		
BP (deg C):				
BP pressure (mm Hg):				

Property/Value	Units	Temp	Data Type	Reference
WS 7.23E+004	mg/L	25	EST	MEYLAN,WM ET AL. (1996)
logP 1.04			EXP	HANSCH,C ET AL. (1995)
VP 1.61E-003	mm Hg	25	EST	NEELY,WB & BLAU,GE (1985)
DC	pKa			
HL 4.20E-007	atm m3/mol	25	EST	MEYLAN,WM & HOWARD,PH (1991)
OH 1.11E-010	cm3/molc sec	25	EST	MEYLAN,WM & HOWARD,PH (1993)

CAS #: 000705-70-4				N-METHYL-4-FLUOROPHENYLCARBAMATE
Formula: $C_8H_8FNO_2$				
Mol Weight: 169.16				
MP (deg C):		FP (deg C):		
BP (deg C):				
BP pressure (mm Hg):				

Property/Value	Units	Temp	Data Type	Reference
WS 5.01E+003	mg/L	25	EST	MEYLAN,WM ET AL. (1996)
logP 1.28			EXP	HANSCH,C & LEO,AJ (1985)
VP 7.45E-002	mm Hg	25	EST	NEELY,WB & BLAU,GE (1985)
DC	pKa			
HL 3.75E-008	atm m3/mol	25	EST	MEYLAN,WM & HOWARD,PH (1991)
OH 7.85E-012	cm3/molc sec	25	EST	MEYLAN,WM & HOWARD,PH (1993)

CAS #: 000706-07-0				4-CHLORO B-NITROSTYRENE
Formula: $C_8H_6ClNO_2$				
Mol Weight: 183.60				
MP (deg C):		FP (deg C):		
BP (deg C):				
BP pressure (mm Hg):				

Property/Value	Units	Temp	Data Type	Reference
WS 1.78E+002	mg/L	25	EST	MEYLAN,WM ET AL. (1996)
logP 2.44			EXP	HANSCH,C & LEO,AJ (1985)
VP 2.53E-003	mm Hg	25	EST	NEELY,WB & BLAU,GE (1985)
DC	pKa			
HL 2.57E-006	atm m3/mol	25	EST	MEYLAN,WM & HOWARD,PH (1991)
OH 1.27E-011	cm3/molc sec	25	EST	MEYLAN,WM & HOWARD,PH (1993)

CAS #: 000706-14-9				GAMMA-DECALACTONE
Formula: $C_{10}H_{18}O_2$				
Mol Weight: 170.25				
MP (deg C):		FP (deg C):		
BP (deg C):				
BP pressure (mm Hg):				

Property/Value	Units	Temp	Data Type	Reference
WS 2.92E+002	mg/L	25	EST	MEYLAN,WM ET AL. (1996)
logP 2.72			EXP	HANSCH,C & LEO,AJ (1985)
VP 5.12E-003	mm Hg	25	EST	NEELY,WB & BLAU,GE (1985)
DC	pKa			
HL 5.62E-004	atm m3/mol	25	EST	MEYLAN,WM & HOWARD,PH (1991)
OH 1.25E-011	cm3/molc sec	25	EST	MEYLAN,WM & HOWARD,PH (1993)

CAS #: 000706-78-5				OCTACHLOROCYCLOPENTENE
Formula: C_5Cl_8				
Mol Weight: 343.68				
MP (deg C): 40		FP (deg C):		
BP (deg C): 283				
BP pressure (mm Hg):				

Property/Value	Units	Temp	Data Type	Reference
WS 1.00E-001	mg/L		EXP	TALIAN,SF ET AL. (1986)
logP 4.66			EST	MEYLAN,WM & HOWARD,PH (1995)
VP 2.18E-003	mm Hg	25	EST	NEELY,WB & BLAU,GE (1985)
DC	pKa			
HL 8.77E-005	atm m3/mol	25	EST	MEYLAN,WM & HOWARD,PH (1991)
OH 2.80E-012	cm3/molc sec	25	EST	MEYLAN,WM & HOWARD,PH (1993)

CAS #: 000707-07-3				TRIMETHOXYMETHYLBENZENE

Formula: $C_{10}H_{14}O_3$

Mol Weight: 182.22

MP (deg C): | FP (deg C):

BP (deg C): 87-88

BP pressure (mm Hg): 7.00E+000

Property/Value	Units	Temp	Data Type	Reference
WS 6.19E+002	mg/L	25	EST	MEYLAN,WM ET AL. (1996)
logP 2.27			EXP	HANSCH,C & LEO,AJ (1985)
VP 7.13E-002	mm Hg	25	EST	NEELY,WB & BLAU,GE (1985)
DC	pKa			
HL 4.78E-008	atm m3/mol	25	EST	MEYLAN,WM & HOWARD,PH (1991)
OH 6.78E-012	cm3/molc sec	25	EST	MEYLAN,WM & HOWARD,PH (1993)

CAS #: 000707-98-2				9-PROPYL ADENINE

Formula: $C_8H_{13}N_5$

Mol Weight: 179.23

MP (deg C): | FP (deg C):

BP (deg C):

BP pressure (mm Hg):

Property/Value	Units	Temp	Data Type	Reference
WS 1.33E+004	mg/L	25	EST	MEYLAN,WM ET AL. (1996)
logP 0.74			EXP	HANSCH,C & LEO,AJ (1985)
VP 1.04E-006	mm Hg	25	EST	NEELY,WB & BLAU,GE (1985)
DC	pKa			
HL 2.64E-012	atm m3/mol	25	EST	MEYLAN,WM & HOWARD,PH (1991)
OH 2.02E-010	cm3/molc sec	25	EST	MEYLAN,WM & HOWARD,PH (1993)

CAS #: 000708-43-0				1-BUTYLBENZOTRIAZOLE

Formula: $C_{10}H_{13}N_3$

Mol Weight: 175.24

MP (deg C): | FP (deg C):

BP (deg C):

BP pressure (mm Hg):

Property/Value	Units	Temp	Data Type	Reference
WS 3.71E+002	mg/L	25	EST	MEYLAN,WM ET AL. (1996)
logP 2.57			EXP	HANSCH,C & LEO,AJ (1985)
VP 4.22E-004	mm Hg	25	EST	NEELY,WB & BLAU,GE (1985)
DC	pKa			
HL 7.32E-006	atm m3/mol	25	EST	MEYLAN,WM & HOWARD,PH (1991)
OH 4.88E-012	cm3/molc sec	25	EST	MEYLAN,WM & HOWARD,PH (1993)

CAS #: 000708-79-2				1-METHYL URIC ACID

Formula: $C_6H_6N_4O_3$

Mol Weight: 182.14

MP (deg C): | FP (deg C):

BP (deg C):

BP pressure (mm Hg):

Property/Value	Units	Temp	Data Type	Reference
WS 5.00E+003	mg/L		EXP	YALKOWSKY,SH & DANNENFELSER (1992)
logP -0.57			EXP	GASPARI,F & BONATI,M (1987)
VP 3.47E-009	mm Hg	25	EST	NEELY,WB & BLAU,GE (1985)
DC	pKa			
HL 2.63E-017	atm m3/mol	25	EST	MEYLAN,WM & HOWARD,PH (1991)
OH 1.29E-011	cm3/molc sec	25	EST	MEYLAN,WM & HOWARD,PH (1993)

CAS #: 000709-79-5				BENZALCYANOACETAMIDE

Formula: $C_{10}H_8N_2O$

Mol Weight: 172.19

MP (deg C): | FP (deg C):

BP (deg C):

BP pressure (mm Hg):

Property/Value	Units	Temp	Data Type	Reference
WS 1.52E+003	mg/L	25	EST	MEYLAN,WM ET AL. (1996)
logP 1.56			EXP	HANSCH,C ET AL. (1995)
VP 1.77E-006	mm Hg	25	EST	NEELY,WB & BLAU,GE (1985)
DC	pKa			
HL 5.89E-013	atm m3/mol	25	EST	MEYLAN,WM & HOWARD,PH (1991)
OH 8.70E-012	cm3/molc sec	25	EST	MEYLAN,WM & HOWARD,PH (1993)

CAS #: 000709-98-8				PROPANIL

Formula: $C_9H_9Cl_2NO$

Mol Weight: 218.08

MP (deg C): 91-93 | FP (deg C):

BP (deg C):

BP pressure (mm Hg):

Property/Value	Units	Temp	Data Type	Reference
WS 2.25E+002	mg/L	25	EXP	WORTHING,CR & WALKER,SB (1987)
logP 3.07			EXP	HANSCH,C & LEO,AJ (1985)
VP 4.00E-005	mm Hg	22	22.5 TYPE NR;	WAUCHOPE,RD ET AL. (1991A)
DC	pKa			
HL 4.50E-009	atm m3/mol	25	EST	MEYLAN,WM & HOWARD,PH (1991)
OH 1.57E-011	cm3/molc sec	25	EST	ATKINSON,R (1987)

CAS #: 000710-15-6				PHENETHYL SULFAMIDE

Formula: $C_8H_{12}N_2O_2S$

Mol Weight: 200.26

MP (deg C): | FP (deg C):

BP (deg C):

BP pressure (mm Hg):

Property/Value	Units	Temp	Data Type	Reference
WS 8.87E+003	mg/L	25	EST	MEYLAN,WM ET AL. (1996)
logP 0.81			EXP	HANSCH,C & LEO,AJ (1985)
VP 2.04E-005	mm Hg	25	EST	NEELY,WB & BLAU,GE (1985)
DC	pKa			
HL 7.52E-009	atm m3/mol	25	EST	MEYLAN,WM & HOWARD,PH (1991)
OH 1.67E-011	cm3/molc sec	25	EST	MEYLAN,WM & HOWARD,PH (1993)

CAS #: 000710-25-8				2-PROPENAMIDE, 3-(5-NITRO-2-FURANYL)-

Formula: $C_7H_6N_2O_4$

Mol Weight: 182.14

MP (deg C): | FP (deg C):

BP (deg C):

BP pressure (mm Hg):

Property/Value	Units	Temp	Data Type	Reference
WS 6.10E+003	mg/L	25	EST	MEYLAN,WM ET AL. (1996)
logP 0.65			EXP	BALAZ,S ET AL. (1985)
VP 4.59E-006	mm Hg	25	EST	NEELY,WB & BLAU,GE (1985)
DC	pKa			
HL 1.04E-012	atm m3/mol	25	EST	MEYLAN,WM & HOWARD,PH (1991)
OH 2.65E-011	cm3/molc sec	25	EST	MEYLAN,WM & HOWARD,PH (1993)

CAS #: 000711-82-0				ETHANOL, 2-(1-NAPHTHALENYLOXY)-

Formula: $C_{12}H_{12}O_2$

Mol Weight: 188.23

MP (deg C):   FP (deg C):

BP (deg C):

BP pressure (mm Hg):

Property/Value	Units	Temp	Data Type	Reference
WS 1.19E+003	mg/L	25	EST	MEYLAN,WM ET AL. (1996)
logP 2.50			EXP	CHAMBERLAIN,K ET AL. (1986)
VP 2.82E-006	mm Hg	25	EST	NEELY,WB & BLAU,GE (1985)
DC	pKa			
HL 1.51E-009	atm m3/mol	25	EST	MEYLAN,WM & HOWARD,PH (1991)
OH 2.11E-010	cm3/molc sec	25	EST	MEYLAN,WM & HOWARD,PH (1993)

CAS #: 000713-05-3				3-PYRIDINEBUTANAMIDE, N-METHYL-·-OXO-

Formula: $C_{10}H_{12}N_2O_2$

Mol Weight: 192.22

MP (deg C):   FP (deg C):

BP (deg C):

BP pressure (mm Hg):

Property/Value	Units	Temp	Data Type	Reference
WS 1.86E+005	mg/L	25	EST	MEYLAN,WM ET AL. (1996)
logP -0.69			EXP	LI,NY & GORROD,JW (1992)
VP 3.28E-006	mm Hg	25	EST	NEELY,WB & BLAU,GE (1985)
DC	pKa			
HL 5.68E-015	atm m3/mol	25	EST	MEYLAN,WM & HOWARD,PH (1991)
OH 1.19E-011	cm3/molc sec	25	EST	MEYLAN,WM & HOWARD,PH (1993)

CAS #: 000713-68-8				3-PHENOXYPHENOL

Formula: $C_{12}H_{10}O_2$

Mol Weight: 186.21

MP (deg C):   FP (deg C):

BP (deg C):

BP pressure (mm Hg):

Property/Value	Units	Temp	Data Type	Reference
WS 1.75E+002	mg/L	25	EST	MEYLAN,WM ET AL. (1996)
logP 3.57			EST	MEYLAN,WM & HOWARD,PH (1995)
VP 6.05E-005	mm Hg	25	EST	NEELY,WB & BLAU,GE (1985)
DC	pKa			
HL 1.23E-008	atm m3/mol	25	EST	MEYLAN,WM & HOWARD,PH (1991)
OH 8.92E-011	cm3/molc sec	25	EST	MEYLAN,WM & HOWARD,PH (1993)

CAS #: 000715-48-0				6-METHYL-4-NITROQUINOLINE-1-OXIDE

Formula: $C_{10}H_{10}N_2O_3$

Mol Weight: 206.20

MP (deg C):   FP (deg C):

BP (deg C):

BP pressure (mm Hg):

Property/Value	Units	Temp	Data Type	Reference
WS 1.04E+003	mg/L	25	EST	MEYLAN,WM ET AL. (1996)
logP 1.43			EXP	HANSCH,C & LEO,AJ (1985)
VP 1.06E-006	mm Hg	25	EST	NEELY,WB & BLAU,GE (1985)
DC	pKa			
HL 3.00E-014	atm m3/mol	25	EST	MEYLAN,WM & HOWARD,PH (1991)
OH 2.83E-012	cm3/molc sec	25	EST	MEYLAN,WM & HOWARD,PH (1993)

CAS #: 000715-99-1				1-PH-3,5-DIME-4-NITROSOPYRAZOLE

Formula: $C_{11}H_{11}N_3O$

Mol Weight: 201.23

MP (deg C):   FP (deg C):

BP (deg C):

BP pressure (mm Hg):

Property/Value	Units	Temp	Data Type	Reference
WS 3.70E+001	mg/L	25	EXP	GUNTHER,FA ET AL. (1968)
logP 2.28			EXP	HANSCH,C & LEO,AJ (1985)
VP 1.93E-004	mm Hg	25	EST	NEELY,WB & BLAU,GE (1985)
DC	pKa			
HL 1.71E-010	atm m3/mol	25	EST	MEYLAN,WM & HOWARD,PH (1991)
OH 3.97E-011	cm3/molc sec	25	EST	MEYLAN,WM & HOWARD,PH (1993)

CAS #: 000716-79-0				2-PHENYLBENZIMIDAZOLE

Formula: $C_{13}H_{10}N_2$

Mol Weight: 194.24

MP (deg C): 291   FP (deg C):

BP (deg C):

BP pressure (mm Hg):

Property/Value	Units	Temp	Data Type	Reference
WS 8.00E+001	mg/L	25	EST	MEYLAN,WM ET AL. (1996)
logP 3.24			EXP	HANSCH,C & LEO,AJ (1985)
VP 5.47E-008	mm Hg	25	EST	NEELY,WB & BLAU,GE (1985)
DC	pKa			
HL 2.82E-008	atm m3/mol	25	EST	MEYLAN,WM & HOWARD,PH (1991)
OH 6.59E-011	cm3/molc sec	25	EST	MEYLAN,WM & HOWARD,PH (1993)

CAS #: 000717-74-8				1,3,5-TRIISOPROPYLBENZENE

Formula: $C_{15}H_{24}$

Mol Weight: 204.36

MP (deg C): -7.4   FP (deg C):

BP (deg C): 238

BP pressure (mm Hg):

Property/Value	Units	Temp	Data Type	Reference
WS 1.54E-001	mg/L	25	EST	MEYLAN,WM ET AL. (1996)
logP 6.36			EST	MEYLAN,WM & HOWARD,PH (1995)
VP 2.58E-002	mm Hg	25	EST	NEELY,WB & BLAU,GE (1985)
DC	pKa			
HL 3.96E-002	atm m3/mol	25	EST	MEYLAN,WM & HOWARD,PH (1991)
OH 3.29E-011	cm3/molc sec	25	EST	MEYLAN,WM & HOWARD,PH (1993)

CAS #: 000718-64-9				2-PROPANOL,2-PHHENYL-1,1,1,3,3,3-HEXAFLUORO

Formula: $C_9H_6F_6O$

Mol Weight: 244.14

MP (deg C):   FP (deg C):

BP (deg C): 160

BP pressure (mm Hg):

Property/Value	Units	Temp	Data Type	Reference
WS 1.01E+002	mg/L	25	EST	MEYLAN,WM ET AL. (1996)
logP 3.41			EXP	HANSCH,C & LEO,AJ (1985)
VP 2.23E-001	mm Hg	25	EST	NEELY,WB & BLAU,GE (1985)
DC	pKa			
HL 2.38E-005	atm m3/mol	25	EST	MEYLAN,WM & HOWARD,PH (1991)
OH 4.44E-012	cm3/molc sec	25	EST	MEYLAN,WM & HOWARD,PH (1993)

CAS #: 000718-67-2	BARBITURIC ACID,5ME-5CYHXENE

Formula: $C_{11}H_{14}N_2O_3$
Mol Weight: 222.25
MP (deg C):
FP (deg C):
BP (deg C):
BP pressure (mm Hg):

Property/Value	Units	Temp	Data Type	Reference
WS 1.44E+003	mg/L	25	EST	MEYLAN,WM ET AL. (1996)
logP 1.60			EXP	YIH,TD & VAN ROSSUM,JM (1977)
VP 4.38E-012	mm Hg	25	EST	NEELY,WB & BLAU,GE (1985)
DC	pKa			
HL 3.86E-013	atm m3/mol	25	EST	MEYLAN,WM & HOWARD,PH (1991)
OH 9.90E-011	cm3/molc sec	25	EST	MEYLAN,WM & HOWARD,PH (1993)

CAS #: 000719-22-2	2,6-BIS(T-BUTYL)-2,5-CYCLOHEXADIENE-1,4-DIONE

Formula: $C_{14}H_{20}O_2$
Mol Weight: 220.31
MP (deg C): 65-67
FP (deg C):
BP (deg C):
BP pressure (mm Hg):

Property/Value	Units	Temp	Data Type	Reference
WS 1.16E+001	mg/L	25	EST	MEYLAN,WM ET AL (1996)
logP 4.42			EXP	BARBER,LB II ET AL (1988)
VP 4.11E-004	mm Hg	25	EST	NEELY,WB & BLAU,GE (1985)
DC	pKa			
HL 1.64E-008	atm m3/mol	25	EST	MEYLAN,WM & HOWARD,PH (1991)
OH 2.23E-011	cm3/molc sec	25	EST	MEYLAN,WM & HOWARD,PH (1993)

CAS #: 000721-50-6	PRILOCAINE

Formula: $C_{13}H_{20}N_2O$
Mol Weight: 220.32
MP (deg C): 37-38
FP (deg C):
BP (deg C): 159-162
BP pressure (mm Hg): 1.00E-001

Property/Value	Units	Temp	Data Type	Reference
WS 5.41E+002	mg/L	25	EST	MEYLAN,WM ET AL. (1996)
logP 2.11			EXP	HANSCH,C ET AL. (1995)
VP 2.05E-006	mm Hg	25	EST	NEELY,WB & BLAU,GE (1985)
DC	pKa			
HL 7.18E-011	atm m3/mol	25	EST	MEYLAN,WM & HOWARD,PH (1991)
OH 1.04E-010	cm3/molc sec	25	EST	MEYLAN,WM & HOWARD,PH (1993)

CAS #: 000723-46-6	SULFAMETHOXAZOLE

Formula: $C_{10}H_{11}N_3O_3S$
Mol Weight: 253.28
MP (deg C): 167
FP (deg C):
BP (deg C):
BP pressure (mm Hg):

Property/Value	Units	Temp	Data Type	Reference
WS 6.10E+002	mg/L	37	EXP	YALKOWSKY,SH & DANNENFELSER,RM (1992)
logP 0.89			EXP	HANSCH,C & LEO,AJ (1985)
VP 6.93E-008	mm Hg	25	EST	NEELY,WB & BLAU,GE (1985)
DC	pKa			
HL 6.42E-013	atm m3/mol	25	EST	MEYLAN,WM & HOWARD,PH (1991)
OH 1.63E-010	cm3/molc sec	25	EST	MEYLAN,WM & HOWARD,PH (1993)

CAS #: 000723-57-9	124-BENZTHIADIZ-1-O2-3-ME-6-CF3

Formula: $C_9H_7F_3N_2O_2S$
Mol Weight: 264.23
MP (deg C):
FP (deg C):
BP (deg C):
BP pressure (mm Hg):

Property/Value	Units	Temp	Data Type	Reference
WS 8.64E+002	mg/L	25	EST	MEYLAN,WM ET AL. (1996)
logP 1.59			EXP	HANSCH,C & LEO,AJ (1985)
VP 8.89E-008	mm Hg	25	EST	NEELY,WB & BLAU,GE (1985)
DC	pKa			
HL 8.04E-007	atm m3/mol	25	EST	MEYLAN,WM & HOWARD,PH (1991)
OH 2.59E-013	cm3/molc sec	25	EST	MEYLAN,WM & HOWARD,PH (1993)

CAS #: 000723-62-6	ANTHRACENE-9-CARBOXYLIC ACID

Formula: $C_{15}H_{10}O_2$
Mol Weight: 222.25
MP (deg C): 217 dec
FP (deg C):
BP (deg C):
BP pressure (mm Hg):

Property/Value	Units	Temp	Data Type	Reference
WS 8.50E-002	mg/L	25	EXP	MEANS,JC ET AL. (1982)
logP 3.85			EXP	HANSCH,C ET AL. (1995)
VP 2.22E-007	mm Hg	25	EST	NEELY,WB & BLAU,GE (1985)
DC 3.65	pKa	20	EXP	KORTUM,G ET AL (1961)
HL 1.03E-009	atm m3/mol	25	EST	MEYLAN,WM & HOWARD,PH (1991)
OH 1.53E-011	cm3/molc sec	25	EST	MEYLAN,WM & HOWARD,PH (1993)

CAS #: 000726-78-3	N-METHYLCYCLOBARBITAL

Formula: $C_{13}H_{18}N_2O_3$
Mol Weight: 250.30
MP (deg C):
FP (deg C):
BP (deg C):
BP pressure (mm Hg):

Property/Value	Units	Temp	Data Type	Reference
WS 2.82E+002	mg/L	25	EST	MEYLAN,WM ET AL. (1996)
logP 2.25			EXP	HANSCH,C & LEO,AJ (1985)
VP 1.35E-011	mm Hg	25	EST	NEELY,WB & BLAU,GE (1985)
DC	pKa			
HL 1.13E-012	atm m3/mol	25	EST	MEYLAN,WM & HOWARD,PH (1991)
OH 9.84E-011	cm3/molc sec	25	EST	MEYLAN,WM & HOWARD,PH (1993)

CAS #: 000728-88-1	1-PROPANONE, 2-METHYL-1-(4-METHYLPHENYL)-3-(1-PI

Formula: $C_{16}H_{23}NO$
Mol Weight: 245.37
MP (deg C): 176-177
FP (deg C):
BP (deg C):
BP pressure (mm Hg):

Property/Value	Units	Temp	Data Type	Reference
WS 1.95E+002	mg/L	25	EST	MEYLAN,WM ET AL. (1996)
logP 3.65			EXP	SANGSTER,J (1994)
VP 3.45E-005	mm Hg	25	EST	NEELY,WB & BLAU,GE (1985)
DC	pKa			
HL 9.76E-009	atm m3/mol	25	EST	MEYLAN,WM & HOWARD,PH (1991)
OH 1.33E-010	cm3/molc sec	25	EST	MEYLAN,WM & HOWARD,PH (1993)

000730-40-5 — 4-((4-NITROPHENYL)AZO)BENZENAMINE

Formula: $C_{12}H_{10}N_4O_2$

Mol Weight: 242.24

MP (deg C):
FP (deg C):
BP (deg C):
BP pressure (mm Hg):

Property/Value	Units	Temp	Data Type	Reference
WS 3.40E-001	mg/L	25	EXP	BAUGHMAN,GL & PERENICH,TA (1988)
logP 3.59			EST	MEYLAN,WM & HOWARD,PH (1995)
VP 3.00E-010	mm Hg	25	EXP	BAUGHMAN,GL & PERENICH,TA (1988)
DC	pKa			
HL 2.81E-010	atm m3/mol	25	EST	VP/WSOL
OH 4.27E-011	cm3/molc sec	25	EST	MEYLAN,WM & HOWARD,PH (1993)

000731-27-1 — TOLYFLUANIDE

Formula: $C_{10}H_{13}Cl_2FN_2O_2S_2$

Mol Weight: 347.26

MP (deg C): 96
FP (deg C):
BP (deg C):
BP pressure (mm Hg):

Property/Value	Units	Temp	Data Type	Reference
WS 9.00E-001	mg/L	25	EXP	TOMLIN,C (1994)
logP 3.90			EXP	TOMLIN,C (1994)
VP 5.06E-007	mm Hg	25	EST	NEELY,WB & BLAU,GE (1985)
DC	pKa			
HL 7.44E-007	atm m3/mol	25	EST	MEYLAN,WM & HOWARD,PH (1991)
OH 1.79E-011	cm3/molc sec	25	EST	MEYLAN,WM & HOWARD,PH (1993)

000731-92-0 — 2,4-DINITRO-6-PHENYLPHENOL

Formula: $C_{12}H_8N_2O_5$

Mol Weight: 260.21

MP (deg C):
FP (deg C):
BP (deg C):
BP pressure (mm Hg):

Property/Value	Units	Temp	Data Type	Reference
WS 2.17E+001	mg/L	25	EST	MEYLAN,WM ET AL. (1996)
logP 3.49			EST	MEYLAN,WM & HOWARD,PH (1995)
VP 6.75E-009	mm Hg	25	EST	NEELY,WB & BLAU,GE (1985)
DC 3.85	pKa	25	EXP	SERJEANT,EP & DEMPSEY,B (1979)
HL 2.12E-009	atm m3/mol	25	EST	MEYLAN,WM & HOWARD,PH (1991)
OH 1.94E-012	cm3/molc sec	25	EST	MEYLAN,WM & HOWARD,PH (1993)

000732-11-6 — PHOSMET

Formula: $C_{11}H_{12}NO_4PS_2$

Mol Weight: 317.32

MP (deg C): 72.5
FP (deg C):
BP (deg C):
BP pressure (mm Hg):

Property/Value	Units	Temp	Data Type	Reference
WS 2.44E+001	mg/L	20	EXP	BOWMAN,BT & SANS,WW (1983)
logP 2.78			EXP	HANSCH,C ET AL. (1995)
VP 4.90E-007	mm Hg	20	EXP	WAUCHOPE,RD ET AL. (1991)
DC	pKa			
HL 8.38E-009	atm m3/mol	20	EST	VP/WSOL
OH 1.45E-010	cm3/molc sec	25	EST	MEYLAN,WM & HOWARD,PH (1993)

000732-26-3 — 2,4,6-TRI(TERT-BUTYL)PHENOL

Formula: $C_{18}H_{30}O$

Mol Weight: 262.44

MP (deg C): 128
FP (deg C):
BP (deg C):
BP pressure (mm Hg):

Property/Value	Units	Temp	Data Type	Reference
WS 3.50E+001	mg/L	25	EXP	CHEM INSPECT TEST INST (1992)
logP 6.06			EXP	CHEM INSPECT TEST INST (1992)
VP 6.61E-004	mm Hg	25	EXT	PERRY,RH & GREEN,D (1984)
DC 12.19	pKa	25	EXP	SERJEANT,EP & DEMPSEY,B (1979)
HL 9.65E-006	atm m3/mol	25	EST	MEYLAN,WM & HOWARD,PH (1991)
OH 1.60E-011	cm3/molc sec	25	EST	MEYLAN,WM & HOWARD,PH (1993)

000738-70-5 — TRIMETHOPRIM

Formula: $C_{14}H_{18}N_4O_3$

Mol Weight: 290.32

MP (deg C): 199-203
FP (deg C):
BP (deg C):
BP pressure (mm Hg):

Property/Value	Units	Temp	Data Type	Reference
WS 4.00E+002	mg/L	25	EXP	YALKOWSKY,SH & DANNENFELSER,RM (1992)
logP 0.91			EXP	HANSCH,C & LEO,AJ (1985)
VP 9.88E-009	mm Hg	25	EST	NEELY,WB & BLAU,GE (1985)
DC 7.12	pKa	20	EXP	PERRIN,DD (1972)
HL 2.39E-014	atm m3/mol	25	EST	MEYLAN,WM & HOWARD,PH (1991)
OH 2.03E-010	cm3/molc sec	25	EST	MEYLAN,WM & HOWARD,PH (1993)

000741-58-2 — BENSULIDE

Formula: $C_8H_{11}NO_2S_2$

Mol Weight: 217.31

MP (deg C): 34.4
FP (deg C):
BP (deg C):
BP pressure (mm Hg):

Property/Value	Units	Temp	Data Type	Reference
WS 2.50E+001	mg/L	20	EXP	YALKOWSKY,SH & DANNENFELSER,RM (1992)
logP 4.20			EXP	TOMLIN,C (1994)
VP 8.00E-007	mm Hg	25	EXP	WAUCHOPE,RD ET AL. (1991A)
DC	pKa			
HL 9.15E-009	atm m3/mol	25	EST	VP/WSOL
OH 1.68E-010	cm3/molc sec	25	EST	MEYLAN,WM & HOWARD,PH (1993)

000746-53-2 — 9H-FLUOREN-9-ONE, 2,4,5,7-TETRANITRO-

Formula: $C_{13}H_4N_4O_9$

Mol Weight: 360.20

MP (deg C):
FP (deg C):
BP (deg C):
BP pressure (mm Hg):

Property/Value	Units	Temp	Data Type	Reference
WS 1.52E+000	mg/L	25	EST	MEYLAN,WM ET AL. (1996)
logP 2.40			EXP	DEBNATH,AK & HANSCH,C (1992)
VP 5.38E-012	mm Hg	25	EST	NEELY,WB & BLAU,GE (1985)
DC	pKa			
HL 1.64E-016	atm m3/mol	25	EST	MEYLAN,WM & HOWARD,PH (1991)
OH 5.29E-015	cm3/molc sec	25	EST	MEYLAN,WM & HOWARD,PH (1993)

CAS #: 000749-02-0				SPIROPERIDOL
Formula: $C_{23}H_{26}FN_3O_2$				
Mol Weight: 395.48				
MP (deg C): 190-93.6		FP (deg C):		
BP (deg C):				
BP pressure (mm Hg):				

Property/Value	Units	Temp	Data Type	Reference
WS 8.44E+000	mg/L	25	EST	MEYLAN,WM ET AL. (1996)
logP 3.03			EXP	HANSCH,C & LEO,AJ (1985)
VP 5.36E-012	mm Hg	25	EST	NEELY,WB & BLAU,GE (1985)
DC	pKa			
HL 8.39E-015	atm m3/mol	25	EST	MEYLAN,WM & HOWARD,PH (1991)
OH 2.89E-010	cm3/molc sec	25	EST	MEYLAN,WM & HOWARD,PH (1993)

CAS #: 000753-90-2				2,2,2-TRIFLUOROETHYLAMINE
Formula: $C_2H_4F_3N$				
Mol Weight: 99.06				
MP (deg C):		FP (deg C):		
BP (deg C): 37-38				
BP pressure (mm Hg):				

Property/Value	Units	Temp	Data Type	Reference
WS 7.48E+005	mg/L	25	EST	MEYLAN,WM ET AL. (1996)
logP 0.24			EXP	HANSCH,C & LEO,AJ (1985)
VP 9.78E+002	mm Hg	25	EST	NEELY,WB & BLAU,GE (1985)
DC 5.70	pKa		EXP	PERRIN,DD (1965)
HL 7.92E-005	atm m3/mol	25	EST	MEYLAN,WM & HOWARD,PH (1991)
OH 6.17E-013	cm3/molc sec	25	EST	MEYLAN,WM & HOWARD,PH (1993)

CAS #: 000756-79-6				Phosphonic acid, methyl-, dimethyl ester
Formula: $C_3H_9O_3P$				
Mol Weight: 124.08				
MP (deg C):		FP (deg C):		
BP (deg C):				
BP pressure (mm Hg): 1.81E+002				

Property/Value	Units	Temp	Data Type	Reference
WS 1.00E+006	mg/L	25	EXP	BENNETT,SR ET AL. (1984)
logP -0.61			EXP	KRIKORIAN,SE ET AL. (1987)
VP 9.62E-001	mm Hg	25	EXT	BOUBLIK,T ET AL. (1984)
DC	pKa			
HL 1.25E-006	atm m3/mol	25	EST	MEYLAN,WM & HOWARD,PH (1991)
OH 5.71E-012	cm3/molc sec	25	EST	MEYLAN,WM & HOWARD,PH (1993)

CAS #: 000757-58-4				HEXAETHYL TETRAPHOSPHATE
Formula: $C_{12}H_{30}O_{13}P_4$				
Mol Weight: 506.26				
MP (deg C): -40		FP (deg C):		
BP (deg C):				
BP pressure (mm Hg):				

Property/Value	Units	Temp	Data Type	Reference
WS 1.00E+006	mg/L		EXP	GUNTHER,FA ET AL. (1968)
logP -0.39			EST	MEYLAN,WM & HOWARD,PH (1995)
VP 2.06E-008	mm Hg	25	EST	NEELY,WB & BLAU,GE (1985)
DC	pKa			
HL 3.30E-017	atm m3/mol	25	EST	MEYLAN,WM & HOWARD,PH (1991)
OH 1.16E-010	cm3/molc sec	25	EST	MEYLAN,WM & HOWARD,PH (1993)

CAS #: 000758-96-3				PROPANAMIDE, N,N-DIMETHYL-
Formula: $C_5H_{11}NO$				
Mol Weight: 101.15				
MP (deg C): -45		FP (deg C):		
BP (deg C): 174-175				
BP pressure (mm Hg):				

Property/Value	Units	Temp	Data Type	Reference
WS 1.44E+005	mg/L	25	EST	MEYLAN,WM ET AL. (1996)
logP -0.11			EXP	HANSCH,C ET AL. (1995)
VP 2.47E+000	mm Hg	25	EST	NEELY,WB & BLAU,GE (1985)
DC	pKa			
HL 7.16E-008	atm m3/mol	25	EST	MEYLAN,WM & HOWARD,PH (1991)
OH 1.73E-011	cm3/molc sec	25	EST	MEYLAN,WM & HOWARD,PH (1993)

CAS #: 000759-73-9				1-ETHYL-1-NITROSOUREA
Formula: $C_3H_7N_3O_2$				
Mol Weight: 117.11				
MP (deg C): 103-104		FP (deg C):		
BP (deg C):				
BP pressure (mm Hg):				

Property/Value	Units	Temp	Data Type	Reference
WS 1.30E+004	mg/L	25	EXP	IARC (1978)
logP 0.23			EXP	HANSCH,C & LEO,AJ (1985)
VP 5.34E-002	mm Hg	25	EST	NEELY,WB & BLAU,GE (1985)
DC	pKa			
HL 1.32E-010	atm m3/mol	25	EST	MEYLAN,WM & HOWARD,PH (1991)
OH 5.00E-012	cm3/molc sec	25	EST	MEYLAN,WM & HOWARD,PH (1993)

CAS #: 000759-94-4				EPTAM (EPTC)
Formula: $C_9H_{19}NOS$				
Mol Weight: 189.32				
MP (deg C):		FP (deg C):		
BP (deg C): 127				
BP pressure (mm Hg): 2.00E+001				

Property/Value	Units	Temp	Data Type	Reference
WS 3.75E+002	mg/L	25	EXP	YAKOWSKY, SH (1992)
logP 3.21			EXP	HANSCH,C & LEO,AJ (1985)
VP 3.40E-002	mm Hg	25	EXP	WAUCHOPE,RD ET AL. (1991A)
DC	pKa			
HL 2.26E-005	atm m3/mol	25	EST	VP/WSOL
OH 2.71E-011	cm3/molc sec	25	EST	ATKINSON, R (1988)

CAS #: 000760-20-3				3-METHYL-1-PENTENE
Formula: C_6H_{12}				
Mol Weight: 84.16				
MP (deg C): -153		FP (deg C):		
BP (deg C): 54.2				
BP pressure (mm Hg):				

Property/Value	Units	Temp	Data Type	Reference
WS 8.76E+001	mg/L	25	EST	MEYLAN,WM ET AL. (1996)
logP 3.08			EST	MEYLAN,WM & HOWARD,PH (1995)
VP 2.69E+002	mm Hg	25	EXP	YAWS,CL (1994A)
DC	pKa			
HL 3.59E-001	atm m3/mol	25	EST	MEYLAN,WM & HOWARD,PH (1991)
OH 3.02E-011	cm3/molc sec	25	EST	MEYLAN,WM & HOWARD,PH (1993)

CAS #: 000760-21-4				2-ETHYL-1-BUTENE

Formula: C6H12			
Mol Weight: 84.16			
MP (deg C): -131.5		FP (deg C):	
BP (deg C): 64.7			
BP pressure (mm Hg):			

Property/Value	Units	Temp	Data Type	Reference
WS 6.79E+001	mg/L	25	EST	MEYLAN,WM ET AL. (1996)
logP 3.21			EST	MEYLAN,WM & HOWARD,PH (1995)
VP 1.75E+002	mm Hg	25	EXP	YAWS,CL (1994A)
DC	pKa			
HL 4.23E-001	atm m3/mol	25	EST	MEYLAN,WM & HOWARD,PH (1991)
OH 5.36E-011	cm3/molc sec	25	EST	MEYLAN,WM & HOWARD,PH (1993)

CAS #: 000760-23-6				3,4-DICHLORO-1-BUTENE

Formula: C4H6Cl2			
Mol Weight: 125.00			
MP (deg C): -61		FP (deg C):	
BP (deg C): 118.6			
BP pressure (mm Hg):			

Property/Value	Units	Temp	Data Type	Reference
WS 5.74E+002	mg/L	25	EST	MEYLAN,WM ET AL. (1996)
logP 2.60			EST	MEYLAN,WM & HOWARD,PH (1995)
VP 2.19E+001	mm Hg	25	EXP	DAUBERT,TE & DANNER,RP (1991)
DC	pKa			
HL 8.56E-003	atm m3/mol	25	EST	VP/WSOL
OH 2.69E-011	cm3/molc sec	25	EST	MEYLAN,WM & HOWARD,PH (1993)

CAS #: 000760-55-4				OCTYLMALONIC ACID

Formula: C11H20O4			
Mol Weight: 216.28			
MP (deg C): 116		FP (deg C):	
BP (deg C):			
BP pressure (mm Hg):			

Property/Value	Units	Temp	Data Type	Reference
WS 2.60E+002	mg/L	25	EXP	STEPHEN,H & STEPHEN,T (1963)
logP 2.61			EST	MEYLAN,WM & HOWARD,PH (1995)
VP 4.50E-006	mm Hg	25	EST	NEELY,WB & BLAU,GE (1985)
DC	pKa			
HL 3.93E-011	atm m3/mol	25	EST	MEYLAN,WM & HOWARD,PH (1991)
OH 1.22E-011	cm3/molc sec	25	EST	MEYLAN,WM & HOWARD,PH (1993)

CAS #: 000760-60-1				UREA, N-(2-METHYLPROPYL)-N-NITROSO-

Formula: C5H11N3O2			
Mol Weight: 145.16			
MP (deg C):		FP (deg C):	
BP (deg C):			
BP pressure (mm Hg):			

Property/Value	Units	Temp	Data Type	Reference
WS 1.03E+004	mg/L	25	EST	MEYLAN,WM ET AL. (1996)
logP 1.04			EXP	HANSCH,C & LEO,AJ (1985)
VP 1.09E-002	mm Hg	25	EST	NEELY,WB & BLAU,GE (1985)
DC	pKa			
HL 2.32E-010	atm m3/mol	25	EST	MEYLAN,WM & HOWARD,PH (1991)
OH 8.43E-012	cm3/molc sec	25	EST	MEYLAN,WM & HOWARD,PH (1993)

CAS #: 000762-04-9				DIETHYL PHOSPHITE

Formula: C4H11O3P			
Mol Weight: 138.10			
MP (deg C):		FP (deg C):	
BP (deg C):			
BP pressure (mm Hg):			

Property/Value	Units	Temp	Data Type	Reference
WS 1.15E+005	mg/L	25	EST	MEYLAN,WM ET AL. (1996)
logP -0.15			EST	MEYLAN,WM & HOWARD,PH (1995)
VP 1.12E+001	mm Hg	25	EXP	PAGE,FM & PURNELL,JH (1958)
DC	pKa			
HL 5.78E-006	atm m3/mol	25	EST	MEYLAN,WM & HOWARD,PH (1991)
OH 3.86E-011	cm3/molc sec	25	EST	MEYLAN,WM & HOWARD,PH (1993)

CAS #: 000763-29-1				2-METHYL-1-PENTENE

Formula: C6H12			
Mol Weight: 84.16			
MP (deg C): -135.7		FP (deg C):	
BP (deg C): 62.1			
BP pressure (mm Hg):			

Property/Value	Units	Temp	Data Type	Reference
WS 7.80E+001	mg/L	25	EXP	YALKOWSKY,SH & DANNENFELSER,RM (1992)
logP 3.21			EST	MEYLAN,WM & HOWARD,PH (1995)
VP 1.95E+002	mm Hg	25	EXP	DAUBERT,TE & DANNER,RP (1989)
DC	pKa			
HL 2.77E-001	atm m3/mol	25	EST	VP/WSOL
OH 6.26E-011	cm3/molc sec	25	EXP	ATKINSON,R (1989)

CAS #: 000763-30-4				2-METHYL-1,4-PENTADIENE

Formula: C6H10			
Mol Weight: 82.15			
MP (deg C):		FP (deg C):	
BP (deg C): 56			
BP pressure (mm Hg):			

Property/Value	Units	Temp	Data Type	Reference
WS 8.97E+001	mg/L	25	EST	MEYLAN,WM ET AL. (1996)
logP 3.07			EST	MEYLAN,WM & HOWARD,PH (1995)
VP 2.03E+002	mm Hg	25	EST	NEELY,WB & BLAU,GE (1985)
DC	pKa			
HL 1.99E-001	atm m3/mol	25	EST	MEYLAN,WM & HOWARD,PH (1991)
OH 7.88E-011	cm3/molc sec	25	EXP	ATKINSON,R (1989)

CAS #: 000763-32-6				3-METHYL-3-BUTEN-1-OL

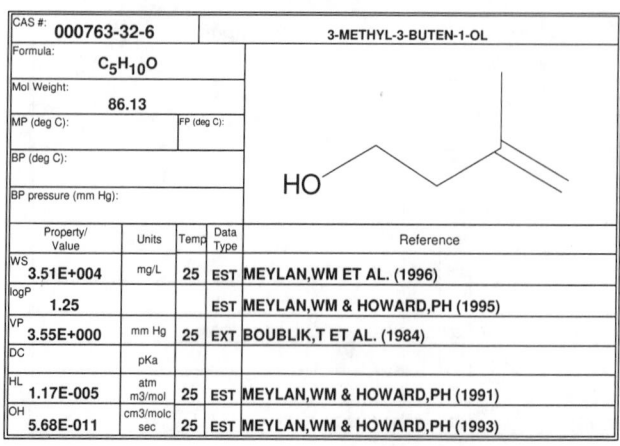

Formula: C5H10O			
Mol Weight: 86.13			
MP (deg C):		FP (deg C):	
BP (deg C):			
BP pressure (mm Hg):			

Property/Value	Units	Temp	Data Type	Reference
WS 3.51E+004	mg/L	25	EST	MEYLAN,WM ET AL. (1996)
logP 1.25			EST	MEYLAN,WM & HOWARD,PH (1995)
VP 3.55E+000	mm Hg	25	EXT	BOUBLIK,T ET AL. (1984)
DC	pKa			
HL 1.17E-005	atm m3/mol	25	EST	MEYLAN,WM & HOWARD,PH (1991)
OH 5.68E-011	cm3/molc sec	25	EST	MEYLAN,WM & HOWARD,PH (1993)

CAS #: 000764-13-6 — 2,5-DIMETHYL-2,4-HEXADIENE

Formula: C_8H_{14}
Mol Weight: 110.20
MP (deg C): 14
FP (deg C):
BP (deg C): 134.5
BP pressure (mm Hg):

Property/Value	Units	Temp	Data Type	Reference
WS 3.20E+001	mg/L	25	EXP	CHEM INSPECT TEST INST (1992)
logP 3.50			EXP	CHEM INSPECT TEST INST (1992)
VP 1.76E+001	mm Hg	25	EST	NEELY,WB & BLAU,GE (1985)
DC	pKa			
HL 4.70E-001	atm m3/mol	25	EST	MEYLAN,WM & HOWARD,PH (1991)
OH 2.10E-010	cm3/molc sec	25	EXP	ATKINSON,R (1989)

CAS #: 000764-41-0 — 1,4-DICHLORO-2-BUTENE

Formula: $C_4H_6Cl_2$
Mol Weight: 125.00
MP (deg C):
FP (deg C):
BP (deg C): 72-75
BP pressure (mm Hg): 4.00E+001

Property/Value	Units	Temp	Data Type	Reference
WS 5.80E+002	mg/L	25	EXP	ALBANESE,V ET AL. (1987)
logP 2.60			EST	MEYLAN,WM & HOWARD,PH (1995)
VP 3.00E+000	mm Hg		EXP	ALBANESE,V ET AL. (1987)
DC	pKa			
HL 8.51E-003	atm m3/mol	25	EST	VP/WSOL
OH 3.33E-011	cm3/molc sec	25	EST	MEYLAN,WM & HOWARD,PH (1993)

CAS #: 000764-42-1 — 2-BUTENEDINITRILE, (E)-

Formula: $C_4H_2N_2$
Mol Weight: 78.07
MP (deg C): 96.8
FP (deg C):
BP (deg C): 186
BP pressure (mm Hg):

Property/Value	Units	Temp	Data Type	Reference
WS 1.17E+005	mg/L	25	EST	MEYLAN,WM ET AL. (1996)
logP -0.25			EXP	HANSCH,C ET AL. (1995)
VP 1.95E-001	mm Hg	25	EST	NEELY,WB & BLAU,GE (1985)
DC	pKa			
HL 3.07E-007	atm m3/mol	25	EST	MEYLAN,WM & HOWARD,PH (1991)
OH 1.44E-012	cm3/molc sec	25	EST	MEYLAN,WM & HOWARD,PH (1993)

CAS #: 000765-09-3 — 1-BROMO-N-TRIDECANE

Formula: $C_{13}H_{27}Br$
Mol Weight: 263.27
MP (deg C): 6.2
FP (deg C):
BP (deg C): 296
BP pressure (mm Hg):

Property/Value	Units	Temp	Data Type	Reference
WS 1.83E-002	mg/L	25	EST	MEYLAN,WM ET AL. (1996)
logP 7.07			EST	MEYLAN,WM & HOWARD,PH (1995)
VP 1.36E-003	mm Hg	25	EXT	LI,JCM & ROSSINI,FD (1953)
DC	pKa			
HL 2.54E-001	atm m3/mol	25	EST	MEYLAN,WM & HOWARD,PH (1991)
OH 1.49E-011	cm3/molc sec	25	EST	MEYLAN,WM & HOWARD,PH (1993)

CAS #: 000765-30-0 — CYCLOPROPANAMINE

Formula: C_3H_7N
Mol Weight: 57.10
MP (deg C): -35.4
FP (deg C):
BP (deg C): 50.5
BP pressure (mm Hg):

Property/Value	Units	Temp	Data Type	Reference
WS 1.00E+006	mg/L	25	EST	MEYLAN,WM ET AL. (1996)
logP 0.07			EXP	HANSCH,C ET AL. (1995)
VP 1.48E+002	mm Hg	25	EST	NEELY,WB & BLAU,GE (1985)
DC	pKa			
HL 5.89E-006	atm m3/mol	25	EST	MEYLAN,WM & HOWARD,PH (1991)
OH 2.16E-011	cm3/molc sec	25	EST	MEYLAN,WM & HOWARD,PH (1993)

CAS #: 000765-34-4 — GLYCIDYLALDEHYDE

Formula: $C_3H_4O_2$
Mol Weight: 72.06
MP (deg C): -62
FP (deg C):
BP (deg C): 112-113
BP pressure (mm Hg):

Property/Value	Units	Temp	Data Type	Reference
WS 1.00E+006	mg/L		EXP	IARC (1976)
logP -0.12			EST	MEYLAN,WM & HOWARD,PH (1995)
VP 4.53E+001	mm Hg	25	EST	NEELY,WB & BLAU,GE (1985)
DC	pKa			
HL 5.11E-007	atm m3/mol	25	EST	MEYLAN,WM & HOWARD,PH (1991)
OH 2.13E-011	cm3/molc sec	25	EST	MEYLAN,WM & HOWARD,PH (1993)

CAS #: 000765-43-5 — CYCLOPROPYL METHYL KETONE

Formula: C_5H_8O
Mol Weight: 84.12
MP (deg C): -68.3
FP (deg C):
BP (deg C): 111.3
BP pressure (mm Hg):

Property/Value	Units	Temp	Data Type	Reference
WS 4.90E+004	mg/L	25	EST	MEYLAN,WM ET AL. (1996)
logP 0.49			EXP	HANSCH,C ET AL. (1995)
VP 4.47E+001	mm Hg	25	EST	NEELY,WB & BLAU,GE (1985)
DC	pKa			
HL 3.85E-005	atm m3/mol	25	EST	MEYLAN,WM & HOWARD,PH (1991)
OH 3.25E-013	cm3/molc sec	25	EST	MEYLAN,WM & HOWARD,PH (1993)

CAS #: 000766-09-6 — PIPERIDINE, 1-ETHYL-

Formula: $C_7H_{15}N$
Mol Weight: 113.20
MP (deg C):
FP (deg C):
BP (deg C): 130.8
BP pressure (mm Hg):

Property/Value	Units	Temp	Data Type	Reference
WS 3.46E+004	mg/L	25	EST	MEYLAN,WM ET AL. (1996)
logP 1.75			EXP	SANGSTER,J (1993)
VP 6.20E+000	mm Hg	25	EST	NEELY,WB & BLAU,GE (1985)
DC	pKa			
HL 5.00E-005	atm m3/mol	25	EST	MEYLAN,WM & HOWARD,PH (1991)
OH 1.00E-010	cm3/molc sec	25	EST	MEYLAN,WM & HOWARD,PH (1993)

CAS #: 000766-17-6				2,6-DIMETHYLPIPERIDINE (CIS-)

Formula: $C_7H_{15}N$

Mol Weight: 113.20

MP (deg C): | FP (deg C):

BP (deg C): 127-128

BP pressure (mm Hg): 7.68E+002

Property/Value	Units	Temp	Data Type	Reference
WS 5.34E+004	mg/L	25	EST	MEYLAN,WM ET AL. (1996)
logP 1.53			EXP	HANSCH,C ET AL. (1995)
VP 2.56E+000	mm Hg	25	EST	NEELY,WB & BLAU,GE (1985)
DC	pKa			
HL 3.02E-005	atm m3/mol	25	EST	MEYLAN,WM & HOWARD,PH (1991)
OH 1.12E-010	cm3/molc sec	25	EST	MEYLAN,WM & HOWARD,PH (1993)

CAS #: 000766-51-8				O-CHLOROANISOLE

Formula: C_7H_7ClO

Mol Weight: 142.59

MP (deg C): -26.8 | FP (deg C):

BP (deg C): 198.5

BP pressure (mm Hg):

Property/Value	Units	Temp	Data Type	Reference
WS 4.20E+002	mg/L	25	EST	MEYLAN,WM ET AL. (1996)
logP 2.68			EXP	NAKAGAWA,Y ET AL. (1992)
VP 3.04E+000	mm Hg	25	EXT	BOUBLIK,T ET AL. (1984)
DC	pKa			
HL 2.36E-004	atm m3/mol	25	EST	MEYLAN,WM & HOWARD,PH (1991)
OH 7.11E-012	cm3/molc sec	25	EST	MEYLAN,WM & HOWARD,PH (1993)

CAS #: 000766-77-8				SILANE, DIMETHYL PHENYL

Formula: $C_8H_{12}Si$

Mol Weight: 136.27

MP (deg C): | FP (deg C):

BP (deg C): 156.5

BP pressure (mm Hg):

Property/Value	Units	Temp	Data Type	Reference
WS 3.40E+001	mg/L	25	EST	MEYLAN,WM ET AL. (1996)
logP 3.99			EXP	HANSCH,C & LEO,AJ (1985)
VP 3.67E+000	mm Hg	25	EXP	OHE,S (1976)
DC	pKa			
HL 9.89E-003	atm m3/mol	25	EST	MEYLAN,WM & HOWARD,PH (1991)
OH 2.25E-012	cm3/molc sec	25	EST	MEYLAN,WM & HOWARD,PH (1993)

CAS #: 000767-00-0				P-CYANOPHENOL

Formula: C_7H_5NO

Mol Weight: 119.12

MP (deg C): 110-113 | FP (deg C):

BP (deg C):

BP pressure (mm Hg):

Property/Value	Units	Temp	Data Type	Reference
WS 8.98E+003	mg/L	25	EST	MEYLAN,WM ET AL. (1996)
logP 1.60			EXP	HANSCH,C & LEO,AJ (1985)
VP 6.65E-003	mm Hg	25	EST	NEELY,WB & BLAU,GE (1985)
DC 7.97	pKa	25	EXP	KORTUM,G ET AL (1961)
HL 5.42E-009	atm m3/mol	25	EST	MEYLAN,WM & HOWARD,PH (1991)
OH 6.02E-012	cm3/molc sec	25	EST	MEYLAN,WM & HOWARD,PH (1993)

CAS #: 000767-15-7				2-AMINO-4,6-DIMETHYLPYRIMIDINE

Formula: $C_6H_9N_3$

Mol Weight: 123.16

MP (deg C): 151-153 | FP (deg C):

BP (deg C):

BP pressure (mm Hg):

Property/Value	Units	Temp	Data Type	Reference
WS 1.44E+004	mg/L	25	EST	MEYLAN,WM ET AL. (1996)
logP 0.97			EST	MEYLAN,WM & HOWARD,PH (1995)
VP 2.00E-002	mm Hg	25	EST	NEELY,WB & BLAU,GE (1985)
DC 4.82	pKa	20	EXP	PERRIN,DD (1965)
HL 4.46E-006	atm m3/mol	25	EST	MEYLAN,WM & HOWARD,PH (1991)
OH 8.68E-011	cm3/molc sec	25	EST	MEYLAN,WM & HOWARD,PH (1993)

CAS #: 000768-03-6				ACRYLOPHENONE

Formula: C_9H_8O

Mol Weight: 132.16

MP (deg C): | FP (deg C):

BP (deg C):

BP pressure (mm Hg):

Property/Value	Units	Temp	Data Type	Reference
WS 2.24E+003	mg/L	25	EST	MEYLAN,WM ET AL. (1996)
logP 1.88			EXP	HANSCH,C & LEO,AJ (1985)
VP 2.38E-001	mm Hg	25	EST	NEELY,WB & BLAU,GE (1985)
DC	pKa			
HL 5.17E-006	atm m3/mol	25	EST	MEYLAN,WM & HOWARD,PH (1991)
OH 1.10E-011	cm3/molc sec	25	EST	MEYLAN,WM & HOWARD,PH (1993)

CAS #: 000768-32-1				SILANE, TRIMETHYL PHENYL

Formula: $C_9H_{14}Si$

Mol Weight: 150.30

MP (deg C): | FP (deg C):

BP (deg C): 169.5

BP pressure (mm Hg):

Property/Value	Units	Temp	Data Type	Reference
WS 7.05E+000	mg/L	25	EST	MEYLAN,WM ET AL. (1996)
logP 4.72			EXP	HANSCH,C & LEO,AJ (1985)
VP 1.60E+000	mm Hg	25	EST	NEELY,WB & BLAU,GE (1985)
DC	pKa			
HL 1.36E-002	atm m3/mol	25	EST	MEYLAN,WM & HOWARD,PH (1991)
OH 2.40E-012	cm3/molc sec	25	EST	MEYLAN,WM & HOWARD,PH (1993)

CAS #: 000768-33-2				CHLORODIMETHYLPHENYLSILANE

Formula: $C_8H_{11}ClSi$

Mol Weight: 170.72

MP (deg C): | FP (deg C):

BP (deg C): 80-84

BP pressure (mm Hg): 1.60E+001

Property/Value	Units	Temp	Data Type	Reference
WS 2.01E+001	mg/L	25	EST	MEYLAN,WM ET AL. (1996)
logP 4.08			EST	MEYLAN,WM & HOWARD,PH (1995)
VP 7.24E-001	mm Hg	25	EXT	PERRY,RH & GREEN,D (1984)
DC	pKa			
HL 3.94E-003	atm m3/mol	25	EST	MEYLAN,WM & HOWARD,PH (1991)
OH 2.25E-012	cm3/molc sec	25	EST	MEYLAN,WM & HOWARD,PH (1993)

CAS #: 000768-49-0				(2-METHYL-1-PROPENYL)BENZENE

Formula: $C_{10}H_{12}$
Mol Weight: 132.21
MP (deg C):   FP (deg C):
BP (deg C): 99
BP pressure (mm Hg): 4.35E+001

Property/Value	Units	Temp	Data Type	Reference
WS 4.60E+001	mg/L	25	EST	MEYLAN,WM ET AL. (1996)
logP 3.85			EST	MEYLAN,WM & HOWARD,PH (1995)
VP 5.51E-001	mm Hg	25	EST	NEELY,WB & BLAU,GE (1985)
DC	pKa			
HL 6.79E-003	atm m3/mol	25	EST	MEYLAN,WM & HOWARD,PH (1991)
OH 3.30E-011	cm3/molc sec	25	EXP	ATKINSON,R (1989)

CAS #: 000768-66-1				PIPERIDINE, 2,2,6,6-TETRAMETHYL-

Formula: $C_9H_{19}N$
Mol Weight: 141.26
MP (deg C): 28   FP (deg C):
BP (deg C): 156
BP pressure (mm Hg):

Property/Value	Units	Temp	Data Type	Reference
WS 1.23E+004	mg/L	25	EST	MEYLAN,WM ET AL. (1996)
logP 2.15			EXP	HANSCH,C ET AL. (1995)
VP 9.81E-001	mm Hg	25	EST	NEELY,WB & BLAU,GE (1985)
DC 11.07	pKa	25	EXP	PERRIN,DD (1965)
HL 5.33E-005	atm m3/mol	25	EST	MEYLAN,WM & HOWARD,PH (1991)
OH 6.79E-011	cm3/molc sec	25	EST	MEYLAN,WM & HOWARD,PH (1993)

CAS #: 000768-94-5				AMANTADINE

Formula: $C_{10}H_{17}N$
Mol Weight: 151.25
MP (deg C): 180   FP (deg C):
BP (deg C):
BP pressure (mm Hg):

Property/Value	Units	Temp	Data Type	Reference
WS 6.29E+003	mg/L	25	EST	MEYLAN,WM ET AL. (1996)
logP 2.44			EXP	HANSCH,C & LEO,AJ (1985)
VP 1.32E-001	mm Hg	25	EST	NEELY,WB & BLAU,GE (1985)
DC	pKa			
HL 8.32E-006	atm m3/mol	25	EST	MEYLAN,WM & HOWARD,PH (1991)
OH 4.03E-011	cm3/molc sec	25	EST	MEYLAN,WM & HOWARD,PH (1993)

CAS #: 000768-95-6				1-HYDROXYADAMANTANE

Formula: $C_{10}H_{16}O$
Mol Weight: 152.24
MP (deg C): 247 sub   FP (deg C):
BP (deg C):
BP pressure (mm Hg):

Property/Value	Units	Temp	Data Type	Reference
WS 3.57E+003	mg/L	25	EST	MEYLAN,WM ET AL. (1996)
logP 2.14			EXP	HANSCH,C & LEO,AJ (1985)
VP 1.80E-002	mm Hg	25	EST	NEELY,WB & BLAU,GE (1985)
DC	pKa			
HL 2.96E-006	atm m3/mol	25	EST	MEYLAN,WM & HOWARD,PH (1991)
OH 2.83E-011	cm3/molc sec	25	EST	MEYLAN,WM & HOWARD,PH (1993)

CAS #: 000769-92-6				BENZENAMINE, 4-(1,1-DIMETHYLETHYL)

Formula: $C_{10}H_{15}N$
Mol Weight: 149.24
MP (deg C): 17   FP (deg C):
BP (deg C): 241
BP pressure (mm Hg):

Property/Value	Units	Temp	Data Type	Reference
WS 3.78E+002	mg/L	25	EST	MEYLAN,WM ET AL. (1996)
logP 2.70			EXP	HANSCH,C ET AL. (1995)
VP 2.92E-002	mm Hg	25	EST	NEELY,WB & BLAU,GE (1985)
DC 4.95	pKa	25	EXP	PERRIN,DD (1965)
HL 4.92E-006	atm m3/mol	25	EST	MEYLAN,WM & HOWARD,PH (1991)
OH 1.30E-010	cm3/molc sec	25	EST	MEYLAN,WM & HOWARD,PH (1993)

CAS #: 000770-09-2				BENZYLTRIMETHYLSILANE

Formula: $C_{10}H_{16}Si$
Mol Weight: 164.33
MP (deg C):   FP (deg C):
BP (deg C): 190.5
BP pressure (mm Hg):

Property/Value	Units	Temp	Data Type	Reference
WS 1.94E+001	mg/L	25	EST	MEYLAN,WM ET AL. (1996)
logP 4.13			EXP	HANSCH,C ET AL. (1995)
VP 5.84E-001	mm Hg	25	EST	NEELY,WB & BLAU,GE (1985)
DC	pKa			
HL 1.78E-002	atm m3/mol	25	EST	MEYLAN,WM & HOWARD,PH (1991)
OH 6.32E-012	cm3/molc sec	25	EST	MEYLAN,WM & HOWARD,PH (1993)

CAS #: 000770-12-7				PHENYL DICHLOROPHOSPHATE

Formula: $C_6H_5Cl_2O_2P$
Mol Weight: 210.99
MP (deg C):   FP (deg C):
BP (deg C): 241-243
BP pressure (mm Hg):

Property/Value	Units	Temp	Data Type	Reference
WS 1.58E+003	mg/L	25	EST	MEYLAN,WM ET AL. (1996)
logP 1.62			EST	MEYLAN,WM & HOWARD,PH (1995)
VP 5.25E-002	mm Hg	25	EXT	PERRY,RH & GREEN,D (1984)
DC	pKa			
HL 1.01E-006	atm m3/mol	25	EST	MEYLAN,WM & HOWARD,PH (1991)
OH 3.61E-012	cm3/molc sec	25	EST	MEYLAN,WM & HOWARD,PH (1993)

CAS #: 000770-19-4				M-FLUOROPHENYLUREA

Formula: $C_7H_7FN_2O$
Mol Weight: 154.15
MP (deg C):   FP (deg C):
BP (deg C):
BP pressure (mm Hg):

Property/Value	Units	Temp	Data Type	Reference
WS 5.76E+003	mg/L	25	EST	MEYLAN,WM ET AL. (1996)
logP 1.29			EXP	HANSCH,C & LEO,AJ (1985)
VP 1.09E-003	mm Hg	25	EST	NEELY,WB & BLAU,GE (1985)
DC	pKa			
HL 2.35E-010	atm m3/mol	25	EST	MEYLAN,WM & HOWARD,PH (1991)
OH 5.44E-011	cm3/molc sec	25	EST	MEYLAN,WM & HOWARD,PH (1993)

CAS #: 000771-39-1 — 2-NH2-1,3-BENZOXAZIN-4-ONE

Formula: $C_8H_6N_2O_2$

Mol Weight: 162.15

MP (deg C):   FP (deg C):

BP (deg C):

BP pressure (mm Hg):

Property/Value	Units	Temp	Data Type	Reference
WS 3.94E+004	mg/L	25	EST	MEYLAN,WM ET AL. (1996)
logP 0.27			EXP	HANSCH,C & LEO,AJ (1985)
VP 3.94E-005	mm Hg	25	EST	NEELY,WB & BLAU,GE (1985)
DC	pKa			
HL 2.87E-012	atm m3/mol	25	EST	MEYLAN,WM & HOWARD,PH (1991)
OH 4.06E-011	cm3/molc sec	25	EST	MEYLAN,WM & HOWARD,PH (1993)

CAS #: 000771-50-6 — INDOLE, 3-CARBOXYLIC ACID

Formula: $C_9H_7NO_2$

Mol Weight: 161.16

MP (deg C): 205-208   FP (deg C):

BP (deg C):

BP pressure (mm Hg):

Property/Value	Units	Temp	Data Type	Reference
WS 1.35E+003	mg/L	25	EST	MEYLAN,WM ET AL. (1996)
logP 1.99			EXP	HANSCH,C ET AL. (1995)
VP 1.21E-005	mm Hg	25	EST	NEELY,WB & BLAU,GE (1985)
DC	pKa			
HL 1.78E-011	atm m3/mol	25	EST	MEYLAN,WM & HOWARD,PH (1991)
OH 4.12E-011	cm3/molc sec	25	EST	MEYLAN,WM & HOWARD,PH (1993)

CAS #: 000771-56-2 — N-PROPYLPENTAFLUOROBENZENE

Formula: $C_7H_3F_5$

Mol Weight: 182.09

MP (deg C): -29.8   FP (deg C):

BP (deg C): 117.5

BP pressure (mm Hg):

Property/Value	Units	Temp	Data Type	Reference
WS 5.08E+001	mg/L	25	EST	MEYLAN,WM ET AL. (1996)
logP 3.54			EST	MEYLAN,WM & HOWARD,PH (1995)
VP 3.98E+001	mm Hg	25	EST	NEELY,WB & BLAU,GE (1985)
DC	pKa			
HL 1.29E-002	atm m3/mol	25	EST	MEYLAN,WM & HOWARD,PH (1991)
OH 3.06E-012	cm3/molc sec	25	EXP	ATKINSON,R (1989)

CAS #: 000771-60-8 — 2,3,4,5,6-PENTAFLUOROANILINE

Formula: $C_6H_2F_5N$

Mol Weight: 183.08

MP (deg C): 35-36   FP (deg C):

BP (deg C): 153

BP pressure (mm Hg):

Property/Value	Units	Temp	Data Type	Reference
WS 8.94E+002	mg/L	25	EST	MEYLAN,WM ET AL. (1996)
logP 2.08			EST	MEYLAN,WM & HOWARD,PH (1995)
VP 2.56E+000	mm Hg	25	EST	NEELY,WB & BLAU,GE (1985)
DC -0.28	pKa	25	EXP	PERRIN,DD (1972)
HL 4.12E-006	atm m3/mol	25	EST	MEYLAN,WM & HOWARD,PH (1991)
OH 6.50E-012	cm3/molc sec	25	EST	MEYLAN,WM & HOWARD,PH (1993)

CAS #: 000771-61-9 — PENTAFLUOROPHENOL

Formula: C_6HF_5O

Mol Weight: 184.07

MP (deg C): 32.8   FP (deg C):

BP (deg C): 145.6

BP pressure (mm Hg):

Property/Value	Units	Temp	Data Type	Reference
WS 3.49E+002	mg/L	25	EST	MEYLAN,WM ET AL. (1996)
logP 3.23			EXP	HANSCH,C & LEO,AJ (1985)
VP 4.61E+001	mm Hg	25	EXT	BOUBLIK,T ET AL. (1984)
DC 5.53	pKa	25	EXP	SERJEANT,EP & DEMPSEY,B (1979)
HL 1.21E-006	atm m3/mol	25	EST	MEYLAN,WM & HOWARD,PH (1991)
OH 2.15E-012	cm3/molc sec	25	EST	MEYLAN,WM & HOWARD,PH (1993)

CAS #: 000771-97-1 — 2,3-NAPHTHALENEDIAMINE

Formula: $C_{10}H_{10}N_2$

Mol Weight: 158.20

MP (deg C): 199   FP (deg C):

BP (deg C):

BP pressure (mm Hg):

Property/Value	Units	Temp	Data Type	Reference
WS 3.38E+003	mg/L	25	EST	MEYLAN,WM ET AL. (1996)
logP 1.54			EXP	SCHULTZ,TW & APPLEHANS,FM (1985)
VP 2.93E-005	mm Hg	25	EST	NEELY,WB & BLAU,GE (1985)
DC	pKa			
HL 6.57E-011	atm m3/mol	25	EST	MEYLAN,WM & HOWARD,PH (1991)
OH 2.00E-010	cm3/molc sec	25	EST	MEYLAN,WM & HOWARD,PH (1993)

CAS #: 000771-99-3 — PIPERIDINE, 4-PHENYL-

Formula: $C_{11}H_{15}N$

Mol Weight: 161.25

MP (deg C): 60.5   FP (deg C):

BP (deg C): 257

BP pressure (mm Hg):

Property/Value	Units	Temp	Data Type	Reference
WS 6.26E+003	mg/L	25	EST	MEYLAN,WM ET AL. (1996)
logP 2.39			EXP	SANGSTER,J (1993)
VP 5.50E-003	mm Hg	25	EST	NEELY,WB & BLAU,GE (1985)
DC	pKa			
HL 1.39E-006	atm m3/mol	25	EST	MEYLAN,WM & HOWARD,PH (1991)
OH 9.48E-011	cm3/molc sec	25	EST	MEYLAN,WM & HOWARD,PH (1993)

CAS #: 000774-40-3 — MANDELIC ACID, ETHYL ESTER

Formula: $C_{10}H_{12}O_3$

Mol Weight: 180.21

MP (deg C): 35   FP (deg C):

BP (deg C): 150

BP pressure (mm Hg): 2.10E+001

Property/Value	Units	Temp	Data Type	Reference
WS 8.78E+003	mg/L	25	EST	MEYLAN,WM ET AL. (1996)
logP 1.53			EXP	HANSCH,C & LEO,AJ (1985)
VP 2.60E-004	mm Hg	25	EST	NEELY,WB & BLAU,GE (1985)
DC	pKa			
HL 2.93E-006	atm m3/mol	25	EST	MEYLAN,WM & HOWARD,PH (1991)
OH 1.15E-011	cm3/molc sec	25	EST	MEYLAN,WM & HOWARD,PH (1993)

CAS #: 000776-35-2			9,10-DIHYDROPHENANTHRENE	

Formula: $C_{14}H_{12}$
Mol Weight: 180.25
MP (deg C): 34.5 FP (deg C):
BP (deg C): 168-169
BP pressure (mm Hg): 1.50E+001

Property/Value	Units	Temp	Data Type	Reference
WS 5.88E-001	mg/L	25	EST	MEYLAN,WM ET AL. (1996)
logP 4.52			EXP	HANSCH,C ET AL. (1995)
VP 4.43E-004	mm Hg	25	EST	NEELY,WB & BLAU,GE (1985)
DC	pKa			
HL 2.22E-004	atm m3/mol	25	EST	MEYLAN,WM & HOWARD,PH (1991)
OH 1.06E-011	cm3/molc sec	25	EST	MEYLAN,WM & HOWARD,PH (1993)

CAS #: 000776-75-0			PIPERIDINE, 1-BENZOYL-	

Formula: $C_{12}H_{15}NO$
Mol Weight: 189.26
MP (deg C): 49 FP (deg C):
BP (deg C): 320.5
BP pressure (mm Hg):

Property/Value	Units	Temp	Data Type	Reference
WS 6.55E+002	mg/L	25	EST	MEYLAN,WM ET AL. (1996)
logP 2.20			EXP	SURYANARAYANA,MVS ET AL. (1991)
VP 1.88E-004	mm Hg	25	EST	NEELY,WB & BLAU,GE (1985)
DC	pKa			
HL 1.10E-008	atm m3/mol	25	EST	MEYLAN,WM & HOWARD,PH (1991)
OH 3.04E-011	cm3/molc sec	25	EST	MEYLAN,WM & HOWARD,PH (1993)

CAS #: 000777-59-3			1-ME-1-NITROSO-3-(P-F PH)UREA	

Formula: $C_8H_8FN_3O_2$
Mol Weight: 197.17
MP (deg C): FP (deg C):
BP (deg C):
BP pressure (mm Hg):

Property/Value	Units	Temp	Data Type	Reference
WS 1.12E+003	mg/L	25	EST	MEYLAN,WM ET AL. (1996)
logP 1.88			EXP	HANSCH,C & LEO,AJ (1985)
VP 1.32E-004	mm Hg	25	EST	NEELY,WB & BLAU,GE (1985)
DC	pKa			
HL 6.38E-011	atm m3/mol	25	EST	MEYLAN,WM & HOWARD,PH (1991)
OH 1.50E-011	cm3/molc sec	25	EST	MEYLAN,WM & HOWARD,PH (1993)

CAS #: 000779-02-2			9-METHYLANTHRACENE	

Formula: $C_{15}H_{12}$
Mol Weight: 192.26
MP (deg C): 81.5 FP (deg C):
BP (deg C): 196-197
BP pressure (mm Hg): 1.20E+001

Property/Value	Units	Temp	Data Type	Reference
WS 2.61E-001	mg/L	25	EXP	YALKOWSKY,SH & DANNENFELSER,RM (1992)
logP 5.07			EXP	SANGSTER,J (1993)
VP 5.01E-005	mm Hg	25	EST	NEELY,WB & BLAU,GE (1985)
DC	pKa			
HL 5.67E-005	atm m3/mol	25	EST	MEYLAN,WM & HOWARD,PH (1991)
OH 1.05E-010	cm3/molc sec	25	EST	MEYLAN,WM & HOWARD,PH (1993)

CAS #: 000779-84-0			2-HYDROXYBENZALANILINE	

Formula: $C_{13}H_{11}NO$
Mol Weight: 197.24
MP (deg C): 49.5 FP (deg C):
BP (deg C):
BP pressure (mm Hg):

Property/Value	Units	Temp	Data Type	Reference
WS 3.95E+002	mg/L	25	EST	MEYLAN,WM ET AL. (1996)
logP 3.09			EXP	HANSCH,C & LEO,AJ (1985)
VP 1.93E-005	mm Hg	25	EST	NEELY,WB & BLAU,GE (1985)
DC	pKa			
HL 1.02E-008	atm m3/mol	25	EST	MEYLAN,WM & HOWARD,PH (1991)
OH 4.43E-011	cm3/molc sec	25	EST	MEYLAN,WM & HOWARD,PH (1993)

CAS #: 000780-11-0			M-T-BUTYLPHENYL N-METHYLCARBAMATE	

Formula: $C_{12}H_{17}NO_2$
Mol Weight: 207.27
MP (deg C): FP (deg C):
BP (deg C):
BP pressure (mm Hg):

Property/Value	Units	Temp	Data Type	Reference
WS 1.26E+002	mg/L	25	EST	MEYLAN,WM ET AL. (1996)
logP 2.93			EXP	HANSCH,C & LEO,AJ (1985)
VP 1.80E-003	mm Hg	25	EST	NEELY,WB & BLAU,GE (1985)
DC	pKa			
HL 8.29E-008	atm m3/mol	25	EST	MEYLAN,WM & HOWARD,PH (1991)
OH 1.10E-011	cm3/molc sec	25	EST	MEYLAN,WM & HOWARD,PH (1993)

CAS #: 000780-59-6			N-METHYLALLOBARBITAL	

Formula: $C_{11}H_{14}N_2O_3$
Mol Weight: 222.25
MP (deg C): FP (deg C):
BP (deg C):
BP pressure (mm Hg):

Property/Value	Units	Temp	Data Type	Reference
WS 1.59E+003	mg/L	25	EST	MEYLAN,WM ET AL. (1996)
logP 1.55			EXP	HANSCH,C & LEO,AJ (1985)
VP 1.54E-010	mm Hg	25	EST	NEELY,WB & BLAU,GE (1985)
DC	pKa			
HL 7.75E-013	atm m3/mol	25	EST	MEYLAN,WM & HOWARD,PH (1991)
OH 6.24E-011	cm3/molc sec	25	EST	MEYLAN,WM & HOWARD,PH (1993)

CAS #: 000780-69-8			SILANE, TRIETHOXYPHENYL-	

Formula: $C_{12}H_{20}O_3Si$
Mol Weight: 240.38
MP (deg C): FP (deg C):
BP (deg C): 112-113
BP pressure (mm Hg): 1.00E+001

Property/Value	Units	Temp	Data Type	Reference
WS 7.47E+001	mg/L	25	EST	MEYLAN,WM ET AL. (1996)
logP 2.99			EXP	HANSCH,C ET AL. (1995)
VP 2.76E-003	mm Hg	25	EST	NEELY,WB & BLAU,GE (1985)
DC	pKa			
HL 4.04E-008	atm m3/mol	25	EST	MEYLAN,WM & HOWARD,PH (1991)
OH 2.05E-011	cm3/molc sec	25	EST	MEYLAN,WM & HOWARD,PH (1993)

000781-43-1 — 9,10-DIMETHYLANTHRACENE

Formula: $C_{16}H_{14}$

Mol Weight: 206.29

MP (deg C): 182-184

FP (deg C):

BP (deg C):

BP pressure (mm Hg):

Property/Value	Units	Temp	Data Type	Reference
WS 5.60E-002	mg/L	25	EXP	YALKOWSKY,SH & DANNENFELSER,RM (1992)
logP 5.69			EXP	SANGSTER,J (1993)
VP 1.82E-005	mm Hg	25	EST	NEELY,WB & BLAU,GE (1985)
DC	pKa			
HL 6.25E-005	atm m3/mol	25	EST	MEYLAN,WM & HOWARD,PH (1991)
OH 1.28E-010	cm3/molc sec	25	EST	MEYLAN,WM & HOWARD,PH (1993)

000786-19-6 — CARBOPHENTHION

Formula: $C_{11}H_{16}ClO_2PS_3$

Mol Weight: 342.87

MP (deg C):

FP (deg C):

BP (deg C): 82

BP pressure (mm Hg): 1.00E-002

Property/Value	Units	Temp	Data Type	Reference
WS 6.30E-001	mg/L	20	EXP	BOWMAN,BT & SANS,WW (1983)
logP 5.33			EXP	HANSCH,C & LEO,AJ (1985)
VP 3.00E-007	mm Hg	20	EXP	SUNTIO,LR ET AL. (1988)
DC	pKa			
HL 2.15E-007	atm m3/mol	20	EST	VP/WSOL
OH 2.48E-010	cm3/molc sec	25	EST	ATKINSON,R (1987)

000789-02-6 — O,P'-DDT

Formula: $C_{14}H_9Cl_5$

Mol Weight: 354.49

MP (deg C):

FP (deg C):

BP (deg C):

BP pressure (mm Hg):

Property/Value	Units	Temp	Data Type	Reference
WS 8.50E-002	mg/L	25	EXP	YALKOWSKY,SH & DANNENFELSER,RM (1992)
logP 6.79			EST	MEYLAN,WM & HOWARD,PH (1995)
VP 1.10E-007	mm Hg	25	EXP	MACKAY,D & SHIU,WY (1981)
DC	pKa			
HL 5.88E-007	atm m3/mol	25	EST	VP/WSOL
OH 3.44E-012	cm3/molc sec	25	EST	MEYLAN,WM & HOWARD,PH (1993)

000789-24-2 — 9-PHENYL-9H-FLUORENE

Formula: $C_{19}H_{14}$

Mol Weight: 242.32

MP (deg C): 146-148

FP (deg C):

BP (deg C):

BP pressure (mm Hg):

Property/Value	Units	Temp	Data Type	Reference
WS 5.29E-002	mg/L	25	EST	MEYLAN,WM ET AL. (1996)
logP 5.37			EST	MEYLAN,WM & HOWARD,PH (1995)
VP 2.47E-006	mm Hg	25	EST	NEELY,WB & BLAU,GE (1985)
DC	pKa			
HL 1.35E-005	atm m3/mol	25	EST	MEYLAN,WM & HOWARD,PH (1991)
OH 1.43E-011	cm3/molc sec	25	EST	MEYLAN,WM & HOWARD,PH (1993)

000789-61-7 — B-2'-DEOXYTHIOGUANOSINE

Formula: $C_{10}H_{13}N_5O_3S$

Mol Weight: 283.31

MP (deg C):

FP (deg C):

BP (deg C):

BP pressure (mm Hg):

Property/Value	Units	Temp	Data Type	Reference
WS 2.30E+004	mg/L	25	EST	MEYLAN,WM ET AL. (1996)
logP -0.56			EXP	HANSCH,C & LEO,AJ (1985)
VP 8.02E-018	mm Hg	25	EST	NEELY,WB & BLAU,GE (1985)
DC	pKa			
HL 2.01E-022	atm m3/mol	25	EST	MEYLAN,WM & HOWARD,PH (1991)
OH 1.71E-010	cm3/molc sec	25	EST	MEYLAN,WM & HOWARD,PH (1993)

000790-53-4 — DIETHYLMALONATE,4-F BENZAL

Formula: $C_{14}H_{15}FO_4$

Mol Weight: 266.27

MP (deg C):

FP (deg C):

BP (deg C):

BP pressure (mm Hg):

Property/Value	Units	Temp	Data Type	Reference
WS 4.32E+001	mg/L	25	EST	MEYLAN,WM ET AL. (1996)
logP 3.10			EXP	HANSCH,C & LEO,AJ (1985)
VP 1.20E-003	mm Hg	25	EST	NEELY,WB & BLAU,GE (1985)
DC	pKa			
HL 1.28E-008	atm m3/mol	25	EST	MEYLAN,WM & HOWARD,PH (1991)
OH 1.63E-011	cm3/molc sec	25	EST	MEYLAN,WM & HOWARD,PH (1993)

000791-28-6 — TRIPHENYLPHOSPHINE OXIDE

Formula: $C_{18}H_{15}OP$

Mol Weight: 278.29

MP (deg C): 156.5

FP (deg C):

BP (deg C): > 360

BP pressure (mm Hg):

Property/Value	Units	Temp	Data Type	Reference
WS 6.28E+001	mg/L	25	EST	MEYLAN,WM ET AL. (1996)
logP 2.83			EXP	HANSCH,C & LEO,AJ (1985)
VP 2.60E-009	mm Hg	25	EXT	DAUBERT,TE & DANNER,RP (1993)
DC	pKa			
HL 5.26E-010	atm m3/mol	25	EST	MEYLAN,WM & HOWARD,PH (1991)
OH 5.85E-012	cm3/molc sec	25	EST	MEYLAN,WM & HOWARD,PH (1993)

000800-79-3 — AMPICILLIN

Formula: $C_{16}H_{19}N_3O_4S$

Mol Weight: 349.41

MP (deg C):

FP (deg C):

BP (deg C):

BP pressure (mm Hg):

Property/Value	Units	Temp	Data Type	Reference
WS 4.39E+002	mg/L	25	EST	MEYLAN,WM ET AL. (1996)
logP 1.35			EXP	HANSCH,C & LEO,AJ (1985)
VP 7.75E-015	mm Hg	25	EST	NEELY,WB & BLAU,GE (1985)
DC	pKa			
HL 2.39E-017	atm m3/mol	25	EST	MEYLAN,WM & HOWARD,PH (1991)
OH 1.25E-010	cm3/molc sec	25	EST	MEYLAN,WM & HOWARD,PH (1993)

CAS #: 000807-28-3 — 1,3-DIMETHYL-1,1,3,3-TETRAPHENYLDISILOXANE

Formula: $C_{26}H_{26}OSi_2$

Mol Weight: 410.67

MP (deg C): | FP (deg C):

BP (deg C):

BP pressure (mm Hg):

Property/Value	Units	Temp	Data Type	Reference
WS 2.62E-006	mg/L	25	EST	MEYLAN,WM ET AL. (1996)
logP 9.63			EST	MEYLAN,WM & HOWARD,PH (1995)
VP 3.35E-009	mm Hg	25	EST	NEELY,WB & BLAU,GE (1985)
DC	pKa			
HL 9.57E-008	atm m3/mol	25	EST	MEYLAN,WM & HOWARD,PH (1991)
OH 8.10E-012	cm3/molc sec	25	EST	MEYLAN,WM & HOWARD,PH (1993)

CAS #: 000811-97-2 — 1,1,1,2-TETRAFLUOROETHANE

Formula: $C_2H_2F_4$

Mol Weight: 102.03

MP (deg C): -101 | FP (deg C):

BP (deg C): -26.5

BP pressure (mm Hg):

Property/Value	Units	Temp	Data Type	Reference
WS 6.70E+001	mg/L	25	EST	VP/HL
logP 1.68			EST	MEYLAN,WM & HOWARD,PH (1995)
VP 4.30E+002	mm Hg	25	EST	PCHEM (1987)
DC	pKa			
HL 1.53E+000	atm m3/mol	25	EST	MEYLAN,WM & HOWARD,PH (1991)
OH 6.00E-015	cm3/molc sec	25	EXP	ATKINSON,R (1989)

CAS #: 000814-75-5 — 3-BROMO-2-BUTANONE

Formula: C_4H_7BrO

Mol Weight: 151.01

MP (deg C): | FP (deg C):

BP (deg C):

BP pressure (mm Hg):

Property/Value	Units	Temp	Data Type	Reference
WS 2.67E+004	mg/L	25	EST	MEYLAN,WM ET AL. (1996)
logP 0.53			EST	MEYLAN,WM & HOWARD,PH (1995)
VP 1.06E+001	mm Hg	25	EST	NEELY,WB & BLAU,GE (1985)
DC	pKa			
HL 7.58E-006	atm m3/mol	25	EST	MEYLAN,WM & HOWARD,PH (1991)
OH 9.31E-013	cm3/molc sec	25	EST	MEYLAN,WM & HOWARD,PH (1993)

CAS #: 000815-24-7 — 3-PENTANONE, 2,2,4,4-TETRAMETHYL-

Formula: $C_9H_{18}O$

Mol Weight: 142.24

MP (deg C): -25.2 | FP (deg C):

BP (deg C): 152

BP pressure (mm Hg):

Property/Value	Units	Temp	Data Type	Reference
WS 2.25E+002	mg/L	25	EST	MEYLAN,WM ET AL. (1996)
logP 3.00			EXP	HANSCH,C ET AL. (1995)
VP 7.40E+000	mm Hg	25	EST	NEELY,WB & BLAU,GE (1985)
DC	pKa			
HL 2.71E-004	atm m3/mol	25	EST	MEYLAN,WM & HOWARD,PH (1991)
OH 3.18E-012	cm3/molc sec	25	EST	MEYLAN,WM & HOWARD,PH (1993)

CAS #: 000816-40-0 — 1-BROMO-2-BUTANONE

Formula: C_4H_7BrO

Mol Weight: 151.01

MP (deg C): | FP (deg C):

BP (deg C): 105

BP pressure (mm Hg): 1.50E+002

Property/Value	Units	Temp	Data Type	Reference
WS 2.31E+004	mg/L	25	EST	MEYLAN,WM ET AL. (1996)
logP 0.60			EST	MEYLAN,WM & HOWARD,PH (1995)
VP 3.63E+000	mm Hg	25	EXP	PERRY,RH & GREEN,D (1984)
DC	pKa			
HL 7.58E-006	atm m3/mol	25	EST	MEYLAN,WM & HOWARD,PH (1991)
OH 1.43E-012	cm3/molc sec	25	EST	MEYLAN,WM & HOWARD,PH (1993)

CAS #: 000816-57-9 — 1-PROPYL-1-NITROSOUREA

Formula: $C_4H_9N_3O_2$

Mol Weight: 131.14

MP (deg C): | FP (deg C):

BP (deg C):

BP pressure (mm Hg):

Property/Value	Units	Temp	Data Type	Reference
WS 2.47E+003	mg/L	25	EST	MEYLAN,WM ET AL. (1996)
logP 0.30			EXP	HANSCH,C & LEO,AJ (1985)
VP 1.65E-002	mm Hg	25	EST	NEELY,WB & BLAU,GE (1985)
DC	pKa			
HL 1.75E-010	atm m3/mol	25	EST	MEYLAN,WM & HOWARD,PH (1991)
OH 7.03E-012	cm3/molc sec	25	EST	MEYLAN,WM & HOWARD,PH (1993)

CAS #: 000816-79-5 — 3-ETHYL-2-PENTENE

Formula: C_7H_{14}

Mol Weight: 98.19

MP (deg C): | FP (deg C):

BP (deg C): 96

BP pressure (mm Hg):

Property/Value	Units	Temp	Data Type	Reference
WS 2.78E+001	mg/L	25	EST	MEYLAN,WM ET AL. (1996)
logP 3.62			EST	MEYLAN,WM & HOWARD,PH (1995)
VP 5.28E+001	mm Hg	25	EXP	DYKYJ,J & REPAS,M (1973)
DC	pKa			
HL 6.63E-001	atm m3/mol	25	EST	MEYLAN,WM & HOWARD,PH (1991)
OH 8.92E-011	cm3/molc sec	25	EST	MEYLAN,WM & HOWARD,PH (1993)

CAS #: 000817-99-2 — ACETAMIDE, N-(2-AMINO-2-OXOETHYL)-2-DIAZO-

Formula: $C_4H_6N_4O_2$

Mol Weight: 142.12

MP (deg C): | FP (deg C):

BP (deg C):

BP pressure (mm Hg):

Property/Value	Units	Temp	Data Type	Reference
WS 4.79E+004	mg/L	25	EST	MEYLAN,WM ET AL. (1996)
logP -1.26			EXP	HANSCH,C & LEO,AJ (1985)
VP 1.05E-011	mm Hg	25	EST	NEELY,WB & BLAU,GE (1985)
DC	pKa			
HL 1.08E-015	atm m3/mol	25	EST	MEYLAN,WM & HOWARD,PH (1991)
OH 1.18E-011	cm3/molc sec	25	EST	MEYLAN,WM & HOWARD,PH (1993)

CAS #: 000818-23-5				8-PENTADECANONE

Formula: $C_{15}H_{30}O$

Mol Weight: 226.41

MP (deg C): 43   FP (deg C):

BP (deg C): 291

BP pressure (mm Hg):

Property/Value	Units	Temp	Data Type	Reference
WS 4.68E-001	mg/L	25	EST	MEYLAN,WM ET AL. (1996)
logP 5.66			EST	MEYLAN,WM & HOWARD,PH (1995)
VP 4.04E-004	mm Hg	25	EXT	ENGINEERING SCIENCES DATA UNIT (1975)
DC	pKa			
HL 1.48E-003	atm m3/mol	25	EST	MEYLAN,WM & HOWARD,PH (1991)
OH 2.18E-011	cm3/molc sec	25	EST	MEYLAN,WM & HOWARD,PH (1993)

CAS #: 000818-38-2				DIETHYL GLUTARATE

Formula: $C_9H_{16}O_4$

Mol Weight: 188.23

MP (deg C): -24.1   FP (deg C):

BP (deg C): 236.5

BP pressure (mm Hg):

Property/Value	Units	Temp	Data Type	Reference
WS 1.25E+003	mg/L	25	EST	MEYLAN,WM ET AL. (1996)
logP 1.88			EST	MEYLAN,WM & HOWARD,PH (1995)
VP 1.61E-001	mm Hg	25	EXT	OHE,S (1976)
DC	pKa			
HL 1.30E-006	atm m3/mol	25	EST	MEYLAN,WM & HOWARD,PH (1991)
OH 6.44E-012	cm3/molc sec	25	EST	MEYLAN,WM & HOWARD,PH (1993)

CAS #: 000818-61-1				2-HYDROXYETHYL ACRYLATE

Formula: $C_5H_8O_3$

Mol Weight: 116.12

MP (deg C):   FP (deg C):

BP (deg C): 90-92

BP pressure (mm Hg): 1.20E+001

Property/Value	Units	Temp	Data Type	Reference
WS 5.07E+005	mg/L	25	EST	MEYLAN,WM ET AL. (1996)
logP -0.21			EXP	HANSCH,C & LEO,AJ (1985)
VP 5.24E-002	mm Hg	25	EXP	YAWS,CL (1994A)
DC	pKa			
HL 4.49E-009	atm m3/mol	25	EST	MEYLAN,WM & HOWARD,PH (1991)
OH 1.52E-011	cm3/molc sec	25	EST	MEYLAN,WM & HOWARD,PH (1993)

CAS #: 000819-06-7				2,2,2-TRIF N,N-DIME ETHYLAMINE

Formula: $C_4H_8F_3N$

Mol Weight: 127.11

MP (deg C):   FP (deg C):

BP (deg C):

BP pressure (mm Hg):

Property/Value	Units	Temp	Data Type	Reference
WS 1.20E+005	mg/L	25	EST	MEYLAN,WM ET AL. (1996)
logP 1.06			EXP	HANSCH,C & LEO,AJ (1985)
VP 7.13E+002	mm Hg	25	EST	NEELY,WB & BLAU,GE (1985)
DC	pKa			
HL 3.82E-004	atm m3/mol	25	EST	MEYLAN,WM & HOWARD,PH (1991)
OH 3.15E-012	cm3/molc sec	25	EST	MEYLAN,WM & HOWARD,PH (1993)

CAS #: 000820-69-9				3-HEXENE-2,5-DIONE (TRANS)

Formula: $C_6H_8O_2$

Mol Weight: 112.13

MP (deg C):   FP (deg C):

BP (deg C):

BP pressure (mm Hg):

Property/Value	Units	Temp	Data Type	Reference
WS 3.46E+004	mg/L	25	EST	MEYLAN,WM ET AL. (1996)
logP 0.57			EST	MEYLAN,WM & HOWARD,PH (1995)
VP 2.17E+000	mm Hg	25	EST	NEELY,WB & BLAU,GE (1985)
DC	pKa			
HL 1.11E-008	atm m3/mol	25	EST	MEYLAN,WM & HOWARD,PH (1991)
OH 5.31E-011	cm3/molc sec	25	EXP	ATKINSON,R (1989)

CAS #: 000821-07-8				TRANS-1,3,5-HEXATRIENE

Formula: C_6H_8

Mol Weight: 80.13

MP (deg C): -12   FP (deg C):

BP (deg C): 78.5

BP pressure (mm Hg):

Property/Value	Units	Temp	Data Type	Reference
WS 1.54E+002	mg/L	25	EST	MEYLAN,WM ET AL. (1996)
logP 2.80			EST	MEYLAN,WM & HOWARD,PH (1995)
VP 1.15E+002	mm Hg	25	EST	NEELY,WB & BLAU,GE (1985)
DC	pKa			
HL 9.84E-002	atm m3/mol	25	EST	MEYLAN,WM & HOWARD,PH (1991)
OH 1.11E-010	cm3/molc sec	25	EXP	ATKINSON,R (1989)

CAS #: 000821-09-0				4-PENTEN-1-OL

Formula: $C_5H_{10}O$

Mol Weight: 86.13

MP (deg C):   FP (deg C):

BP (deg C): 141

BP pressure (mm Hg):

Property/Value	Units	Temp	Data Type	Reference
WS 5.70E+004	mg/L	25	EXP	YALKOWSKY,SH & DANNENFELSER,RM (1992)
logP 1.20			EST	MEYLAN,WM & HOWARD,PH (1995)
VP 3.06E+000	mm Hg	25	EST	NEELY,WB & BLAU,GE (1985)
DC	pKa			
HL 9.88E-006	atm m3/mol	25	EST	MEYLAN,WM & HOWARD,PH (1991)
OH 3.30E-011	cm3/molc sec	25	EST	MEYLAN,WM & HOWARD,PH (1993)

CAS #: 000821-10-3				2-BUTYNE, 1,4-DICHLORO-

Formula: $C_4H_4Cl_2$

Mol Weight: 122.98

MP (deg C):   FP (deg C):

BP (deg C): 165.5

BP pressure (mm Hg):

Property/Value	Units	Temp	Data Type	Reference
WS 1.88E+003	mg/L	25	EST	MEYLAN,WM ET AL. (1996)
logP 2.01			EXP	HANSCH,C ET AL. (1995)
VP 1.86E+000	mm Hg	25	EST	NEELY,WB & BLAU,GE (1985)
DC	pKa			
HL 1.33E-003	atm m3/mol	25	EST	MEYLAN,WM & HOWARD,PH (1991)
OH 1.63E-011	cm3/molc sec	25	EST	MEYLAN,WM & HOWARD,PH (1993)

CAS #: 000821-11-4				2-BUTENE-1,4-DIOL (TRANS)
Formula: $C_4H_8O_2$				
Mol Weight: 88.11				
MP (deg C): 25		FP (deg C):		
BP (deg C): 131				
BP pressure (mm Hg): 1.30E+001				

Property/ Value	Units	Temp	Data Type	Reference
WS 6.19E+005	mg/L	25	EST	MEYLAN,WM ET AL. (1996)
logP -0.43			EST	MEYLAN,WM & HOWARD,PH (1995)
VP 6.99E-003	mm Hg	25	EXT	YAWS,CL (1994)
DC	pKa			
HL 2.03E-007	atm m3/mol	25	EST	MEYLAN,WM & HOWARD,PH (1991)
OH 6.32E-011	cm3/molc sec	25	EST	MEYLAN,WM & HOWARD,PH (1993)

CAS #: 000821-55-6				2-NONANONE
Formula: $C_9H_{18}O$				
Mol Weight: 142.24				
MP (deg C): -7.5		FP (deg C):		
BP (deg C): 195.3				
BP pressure (mm Hg):				

Property/ Value	Units	Temp	Data Type	Reference
WS 3.80E+002	mg/L	25	EXP	ISNARD,P & LAMBERT,S (1989)
logP 3.14			EXP	HANSCH,C ET AL. (1995)
VP 6.24E-001	mm Hg	25	EXP	OHE,S (1976)
DC	pKa			
HL 3.67E-004	atm m3/mol	25	EXP	BUTTERY,RG ET AL. (1969)
OH 1.22E-011	cm3/molc sec	25	EXP	ATKINSON,R (1989)

CAS #: 000821-95-4				1-UNDECENE
Formula: $C_{11}H_{22}$				
Mol Weight: 154.30				
MP (deg C): -49.2		FP (deg C):		
BP (deg C): 192.7				
BP pressure (mm Hg):				

Property/ Value	Units	Temp	Data Type	Reference
WS 3.43E-001	mg/L	25	EST	MEYLAN,WM ET AL. (1996)
logP 5.61			EST	MEYLAN,WM & HOWARD,PH (1995)
VP 4.93E-001	mm Hg	25	EXP	DAUBERT,TE & DANNER,RP (1989)
DC	pKa			
HL 1.48E+000	atm m3/mol	25	EST	MEYLAN,WM & HOWARD,PH (1991)
OH 3.72E-011	cm3/molc sec	25	EST	MEYLAN,WM & HOWARD,PH (1993)

CAS #: 000822-06-0				HEXAMETHYLENE DIISOCYANATE
Formula: $C_8H_{12}N_2O_2$				
Mol Weight: 168.20				
MP (deg C): -67		FP (deg C):		
BP (deg C): 255				
BP pressure (mm Hg):				

Property/ Value	Units	Temp	Data Type	Reference
WS 1.17E+002	mg/L	25	EST	MEYLAN,WM ET AL. (1996)
logP 3.20			EST	MEYLAN,WM & HOWARD,PH (1995)
VP 3.00E-002	mm Hg	25	EST	GOLDBERG,NA & KUCHERYAVYI,VI (1960)
DC	pKa			
HL 4.80E-005	atm m3/mol	25	EST	MEYLAN,WM & HOWARD,PH (1991)
OH 7.74E-012	cm3/molc sec	25	EST	ATKINSON,R (1988)

CAS #: 000822-36-6				1H-IMIDAZOLE, 4-METHYL-
Formula: $C_4H_6N_2$				
Mol Weight: 82.11				
MP (deg C): 56		FP (deg C):		
BP (deg C): 263				
BP pressure (mm Hg):				

Property/ Value	Units	Temp	Data Type	Reference
WS 8.25E+004	mg/L	25	EST	MEYLAN,WM ET AL. (1996)
logP 0.23			EXP	HANSCH,C ET AL. (1995)
VP 1.45E-002	mm Hg	25	EST	NEELY,WB & BLAU,GE (1985)
DC	pKa			
HL 4.14E-006	atm m3/mol	25	EST	MEYLAN,WM & HOWARD,PH (1991)
OH 9.41E-011	cm3/molc sec	25	EST	MEYLAN,WM & HOWARD,PH (1993)

CAS #: 000822-50-4				1,2-DIMETHYLCYCLOPENTANE (TRANS)
Formula: C_7H_{14}				
Mol Weight: 98.19				
MP (deg C): -117.6		FP (deg C):		
BP (deg C): 91.9				
BP pressure (mm Hg):				

Property/ Value	Units	Temp	Data Type	Reference
WS 3.39E+001	mg/L	25	EST	MEYLAN,WM ET AL. (1996)
logP 3.52			EST	MEYLAN,WM & HOWARD,PH (1995)
VP 6.40E+001	mm Hg	25	EXP	DAUBERT,TE & DANNER,RP (1989)
DC	pKa			
HL 3.39E-001	atm m3/mol	25	EST	MEYLAN,WM & HOWARD,PH (1991)
OH 8.42E-012	cm3/molc sec	25	EST	MEYLAN,WM & HOWARD,PH (1993)

CAS #: 000822-84-4				3-NITROTHIOPHENE
Formula: $C_4H_5NO_2S$				
Mol Weight: 131.15				
MP (deg C): 78.5		FP (deg C):		
BP (deg C): 225				
BP pressure (mm Hg):				

Property/ Value	Units	Temp	Data Type	Reference
WS 1.79E+003	mg/L	25	EST	MEYLAN,WM ET AL. (1996)
logP 1.55			EXP	HANSCH,C & LEO,AJ (1985)
VP 1.14E-001	mm Hg	25	EST	NEELY,WB & BLAU,GE (1985)
DC	pKa			
HL 1.16E-005	atm m3/mol	25	EST	MEYLAN,WM & HOWARD,PH (1991)
OH 1.19E-012	cm3/molc sec	25	EST	MEYLAN,WM & HOWARD,PH (1993)

CAS #: 000822-86-6				1,2-DICHLOROCYCLOHEXANE -TRANS
Formula: $C_6H_{10}Cl_2$				
Mol Weight: 153.05				
MP (deg C):		FP (deg C):		
BP (deg C): 193-194				
BP pressure (mm Hg):				

Property/ Value	Units	Temp	Data Type	Reference
WS 1.42E+002	mg/L	25	EST	MEYLAN,WM ET AL. (1996)
logP 3.21			EXP	HANSCH,C & LEO,AJ (1985)
VP 1.32E+000	mm Hg	25	EST	NEELY,WB & BLAU,GE (1985)
DC	pKa			
HL 1.66E-002	atm m3/mol	25	EST	MEYLAN,WM & HOWARD,PH (1991)
OH 4.31E-012	cm3/molc sec	25	EST	MEYLAN,WM & HOWARD,PH (1993)

CAS #: 000823-09-6	PYRIMIDINE, 2-(METHYLTHIO)-

Formula: $C_5H_6N_2S$

Mol Weight: 126.18

MP (deg C):

FP (deg C):

BP (deg C):

BP pressure (mm Hg):

Property/ Value	Units	Temp	Data Type	Reference
WS 1.31E+004	mg/L	25	EST	MEYLAN,WM ET AL. (1996)
logP 1.01			EXP	YAMAGAMI,C ET AL. (1990)
VP 2.11E-001	mm Hg	25	EST	NEELY,WB & BLAU,GE (1985)
DC 0.59	pKa	20	EXP	PERRIN,DD (1965)
HL 3.31E-006	atm m3/mol	25	EST	MEYLAN,WM & HOWARD,PH (1991)
OH 2.54E-012	cm3/molc sec	25	EST	MEYLAN,WM & HOWARD,PH (1993)

CAS #: 000823-40-5	2,6-DIAMINOTOLUENE

Formula: $C_7H_{10}N_2$

Mol Weight: 122.17

MP (deg C): 105

FP (deg C):

BP (deg C): 260

BP pressure (mm Hg):

Property/ Value	Units	Temp	Data Type	Reference
WS 7.25E+004	mg/L	25	EST	MEYLAN,WM ET AL. (1996)
logP 0.16			EST	MEYLAN,WM & HOWARD,PH (1995)
VP 2.46E-003	mm Hg	25	EXT	MILLIGAN,B & GILBERT,KE (1978)
DC	pKa			
HL 7.49E-009	atm m3/mol	25	EST	MEYLAN,WM & HOWARD,PH (1991)
OH 2.40E-010	cm3/molc sec	25	EST	MEYLAN,WM & HOWARD,PH (1993)

CAS #: 000824-46-4	1,4-BENZENEDIOL, 2-METHOXY-

Formula: $C_7H_8O_3$

Mol Weight: 140.14

MP (deg C): 89-91

FP (deg C):

BP (deg C):

BP pressure (mm Hg):

Property/ Value	Units	Temp	Data Type	Reference
WS 3.32E+004	mg/L	25	EST	MEYLAN,WM ET AL. (1996)
logP 0.47			EXP	HANSCH,C ET AL. (1995)
VP 7.12E-004	mm Hg	25	EST	NEELY,WB & BLAU,GE (1985)
DC	pKa			
HL 3.45E-012	atm m3/mol	25	EST	MEYLAN,WM & HOWARD,PH (1991)
OH 2.01E-010	cm3/molc sec	25	EST	MEYLAN,WM & HOWARD,PH (1993)

CAS #: 000824-75-9	P-FLUOROBENZAMIDE

Formula: C_7H_6FNO

Mol Weight: 139.13

MP (deg C): 154-157

FP (deg C):

BP (deg C):

BP pressure (mm Hg):

Property/ Value	Units	Temp	Data Type	Reference
WS 1.41E+004	mg/L	25	EST	MEYLAN,WM ET AL. (1996)
logP 0.91			EXP	HANSCH,C & LEO,AJ (1985)
VP 1.09E-003	mm Hg	25	EST	NEELY,WB & BLAU,GE (1985)
DC	pKa			
HL 2.58E-009	atm m3/mol	25	EST	MEYLAN,WM & HOWARD,PH (1991)
OH 4.23E-012	cm3/molc sec	25	EST	MEYLAN,WM & HOWARD,PH (1993)

CAS #: 000825-44-5	BENZO(B)THIOPHENE S,S-DIOXIDE

Formula: $C_8H_6O_2S$

Mol Weight: 166.20

MP (deg C):

FP (deg C):

BP (deg C):

BP pressure (mm Hg):

Property/ Value	Units	Temp	Data Type	Reference
WS 1.14E+004	mg/L	25	EST	MEYLAN,WM ET AL. (1996)
logP 0.88			EXP	HANSCH,C & LEO,AJ (1985)
VP 1.03E-003	mm Hg	25	EST	NEELY,WB & BLAU,GE (1985)
DC	pKa			
HL 7.11E-007	atm m3/mol	25	EST	MEYLAN,WM & HOWARD,PH (1991)
OH 5.68E-011	cm3/molc sec	25	EST	MEYLAN,WM & HOWARD,PH (1993)

CAS #: 000825-51-4	DECAHYDRO-2-NAPHTHOL

Formula: $C_{10}H_{18}O$

Mol Weight: 154.25

MP (deg C):

FP (deg C):

BP (deg C): 109

BP pressure (mm Hg): 1.40E+001

Property/ Value	Units	Temp	Data Type	Reference
WS 1.26E+003	mg/L	25	EST	MEYLAN,WM ET AL. (1996)
logP 2.66			EST	MEYLAN,WM & HOWARD,PH (1995)
VP 6.06E-003	mm Hg	25	EST	NEELY,WB & BLAU,GE (1985)
DC	pKa			
HL 6.70E-006	atm m3/mol	25	EST	MEYLAN,WM & HOWARD,PH (1991)
OH 2.75E-011	cm3/molc sec	25	EST	MEYLAN,WM & HOWARD,PH (1993)

CAS #: 000825-55-8	2-PHENYLTHIOPHENE

Formula: $C_{10}H_8S$

Mol Weight: 160.24

MP (deg C):

FP (deg C):

BP (deg C):

BP pressure (mm Hg):

Property/ Value	Units	Temp	Data Type	Reference
WS 4.37E+001	mg/L	25	EST	MEYLAN,WM ET AL. (1996)
logP 3.74			EXP	HANSCH,C & LEO,AJ (1985)
VP 2.82E-003	mm Hg	25	EST	NEELY,WB & BLAU,GE (1985)
DC	pKa			
HL 2.25E-004	atm m3/mol	25	EST	MEYLAN,WM & HOWARD,PH (1991)
OH 1.99E-011	cm3/molc sec	25	EST	MEYLAN,WM & HOWARD,PH (1993)

CAS #: 000825-56-9	1,3,4-OXADIAZOLE, 2-PHENYL-

Formula: $C_8H_6N_2O$

Mol Weight: 146.15

MP (deg C):

FP (deg C):

BP (deg C):

BP pressure (mm Hg):

Property/ Value	Units	Temp	Data Type	Reference
WS 6.00E+003	mg/L	25	EST	MEYLAN,WM ET AL. (1996)
logP 1.31			EXP	HANSCH,C ET AL. (1995)
VP 6.86E-004	mm Hg	25	EST	NEELY,WB & BLAU,GE (1985)
DC	pKa			
HL 2.20E-007	atm m3/mol	25	EST	MEYLAN,WM & HOWARD,PH (1991)
OH 1.03E-011	cm3/molc sec	25	EST	MEYLAN,WM & HOWARD,PH (1993)

CAS #: 000825-86-5	4-IODOBENZENESULFONAMIDE

Formula: $C_6H_6INO_2S$
Mol Weight: 283.09
MP (deg C):　FP (deg C):
BP (deg C):
BP pressure (mm Hg):

Property/Value	Units	Temp	Data Type	Reference
WS 6.75E+002	mg/L	25	EST	MEYLAN,WM ET AL. (1996)
logP 1.59			EXP	HANSCH,C ET AL. (1995)
VP 1.61E-005	mm Hg	25	EST	NEELY,WB & BLAU,GE (1985)
DC	pKa			
HL 9.79E-008	atm m3/mol	25	EST	MEYLAN,WM & HOWARD,PH (1991)
OH 2.75E-013	cm3/molc sec	25	EST	MEYLAN,WM & HOWARD,PH (1993)

CAS #: 000826-81-3	2-METHYL 8-QUINOLONOL

Formula: $C_{10}H_9NO$
Mol Weight: 159.19
MP (deg C): 73.8　FP (deg C):
BP (deg C): 267
BP pressure (mm Hg):

Property/Value	Units	Temp	Data Type	Reference
WS 2.69E+003	mg/L	25	EST	MEYLAN,WM ET AL. (1996)
logP 2.33			EXP	HANSCH,C & LEO,AJ (1985)
VP 1.05E-004	mm Hg	25	EST	NEELY,WB & BLAU,GE (1985)
DC 5.55	pKa		EXP	KORTUM,G ET AL (1961)
HL 7.90E-011	atm m3/mol	25	EST	MEYLAN,WM & HOWARD,PH (1991)
OH 8.86E-011	cm3/molc sec	25	EST	MEYLAN,WM & HOWARD,PH (1993)

CAS #: 000827-52-1	PHENYL CYCLOHEXANE

Formula: $C_{12}H_{16}$
Mol Weight: 160.26
MP (deg C): 7.3　FP (deg C):
BP (deg C): 240.1
BP pressure (mm Hg):

Property/Value	Units	Temp	Data Type	Reference
WS 5.33E+000	mg/L	25	EST	MEYLAN,WM ET AL. (1996)
logP 4.81			EST	MEYLAN,WM & HOWARD,PH (1995)
VP 3.99E-002	mm Hg	25	EXP	DAUBERT,TE & DANNER,RP (1989)
DC	pKa			
HL 1.08E-002	atm m3/mol	25	EST	MEYLAN,WM & HOWARD,PH (1991)
OH 1.46E-011	cm3/molc sec	25	EST	MEYLAN,WM & HOWARD,PH (1993)

CAS #: 000828-00-2	2,6-DIMETHYL-1,3-DIOXAN-4-OL ACETATE

Formula: $C_8H_{14}O_4$
Mol Weight: 174.20
MP (deg C):　FP (deg C):
BP (deg C): 74-75
BP pressure (mm Hg): 6.00E+000

Property/Value	Units	Temp	Data Type	Reference
WS 1.00E+006	mg/L		EXP	BUDAVARI,S (1989)
logP 0.49			EST	MEYLAN,WM & HOWARD,PH (1995)
VP 1.42E-001	mm Hg	25	EST	MPBPVP
DC	pKa			
HL 1.24E-007	atm m3/mol	25	EST	MEYLAN,WM & HOWARD,PH (1991)
OH 4.89E-011	cm3/molc sec	25	EST	MEYLAN,WM & HOWARD,PH (1993)

CAS #: 000829-26-5	2,3,6-TRIMETHYLNAPHTHALENE

Formula: $C_{13}H_{14}$
Mol Weight: 170.26
MP (deg C):　FP (deg C):
BP (deg C):
BP pressure (mm Hg):

Property/Value	Units	Temp	Data Type	Reference
WS 5.60E+000	mg/L	25	EST	MEYLAN,WM ET AL. (1996)
logP 4.73			EXP	HANSCH,C & LEO,AJ (1985)
VP 2.52E-003	mm Hg	25	EST	NEELY,WB & BLAU,GE (1985)
DC	pKa			
HL 7.07E-004	atm m3/mol	25	EST	MEYLAN,WM & HOWARD,PH (1991)
OH 1.24E-010	cm3/molc sec	25	EST	MEYLAN,WM & HOWARD,PH (1993)

CAS #: 000830-03-5	P-NITROPHENYLACETATE

Formula: $C_8H_7NO_4$
Mol Weight: 181.15
MP (deg C): 77-79　FP (deg C):
BP (deg C):
BP pressure (mm Hg):

Property/Value	Units	Temp	Data Type	Reference
WS 1.16E+003	mg/L	25	EST	MEYLAN,WM ET AL. (1996)
logP 1.50			EXP	HANSCH,C & LEO,AJ (1985)
VP 1.86E-003	mm Hg	25	EST	NEELY,WB & BLAU,GE (1985)
DC	pKa			
HL 2.56E-007	atm m3/mol	25	EST	MEYLAN,WM & HOWARD,PH (1991)
OH 3.52E-013	cm3/molc sec	25	EST	MEYLAN,WM & HOWARD,PH (1993)

CAS #: 000830-09-1	CINNAMIC ACID, P-METHOXY-

Formula: $C_{10}H_{10}O_3$
Mol Weight: 178.19
MP (deg C): 173.5-190　FP (deg C):
BP (deg C):
BP pressure (mm Hg):

Property/Value	Units	Temp	Data Type	Reference
WS 7.12E+001	mg/L	25	EXP	STEPHEN,H & STEPHEN,T (1963)
logP 2.68			EXP	HANSCH,C & LEO,AJ (1985)
VP 1.59E-004	mm Hg	25	EST	NEELY,WB & BLAU,GE (1985)
DC 4.54	pKa	25	EXP	KORTUM,G ET AL (1961)
HL 7.65E-010	atm m3/mol	25	EST	MEYLAN,WM & HOWARD,PH (1991)
OH 4.13E-011	cm3/molc sec	25	EST	MEYLAN,WM & HOWARD,PH (1993)

CAS #: 000830-13-7	CYCLODODECANONE

Formula: $C_{12}H_{24}$
Mol Weight: 168.33
MP (deg C): 59　FP (deg C):
BP (deg C): 126-128
BP pressure (mm Hg): 1.20E+001

Property/Value	Units	Temp	Data Type	Reference
WS 1.69E+001	mg/L	25	EST	MEYLAN,WM ET AL. (1996)
logP 4.10			EXP	HANSCH,C & LEO,AJ (1985)
VP 4.40E-003	mm Hg	25	EXT	BOUBLIK,T ET AL. (1984)
DC	pKa			
HL 2.80E-004	atm m3/mol	25	EST	MEYLAN,WM & HOWARD,PH (1991)
OH 2.06E-011	cm3/molc sec	25	EST	MEYLAN,WM & HOWARD,PH (1993)

CAS #: 000830-81-9	1-ACETOXYNAPHTHALENE

Formula: $C_{12}H_{10}O_2$

Mol Weight: 186.21

MP (deg C): 43-46   FP (deg C):

BP (deg C):

BP pressure (mm Hg):

Property/Value	Units	Temp	Data Type	Reference
WS 2.17E+002	mg/L	25	EST	MEYLAN,WM ET AL. (1996)
logP 2.78			EXP	HANSCH,C & LEO,AJ (1985)
VP 7.61E-004	mm Hg	25	EST	NEELY,WB & BLAU,GE (1985)
DC	pKa			
HL 6.33E-006	atm m3/mol	25	EST	MEYLAN,WM & HOWARD,PH (1991)
OH 2.24E-011	cm3/molc sec	25	EST	MEYLAN,WM & HOWARD,PH (1993)

CAS #: 000830-96-6	1H-INDOLE-3-PROPANOIC ACID

Formula: $C_{11}H_{11}NO_2$

Mol Weight: 189.22

MP (deg C): 134-135   FP (deg C):

BP (deg C):

BP pressure (mm Hg):

Property/Value	Units	Temp	Data Type	Reference
WS 3.94E+003	mg/L	25	EST	MEYLAN,WM ET AL. (1996)
logP 1.75			EXP	HANSCH,C & LEO,AJ (1985)
VP 2.27E-006	mm Hg	25	EST	NEELY,WB & BLAU,GE (1985)
DC	pKa			
HL 9.65E-012	atm m3/mol	25	EST	MEYLAN,WM & HOWARD,PH (1991)
OH 2.03E-010	cm3/molc sec	25	EST	MEYLAN,WM & HOWARD,PH (1993)

CAS #: 000831-25-4	THIAZOLIDINE,2-(P-NITROPHENYL)

Formula: $C_9H_{10}N_2O_2S$

Mol Weight: 210.26

MP (deg C):   FP (deg C):

BP (deg C):

BP pressure (mm Hg):

Property/Value	Units	Temp	Data Type	Reference
WS 5.91E+003	mg/L	25	EST	MEYLAN,WM ET AL. (1996)
logP 1.68			EXP	HANSCH,C & LEO,AJ (1985)
VP 1.77E-005	mm Hg	25	EST	NEELY,WB & BLAU,GE (1985)
DC	pKa			
HL 1.17E-008	atm m3/mol	25	EST	MEYLAN,WM & HOWARD,PH (1991)
OH 1.44E-010	cm3/molc sec	25	EST	MEYLAN,WM & HOWARD,PH (1993)

CAS #: 000831-61-8	ETHYL GALLATE

Formula: $C_9H_{10}O_5$

Mol Weight: 198.18

MP (deg C):   FP (deg C):

BP (deg C):

BP pressure (mm Hg):

Property/Value	Units	Temp	Data Type	Reference
WS 1.32E+004	mg/L	25	EST	MEYLAN,WM ET AL. (1996)
logP 1.30			EXP	HANSCH,C & LEO,AJ (1985)
VP 6.99E-007	mm Hg	25	EST	NEELY,WB & BLAU,GE (1985)
DC	pKa			
HL 5.19E-017	atm m3/mol	25	EST	MEYLAN,WM & HOWARD,PH (1991)
OH 9.07E-011	cm3/molc sec	25	EST	MEYLAN,WM & HOWARD,PH (1993)

CAS #: 000831-71-0	HYDRAZINECARBOTHIAMIDE, 2-[(5-NITRO-2-FURANYL)ME

Formula: $C_6H_6N_4O_3S$

Mol Weight: 214.20

MP (deg C):   FP (deg C):

BP (deg C):

BP pressure (mm Hg):

Property/Value	Units	Temp	Data Type	Reference
WS 5.98E+002	mg/L	25	EST	MEYLAN,WM ET AL. (1996)
logP 1.64			EXP	DE,AU ET AL. (1983)
VP 8.22E-006	mm Hg	25	EST	NEELY,WB & BLAU,GE (1985)
DC	pKa			
HL 1.34E-010	atm m3/mol	25	EST	MEYLAN,WM & HOWARD,PH (1991)
OH 9.79E-011	cm3/molc sec	25	EST	MEYLAN,WM & HOWARD,PH (1993)

CAS #: 000831-82-3	PHENOL, 4-PHENOXY-

Formula: $C_{12}H_{10}O_2$

Mol Weight: 186.21

MP (deg C): 83-85   FP (deg C):

BP (deg C):

BP pressure (mm Hg):

Property/Value	Units	Temp	Data Type	Reference
WS 2.69E+002	mg/L	25	EST	MEYLAN,WM ET AL. (1996)
logP 3.35			EXP	HANSCH,C ET AL. (1995)
VP 6.05E-005	mm Hg	25	EST	NEELY,WB & BLAU,GE (1985)
DC	pKa			
HL 1.23E-008	atm m3/mol	25	EST	MEYLAN,WM & HOWARD,PH (1991)
OH 3.84E-011	cm3/molc sec	25	EST	MEYLAN,WM & HOWARD,PH (1993)

CAS #: 000832-69-9	1-METHYLPHENANTHRENE

Formula: $C_{15}H_{12}$

Mol Weight: 192.26

MP (deg C):   FP (deg C):

BP (deg C):

BP pressure (mm Hg):

Property/Value	Units	Temp	Data Type	Reference
WS 2.69E-001	mg/L	25	EXP	YALKOWSKY,SH & DANNENFELSER,RM (1992)
logP 5.08			EXP	WANG,L ET AL. (1986)
VP 5.01E-005	mm Hg	25	EST	NEELY,WB & BLAU,GE (1985)
DC	pKa			
HL 5.67E-005	atm m3/mol	25	EST	MEYLAN,WM & HOWARD,PH (1991)
OH 3.41E-011	cm3/molc sec	25	EST	MEYLAN,WM & HOWARD,PH (1993)

CAS #: 000832-71-3	3-METHYLPHENANTHRENE

Formula: $C_{15}H_{12}$

Mol Weight: 192.26

MP (deg C): 65   FP (deg C):

BP (deg C): 350

BP pressure (mm Hg):

Property/Value	Units	Temp	Data Type	Reference
WS 2.63E-001	mg/L	25	EST	MEYLAN,WM ET AL. (1996)
logP 5.15			EXP	WANG,L ET AL. (1986)
VP 5.01E-005	mm Hg	25	EST	NEELY,WB & BLAU,GE (1985)
DC	pKa			
HL 5.67E-005	atm m3/mol	25	EST	MEYLAN,WM & HOWARD,PH (1991)
OH 3.41E-011	cm3/molc sec	25	EST	MEYLAN,WM & HOWARD,PH (1993)

CAS #: 000833-50-1				BENZOXAZOLE, 2-PHENYL-

Formula: $C_{13}H_9NO$

Mol Weight: 195.22

MP (deg C): 102-104   **FP (deg C):**

BP (deg C):

BP pressure (mm Hg):

Property/Value	Units	Temp	Data Type	Reference
WS 3.40E+001	mg/L	25	EST	MEYLAN,WM ET AL. (1996)
logP 3.67			EXP	SANGSTER,J (1993)
VP 3.11E-005	mm Hg	25	EST	NEELY,WB & BLAU,GE (1985)
DC	pKa			
HL 5.27E-008	atm m3/mol	25	EST	MEYLAN,WM & HOWARD,PH (1991)
OH 1.90E-011	cm3/molc sec	25	EST	MEYLAN,WM & HOWARD,PH (1993)

CAS #: 000834-12-8				AMETRYNE

Formula: $C_9H_{17}N_5S$

Mol Weight: 227.33

MP (deg C): 88-89   **FP (deg C):**

BP (deg C):

BP pressure (mm Hg):

Property/Value	Units	Temp	Data Type	Reference
WS 1.85E+002	mg/L	20	EXP	YALKOWSKY,SH & DANNENFELSER,RM (1992)
logP 2.98			EXP	HANSCH,C & LEO,AJ (1985)
VP 8.47E-007	mm Hg	20	EXP	JORDAN,LS ET AL. (1970)
DC 4.10	pKa	20	EXP	TOMLIN,C (1994)
HL 4.58E-009	atm m3/mol	20	EST	VP/WSOL
OH 2.85E-011	cm3/molc sec	25	EST	MEYLAN,WM & HOWARD,PH (1993)

CAS #: 000834-91-3				2,4(1H,3H)-PYRIMIDINEDIONE, 5-[(2-CHLOROETHYL)(2

Formula: $C_9H_{13}ClFN_3O_2$

Mol Weight: 249.67

MP (deg C):   **FP (deg C):**

BP (deg C):

BP pressure (mm Hg):

Property/Value	Units	Temp	Data Type	Reference
WS 4.47E+003	mg/L	25	EST	MEYLAN,WM ET AL. (1996)
logP 0.85			EXP	HANSCH,C ET AL. (1995)
VP 2.91E-009	mm Hg	25	EST	NEELY,WB & BLAU,GE (1985)
DC	pKa			
HL 3.47E-012	atm m3/mol	25	EST	MEYLAN,WM & HOWARD,PH (1991)
OH 2.50E-011	cm3/molc sec	25	EST	MEYLAN,WM & HOWARD,PH (1993)

CAS #: 000835-31-4				NAPHAZOLINE

Formula: $C_{14}H_{14}N_2$

Mol Weight: 210.28

MP (deg C):   **FP (deg C):**

BP (deg C):

BP pressure (mm Hg):

Property/Value	Units	Temp	Data Type	Reference
WS 1.18E+001	mg/L	25	EST	MEYLAN,WM ET AL. (1996)
logP 4.12			EST	MEYLAN,WM & HOWARD,PH (1995)
VP 6.67E-008	mm Hg	25	EST	NEELY,WB & BLAU,GE (1985)
DC	pKa			
HL 5.05E-009	atm m3/mol	25	EST	MEYLAN,WM & HOWARD,PH (1991)
OH 1.27E-010	cm3/molc sec	25	EST	MEYLAN,WM & HOWARD,PH (1993)

CAS #: 000835-64-3				2-(2-HYDROXYPHENYL)BENZOXAZOLE

Formula: $C_{13}H_9NO_2$

Mol Weight: 211.22

MP (deg C): 123.5   **FP (deg C):**

BP (deg C): 338

BP pressure (mm Hg):

Property/Value	Units	Temp	Data Type	Reference
WS 3.97E+002	mg/L	25	EST	MEYLAN,WM ET AL. (1996)
logP 3.00			EST	MEYLAN,WM & HOWARD,PH (1995)
VP 2.30E-008	mm Hg	20	EXP	ARTHURDLITTLE,INC (1982)
DC	pKa			
HL 5.49E-012	atm m3/mol	25	EST	MEYLAN,WM & HOWARD,PH (1991)
OH 3.96E-011	cm3/molc sec	25	EST	MEYLAN,WM & HOWARD,PH (1993)

CAS #: 000836-30-6				4-NITRO-N-PHENYLBENZENAMINE

Formula: $C_{12}H_{10}N_2O_2$

Mol Weight: 214.23

MP (deg C): 135.3   **FP (deg C):**

BP (deg C): 211

BP pressure (mm Hg): 3.00E+001

Property/Value	Units	Temp	Data Type	Reference
WS 2.36E+001	mg/L	25	EST	MEYLAN,WM ET AL. (1996)
logP 3.74			EXP	HANSCH,C ET AL. (1995)
VP 1.66E-005	mm Hg	25	EST	NEELY,WB & BLAU,GE (1985)
DC -2.50	pKa	25	EXP	PERRIN,DD (1965)
HL 4.14E-009	atm m3/mol	25	EST	MEYLAN,WM & HOWARD,PH (1991)
OH 1.65E-010	cm3/molc sec	25	EST	MEYLAN,WM & HOWARD,PH (1993)

CAS #: 000838-85-7				DIPHENYL PHOSPHATE

Formula: $C_{12}H_{11}O_4P$

Mol Weight: 250.19

MP (deg C): 67-68   **FP (deg C):**

BP (deg C):

BP pressure (mm Hg):

Property/Value	Units	Temp	Data Type	Reference
WS 8.24E+001	mg/L	25	EST	MEYLAN,WM ET AL. (1996)
logP 2.88			EST	MEYLAN,WM & HOWARD,PH (1995)
VP 5.90E-008	mm Hg	25	EST	NEELY,WB & BLAU,GE (1985)
DC	pKa			
HL 1.06E-010	atm m3/mol	25	EST	MEYLAN,WM & HOWARD,PH (1991)
OH 7.37E-012	cm3/molc sec	25	EST	MEYLAN,WM & HOWARD,PH (1993)

CAS #: 000838-89-1				1-MEO-1-ME-3-(CF3-PHENYL)UREA

Formula: $C_{10}H_{11}F_3N_2O_2$

Mol Weight: 248.21

MP (deg C):   **FP (deg C):**

BP (deg C):

BP pressure (mm Hg):

Property/Value	Units	Temp	Data Type	Reference
WS 1.35E+002	mg/L	25	EST	MEYLAN,WM ET AL. (1996)
logP 2.64			EXP	HANSCH,C & LEO,AJ (1985)
VP 5.61E-005	mm Hg	25	EST	NEELY,WB & BLAU,GE (1985)
DC	pKa			
HL 1.83E-007	atm m3/mol	25	EST	MEYLAN,WM & HOWARD,PH (1991)
OH 7.04E-012	cm3/molc sec	25	EST	MEYLAN,WM & HOWARD,PH (1993)

000841-06-5 — METHOPROPTRYNE

Formula: $C_{11}H_{21}N_5OS$

Mol Weight: 271.39

MP (deg C): **FP (deg C):**

BP (deg C):

BP pressure (mm Hg):

Property/Value	Units	Temp	Data Type	Reference
WS 3.20E+002	mg/L	20	EXP	YALKOWSKY,SH & DANNENFELSER,RM (1992)
logP 3.04			EST	MEYLAN,WM & HOWARD,PH (1995)
VP 2.85E-007	mm Hg	20	EXP	KEARNEY,PC & KAUFMAN,DD (1975A)
DC	pKa			
HL 3.18E-010	atm m3/mol	20	EST	VP/WSOL
OH 4.22E-011	cm3/molc sec	25	EST	MEYLAN,WM & HOWARD,PH (1993)

000842-00-2 — ET-4-SULFONAMIDONAPHTHYLSULFONE

Formula: $C_{12}H_{13}NO_4S_2$

Mol Weight: 299.37

MP (deg C): **FP (deg C):**

BP (deg C):

BP pressure (mm Hg):

Property/Value	Units	Temp	Data Type	Reference
WS 9.61E+001	mg/L	25	EST	MEYLAN,WM ET AL. (1996)
logP 1.24			EXP	HANSCH,C & LEO,AJ (1985)
VP 1.54E-009	mm Hg	25	EST	NEELY,WB & BLAU,GE (1985)
DC	pKa			
HL 1.27E-011	atm m3/mol	25	EST	MEYLAN,WM & HOWARD,PH (1991)
OH 8.44E-012	cm3/molc sec	25	EST	MEYLAN,WM & HOWARD,PH (1993)

000842-07-9 — 1-(PHENYLAZO)-2-NAPHTHALENOL

Formula: $C_{16}H_{12}N_2O$

Mol Weight: 248.29

MP (deg C): 134 **FP (deg C):**

BP (deg C):

BP pressure (mm Hg):

Property/Value	Units	Temp	Data Type	Reference
WS 6.74E-001	mg/L	25	EST	MEYLAN,WM ET AL. (1996)
logP 5.51			EST	MEYLAN,WM & HOWARD,PH (1995)
VP 7.34E-008	mm Hg	25	EST	NEELY,WB & BLAU,GE (1985)
DC	pKa			
HL 2.62E-011	atm m3/mol	25	EST	MEYLAN,WM & HOWARD,PH (1991)
OH 1.82E-011	cm3/molc sec	25	EST	MEYLAN,WM & HOWARD,PH (1993)

000846-49-1 — LORAZEPAM

Formula: $C_{15}H_{10}Cl_2N_2O_2$

Mol Weight: 321.17

MP (deg C): 166-168 **FP (deg C):**

BP (deg C):

BP pressure (mm Hg):

Property/Value	Units	Temp	Data Type	Reference
WS 8.39E+001	mg/L	25	EST	MEYLAN,WM ET AL. (1996)
logP 2.39			EXP	HANSCH,C ET AL. (1995)
VP 7.95E-013	mm Hg	25	EST	NEELY,WB & BLAU,GE (1985)
DC	pKa			
HL 4.10E-010	atm m3/mol	25	EST	MEYLAN,WM & HOWARD,PH (1991)
OH 1.12E-011	cm3/molc sec	25	EST	MEYLAN,WM & HOWARD,PH (1993)

000846-50-4 — TEMAZEPAM

Formula: $C_{16}H_{13}ClN_2O_2$

Mol Weight: 300.75

MP (deg C): 119-121 **FP (deg C):**

BP (deg C):

BP pressure (mm Hg):

Property/Value	Units	Temp	Data Type	Reference
WS 1.64E+002	mg/L	25	EST	MEYLAN,WM ET AL. (1996)
logP 2.19			EXP	HANSCH,C & LEO,AJ (1985)
VP 2.14E-011	mm Hg	25	EST	NEELY,WB & BLAU,GE (1985)
DC	pKa			
HL 1.13E-008	atm m3/mol	25	EST	MEYLAN,WM & HOWARD,PH (1991)
OH 1.44E-011	cm3/molc sec	25	EST	MEYLAN,WM & HOWARD,PH (1993)

000846-53-7 — 1,4-BENZODIAZPN2-ON,5(2-MES PH)7CL

Formula: $C_{16}H_{13}ClN_2OS$

Mol Weight: 316.81

MP (deg C): **FP (deg C):**

BP (deg C):

BP pressure (mm Hg):

Property/Value	Units	Temp	Data Type	Reference
WS 2.74E+001	mg/L	25	EST	MEYLAN,WM ET AL. (1996)
logP 2.99			EXP	HANSCH,C & LEO,AJ (1985)
VP 2.73E-010	mm Hg	25	EST	NEELY,WB & BLAU,GE (1985)
DC	pKa			
HL 5.18E-012	atm m3/mol	25	EST	MEYLAN,WM & HOWARD,PH (1991)
OH 2.12E-011	cm3/molc sec	25	EST	MEYLAN,WM & HOWARD,PH (1993)

000847-25-6 — P-METHYLSULFONYLAMPHENICOL

Formula: $C_{12}H_{15}Cl_2NO_5S$

Mol Weight: 356.23

MP (deg C): 164.3-6.3 **FP (deg C):**

BP (deg C):

BP pressure (mm Hg):

Property/Value	Units	Temp	Data Type	Reference
WS 9.66E+003	mg/L	25	EST	MEYLAN,WM ET AL. (1996)
logP -0.27			EXP	HANSCH,C & LEO,AJ (1985)
VP 1.11E-014	mm Hg	25	EST	NEELY,WB & BLAU,GE (1985)
DC	pKa			
HL 1.34E-019	atm m3/mol	25	EST	MEYLAN,WM & HOWARD,PH (1991)
OH 3.24E-011	cm3/molc sec	25	EST	MEYLAN,WM & HOWARD,PH (1993)

000862-26-0 — L-TYROSINE, N-(N-CARBOXY-L-VALYL)-, N-BENZYL EST

Formula: $C_{22}H_{26}N_2O_6$

Mol Weight: 414.46

MP (deg C): **FP (deg C):**

BP (deg C):

BP pressure (mm Hg):

Property/Value	Units	Temp	Data Type	Reference
WS 7.96E+001	mg/L	25	EST	MEYLAN,WM ET AL. (1996)
logP 2.43			EXP	HANSCH,C ET AL. (1995)
VP 4.71E-015	mm Hg	25	EST	NEELY,WB & BLAU,GE (1985)
DC	pKa			
HL 2.86E-019	atm m3/mol	25	EST	MEYLAN,WM & HOWARD,PH (1991)
OH 8.23E-011	cm3/molc sec	25	EST	MEYLAN,WM & HOWARD,PH (1993)

000868-54-2 — 1-PROPENE-1,1,3-TRICARBONITRILE, 2-AMINO-

CAS #: 000868-54-2				1-PROPENE-1,1,3-TRICARBONITRILE, 2-AMINO-
Formula: $C_6H_4N_4$				
Mol Weight: 132.13				
MP (deg C): 170-173		**FP (deg C):**		
BP (deg C):				
BP pressure (mm Hg):				

Property/Value	Units	Temp	Data Type	Reference
WS 6.44E+005	mg/L	25	EST	MEYLAN,WM ET AL. (1996)
logP -0.13			EXP	HANSCH,C ET AL. (1995)
VP 4.41E-005	mm Hg	25	EST	NEELY,WB & BLAU,GE (1985)
DC	pKa			
HL 4.29E-014	atm m3/mol	25	EST	MEYLAN,WM & HOWARD,PH (1991)
OH 1.02E-012	cm3/molc sec	25	EST	MEYLAN,WM & HOWARD,PH (1993)

000868-77-9 — 2-HYDROXYETHYL METHACRYLATE

CAS #: 000868-77-9				2-HYDROXYETHYL METHACRYLATE
Formula: $C_6H_{10}O_3$				
Mol Weight: 130.14				
MP (deg C):		**FP (deg C):**		
BP (deg C): 67				
BP pressure (mm Hg): 3.50E+000				

Property/Value	Units	Temp	Data Type	Reference
WS 1.18E+005	mg/L	25	EST	MEYLAN,WM ET AL. (1996)
logP 0.47			EXP	HANSCH,C & LEO,AJ (1985)
VP 7.91E-002	mm Hg	25	EST	NEELY,WB & BLAU,GE (1985)
DC	pKa			
HL 7.03E-009	atm m3/mol	25	EST	MEYLAN,WM & HOWARD,PH (1991)
OH 2.41E-011	cm3/molc sec	25	EST	MEYLAN,WM & HOWARD,PH (1993)

000868-85-9 — DIMETHYL PHOSPHITE

CAS #: 000868-85-9				DIMETHYL PHOSPHITE
Formula: $C_2H_7O_3P$				
Mol Weight: 110.05				
MP (deg C):		**FP (deg C):**		
BP (deg C): 170.5				
BP pressure (mm Hg):				

Property/Value	Units	Temp	Data Type	Reference
WS 1.00E+006	mg/L	25	EST	MEYLAN,WM ET AL. (1996)
logP -1.13			EST	MEYLAN,WM & HOWARD,PH (1995)
VP 4.52E+000	mm Hg	25	EXP	BENNETT,SR ET AL. (1984)
DC	pKa			
HL 3.28E-006	atm m3/mol	25	EST	MEYLAN,WM & HOWARD,PH (1991)
OH 5.58E-012	cm3/molc sec	25	EST	MEYLAN,WM & HOWARD,PH (1993)

000869-01-2 — 1-BUTYL-1-NITROSOUREA

CAS #: 000869-01-2				1-BUTYL-1-NITROSOUREA
Formula: $C_5H_{11}N_3O_2$				
Mol Weight: 145.16				
MP (deg C):		**FP (deg C):**		
BP (deg C):				
BP pressure (mm Hg):				

Property/Value	Units	Temp	Data Type	Reference
WS 1.03E+004	mg/L	25	EST	MEYLAN,WM ET AL. (1996)
logP 1.04			EXP	HANSCH,C & LEO,AJ (1985)
VP 5.27E-003	mm Hg	25	EST	NEELY,WB & BLAU,GE (1985)
DC	pKa			
HL 2.32E-010	atm m3/mol	25	EST	MEYLAN,WM & HOWARD,PH (1991)
OH 8.44E-012	cm3/molc sec	25	EST	MEYLAN,WM & HOWARD,PH (1993)

000869-29-4 — ALLYLIDENEDIACETATE

CAS #: 000869-29-4				ALLYLIDENEDIACETATE
Formula: $C_7H_{10}O_4$				
Mol Weight: 158.16				
MP (deg C): -37.6		**FP (deg C):**		
BP (deg C): 180				
BP pressure (mm Hg):				

Property/Value	Units	Temp	Data Type	Reference
WS 9.00E+003	mg/L	25	EXP	CHEM INSPECT TEST INST (1992)
logP 0.78			EXP	CHEM INSPECT TEST INST (1992)
VP 3.77E+000	mm Hg	25	EST	NEELY,WB & BLAU,GE (1985)
DC	pKa			
HL 5.48E-007	atm m3/mol	25	EST	MEYLAN,WM & HOWARD,PH (1991)
OH 3.15E-011	cm3/molc sec	25	EST	MEYLAN,WM & HOWARD,PH (1993)

000869-79-4 — DI-(SEC-BUTYL)UREA

CAS #: 000869-79-4				DI-(SEC-BUTYL)UREA
Formula: $C_9H_{20}N_2O$				
Mol Weight: 172.27				
MP (deg C):		**FP (deg C):**		
BP (deg C):				
BP pressure (mm Hg):				

Property/Value	Units	Temp	Data Type	Reference
WS 8.31E+002	mg/L	25	EST	MEYLAN,WM ET AL. (1996)
logP 2.18			EST	MEYLAN,WM & HOWARD,PH (1995)
VP 2.47E-003	mm Hg	25	EST	NEELY,WB & BLAU,GE (1985)
DC	pKa			
HL 9.63E-009	atm m3/mol	25	EST	MEYLAN,WM & HOWARD,PH (1991)
OH 2.45E-011	cm3/molc sec	25	EST	MEYLAN,WM & HOWARD,PH (1993)

000870-23-5 — ALLYL MERCAPTAN

CAS #: 000870-23-5				ALLYL MERCAPTAN
Formula: C_3H_6S				
Mol Weight: 74.15				
MP (deg C):		**FP (deg C):**		
BP (deg C): 65				
BP pressure (mm Hg):				

Property/Value	Units	Temp	Data Type	Reference
WS 5.51E+003	mg/L	25	EST	MEYLAN,WM ET AL. (1996)
logP 1.62			EST	MEYLAN,WM & HOWARD,PH (1995)
VP 5.24E+002	mm Hg	25	EXP	PATTE,F ET AL. (1982)
DC	pKa			
HL 3.43E-003	atm m3/mol	25	EST	MEYLAN,WM & HOWARD,PH (1991)
OH 6.61E-011	cm3/molc sec	25	EST	MEYLAN,WM & HOWARD,PH (1993)

000871-83-0 — 2-METHYLNONANE

CAS #: 000871-83-0				2-METHYLNONANE
Formula: $C_{10}H_{22}$				
Mol Weight: 142.29				
MP (deg C): -74.6		**FP (deg C):**		
BP (deg C): 167.1				
BP pressure (mm Hg):				

Property/Value	Units	Temp	Data Type	Reference
WS 8.99E-001	mg/L	25	EST	MEYLAN,WM ET AL. (1996)
logP 5.18			EST	MEYLAN,WM & HOWARD,PH (1995)
VP 1.89E+000	mm Hg	25	EXP	YAWS,CL (1994B)
DC	pKa			
HL 5.30E+000	atm m3/mol	25	EST	MEYLAN,WM & HOWARD,PH (1991)
OH 1.11E-011	cm3/molc sec	25	EST	MEYLAN,WM & HOWARD,PH (1993)

000872-05-9 — 1-DECENE

Formula: $C_{10}H_{20}$
Mol Weight: 140.27
MP (deg C): -66.3
FP (deg C):
BP (deg C): 170.5
BP pressure (mm Hg):

Property/Value	Units	Temp	Data Type	Reference
WS 1.15E-001	mg/L	25	EXP	SHAW,DG (1989)
logP 5.12			EST	MEYLAN,WM & HOWARD,PH (1995)
VP 1.67E+000	mm Hg	25	EXP	DAUBERT,TE & DANNER,RP (1989)
DC	pKa			
HL 2.68E+000	atm m3/mol	25	EST	VP/WSOL
OH 3.58E-011	cm3/molc sec	25	EST	MEYLAN,WM & HOWARD,PH (1993)

000872-31-1 — 3-BROMOTHIOPHENE

Formula: C_4H_3BrS
Mol Weight: 163.04
MP (deg C):
FP (deg C):
BP (deg C): 159.5
BP pressure (mm Hg):

Property/Value	Units	Temp	Data Type	Reference
WS 3.84E+002	mg/L	25	EST	MEYLAN,WM ET AL. (1996)
logP 2.62			EXP	HANSCH,C & LEO,AJ (1985)
VP 1.66E+001	mm Hg	25	EXT	BOUBLIK,T ET AL. (1984)
DC	pKa			
HL 1.17E-003	atm m3/mol	25	EST	MEYLAN,WM & HOWARD,PH (1991)
OH 6.00E-012	cm3/molc sec	25	EST	MEYLAN,WM & HOWARD,PH (1993)

000872-50-4 — N-METHYLPYRROLIDONE

Formula: C_5H_9NO
Mol Weight: 99.13
MP (deg C): -90
FP (deg C):
BP (deg C): 202
BP pressure (mm Hg):

Property/Value	Units	Temp	Data Type	Reference
WS 1.00E+006	mg/L	25	EXP	RIDDICK,JA ET AL. (1986)
logP -0.38			EXP	SASAKI,H ET AL. (1988)
VP 3.42E-001	mm Hg	25	EXP	DAUBERT,TE & DANNER,RP (1989)
DC	pKa			
HL 4.46E-008	atm m3/mol	25	EST	VP/WSOL
OH 7.40E-011	cm3/molc sec	25	EST	ATKINSON,R (1987)

000872-55-9 — 2-ETHYLTHIOPHENE

Formula: C_6H_8S
Mol Weight: 112.19
MP (deg C):
FP (deg C):
BP (deg C): 134
BP pressure (mm Hg):

Property/Value	Units	Temp	Data Type	Reference
WS 2.92E+002	mg/L	25	EXP	YALKOWSKY,SH & DANNENFELSER,RM (1992)
logP 2.87			EXP	HANSCH,C & LEO,AJ (1985)
VP 1.12E+001	mm Hg	25	EXT	CHAO,J ET AL. (1983)
DC	pKa			
HL 4.29E-003	atm m3/mol	25	EST	MEYLAN,WM & HOWARD,PH (1991)
OH 2.48E-011	cm3/molc sec	25	EST	MEYLAN,WM & HOWARD,PH (1993)

000872-85-5 — 4-FORMYLPYRIDINE

Formula: C_6H_5NO
Mol Weight: 107.11
MP (deg C):
FP (deg C):
BP (deg C): 77-78
BP pressure (mm Hg): 1.20E+001

Property/Value	Units	Temp	Data Type	Reference
WS 4.77E+004	mg/L	25	EST	MEYLAN,WM ET AL. (1996)
logP 0.43			EXP	HANSCH,C ET AL. (1995)
VP 5.68E-001	mm Hg	25	EST	NEELY,WB & BLAU,GE (1985)
DC 4.77	pKa	20	EXP	PERRIN,DD (1965)
HL 1.76E-008	atm m3/mol	25	EST	MEYLAN,WM & HOWARD,PH (1991)
OH 1.71E-011	cm3/molc sec	25	EST	MEYLAN,WM & HOWARD,PH (1993)

000872-93-5 — 3-METHYL SULFOLANE

Formula: $C_5H_{10}O_2S$
Mol Weight: 134.20
MP (deg C): 1
FP (deg C):
BP (deg C): 276
BP pressure (mm Hg):

Property/Value	Units	Temp	Data Type	Reference
WS 6.19E+004	mg/L	25	EST	MEYLAN,WM ET AL. (1996)
logP 0.18			EST	MEYLAN,WM & HOWARD,PH (1995)
VP 7.26E-003	mm Hg	25	EXP	DAUBERT,TE & DANNER,RP (1989)
DC	pKa			
HL 6.44E-006	atm m3/mol	25	EST	MEYLAN,WM & HOWARD,PH (1991)
OH 1.80E-011	cm3/molc sec	25	EST	MEYLAN,WM & HOWARD,PH (1993)

000873-49-4 — CYCLOPROPYLBENZENE

Formula: C_9H_{10}
Mol Weight: 118.18
MP (deg C): -31
FP (deg C):
BP (deg C): 173.6
BP pressure (mm Hg):

Property/Value	Units	Temp	Data Type	Reference
WS 1.64E+002	mg/L	25	EST	MEYLAN,WM ET AL. (1996)
logP 3.27			EXP	HANSCH,C & LEO,AJ (1985)
VP 1.32E+000	mm Hg	25	EST	NEELY,WB & BLAU,GE (1985)
DC	pKa			
HL 4.62E-003	atm m3/mol	25	EST	MEYLAN,WM & HOWARD,PH (1991)
OH 4.74E-012	cm3/molc sec	25	EST	MEYLAN,WM & HOWARD,PH (1993)

000873-55-2 — BENZENESULFINIC ACID, SODIUM SALT

Formula: $C_6H_5NaO_2S$
Mol Weight: 164.16
MP (deg C): > 300
FP (deg C):
BP (deg C):
BP pressure (mm Hg):

Property/Value	Units	Temp	Data Type	Reference
WS 1.00E+006	mg/L	25	EST	MEYLAN,WM ET AL. (1996)
logP -3.54			EXP	HANSCH,C ET AL. (1995)
VP 1.09E-010	mm Hg	25	EST	NEELY,WB & BLAU,GE (1985)
DC	pKa			
HL	atm m3/mol			
OH 4.17E-013	cm3/molc sec	25	EST	MEYLAN,WM & HOWARD,PH (1993)

528

CAS #: 000873-62-1				M-CYANOPHENOL

Formula: C_7H_5NO

Mol Weight: 119.12

MP (deg C): 79-81   FP (deg C):

BP (deg C):

BP pressure (mm Hg):

Property/ Value	Units	Temp	Data Type	Reference
WS 7.38E+003	mg/L	25	EST	MEYLAN,WM ET AL. (1996)
logP 1.70			EXP	HANSCH,C & LEO,AJ (1985)
VP 6.65E-003	mm Hg	25	EST	NEELY,WB & BLAU,GE (1985)
DC 8.61	pKa	25	EXP	SERJEANT,EP & DEMPSEY,B (1979)
HL 5.42E-009	atm m3/mol	25	EST	MEYLAN,WM & HOWARD,PH (1991)
OH 4.50E-012	cm3/molc sec	25	EST	MEYLAN,WM & HOWARD,PH (1993)

CAS #: 000873-63-2				M-CHLOROBENZYL ALCOHOL

Formula: C_7H_7ClO

Mol Weight: 142.59

MP (deg C):   FP (deg C):

BP (deg C): 237

BP pressure (mm Hg):

Property/ Value	Units	Temp	Data Type	Reference
WS 5.83E+003	mg/L	25	EST	MEYLAN,WM ET AL. (1996)
logP 1.94			EXP	HANSCH,C & LEO,AJ (1985)
VP 7.89E-003	mm Hg	25	EST	NEELY,WB & BLAU,GE (1985)
DC	pKa			
HL 1.61E-007	atm m3/mol	25	EST	MEYLAN,WM & HOWARD,PH (1991)
OH 6.82E-012	cm3/molc sec	25	EST	MEYLAN,WM & HOWARD,PH (1993)

CAS #: 000873-66-5				TRANS-1-PHENYL-1-PROPENE

Formula: C_9H_{10}

Mol Weight: 118.18

MP (deg C): -29.3   FP (deg C):

BP (deg C): 178.3

BP pressure (mm Hg):

Property/ Value	Units	Temp	Data Type	Reference
WS 1.40E+002	mg/L	25	EST	MEYLAN,WM ET AL. (1996)
logP 3.31			EST	MEYLAN,WM & HOWARD,PH (1995)
VP 1.14E+000	mm Hg	25	EST	NEELY,WB & BLAU,GE (1985)
DC	pKa			
HL 4.33E-003	atm m3/mol	25	EST	MEYLAN,WM & HOWARD,PH (1991)
OH 5.90E-011	cm3/molc sec	25	EXP	ATKINSON,R (1989)

CAS #: 000873-69-8				2-PYRIDINEALDOXIME

Formula: $C_6H_6N_2O$

Mol Weight: 122.13

MP (deg C):   FP (deg C):

BP (deg C):

BP pressure (mm Hg):

Property/ Value	Units	Temp	Data Type	Reference
WS 5.19E+005	mg/L	25	EST	MEYLAN,WM ET AL. (1996)
logP 0.68			EST	MEYLAN,WM & HOWARD,PH (1995)
VP 1.83E-003	mm Hg	25	EST	NEELY,WB & BLAU,GE (1985)
DC 3.59	pKa	20	EXP	PERRIN,DD (1965)
HL 4.68E-010	atm m3/mol	25	EST	MEYLAN,WM & HOWARD,PH (1991)
OH 2.82E-012	cm3/molc sec	25	EST	MEYLAN,WM & HOWARD,PH (1993)

CAS #: 000873-74-5				P-AMINOBENZONITRILE

Formula: $C_7H_6N_2$

Mol Weight: 118.14

MP (deg C): 83-85   FP (deg C):

BP (deg C):

BP pressure (mm Hg):

Property/ Value	Units	Temp	Data Type	Reference
WS 5.51E+003	mg/L	25	EST	MEYLAN,WM ET AL. (1996)
logP 1.17			EST	MEYLAN,WM & HOWARD,PH (1995)
VP 8.87E-003	mm Hg	25	EST	NEELY,WB & BLAU,GE (1985)
DC 1.74	pKa	25	EXP	PERRIN,DD (1965)
HL 1.84E-008	atm m3/mol	25	EST	MEYLAN,WM & HOWARD,PH (1991)
OH 1.90E-011	cm3/molc sec	25	EST	MEYLAN,WM & HOWARD,PH (1993)

CAS #: 000873-76-7				P-CHLOROBENZYL ALCOHOL

Formula: C_7H_7ClO

Mol Weight: 142.59

MP (deg C): 75   FP (deg C):

BP (deg C): 235

BP pressure (mm Hg):

Property/ Value	Units	Temp	Data Type	Reference
WS 5.60E+003	mg/L	25	EST	MEYLAN,WM ET AL. (1996)
logP 1.96			EXP	HANSCH,C & LEO,AJ (1985)
VP 7.89E-003	mm Hg	25	EST	NEELY,WB & BLAU,GE (1985)
DC	pKa			
HL 1.61E-007	atm m3/mol	25	EST	MEYLAN,WM & HOWARD,PH (1991)
OH 5.08E-012	cm3/molc sec	25	EST	MEYLAN,WM & HOWARD,PH (1993)

CAS #: 000874-05-5				3-AMINOINDAZOLE

Formula: $C_7H_7N_3$

Mol Weight: 133.15

MP (deg C):   FP (deg C):

BP (deg C):

BP pressure (mm Hg):

Property/ Value	Units	Temp	Data Type	Reference
WS 1.57E+004	mg/L	25	EST	MEYLAN,WM ET AL. (1996)
logP -0.72			EST	MEYLAN,WM & HOWARD,PH (1995)
VP 5.31E-004	mm Hg	25	EST	NEELY,WB & BLAU,GE (1985)
DC	pKa			
HL 1.92E-010	atm m3/mol	25	EST	MEYLAN,WM & HOWARD,PH (1991)
OH 1.14E-010	cm3/molc sec	25	EST	MEYLAN,WM & HOWARD,PH (1993)

CAS #: 000874-14-6				2,4(1H,3H)-PYRIMIDINEDIONE, 1,3-DIMETHYL-

Formula: $C_6H_8N_2O_2$

Mol Weight: 140.14

MP (deg C):   FP (deg C):

BP (deg C):

BP pressure (mm Hg):

Property/ Value	Units	Temp	Data Type	Reference
WS 1.18E+004	mg/L	25	EST	MEYLAN,WM ET AL. (1996)
logP -0.54			EXP	HANSCH,C ET AL. (1995)
VP 7.95E-005	mm Hg	25	EST	NEELY,WB & BLAU,GE (1985)
DC	pKa			
HL 4.18E-010	atm m3/mol	25	EST	MEYLAN,WM & HOWARD,PH (1991)
OH 9.88E-012	cm3/molc sec	25	EST	MEYLAN,WM & HOWARD,PH (1993)

529

CAS #: 000874-41-9				1,3-DIMETHYL-4-ETHYLBENZENE

Formula: $C_{10}H_{14}$

Mol Weight: 134.22

MP (deg C): -62.9   FP (deg C):

BP (deg C): 188.4

BP pressure (mm Hg):

Property/ Value	Units	Temp	Data Type	Reference
WS 1.35E+001	mg/L	25	EST	MEYLAN,WM ET AL. (1996)
logP 4.47			EXP	SHERBLOM,PM & EGANHOUSE,RP (1988)
VP 8.54E-001	mm Hg	25	EXT	CHAO,J ET AL. (1983)
DC	pKa			
HL 9.61E-003	atm m3/mol	25	EST	MEYLAN,WM & HOWARD,PH (1991)
OH 1.76E-011	cm3/molc sec	25	EST	MEYLAN,WM & HOWARD,PH (1993)

CAS #: 000874-80-6				BUTYLPYRIDINIUM BROMIDE

Formula: $C_9H_{14}BrN$

Mol Weight: 216.12

MP (deg C):   FP (deg C):

BP (deg C):

BP pressure (mm Hg):

Property/ Value	Units	Temp	Data Type	Reference
WS 1.00E+006	mg/L	25	EST	MEYLAN,WM ET AL. (1996)
logP -2.69			EXP	HANSCH,C & LEO,AJ (1985)
VP 2.86E-002	mm Hg	25	EST	NEELY,WB & BLAU,GE (1985)
DC	pKa			
HL 3.52E-010	atm m3/mol	25	EST	MEYLAN,WM & HOWARD,PH (1991)
OH 1.42E-011	cm3/molc sec	25	EST	MEYLAN,WM & HOWARD,PH (1993)

CAS #: 000874-84-0				2-(B-NITROVINYL)THIOPHENE

Formula: $C_6H_7NO_2S$

Mol Weight: 157.19

MP (deg C):   FP (deg C):

BP (deg C):

BP pressure (mm Hg):

Property/ Value	Units	Temp	Data Type	Reference
WS 6.21E+002	mg/L	25	EST	MEYLAN,WM ET AL. (1996)
logP 1.96			EXP	HANSCH,C & LEO,AJ (1985)
VP 8.27E-003	mm Hg	25	EST	NEELY,WB & BLAU,GE (1985)
DC	pKa			
HL 1.89E-006	atm m3/mol	25	EST	MEYLAN,WM & HOWARD,PH (1991)
OH 2.04E-011	cm3/molc sec	25	EST	MEYLAN,WM & HOWARD,PH (1993)

CAS #: 000874-90-8				BENZONITRILE, 4-METHOXY-

Formula: C_8H_7NO

Mol Weight: 133.15

MP (deg C): 61.5   FP (deg C):

BP (deg C): 256.5

BP pressure (mm Hg):

Property/ Value	Units	Temp	Data Type	Reference
WS 1.72E+003	mg/L	25	EST	MEYLAN,WM ET AL. (1996)
logP 1.70			EXP	WANG,W ET AL. (1987)
VP 6.75E-002	mm Hg	25	EST	NEELY,WB & BLAU,GE (1985)
DC	pKa			
HL 3.08E-006	atm m3/mol	25	EST	MEYLAN,WM & HOWARD,PH (1991)
OH 4.63E-012	cm3/molc sec	25	EST	MEYLAN,WM & HOWARD,PH (1993)

CAS #: 000875-79-6				1,2-DIMETHYLINDOLE

Formula: $C_{10}H_{11}N$

Mol Weight: 145.21

MP (deg C): 55-58   FP (deg C):

BP (deg C):

BP pressure (mm Hg):

Property/ Value	Units	Temp	Data Type	Reference
WS 3.11E+002	mg/L	25	EST	MEYLAN,WM ET AL. (1996)
logP 2.82			EXP	HANSCH,C & LEO,AJ (1985)
VP 1.44E-002	mm Hg	25	EST	NEELY,WB & BLAU,GE (1985)
DC	pKa			
HL 2.09E-005	atm m3/mol	25	EST	MEYLAN,WM & HOWARD,PH (1991)
OH 2.00E-010	cm3/molc sec	25	EST	MEYLAN,WM & HOWARD,PH (1993)

CAS #: 000876-98-2				PHENOL, 2,6-DIMETHYL-, ACETATE

Formula: $C_{10}H_{12}O_2$

Mol Weight: 164.21

MP (deg C):   FP (deg C):

BP (deg C):

BP pressure (mm Hg):

Property/ Value	Units	Temp	Data Type	Reference
WS 6.83E+002	mg/L	25	EST	MEYLAN,WM ET AL. (1996)
logP 2.32			EXP	SOTOMATSU,T ET AL. (1993)
VP 6.25E-002	mm Hg	25	EST	NEELY,WB & BLAU,GE (1985)
DC	pKa			
HL 7.89E-005	atm m3/mol	25	EST	MEYLAN,WM & HOWARD,PH (1991)
OH 1.01E-011	cm3/molc sec	25	EST	MEYLAN,WM & HOWARD,PH (1993)

CAS #: 000877-10-1				1,2,4,5-TETRACHLORO-3,6-DIMETHYLBENZENE

Formula: $C_8H_6Cl_4$

Mol Weight: 243.95

MP (deg C):   FP (deg C):

BP (deg C):

BP pressure (mm Hg):

Property/ Value	Units	Temp	Data Type	Reference
WS 3.71E-001	mg/L	25	EST	MEYLAN,WM ET AL. (1996)
logP 5.67			EST	MEYLAN,WM & HOWARD,PH (1995)
VP 3.28E-003	mm Hg	25	EST	NEELY,WB & BLAU,GE (1985)
DC	pKa			
HL 1.98E-003	atm m3/mol	25	EST	MEYLAN,WM & HOWARD,PH (1991)
OH 5.35E-013	cm3/molc sec	25	EST	MEYLAN,WM & HOWARD,PH (1993)

CAS #: 000877-11-2				2,3,4,5,6-PENTACHLOROTOLUENE

Formula: $C_7H_3Cl_5$

Mol Weight: 264.37

MP (deg C): 224.8   FP (deg C):

BP (deg C): 301

BP pressure (mm Hg):

Property/ Value	Units	Temp	Data Type	Reference
WS 2.36E-001	mg/L	25	EST	MEYLAN,WM ET AL. (1996)
logP 5.76			EST	MEYLAN,WM & HOWARD,PH (1995)
VP 1.79E+001	mm Hg	25	EXP	DYKYJ,J & REPAS,M (1973)
DC	pKa			
HL 7.70E-004	atm m3/mol	20	EXP	OLIVER,BG (1985)
OH 2.07E-013	cm3/molc sec	25	EST	MEYLAN,WM & HOWARD,PH (1993)

000877-22-5 — 3-METHOXYSALICYCLIC ACID

Formula: $C_8H_8O_4$

Mol Weight: 168.15

MP (deg C): 150-152

FP (deg C):

BP (deg C):

BP pressure (mm Hg):

Property/Value	Units	Temp	Data Type	Reference
WS 1.38E+003	mg/L	25	EST	MEYLAN,WM ET AL. (1996)
logP 1.94			EXP	HANSCH,C & LEO,AJ (1985)
VP 1.71E-005	mm Hg	25	EST	NEELY,WB & BLAU,GE (1985)
DC	pKa			
HL 8.40E-010	atm m3/mol	25	EST	MEYLAN,WM & HOWARD,PH (1991)
OH 1.22E-011	cm3/molc sec	25	EST	MEYLAN,WM & HOWARD,PH (1993)

000877-43-0 — 2,6-DIMETHYLQUINOLINE

Formula: $C_{11}H_{11}N$

Mol Weight: 157.22

MP (deg C): 60

FP (deg C):

BP (deg C): 266.5

BP pressure (mm Hg):

Property/Value	Units	Temp	Data Type	Reference
WS 1.21E+002	mg/L	25	EST	MEYLAN,WM ET AL. (1996)
logP 3.24			EST	MEYLAN,WM & HOWARD,PH (1995)
VP 4.34E-003	mm Hg	24	EXT	CHAO,J ET AL. (1983)
DC 5.02	pKa	25	EXP	PERRIN,DD (1965)
HL 8.38E-007	atm m3/mol	25	EST	MEYLAN,WM & HOWARD,PH (1991)
OH 4.64E-011	cm3/molc sec	25	EST	MEYLAN,WM & HOWARD,PH (1993)

000877-65-6 — 4-TERT-BUTYLBENZYL ALCOHOL

Formula: $C_{11}H_{16}O$

Mol Weight: 164.25

MP (deg C):

FP (deg C):

BP (deg C): 140

BP pressure (mm Hg): 2.00E+001

Property/Value	Units	Temp	Data Type	Reference
WS 5.98E+002	mg/L	25	EST	MEYLAN,WM ET AL. (1996)
logP 2.99			EST	MEYLAN,WM & HOWARD,PH (1995)
VP 1.18E-003	mm Hg	25	EST	NEELY,WB & BLAU,GE (1985)
DC	pKa			
HL 5.61E-007	atm m3/mol	25	EST	MEYLAN,WM & HOWARD,PH (1991)
OH 9.72E-012	cm3/molc sec	25	EST	MEYLAN,WM & HOWARD,PH (1993)

000877-95-2 — ACETAMIDE, N-(2-PHENYLETHYL)-

Formula: $C_{10}H_{13}NO$

Mol Weight: 163.22

MP (deg C):

FP (deg C):

BP (deg C):

BP pressure (mm Hg):

Property/Value	Units	Temp	Data Type	Reference
WS 6.37E+003	mg/L	25	EST	MEYLAN,WM ET AL. (1996)
logP 1.19			EXP	RADZICKA,A & WOLFENDEN,R (1988)
VP 7.84E-005	mm Hg	25	EST	NEELY,WB & BLAU,GE (1985)
DC	pKa			
HL 2.63E-009	atm m3/mol	25	EST	MEYLAN,WM & HOWARD,PH (1991)
OH 1.63E-011	cm3/molc sec	25	EST	MEYLAN,WM & HOWARD,PH (1993)

000878-13-7 — CYCLOUNDECANONE

Formula: $C_{11}H_{20}O$

Mol Weight: 168.28

MP (deg C):

FP (deg C):

BP (deg C): 106

BP pressure (mm Hg): 4.00E+000

Property/Value	Units	Temp	Data Type	Reference
WS 5.48E+001	mg/L	25	EST	MEYLAN,WM ET AL. (1996)
logP 3.58			EST	MEYLAN,WM & HOWARD,PH (1995)
VP 4.94E-002	mm Hg	25	EXT	BOUBLIK,T ET AL. (1984)
DC	pKa			
HL 2.11E-004	atm m3/mol	25	EST	MEYLAN,WM & HOWARD,PH (1991)
OH 1.92E-011	cm3/molc sec	25	EST	MEYLAN,WM & HOWARD,PH (1993)

000879-39-0 — 2,3,4,5-TETRACHLORONITROBENZENE

Formula: $C_6HCl_4NO_2$

Mol Weight: 260.89

MP (deg C):

FP (deg C):

BP (deg C):

BP pressure (mm Hg):

Property/Value	Units	Temp	Data Type	Reference
WS 7.31E+000	mg/L	20	EXP	YALKOWSKY,SH & DANNENFELSER,RM (1992)
logP 3.93			EXP	NIIMI,AJ ET AL. (1989)
VP 1.38E-004	mm Hg	25	EST	NEELY,WB & BLAU,GE (1985)
DC	pKa			
HL 6.41E-006	atm m3/mol	25	EST	MEYLAN,WM & HOWARD,PH (1991)
OH 7.19E-015	cm3/molc sec	25	EST	MEYLAN,WM & HOWARD,PH (1993)

000881-07-2 — QUINOLINE, 2-METHYL-8-NITRO-

Formula: $C_{10}H_8N_2O_2$

Mol Weight: 188.19

MP (deg C): 139-141

FP (deg C):

BP (deg C):

BP pressure (mm Hg):

Property/Value	Units	Temp	Data Type	Reference
WS 4.08E+002	mg/L	25	EST	MEYLAN,WM ET AL. (1996)
logP 1.99			EXP	DEBNATH,AK ET AL. (1992)
VP 5.10E-005	mm Hg	25	EST	NEELY,WB & BLAU,GE (1985)
DC	pKa			
HL 3.00E-009	atm m3/mol	25	EST	MEYLAN,WM & HOWARD,PH (1991)
OH 2.83E-012	cm3/molc sec	25	EST	MEYLAN,WM & HOWARD,PH (1993)

000881-51-6 — 2-NITRO-4-CHLOROACETANILIDE

Formula: $C_8H_7ClN_2O_3$

Mol Weight: 214.61

MP (deg C):

FP (deg C):

BP (deg C):

BP pressure (mm Hg):

Property/Value	Units	Temp	Data Type	Reference
WS 4.52E+002	mg/L	25	EST	MEYLAN,WM ET AL. (1996)
logP 1.78			EXP	NAKAGAWA,Y ET AL. (1992)
VP 1.14E-006	mm Hg	25	EST	NEELY,WB & BLAU,GE (1985)
DC	pKa			
HL 3.95E-010	atm m3/mol	25	EST	MEYLAN,WM & HOWARD,PH (1991)
OH 5.55E-013	cm3/molc sec	25	EST	MEYLAN,WM & HOWARD,PH (1993)

000882-06-4 — TRANS-P-NITROCINNAMIC ACID

Formula: $C_9H_7NO_4$

Mol Weight: 193.16

MP (deg C):
FP (deg C):
BP (deg C):
BP pressure (mm Hg):

Property/Value	Units	Temp	Data Type	Reference
WS 7.42E+002	mg/L	25	EST	MEYLAN,WM ET AL. (1996)
logP 2.12			EXP	HANSCH,C ET AL. (1995)
VP 1.25E-005	mm Hg	25	EST	NEELY,WB & BLAU,GE (1985)
DC 4.05	pKa		EXP	KORTUM,G ET AL (1961)
HL 5.10E-011	atm m3/mol	25	EST	MEYLAN,WM & HOWARD,PH (1991)
OH 2.05E-011	cm3/molc sec	25	EST	MEYLAN,WM & HOWARD,PH (1993)

000882-09-7 — 2-(P-CLPHENOXY)2-ME-PROPIONIC ACID

Formula: $C_{10}H_{11}ClO_3$

Mol Weight: 214.65

MP (deg C): 118-119
FP (deg C):
BP (deg C):
BP pressure (mm Hg):

Property/Value	Units	Temp	Data Type	Reference
WS 5.83E+002	mg/L	25	EST	MEYLAN,WM ET AL. (1996)
logP 2.57			EXP	HANSCH,C & LEO,AJ (1985)
VP 1.13E-004	mm Hg	25	EST	NEELY,WB & BLAU,GE (1985)
DC	pKa			
HL 2.19E-008	atm m3/mol	25	EST	MEYLAN,WM & HOWARD,PH (1991)
OH 7.75E-012	cm3/molc sec	25	EST	MEYLAN,WM & HOWARD,PH (1993)

000882-14-4 — PROPANAMIDE, N-(3,4-DICHLOROPHENYL)-

Formula: $C_{10}H_{11}Cl_2NO$

Mol Weight: 232.11

MP (deg C):
FP (deg C):
BP (deg C):
BP pressure (mm Hg):

Property/Value	Units	Temp	Data Type	Reference
WS 1.97E+001	mg/L	25	EST	MEYLAN,WM ET AL. (1996)
logP 3.72			EXP	HANSCH,C ET AL. (1995)
VP 6.62E-006	mm Hg	25	EST	NEELY,WB & BLAU,GE (1985)
DC	pKa			
HL 5.97E-009	atm m3/mol	25	EST	MEYLAN,WM & HOWARD,PH (1991)
OH 5.07E-012	cm3/molc sec	25	EST	MEYLAN,WM & HOWARD,PH (1993)

000882-26-8 — 3-NITRO-B-NITROSTYRENE

Formula: $C_8H_6N_2O_4$

Mol Weight: 194.15

MP (deg C):
FP (deg C):
BP (deg C):
BP pressure (mm Hg):

Property/Value	Units	Temp	Data Type	Reference
WS 5.33E+002	mg/L	25	EST	MEYLAN,WM ET AL. (1996)
logP 1.82			EXP	HANSCH,C & LEO,AJ (1985)
VP 1.15E-004	mm Hg	25	EST	NEELY,WB & BLAU,GE (1985)
DC	pKa			
HL 1.37E-008	atm m3/mol	25	EST	MEYLAN,WM & HOWARD,PH (1991)
OH 1.16E-011	cm3/molc sec	25	EST	MEYLAN,WM & HOWARD,PH (1993)

000882-33-7 — DIPHENYL DISULFIDE

Formula: $C_{12}H_{10}S_2$

Mol Weight: 218.34

MP (deg C): 62
FP (deg C):
BP (deg C): 310
BP pressure (mm Hg):

Property/Value	Units	Temp	Data Type	Reference
WS 6.02E+000	mg/L	25	EST	MEYLAN,WM ET AL. (1996)
logP 4.41			EXP	HANSCH,C & LEO,AJ (1985)
VP 2.20E-004	mm Hg	25	EXT	OHE,S (1976)
DC	pKa			
HL 4.84E-005	atm m3/mol	25	EST	MEYLAN,WM & HOWARD,PH (1991)
OH 2.50E-010	cm3/molc sec	25	EST	MEYLAN,WM & HOWARD,PH (1993)

000883-57-8 — 1-HEXYLBENZOTRIAZOLE

Formula: $C_{12}H_{17}N_3$

Mol Weight: 203.29

MP (deg C):
FP (deg C):
BP (deg C):
BP pressure (mm Hg):

Property/Value	Units	Temp	Data Type	Reference
WS 2.91E+001	mg/L	25	EST	MEYLAN,WM ET AL. (1996)
logP 3.70			EXP	HANSCH,C & LEO,AJ (1985)
VP 6.29E-005	mm Hg	25	EST	NEELY,WB & BLAU,GE (1985)
DC	pKa			
HL 1.29E-005	atm m3/mol	25	EST	MEYLAN,WM & HOWARD,PH (1991)
OH 7.70E-012	cm3/molc sec	25	EST	MEYLAN,WM & HOWARD,PH (1993)

000883-93-2 — 2-PHENYLBENZTHIAZOLE

Formula: $C_{13}H_9NS$

Mol Weight: 211.29

MP (deg C): 115
FP (deg C):
BP (deg C): 371
BP pressure (mm Hg):

Property/Value	Units	Temp	Data Type	Reference
WS 8.80E+000	mg/L	25	EST	MEYLAN,WM ET AL. (1996)
logP 4.26			EXP	HANSCH,C & LEO,AJ (1985)
VP 4.84E-006	mm Hg	25	EST	NEELY,WB & BLAU,GE (1985)
DC	pKa			
HL 2.87E-008	atm m3/mol	25	EST	MEYLAN,WM & HOWARD,PH (1991)
OH 1.55E-011	cm3/molc sec	25	EST	MEYLAN,WM & HOWARD,PH (1993)

000885-82-5 — 4-PHENYL-2-NITROPHENOL

Formula: $C_{12}H_9NO_3$

Mol Weight: 215.21

MP (deg C):
FP (deg C):
BP (deg C):
BP pressure (mm Hg):

Property/Value	Units	Temp	Data Type	Reference
WS 8.36E+000	mg/L	20	EXP	SCHWARZENBACH,RP ET AL. (1988)
logP 3.71			EXP	SCHWARZENBACH,RP ET AL. (1988)
VP 5.55E-007	mm Hg	25	EST	NEELY,WB & BLAU,GE (1985)
DC 6.73	pKa	25	EXP	SERJEANT,EP & DEMPSEY,B (1979)
HL 5.37E-007	atm m3/mol	25	EST	MEYLAN,WM & HOWARD,PH (1991)
OH 5.61E-012	cm3/molc sec	25	EST	MEYLAN,WM & HOWARD,PH (1993)

CAS #: 000886-50-0				TERBUTRYN

Formula: $C_{10}H_{19}N_5S$

Mol Weight: 241.36

MP (deg C): 104 FP (deg C):

BP (deg C): 157

BP pressure (mm Hg): 6.00E-002

Property/Value	Units	Temp	Data Type	Reference
WS 5.80E+001	mg/L	20	EXP	YALKOWSKY,SH & DANNENFELSER,RM (1992)
logP 3.74			EXP	HANSCH,C & LEO,AJ (1985)
VP 2.10E-006	mm Hg	25	EXP	WEBER,JB (1994)
DC	pKa			
HL 1.15E-008	atm m3/mol	25	EST	VP/WSOL
OH 1.07E-011	cm3/molc sec	25	EST	MEYLAN,WM & HOWARD,PH (1993)

CAS #: 000886-59-9				3-PHENYL-1-CYCLOHEXYLUREA

Formula: $C_{13}H_{18}N_2O$

Mol Weight: 218.30

MP (deg C): FP (deg C):

BP (deg C):

BP pressure (mm Hg):

Property/Value	Units	Temp	Data Type	Reference
WS 1.51E+002	mg/L	25	EST	MEYLAN,WM ET AL. (1996)
logP 2.77			EXP	HANSCH,C & LEO,AJ (1985)
VP 4.21E-006	mm Hg	25	EST	NEELY,WB & BLAU,GE (1985)
DC	pKa			
HL 8.04E-010	atm m3/mol	25	EST	MEYLAN,WM & HOWARD,PH (1991)
OH 6.28E-011	cm3/molc sec	25	EST	MEYLAN,WM & HOWARD,PH (1993)

CAS #: 000886-65-7				(1,4-DIPHENYL)-1,3-BUTADIENE

Formula: $C_{16}H_{14}$

Mol Weight: 206.29

MP (deg C): 151-153 FP (deg C):

BP (deg C): 350

BP pressure (mm Hg):

Property/Value	Units	Temp	Data Type	Reference
WS 1.23E+000	mg/L	25	EST	MEYLAN,WM ET AL. (1996)
logP 5.29			EST	MEYLAN,WM & HOWARD,PH (1995)
VP 3.54E-004	mm Hg	25	EST	NEELY,WB & BLAU,GE (1985)
DC	pKa			
HL 1.56E-004	atm m3/mol	25	EST	MEYLAN,WM & HOWARD,PH (1991)
OH 1.46E-010	cm3/molc sec	25	EST	MEYLAN,WM & HOWARD,PH (1993)

CAS #: 000890-38-0				INOSINE, 2'-DEOXY-

Formula: $C_{10}H_{12}N_4O_4$

Mol Weight: 252.23

MP (deg C): > 250 dec FP (deg C):

BP (deg C):

BP pressure (mm Hg):

Property/Value	Units	Temp	Data Type	Reference
WS 3.25E+004	mg/L	25	EST	MEYLAN,WM ET AL. (1996)
logP -1.71			EXP	FORD,H ET AL. (1991)
VP 1.08E-012	mm Hg	25	EST	NEELY,WB & BLAU,GE (1985)
DC	pKa			
HL 8.96E-022	atm m3/mol	25	EST	MEYLAN,WM & HOWARD,PH (1991)
OH 2.40E-010	cm3/molc sec	25	EST	MEYLAN,WM & HOWARD,PH (1993)

CAS #: 000890-67-5				3-METHYLHYDROCHLOROTHIAZIDE

Formula: $C_8H_{10}ClN_3O_4S_2$

Mol Weight: 311.77

MP (deg C): FP (deg C):

BP (deg C):

BP pressure (mm Hg):

Property/Value	Units	Temp	Data Type	Reference
WS 7.96E+002	mg/L	25	EST	MEYLAN,WM ET AL. (1996)
logP 0.08			EXP	HANSCH,C & LEO,AJ (1985)
VP 7.40E-010	mm Hg	25	EST	NEELY,WB & BLAU,GE (1985)
DC	pKa			
HL 5.82E-012	atm m3/mol	25	EST	MEYLAN,WM & HOWARD,PH (1991)
OH 1.69E-010	cm3/molc sec	25	EST	MEYLAN,WM & HOWARD,PH (1993)

CAS #: 000891-60-1				BENZAMIDE, 4-AMINO-3-CHLORO-N-[2-(DIETHYLAMINO)E

Formula: $C_{13}H_{20}ClN_3O$

Mol Weight: 269.78

MP (deg C): FP (deg C):

BP (deg C):

BP pressure (mm Hg):

Property/Value	Units	Temp	Data Type	Reference
WS 6.48E+002	mg/L	25	EST	MEYLAN,WM ET AL. (1996)
logP 1.70			EXP	VAN DAMME,M ET AL. (1984)
VP 2.39E-008	mm Hg	25	EST	NEELY,WB & BLAU,GE (1985)
DC	pKa			
HL 1.48E-015	atm m3/mol	25	EST	MEYLAN,WM & HOWARD,PH (1991)
OH 1.33E-010	cm3/molc sec	25	EST	MEYLAN,WM & HOWARD,PH (1993)

CAS #: 000915-67-3				AMARANTH

Formula: $C_{20}H_{11}N_2Na_3O_{10}S_3$

Mol Weight: 604.48

MP (deg C): FP (deg C):

BP (deg C):

BP pressure (mm Hg):

Property/Value	Units	Temp	Data Type	Reference
WS 6.00E+004	mg/L	25	EXP	GREEN,FJ (1990)
logP -5.13			EST	MEYLAN,WM & HOWARD,PH (1995)
VP 5.82E-025	mm Hg	25	EST	NEELY,WB & BLAU,GE (1985)
DC	pKa			
HL	atm m3/mol			
OH 1.80E-012	cm3/molc sec	25	EST	MEYLAN,WM & HOWARD,PH (1993)

CAS #: 000918-00-3				1,1,1-TRICHLOROACETONE

Formula: $C_3H_3Cl_3O$

Mol Weight: 161.42

MP (deg C): FP (deg C):

BP (deg C): 149

BP pressure (mm Hg):

Property/Value	Units	Temp	Data Type	Reference
WS 7.45E+003	mg/L	25	EST	MEYLAN,WM ET AL. (1996)
logP 1.12			EST	MEYLAN,WM & HOWARD,PH (1995)
VP 7.50E+000	mm Hg	25	EST	NEELY,WB & BLAU,GE (1985)
DC	pKa			
HL 2.17E-006	atm m3/mol	25	EST	MEYLAN,WM & HOWARD,PH (1991)
OH 1.50E-014	cm3/molc sec	25	EST	MEYLAN,WM & HOWARD,PH (1993)

533

CAS #: 000918-20-7 — 2-CHLORO-2-METHYLPROPANE

Formula: C_4H_9Cl

Mol Weight: 92.57

MP (deg C):

FP (deg C):

BP (deg C): 50

BP pressure (mm Hg):

Property/Value	Units	Temp	Data Type	Reference
WS 9.94E+002	mg/L	25	EST	MEYLAN,WM ET AL. (1996)
logP 2.45			EST	MEYLAN,WM & HOWARD,PH (1995)
VP 3.07E+002	mm Hg	25	EXP	BOUBLIK,T ET AL. (1984)
DC	pKa			
HL 1.92E-002	atm m3/mol	25	EST	MEYLAN,WM & HOWARD,PH (1991)
OH 4.08E-013	cm3/molc sec	25	EST	MEYLAN,WM & HOWARD,PH (1993)

CAS #: 000919-76-6 — AMIDOTHION

Formula: $C_7H_{16}NO_4PS_2$

Mol Weight: 273.31

MP (deg C):

FP (deg C):

BP (deg C):

BP pressure (mm Hg):

Property/Value	Units	Temp	Data Type	Reference
WS 2.00E+004	mg/L		EXP	GUNTHER,FA ET AL. (1968)
logP 0.00			EST	MEYLAN,WM & HOWARD,PH (1995)
VP 2.72E-006	mm Hg	25	EST	NEELY,WB & BLAU,GE (1985)
DC	pKa			
HL 3.28E-013	atm m3/mol	25	EST	MEYLAN,WM & HOWARD,PH (1991)
OH 9.97E-011	cm3/molc sec	25	EST	MEYLAN,WM & HOWARD,PH (1993)

CAS #: 000919-86-8 — DEMETON-S-METHYL

Formula: $C_6H_{15}O_3PS_2$

Mol Weight: 230.29

MP (deg C):

FP (deg C):

BP (deg C): 118

BP pressure (mm Hg): 1.00E+000

Property/Value	Units	Temp	Data Type	Reference
WS 3.30E+003	mg/L	20	EXP	SHIU,WY ET AL. (1990)
logP 1.02			EXP	HANSCH,C & LEO,AJ (1985)
VP 3.00E-004	mm Hg	20	EXP	TOMLIN,C (1994)
DC	pKa			
HL 2.75E-008	atm m3/mol	25	EST	VP/WSOL
OH 4.72E-011	cm3/molc sec	25	EST	MEYLAN,WM & HOWARD,PH (1993)

CAS #: 000919-94-8 — ETHYL(TERT-AMYL) ETHER

Formula: $C_7H_{16}O$

Mol Weight: 116.20

MP (deg C):

FP (deg C):

BP (deg C): 102

BP pressure (mm Hg):

Property/Value	Units	Temp	Data Type	Reference
WS 4.00E+002	mg/L	20	EXP	EVANS,TW & EDLUND,KR (1936)
logP 2.41			EST	MEYLAN,WM & HOWARD,PH (1995)
VP 5.00E+001	mm Hg	25	EXP	EVANS,TW & EDLUND,KR (1936)
DC	pKa			
HL 1.91E-002	atm m3/mol	25	EST	VP/WSOL
OH 1.06E-011	cm3/molc sec	25	EST	MEYLAN,WM & HOWARD,PH (1993)

CAS #: 000920-66-1 — 1,1,1,3,3,3-HEXAFLUOROPROPAN-2-OL

Formula: $C_3H_2F_6O$

Mol Weight: 168.04

MP (deg C): -4

FP (deg C):

BP (deg C): 59

BP pressure (mm Hg):

Property/Value	Units	Temp	Data Type	Reference
WS 7.77E+003	mg/L	25	EST	MEYLAN,WM ET AL. (1996)
logP 1.66			EXP	HANSCH,C ET AL. (1995)
VP 1.59E+002	mm Hg	25	EXP	BOUBLIK,T ET AL. (1984)
DC 9.30	pKa	25	EXP	SERJEANT,EP & DEMPSEY,B (1979)
HL 4.25E-005	atm m3/mol	25	EXP	ROCHESTER,CH & SYMONDS,JR (1973)
OH 7.10E-014	cm3/molc sec	25	EST	MEYLAN,WM & HOWARD,PH (1993)

CAS #: 000921-09-5 — 1,1,2,3-TETRACHLOROBUTA-1,3-DIENE

Formula: $C_4H_2Cl_4$

Mol Weight: 191.87

MP (deg C):

FP (deg C):

BP (deg C):

BP pressure (mm Hg):

Property/Value	Units	Temp	Data Type	Reference
WS 2.40E+001	mg/L	25	EST	MEYLAN,WM ET AL. (1996)
logP 3.87			EST	MEYLAN,WM & HOWARD,PH (1995)
VP 4.73E+000	mm Hg	25	EST	NEELY,WB & BLAU,GE (1985)
DC	pKa			
HL 2.09E-002	atm m3/mol	25	EST	MEYLAN,WM & HOWARD,PH (1991)
OH 5.06E-013	cm3/molc sec	25	EST	MEYLAN,WM & HOWARD,PH (1993)

CAS #: 000921-47-1 — 2,3,4-TRIMETHYLHEXANE

Formula: C_9H_{20}

Mol Weight: 128.26

MP (deg C):

FP (deg C):

BP (deg C): 139.1

BP pressure (mm Hg):

Property/Value	Units	Temp	Data Type	Reference
WS 3.59E+000	mg/L	25	EST	MEYLAN,WM ET AL. (1996)
logP 4.54			EST	MEYLAN,WM & HOWARD,PH (1995)
VP 9.02E+000	mm Hg	25	EXT	ZWOLINSKI,BJ & WILHOIT,RC (1971)
DC	pKa			
HL 4.00E+000	atm m3/mol	25	EST	MEYLAN,WM & HOWARD,PH (1991)
OH 1.02E-011	cm3/molc sec	25	EST	MEYLAN,WM & HOWARD,PH (1993)

CAS #: 000922-61-2 — 3-METHYL-2-PENTENE

Formula: C_6H_{12}

Mol Weight: 84.16

MP (deg C):

FP (deg C):

BP (deg C): 69

BP pressure (mm Hg):

Property/Value	Units	Temp	Data Type	Reference
WS 7.93E+001	mg/L	25	EST	MEYLAN,WM ET AL. (1996)
logP 3.13			EST	MEYLAN,WM & HOWARD,PH (1995)
VP 1.57E+002	mm Hg	25	EXP	ZWOLINSKI,BJ & WILHOIT,RC (1971)
DC	pKa			
HL 5.00E-001	atm m3/mol	25	EST	MEYLAN,WM & HOWARD,PH (1991)
OH 8.83E-011	cm3/molc sec	25	EST	MEYLAN,WM & HOWARD,PH (1993)

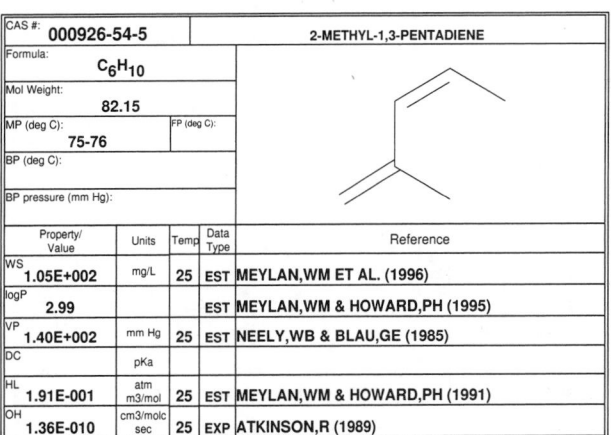

2-ETHYLPROPENAL

CAS #:	000922-63-4			
Formula:	$C_5H_{10}O$			
Mol Weight:	86.13			
MP (deg C):		FP (deg C):		
BP (deg C):	92-93			
BP pressure (mm Hg):				

Property/ Value	Units	Temp	Data Type	Reference
WS 1.12E+004	mg/L	25	EST	MEYLAN,WM ET AL. (1996)
logP 1.24			EXP	HANSCH,C & LEO,AJ (1985)
VP 3.64E+001	mm Hg	25	EST	NEELY,WB & BLAU,GE (1985)
DC	pKa			
HL 7.44E-005	atm m3/mol	25	EST	MEYLAN,WM & HOWARD,PH (1991)
OH 3.55E-011	cm3/molc sec	25	EST	MEYLAN,WM & HOWARD,PH (1993)

2-HYDROXYPROPYL METHACRYLATE

CAS #:	000923-26-2			
Formula:	$C_7H_{12}O_3$			
Mol Weight:	144.17			
MP (deg C):		FP (deg C):		
BP (deg C):	90			
BP pressure (mm Hg):	9.00E+000			

Property/ Value	Units	Temp	Data Type	Reference
WS 3.86E+004	mg/L	25	EST	MEYLAN,WM ET AL. (1996)
logP 0.97			EXP	HANSCH,C & LEO,AJ (1985)
VP 7.24E-002	mm Hg	25	EST	NEELY,WB & BLAU,GE (1985)
DC	pKa			
HL 9.34E-009	atm m3/mol	25	EST	MEYLAN,WM & HOWARD,PH (1991)
OH 2.86E-011	cm3/molc sec	25	EST	MEYLAN,WM & HOWARD,PH (1993)

N-NITROSODIBUTYLAMINE

CAS #:	000924-16-3			
Formula:	$C_8H_{18}N_2O$			
Mol Weight:	158.25			
MP (deg C):	2.13 EST	FP (deg C):		
BP (deg C):	116			
BP pressure (mm Hg):				

Property/ Value	Units	Temp	Data Type	Reference
WS 1.20E+003	mg/L	25	EXP	IARC (1978)
logP 1.92			EXP	HANSCH,C & LEO,AJ (1985)
VP 3.00E-002	mm Hg	25	EXP	KLEIN,RG (1982)
DC	pKa			
HL 1.32E-005	atm m3/mol	37	EXP	MIRVISH,SS ET AL. (1976)
OH 2.69E-011	cm3/molc sec	25	EST	MEYLAN,WM & HOWARD,PH (1993)

N-NITROSOMETHYLPROPYLAMINE

CAS #:	000924-46-9			
Formula:	$C_4H_{10}N_2O$			
Mol Weight:	102.14			
MP (deg C):		FP (deg C):		
BP (deg C):				
BP pressure (mm Hg):				

Property/ Value	Units	Temp	Data Type	Reference
WS 3.83E+004	mg/L	25	EST	MEYLAN,WM ET AL. (1996)
logP 0.56			EXP	VERA,A ET AL. (1992)
VP 5.62E-001	mm Hg	25	EST	NEELY,WB & BLAU,GE (1985)
DC	pKa			
HL 3.63E-006	atm m3/mol	25	EST	MEYLAN,WM & HOWARD,PH (1991)
OH 1.33E-011	cm3/molc sec	25	EST	MEYLAN,WM & HOWARD,PH (1993)

BUTANEDIOIC ACID, DIPROPYL ESTER

CAS #:	000925-15-5			
Formula:	$C_{10}H_{18}O_4$			
Mol Weight:	202.25			
MP (deg C):	-5.9	FP (deg C):		
BP (deg C):	250.8			
BP pressure (mm Hg):				

Property/ Value	Units	Temp	Data Type	Reference
WS 6.09E+002	mg/L	25	EST	MEYLAN,WM ET AL. (1996)
logP 2.16			EXP	FUNASAKI,N ET AL. (1984)
VP 9.55E-002	mm Hg	25	EST	NEELY,WB & BLAU,GE (1985)
DC	pKa			
HL 1.72E-006	atm m3/mol	25	EST	MEYLAN,WM & HOWARD,PH (1991)
OH 8.01E-012	cm3/molc sec	25	EST	MEYLAN,WM & HOWARD,PH (1993)

METHYL-4-HYDROXYBUTYRATE

CAS #:	000925-57-5			
Formula:	$C_5H_{10}O_3$			
Mol Weight:	118.13			
MP (deg C):		FP (deg C):		
BP (deg C):				
BP pressure (mm Hg):				

Property/ Value	Units	Temp	Data Type	Reference
WS 3.71E+005	mg/L	25	EST	MEYLAN,WM ET AL. (1996)
logP -0.06			EXP	HANSCH,C & LEO,AJ (1985)
VP 1.74E-001	mm Hg	25	EST	NEELY,WB & BLAU,GE (1985)
DC	pKa			
HL 1.13E-008	atm m3/mol	25	EST	MEYLAN,WM & HOWARD,PH (1991)
OH 6.64E-012	cm3/molc sec	25	EST	MEYLAN,WM & HOWARD,PH (1993)

2-PROPENOIC ACID, PROPYL ESTER

CAS #:	000925-60-0			
Formula:	$C_6H_{10}O_2$			
Mol Weight:	114.15			
MP (deg C):		FP (deg C):		
BP (deg C):				
BP pressure (mm Hg):				

Property/ Value	Units	Temp	Data Type	Reference
WS 3.65E+003	mg/L	25	EST	MEYLAN,WM ET AL. (1996)
logP 1.71			EST	MEYLAN,WM & HOWARD,PH (1995)
VP 1.45E+001	mm Hg	25	EXP	DAUBERT,TE & DANNER,RP (1989)
DC	pKa			
HL 1.63E-004	atm m3/mol	25	EST	MEYLAN,WM & HOWARD,PH (1991)
OH 1.24E-011	cm3/molc sec	25	EST	MEYLAN,WM & HOWARD,PH (1993)

2-METHYL-1,3-PENTADIENE

CAS #:	000926-54-5			
Formula:	C_6H_{10}			
Mol Weight:	82.15			
MP (deg C):	75-76	FP (deg C):		
BP (deg C):				
BP pressure (mm Hg):				

Property/ Value	Units	Temp	Data Type	Reference
WS 1.05E+002	mg/L	25	EST	MEYLAN,WM ET AL. (1996)
logP 2.99			EST	MEYLAN,WM & HOWARD,PH (1995)
VP 1.40E+002	mm Hg	25	EST	NEELY,WB & BLAU,GE (1985)
DC	pKa			
HL 1.91E-001	atm m3/mol	25	EST	MEYLAN,WM & HOWARD,PH (1991)
OH 1.36E-010	cm3/molc sec	25	EXP	ATKINSON,R (1989)

CAS #: 000926-56-7 — 4-METHYL-1,3-PENTADIENE

Formula: C_6H_{10}

Mol Weight: 82.15

MP (deg C):

FP (deg C):

BP (deg C): 76.5

BP pressure (mm Hg):

Property/Value	Units	Temp	Data Type	Reference
WS 1.05E+002	mg/L	25	EST	MEYLAN,WM ET AL. (1996)
logP 2.99			EST	MEYLAN,WM & HOWARD,PH (1995)
VP 1.40E+002	mm Hg	25	EST	NEELY,WB & BLAU,GE (1985)
DC	pKa			
HL 1.91E-001	atm m3/mol	25	EST	MEYLAN,WM & HOWARD,PH (1991)
OH 1.31E-010	cm3/molc sec	25	EXP	ATKINSON,R (1989)

CAS #: 000926-57-8 — 1,3-DICHLORO-2-BUTENE

Formula: $C_4H_6Cl_2$

Mol Weight: 125.00

MP (deg C):

FP (deg C):

BP (deg C): 131

BP pressure (mm Hg):

Property/Value	Units	Temp	Data Type	Reference
WS 3.63E+002	mg/L	25	EST	MEYLAN,WM ET AL. (1996)
logP 2.84			EST	MEYLAN,WM & HOWARD,PH (1995)
VP 1.26E+001	mm Hg	25	EST	NEELY,WB & BLAU,GE (1985)
DC	pKa			
HL 6.64E-004	atm m3/mol	25	EST	MEYLAN,WM & HOWARD,PH (1991)
OH 1.37E-011	cm3/molc sec	25	EST	MEYLAN,WM & HOWARD,PH (1993)

CAS #: 000927-49-1 — 6-UNDECANONE

Formula: $C_{11}H_{22}O$

Mol Weight: 170.30

MP (deg C): 14-15

FP (deg C):

BP (deg C): 226

BP pressure (mm Hg):

Property/Value	Units	Temp	Data Type	Reference
WS 5.00E+001	mg/L	25	EXP	CHEM INSPECT TEST INST (1992)
logP 3.69			EST	MEYLAN,WM & HOWARD,PH (1995)
VP 5.00E-002	mm Hg	20	EXP	ENGINEERING SCIENCES UNIT (1975)
DC	pKa			
HL 2.24E-004	atm m3/mol	20	EST	VP/WSOL
OH 1.61E-011	cm3/molc sec	25	EST	MEYLAN,WM & HOWARD,PH (1993)

CAS #: 000927-62-8 — DIMETHYLBUTYLAMINE

Formula: $C_6H_{15}N$

Mol Weight: 101.19

MP (deg C):

FP (deg C):

BP (deg C): 95

BP pressure (mm Hg):

Property/Value	Units	Temp	Data Type	Reference
WS 4.18E+004	mg/L	25	EST	MEYLAN,WM ET AL. (1996)
logP 1.70			EXP	HANSCH,C & LEO,AJ (1985)
VP 4.55E+001	mm Hg	25	EST	NEELY,WB & BLAU,GE (1985)
DC 10.19	pKa	20	EXP	PERRIN,DD (1965)
HL 8.54E-005	atm m3/mol	25	EST	MEYLAN,WM & HOWARD,PH (1991)
OH 8.19E-011	cm3/molc sec	25	EST	MEYLAN,WM & HOWARD,PH (1993)

CAS #: 000927-68-4 — ETHANOL, 2-BROMO-, ACETATE

Formula: $C_4H_7BrO_2$

Mol Weight: 167.01

MP (deg C): -13.8

FP (deg C):

BP (deg C): 162.5

BP pressure (mm Hg):

Property/Value	Units	Temp	Data Type	Reference
WS 3.54E+004	mg/L	25	EXP	WASIK,SP ET AL. (1981)
logP 1.11			EXP	HANSCH,C & LEO,AJ (1985)
VP 3.37E+000	mm Hg	25	EST	NEELY,WB & BLAU,GE (1985)
DC	pKa			
HL 2.68E-005	atm m3/mol	25	EST	MEYLAN,WM & HOWARD,PH (1991)
OH 1.11E-012	cm3/molc sec	25	EST	MEYLAN,WM & HOWARD,PH (1993)

CAS #: 000927-73-1 — 4-CHLORO-1-BUTENE

Formula: C_4H_7Cl

Mol Weight: 90.55

MP (deg C):

FP (deg C):

BP (deg C): 63.5

BP pressure (mm Hg):

Property/Value	Units	Temp	Data Type	Reference
WS 1.06E+003	mg/L	25	EST	MEYLAN,WM ET AL. (1996)
logP 2.42			EST	MEYLAN,WM & HOWARD,PH (1995)
VP 6.64E+001	mm Hg	25	EST	NEELY,WB & BLAU,GE (1985)
DC	pKa			
HL 4.52E-002	atm m3/mol	25	EST	MEYLAN,WM & HOWARD,PH (1991)
OH 2.71E-011	cm3/molc sec	25	EST	MEYLAN,WM & HOWARD,PH (1993)

CAS #: 000928-45-0 — BUTYL NITRATE

Formula: $C_4H_9NO_3$

Mol Weight: 119.12

MP (deg C):

FP (deg C):

BP (deg C): 133

BP pressure (mm Hg):

Property/Value	Units	Temp	Data Type	Reference
WS 1.48E+003	mg/L	25	EST	MEYLAN,WM ET AL. (1996)
logP 2.15			EXP	HANSCH,C & LEO,AJ (1985)
VP 8.19E+000	mm Hg	25	EST	NEELY,WB & BLAU,GE (1985)
DC	pKa			
HL 4.64E-004	atm m3/mol	25	EST	MEYLAN,WM & HOWARD,PH (1991)
OH 1.78E-012	cm3/molc sec	25	EXP	ATKINSON,R (1989)

CAS #: 000928-51-8 — 1-BUTANOL, 4-CHLORO-

Formula: C_4H_9ClO

Mol Weight: 108.57

MP (deg C):

FP (deg C):

BP (deg C): 84-85

BP pressure (mm Hg): 1.60E+001

Property/Value	Units	Temp	Data Type	Reference
WS 6.69E+004	mg/L	25	EST	MEYLAN,WM ET AL. (1996)
logP 0.85			EXP	HANSCH,C ET AL. (1995)
VP 3.52E-001	mm Hg	25	EST	NEELY,WB & BLAU,GE (1985)
DC	pKa			
HL 2.22E-006	atm m3/mol	25	EST	MEYLAN,WM & HOWARD,PH (1991)
OH 6.42E-012	cm3/molc sec	25	EST	MEYLAN,WM & HOWARD,PH (1993)

536

CAS #: 000928-65-4				HEXYLTRICHLOROSILANE

Formula: $C_6H_{13}Cl_3Si$

Mol Weight: 219.62

MP (deg C): -140 FP (deg C):

BP (deg C): 55.5

BP pressure (mm Hg):

Property/Value	Units	Temp	Data Type	Reference
WS 5.34E+000	mg/L	25	EST	MEYLAN,WM ET AL. (1996)
logP 4.46			EST	MEYLAN,WM & HOWARD,PH (1995)
VP 1.25E+000	mm Hg	25	EST	NEELY,WB & BLAU,GE (1985)
DC	pKa			
HL 2.20E-002	atm m3/mol	25	EST	MEYLAN,WM & HOWARD,PH (1991)
OH 6.82E-012	cm3/molc sec	25	EST	MEYLAN,WM & HOWARD,PH (1993)

CAS #: 000929-73-7				1-DODECANAMINE, HYDROCHLORIDE

Formula: $C_{12}H_{28}ClN$

Mol Weight: 221.82

MP (deg C): FP (deg C):

BP (deg C):

BP pressure (mm Hg):

Property/Value	Units	Temp	Data Type	Reference
WS 8.86E+002	mg/L	25	EST	MEYLAN,WM ET AL. (1996)
logP 1.85			EXP	HANSCH,C ET AL. (1995)
VP 7.62E-008	mm Hg	25	EST	NEELY,WB & BLAU,GE (1985)
DC	pKa			
HL 1.59E-011	atm m3/mol	25	EST	MEYLAN,WM & HOWARD,PH (1991)
OH 2.47E-011	cm3/molc sec	25	EST	MEYLAN,WM & HOWARD,PH (1993)

CAS #: 000930-27-8				3-METHYLFURAN

Formula: C_5H_6O

Mol Weight: 82.10

MP (deg C): FP (deg C):

BP (deg C): 66

BP pressure (mm Hg):

Property/Value	Units	Temp	Data Type	Reference
WS 3.03E+003	mg/L	25	EST	MEYLAN,WM ET AL. (1996)
logP 1.91			EST	MEYLAN,WM & HOWARD,PH (1995)
VP 7.27E+001	mm Hg	25	EST	NEELY,WB & BLAU,GE (1985)
DC	pKa			
HL 5.93E-003	atm m3/mol	25	EST	MEYLAN,WM & HOWARD,PH (1991)
OH 9.35E-011	cm3/molc sec	25	EXP	ATKINSON,R (1989)

CAS #: 000930-36-9				1-METHYLPYRAZOLE

Formula: $C_4H_6N_2$

Mol Weight: 82.11

MP (deg C): FP (deg C):

BP (deg C): 127

BP pressure (mm Hg):

Property/Value	Units	Temp	Data Type	Reference
WS 8.25E+004	mg/L	25	EST	MEYLAN,WM ET AL. (1996)
logP 0.23			EXP	HANSCH,C ET AL. (1995)
VP 1.71E+001	mm Hg	25	EST	NEELY,WB & BLAU,GE (1985)
DC 2.04	pKa	25	EXP	PERRIN,DD (1965)
HL 7.88E-005	atm m3/mol	25	EST	MEYLAN,WM & HOWARD,PH (1991)
OH 3.61E-011	cm3/molc sec	25	EST	MEYLAN,WM & HOWARD,PH (1993)

CAS #: 000930-55-2				N-NITROSOPYRROLIDINE

Formula: $C_4H_8N_2O$

Mol Weight: 100.12

MP (deg C): FP (deg C):

BP (deg C): 104-106

BP pressure (mm Hg): 2.00E+001

Property/Value	Units	Temp	Data Type	Reference
WS 1.70E+005	mg/L	25	EST	MEYLAN,WM ET AL. (1996)
logP -0.19			EXP	HANSCH,C & LEO,AJ (1985)
VP 9.00E-002	mm Hg	25	EXT	KLEIN,RG (1982)
DC	pKa			
HL 4.89E-008	atm m3/mol	37	EXP	MIRVISH,SS ET AL. (1976)
OH 1.95E-011	cm3/molc sec	25	EST	MEYLAN,WM & HOWARD,PH (1993)

CAS #: 000930-68-7				2-CYCLOHEXENE-1-ONE

Formula: C_6H_8O

Mol Weight: 96.13

MP (deg C): -53 FP (deg C):

BP (deg C): 170

BP pressure (mm Hg):

Property/Value	Units	Temp	Data Type	Reference
WS 3.62E+004	mg/L	25	EST	MEYLAN,WM ET AL. (1996)
logP 0.61			EXP	HANSCH,C & LEO,AJ (1985)
VP 3.62E+000	mm Hg	25	EST	NEELY,WB & BLAU,GE (1985)
DC	pKa			
HL 2.40E-005	atm m3/mol	25	EST	MEYLAN,WM & HOWARD,PH (1991)
OH 5.73E-011	cm3/molc sec	25	EST	MEYLAN,WM & HOWARD,PH (1993)

CAS #: 000930-73-4				N-METHYLPYRIDINIUM IODIDE

Formula: C_6H_8IN

Mol Weight: 221.04

MP (deg C): FP (deg C):

BP (deg C):

BP pressure (mm Hg):

Property/Value	Units	Temp	Data Type	Reference
WS 1.00E+006	mg/L	25	EST	MEYLAN,WM ET AL. (1996)
logP -3.30			EXP	HANSCH,C & LEO,AJ (1985)
VP 1.29E-001	mm Hg	25	EST	NEELY,WB & BLAU,GE (1985)
DC	pKa			
HL 1.50E-010	atm m3/mol	25	EST	MEYLAN,WM & HOWARD,PH (1991)
OH 1.05E-011	cm3/molc sec	25	EST	MEYLAN,WM & HOWARD,PH (1993)

CAS #: 000930-90-5				1-METHYL-2-ETHYLCYCLOPENTANE (TRANS)

Formula: C_8H_{16}

Mol Weight: 112.22

MP (deg C): -105.9 FP (deg C):

BP (deg C): 121.2

BP pressure (mm Hg):

Property/Value	Units	Temp	Data Type	Reference
WS 1.17E+001	mg/L	25	EST	MEYLAN,WM ET AL. (1996)
logP 4.01			EST	MEYLAN,WM & HOWARD,PH (1995)
VP 2.00E+001	mm Hg	25	EXP	ZWOLINSKI,BJ & WILHOIT,RC (1971)
DC	pKa			
HL 4.50E-001	atm m3/mol	25	EST	MEYLAN,WM & HOWARD,PH (1991)
OH 1.01E-011	cm3/molc sec	25	EST	MEYLAN,WM & HOWARD,PH (1993)

CAS #: 000931-17-9 — 1,2-CYCLOHEXANEDIOL

Formula: $C_6H_{12}O_2$
Mol Weight: 116.16
MP (deg C):
FP (deg C):
BP (deg C):
BP pressure (mm Hg):

Property/Value	Units	Temp	Data Type	Reference
WS 6.59E+004	mg/L	25	EST	MEYLAN,WM ET AL. (1996)
logP 0.51			EST	MEYLAN,WM & HOWARD,PH (1995)
VP 1.05E-002	mm Hg	25	EST	NEELY,WB & BLAU,GE (1985)
DC	pKa			
HL 1.79E-007	atm m3/mol	25	EST	MEYLAN,WM & HOWARD,PH (1991)
OH 2.65E-011	cm3/molc sec	25	EST	MEYLAN,WM & HOWARD,PH (1993)

CAS #: 000931-19-1 — 2-METHYL-N-OXIDEPYRIDINE

Formula: C_6H_7NO
Mol Weight: 109.13
MP (deg C): 48-50
FP (deg C):
BP (deg C):
BP pressure (mm Hg):

Property/Value	Units	Temp	Data Type	Reference
WS 5.17E+005	mg/L	25	EST	MEYLAN,WM ET AL. (1996)
logP -0.79			EST	MEYLAN,WM & HOWARD,PH (1995)
VP 5.33E-002	mm Hg	25	EST	NEELY,WB & BLAU,GE (1985)
DC	pKa			
HL 7.78E-011	atm m3/mol	25	EST	MEYLAN,WM & HOWARD,PH (1991)
OH 1.10E-012	cm3/molc sec	25	EST	MEYLAN,WM & HOWARD,PH (1993)

CAS #: 000931-63-5 — PYRIMIDINE, 2-METHOXY-

Formula: $C_5H_6N_2O$
Mol Weight: 110.12
MP (deg C):
FP (deg C):
BP (deg C):
BP pressure (mm Hg):

Property/Value	Units	Temp	Data Type	Reference
WS 6.92E+004	mg/L	25	EST	MEYLAN,WM ET AL. (1996)
logP 0.23			EXP	HANSCH,C ET AL. (1995)
VP 1.60E+000	mm Hg	25	EST	NEELY,WB & BLAU,GE (1985)
DC	pKa			
HL 1.57E-006	atm m3/mol	25	EST	MEYLAN,WM & HOWARD,PH (1991)
OH 3.37E-012	cm3/molc sec	25	EST	MEYLAN,WM & HOWARD,PH (1993)

CAS #: 000931-64-6 — BICYCLO(2.2.2)OCT-2-ENE

Formula: C_8H_{12}
Mol Weight: 108.18
MP (deg C):
FP (deg C):
BP (deg C):
BP pressure (mm Hg):

Property/Value	Units	Temp	Data Type	Reference
WS 4.49E+001	mg/L	25	EST	MEYLAN,WM ET AL. (1996)
logP 3.34			EST	MEYLAN,WM & HOWARD,PH (1995)
VP 9.93E+000	mm Hg	25	EST	NEELY,WB & BLAU,GE (1985)
DC	pKa			
HL 5.15E-002	atm m3/mol	25	EST	MEYLAN,WM & HOWARD,PH (1991)
OH 4.08E-011	cm3/molc sec	25	EXP	ATKINSON,R (1989)

CAS #: 000932-17-2 — 2-PYRROLIDINONE, 1-ACETYL-

Formula: $C_6H_9NO_2$
Mol Weight: 127.14
MP (deg C):
FP (deg C):
BP (deg C):
BP pressure (mm Hg):

Property/Value	Units	Temp	Data Type	Reference
WS 1.04E+005	mg/L	25	EST	MEYLAN,WM ET AL. (1996)
logP -0.05			EXP	SANGSTER,J (1994)
VP 3.49E-004	mm Hg	25	EST	NEELY,WB & BLAU,GE (1985)
DC	pKa			
HL 7.38E-008	atm m3/mol	25	EST	MEYLAN,WM & HOWARD,PH (1991)
OH 2.31E-011	cm3/molc sec	25	EST	MEYLAN,WM & HOWARD,PH (1993)

CAS #: 000932-83-2 — N-NITROSOHEXAMETHYLENEIMINE

Formula: $C_6H_{12}N_2O$
Mol Weight: 128.18
MP (deg C):
FP (deg C):
BP (deg C):
BP pressure (mm Hg):

Property/Value	Units	Temp	Data Type	Reference
WS 1.53E+004	mg/L	25	EST	MEYLAN,WM ET AL. (1996)
logP 0.92			EXP	HANSCH,C & LEO,AJ (1985)
VP 4.51E-002	mm Hg	25	EST	NEELY,WB & BLAU,GE (1985)
DC	pKa			
HL 2.82E-006	atm m3/mol	25	EST	MEYLAN,WM & HOWARD,PH (1991)
OH 2.70E-011	cm3/molc sec	25	EST	MEYLAN,WM & HOWARD,PH (1993)

CAS #: 000932-90-1 — BENZALDOXIME

Formula: C_7H_7NO
Mol Weight: 121.14
MP (deg C):
FP (deg C):
BP (deg C):
BP pressure (mm Hg):

Property/Value	Units	Temp	Data Type	Reference
WS 3.18E+003	mg/L	25	EST	MEYLAN,WM ET AL. (1996)
logP 1.75			EXP	HANSCH,C & LEO,AJ (1985)
VP 4.10E-003	mm Hg	25	EST	NEELY,WB & BLAU,GE (1985)
DC	pKa			
HL 3.57E-007	atm m3/mol	25	EST	MEYLAN,WM & HOWARD,PH (1991)
OH 6.57E-012	cm3/molc sec	25	EST	MEYLAN,WM & HOWARD,PH (1993)

CAS #: 000933-67-5 — 7-METHYLINDOLE

Formula: C_9H_9N
Mol Weight: 131.18
MP (deg C): 85
FP (deg C):
BP (deg C): 266
BP pressure (mm Hg):

Property/Value	Units	Temp	Data Type	Reference
WS 5.92E+002	mg/L	25	EST	MEYLAN,WM ET AL. (1996)
logP 2.56			EXP	HANSCH,C ET AL. (1995)
VP 6.03E-003	mm Hg	25	EST	NEELY,WB & BLAU,GE (1985)
DC	pKa			
HL 9.78E-007	atm m3/mol	25	EST	MEYLAN,WM & HOWARD,PH (1991)
OH 2.00E-010	cm3/molc sec	25	EST	MEYLAN,WM & HOWARD,PH (1993)

539

CAS #: 000934-80-5	4-ETHYL-1,2-DIMETHYLBENZENE

Formula: $C_{10}H_{14}$

Mol Weight: 134.22

MP (deg C): -66.9

FP (deg C):

BP (deg C): 189.5

BP pressure (mm Hg):

Property/ Value	Units	Temp	Data Type	Reference
WS 1.27E+001	mg/L	25	EST	MEYLAN,WM ET AL. (1996)
logP 4.50			EXP	SHERBLOM,PM & EGANHOUSE,RP (1988)
VP 7.48E-001	mm Hg	25	EXP	YAWS,CL (1994B)
DC	pKa			
HL 9.61E-003	atm m3/mol	25	EST	MEYLAN,WM & HOWARD,PH (1991)
OH 1.69E-011	cm3/molc sec	25	EST	MEYLAN,WM & HOWARD,PH (1993)

CAS #: 000934-87-2	THIACETIC ACID, S-PHENYL ESTER

Formula: C_8H_8OS

Mol Weight: 152.22

MP (deg C):

FP (deg C):

BP (deg C): 99-100

BP pressure (mm Hg): 6.00E+000

Property/ Value	Units	Temp	Data Type	Reference
WS 9.25E+002	mg/L	25	EST	MEYLAN,WM ET AL. (1996)
logP 2.23			EXP	HANSCH,C & LEO,AJ (1985)
VP 3.65E-002	mm Hg	25	EST	NEELY,WB & BLAU,GE (1985)
DC	pKa			
HL 1.25E-005	atm m3/mol	25	EST	MEYLAN,WM & HOWARD,PH (1991)
OH 2.33E-012	cm3/molc sec	25	EST	MEYLAN,WM & HOWARD,PH (1993)

CAS #: 000935-30-8	2-AZACYCLO-NONANONE

Formula: $C_8H_{15}NO$

Mol Weight: 141.21

MP (deg C): 77-79

FP (deg C):

BP (deg C):

BP pressure (mm Hg):

Property/ Value	Units	Temp	Data Type	Reference
WS 2.22E+004	mg/L	25	EST	MEYLAN,WM ET AL. (1996)
logP 0.67			EXP	HANSCH,C & LEO,AJ (1985)
VP 3.03E-004	mm Hg	25	EST	NEELY,WB & BLAU,GE (1985)
DC	pKa			
HL 4.46E-008	atm m3/mol	25	EST	MEYLAN,WM & HOWARD,PH (1991)
OH 2.12E-011	cm3/molc sec	25	EST	MEYLAN,WM & HOWARD,PH (1993)

CAS #: 000935-92-2	2,5-CYCLOHEXADIENE-1,4-DIONE, 2,3,5-TRIMETHYL-

Formula: $C_9H_{10}O_2$

Mol Weight: 150.18

MP (deg C):

FP (deg C):

BP (deg C):

BP pressure (mm Hg):

Property/ Value	Units	Temp	Data Type	Reference
WS 2.11E+003	mg/L	25	EST	MEYLAN,WM ET AL. (1996)
logP 1.82			EXP	RICH,PR (1990)
VP 8.22E-003	mm Hg	25	EST	NEELY,WB & BLAU,GE (1985)
DC	pKa			
HL 4.69E-009	atm m3/mol	25	EST	MEYLAN,WM & HOWARD,PH (1991)
OH 2.45E-011	cm3/molc sec	25	EST	MEYLAN,WM & HOWARD,PH (1993)

CAS #: 000935-95-5	2,3,5,6-TETRACHLOROPHENOL

Formula: $C_6H_2Cl_4O$

Mol Weight: 231.89

MP (deg C): 115

FP (deg C):

BP (deg C):

BP pressure (mm Hg):

Property/ Value	Units	Temp	Data Type	Reference
WS 5.49E+001	mg/L	25	EST	MEYLAN,WM ET AL. (1996)
logP 3.88			EXP	HANSCH,C & LEO,AJ (1985)
VP 3.39E-004	mm Hg	25	EST	NEELY,WB & BLAU,GE (1985)
DC	pKa			
HL 1.69E-007	atm m3/mol	25	EST	MEYLAN,WM & HOWARD,PH (1991)
OH 1.55E-012	cm3/molc sec	25	EST	MEYLAN,WM & HOWARD,PH (1993)

CAS #: 000936-02-7	BENZOYLHYDRAZINE, O-HYDROXY

Formula: $C_7H_8N_2O_2$

Mol Weight: 152.15

MP (deg C): 147-150

FP (deg C):

BP (deg C):

BP pressure (mm Hg):

Property/ Value	Units	Temp	Data Type	Reference
WS 8.67E+004	mg/L	25	EST	MEYLAN,WM ET AL. (1996)
logP 0.60			EXP	HANSCH,C & LEO,AJ (1985)
VP 1.24E-006	mm Hg	25	EST	NEELY,WB & BLAU,GE (1985)
DC	pKa			
HL 1.21E-012	atm m3/mol	25	EST	MEYLAN,WM & HOWARD,PH (1991)
OH 3.60E-011	cm3/molc sec	25	EST	MEYLAN,WM & HOWARD,PH (1993)

CAS #: 000936-05-0	1-ME-2-NO2-5-HYDROXYME IMIDAZOLE

Formula: $C_5H_9N_3O_3$

Mol Weight: 159.15

MP (deg C):

FP (deg C):

BP (deg C):

BP pressure (mm Hg):

Property/ Value	Units	Temp	Data Type	Reference
WS 5.72E+004	mg/L	25	EST	MEYLAN,WM ET AL. (1996)
logP -0.35			EXP	HANSCH,C & LEO,AJ (1985)
VP 6.53E-007	mm Hg	25	EST	NEELY,WB & BLAU,GE (1985)
DC	pKa			
HL 1.28E-011	atm m3/mol	25	EST	MEYLAN,WM & HOWARD,PH (1991)
OH 1.47E-011	cm3/molc sec	25	EST	MEYLAN,WM & HOWARD,PH (1993)

CAS #: 000936-98-1	TR-2-PHENYLCYCLOPROPYLCARBINOL

Formula: $C_{10}H_{12}O$

Mol Weight: 148.21

MP (deg C):

FP (deg C):

BP (deg C):

BP pressure (mm Hg):

Property/ Value	Units	Temp	Data Type	Reference
WS 5.40E+003	mg/L	25	EST	MEYLAN,WM ET AL. (1996)
logP 1.95			EXP	HANSCH,C & LEO,AJ (1985)
VP 1.55E-003	mm Hg	25	EST	NEELY,WB & BLAU,GE (1985)
DC	pKa			
HL 2.24E-007	atm m3/mol	25	EST	MEYLAN,WM & HOWARD,PH (1991)
OH 8.95E-012	cm3/molc sec	25	EST	MEYLAN,WM & HOWARD,PH (1993)

000937-05-3 — CYCLOHEXANOL, 4-(1,1-DIMETHYLETHYL)-, CIS-

Formula:	$C_{10}H_{20}O$		
Mol Weight:	156.27		
MP (deg C):		FP (deg C):	
BP (deg C):			
BP pressure (mm Hg):			

Property/Value	Units	Temp	Data Type	Reference
WS 5.29E+002	mg/L	25	EST	MEYLAN,WM ET AL. (1996)
logP 3.02			EXP	FUNASAKI,N ET AL. (1986)
VP 2.63E-002	mm Hg	25	EST	NEELY,WB & BLAU,GE (1985)
DC	pKa			
HL 1.52E-005	atm m3/mol	25	EST	MEYLAN,WM & HOWARD,PH (1991)
OH 2.02E-011	cm3/molc sec	25	EST	MEYLAN,WM & HOWARD,PH (1993)

000937-30-4 — 4-ETHYLACETOPHENONE

Formula:	$C_{10}H_{12}O$		
Mol Weight:	148.21		
MP (deg C):	-20.6	FP (deg C):	
BP (deg C):	125		
BP pressure (mm Hg):	2.00E+001		

Property/Value	Units	Temp	Data Type	Reference
WS 3.73E+002	mg/L	25	EST	MEYLAN,WM ET AL. (1996)
logP 2.71			EST	MEYLAN,WM & HOWARD,PH (1995)
VP 9.14E-001	mm Hg	25	EXP	BOUBLIK,T ET AL. (1984)
DC	pKa			
HL 1.44E-005	atm m3/mol	25	EST	MEYLAN,WM & HOWARD,PH (1991)
OH 5.62E-012	cm3/molc sec	25	EST	MEYLAN,WM & HOWARD,PH (1993)

000937-40-6 — N-BENZYL-N-METHYLNITROSOAMINE

Formula:	$C_8H_{10}N_2O$		
Mol Weight:	150.18		
MP (deg C):		FP (deg C):	
BP (deg C):			
BP pressure (mm Hg):			

Property/Value	Units	Temp	Data Type	Reference
WS 3.59E+003	mg/L	25	EST	MEYLAN,WM ET AL. (1996)
logP 1.55			EXP	SINGER,GM ET AL. (1986)
VP 3.57E-003	mm Hg	25	EST	NEELY,WB & BLAU,GE (1985)
DC	pKa			
HL 1.20E-005	atm m3/mol	37	EXP	MIRVISH,SS ET AL. (1976)
OH 1.49E-011	cm3/molc sec	25	EST	MEYLAN,WM & HOWARD,PH (1993)

000938-16-9 — 2,2-DIMETHYLPROPIOPHENONE

Formula:	$C_{11}H_{14}O$		
Mol Weight:	162.23		
MP (deg C):		FP (deg C):	
BP (deg C):	220		
BP pressure (mm Hg):			

Property/Value	Units	Temp	Data Type	Reference
WS 1.83E+002	mg/L	25	EST	MEYLAN,WM ET AL. (1996)
logP 3.00			EXP	HANSCH,C ET AL. (1995)
VP 8.34E-002	mm Hg	25	EST	NEELY,WB & BLAU,GE (1985)
DC	pKa			
HL 2.30E-005	atm m3/mol	25	EST	MEYLAN,WM & HOWARD,PH (1991)
OH 3.37E-012	cm3/molc sec	25	EST	MEYLAN,WM & HOWARD,PH (1993)

000938-22-7 — 2,3,4,6-TETRACHLOROANISOLE

Formula:	$C_7H_4Cl_4O$		
Mol Weight:	245.92		
MP (deg C):		FP (deg C):	
BP (deg C):			
BP pressure (mm Hg):			

Property/Value	Units	Temp	Data Type	Reference
WS 2.19E+000	mg/L	25	EST	MEYLAN,WM ET AL. (1996)
logP 4.75			EXP	OPPERHUIZEN,A & VOORS,PI (1987)
VP 3.19E-003	mm Hg	25	EST	NEELY,WB & BLAU,GE (1985)
DC	pKa			
HL 9.61E-005	atm m3/mol	25	EST	MEYLAN,WM & HOWARD,PH (1991)
OH 1.00E-012	cm3/molc sec	25	EST	MEYLAN,WM & HOWARD,PH (1993)

000938-33-0 — 8-METHOXYQUINOLINE

Formula:	$C_{10}H_9NO$		
Mol Weight:	159.19		
MP (deg C):	49.5	FP (deg C):	
BP (deg C):	283		
BP pressure (mm Hg):			

Property/Value	Units	Temp	Data Type	Reference
WS 1.85E+003	mg/L	25	EST	MEYLAN,WM ET AL. (1996)
logP 1.84			EXP	HANSCH,C & LEO,AJ (1985)
VP 3.23E-003	mm Hg	25	EST	NEELY,WB & BLAU,GE (1985)
DC 5.01	pKa		EXP	PERRIN,DD (1965)
HL 4.07E-008	atm m3/mol	25	EST	MEYLAN,WM & HOWARD,PH (1991)
OH 1.60E-010	cm3/molc sec	25	EST	MEYLAN,WM & HOWARD,PH (1993)

000938-56-7 — 1-(2-HYDROXYETHYL)BENZOTRIAZOLE

Formula:	$C_8H_{11}N_3O$		
Mol Weight:	165.20		
MP (deg C):		FP (deg C):	
BP (deg C):			
BP pressure (mm Hg):			

Property/Value	Units	Temp	Data Type	Reference
WS 8.01E+004	mg/L	25	EST	MEYLAN,WM ET AL. (1996)
logP 0.50			EXP	HANSCH,C & LEO,AJ (1985)
VP 3.73E-006	mm Hg	25	EST	NEELY,WB & BLAU,GE (1985)
DC	pKa			
HL 1.52E-010	atm m3/mol	25	EST	MEYLAN,WM & HOWARD,PH (1991)
OH 6.31E-012	cm3/molc sec	25	EST	MEYLAN,WM & HOWARD,PH (1993)

000938-86-3 — 2,3,4,5-TETRACHLOROANISOLE

Formula:	$C_7H_4Cl_4O$		
Mol Weight:	245.92		
MP (deg C):	83-85	FP (deg C):	
BP (deg C):			
BP pressure (mm Hg):			

Property/Value	Units	Temp	Data Type	Reference
WS 3.51E+000	mg/L	25	EST	MEYLAN,WM ET AL. (1996)
logP 4.51			EXP	OPPERHUIZEN,A & VOORS,PI (1987)
VP 3.19E-003	mm Hg	25	EST	NEELY,WB & BLAU,GE (1985)
DC	pKa			
HL 9.61E-005	atm m3/mol	25	EST	MEYLAN,WM & HOWARD,PH (1991)
OH 1.74E-012	cm3/molc sec	25	EST	MEYLAN,WM & HOWARD,PH (1993)

CAS #: 000938-91-0	1,3-DIMETHYL PHENYL UREA

Formula: $C_9H_{12}N_2O$

Mol Weight: 164.21

MP (deg C): FP (deg C):

BP (deg C):

BP pressure (mm Hg):

Property/Value	Units	Temp	Data Type	Reference
WS 8.81E+003	mg/L	25	EST	MEYLAN,WM ET AL. (1996)
logP 1.02			EXP	HANSCH,C & LEO,AJ (1985)
VP 2.04E-004	mm Hg	25	EST	NEELY,WB & BLAU,GE (1985)
DC	pKa			
HL 9.06E-009	atm m3/mol	25	EST	MEYLAN,WM & HOWARD,PH (1991)
OH 1.45E-011	cm3/molc sec	25	EST	MEYLAN,WM & HOWARD,PH (1993)

CAS #: 000939-05-9	3-OH-5-PHENYLISOXAZOLE

Formula: $C_9H_8NO_2$

Mol Weight: 162.17

MP (deg C): FP (deg C):

BP (deg C):

BP pressure (mm Hg):

Property/Value	Units	Temp	Data Type	Reference
WS 4.49E+002	mg/L	25	EST	MEYLAN,WM ET AL. (1996)
logP 2.55			EXP	HANSCH,C & LEO,AJ (1985)
VP 1.54E-004	mm Hg	25	EST	NEELY,WB & BLAU,GE (1985)
DC	pKa			
HL 4.53E-010	atm m3/mol	25	EST	MEYLAN,WM & HOWARD,PH (1991)
OH 2.00E-010	cm3/molc sec	25	EST	MEYLAN,WM & HOWARD,PH (1993)

CAS #: 000939-23-1	4-PHENYLPYRIDINE

Formula: $C_{11}H_9N$

Mol Weight: 155.20

MP (deg C): 77.5 FP (deg C):

BP (deg C): 281

BP pressure (mm Hg):

Property/Value	Units	Temp	Data Type	Reference
WS 4.42E+002	mg/L	25	EST	MEYLAN,WM ET AL. (1996)
logP 2.59			EXP	HANSCH,C & LEO,AJ (1985)
VP 2.80E-003	mm Hg	25	EST	NEELY,WB & BLAU,GE (1985)
DC 5.55	pKa		EXP	PERRIN,DD (1965)
HL 5.41E-007	atm m3/mol	25	EST	MEYLAN,WM & HOWARD,PH (1991)
OH 4.03E-012	cm3/molc sec	25	EST	MEYLAN,WM & HOWARD,PH (1993)

CAS #: 000939-27-5	2-ETHYLNAPHTHALENE

Formula: $C_{12}H_{12}$

Mol Weight: 156.23

MP (deg C): -70 FP (deg C):

BP (deg C):

BP pressure (mm Hg):

Property/Value	Units	Temp	Data Type	Reference
WS 7.97E+000	mg/L	25	EXP	SHAW,DG (1989)
logP 4.38			EXP	HANSCH,C & LEO,AJ (1985)
VP 3.16E-002	mm Hg	25	EXP	CHAO,J ET AL. (1983)
DC	pKa			
HL 8.16E-004	atm m3/mol		EST	VP/WSOL
OH 5.48E-011	cm3/molc sec	25	EST	MEYLAN,WM & HOWARD,PH (1993)

CAS #: 000939-48-0	BENZOIC ACID, 1-METHYLETHYL ESTER

Formula: $C_{10}H_{12}O_2$

Mol Weight: 164.21

MP (deg C): FP (deg C):

BP (deg C): 216

BP pressure (mm Hg):

Property/Value	Units	Temp	Data Type	Reference
WS 1.26E+002	mg/L	25	EST	MEYLAN,WM ET AL. (1996)
logP 3.18			EXP	HANSCH,C ET AL. (1995)
VP 1.12E-001	mm Hg	25	EST	NEELY,WB & BLAU,GE (1985)
DC	pKa			
HL 6.12E-005	atm m3/mol	25	EST	MEYLAN,WM & HOWARD,PH (1991)
OH 5.22E-012	cm3/molc sec	25	EST	MEYLAN,WM & HOWARD,PH (1993)

CAS #: 000939-72-0	2-(2-HYDROXYETHYL)BENZOTRIAZOLE

Formula: $C_8H_{11}N_3O$

Mol Weight: 165.20

MP (deg C): FP (deg C):

BP (deg C):

BP pressure (mm Hg):

Property/Value	Units	Temp	Data Type	Reference
WS 3.51E+004	mg/L	25	EST	MEYLAN,WM ET AL. (1996)
logP 0.92			EXP	HANSCH,C & LEO,AJ (1985)
VP 3.73E-006	mm Hg	25	EST	NEELY,WB & BLAU,GE (1985)
DC	pKa			
HL 1.52E-010	atm m3/mol	25	EST	MEYLAN,WM & HOWARD,PH (1991)
OH 6.31E-012	cm3/molc sec	25	EST	MEYLAN,WM & HOWARD,PH (1993)

CAS #: 000940-36-3	CARBAMIC ACID, (4-CHLOROPHENYL)-, METHYL ESTER

Formula: $C_8H_8ClNO_2$

Mol Weight: 185.61

MP (deg C): FP (deg C):

BP (deg C):

BP pressure (mm Hg):

Property/Value	Units	Temp	Data Type	Reference
WS 3.43E+002	mg/L	25	EST	MEYLAN,WM ET AL. (1996)
logP 2.55			EXP	TAKAHASHI,J ET AL. (1988)
VP 8.31E-003	mm Hg	25	EST	NEELY,WB & BLAU,GE (1985)
DC	pKa			
HL 1.62E-008	atm m3/mol	25	EST	MEYLAN,WM & HOWARD,PH (1991)
OH 1.27E-011	cm3/molc sec	25	EST	MEYLAN,WM & HOWARD,PH (1993)

CAS #: 000940-43-2	BENZENEPROPANAMIDE, N-METHYL-

Formula: $C_{10}H_{13}NO$

Mol Weight: 163.22

MP (deg C): FP (deg C):

BP (deg C):

BP pressure (mm Hg):

Property/Value	Units	Temp	Data Type	Reference
WS 6.63E+003	mg/L	25	EST	MEYLAN,WM ET AL. (1996)
logP 1.17			EXP	RADZICKA,A & WOLFENDEN,R (1988)
VP 7.84E-005	mm Hg	25	EST	NEELY,WB & BLAU,GE (1985)
DC	pKa			
HL 2.63E-009	atm m3/mol	25	EST	MEYLAN,WM & HOWARD,PH (1991)
OH 1.54E-011	cm3/molc sec	25	EST	MEYLAN,WM & HOWARD,PH (1993)

CAS #: 000940-64-7				P-METHYLPHENOXYACETIC ACID

Formula: $C_9H_{10}O_3$

Mol Weight: 166.18

MP (deg C): 140-142   FP (deg C):

BP (deg C):

BP pressure (mm Hg):

Property/Value	Units	Temp	Data Type	Reference
WS 4.11E+003	mg/L	25	EST	MEYLAN,WM ET AL. (1996)
logP 1.86			EXP	HANSCH,C & LEO,AJ (1985)
VP 5.34E-004	mm Hg	25	EST	NEELY,WB & BLAU,GE (1985)
DC 3.21	pKa	25	EXP	KORTUM,G ET AL (1961)
HL 1.85E-008	atm m3/mol	25	EST	MEYLAN,WM & HOWARD,PH (1991)
OH 3.13E-011	cm3/molc sec	25	EST	MEYLAN,WM & HOWARD,PH (1993)

CAS #: 000941-69-5				1H-PYRROLE-2,5-DIONE, 1-PHENYL-

Formula: $C_{10}H_7NO_2$

Mol Weight: 173.17

MP (deg C): 89-89.8   FP (deg C):

BP (deg C):

BP pressure (mm Hg):

Property/Value	Units	Temp	Data Type	Reference
WS 6.97E+003	mg/L	25	EST	MEYLAN,WM ET AL. (1996)
logP 1.09			EXP	HANSCH,C ET AL. (1995)
VP 8.99E-007	mm Hg	25	EST	NEELY,WB & BLAU,GE (1985)
DC	pKa			
HL 3.26E-008	atm m3/mol	25	EST	MEYLAN,WM & HOWARD,PH (1991)
OH 1.47E-011	cm3/molc sec	25	EST	MEYLAN,WM & HOWARD,PH (1993)

CAS #: 000941-98-0				ALPHA-NAPHTHYL METHYL KETONE

Formula: $C_{12}H_{10}O$

Mol Weight: 170.21

MP (deg C): 34   FP (deg C):

BP (deg C): 297

BP pressure (mm Hg):

Property/Value	Units	Temp	Data Type	Reference
WS 2.26E+002	mg/L	25	EST	MEYLAN,WM ET AL. (1996)
logP 2.85			EST	MEYLAN,WM & HOWARD,PH (1995)
VP 3.92E-004	mm Hg	25	EXT	PERRY,RH & GREEN,D (1984)
DC	pKa			
HL 9.58E-007	atm m3/mol	25	EST	MEYLAN,WM & HOWARD,PH (1991)
OH 1.98E-011	cm3/molc sec	25	EST	MEYLAN,WM & HOWARD,PH (1993)

CAS #: 000942-24-5				1H-INDOLE-3-CARBOXYLIC ACID, METHYL ESTER

Formula: $C_{10}H_9NO_2$

Mol Weight: 175.19

MP (deg C): 149-152   FP (deg C):

BP (deg C):

BP pressure (mm Hg):

Property/Value	Units	Temp	Data Type	Reference
WS 3.71E+002	mg/L	25	EST	MEYLAN,WM ET AL. (1996)
logP 2.57			EXP	YAMAGAMI,C ET AL. (1992)
VP 2.09E-004	mm Hg	25	EST	NEELY,WB & BLAU,GE (1985)
DC	pKa			
HL 5.71E-009	atm m3/mol	25	EST	MEYLAN,WM & HOWARD,PH (1991)
OH 3.56E-011	cm3/molc sec	25	EST	MEYLAN,WM & HOWARD,PH (1993)

CAS #: 000942-79-0				N-METHYL-2-CYANOPHENYLCARBAMATE

Formula: $C_9H_8N_2O_2$

Mol Weight: 176.18

MP (deg C):   FP (deg C):

BP (deg C):

BP pressure (mm Hg):

Property/Value	Units	Temp	Data Type	Reference
WS 5.75E+003	mg/L	25	EST	MEYLAN,WM ET AL. (1996)
logP 0.86			EXP	HANSCH,C & LEO,AJ (1985)
VP 6.30E-004	mm Hg	25	EST	NEELY,WB & BLAU,GE (1985)
DC	pKa			
HL 3.10E-010	atm m3/mol	25	EST	MEYLAN,WM & HOWARD,PH (1991)
OH 6.41E-012	cm3/molc sec	25	EST	MEYLAN,WM & HOWARD,PH (1993)

CAS #: 000943-49-7				N-METHYL-3-CYANOPHENYLCARBAMATE

Formula: $C_9H_8N_2O_2$

Mol Weight: 176.18

MP (deg C):   FP (deg C):

BP (deg C):

BP pressure (mm Hg):

Property/Value	Units	Temp	Data Type	Reference
WS 4.64E+003	mg/L	25	EST	MEYLAN,WM ET AL. (1996)
logP 0.97			EXP	HANSCH,C & LEO,AJ (1985)
VP 6.30E-004	mm Hg	25	EST	NEELY,WB & BLAU,GE (1985)
DC	pKa			
HL 3.10E-010	atm m3/mol	25	EST	MEYLAN,WM & HOWARD,PH (1991)
OH 6.32E-012	cm3/molc sec	25	EST	MEYLAN,WM & HOWARD,PH (1993)

CAS #: 000943-89-5				P-METHOXYCINNAMIC ACID (TRANS)

Formula: $C_{10}H_{10}O_3$

Mol Weight: 178.19

MP (deg C):   FP (deg C):

BP (deg C):

BP pressure (mm Hg):

Property/Value	Units	Temp	Data Type	Reference
WS 1.97E+003	mg/L	25	EXP	STEPHEN,H & STEPHEN,T (1963)
logP 2.15			EST	MEYLAN,WM & HOWARD,PH (1995)
VP 1.59E-004	mm Hg	25	EST	NEELY,WB & BLAU,GE (1985)
DC 4.54	pKa	25	EXP	KORTUM,G ET AL (1961)
HL 7.65E-010	atm m3/mol	25	EST	MEYLAN,WM & HOWARD,PH (1991)
OH 4.13E-011	cm3/molc sec	25	EST	MEYLAN,WM & HOWARD,PH (1993)

CAS #: 000944-21-8				FONOPHOS, O-ANALOG

Formula: $C_{10}H_{15}O_2PS$

Mol Weight: 230.27

MP (deg C):   FP (deg C):

BP (deg C):

BP pressure (mm Hg):

Property/Value	Units	Temp	Data Type	Reference
WS 4.79E+002	mg/L	25	EST	MEYLAN,WM ET AL. (1996)
logP 2.11			EXP	HANSCH,C & LEO,AJ (1985)
VP 2.42E-004	mm Hg	25	EST	NEELY,WB & BLAU,GE (1985)
DC	pKa			
HL 2.41E-007	atm m3/mol	25	EST	MEYLAN,WM & HOWARD,PH (1991)
OH 3.30E-011	cm3/molc sec	25	EST	MEYLAN,WM & HOWARD,PH (1993)

CAS #: 000944-22-9				FONOPHOS

Formula: $C_{10}H_{15}OPS_2$

Mol Weight: 246.33

MP (deg C):

FP (deg C):

BP (deg C): 130

BP pressure (mm Hg): 1.00E-001

Property/Value	Units	Temp	Data Type	Reference
WS 1.60E+001	mg/L	25	EXP	RACKE,KD (1992)
logP 3.94			EXP	HANSCH,C & LEO,AJ (1985)
VP 2.00E-004	mm Hg	25	EXP	RACKE,KD (1992)
DC	pKa			
HL 5.40E-006	atm m3/mol	25	EST	RAO,PSC ET AL. (1985A)
OH 8.88E-011	cm3/molc sec	25	EST	ATKINSON,R (1988)

CAS #: 000944-73-0				1,3-DIMETHYL URIC ACID

Formula: $C_7H_8N_4O_3$

Mol Weight: 196.17

MP (deg C):

FP (deg C):

BP (deg C):

BP pressure (mm Hg):

Property/Value	Units	Temp	Data Type	Reference
WS 6.23E+003	mg/L	25	EST	MEYLAN,WM ET AL. (1996)
logP -0.52			EXP	GASPARI,F & BONATI,M (1987)
VP 1.51E-008	mm Hg	25	EST	NEELY,WB & BLAU,GE (1985)
DC	pKa			
HL 5.77E-017	atm m3/mol	25	EST	MEYLAN,WM & HOWARD,PH (1991)
OH 1.34E-011	cm3/molc sec	25	EST	MEYLAN,WM & HOWARD,PH (1993)

CAS #: 000945-51-7				DIPHENYL SULFOXIDE

Formula: $C_{12}H_{10}OS$

Mol Weight: 202.28

MP (deg C): 71.2

FP (deg C):

BP (deg C): 340

BP pressure (mm Hg): 1.60E+001

Property/Value	Units	Temp	Data Type	Reference
WS 7.42E+002	mg/L	25	EST	MEYLAN,WM ET AL. (1996)
logP 2.06			EXP	HANSCH,C & LEO,AJ (1985)
VP 3.62E-005	mm Hg	25	EST	NEELY,WB & BLAU,GE (1985)
DC	pKa			
HL 1.98E-009	atm m3/mol	25	EST	MEYLAN,WM & HOWARD,PH (1991)
OH 6.08E-011	cm3/molc sec	25	EST	MEYLAN,WM & HOWARD,PH (1993)

CAS #: 000946-31-6				6-CHLORO-2,4-DINITROPHENOL

Formula: $C_6H_3ClN_2O_5$

Mol Weight: 218.55

MP (deg C):

FP (deg C):

BP (deg C):

BP pressure (mm Hg):

Property/Value	Units	Temp	Data Type	Reference
WS 3.31E+002	mg/L	25	EST	MEYLAN,WM ET AL. (1996)
logP 2.37			EST	MEYLAN,WM & HOWARD,PH (1995)
VP 3.01E-006	mm Hg	25	EST	NEELY,WB & BLAU,GE (1985)
DC 2.10	pKa	25	EXP	SERJEANT,EP & DEMPSEY,B (1979)
HL 2.05E-008	atm m3/mol	25	EST	MEYLAN,WM & HOWARD,PH (1991)
OH 1.47E-013	cm3/molc sec	25	EST	MEYLAN,WM & HOWARD,PH (1993)

CAS #: 000946-76-9				PROPANEDINITRILE, [(4-CHLOROPHENYL)HYDRAZONO]-

Formula: $C_9H_5ClN_4$

Mol Weight: 204.62

MP (deg C):

FP (deg C):

BP (deg C):

BP pressure (mm Hg):

Property/Value	Units	Temp	Data Type	Reference
WS 5.28E+001	mg/L	25	EST	MEYLAN,WM ET AL. (1996)
logP 3.39			EXP	STURDIK,E ET AL. (1985)
VP 5.67E-006	mm Hg	25	EST	NEELY,WB & BLAU,GE (1985)
DC	pKa			
HL 4.29E-009	atm m3/mol	25	EST	MEYLAN,WM & HOWARD,PH (1991)
OH 1.25E-011	cm3/molc sec	25	EST	MEYLAN,WM & HOWARD,PH (1993)

CAS #: 000946-80-5				BENZYL PHENYL ETHER

Formula: $C_{13}H_{12}O$

Mol Weight: 184.24

MP (deg C): 39-41

FP (deg C):

BP (deg C): 286-287

BP pressure (mm Hg):

Property/Value	Units	Temp	Data Type	Reference
WS 3.04E+001	mg/L	25	EST	MEYLAN,WM ET AL. (1996)
logP 3.79			EXP	HANSCH,C & LEO,AJ (1985)
VP 6.30E-003	mm Hg	25	EXT	JORDAN,TE (1954)
DC	pKa			
HL 2.57E-005	atm m3/mol	25	EST	MEYLAN,WM & HOWARD,PH (1991)
OH 3.20E-011	cm3/molc sec	25	EST	MEYLAN,WM & HOWARD,PH (1993)

CAS #: 000947-02-4				CYOLANE

Formula: $C_7H_{14}NO_3PS_2$

Mol Weight: 255.30

MP (deg C): 37-45

FP (deg C):

BP (deg C): 115-118

BP pressure (mm Hg): 0.00E+000

Property/Value	Units	Temp	Data Type	Reference
WS 6.50E+002	mg/L	25	EXP	WORTHING,CR & WALKER,SB (1987)
logP -1.77			EST	LYMAN ET AL. (1982)
VP 5.30E-005	mm Hg	25	EST	NEELY,WB & BLAU,GE (1985)
DC	pKa			
HL 8.96E-011	atm m3/mol	25	EST	MEYLAN,WM & HOWARD,PH (1991)
OH 5.40E-011	cm3/molc sec	25	EST	ATKINSON,R (1988)

CAS #: 000947-04-6				AZACYCLOTRIDECAN-2-ONE

Formula: $C_{12}H_{23}NO$

Mol Weight: 197.32

MP (deg C): 150-153

FP (deg C):

BP (deg C):

BP pressure (mm Hg):

Property/Value	Units	Temp	Data Type	Reference
WS 2.90E+002	mg/L	25	EXP	CHEM INSPECT TEST INST (1992)
logP 2.92			EXP	CHEM INSPECT TEST INST (1992)
VP 7.31E-006	mm Hg	25	EST	NEELY,WB & BLAU,GE (1985)
DC	pKa			
HL 1.39E-007	atm m3/mol	25	EST	MEYLAN,WM & HOWARD,PH (1991)
OH 2.69E-011	cm3/molc sec	25	EST	MEYLAN,WM & HOWARD,PH (1993)

000947-73-9 — 9-PHENANTHRENAMINE

Formula: $C_{14}H_{11}N$

Mol Weight: 193.25

MP (deg C): 138.3

FP (deg C):

BP (deg C):

BP pressure (mm Hg):

Property/Value	Units	Temp	Data Type	Reference
WS 3.35E+000	mg/L	25	EST	MEYLAN,WM ET AL. (1996)
logP 3.56			EXP	DEBNATH,AK ET AL. (1992)
VP 4.05E-006	mm Hg	25	EST	NEELY,WB & BLAU,GE (1985)
DC	pKa			
HL 1.81E-008	atm m3/mol	25	EST	MEYLAN,WM & HOWARD,PH (1991)
OH 2.00E-010	cm3/molc sec	25	EST	MEYLAN,WM & HOWARD,PH (1993)

000947-84-2 — [1,1'-BIPHENYL]-2-CARBOXYLIC ACID

Formula: $C_{13}H_{10}O_2$

Mol Weight: 198.22

MP (deg C): 114.3

FP (deg C):

BP (deg C): 343

BP pressure (mm Hg):

Property/Value	Units	Temp	Data Type	Reference
WS 1.52E+002	mg/L	25	EST	MEYLAN,WM ET AL. (1996)
logP 2.89			EXP	HANSCH,C ET AL. (1995)
VP 5.48E-006	mm Hg	25	EST	NEELY,WB & BLAU,GE (1985)
DC 3.46	pKa		EXP	KORTUM,G ET AL (1961)
HL 8.33E-009	atm m3/mol	25	EST	MEYLAN,WM & HOWARD,PH (1991)
OH 5.16E-012	cm3/molc sec	25	EST	MEYLAN,WM & HOWARD,PH (1993)

000949-67-7 — L-TYROSINE, ETHYL ESTER

Formula: $C_{11}H_{15}NO_3$

Mol Weight: 209.25

MP (deg C):

FP (deg C):

BP (deg C):

BP pressure (mm Hg):

Property/Value	Units	Temp	Data Type	Reference
WS 3.69E+005	mg/L	25	EST	MEYLAN,WM ET AL. (1996)
logP 0.72			EXP	HUANG,CH ET AL. (1985)
VP 2.03E-005	mm Hg	25	EST	NEELY,WB & BLAU,GE (1985)
DC	pKa			
HL 5.33E-012	atm m3/mol	25	EST	MEYLAN,WM & HOWARD,PH (1991)
OH 8.10E-011	cm3/molc sec	25	EST	MEYLAN,WM & HOWARD,PH (1993)

000949-87-1 — (4-METHYLPHENYL)PHENYLDIAZENE

Formula: $C_{13}H_{12}N_2$

Mol Weight: 196.25

MP (deg C): 71.5

FP (deg C):

BP (deg C): 312

BP pressure (mm Hg):

Property/Value	Units	Temp	Data Type	Reference
WS 1.78E+000	mg/L	25	EST	MEYLAN,WM ET AL. (1996)
logP 4.66			EST	MEYLAN,WM & HOWARD,PH (1995)
VP 9.42E-004	mm Hg	25	EST	NEELY,WB & BLAU,GE (1985)
DC	pKa			
HL 1.62E-005	atm m3/mol	25	EST	MEYLAN,WM & HOWARD,PH (1991)
OH 2.93E-012	cm3/molc sec	25	EST	MEYLAN,WM & HOWARD,PH (1993)

000950-10-7 — PHOSPHORAMIDIC ACID, (4-METHYL-1,3-DITHIOLAN-2-Y

Formula: $C_8H_{16}NO_3PS_2$

Mol Weight: 269.32

MP (deg C):

FP (deg C):

BP (deg C): 120

BP pressure (mm Hg): 1.00E-003

Property/Value	Units	Temp	Data Type	Reference
WS 5.70E+001	mg/L	25	EXP	SHIU,WY ET AL. (1990)
logP 1.04			EXP	HANSCH,C ET AL. (1995)
VP 3.18E-005	mm Hg	25	EST	NEELY,WB & BLAU,GE (1985)
DC	pKa			
HL 1.19E-010	atm m3/mol	25	EST	MEYLAN,WM & HOWARD,PH (1991)
OH 6.43E-011	cm3/molc sec	25	EST	MEYLAN,WM & HOWARD,PH (1993)

000950-35-6 — METHYL PARAOXON

Formula: $C_8H_{10}NO_6P$

Mol Weight: 247.15

MP (deg C):

FP (deg C):

BP (deg C):

BP pressure (mm Hg):

Property/Value	Units	Temp	Data Type	Reference
WS 7.31E+002	mg/L	25	EST	MEYLAN,WM ET AL. (1996)
logP 1.33			EXP	HANSCH,C & LEO,AJ (1985)
VP 3.30E-005	mm Hg	25	EST	NEELY,WB & BLAU,GE (1985)
DC	pKa			
HL 3.64E-010	atm m3/mol	25	EST	MEYLAN,WM & HOWARD,PH (1991)
OH 6.03E-012	cm3/molc sec	25	EST	MEYLAN,WM & HOWARD,PH (1993)

000950-37-8 — METHIDATHION

Formula: $C_6H_{11}N_2O_4PS_3$

Mol Weight: 302.33

MP (deg C): 39.5

FP (deg C):

BP (deg C):

BP pressure (mm Hg):

Property/Value	Units	Temp	Data Type	Reference
WS 1.87E+002	mg/L	20	EXP	BOWMAN,BT & SANS,WW (1983)
logP 2.42			EXP	HANSCH,C & LEO,AJ (1985)
VP 3.37E-006	mm Hg	25	EXP	WAUCHOPE,RD ET AL. (1991A)
DC	pKa			
HL 7.17E-009	atm m3/mol	25	EST	VP/WSOL
OH 1.46E-010	cm3/molc sec	25	EST	MEYLAN,WM & HOWARD,PH (1993)

000951-77-9 — CYTIDINE, 2'-DEOXY-

Formula: $C_9H_{13}N_3O_4$

Mol Weight: 227.22

MP (deg C):

FP (deg C):

BP (deg C):

BP pressure (mm Hg):

Property/Value	Units	Temp	Data Type	Reference
WS 6.34E+004	mg/L	25	EST	MEYLAN,WM ET AL. (1996)
logP -1.89			EXP	SANGSTER,J (1993)
VP 4.22E-011	mm Hg	25	EST	NEELY,WB & BLAU,GE (1985)
DC	pKa			
HL 4.28E-018	atm m3/mol	25	EST	MEYLAN,WM & HOWARD,PH (1991)
OH 1.12E-010	cm3/molc sec	25	EST	MEYLAN,WM & HOWARD,PH (1993)

CAS #: 000951-78-0				URIDINE, 2'-DEOXY-

Formula: $C_9H_{12}N_2O_5$

Mol Weight: 228.21

MP (deg C): 163   FP (deg C):

BP (deg C):

BP pressure (mm Hg):

Property/ Value	Units	Temp	Data Type	Reference
WS 2.97E+004	mg/L	25	EST	MEYLAN,WM ET AL. (1996)
logP -1.51			EXP	BALZARINI,J ET AL. (1989)
VP 1.14E-013	mm Hg	25	EST	NEELY,WB & BLAU,GE (1985)
DC	pKa			
HL 7.72E-018	atm m3/mol	25	EST	MEYLAN,WM & HOWARD,PH (1991)
OH 8.45E-011	cm3/molc sec	25	EST	MEYLAN,WM & HOWARD,PH (1993)

CAS #: 000951-86-0				1,2-DICHLORO-1,2-DIPHENYLETHENE (TRANS)

Formula: $C_{14}H_{10}Cl_2$

Mol Weight: 249.14

MP (deg C): 143.5   FP (deg C):

BP (deg C): 183

BP pressure (mm Hg): 1.80E+001

Property/ Value	Units	Temp	Data Type	Reference
WS 4.72E-001	mg/L	25	EST	MEYLAN,WM ET AL. (1996)
logP 5.51			EST	MEYLAN,WM & HOWARD,PH (1995)
VP 2.21E-004	mm Hg	25	EST	NEELY,WB & BLAU,GE (1985)
DC	pKa			
HL 6.40E-005	atm m3/mol	25	EST	MEYLAN,WM & HOWARD,PH (1991)
OH 8.52E-012	cm3/molc sec	25	EST	MEYLAN,WM & HOWARD,PH (1993)

CAS #: 000953-17-3				CARBOPHENOTHION-METHYL

Formula: $C_9H_{12}ClO_2PS_3$

Mol Weight: 314.81

MP (deg C):   FP (deg C):

BP (deg C):

BP pressure (mm Hg):

Property/ Value	Units	Temp	Data Type	Reference
WS 1.00E+000	mg/L		EXP	YALKOWSKY,SH & DANNENFELSER,RM (1992)
logP 4.82			EXP	HANSCH,C & LEO,AJ (1985)
VP 2.67E-005	mm Hg	25	EST	NEELY,WB & BLAU,GE (1985)
DC	pKa			
HL 1.00E-007	atm m3/mol	25	EST	MEYLAN,WM & HOWARD,PH (1991)
OH 2.12E-010	cm3/molc sec	25	EST	MEYLAN,WM & HOWARD,PH (1993)

CAS #: 000957-51-7				DIPHENAMID

Formula: $C_{16}H_{17}NO$

Mol Weight: 239.32

MP (deg C): 135   FP (deg C):

BP (deg C):

BP pressure (mm Hg):

Property/ Value	Units	Temp	Data Type	Reference
WS 2.60E+002	mg/L	27	EXP	WAUCHOPE,RD ET AL. (1992)
logP 2.17			EST	GERSTL,Z & HELLING,CS (1987)
VP 3.00E-008	mm Hg	25	EXP	WAUCHOPE,RD ET AL. (1992)
DC	pKa			
HL 3.63E-011	atm m3/mol	25	EST	VP/WSOL
OH 2.42E-011	cm3/molc sec	25	EST	ATKINSON,R (1988)

CAS #: 000958-09-8				ADENOSINE, 2'-DEOXY-

Formula: $C_{10}H_{13}N_5O_3$

Mol Weight: 251.25

MP (deg C):   FP (deg C):

BP (deg C):

BP pressure (mm Hg):

Property/ Value	Units	Temp	Data Type	Reference
WS 3.36E+003	mg/L	25	EST	MEYLAN,WM ET AL. (1996)
logP -0.55			EXP	HANSCH,C ET AL. (1995)
VP 5.58E-013	mm Hg	25	EST	NEELY,WB & BLAU,GE (1985)
DC	pKa			
HL 3.04E-021	atm m3/mol	25	EST	MEYLAN,WM & HOWARD,PH (1991)
OH 2.40E-010	cm3/molc sec	25	EST	MEYLAN,WM & HOWARD,PH (1993)

CAS #: 000959-24-0				SOTALOL

Formula: $C_{12}H_{19}NO_3S$

Mol Weight: 257.35

MP (deg C): 206.5-207   FP (deg C):

BP (deg C):

BP pressure (mm Hg):

Property/ Value	Units	Temp	Data Type	Reference
WS 1.37E+005	mg/L	25	EST	MEYLAN,WM ET AL. (1996)
logP 0.24			EXP	HANSCH,C & LEO,AJ (1985)
VP 3.37E-008	mm Hg	25	EST	NEELY,WB & BLAU,GE (1985)
DC	pKa			
HL 2.66E-014	atm m3/mol	25	EST	MEYLAN,WM & HOWARD,PH (1991)
OH 1.03E-010	cm3/molc sec	25	EST	MEYLAN,WM & HOWARD,PH (1993)

CAS #: 000959-26-2				BIS(2-HYDROXYETHYL)TEREPHTHALATE

Formula: $C_{12}H_{14}O_6$

Mol Weight: 254.24

MP (deg C):   FP (deg C):

BP (deg C):

BP pressure (mm Hg):

Property/ Value	Units	Temp	Data Type	Reference
WS 1.76E+004	mg/L	25	EST	MEYLAN,WM ET AL. (1996)
logP 0.12			EST	MEYLAN,WM & HOWARD,PH (1995)
VP 1.11E-007	mm Hg	25	EST	NEELY,WB & BLAU,GE (1985)
DC	pKa			
HL 5.27E-013	atm m3/mol	25	EST	MEYLAN,WM & HOWARD,PH (1991)
OH 1.21E-011	cm3/molc sec	25	EST	MEYLAN,WM & HOWARD,PH (1993)

CAS #: 000959-98-8				A-ENDOSULFAN

Formula: $C_9H_6Cl_6O_3S$

Mol Weight: 406.93

MP (deg C):   FP (deg C):

BP (deg C):

BP pressure (mm Hg):

Property/ Value	Units	Temp	Data Type	Reference
WS 5.10E-001	mg/L	20	EXP	BOWMAN,BT & SANS,WW (1983)
logP 3.83			EXP	HANSCH,C & LEO,AJ (1985)
VP 1.00E-005	mm Hg	25	EXP	SUNTIO,LR ET AL. (1988)
DC	pKa			
HL 1.05E-005	atm m3/mol	25	EST	VP/WSOL
OH 8.25E-011	cm3/molc sec	25	EST	MEYLAN,WM & HOWARD,PH (1993)

CAS #: 000961-07-9 — GUANOSINE, 2'-DEOXY-

Formula:	$C_{10}H_{15}N_5O_4$
Mol Weight:	269.26
MP (deg C):	FP (deg C):
BP (deg C):	
BP pressure (mm Hg):	

Property/Value	Units	Temp	Data Type	Reference
WS 1.20E+004	mg/L	25	EST	MEYLAN,WM ET AL. (1996)
logP -1.30			EXP	BALZARINI,JM ET AL. (1989)
VP 2.75E-014	mm Hg	25	EST	NEELY,WB & BLAU,GE (1985)
DC	pKa			
HL 1.12E-021	atm m3/mol	25	EST	MEYLAN,WM & HOWARD,PH (1991)
OH 2.40E-010	cm3/molc sec	25	EST	MEYLAN,WM & HOWARD,PH (1993)

CAS #: 000961-11-5 — STIROFOS

Formula:	$C_{10}H_9Cl_4O_4P$
Mol Weight:	365.97
MP (deg C): 97-98	FP (deg C):
BP (deg C):	
BP pressure (mm Hg):	

Property/Value	Units	Temp	Data Type	Reference
WS 1.10E+001	mg/L	20	EXP	MERCK INDEX (1983)
logP 3.53			EXP	HANSCH,C & LEO,AJ (1985)
VP 4.20E-008	mm Hg	20	EXP	MERCK INDEX (1983)
DC	pKa			
HL 1.84E-009	atm m3/mol	20	EST	VP/WSOL
OH 2.34E-011	cm3/molc sec	25	EST	MEYLAN,WM & HOWARD,PH (1993)

CAS #: 000961-22-8 — METHYL AZINPHOS O-ANALOG

Formula:	$C_{10}H_{12}N_3O_4PS$
Mol Weight:	301.26
MP (deg C):	FP (deg C):
BP (deg C):	
BP pressure (mm Hg):	

Property/Value	Units	Temp	Data Type	Reference
WS 2.60E+003	mg/L	25	EST	MEYLAN,WM ET AL. (1996)
logP 0.78			EXP	HANSCH,C & LEO,AJ (1985)
VP 7.64E-008	mm Hg	25	EST	NEELY,WB & BLAU,GE (1985)
DC	pKa			
HL 6.19E-013	atm m3/mol	25	EST	MEYLAN,WM & HOWARD,PH (1991)
OH 8.54E-011	cm3/molc sec	25	EST	MEYLAN,WM & HOWARD,PH (1993)

CAS #: 000961-68-2 — 2,4-DINITRO-N-PHENYLBENZENAMINE

Formula:	$C_{12}H_9N_3O_4$
Mol Weight:	259.22
MP (deg C): 159-161	FP (deg C):
BP (deg C):	
BP pressure (mm Hg):	

Property/Value	Units	Temp	Data Type	Reference
WS 1.32E+000	mg/L	25	EXP	BAUGHMAN,GL & PERENICH,TA (1988)
logP 3.50			EST	MEYLAN,WM & HOWARD,PH (1995)
VP 3.00E-009	mm Hg	25	EXP	BAUGHMAN,GL & PERENICH,TA (1988)
DC	pKa			
HL 7.75E-010	atm m3/mol	25	EST	VP/WSOL
OH 1.49E-010	cm3/molc sec	25	EST	MEYLAN,WM & HOWARD,PH (1993)

CAS #: 000962-58-3 — DIAZINON, O-ANALOG

Formula:	$C_{12}H_{21}N_2O_4P$
Mol Weight:	288.29
MP (deg C):	FP (deg C):
BP (deg C):	
BP pressure (mm Hg):	

Property/Value	Units	Temp	Data Type	Reference
WS 2.45E+002	mg/L	25	EST	MEYLAN,WM ET AL. (1996)
logP 2.07			EXP	HANSCH,C & LEO,AJ (1985)
VP 1.10E-005	mm Hg	25	EST	NEELY,WB & BLAU,GE (1985)
DC	pKa			
HL 1.89E-010	atm m3/mol	25	EST	MEYLAN,WM & HOWARD,PH (1991)
OH 4.37E-011	cm3/molc sec	25	EST	MEYLAN,WM & HOWARD,PH (1993)

CAS #: 000963-03-1 — BENZENESULFONAMIDE, 4-CHLORO-N-[(CYCLOHEXYLAMINO

Formula:	$C_{13}H_{17}ClN_2O_3S$
Mol Weight:	316.81
MP (deg C):	FP (deg C):
BP (deg C):	
BP pressure (mm Hg):	

Property/Value	Units	Temp	Data Type	Reference
WS 1.27E+001	mg/L	25	EST	MEYLAN,WM ET AL. (1996)
logP 3.38			EXP	SANGSTER,J (1993)
VP 2.31E-009	mm Hg	25	EST	NEELY,WB & BLAU,GE (1985)
DC	pKa			
HL 1.32E-010	atm m3/mol	25	EST	MEYLAN,WM & HOWARD,PH (1991)
OH 2.14E-011	cm3/molc sec	25	EST	MEYLAN,WM & HOWARD,PH (1993)

CAS #: 000963-39-3 — 2H-1,4-BENZODIAZEPIN-2-ONE, 7-CHLORO-1,3-DIHYDRO

Formula:	$C_{15}H_{11}ClN_2O_2$
Mol Weight:	286.72
MP (deg C):	FP (deg C):
BP (deg C):	
BP pressure (mm Hg):	

Property/Value	Units	Temp	Data Type	Reference
WS 7.83E+002	mg/L	25	EST	MEYLAN,WM ET AL. (1996)
logP 1.49			EXP	SANGSTER,J (1993)
VP 2.45E-013	mm Hg	25	EST	NEELY,WB & BLAU,GE (1985)
DC	pKa			
HL 1.43E-017	atm m3/mol	25	EST	MEYLAN,WM & HOWARD,PH (1991)
OH 1.52E-011	cm3/molc sec	25	EST	MEYLAN,WM & HOWARD,PH (1993)

CAS #: 000963-89-3 — 7,9-DIMETHYLBENZ(C)ACRIDINE

Formula:	$C_{19}H_{15}N$
Mol Weight:	257.34
MP (deg C): 157-161	FP (deg C):
BP (deg C): 300	
BP pressure (mm Hg): 1.00E+001	

Property/Value	Units	Temp	Data Type	Reference
WS 2.82E-002	mg/L	25	EST	MEYLAN,WM ET AL. (1996)
logP 5.59			EST	MEYLAN,WM & HOWARD,PH (1995)
VP 6.40E-008	mm Hg	25	EST	NEELY,WB & BLAU,GE (1985)
DC	pKa			
HL 7.99E-009	atm m3/mol	25	EST	MEYLAN,WM & HOWARD,PH (1991)
OH 8.99E-011	cm3/molc sec	25	EST	MEYLAN,WM & HOWARD,PH (1993)

CAS #: 000964-79-4 — N-(2,4-DINITROPHENYL)-M-TOLUIDINE

Formula: $C_{13}H_{11}N_3O_4$

Mol Weight: 273.25

MP (deg C):
FP (deg C):
BP (deg C):
BP pressure (mm Hg):

	Property/Value	Units	Temp	Data Type	Reference
WS	6.07E+000	mg/L	25	EST	MEYLAN,WM ET AL. (1996)
logP	4.05			EST	MEYLAN,WM & HOWARD,PH (1995)
VP	1.31E-007	mm Hg	25	EST	NEELY,WB & BLAU,GE (1985)
DC		pKa			
HL	3.95E-010	atm m3/mol	25	EST	MEYLAN,WM & HOWARD,PH (1991)
OH	2.00E-010	cm3/molc sec	25	EST	MEYLAN,WM & HOWARD,PH (1993)

CAS #: 000968-81-0 — DIMELIN

Formula: $C_{15}H_{20}N_2O_4S$

Mol Weight: 324.40

MP (deg C):
FP (deg C):
BP (deg C):
BP pressure (mm Hg):

	Property/Value	Units	Temp	Data Type	Reference
WS	3.43E+003	mg/L	37	EXP	YALKOWSKY,SH & DANNENFELSER,RM (1992)
logP	2.44			EXP	SANGSTER,J (1993)
VP	3.39E-010	mm Hg	25	EST	NEELY,WB & BLAU,GE (1985)
DC		pKa			
HL	3.25E-013	atm m3/mol	25	EST	MEYLAN,WM & HOWARD,PH (1991)
OH	2.16E-011	cm3/molc sec	25	EST	MEYLAN,WM & HOWARD,PH (1993)

CAS #: 000969-99-3 — OPROMAZINE

Formula: $C_{17}H_{19}ClN_2OS$

Mol Weight: 334.87

MP (deg C): 115
FP (deg C):
BP (deg C):
BP pressure (mm Hg):

	Property/Value	Units	Temp	Data Type	Reference
WS	6.31E+001	mg/L	25	EST	MEYLAN,WM ET AL. (1996)
logP	2.32			EXP	WHELPTON,R (1989)
VP	9.75E-009	mm Hg	25	EST	NEELY,WB & BLAU,GE (1985)
DC		pKa			
HL	2.33E-014	atm m3/mol	25	EST	MEYLAN,WM & HOWARD,PH (1991)
OH	2.05E-010	cm3/molc sec	25	EST	MEYLAN,WM & HOWARD,PH (1993)

CAS #: 000970-76-3 — 2,4-DINITRO-N-(4-NITROPHENYL)BENZENEAMINE

Formula: $C_{12}H_8N_4O_6$

Mol Weight: 304.22

MP (deg C):
FP (deg C):
BP (deg C):
BP pressure (mm Hg):

	Property/Value	Units	Temp	Data Type	Reference
WS	5.42E+000	mg/L	25	EST	MEYLAN,WM ET AL. (1996)
logP	3.90			EST	MEYLAN,WM & HOWARD,PH (1995)
VP	6.04E-009	mm Hg	25	EST	NEELY,WB & BLAU,GE (1985)
DC		pKa			
HL	1.41E-012	atm m3/mol	25	EST	MEYLAN,WM & HOWARD,PH (1991)
OH	2.06E-011	cm3/molc sec	25	EST	MEYLAN,WM & HOWARD,PH (1993)

CAS #: 000970-91-2 — 2,4-DINITRO-N-(3-NITROPHENYL)BENZENAMINE

Formula: $C_{12}H_8N_4O_6$

Mol Weight: 304.22

MP (deg C):
FP (deg C):
BP (deg C):
BP pressure (mm Hg):

	Property/Value	Units	Temp	Data Type	Reference
WS	5.42E+000	mg/L	25	EST	MEYLAN,WM ET AL. (1996)
logP	3.90			EST	MEYLAN,WM & HOWARD,PH (1995)
VP	6.04E-009	mm Hg	25	EST	NEELY,WB & BLAU,GE (1985)
DC		pKa			
HL	1.41E-012	atm m3/mol	25	EST	MEYLAN,WM & HOWARD,PH (1991)
OH	1.51E-011	cm3/molc sec	25	EST	MEYLAN,WM & HOWARD,PH (1993)

CAS #: 000973-21-7 — DINOBUTON

Formula: $C_{14}H_{18}N_2O_7$

Mol Weight: 326.31

MP (deg C): 56-57
FP (deg C):
BP (deg C):
BP pressure (mm Hg):

	Property/Value	Units	Temp	Data Type	Reference
WS	3.97E+000	mg/L	25	EST	MEYLAN,WM ET AL. (1996)
logP	3.94			EST	MEYLAN,WM & HOWARD,PH (1995)
VP	1.10E-006	mm Hg	25	EST	NEELY,WB & BLAU,GE (1985)
DC		pKa			
HL	1.62E-008	atm m3/mol	25	EST	MEYLAN,WM & HOWARD,PH (1991)
OH	7.33E-012	cm3/molc sec	25	EST	MEYLAN,WM & HOWARD,PH (1993)

CAS #: 000976-70-5 — 6,7-DIHYDROCANRENONE

Formula: $C_{22}H_{30}O_3$

Mol Weight: 342.48

MP (deg C): 156-159
FP (deg C):
BP (deg C):
BP pressure (mm Hg):

	Property/Value	Units	Temp	Data Type	Reference
WS	1.96E+001	mg/L	25	EST	MEYLAN,WM ET AL. (1996)
logP	2.98			EXP	HANSCH,C & LEO,AJ (1985)
VP	7.65E-009	mm Hg	25	EST	NEELY,WB & BLAU,GE (1985)
DC		pKa			
HL	1.34E-007	atm m3/mol	25	EST	MEYLAN,WM & HOWARD,PH (1991)
OH	9.56E-011	cm3/molc sec	25	EST	MEYLAN,WM & HOWARD,PH (1993)

CAS #: 000976-71-6 — CANRENONE

Formula: $C_{22}H_{28}O_3$

Mol Weight: 340.47

MP (deg C): 149-151
FP (deg C):
BP (deg C):
BP pressure (mm Hg):

	Property/Value	Units	Temp	Data Type	Reference
WS	3.64E+001	mg/L	25	EST	MEYLAN,WM ET AL. (1996)
logP	2.68			EXP	HANSCH,C & LEO,AJ (1985)
VP	6.85E-009	mm Hg	25	EST	NEELY,WB & BLAU,GE (1985)
DC		pKa			
HL	9.63E-008	atm m3/mol	25	EST	MEYLAN,WM & HOWARD,PH (1991)
OH	1.86E-010	cm3/molc sec	25	EST	MEYLAN,WM & HOWARD,PH (1993)

BETAMETHASONE-17-ACETATE

CAS #:	000987-24-6	
Formula:	$C_{24}H_{31}FO_6$	
Mol Weight:	434.51	
MP (deg C): 231-234		FP (deg C):
BP (deg C):		
BP pressure (mm Hg):		

Property/Value	Units	Temp	Data Type	Reference
WS 8.04E+000	mg/L	25	EST	MEYLAN,WM ET AL. (1996)
logP 2.77			EXP	HANSCH,C & LEO,AJ (1985)
VP 7.66E-014	mm Hg	25	EST	NEELY,WB & BLAU,GE (1985)
DC	pKa			
HL 7.73E-012	atm m3/mol	25	EST	MEYLAN,WM & HOWARD,PH (1991)
OH 6.87E-011	cm3/molc sec	25	EST	MEYLAN,WM & HOWARD,PH (1993)

ERYTHROMYCIN, N-DEMETHYL-

CAS #:	000992-62-1	
Formula:	$C_{36}H_{65}NO_{13}$	
Mol Weight:	719.92	
MP (deg C):		FP (deg C):
BP (deg C):		
BP pressure (mm Hg):		

Property/Value	Units	Temp	Data Type	Reference
WS 3.41E+000	mg/L	25	EST	MEYLAN,WM ET AL. (1996)
logP 2.21			EXP	HANSCH,C ET AL. (1995)
VP 2.52E-027	mm Hg	25	EST	NEELY,WB & BLAU,GE (1985)
DC	pKa			
HL 2.47E-029	atm m3/mol	25	EST	MEYLAN,WM & HOWARD,PH (1991)
OH 4.00E-010	cm3/molc sec	25	EST	MEYLAN,WM & HOWARD,PH (1993)

METHYLPHOPHONIC ACID

CAS #:	000993-13-5	
Formula:	CH_5O_3P	
Mol Weight:	96.02	
MP (deg C): 108.5		FP (deg C):
BP (deg C):		
BP pressure (mm Hg):		

Property/Value	Units	Temp	Data Type	Reference
WS 2.00E+004	mg/L		EXP	ROSENBLATT,DH ET AL. (1975); >
logP -0.70			EST	MEYLAN,WM & HOWARD,PH (1995)
VP 2.00E-006	mm Hg		EST	HL X WSOL
DC 2.38	pKa	25	EXP	KORTUM,G ET AL (1961)
HL 1.22E-011	atm m3/mol	25	EST	MEYLAN,WM & HOWARD,PH (1991)
OH 2.16E-013	cm3/molc sec	25	EST	ATKINSON,R (1987)

METHYL-T-AMYL ETHER

CAS #:	000994-05-8	
Formula:	$C_6H_{14}O$	
Mol Weight:	102.18	
MP (deg C):		FP (deg C):
BP (deg C): 86.3		
BP pressure (mm Hg):		

Property/Value	Units	Temp	Data Type	Reference
WS 2.64E+003	mg/L	25	EST	MEYLAN,WM ET AL. (1996)
logP 1.92			EST	MEYLAN,WM & HOWARD,PH (1995)
VP 7.52E+001	mm Hg	25	EXP	DAUBERT,TE & DANNER,RP (1989)
DC	pKa			
HL 2.68E-003	atm m3/mol	25	EST	MEYLAN,WM & HOWARD,PH (1991)
OH 7.91E-012	cm3/molc sec	25	EXP	ATKINSON,R (1989)

CHLOROTRIETHYLSILANE

CAS #:	000994-30-9	
Formula:	$C_6H_{15}ClSi$	
Mol Weight:	150.73	
MP (deg C):		FP (deg C):
BP (deg C): 144.5		
BP pressure (mm Hg):		

Property/Value	Units	Temp	Data Type	Reference
WS 3.16E+001	mg/L	25	EST	MEYLAN,WM ET AL. (1996)
logP 3.95			EST	MEYLAN,WM & HOWARD,PH (1995)
VP 6.77E+000	mm Hg	25	EXP	PERRY,RH & GREEN,D (1984)
DC	pKa			
HL 1.50E-001	atm m3/mol	25	EST	MEYLAN,WM & HOWARD,PH (1991)
OH 3.58E-012	cm3/molc sec	25	EST	MEYLAN,WM & HOWARD,PH (1993)

N-NITROSO-DIISOBUTYLAMINE

CAS #:	000997-95-5	
Formula:	$C_8H_{18}N_2O$	
Mol Weight:	158.25	
MP (deg C):		FP (deg C):
BP (deg C):		
BP pressure (mm Hg):		

Property/Value	Units	Temp	Data Type	Reference
WS 7.00E+002	mg/L	25	EST	MEYLAN,WM ET AL. (1996)
logP 2.34			EXP	VERA,A ET AL. (1992)
VP 4.28E-002	mm Hg	25	EST	NEELY,WB & BLAU,GE (1985)
DC	pKa			
HL 1.13E-005	atm m3/mol	25	EST	MEYLAN,WM & HOWARD,PH (1991)
OH 2.68E-011	cm3/molc sec	25	EST	MEYLAN,WM & HOWARD,PH (1993)

BUTANOIC ACID, 4-HYDROXY-, ETHYL ESTER

CAS #:	000999-10-0	
Formula:	$C_6H_{12}O_3$	
Mol Weight:	132.16	
MP (deg C):		FP (deg C):
BP (deg C):		
BP pressure (mm Hg):		

Property/Value	Units	Temp	Data Type	Reference
WS 1.25E+005	mg/L	25	EST	MEYLAN,WM ET AL. (1996)
logP 0.43			EXP	SANGSTER,J (1994)
VP 5.22E-002	mm Hg	25	EST	NEELY,WB & BLAU,GE (1985)
DC	pKa			
HL 1.50E-008	atm m3/mol	25	EST	MEYLAN,WM & HOWARD,PH (1991)
OH 8.09E-012	cm3/molc sec	25	EST	MEYLAN,WM & HOWARD,PH (1993)

DIALLYL MALEATE

CAS #:	000999-21-3	
Formula:	$C_{10}H_{12}O_4$	
Mol Weight:	196.20	
MP (deg C):		FP (deg C):
BP (deg C): 108-110		
BP pressure (mm Hg): 3.00E+000		

Property/Value	Units	Temp	Data Type	Reference
WS 1.51E+002	mg/L	25	EST	MEYLAN,WM ET AL. (1996)
logP 2.90			EST	MEYLAN,WM & HOWARD,PH (1995)
VP 6.82E-003	mm Hg	25	EXP	DAUBERT,TE & DANNER,RP (1989)
DC	pKa			
HL 2.39E-007	atm m3/mol	25	EST	MEYLAN,WM & HOWARD,PH (1991)
OH 6.25E-011	cm3/molc sec	25	EST	MEYLAN,WM & HOWARD,PH (1993)

CAS #: 000999-29-1 — N-DIAZOACETYLGLYCINE,ETHYL ESTER

Formula: $C_6H_9N_3O_3$

Mol Weight: 171.16

MP (deg C):   FP (deg C):

BP (deg C):

BP pressure (mm Hg):

Property/Value	Units	Temp	Data Type	Reference
WS 3.48E+003	mg/L	25	EST	MEYLAN,WM ET AL. (1996)
logP -0.08			EXP	HANSCH,C & LEO,AJ (1985)
VP 3.88E-010	mm Hg	25	EST	NEELY,WB & BLAU,GE (1985)
DC	pKa			
HL 1.41E-014	atm m3/mol	25	EST	MEYLAN,WM & HOWARD,PH (1991)
OH 1.15E-011	cm3/molc sec	25	EST	MEYLAN,WM & HOWARD,PH (1993)

CAS #: 000999-33-7 — N-BUTYL-N-CHLORO-1-BUTANAMINE

Formula: $C_8H_{18}ClN$

Mol Weight: 163.69

MP (deg C):   FP (deg C):

BP (deg C):

BP pressure (mm Hg):

Property/Value	Units	Temp	Data Type	Reference
WS 5.45E+002	mg/L	25	EST	MEYLAN,WM ET AL. (1996)
logP 2.44				MEYLAN,WM & HOWARD,PH (1995)
VP 1.25E+000	mm Hg	25	EST	NEELY,WB & BLAU,GE (1985)
DC	pKa			
HL 1.75E-003	atm m3/mol	25	EST	MEYLAN,WM & HOWARD,PH (1991)
OH 2.68E-011	cm3/molc sec	25	EST	MEYLAN,WM & HOWARD,PH (1993)

CAS #: 000999-61-1 — 2-HYDROXYPROPYL ACRYLATE

Formula: $C_6H_{10}O_3$

Mol Weight: 130.14

MP (deg C):   FP (deg C):

BP (deg C): 77

BP pressure (mm Hg): 5.00E+000

Property/Value	Units	Temp	Data Type	Reference
WS 1.49E+005	mg/L	25	EST	MEYLAN,WM ET AL. (1996)
logP 0.35			EXP	HANSCH,C & LEO,AJ (1985)
VP 1.74E-001	mm Hg	25	EST	NEELY,WB & BLAU,GE (1985)
DC	pKa			
HL 5.96E-009	atm m3/mol	25	EST	MEYLAN,WM & HOWARD,PH (1991)
OH 1.97E-011	cm3/molc sec	25	EST	MEYLAN,WM & HOWARD,PH (1993)

CAS #: 000999-81-5 — CHLORMEQUAT CHLORIDE

Formula: $C_5H_{13}Cl_2N$

Mol Weight: 158.07

MP (deg C): 245 dec   FP (deg C):

BP (deg C):

BP pressure (mm Hg):

Property/Value	Units	Temp	Data Type	Reference
WS 9.96E+005	mg/L	25	EXP	BAKER,EA ET AL. (1992)
logP -3.80			EXP	BAKER,EA ET AL. (1992)
VP 7.50E-008	mm Hg	20	EXP	WORTHING,CR & WALKER,SB (1987)
DC	pKa			
HL 1.60E-014	atm m3/mol	20	EST	VP/WSOL
OH 1.32E-012	cm3/molc sec	25	EST	MEYLAN,WM & HOWARD,PH (1993)

CAS #: 000999-97-3 — HEXAMETHYLDISILIZANE

Formula: $C_6H_{19}NSi_2$

Mol Weight: 161.40

MP (deg C):   FP (deg C):

BP (deg C): 125

BP pressure (mm Hg):

Property/Value	Units	Temp	Data Type	Reference
WS 3.92E+002	mg/L	25	EST	MEYLAN,WM ET AL. (1996)
logP 2.62			EST	MEYLAN,WM & HOWARD,PH (1995)
VP 1.38E+001	mm Hg	25	EXP	YAWS,CL (1994A)
DC	pKa			
HL 8.69E-005	atm m3/mol	25	EST	MEYLAN,WM & HOWARD,PH (1991)
OH 8.98E-013	cm3/molc sec	25	EST	MEYLAN,WM & HOWARD,PH (1993)

CAS #: 001000-36-8 — Phosphonic acid, methyl- dipentyl ester

Formula: $C_{11}H_{25}O_3P$

Mol Weight: 236.29

MP (deg C):   FP (deg C):

BP (deg C):

BP pressure (mm Hg):

Property/Value	Units	Temp	Data Type	Reference
WS 1.80E+001	mg/L	25	EST	MEYLAN,WM ET AL. (1996)
logP 3.74			EXP	KRIKORIAN,SE ET AL. (1987)
VP 1.91E-003	mm Hg	25	EST	NEELY,WB & BLAU,GE (1985)
DC	pKa			
HL 1.20E-005	atm m3/mol	25	EST	MEYLAN,WM & HOWARD,PH (1991)
OH 5.55E-011	cm3/molc sec	25	EST	MEYLAN,WM & HOWARD,PH (1993)

CAS #: 001000-49-3 — SILANE, BUTYL TRIMETHYL

Formula: $C_7H_{18}Si$

Mol Weight: 130.31

MP (deg C):   FP (deg C):

BP (deg C): 115

BP pressure (mm Hg):

Property/Value	Units	Temp	Data Type	Reference
WS 2.37E+001	mg/L	25	EST	MEYLAN,WM ET AL. (1996)
logP 4.20			EXP	HANSCH,C & LEO,AJ (1985)
VP 4.52E+001	mm Hg	25	EST	NEELY,WB & BLAU,GE (1985)
DC	pKa			
HL 5.17E-001	atm m3/mol	25	EST	MEYLAN,WM & HOWARD,PH (1991)
OH 4.44E-012	cm3/molc sec	25	EST	MEYLAN,WM & HOWARD,PH (1993)

CAS #: 001001-52-1 — SILANE, BUTYL-DIMETHYL

Formula: $C_6H_{16}Si$

Mol Weight: 116.28

MP (deg C):   FP (deg C):

BP (deg C):

BP pressure (mm Hg):

Property/Value	Units	Temp	Data Type	Reference
WS 9.25E+001	mg/L	25	EST	MEYLAN,WM ET AL. (1996)
logP 3.57			EXP	HANSCH,C & LEO,AJ (1985)
VP 4.29E+001	mm Hg	25	EST	NEELY,WB & BLAU,GE (1985)
DC	pKa			
HL 3.76E-001	atm m3/mol	25	EST	MEYLAN,WM & HOWARD,PH (1991)
OH 4.29E-012	cm3/molc sec	25	EST	MEYLAN,WM & HOWARD,PH (1993)

CAS #: 001002-69-3	1-CHLORODECANE
Formula: $C_{10}H_{21}Cl$	
Mol Weight: 176.73	
MP (deg C): -31.3	FP (deg C):
BP (deg C): 223.4	
BP pressure (mm Hg):	

Property/Value	Units	Temp	Data Type	Reference
WS 1.13E+000	mg/L	25	EST	MEYLAN,WM ET AL. (1996)
logP 5.51			EST	MEYLAN,WM & HOWARD,PH (1995)
VP 9.84E-003	mm Hg	25	EXT	BOUBLIK,T ET AL. (1984)
DC	pKa			
HL 1.05E-001	atm m3/mol	25	EST	MEYLAN,WM & HOWARD,PH (1991)
OH 1.06E-011	cm3/molc sec	25	EST	MEYLAN,WM & HOWARD,PH (1993)

CAS #: 001002-84-2	PENTADECANOIC ACID
Formula: $C_{15}H_{30}O_2$	
Mol Weight: 242.41	
MP (deg C): 52.3	FP (deg C):
BP (deg C): 257	
BP pressure (mm Hg): 1.00E+002	

Property/Value	Units	Temp	Data Type	Reference
WS 1.20E+001	mg/L	20	EXP	YALKOWSKY,SH & DANNENFELSER,RM (1992)
logP 6.47			EST	MEYLAN,WM & HOWARD,PH (1995)
VP 2.45E-008	mm Hg	25	EXT	YAWS,CL (1994B)
DC	pKa			
HL 2.18E-005	atm m3/mol	25	EST	MEYLAN,WM & HOWARD,PH (1991)
OH 1.82E-011	cm3/molc sec	25	EST	MEYLAN,WM & HOWARD,PH (1993)

CAS #: 001003-09-4	2-BROMOTHIOPHENE
Formula: C_4H_3BrS	
Mol Weight: 163.04	
MP (deg C):	FP (deg C):
BP (deg C): 150	
BP pressure (mm Hg):	

Property/Value	Units	Temp	Data Type	Reference
WS 2.97E+002	mg/L	25	EST	MEYLAN,WM ET AL. (1996)
logP 2.75			EXP	HANSCH,C & LEO,AJ (1985)
VP 2.34E+001	mm Hg	25	EXT	BOUBLIK,T ET AL. (1984)
DC	pKa			
HL 1.17E-003	atm m3/mol	25	EST	MEYLAN,WM & HOWARD,PH (1991)
OH 6.00E-012	cm3/molc sec	25	EST	MEYLAN,WM & HOWARD,PH (1993)

CAS #: 001003-29-8	PYRROLE-2-CARBOXALDEHYDE
Formula: C_5H_5NO	
Mol Weight: 95.10	
MP (deg C): 46.5	FP (deg C):
BP (deg C): 218	
BP pressure (mm Hg):	

Property/Value	Units	Temp	Data Type	Reference
WS 3.43E+004	mg/L	25	EST	MEYLAN,WM ET AL. (1996)
logP 0.64			EXP	HANSCH,C & LEO,AJ (1985)
VP 3.26E-001	mm Hg	25	EST	NEELY,WB & BLAU,GE (1985)
DC	pKa			
HL 2.26E-008	atm m3/mol	25	EST	MEYLAN,WM & HOWARD,PH (1991)
OH 7.27E-011	cm3/molc sec	25	EST	MEYLAN,WM & HOWARD,PH (1993)

CAS #: 001003-31-2	2-CYANOTHIOPHENE
Formula: C_5H_5NS	
Mol Weight: 111.17	
MP (deg C):	FP (deg C):
BP (deg C): 192	
BP pressure (mm Hg):	

Property/Value	Units	Temp	Data Type	Reference
WS 4.90E+003	mg/L	25	EST	MEYLAN,WM ET AL. (1996)
logP 1.27			EXP	HANSCH,C & LEO,AJ (1985)
VP 3.38E-001	mm Hg	25	EST	NEELY,WB & BLAU,GE (1985)
DC	pKa			
HL 2.83E-005	atm m3/mol	25	EST	MEYLAN,WM & HOWARD,PH (1991)
OH 1.68E-012	cm3/molc sec	25	EST	MEYLAN,WM & HOWARD,PH (1993)

CAS #: 001003-67-4	4-METHYL-N-OXIDEPYRIDINE
Formula: C_6H_7NO	
Mol Weight: 109.13	
MP (deg C): 38-40	FP (deg C):
BP (deg C): 150	
BP pressure (mm Hg): 1.50E+001	

Property/Value	Units	Temp	Data Type	Reference
WS 1.00E+006	mg/L	20	EXP	GOE,GL (1982)
logP -0.79			EST	MEYLAN,WM & HOWARD,PH (1995)
VP 5.33E-002	mm Hg	25	EST	NEELY,WB & BLAU,GE (1985)
DC 1.29	pKa	24	EXP	PERRIN,DD (1965)
HL 7.78E-011	atm m3/mol	25	EST	MEYLAN,WM & HOWARD,PH (1991)
OH 1.10E-012	cm3/molc sec	25	EST	MEYLAN,WM & HOWARD,PH (1993)

CAS #: 001003-73-2	3-METHYL-N-OXIDEPYRIDINE
Formula: C_6H_7NO	
Mol Weight: 109.13	
MP (deg C): 39	FP (deg C):
BP (deg C): 148	
BP pressure (mm Hg): 1.50E+001	

Property/Value	Units	Temp	Data Type	Reference
WS 1.00E+006	mg/L	25	EXP	GOE,GL (1982)
logP -0.79			EST	MEYLAN,WM & HOWARD,PH (1995)
VP 5.33E-002	mm Hg	25	EST	NEELY,WB & BLAU,GE (1985)
DC 1.08	pKa	24	EXP	PERRIN,DD (1965)
HL 7.78E-011	atm m3/mol	25	EST	MEYLAN,WM & HOWARD,PH (1991)
OH 1.10E-012	cm3/molc sec	25	EST	MEYLAN,WM & HOWARD,PH (1993)

CAS #: 001004-36-0	GAMMA-2,6-DIMETHYLPYRONE
Formula: $C_7H_8O_2$	
Mol Weight: 124.14	
MP (deg C): 132	FP (deg C):
BP (deg C): 251	
BP pressure (mm Hg):	

Property/Value	Units	Temp	Data Type	Reference
WS 7.96E+004	mg/L	25	EST	MEYLAN,WM ET AL. (1996)
logP 0.10			EXP	HANSCH,C ET AL. (1995)
VP 4.26E-001	mm Hg	25	EST	NEELY,WB & BLAU,GE (1985)
DC	pKa			
HL 8.28E-006	atm m3/mol	25	EST	MEYLAN,WM & HOWARD,PH (1991)
OH 6.11E-011	cm3/molc sec	25	EST	MEYLAN,WM & HOWARD,PH (1993)

CAS #: 001004-38-2				2,4,6-PYRIMIDINETRIAMINE
Formula: $C_4H_7N_5$				
Mol Weight: 125.13				
MP (deg C): 249-251		FP (deg C):		
BP (deg C):				
BP pressure (mm Hg):				

Property/Value	Units	Temp	Data Type	Reference
WS 1.00E+006	mg/L	25	EST	MEYLAN,WM ET AL. (1996)
logP -1.95			EST	MEYLAN,WM & HOWARD,PH (1995)
VP 6.04E-005	mm Hg	25	EST	NEELY,WB & BLAU,GE (1985)
DC 6.81	pKa	20	EXP	PERRIN,DD (1965)
HL 4.57E-013	atm m3/mol	25	EST	MEYLAN,WM & HOWARD,PH (1991)
OH 2.00E-010	cm3/molc sec	25	EST	MEYLAN,WM & HOWARD,PH (1993)

CAS #: 001004-66-6				2,6-DIMETHYLANISOLE
Formula: $C_9H_{12}O$				
Mol Weight: 136.20				
MP (deg C):		FP (deg C):		
BP (deg C): 182.5				
BP pressure (mm Hg):				

Property/Value	Units	Temp	Data Type	Reference
WS 2.29E+002	mg/L	25	EST	MEYLAN,WM ET AL. (1996)
logP 3.02			EXP	HANSCH,C ET AL. (1995)
VP 5.49E-001	mm Hg	25	EST	NEELY,WB & BLAU,GE (1985)
DC	pKa			
HL 3.88E-004	atm m3/mol	25	EST	MEYLAN,WM & HOWARD,PH (1991)
OH 3.34E-011	cm3/molc sec	25	EST	MEYLAN,WM & HOWARD,PH (1993)

CAS #: 001006-31-1				2,3,5,6-TETRACHLOROTOLUENE
Formula: $C_7H_4Cl_4$				
Mol Weight: 229.92				
MP (deg C): 93.5		FP (deg C):		
BP (deg C):				
BP pressure (mm Hg):				

Property/Value	Units	Temp	Data Type	Reference
WS 1.30E+000	mg/L	25	EST	MEYLAN,WM ET AL. (1996)
logP 5.12			EST	MEYLAN,WM & HOWARD,PH (1995)
VP 1.08E-002	mm Hg	25	EST	NEELY,WB & BLAU,GE (1985)
DC	pKa			
HL 1.79E-003	atm m3/mol	25	EST	MEYLAN,WM & HOWARD,PH (1991)
OH 3.51E-013	cm3/molc sec	25	EST	MEYLAN,WM & HOWARD,PH (1993)

CAS #: 001006-32-2				2,3,4,5-TETRACHLOROTOLUENE
Formula: $C_7H_4Cl_4$				
Mol Weight: 229.92				
MP (deg C):		FP (deg C):		
BP (deg C):				
BP pressure (mm Hg):				

Property/Value	Units	Temp	Data Type	Reference
WS 1.30E+000	mg/L	25	EST	MEYLAN,WM ET AL. (1996)
logP 5.12			EST	MEYLAN,WM & HOWARD,PH (1995)
VP 1.08E-002	mm Hg	25	EST	NEELY,WB & BLAU,GE (1985)
DC	pKa			
HL 1.79E-003	atm m3/mol	25	EST	MEYLAN,WM & HOWARD,PH (1991)
OH 3.51E-013	cm3/molc sec	25	EST	MEYLAN,WM & HOWARD,PH (1993)

CAS #: 001006-59-3				PHENOL, 2,6-DIETHYL-
Formula: $C_{10}H_{14}O$				
Mol Weight: 150.22				
MP (deg C):		FP (deg C):		
BP (deg C):				
BP pressure (mm Hg):				

Property/Value	Units	Temp	Data Type	Reference
WS 7.44E+002	mg/L	25	EST	MEYLAN,WM ET AL. (1996)
logP 3.03			EXP	SOTOMATSU,T ET AL. (1993)
VP 7.91E-003	mm Hg	25	EST	NEELY,WB & BLAU,GE (1985)
DC	pKa			
HL 1.20E-006	atm m3/mol	25	EST	MEYLAN,WM & HOWARD,PH (1991)
OH 5.18E-011	cm3/molc sec	25	EST	MEYLAN,WM & HOWARD,PH (1993)

CAS #: 001006-64-0				PYRROLIDINE, 2-PHENYL-
Formula: $C_{10}H_{13}N$				
Mol Weight: 147.22				
MP (deg C):		FP (deg C):		
BP (deg C):				
BP pressure (mm Hg):				

Property/Value	Units	Temp	Data Type	Reference
WS 1.93E+004	mg/L	25	EST	MEYLAN,WM ET AL. (1996)
logP 1.89			EXP	SANGSTER,J (1993)
VP 2.06E-002	mm Hg	25	EST	NEELY,WB & BLAU,GE (1985)
DC 4.40	pKa		EXP	PERRIN,DD (1965)
HL 1.04E-006	atm m3/mol	25	EST	MEYLAN,WM & HOWARD,PH (1991)
OH 9.62E-011	cm3/molc sec	25	EST	MEYLAN,WM & HOWARD,PH (1993)

CAS #: 001006-94-6				5-METHOXYINDOLE
Formula: C_9H_9NO				
Mol Weight: 147.18				
MP (deg C): 56-58		FP (deg C):		
BP (deg C): 176-178				
BP pressure (mm Hg): 1.70E+001				

Property/Value	Units	Temp	Data Type	Reference
WS 1.36E+003	mg/L	25	EST	MEYLAN,WM ET AL. (1996)
logP 2.06			EXP	HANSCH,C & LEO,AJ (1985)
VP 1.76E-003	mm Hg	25	EST	NEELY,WB & BLAU,GE (1985)
DC	pKa			
HL 5.24E-008	atm m3/mol	25	EST	MEYLAN,WM & HOWARD,PH (1991)
OH 2.01E-010	cm3/molc sec	25	EST	MEYLAN,WM & HOWARD,PH (1993)

CAS #: 001007-28-9				DESISOPROPYLATRAZINE
Formula: $C_5H_8ClN_5$				
Mol Weight: 173.61				
MP (deg C):		FP (deg C):		
BP (deg C):				
BP pressure (mm Hg):				

Property/Value	Units	Temp	Data Type	Reference
WS 6.70E+002	mg/L	22	EXP	MILLS,MS & THURMAN,EM (1994A)
logP 1.15			EXP	FINIZIO,A ET AL. (1991)
VP 2.11E-004	mm Hg	25	EST	NEELY,WB & BLAU,GE (1985)
DC	pKa			
HL 1.16E-009	atm m3/mol	25	EST	MEYLAN,WM & HOWARD,PH (1991)
OH 8.97E-012	cm3/molc sec	25	EST	MEYLAN,WM & HOWARD,PH (1993)

001013-51-0 — ACETIC ACID,2-METHYL-5-NITROIMIDAZOL-2-YL,METHYL

Formula: $C_7H_9N_3O_4$

Mol Weight: 199.17

MP (deg C):
FP (deg C):
BP (deg C):
BP pressure (mm Hg):

Property/Value	Units	Temp	Data Type	Reference
WS 1.29E+004	mg/L	25	EST	MEYLAN,WM ET AL. (1996)
logP 0.17			EXP	HANSCH,C ET AL. (1995)
VP 1.44E-006	mm Hg	25	EST	NEELY,WB & BLAU,GE (1985)
DC	pKa			
HL 5.17E-011	atm m3/mol	25	EST	MEYLAN,WM & HOWARD,PH (1991)
OH 5.61E-012	cm3/molc sec	25	EST	MEYLAN,WM & HOWARD,PH (1993)

001014-69-3 — DESMETRYNE

Formula: $C_8H_{15}N_5S$

Mol Weight: 213.31

MP (deg C):
FP (deg C):
BP (deg C):
BP pressure (mm Hg):

Property/Value	Units	Temp	Data Type	Reference
WS 5.80E+002	mg/L	20	EXP	YALKOWSKY,SH & DANNENFELSER,RM (1992)
logP 2.82			EST	MEYLAN,WM & HOWARD,PH (1995)
VP 9.98E-008	mm Hg	20	EXP	TOMLIN,C (1994)
DC	pKa			
HL 4.83E-011	atm m3/mol	20	EST	VP/WSOL
OH 2.09E-011	cm3/molc sec	25	EST	MEYLAN,WM & HOWARD,PH (1993)

001014-70-6 — SIMETRYN

Formula: $C_8H_{15}N_5S$

Mol Weight: 213.31

MP (deg C): 82-83
FP (deg C):
BP (deg C):
BP pressure (mm Hg):

Property/Value	Units	Temp	Data Type	Reference
WS 4.50E+002	mg/L	25	EXP	SHIU,WY ET AL. (1990)
logP 2.80			EXP	FINIZIO,A ET AL. (1991)
VP 7.10E-007	mm Hg	25	EXP	AUGUSTIJN-BECKERS,PWM ET AL. (1994)
DC	pKa			
HL 4.43E-010	atm m3/mol	25	EST	VP/WSOL
OH 1.90E-011	cm3/molc sec	25	EST	MEYLAN,WM & HOWARD,PH (1993)

001016-47-3 — ACETAMIDE, N-[2-(1H-INDOL-3-YL)ETHYL]-

Formula: $C_{12}H_{14}N_2O$

Mol Weight: 202.26

MP (deg C):
FP (deg C):
BP (deg C):
BP pressure (mm Hg):

Property/Value	Units	Temp	Data Type	Reference
WS 2.66E+003	mg/L	25	EST	MEYLAN,WM ET AL. (1996)
logP 1.41			EXP	RADZICKA,A & WOLFENDEN,R (1988)
VP 1.59E-007	mm Hg	25	EST	NEELY,WB & BLAU,GE (1985)
DC	pKa			
HL 4.33E-013	atm m3/mol	25	EST	MEYLAN,WM & HOWARD,PH (1991)
OH 2.11E-010	cm3/molc sec	25	EST	MEYLAN,WM & HOWARD,PH (1993)

001017-24-9 — BENZYL-4-PYRIDYL KETONE

Formula: $C_{13}H_{11}NO$

Mol Weight: 197.24

MP (deg C):
FP (deg C):
BP (deg C):
BP pressure (mm Hg):

Property/Value	Units	Temp	Data Type	Reference
WS 3.18E+003	mg/L	25	EST	MEYLAN,WM ET AL. (1996)
logP 1.35			EXP	HANSCH,C & LEO,AJ (1985)
VP 1.41E-004	mm Hg	25	EST	NEELY,WB & BLAU,GE (1985)
DC	pKa			
HL 1.04E-009	atm m3/mol	25	EST	MEYLAN,WM & HOWARD,PH (1991)
OH 5.88E-012	cm3/molc sec	25	EST	MEYLAN,WM & HOWARD,PH (1993)

001022-22-6 — 1,1-BIS(P-CHLOROPHENYL)-2-CHLOROETHYLENE

Formula: $C_{14}H_9Cl_3$

Mol Weight: 283.59

MP (deg C):
FP (deg C):
BP (deg C):
BP pressure (mm Hg):

Property/Value	Units	Temp	Data Type	Reference
WS 3.05E-001	mg/L	25	EST	MEYLAN,WM ET AL. (1996)
logP 5.50			EST	MEYLAN,WM & HOWARD,PH (1995)
VP 2.21E-005	mm Hg	25	EST	NEELY,WB & BLAU,GE (1985)
DC	pKa			
HL 4.89E-005	atm m3/mol	25	EST	MEYLAN,WM & HOWARD,PH (1991)
OH 2.08E-011	cm3/molc sec	25	EST	MEYLAN,WM & HOWARD,PH (1993)

001022-45-3 — 4(1H)-QUINAZOLINONE, 2-PHENYL-

Formula: $C_{14}H_{10}N_2O$

Mol Weight: 222.25

MP (deg C):
FP (deg C):
BP (deg C):
BP pressure (mm Hg):

Property/Value	Units	Temp	Data Type	Reference
WS 2.31E+002	mg/L	25	EST	MEYLAN,WM ET AL. (1996)
logP 2.53			EXP	SANGSTER,J (1993)
VP 2.16E-009	mm Hg	25	EST	NEELY,WB & BLAU,GE (1985)
DC	pKa			
HL 2.95E-011	atm m3/mol	25	EST	MEYLAN,WM & HOWARD,PH (1991)
OH 4.32E-011	cm3/molc sec	25	EST	MEYLAN,WM & HOWARD,PH (1993)

001022-66-8 — 2(1H)-QUINOLINONE, 3,4-DIHYDRO-3-PHENYL-

Formula: $C_{15}H_{13}NO$

Mol Weight: 223.28

MP (deg C):
FP (deg C):
BP (deg C):
BP pressure (mm Hg):

Property/Value	Units	Temp	Data Type	Reference
WS 1.04E+002	mg/L	25	EST	MEYLAN,WM ET AL. (1996)
logP 2.93			EXP	SANGSTER,J (1993)
VP 1.59E-007	mm Hg	25	EST	NEELY,WB & BLAU,GE (1985)
DC	pKa			
HL 2.42E-010	atm m3/mol	25	EST	MEYLAN,WM & HOWARD,PH (1991)
OH 2.52E-011	cm3/molc sec	25	EST	MEYLAN,WM & HOWARD,PH (1993)

001024-57-3 — HEPTACHLOR EPOXIDE

CAS #:	001024-57-3
Formula:	$C_{10}H_5Cl_7O$
Mol Weight:	389.32
MP (deg C):	160-161.5
FP (deg C):	
BP (deg C):	
BP pressure (mm Hg):	

Property/Value	Units	Temp	Data Type	Reference
WS 2.00E-001	mg/L	25	EXP	BIGGAR,JW & RIGGS,RI (1974)
logP 4.98			EXP	SANGSTER,J (1993)
VP 1.95E-005	mm Hg	30	EXP	NASH,RG (1983A)
DC	pKa			
HL 3.20E-005	atm m3/mol	30	EXP	WARNER,HP ET AL. (1987)
OH 4.56E+000	cm3/molc sec	25	EST	MEYLAN,WM & HOWARD,PH (1993)

001025-75-8 — CHLOROTHIAZIDE-3-METHYL

CAS #:	001025-75-8
Formula:	$C_8H_8ClN_3O_4S_2$
Mol Weight:	309.75
MP (deg C):	
FP (deg C):	
BP (deg C):	
BP pressure (mm Hg):	

Property/Value	Units	Temp	Data Type	Reference
WS 9.03E+002	mg/L	25	EST	MEYLAN,WM ET AL. (1996)
logP 0.03			EXP	SANGSTER,J (1994)
VP 9.72E-012	mm Hg	25	EST	NEELY,WB & BLAU,GE (1985)
DC	pKa			
HL 5.37E-012	atm m3/mol	25	EST	MEYLAN,WM & HOWARD,PH (1991)
OH 2.07E-013	cm3/molc sec	25	EST	MEYLAN,WM & HOWARD,PH (1993)

001028-33-7 — 1-HEXYL-3,7-DIMETHYL XANTHINE

CAS #:	001028-33-7
Formula:	$C_{13}H_{22}N_4O_2$
Mol Weight:	266.35
MP (deg C):	82-83
FP (deg C):	
BP (deg C):	
BP pressure (mm Hg):	

Property/Value	Units	Temp	Data Type	Reference
WS 1.16E+002	mg/L	25	EST	MEYLAN,WM ET AL. (1996)
logP 1.20			EXP	HANSCH,C & LEO,AJ (1985)
VP 6.11E-010	mm Hg	25	EST	NEELY,WB & BLAU,GE (1985)
DC	pKa			
HL 1.48E-010	atm m3/mol	25	EST	MEYLAN,WM & HOWARD,PH (1991)
OH 2.91E-011	cm3/molc sec	25	EST	MEYLAN,WM & HOWARD,PH (1993)

001031-07-8 — ENDOSULFAN SULFATE

CAS #:	001031-07-8
Formula:	$C_9H_6Cl_6O_4S$
Mol Weight:	422.93
MP (deg C):	198-201
FP (deg C):	
BP (deg C):	
BP pressure (mm Hg):	

Property/Value	Units	Temp	Data Type	Reference
WS 4.80E-001	mg/L	20	EXP	SHIU,WY ET AL. (1990)
logP 3.66			EXP	HANSCH,C ET AL. (1995)
VP 1.00E-005	mm Hg	25	EXP	MABEY,WR ET AL. (1981)
DC	pKa			
HL 1.16E-005	atm m3/mol	25	EST	VP/WSOL
OH 7.59E-012	cm3/molc sec	25	EST	MEYLAN,WM & HOWARD,PH (1993)

001034-01-1 — OCTYL GALLATE

CAS #:	001034-01-1
Formula:	$C_{15}H_{22}O_5$
Mol Weight:	282.34
MP (deg C):	101-104
FP (deg C):	
BP (deg C):	
BP pressure (mm Hg):	

Property/Value	Units	Temp	Data Type	Reference
WS 4.42E+001	mg/L	25	EST	MEYLAN,WM ET AL. (1996)
logP 3.66			EXP	HANSCH,C & LEO,AJ (1985)
VP 2.91E-009	mm Hg	25	EST	NEELY,WB & BLAU,GE (1985)
DC	pKa			
HL 2.84E-016	atm m3/mol	25	EST	MEYLAN,WM & HOWARD,PH (1991)
OH 9.93E-011	cm3/molc sec	25	EST	MEYLAN,WM & HOWARD,PH (1993)

001058-92-0 — MORDANT BLUE 13

CAS #:	001058-92-0
Formula:	$C_{16}H_9ClN_2Na_2O_9S_2$
Mol Weight:	518.82
MP (deg C):	
FP (deg C):	
BP (deg C):	
BP pressure (mm Hg):	

Property/Value	Units	Temp	Data Type	Reference
WS 4.35E+001	mg/L	25	EST	MEYLAN,WM ET AL. (1996)
logP 1.79			EST	MEYLAN,WM & HOWARD,PH (1995)
VP 3.75E-023	mm Hg	25	EST	NEELY,WB & BLAU,GE (1985)
DC	pKa			
HL 2.61E-031	atm m3/mol	25	EST	MEYLAN,WM & HOWARD,PH (1991)
OH 7.28E-012	cm3/molc sec	25	EST	MEYLAN,WM & HOWARD,PH (1993)

001067-08-9 — 3-METHYL-3-ETHYLPENTANE

CAS #:	001067-08-9
Formula:	C_8H_{18}
Mol Weight:	114.23
MP (deg C):	-90.9
FP (deg C):	
BP (deg C):	118.2
BP pressure (mm Hg):	

Property/Value	Units	Temp	Data Type	Reference
WS 8.58E+000	mg/L	25	EST	MEYLAN,WM ET AL. (1996)
logP 4.16			EST	MEYLAN,WM & HOWARD,PH (1995)
VP 2.30E+001	mm Hg	25	EXP	YAWS,CL (1994B)
DC	pKa			
HL 3.01E+000	atm m3/mol	25	EST	MEYLAN,WM & HOWARD,PH (1991)
OH 4.12E-012	cm3/molc sec	25	EST	MEYLAN,WM & HOWARD,PH (1993)

001067-20-5 — 3,3-DIETHYLPENTANE

CAS #:	001067-20-5
Formula:	C_9H_{20}
Mol Weight:	128.26
MP (deg C):	-33.1
FP (deg C):	
BP (deg C):	146.3
BP pressure (mm Hg):	

Property/Value	Units	Temp	Data Type	Reference
WS 2.90E+000	mg/L	25	EST	MEYLAN,WM ET AL. (1996)
logP 4.65			EST	MEYLAN,WM & HOWARD,PH (1995)
VP 7.30E+000	mm Hg	25	EXP	DAUBERT,TE & DANNER,RP (1989)
DC	pKa			
HL 4.00E+000	atm m3/mol	25	EST	MEYLAN,WM & HOWARD,PH (1991)
OH 5.26E-012	cm3/molc sec	25	EST	MEYLAN,WM & HOWARD,PH (1993)

001068-57-1 — ACETIC ACID HYDRAZIDE

Formula: $C_2H_6N_2O$

Mol Weight: 74.08

MP (deg C): 67 FP (deg C):

BP (deg C): 137

BP pressure (mm Hg): 2.40E+001

Property/Value	Units	Temp	Data Type	Reference
WS 1.47E+005	mg/L	25	EST	MEYLAN,WM ET AL. (1996)
logP -1.58			EXP	HANSCH,C & LEO,AJ (1985)
VP 6.34E-002	mm Hg	25	EST	NEELY,WB & BLAU,GE (1985)
DC 3.24	pKa		EXP	SERJEANT,EP & DEMPSEY,B (1979)
HL 4.67E-011	atm m3/mol	25	EST	MEYLAN,WM & HOWARD,PH (1991)
OH 5.60E-012	cm3/molc sec	25	EST	MEYLAN,WM & HOWARD,PH (1993)

001068-87-7 — 2,4-DIMETHYL-3-ETHYLPENTANE

Formula: C_9H_{20}

Mol Weight: 128.26

MP (deg C): -122.4 FP (deg C):

BP (deg C): 136.7

BP pressure (mm Hg):

Property/Value	Units	Temp	Data Type	Reference
WS 3.59E+000	mg/L	25	EST	MEYLAN,WM ET AL. (1996)
logP 4.54			EST	MEYLAN,WM & HOWARD,PH (1995)
VP 1.01E+001	mm Hg	25	EXP	YAWS,CL (1994B)
DC	pKa			
HL 4.00E+000	atm m3/mol	25	EST	MEYLAN,WM & HOWARD,PH (1991)
OH 1.04E-011	cm3/molc sec	25	EST	MEYLAN,WM & HOWARD,PH (1993)

001069-53-0 — 2,3,5-TRIMETHYLHEXANE

Formula: C_9H_{20}

Mol Weight: 128.26

MP (deg C): -127.9 FP (deg C):

BP (deg C): 131.4

BP pressure (mm Hg):

Property/Value	Units	Temp	Data Type	Reference
WS 3.59E+000	mg/L	25	EST	MEYLAN,WM ET AL. (1996)
logP 4.54			EST	MEYLAN,WM & HOWARD,PH (1995)
VP 1.17E+001	mm Hg	25	EXP	ZWOLINSKI,BJ & WILHOIT,RC (1971)
DC	pKa			
HL 4.00E+000	atm m3/mol	25	EST	MEYLAN,WM & HOWARD,PH (1991)
OH 7.88E-012	cm3/molc sec	25	EXP	ATKINSON,R (1989)

001070-87-7 — 2,2,4,4-TETRAMETHYLPENTANE

Formula: C_9H_{20}

Mol Weight: 128.26

MP (deg C): -66.5 FP (deg C):

BP (deg C): 122.2

BP pressure (mm Hg):

Property/Value	Units	Temp	Data Type	Reference
WS 3.61E+000	mg/L	25	EST	MEYLAN,WM ET AL. (1996)
logP 4.54			EST	MEYLAN,WM & HOWARD,PH (1995)
VP 2.00E+001	mm Hg	25	EXP	DAUBERT,TE & DANNER,RP (1989)
DC	pKa			
HL 4.00E+000	atm m3/mol	25	EST	MEYLAN,WM & HOWARD,PH (1991)
OH 2.42E-012	cm3/molc sec	25	EST	MEYLAN,WM & HOWARD,PH (1993)

001071-26-7 — 2,2-DIMETHYLHEPTANE

Formula: C_9H_{20}

Mol Weight: 128.26

MP (deg C): -113 FP (deg C):

BP (deg C): 132.7

BP pressure (mm Hg):

Property/Value	Units	Temp	Data Type	Reference
WS 2.90E+000	mg/L	25	EST	MEYLAN,WM ET AL. (1996)
logP 4.65			EST	MEYLAN,WM & HOWARD,PH (1995)
VP 1.08E+001	mm Hg	25	EXP	YAWS,CL (1994B)
DC	pKa			
HL 4.00E+000	atm m3/mol	25	EST	MEYLAN,WM & HOWARD,PH (1991)
OH 6.06E-012	cm3/molc sec	25	EST	MEYLAN,WM & HOWARD,PH (1993)

001071-81-4 — 2,2,5,5-TETRAMETHYLHEXANE

Formula: $C_{10}H_{22}$

Mol Weight: 142.29

MP (deg C): -12.6 FP (deg C):

BP (deg C): 137.4

BP pressure (mm Hg):

Property/Value	Units	Temp	Data Type	Reference
WS 1.20E+000	mg/L	25	EST	MEYLAN,WM ET AL. (1996)
logP 5.03			EST	MEYLAN,WM & HOWARD,PH (1995)
VP 8.74E+000	mm Hg	25	EXP	ZWOLINSKI,BJ & WILHOIT,RC (1971)
DC	pKa			
HL 5.30E+000	atm m3/mol	25	EST	MEYLAN,WM & HOWARD,PH (1991)
OH 3.83E-012	cm3/molc sec	25	EST	MEYLAN,WM & HOWARD,PH (1993)

001071-83-6 — GLYPHOSPHATE

Formula: $C_3H_8NO_5P$

Mol Weight: 169.07

MP (deg C): 200 FP (deg C):

BP (deg C):

BP pressure (mm Hg):

Property/Value	Units	Temp	Data Type	Reference
WS 1.20E+004	mg/L	25	EXP	WORTHING,CR & WALKER,SB (1987)
logP -4.47			EST	MEYLAN,WM & HOWARD,PH (1995)
VP 2.89E-010	mm Hg	25	EST	NEELY,WB & BLAU,GE (1985)
DC	pKa			
HL 4.08E-019	atm m3/mol	25	EST	MEYLAN,WM & HOWARD,PH (1991)
OH 7.90E-011	cm3/molc sec	25	EST	MEYLAN,WM & HOWARD,PH (1993)

001072-33-9 — TRIDECYL ACETATE

Formula: $C_{15}H_{30}O_2$

Mol Weight: 242.41

MP (deg C): FP (deg C):

BP (deg C):

BP pressure (mm Hg):

Property/Value	Units	Temp	Data Type	Reference
WS 1.16E-001	mg/L	25	EST	MEYLAN,WM ET AL. (1996)
logP 6.27			EST	MEYLAN,WM & HOWARD,PH (1995)
VP 1.88E-003	mm Hg	25	EST	NEELY,WB & BLAU,GE (1985)
DC	pKa			
HL 5.25E-003	atm m3/mol	25	EST	MEYLAN,WM & HOWARD,PH (1991)
OH 1.73E-011	cm3/molc sec	25	EST	MEYLAN,WM & HOWARD,PH (1993)

2-ACETYLPYRROLE

CAS #:	001072-83-9			
Formula:	C_6H_7NO			
Mol Weight:	109.13			
MP (deg C):	90	FP (deg C):		
BP (deg C):	220			
BP pressure (mm Hg):				

Property/Value	Units	Temp	Data Type	Reference
WS 1.76E+004	mg/L	25	EST	MEYLAN,WM ET AL. (1996)
logP 0.93			EXP	HANSCH,C & LEO,AJ (1985)
VP 2.16E-001	mm Hg	25	EST	NEELY,WB & BLAU,GE (1985)
DC	pKa			
HL 1.65E-008	atm m3/mol	25	EST	MEYLAN,WM & HOWARD,PH (1991)
OH 1.00E-010	cm3/molc sec	25	EST	MEYLAN,WM & HOWARD,PH (1993)

M-BROMOFLUOROBENZENE

CAS #:	001073-06-9			
Formula:	C_6H_4BrF			
Mol Weight:	175.01			
MP (deg C):		FP (deg C):		
BP (deg C):	150			
BP pressure (mm Hg):				

Property/Value	Units	Temp	Data Type	Reference
WS 3.78E+002	mg/L	25	EXP	YALKOWSKY,SH & DANNENFELSER,RM (1992)
logP 2.92			EXP	DUNN,WJ ET AL. (1983)
VP 2.84E+000	mm Hg	25	EST	NEELY,WB & BLAU,GE (1985)
DC	pKa			
HL 2.50E-003	atm m3/mol	25	EST	MEYLAN,WM & HOWARD,PH (1991)
OH 1.54E-012	cm3/molc sec	25	EST	MEYLAN,WM & HOWARD,PH (1993)

2,6-DIMETHYL-N-OXIDEPYRIDINE

CAS #:	001073-23-0			
Formula:	C_7H_9NO			
Mol Weight:	123.16			
MP (deg C):		FP (deg C):		
BP (deg C):	115-119			
BP pressure (mm Hg):	1.80E+001			

Property/Value	Units	Temp	Data Type	Reference
WS 1.57E+005	mg/L	25	EST	MEYLAN,WM ET AL. (1996)
logP -0.24			EST	MEYLAN,WM & HOWARD,PH (1995)
VP 1.53E-002	mm Hg	25	EST	NEELY,WB & BLAU,GE (1985)
DC	pKa			
HL 8.59E-011	atm m3/mol	25	EST	MEYLAN,WM & HOWARD,PH (1991)
OH 2.79E-012	cm3/molc sec	25	EST	MEYLAN,WM & HOWARD,PH (1993)

P-CHLOROSTYRENE

CAS #:	001073-67-2			
Formula:	C_8H_7Cl			
Mol Weight:	138.60			
MP (deg C):	15.9	FP (deg C):		
BP (deg C):	192			
BP pressure (mm Hg):				

Property/Value	Units	Temp	Data Type	Reference
WS 8.05E+001	mg/L	25	EST	MEYLAN,WM ET AL. (1996)
logP 3.54			EST	MEYLAN,WM & HOWARD,PH (1995)
VP 8.16E-001	mm Hg	25	EXT	PERRY,RH & GREEN,D (1984)
DC	pKa			
HL 2.05E-003	atm m3/mol	25	EST	MEYLAN,WM & HOWARD,PH (1991)
OH 2.76E-011	cm3/molc sec	25	EST	MEYLAN,WM & HOWARD,PH (1993)

P-METHIOPHENOL

CAS #:	001073-72-9			
Formula:	C_7H_8OS			
Mol Weight:	140.21			
MP (deg C):		FP (deg C):		
BP (deg C):				
BP pressure (mm Hg):				

Property/Value	Units	Temp	Data Type	Reference
WS 9.59E+003	mg/L	25	EST	MEYLAN,WM ET AL. (1996)
logP 1.78			EXP	HANSCH,C & LEO,AJ (1985)
VP 8.72E-003	mm Hg	25	EST	NEELY,WB & BLAU,GE (1985)
DC 9.53	pKa	25	EXP	KORTUM,G ET AL (1961)
HL 1.63E-008	atm m3/mol	25	EST	MEYLAN,WM & HOWARD,PH (1991)
OH 2.17E-011	cm3/molc sec	25	EST	MEYLAN,WM & HOWARD,PH (1993)

1-METHYL-2-PROPYLBENZENE

CAS #:	001074-17-5			
Formula:	$C_{10}H_{14}$			
Mol Weight:	134.22			
MP (deg C):	-60.3	FP (deg C):		
BP (deg C):	185			
BP pressure (mm Hg):				

Property/Value	Units	Temp	Data Type	Reference
WS 1.27E+001	mg/L	25	EST	MEYLAN,WM ET AL. (1996)
logP 4.50			EXP	SHERBLOM,PM & EGANHOUSE,RP (1988)
VP 9.94E-001	mm Hg	25	EXP	CHAO,J ET AL. (1983)
DC	pKa			
HL 1.16E-002	atm m3/mol	25	EST	MEYLAN,WM & HOWARD,PH (1991)
OH 8.80E-012	cm3/molc sec	25	EST	MEYLAN,WM & HOWARD,PH (1993)

1-METHYL-3-PROPYLBENZENE

CAS #:	001074-43-7			
Formula:	$C_{10}H_{14}$			
Mol Weight:	134.22			
MP (deg C):	-82.5	FP (deg C):		
BP (deg C):	182			
BP pressure (mm Hg):				

Property/Value	Units	Temp	Data Type	Reference
WS 9.09E+000	mg/L	25	EST	MEYLAN,WM ET AL. (1996)
logP 4.67			EXP	SHERBLOM,PM & EGANHOUSE,RP (1988)
VP 2.66E+000	mm Hg	25	EXP	CHAO,J ET AL. (1983)
DC	pKa			
HL 1.16E-002	atm m3/mol	25	EST	MEYLAN,WM & HOWARD,PH (1991)
OH 1.52E-011	cm3/molc sec	25	EST	MEYLAN,WM & HOWARD,PH (1993)

1-METHYL-4-PROPYLBENZENE

CAS #:	001074-55-1			
Formula:	$C_{10}H_{14}$			
Mol Weight:	134.22			
MP (deg C):	-63.6	FP (deg C):		
BP (deg C):	183.4			
BP pressure (mm Hg):				

Property/Value	Units	Temp	Data Type	Reference
WS 1.04E+001	mg/L	25	EST	MEYLAN,WM ET AL. (1996)
logP 4.60			EXP	SHERBLOM,PM & EGANHOUSE,RP (1988)
VP 1.10E+000	mm Hg	25	EXP	CHAO,J ET AL. (1983)
DC	pKa			
HL 1.16E-002	atm m3/mol	25	EST	MEYLAN,WM & HOWARD,PH (1991)
OH 8.80E-012	cm3/molc sec	25	EST	MEYLAN,WM & HOWARD,PH (1993)

CAS #: 001075-49-6				BENZOIC ACID, 4-ETHENYL-
Formula: $C_9H_8O_2$				
Mol Weight: 148.16				
MP (deg C): 143-144		FP (deg C):		
BP (deg C):				
BP pressure (mm Hg):				

Property/Value	Units	Temp	Data Type	Reference
WS 1.06E+003	mg/L	25	EST	MEYLAN,WM ET AL. (1996)
logP 2.18			EXP	SANGSTER,J (1993)
VP 1.84E-003	mm Hg	25	EST	NEELY,WB & BLAU,GE (1985)
DC	pKa			
HL 5.56E-008	atm m3/mol	25	EST	MEYLAN,WM & HOWARD,PH (1991)
OH 2.75E-011	cm3/molc sec	25	EST	MEYLAN,WM & HOWARD,PH (1993)

CAS #: 001076-22-8				3-METHYLXANTHINE
Formula: $C_6H_6N_4O_2$				
Mol Weight: 166.14				
MP (deg C): > 300		FP (deg C):		
BP (deg C):				
BP pressure (mm Hg):				

Property/Value	Units	Temp	Data Type	Reference
WS 8.39E+003	mg/L	25	EST	MEYLAN,WM ET AL. (1996)
logP -0.50			EXP	GASPARI,F & BONATI,M (1987)
VP 1.25E-010	mm Hg	25	EST	NEELY,WB & BLAU,GE (1985)
DC 8.10	pKa	25	EXP	KORTUM,G ET AL (1961)
HL 7.64E-013	atm m3/mol	25	EST	MEYLAN,WM & HOWARD,PH (1991)
OH 1.87E-011	cm3/molc sec	25	EST	MEYLAN,WM & HOWARD,PH (1993)

CAS #: 001077-16-3				N-HEXYLBENZENE
Formula: $C_{12}H_{18}$				
Mol Weight: 162.28				
MP (deg C): -61		FP (deg C):		
BP (deg C): 226.1				
BP pressure (mm Hg):				

Property/Value	Units	Temp	Data Type	Reference
WS 9.02E-001	mg/L	25	EXP	YALKOWSKY,SH & DANNENFELSER,RM (1992)
logP 5.52			EXP	HANSCH,C & LEO,AJ (1985)
VP 1.21E-001	mm Hg	25	EXP	YAWS,CL (1994B)
DC	pKa			
HL 2.86E-002	atm m3/mol	25	EST	VP/WSOL
OH 1.15E-011	cm3/molc sec	25	EST	MEYLAN,WM & HOWARD,PH (1993)

CAS #: 001077-74-3				QUINOLINE, 4,7-DICHLORO-, 1-OXIDE
Formula: $C_9H_7Cl_2NO$				
Mol Weight: 216.07				
MP (deg C):		FP (deg C):		
BP (deg C):				
BP pressure (mm Hg):				

Property/Value	Units	Temp	Data Type	Reference
WS 9.74E+002	mg/L	25	EST	MEYLAN,WM ET AL. (1996)
logP 1.85			EXP	GO,ML & NGIAM,TL (1988)
VP 1.06E-005	mm Hg	25	EST	NEELY,WB & BLAU,GE (1985)
DC	pKa			
HL 3.78E-012	atm m3/mol	25	EST	MEYLAN,WM & HOWARD,PH (1991)
OH 2.96E-012	cm3/molc sec	25	EST	MEYLAN,WM & HOWARD,PH (1993)

CAS #: 001078-71-3				1-PHENYLHEPTANE
Formula: $C_{13}H_{20}$				
Mol Weight: 176.30				
MP (deg C): -48		FP (deg C):		
BP (deg C): 240				
BP pressure (mm Hg):				

Property/Value	Units	Temp	Data Type	Reference
WS 4.47E-001	mg/L	25	EXP	WASIK,SP ET AL. (1981)
logP 5.49			EST	MEYLAN,WM & HOWARD,PH (1995)
VP 3.10E-002	mm Hg	25	EXP	YAWS,CL (1994B)
DC	pKa			
HL 1.51E-002	atm m3/mol	25	EXP	VP/WSOL
OH 1.30E-011	cm3/molc sec	25	EST	MEYLAN,WM & HOWARD,PH (1993)

CAS #: 001080-06-4				L-TYROSINE, METHYL ESTER
Formula: $C_{10}H_{13}NO_3$				
Mol Weight: 195.22				
MP (deg C): 135-139		FP (deg C):		
BP (deg C):				
BP pressure (mm Hg):				

Property/Value	Units	Temp	Data Type	Reference
WS 1.00E+006	mg/L	25	EST	MEYLAN,WM ET AL. (1996)
logP 0.29			EXP	HUANG,CH ET AL. (1985)
VP 5.41E-005	mm Hg	25	EST	NEELY,WB & BLAU,GE (1985)
DC	pKa			
HL 4.01E-012	atm m3/mol	25	EST	MEYLAN,WM & HOWARD,PH (1991)
OH 7.95E-011	cm3/molc sec	25	EST	MEYLAN,WM & HOWARD,PH (1993)

CAS #: 001081-15-8				FORMALDEHYDE, (2,4-DINITROPHENYL)HYDRAZONE
Formula: $C_7H_6N_4O_4$				
Mol Weight: 210.15				
MP (deg C):		FP (deg C):		
BP (deg C):				
BP pressure (mm Hg):				

Property/Value	Units	Temp	Data Type	Reference
WS 1.62E+002	mg/L	25	EST	MEYLAN,WM ET AL. (1996)
logP 2.33			EXP	SANGSTER,J (1993)
VP 2.45E-005	mm Hg	25	EST	NEELY,WB & BLAU,GE (1985)
DC	pKa			
HL 1.73E-009	atm m3/mol	25	EST	MEYLAN,WM & HOWARD,PH (1991)
OH 1.60E-012	cm3/molc sec	25	EST	MEYLAN,WM & HOWARD,PH (1993)

CAS #: 001081-34-1				2,5-THIOPHENYL-THIOPHENE
Formula: $C_{12}H_{14}S_3$				
Mol Weight: 254.44				
MP (deg C): 93-94		FP (deg C):		
BP (deg C):				
BP pressure (mm Hg):				

Property/Value	Units	Temp	Data Type	Reference
WS 4.23E-001	mg/L	25	EST	MEYLAN,WM ET AL. (1996)
logP 5.57			EXP	MCLACHLAN,D ET AL. (1986)
VP 4.10E-007	mm Hg	25	EST	NEELY,WB & BLAU,GE (1985)
DC	pKa			
HL 5.10E-006	atm m3/mol	25	EST	MEYLAN,WM & HOWARD,PH (1991)
OH 4.49E-011	cm3/molc sec	25	EST	MEYLAN,WM & HOWARD,PH (1993)

CAS #: 001081-77-2				N-NONYLBENZENE

Formula: $C_{15}H_{24}$
Mol Weight: 204.36
MP (deg C): -24   FP (deg C):
BP (deg C): 280.5
BP pressure (mm Hg):

Property/ Value	Units	Temp	Data Type	Reference
WS 3.52E-002	mg/L	25	EST	MEYLAN,WM ET AL. (1996)
logP 7.11			EXP	SHERBLOM,PM & EGANHOUSE,RP (1988)
VP 5.71E-003	mm Hg	25	EXP	YAWS,CL (1994B)
DC	pKa			
HL 5.74E-002	atm m3/mol	25	EST	MEYLAN,WM & HOWARD,PH (1991)
OH 1.58E-011	cm3/molc sec	25	EST	MEYLAN,WM & HOWARD,PH (1993)

CAS #: 001082-23-1				2,3,4-TRIMEO AMPHETAMINE

Formula: $C_{12}H_{19}NO_3$
Mol Weight: 225.29
MP (deg C):   FP (deg C):
BP (deg C):
BP pressure (mm Hg):

Property/ Value	Units	Temp	Data Type	Reference
WS 1.69E+004	mg/L	25	EST	MEYLAN,WM ET AL. (1996)
logP 1.51			EXP	HANSCH,C & LEO,AJ (1985)
VP 2.31E-004	mm Hg	25	EST	NEELY,WB & BLAU,GE (1985)
DC	pKa			
HL 2.23E-010	atm m3/mol	25	EST	MEYLAN,WM & HOWARD,PH (1991)
OH 2.47E-010	cm3/molc sec	25	EST	MEYLAN,WM & HOWARD,PH (1993)

CAS #: 001082-88-8				3,4,5-TRIMEO AMPHETAMINE

Formula: $C_{12}H_{19}NO_3$
Mol Weight: 225.29
MP (deg C):   FP (deg C):
BP (deg C):
BP pressure (mm Hg):

Property/ Value	Units	Temp	Data Type	Reference
WS 3.04E+004	mg/L	25	EST	MEYLAN,WM ET AL. (1996)
logP 1.21			EXP	HANSCH,C & LEO,AJ (1985)
VP 2.31E-004	mm Hg	25	EST	NEELY,WB & BLAU,GE (1985)
DC	pKa			
HL 2.23E-010	atm m3/mol	25	EST	MEYLAN,WM & HOWARD,PH (1991)
OH 2.47E-010	cm3/molc sec	25	EST	MEYLAN,WM & HOWARD,PH (1993)

CAS #: 001083-09-6				2,4,5-TRIMEO AMPHETAMINE

Formula: $C_{12}H_{19}NO_3$
Mol Weight: 225.29
MP (deg C):   FP (deg C):
BP (deg C):
BP pressure (mm Hg):

Property/ Value	Units	Temp	Data Type	Reference
WS 1.07E+004	mg/L	25	EST	MEYLAN,WM ET AL. (1996)
logP 1.74			EXP	HANSCH,C & LEO,AJ (1985)
VP 2.31E-004	mm Hg	25	EST	NEELY,WB & BLAU,GE (1985)
DC	pKa			
HL 2.23E-010	atm m3/mol	25	EST	MEYLAN,WM & HOWARD,PH (1991)
OH 2.47E-010	cm3/molc sec	25	EST	MEYLAN,WM & HOWARD,PH (1993)

CAS #: 001083-27-8				HEXYL P-HYDROXYBENZOATE

Formula: $C_{13}H_{18}O_3$
Mol Weight: 222.29
MP (deg C):   FP (deg C):
BP (deg C):
BP pressure (mm Hg):

Property/ Value	Units	Temp	Data Type	Reference
WS 2.45E+001	mg/L	25	EST	MEYLAN,WM ET AL. (1996)
logP 4.35			EXP	HANSCH,C & LEO,AJ (1985)
VP 2.44E-005	mm Hg	25	EST	NEELY,WB & BLAU,GE (1985)
DC	pKa			
HL 1.49E-008	atm m3/mol	25	EST	MEYLAN,WM & HOWARD,PH (1991)
OH 1.83E-011	cm3/molc sec	25	EST	MEYLAN,WM & HOWARD,PH (1993)

CAS #: 001083-41-6				BUTYL GALLATE

Formula: $C_{11}H_{14}O_5$
Mol Weight: 226.23
MP (deg C):   FP (deg C):
BP (deg C):
BP pressure (mm Hg):

Property/ Value	Units	Temp	Data Type	Reference
WS 1.06E+003	mg/L	25	EST	MEYLAN,WM ET AL. (1996)
logP 2.41			EXP	HANSCH,C & LEO,AJ (1985)
VP 1.07E-007	mm Hg	25	EST	NEELY,WB & BLAU,GE (1985)
DC	pKa			
HL 9.15E-017	atm m3/mol	25	EST	MEYLAN,WM & HOWARD,PH (1991)
OH 9.36E-011	cm3/molc sec	25	EST	MEYLAN,WM & HOWARD,PH (1993)

CAS #: 001083-48-3				PYRIDINE, 4- (4-NITROPHENYL)METHYL -

Formula: $C_{12}H_{10}N_2O_2$
Mol Weight: 214.23
MP (deg C): 70-72   FP (deg C):
BP (deg C):
BP pressure (mm Hg):

Property/ Value	Units	Temp	Data Type	Reference
WS 1.04E+003	mg/L	25	EST	MEYLAN,WM ET AL. (1996)
logP 2.88			EXP	SANGSTER,J (1993)
VP 2.01E-005	mm Hg	25	EST	NEELY,WB & BLAU,GE (1985)
DC	pKa			
HL 2.48E-009	atm m3/mol	25	EST	MEYLAN,WM & HOWARD,PH (1991)
OH 2.46E-012	cm3/molc sec	25	EST	MEYLAN,WM & HOWARD,PH (1993)

CAS #: 001083-59-6				2(5-NO2-2-FURFURYLIDENE)PYRIMIDINE

Formula: $C_{10}H_8N_3O_3$
Mol Weight: 218.19
MP (deg C):   FP (deg C):
BP (deg C):
BP pressure (mm Hg):

Property/ Value	Units	Temp	Data Type	Reference
WS 1.04E+003	mg/L	25	EST	MEYLAN,WM ET AL. (1996)
logP 1.34			EXP	HANSCH,C & LEO,AJ (1985)
VP 1.62E-005	mm Hg	25	EST	NEELY,WB & BLAU,GE (1985)
DC	pKa			
HL 2.64E-010	atm m3/mol	25	EST	MEYLAN,WM & HOWARD,PH (1991)
OH 6.14E-011	cm3/molc sec	25	EST	MEYLAN,WM & HOWARD,PH (1993)

559

CAS #: 001085-12-7			HEPTYL P-HYDROXYBENZOATE

Formula: $C_{14}H_{20}O_3$

Mol Weight: 236.31

MP (deg C): | FP (deg C):

BP (deg C):

BP pressure (mm Hg):

Property/Value	Units	Temp	Data Type	Reference
WS 8.02E+000	mg/L	25	EST	MEYLAN,WM ET AL. (1996)
logP 4.83			EXP	HANSCH,C & LEO,AJ (1985)
VP 9.57E-006	mm Hg	25	EST	NEELY,WB & BLAU,GE (1985)
DC	pKa			
HL 1.98E-008	atm m3/mol	25	EST	MEYLAN,WM & HOWARD,PH (1991)
OH 1.97E-011	cm3/molc sec	25	EST	MEYLAN,WM & HOWARD,PH (1993)

CAS #: 001085-98-9			DICHLOFLUANID

Formula: $C_9H_{11}Cl_2FN_2O_2S_2$

Mol Weight: 333.23

MP (deg C): 105-105.6 | FP (deg C):

BP (deg C):

BP pressure (mm Hg):

Property/Value	Units	Temp	Data Type	Reference
WS 6.67E+001	mg/L	25	EST	MEYLAN,WM ET AL. (1996)
logP 2.72			EST	MEYLAN,WM & HOWARD,PH (1995)
VP 1.57E-007	mm Hg	25	EXP	TOMLIN,C (1994)
DC	pKa			
HL 6.70E-007	atm m3/mol	25	EST	MEYLAN,WM & HOWARD,PH (1991)
OH 1.49E-011	cm3/molc sec	25	EST	MEYLAN,WM & HOWARD,PH (1993)

CAS #: 001087-26-9			HEXYL GALLATE

Formula: $C_{13}H_{18}O_5$

Mol Weight: 254.29

MP (deg C): | FP (deg C):

BP (deg C):

BP pressure (mm Hg):

Property/Value	Units	Temp	Data Type	Reference
WS 1.67E+002	mg/L	25	EST	MEYLAN,WM ET AL. (1996)
logP 3.17			EXP	HANSCH,C & LEO,AJ (1985)
VP 1.76E-008	mm Hg	25	EST	NEELY,WB & BLAU,GE (1985)
DC	pKa			
HL 1.61E-016	atm m3/mol	25	EST	MEYLAN,WM & HOWARD,PH (1991)
OH 9.64E-011	cm3/molc sec	25	EST	MEYLAN,WM & HOWARD,PH (1993)

CAS #: 001088-11-5			1,4-BENZDIAZEPIN-2-ONE-5-PH-7-CL

Formula: $C_{15}H_{11}ClN_2O$

Mol Weight: 270.72

MP (deg C): 216.5 | FP (deg C):

BP (deg C):

BP pressure (mm Hg):

Property/Value	Units	Temp	Data Type	Reference
WS 5.70E+001	mg/L	25	EST	MEYLAN,WM ET AL. (1996)
logP 2.93			EXP	HANSCH,C & LEO,AJ (1985)
VP 6.44E-009	mm Hg	25	EST	NEELY,WB & BLAU,GE (1985)
DC	pKa			
HL 1.78E-010	atm m3/mol	25	EST	MEYLAN,WM & HOWARD,PH (1991)
OH 9.35E-012	cm3/molc sec	25	EST	MEYLAN,WM & HOWARD,PH (1993)

CAS #: 001090-16-0			1,4-NAPHTHOQUINONE,3-ANILINO-2-CHLORO

Formula: $C_{16}H_{10}ClNO_2$

Mol Weight: 283.72

MP (deg C): 214-216 | FP (deg C):

BP (deg C):

BP pressure (mm Hg):

Property/Value	Units	Temp	Data Type	Reference
WS 3.80E+000	mg/L	25	EST	MEYLAN,WM ET AL. (1996)
logP 4.22			EXP	HANSCH,C & LEO,AJ (1985)
VP 5.05E-008	mm Hg	25	EST	NEELY,WB & BLAU,GE (1985)
DC	pKa			
HL 3.35E-012	atm m3/mol	25	EST	MEYLAN,WM & HOWARD,PH (1991)
OH 4.51E-011	cm3/molc sec	25	EST	MEYLAN,WM & HOWARD,PH (1993)

CAS #: 001113-02-6			DIMETHOXON

Formula: $C_5H_{12}NO_4PS$

Mol Weight: 213.19

MP (deg C): | FP (deg C):

BP (deg C):

BP pressure (mm Hg):

Property/Value	Units	Temp	Data Type	Reference
WS 1.00E+006	mg/L	25	EXP	HARTLEY,D & KIDD,H (1991)
logP -0.75			EXP	TOMLIN,C (1994)
VP 2.41E-005	mm Hg	20	EST	HARTLEY,D & KIDD,H (1983)
DC	pKa			
HL 4.56E-014	atm m3/mol	25	EST	MEYLAN,WM & HOWARD,PH (1991)
OH 8.19E-011	cm3/molc sec	25	EST	ATKINSON,R (1986)

CAS #: 001114-71-2			PEBULATE

Formula: $C_{10}H_{21}NOS$

Mol Weight: 203.35

MP (deg C): | FP (deg C):

BP (deg C): 142

BP pressure (mm Hg): 2.00E+001

Property/Value	Units	Temp	Data Type	Reference
WS 6.00E+001	mg/L	20	EXP	TOMLIN,C (1994)
logP 3.83			EXP	TOMLIN,C (1994)
VP 3.50E-002	mm Hg	25	EXP	TOMLIN,C (1994)
DC	pKa			
HL 1.56E-004	atm m3/mol	25	EST	MEYLAN,WM & HOWARD,PH (1991)
OH 2.85E-011	cm3/molc sec	25	EST	MEYLAN,WM & HOWARD,PH (1993)

CAS #: 001115-15-7			1,1'-SULFINYLBISETHENE

Formula: C_4H_6OS

Mol Weight: 102.16

MP (deg C): | FP (deg C):

BP (deg C):

BP pressure (mm Hg):

Property/Value	Units	Temp	Data Type	Reference
WS 3.16E+005	mg/L	25	EST	MEYLAN,WM ET AL. (1996)
logP -0.51			EST	MEYLAN,WM & HOWARD,PH (1995)
VP 1.60E+000	mm Hg	25	EST	NEELY,WB & BLAU,GE (1985)
DC	pKa			
HL 3.90E-007	atm m3/mol	25	EST	MEYLAN,WM & HOWARD,PH (1991)
OH 1.13E-010	cm3/molc sec	25	EST	MEYLAN,WM & HOWARD,PH (1993)

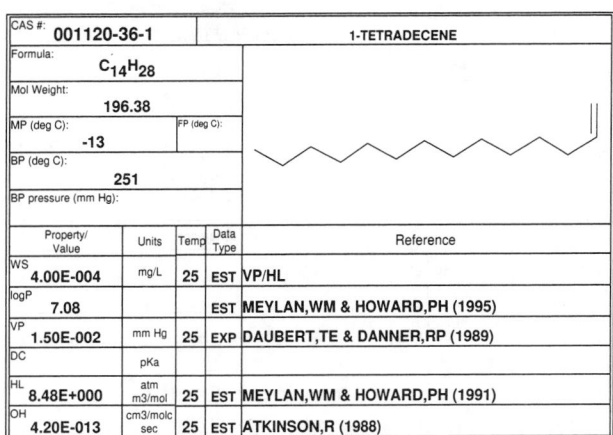

CAS #: 001116-54-7				N-NITROSODIETHANOLAMINE
Formula: $C_4H_{10}N_2O_3$				
Mol Weight: 134.14				
MP (deg C):		FP (deg C):		
BP (deg C): 125				
BP pressure (mm Hg): 1.00E-002				

Property/Value	Units	Temp	Data Type	Reference
WS 1.00E+006	mg/L		EXP	IARC (1978)
logP -1.28			EST	MEYLAN,WM & HOWARD,PH (1995)
VP 5.00E-004	mm Hg	20	EXT	KLEIN,RG (1982)
DC	pKa			
HL 4.85E-012	atm m3/mol	25	EST	MEYLAN,WM & HOWARD,PH (1991)
OH 2.97E-011	cm3/molc sec	25	EST	MEYLAN,WM & HOWARD,PH (1993)

CAS #: 001116-76-3				TRI(N-OCTYL)AMINE
Formula: $C_{24}H_{51}N$				
Mol Weight: 353.68				
MP (deg C):		FP (deg C):		
BP (deg C): 365-367				
BP pressure (mm Hg):				

Property/Value	Units	Temp	Data Type	Reference
WS 8.20E+001	mg/L	25	EXP	NIKOLAEV,AV ET AL. (1969)
logP 10.35			EST	MEYLAN,WM & HOWARD,PH (1995)
VP 1.57E-006	mm Hg	25	EST	NEELY,WB & BLAU,GE (1985)
DC	pKa			
HL 1.40E-002	atm m3/mol	25	EST	MEYLAN,WM & HOWARD,PH (1991)
OH 1.23E-010	cm3/molc sec	25	EST	MEYLAN,WM & HOWARD,PH (1993)

CAS #: 001116-82-1				N-NITROSO-BIS(2-CYANOET)AMINE
Formula: $C_6H_8N_4O$				
Mol Weight: 152.16				
MP (deg C):		FP (deg C):		
BP (deg C):				
BP pressure (mm Hg):				

Property/Value	Units	Temp	Data Type	Reference
WS 3.45E+004	mg/L	25	EST	MEYLAN,WM ET AL. (1996)
logP 0.08			EXP	HANSCH,C & LEO,AJ (1985)
VP 3.68E-005	mm Hg	25	EST	NEELY,WB & BLAU,GE (1985)
DC	pKa			
HL 6.24E-013	atm m3/mol	25	EST	MEYLAN,WM & HOWARD,PH (1991)
OH 2.52E-012	cm3/molc sec	25	EST	MEYLAN,WM & HOWARD,PH (1993)

CAS #: 001117-97-1				O,N-DIMETHYLHYDROXYLAMINE
Formula: C_2H_7NO				
Mol Weight: 61.08				
MP (deg C):		FP (deg C):		
BP (deg C):				
BP pressure (mm Hg):				

Property/Value	Units	Temp	Data Type	Reference
WS 1.21E+005	mg/L	25	EST	MEYLAN,WM ET AL. (1996)
logP 0.06			EST	MEYLAN,WM & HOWARD,PH (1995)
VP 3.79E+002	mm Hg	25	EXP	BOUBLIK,T ET AL. (1983)
DC 4.75	pKa	25	EXP	PERRIN,DD (1965)
HL 4.86E-006	atm m3/mol	25	EST	MEYLAN,WM & HOWARD,PH (1991)
OH 6.43E-011	cm3/molc sec	25	EST	MEYLAN,WM & HOWARD,PH (1993)

CAS #: 001118-68-9				N,N-DIMETHYLGLYCINE
Formula: $C_4H_9NO_2$				
Mol Weight: 103.12				
MP (deg C): -58		FP (deg C):		
BP (deg C): 63.9				
BP pressure (mm Hg):				

Property/Value	Units	Temp	Data Type	Reference
WS 2.98E+005	mg/L	25	EST	MEYLAN,WM ET AL. (1996)
logP -2.91			EXP	TSAI,RS ET AL. (1991)
VP 1.86E-007	mm Hg	25	EST	NEELY,WB & BLAU,GE (1985)
DC	pKa			
HL 5.42E-009	atm m3/mol	25	EST	MEYLAN,WM & HOWARD,PH (1991)
OH 7.56E-011	cm3/molc sec	25	EST	MEYLAN,WM & HOWARD,PH (1993)

CAS #: 001119-40-0				PENTANEDIOIC ACID, DIMETHYL ESTER
Formula: $C_7H_{12}O_4$				
Mol Weight: 160.17				
MP (deg C): -42.5		FP (deg C):		
BP (deg C): 214				
BP pressure (mm Hg):				

Property/Value	Units	Temp	Data Type	Reference
WS 2.02E+004	mg/L	25	EST	MEYLAN,WM ET AL. (1996)
logP 0.62			EXP	HANSCH,C ET AL. (1995)
VP 1.89E+000	mm Hg	25	EST	NEELY,WB & BLAU,GE (1985)
DC	pKa			
HL 7.36E-007	atm m3/mol	25	EST	MEYLAN,WM & HOWARD,PH (1991)
OH 3.55E-012	cm3/molc sec	25	EST	MEYLAN,WM & HOWARD,PH (1993)

CAS #: 001120-21-4				N-UNDECANE
Formula: $C_{11}H_{24}$				
Mol Weight: 156.31				
MP (deg C): -25.6		FP (deg C):		
BP (deg C): 195.5				
BP pressure (mm Hg):				

Property/Value	Units	Temp	Data Type	Reference
WS 6.90E+001	mg/L	25	EXP	YALKOWSKY,SH & DANNENFELSER,RM (1992)
logP 5.74			EST	MEYLAN,WM & HOWARD,PH (1995)
VP 4.12E-001	mm Hg	25	EXP	DAUBERT,TE & DANNER,RP (1989)
DC	pKa			
HL 1.23E-003	atm m3/mol	25	EST	VP/WSOL
OH 1.32E-011	cm3/molc sec	25	EXP	ATKINSON,R (1989)

CAS #: 001120-36-1				1-TETRADECENE
Formula: $C_{14}H_{28}$				
Mol Weight: 196.38				
MP (deg C): -13		FP (deg C):		
BP (deg C): 251				
BP pressure (mm Hg):				

Property/Value	Units	Temp	Data Type	Reference
WS 4.00E-004	mg/L	25	EST	VP/HL
logP 7.08			EST	MEYLAN,WM & HOWARD,PH (1995)
VP 1.50E-002	mm Hg	25	EXP	DAUBERT,TE & DANNER,RP (1989)
DC	pKa			
HL 8.48E+000	atm m3/mol	25	EST	MEYLAN,WM & HOWARD,PH (1991)
OH 4.20E-013	cm3/molc sec	25	EST	ATKINSON,R (1988)

CAS #:	001120-71-4	2,2-DIOXIDE-1,2-OXATHIOLANE

Formula:	$C_3H_6O_3S$
Mol Weight:	122.14
MP (deg C): 31 EST	FP (deg C):
BP (deg C): 112	
BP pressure (mm Hg): 3.00E+001	

Property/Value	Units	Temp	Data Type	Reference
WS 1.71E+005	mg/L	25	EST	MEYLAN,WM ET AL. (1996)
logP -0.28			EST	MEYLAN,WM & HOWARD,PH (1995)
VP 2.70E-001	mm Hg	25	EST	NEELY,WB & BLAU,GE (1985)
DC	pKa			
HL 2.36E-006	atm m3/mol	25	EST	MEYLAN,WM & HOWARD,PH (1991)
OH 2.69E-012	cm3/molc sec	25	EST	MEYLAN,WM & HOWARD,PH (1993)

CAS #:	001120-85-0	1-BUTYLAZIRIDINE

Formula:	$C_6H_{13}N$
Mol Weight:	99.18
MP (deg C):	FP (deg C):
BP (deg C):	
BP pressure (mm Hg):	

Property/Value	Units	Temp	Data Type	Reference
WS 7.64E+004	mg/L	25	EST	MEYLAN,WM ET AL. (1996)
logP 1.40			EST	MEYLAN,WM & HOWARD,PH (1995)
VP 2.31E+001	mm Hg	25	EST	NEELY,WB & BLAU,GE (1985)
DC 7.86	pKa	25	EXP	PERRIN,DD (1965)
HL 3.77E-005	atm m3/mol	25	EST	MEYLAN,WM & HOWARD,PH (1991)
OH 1.98E-011	cm3/molc sec	25	EST	MEYLAN,WM & HOWARD,PH (1993)

CAS #:	001120-87-2	4-BROMOPYRIDINE

Formula:	C_5H_4BrN
Mol Weight:	158.00
MP (deg C): 0.5	FP (deg C):
BP (deg C): 28-30	
BP pressure (mm Hg): 4.00E-001	

Property/Value	Units	Temp	Data Type	Reference
WS 3.38E+003	mg/L	25	EST	MEYLAN,WM ET AL. (1996)
logP 1.54			EXP	HANSCH,C & LEO,AJ (1985)
VP 1.39E+000	mm Hg	25	EST	NEELY,WB & BLAU,GE (1985)
DC 3.78	pKa	20	EXP	PERRIN,DD (1965)
HL 2.81E-006	atm m3/mol	25	EST	MEYLAN,WM & HOWARD,PH (1991)
OH 2.33E-013	cm3/molc sec	25	EST	MEYLAN,WM & HOWARD,PH (1993)

CAS #:	001120-88-3	PYRIDAZINE, 4-METHYL-

Formula:	$C_5H_6N_2$
Mol Weight:	94.12
MP (deg C):	FP (deg C):
BP (deg C): 98-100	
BP pressure (mm Hg): 1.10E+001	

Property/Value	Units	Temp	Data Type	Reference
WS 2.28E+005	mg/L	25	EST	MEYLAN,WM ET AL. (1996)
logP -0.32			EXP	YAMAGAMI,C ET AL. (1990)
VP 1.48E-001	mm Hg	25	EST	NEELY,WB & BLAU,GE (1985)
DC 2.92	pKa		EXP	PERRIN,DD (1965)
HL 3.17E-006	atm m3/mol	25	EST	MEYLAN,WM & HOWARD,PH (1991)
OH 7.36E-013	cm3/molc sec	25	EST	MEYLAN,WM & HOWARD,PH (1993)

CAS #:	001120-90-7	3-IODOPYRIDINE

Formula:	C_5H_4IN
Mol Weight:	205.00
MP (deg C):	FP (deg C):
BP (deg C):	
BP pressure (mm Hg):	

Property/Value	Units	Temp	Data Type	Reference
WS 1.20E+003	mg/L	25	EST	MEYLAN,WM ET AL. (1996)
logP 1.80			EXP	HANSCH,C & LEO,AJ (1985)
VP 2.88E-001	mm Hg	25	EST	NEELY,WB & BLAU,GE (1985)
DC 3.25	pKa	25	EXP	PERRIN,DD (1965)
HL 1.63E-006	atm m3/mol	25	EST	MEYLAN,WM & HOWARD,PH (1991)
OH 2.44E-013	cm3/molc sec	25	EST	MEYLAN,WM & HOWARD,PH (1993)

CAS #:	001120-95-2	PYRIDAZINE, 3-CHLORO-

Formula:	$C_4H_3ClN_2$
Mol Weight:	114.53
MP (deg C):	FP (deg C):
BP (deg C):	
BP pressure (mm Hg):	

Property/Value	Units	Temp	Data Type	Reference
WS 8.62E+004	mg/L	25	EST	MEYLAN,WM ET AL. (1996)
logP 0.10			EXP	YAMAGAMI,C ET AL. (1990)
VP 7.35E-002	mm Hg	25	EST	NEELY,WB & BLAU,GE (1985)
DC	pKa			
HL 1.34E-004	atm m3/mol	25	EST	MEYLAN,WM & HOWARD,PH (1991)
OH 1.62E-013	cm3/molc sec	25	EST	MEYLAN,WM & HOWARD,PH (1993)

CAS #:	001120-97-4	4-METHYLDIOXANE

Formula:	$C_5H_{10}O_2$
Mol Weight:	102.13
MP (deg C): -44.5	FP (deg C):
BP (deg C): 114	
BP pressure (mm Hg):	

Property/Value	Units	Temp	Data Type	Reference
WS 3.53E+004	mg/L	25	EST	MEYLAN,WM ET AL. (1996)
logP 0.60			EST	MEYLAN,WM & HOWARD,PH (1995)
VP 1.92E+001	mm Hg	25	EST	NEELY,WB & BLAU,GE (1985)
DC	pKa			
HL 3.93E-005	atm m3/mol	25	EST	MEYLAN,WM & HOWARD,PH (1991)
OH 1.13E-011	cm3/molc sec	25	EXP	ATKINSON,R (1989)

CAS #:	001121-07-9	N-METHYLSUCCINIMIDE

Formula:	$C_5H_7NO_2$
Mol Weight:	113.12
MP (deg C): 71	FP (deg C):
BP (deg C): 234	
BP pressure (mm Hg):	

Property/Value	Units	Temp	Data Type	Reference
WS 6.88E+005	mg/L	25	EST	MEYLAN,WM ET AL. (1996)
logP -0.95			EXP	HANSCH,C ET AL. (1995)
VP 1.92E-004	mm Hg	25	EST	NEELY,WB & BLAU,GE (1985)
DC	pKa			
HL 5.56E-008	atm m3/mol	25	EST	MEYLAN,WM & HOWARD,PH (1991)
OH 1.99E-011	cm3/molc sec	25	EST	MEYLAN,WM & HOWARD,PH (1993)

CAS #: 001121-21-7				1,2-DICHLOROCYCLOHEXANE -CIS

Formula: $C_6H_{10}Cl_2$

Mol Weight: 153.05

MP (deg C):
FP (deg C):

BP (deg C):

BP pressure (mm Hg):

Property/Value	Units	Temp	Data Type	Reference
WS 1.42E+002	mg/L	25	EST	MEYLAN,WM ET AL. (1996)
logP 3.18			EXP	HANSCH,C & LEO,AJ (1985)
VP 1.32E+000	mm Hg	25	EST	NEELY,WB & BLAU,GE (1985)
DC	pKa			
HL 1.66E-002	atm m3/mol	25	EST	MEYLAN,WM & HOWARD,PH (1991)
OH 4.31E-012	cm3/molc sec	25	EST	MEYLAN,WM & HOWARD,PH (1993)

CAS #: 001121-60-4				2-PYRIDINECARBOXYALDEHYDE

Formula: C_6H_5NO

Mol Weight: 107.11

MP (deg C):
FP (deg C):

BP (deg C): 180

BP pressure (mm Hg):

Property/Value	Units	Temp	Data Type	Reference
WS 4.68E+004	mg/L	25	EST	MEYLAN,WM ET AL. (1996)
logP 0.44			EXP	HANSCH,C ET AL. (1995)
VP 5.68E-001	mm Hg	25	EST	NEELY,WB & BLAU,GE (1985)
DC 3.80	pKa	20	EXP	PERRIN,DD (1965)
HL 1.76E-008	atm m3/mol	25	EST	MEYLAN,WM & HOWARD,PH (1991)
OH 1.71E-011	cm3/molc sec	25	EST	MEYLAN,WM & HOWARD,PH (1993)

CAS #: 001122-54-9				4-ACETYLPYRIDINE

Formula: C_7H_7NO

Mol Weight: 121.14

MP (deg C): 16
FP (deg C):

BP (deg C): 212

BP pressure (mm Hg):

Property/Value	Units	Temp	Data Type	Reference
WS 3.87E+004	mg/L	25	EST	MEYLAN,WM ET AL. (1996)
logP 0.48			EXP	HANSCH,C & LEO,AJ (1985)
VP 3.74E-001	mm Hg	25	EST	NEELY,WB & BLAU,GE (1985)
DC 3.59	pKa	17	EXP	PERRIN,DD (1972)
HL 1.28E-008	atm m3/mol	25	EST	MEYLAN,WM & HOWARD,PH (1991)
OH 4.39E-013	cm3/molc sec	25	EST	MEYLAN,WM & HOWARD,PH (1993)

CAS #: 001122-58-3				4-DIMETHYLAMINOPYRIDINE

Formula: $C_7H_{10}N_2$

Mol Weight: 122.17

MP (deg C): 112-114
FP (deg C):

BP (deg C):

BP pressure (mm Hg):

Property/Value	Units	Temp	Data Type	Reference
WS 7.07E+003	mg/L	25	EST	MEYLAN,WM ET AL. (1996)
logP 1.34			EXP	HANSCH,C & LEO,AJ (1985)
VP 1.00E+000	mm Hg	25	EST	NEELY,WB & BLAU,GE (1985)
DC 10.14	pKa	5	EXP	PERRIN,DD (1965)
HL 1.12E-007	atm m3/mol	25	EST	MEYLAN,WM & HOWARD,PH (1991)
OH 7.27E-011	cm3/molc sec	25	EST	MEYLAN,WM & HOWARD,PH (1993)

CAS #: 001122-61-8				4-NITROPYRIDINE

Formula: $C_5H_4N_2O_2$

Mol Weight: 124.10

MP (deg C):
FP (deg C):

BP (deg C):

BP pressure (mm Hg):

Property/Value	Units	Temp	Data Type	Reference
WS 2.06E+004	mg/L	25	EST	MEYLAN,WM ET AL. (1996)
logP 0.33			EXP	HANSCH,C & LEO,AJ (1985)
VP 1.15E-001	mm Hg	25	EST	NEELY,WB & BLAU,GE (1985)
DC	pKa			
HL 2.78E-008	atm m3/mol	25	EST	MEYLAN,WM & HOWARD,PH (1991)
OH 4.62E-014	cm3/molc sec	25	EST	MEYLAN,WM & HOWARD,PH (1993)

CAS #: 001122-62-9				2-ACETYLPYRIDINE

Formula: C_7H_7NO

Mol Weight: 121.14

MP (deg C):
FP (deg C):

BP (deg C): 192

BP pressure (mm Hg):

Property/Value	Units	Temp	Data Type	Reference
WS 1.87E+004	mg/L	25	EST	MEYLAN,WM ET AL. (1996)
logP 0.85			EXP	HANSCH,C & LEO,AJ (1985)
VP 3.74E-001	mm Hg	25	EST	NEELY,WB & BLAU,GE (1985)
DC 2.73	pKa	17	EXP	PERRIN,DD (1972)
HL 1.28E-008	atm m3/mol	25	EST	MEYLAN,WM & HOWARD,PH (1991)
OH 4.39E-013	cm3/molc sec	25	EST	MEYLAN,WM & HOWARD,PH (1993)

CAS #: 001122-81-2				4-PROPYLPYRIDINE

Formula: $C_8H_{11}N$

Mol Weight: 121.18

MP (deg C):
FP (deg C):

BP (deg C): 185

BP pressure (mm Hg):

Property/Value	Units	Temp	Data Type	Reference
WS 3.19E+004	mg/L	25	EST	MEYLAN,WM ET AL. (1996)
logP 2.10			EXP	HANSCH,C & LEO,AJ (1985)
VP 9.73E-001	mm Hg	25	EST	NEELY,WB & BLAU,GE (1985)
DC 6.05	pKa	25	EXP	PERRIN,DD (1972)
HL 1.37E-005	atm m3/mol	25	EST	MEYLAN,WM & HOWARD,PH (1991)
OH 3.38E-012	cm3/molc sec	25	EST	MEYLAN,WM & HOWARD,PH (1993)

CAS #: 001123-54-2				8-AZAADENINE

Formula: $C_4H_4N_6$

Mol Weight: 136.12

MP (deg C): > 300
FP (deg C):

BP (deg C):

BP pressure (mm Hg):

Property/Value	Units	Temp	Data Type	Reference
WS 2.81E+004	mg/L	25	EST	MEYLAN,WM ET AL. (1996)
logP -0.96			EXP	HANSCH,C & LEO,AJ (1985)
VP 5.50E-006	mm Hg	25	EST	NEELY,WB & BLAU,GE (1985)
DC	pKa			
HL 2.81E-014	atm m3/mol	25	EST	MEYLAN,WM & HOWARD,PH (1991)
OH 5.52E-011	cm3/molc sec	25	EST	MEYLAN,WM & HOWARD,PH (1993)

CAS #: 001124-11-4 — PYRAZINE, TETRAMETHYL-

Formula: $C_8H_{12}N_2$

Mol Weight: 136.20

MP (deg C): 84-86

FP (deg C):

BP (deg C): 190

BP pressure (mm Hg):

Property/Value	Units	Temp	Data Type	Reference
WS 7.01E+003	mg/L	25	EST	MEYLAN,WM ET AL. (1996)
logP 1.28			EXP	YAMAGAMI,C ET AL. (1991)
VP 1.69E-001	mm Hg	25	EST	NEELY,WB & BLAU,GE (1985)
DC 3.55	pKa	27	EXP	PERRIN,DD (1965)
HL 4.33E-006	atm m3/mol	25	EST	MEYLAN,WM & HOWARD,PH (1991)
OH 2.90E-012	cm3/molc sec	25	EST	MEYLAN,WM & HOWARD,PH (1993)

CAS #: 001124-14-7 — 1-DODECENE

Formula: $C_{12}H_{24}$

Mol Weight: 168.33

MP (deg C):

FP (deg C):

BP (deg C):

BP pressure (mm Hg):

Property/Value	Units	Temp	Data Type	Reference
WS 1.13E-001	mg/L	25	EST	MEYLAN,WM ET AL. (1996)
logP 6.10			EST	MEYLAN,WM & HOWARD,PH (1995)
VP 4.23E-001	mm Hg	25	EST	NEELY,WB & BLAU,GE (1985)
DC	pKa			
HL 1.96E+000	atm m3/mol	25	EST	MEYLAN,WM & HOWARD,PH (1991)
OH 3.87E-011	cm3/molc sec	25	EST	MEYLAN,WM & HOWARD,PH (1993)

CAS #: 001124-33-0 — 4-NITROPYRIDINE-1-OXIDE

Formula: $C_5H_6N_2O_3$

Mol Weight: 142.12

MP (deg C): 160.5

FP (deg C):

BP (deg C):

BP pressure (mm Hg):

Property/Value	Units	Temp	Data Type	Reference
WS 1.01E+005	mg/L	25	EST	MEYLAN,WM ET AL. (1996)
logP -0.55			EXP	HANSCH,C ET AL. (1995)
VP 5.49E-004	mm Hg	25	EST	NEELY,WB & BLAU,GE (1985)
DC	pKa			
HL 2.78E-013	atm m3/mol	25	EST	MEYLAN,WM & HOWARD,PH (1991)
OH 4.62E-014	cm3/molc sec	25	EST	MEYLAN,WM & HOWARD,PH (1993)

CAS #: 001125-78-6 — 2-NAPHTHALENOL, 5,6,7,8-TETRAHYDRO-

Formula: $C_{10}H_{12}O$

Mol Weight: 148.21

MP (deg C): 57

FP (deg C):

BP (deg C): 275.5

BP pressure (mm Hg):

Property/Value	Units	Temp	Data Type	Reference
WS 1.50E+003	mg/L	20	EXP	YALKOWSKY,SH & DANNENFELSER,RM (1992)
logP 2.90			EXP	SANGSTER,J (1993)
VP 3.72E-003	mm Hg	25	EST	NEELY,WB & BLAU,GE (1985)
DC 10.48	pKa	25	EXP	KORTUM,G ET AL (1961)
HL 5.31E-007	atm m3/mol	25	EST	MEYLAN,WM & HOWARD,PH (1991)
OH 1.06E-010	cm3/molc sec	25	EST	MEYLAN,WM & HOWARD,PH (1993)

CAS #: 001125-80-0 — ISOQUINOLONE-3-METHYL

Formula: $C_{10}H_9N$

Mol Weight: 143.19

MP (deg C): 68

FP (deg C):

BP (deg C): 249

BP pressure (mm Hg):

Property/Value	Units	Temp	Data Type	Reference
WS 4.18E+002	mg/L	25	EST	MEYLAN,WM ET AL. (1996)
logP 2.68			EXP	HANSCH,C ET AL. (1995)
VP 9.52E-003	mm Hg	25	EST	NEELY,WB & BLAU,GE (1985)
DC 5.64	pKa	25	EXP	PERRIN,DD (1965)
HL 7.60E-007	atm m3/mol	25	EST	MEYLAN,WM & HOWARD,PH (1991)
OH 3.77E-011	cm3/molc sec	25	EST	MEYLAN,WM & HOWARD,PH (1993)

CAS #: 001126-00-7 — 1H-PYRAZOLE, 1-PHENYL-

Formula: $C_9H_8N_2$

Mol Weight: 144.18

MP (deg C):

FP (deg C):

BP (deg C): 141-142

BP pressure (mm Hg): 3.00E+001

Property/Value	Units	Temp	Data Type	Reference
WS 1.06E+003	mg/L	25	EST	MEYLAN,WM ET AL. (1996)
logP 2.20			EXP	HANSCH,C ET AL. (1995)
VP 1.02E-002	mm Hg	25	EST	NEELY,WB & BLAU,GE (1985)
DC	pKa			
HL 1.34E-008	atm m3/mol	25	EST	MEYLAN,WM & HOWARD,PH (1991)
OH 3.94E-011	cm3/molc sec	25	EST	MEYLAN,WM & HOWARD,PH (1993)

CAS #: 001126-34-7 — BENZENESULFONIC ACID, 3-AMINO-, MONOSODIUM SALT

Formula: $C_6H_6NNaO_3S$

Mol Weight: 195.17

MP (deg C):

FP (deg C):

BP (deg C):

BP pressure (mm Hg):

Property/Value	Units	Temp	Data Type	Reference
WS 1.00E+006	mg/L	25	EST	MEYLAN,WM ET AL. (1996)
logP -3.40			EXP	SANGSTER,J (1994)
VP 2.81E-012	mm Hg	25	EST	NEELY,WB & BLAU,GE (1985)
DC	pKa			
HL	atm m3/mol			
OH 2.30E-011	cm3/molc sec	25	EST	MEYLAN,WM & HOWARD,PH (1993)

CAS #: 001126-46-1 — METHYL P-CHLOROBENZOATE

Formula: $C_8H_7ClO_2$

Mol Weight: 170.60

MP (deg C): 43.5

FP (deg C):

BP (deg C):

BP pressure (mm Hg):

Property/Value	Units	Temp	Data Type	Reference
WS 2.16E+002	mg/L	25	EST	MEYLAN,WM ET AL. (1996)
logP 2.87			EXP	PAGOU,M & KOUTSELINIS,A (1993)
VP 9.07E-002	mm Hg	25	EST	NEELY,WB & BLAU,GE (1985)
DC	pKa			
HL 2.57E-005	atm m3/mol	25	EST	MEYLAN,WM & HOWARD,PH (1991)
OH 6.58E-013	cm3/molc sec	25	EST	MEYLAN,WM & HOWARD,PH (1993)

CAS #: 001126-78-9	N-BUTYLANILINE

Formula: $C_{10}H_{15}N$
Mol Weight: 149.24
MP (deg C): -14.4
FP (deg C):
BP (deg C): 243.5
BP pressure (mm Hg):

Property/Value	Units	Temp	Data Type	Reference
WS 6.70E+001	mg/L	25	EST	MEYLAN,WM ET AL. (1996)
logP 3.58			EXP	HANSCH,C & LEO,AJ (1985)
VP 9.05E-002	mm Hg	25	EST	NEELY,WB & BLAU,GE (1985)
DC 5.12	pKa	25	EXP	PERRIN,DD (1965)
HL 9.78E-006	atm m3/mol	25	EST	MEYLAN,WM & HOWARD,PH (1991)
OH 5.61E-011	cm3/molc sec	25	EST	MEYLAN,WM & HOWARD,PH (1993)

CAS #: 001126-79-0	BUTYL PHENYL ETHER

Formula: $C_{10}H_{14}O$
Mol Weight: 150.22
MP (deg C): -19.4
FP (deg C):
BP (deg C): 210
BP pressure (mm Hg):

Property/Value	Units	Temp	Data Type	Reference
WS 7.08E+001	mg/L	25	EST	MEYLAN,WM ET AL. (1996)
logP 3.55			EST	MEYLAN,WM & HOWARD,PH (1995)
VP 2.96E-001	mm Hg	25	EXT	BOUBLIK,T ET AL. (1983)
DC	pKa			
HL 7.46E-004	atm m3/mol	25	EST	MEYLAN,WM & HOWARD,PH (1991)
OH 3.38E-011	cm3/molc sec	25	EST	MEYLAN,WM & HOWARD,PH (1993)

CAS #: 001127-76-0	1-ETHYLNAPHTHALENE

Formula: $C_{12}H_{12}$
Mol Weight: 156.23
MP (deg C): -13.9
FP (deg C):
BP (deg C): 258.6
BP pressure (mm Hg):

Property/Value	Units	Temp	Data Type	Reference
WS 1.07E+001	mg/L	25	EXP	YALKOWSKY,SH & DANNENFELSER,RM (1992)
logP 4.40			EXP	SANGSTER,J (1993)
VP 2.52E-002	mm Hg	25	EXP	YAWS,CL (1994B)
DC	pKa			
HL 4.62E-004	atm m3/mol	25	EXP	SCHWARZ AND WASIK (1977)
OH 5.51E-011	cm3/molc sec	25	EST	MEYLAN,WM & HOWARD,PH (1993)

CAS #: 001128-05-8	3-ACETYL BENZOTHIOPHENE

Formula: $C_{10}H_8OS$
Mol Weight: 176.24
MP (deg C):
FP (deg C):
BP (deg C):
BP pressure (mm Hg):

Property/Value	Units	Temp	Data Type	Reference
WS 2.43E+002	mg/L	25	EST	MEYLAN,WM ET AL. (1996)
logP 2.78			EXP	HANSCH,C & LEO,AJ (1985)
VP 3.93E-004	mm Hg	25	EST	NEELY,WB & BLAU,GE (1985)
DC	pKa			
HL 5.20E-007	atm m3/mol	25	EST	MEYLAN,WM & HOWARD,PH (1991)
OH 2.74E-011	cm3/molc sec	25	EST	MEYLAN,WM & HOWARD,PH (1993)

CAS #: 001128-16-1	3,5-DICHLORO-2-(TRICHLOROMETHYL)PYRIDINE

Formula: $C_6H_2Cl_5N$
Mol Weight: 265.35
MP (deg C):
FP (deg C):
BP (deg C):
BP pressure (mm Hg):

Property/Value	Units	Temp	Data Type	Reference
WS 7.50E+000	mg/L	25	EST	MEYLAN,WM ET AL. (1996)
logP 4.00			EST	MEYLAN,WM & HOWARD,PH (1995)
VP 2.88E-003	mm Hg	25	EST	NEELY,WB & BLAU,GE (1985)
DC	pKa			
HL 1.87E-007	atm m3/mol	25	EST	MEYLAN,WM & HOWARD,PH (1991)
OH 3.36E-014	cm3/molc sec	25	EST	MEYLAN,WM & HOWARD,PH (1993)

CAS #: 001128-78-5	N-METHYL-O-TOLYLCARBAMATE

Formula: $C_9H_{11}NO_2$
Mol Weight: 165.19
MP (deg C):
FP (deg C):
BP (deg C):
BP pressure (mm Hg):

Property/Value	Units	Temp	Data Type	Reference
WS 3.67E+003	mg/L	25	EST	MEYLAN,WM ET AL. (1996)
logP 1.46			EXP	HANSCH,C & LEO,AJ (1985)
VP 1.87E-002	mm Hg	25	EST	NEELY,WB & BLAU,GE (1985)
DC	pKa			
HL 3.54E-008	atm m3/mol	25	EST	MEYLAN,WM & HOWARD,PH (1991)
OH 9.93E-012	cm3/molc sec	25	EST	MEYLAN,WM & HOWARD,PH (1993)

CAS #: 001129-26-6	P-METHOXYBENZENESULFONAMIDE

Formula: $C_7H_9NO_3S$
Mol Weight: 187.22
MP (deg C):
FP (deg C):
BP (deg C):
BP pressure (mm Hg):

Property/Value	Units	Temp	Data Type	Reference
WS 2.01E+004	mg/L	25	EST	MEYLAN,WM ET AL. (1996)
logP 0.47			EXP	HANSCH,C & LEO,AJ (1985)
VP 1.07E-004	mm Hg	25	EST	NEELY,WB & BLAU,GE (1985)
DC	pKa			
HL 2.50E-008	atm m3/mol	25	EST	MEYLAN,WM & HOWARD,PH (1991)
OH 5.43E-012	cm3/molc sec	25	EST	MEYLAN,WM & HOWARD,PH (1993)

CAS #: 001129-37-9	P-NITROBENZALDOXIME

Formula: $C_7H_6N_2O_3$
Mol Weight: 166.14
MP (deg C): 126-130
FP (deg C):
BP (deg C):
BP pressure (mm Hg):

Property/Value	Units	Temp	Data Type	Reference
WS 5.65E+002	mg/L	25	EST	MEYLAN,WM ET AL. (1996)
logP 1.95			EXP	HANSCH,C ET AL. (1995)
VP 7.47E-006	mm Hg	25	EST	NEELY,WB & BLAU,GE (1985)
DC	pKa			
HL 1.41E-009	atm m3/mol	25	EST	MEYLAN,WM & HOWARD,PH (1991)
OH 2.52E-012	cm3/molc sec	25	EST	MEYLAN,WM & HOWARD,PH (1993)

001129-41-5 — N-METHYL-M-TOLYLCARBAMATE

Formula: $C_9H_{11}NO_2$

Mol Weight: 165.19

MP (deg C): FP (deg C):

BP (deg C):

BP pressure (mm Hg):

Property/Value	Units	Temp	Data Type	Reference
WS 2.60E+003	mg/L	30	EXP	YALKOWSKY,SH & DANNENFELSER,RM (1992)
logP 1.70			EXP	HANSCH,C & LEO,AJ (1985)
VP 1.87E-002	mm Hg	25	EST	NEELY,WB & BLAU,GE (1985)
DC	pKa			
HL 3.54E-008	atm m3/mol	25	EST	MEYLAN,WM & HOWARD,PH (1991)
OH 1.14E-011	cm3/molc sec	25	EST	MEYLAN,WM & HOWARD,PH (1993)

001129-48-2 — N-METHYL-P-TOLYLCARBAMATE

Formula: $C_9H_{11}NO_2$

Mol Weight: 165.19

MP (deg C): FP (deg C):

BP (deg C):

BP pressure (mm Hg):

Property/Value	Units	Temp	Data Type	Reference
WS 2.48E+003	mg/L	25	EST	MEYLAN,WM ET AL. (1996)
logP 1.66			EXP	HANSCH,C & LEO,AJ (1985)
VP 1.87E-002	mm Hg	25	EST	NEELY,WB & BLAU,GE (1985)
DC	pKa			
HL 3.54E-008	atm m3/mol	25	EST	MEYLAN,WM & HOWARD,PH (1991)
OH 9.93E-012	cm3/molc sec	25	EST	MEYLAN,WM & HOWARD,PH (1993)

001129-50-6 — BUTYRANILIDE

Formula: $C_{10}H_{13}NO$

Mol Weight: 163.22

MP (deg C): 97 FP (deg C):

BP (deg C): 189

BP pressure (mm Hg): 1.50E+001

Property/Value	Units	Temp	Data Type	Reference
WS 1.17E+003	mg/L	25	EST	MEYLAN,WM ET AL. (1996)
logP 2.05			EXP	HANSCH,C & LEO,AJ (1985)
VP 7.84E-005	mm Hg	25	EST	NEELY,WB & BLAU,GE (1985)
DC	pKa			
HL 1.09E-008	atm m3/mol	25	EST	MEYLAN,WM & HOWARD,PH (1991)
OH 1.71E-011	cm3/molc sec	25	EST	MEYLAN,WM & HOWARD,PH (1993)

001131-60-8 — P-CYCLOHEXYLPHENOL

Formula: $C_{12}H_{16}O$

Mol Weight: 176.26

MP (deg C): 133 FP (deg C):

BP (deg C): 294

BP pressure (mm Hg):

Property/Value	Units	Temp	Data Type	Reference
WS 5.43E+001	mg/L	25	EST	MEYLAN,WM ET AL. (1996)
logP 4.22			EXP	HANSCH,C & LEO,AJ (1985)
VP 5.02E-004	mm Hg	25	EST	NEELY,WB & BLAU,GE (1985)
DC	pKa			
HL 1.12E-006	atm m3/mol	25	EST	MEYLAN,WM & HOWARD,PH (1991)
OH 5.02E-011	cm3/molc sec	25	EST	MEYLAN,WM & HOWARD,PH (1993)

001131-64-2 — 2(1H)-ISOQUINOLINECARBOXIMIDAMIDE, 3,4-DIHYDRO-

Formula: $C_{10}H_{13}N_3$

Mol Weight: 175.24

MP (deg C): 278-280 FP (deg C):

BP (deg C):

BP pressure (mm Hg):

Property/Value	Units	Temp	Data Type	Reference
WS 1.33E+004	mg/L	25	EST	MEYLAN,WM ET AL. (1996)
logP 0.75			EXP	SANGSTER,J (1994)
VP 1.33E-004	mm Hg	25	EST	NEELY,WB & BLAU,GE (1985)
DC	pKa			
HL 4.43E-012	atm m3/mol	25	EST	MEYLAN,WM & HOWARD,PH (1991)
OH 1.13E-010	cm3/molc sec	25	EST	MEYLAN,WM & HOWARD,PH (1993)

001132-18-9 — P-PROPYLBENZENESULFONAMIDE

Formula: $C_9H_{13}NO_2S$

Mol Weight: 199.27

MP (deg C): FP (deg C):

BP (deg C):

BP pressure (mm Hg):

Property/Value	Units	Temp	Data Type	Reference
WS 1.76E+003	mg/L	25	EST	MEYLAN,WM ET AL. (1996)
logP 1.64			EXP	HANSCH,C & LEO,AJ (1985)
VP 4.82E-005	mm Hg	25	EST	NEELY,WB & BLAU,GE (1985)
DC	pKa			
HL 8.21E-007	atm m3/mol	25	EST	MEYLAN,WM & HOWARD,PH (1991)
OH 3.50E-012	cm3/molc sec	25	EST	MEYLAN,WM & HOWARD,PH (1993)

001132-21-4 — 3,5-DIMETHOXYBENZOIC ACID

Formula: $C_9H_{10}O_4$

Mol Weight: 182.18

MP (deg C): 185.5 FP (deg C):

BP (deg C):

BP pressure (mm Hg):

Property/Value	Units	Temp	Data Type	Reference
WS 7.24E+002	mg/L	25	EST	MEYLAN,WM ET AL. (1996)
logP 2.19			EXP	HANSCH,C & LEO,AJ (1985)
VP 1.38E-007	mm Hg	25	EXT	COLOMINA,M ET AL. (1985)
DC 3.97	pKa	25	EXP	SERJEANT,EP & DEMPSEY,B (1979)
HL 3.79E-010	atm m3/mol	25	EST	MEYLAN,WM & HOWARD,PH (1991)
OH 6.69E-011	cm3/molc sec	25	EST	MEYLAN,WM & HOWARD,PH (1993)

001133-80-8 — 2-BROMO-9H-FLUORENE

Formula: $C_{13}H_9Br$

Mol Weight: 245.13

MP (deg C): 113.5 FP (deg C):

BP (deg C): 185

BP pressure (mm Hg): 1.35E+002

Property/Value	Units	Temp	Data Type	Reference
WS 1.26E-001	mg/L	25	EST	MEYLAN,WM ET AL. (1996)
logP 4.91			EST	MEYLAN,WM & HOWARD,PH (1995)
VP 7.51E-005	mm Hg	25	EST	NEELY,WB & BLAU,GE (1985)
DC	pKa			
HL 6.67E-005	atm m3/mol	25	EST	MEYLAN,WM & HOWARD,PH (1991)
OH 5.98E-012	cm3/molc sec	25	EST	MEYLAN,WM & HOWARD,PH (1993)

CAS #: 001134-23-2 — S-ETHYL CYCLOHEXYLETHYLCARBAMOTHIOATE

Formula: $C_{11}H_{21}NOS$

Mol Weight: 215.36

MP (deg C): 11.5
FP (deg C):
BP (deg C): 145
BP pressure (mm Hg):

Property/Value	Units	Temp	Data Type	Reference
WS 8.50E+001	mg/L	22	EXP	SHIU,WY ET AL. (1990)
logP 3.88			EXP	TOMLIN,C (1994)
VP 2.00E-003	mm Hg	25	EXP	KOREN,E ET AL. (1980)
DC	pKa			
HL 6.70E-006	atm m3/mol	25	EST	VP/WSOL
OH 3.90E-011	cm3/molc sec	25	EST	MEYLAN,WM & HOWARD,PH (1993)

CAS #: 001134-47-0 — BENZENEPROPANOIC ACID, á-(AMINOMETHYL)-4-CHLORO-

Formula: $C_{10}H_{12}ClNO_2$

Mol Weight: 213.67

MP (deg C): 206-208
FP (deg C):
BP (deg C):
BP pressure (mm Hg):

Property/Value	Units	Temp	Data Type	Reference
WS 2.09E+003	mg/L	25	EST	MEYLAN,WM ET AL. (1996)
logP -0.96			EXP	HANSCH,C ET AL. (1995)
VP 2.36E-009	mm Hg	25	EST	NEELY,WB & BLAU,GE (1985)
DC	pKa			
HL 5.94E-012	atm m3/mol	25	EST	MEYLAN,WM & HOWARD,PH (1991)
OH 3.77E-011	cm3/molc sec	25	EST	MEYLAN,WM & HOWARD,PH (1993)

CAS #: 001134-61-8 — 2,2'-BITHIOPHENE, 5-(3-BUTEN-1-YNYL)-

Formula: $C_{12}H_{12}S_2$

Mol Weight: 220.36

MP (deg C):
FP (deg C):
BP (deg C):
BP pressure (mm Hg):

Property/Value	Units	Temp	Data Type	Reference
WS 1.44E+000	mg/L	25	EST	MEYLAN,WM ET AL. (1996)
logP 5.15			EXP	MCLACHLAN,D ET AL. (1986)
VP 1.20E-005	mm Hg	25	EST	NEELY,WB & BLAU,GE (1985)
DC	pKa			
HL 2.16E-005	atm m3/mol	25	EST	MEYLAN,WM & HOWARD,PH (1991)
OH 8.24E-011	cm3/molc sec	25	EST	MEYLAN,WM & HOWARD,PH (1993)

CAS #: 001134-87-8 — 3-BENZYL-2,4-PENTANEDIONE

Formula: $C_{12}H_{14}O_2$

Mol Weight: 190.24

MP (deg C):
FP (deg C):
BP (deg C):
BP pressure (mm Hg):

Property/Value	Units	Temp	Data Type	Reference
WS 1.19E+003	mg/L	25	EST	MEYLAN,WM ET AL. (1996)
logP 1.89			EXP	HANSCH,C & LEO,AJ (1985)
VP 2.03E-003	mm Hg	25	EST	NEELY,WB & BLAU,GE (1985)
DC	pKa			
HL 3.57E-009	atm m3/mol	25	EST	MEYLAN,WM & HOWARD,PH (1991)
OH 1.00E-011	cm3/molc sec	25	EST	MEYLAN,WM & HOWARD,PH (1993)

CAS #: 001135-00-8 — P-BUTYLBENZENESULFONAMIDE

Formula: $C_{10}H_{15}NO_2S$

Mol Weight: 213.30

MP (deg C):
FP (deg C):
BP (deg C):
BP pressure (mm Hg):

Property/Value	Units	Temp	Data Type	Reference
WS 3.02E+002	mg/L	25	EST	MEYLAN,WM ET AL. (1996)
logP 2.45			EXP	HANSCH,C & LEO,AJ (1985)
VP 2.05E-005	mm Hg	25	EST	NEELY,WB & BLAU,GE (1985)
DC	pKa			
HL 1.09E-006	atm m3/mol	25	EST	MEYLAN,WM & HOWARD,PH (1991)
OH 4.91E-012	cm3/molc sec	25	EST	MEYLAN,WM & HOWARD,PH (1993)

CAS #: 001135-24-6 — CINNAMIC ACID, 3-METHOXY-4-HYDROXY

Formula: $C_{10}H_{10}O_4$

Mol Weight: 194.19

MP (deg C):
FP (deg C):
BP (deg C):
BP pressure (mm Hg):

Property/Value	Units	Temp	Data Type	Reference
WS 5.97E+003	mg/L	25	EST	MEYLAN,WM ET AL. (1996)
logP 1.51			EXP	HANSCH,C ET AL. (1995)
VP 2.69E-006	mm Hg	25	EST	NEELY,WB & BLAU,GE (1985)
DC 4.58	pKa	25	EXP	SERJEANT,EP & DEMPSEY,B (1979)
HL 7.96E-014	atm m3/mol	25	EST	MEYLAN,WM & HOWARD,PH (1991)
OH 4.83E-011	cm3/molc sec	25	EST	MEYLAN,WM & HOWARD,PH (1993)

CAS #: 001135-43-9 — N-METHYL-4-ACETYLPHENYLCARBAMATE

Formula: $C_{10}H_{11}NO_3$

Mol Weight: 193.20

MP (deg C):
FP (deg C):
BP (deg C):
BP pressure (mm Hg):

Property/Value	Units	Temp	Data Type	Reference
WS 6.50E+003	mg/L	25	EST	MEYLAN,WM ET AL. (1996)
logP 1.01			EXP	HANSCH,C & LEO,AJ (1985)
VP 6.98E-004	mm Hg	25	EST	NEELY,WB & BLAU,GE (1985)
DC	pKa			
HL 5.85E-011	atm m3/mol	25	EST	MEYLAN,WM & HOWARD,PH (1991)
OH 7.99E-012	cm3/molc sec	25	EST	MEYLAN,WM & HOWARD,PH (1993)

CAS #: 001135-58-6 — UREA, N,N''-1,3-PHENYLENEBIS-

Formula: $C_8H_{10}N_4O_2$

Mol Weight: 194.19

MP (deg C):
FP (deg C):
BP (deg C):
BP pressure (mm Hg):

Property/Value	Units	Temp	Data Type	Reference
WS 1.50E+004	mg/L	25	EST	MEYLAN,WM ET AL. (1996)
logP 0.58			EXP	HANSCH,C ET AL. (1995)
VP 1.55E-007	mm Hg	25	EST	NEELY,WB & BLAU,GE (1985)
DC	pKa			
HL 4.75E-021	atm m3/mol	25	EST	MEYLAN,WM & HOWARD,PH (1991)
OH 2.02E-010	cm3/molc sec	25	EST	MEYLAN,WM & HOWARD,PH (1993)

CAS #: 001135-61-1 — BARBITURIC ACID,5ET-5IBU

Formula: $C_{10}H_{16}N_2O_3$

Mol Weight: 212.25

MP (deg C): FP (deg C):

BP (deg C):

BP pressure (mm Hg):

Property/Value	Units	Temp	Data Type	Reference
WS 1.36E+003	mg/L	25	EST	MEYLAN,WM ET AL. (1996)
logP 1.69			EXP	SANGSTER,J (1993)
VP 4.99E-011	mm Hg	25	EST	NEELY,WB & BLAU,GE (1985)
DC	pKa			
HL 6.36E-013	atm m3/mol	25	EST	MEYLAN,WM & HOWARD,PH (1991)
OH 1.66E-011	cm3/molc sec	25	EST	MEYLAN,WM & HOWARD,PH (1993)

CAS #: 001137-41-3 — 4-BENZOYLANILINE

Formula: $C_{13}H_{11}NO$

Mol Weight: 197.24

MP (deg C): 121-124 FP (deg C):

BP (deg C):

BP pressure (mm Hg):

Property/Value	Units	Temp	Data Type	Reference
WS 5.63E+002	mg/L	25	EST	MEYLAN,WM ET AL. (1996)
logP 2.23			EST	MEYLAN,WM & HOWARD,PH (1995)
VP 2.04E-005	mm Hg	25	EST	NEELY,WB & BLAU,GE (1985)
DC 2.24	pKa	20	EXP	PERRIN,DD (1965)
HL 6.86E-010	atm m3/mol	25	EST	MEYLAN,WM & HOWARD,PH (1991)
OH 9.99E-011	cm3/molc sec	25	EST	MEYLAN,WM & HOWARD,PH (1993)

CAS #: 001137-42-4 — P-HYDROXYBENZOPHENONE

Formula: $C_{13}H_{10}O_2$

Mol Weight: 198.22

MP (deg C): 132-135 FP (deg C):

BP (deg C):

BP pressure (mm Hg):

Property/Value	Units	Temp	Data Type	Reference
WS 4.06E+002	mg/L	25	EST	MEYLAN,WM ET AL. (1996)
logP 3.07			EXP	HANSCH,C & LEO,AJ (1985)
VP 1.00E-005	mm Hg	25	EST	NEELY,WB & BLAU,GE (1985)
DC	pKa			
HL 2.02E-010	atm m3/mol	25	EST	MEYLAN,WM & HOWARD,PH (1991)
OH 3.23E-011	cm3/molc sec	25	EST	MEYLAN,WM & HOWARD,PH (1993)

CAS #: 001137-96-8 — ALPHA-PHENYL-N-PHENYLNITRONE

Formula: $C_{13}H_{11}NO$

Mol Weight: 197.24

MP (deg C): FP (deg C):

BP (deg C):

BP pressure (mm Hg):

Property/Value	Units	Temp	Data Type	Reference
WS 1.12E+003	mg/L	25	EST	MEYLAN,WM ET AL. (1996)
logP 1.88			EXP	KIRCHNER,JJ ET AL. (1985)
VP 1.16E-008	mm Hg	25	EST	NEELY,WB & BLAU,GE (1985)
DC	pKa			
HL 7.90E-012	atm m3/mol	25	EST	MEYLAN,WM & HOWARD,PH (1991)
OH 6.98E-012	cm3/molc sec	25	EST	MEYLAN,WM & HOWARD,PH (1993)

CAS #: 001138-58-5 — BENZENESULFONAMIDE, 4-BUTOXY-

Formula: $C_{10}H_{15}NO_3S$

Mol Weight: 229.30

MP (deg C): FP (deg C):

BP (deg C):

BP pressure (mm Hg):

Property/Value	Units	Temp	Data Type	Reference
WS 5.45E+002	mg/L	25	EST	MEYLAN,WM ET AL. (1996)
logP 2.05			EXP	CAROTTI,A ET AL. (1989)
VP 8.11E-006	mm Hg	25	EST	NEELY,WB & BLAU,GE (1985)
DC	pKa			
HL 5.84E-008	atm m3/mol	25	EST	MEYLAN,WM & HOWARD,PH (1991)
OH 1.69E-011	cm3/molc sec	25	EST	MEYLAN,WM & HOWARD,PH (1993)

CAS #: 001138-60-9 — ISOPROPYL GALLATE

Formula: $C_{10}H_{12}O_5$

Mol Weight: 212.20

MP (deg C): FP (deg C):

BP (deg C):

BP pressure (mm Hg):

Property/Value	Units	Temp	Data Type	Reference
WS 6.96E+003	mg/L	25	EST	MEYLAN,WM ET AL. (1996)
logP 1.54			EXP	HANSCH,C & LEO,AJ (1985)
VP 5.03E-007	mm Hg	25	EST	NEELY,WB & BLAU,GE (1985)
DC	pKa			
HL 6.89E-017	atm m3/mol	25	EST	MEYLAN,WM & HOWARD,PH (1991)
OH 1.69E-010	cm3/molc sec	25	EST	MEYLAN,WM & HOWARD,PH (1993)

CAS #: 001141-88-4 — BENZENAMINE, 2,2'-DITHIOBIS-

Formula: $C_{12}H_{12}N_2S_2$

Mol Weight: 248.37

MP (deg C): FP (deg C):

BP (deg C):

BP pressure (mm Hg):

Property/Value	Units	Temp	Data Type	Reference
WS 6.00E+001	mg/L	25	EST	MEYLAN,WM ET AL. (1996)
logP 3.05			EXP	HANSCH,C ET AL. (1995)
VP 9.95E-008	mm Hg	25	EST	NEELY,WB & BLAU,GE (1985)
DC	pKa			
HL 6.05E-012	atm m3/mol	25	EST	MEYLAN,WM & HOWARD,PH (1991)
OH 3.57E-010	cm3/molc sec	25	EST	MEYLAN,WM & HOWARD,PH (1993)

CAS #: 001142-70-7 — BARBITURIC ACID,5-IPR-5(2BR2ALLYL)

Formula: $C_{11}H_{15}BrN_2O_3$

Mol Weight: 303.16

MP (deg C): 130-133 FP (deg C):

BP (deg C):

BP pressure (mm Hg):

Property/Value	Units	Temp	Data Type	Reference
WS 1.56E+002	mg/L	25	EST	MEYLAN,WM ET AL. (1996)
logP 2.20			EXP	YIH,TD & VAN ROSSUM,JM (1977)
VP 1.91E-012	mm Hg	25	EST	NEELY,WB & BLAU,GE (1985)
DC	pKa			
HL 1.25E-013	atm m3/mol	25	EST	MEYLAN,WM & HOWARD,PH (1991)
OH 2.98E-011	cm3/molc sec	25	EST	MEYLAN,WM & HOWARD,PH (1993)

CAS #: 001143-78-8				1H-BENZOTRIAZOLE-1-PROPANOIC ACID, á-METHYL-, ET

Formula: $C_{12}H_{15}N_3O_2$

Mol Weight: 233.27

MP (deg C):

FP (deg C):

BP (deg C):

BP pressure (mm Hg):

Property/Value	Units	Temp	Data Type	Reference
WS 3.37E+002	mg/L	25	EST	MEYLAN,WM ET AL. (1996)
logP 2.27			EXP	SPARATORE,F ET AL. (1988)
VP 1.43E-005	mm Hg	25	EST	NEELY,WB & BLAU,GE (1985)
DC	pKa			
HL 1.74E-008	atm m3/mol	25	EST	MEYLAN,WM & HOWARD,PH (1991)
OH 6.07E-012	cm3/molc sec	25	EST	MEYLAN,WM & HOWARD,PH (1993)

CAS #: 001145-46-6				BENZENESULFONAMIDE, 4-(HEXYLOXY)-

Formula: $C_{12}H_{19}NO_3S$

Mol Weight: 257.35

MP (deg C):

FP (deg C):

BP (deg C):

BP pressure (mm Hg):

Property/Value	Units	Temp	Data Type	Reference
WS 6.77E+001	mg/L	25	EST	MEYLAN,WM ET AL. (1996)
logP 2.93			EXP	HANSCH,C ET AL. (1995)
VP 1.51E-006	mm Hg	25	EST	NEELY,WB & BLAU,GE (1985)
DC	pKa			
HL 1.03E-007	atm m3/mol	25	EST	MEYLAN,WM & HOWARD,PH (1991)
OH 1.98E-011	cm3/molc sec	25	EST	MEYLAN,WM & HOWARD,PH (1993)

CAS #: 001146-98-1				1H-INDENE-1,3(2H)-DIONE, 2-(4-BROMOPHENYL)-

Formula: $C_{15}H_9BrO_2$

Mol Weight: 301.14

MP (deg C): 137-139

FP (deg C):

BP (deg C):

BP pressure (mm Hg):

Property/Value	Units	Temp	Data Type	Reference
WS 7.44E+000	mg/L	25	EST	MEYLAN,WM ET AL. (1996)
logP 3.76			EXP	SANGSTER,J (1993)
VP 3.35E-007	mm Hg	25	EST	NEELY,WB & BLAU,GE (1985)
DC	pKa			
HL 1.91E-010	atm m3/mol	25	EST	MEYLAN,WM & HOWARD,PH (1991)
OH 3.10E-012	cm3/molc sec	25	EST	MEYLAN,WM & HOWARD,PH (1993)

CAS #: 001146-99-2				1H-INDENE-1,3(2H)-DIONE, 2-(4-CHLOROPHENYL)-

Formula: $C_{15}H_9ClO_2$

Mol Weight: 256.69

MP (deg C): 145-146

FP (deg C):

BP (deg C):

BP pressure (mm Hg):

Property/Value	Units	Temp	Data Type	Reference
WS 1.79E+001	mg/L	25	EST	MEYLAN,WM ET AL. (1996)
logP 3.61			EXP	SANGSTER,J (1993)
VP 7.45E-007	mm Hg	25	EST	NEELY,WB & BLAU,GE (1985)
DC	pKa			
HL 3.55E-010	atm m3/mol	25	EST	MEYLAN,WM & HOWARD,PH (1991)
OH 3.27E-012	cm3/molc sec	25	EST	MEYLAN,WM & HOWARD,PH (1993)

CAS #: 001148-79-4				2,2':6',2"-TERPYRIDINE

Formula: $C_{15}H_{11}N_3$

Mol Weight: 233.28

MP (deg C): 89-91

FP (deg C):

BP (deg C):

BP pressure (mm Hg):

Property/Value	Units	Temp	Data Type	Reference
WS 6.19E+002	mg/L	25	EST	MEYLAN,WM ET AL. (1996)
logP 1.96			EXP	DEVOOGT,P ET AL. (1990)
VP 4.17E-007	mm Hg	25	EST	NEELY,WB & BLAU,GE (1985)
DC 4.33	pKa	23	EXP	PERRIN,DD (1965)
HL 7.12E-014	atm m3/mol	25	EST	MEYLAN,WM & HOWARD,PH (1991)
OH 2.40E-012	cm3/molc sec	25	EST	MEYLAN,WM & HOWARD,PH (1993)

CAS #: 001151-51-5				3,5,4'-TRICL SALICYLANILIDE

Formula: $C_{13}H_8Cl_3NO_2$

Mol Weight: 316.57

MP (deg C): 142.5-43

FP (deg C):

BP (deg C):

BP pressure (mm Hg):

Property/Value	Units	Temp	Data Type	Reference
WS 1.49E+000	mg/L	25	EST	MEYLAN,WM ET AL. (1996)
logP 5.15			EXP	TERADA,H ET AL. (1988)
VP 3.64E-010	mm Hg	25	EST	NEELY,WB & BLAU,GE (1985)
DC	pKa			
HL 6.51E-011	atm m3/mol	25	EST	MEYLAN,WM & HOWARD,PH (1991)
OH 4.05E-012	cm3/molc sec	25	EST	MEYLAN,WM & HOWARD,PH (1993)

CAS #: 001152-72-3				3-BUTYN-1-OL, 2-CHLORO-4[5-(1,3-PENTADIYNYL)-2-

Formula: $C_{15}H_{11}ClO_2S$

Mol Weight: 290.77

MP (deg C):

FP (deg C):

BP (deg C):

BP pressure (mm Hg):

Property/Value	Units	Temp	Data Type	Reference
WS 1.64E+000	mg/L	25	EST	MEYLAN,WM ET AL. (1996)
logP 4.60			EXP	MCLACHLAN,D ET AL. (1986)
VP 6.22E-007	mm Hg	25	EST	NEELY,WB & BLAU,GE (1985)
DC	pKa			
HL 5.59E-008	atm m3/mol	25	EST	MEYLAN,WM & HOWARD,PH (1991)
OH 1.01E-010	cm3/molc sec	25	EST	MEYLAN,WM & HOWARD,PH (1993)

CAS #: 001155-00-6				DISULFIDE, BIS(2-NITROPHENYL)

Formula: $C_{12}H_8N_2O_4S_2$

Mol Weight: 308.34

MP (deg C): 196-197

FP (deg C):

BP (deg C):

BP pressure (mm Hg):

Property/Value	Units	Temp	Data Type	Reference
WS 5.92E+000	mg/L	25	EST	MEYLAN,WM ET AL. (1996)
logP 3.37			EXP	HANSCH,C ET AL. (1995)
VP 1.38E-008	mm Hg	25	EST	NEELY,WB & BLAU,GE (1985)
DC	pKa			
HL 7.54E-010	atm m3/mol	25	EST	MEYLAN,WM & HOWARD,PH (1991)
OH 2.28E-010	cm3/molc sec	25	EST	MEYLAN,WM & HOWARD,PH (1993)

TETRADECYLPYRIDINIUM BROMIDE

CAS #: 001155-74-4

Formula: $C_{19}H_{34}BrN$

Mol Weight: 356.39

MP (deg C):

FP (deg C):

BP (deg C):

BP pressure (mm Hg):

Property/Value	Units	Temp	Data Type	Reference
WS 4.23E+002	mg/L	25	EST	MEYLAN,WM ET AL. (1996)
logP 1.32			EXP	HANSCH,C & LEO,AJ (1985)
VP 2.08E-006	mm Hg	25	EST	NEELY,WB & BLAU,GE (1985)
DC	pKa			
HL 5.98E-009	atm m3/mol	25	EST	MEYLAN,WM & HOWARD,PH (1991)
OH 2.84E-011	cm3/molc sec	25	EST	MEYLAN,WM & HOWARD,PH (1993)

TOLAZAMIDE

CAS #: 001156-19-0

Formula: $C_{14}H_{21}N_3O_3S$

Mol Weight: 311.41

MP (deg C): 170-173

FP (deg C):

BP (deg C):

BP pressure (mm Hg):

Property/Value	Units	Temp	Data Type	Reference
WS 6.54E+001	mg/L	30	EXP	YALKOWSKY,SH & DANNENFELSER,RM (1992)
logP 2.69			EXP	SANGSTER,J (1993)
VP 1.07E-009	mm Hg	25	EST	NEELY,WB & BLAU,GE (1985)
DC	pKa			
HL 1.36E-012	atm m3/mol	25	EST	MEYLAN,WM & HOWARD,PH (1991)
OH 3.02E-011	cm3/molc sec	25	EST	MEYLAN,WM & HOWARD,PH (1993)

CLOBENZEPAM

CAS #: 001159-93-9

Formula: $C_{17}H_{18}ClN_3O$

Mol Weight: 315.81

MP (deg C): 165-166

FP (deg C):

BP (deg C):

BP pressure (mm Hg):

Property/Value	Units	Temp	Data Type	Reference
WS 5.02E-001	mg/L	25	EST	MEYLAN,WM ET AL. (1996)
logP 3.73			EXP	SANGSTER,J (1994)
VP 3.10E-009	mm Hg	25	EST	NEELY,WB & BLAU,GE (1985)
DC	pKa			
HL 1.08E-014	atm m3/mol	25	EST	MEYLAN,WM & HOWARD,PH (1991)
OH 2.71E-010	cm3/molc sec	25	EST	MEYLAN,WM & HOWARD,PH (1993)

DECABROMODIPHENYL ETHER

CAS #: 001163-19-5

Formula: $C_{12}Br_{10}O$

Mol Weight: 959.17

MP (deg C): 294-296

FP (deg C):

BP (deg C): 425

BP pressure (mm Hg):

Property/Value	Units	Temp	Data Type	Reference
WS 2.50E-002	mg/L	25	EXP	YALKOWSKY,SH & DANNENFELSER,RM (1992)
logP 5.24			EXP	NORRIS,JM ET AL. (1973)
VP 4.31E-013	mm Hg	25	EST	NEELY,WB & BLAU,GE (1985)
DC	pKa			
HL 1.19E-008	atm m3/mol	25	EST	MEYLAN,WM & HOWARD,PH (1991)
OH 1.74E-013	cm3/molc sec	25	EST	MEYLAN,WM & HOWARD,PH (1993)

CHLOROTHIAZIDE-3-PHENYL

CAS #: 001163-51-5

Formula: $C_{13}H_{10}ClN_3O_4S_2$

Mol Weight: 371.82

MP (deg C):

FP (deg C):

BP (deg C):

BP pressure (mm Hg):

Property/Value	Units	Temp	Data Type	Reference
WS 2.59E+001	mg/L	25	EST	MEYLAN,WM ET AL. (1996)
logP 1.40			EXP	SANGSTER,J (1994)
VP 4.88E-014	mm Hg	25	EST	NEELY,WB & BLAU,GE (1985)
DC	pKa			
HL 3.27E-013	atm m3/mol	25	EST	MEYLAN,WM & HOWARD,PH (1991)
OH 4.34E-012	cm3/molc sec	25	EST	MEYLAN,WM & HOWARD,PH (1993)

2,6-ETO-4-SULFANILAMIDOPYRIMIDINE

CAS #: 001164-13-2

Formula: $C_{14}H_{18}N_4O_4S$

Mol Weight: 338.39

MP (deg C):

FP (deg C):

BP (deg C):

BP pressure (mm Hg):

Property/Value	Units	Temp	Data Type	Reference
WS 6.49E+001	mg/L	25	EST	MEYLAN,WM ET AL. (1996)
logP 2.40			EXP	HANSCH,C & LEO,AJ (1985)
VP 7.12E-010	mm Hg	25	EST	NEELY,WB & BLAU,GE (1985)
DC	pKa			
HL 2.29E-014	atm m3/mol	25	EST	MEYLAN,WM & HOWARD,PH (1991)
OH 2.12E-010	cm3/molc sec	25	EST	MEYLAN,WM & HOWARD,PH (1993)

DIMEFLINE

CAS #: 001165-48-6

Formula: $C_{20}H_{21}NO_3$

Mol Weight: 323.40

MP (deg C): 109.5

FP (deg C):

BP (deg C):

BP pressure (mm Hg):

Property/Value	Units	Temp	Data Type	Reference
WS 6.96E+001	mg/L	25	EST	MEYLAN,WM ET AL. (1996)
logP 3.65			EXP	SANGSTER,J (1994)
VP 2.10E-008	mm Hg	25	EST	NEELY,WB & BLAU,GE (1985)
DC	pKa			
HL 1.83E-012	atm m3/mol	25	EST	MEYLAN,WM & HOWARD,PH (1991)
OH 3.17E-010	cm3/molc sec	25	EST	MEYLAN,WM & HOWARD,PH (1993)

DEXAMETHASONE-17-ACETATE

CAS #: 001177-87-3

Formula: $C_{24}H_{31}FO_6$

Mol Weight: 434.51

MP (deg C): 262-264

FP (deg C):

BP (deg C):

BP pressure (mm Hg):

Property/Value	Units	Temp	Data Type	Reference
WS 6.10E+000	mg/L	25	EST	MEYLAN,WM ET AL. (1996)
logP 2.91			EXP	HANSCH,C & LEO,AJ (1985)
VP 1.69E-013	mm Hg	25	EST	NEELY,WB & BLAU,GE (1985)
DC	pKa			
HL 7.73E-012	atm m3/mol	25	EST	MEYLAN,WM & HOWARD,PH (1991)
OH 6.74E-011	cm3/molc sec	25	EST	MEYLAN,WM & HOWARD,PH (1993)

CAS #: 001182-87-2				PERUVOSIDE

Formula: $C_{30}H_{44}O_9$
Mol Weight: 548.68
MP (deg C): 161-164
FP (deg C):
BP (deg C):
BP pressure (mm Hg):

Property/Value	Units	Temp	Data Type	Reference
WS 4.41E+001	mg/L	25	EST	MEYLAN,WM ET AL. (1996)
logP 1.05			EXP	HANSCH,C & LEO,AJ (1985)
VP 1.16E-020	mm Hg	25	EST	NEELY,WB & BLAU,GE (1985)
DC	pKa			
HL 2.20E-020	atm m3/mol	25	EST	MEYLAN,WM & HOWARD,PH (1991)
OH 1.40E-010	cm3/molc sec	25	EST	MEYLAN,WM & HOWARD,PH (1993)

CAS #: 001185-33-7				2,2-DIMETHYL-1-BUTANOL

Formula: $C_6H_{14}O$
Mol Weight: 102.18
MP (deg C): < -15
FP (deg C):
BP (deg C): 136.5
BP pressure (mm Hg):

Property/Value	Units	Temp	Data Type	Reference
WS 7.60E+003	mg/L	25	EXP	YALKOWSKY,SH & DANNENFELSER,RM (1992)
logP 1.71			EST	MEYLAN,WM & HOWARD,PH (1995)
VP 2.78E+000	mm Hg	25	EST	NEELY,WB & BLAU,GE (1985)
DC	pKa			
HL 1.76E-005	atm m3/mol	25	EST	MEYLAN,WM & HOWARD,PH (1991)
OH 5.81E-012	cm3/molc sec	25	EST	MEYLAN,WM & HOWARD,PH (1993)

CAS #: 001185-97-3				OO-DIME DICHLOROVINYLPHOSPHONATE

Formula: $C_4H_7Cl_2O_3P$
Mol Weight: 204.98
MP (deg C):
FP (deg C):
BP (deg C):
BP pressure (mm Hg):

Property/Value	Units	Temp	Data Type	Reference
WS 2.63E+003	mg/L	25	EST	MEYLAN,WM ET AL. (1996)
logP 1.40			EXP	HANSCH,C & LEO,AJ (1985)
VP 6.62E-002	mm Hg	25	EST	NEELY,WB & BLAU,GE (1985)
DC	pKa			
HL 1.35E-006	atm m3/mol	25	EST	MEYLAN,WM & HOWARD,PH (1991)
OH 9.41E-012	cm3/molc sec	25	EST	MEYLAN,WM & HOWARD,PH (1993)

CAS #: 001186-53-4				2,2,3,4-TETRAMETHYLPENTANE

Formula: C_9H_{20}
Mol Weight: 128.26
MP (deg C): -121
FP (deg C):
BP (deg C): 133
BP pressure (mm Hg):

Property/Value	Units	Temp	Data Type	Reference
WS 3.87E+000	mg/L	25	EST	MEYLAN,WM ET AL. (1996)
logP 4.50			EST	MEYLAN,WM & HOWARD,PH (1995)
VP 1.26E+001	mm Hg	25	EXP	DAUBERT,TE & DANNER,RP (1989)
DC	pKa			
HL 4.00E+000	atm m3/mol	25	EST	MEYLAN,WM & HOWARD,PH (1991)
OH 6.32E-012	cm3/molc sec	25	EST	MEYLAN,WM & HOWARD,PH (1993)

CAS #: 001187-58-2				N-METHYL PROPIONAMIDE

Formula: C_4H_9NO
Mol Weight: 87.12
MP (deg C): -30.9
FP (deg C):
BP (deg C): 148
BP pressure (mm Hg):

Property/Value	Units	Temp	Data Type	Reference
WS 1.89E+005	mg/L	25	EST	MEYLAN,WM ET AL. (1996)
logP -0.21			EST	MEYLAN,WM & HOWARD,PH (1995)
VP 1.15E+000	mm Hg	25	EXT	BOUBLIK,T ET AL. (1984)
DC	pKa			
HL 3.26E-008	atm m3/mol	25	EST	MEYLAN,WM & HOWARD,PH (1991)
OH 7.29E-012	cm3/molc sec	25	EST	MEYLAN,WM & HOWARD,PH (1993)

CAS #: 001188-09-6				2-PROPENOIC ACID, 2-METHYL-, 1,3-PROPANEDIYL EST

Formula: $C_{11}H_{16}O_4$
Mol Weight: 212.25
MP (deg C):
FP (deg C):
BP (deg C):
BP pressure (mm Hg):

Property/Value	Units	Temp	Data Type	Reference
WS 9.76E+002	mg/L	25	EST	MEYLAN,WM ET AL. (1996)
logP 1.86			EXP	SANGSTER,J (1993)
VP 7.12E-002	mm Hg	25	EST	NEELY,WB & BLAU,GE (1985)
DC	pKa			
HL 5.02E-007	atm m3/mol	25	EST	MEYLAN,WM & HOWARD,PH (1991)
OH 4.13E-011	cm3/molc sec	25	EST	MEYLAN,WM & HOWARD,PH (1993)

CAS #: 001190-48-3				L-LYSINE, N-FORMYL-

Formula: $C_7H_{14}N_2O_3$
Mol Weight: 174.20
MP (deg C):
FP (deg C):
BP (deg C):
BP pressure (mm Hg):

Property/Value	Units	Temp	Data Type	Reference
WS 1.33E+005	mg/L	25	EST	MEYLAN,WM ET AL. (1996)
logP -2.84			EXP	HANSCH,C & LEO,AJ (1985)
VP 5.14E-009	mm Hg	25	EST	NEELY,WB & BLAU,GE (1985)
DC	pKa			
HL 1.59E-015	atm m3/mol	25	EST	MEYLAN,WM & HOWARD,PH (1991)
OH 5.86E-011	cm3/molc sec	25	EST	MEYLAN,WM & HOWARD,PH (1993)

CAS #: 001191-95-3				CYCLOBUTANONE

Formula: C_4H_6O
Mol Weight: 70.09
MP (deg C): -50.9
FP (deg C):
BP (deg C): 99
BP pressure (mm Hg):

Property/Value	Units	Temp	Data Type	Reference
WS 1.02E+005	mg/L	25	EST	MEYLAN,WM ET AL. (1996)
logP 0.14			EST	MEYLAN,WM & HOWARD,PH (1995)
VP 4.30E+001	mm Hg	25	EXP	CHAO,J ET AL. (1983)
DC	pKa			
HL 2.90E-005	atm m3/mol	25	EST	MEYLAN,WM & HOWARD,PH (1991)
OH 8.70E-013	cm3/molc sec	25	EXP	ATKINSON,R (1989)

CAS #: 001192-18-3				1,2-DIMETHYLCYCLOPENTANE (CIS)
Formula: C7H14				
Mol Weight: 98.19				
MP (deg C): -54		FP (deg C):		
BP (deg C): 99.5				
BP pressure (mm Hg):				

Property/Value	Units	Temp	Data Type	Reference
WS 3.39E+001	mg/L	25	EST	MEYLAN,WM ET AL. (1996)
logP 3.52			EST	MEYLAN,WM & HOWARD,PH (1995)
VP 4.72E+001	mm Hg	25	EXP	DAUBERT,TE & DANNER,RP (1989)
DC	pKa			
HL 3.39E-001	atm m3/mol	25	EST	MEYLAN,WM & HOWARD,PH (1991)
OH 8.42E-012	cm3/molc sec	25	EST	MEYLAN,WM & HOWARD,PH (1993)

CAS #: 001192-37-6				METHYLENE CYCLOHEXANE
Formula: C7H12				
Mol Weight: 96.17				
MP (deg C): -106.7		FP (deg C):		
BP (deg C): 102.5				
BP pressure (mm Hg):				

Property/Value	Units	Temp	Data Type	Reference
WS 3.01E+001	mg/L	25	EST	MEYLAN,WM ET AL. (1996)
logP 3.59			EST	MEYLAN,WM & HOWARD,PH (1995)
VP 4.02E+001	mm Hg	25	EXT	BOUBLIK,T ET AL. (1983)
DC	pKa			
HL 1.56E-001	atm m3/mol	25	EST	MEYLAN,WM & HOWARD,PH (1991)
OH 5.79E-011	cm3/molc sec	25	EST	MEYLAN,WM & HOWARD,PH (1993)

CAS #: 001192-62-7				2-ACETYLFURAN
Formula: C6H6O2				
Mol Weight: 110.11				
MP (deg C): 33		FP (deg C):		
BP (deg C): 175				
BP pressure (mm Hg):				

Property/Value	Units	Temp	Data Type	Reference
WS 3.91E+004	mg/L	25	EST	MEYLAN,WM ET AL. (1996)
logP 0.52			EXP	HANSCH,C & LEO,AJ (1985)
VP 3.47E+000	mm Hg	25	EST	NEELY,WB & BLAU,GE (1985)
DC	pKa			
HL 9.79E-006	atm m3/mol	25	EST	MEYLAN,WM & HOWARD,PH (1991)
OH 3.70E-011	cm3/molc sec	25	EST	MEYLAN,WM & HOWARD,PH (1993)

CAS #: 001193-02-8				4-AMINOBENZENETHIOL
Formula: C6H7NS				
Mol Weight: 125.19				
MP (deg C): 46		FP (deg C):		
BP (deg C): 143				
BP pressure (mm Hg): 1.70E+001				

Property/Value	Units	Temp	Data Type	Reference
WS 7.30E+003	mg/L	25	EST	MEYLAN,WM ET AL. (1996)
logP 1.31			EXP	HANSCH,C ET AL. (1995)
VP 2.44E-002	mm Hg	25	EST	NEELY,WB & BLAU,GE (1985)
DC	pKa			
HL 1.85E-007	atm m3/mol	25	EST	MEYLAN,WM & HOWARD,PH (1991)
OH 5.81E-011	cm3/molc sec	25	EST	MEYLAN,WM & HOWARD,PH (1993)

CAS #: 001193-62-0				METHYL PYRROLE-2-CARBOXYLATE
Formula: C6H7NO2				
Mol Weight: 125.13				
MP (deg C):		FP (deg C):		
BP (deg C):				
BP pressure (mm Hg):				

Property/Value	Units	Temp	Data Type	Reference
WS 9.62E+003	mg/L	25	EST	MEYLAN,WM ET AL. (1996)
logP 1.17			EXP	HANSCH,C & LEO,AJ (1985)
VP 1.60E-001	mm Hg	25	EST	NEELY,WB & BLAU,GE (1985)
DC	pKa			
HL 5.85E-008	atm m3/mol	25	EST	MEYLAN,WM & HOWARD,PH (1991)
OH 3.56E-011	cm3/molc sec	25	EST	MEYLAN,WM & HOWARD,PH (1993)

CAS #: 001193-82-4				METHYL PHENYL SULFOXIDE
Formula: C7H8OS				
Mol Weight: 140.21				
MP (deg C): 32		FP (deg C):		
BP (deg C): 263.5				
BP pressure (mm Hg):				

Property/Value	Units	Temp	Data Type	Reference
WS 2.83E+004	mg/L	25	EST	MEYLAN,WM ET AL. (1996)
logP 0.55			EXP	HANSCH,C & LEO,AJ (1985)
VP 1.99E-002	mm Hg	25	EST	NEELY,WB & BLAU,GE (1985)
DC	pKa			
HL 9.91E-009	atm m3/mol	25	EST	MEYLAN,WM & HOWARD,PH (1991)
OH 6.15E-011	cm3/molc sec	25	EST	MEYLAN,WM & HOWARD,PH (1993)

CAS #: 001194-65-6				2,6-DICHLOROBENZONITRILE
Formula: C7H3Cl2N				
Mol Weight: 172.01				
MP (deg C): 144-145		FP (deg C):		
BP (deg C): 270-270.1				
BP pressure (mm Hg):				

Property/Value	Units	Temp	Data Type	Reference
WS 1.80E+001	mg/L	20	EXP	WSSA (1983)
logP 2.74			EXP	HANSCH,C ET AL. (1995)
VP 5.50E-004	mm Hg	20	EXP	WSSA (1983)
DC	pKa			
HL 6.92E-006	atm m3/mol	20	EST	VP/WSOL
OH 1.75E-013	cm3/molc sec	25	EST	MEYLAN,WM & HOWARD,PH (1993)

CAS #: 001195-08-0				5-FORMYLURACIL
Formula: C5H4N2O3				
Mol Weight: 140.10				
MP (deg C): > 300		FP (deg C):		
BP (deg C):				
BP pressure (mm Hg):				

Property/Value	Units	Temp	Data Type	Reference
WS 3.11E+004	mg/L	25	EST	MEYLAN,WM ET AL. (1996)
logP -1.03			EXP	HANSCH,C & LEO,AJ (1985)
VP 1.80E-007	mm Hg	25	EST	NEELY,WB & BLAU,GE (1985)
DC	pKa			
HL 1.00E-012	atm m3/mol	25	EST	MEYLAN,WM & HOWARD,PH (1991)
OH 2.51E-011	cm3/molc sec	25	EST	MEYLAN,WM & HOWARD,PH (1993)

CAS #: 001195-16-0 — N-(TETRAHYDRO-2-OXO-3-THIENYL)ACETAMIDE

Formula: $C_6H_9NO_3S$
Mol Weight: 175.21
MP (deg C): 112
FP (deg C):
BP (deg C):
BP pressure (mm Hg):

Property/Value	Units	Temp	Data Type	Reference
WS 1.37E+005	mg/L	25	EST	MEYLAN,WM ET AL. (1996)
logP -0.35			EXP	HANSCH,C ET AL. (1995)
VP 3.31E-005	mm Hg	25	EST	NEELY,WB & BLAU,GE (1985)
DC	pKa			
HL 2.43E-010	atm m3/mol	25	EST	MEYLAN,WM & HOWARD,PH (1991)
OH 1.98E-011	cm3/molc sec	25	EST	MEYLAN,WM & HOWARD,PH (1993)

CAS #: 001195-79-5 — FENCHONE

Formula: $C_{10}H_{16}O$
Mol Weight: 152.24
MP (deg C): 5
FP (deg C):
BP (deg C):
BP pressure (mm Hg):

Property/Value	Units	Temp	Data Type	Reference
WS 7.44E+002	mg/L	25	EST	MEYLAN,WM ET AL. (1996)
logP 2.34			EST	MEYLAN,WM & HOWARD,PH (1995)
VP 8.17E-001	mm Hg	25	EXT	PERRY,RH & GREEN,D (1984)
DC	pKa			
HL 7.00E-005	atm m3/mol	25	EST	MEYLAN,WM & HOWARD,PH (1991)
OH 1.56E-011	cm3/molc sec	25	EST	MEYLAN,WM & HOWARD,PH (1993)

CAS #: 001196-57-2 — QUINOXALINE-2-ONE

Formula: $C_8H_6N_2O$
Mol Weight: 146.15
MP (deg C):
FP (deg C):
BP (deg C):
BP pressure (mm Hg):

Property/Value	Units	Temp	Data Type	Reference
WS 1.64E+004	mg/L	25	EST	MEYLAN,WM ET AL. (1996)
logP 0.80			EXP	HANSCH,C & LEO,AJ (1985)
VP 2.33E-005	mm Hg	25	EST	NEELY,WB & BLAU,GE (1985)
DC	pKa			
HL 9.25E-009	atm m3/mol	25	EST	MEYLAN,WM & HOWARD,PH (1991)
OH 1.51E-011	cm3/molc sec	25	EST	MEYLAN,WM & HOWARD,PH (1993)

CAS #: 001197-19-9 — P-CYANO-N,N-DIMETHYLANILINE

Formula: $C_9H_{10}N_2$
Mol Weight: 146.19
MP (deg C): 73-75
FP (deg C):
BP (deg C): 318
BP pressure (mm Hg):

Property/Value	Units	Temp	Data Type	Reference
WS 5.66E+002	mg/L	25	EST	MEYLAN,WM ET AL. (1996)
logP 2.20			EXP	HANSCH,C ET AL. (1995)
VP 2.06E-002	mm Hg	25	EST	NEELY,WB & BLAU,GE (1985)
DC 1.78	pKa	25	EXP	PERRIN,DD (1965)
HL 8.28E-007	atm m3/mol	25	EST	MEYLAN,WM & HOWARD,PH (1991)
OH 6.78E-011	cm3/molc sec	25	EST	MEYLAN,WM & HOWARD,PH (1993)

CAS #: 001197-22-4 — N-PHENYL METHANESULFONAMIDE

Formula: $C_7H_9NO_2S$
Mol Weight: 171.22
MP (deg C):
FP (deg C):
BP (deg C):
BP pressure (mm Hg):

Property/Value	Units	Temp	Data Type	Reference
WS 9.37E+003	mg/L	25	EST	MEYLAN,WM ET AL. (1996)
logP 0.95			EXP	HANSCH,C & LEO,AJ (1985)
VP 7.16E-004	mm Hg	25	EST	NEELY,WB & BLAU,GE (1985)
DC	pKa			
HL 1.17E-006	atm m3/mol	25	EST	MEYLAN,WM & HOWARD,PH (1991)
OH 4.37E-011	cm3/molc sec	25	EST	MEYLAN,WM & HOWARD,PH (1993)

CAS #: 001197-50-8 — W-PHENYLPROPIONALDEHYDE OXIME

Formula: $C_9H_{11}NO$
Mol Weight: 149.19
MP (deg C):
FP (deg C):
BP (deg C):
BP pressure (mm Hg):

Property/Value	Units	Temp	Data Type	Reference
WS 3.56E+003	mg/L	25	EST	MEYLAN,WM ET AL. (1996)
logP 1.56			EXP	HANSCH,C & LEO,AJ (1985)
VP 4.72E-004	mm Hg	25	EST	NEELY,WB & BLAU,GE (1985)
DC	pKa			
HL 6.30E-007	atm m3/mol	25	EST	MEYLAN,WM & HOWARD,PH (1991)
OH 9.79E-012	cm3/molc sec	25	EST	MEYLAN,WM & HOWARD,PH (1993)

CAS #: 001197-60-0 — 2-FURANMETHANAMINIUM, N,N,N,5-TETRAMETHYL-, IODI

Formula: $C_9H_{16}INO$
Mol Weight: 281.14
MP (deg C):
FP (deg C):
BP (deg C):
BP pressure (mm Hg):

Property/Value	Units	Temp	Data Type	Reference
WS 6.00E+005	mg/L	25	EST	MEYLAN,WM ET AL. (1996)
logP -1.85			EXP	PRATESI,P ET AL. (1986)
VP 2.23E-008	mm Hg	25	EST	NEELY,WB & BLAU,GE (1985)
DC	pKa			
HL 2.34E-013	atm m3/mol	25	EST	MEYLAN,WM & HOWARD,PH (1991)
OH 1.41E-010	cm3/molc sec	25	EST	MEYLAN,WM & HOWARD,PH (1993)

CAS #: 001198-37-4 — 2,4-DIMETHYLQUINOLINE

Formula: $C_{11}H_{11}N$
Mol Weight: 157.22
MP (deg C):
FP (deg C):
BP (deg C): 265
BP pressure (mm Hg):

Property/Value	Units	Temp	Data Type	Reference
WS 1.80E+003	mg/L	25	EXP	YALKOWSKY,SH & DANNENFELSER,RM (1992)
logP 3.24			EST	MEYLAN,WM & HOWARD,PH (1995)
VP 7.94E-003	mm Hg	25	EXT	CHAO,J ET AL. (1983)
DC 5.12	pKa	25	EXP	PERRIN,DD (1965)
HL 8.38E-007	atm m3/mol	25	EST	MEYLAN,WM & HOWARD,PH (1991)
OH 4.64E-011	cm3/molc sec	25	EST	MEYLAN,WM & HOWARD,PH (1993)

CAS #: 001198-55-6				TETRACHLORO-1,2-BENZENEDIOL

Formula: $C_6H_2Cl_4O_2$

Mol Weight: 247.89

MP (deg C): | FP (deg C):

BP (deg C):

BP pressure (mm Hg):

Property/ Value	Units	Temp	Data Type	Reference
WS 2.70E+000	mg/L		EXP	KOLSET,K & HEIBERG,A (1988)
logP 4.29			EXP	HANSCH,C & LEO,AJ (1985)
VP 4.15E-006	mm Hg	25	EST	NEELY,WB & BLAU,GE (1985)
DC	pKa			
HL 1.76E-011	atm m3/mol	25	EST	MEYLAN,WM & HOWARD,PH (1991)
OH 1.25E-012	cm3/molc sec	25	EST	MEYLAN,WM & HOWARD,PH (1993)

CAS #: 001199-77-5				ALPHA-METHYLCINNAMIC ACID

Formula: $C_{10}H_{10}O_2$

Mol Weight: 162.19

MP (deg C): 79-81 | FP (deg C):

BP (deg C):

BP pressure (mm Hg):

Property/ Value	Units	Temp	Data Type	Reference
WS 1.00E+003	mg/L	25	EST	MEYLAN,WM ET AL. (1996)
logP 2.60			EXP	SANGSTER,J (1993)
VP 9.11E-004	mm Hg	25	EXT	PERRY,RH & GREEN,D (1984)
DC	pKa			
HL 2.03E-008	atm m3/mol	25	EST	MEYLAN,WM & HOWARD,PH (1991)
OH 3.29E-011	cm3/molc sec	25	EST	MEYLAN,WM & HOWARD,PH (1993)

CAS #: 001199-98-0				BUTRAMIDE, 4-PHENYL

Formula: $C_{10}H_{13}NO$

Mol Weight: 163.22

MP (deg C): | FP (deg C):

BP (deg C):

BP pressure (mm Hg):

Property/ Value	Units	Temp	Data Type	Reference
WS 4.13E+003	mg/L	25	EST	MEYLAN,WM ET AL. (1996)
logP 1.41			EXP	HANSCH,C & LEO,AJ (1985)
VP 5.34E-005	mm Hg	25	EST	NEELY,WB & BLAU,GE (1985)
DC	pKa			
HL 1.59E-009	atm m3/mol	25	EST	MEYLAN,WM & HOWARD,PH (1991)
OH 1.33E-011	cm3/molc sec	25	EST	MEYLAN,WM & HOWARD,PH (1993)

CAS #: 001199-99-1				N,N-DIME-N-G-PHENYLPROPYLAMINE

Formula: $C_{11}H_{17}N$

Mol Weight: 163.26

MP (deg C): | FP (deg C):

BP (deg C):

BP pressure (mm Hg):

Property/ Value	Units	Temp	Data Type	Reference
WS 3.14E+003	mg/L	25	EST	MEYLAN,WM ET AL. (1996)
logP 2.73			EXP	HANSCH,C & LEO,AJ (1985)
VP 8.20E-002	mm Hg	25	EST	NEELY,WB & BLAU,GE (1985)
DC	pKa			
HL 5.20E-006	atm m3/mol	25	EST	MEYLAN,WM & HOWARD,PH (1991)
OH 8.66E-011	cm3/molc sec	25	EST	MEYLAN,WM & HOWARD,PH (1993)

CAS #: 001200-06-2				4-METHOXYPHENYL ACETATE

Formula: $C_9H_{10}O_3$

Mol Weight: 166.18

MP (deg C): | FP (deg C):

BP (deg C):

BP pressure (mm Hg):

Property/ Value	Units	Temp	Data Type	Reference
WS 3.10E+003	mg/L	25	EST	MEYLAN,WM ET AL. (1996)
logP 1.54			EXP	HANSCH,C & LEO,AJ (1985)
VP 6.02E-002	mm Hg	25	EST	NEELY,WB & BLAU,GE (1985)
DC	pKa			
HL 3.83E-006	atm m3/mol	25	EST	MEYLAN,WM & HOWARD,PH (1991)
OH 1.67E-011	cm3/molc sec	25	EST	MEYLAN,WM & HOWARD,PH (1993)

CAS #: 001201-25-8				1-AZULENE CARBOXYLIC ACID

Formula: $C_{11}H_8O_2$

Mol Weight: 172.19

MP (deg C): | FP (deg C):

BP (deg C):

BP pressure (mm Hg):

Property/ Value	Units	Temp	Data Type	Reference
WS 5.18E+002	mg/L	25	EST	MEYLAN,WM ET AL. (1996)
logP 2.88			EXP	HANSCH,C & LEO,AJ (1985)
VP 1.46E-004	mm Hg	25	EST	NEELY,WB & BLAU,GE (1985)
DC	pKa			
HL 1.84E-007	atm m3/mol	25	EST	MEYLAN,WM & HOWARD,PH (1991)
OH 2.26E-010	cm3/molc sec	25	EST	MEYLAN,WM & HOWARD,PH (1993)

CAS #: 001201-56-5				BENZENEMETHANOL, _-[1-(DIMETHYLAMINO)ETHYL]-, (R

Formula: $C_{11}H_{17}NO$

Mol Weight: 179.26

MP (deg C): | FP (deg C):

BP (deg C):

BP pressure (mm Hg):

Property/ Value	Units	Temp	Data Type	Reference
WS 2.00E+004	mg/L	25	EST	MEYLAN,WM ET AL. (1996)
logP 1.70			EXP	SANGSTER,J (1994)
VP 8.12E-004	mm Hg	25	EST	NEELY,WB & BLAU,GE (1985)
DC 9.20	pKa	26	EXP	PERRIN,DD (1965)
HL 1.90E-010	atm m3/mol	25	EST	MEYLAN,WM & HOWARD,PH (1991)
OH 1.04E-010	cm3/molc sec	25	EST	MEYLAN,WM & HOWARD,PH (1993)

CAS #: 001202-04-6				1H-INDOLE-2-CARBOXYLIC ACID, METHYL ESTER

Formula: $C_{10}H_9NO_2$

Mol Weight: 175.19

MP (deg C): | FP (deg C):

BP (deg C):

BP pressure (mm Hg):

Property/ Value	Units	Temp	Data Type	Reference
WS 4.70E+002	mg/L	25	EST	MEYLAN,WM ET AL. (1996)
logP 2.45			EXP	YAMAGAMI,C ET AL. (1992)
VP 2.09E-004	mm Hg	25	EST	NEELY,WB & BLAU,GE (1985)
DC	pKa			
HL 5.71E-009	atm m3/mol	25	EST	MEYLAN,WM & HOWARD,PH (1991)
OH 3.56E-011	cm3/molc sec	25	EST	MEYLAN,WM & HOWARD,PH (1993)

CAS #: 001202-32-0	B-ETHYL-B-NITROSTYRENE

Formula: $C_{10}H_{11}NO_2$

Mol Weight: 177.20

MP (deg C): 90.5 FP (deg C):

BP (deg C): 307.5

BP pressure (mm Hg):

Property/Value	Units	Temp	Data Type	Reference
WS 8.36E+001	mg/L	25	EST	MEYLAN,WM ET AL. (1996)
logP 2.86			EXP	HANSCH,C & LEO,AJ (1985)
VP 2.59E-003	mm Hg	25	EST	NEELY,WB & BLAU,GE (1985)
DC	pKa			
HL 7.23E-006	atm m3/mol	25	EST	MEYLAN,WM & HOWARD,PH (1991)
OH 2.04E-011	cm3/molc sec	25	EST	MEYLAN,WM & HOWARD,PH (1993)

CAS #: 001202-34-2	N-2-PYRIDINYL-2-PYRIDINAMINE

Formula: $C_{10}H_9N_3$

Mol Weight: 171.20

MP (deg C): 90.5 FP (deg C):

BP (deg C): 307.5

BP pressure (mm Hg):

Property/Value	Units	Temp	Data Type	Reference
WS 8.05E+002	mg/L	25	EST	MEYLAN,WM ET AL. (1996)
logP 2.20			EST	MEYLAN,WM & HOWARD,PH (1995)
VP 5.39E-004	mm Hg	25	EST	NEELY,WB & BLAU,GE (1985)
DC	pKa			
HL 1.80E-012	atm m3/mol	25	EST	MEYLAN,WM & HOWARD,PH (1991)
OH 5.56E-011	cm3/molc sec	25	EST	MEYLAN,WM & HOWARD,PH (1993)

CAS #: 001205-08-9	METHYL N-BENZOGLYCINE

Formula: $C_{10}H_{11}NO_3$

Mol Weight: 193.20

MP (deg C): FP (deg C):

BP (deg C):

BP pressure (mm Hg):

Property/Value	Units	Temp	Data Type	Reference
WS 9.45E+003	mg/L	25	EST	MEYLAN,WM ET AL. (1996)
logP 0.82			EXP	HANSCH,C & LEO,AJ (1985)
VP 1.04E-005	mm Hg	25	EST	NEELY,WB & BLAU,GE (1985)
DC	pKa			
HL 2.31E-010	atm m3/mol	25	EST	MEYLAN,WM & HOWARD,PH (1991)
OH 1.03E-011	cm3/molc sec	25	EST	MEYLAN,WM & HOWARD,PH (1993)

CAS #: 001205-64-7	3-METHYL DIPHENYL AMINE

Formula: $C_{13}H_{13}N$

Mol Weight: 183.26

MP (deg C): 30 FP (deg C):

BP (deg C): 316

BP pressure (mm Hg):

Property/Value	Units	Temp	Data Type	Reference
WS 2.80E+001	mg/L	25	EST	MEYLAN,WM ET AL. (1996)
logP 3.84			EST	MEYLAN,WM & HOWARD,PH (1995)
VP 1.49E-004	mm Hg	25	EXP	WAYAKU,M (1982)
DC	pKa			
HL 1.16E-006	atm m3/mol	25	EST	MEYLAN,WM & HOWARD,PH (1991)
OH 2.00E-010	cm3/molc sec	25	EST	MEYLAN,WM & HOWARD,PH (1993)

CAS #: 001205-71-6	4-CHLORO-N-PHENYLBENZENAMINE

Formula: $C_{12}H_{10}ClN$

Mol Weight: 203.67

MP (deg C): 74 FP (deg C):

BP (deg C): 335

BP pressure (mm Hg):

Property/Value	Units	Temp	Data Type	Reference
WS 1.82E+001	mg/L	25	EST	MEYLAN,WM ET AL. (1996)
logP 3.94			EST	MEYLAN,WM & HOWARD,PH (1995)
VP 3.75E-004	mm Hg	25	EST	NEELY,WB & BLAU,GE (1985)
DC	pKa			
HL 7.78E-007	atm m3/mol	25	EST	MEYLAN,WM & HOWARD,PH (1991)
OH 2.00E-010	cm3/molc sec	25	EST	MEYLAN,WM & HOWARD,PH (1993)

CAS #: 001206-88-8	1-ETHYL AZULENE CARBOXYLATE

Formula: $C_{13}H_{12}O_2$

Mol Weight: 200.24

MP (deg C): FP (deg C):

BP (deg C):

BP pressure (mm Hg):

Property/Value	Units	Temp	Data Type	Reference
WS 2.34E+001	mg/L	25	EST	MEYLAN,WM ET AL. (1996)
logP 3.83			EXP	HANSCH,C & LEO,AJ (1985)
VP 1.22E-003	mm Hg	25	EST	NEELY,WB & BLAU,GE (1985)
DC	pKa			
HL 7.82E-005	atm m3/mol	25	EST	MEYLAN,WM & HOWARD,PH (1991)
OH 2.27E-010	cm3/molc sec	25	EST	MEYLAN,WM & HOWARD,PH (1993)

CAS #: 001207-63-2	124-BENZTHIADIZ-1-O2-3-ET-7-CL

Formula: $C_9H_9ClN_2O_2S$

Mol Weight: 244.70

MP (deg C): FP (deg C):

BP (deg C):

BP pressure (mm Hg):

Property/Value	Units	Temp	Data Type	Reference
WS 1.05E+003	mg/L	25	EST	MEYLAN,WM ET AL. (1996)
logP 1.62			EXP	HANSCH,C & LEO,AJ (1985)
VP 1.29E-008	mm Hg	25	EST	NEELY,WB & BLAU,GE (1985)
DC	pKa			
HL 9.10E-008	atm m3/mol	25	EST	MEYLAN,WM & HOWARD,PH (1991)
OH 1.64E-012	cm3/molc sec	25	EST	MEYLAN,WM & HOWARD,PH (1993)

CAS #: 001208-86-2	4-METHOXY-N-PHENYLBENZENAMINE

Formula: $C_{13}H_{13}NO$

Mol Weight: 199.25

MP (deg C): FP (deg C):

BP (deg C):

BP pressure (mm Hg):

Property/Value	Units	Temp	Data Type	Reference
WS 5.82E+001	mg/L	25	EST	MEYLAN,WM ET AL. (1996)
logP 3.37			EST	MEYLAN,WM & HOWARD,PH (1995)
VP 2.50E-004	mm Hg	25	EST	NEELY,WB & BLAU,GE (1985)
DC	pKa			
HL 6.21E-008	atm m3/mol	25	EST	MEYLAN,WM & HOWARD,PH (1991)
OH 2.01E-010	cm3/molc sec	25	EST	MEYLAN,WM & HOWARD,PH (1993)

CAS #: 001209-98-9				FENCAMFAMINE

Formula: $C_{15}H_{21}N$

Mol Weight: 215.34

MP (deg C): FP (deg C):

BP (deg C): 128-131

BP pressure (mm Hg): 1.00E-001

Property/Value	Units	Temp	Data Type	Reference
WS 6.87E+002	mg/L	25	EST	MEYLAN,WM ET AL. (1996)
logP 3.20			EXP	SANGSTER,J (1994)
VP 4.31E-004	mm Hg	25	EST	NEELY,WB & BLAU,GE (1985)
DC	pKa			
HL 1.90E-006	atm m3/mol	25	EST	MEYLAN,WM & HOWARD,PH (1991)
OH 1.03E-010	cm3/molc sec	25	EST	MEYLAN,WM & HOWARD,PH (1993)

CAS #: 001210-12-4				9-ANTHRACENECARBONITRILE

Formula: $C_{15}H_9N$

Mol Weight: 203.25

MP (deg C): 173-177 FP (deg C):

BP (deg C):

BP pressure (mm Hg):

Property/Value	Units	Temp	Data Type	Reference
WS 4.09E-001	mg/L	25	EST	MEYLAN,WM ET AL. (1996)
logP 4.26			EXP	HANSCH,C ET AL. (1995)
VP 3.26E-006	mm Hg	25	EST	NEELY,WB & BLAU,GE (1985)
DC	pKa			
HL 4.96E-007	atm m3/mol	25	EST	MEYLAN,WM & HOWARD,PH (1991)
OH 7.06E-012	cm3/molc sec	25	EST	MEYLAN,WM & HOWARD,PH (1993)

CAS #: 001212-29-9				THIOUREA, N,N'-DICYCLOHEXYL-

Formula: $C_{13}H_{24}N_2S$

Mol Weight: 240.41

MP (deg C): FP (deg C):

BP (deg C):

BP pressure (mm Hg):

Property/Value	Units	Temp	Data Type	Reference
WS 1.92E+002	mg/L	25	EST	MEYLAN,WM ET AL. (1996)
logP 3.69			EXP	CHEM INSPECT TEST INST (1992)
VP 2.21E-005	mm Hg	25	EST	NEELY,WB & BLAU,GE (1985)
DC	pKa			
HL 2.52E-006	atm m3/mol	25	EST	MEYLAN,WM & HOWARD,PH (1991)
OH 1.95E-010	cm3/molc sec	25	EST	MEYLAN,WM & HOWARD,PH (1993)

CAS #: 001214-32-0				9H-FLUOREN-2-AMINE, 7-NITRO-

Formula: $C_{13}H_{10}N_2O_2$

Mol Weight: 226.24

MP (deg C): FP (deg C):

BP (deg C):

BP pressure (mm Hg):

Property/Value	Units	Temp	Data Type	Reference
WS 2.46E+000	mg/L	25	EST	MEYLAN,WM ET AL. (1996)
logP 3.06			EXP	DEBNATH,AK ET AL. (1991)
VP 3.95E-007	mm Hg	25	EST	NEELY,WB & BLAU,GE (1985)
DC	pKa			
HL 2.33E-010	atm m3/mol	25	EST	MEYLAN,WM & HOWARD,PH (1991)
OH 1.70E-010	cm3/molc sec	25	EST	MEYLAN,WM & HOWARD,PH (1993)

CAS #: 001214-39-7				1H-PURIN-6-AMINE, N-(PHENYLMETHYL)-

Formula: $C_{12}H_{11}N_5$

Mol Weight: 225.25

MP (deg C): 230-233 FP (deg C):

BP (deg C):

BP pressure (mm Hg):

Property/Value	Units	Temp	Data Type	Reference
WS 1.47E+003	mg/L	25	EST	MEYLAN,WM ET AL. (1996)
logP 1.57			EXP	SHAFER,WE (1990)
VP 4.06E-009	mm Hg	25	EST	NEELY,WB & BLAU,GE (1985)
DC	pKa			
HL 1.24E-014	atm m3/mol	25	EST	MEYLAN,WM & HOWARD,PH (1991)
OH 2.14E-010	cm3/molc sec	25	EST	MEYLAN,WM & HOWARD,PH (1993)

CAS #: 001216-40-6				BARBITURIC ACID,5(1-MEBU),5-B-BROMALLYL

Formula: $C_{12}H_{17}BrN_2O_3$

Mol Weight: 317.19

MP (deg C): FP (deg C):

BP (deg C):

BP pressure (mm Hg):

Property/Value	Units	Temp	Data Type	Reference
WS 1.60E+002	mg/L	25	EST	MEYLAN,WM ET AL. (1996)
logP 2.09			EXP	HANSCH,C & LEO,AJ (1985)
VP 8.23E-013	mm Hg	25	EST	NEELY,WB & BLAU,GE (1985)
DC	pKa			
HL 1.66E-013	atm m3/mol	25	EST	MEYLAN,WM & HOWARD,PH (1991)
OH 3.12E-011	cm3/molc sec	25	EST	MEYLAN,WM & HOWARD,PH (1993)

CAS #: 001220-55-9				UREA, N-(1-PYRROLIDINYL)-N'-[(4-METHYLPHENYL)SUL

Formula: $C_{12}H_{17}N_3O_3S$

Mol Weight: 283.35

MP (deg C): FP (deg C):

BP (deg C):

BP pressure (mm Hg):

Property/Value	Units	Temp	Data Type	Reference
WS 4.27E+003	mg/L	25	EST	MEYLAN,WM ET AL. (1996)
logP 0.65			EXP	SANGSTER,J (1993)
VP 6.34E-009	mm Hg	25	EST	NEELY,WB & BLAU,GE (1985)
DC	pKa			
HL 7.71E-013	atm m3/mol	25	EST	MEYLAN,WM & HOWARD,PH (1991)
OH 2.26E-011	cm3/molc sec	25	EST	MEYLAN,WM & HOWARD,PH (1993)

CAS #: 001220-83-3				BENZENESULFONAMIDE, 4-AMINO-N-(6-METHOXY-4-PYRIM

Formula: $C_{11}H_{12}N_4O_3S$

Mol Weight: 280.31

MP (deg C): FP (deg C):

BP (deg C):

BP pressure (mm Hg):

Property/Value	Units	Temp	Data Type	Reference
WS 4.03E+003	mg/L	25	EST	MEYLAN,WM ET AL. (1996)
logP 0.70			EXP	SANGSTER,J (1994)
VP 8.35E-009	mm Hg	25	EST	NEELY,WB & BLAU,GE (1985)
DC	pKa			
HL 2.41E-014	atm m3/mol	25	EST	MEYLAN,WM & HOWARD,PH (1991)
OH 7.93E-011	cm3/molc sec	25	EST	MEYLAN,WM & HOWARD,PH (1993)

CAS #: 001220-94-6 — C.I. DISPERSE VIOLET 4

Formula: $C_{15}H_{12}N_2O_2$

Mol Weight: 252.28

MP (deg C): 193

FP (deg C):

BP (deg C):

BP pressure (mm Hg):

Property/Value	Units	Temp	Data Type	Reference
WS 5.80E-001	mg/L	25	EST	BAUGHMAN,GL & PERENICH,TA (1988)
logP 3.71			EST	MEYLAN,WM & HOWARD,PH (1995)
VP 1.95E-008	mm Hg	25	EST	NEELY,WB & BLAU,GE (1985)
DC	pKa			
HL 8.72E-016	atm m3/mol	25	EST	MEYLAN,WM & HOWARD,PH (1991)
OH 6.98E-011	cm3/molc sec	25	EST	MEYLAN,WM & HOWARD,PH (1993)

CAS #: 001225-64-5 — 10H-PHENOTHIAZINE-10-PROPANAMINE, 2-CHLORO-N-MET

Formula: $C_{16}H_{17}ClN_2S$

Mol Weight: 304.84

MP (deg C):

FP (deg C):

BP (deg C):

BP pressure (mm Hg):

Property/Value	Units	Temp	Data Type	Reference
WS 5.50E-001	mg/L	25	EST	MEYLAN,WM ET AL. (1996)
logP 4.94			EXP	SANGSTER,J (1993)
VP 1.83E-007	mm Hg	25	EST	NEELY,WB & BLAU,GE (1985)
DC	pKa			
HL 1.68E-010	atm m3/mol	25	EST	MEYLAN,WM & HOWARD,PH (1991)
OH 2.40E-010	cm3/molc sec	25	EST	MEYLAN,WM & HOWARD,PH (1993)

CAS #: 001226-52-4 — 2,4,7-PTERIDINETRIAMINE, 6-(4-HYDROXYPHENYL)-

Formula: $C_{12}H_{11}N_7O$

Mol Weight: 269.27

MP (deg C):

FP (deg C):

BP (deg C):

BP pressure (mm Hg):

Property/Value	Units	Temp	Data Type	Reference
WS 1.15E+004	mg/L	25	EST	MEYLAN,WM ET AL. (1996)
logP 0.92			EXP	SANGSTER,J (1993)
VP 1.70E-012	mm Hg	25	EST	NEELY,WB & BLAU,GE (1985)
DC	pKa			
HL 1.93E-022	atm m3/mol	25	EST	MEYLAN,WM & HOWARD,PH (1991)
OH 2.24E-010	cm3/molc sec	25	EST	MEYLAN,WM & HOWARD,PH (1993)

CAS #: 001229-55-6 — SOLVENT RED 1

Formula: $C_{17}H_{14}N_2O_2$

Mol Weight: 278.31

MP (deg C): 183

FP (deg C):

BP (deg C):

BP pressure (mm Hg):

Property/Value	Units	Temp	Data Type	Reference
WS 3.30E-004	mg/L	25	EXP	BAUGHMAN,GL & WEBER,EJ (1991)
logP 7.50			EXP	HOU,M ET AL. (1991)
VP 6.29E-009	mm Hg	25	EST	NEELY,GM & BLAU,GE (1986)
DC	pKa			
HL 1.07E-010	atm m3/mol	25	EST	MEYLAN,WM & HOWARD,PH (1991)
OH 2.68E-011	cm3/molc sec	25	EST	MEYLAN,WM & HOWARD,PH (1993)

CAS #: 001241-94-7 — 2-ETHYLHEXYL DIPHENYL PHOSPHATE

Formula: $C_{20}H_{27}O_4P$

Mol Weight: 362.41

MP (deg C): -80

FP (deg C): 200

BP (deg C): 239

BP pressure (mm Hg):

Property/Value	Units	Temp	Data Type	Reference
WS 1.90E+000	mg/L	25	EXP	SAEGER,VW ET AL. (1979)
logP 5.73			EXP	HANSCH,C & LEO,AJ (1985)
VP 6.29E-005	mm Hg	25	EST	HL X WSOL
DC	pKa			
HL 1.58E-005	atm m3/mol	25	EXP	MUIR,DCG (1984)
OH 4.82E-011	cm3/molc sec	25	EST	MEYLAN,WM & HOWARD,PH (1993)

CAS #: 001260-17-9 — CARMINIC ACID

Formula: $C_{22}H_{20}O_{13}$

Mol Weight: 492.40

MP (deg C): 136 dec

FP (deg C):

BP (deg C):

BP pressure (mm Hg):

Property/Value	Units	Temp	Data Type	Reference
WS 1.30E+003	mg/L	25	EXP	YALKOWSKY,SH & DANNENFELSER,RM (1992)
logP 0.97			EST	MEYLAN,WM & HOWARD,PH (1995)
VP 9.71E-027	mm Hg	25	EST	NEELY,WB & BLAU,GE (1985)
DC	pKa			
HL 6.57E-032	atm m3/mol	25	EST	MEYLAN,WM & HOWARD,PH (1991)
OH 2.63E-010	cm3/molc sec	25	EST	MEYLAN,WM & HOWARD,PH (1993)

CAS #: 001320-98-5 — 4-METHYLPENTANOL

Formula: $C_6H_{14}O$

Mol Weight: 102.18

MP (deg C):

FP (deg C):

BP (deg C):

BP pressure (mm Hg):

Property/Value	Units	Temp	Data Type	Reference
WS 1.20E+004	mg/L	25	EST	MEYLAN,WM ET AL. (1996)
logP 1.75			EST	MEYLAN,WM & HOWARD,PH (1995)
VP 1.70E+000	mm Hg	25	EST	NEELY,WB & BLAU,GE (1985)
DC	pKa			
HL 1.76E-005	atm m3/mol	25	EST	MEYLAN,WM & HOWARD,PH (1991)
OH 9.71E-012	cm3/molc sec	25	EST	MEYLAN,WM & HOWARD,PH (1993)

CAS #: 001321-89-7 — ISODECYLADEHYDE

Formula: $C_{10}H_{20}O$

Mol Weight: 156.27

MP (deg C):

FP (deg C):

BP (deg C): 197

BP pressure (mm Hg):

Property/Value	Units	Temp	Data Type	Reference
WS 5.03E+001	mg/L	25	EST	MEYLAN,WM ET AL. (1996)
logP 3.69			EST	MEYLAN,WM & HOWARD,PH (1995)
VP 2.90E-001	mm Hg	25	EST	NEELY,WB & BLAU,GE (1985)
DC	pKa			
HL 6.54E-004	atm m3/mol	25	EST	MEYLAN,WM & HOWARD,PH (1991)
OH 3.45E-011	cm3/molc sec	25	EST	MEYLAN,WM & HOWARD,PH (1993)

CAS #: 001330-61-6 — ISODECYL ACRYLATE

Formula: $C_{13}H_{24}O_2$

Mol Weight: 212.34

MP (deg C): -100

FP (deg C):

BP (deg C): 158

BP pressure (mm Hg): 5.00E+001

Property/Value	Units	Temp	Data Type	Reference
WS 1.75E+000	mg/L	25	EST	MEYLAN,WM ET AL. (1996)
logP 5.07			EST	MEYLAN,WM & HOWARD,PH (1995)
VP 2.27E-002	mm Hg	25	EST	NEELY,WB & BLAU,GE (1985)
DC	pKa			
HL 1.18E-003	atm m3/mol	25	EST	MEYLAN,WM & HOWARD,PH (1991)
OH 2.22E-011	cm3/molc sec	25	EST	MEYLAN,WM & HOWARD,PH (1993)

CAS #: 001330-78-5 — TRICRESYL PHOSPHATE

Formula: $C_{21}H_{21}O_4P$

Mol Weight: 368.37

MP (deg C): -33

FP (deg C):

BP (deg C): 265

BP pressure (mm Hg):

MIXTURE OF CRESOL ISOMERS

Property/Value	Units	Temp	Data Type	Reference
WS 3.60E-001	mg/L	25	EXP	SAEGER,VW ET AL. (1979)
logP 5.11			EXP	SAEGER,VW ET AL. (1979)
VP 6.00E-007	mm Hg	25	EXT	BOETHLING,RS & COOPER,JC (1985)
DC	pKa			
HL 8.08E-007	atm m3/mol	25	EST	VP/WSOL
OH 1.44E-011	cm3/molc sec	25	EST	ATKINSON,R (1987)

CAS #: 001333-07-9 — M-TOLUENESULFONAMIDE

Formula: $C_7H_9NO_2S$

Mol Weight: 171.22

MP (deg C):

FP (deg C):

BP (deg C):

BP pressure (mm Hg):

Property/Value	Units	Temp	Data Type	Reference
WS 1.14E+004	mg/L	25	EST	MEYLAN,WM ET AL. (1996)
logP 0.92			EST	MEYLAN,WM & HOWARD,PH (1995)
VP 3.06E-004	mm Hg	25	EST	NEELY,WB & BLAU,GE (1985)
DC	pKa			
HL 4.66E-007	atm m3/mol	25	EST	MEYLAN,WM & HOWARD,PH (1991)
OH 1.22E-012	cm3/molc sec	25	EST	MEYLAN,WM & HOWARD,PH (1993)

CAS #: 001335-40-6 — ETHYL 2-FUROATE

Formula: $C_7H_8O_3$

Mol Weight: 140.14

MP (deg C):

FP (deg C):

BP (deg C):

BP pressure (mm Hg):

Property/Value	Units	Temp	Data Type	Reference
WS 4.21E+003	mg/L	25	EST	MEYLAN,WM ET AL. (1996)
logP 1.44			EST	MEYLAN,WM & HOWARD,PH (1995)
VP 4.05E-001	mm Hg	25	EXT	PERRY,RH & GREEN,D (1984)
DC	pKa			
HL 4.60E-005	atm m3/mol	25	EST	MEYLAN,WM & HOWARD,PH (1991)
OH 1.48E-011	cm3/molc sec	25	EST	MEYLAN,WM & HOWARD,PH (1993)

CAS #: 001338-23-4 — PEROXIDE-2-BUTANONE

Formula: $C_4H_8O_2$

Mol Weight: 88.11

MP (deg C):

FP (deg C):

BP (deg C):

BP pressure (mm Hg):

Property/Value	Units	Temp	Data Type	Reference
WS 1.37E+005	mg/L	25	EST	MEYLAN,WM ET AL. (1996)
logP -0.43			EST	MEYLAN,WM & HOWARD,PH (1995)
VP 4.50E-002	mm Hg	25	EST	NEELY,WB & BLAU,GE (1985)
DC	pKa			
HL 1.74E-009	atm m3/mol	25	EST	MEYLAN,WM & HOWARD,PH (1991)
OH 9.21E-012	cm3/molc sec	25	EST	MEYLAN,WM & HOWARD,PH (1993)

CAS #: 001404-64-4 — SPARSOMYCIN

Formula: $C_{13}H_{19}N_3O_5S_2$

Mol Weight: 361.44

MP (deg C):

FP (deg C):

BP (deg C):

BP pressure (mm Hg):

Property/Value	Units	Temp	Data Type	Reference
WS 7.47E+003	mg/L	25	EST	MEYLAN,WM ET AL. (1996)
logP -1.71			EXP	HANSCH,C ET AL. (1995)
VP 2.10E-020	mm Hg	25	EST	NEELY,WB & BLAU,GE (1985)
DC	pKa			
HL 1.67E-029	atm m3/mol	25	EST	MEYLAN,WM & HOWARD,PH (1991)
OH 1.71E-010	cm3/molc sec	25	EST	MEYLAN,WM & HOWARD,PH (1993)

CAS #: 001405-86-3 — GLYCYRRHIZIC ACID

Formula: $C_{42}H_{62}O_{16}$

Mol Weight: 822.95

MP (deg C): 220 dec

FP (deg C):

BP (deg C):

BP pressure (mm Hg):

Property/Value	Units	Temp	Data Type	Reference
WS 5.30E-002	mg/L	25	EST	MEYLAN,WM ET AL. (1996)
logP 2.80			EXP	SANGSTER,J (1993)
VP 5.22E-034	mm Hg	25	EST	NEELY,WB & BLAU,GE (1985)
DC	pKa			
HL 1.23E-035	atm m3/mol	25	EST	MEYLAN,WM & HOWARD,PH (1991)
OH 2.30E-010	cm3/molc sec	25	EST	MEYLAN,WM & HOWARD,PH (1993)

CAS #: 001415-73-2 — ALOIN

Formula: $C_{21}H_{22}O_9$

Mol Weight: 418.40

MP (deg C):

FP (deg C):

BP (deg C):

BP pressure (mm Hg):

Property/Value	Units	Temp	Data Type	Reference
WS 8.30E+003	mg/L	20	EXP	SEIDELL,A (1941)
logP 0.55			EST	MEYLAN,WM & HOWARD,PH (1995)
VP 5.28E-021	mm Hg	25	EST	NEELY,WB & BLAU,GE (1985)
DC	pKa			
HL 2.61E-021	atm m3/mol	25	EST	MEYLAN,WM & HOWARD,PH (1991)
OH 2.77E-010	cm3/molc sec	25	EST	MEYLAN,WM & HOWARD,PH (1993)

CAS #: 001420-06-0 — TRIFENMORPH

Formula: $C_{23}H_{23}NO$

Mol Weight: 329.45

MP (deg C): 174-176

Property/Value	Units	Temp	Data Type	Reference
WS 2.00E-002	mg/L	20	EXP	YALKOWSKY,SH & DANNENFELSER,RM (1992)
logP 4.21			EST	MEYLAN,WM & HOWARD,PH (1995)
VP 3.31E-008	mm Hg	25	EST	NEELY,WB & BLAU,GE (1985)
DC	pKa			
HL 1.32E-010	atm m3/mol	25	EST	MEYLAN,WM & HOWARD,PH (1991)
OH 1.54E-010	cm3/molc sec	25	EST	MEYLAN,WM & HOWARD,PH (1993)

CAS #: 001420-55-9 — ETHYLTHIOPERAZINE

Formula: $C_{22}H_{29}N_3S_2$

Mol Weight: 399.62

MP (deg C): 62-64

BP (deg C): 227

BP pressure (mm Hg): 1.00E-002

Property/Value	Units	Temp	Data Type	Reference
WS 5.84E-002	mg/L	25	EST	MEYLAN,WM ET AL. (1996)
logP 5.41			EXP	HANSCH,C & LEO,AJ (1985)
VP 1.89E-010	mm Hg	25	EST	NEELY,WB & BLAU,GE (1985)
DC	pKa			
HL 5.59E-015	atm m3/mol	25	EST	MEYLAN,WM & HOWARD,PH (1991)
OH 4.06E-010	cm3/molc sec	25	EST	MEYLAN,WM & HOWARD,PH (1993)

CAS #: 001422-07-7 — CODEINE HYDROCHLORIDE

Formula: $C_{18}H_{22}ClNO_3$

Mol Weight: 335.83

MP (deg C): 280

Property/Value	Units	Temp	Data Type	Reference
WS 3.39E+004	mg/L	16	EXP	STEPHEN,H & STEPHEN,T (1963)
logP 1.28			EST	MEYLAN,WM & HOWARD,PH (1995)
VP 4.15E-009	mm Hg	25	EST	NEELY,WB & BLAU,GE (1985)
DC	pKa			
HL 7.58E-014	atm m3/mol	25	EST	MEYLAN,WM & HOWARD,PH (1991)
OH 2.36E-010	cm3/molc sec	25	EST	MEYLAN,WM & HOWARD,PH (1993)

CAS #: 001427-45-8 — 1,4-BENZODIAZPN2-ON,5,(2CF3PH)7NO2

Formula: $C_{16}H_{10}F_3N_3O_3$

Mol Weight: 349.27

Property/Value	Units	Temp	Data Type	Reference
WS 1.83E+001	mg/L	25	EST	MEYLAN,WM ET AL. (1996)
logP 2.51			EXP	HANSCH,C & LEO,AJ (1985)
VP 3.75E-010	mm Hg	25	EST	NEELY,WB & BLAU,GE (1985)
DC	pKa			
HL 8.23E-012	atm m3/mol	25	EST	MEYLAN,WM & HOWARD,PH (1991)
OH 3.43E-012	cm3/molc sec	25	EST	MEYLAN,WM & HOWARD,PH (1993)

CAS #: 001431-39-6 — 1-DIME AMINO-5-SO2NH2-NAPHTHALENE

Formula: $C_{12}H_{14}N_2O_2S$

Mol Weight: 250.32

MP (deg C): 218-221

Property/Value	Units	Temp	Data Type	Reference
WS 4.53E+002	mg/L	25	EST	MEYLAN,WM ET AL. (1996)
logP 2.01			EXP	HANSCH,C & LEO,AJ (1985)
VP 2.72E-007	mm Hg	25	EST	NEELY,WB & BLAU,GE (1985)
DC	pKa			
HL 6.55E-010	atm m3/mol	25	EST	MEYLAN,WM & HOWARD,PH (1991)
OH 8.39E-011	cm3/molc sec	25	EST	MEYLAN,WM & HOWARD,PH (1993)

CAS #: 001435-60-5 — 4-(4-NITROPHENYLAZO)PHENOL

Formula: $C_{12}H_9N_3O_3$

Mol Weight: 243.22

Property/Value	Units	Temp	Data Type	Reference
WS 3.89E+000	mg/L	25	EXP	BAUGHMAN,GL & PERENICH,TA (1988)
logP 4.03			EST	MEYLAN,WM & HOWARD,PH (1995)
VP 3.00E-010	mm Hg	25	EXP	BAUGHMAN,GL & PERENICH,TA (1988)
DC	pKa			
HL 2.47E-011	atm m3/mol	25	EST	VP/WSOL
OH 1.34E-011	cm3/molc sec	25	EST	MEYLAN,WM & HOWARD,PH (1993)

CAS #: 001436-34-6 — OXIRANE, BUTYL-

Formula: $C_6H_{12}O$

Mol Weight: 100.16

MP (deg C): 93-95

Property/Value	Units	Temp	Data Type	Reference
WS 2.53E+003	mg/L	25	EST	MEYLAN,WM ET AL. (1996)
logP 1.95			EXP	DENEER,JW ET AL. (1988)
VP 3.28E+001	mm Hg	25	EST	NEELY,WB & BLAU,GE (1985)
DC	pKa			
HL 3.73E-004	atm m3/mol	25	EST	MEYLAN,WM & HOWARD,PH (1991)
OH 4.64E-012	cm3/molc sec	25	EST	MEYLAN,WM & HOWARD,PH (1993)

CAS #: 001436-43-7 — 2-CYANOQUINOLINE

Formula: $C_{10}H_6N_2$

Mol Weight: 154.17

MP (deg C): 95

BP (deg C): 160

BP pressure (mm Hg): 2.30E+001

Property/Value	Units	Temp	Data Type	Reference
WS 1.70E+003	mg/L	25	EST	MEYLAN,WM ET AL. (1996)
logP 1.60			EXP	HANSCH,C & LEO,AJ (1985)
VP 3.56E-004	mm Hg	25	EST	NEELY,WB & BLAU,GE (1985)
DC	pKa			
HL 6.65E-009	atm m3/mol	25	EST	MEYLAN,WM & HOWARD,PH (1991)
OH 2.54E-012	cm3/molc sec	25	EST	MEYLAN,WM & HOWARD,PH (1993)

001437-15-6 — 1,2-(DI-(A-PYRIDYL)ETHYLENE

Formula: $C_{12}H_{10}N_2$

Mol Weight: 182.23

MP (deg C): 118-119

BP (deg C): 150-160

BP pressure (mm Hg): 2.00E+000

Property/Value	Units	Temp	Data Type	Reference
WS 8.47E+002	mg/L	25	EST	MEYLAN,WM ET AL. (1996)
logP 2.11			EXP	HANSCH,C & LEO,AJ (1985)
VP 3.78E-004	mm Hg	25	EST	NEELY,WB & BLAU,GE (1985)
DC	pKa			
HL 2.12E-010	atm m3/mol	25	EST	MEYLAN,WM & HOWARD,PH (1991)
OH 5.71E-011	cm3/molc sec	25	EST	MEYLAN,WM & HOWARD,PH (1993)

001441-44-7 — 4-(N,N-DIME)-3-PYRIDYLBUTYLAMINE

Formula: $C_{11}H_{18}N_2$

Mol Weight: 178.28

MP (deg C):

BP (deg C):

BP pressure (mm Hg):

Property/Value	Units	Temp	Data Type	Reference
WS 6.09E+005	mg/L	25	EST	MEYLAN,WM ET AL. (1996)
logP 1.49			EXP	HANSCH,C & LEO,AJ (1985)
VP 1.11E-002	mm Hg	25	EST	NEELY,WB & BLAU,GE (1985)
DC	pKa			
HL 9.03E-009	atm m3/mol	25	EST	MEYLAN,WM & HOWARD,PH (1991)
OH 8.41E-011	cm3/molc sec	25	EST	MEYLAN,WM & HOWARD,PH (1993)

001443-80-7 — P-CYANOACETOPHENONE

Formula: C_9H_7NO

Mol Weight: 145.16

MP (deg C): 57-58

BP (deg C):

BP pressure (mm Hg):

Property/Value	Units	Temp	Data Type	Reference
WS 3.93E+003	mg/L	25	EST	MEYLAN,WM ET AL. (1996)
logP 1.22			EXP	HANSCH,C & LEO,AJ (1985)
VP 6.42E-003	mm Hg	25	EST	NEELY,WB & BLAU,GE (1985)
DC	pKa			
HL 9.48E-008	atm m3/mol	25	EST	MEYLAN,WM & HOWARD,PH (1991)
OH 3.35E-013	cm3/molc sec	25	EST	MEYLAN,WM & HOWARD,PH (1993)

001443-94-3 — UREA, N-(1-PIPERIDINYL)-N'-[(4-METHYLPHENYL)SULF

Formula: $C_{13}H_{19}N_3O_3S$

Mol Weight: 297.38

MP (deg C):

BP (deg C):

BP pressure (mm Hg):

Property/Value	Units	Temp	Data Type	Reference
WS 9.12E+002	mg/L	25	EST	MEYLAN,WM ET AL. (1996)
logP 1.34			EXP	SANGSTER,J (1993)
VP 2.61E-009	mm Hg	25	EST	NEELY,WB & BLAU,GE (1985)
DC	pKa			
HL 1.02E-012	atm m3/mol	25	EST	MEYLAN,WM & HOWARD,PH (1991)
OH 2.88E-011	cm3/molc sec	25	EST	MEYLAN,WM & HOWARD,PH (1993)

001445-45-0 — 1,1,1-TRIMETHOXYETHANE

Formula: $C_5H_{12}O_3$

Mol Weight: 120.15

MP (deg C):

BP (deg C): 108

BP pressure (mm Hg):

Property/Value	Units	Temp	Data Type	Reference
WS 1.08E+004	mg/L	25	EST	MEYLAN,WM ET AL. (1996)
logP 1.13			EST	MEYLAN,WM & HOWARD,PH (1995)
VP 4.50E+001	mm Hg	25	EST	NEELY,WB & BLAU,GE (1985)
DC	pKa			
HL 7.86E-007	atm m3/mol	25	EST	MEYLAN,WM & HOWARD,PH (1991)
OH 2.96E-012	cm3/molc sec	25	EST	MEYLAN,WM & HOWARD,PH (1993)

001445-69-8 — 1,4-PHTHALAZINEDIONE, 2,3-DIHYDRO

Formula: $C_8H_6N_2O_2$

Mol Weight: 162.15

MP (deg C): > 300

BP (deg C):

BP pressure (mm Hg):

Property/Value	Units	Temp	Data Type	Reference
WS 2.64E+003	mg/L	25	EST	MEYLAN,WM ET AL. (1996)
logP 0.11			EXP	HALL,IH ET AL. (1989)
VP 5.49E-008	mm Hg	25	EST	NEELY,WB & BLAU,GE (1985)
DC	pKa			
HL 4.27E-011	atm m3/mol	25	EST	MEYLAN,WM & HOWARD,PH (1991)
OH 1.17E-011	cm3/molc sec	25	EST	MEYLAN,WM & HOWARD,PH (1993)

001445-75-6 — PHOSPHONIC ACID, METHYL-, DIISOPROPYL ESTER

Formula: $C_7H_{17}O_3P$

Mol Weight: 180.19

MP (deg C):

BP (deg C):

BP pressure (mm Hg):

Property/Value	Units	Temp	Data Type	Reference
WS 7.25E+003	mg/L	25	EST	MEYLAN,WM ET AL. (1996)
logP 1.03			EXP	KRIKORIAN,SE ET AL. (1987)
VP 1.16E-003	mm Hg	25	EXT	BOUBLIK,T ET AL. (1984)
DC	pKa			
HL 3.88E-006	atm m3/mol	25	EST	MEYLAN,WM & HOWARD,PH (1991)
OH 8.03E-011	cm3/molc sec	25	EST	MEYLAN,WM & HOWARD,PH (1993)

001448-87-9 — 2-CHLOROQUINOXALINE

Formula: $C_8H_5ClN_2$

Mol Weight: 164.60

MP (deg C):

BP (deg C):

BP pressure (mm Hg):

Property/Value	Units	Temp	Data Type	Reference
WS 8.12E+002	mg/L	25	EST	MEYLAN,WM ET AL. (1996)
logP 2.23			EXP	HANSCH,C & LEO,AJ (1985)
VP 2.46E-003	mm Hg	25	EST	NEELY,WB & BLAU,GE (1985)
DC -1.60	pKa	20	EXP	PERRIN,DD (1965)
HL 2.11E-007	atm m3/mol	25	EST	MEYLAN,WM & HOWARD,PH (1991)
OH 1.41E-012	cm3/molc sec	25	EST	MEYLAN,WM & HOWARD,PH (1993)

CAS #: 001450-72-2				2-ACETYL-4-METHYLPHENOL

Formula: $C_9H_{10}O_2$

Mol Weight: 150.18

MP (deg C): 50   FP (deg C):

BP (deg C): 210

BP pressure (mm Hg):

Property/Value	Units	Temp	Data Type	Reference
WS 2.04E+003	mg/L	25	EST	MEYLAN,WM ET AL. (1996)
logP 2.52			EST	MEYLAN,WM & HOWARD,PH (1995)
VP 2.06E-003	mm Hg	25	EST	NEELY,WB & BLAU,GE (1985)
DC	pKa			
HL 1.42E-006	atm m3/mol	25	EST	MEYLAN,WM & HOWARD,PH (1991)
OH 3.76E-011	cm3/molc sec	25	EST	MEYLAN,WM & HOWARD,PH (1993)

CAS #: 001451-20-3				6H-DIBENZO[B,D]PYRAN, 6A,7,10,10A-TETRAHYDRO-1-M

Formula: $C_{22}H_{32}O_2$

Mol Weight: 328.50

MP (deg C):   FP (deg C):

BP (deg C):

BP pressure (mm Hg):

Property/Value	Units	Temp	Data Type	Reference
WS 1.09E-003	mg/L	25	EST	MEYLAN,WM ET AL. (1996)
logP 8.06			EXP	SANGSTER,J (1993)
VP 5.84E-007	mm Hg	25	EST	NEELY,WB & BLAU,GE (1985)
DC	pKa			
HL 1.39E-004	atm m3/mol	25	EST	MEYLAN,WM & HOWARD,PH (1991)
OH 3.10E-010	cm3/molc sec	25	EST	MEYLAN,WM & HOWARD,PH (1993)

CAS #: 001452-77-3				2-PYRIDINECARBOXAMIDE

Formula: $C_6H_6N_2O$

Mol Weight: 122.13

MP (deg C): 110 dec   FP (deg C):

BP (deg C):

BP pressure (mm Hg):

Property/Value	Units	Temp	Data Type	Reference
WS 7.34E+004	mg/L	25	EST	MEYLAN,WM ET AL. (1996)
logP 0.29			EXP	YAMAGAMI,C ET AL. (1990A)
VP 4.20E-004	mm Hg	25	EST	NEELY,WB & BLAU,GE (1985)
DC 2.10	pKa	20	EXP	PERRIN,DD (1965)
HL 2.90E-012	atm m3/mol	25	EST	MEYLAN,WM & HOWARD,PH (1991)
OH 2.34E-012	cm3/molc sec	25	EST	MEYLAN,WM & HOWARD,PH (1993)

CAS #: 001453-82-3				I-NICOTINAMIDE

Formula: $C_6H_6N_2O$

Mol Weight: 122.13

MP (deg C): 155-157   FP (deg C):

BP (deg C):

BP pressure (mm Hg):

Property/Value	Units	Temp	Data Type	Reference
WS 1.71E+005	mg/L	25	EST	MEYLAN,WM ET AL. (1996)
logP -0.28			EXP	HANSCH,C & LEO,AJ (1985)
VP 4.20E-004	mm Hg	25	EST	NEELY,WB & BLAU,GE (1985)
DC 3.61	pKa	20	EXP	PERRIN,DD (1965)
HL 2.90E-012	atm m3/mol	25	EST	MEYLAN,WM & HOWARD,PH (1991)
OH 2.34E-012	cm3/molc sec	25	EST	MEYLAN,WM & HOWARD,PH (1993)

CAS #: 001454-80-4				[1,1'-BIPHENYL]-2,2'-DIAMINE

Formula: $C_{12}H_{12}N_2$

Mol Weight: 184.24

MP (deg C): 81   FP (deg C):

BP (deg C): 162

BP pressure (mm Hg): 4.00E+000

Property/Value	Units	Temp	Data Type	Reference
WS 3.77E+003	mg/L	25	EST	MEYLAN,WM ET AL. (1996)
logP 1.34			EXP	DEBNATH,AK ET AL. (1992)
VP 2.69E-006	mm Hg	25	EXT	YAWS,CL (1994B)
DC	pKa			
HL 5.17E-011	atm m3/mol	25	EST	MEYLAN,WM & HOWARD,PH (1991)
OH 1.54E-010	cm3/molc sec	25	EST	MEYLAN,WM & HOWARD,PH (1993)

CAS #: 001454-85-9				HEPTADECYL ALCOHOL

Formula: $C_{17}H_{36}O$

Mol Weight: 256.48

MP (deg C): 53.8   FP (deg C):

BP (deg C): 333

BP pressure (mm Hg):

Property/Value	Units	Temp	Data Type	Reference
WS 4.76E-002	mg/L	25	EST	MEYLAN,WM ET AL. (1996)
logP 7.23			EST	MEYLAN,WM & HOWARD,PH (1995)
VP 2.61E-006	mm Hg	25	EST	NEELY,WB & BLAU,GE (1985)
DC	pKa			
HL 3.97E-004	atm m3/mol	25	EST	MEYLAN,WM & HOWARD,PH (1991)
OH 2.53E-011	cm3/molc sec	25	EST	MEYLAN,WM & HOWARD,PH (1993)

CAS #: 001455-21-6				1-NONANETHIOL

Formula: $C_9H_{20}S$

Mol Weight: 160.32

MP (deg C): -20.1   FP (deg C):

BP (deg C): 220

BP pressure (mm Hg):

Property/Value	Units	Temp	Data Type	Reference
WS 6.54E+000	mg/L	25	EST	MEYLAN,WM ET AL. (1996)
logP 4.70			EST	MEYLAN,WM & HOWARD,PH (1995)
VP 1.32E-001	mm Hg	25	EXP	YAWS,CL (1994B)
DC	pKa			
HL 2.52E-002	atm m3/mol	25	EST	MEYLAN,WM & HOWARD,PH (1991)
OH 5.13E-011	cm3/molc sec	25	EST	MEYLAN,WM & HOWARD,PH (1993)

CAS #: 001455-77-2				1H-1,2,4-TRIAZOLE-3,5-DIAMINE

Formula: $C_2H_5N_5$

Mol Weight: 99.10

MP (deg C): 204-206 de   FP (deg C):

BP (deg C):

BP pressure (mm Hg):

Property/Value	Units	Temp	Data Type	Reference
WS 1.00E+006	mg/L	25	EST	MEYLAN,WM ET AL. (1996)
logP -1.61			EXP	HANSCH,C ET AL. (1995)
VP 1.34E-004	mm Hg	25	EST	NEELY,WB & BLAU,GE (1985)
DC	pKa			
HL 1.91E-013	atm m3/mol	25	EST	MEYLAN,WM & HOWARD,PH (1991)
OH 1.00E-013	cm3/molc sec	25	EST	MEYLAN,WM & HOWARD,PH (1993)

CAS #: 001455-84-1	5(2-PYRIDYL)-3-ME-1,2,4-OXADIAZOLE

Formula: $C_8H_8N_3O$

Mol Weight: 162.17

MP (deg C): | FP (deg C):

BP (deg C):

BP pressure (mm Hg):

Property/Value	Units	Temp	Data Type	Reference
WS 1.18E+004	mg/L	25	EST	MEYLAN,WM ET AL. (1996)
logP 0.89			EXP	HANSCH,C & LEO,AJ (1985)
VP 1.05E-003	mm Hg	25	EST	NEELY,WB & BLAU,GE (1985)
DC	pKa			
HL 2.60E-009	atm m3/mol	25	EST	MEYLAN,WM & HOWARD,PH (1991)
OH 4.78E-012	cm3/molc sec	25	EST	MEYLAN,WM & HOWARD,PH (1993)

CAS #: 001456-28-6	2,6-DIMETHYL-N-NITROSOMORPHOLINE

Formula: $C_6H_{12}N_2O_2$

Mol Weight: 144.17

MP (deg C): | FP (deg C):

BP (deg C):

BP pressure (mm Hg):

Property/Value	Units	Temp	Data Type	Reference
WS 1.24E+005	mg/L	25	EXP	MIRVISH,SS ET AL. (1976)
logP 0.32			EXP	HANSCH,C & LEO,AJ (1985)
VP 2.38E-002	mm Hg	25	EST	NEELY,WB & BLAU,GE (1985)
DC	pKa			
HL 2.48E-008	atm m3/mol	25	EST	MEYLAN,WM & HOWARD,PH (1991)
OH 5.08E-011	cm3/molc sec	25	EST	MEYLAN,WM & HOWARD,PH (1993)

CAS #: 001457-85-8	OXANILIC ACID,ETHYL ESTER

Formula: $C_{10}H_{11}NO_3$

Mol Weight: 193.20

MP (deg C): 65.5 | FP (deg C):

BP (deg C): 280

BP pressure (mm Hg):

Property/Value	Units	Temp	Data Type	Reference
WS 2.58E+003	mg/L	25	EST	MEYLAN,WM ET AL. (1996)
logP 1.48			EXP	HANSCH,C ET AL. (1995)
VP 1.04E-005	mm Hg	25	EST	NEELY,WB & BLAU,GE (1985)
DC	pKa			
HL 1.15E-010	atm m3/mol	25	EST	MEYLAN,WM & HOWARD,PH (1991)
OH 4.43E-011	cm3/molc sec	25	EST	MEYLAN,WM & HOWARD,PH (1993)

CAS #: 001459-09-2	N-HEXADECYLBENZENE

Formula: $C_{22}H_{38}$

Mol Weight: 302.55

MP (deg C): 27 | FP (deg C):

BP (deg C): 385

BP pressure (mm Hg):

Property/Value	Units	Temp	Data Type	Reference
WS 4.11E-005	mg/L	25	EST	MEYLAN,WM ET AL. (1996)
logP 9.91			EST	MEYLAN,WM & HOWARD,PH (1995)
VP 1.12E-008	mm Hg	25	EXT	ZWOLINSKI,BJ & WILHOIT,RC (1971)
DC	pKa			
HL 4.17E-001	atm m3/mol	25	EST	MEYLAN,WM & HOWARD,PH (1991)
OH 2.57E-011	cm3/molc sec	25	EST	MEYLAN,WM & HOWARD,PH (1993)

CAS #: 001459-10-5	1-PHENYLTETRADECANE

Formula: $C_{20}H_{34}$

Mol Weight: 274.49

MP (deg C): 16 | FP (deg C):

BP (deg C): 359

BP pressure (mm Hg):

Property/Value	Units	Temp	Data Type	Reference
WS 4.12E-004	mg/L	25	EST	MEYLAN,WM ET AL. (1996)
logP 8.92			EST	MEYLAN,WM & HOWARD,PH (1995)
VP 8.45E-007	mm Hg	25	EXT	ZWOLINSKI,BJ & WILHOIT,RC (1971)
DC	pKa			
HL 2.37E-001	atm m3/mol	25	EST	MEYLAN,WM & HOWARD,PH (1991)
OH 2.29E-011	cm3/molc sec	25	EST	MEYLAN,WM & HOWARD,PH (1993)

CAS #: 001460-06-6	1-CHLORO-1,2-DIPHENYLETHENE

Formula: $C_{14}H_{11}Cl$

Mol Weight: 214.70

MP (deg C): | FP (deg C):

BP (deg C):

BP pressure (mm Hg):

Property/Value	Units	Temp	Data Type	Reference
WS 1.91E+000	mg/L	25	EST	MEYLAN,WM ET AL. (1996)
logP 5.02			EST	MEYLAN,WM & HOWARD,PH (1995)
VP 7.22E-004	mm Hg	25	EST	NEELY,WB & BLAU,GE (1985)
DC	pKa			
HL 8.90E-005	atm m3/mol	25	EST	MEYLAN,WM & HOWARD,PH (1991)
OH 2.19E-011	cm3/molc sec	25	EST	MEYLAN,WM & HOWARD,PH (1993)

CAS #: 001460-57-7	1,2-CYCLOHEXANEDIOL-TRANS

Formula: $C_6H_{12}O_2$

Mol Weight: 116.16

MP (deg C): 101-104 | FP (deg C):

BP (deg C):

BP pressure (mm Hg):

Property/Value	Units	Temp	Data Type	Reference
WS 6.59E+004	mg/L	25	EST	MEYLAN,WM ET AL. (1996)
logP 0.08			EXP	HANSCH,C & LEO,AJ (1985)
VP 1.05E-002	mm Hg	25	EST	NEELY,WB & BLAU,GE (1985)
DC	pKa			
HL 1.79E-007	atm m3/mol	25	EST	MEYLAN,WM & HOWARD,PH (1991)
OH 2.65E-011	cm3/molc sec	25	EST	MEYLAN,WM & HOWARD,PH (1993)

CAS #: 001464-44-4	BETA-D-GLUCOPYRANOSIDE, PHENYL

Formula: $C_{12}H_{16}O_6$

Mol Weight: 256.26

MP (deg C): 176-178 | FP (deg C):

BP (deg C):

BP pressure (mm Hg):

Property/Value	Units	Temp	Data Type	Reference
WS 1.26E+005	mg/L	25	EST	MEYLAN,WM ET AL. (1996)
logP -0.75			EXP	HANSCH,C & LEO,AJ (1985)
VP 7.66E-011	mm Hg	25	EST	NEELY,WB & BLAU,GE (1985)
DC	pKa			
HL 1.15E-015	atm m3/mol	25	EST	MEYLAN,WM & HOWARD,PH (1991)
OH 8.35E-011	cm3/molc sec	25	EST	MEYLAN,WM & HOWARD,PH (1993)

CAS #: 001464-53-5				DIEPOXYBUTANE
Formula: $C_4H_6O_2$				
Mol Weight: 86.09				
MP (deg C): 4 EST		FP (deg C):		
BP (deg C): 144				
BP pressure (mm Hg):				

Property/Value	Units	Temp	Data Type	Reference
WS 1.00E+006	mg/L		EXP	DEAN,JA (1985)
logP -0.28			EXP	DENEER,JW ET AL. (1988)
VP 5.68E+001	mm Hg	25	EST	NEELY,WB & BLAU,GE (1985)
DC	pKa			
HL 3.54E-008	atm m3/mol	25	EST	MEYLAN,WM & HOWARD,PH (1991)
OH 8.56E-013	cm3/molc sec	25	EST	MEYLAN,WM & HOWARD,PH (1993)

CAS #: 001466-76-8				2,6-DIMETHOXYBENZOIC ACID
Formula: $C_9H_{10}O_4$				
Mol Weight: 182.18				
MP (deg C): 186-187 de		FP (deg C):		
BP (deg C):				
BP pressure (mm Hg):				

Property/Value	Units	Temp	Data Type	Reference
WS 1.47E+004	mg/L	25	EST	MEYLAN,WM ET AL. (1996)
logP 0.66			EXP	HANSCH,C & LEO,AJ (1985)
VP 1.90E-007	mm Hg	25	EXT	COLOMINA,M ET AL. (1985)
DC 3.44	pKa	25	EXP	SERJEANT,EP & DEMPSEY,B (1979)
HL 3.79E-010	atm m3/mol	25	EST	MEYLAN,WM & HOWARD,PH (1991)
OH 9.00E-011	cm3/molc sec	25	EST	MEYLAN,WM & HOWARD,PH (1993)

CAS #: 001467-23-8				BENZENESULFONAMIDE, N-[(ETHYLAMINOCARBONYL)-4-ME
Formula: $C_{10}H_{14}N_2O_3S$				
Mol Weight: 242.30				
MP (deg C):		FP (deg C):		
BP (deg C):				
BP pressure (mm Hg):				

Property/Value	Units	Temp	Data Type	Reference
WS 2.77E+003	mg/L	25	EST	MEYLAN,WM ET AL. (1996)
logP 1.14			EXP	SANGSTER,J (1993)
VP 1.49E-007	mm Hg	25	EST	NEELY,WB & BLAU,GE (1985)
DC	pKa			
HL 1.44E-010	atm m3/mol	25	EST	MEYLAN,WM & HOWARD,PH (1991)
OH 7.22E-012	cm3/molc sec	25	EST	MEYLAN,WM & HOWARD,PH (1993)

CAS #: 001467-34-1				3,4,5-TRIMETHYLACETANILIDE
Formula: $C_{11}H_{15}NO$				
Mol Weight: 177.25				
MP (deg C):		FP (deg C):		
BP (deg C):				
BP pressure (mm Hg):				

Property/Value	Units	Temp	Data Type	Reference
WS 4.16E+002	mg/L	25	EST	MEYLAN,WM ET AL. (1996)
logP 2.50			EXP	NAKAGAWA,Y ET AL. (1992)
VP 2.61E-005	mm Hg	25	EST	NEELY,WB & BLAU,GE (1985)
DC	pKa			
HL 8.29E-009	atm m3/mol	25	EST	MEYLAN,WM & HOWARD,PH (1991)
OH 1.04E-010	cm3/molc sec	25	EST	MEYLAN,WM & HOWARD,PH (1993)

CAS #: 001467-79-4				DIMETHYL CYANAMIDE
Formula: $C_3H_6N_2$				
Mol Weight: 70.09				
MP (deg C):		FP (deg C):		
BP (deg C): 163.5				
BP pressure (mm Hg):				

Property/Value	Units	Temp	Data Type	Reference
WS 1.82E+005	mg/L	25	EST	MEYLAN,WM ET AL. (1996)
logP -0.15			EXP	HANSCH,C & LEO,AJ (1985)
VP 1.60E+001	mm Hg	25	EST	NEELY,WB & BLAU,GE (1985)
DC	pKa			
HL 1.60E-004	atm m3/mol	25	EST	MEYLAN,WM & HOWARD,PH (1991)
OH 2.53E-012	cm3/molc sec	25	EST	MEYLAN,WM & HOWARD,PH (1993)

CAS #: 001468-83-3				3-ACETYLTHIOPHENE
Formula: C_6H_6OS				
Mol Weight: 126.18				
MP (deg C): 59-63		FP (deg C):		
BP (deg C): 208-210				
BP pressure (mm Hg): 7.48E+002				

Property/Value	Units	Temp	Data Type	Reference
WS 8.31E+003	mg/L	25	EST	MEYLAN,WM ET AL. (1996)
logP 1.24			EXP	HANSCH,C & LEO,AJ (1985)
VP 3.65E-001	mm Hg	25	EST	NEELY,WB & BLAU,GE (1985)
DC	pKa			
HL 5.33E-006	atm m3/mol	25	EST	MEYLAN,WM & HOWARD,PH (1991)
OH 8.79E-012	cm3/molc sec	25	EST	MEYLAN,WM & HOWARD,PH (1993)

CAS #: 001470-44-6				1H-INDENE-1,3(2H)-DIONE, 2-(3-CHLOROPHENYL)-
Formula: $C_{15}H_9ClO_2$				
Mol Weight: 256.69				
MP (deg C):		FP (deg C):		
BP (deg C):				
BP pressure (mm Hg):				

Property/Value	Units	Temp	Data Type	Reference
WS 1.79E+001	mg/L	25	EST	MEYLAN,WM ET AL. (1996)
logP 3.61			EXP	SANGSTER,J (1994)
VP 7.45E-007	mm Hg	25	EST	NEELY,WB & BLAU,GE (1985)
DC	pKa			
HL 3.55E-010	atm m3/mol	25	EST	MEYLAN,WM & HOWARD,PH (1991)
OH 4.88E-012	cm3/molc sec	25	EST	MEYLAN,WM & HOWARD,PH (1993)

CAS #: 001470-94-6				5-INDANOL
Formula: $C_9H_{10}O$				
Mol Weight: 134.18				
MP (deg C): 58		FP (deg C):		
BP (deg C): 253				
BP pressure (mm Hg):				

Property/Value	Units	Temp	Data Type	Reference
WS 6.30E+003	mg/L	25	EXP	SOUTHWORTH,GR & KELLER,JL (1986)
logP 2.99			EST	MEYLAN,WM & HOWARD,PH (1995)
VP 1.35E-002	mm Hg	25	EST	NEELY,WB & BLAU,GE (1985)
DC 10.32	pKa	25	EXP	KORTUM,G ET AL (1961)
HL 4.00E-007	atm m3/mol	25	EST	MEYLAN,WM & HOWARD,PH (1991)
OH 1.04E-010	cm3/molc sec	25	EST	MEYLAN,WM & HOWARD,PH (1993)

CAS #: 001471-03-0				ALLYL PROPYL ETHER

Formula: $C_6H_{12}O$
Mol Weight: 100.16
MP (deg C):   FP (deg C):
BP (deg C): 91
BP pressure (mm Hg):

Property/ Value	Units	Temp	Data Type	Reference
WS 2.81E+003	mg/L	25	EST	MEYLAN,WM ET AL. (1996)
logP 1.89			EST	MEYLAN,WM & HOWARD,PH (1995)
VP 6.08E+001	mm Hg	25	EXP	OHE,S (1976)
DC	pKa			
HL 2.00E-003	atm m3/mol	25	EST	MEYLAN,WM & HOWARD,PH (1991)
OH 4.24E-011	cm3/molc sec	25	EST	MEYLAN,WM & HOWARD,PH (1993)

CAS #: 001472-87-3				DIMETHYL TRIDECANEDIOATE

Formula: $C_{15}H_{28}O_4$
Mol Weight: 272.39
MP (deg C): 35-37   FP (deg C):
BP (deg C): 326-328
BP pressure (mm Hg):

Property/ Value	Units	Temp	Data Type	Reference
WS 1.34E+000	mg/L	25	EST	MEYLAN,WM ET AL. (1996)
logP 4.82			EST	MEYLAN,WM & HOWARD,PH (1995)
VP 2.36E-007	mm Hg	25	EXT	BOUBLIK,T ET AL. (1984)
DC	pKa			
HL 7.10E-006	atm m3/mol	25	EST	MEYLAN,WM & HOWARD,PH (1991)
OH 1.49E-011	cm3/molc sec	25	EST	MEYLAN,WM & HOWARD,PH (1993)

CAS #: 001476-11-5				1,4-DICHLORO-2-BUTENE(CIS)

Formula: $C_4H_6Cl_2$
Mol Weight: 125.00
MP (deg C): -48   FP (deg C):
BP (deg C): 152.5
BP pressure (mm Hg):

Property/ Value	Units	Temp	Data Type	Reference
WS 5.80E+002	mg/L	25	EST	MEYLAN,WM ET AL. (1996)
logP 2.60			EST	MEYLAN,WM & HOWARD,PH (1995)
VP 4.09E+000	mm Hg	25	EXP	YAWS,CL (1994)
DC	pKa			
HL 1.88E-002	atm m3/mol	25	EST	MEYLAN,WM & HOWARD,PH (1991)
OH 3.33E-011	cm3/molc sec	25	EST	MEYLAN,WM & HOWARD,PH (1993)

CAS #: 001477-50-5				INDOLE-2-CARBOXYLIC ACID

Formula: $C_9H_7NO_2$
Mol Weight: 161.16
MP (deg C): 205-208   FP (deg C):
BP (deg C):
BP pressure (mm Hg):

Property/ Value	Units	Temp	Data Type	Reference
WS 7.20E+002	mg/L	25	EST	MEYLAN,WM ET AL. (1996)
logP 2.31			EXP	HANSCH,C & LEO,AJ (1985)
VP 1.21E-005	mm Hg	25	EST	NEELY,WB & BLAU,GE (1985)
DC	pKa			
HL 1.78E-011	atm m3/mol	25	EST	MEYLAN,WM & HOWARD,PH (1991)
OH 4.12E-011	cm3/molc sec	25	EST	MEYLAN,WM & HOWARD,PH (1993)

CAS #: 001492-96-2				1,4-BENZODIAZEPIN-2-ONE,5(4FPH..

Formula: $C_{15}H_{10}ClFN_2O$
Mol Weight: 288.71
MP (deg C):   FP (deg C):
BP (deg C):
BP pressure (mm Hg):

Property/ Value	Units	Temp	Data Type	Reference
WS 4.77E+001	mg/L	25	EST	MEYLAN,WM ET AL. (1996)
logP 2.90			EXP	HANSCH,C & LEO,AJ (1985)
VP 7.76E-009	mm Hg	25	EST	NEELY,WB & BLAU,GE (1985)
DC	pKa			
HL 2.08E-010	atm m3/mol	25	EST	MEYLAN,WM & HOWARD,PH (1991)
OH 7.98E-012	cm3/molc sec	25	EST	MEYLAN,WM & HOWARD,PH (1993)

CAS #: 001493-02-3				FORMYL FLUORIDE

Formula: CHFO
Mol Weight: 48.02
MP (deg C): -142   FP (deg C):
BP (deg C): -29
BP pressure (mm Hg):

Property/ Value	Units	Temp	Data Type	Reference
WS 5.49E+005	mg/L	25	EST	MEYLAN,WM ET AL. (1996)
logP -0.72			EST	MEYLAN,WM & HOWARD,PH (1995)
VP 2.89E+003	mm Hg	25	EXT	BOUBLIK,T ET AL. (1984)
DC	pKa			
HL 2.13E-003	atm m3/mol	25	EST	MEYLAN,WM & HOWARD,PH (1991)
OH 1.59E-012	cm3/molc sec	25	EST	MEYLAN,WM & HOWARD,PH (1993)

CAS #: 001493-27-2				O-FLUORONITROBENZENE

Formula: $C_6H_4FNO_2$
Mol Weight: 141.10
MP (deg C): -6   FP (deg C):
BP (deg C):
BP pressure (mm Hg):

Property/ Value	Units	Temp	Data Type	Reference
WS 1.22E+003	mg/L	25	EST	MEYLAN,WM ET AL. (1996)
logP 1.69			EXP	HANSCH,C & LEO,AJ (1985)
VP 2.94E-001	mm Hg	25	EST	NEELY,WB & BLAU,GE (1985)
DC	pKa			
HL 2.48E-005	atm m3/mol	25	EST	MEYLAN,WM & HOWARD,PH (1991)
OH 3.05E-013	cm3/molc sec	25	EST	MEYLAN,WM & HOWARD,PH (1993)

CAS #: 001498-96-0				P-BUTOXYBENZOIC ACID

Formula: $C_{11}H_{14}O_3$
Mol Weight: 194.23
MP (deg C): 147-150   FP (deg C):
BP (deg C):
BP pressure (mm Hg):

Property/ Value	Units	Temp	Data Type	Reference
WS 4.18E+001	mg/L	25	EST	MEYLAN,WM ET AL. (1996)
logP 3.57			EXP	HANSCH,C & LEO,AJ (1985)
VP 9.28E-005	mm Hg	25	EST	NEELY,WB & BLAU,GE (1985)
DC	pKa			
HL 1.50E-008	atm m3/mol	25	EST	MEYLAN,WM & HOWARD,PH (1991)
OH 2.08E-011	cm3/molc sec	25	EST	MEYLAN,WM & HOWARD,PH (1993)

001499-10-1 — 9,10-DIPHENYLANTHRACENE

Formula: $C_{26}H_{18}$
Mol Weight: 330.43
MP (deg C): 245-248
FP (deg C):
BP (deg C):
BP pressure (mm Hg):

Property/Value	Units	Temp	Data Type	Reference
WS 3.76E-005	mg/L	25	EST	MEYLAN,WM ET AL. (1996)
logP 8.46			EXP	DEVOOGT,P ET AL. (1990)
VP 2.12E-014	mm Hg	25	EXT	JONES,AH (1960)
DC	pKa			
HL 3.03E-007	atm m3/mol	25	EST	MEYLAN,WM & HOWARD,PH (1991)
OH 5.64E-011	cm3/molc sec	25	EST	MEYLAN,WM & HOWARD,PH (1993)

001501-05-9 — BUTANOIC ACID, 4-BENZOYL-

Formula: $C_{11}H_{12}O_3$
Mol Weight: 192.22
MP (deg C): 126.5-128
FP (deg C):
BP (deg C):
BP pressure (mm Hg):

Property/Value	Units	Temp	Data Type	Reference
WS 6.35E+003	mg/L	25	EST	MEYLAN,WM ET AL. (1996)
logP 1.49			EXP	SANGSTER,J (1994)
VP 4.03E-005	mm Hg	25	EST	NEELY,WB & BLAU,GE (1985)
DC	pKa			
HL 1.29E-010	atm m3/mol	25	EST	MEYLAN,WM & HOWARD,PH (1991)
OH 8.50E-012	cm3/molc sec	25	EST	MEYLAN,WM & HOWARD,PH (1993)

001502-22-3 — CYCLOHEXANONE, 2-(1-CYCLOHEXEN-1-YL)-

Formula: $C_{12}H_{18}O$
Mol Weight: 178.28
MP (deg C): -78
FP (deg C):
BP (deg C): 265
BP pressure (mm Hg):

Property/Value	Units	Temp	Data Type	Reference
WS 3.00E+002	mg/L	25	EXP	CHEM INSPECT TEST INST (1992)
logP 3.17			EXP	CHEM INSPECT TEST INST (1992)
VP 6.37E-003	mm Hg	25	EST	NEELY,WB & BLAU,GE (1985)
DC	pKa			
HL 1.28E-004	atm m3/mol	25	EST	MEYLAN,WM & HOWARD,PH (1991)
OH 1.05E-010	cm3/molc sec	25	EST	MEYLAN,WM & HOWARD,PH (1993)

001502-95-0 — DIACETYLMORPHINE, HYDROCHLORIDE

Formula: $C_{21}H_{24}ClNO_5$
Mol Weight: 405.88
MP (deg C):
FP (deg C):
BP (deg C):
BP pressure (mm Hg):

Property/Value	Units	Temp	Data Type	Reference
WS 1.05E+003	mg/L	25	EST	MEYLAN,WM ET AL. (1996)
logP 0.50			EST	MEYLAN,WM & HOWARD,PH (1995)
VP 4.93E-013	mm Hg	25	EST	NEELY,WB & BLAU,GE (1985)
DC	pKa			
HL 6.33E-020	atm m3/mol	25	EST	MEYLAN,WM & HOWARD,PH (1991)
OH 1.57E-010	cm3/molc sec	25	EST	MEYLAN,WM & HOWARD,PH (1993)

001504-26-3 — BENZENEACETAMIDE, N-(2-NITROPHENYL)-

Formula: $C_{14}H_{12}N_2O_3$
Mol Weight: 256.26
MP (deg C):
FP (deg C):
BP (deg C):
BP pressure (mm Hg):

Property/Value	Units	Temp	Data Type	Reference
WS 4.15E+001	mg/L	25	EST	MEYLAN,WM ET AL. (1996)
logP 2.73			EXP	YAMAGAMI,C ET AL. (1984)
VP 1.26E-008	mm Hg	25	EST	NEELY,WB & BLAU,GE (1985)
DC	pKa			
HL 4.30E-011	atm m3/mol	25	EST	MEYLAN,WM & HOWARD,PH (1991)
OH 7.10E-012	cm3/molc sec	25	EST	MEYLAN,WM & HOWARD,PH (1993)

001510-24-3 — 4-BIPHENYLISOTHIOCYANATE

Formula: $C_{13}H_9NS$
Mol Weight: 211.29
MP (deg C):
FP (deg C):
BP (deg C):
BP pressure (mm Hg):

Property/Value	Units	Temp	Data Type	Reference
WS 4.01E+000	mg/L	25	EST	MEYLAN,WM ET AL. (1996)
logP 4.66			EXP	HANSCH,C & LEO,AJ (1985)
VP 1.86E-005	mm Hg	25	EST	NEELY,WB & BLAU,GE (1985)
DC	pKa			
HL 6.00E-005	atm m3/mol	25	EST	MEYLAN,WM & HOWARD,PH (1991)
OH 7.10E-012	cm3/molc sec	25	EST	MEYLAN,WM & HOWARD,PH (1993)

001513-31-1 — DIFLUOROMETHANESULFONANILIDE,P-CL

Formula: $C_7H_6ClF_2NO_2S$
Mol Weight: 241.65
MP (deg C):
FP (deg C):
BP (deg C):
BP pressure (mm Hg):

Property/Value	Units	Temp	Data Type	Reference
WS 9.88E+001	mg/L	25	EST	MEYLAN,WM ET AL. (1996)
logP 2.84			EXP	HANSCH,C & LEO,AJ (1985)
VP 2.29E-004	mm Hg	25	EST	NEELY,WB & BLAU,GE (1985)
DC	pKa			
HL 3.42E-006	atm m3/mol	25	EST	MEYLAN,WM & HOWARD,PH (1991)
OH 1.26E-011	cm3/molc sec	25	EST	MEYLAN,WM & HOWARD,PH (1993)

001513-65-1 — 2,6-DIFLUOROPYRIDINE

Formula: $C_5H_3F_2N$
Mol Weight: 115.08
MP (deg C):
FP (deg C):
BP (deg C): 124.5
BP pressure (mm Hg): 7.43E+002

Property/Value	Units	Temp	Data Type	Reference
WS 1.05E+004	mg/L	25	EST	MEYLAN,WM ET AL. (1996)
logP 1.17			EXP	HANSCH,C ET AL. (1995)
VP 3.38E+001	mm Hg	25	EST	NEELY,WB & BLAU,GE (1985)
DC	pKa			
HL 9.60E-006	atm m3/mol	25	EST	MEYLAN,WM & HOWARD,PH (1991)
OH 5.81E-013	cm3/molc sec	25	EST	MEYLAN,WM & HOWARD,PH (1993)

CAS #: 001515-85-1 — BENZONITRILE, 4-(METHOXYMETHYL)-

Formula: C_9H_9NO

Mol Weight: 147.18

MP (deg C):
FP (deg C):
BP (deg C):
BP pressure (mm Hg):

Property/Value	Units	Temp	Data Type	Reference
WS 3.78E+003	mg/L	25	EST	MEYLAN,WM ET AL. (1996)
logP 1.23			EXP	SANGSTER,J (1993)
VP 2.15E-002	mm Hg	25	EST	NEELY,WB & BLAU,GE (1985)
DC	pKa			
HL 6.72E-007	atm m3/mol	25	EST	MEYLAN,WM & HOWARD,PH (1991)
OH 7.38E-012	cm3/molc sec	25	EST	MEYLAN,WM & HOWARD,PH (1993)

CAS #: 001517-63-1 — PHNEYL-P-TOLYLCARBINOL

Formula: $C_{14}H_{14}O$

Mol Weight: 198.27

MP (deg C): 50-54
FP (deg C):
BP (deg C):
BP pressure (mm Hg):

Property/Value	Units	Temp	Data Type	Reference
WS 3.07E+002	mg/L	25	EST	MEYLAN,WM ET AL. (1996)
logP 3.13			EXP	HANSCH,C & LEO,AJ (1985)
VP 7.67E-006	mm Hg	25	EST	NEELY,WB & BLAU,GE (1985)
DC	pKa			
HL 1.94E-008	atm m3/mol	25	EST	MEYLAN,WM & HOWARD,PH (1991)
OH 1.78E-011	cm3/molc sec	25	EST	MEYLAN,WM & HOWARD,PH (1993)

CAS #: 001517-69-7 — 1-PHENYLETHANOL

Formula: $C_8H_{10}O$

Mol Weight: 122.17

MP (deg C): 9-11
FP (deg C):
BP (deg C): 98
BP pressure (mm Hg): 2.00E+001

Property/Value	Units	Temp	Data Type	Reference
WS 1.47E+004	mg/L	25	EXP	KUHNE,R ET AL. (1995)
logP 1.42			EXP	HANSCH,C & LEO,AJ (1985)
VP 4.80E-002	mm Hg	25	EST	NEELY,WB & BLAU,GE (1985)
DC	pKa			
HL 2.89E-007	atm m3/mol	25	EST	MEYLAN,WM & HOWARD,PH (1991)
OH 1.17E-011	cm3/molc sec	25	EST	MEYLAN,WM & HOWARD,PH (1993)

CAS #: 001518-72-5 — DI-TERT-BUTYLDISULFIDE

Formula: $C_8H_{18}S_2$

Mol Weight: 178.36

MP (deg C):
FP (deg C):
BP (deg C):
BP pressure (mm Hg):

Property/Value	Units	Temp	Data Type	Reference
WS 6.64E+000	mg/L	25	EST	MEYLAN,WM ET AL. (1996)
logP 4.60				MEYLAN,WM & HOWARD,PH (1995)
VP 3.53E-001	mm Hg	25	EST	NEELY,WB & BLAU,GE (1985)
DC	pKa			
HL 6.64E-003	atm m3/mol	25	EST	MEYLAN,WM & HOWARD,PH (1991)
OH 7.51E-012	cm3/molc sec	25	EXP	ATKINSON,R (1989)

CAS #: 001518-83-8 — 4-CYCLOPENTYLPHENOL

Formula: $C_{11}H_{14}O$

Mol Weight: 162.23

MP (deg C): 66-64
FP (deg C):
BP (deg C): 155
BP pressure (mm Hg): 1.20E+001

Property/Value	Units	Temp	Data Type	Reference
WS 1.34E+002	mg/L	25	EST	MEYLAN,WM ET AL. (1996)
logP 3.84				MEYLAN,WM & HOWARD,PH (1995)
VP 1.66E-003	mm Hg	25	EST	NEELY,WB & BLAU,GE (1985)
DC	pKa			
HL 8.47E-007	atm m3/mol	25	EST	MEYLAN,WM & HOWARD,PH (1991)
OH 4.71E-011	cm3/molc sec	25	EST	MEYLAN,WM & HOWARD,PH (1993)

CAS #: 001521-38-6 — 2,3-DIMETHOXYBENZOIC ACID

Formula: $C_9H_{10}O_4$

Mol Weight: 182.18

MP (deg C): 122-124
FP (deg C):
BP (deg C):
BP pressure (mm Hg):

Property/Value	Units	Temp	Data Type	Reference
WS 5.31E+003	mg/L	25	EST	MEYLAN,WM ET AL. (1996)
logP 1.18				MEYLAN,WM & HOWARD,PH (1995)
VP 1.86E-006	mm Hg	25	EXT	COLOMINA,M ET AL. (1985)
DC 3.98	pKa	25	EXP	SERJEANT,EP & DEMPSEY,B (1979)
HL 3.79E-010	atm m3/mol	25	EST	MEYLAN,WM & HOWARD,PH (1991)
OH 9.07E-012	cm3/molc sec	25	EST	MEYLAN,WM & HOWARD,PH (1993)

CAS #: 001522-00-5 — N,N-DIETHYL SUCCINAMIC ACID

Formula: $C_8H_{15}NO_3$

Mol Weight: 173.21

MP (deg C):
FP (deg C):
BP (deg C):
BP pressure (mm Hg):

Property/Value	Units	Temp	Data Type	Reference
WS 2.79E+004	mg/L	25	EXP	SEIDELL,A (1941)
logP 0.24			EST	MEYLAN,WM & HOWARD,PH (1995)
VP 1.25E-004	mm Hg	25	EST	NEELY,WB & BLAU,GE (1985)
DC	pKa			
HL 9.38E-013	atm m3/mol	25	EST	MEYLAN,WM & HOWARD,PH (1991)
OH 2.71E-011	cm3/molc sec	25	EST	MEYLAN,WM & HOWARD,PH (1993)

CAS #: 001523-06-4 — M-NITROPHENYL ACETATE

Formula: $C_8H_7NO_4$

Mol Weight: 181.15

MP (deg C):
FP (deg C):
BP (deg C):
BP pressure (mm Hg):

Property/Value	Units	Temp	Data Type	Reference
WS 6.18E+002	mg/L	25	EST	MEYLAN,WM ET AL. (1996)
logP 1.82			EXP	HANSCH,C & LEO,AJ (1985)
VP 1.86E-003	mm Hg	25	EST	NEELY,WB & BLAU,GE (1985)
DC	pKa			
HL 2.56E-007	atm m3/mol	25	EST	MEYLAN,WM & HOWARD,PH (1991)
OH 2.80E-013	cm3/molc sec	25	EST	MEYLAN,WM & HOWARD,PH (1993)

CAS #: 001528-74-1 — 4,4'-DINITROBIPHENYL

Formula: $C_{12}H_8N_2O_4$

Mol Weight: 244.21

MP (deg C):

FP (deg C):

BP (deg C):

BP pressure (mm Hg):

Property/Value	Units	Temp	Data Type	Reference
WS 1.32E+001	mg/L	25	EST	MEYLAN,WM ET AL. (1996)
logP 3.39			EST	MEYLAN,WM & HOWARD,PH (1995)
VP 2.74E-008	mm Hg	25	EXT	JONES,AH (1960)
DC	pKa			
HL 6.44E-009	atm m3/mol	25	EST	MEYLAN,WM & HOWARD,PH (1991)
OH 6.24E-013	cm3/molc sec	25	EST	MEYLAN,WM & HOWARD,PH (1993)

CAS #: 001529-68-6 — 1,2,3,4-TETRABROMOBUTANE

Formula: $C_4H_6Br_4$

Mol Weight: 373.73

MP (deg C):

FP (deg C):

BP (deg C):

BP pressure (mm Hg):

Property/Value	Units	Temp	Data Type	Reference
WS 4.28E+000	mg/L	25	EST	MEYLAN,WM ET AL. (1996)
logP 3.53			EST	MEYLAN,WM & HOWARD,PH (1995)
VP 6.72E-003	mm Hg	25	EST	NEELY,WB & BLAU,GE (1985)
DC	pKa			
HL 3.04E-005	atm m3/mol	25	EST	MEYLAN,WM & HOWARD,PH (1991)
OH 4.70E-013	cm3/molc sec	25	EST	MEYLAN,WM & HOWARD,PH (1993)

CAS #: 001532-72-5 — 1-ISOQUINOLINAMINE

Formula: $C_9H_8N_2$

Mol Weight: 144.18

MP (deg C): 103-105

FP (deg C):

BP (deg C):

BP pressure (mm Hg):

Property/Value	Units	Temp	Data Type	Reference
WS 4.87E+004	mg/L	25	EST	MEYLAN,WM ET AL. (1996)
logP 0.25			EXP	HANSCH,C ET AL. (1995)
VP 2.07E-004	mm Hg	25	EST	NEELY,WB & BLAU,GE (1985)
DC	pKa			
HL 6.88E-012	atm m3/mol	25	EST	MEYLAN,WM & HOWARD,PH (1991)
OH 1.44E-011	cm3/molc sec	25	EST	MEYLAN,WM & HOWARD,PH (1993)

CAS #: 001532-84-9 — 1-ISOQUINOLINAMINE

Formula: $C_9H_8N_2$

Mol Weight: 144.18

MP (deg C): 122-124

FP (deg C):

BP (deg C):

BP pressure (mm Hg):

Property/Value	Units	Temp	Data Type	Reference
WS 1.92E+003	mg/L	25	EST	MEYLAN,WM ET AL. (1996)
logP 1.90			EXP	HANSCH,C & LEO,AJ (1985)
VP 5.50E-004	mm Hg	25	EST	NEELY,WB & BLAU,GE (1985)
DC 7.59	pKa	20	EXP	PERRIN,DD (1965)
HL 2.43E-010	atm m3/mol	25	EST	MEYLAN,WM & HOWARD,PH (1991)
OH 2.00E-010	cm3/molc sec	25	EST	MEYLAN,WM & HOWARD,PH (1993)

CAS #: 001535-61-1 — FLUOROTHIAZIDE

Formula: $C_7H_6FN_3O_4S_2$

Mol Weight: 279.27

MP (deg C):

FP (deg C):

BP (deg C):

BP pressure (mm Hg):

Property/Value	Units	Temp	Data Type	Reference
WS 2.55E+003	mg/L	25	EST	MEYLAN,WM ET AL. (1996)
logP -0.29			EXP	SANGSTER,J (1994)
VP 9.65E-011	mm Hg	25	EST	NEELY,WB & BLAU,GE (1985)
DC	pKa			
HL 6.37E-012	atm m3/mol	25	EST	MEYLAN,WM & HOWARD,PH (1991)
OH 1.81E-011	cm3/molc sec	25	EST	MEYLAN,WM & HOWARD,PH (1993)

CAS #: 001538-74-5 — BUTYL N-PHENYLCARBAMATE

Formula: $C_{11}H_{15}NO_2$

Mol Weight: 193.25

MP (deg C):

FP (deg C):

BP (deg C):

BP pressure (mm Hg):

Property/Value	Units	Temp	Data Type	Reference
WS 7.20E+001	mg/L	25	EST	MEYLAN,WM ET AL. (1996)
logP 3.30			EXP	BRIGGS,GG (1981)
VP 2.42E-003	mm Hg	25	EST	NEELY,WB & BLAU,GE (1985)
DC	pKa			
HL 5.11E-008	atm m3/mol	25	EST	MEYLAN,WM & HOWARD,PH (1991)
OH 4.72E-011	cm3/molc sec	25	EST	MEYLAN,WM & HOWARD,PH (1993)

CAS #: 001542-59-2 — 4,6-NH2-2,2-DIME-1(4FPH)S-TRIAZINE

Formula: $C_{11}H_{14}FN_5$

Mol Weight: 235.27

MP (deg C):

FP (deg C):

BP (deg C):

BP pressure (mm Hg):

Property/Value	Units	Temp	Data Type	Reference
WS 1.38E+004	mg/L	25	EST	MEYLAN,WM ET AL. (1996)
logP 0.37			EXP	HANSCH,C & LEO,AJ (1985)
VP 6.08E-006	mm Hg	25	EST	NEELY,WB & BLAU,GE (1985)
DC	pKa			
HL 8.50E-013	atm m3/mol	25	EST	MEYLAN,WM & HOWARD,PH (1991)
OH 9.18E-011	cm3/molc sec	25	EST	MEYLAN,WM & HOWARD,PH (1993)

CAS #: 001547-10-0 — 3-CF3 HYDROCHLOROTHIAZIDE

Formula: $C_8H_7ClF_3N_3O_4S_2$

Mol Weight: 365.74

MP (deg C):

FP (deg C):

BP (deg C):

BP pressure (mm Hg):

Property/Value	Units	Temp	Data Type	Reference
WS 1.69E+002	mg/L	25	EST	MEYLAN,WM ET AL. (1996)
logP 0.49			EXP	SANGSTER,J (1994)
VP 1.32E-009	mm Hg	25	EST	NEELY,WB & BLAU,GE (1985)
DC	pKa			
HL 4.59E-011	atm m3/mol	25	EST	MEYLAN,WM & HOWARD,PH (1991)
OH 1.33E-011	cm3/molc sec	25	EST	MEYLAN,WM & HOWARD,PH (1993)

CAS #: 001551-21-9				3-METHYL-2-THIABUTANE

Formula: $C_4H_{10}S$
Mol Weight: 90.19
MP (deg C): -101.5
FP (deg C):
BP (deg C): 84.8
BP pressure (mm Hg):

Property/Value	Units	Temp	Data Type	Reference
WS 3.42E+003	mg/L	25	EST	MEYLAN,WM ET AL. (1996)
logP 1.83			EST	MEYLAN,WM & HOWARD,PH (1995)
VP 8.00E+001	mm Hg	25	EXP	ZWOLINSKI,BJ & WILHOIT,RC (1971)
DC	pKa			
HL 1.38E-003	atm m3/mol	25	EST	MEYLAN,WM & HOWARD,PH (1991)
OH 1.82E-011	cm3/molc sec	25	EST	MEYLAN,WM & HOWARD,PH (1993)

CAS #: 001552-94-9				2,4-PENTADIENOIC ACID, 5-PHENYL-

Formula: $C_{11}H_{10}O_2$
Mol Weight: 174.20
MP (deg C):
FP (deg C):
BP (deg C):
BP pressure (mm Hg):

Property/Value	Units	Temp	Data Type	Reference
WS 2.98E+002	mg/L	25	EST	MEYLAN,WM ET AL. (1996)
logP 3.15			EXP	HANSCH,C & LEO,AJ (1985)
VP 1.59E-004	mm Hg	25	EST	NEELY,WB & BLAU,GE (1985)
DC	pKa			
HL 1.63E-008	atm m3/mol	25	EST	MEYLAN,WM & HOWARD,PH (1991)
OH 5.21E-011	cm3/molc sec	25	EST	MEYLAN,WM & HOWARD,PH (1993)

CAS #: 001553-60-2				BENZENEACETIC ACID, 4-(2-METHYLPROPYL)-

Formula: $C_{12}H_{16}O_2$
Mol Weight: 192.26
MP (deg C): 85-87
FP (deg C):
BP (deg C):
BP pressure (mm Hg):

Property/Value	Units	Temp	Data Type	Reference
WS 1.64E+002	mg/L	25	EST	MEYLAN,WM ET AL. (1996)
logP 3.35			EXP	SANGSTER,J (1994)
VP 1.65E-004	mm Hg	25	EST	NEELY,WB & BLAU,GE (1985)
DC	pKa			
HL 1.14E-007	atm m3/mol	25	EST	MEYLAN,WM & HOWARD,PH (1991)
OH 1.01E-011	cm3/molc sec	25	EST	MEYLAN,WM & HOWARD,PH (1993)

CAS #: 001558-17-4				4,6-DIMETHYLPYRIMIDINE

Formula: $C_6H_8N_2$
Mol Weight: 108.14
MP (deg C): 25
FP (deg C):
BP (deg C): 159
BP pressure (mm Hg):

Property/Value	Units	Temp	Data Type	Reference
WS 3.26E+004	mg/L	25	EST	MEYLAN,WM ET AL. (1996)
logP 0.62			EXP	HANSCH,C & LEO,AJ (1985)
VP 1.50E+000	mm Hg	25	EST	NEELY,WB & BLAU,GE (1985)
DC 2.70	pKa		EXP	PERRIN,DD (1965)
HL 3.55E-006	atm m3/mol	25	EST	MEYLAN,WM & HOWARD,PH (1991)
OH 1.84E-012	cm3/molc sec	25	EST	MEYLAN,WM & HOWARD,PH (1993)

CAS #: 001562-94-3				4,4'-DIMETHOXYAZOXYBENZENE

Formula: $C_{14}H_{14}N_2O_3$
Mol Weight: 258.28
MP (deg C): 118-136
FP (deg C):
BP (deg C):
BP pressure (mm Hg):

Property/Value	Units	Temp	Data Type	Reference
WS 1.26E+001	mg/L	25	EST	MEYLAN,WM ET AL. (1996)
logP 3.27			EST	MEYLAN,WM & HOWARD,PH (1995)
VP 3.73E-010	mm Hg	25	EST	NEELY,WB & BLAU,GE (1985)
DC	pKa			
HL 4.13E-015	atm m3/mol	25	EST	MEYLAN,WM & HOWARD,PH (1991)
OH 1.48E-011	cm3/molc sec	25	EST	MEYLAN,WM & HOWARD,PH (1993)

CAS #: 001563-38-8				CARBOFURAN PHENOL

Formula: $C_{10}H_{12}O_2$
Mol Weight: 164.21
MP (deg C):
FP (deg C):
BP (deg C):
BP pressure (mm Hg):

Property/Value	Units	Temp	Data Type	Reference
WS 1.10E+003	mg/L	25	EST	MEYLAN,WM ET AL. (1996)
logP 2.08			EXP	HANSCH,C & LEO,AJ (1985)
VP 2.11E-003	mm Hg	25	EST	NEELY,WB & BLAU,GE (1985)
DC	pKa			
HL 2.84E-008	atm m3/mol	25	EST	MEYLAN,WM & HOWARD,PH (1991)
OH 7.53E-011	cm3/molc sec	25	EST	MEYLAN,WM & HOWARD,PH (1993)

CAS #: 001563-66-2				CARBOFURAN

Formula: $C_{12}H_{15}NO_3$
Mol Weight: 221.26
MP (deg C): 151
FP (deg C):
BP (deg C):
BP pressure (mm Hg):

Property/Value	Units	Temp	Data Type	Reference
WS 3.20E+002	mg/L	19	EXP	SHAROM,MS ET AL. (1980A)
logP 2.32			EXP	HANSCH,C & LEO,AJ (1985)
VP 4.85E-006	mm Hg	19	EXP	FERREIRA,GAL & SEIBER,JN (1981)
DC	pKa			
HL 3.09E-009	atm m3/mol	19	EST	VP/WSOL
OH 2.89E-011	cm3/molc sec	25	EST	MEYLAN,WM & HOWARD,PH (1993)

CAS #: 001565-17-9				P-ACETYLBENZENESULFONAMIDE

Formula: $C_8H_9NO_3S$
Mol Weight: 199.23
MP (deg C):
FP (deg C):
BP (deg C):
BP pressure (mm Hg):

Property/Value	Units	Temp	Data Type	Reference
WS 2.65E+003	mg/L	25	EST	MEYLAN,WM ET AL. (1996)
logP 0.20			EXP	HANSCH,C & LEO,AJ (1985)
VP 1.79E-005	mm Hg	25	EST	NEELY,WB & BLAU,GE (1985)
DC	pKa			
HL 7.69E-010	atm m3/mol	25	EST	MEYLAN,WM & HOWARD,PH (1991)
OH 4.82E-013	cm3/molc sec	25	EST	MEYLAN,WM & HOWARD,PH (1993)

001567-89-1 — 2-PROPANONE, (2,4-DINITROPHENYL)HYDRAZONE

Formula: $C_9H_{10}N_4O_4$

Mol Weight: 238.20

MP (deg C):
FP (deg C):
BP (deg C):
BP pressure (mm Hg):

Property/Value	Units	Temp	Data Type	Reference
WS 2.78E+001	mg/L	25	EST	MEYLAN,WM ET AL. (1996)
logP 3.05			EXP	PARKIN,JE (1986)
VP 4.74E-006	mm Hg	25	EST	NEELY,WB & BLAU,GE (1985)
DC	pKa			
HL 3.06E-009	atm m3/mol	25	EST	MEYLAN,WM & HOWARD,PH (1991)
OH 1.00E-012	cm3/molc sec	25	EST	MEYLAN,WM & HOWARD,PH (1993)

001568-70-3 — 4-METHOXY-2-NITROPHENOL

Formula: $C_7H_7NO_4$

Mol Weight: 169.14

MP (deg C): 78-80
FP (deg C):
BP (deg C):
BP pressure (mm Hg):

Property/Value	Units	Temp	Data Type	Reference
WS 2.44E+002	mg/L	20	EXP	SCHWARZENBACH,RP ET AL. (1988)
logP 2.02			EXP	SCHWARZENBACH,RP ET AL. (1988)
VP 2.71E-001	mm Hg	20	EXP	SCHWARZENBACH,RP ET AL. (1988)
DC 7.31	pKa	25	EXP	SERJEANT,EP & DEMPSEY,B (1979)
HL 2.47E-004	atm m3/mol	20	EST	VP/WSOL
OH 4.57E-012	cm3/molc sec	25	EST	MEYLAN,WM & HOWARD,PH (1993)

001569-01-3 — 1-PROPOXY-2-PROPANOL

Formula: $C_6H_{14}O_2$

Mol Weight: 118.18

MP (deg C): -80
FP (deg C):
BP (deg C): 149.8
BP pressure (mm Hg):

Property/Value	Units	Temp	Data Type	Reference
WS 1.00E+006	mg/L	25	EXP	BROWN,ES ET AL. (1980)
logP 0.49			EST	MEYLAN,WM & HOWARD,PH (1995)
VP 1.70E+000	mm Hg	20	EXP	BROWN,ES ET AL. (1980)
DC	pKa			
HL 3.46E-008	atm m3/mol	25	EST	MEYLAN,WM & HOWARD,PH (1991)
OH 2.54E-011	cm3/molc sec	25	EST	ATKINSON,R (1988)

001569-50-2 — 3-PENTEN-2-OL

Formula: $C_5H_{10}O$

Mol Weight: 86.13

MP (deg C):
FP (deg C):
BP (deg C):
BP pressure (mm Hg):

Property/Value	Units	Temp	Data Type	Reference
WS 8.92E+004	mg/L	25	EXP	YALKOWSKY,SH & DANNENFELSER,RM (1992)
logP 1.04			EST	MEYLAN,WM & HOWARD,PH (1995)
VP 6.01E+000	mm Hg	25	EST	NEELY,WB & BLAU,GE (1985)
DC	pKa			
HL 1.17E-005	atm m3/mol	25	EST	MEYLAN,WM & HOWARD,PH (1991)
OH 6.36E-011	cm3/molc sec	25	EST	MEYLAN,WM & HOWARD,PH (1993)

001569-69-3 — CYCLOHEXANETHIOL

Formula: $C_6H_{12}S$

Mol Weight: 116.23

MP (deg C):
FP (deg C):
BP (deg C): 158.9
BP pressure (mm Hg):

Property/Value	Units	Temp	Data Type	Reference
WS 2.60E+002	mg/L	25	EST	MEYLAN,WM ET AL. (1996)
logP 3.05			EST	MEYLAN,WM & HOWARD,PH (1995)
VP 3.97E+000	mm Hg	25	EXT	BOUBLIK,T ET AL. (1984)
DC	pKa			
HL 4.75E-003	atm m3/mol	25	EST	MEYLAN,WM & HOWARD,PH (1991)
OH 6.25E-011	cm3/molc sec	25	EST	MEYLAN,WM & HOWARD,PH (1993)

001570-45-2 — ISONICOTINIC ACID, ETHYL ESTER

Formula: $C_8H_9NO_2$

Mol Weight: 151.17

MP (deg C): 23
FP (deg C):
BP (deg C): 219.5
BP pressure (mm Hg):

Property/Value	Units	Temp	Data Type	Reference
WS 4.51E+003	mg/L	25	EST	MEYLAN,WM ET AL. (1996)
logP 1.43			EXP	HANSCH,C & LEO,AJ (1985)
VP 8.40E-002	mm Hg	25	EST	NEELY,WB & BLAU,GE (1985)
DC 3.45	pKa	20	EXP	PERRIN,DD (1965)
HL 6.03E-008	atm m3/mol	25	EST	MEYLAN,WM & HOWARD,PH (1991)
OH 1.78E-012	cm3/molc sec	25	EST	MEYLAN,WM & HOWARD,PH (1993)

001570-64-5 — 2-METHYL-4-CHLOROPHENOL

Formula: C_7H_7ClO

Mol Weight: 142.59

MP (deg C): 51
FP (deg C):
BP (deg C): 223
BP pressure (mm Hg):

Property/Value	Units	Temp	Data Type	Reference
WS 4.00E+003	mg/L	25	EXP	WOODROW,JE ET AL. (1990)
logP 2.63			EXP	HANSCH,C ET AL. (1995)
VP 2.40E-003	mm Hg	25	EXP	WOODROW,JE ET AL. (1990)
DC 9.71	pKa	25	EXP	SERJEANT,EP & DEMPSEY,B (1979)
HL 1.13E-007	atm m3/mol	25	EST	VP/WSOL
OH 1.22E-011	cm3/molc sec	25	EST	MEYLAN,WM & HOWARD,PH (1993)

001571-33-1 — PHENYLPHOSPHONIC ACID

Formula: $C_6H_7O_3P$

Mol Weight: 158.09

MP (deg C): 163-166
FP (deg C):
BP (deg C):
BP pressure (mm Hg):

Property/Value	Units	Temp	Data Type	Reference
WS 1.90E+005	mg/L	15	EXP	YALKOWSKY,SH & DANNENFELSER,RM (1992)
logP 0.54			EXP	HANSCH,C & LEO,AJ (1985)
VP 1.57E-006	mm Hg	25	EST	NEELY,WB & BLAU,GE (1985)
DC 1.83	pKa		EXP	KORTUM,G ET AL (1961)
HL 3.12E-013	atm m3/mol	25	EST	MEYLAN,WM & HOWARD,PH (1991)
OH 2.23E-012	cm3/molc sec	25	EST	MEYLAN,WM & HOWARD,PH (1993)

CAS #: 001572-46-9				9-(PHENYLMETHYL)-9H-FLUORENE
Formula: $C_{20}H_{16}$				
Mol Weight: 256.35				
MP (deg C):		FP (deg C):		
BP (deg C):				
BP pressure (mm Hg):				

Property/Value	Units	Temp	Data Type	Reference
WS 1.68E-002	mg/L	25	EST	MEYLAN,WM ET AL. (1996)
logP 5.86			EST	MEYLAN,WM & HOWARD,PH (1995)
VP 1.05E-006	mm Hg	25	EST	NEELY,WB & BLAU,GE (1985)
DC	pKa			
HL 1.79E-005	atm m3/mol	25	EST	MEYLAN,WM & HOWARD,PH (1991)
OH 1.61E-011	cm3/molc sec	25	EST	MEYLAN,WM & HOWARD,PH (1993)

CAS #: 001573-57-5				1,2,2,3,3,4-HEXACHLOROBUTANE
Formula: $C_4H_4Cl_6$				
Mol Weight: 264.79				
MP (deg C):		FP (deg C):		
BP (deg C):				
BP pressure (mm Hg):				

Property/Value	Units	Temp	Data Type	Reference
WS 1.02E+000	mg/L	25	EST	MEYLAN,WM ET AL. (1996)
logP 5.02			EST	MEYLAN,WM & HOWARD,PH (1995)
VP 3.93E-002	mm Hg	25	EST	NEELY,WB & BLAU,GE (1985)
DC	pKa			
HL 3.29E-004	atm m3/mol	25	EST	MEYLAN,WM & HOWARD,PH (1991)
OH 1.77E-013	cm3/molc sec	25	EST	MEYLAN,WM & HOWARD,PH (1993)

CAS #: 001573-58-6				1,2,3-TRICHLOROBUTA-1,3-DIENE
Formula: $C_4H_3Cl_3$				
Mol Weight: 157.43				
MP (deg C):		FP (deg C):		
BP (deg C): 33-34				
BP pressure (mm Hg): 7.00E+000				

Property/Value	Units	Temp	Data Type	Reference
WS 9.22E+001	mg/L	25	EST	MEYLAN,WM ET AL. (1996)
logP 3.38			EST	MEYLAN,WM & HOWARD,PH (1995)
VP 1.57E+001	mm Hg	25	EST	NEELY,WB & BLAU,GE (1985)
DC	pKa			
HL 2.90E-002	atm m3/mol	25	EST	MEYLAN,WM & HOWARD,PH (1991)
OH 1.76E-012	cm3/molc sec	25	EST	MEYLAN,WM & HOWARD,PH (1993)

CAS #: 001574-10-3				HYDRAZINECARBOXAMIDE, 2-(PHENYLMETHYLENE)-
Formula: $C_8H_9N_3O$				
Mol Weight: 163.18				
MP (deg C):		FP (deg C):		
BP (deg C):				
BP pressure (mm Hg):				

Property/Value	Units	Temp	Data Type	Reference
WS 5.45E+003	mg/L	25	EST	MEYLAN,WM ET AL. (1996)
logP 1.27			EXP	HANSCH,C ET AL. (1995)
VP 9.81E-005	mm Hg	25	EST	NEELY,WB & BLAU,GE (1985)
DC -1.05	pKa		EXP	PERRIN,DD (1965)
HL 7.88E-011	atm m3/mol	25	EST	MEYLAN,WM & HOWARD,PH (1991)
OH 8.57E-012	cm3/molc sec	25	EST	MEYLAN,WM & HOWARD,PH (1993)

CAS #: 001574-41-0				1,3-PENTADIENE (Z)
Formula: C_5H_8				
Mol Weight: 68.12				
MP (deg C): -140.8		FP (deg C):		
BP (deg C): 44.1				
BP pressure (mm Hg):				

Property/Value	Units	Temp	Data Type	Reference
WS 3.26E+002	mg/L	25	EST	MEYLAN,WM ET AL. (1996)
logP 2.40			EXP	SANGSTER,J (1994)
VP 3.80E+002	mm Hg	25	EXP	YAWS,CL (1994A)
DC	pKa			
HL 1.22E-001	atm m3/mol	25	EST	MEYLAN,WM & HOWARD,PH (1991)
OH 1.01E-010	cm3/molc sec	25	EXP	ATKINSON,R (1989)

CAS #: 001576-17-6				3-PHENYL-1-METHYL-1-METHOXYUREA
Formula: $C_9H_{12}N_2O_2$				
Mol Weight: 180.21				
MP (deg C):		FP (deg C):		
BP (deg C):				
BP pressure (mm Hg):				

Property/Value	Units	Temp	Data Type	Reference
WS 4.35E+003	mg/L	25	EST	MEYLAN,WM ET AL. (1996)
logP 1.29			EXP	HANSCH,C & LEO,AJ (1985)
VP 8.47E-005	mm Hg	25	EST	NEELY,WB & BLAU,GE (1985)
DC	pKa			
HL 2.10E-008	atm m3/mol	25	EST	MEYLAN,WM & HOWARD,PH (1991)
OH 4.42E-011	cm3/molc sec	25	EST	MEYLAN,WM & HOWARD,PH (1993)

CAS #: 001576-42-7				BENZAMIDE, 3-(AMINOSULFONYL)-
Formula: $C_7H_8N_2O_3S$				
Mol Weight: 200.22				
MP (deg C):		FP (deg C):		
BP (deg C):				
BP pressure (mm Hg):				

Property/Value	Units	Temp	Data Type	Reference
WS 1.12E+004	mg/L	25	EST	MEYLAN,WM ET AL. (1996)
logP -0.84			EXP	CAROTTI,A ET AL. (1989)
VP 1.39E-007	mm Hg	25	EST	NEELY,WB & BLAU,GE (1985)
DC	pKa			
HL 1.74E-013	atm m3/mol	25	EST	MEYLAN,WM & HOWARD,PH (1991)
OH 2.38E-012	cm3/molc sec	25	EST	MEYLAN,WM & HOWARD,PH (1993)

CAS #: 001576-43-8				P-HYDROXYBENZENESULFONAMIDE
Formula: $C_6H_7NO_3S$				
Mol Weight: 173.19				
MP (deg C):		FP (deg C):		
BP (deg C):				
BP pressure (mm Hg):				

Property/Value	Units	Temp	Data Type	Reference
WS 2.01E+005	mg/L	25	EST	MEYLAN,WM ET AL. (1996)
logP 0.06			EXP	HANSCH,C & LEO,AJ (1985)
VP 1.15E-005	mm Hg	25	EST	NEELY,WB & BLAU,GE (1985)
DC	pKa			
HL 4.39E-011	atm m3/mol	25	EST	MEYLAN,WM & HOWARD,PH (1991)
OH 7.27E-012	cm3/molc sec	25	EST	MEYLAN,WM & HOWARD,PH (1993)

CAS #: 001576-67-6				3,6-DIMETHYLPHENANTHRENE

Formula: $C_{16}H_{14}$
Mol Weight: 206.29
MP (deg C): 144-146
FP (deg C):
BP (deg C):
BP pressure (mm Hg):

Property/ Value	Units	Temp	Data Type	Reference
WS 7.13E-002	mg/L	25	EST	MEYLAN,WM ET AL. (1996)
logP 5.44			EST	MEYLAN,WM & HOWARD,PH (1995)
VP 1.82E-005	mm Hg	25	EST	NEELY,WB & BLAU,GE (1985)
DC	pKa			
HL 6.25E-005	atm m3/mol	25	EST	MEYLAN,WM & HOWARD,PH (1991)
OH 4.19E-011	cm3/molc sec	25	EST	MEYLAN,WM & HOWARD,PH (1993)

CAS #: 001582-09-8				TRIFLURALIN

Formula: $C_{13}H_{16}F_3N_3O_4$
Mol Weight: 335.29
MP (deg C): 48.5-49
FP (deg C):
BP (deg C): 139-140
BP pressure (mm Hg): 4.20E+000

Property/ Value	Units	Temp	Data Type	Reference
WS 8.11E+000	mg/L	20	EXP	KANAZAWA,J (1983)
logP 5.34			EXP	HANSCH,C & LEO,AJ (1985)
VP 4.85E-005	mm Hg	20	EXP	SPENCER,WF & CLIATH,MM (1974)
DC	pKa			
HL 2.64E-005	atm m3/mol	20	EST	VP/WSOL
OH 8.40E-011	cm3/molc sec	25	EST	ATKINSON,R (1988)

CAS #: 001582-75-8				PREGN-4-ENE-21-CARBOXYLIC ACID, 9-FLUORO-11,17-D

Formula: $C_{22}H_{29}FO_4$
Mol Weight: 376.47
MP (deg C):
FP (deg C):
BP (deg C):
BP pressure (mm Hg):

Property/ Value	Units	Temp	Data Type	Reference
WS 1.23E+002	mg/L	25	EST	MEYLAN,WM ET AL. (1996)
logP 2.40			EXP	SANGSTER,J (1993)
VP 6.20E-012	mm Hg	25	EST	NEELY,WB & BLAU,GE (1985)
DC	pKa			
HL 9.78E-012	atm m3/mol	25	EST	MEYLAN,WM & HOWARD,PH (1991)
OH 9.74E-011	cm3/molc sec	25	EST	MEYLAN,WM & HOWARD,PH (1993)

CAS #: 001585-07-5				1-BROMO-4-ETHYLBENZENE

Formula: C_8H_9Br
Mol Weight: 185.07
MP (deg C): -43.5
FP (deg C):
BP (deg C): 204
BP pressure (mm Hg):

Property/ Value	Units	Temp	Data Type	Reference
WS 2.33E+001	mg/L	25	EST	MEYLAN,WM ET AL. (1996)
logP 3.92			EST	MEYLAN,WM & HOWARD,PH (1995)
VP 1.00E+000	mm Hg	30	EXP	PERRY,RH & GREEN,D (1984)
DC	pKa			
HL 3.14E-003	atm m3/mol	25	EST	MEYLAN,WM & HOWARD,PH (1991)
OH 2.60E-012	cm3/molc sec	25	EST	MEYLAN,WM & HOWARD,PH (1993)

CAS #: 001585-40-6				BENZENEPENTACARBOXYLIC ACID

Formula: $C_{11}H_6O_{10}$
Mol Weight: 298.16
MP (deg C):
FP (deg C):
BP (deg C):
BP pressure (mm Hg):

Property/ Value	Units	Temp	Data Type	Reference
WS 2.33E+004	mg/L	25	EST	MEYLAN,WM ET AL. (1996)
logP -0.31			EST	MEYLAN,WM & HOWARD,PH (1995)
VP 2.23E-013	mm Hg	25	EST	NEELY,WB & BLAU,GE (1985)
DC 1.80	pKa		EXP	KORTUM,G ET AL (1961)
HL 1.78E-026	atm m3/mol	25	EST	MEYLAN,WM & HOWARD,PH (1991)
OH 2.61E-012	cm3/molc sec	25	EST	MEYLAN,WM & HOWARD,PH (1993)

CAS #: 001585-79-1				N-MALEOYLGLYCINE, ETHYL ESTER

Formula: $C_8H_9NO_4$
Mol Weight: 183.17
MP (deg C):
FP (deg C):
BP (deg C):
BP pressure (mm Hg):

Property/ Value	Units	Temp	Data Type	Reference
WS 2.03E+004	mg/L	25	EST	MEYLAN,WM ET AL. (1996)
logP 0.49			EXP	HANSCH,C & LEO,AJ (1985)
VP 2.03E-006	mm Hg	25	EST	NEELY,WB & BLAU,GE (1985)
DC	pKa			
HL 4.39E-011	atm m3/mol	25	EST	MEYLAN,WM & HOWARD,PH (1991)
OH 2.39E-011	cm3/molc sec	25	EST	MEYLAN,WM & HOWARD,PH (1993)

CAS #: 001592-38-7				2-NAPHTHALENEMETHANOL

Formula: $C_{11}H_{10}O$
Mol Weight: 158.20
MP (deg C): 81.3
FP (deg C):
BP (deg C): 178
BP pressure (mm Hg): 1.20E+001

Property/ Value	Units	Temp	Data Type	Reference
WS 1.83E+003	mg/L	25	EST	MEYLAN,WM ET AL. (1996)
logP 2.45			EXP	SANGSTER,J (1993)
VP 2.89E-005	mm Hg	25	EST	NEELY,WB & BLAU,GE (1985)
DC	pKa			
HL 2.12E-008	atm m3/mol	25	EST	MEYLAN,WM & HOWARD,PH (1991)
OH 5.71E-011	cm3/molc sec	25	EST	MEYLAN,WM & HOWARD,PH (1993)

CAS #: 001592-82-1				3-ISOPROPYL DIMETHYL PARATHION

Formula: $C_{11}H_{16}NO_5PS$
Mol Weight: 305.29
MP (deg C):
FP (deg C):
BP (deg C):
BP pressure (mm Hg):

Property/ Value	Units	Temp	Data Type	Reference
WS 1.62E+000	mg/L	25	EST	MEYLAN,WM ET AL. (1996)
logP 4.05			EXP	HANSCH,C & LEO,AJ (1985)
VP 8.77E-006	mm Hg	25	EST	NEELY,WB & BLAU,GE (1985)
DC	pKa			
HL 3.27E-007	atm m3/mol	25	EST	MEYLAN,WM & HOWARD,PH (1991)
OH 6.19E-011	cm3/molc sec	25	EST	MEYLAN,WM & HOWARD,PH (1993)

CAS #: 001594-56-5				2,4-DINITROPHENYL THIOCYANATE
Formula: $C_7H_3N_3O_4S$				
Mol Weight: 225.18				
MP (deg C):		FP (deg C):		
BP (deg C):				
BP pressure (mm Hg):				

Property/Value	Units	Temp	Data Type	Reference
WS 2.40E+002	mg/L	25	EST	MEYLAN,WM ET AL. (1996)
logP 2.04			EST	MEYLAN,WM & HOWARD,PH (1995)
VP 8.88E-007	mm Hg	25	EST	NEELY,WB & BLAU,GE (1985)
DC	pKa			
HL 1.36E-010	atm m3/mol	25	EST	MEYLAN,WM & HOWARD,PH (1991)
OH 1.41E-014	cm3/molc sec	25	EST	MEYLAN,WM & HOWARD,PH (1993)

CAS #: 001596-52-7				4,6-DINITROQUINOLINE-1-OXIDE
Formula: $C_9H_7N_3O_5$				
Mol Weight: 237.17				
MP (deg C):		FP (deg C):		
BP (deg C):				
BP pressure (mm Hg):				

Property/Value	Units	Temp	Data Type	Reference
WS 1.98E+003	mg/L	25	EST	MEYLAN,WM ET AL. (1996)
logP 0.90			EXP	HANSCH,C & LEO,AJ (1985)
VP 4.65E-008	mm Hg	25	EST	NEELY,WB & BLAU,GE (1985)
DC	pKa			
HL 1.07E-016	atm m3/mol	25	EST	MEYLAN,WM & HOWARD,PH (1991)
OH 1.57E-013	cm3/molc sec	25	EST	MEYLAN,WM & HOWARD,PH (1993)

CAS #: 001596-84-5				DAMINOZIDE
Formula: $C_6H_{12}N_2O_3$				
Mol Weight: 160.17				
MP (deg C): 157-164		FP (deg C):		
BP (deg C):				
BP pressure (mm Hg):				

Property/Value	Units	Temp	Data Type	Reference
WS 1.00E+005	mg/L	25	EXP	WORTHING,CR & WALKER,SB (1987)
logP -1.51			EST	MEYLAN,WM & HOWARD,PH (1995)
VP 2.00E-004	mm Hg	23	EXP	AUGUSTIJN-BECKERS,PWM ET AL. (1994)
DC	pKa			
HL 4.23E-010	atm m3/mol	25	EST	VP/WSOL
OH 7.87E-011	cm3/molc sec	25	EST	MEYLAN,WM & HOWARD,PH (1993)

CAS #: 001600-27-7				MERCURIC ACETATE
Formula: $C_4H_6HgO_4$				
Mol Weight: 318.68				
MP (deg C): 178-180		FP (deg C):		
BP (deg C):				
BP pressure (mm Hg):				

Property/Value	Units	Temp	Data Type	Reference
WS 2.50E+005	mg/L	10	EXP	ROSENBLATT,DH ET AL. (1975)
logP -1.28			EST	MEYLAN,WM & HOWARD,PH (1995)
VP 1.77E-001	mm Hg	25	EST	NEELY,WB & BLAU,GE (1985)
DC	pKa			
HL	atm m3/mol			
OH 8.43E-014	cm3/molc sec	25	EST	MEYLAN,WM & HOWARD,PH (1993)

CAS #: 001603-41-4				2-AMINO-5-METHYLPYRIDINE
Formula: $C_6H_8N_2$				
Mol Weight: 108.14				
MP (deg C): 76-77		FP (deg C):		
BP (deg C): 227				
BP pressure (mm Hg):				

Property/Value	Units	Temp	Data Type	Reference
WS 1.49E+004	mg/L	25	EST	MEYLAN,WM ET AL. (1996)
logP 1.02			EXP	HANSCH,C & LEO,AJ (1985)
VP 1.44E-001	mm Hg	25	EST	NEELY,WB & BLAU,GE (1985)
DC 7.22	pKa		EXP	PERRIN,DD (1965)
HL 2.75E-009	atm m3/mol	25	EST	MEYLAN,WM & HOWARD,PH (1991)
OH 2.52E-011	cm3/molc sec	25	EST	MEYLAN,WM & HOWARD,PH (1993)

CAS #: 001606-67-3				1-PYRENAMINE
Formula: $C_{16}H_{13}N$				
Mol Weight: 219.29				
MP (deg C): 115-117		FP (deg C):		
BP (deg C):				
BP pressure (mm Hg):				

Property/Value	Units	Temp	Data Type	Reference
WS 5.76E-001	mg/L	25	EST	MEYLAN,WM ET AL. (1996)
logP 4.31			EXP	DEBNATH,AK ET AL. (1992)
VP 1.59E-007	mm Hg	25	EST	NEELY,WB & BLAU,GE (1985)
DC	pKa			
HL 2.93E-009	atm m3/mol	25	EST	MEYLAN,WM & HOWARD,PH (1991)
OH 2.00E-010	cm3/molc sec	25	EST	MEYLAN,WM & HOWARD,PH (1993)

CAS #: 001608-68-0				2-ISOPROPYLPHENYL ACETATE
Formula: $C_{11}H_{14}O_2$				
Mol Weight: 178.23				
MP (deg C):		FP (deg C):		
BP (deg C):				
BP pressure (mm Hg):				

Property/Value	Units	Temp	Data Type	Reference
WS 2.37E+002	mg/L	25	EST	MEYLAN,WM ET AL. (1996)
logP 2.78			EXP	HANSCH,C & LEO,AJ (1985)
VP 4.28E-002	mm Hg	25	EST	NEELY,WB & BLAU,GE (1985)
DC	pKa			
HL 1.26E-004	atm m3/mol	25	EST	MEYLAN,WM & HOWARD,PH (1991)
OH 5.77E-012	cm3/molc sec	25	EST	MEYLAN,WM & HOWARD,PH (1993)

CAS #: 001609-07-0				1-METHYL-2-CYANOGUANIDINE
Formula: $C_3H_6N_4$				
Mol Weight: 98.11				
MP (deg C):		FP (deg C):		
BP (deg C):				
BP pressure (mm Hg):				

Property/Value	Units	Temp	Data Type	Reference
WS 5.38E+005	mg/L	25	EST	MEYLAN,WM ET AL. (1996)
logP -0.77			EXP	HANSCH,C & LEO,AJ (1985)
VP 1.35E-001	mm Hg	25	EST	NEELY,WB & BLAU,GE (1985)
DC	pKa			
HL 4.94E-010	atm m3/mol	25	EST	MEYLAN,WM & HOWARD,PH (1991)
OH 8.53E-011	cm3/molc sec	25	EST	MEYLAN,WM & HOWARD,PH (1993)

592

CAS #: 001610-17-9 — ATRATONE

Formula: $C_9H_{17}N_5O$

Mol Weight: 211.27

MP (deg C): FP (deg C):

BP (deg C):

BP pressure (mm Hg):

Property/Value	Units	Temp	Data Type	Reference
WS 1.80E+003	mg/L	20	EXP	BAILEY,GW ET AL. (1968)
logP 2.69			EXP	FINIZIO,A ET AL. (1991)
VP 2.90E-006	mm Hg	20	EXP	KEARNEY,PC & KAUFMAN,DD (1975A)
DC		pKa		
HL 4.48E-010	atm m3/mol	20	EST	VP/WSOL
OH 2.84E-011	cm3/molc sec	25	EST	MEYLAN,WM & HOWARD,PH (1993)

CAS #: 001610-18-0 — PROMETON

Formula: $C_{10}H_{19}N_5O$

Mol Weight: 225.30

MP (deg C): 91-92 FP (deg C):

BP (deg C):

BP pressure (mm Hg):

Property/Value	Units	Temp	Data Type	Reference
WS 7.50E+002	mg/L	25	EXP	BAILEY,GW ET AL. (1968)
logP 2.99			EXP	FINIZIO,A ET AL. (1991)
VP 5.00E-006	mm Hg	25	EXP	WEBER,JB (1994)
DC		pKa		
HL 1.98E-009	atm m3/mol	25	EST	VP/WSOL
OH 3.79E-011	cm3/molc sec	2525	EST	MEYLAN,WM & HOWARD,PH (1993)

CAS #: 001611-63-8 — UREA, N-(1-METHYLCYCLOHEXYL)-N'-PHENYL-

Formula: $C_{14}H_{20}N_2O$

Mol Weight: 232.33

MP (deg C): FP (deg C):

BP (deg C):

BP pressure (mm Hg):

Property/Value	Units	Temp	Data Type	Reference
WS 2.49E+001	mg/L	25	EST	MEYLAN,WM ET AL. (1996)
logP 3.60			EXP	MITSUTAKE,KI ET AL. (1986)
VP 2.34E-006	mm Hg	25	EST	NEELY,WB & BLAU,GE (1985)
DC		pKa		
HL 1.07E-009	atm m3/mol	25	EST	MEYLAN,WM & HOWARD,PH (1991)
OH 5.09E-011	cm3/molc sec	25	EST	MEYLAN,WM & HOWARD,PH (1993)

CAS #: 001612-15-3 — 2H-BENZOTRIAZOLE-2-PROPANOIC ACID, â-METHYL-, ET

Formula: $C_{12}H_{15}N_3O_2$

Mol Weight: 233.27

MP (deg C): FP (deg C):

BP (deg C):

BP pressure (mm Hg):

Property/Value	Units	Temp	Data Type	Reference
WS 1.45E+002	mg/L	25	EST	MEYLAN,WM ET AL. (1996)
logP 2.70			EXP	SPARATORE,F ET AL. (1988)
VP 1.43E-005	mm Hg	25	EST	NEELY,WB & BLAU,GE (1985)
DC		pKa		
HL 1.74E-008	atm m3/mol	25	EST	MEYLAN,WM & HOWARD,PH (1991)
OH 6.07E-012	cm3/molc sec	25	EST	MEYLAN,WM & HOWARD,PH (1993)

CAS #: 001613-37-2 — QUINOLINE-1-OXIDE

Formula: C_9H_9NO

Mol Weight: 147.18

MP (deg C): FP (deg C):

BP (deg C):

BP pressure (mm Hg):

Property/Value	Units	Temp	Data Type	Reference
WS 3.92E+004	mg/L	25	EST	MEYLAN,WM ET AL. (1996)
logP 0.36			EXP	HANSCH,C & LEO,AJ (1985)
VP 2.07E-004	mm Hg	25	EST	NEELY,WB & BLAU,GE (1985)
DC		pKa		
HL 6.88E-012	atm m3/mol	25	EST	MEYLAN,WM & HOWARD,PH (1991)
OH 1.44E-011	cm3/molc sec	25	EST	MEYLAN,WM & HOWARD,PH (1993)

CAS #: 001613-51-0 — TETRAHYDRO-2H-THIOPYRAN

Formula: $C_5H_{10}S$

Mol Weight: 102.20

MP (deg C): 19 FP (deg C):

BP (deg C): 141.8

BP pressure (mm Hg):

Property/Value	Units	Temp	Data Type	Reference
WS 1.30E+003	mg/L	25	EST	MEYLAN,WM ET AL. (1996)
logP 2.28			EST	MEYLAN,WM & HOWARD,PH (1995)
VP 7.86E+000	mm Hg	25	EXT	BOUBLIK,T ET AL. (1984)
DC		pKa		
HL 8.11E-004	atm m3/mol	25	EST	MEYLAN,WM & HOWARD,PH (1991)
OH 2.39E-011	cm3/molc sec	25	EST	MEYLAN,WM & HOWARD,PH (1993)

CAS #: 001615-06-1 — 1,3-DIHYDRO-BENZOTHIADIAZOL,22-O2

Formula: $C_6H_6N_2O_2S$

Mol Weight: 170.19

MP (deg C): FP (deg C):

BP (deg C):

BP pressure (mm Hg):

Property/Value	Units	Temp	Data Type	Reference
WS 6.78E+003	mg/L	25	EST	MEYLAN,WM ET AL. (1996)
logP 1.12			EXP	HANSCH,C & LEO,AJ (1985)
VP 3.21E-005	mm Hg	25	EST	NEELY,WB & BLAU,GE (1985)
DC		pKa		
HL 4.42E-008	atm m3/mol	25	EST	MEYLAN,WM & HOWARD,PH (1991)
OH 6.99E-011	cm3/molc sec	25	EST	MEYLAN,WM & HOWARD,PH (1993)

CAS #: 001615-80-1 — N,N'-DIETHYLHYDRAZINE

Formula: $C_4H_{12}N_2$

Mol Weight: 88.15

MP (deg C): -53 EST FP (deg C):

BP (deg C): 85.5

BP pressure (mm Hg):

Property/Value	Units	Temp	Data Type	Reference
WS 5.23E+004	mg/L	25	EST	MEYLAN,WM ET AL. (1996)
logP 0.45			EST	MEYLAN,WM & HOWARD,PH (1995)
VP 6.93E+001	mm Hg	25	EST	NEELY,WB & BLAU,GE (1985)
DC 7.71		pKa	25	EXP	PERRIN,DD (1965)
HL 1.22E-007	atm m3/mol	25	EST	MEYLAN,WM & HOWARD,PH (1991)
OH 1.43E-012	cm3/molc sec	25	EST	MEYLAN,WM & HOWARD,PH (1993)

594

CAS #: 001628-89-3				2-METHOXYPYRIDINE
Formula: C6H7NO				
Mol Weight: 109.13				
MP (deg C):		FP (deg C):		
BP (deg C): 142				
BP pressure (mm Hg):				

Property/ Value	Units	Temp	Data Type	Reference
WS 7.86E+003	mg/L	25	EST	MEYLAN,WM ET AL. (1996)
logP 1.34			EXP	YAMAGAMI,C ET AL. (1990A)
VP 2.58E+000	mm Hg	25	EST	NEELY,WB & BLAU,GE (1985)
DC 3.06	pKa	25	EXP	PERRIN,DD (1965)
HL 3.80E-006	atm m3/mol	25	EST	MEYLAN,WM & HOWARD,PH (1991)
OH 4.91E-012	cm3/molc sec	25	EST	MEYLAN,WM & HOWARD,PH (1993)

CAS #: 001632-16-2				2-ETHYL-1-HEXENE
Formula: C8H16				
Mol Weight: 112.22				
MP (deg C):		FP (deg C):		
BP (deg C): 120				
BP pressure (mm Hg):				

Property/ Value	Units	Temp	Data Type	Reference
WS 8.20E+000	mg/L	25	EST	MEYLAN,WM ET AL. (1996)
logP 4.19			EST	MEYLAN,WM & HOWARD,PH (1995)
VP 1.97E+001	mm Hg	25	EXP	DAUBERT,TE & DANNER,RP (1989)
DC	pKa			
HL 7.46E-001	atm m3/mol	25	EST	MEYLAN,WM & HOWARD,PH (1991)
OH 5.64E-011	cm3/molc sec	25	EST	MEYLAN,WM & HOWARD,PH (1993)

CAS #: 001632-73-1				FENCHYL ALCOHOL
Formula: C10H18O				
Mol Weight: 154.25				
MP (deg C):		FP (deg C):		
BP (deg C):				
BP pressure (mm Hg):				

Property/ Value	Units	Temp	Data Type	Reference
WS 8.63E+002	mg/L	25	EST	MEYLAN,WM ET AL. (1996)
logP 2.85			EST	MEYLAN,WM & HOWARD,PH (1995)
VP 3.56E-002	mm Hg	25	EXT	BOUBLIK,T ET AL. (1984)
DC	pKa			
HL 6.70E-006	atm m3/mol	25	EST	MEYLAN,WM & HOWARD,PH (1991)
OH 1.22E-011	cm3/molc sec	25	EST	MEYLAN,WM & HOWARD,PH (1993)

CAS #: 001632-76-4				PYRIDAZINE, 3-METHYL-
Formula: C5H6N2				
Mol Weight: 94.12				
MP (deg C): 184		FP (deg C):		
BP (deg C): 214				
BP pressure (mm Hg):				

Property/ Value	Units	Temp	Data Type	Reference
WS 2.42E+005	mg/L	25	EST	MEYLAN,WM ET AL. (1996)
logP -0.35			EXP	YAMAGAMI,C ET AL. (1990)
VP 1.48E-001	mm Hg	25	EST	NEELY,WB & BLAU,GE (1985)
DC 3.46	pKa	17	EXP	PERRIN,DD (1965)
HL 3.17E-006	atm m3/mol	25	EST	MEYLAN,WM & HOWARD,PH (1991)
OH 7.36E-013	cm3/molc sec	25	EST	MEYLAN,WM & HOWARD,PH (1993)

CAS #: 001633-44-9				3,4-PYRIDINEDICARBONITRILE
Formula: C7H3N3				
Mol Weight: 129.12				
MP (deg C): 79-81		FP (deg C):		
BP (deg C):				
BP pressure (mm Hg):				

Property/ Value	Units	Temp	Data Type	Reference
WS 3.14E+004	mg/L	25	EST	MEYLAN,WM ET AL. (1996)
logP 0.24			EXP	HANSCH,C ET AL. (1995)
VP 3.05E-003	mm Hg	25	EST	NEELY,WB & BLAU,GE (1985)
DC	pKa			
HL 6.58E-010	atm m3/mol	25	EST	MEYLAN,WM & HOWARD,PH (1991)
OH 8.55E-015	cm3/molc sec	25	EST	MEYLAN,WM & HOWARD,PH (1993)

CAS #: 001634-04-4				METHYL T-BUTYL ETHER
Formula: C5H12O				
Mol Weight: 88.15				
MP (deg C): -109		FP (deg C):		
BP (deg C): 55.2				
BP pressure (mm Hg):				

Property/ Value	Units	Temp	Data Type	Reference
WS 5.10E+004	mg/L	25	EXP	BENNETT,GM & PHILIP,WG (1928)
logP 0.94			EXP	HANSCH,C ET AL. (1995)
VP 2.49E+002	mm Hg	25	EXP	DAUBERT,TE & DANNER,RP (1985)
DC -3.70	pKa	23	EXP	PLETCHER,T & CORDES,EH (1967)
HL 5.87E-004	atm m3/mol	25	EXP	HINE,J & MOOKERJEE,PK (1975)
OH 2.83E-012	cm3/molc sec	25	EXP	ATKINSON,R (1989)

CAS #: 001636-33-5				2-ISOTHIOCYANONAPHTHALENE
Formula: C11H7NS				
Mol Weight: 185.25				
MP (deg C):		FP (deg C):		
BP (deg C):				
BP pressure (mm Hg):				

Property/ Value	Units	Temp	Data Type	Reference
WS 1.02E+001	mg/L	25	EST	MEYLAN,WM ET AL. (1996)
logP 4.34			EXP	HANSCH,C & LEO,AJ (1985)
VP 1.32E-004	mm Hg	25	EST	NEELY,WB & BLAU,GE (1985)
DC	pKa			
HL 7.62E-005	atm m3/mol	25	EST	MEYLAN,WM & HOWARD,PH (1991)
OH 2.37E-011	cm3/molc sec	25	EST	MEYLAN,WM & HOWARD,PH (1993)

CAS #: 001637-31-6				1,2,3,4-TETRACHLOROBUTA-1,3-DIENE
Formula: C4H2Cl4				
Mol Weight: 191.87				
MP (deg C): -4		FP (deg C):		
BP (deg C): 188				
BP pressure (mm Hg):				

Property/ Value	Units	Temp	Data Type	Reference
WS 3.12E+001	mg/L	25	EST	MEYLAN,WM ET AL. (1996)
logP 3.73			EST	MEYLAN,WM & HOWARD,PH (1995)
VP 2.50E+000	mm Hg	25	EST	NEELY,WB & BLAU,GE (1985)
DC	pKa			
HL 2.09E-002	atm m3/mol	25	EST	MEYLAN,WM & HOWARD,PH (1991)
OH 5.06E-013	cm3/molc sec	25	EST	MEYLAN,WM & HOWARD,PH (1993)

CAS #: 001638-22-8 — P-BUTYLPHENOL

Formula: $C_{10}H_{14}O$
Mol Weight: 150.22
MP (deg C): 22
FP (deg C):
BP (deg C): 248
BP pressure (mm Hg):

Property/Value	Units	Temp	Data Type	Reference
WS 2.20E+002	mg/L	25	EST	MEYLAN,WM ET AL. (1996)
logP 3.65			EXP	SANGSTER,J (1993)
VP 2.22E-003	mm Hg	25	EXT	JORDAN,TE (1954)
DC	pKa			
HL 1.45E-006	atm m3/mol	25	EST	MEYLAN,WM & HOWARD,PH (1991)
OH 4.46E-011	cm3/molc sec	25	EST	MEYLAN,WM & HOWARD,PH (1993)

CAS #: 001638-26-2 — 1,1-DIMETHYLCYCLOPENTANE

Formula: C_7H_{14}
Mol Weight: 98.19
MP (deg C): -69.8
FP (deg C):
BP (deg C): 87.5
BP pressure (mm Hg):

Property/Value	Units	Temp	Data Type	Reference
WS 3.16E+001	mg/L	25	EST	MEYLAN,WM ET AL. (1996)
logP 3.56			EST	MEYLAN,WM & HOWARD,PH (1995)
VP 7.62E+001	mm Hg	25	EXP	YAWS,CL (1994A)
DC	pKa			
HL 3.39E-001	atm m3/mol	25	EST	MEYLAN,WM & HOWARD,PH (1991)
OH 4.86E-012	cm3/molc sec	25	EST	MEYLAN,WM & HOWARD,PH (1993)

CAS #: 001639-09-4 — 1-HEPTANETHIOL

Formula: $C_7H_{16}S$
Mol Weight: 132.27
MP (deg C): -43
FP (deg C):
BP (deg C): 177
BP pressure (mm Hg):

Property/Value	Units	Temp	Data Type	Reference
WS 5.96E+001	mg/L	25	EST	MEYLAN,WM ET AL. (1996)
logP 3.72			EST	MEYLAN,WM & HOWARD,PH (1995)
VP 1.31E+000	mm Hg	25	EXP	YAWS,CL (1994A)
DC	pKa			
HL 1.43E-002	atm m3/mol	25	EST	MEYLAN,WM & HOWARD,PH (1991)
OH 4.84E-011	cm3/molc sec	25	EST	MEYLAN,WM & HOWARD,PH (1993)

CAS #: 001639-66-3 — DI-N-OCTYL SODIUM SULFOSUCCINATE

Formula: $C_{20}H_{37}NaO_7S$
Mol Weight: 444.57
MP (deg C):
FP (deg C):
BP (deg C):
BP pressure (mm Hg):

Property/Value	Units	Temp	Data Type	Reference
WS 4.56E-002	mg/L	25	EST	MEYLAN,WM ET AL. (1996)
logP 4.10			EST	MEYLAN,WM & HOWARD,PH (1995)
VP 1.40E-016	mm Hg	25	EST	NEELY,WB & BLAU,GE (1985)
DC	pKa			
HL	atm m3/mol			
OH 2.31E-011	cm3/molc sec	25	EST	MEYLAN,WM & HOWARD,PH (1993)

CAS #: 001640-89-7 — ETHYLCYCLOPENTANE

Formula: C_7H_{14}
Mol Weight: 98.19
MP (deg C): -138.4
FP (deg C):
BP (deg C): 103.5
BP pressure (mm Hg):

Property/Value	Units	Temp	Data Type	Reference
WS 2.94E+001	mg/L	25	EST	MEYLAN,WM ET AL. (1996)
logP 3.59			EST	MEYLAN,WM & HOWARD,PH (1995)
VP 3.98E+001	mm Hg	25	EXP	DAUBERT,TE & DANNER,RP (1989)
DC	pKa			
HL 3.39E-001	atm m3/mol	25	EST	MEYLAN,WM & HOWARD,PH (1991)
OH 8.73E-012	cm3/molc sec	25	EST	MEYLAN,WM & HOWARD,PH (1993)

CAS #: 001643-15-8 — M-METHYLPHENOXYACETIC

Formula: $C_9H_{10}O_3$
Mol Weight: 166.18
MP (deg C):
FP (deg C):
BP (deg C):
BP pressure (mm Hg):

Property/Value	Units	Temp	Data Type	Reference
WS 4.81E+003	mg/L	25	EST	MEYLAN,WM ET AL. (1996)
logP 1.78			EXP	HANSCH,C & LEO,AJ (1985)
VP 5.34E-004	mm Hg	25	EST	NEELY,WB & BLAU,GE (1985)
DC 3.20	pKa	25	EXP	KORTUM,G ET AL (1961)
HL 1.85E-008	atm m3/mol	25	EST	MEYLAN,WM & HOWARD,PH (1991)
OH 6.11E-011	cm3/molc sec	25	EST	MEYLAN,WM & HOWARD,PH (1993)

CAS #: 001643-16-9 — P-I-PROPYLPHENOXYACETIC ACID

Formula: $C_{11}H_{14}O_3$
Mol Weight: 194.23
MP (deg C):
FP (deg C):
BP (deg C):
BP pressure (mm Hg):

Property/Value	Units	Temp	Data Type	Reference
WS 5.86E+002	mg/L	25	EST	MEYLAN,WM ET AL. (1996)
logP 2.69			EXP	HANSCH,C & LEO,AJ (1985)
VP 1.46E-004	mm Hg	25	EST	NEELY,WB & BLAU,GE (1985)
DC	pKa			
HL 3.26E-008	atm m3/mol	25	EST	MEYLAN,WM & HOWARD,PH (1991)
OH 3.29E-011	cm3/molc sec	25	EST	MEYLAN,WM & HOWARD,PH (1993)

CAS #: 001643-20-5 — N,N-DIMETHYLDODECYLAMINE OXIDE

Formula: $C_{14}H_{31}NO$
Mol Weight: 229.41
MP (deg C): 132-133
FP (deg C):
BP (deg C):
BP pressure (mm Hg):

Property/Value	Units	Temp	Data Type	Reference
WS 1.90E+005	mg/L	25	EXP	BROWN,SL ET AL. (1975C)
logP 6.27			EST	MEYLAN,WM & HOWARD,PH (1995)
VP 6.23E-008	mm Hg	25	EST	NEELY,WB & BLAU,GE (1985)
DC	pKa			
HL 6.61E-011	atm m3/mol	25	EST	MEYLAN,WM & HOWARD,PH (1991)
OH 2.72E-011	cm3/molc sec	25	EST	MEYLAN,WM & HOWARD,PH (1993)

CAS #: 001646-27-1				2-BENZOFURANCARBOXYLIC ACID, METHYL ESTER

Formula: $C_{10}H_8O_3$

Mol Weight: 176.17

MP (deg C): FP (deg C):

BP (deg C):

BP pressure (mm Hg):

Property/Value	Units	Temp	Data Type	Reference
WS 3.97E+002	mg/L	25	EST	MEYLAN,WM ET AL. (1996)
logP 2.53			EXP	YAMAGAMI,C ET AL. (1992)
VP 3.43E-003	mm Hg	25	EST	NEELY,WB & BLAU,GE (1985)
DC	pKa			
HL 3.38E-006	atm m3/mol	25	EST	MEYLAN,WM & HOWARD,PH (1991)
OH 1.22E-011	cm3/molc sec	25	EST	MEYLAN,WM & HOWARD,PH (1993)

CAS #: 001646-87-3				ALDICARB SULFOXIDE

Formula: $C_7H_{14}N_2O_3S$

Mol Weight: 206.27

MP (deg C): FP (deg C):

BP (deg C):

BP pressure (mm Hg):

Property/Value	Units	Temp	Data Type	Reference
WS 2.80E+004	mg/L	25	EXP	CARSEL,RF ET AL. (1985)
logP -0.78			EST	MEYLAN,WM & HOWARD,PH (1995)
VP 1.00E-004	mm Hg	25	EXP	DIERBERG,FE & GIVEN,CJ (1986)
DC	pKa			
HL 9.69E-010	atm m3/mol	25	EST	VP/WSOL
OH 6.98E-011	cm3/molc sec	25	EST	MEYLAN,WM & HOWARD,PH (1993)

CAS #: 001646-88-4				ALDICARB SULFONE

Formula: $C_7H_{14}N_2O_4S$

Mol Weight: 222.26

MP (deg C): 140-142 FP (deg C):

BP (deg C):

BP pressure (mm Hg):

Property/Value	Units	Temp	Data Type	Reference
WS 7.80E+003	mg/L	25	EXP	CARSEL,RF ET AL. (1985)
logP -0.57			EXP	HANSCH,C & LEO,AJ (1985)
VP 9.00E-005	mm Hg	25	EXP	WAUCHOPE,RD ET AL. (1991A)
DC	pKa			
HL 3.37E-009	atm m3/mol	25	EST	VP/WSOL
OH 9.84E-012	cm3/molc sec	25	EST	MEYLAN,WM & HOWARD,PH (1993)

CAS #: 001649-08-7				1,2-DICHLORO-1,1-DIFLUOROETHANE

Formula: $C_2H_2Cl_2F_2$

Mol Weight: 134.94

MP (deg C): -101 FP (deg C):

BP (deg C): 46.8

BP pressure (mm Hg):

Property/Value	Units	Temp	Data Type	Reference
WS 9.99E+002	mg/L	25	EST	VP/HL
logP 2.31			EST	MEYLAN,WM & HOWARD,PH (1995)
VP 2.14E+002	mm Hg	25	EST	NEELY,WB & BLAU,GE (1985)
DC	pKa			
HL 4.80E-002	atm m3/mol	25	EST	MEYLAN,WM & HOWARD,PH (1991)
OH 2.60E-012	cm3/molc sec	25	EXP	ATKINSON,R (1989)

CAS #: 001649-18-9				AZAPERONE

Formula: $C_{19}H_{22}FN_3O$

Mol Weight: 327.41

MP (deg C): 73-75 FP (deg C):

BP (deg C):

BP pressure (mm Hg):

Property/Value	Units	Temp	Data Type	Reference
WS 1.31E+002	mg/L	25	EST	MEYLAN,WM ET AL. (1996)
logP 3.30			EXP	HANSCH,C & LEO,AJ (1985)
VP 5.46E-008	mm Hg	25	EST	NEELY,WB & BLAU,GE (1985)
DC	pKa			
HL 8.91E-015	atm m3/mol	25	EST	MEYLAN,WM & HOWARD,PH (1991)
OH 1.55E-010	cm3/molc sec	25	EST	MEYLAN,WM & HOWARD,PH (1993)

CAS #: 001653-19-6				2,3-DICHLOROBUTA-1,3-DIENE

Formula: $C_4H_4Cl_2$

Mol Weight: 122.98

MP (deg C): FP (deg C):

BP (deg C): 98

BP pressure (mm Hg):

Property/Value	Units	Temp	Data Type	Reference
WS 2.59E+002	mg/L	25	EST	MEYLAN,WM ET AL. (1996)
logP 3.02			EST	MEYLAN,WM & HOWARD,PH (1995)
VP 1.00E+002	mm Hg	25	EST	NEELY,WB & BLAU,GE (1985)
DC	pKa			
HL 4.03E-002	atm m3/mol	25	EST	MEYLAN,WM & HOWARD,PH (1991)
OH 6.26E-012	cm3/molc sec	25	EST	MEYLAN,WM & HOWARD,PH (1993)

CAS #: 001653-30-1				2-UNDECANOL

Formula: $C_{11}H_{24}O$

Mol Weight: 172.31

MP (deg C): FP (deg C):

BP (deg C):

BP pressure (mm Hg):

Property/Value	Units	Temp	Data Type	Reference
WS 1.97E+002	mg/L	25	EXP	SUZUKI,T (1991)
logP 4.21			EST	MEYLAN,WM & HOWARD,PH (1995)
VP 3.34E-002	mm Hg	25	EXT	PERRY,RH & GREEN,D (1984)
DC	pKa			
HL 7.26E-005	atm m3/mol	25	EST	MEYLAN,WM & HOWARD,PH (1991)
OH 1.99E-011	cm3/molc sec	25	EST	MEYLAN,WM & HOWARD,PH (1993)

CAS #: 001668-99-1				QUINIDINE HYDROCHLORIDE

Formula: $C_{20}H_{25}ClN_2O_2$

Mol Weight: 360.89

MP (deg C): FP (deg C):

BP (deg C):

BP pressure (mm Hg):

Property/Value	Units	Temp	Data Type	Reference
WS 1.16E+004	mg/L	25	EXP	SEIDELL,A (1941)
logP 1.78			EST	MEYLAN,WM & HOWARD,PH (1995)
VP 1.19E-016	mm Hg	25	EST	NEELY,WB & BLAU,GE (1985)
DC	pKa			
HL 5.41E-023	atm m3/mol	25	EST	MEYLAN,WM & HOWARD,PH (1991)
OH 1.69E-010	cm3/molc sec	25	EST	MEYLAN,WM & HOWARD,PH (1993)

CAS #: 001670-14-0 — BENZAMIDINE HCL

Formula: $C_7H_8N_2$
Mol Weight: 120.16
MP (deg C): 86-88
FP (deg C):
BP (deg C):
BP pressure (mm Hg):

Property/Value	Units	Temp	Data Type	Reference
WS 2.79E+004	mg/L	25	EST	MEYLAN,WM ET AL. (1996)
logP 0.65			EXP	HANSCH,C ET AL. (1995)
VP 2.82E-002	mm Hg	25	EST	NEELY,WB & BLAU,GE (1985)
DC 11.60	pKa	20	EXP	PERRIN,DD (1965)
HL 1.83E-008	atm m3/mol	25	EST	MEYLAN,WM & HOWARD,PH (1991)
OH 2.53E-011	cm3/molc sec	25	EST	MEYLAN,WM & HOWARD,PH (1993)

CAS #: 001670-84-4 — 1H-INDOLE-2-CARBOXAMIDE

Formula: $C_9H_8N_2O$
Mol Weight: 160.18
MP (deg C):
FP (deg C):
BP (deg C):
BP pressure (mm Hg):

Property/Value	Units	Temp	Data Type	Reference
WS 2.88E+003	mg/L	25	EST	MEYLAN,WM ET AL. (1996)
logP 1.61			EXP	SANGSTER,J (1994)
VP 1.41E-006	mm Hg	25	EST	NEELY,WB & BLAU,GE (1985)
DC	pKa			
HL 3.64E-013	atm m3/mol	25	EST	MEYLAN,WM & HOWARD,PH (1991)
OH 1.02E-010	cm3/molc sec	25	EST	MEYLAN,WM & HOWARD,PH (1993)

CAS #: 001670-86-6 — INDOLE-4-CARBOXAMIDE

Formula: $C_9H_8N_2O$
Mol Weight: 160.18
MP (deg C):
FP (deg C):
BP (deg C):
BP pressure (mm Hg):

Property/Value	Units	Temp	Data Type	Reference
WS 1.87E+004	mg/L	25	EST	MEYLAN,WM ET AL. (1996)
logP 0.66			EXP	HANSCH,C ET AL. (1995)
VP 1.41E-006	mm Hg	25	EST	NEELY,WB & BLAU,GE (1985)
DC	pKa			
HL 3.64E-013	atm m3/mol	25	EST	MEYLAN,WM & HOWARD,PH (1991)
OH 1.02E-010	cm3/molc sec	25	EST	MEYLAN,WM & HOWARD,PH (1993)

CAS #: 001670-87-7 — INDOLE-5-CARBOXAMIDE

Formula: $C_9H_8N_2O$
Mol Weight: 160.18
MP (deg C):
FP (deg C):
BP (deg C):
BP pressure (mm Hg):

Property/Value	Units	Temp	Data Type	Reference
WS 1.24E+004	mg/L	25	EST	MEYLAN,WM ET AL. (1996)
logP 0.87			EXP	HANSCH,C ET AL. (1995)
VP 1.41E-006	mm Hg	25	EST	NEELY,WB & BLAU,GE (1985)
DC	pKa			
HL 3.64E-013	atm m3/mol	25	EST	MEYLAN,WM & HOWARD,PH (1991)
OH 1.02E-010	cm3/molc sec	25	EST	MEYLAN,WM & HOWARD,PH (1993)

CAS #: 001671-82-5 — 1-METHYL-2-NITRO-1H-IMIDAZOLE

Formula: $C_4H_5N_3O_2$
Mol Weight: 127.10
MP (deg C):
FP (deg C):
BP (deg C):
BP pressure (mm Hg):

Property/Value	Units	Temp	Data Type	Reference
WS 5.37E+004	mg/L	25	EST	MEYLAN,WM ET AL. (1996)
logP -0.17			EXP	SUWINSKI,J ET AL. (1985)
VP 1.69E-003	mm Hg	25	EST	NEELY,WB & BLAU,GE (1985)
DC	pKa			
HL 3.16E-007	atm m3/mol	25	EST	MEYLAN,WM & HOWARD,PH (1991)
OH 4.64E-012	cm3/molc sec	25	EST	MEYLAN,WM & HOWARD,PH (1993)

CAS #: 001672-46-4 — DIGOXIGENIN

Formula: $C_{23}H_{34}O_5$
Mol Weight: 390.52
MP (deg C): 222
FP (deg C):
BP (deg C):
BP pressure (mm Hg):

Property/Value	Units	Temp	Data Type	Reference
WS 4.03E+002	mg/L	25	EST	MEYLAN,WM ET AL. (1996)
logP 1.10			EXP	HANSCH,C & LEO,AJ (1985)
VP 6.69E-015	mm Hg	25	EST	NEELY,WB & BLAU,GE (1985)
DC	pKa			
HL 2.29E-011	atm m3/mol	25	EST	MEYLAN,WM & HOWARD,PH (1991)
OH 6.80E-011	cm3/molc sec	25	EST	MEYLAN,WM & HOWARD,PH (1993)

CAS #: 001673-47-8 — BENZOYLHYDRAZINE, M-CHLORO

Formula: $C_7H_7ClN_2O$
Mol Weight: 170.60
MP (deg C):
FP (deg C):
BP (deg C):
BP pressure (mm Hg):

Property/Value	Units	Temp	Data Type	Reference
WS 6.00E+003	mg/L	25	EST	MEYLAN,WM ET AL. (1996)
logP 1.18			EXP	HANSCH,C & LEO,AJ (1985)
VP 2.02E-005	mm Hg	25	EST	NEELY,WB & BLAU,GE (1985)
DC	pKa			
HL 6.85E-012	atm m3/mol	25	EST	MEYLAN,WM & HOWARD,PH (1991)
OH 6.08E-012	cm3/molc sec	25	EST	MEYLAN,WM & HOWARD,PH (1993)

CAS #: 001675-02-1 — ERYTHROMYCIN, 3"-O-DEMETHYL-

Formula: $C_{36}H_{65}NO_{13}$
Mol Weight: 719.92
MP (deg C):
FP (deg C):
BP (deg C):
BP pressure (mm Hg):

Property/Value	Units	Temp	Data Type	Reference
WS 2.21E+001	mg/L	25	EST	MEYLAN,WM ET AL. (1996)
logP 1.26			EXP	HANSCH,C ET AL. (1995)
VP 9.44E-028	mm Hg	25	EST	NEELY,WB & BLAU,GE (1985)
DC	pKa			
HL 1.69E-028	atm m3/mol	25	EST	MEYLAN,WM & HOWARD,PH (1991)
OH 4.06E-010	cm3/molc sec	25	EST	MEYLAN,WM & HOWARD,PH (1993)

598

CAS #: 001678-25-7			BENZENESULFANILIDE
Formula: $C_{12}H_{11}NO_2S$			
Mol Weight: 233.29			
MP (deg C):	FP (deg C):		
BP (deg C):			
BP pressure (mm Hg):			

Property/Value	Units	Temp	Data Type	Reference
WS 1.83E+002	mg/L	25	EST	MEYLAN,WM ET AL. (1996)
logP 2.58			EXP	HANSCH,C & LEO,AJ (1985)
VP 3.00E-006	mm Hg	25	EST	NEELY,WB & BLAU,GE (1985)
DC	pKa			
HL 2.33E-007	atm m3/mol	25	EST	MEYLAN,WM & HOWARD,PH (1991)
OH 4.31E-011	cm3/molc sec	25	EST	MEYLAN,WM & HOWARD,PH (1993)

CAS #: 001678-91-7			ETHYLCYCLOHEXANE
Formula: C_8H_{16}			
Mol Weight: 112.22			
MP (deg C): -111.3	FP (deg C):		
BP (deg C): 131.9			
BP pressure (mm Hg):			

Property/Value	Units	Temp	Data Type	Reference
WS 6.30E+000	mg/L	20	EXP	YALKOWSKY,SH & DANNENFELSER,RM (1992)
logP 4.08			EST	MEYLAN,WM & HOWARD,PH (1995)
VP 1.28E+001	mm Hg	25	EXP	DAUBERT,TE & DANNER,RP (1989)
DC 10.45	pKa	23	EXP	PERRIN,DD (1965)
HL 3.00E-001	atm m3/mol	25	EST	VP/WSOL
OH 1.20E-011	cm3/molc sec	25	EST	MEYLAN,WM & HOWARD,PH (1993)

CAS #: 001678-92-8			N-PROPYLCYCLOHEXANE
Formula: C_9H_{18}			
Mol Weight: 126.24			
MP (deg C): -94.9	FP (deg C):		
BP (deg C): 156.7			
BP pressure (mm Hg):			

Property/Value	Units	Temp	Data Type	Reference
WS 3.42E+000	mg/L	25	EST	MEYLAN,WM ET AL. (1996)
logP 4.58			EST	MEYLAN,WM & HOWARD,PH (1995)
VP 4.19E+000	mm Hg	25	EXP	DAUBERT,TE & DANNER,RP (1989)
DC	pKa			
HL 5.97E-001	atm m3/mol	25	EST	MEYLAN,WM & HOWARD,PH (1991)
OH 1.34E-011	cm3/molc sec	25	EST	MEYLAN,WM & HOWARD,PH (1993)

CAS #: 001678-93-9			BUTYLCYCLOHEXANE
Formula: $C_{10}H_{20}$			
Mol Weight: 140.27			
MP (deg C): -74.7	FP (deg C):		
BP (deg C): 180.9			
BP pressure (mm Hg):			

Property/Value	Units	Temp	Data Type	Reference
WS 1.14E+000	mg/L	25	EST	MEYLAN,WM ET AL. (1996)
logP 5.07			EST	MEYLAN,WM & HOWARD,PH (1995)
VP 1.31E+000	mm Hg	25	EXP	DAUBERT,TE & DANNER,RP (1989)
DC	pKa			
HL 7.92E-001	atm m3/mol	25	EST	MEYLAN,WM & HOWARD,PH (1991)
OH 1.48E-011	cm3/molc sec	25	EST	MEYLAN,WM & HOWARD,PH (1993)

CAS #: 001678-98-4			ISOBUTYLCYCLOHEXANE
Formula: $C_{10}H_{20}$			
Mol Weight: 140.27			
MP (deg C): -95	FP (deg C):		
BP (deg C): 171.3			
BP pressure (mm Hg):			

Property/Value	Units	Temp	Data Type	Reference
WS 1.32E+000	mg/L	25	EST	MEYLAN,WM ET AL. (1996)
logP 4.99			EST	MEYLAN,WM & HOWARD,PH (1995)
VP 4.83E+000	mm Hg	25	EXT	CHAO,J ET AL. (1983)
DC	pKa			
HL 7.92E-001	atm m3/mol	25	EST	MEYLAN,WM & HOWARD,PH (1991)
OH 1.48E-011	cm3/molc sec	25	EST	MEYLAN,WM & HOWARD,PH (1993)

CAS #: 001679-07-8			CYCLOPENTANETHIOL
Formula: $C_5H_{10}S$			
Mol Weight: 102.20			
MP (deg C):	FP (deg C):		
BP (deg C): 132.1			
BP pressure (mm Hg):			

Property/Value	Units	Temp	Data Type	Reference
WS 7.58E+002	mg/L	25	EST	MEYLAN,WM ET AL. (1996)
logP 2.55			EST	MEYLAN,WM & HOWARD,PH (1995)
VP 1.14E+001	mm Hg	25	EXT	BOUBLIK,T ET AL. (1984)
DC	pKa			
HL 3.58E-003	atm m3/mol	25	EST	MEYLAN,WM & HOWARD,PH (1991)
OH 5.53E-011	cm3/molc sec	25	EST	MEYLAN,WM & HOWARD,PH (1993)

CAS #: 001679-09-0			2-METHYL-2-BUTANETHIOL
Formula: $C_5H_{12}S$			
Mol Weight: 104.22			
MP (deg C):	FP (deg C):		
BP (deg C): 99.1			
BP pressure (mm Hg):			

Property/Value	Units	Temp	Data Type	Reference
WS 6.45E+002	mg/L	25	EST	MEYLAN,WM ET AL. (1996)
logP 2.63			EST	MEYLAN,WM & HOWARD,PH (1995)
VP 4.79E+001	mm Hg	25	EXT	BOUBLIK,T ET AL. (1984)
DC	pKa			
HL 8.12E-003	atm m3/mol	25	EST	MEYLAN,WM & HOWARD,PH (1991)
OH 3.42E-011	cm3/molc sec	25	EST	MEYLAN,WM & HOWARD,PH (1993)

CAS #: 001679-64-7			METHYL TEREPHTHALATE
Formula: $C_9H_8O_4$			
Mol Weight: 180.16			
MP (deg C): 221-223	FP (deg C):		
BP (deg C):			
BP pressure (mm Hg):			

Property/Value	Units	Temp	Data Type	Reference
WS 1.91E+003	mg/L	25	EST	MEYLAN,WM ET AL. (1996)
logP 1.71			EST	MEYLAN,WM & HOWARD,PH (1995)
VP 1.84E-004	mm Hg	25	EST	NEELY,WB & BLAU,GE (1985)
DC	pKa			
HL 6.99E-010	atm m3/mol	25	EST	MEYLAN,WM & HOWARD,PH (1991)
OH 9.09E-013	cm3/molc sec	25	EST	MEYLAN,WM & HOWARD,PH (1993)

CAS #: 001680-21-3 — 2-PROPENOIC ACID, 1,2-ETHANEDIYLBIS(OXY-2,1-ETHA

Formula: $C_{12}H_{18}O_6$

Mol Weight: 258.27

MP (deg C):   FP (deg C):

BP (deg C):

BP pressure (mm Hg):

Property/ Value	Units	Temp	Data Type	Reference
WS 3.22E+003	mg/L	25	EST	MEYLAN,WM ET AL. (1996)
logP 0.96			EXP	SANGSTER,J (1993)
VP 2.89E-003	mm Hg	25	EST	NEELY,WB & BLAU,GE (1985)
DC	pKa			
HL 3.72E-011	atm m3/mol	25	EST	MEYLAN,WM & HOWARD,PH (1991)
OH 5.01E-011	cm3/molc sec	25	EST	MEYLAN,WM & HOWARD,PH (1993)

CAS #: 001680-44-0 — 1H-1,2,3-TRIAZOLE, 4-PHENYL-

Formula: $C_8H_7N_3$

Mol Weight: 145.17

MP (deg C):   FP (deg C):

BP (deg C):

BP pressure (mm Hg):

Property/ Value	Units	Temp	Data Type	Reference
WS 1.86E+003	mg/L	25	EST	MEYLAN,WM ET AL. (1996)
logP 1.91			EXP	SANGSTER,J (1993)
VP 6.79E-005	mm Hg	25	EST	NEELY,WB & BLAU,GE (1985)
DC	pKa			
HL 1.15E-007	atm m3/mol	25	EST	MEYLAN,WM & HOWARD,PH (1991)
OH 3.56E-012	cm3/molc sec	25	EST	MEYLAN,WM & HOWARD,PH (1993)

CAS #: 001689-09-4 — 1(3H)ISOBENZOFURANONE-3,3-DIMETHYL

Formula: $C_{10}H_{10}O_2$

Mol Weight: 162.19

MP (deg C):   FP (deg C):

BP (deg C):

BP pressure (mm Hg):

Property/ Value	Units	Temp	Data Type	Reference
WS 2.23E+003	mg/L	25	EST	MEYLAN,WM ET AL. (1996)
logP 1.73			EXP	HANSCH,C & LEO,AJ (1985)
VP 1.64E-003	mm Hg	25	EST	NEELY,WB & BLAU,GE (1985)
DC	pKa			
HL 2.24E-005	atm m3/mol	25	EST	MEYLAN,WM & HOWARD,PH (1991)
OH 4.25E-012	cm3/molc sec	25	EST	MEYLAN,WM & HOWARD,PH (1993)

CAS #: 001689-82-3 — P-PHENYLAZOPHENOL

Formula: $C_{12}H_{10}N_2O$

Mol Weight: 198.23

MP (deg C): 155   FP (deg C):

BP (deg C): 225

BP pressure (mm Hg): 2.00E+001

Property/ Value	Units	Temp	Data Type	Reference
WS 9.00E+001	mg/L	20	EXP	YALKOWSKY,SH & DANNENFELSER,RM (1992)
logP 3.63			EST	MEYLAN,WM & HOWARD,PH (1995)
VP 2.31E-007	mm Hg	25	EXP	SHIMIZU,T ET AL. (1987)
DC	pKa			
HL 6.70E-010	atm m3/mol	25	EST	VP/WSOL
OH 1.41E-011	cm3/molc sec	25	EST	MEYLAN,WM & HOWARD,PH (1993)

CAS #: 001689-83-4 — IOXYNIL

Formula: $C_7H_3I_2NO$

Mol Weight: 370.92

MP (deg C): 212-213   FP (deg C):

BP (deg C):

BP pressure (mm Hg):

Property/ Value	Units	Temp	Data Type	Reference
WS 5.00E+001	mg/L	25	EXP	BUIKEMA,ALJR ET AL. (1979)
logP 0.90			EXP	HANSCH,C ET AL. (1995)
VP 1.38E-007	mm Hg	25	EST	NEELY,WB & BLAU,GE (1985)
DC 3.96	pKa	25	EXP	BUIKEMA,ALJR ET AL. (1979)
HL 5.48E-010	atm m3/mol	25	EST	MEYLAN,WM & HOWARD,PH (1991)
OH 2.16E-013	cm3/molc sec	25	EST	MEYLAN,WM & HOWARD,PH (1993)

CAS #: 001689-84-5 — BROMOXYNIL

Formula: $C_7H_3Br_2NO$

Mol Weight: 276.93

MP (deg C): 194-195   FP (deg C):

BP (deg C):

BP pressure (mm Hg):

Property/ Value	Units	Temp	Data Type	Reference
WS 1.30E+002	mg/L	25	EXP	WORTHING,CR & WALKER,SB (1987)
logP 3.39			EST	MEYLAN,WM & HOWARD,PH (1995)
VP 5.00E-007	mm Hg			WAUCHOPE,RD ET AL. (1991A); <5e-7
DC 4.10	pKa			WAUCHOPE,RD ET AL. (1991A)
HL 9.52E-010	atm m3/mol	25	EST	MEYLAN,WM & HOWARD,PH (1991)
OH 2.10E-013	cm3/molc sec	25	EST	MEYLAN,WM & HOWARD,PH (1993)

CAS #: 001689-99-2 — BROMOXYNIL OCTANOATE

Formula: $C_{15}H_{17}Br_2NO_2$

Mol Weight: 403.13

MP (deg C): 45.5   FP (deg C):

BP (deg C):

BP pressure (mm Hg):

Property/ Value	Units	Temp	Data Type	Reference
WS 9.83E-003	mg/L	25	EST	MEYLAN,WM ET AL. (1996)
logP 6.10			EXP	KRAWCHUK,BP & WEBSTER,GRB (1987)
VP 1.69E-007	mm Hg	25	EST	NEELY,WB & BLAU,GE (1985)
DC	pKa			
HL 5.44E-007	atm m3/mol	25	EST	MEYLAN,WM & HOWARD,PH (1991)
OH 7.89E-012	cm3/molc sec	25	EST	MEYLAN,WM & HOWARD,PH (1993)

CAS #: 001693-94-3 — 4H-PYRIDO(12A)PYRIMIDIN-4-ONE,2ME

Formula: $C_9H_{10}N_2O$

Mol Weight: 162.19

MP (deg C):   FP (deg C):

BP (deg C):

BP pressure (mm Hg):

Property/ Value	Units	Temp	Data Type	Reference
WS 2.23E+004	mg/L	25	EST	MEYLAN,WM ET AL. (1996)
logP 0.57			EXP	HANSCH,C & LEO,AJ (1985)
VP 1.60E-004	mm Hg	25	EST	NEELY,WB & BLAU,GE (1985)
DC	pKa			
HL 2.35E-009	atm m3/mol	25	EST	MEYLAN,WM & HOWARD,PH (1991)
OH 1.10E-010	cm3/molc sec	25	EST	MEYLAN,WM & HOWARD,PH (1993)

001695-04-1 — O-PHENOXYANISOLE

CAS #: 001695-04-1

Formula: $C_{13}H_{12}O_2$

Mol Weight: 200.24

MP (deg C): 79

FP (deg C):

BP (deg C): 288

BP pressure (mm Hg):

Property/Value	Units	Temp	Data Type	Reference
WS 1.40E+002	mg/L	25	EST	MEYLAN,WM ET AL. (1996)
logP 2.92			EXP	HANSCH,C & LEO,AJ (1985)
VP 7.25E-004	mm Hg	25	EST	NEELY,WB & BLAU,GE (1985)
DC	pKa			
HL 6.97E-006	atm m3/mol	25	EST	MEYLAN,WM & HOWARD,PH (1991)
OH 2.73E-011	cm3/molc sec	25	EST	MEYLAN,WM & HOWARD,PH (1993)

001696-17-9 — N,N-DIETHYLBENZAMIDE

CAS #: 001696-17-9

Formula: $C_{11}H_{15}NO$

Mol Weight: 177.25

MP (deg C):

FP (deg C):

BP (deg C):

BP pressure (mm Hg):

Property/Value	Units	Temp	Data Type	Reference
WS 4.86E+003	mg/L	25	EST	MEYLAN,WM ET AL. (1996)
logP 1.25			EXP	HANSCH,C & LEO,AJ (1985)
VP 1.05E-003	mm Hg	25	EST	NEELY,WB & BLAU,GE (1985)
DC	pKa			
HL 1.88E-008	atm m3/mol	25	EST	MEYLAN,WM & HOWARD,PH (1991)
OH 2.48E-011	cm3/molc sec	25	EST	MEYLAN,WM & HOWARD,PH (1993)

001697-18-3 — 2'-CL SALICYLANILIDE

CAS #: 001697-18-3

Formula: $C_{13}H_{10}ClNO_2$

Mol Weight: 247.68

MP (deg C):

FP (deg C):

BP (deg C):

BP pressure (mm Hg):

Property/Value	Units	Temp	Data Type	Reference
WS 1.21E+002	mg/L	25	EST	MEYLAN,WM ET AL. (1996)
logP 4.35			EXP	TERADA,H ET AL. (1988)
VP 5.90E-009	mm Hg	25	EST	NEELY,WB & BLAU,GE (1985)
DC	pKa			
HL 1.19E-010	atm m3/mol	25	EST	MEYLAN,WM & HOWARD,PH (1991)
OH 3.41E-011	cm3/molc sec	25	EST	MEYLAN,WM & HOWARD,PH (1993)

001698-60-8 — CHLORIDAZON

CAS #: 001698-60-8

Formula: $C_{10}H_8ClN_3O$

Mol Weight: 221.65

MP (deg C): 206

FP (deg C):

BP (deg C):

BP pressure (mm Hg):

Property/Value	Units	Temp	Data Type	Reference
WS 3.40E+002	mg/L	20	EXP	TOMLIN,C (1994)
logP 1.14			EXP	BRAUMANN,T & GRIMME,LH (1981)
VP 5.00E-002	mm Hg	25	EXP	WAUCHOPE,RD ET AL. (1991A)
DC 3.38	pKa		EXP	GIOCOLEA,A ET AL. (1991)
HL 4.29E-005	atm m3/mol	25	EST	VP/WSOL
OH 4.02E-011	cm3/molc sec	25	EST	MEYLAN,WM & HOWARD,PH (1993)

001698-62-0 — 3(2H)-PYRIDAZINONE, 4-CHLORO-5-(METHYLAMINO)-2-P

CAS #: 001698-62-0

Formula: $C_{11}H_{10}ClN_3O$

Mol Weight: 235.67

MP (deg C):

FP (deg C):

BP (deg C):

BP pressure (mm Hg):

Property/Value	Units	Temp	Data Type	Reference
WS 2.73E+003	mg/L	25	EST	MEYLAN,WM ET AL. (1996)
logP 1.19			EXP	SANGSTER,J (1994)
VP 1.04E-006	mm Hg	25	EST	NEELY,WB & BLAU,GE (1985)
DC	pKa			
HL 1.44E-011	atm m3/mol	25	EST	MEYLAN,WM & HOWARD,PH (1991)
OH 8.35E-011	cm3/molc sec	25	EST	MEYLAN,WM & HOWARD,PH (1993)

001701-69-5 — ETHYL-4-PYRIDYL KETONE

CAS #: 001701-69-5

Formula: C_8H_9NO

Mol Weight: 135.17

MP (deg C):

FP (deg C):

BP (deg C):

BP pressure (mm Hg):

Property/Value	Units	Temp	Data Type	Reference
WS 1.93E+004	mg/L	25	EST	MEYLAN,WM ET AL. (1996)
logP 0.77			EXP	HANSCH,C & LEO,AJ (1985)
VP 1.23E-001	mm Hg	25	EST	NEELY,WB & BLAU,GE (1985)
DC	pKa			
HL 1.70E-008	atm m3/mol	25	EST	MEYLAN,WM & HOWARD,PH (1991)
OH 1.57E-012	cm3/molc sec	25	EST	MEYLAN,WM & HOWARD,PH (1993)

001702-17-6 — 3,6-DICHLOROPICOLINIC ACID

CAS #: 001702-17-6

Formula: $C_6H_3Cl_2NO_2$

Mol Weight: 192.00

MP (deg C): 151-2

FP (deg C):

BP (deg C):

BP pressure (mm Hg):

Property/Value	Units	Temp	Data Type	Reference
WS 9.00E+003	mg/L	20	EXP	WORTHING,CR & WALKER,SB (1987)
logP -0.68			EXP	GARTEN,CTJR & TRABALKA,JR (1983)
VP 1.20E-005	mm Hg	25	EXP	WORTHING,CR & WALKER,SB (1987)
DC 2.33	pKa		EXP	WORTHING,CR & WALKER,SB (1987)
HL 3.03E-009	atm m3/mol	25	EST	VP/WSOL
OH 6.36E-014	cm3/molc sec	25	EST	ATKINSON,R (1988)

001705-85-7 — 6-METHYLCHRYSENE

CAS #: 001705-85-7

Formula: $C_{19}H_{14}$

Mol Weight: 242.32

MP (deg C):

FP (deg C):

BP (deg C):

BP pressure (mm Hg):

Property/Value	Units	Temp	Data Type	Reference
WS 6.50E-002	mg/L	27	EXP	YALKOWSKY,SH & DANNENFELSER,RM (1992)
logP 6.07			EST	MEYLAN,WM & HOWARD,PH (1995)
VP 2.53E-007	mm Hg	25	EST	NEELY,WB & BLAU,GE (1985)
DC	pKa			
HL 5.53E-006	atm m3/mol	25	EST	MEYLAN,WM & HOWARD,PH (1991)
OH 1.31E-010	cm3/molc sec	25	EST	MEYLAN,WM & HOWARD,PH (1993)

CAS #: 001708-29-8				2,5-DIHYDROFURAN

Formula: C_4H_6O

Mol Weight: 70.09

MP (deg C):   **FP (deg C):**

BP (deg C): 66-67

BP pressure (mm Hg):

Property/Value	Units	Temp	Data Type	Reference
WS 5.48E+004	mg/L	25	EST	MEYLAN,WM ET AL. (1996)
logP 0.46			EXP	HANSCH,C & LEO,AJ (1985)
VP 1.58E+002	mm Hg	25	EXT	YAWS,CL (1994)
DC	pKa			
HL 5.89E-004	atm m3/mol	25	EST	MEYLAN,WM & HOWARD,PH (1991)
OH 6.55E-011	cm3/molc sec	25	EST	MEYLAN,WM & HOWARD,PH (1993)

CAS #: 001709-52-0				N1-METHYLSULFANILAMIDE

Formula: $C_7H_{10}N_2O_2S$

Mol Weight: 186.23

MP (deg C):   **FP (deg C):**

BP (deg C):

BP pressure (mm Hg):

Property/Value	Units	Temp	Data Type	Reference
WS 5.23E+004	mg/L	25	EST	MEYLAN,WM ET AL. (1996)
logP -0.01			EXP	HANSCH,C & LEO,AJ (1985)
VP 2.00E-005	mm Hg	25	EST	NEELY,WB & BLAU,GE (1985)
DC	pKa			
HL 3.28E-010	atm m3/mol	25	EST	MEYLAN,WM & HOWARD,PH (1991)
OH 2.43E-011	cm3/molc sec	25	EST	MEYLAN,WM & HOWARD,PH (1993)

CAS #: 001709-59-7				N1-DIMETHYLSULFANILAMIDE

Formula: $C_8H_{12}N_2O_2S$

Mol Weight: 200.26

MP (deg C):   **FP (deg C):**

BP (deg C):

BP pressure (mm Hg):

Property/Value	Units	Temp	Data Type	Reference
WS 1.17E+004	mg/L	25	EST	MEYLAN,WM ET AL. (1996)
logP 0.67			EXP	HANSCH,C & LEO,AJ (1985)
VP 1.95E-005	mm Hg	25	EST	NEELY,WB & BLAU,GE (1985)
DC	pKa			
HL 7.19E-010	atm m3/mol	25	EST	MEYLAN,WM & HOWARD,PH (1991)
OH 2.55E-011	cm3/molc sec	25	EST	MEYLAN,WM & HOWARD,PH (1993)

CAS #: 001712-64-7				ISOPROPYL NITRATE

Formula: $C_3H_7NO_3$

Mol Weight: 105.09

MP (deg C):   **FP (deg C):**

BP (deg C): 100

BP pressure (mm Hg):

Property/Value	Units	Temp	Data Type	Reference
WS 4.27E+003	mg/L	25	EST	MEYLAN,WM ET AL. (1996)
logP 1.66			EST	MEYLAN,WM & HOWARD,PH (1995)
VP 3.41E+001	mm Hg	25	EXP	BOUBLIK,T ET AL. (1984)
DC	pKa			
HL 3.50E-004	atm m3/mol	25	EST	MEYLAN,WM & HOWARD,PH (1991)
OH 4.92E-013	cm3/molc sec	25	EST	MEYLAN,WM & HOWARD,PH (1993)

CAS #: 001717-00-6				1,1-DICHLOROFLUOROETHANE

Formula: $C_2H_3Cl_2F$

Mol Weight: 116.95

MP (deg C): -103.5   **FP (deg C):**

BP (deg C): 32

BP pressure (mm Hg):

Property/Value	Units	Temp	Data Type	Reference
WS 2.63E+003	mg/L	25	EST	VP/HL
logP 2.37			EST	MEYLAN,WM & HOWARD,PH (1995)
VP 7.07E+002	mm Hg	25	EST	NEELY,WB & BLAU,GE (1985)
DC	pKa			
HL 2.41E-002	atm m3/mol	25	EST	MEYLAN,WM & HOWARD,PH (1991)
OH 1.61E-014	cm3/molc sec	25	EXP	BROWN,AC ET AL. (1990A)

CAS #: 001719-53-5				DICHLORODIETHYLSILANE

Formula: $C_4H_{10}Cl_2Si$

Mol Weight: 157.12

MP (deg C): -96.5   **FP (deg C):**

BP (deg C):

BP pressure (mm Hg):

Property/Value	Units	Temp	Data Type	Reference
WS 1.24E+002	mg/L	25	EST	MEYLAN,WM ET AL. (1996)
logP 3.23			EST	MEYLAN,WM & HOWARD,PH (1995)
VP 4.53E+001	mm Hg	25	EST	NEELY,WB & BLAU,GE (1985)
DC	pKa			
HL 3.26E-002	atm m3/mol	25	EST	MEYLAN,WM & HOWARD,PH (1991)
OH 2.39E-012	cm3/molc sec	25	EST	MEYLAN,WM & HOWARD,PH (1993)

CAS #: 001722-10-7				3-CHLORO-6-METHOXYPYRIDAZINE

Formula: $C_5H_5ClN_2O$

Mol Weight: 144.56

MP (deg C): 84-85   **FP (deg C):**

BP (deg C):

BP pressure (mm Hg):

Property/Value	Units	Temp	Data Type	Reference
WS 1.39E+004	mg/L	25	EST	MEYLAN,WM ET AL. (1996)
logP 0.89			EXP	HANSCH,C & LEO,AJ (1985)
VP 6.10E-003	mm Hg	25	EST	NEELY,WB & BLAU,GE (1985)
DC	pKa			
HL 7.24E-005	atm m3/mol	25	EST	MEYLAN,WM & HOWARD,PH (1991)
OH 1.57E-012	cm3/molc sec	25	EST	MEYLAN,WM & HOWARD,PH (1993)

CAS #: 001722-12-9				PYRIMIDINE, 2-CHLORO-

Formula: $C_4H_3ClN_2$

Mol Weight: 114.53

MP (deg C): 66-68   **FP (deg C):**

BP (deg C): 75-76

BP pressure (mm Hg): 1.00E+001

Property/Value	Units	Temp	Data Type	Reference
WS 5.17E+004	mg/L	25	EST	MEYLAN,WM ET AL. (1996)
logP 0.36			EXP	YAMAGAMI,C ET AL. (1990)
VP 2.48E+000	mm Hg	25	EST	NEELY,WB & BLAU,GE (1985)
DC	pKa			
HL 2.16E-006	atm m3/mol	25	EST	MEYLAN,WM & HOWARD,PH (1991)
OH 1.62E-013	cm3/molc sec	25	EST	MEYLAN,WM & HOWARD,PH (1993)

CAS #: 001722-18-5 — 2-PHENYL-S-TRIAZINE

Formula: $C_9H_7N_3$
Mol Weight: 157.18
MP (deg C):
FP (deg C):
BP (deg C):
BP pressure (mm Hg):

Property/Value	Units	Temp	Data Type	Reference
WS 1.59E+003	mg/L	25	EST	MEYLAN,WM ET AL. (1996)
logP 1.93			EXP	HANSCH,C & LEO,AJ (1985)
VP 6.15E-004	mm Hg	25	EST	NEELY,WB & BLAU,GE (1985)
DC	pKa			
HL 9.27E-008	atm m3/mol	25	EST	MEYLAN,WM & HOWARD,PH (1991)
OH 3.65E-012	cm3/molc sec	25	EST	MEYLAN,WM & HOWARD,PH (1993)

CAS #: 001724-39-6 — CYCLODODECANOL

Formula: $C_{12}H_{24}O$
Mol Weight: 184.32
MP (deg C): 76-79
FP (deg C):
BP (deg C):
BP pressure (mm Hg):

Property/Value	Units	Temp	Data Type	Reference
WS 2.07E+001	mg/L	25	EST	MEYLAN,WM ET AL. (1996)
logP 4.58			EST	MEYLAN,WM & HOWARD,PH (1995)
VP 3.79E-004	mm Hg	25	EST	NEELY,WB & BLAU,GE (1985)
DC	pKa			
HL 2.68E-005	atm m3/mol	25	EST	MEYLAN,WM & HOWARD,PH (1991)
OH 2.60E-011	cm3/molc sec	25	EST	MEYLAN,WM & HOWARD,PH (1993)

CAS #: 001725-74-2 — 1,2,3,4,5,6-HEXACHLORO-3-HEXENE

Formula: $C_6H_6Cl_6$
Mol Weight: 290.83
MP (deg C): 58.5
FP (deg C):
BP (deg C): 112
BP pressure (mm Hg): 2.00E+000

Property/Value	Units	Temp	Data Type	Reference
WS 8.61E-001	mg/L	25	EST	MEYLAN,WM ET AL. (1996)
logP 4.93			EST	MEYLAN,WM & HOWARD,PH (1995)
VP 2.27E-003	mm Hg	25	EST	NEELY,WB & BLAU,GE (1985)
DC	pKa			
HL 2.13E-003	atm m3/mol	25	EST	MEYLAN,WM & HOWARD,PH (1991)
OH 3.59E-012	cm3/molc sec	25	EST	MEYLAN,WM & HOWARD,PH (1993)

CAS #: 001726-94-9 — TRIPHENYLMETHYLISOTHIOCYANATE

Formula: $C_{20}H_{15}NS$
Mol Weight: 301.41
MP (deg C):
FP (deg C):
BP (deg C):
BP pressure (mm Hg):

Property/Value	Units	Temp	Data Type	Reference
WS 1.22E-002	mg/L	25	EST	MEYLAN,WM ET AL. (1996)
logP 7.02			EXP	HANSCH,C & LEO,AJ (1985)
VP 1.17E-007	mm Hg	25	EST	NEELY,WB & BLAU,GE (1985)
DC	pKa			
HL 1.64E-006	atm m3/mol	25	EST	MEYLAN,WM & HOWARD,PH (1991)
OH 1.29E-011	cm3/molc sec	25	EST	MEYLAN,WM & HOWARD,PH (1993)

CAS #: 001730-37-6 — 1-METHYLFLUORENE

Formula: $C_{14}H_{12}$
Mol Weight: 180.25
MP (deg C): 84-86
FP (deg C):
BP (deg C):
BP pressure (mm Hg):

Property/Value	Units	Temp	Data Type	Reference
WS 1.09E+000	mg/L	25	EXP	YALKOWSKY,SH & DANNENFELSER,RM (1992)
logP 4.97			EXP	MILLER,MM ET AL. (1985)
VP 3.98E-004	mm Hg	25	EST	NEELY,WB & BLAU,GE (1985)
DC	pKa			
HL 1.85E-004	atm m3/mol	25	EST	MEYLAN,WM & HOWARD,PH (1991)
OH 1.58E-011	cm3/molc sec	25	EST	MEYLAN,WM & HOWARD,PH (1993)

CAS #: 001731-81-3 — UNDECYL ALCOHOL ACETATE

Formula: $C_{13}H_{26}O_2$
Mol Weight: 214.35
MP (deg C):
FP (deg C):
BP (deg C):
BP pressure (mm Hg):

Property/Value	Units	Temp	Data Type	Reference
WS 1.13E+000	mg/L	25	EST	MEYLAN,WM ET AL. (1996)
logP 5.28			EST	MEYLAN,WM & HOWARD,PH (1995)
VP 1.22E-002	mm Hg	25	EST	NEELY,WB & BLAU,GE (1985)
DC	pKa			
HL 2.98E-003	atm m3/mol	25	EST	MEYLAN,WM & HOWARD,PH (1991)
OH 1.45E-011	cm3/molc sec	25	EST	MEYLAN,WM & HOWARD,PH (1993)

CAS #: 001731-84-6 — METHYL NONANOATE

Formula: $C_{10}H_{20}O_2$
Mol Weight: 172.27
MP (deg C):
FP (deg C):
BP (deg C): 213.5
BP pressure (mm Hg):

Property/Value	Units	Temp	Data Type	Reference
WS 2.24E+002	mg/L	25	EXP	TEWARI,YB ET AL. (1982A)
logP 4.32			EXP	TEWARI,YB ET AL. (1982)
VP 2.11E-001	mm Hg	25	EST	NEELY,WB & BLAU,GE (1985)
DC	pKa			
HL 1.27E-003	atm m3/mol	25	EST	MEYLAN,WM & HOWARD,PH (1991)
OH 9.45E-012	cm3/molc sec	25	EST	MEYLAN,WM & HOWARD,PH (1993)

CAS #: 001731-88-0 — METHYL TRIDECANOATE

Formula: $C_{14}H_{28}O_2$
Mol Weight: 228.38
MP (deg C): 6.5
FP (deg C):
BP (deg C): 92
BP pressure (mm Hg): 1.00E+000

Property/Value	Units	Temp	Data Type	Reference
WS 3.63E-001	mg/L	25	EST	MEYLAN,WM ET AL. (1996)
logP 5.78			EST	MEYLAN,WM & HOWARD,PH (1995)
VP 1.29E-003	mm Hg	26	EXP	BACCANARI,DP ET AL. (1968)
DC	pKa			
HL 3.96E-003	atm m3/mol	25	EST	MEYLAN,WM & HOWARD,PH (1991)
OH 1.51E-011	cm3/molc sec	25	EST	MEYLAN,WM & HOWARD,PH (1993)

CAS #: 001732-08-7				HEPTANEDIOIC ACID, DIMETHYL ESTER

Formula: $C_9H_{16}O_4$

Mol Weight: 188.23

MP (deg C): -21 FP (deg C):

BP (deg C): 120

BP pressure (mm Hg): 1.00E+001

Property/Value	Units	Temp	Data Type	Reference
WS 3.33E+003	mg/L	25	EST	MEYLAN,WM ET AL. (1996)
logP 1.38			EXP	HANSCH,C ET AL. (1995)
VP 2.54E-001	mm Hg	25	EST	NEELY,WB & BLAU,GE (1985)
DC	pKa			
HL 1.30E-006	atm m3/mol	25	EST	MEYLAN,WM & HOWARD,PH (1991)
OH 6.37E-012	cm3/molc sec	25	EST	MEYLAN,WM & HOWARD,PH (1993)

CAS #: 001737-06-0				4-OCF3 ACETANILIDE

Formula: $C_9H_8F_3NO_2$

Mol Weight: 219.16

MP (deg C): FP (deg C):

BP (deg C):

BP pressure (mm Hg):

Property/Value	Units	Temp	Data Type	Reference
WS 2.01E+002	mg/L	25	EST	MEYLAN,WM ET AL. (1996)
logP 2.62			EXP	NAKAGAWA,Y ET AL. (1992)
VP 1.17E-004	mm Hg	25	EST	NEELY,WB & BLAU,GE (1985)
DC	pKa			
HL 2.88E-009	atm m3/mol	25	EST	MEYLAN,WM & HOWARD,PH (1991)
OH 1.14E-011	cm3/molc sec	25	EST	MEYLAN,WM & HOWARD,PH (1993)

CAS #: 001740-57-4				ISO-PHTHALAMIDE

Formula: $C_8H_8N_2O_2$

Mol Weight: 164.17

MP (deg C): FP (deg C):

BP (deg C):

BP pressure (mm Hg):

Property/Value	Units	Temp	Data Type	Reference
WS 4.85E+003	mg/L	25	EST	MEYLAN,WM ET AL. (1996)
logP -0.21			EXP	HANSCH,C & LEO,AJ (1985)
VP 1.55E-007	mm Hg	25	EST	NEELY,WB & BLAU,GE (1985)
DC	pKa			
HL 1.44E-012	atm m3/mol	25	EST	MEYLAN,WM & HOWARD,PH (1991)
OH 5.62E-012	cm3/molc sec	25	EST	MEYLAN,WM & HOWARD,PH (1993)

CAS #: 001741-83-9				METHYL PENTYL SULFIDE

Formula: $C_6H_{14}S$

Mol Weight: 118.24

MP (deg C): -94 FP (deg C):

BP (deg C): 145.1

BP pressure (mm Hg):

Property/Value	Units	Temp	Data Type	Reference
WS 3.51E+002	mg/L	25	EST	MEYLAN,WM ET AL. (1996)
logP 2.88			EST	MEYLAN,WM & HOWARD,PH (1995)
VP 6.45E+000	mm Hg	25	EST	NEELY,WB & BLAU,GE (1985)
DC	pKa			
HL 2.44E-003	atm m3/mol	25	EST	MEYLAN,WM & HOWARD,PH (1991)
OH 1.59E-011	cm3/molc sec	25	EST	MEYLAN,WM & HOWARD,PH (1993)

CAS #: 001746-01-6				2,3,7,8-TETRACHLORODIBENZO-P-DIOXIN

Formula: $C_{12}H_4Cl_4O_2$

Mol Weight: 321.98

MP (deg C): 305 FP (deg C):

BP (deg C):

BP pressure (mm Hg):

Property/Value	Units	Temp	Data Type	Reference
WS 1.26E-005	mg/L	25	EXP	MARPLE,L ET AL. (1986)
logP 6.80			EXP	SHIU,WY ET AL. (1988)
VP 1.50E-009	mm Hg	25	EXP	RORDORF,BF (1987)
DC	pKa			
HL 5.00E-005	atm m3/mol	25	EST	VP/WSOL
OH 2.87E-012	cm3/molc sec	25	EST	MEYLAN,WM & HOWARD,PH (1993)

CAS #: 001746-13-0				ALLYLPHENYL ETHER

Formula: $C_9H_{10}O$

Mol Weight: 134.18

MP (deg C): FP (deg C):

BP (deg C): 191.7

BP pressure (mm Hg):

Property/Value	Units	Temp	Data Type	Reference
WS 2.73E+002	mg/L	25	EST	MEYLAN,WM ET AL. (1996)
logP 2.94			EXP	HANSCH,C & LEO,AJ (1985)
VP 5.91E-001	mm Hg	25	EST	NEELY,WB & BLAU,GE (1985)
DC	pKa			
HL 4.18E-004	atm m3/mol	25	EST	MEYLAN,WM & HOWARD,PH (1991)
OH 5.35E-011	cm3/molc sec	25	EST	MEYLAN,WM & HOWARD,PH (1993)

CAS #: 001746-25-4				2,4,5-TRIIODOIMIDAZOLE

Formula: $C_3HI_3N_2$

Mol Weight: 445.77

MP (deg C): FP (deg C):

BP (deg C):

BP pressure (mm Hg):

Property/Value	Units	Temp	Data Type	Reference
WS 6.70E+000	mg/L	25	EST	MEYLAN,WM ET AL. (1996)
logP 2.78			EXP	HANSCH,C & LEO,AJ (1985)
VP 2.62E-007	mm Hg	25	EST	NEELY,WB & BLAU,GE (1985)
DC	pKa			
HL 4.68E-008	atm m3/mol	25	EST	MEYLAN,WM & HOWARD,PH (1991)
OH 3.60E-011	cm3/molc sec	25	EST	MEYLAN,WM & HOWARD,PH (1993)

CAS #: 001746-81-2				MONOLINURON

Formula: $C_9H_{11}ClN_2O_2$

Mol Weight: 214.65

MP (deg C): 80-83 FP (deg C):

BP (deg C):

BP pressure (mm Hg):

Property/Value	Units	Temp	Data Type	Reference
WS 9.30E+002	mg/L	20	EXP	YALKOWSKY,SH & DANNENFELSER,RM (1992)
logP 2.30			EXP	HANSCH,C & LEO,AJ (1985)
VP 1.50E-004	mm Hg	20	EXP	AUGUSTIJN-BECKERS.PWM ET AL. (1994)
DC	pKa			
HL 4.60E-008	atm m3/mol	20	EST	VP/WSOL
OH 1.40E-011	cm3/molc sec	25	EST	MEYLAN,WM & HOWARD,PH (1993)

CAS #: 001747-53-1 — 4-PYRIMIDINECARBOXYLIC ACID, 1,2,3,6-TETRAHYDRO-

Formula:	$C_7H_8N_2O_4$		
Mol Weight:	184.15		
MP (deg C):		**FP (deg C):**	
BP (deg C):			
BP pressure (mm Hg):			

Property/Value	Units	Temp	Data Type	Reference
WS 1.38E+004	mg/L	25	EST	MEYLAN,WM ET AL. (1996)
logP 0.68			EXP	SANGSTER,J (1994)
VP 4.56E-008	mm Hg	25	EST	NEELY,WB & BLAU,GE (1985)
DC	pKa			
HL 3.44E-012	atm m3/mol	25	EST	MEYLAN,WM & HOWARD,PH (1991)
OH 1.00E-011	cm3/molc sec	25	EST	MEYLAN,WM & HOWARD,PH (1993)

CAS #: 001752-26-7 — PENICILLIN, A-PHENOXYETHYL

Formula:	$C_{17}H_{20}N_2O_5S$		
Mol Weight:	364.42		
MP (deg C):		**FP (deg C):**	
BP (deg C):			
BP pressure (mm Hg):			

Property/Value	Units	Temp	Data Type	Reference
WS 5.72E+001	mg/L	25	EST	MEYLAN,WM ET AL. (1996)
logP 2.28			EXP	HANSCH,C & LEO,AJ (1985)
VP 7.59E-013	mm Hg	25	EST	NEELY,WB & BLAU,GE (1985)
DC	pKa			
HL 5.87E-015	atm m3/mol	25	EST	MEYLAN,WM & HOWARD,PH (1991)
OH 1.17E-010	cm3/molc sec	25	EST	MEYLAN,WM & HOWARD,PH (1993)

CAS #: 001752-96-1 — NICOTINANILIDE

Formula:	$C_{12}H_{10}N_2O$		
Mol Weight:	198.23		
MP (deg C):		**FP (deg C):**	
BP (deg C):			
BP pressure (mm Hg):			

Property/Value	Units	Temp	Data Type	Reference
WS 1.49E+003	mg/L	25	EST	MEYLAN,WM ET AL. (1996)
logP 1.73			EXP	HANSCH,C & LEO,AJ (1985)
VP 1.19E-006	mm Hg	25	EST	NEELY,WB & BLAU,GE (1985)
DC	pKa			
HL 1.60E-012	atm m3/mol	25	EST	MEYLAN,WM & HOWARD,PH (1991)
OH 1.28E-011	cm3/molc sec	25	EST	MEYLAN,WM & HOWARD,PH (1993)

CAS #: 001754-47-8 — DIOCTYL PHENYL PHOSPHONATE

Formula:	$C_{22}H_{39}O_3P$		
Mol Weight:	382.53		
MP (deg C):		**FP (deg C):**	
BP (deg C):	207		
BP pressure (mm Hg):	4.00E+000		

Property/Value	Units	Temp	Data Type	Reference
WS 1.53E-003	mg/L	25	EST	MEYLAN,WM ET AL. (1996)
logP 7.51			EST	MEYLAN,WM & HOWARD,PH (1995)
VP 1.95E-006	mm Hg	25	EXP	DOBRY,A & KELLER,R (1957)
DC	pKa			
HL 1.69E-006	atm m3/mol	25	EST	MEYLAN,WM & HOWARD,PH (1991)
OH 6.58E-011	cm3/molc sec	25	EST	MEYLAN,WM & HOWARD,PH (1993)

CAS #: 001754-58-1 — o-PHENYL N,N'-DIMETHYL PHOSPHORODIAMIDATE

Formula:	$C_8H_{13}N_2O_2P$		
Mol Weight:	200.18		
MP (deg C):	101-103	**FP (deg C):**	
BP (deg C):			
BP pressure (mm Hg):			

Property/Value	Units	Temp	Data Type	Reference
WS 5.00E+001	mg/L			SHIU,WY ET AL. (1990)
logP 0.32			EST	MEYLAN,WM & HOWARD,PH (1995)
VP 3.62E-004	mm Hg	25	EST	NEELY,WB & BLAU,GE (1985)
DC	pKa			
HL 1.79E-011	atm m3/mol	25	EST	MEYLAN,WM & HOWARD,PH (1991)
OH 6.11E-011	cm3/molc sec	25	EST	MEYLAN,WM & HOWARD,PH (1993)

CAS #: 001758-88-9 — 2-ETHYL-1,4-DIMETHYLBENZENE

Formula:	$C_{10}H_{14}$		
Mol Weight:	134.22		
MP (deg C):	-53.7	**FP (deg C):**	
BP (deg C):	186.9		
BP pressure (mm Hg):			

Property/Value	Units	Temp	Data Type	Reference
WS 1.46E+001	mg/L	25	EST	MEYLAN,WM ET AL. (1996)
logP 4.43			EXP	SHERBLOM,PM & EGANHOUSE,RP (1988)
VP 9.39E-001	mm Hg	25	EXT	CHAO,J ET AL. (1983)
DC	pKa			
HL 9.61E-003	atm m3/mol	25	EST	MEYLAN,WM & HOWARD,PH (1991)
OH 1.69E-011	cm3/molc sec	25	EST	MEYLAN,WM & HOWARD,PH (1993)

CAS #: 001759-02-0 — RH-1911 O-PH(4-CL) N-BUTYLENE CARBAMATE

Formula:	$C_{11}H_{12}ClNO_2$		
Mol Weight:	225.68		
MP (deg C):		**FP (deg C):**	
BP (deg C):			
BP pressure (mm Hg):			

Property/Value	Units	Temp	Data Type	Reference
WS 4.17E+001	mg/L	25	EST	MEYLAN,WM ET AL. (1996)
logP 3.38			EXP	NANDIHALLI,UB ET AL. (1993)
VP 3.30E-004	mm Hg	25	EST	NEELY,WB & BLAU,GE (1985)
DC	pKa			
HL 4.06E-008	atm m3/mol	25	EST	MEYLAN,WM & HOWARD,PH (1991)
OH 2.58E-011	cm3/molc sec	25	EST	MEYLAN,WM & HOWARD,PH (1993)

CAS #: 001759-53-1 — CYCLOPROPANECARBOXYLIC ACID

Formula:	$C_4H_6O_2$		
Mol Weight:	86.09		
MP (deg C):	18.5	**FP (deg C):**	
BP (deg C):	183		
BP pressure (mm Hg):			

Property/Value	Units	Temp	Data Type	Reference
WS 9.15E+004	mg/L	25	EST	MEYLAN,WM ET AL. (1996)
logP 0.63			EXP	HANSCH,C ET AL. (1995)
VP 1.53E+000	mm Hg	25	EST	NEELY,WB & BLAU,GE (1985)
DC 4.83	pKa	25	EXP	KORTUM,G ET AL. (1961)
HL 4.26E-007	atm m3/mol	25	EST	MEYLAN,WM & HOWARD,PH (1991)
OH 6.21E-013	cm3/molc sec	25	EST	MEYLAN,WM & HOWARD,PH (1993)

606

CAS #: 001795-15-9				1-CYCLOHEXYLOCTANE

Formula: $C_{14}H_{28}$
Mol Weight: 196.38
MP (deg C): -20   FP (deg C):
BP (deg C): 264
BP pressure (mm Hg):

Property/Value	Units	Temp	Data Type	Reference
WS 1.31E-002	mg/L	25	EST	MEYLAN,WM ET AL. (1996)
logP 7.03			EST	MEYLAN,WM & HOWARD,PH (1995)
VP 9.15E-003	mm Hg	25	EXT	ZWOLINSKI,BJ & WILHOIT,RC (1971)
DC	pKa			
HL 2.46E+000	atm m3/mol	25	EST	MEYLAN,WM & HOWARD,PH (1991)
OH 2.05E-011	cm3/molc sec	25	EST	MEYLAN,WM & HOWARD,PH (1993)

CAS #: 001795-16-0				N-DECYLCYCLOHEXANE

Formula: $C_{16}H_{32}$
Mol Weight: 224.43
MP (deg C): -1.7   FP (deg C):
BP (deg C): 299
BP pressure (mm Hg):

Property/Value	Units	Temp	Data Type	Reference
WS 1.36E-003	mg/L	25	EST	MEYLAN,WM ET AL. (1996)
logP 8.01			EST	MEYLAN,WM & HOWARD,PH (1995)
VP 1.03E-003	mm Hg	25	EXP	YAWS,CL (1994B)
DC	pKa			
HL 4.34E+000	atm m3/mol	25	EST	MEYLAN,WM & HOWARD,PH (1991)
OH 2.33E-011	cm3/molc sec	25	EST	MEYLAN,WM & HOWARD,PH (1993)

CAS #: 001795-17-1				DODECYLCYCLOHEXANE

Formula: $C_{18}H_{36}$
Mol Weight: 252.49
MP (deg C): 12.5   FP (deg C):
BP (deg C): 331
BP pressure (mm Hg):

Property/Value	Units	Temp	Data Type	Reference
WS 1.38E-004	mg/L	25	EST	MEYLAN,WM ET AL. (1996)
logP 9.00			EST	MEYLAN,WM & HOWARD,PH (1995)
VP 4.08E-005	mm Hg	25	EXT	ZWOLINSKI,BJ & WILHOIT,RC (1971)
DC	pKa			
HL 7.64E+000	atm m3/mol	25	EST	MEYLAN,WM & HOWARD,PH (1991)
OH 2.61E-011	cm3/molc sec	25	EST	MEYLAN,WM & HOWARD,PH (1993)

CAS #: 001795-18-2				N-TETRADECYLCYCLOHEXANE

Formula: $C_{20}H_{40}$
Mol Weight: 280.54
MP (deg C): 24   FP (deg C):
BP (deg C): 360
BP pressure (mm Hg):

Property/Value	Units	Temp	Data Type	Reference
WS 1.39E-005	mg/L	25	EST	MEYLAN,WM ET AL. (1996)
logP 9.98			EST	MEYLAN,WM & HOWARD,PH (1995)
VP 1.88E-006	mm Hg	25	EXT	ZWOLINSKI,BJ & WILHOIT,RC (1971)
DC	pKa			
HL 1.35E+001	atm m3/mol	25	EST	MEYLAN,WM & HOWARD,PH (1991)
OH 2.89E-011	cm3/molc sec	25	EST	MEYLAN,WM & HOWARD,PH (1993)

CAS #: 001798-03-4				O-ETHYL PHENOXYACETIC ACID

Formula: $C_{10}H_{12}O_3$
Mol Weight: 180.21
MP (deg C):   FP (deg C):
BP (deg C):
BP pressure (mm Hg):

Property/Value	Units	Temp	Data Type	Reference
WS 9.42E+002	mg/L	25	EST	MEYLAN,WM ET AL. (1996)
logP 2.53			EXP	HANSCH,C & LEO,AJ (1985)
VP 2.35E-004	mm Hg	25	EST	NEELY,WB & BLAU,GE (1985)
DC	pKa			
HL 2.46E-008	atm m3/mol	25	EST	MEYLAN,WM & HOWARD,PH (1991)
OH 3.21E-011	cm3/molc sec	25	EST	MEYLAN,WM & HOWARD,PH (1993)

CAS #: 001798-06-7				P-IODOPHENYLACETIC ACID

Formula: $C_8H_7IO_2$
Mol Weight: 262.05
MP (deg C):   FP (deg C):
BP (deg C):
BP pressure (mm Hg):

Property/Value	Units	Temp	Data Type	Reference
WS 2.80E+002	mg/L	25	EST	MEYLAN,WM ET AL. (1996)
logP 2.64			EXP	HANSCH,C & LEO,AJ (1985)
VP 7.89E-005	mm Hg	25	EST	NEELY,WB & BLAU,GE (1985)
DC 4.18	pKa	25	EXP	KORTUM,G ET AL (1961)
HL 1.02E-008	atm m3/mol	25	EST	MEYLAN,WM & HOWARD,PH (1991)
OH 2.55E-012	cm3/molc sec	25	EST	MEYLAN,WM & HOWARD,PH (1993)

CAS #: 001798-09-0				M-METHOXYPHENYLACETIC ACID

Formula: $C_9H_{10}O_3$
Mol Weight: 166.18
MP (deg C): 71-73   FP (deg C):
BP (deg C):
BP pressure (mm Hg):

Property/Value	Units	Temp	Data Type	Reference
WS 8.33E+003	mg/L	25	EST	MEYLAN,WM ET AL. (1996)
logP 1.50			EXP	HANSCH,C & LEO,AJ (1985)
VP 5.34E-004	mm Hg	25	EST	NEELY,WB & BLAU,GE (1985)
DC	pKa			
HL 2.62E-009	atm m3/mol	25	EST	MEYLAN,WM & HOWARD,PH (1991)
OH 3.77E-011	cm3/molc sec	25	EST	MEYLAN,WM & HOWARD,PH (1993)

CAS #: 001798-11-4				P-NITROPHENOXYACETIC ACID

Formula: $C_8H_7NO_5$
Mol Weight: 197.15
MP (deg C):   FP (deg C):
BP (deg C):
BP pressure (mm Hg):

Property/Value	Units	Temp	Data Type	Reference
WS 2.49E+003	mg/L	25	EST	MEYLAN,WM ET AL. (1996)
logP 1.48			EXP	HANSCH,C & LEO,AJ (1985)
VP 1.42E-005	mm Hg	25	EST	NEELY,WB & BLAU,GE (1985)
DC 2.89	pKa	25	EXP	KORTUM,G ET AL (1961)
HL 6.62E-011	atm m3/mol	25	EST	MEYLAN,WM & HOWARD,PH (1991)
OH 7.48E-012	cm3/molc sec	25	EST	MEYLAN,WM & HOWARD,PH (1993)

CAS #: 001798-99-8				M-BROMOPHENOXYACETIC ACID

Formula: $C_8H_7BrO_3$

Mol Weight: 231.05

MP (deg C): | FP (deg C):

BP (deg C):

BP pressure (mm Hg):

Property/Value	Units	Temp	Data Type	Reference
WS 9.48E+002	mg/L	25	EST	MEYLAN,WM ET AL. (1996)
logP 2.22			EXP	HANSCH,C & LEO,AJ (1985)
VP 1.07E-004	mm Hg	25	EST	NEELY,WB & BLAU,GE (1985)
DC 3.09	pKa	25	EXP	KORTUM,G ET AL (1961)
HL 6.68E-009	atm m3/mol	25	EST	MEYLAN,WM & HOWARD,PH (1991)
OH 1.83E-011	cm3/molc sec	25	EST	MEYLAN,WM & HOWARD,PH (1993)

CAS #: 001801-72-5				DIALLYL UREA

Formula: $C_7H_{12}N_2O$

Mol Weight: 140.19

MP (deg C): 90-93 | FP (deg C):

BP (deg C):

BP pressure (mm Hg):

Property/Value	Units	Temp	Data Type	Reference
WS 2.38E+004	mg/L	25	EST	MEYLAN,WM ET AL. (1996)
logP 0.64			EXP	HANSCH,C & LEO,AJ (1985)
VP 6.16E-003	mm Hg	25	EST	NEELY,WB & BLAU,GE (1985)
DC	pKa			
HL 3.03E-009	atm m3/mol	25	EST	MEYLAN,WM & HOWARD,PH (1991)
OH 6.23E-011	cm3/molc sec	25	EST	MEYLAN,WM & HOWARD,PH (1993)

CAS #: 001805-32-9				3,4-DICHLOROBENZYL ALCOHOL

Formula: $C_7H_6Cl_2O$

Mol Weight: 177.03

MP (deg C): 35-38 | FP (deg C):

BP (deg C): 148-151

BP pressure (mm Hg):

Property/Value	Units	Temp	Data Type	Reference
WS 1.60E+003	mg/L	25	EXP	CHEM INSPECT TEST INST (1992)
logP 2.74			EXP	CHEM INSPECT TEST INST (1992)
VP 6.99E-004	mm Hg	25	EST	NEELY,WB & BLAU,GE (1985)
DC	pKa			
HL 1.19E-007	atm m3/mol	25	EST	MEYLAN,WM & HOWARD,PH (1991)
OH 4.40E-012	cm3/molc sec	25	EST	MEYLAN,WM & HOWARD,PH (1993)

CAS #: 001806-29-7				2,2'-BIPHENOL

Formula: $C_{12}H_{10}O_2$

Mol Weight: 186.21

MP (deg C): 109 | FP (deg C):

BP (deg C): 320

BP pressure (mm Hg):

Property/Value	Units	Temp	Data Type	Reference
WS 7.98E+002	mg/L	25	EST	MEYLAN,WM ET AL. (1996)
logP 2.80			EST	MEYLAN,WM & HOWARD,PH (1995)
VP 7.89E-007	mm Hg	25	EST	NEELY,WB & BLAU,GE (1985)
DC	pKa			
HL 4.48E-012	atm m3/mol	25	EST	MEYLAN,WM & HOWARD,PH (1991)
OH 4.79E-011	cm3/molc sec	25	EST	MEYLAN,WM & HOWARD,PH (1993)

CAS #: 001806-34-4				2,2'-(1,4-PHENYLENE)BIS(5-PHENYLOXAZOLE)

Formula: $C_{24}H_{16}N_2O_2$

Mol Weight: 364.41

MP (deg C): 244-246 | FP (deg C):

BP (deg C):

BP pressure (mm Hg):

Property/Value	Units	Temp	Data Type	Reference
WS 1.03E-001	mg/L	25	EST	MEYLAN,WM ET AL. (1996)
logP 5.49			EST	MEYLAN,WM & HOWARD,PH (1995)
VP 7.72E-012	mm Hg	25	EST	NEELY,WB & BLAU,GE (1985)
DC	pKa			
HL 3.19E-013	atm m3/mol	25	EST	MEYLAN,WM & HOWARD,PH (1991)
OH 3.18E-011	cm3/molc sec	25	EST	MEYLAN,WM & HOWARD,PH (1993)

CAS #: 001809-19-4				DIBUTYL PHOSPHITE

Formula: $C_8H_{19}O_3P$

Mol Weight: 194.21

MP (deg C): | FP (deg C):

BP (deg C): 95

BP pressure (mm Hg): 1.00E+000

Property/Value	Units	Temp	Data Type	Reference
WS 7.30E+003	mg/L	25	EXP	BEILSTEIN ONLINE (1991)
logP 1.81			EST	MEYLAN,WM & HOWARD,PH (1995)
VP 2.43E-002	mm Hg	25	EST	NEELY,WB & BLAU,GE (1985)
DC	pKa			
HL 1.80E-005	atm m3/mol	25	EST	MEYLAN,WM & HOWARD,PH (1991)
OH 5.26E-011	cm3/molc sec	25	EST	MEYLAN,WM & HOWARD,PH (1993)

CAS #: 001809-20-7				PHOSPHONIC ACID, BIS(1-METHYLETHYL) ESTER

Formula: $C_6H_{15}O_3P$

Mol Weight: 166.16

MP (deg C): | FP (deg C):

BP (deg C):

BP pressure (mm Hg):

Property/Value	Units	Temp	Data Type	Reference
WS 2.65E+004	mg/L	25	EST	MEYLAN,WM ET AL. (1996)
logP 0.45			EXP	HANSCH,C ET AL. (1995)
VP 5.44E-001	mm Hg	25	EST	NEELY,WB & BLAU,GE (1985)
DC	pKa			
HL 1.02E-005	atm m3/mol	25	EST	MEYLAN,WM & HOWARD,PH (1991)
OH 8.02E-011	cm3/molc sec	25	EST	MEYLAN,WM & HOWARD,PH (1993)

CAS #: 001812-30-2				BROMAZEPAM

Formula: $C_{14}H_{10}BrN_3O$

Mol Weight: 316.16

MP (deg C): | FP (deg C):

BP (deg C):

BP pressure (mm Hg):

Property/Value	Units	Temp	Data Type	Reference
WS 1.75E+002	mg/L	25	EST	MEYLAN,WM ET AL. (1996)
logP 2.05			EXP	SANGSTER,J (1994)
VP 1.90E-009	mm Hg	25	EST	NEELY,WB & BLAU,GE (1985)
DC	pKa			
HL 1.25E-013	atm m3/mol	25	EST	MEYLAN,WM & HOWARD,PH (1991)
OH 5.78E-012	cm3/molc sec	25	EST	MEYLAN,WM & HOWARD,PH (1993)

CAS #: 001814-88-6			1,1,1,2,2-PENTAFLUOROPROPANE

Formula: $C_3H_3F_5$

Mol Weight: 134.05

MP (deg C): | FP (deg C):

BP (deg C): -17.4

BP pressure (mm Hg):

Property/Value	Units	Temp	Data Type	Reference
WS 7.85E+001	mg/L	25	EST	MEYLAN,WM ET AL. (1996)
logP 2.71			EST	MEYLAN,WM & HOWARD,PH (1995)
VP 3.45E+003	mm Hg	25	EXP	OHE,S (1976)
DC	pKa			
HL 4.05E+000	atm m3/mol	25	EST	MEYLAN,WM & HOWARD,PH (1991)
OH 2.45E-015	cm3/molc sec	25	EST	MEYLAN,WM & HOWARD,PH (1993)

CAS #: 001817-47-6			4-ISOPROPYLNITROBENZENE

Formula: $C_9H_{11}NO_2$

Mol Weight: 165.19

MP (deg C): | FP (deg C):

BP (deg C): 122

BP pressure (mm Hg): 9.00E+000

Property/Value	Units	Temp	Data Type	Reference
WS 2.99E+001	mg/L	25	EST	MEYLAN,WM ET AL. (1996)
logP 3.45			EXP	HANSCH,C ET AL. (1995)
VP 1.47E-002	mm Hg	25	EST	NEELY,WB & BLAU,GE (1985)
DC	pKa			
HL 4.13E-005	atm m3/mol	25	EST	MEYLAN,WM & HOWARD,PH (1991)
OH 2.85E-012	cm3/molc sec	25	EST	MEYLAN,WM & HOWARD,PH (1993)

CAS #: 001820-81-1			5-CHLOROURACIL

Formula: $C_4H_3ClN_2O_2$

Mol Weight: 146.53

MP (deg C): > 300 | FP (deg C):

BP (deg C):

BP pressure (mm Hg):

Property/Value	Units	Temp	Data Type	Reference
WS 7.66E+003	mg/L	25	EST	MEYLAN,WM ET AL. (1996)
logP -0.35			EXP	HANSCH,C & LEO,AJ (1985)
VP 8.97E-007	mm Hg	25	EST	NEELY,WB & BLAU,GE (1985)
DC	pKa			
HL 6.24E-011	atm m3/mol	25	EST	MEYLAN,WM & HOWARD,PH (1991)
OH 5.83E-012	cm3/molc sec	25	EST	MEYLAN,WM & HOWARD,PH (1993)

CAS #: 001821-12-1			4-PHENYLBUTYRIC ACID

Formula: $C_{10}H_{12}O_2$

Mol Weight: 164.21

MP (deg C): 52 | FP (deg C):

BP (deg C): 290

BP pressure (mm Hg):

Property/Value	Units	Temp	Data Type	Reference
WS 1.39E+003	mg/L	25	EST	MEYLAN,WM ET AL. (1996)
logP 2.42			EXP	HANSCH,C & LEO,AJ (1985)
VP 6.64E-004	mm Hg	25	EST	NEELY,WB & BLAU,GE (1985)
DC 4.76	pKa	25	EXP	KORTUM,G ET AL (1961)
HL 7.79E-008	atm m3/mol	25	EST	MEYLAN,WM & HOWARD,PH (1991)
OH 8.79E-012	cm3/molc sec	25	EST	MEYLAN,WM & HOWARD,PH (1993)

CAS #: 001821-33-6			BENZAMIDE, N-[(PHENYLAMINO)CARBONYL]-

Formula: $C_{14}H_{12}N_2O_2$

Mol Weight: 240.26

MP (deg C): | FP (deg C):

BP (deg C):

BP pressure (mm Hg):

Property/Value	Units	Temp	Data Type	Reference
WS 3.41E+001	mg/L	25	EST	MEYLAN,WM ET AL. (1996)
logP 3.39			EXP	SOTOMATSU,T ET AL. (1987)
VP 7.31E-010	mm Hg	25	EST	NEELY,WB & BLAU,GE (1985)
DC	pKa			
HL 1.18E-011	atm m3/mol	25	EST	MEYLAN,WM & HOWARD,PH (1991)
OH 4.54E-011	cm3/molc sec	25	EST	MEYLAN,WM & HOWARD,PH (1993)

CAS #: 001824-09-5			G 34698

Formula: $C_{11}H_{20}ClN_5O$

Mol Weight: 273.77

MP (deg C): | FP (deg C):

BP (deg C):

BP pressure (mm Hg):

Property/Value	Units	Temp	Data Type	Reference
WS 1.20E+002	mg/L	25	EXP	RAHMAN,A & MATTHEWS,LJ (1979)
logP 2.54			EST	MEYLAN,WM & HOWARD,PH (1995)
VP 8.40E-006	mm Hg	25	EST	NEELY,WB & BLAU,GE (1985)
DC	pKa			
HL 6.95E-011	atm m3/mol	25	EST	MEYLAN,WM & HOWARD,PH (1991)
OH 4.10E-011	cm3/molc sec	25	EST	MEYLAN,WM & HOWARD,PH (1993)

CAS #: 001824-46-0			6-METHYLHYDROTHIAZIDE

Formula: $C_8H_{11}N_3O_4S_2$

Mol Weight: 277.32

MP (deg C): | FP (deg C):

BP (deg C):

BP pressure (mm Hg):

Property/Value	Units	Temp	Data Type	Reference
WS 2.28E+003	mg/L	25	EST	MEYLAN,WM ET AL. (1996)
logP -0.22			EXP	HANSCH,C & LEO,AJ (1985)
VP 2.18E-009	mm Hg	25	EST	NEELY,WB & BLAU,GE (1985)
DC	pKa			
HL 6.53E-012	atm m3/mol	25	EST	MEYLAN,WM & HOWARD,PH (1991)
OH 8.60E-011	cm3/molc sec	25	EST	MEYLAN,WM & HOWARD,PH (1993)

CAS #: 001824-47-1			3-CHLOROMETHYLHYDROCHLOROTHIAZIDE

Formula: $C_8H_9Cl_2N_3O_4S_2$

Mol Weight: 346.21

MP (deg C): | FP (deg C):

BP (deg C):

BP pressure (mm Hg):

Property/Value	Units	Temp	Data Type	Reference
WS 4.68E+002	mg/L	25	EST	MEYLAN,WM ET AL. (1996)
logP 0.11			EXP	HANSCH,C & LEO,AJ (1985)
VP 8.12E-011	mm Hg	25	EST	NEELY,WB & BLAU,GE (1985)
DC	pKa			
HL 2.05E-012	atm m3/mol	25	EST	MEYLAN,WM & HOWARD,PH (1991)
OH 6.22E-011	cm3/molc sec	25	EST	MEYLAN,WM & HOWARD,PH (1993)

CAS #: 001824-50-6	3-BENZYLHYDROCHLOROTHIAZIDE

Formula: $C_{14}H_{14}ClN_3O_4S_2$

Mol Weight: 387.87

MP (deg C): FP (deg C):

BP (deg C):

BP pressure (mm Hg):

Property/ Value	Units	Temp	Data Type	Reference
WS 5.20E+001	mg/L	25	EST	MEYLAN,WM ET AL. (1996)
logP 0.93			EXP	HANSCH,C ET AL. (1995)
VP 1.77E-012	mm Hg	25	EST	NEELY,WB & BLAU,GE (1985)
DC	pKa			
HL 4.70E-013	atm m3/mol	25	EST	MEYLAN,WM & HOWARD,PH (1991)
OH 2.14E-010	cm3/molc sec	25	EST	MEYLAN,WM & HOWARD,PH (1993)

CAS #: 001824-58-4	3-ETHYLHYDROCHLOROTHIAZIDE

Formula: $C_9H_{12}ClN_3O_4S_2$

Mol Weight: 325.79

MP (deg C): 269-270 FP (deg C):

BP (deg C):

BP pressure (mm Hg):

Property/ Value	Units	Temp	Data Type	Reference
WS 4.27E+002	mg/L	25	EST	MEYLAN,WM ET AL. (1996)
logP 0.30			EXP	HANSCH,C & LEO,AJ (1985)
VP 3.25E-010	mm Hg	25	EST	NEELY,WB & BLAU,GE (1985)
DC	pKa			
HL 7.73E-012	atm m3/mol	25	EST	MEYLAN,WM & HOWARD,PH (1991)
OH 2.09E-010	cm3/molc sec	25	EST	MEYLAN,WM & HOWARD,PH (1993)

CAS #: 001825-21-4	PENTACHLOROANISOLE

Formula: $C_7H_3Cl_5O$

Mol Weight: 280.37

MP (deg C): 107-109 FP (deg C):

BP (deg C):

BP pressure (mm Hg):

Property/ Value	Units	Temp	Data Type	Reference
WS 3.54E-001	mg/L	25	EST	MEYLAN,WM ET AL. (1996)
logP 5.45			EXP	OPPERHUIZEN,A & VOORS,PI (1987)
VP 5.92E-004	mm Hg	25	EST	NEELY,WB & BLAU,GE (1985)
DC	pKa			
HL 1.93E-003	atm m3/mol	25	EST	MEYLAN,WM & HOWARD,PH (1991)
OH 1.09E-012	cm3/molc sec	25	EST	MEYLAN,WM & HOWARD,PH (1993)

CAS #: 001825-30-5	1,5-DICHLORONAPHTHALENE

Formula: $C_{10}H_6Cl_2$

Mol Weight: 197.07

MP (deg C): 107 FP (deg C):

BP (deg C):

BP pressure (mm Hg):

Property/ Value	Units	Temp	Data Type	Reference
WS 9.63E+000	mg/L	25	EST	MEYLAN,WM ET AL. (1996)
logP 4.67			EXP	OPPERHUIZEN,A (1987)
VP 1.76E-003	mm Hg	25	EST	NEELY,WB & BLAU,GE (1985)
DC	pKa			
HL 2.89E-004	atm m3/mol	25	EST	MEYLAN,WM & HOWARD,PH (1991)
OH 4.44E-012	cm3/molc sec	25	EST	MEYLAN,WM & HOWARD,PH (1993)

CAS #: 001825-31-6	1,4-DICHLORONAPHTHALENE

Formula: $C_{10}H_6Cl_2$

Mol Weight: 197.07

MP (deg C): 67.5 FP (deg C):

BP (deg C): 288

BP pressure (mm Hg):

Property/ Value	Units	Temp	Data Type	Reference
WS 3.10E-001	mg/L	25	EXP	OPPERHUIZEN,A ET AL. (1985)
logP 4.66			EXP	OPPERHUIZEN,A (1987)
VP 1.76E-003	mm Hg	25	EST	NEELY,WB & BLAU,GE (1985)
DC	pKa			
HL 2.89E-004	atm m3/mol	25	EST	MEYLAN,WM & HOWARD,PH (1991)
OH 5.80E-012	cm3/molc sec	25	EXP	ATKINSON,R (1989)

CAS #: 001825-62-3	ETHOXYTRIMETHYLSILANE

Formula: $C_5H_{14}OSi$

Mol Weight: 118.25

MP (deg C): FP (deg C):

BP (deg C): 76

BP pressure (mm Hg):

Property/ Value	Units	Temp	Data Type	Reference
WS 1.04E+003	mg/L	25	EST	MEYLAN,WM ET AL. (1996)
logP 2.33			EST	MEYLAN,WM & HOWARD,PH (1995)
VP 1.15E+002	mm Hg	25	EXP	OHE,S (1976)
DC	pKa			
HL 3.18E-003	atm m3/mol	25	EST	MEYLAN,WM & HOWARD,PH (1991)
OH 6.62E-012	cm3/molc sec	25	EST	MEYLAN,WM & HOWARD,PH (1993)

CAS #: 001826-11-5	THIAZOLE, 2-PHENYL-

Formula: C_9H_7NS

Mol Weight: 161.23

MP (deg C): FP (deg C):

BP (deg C):

BP pressure (mm Hg):

Property/ Value	Units	Temp	Data Type	Reference
WS 2.91E+002	mg/L	25	EST	MEYLAN,WM ET AL. (1996)
logP 2.77			EXP	SANGSTER,J (1993)
VP 1.19E-003	mm Hg	25	EST	NEELY,WB & BLAU,GE (1985)
DC	pKa			
HL 2.94E-007	atm m3/mol	25	EST	MEYLAN,WM & HOWARD,PH (1991)
OH 5.82E-012	cm3/molc sec	25	EST	MEYLAN,WM & HOWARD,PH (1993)

CAS #: 001828-76-8	1(3H)-ISOBENZOFURANONE, 3-HYDROXY-3-METHYL-

Formula: $C_9H_8O_3$

Mol Weight: 164.16

MP (deg C): FP (deg C):

BP (deg C):

BP pressure (mm Hg):

Property/ Value	Units	Temp	Data Type	Reference
WS 4.31E+004	mg/L	25	EST	MEYLAN,WM ET AL. (1996)
logP 0.81			EXP	HANSCH,C ET AL. (1995)
VP 9.70E-006	mm Hg	25	EST	NEELY,WB & BLAU,GE (1985)
DC	pKa			
HL 6.18E-010	atm m3/mol	25	EST	MEYLAN,WM & HOWARD,PH (1991)
OH 4.57E-012	cm3/molc sec	25	EST	MEYLAN,WM & HOWARD,PH (1993)

CAS #: 001829-37-4				2-BROMOPHENYL ACETATE

Formula: $C_8H_7BrO_2$

Mol Weight: 215.05

MP (deg C):

FP (deg C):

BP (deg C):

BP pressure (mm Hg):

Property/Value	Units	Temp	Data Type	Reference
WS 4.83E+002	mg/L	25	EST	MEYLAN,WM ET AL. (1996)
logP 2.20			EXP	HANSCH,C & LEO,AJ (1985)
VP 2.46E-002	mm Hg	25	EST	NEELY,WB & BLAU,GE (1985)
DC	pKa			
HL 2.58E-005	atm m3/mol	25	EST	MEYLAN,WM & HOWARD,PH (1991)
OH 1.00E-012	cm3/molc sec	25	EST	MEYLAN,WM & HOWARD,PH (1993)

CAS #: 001832-53-7				ETHYL METHYL PHOSPHONATE

Formula: $C_3H_9O_3P$

Mol Weight: 124.08

MP (deg C):

FP (deg C):

BP (deg C): 106-107

BP pressure (mm Hg): 1.00E-001

Property/Value	Units	Temp	Data Type	Reference
WS 1.30E+005	mg/L	25	EST	MEYLAN,WM ET AL. (1996)
logP -0.15			EST	MEYLAN,WM & HOWARD,PH (1995)
VP 1.90E-002	mm Hg	25	EST	NEELY,WB & BLAU,GE (1985)
DC	pKa			
HL 5.18E-009	atm m3/mol	25	EST	MEYLAN,WM & HOWARD,PH (1991)
OH 1.96E-011	cm3/molc sec	25	EST	MEYLAN,WM & HOWARD,PH (1993)

CAS #: 001832-54-8				ISOPROPYL METHYLPHOSPHONATE

Formula: $C_4H_{11}O_3P$

Mol Weight: 138.10

MP (deg C):

FP (deg C):

BP (deg C):

BP pressure (mm Hg):

Property/Value	Units	Temp	Data Type	Reference
WS 5.04E+004	mg/L	25	EST	MEYLAN,WM ET AL. (1996)
logP 0.27			EST	MEYLAN,WM & HOWARD,PH (1995)
VP 1.19E-002	mm Hg	25	EST	NEELY,WB & BLAU,GE (1985)
DC	pKa			
HL 6.88E-009	atm m3/mol	25	EST	MEYLAN,WM & HOWARD,PH (1991)
OH 4.04E-011	cm3/molc sec	25	EST	MEYLAN,WM & HOWARD,PH (1993)

CAS #: 001832-68-4				DIOCTYL METHYLPHOSPHONATE

Formula: $C_{17}H_{37}O_3P$

Mol Weight: 320.46

MP (deg C):

FP (deg C):

BP (deg C):

BP pressure (mm Hg):

Property/Value	Units	Temp	Data Type	Reference
WS 5.42E-002	mg/L	25	EST	MEYLAN,WM ET AL. (1996)
logP 6.13			EXP	KRIKORIAN,SE ET AL. (1987)
VP 8.37E-006	mm Hg	25	EST	NEELY,WB & BLAU,GE (1985)
DC	pKa			
HL 6.60E-005	atm m3/mol	25	EST	MEYLAN,WM & HOWARD,PH (1991)
OH 6.40E-011	cm3/molc sec	25	EST	MEYLAN,WM & HOWARD,PH (1993)

CAS #: 001836-75-5				NITROFEN

Formula: $C_{12}H_7Cl_2NO_3$

Mol Weight: 284.10

MP (deg C): 70-71

FP (deg C):

BP (deg C):

BP pressure (mm Hg):

Property/Value	Units	Temp	Data Type	Reference
WS 1.00E+000	mg/L	22	EXP	YALKOWSKY,SH & DANNENFELSER,RM (1992)
logP 4.64			EXP	SANGSTER,J (1994)
VP 8.00E-006	mm Hg	40	EXP	MARTIN,H & WORTHING,CR (1977)
DC	pKa			
HL 2.55E-007	atm m3/mol	25	EST	MEYLAN,WM & HOWARD,PH (1991)
OH 2.00E-012	cm3/molc sec	25	EST	MEYLAN,WM & HOWARD,PH (1993)

CAS #: 001836-77-7				CHLORNITROFEN

Formula: $C_{12}H_6Cl_3NO_3$

Mol Weight: 318.55

MP (deg C):

FP (deg C):

BP (deg C): 107-108

BP pressure (mm Hg):

Property/Value	Units	Temp	Data Type	Reference
WS 2.50E-001	mg/L	25	EXP	CHEM INSPECT TEST INST (1992)
logP 4.97			EXP	KAWAMOTO,K & URANO,K (1989)
VP 5.27E-007	mm Hg	25	EST	NEELY,WB & BLAU,GE (1985)
DC	pKa			
HL 1.89E-007	atm m3/mol	25	EST	MEYLAN,WM & HOWARD,PH (1991)
OH 1.29E-012	cm3/molc sec	25	EST	MEYLAN,WM & HOWARD,PH (1993)

CAS #: 001848-75-5				2-IMIDAZOLIDINIMINE, N-PHENYL-

Formula: $C_9H_{11}N_3$

Mol Weight: 161.21

MP (deg C):

FP (deg C):

BP (deg C):

BP pressure (mm Hg):

Property/Value	Units	Temp	Data Type	Reference
WS 9.45E+004	mg/L	25	EST	MEYLAN,WM ET AL. (1996)
logP 1.01			EXP	HANSCH,C & LEO,AJ (1985)
VP 4.85E-004	mm Hg	25	EST	NEELY,WB & BLAU,GE (1985)
DC	pKa			
HL 2.74E-011	atm m3/mol	25	EST	MEYLAN,WM & HOWARD,PH (1991)
OH 1.45E-010	cm3/molc sec	25	EST	MEYLAN,WM & HOWARD,PH (1993)

CAS #: 001861-32-1				DIMETHYL TETRACHLOROTEREPHTHALATE

Formula: $C_{10}H_6Cl_4O_4$

Mol Weight: 331.97

MP (deg C): 156

FP (deg C):

BP (deg C): 360-370

BP pressure (mm Hg):

Property/Value	Units	Temp	Data Type	Reference
WS 5.00E-001	mg/L	25	EXP	YALKOWSKY,SH & DANNENFELSER,RM (1992)
logP 4.40			EXP	HANSCH,C ET AL. (1995)
VP 2.50E-006	mm Hg	25	EXP	GLOTFELTY,DE ET AL. (1984A)
DC	pKa			
HL 2.18E-006	atm m3/mol	25	EST	VP/WSOL
OH 4.41E-013	cm3/molc sec	25	EST	MEYLAN,WM & HOWARD,PH (1993)

CAS #: 001861-40-1 — BENEFIN

Formula: $C_{13}H_{16}F_3N_3O_4$

Mol Weight: 335.29

MP (deg C): 65-66.5

BP (deg C): 370

BP pressure (mm Hg):

Property/Value	Units	Temp	Data Type	Reference
WS 1.00E-001	mg/L	25	EXP	WSSA (1983)
logP 5.29			EXP	TOMLIN,C (1994)
VP 6.60E-005	mm Hg	25	EXP	WAUCHOPE,RD ET AL. (1992)
DC	pKa			
HL 2.91E-004	atm m3/mol	25	EST	VP/WSOL
OH 8.20E-011	cm3/molc sec	25	EST	ATKINSON,R (1988)

CAS #: 001861-44-5 — TRITAC

Formula: $C_{10}H_{11}Cl_3O_2$

Mol Weight: 269.56

MP (deg C):

BP (deg C):

BP pressure (mm Hg):

Property/Value	Units	Temp	Data Type	Reference
WS 7.30E+001	mg/L	25	EXP	SHIU,WY ET AL. (1990)
logP 3.15			EST	MEYLAN,WM & HOWARD,PH (1995)
VP 1.00E-004	mm Hg	25	EXP	BAILEY,GW & WHITE,JL (1965)
DC	pKa			
HL 4.86E-007	atm m3/mol	25	EST	VP/WSOL
OH 2.17E-011	cm3/molc sec	25	EST	MEYLAN,WM & HOWARD,PH (1993)

CAS #: 001864-94-4 — PHENYL FORMATE

Formula: $C_7H_6O_2$

Mol Weight: 122.12

MP (deg C):

BP (deg C):

BP pressure (mm Hg):

Property/Value	Units	Temp	Data Type	Reference
WS 8.27E+003	mg/L	25	EST	MEYLAN,WM ET AL. (1996)
logP 1.26			EXP	HANSCH,C & LEO,AJ (1985)
VP 1.10E+000	mm Hg	25	EST	NEELY,WB & BLAU,GE (1985)
DC	pKa			
HL 8.87E-005	atm m3/mol	25	EST	MEYLAN,WM & HOWARD,PH (1991)
OH 2.15E-011	cm3/molc sec	25	EST	MEYLAN,WM & HOWARD,PH (1993)

CAS #: 001867-66-9 — KETAMINE

Formula: $C_{13}H_{17}Cl_2NO$

Mol Weight: 274.19

MP (deg C): 262-263

BP (deg C):

BP pressure (mm Hg):

Property/Value	Units	Temp	Data Type	Reference
WS 3.87E+003	mg/L	25	EST	MEYLAN,WM ET AL. (1996)
logP 2.18			EXP	HANSCH,C & LEO,AJ (1985)
VP 3.41E-005	mm Hg	25	EST	NEELY,WB & BLAU,GE (1985)
DC	pKa			
HL 1.38E-008	atm m3/mol	25	EST	MEYLAN,WM & HOWARD,PH (1991)
OH 7.71E-011	cm3/molc sec	25	EST	MEYLAN,WM & HOWARD,PH (1993)

CAS #: 001869-67-6 — 2,4-DINITRO-N-(3-(TRIFLUOROMETHYL)PHENYL)BENZE*

Formula: $C_{13}H_8F_3N_3O_4$

Mol Weight: 327.22

MP (deg C):

BP (deg C):

BP pressure (mm Hg):

Property/Value	Units	Temp	Data Type	Reference
WS 1.30E+000	mg/L	25	EST	MEYLAN,WM ET AL. (1996)
logP 4.47			EST	MEYLAN,WM & HOWARD,PH (1995)
VP 2.25E-007	mm Hg	25	EST	NEELY,WB & BLAU,GE (1985)
DC	pKa			
HL 3.11E-009	atm m3/mol	25	EST	MEYLAN,WM & HOWARD,PH (1991)
OH 2.45E-011	cm3/molc sec	25	EST	MEYLAN,WM & HOWARD,PH (1993)

CAS #: 001871-57-4 — 2-(CHLOROMETHYL)-3-CHLORO-1-PROPENE

Formula: $C_4H_6Cl_2$

Mol Weight: 125.00

MP (deg C): -14

BP (deg C): 138

BP pressure (mm Hg):

Property/Value	Units	Temp	Data Type	Reference
WS 4.45E+002	mg/L	25	EST	MEYLAN,WM ET AL. (1996)
logP 2.73			EST	MEYLAN,WM & HOWARD,PH (1995)
VP 5.57E+000	mm Hg	25	EST	NEELY,WB & BLAU,GE (1985)
DC	pKa			
HL 1.88E-002	atm m3/mol	25	EST	MEYLAN,WM & HOWARD,PH (1991)
OH 3.35E-011	cm3/molc sec	25	EXP	ATKINSON,R (1989)

CAS #: 001871-96-1 — DECANEDINITRILE

Formula: $C_{10}H_{16}N_2$

Mol Weight: 164.25

MP (deg C):

BP (deg C): 204

BP pressure (mm Hg): 1.60E+001

Property/Value	Units	Temp	Data Type	Reference
WS 1.14E+003	mg/L	25	EST	MEYLAN,WM ET AL. (1996)
logP 1.75			EXP	TANII,H & HASHIMOTO,K (1985)
VP 5.77E-004	mm Hg	25	EST	NEELY,WB & BLAU,GE (1985)
DC	pKa			
HL 9.21E-008	atm m3/mol	25	EST	MEYLAN,WM & HOWARD,PH (1991)
OH 6.36E-012	cm3/molc sec	25	EST	MEYLAN,WM & HOWARD,PH (1993)

CAS #: 001874-22-2 — 2-PROPENAL, 3-(5-NITRO-2-FURANYL)-

Formula: $C_7H_5NO_4$

Mol Weight: 167.12

MP (deg C): 117-120

BP (deg C):

BP pressure (mm Hg):

Property/Value	Units	Temp	Data Type	Reference
WS 3.91E+003	mg/L	25	EST	MEYLAN,WM ET AL. (1996)
logP 0.96			EXP	DEBNATH,AK ET AL. (1991)
VP 2.00E-003	mm Hg	25	EST	NEELY,WB & BLAU,GE (1985)
DC	pKa			
HL 6.31E-009	atm m3/mol	25	EST	MEYLAN,WM & HOWARD,PH (1991)
OH 4.08E-011	cm3/molc sec	25	EST	MEYLAN,WM & HOWARD,PH (1993)

612

CAS #: 001877-71-0 — ISOPHTHALIC ACID, METHYL ESTER

Formula: $C_9H_8O_4$

Mol Weight: 180.16

MP (deg C):

FP (deg C):

BP (deg C):

BP pressure (mm Hg):

Property/Value	Units	Temp	Data Type	Reference
WS 1.50E+003	mg/L	25	EST	MEYLAN,WM ET AL. (1996)
logP 1.83			EXP	HANSCH,C & LEO,AJ (1985)
VP 1.84E-004	mm Hg	25	EST	NEELY,WB & BLAU,GE (1985)
DC	pKa			
HL 6.99E-010	atm m3/mol	25	EST	MEYLAN,WM & HOWARD,PH (1991)
OH 9.70E-013	cm3/molc sec	25	EST	MEYLAN,WM & HOWARD,PH (1993)

CAS #: 001877-72-1 — M-CYANOBENZOIC ACID

Formula: $C_8H_5NO_2$

Mol Weight: 147.13

MP (deg C): 219

FP (deg C):

BP (deg C):

BP pressure (mm Hg):

Property/Value	Units	Temp	Data Type	Reference
WS 2.31E+003	mg/L	25	EST	MEYLAN,WM ET AL. (1996)
logP 1.48			EXP	HANSCH,C & LEO,AJ (1985)
VP 2.09E-004	mm Hg	25	EST	NEELY,WB & BLAU,GE (1985)
DC 3.60	pKa	20	EXP	KORTUM,G ET AL (1961)
HL 1.05E-009	atm m3/mol	25	EST	MEYLAN,WM & HOWARD,PH (1991)
OH 6.47E-013	cm3/molc sec	25	EST	MEYLAN,WM & HOWARD,PH (1993)

CAS #: 001877-73-2 — M-NITROPHENYLACETIC ACID

Formula: $C_8H_7NO_4$

Mol Weight: 181.15

MP (deg C): 117-120

FP (deg C):

BP (deg C):

BP pressure (mm Hg):

Property/Value	Units	Temp	Data Type	Reference
WS 3.18E+003	mg/L	25	EST	MEYLAN,WM ET AL. (1996)
logP 1.45			EXP	HANSCH,C & LEO,AJ (1985)
VP 2.91E-005	mm Hg	25	EST	NEELY,WB & BLAU,GE (1985)
DC 3.97	pKa	25	EXP	KORTUM,G ET AL (1961)
HL 1.74E-010	atm m3/mol	25	EST	MEYLAN,WM & HOWARD,PH (1991)
OH 1.50E-012	cm3/molc sec	25	EST	MEYLAN,WM & HOWARD,PH (1993)

CAS #: 001877-75-4 — P-METHOXYPHENOXYACETIC ACID

Formula: $C_9H_{10}O_4$

Mol Weight: 182.18

MP (deg C):

FP (deg C):

BP (deg C):

BP pressure (mm Hg):

Property/Value	Units	Temp	Data Type	Reference
WS 1.19E+004	mg/L	25	EST	MEYLAN,WM ET AL. (1996)
logP 1.23			EXP	HANSCH,C & LEO,AJ (1985)
VP 2.12E-004	mm Hg	25	EST	NEELY,WB & BLAU,GE (1985)
DC 3.21	pKa	25	EXP	KORTUM,G ET AL (1961)
HL 9.93E-010	atm m3/mol	25	EST	MEYLAN,WM & HOWARD,PH (1991)
OH 2.42E-011	cm3/molc sec	25	EST	MEYLAN,WM & HOWARD,PH (1993)

CAS #: 001877-77-6 — M-AMINOBENZYL ALCOHOL

Formula: C_7H_9NO

Mol Weight: 123.16

MP (deg C): 93-95

FP (deg C):

BP (deg C):

BP pressure (mm Hg):

Property/Value	Units	Temp	Data Type	Reference
WS 1.08E+005	mg/L	25	EST	MEYLAN,WM ET AL. (1996)
logP -0.05			EXP	HANSCH,C & LEO,AJ (1985)
VP 3.71E-004	mm Hg	25	EST	NEELY,WB & BLAU,GE (1985)
DC	pKa			
HL 7.68E-011	atm m3/mol	25	EST	MEYLAN,WM & HOWARD,PH (1991)
OH 1.35E-010	cm3/molc sec	25	EST	MEYLAN,WM & HOWARD,PH (1993)

CAS #: 001878-18-8 — 2-METHYL-1-BUTANETHIOL

Formula: $C_5H_{12}S$

Mol Weight: 104.22

MP (deg C):

FP (deg C):

BP (deg C): 116-117

BP pressure (mm Hg):

Property/Value	Units	Temp	Data Type	Reference
WS 5.99E+002	mg/L	25	EST	MEYLAN,WM ET AL. (1996)
logP 2.67			EST	MEYLAN,WM & HOWARD,PH (1995)
VP 2.00E+001	mm Hg	25	EXT	BOUBLIK,T ET AL. (1984)
DC	pKa			
HL 8.12E-003	atm m3/mol	25	EST	MEYLAN,WM & HOWARD,PH (1991)
OH 5.40E-011	cm3/molc sec	25	EXP	ATKINSON,R (1989)

CAS #: 001878-49-5 — O-METHYLPHENOXYACETIC ACID

Formula: $C_9H_{10}O_3$

Mol Weight: 166.18

MP (deg C):

FP (deg C):

BP (deg C):

BP pressure (mm Hg):

Property/Value	Units	Temp	Data Type	Reference
WS 3.24E+003	mg/L	25	EST	MEYLAN,WM ET AL. (1996)
logP 1.98			EXP	HANSCH,C & LEO,AJ (1985)
VP 5.34E-004	mm Hg	25	EST	NEELY,WB & BLAU,GE (1985)
DC 3.23	pKa	25	EXP	KORTUM,G ET AL (1961)
HL 1.85E-008	atm m3/mol	25	EST	MEYLAN,WM & HOWARD,PH (1991)
OH 3.13E-011	cm3/molc sec	25	EST	MEYLAN,WM & HOWARD,PH (1993)

CAS #: 001878-51-9 — M-ETHYLPHENOXYACETIC ACID

Formula: $C_{10}H_{12}O_3$

Mol Weight: 180.21

MP (deg C):

FP (deg C):

BP (deg C):

BP pressure (mm Hg):

Property/Value	Units	Temp	Data Type	Reference
WS 1.63E+003	mg/L	25	EST	MEYLAN,WM ET AL. (1996)
logP 2.25			EXP	HANSCH,C & LEO,AJ (1985)
VP 2.35E-004	mm Hg	25	EST	NEELY,WB & BLAU,GE (1985)
DC	pKa			
HL 2.46E-008	atm m3/mol	25	EST	MEYLAN,WM & HOWARD,PH (1991)
OH 5.93E-011	cm3/molc sec	25	EST	MEYLAN,WM & HOWARD,PH (1993)

CAS #: 001878-52-0 — M-I-PROPYLPHENOXYACETIC ACID

Formula: $C_{11}H_{14}O_3$

Mol Weight: 194.23

MP (deg C):   FP (deg C):

BP (deg C):

BP pressure (mm Hg):

Property/Value	Units	Temp	Data Type	Reference
WS 7.14E+002	mg/L	25	EST	MEYLAN,WM ET AL. (1996)
logP 2.59			EXP	HANSCH,C & LEO,AJ (1985)
VP 1.46E-004	mm Hg	25	EST	NEELY,WB & BLAU,GE (1985)
DC	pKa			
HL 3.26E-008	atm m3/mol	25	EST	MEYLAN,WM & HOWARD,PH (1991)
OH 5.81E-011	cm3/molc sec	25	EST	MEYLAN,WM & HOWARD,PH (1993)

CAS #: 001878-53-1 — M-BUTYLPHENOXYACETIC ACID

Formula: $C_{12}H_{16}O_3$

Mol Weight: 208.26

MP (deg C):   FP (deg C):

BP (deg C):

BP pressure (mm Hg):

Property/Value	Units	Temp	Data Type	Reference
WS 1.90E+002	mg/L	25	EST	MEYLAN,WM ET AL. (1996)
logP 3.18			EXP	HANSCH,C & LEO,AJ (1985)
VP 3.86E-005	mm Hg	25	EST	NEELY,WB & BLAU,GE (1985)
DC	pKa			
HL 4.33E-008	atm m3/mol	25	EST	MEYLAN,WM & HOWARD,PH (1991)
OH 6.21E-011	cm3/molc sec	25	EST	MEYLAN,WM & HOWARD,PH (1993)

CAS #: 001878-55-3 — M-T-BUTYLPHENOXYACETIC ACID

Formula: $C_{12}H_{16}O_3$

Mol Weight: 208.26

MP (deg C):   FP (deg C):

BP (deg C):

BP pressure (mm Hg):

Property/Value	Units	Temp	Data Type	Reference
WS 2.92E+002	mg/L	25	EST	MEYLAN,WM ET AL. (1996)
logP 2.96			EXP	HANSCH,C & LEO,AJ (1985)
VP 8.54E-005	mm Hg	25	EST	NEELY,WB & BLAU,GE (1985)
DC	pKa			
HL 4.33E-008	atm m3/mol	25	EST	MEYLAN,WM & HOWARD,PH (1991)
OH 5.27E-011	cm3/molc sec	25	EST	MEYLAN,WM & HOWARD,PH (1993)

CAS #: 001878-56-4 — PHENOXYACETIC ACID,4-CYCLOHEXYL

Formula: $C_{14}H_{18}O_3$

Mol Weight: 234.30

MP (deg C):   FP (deg C):

BP (deg C):

BP pressure (mm Hg):

Property/Value	Units	Temp	Data Type	Reference
WS 4.15E+001	mg/L	25	EST	MEYLAN,WM ET AL. (1996)
logP 3.79			EXP	HANSCH,C & LEO,AJ (1985)
VP 4.94E-006	mm Hg	25	EST	NEELY,WB & BLAU,GE (1985)
DC	pKa			
HL 3.37E-008	atm m3/mol	25	EST	MEYLAN,WM & HOWARD,PH (1991)
OH 4.07E-011	cm3/molc sec	25	EST	MEYLAN,WM & HOWARD,PH (1993)

CAS #: 001878-57-5 — M-PHENYLPHENOXYACETIC ACID

Formula: $C_{14}H_{12}O_3$

Mol Weight: 228.25

MP (deg C):   FP (deg C):

BP (deg C):

BP pressure (mm Hg):

Property/Value	Units	Temp	Data Type	Reference
WS 1.49E+002	mg/L	25	EST	MEYLAN,WM ET AL. (1996)
logP 3.18			EXP	HANSCH,C & LEO,AJ (1985)
VP 9.21E-007	mm Hg	25	EST	NEELY,WB & BLAU,GE (1985)
DC	pKa			
HL 1.29E-009	atm m3/mol	25	EST	MEYLAN,WM & HOWARD,PH (1991)
OH 4.55E-011	cm3/molc sec	25	EST	MEYLAN,WM & HOWARD,PH (1993)

CAS #: 001878-58-6 — 5-INDANOXYACETIC ACID

Formula: $C_{11}H_{12}O_3$

Mol Weight: 192.22

MP (deg C):   FP (deg C):

BP (deg C):

BP pressure (mm Hg):

Property/Value	Units	Temp	Data Type	Reference
WS 1.22E+003	mg/L	25	EST	MEYLAN,WM ET AL. (1996)
logP 2.33			EXP	HANSCH,C & LEO,AJ (1985)
VP 5.15E-005	mm Hg	25	EST	NEELY,WB & BLAU,GE (1985)
DC	pKa			
HL 1.20E-008	atm m3/mol	25	EST	MEYLAN,WM & HOWARD,PH (1991)
OH 7.29E-011	cm3/molc sec	25	EST	MEYLAN,WM & HOWARD,PH (1993)

CAS #: 001878-59-7 — 2(5678-TETRAHYDRONAPHTHYLOXY)ACETIC ACID

Formula: $C_{12}H_{14}O_3$

Mol Weight: 206.24

MP (deg C):   FP (deg C):

BP (deg C):

BP pressure (mm Hg):

Property/Value	Units	Temp	Data Type	Reference
WS 5.29E+002	mg/L	25	EST	MEYLAN,WM ET AL. (1996)
logP 2.67			EXP	HANSCH,C & LEO,AJ (1985)
VP 2.08E-005	mm Hg	25	EST	NEELY,WB & BLAU,GE (1985)
DC	pKa			
HL 1.59E-008	atm m3/mol	25	EST	MEYLAN,WM & HOWARD,PH (1991)
OH 7.50E-011	cm3/molc sec	25	EST	MEYLAN,WM & HOWARD,PH (1993)

CAS #: 001878-61-1 — PHENOXYACETIC ACID, M-CARBOXY

Formula: $C_9H_8O_5$

Mol Weight: 196.16

MP (deg C):   FP (deg C):

BP (deg C):

BP pressure (mm Hg):

Property/Value	Units	Temp	Data Type	Reference
WS 1.28E+004	mg/L	25	EST	MEYLAN,WM ET AL. (1996)
logP 1.11			EXP	HANSCH,C & LEO,AJ (1985)
VP 2.06E-006	mm Hg	25	EST	NEELY,WB & BLAU,GE (1985)
DC	pKa			
HL 3.38E-013	atm m3/mol	25	EST	MEYLAN,WM & HOWARD,PH (1991)
OH 1.12E-011	cm3/molc sec	25	EST	MEYLAN,WM & HOWARD,PH (1993)

CAS #: 001878-62-2				O-ACETYLPHENOXYACETIC ACID
Formula: $C_{10}H_{10}O_4$				
Mol Weight: 194.19				
MP (deg C):		FP (deg C):		
BP (deg C):				
BP pressure (mm Hg):				

Property/ Value	Units	Temp	Data Type	Reference
WS 9.96E+003	mg/L	25	EST	MEYLAN,WM ET AL. (1996)
logP 1.25			EXP	HANSCH,C & LEO,AJ (1985)
VP 3.38E-005	mm Hg	25	EST	NEELY,WB & BLAU,GE (1985)
DC	pKa			
HL 3.06E-011	atm m3/mol	25	EST	MEYLAN,WM & HOWARD,PH (1991)
OH 2.45E-011	cm3/molc sec	25	EST	MEYLAN,WM & HOWARD,PH (1993)

CAS #: 001878-65-5				M-CHLOROPHENYLACETIC ACID
Formula: $C_8H_7ClO_2$				
Mol Weight: 170.60				
MP (deg C): 78-79.5		FP (deg C):		
BP (deg C):				
BP pressure (mm Hg):				

Property/ Value	Units	Temp	Data Type	Reference
WS 2.49E+003	mg/L	25	EST	MEYLAN,WM ET AL. (1996)
logP 2.09			EXP	HANSCH,C & LEO,AJ (1985)
VP 8.17E-004	mm Hg	25	EST	NEELY,WB & BLAU,GE (1985)
DC 4.14	pKa	25	EXP	KORTUM,G ET AL (1961)
HL 3.28E-008	atm m3/mol	25	EST	MEYLAN,WM & HOWARD,PH (1991)
OH 3.50E-012	cm3/molc sec	25	EST	MEYLAN,WM & HOWARD,PH (1993)

CAS #: 001878-66-6				P-CHLOROPHENYLACETIC ACID
Formula: $C_8H_7ClO_2$				
Mol Weight: 170.60				
MP (deg C): 105-108		FP (deg C):		
BP (deg C):				
BP pressure (mm Hg):				

Property/ Value	Units	Temp	Data Type	Reference
WS 2.35E+003	mg/L	25	EST	MEYLAN,WM ET AL. (1996)
logP 2.12			EXP	HANSCH,C & LEO,AJ (1985)
VP 8.17E-004	mm Hg	25	EST	NEELY,WB & BLAU,GE (1985)
DC 4.19	pKa	25	EXP	KORTUM,G ET AL (1961)
HL 3.28E-008	atm m3/mol	25	EST	MEYLAN,WM & HOWARD,PH (1991)
OH 2.64E-012	cm3/molc sec	25	EST	MEYLAN,WM & HOWARD,PH (1993)

CAS #: 001878-67-7				M-BROMOPHENYLACETIC ACID
Formula: $C_8H_7BrO_2$				
Mol Weight: 215.05				
MP (deg C): 100-102		FP (deg C):		
BP (deg C):				
BP pressure (mm Hg):				

Property/ Value	Units	Temp	Data Type	Reference
WS 8.59E+002	mg/L	25	EST	MEYLAN,WM ET AL. (1996)
logP 2.37			EXP	HANSCH,C & LEO,AJ (1985)
VP 3.05E-004	mm Hg	25	EST	NEELY,WB & BLAU,GE (1985)
DC	pKa			
HL 1.76E-008	atm m3/mol	25	EST	MEYLAN,WM & HOWARD,PH (1991)
OH 3.26E-012	cm3/molc sec	25	EST	MEYLAN,WM & HOWARD,PH (1993)

CAS #: 001878-68-8				P-BROMOPHENYLACETIC ACID
Formula: $C_8H_7BrO_2$				
Mol Weight: 215.05				
MP (deg C): 117-119		FP (deg C):		
BP (deg C):				
BP pressure (mm Hg):				

Property/ Value	Units	Temp	Data Type	Reference
WS 9.67E+002	mg/L	25	EST	MEYLAN,WM ET AL. (1996)
logP 2.31			EXP	HANSCH,C & LEO,AJ (1985)
VP 3.05E-004	mm Hg	25	EST	NEELY,WB & BLAU,GE (1985)
DC 4.19	pKa	25	EXP	KORTUM,G ET AL (1961)
HL 1.76E-008	atm m3/mol	25	EST	MEYLAN,WM & HOWARD,PH (1991)
OH 2.49E-012	cm3/molc sec	25	EST	MEYLAN,WM & HOWARD,PH (1993)

CAS #: 001878-69-9				M-IODOPHENYLACETIC ACID
Formula: $C_8H_7IO_2$				
Mol Weight: 262.05				
MP (deg C):		FP (deg C):		
BP (deg C):				
BP pressure (mm Hg):				

Property/ Value	Units	Temp	Data Type	Reference
WS 2.91E+002	mg/L	25	EST	MEYLAN,WM ET AL. (1996)
logP 2.62			EXP	HANSCH,C & LEO,AJ (1985)
VP 7.89E-005	mm Hg	25	EST	NEELY,WB & BLAU,GE (1985)
DC 4.16	pKa	25	EXP	KORTUM,G ET AL (1961)
HL 1.02E-008	atm m3/mol	25	EST	MEYLAN,WM & HOWARD,PH (1991)
OH 3.35E-012	cm3/molc sec	25	EST	MEYLAN,WM & HOWARD,PH (1993)

CAS #: 001878-71-3				M-CYANOPHENYLACETIC ACID
Formula: $C_9H_7NO_2$				
Mol Weight: 161.16				
MP (deg C):		FP (deg C):		
BP (deg C):				
BP pressure (mm Hg):				

Property/ Value	Units	Temp	Data Type	Reference
WS 8.96E+003	mg/L	25	EST	MEYLAN,WM ET AL. (1996)
logP 1.18			EXP	HANSCH,C & LEO,AJ (1985)
VP 8.25E-005	mm Hg	25	EST	NEELY,WB & BLAU,GE (1985)
DC	pKa			
HL 4.27E-010	atm m3/mol	25	EST	MEYLAN,WM & HOWARD,PH (1991)
OH 1.64E-012	cm3/molc sec	25	EST	MEYLAN,WM & HOWARD,PH (1993)

CAS #: 001878-80-4				M-ACETYLPHENOXYACETIC ACID
Formula: $C_{10}H_{10}O_4$				
Mol Weight: 194.19				
MP (deg C):		FP (deg C):		
BP (deg C):				
BP pressure (mm Hg):				

Property/ Value	Units	Temp	Data Type	Reference
WS 1.69E+004	mg/L	25	EST	MEYLAN,WM ET AL. (1996)
logP 0.98			EXP	HANSCH,C & LEO,AJ (1985)
VP 3.38E-005	mm Hg	25	EST	NEELY,WB & BLAU,GE (1985)
DC	pKa			
HL 3.06E-011	atm m3/mol	25	EST	MEYLAN,WM & HOWARD,PH (1991)
OH 1.40E-011	cm3/molc sec	25	EST	MEYLAN,WM & HOWARD,PH (1993)

001878-81-5 — P-ACETYLPHENOXYACETIC ACID

CAS #: 001878-81-5

Formula: $C_{10}H_{10}O_4$

Mol Weight: 194.19

MP (deg C): 175-177

FP (deg C):

BP (deg C):

BP pressure (mm Hg):

Property/ Value	Units	Temp	Data Type	Reference
WS 2.10E+004	mg/L	25	EST	MEYLAN,WM ET AL. (1996)
logP 0.87			EXP	HANSCH,C & LEO,AJ (1985)
VP 3.38E-005	mm Hg	25	EST	NEELY,WB & BLAU,GE (1985)
DC	pKa			
HL 3.06E-011	atm m3/mol	25	EST	MEYLAN,WM & HOWARD,PH (1991)
OH 2.45E-011	cm3/molc sec	25	EST	MEYLAN,WM & HOWARD,PH (1993)

001878-82-6 — P-CYANOPHENOXYACETIC ACID

CAS #: 001878-82-6

Formula: $C_9H_7NO_3$

Mol Weight: 177.16

MP (deg C):

FP (deg C):

BP (deg C):

BP pressure (mm Hg):

Property/ Value	Units	Temp	Data Type	Reference
WS 1.23E+004	mg/L	25	EST	MEYLAN,WM ET AL. (1996)
logP 0.93			EXP	HANSCH,C & LEO,AJ (1985)
VP 3.11E-005	mm Hg	25	EST	NEELY,WB & BLAU,GE (1985)
DC 2.93	pKa	25	EXP	KORTUM,G ET AL (1961)
HL 1.62E-010	atm m3/mol	25	EST	MEYLAN,WM & HOWARD,PH (1991)
OH 8.59E-012	cm3/molc sec	25	EST	MEYLAN,WM & HOWARD,PH (1993)

001878-83-7 — M-HYDROXYPHENOXYACETIC ACID

CAS #: 001878-83-7

Formula: $C_8H_8O_4$

Mol Weight: 168.15

MP (deg C):

FP (deg C):

BP (deg C):

BP pressure (mm Hg):

Property/ Value	Units	Temp	Data Type	Reference
WS 1.33E+005	mg/L	25	EST	MEYLAN,WM ET AL. (1996)
logP 0.76			EXP	HANSCH,C & LEO,AJ (1985)
VP 1.85E-005	mm Hg	25	EST	NEELY,WB & BLAU,GE (1985)
DC	pKa			
HL 1.75E-012	atm m3/mol	25	EST	MEYLAN,WM & HOWARD,PH (1991)
OH 2.05E-010	cm3/molc sec	25	EST	MEYLAN,WM & HOWARD,PH (1993)

001878-84-8 — P-HYDROXYPHENOXYACETIC ACID

CAS #: 001878-84-8

Formula: $C_8H_8O_4$

Mol Weight: 168.15

MP (deg C):

FP (deg C):

BP (deg C):

BP pressure (mm Hg):

Property/ Value	Units	Temp	Data Type	Reference
WS 1.65E+005	mg/L	25	EST	MEYLAN,WM ET AL. (1996)
logP 0.65			EXP	HANSCH,C & LEO,AJ (1985)
VP 1.85E-005	mm Hg	25	EST	NEELY,WB & BLAU,GE (1985)
DC	pKa			
HL 1.75E-012	atm m3/mol	25	EST	MEYLAN,WM & HOWARD,PH (1991)
OH 3.38E-011	cm3/molc sec	25	EST	MEYLAN,WM & HOWARD,PH (1993)

001878-85-9 — O-METHOXY PHENOXYACETIC ACID

CAS #: 001878-85-9

Formula: $C_9H_{10}O_4$

Mol Weight: 182.18

MP (deg C):

FP (deg C):

BP (deg C):

BP pressure (mm Hg):

Property/ Value	Units	Temp	Data Type	Reference
WS 1.94E+004	mg/L	25	EST	MEYLAN,WM ET AL. (1996)
logP 0.98			EXP	HANSCH,C & LEO,AJ (1985)
VP 2.12E-004	mm Hg	25	EST	NEELY,WB & BLAU,GE (1985)
DC 3.23	pKa	25	EXP	KORTUM,G ET AL (1961)
HL 9.93E-010	atm m3/mol	25	EST	MEYLAN,WM & HOWARD,PH (1991)
OH 2.42E-011	cm3/molc sec	25	EST	MEYLAN,WM & HOWARD,PH (1993)

001878-87-1 — O-NITROPHENOXYACETIC ACID

CAS #: 001878-87-1

Formula: $C_8H_7NO_5$

Mol Weight: 197.15

MP (deg C):

FP (deg C):

BP (deg C):

BP pressure (mm Hg):

Property/ Value	Units	Temp	Data Type	Reference
WS 4.96E+003	mg/L	25	EST	MEYLAN,WM ET AL. (1996)
logP 1.13			EXP	SANGSTER,J (1994)
VP 1.42E-005	mm Hg	25	EST	NEELY,WB & BLAU,GE (1985)
DC 2.90	pKa	25	EXP	KORTUM,G ET AL (1961)
HL 6.62E-011	atm m3/mol	25	EST	MEYLAN,WM & HOWARD,PH (1991)
OH 7.48E-012	cm3/molc sec	25	EST	MEYLAN,WM & HOWARD,PH (1993)

001878-88-2 — M-NITROPHENOXYACETIC ACID

CAS #: 001878-88-2

Formula: $C_8H_7NO_5$

Mol Weight: 197.15

MP (deg C):

FP (deg C):

BP (deg C):

BP pressure (mm Hg):

Property/ Value	Units	Temp	Data Type	Reference
WS 3.09E+003	mg/L	25	EST	MEYLAN,WM ET AL. (1996)
logP 1.37			EXP	HANSCH,C & LEO,AJ (1985)
VP 1.42E-005	mm Hg	25	EST	NEELY,WB & BLAU,GE (1985)
DC 2.95	pKa	25	EXP	KORTUM,G ET AL (1961)
HL 6.62E-011	atm m3/mol	25	EST	MEYLAN,WM & HOWARD,PH (1991)
OH 6.67E-012	cm3/molc sec	25	EST	MEYLAN,WM & HOWARD,PH (1993)

001878-89-3 — PHENOXYACETIC ACID, M-BENZAMIDO

CAS #: 001878-89-3

Formula: $C_{15}H_{13}NO_4$

Mol Weight: 271.28

MP (deg C):

FP (deg C):

BP (deg C):

BP pressure (mm Hg):

Property/ Value	Units	Temp	Data Type	Reference
WS 3.59E+002	mg/L	25	EST	MEYLAN,WM ET AL. (1996)
logP 1.99			EXP	HANSCH,C & LEO,AJ (1985)
VP 8.27E-010	mm Hg	25	EST	NEELY,WB & BLAU,GE (1985)
DC	pKa			
HL 3.80E-015	atm m3/mol	25	EST	MEYLAN,WM & HOWARD,PH (1991)
OH 1.44E-010	cm3/molc sec	25	EST	MEYLAN,WM & HOWARD,PH (1993)

001878-91-7 — P-BROMOPHENOXYACETIC ACID

Formula: $C_8H_7BrO_3$

Mol Weight: 231.05

MP (deg C): FP (deg C):

BP (deg C):

BP pressure (mm Hg):

Property/Value	Units	Temp	Data Type	Reference
WS 6.03E+002	mg/L	25	EST	MEYLAN,WM ET AL. (1996)
logP 2.45			EXP	HANSCH,C & LEO,AJ (1985)
VP 1.07E-004	mm Hg	25	EST	NEELY,WB & BLAU,GE (1985)
DC 3.13	pKa	25	EXP	KORTUM,G ET AL (1961)
HL 6.68E-009	atm m3/mol	25	EST	MEYLAN,WM & HOWARD,PH (1991)
OH 1.10E-011	cm3/molc sec	25	EST	MEYLAN,WM & HOWARD,PH (1993)

001878-92-8 — O-IODOPHENOXYACETIC ACID

Formula: $C_8H_7IO_3$

Mol Weight: 278.05

MP (deg C): FP (deg C):

BP (deg C):

BP pressure (mm Hg):

Property/Value	Units	Temp	Data Type	Reference
WS 4.18E+002	mg/L	25	EST	MEYLAN,WM ET AL. (1996)
logP 2.33			EXP	HANSCH,C & LEO,AJ (1985)
VP 3.01E-005	mm Hg	25	EST	NEELY,WB & BLAU,GE (1985)
DC 3.17	pKa	25	EXP	KORTUM,G ET AL (1961)
HL 3.89E-009	atm m3/mol	25	EST	MEYLAN,WM & HOWARD,PH (1991)
OH 1.19E-011	cm3/molc sec	25	EST	MEYLAN,WM & HOWARD,PH (1993)

001878-93-9 — M-IODOPHENOXYACETIC ACID

Formula: $C_8H_7IO_3$

Mol Weight: 278.05

MP (deg C): FP (deg C):

BP (deg C):

BP pressure (mm Hg):

Property/Value	Units	Temp	Data Type	Reference
WS 3.37E+002	mg/L	25	EST	MEYLAN,WM ET AL. (1996)
logP 2.44			EXP	HANSCH,C & LEO,AJ (1985)
VP 3.01E-005	mm Hg	25	EST	NEELY,WB & BLAU,GE (1985)
DC 3.13	pKa	25	EXP	KORTUM,G ET AL (1961)
HL 3.89E-009	atm m3/mol	25	EST	MEYLAN,WM & HOWARD,PH (1991)
OH 1.90E-011	cm3/molc sec	25	EST	MEYLAN,WM & HOWARD,PH (1993)

001878-94-0 — P-IODOPHENOXYACETIC ACID

Formula: $C_8H_7IO_3$

Mol Weight: 278.05

MP (deg C): FP (deg C):

BP (deg C):

BP pressure (mm Hg):

Property/Value	Units	Temp	Data Type	Reference
WS 2.06E+002	mg/L	25	EST	MEYLAN,WM ET AL. (1996)
logP 2.69			EXP	HANSCH,C & LEO,AJ (1985)
VP 3.01E-005	mm Hg	25	EST	NEELY,WB & BLAU,GE (1985)
DC 3.16	pKa	25	EXP	KORTUM,G ET AL (1961)
HL 3.89E-009	atm m3/mol	25	EST	MEYLAN,WM & HOWARD,PH (1991)
OH 1.19E-011	cm3/molc sec	25	EST	MEYLAN,WM & HOWARD,PH (1993)

001879-09-0 — 2-(1,1-DIMETHYLETHYL)-4,6-DIMETHYLPHENOL

Formula: $C_{12}H_{18}O$

Mol Weight: 178.28

MP (deg C): 22.3 FP (deg C):

BP (deg C): 249

BP pressure (mm Hg):

Property/Value	Units	Temp	Data Type	Reference
WS 2.96E+001	mg/L	25	EST	MEYLAN,WM ET AL. (1996)
logP 4.52			EST	MEYLAN,WM & HOWARD,PH (1995)
VP 4.48E-002	mm Hg	25	EXT	CHAO,J ET AL. (1983)
DC 12.04	pKa	20	EXP	SERJEANT,EP & DEMPSEY,B (1979)
HL 1.76E-006	atm m3/mol	25	EST	MEYLAN,WM & HOWARD,PH (1991)
OH 2.11E-011	cm3/molc sec	25	EST	MEYLAN,WM & HOWARD,PH (1993)

001879-56-7 — O-BROMOPHENOXYACETIC ACID

Formula: $C_8H_7BrO_3$

Mol Weight: 231.05

MP (deg C): FP (deg C):

BP (deg C):

BP pressure (mm Hg):

Property/Value	Units	Temp	Data Type	Reference
WS 1.35E+003	mg/L	25	EST	MEYLAN,WM ET AL. (1996)
logP 2.04			EXP	HANSCH,C & LEO,AJ (1985)
VP 1.07E-004	mm Hg	25	EST	NEELY,WB & BLAU,GE (1985)
DC 3.13	pKa	25	EXP	KORTUM,G ET AL (1961)
HL 6.68E-009	atm m3/mol	25	EST	MEYLAN,WM & HOWARD,PH (1991)
OH 1.10E-011	cm3/molc sec	25	EST	MEYLAN,WM & HOWARD,PH (1993)

001879-58-9 — M-CYANOPHENOXYACETIC ACID

Formula: $C_9H_7NO_3$

Mol Weight: 177.16

MP (deg C): FP (deg C):

BP (deg C):

BP pressure (mm Hg):

Property/Value	Units	Temp	Data Type	Reference
WS 1.18E+004	mg/L	25	EST	MEYLAN,WM ET AL. (1996)
logP 0.95			EXP	HANSCH,C & LEO,AJ (1985)
VP 3.11E-005	mm Hg	25	EST	NEELY,WB & BLAU,GE (1985)
DC 3.03	pKa	25	EXP	KORTUM,G ET AL (1961)
HL 1.62E-010	atm m3/mol	25	EST	MEYLAN,WM & HOWARD,PH (1991)
OH 7.61E-012	cm3/molc sec	25	EST	MEYLAN,WM & HOWARD,PH (1993)

001882-26-4 — PYRIDINOL CARBAMATE

Formula: $C_{11}H_{15}N_3O_4$

Mol Weight: 253.26

MP (deg C): 136-137 FP (deg C):

BP (deg C):

BP pressure (mm Hg):

Property/Value	Units	Temp	Data Type	Reference
WS 1.38E+004	mg/L	25	EST	MEYLAN,WM ET AL. (1996)
logP 0.24			EXP	HANSCH,C & LEO,AJ (1985)
VP 4.35E-005	mm Hg	25	EST	NEELY,WB & BLAU,GE (1985)
DC	pKa			
HL 1.89E-014	atm m3/mol	25	EST	MEYLAN,WM & HOWARD,PH (1991)
OH 1.74E-011	cm3/molc sec	25	EST	MEYLAN,WM & HOWARD,PH (1993)

CAS #: 001883-15-4	3,4-PYRIDINEDIMETHANOL, 6-METHYL-5-(PHOSPHONOOXY

Formula: $C_7H_{10}NO_6P$

Mol Weight: 235.13

MP (deg C):
FP (deg C):

BP (deg C):

BP pressure (mm Hg):

Property/Value	Units	Temp	Data Type	Reference
WS 3.25E+005	mg/L	25	EST	MEYLAN,WM ET AL. (1996)
logP -0.64			EXP	SANGSTER,J (1994)
VP 9.84E-011	mm Hg	25	EST	NEELY,WB & BLAU,GE (1985)
DC	pKa			
HL 5.45E-024	atm m3/mol	25	EST	MEYLAN,WM & HOWARD,PH (1991)
OH 2.15E-011	cm3/molc sec	25	EST	MEYLAN,WM & HOWARD,PH (1993)

CAS #: 001885-38-7	3-PHENYL-2-PROPENENITRILE

Formula: C_9H_7N

Mol Weight: 129.16

MP (deg C): 22
FP (deg C):

BP (deg C): 263.8

BP pressure (mm Hg):

Property/Value	Units	Temp	Data Type	Reference
WS 1.07E+003	mg/L	25	EST	MEYLAN,WM ET AL. (1996)
logP 1.96			EXP	HANSCH,C & LEO,AJ (1985)
VP 2.29E-002	mm Hg	25	EST	NEELY,WB & BLAU,GE (1985)
DC	pKa			
HL 6.16E-006	atm m3/mol	25	EST	MEYLAN,WM & HOWARD,PH (1991)
OH 1.09E-011	cm3/molc sec	25	EST	MEYLAN,WM & HOWARD,PH (1993)

CAS #: 001885-87-6	N-BENZALAMINOPIPERIDINE

Formula: $C_{12}H_{16}N_2$

Mol Weight: 188.27

MP (deg C):
FP (deg C):

BP (deg C):

BP pressure (mm Hg):

Property/Value	Units	Temp	Data Type	Reference
WS 1.27E+002	mg/L	25	EST	MEYLAN,WM ET AL. (1996)
logP 3.04			EXP	HANSCH,C & LEO,AJ (1985)
VP 2.02E-003	mm Hg	25	EST	NEELY,WB & BLAU,GE (1985)
DC	pKa			
HL 3.71E-006	atm m3/mol	25	EST	MEYLAN,WM & HOWARD,PH (1991)
OH 3.22E-011	cm3/molc sec	25	EST	MEYLAN,WM & HOWARD,PH (1993)

CAS #: 001886-81-3	DODECYL BENZENESULFONATE

Formula: $C_{18}H_{30}O_3S$

Mol Weight: 326.50

MP (deg C):
FP (deg C):

BP (deg C):

BP pressure (mm Hg):

Property/Value	Units	Temp	Data Type	Reference
WS 3.43E-002	mg/L	25	EST	MEYLAN,WM ET AL. (1996)
logP 6.32			EST	MEYLAN,WM & HOWARD,PH (1995)
VP 1.33E-007	mm Hg	25	EST	NEELY,WB & BLAU,GE (1985)
DC	pKa			
HL 1.82E-005	atm m3/mol	25	EST	MEYLAN,WM & HOWARD,PH (1991)
OH 1.53E-011	cm3/molc sec	25	EST	MEYLAN,WM & HOWARD,PH (1993)

CAS #: 001888-71-7	PERCHLOROPROPYLENE

Formula: C_3Cl_6

Mol Weight: 248.75

MP (deg C): -72.9
FP (deg C):

BP (deg C): 209.5

BP pressure (mm Hg):

Property/Value	Units	Temp	Data Type	Reference
WS 1.70E+001	mg/L	25	EXP	OKOUCHI,S ET AL. (1992)
logP 4.38			EST	MEYLAN,WM & HOWARD,PH (1995)
VP 2.44E-001	mm Hg	25	EXT	DYKYJ,J (1970)
DC	pKa			
HL 4.70E-003	atm m3/mol	25	EST	VP/WSOL
OH 7.74E-013	cm3/molc sec	25	EST	MEYLAN,WM & HOWARD,PH (1993)

CAS #: 001891-90-3	P-TRIFLUOROMETHYLBENZAMIDE

Formula: $C_8H_6F_3NO$

Mol Weight: 189.14

MP (deg C): 184-186
FP (deg C):

BP (deg C):

BP pressure (mm Hg):

Property/Value	Units	Temp	Data Type	Reference
WS 1.72E+003	mg/L	25	EST	MEYLAN,WM ET AL. (1996)
logP 1.71			EXP	HANSCH,C & LEO,AJ (1985)
VP 6.01E-004	mm Hg	25	EST	NEELY,WB & BLAU,GE (1985)
DC	pKa			
HL 1.92E-008	atm m3/mol	25	EST	MEYLAN,WM & HOWARD,PH (1991)
OH 2.69E-013	cm3/molc sec	25	EST	MEYLAN,WM & HOWARD,PH (1993)

CAS #: 001893-33-0	PIPAMPERONE

Formula: $C_{21}H_{30}FN_3O_2$

Mol Weight: 375.49

MP (deg C):
FP (deg C):

BP (deg C):

BP pressure (mm Hg):

Property/Value	Units	Temp	Data Type	Reference
WS 8.17E+001	mg/L	25	EST	MEYLAN,WM ET AL. (1996)
logP 2.02			EXP	HANSCH,C ET AL. (1995)
VP 9.48E-011	mm Hg	25	EST	NEELY,WB & BLAU,GE (1985)
DC	pKa			
HL 1.60E-017	atm m3/mol	25	EST	MEYLAN,WM & HOWARD,PH (1991)
OH 2.08E-010	cm3/molc sec	25	EST	MEYLAN,WM & HOWARD,PH (1993)

CAS #: 001897-45-6	CHLOROTHANONIL

Formula: $C_8Cl_4N_2$

Mol Weight: 265.91

MP (deg C): 250-251
FP (deg C):

BP (deg C): 350

BP pressure (mm Hg):

Property/Value	Units	Temp	Data Type	Reference
WS 6.00E-001	mg/L	25	EXP	WORTHING,CR & WALKER,SB (1987)
logP 3.05			EXP	KRAWCHUK,BP & WEBSTER,GRB (1987)
VP 3.40E-007	mm Hg	25	EST	HL X WSOL
DC	pKa			
HL 2.00E-006	atm m3/mol	25	EXP	KAWAMOTO,K & URANO,K (1989)
OH 6.20E-015	cm3/molc sec	25	EST	MEYLAN,WM & HOWARD,PH (1993)

CAS #: 001899-48-5			2,4-DIAMINOQUINAZOLINE

Formula: $C_8H_8N_4$

Mol Weight: 160.18

MP (deg C): FP (deg C):

BP (deg C):

BP pressure (mm Hg):

Property/ Value	Units	Temp	Data Type	Reference
WS 9.56E+003	mg/L	25	EST	MEYLAN,WM ET AL. (1996)
logP 1.00			EXP	HANSCH,C & LEO,AJ (1985)
VP 8.75E-006	mm Hg	25	EST	NEELY,WB & BLAU,GE (1985)
DC 7.96	pKa	20	EXP	PERRIN,DD (1972)
HL 1.26E-010	atm m3/mol	25	EST	MEYLAN,WM & HOWARD,PH (1991)
OH 1.81E-010	cm3/molc sec	25	EST	MEYLAN,WM & HOWARD,PH (1993)

CAS #: 001899-94-1			M-METHYLBENZENESULFONAMIDE

Formula: $C_7H_9NO_2S$

Mol Weight: 171.22

MP (deg C): FP (deg C):

BP (deg C):

BP pressure (mm Hg):

Property/ Value	Units	Temp	Data Type	Reference
WS 7.82E+003	mg/L	25	EXP	SUZUKI,T (1991)
logP 0.85			EXP	HANSCH,C & LEO,AJ (1985)
VP 3.06E-004	mm Hg	25	EST	NEELY,WB & BLAU,GE (1985)
DC	pKa			
HL 4.66E-007	atm m3/mol	25	EST	MEYLAN,WM & HOWARD,PH (1991)
OH 1.22E-012	cm3/molc sec	25	EST	MEYLAN,WM & HOWARD,PH (1993)

CAS #: 001904-58-1			BENZOYLHYDRAZINE, O-AMINO

Formula: $C_7H_9N_3O$

Mol Weight: 151.17

MP (deg C): FP (deg C):

BP (deg C):

BP pressure (mm Hg):

Property/ Value	Units	Temp	Data Type	Reference
WS 5.23E+003	mg/L	25	EST	MEYLAN,WM ET AL. (1996)
logP -0.18			EXP	HANSCH,C & LEO,AJ (1985)
VP 3.14E-006	mm Hg	25	EST	NEELY,WB & BLAU,GE (1985)
DC	pKa			
HL 3.27E-015	atm m3/mol	25	EST	MEYLAN,WM & HOWARD,PH (1991)
OH 1.04E-010	cm3/molc sec	25	EST	MEYLAN,WM & HOWARD,PH (1993)

CAS #: 001906-79-2			1-ETHYLPYRIDINIUM BROMIDE

Formula: $C_7H_{10}BrN$

Mol Weight: 188.07

MP (deg C): FP (deg C):

BP (deg C):

BP pressure (mm Hg):

Property/ Value	Units	Temp	Data Type	Reference
WS 1.00E+006	mg/L	25	EST	MEYLAN,WM ET AL. (1996)
logP -4.07			EST	MEYLAN,WM & HOWARD,PH (1995)
VP 2.26E-001	mm Hg	25	EST	NEELY,WB & BLAU,GE (1985)
DC	pKa			
HL 2.00E-010	atm m3/mol	25	EST	MEYLAN,WM & HOWARD,PH (1991)
OH 1.15E-011	cm3/molc sec	25	EST	MEYLAN,WM & HOWARD,PH (1993)

CAS #: 001912-24-9			ATRAZINE

Formula: $C_8H_{14}ClN_5$

Mol Weight: 215.69

MP (deg C): 171-174 FP (deg C):

BP (deg C):

BP pressure (mm Hg):

Property/ Value	Units	Temp	Data Type	Reference
WS 2.80E+001	mg/L	20	EXP	MARTIN,H & WORTHING,CR (1977)
logP 2.61			EXP	HANSCH,C ET AL. (1995)
VP 3.00E-007	mm Hg	20	EXT	MARTIN,H & WORTHING,CR (1977)
DC 1.68	pKa		EXP	WEBER,JB (1970)
HL 4.47E-009	atm m3/mol	25	EST	MEYLAN,WM & HOWARD,PH (1991)
OH 2.73E-011	cm3/molc sec	25	EST	MEYLAN,WM & HOWARD,PH (1993)

CAS #: 001912-25-0			IPAZINE

Formula: $C_{10}H_{18}ClN_5$

Mol Weight: 243.74

MP (deg C): FP (deg C):

BP (deg C):

BP pressure (mm Hg):

Property/ Value	Units	Temp	Data Type	Reference
WS 4.00E+001	mg/L	21	EXP	YALKOWSKY,SH & DANNENFELSER,RM (1992)
logP 3.94			EXP	HANSCH,C & LEO,AJ (1985)
VP 4.89E-005	mm Hg	25	EST	NEELY,WB & BLAU,GE (1985)
DC	pKa			
HL 1.22E-007	atm m3/mol	25	EST	MEYLAN,WM & HOWARD,PH (1991)
OH 3.62E-011	cm3/molc sec	25	EST	MEYLAN,WM & HOWARD,PH (1993)

CAS #: 001912-26-1			TRIETAZINE

Formula: $C_9H_{16}ClN_5$

Mol Weight: 229.71

MP (deg C): 100-102 FP (deg C):

BP (deg C):

BP pressure (mm Hg):

Property/ Value	Units	Temp	Data Type	Reference
WS 2.00E+001	mg/L	20	EXP	BAILEY,GW ET AL. (1968)
logP 3.34			EXP	HANSCH,C & LEO,AJ (1985)
VP 6.73E-005	mm Hg	25	EST	NEELY,WB & BLAU,GE (1985)
DC 1.88	pKa		EXP	BAILEY,GW ET AL. (1968)
HL 9.17E-008	atm m3/mol	25	EST	MEYLAN,WM & HOWARD,PH (1991)
OH 2.67E-011	cm3/molc sec	25	EST	MEYLAN,WM & HOWARD,PH (1993)

CAS #: 001918-00-9			DICAMBA

Formula: $C_8H_6Cl_2O_3$

Mol Weight: 221.04

MP (deg C): 114-116 FP (deg C):

BP (deg C):

BP pressure (mm Hg):

Property/ Value	Units	Temp	Data Type	Reference
WS 4.50E+003	mg/L	25	EXP	MARTIN,H & WORTHING,CR (1977)
logP 2.21			EXP	HANSCH,C & LEO,AJ (1985)
VP 3.37E-005	mm Hg	25	EXP	WEBER,JB (1994)
DC 1.90	pKa		EXP	CESSNA,AJ & GROVER,R (1978)
HL 2.18E-009	atm m3/mol	25	EST	VP/WSOL
OH 2.98E-012	cm3/molc sec	25	EST	MEYLAN,WM & HOWARD,PH (1993)

001918-02-1 — PICLORAM

CAS #	001918-02-1
Formula	$C_6H_3Cl_3N_2O_2$
Mol Weight	241.46
MP (deg C)	218-219
FP (deg C)	
BP (deg C)	
BP pressure (mm Hg)	

Property/Value	Units	Temp	Data Type	Reference
WS 4.30E+002	mg/L	25	EXP	BAILEY,GW ET AL. (1968)
logP 0.30			EXP	HANSCH,C & LEO,AJ (1985)
VP 7.21E-011	mm Hg	20	EXP	DOBBS, AJ (1982)
DC 3.60	pKa		EXP	MERCK INDEX (1983)
HL 4.05E+000	atm m3/mol	20	EST	VP/WSOL
OH 2.07E-011	cm3/molc sec	25	EST	MEYLAN,WM & HOWARD,PH (1993)

001918-08-7 — DIPROPALIN

CAS #	001918-08-7
Formula	$C_{13}H_{19}N_3O_4$
Mol Weight	281.31
MP (deg C)	80
FP (deg C)	
BP (deg C)	
BP pressure (mm Hg)	

Property/Value	Units	Temp	Data Type	Reference
WS 3.04E+002	mg/L		EXP	GUNTHER,FA ET AL. (1968)
logP 4.90			EST	MEYLAN,WM & HOWARD,PH (1995)
VP 9.12E-007	mm Hg	25	EST	NEELY,WB & BLAU,GE (1985)
DC	pKa			
HL 2.69E-005	atm m3/mol	25	EST	MEYLAN,WM & HOWARD,PH (1991)
OH 2.42E-011	cm3/molc sec	25	EST	MEYLAN,WM & HOWARD,PH (1993)

001918-11-2 — 2,6-DI-T-BUTYL-P-TOLYL METHYLCARBAMATE

CAS #	001918-11-2
Formula	$C_{17}H_{27}NO_2$
Mol Weight	277.41
MP (deg C)	
FP (deg C)	
BP (deg C)	
BP pressure (mm Hg)	

Property/Value	Units	Temp	Data Type	Reference
WS 7.00E+000	mg/L	25	EXP	SHIU,WY ET AL. (1990)
logP 5.28			EST	MEYLAN,WM & HOWARD,PH (1995)
VP 3.95E-005	mm Hg	25	EST	NEELY,WB & BLAU,GE (1985)
DC	pKa			
HL 2.36E-007	atm m3/mol	25	EST	MEYLAN,WM & HOWARD,PH (1991)
OH 2.53E-011	cm3/molc sec	25	EST	MEYLAN,WM & HOWARD,PH (1993)

001918-13-4 — CHLORTHIAMID

CAS #	001918-13-4
Formula	$C_7H_5Cl_2NS$
Mol Weight	206.09
MP (deg C)	
FP (deg C)	
BP (deg C)	
BP pressure (mm Hg)	

Property/Value	Units	Temp	Data Type	Reference
WS 9.50E+002	mg/L	21	EXP	YALKOWSKY,SH & DANNENFELSER,RM (1992)
logP 2.96			EST	MEYLAN,WM & HOWARD,PH (1995)
VP 1.00E-006	mm Hg	20	EXP	KEARNEY,PC & KAUFMAN,DD (1975A)
DC	pKa			
HL 2.86E-011	atm m3/mol	20	EST	VP/WSOL
OH 2.22E-011	cm3/molc sec	25	EST	MEYLAN,WM & HOWARD,PH (1993)

001918-16-7 — PROPACHLOR

CAS #	001918-16-7
Formula	$C_{11}H_{14}ClNO$
Mol Weight	211.69
MP (deg C)	77
FP (deg C)	
BP (deg C)	110
BP pressure (mm Hg)	3.00E-002

Property/Value	Units	Temp	Data Type	Reference
WS 6.13E+002	mg/L	25	EXP	WAUCHOPE,RD ET AL. (1992)
logP 2.18			EXP	HANSCH,C & LEO,AJ (1985)
VP 2.30E-004	mm Hg	25	EXP	WAUCHOPE,RD ET AL. (1992)
DC	pKa			
HL 1.09E-007	atm m3/mol	20	EST	SUNTIO,LR ET AL. (1988)
OH 2.23E-011	cm3/molc sec	25	EST	MEYLAN,WM & HOWARD,PH (1993)

001918-18-9 — METHYL (3,4-DICHLOROPHENYL)CARBAMATE

CAS #	001918-18-9
Formula	$C_8H_7Cl_2NO_2$
Mol Weight	220.06
MP (deg C)	
FP (deg C)	
BP (deg C)	
BP pressure (mm Hg)	

Property/Value	Units	Temp	Data Type	Reference
WS 5.03E+001	mg/L	25	EST	MEYLAN,WM ET AL. (1996)
logP 3.32			EXP	SANGSTER,J (1993)
VP 1.39E-003	mm Hg	25	EST	NEELY,WB & BLAU,GE (1985)
DC	pKa			
HL 1.20E-008	atm m3/mol	25	EST	MEYLAN,WM & HOWARD,PH (1991)
OH 8.98E-012	cm3/molc sec	25	EST	MEYLAN,WM & HOWARD,PH (1993)

001928-37-6 — 2,4,5-T, METHYL ESTER

CAS #	001928-37-6
Formula	$C_9H_7Cl_3O_3$
Mol Weight	269.51
MP (deg C)	
FP (deg C)	
BP (deg C)	
BP pressure (mm Hg)	

Property/Value	Units	Temp	Data Type	Reference
WS 1.72E+001	mg/L	25	EST	MEYLAN,WM ET AL. (1996)
logP 3.55			EST	MEYLAN,WM & HOWARD,PH (1995)
VP 2.27E-004	mm Hg	25	EST	NEELY,WB & BLAU,GE (1985)
DC	pKa			
HL 2.19E-006	atm m3/mol	25	EST	MEYLAN,WM & HOWARD,PH (1991)
OH 5.72E-012	cm3/molc sec	25	EST	MEYLAN,WM & HOWARD,PH (1993)

001928-38-7 — 2,4-D, METHYL ESTER

CAS #	001928-38-7
Formula	$C_9H_8Cl_2O_3$
Mol Weight	235.07
MP (deg C)	38-39
FP (deg C)	
BP (deg C)	
BP pressure (mm Hg)	

Property/Value	Units	Temp	Data Type	Reference
WS 2.58E+002	mg/L	25	EXP	PARIS,DF ET AL. (1984)
logP 2.90			EST	MEYLAN,WM & HOWARD,PH (1995)
VP 2.30E-003	mm Hg	25	EXP	BOSCH,SJ (1983)
DC	pKa			
HL 2.76E-006	atm m3/mol	25	EST	VP/WSOL
OH 6.27E-012	cm3/molc sec	25	EST	MEYLAN,WM & HOWARD,PH (1993)

CAS #: 001928-39-8 — 2,4,5-T, ETHYL ESTER

Formula: $C_{10}H_9Cl_3O_3$
Mol Weight: 283.54
MP (deg C): 64
FP (deg C):
BP (deg C):
BP pressure (mm Hg):

Property/Value	Units	Temp	Data Type	Reference
WS 4.01E+000	mg/L	25	EXP	PARIS,DF ET AL. (1984)
logP 4.04			EST	MEYLAN,WM & HOWARD,PH (1995)
VP 8.91E-005	mm Hg	25	EST	NEELY,WB & BLAU,GE (1985)
DC	pKa			
HL 2.90E-006	atm m3/mol	25	EST	MEYLAN,WM & HOWARD,PH (1991)
OH 4.72E-012	cm3/molc sec	25	EST	MEYLAN,WM & HOWARD,PH (1993)

CAS #: 001928-44-5 — 2,4-D, OCTYL ESTER

Formula: $C_{16}H_{22}Cl_2O_3$
Mol Weight: 333.26
MP (deg C):
FP (deg C):
BP (deg C): 170-172
BP pressure (mm Hg):

Property/Value	Units	Temp	Data Type	Reference
WS 3.17E-002	mg/L	25	EXP	PARIS,DF ET AL. (1984)
logP 6.34			EST	MEYLAN,WM & HOWARD,PH (1995)
VP 2.56E-006	mm Hg	25	EST	NEELY,WB & BLAU,GE (1985)
DC	pKa			
HL 2.14E-005	atm m3/mol	25	EST	MEYLAN,WM & HOWARD,PH (1991)
OH 1.63E-011	cm3/molc sec	25	EST	MEYLAN,WM & HOWARD,PH (1993)

CAS #: 001928-47-8 — 2,4,5-T, 2-ETHYLHEXYL ESTER

Formula: $C_{16}H_{21}Cl_3O_3$
Mol Weight: 367.70
MP (deg C):
FP (deg C):
BP (deg C):
BP pressure (mm Hg):

Property/Value	Units	Temp	Data Type	Reference
WS 6.05E-003	mg/L	25	EST	MEYLAN,WM ET AL. (1996)
logP 6.91			EST	MEYLAN,WM & HOWARD,PH (1995)
VP 1.20E-006	mm Hg	25	EXP	HAMILTON,DJ (1980)
DC	pKa			
HL 1.59E-005	atm m3/mol	25	EST	MEYLAN,WM & HOWARD,PH (1991)
OH 1.40E-011	cm3/molc sec	25	EST	MEYLAN,WM & HOWARD,PH (1993)

CAS #: 001928-61-6 — 2,4-D, PROPYL ESTER

Formula: $C_{11}H_{12}Cl_2O_3$
Mol Weight: 263.12
MP (deg C):
FP (deg C):
BP (deg C): 147
BP pressure (mm Hg): 1.67E+000

Property/Value	Units	Temp	Data Type	Reference
WS 1.48E+001	mg/L	25	EXP	PARIS,DF ET AL. (1984)
logP 3.89			EST	MEYLAN,WM & HOWARD,PH (1995)
VP 1.90E-004	mm Hg	25	EST	NEELY,WB & BLAU,GE (1985)
DC	pKa			
HL 5.20E-006	atm m3/mol	25	EST	MEYLAN,WM & HOWARD,PH (1991)
OH 6.75E-012	cm3/molc sec	25	EST	MEYLAN,WM & HOWARD,PH (1993)

CAS #: 001929-29-9 — P-MEO-B-PHENYLPROPIONIC ACID

Formula: $C_{10}H_{12}O_3$
Mol Weight: 180.21
MP (deg C): 98-100
FP (deg C):
BP (deg C):
BP pressure (mm Hg):

Property/Value	Units	Temp	Data Type	Reference
WS 3.25E+003	mg/L	25	EST	MEYLAN,WM ET AL. (1996)
logP 1.90			EXP	HANSCH,C & LEO,AJ (1985)
VP 3.83E-004	mm Hg	25	EST	NEELY,WB & BLAU,GE (1985)
DC	pKa			
HL 3.47E-009	atm m3/mol	25	EST	MEYLAN,WM & HOWARD,PH (1991)
OH 2.88E-011	cm3/molc sec	25	EST	MEYLAN,WM & HOWARD,PH (1993)

CAS #: 001929-73-3 — 2,4-D, BUTOXYETHYL ESTER

Formula: $C_{14}H_{18}Cl_2O_4$
Mol Weight: 321.20
MP (deg C):
FP (deg C):
BP (deg C): 156-162
BP pressure (mm Hg): 1.00E+000

Property/Value	Units	Temp	Data Type	Reference
WS 1.20E+001	mg/L	25	EXP	BOSCH,SJ (1983)
logP 4.10			EST	MEYLAN,WM & HOWARD,PH (1995)
VP 4.50E-006	mm Hg	25	EXP	BOSCH,SJ (1983)
DC	pKa			
HL 1.59E-007	atm m3/mol	25	EST	VP/WSOL
OH 2.72E-011	cm3/molc sec	25	EST	MEYLAN,WM & HOWARD,PH (1993)

CAS #: 001929-77-7 — VERNOLATE

Formula: $C_{10}H_{21}NOS$
Mol Weight: 203.35
MP (deg C):
FP (deg C):
BP (deg C): 150
BP pressure (mm Hg): 3.00E+001

Property/Value	Units	Temp	Data Type	Reference
WS 9.00E+001	mg/L	20	EXP	SUNTIO,LR ET AL. (1988)
logP 3.84			EXP	TOMLIN,C (1994)
VP 1.04E-002	mm Hg	25	EXP	WEED SCIENCE SOCIETY OF AMERICA (1989)
DC	pKa			
HL 3.09E-005	atm m3/mol	25	EST	VP/WSOL
OH 2.92E-011	cm3/molc sec	25	EST	MEYLAN,WM & HOWARD,PH (1993)

CAS #: 001929-82-4 — NITRAPYRIN

Formula: $C_6H_3Cl_4N$
Mol Weight: 230.91
MP (deg C): 62.5-62.9
FP (deg C):
BP (deg C): 136-137.5
BP pressure (mm Hg): 1.10E+001

Property/Value	Units	Temp	Data Type	Reference
WS 5.40E+001	mg/L	25	EXP	SHIU,WY ET AL. (1990)
logP 3.41			EXP	HANSCH,C & LEO,AJ (1985)
VP 2.80E-003	mm Hg	25	EXP	WAUCHOPE,RD ET AL. (1991A)
DC	pKa			
HL 1.58E-005	atm m3/mol	25	EST	VP/WSOL
OH 3.50E-014	cm3/molc sec	25	EST	MEYLAN,WM & HOWARD,PH (1993)

CAS #: 001929-86-8				MECOPROP, POTASSIUM SALT
Formula: $C_{10}H_{10}ClKO_3$				
Mol Weight: 252.74				
MP (deg C):		FP (deg C):		
BP (deg C):				
BP pressure (mm Hg):				

Property/ Value	Units	Temp	Data Type	Reference
WS 7.95E+005	mg/L	25	EXP	KENAGA,EE (1980)
logP -0.87			EST	MEYLAN,WM & HOWARD,PH (1995)
VP 5.21E-011	mm Hg	25	EST	NEELY,WB & BLAU,GE (1985)
DC	pKa			
HL	atm m3/mol			
OH 1.17E-011	cm3/molc sec	25	EST	MEYLAN,WM & HOWARD,PH (1993)

CAS #: 001934-92-5				N-METHYLBENZANILIDE
Formula: $C_{14}H_{13}NO$				
Mol Weight: 211.27				
MP (deg C):		FP (deg C):		
BP (deg C):				
BP pressure (mm Hg):				

Property/ Value	Units	Temp	Data Type	Reference
WS 3.69E+002	mg/L	25	EST	MEYLAN,WM ET AL. (1996)
logP 2.36			EXP	HANSCH,C & LEO,AJ (1985)
VP 2.10E-005	mm Hg	25	EST	NEELY,WB & BLAU,GE (1985)
DC	pKa			
HL 2.50E-008	atm m3/mol	25	EST	MEYLAN,WM & HOWARD,PH (1991)
OH 1.48E-011	cm3/molc sec	25	EST	MEYLAN,WM & HOWARD,PH (1993)

CAS #: 001936-15-8				C.I. 16230
Formula: $C_{16}H_{10}N_2Na_2O_7S_2$				
Mol Weight: 452.37				
MP (deg C):		FP (deg C):		
BP (deg C):				
BP pressure (mm Hg):				

Property/ Value	Units	Temp	Data Type	Reference
WS 8.00E+004	mg/L	25	EXP	GREEN,FJ (1990)
logP -4.56			EST	MEYLAN,WM & HOWARD,PH (1995)
VP 2.49E-020	mm Hg	25	EST	NEELY,B & BLAU,GE (1985)
DC 11.50	pKa	25	EXP	HAAG,WR & MILL,T (1987)
HL	atm m3/mol			
OH 1.49E-012	cm3/molc sec	25	EST	MEYLAN,WM & HOWARD,PH (1993)

CAS #: 001937-37-7				C.I. DIRECT BLACK 38
Formula: $C_{36}H_{27}N_7Na_2O_7S_2$				
Mol Weight: 779.77				
MP (deg C):		FP (deg C):		
BP (deg C):				
BP pressure (mm Hg):				

Property/ Value	Units	Temp	Data Type	Reference
WS 3.00E+003	mg/L	25	EST	MEYLAN,WM ET AL. (1996)
logP 2.04			EST	MEYLAN,WM & HOWARD,PH (1995)
VP 1.53E-036	mm Hg	25	EST	NEELY,WB & BLAU,GE (1985)
DC	pKa			
HL 8.23E-040	atm m3/mol	25	EST	MEYLAN,WM & HOWARD,PH (1991)
OH 2.63E-010	cm3/molc sec	25	EST	MEYLAN,WM & HOWARD,PH (1993)

CAS #: 001940-18-7				1-ETHYLCYCLOHEXANOL
Formula: $C_8H_{16}O$				
Mol Weight: 128.22				
MP (deg C): 34.5		FP (deg C):		
BP (deg C): 166				
BP pressure (mm Hg):				

Property/ Value	Units	Temp	Data Type	Reference
WS 1.89E+003	mg/L	25	EST	MEYLAN,WM ET AL. (1996)
logP 2.58			EST	MEYLAN,WM & HOWARD,PH (1995)
VP 2.03E-001	mm Hg	25	EST	NEELY,WB & BLAU,GE (1985)
DC	pKa			
HL 8.63E-006	atm m3/mol	25	EST	MEYLAN,WM & HOWARD,PH (1991)
OH 1.68E-011	cm3/molc sec	25	EST	MEYLAN,WM & HOWARD,PH (1993)

CAS #: 001942-45-6				4-OCTYNE
Formula: C_8H_{14}				
Mol Weight: 110.20				
MP (deg C): -101		FP (deg C):		
BP (deg C): 131.6				
BP pressure (mm Hg):				

Property/ Value	Units	Temp	Data Type	Reference
WS 2.91E+001	mg/L	25	EST	MEYLAN,WM ET AL. (1996)
logP 3.55			EST	MEYLAN,WM & HOWARD,PH (1995)
VP 9.67E+000	mm Hg	25	EXT	BOUBLIK,T ET AL. (1984)
DC	pKa			
HL 3.34E-002	atm m3/mol	25	EST	MEYLAN,WM & HOWARD,PH (1991)
OH 3.19E-011	cm3/molc sec	25	EST	MEYLAN,WM & HOWARD,PH (1993)

CAS #: 001942-48-9				5,7-UNDECANEDIONE
Formula: $C_{11}H_{20}O_2$				
Mol Weight: 184.28				
MP (deg C):		FP (deg C):		
BP (deg C):				
BP pressure (mm Hg):				

Property/ Value	Units	Temp	Data Type	Reference
WS 4.78E+001	mg/L	25	EST	MEYLAN,WM ET AL. (1996)
logP 3.56			EXP	SANGSTER,J (1994)
VP 2.07E-002	mm Hg	25	EST	NEELY,WB & BLAU,GE (1985)
DC	pKa			
HL 1.83E-007	atm m3/mol	25	EST	MEYLAN,WM & HOWARD,PH (1991)
OH 1.38E-011	cm3/molc sec	25	EST	MEYLAN,WM & HOWARD,PH (1993)

CAS #: 001943-79-9				N-METHYLPHENYLCARBAMATE
Formula: $C_8H_9NO_2$				
Mol Weight: 151.17				
MP (deg C):		FP (deg C):		
BP (deg C):				
BP pressure (mm Hg):				

Property/ Value	Units	Temp	Data Type	Reference
WS 6.55E+003	mg/L	25	EST	MEYLAN,WM ET AL. (1996)
logP 1.24			EXP	HANSCH,C & LEO,AJ (1985)
VP 7.07E-002	mm Hg	25	EST	NEELY,WB & BLAU,GE (1985)
DC	pKa			
HL 3.21E-008	atm m3/mol	25	EST	MEYLAN,WM & HOWARD,PH (1991)
OH 8.07E-012	cm3/molc sec	25	EST	MEYLAN,WM & HOWARD,PH (1993)

CAS #: 001943-87-9 — CARBAMIC ACID, (4-NITROPHENYL)-, METHYL ESTER

Formula: $C_8H_8N_2O_4$

Mol Weight: 196.16

MP (deg C): FP (deg C):

BP (deg C):

BP pressure (mm Hg):

Property/Value	Units	Temp	Data Type	Reference
WS 7.08E+002	mg/L	25	EST	MEYLAN,WM ET AL. (1996)
logP 2.12			EXP	TAKAHASHI,J ET AL. (1988)
VP 2.34E-004	mm Hg	25	EST	NEELY,WB & BLAU,GE (1985)
DC	pKa			
HL 8.62E-011	atm m3/mol	25	EST	MEYLAN,WM & HOWARD,PH (1991)
OH 5.55E-012	cm3/molc sec	25	EST	MEYLAN,WM & HOWARD,PH (1993)

CAS #: 001948-71-6 — TYROSINE-AMIDE,N-ACETYL

Formula: $C_{11}H_{14}N_2O_3$

Mol Weight: 222.25

MP (deg C): 223-225 FP (deg C):

BP (deg C):

BP pressure (mm Hg):

Property/Value	Units	Temp	Data Type	Reference
WS 3.45E+004	mg/L	25	EST	MEYLAN,WM ET AL. (1996)
logP -0.87			EXP	HANSCH,C & LEO,AJ (1985)
VP 5.04E-011	mm Hg	25	EST	NEELY,WB & BLAU,GE (1985)
DC				
HL 1.32E-015	atm m3/mol	25	EST	MEYLAN,WM & HOWARD,PH (1991)
OH 8.76E-011	cm3/molc sec	25	EST	MEYLAN,WM & HOWARD,PH (1993)

CAS #: 001948-92-1 — BENZENAMINE, 4-[(4-NITROPHENYL)SULFONYL]-

Formula: $C_{12}H_{10}N_2O_4S$

Mol Weight: 278.29

MP (deg C): 169-171 FP (deg C):

BP (deg C):

BP pressure (mm Hg):

Property/Value	Units	Temp	Data Type	Reference
WS 1.21E+002	mg/L	25	EST	MEYLAN,WM ET AL. (1996)
logP 2.04			EXP	ALTOMARE,C ET AL. (1991)
VP 8.67E-009	mm Hg	25	EST	NEELY,WB & BLAU,GE (1985)
DC	pKa			
HL 3.48E-013	atm m3/mol	25	EST	MEYLAN,WM & HOWARD,PH (1991)
OH 2.31E-011	cm3/molc sec	25	EST	MEYLAN,WM & HOWARD,PH (1993)

CAS #: 001951-56-0 — N-IPR-3-(5-NO2-2-FURYL)ACRYLAMIDE

Formula: $C_{10}H_{13}N_2O_4$

Mol Weight: 225.23

MP (deg C): FP (deg C):

BP (deg C):

BP pressure (mm Hg):

Property/Value	Units	Temp	Data Type	Reference
WS 6.08E+002	mg/L	25	EST	MEYLAN,WM ET AL. (1996)
logP 1.57			EXP	SANGSTER,J (1993)
VP 9.14E-007	mm Hg	25	EST	NEELY,WB & BLAU,GE (1985)
DC	pKa			
HL 4.02E-012	atm m3/mol	25	EST	MEYLAN,WM & HOWARD,PH (1991)
OH 3.83E-011	cm3/molc sec	25	EST	MEYLAN,WM & HOWARD,PH (1993)

CAS #: 001953-33-9 — BARBITURIC ACID, 5-BUTYL

Formula: $C_8H_{12}N_2O_3$

Mol Weight: 184.20

MP (deg C): FP (deg C):

BP (deg C):

BP pressure (mm Hg):

Property/Value	Units	Temp	Data Type	Reference
WS 1.16E+004	mg/L	25	EST	MEYLAN,WM ET AL. (1996)
logP 0.77			EXP	HANSCH,C & LEO,AJ (1985)
VP 1.03E-010	mm Hg	25	EST	NEELY,WB & BLAU,GE (1985)
DC	pKa			
HL 3.61E-013	atm m3/mol	25	EST	MEYLAN,WM & HOWARD,PH (1991)
OH 1.06E-011	cm3/molc sec	25	EST	MEYLAN,WM & HOWARD,PH (1993)

CAS #: 001953-54-4 — 1H-INDOL-5-OL

Formula: C_8H_7NO

Mol Weight: 133.15

MP (deg C): 107-108 FP (deg C):

BP (deg C):

BP pressure (mm Hg):

Property/Value	Units	Temp	Data Type	Reference
WS 3.83E+004	mg/L	25	EST	MEYLAN,WM ET AL. (1996)
logP 1.11			EXP	SANGSTER,J (1993)
VP 1.97E-004	mm Hg	25	EST	NEELY,WB & BLAU,GE (1985)
DC	pKa			
HL 9.22E-011	atm m3/mol	25	EST	MEYLAN,WM & HOWARD,PH (1991)
OH 2.00E-010	cm3/molc sec	25	EST	MEYLAN,WM & HOWARD,PH (1993)

CAS #: 001959-97-3 — 1,4-BENZODIAZPN2-ON,1-ME5(4-CLPH)7F

Formula: $C_{16}H_{12}ClFN_2O$

Mol Weight: 302.74

MP (deg C): FP (deg C):

BP (deg C):

BP pressure (mm Hg):

Property/Value	Units	Temp	Data Type	Reference
WS 2.67E+001	mg/L	25	EST	MEYLAN,WM ET AL. (1996)
logP 3.10			EXP	HANSCH,C & LEO,AJ (1985)
VP 3.35E-008	mm Hg	25	EST	NEELY,WB & BLAU,GE (1985)
DC	pKa			
HL 4.25E-009	atm m3/mol	25	EST	MEYLAN,WM & HOWARD,PH (1991)
OH 7.93E-012	cm3/molc sec	25	EST	MEYLAN,WM & HOWARD,PH (1993)

CAS #: 001962-75-0 — P-DIBUTYLPHTHALATE

Formula: $C_{16}H_{22}O_4$

Mol Weight: 278.35

MP (deg C): FP (deg C):

BP (deg C):

BP pressure (mm Hg):

Property/Value	Units	Temp	Data Type	Reference
WS 3.10E-001	mg/L	25	EST	MEYLAN,WM ET AL. (1996)
logP 5.53			EXP	HANSCH,C ET AL. (1995)
VP 2.55E-004	mm Hg	25	EST	NEELY,WB & BLAU,GE (1985)
DC	pKa			
HL 1.22E-006	atm m3/mol	25	EST	MEYLAN,WM & HOWARD,PH (1991)
OH 9.28E-012	cm3/molc sec	25	EST	MEYLAN,WM & HOWARD,PH (1993)

CAS #: 001966-58-1 — 3,4-DICHLOROBENZYL METHYLCARBAMATE

Formula: $C_9H_9Cl_2NO_2$
Mol Weight: 234.08
MP (deg C): | FP (deg C):
BP (deg C):
BP pressure (mm Hg):

Property/Value	Units	Temp	Data Type	Reference
WS 1.70E+002	mg/L	25	EXP	WEBER,JB (1972)
logP 2.95			EST	MEYLAN,WM & HOWARD,PH (1995)
VP 5.00E-004	mm Hg	25	EST	NEELY,WB & BLAU,GE (1985)
DC	pKa			
HL 3.85E-009	atm m3/mol	25	EST	MEYLAN,WM & HOWARD,PH (1991)
OH 8.55E-012	cm3/molc sec	25	EST	MEYLAN,WM & HOWARD,PH (1993)

CAS #: 001967-16-4 — CHLORBUFAM

Formula: $C_{11}H_{10}ClNO_2$
Mol Weight: 223.66
MP (deg C): 45.5 | FP (deg C):
BP (deg C):
BP pressure (mm Hg):

Property/Value	Units	Temp	Data Type	Reference
WS 5.40E+002	mg/L	20	EXP	YALKOWSKY,SH & DANNENFELSER,RM (1992)
logP 3.02			EST	MEYLAN,WM & HOWARD,PH (1995)
VP 1.58E-005	mm Hg	20	EXP	TOMLIN,C (1994)
DC	pKa			
HL 8.61E-009	atm m3/mol	20	EST	VP/WSOL
OH 4.03E-011	cm3/molc sec	25	EST	MEYLAN,WM & HOWARD,PH (1993)

CAS #: 001967-25-5 — P-BROMOPHENYLUREA

Formula: $C_7H_7BrN_2O$
Mol Weight: 215.06
MP (deg C): 226 | FP (deg C):
BP (deg C): 260
BP pressure (mm Hg):

Property/Value	Units	Temp	Data Type	Reference
WS 7.45E+002	mg/L	25	EST	MEYLAN,WM ET AL. (1996)
logP 1.98			EXP	HANSCH,C & LEO,AJ (1985)
VP 5.64E-005	mm Hg	25	EST	NEELY,WB & BLAU,GE (1985)
DC	pKa			
HL 8.02E-011	atm m3/mol	25	EST	MEYLAN,WM & HOWARD,PH (1991)
OH 1.32E-011	cm3/molc sec	25	EST	MEYLAN,WM & HOWARD,PH (1993)

CAS #: 001967-27-7 — M-CHLOROPHENYLUREA

Formula: $C_7H_7ClN_2O$
Mol Weight: 170.60
MP (deg C): | FP (deg C):
BP (deg C):
BP pressure (mm Hg):

Property/Value	Units	Temp	Data Type	Reference
WS 1.71E+003	mg/L	25	EST	MEYLAN,WM ET AL. (1996)
logP 1.82			EXP	HANSCH,C & LEO,AJ (1985)
VP 1.65E-004	mm Hg	25	EST	NEELY,WB & BLAU,GE (1985)
DC	pKa			
HL 1.49E-010	atm m3/mol	25	EST	MEYLAN,WM & HOWARD,PH (1991)
OH 3.10E-011	cm3/molc sec	25	EST	MEYLAN,WM & HOWARD,PH (1993)

CAS #: 001970-40-7 — PYRICLOR

Formula: $C_5H_2Cl_3NO$
Mol Weight: 198.44
MP (deg C): | FP (deg C):
BP (deg C):
BP pressure (mm Hg):

Property/Value	Units	Temp	Data Type	Reference
WS 5.26E+002	mg/L	25	EST	MEYLAN,WM ET AL. (1996)
logP 2.26			EST	MEYLAN,WM & HOWARD,PH (1995)
VP 1.03E-003	mm Hg	25	EST	NEELY,WB & BLAU,GE (1985)
DC	pKa			
HL 1.88E-008	atm m3/mol	25	EST	MEYLAN,WM & HOWARD,PH (1991)
OH 1.77E-013	cm3/molc sec	25	EST	MEYLAN,WM & HOWARD,PH (1993)

CAS #: 001972-71-0 — DISALICYLIMIDE

Formula: $C_{14}H_{11}NO_4$
Mol Weight: 257.25
MP (deg C): | FP (deg C):
BP (deg C):
BP pressure (mm Hg):

Property/Value	Units	Temp	Data Type	Reference
WS 1.88E+003	mg/L	25	EST	MEYLAN,WM ET AL. (1996)
logP 1.92			EXP	HANSCH,C & LEO,AJ (1985)
VP 6.24E-014	mm Hg	25	EST	NEELY,WB & BLAU,GE (1985)
DC	pKa			
HL 3.86E-011	atm m3/mol	25	EST	MEYLAN,WM & HOWARD,PH (1991)
OH 6.65E-011	cm3/molc sec	25	EST	MEYLAN,WM & HOWARD,PH (1993)

CAS #: 001974-04-5 — 2-BROMOHEPTANE

Formula: $C_7H_{15}Br$
Mol Weight: 179.11
MP (deg C): 47 | FP (deg C):
BP (deg C): 166
BP pressure (mm Hg):

Property/Value	Units	Temp	Data Type	Reference
WS 1.94E+001	mg/L	25	EST	MEYLAN,WM ET AL. (1996)
logP 4.05			EST	MEYLAN,WM & HOWARD,PH (1995)
VP 2.28E+000	mm Hg	25	EXP	DYKYJ,J & REAPAS,M (1973)
DC	pKa			
HL 4.65E-002	atm m3/mol	25	EST	MEYLAN,WM & HOWARD,PH (1991)
OH 5.40E-012	cm3/molc sec	25	EST	MEYLAN,WM & HOWARD,PH (1993)

CAS #: 001977-00-0 — BENZOIC ACID, 2-[[3-(TRIFLUOROMETHYL)PHENYL]AMIN

Formula: $C_{14}H_9F_3NNaO_2$
Mol Weight: 303.22
MP (deg C): | FP (deg C):
BP (deg C):
BP pressure (mm Hg):

Property/Value	Units	Temp	Data Type	Reference
WS 9.30E+002	mg/L	25	EST	MEYLAN,WM ET AL. (1996)
logP 1.29			EXP	SANGSTER,J (1993)
VP 1.34E-012	mm Hg	25	EST	NEELY,WB & BLAU,GE (1985)
DC	pKa			
HL	atm m3/mol			
OH 1.56E-010	cm3/molc sec	25	EST	MEYLAN,WM & HOWARD,PH (1993)

CAS #: 001978-59-2		PYRIDINE, 4-(4-FLUOROPHENYL)-1,2,3,6-TETRAHYDRO-

Formula: $C_{11}H_{12}FN$

Mol Weight: 177.22

MP (deg C): FP (deg C):

BP (deg C):

BP pressure (mm Hg):

Property/Value	Units	Temp	Data Type	Reference
WS 5.48E+003	mg/L	25	EST	MEYLAN,WM ET AL. (1996)
logP 2.37			EXP	ALTOMARE,CA ET AL. (1992)
VP 4.53E-003	mm Hg	25	EST	NEELY,WB & BLAU,GE (1985)
DC	pKa			
HL 7.89E-007	atm m3/mol	25	EST	MEYLAN,WM & HOWARD,PH (1991)
OH 1.73E-010	cm3/molc sec	25	EST	MEYLAN,WM & HOWARD,PH (1993)

CAS #: 001982-37-2		METHDILAZINE

Formula: $C_{18}H_{20}N_2S$

Mol Weight: 296.44

MP (deg C): 87-88 FP (deg C):

BP (deg C):

BP pressure (mm Hg):

Property/Value	Units	Temp	Data Type	Reference
WS 3.48E-001	mg/L	25	EST	MEYLAN,WM ET AL. (1996)
logP 5.23			EXP	HANSCH,C & LEO,AJ (1985)
VP 1.73E-007	mm Hg	25	EST	NEELY,WB & BLAU,GE (1985)
DC	pKa			
HL 2.91E-010	atm m3/mol	25	EST	MEYLAN,WM & HOWARD,PH (1991)
OH 2.79E-010	cm3/molc sec	25	EST	MEYLAN,WM & HOWARD,PH (1993)

CAS #: 001982-47-4		CHLOROXURON

Formula: $C_{15}H_{15}ClN_2O_2$

Mol Weight: 290.75

MP (deg C): 151 FP (deg C):

BP (deg C):

BP pressure (mm Hg):

Property/Value	Units	Temp	Data Type	Reference
WS 3.70E+000	mg/L	20	EXP	YALKOWSKY,SH & DANNENFELSER,RM (1992)
logP 4.08			EST	MEYLAN,WM & HOWARD,PH (1995)
VP 3.90E-009	mm Hg	25	EXP	WAUCHOPE,RD ET AL. (1991A)
DC	pKa			
HL 4.03E-010	atm m3/mol	25		VP/WSOL
OH 4.62E-011	cm3/molc sec	25	EST	MEYLAN,WM & HOWARD,PH (1993)

CAS #: 001982-49-6		SIDURON

Formula: $C_{14}H_{20}N_2O$

Mol Weight: 232.33

MP (deg C): 133-138 FP (deg C):

BP (deg C):

BP pressure (mm Hg):

Property/Value	Units	Temp	Data Type	Reference
WS 1.80E+001	mg/L	25	EXP	YALKOWSKY,SH & DANNENFELSER,RM (1992)
logP 3.86			EST	MEYLAN,WM & HOWARD,PH (1995)
VP 4.00E-009	mm Hg	25	EXP	WAUCHOPE,RD ET AL. (1991A)
DC	pKa			
HL 6.79E-011	atm m3/mol	25	EST	VP/WSOL
OH 6.44E-011	cm3/molc sec	25	EST	MEYLAN,WM & HOWARD,PH (1993)

CAS #: 001985-12-2		P-BROMOPHENYL ISOTHIOCYANATE

Formula: C_7H_4BrNS

Mol Weight: 214.09

MP (deg C): 60-61 FP (deg C):

BP (deg C):

BP pressure (mm Hg):

Property/Value	Units	Temp	Data Type	Reference
WS 1.16E+005	mg/L	25	EXP	YALKOWSKY,SH & DANNENFELSER,RM (1992)
logP 4.03			EXP	HANSCH,C & LEO,AJ (1985)
VP 4.11E-003	mm Hg	25	EST	NEELY,WB & BLAU,GE (1985)
DC	pKa			
HL 3.11E-004	atm m3/mol	25	EST	MEYLAN,WM & HOWARD,PH (1991)
OH 1.35E-012	cm3/molc sec	25	EST	MEYLAN,WM & HOWARD,PH (1993)

CAS #: 001985-51-9		NEOPENTYLGLYCOL DIMETHACRYLATE

Formula: $C_{13}H_{20}O_4$

Mol Weight: 240.30

MP (deg C): FP (deg C):

BP (deg C): 87

BP pressure (mm Hg): 6.00E-001

Property/Value	Units	Temp	Data Type	Reference
WS 1.11E+002	mg/L	25	EST	MEYLAN,WM ET AL. (1996)
logP 2.79			EXP	SANGSTER,J (1993)
VP 2.76E-002	mm Hg	25	EST	NEELY,WB & BLAU,GE (1985)
DC	pKa			
HL 8.85E-007	atm m3/mol	25	EST	MEYLAN,WM & HOWARD,PH (1991)
OH 4.03E-011	cm3/molc sec	25	EST	MEYLAN,WM & HOWARD,PH (1993)

CAS #: 001989-33-9		9-CARBOXYFLUORENE

Formula: $C_{14}H_{10}O_2$

Mol Weight: 210.23

MP (deg C): 228-231 FP (deg C):

BP (deg C):

BP pressure (mm Hg):

Property/Value	Units	Temp	Data Type	Reference
WS 3.03E+001	mg/L	25	EST	MEYLAN,WM ET AL. (1996)
logP 2.80			EXP	HANSCH,C & LEO,AJ (1985)
VP 1.89E-006	mm Hg	25	EST	NEELY,WB & BLAU,GE (1985)
DC	pKa			
HL 1.24E-009	atm m3/mol	25	EST	MEYLAN,WM & HOWARD,PH (1991)
OH 9.84E-012	cm3/molc sec	25	EST	MEYLAN,WM & HOWARD,PH (1993)

CAS #: 001991-78-2		P-CHLOROPHENYLALANINE

Formula: $C_9H_{10}ClNO_2$

Mol Weight: 199.64

MP (deg C): FP (deg C):

BP (deg C):

BP pressure (mm Hg):

Property/Value	Units	Temp	Data Type	Reference
WS 9.61E+002	mg/L	25	EST	MEYLAN,WM ET AL. (1996)
logP -0.48			EXP	HANSCH,C & LEO,AJ (1985)
VP 6.09E-009	mm Hg	25	EST	NEELY,WB & BLAU,GE (1985)
DC 2.08	pKa	25	EXP	KORTUM,G ET AL (1961)
HL 8.93E-011	atm m3/mol	25	EST	MEYLAN,WM & HOWARD,PH (1991)
OH 4.10E-011	cm3/molc sec	25	EST	MEYLAN,WM & HOWARD,PH (1993)

002002-16-6 — 2-PHENYLGUANIDINE

CAS #: 002002-16-6	2-PHENYLGUANIDINE
Formula: $C_7H_9N_3$	
Mol Weight: 135.17	
MP (deg C):	FP (deg C):
BP (deg C):	
BP pressure (mm Hg):	

Property/Value	Units	Temp	Data Type	Reference
WS 3.15E+005	mg/L	25	EST	MEYLAN,WM ET AL. (1996)
logP 0.53			EXP	HANSCH,C & LEO,AJ (1985)
VP 1.06E-002	mm Hg	25	EST	NEELY,WB & BLAU,GE (1985)
DC	pKa			
HL 1.29E-011	atm m3/mol	25	EST	MEYLAN,WM & HOWARD,PH (1991)
OH 4.41E-011	cm3/molc sec	25	EST	MEYLAN,WM & HOWARD,PH (1993)

002002-29-1 — FLUMETHASONE-21-PIVALATE

CAS #: 002002-29-1	FLUMETHASONE-21-PIVALATE
Formula: $C_{27}H_{36}F_2O_6$	
Mol Weight: 494.58	
MP (deg C):	FP (deg C):
BP (deg C):	
BP pressure (mm Hg):	

Property/Value	Units	Temp	Data Type	Reference
WS 3.92E-001	mg/L	25	EST	MEYLAN,WM ET AL. (1996)
logP 3.86			EXP	HANSCH,C & LEO,AJ (1985)
VP 2.75E-014	mm Hg	25	EST	NEELY,WB & BLAU,GE (1985)
DC	pKa			
HL 3.60E-011	atm m3/mol	25	EST	MEYLAN,WM & HOWARD,PH (1991)
OH 5.88E-011	cm3/molc sec	25	EST	MEYLAN,WM & HOWARD,PH (1993)

002002-36-0 — 9-PENTYLADENINE

CAS #: 002002-36-0	9-PENTYLADENINE
Formula: $C_{10}H_{17}N_5$	
Mol Weight: 207.28	
MP (deg C):	FP (deg C):
BP (deg C):	
BP pressure (mm Hg):	

Property/Value	Units	Temp	Data Type	Reference
WS 1.22E+003	mg/L	25	EST	MEYLAN,WM ET AL. (1996)
logP 1.79			EXP	HANSCH,C & LEO,AJ (1985)
VP 1.99E-007	mm Hg	25	EST	NEELY,WB & BLAU,GE (1985)
DC	pKa			
HL 4.65E-012	atm m3/mol	25	EST	MEYLAN,WM & HOWARD,PH (1991)
OH 2.05E-010	cm3/molc sec	25	EST	MEYLAN,WM & HOWARD,PH (1993)

002004-70-8 — 1,3-PENTADIENE (E)

CAS #: 002004-70-8	1,3-PENTADIENE (E)
Formula: C_5H_8	
Mol Weight: 68.12	
MP (deg C): -87.4	FP (deg C):
BP (deg C): 42	
BP pressure (mm Hg):	

Property/Value	Units	Temp	Data Type	Reference
WS 3.26E+002	mg/L	25	EST	MEYLAN,WM ET AL. (1996)
logP 2.44			EXP	SANGSTER,J (1994)
VP 4.11E+002	mm Hg	25	EXP	DAUBERT,TE & DANNER,RP (1989)
DC	pKa			
HL 1.22E-001	atm m3/mol	25	EST	MEYLAN,WM & HOWARD,PH (1991)
OH 1.05E-010	cm3/molc sec	25	EST	MEYLAN,WM & HOWARD,PH (1993)

002008-39-1 — 2,4-D, DIMETHYLAMINE SALT

CAS #: 002008-39-1	2,4-D, DIMETHYLAMINE SALT
Formula: $C_{10}H_{13}Cl_2NO_3$	
Mol Weight: 266.13	
MP (deg C):	FP (deg C):
BP (deg C):	
BP pressure (mm Hg):	

Property/Value	Units	Temp	Data Type	Reference
WS 3.00E+006	mg/L	20	EXP	BOSCH,SJ (1983)
logP 0.65			EXP	MOODY,RP ET AL. (1987)
VP 1.00E-009	mm Hg	25	EXP	BOSCH,SJ (1983)
DC	pKa			
HL 1.17E-016	atm m3/mol	25	EST	VP/WSOL
OH 8.58E-012	cm3/molc sec	25	EST	MEYLAN,WM & HOWARD,PH (1993)

002008-41-5 — BUTYLATE

CAS #: 002008-41-5	BUTYLATE
Formula: $C_{11}H_{23}NOS$	
Mol Weight: 217.38	
MP (deg C):	FP (deg C):
BP (deg C): 138	
BP pressure (mm Hg): 2.10E+001	

Property/Value	Units	Temp	Data Type	Reference
WS 4.40E+001	mg/L	25	EST	WAUCHOPE,RD ET AL. (1991A)
logP 4.15			EXP	HANSCH,C ET AL. (1995)
VP 1.30E-003	mm Hg	25	EXP	WORTHING,CR & WALKER,SB (1987)
DC	pKa			
HL 8.45E-006	atm m3/mol	25	EST	VP/WSOL
OH 3.00E-011	cm3/molc sec	25	EST	MEYLAN,WM & HOWARD,PH (1993)

002008-58-4 — 2,6-DICHLOROBENZAMIDE

CAS #: 002008-58-4	2,6-DICHLOROBENZAMIDE
Formula: $C_7H_5Cl_2NO$	
Mol Weight: 190.03	
MP (deg C): 196-199	FP (deg C):
BP (deg C):	
BP pressure (mm Hg):	

Property/Value	Units	Temp	Data Type	Reference
WS 2.70E+003	mg/L	25	EXP	GEYER,H ET AL. (1981)
logP 0.77			EXP	NAKAGAWA,Y ET AL. (1992)
VP 3.26E-005	mm Hg	25	EST	NEELY,WB & BLAU,GE (1985)
DC	pKa			
HL 1.22E-009	atm m3/mol	25	EST	MEYLAN,WM & HOWARD,PH (1991)
OH 2.88E-012	cm3/molc sec	25	EST	MEYLAN,WM & HOWARD,PH (1993)

002008-73-3 — 1-(3,4-DICLPHENYL)-3-PHENYLUREA

CAS #: 002008-73-3	1-(3,4-DICLPHENYL)-3-PHENYLUREA
Formula: $C_{13}H_{10}Cl_2N_2O$	
Mol Weight: 281.14	
MP (deg C):	FP (deg C):
BP (deg C):	
BP pressure (mm Hg):	

Property/Value	Units	Temp	Data Type	Reference
WS 1.53E+000	mg/L	25	EST	MEYLAN,WM ET AL. (1996)
logP 4.70			EXP	HANSCH,C & LEO,AJ (1985)
VP 1.19E-007	mm Hg	25	EST	NEELY,WB & BLAU,GE (1985)
DC	pKa			
HL 6.10E-011	atm m3/mol	25	EST	MEYLAN,WM & HOWARD,PH (1991)
OH 5.14E-011	cm3/molc sec	25	EST	MEYLAN,WM & HOWARD,PH (1993)

002011-57-6 — 2,3-DIME-4-HYDROXYACETANILIDE

CAS #:	002011-57-6		
Formula:	$C_{10}H_{13}NO_2$		
Mol Weight:	179.22		
MP (deg C):		FP (deg C):	
BP (deg C):			
BP pressure (mm Hg):			

Property/Value	Units	Temp	Data Type	Reference
WS 1.81E+004	mg/L	25	EST	MEYLAN,WM ET AL. (1996)
logP 0.57			EXP	HANSCH,C & LEO,AJ (1985)
VP 1.05E-006	mm Hg	25	EST	NEELY,WB & BLAU,GE (1985)
DC	pKa			
HL 7.82E-013	atm m3/mol	25	EST	MEYLAN,WM & HOWARD,PH (1991)
OH 5.63E-011	cm3/molc sec	25	EST	MEYLAN,WM & HOWARD,PH (1993)

002011-67-8 — NIMETAZEPAM

CAS #:	002011-67-8		
Formula:	$C_{16}H_{13}N_3O_3$		
Mol Weight:	295.30		
MP (deg C):	156.5-7.5	FP (deg C):	
BP (deg C):			
BP pressure (mm Hg):			

Property/Value	Units	Temp	Data Type	Reference
WS 7.62E+001	mg/L	25	EST	MEYLAN,WM ET AL. (1996)
logP 2.16			EXP	HANSCH,C & LEO,AJ (1985)
VP 2.11E-009	mm Hg	25	EST	NEELY,WB & BLAU,GE (1985)
DC	pKa			
HL 1.94E-011	atm m3/mol	25	EST	MEYLAN,WM & HOWARD,PH (1991)
OH 7.42E-012	cm3/molc sec	25	EST	MEYLAN,WM & HOWARD,PH (1993)

002012-81-9 — BENZENEACETONITRILE, 4-CHLORO-.ALPHA.-(1-METHYLE

CAS #:	002012-81-9		
Formula:	$C_{11}H_{12}ClN$		
Mol Weight:	193.68		
MP (deg C):		FP (deg C):	
BP (deg C):			
BP pressure (mm Hg):			

Property/Value	Units	Temp	Data Type	Reference
WS 3.13E+001	mg/L	25	EST	MEYLAN,WM ET AL. (1996)
logP 3.41			EXP	WANG,W ET AL. (1987)
VP 2.25E-003	mm Hg	25	EST	NEELY,WB & BLAU,GE (1985)
DC	pKa			
HL 4.28E-006	atm m3/mol	25	EST	MEYLAN,WM & HOWARD,PH (1991)
OH 4.82E-012	cm3/molc sec	25	EST	MEYLAN,WM & HOWARD,PH (1993)

002016-42-4 — 1-TETRADECANAMINE

CAS #:	002016-42-4		
Formula:	$C_{14}H_{31}N$		
Mol Weight:	213.41		
MP (deg C):	83.1	FP (deg C):	
BP (deg C):	291.2		
BP pressure (mm Hg):			

Property/Value	Units	Temp	Data Type	Reference
WS 4.71E+000	mg/L	25	EST	MEYLAN,WM ET AL. (1996)
logP 5.75			EST	MEYLAN,WM & HOWARD,PH (1995)
VP 9.74E-004	mm Hg	25	EXP	YAWS,CL (1994B)
DC 10.62	pKa	25	EXP	PERRIN,DD (1965)
HL 3.01E-004	atm m3/mol	25	EST	MEYLAN,WM & HOWARD,PH (1991)
OH 4.85E-011	cm3/molc sec	25	EST	MEYLAN,WM & HOWARD,PH (1993)

002016-57-1 — N-DECYLAMINE

CAS #:	002016-57-1		
Formula:	$C_{10}H_{23}N$		
Mol Weight:	157.30		
MP (deg C):	17	FP (deg C):	
BP (deg C):	220.5		
BP pressure (mm Hg):			

Property/Value	Units	Temp	Data Type	Reference
WS 5.50E+002	mg/L	25	EXP	CHRISTIE,AO & CRISP,DJ (1967)
logP 3.78			EST	MEYLAN,WM & HOWARD,PH (1995)
VP 1.00E-001	mm Hg	25	EST	DREISBACH,RR (1961)
DC 10.64	pKa	25	EXP	PERRIN,DD (1965)
HL 9.70E-005	atm m3/mol	25	EST	MEYLAN,WM & HOWARD,PH (1991)
OH 4.29E-011	cm3/molc sec	25	EST	MEYLAN,WM & HOWARD,PH (1993)

002018-61-3 — L-PHENYLALANINE, N-ACETYL-

CAS #:	002018-61-3		
Formula:	$C_{11}H_{13}NO_3$		
Mol Weight:	207.23		
MP (deg C):	171-173	FP (deg C):	
BP (deg C):			
BP pressure (mm Hg):			

Property/Value	Units	Temp	Data Type	Reference
WS 6.45E+003	mg/L	25	EST	MEYLAN,WM ET AL. (1996)
logP 0.93			EXP	GREEN,PG ET AL. (1991)
VP 2.43E-007	mm Hg	25	EST	NEELY,WB & BLAU,GE (1985)
DC	pKa			
HL 3.91E-013	atm m3/mol	25	EST	MEYLAN,WM & HOWARD,PH (1991)
OH 1.95E-011	cm3/molc sec	25	EST	MEYLAN,WM & HOWARD,PH (1993)

002021-26-3 — 1H-ISOINDOLE-1,3(2H)-DIONE, 5-PHENYL-

CAS #:	002021-26-3		
Formula:	$C_{14}H_9NO_2$		
Mol Weight:	223.23		
MP (deg C):		FP (deg C):	
BP (deg C):			
BP pressure (mm Hg):			

Property/Value	Units	Temp	Data Type	Reference
WS 9.62E+001	mg/L	25	EST	MEYLAN,WM ET AL. (1996)
logP 2.97			EXP	SANGSTER,J (1993)
VP 5.50E-011	mm Hg	25	EST	NEELY,WB & BLAU,GE (1985)
DC	pKa			
HL 7.85E-010	atm m3/mol	25	EST	MEYLAN,WM & HOWARD,PH (1991)
OH 1.02E-011	cm3/molc sec	25	EST	MEYLAN,WM & HOWARD,PH (1993)

002021-28-5 — ETHYL B-PHENYLPROPIONATE

CAS #:	002021-28-5		
Formula:	$C_{11}H_{14}O_2$		
Mol Weight:	178.23		
MP (deg C):		FP (deg C):	
BP (deg C):	247.2		
BP pressure (mm Hg):			

Property/Value	Units	Temp	Data Type	Reference
WS 2.62E+002	mg/L	25	EST	MEYLAN,WM ET AL. (1996)
logP 2.73			EXP	HANSCH,C & LEO,AJ (1985)
VP 2.43E-002	mm Hg	25	EST	NEELY,WB & BLAU,GE (1985)
DC	pKa			
HL 2.49E-005	atm m3/mol	25	EST	MEYLAN,WM & HOWARD,PH (1991)
OH 8.51E-012	cm3/molc sec	25	EST	MEYLAN,WM & HOWARD,PH (1993)

CAS #: 002024-34-2	1,4-BENZODIAZEPIN-2-ONE,7-FLUORO-5-(2-FLUOROPHEN

Formula: $C_{16}H_{12}F_2N_2O$

Mol Weight: 286.28

MP (deg C):   FP (deg C):

BP (deg C):

BP pressure (mm Hg):

Property/Value	Units	Temp	Data Type	Reference
WS 1.29E+002	mg/L	25	EST	MEYLAN,WM ET AL. (1996)
logP 2.41			EXP	HANSCH,C ET AL. (1995)
VP 1.47E-007	mm Hg	25	EST	NEELY,WB & BLAU,GE (1985)
DC	pKa			
HL 6.70E-009	atm m3/mol	25	EST	MEYLAN,WM & HOWARD,PH (1991)
OH 9.22E-012	cm3/molc sec	25	EST	MEYLAN,WM & HOWARD,PH (1993)

CAS #: 002025-40-3	2-CYANO-3-PHENYL ET-2-PROPENOATE

Formula: $C_{12}H_{11}NO_2$

Mol Weight: 201.23

MP (deg C): 51   FP (deg C):

BP (deg C): 188

BP pressure (mm Hg): 1.50E+001

Property/Value	Units	Temp	Data Type	Reference
WS 1.97E+002	mg/L	25	EST	MEYLAN,WM ET AL. (1996)
logP 2.43			EXP	HANSCH,C & LEO,AJ (1985)
VP 1.53E-004	mm Hg	25	EST	NEELY,WB & BLAU,GE (1985)
DC	pKa			
HL 1.23E-008	atm m3/mol	25	EST	MEYLAN,WM & HOWARD,PH (1991)
OH 8.36E-012	cm3/molc sec	25	EST	MEYLAN,WM & HOWARD,PH (1993)

CAS #: 002027-17-0	2-ISOPROPYLNAPHTHALENE

Formula: $C_{13}H_{14}$

Mol Weight: 170.26

MP (deg C): 225   FP (deg C):

BP (deg C): 268.2

BP pressure (mm Hg):

Property/Value	Units	Temp	Data Type	Reference
WS 6.89E+000	mg/L	25	EST	MEYLAN,WM ET AL. (1996)
logP 4.63			EST	MEYLAN,WM & HOWARD,PH (1995)
VP 5.18E-003	mm Hg	25	EXP	ZWOLINSKI,BJ & WILHOIT,RC (1971)
DC	pKa			
HL 8.48E-004	atm m3/mol	25	EST	MEYLAN,WM & HOWARD,PH (1991)
OH 5.38E-011	cm3/molc sec	25	EST	MEYLAN,WM & HOWARD,PH (1993)

CAS #: 002027-47-6	9-OCTADECENOIC ACID

Formula: $C_{18}H_{34}O_2$

Mol Weight: 282.47

MP (deg C):   FP (deg C):

BP (deg C):

BP pressure (mm Hg):

Property/Value	Units	Temp	Data Type	Reference
WS 1.15E-002	mg/L	25	EST	MEYLAN,WM ET AL. (1996)
logP 7.64			EXP	SANGSTER,J (1993)
VP 1.54E-006	mm Hg	25	EST	NEELY,WB & BLAU,GE (1985)
DC	pKa			
HL 4.48E-005	atm m3/mol	25	EST	MEYLAN,WM & HOWARD,PH (1991)
OH 7.55E-011	cm3/molc sec	25	EST	MEYLAN,WM & HOWARD,PH (1993)

CAS #: 002029-64-3	O-OCTYL CARBAMATE

Formula: $C_9H_{19}NO_2$

Mol Weight: 173.26

MP (deg C):   FP (deg C):

BP (deg C):

BP pressure (mm Hg):

Property/Value	Units	Temp	Data Type	Reference
WS 2.23E+002	mg/L	25	EST	MEYLAN,WM ET AL. (1996)
logP 2.84			EXP	HANSCH,C & LEO,AJ (1985)
VP 3.30E-002	mm Hg	25	EST	NEELY,WB & BLAU,GE (1985)
DC	pKa			
HL 2.88E-007	atm m3/mol	25	EST	MEYLAN,WM & HOWARD,PH (1991)
OH 1.22E-011	cm3/molc sec	25	EST	MEYLAN,WM & HOWARD,PH (1993)

CAS #: 002030-63-9	CLOFAZIMINE

Formula: $C_{27}H_{22}Cl_2N_4$

Mol Weight: 473.41

MP (deg C): 210-212   FP (deg C):

BP (deg C):

BP pressure (mm Hg):

Property/Value	Units	Temp	Data Type	Reference
WS 2.25E-001	mg/L	25	EST	MEYLAN,WM ET AL. (1996)
logP 4.30			EXP	QUIGLEY,JM ET AL. (1990)
VP 1.45E-012	mm Hg	25	EST	NEELY,WB & BLAU,GE (1985)
DC	pKa			
HL 3.60E-011	atm m3/mol	25	EST	MEYLAN,WM & HOWARD,PH (1991)
OH 1.36E-010	cm3/molc sec	25	EST	MEYLAN,WM & HOWARD,PH (1993)

CAS #: 002032-59-9	AMINOCARB

Formula: $C_{11}H_{16}N_2O_2$

Mol Weight: 208.26

MP (deg C): 93-94   FP (deg C):

BP (deg C):

BP pressure (mm Hg):

Property/Value	Units	Temp	Data Type	Reference
WS 9.15E+002	mg/L	20	EXP	BOWMAN,BT & SANS,WW (1983)
logP 1.90			EXP	SANGSTER,J (1993)
VP 1.88E-006	mm Hg	20	EST	HL X WSOL
DC	pKa			
HL 5.64E-010	atm m3/mol	25	EST	MEYLAN,WM & HOWARD,PH (1991)
OH 2.75E-010	cm3/molc sec	25	EST	ATKINSON,R (1988)

CAS #: 002032-65-7	BAYER 37344

Formula: $C_{11}H_{15}NO_2S$

Mol Weight: 225.31

MP (deg C): 121.5   FP (deg C):

BP (deg C):

BP pressure (mm Hg):

Property/Value	Units	Temp	Data Type	Reference
WS 1.04E+002	mg/L	25	EST	MEYLAN,WM ET AL. (1996)
logP 2.92			EXP	HANSCH,C ET AL. (1995)
VP 1.11E-004	mm Hg	25	EST	NEELY,WB & BLAU,GE (1985)
DC	pKa			
HL 1.14E-009	atm m3/mol	25	EST	MEYLAN,WM & HOWARD,PH (1991)
OH 1.58E-011	cm3/molc sec	25	EST	MEYLAN,WM & HOWARD,PH (1993)

CAS #: 002034-22-2 — 2,4,5-TRIBROMOIMIDAZOLE

Formula: $C_3HBr_3N_2$

Mol Weight: 304.78

MP (deg C): 217-220 de

FP (deg C):

BP (deg C):

BP pressure (mm Hg):

Property/ Value	Units	Temp	Data Type	Reference
WS 2.44E+002	mg/L	25	EST	MEYLAN,WM ET AL. (1996)
logP 1.96			EXP	HANSCH,C & LEO,AJ (1985)
VP 9.70E-006	mm Hg	25	EST	NEELY,WB & BLAU,GE (1985)
DC	pKa			
HL 2.37E-007	atm m3/mol	25	EST	MEYLAN,WM & HOWARD,PH (1991)
OH 3.60E-011	cm3/molc sec	25	EST	MEYLAN,WM & HOWARD,PH (1993)

CAS #: 002035-15-6 — 6H-[1,3]DIOXOLO[5,6]BENZOFURO[3,2-C][1]BENZOPYRA

Formula: $C_{16}H_{12}O_5$

Mol Weight: 284.27

MP (deg C):

FP (deg C):

BP (deg C):

BP pressure (mm Hg):

Property/ Value	Units	Temp	Data Type	Reference
WS 3.40E+002	mg/L	25	EST	MEYLAN,WM ET AL. (1996)
logP 2.61			EXP	ARNOLDI,A & MERLINI,L (1990)
VP 1.54E-008	mm Hg	25	EST	NEELY,WB & BLAU,GE (1985)
DC	pKa			
HL 3.95E-014	atm m3/mol	25	EST	MEYLAN,WM & HOWARD,PH (1991)
OH 2.21E-010	cm3/molc sec	25	EST	MEYLAN,WM & HOWARD,PH (1993)

CAS #: 002036-41-1 — PYRIMIDINE, 5-METHYL-

Formula: $C_5H_6N_2$

Mol Weight: 94.12

MP (deg C): 30.5

FP (deg C):

BP (deg C): 153

BP pressure (mm Hg):

Property/ Value	Units	Temp	Data Type	Reference
WS 1.19E+005	mg/L	25	EST	MEYLAN,WM ET AL. (1996)
logP 0.01			EXP	YAMAGAMI,C ET AL. (1990)
VP 4.63E+000	mm Hg	25	EST	NEELY,WB & BLAU,GE (1985)
DC	pKa			
HL 3.22E-006	atm m3/mol	25	EST	MEYLAN,WM & HOWARD,PH (1991)
OH 7.36E-013	cm3/molc sec	25	EST	MEYLAN,WM & HOWARD,PH (1993)

CAS #: 002037-31-2 — 3-CHLOROBENZENETHIOL

Formula: C_6H_5ClS

Mol Weight: 144.62

MP (deg C):

FP (deg C):

BP (deg C): 206

BP pressure (mm Hg):

Property/ Value	Units	Temp	Data Type	Reference
WS 1.15E+002	mg/L	25	EST	MEYLAN,WM ET AL. (1996)
logP 3.33			EST	MEYLAN,WM & HOWARD,PH (1995)
VP 2.32E-001	mm Hg	25	EST	NEELY,WB & BLAU,GE (1985)
DC	pKa			
HL 3.87E-004	atm m3/mol	25	EST	MEYLAN,WM & HOWARD,PH (1991)
OH 7.96E-012	cm3/molc sec	25	EST	MEYLAN,WM & HOWARD,PH (1993)

CAS #: 002038-57-5 — 3-PHENYL PROPYLAMINE

Formula: $C_9H_{13}N$

Mol Weight: 135.21

MP (deg C):

FP (deg C):

BP (deg C): 221

BP pressure (mm Hg):

Property/ Value	Units	Temp	Data Type	Reference
WS 2.44E+004	mg/L	25	EST	MEYLAN,WM ET AL. (1996)
logP 1.83			EXP	HANSCH,C & LEO,AJ (1985)
VP 1.12E-001	mm Hg	25	EST	NEELY,WB & BLAU,GE (1985)
DC 10.16	pKa	25	EXP	PERRIN,DD (1965)
HL 1.08E-006	atm m3/mol	25	EST	MEYLAN,WM & HOWARD,PH (1991)
OH 3.91E-011	cm3/molc sec	25	EST	MEYLAN,WM & HOWARD,PH (1993)

CAS #: 002038-62-2 — PROPYLFLUORIDE, G-PHENYL

Formula: $C_9H_{11}F$

Mol Weight: 138.19

MP (deg C):

FP (deg C):

BP (deg C):

BP pressure (mm Hg):

Property/ Value	Units	Temp	Data Type	Reference
WS 2.58E+002	mg/L	25	EST	MEYLAN,WM ET AL. (1996)
logP 2.95			EXP	HANSCH,C & LEO,AJ (1985)
VP 1.42E+000	mm Hg	25	EST	NEELY,WB & BLAU,GE (1985)
DC	pKa			
HL 2.08E-002	atm m3/mol	25	EST	MEYLAN,WM & HOWARD,PH (1991)
OH 6.80E-012	cm3/molc sec	25	EST	MEYLAN,WM & HOWARD,PH (1993)

CAS #: 002039-82-9 — 4-BROMOSTYRENE

Formula: C_8H_7Br

Mol Weight: 183.05

MP (deg C): 7.7

FP (deg C):

BP (deg C): 212

BP pressure (mm Hg):

Property/ Value	Units	Temp	Data Type	Reference
WS 3.12E+001	mg/L	25	EST	MEYLAN,WM ET AL. (1996)
logP 3.78			EST	MEYLAN,WM & HOWARD,PH (1995)
VP 7.79E-001	mm Hg	25	EXT	BOUBLIK,T ET AL. (1984)
DC	pKa			
HL 1.10E-003	atm m3/mol	25	EST	MEYLAN,WM & HOWARD,PH (1991)
OH 2.75E-011	cm3/molc sec	25	EST	MEYLAN,WM & HOWARD,PH (1993)

CAS #: 002039-85-2 — M-CHLOROSTYRENE

Formula: C_8H_7Cl

Mol Weight: 138.60

MP (deg C):

FP (deg C):

BP (deg C): 62-63

BP pressure (mm Hg): 6.00E+000

Property/ Value	Units	Temp	Data Type	Reference
WS 8.05E+001	mg/L	25	EST	MEYLAN,WM ET AL. (1996)
logP 3.54			EST	MEYLAN,WM & HOWARD,PH (1995)
VP 9.81E-001	mm Hg	25	EXP	PERRY,RH & GREEN,D (1984)
DC	pKa			
HL 2.05E-003	atm m3/mol	25	EST	MEYLAN,WM & HOWARD,PH (1991)
OH 2.76E-011	cm3/molc sec	25	EST	MEYLAN,WM & HOWARD,PH (1993)

CAS #: 002039-87-4				O-CHLOROSTYRENE
Formula: C_8H_7Cl				
Mol Weight: 138.60				
MP (deg C): -63.1		FP (deg C):		
BP (deg C): 188.7				
BP pressure (mm Hg):				

Property/Value	Units	Temp	Data Type	Reference
WS 8.05E+001	mg/L	25	EST	MEYLAN,WM ET AL. (1996)
logP 3.54			EST	MEYLAN,WM & HOWARD,PH (1995)
VP 9.61E-001	mm Hg	25	EXT	BOUBLIK,T ET AL. (1984)
DC	pKa			
HL 2.05E-003	atm m3/mol	25	EST	MEYLAN,WM & HOWARD,PH (1991)
OH 2.76E-011	cm3/molc sec	25	EST	MEYLAN,WM & HOWARD,PH (1993)

CAS #: 002040-96-2				PROPYLCYCLOPENTANE
Formula: C_8H_{16}				
Mol Weight: 112.22				
MP (deg C): -117.3		FP (deg C):		
BP (deg C): 131				
BP pressure (mm Hg):				

Property/Value	Units	Temp	Data Type	Reference
WS 2.04E+000	mg/L	25	EXP	MACKAY,D & SHIU,WY (1981)
logP 4.08			EST	MEYLAN,WM & HOWARD,PH (1995)
VP 1.24E+001	mm Hg	25	EXP	DAUBERT,TE & DANNER,RP (1989)
DC	pKa			
HL 8.98E-001	atm m3/mol	25	EST	VP/WSOL
OH 1.01E-011	cm3/molc sec	25	EST	MEYLAN,WM & HOWARD,PH (1993)

CAS #: 002042-14-0				3-NITRO-P-CRESOL
Formula: $C_7H_7NO_3$				
Mol Weight: 153.14				
MP (deg C):		FP (deg C):		
BP (deg C):				
BP pressure (mm Hg):				

Property/Value	Units	Temp	Data Type	Reference
WS 3.84E+003	mg/L	25	EST	MEYLAN,WM ET AL. (1996)
logP 2.18			EXP	CHEM INSPECT TEST INST (1992)
VP 6.32E-004	mm Hg	25	EST	NEELY,WB & BLAU,GE (1985)
DC 8.62	pKa	25	EXP	SERJEANT,EP & DEMPSEY,B (1979)
HL 2.44E-009	atm m3/mol	25	EST	MEYLAN,WM & HOWARD,PH (1991)
OH 3.85E-012	cm3/molc sec	25	EST	MEYLAN,WM & HOWARD,PH (1993)

CAS #: 002042-37-7				2-BROMOBENZONITRILE
Formula: C_7H_4BrN				
Mol Weight: 182.03				
MP (deg C): 55.5		FP (deg C):		
BP (deg C): 252				
BP pressure (mm Hg):				

Property/Value	Units	Temp	Data Type	Reference
WS 2.46E+002	mg/L	25	EST	MEYLAN,WM ET AL. (1996)
logP 2.43			EST	MEYLAN,WM & HOWARD,PH (1995)
VP 2.73E-002	mm Hg	25	EST	NEELY,WB & BLAU,GE (1985)
DC	pKa			
HL 2.07E-005	atm m3/mol	25	EST	MEYLAN,WM & HOWARD,PH (1991)
OH 2.17E-013	cm3/molc sec	25	EST	MEYLAN,WM & HOWARD,PH (1993)

CAS #: 002046-17-5				METHYL 4-PHENYLBUTYRATE
Formula: $C_{11}H_{14}O_2$				
Mol Weight: 178.23				
MP (deg C):		FP (deg C):		
BP (deg C):				
BP pressure (mm Hg):				

Property/Value	Units	Temp	Data Type	Reference
WS 2.42E+002	mg/L	25	EST	MEYLAN,WM ET AL. (1996)
logP 2.77			EXP	HANSCH,C & LEO,AJ (1985)
VP 2.43E-002	mm Hg	25	EST	NEELY,WB & BLAU,GE (1985)
DC	pKa			
HL 2.49E-005	atm m3/mol	25	EST	MEYLAN,WM & HOWARD,PH (1991)
OH 8.47E-012	cm3/molc sec	25	EST	MEYLAN,WM & HOWARD,PH (1993)

CAS #: 002046-18-6				G-PHENYLPROPYLCYANIDE
Formula: $C_{10}H_{11}N$				
Mol Weight: 145.21				
MP (deg C):		FP (deg C):		
BP (deg C): 97-99				
BP pressure (mm Hg): 1.70E+000				

Property/Value	Units	Temp	Data Type	Reference
WS 5.61E+002	mg/L	25	EST	MEYLAN,WM ET AL. (1996)
logP 2.21			EXP	HANSCH,C & LEO,AJ (1985)
VP 1.08E-002	mm Hg	25	EST	NEELY,WB & BLAU,GE (1985)
DC	pKa			
HL 4.35E-006	atm m3/mol	25	EST	MEYLAN,WM & HOWARD,PH (1991)
OH 6.35E-012	cm3/molc sec	25	EST	MEYLAN,WM & HOWARD,PH (1993)

CAS #: 002046-19-7				2-AMINO-5-PHENYLPENTANOIC ACID
Formula: $C_{11}H_{15}NO_2$				
Mol Weight: 193.25				
MP (deg C):		FP (deg C):		
BP (deg C):				
BP pressure (mm Hg):				

Property/Value	Units	Temp	Data Type	Reference
WS 8.18E+002	mg/L	25	EST	MEYLAN,WM ET AL. (1996)
logP -0.36			EXP	HANSCH,C & LEO,AJ (1985)
VP 7.11E-009	mm Hg	25	EST	NEELY,WB & BLAU,GE (1985)
DC	pKa			
HL 2.12E-010	atm m3/mol	25	EST	MEYLAN,WM & HOWARD,PH (1991)
OH 4.70E-011	cm3/molc sec	25	EST	MEYLAN,WM & HOWARD,PH (1993)

CAS #: 002046-21-1				6-KETO METHYLHEPTANOATE
Formula: $C_8H_{14}O_3$				
Mol Weight: 158.20				
MP (deg C):		FP (deg C):		
BP (deg C):				
BP pressure (mm Hg):				

Property/Value	Units	Temp	Data Type	Reference
WS 2.37E+004	mg/L	25	EST	MEYLAN,WM ET AL. (1996)
logP 0.55			EXP	HANSCH,C & LEO,AJ (1985)
VP 2.20E-001	mm Hg	25	EST	NEELY,WB & BLAU,GE (1985)
DC	pKa			
HL 2.76E-007	atm m3/mol	25	EST	MEYLAN,WM & HOWARD,PH (1991)
OH 7.92E-012	cm3/molc sec	25	EST	MEYLAN,WM & HOWARD,PH (1993)

002046-23-3 — 4-CYCLOPROPYL 2-BUTANONE

Formula: $C_7H_{12}O$

Mol Weight: 112.17

MP (deg C):
FP (deg C):
BP (deg C):
BP pressure (mm Hg):

Property/Value	Units	Temp	Data Type	Reference
WS 5.60E+003	mg/L	25	EST	MEYLAN,WM ET AL. (1996)
logP 1.50			EXP	HANSCH,C & LEO,AJ (1985)
VP 6.28E+000	mm Hg	25	EST	NEELY,WB & BLAU,GE (1985)
DC	pKa			
HL 6.79E-005	atm m3/mol	25	EST	MEYLAN,WM & HOWARD,PH (1991)
OH 5.57E-012	cm3/molc sec	25	EST	MEYLAN,WM & HOWARD,PH (1993)

002046-33-5 — 1-METHOXY-3-PHENYLPROPANE

Formula: $C_{10}H_{14}O$

Mol Weight: 150.22

MP (deg C):
FP (deg C):
BP (deg C):
BP pressure (mm Hg):

Property/Value	Units	Temp	Data Type	Reference
WS 3.74E+002	mg/L	25	EST	MEYLAN,WM ET AL. (1996)
logP 2.70			EXP	HANSCH,C & LEO,AJ (1985)
VP 2.03E-001	mm Hg	25	EST	NEELY,WB & BLAU,GE (1985)
DC	pKa			
HL 1.23E-004	atm m3/mol	25	EST	MEYLAN,WM & HOWARD,PH (1991)
OH 1.79E-011	cm3/molc sec	25	EST	MEYLAN,WM & HOWARD,PH (1993)

002050-20-6 — DIETHYL PIMELATE

Formula: $C_{11}H_{20}O_4$

Mol Weight: 216.28

MP (deg C): -24
FP (deg C):
BP (deg C): 254
BP pressure (mm Hg):

Property/Value	Units	Temp	Data Type	Reference
WS 1.97E+003	mg/L	25	EXP	SUZUKI,T (1991)
logP 2.86			EST	MEYLAN,WM & HOWARD,PH (1995)
VP 3.10E-003	mm Hg	25	EXP	BENNETT,SR ET AL. (1984)
DC	pKa			
HL 4.48E-007	atm m3/mol	25	EST	VP/WSOL
OH 9.26E-012	cm3/molc sec	25	EST	MEYLAN,WM & HOWARD,PH (1993)

002050-23-9 — DIETHYL SUBERATE

Formula: $C_{12}H_{22}O_4$

Mol Weight: 230.31

MP (deg C): 5.9
FP (deg C):
BP (deg C): 282.6
BP pressure (mm Hg):

Property/Value	Units	Temp	Data Type	Reference
WS 6.79E+002	mg/L	25	EXP	SUZUKI,T (1991)
logP 3.35			EST	MEYLAN,WM & HOWARD,PH (1995)
VP 1.45E-002	mm Hg	25	EST	NEELY,WB & BLAU,GE (1985)
DC	pKa			
HL 3.03E-006	atm m3/mol	25	EST	MEYLAN,WM & HOWARD,PH (1991)
OH 1.07E-011	cm3/molc sec	25	EST	MEYLAN,WM & HOWARD,PH (1993)

002050-24-0 — 1,3-DIETHYL-5-METHYLBENZENE

Formula: $C_{11}H_{16}$

Mol Weight: 148.25

MP (deg C): -74.1
FP (deg C):
BP (deg C): 205
BP pressure (mm Hg):

Property/Value	Units	Temp	Data Type	Reference
WS 8.81E+000	mg/L	25	EST	MEYLAN,WM ET AL. (1996)
logP 4.62			EST	MEYLAN,WM & HOWARD,PH (1995)
VP 1.00E+000	mm Hg	25	EXP	CHAO,J ET AL. (1983)
DC	pKa			
HL 1.28E-002	atm m3/mol	25	EST	MEYLAN,WM & HOWARD,PH (1991)
OH 3.38E-011	cm3/molc sec	25	EST	MEYLAN,WM & HOWARD,PH (1993)

002050-43-3 — 2,4-DIMETHYLACETANILIDE

Formula: $C_{10}H_{13}NO$

Mol Weight: 163.22

MP (deg C): 129.3
FP (deg C):
BP (deg C): 170
BP pressure (mm Hg): 1.00E+001

Property/Value	Units	Temp	Data Type	Reference
WS 3.75E+003	mg/L	25	EST	MEYLAN,WM ET AL. (1996)
logP 1.46			EXP	NAKAGAWA,Y ET AL. (1992)
VP 6.74E-005	mm Hg	25	EST	NEELY,WB & BLAU,GE (1985)
DC	pKa			
HL 7.52E-009	atm m3/mol	25	EST	MEYLAN,WM & HOWARD,PH (1991)
OH 1.90E-011	cm3/molc sec	25	EST	MEYLAN,WM & HOWARD,PH (1993)

002050-44-4 — 2,5-DIMETHYLACETANILIDE

Formula: $C_{10}H_{13}NO$

Mol Weight: 163.22

MP (deg C):
FP (deg C):
BP (deg C):
BP pressure (mm Hg):

Property/Value	Units	Temp	Data Type	Reference
WS 3.67E+003	mg/L	25	EST	MEYLAN,WM ET AL. (1996)
logP 1.47			EXP	NAKAGAWA,Y ET AL. (1992)
VP 6.74E-005	mm Hg	25	EST	NEELY,WB & BLAU,GE (1985)
DC	pKa			
HL 7.52E-009	atm m3/mol	25	EST	MEYLAN,WM & HOWARD,PH (1991)
OH 4.01E-011	cm3/molc sec	25	EST	MEYLAN,WM & HOWARD,PH (1993)

002050-45-5 — 3,5-DIMETHYLACETANILIDE

Formula: $C_{10}H_{13}NO$

Mol Weight: 163.22

MP (deg C):
FP (deg C):
BP (deg C):
BP pressure (mm Hg):

Property/Value	Units	Temp	Data Type	Reference
WS 9.28E+002	mg/L	25	EST	MEYLAN,WM ET AL. (1996)
logP 2.17			EXP	NAKAGAWA,Y ET AL. (1992)
VP 6.74E-005	mm Hg	25	EST	NEELY,WB & BLAU,GE (1985)
DC	pKa			
HL 7.52E-009	atm m3/mol	25	EST	MEYLAN,WM & HOWARD,PH (1991)
OH 8.50E-011	cm3/molc sec	25	EST	MEYLAN,WM & HOWARD,PH (1993)

002050-60-4 — DIBUTYL OXALATE

CAS #: 002050-60-4				DIBUTYL OXALATE
Formula: $C_{10}H_{18}O_4$				
Mol Weight: 202.25				
MP (deg C): -30.5	**FP (deg C):**			
BP (deg C): 241				
BP pressure (mm Hg):				

Property/ Value	Units	Temp	Data Type	Reference
WS 4.04E+002	mg/L	25	EST	MEYLAN,WM ET AL. (1996)
logP 2.37			EST	MEYLAN,WM & HOWARD,PH (1995)
VP 9.55E-002	mm Hg	25	EST	NEELY,WB & BLAU,GE (1985)
DC	pKa			
HL 1.35E-005	atm m3/mol	25	EST	MEYLAN,WM & HOWARD,PH (1991)
OH 9.13E-012	cm3/mol sec	25	EST	MEYLAN,WM & HOWARD,PH (1993)

002050-67-1 — 3,3'-DICHLOROBIPHENYL

CAS #: 002050-67-1				3,3'-DICHLOROBIPHENYL
Formula: $C_{12}H_8Cl_2$				
Mol Weight: 223.10				
MP (deg C): 29	**FP (deg C):**			
BP (deg C): 320				
BP pressure (mm Hg):				

Property/ Value	Units	Temp	Data Type	Reference
WS 3.55E-001	mg/L	25	EXP	YALKOWSKY,SH & DANNENFELSER,RM (1992)
logP 5.27			EXP	SANGSTER,J (1993)
VP 6.49E-004	mm Hg	25	EXP	BIDLEMAN,TF (1984)
DC	pKa			
HL 5.37E-004	atm m3/mol	25	EST	VP/WSOL
OH 3.51E-012	cm3/mol sec	25	EST	MEYLAN,WM & HOWARD,PH (1993)

002050-68-2 — 4,4'-DICHLOROBIPHENYL

CAS #: 002050-68-2				4,4'-DICHLOROBIPHENYL
Formula: $C_{12}H_8Cl_2$				
Mol Weight: 223.10				
MP (deg C): 149.3	**FP (deg C):**			
BP (deg C): 317				
BP pressure (mm Hg):				

Property/ Value	Units	Temp	Data Type	Reference
WS 6.20E-002	mg/L	20	EXP	CHIOU,CT ET AL. (1977)
logP 4.82			EXP	HANSCH,C & LEO,AJ (1985)
VP 5.35E-004	mm Hg	25	EXP	BIDLEMAN,TF (1984)
DC	pKa			
HL 2.53E-003	atm m3/mol	25	EST	VP/WSOL
OH 1.73E-012	cm3/mol sec	25	EST	MEYLAN,WM & HOWARD,PH (1993)

002050-69-3 — 1,2-DICHLORONAPHTHALENE

CAS #: 002050-69-3				1,2-DICHLORONAPHTHALENE
Formula: $C_{10}H_6Cl_2$				
Mol Weight: 197.07				
MP (deg C): 36	**FP (deg C):**			
BP (deg C): 296.5				
BP pressure (mm Hg):				

Property/ Value	Units	Temp	Data Type	Reference
WS 7.61E+000	mg/L	25	EST	MEYLAN,WM ET AL. (1996)
logP 4.42			EXP	HANSCH,C & LEO,AJ (1985)
VP 1.76E-003	mm Hg	25	EST	NEELY,WB & BLAU,GE (1985)
DC	pKa			
HL 2.89E-004	atm m3/mol	25	EST	MEYLAN,WM & HOWARD,PH (1991)
OH 4.44E-012	cm3/mol sec	25	EST	MEYLAN,WM & HOWARD,PH (1993)

002050-73-9 — NAPHTHALENE, 1,7-DICHLORO-

CAS #: 002050-73-9				NAPHTHALENE, 1,7-DICHLORO-
Formula: $C_{10}H_6Cl_2$				
Mol Weight: 197.07				
MP (deg C): 63.5	**FP (deg C):**			
BP (deg C): 285.5				
BP pressure (mm Hg):				

Property/ Value	Units	Temp	Data Type	Reference
WS 5.78E+000	mg/L	25	EST	MEYLAN,WM ET AL. (1996)
logP 4.56			EXP	OPPERHUIZEN,A (1987)
VP 1.76E-003	mm Hg	25	EST	NEELY,WB & BLAU,GE (1985)
DC	pKa			
HL 2.89E-004	atm m3/mol	25	EST	MEYLAN,WM & HOWARD,PH (1991)
OH 4.44E-012	cm3/mol sec	25	EST	MEYLAN,WM & HOWARD,PH (1993)

002050-74-0 — 1,8-DICHLORONAPHTHALENE

CAS #: 002050-74-0				1,8-DICHLORONAPHTHALENE
Formula: $C_{10}H_6Cl_2$				
Mol Weight: 197.07				
MP (deg C): 89	**FP (deg C):**			
BP (deg C):				
BP pressure (mm Hg):				

Property/ Value	Units	Temp	Data Type	Reference
WS 3.13E-001	mg/L	25	EXP	OPPERHUIZEN,A ET AL. (1985)
logP 4.42			EXP	HANSCH,C & LEO,AJ (1985)
VP 1.76E-003	mm Hg	25	EST	NEELY,WB & BLAU,GE (1985)
DC	pKa			
HL 2.89E-004	atm m3/mol	25	EST	MEYLAN,WM & HOWARD,PH (1991)
OH 4.44E-012	cm3/mol sec	25	EST	MEYLAN,WM & HOWARD,PH (1993)

002050-75-1 — 2,3-DICHLORONAPHTHALENE

CAS #: 002050-75-1				2,3-DICHLORONAPHTHALENE
Formula: $C_{10}H_6Cl_2$				
Mol Weight: 197.07				
MP (deg C):	**FP (deg C):**			
BP (deg C):				
BP pressure (mm Hg):				

Property/ Value	Units	Temp	Data Type	Reference
WS 8.76E-002	mg/L	22	EXP	OPPERHUIZEN,A ET AL. (1985)
logP 4.42			EXP	HANSCH,C & LEO,AJ (1985)
VP 1.76E-003	mm Hg	25	EST	NEELY,WB & BLAU,GE (1985)
DC	pKa			
HL 2.89E-004	atm m3/mol	25	EST	MEYLAN,WM & HOWARD,PH (1991)
OH 4.44E-012	cm3/mol sec	25	EST	MEYLAN,WM & HOWARD,PH (1993)

002050-76-2 — 2,4-DICHLORO-1-NAPHTHOL

CAS #: 002050-76-2				2,4-DICHLORO-1-NAPHTHOL
Formula: $C_{10}H_6Cl_2O$				
Mol Weight: 213.06				
MP (deg C): 107.5	**FP (deg C):**			
BP (deg C): 180				
BP pressure (mm Hg):				

Property/ Value	Units	Temp	Data Type	Reference
WS 3.48E+001	mg/L	25	EST	MEYLAN,WM ET AL. (1996)
logP 4.23			EXP	CHEM INSPECT TEST INST (1992)
VP 2.03E-005	mm Hg	25	EST	NEELY,WB & BLAU,GE (1985)
DC	pKa			
HL 3.00E-008	atm m3/mol	25	EST	MEYLAN,WM & HOWARD,PH (1991)
OH 1.54E-011	cm3/mol sec	25	EST	MEYLAN,WM & HOWARD,PH (1993)

CAS #: 002050-77-3				1-IODODECANE

Formula: $C_{10}H_{21}I$

Mol Weight: 268.18

MP (deg C): -16.3   FP (deg C):

BP (deg C): 132

BP pressure (mm Hg): 1.50E+001

Property/Value	Units	Temp	Data Type	Reference
WS 1.38E-001	mg/L	25	EST	MEYLAN,WM ET AL. (1996)
logP 6.01			EST	MEYLAN,WM & HOWARD,PH (1995)
VP 1.40E-002	mm Hg	25	EXT	LI,JCM & ROSSINI,FD (1953)
DC	pKa			
HL 7.04E-002	atm m3/mol	25	EST	MEYLAN,WM & HOWARD,PH (1991)
OH 1.09E-011	cm3/molc sec	25	EST	MEYLAN,WM & HOWARD,PH (1993)

CAS #: 002050-85-3				ACETAMIDE, N,N'-(1,2-PHENYLENE)BIS-

Formula: $C_{10}H_{12}N_2O_2$

Mol Weight: 192.22

MP (deg C):   FP (deg C):

BP (deg C):

BP pressure (mm Hg):

Property/Value	Units	Temp	Data Type	Reference
WS 2.49E+003	mg/L	25	EST	MEYLAN,WM ET AL. (1996)
logP -0.03			EXP	HANSCH,C ET AL. (1995)
VP 5.16E-008	mm Hg	25	EST	NEELY,WB & BLAU,GE (1985)
DC	pKa			
HL 1.12E-011	atm m3/mol	25	EST	MEYLAN,WM & HOWARD,PH (1991)
OH 6.70E-012	cm3/molc sec	25	EST	MEYLAN,WM & HOWARD,PH (1993)

CAS #: 002050-92-2				DIPENTYLAMINE

Formula: $C_{10}H_{23}N$

Mol Weight: 157.30

MP (deg C): -7.85   FP (deg C):

BP (deg C): 202

BP pressure (mm Hg):

Property/Value	Units	Temp	Data Type	Reference
WS 4.44E+002	mg/L	25	EST	MEYLAN,WM ET AL. (1996)
logP 3.76			EST	MEYLAN,WM & HOWARD,PH (1995)
VP 1.53E-001	mm Hg	25	EXP	DAUBERT,TE & DANNER,RP (1993)
DC 11.16	pKa	26	EXP	PERRIN,DD (1965)
HL 2.08E-004	atm m3/mol	25	EST	MEYLAN,WM & HOWARD,PH (1991)
OH 8.97E-011	cm3/molc sec	25	EST	MEYLAN,WM & HOWARD,PH (1993)

CAS #: 002051-24-3				DECACHLOROBIPHENYL

Formula: $C_{12}Cl_{10}$

Mol Weight: 498.66

MP (deg C):   FP (deg C):

BP (deg C):

BP pressure (mm Hg):

Property/Value	Units	Temp	Data Type	Reference
WS 7.40E-006	mg/L	25	EXP	YALKOWSKY,SH & DANNENFELSER,RM (1992)
logP 8.27			EXP	DEBRUIJN,J ET AL. (1989)
VP 1.06E-007	mm Hg	25	EXP	HINCKLEY,DA ET AL. (1990)
DC	pKa			
HL 9.40E-003	atm m3/mol	25	EST	VP/WSOL
OH 1.82E-014	cm3/molc sec	25	EST	MEYLAN,WM & HOWARD,PH (1993)

CAS #: 002051-30-1				2,6-DIMETHYLOCTANE

Formula: $C_{10}H_{22}$

Mol Weight: 142.29

MP (deg C):   FP (deg C):

BP (deg C): 160.4

BP pressure (mm Hg):

Property/Value	Units	Temp	Data Type	Reference
WS 1.04E+000	mg/L	25	EST	MEYLAN,WM ET AL. (1996)
logP 5.11			EST	MEYLAN,WM & HOWARD,PH (1995)
VP 2.95E+000	mm Hg	25	EXT	ZWOLINSKI,BJ & WILHOIT,RC (1971)
DC	pKa			
HL 5.30E+000	atm m3/mol	25	EST	MEYLAN,WM & HOWARD,PH (1991)
OH 1.14E-011	cm3/molc sec	25	EST	MEYLAN,WM & HOWARD,PH (1993)

CAS #: 002051-60-7				2-CHLOROBIPHENYL

Formula: $C_{12}H_9Cl$

Mol Weight: 188.66

MP (deg C): 34   FP (deg C):

BP (deg C): 274

BP pressure (mm Hg):

Property/Value	Units	Temp	Data Type	Reference
WS 4.83E+000	mg/L	25	EXP	CHIOU,CT ET AL. (1983)
logP 4.53			EXP	HANSCH,C ET AL. (1995)
VP 1.38E-003	mm Hg	25	EXP	BIDLEMAN,TF (1984)
DC	pKa			
HL 7.09E-005	atm m3/mol	25	EST	VP/WSOL
OH 2.80E-012	cm3/molc sec	25	EXP	ATKINSON,R (1989)

CAS #: 002051-61-8				3-CHLOROBIPHENYL

Formula: $C_{12}H_9Cl$

Mol Weight: 188.66

MP (deg C): 16   FP (deg C):

BP (deg C): 284.5

BP pressure (mm Hg):

Property/Value	Units	Temp	Data Type	Reference
WS 1.30E+000	mg/L	25	EXP	YALKOWSKY,SH & DANNENFELSER,RM (1992)
logP 4.58			EXP	HANSCH,C ET AL. (1995)
VP 7.35E-003	mm Hg	25	EXP	BIDLEMAN,TF (1984)
DC	pKa			
HL 1.40E-003	atm m3/mol	25	EST	VP/WSOL
OH 5.30E-012	cm3/molc sec	25	EXP	ATKINSON,R (1989)

CAS #: 002051-62-9				4-CHLOROBIPHENYL

Formula: $C_{12}H_9Cl$

Mol Weight: 188.66

MP (deg C): 78.8   FP (deg C):

BP (deg C): 292.9

BP pressure (mm Hg):

Property/Value	Units	Temp	Data Type	Reference
WS 1.34E+000	mg/L	25	EXP	YALKOWSKY,SH & DANNENFELSER,RM (1992)
logP 4.61			EXP	HANSCH,C ET AL. (1995)
VP 1.05E-002	mm Hg	25	EXP	BIDLEMAN,TF (1984)
DC	pKa			
HL 1.95E-003	atm m3/mol	25	EST	VP/WSOL
OH 3.90E-012	cm3/molc sec	25	EXP	ATKINSON,R (1989)

CAS #: 002051-95-8 — BENZENEBUTANOIC ACID, GAMMA -OXO-

Formula: $C_{10}H_{10}O_3$

Mol Weight: 178.19

MP (deg C): 117-119

FP (deg C):

BP (deg C):

BP pressure (mm Hg):

Property/Value	Units	Temp	Data Type	Reference
WS 1.08E+004	mg/L	25	EST	MEYLAN,WM ET AL. (1996)
logP 1.30			EXP	SANGSTER,J (1993)
VP 9.73E-005	mm Hg	25	EST	NEELY,WB & BLAU,GE (1985)
DC	pKa			
HL 9.69E-011	atm m3/mol	25	EST	MEYLAN,WM & HOWARD,PH (1991)
OH 5.89E-012	cm3/molc sec	25	EST	MEYLAN,WM & HOWARD,PH (1993)

CAS #: 002052-07-5 — O-BROMOBIPHENYL

Formula: $C_{12}H_9Br$

Mol Weight: 233.11

MP (deg C): 1.5

FP (deg C):

BP (deg C): 297

BP pressure (mm Hg):

Property/Value	Units	Temp	Data Type	Reference
WS 3.52E+000	mg/L	25	EST	MEYLAN,WM ET AL. (1996)
logP 4.59			EXP	DOUCETTE,WJ & ANDREN,AW (1987)
VP 3.54E-004	mm Hg	25	EST	NEELY,WB & BLAU,GE (1985)
DC	pKa			
HL 1.65E-004	atm m3/mol	25	EST	MEYLAN,WM & HOWARD,PH (1991)
OH 3.47E-012	cm3/molc sec	25	EST	MEYLAN,WM & HOWARD,PH (1993)

CAS #: 002052-14-4 — BENZOIC ACID, 2-HYDROXY-, BUTYL ESTER

Formula: $C_{11}H_{14}O_3$

Mol Weight: 194.23

MP (deg C): -5.9

FP (deg C):

BP (deg C): 271

BP pressure (mm Hg):

Property/Value	Units	Temp	Data Type	Reference
WS 1.98E+001	mg/L	25	EST	MEYLAN,WM ET AL. (1996)
logP 4.63			EXP	KORENMAN,YI & DANILOV,VN (1990)
VP 1.86E-004	mm Hg	25	EST	NEELY,WB & BLAU,GE (1985)
DC	pKa			
HL 1.06E-005	atm m3/mol	25	EST	MEYLAN,WM & HOWARD,PH (1991)
OH 1.55E-011	cm3/molc sec	25	EST	MEYLAN,WM & HOWARD,PH (1993)

CAS #: 002055-14-3 — N,N-DIETHYL-3-PYRIDYLMETHYLAMINE

Formula: $C_{10}H_{16}N_2$

Mol Weight: 164.25

MP (deg C):

FP (deg C):

BP (deg C):

BP pressure (mm Hg):

Property/Value	Units	Temp	Data Type	Reference
WS 1.00E+006	mg/L	25	EST	MEYLAN,WM ET AL. (1996)
logP 1.01			EXP	HANSCH,C & LEO,AJ (1985)
VP 4.30E-003	mm Hg	25	EXP	ARTHURDLITTLE,INC (1982)
DC	pKa			
HL 6.80E-009	atm m3/mol	25	EST	MEYLAN,WM & HOWARD,PH (1991)
OH 9.33E-011	cm3/molc sec	25	EST	MEYLAN,WM & HOWARD,PH (1993)

CAS #: 002055-21-2 — N,N-DIMETHYL-3-PYRIDYLMETHYLAMINE

Formula: $C_8H_{12}N_2$

Mol Weight: 136.20

MP (deg C):

FP (deg C):

BP (deg C):

BP pressure (mm Hg):

Property/Value	Units	Temp	Data Type	Reference
WS 1.00E+006	mg/L	25	EST	MEYLAN,WM ET AL. (1996)
logP 0.49			EXP	HANSCH,C & LEO,AJ (1985)
VP 3.67E-001	mm Hg	25	EST	NEELY,WB & BLAU,GE (1985)
DC 8.00	pKa	25	EXP	PERRIN,DD (1965)
HL 3.86E-009	atm m3/mol	25	EST	MEYLAN,WM & HOWARD,PH (1991)
OH 7.81E-011	cm3/molc sec	25	EST	MEYLAN,WM & HOWARD,PH (1993)

CAS #: 002055-46-1 — N,N'-TRIMETHYLENETHIOUREA

Formula: $C_4H_8N_2S$

Mol Weight: 116.19

MP (deg C): 210-212

FP (deg C):

BP (deg C):

BP pressure (mm Hg):

Property/Value	Units	Temp	Data Type	Reference
WS 1.97E+005	mg/L	25	EST	MEYLAN,WM ET AL. (1996)
logP -0.68			EXP	HANSCH,C & LEO,AJ (1985)
VP 1.58E-003	mm Hg	25	EST	NEELY,WB & BLAU,GE (1985)
DC	pKa			
HL 4.46E-007	atm m3/mol	25	EST	MEYLAN,WM & HOWARD,PH (1991)
OH 1.49E-010	cm3/molc sec	25	EST	MEYLAN,WM & HOWARD,PH (1993)

CAS #: 002058-74-4 — N-METHYLINDOL-2,3-DIONE

Formula: $C_9H_7NO_2$

Mol Weight: 161.16

MP (deg C): 130-133

FP (deg C):

BP (deg C):

BP pressure (mm Hg):

Property/Value	Units	Temp	Data Type	Reference
WS 2.16E+004	mg/L	25	EST	MEYLAN,WM ET AL. (1996)
logP 0.58			EXP	HANSCH,C & LEO,AJ (1985)
VP 5.51E-005	mm Hg	25	EST	NEELY,WB & BLAU,GE (1985)
DC	pKa			
HL 4.52E-010	atm m3/mol	25	EST	MEYLAN,WM & HOWARD,PH (1991)
OH 1.19E-011	cm3/molc sec	25	EST	MEYLAN,WM & HOWARD,PH (1993)

CAS #: 002059-76-9 — BENZENE, 1-IODO-4-ISOTHIOCYANATO-

Formula: C_7H_4INS

Mol Weight: 261.09

MP (deg C):

FP (deg C):

BP (deg C):

BP pressure (mm Hg):

Property/Value	Units	Temp	Data Type	Reference
WS 5.11E+000	mg/L	25	EST	MEYLAN,WM ET AL. (1996)
logP 4.22			EXP	AUGUSTIN,J ET AL. (1987)
VP 8.28E-004	mm Hg	25	EST	NEELY,WB & BLAU,GE (1985)
DC	pKa			
HL 1.81E-004	atm m3/mol	25	EST	MEYLAN,WM & HOWARD,PH (1991)
OH 1.41E-012	cm3/molc sec	25	EST	MEYLAN,WM & HOWARD,PH (1993)

CAS #: 002060-58-4		3,11-TRIDECADIENE-5,7,9-TRIYN-1-OL, 2-CHLORO-

Formula: $C_{13}H_{11}ClO$	
Mol Weight: 218.69	
MP (deg C):	FP (deg C):
BP (deg C):	
BP pressure (mm Hg):	

Property/Value	Units	Temp	Data Type	Reference
WS 5.18E+001	mg/L	25	EST	MEYLAN,WM ET AL. (1996)
logP 3.91			EXP	MCLACHLAN,D ET AL. (1986)
VP 4.41E-007	mm Hg	25	EST	NEELY,WB & BLAU,GE (1985)
DC	pKa			
HL 4.12E-007	atm m3/mol	25	EST	MEYLAN,WM & HOWARD,PH (1991)
OH 2.37E-010	cm3/molc sec	25	EST	MEYLAN,WM & HOWARD,PH (1993)

CAS #: 002062-78-4		PIMOZIDE

Formula: $C_{28}H_{29}F_2N_3O$	
Mol Weight: 461.56	
MP (deg C): 214-218	FP (deg C):
BP (deg C):	
BP pressure (mm Hg):	

Property/Value	Units	Temp	Data Type	Reference
WS 5.25E-003	mg/L	25	EST	MEYLAN,WM ET AL. (1996)
logP 6.30			EXP	HANSCH,C ET AL. (1995)
VP 6.25E-013	mm Hg	25	EST	NEELY,WB & BLAU,GE (1985)
DC	pKa			
HL 1.10E-013	atm m3/mol	25	EST	MEYLAN,WM & HOWARD,PH (1991)
OH 1.39E-010	cm3/molc sec	25	EST	MEYLAN,WM & HOWARD,PH (1993)

CAS #: 002062-84-2		BENPERIDOL

Formula: $C_{22}H_{24}FN_3O_2$	
Mol Weight: 381.45	
MP (deg C): 170-171.8	FP (deg C):
BP (deg C):	
BP pressure (mm Hg):	

Property/Value	Units	Temp	Data Type	Reference
WS 1.83E+000	mg/L	25	EST	MEYLAN,WM ET AL. (1996)
logP 3.91			EXP	HANSCH,C ET AL. (1995)
VP 2.67E-011	mm Hg	25	EST	NEELY,WB & BLAU,GE (1985)
DC	pKa			
HL 1.45E-015	atm m3/mol	25	EST	MEYLAN,WM & HOWARD,PH (1991)
OH 1.36E-010	cm3/molc sec	25	EST	MEYLAN,WM & HOWARD,PH (1993)

CAS #: 002065-23-8		METHYL PHENOXYACETATE

Formula: $C_9H_{10}O_3$	
Mol Weight: 166.18	
MP (deg C): 245	FP (deg C):
BP (deg C):	
BP pressure (mm Hg):	

Property/Value	Units	Temp	Data Type	Reference
WS 4.01E+003	mg/L	25	EST	MEYLAN,WM ET AL. (1996)
logP 1.41			EXP	HANSCH,C & LEO,AJ (1985)
VP 6.01E-002	mm Hg	25	EST	NEELY,WB & BLAU,GE (1985)
DC	pKa			
HL 5.37E-006	atm m3/mol	25	EST	MEYLAN,WM & HOWARD,PH (1991)
OH 2.59E-011	cm3/molc sec	25	EST	MEYLAN,WM & HOWARD,PH (1993)

CAS #: 002065-37-4		2-BROMO-1,4-NAPHTHOQUINONE

Formula: $C_{10}H_5BrO_2$	
Mol Weight: 237.06	
MP (deg C):	FP (deg C):
BP (deg C):	
BP pressure (mm Hg):	

Property/Value	Units	Temp	Data Type	Reference
WS 2.30E+002	mg/L	25	EST	MEYLAN,WM ET AL. (1996)
logP 2.44			EXP	HANSCH,C ET AL. (1995)
VP 2.58E-005	mm Hg	25	EST	NEELY,WB & BLAU,GE (1985)
DC	pKa			
HL 3.92E-010	atm m3/mol	25	EST	MEYLAN,WM & HOWARD,PH (1991)
OH 3.52E-012	cm3/molc sec	25	EST	MEYLAN,WM & HOWARD,PH (1993)

CAS #: 002065-70-5		2,6-DICHLORONAPHTHALENE

Formula: $C_{10}H_6Cl_2$	
Mol Weight: 197.07	
MP (deg C): 140.5	FP (deg C):
BP (deg C): 285	
BP pressure (mm Hg):	

Property/Value	Units	Temp	Data Type	Reference
WS 5.40E-002	mg/L	36	EXP	OPPERHUIZEN,A (1987)
logP 4.46			EST	MEYLAN,WM & HOWARD,PH (1995)
VP 1.76E-003	mm Hg	25	EST	NEELY,WB & BLAU,GE (1985)
DC	pKa			
HL 2.89E-004	atm m3/mol	25	EST	MEYLAN,WM & HOWARD,PH (1991)
OH 4.44E-012	cm3/molc sec	25	EST	MEYLAN,WM & HOWARD,PH (1993)

CAS #: 002067-33-6		5-BROMOPENTANOIC ACID

Formula: $C_5H_9BrO_2$	
Mol Weight: 181.04	
MP (deg C): 38-40	FP (deg C):
BP (deg C):	
BP pressure (mm Hg):	

Property/Value	Units	Temp	Data Type	Reference
WS 3.20E+003	mg/L	25	EST	MEYLAN,WM ET AL. (1996)
logP 1.90			EST	MEYLAN,WM & HOWARD,PH (1995)
VP 1.54E-002	mm Hg	25	EST	NEELY,WB & BLAU,GE (1985)
DC	pKa			
HL 1.48E-007	atm m3/mol	25	EST	MEYLAN,WM & HOWARD,PH (1991)
OH 3.64E-012	cm3/molc sec	25	EST	MEYLAN,WM & HOWARD,PH (1993)

CAS #: 002070-61-3		FLUOROMETHANE SULFONANILIDE

Formula: $C_7H_8FNO_2S$	
Mol Weight: 189.21	
MP (deg C):	FP (deg C):
BP (deg C):	
BP pressure (mm Hg):	

Property/Value	Units	Temp	Data Type	Reference
WS 3.49E+003	mg/L	25	EST	MEYLAN,WM ET AL. (1996)
logP 1.35			EXP	HANSCH,C & LEO,AJ (1985)
VP 6.51E-004	mm Hg	25	EST	NEELY,WB & BLAU,GE (1985)
DC	pKa			
HL 2.32E-006	atm m3/mol	25	EST	MEYLAN,WM & HOWARD,PH (1991)
OH 4.33E-011	cm3/molc sec	25	EST	MEYLAN,WM & HOWARD,PH (1993)

CAS #: 002075-46-9	4-NITROPYRAZOLE

Formula: $C_3H_3N_3O_2$

Mol Weight: 113.08

MP (deg C):   FP (deg C):

BP (deg C):

BP pressure (mm Hg):

Property/Value	Units	Temp	Data Type	Reference
WS 1.36E+004	mg/L	25	EST	MEYLAN,WM ET AL. (1996)
logP 0.59			EXP	HANSCH,C & LEO,AJ (1985)
VP 4.11E-003	mm Hg	25	EST	NEELY,WB & BLAU,GE (1985)
DC	pKa			
HL 1.46E-008	atm m3/mol	25	EST	MEYLAN,WM & HOWARD,PH (1991)
OH 4.50E-012	cm3/molc sec	25	EST	MEYLAN,WM & HOWARD,PH (1993)

CAS #: 002077-46-5	2,3,6-TRICHLOROTOLUENE

Formula: $C_7H_5Cl_3$

Mol Weight: 195.48

MP (deg C):   FP (deg C):

BP (deg C):

BP pressure (mm Hg):

Property/Value	Units	Temp	Data Type	Reference
WS 6.97E+000	mg/L	25	EST	MEYLAN,WM ET AL. (1996)
logP 4.47			EST	MEYLAN,WM & HOWARD,PH (1995)
VP 7.32E-002	mm Hg	25	EST	NEELY,WB & BLAU,GE (1985)
DC	pKa			
HL 1.50E-003	atm m3/mol	20	EXP	OLIVER,BG (1985)
OH 5.01E-013	cm3/molc sec	25	EST	MEYLAN,WM & HOWARD,PH (1993)

CAS #: 002077-99-8	26NO2-N-PR-AAA-TRIF-P-TOLUIDINE

Formula: $C_{10}H_{10}F_3N_3O_4$

Mol Weight: 293.20

MP (deg C):   FP (deg C):

BP (deg C):

BP pressure (mm Hg):

Property/Value	Units	Temp	Data Type	Reference
WS 1.03E+001	mg/L	25	EST	MEYLAN,WM ET AL. (1996)
logP 3.65			EXP	HANSCH,C & LEO,AJ (1985)
VP 7.02E-006	mm Hg	25	EST	NEELY,WB & BLAU,GE (1985)
DC	pKa			
HL 5.87E-006	atm m3/mol	25	EST	MEYLAN,WM & HOWARD,PH (1991)
OH 1.20E-011	cm3/molc sec	25	EST	MEYLAN,WM & HOWARD,PH (1993)

CAS #: 002078-54-8	PHENOL, 2,6-BIS(1-METHYLETHYL)-

Formula: $C_{12}H_{18}O$

Mol Weight: 178.28

MP (deg C): 19   FP (deg C):

BP (deg C): 136

BP pressure (mm Hg): 3.00E+001

Property/Value	Units	Temp	Data Type	Reference
WS 1.24E+002	mg/L	25	EST	MEYLAN,WM ET AL. (1996)
logP 3.79			EXP	HANSCH,C ET AL. (1995)
VP 3.05E-003	mm Hg	25	EST	NEELY,WB & BLAU,GE (1985)
DC 11.10	pKa	20	EXP	SERJEANT,EP & DEMPSEY,B (1979)
HL 2.12E-006	atm m3/mol	25	EST	MEYLAN,WM & HOWARD,PH (1991)
OH 5.30E-011	cm3/molc sec	25	EST	MEYLAN,WM & HOWARD,PH (1993)

CAS #: 002082-79-3	OCTADECYL 3,5-BIS(TERT-BUTYL)-4-HYDROXYBENZENEP*

Formula: $C_{35}H_{62}O_3$

Mol Weight: 530.88

MP (deg C):   FP (deg C):

BP (deg C):

BP pressure (mm Hg):

Property/Value	Units	Temp	Data Type	Reference
WS 6.09E-009	mg/L	25	EST	MEYLAN,WM ET AL. (1996)
logP 14.26			EST	MEYLAN,WM & HOWARD,PH (1995)
VP 3.38E-013	mm Hg	25	EST	NEELY,WB & BLAU,GE (1985)
DC	pKa			
HL 1.61E-006	atm m3/mol	25	EST	MEYLAN,WM & HOWARD,PH (1991)
OH 4.37E-011	cm3/molc sec	25	EST	MEYLAN,WM & HOWARD,PH (1993)

CAS #: 002082-81-7	1,4-BUTANEDIYL BIS(2-METHYL-2-PROPENOATE)

Formula: $C_{12}H_{18}O_4$

Mol Weight: 226.27

MP (deg C): 226.28   FP (deg C):

BP (deg C): 132-134

BP pressure (mm Hg): 4.00E+000

Property/Value	Units	Temp	Data Type	Reference
WS 6.00E+001	mg/L	25	EST	MEYLAN,WM ET AL. (1996)
logP 3.19			EST	MEYLAN,WM & HOWARD,PH (1995)
VP 2.76E-002	mm Hg	25	EST	NEELY,WB & BLAU,GE (1985)
DC	pKa			
HL 6.67E-007	atm m3/mol	25	EST	MEYLAN,WM & HOWARD,PH (1991)
OH 4.28E-011	cm3/molc sec	25	EST	MEYLAN,WM & HOWARD,PH (1993)

CAS #: 002084-11-9	M-PROPYLPHENOXYACETIC ACID

Formula: $C_{11}H_{14}O_3$

Mol Weight: 194.23

MP (deg C):   FP (deg C):

BP (deg C):

BP pressure (mm Hg):

Property/Value	Units	Temp	Data Type	Reference
WS 5.64E+002	mg/L	25	EST	MEYLAN,WM ET AL. (1996)
logP 2.71			EXP	HANSCH,C & LEO,AJ (1985)
VP 9.28E-005	mm Hg	25	EST	NEELY,WB & BLAU,GE (1985)
DC	pKa			
HL 3.26E-008	atm m3/mol	25	EST	MEYLAN,WM & HOWARD,PH (1991)
OH 6.07E-011	cm3/molc sec	25	EST	MEYLAN,WM & HOWARD,PH (1993)

CAS #: 002084-13-1	M-CARBOXYPHENYLACETIC ACID

Formula: $C_9H_8O_4$

Mol Weight: 180.16

MP (deg C):   FP (deg C):

BP (deg C):

BP pressure (mm Hg):

Property/Value	Units	Temp	Data Type	Reference
WS 1.45E+004	mg/L	25	EST	MEYLAN,WM ET AL. (1996)
logP 1.14			EXP	HANSCH,C & LEO,AJ (1985)
VP 5.18E-006	mm Hg	25	EST	NEELY,WB & BLAU,GE (1985)
DC	pKa			
HL 8.90E-013	atm m3/mol	25	EST	MEYLAN,WM & HOWARD,PH (1991)
OH 2.62E-012	cm3/molc sec	25	EST	MEYLAN,WM & HOWARD,PH (1993)

002088-07-5 — 2-METHYL-1-PENTEN-3-OL

Formula: $C_6H_{12}O$
Mol Weight: 100.16
MP (deg C): FP (deg C):
BP (deg C):
BP pressure (mm Hg):

Property/Value	Units	Temp	Data Type	Reference
WS 1.42E+004	mg/L	25	EST	MEYLAN,WM ET AL. (1996)
logP 1.67			EST	MEYLAN,WM & HOWARD,PH (1995)
VP 3.92E+000	mm Hg	25	EST	NEELY,WB & BLAU,GE (1985)
DC	pKa			
HL 1.55E-005	atm m3/mol	25	EST	MEYLAN,WM & HOWARD,PH (1991)
OH 6.13E-011	cm3/molc sec	25	EST	MEYLAN,WM & HOWARD,PH (1993)

002088-24-6 — M-METHOXYPHENOXYACETIC ACID

Formula: $C_9H_{10}O_4$
Mol Weight: 182.18
MP (deg C): FP (deg C):
BP (deg C):
BP pressure (mm Hg):

Property/Value	Units	Temp	Data Type	Reference
WS 8.85E+003	mg/L	25	EST	MEYLAN,WM ET AL. (1996)
logP 1.38			EXP	HANSCH,C & LEO,AJ (1985)
VP 2.12E-004	mm Hg	25	EST	NEELY,WB & BLAU,GE (1985)
DC 3.14	pKa	25	EXP	KORTUM,G ET AL (1961)
HL 9.93E-010	atm m3/mol	25	EST	MEYLAN,WM & HOWARD,PH (1991)
OH 2.06E-010	cm3/molc sec	25	EST	MEYLAN,WM & HOWARD,PH (1993)

002094-99-7 — 1-(1-ISOCYANATO-1-METHYLETHYL)-3-(1-METHYLETHE*)

Formula: $C_{13}H_{15}NO$
Mol Weight: 201.27
MP (deg C): FP (deg C):
BP (deg C): 268-271
BP pressure (mm Hg):

Property/Value	Units	Temp	Data Type	Reference
WS 4.69E+000	mg/L	25	EST	MEYLAN,WM ET AL. (1996)
logP 4.64			EST	MEYLAN,WM & HOWARD,PH (1995)
VP 2.07E-004	mm Hg	25	EST	NEELY,WB & BLAU,GE (1985)
DC	pKa			
HL 2.01E-006	atm m3/mol	25	EST	MEYLAN,WM & HOWARD,PH (1991)
OH 5.35E-011	cm3/molc sec	25	EST	MEYLAN,WM & HOWARD,PH (1993)

002095-17-2 — 10H-PHENOTHIAZINE-10-PROPANAMINE,2-CHLORO-

Formula: $C_{15}H_{15}ClN_2S$
Mol Weight: 290.82
MP (deg C): FP (deg C):
BP (deg C):
BP pressure (mm Hg):

Property/Value	Units	Temp	Data Type	Reference
WS 1.67E+000	mg/L	25	EST	MEYLAN,WM ET AL. (1996)
logP 4.47			EXP	WHELPTON,R (1989)
VP 2.13E-007	mm Hg	25	EST	NEELY,WB & BLAU,GE (1985)
DC	pKa			
HL 7.65E-011	atm m3/mol	25	EST	MEYLAN,WM & HOWARD,PH (1991)
OH 1.97E-010	cm3/molc sec	25	EST	MEYLAN,WM & HOWARD,PH (1993)

002100-42-7 — 1,4-DIMETHOXY-2-CHLOROBENZENE

Formula: $C_8H_9ClO_2$
Mol Weight: 172.61
MP (deg C): FP (deg C):
BP (deg C):
BP pressure (mm Hg):

Property/Value	Units	Temp	Data Type	Reference
WS 3.01E+002	mg/L	25	EST	MEYLAN,WM ET AL. (1996)
logP 2.69			EXP	NAKAGAWA,Y ET AL. (1992)
VP 1.08E-001	mm Hg	25	EST	NEELY,WB & BLAU,GE (1985)
DC	pKa			
HL 1.40E-005	atm m3/mol	25	EST	MEYLAN,WM & HOWARD,PH (1991)
OH 1.47E-011	cm3/molc sec	25	EST	MEYLAN,WM & HOWARD,PH (1993)

002104-64-5 — EPN

Formula: $C_{14}H_{14}NO_4PS$
Mol Weight: 323.31
MP (deg C): 36 FP (deg C):
BP (deg C): 100
BP pressure (mm Hg): 3.00E-002

Property/Value	Units	Temp	Data Type	Reference
WS 9.10E-001	mg/L	25	EXP	CHEM INSPECT TEST INST (1992)
logP 4.78			EXP	CHEM INSPECT TEST INST (1992)
VP 9.50E-007	mm Hg	25	EXP	AUGUSTIJN-BECKERS,PWM ET AL. (1994)
DC	pKa			
HL 4.44E-007	atm m3/mol	25	EST	VP/WSOL
OH 7.47E-011	cm3/molc sec	25	EST	MEYLAN,WM & HOWARD,PH (1993)

002104-96-3 — BROMOPHOS

Formula: $C_8H_8BrCl_2O_3PS$
Mol Weight: 366.00
MP (deg C): 53-54 FP (deg C):
BP (deg C): 140-142
BP pressure (mm Hg): 1.00E-002

Property/Value	Units	Temp	Data Type	Reference
WS 3.00E-001	mg/L	20	EXP	BOWMAN,BT & SANS,WW (1983A)
logP 5.21			EXP	HANSCH,C ET AL. (1995)
VP 1.28E-004	mm Hg	20		WORTHING,CR & WALKER,SB (1987)
DC	pKa			
HL 2.05E-004	atm m3/mol	20	EST	VP/WSOL
OH 6.10E-011	cm3/molc sec	25	EST	ATKINSON,R (1988)

002113-47-5 — 2-PHENYLACETANILIDE

Formula: $C_{14}H_{13}NO$
Mol Weight: 211.27
MP (deg C): 121 FP (deg C):
BP (deg C): 355
BP pressure (mm Hg):

Property/Value	Units	Temp	Data Type	Reference
WS 7.21E+002	mg/L	25	EST	MEYLAN,WM ET AL. (1996)
logP 2.02			EXP	NAKAGAWA,Y ET AL. (1992)
VP 3.65E-007	mm Hg	25	EST	NEELY,WB & BLAU,GE (1985)
DC	pKa			
HL 4.74E-010	atm m3/mol	25	EST	MEYLAN,WM & HOWARD,PH (1991)
OH 1.24E-011	cm3/molc sec	25	EST	MEYLAN,WM & HOWARD,PH (1993)

CAS #: 002113-57-7				3-BROMOBIPHENYL

Formula: $C_{12}H_9Br$

Mol Weight: 233.11

MP (deg C):   FP (deg C):

BP (deg C): 300

BP pressure (mm Hg):

Property/ Value	Units	Temp	Data Type	Reference
WS 2.11E+000	mg/L	25	EST	MEYLAN,WM ET AL. (1996)
logP 4.85			EXP	DOUCETTE,WJ & ANDREN,AW (1987)
VP 3.54E-004	mm Hg	25	EST	NEELY,WB & BLAU,GE (1985)
DC	pKa			
HL 1.65E-004	atm m3/mol	25	EST	MEYLAN,WM & HOWARD,PH (1991)
OH 4.63E-012	cm3/molc sec	25	EST	MEYLAN,WM & HOWARD,PH (1993)

CAS #: 002113-58-8				3-NITROBIPHENYL

Formula: $C_{12}H_9NO_2$

Mol Weight: 199.21

MP (deg C): 62   FP (deg C):

BP (deg C): 227

BP pressure (mm Hg): 3.50E+001

Property/ Value	Units	Temp	Data Type	Reference
WS 1.37E+000	mg/L	25	EST	GANG,Y & XIAOBAI,XU (1992)
logP 3.87			EXP	HANSCH,C ET AL. (1995)
VP 3.25E-005	mm Hg	25	EST	NEELY,WB & BLAU,GE (1985)
DC	pKa			
HL 1.63E-006	atm m3/mol	25	EST	MEYLAN,WM & HOWARD,PH (1991)
OH 2.79E-012	cm3/molc sec	25	EST	MEYLAN,WM & HOWARD,PH (1993)

CAS #: 002114-15-0				O-ISOBUTYL CARBAMATE

Formula: $C_5H_{11}NO_2$

Mol Weight: 117.15

MP (deg C):   FP (deg C):

BP (deg C):

BP pressure (mm Hg):

Property/ Value	Units	Temp	Data Type	Reference
WS 2.86E+004	mg/L	25	EST	MEYLAN,WM ET AL. (1996)
logP 0.65			EXP	HANSCH,C & LEO,AJ (1985)
VP 4.11E+000	mm Hg	25	EST	NEELY,WB & BLAU,GE (1985)
DC	pKa			
HL 9.26E-008	atm m3/mol	25	EST	MEYLAN,WM & HOWARD,PH (1991)
OH 6.56E-012	cm3/molc sec	25	EST	MEYLAN,WM & HOWARD,PH (1993)

CAS #: 002114-20-7				O-HEXYL CARBAMATE

Formula: $C_7H_{15}NO_2$

Mol Weight: 145.20

MP (deg C):   FP (deg C):

BP (deg C):

BP pressure (mm Hg):

Property/ Value	Units	Temp	Data Type	Reference
WS 2.09E+003	mg/L	25	EST	MEYLAN,WM ET AL. (1996)
logP 1.85			EXP	HANSCH,C & LEO,AJ (1985)
VP 2.97E-001	mm Hg	25	EST	NEELY,WB & BLAU,GE (1985)
DC	pKa			
HL 1.63E-007	atm m3/mol	25	EST	MEYLAN,WM & HOWARD,PH (1991)
OH 9.39E-012	cm3/molc sec	25	EST	MEYLAN,WM & HOWARD,PH (1993)

CAS #: 002116-65-6				4-BENZYLPYRIDINE

Formula: $C_{12}H_{11}N$

Mol Weight: 169.23

MP (deg C): 12.4   FP (deg C):

BP (deg C): 288

BP pressure (mm Hg):

Property/ Value	Units	Temp	Data Type	Reference
WS 7.16E+003	mg/L	25	EST	MEYLAN,WM ET AL. (1996)
logP 3.06			EST	MEYLAN,WM & HOWARD,PH (1995)
VP 3.02E-003	mm Hg	25	EST	NEELY,WB & BLAU,GE (1985)
DC 5.59	pKa	25	EXP	PERRIN,DD (1972)
HL 6.28E-007	atm m3/mol	25	EST	MEYLAN,WM & HOWARD,PH (1991)
OH 6.70E-012	cm3/molc sec	25	EST	MEYLAN,WM & HOWARD,PH (1993)

CAS #: 002122-63-6				BENZOIC ACID, 4-AMINO-3-IODO-

Formula: $C_7H_6INO_2$

Mol Weight: 263.04

MP (deg C):   FP (deg C):

BP (deg C):

BP pressure (mm Hg):

Property/ Value	Units	Temp	Data Type	Reference
WS 1.93E+002	mg/L	25	EST	MEYLAN,WM ET AL. (1996)
logP 2.36			EXP	SANGSTER,J (1994)
VP 5.87E-006	mm Hg	25	EST	NEELY,WB & BLAU,GE (1985)
DC	pKa			
HL 8.88E-012	atm m3/mol	25	EST	MEYLAN,WM & HOWARD,PH (1991)
OH 1.37E-011	cm3/molc sec	25	EST	MEYLAN,WM & HOWARD,PH (1993)

CAS #: 002124-31-4				P-ACETYL-N,N-DIMETHYLANILINE

Formula: $C_{10}H_{13}NO$

Mol Weight: 163.22

MP (deg C):   FP (deg C):

BP (deg C):

BP pressure (mm Hg):

Property/ Value	Units	Temp	Data Type	Reference
WS 1.06E+003	mg/L	25	EST	MEYLAN,WM ET AL. (1996)
logP 2.10			EXP	HANSCH,C & LEO,AJ (1985)
VP 2.33E-002	mm Hg	25	EST	NEELY,WB & BLAU,GE (1985)
DC	pKa			
HL 1.56E-007	atm m3/mol	25	EST	MEYLAN,WM & HOWARD,PH (1991)
OH 2.03E-010	cm3/molc sec	25	EST	MEYLAN,WM & HOWARD,PH (1993)

CAS #: 002125-48-6				PHENYLPROPYL-TRIME-AMMONIUM IODIDE

Formula: $C_{12}H_{20}IN$

Mol Weight: 305.20

MP (deg C):   FP (deg C):

BP (deg C):

BP pressure (mm Hg):

Property/ Value	Units	Temp	Data Type	Reference
WS 6.08E+005	mg/L	25	EST	MEYLAN,WM ET AL. (1996)
logP -2.02			EXP	HANSCH,C & LEO,AJ (1985)
VP 2.54E-009	mm Hg	25	EST	NEELY,WB & BLAU,GE (1985)
DC	pKa			
HL 3.75E-013	atm m3/mol	25	EST	MEYLAN,WM & HOWARD,PH (1991)
OH 2.19E-011	cm3/molc sec	25	EST	MEYLAN,WM & HOWARD,PH (1993)

CAS #: 002131-18-2				1-PHENYLPENTADECANE

Formula: $C_{21}H_{36}$

Mol Weight: 288.52

MP (deg C): 22   **FP (deg C):**

BP (deg C): 373

BP pressure (mm Hg):

Property/ Value	Units	Temp	Data Type	Reference
WS 1.30E-004	mg/L	25	EST	MEYLAN,WM ET AL. (1996)
logP 9.42			EST	MEYLAN,WM & HOWARD,PH (1995)
VP 1.82E-005	mm Hg	25	EST	NEELY,WB & BLAU,GE (1985)
DC	pKa			
HL 3.14E-001	atm m3/mol	25	EST	MEYLAN,WM & HOWARD,PH (1991)
OH 2.43E-011	cm3/molc sec	25	EST	MEYLAN,WM & HOWARD,PH (1993)

CAS #: 002131-41-1				1,4,5-TRIMETHYLNAPHTHALENE

Formula: $C_{13}H_{14}$

Mol Weight: 170.26

MP (deg C):   **FP (deg C):**

BP (deg C):

BP pressure (mm Hg):

Property/ Value	Units	Temp	Data Type	Reference
WS 2.10E+000	mg/L	25	EXP	YALKOWSKY,SH & DANNENFELSER,RM (1992)
logP 4.90			EXP	MILLER,MM ET AL. (1985)
VP 2.52E-003	mm Hg	25	EST	NEELY,WB & BLAU,GE (1985)
DC	pKa			
HL 7.07E-004	atm m3/mol	25	EST	MEYLAN,WM & HOWARD,PH (1991)
OH 1.24E-010	cm3/molc sec	25	EST	MEYLAN,WM & HOWARD,PH (1993)

CAS #: 002131-55-7				BENZENE, 1-CHLORO-4-ISOTHIOCYANATO-

Formula: C_7H_4ClNS

Mol Weight: 169.63

MP (deg C): 46   **FP (deg C):**

BP (deg C): 249.5

BP pressure (mm Hg):

Property/ Value	Units	Temp	Data Type	Reference
WS 2.83E+001	mg/L	25	EST	MEYLAN,WM ET AL. (1996)
logP 3.91			EXP	AUGUSTIN,J ET AL. (1987)
VP 1.34E-002	mm Hg	25	EST	NEELY,WB & BLAU,GE (1985)
DC	pKa			
HL 5.79E-004	atm m3/mol	25	EST	MEYLAN,WM & HOWARD,PH (1991)
OH 1.50E-012	cm3/molc sec	25	EST	MEYLAN,WM & HOWARD,PH (1993)

CAS #: 002131-59-1				BENZENE, 1-BROMO-3-ISOTHIOCYANATO-

Formula: C_7H_4BrNS

Mol Weight: 214.09

MP (deg C):   **FP (deg C):**

BP (deg C): 256

BP pressure (mm Hg):

Property/ Value	Units	Temp	Data Type	Reference
WS 1.75E+001	mg/L	25	EXP	YALKOWSKY,SH & DANNEDFELSER,RM (1992)
logP 4.12			EXP	HANSCH,C ET AL. (1995)
VP 4.11E-003	mm Hg	25	EST	NEELY,WB & BLAU,GE (1985)
DC	pKa			
HL 3.11E-004	atm m3/mol	25	EST	MEYLAN,WM & HOWARD,PH (1991)
OH 7.73E-013	cm3/molc sec	25	EST	MEYLAN,WM & HOWARD,PH (1993)

CAS #: 002131-60-4				4-HYDROXYPHENYLISOTHIOCYANATE

Formula: C_7H_5NOS

Mol Weight: 151.19

MP (deg C):   **FP (deg C):**

BP (deg C):

BP pressure (mm Hg):

Property/ Value	Units	Temp	Data Type	Reference
WS 1.30E+003	mg/L	25	EST	MEYLAN,WM ET AL. (1996)
logP 2.74			EXP	HANSCH,C & LEO,AJ (1985)
VP 9.26E-004	mm Hg	25	EST	NEELY,WB & BLAU,GE (1985)
DC	pKa			
HL 8.13E-008	atm m3/mol	25	EST	MEYLAN,WM & HOWARD,PH (1991)
OH 3.67E-011	cm3/molc sec	25	EST	MEYLAN,WM & HOWARD,PH (1993)

CAS #: 002131-61-5				4-NITROPHENYLISOTHIOCYANATE

Formula: $C_7H_4N_2O_2S$

Mol Weight: 180.19

MP (deg C): 110-112   **FP (deg C):**

BP (deg C): 137-138

BP pressure (mm Hg): 1.10E+001

Property/ Value	Units	Temp	Data Type	Reference
WS 4.45E+001	mg/L	25	EST	MEYLAN,WM ET AL. (1996)
logP 3.62			EXP	HANSCH,C & LEO,AJ (1985)
VP 3.45E-004	mm Hg	25	EST	NEELY,WB & BLAU,GE (1985)
DC	pKa			
HL 3.08E-006	atm m3/mol	25	EST	MEYLAN,WM & HOWARD,PH (1991)
OH 1.87E-013	cm3/molc sec	25	EST	MEYLAN,WM & HOWARD,PH (1993)

CAS #: 002131-62-6				4-CARBOXYPHENYLISOTHIOCYANATE

Formula: C_8H_5NOS

Mol Weight: 163.20

MP (deg C):   **FP (deg C):**

BP (deg C):

BP pressure (mm Hg):

Property/ Value	Units	Temp	Data Type	Reference
WS 6.53E+001	mg/L	25	EST	MEYLAN,WM ET AL. (1996)
logP 3.52			EXP	HANSCH,C & LEO,AJ (1985)
VP 1.62E-003	mm Hg	25	EST	NEELY,WB & BLAU,GE (1985)
DC	pKa			
HL 1.95E-006	atm m3/mol	25	EST	MEYLAN,WM & HOWARD,PH (1991)
OH 1.80E-011	cm3/molc sec	25	EST	MEYLAN,WM & HOWARD,PH (1993)

CAS #: 002131-64-8				BENZENAMINE, 4-ISOTHIOCYANATO-N,N-DIMETHYL-

Formula: $C_9H_{10}N_2S$

Mol Weight: 178.26

MP (deg C):   **FP (deg C):**

BP (deg C):

BP pressure (mm Hg):

Property/ Value	Units	Temp	Data Type	Reference
WS 2.33E+001	mg/L	25	EST	MEYLAN,WM ET AL. (1996)
logP 3.96			EXP	AUGUSTIN,J ET AL. (1987)
VP 2.69E-003	mm Hg	25	EST	NEELY,WB & BLAU,GE (1985)
DC	pKa			
HL 1.24E-005	atm m3/mol	25	EST	MEYLAN,WM & HOWARD,PH (1991)
OH 2.03E-010	cm3/molc sec	25	EST	MEYLAN,WM & HOWARD,PH (1993)

CAS #: 002133-81-5			A-2'-DEOXYTHIOGUANOSINE	
Formula: $C_{10}H_{15}N_5O_3S$				
Mol Weight: 285.33				
MP (deg C):		FP (deg C):		
BP (deg C):				
BP pressure (mm Hg):				

Property/Value	Units	Temp	Data Type	Reference
WS 3.55E+003	mg/L	25	EST	MEYLAN,WM ET AL. (1996)
logP -0.79			EXP	HANSCH,C & LEO,AJ (1985)
VP 1.61E-018	mm Hg	25	EST	NEELY,WB & BLAU,GE (1985)
DC	pKa			
HL 8.22E-024	atm m3/mol	25	EST	MEYLAN,WM & HOWARD,PH (1991)
OH 1.11E-010	cm3/molc sec	25	EST	MEYLAN,WM & HOWARD,PH (1993)

CAS #: 002135-17-3			FLUMETHASONE	
Formula: $C_{22}H_{28}F_2O_5$				
Mol Weight: 410.46				
MP (deg C):		FP (deg C):		
BP (deg C):				
BP pressure (mm Hg):				

Property/Value	Units	Temp	Data Type	Reference
WS 1.00E+000	mg/L	20	EXP	YALKOWSKY,SH & DANNENFELSER,RM (1992)
logP 1.94			EXP	HANSCH,C & LEO,AJ (1985)
VP 1.08E-013	mm Hg	25	EST	NEELY,WB & BLAU,GE (1985)
DC	pKa			
HL 1.42E-007	atm m3/mol	25	EST	MEYLAN,WM & HOWARD,PH (1991)
OH 5.98E-011	cm3/molc sec	25	EST	MEYLAN,WM & HOWARD,PH (1993)

CAS #: 002136-89-2			2-CHLOROBENZOTRICHLORIDE	
Formula: $C_7H_4Cl_4$				
Mol Weight: 229.92				
MP (deg C): 29.4		FP (deg C):		
BP (deg C): 264.3				
BP pressure (mm Hg):				

Property/Value	Units	Temp	Data Type	Reference
WS 4.04E+000	mg/L	25	EST	MEYLAN,WM ET AL. (1996)
logP 4.54			EST	MEYLAN,WM & HOWARD,PH (1995)
VP 5.04E-002	mm Hg	25	EXT	PERRY,RH & GREEN,D (1984)
DC	pKa			
HL 1.93E-004	atm m3/mol	25	EST	MEYLAN,WM & HOWARD,PH (1991)
OH 2.51E-013	cm3/molc sec	25	EST	MEYLAN,WM & HOWARD,PH (1993)

CAS #: 002136-99-4			2,2',3,3',5,5',6,6'-OCTACHLOROBIPHENYL	
Formula: $C_{12}H_2Cl_8$				
Mol Weight: 429.77				
MP (deg C):		FP (deg C):		
BP (deg C):				
BP pressure (mm Hg):				

Property/Value	Units	Temp	Data Type	Reference
WS 1.47E-004	mg/L	25	EXP	YALKOWSKY,SH & DANNENFELSER,RM (1992)
logP 7.73			EXP	DEBRUIJN,J ET AL. (1989)
VP 3.93E-006	mm Hg	25	EXP	HINCKLEY,DA ET AL. (1990)
DC	pKa			
HL 1.51E-002	atm m3/mol	25	EST	VP/WSOL
OH 5.59E-014	cm3/molc sec	25	EST	MEYLAN,WM & HOWARD,PH (1993)

CAS #: 002140-61-6			5-METHYLCYTIDINE	
Formula: $C_{10}H_{15}N_3O_5$				
Mol Weight: 257.25				
MP (deg C): 212-215 de		FP (deg C):		
BP (deg C):				
BP pressure (mm Hg):				

Property/Value	Units	Temp	Data Type	Reference
WS 5.50E+004	mg/L	25	EST	MEYLAN,WM ET AL. (1996)
logP -2.01			EXP	HANSCH,C ET AL. (1995)
VP 6.56E-012	mm Hg	25	EST	NEELY,WB & BLAU,GE (1985)
DC	pKa			
HL 2.45E-019	atm m3/mol	25	EST	MEYLAN,WM & HOWARD,PH (1991)
OH 1.57E-010	cm3/molc sec	25	EST	MEYLAN,WM & HOWARD,PH (1993)

CAS #: 002142-63-4			M-BROMOACETOPHENONE	
Formula: C_8H_7BrO				
Mol Weight: 199.05				
MP (deg C): 7.5		FP (deg C):		
BP (deg C): 133				
BP pressure (mm Hg): 1.90E+001				

Property/Value	Units	Temp	Data Type	Reference
WS 3.44E+002	mg/L	25	EST	MEYLAN,WM ET AL. (1996)
logP 2.47			EXP	DUNN,WJ ET AL. (1983)
VP 3.08E-002	mm Hg	25	EST	NEELY,WB & BLAU,GE (1985)
DC	pKa			
HL 3.91E-006	atm m3/mol	25	EST	MEYLAN,WM & HOWARD,PH (1991)
OH 6.19E-013	cm3/molc sec	25	EST	MEYLAN,WM & HOWARD,PH (1993)

CAS #: 002142-68-9			O-CHLOROACETOPHENONE	
Formula: C_8H_7ClO				
Mol Weight: 154.60				
MP (deg C):		FP (deg C):		
BP (deg C):				
BP pressure (mm Hg):				

Property/Value	Units	Temp	Data Type	Reference
WS 1.19E+003	mg/L	25	EST	MEYLAN,WM ET AL. (1996)
logP 2.09			EXP	HANSCH,C & LEO,AJ (1985)
VP 1.22E-001	mm Hg	25	EST	NEELY,WB & BLAU,GE (1985)
DC	pKa			
HL 7.27E-006	atm m3/mol	25	EST	MEYLAN,WM & HOWARD,PH (1991)
OH 1.35E-012	cm3/molc sec	25	EST	MEYLAN,WM & HOWARD,PH (1993)

CAS #: 002142-69-0			ETHANONE, 1-(2-BROMOPHENYL)-	
Formula: C_8H_7BrO				
Mol Weight: 199.05				
MP (deg C):		FP (deg C):		
BP (deg C):				
BP pressure (mm Hg):				

Property/Value	Units	Temp	Data Type	Reference
WS 5.00E+002	mg/L	25	EST	MEYLAN,WM ET AL. (1996)
logP 2.28			EXP	SANGSTER,J (1993)
VP 3.08E-002	mm Hg	25	EST	NEELY,WB & BLAU,GE (1985)
DC	pKa			
HL 3.91E-006	atm m3/mol	25	EST	MEYLAN,WM & HOWARD,PH (1991)
OH 1.22E-012	cm3/molc sec	25	EST	MEYLAN,WM & HOWARD,PH (1993)

CAS #: 002144-41-4			CIS-2,5-DIMETHYLTETRAHYDROFURAN

Formula: $C_6H_{12}O$

Mol Weight: 100.16

MP (deg C): | FP (deg C):

BP (deg C):

BP pressure (mm Hg):

Property/ Value	Units	Temp	Data Type	Reference
WS 8.38E+003	mg/L	25	EST	MEYLAN,WM ET AL. (1996)
logP 1.22			EXP	FUNASAKI,N ET AL. (1985)
VP 3.64E+001	mm Hg	25	EST	NEELY,WB & BLAU,GE (1985)
DC	pKa			
HL 1.49E-004	atm m3/mol	25	EST	MEYLAN,WM & HOWARD,PH (1991)
OH 2.59E-011	cm3/molc sec	25	EST	MEYLAN,WM & HOWARD,PH (1993)

CAS #: 002150-47-2			METHYL 2,4-DIHYDROXYBENZOATE

Formula: $C_8H_8O_4$

Mol Weight: 168.15

MP (deg C): 118-121 | FP (deg C):

BP (deg C):

BP pressure (mm Hg):

Property/ Value	Units	Temp	Data Type	Reference
WS 3.67E+003	mg/L	25	EST	MEYLAN,WM ET AL. (1996)
logP 2.12			EST	MEYLAN,WM & HOWARD,PH (1995)
VP 4.99E-005	mm Hg	25	EST	NEELY,WB & BLAU,GE (1985)
DC	pKa			
HL 4.73E-010	atm m3/mol	25	EST	MEYLAN,WM & HOWARD,PH (1991)
OH 1.84E-010	cm3/molc sec	25	EST	MEYLAN,WM & HOWARD,PH (1993)

CAS #: 002150-88-1			ME-N-(M-CHLOROPHENYL)CARBAMATE

Formula: $C_8H_8ClNO_2$

Mol Weight: 185.61

MP (deg C): | FP (deg C):

BP (deg C):

BP pressure (mm Hg):

Property/ Value	Units	Temp	Data Type	Reference
WS 3.24E+002	mg/L	25	EST	MEYLAN,WM ET AL. (1996)
logP 2.58			EXP	HANSCH,C & LEO,AJ (1985)
VP 8.31E-003	mm Hg	25	EST	NEELY,WB & BLAU,GE (1985)
DC	pKa			
HL 1.62E-008	atm m3/mol	25	EST	MEYLAN,WM & HOWARD,PH (1991)
OH 3.02E-011	cm3/molc sec	25	EST	MEYLAN,WM & HOWARD,PH (1993)

CAS #: 002150-93-8			N-(3,4-DICHLOROPHENYL)ACETAMIDE

Formula: $C_8H_7Cl_2NO$

Mol Weight: 204.06

MP (deg C): | FP (deg C):

BP (deg C):

BP pressure (mm Hg):

Property/ Value	Units	Temp	Data Type	Reference
WS 1.14E+002	mg/L	25	EST	MEYLAN,WM ET AL. (1996)
logP 3.00			EXP	NAKAGAWA,Y ET AL. (1992)
VP 2.02E-005	mm Hg	25	EST	NEELY,WB & BLAU,GE (1985)
DC	pKa			
HL 3.39E-009	atm m3/mol	25	EST	MEYLAN,WM & HOWARD,PH (1991)
OH 2.65E-012	cm3/molc sec	25	EST	MEYLAN,WM & HOWARD,PH (1993)

CAS #: 002150-95-0			BUTANAMIDE, N-(3,4-DICHLOROPHENYL)-

Formula: $C_{10}H_{11}Cl_2NO$

Mol Weight: 232.11

MP (deg C): | FP (deg C):

BP (deg C):

BP pressure (mm Hg):

Property/ Value	Units	Temp	Data Type	Reference
WS 1.79E+001	mg/L	25	EST	MEYLAN,WM ET AL. (1996)
logP 3.77			EXP	HANSCH,C ET AL. (1995)
VP 4.60E-006	mm Hg	25	EST	NEELY,WB & BLAU,GE (1985)
DC	pKa			
HL 5.97E-009	atm m3/mol	25	EST	MEYLAN,WM & HOWARD,PH (1991)
OH 7.22E-012	cm3/molc sec	25	EST	MEYLAN,WM & HOWARD,PH (1993)

CAS #: 002150-96-1			PENTANAMIDE, N-(3,4-DICHLOROPHENYL)-

Formula: $C_{11}H_{13}Cl_2NO$

Mol Weight: 246.14

MP (deg C): | FP (deg C):

BP (deg C):

BP pressure (mm Hg):

Property/ Value	Units	Temp	Data Type	Reference
WS 6.83E+000	mg/L	25	EST	MEYLAN,WM ET AL. (1996)
logP 4.17			EXP	HANSCH,C ET AL. (1995)
VP 2.00E-006	mm Hg	25	EST	NEELY,WB & BLAU,GE (1985)
DC	pKa			
HL 7.93E-009	atm m3/mol	25	EST	MEYLAN,WM & HOWARD,PH (1991)
OH 9.21E-012	cm3/molc sec	25	EST	MEYLAN,WM & HOWARD,PH (1993)

CAS #: 002152-44-5			BETAMETHASONE-17-VALERATE

Formula: $C_{27}H_{37}FO_6$

Mol Weight: 476.59

MP (deg C): 183-184 | FP (deg C):

BP (deg C):

BP pressure (mm Hg):

Property/ Value	Units	Temp	Data Type	Reference
WS 8.52E-001	mg/L	25	EST	MEYLAN,WM ET AL. (1996)
logP 3.60			EXP	HANSCH,C & LEO,AJ (1985)
VP 3.70E-015	mm Hg	25	EST	NEELY,WB & BLAU,GE (1985)
DC	pKa			
HL 1.81E-011	atm m3/mol	25	EST	MEYLAN,WM & HOWARD,PH (1991)
OH 7.22E-011	cm3/molc sec	25	EST	MEYLAN,WM & HOWARD,PH (1993)

CAS #: 002153-10-8			4-HYDROXY DICHLOROACETANILIDE

Formula: $C_8H_7Cl_2NO_2$

Mol Weight: 220.06

MP (deg C): | FP (deg C):

BP (deg C):

BP pressure (mm Hg):

Property/ Value	Units	Temp	Data Type	Reference
WS 1.48E+003	mg/L	25	EST	MEYLAN,WM ET AL. (1996)
logP 1.60			EXP	DUTTA,H ET AL. (1988)
VP 2.66E-007	mm Hg	25	EST	NEELY,WB & BLAU,GE (1985)
DC	pKa			
HL 7.97E-014	atm m3/mol	25	EST	MEYLAN,WM & HOWARD,PH (1991)
OH 1.78E-011	cm3/molc sec	25	EST	MEYLAN,WM & HOWARD,PH (1993)

641

002156-14-1

CAS #: 002156-14-1

1H-INDENE-1,3(2H)-DIONE, 2-[1,1'-BIPHENYL]-4-YL-

Formula: $C_{21}H_{14}O_2$

Mol Weight: 298.34

MP (deg C):　**FP (deg C):**

BP (deg C):

BP pressure (mm Hg):

Property/Value	Units	Temp	Data Type	Reference
WS 1.02E+000	mg/L	25	EST	MEYLAN,WM ET AL. (1996)
logP 4.79			EXP	SANGSTER,J (1993)
VP 2.62E-009	mm Hg	25	EST	NEELY,WB & BLAU,GE (1985)
DC	pKa			
HL 3.68E-011	atm m3/mol	25	EST	MEYLAN,WM & HOWARD,PH (1991)
OH 9.09E-012	cm3/molc sec	25	EST	MEYLAN,WM & HOWARD,PH (1993)

002162-92-7

CAS #: 002162-92-7

1,2-DICHLOROHEXANE

Formula: $C_6H_{12}Cl_2$

Mol Weight: 155.07

MP (deg C):　**FP (deg C):**

BP (deg C): 173

BP pressure (mm Hg):

Property/Value	Units	Temp	Data Type	Reference
WS 4.77E+001	mg/L	25	EST	MEYLAN,WM ET AL. (1996)
logP 3.72			EST	MEYLAN,WM & HOWARD,PH (1995)
VP 1.64E+000	mm Hg	25	EST	NEELY,WB & BLAU,GE (1985)
DC	pKa			
HL 3.77E-002	atm m3/mol	25	EST	MEYLAN,WM & HOWARD,PH (1991)
OH 3.60E-012	cm3/molc sec	25	EST	MEYLAN,WM & HOWARD,PH (1993)

002163-68-0

CAS #: 002163-68-0

4-ETHYLAMINO-6-ISOPROPYLAMINO-S-TRIAZIN-2-OL

Formula: $C_8H_{15}N_5O$

Mol Weight: 197.24

MP (deg C):　**FP (deg C):**

BP (deg C):

BP pressure (mm Hg):

Property/Value	Units	Temp	Data Type	Reference
WS 5.92E+000	mg/L	26	EXP	WARD,TM & WEBER,JB (1968) @ pH = 7
logP 2.09			EST	MEYLAN,WM & HOWARD,PH (1995)
VP 8.49E-006	mm Hg	25	EST	NEELY,WB & BLAU,GE (1985)
DC	pKa			
HL 6.28E-013	atm m3/mol	25	EST	MEYLAN,WM & HOWARD,PH (1991)
OH 2.76E-011	cm3/molc sec	25	EST	MEYLAN,WM & HOWARD,PH (1993)

002163-69-1

CAS #: 002163-69-1

CYCLURON

Formula: $C_{11}H_{22}N_2O$

Mol Weight: 198.31

MP (deg C):　**FP (deg C):**

BP (deg C):

BP pressure (mm Hg):

Property/Value	Units	Temp	Data Type	Reference
WS 1.20E+003	mg/L	20	EXP	YALKOWSKY,SH & DANNENFELSER,RM (1992)
logP 2.84			EST	MEYLAN,WM & HOWARD,PH (1995)
VP 6.67E-005	mm Hg	25	EST	NEELY,WB & BLAU,GE (1985)
DC	pKa			
HL 1.24E-008	atm m3/mol	25	EST	MEYLAN,WM & HOWARD,PH (1991)
OH 2.50E-011	cm3/molc sec	25	EST	MEYLAN,WM & HOWARD,PH (1993)

002163-79-3

CAS #: 002163-79-3

3-(HEXAHYDRO-4,7-METHANOINDAN-5-YL)-1,1-DIMETHY*

Formula: $C_{13}H_{22}N_2O$

Mol Weight: 222.33

MP (deg C):　**FP (deg C):**

BP (deg C):

BP pressure (mm Hg):

Property/Value	Units	Temp	Data Type	Reference
WS 1.94E+002	mg/L	25	EST	MEYLAN,WM ET AL. (1996)
logP 2.62			EST	MEYLAN,WM & HOWARD,PH (1995)
VP 1.71E-005	mm Hg	25	EST	NEELY,WB & BLAU,GE (1985)
DC	pKa			
HL 4.24E-009	atm m3/mol	25	EST	MEYLAN,WM & HOWARD,PH (1991)
OH 2.03E-011	cm3/molc sec	25	EST	MEYLAN,WM & HOWARD,PH (1993)

002164-08-1

CAS #: 002164-08-1

LENACIL

Formula: $C_{13}H_{18}N_2O_2$

Mol Weight: 234.30

MP (deg C): 290　**FP (deg C):**

BP (deg C):

BP pressure (mm Hg):

Property/Value	Units	Temp	Data Type	Reference
WS 6.00E+000	mg/L	25	EXP	YALKOWSKY,SH & DANNENFELSER,RM (1992)
logP 3.09			EST	MEYLAN,WM & HOWARD,PH (1995)
VP 1.12E-008	mm Hg	25	EST	NEELY,WB & BLAU,GE (1985)
DC	pKa			
HL 4.97E-010	atm m3/mol	25	EST	MEYLAN,WM & HOWARD,PH (1991)
OH 4.33E-011	cm3/molc sec	25	EST	MEYLAN,WM & HOWARD,PH (1993)

002164-09-2

CAS #: 002164-09-2

DICRYL

Formula: $C_{10}H_9Cl_2NO$

Mol Weight: 230.10

MP (deg C): 128　**FP (deg C):**

BP (deg C):

BP pressure (mm Hg):

Property/Value	Units	Temp	Data Type	Reference
WS 8.50E+000	mg/L		EXP	GUNTHER,FA ET AL. (1968)
logP 3.29			EST	MEYLAN,WM & HOWARD,PH (1995)
VP 5.47E-006	mm Hg	25	EST	NEELY,WB & BLAU,GE (1985)
DC	pKa			
HL 2.80E-009	atm m3/mol	25	EST	MEYLAN,WM & HOWARD,PH (1991)
OH 2.69E-011	cm3/molc sec	25	EST	MEYLAN,WM & HOWARD,PH (1993)

002164-13-8

CAS #: 002164-13-8

BCPC

Formula: $C_{11}H_{14}ClNO_2$

Mol Weight: 227.69

MP (deg C):　**FP (deg C):**

BP (deg C):

BP pressure (mm Hg):

Property/Value	Units	Temp	Data Type	Reference
WS 1.53E+002	mg/L	30	EXP	FREED,VH ET AL. (1967)
logP 3.80			EST	MEYLAN,WM & HOWARD,PH (1995)
VP 7.13E-004	mm Hg	25	EST	NEELY,WB & BLAU,GE (1985)
DC	pKa			
HL 3.79E-008	atm m3/mol	25	EST	MEYLAN,WM & HOWARD,PH (1991)
OH 4.29E-011	cm3/molc sec	25	EST	MEYLAN,WM & HOWARD,PH (1993)

CAS #: 002164-17-2 — FLUOMETURON

Formula: $C_{10}H_{11}F_3N_2O$

Mol Weight: 232.21

MP (deg C): 163-164.5

FP (deg C):

BP (deg C):

BP pressure (mm Hg):

Property/Value	Units	Temp	Data Type	Reference
WS 1.10E+002	mg/L	25	EXP	WAUCHOPE,RD ET AL. (1992)
logP 2.42			EXP	HANSCH,C ET AL. (1995)
VP 5.00E-007	mm Hg	20	NR	MARTIN,H & WORTHING,CR (1977)
DC	pKa			
HL 1.80E-009	atm m3/mol	20	EST	VP/WSOL
OH 3.16E+001	cm3/molc sec	25	EST	MEYLAN,WM & HOWARD,PH (1993)

CAS #: 002169-64-4 — 6-AZAURIDINETRIACETATE

Formula: $C_{14}H_{17}N_3O_9$

Mol Weight: 371.31

MP (deg C): 99-101

FP (deg C):

BP (deg C):

BP pressure (mm Hg):

Property/Value	Units	Temp	Data Type	Reference
WS 3.34E+002	mg/L	25	EST	MEYLAN,WM ET AL. (1996)
logP -0.20			EXP	HANSCH,C ET AL. (1995)
VP 1.60E-012	mm Hg	25	EST	NEELY,WB & BLAU,GE (1985)
DC	pKa			
HL 1.72E-018	atm m3/mol	25	EST	MEYLAN,WM & HOWARD,PH (1991)
OH 8.89E-011	cm3/molc sec	25	EST	MEYLAN,WM & HOWARD,PH (1993)

CAS #: 002170-44-7 — 1,2,4,6,7,9-HEXABROMO-3,8-DICHLORODIBENZO-P-DIO*

Formula: $C_{12}Br_6Cl_2O_2$

Mol Weight: 726.49

MP (deg C):

FP (deg C):

BP (deg C):

BP pressure (mm Hg):

Property/Value	Units	Temp	Data Type	Reference
WS 7.76E-010	mg/L	25	EST	MEYLAN,WM ET AL. (1996)
logP 10.97			EST	MEYLAN,WM & HOWARD,PH (1995)
VP 4.35E-011	mm Hg	25	EST	NEELY,WB & BLAU,GE (1985)
DC	pKa			
HL 2.57E-008	atm m3/mol	25	EST	MEYLAN,WM & HOWARD,PH (1991)
OH 3.58E-013	cm3/molc sec	25	EST	MEYLAN,WM & HOWARD,PH (1993)

CAS #: 002170-45-8 — OCTABROMODIBENZO-P-DIOXIN

Formula: $C_{12}Br_8O_2$

Mol Weight: 815.40

MP (deg C):

FP (deg C):

BP (deg C):

BP pressure (mm Hg):

Property/Value	Units	Temp	Data Type	Reference
WS 7.47E-011	mg/L	25	EST	MEYLAN,WM ET AL. (1996)
logP 11.46			EST	MEYLAN,WM & HOWARD,PH (1995)
VP 7.73E-012	mm Hg	25	EST	NEELY,WB & BLAU,GE (1985)
DC	pKa			
HL 7.43E-009	atm m3/mol	25	EST	MEYLAN,WM & HOWARD,PH (1991)
OH 3.20E-013	cm3/molc sec	25	EST	MEYLAN,WM & HOWARD,PH (1993)

CAS #: 002176-16-1 — BENZENEETHANAMINE, 3,5-DIMETHOXY-4-(PHENYLMETHOX

Formula: $C_{17}H_{21}NO_3$

Mol Weight: 287.36

MP (deg C):

FP (deg C):

BP (deg C):

BP pressure (mm Hg):

Property/Value	Units	Temp	Data Type	Reference
WS 1.32E+003	mg/L	25	EST	MEYLAN,WM ET AL. (1996)
logP 2.40			EXP	SANGSTER,J (1993)
VP 5.67E-007	mm Hg	25	EST	NEELY,WB & BLAU,GE (1985)
DC	pKa			
HL 1.36E-011	atm m3/mol	25	EST	MEYLAN,WM & HOWARD,PH (1991)
OH 2.40E-010	cm3/molc sec	25	EST	MEYLAN,WM & HOWARD,PH (1993)

CAS #: 002176-62-7 — 2,3,4,5,6-PENTACHLORPYRIDINE

Formula: C_5Cl_5N

Mol Weight: 251.33

MP (deg C): 123

FP (deg C):

BP (deg C): 94

BP pressure (mm Hg):

Property/Value	Units	Temp	Data Type	Reference
WS 8.50E+001	mg/L	25	EST	MEYLAN,WM ET AL. (1996)
logP 3.53			EXP	HANSCH,C & LEO,AJ (1985)
VP 1.40E-002	mm Hg	25	EXT	GEHRING,J ET AL. (1967)
DC -1.00	pKa		EXP	GEHRING,J ET AL. (1967)
HL 3.08E-006	atm m3/mol	25	EST	MEYLAN,WM & HOWARD,PH (1991)
OH 1.42E-014	cm3/molc sec			ATKINSON,R (1987)

CAS #: 002180-92-9 — BUPIVACAINE

Formula: $C_{18}H_{28}N_2O$

Mol Weight: 288.44

MP (deg C): 107-108

FP (deg C):

BP (deg C):

BP pressure (mm Hg):

Property/Value	Units	Temp	Data Type	Reference
WS 2.40E+003	mg/L	25	EXP	YALKOWSKY,SH & DANNENFELSER,RM (1992)
logP 3.41			EXP	HANSCH,C ET AL. (1995)
VP 1.79E-008	mm Hg	25	EST	NEELY,WB & BLAU,GE (1985)
DC	pKa			
HL 1.79E-010	atm m3/mol	25	EST	MEYLAN,WM & HOWARD,PH (1991)
OH 1.33E-010	cm3/molc sec	25	EST	MEYLAN,WM & HOWARD,PH (1993)

CAS #: 002183-54-2 — BENZENEMETHANOL, 3,5-DIBROMO-2-HYDROXY-

Formula: $C_7H_6Br_2O_2$

Mol Weight: 281.94

MP (deg C):

FP (deg C):

BP (deg C):

BP pressure (mm Hg):

Property/Value	Units	Temp	Data Type	Reference
WS 9.14E+002	mg/L	25	EST	MEYLAN,WM ET AL. (1996)
logP 2.72			EXP	TAKAGI,T ET AL. (1992)
VP 1.27E-006	mm Hg	25	EST	NEELY,WB & BLAU,GE (1985)
DC	pKa			
HL 3.59E-012	atm m3/mol	25	EST	MEYLAN,WM & HOWARD,PH (1991)
OH 4.87E-012	cm3/molc sec	25	EST	MEYLAN,WM & HOWARD,PH (1993)

CAS #: 002189-60-8				OCTYLBENZENE

Formula: $C_{14}H_{22}$

Mol Weight: 190.33

MP (deg C): -36   FP (deg C):

BP (deg C): 264.5

BP pressure (mm Hg):

Property/Value	Units	Temp	Data Type	Reference
WS 6.60E-002	mg/L	25	EXP	WASIK,SP ET AL. (1981)
logP 6.30			EXP	SANGSTER,J (1993)
VP 1.13E-002	mm Hg	25	EXP	YAWS,CL (1994B)
DC	pKa			
HL 4.29E-002	atm m3/mol	25	EST	VP/WSOL
OH 1.44E-011	cm3/molc sec	25	EST	MEYLAN,WM & HOWARD,PH (1993)

CAS #: 002189-61-9				CARBAMIC ACID, (3-NITROPHENYL)-, METHYL ESTER

Formula: $C_8H_8N_2O_4$

Mol Weight: 196.16

MP (deg C):   FP (deg C):

BP (deg C):

BP pressure (mm Hg):

Property/Value	Units	Temp	Data Type	Reference
WS 8.97E+002	mg/L	25	EST	MEYLAN,WM ET AL. (1996)
logP 2.00			EXP	TAKAHASHI,J ET AL. (1988)
VP 2.34E-004	mm Hg	25	EST	NEELY,WB & BLAU,GE (1985)
DC	pKa			
HL 8.62E-011	atm m3/mol	25	EST	MEYLAN,WM & HOWARD,PH (1991)
OH 3.94E-012	cm3/molc sec	25	EST	MEYLAN,WM & HOWARD,PH (1993)

CAS #: 002198-53-0				2,6-DIMETHYLACETANILIDE

Formula: $C_{10}H_{13}NO$

Mol Weight: 163.22

MP (deg C): 182-184   FP (deg C):

BP (deg C):

BP pressure (mm Hg):

Property/Value	Units	Temp	Data Type	Reference
WS 8.07E+003	mg/L	25	EST	MEYLAN,WM ET AL. (1996)
logP 1.07			EXP	NAKAGAWA,Y ET AL. (1992)
VP 6.74E-005	mm Hg	25	EST	NEELY,WB & BLAU,GE (1985)
DC	pKa			
HL 7.52E-009	atm m3/mol	25	EST	MEYLAN,WM & HOWARD,PH (1991)
OH 1.90E-011	cm3/molc sec	25	EST	MEYLAN,WM & HOWARD,PH (1993)

CAS #: 002198-54-1				3,4-DIMETHYLACETANILIDE

Formula: $C_{10}H_{13}NO$

Mol Weight: 163.22

MP (deg C):   FP (deg C):

BP (deg C):

BP pressure (mm Hg):

Property/Value	Units	Temp	Data Type	Reference
WS 1.06E+003	mg/L	25	EST	MEYLAN,WM ET AL. (1996)
logP 2.10			EXP	NAKAGAWA,Y ET AL. (1992)
VP 6.74E-005	mm Hg	25	EST	NEELY,WB & BLAU,GE (1985)
DC	pKa			
HL 7.52E-009	atm m3/mol	25	EST	MEYLAN,WM & HOWARD,PH (1991)
OH 4.01E-011	cm3/molc sec	25	EST	MEYLAN,WM & HOWARD,PH (1993)

CAS #: 002198-77-8				2,7-DICHLORONAPHTHALENE

Formula: $C_{10}H_6Cl_2$

Mol Weight: 197.07

MP (deg C):   FP (deg C):

BP (deg C):

BP pressure (mm Hg):

Property/Value	Units	Temp	Data Type	Reference
WS 2.36E-001	mg/L	22	EXP	OPPERHUIZEN,A ET AL. (1985)
logP 4.81			EXP	OPPERHUIZEN,A ET AL. (1985)
VP 1.76E-003	mm Hg	25	EST	NEELY,WB & BLAU,GE (1985)
DC	pKa			
HL 2.89E-004	atm m3/mol	25	EST	MEYLAN,WM & HOWARD,PH (1991)
OH 4.44E-012	cm3/molc sec	25	EST	MEYLAN,WM & HOWARD,PH (1993)

CAS #: 002199-43-1				1H-PYRROLE-2-CARBOXYLIC ACID, ETHYL ESTER

Formula: $C_7H_9NO_2$

Mol Weight: 139.16

MP (deg C):   FP (deg C):

BP (deg C):

BP pressure (mm Hg):

Property/Value	Units	Temp	Data Type	Reference
WS 2.81E+003	mg/L	25	EST	MEYLAN,WM ET AL. (1996)
logP 1.73			EXP	YAMAGAMI,C ET AL. (1994)
VP 5.10E-002	mm Hg	25	EST	NEELY,WB & BLAU,GE (1985)
DC	pKa			
HL 7.76E-008	atm m3/mol	25	EST	MEYLAN,WM & HOWARD,PH (1991)
OH 3.72E-011	cm3/molc sec	25	EST	MEYLAN,WM & HOWARD,PH (1993)

CAS #: 002199-49-7				PYRROLE-3-CARBOXYLIC ACID-4-METHYL,ETHYL ESTER

Formula: $C_8H_{11}NO_2$

Mol Weight: 153.18

MP (deg C):   FP (deg C):

BP (deg C):

BP pressure (mm Hg):

Property/Value	Units	Temp	Data Type	Reference
WS 1.09E+003	mg/L	25	EST	MEYLAN,WM ET AL. (1996)
logP 2.14			EXP	HANSCH,C ET AL. (1995)
VP 1.43E-002	mm Hg	25	EST	NEELY,WB & BLAU,GE (1985)
DC	pKa			
HL 8.56E-008	atm m3/mol	25	EST	MEYLAN,WM & HOWARD,PH (1991)
OH 9.46E-011	cm3/molc sec	25	EST	MEYLAN,WM & HOWARD,PH (1993)

CAS #: 002202-79-1				1,4-NAPHTHOQUINONE,6,7-DIMETHYL

Formula: $C_{12}H_{10}O_2$

Mol Weight: 186.21

MP (deg C):   FP (deg C):

BP (deg C):

BP pressure (mm Hg):

Property/Value	Units	Temp	Data Type	Reference
WS 3.84E+002	mg/L	25	EST	MEYLAN,WM ET AL. (1996)
logP 2.49			EXP	HANSCH,C & LEO,AJ (1985)
VP 7.55E-005	mm Hg	25	EST	NEELY,WB & BLAU,GE (1985)
DC	pKa			
HL 2.40E-009	atm m3/mol	25	EST	MEYLAN,WM & HOWARD,PH (1991)
OH 4.93E-012	cm3/molc sec	25	EST	MEYLAN,WM & HOWARD,PH (1993)

CAS #: 002207-01-4	1,2-DIMETHYLCYCLOHEXANE (CIS)

Formula: C_8H_{16}
Mol Weight: 112.22
MP (deg C): -49.9   FP (deg C):
BP (deg C): 129.8
BP pressure (mm Hg):

Property/Value	Units	Temp	Data Type	Reference
WS 6.00E+000	mg/L	25	EXP	RIDDICK,JA ET AL. (1986)
logP 4.01			EST	MEYLAN,WM & HOWARD,PH (1995)
VP 1.45E+001	mm Hg	25	EXP	DAUBERT,TE & DANNER,RP (1989)
DC	pKa			
HL 3.57E-001	atm m3/mol	25	EST	VP/WSOL
OH 1.19E-011	cm3/molc sec	25	EST	MEYLAN,WM & HOWARD,PH (1993)

CAS #: 002207-03-6	1,3-DIMETHYLCYCLOHEXANE (TRANS)

Formula: C_8H_{16}
Mol Weight: 112.22
MP (deg C): -90.1   FP (deg C):
BP (deg C): 124.5
BP pressure (mm Hg):

Property/Value	Units	Temp	Data Type	Reference
WS 1.17E+001	mg/L	25	EST	MEYLAN,WM ET AL. (1996)
logP 4.01			EST	MEYLAN,WM & HOWARD,PH (1995)
VP 1.76E+001	mm Hg	25	EXP	DAUBERT,TE & DANNER,RP (1989)
DC	pKa			
HL 4.50E-001	atm m3/mol	25	EST	MEYLAN,WM & HOWARD,PH (1991)
OH 1.19E-011	cm3/molc sec	25	EST	MEYLAN,WM & HOWARD,PH (1993)

CAS #: 002207-04-7	1,4-DIMETHYLCYCLOHEXANE (TRANS)

Formula: C_8H_{16}
Mol Weight: 112.22
MP (deg C): -36.9   FP (deg C):
BP (deg C): 119.4
BP pressure (mm Hg):

Property/Value	Units	Temp	Data Type	Reference
WS 1.17E+001	mg/L	25	EST	MEYLAN,WM ET AL. (1996)
logP 4.01			EST	MEYLAN,WM & HOWARD,PH (1995)
VP 2.27E+001	mm Hg	25	EXP	DAUBERT,TE & DANNER,RP (1989)
DC	pKa			
HL 4.50E-001	atm m3/mol	25	EST	MEYLAN,WM & HOWARD,PH (1991)
OH 1.19E-011	cm3/molc sec	25	EST	MEYLAN,WM & HOWARD,PH (1993)

CAS #: 002209-02-1	2-FURANMETHANAMINIUM, TETRAHYDRO-4-HYDROXY-N,N,N

Formula: $C_8H_{18}INO_2$
Mol Weight: 287.14
MP (deg C):   FP (deg C):
BP (deg C):
BP pressure (mm Hg):

Property/Value	Units	Temp	Data Type	Reference
WS 1.00E+006	mg/L	25	EST	MEYLAN,WM ET AL. (1996)
logP -2.95			EXP	PRATESI,P ET AL. (1986)
VP 4.74E-011	mm Hg	25	EST	NEELY,WB & BLAU,GE (1985)
DC	pKa			
HL 2.92E-018	atm m3/mol	25	EST	MEYLAN,WM & HOWARD,PH (1991)
OH 1.33E-010	cm3/molc sec	25	EST	MEYLAN,WM & HOWARD,PH (1993)

CAS #: 002210-28-8	N-PROPYL METHACRYLATE

Formula: $C_7H_{12}O_2$
Mol Weight: 128.17
MP (deg C):   FP (deg C):
BP (deg C): 141
BP pressure (mm Hg):

Property/Value	Units	Temp	Data Type	Reference
WS 1.10E+003	mg/L	25	EST	MEYLAN,WM ET AL. (1996)
logP 2.26			EST	MEYLAN,WM & HOWARD,PH (1995)
VP 6.38E+000	mm Hg	25	EXP	DAUBERT,TE & DANNER,RP (1989)
DC	pKa			
HL 2.55E-004	atm m3/mol	25	EST	MEYLAN,WM & HOWARD,PH (1991)
OH 2.13E-011	cm3/molc sec	25	EST	MEYLAN,WM & HOWARD,PH (1993)

CAS #: 002211-33-8	N-PHENYL-2,4,5-IMIDAZOLITRIONE

Formula: $C_9H_6N_2O_3$
Mol Weight: 190.16
MP (deg C):   FP (deg C):
BP (deg C):
BP pressure (mm Hg):

Property/Value	Units	Temp	Data Type	Reference
WS 1.51E+004	mg/L	25	EST	MEYLAN,WM ET AL. (1996)
logP 0.60			EXP	LIPINSKI,CA ET AL. (1991)
VP 1.39E-010	mm Hg	25	EST	NEELY,WB & BLAU,GE (1985)
DC	pKa			
HL 3.54E-012	atm m3/mol	25	EST	MEYLAN,WM & HOWARD,PH (1991)
OH 1.34E-011	cm3/molc sec	25	EST	MEYLAN,WM & HOWARD,PH (1993)

CAS #: 002211-99-6	4-(1-METHYLUNDECYL)BENZENESULFONIC ACID, SODIUM*

Formula: $C_{18}H_{29}NaO_3S$
Mol Weight: 348.48
MP (deg C):   FP (deg C):
BP (deg C):
BP pressure (mm Hg):

Property/Value	Units	Temp	Data Type	Reference
WS 2.02E+001	mg/L	25	EST	MEYLAN,WM ET AL. (1996)
logP 2.92			EST	MEYLAN,WM & HOWARD,PH (1995)
VP 3.85E-015	mm Hg	25	EST	NEELY,WB & BLAU,GE (1985)
DC	pKa			
HL	atm m3/mol			
OH 1.62E-011	cm3/molc sec	25	EST	MEYLAN,WM & HOWARD,PH (1993)

CAS #: 002212-52-4	4-(1-PENTYLHEPTYL)BENZENESULFONIC ACID, SODIUM *

Formula: $C_{18}H_{29}NaO_3S$
Mol Weight: 348.48
MP (deg C):   FP (deg C):
BP (deg C):
BP pressure (mm Hg):

Property/Value	Units	Temp	Data Type	Reference
WS 2.02E+001	mg/L	25	EST	MEYLAN,WM ET AL. (1996)
logP 2.92			EST	MEYLAN,WM & HOWARD,PH (1995)
VP 3.85E-015	mm Hg	25	EST	NEELY,WB & BLAU,GE (1985)
DC	pKa			
HL	atm m3/mol			
OH 1.64E-011	cm3/molc sec	25	EST	MEYLAN,WM & HOWARD,PH (1993)

CAS #: 002212-67-1				MOLINATE

Formula: $C_9H_{17}NOS$

Mol Weight: 187.31

MP (deg C): | **FP (deg C):**

BP (deg C): 202

BP pressure (mm Hg): 1.00E+001

Property/Value	Units	Temp	Data Type	Reference
WS 9.70E+002	mg/L	25	EXP	WAUCHOPE,RD ET AL. (1991A)
logP 3.21			EXP	HANSCH,C & LEO,AJ (1985)
VP 5.60E-003	mm Hg	25	EXP	WAUCHOPE,RD ET AL. (1991A)
DC	pKa			
HL 4.10E-006	atm m3/mol	20	EXP	SAGEBIEL,JC ET AL. (1992)
OH 3.01E-011	cm3/molc sec	25	EST	ATKINSON,R (1988)

CAS #: 002213-32-3				2,4-DIMETHYL-1-PENTENE

Formula: C_7H_{14}

Mol Weight: 98.19

MP (deg C): -124.1 | **FP (deg C):**

BP (deg C): 81.6

BP pressure (mm Hg):

Property/Value	Units	Temp	Data Type	Reference
WS 2.76E+001	mg/L	25	EST	MEYLAN,WM ET AL. (1996)
logP 3.63			EST	MEYLAN,WM & HOWARD,PH (1995)
VP 9.45E+001	mm Hg	25	EXP	ZWOLINSKI,BJ & WILHOIT,RC (1971)
DC	pKa			
HL 5.62E-001	atm m3/mol	25	EST	MEYLAN,WM & HOWARD,PH (1991)
OH 5.54E-011	cm3/molc sec	25	EST	MEYLAN,WM & HOWARD,PH (1993)

CAS #: 002213-43-6				N-AMINOPIPERIDINE

Formula: $C_5H_{12}N_2$

Mol Weight: 100.16

MP (deg C): | **FP (deg C):**

BP (deg C): 147

BP pressure (mm Hg):

Property/Value	Units	Temp	Data Type	Reference
WS 2.03E+005	mg/L	25	EST	MEYLAN,WM ET AL. (1996)
logP -0.28			EXP	HANSCH,C & LEO,AJ (1985)
VP 3.18E+000	mm Hg	25	EST	NEELY,WB & BLAU,GE (1985)
DC	pKa			
HL 7.17E-008	atm m3/mol	25	EST	MEYLAN,WM & HOWARD,PH (1991)
OH 2.56E-011	cm3/molc sec	25	EST	MEYLAN,WM & HOWARD,PH (1993)

CAS #: 002213-70-9				O=P(OME)(NME)O-2,4,5-CLPHENYL

Formula: $C_8H_9Cl_3NO_3P$

Mol Weight: 304.50

MP (deg C): | **FP (deg C):**

BP (deg C):

BP pressure (mm Hg):

Property/Value	Units	Temp	Data Type	Reference
WS 2.82E+001	mg/L	25	EST	MEYLAN,WM ET AL. (1996)
logP 3.06			EXP	HANSCH,C & LEO,AJ (1985)
VP 1.78E-005	mm Hg	25	EST	NEELY,WB & BLAU,GE (1985)
DC	pKa			
HL 5.22E-010	atm m3/mol	25	EST	MEYLAN,WM & HOWARD,PH (1991)
OH 3.18E-011	cm3/molc sec	25	EST	MEYLAN,WM & HOWARD,PH (1993)

CAS #: 002213-84-5				O=P(OME)(NET)O-2,4,5-CLPHENYL

Formula: $C_9H_{11}Cl_3NO_3P$

Mol Weight: 318.53

MP (deg C): | **FP (deg C):**

BP (deg C):

BP pressure (mm Hg):

Property/Value	Units	Temp	Data Type	Reference
WS 1.84E+001	mg/L	25	EST	MEYLAN,WM ET AL. (1996)
logP 3.18			EXP	HANSCH,C & LEO,AJ (1985)
VP 8.65E-006	mm Hg	25	EST	NEELY,WB & BLAU,GE (1985)
DC	pKa			
HL 6.93E-010	atm m3/mol	25	EST	MEYLAN,WM & HOWARD,PH (1991)
OH 3.94E-011	cm3/molc sec	25	EST	MEYLAN,WM & HOWARD,PH (1993)

CAS #: 002213-85-6				O=P(OME)(NPR)O-2,4,5-CLPHENYL

Formula: $C_{10}H_{13}Cl_3NO_3P$

Mol Weight: 332.55

MP (deg C): | **FP (deg C):**

BP (deg C):

BP pressure (mm Hg):

Property/Value	Units	Temp	Data Type	Reference
WS 5.46E+000	mg/L	25	EST	MEYLAN,WM ET AL. (1996)
logP 3.70			EXP	HANSCH,C & LEO,AJ (1985)
VP 5.19E-006	mm Hg	25	EST	NEELY,WB & BLAU,GE (1985)
DC	pKa			
HL 9.20E-010	atm m3/mol	25	EST	MEYLAN,WM & HOWARD,PH (1991)
OH 4.25E-011	cm3/molc sec	25	EST	MEYLAN,WM & HOWARD,PH (1993)

CAS #: 002213-87-8				PHOSPHORAMIDIC ACID, N-BUTYL-, O-METHYL O-(2,4,5

Formula: $C_{11}H_{15}Cl_3NO_3P$

Mol Weight: 346.58

MP (deg C): | **FP (deg C):**

BP (deg C):

BP pressure (mm Hg):

Property/Value	Units	Temp	Data Type	Reference
WS 1.75E+000	mg/L	25	EST	MEYLAN,WM ET AL. (1996)
logP 4.18			EXP	HANSCH,C & LEO,AJ (1985)
VP 2.61E-006	mm Hg	25	EST	NEELY,WB & BLAU,GE (1985)
DC	pKa			
HL 1.22E-009	atm m3/mol	25	EST	MEYLAN,WM & HOWARD,PH (1991)
OH 4.39E-011	cm3/molc sec	25	EST	MEYLAN,WM & HOWARD,PH (1993)

CAS #: 002213-88-9				PHOSPHORAMIDIC ACID, N-(2-METHYLPROPYL)-, O-METH

Formula: $C_{11}H_{15}Cl_3NO_3P$

Mol Weight: 346.58

MP (deg C): | **FP (deg C):**

BP (deg C):

BP pressure (mm Hg):

Property/Value	Units	Temp	Data Type	Reference
WS 3.28E+000	mg/L	25	EST	MEYLAN,WM ET AL. (1996)
logP 3.86			EXP	HANSCH,C & LEO,AJ (1985)
VP 3.96E-006	mm Hg	25	EST	NEELY,WB & BLAU,GE (1985)
DC	pKa			
HL 1.22E-009	atm m3/mol	25	EST	MEYLAN,WM & HOWARD,PH (1991)
OH 4.39E-011	cm3/molc sec	25	EST	MEYLAN,WM & HOWARD,PH (1993)

CAS #: 002214-33-7 — PHOSPHORAMIDIC ACID, N-(1,1-DIMETHYLETHYL)-, O-

Formula: $C_{11}H_{15}Cl_3NO_3P$

Mol Weight: 346.58

MP (deg C): FP (deg C):

BP (deg C):

BP pressure (mm Hg):

Property/ Value	Units	Temp	Data Type	Reference
WS 3.48E+000	mg/L	25	EST	MEYLAN,WM ET AL. (1996)
logP 3.83			EXP	HANSCH,C & LEO,AJ (1985)
VP 5.19E-006	mm Hg	25	EST	NEELY,WB & BLAU,GE (1985)
DC	pKa			
HL 1.22E-009	atm m3/mol	25	EST	MEYLAN,WM & HOWARD,PH (1991)
OH 3.10E-011	cm3/molc sec	25	EST	MEYLAN,WM & HOWARD,PH (1993)

CAS #: 002214-34-8 — O=P(OME)(N)O-2,4,5-CLPHENYL

Formula: $C_7H_7Cl_3NO_3P$

Mol Weight: 290.47

MP (deg C): FP (deg C):

BP (deg C):

BP pressure (mm Hg):

Property/ Value	Units	Temp	Data Type	Reference
WS 9.63E+001	mg/L	25	EST	MEYLAN,WM ET AL. (1996)
logP 2.53			EXP	HANSCH,C & LEO,AJ (1985)
VP 2.08E-005	mm Hg	25	EST	NEELY,WB & BLAU,GE (1985)
DC	pKa			
HL 2.38E-010	atm m3/mol	25	EST	MEYLAN,WM & HOWARD,PH (1991)
OH 3.05E-011	cm3/molc sec	25	EST	MEYLAN,WM & HOWARD,PH (1993)

CAS #: 002214-53-1 — 4-CYANO-2-METHYLPYRIDINE

Formula: $C_7H_6N_2$

Mol Weight: 118.14

MP (deg C): FP (deg C):

BP (deg C):

BP pressure (mm Hg):

Property/ Value	Units	Temp	Data Type	Reference
WS 1.13E+004	mg/L	25	EST	MEYLAN,WM ET AL. (1996)
logP 0.81			EXP	HANSCH,C & LEO,AJ (1985)
VP 1.01E-001	mm Hg	25	EST	NEELY,WB & BLAU,GE (1985)
DC	pKa			
HL 7.52E-008	atm m3/mol	25	EST	MEYLAN,WM & HOWARD,PH (1991)
OH 2.62E-013	cm3/molc sec	25	EST	MEYLAN,WM & HOWARD,PH (1993)

CAS #: 002215-77-2 — P-PHENOXYBENZOIC ACID

Formula: $C_{13}H_{10}O_3$

Mol Weight: 214.22

MP (deg C): FP (deg C):

BP (deg C):

BP pressure (mm Hg):

Property/ Value	Units	Temp	Data Type	Reference
WS 1.69E+001	mg/L	25	EST	MEYLAN,WM ET AL. (1996)
logP 3.91			EXP	HANSCH,C & LEO,AJ (1985)
VP 5.45E-006	mm Hg	25	EST	NEELY,WB & BLAU,GE (1985)
DC 4.52	pKa		EXP	KORTUM,G ET AL (1961)
HL 2.37E-009	atm m3/mol	25	EST	MEYLAN,WM & HOWARD,PH (1991)
OH 7.26E-012	cm3/molc sec	25	EST	MEYLAN,WM & HOWARD,PH (1993)

CAS #: 002216-33-3 — 3-METHYLOCTANE

Formula: C_9H_{20}

Mol Weight: 128.26

MP (deg C): -107.6 FP (deg C):

BP (deg C): 144.2

BP pressure (mm Hg):

Property/ Value	Units	Temp	Data Type	Reference
WS 8.70E+001	mg/L	20	EXP	KERTES,AS (1989)
logP 4.69			EST	MEYLAN,WM & HOWARD,PH (1995)
VP 6.25E+000	mm Hg	25	EXP	DAUBERT,TE & DANNER,RP (1989)
DC	pKa			
HL 1.21E-002	atm m3/mol	25	EST	VP/WSOL
OH 9.97E-012	cm3/molc sec	25	EST	MEYLAN,WM & HOWARD,PH (1993)

CAS #: 002216-34-4 — 4-METHYLOCTANE

Formula: C_9H_{20}

Mol Weight: 128.26

MP (deg C): -113.3 FP (deg C):

BP (deg C): 142.4

BP pressure (mm Hg):

Property/ Value	Units	Temp	Data Type	Reference
WS 1.15E-001	mg/L	25	EXP	KERTES,AS (1989)
logP 4.69			EST	MEYLAN,WM & HOWARD,PH (1995)
VP 6.83E+000	mm Hg	25	EXP	DAUBERT,TE & DANNER,RP (1989)
DC	pKa			
HL 1.00E+001	atm m3/mol	25	EST	VP/WSOL
OH 9.72E-012	cm3/molc sec	25	EXP	ATKINSON,R (1989)

CAS #: 002216-45-7 — P-METHYLBENZYL ACETATE

Formula: $C_{10}H_{12}O_2$

Mol Weight: 164.21

MP (deg C): FP (deg C):

BP (deg C):

BP pressure (mm Hg):

Property/ Value	Units	Temp	Data Type	Reference
WS 5.40E+002	mg/L	25	EST	MEYLAN,WM ET AL. (1996)
logP 2.44			EXP	HANSCH,C & LEO,AJ (1985)
VP 6.24E-002	mm Hg	25	EST	NEELY,WB & BLAU,GE (1985)
DC	pKa			
HL 1.56E-005	atm m3/mol	25	EST	MEYLAN,WM & HOWARD,PH (1991)
OH 7.93E-012	cm3/molc sec	25	EST	MEYLAN,WM & HOWARD,PH (1993)

CAS #: 002216-51-5 — MENTHOL (L)

Formula: $C_{10}H_{20}O$

Mol Weight: 156.27

MP (deg C): 35-36 FP (deg C):

BP (deg C): 103-105

BP pressure (mm Hg): 1.60E+001

Property/ Value	Units	Temp	Data Type	Reference
WS 4.90E+002	mg/L	25	EXP	CHEM INSPECT TEST INST (1992)
logP 3.30			EXP	CHEM INSPECT TEST INST (1992)
VP 6.37E-002	mm Hg	25	EXT	PERRY,RH & GREEN,D (1984)
DC	pKa			
HL 1.52E-005	atm m3/mol	25	EST	MEYLAN,WM & HOWARD,PH (1991)
OH 2.41E-011	cm3/molc sec	25	EST	MEYLAN,WM & HOWARD,PH (1993)

CAS #: 002216-68-4 — 1-NAPHTHALENAMINE, N-METHYL-

Formula: $C_{11}H_{11}N$

Mol Weight: 157.22

MP (deg C): 174

BP (deg C): 294.5

BP pressure (mm Hg):

	Property/Value	Units	Temp	Data Type	Reference
WS	2.45E+002	mg/L	25	EST	MEYLAN,WM ET AL. (1996)
logP	2.88			EXP	HANSCH,C ET AL. (1995)
VP	2.69E-003	mm Hg	25	EST	NEELY,WB & BLAU,GE (1985)
DC	3.67	pKa	27	EXP	PERRIN,DD (1965)
HL	4.08E-007	atm m3/mol	25	EST	MEYLAN,WM & HOWARD,PH (1991)
OH	2.01E-010	cm3/molc sec	25	EST	MEYLAN,WM & HOWARD,PH (1993)

CAS #: 002216-69-5 — NAPHTHALENE, 1-METHOXY-

Formula: $C_{11}H_{10}O$

Mol Weight: 158.20

MP (deg C): < -10

BP (deg C): 269

BP pressure (mm Hg):

	Property/Value	Units	Temp	Data Type	Reference
WS	1.00E+001	mg/L	25	EXP	CHEM INSPECT TEST INST (1992)
logP	3.45			EXP	CHEM INSPECT TEST INST (1992)
VP	8.23E-003	mm Hg	25	EST	NEELY,WB & BLAU,GE (1985)
DC		pKa			
HL	3.11E-005	atm m3/mol	25	EST	MEYLAN,WM & HOWARD,PH (1991)
OH	2.01E-010	cm3/molc sec	25	EST	MEYLAN,WM & HOWARD,PH (1993)

CAS #: 002217-08-5 — BARBITURIC ACID,5,5-DIPR

Formula: $C_{10}H_{16}N_2O_3$

Mol Weight: 212.25

MP (deg C):

BP (deg C):

BP pressure (mm Hg):

	Property/Value	Units	Temp	Data Type	Reference
WS	6.00E+002	mg/L	20	EXP	YALSOWSKY,SH & DANNENFELSER,RM (1992)
logP	1.75			EXP	WONG,O & MCKEOWN,RH (1988)
VP	3.02E-011	mm Hg	25	EST	NEELY,WB & BLAU,GE (1985)
DC		pKa			
HL	6.36E-013	atm m3/mol	25	EST	MEYLAN,WM & HOWARD,PH (1991)
OH	1.36E-011	cm3/molc sec	25	EST	MEYLAN,WM & HOWARD,PH (1993)

CAS #: 002217-82-5 — P-BIPHENYLSULFONIC ACID, SODIUM SALT

Formula: $C_{12}H_9NaO_3S$

Mol Weight: 256.26

MP (deg C):

BP (deg C):

BP pressure (mm Hg):

	Property/Value	Units	Temp	Data Type	Reference
WS	8.15E+004	mg/L	25	EST	MEYLAN,WM ET AL. (1996)
logP	-0.67			EXP	HANSCH,C & LEO,AJ (1985)
VP	4.79E-014	mm Hg	25	EST	NEELY,WB & BLAU,GE (1985)
DC		pKa			
HL		atm m3/mol			
OH	4.11E-012	cm3/molc sec	25	EST	MEYLAN,WM & HOWARD,PH (1993)

CAS #: 002221-11-6 — 2,4-IMIDAZOLIDINEDIONE, 3,5-BIS(PHENYLMETHYL)-

Formula: $C_{17}H_{16}N_2O_2$

Mol Weight: 280.33

MP (deg C):

BP (deg C):

BP pressure (mm Hg):

	Property/Value	Units	Temp	Data Type	Reference
WS	5.99E+001	mg/L	25	EST	MEYLAN,WM ET AL. (1996)
logP	2.84			EXP	SANGSTER,J (1994)
VP	1.52E-010	mm Hg	25	EST	NEELY,WB & BLAU,GE (1985)
DC		pKa			
HL	2.98E-011	atm m3/mol	25	EST	MEYLAN,WM & HOWARD,PH (1991)
OH	2.50E-011	cm3/molc sec	25	EST	MEYLAN,WM & HOWARD,PH (1993)

CAS #: 002221-12-7 — 2,4-IMIDAZOLIDINEDIONE, 1-METHYL-3-PHENYL-

Formula: $C_{10}H_{10}N_2O_2$

Mol Weight: 190.20

MP (deg C):

BP (deg C):

BP pressure (mm Hg):

	Property/Value	Units	Temp	Data Type	Reference
WS	1.82E+003	mg/L	25	EST	MEYLAN,WM ET AL. (1996)
logP	0.14			EXP	SANGSTER,J (1994)
VP	1.24E-006	mm Hg	25	EST	NEELY,WB & BLAU,GE (1985)
DC		pKa			
HL	1.77E-008	atm m3/mol	25	EST	MEYLAN,WM & HOWARD,PH (1991)
OH	1.63E-011	cm3/molc sec	25	EST	MEYLAN,WM & HOWARD,PH (1993)

CAS #: 002221-13-8 — N-PHENYL-2,5-IMIDAZOLIDIONE

Formula: $C_9H_8N_2O_2$

Mol Weight: 176.18

MP (deg C):

BP (deg C):

BP pressure (mm Hg):

	Property/Value	Units	Temp	Data Type	Reference
WS	2.60E+003	mg/L	25	EST	MEYLAN,WM ET AL. (1996)
logP	0.04			EXP	LIPINSKI,CA ET AL. (1991)
VP	2.16E-007	mm Hg	25	EST	NEELY,WB & BLAU,GE (1985)
DC		pKa			
HL	8.08E-009	atm m3/mol	25	EST	MEYLAN,WM & HOWARD,PH (1991)
OH	1.57E-011	cm3/molc sec	25	EST	MEYLAN,WM & HOWARD,PH (1993)

CAS #: 002227-13-6 — TETRASUL

Formula: $C_{12}H_6Cl_4S$

Mol Weight: 324.06

MP (deg C):

BP (deg C):

BP pressure (mm Hg):

	Property/Value	Units	Temp	Data Type	Reference
WS	1.21E-002	mg/L	25	EST	MEYLAN,WM ET AL. (1996)
logP	6.87			EST	MEYLAN,WM & HOWARD,PH (1995)
VP	2.26E-006	mm Hg	25	EST	NEELY,WB & BLAU,GE (1985)
DC		pKa			
HL	9.46E-006	atm m3/mol	25	EST	MEYLAN,WM & HOWARD,PH (1991)
OH	4.42E-012	cm3/molc sec	25	EST	MEYLAN,WM & HOWARD,PH (1993)

CAS #: 002227-17-0				DIENOCHLOR
Formula: C$_{10}$Cl$_{10}$				
Mol Weight: 474.64				
MP (deg C): 121.5-122		FP (deg C):		
BP (deg C):				
BP pressure (mm Hg):				

Property/Value	Units	Temp	Data Type	Reference
WS 2.50E+001	mg/L	20	EXP	AUGUSTIJN-BECKERS,PWM ET AL. (1994)
logP 8.39			EST	MEYLAN,WM & HOWARD,PH (1995)
VP 1.00E-005	mm Hg	25	EXP	AUGUSTIJN-BECKERS,PWM ET AL. (1994)
DC	pKa			
HL 2.50E-007	atm m3/mol	25	EST	VP/WSOL
OH 7.86E-013	cm3/molc sec	25	EST	MEYLAN,WM & HOWARD,PH (1993)

CAS #: 002227-79-4				THIOBENZAMIDE
Formula: C$_7$H$_7$NS				
Mol Weight: 137.20				
MP (deg C):		FP (deg C):		
BP (deg C):				
BP pressure (mm Hg):				

Property/Value	Units	Temp	Data Type	Reference
WS 4.68E+004	mg/L	25	EST	MEYLAN,WM ET AL. (1996)
logP 1.49			EXP	HANSCH,C & LEO,AJ (1985)
VP 1.75E-002	mm Hg	25	EST	NEELY,WB & BLAU,GE (1985)
DC	pKa			
HL 3.92E-007	atm m3/mol	25	EST	MEYLAN,WM & HOWARD,PH (1991)
OH 2.53E-011	cm3/molc sec	25	EST	MEYLAN,WM & HOWARD,PH (1993)

CAS #: 002234-13-1				OCTACHLORONAPHTHALENE
Formula: C$_{10}$Cl$_8$				
Mol Weight: 403.74				
MP (deg C): 197-198		FP (deg C):		
BP (deg C): 440-442				
BP pressure (mm Hg): 7.40E+000				

Property/Value	Units	Temp	Data Type	Reference
WS 8.00E-005	mg/L	22	EXP	OPPERHUIZEN,A (1986)
logP 8.24			EXP	OPPERHUIZEN,A ET AL. (1985)
VP 1.52E-008	mm Hg	22	EXT	WEAST,RC ET AL. (1985)
DC	pKa			
HL 1.02E-004	atm m3/mol	25	EST	MEYLAN,WM & HOWARD,PH (1991)
OH 3.67E-014	cm3/molc sec	25	EST	MEYLAN,WM & HOWARD,PH (1993)

CAS #: 002235-12-3				1,3,5-HEXATRIENE
Formula: C$_6$H$_8$				
Mol Weight: 80.13				
MP (deg C):		FP (deg C):		
BP (deg C): 76-79				
BP pressure (mm Hg):				

Property/Value	Units	Temp	Data Type	Reference
WS 1.54E+002	mg/L	25	EST	MEYLAN,WM ET AL. (1996)
logP 2.80			EST	MEYLAN,WM & HOWARD,PH (1995)
VP 8.99E+001	mm Hg	25	EXT	BOUBLIK,T ET AL. (1984)
DC	pKa			
HL 9.84E-002	atm m3/mol	25	EST	MEYLAN,WM & HOWARD,PH (1991)
OH 1.12E-010	cm3/molc sec	25	EST	MEYLAN,WM & HOWARD,PH (1993)

CAS #: 002235-25-8				ETHYL MERCURIC PHOSPHATE
Formula: C$_2$H$_7$HgO$_4$P				
Mol Weight: 326.64				
MP (deg C):		FP (deg C):		
BP (deg C):				
BP pressure (mm Hg):				

Property/Value	Units	Temp	Data Type	Reference
WS 5.25E+004	mg/L	25	EST	MEYLAN,WM ET AL. (1996)
logP -0.92			EST	MEYLAN,WM & HOWARD,PH (1995)
VP 5.12E-007	mm Hg	25	EST	NEELY,WB & BLAU,GE (1985)
DC	pKa			
HL	atm m3/mol			
OH 5.65E-011	cm3/molc sec	25	EST	MEYLAN,WM & HOWARD,PH (1993)

CAS #: 002235-83-8				5-PHENYL-2-PENTANONE
Formula: C$_{11}$H$_{14}$O				
Mol Weight: 162.23				
MP (deg C):		FP (deg C):		
BP (deg C):				
BP pressure (mm Hg):				

Property/Value	Units	Temp	Data Type	Reference
WS 5.73E+002	mg/L	25	EST	MEYLAN,WM ET AL. (1996)
logP 2.42			EXP	HANSCH,C & LEO,AJ (1985)
VP 3.22E-002	mm Hg	25	EST	NEELY,WB & BLAU,GE (1985)
DC	pKa			
HL 7.05E-006	atm m3/mol	25	EST	MEYLAN,WM & HOWARD,PH (1991)
OH 1.14E-011	cm3/molc sec	25	EST	MEYLAN,WM & HOWARD,PH (1993)

CAS #: 002237-30-1				M-AMINOBENZONITRILE
Formula: C$_7$H$_6$N$_2$				
Mol Weight: 118.14				
MP (deg C): 54.3		FP (deg C):		
BP (deg C): 289				
BP pressure (mm Hg):				

Property/Value	Units	Temp	Data Type	Reference
WS 6.75E+003	mg/L	25	EST	MEYLAN,WM ET AL. (1996)
logP 1.07			EXP	HANSCH,C & LEO,AJ (1985)
VP 8.87E-003	mm Hg	25	EST	NEELY,WB & BLAU,GE (1985)
DC 2.75	pKa	25	EXP	PERRIN,DD (1965)
HL 1.84E-008	atm m3/mol	25	EST	MEYLAN,WM & HOWARD,PH (1991)
OH 1.41E-011	cm3/molc sec	25	EST	MEYLAN,WM & HOWARD,PH (1993)

CAS #: 002240-28-0				BETAMETHASONE-21-VALERATE
Formula: C$_{27}$H$_{37}$FO$_6$				
Mol Weight: 476.59				
MP (deg C):		FP (deg C):		
BP (deg C):				
BP pressure (mm Hg):				

Property/Value	Units	Temp	Data Type	Reference
WS 5.01E-001	mg/L	25	EST	MEYLAN,WM ET AL. (1996)
logP 3.87			EXP	HANSCH,C & LEO,AJ (1985)
VP 8.25E-015	mm Hg	25	EST	NEELY,WB & BLAU,GE (1985)
DC	pKa			
HL 1.81E-011	atm m3/mol	25	EST	MEYLAN,WM & HOWARD,PH (1991)
OH 7.08E-011	cm3/molc sec	25	EST	MEYLAN,WM & HOWARD,PH (1993)

CAS #: 002240-41-7				O,O-DIMETHYLPHENYLPHOSPHONATE

Formula: $C_8H_{11}O_3P$

Mol Weight: 186.15

MP (deg C): FP (deg C):

BP (deg C):

BP pressure (mm Hg):

Property/Value	Units	Temp	Data Type	Reference
WS 7.93E+003	mg/L	25	EST	MEYLAN,WM ET AL. (1996)
logP 0.95			EXP	HANSCH,C & LEO,AJ (1985)
VP 8.14E-003	mm Hg	25	EST	NEELY,WB & BLAU,GE (1985)
DC	pKa			
HL 3.20E-008	atm m3/mol	25	EST	MEYLAN,WM & HOWARD,PH (1991)
OH 7.53E-012	cm3/molc sec	25	EST	MEYLAN,WM & HOWARD,PH (1993)

CAS #: 002240-88-2				3,3,3-TRIFLUORO-1-PROPANOL

Formula: $C_3H_5F_3O$

Mol Weight: 114.07

MP (deg C): FP (deg C):

BP (deg C):

BP pressure (mm Hg):

Property/Value	Units	Temp	Data Type	Reference
WS 1.58E+005	mg/L	25	EST	MEYLAN,WM ET AL. (1996)
logP 0.39			EXP	MULLER,N (1986)
VP 8.07E+001	mm Hg	25	EST	NEELY,WB & BLAU,GE (1985)
DC	pKa			
HL 3.74E-005	atm m3/mol	25	EST	MEYLAN,WM & HOWARD,PH (1991)
OH 4.24E-012	cm3/molc sec	25	EST	MEYLAN,WM & HOWARD,PH (1993)

CAS #: 002243-27-8				NONANONITRILE

Formula: $C_9H_{17}N$

Mol Weight: 139.24

MP (deg C): -34.2 FP (deg C):

BP (deg C): 224.4

BP pressure (mm Hg):

Property/Value	Units	Temp	Data Type	Reference
WS 7.00E+001	mg/L	25	EXP	CHEM INSPECT TEST INST (1992)
logP 3.12			EXP	TANII,H & HASHIMOTO,K (1984)
VP 1.01E-001	mm Hg	25	EST	NEELY,WB & BLAU,GE (1985)
DC	pKa			
HL 2.22E-004	atm m3/mol	25	EST	MEYLAN,WM & HOWARD,PH (1991)
OH 7.32E-012	cm3/molc sec	25	EST	MEYLAN,WM & HOWARD,PH (1993)

CAS #: 002243-42-7				O-PHENOXYBENZOIC ACID

Formula: $C_{13}H_{10}O_3$

Mol Weight: 214.22

MP (deg C): 113 FP (deg C):

BP (deg C): 355

BP pressure (mm Hg):

Property/Value	Units	Temp	Data Type	Reference
WS 8.15E+001	mg/L	25	EST	MEYLAN,WM ET AL. (1996)
logP 3.11			EXP	HANSCH,C & LEO,AJ (1985)
VP 5.45E-006	mm Hg	25	EST	NEELY,WB & BLAU,GE (1985)
DC 3.53	pKa		EXP	KORTUM,G ET AL (1961)
HL 2.37E-009	atm m3/mol	25	EST	MEYLAN,WM & HOWARD,PH (1991)
OH 7.26E-012	cm3/molc sec	25	EST	MEYLAN,WM & HOWARD,PH (1993)

CAS #: 002243-62-1				1,5-DIAMINONAPHTHALENE

Formula: $C_{10}H_{10}N_2$

Mol Weight: 158.20

MP (deg C): 190 FP (deg C):

BP (deg C):

BP pressure (mm Hg):

Property/Value	Units	Temp	Data Type	Reference
WS 3.80E+002	mg/L	25	EXP	CHEM INSPECT TEST INST (1992)
logP 0.89			EXP	SCHULTZ,TW & APPLEHANS,FM (1985)
VP 2.93E-005	mm Hg	25	EST	NEELY,WB & BLAU,GE (1985)
DC 4.44	pKa	25	EXP	PERRIN,DD (1972)
HL 6.57E-011	atm m3/mol	25	EST	MEYLAN,WM & HOWARD,PH (1991)
OH 2.00E-010	cm3/molc sec	25	EST	MEYLAN,WM & HOWARD,PH (1993)

CAS #: 002243-81-4				1-NAPHTHALENECARBOXAMIDE

Formula: $C_{11}H_9NO$

Mol Weight: 171.20

MP (deg C): 205.8 FP (deg C):

BP (deg C):

BP pressure (mm Hg):

Property/Value	Units	Temp	Data Type	Reference
WS 1.51E+003	mg/L	25	EST	MEYLAN,WM ET AL. (1996)
logP 1.88			EXP	ALCORN,CJ ET AL. (1993)
VP 3.40E-006	mm Hg	25	EST	NEELY,WB & BLAU,GE (1985)
DC	pKa			
HL 2.16E-010	atm m3/mol	25	EST	MEYLAN,WM & HOWARD,PH (1991)
OH 2.17E-011	cm3/molc sec	25	EST	MEYLAN,WM & HOWARD,PH (1993)

CAS #: 002245-38-7				1,6,7-TRIMETHYLNAPHTHALENE

Formula: $C_{13}H_{14}$

Mol Weight: 170.26

MP (deg C): FP (deg C):

BP (deg C): 285

BP pressure (mm Hg): 7.62E+002

Property/Value	Units	Temp	Data Type	Reference
WS 4.78E+000	mg/L	25	EST	MEYLAN,WM ET AL. (1996)
logP 4.81			EST	MEYLAN,WM & HOWARD,PH (1995)
VP 2.52E-003	mm Hg	25	EST	NEELY,WB & BLAU,GE (1985)
DC	pKa			
HL 7.07E-004	atm m3/mol	25	EST	MEYLAN,WM & HOWARD,PH (1991)
OH 1.24E-010	cm3/molc sec	25	EST	MEYLAN,WM & HOWARD,PH (1993)

CAS #: 002251-64-1				124-BENZTHIADIZ-1-O2-3-CF3-6-CL

Formula: $C_8H_4ClF_3N_2O_2S$

Mol Weight: 284.65

MP (deg C): FP (deg C):

BP (deg C):

BP pressure (mm Hg):

Property/Value	Units	Temp	Data Type	Reference
WS 5.88E+002	mg/L	25	EST	MEYLAN,WM ET AL. (1996)
logP 1.65			EXP	HANSCH,C & LEO,AJ (1985)
VP 5.06E-008	mm Hg	25	EST	NEELY,WB & BLAU,GE (1985)
DC	pKa			
HL 5.40E-007	atm m3/mol	25	EST	MEYLAN,WM & HOWARD,PH (1991)
OH 1.85E-013	cm3/molc sec	25	EST	MEYLAN,WM & HOWARD,PH (1993)

CAS #: 002262-52-4				DIETHYLMALONATE,2-F BENZAL

Formula: $C_{14}H_{15}FO_4$

Mol Weight: 266.27

MP (deg C): | FP (deg C):

BP (deg C):

BP pressure (mm Hg):

Property/Value	Units	Temp	Data Type	Reference
WS 4.86E+001	mg/L	25	EST	MEYLAN,WM ET AL. (1996)
logP 3.04			EXP	HANSCH,C & LEO,AJ (1985)
VP 1.20E-003	mm Hg	25	EST	NEELY,WB & BLAU,GE (1985)
DC	pKa			
HL 1.28E-008	atm m3/mol	25	EST	MEYLAN,WM & HOWARD,PH (1991)
OH 1.63E-011	cm3/molc sec	25	EST	MEYLAN,WM & HOWARD,PH (1993)

CAS #: 002265-94-3				BENZENE, 1,3-DIFLUORO-5-NITRO-

Formula: $C_6H_3F_2NO_2$

Mol Weight: 159.09

MP (deg C): 17 | FP (deg C):

BP (deg C): 176-177

BP pressure (mm Hg):

Property/Value	Units	Temp	Data Type	Reference
WS 5.20E+002	mg/L	25	EST	MEYLAN,WM ET AL. (1996)
logP 2.03			EXP	SANGSTER,J (1993)
VP 3.50E-001	mm Hg	25	EST	NEELY,WB & BLAU,GE (1985)
DC	pKa			
HL 2.90E-005	atm m3/mol	25	EST	MEYLAN,WM & HOWARD,PH (1991)
OH 2.67E-013	cm3/molc sec	25	EST	MEYLAN,WM & HOWARD,PH (1993)

CAS #: 002270-20-4				5-PHENYLPENTANOIC ACID

Formula: $C_{11}H_{14}O_2$

Mol Weight: 178.23

MP (deg C): 57.5 | FP (deg C):

BP (deg C): 190

BP pressure (mm Hg): 3.00E+001

Property/Value	Units	Temp	Data Type	Reference
WS 6.90E+002	mg/L	25	EST	MEYLAN,WM ET AL. (1996)
logP 2.70			EXP	HANSCH,C ET AL. (1995)
VP 2.48E-004	mm Hg	25	EST	NEELY,WB & BLAU,GE (1985)
DC 4.88	pKa	20	EXP	SERJEANT,EP & DEMPSEY,B (1979)
HL 1.03E-007	atm m3/mol	25	EST	MEYLAN,WM & HOWARD,PH (1991)
OH 1.02E-011	cm3/molc sec	25	EST	MEYLAN,WM & HOWARD,PH (1993)

CAS #: 002274-01-3				3-ETHYLBUTANOYLUREA

Formula: $C_7H_{14}N_2O_2$

Mol Weight: 158.20

MP (deg C): | FP (deg C):

BP (deg C):

BP pressure (mm Hg):

Property/Value	Units	Temp	Data Type	Reference
WS 1.17E+004	mg/L	25	EST	MEYLAN,WM ET AL. (1996)
logP 0.91			EXP	HANSCH,C & LEO,AJ (1985)
VP 2.09E-006	mm Hg	25	EST	NEELY,WB & BLAU,GE (1985)
DC	pKa			
HL 3.35E-010	atm m3/mol	25	EST	MEYLAN,WM & HOWARD,PH (1991)
OH 1.37E-011	cm3/molc sec	25	EST	MEYLAN,WM & HOWARD,PH (1993)

CAS #: 002274-08-0				3-METHYLBUTANOYLUREA

Formula: $C_6H_{12}N_2O_2$

Mol Weight: 144.17

MP (deg C): | FP (deg C):

BP (deg C):

BP pressure (mm Hg):

Property/Value	Units	Temp	Data Type	Reference
WS 3.32E+004	mg/L	25	EST	MEYLAN,WM ET AL. (1996)
logP 0.45			EXP	HANSCH,C & LEO,AJ (1985)
VP 4.81E-006	mm Hg	25	EST	NEELY,WB & BLAU,GE (1985)
DC	pKa			
HL 2.52E-010	atm m3/mol	25	EST	MEYLAN,WM & HOWARD,PH (1991)
OH 8.13E-012	cm3/molc sec	25	EST	MEYLAN,WM & HOWARD,PH (1993)

CAS #: 002274-42-2				METHYLSULFONYLACETONITRILE

Formula: $C_3H_5NO_2S$

Mol Weight: 119.14

MP (deg C): | FP (deg C):

BP (deg C):

BP pressure (mm Hg):

Property/Value	Units	Temp	Data Type	Reference
WS 8.96E+005	mg/L	25	EST	MEYLAN,WM ET AL. (1996)
logP -1.42			EXP	HANSCH,C ET AL. (1995)
VP 2.15E-002	mm Hg	25	EST	NEELY,WB & BLAU,GE (1985)
DC	pKa			
HL 2.59E-009	atm m3/mol	25	EST	MEYLAN,WM & HOWARD,PH (1991)
OH 2.44E-012	cm3/molc sec	25	EST	MEYLAN,WM & HOWARD,PH (1993)

CAS #: 002274-67-1				PHOSPHORIC ACID,2-CHLORO-(2,4-DICHLOROPHENYL)VIN

Formula: $C_{10}H_{10}Cl_3O_4P$

Mol Weight: 331.52

MP (deg C): | FP (deg C):

BP (deg C):

BP pressure (mm Hg):

Property/Value	Units	Temp	Data Type	Reference
WS 1.70E+001	mg/L	25	EST	MEYLAN,WM ET AL. (1996)
logP 3.13			EXP	TOMLIN,C (1994)
VP 9.65E-006	mm Hg	25	EST	NEELY,WB & BLAU,GE (1985)
DC	pKa			
HL 2.93E-008	atm m3/mol	25	EST	MEYLAN,WM & HOWARD,PH (1991)
OH 2.47E-011	cm3/molc sec	25	EST	MEYLAN,WM & HOWARD,PH (1993)

CAS #: 002275-14-1				PHENKAPTON

Formula: $C_{11}H_{15}Cl_2O_2PS_3$

Mol Weight: 377.31

MP (deg C): | FP (deg C):

BP (deg C):

BP pressure (mm Hg):

Property/Value	Units	Temp	Data Type	Reference
WS 4.36E-002	mg/L	25	EST	MEYLAN,WM ET AL. (1996)
logP 5.84			EST	MEYLAN,WM & HOWARD,PH (1995)
VP 7.56E-007	mm Hg	25	EST	NEELY,WB & BLAU,GE (1985)
DC	pKa			
HL 1.31E-007	atm m3/mol	25	EST	MEYLAN,WM & HOWARD,PH (1991)
OH 2.44E-010	cm3/molc sec	25	EST	MEYLAN,WM & HOWARD,PH (1993)

CAS #: 002275-18-5 — PROTHOATE

Formula: $C_9H_{20}NO_3PS_2$

Mol Weight: 285.37

MP (deg C):
FP (deg C):
BP (deg C):
BP pressure (mm Hg):

Property/Value	Units	Temp	Data Type	Reference
WS 2.50E+003	mg/L	20	EXP	YALKOWSKY,SH & DANNENFELSER,RM (1992)
logP 2.17			EST	MEYLAN,WM & HOWARD,PH (1995)
VP 2.12E-006	mm Hg	25	EST	NEELY,WB & BLAU,GE (1985)
DC	pKa			
HL 6.55E-011	atm m3/mol	25	EST	MEYLAN,WM & HOWARD,PH (1991)
OH 1.20E-010	cm3/molc sec	25	EST	MEYLAN,WM & HOWARD,PH (1993)

CAS #: 002275-23-2 — VAMIDOTHION

Formula: $C_8H_{18}NO_4PS_2$

Mol Weight: 287.34

MP (deg C): 46-48
FP (deg C):
BP (deg C):
BP pressure (mm Hg):

Property/Value	Units	Temp	Data Type	Reference
WS 4.00E+006	mg/L	20	EXP	WORTHING,CR & WALKER,SB (1987)
logP 0.16			EST	MEYLAN,WM & HOWARD,PH (1995)
VP 9.00E-009	mm Hg	25	EST	HL X WSOL
DC	pKa			
HL 8.60E-016	atm m3/mol	25	EST	MEYLAN,WM & HOWARD,PH (1991)
OH 1.16E-010	cm3/molc sec	25	EST	ATKINSON,R (1988)

CAS #: 002278-22-0 — PEROXYACETYLNITRATE

Formula: $C_2H_3NO_5$

Mol Weight: 121.05

MP (deg C):
FP (deg C):
BP (deg C):
BP pressure (mm Hg):

Property/Value	Units	Temp	Data Type	Reference
WS 1.46E+005	mg/L	25	EST	MEYLAN,WM ET AL. (1996)
logP -0.19			EST	MEYLAN,WM & HOWARD,PH (1995)
VP 2.92E+001	mm Hg	25	EXP	BRUCKMANN,PW & WILLNER,H (1983)
DC	pKa			
HL 2.78E-004	atm m3/mol	25	EXP	GAFFNEY,JS ET AL. (1987)
OH 1.00E-013	cm3/molc sec	25	EXP	ATKINSON,R (1989)

CAS #: 002279-64-3 — AGROX

Formula: $C_7H_8HgN_2O$

Mol Weight: 336.74

MP (deg C):
FP (deg C):
BP (deg C):
BP pressure (mm Hg):

Property/Value	Units	Temp	Data Type	Reference
WS 1.12E+003	mg/L	25	EST	MEYLAN,WM ET AL. (1996)
logP -0.57			EST	MEYLAN,WM & HOWARD,PH (1995)
VP 6.92E-006	mm Hg	25	EST	NEELY,WB & BLAU,GE (1985)
DC	pKa			
HL	atm m3/mol			
OH 3.95E-012	cm3/molc sec	25	EST	MEYLAN,WM & HOWARD,PH (1993)

CAS #: 002284-20-0 — BENZENE, 1-ISOTHIOCYANATO-4-METHOXY

Formula: C_8H_7NOS

Mol Weight: 165.22

MP (deg C): 18
FP (deg C):
BP (deg C): 280-281
BP pressure (mm Hg):

Property/Value	Units	Temp	Data Type	Reference
WS 5.68E+001	mg/L	25	EST	MEYLAN,WM ET AL. (1996)
logP 3.58			EXP	AUGUSTIN,J ET AL. (1987)
VP 8.41E-003	mm Hg	25	EST	NEELY,WB & BLAU,GE (1985)
DC	pKa			
HL 4.62E-005	atm m3/mol	25	EST	MEYLAN,WM & HOWARD,PH (1991)
OH 2.44E-011	cm3/molc sec	25	EST	MEYLAN,WM & HOWARD,PH (1993)

CAS #: 002285-16-7 — 1,4-BENZODIAZEPIN-2-ONE,5(CF3)PH

Formula: $C_{16}H_{11}F_3N_2O$

Mol Weight: 304.27

MP (deg C):
FP (deg C):
BP (deg C):
BP pressure (mm Hg):

Property/Value	Units	Temp	Data Type	Reference
WS 2.37E+001	mg/L	25	EST	MEYLAN,WM ET AL. (1996)
logP 3.15			EXP	HANSCH,C & LEO,AJ (1985)
VP 1.87E-008	mm Hg	25	EST	NEELY,WB & BLAU,GE (1985)
DC	pKa			
HL 2.09E-009	atm m3/mol	25	EST	MEYLAN,WM & HOWARD,PH (1991)
OH 1.62E-011	cm3/molc sec	25	EST	MEYLAN,WM & HOWARD,PH (1993)

CAS #: 002294-82-8 — 9-ETHYL-9H-FLUORENE

Formula: $C_{15}H_{14}$

Mol Weight: 194.28

MP (deg C):
FP (deg C):
BP (deg C): 123-124
BP pressure (mm Hg): 1.00E+000

Property/Value	Units	Temp	Data Type	Reference
WS 3.95E-001	mg/L	25	EST	MEYLAN,WM ET AL. (1996)
logP 4.64			EST	MEYLAN,WM & HOWARD,PH (1995)
VP 2.34E-004	mm Hg	25	EST	NEELY,WB & BLAU,GE (1985)
DC	pKa			
HL 2.95E-004	atm m3/mol	25	EST	MEYLAN,WM & HOWARD,PH (1991)
OH 1.14E-011	cm3/molc sec	25	EST	MEYLAN,WM & HOWARD,PH (1993)

CAS #: 002295-35-4 — 2-AZACYCLOPENTANTHIONE

Formula: C_4H_7NS

Mol Weight: 101.17

MP (deg C):
FP (deg C):
BP (deg C):
BP pressure (mm Hg):

Property/Value	Units	Temp	Data Type	Reference
WS 1.28E+005	mg/L	25	EST	MEYLAN,WM ET AL. (1996)
logP -0.05			EXP	HANSCH,C & LEO,AJ (1985)
VP 4.36E-002	mm Hg	25	EST	NEELY,WB & BLAU,GE (1985)
DC	pKa			
HL 8.28E-006	atm m3/mol	25	EST	MEYLAN,WM & HOWARD,PH (1991)
OH 7.38E-011	cm3/molc sec	25	EST	MEYLAN,WM & HOWARD,PH (1993)

CAS #: 002297-94-1				2,4,6-CYCLOHEPTATRIENE-1-ONE, 2-HYDROXY-5-NITROS

Formula: $C_7H_5NO_3$
Mol Weight: 151.12
MP (deg C): | FP (deg C):
BP (deg C):
BP pressure (mm Hg):

Property/Value	Units	Temp	Data Type	Reference
WS 1.47E+004	mg/L	25	EST	MEYLAN,WM ET AL. (1996)
logP 0.83			EXP	SANGSTER,J (1993)
VP 7.76E-005	mm Hg	25	EST	NEELY,WB & BLAU,GE (1985)
DC	pKa			
HL 6.86E-008	atm m3/mol	25	EST	MEYLAN,WM & HOWARD,PH (1991)
OH 5.99E-011	cm3/molc sec	25	EST	MEYLAN,WM & HOWARD,PH (1993)

CAS #: 002301-40-8				2,4-IMIDAZOLIDINEDIONE, 3-(PHENYLMETHYL)-

Formula: $C_{10}H_{10}N_2O_2$
Mol Weight: 190.20
MP (deg C): | FP (deg C):
BP (deg C):
BP pressure (mm Hg):

Property/Value	Units	Temp	Data Type	Reference
WS 9.78E+003	mg/L	25	EST	MEYLAN,WM ET AL. (1996)
logP 0.82			EXP	SANGSTER,J (1994)
VP 9.77E-008	mm Hg	25	EST	NEELY,WB & BLAU,GE (1985)
DC	pKa			
HL 2.78E-010	atm m3/mol	25	EST	MEYLAN,WM & HOWARD,PH (1991)
OH 1.30E-011	cm3/molc sec	25	EST	MEYLAN,WM & HOWARD,PH (1993)

CAS #: 002303-16-4				DIALLATE

Formula: $C_{10}H_{17}Cl_2NOS$
Mol Weight: 270.22
MP (deg C): 25-30 | FP (deg C):
BP (deg C): 150 9
BP pressure (mm Hg):

Property/Value	Units	Temp	Data Type	Reference
WS 1.40E+001	mg/L	25	EXP	WORTHING,CR & WALKER,SB (1983)
logP 4.80			EST	MILL,T & MABEY,W (1985)
VP 1.50E-004	mm Hg	25	EXP	WORTHING,CR & WALKER,SB (1983)
DC	pKa			
HL 3.80E-006	atm m3/mol	25	EST	VP/WSOL
OH 3.49E-011	cm3/molc sec	25	EST	MEYLAN,WM & HOWARD,PH (1993)

CAS #: 002303-17-5				DIISOPROPYL CARBAMATE

Formula: $C_{10}H_{16}Cl_3NOS$
Mol Weight: 304.67
MP (deg C): 29-20 | FP (deg C):
BP (deg C): 117
BP pressure (mm Hg): 4.00E-001

Property/Value	Units	Temp	Data Type	Reference
WS 4.00E+000	mg/L	25	EXP	KENAGA, EE (1980)
logP 4.57			EST	MEYLAN,WM & HOWARD, PH (1952)
VP 1.93E-004	mm Hg	25	EST	GROVER, R ET AL. (1988)
DC	pKa			
HL 1.93E-005	atm m3/mol	25	EST	VP/WSOL
OH 3.23E-011	cm3/molc sec	25	EST	ATKINSON, R (1988)

CAS #: 002303-25-5				3-ME-4'-NO2-DIPHENYL ETHER

Formula: $C_{13}H_{11}NO_3$
Mol Weight: 229.24
MP (deg C): | FP (deg C):
BP (deg C):
BP pressure (mm Hg):

Property/Value	Units	Temp	Data Type	Reference
WS 5.00E+000	mg/L	25	EXP	KEARNEY,PC & KAUFMAN,DD (1975A)
logP 4.40			EXP	NANDIHALLI,UB ET AL. (1993)
VP 1.52E-005	mm Hg	25	EST	NEELY,WB & BLAU,GE (1985)
DC	pKa			
HL 5.13E-007	atm m3/mol	25	EST	MEYLAN,WM & HOWARD,PH (1991)
OH 1.36E-011	cm3/molc sec	25	EST	MEYLAN,WM & HOWARD,PH (1993)

CAS #: 002307-49-5				TRICAMBA

Formula: $C_8H_5Cl_3O_3$
Mol Weight: 255.49
MP (deg C): | FP (deg C):
BP (deg C):
BP pressure (mm Hg):

Property/Value	Units	Temp	Data Type	Reference
WS 9.16E+001	mg/L	25	EST	MEYLAN,WM ET AL. (1996)
logP 2.79			EST	MEYLAN,WM & HOWARD,PH (1995)
VP 1.50E-005	mm Hg	25	EST	NEELY,WB & BLAU,GE (1985)
DC	pKa			
HL 2.61E-009	atm m3/mol	25	EST	MEYLAN,WM & HOWARD,PH (1991)
OH 1.42E-012	cm3/molc sec	25	EST	MEYLAN,WM & HOWARD,PH (1993)

CAS #: 002307-55-3				2,4-D, AMINE SALT

Formula: $C_8H_9Cl_2NO_3$
Mol Weight: 238.07
MP (deg C): | FP (deg C):
BP (deg C):
BP pressure (mm Hg):

Property/Value	Units	Temp	Data Type	Reference
WS 4.58E+004	mg/L	25	EST	MEYLAN,WM ET AL. (1996)
logP -0.26			EST	MEYLAN,WM & HOWARD,PH (1995)
VP 1.00E-010	mm Hg	38	EXP	DAVIDSON,JM ET AL. (1978)
DC	pKa			
HL 4.43E-017	atm m3/mol	25	EST	MEYLAN,WM & HOWARD,PH (1991)
OH 3.60E-012	cm3/molc sec	25	EST	MEYLAN,WM & HOWARD,PH (1993)

CAS #: 002307-68-8				SOLAN

Formula: $C_{13}H_{18}ClNO$
Mol Weight: 239.75
MP (deg C): 79-80 | FP (deg C):
BP (deg C):
BP pressure (mm Hg):

Property/Value	Units	Temp	Data Type	Reference
WS 8.50E+000	mg/L		EXP	YALKOWSKY,SH & DANNENFELSER,RM (1992)
logP 4.31			EXP	MITSUTAKE,KI ET AL. (1986)
VP 2.22E-006	mm Hg	25	EST	NEELY,WB & BLAU,GE (1985)
DC	pKa			
HL 1.57E-008	atm m3/mol	25	EST	MEYLAN,WM & HOWARD,PH (1991)
OH 1.90E-011	cm3/molc sec	25	EST	MEYLAN,WM & HOWARD,PH (1993)

CAS #: 002310-17-0 — PHOSALONE

Formula: $C_{12}H_{15}ClNO_4PS_2$

Mol Weight: 367.81

MP (deg C): 46

FP (deg C):

BP (deg C):

BP pressure (mm Hg):

Property/ Value	Units	Temp	Data Type	Reference
WS 2.15E+000	mg/L	20	EXP	CHIOU,CT ET AL. (1977)
logP 4.38			EXP	HANSCH,C ET AL. (1995)
VP 4.54E-008	mm Hg	25	EST	NEELY,WB & BLAU,GE (1985)
DC	pKa			
HL 3.94E-007	atm m3/mol	25	EST	MEYLAN,WM & HOWARD,PH (1991)
OH 1.73E-010	cm3/molc sec	25	EST	MEYLAN,WM & HOWARD,PH (1993)

CAS #: 002311-46-8 — ISOPROPYL HEXANOATE

Formula: $C_9H_{18}O_2$

Mol Weight: 158.24

MP (deg C):

FP (deg C):

BP (deg C):

BP pressure (mm Hg):

Property/ Value	Units	Temp	Data Type	Reference
WS 1.18E+002	mg/L	25	EST	MEYLAN,WM ET AL. (1996)
logP 3.25			EST	MEYLAN,WM & HOWARD,PH (1995)
VP 1.04E+000	mm Hg	25	EST	NEELY,WB & BLAU,GE (1985)
DC	pKa			
HL 9.60E-004	atm m3/mol	25	EST	MEYLAN,WM & HOWARD,PH (1991)
OH 8.43E-012	cm3/molc sec	25	EST	MEYLAN,WM & HOWARD,PH (1993)

CAS #: 002312-35-8 — PROPARGITE

Formula: $C_{19}H_{26}O_4S$

Mol Weight: 350.48

MP (deg C):

FP (deg C):

BP (deg C):

BP pressure (mm Hg):

Property/ Value	Units	Temp	Data Type	Reference
WS 5.00E-001	mg/L	25	EXP	NOYES DATA CORP (1988)
logP 5.00			EXP	SAITO,H ET AL. (1993)
VP 3.00E+000	mm Hg	20	EXP	WORTHING,CR & WALKER,SB (1987)
DC	pKa			
HL 2.77E+000	atm m3/mol	25	EST	VP/WSOL
OH 1.20E-010	cm3/molc sec	25	EST	MEYLAN,WM & HOWARD,PH (1993)

CAS #: 002314-79-6 — N-(P-TOLYL)SUCCINIMIDE

Formula: $C_{10}H_9NO_2$

Mol Weight: 175.19

MP (deg C):

FP (deg C):

BP (deg C):

BP pressure (mm Hg):

Property/ Value	Units	Temp	Data Type	Reference
WS 1.59E+004	mg/L	25	EST	MEYLAN,WM ET AL. (1996)
logP 0.58			EXP	HANSCH,C & LEO,AJ (1985)
VP 4.03E-007	mm Hg	25	EST	NEELY,WB & BLAU,GE (1985)
DC	pKa			
HL 1.44E-007	atm m3/mol	25	EST	MEYLAN,WM & HOWARD,PH (1991)
OH 1.97E-011	cm3/molc sec	25	EST	MEYLAN,WM & HOWARD,PH (1993)

CAS #: 002314-80-9 — P-METHOXY-N-PHENYLSUCCINIMIDE

Formula: $C_{11}H_{11}NO_3$

Mol Weight: 205.22

MP (deg C):

FP (deg C):

BP (deg C):

BP pressure (mm Hg):

Property/ Value	Units	Temp	Data Type	Reference
WS 3.06E+004	mg/L	25	EST	MEYLAN,WM ET AL. (1996)
logP 0.15			EXP	HANSCH,C & LEO,AJ (1985)
VP 1.58E-007	mm Hg	25	EST	NEELY,WB & BLAU,GE (1985)
DC	pKa			
HL 7.72E-009	atm m3/mol	25	EST	MEYLAN,WM & HOWARD,PH (1991)
OH 1.64E-011	cm3/molc sec	25	EST	MEYLAN,WM & HOWARD,PH (1993)

CAS #: 002315-36-8 — CDEA

Formula: $C_6H_{12}ClNO$

Mol Weight: 149.62

MP (deg C):

FP (deg C):

BP (deg C): 148-150

BP pressure (mm Hg): 5.50E+001

Property/ Value	Units	Temp	Data Type	Reference
WS 8.55E+004	mg/L	25	EXP	SHIU,WY ET AL. (1990)
logP 1.08			EST	MEYLAN,WM & HOWARD,PH (1995)
VP 5.99E-002	mm Hg	25	EST	NEELY,WB & BLAU,GE (1985)
DC	pKa			
HL 3.35E-008	atm m3/mol	25	EST	MEYLAN,WM & HOWARD,PH (1991)
OH 2.33E-011	cm3/molc sec	25	EST	MEYLAN,WM & HOWARD,PH (1993)

CAS #: 002315-68-6 — PROPYL BENZOATE

Formula: $C_{10}H_{12}O_2$

Mol Weight: 164.21

MP (deg C): -51.6

FP (deg C):

BP (deg C): 211

BP pressure (mm Hg):

Property/ Value	Units	Temp	Data Type	Reference
WS 3.51E+002	mg/L	25	EXP	KUHNE,R ET AL. (1995)
logP 3.01			EXP	HANSCH,C ET AL. (1995)
VP 1.36E-001	mm Hg	25	EXT	OHE,S (1976)
DC	pKa			
HL 6.12E-005	atm m3/mol	25	EST	MEYLAN,WM & HOWARD,PH (1991)
OH 3.78E-012	cm3/molc sec	25	EST	MEYLAN,WM & HOWARD,PH (1993)

CAS #: 002316-26-9 — CINNAMIC ACID,3,4-DIMETHOXY

Formula: $C_{11}H_{12}O_4$

Mol Weight: 208.22

MP (deg C): 181-183

FP (deg C):

BP (deg C):

BP pressure (mm Hg):

Property/ Value	Units	Temp	Data Type	Reference
WS 9.89E+002	mg/L	25	EST	MEYLAN,WM ET AL. (1996)
logP 2.34			EXP	HANSCH,C & LEO,AJ (1985)
VP 2.64E-005	mm Hg	25	EST	NEELY,WB & BLAU,GE (1985)
DC 4.53	pKa	25	EXP	SERJEANT,EP & DEMPSEY,B (1979)
HL 4.53E-011	atm m3/mol	25	EST	MEYLAN,WM & HOWARD,PH (1991)
OH 3.94E-011	cm3/molc sec	25	EST	MEYLAN,WM & HOWARD,PH (1993)

655

CAS #: 002316-64-5				BENZENEMETHANOL, 5-BROMO-2-HYDROXY-

Formula: $C_7H_7BrO_2$

Mol Weight: 203.04

MP (deg C): 113   FP (deg C):

BP (deg C):

BP pressure (mm Hg):

Property/Value	Units	Temp	Data Type	Reference
WS 1.76E+004	mg/L	25	EST	MEYLAN,WM ET AL. (1996)
logP 1.72			EXP	SANGSTER,J (1993)
VP 2.85E-005	mm Hg	25	EST	NEELY,WB & BLAU,GE (1985)
DC	pKa			
HL 9.01E-012	atm m3/mol	25	EST	MEYLAN,WM & HOWARD,PH (1991)
OH 1.52E-011	cm3/molc sec	25	EST	MEYLAN,WM & HOWARD,PH (1993)

CAS #: 002327-02-8				(3,4-DICHLOROPHENYL)UREA

Formula: $C_7H_6Cl_2N_2O$

Mol Weight: 205.04

MP (deg C):   FP (deg C):

BP (deg C):

BP pressure (mm Hg):

Property/Value	Units	Temp	Data Type	Reference
WS 2.25E+002	mg/L	25	EST	MEYLAN,WM ET AL. (1996)
logP 2.65			EXP	SANGSTER,J (1994)
VP 3.26E-005	mm Hg	25	EST	NEELY,WB & BLAU,GE (1985)
DC	pKa			
HL 1.11E-010	atm m3/mol	25	EST	MEYLAN,WM & HOWARD,PH (1991)
OH 9.76E-012	cm3/molc sec	25	EST	MEYLAN,WM & HOWARD,PH (1993)

CAS #: 002338-10-5				4,5,6,7-TETRACHLOROBENZOTRIAZOLE

Formula: $C_6HCl_4N_3$

Mol Weight: 256.91

MP (deg C):   FP (deg C):

BP (deg C):

BP pressure (mm Hg):

Property/Value	Units	Temp	Data Type	Reference
WS 5.95E+000	mg/L	25	EST	MEYLAN,WM ET AL. (1996)
logP 4.17			EXP	HANSCH,C & LEO,AJ (1985)
VP 1.45E-006	mm Hg	25	EST	NEELY,WB & BLAU,GE (1985)
DC	pKa			
HL 4.42E-008	atm m3/mol	25	EST	MEYLAN,WM & HOWARD,PH (1991)
OH 4.22E-014	cm3/molc sec	25	EST	MEYLAN,WM & HOWARD,PH (1993)

CAS #: 002338-12-7				5-NITROBENZOTRIAZOLE

Formula: $C_6H_{10}N_4O_2$

Mol Weight: 170.17

MP (deg C): 217   FP (deg C):

BP (deg C):

BP pressure (mm Hg):

Property/Value	Units	Temp	Data Type	Reference
WS 5.77E+002	mg/L	25	EST	MEYLAN,WM ET AL. (1996)
logP 1.95			EXP	HANSCH,C & LEO,AJ (1985)
VP 5.60E-006	mm Hg	25	EST	NEELY,WB & BLAU,GE (1985)
DC	pKa			
HL 5.79E-010	atm m3/mol	25	EST	MEYLAN,WM & HOWARD,PH (1991)
OH 1.25E-013	cm3/molc sec	25	EST	MEYLAN,WM & HOWARD,PH (1993)

CAS #: 002338-25-2				2-CF3-5,6-DICHLOROBENZIMIDAZOLE

Formula: $C_8H_3Cl_2F_3N_2$

Mol Weight: 255.03

MP (deg C):   FP (deg C):

BP (deg C):

BP pressure (mm Hg):

Property/Value	Units	Temp	Data Type	Reference
WS 2.32E+001	mg/L	25	EST	MEYLAN,WM ET AL. (1996)
logP 3.99			EXP	HANSCH,C & LEO,AJ (1985)
VP 2.90E-006	mm Hg	25	EST	NEELY,WB & BLAU,GE (1985)
DC	pKa			
HL 1.75E-006	atm m3/mol	25	EST	MEYLAN,WM & HOWARD,PH (1991)
OH 1.29E-012	cm3/molc sec	25	EST	MEYLAN,WM & HOWARD,PH (1993)

CAS #: 002338-27-4				1H-BENZIMIDAZOLE, 4,5,6-TRICHLORO-2-(TRIFLUOROME

Formula: $C_8H_2Cl_3F_3N_2$

Mol Weight: 289.47

MP (deg C):   FP (deg C):

BP (deg C):

BP pressure (mm Hg):

Property/Value	Units	Temp	Data Type	Reference
WS 7.00E+000	mg/L	25	EST	MEYLAN,WM ET AL. (1996)
logP 3.87			EXP	SANGSTER,J (1994)
VP 7.61E-007	mm Hg	25	EST	NEELY,WB & BLAU,GE (1985)
DC	pKa			
HL 1.30E-006	atm m3/mol	25	EST	MEYLAN,WM & HOWARD,PH (1991)
OH 5.84E-013	cm3/molc sec	25	EST	MEYLAN,WM & HOWARD,PH (1993)

CAS #: 002338-30-9				4567-TETRABR 2-CF3 BENZIMIDAZOLE

Formula: $C_8HBr_4F_3N_2$

Mol Weight: 501.74

MP (deg C):   FP (deg C):

BP (deg C):

BP pressure (mm Hg):

Property/Value	Units	Temp	Data Type	Reference
WS 5.45E-002	mg/L	25	EST	MEYLAN,WM ET AL. (1996)
logP 4.81			EXP	HANSCH,C & LEO,AJ (1985)
VP 4.70E-009	mm Hg	25	EST	NEELY,WB & BLAU,GE (1985)
DC	pKa			
HL 8.03E-008	atm m3/mol	25	EST	MEYLAN,WM & HOWARD,PH (1991)
OH 2.04E-013	cm3/molc sec	25	EST	MEYLAN,WM & HOWARD,PH (1993)

CAS #: 002338-31-0				2-CF3-4-CHLOROBENZIMIDAZOLE

Formula: $C_8H_4ClF_3N_2$

Mol Weight: 220.58

MP (deg C):   FP (deg C):

BP (deg C):

BP pressure (mm Hg):

Property/Value	Units	Temp	Data Type	Reference
WS 1.08E+002	mg/L	25	EST	MEYLAN,WM ET AL. (1996)
logP 2.93			EXP	HANSCH,C & LEO,AJ (1985)
VP 1.11E-005	mm Hg	25	EST	NEELY,WB & BLAU,GE (1985)
DC	pKa			
HL 2.36E-006	atm m3/mol	25	EST	MEYLAN,WM & HOWARD,PH (1991)
OH 2.85E-012	cm3/molc sec	25	EST	MEYLAN,WM & HOWARD,PH (1993)

CAS #: 002345-34-8				4-ACETOXYBENZOIC ACID

Formula: $C_9H_8O_4$

Mol Weight: 180.16

MP (deg C): 191-194 FP (deg C):

BP (deg C):

BP pressure (mm Hg):

Property/ Value	Units	Temp	Data Type	Reference
WS 3.05E+003	mg/L	25	EST	MEYLAN,WM ET AL. (1996)
logP 1.47			EST	MEYLAN,WM & HOWARD,PH (1995)
VP 1.84E-004	mm Hg	25	EST	NEELY,WB & BLAU,GE (1985)
DC	pKa			
HL 1.30E-009	atm m3/mol	25	EST	MEYLAN,WM & HOWARD,PH (1991)
OH 1.37E-012	cm3/molc sec	25	EST	MEYLAN,WM & HOWARD,PH (1993)

CAS #: 002348-74-5				1,4-NAPHTHOQUINONE,2-ACETAMIDO

Formula: $C_{12}H_9NO_3$

Mol Weight: 215.21

MP (deg C): FP (deg C):

BP (deg C):

BP pressure (mm Hg):

Property/ Value	Units	Temp	Data Type	Reference
WS 2.89E+003	mg/L	25	EST	MEYLAN,WM ET AL. (1996)
logP 1.29			EXP	HANSCH,C & LEO,AJ (1985)
VP 5.39E-008	mm Hg	25	EST	NEELY,WB & BLAU,GE (1985)
DC	pKa			
HL 2.74E-014	atm m3/mol	25	EST	MEYLAN,WM & HOWARD,PH (1991)
OH 1.27E-011	cm3/molc sec	25	EST	MEYLAN,WM & HOWARD,PH (1993)

CAS #: 002348-81-4				1,4-NAPHTHALENEDIONE, 2-AMINO-

Formula: $C_{10}H_7NO_2$

Mol Weight: 173.17

MP (deg C): FP (deg C):

BP (deg C):

BP pressure (mm Hg):

Property/ Value	Units	Temp	Data Type	Reference
WS 1.86E+004	mg/L	25	EST	MEYLAN,WM ET AL. (1996)
logP 1.77			EXP	HANSCH,C ET AL. (1995)
VP 3.55E-005	mm Hg	25	EST	NEELY,WB & BLAU,GE (1985)
DC	pKa			
HL 8.45E-012	atm m3/mol	25	EST	MEYLAN,WM & HOWARD,PH (1991)
OH 2.81E-011	cm3/molc sec	25	EST	MEYLAN,WM & HOWARD,PH (1993)

CAS #: 002348-82-5				2-METHOXY-1,4-NAPHTHOQUINONE

Formula: $C_{11}H_8O_3$

Mol Weight: 188.18

MP (deg C): FP (deg C):

BP (deg C):

BP pressure (mm Hg):

Property/ Value	Units	Temp	Data Type	Reference
WS 3.53E+003	mg/L	25	EST	MEYLAN,WM ET AL. (1996)
logP 1.35			EXP	HANSCH,C & LEO,AJ (1985)
VP 7.92E-005	mm Hg	25	EST	NEELY,WB & BLAU,GE (1985)
DC	pKa			
HL 1.29E-009	atm m3/mol	25	EST	MEYLAN,WM & HOWARD,PH (1991)
OH 1.54E-011	cm3/molc sec	25	EST	MEYLAN,WM & HOWARD,PH (1993)

CAS #: 002358-84-1				DIETHYLENE GLYCOL DIMETHACRYLATE

Formula: $C_{12}H_{18}O_5$

Mol Weight: 242.27

MP (deg C): FP (deg C):

BP (deg C): 120-125

BP pressure (mm Hg): 2.00E+000

Property/ Value	Units	Temp	Data Type	Reference
WS 5.82E+002	mg/L	25	EST	MEYLAN,WM ET AL. (1996)
logP 1.93			EST	MEYLAN,WM & HOWARD,PH (1995)
VP 1.06E-002	mm Hg	25	EST	NEELY,WB & BLAU,GE (1985)
DC	pKa			
HL 7.92E-010	atm m3/mol	25	EST	MEYLAN,WM & HOWARD,PH (1991)
OH 5.39E-011	cm3/molc sec	25	EST	MEYLAN,WM & HOWARD,PH (1993)

CAS #: 002362-57-4				2-(PHENYLAZO)PHENOL

Formula: $C_{12}H_{10}N_2O$

Mol Weight: 198.23

MP (deg C): FP (deg C):

BP (deg C):

BP pressure (mm Hg):

Property/ Value	Units	Temp	Data Type	Reference
WS 1.26E+001	mg/L	25	EST	MEYLAN,WM ET AL. (1996)
logP 4.33			EST	MEYLAN,WM & HOWARD,PH (1995)
VP 1.61E-005	mm Hg	25	EST	NEELY,WB & BLAU,GE (1985)
DC	pKa			
HL 1.53E-009	atm m3/mol	25	EST	MEYLAN,WM & HOWARD,PH (1991)
OH 1.41E-011	cm3/molc sec	25	EST	MEYLAN,WM & HOWARD,PH (1993)

CAS #: 002362-62-1				BENZENECARBOTHIAMIDE, 4-METHYL-

Formula: C_8H_9NS

Mol Weight: 151.23

MP (deg C): FP (deg C):

BP (deg C):

BP pressure (mm Hg):

Property/ Value	Units	Temp	Data Type	Reference
WS 1.47E+004	mg/L	25	EST	MEYLAN,WM ET AL. (1996)
logP 2.01			EXP	SANGSTER,J (1994)
VP 5.24E-003	mm Hg	25	EST	NEELY,WB & BLAU,GE (1985)
DC	pKa			
HL 4.32E-007	atm m3/mol	25	EST	MEYLAN,WM & HOWARD,PH (1991)
OH 2.72E-011	cm3/molc sec	25	EST	MEYLAN,WM & HOWARD,PH (1993)

CAS #: 002362-64-3				BENZENECARBOTHIAMIDE, 4-METHOXY-

Formula: C_8H_9NOS

Mol Weight: 167.23

MP (deg C): FP (deg C):

BP (deg C):

BP pressure (mm Hg):

Property/ Value	Units	Temp	Data Type	Reference
WS 2.62E+004	mg/L	25	EST	MEYLAN,WM ET AL. (1996)
logP 1.63			EXP	POLASEK,M ET AL. (1988)
VP 1.58E-003	mm Hg	25	EST	NEELY,WB & BLAU,GE (1985)
DC	pKa			
HL 2.32E-008	atm m3/mol	25	EST	MEYLAN,WM & HOWARD,PH (1991)
OH 4.76E-011	cm3/molc sec	25	EST	MEYLAN,WM & HOWARD,PH (1993)

CAS #: 002363-89-5 — 2-OCTENAL

Formula: $C_8H_{14}O$
Mol Weight: 126.20
MP (deg C):
FP (deg C):
BP (deg C):
BP pressure (mm Hg):

Property/Value	Units	Temp	Data Type	Reference
WS 6.13E+002	mg/L	25	EST	MEYLAN,WM ET AL. (1996)
logP 2.57			EST	MEYLAN,WM & HOWARD,PH (1995)
VP 8.60E-001	mm Hg	25	EST	NEELY,WB & BLAU,GE (1985)
DC	pKa			
HL 7.34E-005	atm m3/mol	25	EXP	BUTTERY,RG ET AL. (1971)
OH 3.68E-011	cm3/molc sec	25	EXP	ATKINSON,R (1989)

CAS #: 002364-46-7 — NAPHTHALENE, 1,3,8-TRINITRO-

Formula: $C_{10}H_5N_3O_6$
Mol Weight: 263.17
MP (deg C):
FP (deg C):
BP (deg C):
BP pressure (mm Hg):

Property/Value	Units	Temp	Data Type	Reference
WS 8.83E+001	mg/L	25	EST	MEYLAN,WM ET AL. (1996)
logP 2.30			EXP	DEBNATH,AK ET AL. (1992)
VP 8.51E-008	mm Hg	25	EST	NEELY,WB & BLAU,GE (1985)
DC	pKa			
HL 3.23E-011	atm m3/mol	25	EST	MEYLAN,WM & HOWARD,PH (1991)
OH 2.47E-014	cm3/molc sec	25	EST	MEYLAN,WM & HOWARD,PH (1993)

CAS #: 002364-50-3 — VALERANILIDE

Formula: $C_{11}H_{15}NO$
Mol Weight: 177.25
MP (deg C):
FP (deg C):
BP (deg C):
BP pressure (mm Hg):

Property/Value	Units	Temp	Data Type	Reference
WS 3.35E+002	mg/L	25	EST	MEYLAN,WM ET AL. (1996)
logP 2.51			EXP	HANSCH,C & LEO,AJ (1985)
VP 3.28E-005	mm Hg	25	EST	NEELY,WB & BLAU,GE (1985)
DC	pKa			
HL 1.44E-008	atm m3/mol	25	EST	MEYLAN,WM & HOWARD,PH (1991)
OH 1.91E-011	cm3/molc sec	25	EST	MEYLAN,WM & HOWARD,PH (1993)

CAS #: 002365-30-2 — 1-(2-CHLOROETHYL)-1-NITROSOUREA

Formula: $C_3H_6ClN_3O_2$
Mol Weight: 151.55
MP (deg C):
FP (deg C):
BP (deg C):
BP pressure (mm Hg):

Property/Value	Units	Temp	Data Type	Reference
WS 2.44E+004	mg/L	25	EST	MEYLAN,WM ET AL. (1996)
logP 0.57			EXP	HANSCH,C & LEO,AJ (1985)
VP 2.49E-003	mm Hg	25	EST	NEELY,WB & BLAU,GE (1985)
DC	pKa			
HL 4.63E-011	atm m3/mol	25	EST	MEYLAN,WM & HOWARD,PH (1991)
OH 2.82E-012	cm3/molc sec	25	EST	MEYLAN,WM & HOWARD,PH (1993)

CAS #: 002366-52-1 — 1-FLUOROBUTANE

Formula: C_4H_9F
Mol Weight: 76.11
MP (deg C): -134
FP (deg C):
BP (deg C): 32.5
BP pressure (mm Hg):

Property/Value	Units	Temp	Data Type	Reference
WS 8.32E+002	mg/L	25	EST	MEYLAN,WM ET AL. (1996)
logP 2.58			EXP	HANSCH,C ET AL. (1995)
VP 5.78E+002	mm Hg	25	EXP	LI,JCM & ROSSINI,FD (1953)
DC	pKa			
HL 3.85E-002	atm m3/mol	25	EST	MEYLAN,WM & HOWARD,PH (1991)
OH 2.12E-012	cm3/molc sec	25	EST	MEYLAN,WM & HOWARD,PH (1993)

CAS #: 002367-82-0 — 1,2,3,5-TETRAFLUOROBENZENE

Formula: $C_6H_2F_4$
Mol Weight: 150.08
MP (deg C): -48
FP (deg C):
BP (deg C): 84.4
BP pressure (mm Hg):

Property/Value	Units	Temp	Data Type	Reference
WS 7.42E+002	mg/L	25	EXP	YALKOWSKY,SH & DANNENFELSER,RM (1992)
logP 2.79			EST	MEYLAN,WM & HOWARD,PH (1995)
VP 7.33E+001	mm Hg	25	EXP	BOUBLIK,T ET AL. (1984)
DC	pKa			
HL 1.95E-002	atm m3/mol	25	EST	VP/WSOL
OH 1.29E-012	cm3/molc sec	25	EST	MEYLAN,WM & HOWARD,PH (1993)

CAS #: 002370-12-9 — 2,2-DIMETHYL-1-PENTANOL

Formula: $C_7H_{16}O$
Mol Weight: 116.20
MP (deg C):
FP (deg C):
BP (deg C):
BP pressure (mm Hg):

Property/Value	Units	Temp	Data Type	Reference
WS 4.23E+003	mg/L	25	EXP	WAKITA,K ET AL. (1986)
logP 2.20			EST	MEYLAN,WM & HOWARD,PH (1995)
VP 8.08E-001	mm Hg	25	EST	NEELY,WB & BLAU,GE (1985)
DC	pKa			
HL 2.34E-005	atm m3/mol	25	EST	MEYLAN,WM & HOWARD,PH (1991)
OH 7.22E-012	cm3/molc sec	25	EST	MEYLAN,WM & HOWARD,PH (1993)

CAS #: 002371-42-8 — 2-METHYLISOBORNEOL

Formula: $C_{11}H_{20}O$
Mol Weight: 168.28
MP (deg C):
FP (deg C):
BP (deg C):
BP pressure (mm Hg):

Property/Value	Units	Temp	Data Type	Reference
WS 3.05E+002	mg/L	25	EST	MEYLAN,WM ET AL. (1996)
logP 3.31			EST	MEYLAN,WM & HOWARD,PH (1995)
VP 3.20E-002	mm Hg	25	EST	NEELY,WB & BLAU,GE (1985)
DC	pKa			
HL 8.90E-006	atm m3/mol	25	EST	MEYLAN,WM & HOWARD,PH (1991)
OH 8.07E-012	cm3/molc sec	25	EST	MEYLAN,WM & HOWARD,PH (1993)

CAS #: 002373-41-3				BENZENAMINIUM, 3-CHLORO-N,N,N-TRIMETHYL-, IODIDE
Formula: $C_9H_{13}ClIN$				
Mol Weight: 297.57				
MP (deg C):		**FP (deg C):**		
BP (deg C):				
BP pressure (mm Hg):				

Property/ Value	Units	Temp	Data Type	Reference
WS 1.00E+006	mg/L	25	EST	MEYLAN,WM ET AL. (1996)
logP -2.30			EXP	HANSCH,C ET AL. (1995)
VP 7.50E-009	mm Hg	25	EST	NEELY,WB & BLAU,GE (1985)
DC	pKa			
HL 4.90E-013	atm m3/mol	25	EST	MEYLAN,WM & HOWARD,PH (1991)
OH 4.09E-012	cm3/molc sec	25	EST	MEYLAN,WM & HOWARD,PH (1993)

CAS #: 002373-80-0				3,4-(METHYLENEDIOXY)CINNAMIC ACID
Formula: $C_{10}H_8O_4$				
Mol Weight: 192.17				
MP (deg C): 242-244 de		**FP (deg C):**		
BP (deg C):				
BP pressure (mm Hg):				

Property/ Value	Units	Temp	Data Type	Reference
WS 2.70E+001	mg/L	25	EXP	STEPHEN,H & STEPHEN,T (1963)
logP 2.59			EXP	SANGSTER,J (1993)
VP 2.48E-005	mm Hg	25	EST	NEELY,WB & BLAU,GE (1985)
DC	pKa			
HL 1.50E-011	atm m3/mol	25	EST	MEYLAN,WM & HOWARD,PH (1991)
OH 4.05E-011	cm3/molc sec	25	EST	MEYLAN,WM & HOWARD,PH (1993)

CAS #: 002373-84-4				5-ALLYL-5-ETHYL BARBITURIC ACID
Formula: $C_9H_{12}N_2O_3$				
Mol Weight: 196.21				
MP (deg C):		**FP (deg C):**		
BP (deg C):				
BP pressure (mm Hg):				

Property/ Value	Units	Temp	Data Type	Reference
WS 7.06E+003	mg/L	25	EST	MEYLAN,WM ET AL. (1996)
logP 0.95			EXP	HANSCH,C & LEO,AJ (1985)
VP 7.39E-011	mm Hg	25	EST	NEELY,WB & BLAU,GE (1985)
DC	pKa			
HL 3.57E-013	atm m3/mol	25	EST	MEYLAN,WM & HOWARD,PH (1991)
OH 3.58E-011	cm3/molc sec	25	EST	MEYLAN,WM & HOWARD,PH (1993)

CAS #: 002374-03-0				3-HYDROXY-4-AMINOBENZOIC ACID
Formula: $C_7H_7NO_3$				
Mol Weight: 153.14				
MP (deg C): 211-215		**FP (deg C):**		
BP (deg C):				
BP pressure (mm Hg):				

Property/ Value	Units	Temp	Data Type	Reference
WS 2.75E+004	mg/L	25	EST	MEYLAN,WM ET AL. (1996)
logP 0.50			EXP	HANSCH,C & LEO,AJ (1985)
VP 4.00E-006	mm Hg	25	EST	NEELY,WB & BLAU,GE (1985)
DC	pKa			
HL 3.99E-015	atm m3/mol	25	EST	MEYLAN,WM & HOWARD,PH (1991)
OH 2.81E-011	cm3/molc sec	25	EST	MEYLAN,WM & HOWARD,PH (1993)

CAS #: 002379-74-0				C.I. VAT RED 1
Formula: $C_{18}H_{10}Cl_2O_2S_2$				
Mol Weight: 393.31				
MP (deg C):		**FP (deg C):**		
BP (deg C):				
BP pressure (mm Hg):				

Property/ Value	Units	Temp	Data Type	Reference
WS 2.99E-002	mg/L	25	EST	MEYLAN,WM ET AL. (1996)
logP 5.92			EST	MEYLAN,WM & HOWARD,PH (1995)
VP 3.00E-014	mm Hg	25	EXP	BAUGHMAN,GL & PERENICH,TA (1988)
DC	pKa			
HL 3.05E-013	atm m3/mol	25	EST	MEYLAN,WM & HOWARD,PH (1991)
OH 5.58E-011	cm3/molc sec	25	EST	MEYLAN,WM & HOWARD,PH (1993)

CAS #: 002380-78-1				3-METHOXY-4-HYDROXYPHENYLETHANOL
Formula: $C_9H_{12}O_3$				
Mol Weight: 168.19				
MP (deg C): 40-42		**FP (deg C):**		
BP (deg C):				
BP pressure (mm Hg):				

Property/ Value	Units	Temp	Data Type	Reference
WS 8.05E+004	mg/L	25	EST	MEYLAN,WM ET AL. (1996)
logP 0.47			EXP	HANSCH,C & LEO,AJ (1985)
VP 2.04E-005	mm Hg	25	EST	NEELY,WB & BLAU,GE (1985)
DC	pKa			
HL 1.78E-012	atm m3/mol	25	EST	MEYLAN,WM & HOWARD,PH (1991)
OH 4.31E-011	cm3/molc sec	25	EST	MEYLAN,WM & HOWARD,PH (1993)

CAS #: 002381-15-9				10-METHYL-BENZ(A)ANTHRACENE
Formula: $C_{19}H_{14}$				
Mol Weight: 242.32				
MP (deg C):		**FP (deg C):**		
BP (deg C):				
BP pressure (mm Hg):				

Property/ Value	Units	Temp	Data Type	Reference
WS 5.50E-002	mg/L	25	EXP	SHAW,DG (1989)
logP 6.07			EST	MEYLAN,WM & HOWARD,PH (1995)
VP 2.53E-007	mm Hg	25	EST	NEELY,WB & BLAU,GE (1985)
DC	pKa			
HL 5.53E-006	atm m3/mol	25	EST	MEYLAN,WM & HOWARD,PH (1991)
OH 1.31E-010	cm3/molc sec	25	EST	MEYLAN,WM & HOWARD,PH (1993)

CAS #: 002381-39-7				6-METHYLBENZO(A)PYRENE
Formula: $C_{21}H_{14}$				
Mol Weight: 266.35				
MP (deg C):		**FP (deg C):**		
BP (deg C):				
BP pressure (mm Hg):				

Property/ Value	Units	Temp	Data Type	Reference
WS 7.98E-004	mg/L	25	EXP	PEARLMAN,RS ET AL. (1984)
logP 6.66			EST	MEYLAN,WM & HOWARD,PH (1995)
VP 1.08E-008	mm Hg	25	EST	NEELY,WB & BLAU,GE (1985)
DC	pKa			
HL 8.94E-007	atm m3/mol	25	EST	MEYLAN,WM & HOWARD,PH (1991)
OH 1.31E-010	cm3/molc sec	25	EST	MEYLAN,WM & HOWARD,PH (1993)

CAS #: 002385-85-5 — MIREX

Formula: $C_{10}Cl_{12}$

Mol Weight: 545.55

MP (deg C): 485 dec

FP (deg C):

BP (deg C):

BP pressure (mm Hg):

Property/Value	Units	Temp	Data Type	Reference
WS 8.50E-002	mg/L	25	EXP	YALKOWSKY,SH & DANNENFELSER,RM (1992)
logP 6.89			EXP	VEITH,GD ET AL. (1979)
VP 8.00E-007	mm Hg	25	EXP	AUGUSTIJN-BECKERS,PWM ET AL. (1994)
DC	pKa			
HL 5.16E-004	atm m3/mol	22	EXP	YIN,C & HASSETT,JP (1986)
OH 0.00E+000	cm3/molc sec	25	EST	MEYLAN,WM & HOWARD,PH (1993)

CAS #: 002386-53-0 — DODECYL SULFONATE, SODIUM SALT

Formula: $C_{12}H_{25}NaO_3S$

Mol Weight: 272.39

MP (deg C): > 300

FP (deg C):

BP (deg C):

BP pressure (mm Hg):

Property/Value	Units	Temp	Data Type	Reference
WS 2.68E+003	mg/L	25	EST	MEYLAN,WM ET AL. (1996)
logP 0.96			EXP	HANSCH,C & LEO,AJ (1985)
VP 1.13E-012	mm Hg	25	EST	NEELY,WB & BLAU,GE (1985)
DC	pKa			
HL	atm m3/mol			
OH 1.52E-011	cm3/molc sec	25	EST	MEYLAN,WM & HOWARD,PH (1993)

CAS #: 002387-16-8 — 1,1-DICHLORO-2,2-DIPHENYLETHANE

Formula: $C_{14}H_{12}Cl_2$

Mol Weight: 251.16

MP (deg C):

FP (deg C):

BP (deg C):

BP pressure (mm Hg):

Property/Value	Units	Temp	Data Type	Reference
WS 3.28E+000	mg/L	25	EST	MEYLAN,WM ET AL. (1996)
logP 4.51			EXP	HANSCH,C & LEO,AJ (1985)
VP 1.69E-004	mm Hg	25	EST	NEELY,WB & BLAU,GE (1985)
DC	pKa			
HL 7.91E-005	atm m3/mol	25	EST	MEYLAN,WM & HOWARD,PH (1991)
OH 1.03E-011	cm3/molc sec	25	EST	MEYLAN,WM & HOWARD,PH (1993)

CAS #: 002388-12-7 — DODECANEPEROXOIC ACID

Formula: $C_{12}H_{24}O_3$

Mol Weight: 216.32

MP (deg C):

FP (deg C):

BP (deg C):

BP pressure (mm Hg):

Property/Value	Units	Temp	Data Type	Reference
WS 4.82E+000	mg/L	25	EST	MEYLAN,WM ET AL. (1996)
logP 5.00			EST	MEYLAN,WM & HOWARD,PH (1995)
VP 2.50E-004	mm Hg	25	EXP	SWAIN,HAJR ET AL. (1980)
DC	pKa			
HL 3.29E-005	atm m3/mol	25	EST	MEYLAN,WM & HOWARD,PH (1991)
OH 1.50E-011	cm3/molc sec	25	EST	MEYLAN,WM & HOWARD,PH (1993)

CAS #: 002388-47-8 — PHOSPHORAMIDIC ACID, N-(1-METHYLPROPYL)-, O-METH

Formula: $C_{11}H_{15}Cl_3NO_3P$

Mol Weight: 346.58

MP (deg C):

FP (deg C):

BP (deg C):

BP pressure (mm Hg):

Property/Value	Units	Temp	Data Type	Reference
WS 2.54E+000	mg/L	25	EST	MEYLAN,WM ET AL. (1996)
logP 3.99			EXP	HANSCH,C & LEO,AJ (1985)
VP 3.96E-006	mm Hg	25	EST	NEELY,WB & BLAU,GE (1985)
DC	pKa			
HL 1.22E-009	atm m3/mol	25	EST	MEYLAN,WM & HOWARD,PH (1991)
OH 5.42E-011	cm3/molc sec	25	EST	MEYLAN,WM & HOWARD,PH (1993)

CAS #: 002389-37-9 — 5-NO2 SALICYLANILIDE

Formula: $C_{13}H_{10}N_2O_4$

Mol Weight: 258.24

MP (deg C):

FP (deg C):

BP (deg C):

BP pressure (mm Hg):

Property/Value	Units	Temp	Data Type	Reference
WS 6.06E+001	mg/L	25	EST	MEYLAN,WM ET AL. (1996)
logP 3.66			EXP	TERADA,H ET AL. (1988)
VP 3.75E-010	mm Hg	25	EST	NEELY,WB & BLAU,GE (1985)
DC	pKa			
HL 6.31E-013	atm m3/mol	25	EST	MEYLAN,WM & HOWARD,PH (1991)
OH 1.64E-011	cm3/molc sec	25	EST	MEYLAN,WM & HOWARD,PH (1993)

CAS #: 002390-94-5 — TRANS-2,5-DIMETHYLTETRAHYDROFURAN

Formula: $C_6H_{12}O$

Mol Weight: 100.16

MP (deg C):

FP (deg C):

BP (deg C):

BP pressure (mm Hg):

Property/Value	Units	Temp	Data Type	Reference
WS 8.38E+003	mg/L	25	EST	MEYLAN,WM ET AL. (1996)
logP 1.34			EXP	FUNASAKI,N ET AL. (1985)
VP 3.64E+001	mm Hg	25	EST	NEELY,WB & BLAU,GE (1985)
DC	pKa			
HL 1.49E-004	atm m3/mol	25	EST	MEYLAN,WM & HOWARD,PH (1991)
OH 2.59E-011	cm3/molc sec	25	EST	MEYLAN,WM & HOWARD,PH (1993)

CAS #: 002392-68-9 — 1-CHLORO-3-ISOTHIOCYANATO-BENZENE

Formula: C_7H_4ClNS

Mol Weight: 169.63

MP (deg C):

FP (deg C):

BP (deg C): 249-250

BP pressure (mm Hg):

Property/Value	Units	Temp	Data Type	Reference
WS 1.90E+001	mg/L	25	EXP	YALKOWSKY,SH & DANNENFELSER,RM (1992)
logP 3.36			EXP	SANGSTER,J (1994)
VP 1.34E-002	mm Hg	25	EST	NEELY,WB & BLAU,GE (1985)
DC	pKa			
HL 5.79E-004	atm m3/mol	25	EST	MEYLAN,WM & HOWARD,PH (1991)
OH 8.63E-013	cm3/molc sec	25	EST	MEYLAN,WM & HOWARD,PH (1993)

CAS #: 002396-60-3				(4-METHOXYPHENYL)PHENYLDIAZENE

Formula: $C_{13}H_{12}N_2O$
Mol Weight: 212.25
MP (deg C): 56   FP (deg C):
BP (deg C): 340
BP pressure (mm Hg):

Property/ Value	Units	Temp	Data Type	Reference
WS 3.68E+000	mg/L	25	EST	MEYLAN,WM.ET AL. (1996)
logP 4.19			EST	MEYLAN,WM & HOWARD,PH (1995)
VP 2.56E-004	mm Hg	25	EST	NEELY,WB & BLAU,GE (1985)
DC	pKa			
HL 8.70E-007	atm m3/mol	25	EST	MEYLAN,WM & HOWARD,PH (1991)
OH 1.01E-011	cm3/molc sec	25	EST	MEYLAN,WM & HOWARD,PH (1993)

CAS #: 002396-63-6				1,6-HEPTADIENE

Formula: C_7H_{12}
Mol Weight: 96.17
MP (deg C): -85   FP (deg C):
BP (deg C): 112
BP pressure (mm Hg):

Property/ Value	Units	Temp	Data Type	Reference
WS 1.65E+003	mg/L	25	EXP	YALKOWSKY,SH & DANNENFELSER,RM (1992)
logP 2.24			EST	MEYLAN,WM & HOWARD,PH (1995)
VP 2.53E+001	mm Hg	25	EST	NEELY,WB & BLAU,GE (1985)
DC	pKa			
HL 6.18E-003	atm m3/mol	25	EST	MEYLAN,WM & HOWARD,PH (1991)
OH 1.77E-011	cm3/molc sec	25	EST	MEYLAN,WM & HOWARD,PH (1993)

CAS #: 002396-65-8				1,8-NONADIYNE

Formula: C_9H_{12}
Mol Weight: 120.20
MP (deg C): -27.3   FP (deg C):
BP (deg C): 162
BP pressure (mm Hg):

Property/ Value	Units	Temp	Data Type	Reference
WS 1.25E+002	mg/L	25	EXP	YALKOWSKY,SH & DANNENFELSER,RM (1992)
logP 3.22			EST	MEYLAN,WM & HOWARD,PH (1995)
VP 3.57E+000	mm Hg	25	EST	NEELY,WB & BLAU,GE (1985)
DC	pKa			
HL 1.09E-002	atm m3/mol	25	EST	MEYLAN,WM & HOWARD,PH (1991)
OH 2.05E-011	cm3/molc sec	25	EST	MEYLAN,WM & HOWARD,PH (1993)

CAS #: 002398-37-0				1-BROMO-3-METHOXYBENZENE

Formula: C_7H_7BrO
Mol Weight: 187.04
MP (deg C):   FP (deg C):
BP (deg C): 211
BP pressure (mm Hg):

Property/ Value	Units	Temp	Data Type	Reference
WS 1.31E+002	mg/L	25	EST	MEYLAN,WM ET AL. (1996)
logP 2.96			EST	MEYLAN,WM & HOWARD,PH (1995)
VP 2.88E-001	mm Hg	25	EST	NEELY,WB & BLAU,GE (1985)
DC	pKa			
HL 1.27E-004	atm m3/mol	25	EST	MEYLAN,WM & HOWARD,PH (1991)
OH 1.44E-011	cm3/molc sec	25	EST	MEYLAN,WM & HOWARD,PH (1993)

CAS #: 002398-81-4				N-OXIDENICOTINIC ACID

Formula: $C_6H_5NO_3$
Mol Weight: 139.11
MP (deg C): 254-255   FP (deg C):
BP (deg C):
BP pressure (mm Hg):

Property/ Value	Units	Temp	Data Type	Reference
WS 7.00E+003	mg/L	25	EXP	GOE,GL (1982)
logP -1.45			EST	MEYLAN,WM & HOWARD,PH (1995)
VP 5.82E-005	mm Hg	25	EST	NEELY,WB & BLAU,GE (1985)
DC 0.74	pKa	24	EXP	PERRIN,DD (1965)
HL 1.42E-015	atm m3/mol	25	EST	MEYLAN,WM & HOWARD,PH (1991)
OH 6.57E-013	cm3/molc sec	25	EST	MEYLAN,WM & HOWARD,PH (1993)

CAS #: 002400-00-2				BENZENE, (1-ETHYLDECYL)-

Formula: $C_{18}H_{30}$
Mol Weight: 246.44
MP (deg C):   FP (deg C):
BP (deg C):
BP pressure (mm Hg):

Property/ Value	Units	Temp	Data Type	Reference
WS 2.99E-003	mg/L	25	EST	MEYLAN,WM ET AL. (1996)
logP 8.10			EXP	SHERBLOM,PM ET AL. (1992)
VP 5.84E-004	mm Hg	25	EST	NEELY,WB & BLAU,GE (1985)
DC	pKa			
HL 1.34E-001	atm m3/mol	25	EST	MEYLAN,WM & HOWARD,PH (1991)
OH 2.01E-011	cm3/molc sec	25	EST	MEYLAN,WM & HOWARD,PH (1993)

CAS #: 002402-77-9				2,3-DICHLOROPYRIDINE

Formula: $C_5H_3Cl_2N$
Mol Weight: 147.99
MP (deg C): 65-67   FP (deg C):
BP (deg C):
BP pressure (mm Hg):

Property/ Value	Units	Temp	Data Type	Reference
WS 1.22E+003	mg/L	25	EST	MEYLAN,WM ET AL. (1996)
logP 2.11			EXP	HANSCH,C & LEO,AJ (1985)
VP 7.78E-001	mm Hg	25	EST	NEELY,WB & BLAU,GE (1985)
DC -0.85	pKa		EXP	PERRIN,DD (1972)
HL 2.44E-004	atm m3/mol	25	EST	MEYLAN,WM & HOWARD,PH (1991)
OH 7.60E-014	cm3/molc sec	25	EST	MEYLAN,WM & HOWARD,PH (1993)

CAS #: 002402-78-0				2,6-DICHLOROPYRIDINE

Formula: $C_5H_3Cl_2N$
Mol Weight: 147.99
MP (deg C): 86-88.5   FP (deg C):
BP (deg C):
BP pressure (mm Hg):

Property/ Value	Units	Temp	Data Type	Reference
WS 1.13E+003	mg/L	25	EST	MEYLAN,WM ET AL. (1996)
logP 2.15			EXP	HANSCH,C & LEO,AJ (1985)
VP 7.78E-001	mm Hg	25	EST	NEELY,WB & BLAU,GE (1985)
DC -2.86	pKa		EXP	PERRIN,DD (1972)
HL 1.54E-002	atm m3/mol	25	EST	MEYLAN,WM & HOWARD,PH (1991)
OH 1.83E-013	cm3/molc sec	25	EST	MEYLAN,WM & HOWARD,PH (1993)

002402-79-1 — 2,3,5,6-TETRACHLORPYRIDINE

Formula: C_5HCl_4N
Mol Weight: 216.88
MP (deg C): 90.5
FP (deg C):
BP (deg C): 250.5
BP pressure (mm Hg):

Property/Value	Units	Temp	Data Type	Reference
WS 1.33E+001	mg/L	25	EST	MEYLAN,WM ET AL. (1996)
logP 3.32			EXP	HANSCH,C & LEO,AJ (1985)
VP 1.69E-004	mm Hg	25	EST	WSOL X HL
DC -0.80	pKa		EXP	HANSCH,C & LEO,AJ (1985)
HL 3.62E-006	atm m3/mol	25	EST	MEYLAN,WM & HOWARD,PH (1991)
OH 1.52E-014	cm3/molc sec	25	EST	MEYLAN,WM & HOWARD,PH (1993)

002403-22-7 — BENZENEMETHANAMINE, N-BUTYL-

Formula: $C_{11}H_{17}N$
Mol Weight: 163.26
MP (deg C):
FP (deg C):
BP (deg C): 87-89
BP pressure (mm Hg): 3.00E+000

Property/Value	Units	Temp	Data Type	Reference
WS 4.14E+003	mg/L	25	EST	MEYLAN,WM ET AL. (1996)
logP 2.59			EXP	KRIL,MB & FUNG,HL (1990)
VP 3.48E-002	mm Hg	25	EST	NEELY,WB & BLAU,GE (1985)
DC	pKa			
HL 3.14E-006	atm m3/mol	25	EST	MEYLAN,WM & HOWARD,PH (1991)
OH 8.99E-011	cm3/molc sec	25	EST	MEYLAN,WM & HOWARD,PH (1993)

002403-66-9 — 2-(G-HYDROXYPROPYL)BENZIMIDAZOLE

Formula: $C_{10}H_{12}N_2O$
Mol Weight: 176.22
MP (deg C):
FP (deg C):
BP (deg C):
BP pressure (mm Hg):

Property/Value	Units	Temp	Data Type	Reference
WS 1.59E+004	mg/L	25	EST	MEYLAN,WM ET AL. (1996)
logP 1.25			EXP	HANSCH,C & LEO,AJ (1985)
VP 4.75E-009	mm Hg	25	EST	NEELY,WB & BLAU,GE (1985)
DC	pKa			
HL 2.61E-011	atm m3/mol	25	EST	MEYLAN,WM & HOWARD,PH (1991)
OH 9.62E-011	cm3/molc sec	25	EST	MEYLAN,WM & HOWARD,PH (1993)

002404-03-7 — DIETHYL DIMETHYLPHOSPHORAMIDATE

Formula: $C_6H_{16}NO_3P$
Mol Weight: 181.17
MP (deg C): 8.30
FP (deg C):
BP (deg C): 232.05
BP pressure (mm Hg):

Property/Value	Units	Temp	Data Type	Reference
WS 2.00E+004	mg/L	25	EST	MEYLAN,WM ET AL. (1996)
logP 0.51			EST	MEYLAN,WM & HOWARD,PH (1995)
VP 7.03E-002	mm Hg	25	EST	MPBPVP
DC	pKa			
HL 1.34E-008	atm m3/mol	25	EST	MEYLAN,WM & HOWARD,PH (1991)
OH 9.77E-011	cm3/molc sec	25	EST	MEYLAN,WM & HOWARD,PH (1993)

002404-73-1 — DIBUTYL METHYLPHOSPHONATE

Formula: $C_9H_{21}O_3P$
Mol Weight: 208.24
MP (deg C):
FP (deg C):
BP (deg C):
BP pressure (mm Hg):

Property/Value	Units	Temp	Data Type	Reference
WS 6.52E+001	mg/L	25	EST	MEYLAN,WM ET AL. (1996)
logP 3.26			EXP	KRIKORIAN,SE ET AL. (1987)
VP 1.06E-002	mm Hg	25	EST	NEELY,WB & BLAU,GE (1985)
DC	pKa			
HL 6.84E-006	atm m3/mol	25	EST	MEYLAN,WM & HOWARD,PH (1991)
OH 5.27E-011	cm3/molc sec	25	EST	MEYLAN,WM & HOWARD,PH (1993)

002409-55-4 — 2-(TERT-BUTYL)-4-METHYLPHENOL

Formula: $C_{11}H_{16}O$
Mol Weight: 164.25
MP (deg C): 55
FP (deg C):
BP (deg C): 237
BP pressure (mm Hg):

Property/Value	Units	Temp	Data Type	Reference
WS 1.84E+002	mg/L	25	EST	MEYLAN,WM ET AL. (1996)
logP 3.97			EST	MEYLAN,WM & HOWARD,PH (1995)
VP 2.50E-002	mm Hg	25	EXT	CHAO,J ET AL. (1983)
DC 11.72	pKa	20	EXP	SERJEANT,EP & DEMPSEY,B (1979)
HL 1.50E-006	atm m3/mol	25	EST	MEYLAN,WM & HOWARD,PH (1991)
OH 4.98E-011	cm3/molc sec	25	EST	MEYLAN,WM & HOWARD,PH (1993)

002410-24-4 — P-AMINOPHENYLALANINE

Formula: $C_9H_{12}N_2O_2$
Mol Weight: 180.21
MP (deg C):
FP (deg C):
BP (deg C):
BP pressure (mm Hg):

Property/Value	Units	Temp	Data Type	Reference
WS 3.54E+004	mg/L	25	EST	MEYLAN,WM ET AL. (1996)
logP -2.20			EXP	HANSCH,C & LEO,AJ (1985)
VP 1.05E-009	mm Hg	25	EST	NEELY,WB & BLAU,GE (1985)
DC	pKa			
HL 4.26E-014	atm m3/mol	25	EST	MEYLAN,WM & HOWARD,PH (1991)
OH 1.70E-010	cm3/molc sec	25	EST	MEYLAN,WM & HOWARD,PH (1993)

002416-94-6 — 2,3,6-TRIMETHYLPHENOL

Formula: $C_9H_{12}O$
Mol Weight: 136.20
MP (deg C): 62-64
FP (deg C):
BP (deg C):
BP pressure (mm Hg):

Property/Value	Units	Temp	Data Type	Reference
WS 1.58E+003	mg/L	25	EXP	WASIK,SP ET AL. (1981)
logP 2.67			EXP	SANGSTER,J (1993)
VP 3.47E-002	mm Hg	25	EXT	BAGLAY,AK ET AL. (1988)
DC	pKa			
HL 7.53E-007	atm m3/mol	25	EST	MEYLAN,WM & HOWARD,PH (1991)
OH 1.31E-010	cm3/molc sec	25	EST	MEYLAN,WM & HOWARD,PH (1993)

CAS #: 002419-74-1	1,4-DICHLORO-2-HYDROXYBUTANE

Formula: $C_4H_8Cl_2O$

Mol Weight: 143.01

MP (deg C): FP (deg C):

BP (deg C):

BP pressure (mm Hg):

Property/Value	Units	Temp	Data Type	Reference
WS 2.14E+004	mg/L	25	EST	MEYLAN,WM ET AL. (1996)
logP 1.28			EST	MEYLAN,WM & HOWARD,PH (1995)
VP 4.23E-002	mm Hg	25	EST	NEELY,WB & BLAU,GE (1985)
DC	pKa			
HL 7.82E-007	atm m3/mol	25	EST	MEYLAN,WM & HOWARD,PH (1991)
OH 4.43E-012	cm3/molc sec	25	EST	MEYLAN,WM & HOWARD,PH (1993)

CAS #: 002422-79-9	9-METHYLBENZ(A)ANTHRACENE

Formula: $C_{19}H_{14}$

Mol Weight: 242.32

MP (deg C): FP (deg C):

BP (deg C):

BP pressure (mm Hg):

Property/Value	Units	Temp	Data Type	Reference
WS 6.60E-002	mg/L	27	EXP	KERTES,AS (1989)
logP 6.07			EST	MEYLAN,WM & HOWARD,PH (1995)
VP 2.53E-007	mm Hg	25	EST	NEELY,WB & BLAU,GE (1985)
DC	pKa			
HL 5.53E-006	atm m3/mol	25	EST	MEYLAN,WM & HOWARD,PH (1991)
OH 1.31E-010	cm3/molc sec	25	EST	MEYLAN,WM & HOWARD,PH (1993)

CAS #: 002422-85-7	1,2-DIVINYLCYCLOBUTANE

Formula: C_8H_{12}

Mol Weight: 108.18

MP (deg C): FP (deg C):

BP (deg C):

BP pressure (mm Hg):

Property/Value	Units	Temp	Data Type	Reference
WS 2.06E+001	mg/L	25	EST	MEYLAN,WM ET AL. (1996)
logP 3.74			EST	MEYLAN,WM & HOWARD,PH (1995)
VP 2.25E+001	mm Hg	25	EST	NEELY,WB & BLAU,GE (1985)
DC	pKa			
HL 1.31E-001	atm m3/mol	25	EST	MEYLAN,WM & HOWARD,PH (1991)
OH 5.50E-011	cm3/molc sec	25	EST	MEYLAN,WM & HOWARD,PH (1993)

CAS #: 002423-71-4	2,6-DIMETHYL-4-NITROPHENOL

Formula: $C_8H_9NO_3$

Mol Weight: 167.17

MP (deg C): 168 dec FP (deg C):

BP (deg C):

BP pressure (mm Hg):

Property/Value	Units	Temp	Data Type	Reference
WS 6.57E+002	mg/L	25	EST	MEYLAN,WM ET AL. (1996)
logP 3.00			EST	MEYLAN,WM & HOWARD,PH (1995)
VP 1.92E-004	mm Hg	25	EST	NEELY,WB & BLAU,GE (1985)
DC 7.07	pKa	25	EXP	KORTUM,G ET AL (1961)
HL 2.69E-009	atm m3/mol	25	EST	MEYLAN,WM & HOWARD,PH (1991)
OH 1.21E-012	cm3/molc sec	25	EST	MEYLAN,WM & HOWARD,PH (1993)

CAS #: 002424-02-4	ALLYLETHYLAMINE

Formula: $C_5H_{11}N$

Mol Weight: 85.15

MP (deg C): FP (deg C):

BP (deg C):

BP pressure (mm Hg):

Property/Value	Units	Temp	Data Type	Reference
WS 2.65E+005	mg/L	25	EST	MEYLAN,WM ET AL. (1996)
logP 0.81			EXP	HANSCH,C & LEO,AJ (1985)
VP 5.46E+001	mm Hg	25	EST	NEELY,WB & BLAU,GE (1985)
DC	pKa			
HL 2.90E-005	atm m3/mol	25	EST	MEYLAN,WM & HOWARD,PH (1991)
OH 1.07E-010	cm3/molc sec	25	EST	MEYLAN,WM & HOWARD,PH (1993)

CAS #: 002425-06-1	CAPTAFOL

Formula: $C_{10}H_9Cl_4NO_2S$

Mol Weight: 349.06

MP (deg C): 160-161 FP (deg C):

BP (deg C):

BP pressure (mm Hg):

Property/Value	Units	Temp	Data Type	Reference
WS 1.40E+000	mg/L			WORTHING,CR & WALKER,SB (1987)
logP 2.51			EXP	HANSCH,C & LEO,AJ (1985)
VP 6.55E-009	mm Hg		EST	HL X WSOL
DC	pKa			
HL 2.15E-009	atm m3/mol	25	EST	MEYLAN,WM & HOWARD,PH (1991)
OH 8.88E-011	cm3/molc sec	25	EST	MEYLAN,WM & HOWARD,PH (1993)

CAS #: 002425-10-7	3,4-DIMEPHENYL N-METHYLCARBAMATE

Formula: $C_{10}H_{13}NO_2$

Mol Weight: 179.22

MP (deg C): FP (deg C):

BP (deg C):

BP pressure (mm Hg):

Property/Value	Units	Temp	Data Type	Reference
WS 1.30E+003	mg/L	30	EXP	YALKOWSKY,SH & DANNENFELSER,RM (1992)
logP 2.09			EXP	HANSCH,C & LEO,AJ (1985)
VP 5.78E-004	mm Hg	25	EXP	WATANABE,T (1993)
DC	pKa			
HL 1.05E-007	atm m3/mol	25	EST	VP/WSOL
OH 1.28E-011	cm3/molc sec	25	EST	MEYLAN,WM & HOWARD,PH (1993)

CAS #: 002425-54-9	1-CHLOROTETRADECANE

Formula: $C_{14}H_{29}Cl$

Mol Weight: 232.84

MP (deg C): 4.9 FP (deg C):

BP (deg C): 292

BP pressure (mm Hg):

Property/Value	Units	Temp	Data Type	Reference
WS 1.22E-002	mg/L	25	EST	MEYLAN,WM ET AL. (1996)
logP 7.47			EST	MEYLAN,WM & HOWARD,PH (1995)
VP 7.26E-004	mm Hg	25	EXT	BOUBLIK,T ET AL. (1984)
DC	pKa			
HL 3.26E-001	atm m3/mol	25	EST	MEYLAN,WM & HOWARD,PH (1991)
OH 1.63E-011	cm3/molc sec	25	EST	MEYLAN,WM & HOWARD,PH (1993)

CAS #: 002426-08-6				GLYCIDYL N-BUTYL ETHER

Formula: $C_7H_{14}O_2$
Mol Weight: 130.19
MP (deg C): | FP (deg C):
BP (deg C): 169
BP pressure (mm Hg):

Property/Value	Units	Temp	Data Type	Reference
WS 2.66E+004	mg/L	25	EST	MEYLAN,WM ET AL. (1996)
logP 0.63			EXP	HANSCH,C ET AL. (1995)
VP 4.44E+000	mm Hg	25	EST	NEELY,WB & BLAU,GE (1985)
DC	pKa			
HL 4.37E-006	atm m3/mol	25	EST	MEYLAN,WM & HOWARD,PH (1991)
OH 1.99E-011	cm3/molc sec	25	EST	MEYLAN,WM & HOWARD,PH (1993)

CAS #: 002426-12-2				BENZALDEHYDE,O-((MEAMINO)CO)OXIME

Formula: $C_9H_{10}N_2O_2$
Mol Weight: 178.19
MP (deg C): | FP (deg C):
BP (deg C):
BP pressure (mm Hg):

Property/Value	Units	Temp	Data Type	Reference
WS 3.00E+003	mg/L	25	EST	MEYLAN,WM ET AL. (1996)
logP 1.49			EXP	HANSCH,C & LEO,AJ (1985)
VP 5.35E-003	mm Hg	25	EST	NEELY,WB & BLAU,GE (1985)
DC	pKa			
HL 1.15E-008	atm m3/mol	25	EST	MEYLAN,WM & HOWARD,PH (1991)
OH 1.26E-011	cm3/molc sec	25	EST	MEYLAN,WM & HOWARD,PH (1993)

CAS #: 002427-90-9				2,3-PYRAZINEDICARBOXYLIC ACID, DIETHYL ESTER

Formula: $C_{10}H_{12}N_2O_4$
Mol Weight: 224.22
MP (deg C): | FP (deg C):
BP (deg C):
BP pressure (mm Hg):

Property/Value	Units	Temp	Data Type	Reference
WS 9.11E+003	mg/L	25	EST	MEYLAN,WM ET AL. (1996)
logP 0.65			EXP	YAMAGAMI,C ET AL. (1991)
VP 1.04E-003	mm Hg	25	EST	NEELY,WB & BLAU,GE (1985)
DC	pKa			
HL 2.14E-010	atm m3/mol	25	EST	MEYLAN,WM & HOWARD,PH (1991)
OH 3.34E-012	cm3/molc sec	25	EST	MEYLAN,WM & HOWARD,PH (1993)

CAS #: 002430-22-0				7-METHYL-1-OCTANOL

Formula: $C_9H_{20}O$
Mol Weight: 144.26
MP (deg C): 64.5 | FP (deg C):
BP (deg C): 206
BP pressure (mm Hg):

Property/Value	Units	Temp	Data Type	Reference
WS 4.67E+002	mg/L	20	EXP	SUZUKI,T (1991)
logP 3.22				EST MEYLAN,WM & HOWARD,PH (1995)
VP 4.41E-002	mm Hg	25	EST	NEELY,WB & BLAU,GE (1985)
DC	pKa			
HL 4.12E-005	atm m3/mol	25	EST	MEYLAN,WM & HOWARD,PH (1991)
OH 1.39E-011	cm3/molc sec	25	EST	MEYLAN,WM & HOWARD,PH (1993)

CAS #: 002431-54-1				1,2,4-TRICHLORO-2-BUTENE

Formula: $C_4H_5Cl_3$
Mol Weight: 159.44
MP (deg C): | FP (deg C):
BP (deg C): 67-69
BP pressure (mm Hg): 1.00E+001

Property/Value	Units	Temp	Data Type	Reference
WS 1.58E+002	mg/L	25	EST	MEYLAN,WM ET AL. (1996)
logP 3.09				EST MEYLAN,WM & HOWARD,PH (1995)
VP 8.19E-001	mm Hg	25	EST	NEELY,WB & BLAU,GE (1985)
DC	pKa			
HL 1.35E-002	atm m3/mol	25	EST	MEYLAN,WM & HOWARD,PH (1991)
OH 1.13E-011	cm3/molc sec	25	EST	MEYLAN,WM & HOWARD,PH (1993)

CAS #: 002431-55-2				1,1,2,2,3,4-HEXACHLOROBUTANE

Formula: $C_4H_4Cl_6$
Mol Weight: 264.79
MP (deg C): | FP (deg C):
BP (deg C):
BP pressure (mm Hg):

Property/Value	Units	Temp	Data Type	Reference
WS 4.36E+000	mg/L	25	EST	MEYLAN,WM ET AL. (1996)
logP 4.28				EST MEYLAN,WM & HOWARD,PH (1995)
VP 3.74E-002	mm Hg	25	EST	NEELY,WB & BLAU,GE (1985)
DC	pKa			
HL 3.29E-004	atm m3/mol	25	EST	MEYLAN,WM & HOWARD,PH (1991)
OH 2.64E-013	cm3/molc sec	25	EST	MEYLAN,WM & HOWARD,PH (1993)

CAS #: 002431-96-1				BENZENEACETAMIDE, N,N-DIETHYL-

Formula: $C_{12}H_{17}NO$
Mol Weight: 191.28
MP (deg C): | FP (deg C):
BP (deg C): 169-171
BP pressure (mm Hg): 1.80E+001

Property/Value	Units	Temp	Data Type	Reference
WS 4.32E+002	mg/L	25	EST	MEYLAN,WM ET AL. (1996)
logP 2.40			EXP	SURYANARAYANA,MVS ET AL. (1991)
VP 1.04E+000	mm Hg	30	EXP	SURYANARAYANA,MVS ET AL. (1991)
DC	pKa			
HL 7.67E-009	atm m3/mol	25	EST	MEYLAN,WM & HOWARD,PH (1991)
OH 2.85E-011	cm3/molc sec	25	EST	MEYLAN,WM & HOWARD,PH (1993)

CAS #: 002432-12-4				4-METHYL-2,6-DICHLOROPHENOL

Formula: $C_7H_6Cl_2O$
Mol Weight: 177.03
MP (deg C): 39 | FP (deg C):
BP (deg C): 231
BP pressure (mm Hg):

Property/Value	Units	Temp	Data Type	Reference
WS 2.99E+002	mg/L	25	EST	MEYLAN,WM ET AL. (1996)
logP 3.35				EST MEYLAN,WM & HOWARD,PH (1995)
VP 5.43E-003	mm Hg	25	EST	NEELY,WB & BLAU,GE (1985)
DC 7.19	pKa	25	EXP	SERJEANT,EP & DEMPSEY,B (1979)
HL 3.40E-007	atm m3/mol	25	EST	MEYLAN,WM & HOWARD,PH (1991)
OH 2.01E-012	cm3/molc sec	25	EST	MEYLAN,WM & HOWARD,PH (1993)

CAS #: 002432-90-8				DIDODECYLPHTHALATE

Formula: $C_{32}H_{54}O_4$
Mol Weight: 502.78
MP (deg C): | FP (deg C):
BP (deg C):
BP pressure (mm Hg):

Property/ Value	Units	Temp	Data Type	Reference
WS 1.40E-001	mg/L	24	EXP	YALKOWSKY,SH & DANNENFELSER,RM (1992)
logP 12.47			EST	MEYLAN,WM & HOWARD,PH (1995)
VP 1.60E-010	mm Hg	25	EST	NEELY,WB & BLAU,GE (1985)
DC	pKa			
HL 1.14E-004	atm m3/mol	25	EST	MEYLAN,WM & HOWARD,PH (1991)
OH 3.19E-011	cm3/molc sec	25	EST	MEYLAN,WM & HOWARD,PH (1993)

CAS #: 002436-66-0				4(3H)-QUINAZOLINONE, 3-METHYL-

Formula: $C_9H_8N_2O$
Mol Weight: 160.18
MP (deg C): | FP (deg C):
BP (deg C):
BP pressure (mm Hg):

Property/ Value	Units	Temp	Data Type	Reference
WS 1.76E+004	mg/L	25	EST	MEYLAN,WM ET AL. (1996)
logP 0.69			EXP	HANSCH,C ET AL. (1995)
VP 1.24E-004	mm Hg	25	EST	NEELY,WB & BLAU,GE (1985)
DC	pKa			
HL 8.02E-010	atm m3/mol	25	EST	MEYLAN,WM & HOWARD,PH (1991)
OH 2.46E-011	cm3/molc sec	25	EST	MEYLAN,WM & HOWARD,PH (1993)

CAS #: 002437-25-4				LAURONITRILE

Formula: $C_{12}H_{23}N$
Mol Weight: 181.32
MP (deg C): 4 | FP (deg C):
BP (deg C): 277
BP pressure (mm Hg):

Property/ Value	Units	Temp	Data Type	Reference
WS 2.51E+000	mg/L	25	EST	MEYLAN,WM ET AL. (1996)
logP 4.77			EST	MEYLAN,WM & HOWARD,PH (1995)
VP 2.36E-003	mm Hg	25	EXT	BOUBLIK,T ET AL. (1984)
DC	pKa			
HL 5.19E-004	atm m3/mol	25	EST	MEYLAN,WM & HOWARD,PH (1991)
OH 1.16E-011	cm3/molc sec	25	EST	MEYLAN,WM & HOWARD,PH (1993)

CAS #: 002437-49-2				2,4,6-TRIBROMORESORCINOL

Formula: $C_6H_3Br_3O_2$
Mol Weight: 346.82
MP (deg C): 114-116 | FP (deg C):
BP (deg C):
BP pressure (mm Hg):

Property/ Value	Units	Temp	Data Type	Reference
WS 4.56E+000	mg/L	25	EST	MEYLAN,WM ET AL. (1996)
logP 4.37			EXP	HANSCH,C & LEO,AJ (1985)
VP 8.49E-007	mm Hg	25	EST	NEELY,WB & BLAU,GE (1985)
DC	pKa			
HL 3.69E-012	atm m3/mol	25	EST	MEYLAN,WM & HOWARD,PH (1991)
OH 2.97E-011	cm3/molc sec	25	EST	MEYLAN,WM & HOWARD,PH (1993)

CAS #: 002437-56-1				1-TRIDECENE

Formula: $C_{13}H_{26}$
Mol Weight: 182.35
MP (deg C): -13 | FP (deg C):
BP (deg C): 232.8
BP pressure (mm Hg):

Property/ Value	Units	Temp	Data Type	Reference
WS 3.67E-002	mg/L	25	EST	MEYLAN,WM ET AL. (1996)
logP 6.59			EST	MEYLAN,WM & HOWARD,PH (1995)
VP 6.38E-002	mm Hg	25	EXP	DAUBERT,TE & DANNER,RP (1989)
DC	pKa			
HL 2.61E+000	atm m3/mol	25	EST	MEYLAN,WM & HOWARD,PH (1991)
OH 4.01E-011	cm3/molc sec	25	EST	MEYLAN,WM & HOWARD,PH (1993)

CAS #: 002437-79-8				2,2',4,4'-TETRACHLOROBIPHENYL

Formula: $C_{12}H_6Cl_4$
Mol Weight: 291.99
MP (deg C): | FP (deg C):
BP (deg C):
BP pressure (mm Hg):

Property/ Value	Units	Temp	Data Type	Reference
WS 5.41E-002	mg/L	22	EXP	YALKOWSKY,SH & DANNENFELSER,RM (1992)
logP 6.29			EXP	HANSCH,C & LEO,AJ (1985)
VP 8.63E-005	mm Hg	25	EXP	SHIU,WY & MACKAY,D (1986)
DC	pKa			
HL 6.13E-004	atm m3/mol	25	EST	VP/WSOL
OH 8.13E-013	cm3/molc sec	25	EST	MEYLAN,WM & HOWARD,PH (1993)

CAS #: 002438-04-2				O-ISOPROPYLBENZOIC ACID

Formula: $C_{10}H_{12}O_2$
Mol Weight: 164.21
MP (deg C): 64 | FP (deg C):
BP (deg C): 160-161
BP pressure (mm Hg): 2.50E+001

Property/ Value	Units	Temp	Data Type	Reference
WS 1.84E+002	mg/L	25	EST	MEYLAN,WM ET AL. (1996)
logP 2.99			EST	MEYLAN,WM & HOWARD,PH (1995)
VP 3.47E-004	mm Hg	25	EXT	COLOMINA,M ET AL. (1987)
DC 3.63	pKa	25	EXP	KORTUM,G ET AL (1961)
HL 2.11E-007	atm m3/mol	25	EST	MEYLAN,WM & HOWARD,PH (1991)
OH 4.51E-012	cm3/molc sec	25	EST	MEYLAN,WM & HOWARD,PH (1993)

CAS #: 002438-05-3				4-PROPYLBENZOIC ACID

Formula: $C_{10}H_{12}O_2$
Mol Weight: 164.21
MP (deg C): 142-144 | FP (deg C):
BP (deg C):
BP pressure (mm Hg):

Property/ Value	Units	Temp	Data Type	Reference
WS 7.86E+001	mg/L	25	EST	MEYLAN,WM ET AL. (1996)
logP 3.42			EXP	DA,YZ ET AL. (1992)
VP 6.03E-004	mm Hg	25	EST	NEELY,WB & BLAU,GE (1985)
DC	pKa			
HL 2.11E-007	atm m3/mol	25	EST	MEYLAN,WM & HOWARD,PH (1991)
OH 4.78E-012	cm3/molc sec	25	EST	MEYLAN,WM & HOWARD,PH (1993)

665

CAS #: 002438-72-4				BUFEXAMIC ACID

Formula: $C_{12}H_{17}NO_3$

Mol Weight: 223.27

MP (deg C): 153-155

FP (deg C):

BP (deg C):

BP pressure (mm Hg):

Property/Value	Units	Temp	Data Type	Reference
WS 5.54E+002	mg/L	25	EST	MEYLAN,WM ET AL. (1996)
logP 2.08			EXP	HANSCH,C & LEO,AJ (1985)
VP 2.44E-009	mm Hg	25	EST	NEELY,WB & BLAU,GE (1985)
DC	pKa			
HL 2.50E-013	atm m3/mol	25	EST	MEYLAN,WM & HOWARD,PH (1991)
OH 4.47E-011	cm3/molc sec	25	EST	MEYLAN,WM & HOWARD,PH (1993)

CAS #: 002438-76-8				3-SULFONAMIDOPYRIDINE

Formula: $C_5H_6N_2O_2S$

Mol Weight: 158.18

MP (deg C):

FP (deg C):

BP (deg C):

BP pressure (mm Hg):

Property/Value	Units	Temp	Data Type	Reference
WS 1.83E+005	mg/L	25	EST	MEYLAN,WM ET AL. (1996)
logP -0.49			EXP	HANSCH,C & LEO,AJ (1985)
VP 4.02E-004	mm Hg	25	EST	NEELY,WB & BLAU,GE (1985)
DC	pKa			
HL 5.53E-010	atm m3/mol	25	EST	MEYLAN,WM & HOWARD,PH (1991)
OH 7.91E-014	cm3/molc sec	25	EST	MEYLAN,WM & HOWARD,PH (1993)

CAS #: 002438-88-2				2,3,5,6-TETRACHLORO-4-NITROANILINE

Formula: $C_7H_3Cl_4NO_3$

Mol Weight: 290.92

MP (deg C): 101-105

FP (deg C):

BP (deg C):

BP pressure (mm Hg):

Property/Value	Units	Temp	Data Type	Reference
WS 5.40E-001	mg/L	25	EST	MEYLAN,WM ET AL. (1996)
logP 4.47			EST	MEYLAN,WM & HOWARD,PH (1995)
VP 3.80E-005	mm Hg	25	EST	NEELY,WB & BLAU,GE (1985)
DC	pKa			
HL 1.90E-005	atm m3/mol	25	EST	MEYLAN,WM & HOWARD,PH (1991)
OH 9.40E-013	cm3/molc sec	25	EST	MEYLAN,WM & HOWARD,PH (1993)

CAS #: 002439-00-1				2,3,6-TRICHLOROPHENYLACETIC ACID, SODIUM SALT

Formula: $C_8H_4Cl_3NaO_2$

Mol Weight: 261.47

MP (deg C):

FP (deg C):

BP (deg C):

BP pressure (mm Hg):

Property/Value	Units	Temp	Data Type	Reference
WS 2.40E+004	mg/L	25	EST	MEYLAN,WM ET AL. (1996)
logP -0.08			EST	MEYLAN,WM & HOWARD,PH (1995)
VP 2.71E-011	mm Hg	25	EST	NEELY,WB & BLAU,GE (1985)
DC	pKa			
HL	atm m3/mol			
OH 5.80E-013	cm3/molc sec	25	EST	MEYLAN,WM & HOWARD,PH (1993)

CAS #: 002439-01-2				OXYTHIOQUINOX

Formula: $C_{10}H_6N_2OS_2$

Mol Weight: 234.30

MP (deg C): 172

FP (deg C):

BP (deg C):

BP pressure (mm Hg):

Property/Value	Units	Temp	Data Type	Reference
WS 1.00E+000	mg/L	25	EXP	WAUCHOPE,RD ET AL. (1991A)
logP 3.78			EXP	TOMLIN,C (1994)
VP 2.00E-007	mm Hg	20	EXP	WAUCHOPE,RD ET AL. (1991A)
DC	pKa			
HL 6.17E-008	atm m3/mol	20	EST	VP/WSOL
OH 3.35E-012	cm3/molc sec	25	EST	MEYLAN,WM & HOWARD,PH (1993)

CAS #: 002439-10-3				DODINE

Formula: $C_{15}H_{33}N_3O_2$

Mol Weight: 287.45

MP (deg C): 136

FP (deg C):

BP (deg C):

BP pressure (mm Hg):

Property/Value	Units	Temp	Data Type	Reference
WS 6.30E+002	mg/L	25	EXP	GUNTHER,FA ET AL. (1968)
logP 2.64			EST	MEYLAN,WM & HOWARD,PH (1995)
VP 1.50E-007	mm Hg	25	EXP	WEBER,JB (1994)
DC 9.00	pKa	25	EXP	WEBER,JB (1994)
HL 9.01E-011	atm m3/mol	25	EST	VP/WSOL
OH 8.40E-011	cm3/molc sec	25	EST	MEYLAN,WM & HOWARD,PH (1993)

CAS #: 002439-77-2				O-METHOXYBENZAMIDE

Formula: $C_8H_9NO_2$

Mol Weight: 151.17

MP (deg C):

FP (deg C):

BP (deg C):

BP pressure (mm Hg):

Property/Value	Units	Temp	Data Type	Reference
WS 1.38E+004	mg/L	25	EST	MEYLAN,WM ET AL. (1996)
logP 0.86			EXP	HANSCH,C ET AL. (1995)
VP 1.11E-004	mm Hg	25	EST	NEELY,WB & BLAU,GE (1985)
DC	pKa			
HL 1.31E-010	atm m3/mol	25	EST	MEYLAN,WM & HOWARD,PH (1991)
OH 2.24E-011	cm3/molc sec	25	EST	MEYLAN,WM & HOWARD,PH (1993)

CAS #: 002439-99-8				GLYPHOSINE

Formula: $C_4H_{11}NO_8P_2$

Mol Weight: 263.08

MP (deg C):

FP (deg C):

BP (deg C):

BP pressure (mm Hg):

Property/Value	Units	Temp	Data Type	Reference
WS 2.48E+005	mg/L	20	EXP	YALKOWSKY,SH & DANNENFELSER,RM (1992)
logP -5.79			EST	MEYLAN,WM & HOWARD,PH (1995)
VP 6.63E-014	mm Hg	25	EST	NEELY,WB & BLAU,GE (1985)
DC	pKa			
HL 1.48E-028	atm m3/mol	25	EST	MEYLAN,WM & HOWARD,PH (1991)
OH 9.10E-011	cm3/molc sec	25	EST	MEYLAN,WM & HOWARD,PH (1993)

666

002440-22-4 — 2-(2-HYDROXY-5-METHYLPHENYL)BENZOTRIAZOLE

Formula: $C_{13}H_{11}N_3O$

Mol Weight: 225.25

MP (deg C): 131-133

FP (deg C):

BP (deg C):

BP pressure (mm Hg):

Property/Value	Units	Temp	Data Type	Reference
WS 2.56E+001	mg/L	25	EST	MEYLAN,WM ET AL. (1996)
logP 3.00			EST	MEYLAN,WM & HOWARD,PH (1995)
VP 7.95E-008	mm Hg	25	EST	NEELY,WB & BLAU,GE (1985)
DC	pKa			
HL 6.12E-014	atm m3/mol	25	EST	MEYLAN,WM & HOWARD,PH (1991)
OH 3.05E-011	cm3/molc sec	25	EST	MEYLAN,WM & HOWARD,PH (1993)

002440-79-1 — TYROSINE-N-ACETYL,METHYL ESTER

Formula: $C_{12}H_{15}NO_4$

Mol Weight: 237.26

MP (deg C):

FP (deg C):

BP (deg C):

BP pressure (mm Hg):

Property/Value	Units	Temp	Data Type	Reference
WS 5.52E+004	mg/L	25	EST	MEYLAN,WM ET AL. (1996)
logP 0.33			EXP	HANSCH,C ET AL. (1995)
VP 4.30E-008	mm Hg	25	EST	NEELY,WB & BLAU,GE (1985)
DC	pKa			
HL 1.30E-014	atm m3/mol	25	EST	MEYLAN,WM & HOWARD,PH (1991)
OH 5.50E-011	cm3/molc sec	25	EST	MEYLAN,WM & HOWARD,PH (1993)

002443-62-1 — HEXANOIC ACID HYDRAZIDE

Formula: $C_6H_{14}N_2O$

Mol Weight: 130.19

MP (deg C):

FP (deg C):

BP (deg C):

BP pressure (mm Hg):

Property/Value	Units	Temp	Data Type	Reference
WS 3.57E+004	mg/L	25	EST	MEYLAN,WM ET AL. (1996)
logP 0.48			EXP	HANSCH,C & LEO,AJ (1985)
VP 7.49E-004	mm Hg	25	EST	NEELY,WB & BLAU,GE (1985)
DC	pKa			
HL 1.45E-010	atm m3/mol	25	EST	MEYLAN,WM & HOWARD,PH (1991)
OH 1.36E-011	cm3/molc sec	25	EST	MEYLAN,WM & HOWARD,PH (1993)

002446-62-0 — 11H-PYRIDO[2,1-B]QUINAZOLIN-11-ONE, 6,7,8,9-TETR

Formula: $C_{12}H_{12}N_2O$

Mol Weight: 200.24

MP (deg C):

FP (deg C):

BP (deg C):

BP pressure (mm Hg):

Property/Value	Units	Temp	Data Type	Reference
WS 2.20E+003	mg/L	25	EST	MEYLAN,WM ET AL. (1996)
logP 1.52			EXP	SANGSTER,J (1994)
VP 6.50E-006	mm Hg	25	EST	NEELY,WB & BLAU,GE (1985)
DC	pKa			
HL 8.27E-010	atm m3/mol	25	EST	MEYLAN,WM & HOWARD,PH (1991)
OH 1.61E-011	cm3/molc sec	25	EST	MEYLAN,WM & HOWARD,PH (1993)

002447-57-6 — BENZENESULFONAMIDE, 4-AMINO-N-(5,6-DIMETHOXY-4-P

Formula: $C_{12}H_{14}N_4O_4S$

Mol Weight: 310.33

MP (deg C): 190-194

FP (deg C):

BP (deg C):

BP pressure (mm Hg):

Property/Value	Units	Temp	Data Type	Reference
WS 2.70E+003	mg/L	25	EST	MEYLAN,WM ET AL. (1996)
logP 0.70			EXP	SANGSTER,J (1994)
VP 1.59E-009	mm Hg	25	EST	NEELY,WB & BLAU,GE (1985)
DC	pKa			
HL 1.42E-015	atm m3/mol	25	EST	MEYLAN,WM & HOWARD,PH (1991)
OH 2.83E-011	cm3/molc sec	25	EST	MEYLAN,WM & HOWARD,PH (1993)

002450-08-0 — 4-PYRIMIDINECARBOXYLIC ACID, METHYL ESTER

Formula: $C_6H_6N_2O_2$

Mol Weight: 138.13

MP (deg C):

FP (deg C):

BP (deg C):

BP pressure (mm Hg):

Property/Value	Units	Temp	Data Type	Reference
WS 1.45E+005	mg/L	25	EST	MEYLAN,WM ET AL. (1996)
logP -0.27			EXP	YAMAGAMI,C ET AL. (1990)
VP 1.38E-001	mm Hg	25	EST	NEELY,WB & BLAU,GE (1985)
DC	pKa			
HL 1.88E-008	atm m3/mol	25	EST	MEYLAN,WM & HOWARD,PH (1991)
OH 2.91E-013	cm3/molc sec	25	EST	MEYLAN,WM & HOWARD,PH (1993)

002450-71-7 — 2-PROPYNYLAMINE

Formula: C_3H_5N

Mol Weight: 55.08

MP (deg C):

FP (deg C):

BP (deg C): 83

BP pressure (mm Hg):

Property/Value	Units	Temp	Data Type	Reference
WS 1.00E+006	mg/L	25	EST	MEYLAN,WM ET AL. (1996)
logP -0.43			EXP	HANSCH,C ET AL. (1995)
VP 1.23E+002	mm Hg	25	EST	NEELY,WB & BLAU,GE (1985)
DC 8.15	pKa		EXP	PERRIN,DD (1965)
HL 1.65E-006	atm m3/mol	25	EST	MEYLAN,WM & HOWARD,PH (1991)
OH 3.67E-011	cm3/molc sec	25	EST	MEYLAN,WM & HOWARD,PH (1993)

002451-01-6 — TERPIN HYDRATE

Formula: $C_{10}H_{22}O_3$

Mol Weight: 190.29

MP (deg C):

FP (deg C):

BP (deg C):

BP pressure (mm Hg):

Property/Value	Units	Temp	Data Type	Reference
WS 6.25E+002	mg/L	25	EST	MEYLAN,WM ET AL. (1996)
logP 2.32			EST	MEYLAN,WM & HOWARD,PH (1995)
VP 9.81E-004	mm Hg	25	EST	NEELY,WB & BLAU,GE (1985)
DC	pKa			
HL 5.56E-007	atm m3/mol	25	EST	MEYLAN,WM & HOWARD,PH (1991)
OH 2.45E-011	cm3/molc sec	25	EST	MEYLAN,WM & HOWARD,PH (1993)

002452-84-8 — C.I. SOLVENT YELLOW 58

Formula: $C_{16}H_{19}N_3O_2$
Mol Weight: 285.35
MP (deg C): 134
FP (deg C):
BP (deg C):
BP pressure (mm Hg):

Property/Value	Units	Temp	Data Type	Reference
WS 3.14E+001	mg/L	25	EXP	BAUGHMAN,GL & PERENICH,TA (1988)
logP 2.75			EST	MEYLAN,WM & HOWARD,PH (1995)
VP 1.10E-004	mm Hg	25	EXP	BAUGHMAN,GL & PERENICH,TA (1988)
DC	pKa			
HL 1.32E-006	atm m3/mol	25	EST	VP/WSOL
OH 8.85E-011	cm3/molc sec	25	EST	MEYLAN,WM & HOWARD,PH (1993)

002452-99-5 — 1,2-DIMETHYLCYCLOPENTANE

Formula: C_7H_{14}
Mol Weight: 98.19
MP (deg C):
FP (deg C):
BP (deg C):
BP pressure (mm Hg):

Property/Value	Units	Temp	Data Type	Reference
WS 3.39E+001	mg/L	25	EST	MEYLAN,WM ET AL. (1996)
logP 3.52			EST	MEYLAN,WM & HOWARD,PH (1995)
VP 4.72E+001	mm Hg	25	EXP	ZWOLINSKI,BJ & WILHOIT,RC (1971)
DC	pKa			
HL 3.39E-001	atm m3/mol	25	EST	MEYLAN,WM & HOWARD,PH (1991)
OH 8.42E-012	cm3/molc sec	25	EST	MEYLAN,WM & HOWARD,PH (1993)

002454-39-9 — O-AMINOTHIOBENZAMIDE

Formula: $C_7H_8N_2S$
Mol Weight: 152.22
MP (deg C):
FP (deg C):
BP (deg C):
BP pressure (mm Hg):

Property/Value	Units	Temp	Data Type	Reference
WS 1.08E+005	mg/L	25	EST	MEYLAN,WM ET AL. (1996)
logP 0.99			EXP	HANSCH,C & LEO,AJ (1985)
VP 2.93E-004	mm Hg	25	EST	NEELY,WB & BLAU,GE (1985)
DC	pKa			
HL 1.38E-010	atm m3/mol	25	EST	MEYLAN,WM & HOWARD,PH (1991)
OH 1.50E-010	cm3/molc sec	25	EST	MEYLAN,WM & HOWARD,PH (1993)

002455-15-4 — ETHANAMINE, N,N-DIMETHYL-2-(3-METHYLPHENOXY)-

Formula: $C_{11}H_{17}NO$
Mol Weight: 179.26
MP (deg C):
FP (deg C):
BP (deg C):
BP pressure (mm Hg):

Property/Value	Units	Temp	Data Type	Reference
WS 1.00E+004	mg/L	25	EST	MEYLAN,WM ET AL. (1996)
logP 2.05			EXP	AL-SAADI,D ET AL. (1993)
VP 2.54E-002	mm Hg	25	EST	NEELY,WB & BLAU,GE (1985)
DC	pKa			
HL 2.32E-007	atm m3/mol	25	EST	MEYLAN,WM & HOWARD,PH (1991)
OH 1.62E-010	cm3/molc sec	25	EST	MEYLAN,WM & HOWARD,PH (1993)

002456-27-1 — DINONYL ETHER

Formula: $C_{18}H_{38}O$
Mol Weight: 270.50
MP (deg C):
FP (deg C):
BP (deg C):
BP pressure (mm Hg):

Property/Value	Units	Temp	Data Type	Reference
WS 3.10E-003	mg/L	25	EST	MEYLAN,WM ET AL. (1996)
logP 7.92			EST	MEYLAN,WM & HOWARD,PH (1995)
VP 5.26E-005	mm Hg	25	EXP	DAUBERT,TE & DANNER,RP (1989)
DC	pKa			
HL 8.02E-002	atm m3/mol	25	EST	MEYLAN,WM & HOWARD,PH (1991)
OH 3.88E-011	cm3/molc sec	25	EST	MEYLAN,WM & HOWARD,PH (1993)

002457-47-8 — 3,5-DICHLOROPYRIDINE

Formula: $C_5H_3Cl_2N$
Mol Weight: 147.99
MP (deg C): 65-67
FP (deg C):
BP (deg C):
BP pressure (mm Hg):

Property/Value	Units	Temp	Data Type	Reference
WS 5.04E+002	mg/L	25	EST	MEYLAN,WM ET AL. (1996)
logP 2.56			EXP	HANSCH,C & LEO,AJ (1985)
VP 7.78E-001	mm Hg	25	EST	NEELY,WB & BLAU,GE (1985)
DC 0.67	pKa	25	EXP	PERRIN,DD (1972)
HL 3.87E-006	atm m3/mol	25	EST	MEYLAN,WM & HOWARD,PH (1991)
OH 1.83E-013	cm3/molc sec	25	EST	MEYLAN,WM & HOWARD,PH (1993)

002457-76-3 — 2-CHLORO-4-AMINOBENZOIC ACID

Formula: $C_7H_6ClNO_2$
Mol Weight: 171.58
MP (deg C): 211 dec
FP (deg C):
BP (deg C):
BP pressure (mm Hg):

Property/Value	Units	Temp	Data Type	Reference
WS 4.42E+003	mg/L	25	EST	MEYLAN,WM ET AL. (1996)
logP 1.33			EXP	HANSCH,C & LEO,AJ (1985)
VP 5.32E-005	mm Hg	25	EST	NEELY,WB & BLAU,GE (1985)
DC	pKa			
HL 2.84E-011	atm m3/mol	25	EST	MEYLAN,WM & HOWARD,PH (1991)
OH 2.86E-011	cm3/molc sec	25	EST	MEYLAN,WM & HOWARD,PH (1993)

002459-07-6 — PICOLINIC ACID, METHYL ESTER

Formula: $C_7H_7NO_2$
Mol Weight: 137.14
MP (deg C):
FP (deg C):
BP (deg C):
BP pressure (mm Hg):

Property/Value	Units	Temp	Data Type	Reference
WS 4.24E+004	mg/L	25	EST	MEYLAN,WM ET AL. (1996)
logP 0.36			EXP	HANSCH,C ET AL. (1995)
VP 2.77E-001	mm Hg	25	EST	NEELY,WB & BLAU,GE (1985)
DC 2.21	pKa	22	EXP	PERRIN,DD (1965)
HL 4.54E-008	atm m3/mol	25	EST	MEYLAN,WM & HOWARD,PH (1991)
OH 3.36E-013	cm3/molc sec	25	EST	MEYLAN,WM & HOWARD,PH (1993)

CAS #: 002459-09-8	I-NICOTINIC ACID, METHYL ESTER

Formula: $C_7H_7NO_2$

Mol Weight: 137.14

MP (deg C): 16.1   FP (deg C):

BP (deg C): 208

BP pressure (mm Hg):

Property/Value	Units	Temp	Data Type	Reference
WS 1.56E+004	mg/L	25	EST	MEYLAN,WM ET AL. (1996)
logP 0.87			EXP	HANSCH,C & LEO,AJ (1985)
VP 2.77E-001	mm Hg	25	EST	NEELY,WB & BLAU,GE (1985)
DC 3.26	pKa	22	EXP	PERRIN,DD (1965)
HL 4.54E-008	atm m3/mol	25	EST	MEYLAN,WM & HOWARD,PH (1991)
OH 3.36E-013	cm3/molc sec	25	EST	MEYLAN,WM & HOWARD,PH (1993)

CAS #: 002460-49-3	4,5-DICHLORO-2-METHOXYPHENOL

Formula: $C_7H_6Cl_2O_2$

Mol Weight: 193.03

MP (deg C):   FP (deg C):

BP (deg C):

BP pressure (mm Hg):

Property/Value	Units	Temp	Data Type	Reference
WS 7.80E+001	mg/L	25	EST	MEYLAN,WM ET AL. (1996)
logP 3.26			EXP	HANSCH,C & LEO,AJ (1985)
VP 1.16E-002	mm Hg	25	EXP	BIDLEMAN,TF & RENBERG,L (1985)
DC	pKa			
HL 1.82E-008	atm m3/mol	25	EST	MEYLAN,WM & HOWARD,PH (1991)
OH 6.89E-012	cm3/molc sec	25	EST	MEYLAN,WM & HOWARD,PH (1993)

CAS #: 002463-84-5	DICAPTHON

Formula: $C_8H_9ClNO_5PS$

Mol Weight: 297.66

MP (deg C): 52-53   FP (deg C):

BP (deg C):

BP pressure (mm Hg):

Property/Value	Units	Temp	Data Type	Reference
WS 1.47E+001	mg/L	20	EXP	BOWMAN,BT & SANS,WW (1983)
logP 3.58			EXP	HANSCH,C & LEO,AJ (1985)
VP 3.60E-006	mm Hg	20	EXP	FREED,VH ET AL. (1979A)
DC	pKa			
HL 9.59E-008	atm m3/mol	20	EST	VP/WSOL
OH 6.10E-011	cm3/molc sec	25	EST	MEYLAN,WM & HOWARD,PH (1993)

CAS #: 002465-27-2	BASIC YELLOW 2

Formula: $C_{17}H_{22}ClN_3$

Mol Weight: 303.84

MP (deg C):   FP (deg C):

BP (deg C):

BP pressure (mm Hg):

Property/Value	Units	Temp	Data Type	Reference
WS 1.00E+004	mg/L		EXP	GREEN,FJ (1990)
logP 2.98			EST	MEYLAN,WM & HOWARD,PH (1995)
VP 1.29E-006	mm Hg	25	EST	NEELY,WB & BLAU,GE (1985)
DC	pKa			
HL 3.64E-009	atm m3/mol	25	EST	MEYLAN,WM & HOWARD,PH (1991)
OH 2.05E-010	cm3/molc sec	25	EST	MEYLAN,WM & HOWARD,PH (1993)

CAS #: 002466-76-4	1H-IMIDAZOLE, 1-ACETYL-

Formula: $C_5H_6N_2O$

Mol Weight: 110.12

MP (deg C): 99-105   FP (deg C):

BP (deg C):

BP pressure (mm Hg):

Property/Value	Units	Temp	Data Type	Reference
WS 2.12E+005	mg/L	25	EST	MEYLAN,WM ET AL. (1996)
logP -0.34			EXP	HANSCH,C ET AL. (1995)
VP 1.59E-002	mm Hg	25	EST	NEELY,WB & BLAU,GE (1985)
DC 3.60	pKa	25	EXP	PERRIN,DD (1965)
HL 1.69E-006	atm m3/mol	25	EST	MEYLAN,WM & HOWARD,PH (1991)
OH 3.61E-011	cm3/molc sec	25	EST	MEYLAN,WM & HOWARD,PH (1993)

CAS #: 002473-01-0	1-CHLORONONANE

Formula: $C_9H_{19}Cl$

Mol Weight: 162.70

MP (deg C): -39.4   FP (deg C):

BP (deg C): 203.5

BP pressure (mm Hg):

Property/Value	Units	Temp	Data Type	Reference
WS 3.47E+000	mg/L	25	EST	MEYLAN,WM ET AL. (1996)
logP 5.02			EST	MEYLAN,WM & HOWARD,PH (1995)
VP 2.67E-001	mm Hg	25	EXP	BOUBLIK,T ET AL. (1984)
DC	pKa			
HL 7.91E-002	atm m3/mol	25	EST	MEYLAN,WM & HOWARD,PH (1991)
OH 9.23E-012	cm3/molc sec	25	EST	MEYLAN,WM & HOWARD,PH (1993)

CAS #: 002474-50-2	2,3-QUINOXALINEDIONE,1,4-DIHYDRO-6,7-DIMETHYL-

Formula: $C_{10}H_{10}N_2O_2$

Mol Weight: 190.20

MP (deg C):   FP (deg C):

BP (deg C):

BP pressure (mm Hg):

Property/Value	Units	Temp	Data Type	Reference
WS 3.52E+003	mg/L	25	EST	MEYLAN,WM ET AL. (1996)
logP 1.34			EXP	HANSCH,C ET AL. (1995)
VP 1.10E-008	mm Hg	25	EST	NEELY,WB & BLAU,GE (1985)
DC	pKa			
HL 2.68E-011	atm m3/mol	25	EST	MEYLAN,WM & HOWARD,PH (1991)
OH 2.00E-010	cm3/molc sec	25	EST	MEYLAN,WM & HOWARD,PH (1993)

CAS #: 002475-31-2	C.I. VAT BLUE 5

Formula: $C_{16}H_6Br_4N_2O_2$

Mol Weight: 577.85

MP (deg C):   FP (deg C):

BP (deg C):

BP pressure (mm Hg):

Property/Value	Units	Temp	Data Type	Reference
WS 4.48E-004	mg/L	25	EST	MEYLAN,WM ET AL. (1996)
logP 6.67			EST	MEYLAN,WM & HOWARD,PH (1995)
VP 8.00E-012	mm Hg	25	EXP	BAUGHMAN,GL & PERENICH,TA (1988)
DC	pKa			
HL 1.26E-015	atm m3/mol	25	EST	MEYLAN,WM & HOWARD,PH (1991)
OH 5.92E-012	cm3/molc sec	25	EST	MEYLAN,WM & HOWARD,PH (1993)

CAS #: 002475-44-7 — 1,4-BIS(METHYLAMINO)ANTHRAQUINONE

Formula: $C_{16}H_{14}N_2O_2$
Mol Weight: 266.30
MP (deg C): 220-222
FP (deg C):
BP (deg C):
BP pressure (mm Hg):

Property/Value	Units	Temp	Data Type	Reference
WS 3.70E-002	mg/L	25	EXP	BAUGHMAN,GL & PERENICH,TA (1988)
logP 4.25			EST	MEYLAN,WM & HOWARD,PH (1995)
VP 2.00E-011	mm Hg	25	EXP	BAUGHMAN,GL & PERENICH,TA (1988)
DC	pKa			
HL 1.89E-010	atm m3/mol	25	EST	VP/WSOL
OH 3.01E-011	cm3/mol sec	25	EST	MEYLAN,WM & HOWARD,PH (1993)

CAS #: 002475-45-8 — 1,4,5,8-TETRAAMINOANTHRAQUINONE

Formula: $C_{14}H_{12}N_4O_2$
Mol Weight: 268.28
MP (deg C): 332
FP (deg C):
BP (deg C):
BP pressure (mm Hg):

Property/Value	Units	Temp	Data Type	Reference
WS 3.00E-002	mg/L	25	EXP	KUROIWA,S & OGASAWARA,S (1973)
logP 2.98			EST	MEYLAN,WM & HOWARD,PH (1995)
VP 1.80E-008	mm Hg	25	EXP	BAUGHMAN,GL & PERENICH,TA (1988)
DC	pKa			
HL 2.12E-007	atm m3/mol	25	EST	VP/WSOL
OH 1.35E-010	cm3/mol sec	25	EST	MEYLAN,WM & HOWARD,PH (1993)

CAS #: 002475-46-9 — C.I. DISPERSE BLUE 3

Formula: $C_{17}H_{16}N_2O_3$
Mol Weight: 296.33
MP (deg C): 187
FP (deg C):
BP (deg C):
BP pressure (mm Hg):

Property/Value	Units	Temp	Data Type	Reference
WS 3.55E-002	mg/L	25	EXP	BAUGHMAN,GL & PERENICH,TA (1988)
logP 3.28			EST	MEYLAN,WM & HOWARD,PH (1995)
VP 6.43E-012	mm Hg	25	EST	NEELY,WB & BLAU,GE (1985)
DC	pKa			
HL 9.29E-020	atm m3/mol	25	EST	MEYLAN,WM & HOWARD,PH (1991)
OH 4.37E-011	cm3/molc sec	25	EST	MEYLAN,WM & HOWARD,PH (1993)

CAS #: 002481-94-9 — C.I. SOLVENT YELLOW 56

Formula: $C_{16}H_{19}N_3$
Mol Weight: 253.35
MP (deg C): 97-98
FP (deg C):
BP (deg C):
BP pressure (mm Hg):

Property/Value	Units	Temp	Data Type	Reference
WS 2.64E-001	mg/L	25	EST	MEYLAN,WM ET AL. (1996)
logP 5.27			EST	MEYLAN,WM & HOWARD,PH (1995)
VP 8.45E-008	mm Hg	25	EXP	SHIMIZU,T ET AL. (1987)
DC	pKa			
HL 4.12E-007	atm m3/mol	25	EST	MEYLAN,WM & HOWARD,PH (1991)
OH 7.66E-011	cm3/molc sec	25	EST	MEYLAN,WM & HOWARD,PH (1993)

CAS #: 002482-68-0 — 2-PERCHLOROBUTENE

Formula: C_4Cl_8
Mol Weight: 331.67
MP (deg C):
FP (deg C):
BP (deg C):
BP pressure (mm Hg):

Property/Value	Units	Temp	Data Type	Reference
WS 9.13E-002	mg/L	25	EST	MEYLAN,WM ET AL. (1996)
logP 5.79			EST	MEYLAN,WM & HOWARD,PH (1995)
VP 1.89E-002	mm Hg	25	EST	NEELY,WB & BLAU,GE (1985)
DC	pKa			
HL 1.50E-004	atm m3/mol	25	EST	MEYLAN,WM & HOWARD,PH (1991)
OH 2.80E-012	cm3/molc sec	25	EST	MEYLAN,WM & HOWARD,PH (1993)

CAS #: 002486-02-4 — ISOAMYL GALLATE

Formula: $C_{12}H_{16}O_5$
Mol Weight: 240.26
MP (deg C):
FP (deg C):
BP (deg C):
BP pressure (mm Hg):

Property/Value	Units	Temp	Data Type	Reference
WS 4.13E+002	mg/L	25	EST	MEYLAN,WM ET AL. (1996)
logP 2.80			EXP	HANSCH,C & LEO,AJ (1985)
VP 7.69E-008	mm Hg	25	EST	NEELY,WB & BLAU,GE (1985)
DC	pKa			
HL 1.21E-016	atm m3/mol	25	EST	MEYLAN,WM & HOWARD,PH (1991)
OH 9.50E-011	cm3/molc sec	25	EST	MEYLAN,WM & HOWARD,PH (1993)

CAS #: 002486-52-4 — 2-BR-4-AMINOBENZOIC ACID

Formula: $C_7H_6BrNO_2$
Mol Weight: 216.04
MP (deg C):
FP (deg C):
BP (deg C):
BP pressure (mm Hg):

Property/Value	Units	Temp	Data Type	Reference
WS 1.93E+003	mg/L	25	EST	MEYLAN,WM ET AL. (1996)
logP 1.49			EXP	HANSCH,C & LEO,AJ (1985)
VP 2.37E-005	mm Hg	25	EST	NEELY,WB & BLAU,GE (1985)
DC	pKa			
HL 1.53E-011	atm m3/mol	25	EST	MEYLAN,WM & HOWARD,PH (1991)
OH 2.56E-011	cm3/molc sec	25	EST	MEYLAN,WM & HOWARD,PH (1993)

CAS #: 002486-70-6 — 4-AMINO-M-TOLUIC ACID

Formula: $C_8H_9NO_2$
Mol Weight: 151.17
MP (deg C): 169-170
FP (deg C):
BP (deg C):
BP pressure (mm Hg):

Property/Value	Units	Temp	Data Type	Reference
WS 5.07E+003	mg/L	25	EST	MEYLAN,WM ET AL. (1996)
logP 1.37			EXP	SANGSTER,J (1994)
VP 9.98E-005	mm Hg	25	EST	NEELY,WB & BLAU,GE (1985)
DC	pKa			
HL 4.23E-011	atm m3/mol	25	EST	MEYLAN,WM & HOWARD,PH (1991)
OH 4.95E-011	cm3/molc sec	25	EST	MEYLAN,WM & HOWARD,PH (1993)

CAS #: 002486-75-1			BENZOIC ACID, 4-AMINO-2-METHYL-

Formula: $C_8H_9NO_2$

Mol Weight: 151.17

MP (deg C):   **FP (deg C):**

BP (deg C):

BP pressure (mm Hg):

Property/ Value	Units	Temp	Data Type	Reference
WS 7.97E+003	mg/L	25	EST	MEYLAN,WM ET AL. (1996)
logP 1.14			EXP	SANGSTER,J (1994)
VP 9.98E-005	mm Hg	25	EST	NEELY,WB & BLAU,GE (1985)
DC	pKa			
HL 4.23E-011	atm m3/mol	25	EST	MEYLAN,WM & HOWARD,PH (1991)
OH 1.05E-010	cm3/molc sec	25	EST	MEYLAN,WM & HOWARD,PH (1993)

CAS #: 002486-77-3			3-ALLYLOXY-4-AMINOBENZOIC ACID

Formula: $C_{10}H_{11}NO_3$

Mol Weight: 193.20

MP (deg C):   **FP (deg C):**

BP (deg C):

BP pressure (mm Hg):

Property/ Value	Units	Temp	Data Type	Reference
WS 4.06E+003	mg/L	25	EST	MEYLAN,WM ET AL. (1996)
logP 1.25			EXP	HANSCH,C & LEO,AJ (1985)
VP 7.30E-006	mm Hg	25	EST	NEELY,WB & BLAU,GE (1985)
DC	pKa			
HL 2.98E-012	atm m3/mol	25	EST	MEYLAN,WM & HOWARD,PH (1991)
OH 6.70E-011	cm3/molc sec	25	EST	MEYLAN,WM & HOWARD,PH (1993)

CAS #: 002486-79-5			2-PROPOXY-4-AMINOBENZOIC ACID

Formula: $C_{10}H_{13}NO_3$

Mol Weight: 195.22

MP (deg C):   **FP (deg C):**

BP (deg C):

BP pressure (mm Hg):

Property/ Value	Units	Temp	Data Type	Reference
WS 2.28E+003	mg/L	25	EST	MEYLAN,WM ET AL. (1996)
logP 1.53			EXP	HANSCH,C & LEO,AJ (1985)
VP 6.82E-006	mm Hg	25	EST	NEELY,WB & BLAU,GE (1985)
DC	pKa			
HL 3.99E-012	atm m3/mol	25	EST	MEYLAN,WM & HOWARD,PH (1991)
OH 2.11E-010	cm3/molc sec	25	EST	MEYLAN,WM & HOWARD,PH (1993)

CAS #: 002486-80-8			BENZOIC ACID, 4-AMINO-2-METHOXY-

Formula: $C_8H_9NO_3$

Mol Weight: 167.17

MP (deg C):   **FP (deg C):**

BP (deg C):

BP pressure (mm Hg):

Property/ Value	Units	Temp	Data Type	Reference
WS 2.62E+004	mg/L	25	EST	MEYLAN,WM ET AL. (1996)
logP 0.45			EXP	SANGSTER,J (1994)
VP 3.68E-005	mm Hg	25	EST	NEELY,WB & BLAU,GE (1985)
DC	pKa			
HL 2.27E-012	atm m3/mol	25	EST	MEYLAN,WM & HOWARD,PH (1991)
OH 2.01E-010	cm3/molc sec	25	EST	MEYLAN,WM & HOWARD,PH (1993)

CAS #: 002487-90-3			TRIMETHOXY SILANE

Formula: $C_3H_{10}O_3Si$

Mol Weight: 122.20

MP (deg C): -115   **FP (deg C):**

BP (deg C): 81

BP pressure (mm Hg):

Property/ Value	Units	Temp	Data Type	Reference
WS 1.00E+006	mg/L	25	EST	MEYLAN,WM ET AL. (1996)
logP -1.22			EST	MEYLAN,WM & HOWARD,PH (1995)
VP 7.60E+001	mm Hg	25	EXT	DAUBERT,TE ET AL. (1987)
DC	pKa			
HL 2.04E-007	atm m3/mol	25	EST	MEYLAN,WM & HOWARD,PH (1991)
OH 2.49E-012	cm3/molc sec	25	EST	MEYLAN,WM & HOWARD,PH (1993)

CAS #: 002488-78-0			DIGITOXIGENIN-3-A-SULFATE

Formula: $C_{23}H_{34}O_7S$

Mol Weight: 454.59

MP (deg C):   **FP (deg C):**

BP (deg C):

BP pressure (mm Hg):

Property/ Value	Units	Temp	Data Type	Reference
WS 2.11E+003	mg/L	25	EST	MEYLAN,WM ET AL. (1996)
logP -0.14			EXP	HANSCH,C & LEO,AJ (1985)
VP 4.81E-017	mm Hg	25	EST	NEELY,WB & BLAU,GE (1985)
DC	pKa			
HL 1.89E-015	atm m3/mol	25	EST	MEYLAN,WM & HOWARD,PH (1991)
OH 5.70E-011	cm3/molc sec	25	EST	MEYLAN,WM & HOWARD,PH (1993)

CAS #: 002488-80-4			DIGITOXIGENIN-3-B-SULFATE

Formula: $C_{23}H_{34}O_7S$

Mol Weight: 454.59

MP (deg C):   **FP (deg C):**

BP (deg C):

BP pressure (mm Hg):

Property/ Value	Units	Temp	Data Type	Reference
WS 2.11E+003	mg/L	25	EST	MEYLAN,WM ET AL. (1996)
logP -0.21			EXP	HANSCH,C & LEO,AJ (1985)
VP 4.81E-017	mm Hg	25	EST	NEELY,WB & BLAU,GE (1985)
DC	pKa			
HL 1.89E-015	atm m3/mol	25	EST	MEYLAN,WM & HOWARD,PH (1991)
OH 5.70E-011	cm3/molc sec	25	EST	MEYLAN,WM & HOWARD,PH (1993)

CAS #: 002491-52-3			(4-NITROPHENYL)PHENYLDIAZENE

Formula: $C_{12}H_9N_3O_2$

Mol Weight: 227.22

MP (deg C): 132-134   **FP (deg C):**

BP (deg C):

BP pressure (mm Hg):

Property/ Value	Units	Temp	Data Type	Reference
WS 1.65E+000	mg/L	25	EST	MEYLAN,WM ET AL. (1996)
logP 4.51			EST	MEYLAN,WM & HOWARD,PH (1995)
VP 1.17E-006	mm Hg	25	EXP	SHIMIZU,T ETAL. (1987)
DC -3.47	pKa	25	EXP	PERRIN,DD (1965)
HL 5.80E-008	atm m3/mol	25	EST	MEYLAN,WM & HOWARD,PH (1991)
OH 8.40E-013	cm3/molc sec	25	EST	MEYLAN,WM & HOWARD,PH (1993)

CAS #: 002491-71-6 — 4-((4-AMINOPHENYL)AZO)BENZENESULFONIC ACID,MONO*

Formula: $C_{12}H_{10}N_3NaO_3S$

Mol Weight: 299.29

MP (deg C): FP (deg C):

BP (deg C):

BP pressure (mm Hg):

Property/Value	Units	Temp	Data Type	Reference
WS 1.43E+005	mg/L	25	EST	MEYLAN,WM ET AL. (1996)
logP -1.75			EST	MEYLAN,WM & HOWARD,PH (1995)
VP 6.18E-016	mm Hg	25	EST	NEELY,WB & BLAU,GE (1985)
DC	pKa			
HL	atm m3/mol			
OH 4.28E-011	cm3/molc sec	25	EST	MEYLAN,WM & HOWARD,PH (1993)

CAS #: 002491-74-9 — 4-NITRO-((4-(N-DIMETHYL)AMINOPHENYL)AZO)BENZENE

Formula: $C_{14}H_{14}N_4O_2$

Mol Weight: 270.29

MP (deg C): FP (deg C):

BP (deg C):

BP pressure (mm Hg):

Property/Value	Units	Temp	Data Type	Reference
WS 5.40E-004	mg/L	25	EXP	BAUGHMAN,GL & PERENICH,TA (1988)
logP 4.68			EST	MEYLAN,WM & HOWARD,PH (1995)
VP 1.00E-010	mm Hg	25	EXP	BAUGHMAN,GL & PERENICH,TA (1988)
DC 1.81	pKa		EXP	PERRIN,DD (1965)
HL 6.59E-008	atm m3/mol	25	EST	VP/WSOL
OH 1.49E-010	cm3/molc sec	25	EST	MEYLAN,WM & HOWARD,PH (1993)

CAS #: 002492-87-7 — BETA-D-GLUCOPYRANOSIDE, 4-NITROPHENYL

Formula: $C_{12}H_{15}NO_8$

Mol Weight: 301.26

MP (deg C): 165-168 FP (deg C):

BP (deg C):

BP pressure (mm Hg):

Property/Value	Units	Temp	Data Type	Reference
WS 7.01E+003	mg/L	25	EST	MEYLAN,WM ET AL. (1996)
logP -0.44			EXP	HANSCH,C & LEO,AJ (1985)
VP 3.68E-013	mm Hg	25	EST	NEELY,WB & BLAU,GE (1985)
DC	pKa			
HL 4.54E-018	atm m3/mol	25	EST	MEYLAN,WM & HOWARD,PH (1991)
OH 6.47E-011	cm3/molc sec	25	EST	MEYLAN,WM & HOWARD,PH (1993)

CAS #: 002494-55-5 — ETHANAMINIUM, N,N,N-TRIMETHYL-2-(1-OXOPROPOXY)-,

Formula: $C_8H_{18}INO_2$

Mol Weight: 287.14

MP (deg C): FP (deg C):

BP (deg C):

BP pressure (mm Hg):

Property/Value	Units	Temp	Data Type	Reference
WS 1.00E+006	mg/L	25	EST	MEYLAN,WM ET AL. (1996)
logP -2.96			EXP	SANGSTER,J (1994)
VP 3.35E-008	mm Hg	25	EST	NEELY,WB & BLAU,GE (1985)
DC	pKa			
HL 1.11E-014	atm m3/mol	25	EST	MEYLAN,WM & HOWARD,PH (1991)
OH 1.72E-011	cm3/molc sec	25	EST	MEYLAN,WM & HOWARD,PH (1993)

CAS #: 002495-37-6 — BENZYL METHACRYLATE

Formula: $C_{11}H_{12}O_2$

Mol Weight: 176.22

MP (deg C): FP (deg C):

BP (deg C): 95-98

BP pressure (mm Hg): 4.00E+000

Property/Value	Units	Temp	Data Type	Reference
WS 3.97E+002	mg/L	25	EST	MEYLAN,WM ET AL. (1996)
logP 2.53			EXP	SANGSTER,J (1993)
VP 3.36E-002	mm Hg	25	EST	NEELY,WB & BLAU,GE (1985)
DC	pKa			
HL 1.17E-005	atm m3/mol	25	EST	MEYLAN,WM & HOWARD,PH (1991)
OH 2.45E-011	cm3/molc sec	25	EST	MEYLAN,WM & HOWARD,PH (1993)

CAS #: 002497-06-5 — DISULFOTON SULFONE

Formula: $C_8H_{19}O_4PS_3$

Mol Weight: 306.40

MP (deg C): FP (deg C):

BP (deg C):

BP pressure (mm Hg):

Property/Value	Units	Temp	Data Type	Reference
WS 8.83E+002	mg/L	20	EXP	SHIU,WY ET AL. (1990)
logP 1.87			EXP	HANSCH,C & LEO,AJ (1985)
VP 1.12E-005	mm Hg	25	EST	NEELY,WB & BLAU,GE (1985)
DC	pKa			
HL 1.67E-008	atm m3/mol	25	EST	MEYLAN,WM & HOWARD,PH (1991)
OH 1.32E-010	cm3/molc sec	25	EST	MEYLAN,WM & HOWARD,PH (1993)

CAS #: 002497-07-6 — DISULFOTON SULFOXIDE

Formula: $C_8H_{19}O_3PS_3$

Mol Weight: 290.40

MP (deg C): FP (deg C):

BP (deg C):

BP pressure (mm Hg):

Property/Value	Units	Temp	Data Type	Reference
WS 1.00E+001	mg/L	20	EXP	SHIU,WY ET AL. (1990)
logP 1.73			EXP	HANSCH,C & LEO,AJ (1985)
VP 3.65E-005	mm Hg	25	EST	NEELY,WB & BLAU,GE (1985)
DC	pKa			
HL 1.32E-010	atm m3/mol	25	EST	MEYLAN,WM & HOWARD,PH (1991)
OH 1.92E-010	cm3/molc sec	25	EST	MEYLAN,WM & HOWARD,PH (1993)

CAS #: 002498-20-6 — TETRAPENTYLAMMONIUM IODIDE

Formula: $C_{20}H_{44}IN$

Mol Weight: 425.48

MP (deg C): FP (deg C):

BP (deg C):

BP pressure (mm Hg):

Property/Value	Units	Temp	Data Type	Reference
WS 7.40E+003	mg/L	25	EXP	PEDDLE,CJ & TURNER,WES (1913)
logP 3.68			EST	MEYLAN,WM & HOWARD,PH (1995)
VP 8.51E-012	mm Hg	25	EST	NEELY,WB & BLAU,GE (1985)
DC	pKa			
HL 2.45E-010	atm m3/mol	25	EST	MEYLAN,WM & HOWARD,PH (1991)
OH 5.93E-011	cm3/molc sec	25	EST	MEYLAN,WM & HOWARD,PH (1993)

CAS #: 002498-27-3 — BENZENEMETHANAMINIUM, 3-HYDROXY-N,N,N-TRIMETHYL-

Formula: $C_9H_{14}INO$

Mol Weight: 279.12

MP (deg C):　　FP (deg C):

BP (deg C):

BP pressure (mm Hg):

Property/ Value	Units	Temp	Data Type	Reference
WS 1.00E+006	mg/L	25	EST	MEYLAN,WM ET AL. (1996)
logP -2.68			EXP	HANSCH,C ET AL. (1995)
VP 3.38E-010	mm Hg	25	EST	NEELY,WB & BLAU,GE (1985)
DC	pKa			
HL 6.89E-017	atm m3/mol	25	EST	MEYLAN,WM & HOWARD,PH (1991)
OH 1.11E-011	cm3/molc sec	25	EST	MEYLAN,WM & HOWARD,PH (1993)

CAS #: 002498-50-2 — 4-AMINOBENZAMIDINE

Formula: $C_7H_9N_3$

Mol Weight: 135.17

MP (deg C): 181-183　　FP (deg C):

BP (deg C):

BP pressure (mm Hg):

Property/ Value	Units	Temp	Data Type	Reference
WS 1.82E+005	mg/L	25	EST	MEYLAN,WM ET AL. (1996)
logP -0.37			EXP	HANSCH,C ET AL. (1995)
VP 4.41E-004	mm Hg	25	EST	NEELY,WB & BLAU,GE (1985)
DC	pKa			
HL 6.48E-012	atm m3/mol	25	EST	MEYLAN,WM & HOWARD,PH (1991)
OH 1.50E-010	cm3/molc sec	25	EST	MEYLAN,WM & HOWARD,PH (1993)

CAS #: 002498-66-0 — BENZ A ANTHRACENE-7,12-DIONE

Formula: $C_{18}H_{10}O_2$

Mol Weight: 258.28

MP (deg C): 169-171　　FP (deg C):

BP (deg C):

BP pressure (mm Hg):

Property/ Value	Units	Temp	Data Type	Reference
WS 2.89E-001	mg/L	25	EST	MEYLAN,WM ET AL. (1996)
logP 4.40			EXP	CHEM INSPECT TEST INST (1992)
VP 3.50E-008	mm Hg	25	EST	NEELY,WB & BLAU,GE (1985)
DC	pKa			
HL 3.10E-010	atm m3/mol	25	EST	MEYLAN,WM & HOWARD,PH (1991)
OH 9.05E-012	cm3/molc sec	25	EST	MEYLAN,WM & HOWARD,PH (1993)

CAS #: 002498-77-3 — 1-METHYLBENZ(A)ANTHRACENE

Formula: $C_{19}H_{14}$

Mol Weight: 242.32

MP (deg C):　　FP (deg C):

BP (deg C):

BP pressure (mm Hg):

Property/ Value	Units	Temp	Data Type	Reference
WS 5.50E-002	mg/L	27	EXP	YALKOWSKY,SH & DANNENFELSER,RM (1992)
logP 6.07			EST	MEYLAN,WM & HOWARD,PH (1995)
VP 2.53E-007	mm Hg	25	EST	NEELY,WB & BLAU,GE (1985)
DC	pKa			
HL 5.53E-006	atm m3/mol	25	EST	MEYLAN,WM & HOWARD,PH (1991)
OH 1.31E-010	cm3/molc sec	25	EST	MEYLAN,WM & HOWARD,PH (1993)

CAS #: 002499-59-4 — OCTYL ACRYLATE

Formula: $C_{11}H_{20}O_2$

Mol Weight: 184.28

MP (deg C):　　FP (deg C):

BP (deg C):

BP pressure (mm Hg):

Property/ Value	Units	Temp	Data Type	Reference
WS 1.45E+001	mg/L	25	EST	MEYLAN,WM ET AL. (1996)
logP 4.17			EST	MEYLAN,WM & HOWARD,PH (1995)
VP 1.02E-001	mm Hg	25	EXT	PERRY,RH & GREEN,D (1984)
DC	pKa			
HL 6.72E-004	atm m3/mol	25	EST	MEYLAN,WM & HOWARD,PH (1991)
OH 1.94E-011	cm3/molc sec	25	EST	MEYLAN,WM & HOWARD,PH (1993)

CAS #: 002504-18-9 — N-NITROSOMETHYL-ISOBUTYLAMINE

Formula: $C_5H_{12}N_2O$

Mol Weight: 116.16

MP (deg C):　　FP (deg C):

BP (deg C):

BP pressure (mm Hg):

Property/ Value	Units	Temp	Data Type	Reference
WS 2.11E+004	mg/L	25	EST	MEYLAN,WM ET AL. (1996)
logP 0.81			EXP	VERA,A ET AL. (1992)
VP 5.61E-001	mm Hg	25	EST	NEELY,WB & BLAU,GE (1985)
DC	pKa			
HL 4.81E-006	atm m3/mol	25	EST	MEYLAN,WM & HOWARD,PH (1991)
OH 1.77E-012	cm3/molc sec	25	EST	MEYLAN,WM & HOWARD,PH (1993)

CAS #: 002508-86-3 — 4-AMINOQUINOLINE-1-OXIDE

Formula: $C_9H_{10}N_2O$

Mol Weight: 162.19

MP (deg C):　　FP (deg C):

BP (deg C):

BP pressure (mm Hg):

Property/ Value	Units	Temp	Data Type	Reference
WS 6.19E+004	mg/L	25	EST	MEYLAN,WM ET AL. (1996)
logP 0.05			EXP	HANSCH,C & LEO,AJ (1985)
VP 7.04E-006	mm Hg	25	EST	NEELY,WB & BLAU,GE (1985)
DC	pKa			
HL 2.43E-015	atm m3/mol	25	EST	MEYLAN,WM & HOWARD,PH (1991)
OH 2.00E-010	cm3/molc sec	25	EST	MEYLAN,WM & HOWARD,PH (1993)

CAS #: 002510-86-3 — O,O-DIETHYL-O-PHENYLPHOSPHATE

Formula: $C_{10}H_{15}O_4P$

Mol Weight: 230.20

MP (deg C):　　FP (deg C):

BP (deg C):

BP pressure (mm Hg):

Property/ Value	Units	Temp	Data Type	Reference
WS 1.21E+003	mg/L	25	EST	MEYLAN,WM ET AL. (1996)
logP 1.64			EXP	HANSCH,C & LEO,AJ (1985)
VP 3.40E-004	mm Hg	25	EST	NEELY,WB & BLAU,GE (1985)
DC	pKa			
HL 1.62E-007	atm m3/mol	25	EST	MEYLAN,WM & HOWARD,PH (1991)
OH 4.22E-011	cm3/molc sec	25	EST	MEYLAN,WM & HOWARD,PH (1993)

CAS #: 002511-10-6	Phosphonothioic acid, methyl-, O,S-diethyl ester

Formula: $C_5H_{13}O_2PS$

Mol Weight: 168.20

MP (deg C): -8.3

FP (deg C):

BP (deg C): 229.4

BP pressure (mm Hg):

Property/Value	Units	Temp	Data Type	Reference
WS 1.55E+004	mg/L	25	EST	MEYLAN,WM ET AL. (1996)
logP 0.71			EXP	KRIKORIAN,SE ET AL. (1987)
VP 8.10E-002	mm Hg	25	EST	MPVPBP
DC	pKa			
HL 1.21E-006	atm m3/mol	25	EST	MEYLAN,WM & HOWARD,PH (1991)
OH 3.65E-011	cm3/molc sec	25	EST	MEYLAN,WM & HOWARD,PH (1993)

CAS #: 002511-17-3	PHOSPHONIC DIAMIDE, PENTAMETHYL-

Formula: $C_5H_{15}N_2OP$

Mol Weight: 150.16

MP (deg C):

FP (deg C):

BP (deg C):

BP pressure (mm Hg):

Property/Value	Units	Temp	Data Type	Reference
WS 2.83E+005	mg/L	25	EST	MEYLAN,WM ET AL. (1996)
logP -0.67			EXP	DEBORD,J & LABADIE,M (1985)
VP 5.55E-001	mm Hg	25	EST	NEELY,WB & BLAU,GE (1985)
DC	pKa			
HL 1.17E-009	atm m3/mol	25	EST	MEYLAN,WM & HOWARD,PH (1991)
OH 6.02E-011	cm3/molc sec	25	EST	MEYLAN,WM & HOWARD,PH (1993)

CAS #: 002513-25-9	2,2-DIMETHYLBENZOPYRAN

Formula: $C_{11}H_{12}O$

Mol Weight: 160.22

MP (deg C):

FP (deg C):

BP (deg C):

BP pressure (mm Hg):

Property/Value	Units	Temp	Data Type	Reference
WS 1.02E+002	mg/L	25	EST	MEYLAN,WM ET AL. (1996)
logP 3.31			EXP	HANSCH,C ET AL. (1995)
VP 5.56E-002	mm Hg	25	EST	NEELY,WB & BLAU,GE (1985)
DC	pKa			
HL 1.50E-004	atm m3/mol	25	EST	MEYLAN,WM & HOWARD,PH (1991)
OH 7.76E-011	cm3/molc sec	25	EST	MEYLAN,WM & HOWARD,PH (1993)

CAS #: 002516-72-5	N-ME QUINOLINIUM BROMIDE

Formula: $C_{10}H_{10}BrN$

Mol Weight: 224.10

MP (deg C):

FP (deg C):

BP (deg C):

BP pressure (mm Hg):

Property/Value	Units	Temp	Data Type	Reference
WS 1.00E+006	mg/L	25	EST	MEYLAN,WM ET AL. (1996)
logP -2.64			EXP	HANSCH,C & LEO,AJ (1985)
VP 7.96E-004	mm Hg	25	EST	NEELY,WB & BLAU,GE (1985)
DC	pKa			
HL 1.47E-011	atm m3/mol	25	EST	MEYLAN,WM & HOWARD,PH (1991)
OH 2.45E-011	cm3/molc sec	25	EST	MEYLAN,WM & HOWARD,PH (1993)

CAS #: 002516-95-2	2-NITRO-5-CHLOROBENZOIC ACID

Formula: $C_7H_4ClNO_4$

Mol Weight: 201.57

MP (deg C): 138-140

FP (deg C):

BP (deg C):

BP pressure (mm Hg):

Property/Value	Units	Temp	Data Type	Reference
WS 9.67E+003	mg/L	25	EXP	YALKOWSKY,SH & DANNENFELSER,RM (1992)
logP 2.13			EXP	HANSCH,C & LEO,AJ (1985)
VP 1.89E-005	mm Hg	25	EST	NEELY,WB & BLAU,GE (1985)
DC	pKa			
HL 3.17E-010	atm m3/mol	25	EST	MEYLAN,WM & HOWARD,PH (1991)
OH 5.67E-013	cm3/molc sec	25	EST	MEYLAN,WM & HOWARD,PH (1993)

CAS #: 002516-96-3	2-CHLORO-5-NITROBENZOIC ACID

Formula: $C_7H_4ClNO_4$

Mol Weight: 201.57

MP (deg C): 166.5

FP (deg C):

BP (deg C):

BP pressure (mm Hg):

Property/Value	Units	Temp	Data Type	Reference
WS 3.23E+002	mg/L	25	EST	MEYLAN,WM ET AL. (1996)
logP 2.03			EXP	HANSCH,C & LEO,AJ (1985)
VP 1.89E-005	mm Hg	25	EST	NEELY,WB & BLAU,GE (1985)
DC 2.17	pKa	25	EXP	KORTUM,G ET AL (1961)
HL 3.17E-010	atm m3/mol	25	EST	MEYLAN,WM & HOWARD,PH (1991)
OH 5.83E-013	cm3/molc sec	25	EST	MEYLAN,WM & HOWARD,PH (1993)

CAS #: 002517-04-6	2-AZETIDINECARBOXYLIC ACID

Formula: $C_4H_7NO_2$

Mol Weight: 101.11

MP (deg C): 217 dec

FP (deg C):

BP (deg C):

BP pressure (mm Hg):

Property/Value	Units	Temp	Data Type	Reference
WS 2.63E+005	mg/L	25	EST	MEYLAN,WM ET AL. (1996)
logP -2.84			EXP	TSAI,RS ET AL. (1991)
VP 8.10E-009	mm Hg	25	EST	NEELY,WB & BLAU,GE (1985)
DC	pKa			
HL 1.44E-009	atm m3/mol	25	EST	MEYLAN,WM & HOWARD,PH (1991)
OH 2.36E-011	cm3/molc sec	25	EST	MEYLAN,WM & HOWARD,PH (1993)

CAS #: 002517-43-3	3-METHOXY-1-BUTANOL

Formula: $C_5H_{12}O_2$

Mol Weight: 104.15

MP (deg C):

FP (deg C):

BP (deg C): 157

BP pressure (mm Hg):

Property/Value	Units	Temp	Data Type	Reference
WS 3.66E+005	mg/L	25	EST	MEYLAN,WM ET AL. (1996)
logP 0.00			EST	MEYLAN,WM & HOWARD,PH (1995)
VP 1.62E+000	mm Hg	25	EST	NEELY,WB & BLAU,GE (1985)
DC	pKa			
HL 7.38E-008	atm m3/mol	25	EST	MEYLAN,WM & HOWARD,PH (1991)
OH 2.36E-011	cm3/molc sec	25	EXP	ATKINSON,R (1989)

CAS #: 002518-42-5 — FURAZAN, DIMETHYL-, 2-OXIDE

Formula: $C_4H_6N_2O_2$

Mol Weight: 114.10

MP (deg C):　　FP (deg C):

BP (deg C):

BP pressure (mm Hg):

Property/Value	Units	Temp	Data Type	Reference
WS 6.70E+004	mg/L	25	EST	MEYLAN,WM ET AL. (1996)
logP 0.23			EXP	CALVINO,R ET AL. (1992)
VP 4.90E-002	mm Hg	25	EST	NEELY,WB & BLAU,GE (1985)
DC	pKa			
HL 2.30E-009	atm m3/mol	25	EST	MEYLAN,WM & HOWARD,PH (1991)
OH 4.27E-012	cm3/molc sec	25	EST	MEYLAN,WM & HOWARD,PH (1993)

CAS #: 002518-72-1 — 5-ETHYLBARBITURIC ACID

Formula: $C_6H_8N_2O_3$

Mol Weight: 156.14

MP (deg C):　　FP (deg C):

BP (deg C):

BP pressure (mm Hg):

Property/Value	Units	Temp	Data Type	Reference
WS 1.42E+005	mg/L	25	EST	MEYLAN,WM ET AL. (1996)
logP -0.35			EXP	HANSCH,C & LEO,AJ (1985)
VP 5.35E-010	mm Hg	25	EST	NEELY,WB & BLAU,GE (1985)
DC	pKa			
HL 2.05E-013	atm m3/mol	25	EST	MEYLAN,WM & HOWARD,PH (1991)
OH 7.15E-012	cm3/molc sec	25	EST	MEYLAN,WM & HOWARD,PH (1993)

CAS #: 002519-75-7 — N-DIETHYLAMINOETANILINE,3-CL-4-ME

Formula: $C_{13}H_{21}ClN_2$

Mol Weight: 240.78

MP (deg C):　　FP (deg C):

BP (deg C):

BP pressure (mm Hg):

Property/Value	Units	Temp	Data Type	Reference
WS 7.89E+001	mg/L	25	EST	MEYLAN,WM ET AL. (1996)
logP 4.14			EXP	HANSCH,C & LEO,AJ (1985)
VP 1.81E-004	mm Hg	25	EST	NEELY,WB & BLAU,GE (1985)
DC	pKa			
HL 3.97E-009	atm m3/mol	25	EST	MEYLAN,WM & HOWARD,PH (1991)
OH 1.42E-010	cm3/molc sec	25	EST	MEYLAN,WM & HOWARD,PH (1993)

CAS #: 002521-24-6 — BENZENECARBOTHIAMIDE, 4-CHLORO-

Formula: C_7H_6ClNS

Mol Weight: 171.65

MP (deg C):　　FP (deg C):

BP (deg C):

BP pressure (mm Hg):

Property/Value	Units	Temp	Data Type	Reference
WS 5.49E+003	mg/L	25	EST	MEYLAN,WM ET AL. (1996)
logP 2.40			EXP	SANGSTER,J (1994)
VP 2.47E-003	mm Hg	25	EST	NEELY,WB & BLAU,GE (1985)
DC	pKa			
HL 2.90E-007	atm m3/mol	25	EST	MEYLAN,WM & HOWARD,PH (1991)
OH 2.26E-011	cm3/molc sec	25	EST	MEYLAN,WM & HOWARD,PH (1993)

CAS #: 002523-37-7 — 9-METHYL-9H-FLUORENE

Formula: $C_{14}H_{12}$

Mol Weight: 180.25

MP (deg C): 46.5　　FP (deg C):

BP (deg C): 154-156

BP pressure (mm Hg): 1.50E+001

Property/Value	Units	Temp	Data Type	Reference
WS 1.22E+000	mg/L	25	EST	MEYLAN,WM ET AL. (1996)
logP 4.15			EST	MEYLAN,WM & HOWARD,PH (1995)
VP 6.20E-004	mm Hg	25	EST	NEELY,WB & BLAU,GE (1985)
DC	pKa			
HL 2.22E-004	atm m3/mol	25	EST	MEYLAN,WM & HOWARD,PH (1991)
OH 9.87E-012	cm3/molc sec	25	EST	MEYLAN,WM & HOWARD,PH (1993)

CAS #: 002523-44-6 — 2-CHLOROFLOURENE

Formula: $C_{13}H_9Cl$

Mol Weight: 200.67

MP (deg C):　　FP (deg C):

BP (deg C):

BP pressure (mm Hg):

Property/Value	Units	Temp	Data Type	Reference
WS 3.53E-001	mg/L	25	EST	MEYLAN,WM ET AL. (1996)
logP 4.66			EST	MEYLAN,WM & HOWARD,PH (1995)
VP 2.25E-004	mm Hg	25	EST	NEELY,WB & BLAU,GE (1985)
DC	pKa			
HL 1.24E-004	atm m3/mol	25	EST	MEYLAN,WM & HOWARD,PH (1991)
OH 6.23E-012	cm3/molc sec	25	EST	MEYLAN,WM & HOWARD,PH (1993)

CAS #: 002523-46-8 — 2-METHOXY-9H-FLUORENE

Formula: $C_{14}H_{12}O$

Mol Weight: 196.25

MP (deg C):　　FP (deg C):

BP (deg C):

BP pressure (mm Hg):

Property/Value	Units	Temp	Data Type	Reference
WS 1.13E+000	mg/L	25	EST	MEYLAN,WM ET AL. (1996)
logP 4.10			EST	MEYLAN,WM & HOWARD,PH (1995)
VP 1.53E-004	mm Hg	25	EST	NEELY,WB & BLAU,GE (1985)
DC	pKa			
HL 9.90E-006	atm m3/mol	25	EST	MEYLAN,WM & HOWARD,PH (1991)
OH 4.39E-011	cm3/molc sec	25	EST	MEYLAN,WM & HOWARD,PH (1993)

CAS #: 002523-48-0 — 2-CYANO-9H-FLUORENE

Formula: $C_{14}H_9N$

Mol Weight: 191.23

MP (deg C):　　FP (deg C):

BP (deg C):

BP pressure (mm Hg):

Property/Value	Units	Temp	Data Type	Reference
WS 1.85E+000	mg/L	25	EST	MEYLAN,WM ET AL. (1996)
logP 3.56			EST	MEYLAN,WM & HOWARD,PH (1995)
VP 2.24E-005	mm Hg	25	EST	NEELY,WB & BLAU,GE (1985)
DC	pKa			
HL 1.62E-006	atm m3/mol	25	EST	MEYLAN,WM & HOWARD,PH (1991)
OH 4.53E-012	cm3/molc sec	25	EST	MEYLAN,WM & HOWARD,PH (1993)

CAS #:	002524-03-0	O,O-DIMETHYL PHOSPHOROCHLORIDOTHIOATE

Formula:	$C_2H_6ClO_2PS$

Mol Weight: 160.56

MP (deg C):		FP (deg C):

BP (deg C): 66-67

BP pressure (mm Hg): 1.60E+001

Property/Value	Units	Temp	Data Type	Reference
WS 4.40E+003	mg/L	25	EST	MEYLAN,WM ET AL. (1996)
logP 1.39			EST	MEYLAN,WM & HOWARD,PH (1995)
VP 5.00E-001	mm Hg	25	EXT	ALDRICH (1988)
DC	pKa			
HL 1.52E-003	atm m3/mol	25	EST	MEYLAN,WM & HOWARD,PH (1991)
OH 5.90E-011	cm3/molc sec	23	EXP	ATKINSON,R ET AL. (1988)

CAS #:	002524-52-9	PICOLINIC ACID, ETHYL ESTER

Formula:	$C_8H_9NO_2$

Mol Weight: 151.17

MP (deg C): 1		FP (deg C):

BP (deg C): 243

BP pressure (mm Hg):

Property/Value	Units	Temp	Data Type	Reference
WS 1.36E+004	mg/L	25	EST	MEYLAN,WM ET AL. (1996)
logP 0.87			EXP	HANSCH,C & LEO,AJ (1985)
VP 8.40E-002	mm Hg	25	EST	NEELY,WB & BLAU,GE (1985)
DC	pKa			
HL 6.03E-008	atm m3/mol	25	EST	MEYLAN,WM & HOWARD,PH (1991)
OH 1.78E-012	cm3/molc sec	25	EST	MEYLAN,WM & HOWARD,PH (1993)

CAS #:	002524-78-9	ACETAMIDE, N-[3-(METHYLTHIO)PHENYL]-

Formula:	$C_9H_{11}NOS$

Mol Weight: 181.26

MP (deg C):		FP (deg C):

BP (deg C):

BP pressure (mm Hg):

Property/Value	Units	Temp	Data Type	Reference
WS 8.08E+002	mg/L	25	EST	MEYLAN,WM ET AL. (1996)
logP 2.14			EXP	HANSCH,C ET AL. (1995)
VP 1.46E-005	mm Hg	25	EST	NEELY,WB & BLAU,GE (1985)
DC	pKa			
HL 1.80E-010	atm m3/mol	25	EST	MEYLAN,WM & HOWARD,PH (1991)
OH 8.12E-011	cm3/molc sec	25	EST	MEYLAN,WM & HOWARD,PH (1993)

CAS #:	002525-05-5	1,2-BENZENEDIOL, 4-BUTYL-

Formula:	$C_{10}H_{14}O_2$

Mol Weight: 166.22

MP (deg C):		FP (deg C):

BP (deg C):

BP pressure (mm Hg):

Property/Value	Units	Temp	Data Type	Reference
WS 7.22E+002	mg/L	25	EST	MEYLAN,WM ET AL. (1996)
logP 2.96			EXP	NAITO,Y ET AL. (1991)
VP 7.36E-005	mm Hg	25	EST	NEELY,WB & BLAU,GE (1985)
DC	pKa			
HL 1.51E-010	atm m3/mol	25	EST	MEYLAN,WM & HOWARD,PH (1991)
OH 6.12E-011	cm3/molc sec	25	EST	MEYLAN,WM & HOWARD,PH (1993)

CAS #:	002525-11-3	1,2-BENZENEDIOL, 4-PENTYL-

Formula:	$C_{11}H_{16}O_2$

Mol Weight: 180.25

MP (deg C):		FP (deg C):

BP (deg C):

BP pressure (mm Hg):

Property/Value	Units	Temp	Data Type	Reference
WS 4.61E+002	mg/L	25	EST	MEYLAN,WM ET AL. (1996)
logP 3.11			EXP	ITOKAWA,H ET AL. (1989)
VP 2.43E-005	mm Hg	25	EST	NEELY,WB & BLAU,GE (1985)
DC	pKa			
HL 2.00E-010	atm m3/mol	25	EST	MEYLAN,WM & HOWARD,PH (1991)
OH 6.26E-011	cm3/molc sec	25	EST	MEYLAN,WM & HOWARD,PH (1993)

CAS #:	002527-99-3	2-FURANCARBOXYLIC ACID, 5-BROMO-, METHYL ESTER

Formula:	$C_6H_6BrO_3$

Mol Weight: 206.02

MP (deg C):		FP (deg C):

BP (deg C):

BP pressure (mm Hg):

Property/Value	Units	Temp	Data Type	Reference
WS 1.11E+003	mg/L	25	EST	MEYLAN,WM ET AL. (1996)
logP 1.84			EXP	YAMAGAMI,C ET AL. (1991)
VP 1.64E-001	mm Hg	25	EST	NEELY,WB & BLAU,GE (1985)
DC	pKa			
HL 1.38E-005	atm m3/mol	25	EST	MEYLAN,WM & HOWARD,PH (1991)
OH 8.41E-012	cm3/molc sec	25	EST	MEYLAN,WM & HOWARD,PH (1993)

CAS #:	002528-36-1	DIBUTYL PHENYL PHOSPHATE

Formula:	$C_{14}H_{23}O_4P$

Mol Weight: 286.31

MP (deg C):		FP (deg C):

BP (deg C): 155

BP pressure (mm Hg): 1.00E+000

Property/Value	Units	Temp	Data Type	Reference
WS 9.60E+001	mg/L	25	EXP	SAEGER,VW ET AL. (1979)
logP 4.27			EXP	HANSCH,C & LEO,AJ (1985)
VP 2.30E-004	mm Hg	25	EST	BOETHLING,RS & COOPER,JC (1985)
DC	pKa			
HL 5.04E-007	atm m3/mol	25	EST	MEYLAN,WM & HOWARD,PH (1991)
OH 5.93E-011	cm3/molc sec	25	EST	ATKINSON,R (1988)

CAS #:	002528-61-2	HEPTANOYL CHLORIDE

Formula:	$C_7H_{13}ClO$

Mol Weight: 148.63

MP (deg C): -83.8		FP (deg C):

BP (deg C): 125.3

BP pressure (mm Hg):

Property/Value	Units	Temp	Data Type	Reference
WS 1.56E+003	mg/L	25	EST	MEYLAN,WM ET AL. (1996)
logP 1.98			EST	MEYLAN,WM & HOWARD,PH (1995)
VP 4.46E-001	mm Hg	25	EXT	PERRY,RH & GREEN,D (1984)
DC	pKa			
HL 2.74E-003	atm m3/mol	25	EST	MEYLAN,WM & HOWARD,PH (1991)
OH 8.70E-012	cm3/molc sec	25	EST	MEYLAN,WM & HOWARD,PH (1993)

CAS #: 002530-26-9	3-NITROPYRIDINE

Formula: $C_5H_4N_2O_2$

Mol Weight: 124.10

MP (deg C): FP (deg C):

BP (deg C):

BP pressure (mm Hg):

Property/ Value	Units	Temp	Data Type	Reference
WS 1.21E+004	mg/L	25	EST	MEYLAN,WM ET AL. (1996)
logP 0.60			EXP	HANSCH,C & LEO,AJ (1985)
VP 1.15E-001	mm Hg	25	EST	NEELY,WB & BLAU,GE (1985)
DC 1.18	pKa	25	EXP	PERRIN,DD (1972)
HL 2.78E-008	atm m3/mol	25	EST	MEYLAN,WM & HOWARD,PH (1991)
OH 4.62E-014	cm3/molc sec	25	EST	MEYLAN,WM & HOWARD,PH (1993)

CAS #: 002531-04-6	PIPERYLONE

Formula: $C_{17}H_{23}N_3O$

Mol Weight: 285.39

MP (deg C): FP (deg C):

BP (deg C):

BP pressure (mm Hg):

Property/ Value	Units	Temp	Data Type	Reference
WS 5.09E+002	mg/L	25	EST	MEYLAN,WM ET AL. (1996)
logP 1.72			EST	MEYLAN,WM & HOWARD,PH (1995)
VP 1.67E-008	mm Hg	25	EST	NEELY,WB & BLAU,GE (1985)
DC	pKa			
HL 3.93E-015	atm m3/mol	25	EST	MEYLAN,WM & HOWARD,PH (1991)
OH 2.00E-010	cm3/molc sec	25	EST	MEYLAN,WM & HOWARD,PH (1993)

CAS #: 002531-84-2	2-METHYLPHENANTHRENE

Formula: $C_{15}H_{12}$

Mol Weight: 192.26

MP (deg C): 57-59 FP (deg C):

BP (deg C): 155-160

BP pressure (mm Hg): 3.00E+000

Property/ Value	Units	Temp	Data Type	Reference
WS 2.80E-001	mg/L	25	EXP	ISNARD,P & LAMBERT,S (1989)
logP 4.86			EXP	VEITH,GD ET AL. (1979)
VP 5.01E-005	mm Hg	25	EST	NEELY,WB & BLAU,GE (1985)
DC	pKa			
HL 5.67E-005	atm m3/mol	25	EST	MEYLAN,WM & HOWARD,PH (1991)
OH 3.41E-011	cm3/molc sec	25	EST	MEYLAN,WM & HOWARD,PH (1993)

CAS #: 002532-58-3	1,3-DIMETHYLCYCLOPENTANE (CIS)

Formula: C_7H_{14}

Mol Weight: 98.19

MP (deg C): -133.7 FP (deg C):

BP (deg C): 90.8

BP pressure (mm Hg):

Property/ Value	Units	Temp	Data Type	Reference
WS 3.39E+001	mg/L	25	EST	MEYLAN,WM ET AL. (1996)
logP 3.52			EST	MEYLAN,WM & HOWARD,PH (1995)
VP 6.61E+001	mm Hg	25	EXP	YAWS,CL (1994A)
DC	pKa			
HL 3.39E-001	atm m3/mol	25	EST	MEYLAN,WM & HOWARD,PH (1991)
OH 8.42E-012	cm3/molc sec	25	EST	MEYLAN,WM & HOWARD,PH (1993)

CAS #: 002533-89-3	KARSIL

Formula: $C_{12}H_{15}Cl_2NO$

Mol Weight: 260.17

MP (deg C): 106-107 FP (deg C):

BP (deg C):

BP pressure (mm Hg):

Property/ Value	Units	Temp	Data Type	Reference
WS 4.60E+000	mg/L	25	EST	MEYLAN,WM ET AL. (1996)
logP 4.28			EST	MEYLAN,WM & HOWARD,PH (1995)
VP 1.49E-006	mm Hg	25	EST	NEELY,WB & BLAU,GE (1985)
DC	pKa			
HL 1.05E-008	atm m3/mol	25	EST	MEYLAN,WM & HOWARD,PH (1991)
OH 1.07E-011	cm3/molc sec	25	EST	MEYLAN,WM & HOWARD,PH (1993)

CAS #: 002534-65-8	DECYLPYRIDINIUM BROMIDE

Formula: $C_{15}H_{26}BrN$

Mol Weight: 300.29

MP (deg C): FP (deg C):

BP (deg C):

BP pressure (mm Hg):

Property/ Value	Units	Temp	Data Type	Reference
WS 5.04E+004	mg/L	25	EST	MEYLAN,WM ET AL. (1996)
logP -0.72			EXP	HANSCH,C & LEO,AJ (1985)
VP 6.54E-005	mm Hg	25	EST	NEELY,WB & BLAU,GE (1985)
DC	pKa			
HL 1.93E-009	atm m3/mol	25	EST	MEYLAN,WM & HOWARD,PH (1991)
OH 2.27E-011	cm3/molc sec	25	EST	MEYLAN,WM & HOWARD,PH (1993)

CAS #: 002534-66-9	OCTYLPYRIDINIUM BROMIDE

Formula: $C_{13}H_{22}BrN$

Mol Weight: 272.23

MP (deg C): FP (deg C):

BP (deg C):

BP pressure (mm Hg):

Property/ Value	Units	Temp	Data Type	Reference
WS 1.15E+005	mg/L	25	EST	MEYLAN,WM ET AL. (1996)
logP -0.95			EXP	HANSCH,C & LEO,AJ (1985)
VP 4.19E-004	mm Hg	25	EST	NEELY,WB & BLAU,GE (1985)
DC	pKa			
HL 1.09E-009	atm m3/mol	25	EST	MEYLAN,WM & HOWARD,PH (1991)
OH 1.99E-011	cm3/molc sec	25	EST	MEYLAN,WM & HOWARD,PH (1993)

CAS #: 002536-31-4	CHLORFLURECOL METHYL

Formula: $C_{15}H_{11}ClO_3$

Mol Weight: 274.71

MP (deg C): 136-142 FP (deg C):

BP (deg C):

BP pressure (mm Hg):

Property/ Value	Units	Temp	Data Type	Reference
WS 1.80E+001	mg/L	25	EXP	SHIU,WY ET AL. (1990)
logP 2.86			EST	MEYLAN,WM & HOWARD,PH (1995)
VP 1.93E-008	mm Hg	25	EST	NEELY,WB & BLAU,GE (1985)
DC	pKa			
HL 4.60E-008	atm m3/mol	25	EST	MEYLAN,WM & HOWARD,PH (1991)
OH 5.51E-012	cm3/molc sec	25	EST	MEYLAN,WM & HOWARD,PH (1993)

CAS #: 002537-29-3 — BARBITURIC ACID,5(2HOPR)5-ALLYL

Formula: $C_{10}H_{14}N_2O_4$

Mol Weight: 226.23

MP (deg C): 157-158

FP (deg C):

BP (deg C):

BP pressure (mm Hg):

Property/Value	Units	Temp	Data Type	Reference
WS 3.52E+004	mg/L	25	EST	MEYLAN,WM ET AL. (1996)
logP -0.05			EXP	WITTEKIND,HH ET AL. (1988)
VP 9.58E-015	mm Hg	25	EST	NEELY,WB & BLAU,GE (1985)
DC	pKa			
HL 1.73E-017	atm m3/mol	25	EST	MEYLAN,WM & HOWARD,PH (1991)
OH 4.51E-011	cm3/molc sec	25	EST	MEYLAN,WM & HOWARD,PH (1993)

CAS #: 002538-85-4 — CALCON

Formula: $C_{20}H_{13}N_2NaO_5S$

Mol Weight: 416.39

MP (deg C):

FP (deg C):

BP (deg C):

BP pressure (mm Hg):

Property/Value	Units	Temp	Data Type	Reference
WS 5.63E+000	mg/L	25	EST	MEYLAN,WM ET AL. (1996)
logP 1.96			EST	MEYLAN,WM & HOWARD,PH (1995)
VP 2.29E-024	mm Hg	25	EST	NEELY,WB & BLAU,GE (1985)
DC	pKa			
HL	atm m3/mol			
OH 2.14E-011	cm3/molc sec	25	EST	MEYLAN,WM & HOWARD,PH (1993)

CAS #: 002539-17-5 — TETRACHLOROGUAIACOL

Formula: $C_7H_4Cl_4O_2$

Mol Weight: 261.92

MP (deg C):

FP (deg C):

BP (deg C):

BP pressure (mm Hg):

Property/Value	Units	Temp	Data Type	Reference
WS 4.20E+000	mg/L	25	EXP	KOLSET,K & HEIBERG,A (1988)
logP 4.59			EXP	SANGSTER,J (1994)
VP 1.04E-003	mm Hg	25	EXP	BIDLEMAN,TF & RENBERG,L (1985)
DC	pKa			
HL 8.53E-005	atm m3/mol	25	EST	VP/WSOL
OH 2.19E-012	cm3/molc sec	25	EST	MEYLAN,WM & HOWARD,PH (1993)

CAS #: 002539-26-6 — 3,4,5-TRICL-2,6-MEO PHENOL

Formula: $C_8H_7Cl_3O_3$

Mol Weight: 257.50

MP (deg C):

FP (deg C):

BP (deg C):

BP pressure (mm Hg):

Property/Value	Units	Temp	Data Type	Reference
WS 1.38E+001	mg/L	25	EST	MEYLAN,WM ET AL. (1996)
logP 3.74			EXP	HANSCH,C & LEO,AJ (1985)
VP 2.26E-005	mm Hg	25	EST	NEELY,WB & BLAU,GE (1985)
DC	pKa			
HL 7.98E-010	atm m3/mol	25	EST	MEYLAN,WM & HOWARD,PH (1991)
OH 1.16E-011	cm3/molc sec	25	EST	MEYLAN,WM & HOWARD,PH (1993)

CAS #: 002540-82-1 — FORMOTHION

Formula: $C_6H_{12}NO_4PS_2$

Mol Weight: 257.27

MP (deg C): 25

FP (deg C):

BP (deg C):

BP pressure (mm Hg):

Property/Value	Units	Temp	Data Type	Reference
WS 2.60E+003	mg/L	24	EXP	YALKOWSKY,SH & DANNENFELSER,RM (1992)
logP -0.48			EST	MEYLAN,WM & HOWARD,PH (1995)
VP 3.94E-006	mm Hg	25	EST	NEELY,WB & BLAU,GE (1985)
DC	pKa			
HL 1.48E-010	atm m3/mol	25	EST	MEYLAN,WM & HOWARD,PH (1991)
OH 8.85E-011	cm3/molc sec	25	EST	MEYLAN,WM & HOWARD,PH (1993)

CAS #: 002541-69-7 — BENZ(A)ANTHRACENE-7-METHYL

Formula: $C_{19}H_{14}$

Mol Weight: 242.32

MP (deg C):

FP (deg C):

BP (deg C):

BP pressure (mm Hg):

Property/Value	Units	Temp	Data Type	Reference
WS 1.10E-002	mg/L	24	EXP	YALKOWSKY,SH & DANNENFELSER,RM (1992)
logP 6.07			EST	MEYLAN,WM & HOWARD,PH (1995)
VP 2.53E-007	mm Hg	25	EST	NEELY,WB & BLAU,GE (1985)
DC	pKa			
HL 5.53E-006	atm m3/mol	25	EST	MEYLAN,WM & HOWARD,PH (1991)
OH 1.31E-010	cm3/molc sec	25	EST	MEYLAN,WM & HOWARD,PH (1993)

CAS #: 002545-59-7 — 2,4,5-T, BUTOXYETHYL ESTER

Formula: $C_{14}H_{17}Cl_3O_4$

Mol Weight: 355.65

MP (deg C):

FP (deg C):

BP (deg C): 163-166

BP pressure (mm Hg): 1.00E+000

Property/Value	Units	Temp	Data Type	Reference
WS 8.80E+000	mg/L	25	EST	VP/HL
logP 4.75			EST	MEYLAN,WM & HOWARD,PH (1995)
VP 1.50E-006	mm Hg	25	EXT	HAMILTON,DJ (1980)
DC	pKa			
HL 7.95E-008	atm m3/mol	25	EST	MEYLAN,WM & HOWARD,PH (1991)
OH 2.23E-011	cm3/molc sec	25	EST	ATKINSON,R (1988)

CAS #: 002549-67-9 — 2-ETHYLAZIRIDINE

Formula: C_4H_9N

Mol Weight: 71.12

MP (deg C):

FP (deg C):

BP (deg C):

BP pressure (mm Hg):

Property/Value	Units	Temp	Data Type	Reference
WS 4.03E+005	mg/L	25	EST	MEYLAN,WM ET AL. (1996)
logP 0.62			EST	MEYLAN,WM & HOWARD,PH (1995)
VP 5.52E+001	mm Hg	25	EST	NEELY,WB & BLAU,GE (1985)
DC 8.31	pKa	25	EXP	PERRIN,DD (1965)
HL 1.29E-005	atm m3/mol	25	EST	MEYLAN,WM & HOWARD,PH (1991)
OH 8.08E-012	cm3/molc sec	25	EST	MEYLAN,WM & HOWARD,PH (1993)

CAS #: 002550-28-9				1-HEXYN-5-ONE

Formula: C_6H_8O

Mol Weight: 96.13

MP (deg C): FP (deg C):

BP (deg C): 149

BP pressure (mm Hg):

Property/Value	Units	Temp	Data Type	Reference
WS 3.84E+004	mg/L	25	EST	MEYLAN,WM ET AL. (1996)
logP 0.58			EXP	HANSCH,C & LEO,AJ (1985)
VP 1.40E+001	mm Hg	25	EST	NEELY,WB & BLAU,GE (1985)
DC	pKa			
HL 1.44E-005	atm m3/mol	25	EST	MEYLAN,WM & HOWARD,PH (1991)
OH 1.16E-011	cm3/molc sec	25	EST	MEYLAN,WM & HOWARD,PH (1993)

CAS #: 002553-19-7				DIPHENYLDIETHOXYSILANE

Formula: $C_{16}H_{20}O_2Si$

Mol Weight: 272.42

MP (deg C): FP (deg C):

BP (deg C): 167

BP pressure (mm Hg): 1.50E+001

Property/Value	Units	Temp	Data Type	Reference
WS 1.11E+000	mg/L	25	EST	MEYLAN,WM ET AL. (1996)
logP 4.92			EXP	HANSCH,C ET AL. (1995)
VP 4.72E-005	mm Hg	25	EST	NEELY,WB & BLAU,GE (1985)
DC	pKa			
HL 1.73E-007	atm m3/mol	25	EST	MEYLAN,WM & HOWARD,PH (1991)
OH 1.62E-011	cm3/molc sec	25	EST	MEYLAN,WM & HOWARD,PH (1993)

CAS #: 002557-81-5				PHENYLSULFUR PENTAFLUORIDE

Formula: $C_6H_5F_5S$

Mol Weight: 204.16

MP (deg C): FP (deg C):

BP (deg C):

BP pressure (mm Hg):

Property/Value	Units	Temp	Data Type	Reference
WS 5.63E+001	mg/L	25	EST	MEYLAN,WM ET AL. (1996)
logP 3.36			EXP	HANSCH,C & LEO,AJ (1985)
VP 1.59E+000	mm Hg	25	EST	NEELY,WB & BLAU,GE (1985)
DC	pKa			
HL 4.93E-002	atm m3/mol	25	EST	MEYLAN,WM & HOWARD,PH (1991)
OH 4.17E-013	cm3/molc sec	25	EST	MEYLAN,WM & HOWARD,PH (1993)

CAS #: 002558-30-7				1,4-BENZODIAZPIN-2-ONE,5-FPH,7NO2

Formula: $C_{15}H_{10}FN_3O_3$

Mol Weight: 299.26

MP (deg C): FP (deg C):

BP (deg C):

BP pressure (mm Hg):

Property/Value	Units	Temp	Data Type	Reference
WS 7.37E+001	mg/L	25	EST	MEYLAN,WM ET AL. (1996)
logP 2.15			EXP	HANSCH,C & LEO,AJ (1985)
VP 5.77E-010	mm Hg	25	EST	NEELY,WB & BLAU,GE (1985)
DC	pKa			
HL 1.11E-012	atm m3/mol	25	EST	MEYLAN,WM & HOWARD,PH (1991)
OH 7.94E-012	cm3/molc sec	25	EST	MEYLAN,WM & HOWARD,PH (1993)

CAS #: 002565-47-1				N-METHYL BARBITURIC ACID

Formula: $C_5H_6N_2O_3$

Mol Weight: 142.12

MP (deg C): FP (deg C):

BP (deg C):

BP pressure (mm Hg):

Property/Value	Units	Temp	Data Type	Reference
WS 4.37E+005	mg/L	25	EST	MEYLAN,WM ET AL. (1996)
logP -0.85			EXP	HANSCH,C & LEO,AJ (1985)
VP 9.45E-009	mm Hg	25	EST	NEELY,WB & BLAU,GE (1985)
DC 4.35	pKa	25	EXP	KORTUM,G ET AL (1961)
HL 2.55E-013	atm m3/mol	25	EST	MEYLAN,WM & HOWARD,PH (1991)
OH 3.08E-012	cm3/molc sec	25	EST	MEYLAN,WM & HOWARD,PH (1993)

CAS #: 002567-14-8				1,1,3-TRICHLOROPROPENE

Formula: $C_3H_3Cl_3$

Mol Weight: 145.42

MP (deg C): FP (deg C):

BP (deg C): 131-132

BP pressure (mm Hg):

Property/Value	Units	Temp	Data Type	Reference
WS 3.34E+002	mg/L	25	EST	MEYLAN,WM ET AL. (1996)
logP 2.78			EST	MEYLAN,WM & HOWARD,PH (1995)
VP 7.55E+000	mm Hg	25	EST	NEELY,WB & BLAU,GE (1985)
DC	pKa			
HL 1.76E-002	atm m3/mol	25	EST	MEYLAN,WM & HOWARD,PH (1991)
OH 3.27E-012	cm3/molc sec	25	EST	MEYLAN,WM & HOWARD,PH (1993)

CAS #: 002571-81-5				AZULENO[4,5-B]FURAN-2,9-DIONE, DECAHYDRO-6A-HYDR

Formula: $C_{15}H_{20}O_4$

Mol Weight: 264.32

MP (deg C): FP (deg C):

BP (deg C):

BP pressure (mm Hg):

Property/Value	Units	Temp	Data Type	Reference
WS 1.25E+004	mg/L	25	EST	MEYLAN,WM ET AL. (1996)
logP 0.83			EXP	SANGSTER,J (1994)
VP 1.01E-008	mm Hg	25	EST	NEELY,WB & BLAU,GE (1985)
DC	pKa			
HL 2.95E-012	atm m3/mol	25	EST	MEYLAN,WM & HOWARD,PH (1991)
OH 4.73E-011	cm3/molc sec	25	EST	MEYLAN,WM & HOWARD,PH (1993)

CAS #: 002581-34-2				3-METHYL-4-NITROPHENOL

Formula: $C_7H_7NO_3$

Mol Weight: 153.14

MP (deg C): 127-129 FP (deg C):

BP (deg C):

BP pressure (mm Hg):

Property/Value	Units	Temp	Data Type	Reference
WS 1.19E+003	mg/L	20	EXP	SCHWARZENBACH,RP ET AL. (1988)
logP 2.48			EXP	SCHWARZENBACH,RP ET AL. (1988)
VP 6.32E-004	mm Hg	25	EST	NEELY,WB & BLAU,GE (1985)
DC	pKa			
HL 2.44E-009	atm m3/mol	25	EST	MEYLAN,WM & HOWARD,PH (1991)
OH 1.11E-011	cm3/molc sec	25	EST	MEYLAN,WM & HOWARD,PH (1993)

C.I. DISPERSE ORANGE 1

CAS #:	002581-69-3
Formula:	$C_{18}H_{14}N_4O_2$
Mol Weight:	318.34
MP (deg C):	
FP (deg C):	
BP (deg C):	
BP pressure (mm Hg):	

Property/Value	Units	Temp	Data Type	Reference
WS 9.54E-003	mg/L	25	EXP	BAUGHMAN,GL & PERENICH,TA (1988)
logP 5.80			EST	MEYLAN,WM & HOWARD,PH (1995)
VP 4.69E-009	mm Hg	25	EST	NEELY,WB & BLAU,GE (1985)
DC	pKa			
HL 1.13E-011	atm m3/mol	25	EST	MEYLAN,WM & HOWARD,PH (1991)
OH 2.00E-010	cm3/molc sec	25	EST	MEYLAN,WM & HOWARD,PH (1993)

DEMEPHION

CAS #:	002587-90-8
Formula:	$C_5H_{13}O_3PS_2$
Mol Weight:	216.26
MP (deg C):	
FP (deg C):	
BP (deg C):	
BP pressure (mm Hg):	

Property/Value	Units	Temp	Data Type	Reference
WS 3.00E+002	mg/L	25	EXP	KENAGA,EE (1980)
logP 0.62			EST	MEYLAN,WM & HOWARD,PH (1995)
VP 2.34E-003	mm Hg	25	EST	NEELY,WB & BLAU,GE (1985)
DC	pKa			
HL 1.94E-009	atm m3/mol	25	EST	MEYLAN,WM & HOWARD,PH (1991)
OH 4.08E-011	cm3/molc sec	25	EST	MEYLAN,WM & HOWARD,PH (1993)

PHORATE SULFONE

CAS #:	002588-04-7
Formula:	$C_7H_{17}O_4PS_3$
Mol Weight:	292.38
MP (deg C):	
FP (deg C):	
BP (deg C):	
BP pressure (mm Hg):	

Property/Value	Units	Temp	Data Type	Reference
WS 8.61E+002	mg/L	20	EXP	BOWMAN,BT & SANS,WW (1983A)
logP 1.34			EST	MEYLAN,WM & HOWARD,PH (1995)
VP 2.67E-005	mm Hg	25	EST	NEELY,WB & BLAU,GE (1985)
DC	pKa			
HL 1.25E-008	atm m3/mol	25	EST	MEYLAN,WM & HOWARD,PH (1991)
OH 2.48E-010	cm3/molc sec	25	EST	MEYLAN,WM & HOWARD,PH (1993)

PHORATE SULFOXIDE

CAS #:	002588-05-8
Formula:	$C_7H_{17}O_4PS_2$
Mol Weight:	260.31
MP (deg C):	
FP (deg C):	
BP (deg C):	
BP pressure (mm Hg):	

Property/Value	Units	Temp	Data Type	Reference
WS 5.08E+002	mg/L	25	EST	MEYLAN,WM ET AL. (1996)
logP 1.78			EXP	HANSCH,C & LEO,AJ (1985)
VP 6.92E-005	mm Hg	25	EST	NEELY,WB & BLAU,GE (1985)
DC	pKa			
HL 9.97E-011	atm m3/mol	25	EST	MEYLAN,WM & HOWARD,PH (1991)
OH 3.08E-010	cm3/molc sec	25	EST	MEYLAN,WM & HOWARD,PH (1993)

PHOSPHOROTHIOIC ACID, O,O-DIETHYL S-[(ETHYLSULFO

CAS #:	002588-06-9
Formula:	$C_7H_{17}O_5PS_2$
Mol Weight:	276.31
MP (deg C):	
FP (deg C):	
BP (deg C):	
BP pressure (mm Hg):	

Property/Value	Units	Temp	Data Type	Reference
WS 3.43E+002	mg/L	25	EST	MEYLAN,WM ET AL. (1996)
logP 1.98			EXP	BOWMAN,BT & SANS,WW (1983)
VP 8.67E-006	mm Hg	25	EST	NEELY,WB & BLAU,GE (1985)
DC	pKa			
HL 2.71E-011	atm m3/mol	25	EST	MEYLAN,WM & HOWARD,PH (1991)
OH 1.95E-010	cm3/molc sec	25	EST	MEYLAN,WM & HOWARD,PH (1993)

BENZYLPYRIDINIUM BROMIDE

CAS #:	002589-31-3
Formula:	$C_{12}H_{12}BrN$
Mol Weight:	250.14
MP (deg C):	
FP (deg C):	
BP (deg C):	
BP pressure (mm Hg):	

Property/Value	Units	Temp	Data Type	Reference
WS 1.00E+006	mg/L	25	EST	MEYLAN,WM ET AL. (1996)
logP -2.62			EXP	HANSCH,C & LEO,AJ (1985)
VP 2.85E-004	mm Hg	25	EST	NEELY,WB & BLAU,GE (1985)
DC	pKa			
HL 1.21E-011	atm m3/mol	25	EST	MEYLAN,WM & HOWARD,PH (1991)
OH 1.61E-011	cm3/molc sec	25	EST	MEYLAN,WM & HOWARD,PH (1993)

3-ME-4-CLPHENYL-N-ME CARBAMATE

CAS #:	002589-65-3
Formula:	$C_9H_{10}ClNO_2$
Mol Weight:	199.64
MP (deg C):	
FP (deg C):	
BP (deg C):	
BP pressure (mm Hg):	

Property/Value	Units	Temp	Data Type	Reference
WS 2.81E+002	mg/L	25	EST	MEYLAN,WM ET AL. (1996)
logP 2.57			EXP	HANSCH,C & LEO,AJ (1985)
VP 2.88E-003	mm Hg	25	EST	NEELY,WB & BLAU,GE (1985)
DC	pKa			
HL 2.63E-008	atm m3/mol	25	EST	MEYLAN,WM & HOWARD,PH (1991)
OH 7.73E-012	cm3/molc sec	25	EST	MEYLAN,WM & HOWARD,PH (1993)

TERRAZOLE

CAS #:	002593-15-9
Formula:	$C_5H_5Cl_3N_2OS$
Mol Weight:	247.53
MP (deg C):	19.9
FP (deg C):	
BP (deg C):	95
BP pressure (mm Hg):	9.70E-001

Property/Value	Units	Temp	Data Type	Reference
WS 5.00E+001	mg/L	25	EXP	WORTHING,CR & WALKER,SB (1987)
logP 3.37			EXP	TOMLIN,C (1994)
VP 9.75E-005	mm Hg	25	EXP	WORTHING,CR & WALKER,SB (1987)
DC	pKa			
HL 6.32E-007	atm m3/mol	25	EST	VP/WSOL
OH 6.46E-012	cm3/molc sec	25	EST	MEYLAN,WM & HOWARD,PH (1993)

002594-17-4

O-PENTYL-N-METHYLCARBAMATE

Formula: $C_7H_{15}NO_2$

Mol Weight: 145.20

MP (deg C):
FP (deg C):

BP (deg C):

BP pressure (mm Hg):

Property/Value	Units	Temp	Data Type	Reference
WS 1.69E+003	mg/L	25	EST	MEYLAN,WM ET AL. (1996)
logP 1.96			EXP	HANSCH,C & LEO,AJ (1985)
VP 6.48E-001	mm Hg	25	EST	NEELY,WB & BLAU,GE (1985)
DC	pKa			
HL 2.70E-007	atm m3/mol	25	EST	MEYLAN,WM & HOWARD,PH (1991)
OH 1.20E-011	cm3/molc sec	25	EST	MEYLAN,WM & HOWARD,PH (1993)

002595-54-2

MECARBAM

Formula: $C_{10}H_{20}NO_5PS_2$

Mol Weight: 329.38

MP (deg C):
FP (deg C):

BP (deg C): 144

BP pressure (mm Hg): 2.00E-002

Property/Value	Units	Temp	Data Type	Reference
WS 1.00E+003	mg/L	21	EXP	YALKOWSKY,SH & DANNENFELSER,RM (1992)
logP 2.29			EST	MEYLAN,WM & HOWARD,PH (1995)
VP 1.30E+000	mm Hg	25	EST	NEELY,WB & BLAU,GE (1985)
DC	pKa			
HL 9.00E-010	atm m3/mol	25	EST	MEYLAN,WM & HOWARD,PH (1991)
OH 1.23E-010	cm3/molc sec	25	EST	MEYLAN,WM & HOWARD,PH (1993)

002597-03-7

FENTHOATE

Formula: $C_{12}H_{17}O_4PS_2$

Mol Weight: 320.37

MP (deg C): 17-18
FP (deg C):

BP (deg C): 70-80

BP pressure (mm Hg): 3.50E-005

Property/Value	Units	Temp	Data Type	Reference
WS 2.00E+002	mg/L	20	EXP	YALKOWSKY,SH & DANNENFELSER,RM (1992)
logP 3.69			EXP	HANSCH,C ET AL. (1995)
VP 2.60E-006	mm Hg	25	EXP	AUGUSTIJN-BECKERS,PWM ET AL. (1994)
DC	pKa			
HL 5.48E-009	atm m3/mol	25	EST	VP/WSOL
OH 7.70E-011	cm3/molc sec	25	EST	MEYLAN,WM & HOWARD,PH (1993)

002597-93-5

50-CS-46

Formula: $C_{11}H_7Cl_6HgNO_2$

Mol Weight: 598.49

MP (deg C):
FP (deg C):

BP (deg C):

BP pressure (mm Hg):

Property/Value	Units	Temp	Data Type	Reference
WS 4.09E-001	mg/L	25	EST	MEYLAN,WM ET AL. (1996)
logP 3.05			EST	MEYLAN,WM & HOWARD,PH (1995)
VP 3.92E-011	mm Hg	25	EST	NEELY,WB & BLAU,GE (1985)
DC	pKa			
HL	atm m3/mol			
OH 8.07E-011	cm3/molc sec	25	EST	MEYLAN,WM & HOWARD,PH (1993)

002597-97-9

CHIPCOTE

Formula: C_2H_3HgN

Mol Weight: 241.64

MP (deg C):
FP (deg C):

BP (deg C):

BP pressure (mm Hg):

Property/Value	Units	Temp	Data Type	Reference
WS 2.00E+005	mg/L		EXP	SHIU,WY ET AL. (1990)
logP -0.85			EST	MEYLAN,WM & HOWARD,PH (1995)
VP 5.02E-001	mm Hg	25	EST	NEELY,WB & BLAU,GE (1985)
DC	pKa			
HL	atm m3/mol			
OH 1.36E-013	cm3/molc sec	25	EST	MEYLAN,WM & HOWARD,PH (1993)

002598-29-0

8-ACETYLOXYQUINOLINE

Formula: $C_{11}H_9NO_2$

Mol Weight: 187.20

MP (deg C):
FP (deg C):

BP (deg C):

BP pressure (mm Hg):

Property/Value	Units	Temp	Data Type	Reference
WS 9.37E+002	mg/L	25	EST	MEYLAN,WM ET AL. (1996)
logP 2.03			EXP	HANSCH,C ET AL. (1995)
VP 3.19E-004	mm Hg	25	EST	NEELY,WB & BLAU,GE (1985)
DC	pKa			
HL 8.28E-009	atm m3/mol	25	EST	MEYLAN,WM & HOWARD,PH (1991)
OH 1.50E-011	cm3/molc sec	25	EST	MEYLAN,WM & HOWARD,PH (1993)

002600-69-3

PHORATE, O-ANALOG

Formula: $C_7H_{17}O_3PS_2$

Mol Weight: 244.31

MP (deg C):
FP (deg C):

BP (deg C):

BP pressure (mm Hg):

Property/Value	Units	Temp	Data Type	Reference
WS 4.34E+002	mg/L	25	EST	MEYLAN,WM ET AL. (1996)
logP 2.07			EXP	HANSCH,C & LEO,AJ (1985)
VP 3.32E-004	mm Hg	25	EST	NEELY,WB & BLAU,GE (1985)
DC	pKa			
HL 3.42E-009	atm m3/mol	25	EST	MEYLAN,WM & HOWARD,PH (1991)
OH 1.97E-010	cm3/molc sec	25	EST	MEYLAN,WM & HOWARD,PH (1993)

002603-10-3

METHYL N-PHENYLCARBAMATE

Formula: $C_8H_9NO_2$

Mol Weight: 151.17

MP (deg C):
FP (deg C):

BP (deg C):

BP pressure (mm Hg):

Property/Value	Units	Temp	Data Type	Reference
WS 2.40E+003	mg/L	25	EST	MEYLAN,WM ET AL. (1996)
logP 1.75			EXP	TAKAHASHI,J ET AL. (1988)
VP 7.07E-002	mm Hg	25	EST	NEELY,WB & BLAU,GE (1985)
DC	pKa			
HL 2.18E-008	atm m3/mol	25	EST	MEYLAN,WM & HOWARD,PH (1991)
OH 4.29E-011	cm3/molc sec	25	EST	MEYLAN,WM & HOWARD,PH (1993)

CAS #: 002604-08-2				B-NITRO-4-DIMETHYLAMINOSTYRENE
Formula: $C_{10}H_{12}N_2O_2$				
Mol Weight: 192.22				
MP (deg C):		FP (deg C):		
BP (deg C):				
BP pressure (mm Hg):				

Property/ Value	Units	Temp	Data Type	Reference
WS 1.02E+002	mg/L	25	EST	MEYLAN,WM ET AL. (1996)
logP 2.67			EXP	HANSCH,C & LEO,AJ (1985)
VP 5.69E-004	mm Hg	25	EST	NEELY,WB & BLAU,GE (1985)
DC	pKa			
HL 5.52E-008	atm m3/mol	25	EST	MEYLAN,WM & HOWARD,PH (1991)
OH 2.14E-010	cm3/molc sec	25	EST	MEYLAN,WM & HOWARD,PH (1993)

CAS #: 002610-11-9				DIRECT RED 81
Formula: $C_{29}H_{19}N_5Na_2O_8S_2$				
Mol Weight: 675.61				
MP (deg C): 240 dec		FP (deg C):		
BP (deg C):				
BP pressure (mm Hg):				

Property/ Value	Units	Temp	Data Type	Reference
WS 2.81E+000	mg/L	25	EST	MEYLAN,WM ET AL. (1996)
logP 1.65			EST	MEYLAN,WM & HOWARD,PH (1995)
VP 5.83E-032	mm Hg	25	EST	NEELY,WB & BLAU,GE (1985)
DC	pKa			
HL	atm m3/mol			
OH 8.97E-012	cm3/molc sec	25	EST	MEYLAN,WM & HOWARD,PH (1993)

CAS #: 002611-82-7				ACID RED 18
Formula: $C_{20}H_{11}N_2Na_3O_{10}S_3$				
Mol Weight: 604.48				
MP (deg C):		FP (deg C):		
BP (deg C):				
BP pressure (mm Hg):				

Property/ Value	Units	Temp	Data Type	Reference
WS 8.00E+004	mg/L	25	EXP	GREEN,FJ (1990)
logP -1.74			EST	MEYLAN,WM & HOWARD,PH (1995)
VP 1.29E-027	mm Hg	25	EST	NEELY,WB & BLAU,GE (1985)
DC	pKa			
HL 1.49E-030	atm m3/mol	25	EST	MEYLAN,WM & HOWARD,PH (1991)
OH 2.44E-012	cm3/molc sec	25	EST	MEYLAN,WM & HOWARD,PH (1993)

CAS #: 002612-46-6				CIS-1,3,5-HEXATRIENE
Formula: C_6H_8				
Mol Weight: 80.13				
MP (deg C): -12		FP (deg C):		
BP (deg C): 78				
BP pressure (mm Hg):				

Property/ Value	Units	Temp	Data Type	Reference
WS 1.54E+002	mg/L	25	EST	MEYLAN,WM ET AL. (1996)
logP 2.80			EST	MEYLAN,WM & HOWARD,PH (1995)
VP 1.15E+002	mm Hg	25	EST	NEELY,WB & BLAU,GE (1985)
DC	pKa			
HL 9.84E-002	atm m3/mol	25	EST	MEYLAN,WM & HOWARD,PH (1991)
OH 1.10E-013	cm3/molc sec	25	EXP	ATKINSON,R (1989)

CAS #: 002615-15-8				HEXAETHYLENE GLYCOL
Formula: $C_{12}H_{26}O_7$				
Mol Weight: 282.34				
MP (deg C): 5-7		FP (deg C):		
BP (deg C): 217				
BP pressure (mm Hg): 4.00E+000				

Property/ Value	Units	Temp	Data Type	Reference
WS 1.00E+006	mg/L	25	EST	MEYLAN,WM ET AL. (1996)
logP -2.57			EST	MEYLAN,WM & HOWARD,PH (1995)
VP 1.27E-008	mm Hg	25	EST	NEELY,WB & BLAU,GE (1985)
DC	pKa			
HL 1.18E-016	atm m3/mol	25	EST	MEYLAN,WM & HOWARD,PH (1991)
OH 7.84E-011	cm3/molc sec	25	EST	MEYLAN,WM & HOWARD,PH (1993)

CAS #: 002617-79-0				P-CHLOROFORMANILIDE
Formula: C_7H_6ClNO				
Mol Weight: 155.58				
MP (deg C):		FP (deg C):		
BP (deg C):				
BP pressure (mm Hg):				

Property/ Value	Units	Temp	Data Type	Reference
WS 1.30E+003	mg/L	25	EST	MEYLAN,WM ET AL. (1996)
logP 2.04			EXP	HANSCH,C & LEO,AJ (1985)
VP 2.06E-004	mm Hg	25	EST	NEELY,WB & BLAU,GE (1985)
DC	pKa			
HL 6.26E-009	atm m3/mol	25	EST	MEYLAN,WM & HOWARD,PH (1991)
OH 1.25E-011	cm3/molc sec	25	EST	MEYLAN,WM & HOWARD,PH (1993)

CAS #: 002618-96-4				BENZENESULFONAMIDE, N-(PHENYLSULFONYL)-
Formula: $C_{12}H_{11}NO_4S_2$				
Mol Weight: 297.35				
MP (deg C):		FP (deg C):		
BP (deg C):				
BP pressure (mm Hg):				

Property/ Value	Units	Temp	Data Type	Reference
WS 4.58E+003	mg/L	25	EST	MEYLAN,WM ET AL. (1996)
logP 0.52			EXP	HANSCH,C ET AL. (1995)
VP 5.85E-009	mm Hg	25	EST	NEELY,WB & BLAU,GE (1985)
DC	pKa			
HL 5.17E-008	atm m3/mol	25	EST	MEYLAN,WM & HOWARD,PH (1991)
OH 8.34E-013	cm3/molc sec	25	EST	MEYLAN,WM & HOWARD,PH (1993)

CAS #: 002620-53-3				N-METHYL-4-CHLOROPHENYLCARBAMATE
Formula: $C_8H_8ClNO_2$				
Mol Weight: 185.61				
MP (deg C):		FP (deg C):		
BP (deg C):				
BP pressure (mm Hg):				

Property/ Value	Units	Temp	Data Type	Reference
WS 9.93E+002	mg/L	25	EST	MEYLAN,WM ET AL. (1996)
logP 2.01			EXP	HANSCH,C & LEO,AJ (1985)
VP 8.31E-003	mm Hg	25	EST	NEELY,WB & BLAU,GE (1985)
DC	pKa			
HL 2.38E-008	atm m3/mol	25	EST	MEYLAN,WM & HOWARD,PH (1991)
OH 7.07E-012	cm3/molc sec	25	EST	MEYLAN,WM & HOWARD,PH (1993)

CAS #: 002620-63-5				GLYCIN-AMIDE,N-ACETYL

Formula: $C_4H_8N_2O_2$

Mol Weight: 116.12

MP (deg C): 140-143　　FP (deg C):

BP (deg C):

BP pressure (mm Hg):

Property/ Value	Units	Temp	Data Type	Reference
WS 1.86E+005	mg/L	25	EST	MEYLAN,WM ET AL. (1996)
logP -1.83			EXP	HANSCH,C & LEO,AJ (1985)
VP 1.96E-006	mm Hg	25	EST	NEELY,WB & BLAU,GE (1985)
DC	pKa			
HL 1.18E-010	atm m3/mol	25	EST	MEYLAN,WM & HOWARD,PH (1991)
OH 3.31E-011	cm3/molc sec	25	EST	MEYLAN,WM & HOWARD,PH (1993)

CAS #: 002621-62-7				2,5-DICHLOROACETANILIDE

Formula: $C_8H_7Cl_2NO$

Mol Weight: 204.06

MP (deg C):　　FP (deg C):

BP (deg C):

BP pressure (mm Hg):

Property/ Value	Units	Temp	Data Type	Reference
WS 5.62E+002	mg/L	25	EST	MEYLAN,WM ET AL. (1996)
logP 2.19			EXP	NAKAGAWA,Y ET AL. (1992)
VP 2.02E-005	mm Hg	25	EST	NEELY,WB & BLAU,GE (1985)
DC	pKa			
HL 3.39E-009	atm m3/mol	25	EST	MEYLAN,WM & HOWARD,PH (1991)
OH 2.65E-012	cm3/molc sec	25	EST	MEYLAN,WM & HOWARD,PH (1993)

CAS #: 002621-80-9				P-CHLOROBENZYLCARBAMATE

Formula: $C_8H_8ClNO_2$

Mol Weight: 185.61

MP (deg C):　　FP (deg C):

BP (deg C):

BP pressure (mm Hg):

Property/ Value	Units	Temp	Data Type	Reference
WS 1.07E+003	mg/L	25	EST	MEYLAN,WM ET AL. (1996)
logP 1.97			EXP	HANSCH,C ET AL. (1995)
VP 3.40E-003	mm Hg	25	EST	NEELY,WB & BLAU,GE (1985)
DC	pKa			
HL 2.37E-009	atm m3/mol	25	EST	MEYLAN,WM & HOWARD,PH (1991)
OH 5.17E-012	cm3/molc sec	25	EST	MEYLAN,WM & HOWARD,PH (1993)

CAS #: 002622-07-3				PHOSPHORIC TRIAMIDE, HEXAETHYL-

Formula: $C_{12}H_{30}N_3OP$

Mol Weight: 263.37

MP (deg C):　　FP (deg C):

BP (deg C):

BP pressure (mm Hg):

Property/ Value	Units	Temp	Data Type	Reference
WS 1.38E+002	mg/L	25	EST	MEYLAN,WM ET AL. (1996)
logP 2.53			EXP	DEBORD,J & LABADIE,M (1985)
VP 1.76E-004	mm Hg	25	EST	NEELY,WB & BLAU,GE (1985)
DC	pKa			
HL 3.90E-011	atm m3/mol	25	EST	MEYLAN,WM & HOWARD,PH (1991)
OH 1.36E-010	cm3/molc sec	25	EST	MEYLAN,WM & HOWARD,PH (1993)

CAS #: 002622-26-6				PERICYAZINE

Formula: $C_{21}H_{23}N_3OS$

Mol Weight: 365.50

MP (deg C): 116-117　　FP (deg C):

BP (deg C):

BP pressure (mm Hg):

Property/ Value	Units	Temp	Data Type	Reference
WS 3.80E+001	mg/L	37	EXP	YALKOWSKY,SH & DANNENFELSER,RM (1992)
logP 3.52			EXP	HANSCH,C & LEO,AJ (1985)
VP 4.03E-013	mm Hg	25	EST	NEELY,WB & BLAU,GE (1985)
DC	pKa			
HL 1.81E-016	atm m3/mol	25	EST	MEYLAN,WM & HOWARD,PH (1991)
OH 2.25E-010	cm3/molc sec	25	EST	MEYLAN,WM & HOWARD,PH (1993)

CAS #: 002623-33-8				P-ACETOXYACETANILIDE

Formula: $C_{10}H_{11}NO_3$

Mol Weight: 193.20

MP (deg C):　　FP (deg C):

BP (deg C):

BP pressure (mm Hg):

Property/ Value	Units	Temp	Data Type	Reference
WS 2.39E+003	mg/L	25	EXP	YALKOWSKY,SH & DANNENFELSER,RM (1992)
logP 1.05			EXP	HANSCH,C & LEO,AJ (1985)
VP 9.60E-006	mm Hg	25	EST	NEELY,WB & BLAU,GE (1985)
DC	pKa			
HL 7.42E-011	atm m3/mol	25	EST	MEYLAN,WM & HOWARD,PH (1991)
OH 9.32E-012	cm3/molc sec	25	EST	MEYLAN,WM & HOWARD,PH (1993)

CAS #: 002623-36-1				4-((4-HYDROXYPHENYL)AZO)BENZENESULFONIC ACID,MO*

Formula: $C_{12}H_9N_2NaO_4S$

Mol Weight: 300.27

MP (deg C):　　FP (deg C):

BP (deg C):

BP pressure (mm Hg):

Property/ Value	Units	Temp	Data Type	Reference
WS 2.28E+005	mg/L	25	EST	MEYLAN,WM ET AL. (1996)
logP -1.31				MEYLAN,WM & HOWARD,PH (1995)
VP 6.03E-017	mm Hg	25	EST	NEELY,WB & BLAU,GE (1985)
DC	pKa			
HL	atm m3/mol			
OH 1.35E-011			25	EST	MEYLAN,WM & HOWARD,PH (1993)

CAS #: 002626-81-5				N-ME-2-T-BUTYLPHENYLCARBAMATE

Formula: $C_{12}H_{17}NO_2$

Mol Weight: 207.27

MP (deg C):　　FP (deg C):

BP (deg C):

BP pressure (mm Hg):

Property/ Value	Units	Temp	Data Type	Reference
WS 2.19E+002	mg/L	25	EST	MEYLAN,WM ET AL. (1996)
logP 2.65			EXP	HANSCH,C & LEO,AJ (1985)
VP 1.80E-003	mm Hg	25	EST	NEELY,WB & BLAU,GE (1985)
DC	pKa			
HL 8.29E-008	atm m3/mol	25	EST	MEYLAN,WM & HOWARD,PH (1991)
OH 9.71E-012	cm3/molc sec	25	EST	MEYLAN,WM & HOWARD,PH (1993)

CAS #:	002626-83-7			N-ME-4-T-BUTYLPHENYLCARBAMATE

Formula: $C_{12}H_{17}NO_2$

Mol Weight: 207.27

MP (deg C): | FP (deg C):

BP (deg C):

BP pressure (mm Hg):

Property/Value	Units	Temp	Data Type	Reference
WS 9.78E+001	mg/L	25	EST	MEYLAN,WM ET AL. (1996)
logP 3.06			EXP	HANSCH,C & LEO,AJ (1985)
VP 1.80E-003	mm Hg	25	EST	NEELY,WB & BLAU,GE (1985)
DC	pKa			
HL 8.29E-008	atm m3/mol	25	EST	MEYLAN,WM & HOWARD,PH (1991)
OH 9.71E-012	cm3/molc sec	25	EST	MEYLAN,WM & HOWARD,PH (1993)

CAS #:	002627-77-2			4'-BR SALICYLANILIDE

Formula: $C_{13}H_{10}BrNO_2$

Mol Weight: 292.14

MP (deg C): | FP (deg C):

BP (deg C):

BP pressure (mm Hg):

Property/Value	Units	Temp	Data Type	Reference
WS 7.30E+000	mg/L	25	EST	MEYLAN,WM ET AL. (1996)
logP 4.51			EXP	TERADA,H ET AL. (1988)
VP 2.42E-009	mm Hg	25	EST	NEELY,WB & BLAU,GE (1985)
DC	pKa			
HL 6.38E-011	atm m3/mol	25	EST	MEYLAN,WM & HOWARD,PH (1991)
OH 3.41E-011	cm3/molc sec	25	EST	MEYLAN,WM & HOWARD,PH (1993)

CAS #:	002628-58-2			N-METHYLTHIOBENZANILIDE

Formula: $C_{14}H_{13}NS$

Mol Weight: 227.33

MP (deg C): | FP (deg C):

BP (deg C):

BP pressure (mm Hg):

Property/Value	Units	Temp	Data Type	Reference
WS 6.81E+001	mg/L	25	EST	MEYLAN,WM ET AL. (1996)
logP 3.12			EXP	HANSCH,C ET AL. (1995)
VP 3.01E-005	mm Hg	25	EST	NEELY,WB & BLAU,GE (1985)
DC	pKa			
HL 4.43E-006	atm m3/mol	25	EST	MEYLAN,WM & HOWARD,PH (1991)
OH 1.52E-010	cm3/molc sec	25	EST	MEYLAN,WM & HOWARD,PH (1993)

CAS #:	002629-72-3			4-PYRIDINEPROPANOL

Formula: $C_8H_{11}NO$

Mol Weight: 137.18

MP (deg C): 35-39 | FP (deg C):

BP (deg C): 289

BP pressure (mm Hg):

Property/Value	Units	Temp	Data Type	Reference
WS 1.00E+006	mg/L	25	EST	MEYLAN,WM ET AL. (1996)
logP 0.58			EXP	HANSCH,C & LEO,AJ (1985)
VP 1.73E-003	mm Hg	25	EST	NEELY,WB & BLAU,GE (1985)
DC 5.84	pKa	25	EXP	PERRIN,DD (1972)
HL 5.01E-010	atm m3/mol	25	EST	MEYLAN,WM & HOWARD,PH (1991)
OH 7.64E-012	cm3/molc sec	25	EST	MEYLAN,WM & HOWARD,PH (1993)

CAS #:	002631-30-3			N-METHYL-4-ETHYLPHENYLCARBAMATE

Formula: $C_{10}H_{13}NO_2$

Mol Weight: 179.22

MP (deg C): | FP (deg C):

BP (deg C):

BP pressure (mm Hg):

Property/Value	Units	Temp	Data Type	Reference
WS 6.92E+002	mg/L	25	EST	MEYLAN,WM ET AL. (1996)
logP 2.23			EXP	HANSCH,C & LEO,AJ (1985)
VP 6.17E-003	mm Hg	25	EST	NEELY,WB & BLAU,GE (1985)
DC	pKa			
HL 4.71E-008	atm m3/mol	25	EST	MEYLAN,WM & HOWARD,PH (1991)
OH 1.07E-011	cm3/molc sec	25	EST	MEYLAN,WM & HOWARD,PH (1993)

CAS #:	002631-37-0			3-I-PR-5-MEPHENYL-N-ME CARBAMATE

Formula: $C_{12}H_{17}NO_2$

Mol Weight: 207.27

MP (deg C): 87-87.5 | FP (deg C):

BP (deg C):

BP pressure (mm Hg):

Property/Value	Units	Temp	Data Type	Reference
WS 9.20E+001	mg/L	25	EXP	YALKOWSKY,SH & DANNENFELSER,RM (1992)
logP 3.10			EXP	HANSCH,C & LEO,AJ (1985)
VP 3.00E-005	mm Hg	25	EXP	AUGUSTIJN-BECKERS,PWM ET AL. (1994)
DC	pKa			
HL 8.89E-008	atm m3/mol	25	EST	VP/WSOL
OH 2.09E-011	cm3/molc sec	25	EST	MEYLAN,WM & HOWARD,PH (1993)

CAS #:	002631-39-2			N-ME-3-DIMEAMINOPHENYL CARBAMATE

Formula: $C_{10}H_{14}N_2O_2$

Mol Weight: 194.24

MP (deg C): | FP (deg C):

BP (deg C):

BP pressure (mm Hg):

Property/Value	Units	Temp	Data Type	Reference
WS 2.81E+003	mg/L	25	EST	MEYLAN,WM ET AL. (1996)
logP 1.43			EXP	HANSCH,C & LEO,AJ (1985)
VP 2.01E-003	mm Hg	25	EST	NEELY,WB & BLAU,GE (1985)
DC	pKa			
HL 5.11E-010	atm m3/mol	25	EST	MEYLAN,WM & HOWARD,PH (1991)
OH 2.09E-010	cm3/molc sec	25	EST	MEYLAN,WM & HOWARD,PH (1993)

CAS #:	002631-40-5			N-METHYL-2-ISOPROPYLPHENYLCARBAM

Formula: $C_{11}H_{15}NO_2$

Mol Weight: 193.25

MP (deg C): 89-91 | FP (deg C):

BP (deg C): 128-129

BP pressure (mm Hg): 2.00E+001

Property/Value	Units	Temp	Data Type	Reference
WS 4.00E+002	mg/L	25	EXP	CHEM INSPECT TEST INST (1992)
logP 2.31			EXP	HANSCH,C & LEO,AJ (1985)
VP 4.39E-003	mm Hg	25	EST	NEELY,WB & BLAU,GE (1985)
DC	pKa			
HL 6.25E-008	atm m3/mol	25	EST	MEYLAN,WM & HOWARD,PH (1991)
OH 1.17E-011	cm3/molc sec	25	EST	MEYLAN,WM & HOWARD,PH (1993)

CAS #: 002631-42-7	N-METHYL-2-ETHYLPHENYLCARBAMATE

Formula: $C_{10}H_{13}NO_2$

Mol Weight: 179.22

MP (deg C): FP (deg C):

BP (deg C):

BP pressure (mm Hg):

Property/Value	Units	Temp	Data Type	Reference
WS 1.25E+003	mg/L	25	EST	MEYLAN,WM ET AL. (1996)
logP 1.93			EXP	HANSCH,C & LEO,AJ (1985)
VP 6.17E-003	mm Hg	25	EST	NEELY,WB & BLAU,GE (1985)
DC	pKa			
HL 4.71E-008	atm m3/mol	25	EST	MEYLAN,WM & HOWARD,PH (1991)
OH 1.07E-011	cm3/molc sec	25	EST	MEYLAN,WM & HOWARD,PH (1993)

CAS #: 002631-68-7	TRICHLORO-2,4,6-TRINITROBENZENE

Formula: $C_6Cl_3N_3O_6$

Mol Weight: 316.44

MP (deg C): FP (deg C):

BP (deg C):

BP pressure (mm Hg):

Property/Value	Units	Temp	Data Type	Reference
WS 5.21E+000	mg/L	25	EST	MEYLAN,WM ET AL. (1996)
logP 3.38			EST	MEYLAN,WM & HOWARD,PH (1995)
VP 2.22E-007	mm Hg	25	EST	NEELY,WB & BLAU,GE (1985)
DC	pKa			
HL 1.35E-010	atm m3/mol	25	EST	MEYLAN,WM & HOWARD,PH (1991)
OH 1.33E-015	cm3/molc sec	25	EST	MEYLAN,WM & HOWARD,PH (1993)

CAS #: 002636-26-2	CYANOPHOS

Formula: $C_9H_{10}NO_3PS$

Mol Weight: 243.22

MP (deg C): 14-15 FP (deg C):

BP (deg C): 119-120

BP pressure (mm Hg): 9.00E-002

Property/Value	Units	Temp	Data Type	Reference
WS 4.60E+001	mg/L	30	EXP	YALKOWSKY,SH & DANNENFELSER,RM (1992)
logP 2.71			EXP	DEBRUIJN,J & HERMENS,J (1991)
VP 2.36E-004	mm Hg	25	EST	NEELY,WB & BLAU,GE (1985)
DC	pKa			
HL 4.12E-007	atm m3/mol	25	EST	MEYLAN,WM & HOWARD,PH (1991)
OH 5.92E-011	cm3/molc sec	25	EST	MEYLAN,WM & HOWARD,PH (1993)

CAS #: 002637-34-5	2(1H)-PYRIDINETHIONE

Formula: C_5H_5NS

Mol Weight: 111.17

MP (deg C): 128-130 FP (deg C):

BP (deg C):

BP pressure (mm Hg):

Property/Value	Units	Temp	Data Type	Reference
WS 1.39E+005	mg/L	25	EST	MEYLAN,WM ET AL. (1996)
logP -0.13			EXP	HANSCH,C ET AL. (1995)
VP 8.63E-003	mm Hg	25	EST	NEELY,WB & BLAU,GE (1985)
DC	pKa			
HL 1.23E-005	atm m3/mol	25	EST	MEYLAN,WM & HOWARD,PH (1991)
OH 1.48E-010	cm3/molc sec	25	EST	MEYLAN,WM & HOWARD,PH (1993)

CAS #: 002642-71-9	AZINPHOS ETHYL

Formula: $C_{12}H_{16}N_3O_3PS_2$

Mol Weight: 345.38

MP (deg C): 53 FP (deg C):

BP (deg C): 111

BP pressure (mm Hg): 0.00E+000

Property/Value	Units	Temp	Data Type	Reference
WS 1.05E+001	mg/L	20	EXP	BOWMAN,BT & SANS,WW (1983)
logP 3.40			EXP	HANSCH,C & LEO,AJ (1985)
VP 2.20E-007	mm Hg	20	EXP	MARTIN,H & WORTHING,CR (1977)
DC	pKa			
HL 9.52E-009	atm m3/mol	20	EST	VP/WSOL
OH 1.71E-010	cm3/molc sec	25	EST	MEYLAN,WM & HOWARD,PH (1993)

CAS #: 002642-98-0	6-AMINOCHRYSENE

Formula: $C_{18}H_{13}N$

Mol Weight: 243.31

MP (deg C): 210-211 FP (deg C):

BP (deg C):

BP pressure (mm Hg):

Property/Value	Units	Temp	Data Type	Reference
WS 1.55E-005	mg/L	25	EXP	MEANS,JC ET AL. (1982)
logP 4.99			EXP	HANSCH,C & LEO,AJ (1985)
VP 2.61E-008	mm Hg	25	EST	NEELY,WB & BLAU,GE (1985)
DC	pKa			
HL 1.77E-009	atm m3/mol	25	EST	MEYLAN,WM & HOWARD,PH (1991)
OH 2.00E-010	cm3/molc sec	25	EST	MEYLAN,WM & HOWARD,PH (1993)

CAS #: 002644-96-4	P-NITRO-METHYLHIPPURATE

Formula: $C_{10}H_{10}N_2O_5$

Mol Weight: 238.20

MP (deg C): FP (deg C):

BP (deg C):

BP pressure (mm Hg):

Property/Value	Units	Temp	Data Type	Reference
WS 1.63E+003	mg/L	25	EST	MEYLAN,WM ET AL. (1996)
logP 0.98			EXP	HANSCH,C & LEO,AJ (1985)
VP 1.39E-007	mm Hg	25	EST	NEELY,WB & BLAU,GE (1985)
DC	pKa			
HL 9.11E-013	atm m3/mol	25	EST	MEYLAN,WM & HOWARD,PH (1991)
OH 8.71E-012	cm3/molc sec	25	EST	MEYLAN,WM & HOWARD,PH (1993)

CAS #: 002646-26-6	HYDRAZINECARBOXAMIDE, N-(4-BROMOPHENYL)

Formula: $C_7H_8BrN_3O$

Mol Weight: 230.07

MP (deg C): FP (deg C):

BP (deg C):

BP pressure (mm Hg):

Property/Value	Units	Temp	Data Type	Reference
WS 4.26E+003	mg/L	25	EST	MEYLAN,WM ET AL. (1996)
logP 1.00			EXP	KRAMER,CR & BECK,L (1981)
VP 7.82E-006	mm Hg	25	EST	NEELY,WB & BLAU,GE (1985)
DC	pKa			
HL 3.35E-013	atm m3/mol	25	EST	MEYLAN,WM & HOWARD,PH (1991)
OH 1.32E-011	cm3/molc sec	25	EST	MEYLAN,WM & HOWARD,PH (1993)

CAS #: 002648-00-2				1,4-BENZODIAZPIN-2-ONE,5-PH,7-F

Formula: $C_{15}H_{11}FN_2O$

Mol Weight: 254.27

MP (deg C): | FP (deg C):

BP (deg C):

BP pressure (mm Hg):

Property/Value	Units	Temp	Data Type	Reference
WS 2.34E+002	mg/L	25	EST	MEYLAN,WM ET AL. (1996)
logP 2.32			EXP	HANSCH,C & LEO,AJ (1985)
VP 2.84E-008	mm Hg	25	EST	NEELY,WB & BLAU,GE (1985)
DC	pKa			
HL 2.80E-010	atm m3/mol	25	EST	MEYLAN,WM & HOWARD,PH (1991)
OH 1.00E-011	cm3/molc sec	25	EST	MEYLAN,WM & HOWARD,PH (1993)

CAS #: 002648-01-3				1,4-BENZODIAZPIN-2-ONE,5-(2-F PH)

Formula: $C_{15}H_{11}FN_2O$

Mol Weight: 254.27

MP (deg C): | FP (deg C):

BP (deg C):

BP pressure (mm Hg):

Property/Value	Units	Temp	Data Type	Reference
WS 2.08E+002	mg/L	25	EST	MEYLAN,WM ET AL. (1996)
logP 2.38			EXP	HANSCH,C & LEO,AJ (1985)
VP 2.84E-008	mm Hg	25	EST	NEELY,WB & BLAU,GE (1985)
DC	pKa			
HL 2.80E-010	atm m3/mol	25	EST	MEYLAN,WM & HOWARD,PH (1991)
OH 1.85E-011	cm3/molc sec	25	EST	MEYLAN,WM & HOWARD,PH (1993)

CAS #: 002652-13-3				NONASILOXANE, EICOSAMETHYL-

Formula: $C_{20}H_{60}O_8Si_9$

Mol Weight: 681.47

MP (deg C): | FP (deg C):

BP (deg C): 173

BP pressure (mm Hg): 4.90E+000

Property/Value	Units	Temp	Data Type	Reference
WS 1.07E-009	mg/L	25	EST	MEYLAN,WM ET AL. (1996)
logP 8.30			EXP	BRUGGEMAN,WA ET AL. (1984)
VP 8.22E-003	mm Hg	25	EST	NEELY,WB & BLAU,GE (1985)
DC	pKa			
HL 1.48E-013	atm m3/mol	25	EST	MEYLAN,WM & HOWARD,PH (1991)
OH 2.99E-012	cm3/molc sec	25	EST	MEYLAN,WM & HOWARD,PH (1993)

CAS #: 002652-77-9				2,5-PYRAZOLIDINDIONE,1,2-DIPHENYL

Formula: $C_{15}H_{12}N_2O_2$

Mol Weight: 252.28

MP (deg C): | FP (deg C):

BP (deg C):

BP pressure (mm Hg):

Property/Value	Units	Temp	Data Type	Reference
WS 1.33E+003	mg/L	25	EST	MEYLAN,WM ET AL. (1996)
logP 1.45			EXP	HANSCH,C & LEO,AJ (1985)
VP 6.16E-009	mm Hg	25	EST	NEELY,WB & BLAU,GE (1985)
DC	pKa			
HL 2.11E-009	atm m3/mol	25	EST	MEYLAN,WM & HOWARD,PH (1991)
OH 2.53E-011	cm3/molc sec	25	EST	MEYLAN,WM & HOWARD,PH (1993)

CAS #: 002655-12-1				N-ME-2,3-DIMETHYLPHENYL CARBAMATE

Formula: $C_{10}H_{13}NO_2$

Mol Weight: 179.22

MP (deg C): | FP (deg C):

BP (deg C):

BP pressure (mm Hg):

Property/Value	Units	Temp	Data Type	Reference
WS 1.20E+003	mg/L	25	EST	MEYLAN,WM ET AL. (1996)
logP 1.95			EXP	HANSCH,C & LEO,AJ (1985)
VP 5.37E-003	mm Hg	25	EST	NEELY,WB & BLAU,GE (1985)
DC	pKa			
HL 3.91E-008	atm m3/mol	25	EST	MEYLAN,WM & HOWARD,PH (1991)
OH 1.28E-011	cm3/molc sec	25	EST	MEYLAN,WM & HOWARD,PH (1993)

CAS #: 002655-14-3				N-ME-3,5-DIMETHYLPHENYL CARBAMATE

Formula: $C_{10}H_{13}NO_2$

Mol Weight: 179.22

MP (deg C): 99 | FP (deg C):

BP (deg C):

BP pressure (mm Hg):

Property/Value	Units	Temp	Data Type	Reference
WS 1.35E+002	mg/L	20	EST	MEYLAN,WM ET AL (1996)
logP 2.23			EXP	HANSCH,C & LEO,AJ (1985)
VP 3.60E-004	mm Hg	25	EXP	WATANABE,T (1993)
DC	pKa			
HL 3.91E-008	atm m3/mol	25	EST	MEYLAN,WM & HOWARD,PH (1991)
OH 6.85E-011	cm3/molc sec	25	EST	ATKINSON,R (1988)

CAS #: 002655-19-8				3,5-BIS(TERT-BUTYL)PHENOL METHYLCARBAMATE

Formula: $C_{16}H_{25}NO_2$

Mol Weight: 263.38

MP (deg C): | FP (deg C):

BP (deg C):

BP pressure (mm Hg):

Property/Value	Units	Temp	Data Type	Reference
WS 1.50E+001	mg/L	20	EXP	YALKOWSKY,SH & DANNENFELSER,RM (1992)
logP 4.99			EST	MEYLAN,WM & HOWARD,PH (1995)
VP 8.88E-005	mm Hg	25	EST	NEELY,WB & BLAU,GE (1985)
DC	pKa			
HL 2.14E-007	atm m3/mol	25	EST	MEYLAN,WM & HOWARD,PH (1991)
OH 1.68E-011	cm3/molc sec	25	EST	MEYLAN,WM & HOWARD,PH (1993)

CAS #: 002658-24-4				2,2-DIMETHYLETHYLENIMINE

Formula: C_4H_9N

Mol Weight: 71.12

MP (deg C): | FP (deg C):

BP (deg C):

BP pressure (mm Hg):

Property/Value	Units	Temp	Data Type	Reference
WS 1.00E+006	mg/L	25	EST	MEYLAN,WM ET AL. (1996)
logP 0.14			EST	MEYLAN,WM & HOWARD,PH (1995)
VP 9.51E+001	mm Hg	25	EST	NEELY,WB & BLAU,GE (1985)
DC 8.64	pKa	25	EXP	PERRIN,DD (1965)
HL 1.29E-005	atm m3/mol	25	EST	MEYLAN,WM & HOWARD,PH (1991)
OH 6.55E-012	cm3/molc sec	25	EST	MEYLAN,WM & HOWARD,PH (1993)

CAS #: 002664-63-3 — PHENOL, 4,4'-THIOBIS-

Formula: $C_{12}H_{10}O_2S$

Mol Weight: 218.28

MP (deg C): 154-156

FP (deg C):

BP (deg C):

BP pressure (mm Hg):

Property/Value	Units	Temp	Data Type	Reference
WS 1.88E+002	mg/L	25	EST	MEYLAN,WM ET AL. (1996)
logP 3.34			EXP	HANSCH,C ET AL. (1995)
VP 1.85E-007	mm Hg	25	EST	NEELY,WB & BLAU,GE (1985)
DC	pKa			
HL 3.40E-013	atm m3/mol	25	EST	MEYLAN,WM & HOWARD,PH (1991)
OH 4.12E-011	cm3/molc sec	25	EST	MEYLAN,WM & HOWARD,PH (1993)

CAS #: 002668-24-8 — 4,5,6-TRICHLOROGUAIACOL

Formula: $C_7H_5Cl_3O_2$

Mol Weight: 227.48

MP (deg C):

FP (deg C):

BP (deg C):

BP pressure (mm Hg):

Property/Value	Units	Temp	Data Type	Reference
WS 2.09E+001	mg/L	25	EST	MEYLAN,WM ET AL. (1996)
logP 3.72			EXP	XIE,TM ET AL. (1984)
VP 1.87E-003	mm Hg	25	EXP	BIDLEMAN,TF & RENBERG,L (1985)
DC	pKa			
HL 1.35E-008	atm m3/mol	25	EST	MEYLAN,WM & HOWARD,PH (1991)
OH 3.11E-012	cm3/molc sec	25	EST	MEYLAN,WM & HOWARD,PH (1993)

CAS #: 002670-77-1 — OO-DIET O-2-METHYLTHIOPH P=S

Formula: $C_{12}H_{19}O_3PS_2$

Mol Weight: 306.39

MP (deg C):

FP (deg C):

BP (deg C):

BP pressure (mm Hg):

Property/Value	Units	Temp	Data Type	Reference
WS 3.16E+000	mg/L	25	EST	MEYLAN,WM ET AL. (1996)
logP 4.16			EXP	HANSCH,C & LEO,AJ (1985)
VP 3.32E-005	mm Hg	25	EST	NEELY,WB & BLAU,GE (1985)
DC	pKa			
HL 2.42E-006	atm m3/mol	25	EST	MEYLAN,WM & HOWARD,PH (1991)
OH 1.17E-010	cm3/molc sec	25	EST	MEYLAN,WM & HOWARD,PH (1993)

CAS #: 002675-77-6 — 1,4-DICHLORO-2,5-DIMETHOXYBENZENE

Formula: $C_8H_8Cl_2O_2$

Mol Weight: 207.06

MP (deg C): 134

FP (deg C):

BP (deg C): 268

BP pressure (mm Hg):

Property/Value	Units	Temp	Data Type	Reference
WS 8.00E+000	mg/L	25	EXP	YALKOWSKY,SH & DANNENFELSER,RM (1992)
logP 3.44			EST	MEYLAN,WM & HOWARD,PH (1995)
VP 1.42E-002	mm Hg	25	EST	NEELY,WB & BLAU,GE (1985)
DC	pKa			
HL 1.04E-005	atm m3/mol	25	EST	MEYLAN,WM & HOWARD,PH (1991)
OH 5.48E-012	cm3/molc sec	25	EST	MEYLAN,WM & HOWARD,PH (1993)

CAS #: 002683-90-1 — 8-AZAHYPOXANTHINE

Formula: $C_4H_3N_5O$

Mol Weight: 137.10

MP (deg C):

FP (deg C):

BP (deg C):

BP pressure (mm Hg):

Property/Value	Units	Temp	Data Type	Reference
WS 2.03E+005	mg/L	25	EST	MEYLAN,WM ET AL. (1996)
logP -1.97			EXP	HANSCH,C & LEO,AJ (1985)
VP 6.77E-009	mm Hg	25	EST	NEELY,WB & BLAU,GE (1985)
DC	pKa			
HL 1.02E-013	atm m3/mol	25	EST	MEYLAN,WM & HOWARD,PH (1991)
OH 1.36E-011	cm3/molc sec	25	EST	MEYLAN,WM & HOWARD,PH (1993)

CAS #: 002686-99-9 — N-ME-3,4,5-TRIMEPHENYL CARBAMATE

Formula: $C_{11}H_{15}NO_2$

Mol Weight: 193.25

MP (deg C):

FP (deg C):

BP (deg C):

BP pressure (mm Hg):

Property/Value	Units	Temp	Data Type	Reference
WS 2.53E+002	mg/L	25	EST	MEYLAN,WM ET AL. (1996)
logP 2.66			EXP	HANSCH,C & LEO,AJ (1985)
VP 1.90E-003	mm Hg	25	EST	NEELY,WB & BLAU,GE (1985)
DC	pKa			
HL 4.32E-008	atm m3/mol	25	EST	MEYLAN,WM & HOWARD,PH (1991)
OH 2.33E-011	cm3/molc sec	25	EST	MEYLAN,WM & HOWARD,PH (1993)

CAS #: 002687-25-4 — 2,3-DIAMINOTOLUENE

Formula: $C_7H_{10}N_2$

Mol Weight: 122.17

MP (deg C): 63-64

FP (deg C):

BP (deg C): 255

BP pressure (mm Hg):

Property/Value	Units	Temp	Data Type	Reference
WS 2.46E+004	mg/L	25	EST	MEYLAN,WM ET AL. (1996)
logP 0.71			EST	MEYLAN,WM & HOWARD,PH (1995)
VP 5.53E-004	mm Hg	25	EXT	MILLIGAN,B & GILBERT,KE (1978)
DC	pKa			
HL 7.43E-009	atm m3/mol	25	EST	MEYLAN,WM & HOWARD,PH (1991)
OH 2.40E-010	cm3/molc sec	25	EST	MEYLAN,WM & HOWARD,PH (1993)

CAS #: 002687-91-4 — 2-PYRROLIDINONE, 1-ETHYL-

Formula: $C_6H_{11}NO$

Mol Weight: 113.16

MP (deg C):

FP (deg C):

BP (deg C): 97

BP pressure (mm Hg): 2.00E+001

Property/Value	Units	Temp	Data Type	Reference
WS 1.15E+005	mg/L	25	EST	MEYLAN,WM ET AL. (1996)
logP -0.04			EXP	KIM,KH ET AL. (1993)
VP 1.03E-001	mm Hg	25	EST	NEELY,WB & BLAU,GE (1985)
DC	pKa			
HL 4.19E-008	atm m3/mol	25	EST	MEYLAN,WM & HOWARD,PH (1991)
OH 2.70E-011	cm3/molc sec	25	EST	MEYLAN,WM & HOWARD,PH (1993)

002687-96-9 — 1-LAURYL-2-PYRROLIDONE

Formula: $C_{16}H_{31}NO$
Mol Weight: 253.43
MP (deg C): —   FP (deg C):
BP (deg C): 202-205
BP pressure (mm Hg): 1.10E+001

Property/Value	Units	Temp	Data Type	Reference
WS 5.86E+000	mg/L	25	EST	MEYLAN,WM ET AL. (1996)
logP 4.20			EXP	SASAKI,H ET AL. (1988)
VP 5.34E-006	mm Hg	25	EST	NEELY,WB & BLAU,GE (1985)
DC	pKa			
HL 7.12E-007	atm m3/mol	25	EST	MEYLAN,WM & HOWARD,PH (1991)
OH 4.18E-011	cm3/molc sec	25	EST	MEYLAN,WM & HOWARD,PH (1993)

002688-84-8 — O-PHENOXYANILINE

Formula: $C_{12}H_{11}NO$
Mol Weight: 185.23
MP (deg C): 47-49   FP (deg C):
BP (deg C): 170
BP pressure (mm Hg): 1.80E+001

Property/Value	Units	Temp	Data Type	Reference
WS 4.12E+002	mg/L	25	EST	MEYLAN,WM ET AL. (1996)
logP 2.46			EXP	HANSCH,C & LEO,AJ (1985)
VP 1.25E-004	mm Hg	25	EST	NEELY,WB & BLAU,GE (1985)
DC	pKa			
HL 4.16E-008	atm m3/mol	25	EST	MEYLAN,WM & HOWARD,PH (1991)
OH 1.13E-010	cm3/molc sec	25	EST	MEYLAN,WM & HOWARD,PH (1993)

002689-47-6 — 3-ACETYLPHENYLDIMETHYLCARBAMATE

Formula: $C_{11}H_{13}NO_3$
Mol Weight: 207.23
MP (deg C): —   FP (deg C):
BP (deg C):
BP pressure (mm Hg):

Property/Value	Units	Temp	Data Type	Reference
WS 3.95E+003	mg/L	25	EST	MEYLAN,WM ET AL. (1996)
logP 1.18			EXP	HANSCH,C & LEO,AJ (1985)
VP 6.80E-004	mm Hg	25	EST	NEELY,WB & BLAU,GE (1985)
DC	pKa			
HL 1.28E-010	atm m3/mol	25	EST	MEYLAN,WM & HOWARD,PH (1991)
OH 1.75E-011	cm3/molc sec	25	EST	MEYLAN,WM & HOWARD,PH (1993)

002691-41-0 — OCTAHYDRO-1,3,5,7-TETRANITRO-1,3,5,7-TETRAZOC*

Formula: $C_4H_8N_8O_8$
Mol Weight: 296.16
MP (deg C): —   FP (deg C):
BP (deg C):
BP pressure (mm Hg):

Property/Value	Units	Temp	Data Type	Reference
WS 1.40E+002	mg/L	83	EXP	RYON,MG ET AL. (1984)
logP -5.18			EST	MEYLAN,WM & HOWARD,PH (1995)
VP 2.41E-008	mm Hg	25	EST	NEELY,WB & BLAU,GE (1985)
DC	pKa			
HL 8.67E-010	atm m3/mol	25	EST	MEYLAN,WM & HOWARD,PH (1991)
OH 3.28E-010	cm3/molc sec	25	EST	MEYLAN,WM & HOWARD,PH (1993)

002693-46-1 — 3-FLUORANTHENAMINE

Formula: $C_{16}H_{11}N$
Mol Weight: 217.27
MP (deg C): 115-117   FP (deg C):
BP (deg C):
BP pressure (mm Hg):

Property/Value	Units	Temp	Data Type	Reference
WS 7.15E-001	mg/L	25	EST	MEYLAN,WM ET AL. (1996)
logP 4.20			EXP	DEBNATH,AK ET AL. (1992)
VP 1.59E-007	mm Hg	25	EST	NEELY,WB & BLAU,GE (1985)
DC	pKa			
HL 2.93E-009	atm m3/mol	25	EST	MEYLAN,WM & HOWARD,PH (1991)
OH 2.02E-010	cm3/molc sec	25	EST	MEYLAN,WM & HOWARD,PH (1993)

002696-84-6 — BENZENAMINE, 4-PROPYL-

Formula: $C_9H_{13}N$
Mol Weight: 135.21
MP (deg C): —   FP (deg C):
BP (deg C): 224-226
BP pressure (mm Hg):

Property/Value	Units	Temp	Data Type	Reference
WS 7.82E+002	mg/L	25	EST	MEYLAN,WM ET AL. (1996)
logP 2.40			EXP	SANGSTER,J (1993)
VP 3.37E-002	mm Hg	25	EST	NEELY,WB & BLAU,GE (1985)
DC	pKa			
HL 3.70E-006	atm m3/mol	25	EST	MEYLAN,WM & HOWARD,PH (1991)
OH 1.34E-010	cm3/molc sec	25	EST	MEYLAN,WM & HOWARD,PH (1993)

002700-22-3 — MALONONITRILE, BENZAL

Formula: $C_{10}H_6N_2$
Mol Weight: 154.17
MP (deg C): 83-85   FP (deg C):
BP (deg C):
BP pressure (mm Hg):

Property/Value	Units	Temp	Data Type	Reference
WS 5.43E+002	mg/L	25	EST	MEYLAN,WM ET AL. (1996)
logP 2.18			EXP	HANSCH,C & LEO,AJ (1985)
VP 1.84E-004	mm Hg	25	EST	NEELY,WB & BLAU,GE (1985)
DC	pKa			
HL 1.37E-008	atm m3/mol	25	EST	MEYLAN,WM & HOWARD,PH (1991)
OH 4.06E-012	cm3/molc sec	25	EST	MEYLAN,WM & HOWARD,PH (1993)

002702-58-1 — BENZOIC ACID, 3,5-DINITRO-, METHYL ESTER

Formula: $C_8H_6N_2O_6$
Mol Weight: 226.15
MP (deg C): 107-109   FP (deg C):
BP (deg C):
BP pressure (mm Hg):

Property/Value	Units	Temp	Data Type	Reference
WS 3.49E+002	mg/L	25	EST	MEYLAN,WM ET AL. (1996)
logP 1.84			EXP	PARKIN,JE (1986)
VP 1.91E-005	mm Hg	25	EST	NEELY,WB & BLAU,GE (1985)
DC	pKa			
HL 5.40E-010	atm m3/mol	25	EST	MEYLAN,WM & HOWARD,PH (1991)
OH 2.24E-013	cm3/molc sec	25	EST	MEYLAN,WM & HOWARD,PH (1993)

CAS #: 002702-72-9	2,4-D, SODIUM SALT

Formula: $C_8H_5Cl_2NaO_3$
Mol Weight: 243.02
MP (deg C): FP (deg C):
BP (deg C):
BP pressure (mm Hg):

Property/Value	Units	Temp	Data Type	Reference
WS 3.35E+005	mg/L	20	EXP	BOSCH,SJ (1983)
logP -1.19			EST	MEYLAN,WM & HOWARD,PH (1995)
VP 4.36E-011	mm Hg	25	EST	NEELY,WB & BLAU,GE (1985)
DC	pKa			
HL	atm m3/mol			
OH 6.05E-012	cm3/molc sec	25	EST	MEYLAN,WM & HOWARD,PH (1993)

CAS #: 002706-22-1	1-ACETYL-4-(4-BR PHENYL) SEMICARBAZIDE

Formula: $C_9H_{10}BrN_3O_2$
Mol Weight: 272.11
MP (deg C): FP (deg C):
BP (deg C):
BP pressure (mm Hg):

Property/Value	Units	Temp	Data Type	Reference
WS 1.18E+003	mg/L	25	EST	MEYLAN,WM ET AL. (1996)
logP 1.38			EXP	STEIN,J ET AL. (1986)
VP 1.37E-008	mm Hg	25	EST	NEELY,WB & BLAU,GE (1985)
DC	pKa			
HL 1.09E-015	atm m3/mol	25	EST	MEYLAN,WM & HOWARD,PH (1991)
OH 1.88E-011	cm3/molc sec	25	EST	MEYLAN,WM & HOWARD,PH (1993)

CAS #: 002706-56-1	3-PYRIDINE ETHANEAMINE

Formula: $C_7H_{10}N_2$
Mol Weight: 122.17
MP (deg C): FP (deg C):
BP (deg C): 131.5
BP pressure (mm Hg): 5.00E+001

Property/Value	Units	Temp	Data Type	Reference
WS 1.00E+006	mg/L	25	EST	MEYLAN,WM ET AL. (1996)
logP 0.08			EXP	HANSCH,C & LEO,AJ (1985)
VP 1.48E-001	mm Hg	25	EST	NEELY,WB & BLAU,GE (1985)
DC 10.03	pKa	10	EXP	PERRIN,DD (1965)
HL 1.06E-009	atm m3/mol	25	EST	MEYLAN,WM & HOWARD,PH (1991)
OH 3.38E-011	cm3/molc sec	25	EST	MEYLAN,WM & HOWARD,PH (1993)

CAS #: 002709-56-0	FLUPENTIXOL

Formula: $C_{23}H_{25}F_3N_2OS$
Mol Weight: 434.53
MP (deg C): FP (deg C):
BP (deg C):
BP pressure (mm Hg):

Property/Value	Units	Temp	Data Type	Reference
WS 3.46E-001	mg/L	25	EST	MEYLAN,WM ET AL. (1996)
logP 4.51			EXP	HANSCH,C & LEO,AJ (1985)
VP 1.05E-012	mm Hg	25	EST	NEELY,WB & BLAU,GE (1985)
DC	pKa			
HL 4.13E-016	atm m3/mol	25	EST	MEYLAN,WM & HOWARD,PH (1991)
OH 1.71E-010	cm3/molc sec	25	EST	MEYLAN,WM & HOWARD,PH (1993)

CAS #: 002715-70-0	9-BUTYL ADENINE

Formula: $C_9H_{15}N_5$
Mol Weight: 193.25
MP (deg C): FP (deg C):
BP (deg C):
BP pressure (mm Hg):

Property/Value	Units	Temp	Data Type	Reference
WS 4.15E+003	mg/L	25	EST	MEYLAN,WM ET AL. (1996)
logP 1.25			EXP	HANSCH,C & LEO,AJ (1985)
VP 4.48E-007	mm Hg	25	EST	NEELY,WB & BLAU,GE (1985)
DC	pKa			
HL 3.50E-012	atm m3/mol	25	EST	MEYLAN,WM & HOWARD,PH (1991)
OH 2.04E-010	cm3/molc sec	25	EST	MEYLAN,WM & HOWARD,PH (1993)

CAS #: 002719-08-6	BENZOIC ACID, 2-(ACETYLAMINO)-, METHYL ESTER

Formula: $C_{10}H_{11}NO_3$
Mol Weight: 193.20
MP (deg C): FP (deg C):
BP (deg C):
BP pressure (mm Hg):

Property/Value	Units	Temp	Data Type	Reference
WS 1.85E+003	mg/L	25	EST	MEYLAN,WM ET AL. (1996)
logP 1.65			EXP	HANSCH,C ET AL. (1995)
VP 9.60E-006	mm Hg	25	EST	NEELY,WB & BLAU,GE (1985)
DC	pKa			
HL 3.98E-011	atm m3/mol	25	EST	MEYLAN,WM & HOWARD,PH (1991)
OH 4.31E-012	cm3/molc sec	25	EST	MEYLAN,WM & HOWARD,PH (1993)

CAS #: 002719-21-3	ACETAMIDE, N-(4-ACETYLPHENYL)-

Formula: $C_{10}H_{11}NO_2$
Mol Weight: 177.20
MP (deg C): FP (deg C):
BP (deg C):
BP pressure (mm Hg):

Property/Value	Units	Temp	Data Type	Reference
WS 4.96E+003	mg/L	25	EST	MEYLAN,WM ET AL. (1996)
logP 1.24			EXP	HANSCH,C ET AL. (1995)
VP 1.15E-005	mm Hg	25	EST	NEELY,WB & BLAU,GE (1985)
DC	pKa			
HL 1.12E-011	atm m3/mol	25	EST	MEYLAN,WM & HOWARD,PH (1991)
OH 1.15E-011	cm3/molc sec	25	EST	MEYLAN,WM & HOWARD,PH (1993)

CAS #: 002719-23-5	ACETAMIDE, N-2-THIAZOLYL-

Formula: $C_5H_8N_2OS$
Mol Weight: 144.20
MP (deg C): FP (deg C):
BP (deg C):
BP pressure (mm Hg):

Property/Value	Units	Temp	Data Type	Reference
WS 1.84E+004	mg/L	25	EST	MEYLAN,WM ET AL. (1996)
logP 0.76			EXP	HANSCH,C ET AL. (1995)
VP 1.35E-004	mm Hg	25	EST	NEELY,WB & BLAU,GE (1985)
DC	pKa			
HL 4.39E-012	atm m3/mol	25	EST	MEYLAN,WM & HOWARD,PH (1991)
OH 9.02E-012	cm3/molc sec	25	EST	MEYLAN,WM & HOWARD,PH (1993)

CAS #: 002719-61-1 — BENZENE, (1-METHYLUNDECYL)-

Formula: $C_{18}H_{30}$
Mol Weight: 246.44
MP (deg C):   FP (deg C):
BP (deg C):
BP pressure (mm Hg):

Property/Value	Units	Temp	Data Type	Reference
WS 2.51E-003	mg/L	25	EST	MEYLAN,WM ET AL. (1996)
logP 8.19			EXP	SHERBLOM,PM ET AL. (1992)
VP 5.84E-004	mm Hg	25	EST	NEELY,WB & BLAU,GE (1985)
DC	pKa			
HL 1.34E-001	atm m3/mol	25	EST	MEYLAN,WM & HOWARD,PH (1991)
OH 1.98E-011	cm3/molc sec	25	EST	MEYLAN,WM & HOWARD,PH (1993)

CAS #: 002719-62-2 — BENZENE, (1-PENTYLHEPTYL)-

Formula: $C_{18}H_{30}$
Mol Weight: 246.44
MP (deg C):   FP (deg C):
BP (deg C):
BP pressure (mm Hg):

Property/Value	Units	Temp	Data Type	Reference
WS 3.18E-003	mg/L	25	EST	MEYLAN,WM ET AL. (1996)
logP 8.01			EXP	SHERBLOM,PM ET AL. (1992)
VP 5.84E-004	mm Hg	25	EST	NEELY,WB & BLAU,GE (1985)
DC	pKa			
HL 1.34E-001	atm m3/mol	25	EST	MEYLAN,WM & HOWARD,PH (1991)
OH 2.01E-011	cm3/molc sec	25	EST	MEYLAN,WM & HOWARD,PH (1993)

CAS #: 002719-63-3 — BENZENE, (1-BUTYLOCTYL)-

Formula: $C_{18}H_{30}$
Mol Weight: 246.44
MP (deg C):   FP (deg C):
BP (deg C):
BP pressure (mm Hg):

Property/Value	Units	Temp	Data Type	Reference
WS 3.18E-003	mg/L	25	EST	MEYLAN,WM ET AL. (1996)
logP 8.01			EXP	SHERBLOM,PM ET AL. (1992)
VP 5.84E-004	mm Hg	25	EST	NEELY,WB & BLAU,GE (1985)
DC	pKa			
HL 1.34E-001	atm m3/mol	25	EST	MEYLAN,WM & HOWARD,PH (1991)
OH 2.01E-011	cm3/molc sec	25	EST	MEYLAN,WM & HOWARD,PH (1993)

CAS #: 002719-64-4 — (1-PROPYLNONYL)BENZENE

Formula: $C_{18}H_{30}$
Mol Weight: 246.44
MP (deg C):   FP (deg C):
BP (deg C):
BP pressure (mm Hg):

Property/Value	Units	Temp	Data Type	Reference
WS 3.18E-003	mg/L	25	EST	MEYLAN,WM ET AL. (1996)
logP 8.07			EXP	SHERBLOM,PM ET AL. (1992)
VP 5.84E-004	mm Hg	25	EST	NEELY,WB & BLAU,GE (1985)
DC	pKa			
HL 1.34E-001	atm m3/mol	25	EST	MEYLAN,WM & HOWARD,PH (1991)
OH 2.01E-011	cm3/molc sec	25	EST	MEYLAN,WM & HOWARD,PH (1993)

CAS #: 002724-69-8 — 1-METHYL-3-PHENYLTHIOUREA

Formula: $C_8H_{10}N_2S$
Mol Weight: 166.25
MP (deg C):   FP (deg C):
BP (deg C):
BP pressure (mm Hg):

Property/Value	Units	Temp	Data Type	Reference
WS 1.23E+005	mg/L	25	EST	MEYLAN,WM ET AL. (1996)
logP 0.85			EXP	HANSCH,C ET AL. (1995)
VP 1.66E-003	mm Hg	25	EST	NEELY,WB & BLAU,GE (1985)
DC	pKa			
HL 1.91E-007	atm m3/mol	25	EST	MEYLAN,WM & HOWARD,PH (1991)
OH 1.07E-010	cm3/molc sec	25	EST	MEYLAN,WM & HOWARD,PH (1993)

CAS #: 002728-04-3 — BENZAMIDE, N,N-DIETHYL-2-METHYL-

Formula: $C_{12}H_{17}NO$
Mol Weight: 191.28
MP (deg C):   FP (deg C):
BP (deg C):
BP pressure (mm Hg):

Property/Value	Units	Temp	Data Type	Reference
WS 9.12E+002	mg/L	25	EST	MEYLAN,WM ET AL. (1996)
logP 2.02			EXP	SANGSTER,J (1994)
VP 3.32E-004	mm Hg	25	EST	NEELY,WB & BLAU,GE (1985)
DC	pKa			
HL 2.08E-008	atm m3/mol	25	EST	MEYLAN,WM & HOWARD,PH (1991)
OH 2.78E-011	cm3/molc sec	25	EST	MEYLAN,WM & HOWARD,PH (1993)

CAS #: 002728-05-4 — BENZAMIDE, N,N-DIETHYL-4-METHYL-

Formula: $C_{12}H_{17}NO$
Mol Weight: 191.28
MP (deg C):   FP (deg C):
BP (deg C):
BP pressure (mm Hg):

Property/Value	Units	Temp	Data Type	Reference
WS 4.49E+002	mg/L	25	EST	MEYLAN,WM ET AL. (1996)
logP 2.38			EXP	SURYANARAYANA,MVS ET AL. (1991)
VP 2.44E-001	mm Hg	25	EXP	SURYANARAYANA,MVS ET AL. (1991)
DC	pKa			
HL 2.08E-008	atm m3/mol	25	EST	MEYLAN,WM & HOWARD,PH (1991)
OH 2.78E-011	cm3/molc sec	25	EST	MEYLAN,WM & HOWARD,PH (1993)

CAS #: 002730-05-4 — 1,4-BENZODIAZPIN-2-ONE,5-(2CF3PH)

Formula: $C_{16}H_{11}F_3N_2O$
Mol Weight: 304.27
MP (deg C):   FP (deg C):
BP (deg C):
BP pressure (mm Hg):

Property/Value	Units	Temp	Data Type	Reference
WS 9.01E+001	mg/L	25	EST	MEYLAN,WM ET AL. (1996)
logP 2.47			EXP	HANSCH,C & LEO,AJ (1985)
VP 1.87E-008	mm Hg	25	EST	NEELY,WB & BLAU,GE (1985)
DC	pKa			
HL 2.09E-009	atm m3/mol	25	EST	MEYLAN,WM & HOWARD,PH (1991)
OH 1.65E-011	cm3/molc sec	25	EST	MEYLAN,WM & HOWARD,PH (1993)

CAS #: 002730-67-8				N-ME ETHYLAMINE,2,2,2-TRIFLUORO
Formula: $C_3H_6F_3N$				
Mol Weight: 113.08				
MP (deg C):		FP (deg C):		
BP (deg C):				
BP pressure (mm Hg):				

Property/ Value	Units	Temp	Data Type	Reference
WS 3.67E+005	mg/L	25	EST	MEYLAN,WM ET AL. (1996)
logP 0.55			EXP	HANSCH,C & LEO,AJ (1985)
VP 7.89E+002	mm Hg	25	EST	NEELY,WB & BLAU,GE (1985)
DC	pKa			
HL 1.74E-004	atm m3/mol	25	EST	MEYLAN,WM & HOWARD,PH (1991)
OH 1.88E-012	cm3/molc sec	25	EST	MEYLAN,WM & HOWARD,PH (1993)

CAS #: 002734-52-3				C.I. DISPERSE RED 19
Formula: $C_{16}H_{18}N_4O_4$				
Mol Weight: 330.35				
MP (deg C):		FP (deg C):		
BP (deg C):				
BP pressure (mm Hg):				

Property/ Value	Units	Temp	Data Type	Reference
WS 3.87E+000	mg/L	60	EXP	YALKOWSKY,SH & DANNENFELSER,RM (1992)
logP 3.14			EST	MEYLAN,WM & HOWARD,PH (1995)
VP 5.61E-013	mm Hg	25	EST	NEELY,WB & BLAU,GE (1985)
DC	pKa			
HL 2.17E-015	atm m3/mol	25	EST	MEYLAN,WM & HOWARD,PH (1991)
OH 8.78E-011	cm3/molc sec	25	EST	MEYLAN,WM & HOWARD,PH (1993)

CAS #: 002738-19-4				2-METHYL-2-HEXENE
Formula: C_7H_{14}				
Mol Weight: 98.19				
MP (deg C): -130.4		FP (deg C):		
BP (deg C): 95.4				
BP pressure (mm Hg):				

Property/ Value	Units	Temp	Data Type	Reference
WS 2.78E+001	mg/L	25	EST	MEYLAN,WM ET AL. (1996)
logP 3.62			EST	MEYLAN,WM & HOWARD,PH (1995)
VP 5.40E+001	mm Hg	25	EXT	DYKYJ,J & REPAS,M (1973)
DC	pKa			
HL 6.63E-001	atm m3/mol	25	EST	MEYLAN,WM & HOWARD,PH (1991)
OH 8.96E-011	cm3/molc sec	25	EST	MEYLAN,WM & HOWARD,PH (1993)

CAS #: 002739-57-3				2-DECENE-4,6,8-TRIYNOIC ACID, METHYL ESTER, (Z)-
Formula: $C_{11}H_8O_2$				
Mol Weight: 172.19				
MP (deg C):		FP (deg C):		
BP (deg C):				
BP pressure (mm Hg):				

Property/ Value	Units	Temp	Data Type	Reference
WS 5.15E+002	mg/L	25	EST	MEYLAN,WM ET AL. (1996)
logP 2.42			EXP	MCLACHLAN,D ET AL. (1986)
VP 1.68E-003	mm Hg	25	EST	NEELY,WB & BLAU,GE (1985)
DC	pKa			
HL 1.22E-005	atm m3/mol	25	EST	MEYLAN,WM & HOWARD,PH (1991)
OH 1.37E-010	cm3/molc sec	25	EST	MEYLAN,WM & HOWARD,PH (1993)

CAS #: 002740-94-5				1-METHYL-3-BENZYL THIOUREA
Formula: $C_9H_{12}N_2S$				
Mol Weight: 180.27				
MP (deg C): 74-76		FP (deg C):		
BP (deg C):				
BP pressure (mm Hg):				

Property/ Value	Units	Temp	Data Type	Reference
WS 3.93E+004	mg/L	25	EST	MEYLAN,WM ET AL. (1996)
logP 1.35			EXP	HANSCH,C & LEO,AJ (1985)
VP 5.94E-004	mm Hg	25	EST	NEELY,WB & BLAU,GE (1985)
DC	pKa			
HL 6.15E-008	atm m3/mol	25	EST	MEYLAN,WM & HOWARD,PH (1991)
OH 1.41E-010	cm3/molc sec	25	EST	MEYLAN,WM & HOWARD,PH (1993)

CAS #: 002741-06-2				1-PHENYL-3-ETHYL THIOUREA
Formula: $C_9H_{12}N_2S$				
Mol Weight: 180.27				
MP (deg C):		FP (deg C):		
BP (deg C):				
BP pressure (mm Hg):				

Property/ Value	Units	Temp	Data Type	Reference
WS 3.43E+004	mg/L	25	EST	MEYLAN,WM ET AL. (1996)
logP 1.42			EXP	HANSCH,C & LEO,AJ (1985)
VP 5.94E-004	mm Hg	25	EST	NEELY,WB & BLAU,GE (1985)
DC	pKa			
HL 2.54E-007	atm m3/mol	25	EST	MEYLAN,WM & HOWARD,PH (1991)
OH 1.15E-010	cm3/molc sec	25	EST	MEYLAN,WM & HOWARD,PH (1993)

CAS #: 002741-14-2				THIOUREA, N'-(PHENYLMETHYL)-N,N-DIMETHYL
Formula: $C_{10}H_{14}N_2S$				
Mol Weight: 194.30				
MP (deg C):		FP (deg C):		
BP (deg C):				
BP pressure (mm Hg):				

Property/ Value	Units	Temp	Data Type	Reference
WS 2.09E+004	mg/L	25	EST	MEYLAN,WM ET AL. (1996)
logP 1.59			EXP	HANSCH,C ET AL. (1995)
VP 5.79E-004	mm Hg	25	EST	NEELY,WB & BLAU,GE (1985)
DC	pKa			
HL 1.35E-007	atm m3/mol	25	EST	MEYLAN,WM & HOWARD,PH (1991)
OH 1.45E-010	cm3/molc sec	25	EST	MEYLAN,WM & HOWARD,PH (1993)

CAS #: 002745-25-7				2-CYANOMETHYLFURAN
Formula: C_6H_6NO				
Mol Weight: 108.12				
MP (deg C):		FP (deg C):		
BP (deg C): 75-80				
BP pressure (mm Hg): 2.00E+001				

Property/ Value	Units	Temp	Data Type	Reference
WS 1.14E+004	mg/L	25	EST	MEYLAN,WM ET AL. (1996)
logP 0.85			EXP	HANSCH,C & LEO,AJ (1985)
VP 4.33E-001	mm Hg	25	EST	NEELY,WB & BLAU,GE (1985)
DC	pKa			
HL 2.46E-006	atm m3/mol	25	EST	MEYLAN,WM & HOWARD,PH (1991)
OH 3.94E-011	cm3/molc sec	25	EST	MEYLAN,WM & HOWARD,PH (1993)

CAS #: 002752-68-3 — 2,4-DICL-5-ME-N-ME PHENYLCARBAMATE

Formula: $C_9H_9Cl_2NO_2$

Mol Weight: 234.08

MP (deg C): FP (deg C):

BP (deg C):

BP pressure (mm Hg):

Property/Value	Units	Temp	Data Type	Reference
WS 7.93E+001	mg/L	25	EST	MEYLAN,WM ET AL. (1996)
logP 3.00			EXP	HANSCH,C & LEO,AJ (1985)
VP 4.54E-004	mm Hg	25	EST	NEELY,WB & BLAU,GE (1985)
DC	pKa			
HL 1.95E-008	atm m3/mol	25	EST	MEYLAN,WM & HOWARD,PH (1991)
OH 7.06E-012	cm3/molc sec	25	EST	MEYLAN,WM & HOWARD,PH (1993)

CAS #: 002757-10-0 — N-METHYLCINNAMAMIDE

Formula: $C_{10}H_{11}NO$

Mol Weight: 161.21

MP (deg C): FP (deg C):

BP (deg C):

BP pressure (mm Hg):

Property/Value	Units	Temp	Data Type	Reference
WS 1.92E+003	mg/L	25	EST	MEYLAN,WM ET AL. (1996)
logP 1.81			EXP	HANSCH,C & LEO,AJ (1985)
VP 6.28E-005	mm Hg	25	EST	NEELY,WB & BLAU,GE (1985)
DC	pKa			
HL 5.80E-010	atm m3/mol	25	EST	MEYLAN,WM & HOWARD,PH (1991)
OH 2.76E-011	cm3/molc sec	25	EST	MEYLAN,WM & HOWARD,PH (1993)

CAS #: 002759-71-9 — CYPROMIDE

Formula: $C_{10}H_9Cl_2NO$

Mol Weight: 230.10

MP (deg C): FP (deg C):

BP (deg C):

BP pressure (mm Hg):

Property/Value	Units	Temp	Data Type	Reference
WS 1.63E+001	mg/L	25	EST	MEYLAN,WM ET AL. (1996)
logP 3.83			EXP	HANSCH,C ET AL. (1995)
VP 3.69E-006	mm Hg	25	EST	NEELY,WB & BLAU,GE (1985)
DC	pKa			
HL 2.63E-009	atm m3/mol	25	EST	MEYLAN,WM & HOWARD,PH (1991)
OH 2.77E-012	cm3/molc sec	25	EST	MEYLAN,WM & HOWARD,PH (1993)

CAS #: 002763-96-4 — MUSCIMOL

Formula: $C_4H_6N_2O_2$

Mol Weight: 114.10

MP (deg C): 175 FP (deg C):

BP (deg C):

BP pressure (mm Hg):

Property/Value	Units	Temp	Data Type	Reference
WS 5.67E+005	mg/L	25	EST	MEYLAN,WM ET AL. (1996)
logP -2.39			EXP	HANSCH,C & LEO,AJ (1985)
VP 2.94E-004	mm Hg	25	EST	NEELY,WB & BLAU,GE (1985)
DC	pKa			
HL 1.61E-012	atm m3/mol	25	EST	MEYLAN,WM & HOWARD,PH (1991)
OH 6.56E-011	cm3/molc sec	25	EST	MEYLAN,WM & HOWARD,PH (1993)

CAS #: 002764-72-9 — DIQUAT

Formula: $C_{12}H_{14}N_2$

Mol Weight: 186.26

MP (deg C): FP (deg C):

BP (deg C):

BP pressure (mm Hg):

Property/Value	Units	Temp	Data Type	Reference
WS 7.00E+001	mg/L	20	EXP	GUNTHER,FA ET AL. (1968)
logP 2.36			EST	MEYLAN,WM & HOWARD,PH (1995)
VP 1.00E-008	mm Hg	25	EXP	WEBER,JB (1994)
DC	pKa			
HL 3.50E-011	atm m3/mol	25	EST	VP/WSOL
OH 3.58E-012	cm3/molc sec	25	EST	MEYLAN,WM & HOWARD,PH (1993)

CAS #: 002777-05-1 — CHLOROAZOBENZENE-4'-SULFONIC ACID, NA SALT

Formula: $C_{12}H_8ClN_2NaO_2S$

Mol Weight: 302.72

MP (deg C): FP (deg C):

BP (deg C):

BP pressure (mm Hg):

Property/Value	Units	Temp	Data Type	Reference
WS 7.51E+003	mg/L	25	EST	MEYLAN,WM ET AL. (1996)
logP -0.27			EST	MEYLAN,WM & HOWARD,PH (1995)
VP 6.76E-015	mm Hg	25	EST	NEELY,WB & BLAU,GE (1985)
DC	pKa			
HL	atm m3/mol			
OH 7.09E-013	cm3/molc sec	25	EST	MEYLAN,WM & HOWARD,PH (1993)

CAS #: 002778-04-3 — ENDOTHION

Formula: $C_9H_{13}O_6PS$

Mol Weight: 280.24

MP (deg C): FP (deg C):

BP (deg C):

BP pressure (mm Hg):

Property/Value	Units	Temp	Data Type	Reference
WS 1.50E+006	mg/L		EXP	YALKOWSKY,SH & DANNENFELSER,RM (1992)
logP -1.22			EST	MEYLAN,WM & HOWARD,PH (1995)
VP 5.69E-006	mm Hg	25	EST	NEELY,WB & BLAU,GE (1985)
DC	pKa			
HL 6.46E-012	atm m3/mol	25	EST	MEYLAN,WM & HOWARD,PH (1991)
OH 9.55E-011	cm3/molc sec	25	EST	MEYLAN,WM & HOWARD,PH (1993)

CAS #: 002778-41-8 — 1,4-BIS(1-ISOCYANATO-1-METHYLETHYL)BENZENE

Formula: $C_{14}H_{16}N_2O_2$

Mol Weight: 244.30

MP (deg C): FP (deg C):

BP (deg C):

BP pressure (mm Hg):

Property/Value	Units	Temp	Data Type	Reference
WS 2.29E+000	mg/L	25	EST	MEYLAN,WM ET AL. (1996)
logP 4.74			EST	MEYLAN,WM & HOWARD,PH (1995)
VP 7.58E-004	mm Hg	25	EXT	ACHORN,PJ ET AL. (1986)
DC	pKa			
HL 3.22E-006	atm m3/mol	25	EST	MEYLAN,WM & HOWARD,PH (1991)
OH 5.82E-012	cm3/molc sec	25	EST	MEYLAN,WM & HOWARD,PH (1993)

CAS #: 002778-42-9	1,3-BIS(1-ISOCYANATO-1-METHYLETHYL)BENZENE

Formula: $C_{14}H_{16}N_2O_2$

Mol Weight: 244.30

MP (deg C): | FP (deg C):

BP (deg C): 106

BP pressure (mm Hg): 9.00E-001

Property/ Value	Units	Temp	Data Type	Reference
WS 2.29E+000	mg/L	25	EST	MEYLAN,WM ET AL. (1996)
logP 4.74			EST	MEYLAN,WM & HOWARD,PH (1995)
VP 3.20E-003	mm Hg	25	EXP	ACHORN,PJ ET AL. (1986)
DC	pKa			
HL 3.22E-006	atm m3/mol	25	EST	MEYLAN,WM & HOWARD,PH (1991)
OH 1.01E-011	cm3/molc sec	25	EST	MEYLAN,WM & HOWARD,PH (1993)

CAS #: 002782-91-4	TETRAMETHYLTHIOUREA

Formula: $C_5H_{12}N_2S$

Mol Weight: 132.23

MP (deg C): 78-79 | FP (deg C):

BP (deg C): 245

BP pressure (mm Hg):

Property/ Value	Units	Temp	Data Type	Reference
WS 5.40E+003	mg/L	25	EST	MEYLAN,WM ET AL. (1996)
logP 0.49			EXP	GOVERS,H ET AL. (1986)
VP 1.45E+000	mm Hg	25	EST	NEELY,WB & BLAU,GE (1985)
DC	pKa			
HL 1.16E-008	atm m3/mol	25	EST	MEYLAN,WM & HOWARD,PH (1991)
OH 1.26E-010	cm3/molc sec	25	EST	ATKINSON,R (1988)

CAS #: 002784-27-2	HYDANTOIN,5-PH-5-(P-HYDROXY)PH

Formula: $C_{15}H_{12}N_2O_3$

Mol Weight: 268.27

MP (deg C): | FP (deg C):

BP (deg C):

BP pressure (mm Hg):

Property/ Value	Units	Temp	Data Type	Reference
WS 2.51E+003	mg/L	25	EST	MEYLAN,WM ET AL. (1996)
logP 1.70			EXP	HANSCH,C & LEO,AJ (1985)
VP 1.01E-012	mm Hg	25	EST	NEELY,WB & BLAU,GE (1985)
DC	pKa			
HL 1.06E-015	atm m3/mol	25	EST	MEYLAN,WM & HOWARD,PH (1991)
OH 4.64E-011	cm3/molc sec	25	EST	MEYLAN,WM & HOWARD,PH (1993)

CAS #: 002786-62-1	3-METHYLBENZOTHIAZOL-2-ONE

Formula: C_8H_7NOS

Mol Weight: 165.22

MP (deg C): | FP (deg C):

BP (deg C):

BP pressure (mm Hg):

Property/ Value	Units	Temp	Data Type	Reference
WS 1.32E+003	mg/L	25	EST	MEYLAN,WM ET AL. (1996)
logP 1.98			EXP	HANSCH,C ET AL. (1995)
VP 1.38E-004	mm Hg	25	EST	NEELY,WB & BLAU,GE (1985)
DC	pKa			
HL 7.94E-006	atm m3/mol	25	EST	MEYLAN,WM & HOWARD,PH (1991)
OH 7.26E-012	cm3/molc sec	25	EST	MEYLAN,WM & HOWARD,PH (1993)

CAS #: 002790-16-1	CYCLOPROPANECARBOXAMIDE, N-(3,4-DICHLOROPHENYL)-

Formula: $C_{11}H_{11}Cl_2NO$

Mol Weight: 244.12

MP (deg C): | FP (deg C):

BP (deg C):

BP pressure (mm Hg):

Property/ Value	Units	Temp	Data Type	Reference
WS 2.03E+001	mg/L	25	EST	MEYLAN,WM ET AL. (1996)
logP 3.63			EXP	HANSCH,C ET AL. (1995)
VP 2.13E-006	mm Hg	25	EST	NEELY,WB & BLAU,GE (1985)
DC	pKa			
HL 3.50E-009	atm m3/mol	25	EST	MEYLAN,WM & HOWARD,PH (1991)
OH 3.26E-012	cm3/molc sec	25	EST	MEYLAN,WM & HOWARD,PH (1993)

CAS #: 002797-51-5	2-AMINO-3-CHLORO-1,4-NAPHTHOQUINONE

Formula: $C_{10}H_6ClNO_2$

Mol Weight: 207.62

MP (deg C): 198-200 | FP (deg C):

BP (deg C):

BP pressure (mm Hg):

Property/ Value	Units	Temp	Data Type	Reference
WS 6.30E+003	mg/L	25	EST	MEYLAN,WM ET AL. (1996)
logP 2.12			EXP	HANSCH,C & LEO,AJ (1985)
VP 1.13E-005	mm Hg	25	EST	NEELY,WB & BLAU,GE (1985)
DC	pKa			
HL 6.08E-012	atm m3/mol	25	EST	MEYLAN,WM & HOWARD,PH (1991)
OH 2.34E-011	cm3/molc sec	25	EST	MEYLAN,WM & HOWARD,PH (1993)

CAS #: 002798-54-1	N-ME-3-ME-6-T-BU PHENYLCARBAMATE

Formula: $C_{13}H_{19}NO_2$

Mol Weight: 221.30

MP (deg C): | FP (deg C):

BP (deg C):

BP pressure (mm Hg):

Property/ Value	Units	Temp	Data Type	Reference
WS 7.05E+001	mg/L	25	EST	MEYLAN,WM ET AL. (1996)
logP 3.14			EXP	HANSCH,C & LEO,AJ (1985)
VP 6.74E-004	mm Hg	25	EST	NEELY,WB & BLAU,GE (1985)
DC	pKa			
HL 9.15E-008	atm m3/mol	25	EST	MEYLAN,WM & HOWARD,PH (1991)
OH 1.30E-011	cm3/molc sec	25	EST	MEYLAN,WM & HOWARD,PH (1993)

CAS #: 002799-07-7	L-CYSTEINE, S-(TRIPHENYLMETHYL)-

Formula: $C_{22}H_{21}NO_2S$

Mol Weight: 363.48

MP (deg C): 182-183 de | FP (deg C):

BP (deg C):

BP pressure (mm Hg):

Property/ Value	Units	Temp	Data Type	Reference
WS 7.61E-001	mg/L	25	EST	MEYLAN,WM ET AL. (1996)
logP 2.06			EXP	HANSCH,C ET AL. (1995)
VP 4.87E-014	mm Hg	25	EST	NEELY,WB & BLAU,GE (1985)
DC	pKa			
HL 8.38E-015	atm m3/mol	25	EST	MEYLAN,WM & HOWARD,PH (1991)
OH 6.17E-011	cm3/molc sec	25	EST	MEYLAN,WM & HOWARD,PH (1993)

CAS #: 002801-68-5				2,5-DIMETHOXYAMPHETAMINE

Formula: $C_{11}H_{17}NO_2$

Mol Weight: 195.26

MP (deg C): | FP (deg C):

BP (deg C):

BP pressure (mm Hg):

Property/Value	Units	Temp	Data Type	Reference
WS 1.60E+004	mg/L	25	EST	MEYLAN,WM ET AL. (1996)
logP 1.72			EXP	HANSCH,C & LEO,AJ (1985)
VP 1.83E-003	mm Hg	25	EST	NEELY,WB & BLAU,GE (1985)
DC	pKa			
HL 3.77E-009	atm m3/mol	25	EST	MEYLAN,WM & HOWARD,PH (1991)
OH 9.24E-011	cm3/molc sec	25	EST	MEYLAN,WM & HOWARD,PH (1993)

CAS #: 002807-30-9				ETHYLENE GLYCOL MONOPROPYL ETHER

Formula: $C_5H_{12}O_2$

Mol Weight: 104.15

MP (deg C): | FP (deg C):

BP (deg C): 151

BP pressure (mm Hg):

Property/Value	Units	Temp	Data Type	Reference
WS 3.17E+005	mg/L	25	EST	MEYLAN,WM ET AL. (1996)
logP 0.08			EST	MEYLAN,WM & HOWARD,PH (1995)
VP 3.12E+000	mm Hg	25	EXP	YAWS,CL (1994A)
DC	pKa			
HL 1.50E-008	atm m3/mol	25	EST	MEYLAN,WM & HOWARD,PH (1991)
OH 2.16E-011	cm3/molc sec	25	EST	MEYLAN,WM & HOWARD,PH (1993)

CAS #: 002809-67-8				2-OCTYNE

Formula: C_8H_{14}

Mol Weight: 110.20

MP (deg C): -79.3 | FP (deg C):

BP (deg C): 126.3

BP pressure (mm Hg):

Property/Value	Units	Temp	Data Type	Reference
WS 2.91E+001	mg/L	25	EST	MEYLAN,WM ET AL. (1996)
logP 3.55			EST	MEYLAN,WM & HOWARD,PH (1995)
VP 7.50E+000	mm Hg	25	EXT	BOUBLIK,T ET AL. (1984)
DC	pKa			
HL 3.34E-002	atm m3/mol	25	EST	MEYLAN,WM & HOWARD,PH (1991)
OH 3.24E-011	cm3/molc sec	25	EST	MEYLAN,WM & HOWARD,PH (1993)

CAS #: 002809-69-0				2,4-HEXADIYNE

Formula: C_6H_6

Mol Weight: 78.11

MP (deg C): 67.8 | FP (deg C):

BP (deg C): 129.5

BP pressure (mm Hg):

Property/Value	Units	Temp	Data Type	Reference
WS 4.68E+002	mg/L	25	EST	MEYLAN,WM ET AL. (1996)
logP 2.24			EXP	HANSCH,C ET AL. (1995)
VP 3.25E+001	mm Hg	25	EST	NEELY,WB & BLAU,GE (1985)
DC	pKa			
HL 8.54E-003	atm m3/mol	25	EST	MEYLAN,WM & HOWARD,PH (1991)
OH 7.05E-011	cm3/molc sec	25	EST	MEYLAN,WM & HOWARD,PH (1993)

CAS #: 002810-04-0				THIOPHENE, 2-ETHYL CARBOXYLATE

Formula: $C_7H_{10}O_2S$

Mol Weight: 158.22

MP (deg C): | FP (deg C):

BP (deg C): 218

BP pressure (mm Hg):

Property/Value	Units	Temp	Data Type	Reference
WS 7.29E+002	mg/L	25	EST	MEYLAN,WM ET AL. (1996)
logP 2.33			EXP	HANSCH,C & LEO,AJ (1985)
VP 8.41E-002	mm Hg	25	EST	NEELY,WB & BLAU,GE (1985)
DC	pKa			
HL 2.50E-005	atm m3/mol	25	EST	MEYLAN,WM & HOWARD,PH (1991)
OH 4.74E-012	cm3/molc sec	25	EST	MEYLAN,WM & HOWARD,PH (1993)

CAS #: 002813-95-8				DINOSEB ACETATE

Formula: $C_{12}H_{14}N_2O_6$

Mol Weight: 282.26

MP (deg C): | FP (deg C):

BP (deg C):

BP pressure (mm Hg):

Property/Value	Units	Temp	Data Type	Reference
WS 2.20E+003	mg/L	25	EXP	YALKOWSKY,SH & DANNENFELSER,RM (1992)
logP 3.17			EST	MEYLAN,WM & HOWARD,PH (1995)
VP 6.00E-004	mm Hg	20	EXP	PLIMMER,JR (1976)
DC	pKa			
HL 1.01E-007	atm m3/mol	20	EST	VP/WSOL
OH 4.00E-012	cm3/molc sec	25	EST	MEYLAN,WM & HOWARD,PH (1993)

CAS #: 002816-24-2				beta-D-GLUCOPYRANOSIDE, 2-NITROPHENYL

Formula: $C_{12}H_{15}NO_8$

Mol Weight: 301.26

MP (deg C): | FP (deg C):

BP (deg C):

BP pressure (mm Hg):

Property/Value	Units	Temp	Data Type	Reference
WS 2.28E+004	mg/L	25	EST	MEYLAN,WM ET AL. (1996)
logP -0.78			EXP	SANGSTER,J (1994)
VP 3.68E-013	mm Hg	25	EST	NEELY,WB & BLAU,GE (1985)
DC	pKa			
HL 4.54E-018	atm m3/mol	25	EST	MEYLAN,WM & HOWARD,PH (1991)
OH 6.47E-011	cm3/molc sec	25	EST	MEYLAN,WM & HOWARD,PH (1993)

CAS #: 002818-58-8				BETA-D-GALACTOPYRANOSIDE, PHENYL

Formula: $C_{12}H_{16}O_6$

Mol Weight: 256.26

MP (deg C): 153-155 | FP (deg C):

BP (deg C):

BP pressure (mm Hg):

Property/Value	Units	Temp	Data Type	Reference
WS 1.26E+005	mg/L	25	EST	MEYLAN,WM ET AL. (1996)
logP -0.89			EXP	SANGSTER,J (1994)
VP 7.66E-011	mm Hg	25	EST	NEELY,WB & BLAU,GE (1985)
DC	pKa			
HL 1.15E-015	atm m3/mol	25	EST	MEYLAN,WM & HOWARD,PH (1991)
OH 8.35E-011	cm3/molc sec	25	EST	MEYLAN,WM & HOWARD,PH (1993)

CAS #: 002820-55-5 — PYRIDINE, 3-(1-METHYL-2-PYRROLIDINYL)-, N-OXIDE,

Formula: $C_{10}H_{16}N_2O$

Mol Weight: 180.25

MP (deg C): FP (deg C):

BP (deg C):

BP pressure (mm Hg):

Property/Value	Units	Temp	Data Type	Reference
WS 1.00E+006	mg/L	25	EST	MEYLAN,WM ET AL. (1996)
logP -1.80			EXP	LI,NY & GORROD,JW (1992)
VP 1.17E-004	mm Hg	25	EST	NEELY,WB & BLAU,GE (1985)
DC	pKa			
HL 3.00E-014	atm m3/mol	25	EST	MEYLAN,WM & HOWARD,PH (1991)
OH 9.67E-011	cm3/molc sec	25	EST	MEYLAN,WM & HOWARD,PH (1993)

CAS #: 002826-25-7 — MALONONITRILE, 4-METHYLBENZAL

Formula: $C_{11}H_8N_2$

Mol Weight: 168.20

MP (deg C): FP (deg C):

BP (deg C):

BP pressure (mm Hg):

Property/Value	Units	Temp	Data Type	Reference
WS 3.04E+002	mg/L	25	EST	MEYLAN,WM ET AL. (1996)
logP 2.40			EXP	HANSCH,C & LEO,AJ (1985)
VP 6.50E-005	mm Hg	25	EST	NEELY,WB & BLAU,GE (1985)
DC	pKa			
HL 1.51E-008	atm m3/mol	25	EST	MEYLAN,WM & HOWARD,PH (1991)
OH 7.15E-012	cm3/molc sec	25	EST	MEYLAN,WM & HOWARD,PH (1993)

CAS #: 002827-47-6 — TETRAMETHYLMELAMINE

Formula: $C_7H_{14}N_6$

Mol Weight: 182.23

MP (deg C): FP (deg C):

BP (deg C):

BP pressure (mm Hg):

Property/Value	Units	Temp	Data Type	Reference
WS 5.07E+003	mg/L	25	EST	MEYLAN,WM ET AL. (1996)
logP 1.20			EXP	HANSCH,C & LEO,AJ (1985)
VP 1.35E-004	mm Hg	25	EST	NEELY,WB & BLAU,GE (1985)
DC	pKa			
HL 3.83E-010	atm m3/mol	25	EST	MEYLAN,WM & HOWARD,PH (1991)
OH 5.72E-012	cm3/molc sec	25	EST	MEYLAN,WM & HOWARD,PH (1993)

CAS #: 002832-40-8 — 4'-((6-HYDROXY-M-TOLYL)AZO)ACETANILIDE

Formula: $C_{15}H_{15}N_3O_2$

Mol Weight: 269.31

MP (deg C): 195 FP (deg C):

BP (deg C):

BP pressure (mm Hg):

Property/Value	Units	Temp	Data Type	Reference
WS 1.18E+000	mg/L	25	EXP	BAUGHMAN,GL & PERENICH,TA (1988)
logP 3.98			EST	MEYLAN,WM & HOWARD,PH (1995)
VP 5.00E-011	mm Hg	25	EXP	BAUGHMAN,GL & PERENICH,TA (1988)
DC	pKa			
HL 1.50E-011	atm m3/mol	25	EST	VP/WSOL
OH 3.45E-011	cm3/molc sec	25	EST	MEYLAN,WM & HOWARD,PH (1993)

CAS #: 002835-06-5 — BENZENEACETIC ACID, .ALPHA.-AMINO-, (.+-.)-

Formula: $C_8H_9NO_2$

Mol Weight: 151.17

MP (deg C): 290 sub FP (deg C):

BP (deg C):

BP pressure (mm Hg):

Property/Value	Units	Temp	Data Type	Reference
WS 1.15E+005	mg/L	100	EXP	YALKOWSKY,SH & DANNENFELSER,RM (1992)
logP -2.07			EXP	HANSCH,C ET AL. (1995)
VP 1.18E-007	mm Hg	25	EST	NEELY,WB & BLAU,GE (1985)
DC	pKa			
HL 9.08E-011	atm m3/mol	25	EST	MEYLAN,WM & HOWARD,PH (1991)
OH 3.97E-011	cm3/molc sec	25	EST	MEYLAN,WM & HOWARD,PH (1993)

CAS #: 002835-68-9 — 4-AMINOBENZAMIDE

Formula: $C_7H_8N_2O$

Mol Weight: 136.15

MP (deg C): 181-183 FP (deg C):

BP (deg C):

BP pressure (mm Hg):

Property/Value	Units	Temp	Data Type	Reference
WS 9.53E+003	mg/L	25	EST	MEYLAN,WM ET AL. (1996)
logP -0.41			EXP	HANSCH,C & LEO,AJ (1985)
VP 2.63E-005	mm Hg	25	EST	NEELY,WB & BLAU,GE (1985)
DC	pKa			
HL 7.83E-013	atm m3/mol	25	EST	MEYLAN,WM & HOWARD,PH (1991)
OH 1.00E-010	cm3/molc sec	25	EST	MEYLAN,WM & HOWARD,PH (1993)

CAS #: 002836-32-0 — GLYCOLIC ACID, SODIUM SALT

Formula: $C_2H_3NaO_3$

Mol Weight: 98.03

MP (deg C): FP (deg C):

BP (deg C):

BP pressure (mm Hg):

Property/Value	Units	Temp	Data Type	Reference
WS 1.00E+006	mg/L	25	EST	MEYLAN,WM ET AL. (1996)
logP -5.19			EST	MEYLAN,WM & HOWARD,PH (1995)
VP 4.58E-010	mm Hg	25	EST	NEELY,WB & BLAU,GE (1985)
DC	pKa			
HL	atm m3/mol			
OH 1.15E-012	cm3/molc sec	25	EST	MEYLAN,WM & HOWARD,PH (1993)

CAS #: 002845-79-6 — 1-BENZOYL-2-CARBAMYLHYDRAZINE

Formula: $C_8H_9N_3O_2$

Mol Weight: 179.18

MP (deg C): FP (deg C):

BP (deg C):

BP pressure (mm Hg):

Property/Value	Units	Temp	Data Type	Reference
WS 7.28E+003	mg/L	25	EST	MEYLAN,WM ET AL. (1996)
logP -0.50			EXP	HANSCH,C & LEO,AJ (1985)
VP 1.98E-007	mm Hg	25	EST	NEELY,WB & BLAU,GE (1985)
DC	pKa			
HL 9.78E-016	atm m3/mol	25	EST	MEYLAN,WM & HOWARD,PH (1991)
OH 9.28E-012	cm3/molc sec	25	EST	MEYLAN,WM & HOWARD,PH (1993)

CAS #: 002845-89-8				M-CHLOROANISOLE

Formula: C_7H_7ClO
Mol Weight: 142.59
MP (deg C):   **FP (deg C):**
BP (deg C): 193.5
BP pressure (mm Hg):

Property/ Value	Units	Temp	Data Type	Reference
WS 2.33E+002	mg/L	25	EST	MEYLAN,WM ET AL. (1996)
logP 2.98			EXP	HANSCH,C ET AL. (1995)
VP 8.11E-001	mm Hg	25	EST	NEELY,WB & BLAU,GE (1985)
DC	pKa			
HL 2.36E-004	atm m3/mol	25	EST	MEYLAN,WM & HOWARD,PH (1991)
OH 1.60E-011	cm3/molc sec	25	EST	MEYLAN,WM & HOWARD,PH (1993)

CAS #: 002847-30-5				PYRAZINE, 2-METHOXY-3-METHYL-

Formula: $C_6H_8N_2O$
Mol Weight: 124.14
MP (deg C):   **FP (deg C):**
BP (deg C):
BP pressure (mm Hg):

Property/ Value	Units	Temp	Data Type	Reference
WS 8.46E+003	mg/L	25	EST	MEYLAN,WM ET AL. (1996)
logP 1.24			EXP	YAMAGAMI,C ET AL. (1991)
VP 5.84E-001	mm Hg	25	EST	NEELY,WB & BLAU,GE (1985)
DC	pKa			
HL 1.74E-006	atm m3/mol	25	EST	MEYLAN,WM & HOWARD,PH (1991)
OH 4.08E-012	cm3/molc sec	25	EST	MEYLAN,WM & HOWARD,PH (1993)

CAS #: 002852-07-5				1,1,2-TRICHLORO-1,3-BUTADIENE

Formula: $C_4H_3Cl_3$
Mol Weight: 157.43
MP (deg C):   **FP (deg C):**
BP (deg C):
BP pressure (mm Hg):

Property/ Value	Units	Temp	Data Type	Reference
WS 9.22E+001	mg/L	25	EST	MEYLAN,WM ET AL. (1996)
logP 3.38			EST	MEYLAN,WM & HOWARD,PH (1995)
VP 1.57E+001	mm Hg	25	EST	NEELY,WB & BLAU,GE (1985)
DC	pKa			
HL 2.90E-002	atm m3/mol	25	EST	MEYLAN,WM & HOWARD,PH (1991)
OH 1.76E-012	cm3/molc sec	25	EST	MEYLAN,WM & HOWARD,PH (1993)

CAS #: 002854-98-0				CHLOROTHIAZIDE-3-AMYL

Formula: $C_{12}H_{16}ClN_3O_4S_2$
Mol Weight: 365.86
MP (deg C):   **FP (deg C):**
BP (deg C):
BP pressure (mm Hg):

Property/ Value	Units	Temp	Data Type	Reference
WS 8.48E+000	mg/L	25	EST	MEYLAN,WM ET AL. (1996)
logP 2.01			EXP	HANSCH,C ET AL. (1995)
VP 3.33E-013	mm Hg	25	EST	NEELY,WB & BLAU,GE (1985)
DC	pKa			
HL 1.67E-011	atm m3/mol	25	EST	MEYLAN,WM & HOWARD,PH (1991)
OH 5.59E-012	cm3/molc sec	25	EST	MEYLAN,WM & HOWARD,PH (1993)

CAS #: 002854-99-1				3-PROPYLCHLOROTHIAZIDE

Formula: $C_{10}H_{12}ClN_3O_4S_2$
Mol Weight: 337.81
MP (deg C):   **FP (deg C):**
BP (deg C):
BP pressure (mm Hg):

Property/ Value	Units	Temp	Data Type	Reference
WS 6.04E+001	mg/L	25	EST	MEYLAN,WM ET AL. (1996)
logP 1.21			EXP	SANGSTER,J (1994)
VP 1.81E-012	mm Hg	25	EST	NEELY,WB & BLAU,GE (1985)
DC	pKa			
HL 9.47E-012	atm m3/mol	25	EST	MEYLAN,WM & HOWARD,PH (1991)
OH 2.77E-012	cm3/molc sec	25	EST	MEYLAN,WM & HOWARD,PH (1993)

CAS #: 002859-67-8				3-PYRIDINEPROPANOL

Formula: $C_8H_{11}NO$
Mol Weight: 137.18
MP (deg C):   **FP (deg C):**
BP (deg C): 284
BP pressure (mm Hg):

Property/ Value	Units	Temp	Data Type	Reference
WS 1.00E+006	mg/L	25	EST	MEYLAN,WM ET AL. (1996)
logP 0.60			EXP	HANSCH,C & LEO,AJ (1985)
VP 1.73E-003	mm Hg	25	EST	NEELY,WB & BLAU,GE (1985)
DC 5.47	pKa	25	EXP	PERRIN,DD (1972)
HL 5.01E-010	atm m3/mol	25	EST	MEYLAN,WM & HOWARD,PH (1991)
OH 7.64E-012	cm3/molc sec	25	EST	MEYLAN,WM & HOWARD,PH (1993)

CAS #: 002859-68-9				2-PYRIDINEPROPANOL

Formula: $C_8H_{11}NO$
Mol Weight: 137.18
MP (deg C): 34   **FP (deg C):**
BP (deg C): 260.2
BP pressure (mm Hg):

Property/ Value	Units	Temp	Data Type	Reference
WS 1.00E+006	mg/L	25	EST	MEYLAN,WM ET AL. (1996)
logP 0.58			EXP	HANSCH,C & LEO,AJ (1985)
VP 1.73E-003	mm Hg	25	EST	NEELY,WB & BLAU,GE (1985)
DC 5.61	pKa	25	EXP	PERRIN,DD (1972)
HL 5.01E-010	atm m3/mol	25	EST	MEYLAN,WM & HOWARD,PH (1991)
OH 7.64E-012	cm3/molc sec	25	EST	MEYLAN,WM & HOWARD,PH (1993)

CAS #: 002866-43-5				DISPERSE BRIGHTENER

Formula: $C_{18}H_{10}N_2O_2S$
Mol Weight: 318.36
MP (deg C): 219   **FP (deg C):**
BP (deg C):
BP pressure (mm Hg):

Property/ Value	Units	Temp	Data Type	Reference
WS 9.54E-003	mg/L	25	EXP	BAUGHMAN,GL & PERENICH,TA (1988)
logP 4.79			EST	MEYLAN,WM & HOWARD,PH (1995)
VP 2.58E-010	mm Hg	25	EST	NEELY,WB & BLAU,GE (1985)
DC	pKa			
HL 2.81E-013	atm m3/mol	25	EST	MEYLAN,WM & HOWARD,PH (1991)
OH 4.31E-011	cm3/molc sec	25	EST	MEYLAN,WM & HOWARD,PH (1993)

CAS #: 002867-47-2	2-(DIMETHYLAMINO)ETHYL 2-METHYL-2-PROPENOATE

Formula: $C_8H_{15}NO_2$

Mol Weight: 157.21

MP (deg C): | FP (deg C):

BP (deg C): 62-65

BP pressure (mm Hg): 6.00E+000

Property/ Value	Units	Temp	Data Type	Reference
WS 1.06E+005	mg/L	25	EST	MEYLAN,WM ET AL. (1996)
logP 0.97			EST	MEYLAN,WM & HOWARD,PH (1995)
VP 8.29E-001	mm Hg	25	EST	NEELY,WB & BLAU,GE (1985)
DC	pKa			
HL 9.54E-008	atm m3/mol	25	EST	MEYLAN,WM & HOWARD,PH (1991)
OH 9.92E-011	cm3/molc sec	25	EST	MEYLAN,WM & HOWARD,PH (1993)

CAS #: 002870-04-4	1,3-DIMETHYL-2-ETHYLBENZENE

Formula: $C_{10}H_{14}$

Mol Weight: 134.22

MP (deg C): -16.2 | FP (deg C):

BP (deg C): 190

BP pressure (mm Hg):

Property/ Value	Units	Temp	Data Type	Reference
WS 1.96E+001	mg/L	25	EST	MEYLAN,WM ET AL. (1996)
logP 4.28			EXP	SHERBLOM,PM & EGANHOUSE,RP (1988)
VP 7.39E-001	mm Hg	25	EXT	CHAO,J ET AL. (1983)
DC	pKa			
HL 9.61E-003	atm m3/mol	25	EST	MEYLAN,WM & HOWARD,PH (1991)
OH 1.76E-011	cm3/molc sec	25	EST	MEYLAN,WM & HOWARD,PH (1993)

CAS #: 002870-71-5	ATROPINE METHYL BROMIDE

Formula: $C_{18}H_{26}BrNO_3$

Mol Weight: 384.32

MP (deg C): | FP (deg C):

BP (deg C):

BP pressure (mm Hg):

Property/ Value	Units	Temp	Data Type	Reference
WS 5.00E+005	mg/L	20	EXP	STEPHEN,H & STEPHEN,T (1963)
logP -1.61			EST	MEYLAN,WM & HOWARD,PH (1995)
VP 1.20E-015	mm Hg	25	EST	NEELY,WB & BLAU,GE (1985)
DC	pKa			
HL 2.48E-020	atm m3/mol	25	EST	MEYLAN,WM & HOWARD,PH (1991)
OH 6.50E-011	cm3/molc sec	25	EST	MEYLAN,WM & HOWARD,PH (1993)

CAS #: 002872-48-2	DISPERSE RED 11

Formula: $C_{15}H_{12}N_2O_3$

Mol Weight: 268.27

MP (deg C): 242 | FP (deg C):

BP (deg C):

BP pressure (mm Hg):

Property/ Value	Units	Temp	Data Type	Reference
WS 1.50E+000	mg/L	25	EXP	BAUGHMAN,GL & PERENICH,TA (1988)
logP 3.50			EXP	BAUGHMAN,GL & WEBER,EJ (1991)
VP 1.95E-009	mm Hg	25	EST	NEELY,WB & BLAU,GE (1985)
DC	pKa			
HL 2.35E-017	atm m3/mol	25	EST	MEYLAN,WM & HOWARD,PH (1991)
OH 2.01E-010	cm3/molc sec	25	EST	MEYLAN,WM & HOWARD,PH (1993)

CAS #: 002872-52-8	DISPERSE RED 1

Formula: $C_{16}H_{18}N_4O_3$

Mol Weight: 314.35

MP (deg C): 160-162 | FP (deg C):

BP (deg C):

BP pressure (mm Hg):

Property/ Value	Units	Temp	Data Type	Reference
WS 1.98E-001	mg/L	25	EXP	BAUGHMAN,GL & PERENICH,TA (1988)
logP 4.30			EXP	BAUGHMAN,GL & WEBER,EJ (1991)
VP 4.00E-014	mm Hg	25	EXP	BAUGHMAN,GL & PERENICH,TA (1988)
DC	pKa			
HL 8.36E-014	atm m3/mol	25	EST	VP/WSOL
OH 8.18E-011	cm3/molc sec	25	EST	MEYLAN,WM & HOWARD,PH (1993)

CAS #: 002873-74-7	GLUTARYL CHLORIDE

Formula: $C_5H_6Cl_2O_2$

Mol Weight: 169.01

MP (deg C): | FP (deg C):

BP (deg C): 217

BP pressure (mm Hg):

Property/ Value	Units	Temp	Data Type	Reference
WS 2.97E+005	mg/L	25	EST	MEYLAN,WM ET AL. (1996)
logP -0.79			EST	MEYLAN,WM & HOWARD,PH (1995)
VP 1.17E-001	mm Hg	25	EXT	PERRY,RH & GREEN,D (1984)
DC	pKa			
HL 1.06E-005	atm m3/mol	25	EST	MEYLAN,WM & HOWARD,PH (1991)
OH 3.80E-012	cm3/molc sec	25	EST	MEYLAN,WM & HOWARD,PH (1993)

CAS #: 002882-20-4	PYRAZINE, 2-METHYL-3-(METHYLTHIO)-

Formula: $C_6H_8N_2S$

Mol Weight: 140.21

MP (deg C): | FP (deg C):

BP (deg C):

BP pressure (mm Hg):

Property/ Value	Units	Temp	Data Type	Reference
WS 2.38E+003	mg/L	25	EST	MEYLAN,WM ET AL. (1996)
logP 1.81			EXP	YAMAGAMI,C ET AL. (1991)
VP 6.78E-002	mm Hg	25	EST	NEELY,WB & BLAU,GE (1985)
DC	pKa			
HL 3.65E-006	atm m3/mol	25	EST	MEYLAN,WM & HOWARD,PH (1991)
OH 3.01E-012	cm3/molc sec	25	EST	MEYLAN,WM & HOWARD,PH (1993)

CAS #: 002882-21-5	PYRAZINE, 2-METHOXY-6-METHYL-

Formula: $C_6H_8N_2O$

Mol Weight: 124.14

MP (deg C): | FP (deg C):

BP (deg C):

BP pressure (mm Hg):

Property/ Value	Units	Temp	Data Type	Reference
WS 7.67E+003	mg/L	25	EST	MEYLAN,WM ET AL. (1996)
logP 1.29			EXP	YAMAGAMI,C ET AL. (1991)
VP 5.84E-001	mm Hg	25	EST	NEELY,WB & BLAU,GE (1985)
DC	pKa			
HL 1.74E-006	atm m3/mol	25	EST	MEYLAN,WM & HOWARD,PH (1991)
OH 7.59E-012	cm3/molc sec	25	EST	MEYLAN,WM & HOWARD,PH (1993)

CAS #: 002883-02-5				N-NONYLCYCLOHEXANE

Formula: $C_{15}H_{30}$

Mol Weight: 210.41

MP (deg C): -10   FP (deg C):

BP (deg C): 282

BP pressure (mm Hg):

Property/Value	Units	Temp	Data Type	Reference
WS 4.23E-003	mg/L	25	EST	MEYLAN,WM ET AL. (1996)
logP 7.52			EST	MEYLAN,WM & HOWARD,PH (1995)
VP 2.48E-003	mm Hg	25	EXT	ZWOLINSKI,BJ & WILHOIT,RC (1971)
DC	pKa			
HL 3.27E+000	atm m3/mol	25	EST	MEYLAN,WM & HOWARD,PH (1991)
OH 2.19E-011	cm3/molc sec	25	EST	MEYLAN,WM & HOWARD,PH (1993)

CAS #: 002885-00-9				1-OCTADECANETHIOL

Formula: $C_{18}H_{38}S$

Mol Weight: 286.57

MP (deg C): 30   FP (deg C):

BP (deg C): 204-210

BP pressure (mm Hg): 1.10E+001

Property/Value	Units	Temp	Data Type	Reference
WS 2.37E-004	mg/L	25	EST	MEYLAN,WM ET AL. (1996)
logP 9.12			EST	MEYLAN,WM & HOWARD,PH (1995)
VP 3.97E-007	mm Hg	25	EXT	ZWOLINSKI,BJ & WILHOIT,RC (1971)
DC	pKa			
HL 3.23E-001	atm m3/mol	25	EST	MEYLAN,WM & HOWARD,PH (1991)
OH 6.40E-011	cm3/molc sec	25	EST	MEYLAN,WM & HOWARD,PH (1993)

CAS #: 002886-65-9				1,4-BENZODIAZPIN-2-ONE,5(2FPH)7CL

Formula: $C_{15}H_{10}ClFN_2O$

Mol Weight: 288.71

MP (deg C):   FP (deg C):

BP (deg C):

BP pressure (mm Hg):

Property/Value	Units	Temp	Data Type	Reference
WS 7.06E+001	mg/L	25	EST	MEYLAN,WM ET AL. (1996)
logP 2.70			EXP	HANSCH,C & LEO,AJ (1985)
VP 7.76E-009	mm Hg	25	EST	NEELY,WB & BLAU,GE (1985)
DC	pKa			
HL 2.08E-010	atm m3/mol	25	EST	MEYLAN,WM & HOWARD,PH (1991)
OH 7.98E-012	cm3/molc sec	25	EST	MEYLAN,WM & HOWARD,PH (1993)

CAS #: 002889-58-9				1-(1-ISOCYANATO-1-METHYLETHYL)-4-(1-METHYLETHE*)

Formula: $C_{13}H_{15}NO$

Mol Weight: 201.27

MP (deg C):   FP (deg C):

BP (deg C):

BP pressure (mm Hg):

Property/Value	Units	Temp	Data Type	Reference
WS 3.33E+000	mg/L	25	EST	MEYLAN,WM ET AL. (1996)
logP 4.81			EST	MEYLAN,WM & HOWARD,PH (1995)
VP 4.67E-003	mm Hg	25	EXP	ACHORN,PJ ET AL. (1986)
DC	pKa			
HL 1.06E-004	atm m3/mol	25	EST	MEYLAN,WM & HOWARD,PH (1991)
OH 5.59E-011	cm3/molc sec	25	EST	MEYLAN,WM & HOWARD,PH (1993)

CAS #: 002891-09-0				1,4-BENZDIAZ-2-ON,1ME5(PH)7DIMEAM

Formula: $C_{18}H_{19}N_3O$

Mol Weight: 293.37

MP (deg C):   FP (deg C):

BP (deg C):

BP pressure (mm Hg):

Property/Value	Units	Temp	Data Type	Reference
WS 7.62E+001	mg/L	25	EST	MEYLAN,WM ET AL. (1996)
logP 2.63			EXP	HANSCH,C & LEO,AJ (1985)
VP 9.29E-009	mm Hg	25	EST	NEELY,WB & BLAU,GE (1985)
DC	pKa			
HL 7.82E-011	atm m3/mol	25	EST	MEYLAN,WM & HOWARD,PH (1991)
OH 2.04E-010	cm3/molc sec	25	EST	MEYLAN,WM & HOWARD,PH (1993)

CAS #: 002891-12-5				1,4-BENZODIAZPIN-2-ONE,5(PH)7MES

Formula: $C_{16}H_{14}N_2OS$

Mol Weight: 282.37

MP (deg C):   FP (deg C):

BP (deg C):

BP pressure (mm Hg):

Property/Value	Units	Temp	Data Type	Reference
WS 4.98E+001	mg/L	25	EST	MEYLAN,WM ET AL. (1996)
logP 2.92			EXP	HANSCH,C & LEO,AJ (1985)
VP 1.03E-009	mm Hg	25	EST	NEELY,WB & BLAU,GE (1985)
DC	pKa			
HL 7.00E-012	atm m3/mol	25	EST	MEYLAN,WM & HOWARD,PH (1991)
OH 2.05E-011	cm3/molc sec	25	EST	MEYLAN,WM & HOWARD,PH (1993)

CAS #: 002894-61-3				2H-1,4-BENZODIAZEPIN-2-ONE, 7-BROMO-1,3-DIHYDRO-

Formula: $C_{15}H_{11}BrN_2O$

Mol Weight: 315.17

MP (deg C):   FP (deg C):

BP (deg C):

BP pressure (mm Hg):

Property/Value	Units	Temp	Data Type	Reference
WS 2.21E+001	mg/L	25	EST	MEYLAN,WM ET AL. (1996)
logP 3.11			EXP	SANGSTER,J (1993)
VP 2.78E-009	mm Hg	25	EST	NEELY,WB & BLAU,GE (1985)
DC	pKa			
HL 9.56E-011	atm m3/mol	25	EST	MEYLAN,WM & HOWARD,PH (1991)
OH 9.27E-012	cm3/molc sec	25	EST	MEYLAN,WM & HOWARD,PH (1993)

CAS #: 002894-67-9				1,4-BENZDIAZEPIN-2-ONE-5(2-CLPH)-7-CL

Formula: $C_{15}H_{10}Cl_2N_2O$

Mol Weight: 305.17

MP (deg C):   FP (deg C):

BP (deg C):

BP pressure (mm Hg):

Property/Value	Units	Temp	Data Type	Reference
WS 2.34E+001	mg/L	25	EST	MEYLAN,WM ET AL. (1996)
logP 3.15			EXP	HANSCH,C & LEO,AJ (1985)
VP 1.74E-009	mm Hg	25	EST	NEELY,WB & BLAU,GE (1985)
DC	pKa			
HL 1.32E-010	atm m3/mol	25	EST	MEYLAN,WM & HOWARD,PH (1991)
OH 6.70E-012	cm3/molc sec	25	EST	MEYLAN,WM & HOWARD,PH (1993)

CAS #: 002894-68-0 — 2H-1,4-BENZODIAZEPIN-2-ONE, 7-CHLORO-5-(2-CHLORO

Formula: $C_{16}H_{13}ClN_2O$

Mol Weight: 284.75

MP (deg C): FP (deg C):

BP (deg C):

BP pressure (mm Hg):

Property/Value	Units	Temp	Data Type	Reference
WS 2.05E+001	mg/L	25	EST	MEYLAN,WM ET AL. (1996)
logP 3.12			EXP	SANGSTER,J (1994)
VP 7.59E-009	mm Hg	25	EST	NEELY,WB & BLAU,GE (1985)
DC	pKa			
HL 2.70E-009	atm m3/mol	25	EST	MEYLAN,WM & HOWARD,PH (1991)
OH 7.25E-012	cm3/molc sec	25	EST	MEYLAN,WM & HOWARD,PH (1993)

CAS #: 002896-70-0 — 1-PIPERIDINYLOXY, 2,2,6,6-TETRAMETHYL-4-OXO-

Formula: $C_9H_{17}NO_2$

Mol Weight: 171.24

MP (deg C): FP (deg C):

BP (deg C):

BP pressure (mm Hg):

Property/Value	Units	Temp	Data Type	Reference
WS 4.79E+004	mg/L	25	EST	MEYLAN,WM ET AL. (1996)
logP 0.12			EXP	SANGSTER,J (1993)
VP 3.71E-006	mm Hg	25	EST	NEELY,WB & BLAU,GE (1985)
DC	pKa			
HL 4.08E-011	atm m3/mol	25	EST	MEYLAN,WM & HOWARD,PH (1991)
OH 6.84E-011	cm3/molc sec	25	EST	MEYLAN,WM & HOWARD,PH (1993)

CAS #: 002898-08-0 — 1,4-BENZDIAZEPIN-2-ONE-5-PHENYL

Formula: $C_{15}H_{12}N_2O$

Mol Weight: 236.28

MP (deg C): FP (deg C):

BP (deg C):

BP pressure (mm Hg):

Property/Value	Units	Temp	Data Type	Reference
WS 3.87E+002	mg/L	25	EST	MEYLAN,WM ET AL. (1996)
logP 2.18			EXP	HANSCH,C & LEO,AJ (1985)
VP 2.36E-008	mm Hg	25	EST	NEELY,WB & BLAU,GE (1985)
DC	pKa			
HL 2.40E-010	atm m3/mol	25	EST	MEYLAN,WM & HOWARD,PH (1991)
OH 1.99E-011	cm3/molc sec	25	EST	MEYLAN,WM & HOWARD,PH (1993)

CAS #: 002898-12-6 — MEDAZEPAM

Formula: $C_{16}H_{15}ClN_2$

Mol Weight: 270.76

MP (deg C): FP (deg C):

BP (deg C):

BP pressure (mm Hg):

Property/Value	Units	Temp	Data Type	Reference
WS 3.10E+000	mg/L	25	EST	MEYLAN,WM ET AL. (1996)
logP 4.41			EXP	HANSCH,C & LEO,AJ (1985)
VP 2.13E-006	mm Hg	25	EST	NEELY,WB & BLAU,GE (1985)
DC	pKa			
HL 1.64E-007	atm m3/mol	25	EST	MEYLAN,WM & HOWARD,PH (1991)
OH 6.87E-011	cm3/molc sec	25	EST	MEYLAN,WM & HOWARD,PH (1993)

CAS #: 002904-53-2 — 135-TRIAZINE,N-ETOH,N'-IPR,4-CL-

Formula: $C_8H_{14}ClN_5O$

Mol Weight: 231.69

MP (deg C): FP (deg C):

BP (deg C):

BP pressure (mm Hg):

Property/Value	Units	Temp	Data Type	Reference
WS 6.43E+003	mg/L	25	EST	MEYLAN,WM ET AL. (1996)
logP 0.78			EXP	BALKE,NE & PRICE,TP (1988)
VP 1.63E-007	mm Hg	25	EST	NEELY,WB & BLAU,GE (1985)
DC	pKa			
HL 1.64E-013	atm m3/mol	25	EST	MEYLAN,WM & HOWARD,PH (1991)
OH 3.33E-011	cm3/molc sec	25	EST	MEYLAN,WM & HOWARD,PH (1993)

CAS #: 002905-65-9 — BENZOIC ACID, 3-CHLORO-, METHYL ESTER

Formula: $C_8H_7ClO_2$

Mol Weight: 170.60

MP (deg C): 21 FP (deg C):

BP (deg C): 99-101

BP pressure (mm Hg): 1.20E+001

Property/Value	Units	Temp	Data Type	Reference
WS 2.25E+002	mg/L	25	EST	MEYLAN,WM ET AL. (1996)
logP 2.85			EXP	SOTOMATSU,T ET AL. (1993)
VP 9.07E-002	mm Hg	25	EST	NEELY,WB & BLAU,GE (1985)
DC	pKa			
HL 2.57E-005	atm m3/mol	25	EST	MEYLAN,WM & HOWARD,PH (1991)
OH 5.21E-013	cm3/molc sec	25	EST	MEYLAN,WM & HOWARD,PH (1993)

CAS #: 002905-69-3 — METHYL 2,5-DICHLOROBENZOATE

Formula: $C_8H_6Cl_2O_2$

Mol Weight: 205.04

MP (deg C): 37-40 FP (deg C):

BP (deg C):

BP pressure (mm Hg):

Property/Value	Units	Temp	Data Type	Reference
WS 8.98E+001	mg/L	25	EST	MEYLAN,WM ET AL. (1996)
logP 3.12			EST	MEYLAN,WM & HOWARD,PH (1995)
VP 1.20E-002	mm Hg	25	EST	NEELY,WB & BLAU,GE (1985)
DC	pKa			
HL 1.91E-005	atm m3/mol	25	EST	MEYLAN,WM & HOWARD,PH (1991)
OH 3.46E-013	cm3/molc sec	25	EST	MEYLAN,WM & HOWARD,PH (1993)

CAS #: 002909-38-8 — 3-CHLOROPHENYL ISOCYANATE

Formula: C_7H_4ClNO

Mol Weight: 153.57

MP (deg C): -4 FP (deg C):

BP (deg C): 113-114

BP pressure (mm Hg): 4.30E+001

Property/Value	Units	Temp	Data Type	Reference
WS 1.26E+002	mg/L	25	EST	MEYLAN,WM ET AL. (1996)
logP 3.24			EST	MEYLAN,WM & HOWARD,PH (1995)
VP 1.94E-001	mm Hg	25	EXT	BOUBLIK,T ET AL. (1984)
DC	pKa			
HL 1.72E-004	atm m3/mol	25	EST	MEYLAN,WM & HOWARD,PH (1991)
OH 8.63E-013	cm3/molc sec	25	EST	MEYLAN,WM & HOWARD,PH (1993)

002916-31-6

1,3-DIOXOLANE, 2,2-DIMETHYL-

Formula: $C_5H_{10}O_2$

Mol Weight: 102.13

MP (deg C):

FP (deg C):

BP (deg C): 92-93

BP pressure (mm Hg):

Property/Value	Units	Temp	Data Type	Reference
WS 4.66E+004	mg/L	25	EST	MEYLAN,WM ET AL. (1996)
logP 0.46			EXP	HANSCH,C ET AL. (1995)
VP 3.52E+001	mm Hg	25	EST	NEELY,WB & BLAU,GE (1985)
DC	pKa			
HL 3.93E-005	atm m3/mol	25	EST	MEYLAN,WM & HOWARD,PH (1991)
OH 1.22E-011	cm3/molc sec	25	EST	MEYLAN,WM & HOWARD,PH (1993)

002917-26-2

1-HEXADECANETHIOL

Formula: $C_{16}H_{34}S$

Mol Weight: 258.51

MP (deg C): 19

FP (deg C):

BP (deg C): 123-128

BP pressure (mm Hg): 5.00E-001

Property/Value	Units	Temp	Data Type	Reference
WS 2.36E-003	mg/L	25	EST	MEYLAN,WM ET AL. (1996)
logP 8.14			EST	MEYLAN,WM & HOWARD,PH (1995)
VP 1.03E-005	mm Hg	25	EXT	ZWOLINSKI,BJ & WILHOIT,RC (1971)
DC	pKa			
HL 1.83E-001	atm m3/mol	25	EST	MEYLAN,WM & HOWARD,PH (1991)
OH 6.11E-011	cm3/molc sec	25	EST	MEYLAN,WM & HOWARD,PH (1993)

002917-98-8

PIPERAZINE,1,4-BIS(2,3-EPOXYPROPYL)

Formula: $C_{10}H_{18}N_2O_2$

Mol Weight: 198.27

MP (deg C):

FP (deg C):

BP (deg C):

BP pressure (mm Hg):

Property/Value	Units	Temp	Data Type	Reference
WS 1.00E+006	mg/L	25	EST	MEYLAN,WM ET AL. (1996)
logP -0.84			EXP	HANSCH,C & LEO,AJ (1985)
VP 2.07E-003	mm Hg	25	EST	NEELY,WB & BLAU,GE (1985)
DC	pKa			
HL 1.57E-013	atm m3/mol	25	EST	MEYLAN,WM & HOWARD,PH (1991)
OH 2.25E-010	cm3/molc sec	25	EST	MEYLAN,WM & HOWARD,PH (1993)

002921-88-2

CHLORPYRIFOS

Formula: $C_9H_{11}Cl_3NO_3PS$

Mol Weight: 350.59

MP (deg C): 41-42

FP (deg C):

BP (deg C):

BP pressure (mm Hg):

Property/Value	Units	Temp	Data Type	Reference
WS 4.00E-001	mg/L	23	EXP	CHIOU,CT ET AL. (1977)
logP 4.96			EXP	SANGSTER,J (1994)
VP 1.70E-005	mm Hg	25	EXP	WAUCHOPE,RD ET AL. (1991A)
DC	pKa			
HL 2.93E-006	atm m3/mol	20	EXP	RICE,CP & CHERNYAK,SM (1995)
OH 9.17E-011	cm3/molc sec	25	EST	MEYLAN,WM & HOWARD,PH (1993)

002922-40-9

P-NITROPHENYLALANINE

Formula: $C_9H_{10}N_2O_4$

Mol Weight: 210.19

MP (deg C):

FP (deg C):

BP (deg C):

BP pressure (mm Hg):

Property/Value	Units	Temp	Data Type	Reference
WS 1.57E+003	mg/L	25	EST	MEYLAN,WM ET AL. (1996)
logP -1.25			EXP	HANSCH,C & LEO,AJ (1985)
VP 4.71E-010	mm Hg	25	EST	NEELY,WB & BLAU,GE (1985)
DC	pKa			
HL 4.75E-013	atm m3/mol	25	EST	MEYLAN,WM & HOWARD,PH (1991)
OH 3.99E-011	cm3/molc sec	25	EST	MEYLAN,WM & HOWARD,PH (1993)

002924-27-8

BENZENAMINE, N,N-BIS(2,2,2-TRIFLUOROETHYL)-

Formula: $C_{10}H_9F_6N$

Mol Weight: 257.18

MP (deg C):

FP (deg C):

BP (deg C):

BP pressure (mm Hg):

Property/Value	Units	Temp	Data Type	Reference
WS 1.52E+001	mg/L	25	EST	MEYLAN,WM ET AL. (1996)
logP 3.69			EXP	HANSCH,C ET AL. (1995)
VP 8.87E-001	mm Hg	25	EST	NEELY,WB & BLAU,GE (1985)
DC	pKa			
HL 9.36E-003	atm m3/mol	25	EST	MEYLAN,WM & HOWARD,PH (1991)
OH 1.48E-010	cm3/molc sec	25	EST	MEYLAN,WM & HOWARD,PH (1993)

002933-44-0

DOBEROL

Formula: $C_{13}H_{21}NO_2$

Mol Weight: 223.32

MP (deg C):

FP (deg C):

BP (deg C):

BP pressure (mm Hg):

Property/Value	Units	Temp	Data Type	Reference
WS 7.57E+003	mg/L	25	EST	MEYLAN,WM ET AL. (1996)
logP 1.93			EXP	HANSCH,C & LEO,AJ (1985)
VP 5.62E-006	mm Hg	25	EST	NEELY,WB & BLAU,GE (1985)
DC	pKa			
HL 9.02E-012	atm m3/mol	25	EST	MEYLAN,WM & HOWARD,PH (1991)
OH 1.66E-010	cm3/molc sec	25	EST	MEYLAN,WM & HOWARD,PH (1993)

002933-75-7

ANISOLE,O-(2-HYDROXYETHYL)AMINO

Formula: $C_9H_{13}NO_2$

Mol Weight: 167.21

MP (deg C):

FP (deg C):

BP (deg C):

BP pressure (mm Hg):

Property/Value	Units	Temp	Data Type	Reference
WS 3.32E+003	mg/L	25	EST	MEYLAN,WM ET AL. (1996)
logP 1.50			EXP	HANSCH,C & LEO,AJ (1985)
VP 8.03E-005	mm Hg	25	EST	NEELY,WB & BLAU,GE (1985)
DC	pKa			
HL 1.20E-011	atm m3/mol	25	EST	MEYLAN,WM & HOWARD,PH (1991)
OH 5.26E-011	cm3/molc sec	25	EST	MEYLAN,WM & HOWARD,PH (1993)

CAS #: 002937-50-0 — ALLYL CHLOROCARBONATE

Formula:	$C_4H_5ClO_2$		
Mol Weight:	120.54		
MP (deg C):		FP (deg C):	
BP (deg C):	109-110		
BP pressure (mm Hg):			

Property/Value	Units	Temp	Data Type	Reference
WS 1.45E+004	mg/L	25	EST	MEYLAN,WM ET AL. (1996)
logP 0.98			EST	MEYLAN,WM & HOWARD,PH (1995)
VP 2.19E+001	mm Hg	25	EST	NEELY,WB & BLAU,GE (1985)
DC	pKa			
HL 3.09E-003	atm m3/mol	25	EST	MEYLAN,WM & HOWARD,PH (1991)
OH 2.78E-011	cm3/molc sec	25	EST	MEYLAN,WM & HOWARD,PH (1993)

CAS #: 002947-61-7 — BENZENEACETONITRILE, 4-METHYL-

Formula:	C_9H_9N		
Mol Weight:	131.18		
MP (deg C):	18	FP (deg C):	
BP (deg C):	242.5		
BP pressure (mm Hg):			

Property/Value	Units	Temp	Data Type	Reference
WS 1.85E+003	mg/L	25	EST	MEYLAN,WM ET AL. (1996)
logP 1.67			EXP	SANGSTER,J (1994)
VP 2.87E-002	mm Hg	25	EST	NEELY,WB & BLAU,GE (1985)
DC	pKa			
HL 2.72E-006	atm m3/mol	25	EST	MEYLAN,WM & HOWARD,PH (1991)
OH 3.33E-012	cm3/molc sec	25	EST	MEYLAN,WM & HOWARD,PH (1993)

CAS #: 002953-29-9 — PHOSPHORODITHIOIC ACID, O,O,S-TRIMETHYL ESTER

Formula:	$C_3H_9O_2PS_2$		
Mol Weight:	172.21		
MP (deg C):		FP (deg C):	
BP (deg C):			
BP pressure (mm Hg):			

Property/Value	Units	Temp	Data Type	Reference
WS 1.72E+003	mg/L	25	EST	MEYLAN,WM ET AL. (1996)
logP 1.81			EST	MEYLAN,WM & HOWARD,PH (1995)
VP 2.76E-001	mm Hg	25	EST	NEELY,WB & BLAU,GE (1985)
DC	pKa			
HL 6.33E-005	atm m3/mol	25	EST	MEYLAN,WM & HOWARD,PH (1991)
OH 5.60E-011	cm3/molc sec	25	EXP	ATKINSON,R (1989)

CAS #: 002954-50-9 — 2-NAPHTHALENAMINE, 1,2,3,4-TETRAHYDRO-

Formula:	$C_{10}H_{13}N$		
Mol Weight:	147.22		
MP (deg C):		FP (deg C):	
BP (deg C):			
BP pressure (mm Hg):			

Property/Value	Units	Temp	Data Type	Reference
WS 1.33E+004	mg/L	25	EST	MEYLAN,WM ET AL. (1996)
logP 2.08			EXP	SANGSTER,J (1993)
VP 2.00E-002	mm Hg	25	EST	NEELY,WB & BLAU,GE (1985)
DC 9.93	pKa	17	EXP	PERRIN,DD (1965)
HL 5.24E-007	atm m3/mol	25	EST	MEYLAN,WM & HOWARD,PH (1991)
OH 5.79E-011	cm3/molc sec	25	EST	MEYLAN,WM & HOWARD,PH (1993)

CAS #: 002955-38-6 — PRAZEPAM

Formula:	$C_{19}H_{17}ClN_2O$		
Mol Weight:	324.81		
MP (deg C):	145-146	FP (deg C):	
BP (deg C):			
BP pressure (mm Hg):			

Property/Value	Units	Temp	Data Type	Reference
WS 5.72E+000	mg/L	25	EST	MEYLAN,WM ET AL. (1996)
logP 3.73			EXP	HANSCH,C & LEO,AJ (1985)
VP 2.14E-009	mm Hg	25	EST	NEELY,WB & BLAU,GE (1985)
DC	pKa			
HL 3.76E-009	atm m3/mol	25	EST	MEYLAN,WM & HOWARD,PH (1991)
OH 2.32E-011	cm3/molc sec	25	EST	MEYLAN,WM & HOWARD,PH (1993)

CAS #: 002961-50-4 — P-AMYLPYRIDINE

Formula:	$C_{10}H_{15}N$		
Mol Weight:	149.24		
MP (deg C):		FP (deg C):	
BP (deg C):			
BP pressure (mm Hg):			

Property/Value	Units	Temp	Data Type	Reference
WS 9.57E+002	mg/L	25	EST	MEYLAN,WM ET AL. (1996)
logP 3.75			EXP	HANSCH,C & LEO,AJ (1985)
VP 1.34E-001	mm Hg	25	EST	NEELY,WB & BLAU,GE (1985)
DC	pKa			
HL 2.42E-005	atm m3/mol	25	EST	MEYLAN,WM & HOWARD,PH (1991)
OH 6.21E-012	cm3/molc sec	25	EST	MEYLAN,WM & HOWARD,PH (1993)

CAS #: 002963-66-8 — 2-(2-HYDROXYPHENYL)BENZOIMIDAZOLE

Formula:	$C_{13}H_{10}N_2O$		
Mol Weight:	210.24		
MP (deg C):		FP (deg C):	
BP (deg C):			
BP pressure (mm Hg):			

Property/Value	Units	Temp	Data Type	Reference
WS 1.04E+003	mg/L	25	EST	MEYLAN,WM ET AL. (1996)
logP 2.52			EST	MEYLAN,WM & HOWARD,PH (1995)
VP 5.25E-010	mm Hg	25	EST	NEELY,WB & BLAU,GE (1985)
DC	pKa			
HL 2.93E-012	atm m3/mol	25	EST	MEYLAN,WM & HOWARD,PH (1991)
OH 8.65E-011	cm3/molc sec	25	EST	MEYLAN,WM & HOWARD,PH (1993)

CAS #: 002971-22-4 — 2,2-DIPHENYL-1,1,1-TRICL ETHANE

Formula:	$C_{14}H_{11}Cl_3$		
Mol Weight:	285.60		
MP (deg C):		FP (deg C):	
BP (deg C):			
BP pressure (mm Hg):			

Property/Value	Units	Temp	Data Type	Reference
WS 1.03E+000	mg/L	25	EST	MEYLAN,WM ET AL. (1996)
logP 4.87			EXP	HANSCH,C & LEO,AJ (1985)
VP 8.99E-005	mm Hg	25	EST	NEELY,WB & BLAU,GE (1985)
DC	pKa			
HL 2.79E-005	atm m3/mol	25	EST	MEYLAN,WM & HOWARD,PH (1991)
OH 9.39E-012	cm3/molc sec	25	EST	MEYLAN,WM & HOWARD,PH (1993)

CAS #: 002971-90-6 — 3,5-DICHLORO-2,6-DIMETHYL-4-PYRIDINOL

Formula: $C_7H_7Cl_2NO$
Mol Weight: 192.05
MP (deg C): 320
FP (deg C):
BP (deg C):
BP pressure (mm Hg):

Property/Value	Units	Temp	Data Type	Reference
WS 1.00E+001	mg/L	25	EXP	SHIU,WY ET AL. (1990)
logP 2.71			EST	MEYLAN,WM & HOWARD,PH (1995)
VP 6.58E-004	mm Hg	25	EST	NEELY,WB & BLAU,GE (1985)
DC	pKa			
HL 4.91E-010	atm m3/mol	25	EST	MEYLAN,WM & HOWARD,PH (1991)
OH 4.09E-012	cm3/molc sec	25	EST	MEYLAN,WM & HOWARD,PH (1993)

CAS #: 002973-27-5 — 4-CYANOQUINOLINE

Formula: $C_{10}H_6N_2$
Mol Weight: 154.17
MP (deg C): 103.5
FP (deg C):
BP (deg C): 242.5
BP pressure (mm Hg):

Property/Value	Units	Temp	Data Type	Reference
WS 8.05E+002	mg/L	25	EST	MEYLAN,WM ET AL. (1996)
logP 1.98			EXP	HANSCH,C & LEO,AJ (1985)
VP 3.56E-004	mm Hg	25	EST	NEELY,WB & BLAU,GE (1985)
DC	pKa			
HL 6.65E-009	atm m3/mol	25	EST	MEYLAN,WM & HOWARD,PH (1991)
OH 2.54E-012	cm3/molc sec	25	EST	MEYLAN,WM & HOWARD,PH (1993)

CAS #: 002974-90-5 — BIPHENYL, 3,4'-DICHLORO-

Formula: $C_{12}H_8Cl_2$
Mol Weight: 223.10
MP (deg C):
FP (deg C):
BP (deg C):
BP pressure (mm Hg):

Property/Value	Units	Temp	Data Type	Reference
WS 1.33E+000	mg/L	25	EST	MEYLAN,WM ET AL. (1996)
logP 5.15			EXP	SANGSTER,J (1993)
VP 1.91E-004	mm Hg	25	EST	NEELY,WB & BLAU,GE (1985)
DC	pKa			
HL 2.27E-004	atm m3/mol	25	EST	MEYLAN,WM & HOWARD,PH (1991)
OH 2.62E-012	cm3/molc sec	25	EST	MEYLAN,WM & HOWARD,PH (1993)

CAS #: 002974-92-7 — 3,4-DICHLOROBIPHENYL

Formula: $C_{12}H_8Cl_2$
Mol Weight: 223.10
MP (deg C):
FP (deg C):
BP (deg C):
BP pressure (mm Hg):

Property/Value	Units	Temp	Data Type	Reference
WS 7.92E-003	mg/L	25	EXP	YALKOWSKY,SH & DANNENFELSER,RM (1992)
logP 5.29			EXP	SANGSTER,J (1993)
VP 1.91E-004	mm Hg	25	EST	NEELY,WB & BLAU,GE (1985)
DC	pKa			
HL 2.27E-004	atm m3/mol	25	EST	MEYLAN,WM & HOWARD,PH (1991)
OH 2.47E-012	cm3/molc sec	25	EST	MEYLAN,WM & HOWARD,PH (1993)

CAS #: 002976-32-1 — 3-METHYLSUFONYLNITROBENZENE

Formula: $C_7H_7NO_4S$
Mol Weight: 201.20
MP (deg C):
FP (deg C):
BP (deg C):
BP pressure (mm Hg):

Property/Value	Units	Temp	Data Type	Reference
WS 6.58E+003	mg/L	25	EST	MEYLAN,WM ET AL. (1996)
logP 0.50			EXP	HANSCH,C & LEO,AJ (1985)
VP 3.39E-005	mm Hg	25	EST	NEELY,WB & BLAU,GE (1985)
DC	pKa			
HL 4.92E-009	atm m3/mol	25	EST	MEYLAN,WM & HOWARD,PH (1991)
OH 1.11E-012	cm3/molc sec	25	EST	MEYLAN,WM & HOWARD,PH (1993)

CAS #: 002976-74-1 — 2,3-DICHLOROPHENOXYACETIC ACID

Formula: $C_8H_6Cl_2O_3$
Mol Weight: 221.04
MP (deg C): 173-175
FP (deg C):
BP (deg C):
BP pressure (mm Hg):

Property/Value	Units	Temp	Data Type	Reference
WS 3.43E+002	mg/L	25	EXP	YALKOWSKY,SH & DANNENFELSER,RM (1992)
logP 2.62			EST	MEYLAN,WM & HOWARD,PH (1995)
VP 6.11E-005	mm Hg	25	EST	NEELY,WB & BLAU,GE (1985)
DC	pKa			
HL 9.21E-009	atm m3/mol	25	EST	MEYLAN,WM & HOWARD,PH (1991)
OH 9.21E-012	cm3/molc sec	25	EST	MEYLAN,WM & HOWARD,PH (1993)

CAS #: 002976-75-2 — ACETIC ACID, (1-NAPHTHALENYLOXY)-

Formula: $C_{12}H_{10}O_3$
Mol Weight: 202.21
MP (deg C): 195-197
FP (deg C):
BP (deg C):
BP pressure (mm Hg):

Property/Value	Units	Temp	Data Type	Reference
WS 6.37E+002	mg/L	25	EST	MEYLAN,WM ET AL. (1996)
logP 2.60			EXP	CHAMBERLAIN,K ET AL. (1986)
VP 6.27E-006	mm Hg	25	EST	NEELY,WB & BLAU,GE (1985)
DC	pKa			
HL 1.64E-009	atm m3/mol	25	EST	MEYLAN,WM & HOWARD,PH (1991)
OH 2.05E-010	cm3/molc sec	25	EST	MEYLAN,WM & HOWARD,PH (1993)

CAS #: 002979-53-5 — M-NITROPHENYLHIPPURATE

Formula: $C_{15}H_{12}N_2O_5$
Mol Weight: 300.27
MP (deg C):
FP (deg C):
BP (deg C):
BP pressure (mm Hg):

Property/Value	Units	Temp	Data Type	Reference
WS 3.95E+001	mg/L	25	EST	MEYLAN,WM ET AL. (1996)
logP 2.46			EXP	HANSCH,C & LEO,AJ (1985)
VP 8.90E-010	mm Hg	25	EST	NEELY,WB & BLAU,GE (1985)
DC	pKa			
HL 3.37E-013	atm m3/mol	25	EST	MEYLAN,WM & HOWARD,PH (1991)
OH 1.03E-011	cm3/molc sec	25	EST	MEYLAN,WM & HOWARD,PH (1993)

CAS #: 002979-54-6				PHENYLHIPPURATE

Formula: $C_{15}H_{13}NO_3$

Mol Weight: 255.28

MP (deg C):

FP (deg C):

BP (deg C):

BP pressure (mm Hg):

Property/ Value	Units	Temp	Data Type	Reference
WS 2.35E+002	mg/L	25	EST	MEYLAN,WM ET AL. (1996)
logP 2.31			EXP	HANSCH,C & LEO,AJ (1985)
VP 5.82E-008	mm Hg	25	EST	NEELY,WB & BLAU,GE (1985)
DC	pKa			
HL 8.54E-011	atm m3/mol	25	EST	MEYLAN,WM & HOWARD,PH (1991)
OH 1.21E-011	cm3/molc sec	25	EST	MEYLAN,WM & HOWARD,PH (1993)

CAS #: 002984-42-1				1,4-DICHLORO-1,3-BUTADIENE

Formula: $C_4H_4Cl_2$

Mol Weight: 122.98

MP (deg C):

FP (deg C):

BP (deg C):

BP pressure (mm Hg):

Property/ Value	Units	Temp	Data Type	Reference
WS 4.40E+002	mg/L	25	EST	MEYLAN,WM ET AL. (1996)
logP 2.75			EST	MEYLAN,WM & HOWARD,PH (1995)
VP 2.76E+001	mm Hg	25	EST	NEELY,WB & BLAU,GE (1985)
DC	pKa			
HL 4.03E-002	atm m3/mol	25	EST	MEYLAN,WM & HOWARD,PH (1991)
OH 6.26E-012	cm3/molc sec	25	EST	MEYLAN,WM & HOWARD,PH (1993)

CAS #: 002987-53-3				O-METHIOANILINE

Formula: C_7H_9NS

Mol Weight: 139.22

MP (deg C):

FP (deg C):

BP (deg C): 234

BP pressure (mm Hg):

Property/ Value	Units	Temp	Data Type	Reference
WS 7.97E+003	mg/L	25	EST	MEYLAN,WM ET AL. (1996)
logP 1.20			EXP	HANSCH,C & LEO,AJ (1985)
VP 1.24E-002	mm Hg	25	EST	NEELY,WB & BLAU,GE (1985)
DC 3.45	pKa	20	EXP	PERRIN,DD (1972)
HL 5.55E-008	atm m3/mol	25	EST	MEYLAN,WM & HOWARD,PH (1991)
OH 6.72E-011	cm3/molc sec	25	EST	MEYLAN,WM & HOWARD,PH (1993)

CAS #: 002989-17-5				O-ME PHENOXYACETIC ACID,ME ESTER

Formula: $C_{10}H_{12}O_3$

Mol Weight: 180.21

MP (deg C):

FP (deg C):

BP (deg C):

BP pressure (mm Hg):

Property/ Value	Units	Temp	Data Type	Reference
WS 9.20E+002	mg/L	25	EST	MEYLAN,WM ET AL. (1996)
logP 2.08			EXP	HANSCH,C & LEO,AJ (1985)
VP 1.94E-002	mm Hg	25	EST	NEELY,WB & BLAU,GE (1985)
DC	pKa			
HL 5.93E-006	atm m3/mol	25	EST	MEYLAN,WM & HOWARD,PH (1991)
OH 3.09E-011	cm3/molc sec	25	EST	MEYLAN,WM & HOWARD,PH (1993)

CAS #: 002989-98-2				M-BROMOPHENYLUREA

Formula: $C_7H_7BrN_2O$

Mol Weight: 215.06

MP (deg C):

FP (deg C):

BP (deg C):

BP pressure (mm Hg):

Property/ Value	Units	Temp	Data Type	Reference
WS 6.12E+002	mg/L	25	EST	MEYLAN,WM ET AL. (1996)
logP 2.08			EXP	HANSCH,C & LEO,AJ (1985)
VP 5.64E-005	mm Hg	25	EST	NEELY,WB & BLAU,GE (1985)
DC	pKa			
HL 8.02E-011	atm m3/mol	25	EST	MEYLAN,WM & HOWARD,PH (1991)
OH 2.79E-011	cm3/molc sec	25	EST	MEYLAN,WM & HOWARD,PH (1993)

CAS #: 002990-01-4				1,1-DIPHENYL-3,3-DIMETHYLUREA

Formula: $C_{15}H_{16}N_2O$

Mol Weight: 240.31

MP (deg C):

FP (deg C):

BP (deg C):

BP pressure (mm Hg):

Property/ Value	Units	Temp	Data Type	Reference
WS 1.09E+002	mg/L	25	EST	MEYLAN,WM ET AL. (1996)
logP 2.80			EXP	HANSCH,C & LEO,AJ (1985)
VP 9.44E-006	mm Hg	25	EST	NEELY,WB & BLAU,GE (1985)
DC	pKa			
HL 5.00E-009	atm m3/mol	25	EST	MEYLAN,WM & HOWARD,PH (1991)
OH 2.69E-011	cm3/molc sec	25	EST	MEYLAN,WM & HOWARD,PH (1993)

CAS #: 002990-06-9				ACETAMIDE, N-(4-CHLOROPHENYL)-2-PHENYL-

Formula: $C_{14}H_{12}ClNO$

Mol Weight: 245.71

MP (deg C):

FP (deg C):

BP (deg C):

BP pressure (mm Hg):

Property/ Value	Units	Temp	Data Type	Reference
WS 2.37E+001	mg/L	25	EST	MEYLAN,WM ET AL. (1996)
logP 3.54			EXP	YAMAGAMI,C ET AL. (1984)
VP 2.28E-007	mm Hg	25	EST	NEELY,WB & BLAU,GE (1985)
DC	pKa			
HL 3.69E-010	atm m3/mol	25	EST	MEYLAN,WM & HOWARD,PH (1991)
OH 9.17E-012	cm3/molc sec	25	EST	MEYLAN,WM & HOWARD,PH (1993)

CAS #: 003000-74-6				4-(N-ME)-3-PYRIDYLBUTYLAMINE

Formula: $C_{10}H_{16}N_2$

Mol Weight: 164.25

MP (deg C):

FP (deg C):

BP (deg C):

BP pressure (mm Hg):

Property/ Value	Units	Temp	Data Type	Reference
WS 1.00E+006	mg/L	25	EST	MEYLAN,WM ET AL. (1996)
logP 0.91			EXP	HANSCH,C & LEO,AJ (1985)
VP 2.42E-002	mm Hg	25	EST	NEELY,WB & BLAU,GE (1985)
DC	pKa			
HL 4.11E-009	atm m3/mol	25	EST	MEYLAN,WM & HOWARD,PH (1991)
OH 9.01E-011	cm3/molc sec	25	EST	MEYLAN,WM & HOWARD,PH (1993)

CAS #:	003000-75-7			N-ETHYL-3-PYRIDYLMETHYLAMINE	

Formula: $C_8H_{12}N_2$

Mol Weight: 136.20

MP (deg C): | FP (deg C):

BP (deg C):

BP pressure (mm Hg):

Property/Value	Units	Temp	Data Type	Reference
WS 1.00E+006	mg/L	25	EST	MEYLAN,WM ET AL. (1996)
logP 0.76			EXP	HANSCH,C & LEO,AJ (1985)
VP 1.33E-001	mm Hg	25	EST	NEELY,WB & BLAU,GE (1985)
DC	pKa			
HL 2.33E-009	atm m3/mol	25	EST	MEYLAN,WM & HOWARD,PH (1991)
OH 8.15E-011	cm3/molc sec	25	EST	MEYLAN,WM & HOWARD,PH (1993)

CAS #:	003001-66-9			2-OCTANETHIOL	

Formula: $C_8H_{18}S$

Mol Weight: 146.30

MP (deg C): | FP (deg C):

BP (deg C):

BP pressure (mm Hg):

Property/Value	Units	Temp	Data Type	Reference
WS 2.29E+001	mg/L	25	EST	MEYLAN,WM ET AL. (1996)
logP 4.14			EST	MEYLAN,WM & HOWARD,PH (1995)
VP 7.12E-001	mm Hg	25	EXT	ZWOLINSKI,BJ & WILHOIT,RC (1971)
DC	pKa			
HL 1.90E-002	atm m3/mol	25	EST	MEYLAN,WM & HOWARD,PH (1991)
OH 5.82E-011	cm3/molc sec	25	EST	MEYLAN,WM & HOWARD,PH (1993)

CAS #:	003004-70-4			IPATONE	

Formula: $C_{11}H_{21}N_5O$

Mol Weight: 239.32

MP (deg C): | FP (deg C):

BP (deg C):

BP pressure (mm Hg):

Property/Value	Units	Temp	Data Type	Reference
WS 3.09E+000	mg/L	26	EXP	SHIU,WY ET AL. (1990) @ pH=3
logP 4.19			EST	MEYLAN,WM & HOWARD,PH (1995)
VP 3.36E-005	mm Hg	25	EST	NEELY,WB & BLAU,GE (1985)
DC	pKa			
HL 8.86E-008	atm m3/mol	25	EST	MEYLAN,WM & HOWARD,PH (1991)
OH 3.73E-011	cm3/molc sec	25	EST	MEYLAN,WM & HOWARD,PH (1993)

CAS #:	003004-71-5			NORAZINE	

Formula: $C_7H_{12}ClN_5$

Mol Weight: 201.66

MP (deg C): | FP (deg C):

BP (deg C):

BP pressure (mm Hg):

Property/Value	Units	Temp	Data Type	Reference
WS 2.60E+002	mg/L	21	EXP	SHIU,WY ET AL. (1990)
logP 2.33			EST	MEYLAN,WM & HOWARD,PH (1995)
VP 3.45E-004	mm Hg	25	EST	NEELY,WB & BLAU,GE (1985)
DC	pKa			
HL 3.37E-009	atm m3/mol	25	EST	MEYLAN,WM & HOWARD,PH (1991)
OH 1.98E-011	cm3/molc sec	25	EST	MEYLAN,WM & HOWARD,PH (1993)

CAS #:	003010-38-6			DIS. A. 15	

Formula: $C_{14}H_{14}N_4O_2$

Mol Weight: 270.29

MP (deg C): | FP (deg C):

BP (deg C):

BP pressure (mm Hg):

Property/Value	Units	Temp	Data Type	Reference
WS 6.73E-001	mg/L	25	EST	MEYLAN,WM ET AL. (1996)
logP 4.68			EST	MEYLAN,WM & HOWARD,PH (1995)
VP 1.00E-009	mm Hg	25	EXP	BAUGHMAN,GL & PERENICH,TA (1988)
DC	pKa			
HL 9.22E-010	atm m3/mol	25	EST	MEYLAN,WM & HOWARD,PH (1991)
OH 1.49E-010	cm3/molc sec	25	EST	MEYLAN,WM & HOWARD,PH (1993)

CAS #:	003011-34-5			4-FORMYL-2-NITROPHENOL	

Formula: $C_7H_5NO_4$

Mol Weight: 167.12

MP (deg C): 140-142 | FP (deg C):

BP (deg C):

BP pressure (mm Hg):

Property/Value	Units	Temp	Data Type	Reference
WS 1.87E+002	mg/L	20	EXP	SCHWARZENBACH,RP ET AL. (1988)
logP 1.48			EXP	SCHWARZENBACH,RP ET AL. (1988)
VP 8.93E-004	mm Hg	20	EXP	SCHWARZENBACH,RP ET AL. (1988)
DC	pKa			
HL 1.05E-006	atm m3/mol	20	EST	VP/WSOL
OH 1.84E-011	cm3/molc sec	25	EST	MEYLAN,WM & HOWARD,PH (1993)

CAS #:	003012-37-1			BENZYLTHIOCYANATE	

Formula: C_8H_7NS

Mol Weight: 149.22

MP (deg C): 43 | FP (deg C):

BP (deg C): 232

BP pressure (mm Hg):

Property/Value	Units	Temp	Data Type	Reference
WS 1.53E+003	mg/L	25	EST	MEYLAN,WM ET AL. (1996)
logP 1.99			EXP	HANSCH,C & LEO,AJ (1985)
VP 4.20E-003	mm Hg	25	EST	NEELY,WB & BLAU,GE (1985)
DC	pKa			
HL 3.54E-006	atm m3/mol	25	EST	MEYLAN,WM & HOWARD,PH (1991)
OH 1.21E-011	cm3/molc sec	25	EST	MEYLAN,WM & HOWARD,PH (1993)

CAS #:	003017-95-6			2-BROMO-1-CHLOROPROPANE	

Formula: C_3H_6BrCl

Mol Weight: 157.44

MP (deg C): | FP (deg C):

BP (deg C): 118

BP pressure (mm Hg):

Property/Value	Units	Temp	Data Type	Reference
WS 2.24E+003	mg/L	25	EXP	WASIK,SP ET AL. (1981)
logP 2.34			EST	MEYLAN,WM & HOWARD,PH (1995)
VP 1.13E+001	mm Hg	25	EST	NEELY,WB & BLAU,GE (1985)
DC	pKa			
HL 5.27E-003	atm m3/mol	25	EST	MEYLAN,WM & HOWARD,PH (1991)
OH 4.21E-013	cm3/molc sec	25	EST	MEYLAN,WM & HOWARD,PH (1993)

003018-12-0 — DICHLOROACETONITRILE

Formula: C_2HCl_2N
Mol Weight: 109.94
MP (deg C):
FP (deg C):
BP (deg C): 112.5
BP pressure (mm Hg):

Property/Value	Units	Temp	Data Type	Reference
WS 3.35E+004	mg/L	25	EST	MEYLAN,WM ET AL. (1996)
logP 0.29			EST	MEYLAN,WM & HOWARD,PH (1995)
VP 2.82E+000	mm Hg	25	EST	NEELY,WB & BLAU,GE (1985)
DC	pKa			
HL 3.79E-006	atm m3/mol	25	EST	MEYLAN,WM & HOWARD,PH (1991)
OH 5.32E-014	cm3/molc sec	25	EST	MEYLAN,WM & HOWARD,PH (1993)

003021-63-4 — PERFLUORO-2,7-DIMETHYLOCTANE

Formula: $C_{10}F_{22}$
Mol Weight: 538.08
MP (deg C):
FP (deg C):
BP (deg C):
BP pressure (mm Hg):

Property/Value	Units	Temp	Data Type	Reference
WS 4.27E-006	mg/L	25	EST	MEYLAN,WM ET AL. (1996)
logP 8.47			EST	MEYLAN,WM & HOWARD,PH (1995)
VP 4.64E+003	mm Hg	25	EST	NEELY,WB & BLAU,GE (1985)
DC	pKa			
HL 1.40E+007	atm m3/mol	25	EST	MEYLAN,WM & HOWARD,PH (1991)
OH 0.00E+000	cm3/molc sec	25	EST	MEYLAN,WM & HOWARD,PH (1993)

003023-44-7 — 1,4-BENZODIAZP-2-ON,5(2MEOPH)7CL

Formula: $C_{16}H_{13}ClN_2O_2$
Mol Weight: 300.75
MP (deg C):
FP (deg C):
BP (deg C):
BP pressure (mm Hg):

Property/Value	Units	Temp	Data Type	Reference
WS 6.90E+001	mg/L	25	EST	MEYLAN,WM ET AL. (1996)
logP 2.63			EXP	HANSCH,C & LEO,AJ (1985)
VP 1.23E-009	mm Hg	25	EST	NEELY,WB & BLAU,GE (1985)
DC	pKa			
HL 1.05E-011	atm m3/mol	25	EST	MEYLAN,WM & HOWARD,PH (1991)
OH 3.17E-011	cm3/molc sec	25	EST	MEYLAN,WM & HOWARD,PH (1993)

003025-52-3 — DIS. A. 5

Formula: $C_{16}H_{18}N_4O_2$
Mol Weight: 298.35
MP (deg C):
FP (deg C):
BP (deg C):
BP pressure (mm Hg):

Property/Value	Units	Temp	Data Type	Reference
WS 1.19E-005	mg/L	25	EXP	BAUGHMAN,GL & PERENICH,TA (1988)
logP 5.67			EST	MEYLAN,WM & HOWARD,PH (1995)
VP 8.00E-011	mm Hg	25	EXP	BAUGHMAN,GL & PERENICH,TA (1988)
DC	pKa			
HL 2.64E-006	atm m3/mol	25	EST	VP/WSOL
OH 7.59E-011	cm3/molc sec	25	EST	MEYLAN,WM & HOWARD,PH (1993)

003031-73-0 — METHYLHYDROPEROXIDE

Formula: CH_4O_2
Mol Weight: 48.04
MP (deg C):
FP (deg C):
BP (deg C): 38-40
BP pressure (mm Hg): 6.50E+001

Property/Value	Units	Temp	Data Type	Reference
WS 3.09E+005	mg/L	25	EST	MEYLAN,WM ET AL. (1996)
logP -0.42			EST	MEYLAN,WM & HOWARD,PH (1995)
VP 2.96E+002	mm Hg	25	EST	NEELY,WB & BLAU,GE (1985)
DC 11.50	pKa	20	EXP	SERJEANT,EP & DEMPSEY,B (1979)
HL 4.54E-006	atm m3/mol	25	EXP	LIND,JA & KOK,GL (1986)
OH 5.54E-012	cm3/molc sec	25	EXP	ATKINSON,R (1989)

003034-34-2 — P-CYANOBENZAMIDE

Formula: $C_8H_6N_2O$
Mol Weight: 146.15
MP (deg C):
FP (deg C):
BP (deg C):
BP pressure (mm Hg):

Property/Value	Units	Temp	Data Type	Reference
WS 1.67E+004	mg/L	25	EST	MEYLAN,WM ET AL. (1996)
logP 0.48			EXP	NAKAGAWA,Y ET AL. (1992)
VP 1.69E-005	mm Hg	25	EST	NEELY,WB & BLAU,GE (1985)
DC	pKa			
HL 2.14E-011	atm m3/mol	25	EST	MEYLAN,WM & HOWARD,PH (1991)
OH 2.23E-012	cm3/molc sec	25	EST	MEYLAN,WM & HOWARD,PH (1993)

003034-38-6 — 4-NITROIMIDAZOLE

Formula: $C_3H_3N_3O_2$
Mol Weight: 113.08
MP (deg C): 303 dec
FP (deg C):
BP (deg C):
BP pressure (mm Hg):

Property/Value	Units	Temp	Data Type	Reference
WS 4.41E+004	mg/L	25	EST	MEYLAN,WM ET AL. (1996)
logP -0.11			EXP	HANSCH,C & LEO,AJ (1985)
VP 1.75E-004	mm Hg	25	EST	NEELY,WB & BLAU,GE (1985)
DC -0.05	pKa	20	EXP	PERRIN,DD (1965)
HL 1.48E-008	atm m3/mol	25	EST	MEYLAN,WM & HOWARD,PH (1991)
OH 4.50E-012	cm3/molc sec	25	EST	MEYLAN,WM & HOWARD,PH (1993)

003034-41-1 — 1-METHYL-4-NITRO-IH-IMIDAZOLE

Formula: $C_4H_5N_3O_2$
Mol Weight: 127.10
MP (deg C):
FP (deg C):
BP (deg C):
BP pressure (mm Hg):

Property/Value	Units	Temp	Data Type	Reference
WS 9.13E+004	mg/L	25	EST	MEYLAN,WM ET AL. (1996)
logP -0.44			EXP	SUWINSKI,J ET AL. (1985)
VP 1.69E-003	mm Hg	25	EST	NEELY,WB & BLAU,GE (1985)
DC -0.53	pKa	20	EXP	PERRIN,DD (1965)
HL 3.16E-007	atm m3/mol	25	EST	MEYLAN,WM & HOWARD,PH (1991)
OH 4.64E-012	cm3/molc sec	25	EST	MEYLAN,WM & HOWARD,PH (1993)

CAS #: 003034-42-2 — 1-METHYL-5-NITROIMIDAZOLE

Formula: $C_4H_5N_3O_2$

Mol Weight: 127.10

MP (deg C):   FP (deg C):

BP (deg C):

BP pressure (mm Hg):

Property/Value	Units	Temp	Data Type	Reference
WS 2.98E+004	mg/L	25	EST	MEYLAN,WM ET AL. (1996)
logP 0.13			EXP	SUWINSKI,J ET AL. (1985)
VP 1.69E-003	mm Hg	25	EST	NEELY,WB & BLAU,GE (1985)
DC 2.13	pKa	25	EXP	PERRIN,DD (1965)
HL 3.16E-007	atm m3/mol	25	EST	MEYLAN,WM & HOWARD,PH (1991)
OH 4.64E-012	cm3/molc sec	25	EST	MEYLAN,WM & HOWARD,PH (1993)

CAS #: 003035-45-8 — G-32292 (ARATONE)

Formula: $C_8H_{15}N_5O$

Mol Weight: 197.24

MP (deg C):   FP (deg C):

BP (deg C):

BP pressure (mm Hg):

Property/Value	Units	Temp	Data Type	Reference
WS 3.50E+003	mg/L	21	EXP	GUNTHER,FA ET AL. (1968)
logP 2.66			EST	MEYLAN,WM & HOWARD,PH (1995)
VP 2.27E-004	mm Hg	25	EST	NEELY,WB & BLAU,GE (1985)
DC	pKa			
HL 2.45E-009	atm m3/mol	25	EST	MEYLAN,WM & HOWARD,PH (1991)
OH 2.08E-011	cm3/molc sec	25	EST	MEYLAN,WM & HOWARD,PH (1993)

CAS #: 003038-48-0 — ACETIC ACID,O-CF3 PHENYL ESTER

Formula: $C_9H_7F_3O_2$

Mol Weight: 204.15

MP (deg C): 100-102   FP (deg C):

BP (deg C):

BP pressure (mm Hg):

Property/Value	Units	Temp	Data Type	Reference
WS 2.56E+002	mg/L	25	EST	MEYLAN,WM ET AL. (1996)
logP 2.59			EXP	HANSCH,C & LEO,AJ (1985)
VP 3.30E-001	mm Hg	25	EST	NEELY,WB & BLAU,GE (1985)
DC	pKa			
HL 5.63E-004	atm m3/mol	25	EST	MEYLAN,WM & HOWARD,PH (1991)
OH 5.05E-013	cm3/molc sec	25	EST	MEYLAN,WM & HOWARD,PH (1993)

CAS #: 003039-74-5 — 4-HYDROXYQUINOLINE-1-OXIDE

Formula: $C_9H_9NO_2$

Mol Weight: 163.18

MP (deg C):   FP (deg C):

BP (deg C):

BP pressure (mm Hg):

Property/Value	Units	Temp	Data Type	Reference
WS 1.70E+005	mg/L	25	EST	MEYLAN,WM ET AL. (1996)
logP -0.47			EXP	HANSCH,C & LEO,AJ (1985)
VP 3.49E-006	mm Hg	25	EST	NEELY,WB & BLAU,GE (1985)
DC	pKa			
HL 7.16E-016	atm m3/mol	25	EST	MEYLAN,WM & HOWARD,PH (1991)
OH 2.00E-010	cm3/molc sec	25	EST	MEYLAN,WM & HOWARD,PH (1993)

CAS #: 003040-44-6 — 1-PIPERIDINE ETHANOL

Formula: $C_7H_{15}NO$

Mol Weight: 129.20

MP (deg C): 17.9   FP (deg C):

BP (deg C): 202

BP pressure (mm Hg):

Property/Value	Units	Temp	Data Type	Reference
WS 1.43E+005	mg/L	25	EST	MEYLAN,WM ET AL. (1996)
logP 0.96			EXP	HANSCH,C & LEO,AJ (1985)
VP 2.13E-002	mm Hg	25	EST	NEELY,WB & BLAU,GE (1985)
DC	pKa			
HL 1.83E-009	atm m3/mol	25	EST	MEYLAN,WM & HOWARD,PH (1991)
OH 1.06E-010	cm3/molc sec	25	EST	MEYLAN,WM & HOWARD,PH (1993)

CAS #: 003042-22-6 — 1H-PYRROLE, 2-PHENYL-

Formula: $C_{10}H_9N$

Mol Weight: 143.19

MP (deg C): 129   FP (deg C):

BP (deg C): 272

BP pressure (mm Hg):

Property/Value	Units	Temp	Data Type	Reference
WS 3.57E+002	mg/L	25	EST	MEYLAN,WM ET AL. (1996)
logP 2.76			EXP	SANGSTER,J (1993)
VP 1.83E-003	mm Hg	25	EST	NEELY,WB & BLAU,GE (1985)
DC	pKa			
HL 6.97E-007	atm m3/mol	25	EST	MEYLAN,WM & HOWARD,PH (1991)
OH 1.94E-010	cm3/molc sec	25	EST	MEYLAN,WM & HOWARD,PH (1993)

CAS #: 003055-86-5 — B-PHENOXYPROPIONITRILE

Formula: C_9H_9NO

Mol Weight: 147.18

MP (deg C):   FP (deg C):

BP (deg C):

BP pressure (mm Hg):

Property/Value	Units	Temp	Data Type	Reference
WS 1.79E+003	mg/L	25	EST	MEYLAN,WM ET AL. (1996)
logP 1.61			EXP	HANSCH,C & LEO,AJ (1985)
VP 9.31E-003	mm Hg	25	EST	NEELY,WB & BLAU,GE (1985)
DC	pKa			
HL 1.76E-007	atm m3/mol	25	EST	MEYLAN,WM & HOWARD,PH (1991)
OH 2.28E-011	cm3/molc sec	25	EST	MEYLAN,WM & HOWARD,PH (1993)

CAS #: 003056-17-5 — THYMIDINE, 2',3'-DIDEHYDRO-3'-DEOXY-

Formula: $C_{10}H_{12}N_2O_4$

Mol Weight: 224.22

MP (deg C):   FP (deg C):

BP (deg C):

BP pressure (mm Hg):

Property/Value	Units	Temp	Data Type	Reference
WS 6.60E+003	mg/L	25	EST	MEYLAN,WM ET AL. (1996)
logP -0.72			EXP	SANGSTER,J (1993)
VP 2.82E-012	mm Hg	25	EST	NEELY,WB & BLAU,GE (1985)
DC	pKa			
HL 2.31E-015	atm m3/mol	25	EST	MEYLAN,WM & HOWARD,PH (1991)
OH 1.31E-010	cm3/molc sec	25	EST	MEYLAN,WM & HOWARD,PH (1993)

CAS #: 003056-59-5				2-ETHYLPHENYL ACETATE

Formula: $C_{10}H_{12}O_2$

Mol Weight: 164.21

MP (deg C):   FP (deg C):

BP (deg C):

BP pressure (mm Hg):

Property/Value	Units	Temp	Data Type	Reference
WS 5.61E+002	mg/L	25	EST	MEYLAN,WM ET AL. (1996)
logP 2.42			EXP	HANSCH,C & LEO,AJ (1985)
VP 6.24E-002	mm Hg	25	EST	NEELY,WB & BLAU,GE (1985)
DC	pKa			
HL 9.50E-005	atm m3/mol	25	EST	MEYLAN,WM & HOWARD,PH (1991)
OH 4.76E-012	cm3/molc sec	25	EST	MEYLAN,WM & HOWARD,PH (1993)

CAS #: 003056-60-8				ACETIC ACID, 3-ETHYLPHENYL ESTER

Formula: $C_{10}H_{12}O_2$

Mol Weight: 164.21

MP (deg C):   FP (deg C):

BP (deg C):

BP pressure (mm Hg):

Property/Value	Units	Temp	Data Type	Reference
WS 5.29E+002	mg/L	25	EST	MEYLAN,WM ET AL. (1996)
logP 2.45			EXP	SOTOMATSU,T ET AL. (1993)
VP 6.24E-002	mm Hg	25	EST	NEELY,WB & BLAU,GE (1985)
DC	pKa			
HL 9.50E-005	atm m3/mol	25	EST	MEYLAN,WM & HOWARD,PH (1991)
OH 6.20E-012	cm3/molc sec	25	EST	MEYLAN,WM & HOWARD,PH (1993)

CAS #: 003056-73-3				CINNAMANILIDE

Formula: $C_{15}H_{13}NO$

Mol Weight: 223.28

MP (deg C):   FP (deg C):

BP (deg C):

BP pressure (mm Hg):

Property/Value	Units	Temp	Data Type	Reference
WS 2.73E+001	mg/L	25	EST	MEYLAN,WM ET AL. (1996)
logP 3.61			EXP	HANSCH,C & LEO,AJ (1985)
VP 2.98E-007	mm Hg	25	EST	NEELY,WB & BLAU,GE (1985)
DC	pKa			
HL 1.46E-010	atm m3/mol	25	EST	MEYLAN,WM & HOWARD,PH (1991)
OH 6.42E-011	cm3/molc sec	25	EST	MEYLAN,WM & HOWARD,PH (1993)

CAS #: 003060-89-7				METOBROMURON

Formula: $C_9H_{11}BrN_2O_2$

Mol Weight: 259.11

MP (deg C): 95-96   FP (deg C):

BP (deg C):

BP pressure (mm Hg):

Property/Value	Units	Temp	Data Type	Reference
WS 3.30E+002	mg/L	20	EXP	YALKOWSKY,SH & DANNENFELSER,RM (1992)
logP 2.38			EXP	BRIGGS,GG (1981)
VP 3.00E-006	mm Hg	25	EXP	WEBER,JB (1972)
DC	pKa			
HL 3.10E-009	atm m3/mol	25	EST	VP/WSOL
OH 1.38E-011	cm3/molc sec	25	EST	MEYLAN,WM & HOWARD,PH (1993)

CAS #: 003066-90-8				1,4-BENZENEDIOL, 2,3-DIMETHOXY-5-METHYL-

Formula: $C_9H_{12}O_4$

Mol Weight: 184.19

MP (deg C):   FP (deg C):

BP (deg C):

BP pressure (mm Hg):

Property/Value	Units	Temp	Data Type	Reference
WS 6.66E+003	mg/L	25	EST	MEYLAN,WM ET AL. (1996)
logP 1.05			EXP	RICH,PR (1990)
VP 1.69E-005	mm Hg	25	EST	NEELY,WB & BLAU,GE (1985)
DC	pKa			
HL 2.25E-013	atm m3/mol	25	EST	MEYLAN,WM & HOWARD,PH (1991)
OH 2.02E-010	cm3/molc sec	25	EST	MEYLAN,WM & HOWARD,PH (1993)

CAS #: 003068-88-0				4-METHYL-2-OXETANONE

Formula: $C_4H_6O_2$

Mol Weight: 86.09

MP (deg C):   FP (deg C):

BP (deg C):

BP pressure (mm Hg):

Property/Value	Units	Temp	Data Type	Reference
WS 2.68E+005	mg/L	25	EST	MEYLAN,WM ET AL. (1996)
logP -0.38			EST	MEYLAN,WM & HOWARD,PH (1995)
VP 1.65E+000	mm Hg	25	EST	NEELY,WB & BLAU,GE (1985)
DC	pKa			
HL 1.03E-004	atm m3/mol	25	EST	MEYLAN,WM & HOWARD,PH (1991)
OH 1.47E-012	cm3/molc sec	25	EST	MEYLAN,WM & HOWARD,PH (1993)

CAS #: 003070-13-1				O,O-DIET-O-(4-METHIOPH)PHOSPHATE

Formula: $C_{11}H_{17}O_4PS$

Mol Weight: 276.29

MP (deg C):   FP (deg C):

BP (deg C):

BP pressure (mm Hg):

Property/Value	Units	Temp	Data Type	Reference
WS 6.97E+001	mg/L	25	EST	MEYLAN,WM ET AL. (1996)
logP 2.79			EXP	HANSCH,C & LEO,AJ (1985)
VP 1.54E-005	mm Hg	25	EST	NEELY,WB & BLAU,GE (1985)
DC	pKa			
HL 4.73E-009	atm m3/mol	25	EST	MEYLAN,WM & HOWARD,PH (1991)
OH 4.89E-011	cm3/molc sec	25	EST	MEYLAN,WM & HOWARD,PH (1993)

CAS #: 003070-15-3				FENSULFOTHION SULFIDE

Formula: $C_{11}H_{17}O_3PS_2$

Mol Weight: 292.36

MP (deg C):   FP (deg C):

BP (deg C):

BP pressure (mm Hg):

Property/Value	Units	Temp	Data Type	Reference
WS 3.81E+000	mg/L	25	EST	MEYLAN,WM ET AL. (1996)
logP 4.16			EXP	HANSCH,C & LEO,AJ (1985)
VP 7.32E-005	mm Hg	25	EST	NEELY,WB & BLAU,GE (1985)
DC	pKa			
HL 2.19E-006	atm m3/mol	25	EST	MEYLAN,WM & HOWARD,PH (1991)
OH 1.02E-010	cm3/molc sec	25	EST	MEYLAN,WM & HOWARD,PH (1993)

CAS #: 003073-66-3				1,1,3-TRIMETHYLCYCLOHEXANE
Formula: C$_9$H$_{18}$				
Mol Weight: 126.24				
MP (deg C): -65.7		FP (deg C):		
BP (deg C): 136.6				
BP pressure (mm Hg):				

Property/ Value	Units	Temp	Data Type	Reference
WS 1.77E+000	mg/L	25	EXP	YALKOWSKY,SH & DANNENFELSER,RM (1992)
logP 4.46			EST	MEYLAN,WM & HOWARD,PH (1995)
VP 1.08E+001	mm Hg	25	EXT	OHE,S (1976)
DC	pKa			
HL 5.97E-001	atm m3/mol	25	EST	MEYLAN,WM & HOWARD,PH (1991)
OH 8.73E-012	cm3/molc sec	25	EXP	ATKINSON,R (1989)

CAS #: 003077-12-1				ETHANOL, 2,2'- (4-METHYLPHENYL)IMINO BIS-
Formula: C$_{11}$H$_{17}$NO$_2$				
Mol Weight: 195.26				
MP (deg C):		FP (deg C):		
BP (deg C): 338-340				
BP pressure (mm Hg):				

Property/ Value	Units	Temp	Data Type	Reference
WS 5.42E+003	mg/L	25	EST	MEYLAN,WM ET AL. (1996)
logP 1.09			EXP	SANGSTER,J (1993)
VP 7.34E-007	mm Hg	25	EST	NEELY,WB & BLAU,GE (1985)
DC	pKa			
HL 2.23E-010	atm m3/mol	25	EST	MEYLAN,WM & HOWARD,PH (1991)
OH 2.09E-010	cm3/molc sec	25	EST	MEYLAN,WM & HOWARD,PH (1993)

CAS #: 003085-54-9				P-METHYLFORMANILIDE
Formula: C$_8$H$_9$NO				
Mol Weight: 135.17				
MP (deg C):		FP (deg C):		
BP (deg C):				
BP pressure (mm Hg):				

Property/ Value	Units	Temp	Data Type	Reference
WS 3.70E+003	mg/L	25	EST	MEYLAN,WM ET AL. (1996)
logP 1.61			EXP	HANSCH,C & LEO,AJ (1985)
VP 4.02E-004	mm Hg	25	EST	NEELY,WB & BLAU,GE (1985)
DC	pKa			
HL 9.32E-009	atm m3/mol	25	EST	MEYLAN,WM & HOWARD,PH (1991)
OH 5.24E-011	cm3/molc sec	25	EST	MEYLAN,WM & HOWARD,PH (1993)

CAS #: 003085-79-8				1-BUTANAMINIUM, N,N-DIBUTYL-N-METHYL-, IODIDE
Formula: C$_{13}$H$_{30}$IN				
Mol Weight: 327.30				
MP (deg C):		FP (deg C):		
BP (deg C):				
BP pressure (mm Hg):				

Property/ Value	Units	Temp	Data Type	Reference
WS 5.08E+003	mg/L	25	EST	MEYLAN,WM ET AL. (1996)
logP 0.26			EXP	NEEF,C & MEIJER,DKF (1984)
VP 3.39E-009	mm Hg	25	EST	NEELY,WB & BLAU,GE (1985)
DC	pKa			
HL 3.37E-011	atm m3/mol	25	EST	MEYLAN,WM & HOWARD,PH (1991)
OH 4.15E-011	cm3/molc sec	25	EST	MEYLAN,WM & HOWARD,PH (1993)

CAS #: 003088-31-1				DIETHYLENE GLYCOL MOMOLAURYL ETHER SODIUM SULFAT
Formula: C$_{16}$H$_{33}$NaO$_6$S				
Mol Weight: 376.49				
MP (deg C):		FP (deg C):		
BP (deg C):				
BP pressure (mm Hg):				

Property/ Value	Units	Temp	Data Type	Reference
WS 4.50E+001	mg/L	25	EST	MEYLAN,WM ET AL. (1996)
logP 1.14			EST	MEYLAN,WM & HOWARD,PH (1995)
VP 2.57E-015	mm Hg	25	EST	NEELY,WB & BLAU,GE (1985)
DC	pKa			
HL	atm m3/mol			
OH 4.55E-011	cm3/molc sec	25	EST	MEYLAN,WM & HOWARD,PH (1993)

CAS #: 003093-23-0				ETHANOL, 2-[2-[4-[3-[2-(TRIFLUOROMETHYL)PHENOTHI
Formula: C$_{24}$H$_{30}$F$_3$N$_3$O$_2$S				
Mol Weight: 481.58				
MP (deg C):		FP (deg C):		
BP (deg C):				
BP pressure (mm Hg):				

Property/ Value	Units	Temp	Data Type	Reference
WS 1.09E-001	mg/L	25	EST	MEYLAN,WM ET AL. (1996)
logP 4.49			EXP	HANSCH,C ET AL. (1995)
VP 5.30E-014	mm Hg	25	EST	NEELY,WB & BLAU,GE (1985)
DC	pKa			
HL 9.47E-019	atm m3/mol	25	EST	MEYLAN,WM & HOWARD,PH (1991)
OH 2.18E-010	cm3/molc sec	25	EST	MEYLAN,WM & HOWARD,PH (1993)

CAS #: 003097-21-0				2-BENZIMIDAZOLINONE
Formula: C$_9$H$_{10}$N$_2$O				
Mol Weight: 162.19				
MP (deg C):		FP (deg C):		
BP (deg C):				
BP pressure (mm Hg):				

Property/ Value	Units	Temp	Data Type	Reference
WS 3.79E+003	mg/L	25	EST	MEYLAN,WM ET AL. (1996)
logP 1.46			EXP	HANSCH,C & LEO,AJ (1985)
VP 1.15E-003	mm Hg	25	EST	NEELY,WB & BLAU,GE (1985)
DC	pKa			
HL 2.12E-007	atm m3/mol	25	EST	MEYLAN,WM & HOWARD,PH (1991)
OH 7.61E-012	cm3/molc sec	25	EST	MEYLAN,WM & HOWARD,PH (1993)

CAS #: 003112-85-4				METHYL PHENYL SULFONE
Formula: C$_7$H$_8$O$_2$S				
Mol Weight: 156.20				
MP (deg C):		FP (deg C):		
BP (deg C):				
BP pressure (mm Hg):				

Property/ Value	Units	Temp	Data Type	Reference
WS 2.66E+004	mg/L	25	EST	MEYLAN,WM ET AL. (1996)
logP 0.56			EST	MEYLAN,WM & HOWARD,PH (1995)
VP 5.83E-003	mm Hg	25	EST	NEELY,WB & BLAU,GE (1985)
DC	pKa			
HL 1.25E-006	atm m3/mol	25	EST	MEYLAN,WM & HOWARD,PH (1991)
OH 1.48E-012	cm3/molc sec	25	EST	MEYLAN,WM & HOWARD,PH (1993)

CAS #: 003112-88-7 — BENZYL PHENYL SULFONE

Formula: $C_{13}H_{12}O_2S$
Mol Weight: 232.30
MP (deg C): 146
FP (deg C):
BP (deg C):
BP pressure (mm Hg):

Property/Value	Units	Temp	Data Type	Reference
WS 5.25E+002	mg/L	25	EST	MEYLAN,WM ET AL. (1996)
logP 2.05			EXP	HANSCH,C & LEO,AJ (1985)
VP 6.59E-006	mm Hg	25	EST	NEELY,WB & BLAU,GE (1985)
DC	pKa			
HL 1.01E-007	atm m3/mol	25	EST	MEYLAN,WM & HOWARD,PH (1991)
OH 1.25E-011	cm3/molc sec	25	EST	MEYLAN,WM & HOWARD,PH (1993)

CAS #: 003115-28-4 — HEXANOIC ACID, 2-BUTYL-

Formula: $C_{10}H_{20}O_2$
Mol Weight: 172.27
MP (deg C):
FP (deg C):
BP (deg C):
BP pressure (mm Hg):

Property/Value	Units	Temp	Data Type	Reference
WS 2.76E+002	mg/L	25	EST	MEYLAN,WM ET AL. (1996)
logP 3.20			EXP	SANGSTER,J (1993)
VP 4.80E-003	mm Hg	25	EST	NEELY,WB & BLAU,GE (1985)
DC	pKa			
HL 5.28E-006	atm m3/mol	25	EST	MEYLAN,WM & HOWARD,PH (1991)
OH 1.10E-011	cm3/molc sec	25	EST	MEYLAN,WM & HOWARD,PH (1993)

CAS #: 003116-76-5 — DICLOXACILLIN

Formula: $C_{19}H_{17}Cl_2N_3O_5S$
Mol Weight: 470.33
MP (deg C):
FP (deg C):
BP (deg C):
BP pressure (mm Hg):

Property/Value	Units	Temp	Data Type	Reference
WS 3.63E+000	mg/L	25	EST	MEYLAN,WM ET AL. (1996)
logP 2.91			EXP	HANSCH,C & LEO,AJ (1985)
VP 3.37E-016	mm Hg	25	EST	NEELY,WB & BLAU,GE (1985)
DC	pKa			
HL 1.40E-017	atm m3/mol	25	EST	MEYLAN,WM & HOWARD,PH (1991)
OH 9.58E-011	cm3/molc sec	25	EST	MEYLAN,WM & HOWARD,PH (1993)

CAS #: 003117-02-0 — 2,3-DIMETHOXY-P-BENZOQUINONE

Formula: $C_8H_8O_4$
Mol Weight: 168.15
MP (deg C):
FP (deg C):
BP (deg C):
BP pressure (mm Hg):

Property/Value	Units	Temp	Data Type	Reference
WS 3.22E+004	mg/L	25	EST	MEYLAN,WM ET AL. (1996)
logP 0.34			EXP	HUANG,JX ET AL. (1985)
VP 2.34E-003	mm Hg	25	EST	NEELY,WB & BLAU,GE (1985)
DC	pKa			
HL 5.27E-010	atm m3/mol	25	EST	MEYLAN,WM & HOWARD,PH (1991)
OH 2.67E-011	cm3/molc sec	25	EST	MEYLAN,WM & HOWARD,PH (1993)

CAS #: 003117-05-3 — 2,3,5-TRIMETHOXY-P-BENZOQUINONE

Formula: $C_9H_{10}O_5$
Mol Weight: 198.18
MP (deg C):
FP (deg C):
BP (deg C):
BP pressure (mm Hg):

Property/Value	Units	Temp	Data Type	Reference
WS 2.90E+004	mg/L	25	EST	MEYLAN,WM ET AL. (1996)
logP 0.22			EXP	HUANG,JX ET AL. (1985)
VP 3.60E-004	mm Hg	25	EST	NEELY,WB & BLAU,GE (1985)
DC	pKa			
HL 3.47E-010	atm m3/mol	25	EST	MEYLAN,WM & HOWARD,PH (1991)
OH 3.91E-011	cm3/molc sec	25	EST	MEYLAN,WM & HOWARD,PH (1993)

CAS #: 003117-06-4 — 2,3,5,6-TETRAMETHOXY-P-BENZOQUINONE

Formula: $C_{10}H_{12}O_6$
Mol Weight: 228.20
MP (deg C):
FP (deg C):
BP (deg C):
BP pressure (mm Hg):

Property/Value	Units	Temp	Data Type	Reference
WS 1.39E+004	mg/L	25	EST	MEYLAN,WM ET AL. (1996)
logP 0.41			EXP	HUANG,JX ET AL. (1985)
VP 5.71E-005	mm Hg	25	EST	NEELY,WB & BLAU,GE (1985)
DC	pKa			
HL 2.28E-010	atm m3/mol	25	EST	MEYLAN,WM & HOWARD,PH (1991)
OH 4.89E-011	cm3/molc sec	25	EST	MEYLAN,WM & HOWARD,PH (1993)

CAS #: 003118-68-1 — BENZENESULFONAMIDE, 3-CYANO-

Formula: $C_7H_6N_2O_2S$
Mol Weight: 182.20
MP (deg C):
FP (deg C):
BP (deg C):
BP pressure (mm Hg):

Property/Value	Units	Temp	Data Type	Reference
WS 1.82E+004	mg/L	25	EST	MEYLAN,WM ET AL. (1996)
logP 0.24			EXP	CAROTTI,A ET AL. (1989)
VP 1.65E-005	mm Hg	25	EST	NEELY,WB & BLAU,GE (1985)
DC	pKa			
HL 4.08E-009	atm m3/mol	25	EST	MEYLAN,WM & HOWARD,PH (1991)
OH 7.36E-014	cm3/molc sec	25	EST	MEYLAN,WM & HOWARD,PH (1993)

CAS #: 003118-97-6 — C.I. SOLVENT ORANGE 7

Formula: $C_{18}H_{16}N_2O$
Mol Weight: 276.34
MP (deg C): 166
FP (deg C):
BP (deg C):
BP pressure (mm Hg):

Property/Value	Units	Temp	Data Type	Reference
WS 5.45E-002	mg/L	25	EST	MEYLAN,WM ET AL. (1996)
logP 6.60			EST	MEYLAN,WM & HOWARD,PH (1995)
VP 1.17E-008	mm Hg	25	EST	NEELY,WB & BLAU,GE (1985)
DC	pKa			
HL 2.74E-011	atm m3/mol	25	EST	MEYLAN,WM & HOWARD,PH (1991)
OH 2.32E-011	cm3/molc sec	25	EST	MEYLAN,WM & HOWARD,PH (1993)

CAS #: 003119-02-6				P-CYANOBENZENESULFONAMIDE

Formula: $C_7H_6N_2O_2S$

Mol Weight: 182.20

MP (deg C):   FP (deg C):

BP (deg C):

BP pressure (mm Hg):

Property/Value	Units	Temp	Data Type	Reference
WS 1.86E+004	mg/L	25	EST	MEYLAN,WM ET AL. (1996)
logP 0.23			EXP	HANSCH,C & LEO,AJ (1985)
VP 1.65E-005	mm Hg	25	EST	NEELY,WB & BLAU,GE (1985)
DC	pKa			
HL 4.08E-009	atm m3/mol	25	EST	MEYLAN,WM & HOWARD,PH (1991)
OH 7.36E-014	cm3/molc sec	25	EST	MEYLAN,WM & HOWARD,PH (1993)

CAS #: 003121-79-7				4,4-DIMETHYL-1-PENTANOL

Formula: $C_7H_{16}O$

Mol Weight: 116.20

MP (deg C):   FP (deg C):

BP (deg C):

BP pressure (mm Hg):

Property/Value	Units	Temp	Data Type	Reference
WS 3.35E+003	mg/L	25	EXP	BARTON,AFM (1984)
logP 2.20			EST	MEYLAN,WM & HOWARD,PH (1995)
VP 8.08E-001	mm Hg	25	EST	NEELY,WB & BLAU,GE (1985)
DC	pKa			
HL 2.34E-005	atm m3/mol	25	EST	MEYLAN,WM & HOWARD,PH (1991)
OH 7.49E-012	cm3/molc sec	25	EST	MEYLAN,WM & HOWARD,PH (1993)

CAS #: 003125-63-1				PHENOL, 3-ISOTHIOCYANATO-

Formula: C_7H_5NOS

Mol Weight: 151.19

MP (deg C):   FP (deg C):

BP (deg C):

BP pressure (mm Hg):

Property/Value	Units	Temp	Data Type	Reference
WS 4.97E+002	mg/L	25	EST	MEYLAN,WM ET AL. (1996)
logP 3.23			EXP	AUGUSTIN,J ET AL. (1987)
VP 9.26E-004	mm Hg	25	EST	NEELY,WB & BLAU,GE (1985)
DC	pKa			
HL 8.13E-008	atm m3/mol	25	EST	MEYLAN,WM & HOWARD,PH (1991)
OH 2.11E-011	cm3/molc sec	25	EST	MEYLAN,WM & HOWARD,PH (1993)

CAS #: 003125-64-2				BENZENE, 1-ISOTHIOCYANATO-3-METHOXY

Formula: C_8H_7NOS

Mol Weight: 165.22

MP (deg C):   FP (deg C):

BP (deg C):

BP pressure (mm Hg):

Property/Value	Units	Temp	Data Type	Reference
WS 4.66E+001	mg/L	25	EST	MEYLAN,WM ET AL. (1996)
logP 3.68			EXP	AUGUSTIN,J ET AL. (1987)
VP 8.41E-003	mm Hg	25	EST	NEELY,WB & BLAU,GE (1985)
DC	pKa			
HL 4.62E-005	atm m3/mol	25	EST	MEYLAN,WM & HOWARD,PH (1991)
OH 1.44E-011	cm3/molc sec	25	EST	MEYLAN,WM & HOWARD,PH (1993)

CAS #: 003125-78-8				BENZONITRILE, 3-ISOTHIOCYANATO-

Formula: $C_8H_4N_2S$

Mol Weight: 160.20

MP (deg C):   FP (deg C):

BP (deg C):

BP pressure (mm Hg):

Property/Value	Units	Temp	Data Type	Reference
WS 7.88E+001	mg/L	25	EST	MEYLAN,WM ET AL. (1996)
logP 3.13			EXP	AUGUSTIN,J ET AL. (1987)
VP 9.53E-004	mm Hg	25	EST	NEELY,WB & BLAU,GE (1985)
DC	pKa			
HL 7.55E-006	atm m3/mol	25	EST	MEYLAN,WM & HOWARD,PH (1991)
OH 3.78E-013	cm3/molc sec	25	EST	MEYLAN,WM & HOWARD,PH (1993)

CAS #: 003126-90-7				1,3-BENZENEDICARBOXYLIC ACID, DIBUTYL ESTER

Formula: $C_{16}H_{22}O_4$

Mol Weight: 278.35

MP (deg C):   FP (deg C):

BP (deg C):

BP pressure (mm Hg):

Property/Value	Units	Temp	Data Type	Reference
WS 3.77E+000	mg/L	25	EST	MEYLAN,WM ET AL. (1996)
logP 4.26			EXP	HANSCH,C ET AL. (1995)
VP 2.55E-004	mm Hg	25	EST	NEELY,WB & BLAU,GE (1985)
DC	pKa			
HL 1.22E-006	atm m3/mol	25	EST	MEYLAN,WM & HOWARD,PH (1991)
OH 9.34E-012	cm3/molc sec	25	EST	MEYLAN,WM & HOWARD,PH (1993)

CAS #: 003132-64-7				ALPHA-EPIBROMOHYDRIN

Formula: C_3H_5BrO

Mol Weight: 136.98

MP (deg C): -40   FP (deg C):

BP (deg C): 134-6

BP pressure (mm Hg):

Property/Value	Units	Temp	Data Type	Reference
WS 1.62E+004	mg/L	25	EST	MEYLAN,WM ET AL. (1996)
logP 0.85			EXP	DENEER,JW ET AL. (1988)
VP 2.00E+000	mm Hg	25	EST	SEE COMMENTS
DC	pKa			
HL 1.84E-005	atm m3/mol	25	EST	MEYLAN,WM & HOWARD,PH (1991)
OH 5.76E-013	cm3/molc sec	25	EST	ATKINSON,R (1988)

CAS #: 003140-93-0				THIOPHENE, 2,3-DIBROMO-

Formula: $C_4H_2Br_2S$

Mol Weight: 241.94

MP (deg C): -17.5   FP (deg C):

BP (deg C): 218.5

BP pressure (mm Hg):

Property/Value	Units	Temp	Data Type	Reference
WS 2.53E+001	mg/L	25	EST	MEYLAN,WM ET AL. (1996)
logP 3.53			EXP	HANSCH,C & LEO,AJ (1985)
VP 6.93E-002	mm Hg	25	EST	NEELY,WB & BLAU,GE (1985)
DC	pKa			
HL 4.65E-004	atm m3/mol	25	EST	MEYLAN,WM & HOWARD,PH (1991)
OH 1.72E-012	cm3/molc sec	25	EST	MEYLAN,WM & HOWARD,PH (1993)

003141-26-2 — 3,4-DIBROMOTHIOPHENE

Formula:	$C_4H_2Br_2S$			
Mol Weight:	241.93			
MP (deg C):	4-5	FP (deg C):		
BP (deg C):	221-222			
BP pressure (mm Hg):				

Property/Value	Units	Temp	Data Type	Reference
WS 2.24E+001	mg/L	25	EST	MEYLAN,WM ET AL. (1996)
logP 3.59			EST	MEYLAN,WM & HOWARD,PH (1995)
VP 1.83E+000	mm Hg	25	EXT	OHE,S (1976)
DC	pKa			
HL 4.65E-004	atm m3/mol	25	EST	MEYLAN,WM & HOWARD,PH (1991)
OH 1.72E-012	cm3/molc sec	25	EST	MEYLAN,WM & HOWARD,PH (1993)

003141-27-3 — 2,5-DIBROMOTHIOPHENE

Formula:	$C_4H_2Br_2S$			
Mol Weight:	241.94			
MP (deg C):	-6	FP (deg C):		
BP (deg C):	210.3			
BP pressure (mm Hg):				

Property/Value	Units	Temp	Data Type	Reference
WS 2.24E+001	mg/L	25	EST	MEYLAN,WM ET AL. (1996)
logP 3.59			EST	MEYLAN,WM & HOWARD,PH (1995)
VP 6.93E-002	mm Hg	25	EST	NEELY,WB & BLAU,GE (1985)
DC	pKa			
HL 4.65E-004	atm m3/mol	25	EST	MEYLAN,WM & HOWARD,PH (1991)
OH 1.72E-012	cm3/molc sec	25	EST	MEYLAN,WM & HOWARD,PH (1993)

003141-42-2 — BENZENE, 1,1'-(2,2-DICHLOROCYCLOPROPYLIDENE)BIS-

Formula:	$C_{15}H_{12}Cl_2$			
Mol Weight:	263.17			
MP (deg C):		FP (deg C):		
BP (deg C):				
BP pressure (mm Hg):				

Property/Value	Units	Temp	Data Type	Reference
WS 1.36E+000	mg/L	25	EST	MEYLAN,WM ET AL. (1996)
logP 4.88			EXP	NISHIMURA,K ET AL. (1986)
VP 8.59E-005	mm Hg	25	EST	NEELY,WB & BLAU,GE (1985)
DC	pKa			
HL 4.63E-005	atm m3/mol	25	EST	MEYLAN,WM & HOWARD,PH (1991)
OH 8.60E-012	cm3/molc sec	25	EST	MEYLAN,WM & HOWARD,PH (1993)

003144-16-9 — CAMPHORSULFONIC ACID

Formula:	$C_{10}H_{16}O_4S$			
Mol Weight:	232.30			
MP (deg C):	195 dec	FP (deg C):		
BP (deg C):				
BP pressure (mm Hg):				

Property/Value	Units	Temp	Data Type	Reference
WS 1.65E+005	mg/L	25	EST	MEYLAN,WM ET AL. (1996)
logP -0.87			EST	MEYLAN,WM & HOWARD,PH (1995)
VP 6.83E-008	mm Hg	25	EST	NEELY,WB & BLAU,GE (1985)
DC	pKa			
HL 1.20E-011	atm m3/mol	25	EST	MEYLAN,WM & HOWARD,PH (1991)
OH 1.58E-011	cm3/molc sec	25	EST	MEYLAN,WM & HOWARD,PH (1993)

003146-66-5 — 2-CL TETRAMETHYLMELAMINE

Formula:	$C_7H_{12}ClN_5$			
Mol Weight:	201.66			
MP (deg C):		FP (deg C):		
BP (deg C):				
BP pressure (mm Hg):				

Property/Value	Units	Temp	Data Type	Reference
WS 4.57E+002	mg/L	25	EST	MEYLAN,WM ET AL. (1996)
logP 2.31			EXP	HANSCH,C & LEO,AJ (1985)
VP 1.25E-003	mm Hg	25	EST	NEELY,WB & BLAU,GE (1985)
DC	pKa			
HL 8.03E-007	atm m3/mol	25	EST	MEYLAN,WM & HOWARD,PH (1991)
OH 5.18E-012	cm3/molc sec	25	EST	MEYLAN,WM & HOWARD,PH (1993)

003149-01-7 — BENZENESULFONAMIDE, N-[[(1-METHYLETHYL)AMINO]CAR

Formula:	$C_{10}H_{14}N_2O_3S$			
Mol Weight:	242.30			
MP (deg C):		FP (deg C):		
BP (deg C):				
BP pressure (mm Hg):				

Property/Value	Units	Temp	Data Type	Reference
WS 1.87E+003	mg/L	25	EST	MEYLAN,WM ET AL. (1996)
logP 1.34			EXP	CLOUX,JL ET AL. (1988)
VP 2.77E-007	mm Hg	25	EST	NEELY,WB & BLAU,GE (1985)
DC	pKa			
HL 1.73E-010	atm m3/mol	25	EST	MEYLAN,WM & HOWARD,PH (1991)
OH 1.07E-011	cm3/molc sec	25	EST	MEYLAN,WM & HOWARD,PH (1993)

003149-28-8 — 2-METHOXYPYRAZINE

Formula:	$C_5H_6N_2O$			
Mol Weight:	110.12			
MP (deg C):		FP (deg C):		
BP (deg C):	61			
BP pressure (mm Hg):	2.90E+001			

Property/Value	Units	Temp	Data Type	Reference
WS 2.59E+004	mg/L	25	EST	MEYLAN,WM ET AL. (1996)
logP 0.73			EXP	YAMAGAMI,C ET AL. (1990A)
VP 1.60E+000	mm Hg	25	EST	NEELY,WB & BLAU,GE (1985)
DC 0.75	pKa	20	EXP	PERRIN,DD (1965)
HL 1.57E-006	atm m3/mol	25	EST	MEYLAN,WM & HOWARD,PH (1991)
OH 3.37E-012	cm3/molc sec	25	EST	MEYLAN,WM & HOWARD,PH (1993)

003150-24-1 — GALACTOPYRANOSIDE,2-NITROPHENYL

Formula:	$C_{12}H_{15}NO_8$			
Mol Weight:	301.26			
MP (deg C):		FP (deg C):		
BP (deg C):				
BP pressure (mm Hg):				

Property/Value	Units	Temp	Data Type	Reference
WS 7.01E+003	mg/L	25	EST	MEYLAN,WM ET AL. (1996)
logP -0.59			EXP	HANSCH,C & LEO,AJ (1985)
VP 3.68E-013	mm Hg	25	EST	NEELY,WB & BLAU,GE (1985)
DC	pKa			
HL 4.54E-018	atm m3/mol	25	EST	MEYLAN,WM & HOWARD,PH (1991)
OH 6.47E-011	cm3/molc sec	25	EST	MEYLAN,WM & HOWARD,PH (1993)

CAS #: 003150-53-6	1-(2-OHET)-5-CYANO-2-NO2 PYRROLE

Formula: $C_7H_9N_3O_3$

Mol Weight: 183.17

MP (deg C): | **FP (deg C):**

BP (deg C):

BP pressure (mm Hg):

Property/Value	Units	Temp	Data Type	Reference
WS 3.70E+003	mg/L	25	EST	MEYLAN,WM ET AL. (1996)
logP 0.60			EXP	HANSCH,C & LEO,AJ (1985)
VP 6.99E-007	mm Hg	25	EST	NEELY,WB & BLAU,GE (1985)
DC	pKa			
HL 3.58E-013	atm m3/mol	25	EST	MEYLAN,WM & HOWARD,PH (1991)
OH 7.11E-012	cm3/molc sec	25	EST	MEYLAN,WM & HOWARD,PH (1993)

CAS #: 003153-44-4	BENZENEBUTANOIC ACID, 4-METHOXY---OXO-

Formula: $C_{11}H_{12}O_4$

Mol Weight: 208.22

MP (deg C): 148-150 | **FP (deg C):**

BP (deg C):

BP pressure (mm Hg):

Property/Value	Units	Temp	Data Type	Reference
WS 6.53E+003	mg/L	25	EST	MEYLAN,WM ET AL. (1996)
logP 1.38			EXP	KUCHAR,M ET AL. (1985)
VP 1.46E-005	mm Hg	25	EST	NEELY,WB & BLAU,GE (1985)
DC	pKa			
HL 5.73E-012	atm m3/mol	25	EST	MEYLAN,WM & HOWARD,PH (1991)
OH 2.45E-011	cm3/molc sec	25	EST	MEYLAN,WM & HOWARD,PH (1993)

CAS #: 003156-34-1	2-CHLORO-B-NITROSTYRENE

Formula: $C_8H_6ClNO_2$

Mol Weight: 183.60

MP (deg C): | **FP (deg C):**

BP (deg C):

BP pressure (mm Hg):

Property/Value	Units	Temp	Data Type	Reference
WS 7.93E+001	mg/L	25	EST	MEYLAN,WM ET AL. (1996)
logP 2.85			EXP	HANSCH,C & LEO,AJ (1985)
VP 2.53E-003	mm Hg	25	EST	NEELY,WB & BLAU,GE (1985)
DC	pKa			
HL 2.57E-006	atm m3/mol	25	EST	MEYLAN,WM & HOWARD,PH (1991)
OH 1.27E-011	cm3/molc sec	25	EST	MEYLAN,WM & HOWARD,PH (1993)

CAS #: 003156-35-2	3-CHLORO-B-NITROSTYRENE

Formula: $C_8H_6ClNO_2$

Mol Weight: 183.60

MP (deg C): | **FP (deg C):**

BP (deg C):

BP pressure (mm Hg):

Property/Value	Units	Temp	Data Type	Reference
WS 1.38E+002	mg/L	25	EST	MEYLAN,WM ET AL. (1996)
logP 2.57			EXP	HANSCH,C & LEO,AJ (1985)
VP 2.53E-003	mm Hg	25	EST	NEELY,WB & BLAU,GE (1985)
DC	pKa			
HL 2.57E-006	atm m3/mol	25	EST	MEYLAN,WM & HOWARD,PH (1991)
OH 1.27E-011	cm3/molc sec	25	EST	MEYLAN,WM & HOWARD,PH (1993)

CAS #: 003156-39-6	2-NITRO-B-NITROSTYRENE

Formula: $C_8H_6N_2O_4$

Mol Weight: 194.15

MP (deg C): | **FP (deg C):**

BP (deg C):

BP pressure (mm Hg):

Property/Value	Units	Temp	Data Type	Reference
WS 5.54E+002	mg/L	25	EST	MEYLAN,WM ET AL. (1996)
logP 1.80			EXP	HANSCH,C & LEO,AJ (1985)
VP 1.15E-004	mm Hg	25	EST	NEELY,WB & BLAU,GE (1985)
DC	pKa			
HL 1.37E-008	atm m3/mol	25	EST	MEYLAN,WM & HOWARD,PH (1991)
OH 1.16E-011	cm3/molc sec	25	EST	MEYLAN,WM & HOWARD,PH (1993)

CAS #: 003156-41-0	4-NITRO-B-NITROSTYRENE

Formula: $C_8H_6N_2O_4$

Mol Weight: 194.15

MP (deg C): | **FP (deg C):**

BP (deg C):

BP pressure (mm Hg):

Property/Value	Units	Temp	Data Type	Reference
WS 4.64E+002	mg/L	25	EST	MEYLAN,WM ET AL. (1996)
logP 1.89			EXP	HANSCH,C & LEO,AJ (1985)
VP 1.15E-004	mm Hg	25	EST	NEELY,WB & BLAU,GE (1985)
DC	pKa			
HL 1.37E-008	atm m3/mol	25	EST	MEYLAN,WM & HOWARD,PH (1991)
OH 1.16E-011	cm3/molc sec	25	EST	MEYLAN,WM & HOWARD,PH (1993)

CAS #: 003156-44-3	3-HYDROXY-B-NITROSTYRENE

Formula: $C_8H_7NO_3$

Mol Weight: 165.15

MP (deg C): | **FP (deg C):**

BP (deg C):

BP pressure (mm Hg):

Property/Value	Units	Temp	Data Type	Reference
WS 1.71E+003	mg/L	25	EST	MEYLAN,WM ET AL. (1996)
logP 2.07			EXP	HANSCH,C & LEO,AJ (1985)
VP 1.68E-004	mm Hg	25	EST	NEELY,WB & BLAU,GE (1985)
DC	pKa			
HL 3.62E-010	atm m3/mol	25	EST	MEYLAN,WM & HOWARD,PH (1991)
OH 4.29E-011	cm3/molc sec	25	EST	MEYLAN,WM & HOWARD,PH (1993)

CAS #: 003160-37-0	PIPERONYL ACETONE

Formula: $C_{11}H_{10}O_3$

Mol Weight: 190.20

MP (deg C): 111 | **FP (deg C):**

BP (deg C):

BP pressure (mm Hg):

Property/Value	Units	Temp	Data Type	Reference
WS 7.90E+002	mg/L	25	EST	MEYLAN,WM ET AL. (1996)
logP 2.10			EST	MEYLAN,WM & HOWARD,PH (1995)
VP 5.95E-004	mm Hg	25	EST	NEELY,WB & BLAU,GE (1985)
DC	pKa			
HL 1.36E-009	atm m3/mol	25	EST	MEYLAN,WM & HOWARD,PH (1991)
OH 7.11E-011	cm3/molc sec	25	EST	MEYLAN,WM & HOWARD,PH (1993)

CAS #: 003163-07-3	1,3-BENZENEDIOL, 4-NITRO-

Formula: C6H5NO4	
Mol Weight: 155.11	
MP (deg C): 122	FP (deg C):
BP (deg C): 178	
BP pressure (mm Hg): 1.10E+001	

Property/Value	Units	Temp	Data Type	Reference
WS 9.14E+003	mg/L	25	EST	MEYLAN,WM ET AL. (1996)
logP 1.05			EXP	SANGSTER,J (1993)
VP 2.27E-005	mm Hg	25	EST	NEELY,WB & BLAU,GE (1985)
DC	pKa			
HL 7.28E-010	atm m3/mol	25	EST	MEYLAN,WM & HOWARD,PH (1991)
OH 7.15E-011	cm3/molc sec	25	EST	MEYLAN,WM & HOWARD,PH (1993)

CAS #: 003165-93-3	4-CHLORO-2-METHYLBENZENAMINE HYDROCHLORIDE

Formula: C7H9Cl2N	
Mol Weight: 178.06	
MP (deg C):	FP (deg C):
BP (deg C):	
BP pressure (mm Hg):	

Property/Value	Units	Temp	Data Type	Reference
WS 9.54E+002	mg/L	25	EST	MEYLAN,WM ET AL. (1996)
logP 2.27			EST	MEYLAN,WM & HOWARD,PH (1995)
VP 4.08E-002	mm Hg	25	EST	NEELY,WB & BLAU,GE (1985)
DC 3.85	pKa	25	EXP	PERRIN,DD (1972)
HL 1.56E-006	atm m3/mol	25	EST	MEYLAN,WM & HOWARD,PH (1991)
OH 3.87E-011	cm3/molc sec	25	EST	MEYLAN,WM & HOWARD,PH (1993)

CAS #: 003172-52-9	2,5-DICHLOROTHIOPHENE

Formula: C4H2Cl2S	
Mol Weight: 153.03	
MP (deg C): -40.5	FP (deg C):
BP (deg C): 162	
BP pressure (mm Hg):	

Property/Value	Units	Temp	Data Type	Reference
WS 1.65E+002	mg/L	25	EST	MEYLAN,WM ET AL. (1996)
logP 3.10			EST	MEYLAN,WM & HOWARD,PH (1995)
VP 8.59E+000	mm Hg	25	EXT	BOUBLIK,T ET AL. (1984)
DC	pKa			
HL 1.61E-003	atm m3/mol	25	EST	MEYLAN,WM & HOWARD,PH (1991)
OH 1.96E-012	cm3/molc sec	25	EST	MEYLAN,WM & HOWARD,PH (1993)

CAS #: 003173-53-3	ISOCYANATOCYCLOHEXANE

Formula: C7H11NO	
Mol Weight: 125.17	
MP (deg C):	FP (deg C):
BP (deg C): 172	
BP pressure (mm Hg):	

Property/Value	Units	Temp	Data Type	Reference
WS 2.36E+002	mg/L	25	EST	MEYLAN,WM ET AL. (1996)
logP 3.06			EST	MEYLAN,WM & HOWARD,PH (1995)
VP 1.02E+003	mm Hg	25	EXP	DAUBERT,TE & DANNER,RP (1989)
DC	pKa			
HL 1.68E-003	atm m3/mol	25	EST	MEYLAN,WM & HOWARD,PH (1991)
OH 1.00E-011	cm3/molc sec	25	EST	MEYLAN,WM & HOWARD,PH (1993)

CAS #: 003175-23-3	1,2,2-TRICHLOROPROPANE

Formula: C3H5Cl3	
Mol Weight: 147.43	
MP (deg C):	FP (deg C):
BP (deg C): 124	
BP pressure (mm Hg):	

Property/Value	Units	Temp	Data Type	Reference
WS 1.53E+002	mg/L	25	EST	MEYLAN,WM ET AL. (1996)
logP 3.17			EST	MEYLAN,WM & HOWARD,PH (1995)
VP 1.83E+001	mm Hg	25	EST	NEELY,WB & BLAU,GE (1985)
DC	pKa			
HL 5.67E-003	atm m3/mol	25	EST	MEYLAN,WM & HOWARD,PH (1991)
OH 1.23E-013	cm3/molc sec	25	EST	MEYLAN,WM & HOWARD,PH (1993)

CAS #: 003178-22-1	TERT-BUTYLCYCLOHEXANE

Formula: C10H20	
Mol Weight: 140.27	
MP (deg C): -41.2	FP (deg C):
BP (deg C): 171.5	
BP pressure (mm Hg):	

Property/Value	Units	Temp	Data Type	Reference
WS 1.42E+000	mg/L	25	EST	MEYLAN,WM ET AL. (1996)
logP 4.96			EST	MEYLAN,WM & HOWARD,PH (1995)
VP 2.33E+000	mm Hg	25	EXT	CHAO,J ET AL. (1983)
DC	pKa			
HL 7.92E-001	atm m3/mol	25	EST	MEYLAN,WM & HOWARD,PH (1991)
OH 1.12E-011	cm3/molc sec	25	EST	MEYLAN,WM & HOWARD,PH (1993)

CAS #: 003179-08-6	4-HYDROXY-B-NITROSTYRENE

Formula: C8H7NO3	
Mol Weight: 165.15	
MP (deg C):	FP (deg C):
BP (deg C):	
BP pressure (mm Hg):	

Property/Value	Units	Temp	Data Type	Reference
WS 1.55E+003	mg/L	25	EST	MEYLAN,WM ET AL. (1996)
logP 2.12			EXP	HANSCH,C & LEO,AJ (1985)
VP 1.68E-004	mm Hg	25	EST	NEELY,WB & BLAU,GE (1985)
DC	pKa			
HL 3.62E-010	atm m3/mol	25	EST	MEYLAN,WM & HOWARD,PH (1991)
OH 4.29E-011	cm3/molc sec	25	EST	MEYLAN,WM & HOWARD,PH (1993)

CAS #: 003179-09-7	3-METHOXY B-NITROSTYRENE

Formula: C9H9NO3	
Mol Weight: 179.18	
MP (deg C):	FP (deg C):
BP (deg C):	
BP pressure (mm Hg):	

Property/Value	Units	Temp	Data Type	Reference
WS 2.14E+002	mg/L	25	EST	MEYLAN,WM ET AL. (1996)
logP 2.37			EXP	HANSCH,C & LEO,AJ (1985)
VP 1.62E-003	mm Hg	25	EST	NEELY,WB & BLAU,GE (1985)
DC	pKa			
HL 2.06E-007	atm m3/mol	25	EST	MEYLAN,WM & HOWARD,PH (1991)
OH 3.25E-011	cm3/molc sec	25	EST	MEYLAN,WM & HOWARD,PH (1993)

CAS #: 003179-10-0				4-METHOXY-B-NITROSTYRENE

Formula: $C_9H_9NO_3$
Mol Weight: 179.18
MP (deg C): | FP (deg C):
BP (deg C):
BP pressure (mm Hg):

Property/ Value	Units	Temp	Data Type	Reference
WS 2.99E+002	mg/L	25	EST	MEYLAN,WM ET AL. (1996)
logP 2.20			EXP	HANSCH,C & LEO,AJ (1985)
VP 1.62E-003	mm Hg	25	EST	NEELY,WB & BLAU,GE (1985)
DC	pKa			
HL 2.06E-007	atm m3/mol	25	EST	MEYLAN,WM & HOWARD,PH (1991)
OH 3.25E-011	cm3/molc sec	25	EST	MEYLAN,WM & HOWARD,PH (1993)

CAS #: 003179-89-3				C.I. DISPERSE RED 17

Formula: $C_{17}H_{20}N_4O_4$
Mol Weight: 344.37
MP (deg C): 160 | FP (deg C):
BP (deg C):
BP pressure (mm Hg):

Property/ Value	Units	Temp	Data Type	Reference
WS 7.57E-001	mg/L	25	EXP	BAUGHMAN,GL & PERENICH,TA (1988)
logP 3.69			EST	MEYLAN,WM & HOWARD,PH (1995)
VP 2.07E-013	mm Hg	25	EST	NEELY,WB & BLAU,GE (1985)
DC	pKa			
HL 2.40E-015	atm m3/mol	25	EST	MEYLAN,WM & HOWARD,PH (1991)
OH 1.82E-010	cm3/molc sec	25	EST	MEYLAN,WM & HOWARD,PH (1993)

CAS #: 003180-09-4				2-BUTYLPHENOL

Formula: $C_{10}H_{14}O$
Mol Weight: 150.22
MP (deg C): -20 | FP (deg C):
BP (deg C): 235
BP pressure (mm Hg):

Property/ Value	Units	Temp	Data Type	Reference
WS 2.76E+002	mg/L	25	EST	MEYLAN,WM ET AL. (1996)
logP 3.53			EST	MEYLAN,WM & HOWARD,PH (1995)
VP 1.03E-002	mm Hg	25	EXT	JORDAN,TE (1954)
DC 10.58	pKa	20	EXP	SERJEANT,EP & DEMPSEY,B (1979)
HL 1.45E-006	atm m3/mol	25	EST	MEYLAN,WM & HOWARD,PH (1991)
OH 4.46E-011	cm3/molc sec	25	EST	MEYLAN,WM & HOWARD,PH (1993)

CAS #: 003180-81-2				C.I. 11115

Formula: $C_{16}H_{17}ClN_4O_3$
Mol Weight: 348.79
MP (deg C): 133 | FP (deg C):
BP (deg C):
BP pressure (mm Hg):

Property/ Value	Units	Temp	Data Type	Reference
WS 1.20E-002	mg/L	25	EXP	BAUGHMAN,GL & PERENICH,TA (1988)
logP 4.85			EST	MEYLAN,WM & HOWARD,PH (1995)
VP 1.36E-011	mm Hg	25	EST	NEELY,WB & BLAU,GE (1985)
DC	pKa			
HL 4.40E-014	atm m3/mol	25	EST	MEYLAN,WM & HOWARD,PH (1991)
OH 8.18E-011	cm3/molc sec	25	EST	MEYLAN,WM & HOWARD,PH (1993)

CAS #: 003188-13-4				CHLOROMETHYL ETHYL ETHER

Formula: C_3H_7ClO
Mol Weight: 94.54
MP (deg C): | FP (deg C):
BP (deg C): 83
BP pressure (mm Hg):

Property/ Value	Units	Temp	Data Type	Reference
WS 2.45E+004	mg/L	25	EST	MEYLAN,WM ET AL. (1996)
logP 0.81			EST	MEYLAN,WM & HOWARD,PH (1995)
VP 5.91E+001	mm Hg	25	EST	NEELY,WB & BLAU,GE (1985)
DC	pKa			
HL 4.03E-004	atm m3/mol	25	EST	MEYLAN,WM & HOWARD,PH (1991)
OH 3.58E-012	cm3/molc sec	25	EST	MEYLAN,WM & HOWARD,PH (1993)

CAS #: 003194-55-6				1,2,5,6,9,10-HEXABROMOCYCLODODECANE

Formula: $C_{12}H_{18}Br_6$
Mol Weight: 641.70
MP (deg C): | FP (deg C):
BP (deg C):
BP pressure (mm Hg):

Property/ Value	Units	Temp	Data Type	Reference
WS 2.09E-005	mg/L	25	EST	MEYLAN,WM ET AL. (1996)
logP 7.74			EST	MEYLAN,WM & HOWARD,PH (1995)
VP 1.68E-008	mm Hg	25	EST	NEELY,WB & BLAU,GE (1985)
DC	pKa			
HL 1.72E-006	atm m3/mol	25	EST	MEYLAN,WM & HOWARD,PH (1991)
OH 5.01E-012	cm3/molc sec	25	EST	MEYLAN,WM & HOWARD,PH (1993)

CAS #: 003199-61-9				2-BENZOFURANCARBOXYLIC ACID, ETHYL ESTER

Formula: $C_{11}H_{10}O_3$
Mol Weight: 190.20
MP (deg C): 30.5 | FP (deg C):
BP (deg C): 276
BP pressure (mm Hg):

Property/ Value	Units	Temp	Data Type	Reference
WS 1.22E+002	mg/L	25	EST	MEYLAN,WM ET AL. (1996)
logP 3.05			EXP	YAMAGAMI,C ET AL. (1994)
VP 1.19E-003	mm Hg	25	EST	NEELY,WB & BLAU,GE (1985)
DC	pKa			
HL 4.49E-006	atm m3/mol	25	EST	MEYLAN,WM & HOWARD,PH (1991)
OH 1.37E-011	cm3/molc sec	25	EST	MEYLAN,WM & HOWARD,PH (1993)

CAS #: 003201-53-4				BENZAMIDE, N-[(METHYLAMINO)CARBONYL]

Formula: $C_9H_{10}N_2O_2$
Mol Weight: 178.19
MP (deg C): | FP (deg C):
BP (deg C):
BP pressure (mm Hg):

Property/ Value	Units	Temp	Data Type	Reference
WS 4.63E+003	mg/L	25	EST	MEYLAN,WM ET AL. (1996)
logP 1.27			EXP	SOTOMATSU,T ET AL. (1987)
VP 1.15E-007	mm Hg	25	EST	NEELY,WB & BLAU,GE (1985)
DC	pKa			
HL 4.69E-011	atm m3/mol	25	EST	MEYLAN,WM & HOWARD,PH (1991)
OH 4.34E-012	cm3/molc sec	25	EST	MEYLAN,WM & HOWARD,PH (1993)

CAS #: 003208-16-0				2-ETHYLFURAN

Formula: C_6H_8O

Mol Weight: 96.13

MP (deg C):   FP (deg C):

BP (deg C): 92

BP pressure (mm Hg):

Property/Value	Units	Temp	Data Type	Reference
WS 1.07E+003	mg/L	25	EST	MEYLAN,WM ET AL. (1996)
logP 2.40			EXP	HANSCH,C & LEO,AJ (1985)
VP 2.59E+001	mm Hg	25	EST	NEELY,WB & BLAU,GE (1985)
DC	pKa			
HL 7.88E-003	atm m3/mol	25	EST	MEYLAN,WM & HOWARD,PH (1991)
OH 1.02E-010	cm3/molc sec	25	EST	MEYLAN,WM & HOWARD,PH (1993)

CAS #: 003209-22-1				2,3-DICHLORONITROBENZENE

Formula: $C_6H_3Cl_2NO_2$

Mol Weight: 192.00

MP (deg C): 61.5   FP (deg C):

BP (deg C): 257.5

BP pressure (mm Hg):

Property/Value	Units	Temp	Data Type	Reference
WS 6.24E+001	mg/L	20	EXP	YALKOWSKY,SH & DANNENFELSER,RM (1992)
logP 3.05			EXP	HANSCH,C & LEO,AJ (1985)
VP 5.05E-003	mm Hg	25	EST	NEELY,WB & BLAU,GE (1985)
DC	pKa			
HL 1.17E-005	atm m3/mol	25	EST	MEYLAN,WM & HOWARD,PH (1991)
OH 5.01E-014	cm3/molc sec	25	EST	MEYLAN,WM & HOWARD,PH (1993)

CAS #: 003211-40-3				CHLOROTHIAZIDE-3-BENZYL

Formula: $C_{14}H_{12}ClN_3O_4S_2$

Mol Weight: 385.85

MP (deg C):   FP (deg C):

BP (deg C):

BP pressure (mm Hg):

Property/Value	Units	Temp	Data Type	Reference
WS 1.18E+001	mg/L	25	EST	MEYLAN,WM ET AL. (1996)
logP 1.70			EXP	SANGSTER,J (1994)
VP 2.07E-014	mm Hg	25	EST	NEELY,WB & BLAU,GE (1985)
DC	pKa			
HL 4.34E-013	atm m3/mol	25	EST	MEYLAN,WM & HOWARD,PH (1991)
OH 6.03E-012	cm3/molc sec	25	EST	MEYLAN,WM & HOWARD,PH (1993)

CAS #: 003215-24-5				2-AMINOSTRYCHNINE

Formula: $C_{21}H_{23}N_3O_2$

Mol Weight: 349.44

MP (deg C):   FP (deg C):

BP (deg C):

BP pressure (mm Hg):

Property/Value	Units	Temp	Data Type	Reference
WS 2.16E+003	mg/L	25	EST	MEYLAN,WM ET AL. (1996)
logP 0.54			EXP	HANSCH,C & LEO,AJ (1985)
VP 1.44E-010	mm Hg	25	EST	NEELY,WB & BLAU,GE (1985)
DC	pKa			
HL 2.11E-017	atm m3/mol	25	EST	MEYLAN,WM & HOWARD,PH (1991)
OH 3.77E-010	cm3/molc sec	25	EST	MEYLAN,WM & HOWARD,PH (1993)

CAS #: 003217-09-2				N-BENZYL-2-AMINOPROPANOL

Formula: $C_{10}H_{15}NO$

Mol Weight: 165.24

MP (deg C):   FP (deg C):

BP (deg C):

BP pressure (mm Hg):

Property/Value	Units	Temp	Data Type	Reference
WS 5.32E+004	mg/L	25	EST	MEYLAN,WM ET AL. (1996)
logP 1.28			EXP	HANSCH,C ET AL. (1995)
VP 2.49E-004	mm Hg	25	EST	NEELY,WB & BLAU,GE (1985)
DC	pKa			
HL 8.65E-011	atm m3/mol	25	EST	MEYLAN,WM & HOWARD,PH (1991)
OH 1.03E-010	cm3/molc sec	25	EST	MEYLAN,WM & HOWARD,PH (1993)

CAS #: 003217-15-0				4-BROMO-2,6-DICHLOROPHENOL

Formula: $C_6H_3BrCl_2O$

Mol Weight: 241.91

MP (deg C):   FP (deg C):

BP (deg C):

BP pressure (mm Hg):

Property/Value	Units	Temp	Data Type	Reference
WS 7.01E+001	mg/L	25	EST	MEYLAN,WM ET AL. (1996)
logP 3.69			EST	MEYLAN,WM & HOWARD,PH (1995)
VP 6.69E-004	mm Hg	25	EST	NEELY,WB & BLAU,GE (1985)
DC 6.21	pKa	25	EXP	SERJEANT,EP & DEMPSEY,B (1979)
HL 1.23E-007	atm m3/mol	25	EST	MEYLAN,WM & HOWARD,PH (1991)
OH 5.58E-013	cm3/molc sec	25	EST	MEYLAN,WM & HOWARD,PH (1993)

CAS #: 003221-61-2				2-METHYLOCTANE

Formula: C_9H_{20}

Mol Weight: 128.26

MP (deg C): -80.3   FP (deg C):

BP (deg C): 143.2

BP pressure (mm Hg):

Property/Value	Units	Temp	Data Type	Reference
WS 9.00E+001	mg/L	20	EXP	KERTES,AS (1989)
logP 4.69			EST	MEYLAN,WM & HOWARD,PH (1995)
VP 6.21E+000	mm Hg	25	EXP	DAUBERT,TE & DANNER,RP (1989)
DC	pKa			
HL 1.16E-002	atm m3/mol	25	EST	VP/WSOL
OH 1.01E-011	cm3/molc sec	25	EXP	ATKINSON,R (1989)

CAS #: 003223-70-9				CHLOROALLENE

Formula: C_3H_3Cl

Mol Weight: 74.51

MP (deg C):   FP (deg C):

BP (deg C):

BP pressure (mm Hg):

Property/Value	Units	Temp	Data Type	Reference
WS 2.58E+003	mg/L	25	EST	MEYLAN,WM ET AL. (1996)
logP 2.01			EST	MEYLAN,WM & HOWARD,PH (1995)
VP 5.71E+002	mm Hg	25	EST	NEELY,WB & BLAU,GE (1985)
DC	pKa			
HL 4.44E-002	atm m3/mol	25	EST	MEYLAN,WM & HOWARD,PH (1991)
OH 6.51E-012	cm3/molc sec	25	EST	MEYLAN,WM & HOWARD,PH (1993)

CAS #: 003229-00-3	PROPANE, 1,3-DIBROMO-2,2-BIS(BROMOMETHYL)-

Formula: $C_5H_8Br_4$

Mol Weight: 387.76

MP (deg C): 163 FP (deg C):

BP (deg C): 305.5

BP pressure (mm Hg):

Property/Value	Units	Temp	Data Type	Reference
WS 1.60E+000	mg/L	25	EXP	CHEM INSPECT TEST INST (1992)
logP 3.99			EXP	CHEM INSPECT TEST INST (1992)
VP 1.74E-003	mm Hg	25	EST	NEELY,WB & BLAU,GE (1985)
DC	pKa			
HL 4.04E-005	atm m3/mol	25	EST	MEYLAN,WM & HOWARD,PH (1991)
OH 1.29E-012	cm3/molc sec	25	EST	MEYLAN,WM & HOWARD,PH (1993)

CAS #: 003235-02-7	P-METHYLBENZALDOXIME

Formula: C_8H_9NO

Mol Weight: 135.17

MP (deg C): FP (deg C):

BP (deg C):

BP pressure (mm Hg):

Property/Value	Units	Temp	Data Type	Reference
WS 7.82E+002	mg/L	25	EST	MEYLAN,WM ET AL. (1996)
logP 2.40			EXP	HANSCH,C ET AL. (1995)
VP 1.34E-003	mm Hg	25	EST	NEELY,WB & BLAU,GE (1985)
DC	pKa			
HL 3.94E-007	atm m3/mol	25	EST	MEYLAN,WM & HOWARD,PH (1991)
OH 8.20E-012	cm3/molc sec	25	EST	MEYLAN,WM & HOWARD,PH (1993)

CAS #: 003236-48-4	N-(4-BROMOBUTYL)PHTHALIMIDE

Formula: $C_{12}H_{12}BrNO_2$

Mol Weight: 282.14

MP (deg C): FP (deg C):

BP (deg C):

BP pressure (mm Hg):

Property/Value	Units	Temp	Data Type	Reference
WS 4.62E+001	mg/L	25	EST	MEYLAN,WM ET AL. (1996)
logP 2.96			EXP	HANSCH,C & LEO,AJ (1985)
VP 1.39E-008	mm Hg	25	EST	NEELY,WB & BLAU,GE (1985)
DC	pKa			
HL 6.05E-009	atm m3/mol	25	EST	MEYLAN,WM & HOWARD,PH (1991)
OH 2.27E-011	cm3/molc sec	25	EST	MEYLAN,WM & HOWARD,PH (1993)

CAS #: 003237-22-7	PROPANEDINITRILE, (2-FURANYLMETHYLENE)

Formula: $C_8H_4N_2O$

Mol Weight: 144.13

MP (deg C): FP (deg C):

BP (deg C):

BP pressure (mm Hg):

Property/Value	Units	Temp	Data Type	Reference
WS 2.79E+003	mg/L	25	EST	MEYLAN,WM ET AL. (1996)
logP 1.40			EXP	BALAZ,S ET AL. (1985)
VP 7.24E-004	mm Hg	25	EST	NEELY,WB & BLAU,GE (1985)
DC	pKa			
HL 1.37E-008	atm m3/mol	25	EST	MEYLAN,WM & HOWARD,PH (1991)
OH 4.03E-011	cm3/molc sec	25	EST	MEYLAN,WM & HOWARD,PH (1993)

CAS #: 003244-90-4	TETRAPROPYL THIOPYROPHOSPHORATE

Formula: $C_{12}H_{28}O_5P_2S_2$

Mol Weight: 378.43

MP (deg C): FP (deg C):

BP (deg C):

BP pressure (mm Hg):

Property/Value	Units	Temp	Data Type	Reference
WS 3.00E+001	mg/L	20	EXP	SHIU,WY ET AL. (1990)
logP 4.95			EST	MEYLAN,WM & HOWARD,PH (1995)
VP 1.07E-005	mm Hg	25	EST	NEELY,WB & BLAU,GE (1985)
DC	pKa			
HL 3.82E-006	atm m3/mol	25	EST	MEYLAN,WM & HOWARD,PH (1991)
OH 2.05E-010	cm3/molc sec	25	EST	MEYLAN,WM & HOWARD,PH (1993)

CAS #: 003245-23-6	4-ETHYLPHENYL ACETATE

Formula: $C_{10}H_{12}O_2$

Mol Weight: 164.21

MP (deg C): FP (deg C):

BP (deg C):

BP pressure (mm Hg):

Property/Value	Units	Temp	Data Type	Reference
WS 4.26E+002	mg/L	25	EST	MEYLAN,WM ET AL. (1996)
logP 2.56			EXP	HANSCH,C & LEO,AJ (1985)
VP 6.24E-002	mm Hg	25	EST	NEELY,WB & BLAU,GE (1985)
DC	pKa			
HL 9.50E-005	atm m3/mol	25	EST	MEYLAN,WM & HOWARD,PH (1991)
OH 4.76E-012	cm3/molc sec	25	EST	MEYLAN,WM & HOWARD,PH (1993)

CAS #: 003252-43-5	DIBROMOACETONITRILE

Formula: C_2HBr_2N

Mol Weight: 198.85

MP (deg C): FP (deg C):

BP (deg C): 67-69

BP pressure (mm Hg): 2.40E+001

Property/Value	Units	Temp	Data Type	Reference
WS 9.60E+003	mg/L	25	EST	MEYLAN,WM ET AL. (1996)
logP 0.47			EST	MEYLAN,WM & HOWARD,PH (1995)
VP 3.01E-001	mm Hg	25	EST	NEELY,WB & BLAU,GE (1985)
DC	pKa			
HL 4.06E-007	atm m3/mol	25	EST	MEYLAN,WM & HOWARD,PH (1991)
OH 2.89E-014	cm3/molc sec	25	EST	MEYLAN,WM & HOWARD,PH (1993)

CAS #: 003253-62-1	CONVALLOTOXOL

Formula: $C_{29}H_{44}O_{10}$

Mol Weight: 552.67

MP (deg C): FP (deg C):

BP (deg C):

BP pressure (mm Hg):

Property/Value	Units	Temp	Data Type	Reference
WS 1.11E+003	mg/L	25	EST	MEYLAN,WM ET AL. (1996)
logP -0.62			EXP	HANSCH,C & LEO,AJ (1985)
VP 4.82E-023	mm Hg	25	EST	NEELY,WB & BLAU,GE (1985)
DC	pKa			
HL 1.32E-019	atm m3/mol	25	EST	MEYLAN,WM & HOWARD,PH (1991)
OH 1.08E-010	cm3/molc sec	25	EST	MEYLAN,WM & HOWARD,PH (1993)

CAS #: 003256-99-3	METHYLTHIAZIDE

Formula: $C_8H_9N_3O_4S_2$

Mol Weight: 275.31

MP (deg C): FP (deg C):

BP (deg C):

BP pressure (mm Hg):

Property/Value	Units	Temp	Data Type	Reference
WS 2.84E+003	mg/L	25	EST	MEYLAN,WM ET AL. (1996)
logP -0.32			EXP	SANGSTER,J (1994)
VP 3.48E-011	mm Hg	25	EST	NEELY,WB & BLAU,GE (1985)
DC	pKa			
HL 6.03E-012	atm m3/mol	25	EST	MEYLAN,WM & HOWARD,PH (1991)
OH 1.83E-011	cm3/molc sec	25	EST	MEYLAN,WM & HOWARD,PH (1993)

CAS #: 003261-62-9	BENZENEETHANAMINE, 4-METHYL-

Formula: $C_9H_{13}N$

Mol Weight: 135.21

MP (deg C): FP (deg C):

BP (deg C): 214

BP pressure (mm Hg):

Property/Value	Units	Temp	Data Type	Reference
WS 2.44E+004	mg/L	25	EST	MEYLAN,WM ET AL. (1996)
logP 1.83			EXP	SANGSTER,J (1993)
VP 9.74E-002	mm Hg	25	EST	NEELY,WB & BLAU,GE (1985)
DC	pKa			
HL 8.96E-007	atm m3/mol	25	EST	MEYLAN,WM & HOWARD,PH (1991)
OH 3.92E-011	cm3/molc sec	25	EST	MEYLAN,WM & HOWARD,PH (1993)

CAS #: 003263-31-8	C.I. VAT ORANGE 5

Formula: $C_{20}H_{16}O_4S_2$

Mol Weight: 384.48

MP (deg C): FP (deg C):

BP (deg C):

BP pressure (mm Hg):

Property/Value	Units	Temp	Data Type	Reference
WS 3.88E-001	mg/L	25	EST	MEYLAN,WM ET AL. (1996)
logP 4.68			EST	MEYLAN,WM & HOWARD,PH (1995)
VP 2.00E-007	mm Hg	25	EXP	BAUGHMAN,GL & PERENICH,TA (1988)
DC	pKa			
HL 2.81E-015	atm m3/mol	25	EST	MEYLAN,WM & HOWARD,PH (1991)
OH 2.26E-010	cm3/molc sec	25	EST	MEYLAN,WM & HOWARD,PH (1993)

CAS #: 003268-87-9	1,2,3,4,6,7,8,9-OCTACHLORODIBENZO-P-DIOXIN

Formula: $C_{12}Cl_8O_2$

Mol Weight: 459.76

MP (deg C): 330 FP (deg C):

BP (deg C): 510

BP pressure (mm Hg):

Property/Value	Units	Temp	Data Type	Reference
WS 7.40E-008	mg/L	25	EXP	DOUCETTE,WJ & ANDREN,AW (1988)
logP 8.20			EXP	SHIU,WY ET AL. (1988)
VP 8.25E-013	mm Hg	25	EXP	RORDORF,BF (1989)
DC	pKa			
HL 6.74E-006	atm m3/mol	25	EST	VP/WSOL
OH 1.60E-012	cm3/molc sec	25	EST	ATKINSON,R (1991A)

CAS #: 003271-01-0	2-SULFANILAMIDO-5-ETHYLPYRIMIDINE

Formula: $C_{12}H_{14}N_4O_2S$

Mol Weight: 278.33

MP (deg C): FP (deg C):

BP (deg C):

BP pressure (mm Hg):

Property/Value	Units	Temp	Data Type	Reference
WS 2.79E+003	mg/L	25	EST	MEYLAN,WM ET AL. (1996)
logP 0.90			EXP	HANSCH,C & LEO,AJ (1985)
VP 8.61E-009	mm Hg	25	EST	NEELY,WB & BLAU,GE (1985)
DC	pKa			
HL 2.32E-010	atm m3/mol	25	EST	MEYLAN,WM & HOWARD,PH (1991)
OH 2.51E-011	cm3/molc sec	25	EST	MEYLAN,WM & HOWARD,PH (1993)

CAS #: 003271-76-9	C.I.VAT GREEN 3

Formula: $C_{31}H_{15}NO_3$

Mol Weight: 449.47

MP (deg C): FP (deg C):

BP (deg C):

BP pressure (mm Hg):

Property/Value	Units	Temp	Data Type	Reference
WS 6.73E-006	mg/L	25	EST	MEYLAN,WM ET AL. (1996)
logP 8.48			EST	MEYLAN,WM & HOWARD,PH (1995)
VP 1.00E-010	mm Hg	25	EXP	BAUGHMAN,GL & PERENICH,TA (1988)
DC	pKa			
HL 2.04E-018	atm m3/mol	25	EST	MEYLAN,WM & HOWARD,PH (1991)
OH 2.28E-010	cm3/molc sec	25	EST	MEYLAN,WM & HOWARD,PH (1993)

CAS #: 003274-20-2	2-METHOXY-9,10-ANTHRACENEDIONE

Formula: $C_{15}H_{10}O_3$

Mol Weight: 238.25

MP (deg C): FP (deg C):

BP (deg C):

BP pressure (mm Hg):

Property/Value	Units	Temp	Data Type	Reference
WS 2.53E+000	mg/L	25	EST	MEYLAN,WM ET AL. (1996)
logP 3.43			EST	MEYLAN,WM & HOWARD,PH (1995)
VP 6.85E-011	mm Hg	25	EXP	SHIMIZU,T ET AL. (1987)
DC	pKa			
HL 1.88E-010	atm m3/mol	25	EST	MEYLAN,WM & HOWARD,PH (1991)
OH 9.84E-012	cm3/molc sec	25	EST	MEYLAN,WM & HOWARD,PH (1993)

CAS #: 003274-28-0	HEXANOIC ACID, 2-PROPYL-

Formula: $C_9H_{18}O_2$

Mol Weight: 158.24

MP (deg C): FP (deg C):

BP (deg C):

BP pressure (mm Hg):

Property/Value	Units	Temp	Data Type	Reference
WS 4.65E+002	mg/L	25	EST	MEYLAN,WM ET AL. (1996)
logP 3.01			EXP	SANGSTER,J (1993)
VP 1.46E-002	mm Hg	25	EST	NEELY,WB & BLAU,GE (1985)
DC	pKa			
HL 3.98E-006	atm m3/mol	25	EST	MEYLAN,WM & HOWARD,PH (1991)
OH 9.59E-012	cm3/molc sec	25	EST	MEYLAN,WM & HOWARD,PH (1993)

CAS #:	003278-46-4				2,2,3-TRICHLOROPROPIONIC ACID

Formula: $C_3H_3Cl_3O_2$

Mol Weight: 177.42

MP (deg C): 65.5   FP (deg C):

BP (deg C): 140

BP pressure (mm Hg): 4.00E+001

Property/Value	Units	Temp	Data Type	Reference
WS 3.14E+003	mg/L	25	EST	MEYLAN,WM ET AL. (1996)
logP 1.93			EST	MEYLAN,WM & HOWARD,PH (1995)
VP 2.16E-002	mm Hg	25	EST	NEELY,WB & BLAU,GE (1985)
DC	pKa			
HL 3.18E-008	atm m3/mol	25	EST	MEYLAN,WM & HOWARD,PH (1991)
OH 6.09E-013	cm3/molc sec	25	EST	MEYLAN,WM & HOWARD,PH (1993)

CAS #:	003279-07-0				4-(TERT-BUTYL)-2-NITROPHENOL

Formula: $C_{10}H_{13}NO_3$

Mol Weight: 195.22

MP (deg C):   FP (deg C):

BP (deg C):

BP pressure (mm Hg):

Property/Value	Units	Temp	Data Type	Reference
WS 2.54E+001	mg/L	25	EST	MEYLAN,WM ET AL. (1996)
logP 3.82			EST	MEYLAN,WM & HOWARD,PH (1995)
VP 6.57E-005	mm Hg	25	EST	NEELY,WB & BLAU,GE (1985)
DC	pKa			
HL 1.81E-005	atm m3/mol	25	EST	MEYLAN,WM & HOWARD,PH (1991)
OH 5.64E-012	cm3/molc sec	25	EST	MEYLAN,WM & HOWARD,PH (1993)

CAS #:	003287-99-8				BENZYLAMINE, HYDROCHLORIDE

Formula: $C_7H_{10}ClN$

Mol Weight: 143.62

MP (deg C): 262-263   FP (deg C):

BP (deg C):

BP pressure (mm Hg):

Property/Value	Units	Temp	Data Type	Reference
WS 5.06E+005	mg/L	25	EXP	SEIDELL,A (1941)
logP -1.96			EXP	HANSCH,C ET AL. (1995)
VP 1.72E-006	mm Hg	25	EST	NEELY,WB & BLAU,GE (1985)
DC	pKa			
HL 5.67E-014	atm m3/mol	25	EST	MEYLAN,WM & HOWARD,PH (1991)
OH 1.35E-011	cm3/molc sec	25	EST	MEYLAN,WM & HOWARD,PH (1993)

CAS #:	003290-92-4				TRIMETHYLOLPROPANE TRIMETHACRYLATE

Formula: $C_{18}H_{26}O_6$

Mol Weight: 338.40

MP (deg C): <-10   FP (deg C):

BP (deg C):

BP pressure (mm Hg):

Property/Value	Units	Temp	Data Type	Reference
WS 1.30E+001	mg/L	25	EXP	CHEM INSPECT TEST INST (1992)
logP 4.39			EXP	CHEM INSPECT TEST INST (1992)
VP 1.37E-004	mm Hg	25	EST	NEELY,WB & BLAU,GE (1985)
DC	pKa			
HL 2.31E-009	atm m3/mol	25	EST	MEYLAN,WM & HOWARD,PH (1991)
OH 6.12E-011	cm3/molc sec	25	EST	MEYLAN,WM & HOWARD,PH (1993)

CAS #:	003290-99-1				BENZOYLHYDRAZINE, P-METHOXY

Formula: $C_8H_{10}N_2O_2$

Mol Weight: 166.18

MP (deg C):   FP (deg C):

BP (deg C):

BP pressure (mm Hg):

Property/Value	Units	Temp	Data Type	Reference
WS 1.92E+003	mg/L	25	EST	MEYLAN,WM ET AL. (1996)
logP 0.25			EXP	HANSCH,C & LEO,AJ (1985)
VP 1.43E-005	mm Hg	25	EST	NEELY,WB & BLAU,GE (1985)
DC	pKa			
HL 5.47E-013	atm m3/mol	25	EST	MEYLAN,WM & HOWARD,PH (1991)
OH 2.59E-011	cm3/molc sec	25	EST	MEYLAN,WM & HOWARD,PH (1993)

CAS #:	003296-90-0				1,3-PROPANEDIOL, 2,2-BIS(BRME)-

Formula: $C_5H_{10}Br_2O_2$

Mol Weight: 261.95

MP (deg C): 112-114   FP (deg C):

BP (deg C):

BP pressure (mm Hg):

Property/Value	Units	Temp	Data Type	Reference
WS 3.80E+004	mg/L	25	EXP	CHEM INSPEC TEST INST
logP 1.06			EXP	HANSCH,C & LEO,AJ (1985)
VP 1.30E-005	mm Hg	25	EST	NEELY,WB & BLAU,GE (1985)
DC	pKa			
HL 4.06E-009	atm m3/mol	25	EST	MEYLAN,WM & HOWARD,PH (1991)
OH 8.97E-012	cm3/molc sec	25	EST	MEYLAN,WM & HOWARD,PH (1993)

CAS #:	003299-99-8				9-(1-METHYLETHYL)-9H-FLUORENE

Formula: $C_{16}H_{16}$

Mol Weight: 208.31

MP (deg C):   FP (deg C):

BP (deg C):

BP pressure (mm Hg):

Property/Value	Units	Temp	Data Type	Reference
WS 1.47E-001	mg/L	25	EST	MEYLAN,WM ET AL. (1996)
logP 5.06			EST	MEYLAN,WM & HOWARD,PH (1995)
VP 1.77E-004	mm Hg	25	EST	NEELY,WB & BLAU,GE (1985)
DC	pKa			
HL 3.91E-004	atm m3/mol	25	EST	MEYLAN,WM & HOWARD,PH (1991)
OH 1.28E-011	cm3/molc sec	25	EST	MEYLAN,WM & HOWARD,PH (1993)

CAS #:	003307-39-9				2-(P-CHLOROPHENOXY)PROPIONIC ACID

Formula: $C_9H_9ClO_3$

Mol Weight: 200.62

MP (deg C):   FP (deg C):

BP (deg C):

BP pressure (mm Hg):

Property/Value	Units	Temp	Data Type	Reference
WS 1.48E+003	mg/L	25	EXP	YALKOWSKY,SH & DANNENFELSER,RM (1992)
logP 2.31			EXP	HANSCH,C & LEO,AJ (1985)
VP 2.28E-004	mm Hg	25	EST	NEELY,WB & BLAU,GE (1985)
DC	pKa			
HL 1.65E-008	atm m3/mol	25	EST	MEYLAN,WM & HOWARD,PH (1991)
OH 1.58E-011	cm3/molc sec	25	EST	MEYLAN,WM & HOWARD,PH (1993)

CAS #: 003316-09-4 — 1,2-DIHYDROXY-4-NITROBENZENE

Formula:	$C_6H_5NO_4$
Mol Weight:	155.11
MP (deg C): 174-176	FP (deg C):
BP (deg C):	
BP pressure (mm Hg):	

Property/Value	Units	Temp	Data Type	Reference
WS 1.05E+004	mg/L	25	EST	MEYLAN,WM ET AL. (1996)
logP 1.66			EXP	HANSCH,C ET AL. (1995)
VP 2.27E-005	mm Hg	25	EST	NEELY,WB & BLAU,GE (1985)
DC	pKa			
HL 2.30E-013	atm m3/mol	25	EST	MEYLAN,WM & HOWARD,PH (1991)
OH 3.15E-012	cm3/molc sec	25	EST	MEYLAN,WM & HOWARD,PH (1993)

CAS #: 003319-31-1 — TRIS(2-ETHYLHEXYL) TRIMELLITATE

Formula:	$C_{33}H_{54}O_6$
Mol Weight:	546.79
MP (deg C):	FP (deg C):
BP (deg C): 414	
BP pressure (mm Hg):	

Property/Value	Units	Temp	Data Type	Reference
WS 1.00E+002	mg/L	25	EXP	CHEM INSPECT TEST INST (1992)
logP 11.59			EST	MEYLAN,WM & HOWARD,PH (1995)
VP 3.94E-011	mm Hg	25	EST	NEELY,WB & BLAU,GE (1985)
DC	pKa			
HL 4.45E-007	atm m3/mol	25	EST	MEYLAN,WM & HOWARD,PH (1991)
OH 3.27E-011	cm3/molc sec	25	EST	MEYLAN,WM & HOWARD,PH (1993)

CAS #: 003322-93-8 — 1,2-DIBROMO-4-(1,2-DIBROMOETHYL)CYCLOHEXANE

Formula:	$C_8H_{12}Br_4$
Mol Weight:	427.82
MP (deg C):	FP (deg C):
BP (deg C):	
BP pressure (mm Hg):	

Property/Value	Units	Temp	Data Type	Reference
WS 6.92E-002	mg/L	25	EST	MEYLAN,WM ET AL. (1996)
logP 5.24			EST	MEYLAN,WM & HOWARD,PH (1995)
VP 1.05E-004	mm Hg	25	EST	NEELY,WB & BLAU,GE (1985)
DC	pKa			
HL 5.73E-008	atm m3/mol	25	EST	MEYLAN,WM & HOWARD,PH (1991)
OH 5.72E-012	cm3/molc sec	25	EST	MEYLAN,WM & HOWARD,PH (1993)

CAS #: 003324-71-8 — DIMETHYLGUANIDINE

Formula:	$C_3H_9N_3$
Mol Weight:	87.13
MP (deg C):	FP (deg C):
BP (deg C):	
BP pressure (mm Hg):	

Property/Value	Units	Temp	Data Type	Reference
WS 1.62E+003	mg/L	20	EXP	GREENWALD,I (1926)
logP -0.95			EST	MEYLAN,WM & HOWARD,PH (1995)
VP 2.95E+000	mm Hg	25	EST	NEELY,WB & BLAU,GE (1985)
DC	pKa			
HL 1.13E-010	atm m3/mol	25	EST	MEYLAN,WM & HOWARD,PH (1991)
OH 8.95E-011	cm3/molc sec	25	EST	MEYLAN,WM & HOWARD,PH (1993)

CAS #: 003324-76-3 — 2,4,6-CYCLOHEPTATRIEN-1-ONE, 3-HYDROXY-

Formula:	$C_7H_6O_2$
Mol Weight:	122.12
MP (deg C):	FP (deg C):
BP (deg C):	
BP pressure (mm Hg):	

Property/Value	Units	Temp	Data Type	Reference
WS 2.00E+004	mg/L	25	EST	MEYLAN,WM ET AL. (1996)
logP 0.81			EXP	SANGSTER,J (1993)
VP 6.79E-004	mm Hg	25	EST	NEELY,WB & BLAU,GE (1985)
DC	pKa			
HL 2.53E-009	atm m3/mol	25	EST	MEYLAN,WM & HOWARD,PH (1991)
OH 6.82E-011	cm3/molc sec	25	EST	MEYLAN,WM & HOWARD,PH (1993)

CAS #: 003337-71-1 — ASULAM

Formula:	$C_8H_{10}N_2O_4S$
Mol Weight:	230.24
MP (deg C): 144	FP (deg C):
BP (deg C):	
BP pressure (mm Hg):	

Property/Value	Units	Temp	Data Type	Reference
WS 5.00E+003	mg/L		EXP	YALKOWSKY,SH & DANNENFELSER,RM (1992)
logP -0.27			EXP	HANSCH,C & LEO,AJ (1985)
VP 1.44E-006	mm Hg	25	EST	NEELY,WB & BLAU,GE (1985)
DC	pKa			
HL 1.71E-012	atm m3/mol	25	EST	MEYLAN,WM & HOWARD,PH (1991)
OH 2.87E-011	cm3/molc sec	25	EST	MEYLAN,WM & HOWARD,PH (1993)

CAS #: 003343-19-9 — 3,4-DIHYDROXYPHENYLGLYCOL

Formula:	$C_8H_{10}O_4$
Mol Weight:	170.17
MP (deg C):	FP (deg C):
BP (deg C):	
BP pressure (mm Hg):	

Property/Value	Units	Temp	Data Type	Reference
WS 1.00E+006	mg/L	25	EST	MEYLAN,WM ET AL. (1996)
logP -1.01			EXP	HANSCH,C & LEO,AJ (1985)
VP 1.61E-007	mm Hg	25	EST	NEELY,WB & BLAU,GE (1985)
DC	pKa			
HL 1.14E-016	atm m3/mol	25	EST	MEYLAN,WM & HOWARD,PH (1991)
OH 6.74E-011	cm3/molc sec	25	EST	MEYLAN,WM & HOWARD,PH (1993)

CAS #: 003343-22-4 — URIDINE, 2'-DEOXY-5-FLUORO-, 3',5'-DIBUTANOATE

Formula:	$C_{17}H_{23}FN_2O_7$
Mol Weight:	386.38
MP (deg C):	FP (deg C):
BP (deg C):	
BP pressure (mm Hg):	

Property/Value	Units	Temp	Data Type	Reference
WS 8.86E+001	mg/L	25	EST	MEYLAN,WM ET AL. (1996)
logP 1.90			EXP	SANGSTER,J (1994)
VP 2.65E-012	mm Hg	25	EST	NEELY,WB & BLAU,GE (1985)
DC	pKa			
HL 1.95E-016	atm m3/mol	25	EST	MEYLAN,WM & HOWARD,PH (1991)
OH 7.90E-011	cm3/molc sec	25	EST	MEYLAN,WM & HOWARD,PH (1993)

CAS #: 003347-22-6 — DITHIANONE

Formula: $C_{14}H_4N_2O_2S_2$

Mol Weight: 296.33

MP (deg C): 225

FP (deg C):

BP (deg C):

BP pressure (mm Hg):

Property/Value	Units	Temp	Data Type	Reference
WS 5.00E-001	mg/L	20	EXP	SHIU,WY ET AL. (1990)
logP 2.84			EXP	HANSCH,C ET AL. (1995)
VP 1.04E-010	mm Hg	25	EST	NEELY,WB & BLAU,GE (1985)
DC	pKa			
HL 3.89E-017	atm m3/mol	25	EST	MEYLAN,WM & HOWARD,PH (1991)
OH 2.04E-011	cm3/molc sec	25	EST	MEYLAN,WM & HOWARD,PH (1993)

CAS #: 003351-05-1 — ACID BLUE 113

Formula: $C_{32}H_{21}N_5Na_2O_6S_2$

Mol Weight: 681.66

MP (deg C):

FP (deg C):

BP (deg C):

BP pressure (mm Hg):

Property/Value	Units	Temp	Data Type	Reference
WS 2.49E-003	mg/L	25	EST	MEYLAN,WM ET AL. (1996)
logP 3.20			EST	MEYLAN,WM & HOWARD,PH (1995)
VP 3.87E-028	mm Hg	25	EST	NEELY,WB & BLAU,GE (1985)
DC	pKa			
HL	atm m3/mol			
OH 1.60E-010	cm3/molc sec	25	EST	MEYLAN,WM & HOWARD,PH (1993)

CAS #: 003357-37-7 — HYDRAZINECARBOXIMIDAMIDE, 2-(PHENYLMETHYLENE)-

Formula: $C_8H_{10}N_4$

Mol Weight: 162.20

MP (deg C):

FP (deg C):

BP (deg C):

BP pressure (mm Hg):

Property/Value	Units	Temp	Data Type	Reference
WS 6.57E+003	mg/L	25	EST	MEYLAN,WM ET AL. (1996)
logP 1.18			EXP	SANGSTER,J (1993)
VP 2.62E-004	mm Hg	25	EST	NEELY,WB & BLAU,GE (1985)
DC	pKa			
HL 5.05E-012	atm m3/mol	25	EST	MEYLAN,WM & HOWARD,PH (1991)
OH 9.06E-011	cm3/molc sec	25	EST	MEYLAN,WM & HOWARD,PH (1993)

CAS #: 003357-42-4 — 1H-1,2,4-TRIAZOLE, 3-PHENYL-

Formula: $C_8H_7N_3$

Mol Weight: 145.17

MP (deg C):

FP (deg C):

BP (deg C):

BP pressure (mm Hg):

Property/Value	Units	Temp	Data Type	Reference
WS 3.85E+003	mg/L	25	EST	MEYLAN,WM ET AL. (1996)
logP 1.54			EXP	SANGSTER,J (1993)
VP 6.79E-005	mm Hg	25	EST	NEELY,WB & BLAU,GE (1985)
DC	pKa			
HL 1.17E-007	atm m3/mol	25	EST	MEYLAN,WM & HOWARD,PH (1991)
OH 3.56E-012	cm3/molc sec	25	EST	MEYLAN,WM & HOWARD,PH (1993)

CAS #: 003360-41-6 — 4-PHENYLBUTANOL

Formula: $C_{10}H_{14}O$

Mol Weight: 150.22

MP (deg C):

FP (deg C):

BP (deg C): 140

BP pressure (mm Hg): 1.40E+001

Property/Value	Units	Temp	Data Type	Reference
WS 2.41E+003	mg/L	25	EST	MEYLAN,WM ET AL. (1996)
logP 2.35			EXP	HANSCH,C & LEO,AJ (1985)
VP 1.56E-003	mm Hg	25	EST	NEELY,WB & BLAU,GE (1985)
DC	pKa			
HL 5.09E-007	atm m3/mol	25	EST	MEYLAN,WM & HOWARD,PH (1991)
OH 1.30E-011	cm3/molc sec	25	EST	MEYLAN,WM & HOWARD,PH (1993)

CAS #: 003366-95-8 — 1H-IMIDAZOLE-1-ETHANOL, à,2-DIMETHYL-5-NITRO-

Formula: $C_7H_{11}N_3O_3$

Mol Weight: 185.18

MP (deg C): 76

FP (deg C):

BP (deg C):

BP pressure (mm Hg):

Property/Value	Units	Temp	Data Type	Reference
WS 1.37E+004	mg/L	25	EST	MEYLAN,WM ET AL. (1996)
logP 0.22			EXP	HANSCH,C ET AL. (1995)
VP 2.91E-007	mm Hg	25	EST	NEELY,WB & BLAU,GE (1985)
DC	pKa			
HL 2.25E-011	atm m3/mol	25	EST	MEYLAN,WM & HOWARD,PH (1991)
OH 1.38E-011	cm3/molc sec	25	EST	MEYLAN,WM & HOWARD,PH (1993)

CAS #: 003373-86-2 — O-NITROPHENYLDIMETHYLCARBAMATE

Formula: $C_9H_{10}N_2O_4$

Mol Weight: 210.19

MP (deg C):

FP (deg C):

BP (deg C):

BP pressure (mm Hg):

Property/Value	Units	Temp	Data Type	Reference
WS 1.11E+003	mg/L	25	EST	MEYLAN,WM ET AL. (1996)
logP 1.35			EXP	HANSCH,C & LEO,AJ (1985)
VP 2.25E-004	mm Hg	25	EST	NEELY,WB & BLAU,GE (1985)
DC	pKa			
HL 2.78E-010	atm m3/mol	25	EST	MEYLAN,WM & HOWARD,PH (1991)
OH 1.64E-011	cm3/molc sec	25	EST	MEYLAN,WM & HOWARD,PH (1993)

CAS #: 003375-22-2 — 1,3-DICHLORO-2-METHYLPROPENE

Formula: $C_4H_6Cl_2$

Mol Weight: 125.00

MP (deg C):

FP (deg C):

BP (deg C):

BP pressure (mm Hg):

Property/Value	Units	Temp	Data Type	Reference
WS 3.63E+002	mg/L	25	EST	MEYLAN,WM ET AL. (1996)
logP 2.84			EST	MEYLAN,WM & HOWARD,PH (1995)
VP 1.26E+001	mm Hg	25	EST	NEELY,WB & BLAU,GE (1985)
DC	pKa			
HL 3.84E-002	atm m3/mol	25	EST	MEYLAN,WM & HOWARD,PH (1991)
OH 1.44E-011	cm3/molc sec	25	EST	MEYLAN,WM & HOWARD,PH (1993)

CAS #: 003376-32-7				O-METHYLBENZALDOXIME

Formula: C_8H_9NO

Mol Weight: 135.17

MP (deg C): | FP (deg C):

BP (deg C):

BP pressure (mm Hg):

Property/Value	Units	Temp	Data Type	Reference
WS 6.06E+002	mg/L	25	EST	MEYLAN,WM ET AL. (1996)
logP 2.53			EXP	HANSCH,C & LEO,AJ (1985)
VP 1.34E-003	mm Hg	25	EST	NEELY,WB & BLAU,GE (1985)
DC	pKa			
HL 3.94E-007	atm m3/mol	25	EST	MEYLAN,WM & HOWARD,PH (1991)
OH 8.20E-012	cm3/molc sec	25	EST	MEYLAN,WM & HOWARD,PH (1993)

CAS #: 003377-86-4				2-BROMOHEXANE

Formula: $C_6H_{13}Br$

Mol Weight: 165.08

MP (deg C): | FP (deg C):

BP (deg C): 144

BP pressure (mm Hg):

Property/Value	Units	Temp	Data Type	Reference
WS 5.94E+001	mg/L	25	EST	MEYLAN,WM ET AL. (1996)
logP 3.56			EST	MEYLAN,WM & HOWARD,PH (1995)
VP 7.24E+000	mm Hg	25	EXP	LEVANOVA,SV ET AL. (1967)
DC	pKa			
HL 3.50E-002	atm m3/mol	25	EST	MEYLAN,WM & HOWARD,PH (1991)
OH 3.99E-012	cm3/molc sec	25	EST	MEYLAN,WM & HOWARD,PH (1993)

CAS #: 003380-34-5				5-CHLORO-2-(2,4-DICHLOROPHENOXY)PHENOL

Formula: $C_{12}H_7Cl_3O_2$

Mol Weight: 289.55

MP (deg C): 54-57.3 | FP (deg C):

BP (deg C):

BP pressure (mm Hg):

Property/Value	Units	Temp	Data Type	Reference
WS 1.00E+001	mg/L	20	EXP	YALKOWSKY,SH & DANNENFELSER,RM (1992)
logP 4.76			EXP	CHEM INSPECT TEST INST (1992)
VP 6.45E-007	mm Hg	25	EST	NEELY,WB & BLAU,GE (1985)
DC	pKa			
HL 4.99E-009	atm m3/mol	25	EST	MEYLAN,WM & HOWARD,PH (1991)
OH 2.45E-011	cm3/molc sec	25	EST	MEYLAN,WM & HOWARD,PH (1993)

CAS #: 003382-56-7				4-T-BUTYLNITROBENZENE

Formula: $C_{10}H_{13}NO_2$

Mol Weight: 179.22

MP (deg C): | FP (deg C):

BP (deg C):

BP pressure (mm Hg):

Property/Value	Units	Temp	Data Type	Reference
WS 1.08E+001	mg/L	25	EST	MEYLAN,WM ET AL. (1996)
logP 3.89			EXP	HANSCH,C ET AL. (1995)
VP 6.35E-003	mm Hg	25	EST	NEELY,WB & BLAU,GE (1985)
DC	pKa			
HL 5.49E-005	atm m3/mol	25	EST	MEYLAN,WM & HOWARD,PH (1991)
OH 1.04E-012	cm3/molc sec	25	EST	MEYLAN,WM & HOWARD,PH (1993)

CAS #: 003383-96-8				ABATE

Formula: $C_{16}H_{20}O_6P_2S_3$

Mol Weight: 466.47

MP (deg C): 30 | FP (deg C):

BP (deg C):

BP pressure (mm Hg):

Property/Value	Units	Temp	Data Type	Reference
WS 2.70E-001	mg/L	20	EXP	SHIU,WY ET AL. (1990)
logP 5.96			EXP	HANSCH,C & LEO,AJ (1985)
VP 7.91E-008	mm Hg	25	EST	NEELY,WB & BLAU,GE (1985)
DC	pKa			
HL 1.96E-009	atm m3/mol	25	EST	MEYLAN,WM & HOWARD,PH (1991)
OH 1.43E-010	cm3/molc sec	25	EST	MEYLAN,WM & HOWARD (1993)

CAS #: 003386-18-3				NONAETHYLENE GLYCOL

Formula: $C_{18}H_{38}O_{10}$

Mol Weight: 414.50

MP (deg C): | FP (deg C):

BP (deg C):

BP pressure (mm Hg):

Property/Value	Units	Temp	Data Type	Reference
WS 1.00E+006	mg/L	25	EST	MEYLAN,WM ET AL. (1996)
logP -3.39			EST	MEYLAN,WM & HOWARD,PH (1995)
VP 1.15E-012	mm Hg	25	EST	NEELY,WB & BLAU,GE (1985)
DC	pKa			
HL 4.45E-022	atm m3/mol	25	EST	MEYLAN,WM & HOWARD,PH (1991)
OH 1.20E-010	cm3/molc sec	25	EST	MEYLAN,WM & HOWARD,PH (1993)

CAS #: 003386-33-2				1-CHLOROOCTADECANE

Formula: $C_{18}H_{37}Cl$

Mol Weight: 288.95

MP (deg C): 28.6 | FP (deg C):

BP (deg C): 348

BP pressure (mm Hg):

Property/Value	Units	Temp	Data Type	Reference
WS 1.25E-004	mg/L	25	EST	MEYLAN,WM ET AL. (1996)
logP 9.44			EST	MEYLAN,WM & HOWARD,PH (1995)
VP 1.71E-005	mm Hg	25	EXP	LI,JCM & ROSSINI.FD (1953)
DC	pKa			
HL 1.01E+000	atm m3/mol	25	EST	MEYLAN,WM & HOWARD,PH (1991)
OH 2.19E-011	cm3/molc sec	25	EST	MEYLAN,WM & HOWARD,PH (1993)

CAS #: 003389-71-7				1,2,3,4,7,7-HEXACHLORO-2,5-NORBORNADIENE

Formula: $C_7H_2Cl_6$

Mol Weight: 298.81

MP (deg C): | FP (deg C):

BP (deg C):

BP pressure (mm Hg):

Property/Value	Units	Temp	Data Type	Reference
WS 4.95E-001	mg/L	25	EST	MEYLAN,WM ET AL. (1996)
logP 5.15			EST	MEYLAN,WM & HOWARD,PH (1995)
VP 4.68E-003	mm Hg	25	EST	NEELY,WB & BLAU,GE (1985)
DC	pKa			
HL 4.83E-004	atm m3/mol	25	EST	MEYLAN,WM & HOWARD,PH (1991)
OH 6.13E-011	cm3/molc sec	25	EST	MEYLAN,WM & HOWARD,PH (1993)

CAS #: 003393-59-7				4,5,7-TRICL-2-CF3-BENZIMIDAZOLE
Formula: $C_8H_2Cl_3F_3N_2$				
Mol Weight: 289.47				
MP (deg C):		FP (deg C):		
BP (deg C):				
BP pressure (mm Hg):				

Property/Value	Units	Temp	Data Type	Reference
WS 8.36E+000	mg/L	25	EST	MEYLAN,WM ET AL. (1996)
logP 3.78			EXP	HANSCH,C & LEO,AJ (1985)
VP 7.61E-007	mm Hg	25	EST	NEELY,WB & BLAU,GE (1985)
DC	pKa			
HL 1.30E-006	atm m3/mol	25	EST	MEYLAN,WM & HOWARD,PH (1991)
OH 5.84E-013	cm3/molc sec	25	EST	MEYLAN,WM & HOWARD,PH (1993)

CAS #: 003393-64-4				4-HYDROXY-3-METHYL-2-BUTANONE
Formula: $C_5H_{10}O_2$				
Mol Weight: 102.13				
MP (deg C):		FP (deg C):		
BP (deg C): 90-95				
BP pressure (mm Hg): 1.50E+001				

Property/Value	Units	Temp	Data Type	Reference
WS 1.00E+006	mg/L	25	EST	MEYLAN,WM ET AL. (1996)
logP -0.79			EST	MEYLAN,WM & HOWARD,PH (1995)
VP 2.26E-001	mm Hg	25	EXT	PERRY,RH & GREEN,D (1984)
DC	pKa			
HL 3.19E-009	atm m3/mol	25	EST	MEYLAN,WM & HOWARD,PH (1991)
OH 1.53E-011	cm3/molc sec	25	EST	MEYLAN,WM & HOWARD,PH (1993)

CAS #: 003397-78-2				4-METHOXYCINNOLINE
Formula: $C_9H_8N_2O$				
Mol Weight: 160.18				
MP (deg C):		FP (deg C):		
BP (deg C):				
BP pressure (mm Hg):				

Property/Value	Units	Temp	Data Type	Reference
WS 4.27E+003	mg/L	25	EST	MEYLAN,WM ET AL. (1996)
logP 1.41			EXP	HANSCH,C & LEO,AJ (1985)
VP 6.78E-005	mm Hg	25	EST	NEELY,WB & BLAU,GE (1985)
DC	pKa			
HL 1.66E-008	atm m3/mol	25	EST	MEYLAN,WM & HOWARD,PH (1991)
OH 2.29E-011	cm3/molc sec	25	EST	MEYLAN,WM & HOWARD,PH (1993)

CAS #: 003398-69-4				N-NITROSO-N-ETHYL-(T-BUTYL)AMINE
Formula: $C_6H_{14}N_2O$				
Mol Weight: 130.19				
MP (deg C):		FP (deg C):		
BP (deg C):				
BP pressure (mm Hg):				

Property/Value	Units	Temp	Data Type	Reference
WS 8.42E+003	mg/L	25	EST	MEYLAN,WM ET AL. (1996)
logP 1.21			EST	MEYLAN,WM & HOWARD,PH (1995)
VP 2.19E-001	mm Hg	25	EST	NEELY,WB & BLAU,GE (1985)
DC	pKa			
HL 6.39E-006	atm m3/mol	25	EST	MEYLAN,WM & HOWARD,PH (1991)
OH 9.36E-012	cm3/molc sec	25	EST	MEYLAN,WM & HOWARD,PH (1993)

CAS #: 003404-61-3				3-METHYL-1-HEXENE
Formula: C_7H_{14}				
Mol Weight: 98.19				
MP (deg C):		FP (deg C):		
BP (deg C): 83.9				
BP pressure (mm Hg):				

Property/Value	Units	Temp	Data Type	Reference
WS 3.08E+001	mg/L	25	EST	MEYLAN,WM ET AL. (1996)
logP 3.57			EST	MEYLAN,WM & HOWARD,PH (1995)
VP 8.24E+001	mm Hg	25	EXP	YAWS,CL (1994A)
DC	pKa			
HL 4.76E-001	atm m3/mol	25	EST	MEYLAN,WM & HOWARD,PH (1991)
OH 3.16E-011	cm3/molc sec	25	EST	MEYLAN,WM & HOWARD,PH (1993)

CAS #: 003404-65-7				3-METHYL-3-HEXENE
Formula: C_7H_{14}				
Mol Weight: 98.19				
MP (deg C):		FP (deg C):		
BP (deg C): 112.5				
BP pressure (mm Hg):				

Property/Value	Units	Temp	Data Type	Reference
WS 2.78E+001	mg/L	25	EST	MEYLAN,WM ET AL. (1996)
logP 3.62			EST	MEYLAN,WM & HOWARD,PH (1995)
VP 8.22E+001	mm Hg	25	EXP	DYKYJ,J & REPAS,M (1973)
DC	pKa			
HL 6.63E-001	atm m3/mol	25	EST	MEYLAN,WM & HOWARD,PH (1991)
OH 8.92E-011	cm3/molc sec	25	EST	MEYLAN,WM & HOWARD,PH (1993)

CAS #: 003404-72-6				2,3-DIMETHYL-1-PENTENE
Formula: C_7H_{14}				
Mol Weight: 98.19				
MP (deg C): -134.3		FP (deg C):		
BP (deg C): 84.3				
BP pressure (mm Hg):				

Property/Value	Units	Temp	Data Type	Reference
WS 2.76E+001	mg/L	25	EST	MEYLAN,WM ET AL. (1996)
logP 3.63			EST	MEYLAN,WM & HOWARD,PH (1995)
VP 8.11E+001	mm Hg	25	EXP	DYKYJ,J & REPAS,M (1973)
DC	pKa			
HL 5.62E-001	atm m3/mol	25	EST	MEYLAN,WM & HOWARD,PH (1991)
OH 5.54E-011	cm3/molc sec	25	EST	MEYLAN,WM & HOWARD,PH (1993)

CAS #: 003404-73-7				3,3-DIMETHYL-1-PENTENE
Formula: C_7H_{14}				
Mol Weight: 98.19				
MP (deg C): -134.3		FP (deg C):		
BP (deg C): 77.5				
BP pressure (mm Hg):				

Property/Value	Units	Temp	Data Type	Reference
WS 3.31E+001	mg/L	25	EST	MEYLAN,WM ET AL. (1996)
logP 3.53			EST	MEYLAN,WM & HOWARD,PH (1995)
VP 1.04E+002	mm Hg	25	EXP	DYKYJ,J & REPAS,M (1973)
DC	pKa			
HL 4.76E-001	atm m3/mol	25	EST	MEYLAN,WM & HOWARD,PH (1991)
OH 2.80E-011	cm3/molc sec	25	EST	MEYLAN,WM & HOWARD,PH (1993)

CAS #: 003404-78-2	2,5-DIMETHYL-2-HEXENE

Formula: C_8H_{16}
Mol Weight: 112.22
MP (deg C):   FP (deg C):
BP (deg C): 112.2
BP pressure (mm Hg):

Property/Value	Units	Temp	Data Type	Reference
WS 1.11E+001	mg/L	25	EST	MEYLAN,WM ET AL. (1996)
logP 4.04			EST	MEYLAN,WM & HOWARD,PH (1995)
VP 2.50E+001	mm Hg	25	EXP	ZWOLINSKI,BJ & WILHOIT,RC (1971)
DC	pKa			
HL 8.81E-001	atm m3/mol	25	EST	MEYLAN,WM & HOWARD,PH (1991)
OH 9.10E-011	cm3/molc sec	25	EST	MEYLAN,WM & HOWARD,PH (1993)

CAS #: 003405-32-1	1,2,3,4-TETRACHLOROBUTANE

Formula: $C_4H_6Cl_4$
Mol Weight: 195.90
MP (deg C):   FP (deg C):
BP (deg C):
BP pressure (mm Hg):

Property/Value	Units	Temp	Data Type	Reference
WS 8.92E+001	mg/L	25	EST	MEYLAN,WM ET AL. (1996)
logP 3.17			EST	MEYLAN,WM & HOWARD,PH (1995)
VP 1.80E-001	mm Hg	25	EST	NEELY,WB & BLAU,GE (1985)
DC	pKa			
HL 2.65E-003	atm m3/mol	25	EST	MEYLAN,WM & HOWARD,PH (1991)
OH 4.47E-013	cm3/molc sec	25	EST	MEYLAN,WM & HOWARD,PH (1993)

CAS #: 003411-95-8	2-(2-HYDROXYPHENYL)BENZOXTHIAZOLE

Formula: $C_{13}H_9NOS$
Mol Weight: 227.29
MP (deg C): 131   FP (deg C):
BP (deg C): 175-193
BP pressure (mm Hg): 3.00E+000

Property/Value	Units	Temp	Data Type	Reference
WS 1.35E+002	mg/L	25	EST	MEYLAN,WM ET AL. (1996)
logP 3.45			EST	MEYLAN,WM & HOWARD,PH (1995)
VP 6.29E-008	mm Hg	25	EST	NEELY,WB & BLAU,GE (1985)
DC	pKa			
HL 2.99E-012	atm m3/mol	25	EST	MEYLAN,WM & HOWARD,PH (1991)
OH 3.61E-011	cm3/molc sec	25	EST	MEYLAN,WM & HOWARD,PH (1993)

CAS #: 003416-05-5	THYMIDINE, 3'-DEOXY-

Formula: $C_{10}H_{14}N_2O_4$
Mol Weight: 226.23
MP (deg C): 155-156   FP (deg C):
BP (deg C):
BP pressure (mm Hg):

Property/Value	Units	Temp	Data Type	Reference
WS 5.08E+003	mg/L	25	EST	MEYLAN,WM ET AL. (1996)
logP -0.60			EXP	SANGSTER,J (1993)
VP 3.19E-012	mm Hg	25	EST	NEELY,WB & BLAU,GE (1985)
DC	pKa			
HL 3.31E-016	atm m3/mol	25	EST	MEYLAN,WM & HOWARD,PH (1991)
OH 8.89E-011	cm3/molc sec	25	EST	MEYLAN,WM & HOWARD,PH (1993)

CAS #: 003416-26-0	LIDOFLAZINE

Formula: $C_{29}H_{33}F_2N_3O$
Mol Weight: 477.60
MP (deg C): 159-161   FP (deg C):
BP (deg C):
BP pressure (mm Hg):

Property/Value	Units	Temp	Data Type	Reference
WS 4.39E-002	mg/L	25	EST	MEYLAN,WM ET AL. (1996)
logP 5.10			EXP	SANGSTER,J (1993)
VP 1.75E-013	mm Hg	25	EST	NEELY,WB & BLAU,GE (1985)
DC	pKa			
HL 3.38E-016	atm m3/mol	25	EST	MEYLAN,WM & HOWARD,PH (1991)
OH 2.21E-010	cm3/molc sec	25	EST	MEYLAN,WM & HOWARD,PH (1993)

CAS #: 003424-57-5	2,4,6-TRIMETHYL-2,4,6-TRIPHENYLCYCLOTRISILOXANE

Formula: $C_{21}H_{24}O_3Si_3$
Mol Weight: 408.68
MP (deg C):   FP (deg C):
BP (deg C):
BP pressure (mm Hg):

Property/Value	Units	Temp	Data Type	Reference
WS 9.71E-005	mg/L	25	EST	MEYLAN,WM ET AL. (1996)
logP 8.12			EST	MEYLAN,WM & HOWARD,PH (1995)
VP	mm Hg			
DC	pKa			
HL	atm m3/mol			
OH 6.30E-012	cm3/molc sec	25	EST	MEYLAN,WM & HOWARD,PH (1993)

CAS #: 003424-82-6	O,P'-DDE

Formula: $C_{14}H_8Cl_4$
Mol Weight: 318.03
MP (deg C):   FP (deg C):
BP (deg C):
BP pressure (mm Hg):

Property/Value	Units	Temp	Data Type	Reference
WS 1.40E-001	mg/L	25	EXP	YALKOWSKY,SH & DANNENFELSER,RM (1992)
logP 6.00			EST	MEYLAN,WM & HOWARD,PH (1995)
VP 6.20E-006	mm Hg	25	EXP	SUNTIO,LR ET AL. (1988)
DC	pKa			
HL 1.85E-005	atm m3/mol	25	EST	VP/WSOL
OH 7.43E-012	cm3/molc sec	25	EST	MEYLAN,WM & HOWARD,PH (1993)

CAS #: 003424-93-9	P-METHOXYBENZAMIDE

Formula: $C_8H_9NO_2$
Mol Weight: 151.17
MP (deg C): 166.5   FP (deg C):
BP (deg C): 295
BP pressure (mm Hg):

Property/Value	Units	Temp	Data Type	Reference
WS 1.38E+004	mg/L	25	EST	MEYLAN,WM ET AL. (1996)
logP 0.86			EXP	HANSCH,C & LEO,AJ (1985)
VP 1.11E-004	mm Hg	25	EST	NEELY,WB & BLAU,GE (1985)
DC	pKa			
HL 1.31E-010	atm m3/mol	25	EST	MEYLAN,WM & HOWARD,PH (1991)
OH 2.24E-011	cm3/molc sec	25	EST	MEYLAN,WM & HOWARD,PH (1993)

CAS #: 003435-28-7	6-METHYL-4-PYRIMIDINAMINE

Formula: $C_5H_7N_3$

Mol Weight: 109.13

MP (deg C): 197

FP (deg C):

BP (deg C):

BP pressure (mm Hg):

Property/Value	Units	Temp	Data Type	Reference
WS 7.54E+004	mg/L	25	EST	MEYLAN,WM ET AL. (1996)
logP 0.19			EXP	HANSCH,C & LEO,AJ (1985)
VP 6.92E-002	mm Hg	25	EST	NEELY,WB & BLAU,GE (1985)
DC 6.16	pKa		EXP	PERRIN,DD (1965)
HL 1.14E-009	atm m3/mol	25	EST	MEYLAN,WM & HOWARD,PH (1991)
OH 3.33E-011	cm3/molc sec	25	EST	MEYLAN,WM & HOWARD,PH (1993)

CAS #: 003437-95-4	2-IODOTHIOPHENE

Formula: C_4H_3IS

Mol Weight: 210.04

MP (deg C): -40

FP (deg C):

BP (deg C): 181

BP pressure (mm Hg):

Property/Value	Units	Temp	Data Type	Reference
WS 1.11E+002	mg/L	25	EST	MEYLAN,WM ET AL. (1996)
logP 2.98			EST	MEYLAN,WM & HOWARD,PH (1995)
VP 8.05E+000	mm Hg	25	EXT	BOUBLIK,T ET AL. (1984)
DC	pKa			
HL 6.78E-004	atm m3/mol	25	EST	MEYLAN,WM & HOWARD,PH (1991)
OH 6.28E-012	cm3/molc sec	25	EST	MEYLAN,WM & HOWARD,PH (1993)

CAS #: 003438-46-8	4-METHYLPYRIMIDINE

Formula: $C_5H_6N_2$

Mol Weight: 94.12

MP (deg C): 32

FP (deg C):

BP (deg C): 142

BP pressure (mm Hg):

Property/Value	Units	Temp	Data Type	Reference
WS 8.87E+004	mg/L	25	EST	MEYLAN,WM ET AL. (1996)
logP 0.16			EXP	HANSCH,C & LEO,AJ (1985)
VP 4.63E+000	mm Hg	25	EST	NEELY,WB & BLAU,GE (1985)
DC 1.91	pKa	20	EXP	PERRIN,DD (1965)
HL 3.22E-006	atm m3/mol	25	EST	MEYLAN,WM & HOWARD,PH (1991)
OH 7.36E-013	cm3/molc sec	25	EST	MEYLAN,WM & HOWARD,PH (1993)

CAS #: 003438-48-0	PYRIMIDINE, 4-PHENYL-

Formula: $C_{10}H_8N_2$

Mol Weight: 156.19

MP (deg C): 54-58

FP (deg C):

BP (deg C):

BP pressure (mm Hg):

Property/Value	Units	Temp	Data Type	Reference
WS 1.63E+003	mg/L	25	EST	MEYLAN,WM ET AL. (1996)
logP 1.92			EXP	HANSCH,C ET AL. (1995)
VP 1.20E-003	mm Hg	25	EST	NEELY,WB & BLAU,GE (1985)
DC	pKa			
HL 2.24E-007	atm m3/mol	25	EST	MEYLAN,WM & HOWARD,PH (1991)
OH 3.79E-012	cm3/molc sec	25	EST	MEYLAN,WM & HOWARD,PH (1993)

CAS #: 003441-01-8	M-CYANOBENZAMIDE

Formula: $C_8H_6N_2O$

Mol Weight: 146.15

MP (deg C):

FP (deg C):

BP (deg C):

BP pressure (mm Hg):

Property/Value	Units	Temp	Data Type	Reference
WS 1.54E+004	mg/L	25	EST	MEYLAN,WM ET AL. (1996)
logP 0.52			EXP	NAKAGAWA,Y ET AL. (1992)
VP 1.69E-005	mm Hg	25	EST	NEELY,WB & BLAU,GE (1985)
DC	pKa			
HL 2.14E-011	atm m3/mol	25	EST	MEYLAN,WM & HOWARD,PH (1991)
OH 2.31E-012	cm3/molc sec	25	EST	MEYLAN,WM & HOWARD,PH (1993)

CAS #: 003452-07-1	1-EICOSENE

Formula: $C_{20}H_{40}$

Mol Weight: 280.54

MP (deg C): 28.5

FP (deg C):

BP (deg C): 341

BP pressure (mm Hg):

Property/Value	Units	Temp	Data Type	Reference
WS 1.26E-005	mg/L	25	EST	MEYLAN,WM ET AL. (1996)
logP 10.03			EST	MEYLAN,WM & HOWARD,PH (1995)
VP 1.06E-005	mm Hg	25	EXT	YAWS,CL (1994B)
DC	pKa			
HL 1.89E+001	atm m3/mol	25	EST	MEYLAN,WM & HOWARD,PH (1991)
OH 5.00E-011	cm3/molc sec	25	EST	MEYLAN,WM & HOWARD,PH (1993)

CAS #: 003452-09-3	1-NONYNE

Formula: C_9H_{16}

Mol Weight: 124.23

MP (deg C): -50

FP (deg C):

BP (deg C): 150.8

BP pressure (mm Hg):

Property/Value	Units	Temp	Data Type	Reference
WS 7.14E+000	mg/L	25	EXP	SUZUKI,T (1991)
logP 3.99			EST	MEYLAN,WM & HOWARD,PH (1995)
VP 6.26E+000	mm Hg	25	EXP	HINE,J & MOOKERJEE,PK (1975)
DC	pKa			
HL 1.43E-001	atm m3/mol	25	EST	VP/WSOL
OH 1.51E-011	cm3/molc sec	25	EST	MEYLAN,WM & HOWARD,PH (1993)

CAS #: 003452-97-9	ISONONYL ALCOHOL

Formula: $C_9H_{20}O$

Mol Weight: 144.26

MP (deg C):

FP (deg C):

BP (deg C): 194

BP pressure (mm Hg):

Property/Value	Units	Temp	Data Type	Reference
WS 5.72E+002	mg/L	25	EST	MEYLAN,WM ET AL. (1996)
logP 3.11			EST	MEYLAN,WM & HOWARD,PH (1995)
VP 3.00E-001	mm Hg	20	EXP	FLICK,EW (1991)
DC	pKa			
HL 4.12E-005	atm m3/mol	25	EST	MEYLAN,WM & HOWARD,PH (1991)
OH 1.06E-011	cm3/molc sec	25	EST	MEYLAN,WM & HOWARD,PH (1993)

CAS #: 003453-64-3				3-METHYL-1(3H)-ISOBENZOFURANONE

Formula: $C_9H_8O_2$

Mol Weight: 148.16

MP (deg C):

FP (deg C):

BP (deg C):

BP pressure (mm Hg):

Property/ Value	Units	Temp	Data Type	Reference
WS 4.83E+003	mg/L	25	EST	MEYLAN,WM ET AL. (1996)
logP 1.41			EXP	HANSCH,C & LEO,AJ (1985)
VP 3.71E-003	mm Hg	25	EST	NEELY,WB & BLAU,GE (1985)
DC	pKa			
HL 1.69E-005	atm m3/mol	25	EST	MEYLAN,WM & HOWARD,PH (1991)
OH 6.87E-012	cm3/molc sec	25	EST	MEYLAN,WM & HOWARD,PH (1993)

CAS #: 003454-07-7				P-ETHYLSTYRENE

Formula: $C_{10}H_{12}$

Mol Weight: 132.21

MP (deg C): -49.7

FP (deg C):

BP (deg C): 192.3

BP pressure (mm Hg):

Property/ Value	Units	Temp	Data Type	Reference
WS 2.38E+001	mg/L	25	EST	MEYLAN,WM ET AL. (1996)
logP 4.19			EXP	CHEM INSPECT TEST INST (1992)
VP 9.45E-001	mm Hg	25	EXT	OHE,S (1976)
DC	pKa			
HL 4.05E-003	atm m3/mol	25	EST	MEYLAN,WM & HOWARD,PH (1991)
OH 3.20E-011	cm3/molc sec	25	EST	MEYLAN,WM & HOWARD,PH (1993)

CAS #: 003456-79-9				2H-INDOL-2-ONE, 1,3-DIHYDRO-3-PHENYL-

Formula: $C_{14}H_{11}NO$

Mol Weight: 209.25

MP (deg C):

FP (deg C):

BP (deg C):

BP pressure (mm Hg):

Property/ Value	Units	Temp	Data Type	Reference
WS 2.41E+002	mg/L	25	EST	MEYLAN,WM ET AL. (1996)
logP 2.59			EXP	SANGSTER,J (1993)
VP 3.94E-007	mm Hg	25	EST	NEELY,WB & BLAU,GE (1985)
DC	pKa			
HL 1.83E-010	atm m3/mol	25	EST	MEYLAN,WM & HOWARD,PH (1991)
OH 2.07E-011	cm3/molc sec	25	EST	MEYLAN,WM & HOWARD,PH (1993)

CAS #: 003458-28-4				D-MANNOSE

Formula: $C_6H_{12}O_6$

Mol Weight: 180.16

MP (deg C): 132 dec

FP (deg C):

BP (deg C):

BP pressure (mm Hg):

Property/ Value	Units	Temp	Data Type	Reference
WS 7.13E+005	mg/L	17	EXP	YALKOWSKY,SH & DANNENFELSER,RM (1992)
logP -2.43			EST	MEYLAN,WM & HOWARD,PH (1995)
VP 1.82E-008	mm Hg	25	EST	NEELY,WB & BLAU,GE (1985)
DC 12.92	pKa	0	EXP	KORTUM,G ET AL (1961)
HL 5.88E-011	atm m3/mol	25	EST	MEYLAN,WM & HOWARD,PH (1991)
OH 8.49E-011	cm3/molc sec	25	EST	MEYLAN,WM & HOWARD,PH (1993)

CAS #: 003469-69-0				4-IODOPYRAZOLE

Formula: $C_3H_3IN_2$

Mol Weight: 193.98

MP (deg C): 108-110

FP (deg C):

BP (deg C):

BP pressure (mm Hg):

Property/ Value	Units	Temp	Data Type	Reference
WS 1.66E+003	mg/L	25	EST	MEYLAN,WM ET AL. (1996)
logP 1.70			EXP	HANSCH,C & LEO,AJ (1985)
VP 1.12E-002	mm Hg	25	EST	NEELY,WB & BLAU,GE (1985)
DC	pKa			
HL 8.56E-007	atm m3/mol	25	EST	MEYLAN,WM & HOWARD,PH (1991)
OH 2.37E-011	cm3/molc sec	25	EST	MEYLAN,WM & HOWARD,PH (1993)

CAS #: 003475-63-6				1-NITROSO-TRIMETHYL UREA

Formula: $C_4H_9N_3O_2$

Mol Weight: 131.14

MP (deg C):

FP (deg C):

BP (deg C):

BP pressure (mm Hg):

Property/ Value	Units	Temp	Data Type	Reference
WS 4.48E+004	mg/L	25	EST	MEYLAN,WM ET AL. (1996)
logP 0.36			EXP	HANSCH,C & LEO,AJ (1985)
VP 1.62E+000	mm Hg	25	EST	NEELY,WB & BLAU,GE (1985)
DC	pKa			
HL 4.78E-010	atm m3/mol	25	EST	MEYLAN,WM & HOWARD,PH (1991)
OH 2.67E-012	cm3/molc sec	25	EST	MEYLAN,WM & HOWARD,PH (1993)

CAS #: 003476-50-4				N-DESACETYL-COLCHICINE

Formula: $C_{20}H_{23}NO_5$

Mol Weight: 357.41

MP (deg C):

FP (deg C):

BP (deg C):

BP pressure (mm Hg):

Property/ Value	Units	Temp	Data Type	Reference
WS 6.54E+003	mg/L	25	EST	MEYLAN,WM ET AL. (1996)
logP 1.10			EXP	HANSCH,C ET AL. (1995)
VP 1.77E-009	mm Hg	25	EST	NEELY,WB & BLAU,GE (1985)
DC	pKa			
HL 5.53E-015	atm m3/mol	25	EST	MEYLAN,WM & HOWARD,PH (1991)
OH 3.60E-010	cm3/molc sec	25	EST	MEYLAN,WM & HOWARD,PH (1993)

CAS #: 003481-09-2				N-CHLOROPHTHALIMIDE

Formula: $C_8H_4ClNO_2$

Mol Weight: 181.58

MP (deg C):

FP (deg C):

BP (deg C):

BP pressure (mm Hg):

Property/ Value	Units	Temp	Data Type	Reference
WS 8.16E+003	mg/L	25	EST	MEYLAN,WM ET AL. (1996)
logP 0.96			EST	MEYLAN,WM & HOWARD,PH (1995)
VP 1.49E-006	mm Hg	25	EST	NEELY,WB & BLAU,GE (1985)
DC	pKa			
HL 1.96E-007	atm m3/mol	25	EST	MEYLAN,WM & HOWARD,PH (1991)
OH 1.57E-011	cm3/molc sec	25	EST	MEYLAN,WM & HOWARD,PH (1993)

CAS #: 003481-20-7 — 2,3,5,6-TETRACHLOROANILINE

Formula: $C_6H_3Cl_4N$

Mol Weight: 230.91

MP (deg C): 106-108

FP (deg C):

BP (deg C):

BP pressure (mm Hg):

Property/Value	Units	Temp	Data Type	Reference
WS 9.48E+000	mg/L	25	EST	MEYLAN,WM ET AL. (1996)
logP 4.10			EXP	SANGSTER,J (1994)
VP 5.57E-004	mm Hg	25	EST	NEELY,WB & BLAU,GE (1985)
DC	pKa			
HL 5.74E-007	atm m3/mol	25	EST	MEYLAN,WM & HOWARD,PH (1991)
OH 4.54E-012	cm3/molc sec	25	EST	MEYLAN,WM & HOWARD,PH (1993)

CAS #: 003483-39-4 — OXIRANE, CYCLOHEXYL-

Formula: $C_8H_{14}O$

Mol Weight: 126.20

MP (deg C):

FP (deg C):

BP (deg C):

BP pressure (mm Hg):

Property/Value	Units	Temp	Data Type	Reference
WS 8.00E+002	mg/L	25	EST	MEYLAN,WM ET AL. (1996)
logP 2.43			EXP	SERRENTINO,R ET AL. (1983)
VP 2.99E+000	mm Hg	25	EST	NEELY,WB & BLAU,GE (1985)
DC	pKa			
HL 2.90E-004	atm m3/mol	25	EST	MEYLAN,WM & HOWARD,PH (1991)
OH 4.98E-012	cm3/molc sec	25	EST	MEYLAN,WM & HOWARD,PH (1993)

CAS #: 003485-14-1 — CYCLACILLIN

Formula: $C_{15}H_{23}N_3O_4S$

Mol Weight: 341.43

MP (deg C): 182-183

FP (deg C):

BP (deg C):

BP pressure (mm Hg):

Property/Value	Units	Temp	Data Type	Reference
WS 5.31E+002	mg/L	25	EST	MEYLAN,WM ET AL. (1996)
logP 1.31			EXP	SANGSTER,J (1994)
VP 2.84E-014	mm Hg	25	EST	NEELY,WB & BLAU,GE (1985)
DC	pKa			
HL 5.38E-016	atm m3/mol	25	EST	MEYLAN,WM & HOWARD,PH (1991)
OH 4.14E-010	cm3/molc sec	25	EST	MEYLAN,WM & HOWARD,PH (1993)

CAS #: 003489-59-6 — 2H-1,4-BENZODIAZEPIN-2-ONE,1,3-DIHYDRO-1-METHYL-

Formula: $C_{17}H_{13}N_3O$

Mol Weight: 275.31

MP (deg C):

FP (deg C):

BP (deg C):

BP pressure (mm Hg):

Property/Value	Units	Temp	Data Type	Reference
WS 2.69E+002	mg/L	25	EST	MEYLAN,WM ET AL. (1996)
logP 1.80			EXP	HANSCH,C ET AL. (1995)
VP 8.46E-010	mm Hg	25	EST	NEELY,WB & BLAU,GE (1985)
DC	pKa			
HL 3.22E-013	atm m3/mol	25	EST	MEYLAN,WM & HOWARD,PH (1991)
OH 2.46E-011	cm3/molc sec	25	EST	MEYLAN,WM & HOWARD,PH (1993)

CAS #: 003508-00-7 — 1-BROMOHEPTADECANE

Formula: $C_{17}H_{35}Br$

Mol Weight: 319.38

MP (deg C): 32

FP (deg C):

BP (deg C): 349

BP pressure (mm Hg):

Property/Value	Units	Temp	Data Type	Reference
WS 1.82E-004	mg/L	25	EST	MEYLAN,WM ET AL. (1996)
logP 9.03			EST	MEYLAN,WM & HOWARD,PH (1995)
VP 1.83E-005	mm Hg	25	EXT	LI,JCM & ROSSINI,FD (1953)
DC	pKa			
HL 7.90E-001	atm m3/mol	25	EST	MEYLAN,WM & HOWARD,PH (1991)
OH 2.05E-011	cm3/molc sec	25	EST	MEYLAN,WM & HOWARD,PH (1993)

CAS #: 003510-70-1 — SILANE, TRIMETHYLPROPYL-

Formula: $C_6H_{16}Si$

Mol Weight: 116.28

MP (deg C):

FP (deg C):

BP (deg C): 89

BP pressure (mm Hg):

Property/Value	Units	Temp	Data Type	Reference
WS 5.44E+001	mg/L	25	EST	MEYLAN,WM ET AL. (1996)
logP 3.84			EXP	HANSCH,C ET AL. (1995)
VP 1.27E+002	mm Hg	25	EST	NEELY,WB & BLAU,GE (1985)
DC	pKa			
HL 3.90E-001	atm m3/mol	25	EST	MEYLAN,WM & HOWARD,PH (1991)
OH 3.03E-012	cm3/molc sec	25	EST	MEYLAN,WM & HOWARD,PH (1993)

CAS #: 003522-94-9 — 2,2,5-TRIMETHYLHEXANE

Formula: C_9H_{20}

Mol Weight: 128.26

MP (deg C): -105.7

FP (deg C):

BP (deg C): 124

BP pressure (mm Hg):

Property/Value	Units	Temp	Data Type	Reference
WS 1.15E+000	mg/L	25	EXP	YALKOWSKY,SH & DANNENFELSER,RM (1992)
logP 4.58			EST	MEYLAN,WM & HOWARD,PH (1995)
VP 1.66E+001	mm Hg	25	EXP	DAUBERT,TE & DANNER,RP (1989)
DC	pKa			
HL 2.44E+000	atm m3/mol	25	EST	VP/WSOL
OH 6.05E-012	cm3/molc sec	25	EST	MEYLAN,WM & HOWARD,PH (1993)

CAS #: 003530-82-3 — 5-BENZYL-2,4-IMIDAZOLIDIONE

Formula: $C_{10}H_{10}N_2O_2$

Mol Weight: 190.20

MP (deg C):

FP (deg C):

BP (deg C):

BP pressure (mm Hg):

Property/Value	Units	Temp	Data Type	Reference
WS 1.51E+004	mg/L	25	EST	MEYLAN,WM ET AL. (1996)
logP 0.60			EXP	LIPINSKI,CA ET AL. (1991)
VP 1.29E-008	mm Hg	25	EST	NEELY,WB & BLAU,GE (1985)
DC	pKa			
HL 1.68E-010	atm m3/mol	25	EST	MEYLAN,WM & HOWARD,PH (1991)
OH 1.64E-011	cm3/molc sec	25	EST	MEYLAN,WM & HOWARD,PH (1993)

003531-14-4 — NNN-TRIPR-N-ME AMMONIUM IODIDE

Formula: $C_{10}H_{24}IN$

Mol Weight: 285.21

MP (deg C):
FP (deg C):
BP (deg C):
BP pressure (mm Hg):

Property/Value	Units	Temp	Data Type	Reference
WS 1.03E+005	mg/L	25	EST	MEYLAN,WM ET AL. (1996)
logP -0.98			EXP	NEEF,C & MEIJER,DKF (1984)
VP 4.57E-008	mm Hg	25	EST	NEELY,WB & BLAU,GE (1985)
DC	pKa			
HL 1.44E-011	atm m3/mol	25	EST	MEYLAN,WM & HOWARD,PH (1991)
OH 3.73E-011	cm3/molc sec	25	EST	MEYLAN,WM & HOWARD,PH (1993)

003538-65-6 — BUTYRIC ACID HYDRAZIDE

Formula: $C_4H_{10}N_2O$

Mol Weight: 102.14

MP (deg C): 45.5
FP (deg C):
BP (deg C): 138
BP pressure (mm Hg): 2.00E+001

Property/Value	Units	Temp	Data Type	Reference
WS 1.91E+004	mg/L	25	EST	MEYLAN,WM ET AL. (1996)
logP -0.62			EXP	HANSCH,C & LEO,AJ (1985)
VP 6.36E-003	mm Hg	25	EST	NEELY,WB & BLAU,GE (1985)
DC	pKa			
HL 8.23E-011	atm m3/mol	25	EST	MEYLAN,WM & HOWARD,PH (1991)
OH 1.02E-011	cm3/molc sec	25	EST	MEYLAN,WM & HOWARD,PH (1993)

003544-24-9 — BENZAMIDE, 3-AMINO-

Formula: $C_7H_8N_2O$

Mol Weight: 136.15

MP (deg C): 115-116
FP (deg C):
BP (deg C):
BP pressure (mm Hg):

Property/Value	Units	Temp	Data Type	Reference
WS 8.14E+003	mg/L	25	EST	MEYLAN,WM ET AL. (1996)
logP -0.33			EXP	HANSCH,C ET AL. (1995)
VP 2.63E-005	mm Hg	25	EST	NEELY,WB & BLAU,GE (1985)
DC	pKa			
HL 7.83E-013	atm m3/mol	25	EST	MEYLAN,WM & HOWARD,PH (1991)
OH 4.74E-011	cm3/molc sec	25	EST	MEYLAN,WM & HOWARD,PH (1993)

003544-35-2 — N'-IPR-(4-CLPHENOXY)ACETIC HYDRAZIDE

Formula: $C_{11}H_{15}ClN_2O_2$

Mol Weight: 242.71

MP (deg C): 93-94
FP (deg C):
BP (deg C):
BP pressure (mm Hg):

Property/Value	Units	Temp	Data Type	Reference
WS 4.89E+002	mg/L	25	EST	MEYLAN,WM ET AL. (1996)
logP 2.02			EXP	HANSCH,C & LEO,AJ (1985)
VP 1.06E-006	mm Hg	25	EST	NEELY,WB & BLAU,GE (1985)
DC	pKa			
HL 4.10E-012	atm m3/mol	25	EST	MEYLAN,WM & HOWARD,PH (1991)
OH 9.74E-011	cm3/molc sec	25	EST	MEYLAN,WM & HOWARD,PH (1993)

003547-07-7 — 4-(4-CHLOROPHENOXY) BUTYRIC ACID

Formula: $C_{10}H_{11}ClO_3$

Mol Weight: 214.65

MP (deg C):
FP (deg C):
BP (deg C):
BP pressure (mm Hg):

Property/Value	Units	Temp	Data Type	Reference
WS 2.73E+002	mg/L	25	EST	MEYLAN,WM ET AL. (1996)
logP 2.95			EST	MEYLAN,WM & HOWARD,PH (1995)
VP 5.02E-005	mm Hg	25	EST	NEELY,WB & BLAU,GE (1985)
DC	pKa			
HL 3.10E-009	atm m3/mol	25	EST	MEYLAN,WM & HOWARD,PH (1991)
OH 1.87E-011	cm3/molc sec	25	EST	MEYLAN,WM & HOWARD,PH (1993)

003547-33-9 — ETHANOL, 2-(OCTYLTHIO)-

Formula: $C_{10}H_{22}OS$

Mol Weight: 190.35

MP (deg C):
FP (deg C):
BP (deg C):
BP pressure (mm Hg):

Property/Value	Units	Temp	Data Type	Reference
WS 3.81E+001	mg/L	25	EST	MEYLAN,WM ET AL. (1996)
logP 3.64			EXP	TOMLIN,C (1994)
VP 2.35E-004	mm Hg	25	EST	NEELY,WB & BLAU,GE (1985)
DC	pKa			
HL 2.77E-007	atm m3/mol	25	EST	MEYLAN,WM & HOWARD,PH (1991)
OH 3.22E-011	cm3/molc sec	25	EST	MEYLAN,WM & HOWARD,PH (1993)

003550-21-8 — BENZENE, 1,1'-(ISOTHIOCYANATOMETHYLENE)BIS-

Formula: $C_{14}H_{11}NS$

Mol Weight: 225.31

MP (deg C):
FP (deg C):
BP (deg C):
BP pressure (mm Hg):

Property/Value	Units	Temp	Data Type	Reference
WS 1.45E+000	mg/L	25	EST	MEYLAN,WM ET AL. (1996)
logP 5.09			EXP	SANGSTER,J (1993)
VP 4.09E-005	mm Hg	25	EST	NEELY,WB & BLAU,GE (1985)
DC	pKa			
HL 2.03E-005	atm m3/mol	25	EST	MEYLAN,WM & HOWARD,PH (1991)
OH 1.12E-011	cm3/molc sec	25	EST	MEYLAN,WM & HOWARD,PH (1993)

003555-18-8 — 4-(SEC-BUTYL)-2-NITROPHENOL

Formula: $C_{10}H_{13}NO_3$

Mol Weight: 195.22

MP (deg C):
FP (deg C):
BP (deg C):
BP pressure (mm Hg):

Property/Value	Units	Temp	Data Type	Reference
WS 2.43E+001	mg/L	25	EST	MEYLAN,WM ET AL. (1996)
logP 3.84			EXP	SCHWARZENBACH,RP ET AL. (1988)
VP 4.58E-003	mm Hg	25	EXP	SCHWARZENBACH,RP ET AL. (1988)
DC	pKa			
HL 1.81E-005	atm m3/mol	25	EST	MEYLAN,WM & HOWARD,PH (1991)
OH 9.02E-012	cm3/molc sec	25	EST	MEYLAN,WM & HOWARD,PH (1993)

CAS #: 003555-47-3 — TETRAKIS(TRIMETHYLSILOXY)SILANE

Formula: $C_{12}H_{36}O_4Si_5$

Mol Weight: 384.85

MP (deg C):
FP (deg C):
BP (deg C):
BP pressure (mm Hg):

Property/Value	Units	Temp	Data Type	Reference
WS 1.11E-004	mg/L	25	EST	MEYLAN,WM ET AL. (1996)
logP 6.52			EST	MEYLAN,WM & HOWARD,PH (1995)
VP 6.72E-002	mm Hg	25	EXT	FLANINGAM,OL (1986)
DC	pKa			
HL 1.81E-007	atm m3/mol	25	EST	MEYLAN,WM & HOWARD,PH (1991)
OH 1.80E-012	cm3/molc sec	25	EST	MEYLAN,WM & HOWARD,PH (1993)

CAS #: 003558-69-8 — 2,6-DIPHENYLPYRIDINE

Formula: $C_{17}H_{13}N$

Mol Weight: 231.30

MP (deg C): 82
FP (deg C):
BP (deg C): 397
BP pressure (mm Hg):

Property/Value	Units	Temp	Data Type	Reference
WS 2.29E+000	mg/L	25	EST	MEYLAN,WM ET AL. (1996)
logP 4.82			EXP	HANSCH,C & LEO,AJ (1985)
VP 1.32E-006	mm Hg	25	EST	NEELY,WB & BLAU,GE (1985)
DC	pKa			
HL 4.16E-008	atm m3/mol	25	EST	MEYLAN,WM & HOWARD,PH (1991)
OH 7.23E-012	cm3/molc sec	25	EST	MEYLAN,WM & HOWARD,PH (1993)

CAS #: 003566-00-5 — N-ME CARBAMATE,3-ME-4-MES PHENYL

Formula: $C_{10}H_{13}NO_2S$

Mol Weight: 211.28

MP (deg C): 117-118
FP (deg C):
BP (deg C):
BP pressure (mm Hg):

Property/Value	Units	Temp	Data Type	Reference
WS 3.00E+001	mg/L	25	EXP	SHIU,WY ET AL. (1990)
logP 2.47			EXP	HANSCH,C & LEO,AJ (1985)
VP 3.15E-004	mm Hg	25	EST	NEELY,WB & BLAU,GE (1985)
DC	pKa			
HL 1.03E-009	atm m3/mol	25	EST	MEYLAN,WM & HOWARD,PH (1991)
OH 1.86E-011	cm3/molc sec	25	EST	MEYLAN,WM & HOWARD,PH (1993)

CAS #: 003567-38-2 — 3-PHENYL-CARBAMOYLOXY-1-PROPYNE

Formula: $C_{10}H_9NO_2$

Mol Weight: 175.19

MP (deg C): 86-87
FP (deg C):
BP (deg C):
BP pressure (mm Hg):

Property/Value	Units	Temp	Data Type	Reference
WS 2.55E+003	mg/L	25	EST	MEYLAN,WM ET AL. (1996)
logP 1.59			EXP	HANSCH,C & LEO,AJ (1985)
VP 2.71E-003	mm Hg	25	EST	NEELY,WB & BLAU,GE (1985)
DC	pKa			
HL 6.97E-010	atm m3/mol	25	EST	MEYLAN,WM & HOWARD,PH (1991)
OH 1.67E-011	cm3/molc sec	25	EST	MEYLAN,WM & HOWARD,PH (1993)

CAS #: 003567-62-2 — 3-(3,4-DICHLOROPHENYL)-1-METHYLUREA

Formula: $C_8H_8Cl_2N_2O$

Mol Weight: 219.07

MP (deg C):
FP (deg C):
BP (deg C):
BP pressure (mm Hg):

Property/Value	Units	Temp	Data Type	Reference
WS 1.07E+002	mg/L	25	EST	MEYLAN,WM ET AL. (1996)
logP 2.94			EXP	BRIGGS,GG (1981)
VP 2.02E-005	mm Hg	25	EST	NEELY,WB & BLAU,GE (1985)
DC	pKa			
HL 2.43E-010	atm m3/mol	25	EST	MEYLAN,WM & HOWARD,PH (1991)
OH 1.03E-011	cm3/molc sec	25	EST	MEYLAN,WM & HOWARD,PH (1993)

CAS #: 003567-84-8 — 4,6-NH2-2,2-DIME-1(4BRPH)S-TRIAZINE

Formula: $C_{11}H_{14}BrN_5$

Mol Weight: 296.18

MP (deg C):
FP (deg C):
BP (deg C):
BP pressure (mm Hg):

Property/Value	Units	Temp	Data Type	Reference
WS 1.52E+003	mg/L	25	EST	MEYLAN,WM ET AL. (1996)
logP 1.09			EXP	HANSCH,C & LEO,AJ (1985)
VP 6.44E-007	mm Hg	25	EST	NEELY,WB & BLAU,GE (1985)
DC	pKa			
HL 2.90E-013	atm m3/mol	25	EST	MEYLAN,WM & HOWARD,PH (1991)
OH 8.43E-011	cm3/molc sec	25	EST	MEYLAN,WM & HOWARD,PH (1993)

CAS #: 003570-61-4 — 2,4,5-TES

Formula: $C_8H_6Cl_3NaO_5S$

Mol Weight: 343.55

MP (deg C):
FP (deg C):
BP (deg C):
BP pressure (mm Hg):

Property/Value	Units	Temp	Data Type	Reference
WS 6.00E+004	mg/L	25	EXP	BAILEY,GW & WHITE,JL (1965)
logP -0.04			EST	MEYLAN,WM & HOWARD,PH (1995)
VP 3.27E-014	mm Hg	25	EST	NEELY,WB & BLAU,GE (1985)
DC	pKa			
HL	atm m3/mol			
OH 9.10E-012	cm3/molc sec	25	EST	MEYLAN,WM & HOWARD,PH (1993)

CAS #: 003572-34-7 — BENZENAMINE, 4-[(4-AMINOPHENYL)SULFONYL]-N-ETHYL

Formula: $C_{14}H_{16}N_2O_2S$

Mol Weight: 276.36

MP (deg C):
FP (deg C):
BP (deg C):
BP pressure (mm Hg):

Property/Value	Units	Temp	Data Type	Reference
WS 3.43E+002	mg/L	25	EST	MEYLAN,WM ET AL. (1996)
logP 1.98			EXP	ALTOMARE,C ET AL. (1991)
VP 1.95E-008	mm Hg	25	EST	NEELY,WB & BLAU,GE (1985)
DC	pKa			
HL 9.07E-014	atm m3/mol	25	EST	MEYLAN,WM & HOWARD,PH (1991)
OH 4.10E-011	cm3/molc sec	25	EST	MEYLAN,WM & HOWARD,PH (1993)

CAS #: 003585-93-1			ALPHA-(P-MEO-PHENYL)-N-PHENYLNITRONE	
Formula: $C_{14}H_{13}NO_2$				
Mol Weight: 227.27				
MP (deg C):		FP (deg C):		
BP (deg C):				
BP pressure (mm Hg):				

Property/Value	Units	Temp	Data Type	Reference
WS 6.94E+002	mg/L	25	EST	MEYLAN,WM ET AL. (1996)
logP 1.94			EXP	KIRCHNER,JJ ET AL. (1985)
VP 2.22E-009	mm Hg	25	EST	NEELY,WB & BLAU,GE (1985)
DC	pKa			
HL 4.67E-013	atm m3/mol	25	EST	MEYLAN,WM & HOWARD,PH (1991)
OH 2.91E-011	cm3/molc sec	25	EST	MEYLAN,WM & HOWARD,PH (1993)

CAS #: 003607-78-1			1,1,1,3,3,3-HEXACHLOROPROPANE	
Formula: $C_3H_2Cl_6$				
Mol Weight: 250.77				
MP (deg C): -27		FP (deg C):		
BP (deg C): 206				
BP pressure (mm Hg):				

Property/Value	Units	Temp	Data Type	Reference
WS 3.20E+000	mg/L	25	EST	MEYLAN,WM ET AL. (1996)
logP 4.53			EST	MEYLAN,WM & HOWARD,PH (1995)
VP 1.18E+000	mm Hg	25	EST	NEELY,WB & BLAU,GE (1985)
DC	pKa			
HL 2.48E-004	atm m3/mol	25	EST	MEYLAN,WM & HOWARD,PH (1991)
OH 4.45E-015	cm3/molc sec	25	EST	MEYLAN,WM & HOWARD,PH (1993)

CAS #: 003615-21-2			2-CF3-4,5-DICHLOROBENZIMIDAZOLE	
Formula: $C_8H_3Cl_2F_3N_2$				
Mol Weight: 255.03				
MP (deg C): 213.5		FP (deg C):		
BP (deg C):				
BP pressure (mm Hg):				

Property/Value	Units	Temp	Data Type	Reference
WS 6.90E+001	mg/L	25	EXP	SHIU,WY ET AL. (1990)
logP 3.49			EXP	HANSCH,C & LEO,AJ (1985)
VP 2.90E-006	mm Hg	25	EST	NEELY,WB & BLAU,GE (1985)
DC	pKa			
HL 1.75E-006	atm m3/mol	25	EST	MEYLAN,WM & HOWARD,PH (1991)
OH 1.29E-012	cm3/molc sec	25	EST	MEYLAN,WM & HOWARD,PH (1993)

CAS #: 003619-01-0			BUTANOIC ACID,2-ACETYLAMINO,METHYL ESTER	
Formula: $C_7H_{13}NO_3$				
Mol Weight: 159.19				
MP (deg C):		FP (deg C):		
BP (deg C):				
BP pressure (mm Hg):				

Property/Value	Units	Temp	Data Type	Reference
WS 9.64E+004	mg/L	25	EST	MEYLAN,WM ET AL. (1996)
logP -0.17			EXP	HANSCH,C ET AL. (1995)
VP 8.31E-004	mm Hg	25	EST	NEELY,WB & BLAU,GE (1985)
DC	pKa			
HL 2.06E-009	atm m3/mol	25	EST	MEYLAN,WM & HOWARD,PH (1991)
OH 1.44E-011	cm3/molc sec	25	EST	MEYLAN,WM & HOWARD,PH (1993)

CAS #: 003619-02-1			ALANINE-N-ACETYL,METHYL ESTER	
Formula: $C_6H_{11}NO_3$				
Mol Weight: 145.16				
MP (deg C):		FP (deg C):		
BP (deg C):				
BP pressure (mm Hg):				

Property/Value	Units	Temp	Data Type	Reference
WS 2.44E+005	mg/L	25	EST	MEYLAN,WM ET AL. (1996)
logP -0.57			EXP	HANSCH,C ET AL. (1995)
VP 2.38E-003	mm Hg	25	EST	NEELY,WB & BLAU,GE (1985)
DC	pKa			
HL 1.55E-009	atm m3/mol	25	EST	MEYLAN,WM & HOWARD,PH (1991)
OH 1.19E-011	cm3/molc sec	25	EST	MEYLAN,WM & HOWARD,PH (1993)

CAS #: 003619-17-8			I-BUTRYIC ACID HYDRAZIDE	
Formula: $C_4H_{10}N_2O$				
Mol Weight: 102.14				
MP (deg C):		FP (deg C):		
BP (deg C):				
BP pressure (mm Hg):				

Property/Value	Units	Temp	Data Type	Reference
WS 1.95E+004	mg/L	25	EST	MEYLAN,WM ET AL. (1996)
logP -0.63			EXP	HANSCH,C & LEO,AJ (1985)
VP 1.31E-002	mm Hg	25	EST	NEELY,WB & BLAU,GE (1985)
DC	pKa			
HL 8.23E-011	atm m3/mol	25	EST	MEYLAN,WM & HOWARD,PH (1991)
OH 8.02E-012	cm3/molc sec	25	EST	MEYLAN,WM & HOWARD,PH (1993)

CAS #: 003619-22-5			BENZOYLHYDRAZINE, P-METHYL	
Formula: $C_8H_{10}N_2O$				
Mol Weight: 150.18				
MP (deg C): 116-118		FP (deg C):		
BP (deg C):				
BP pressure (mm Hg):				

Property/Value	Units	Temp	Data Type	Reference
WS 1.84E+004	mg/L	25	EST	MEYLAN,WM ET AL. (1996)
logP 0.72			EXP	HANSCH,C & LEO,AJ (1985)
VP 2.95E-005	mm Hg	25	EST	NEELY,WB & BLAU,GE (1985)
DC	pKa			
HL 1.02E-011	atm m3/mol	25	EST	MEYLAN,WM & HOWARD,PH (1991)
OH 1.03E-011	cm3/molc sec	25	EST	MEYLAN,WM & HOWARD,PH (1993)

CAS #: 003624-68-8			C.I. MORDANT BLUE 9, DISODIUM SALT	
Formula: $C_{16}H_9ClN_2Na_2O_8S_2$				
Mol Weight: 502.82				
MP (deg C):		FP (deg C):		
BP (deg C):				
BP pressure (mm Hg):				

Property/Value	Units	Temp	Data Type	Reference
WS 1.79E+003	mg/L	25	EST	MEYLAN,WM ET AL. (1996)
logP -0.31			EST	MEYLAN,WM & HOWARD,PH (1995)
VP 9.37E-026	mm Hg	25	EST	NEELY,WB & BLAU,GE (1985)
DC	pKa			
HL	atm m3/mol			
OH 4.02E-012	cm3/molc sec	25	EST	MEYLAN,WM & HOWARD,PH (1993)

CAS #: 003625-25-0				REPOSAL

Formula: $C_{14}H_{18}N_2O_3$

Mol Weight: 262.31

MP (deg C): 213 FP (deg C):

BP (deg C):

BP pressure (mm Hg):

Property/Value	Units	Temp	Data Type	Reference
WS 4.46E+002	mg/L	25	EXP	YALKOWSKY,SH & DANNENFELSER,RM (1992)
logP 2.53			EXP	HANSCH,C & LEO,AJ (1985)
VP 6.50E-013	mm Hg	25	EST	NEELY,WB & BLAU,GE (1985)
DC	pKa			
HL 3.99E-013	atm m3/mol	25	EST	MEYLAN,WM & HOWARD,PH (1991)
OH 1.02E-010	cm3/molc sec	25	EST	MEYLAN,WM & HOWARD,PH (1993)

CAS #: 003626-62-8				ETHANONE, 2-PHENYL-1-(1-PIPERIDINYL)-

Formula: $C_{13}H_{17}NO$

Mol Weight: 203.29

MP (deg C): FP (deg C):

BP (deg C):

BP pressure (mm Hg):

Property/Value	Units	Temp	Data Type	Reference
WS 3.47E+002	mg/L	25	EST	MEYLAN,WM ET AL. (1996)
logP 2.44			EXP	SURYANARAYANA,MVS ET AL. (1991)
VP 7.37E-005	mm Hg	25	EST	NEELY,WB & BLAU,GE (1985)
DC	pKa			
HL 4.49E-009	atm m3/mol	25	EST	MEYLAN,WM & HOWARD,PH (1991)
OH 3.42E-011	cm3/molc sec	25	EST	MEYLAN,WM & HOWARD,PH (1993)

CAS #: 003638-35-5				ISOPROPYLCYCLOPROPANE

Formula: C_6H_{12}

Mol Weight: 84.16

MP (deg C): -112.9 FP (deg C):

BP (deg C): 58.3

BP pressure (mm Hg):

Property/Value	Units	Temp	Data Type	Reference
WS 9.66E+001	mg/L	25	EST	MEYLAN,WM ET AL. (1996)
logP 3.03			EST	MEYLAN,WM & HOWARD,PH (1995)
VP 2.05E+002	mm Hg	25	EST	NEELY,WB & BLAU,GE (1985)
DC	pKa			
HL 2.55E-001	atm m3/mol	25	EST	MEYLAN,WM & HOWARD,PH (1991)
OH 2.84E-012	cm3/molc sec	25	EXP	ATKINSON,R (1989)

CAS #: 003642-85-1				PREGN-4-ENE-3,11,20-TRIONE, 6-METHYL-, (6_)-

Formula: $C_{22}H_{30}O_3$

Mol Weight: 342.48

MP (deg C): FP (deg C):

BP (deg C):

BP pressure (mm Hg):

Property/Value	Units	Temp	Data Type	Reference
WS 3.08E+001	mg/L	25	EST	MEYLAN,WM ET AL. (1996)
logP 2.75			EXP	SANGSTER,J (1994)
VP 5.99E-008	mm Hg	25	EST	NEELY,WB & BLAU,GE (1985)
DC	pKa			
HL 3.29E-011	atm m3/mol	25	EST	MEYLAN,WM & HOWARD,PH (1991)
OH 1.07E-010	cm3/molc sec	25	EST	MEYLAN,WM & HOWARD,PH (1993)

CAS #: 003645-61-2				SODIUM P-PHENYLPHENOXIDE

Formula: $C_{12}H_9NaO$

Mol Weight: 192.19

MP (deg C): FP (deg C):

BP (deg C):

BP pressure (mm Hg):

Property/Value	Units	Temp	Data Type	Reference
WS 2.14E+004	mg/L	25	EST	MEYLAN,WM ET AL. (1996)
logP 0.41			EXP	HANSCH,C & LEO,AJ (1985)
VP 1.91E-011	mm Hg	25	EST	NEELY,WB & BLAU,GE (1985)
DC	pKa			
HL	atm m3/mol			
OH 1.87E-011	cm3/molc sec	25	EST	MEYLAN,WM & HOWARD,PH (1993)

CAS #: 003648-20-2				DIUNDECYL PHTHALATE

Formula: $C_{30}H_{50}O_4$

Mol Weight: 474.73

MP (deg C): FP (deg C):

BP (deg C):

BP pressure (mm Hg):

Property/Value	Units	Temp	Data Type	Reference
WS 1.11E+000	mg/L	20	EXP	HOWARD,PH ET AL. (1985)
logP 11.49			EST	MEYLAN,WM & HOWARD,PH (1995)
VP 1.10E-004	mm Hg	25	EST	HL X WSOL
DC	pKa			
HL 5.60E-005	atm m3/mol	25	EST	MEYLAN,WM & HOWARD,PH (1991)
OH 2.82E-011	cm3/molc sec	25	EST	MEYLAN,WM & HOWARD,PH (1993)

CAS #: 003648-21-3				DIHEPTYL PHTHALATE

Formula: $C_{22}H_{34}O_4$

Mol Weight: 362.51

MP (deg C): FP (deg C):

BP (deg C):

BP pressure (mm Hg):

Property/Value	Units	Temp	Data Type	Reference
WS 1.83E-003	mg/L	25	EST	MEYLAN,WM ET AL. (1996)
logP 7.56			EST	MEYLAN,WM & HOWARD,PH (1995)
VP 2.07E-006	mm Hg	25	EST	NEELY,WB & BLAU,GE (1985)
DC	pKa			
HL 6.70E-006	atm m3/mol	25	EST	MEYLAN,WM & HOWARD,PH (1991)
OH 1.78E-011	cm3/molc sec	25	EST	MEYLAN,WM & HOWARD,PH (1993)

CAS #: 003651-23-8				DIALLYLDICHLOROSILANE

Formula: $C_6H_{10}Cl_2Si$

Mol Weight: 181.14

MP (deg C): FP (deg C):

BP (deg C):

BP pressure (mm Hg):

Property/Value	Units	Temp	Data Type	Reference
WS 2.37E+001	mg/L	25	EST	MEYLAN,WM ET AL. (1996)
logP 3.94			EST	MEYLAN,WM & HOWARD,PH (1995)
VP 2.79E+000	mm Hg	25	EXP	PERRY,RH & GREEN,D (1984)
DC	pKa			
HL 3.19E-002	atm m3/mol	25	EST	MEYLAN,WM & HOWARD,PH (1991)
OH 5.47E-011	cm3/molc sec	25	EST	MEYLAN,WM & HOWARD,PH (1993)

003663-20-5 — 4-BUTYLACETANILIDE

Formula: $C_{12}H_{17}NO$

Mol Weight: 191.28

MP (deg C): | FP (deg C):

BP (deg C):

BP pressure (mm Hg):

Property/Value	Units	Temp	Data Type	Reference
WS 7.36E+001	mg/L	25	EST	MEYLAN,WM ET AL. (1996)
logP 3.30			EXP	NAKAGAWA,Y ET AL. (1992)
VP 1.32E-005	mm Hg	25	EST	NEELY,WB & BLAU,GE (1985)
DC	pKa			
HL 1.59E-008	atm m3/mol	25	EST	MEYLAN,WM & HOWARD,PH (1991)
OH 1.91E-011	cm3/molc sec	25	EST	MEYLAN,WM & HOWARD,PH (1993)

003663-21-6 — 2-NITRO-4-BUTYLACETANILIDE

Formula: $C_{12}H_{16}N_2O_3$

Mol Weight: 236.27

MP (deg C): | FP (deg C):

BP (deg C):

BP pressure (mm Hg):

Property/Value	Units	Temp	Data Type	Reference
WS 2.58E+001	mg/L	25	EST	MEYLAN,WM ET AL. (1996)
logP 3.10			EXP	NAKAGAWA,Y ET AL. (1992)
VP 1.70E-007	mm Hg	25	EST	NEELY,WB & BLAU,GE (1985)
DC	pKa			
HL 1.38E-009	atm m3/mol	25	EST	MEYLAN,WM & HOWARD,PH (1991)
OH 5.87E-012	cm3/molc sec	25	EST	MEYLAN,WM & HOWARD,PH (1993)

003663-34-1 — 4-ETHYLACETANILIDE

Formula: $C_{10}H_{13}NO$

Mol Weight: 163.22

MP (deg C): | FP (deg C):

BP (deg C):

BP pressure (mm Hg):

Property/Value	Units	Temp	Data Type	Reference
WS 8.74E+002	mg/L	25	EST	MEYLAN,WM ET AL. (1996)
logP 2.20			EXP	NAKAGAWA,Y ET AL. (1992)
VP 7.27E-005	mm Hg	25	EST	NEELY,WB & BLAU,GE (1985)
DC	pKa			
HL 9.04E-009	atm m3/mol	25	EST	MEYLAN,WM & HOWARD,PH (1991)
OH 1.63E-011	cm3/molc sec	25	EST	MEYLAN,WM & HOWARD,PH (1993)

003665-80-3 — N-ETHYL-4-NITROBENZENAMINE

Formula: $C_8H_{10}N_2O_2$

Mol Weight: 166.18

MP (deg C): 96-98 | FP (deg C):

BP (deg C):

BP pressure (mm Hg):

Property/Value	Units	Temp	Data Type	Reference
WS 4.61E+002	mg/L	25	EST	MEYLAN,WM ET AL. (1996)
logP 2.51			EST	MEYLAN,WM & HOWARD,PH (1995)
VP 2.57E-003	mm Hg	25	EST	NEELY,WB & BLAU,GE (1985)
DC	pKa			
HL 2.19E-008	atm m3/mol	25	EST	MEYLAN,WM & HOWARD,PH (1991)
OH 1.42E-011	cm3/molc sec	25	EST	MEYLAN,WM & HOWARD,PH (1993)

003671-60-1 — 2-CF3-5-BROMOBENZIMIDAZOLE

Formula: $C_8H_4BrF_3N_2$

Mol Weight: 265.04

MP (deg C): | FP (deg C):

BP (deg C):

BP pressure (mm Hg):

Property/Value	Units	Temp	Data Type	Reference
WS 1.74E+001	mg/L	25	EST	MEYLAN,WM ET AL. (1996)
logP 3.57			EXP	HANSCH,C & LEO,AJ (1985)
VP 4.89E-006	mm Hg	25	EST	NEELY,WB & BLAU,GE (1985)
DC	pKa			
HL 1.27E-006	atm m3/mol	25	EST	MEYLAN,WM & HOWARD,PH (1991)
OH 2.67E-012	cm3/molc sec	25	EST	MEYLAN,WM & HOWARD,PH (1993)

003679-63-8 — 4'-CL SALICYLANILIDE

Formula: $C_{13}H_{10}ClNO_2$

Mol Weight: 247.68

MP (deg C): | FP (deg C):

BP (deg C):

BP pressure (mm Hg):

Property/Value	Units	Temp	Data Type	Reference
WS 1.33E+001	mg/L	25	EST	MEYLAN,WM ET AL. (1996)
logP 4.50			EXP	TERADA,H ET AL. (1988)
VP 5.90E-009	mm Hg	25	EST	NEELY,WB & BLAU,GE (1985)
DC	pKa			
HL 1.19E-010	atm m3/mol	25	EST	MEYLAN,WM & HOWARD,PH (1991)
OH 3.41E-011	cm3/molc sec	25	EST	MEYLAN,WM & HOWARD,PH (1993)

003679-64-9 — 4'-CL-5-BR SALICYLANILIDE

Formula: $C_{13}H_9BrClNO_2$

Mol Weight: 326.58

MP (deg C): 238-243 | FP (deg C):

BP (deg C):

BP pressure (mm Hg):

Property/Value	Units	Temp	Data Type	Reference
WS 1.68E+000	mg/L	25	EST	MEYLAN,WM ET AL. (1996)
logP 5.02			EXP	TERADA,H ET AL. (1988)
VP 6.00E-010	mm Hg	25	EST	NEELY,WB & BLAU,GE (1985)
DC	pKa			
HL 4.72E-011	atm m3/mol	25	EST	MEYLAN,WM & HOWARD,PH (1991)
OH 1.25E-011	cm3/molc sec	25	EST	MEYLAN,WM & HOWARD,PH (1993)

003688-53-7 — 2-(2-FURYL)-3-(5-NO2FURYL)ACRYLAMIDE

Formula: $C_{11}H_{10}N_2O_5$

Mol Weight: 250.21

MP (deg C): | FP (deg C):

BP (deg C):

BP pressure (mm Hg):

Property/Value	Units	Temp	Data Type	Reference
WS 3.60E+002	mg/L	25	EST	MEYLAN,WM ET AL. (1996)
logP 0.15			EXP	HANSCH,C & LEO,AJ (1985)
VP 6.93E-008	mm Hg	25	EST	NEELY,WB & BLAU,GE (1985)
DC	pKa			
HL 4.65E-014	atm m3/mol	25	EST	MEYLAN,WM & HOWARD,PH (1991)
OH 7.53E-011	cm3/molc sec	25	EST	MEYLAN,WM & HOWARD,PH (1993)

CAS #: 003688-82-2				BENZAMIDE, 2-ETHOXY-N,N-DIETHYL-
Formula: $C_{13}H_{19}NO_2$				
Mol Weight: 221.30				
MP (deg C):		FP (deg C):		
BP (deg C):				
BP pressure (mm Hg):				

Property/ Value	Units	Temp	Data Type	Reference
WS 2.48E+002	mg/L	25	EST	MEYLAN,WM ET AL. (1996)
logP 2.50			EXP	SURYANARAYANA,MVS ET AL. (1991)
VP 1.20E-002	mm Hg	30	EXP	SURYANARAYANA,MVS ET AL. (1991)
DC	pKa			
HL 1.48E-009	atm m3/mol	25	EST	MEYLAN,WM & HOWARD,PH (1991)
OH 4.88E-011	cm3/molc sec	25	EST	MEYLAN,WM & HOWARD,PH (1993)

CAS #: 003689-24-5				SULFOTEPP
Formula: $C_8H_{20}O_5P_2S_2$				
Mol Weight: 322.32				
MP (deg C):		FP (deg C):		
BP (deg C): 136-139				
BP pressure (mm Hg): 2.00E+000				

Property/ Value	Units	Temp	Data Type	Reference
WS 3.00E+001	mg/L	20	EXP	YALKOWSKY,SH & DANNENFELSER,RM (1992)
logP 3.99			EXP	TOMLIN,C (1994)
VP 1.70E-004	mm Hg	20	EXP	BACCI,E ET AL. (1990)
DC	pKa			
HL 2.40E-006	atm m3/mol	20	EST	VP/WSOL
OH 1.83E-010	cm3/molc sec	25	EST	MEYLAN,WM & HOWARD,PH (1993)

CAS #: 003689-50-7				OXOMEMAZINE
Formula: $C_{18}H_{22}N_2O_2S$				
Mol Weight: 330.45				
MP (deg C): 115		FP (deg C):		
BP (deg C):				
BP pressure (mm Hg):				

Property/ Value	Units	Temp	Data Type	Reference
WS 4.61E+001	mg/L	25	EST	MEYLAN,WM ET AL. (1996)
logP 2.51			EXP	HANSCH,C & LEO,AJ (1985)
VP 1.47E-008	mm Hg	25	EST	NEELY,WB & BLAU,GE (1985)
DC	pKa			
HL 5.25E-012	atm m3/mol	25	EST	MEYLAN,WM & HOWARD,PH (1991)
OH 1.56E-010	cm3/molc sec	25	EST	MEYLAN,WM & HOWARD,PH (1993)

CAS #: 003690-53-7				N-(2-DIETHYLAMINOETHYL)-BENZAMIDE
Formula: $C_{13}H_{20}N_2O$				
Mol Weight: 220.32				
MP (deg C):		FP (deg C):		
BP (deg C):				
BP pressure (mm Hg):				

Property/ Value	Units	Temp	Data Type	Reference
WS 6.85E+002	mg/L	25	EST	MEYLAN,WM ET AL. (1996)
logP 1.99			EXP	HANSCH,C & LEO,AJ (1985)
VP 2.40E-006	mm Hg	25	EST	NEELY,WB & BLAU,GE (1985)
DC	pKa			
HL 5.64E-012	atm m3/mol	25	EST	MEYLAN,WM & HOWARD,PH (1991)
OH 1.06E-010	cm3/molc sec	25	EST	MEYLAN,WM & HOWARD,PH (1993)

CAS #: 003694-52-8				3-NITRO-O-PHENYLENEDIAMINE
Formula: $C_6H_7N_3O_2$				
Mol Weight: 153.14				
MP (deg C): 157-159		FP (deg C):		
BP (deg C):				
BP pressure (mm Hg):				

Property/ Value	Units	Temp	Data Type	Reference
WS 6.05E+003	mg/L	25	EST	MEYLAN,WM ET AL. (1996)
logP 1.27			EXP	HANSCH,C & LEO,AJ (1985)
VP 5.60E-005	mm Hg	25	EST	NEELY,WB & BLAU,GE (1985)
DC	pKa			
HL 5.81E-011	atm m3/mol	25	EST	MEYLAN,WM & HOWARD,PH (1991)
OH 2.20E-011	cm3/molc sec	25	EST	MEYLAN,WM & HOWARD,PH (1993)

CAS #: 003695-86-1				2-PROPENOIC ACID, 2-CYANO-3-(2-FURANYL)-, METHYL
Formula: $C_9H_7NO_3$				
Mol Weight: 177.16				
MP (deg C):		FP (deg C):		
BP (deg C):				
BP pressure (mm Hg):				

Property/ Value	Units	Temp	Data Type	Reference
WS 1.20E+003	mg/L	25	EST	MEYLAN,WM ET AL. (1996)
logP 1.65			EXP	BALAZ,S ET AL. (1985)
VP 1.61E-003	mm Hg	25	EST	NEELY,WB & BLAU,GE (1985)
DC	pKa			
HL 9.21E-009	atm m3/mol	25	EST	MEYLAN,WM & HOWARD,PH (1991)
OH 4.32E-011	cm3/molc sec	25	EST	MEYLAN,WM & HOWARD,PH (1993)

CAS #: 003697-24-3				5-METHYLCHRYSENE
Formula: $C_{19}H_{14}$				
Mol Weight: 242.32				
MP (deg C):		FP (deg C):		
BP (deg C):				
BP pressure (mm Hg):				

Property/ Value	Units	Temp	Data Type	Reference
WS 6.20E-002	mg/L	27	EXP	YALKOWSKY,SH & DANNENFELSER,RM (1992)
logP 6.07			EST	MEYLAN,WM & HOWARD,PH (1995)
VP 2.53E-007	mm Hg	25	EST	NEELY,WB & BLAU,GE (1985)
DC	pKa			
HL 5.53E-006	atm m3/mol	25	EST	MEYLAN,WM & HOWARD,PH (1991)
OH 1.31E-010	cm3/molc sec	25	EST	MEYLAN,WM & HOWARD,PH (1993)

CAS #: 003697-27-6				5,6-DIMETHYLCHRYSENE
Formula: $C_{20}H_{16}$				
Mol Weight: 256.35				
MP (deg C): 129.3		FP (deg C):		
BP (deg C): 200				
BP pressure (mm Hg): 5.00E-001				

Property/ Value	Units	Temp	Data Type	Reference
WS 2.50E-002	mg/L	27	EXP	YALKOWSKY,SH & DANNENFELSER,RM (1992)
logP 6.62			EST	MEYLAN,WM & HOWARD,PH (1995)
VP 1.14E-007	mm Hg	25	EST	NEELY,WB & BLAU,GE (1985)
DC	pKa			
HL 6.10E-006	atm m3/mol	25	EST	MEYLAN,WM & HOWARD,PH (1991)
OH 1.60E-010	cm3/molc sec	25	EST	MEYLAN,WM & HOWARD,PH (1993)

CAS #: 003697-30-1 — 7-ETHYLBENZ(A)ANTHRACENE

Formula: $C_{20}H_{16}$

Mol Weight: 256.35

MP (deg C):   FP (deg C):

BP (deg C):

BP pressure (mm Hg):

Property/Value	Units	Temp	Data Type	Reference
WS 4.10E-002	mg/L	27	EXP	PEARLMAN,RS ET AL. (1984)
logP 6.56			EST	MEYLAN,WM & HOWARD,PH (1995)
VP 1.06E-007	mm Hg	25	EST	NEELY,WB & BLAU,GE (1985)
DC	pKa			
HL 7.34E-006	atm m3/mol	25	EST	MEYLAN,WM & HOWARD,PH (1991)
OH 1.25E-010	cm3/molc sec	25	EST	MEYLAN,WM & HOWARD,PH (1993)

CAS #: 003701-01-7 — 1,3-BENZENEDISULFONAMIDE

Formula: $C_6H_8N_2O_4S_2$

Mol Weight: 236.27

MP (deg C):   FP (deg C):

BP (deg C):

BP pressure (mm Hg):

Property/Value	Units	Temp	Data Type	Reference
WS 7.38E+003	mg/L	25	EST	MEYLAN,WM ET AL. (1996)
logP -0.55			EXP	HANSCH,C & LEO,AJ (1985)
VP 1.26E-007	mm Hg	25	EST	NEELY,WB & BLAU,GE (1985)
DC	pKa			
HL 3.31E-011	atm m3/mol	25	EST	MEYLAN,WM & HOWARD,PH (1991)
OH 8.91E-014	cm3/molc sec	25	EST	MEYLAN,WM & HOWARD,PH (1993)

CAS #: 003704-09-4 — ESTR-4-EN-3-ONE, 17-HYDROXY-7,17-DIMETHYL-, (7AL

Formula: $C_{20}H_{30}O_2$

Mol Weight: 302.46

MP (deg C):   FP (deg C):

BP (deg C):

BP pressure (mm Hg):

Property/Value	Units	Temp	Data Type	Reference
WS 7.25E+001	mg/L	25	EST	MEYLAN,WM ET AL. (1996)
logP 3.19			EXP	SANGSTER,J (1993)
VP 1.72E-008	mm Hg	25	EST	NEELY,WB & BLAU,GE (1985)
DC	pKa			
HL 4.68E-009	atm m3/mol	25	EST	MEYLAN,WM & HOWARD,PH (1991)
OH 1.05E-010	cm3/molc sec	25	EST	MEYLAN,WM & HOWARD,PH (1993)

CAS #: 003707-98-0 — 3(2H)-PYRIDAZINONE, 4-CHLORO-5-(DIMETHYLAMINO)-2

Formula: $C_{12}H_{12}ClN_3O$

Mol Weight: 249.70

MP (deg C):   FP (deg C):

BP (deg C):

BP pressure (mm Hg):

Property/Value	Units	Temp	Data Type	Reference
WS 1.13E+003	mg/L	25	EST	MEYLAN,WM ET AL. (1996)
logP 1.55			EXP	BRAUMANN,T & GRIMME,LH (1981)
VP 1.01E-006	mm Hg	25	EST	NEELY,WB & BLAU,GE (1985)
DC	pKa			
HL 3.15E-011	atm m3/mol	25	EST	MEYLAN,WM & HOWARD,PH (1991)
OH 8.77E-011	cm3/molc sec	25	EST	MEYLAN,WM & HOWARD,PH (1993)

CAS #: 003710-84-7 — DIETHYLHYDROXYLAMINE

Formula: $C_4H_{11}NO$

Mol Weight: 89.14

MP (deg C): 10   FP (deg C):

BP (deg C): 133

BP pressure (mm Hg):

Property/Value	Units	Temp	Data Type	Reference
WS 5.36E+004	mg/L	25	EST	MEYLAN,WM ET AL. (1996)
logP 0.43			EST	MEYLAN,WM & HOWARD,PH (1995)
VP 2.23E-002	mm Hg	25	EST	NEELY,WB & BLAU,GE (1985)
DC	pKa			
HL 5.87E-008	atm m3/mol	25	EST	MEYLAN,WM & HOWARD,PH (1991)
OH 1.01E-010	cm3/molc sec	25	EXP	ATKINSON,R (1989)

CAS #: 003718-65-8 — 3,5-DIMETHYLPYRIDINE-N-OXIDE

Formula: C_7H_9NO

Mol Weight: 123.16

MP (deg C):   FP (deg C):

BP (deg C):

BP pressure (mm Hg):

Property/Value	Units	Temp	Data Type	Reference
WS 1.00E+006	mg/L	25	EXP	GOE,GL (1982)
logP -0.24			EST	MEYLAN,WM & HOWARD,PH (1995)
VP 1.53E-002	mm Hg	25	EST	NEELY,WB & BLAU,GE (1985)
DC	pKa			
HL 8.59E-011	atm m3/mol	25	EST	MEYLAN,WM & HOWARD,PH (1991)
OH 2.79E-012	cm3/molc sec	25	EST	MEYLAN,WM & HOWARD,PH (1993)

CAS #: 003722-12-1 — PROPANEDINITRILE, [(4-NITROPHENYL)HYDRAZONO]-

Formula: $C_9H_6N_5O_2$

Mol Weight: 216.18

MP (deg C):   FP (deg C):

BP (deg C):

BP pressure (mm Hg):

Property/Value	Units	Temp	Data Type	Reference
WS 1.41E+002	mg/L	25	EST	MEYLAN,WM ET AL. (1996)
logP 2.37			EXP	STURDIK,E ET AL. (1985)
VP 3.13E-007	mm Hg	25	EST	NEELY,WB & BLAU,GE (1985)
DC	pKa			
HL 2.28E-011	atm m3/mol	25	EST	MEYLAN,WM & HOWARD,PH (1991)
OH 5.33E-012	cm3/molc sec	25	EST	MEYLAN,WM & HOWARD,PH (1993)

CAS #: 003724-16-1 — 3-PYRIDINE ACETAMIDE

Formula: $C_7H_8N_2O$

Mol Weight: 136.15

MP (deg C):   FP (deg C):

BP (deg C):

BP pressure (mm Hg):

Property/Value	Units	Temp	Data Type	Reference
WS 1.00E+006	mg/L	25	EST	MEYLAN,WM ET AL. (1996)
logP -0.71			EXP	HANSCH,C & LEO,AJ (1985)
VP 1.57E-004	mm Hg	25	EST	NEELY,WB & BLAU,GE (1985)
DC	pKa			
HL 1.18E-012	atm m3/mol	25	EST	MEYLAN,WM & HOWARD,PH (1991)
OH 3.62E-012	cm3/molc sec	25	EST	MEYLAN,WM & HOWARD,PH (1993)

CAS #: 003724-65-0 — CROTONIC ACID

Formula: $C_4H_6O_2$

Mol Weight: 86.09

MP (deg C): 15.5

FP (deg C):

BP (deg C): 185

BP pressure (mm Hg):

Property/Value	Units	Temp	Data Type	Reference
WS 8.60E+004	mg/L	25	EXP	YALKOWSKY,SH & DANNENFELSER,RM (1992)
logP 0.72			EXP	HANSCH,C & LEO,AJ (1985)
VP 5.20E-001	mm Hg	25	EXT	PERRY,RH & GREEN,D (1984)
DC 4.17	pKa	18	EXP	KORTUM,G ET AL (1961)
HL 4.53E-007	atm m3/mol	25	EST	MEYLAN,WM & HOWARD,PH (1991)
OH 2.04E-011	cm3/molc sec	25	EST	MEYLAN,WM & HOWARD,PH (1993)

CAS #: 003726-47-4 — 1-METHYL-3-ETHYLCYCLOPENTANE

Formula: C_8H_{16}

Mol Weight: 112.22

MP (deg C):

FP (deg C):

BP (deg C):

BP pressure (mm Hg):

Property/Value	Units	Temp	Data Type	Reference
WS 1.17E+001	mg/L	25	EST	MEYLAN,WM ET AL. (1996)
logP 4.01			EST	MEYLAN,WM & HOWARD,PH (1995)
VP 6.29E+000	mm Hg	25	EXT	CHAO,J ET AL. (1983)
DC	pKa			
HL 4.50E-001	atm m3/mol	25	EST	MEYLAN,WM & HOWARD,PH (1991)
OH 1.01E-011	cm3/molc sec	25	EST	MEYLAN,WM & HOWARD,PH (1993)

CAS #: 003731-51-9 — 2-PYRIDINEMETHANEAMINE

Formula: $C_6H_8N_2$

Mol Weight: 108.14

MP (deg C):

FP (deg C):

BP (deg C): 203

BP pressure (mm Hg):

Property/Value	Units	Temp	Data Type	Reference
WS 1.00E+006	mg/L	25	EST	MEYLAN,WM ET AL. (1996)
logP -0.21			EXP	HANSCH,C & LEO,AJ (1985)
VP 5.06E-001	mm Hg	25	EST	NEELY,WB & BLAU,GE (1985)
DC 9.09	pKa	10	EXP	PERRIN,DD (1965)
HL 8.01E-010	atm m3/mol	25	EST	MEYLAN,WM & HOWARD,PH (1991)
OH 3.06E-011	cm3/molc sec	25	EST	MEYLAN,WM & HOWARD,PH (1993)

CAS #: 003731-52-0 — 3-PYRIDINEMETHANEAMINE

Formula: $C_6H_8N_2$

Mol Weight: 108.14

MP (deg C): -21.1

FP (deg C):

BP (deg C): 226

BP pressure (mm Hg):

Property/Value	Units	Temp	Data Type	Reference
WS 1.00E+006	mg/L	25	EST	MEYLAN,WM ET AL. (1996)
logP -0.32			EXP	HANSCH,C & LEO,AJ (1985)
VP 5.06E-001	mm Hg	25	EST	NEELY,WB & BLAU,GE (1985)
DC 5.96	pKa		EXP	PERRIN,DD (1965)
HL 8.01E-010	atm m3/mol	25	EST	MEYLAN,WM & HOWARD,PH (1991)
OH 3.06E-011	cm3/molc sec	25	EST	MEYLAN,WM & HOWARD,PH (1993)

CAS #: 003731-53-1 — 4-PYRIDINEMETHANEAMINE

Formula: $C_6H_8N_2$

Mol Weight: 108.14

MP (deg C): -7.6

FP (deg C):

BP (deg C): 230

BP pressure (mm Hg):

Property/Value	Units	Temp	Data Type	Reference
WS 1.00E+006	mg/L	25	EST	MEYLAN,WM ET AL. (1996)
logP -0.38			EXP	HANSCH,C & LEO,AJ (1985)
VP 5.06E-001	mm Hg	25	EST	NEELY,WB & BLAU,GE (1985)
DC	pKa			
HL 8.01E-010	atm m3/mol	25	EST	MEYLAN,WM & HOWARD,PH (1991)
OH 3.06E-011	cm3/molc sec	25	EST	MEYLAN,WM & HOWARD,PH (1993)

CAS #: 003734-48-3 — CHLORDENE

Formula: $C_{10}H_6Cl_6$

Mol Weight: 338.88

MP (deg C):

FP (deg C):

BP (deg C):

BP pressure (mm Hg):

Property/Value	Units	Temp	Data Type	Reference
WS 4.62E-004	mg/L	25	EXT	METCALF,RL (1977)
logP 6.03			EST	MEYLAN,WM & HOWARD,PH (1995)
VP 2.07E-004	mm Hg	25	EST	NEELY,WB & BLAU,GE (1985)
DC	pKa			
HL 4.98E-004	atm m3/mol	25	EST	MEYLAN,WM & HOWARD,PH (1991)
OH 6.46E-011	cm3/molc sec	25	EST	MEYLAN,WM & HOWARD,PH (1993)

CAS #: 003735-01-1 — PARATHION-AMINO

Formula: $C_{10}H_{16}NO_3PS$

Mol Weight: 261.28

MP (deg C):

FP (deg C):

BP (deg C):

BP pressure (mm Hg):

Property/Value	Units	Temp	Data Type	Reference
WS 3.90E+002	mg/L		EXP	REINBOLD,KA ET AL. (1979)
logP 2.60			EXP	HANSCH,C & LEO,AJ (1985)
VP 5.76E-005	mm Hg	25	EST	NEELY,WB & BLAU,GE (1985)
DC	pKa			
HL 2.65E-008	atm m3/mol	25	EST	MEYLAN,WM & HOWARD,PH (1991)
OH 1.71E-010	cm3/molc sec	25	EST	MEYLAN,WM & HOWARD,PH (1993)

CAS #: 003736-81-0 — DILOXANIDE FUROATE

Formula: $C_{14}H_{11}Cl_2NO_4$

Mol Weight: 328.15

MP (deg C):

FP (deg C):

BP (deg C):

BP pressure (mm Hg):

Property/Value	Units	Temp	Data Type	Reference
WS 1.02E+002	mg/L	25	EST	MEYLAN,WM ET AL. (1996)
logP 2.24			EXP	DUTTA,H ET AL. (1988)
VP 1.04E-007	mm Hg	25	EST	NEELY,WB & BLAU,GE (1985)
DC	pKa			
HL 3.73E-011	atm m3/mol	25	EST	MEYLAN,WM & HOWARD,PH (1991)
OH 4.68E-011	cm3/molc sec	25	EST	MEYLAN,WM & HOWARD,PH (1993)

003737-00-6 — CHLOROBROMOPROPENE

CAS #: 003737-00-6

Formula: C_3H_4BrCl

Mol Weight: 155.43

MP (deg C):

FP (deg C):

BP (deg C):

BP pressure (mm Hg):

Property/Value	Units	Temp	Data Type	Reference
WS 6.67E+002	mg/L	25	EST	MEYLAN,WM ET AL. (1996)
logP 2.38			EST	MEYLAN,WM & HOWARD,PH (1995)
VP 1.80E+001	mm Hg	25	EST	NEELY,WB & BLAU,GE (1985)
DC	pKa			
HL 8.02E-003	atm m3/mol	25	EST	MEYLAN,WM & HOWARD,PH (1991)
OH 9.26E-012	cm3/molc sec	25	EST	MEYLAN,WM & HOWARD,PH (1993)

003737-09-5 — DISOPYRAMIDE

CAS #: 003737-09-5

Formula: $C_{21}H_{29}N_3O$

Mol Weight: 339.48

MP (deg C): 94.5-95

FP (deg C):

BP (deg C):

BP pressure (mm Hg):

Property/Value	Units	Temp	Data Type	Reference
WS 4.49E+001	mg/L	25	EST	MEYLAN,WM ET AL. (1996)
logP 2.58			EXP	MANNHOLD,R ET AL. (1993)
VP 1.35E-009	mm Hg	25	EST	NEELY,WB & BLAU,GE (1985)
DC	pKa			
HL 2.59E-016	atm m3/mol	25	EST	MEYLAN,WM & HOWARD,PH (1991)
OH 1.25E-010	cm3/molc sec	25	EST	MEYLAN,WM & HOWARD,PH (1993)

003739-38-6 — M-PHENOXYBENZOIC ACID

CAS #: 003739-38-6

Formula: $C_{13}H_{10}O_3$

Mol Weight: 214.22

MP (deg C): 149-150

FP (deg C):

BP (deg C):

BP pressure (mm Hg):

Property/Value	Units	Temp	Data Type	Reference
WS 1.69E+001	mg/L	25	EST	MEYLAN,WM ET AL. (1996)
logP 3.91			EXP	HANSCH,C & LEO,AJ (1985)
VP 5.45E-006	mm Hg	25	EST	NEELY,WB & BLAU,GE (1985)
DC 3.92	pKa		EXP	KORTUM,G ET AL (1961)
HL 2.37E-009	atm m3/mol	25	EST	MEYLAN,WM & HOWARD,PH (1991)
OH 6.78E-012	cm3/molc sec	25	EST	MEYLAN,WM & HOWARD,PH (1993)

003739-82-0 — PYRIMIDINE, 2-ETHOXY-

CAS #: 003739-82-0

Formula: $C_6H_8N_2O$

Mol Weight: 124.14

MP (deg C):

FP (deg C):

BP (deg C):

BP pressure (mm Hg):

Property/Value	Units	Temp	Data Type	Reference
WS 2.26E+004	mg/L	25	EST	MEYLAN,WM ET AL. (1996)
logP 0.74			EXP	YAMAGAMI,C ET AL. (1990)
VP 5.84E-001	mm Hg	25	EST	NEELY,WB & BLAU,GE (1985)
DC 1.27	pKa	20	EXP	PERRIN,DD (1972)
HL 2.09E-006	atm m3/mol	25	EST	MEYLAN,WM & HOWARD,PH (1991)
OH 8.71E-012	cm3/molc sec	25	EST	MEYLAN,WM & HOWARD,PH (1993)

003740-92-9 — FENCLORIM

CAS #: 003740-92-9

Formula: $C_{10}H_6Cl_2N_2$

Mol Weight: 225.08

MP (deg C):

FP (deg C):

BP (deg C):

BP pressure (mm Hg):

Property/Value	Units	Temp	Data Type	Reference
WS 8.88E+000	mg/L	25	EST	MEYLAN,WM ET AL. (1996)
logP 4.17			EXP	HANSCH,C ET AL. (1995)
VP 4.66E-005	mm Hg	25	EST	NEELY,WB & BLAU,GE (1985)
DC	pKa			
HL 4.90E-004	atm m3/mol	25	EST	MEYLAN,WM & HOWARD,PH (1991)
OH 1.98E-012	cm3/molc sec	25	EST	MEYLAN,WM & HOWARD,PH (1993)

003741-12-6 — 6-CHLORO-4-NITROQUINOLINE-1-OXIDE

CAS #: 003741-12-6

Formula: $C_9H_7ClN_2O_3$

Mol Weight: 226.62

MP (deg C):

FP (deg C):

BP (deg C):

BP pressure (mm Hg):

Property/Value	Units	Temp	Data Type	Reference
WS 8.28E+002	mg/L	25	EST	MEYLAN,WM ET AL. (1996)
logP 1.41			EXP	HANSCH,C & LEO,AJ (1985)
VP 5.92E-007	mm Hg	25	EST	NEELY,WB & BLAU,GE (1985)
DC	pKa			
HL 2.01E-014	atm m3/mol	25	EST	MEYLAN,WM & HOWARD,PH (1991)
OH 6.82E-013	cm3/molc sec	25	EST	MEYLAN,WM & HOWARD,PH (1993)

003741-15-9 — 4-NITROQUINOLINE

CAS #: 003741-15-9

Formula: $C_9H_6N_2O_2$

Mol Weight: 174.16

MP (deg C):

FP (deg C):

BP (deg C):

BP pressure (mm Hg):

Property/Value	Units	Temp	Data Type	Reference
WS 4.17E+002	mg/L	25	EST	MEYLAN,WM ET AL. (1996)
logP 2.06			EXP	HANSCH,C & LEO,AJ (1985)
VP 1.38E-004	mm Hg	25	EST	NEELY,WB & BLAU,GE (1985)
DC	pKa			
HL 2.72E-009	atm m3/mol	25	EST	MEYLAN,WM & HOWARD,PH (1991)
OH 1.80E-012	cm3/molc sec	25	EST	MEYLAN,WM & HOWARD,PH (1993)

003743-11-1 — 2-BENZAMIDO-5-METHYLPHENOL

CAS #: 003743-11-1

Formula: $C_{14}H_{13}NO_2$

Mol Weight: 227.27

MP (deg C):

FP (deg C):

BP (deg C):

BP pressure (mm Hg):

Property/Value	Units	Temp	Data Type	Reference
WS 7.37E+001	mg/L	25	EST	MEYLAN,WM ET AL. (1996)
logP 3.08			EXP	HANSCH,C & LEO,AJ (1985)
VP 9.97E-009	mm Hg	25	EST	NEELY,WB & BLAU,GE (1985)
DC	pKa			
HL 1.40E-013	atm m3/mol	25	EST	MEYLAN,WM & HOWARD,PH (1991)
OH 4.76E-011	cm3/molc sec	25	EST	MEYLAN,WM & HOWARD,PH (1993)

003766-60-7 — BUTURON

Formula: $C_{12}H_{13}ClN_2O$

Mol Weight: 236.70

MP (deg C): — FP (deg C): —

BP (deg C): —

BP pressure (mm Hg): —

Property/Value	Units	Temp	Data Type	Reference
WS 3.00E+001	mg/L	20	EXP	YALKOWSKY,SH & DANNENFELSER,RM (1992)
logP 3.00			EXP	HANSCH,C & LEO,AJ (1985)
VP 4.55E-006	mm Hg	25	EST	NEELY,WB & BLAU,GE (1985)
DC	pKa			
HL 2.08E-010	atm m3/mol	25	EST	MEYLAN,WM & HOWARD,PH (1991)
OH 2.91E-011	cm3/molc sec	25	EST	MEYLAN,WM & HOWARD,PH (1993)

003766-81-2 — N-METHYL O-SEC-BUTYL PHENYL CARBAMATE

Formula: $C_{12}H_{17}NO_2$

Mol Weight: 207.27

MP (deg C): — FP (deg C): —

BP (deg C): —

BP pressure (mm Hg): —

Property/Value	Units	Temp	Data Type	Reference
WS 6.60E+002	mg/L	30	EXP	YALKOWSKY,SH & DANNENFELSER,RM (1992)
logP 2.78			EXP	HANSCH,C & LEO,AJ (1985)
VP 1.43E-004	mm Hg	25	EXP	WATANABE,T (1993)
DC	pKa			
HL 5.91E-008	atm m3/mol	25	EST	VP/WSOL
OH 1.33E-011	cm3/molc sec	25	EST	MEYLAN,WM & HOWARD,PH (1993)

003766-82-3 — N-ME-3-ME-4-IPRPHENYL CARBAMATE

Formula: $C_{12}H_{17}NO_2$

Mol Weight: 207.27

MP (deg C): — FP (deg C): —

BP (deg C): —

BP pressure (mm Hg): —

Property/Value	Units	Temp	Data Type	Reference
WS 8.86E+001	mg/L	25	EST	MEYLAN,WM ET AL. (1996)
logP 3.11			EXP	HANSCH,C & LEO,AJ (1985)
VP 1.32E-003	mm Hg	25	EST	NEELY,WB & BLAU,GE (1985)
DC	pKa			
HL 6.89E-008	atm m3/mol	25	EST	MEYLAN,WM & HOWARD,PH (1991)
OH 1.48E-011	cm3/molc sec	25	EST	MEYLAN,WM & HOWARD,PH (1993)

003767-28-0 — à-D-GLUCOPYRANOSIDE, 4-NITROPHENYL

Formula: $C_{12}H_{15}NO_8$

Mol Weight: 301.26

MP (deg C): 209-213 FP (deg C): —

BP (deg C): —

BP pressure (mm Hg): —

Property/Value	Units	Temp	Data Type	Reference
WS 7.01E+003	mg/L	25	EST	MEYLAN,WM ET AL. (1996)
logP -0.39			EXP	SANGSTER,J (1993)
VP 3.68E-013	mm Hg	25	EST	NEELY,WB & BLAU,GE (1985)
DC	pKa			
HL 4.54E-018	atm m3/mol	25	EST	MEYLAN,WM & HOWARD,PH (1991)
OH 6.47E-011	cm3/molc sec	25	EST	MEYLAN,WM & HOWARD,PH (1993)

003769-23-1 — 4-METHYL-1-HEXENE

Formula: C_7H_{14}

Mol Weight: 98.19

MP (deg C): -141.5 FP (deg C): —

BP (deg C): 86.7

BP pressure (mm Hg): —

Property/Value	Units	Temp	Data Type	Reference
WS 3.08E+001	mg/L	25	EST	MEYLAN,WM ET AL. (1996)
logP 3.57			EST	MEYLAN,WM & HOWARD,PH (1995)
VP 7.35E+001	mm Hg	25	EXP	YAWS,CL (1994A)
DC	pKa			
HL 4.76E-001	atm m3/mol	25	EST	MEYLAN,WM & HOWARD,PH (1991)
OH 3.19E-011	cm3/molc sec	25	EST	MEYLAN,WM & HOWARD,PH (1993)

003769-57-1 — DISPERSE RED 5

Formula: $C_{17}H_{19}ClN_4O_4$

Mol Weight: 378.82

MP (deg C): 192 FP (deg C): —

BP (deg C): —

BP pressure (mm Hg): —

Property/Value	Units	Temp	Data Type	Reference
WS 9.84E-002	mg/L	25	EXP	BAUGHMAN,GL & PERENICH,TA (1988)
logP 4.30			EXP	BAUGHMAN,GL & WEBER,EJ (1991)
VP 4.10E-014	mm Hg	25	EST	NEELY,WB & BLAU,GE (1985)
DC	pKa			
HL 1.78E-015	atm m3/mol	25	EST	MEYLAN,WM & HOWARD,PH (1991)
OH 1.81E-010	cm3/molc sec	25	EST	MEYLAN,WM & HOWARD,PH (1993)

003770-50-1 — 1H-INDOLE-2-CARBOXYLIC ACID, ETHYL ESTER

Formula: $C_{11}H_{11}NO_2$

Mol Weight: 189.22

MP (deg C): 122-125 FP (deg C): —

BP (deg C): —

BP pressure (mm Hg): —

Property/Value	Units	Temp	Data Type	Reference
WS 8.82E+001	mg/L	25	EST	MEYLAN,WM ET AL. (1996)
logP 3.22			EXP	YAMAGAMI,C ET AL. (1994)
VP 8.33E-005	mm Hg	25	EST	NEELY,WB & BLAU,GE (1985)
DC	pKa			
HL 7.58E-009	atm m3/mol	25	EST	MEYLAN,WM & HOWARD,PH (1991)
OH 3.72E-011	cm3/molc sec	25	EST	MEYLAN,WM & HOWARD,PH (1993)

003771-14-0 — 1-NAPHTHOL-3-SULFONIC ACID

Formula: $C_{10}H_8O_4S$

Mol Weight: 224.24

MP (deg C): — FP (deg C): —

BP (deg C): —

BP pressure (mm Hg): —

Property/Value	Units	Temp	Data Type	Reference
WS 1.77E+005	mg/L	25	EST	MEYLAN,WM ET AL. (1996)
logP -0.18			EXP	HANSCH,C & LEO,AJ (1985)
VP 9.03E-010	mm Hg	25	EST	NEELY,WB & BLAU,GE (1985)
DC	pKa			
HL 2.56E-014	atm m3/mol	25	EST	MEYLAN,WM & HOWARD,PH (1991)
OH 1.61E-011	cm3/molc sec	25	EST	MEYLAN,WM & HOWARD,PH (1993)

CAS #: 003771-38-8	DIS. A. 2

Formula: $C_{17}H_{21}N_3O_2$

Mol Weight: 299.38

MP (deg C): | FP (deg C):

BP (deg C):

BP pressure (mm Hg):

Property/Value	Units	Temp	Data Type	Reference
WS 2.24E+001	mg/L	25	EXP	BAUGHMAN,GL & PERENICH,TA (1988)
logP 3.29			EST	MEYLAN,WM & HOWARD,PH (1995)
VP 3.83E-011	mm Hg	25	EST	NEELY,WB & BLAU,GE (1985)
DC	pKa			
HL 6.08E-013	atm m3/mol	25	EST	MEYLAN,WM & HOWARD,PH (1991)
OH 1.82E-010	cm3/molc sec	25	EST	MEYLAN,WM & HOWARD,PH (1993)

CAS #: 003772-76-7	SULFAMETHOMIDINE

Formula: $C_{12}H_{14}N_4O_3S$

Mol Weight: 294.33

MP (deg C): 146 | FP (deg C):

BP (deg C):

BP pressure (mm Hg):

Property/Value	Units	Temp	Data Type	Reference
WS 8.43E+002	mg/L		EXP	YALKOWSKY,SH & DANNENFELSER,RM (1992)
logP 0.61			EXP	HANSCH,C & LEO,AJ (1985)
VP 3.71E-009	mm Hg	25	EST	NEELY,WB & BLAU,GE (1985)
DC	pKa			
HL 2.66E-014	atm m3/mol	25	EST	MEYLAN,WM & HOWARD,PH (1991)
OH 1.69E-010	cm3/molc sec	25	EST	MEYLAN,WM & HOWARD,PH (1993)

CAS #: 003775-60-8	2-AMINO-1,3,4-OXADIAZOLE

Formula: $C_2H_4N_3O$

Mol Weight: 86.07

MP (deg C): | FP (deg C):

BP (deg C):

BP pressure (mm Hg):

Property/Value	Units	Temp	Data Type	Reference
WS 1.00E+006	mg/L	25	EST	MEYLAN,WM ET AL. (1996)
logP -1.44			EXP	HANSCH,C & LEO,AJ (1985)
VP 1.28E-001	mm Hg	25	EST	NEELY,WB & BLAU,GE (1985)
DC	pKa			
HL 1.01E-009	atm m3/mol	25	EST	MEYLAN,WM & HOWARD,PH (1991)
OH 2.00E-010	cm3/molc sec	25	EST	MEYLAN,WM & HOWARD,PH (1993)

CAS #: 003775-61-9	2-NH2-5-ET-1,3,4-OXADIAZOLE

Formula: $C_4H_8N_3O$

Mol Weight: 114.13

MP (deg C): | FP (deg C):

BP (deg C):

BP pressure (mm Hg):

Property/Value	Units	Temp	Data Type	Reference
WS 1.92E+005	mg/L	25	EST	MEYLAN,WM ET AL. (1996)
logP -0.30			EXP	HANSCH,C & LEO,AJ (1985)
VP 1.16E-002	mm Hg	25	EST	NEELY,WB & BLAU,GE (1985)
DC	pKa			
HL 1.48E-009	atm m3/mol	25	EST	MEYLAN,WM & HOWARD,PH (1991)
OH 5.10E-012	cm3/molc sec	25	EST	MEYLAN,WM & HOWARD,PH (1993)

CAS #: 003778-73-2	IFOSFAMIDE

Formula: $C_7H_{15}Cl_2N_2O_2P$

Mol Weight: 261.09

MP (deg C): 39-41 | FP (deg C):

BP (deg C):

BP pressure (mm Hg):

Property/Value	Units	Temp	Data Type	Reference
WS 3.78E+003	mg/L	25	EST	MEYLAN,WM ET AL. (1996)
logP 0.86			EXP	HANSCH,C ET AL. (1995)
VP 2.98E-005	mm Hg	25	EST	NEELY,WB & BLAU,GE (1985)
DC	pKa			
HL 1.36E-011	atm m3/mol	25	EST	MEYLAN,WM & HOWARD,PH (1991)
OH 4.28E-011	cm3/molc sec	25	EST	MEYLAN,WM & HOWARD,PH (1993)

CAS #: 003786-76-3	DIGITOXIN, DIHYDRO-

Formula: $C_{41}H_{66}O_{13}$

Mol Weight: 766.98

MP (deg C): | FP (deg C):

BP (deg C):

BP pressure (mm Hg):

Property/Value	Units	Temp	Data Type	Reference
WS 4.17E-002	mg/L	25	EST	MEYLAN,WM ET AL. (1996)
logP 2.90			EXP	HANSCH,C ET AL. (1995)
VP 9.59E-029	mm Hg	25	EST	NEELY,WB & BLAU,GE (1985)
DC	pKa			
HL 2.30E-025	atm m3/mol	25	EST	MEYLAN,WM & HOWARD,PH (1991)
OH 1.38E-010	cm3/molc sec	25	EST	MEYLAN,WM & HOWARD,PH (1993)

CAS #: 003801-06-7	FLUOROMETHOLONE ACETATE

Formula: $C_{24}H_{31}FO_5$

Mol Weight: 418.51

MP (deg C): 292-303 | FP (deg C):

BP (deg C):

BP pressure (mm Hg):

Property/Value	Units	Temp	Data Type	Reference
WS 3.98E+002	mg/L	25	EST	MEYLAN,WM ET AL. (1996)
logP 1.50			EXP	HANSCH,C & LEO,AJ (1985)
VP 9.45E-012	mm Hg	25	EST	NEELY,WB & BLAU,GE (1985)
DC	pKa			
HL 4.95E-014	atm m3/mol	25	EST	MEYLAN,WM & HOWARD,PH (1991)
OH 6.33E-011	cm3/molc sec	25	EST	MEYLAN,WM & HOWARD,PH (1993)

CAS #: 003811-49-2	SALITHION

Formula: $C_8H_9O_3PS$

Mol Weight: 216.20

MP (deg C): | FP (deg C):

BP (deg C):

BP pressure (mm Hg):

Property/Value	Units	Temp	Data Type	Reference
WS 5.80E+001	mg/L	30	EXP	YALKOWSKY,SH & DANNENFELSER,RM (1992)
logP 2.67			EXP	HANSCH,C & LEO,AJ (1985)
VP 2.22E-003	mm Hg	25	EST	NEELY,WB & BLAU,GE (1985)
DC	pKa			
HL 1.56E-005	atm m3/mol	25	EST	MEYLAN,WM & HOWARD,PH (1991)
OH 7.93E-011	cm3/molc sec	25	EST	MEYLAN,WM & HOWARD,PH (1993)

BENAZOLIN

CAS #: 003813-05-6				BENAZOLIN
Formula: $C_9H_6ClNO_3S$				
Mol Weight: 243.67				
MP (deg C):			**FP (deg C):**	
BP (deg C):				
BP pressure (mm Hg):				

Property/Value	Units	Temp	Data Type	Reference
WS 6.00E+002	mg/L	20	EXP	YALKOWSKY,SH & DANNENFELSER,RM (1992)
logP 2.08			EST	MEYLAN,WM & HOWARD,PH (1995)
VP 7.63E-008	mm Hg	25	EST	NEELY,WB & BLAU,GE (1985)
DC	pKa			
HL 8.73E-010	atm m3/mol	25	EST	MEYLAN,WM & HOWARD,PH (1991)
OH 5.35E-012	cm3/molc sec	25	EST	MEYLAN,WM & HOWARD,PH (1993)

ALLYL-ISOPROPYL ACETAMIDE

CAS #: 003829-78-5				ALLYL-ISOPROPYL ACETAMIDE
Formula: $C_8H_{15}NO$				
Mol Weight: 141.21				
MP (deg C):			**FP (deg C):**	
BP (deg C):				
BP pressure (mm Hg):				

Property/Value	Units	Temp	Data Type	Reference
WS 8.80E+003	mg/L	25	EST	MEYLAN,WM ET AL. (1996)
logP 1.14			EXP	HANSCH,C & LEO,AJ (1985)
VP 2.40E-001	mm Hg	25	EST	NEELY,WB & BLAU,GE (1985)
DC	pKa			
HL 1.25E-007	atm m3/mol	25	EST	MEYLAN,WM & HOWARD,PH (1991)
OH 5.35E-011	cm3/molc sec	25	EST	MEYLAN,WM & HOWARD,PH (1993)

4-METHYL-8-QUINOLINOL

CAS #: 003846-73-9				4-METHYL-8-QUINOLINOL
Formula: $C_{10}H_9NO$				
Mol Weight: 159.19				
MP (deg C):			**FP (deg C):**	
BP (deg C):				
BP pressure (mm Hg):				

Property/Value	Units	Temp	Data Type	Reference
WS 2.30E+003	mg/L	25	EST	MEYLAN,WM ET AL. (1996)
logP 2.41			EXP	HANSCH,C ET AL. (1995)
VP 1.05E-004	mm Hg	25	EST	NEELY,WB & BLAU,GE (1985)
DC 5.56	pKa		EXP	KORTUM,G ET AL (1961)
HL 7.90E-011	atm m3/mol	25	EST	MEYLAN,WM & HOWARD,PH (1991)
OH 8.86E-011	cm3/molc sec	25	EST	MEYLAN,WM & HOWARD,PH (1993)

P-CHLOROBENZALDOXIME

CAS #: 003848-36-0				P-CHLOROBENZALDOXIME
Formula: C_7H_6ClNO				
Mol Weight: 155.58				
MP (deg C):			**FP (deg C):**	
BP (deg C):				
BP pressure (mm Hg):				

Property/Value	Units	Temp	Data Type	Reference
WS 3.76E+002	mg/L	25	EST	MEYLAN,WM ET AL. (1996)
logP 2.67			EXP	HANSCH,C ET AL. (1995)
VP 5.92E-004	mm Hg	25	EST	NEELY,WB & BLAU,GE (1985)
DC	pKa			
HL 2.65E-007	atm m3/mol	25	EST	MEYLAN,WM & HOWARD,PH (1991)
OH 3.59E-012	cm3/molc sec	25	EST	MEYLAN,WM & HOWARD,PH (1993)

ISOBUTYL GALLATE

CAS #: 003856-05-1				ISOBUTYL GALLATE
Formula: $C_{11}H_{14}O_5$				
Mol Weight: 226.23				
MP (deg C):			**FP (deg C):**	
BP (deg C):				
BP pressure (mm Hg):				

Property/Value	Units	Temp	Data Type	Reference
WS 9.61E+002	mg/L	25	EST	MEYLAN,WM ET AL. (1996)
logP 2.46			EXP	HANSCH,C & LEO,AJ (1985)
VP 1.97E-007	mm Hg	25	EST	NEELY,WB & BLAU,GE (1985)
DC	pKa			
HL 9.15E-017	atm m3/mol	25	EST	MEYLAN,WM & HOWARD,PH (1991)
OH 9.36E-011	cm3/molc sec	25	EST	MEYLAN,WM & HOWARD,PH (1993)

C.I. DISPERSE BLUE 26

CAS #: 003860-63-7				C.I. DISPERSE BLUE 26
Formula: $C_{16}H_{14}N_2O_4$				
Mol Weight: 298.30				
MP (deg C): 217			**FP (deg C):**	
BP (deg C):				
BP pressure (mm Hg):				

Property/Value	Units	Temp	Data Type	Reference
WS 2.03E-002	mg/L	25	EXP	BAUGHMAN,G & PERENICH,TA (1988)
logP 4.70			EST	MEYLAN,WM & HOWARD,PH (1995)
VP 5.95E-012	mm Hg	25	EST	NEELY,WB & BLAU,GE (1985)
DC	pKa			
HL 3.28E-017	atm m3/mol	25	EST	MEYLAN,WM & HOWARD,PH (1991)
OH 4.48E-011	cm3/molc sec	25	EST	MEYLAN,WM & HOWARD,PH (1993)

1,4-BENZODIAZPN2-ON,5(2CF3PH)7CL

CAS #: 003864-49-1				1,4-BENZODIAZPN2-ON,5(2CF3PH)7CL
Formula: $C_{16}H_{10}ClF_3N_2O$				
Mol Weight: 338.72				
MP (deg C):			**FP (deg C):**	
BP (deg C):				
BP pressure (mm Hg):				

Property/Value	Units	Temp	Data Type	Reference
WS 1.37E+001	mg/L	25	EST	MEYLAN,WM ET AL. (1996)
logP 3.19			EXP	HANSCH,C & LEO,AJ (1985)
VP 5.07E-009	mm Hg	25	EST	NEELY,WB & BLAU,GE (1985)
DC	pKa			
HL 1.55E-009	atm m3/mol	25	EST	MEYLAN,WM & HOWARD,PH (1991)
OH 5.91E-012	cm3/molc sec	25	EST	MEYLAN,WM & HOWARD,PH (1993)

ISOPROPYLCYCLOPENTANE

CAS #: 003875-51-2				ISOPROPYLCYCLOPENTANE
Formula: C_8H_{16}				
Mol Weight: 112.22				
MP (deg C): -111.4			**FP (deg C):**	
BP (deg C): 126.5				
BP pressure (mm Hg):				

Property/Value	Units	Temp	Data Type	Reference
WS 1.02E+002	mg/L	20	EXP	SHAW,DG (1989)
logP 4.01			EST	MEYLAN,WM & HOWARD,PH (1995)
VP 1.62E+001	mm Hg	25	EXP	CHAO,J ET AL. (1983)
DC	pKa			
HL 2.35E-002	atm m3/mol	25	EST	VP/WSOL
OH 1.01E-011	cm3/molc sec	25	EST	MEYLAN,WM & HOWARD,PH (1993)

CAS #:	003877-15-4				N-PROPYLMETHYLSULFIDE
Formula:	$C_4H_{10}S$				
Mol Weight:	90.19				
MP (deg C): -113		FP (deg C):			
BP (deg C): 95.6					
BP pressure (mm Hg):					

Property/ Value	Units	Temp	Data Type	Reference
WS 2.96E+003	mg/L	25	EST	MEYLAN,WM ET AL. (1996)
logP 1.90			EST	MEYLAN,WM & HOWARD,PH (1995)
VP 5.08E+001	mm Hg	25	EXP	ZWOLINSKI,BJ & WILHOIT,RC (1971)
DC	pKa			
HL 1.38E-003	atm m3/mol	25	EST	MEYLAN,WM & HOWARD,PH (1991)
OH 1.30E-011	cm3/molc sec	25	EST	MEYLAN,WM & HOWARD,PH (1993)

CAS #:	003878-19-1				FUBERIDAZOLE
Formula:	$C_{11}H_8N_2O$				
Mol Weight:	184.20				
MP (deg C): -43.1		FP (deg C):			
BP (deg C): 221					
BP pressure (mm Hg):					

Property/ Value	Units	Temp	Data Type	Reference
WS 7.80E+001	mg/L	25	EXP	SHIU,WY ET AL. (1990)
logP 2.67			EXP	TOMLIN,C (1994)
VP 2.38E-007	mm Hg	25	EST	NEELY,WB & BLAU,GE (1985)
DC	pKa			
HL 2.81E-008	atm m3/mol	25	EST	MEYLAN,WM & HOWARD,PH (1991)
OH 1.33E-010	cm3/molc sec	25	EST	MEYLAN,WM & HOWARD,PH (1993)

CAS #:	003891-59-6				D-GLUCOSE, 2,3,4,5,6-PENTAACETATE
Formula:	$C_{16}H_{22}O_{11}$				
Mol Weight:	390.35				
MP (deg C):		FP (deg C):			
BP (deg C):					
BP pressure (mm Hg):					

Property/ Value	Units	Temp	Data Type	Reference
WS 1.50E+003	mg/L	18	EXP	YALKOWSKY,SH & DANNENFELSER,RM (1992)
logP 0.63			EXP	HANSCH,C ET AL. (1995)
VP 2.52E-006	mm Hg	25	EST	NEELY,WB & BLAU,GE (1985)
DC	pKa			
HL 1.14E-016	atm m3/mol	25	EST	MEYLAN,WM & HOWARD,PH (1991)
OH 5.02E-011	cm3/molc sec	25	EST	MEYLAN,WM & HOWARD,PH (1993)

CAS #:	003894-63-1				1,4-BENZODIAZPIN2-ONE,5(4CF3PH)
Formula:	$C_{16}H_{11}F_3N_2O$				
Mol Weight:	304.27				
MP (deg C):		FP (deg C):			
BP (deg C):					
BP pressure (mm Hg):					

Property/ Value	Units	Temp	Data Type	Reference
WS 1.63E+001	mg/L	25	EST	MEYLAN,WM ET AL. (1996)
logP 3.34			EXP	HANSCH,C & LEO,AJ (1985)
VP 1.87E-008	mm Hg	25	EST	NEELY,WB & BLAU,GE (1985)
DC	pKa			
HL 2.09E-009	atm m3/mol	25	EST	MEYLAN,WM & HOWARD,PH (1991)
OH 1.65E-011	cm3/molc sec	25	EST	MEYLAN,WM & HOWARD,PH (1993)

CAS #:	003896-29-5				2,6-DIFLUOROACETANILIDE
Formula:	$C_8H_7F_2NO$				
Mol Weight:	171.15				
MP (deg C):		FP (deg C):			
BP (deg C):					
BP pressure (mm Hg):					

Property/ Value	Units	Temp	Data Type	Reference
WS 1.56E+004	mg/L	25	EST	MEYLAN,WM ET AL. (1996)
logP 0.69			EXP	NAKAGAWA,Y ET AL. (1992)
VP 6.87E-004	mm Hg	25	EST	NEELY,WB & BLAU,GE (1985)
DC	pKa			
HL 8.40E-009	atm m3/mol	25	EST	MEYLAN,WM & HOWARD,PH (1991)
OH 1.70E-012	cm3/molc sec	25	EST	MEYLAN,WM & HOWARD,PH (1993)

CAS #:	003900-31-0				FLUDIAZEPAM
Formula:	$C_{16}H_{12}ClFN_2O$				
Mol Weight:	302.74				
MP (deg C): 295-297		FP (deg C):			
BP (deg C):					
BP pressure (mm Hg):					

Property/ Value	Units	Temp	Data Type	Reference
WS 5.30E+001	mg/L	25	EST	MEYLAN,WM ET AL. (1996)
logP 2.75			EXP	HANSCH,C & LEO,AJ (1985)
VP 3.35E-008	mm Hg	25	EST	NEELY,WB & BLAU,GE (1985)
DC	pKa			
HL 4.25E-009	atm m3/mol	25	EST	MEYLAN,WM & HOWARD,PH (1991)
OH 8.54E-012	cm3/molc sec	25	EST	MEYLAN,WM & HOWARD,PH (1993)

CAS #:	003915-60-4				2-THIOBENZYLPYRIDINE OXIDE
Formula:	$C_{12}H_{13}NOS$				
Mol Weight:	219.31				
MP (deg C):		FP (deg C):			
BP (deg C):					
BP pressure (mm Hg):					

Property/ Value	Units	Temp	Data Type	Reference
WS 3.11E+003	mg/L	25	EST	MEYLAN,WM ET AL. (1996)
logP 1.24			EXP	HANSCH,C & LEO,AJ (1985)
VP 3.12E-006	mm Hg	25	EST	NEELY,WB & BLAU,GE (1985)
DC -0.23	pKa		EXP	PERRIN,DD (1965)
HL 6.46E-012	atm m3/mol	25	EST	MEYLAN,WM & HOWARD,PH (1991)
OH 1.45E-011	cm3/molc sec	25	EST	MEYLAN,WM & HOWARD,PH (1993)

CAS #:	003926-67-8				10H-PHENOTHIAZIN-2-OL, 8-CHLORO-10-[3-(DIMETHYLA
Formula:	$C_{17}H_{19}ClN_2OS$				
Mol Weight:	334.87				
MP (deg C):		FP (deg C):			
BP (deg C):					
BP pressure (mm Hg):					

Property/ Value	Units	Temp	Data Type	Reference
WS 6.72E-001	mg/L	25	EST	MEYLAN,WM ET AL. (1996)
logP 4.63			EXP	WHELPTON,R (1989)
VP 2.56E-009	mm Hg	25	EST	NEELY,WB & BLAU,GE (1985)
DC	pKa			
HL 3.84E-014	atm m3/mol	25	EST	MEYLAN,WM & HOWARD,PH (1991)
OH 2.91E-010	cm3/molc sec	25	EST	MEYLAN,WM & HOWARD,PH (1993)

CAS #: 003930-20-9 — SOTALOL

Formula:	$C_{12}H_{20}N_2O_3S$
Mol Weight:	272.37
MP (deg C):	206.5-207
FP (deg C):	
BP (deg C):	
BP pressure (mm Hg):	

Property/Value	Units	Temp	Data Type	Reference
WS 5.51E+003	mg/L	25	EST	MEYLAN,WM ET AL. (1996)
logP 0.24			EXP	BURGOT,G ET AL. (1990)
VP 5.30E-009	mm Hg	25	EST	NEELY,WB & BLAU,GE (1985)
DC	pKa			
HL 2.49E-014	atm m3/mol	25	EST	MEYLAN,WM & HOWARD,PH (1991)
OH 1.53E-010	cm3/molc sec	25	EST	MEYLAN,WM & HOWARD,PH (1993)

CAS #: 003930-83-4 — 2-IODOBENZAMIDE

Formula:	C_7H_6INO
Mol Weight:	247.04
MP (deg C):	
FP (deg C):	
BP (deg C):	
BP pressure (mm Hg):	

Property/Value	Units	Temp	Data Type	Reference
WS 3.95E+003	mg/L	25	EST	MEYLAN,WM ET AL. (1996)
logP 0.93			EXP	NAKAGAWA,Y ET AL. (1992)
VP 1.63E-005	mm Hg	25	EST	NEELY,WB & BLAU,GE (1985)
DC	pKa			
HL 5.13E-010	atm m3/mol	25	EST	MEYLAN,WM & HOWARD,PH (1991)
OH 3.17E-012	cm3/molc sec	25	EST	MEYLAN,WM & HOWARD,PH (1993)

CAS #: 003938-20-3 — N-ME-3-I-PROPOXYPHENYLCARBAMATE

Formula:	$C_{11}H_{15}NO_3$
Mol Weight:	209.25
MP (deg C):	
FP (deg C):	
BP (deg C):	
BP pressure (mm Hg):	

Property/Value	Units	Temp	Data Type	Reference
WS 8.31E+002	mg/L	25	EST	MEYLAN,WM ET AL. (1996)
logP 1.96			EXP	HANSCH,C & LEO,AJ (1985)
VP 1.30E-003	mm Hg	25	EST	NEELY,WB & BLAU,GE (1985)
DC	pKa			
HL 3.35E-009	atm m3/mol	25	EST	MEYLAN,WM & HOWARD,PH (1991)
OH 4.04E-011	cm3/molc sec	25	EST	MEYLAN,WM & HOWARD,PH (1993)

CAS #: 003938-24-7 — N-ME-2-METHOXYPHENYLCARBAMATE

Formula:	$C_9H_{11}NO_3$
Mol Weight:	181.19
MP (deg C):	
FP (deg C):	
BP (deg C):	
BP pressure (mm Hg):	

Property/Value	Units	Temp	Data Type	Reference
WS 1.11E+004	mg/L	25	EST	MEYLAN,WM ET AL. (1996)
logP 0.81			EXP	HANSCH,C & LEO,AJ (1985)
VP 5.26E-003	mm Hg	25	EST	NEELY,WB & BLAU,GE (1985)
DC	pKa			
HL 1.90E-009	atm m3/mol	25	EST	MEYLAN,WM & HOWARD,PH (1991)
OH 2.27E-011	cm3/molc sec	25	EST	MEYLAN,WM & HOWARD,PH (1993)

CAS #: 003938-28-1 — N-ME-3-METHOXYPHENYLCARBAMATE

Formula:	$C_9H_{11}NO_3$
Mol Weight:	181.19
MP (deg C):	
FP (deg C):	
BP (deg C):	
BP pressure (mm Hg):	

Property/Value	Units	Temp	Data Type	Reference
WS 4.22E+003	mg/L	25	EST	MEYLAN,WM ET AL. (1996)
logP 1.30			EXP	HANSCH,C & LEO,AJ (1985)
VP 5.26E-003	mm Hg	25	EST	NEELY,WB & BLAU,GE (1985)
DC	pKa			
HL 1.90E-009	atm m3/mol	25	EST	MEYLAN,WM & HOWARD,PH (1991)
OH 2.91E-011	cm3/molc sec	25	EST	MEYLAN,WM & HOWARD,PH (1993)

CAS #: 003938-29-2 — N-METHYL-4-METHOXYPHENYLCARBATE

Formula:	$C_9H_{11}NO_3$
Mol Weight:	181.19
MP (deg C):	
FP (deg C):	
BP (deg C):	
BP pressure (mm Hg):	

Property/Value	Units	Temp	Data Type	Reference
WS 5.13E+003	mg/L	25	EST	MEYLAN,WM ET AL. (1996)
logP 1.20			EXP	HANSCH,C & LEO,AJ (1985)
VP 5.26E-003	mm Hg	25	EST	NEELY,WB & BLAU,GE (1985)
DC	pKa			
HL 1.90E-009	atm m3/mol	25	EST	MEYLAN,WM & HOWARD,PH (1991)
OH 2.27E-011	cm3/molc sec	25	EST	MEYLAN,WM & HOWARD,PH (1993)

CAS #: 003938-34-9 — N-ME 4-METHYLTHIOPHENYLCARBAMATE

Formula:	$C_9H_{11}NO_2S$
Mol Weight:	197.26
MP (deg C):	
FP (deg C):	
BP (deg C):	
BP pressure (mm Hg):	

Property/Value	Units	Temp	Data Type	Reference
WS 1.04E+003	mg/L	25	EST	MEYLAN,WM ET AL. (1996)
logP 1.92			EXP	HANSCH,C & LEO,AJ (1985)
VP 8.37E-004	mm Hg	25	EST	NEELY,WB & BLAU,GE (1985)
DC	pKa			
HL 9.36E-010	atm m3/mol	25	EST	MEYLAN,WM & HOWARD,PH (1991)
OH 1.64E-011	cm3/molc sec	25	EST	MEYLAN,WM & HOWARD,PH (1993)

CAS #: 003942-51-6 — N-ME-4-SEC-BUTYLPHENYLCARBAMATE

Formula:	$C_{12}H_{17}NO_2$
Mol Weight:	207.27
MP (deg C):	
FP (deg C):	
BP (deg C):	
BP pressure (mm Hg):	

Property/Value	Units	Temp	Data Type	Reference
WS 7.43E+001	mg/L	25	EST	MEYLAN,WM ET AL. (1996)
logP 3.20			EXP	HANSCH,C & LEO,AJ (1985)
VP 1.53E-003	mm Hg	25	EST	NEELY,WB & BLAU,GE (1985)
DC	pKa			
HL 8.29E-008	atm m3/mol	25	EST	MEYLAN,WM & HOWARD,PH (1991)
OH 1.33E-011	cm3/molc sec	25	EST	MEYLAN,WM & HOWARD,PH (1993)

CAS #: 003942-54-9 — N-ME-2-CHLOROPHENYLCARBAMATE

Formula: $C_8H_8ClNO_2$
Mol Weight: 185.61
MP (deg C):
FP (deg C):
BP (deg C):
BP pressure (mm Hg):

Property/Value	Units	Temp	Data Type	Reference
WS 2.06E+003	mg/L	25	EST	MEYLAN,WM ET AL. (1996)
logP 1.64			EXP	HANSCH,C & LEO,AJ (1985)
VP 8.31E-003	mm Hg	25	EST	NEELY,WB & BLAU,GE (1985)
DC	pKa			
HL 2.38E-008	atm m3/mol	25	EST	MEYLAN,WM & HOWARD,PH (1991)
OH 7.07E-012	cm3/molc sec	25	EST	MEYLAN,WM & HOWARD,PH (1993)

CAS #: 003956-07-8 — 4-IODOBENZAMIDE

Formula: C_7H_6INO
Mol Weight: 247.04
MP (deg C):
FP (deg C):
BP (deg C):
BP pressure (mm Hg):

Property/Value	Units	Temp	Data Type	Reference
WS 4.91E+002	mg/L	25	EST	MEYLAN,WM ET AL. (1996)
logP 1.99			EXP	HANSCH,C & LEO,AJ (1985)
VP 1.63E-005	mm Hg	25	EST	NEELY,WB & BLAU,GE (1985)
DC	pKa			
HL 5.13E-010	atm m3/mol	25	EST	MEYLAN,WM & HOWARD,PH (1991)
OH 3.17E-012	cm3/molc sec	25	EST	MEYLAN,WM & HOWARD,PH (1993)

CAS #: 003956-55-6 — DISPERSE BLUE 79

Formula: $C_{23}H_{25}BrN_6O_{10}$
Mol Weight: 625.39
MP (deg C):
FP (deg C):
BP (deg C):
BP pressure (mm Hg):

Property/Value	Units	Temp	Data Type	Reference
WS 3.22E-003	mg/L	25	EST	MEYLAN,WM ET AL. (1996)
logP 4.80			EXP	BAUGHMAN,GL & WEBER,EJ (1991)
VP 1.74E-017	mm Hg	25	EST	NEELY,WB & BLAU,GE (1985)
DC	pKa			
HL 9.82E-025	atm m3/mol	25	EST	MEYLAN,WM & HOWARD,PH (1991)
OH 2.26E-010	cm3/molc sec	25	EST	MEYLAN,WM & HOWARD,PH (1993)

CAS #: 003957-74-2 — 4-ACETOXY DICHLOROACETANILIDE

Formula: $C_{10}H_9Cl_2NO_3$
Mol Weight: 262.09
MP (deg C):
FP (deg C):
BP (deg C):
BP pressure (mm Hg):

Property/Value	Units	Temp	Data Type	Reference
WS 4.83E+002	mg/L	25	EST	MEYLAN,WM ET AL. (1996)
logP 1.90			EXP	DUTTA,H ET AL. (1988)
VP 5.19E-007	mm Hg	25	EST	NEELY,WB & BLAU,GE (1985)
DC	pKa			
HL 9.21E-012	atm m3/mol	25	EST	MEYLAN,WM & HOWARD,PH (1991)
OH 9.43E-012	cm3/molc sec	25	EST	MEYLAN,WM & HOWARD,PH (1993)

CAS #: 003964-58-7 — 3-CHLORO-4-HYDROXYBENZOIC ACID

Formula: $C_7H_5ClO_3$
Mol Weight: 172.57
MP (deg C): 171
FP (deg C):
BP (deg C):
BP pressure (mm Hg):

Property/Value	Units	Temp	Data Type	Reference
WS 1.99E+003	mg/L	25	EST	MEYLAN,WM ET AL. (1996)
logP 2.41			EXP	SANGSTER,J (1993)
VP 2.58E-005	mm Hg	25	EST	NEELY,WB & BLAU,GE (1985)
DC 7.52	pKa	25	EXP	SERJEANT,EP & DEMPSEY,B (1979)
HL 8.36E-012	atm m3/mol	25	EST	MEYLAN,WM & HOWARD,PH (1991)
OH 4.26E-012	cm3/molc sec	25	EST	MEYLAN,WM & HOWARD,PH (1993)

CAS #: 003970-35-2 — 2-CHLORO-3-NITROBENZOIC ACID

Formula: $C_7H_4ClNO_4$
Mol Weight: 201.57
MP (deg C): 183.5
FP (deg C):
BP (deg C):
BP pressure (mm Hg):

Property/Value	Units	Temp	Data Type	Reference
WS 3.47E+002	mg/L	25	EST	MEYLAN,WM ET AL. (1996)
logP 1.99			EST	MEYLAN,WM & HOWARD,PH (1995)
VP 1.89E-005	mm Hg	25	EST	NEELY,WB & BLAU,GE (1985)
DC 2.20	pKa	25	EXP	KORTUM,G ET AL (1961)
HL 3.17E-010	atm m3/mol	25	EST	MEYLAN,WM & HOWARD,PH (1991)
OH 5.83E-013	cm3/molc sec	25	EST	MEYLAN,WM & HOWARD,PH (1993)

CAS #: 003970-62-5 — 2,2-DIMETHYL-3-PENTANOL

Formula: $C_7H_{16}O$
Mol Weight: 116.20
MP (deg C): -5
FP (deg C):
BP (deg C): 135
BP pressure (mm Hg):

Property/Value	Units	Temp	Data Type	Reference
WS 8.20E+003	mg/L	25	EXP	YALKOWSKY,SH & DANNENFELSER,RM (1992)
logP 2.13			EST	MEYLAN,WM & HOWARD,PH (1995)
VP 2.54E+000	mm Hg	25	EST	NEELY,WB & BLAU,GE (1985)
DC	pKa			
HL 2.34E-005	atm m3/mol	25	EST	MEYLAN,WM & HOWARD,PH (1991)
OH 1.22E-011	cm3/molc sec	25	EST	MEYLAN,WM & HOWARD,PH (1993)

CAS #: 003971-99-1 — N-ME-2,5-DIMETHYLPHENYLCARBAMATE

Formula: $C_{10}H_{13}NO_2$
Mol Weight: 179.22
MP (deg C):
FP (deg C):
BP (deg C):
BP pressure (mm Hg):

Property/Value	Units	Temp	Data Type	Reference
WS 1.03E+003	mg/L	25	EST	MEYLAN,WM ET AL. (1996)
logP 2.03			EXP	HANSCH,C & LEO,AJ (1985)
VP 5.37E-003	mm Hg	25	EST	NEELY,WB & BLAU,GE (1985)
DC	pKa			
HL 3.91E-008	atm m3/mol	25	EST	MEYLAN,WM & HOWARD,PH (1991)
OH 1.28E-011	cm3/molc sec	25	EST	MEYLAN,WM & HOWARD,PH (1993)

CAS #:	003978-68-5	N-ME-3-BUTOXYPHENYLCARBAMATE

Formula: $C_{12}H_{17}NO_3$

Mol Weight: 223.27

MP (deg C):　　FP (deg C):

BP (deg C):

BP pressure (mm Hg):

Property/Value	Units	Temp	Data Type	Reference
WS 9.81E+001	mg/L	25	EST	MEYLAN,WM ET AL. (1996)
logP 2.96			EXP	HANSCH,C & LEO,AJ (1985)
VP 2.36E-004	mm Hg	25	EST	NEELY,WB & BLAU,GE (1985)
DC	pKa			
HL 4.45E-009	atm m3/mol	25	EST	MEYLAN,WM & HOWARD,PH (1991)
OH 4.06E-011	cm3/molc sec	25	EST	MEYLAN,WM & HOWARD,PH (1993)

CAS #:	003978-69-6	N-ME-4-BUTOXYPHENYLCARBAMATE

Formula: $C_{12}H_{17}NO_3$

Mol Weight: 223.27

MP (deg C):　　FP (deg C):

BP (deg C):

BP pressure (mm Hg):

Property/Value	Units	Temp	Data Type	Reference
WS 1.19E+002	mg/L	25	EST	MEYLAN,WM ET AL. (1996)
logP 2.86			EXP	HANSCH,C & LEO,AJ (1985)
VP 2.36E-004	mm Hg	25	EST	NEELY,WB & BLAU,GE (1985)
DC	pKa			
HL 4.45E-009	atm m3/mol	25	EST	MEYLAN,WM & HOWARD,PH (1991)
OH 3.42E-011	cm3/molc sec	25	EST	MEYLAN,WM & HOWARD,PH (1993)

CAS #:	003978-81-2	4-(TERT-BUTYL)PYRIDINE

Formula: $C_9H_{13}N$

Mol Weight: 135.21

MP (deg C): -41　　FP (deg C):

BP (deg C): 196.5

BP pressure (mm Hg):

Property/Value	Units	Temp	Data Type	Reference
WS 4.22E+002	mg/L	25	EST	MEYLAN,WM ET AL. (1996)
logP 2.71			EST	MEYLAN,WM & HOWARD,PH (1995)
VP 9.72E-001	mm Hg	25	EST	NEELY,WB & BLAU,GE (1985)
DC 5.99	pKa	25	EXP	PERRIN,DD (1965)
HL 1.82E-005	atm m3/mol	25	EST	MEYLAN,WM & HOWARD,PH (1991)
OH 1.32E-012	cm3/molc sec	25	EST	MEYLAN,WM & HOWARD,PH (1993)

CAS #:	003982-82-9	1,3,3,5-TETRAMETHYL-1,1,5,5-TETRAPHENYLTRISILO*

Formula: $C_{28}H_{32}O_2Si_3$

Mol Weight: 484.82

MP (deg C):　　FP (deg C):

BP (deg C):

BP pressure (mm Hg):

Property/Value	Units	Temp	Data Type	Reference
WS 1.14E-007	mg/L	25	EST	MEYLAN,WM ET AL. (1996)
logP 10.21			EST	MEYLAN,WM & HOWARD,PH (1995)
VP 5.86E-010	mm Hg	25	EST	NEELY,WB & BLAU,GE (1985)
DC	pKa			
HL 2.88E-009	atm m3/mol	25	EST	MEYLAN,WM & HOWARD,PH (1991)
OH 8.40E-012	cm3/molc sec	25	EST	MEYLAN,WM & HOWARD,PH (1993)

CAS #:	003988-76-9	2-PROPEN-1-ONE, 1,3-BIS(2-FURANYL)-

Formula: $C_{11}H_{10}O_3$

Mol Weight: 190.20

MP (deg C):　　FP (deg C):

BP (deg C):

BP pressure (mm Hg):

Property/Value	Units	Temp	Data Type	Reference
WS 5.04E+002	mg/L	25	EST	MEYLAN,WM ET AL. (1996)
logP 2.34			EXP	HANSCH,C & LEO,AJ (1985)
VP 2.10E-003	mm Hg	25	EST	NEELY,WB & BLAU,GE (1985)
DC	pKa			
HL 2.31E-007	atm m3/mol	25	EST	MEYLAN,WM & HOWARD,PH (1991)
OH 9.47E-011	cm3/molc sec	25	EST	MEYLAN,WM & HOWARD,PH (1993)

CAS #:	003994-23-8	2-CL-C6H4NHN=C(CN)COOET

Formula: $C_{11}H_{10}ClN_3O_2$

Mol Weight: 251.67

MP (deg C):　　FP (deg C):

BP (deg C):

BP pressure (mm Hg):

Property/Value	Units	Temp	Data Type	Reference
WS 3.01E+001	mg/L	25	EST	MEYLAN,WM ET AL. (1996)
logP 3.38			EXP	HANSCH,C & LEO,AJ (1985)
VP 4.72E-006	mm Hg	25	EST	NEELY,WB & BLAU,GE (1985)
DC	pKa			
HL 6.52E-007	atm m3/mol	25	EST	MEYLAN,WM & HOWARD,PH (1991)
OH 1.41E-011	cm3/molc sec	25	EST	MEYLAN,WM & HOWARD,PH (1993)

CAS #:	004001-73-4	2-BROMOBENZAMIDE

Formula: C_7H_6BrNO

Mol Weight: 200.04

MP (deg C): 160.5　　FP (deg C):

BP (deg C):

BP pressure (mm Hg):

Property/Value	Units	Temp	Data Type	Reference
WS 9.62E+003	mg/L	25	EST	MEYLAN,WM ET AL. (1996)
logP 0.77			EXP	NAKAGAWA,Y ET AL. (1992)
VP 5.64E-005	mm Hg	25	EST	NEELY,WB & BLAU,GE (1985)
DC	pKa			
HL 8.82E-010	atm m3/mol	25	EST	MEYLAN,WM & HOWARD,PH (1991)
OH 3.12E-012	cm3/molc sec	25	EST	MEYLAN,WM & HOWARD,PH (1993)

CAS #:	004005-51-0	2-AMINO-1,3,4-THIADIAZOLE

Formula: $C_2H_5N_3S$

Mol Weight: 103.15

MP (deg C): 190-192　　FP (deg C):

BP (deg C):

BP pressure (mm Hg):

Property/Value	Units	Temp	Data Type	Reference
WS 3.56E+005	mg/L	25	EST	MEYLAN,WM ET AL. (1996)
logP -0.57			EXP	HANSCH,C & LEO,AJ (1985)
VP 8.28E-003	mm Hg	25	EST	NEELY,WB & BLAU,GE (1985)
DC	pKa			
HL 5.51E-010	atm m3/mol	25	EST	MEYLAN,WM & HOWARD,PH (1991)
OH 3.86E-011	cm3/molc sec	25	EST	MEYLAN,WM & HOWARD,PH (1993)

004008-48-4 — NITROXOLINE

Formula: $C_9H_6N_2O_3$
Mol Weight: 190.16
MP (deg C): 179-181.5
FP (deg C):
BP (deg C):
BP pressure (mm Hg):

Property/Value	Units	Temp	Data Type	Reference
WS 3.73E+003	mg/L	25	EST	MEYLAN,WM ET AL. (1996)
logP 1.99			EXP	HANSCH,C ET AL. (1995)
VP 2.13E-006	mm Hg	25	EST	NEELY,WB & BLAU,GE (1985)
DC	pKa			
HL 2.83E-013	atm m3/mol	25	EST	MEYLAN,WM & HOWARD,PH (1991)
OH 5.30E-012	cm3/molc sec	25	EST	MEYLAN,WM & HOWARD,PH (1993)

004022-58-6 — 4,6-NH2 1,2-H 2,2-DIME-1PH TRIAZENE

Formula: $C_{11}H_{15}N_5$
Mol Weight: 217.28
MP (deg C):
FP (deg C):
BP (deg C):
BP pressure (mm Hg):

Property/Value	Units	Temp	Data Type	Reference
WS 2.05E+004	mg/L	25	EST	MEYLAN,WM ET AL. (1996)
logP 0.28			EXP	HANSCH,C & LEO,AJ (1985)
VP 5.09E-006	mm Hg	25	EST	NEELY,WB & BLAU,GE (1985)
DC	pKa			
HL 7.28E-013	atm m3/mol	25	EST	MEYLAN,WM & HOWARD,PH (1991)
OH 1.89E-010	cm3/molc sec	25	EST	MEYLAN,WM & HOWARD,PH (1993)

004024-81-1 — OCH

Formula: C_6Cl_8O
Mol Weight: 371.69
MP (deg C):
FP (deg C):
BP (deg C):
BP pressure (mm Hg):

Property/Value	Units	Temp	Data Type	Reference
WS 5.84E+000	mg/L	25	EST	MEYLAN,WM ET AL. (1996)
logP 3.39			EST	MEYLAN,WM & HOWARD,PH (1995)
VP 4.72E-005	mm Hg	25	EST	NEELY,WB & BLAU,GE (1985)
DC	pKa			
HL 2.37E-008	atm m3/mol	25	EST	MEYLAN,WM & HOWARD,PH (1991)
OH 3.32E-012	cm3/molc sec	25	EST	MEYLAN,WM & HOWARD,PH (1993)

004032-26-2 — DIQUAT DICHLORIDE

Formula: $C_{12}H_{12}Cl_2N_2$
Mol Weight: 255.15
MP (deg C):
FP (deg C):
BP (deg C):
BP pressure (mm Hg):

Property/Value	Units	Temp	Data Type	Reference
WS 4.43E+005	mg/L	25	EST	MEYLAN,WM ET AL. (1996)
logP -2.82			EST	MEYLAN,WM & HOWARD,PH (1995)
VP 1.11E-005	mm Hg	25	EST	NEELY,WB & BLAU,GE (1985)
DC	pKa			
HL 1.42E-013	atm m3/mol	25	EST	MEYLAN,WM & HOWARD,PH (1991)
OH 2.32E-011	cm3/molc sec	25	EST	MEYLAN,WM & HOWARD,PH (1993)

004033-40-3 — L-ASPARAGINE, N-ACETYL-

Formula: $C_6H_{10}N_2O_4$
Mol Weight: 174.16
MP (deg C):
FP (deg C):
BP (deg C):
BP pressure (mm Hg):

Property/Value	Units	Temp	Data Type	Reference
WS 4.78E+005	mg/L	25	EST	MEYLAN,WM ET AL. (1996)
logP -2.60			EXP	HANSCH,C & LEO,AJ (1985)
VP 3.29E-008	mm Hg	25	EST	NEELY,WB & BLAU,GE (1985)
DC	pKa			
HL 1.16E-015	atm m3/mol	25	EST	MEYLAN,WM & HOWARD,PH (1991)
OH 1.82E-011	cm3/molc sec	25	EST	MEYLAN,WM & HOWARD,PH (1993)

004038-04-4 — 3-ETHYL-1-PENTENE

Formula: C_7H_{14}
Mol Weight: 98.19
MP (deg C): -127.5
FP (deg C):
BP (deg C): 84.1
BP pressure (mm Hg):

Property/Value	Units	Temp	Data Type	Reference
WS 3.08E+001	mg/L	25	EST	MEYLAN,WM ET AL. (1996)
logP 3.57			EST	MEYLAN,WM & HOWARD,PH (1995)
VP 8.15E+001	mm Hg	27	EXP	YAWS,CL (1994A)
DC	pKa			
HL 4.76E-001	atm m3/mol	25	EST	MEYLAN,WM & HOWARD,PH (1991)
OH 3.19E-011	cm3/molc sec	25	EST	MEYLAN,WM & HOWARD,PH (1993)

004043-23-6 — N-ME-M-ETHYLPHENYLCARBAMATE

Formula: $C_{10}H_{13}NO_2$
Mol Weight: 179.22
MP (deg C):
FP (deg C):
BP (deg C):
BP pressure (mm Hg):

Property/Value	Units	Temp	Data Type	Reference
WS 7.34E+002	mg/L	25	EST	MEYLAN,WM ET AL. (1996)
logP 2.20			EXP	HANSCH,C & LEO,AJ (1985)
VP 6.17E-003	mm Hg	25	EST	NEELY,WB & BLAU,GE (1985)
DC	pKa			
HL 4.71E-008	atm m3/mol	25	EST	MEYLAN,WM & HOWARD,PH (1991)
OH 1.22E-011	cm3/molc sec	25	EST	MEYLAN,WM & HOWARD,PH (1993)

004049-81-4 — 2-METHYL-1,5-HEXADIENE

Formula: C_7H_{12}
Mol Weight: 96.17
MP (deg C): 128.8
FP (deg C):
BP (deg C): 88.1
BP pressure (mm Hg):

Property/Value	Units	Temp	Data Type	Reference
WS 3.16E+001	mg/L	25	EST	MEYLAN,WM ET AL. (1996)
logP 3.56			EST	MEYLAN,WM & HOWARD,PH (1995)
VP 7.65E+001	mm Hg	25	EST	NEELY,WB & BLAU,GE (1985)
DC	pKa			
HL 2.64E-001	atm m3/mol	25	EST	MEYLAN,WM & HOWARD,PH (1991)
OH 9.61E-011	cm3/molc sec	25	EXP	ATKINSON,R (1989)

CAS #: 004050-45-7				2-HEXENE (TRANS)

Formula: C_6H_{12}

Mol Weight: 84.16

MP (deg C): -133 FP (deg C):

BP (deg C): 67.9

BP pressure (mm Hg):

Property/Value	Units	Temp	Data Type	Reference
WS 8.85E+001	mg/L	25	EST	MEYLAN,WM ET AL. (1996)
logP 3.07			EST	MEYLAN,WM & HOWARD,PH (1995)
VP 1.55E+002	mm Hg	25	EXP	DAUBERT,TE & DANNER,RP (1989)
DC	pKa			
HL 4.23E-001	atm m3/mol	25	EST	MEYLAN,WM & HOWARD,PH (1991)
OH 5.90E-011	cm3/molc sec	25	EST	MEYLAN,WM & HOWARD,PH (1993)

CAS #: 004052-30-6				P-METHYLSULFONYLBENZOIC ACID

Formula: $C_8H_8O_4S$

Mol Weight: 200.21

MP (deg C): 268-271 FP (deg C):

BP (deg C):

BP pressure (mm Hg):

Property/Value	Units	Temp	Data Type	Reference
WS 1.17E+004	mg/L	25	EST	MEYLAN,WM ET AL. (1996)
logP 0.67			EXP	HANSCH,C & LEO,AJ (1985)
VP 5.88E-006	mm Hg	25	EST	NEELY,WB & BLAU,GE (1985)
DC 3.64	pKa		EXP	KORTUM,G ET AL (1961)
HL 2.51E-011	atm m3/mol	25	EST	MEYLAN,WM & HOWARD,PH (1991)
OH 1.74E-012	cm3/molc sec	25	EST	MEYLAN,WM & HOWARD,PH (1993)

CAS #: 004053-42-3				6-METHYLQUINOLINE, 1-OXIDE

Formula: $C_{10}H_{11}NO$

Mol Weight: 161.21

MP (deg C): FP (deg C):

BP (deg C):

BP pressure (mm Hg):

Property/Value	Units	Temp	Data Type	Reference
WS 1.07E+004	mg/L	25	EST	MEYLAN,WM ET AL. (1996)
logP 0.95			EXP	HANSCH,C & LEO,AJ (1985)
VP 8.90E-005	mm Hg	25	EST	NEELY,WB & BLAU,GE (1985)
DC	pKa			
HL 7.60E-012	atm m3/mol	25	EST	MEYLAN,WM & HOWARD,PH (1991)
OH 3.77E-011	cm3/molc sec	25	EST	MEYLAN,WM & HOWARD,PH (1993)

CAS #: 004054-38-0				1,3-CYCLOHEPTADIENE

Formula: C_5H_6

Mol Weight: 66.10

MP (deg C): -110.4 FP (deg C):

BP (deg C): 120.5

BP pressure (mm Hg):

Property/Value	Units	Temp	Data Type	Reference
WS 6.07E+001	mg/L	25	EST	MEYLAN,WM ET AL. (1996)
logP 3.24			EST	MEYLAN,WM & HOWARD,PH (1995)
VP 4.40E+002	mm Hg	25	EXP	CHAO,J ET AL. (1983)
DC	pKa			
HL 1.12E-001	atm m3/mol	25	EST	MEYLAN,WM & HOWARD,PH (1991)
OH 1.39E-010	cm3/molc sec	25	EXP	ATKINSON,R (1989)

CAS #: 004055-39-4				MITOMYCIN A

Formula: $C_{16}H_{19}N_3O_6$

Mol Weight: 349.35

MP (deg C): FP (deg C):

BP (deg C):

BP pressure (mm Hg):

Property/Value	Units	Temp	Data Type	Reference
WS 1.32E+004	mg/L	25	EST	MEYLAN,WM ET AL. (1996)
logP -0.73			EST	MEYLAN,WM & HOWARD,PH (1995)
VP 1.12E-009	mm Hg	25	EST	NEELY,WB & BLAU,GE (1985)
DC	pKa			
HL 1.74E-022	atm m3/mol	25	EST	MEYLAN,WM & HOWARD,PH (1991)
OH 6.97E-013	cm3/molc sec	25	EST	MEYLAN,WM & HOWARD,PH (1993)

CAS #: 004055-40-7				MITOMYCIN B

Formula: $C_{16}H_{19}N_3O_6$

Mol Weight: 349.35

MP (deg C): FP (deg C):

BP (deg C):

BP pressure (mm Hg):

Property/Value	Units	Temp	Data Type	Reference
WS 1.01E+005	mg/L	25	EST	MEYLAN,WM ET AL. (1996)
logP -1.77			EST	MEYLAN,WM & HOWARD,PH (1995)
VP 6.15E-012	mm Hg	25	EST	NEELY,WB & BLAU,GE (1985)
DC	pKa			
HL 1.20E-024	atm m3/mol	25	EST	MEYLAN,WM & HOWARD,PH (1991)
OH 7.16E-013	cm3/molc sec	25	EST	MEYLAN,WM & HOWARD,PH (1993)

CAS #: 004070-75-1				1,2-DIBENZOYLETHYLENE

Formula: $C_{16}H_{12}O_2$

Mol Weight: 236.27

MP (deg C): FP (deg C):

BP (deg C):

BP pressure (mm Hg):

Property/Value	Units	Temp	Data Type	Reference
WS 5.63E+001	mg/L	25	EST	MEYLAN,WM ET AL. (1996)
logP 3.16			EXP	HANSCH,C & LEO,AJ (1985)
VP 7.13E-006	mm Hg	25	EST	NEELY,WB & BLAU,GE (1985)
DC	pKa			
HL 4.34E-010	atm m3/mol	25	EST	MEYLAN,WM & HOWARD,PH (1991)
OH 1.05E-011	cm3/molc sec	25	EST	MEYLAN,WM & HOWARD,PH (1993)

CAS #: 004074-88-8				DIETHYLENE GLYCOL DIACRYLATE

Formula: $C_{10}H_{14}O_5$

Mol Weight: 214.22

MP (deg C): FP (deg C):

BP (deg C): 94

BP pressure (mm Hg): 2.25E-001

Property/Value	Units	Temp	Data Type	Reference
WS 7.08E+003	mg/L	25	EST	MEYLAN,WM ET AL. (1996)
logP 0.84			EST	MEYLAN,WM & HOWARD,PH (1995)
VP 4.08E-002	mm Hg	25	EST	NEELY,WB & BLAU,GE (1985)
DC	pKa			
HL 9.52E-010	atm m3/mol	25	EST	MEYLAN,WM & HOWARD,PH (1991)
OH 5.00E-011	cm3/molc sec	25	EST	MEYLAN,WM & HOWARD,PH (1993)

004075-58-5

CAS #: 004075-58-5

THIOPHEN-2-COCO-O-ETHYL

Formula: $C_8H_{10}O_3S$

Mol Weight: 186.23

MP (deg C):   **FP (deg C):**

BP (deg C):

BP pressure (mm Hg):

Property/ Value	Units	Temp	Data Type	Reference
WS 1.62E+003	mg/L	25	EST	MEYLAN,WM ET AL. (1996)
logP 1.77			EXP	HANSCH,C & LEO,AJ (1985)
VP 2.74E-003	mm Hg	25	EST	NEELY,WB & BLAU,GE (1985)
DC	pKa			
HL 9.97E-008	atm m3/mol	25	EST	MEYLAN,WM & HOWARD,PH (1991)
OH 1.03E-011	cm3/molc sec	25	EST	MEYLAN,WM & HOWARD,PH (1993)

004080-31-3

CAS #: 004080-31-3

N-(3-CHLORALLYL) HEXAMINIUM CHLORIDE

Formula: $C_9H_{16}Cl_2N_4$

Mol Weight: 251.16

MP (deg C):   **FP (deg C):**

BP (deg C):

BP pressure (mm Hg):

Property/ Value	Units	Temp	Data Type	Reference
WS 1.00E+006	mg/L	25	EST	MEYLAN,WM ET AL. (1996)
logP -5.92			EST	MEYLAN,WM & HOWARD,PH (1995)
VP 4.19E-009	mm Hg	25	EST	NEELY,WB & BLAU,GE (1985)
DC	pKa			
HL 1.76E-008	atm m3/mol	25	EST	MEYLAN,WM & HOWARD,PH (1991)
OH 5.23E-010	cm3/molc sec	25	EST	MEYLAN,WM & HOWARD,PH (1993)

004081-02-1

CAS #: 004081-02-1

4,4'-DIHYDROXYTRIPHENYLMETHANE

Formula: $C_{19}H_{16}O_2$

Mol Weight: 276.34

MP (deg C):   **FP (deg C):**

BP (deg C):

BP pressure (mm Hg):

Property/ Value	Units	Temp	Data Type	Reference
WS 1.36E+001	mg/L	25	EST	MEYLAN,WM ET AL. (1996)
logP 4.30			EXP	HANSCH,C & LEO,AJ (1985)
VP 1.92E-009	mm Hg	25	EST	NEELY,WB & BLAU,GE (1985)
DC	pKa			
HL 4.20E-013	atm m3/mol	25	EST	MEYLAN,WM & HOWARD,PH (1991)
OH 8.71E-011	cm3/molc sec	25	EST	MEYLAN,WM & HOWARD,PH (1993)

004088-60-2

CAS #: 004088-60-2

2-BUTEN-1-OL (CIS)

Formula: C_4H_8O

Mol Weight: 72.11

MP (deg C):   **FP (deg C):**

BP (deg C): 123

BP pressure (mm Hg):

Property/ Value	Units	Temp	Data Type	Reference
WS 1.66E+005	mg/L	25	EXP	RIDDICK,JA ET AL. (1986)
logP 0.63			EST	MEYLAN,WM & HOWARD,PH (1995)
VP 6.00E+001	mm Hg	63	25	RIDDICK,JA ET AL. (1986)
DC	pKa			
HL 8.78E-006	atm m3/mol	25	EST	MEYLAN,WM & HOWARD,PH (1991)
OH 5.99E-011	cm3/molc sec	25	EST	MEYLAN,WM & HOWARD,PH (1993)

004089-04-7

CAS #: 004089-04-7

N-METHYL-2-NAPHTHYLCARBAMATE

Formula: $C_{12}H_{11}NO_2$

Mol Weight: 201.23

MP (deg C):   **FP (deg C):**

BP (deg C):

BP pressure (mm Hg):

Property/ Value	Units	Temp	Data Type	Reference
WS 2.81E+002	mg/L	25	EST	MEYLAN,WM ET AL. (1996)
logP 2.56			EXP	HANSCH,C & LEO,AJ (1985)
VP 1.03E-004	mm Hg	25	EST	NEELY,WB & BLAU,GE (1985)
DC	pKa			
HL 3.14E-009	atm m3/mol	25	EST	MEYLAN,WM & HOWARD,PH (1991)
OH 2.83E-011	cm3/molc sec	25	EST	MEYLAN,WM & HOWARD,PH (1993)

004089-99-0

CAS #: 004089-99-0

N-ME-4-I-PROPYLPHENYLCARBAMATE

Formula: $C_{11}H_{15}NO_2$

Mol Weight: 193.25

MP (deg C):   **FP (deg C):**

BP (deg C):

BP pressure (mm Hg):

Property/ Value	Units	Temp	Data Type	Reference
WS 1.92E+002	mg/L	25	EST	MEYLAN,WM ET AL. (1996)
logP 2.80			EXP	HANSCH,C & LEO,AJ (1985)
VP 4.39E-003	mm Hg	25	EST	NEELY,WB & BLAU,GE (1985)
DC	pKa			
HL 6.25E-008	atm m3/mol	25	EST	MEYLAN,WM & HOWARD,PH (1991)
OH 1.17E-011	cm3/molc sec	25	EST	MEYLAN,WM & HOWARD,PH (1993)

004090-00-0

CAS #: 004090-00-0

N-METHYL-3-CHLOROPHENYLCARBAMATE

Formula: $C_8H_8ClNO_2$

Mol Weight: 185.61

MP (deg C):   **FP (deg C):**

BP (deg C):

BP pressure (mm Hg):

Property/ Value	Units	Temp	Data Type	Reference
WS 9.54E+002	mg/L	25	EST	MEYLAN,WM ET AL. (1996)
logP 2.03			EXP	HANSCH,C & LEO,AJ (1985)
VP 8.31E-003	mm Hg	25	EST	NEELY,WB & BLAU,GE (1985)
DC	pKa			
HL 2.38E-008	atm m3/mol	25	EST	MEYLAN,WM & HOWARD,PH (1991)
OH 7.47E-012	cm3/molc sec	25	EST	MEYLAN,WM & HOWARD,PH (1993)

004091-39-8

CAS #: 004091-39-8

3-CHLORO-2-BUTANONE

Formula: C_4H_7ClO

Mol Weight: 106.55

MP (deg C):   **FP (deg C):**

BP (deg C): 115

BP pressure (mm Hg):

Property/ Value	Units	Temp	Data Type	Reference
WS 4.73E+004	mg/L	25	EST	MEYLAN,WM ET AL. (1996)
logP 0.44			EST	MEYLAN,WM & HOWARD,PH (1995)
VP 2.68E+001	mm Hg	25	EST	NEELY,WB & BLAU,GE (1985)
DC	pKa			
HL 2.32E-005	atm m3/mol	25	EST	MEYLAN,WM & HOWARD,PH (1991)
OH 1.08E-012	cm3/molc sec	25	EST	MEYLAN,WM & HOWARD,PH (1993)

CAS #: 004093-35-0 — BROMOPRIDE

Formula: $C_{14}H_{22}BrN_3O_2$

Mol Weight: 344.26

MP (deg C):
FP (deg C):
BP (deg C):
BP pressure (mm Hg):

Property/ Value	Units	Temp	Data Type	Reference
WS 2.57E+001	mg/L	25	EST	MEYLAN,WM ET AL. (1996)
logP 2.83			EXP	SANGSTER,J (1993)
VP 1.98E-009	mm Hg	25	EST	NEELY,WB & BLAU,GE (1985)
DC	pKa			
HL 4.70E-017	atm m3/mol	25	EST	MEYLAN,WM & HOWARD,PH (1991)
OH 3.05E-010	cm3/molc sec	25	EST	MEYLAN,WM & HOWARD,PH (1993)

CAS #: 004093-42-9 — 2MEO-4NH2-5NO2 BENZAMIDE,N(DIETAMET)

Formula: $C_{14}H_{22}N_4O_4$

Mol Weight: 310.36

MP (deg C):
FP (deg C):
BP (deg C):
BP pressure (mm Hg):

Property/ Value	Units	Temp	Data Type	Reference
WS 1.69E+002	mg/L	25	EST	MEYLAN,WM ET AL. (1996)
logP 2.11			EXP	HANSCH,C & LEO,AJ (1985)
VP 3.40E-010	mm Hg	25	EST	NEELY,WB & BLAU,GE (1985)
DC	pKa			
HL 1.02E-017	atm m3/mol	25	EST	MEYLAN,WM & HOWARD,PH (1991)
OH 2.41E-010	cm3/molc sec	25	EST	MEYLAN,WM & HOWARD,PH (1993)

CAS #: 004094-38-6 — BENZENAMINE, 4-[(4-METHYLPHENYL)SULFONYL]-

Formula: $C_{13}H_{13}NO_2S$

Mol Weight: 247.32

MP (deg C):
FP (deg C):
BP (deg C):
BP pressure (mm Hg):

Property/ Value	Units	Temp	Data Type	Reference
WS 2.18E+002	mg/L	25	EST	MEYLAN,WM ET AL. (1996)
logP 2.40			EXP	ALTOMARE,C ET AL. (1991)
VP 2.70E-007	mm Hg	25	EST	NEELY,WB & BLAU,GE (1985)
DC	pKa			
HL 9.72E-011	atm m3/mol	25	EST	MEYLAN,WM & HOWARD,PH (1991)
OH 2.42E-011	cm3/molc sec	25	EST	MEYLAN,WM & HOWARD,PH (1993)

CAS #: 004096-20-2 — N-PHENYLPIPERIDINE

Formula: $C_{11}H_{15}N$

Mol Weight: 161.25

MP (deg C): 4.7
FP (deg C):
BP (deg C): 258
BP pressure (mm Hg):

Property/ Value	Units	Temp	Data Type	Reference
WS 1.93E+002	mg/L	25	EST	MEYLAN,WM ET AL. (1996)
logP 2.98			EXP	HANSCH,C & LEO,AJ (1985)
VP 2.37E-002	mm Hg	25	EST	NEELY,WB & BLAU,GE (1985)
DC	pKa			
HL 8.84E-005	atm m3/mol	25	EST	MEYLAN,WM & HOWARD,PH (1991)
OH 1.72E-010	cm3/molc sec	25	EST	MEYLAN,WM & HOWARD,PH (1993)

CAS #: 004096-21-3 — 1-PHENYLPYRROLIDINE

Formula: $C_{10}H_{13}N$

Mol Weight: 147.22

MP (deg C):
FP (deg C):
BP (deg C): 119-120
BP pressure (mm Hg): 1.20E+001

Property/ Value	Units	Temp	Data Type	Reference
WS 7.11E+001	mg/L	25	EST	MEYLAN,WM ET AL. (1996)
logP 3.56			EXP	HANSCH,C ET AL. (1995)
VP 7.98E-002	mm Hg	25	EST	NEELY,WB & BLAU,GE (1985)
DC 9.70	pKa		EXP	PERRIN,DD (1965)
HL 6.66E-005	atm m3/mol	25	EST	MEYLAN,WM & HOWARD,PH (1991)
OH 1.66E-010	cm3/molc sec	25	EST	MEYLAN,WM & HOWARD,PH (1993)

CAS #: 004097-22-7 — ADENOSINE, 2',3'-DIDEOXY-

Formula: $C_{10}H_{13}N_5O_2$

Mol Weight: 235.25

MP (deg C): 184-186
FP (deg C):
BP (deg C):
BP pressure (mm Hg):

Property/ Value	Units	Temp	Data Type	Reference
WS 2.28E+003	mg/L	25	EST	MEYLAN,WM ET AL. (1996)
logP -0.25			EXP	SANGSTER,J (1993)
VP 3.50E-011	mm Hg	25	EST	NEELY,WB & BLAU,GE (1985)
DC	pKa			
HL 8.32E-020	atm m3/mol	25	EST	MEYLAN,WM & HOWARD,PH (1991)
OH 2.32E-010	cm3/molc sec	25	EST	MEYLAN,WM & HOWARD,PH (1993)

CAS #: 004097-36-3 — DNAP

Formula: $C_{11}H_{14}N_2O_5$

Mol Weight: 254.24

MP (deg C):
FP (deg C):
BP (deg C):
BP pressure (mm Hg):

Property/ Value	Units	Temp	Data Type	Reference
WS 6.23E+000	mg/L	25	EST	MEYLAN,WM ET AL. (1996)
logP 4.16			EST	MEYLAN,WM & HOWARD,PH (1995)
VP 2.80E-007	mm Hg	25	EST	NEELY,WB & BLAU,GE (1985)
DC	pKa			
HL 9.46E-008	atm m3/mol	25	EST	MEYLAN,WM & HOWARD,PH (1991)
OH 5.45E-012	cm3/molc sec	25	EST	MEYLAN,WM & HOWARD,PH (1993)

CAS #: 004097-49-8 — 4-(TERT-BUTYL)-2,6-DINITROPHENOL

Formula: $C_{10}H_{12}N_2O_5$

Mol Weight: 240.22

MP (deg C): 94-96
FP (deg C):
BP (deg C):
BP pressure (mm Hg):

Property/ Value	Units	Temp	Data Type	Reference
WS 2.11E+001	mg/L	25	EST	MEYLAN,WM ET AL. (1996)
logP 3.64			EST	MEYLAN,WM & HOWARD,PH (1995)
VP 8.72E-007	mm Hg	25	EST	NEELY,WB & BLAU,GE (1985)
DC	pKa			
HL 7.13E-008	atm m3/mol	25	EST	MEYLAN,WM & HOWARD,PH (1991)
OH 6.64E-013	cm3/molc sec	25	EST	MEYLAN,WM & HOWARD,PH (1993)

CAS #: 004097-58-9				4-CYCLOHEXYL-2,6-DINITROPHENOL

Formula: $C_{12}H_{14}N_2O_5$

Mol Weight: 266.26

MP (deg C): | FP (deg C):

BP (deg C):

BP pressure (mm Hg):

Property/Value	Units	Temp	Data Type	Reference
WS 2.53E+000	mg/L	25	EST	MEYLAN,WM ET AL. (1996)
logP 4.54			EST	MEYLAN,WM & HOWARD,PH (1995)
VP 1.75E-002	mm Hg	25	EXT	OHE,S (1976)
DC	pKa			
HL 5.54E-008	atm m3/mol	25	EST	MEYLAN,WM & HOWARD,PH (1991)
OH 1.02E-011	cm3/molc sec	25	EST	MEYLAN,WM & HOWARD,PH (1993)

CAS #: 004099-65-4				2,4-DINITRO-6-HEXYLPHENOL

Formula: $C_{12}H_{16}N_2O_5$

Mol Weight: 268.27

MP (deg C): | FP (deg C):

BP (deg C):

BP pressure (mm Hg):

Property/Value	Units	Temp	Data Type	Reference
WS 1.71E+000	mg/L	25	EST	MEYLAN,WM ET AL. (1996)
logP 4.73			EST	MEYLAN,WM & HOWARD,PH (1995)
VP 6.35E-008	mm Hg	25	EST	NEELY,WB & BLAU,GE (1985)
DC	pKa			
HL 1.26E-007	atm m3/mol	25	EST	MEYLAN,WM & HOWARD,PH (1991)
OH 6.87E-012	cm3/molc sec	25	EST	MEYLAN,WM & HOWARD,PH (1993)

CAS #: 004100-80-5				PYROTATARIC ANHYDRIDE

Formula: $C_5H_6O_3$

Mol Weight: 114.10

MP (deg C): 34 | FP (deg C):

BP (deg C): 239

BP pressure (mm Hg):

Property/Value	Units	Temp	Data Type	Reference
WS 9.42E+003	mg/L	25	EST	MEYLAN,WM ET AL. (1996)
logP 1.23			EST	MEYLAN,WM & HOWARD,PH (1995)
VP 3.09E-002	mm Hg	25	EXT	PERRY,RH & GREEN,D (1984)
DC	pKa			
HL 2.09E-005	atm m3/mol	25	EST	MEYLAN,WM & HOWARD,PH (1991)
OH 2.26E-012	cm3/molc sec	25	EST	MEYLAN,WM & HOWARD,PH (1993)

CAS #: 004101-68-2				1,2-DIBROMODECANE

Formula: $C_{10}H_{20}Br_2$

Mol Weight: 300.09

MP (deg C): 28 | FP (deg C):

BP (deg C): 161

BP pressure (mm Hg): 9.00E+000

Property/Value	Units	Temp	Data Type	Reference
WS 7.28E-002	mg/L	25	EST	MEYLAN,WM ET AL. (1996)
logP 6.12			EXP	CHEM INSPECT TEST INST (1992)
VP 3.73E-003	mm Hg	25	EXT	PERRY,RH & GREEN,D (1984)
DC	pKa			
HL 1.25E-002	atm m3/mol	25	EST	MEYLAN,WM & HOWARD,PH (1991)
OH 1.02E-011	cm3/molc sec	25	EST	MEYLAN,WM & HOWARD,PH (1993)

CAS #: 004104-75-0				UREA,1-METHYL-1-PHENYL-2-THIO

Formula: $C_8H_{10}N_2S$

Mol Weight: 166.25

MP (deg C): | FP (deg C):

BP (deg C):

BP pressure (mm Hg):

Property/Value	Units	Temp	Data Type	Reference
WS 1.23E+005	mg/L	25	EST	MEYLAN,WM ET AL. (1996)
logP 0.85			EXP	HANSCH,C & LEO,AJ (1985)
VP 1.72E-003	mm Hg	25	EST	NEELY,WB & BLAU,GE (1985)
DC	pKa			
HL 1.79E-006	atm m3/mol	25	EST	MEYLAN,WM & HOWARD,PH (1991)
OH 1.69E-010	cm3/molc sec	25	EST	MEYLAN,WM & HOWARD,PH (1993)

CAS #: 004110-50-3				ETHYLPROPYLSULFIDE

Formula: $C_5H_{12}S$

Mol Weight: 104.22

MP (deg C): -117 | FP (deg C):

BP (deg C): 118.6

BP pressure (mm Hg):

Property/Value	Units	Temp	Data Type	Reference
WS 1.03E+003	mg/L	25	EST	MEYLAN,WM ET AL. (1996)
logP 2.39			EST	MEYLAN,WM & HOWARD,PH (1995)
VP 1.88E+001	mm Hg	25	EXP	ZWOLINSKI,BJ & WILHOIT,RC (1971)
DC	pKa			
HL 1.84E-003	atm m3/mol	25	EST	MEYLAN,WM & HOWARD,PH (1991)
OH 1.94E-011	cm3/molc sec	25	EST	MEYLAN,WM & HOWARD,PH (1993)

CAS #: 004119-41-9				G-PHENYL PROPYLIODINE

Formula: $C_9H_{11}I$

Mol Weight: 246.09

MP (deg C): | FP (deg C):

BP (deg C):

BP pressure (mm Hg):

Property/Value	Units	Temp	Data Type	Reference
WS 1.16E+001	mg/L	25	EST	MEYLAN,WM ET AL. (1996)
logP 3.90			EXP	HANSCH,C & LEO,AJ (1985)
VP 1.60E-002	mm Hg	25	EST	NEELY,WB & BLAU,GE (1985)
DC	pKa			
HL 7.82E-004	atm m3/mol	25	EST	MEYLAN,WM & HOWARD,PH (1991)
OH 7.13E-012	cm3/molc sec	25	EST	MEYLAN,WM & HOWARD,PH (1993)

CAS #: 004122-57-0				1(3H)-ISOBENZOFURANONE-3-METHOXYL

Formula: $C_9H_8O_3$

Mol Weight: 164.16

MP (deg C): | FP (deg C):

BP (deg C):

BP pressure (mm Hg):

Property/Value	Units	Temp	Data Type	Reference
WS 9.17E+003	mg/L	25	EST	MEYLAN,WM ET AL. (1996)
logP 1.00			EXP	HANSCH,C & LEO,AJ (1985)
VP 1.14E-003	mm Hg	25	EST	NEELY,WB & BLAU,GE (1985)
DC	pKa			
HL 1.49E-007	atm m3/mol	25	EST	MEYLAN,WM & HOWARD,PH (1991)
OH 2.02E-011	cm3/molc sec	25	EST	MEYLAN,WM & HOWARD,PH (1993)

CAS #: 004124-31-6				TRICHLOROACETIC ANHYDRIDE

Formula: $C_4Cl_6O_3$

Mol Weight: 308.76

MP (deg C): | FP (deg C):

BP (deg C):

BP pressure (mm Hg):

Property/Value	Units	Temp	Data Type	Reference
WS 1.64E+002	mg/L	25	EST	MEYLAN,WM ET AL. (1996)
logP 2.13			EST	MEYLAN,WM & HOWARD,PH (1995)
VP 1.24E-001	mm Hg	25	EXT	PERRY,RH & GREEN,D (1984)
DC	pKa			
HL 6.81E-008	atm m3/mol	25	EST	MEYLAN,WM & HOWARD,PH (1991)
OH 0.00E+000	cm3/molc sec	25	EST	MEYLAN,WM & HOWARD,PH (1993)

CAS #: 004128-37-4				N,N'-BIS(1-METHYLETHYL)UREA

Formula: $C_7H_{16}N_2O$

Mol Weight: 144.22

MP (deg C): | FP (deg C):

BP (deg C):

BP pressure (mm Hg):

Property/Value	Units	Temp	Data Type	Reference
WS 7.68E+003	mg/L	25	EST	MEYLAN,WM ET AL. (1996)
logP 1.19			EST	MEYLAN,WM & HOWARD,PH (1995)
VP 2.24E-002	mm Hg	25	EST	NEELY,WB & BLAU,GE (1985)
DC	pKa			
HL 5.46E-009	atm m3/mol	25	EST	MEYLAN,WM & HOWARD,PH (1991)
OH 1.86E-011	cm3/molc sec	25	EST	MEYLAN,WM & HOWARD,PH (1993)

CAS #: 004128-71-6				N-(4-(PHENYLAZO)PHENYL)ACETAMIDE

Formula: $C_{14}H_{13}N_3O$

Mol Weight: 239.28

MP (deg C): | FP (deg C):

BP (deg C):

BP pressure (mm Hg):

Property/Value	Units	Temp	Data Type	Reference
WS 1.79E+001	mg/L	25	EST	MEYLAN,WM ET AL. (1996)
logP 3.22			EST	MEYLAN,WM & HOWARD,PH (1995)
VP 1.23E-007	mm Hg	25	EST	NEELY,WB & BLAU,GE (1985)
DC	pKa			
HL 1.68E-011	atm m3/mol	25	EST	MEYLAN,WM & HOWARD,PH (1991)
OH 5.80E-012	cm3/molc sec	25	EST	MEYLAN,WM & HOWARD,PH (1993)

CAS #: 004130-92-1				2,6-DI(TERT-BUTYL)-4-ETHYLPHENOL

Formula: $C_{16}H_{26}O$

Mol Weight: 234.39

MP (deg C): | FP (deg C):

BP (deg C):

BP pressure (mm Hg):

Property/Value	Units	Temp	Data Type	Reference
WS 2.12E+000	mg/L	25	EST	MEYLAN,WM ET. AL. (1996)
logP 5.52			EST	MEYLAN,WM & HOWARD,PH (1995)
VP 7.81E-005	mm Hg	25	EST	NEELY,WB & BLAU,GE (1985)
DC	pKa			
HL 5.48E-006	atm m3/mol	25	EST	MEYLAN,WM & HOWARD,PH (1991)
OH 1.84E-011	cm3/molc sec	25	EST	MEYLAN,WM & HOWARD,PH (1993)

CAS #: 004138-38-9				N-(1,1-DIMETHYLETHYL)-4-NITROBENZENAMINE

Formula: $C_{10}H_{14}N_2O_2$

Mol Weight: 194.24

MP (deg C): | FP (deg C):

BP (deg C):

BP pressure (mm Hg):

Property/Value	Units	Temp	Data Type	Reference
WS 6.07E+001	mg/L	25	EST	MEYLAN,WM ET AL. (1996)
logP 3.38			EST	MEYLAN,WM & HOWARD,PH (1995)
VP 8.33E-004	mm Hg	25	EST	NEELY,WB & BLAU,GE (1985)
DC	pKa			
HL 3.86E-008	atm m3/mol	25	EST	MEYLAN,WM & HOWARD,PH (1991)
OH 5.83E-012	cm3/molc sec	25	EST	MEYLAN,WM & HOWARD,PH (1993)

CAS #: 004143-63-9				2H-1-BENZOPYRAN-2-ONE, 2-(4-HYDROXYPHENYL)-

Formula: $C_{15}H_{10}O_3$

Mol Weight: 238.25

MP (deg C): | FP (deg C):

BP (deg C):

BP pressure (mm Hg):

Property/Value	Units	Temp	Data Type	Reference
WS 1.93E+002	mg/L	25	EST	MEYLAN,WM ET AL. (1996)
logP 3.20			EXP	HANSCH,C ET AL. (1995)
VP 1.25E-007	mm Hg	25	EST	NEELY,WB & BLAU,GE (1985)
DC	pKa			
HL 3.76E-012	atm m3/mol	25	EST	MEYLAN,WM & HOWARD,PH (1991)
OH 8.15E-011	cm3/molc sec	25	EST	MEYLAN,WM & HOWARD,PH (1993)

CAS #: 004144-64-3				1H-BENZOTRIAZOLE-1-ACETIC ACID

Formula: $C_8H_7N_3O_2$

Mol Weight: 177.16

MP (deg C): | FP (deg C):

BP (deg C):

BP pressure (mm Hg):

Property/Value	Units	Temp	Data Type	Reference
WS 1.00E+006	mg/L	25	EST	MEYLAN,WM ET AL. (1996)
logP -1.88			EXP	SANGSTER,J (1994)
VP 7.82E-006	mm Hg	25	EST	NEELY,WB & BLAU,GE (1985)
DC	pKa			
HL 2.33E-011	atm m3/mol	25	EST	MEYLAN,WM & HOWARD,PH (1991)
OH 2.22E-012	cm3/molc sec	25	EST	MEYLAN,WM & HOWARD,PH (1993)

CAS #: 004144-68-7				2-CARBOXYMETHYLBENZOTRIAZOLE

Formula: $C_8H_9N_3O_2$

Mol Weight: 179.18

MP (deg C): | FP (deg C):

BP (deg C):

BP pressure (mm Hg):

Property/Value	Units	Temp	Data Type	Reference
WS 1.00E+006	mg/L	25	EST	MEYLAN,WM ET AL. (1996)
logP -1.64			EXP	HANSCH,C & LEO,AJ (1985)
VP 7.82E-006	mm Hg	25	EST	NEELY,WB & BLAU,GE (1985)
DC	pKa			
HL 2.33E-011	atm m3/mol	25	EST	MEYLAN,WM & HOWARD,PH (1991)
OH 2.22E-012	cm3/molc sec	25	EST	MEYLAN,WM & HOWARD,PH (1993)

CAS #: 004144-70-1	2-(3-CARBOXYPROPYL)BENZOTRIAZOLE

Formula: $C_{10}H_{13}N_3O_2$

Mol Weight: 207.23

MP (deg C): FP (deg C):

BP (deg C):

BP pressure (mm Hg):

Property/Value	Units	Temp	Data Type	Reference
WS 3.14E+004	mg/L	25	EST	MEYLAN,WM ET AL. (1996)
logP 0.60			EXP	HANSCH,C & LEO,AJ (1985)
VP 1.48E-006	mm Hg	25	EST	NEELY,WB & BLAU,GE (1985)
DC	pKa			
HL 4.10E-011	atm m3/mol	25	EST	MEYLAN,WM & HOWARD,PH (1991)
OH 4.94E-012	cm3/molc sec	25	EST	MEYLAN,WM & HOWARD,PH (1993)

CAS #: 004147-51-7	DIPROPETRYNE

Formula: $C_{11}H_{21}N_5S$

Mol Weight: 255.39

MP (deg C): 104-106 FP (deg C):

BP (deg C):

BP pressure (mm Hg):

Property/Value	Units	Temp	Data Type	Reference
WS 1.60E+001	mg/L	20	EXP	YALKOWSKY,SH & DANNENFELSER,RM (1992)
logP 4.22			EST	MEYLAN,WM & HOWARD,PH (1995)
VP 1.60E-006	mm Hg	25	EXP	WEBER,JB (1994)
DC	pKa			
HL 3.36E-008	atm m3/mol	25	EST	VP/WSOL
OH 4.45E-011	cm3/molc sec	25	EST	MEYLAN,WM & HOWARD,PH (1993)

CAS #: 004151-47-7	3-T-BUTYL-4-HYDROXYACETANILIDE

Formula: $C_{12}H_{17}NO_2$

Mol Weight: 207.27

MP (deg C): FP (deg C):

BP (deg C):

BP pressure (mm Hg):

Property/Value	Units	Temp	Data Type	Reference
WS 3.87E+002	mg/L	25	EST	MEYLAN,WM ET AL. (1996)
logP 2.36			EXP	HANSCH,C & LEO,AJ (1985)
VP 4.28E-007	mm Hg	25	EST	NEELY,WB & BLAU,GE (1985)
DC	pKa			
HL 1.66E-012	atm m3/mol	25	EST	MEYLAN,WM & HOWARD,PH (1991)
OH 2.17E-011	cm3/molc sec	25	EST	MEYLAN,WM & HOWARD,PH (1993)

CAS #: 004153-42-8	ETHANAMINIUM, 1,1-DIMETHYL-N,N,N-TRIMETHYL-, IOD

Formula: $C_7H_{18}IN$

Mol Weight: 243.13

MP (deg C): FP (deg C):

BP (deg C):

BP pressure (mm Hg):

Property/Value	Units	Temp	Data Type	Reference
WS 1.00E+006	mg/L	25	EST	MEYLAN,WM ET AL. (1996)
logP -2.08			EXP	NEEF,C & MEIJER,DKF (1984)
VP 1.07E-006	mm Hg	25	EST	NEELY,WB & BLAU,GE (1985)
DC	pKa			
HL 6.16E-012	atm m3/mol	25	EST	MEYLAN,WM & HOWARD,PH (1991)
OH 4.30E-012	cm3/molc sec	25	EST	MEYLAN,WM & HOWARD,PH (1993)

CAS #: 004162-45-2	ETHOXYLATED TETRABROMOBISPHENOL A

Formula: $C_{19}H_{20}Br_4O_4$

Mol Weight: 632.00

MP (deg C): 107 FP (deg C):

BP (deg C):

BP pressure (mm Hg):

Property/Value	Units	Temp	Data Type	Reference
WS 1.59E-004	mg/L	25	EST	MEYLAN,WM ET AL. (1996)
logP 6.78			EST	MEYLAN,WM & HOWARD,PH (1995)
VP 9.68E-016	mm Hg	25	EST	NEELY,WB & BLAU,GE (1985)
DC	pKa			
HL 1.76E-013	atm m3/mol	25	EST	MEYLAN,WM & HOWARD,PH (1991)
OH 2.56E-011	cm3/molc sec	25	EST	MEYLAN,WM & HOWARD,PH (1993)

CAS #: 004164-28-7	DIMETHYLNITRAMINE

Formula: $C_2H_6N_2O_2$

Mol Weight: 90.08

MP (deg C): 58 FP (deg C):

BP (deg C): 187

BP pressure (mm Hg):

Property/Value	Units	Temp	Data Type	Reference
WS 3.45E+005	mg/L	25	EST	MEYLAN,WM ET AL. (1996)
logP -0.52			EST	MEYLAN,WM & HOWARD,PH (1995)
VP 3.61E-001	mm Hg	25	EXP	BRADLEY,RS ET AL. (1952)
DC	pKa			
HL 3.20E-006	atm m3/mol	25	EST	MEYLAN,WM & HOWARD,PH (1991)
OH 3.84E-012	cm3/molc sec	25	EXP	ATKINSON,R (1989)

CAS #: 004169-04-4	PROPYLENE GLYCOL MONOPHENYL ETHER

Formula: $C_9H_{12}O_2$

Mol Weight: 152.19

MP (deg C): FP (deg C):

BP (deg C): 244

BP pressure (mm Hg):

Property/Value	Units	Temp	Data Type	Reference
WS 1.21E+004	mg/L	25	EST	MEYLAN,WM ET AL. (1996)
logP 1.52			EST	MEYLAN,WM & HOWARD,PH (1995)
VP 3.08E-003	mm Hg	25	EST	NEELY,WB & BLAU,GE (1985)
DC	pKa			
HL 2.05E-008	atm m3/mol	25	EST	MEYLAN,WM & HOWARD,PH (1991)
OH 4.04E-011	cm3/molc sec	25	EST	MEYLAN,WM & HOWARD,PH (1993)

CAS #: 004170-30-3	CROTONALDEHYDE

Formula: C_4H_6O

Mol Weight: 70.09

MP (deg C): -76.5 FP (deg C):

BP (deg C): 104

BP pressure (mm Hg):

Property/Value	Units	Temp	Data Type	Reference
WS 1.81E+005	mg/L	20	EXP	BAXTER,WFJR (1979)
logP 0.60			EST	MEYLAN,WM & HOWARD,PH (1995)
VP 3.00E+001	mm Hg	20	EXP	WEBER,RC ET AL. (1981)
DC	pKa			
HL 9.68E-006	atm m3/mol	20	EST	VP/WSOL
OH 3.62E-011	cm3/molc sec	25	EST	MEYLAN,WM & HOWARD,PH (1993)

CAS #: 004180-23-8	ANETHOLE (TRANS)

Formula: $C_{10}H_{12}O$
Mol Weight: 148.21
MP (deg C): 21.35 FP (deg C):
BP (deg C): 234
BP pressure (mm Hg):

Property/Value	Units	Temp	Data Type	Reference
WS 9.87E+001	mg/L	25	EST	MEYLAN,WM ET AL. (1996)
logP 3.39			EST	MEYLAN,WM & HOWARD,PH (1995)
VP 7.05E-002	mm Hg	25	EXT	PERRY,RH & GREEN,D (1984)
DC	pKa			
HL 2.56E-004	atm m3/mol	25	EST	MEYLAN,WM & HOWARD,PH (1991)
OH 7.76E-011	cm3/molc sec	25	EST	MEYLAN,WM & HOWARD,PH (1993)

CAS #: 004181-95-7	TETRACONTANE

Formula: $C_{40}H_{82}$
Mol Weight: 563.10
MP (deg C): 81-83 FP (deg C):
BP (deg C):
BP pressure (mm Hg):

Property/Value	Units	Temp	Data Type	Reference
WS 6.96E-016	mg/L	25	EST	MEYLAN,WM ET AL. (1996)
logP 19.99			EST	MEYLAN,WM & HOWARD,PH (1995)
VP 2.49E-031	mm Hg	25	EXT	ZWOLINSKI,BJ & WILHOIT,RC (1971)
DC	pKa			
HL 2.61E+004	atm m3/mol	25	EST	MEYLAN,WM & HOWARD,PH (1991)
OH 5.35E-011	cm3/molc sec	25	EST	MEYLAN,WM & HOWARD,PH (1993)

CAS #: 004201-22-3	2-(2-CL-4-ME-PHIMINO)IMIDAZOLIDINE

Formula: $C_{10}H_{12}ClN_3$
Mol Weight: 209.68
MP (deg C): 148-150 FP (deg C):
BP (deg C):
BP pressure (mm Hg):

Property/Value	Units	Temp	Data Type	Reference
WS 1.15E+004	mg/L	25	EST	MEYLAN,WM ET AL. (1996)
logP 1.80			EXP	HANSCH,C & LEO,AJ (1985)
VP 3.31E-005	mm Hg	25	EST	NEELY,WB & BLAU,GE (1985)
DC	pKa			
HL 2.24E-011	atm m3/mol	25	EST	MEYLAN,WM & HOWARD,PH (1991)
OH 1.47E-010	cm3/molc sec	25	EST	MEYLAN,WM & HOWARD,PH (1993)

CAS #: 004201-24-5	2-(2-CL-6-ME-PHIMINO)IMIDAZOLIDINE

Formula: $C_{10}H_{12}ClN_3$
Mol Weight: 209.68
MP (deg C): FP (deg C):
BP (deg C):
BP pressure (mm Hg):

Property/Value	Units	Temp	Data Type	Reference
WS 1.38E+004	mg/L	25	EST	MEYLAN,WM ET AL. (1996)
logP 1.71			EXP	HANSCH,C & LEO,AJ (1985)
VP 3.31E-005	mm Hg	25	EST	NEELY,WB & BLAU,GE (1985)
DC	pKa			
HL 2.24E-011	atm m3/mol	25	EST	MEYLAN,WM & HOWARD,PH (1991)
OH 1.47E-010	cm3/molc sec	25	EST	MEYLAN,WM & HOWARD,PH (1993)

CAS #: 004201-26-7	2-(4-CL-2-ME-PHIMINO)IMIDAZOLIDIN

Formula: $C_{10}H_{12}ClN_3$
Mol Weight: 209.68
MP (deg C): FP (deg C):
BP (deg C):
BP pressure (mm Hg):

Property/Value	Units	Temp	Data Type	Reference
WS 1.13E+003	mg/L	25	EST	MEYLAN,WM ET AL. (1996)
logP 1.80			EXP	HANSCH,C & LEO,AJ (1985)
VP 4.92E-007	mm Hg	25	EST	NEELY,WB & BLAU,GE (1985)
DC	pKa			
HL 2.24E-011	atm m3/mol	25	EST	MEYLAN,WM & HOWARD,PH (1991)
OH 8.79E-011	cm3/molc sec	25	EST	MEYLAN,WM & HOWARD,PH (1993)

CAS #: 004201-33-6	2(2,6-DICL-4MEPH IMINO)IMIDAZOLIDINE

Formula: $C_{10}H_{11}Cl_2N_3$
Mol Weight: 244.13
MP (deg C): FP (deg C):
BP (deg C):
BP pressure (mm Hg):

Property/Value	Units	Temp	Data Type	Reference
WS 5.84E+002	mg/L	25	EST	MEYLAN,WM ET AL. (1996)
logP 1.92			EXP	HANSCH,C & LEO,AJ (1985)
VP 1.10E-007	mm Hg	25	EST	NEELY,WB & BLAU,GE (1985)
DC	pKa			
HL 1.66E-011	atm m3/mol	25	EST	MEYLAN,WM & HOWARD,PH (1991)
OH 7.67E-011	cm3/molc sec	25	EST	MEYLAN,WM & HOWARD,PH (1993)

CAS #: 004201-34-7	2(2,4-DICL-6MEPH IMINO)IMIDAZOLIDINE

Formula: $C_{10}H_{11}Cl_2N_3$
Mol Weight: 244.13
MP (deg C): FP (deg C):
BP (deg C):
BP pressure (mm Hg):

Property/Value	Units	Temp	Data Type	Reference
WS 2.37E+002	mg/L	25	EST	MEYLAN,WM ET AL. (1996)
logP 2.38			EXP	HANSCH,C & LEO,AJ (1985)
VP 1.10E-007	mm Hg	25	EST	NEELY,WB & BLAU,GE (1985)
DC	pKa			
HL 1.66E-011	atm m3/mol	25	EST	MEYLAN,WM & HOWARD,PH (1991)
OH 7.67E-011	cm3/molc sec	25	EST	MEYLAN,WM & HOWARD,PH (1993)

CAS #: 004201-36-9	2(24-DIME-6-CL-PH IMINO)IMIDAZOLINE

Formula: $C_{11}H_{14}ClN_3$
Mol Weight: 223.71
MP (deg C): FP (deg C):
BP (deg C):
BP pressure (mm Hg):

Property/Value	Units	Temp	Data Type	Reference
WS 5.39E+003	mg/L	25	EST	MEYLAN,WM ET AL. (1996)
logP 2.10			EXP	HANSCH,C & LEO,AJ (1985)
VP 1.64E-005	mm Hg	25	EST	NEELY,WB & BLAU,GE (1985)
DC	pKa			
HL 2.47E-011	atm m3/mol	25	EST	MEYLAN,WM & HOWARD,PH (1991)
OH 1.54E-010	cm3/molc sec	25	EST	MEYLAN,WM & HOWARD,PH (1993)

CAS #: 004201-38-1

2-IMIDAZOLIDINIMINE, N-(4-CHLORO-2,6-DIMETHYLPHE

Formula: $C_{11}H_{14}ClN_3$

Mol Weight: 223.71

MP (deg C): FP (deg C):

BP (deg C):

BP pressure (mm Hg):

Property/Value	Units	Temp	Data Type	Reference
WS 2.46E+003	mg/L	25	EST	MEYLAN,WM ET AL. (1996)
logP 2.50			EXP	SANGSTER,J (1993)
VP 1.64E-005	mm Hg	25	EST	NEELY,WB & BLAU,GE (1985)
DC	pKa			
HL 2.47E-011	atm m3/mol	25	EST	MEYLAN,WM & HOWARD,PH (1991)
OH 1.54E-010	cm3/molc sec	25	EST	MEYLAN,WM & HOWARD,PH (1993)

CAS #: 004201-40-5

2-(246-TRIME PH IMINO)IMIDAZOLIDINE

Formula: $C_{12}H_{17}N_3$

Mol Weight: 203.29

MP (deg C): FP (deg C):

BP (deg C):

BP pressure (mm Hg):

Property/Value	Units	Temp	Data Type	Reference
WS 4.48E+003	mg/L	25	EST	MEYLAN,WM ET AL. (1996)
logP 2.32			EXP	HANSCH,C & LEO,AJ (1985)
VP 2.91E-005	mm Hg	25	EST	NEELY,WB & BLAU,GE (1985)
DC	pKa			
HL 3.69E-011	atm m3/mol	25	EST	MEYLAN,WM & HOWARD,PH (1991)
OH 1.82E-010	cm3/molc sec	25	EST	MEYLAN,WM & HOWARD,PH (1993)

CAS #: 004202-14-6

PHOSPHONIC ACID, (2-OXOPROPYL)-, DIMETHYL ESTER

Formula: $C_5H_{11}O_4P$

Mol Weight: 166.11

MP (deg C): FP (deg C):

BP (deg C): 76-79

BP pressure (mm Hg): 3.00E+000

Property/Value	Units	Temp	Data Type	Reference
WS 1.18E+004	mg/L	25	EST	MEYLAN,WM ET AL. (1996)
logP 0.86			EXP	HANSCH,C ET AL. (1995)
VP 7.15E-002	mm Hg	25	EST	NEELY,WB & BLAU,GE (1985)
DC	pKa			
HL 8.40E-010	atm m3/mol	25	EST	MEYLAN,WM & HOWARD,PH (1991)
OH 6.38E-012	cm3/molc sec	25	EST	MEYLAN,WM & HOWARD,PH (1993)

CAS #: 004205-90-7

2-(2,6-DICL PH IMINO)IMIDAZOLIDINE

Formula: $C_9H_9Cl_2N_3$

Mol Weight: 230.10

MP (deg C): 130 FP (deg C):

BP (deg C):

BP pressure (mm Hg):

Property/Value	Units	Temp	Data Type	Reference
WS 1.36E+004	mg/L	25	EST	MEYLAN,WM ET AL. (1996)
logP 1.59			EXP	HANSCH,C & LEO,AJ (1985)
VP 2.28E-005	mm Hg	25	EST	NEELY,WB & BLAU,GE (1985)
DC	pKa			
HL 1.51E-011	atm m3/mol	25	EST	MEYLAN,WM & HOWARD,PH (1991)
OH 1.44E-010	cm3/molc sec	25	EST	MEYLAN,WM & HOWARD,PH (1993)

CAS #: 004205-93-0

2-(2,6-DIBR PH IMINO)IMIDAZOLIDINE

Formula: $C_9H_9Br_2N_3$

Mol Weight: 319.01

MP (deg C): FP (deg C):

BP (deg C):

BP pressure (mm Hg):

Property/Value	Units	Temp	Data Type	Reference
WS 2.05E+003	mg/L	25	EST	MEYLAN,WM ET AL. (1996)
logP 1.96			EXP	HANSCH,C & LEO,AJ (1985)
VP 3.39E-006	mm Hg	25	EST	NEELY,WB & BLAU,GE (1985)
DC	pKa			
HL 4.35E-012	atm m3/mol	25	EST	MEYLAN,WM & HOWARD,PH (1991)
OH 1.44E-010	cm3/molc sec	25	EST	MEYLAN,WM & HOWARD,PH (1993)

CAS #: 004206-94-4

METHYLPHOSPHINIC ACID

Formula: CH_5O_2P

Mol Weight: 80.02

MP (deg C): -40.01 FP (deg C):

BP (deg C): 162.6

BP pressure (mm Hg):

Property/Value	Units	Temp	Data Type	Reference
WS 1.00E+006	mg/L	25	EST	MEYLAN,WM ET AL. (1996)
logP -1.17			EST	MEYLAN,WM & HOWARD,PH (1995)
VP 6.62E-001	mm Hg	25	EST	MPBPVP
DC -4.20	pKa		EXP	COOK,AG & MASON,GW (1973)
HL 5.14E-008	atm m3/mol	25	EST	MEYLAN,WM & HOWARD,PH (1991)
OH 1.80E-013	cm3/molc sec	25	EST	MEYLAN,WM & HOWARD,PH (1993)

CAS #: 004214-48-6

3,5-DICL SALICYLANILIDE

Formula: $C_{13}H_9Cl_2NO_2$

Mol Weight: 282.13

MP (deg C): FP (deg C):

BP (deg C):

BP pressure (mm Hg):

Property/Value	Units	Temp	Data Type	Reference
WS 2.23E+001	mg/L	25	EST	MEYLAN,WM ET AL. (1996)
logP 4.01			EXP	TERADA,H ET AL. (1988)
VP 1.47E-009	mm Hg	25	EST	NEELY,WB & BLAU,GE (1985)
DC	pKa			
HL 8.79E-011	atm m3/mol	25	EST	MEYLAN,WM & HOWARD,PH (1991)
OH 1.28E-011	cm3/molc sec	25	EST	MEYLAN,WM & HOWARD,PH (1993)

CAS #: 004214-76-0

2-PYRIDINAMINE, 5-NITRO-

Formula: $C_5H_5N_3O_2$

Mol Weight: 139.11

MP (deg C): 186-188 FP (deg C):

BP (deg C):

BP pressure (mm Hg):

Property/Value	Units	Temp	Data Type	Reference
WS 2.35E+004	mg/L	25	EST	MEYLAN,WM ET AL. (1996)
logP 0.65			EXP	SCHULTZ,TW & MOULTON,BA (1985)
VP 1.32E-003	mm Hg	25	EST	NEELY,WB & BLAU,GE (1985)
DC 2.78	pKa	20	EXP	PERRIN,DD (1972)
HL 9.83E-012	atm m3/mol	25	EST	MEYLAN,WM & HOWARD,PH (1991)
OH 2.55E-012	cm3/molc sec	25	EST	MEYLAN,WM & HOWARD,PH (1993)

CAS #: 004224-00-4				2,4-IMIDAZOLIDINEDIONE, 5,5-DIPHENYL-3-METHYL-
Formula: $C_{16}H_{14}N_2O_2$				
Mol Weight: 266.30				
MP (deg C):		FP (deg C):		
BP (deg C):				
BP pressure (mm Hg):				

Property/Value	Units	Temp	Data Type	Reference
WS 2.17E+002	mg/L	25	EST	MEYLAN,WM ET AL. (1996)
logP 2.28			EXP	SANGSTER,J (1994)
VP 5.34E-010	mm Hg	25	EST	NEELY,WB & BLAU,GE (1985)
DC	pKa			
HL 2.25E-011	atm m3/mol	25	EST	MEYLAN,WM & HOWARD,PH (1991)
OH 1.11E-011	cm3/molc sec	25	EST	MEYLAN,WM & HOWARD,PH (1993)

CAS #: 004228-88-0				2-CF3-4,6-DICHLOROBENZIMIDAZOLE
Formula: $C_8H_3Cl_2F_3N_2$				
Mol Weight: 255.03				
MP (deg C):		FP (deg C):		
BP (deg C):				
BP pressure (mm Hg):				

Property/Value	Units	Temp	Data Type	Reference
WS 2.32E+001	mg/L	25	EST	MEYLAN,WM ET AL. (1996)
logP 3.49			EXP	HANSCH,C & LEO,AJ (1985)
VP 2.90E-006	mm Hg	25	EST	NEELY,WB & BLAU,GE (1985)
DC	pKa			
HL 1.75E-006	atm m3/mol	25	EST	MEYLAN,WM & HOWARD,PH (1991)
OH 1.29E-012	cm3/molc sec	25	EST	MEYLAN,WM & HOWARD,PH (1993)

CAS #: 004228-89-1				2-CF3-4,7-DICHLOROBENZIMIDAZOLE
Formula: $C_8H_3Cl_2F_3N_2$				
Mol Weight: 255.03				
MP (deg C):		FP (deg C):		
BP (deg C):				
BP pressure (mm Hg):				

Property/Value	Units	Temp	Data Type	Reference
WS 7.85E+001	mg/L	25	EST	MEYLAN,WM ET AL. (1996)
logP 2.87			EXP	HANSCH,C & LEO,AJ (1985)
VP 2.90E-006	mm Hg	25	EST	NEELY,WB & BLAU,GE (1985)
DC	pKa			
HL 1.75E-006	atm m3/mol	25	EST	MEYLAN,WM & HOWARD,PH (1991)
OH 1.29E-012	cm3/molc sec	25	EST	MEYLAN,WM & HOWARD,PH (1993)

CAS #: 004233-61-8				1H-BENZOTRIAZOLE-1-ACETIC ACID, à,à-DIMETHYL-
Formula: $C_{10}H_{11}N_3O_2$				
Mol Weight: 205.22				
MP (deg C):		FP (deg C):		
BP (deg C):				
BP pressure (mm Hg):				

Property/Value	Units	Temp	Data Type	Reference
WS 1.51E+005	mg/L	25	EST	MEYLAN,WM ET AL. (1996)
logP -0.20			EXP	SPARATORE,F ET AL. (1988)
VP 3.22E-006	mm Hg	25	EST	NEELY,WB & BLAU,GE (1985)
DC	pKa			
HL 4.10E-011	atm m3/mol	25	EST	MEYLAN,WM & HOWARD,PH (1991)
OH 1.85E-012	cm3/molc sec	25	EST	MEYLAN,WM & HOWARD,PH (1993)

CAS #: 004238-50-0				2-METHOXYETHYL-TRIME AMMONIUM IODIDE
Formula: $C_6H_{16}INO$				
Mol Weight: 245.10				
MP (deg C):		FP (deg C):		
BP (deg C):				
BP pressure (mm Hg):				

Property/Value	Units	Temp	Data Type	Reference
WS 1.00E+006	mg/L	25	EST	MEYLAN,WM ET AL. (1996)
logP -2.62			EXP	HANSCH,C & LEO,AJ (1985)
VP 5.40E-007	mm Hg	25	EST	NEELY,WB & BLAU,GE (1985)
DC	pKa			
HL 4.09E-014	atm m3/mol	25	EST	MEYLAN,WM & HOWARD,PH (1991)
OH 4.20E-011	cm3/molc sec	25	EST	MEYLAN,WM & HOWARD,PH (1993)

CAS #: 004238-66-8				6H-PYRIDO[4,3-B]CARBAZOLE, 5-METHYL-
Formula: $C_{16}H_{12}N_2$				
Mol Weight: 232.29				
MP (deg C):		FP (deg C):		
BP (deg C):				
BP pressure (mm Hg):				

Property/Value	Units	Temp	Data Type	Reference
WS 7.83E-001	mg/L	25	EST	MEYLAN,WM ET AL. (1996)
logP 4.06			EXP	HANSCH,C ET AL. (1995)
VP 2.05E-007	mm Hg	25	EST	NEELY,WB & BLAU,GE (1985)
DC	pKa			
HL 5.16E-011	atm m3/mol	25	EST	MEYLAN,WM & HOWARD,PH (1991)
OH 3.05E-010	cm3/molc sec	25	EST	MEYLAN,WM & HOWARD,PH (1993)

CAS #: 004238-71-5				1H-IMIDAZOLE, 1-(PHENYLMETHYL)-
Formula: $C_{10}H_{10}N_2$				
Mol Weight: 158.20				
MP (deg C): 68-70		FP (deg C):		
BP (deg C): 310				
BP pressure (mm Hg):				

Property/Value	Units	Temp	Data Type	Reference
WS 3.00E+003	mg/L	25	EST	MEYLAN,WM ET AL. (1996)
logP 1.60			EXP	AVDEEF,A (1993)
VP 2.10E-004	mm Hg	25	EST	NEELY,WB & BLAU,GE (1985)
DC	pKa			
HL 6.47E-006	atm m3/mol	25	EST	MEYLAN,WM & HOWARD,PH (1991)
OH 4.18E-011	cm3/molc sec	25	EST	MEYLAN,WM & HOWARD,PH (1993)

CAS #: 004241-40-1				BARBITURIC ACID,5-ET,5-(3-OH-1-ME)BUTYL
Formula: $C_{11}H_{18}N_2O_4$				
Mol Weight: 242.28				
MP (deg C):		FP (deg C):		
BP (deg C):				
BP pressure (mm Hg):				

Property/Value	Units	Temp	Data Type	Reference
WS 1.31E+004	mg/L	25	EST	MEYLAN,WM ET AL. (1996)
logP 0.35			EXP	HANSCH,C & LEO,AJ (1985)
VP 6.02E-015	mm Hg	25	EST	NEELY,WB & BLAU,GE (1985)
DC	pKa			
HL 3.09E-017	atm m3/mol	25	EST	MEYLAN,WM & HOWARD,PH (1991)
OH 2.54E-011	cm3/molc sec	25	EST	MEYLAN,WM & HOWARD,PH (1993)

CAS #: 004248-18-4 — 2H-BENZOTRIAZOLE-2-PROPANOIC ACID, á-METHYL-

Formula: $C_{10}H_{11}N_3O_2$

Mol Weight: 205.22

MP (deg C):   **FP (deg C):**

BP (deg C):

BP pressure (mm Hg):

Property/Value	Units	Temp	Data Type	Reference
WS 5.23E+004	mg/L	25	EST	MEYLAN,WM ET AL. (1996)
logP 0.34			EXP	SPARATORE,F ET AL. (1988)
VP 2.55E-006	mm Hg	25	EST	NEELY,WB & BLAU,GE (1985)
DC	pKa			
HL 4.10E-011	atm m3/mol	25	EST	MEYLAN,WM & HOWARD,PH (1991)
OH 4.94E-012	cm3/molc sec	25	EST	MEYLAN,WM & HOWARD,PH (1993)

CAS #: 004248-19-5 — O-T-BUTYL CARBAMATE

Formula: $C_5H_{11}NO_2$

Mol Weight: 117.15

MP (deg C): 105-108   **FP (deg C):**

BP (deg C):

BP pressure (mm Hg):

Property/Value	Units	Temp	Data Type	Reference
WS 1.46E+005	mg/L	37	EXP	YALKOWSKY,SH & DANNENFELSER,RM (1992)
logP 0.48			EXP	HANSCH,C & LEO,AJ (1985)
VP 6.18E+000	mm Hg	25	EST	NEELY,WB & BLAU,GE (1985)
DC	pKa			
HL 9.26E-008	atm m3/mol	25	EST	MEYLAN,WM & HOWARD,PH (1991)
OH 2.50E-012	cm3/molc sec	25	EST	MEYLAN,WM & HOWARD,PH (1993)

CAS #: 004248-20-8 — O-HEPTYL CARBAMATE

Formula: $C_8H_{17}NO_2$

Mol Weight: 159.23

MP (deg C):   **FP (deg C):**

BP (deg C):

BP pressure (mm Hg):

Property/Value	Units	Temp	Data Type	Reference
WS 6.66E+002	mg/L	25	EST	MEYLAN,WM ET AL. (1996)
logP 2.36			EXP	HANSCH,C & LEO,AJ (1985)
VP 1.11E-001	mm Hg	25	EST	NEELY,WB & BLAU,GE (1985)
DC	pKa			
HL 2.17E-007	atm m3/mol	25	EST	MEYLAN,WM & HOWARD,PH (1991)
OH 1.08E-011	cm3/molc sec	25	EST	MEYLAN,WM & HOWARD,PH (1993)

CAS #: 004249-64-3 — 2-N,N-DIETAMINO-1-PHENYLETHANOL

Formula: $C_{12}H_{19}NO$

Mol Weight: 193.29

MP (deg C):   **FP (deg C):**

BP (deg C):

BP pressure (mm Hg):

Property/Value	Units	Temp	Data Type	Reference
WS 9.08E+003	mg/L	25	EST	MEYLAN,WM ET AL. (1996)
logP 2.02			EXP	HANSCH,C & LEO,AJ (1985)
VP 1.04E-004	mm Hg	25	EST	NEELY,WB & BLAU,GE (1985)
DC	pKa			
HL 2.52E-010	atm m3/mol	25	EST	MEYLAN,WM & HOWARD,PH (1991)
OH 1.08E-010	cm3/molc sec	25	EST	MEYLAN,WM & HOWARD,PH (1993)

CAS #: 004261-14-7 — 9H-PURIN-6-AMINE, 9-(PHENYLMETHYL)-

Formula: $C_{12}H_{11}N_5$

Mol Weight: 225.25

MP (deg C):   **FP (deg C):**

BP (deg C):

BP pressure (mm Hg):

Property/Value	Units	Temp	Data Type	Reference
WS 1.14E+003	mg/L	25	EST	MEYLAN,WM ET AL. (1996)
logP 1.70			EXP	LAM,SP ET AL. (1989)
VP 1.48E-008	mm Hg	25	EST	NEELY,WB & BLAU,GE (1985)
DC	pKa			
HL 1.21E-013	atm m3/mol	25	EST	MEYLAN,WM & HOWARD,PH (1991)
OH 2.06E-010	cm3/molc sec	25	EST	MEYLAN,WM & HOWARD,PH (1993)

CAS #: 004265-25-2 — 2-METHYLBENZOFURAN

Formula: C_9H_8O

Mol Weight: 132.16

MP (deg C):   **FP (deg C):**

BP (deg C): 197.5

BP pressure (mm Hg):

Property/Value	Units	Temp	Data Type	Reference
WS 1.60E+002	mg/L	25	EST	MEYLAN,WM ET AL. (1996)
logP 3.22			EXP	SANGSTER,J (1993)
VP 4.87E-001	mm Hg	25	EXP	YAWS,CL (1994B)
DC	pKa			
HL 5.79E-004	atm m3/mol	25	EST	MEYLAN,WM & HOWARD,PH (1991)
OH 9.75E-011	cm3/molc sec	25	EST	MEYLAN,WM & HOWARD,PH (1993)

CAS #: 004267-05-4 — 3-CCL3 HYDROCHLOROTHIAZIDE

Formula: $C_8H_7Cl_4N_3O_4S_2$

Mol Weight: 415.10

MP (deg C): 300-303   **FP (deg C):**

BP (deg C):

BP pressure (mm Hg):

Property/Value	Units	Temp	Data Type	Reference
WS 4.38E+001	mg/L	25	EST	MEYLAN,WM ET AL. (1996)
logP 0.82			EXP	SANGSTER,J (1994)
VP 3.29E-011	mm Hg	25	EST	NEELY,WB & BLAU,GE (1985)
DC	pKa			
HL 2.55E-013	atm m3/mol	25	EST	MEYLAN,WM & HOWARD,PH (1991)
OH 1.29E-011	cm3/molc sec	25	EST	MEYLAN,WM & HOWARD,PH (1993)

CAS #: 004275-43-8 — A-BENYZL BENZENEETHANEAMINE

Formula: $C_{15}H_{17}N$

Mol Weight: 211.31

MP (deg C):   **FP (deg C):**

BP (deg C):

BP pressure (mm Hg):

Property/Value	Units	Temp	Data Type	Reference
WS 5.37E+002	mg/L	25	EST	MEYLAN,WM ET AL. (1996)
logP 3.35			EXP	HANSCH,C & LEO,AJ (1985)
VP 9.55E-005	mm Hg	25	EST	NEELY,WB & BLAU,GE (1985)
DC	pKa			
HL 8.70E-008	atm m3/mol	25	EST	MEYLAN,WM & HOWARD,PH (1991)
OH 6.03E-011	cm3/molc sec	25	EST	MEYLAN,WM & HOWARD,PH (1993)

004276-49-7 — 1-BROMOEICOSANE

Formula:	$C_{20}H_{41}Br$
Mol Weight:	361.46
MP (deg C):	37-38
FP (deg C):	
BP (deg C):	
BP pressure (mm Hg):	

Property/Value	Units	Temp	Data Type	Reference
WS 5.62E-006	mg/L	25	EST	MEYLAN,WM ET AL. (1996)
logP 10.51			EST	MEYLAN,WM & HOWARD,PH (1995)
VP 8.26E-007	mm Hg	25	EXT	LI,JCM & ROSSINI,FD (1953)
DC	pKa			
HL 1.85E+000	atm m3/mol	25	EST	MEYLAN,WM & HOWARD,PH (1991)
OH 2.48E-011	cm3/molc sec	25	EST	MEYLAN,WM & HOWARD,PH (1993)

004282-40-0 — 1-IODOHEPTANE

Formula:	$C_7H_{15}I$
Mol Weight:	226.10
MP (deg C):	-48.2
FP (deg C):	
BP (deg C):	204
BP pressure (mm Hg):	

Property/Value	Units	Temp	Data Type	Reference
WS 3.51E+000	mg/L	25	EXP	YALKOWSKY,SH & DANNENFELSER,RM (1992)
logP 4.70			EXP	HANSCH,C & LEO,AJ (1985)
VP 4.46E-001	mm Hg	25	EXT	DYKYJ,J & REPAS,M (1973)
DC	pKa			
HL 3.01E-002	atm m3/mol	25	EST	MEYLAN,WM & HOWARD,PH (1991)
OH 6.69E-012	cm3/molc sec	25	EST	MEYLAN,WM & HOWARD,PH (1993)

004282-42-2 — IODONONANE

Formula:	$C_9H_{19}I$
Mol Weight:	254.16
MP (deg C):	
FP (deg C):	
BP (deg C):	107-108
BP pressure (mm Hg):	8.00E+000

Property/Value	Units	Temp	Data Type	Reference
WS 4.34E-001	mg/L	25	EST	MEYLAN,WM ET AL. (1996)
logP 5.52			EST	MEYLAN,WM & HOWARD,PH (1995)
VP 2.73E-002	mm Hg	25	EXT	PERRY,RH & GREEN,D (1984)
DC	pKa			
HL 5.30E-002	atm m3/mol	25	EST	MEYLAN,WM & HOWARD,PH (1991)
OH 9.52E-012	cm3/molc sec	25	EST	MEYLAN,WM & HOWARD,PH (1993)

004282-44-4 — 1-IODOUNDECANE

Formula:	$C_{11}H_{23}I$
Mol Weight:	282.21
MP (deg C):	
FP (deg C):	
BP (deg C):	125-130
BP pressure (mm Hg):	5.00E+000

Property/Value	Units	Temp	Data Type	Reference
WS 4.36E-002	mg/L	25	EST	MEYLAN,WM ET AL. (1996)
logP 6.50			EST	MEYLAN,WM & HOWARD,PH (1995)
VP 4.43E-003	mm Hg	25	EXT	LI,JCM & ROSSINI,FD (1953)
DC	pKa			
HL 9.35E-002	atm m3/mol	25	EST	MEYLAN,WM & HOWARD,PH (1991)
OH 1.23E-011	cm3/molc sec	25	EST	MEYLAN,WM & HOWARD,PH (1993)

004284-51-9 — P-CHLORO METHANESULFONANILIDE

Formula:	$C_7H_8ClNO_2S$
Mol Weight:	205.66
MP (deg C):	
FP (deg C):	
BP (deg C):	
BP pressure (mm Hg):	

Property/Value	Units	Temp	Data Type	Reference
WS 1.08E+003	mg/L	25	EST	MEYLAN,WM ET AL. (1996)
logP 1.85			EXP	HANSCH,C & LEO,AJ (1985)
VP 1.47E-004	mm Hg	25	EST	NEELY,WB & BLAU,GE (1985)
DC	pKa			
HL 8.64E-007	atm m3/mol	25	EST	MEYLAN,WM & HOWARD,PH (1991)
OH 1.35E-011	cm3/molc sec	25	EST	MEYLAN,WM & HOWARD,PH (1993)

004290-62-4 — 2-PROPANOL, 1-[(1-METHYLETHYL)AMINO]-3-(1-NAPHTH

Formula:	$C_{18}H_{23}NO_3$
Mol Weight:	301.39
MP (deg C):	
FP (deg C):	
BP (deg C):	
BP pressure (mm Hg):	

Property/Value	Units	Temp	Data Type	Reference
WS 2.41E+001	mg/L	25	EST	MEYLAN,WM ET AL. (1996)
logP 4.34			EXP	SANGSTER,J (1994)
VP 7.51E-007	mm Hg	25	EST	NEELY,WB & BLAU,GE (1985)
DC	pKa			
HL 5.20E-011	atm m3/mol	25	EST	MEYLAN,WM & HOWARD,PH (1991)
OH 3.04E-010	cm3/molc sec	25	EST	MEYLAN,WM & HOWARD,PH (1993)

004292-19-7 — 1-IODODODECANE

Formula:	$C_{12}H_{25}I$
Mol Weight:	296.24
MP (deg C):	0.3
FP (deg C):	
BP (deg C):	298.2
BP pressure (mm Hg):	

Property/Value	Units	Temp	Data Type	Reference
WS 1.38E-002	mg/L	25	EST	MEYLAN,WM ET AL. (1996)
logP 6.99			EST	MEYLAN,WM & HOWARD,PH (1995)
VP 1.40E-002	mm Hg	25	EXT	LI,JCM & ROSSINI,FD (1953)
DC	pKa			
HL 1.24E-001	atm m3/mol	25	EST	MEYLAN,WM & HOWARD,PH (1991)
OH 1.38E-011	cm3/molc sec	25	EST	MEYLAN,WM & HOWARD,PH (1993)

004292-75-5 — 1-CYCLOHEXYLHEXANE

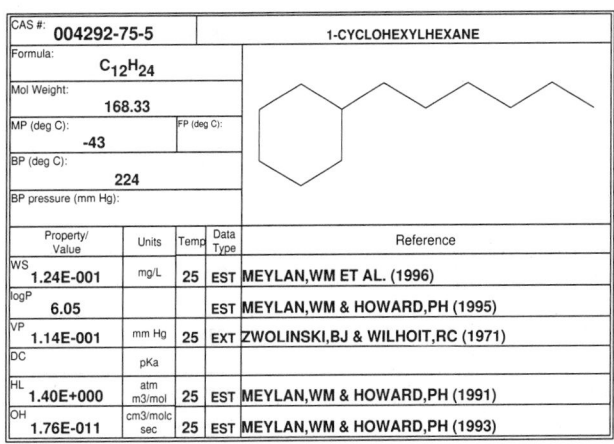

Formula:	$C_{12}H_{24}$
Mol Weight:	168.33
MP (deg C):	-43
FP (deg C):	
BP (deg C):	224
BP pressure (mm Hg):	

Property/Value	Units	Temp	Data Type	Reference
WS 1.24E-001	mg/L	25	EST	MEYLAN,WM ET AL. (1996)
logP 6.05			EST	MEYLAN,WM & HOWARD,PH (1995)
VP 1.14E-001	mm Hg	25	EXT	ZWOLINSKI,BJ & WILHOIT,RC (1971)
DC	pKa			
HL 1.40E+000	atm m3/mol	25	EST	MEYLAN,WM & HOWARD,PH (1991)
OH 1.76E-011	cm3/molc sec	25	EST	MEYLAN,WM & HOWARD,PH (1993)

CAS #: 004292-92-6				PENTYLCYCLOHEXANE

Formula: $C_{11}H_{22}$

Mol Weight: 154.30

MP (deg C): -57.5

FP (deg C):

BP (deg C): 203.7

BP pressure (mm Hg):

Property/Value	Units	Temp	Data Type	Reference
WS 3.78E-001	mg/L	25	EST	MEYLAN,WM ET AL. (1996)
logP 5.56			EST	MEYLAN,WM & HOWARD,PH (1995)
VP 3.72E-001	mm Hg	25	EXT	ZWOLINSKI,BJ & WILHOIT,RC (1971)
DC	pKa			
HL 1.05E+000	atm m3/mol	25	EST	MEYLAN,WM & HOWARD,PH (1991)
OH 1.62E-011	cm3/molc sec	25	EST	MEYLAN,WM & HOWARD,PH (1993)

CAS #: 004300-27-0				1-PHENYL-1,3,5-HEPTATRIYNE

Formula: $C_{13}H_8$

Mol Weight: 164.21

MP (deg C):

FP (deg C):

BP (deg C):

BP pressure (mm Hg):

Property/Value	Units	Temp	Data Type	Reference
WS 4.20E+000	mg/L	25	EST	MEYLAN,WM ET AL. (1996)
logP 4.91			EXP	MCLACHLAN,D ET AL. (1986)
VP 7.91E-004	mm Hg	25	EST	NEELY,WB & BLAU,GE (1985)
DC	pKa			
HL 5.00E-004	atm m3/mol	25	EST	MEYLAN,WM & HOWARD,PH (1991)
OH 1.17E-010	cm3/molc sec	25	EST	MEYLAN,WM & HOWARD,PH (1993)

CAS #: 004304-12-5				BENZYL GLUCOPYRANOSIDE

Formula: $C_{13}H_{18}O_6$

Mol Weight: 270.28

MP (deg C):

FP (deg C):

BP (deg C):

BP pressure (mm Hg):

Property/Value	Units	Temp	Data Type	Reference
WS 7.21E+004	mg/L	25	EST	MEYLAN,WM ET AL. (1996)
logP -0.70			EXP	HANSCH,C & LEO,AJ (1985)
VP 2.65E-011	mm Hg	25	EST	NEELY,WB & BLAU,GE (1985)
DC	pKa			
HL 1.26E-015	atm m3/mol	25	EST	MEYLAN,WM & HOWARD,PH (1991)
OH 7.26E-011	cm3/molc sec	25	EST	MEYLAN,WM & HOWARD,PH (1993)

CAS #: 004313-13-7				DIS. A. 16

Formula: $C_{15}H_{16}N_4O_2$

Mol Weight: 284.32

MP (deg C):

FP (deg C):

BP (deg C):

BP pressure (mm Hg):

Property/Value	Units	Temp	Data Type	Reference
WS 1.91E-001	mg/L	25	EST	MEYLAN,WM ET AL. (1996)
logP 5.23			EST	MEYLAN,WM & HOWARD,PH (1995)
VP 1.00E-008	mm Hg	25	EXP	BAUGHMAN,GL & PERENICH,TA (1988)
DC	pKa			
HL 1.02E-009	atm m3/mol	25	EST	MEYLAN,WM & HOWARD,PH (1991)
OH 1.82E-010	cm3/molc sec	25	EST	MEYLAN,WM & HOWARD,PH (1993)

CAS #: 004313-14-8				DIS. A. 17

Formula: $C_{15}H_{16}N_4O_2$

Mol Weight: 284.32

MP (deg C):

FP (deg C):

BP (deg C):

BP pressure (mm Hg):

Property/Value	Units	Temp	Data Type	Reference
WS 1.91E-001	mg/L	25	EST	MEYLAN,WM ET AL. (1996)
logP 5.23			EST	MEYLAN,WM & HOWARD,PH (1995)
VP 2.00E-007	mm Hg	25	EXP	BAUGHMAN,GL & PERENICH,TA (1988)
DC	pKa			
HL 1.02E-009	atm m3/mol	25	EST	MEYLAN,WM & HOWARD,PH (1991)
OH 1.82E-010	cm3/molc sec	25	EST	MEYLAN,WM & HOWARD,PH (1993)

CAS #: 004314-66-3				N-ETHYLNICOTINAMIDE

Formula: $C_8H_{10}N_2O$

Mol Weight: 150.18

MP (deg C):

FP (deg C):

BP (deg C):

BP pressure (mm Hg):

Property/Value	Units	Temp	Data Type	Reference
WS 4.12E+004	mg/L	25	EST	MEYLAN,WM ET AL. (1996)
logP 0.31			EXP	HANSCH,C & LEO,AJ (1985)
VP 9.11E-005	mm Hg	25	EST	NEELY,WB & BLAU,GE (1985)
DC	pKa			
HL 8.45E-012	atm m3/mol	25	EST	MEYLAN,WM & HOWARD,PH (1991)
OH 9.83E-012	cm3/molc sec	25	EST	MEYLAN,WM & HOWARD,PH (1993)

CAS #: 004329-03-7				ETHOXYCHLOR

Formula: $C_{18}H_{19}Cl_3O_2$

Mol Weight: 373.71

MP (deg C):

FP (deg C):

BP (deg C):

BP pressure (mm Hg):

Property/Value	Units	Temp	Data Type	Reference
WS 9.32E-003	mg/L	25	EST	MEYLAN,WM ET AL. (1996)
logP 6.65			EST	MEYLAN,WM & HOWARD,PH (1995)
VP 4.74E-007	mm Hg	25	EST	NEELY,WB & BLAU,GE (1985)
DC	pKa			
HL 1.72E-007	atm m3/mol	25	EST	MEYLAN,WM & HOWARD,PH (1991)
OH 6.42E-011	cm3/molc sec	25	EST	MEYLAN,WM & HOWARD,PH (1993)

CAS #: 004335-90-4				BENZAL ACETYLACETONE

Formula: $C_{12}H_{12}O_2$

Mol Weight: 188.23

MP (deg C):

FP (deg C):

BP (deg C):

BP pressure (mm Hg):

Property/Value	Units	Temp	Data Type	Reference
WS 2.20E+003	mg/L	25	EST	MEYLAN,WM ET AL. (1996)
logP 1.59			EXP	HANSCH,C & LEO,AJ (1985)
VP 1.19E-003	mm Hg	25	EST	NEELY,WB & BLAU,GE (1985)
DC	pKa			
HL 4.95E-010	atm m3/mol	25	EST	MEYLAN,WM & HOWARD,PH (1991)
OH 7.24E-011	cm3/molc sec	25	EST	MEYLAN,WM & HOWARD,PH (1993)

CAS #: 004340-77-6				4-CHLOROAZOBENZENE

Formula: $C_{12}H_9ClN_2$

Mol Weight: 216.67

MP (deg C): FP (deg C):

BP (deg C):

BP pressure (mm Hg):

Property/ Value	Units	Temp	Data Type	Reference
WS 4.30E-001	mg/L	25	EXP	BAUGHMAN,GL & PERENICH,TA (1988)
logP 4.76			EST	MEYLAN,WM & HOWARD,PH (1995)
VP 2.00E-006	mm Hg	25	EXP	BAUGHMAN,GL & PERENICH,TA (1988)
DC	pKa			
HL 1.33E-006	atm m3/mol	25	EST	VP/WSOL
OH 1.32E-012	cm3/molc sec		EST	MEYLAN,WM & HOWARD,PH (1993)

CAS #: 004342-03-4				IMIDAZOL-4-CONH2-5(33DIME-1-TRIAZENO)

Formula: $C_6H_{10}N_6O$

Mol Weight: 182.19

MP (deg C): 205 FP (deg C):

BP (deg C):

BP pressure (mm Hg):

Property/ Value	Units	Temp	Data Type	Reference
WS 4.22E+003	mg/L	25	EST	MEYLAN,WM ET AL. (1996)
logP -0.24			EXP	HANSCH,C & LEO,AJ (1985)
VP 2.20E-008	mm Hg	25	EST	NEELY,WB & BLAU,GE (1985)
DC	pKa			
HL 1.54E-016	atm m3/mol	25	EST	MEYLAN,WM & HOWARD,PH (1991)
OH 1.75E-011	cm3/molc sec	25	EST	MEYLAN,WM & HOWARD,PH (1993)

CAS #: 004343-96-8				4-PYRIDINEPENTANOL

Formula: $C_{10}H_{15}NO$

Mol Weight: 165.24

MP (deg C): FP (deg C):

BP (deg C):

BP pressure (mm Hg):

Property/ Value	Units	Temp	Data Type	Reference
WS 2.72E+005	mg/L	25	EST	MEYLAN,WM ET AL. (1996)
logP 1.39			EXP	HANSCH,C & LEO,AJ (1985)
VP 1.36E-004	mm Hg	25	EST	NEELY,WB & BLAU,GE (1985)
DC	pKa			
HL 8.84E-010	atm m3/mol	25	EST	MEYLAN,WM & HOWARD,PH (1991)
OH 1.05E-011	cm3/molc sec	25	EST	MEYLAN,WM & HOWARD,PH (1993)

CAS #: 004346-64-9				PHENYL METHYL SELENIDE

Formula: C_7H_8Se

Mol Weight: 171.10

MP (deg C): FP (deg C):

BP (deg C):

BP pressure (mm Hg):

Property/ Value	Units	Temp	Data Type	Reference
WS 2.15E+002	mg/L	25	EST	MEYLAN,WM ET AL. (1996)
logP 2.87			EXP	HANSCH,C & LEO,AJ (1985)
VP 2.57E-001	mm Hg	25	EST	NEELY,WB & BLAU,GE (1985)
DC	pKa			
HL 2.17E-004	atm m3/mol	25	EST	MEYLAN,WM & HOWARD,PH (1991)
OH 6.80E-011	cm3/molc sec	25	EST	MEYLAN,WM & HOWARD,PH (1993)

CAS #: 004351-48-8				BETAMETHASONE-17-PROPIONATE-21-DESOXY

Formula: $C_{25}H_{33}FO_5$

Mol Weight: 432.54

MP (deg C): FP (deg C):

BP (deg C):

BP pressure (mm Hg):

Property/ Value	Units	Temp	Data Type	Reference
WS 1.43E+001	mg/L	25	EST	MEYLAN,WM ET AL. (1996)
logP 3.09			EXP	HANSCH,C & LEO,AJ (1985)
VP 3.62E-012	mm Hg	25	EST	NEELY,WB & BLAU,GE (1985)
DC	pKa			
HL 6.58E-014	atm m3/mol	25	EST	MEYLAN,WM & HOWARD,PH (1991)
OH 6.70E-011	cm3/molc sec	25	EST	MEYLAN,WM & HOWARD,PH (1993)

CAS #: 004351-59-1				CLOBETASONE-17-PROPIONATE-DES-CL

Formula: $C_{25}H_{31}FO_5$

Mol Weight: 430.52

MP (deg C): FP (deg C):

BP (deg C):

BP pressure (mm Hg):

Property/ Value	Units	Temp	Data Type	Reference
WS 4.54E+000	mg/L	25	EST	MEYLAN,WM ET AL. (1996)
logP 3.09			EXP	HANSCH,C & LEO,AJ (1985)
VP 6.25E-010	mm Hg	25	EST	NEELY,WB & BLAU,GE (1985)
DC	pKa			
HL 6.87E-013	atm m3/mol	25	EST	MEYLAN,WM & HOWARD,PH (1991)
OH 6.02E-011	cm3/molc sec	25	EST	MEYLAN,WM & HOWARD,PH (1993)

CAS #: 004358-63-8				PHENYLBENZENESULFONATE

Formula: $C_{12}H_{10}O_3S$

Mol Weight: 234.28

MP (deg C): FP (deg C):

BP (deg C):

BP pressure (mm Hg):

Property/ Value	Units	Temp	Data Type	Reference
WS 7.03E+001	mg/L	25	EST	MEYLAN,WM ET AL. (1996)
logP 3.06			EXP	HANSCH,C & LEO,AJ (1985)
VP 5.84E-006	mm Hg	25	EST	NEELY,WB & BLAU,GE (1985)
DC	pKa			
HL 2.98E-007	atm m3/mol	25	EST	MEYLAN,WM & HOWARD,PH (1991)
OH 2.19E-011	cm3/molc sec	25	EST	MEYLAN,WM & HOWARD,PH (1993)

CAS #: 004360-12-7				AJMALINE

Formula: $C_{20}H_{28}N_2O_2$

Mol Weight: 328.46

MP (deg C): 158-160 FP (deg C):

BP (deg C):

BP pressure (mm Hg):

Property/ Value	Units	Temp	Data Type	Reference
WS 4.90E+002	mg/L	30	EXP	YALKOWSKY,SH & DANNENFELSER,RM (1992)
logP 1.81			EXP	HANSCH,C ET AL. (1995)
VP 7.29E-011	mm Hg	25	EST	NEELY,WB & BLAU,GE (1985)
DC	pKa			
HL 2.67E-013	atm m3/mol	25	EST	MEYLAN,WM & HOWARD,PH (1991)
OH 3.02E-010	cm3/molc sec	25	EST	MEYLAN,WM & HOWARD,PH (1993)

004368-68-7 — 1-BENZYL-1,2,3-TRIAZOLE

Formula: $C_9H_{11}N_3$
Mol Weight: 161.21
MP (deg C):
FP (deg C):
BP (deg C):
BP pressure (mm Hg):

Property/Value	Units	Temp	Data Type	Reference
WS 5.68E+003	mg/L	25	EST	MEYLAN,WM ET AL. (1996)
logP 1.27			EXP	HANSCH,C ET AL. (1995)
VP 3.08E-003	mm Hg	25	EST	NEELY,WB & BLAU,GE (1985)
DC	pKa			
HL 2.59E-006	atm m3/mol	25	EST	MEYLAN,WM & HOWARD,PH (1991)
OH 5.88E-012	cm3/molc sec	25	EST	MEYLAN,WM & HOWARD,PH (1993)

004371-23-7 — [1,1'-BIPHENYL]-4-SULFONAMIDE

Formula: $C_{12}H_{11}NO_2S$
Mol Weight: 233.29
MP (deg C):
FP (deg C):
BP (deg C):
BP pressure (mm Hg):

Property/Value	Units	Temp	Data Type	Reference
WS 3.57E+002	mg/L	25	EST	MEYLAN,WM ET AL. (1996)
logP 2.24			EXP	CAROTTI,A ET AL. (1989)
VP 5.76E-007	mm Hg	25	EST	NEELY,WB & BLAU,GE (1985)
DC	pKa			
HL 3.24E-008	atm m3/mol	25	EST	MEYLAN,WM & HOWARD,PH (1991)
OH 4.11E-012	cm3/molc sec	25	EST	MEYLAN,WM & HOWARD,PH (1993)

004372-29-6 — DECYLPROPANEDIOIC ACID

Formula: $C_{13}H_{24}O_4$
Mol Weight: 244.33
MP (deg C):
FP (deg C):
BP (deg C):
BP pressure (mm Hg):

Property/Value	Units	Temp	Data Type	Reference
WS 2.50E+001	mg/L	25	EXP	SEIDELL,A (1941)
logP 3.59			EST	MEYLAN,WM & HOWARD,PH (1995)
VP 1.02E-006	mm Hg	25	EST	NEELY,WB & BLAU,GE (1985)
DC	pKa			
HL 6.93E-011	atm m3/mol	25	EST	MEYLAN,WM & HOWARD,PH (1991)
OH 1.50E-011	cm3/molc sec	25	EST	MEYLAN,WM & HOWARD,PH (1993)

004376-18-5 — MONOMETHYLPHTHALATE

Formula: $C_9H_8O_4$
Mol Weight: 180.16
MP (deg C): 82-84
FP (deg C):
BP (deg C):
BP pressure (mm Hg):

Property/Value	Units	Temp	Data Type	Reference
WS 5.96E+003	mg/L	25	EST	MEYLAN,WM ET AL. (1996)
logP 1.13			EXP	HANSCH,C & LEO,AJ (1985)
VP 1.84E-004	mm Hg	25	EST	NEELY,WB & BLAU,GE (1985)
DC	pKa			
HL 6.99E-010	atm m3/mol	25	EST	MEYLAN,WM & HOWARD,PH (1991)
OH 9.09E-013	cm3/molc sec	25	EST	MEYLAN,WM & HOWARD,PH (1993)

004383-07-7 — BENZENEMETHANOL, 2-HYDROXY-5-METHYL-

Formula: $C_8H_{10}O_2$
Mol Weight: 138.17
MP (deg C):
FP (deg C):
BP (deg C):
BP pressure (mm Hg):

Property/Value	Units	Temp	Data Type	Reference
WS 1.56E+005	mg/L	25	EST	MEYLAN,WM ET AL. (1996)
logP 0.97			EXP	SANGSTER,J (1993)
VP 2.12E-004	mm Hg	25	EST	NEELY,WB & BLAU,GE (1985)
DC 10.15	pKa	25	EXP	KORTUM,G ET AL (1961)
HL 2.50E-011	atm m3/mol	25	EST	MEYLAN,WM & HOWARD,PH (1991)
OH 5.35E-011	cm3/molc sec	25	EST	MEYLAN,WM & HOWARD,PH (1993)

004388-79-8 — BARBITURIC ACID,5-ET-5-IPE-2-THIO

Formula: $C_{11}H_{18}N_2O_2S$
Mol Weight: 242.34
MP (deg C):
FP (deg C):
BP (deg C):
BP pressure (mm Hg):

Property/Value	Units	Temp	Data Type	Reference
WS 7.43E+001	mg/L	25	EST	MEYLAN,WM ET AL. (1996)
logP 2.98			EXP	HANSCH,C & LEO,AJ (1985)
VP 2.15E-010	mm Hg	25	EST	NEELY,WB & BLAU,GE (1985)
DC	pKa			
HL 6.35E-009	atm m3/mol	25	EST	MEYLAN,WM & HOWARD,PH (1991)
OH 2.34E-011	cm3/molc sec	25	EST	MEYLAN,WM & HOWARD,PH (1993)

004394-85-8 — N-FORMYLMORPHOLINE

Formula: $C_5H_9NO_2$
Mol Weight: 115.13
MP (deg C): 21
FP (deg C):
BP (deg C): 239
BP pressure (mm Hg):

Property/Value	Units	Temp	Data Type	Reference
WS 1.00E+006	mg/L	25	EST	MEYLAN,WM ET AL. (1996)
logP -1.32			EST	MEYLAN,WM & HOWARD,PH (1995)
VP 3.28E-001	mm Hg	25	EST	NEELY,WB & BLAU,GE (1985)
DC	pKa			
HL 5.06E-010	atm m3/mol	25	EST	MEYLAN,WM & HOWARD,PH (1991)
OH 8.98E-011	cm3/molc sec	25	EST	MEYLAN,WM & HOWARD,PH (1993)

004395-65-7 — C.I. DISPERSE BLUE 19

Formula: $C_{20}H_{14}N_2O_2$
Mol Weight: 314.35
MP (deg C): 192
FP (deg C):
BP (deg C):
BP pressure (mm Hg):

Property/Value	Units	Temp	Data Type	Reference
WS 1.92E-004	mg/L	25	EXP	BAUGHMAN,GL & PERENICH,TA (1988)
logP 5.38			EST	MEYLAN,WM & HOWARD,PH (1995)
VP 1.18E-010	mm Hg	25	EST	NEELY,WB & BLAU,GE (1985)
DC	pKa			
HL 2.19E-016	atm m3/mol	25	EST	MEYLAN,WM & HOWARD,PH (1991)
OH 2.00E-010	cm3/molc sec	25	EST	MEYLAN,WM & HOWARD,PH (1993)

CAS #: 004406-41-1				I-BUTYRANILIDE

Formula: $C_{10}H_{13}NO$

Mol Weight: 163.22

MP (deg C): | FP (deg C):

BP (deg C):

BP pressure (mm Hg):

Property/ Value	Units	Temp	Data Type	Reference
WS 1.43E+003	mg/L	25	EST	MEYLAN,WM ET AL. (1996)
logP 1.95			EXP	HANSCH,C & LEO,AJ (1985)
VP 1.43E-004	mm Hg	25	EST	NEELY,WB & BLAU,GE (1985)
DC	pKa			
HL 1.09E-008	atm m3/mol	25	EST	MEYLAN,WM & HOWARD,PH (1991)
OH 1.49E-011	cm3/molc sec	25	EST	MEYLAN,WM & HOWARD,PH (1993)

CAS #: 004408-78-0				PHOSPHONOACETIC ACID

Formula: $C_2H_5O_5P$

Mol Weight: 140.03

MP (deg C): 143-146 | FP (deg C):

BP (deg C):

BP pressure (mm Hg):

Property/ Value	Units	Temp	Data Type	Reference
WS 3.92E+005	mg/L	0	EXP	STEPHEN,H & STEPHEN,T (1963)
logP 0.86			EST	MEYLAN,WM & HOWARD,PH (1995)
VP 6.67E-007	mm Hg	25	EST	NEELY,WB & BLAU,GE (1985)
DC	pKa			
HL 9.06E-017	atm m3/mol	25	EST	MEYLAN,WM & HOWARD,PH (1991)
OH 1.50E-012	cm3/molc sec	25	EST	MEYLAN,WM & HOWARD,PH (1993)

CAS #: 004412-35-5				4H-1-BENZOPYRAN-4-ONE, 3-METHYL-2-PHENYL-8-(2-PR

Formula: $C_{19}H_{16}O_2$

Mol Weight: 276.34

MP (deg C): | FP (deg C):

BP (deg C):

BP pressure (mm Hg):

Property/ Value	Units	Temp	Data Type	Reference
WS 1.73E+000	mg/L	25	EST	MEYLAN,WM ET AL. (1996)
logP 4.67			EXP	HANSCH,C ET AL. (1995)
VP 3.18E-007	mm Hg	25	EST	NEELY,WB & BLAU,GE (1985)
DC	pKa			
HL 8.20E-008	atm m3/mol	25	EST	MEYLAN,WM & HOWARD,PH (1991)
OH 9.16E-011	cm3/molc sec	25	EST	MEYLAN,WM & HOWARD,PH (1993)

CAS #: 004412-91-3				3-HYDROXYMETHYLFURAN

Formula: $C_5H_6O_2$

Mol Weight: 98.10

MP (deg C): | FP (deg C):

BP (deg C): 79-80

BP pressure (mm Hg): 1.70E+001

Property/ Value	Units	Temp	Data Type	Reference
WS 2.13E+005	mg/L	25	EST	MEYLAN,WM ET AL. (1996)
logP 0.30			EXP	HANSCH,C & LEO,AJ (1985)
VP 4.29E-001	mm Hg	25	EST	NEELY,WB & BLAU,GE (1985)
DC	pKa			
HL 2.17E-007	atm m3/mol	25	EST	MEYLAN,WM & HOWARD,PH (1991)
OH 1.04E-010	cm3/molc sec	25	EST	MEYLAN,WM & HOWARD,PH (1993)

CAS #: 004413-31-4				1,1'-(2,2,2-TRICHLOROETHYLIDENE)BIS(4-METHYLB*)

Formula: $C_{16}H_{15}Cl_3$

Mol Weight: 313.66

MP (deg C): | FP (deg C):

BP (deg C):

BP pressure (mm Hg):

Property/ Value	Units	Temp	Data Type	Reference
WS 2.36E-002	mg/L	25	EST	MEYLAN,WM ET AL. (1996)
logP 6.60			EST	MEYLAN,WM & HOWARD,PH (1995)
VP 1.47E-005	mm Hg	25	EST	NEELY,WB & BLAU,GE (1985)
DC	pKa			
HL 3.39E-005	atm m3/mol	25	EST	MEYLAN,WM & HOWARD,PH (1991)
OH 1.27E-011	cm3/molc sec	25	EST	MEYLAN,WM & HOWARD,PH (1993)

CAS #: 004418-61-5				5-AMINO-1H-TETRAZOLE

Formula: CH_3N_5

Mol Weight: 85.07

MP (deg C): 200-204 de | FP (deg C):

BP (deg C):

BP pressure (mm Hg):

Property/ Value	Units	Temp	Data Type	Reference
WS 1.20E+004	mg/L	18	EXP	YALKOWSKY,SH & DANNENFELSER,RM (1992)
logP -3.41			EST	MEYLAN,WM & HOWARD,PH (1995)
VP 5.26E-003	mm Hg	25	EST	NEELY,WB & BLAU,GE (1985)
DC	pKa			
HL 1.31E-014	atm m3/mol	25	EST	MEYLAN,WM & HOWARD,PH (1991)
OH 1.52E-009	cm3/molc sec	25	EST	MEYLAN,WM & HOWARD,PH (1993)

CAS #: 004423-58-9				AMPHENICOL

Formula: $C_{11}H_{14}N_2O_5$

Mol Weight: 254.24

MP (deg C): | FP (deg C):

BP (deg C):

BP pressure (mm Hg):

Property/ Value	Units	Temp	Data Type	Reference
WS 4.74E+002	mg/L	25	EST	MEYLAN,WM ET AL. (1996)
logP -0.03			EXP	HANSCH,C & LEO,AJ (1985)
VP 1.14E-011	mm Hg	25	EST	NEELY,WB & BLAU,GE (1985)
DC	pKa			
HL 1.84E-017	atm m3/mol	25	EST	MEYLAN,WM & HOWARD,PH (1991)
OH 3.09E-011	cm3/molc sec	25	EST	MEYLAN,WM & HOWARD,PH (1993)

CAS #: 004425-23-4				AZEPINO[2,1-B]QUINAZOLIN-12(6H)-ONE, 7,8,9,10-TE

Formula: $C_{13}H_{14}N_2O$

Mol Weight: 214.27

MP (deg C): | FP (deg C):

BP (deg C):

BP pressure (mm Hg):

Property/ Value	Units	Temp	Data Type	Reference
WS 7.23E+002	mg/L	25	EST	MEYLAN,WM ET AL. (1996)
logP 2.00			EXP	SANGSTER,J (1994)
VP 2.68E-006	mm Hg	25	EST	NEELY,WB & BLAU,GE (1985)
DC	pKa			
HL 1.10E-009	atm m3/mol	25	EST	MEYLAN,WM & HOWARD,PH (1991)
OH 1.75E-011	cm3/molc sec	25	EST	MEYLAN,WM & HOWARD,PH (1993)

CAS #: 004425-56-3				5-CYANOURACIL
Formula: $C_5H_3N_3O_2$				
Mol Weight: 137.10				
MP (deg C):		FP (deg C):		
BP (deg C):				
BP pressure (mm Hg):				

Property/ Value	Units	Temp	Data Type	Reference
WS 1.43E+004	mg/L	25	EST	MEYLAN,WM ET AL. (1996)
logP -0.93			EXP	HANSCH,C & LEO,AJ (1985)
VP 5.54E-008	mm Hg	25	EST	NEELY,WB & BLAU,GE (1985)
DC	pKa			
HL 1.93E-013	atm m3/mol	25	EST	MEYLAN,WM & HOWARD,PH (1991)
OH 4.92E-012	cm3/molc sec	25	EST	MEYLAN,WM & HOWARD,PH (1993)

CAS #: 004426-79-3				BUTANE, 2-ISOTHIOCYANATO-
Formula: C_5H_9NS				
Mol Weight: 115.20				
MP (deg C):		FP (deg C):		
BP (deg C):				
BP pressure (mm Hg):				

Property/ Value	Units	Temp	Data Type	Reference
WS 4.16E+002	mg/L	25	EST	MEYLAN,WM ET AL. (1996)
logP 2.81			EXP	HANSCH,C ET AL. (1995)
VP 4.73E+000	mm Hg	25	EST	NEELY,WB & BLAU,GE (1985)
DC	pKa			
HL 7.27E-003	atm m3/mol	25	EST	MEYLAN,WM & HOWARD,PH (1991)
OH 3.87E-012	cm3/molc sec	25	EST	MEYLAN,WM & HOWARD,PH (1993)

CAS #: 004426-90-8				BENZENESULFONAMIDE, 4-[(METHYLSULFONYL)AMINO]-
Formula: $C_7H_{10}N_2O_4S_2$				
Mol Weight: 250.30				
MP (deg C):		FP (deg C):		
BP (deg C):				
BP pressure (mm Hg):				

Property/ Value	Units	Temp	Data Type	Reference
WS 2.54E+003	mg/L	25	EST	MEYLAN,WM ET AL. (1996)
logP -0.40			EXP	CAROTTI,A ET AL. (1989)
VP 1.06E-007	mm Hg	25	EST	NEELY,WB & BLAU,GE (1985)
DC	pKa			
HL 9.14E-011	atm m3/mol	25	EST	MEYLAN,WM & HOWARD,PH (1991)
OH 1.02E-011	cm3/molc sec	25	EST	MEYLAN,WM & HOWARD,PH (1993)

CAS #: 004434-66-6				1-BROMONONADECANE
Formula: $C_{19}H_{39}Br$				
Mol Weight: 347.43				
MP (deg C):		FP (deg C):		
BP (deg C):				
BP pressure (mm Hg):				

Property/ Value	Units	Temp	Data Type	Reference
WS 1.80E-005	mg/L	25	EST	MEYLAN,WM ET AL. (1996)
logP 10.02			EST	MEYLAN,WM & HOWARD,PH (1995)
VP 2.23E-006	mm Hg	25	EXT	LI,JCM & ROSSINI,FD (1953)
DC	pKa			
HL 1.39E+000	atm m3/mol	25	EST	MEYLAN,WM & HOWARD,PH (1991)
OH 2.34E-011	cm3/molc sec	25	EST	MEYLAN,WM & HOWARD,PH (1993)

CAS #: 004435-50-1				1,2,3-TRIHYDROXYBUTANE
Formula: $C_4H_{10}O_3$				
Mol Weight: 106.12				
MP (deg C):		FP (deg C):		
BP (deg C): 170				
BP pressure (mm Hg): 2.00E+001				

Property/ Value	Units	Temp	Data Type	Reference
WS 1.00E+006	mg/L	25	EST	MEYLAN,WM ET AL. (1996)
logP -1.23			EST	MEYLAN,WM & HOWARD,PH (1995)
VP 3.36E-003	mm Hg	25	EXT	JORDAN,TE (1954)
DC	pKa			
HL 8.42E-009	atm m3/mol	25	EST	MEYLAN,WM & HOWARD,PH (1991)
OH 2.32E-011	cm3/molc sec	25	EST	MEYLAN,WM & HOWARD,PH (1993)

CAS #: 004435-53-4				3-METHOXYBUTYL ACETATE
Formula: $C_7H_{14}O_3$				
Mol Weight: 146.19				
MP (deg C):		FP (deg C):		
BP (deg C):				
BP pressure (mm Hg):				

Property/ Value	Units	Temp	Data Type	Reference
WS 4.60E+003	mg/L	20	EXP	CHEM INSPECT TEST INST (1992)
logP 1.01			EST	MEYLAN,WM & HOWARD,PH (1995)
VP 2.77E+000	mm Hg	25	EST	NEELY,WB & BLAU,GE (1985)
DC	pKa			
HL 4.80E-006	atm m3/mol	25	EST	MEYLAN,WM & HOWARD,PH (1991)
OH 1.89E-011	cm3/molc sec	25	EST	MEYLAN,WM & HOWARD,PH (1993)

CAS #: 004439-24-1				ETHANOL, 2-(2-METHYLPROPOXY)-
Formula: $C_6H_{14}O_2$				
Mol Weight: 118.18				
MP (deg C):		FP (deg C):		
BP (deg C): 160				
BP pressure (mm Hg):				

Property/ Value	Units	Temp	Data Type	Reference
WS 7.55E+004	mg/L	25	EST	MEYLAN,WM ET AL. (1996)
logP 0.75			EXP	TANII,H ET AL. (1992)
VP 4.73E-001	mm Hg	25	EST	NEELY,WB & BLAU,GE (1985)
DC	pKa			
HL 9.79E-008	atm m3/mol	25	EST	MEYLAN,WM & HOWARD,PH (1991)
OH 2.53E-011	cm3/molc sec	25	EST	MEYLAN,WM & HOWARD,PH (1993)

CAS #: 004442-79-9				CYCLOHEXANEETHANOL
Formula: $C_8H_{16}O$				
Mol Weight: 128.22				
MP (deg C):		FP (deg C):		
BP (deg C): 208				
BP pressure (mm Hg):				

Property/ Value	Units	Temp	Data Type	Reference
WS 1.75E+003	mg/L	25	EST	MEYLAN,WM ET AL. (1996)
logP 2.62			EST	MEYLAN,WM & HOWARD,PH (1995)
VP 1.66E-001	mm Hg	25	EXT	PERRY,RH & GREEN,D (1984)
DC	pKa			
HL 8.63E-006	atm m3/mol	25	EST	MEYLAN,WM & HOWARD,PH (1991)
OH 1.62E-011	cm3/molc sec	25	EST	MEYLAN,WM & HOWARD,PH (1993)

CAS #: 004450-00-4 — 4-PYRIMIDINECARBOXYLIC ACID, 1,2,3,6-TETRAHYDRO-

Formula: $C_{10}H_{14}N_2O_4$

Mol Weight: 226.23

MP (deg C): | FP (deg C):

BP (deg C):

BP pressure (mm Hg):

Property/Value	Units	Temp	Data Type	Reference
WS 1.32E+003	mg/L	25	EST	MEYLAN,WM ET AL. (1996)
logP 1.62			EXP	FURST,W ET AL. (1990)
VP 4.05E-009	mm Hg	25	EST	NEELY,WB & BLAU,GE (1985)
DC	pKa			
HL 8.05E-012	atm m3/mol	25	EST	MEYLAN,WM & HOWARD,PH (1991)
OH 1.44E-011	cm3/molc sec	25	EST	MEYLAN,WM & HOWARD,PH (1993)

CAS #: 004450-01-5 — 4-PYRIMIDINECARBOXYLIC ACID, 1,2,3,6-TETRAHYDRO-

Formula: $C_{12}H_{16}N_2O_4$

Mol Weight: 252.27

MP (deg C): | FP (deg C):

BP (deg C):

BP pressure (mm Hg):

Property/Value	Units	Temp	Data Type	Reference
WS 7.22E+002	mg/L	25	EST	MEYLAN,WM ET AL. (1996)
logP 1.76			EXP	FURST,W ET AL. (1990)
VP 7.42E-010	mm Hg	25	EST	NEELY,WB & BLAU,GE (1985)
DC	pKa			
HL 1.07E-011	atm m3/mol	25	EST	MEYLAN,WM & HOWARD,PH (1991)
OH 1.56E-011	cm3/molc sec	25	EST	MEYLAN,WM & HOWARD,PH (1993)

CAS #: 004450-03-7 — 4-PYRIMIDINECARBOXYLIC ACID, 1,2,3,6-TETRAHYDRO-

Formula: $C_8H_{10}N_2O_4$

Mol Weight: 198.18

MP (deg C): | FP (deg C):

BP (deg C):

BP pressure (mm Hg):

Property/Value	Units	Temp	Data Type	Reference
WS 9.09E+003	mg/L	25	EST	MEYLAN,WM ET AL. (1996)
logP 0.81			EXP	FURST,W ET AL. (1990)
VP 3.32E-008	mm Hg	25	EST	NEELY,WB & BLAU,GE (1985)
DC	pKa			
HL 4.57E-012	atm m3/mol	25	EST	MEYLAN,WM & HOWARD,PH (1991)
OH 1.18E-011	cm3/molc sec	25	EST	MEYLAN,WM & HOWARD,PH (1993)

CAS #: 004450-04-8 — 4-PYRIMIDINECARBOXYLIC ACID, 1,2,3,6-TETRAHYDRO-

Formula: $C_{10}H_{14}N_2O_4$

Mol Weight: 226.23

MP (deg C): | FP (deg C):

BP (deg C):

BP pressure (mm Hg):

Property/Value	Units	Temp	Data Type	Reference
WS 1.74E+003	mg/L	25	EST	MEYLAN,WM ET AL. (1996)
logP 1.48			EXP	FURST,W ET AL. (1990)
VP 6.61E-009	mm Hg	25	EST	NEELY,WB & BLAU,GE (1985)
DC	pKa			
HL 8.05E-012	atm m3/mol	25	EST	MEYLAN,WM & HOWARD,PH (1991)
OH 1.44E-011	cm3/molc sec	25	EST	MEYLAN,WM & HOWARD,PH (1993)

CAS #: 004450-26-4 — TEH-PYRIMIDIN26DIONE,4-IPR CARBOXYLATE

Formula: $C_8H_{10}N_2O_4$

Mol Weight: 198.18

MP (deg C): | FP (deg C):

BP (deg C):

BP pressure (mm Hg):

Property/Value	Units	Temp	Data Type	Reference
WS 5.14E+003	mg/L	25	EST	MEYLAN,WM ET AL. (1996)
logP 1.10			EXP	FURST,W ET AL. (1990)
VP 2.04E-008	mm Hg	25	EST	NEELY,WB & BLAU,GE (1985)
DC	pKa			
HL 4.57E-012	atm m3/mol	25	EST	MEYLAN,WM & HOWARD,PH (1991)
OH 1.15E-011	cm3/molc sec	25	EST	MEYLAN,WM & HOWARD,PH (1993)

CAS #: 004461-29-4 — THIOPHEN-2-YL-ACETAMIDE

Formula: C_6H_9NOS

Mol Weight: 143.21

MP (deg C): | FP (deg C):

BP (deg C):

BP pressure (mm Hg):

Property/Value	Units	Temp	Data Type	Reference
WS 4.33E+004	mg/L	25	EST	MEYLAN,WM ET AL. (1996)
logP 0.33			EXP	HANSCH,C & LEO,AJ (1985)
VP 1.56E-004	mm Hg	25	EST	NEELY,WB & BLAU,GE (1985)
DC	pKa			
HL 4.91E-010	atm m3/mol	25	EST	MEYLAN,WM & HOWARD,PH (1991)
OH 2.64E-011	cm3/molc sec	25	EST	MEYLAN,WM & HOWARD,PH (1993)

CAS #: 004461-42-1 — 1-CHLORO-1-BUTENE

Formula: C_4H_7Cl

Mol Weight: 90.55

MP (deg C): | FP (deg C):

BP (deg C):

BP pressure (mm Hg):

Property/Value	Units	Temp	Data Type	Reference
WS 8.62E+002	mg/L	25	EST	MEYLAN,WM ET AL. (1996)
logP 2.53			EST	MEYLAN,WM & HOWARD,PH (1995)
VP 1.62E+002	mm Hg	25	EST	NEELY,WB & BLAU,GE (1985)
DC	pKa			
HL 9.24E-002	atm m3/mol	25	EST	MEYLAN,WM & HOWARD,PH (1991)
OH 1.29E-011	cm3/molc sec	25	EST	MEYLAN,WM & HOWARD,PH (1993)

CAS #: 004461-48-7 — 4-METHYL-2-PENTENE

Formula: C_6H_{12}

Mol Weight: 84.16

MP (deg C): | FP (deg C):

BP (deg C):

BP pressure (mm Hg):

Property/Value	Units	Temp	Data Type	Reference
WS 1.02E+002	mg/L	25	EST	MEYLAN,WM ET AL. (1996)
logP 3.00			EST	MEYLAN,WM & HOWARD,PH (1995)
VP 2.07E+002	mm Hg	21	EXP	FLICK,EW (1991)
DC	pKa			
HL 4.23E-001	atm m3/mol	25	EST	MEYLAN,WM & HOWARD,PH (1991)
OH 5.88E-011	cm3/molc sec	25	EST	MEYLAN,WM & HOWARD,PH (1993)

CAS #: 004465-58-1				C.I. DISPERSE RED 3

Formula: $C_{16}H_{13}NO_3$

Mol Weight: 267.29

MP (deg C):   **FP (deg C):**

BP (deg C):

BP pressure (mm Hg):

Property/Value	Units	Temp	Data Type	Reference
WS 3.30E+000	mg/L	25	EST	MEYLAN,WM ET AL. (1996)
logP 3.10			EST	MEYLAN,WM & HOWARD,PH (1995)
VP 1.00E-011	mm Hg	25	EXP	BAUGHMAN,GL & PERENICH,TA (1988)
DC	pKa			
HL 1.20E-016	atm m3/mol	25	EST	MEYLAN,WM & HOWARD,PH (1991)
OH 3.20E-011	cm3/molc sec	25	EST	MEYLAN,WM & HOWARD,PH (1993)

CAS #: 004465-61-6				4-MEO PHENYL 4-AMINOSALICYLATE

Formula: $C_{14}H_{13}NO_4$

Mol Weight: 259.26

MP (deg C):   **FP (deg C):**

BP (deg C):

BP pressure (mm Hg):

Property/Value	Units	Temp	Data Type	Reference
WS 5.02E+001	mg/L	25	EST	MEYLAN,WM ET AL. (1996)
logP 3.07			EXP	HANSCH,C & LEO,AJ (1985)
VP 3.79E-008	mm Hg	25	EST	NEELY,WB & BLAU,GE (1985)
DC	pKa			
HL 3.51E-011	atm m3/mol	25	EST	MEYLAN,WM & HOWARD,PH (1991)
OH 2.01E-010	cm3/molc sec	25	EST	MEYLAN,WM & HOWARD,PH (1993)

CAS #: 004471-41-4				C.I. DISPERSE BLUE 23

Formula: $C_{18}H_{18}N_2O_4$

Mol Weight: 326.36

MP (deg C): 248   **FP (deg C):**

BP (deg C):

BP pressure (mm Hg):

Property/Value	Units	Temp	Data Type	Reference
WS 5.54E-001	mg/L	25	EXP	BAUGHMAN,GL & PERENICH,TA (1988)
logP 2.71			EST	MEYLAN,WM & HOWARD,PH (1995)
VP 1.87E-014	mm Hg	25	EST	NEELY,WB & BLAU,GE (1985)
DC	pKa			
HL 4.51E-021	atm m3/mol	25	EST	MEYLAN,WM & HOWARD,PH (1991)
OH 5.73E-011	cm3/molc sec	25	EST	MEYLAN,WM & HOWARD,PH (1993)

CAS #: 004478-92-6				1-PHENETHYLISOTHIOCYANATE

Formula: C_9H_9NS

Mol Weight: 163.24

MP (deg C):   **FP (deg C):**

BP (deg C):

BP pressure (mm Hg):

Property/Value	Units	Temp	Data Type	Reference
WS 7.34E+001	mg/L	25	EST	MEYLAN,WM ET AL. (1996)
logP 3.46			EXP	HANSCH,C & LEO,AJ (1985)
VP 1.13E-002	mm Hg	25	EST	NEELY,WB & BLAU,GE (1985)
DC	pKa			
HL 3.33E-004	atm m3/mol	25	EST	MEYLAN,WM & HOWARD,PH (1991)
OH 7.14E-012	cm3/molc sec	25	EST	MEYLAN,WM & HOWARD,PH (1993)

CAS #: 004482-55-7				FENURON TCA

Formula: $C_{11}H_{13}Cl_3N_2O_3$

Mol Weight: 327.60

MP (deg C):   **FP (deg C):**

BP (deg C):

BP pressure (mm Hg):

Property/Value	Units	Temp	Data Type	Reference
WS 2.41E+003	mg/L	25	EST	MEYLAN,WM ET AL. (1996)
logP 0.64			EST	MEYLAN,WM & HOWARD,PH (1995)
VP 2.84E-011	mm Hg	25	EST	NEELY,WB & BLAU,GE (1985)
DC	pKa			
HL 1.63E-018	atm m3/mol	25	EST	MEYLAN,WM & HOWARD,PH (1991)
OH 2.53E-012	cm3/molc sec	25	EST	MEYLAN,WM & HOWARD,PH (1993)

CAS #: 004493-18-9				6-(N-HEXANOYLAMINO)PENCILLANIC ACID

Formula: $C_{14}H_{22}N_2O_4S$

Mol Weight: 314.41

MP (deg C):   **FP (deg C):**

BP (deg C):

BP pressure (mm Hg):

Property/Value	Units	Temp	Data Type	Reference
WS 9.56E+001	mg/L	25	EST	MEYLAN,WM ET AL. (1996)
logP 2.37			EXP	HANSCH,C & LEO,AJ (1985)
VP 3.88E-011	mm Hg	25	EST	NEELY,WB & BLAU,GE (1985)
DC	pKa			
HL 4.48E-013	atm m3/mol	25	EST	MEYLAN,WM & HOWARD,PH (1991)
OH 9.40E-011	cm3/molc sec	25	EST	MEYLAN,WM & HOWARD,PH (1993)

CAS #: 004513-94-4				2-CYANOPYRROLE

Formula: $C_5H_4N_2$

Mol Weight: 92.10

MP (deg C):   **FP (deg C):**

BP (deg C):

BP pressure (mm Hg):

Property/Value	Units	Temp	Data Type	Reference
WS 7.24E+003	mg/L	25	EST	MEYLAN,WM ET AL. (1996)
logP 1.13			EXP	HANSCH,C ET AL. (1995)
VP 2.00E-001	mm Hg	25	EST	NEELY,WB & BLAU,GE (1985)
DC	pKa			
HL 8.77E-008	atm m3/mol	25	EST	MEYLAN,WM & HOWARD,PH (1991)
OH 1.94E-011	cm3/molc sec	25	EST	MEYLAN,WM & HOWARD,PH (1993)

CAS #: 004514-53-8				4'-CHLOROBENZOGUANAMINE

Formula: $C_9H_8ClN_5$

Mol Weight: 221.65

MP (deg C):   **FP (deg C):**

BP (deg C):

BP pressure (mm Hg):

Property/Value	Units	Temp	Data Type	Reference
WS 3.45E+002	mg/L	25	EST	MEYLAN,WM ET AL. (1996)
logP 2.33			EXP	HANSCH,C & LEO,AJ (1985)
VP 2.11E-007	mm Hg	25	EST	NEELY,WB & BLAU,GE (1985)
DC 3.97	pKa	20	EXP	PERRIN,DD (1972)
HL 3.05E-011	atm m3/mol	25	EST	MEYLAN,WM & HOWARD,PH (1991)
OH 1.24E-012	cm3/molc sec	25	EST	MEYLAN,WM & HOWARD,PH (1993)

CAS #: 004514-54-9	3'-CHLOROBENZOGUANAMINE		
Formula: $C_9H_8ClN_5$			
Mol Weight: 221.65			
MP (deg C):	FP (deg C):		
BP (deg C):			
BP pressure (mm Hg):			

Property/Value	Units	Temp	Data Type	Reference
WS 3.45E+002	mg/L	25	EST	MEYLAN,WM ET AL. (1996)
logP 2.33			EXP	HANSCH,C & LEO,AJ (1985)
VP 2.11E-007	mm Hg	25	EST	NEELY,WB & BLAU,GE (1985)
DC	pKa			
HL 3.05E-011	atm m3/mol	25	EST	MEYLAN,WM & HOWARD,PH (1991)
OH 2.64E-012	cm3/molc sec	25	EST	MEYLAN,WM & HOWARD,PH (1993)

CAS #: 004516-69-2	1,1,3-TRIMETHYLCYCLOPENTANE		
Formula: C_8H_{16}			
Mol Weight: 112.22			
MP (deg C): -142.4	FP (deg C):		
BP (deg C): 104.9			
BP pressure (mm Hg):			

Property/Value	Units	Temp	Data Type	Reference
WS 3.73E+000	mg/L	25	EXP	YALKOWSKY,SH & DANNENFELSER,RM (1992)
logP 3.97			EST	MEYLAN,WM & HOWARD,PH (1995)
VP 1.11E+001	mm Hg	25	EXT	CHAO,J ET AL. (1983)
DC	pKa			
HL 4.50E-001	atm m3/mol	25	EST	MEYLAN,WM & HOWARD,PH (1991)
OH 6.24E-012	cm3/molc sec	25	EST	MEYLAN,WM & HOWARD,PH (1993)

CAS #: 004521-28-2	P-METHOXY-G-PHENYLBUTYRIC ACID		
Formula: $C_{11}H_{14}O_3$			
Mol Weight: 194.23			
MP (deg C): 56-59	FP (deg C):		
BP (deg C):			
BP pressure (mm Hg):			

Property/Value	Units	Temp	Data Type	Reference
WS 1.19E+003	mg/L	25	EST	MEYLAN,WM ET AL. (1996)
logP 2.33			EXP	HANSCH,C & LEO,AJ (1985)
VP 9.28E-005	mm Hg	25	EST	NEELY,WB & BLAU,GE (1985)
DC	pKa			
HL 4.61E-009	atm m3/mol	25	EST	MEYLAN,WM & HOWARD,PH (1991)
OH 3.10E-011	cm3/molc sec	25	EST	MEYLAN,WM & HOWARD,PH (1993)

CAS #: 004525-75-1	2-CHLOROPHENYL ACETATE		
Formula: $C_8H_7ClO_2$			
Mol Weight: 170.60			
MP (deg C):	FP (deg C):		
BP (deg C):			
BP pressure (mm Hg):			

Property/Value	Units	Temp	Data Type	Reference
WS 8.40E+002	mg/L	25	EST	MEYLAN,WM ET AL. (1996)
logP 2.18			EXP	HANSCH,C & LEO,AJ (1985)
VP 9.07E-002	mm Hg	25	EST	NEELY,WB & BLAU,GE (1985)
DC	pKa			
HL 4.80E-005	atm m3/mol	25	EST	MEYLAN,WM & HOWARD,PH (1991)
OH 1.11E-012	cm3/molc sec	25	EST	MEYLAN,WM & HOWARD,PH (1993)

CAS #: 004531-54-8	1-METHYL-4-NITRO-1H-IMIDAZOL-5-AMINE		
Formula: $C_4H_6N_4O_2$			
Mol Weight: 142.12			
MP (deg C):	FP (deg C):		
BP (deg C):			
BP pressure (mm Hg):			

Property/Value	Units	Temp	Data Type	Reference
WS 2.39E+005	mg/L	25	EST	MEYLAN,WM ET AL. (1996)
logP -1.00			EXP	SUWINSKI,J ET AL. (1985)
VP 4.27E-005	mm Hg	25	EST	NEELY,WB & BLAU,GE (1985)
DC	pKa			
HL 1.12E-010	atm m3/mol	25	EST	MEYLAN,WM & HOWARD,PH (1991)
OH 7.51E-012	cm3/molc sec	25	EST	MEYLAN,WM & HOWARD,PH (1993)

CAS #: 004531-79-7	3-NITRO-N-PHENYLBENZENAMINE		
Formula: $C_{12}H_{10}N_2O_2$			
Mol Weight: 214.23			
MP (deg C):	FP (deg C):		
BP (deg C):			
BP pressure (mm Hg):			

Property/Value	Units	Temp	Data Type	Reference
WS 2.62E+001	mg/L	25	EST	MEYLAN,WM ET AL. (1996)
logP 3.69			EST	MEYLAN,WM & HOWARD,PH (1995)
VP 1.66E-005	mm Hg	25	EST	NEELY,WB & BLAU,GE (1985)
DC	pKa			
HL 4.14E-009	atm m3/mol	25	EST	MEYLAN,WM & HOWARD,PH (1991)
OH 1.59E-010	cm3/molc sec	25	EST	MEYLAN,WM & HOWARD,PH (1993)

CAS #: 004534-49-0	BENZENE, (1-PENTYLOCTYL)-		
Formula: $C_{19}H_{32}$			
Mol Weight: 260.47			
MP (deg C):	FP (deg C):		
BP (deg C):			
BP pressure (mm Hg):			

Property/Value	Units	Temp	Data Type	Reference
WS 9.00E-004	mg/L	25	EST	MEYLAN,WM ET AL. (1996)
logP 8.56			EXP	SHERBLOM,PM ET AL. (1992)
VP 2.29E-004	mm Hg	25	EST	NEELY,WB & BLAU,GE (1985)
DC	pKa			
HL 1.78E-001	atm m3/mol	25	EST	MEYLAN,WM & HOWARD,PH (1991)
OH 2.15E-011	cm3/molc sec	25	EST	MEYLAN,WM & HOWARD,PH (1993)

CAS #: 004534-50-3	BENZENE, (1-BUTYLNONYL)-		
Formula: $C_{19}H_{32}$			
Mol Weight: 260.47			
MP (deg C):	FP (deg C):		
BP (deg C):			
BP pressure (mm Hg):			

Property/Value	Units	Temp	Data Type	Reference
WS 9.00E-004	mg/L	25	EST	MEYLAN,WM ET AL. (1996)
logP 8.56			EXP	SHERBLOM,PM ET AL. (1992)
VP 2.29E-004	mm Hg	25	EST	NEELY,WB & BLAU,GE (1985)
DC	pKa			
HL 1.78E-001	atm m3/mol	25	EST	MEYLAN,WM & HOWARD,PH (1991)
OH 2.15E-011	cm3/molc sec	25	EST	MEYLAN,WM & HOWARD,PH (1993)

CAS #: 004534-51-4 — BENZENE, (1-PROPYLDECYL)-

Formula: $C_{19}H_{32}$
Mol Weight: 260.47
MP (deg C):   FP (deg C):
BP (deg C):
BP pressure (mm Hg):

Property/Value	Units	Temp	Data Type	Reference
WS 9.00E-004	mg/L	25	EST	MEYLAN,WM ET AL. (1996)
logP 8.62			EXP	SHERBLOM,PM ET AL. (1992)
VP 2.29E-004	mm Hg	25	EST	NEELY,WB & BLAU,GE (1985)
DC	pKa			
HL 1.78E-001	atm m3/mol	25	EST	MEYLAN,WM & HOWARD,PH (1991)
OH 2.15E-011	cm3/molc sec	25	EST	MEYLAN,WM & HOWARD,PH (1993)

CAS #: 004534-52-5 — BENZENE, (1-ETHYLUNDECYL)-

Formula: $C_{19}H_{32}$
Mol Weight: 260.47
MP (deg C):   FP (deg C):
BP (deg C):
BP pressure (mm Hg):

Property/Value	Units	Temp	Data Type	Reference
WS 8.48E-004	mg/L	25	EST	MEYLAN,WM ET AL. (1996)
logP 8.65			EXP	SHERBLOM,PM ET AL. (1992)
VP 2.29E-004	mm Hg	25	EST	NEELY,WB & BLAU,GE (1985)
DC	pKa			
HL 1.78E-001	atm m3/mol	25	EST	MEYLAN,WM & HOWARD,PH (1991)
OH 2.15E-011	cm3/molc sec	25	EST	MEYLAN,WM & HOWARD,PH (1993)

CAS #: 004534-53-6 — BENZENE, (1-METHYLDODECYL)-

Formula: $C_{19}H_{32}$
Mol Weight: 260.47
MP (deg C):   FP (deg C):
BP (deg C):
BP pressure (mm Hg):

Property/Value	Units	Temp	Data Type	Reference
WS 6.70E-004	mg/L	25	EST	MEYLAN,WM ET AL. (1996)
logP 8.77			EXP	SHERBLOM,PM ET AL. (1992)
VP 2.29E-004	mm Hg	25	EST	NEELY,WB & BLAU,GE (1985)
DC	pKa			
HL 1.78E-001	atm m3/mol	25	EST	MEYLAN,WM & HOWARD,PH (1991)
OH 2.12E-011	cm3/molc sec	25	EST	MEYLAN,WM & HOWARD,PH (1993)

CAS #: 004534-55-8 — BENZENE, (1-PENTYLNONYL)-

Formula: $C_{20}H_{34}$
Mol Weight: 274.49
MP (deg C):   FP (deg C):
BP (deg C):
BP pressure (mm Hg):

Property/Value	Units	Temp	Data Type	Reference
WS 2.17E-004	mg/L	25	EST	MEYLAN,WM ET AL. (1996)
logP 9.16			EXP	SHERBLOM,PM ET AL. (1992)
VP 9.39E-005	mm Hg	25	EST	NEELY,WB & BLAU,GE (1985)
DC	pKa			
HL 2.37E-001	atm m3/mol	25	EST	MEYLAN,WM & HOWARD,PH (1991)
OH 2.29E-011	cm3/molc sec	25	EST	MEYLAN,WM & HOWARD,PH (1993)

CAS #: 004534-56-9 — BENZENE, (1-BUTYLDECYL)-

Formula: $C_{20}H_{34}$
Mol Weight: 274.49
MP (deg C):   FP (deg C):
BP (deg C):
BP pressure (mm Hg):

Property/Value	Units	Temp	Data Type	Reference
WS 2.17E-004	mg/L	25	EST	MEYLAN,WM ET AL. (1996)
logP 9.16			EXP	SHERBLOM,PM ET AL. (1992)
VP 9.39E-005	mm Hg	25	EST	NEELY,WB & BLAU,GE (1985)
DC	pKa			
HL 2.37E-001	atm m3/mol	25	EST	MEYLAN,WM & HOWARD,PH (1991)
OH 2.29E-011	cm3/molc sec	25	EST	MEYLAN,WM & HOWARD,PH (1993)

CAS #: 004534-57-0 — BENZENE, (1-PROPYLUNDECYL)-

Formula: $C_{20}H_{34}$
Mol Weight: 274.49
MP (deg C):   FP (deg C):
BP (deg C):
BP pressure (mm Hg):

Property/Value	Units	Temp	Data Type	Reference
WS 2.17E-004	mg/L	25	EST	MEYLAN,WM ET AL. (1996)
logP 9.25			EXP	SHERBLOM,PM ET AL. (1992)
VP 9.39E-005	mm Hg	25	EST	NEELY,WB & BLAU,GE (1985)
DC	pKa			
HL 2.37E-001	atm m3/mol	25	EST	MEYLAN,WM & HOWARD,PH (1991)
OH 2.29E-011	cm3/molc sec	25	EST	MEYLAN,WM & HOWARD,PH (1993)

CAS #: 004534-58-1 — BENZENE, (1-ETHYLDODECYL)-

Formula: $C_{20}H_{34}$
Mol Weight: 274.49
MP (deg C):   FP (deg C):
BP (deg C):
BP pressure (mm Hg):

Property/Value	Units	Temp	Data Type	Reference
WS 2.01E-004	mg/L	25	EST	MEYLAN,WM ET AL. (1996)
logP 9.29			EXP	SHERBLOM,PM ET AL. (1992)
VP 9.39E-005	mm Hg	25	EST	NEELY,WB & BLAU,GE (1985)
DC	pKa			
HL 2.37E-001	atm m3/mol	25	EST	MEYLAN,WM & HOWARD,PH (1991)
OH 2.29E-011	cm3/molc sec	25	EST	MEYLAN,WM & HOWARD,PH (1993)

CAS #: 004534-59-2 — BENZENE, (1-METHYLTRIDECYL)-

Formula: $C_{20}H_{34}$
Mol Weight: 274.49
MP (deg C):   FP (deg C):
BP (deg C):
BP pressure (mm Hg):

Property/Value	Units	Temp	Data Type	Reference
WS 2.01E-004	mg/L	25	EST	MEYLAN,WM ET AL. (1996)
logP 9.29			EXP	SHERBLOM,PM ET AL. (1992)
VP 9.39E-005	mm Hg	25	EST	NEELY,WB & BLAU,GE (1985)
DC	pKa			
HL 2.37E-001	atm m3/mol	25	EST	MEYLAN,WM & HOWARD,PH (1991)
OH 2.26E-011	cm3/molc sec	25	EST	MEYLAN,WM & HOWARD,PH (1993)

CAS #: 004536-86-1				BENZENE, (1-PROPYLOCTYL)-

Formula: $C_{17}H_{28}$
Mol Weight: 232.41
MP (deg C):
FP (deg C):
BP (deg C):
BP pressure (mm Hg):

Property/Value	Units	Temp	Data Type	Reference
WS 1.28E-002	mg/L	25	EST	MEYLAN,WM ET AL. (1996)
logP 7.59			EXP	SHERBLOM,PM ET AL. (1992)
VP 1.55E-003	mm Hg	25	EST	NEELY,WB & BLAU,GE (1985)
DC	pKa			
HL 1.01E-001	atm m3/mol	25	EST	MEYLAN,WM & HOWARD,PH (1991)
OH 1.87E-011	cm3/molc sec	25	EST	MEYLAN,WM & HOWARD,PH (1993)

CAS #: 004536-87-2				BENZENE, (1-ETHYLNONYL)-

Formula: $C_{17}H_{28}$
Mol Weight: 232.41
MP (deg C):
FP (deg C):
BP (deg C):
BP pressure (mm Hg):

Property/Value	Units	Temp	Data Type	Reference
WS 1.10E-002	mg/L	25	EST	MEYLAN,WM ET AL. (1996)
logP 7.53			EXP	SHERBLOM,PM ET AL. (1992)
VP 1.55E-003	mm Hg	25	EST	NEELY,WB & BLAU,GE (1985)
DC	pKa			
HL 1.01E-001	atm m3/mol	25	EST	MEYLAN,WM & HOWARD,PH (1991)
OH 1.87E-011	cm3/molc sec	25	EST	MEYLAN,WM & HOWARD,PH (1993)

CAS #: 004536-88-3				BENZENE, (1-METHYLDECYL)-

Formula: $C_{17}H_{28}$
Mol Weight: 232.41
MP (deg C):
FP (deg C):
BP (deg C):
BP pressure (mm Hg):

Property/Value	Units	Temp	Data Type	Reference
WS 9.00E-003	mg/L	25	EST	MEYLAN,WM ET AL. (1996)
logP 7.63			EXP	SHERBLOM,PM ET AL. (1992)
VP 1.55E-003	mm Hg	25	EST	NEELY,WB & BLAU,GE (1985)
DC	pKa			
HL 1.01E-001	atm m3/mol	25	EST	MEYLAN,WM & HOWARD,PH (1991)
OH 1.84E-011	cm3/molc sec	25	EST	MEYLAN,WM & HOWARD,PH (1993)

CAS #: 004537-11-5				BENZENE, (1-BUTYLHEXYL)-

Formula: $C_{16}H_{26}$
Mol Weight: 218.39
MP (deg C):
FP (deg C):
BP (deg C):
BP pressure (mm Hg):

Property/Value	Units	Temp	Data Type	Reference
WS 3.69E-002	mg/L	25	EST	MEYLAN,WM ET AL. (1996)
logP 6.90			EXP	SHERBLOM,PM ET AL. (1992)
VP 3.82E-003	mm Hg	25	EST	NEELY,WB & BLAU,GE (1985)
DC	pKa			
HL 7.61E-002	atm m3/mol	25	EST	MEYLAN,WM & HOWARD,PH (1991)
OH 1.73E-011	cm3/molc sec	25	EST	MEYLAN,WM & HOWARD,PH (1993)

CAS #: 004537-12-6				BENZENE, (1-PROPYLHEPTYL)-

Formula: $C_{16}H_{26}$
Mol Weight: 218.39
MP (deg C):
FP (deg C):
BP (deg C):
BP pressure (mm Hg):

Property/Value	Units	Temp	Data Type	Reference
WS 3.69E-002	mg/L	25	EST	MEYLAN,WM ET AL. (1996)
logP 7.00			EXP	SHERBLOM,PM ET AL. (1992)
VP 3.82E-003	mm Hg	25	EST	NEELY,WB & BLAU,GE (1985)
DC	pKa			
HL 7.61E-002	atm m3/mol	25	EST	MEYLAN,WM & HOWARD,PH (1991)
OH 1.73E-011	cm3/molc sec	25	EST	MEYLAN,WM & HOWARD,PH (1993)

CAS #: 004537-13-7				BENZENE, (1-METHYLNONYL)-

Formula: $C_{16}H_{26}$
Mol Weight: 218.39
MP (deg C):
FP (deg C):
BP (deg C):
BP pressure (mm Hg):

Property/Value	Units	Temp	Data Type	Reference
WS 3.28E-002	mg/L	25	EST	MEYLAN,WM ET AL. (1996)
logP 7.06			EXP	SHERBLOM,PM ET AL. (1992)
VP 3.82E-003	mm Hg	25	EST	NEELY,WB & BLAU,GE (1985)
DC	pKa			
HL 7.61E-002	atm m3/mol	25	EST	MEYLAN,WM & HOWARD,PH (1991)
OH 1.70E-011	cm3/molc sec	25	EST	MEYLAN,WM & HOWARD,PH (1993)

CAS #: 004537-14-8				BENZENE, (1-PENTYLHEXYL)-

Formula: $C_{17}H_{28}$
Mol Weight: 232.41
MP (deg C):
FP (deg C):
BP (deg C):
BP pressure (mm Hg):

Property/Value	Units	Temp	Data Type	Reference
WS 1.28E-002	mg/L	25	EST	MEYLAN,WM ET AL. (1996)
logP 7.45			EXP	SHERBLOM,PM ET AL. (1992)
VP 1.55E-003	mm Hg	25	EST	NEELY,WB & BLAU,GE (1985)
DC	pKa			
HL 1.01E-001	atm m3/mol	25	EST	MEYLAN,WM & HOWARD,PH (1991)
OH 1.87E-011	cm3/molc sec	25	EST	MEYLAN,WM & HOWARD,PH (1993)

CAS #: 004537-15-9				BENZENE, (1-BUTYLHEPTYL)-

Formula: $C_{17}H_{28}$
Mol Weight: 232.41
MP (deg C):
FP (deg C):
BP (deg C):
BP pressure (mm Hg):

Property/Value	Units	Temp	Data Type	Reference
WS 1.28E-002	mg/L	25	EST	MEYLAN,WM ET AL. (1996)
logP 7.45			EXP	SHERBLOM,PM ET AL. (1992)
VP 1.55E-003	mm Hg	25	EST	NEELY,WB & BLAU,GE (1985)
DC	pKa			
HL 1.01E-001	atm m3/mol	25	EST	MEYLAN,WM & HOWARD,PH (1991)
OH 1.87E-011	cm3/molc sec	25	EST	MEYLAN,WM & HOWARD,PH (1993)

CAS #: 004537-73-9				PYRIMIDINE-2-THIOCARBOXAMIDE

Formula: $C_5H_5N_3S$

Mol Weight: 139.18

MP (deg C):   FP (deg C):

BP (deg C):

BP pressure (mm Hg):

Property/Value	Units	Temp	Data Type	Reference
WS 8.22E+004	mg/L	25	EST	MEYLAN,WM ET AL. (1996)
logP -0.34			EXP	HANSCH,C & LEO,AJ (1985)
VP 3.64E-003	mm Hg	25	EST	NEELY,WB & BLAU,GE (1985)
DC	pKa			
HL 2.12E-010	atm m3/mol	25	EST	MEYLAN,WM & HOWARD,PH (1991)
OH 2.15E-011	cm3/molc sec	25	EST	MEYLAN,WM & HOWARD,PH (1993)

CAS #: 004540-00-5				C.I. DISPERSE RED 7

Formula: $C_{16}H_{17}ClN_4O_4$

Mol Weight: 364.79

MP (deg C): 190   FP (deg C):

BP (deg C):

BP pressure (mm Hg):

Property/Value	Units	Temp	Data Type	Reference
WS 4.01E-001	mg/L	25	EXP	BAUGHMAN,GL & PERENICH,TA (1988)
logP 3.79			EST	MEYLAN,WM & HOWARD,PH (1995)
VP 1.12E-013	mm Hg	25	EST	NEELY,WB & BLAU,GE (1985)
DC	pKa			
HL 1.61E-015	atm m3/mol	25	EST	MEYLAN,WM & HOWARD,PH (1991)
OH 7.06E-011	cm3/molc sec	25	EST	MEYLAN,WM & HOWARD,PH (1993)

CAS #: 004546-55-8				ADENOSINE, N-BENZOYL-

Formula: $C_{17}H_{17}N_5O_5$

Mol Weight: 371.36

MP (deg C):   FP (deg C):

BP (deg C):

BP pressure (mm Hg):

Property/Value	Units	Temp	Data Type	Reference
WS 2.74E+002	mg/L	25	EST	MEYLAN,WM ET AL. (1996)
logP -0.10			EXP	HANSCH,C ET AL. (1995)
VP 7.76E-021	mm Hg	25	EST	NEELY,WB & BLAU,GE (1985)
DC	pKa			
HL 7.14E-026	atm m3/mol	25	EST	MEYLAN,WM & HOWARD,PH (1991)
OH 2.49E-010	cm3/molc sec	25	EST	MEYLAN,WM & HOWARD,PH (1993)

CAS #: 004546-70-7				ADENOSINE, 2-AMINO-2'-DEOXY-

Formula: $C_{10}H_{14}N_6O_3$

Mol Weight: 266.26

MP (deg C):   FP (deg C):

BP (deg C):

BP pressure (mm Hg):

Property/Value	Units	Temp	Data Type	Reference
WS 2.61E+003	mg/L	25	EST	MEYLAN,WM ET AL. (1996)
logP -0.52			EXP	BALZARINI,J ET AL. (1989)
VP 1.40E-014	mm Hg	25	EST	NEELY,WB & BLAU,GE (1985)
DC	pKa			
HL 3.82E-021	atm m3/mol	25	EST	MEYLAN,WM & HOWARD,PH (1991)
OH 2.40E-010	cm3/molc sec	25	EST	MEYLAN,WM & HOWARD,PH (1993)

CAS #: 004546-72-9				ADENOSINE, N-BENZOYL-2'-DEOXY-

Formula: $C_{17}H_{17}N_5O_4$

Mol Weight: 355.36

MP (deg C):   FP (deg C):

BP (deg C):

BP pressure (mm Hg):

Property/Value	Units	Temp	Data Type	Reference
WS 1.83E+002	mg/L	25	EST	MEYLAN,WM ET AL. (1996)
logP 0.22			EXP	HANSCH,C ET AL. (1995)
VP 1.09E-018	mm Hg	25	EST	NEELY,WB & BLAU,GE (1985)
DC	pKa			
HL 1.95E-024	atm m3/mol	25	EST	MEYLAN,WM & HOWARD,PH (1991)
OH 2.41E-010	cm3/molc sec	25	EST	MEYLAN,WM & HOWARD,PH (1993)

CAS #: 004548-15-6				1H-IMIDAZOL-1-ETHANOL, 2-(4-FLUOROPHENYL)-5-NITR

Formula: $C_{11}H_{10}FN_3O_3$

Mol Weight: 251.22

MP (deg C):   FP (deg C):

BP (deg C):

BP pressure (mm Hg):

Property/Value	Units	Temp	Data Type	Reference
WS 6.94E+002	mg/L	25	EST	MEYLAN,WM ET AL. (1996)
logP 1.33			EXP	SANGSTER,J (1994)
VP 2.36E-010	mm Hg	25	EST	NEELY,WB & BLAU,GE (1985)
DC	pKa			
HL 1.38E-012	atm m3/mol	25	EST	MEYLAN,WM & HOWARD,PH (1991)
OH 8.83E-012	cm3/molc sec	25	EST	MEYLAN,WM & HOWARD,PH (1993)

CAS #: 004549-40-0				N-NITROSOMETHYLVINYLAMINE

Formula: $C_3H_6N_2O$

Mol Weight: 86.09

MP (deg C):   FP (deg C):

BP (deg C): 138

BP pressure (mm Hg):

Property/Value	Units	Temp	Data Type	Reference
WS 2.22E+005	mg/L	25	EST	MEYLAN,WM ET AL. (1996)
logP -0.28			EST	MEYLAN,WM & HOWARD,PH (1995)
VP 8.96E+000	mm Hg	25	EXT	KLEIN,RG (1982)
DC	pKa			
HL 3.60E-006	atm m3/mol	25	EST	MEYLAN,WM & HOWARD,PH (1991)
OH 1.72E+001	cm3/molc sec	25	EST	MEYLAN,WM & HOWARD,PH (1993)

CAS #: 004549-43-3				N-NITROSOMETHYLALLYLAMINE

Formula: $C_4H_8N_2O$

Mol Weight: 100.12

MP (deg C):   FP (deg C):

BP (deg C):

BP pressure (mm Hg):

Property/Value	Units	Temp	Data Type	Reference
WS 7.78E+004	mg/L	25	EST	MEYLAN,WM ET AL. (1996)
logP 0.21			EST	MEYLAN,WM & HOWARD,PH (1995)
VP 6.03E-001	mm Hg	25	EST	NEELY,WB & BLAU,GE (1985)
DC	pKa			
HL 2.70E-006	atm m3/mol	25	EST	MEYLAN,WM & HOWARD,PH (1991)
OH 3.63E-011	cm3/molc sec	25	EST	MEYLAN,WM & HOWARD,PH (1993)

CAS #: 004549-44-4				N-NITROSOETHYL-N-BUTYLAMINE

Formula: $C_6H_{14}N_2O$

Mol Weight: 130.19

MP (deg C): | FP (deg C):

BP (deg C):

BP pressure (mm Hg):

Property/Value	Units	Temp	Data Type	Reference
WS 3.72E+003	mg/L	25	EST	MEYLAN,WM ET AL. (1996)
logP 1.63			EXP	VERA,A ET AL. (1992)
VP 8.74E-002	mm Hg	25	EST	NEELY,WB & BLAU,GE (1985)
DC	pKa			
HL 6.39E-006	atm m3/mol	25	EST	MEYLAN,WM & HOWARD,PH (1991)
OH 2.23E-011	cm3/molc sec	25	EST	MEYLAN,WM & HOWARD,PH (1993)

CAS #: 004549-74-0				3-METHYL-1,3-PENTADIENE

Formula: C_6H_{10}

Mol Weight: 82.15

MP (deg C): | FP (deg C):

BP (deg C): 77

BP pressure (mm Hg):

Property/Value	Units	Temp	Data Type	Reference
WS 1.05E+002	mg/L	25	EST	MEYLAN,WM ET AL. (1996)
logP 2.99			EST	MEYLAN,WM & HOWARD,PH (1995)
VP 1.40E+002	mm Hg	25	EST	NEELY,WB & BLAU,GE (1985)
DC	pKa			
HL 1.91E-001	atm m3/mol	25	EST	MEYLAN,WM & HOWARD,PH (1991)
OH 1.36E-010	cm3/molc sec	25	EXP	ATKINSON,R (1989)

CAS #: 004551-72-8				1H-PYRROLE-2-CARBOXAMIDE

Formula: $C_5H_6N_2O$

Mol Weight: 110.12

MP (deg C): | FP (deg C):

BP (deg C):

BP pressure (mm Hg):

Property/Value	Units	Temp	Data Type	Reference
WS 4.46E+003	mg/L	25	EST	MEYLAN,WM ET AL. (1996)
logP 0.09			EXP	YAMAGAMI,C ET AL. (1994)
VP 2.63E-004	mm Hg	25	EST	NEELY,WB & BLAU,GE (1985)
DC	pKa			
HL 3.73E-012	atm m3/mol	25	EST	MEYLAN,WM & HOWARD,PH (1991)
OH 1.02E-010	cm3/molc sec	25	EST	MEYLAN,WM & HOWARD,PH (1993)

CAS #: 004551-76-2				3,5-DINITRO-4-METHYLBENZAMIDE

Formula: $C_8H_7N_3O_5$

Mol Weight: 225.16

MP (deg C): | FP (deg C):

BP (deg C):

BP pressure (mm Hg):

Property/Value	Units	Temp	Data Type	Reference
WS 3.46E+003	mg/L	25	EST	MEYLAN,WM ET AL. (1996)
logP 0.68			EXP	HANSCH,C & LEO,AJ (1985)
VP 7.02E-008	mm Hg	25	EST	NEELY,WB & BLAU,GE (1985)
DC	pKa			
HL 3.80E-014	atm m3/mol	25	EST	MEYLAN,WM & HOWARD,PH (1991)
OH 2.14E-012	cm3/molc sec	25	EST	MEYLAN,WM & HOWARD,PH (1993)

CAS #: 004553-62-2				2-METHYLPENTANEDINITRILE

Formula: $C_6H_8N_2$

Mol Weight: 108.14

MP (deg C): -45 | FP (deg C):

BP (deg C): 263

BP pressure (mm Hg):

Property/Value	Units	Temp	Data Type	Reference
WS 3.48E+004	mg/L	25	EST	MEYLAN,WM ET AL. (1996)
logP 0.28			EST	MEYLAN,WM & HOWARD,PH (1995)
VP 5.10E-003	mm Hg	25	EXP	DAUBERT,TE & DANNER,RP (1991)
DC	pKa			
HL 2.97E-008	atm m3/mol	25	EST	MEYLAN,WM & HOWARD,PH (1991)
OH 1.21E-012	cm3/molc sec	25	EST	ATKINSON,R (1988)

CAS #: 004559-87-9				1-METHYL-1-PHENYL UREA

Formula: $C_8H_{10}N_2O$

Mol Weight: 150.18

MP (deg C): | FP (deg C):

BP (deg C):

BP pressure (mm Hg):

Property/Value	Units	Temp	Data Type	Reference
WS 3.32E+004	mg/L	25	EST	MEYLAN,WM ET AL. (1996)
logP 0.42			EXP	HANSCH,C & LEO,AJ (1985)
VP 4.12E-004	mm Hg	25	EST	NEELY,WB & BLAU,GE (1985)
DC	pKa			
HL 4.13E-009	atm m3/mol	25	EST	MEYLAN,WM & HOWARD,PH (1991)
OH 1.40E-011	cm3/molc sec	25	EST	MEYLAN,WM & HOWARD,PH (1993)

CAS #: 004562-27-0				4-PYRIMIDONE

Formula: $C_4H_4N_2O$

Mol Weight: 96.09

MP (deg C): | FP (deg C):

BP (deg C):

BP pressure (mm Hg):

Property/Value	Units	Temp	Data Type	Reference
WS 1.00E+006	mg/L	25	EST	MEYLAN,WM ET AL. (1996)
logP -1.38			EXP	HANSCH,C ET AL. (1995)
VP 6.64E-005	mm Hg	25	EST	NEELY,WB & BLAU,GE (1985)
DC	pKa			
HL 5.92E-010	atm m3/mol	25	EST	MEYLAN,WM & HOWARD,PH (1991)
OH 8.78E-011	cm3/molc sec	25	EST	MEYLAN,WM & HOWARD,PH (1993)

CAS #: 004562-36-1				GITOXIN

Formula: $C_{41}H_{64}O_{14}$

Mol Weight: 780.96

MP (deg C): 285 dec | FP (deg C):

BP (deg C):

BP pressure (mm Hg):

Property/Value	Units	Temp	Data Type	Reference
WS 2.34E+000	mg/L		EXP	YALKOWSKY,SH & DANNENFELSER,RM (1992)
logP 1.67			EXP	SANGSTER,J (1993)
VP 3.30E-030	mm Hg	25	EST	NEELY,WB & BLAU,GE (1985)
DC	pKa			
HL 4.66E-027	atm m3/mol	25	EST	MEYLAN,WM & HOWARD,PH (1991)
OH 1.70E-010	cm3/molc sec	25	EST	MEYLAN,WM & HOWARD,PH (1993)

CAS #: 004563-33-1 — BENZENEMETHANESULFONAMIDE

Formula: $C_7H_9NO_2S$

Mol Weight: 171.22

MP (deg C): FP (deg C):

BP (deg C):

BP pressure (mm Hg):

Property/Value	Units	Temp	Data Type	Reference
WS 4.70E+004	mg/L	25	EST	MEYLAN,WM ET AL. (1996)
logP 0.13			EXP	HANSCH,C ET AL. (1995)
VP 3.30E-004	mm Hg	25	EST	NEELY,WB & BLAU,GE (1985)
DC	pKa			
HL 1.71E-007	atm m3/mol	25	EST	MEYLAN,WM & HOWARD,PH (1991)
OH 1.21E-011	cm3/molc sec	25	EST	MEYLAN,WM & HOWARD,PH (1993)

CAS #: 004568-93-8 — AMYL GALLATE

Formula: $C_{12}H_{16}O_5$

Mol Weight: 240.26

MP (deg C): FP (deg C):

BP (deg C):

BP pressure (mm Hg):

Property/Value	Units	Temp	Data Type	Reference
WS 5.34E+002	mg/L	25	EST	MEYLAN,WM ET AL. (1996)
logP 2.67			EXP	HANSCH,C & LEO,AJ (1985)
VP 4.30E-008	mm Hg	25	EST	NEELY,WB & BLAU,GE (1985)
DC	pKa			
HL 1.21E-016	atm m3/mol	25	EST	MEYLAN,WM & HOWARD,PH (1991)
OH 9.50E-011	cm3/molc sec	25	EST	MEYLAN,WM & HOWARD,PH (1993)

CAS #: 004570-41-6 — BENZOXAZOLE, 2-AMINO

Formula: $C_7H_6N_2O$

Mol Weight: 134.14

MP (deg C): FP (deg C):

BP (deg C):

BP pressure (mm Hg):

Property/Value	Units	Temp	Data Type	Reference
WS 4.37E+003	mg/L	25	EST	MEYLAN,WM ET AL. (1996)
logP 1.53			EXP	HANSCH,C & LEO,AJ (1985)
VP 2.86E-003	mm Hg	25	EST	NEELY,WB & BLAU,GE (1985)
DC	pKa			
HL 2.43E-010	atm m3/mol	25	EST	MEYLAN,WM & HOWARD,PH (1991)
OH 2.00E-010	cm3/molc sec	25	EST	MEYLAN,WM & HOWARD,PH (1993)

CAS #: 004570-45-0 — OXAZOLE-2-AMINE

Formula: $C_3H_5N_2O$

Mol Weight: 85.09

MP (deg C): FP (deg C):

BP (deg C):

BP pressure (mm Hg):

Property/Value	Units	Temp	Data Type	Reference
WS 1.83E+005	mg/L	25	EST	MEYLAN,WM ET AL. (1996)
logP -0.18			EXP	HANSCH,C & LEO,AJ (1985)
VP 2.88E+000	mm Hg	25	EST	NEELY,WB & BLAU,GE (1985)
DC	pKa			
HL 2.49E-009	atm m3/mol	25	EST	MEYLAN,WM & HOWARD,PH (1991)
OH 2.00E-010	cm3/molc sec	25	EST	MEYLAN,WM & HOWARD,PH (1993)

CAS #: 004574-37-2 — IMIDAZOL-5-CONH2-4-(3,3-DIET-1-NNN)

Formula: $C_8H_{14}N_6O$

Mol Weight: 210.24

MP (deg C): FP (deg C):

BP (deg C):

BP pressure (mm Hg):

Property/Value	Units	Temp	Data Type	Reference
WS 8.36E+003	mg/L	25	EST	MEYLAN,WM ET AL. (1996)
logP 0.78			EXP	HANSCH,C & LEO,AJ (1985)
VP 4.08E-009	mm Hg	25	EST	NEELY,WB & BLAU,GE (1985)
DC	pKa			
HL 2.71E-016	atm m3/mol	25	EST	MEYLAN,WM & HOWARD,PH (1991)
OH 3.27E-011	cm3/molc sec	25	EST	MEYLAN,WM & HOWARD,PH (1993)

CAS #: 004593-90-2 — 3-PHENYL-N-BUTYRIC ACID

Formula: $C_{10}H_{12}O_2$

Mol Weight: 164.21

MP (deg C): 39-37 FP (deg C):

BP (deg C): 710-172

BP pressure (mm Hg): 2.00E+001

Property/Value	Units	Temp	Data Type	Reference
WS 2.24E+003	mg/L	25	EST	MEYLAN,WM ET AL. (1996)
logP 2.18			EXP	HANSCH,C & LEO,AJ (1985)
VP 1.31E-003	mm Hg	25	EST	NEELY,WB & BLAU,GE (1985)
DC	pKa			
HL 7.79E-008	atm m3/mol	25	EST	MEYLAN,WM & HOWARD,PH (1991)
OH 8.25E-012	cm3/molc sec	25	EST	MEYLAN,WM & HOWARD,PH (1993)

CAS #: 004595-59-9 — 5-BROMOPYRIMIDINE

Formula: $C_4H_3BrN_2$

Mol Weight: 158.99

MP (deg C): 71-73 FP (deg C):

BP (deg C):

BP pressure (mm Hg):

Property/Value	Units	Temp	Data Type	Reference
WS 1.68E+004	mg/L	25	EST	MEYLAN,WM ET AL. (1996)
logP 0.72			EXP	HANSCH,C & LEO,AJ (1985)
VP 8.67E-001	mm Hg	25	EST	NEELY,WB & BLAU,GE (1985)
DC	pKa			
HL 1.16E-006	atm m3/mol	25	EST	MEYLAN,WM & HOWARD,PH (1991)
OH 1.45E-013	cm3/molc sec	25	EST	MEYLAN,WM & HOWARD,PH (1993)

CAS #: 004595-60-2 — PYRIMIDINE, 2-BROMO-

Formula: $C_4H_3BrN_2$

Mol Weight: 158.99

MP (deg C): FP (deg C):

BP (deg C):

BP pressure (mm Hg):

Property/Value	Units	Temp	Data Type	Reference
WS 2.59E+004	mg/L	25	EST	MEYLAN,WM ET AL. (1996)
logP 0.50			EXP	YAMAGAMI,C ET AL. (1990)
VP 8.67E-001	mm Hg	25	EST	NEELY,WB & BLAU,GE (1985)
DC -1.63	pKa	20	EXP	PERRIN,DD (1972)
HL 1.16E-006	atm m3/mol	25	EST	MEYLAN,WM & HOWARD,PH (1991)
OH 1.45E-013	cm3/molc sec	25	EST	MEYLAN,WM & HOWARD,PH (1993)

CAS #: 004597-87-9				2-METHYLAMINOPYRIDINE

Formula: $C_6H_8N_2$

Mol Weight: 108.14

MP (deg C): 15   FP (deg C):

BP (deg C): 200.5

BP pressure (mm Hg):

Property/ Value	Units	Temp	Data Type	Reference
WS 1.22E+004	mg/L	25	EST	MEYLAN,WM ET AL. (1996)
logP 1.12			EXP	YAMAGAMI,C ET AL. (1990A)
VP 1.11E+000	mm Hg	25	EST	NEELY,WB & BLAU,GE (1985)
DC	pKa			
HL 5.47E-009	atm m3/mol	25	EST	MEYLAN,WM & HOWARD,PH (1991)
OH 9.36E-012	cm3/molc sec	25	EST	MEYLAN,WM & HOWARD,PH (1993)

CAS #: 004598-39-4				4-PYRIMIDINECARBOXYLIC ACID, 1,2,3,6-TETRAHYDRO-

Formula: $C_9H_{12}N_2O_4$

Mol Weight: 212.21

MP (deg C):   FP (deg C):

BP (deg C):

BP pressure (mm Hg):

Property/ Value	Units	Temp	Data Type	Reference
WS 5.19E+003	mg/L	25	EST	MEYLAN,WM ET AL. (1996)
logP 1.01			EXP	FURST,W ET AL. (1990)
VP 1.48E-008	mm Hg	25	EST	NEELY,WB & BLAU,GE (1985)
DC	pKa			
HL 6.06E-012	atm m3/mol	25	EST	MEYLAN,WM & HOWARD,PH (1991)
OH 1.29E-011	cm3/molc sec	25	EST	MEYLAN,WM & HOWARD,PH (1993)

CAS #: 004608-25-7				2-ETHYLISONIAZID

Formula: $C_8H_{11}N_3O$

Mol Weight: 165.20

MP (deg C):   FP (deg C):

BP (deg C):

BP pressure (mm Hg):

Property/ Value	Units	Temp	Data Type	Reference
WS 2.51E+003	mg/L	25	EST	MEYLAN,WM ET AL. (1996)
logP 0.12			EXP	HANSCH,C & LEO,AJ (1985)
VP 7.99E-006	mm Hg	25	EST	NEELY,WB & BLAU,GE (1985)
DC	pKa			
HL 1.77E-014	atm m3/mol	25	EST	MEYLAN,WM & HOWARD,PH (1991)
OH 7.01E-012	cm3/molc sec	25	EST	MEYLAN,WM & HOWARD,PH (1993)

CAS #: 004609-10-3				BUTANOIC ACID, 4-(4-METHOXYBENZOYL)-

Formula: $C_{12}H_{14}O_4$

Mol Weight: 222.24

MP (deg C):   FP (deg C):

BP (deg C):

BP pressure (mm Hg):

Property/ Value	Units	Temp	Data Type	Reference
WS 3.37E+003	mg/L	25	EST	MEYLAN,WM ET AL. (1996)
logP 1.63			EXP	SANGSTER,J (1994)
VP 6.35E-006	mm Hg	25	EST	NEELY,WB & BLAU,GE (1985)
DC	pKa			
HL 7.61E-012	atm m3/mol	25	EST	MEYLAN,WM & HOWARD,PH (1991)
OH 2.72E-011	cm3/molc sec	25	EST	MEYLAN,WM & HOWARD,PH (1993)

CAS #: 004620-70-6				ETHANOL, 2- (1,1-DIMETHYLETHYL)AMINO -

Formula: $C_6H_{15}NO$

Mol Weight: 117.19

MP (deg C): 44   FP (deg C):

BP (deg C): 176.5

BP pressure (mm Hg):

Property/ Value	Units	Temp	Data Type	Reference
WS 4.67E+005	mg/L	25	EST	MEYLAN,WM ET AL. (1996)
logP 0.41			EXP	HANSCH,C ET AL. (1995)
VP 2.78E-001	mm Hg	25	EST	NEELY,WB & BLAU,GE (1985)
DC	pKa			
HL 1.89E-009	atm m3/mol	25	EST	MEYLAN,WM & HOWARD,PH (1991)
OH 7.83E-011	cm3/molc sec	25	EST	MEYLAN,WM & HOWARD,PH (1993)

CAS #: 004621-36-7				BENZENE, (1-ETHYLOCTYL)-

Formula: $C_{16}H_{26}$

Mol Weight: 218.39

MP (deg C):   FP (deg C):

BP (deg C):

BP pressure (mm Hg):

Property/ Value	Units	Temp	Data Type	Reference
WS 3.69E-002	mg/L	25	EST	MEYLAN,WM ET AL. (1996)
logP 7.00			EXP	SHERBLOM,PM ET AL. (1992)
VP 3.82E-003	mm Hg	25	EST	NEELY,WB & BLAU,GE (1985)
DC	pKa			
HL 7.61E-002	atm m3/mol	25	EST	MEYLAN,WM & HOWARD,PH (1991)
OH 1.73E-011	cm3/molc sec	25	EST	MEYLAN,WM & HOWARD,PH (1993)

CAS #: 004621-66-3				THIONICOTINAMIDE

Formula: $C_6H_6N_2S$

Mol Weight: 138.19

MP (deg C):   FP (deg C):

BP (deg C):

BP pressure (mm Hg):

Property/ Value	Units	Temp	Data Type	Reference
WS 2.32E+005	mg/L	25	EST	MEYLAN,WM ET AL. (1996)
logP 0.67			EXP	HANSCH,C & LEO,AJ (1985)
VP 8.41E-003	mm Hg	25	EST	NEELY,WB & BLAU,GE (1985)
DC	pKa			
HL 5.13E-010	atm m3/mol	25	EST	MEYLAN,WM & HOWARD,PH (1991)
OH 2.18E-011	cm3/molc sec	25	EST	MEYLAN,WM & HOWARD,PH (1993)

CAS #: 004630-06-2				6-METHYL-5-HEPTEN-2-OL

Formula: $C_8H_{16}O$

Mol Weight: 128.22

MP (deg C):   FP (deg C):

BP (deg C): 78

BP pressure (mm Hg): 1.40E+001

Property/ Value	Units	Temp	Data Type	Reference
WS 2.48E+003	mg/L	25	EST	MEYLAN,WM ET AL. (1996)
logP 2.44			EST	MEYLAN,WM & HOWARD,PH (1995)
VP 2.72E-001	mm Hg	25	EXT	PERRY,RH & GREEN,D (1984)
DC	pKa			
HL 2.73E-005	atm m3/mol	25	EST	MEYLAN,WM & HOWARD,PH (1991)
OH 6.74E-011	cm3/molc sec	25	EST	MEYLAN,WM & HOWARD,PH (1993)

CAS #: 004637-56-3				4-(HYDROXYAMINO)QUINOLINE 1-OXIDE
Formula: $C_9H_{10}N_2O_2$				
Mol Weight: 178.19				
MP (deg C):		FP (deg C):		
BP (deg C):				
BP pressure (mm Hg):				

Property/ Value	Units	Temp	Data Type	Reference
WS 5.75E+004	mg/L	25	EST	MEYLAN,WM ET AL. (1996)
logP 0.00			EXP	SANGSTER,J (1993)
VP 2.28E-008	mm Hg	25	EST	NEELY,WB & BLAU,GE (1985)
DC	pKa			
HL 4.87E-018	atm m3/mol	25	EST	MEYLAN,WM & HOWARD,PH (1991)
OH 2.00E-010	cm3/molc sec	25	EST	MEYLAN,WM & HOWARD,PH (1993)

CAS #: 004637-59-6				4-CHLOROQUINOLINE-1-OXIDE
Formula: C_9H_8ClNO				
Mol Weight: 181.62				
MP (deg C):		FP (deg C):		
BP (deg C):				
BP pressure (mm Hg):				

Property/ Value	Units	Temp	Data Type	Reference
WS 6.61E+003	mg/L	25	EST	MEYLAN,WM ET AL. (1996)
logP 1.08			EXP	HANSCH,C & LEO,AJ (1985)
VP 4.81E-005	mm Hg	25	EST	NEELY,WB & BLAU,GE (1985)
DC	pKa			
HL 5.10E-012	atm m3/mol	25	EST	MEYLAN,WM & HOWARD,PH (1991)
OH 1.01E-011	cm3/molc sec	25	EST	MEYLAN,WM & HOWARD,PH (1993)

CAS #: 004638-48-6				5-CHLOROSALICYLANILIDE
Formula: $C_{13}H_{10}ClNO_2$				
Mol Weight: 247.68				
MP (deg C): 210-212		FP (deg C):		
BP (deg C):				
BP pressure (mm Hg):				

Property/ Value	Units	Temp	Data Type	Reference
WS 3.22E+001	mg/L	25	EST	MEYLAN,WM ET AL. (1996)
logP 4.05			EXP	TERADA,H ET AL. (1988)
VP 5.90E-009	mm Hg	25	EST	NEELY,WB & BLAU,GE (1985)
DC	pKa			
HL 1.19E-010	atm m3/mol	25	EST	MEYLAN,WM & HOWARD,PH (1991)
OH 2.14E-011	cm3/molc sec	25	EST	MEYLAN,WM & HOWARD,PH (1993)

CAS #: 004649-27-8				4-CARBOXY-N-PHENYLPHATHALIMIDE
Formula: $C_{15}H_9NO_4$				
Mol Weight: 267.24				
MP (deg C):		FP (deg C):		
BP (deg C):				
BP pressure (mm Hg):				

Property/ Value	Units	Temp	Data Type	Reference
WS 4.62E+002	mg/L	25	EST	MEYLAN,WM ET AL. (1996)
logP 1.89			EST	MEYLAN,WM & HOWARD,PH (1995)
VP 4.09E-011	mm Hg	25	EST	NEELY,WB & BLAU,GE (1985)
DC	pKa			
HL 1.06E-012	atm m3/mol	25	EST	MEYLAN,WM & HOWARD,PH (1991)
OH 1.32E-011	cm3/molc sec	25	EST	MEYLAN,WM & HOWARD,PH (1993)

CAS #: 004653-73-0				1,3,5-TRIAZINE-2,4-DIAMINE, 1-(4-BUTYLPHENYL)-1,
Formula: $C_{15}H_{23}N_5$				
Mol Weight: 273.38				
MP (deg C):		FP (deg C):		
BP (deg C):				
BP pressure (mm Hg):				

Property/ Value	Units	Temp	Data Type	Reference
WS 9.35E+001	mg/L	25	EST	MEYLAN,WM ET AL. (1996)
logP 2.66			EXP	SANGSTER,J (1993)
VP 2.21E-007	mm Hg	25	EST	NEELY,WB & BLAU,GE (1985)
DC	pKa			
HL 1.88E-012	atm m3/mol	25	EST	MEYLAN,WM & HOWARD,PH (1991)
OH 2.25E-010	cm3/molc sec	25	EST	MEYLAN,WM & HOWARD,PH (1993)

CAS #: 004655-34-9				ISOPROPYL METHACRYLATE
Formula: $C_7H_{12}O_2$				
Mol Weight: 128.17				
MP (deg C):		FP (deg C):		
BP (deg C): 125				
BP pressure (mm Hg):				

Property/ Value	Units	Temp	Data Type	Reference
WS 1.12E+003	mg/L	25	EST	MEYLAN,WM ET AL. (1996)
logP 2.25			EXP	HANSCH,C & LEO,AJ (1985)
VP 1.15E+001	mm Hg	25	EST	NEELY,WB & BLAU,GE (1985)
DC	pKa			
HL 2.55E-004	atm m3/mol	25	EST	MEYLAN,WM & HOWARD,PH (1991)
OH 2.16E-011	cm3/molc sec	25	EST	MEYLAN,WM & HOWARD,PH (1993)

CAS #: 004657-20-9				2,2,4,6,6,8-HEXAMETHYL-4,8-DIPHENYLCYCLOTETRAS*
Formula: $C_{18}H_{28}O_4Si_4$				
Mol Weight: 420.77				
MP (deg C):		FP (deg C):		
BP (deg C):				
BP pressure (mm Hg):				

Property/ Value	Units	Temp	Data Type	Reference
WS 8.30E-005	mg/L	25	EST	MEYLAN,WM ET AL. (1996)
logP 7.52			EST	MEYLAN,WM & HOWARD,PH (1995)
VP 8.37E-006	mm Hg	25	EST	NEELY,WB & BLAU,GE (1985)
DC	pKa			
HL 7.58E-011	atm m3/mol	25	EST	MEYLAN,WM & HOWARD,PH (1991)
OH 4.80E-012	cm3/molc sec	25	EST	MEYLAN,WM & HOWARD,PH (1993)

CAS #: 004658-28-0				AZIPROTRYNE
Formula: $C_7H_{11}N_7S$				
Mol Weight: 225.28				
MP (deg C):		FP (deg C):		
BP (deg C):				
BP pressure (mm Hg):				

Property/ Value	Units	Temp	Data Type	Reference
WS 5.50E+001	mg/L	20	EXP	YALKOWSKY,SH & DANNENFELSER,RM (1992)
logP -4.01			EST	MEYLAN,WM & HOWARD,PH (1995)
VP 1.24E-010	mm Hg	25	EST	NEELY,WB & BLAU,GE (1985)
DC	pKa			
HL 5.02E-013	atm m3/mol	25	EST	MEYLAN,WM & HOWARD,PH (1991)
OH 1.95E-011	cm3/molc sec	25	EST	MEYLAN,WM & HOWARD,PH (1993)

CAS #: 004671-97-0			N-PHENYL-3-N'-PIPERIDINOACETAMIDE	

Formula: $C_{13}H_{18}N_2O$

Mol Weight: 218.30

MP (deg C): | FP (deg C):

BP (deg C):

BP pressure (mm Hg):

Property/Value	Units	Temp	Data Type	Reference
WS 1.70E+002	mg/L	25	EST	MEYLAN,WM ET AL. (1996)
logP 2.71			EXP	HANSCH,C ET AL. (1995)
VP 1.05E-006	mm Hg	25	EST	NEELY,WB & BLAU,GE (1985)
DC	pKa			
HL 6.30E-011	atm m3/mol	25	EST	MEYLAN,WM & HOWARD,PH (1991)
OH 1.11E-010	cm3/molc sec	25	EST	MEYLAN,WM & HOWARD,PH (1993)

CAS #: 004678-45-9			GRAYANOTOXIN III	

Formula: $C_{20}H_{34}O_6$

Mol Weight: 370.49

MP (deg C): | FP (deg C):

BP (deg C):

BP pressure (mm Hg):

Property/Value	Units	Temp	Data Type	Reference
WS 2.74E+003	mg/L	25	EST	MEYLAN,WM ET AL. (1996)
logP 0.27			EXP	HANSCH,C & LEO,AJ (1985)
VP 5.45E-014	mm Hg	25	EST	NEELY,WB & BLAU,GE (1985)
DC	pKa			
HL 1.45E-012	atm m3/mol	25	EST	MEYLAN,WM & HOWARD,PH (1991)
OH 7.84E-011	cm3/molc sec	25	EST	MEYLAN,WM & HOWARD,PH (1993)

CAS #: 004684-94-0			6-CHLOROPICOLINIC ACID	

Formula: $C_6H_4ClNO_2$

Mol Weight: 157.56

MP (deg C): | FP (deg C):

BP (deg C):

BP pressure (mm Hg):

Property/Value	Units	Temp	Data Type	Reference
WS 3.68E+003	mg/L	25	EST	MEYLAN,WM ET AL. (1996)
logP 1.50			EXP	SANGSTER,J (1994)
VP 1.20E-003	mm Hg	25	EST	NEELY,WB & BLAU,GE (1985)
DC	pKa			
HL 6.63E-009	atm m3/mol	25	EST	MEYLAN,WM & HOWARD,PH (1991)
OH 5.91E-013	cm3/molc sec	25	EST	MEYLAN,WM & HOWARD,PH (1993)

CAS #: 004685-14-7			PARAQUAT	

Formula: $C_{12}H_{14}Cl_2N_2$

Mol Weight: 257.16

MP (deg C): | FP (deg C):

BP (deg C):

BP pressure (mm Hg):

Property/Value	Units	Temp	Data Type	Reference
WS 6.20E+005	mg/L	25	EXP	WAUCHOPE,RD ET AL. (1992)
logP -4.22			EXP	PLATFORD,RF (1983)
VP 1.00E-007	mm Hg	25	EST	<1E-7; SEIBER,JN & WOODROW,JE (1984)
DC	pKa			
HL 1.00E-009	atm m3/mol	25	EST	<1E-9; SEIBER,JN & WOODROW,JE (1984)
OH 2.12E-011	cm3/molc sec	25	EST	MEYLAN,WM & HOWARD,PH (1993)

CAS #: 004697-36-3			CARBENICILLIN	

Formula: $C_{17}H_{18}N_2O_6S$

Mol Weight: 378.41

MP (deg C): | FP (deg C):

BP (deg C):

BP pressure (mm Hg):

Property/Value	Units	Temp	Data Type	Reference
WS 4.51E+002	mg/L	25	EST	MEYLAN,WM ET AL. (1996)
logP 1.13			EXP	HANSCH,C & LEO,AJ (1985)
VP 1.00E-014	mm Hg	25	EST	NEELY,WB & BLAU,GE (1985)
DC	pKa			
HL 8.66E-020	atm m3/mol	25	EST	MEYLAN,WM & HOWARD,PH (1991)
OH 9.21E-011	cm3/molc sec	25	EST	MEYLAN,WM & HOWARD,PH (1993)

CAS #: 004699-82-5			2H-1,4-BENZODIAZEPIN-2-ONE, 7-CHLORO-1,3-DIHYDRO	

Formula: $C_{16}H_{13}ClN_2O$

Mol Weight: 284.75

MP (deg C): | FP (deg C):

BP (deg C):

BP pressure (mm Hg):

Property/Value	Units	Temp	Data Type	Reference
WS 2.16E+001	mg/L	25	EST	MEYLAN,WM ET AL. (1996)
logP 3.33			EXP	SANGSTER,J (1994)
VP 2.86E-009	mm Hg	25	EST	NEELY,WB & BLAU,GE (1985)
DC	pKa			
HL 1.96E-010	atm m3/mol	25	EST	MEYLAN,WM & HOWARD,PH (1991)
OH 1.64E-011	cm3/molc sec	25	EST	MEYLAN,WM & HOWARD,PH (1993)

CAS #: 004712-38-3			1-BUTANOL (D)	

Formula: $C_4H_{10}O$

Mol Weight: 76.12

MP (deg C): | FP (deg C):

BP (deg C): 118

BP pressure (mm Hg):

Property/Value	Units	Temp	Data Type	Reference
WS 7.67E+004	mg/L	25	EST	MEYLAN,WM ET AL. (1996)
logP 0.84			EST	MEYLAN,WM & HOWARD,PH (1995)
VP 9.67E+000	mm Hg	25	EST	NEELY,WB & BLAU,GE (1985)
DC 16.10	pKa		EXP	SERJEANT,EP & DEMPSEY,B (1979)
HL 9.99E-006	atm m3/mol	25	EST	MEYLAN,WM & HOWARD,PH (1991)
OH 6.89E-012	cm3/molc sec	25	EST	MEYLAN,WM & HOWARD,PH (1993)

CAS #: 004712-39-4			2-BUTANOL (D)	

Formula: $C_4H_{10}O$

Mol Weight: 76.12

MP (deg C): | FP (deg C):

BP (deg C):

BP pressure (mm Hg):

Property/Value	Units	Temp	Data Type	Reference
WS 1.30E+005	mg/L	25	EST	MEYLAN,WM ET AL. (1996)
logP 0.77			EST	MEYLAN,WM & HOWARD,PH (1995)
VP 3.07E+001	mm Hg	25	EST	NEELY,WB & BLAU,GE (1985)
DC 17.60	pKa	25	EXP	SERJEANT,EP & DEMPSEY,B (1979)
HL 9.99E-006	atm m3/mol	25	EST	MEYLAN,WM & HOWARD,PH (1991)
OH 9.98E-012	cm3/molc sec	25	EST	MEYLAN,WM & HOWARD,PH (1993)

CAS #: 004720-09-6				ANDROMEDOTOXIN

Formula: $C_{22}H_{36}O_7$

Mol Weight: 412.53

MP (deg C): **FP (deg C):**

BP (deg C):

BP pressure (mm Hg):

Property/ Value	Units	Temp	Data Type	Reference
WS 2.81E+004	mg/L	12	EXP	SEIDELL,A (1941)
logP -0.03			EST	MEYLAN,WM & HOWARD,PH (1995)
VP 2.30E-014	mm Hg	25	EST	NEELY,WB & BLAU,GE (1985)
DC	pKa			
HL 9.43E-014	atm m3/mol	25	EST	MEYLAN,WM & HOWARD,PH (1991)
OH 7.47E-011	cm3/molc sec	25	EST	MEYLAN,WM & HOWARD,PH (1993)

CAS #: 004726-14-1				NITRALIN

Formula: $C_{13}H_{19}N_3O_6S$

Mol Weight: 345.38

MP (deg C): 151-152 **FP (deg C):**

BP (deg C):

BP pressure (mm Hg):

Property/ Value	Units	Temp	Data Type	Reference
WS 6.00E-001	mg/L	25	EXP	YALKOWSKY,SH & DANNENFELSER,RM (1992)
logP 2.92			EST	MEYLAN,WM & HOWARD,PH (1995)
VP 9.30E-009	mm Hg	20	EXP	MERCK INDEX (1989)
DC	pKa			
HL 7.04E-009	atm m3/mol	20	EST	VP/WSOL
OH 2.51E-011	cm3/molc sec	25	EST	MEYLAN,WM & HOWARD,PH (1993)

CAS #: 004741-74-6				1,1-BIS(P-METHOXYPHENYL)-2,2-DIMETHYLPROPANE

Formula: $C_{19}H_{24}O_2$

Mol Weight: 284.40

MP (deg C): **FP (deg C):**

BP (deg C):

BP pressure (mm Hg):

Property/ Value	Units	Temp	Data Type	Reference
WS 2.16E-001	mg/L	25	EST	MEYLAN,WM ET AL. (1996)
logP 5.67			EST	MEYLAN,WM & HOWARD,PH (1995)
VP 1.28E-005	mm Hg	25	EST	NEELY,WB & BLAU,GE (1985)
DC	pKa			
HL 5.22E-006	atm m3/mol	25	EST	MEYLAN,WM & HOWARD,PH (1991)
OH 5.63E-011	cm3/molc sec	25	EST	MEYLAN,WM & HOWARD,PH (1993)

CAS #: 004746-61-6				O-HYDROXY PHENYLACETAMIDE

Formula: $C_8H_9NO_2$

Mol Weight: 151.17

MP (deg C): **FP (deg C):**

BP (deg C):

BP pressure (mm Hg):

Property/ Value	Units	Temp	Data Type	Reference
WS 1.86E+004	mg/L	25	EST	MEYLAN,WM ET AL. (1996)
logP 0.71			EXP	HANSCH,C & LEO,AJ (1985)
VP 7.00E-006	mm Hg	25	EST	NEELY,WB & BLAU,GE (1985)
DC	pKa			
HL 6.42E-013	atm m3/mol	25	EST	MEYLAN,WM & HOWARD,PH (1991)
OH 1.77E-011	cm3/molc sec	25	EST	MEYLAN,WM & HOWARD,PH (1993)

CAS #: 004749-68-2				2-(2-CHLOROPHENYLIMINO)IMIDAZOLINE

Formula: $C_9H_{10}ClN_3$

Mol Weight: 195.65

MP (deg C): **FP (deg C):**

BP (deg C):

BP pressure (mm Hg):

Property/ Value	Units	Temp	Data Type	Reference
WS 3.18E+003	mg/L	25	EST	MEYLAN,WM ET AL. (1996)
logP 1.36			EXP	HANSCH,C & LEO,AJ (1985)
VP 1.03E-006	mm Hg	25	EST	NEELY,WB & BLAU,GE (1985)
DC	pKa			
HL 2.03E-011	atm m3/mol	25	EST	MEYLAN,WM & HOWARD,PH (1991)
OH 8.49E-011	cm3/molc sec	25	EST	MEYLAN,WM & HOWARD,PH (1993)

CAS #: 004751-48-8				2-(2,6-DIETPHENYLIMINO)IMIDAZOLIDINE

Formula: $C_{13}H_{19}N_3$

Mol Weight: 217.32

MP (deg C): **FP (deg C):**

BP (deg C):

BP pressure (mm Hg):

Property/ Value	Units	Temp	Data Type	Reference
WS 2.76E+003	mg/L	25	EST	MEYLAN,WM ET AL. (1996)
logP 2.48			EXP	HANSCH,C & LEO,AJ (1985)
VP 1.34E-005	mm Hg	25	EST	NEELY,WB & BLAU,GE (1985)
DC	pKa			
HL 5.89E-011	atm m3/mol	25	EST	MEYLAN,WM & HOWARD,PH (1991)
OH 1.59E-010	cm3/molc sec	25	EST	MEYLAN,WM & HOWARD,PH (1993)

CAS #: 004755-77-5				ETHYL CHLOROGLYOXYLATE

Formula: $C_4H_5ClO_3$

Mol Weight: 136.54

MP (deg C): **FP (deg C):**

BP (deg C): 137

BP pressure (mm Hg):

Property/ Value	Units	Temp	Data Type	Reference
WS 3.45E+005	mg/L	25	EST	MEYLAN,WM ET AL. (1996)
logP -0.70			EST	MEYLAN,WM & HOWARD,PH (1995)
VP 7.61E+000	mm Hg	25	EXP	PERRY,RH & GREEN,D (1984)
DC	pKa			
HL 1.24E-005	atm m3/mol	25	EST	MEYLAN,WM & HOWARD,PH (1991)
OH 1.66E-012	cm3/molc sec	25	EST	MEYLAN,WM & HOWARD,PH (1993)

CAS #: 004756-30-3				4-CHLOROPHENYL GLUCOPYRANOSIDE

Formula: $C_{12}H_{15}ClO_6$

Mol Weight: 290.70

MP (deg C): **FP (deg C):**

BP (deg C):

BP pressure (mm Hg):

Property/ Value	Units	Temp	Data Type	Reference
WS 8.34E+003	mg/L	25	EST	MEYLAN,WM ET AL. (1996)
logP 0.26			EXP	HANSCH,C & LEO,AJ (1985)
VP 1.21E-011	mm Hg	25	EST	NEELY,WB & BLAU,GE (1985)
DC	pKa			
HL 8.52E-016	atm m3/mol	25	EST	MEYLAN,WM & HOWARD,PH (1991)
OH 6.83E-011	cm3/molc sec	25	EST	MEYLAN,WM & HOWARD,PH (1993)

CAS #: 004759-24-4 — ALPHA-HYDROXYBENZYLPENCILLIN

Formula: $C_{16}H_{18}N_2O_5S$

Mol Weight: 350.40

MP (deg C): FP (deg C):

BP (deg C):

BP pressure (mm Hg):

Property/Value	Units	Temp	Data Type	Reference
WS 3.93E+002	mg/L	25	EST	MEYLAN,WM ET AL. (1996)
logP 1.40			EXP	HANSCH,C & LEO,AJ (1985)
VP 4.79E-016	mm Hg	25	EST	NEELY,WB & BLAU,GE (1985)
DC	pKa			
HL 1.82E-015	atm m3/mol	25	EST	MEYLAN,WM & HOWARD,PH (1991)
OH 9.58E-011	cm3/molc sec	25	EST	MEYLAN,WM & HOWARD,PH (1993)

CAS #: 004764-17-4 — 3,4-METHYLENEDIOXYAMPHETAMINE

Formula: $C_{10}H_{13}NO_2$

Mol Weight: 179.22

MP (deg C): FP (deg C):

BP (deg C):

BP pressure (mm Hg):

Property/Value	Units	Temp	Data Type	Reference
WS 2.25E+004	mg/L	25	EST	MEYLAN,WM ET AL. (1996)
logP 1.64			EXP	HANSCH,C & LEO,AJ (1985)
VP 1.69E-003	mm Hg	25	EST	NEELY,WB & BLAU,GE (1985)
DC 9.67	pKa	25	EXP	PERRIN,DD (1965)
HL 1.25E-009	atm m3/mol	25	EST	MEYLAN,WM & HOWARD,PH (1991)
OH 9.35E-011	cm3/molc sec	25	EST	MEYLAN,WM & HOWARD,PH (1993)

CAS #: 004770-31-4 — 1(3H)-ISOBENZOFURANONE,3,3-DIETHYL

Formula: $C_{12}H_{14}O_2$

Mol Weight: 190.24

MP (deg C): FP (deg C):

BP (deg C):

BP pressure (mm Hg):

Property/Value	Units	Temp	Data Type	Reference
WS 2.47E+002	mg/L	25	EST	MEYLAN,WM ET AL. (1996)
logP 2.69			EXP	HANSCH,C & LEO,AJ (1985)
VP 2.27E-004	mm Hg	25	EST	NEELY,WB & BLAU,GE (1985)
DC	pKa			
HL 3.95E-005	atm m3/mol	25	EST	MEYLAN,WM & HOWARD,PH (1991)
OH 6.55E-012	cm3/molc sec	25	EST	MEYLAN,WM & HOWARD,PH (1993)

CAS #: 004774-14-5 — 2-CHLORO-6-CHLORO-PYRAZINE

Formula: $C_4H_2Cl_2N_2$

Mol Weight: 148.98

MP (deg C): 55-58 FP (deg C):

BP (deg C):

BP pressure (mm Hg):

Property/Value	Units	Temp	Data Type	Reference
WS 3.78E+003	mg/L	25	EST	MEYLAN,WM ET AL. (1996)
logP 1.53			EXP	YAMAGAMI,C & TAKAO,N (1991)
VP 4.38E-001	mm Hg	25	EST	NEELY,WB & BLAU,GE (1985)
DC	pKa			
HL 6.38E-003	atm m3/mol	25	EST	MEYLAN,WM & HOWARD,PH (1991)
OH 1.14E-013	cm3/molc sec	25	EST	MEYLAN,WM & HOWARD,PH (1993)

CAS #: 004774-15-6 — PYRAZINE, 2,6-DIMETHOXY-

Formula: $C_6H_8N_2O_2$

Mol Weight: 140.14

MP (deg C): FP (deg C):

BP (deg C):

BP pressure (mm Hg):

Property/Value	Units	Temp	Data Type	Reference
WS 3.74E+003	mg/L	25	EST	MEYLAN,WM ET AL. (1996)
logP 1.58			EXP	YAMAGAMI,C ET AL. (1991)
VP 1.57E-001	mm Hg	25	EST	NEELY,WB & BLAU,GE (1985)
DC	pKa			
HL 8.49E-007	atm m3/mol	25	EST	MEYLAN,WM & HOWARD,PH (1991)
OH 2.96E-011	cm3/molc sec	25	EST	MEYLAN,WM & HOWARD,PH (1993)

CAS #: 004774-24-7 — QUINOLINE, 2-(1-PIPERAZINYL)-

Formula: $C_{13}H_{15}N_3$

Mol Weight: 213.28

MP (deg C): FP (deg C):

BP (deg C):

BP pressure (mm Hg):

Property/Value	Units	Temp	Data Type	Reference
WS 6.89E+003	mg/L	25	EST	MEYLAN,WM ET AL. (1996)
logP 2.04			EXP	CACCIA,S ET AL. (1985)
VP 5.19E-006	mm Hg	25	EST	NEELY,WB & BLAU,GE (1985)
DC	pKa			
HL 1.45E-012	atm m3/mol	25	EST	MEYLAN,WM & HOWARD,PH (1991)
OH 3.06E-010	cm3/molc sec	25	EST	MEYLAN,WM & HOWARD,PH (1993)

CAS #: 004780-24-9 — ISOPROPICILLIN

Formula: $C_{18}H_{22}N_2O_5S$

Mol Weight: 378.45

MP (deg C): FP (deg C):

BP (deg C):

BP pressure (mm Hg):

Property/Value	Units	Temp	Data Type	Reference
WS 1.83E+001	mg/L	25	EST	MEYLAN,WM ET AL. (1996)
logP 2.76			EXP	HANSCH,C & LEO,AJ (1985)
VP 4.55E-013	mm Hg	25	EST	NEELY,WB & BLAU,GE (1985)
DC	pKa			
HL 7.79E-015	atm m3/mol	25	EST	MEYLAN,WM & HOWARD,PH (1991)
OH 1.08E-010	cm3/molc sec	25	EST	MEYLAN,WM & HOWARD,PH (1993)

CAS #: 004780-79-4 — 1-NAPHTHALENEMETHANOL

Formula: $C_{11}H_{10}O$

Mol Weight: 158.20

MP (deg C): FP (deg C):

BP (deg C):

BP pressure (mm Hg):

Property/Value	Units	Temp	Data Type	Reference
WS 2.84E+003	mg/L	25	EXP	SOUTHWORTH,GR & KELLER,JL (1986)
logP 2.37			EXP	SANGSTER,J (1993)
VP 2.89E-005	mm Hg	25	EST	NEELY,WB & BLAU,GE (1985)
DC	pKa			
HL 2.12E-008	atm m3/mol	25	EST	MEYLAN,WM & HOWARD,PH (1991)
OH 5.71E-011	cm3/molc sec	25	EST	MEYLAN,WM & HOWARD,PH (1993)

CAS #: 004783-68-0 — 2-PHENOXYPYRIDINE

Formula: $C_{11}H_9NO$
Mol Weight: 171.20

Property/Value	Units	Temp	Data Type	Reference
WS 5.52E+002	mg/L	25	EST	MEYLAN,WM ET AL. (1996)
logP 2.39			EXP	HANSCH,C & LEO,AJ (1985)
VP 2.57E-003	mm Hg	25	EST	NEELY,WB & BLAU,GE (1985)
DC	pKa			
HL 1.41E-006	atm m3/mol	25	EST	MEYLAN,WM & HOWARD,PH (1991)
OH 5.85E-012	cm3/molc sec	25	EST	MEYLAN,WM & HOWARD,PH (1993)

CAS #: 004784-77-4 — 2-BROMO-2-BUTENE

Formula: C_4H_7Br
Mol Weight: 135.01
BP (deg C): 104.5

Property/Value	Units	Temp	Data Type	Reference
WS 7.32E+002	mg/L	25	EST	MEYLAN,WM ET AL. (1996)
logP 2.43			EST	MEYLAN,WM & HOWARD,PH (1995)
VP 8.04E+001	mm Hg	25	EXP	PERRY,RH & GREEN,D (1984)
DC	pKa			
HL 1.75E-002	atm m3/mol	25	EST	MEYLAN,WM & HOWARD,PH (1991)
OH 4.33E-011	cm3/molc sec	25	EST	MEYLAN,WM & HOWARD,PH (1993)

CAS #: 004786-20-3 — CROTONONITRILE

Formula: C_4H_5N
Mol Weight: 67.09
BP (deg C): 120-121

Property/Value	Units	Temp	Data Type	Reference
WS 2.18E+004	mg/L	25	EST	MEYLAN,WM ET AL. (1996)
logP 0.62			EST	MEYLAN,WM & HOWARD,PH (1995)
VP 3.20E+001	mm Hg	25	EXP	YAWS,CL (1994)
DC	pKa			
HL 2.16E-004	atm m3/mol	25	EST	MEYLAN,WM & HOWARD,PH (1991)
OH 9.16E-012	cm3/molc sec	25	EST	MEYLAN,WM & HOWARD,PH (1993)

CAS #: 004794-83-6 — 2-(2,4-DIME PHENYLIMINO)IMIDAZOLINE

Formula: $C_{11}H_{15}N_3$
Mol Weight: 189.26

Property/Value	Units	Temp	Data Type	Reference
WS 1.47E+003	mg/L	25	EST	MEYLAN,WM ET AL. (1996)
logP 1.79			EXP	HANSCH,C & LEO,AJ (1985)
VP 7.37E-007	mm Hg	25	EST	NEELY,WB & BLAU,GE (1985)
DC	pKa			
HL 3.34E-011	atm m3/mol	25	EST	MEYLAN,WM & HOWARD,PH (1991)
OH 1.37E-010	cm3/molc sec	25	EST	MEYLAN,WM & HOWARD,PH (1993)

CAS #: 004796-68-3 — BICYCLO[2.2.1]HEPTA-2,5-DIEN-7-OL, BENZOATE

Formula: $C_{14}H_{12}O_2$
Mol Weight: 212.25

Property/Value	Units	Temp	Data Type	Reference
WS 9.77E+001	mg/L	25	EST	MEYLAN,WM ET AL. (1996)
logP 3.03			EXP	LEAHY,DE ET AL. (1989)
VP 5.45E-004	mm Hg	25	EST	NEELY,WB & BLAU,GE (1985)
DC	pKa			
HL 2.86E-005	atm m3/mol	25	EST	MEYLAN,WM & HOWARD,PH (1991)
OH 1.21E-010	cm3/molc sec	25	EST	MEYLAN,WM & HOWARD,PH (1993)

CAS #: 004798-44-1 — 1-HEXEN-3-OL

Formula: $C_6H_{12}O$
Mol Weight: 100.16
BP (deg C): 134

Property/Value	Units	Temp	Data Type	Reference
WS 2.52E+004	mg/L	25	EXP	YALKOWSKY,SH & DANNENFELSER,RM (1992)
logP 1.61			EST	MEYLAN,WM & HOWARD,PH (1995)
VP 2.79E+000	mm Hg	25	EST	NEELY,WB & BLAU,GE (1985)
DC	pKa			
HL 1.31E-005	atm m3/mol	25	EST	MEYLAN,WM & HOWARD,PH (1991)
OH 3.75E-011	cm3/molc sec	25	EST	MEYLAN,WM & HOWARD,PH (1993)

CAS #: 004798-45-2 — 4-METHYL-1-PENTEN-3-OL

Formula: $C_6H_{12}O$
Mol Weight: 100.16

Property/Value	Units	Temp	Data Type	Reference
WS 3.06E+004	mg/L	25	EXP	BARTON,AFM (1984)
logP 1.54			EST	MEYLAN,WM & HOWARD,PH (1995)
VP 5.90E+000	mm Hg	25	EST	NEELY,WB & BLAU,GE (1985)
DC	pKa			
HL 1.31E-005	atm m3/mol	25	EST	MEYLAN,WM & HOWARD,PH (1991)
OH 3.75E-011	cm3/molc sec	25	EST	MEYLAN,WM & HOWARD,PH (1993)

CAS #: 004798-58-7 — 2-HEXENE-4-OL

Formula: $C_6H_{12}O$
Mol Weight: 100.16

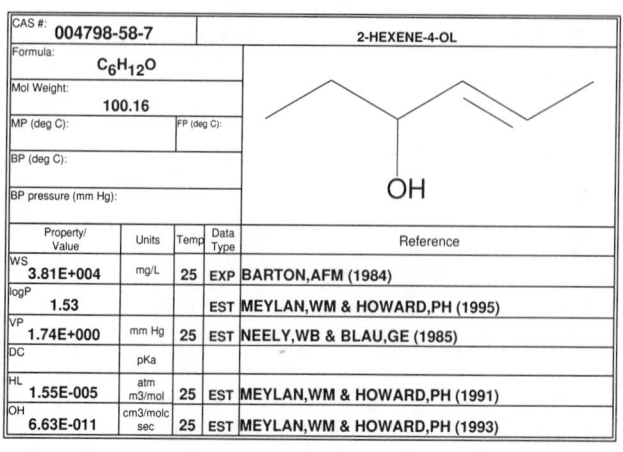

Property/Value	Units	Temp	Data Type	Reference
WS 3.81E+004	mg/L	25	EXP	BARTON,AFM (1984)
logP 1.53			EST	MEYLAN,WM & HOWARD,PH (1995)
VP 1.74E+000	mm Hg	25	EST	NEELY,WB & BLAU,GE (1985)
DC	pKa			
HL 1.55E-005	atm m3/mol	25	EST	MEYLAN,WM & HOWARD,PH (1991)
OH 6.63E-011	cm3/molc sec	25	EST	MEYLAN,WM & HOWARD,PH (1993)

CAS #: 004799-59-1 — O,O-DIET-O-(3-ME-4-SMEPH)P=O

Formula: $C_{12}H_{19}O_4PS$

Mol Weight: 290.32

MP (deg C): | FP (deg C):

BP (deg C):

BP pressure (mm Hg):

Property/ Value	Units	Temp	Data Type	Reference
WS 2.21E+001	mg/L	25	EST	MEYLAN,WM ET AL. (1996)
logP 3.28			EXP	HANSCH,C & LEO,AJ (1985)
VP 8.96E-006	mm Hg	25	EST	NEELY,WB & BLAU,GE (1985)
DC	pKa			
HL 5.22E-009	atm m3/mol	25	EST	MEYLAN,WM & HOWARD,PH (1991)
OH 5.11E-011	cm3/molc sec	25	EST	MEYLAN,WM & HOWARD,PH (1993)

CAS #: 004812-20-8 — O-ISOPROPOXYPHENOL

Formula: $C_9H_{12}O_2$

Mol Weight: 152.19

MP (deg C): | FP (deg C):

BP (deg C): 100-102

BP pressure (mm Hg): 1.10E+001

Property/ Value	Units	Temp	Data Type	Reference
WS 1.22E+003	mg/L	25	EST	MEYLAN,WM ET AL. (1996)
logP 2.09			EXP	HANSCH,C & LEO,AJ (1985)
VP 1.48E-002	mm Hg	25	EST	NEELY,WB & BLAU,GE (1985)
DC	pKa			
HL 5.84E-008	atm m3/mol	25	EST	MEYLAN,WM & HOWARD,PH (1991)
OH 4.11E-011	cm3/molc sec	25	EST	MEYLAN,WM & HOWARD,PH (1993)

CAS #: 004824-72-0 — 4-NITRO-2,3,5,6-TETRACHLOROPHENOL

Formula: $C_6HCl_4NO_3$

Mol Weight: 276.89

MP (deg C): | FP (deg C):

BP (deg C):

BP pressure (mm Hg):

Property/ Value	Units	Temp	Data Type	Reference
WS 9.36E+000	mg/L	25	EST	MEYLAN,WM ET AL. (1996)
logP 4.49			EST	MEYLAN,WM & HOWARD,PH (1995)
VP 2.16E-006	mm Hg	25	EST	NEELY,WB & BLAU,GE (1985)
DC	pKa			
HL 6.67E-010	atm m3/mol	25	EST	MEYLAN,WM & HOWARD,PH (1991)
OH 3.16E-013	cm3/molc sec	25	EST	MEYLAN,WM & HOWARD,PH (1993)

CAS #: 004824-78-6 — BROMOPHOS ETHYL

Formula: $C_{10}H_{12}BrCl_2O_3PS$

Mol Weight: 394.06

MP (deg C): | FP (deg C):

BP (deg C): 122-33

BP pressure (mm Hg): 0.00E+000

Property/ Value	Units	Temp	Data Type	Reference
WS 4.40E-001	mg/L	20	EXP	BOWMAN,BT & SANS,WW (1983A)
logP 6.15			EXP	DEBRUIJN,J ET AL. (1989)
VP 4.60E-005	mm Hg	30		WORTHING,CR & WALKER,SB (1987)
DC	pKa			
HL 1.64E-005	atm m3/mol	25	EST	MEYLAN,WM & HOWARD,PH (1991)
OH 9.00E-011	cm3/molc sec	25	EST	ATKINSON,R (1988)

CAS #: 004826-22-6 — PROPANOIC ACID, 3-(4H-1,2,4-BENZOTHIADIAZIN-3-YL

Formula: $C_{10}H_{10}N_2O_4S$

Mol Weight: 254.27

MP (deg C): | FP (deg C):

BP (deg C):

BP pressure (mm Hg):

Property/ Value	Units	Temp	Data Type	Reference
WS 7.93E+004	mg/L	25	EST	MEYLAN,WM ET AL. (1996)
logP -0.18			EXP	GAMBERINI,G ET AL. (1989)
VP 1.44E-010	mm Hg	25	EST	NEELY,WB & BLAU,GE (1985)
DC	pKa			
HL 9.13E-013	atm m3/mol	25	EST	MEYLAN,WM & HOWARD,PH (1991)
OH 1.19E-011	cm3/molc sec	25	EST	MEYLAN,WM & HOWARD,PH (1993)

CAS #: 004826-24-8 — BUTANOIC ACID, 4-(4H-1,2,4-BENZOTHIADIAZIN-3-YL)

Formula: $C_{11}H_{12}N_2O_4S$

Mol Weight: 268.29

MP (deg C): | FP (deg C):

BP (deg C):

BP pressure (mm Hg):

Property/ Value	Units	Temp	Data Type	Reference
WS 6.23E+004	mg/L	25	EST	MEYLAN,WM ET AL. (1996)
logP -0.15			EXP	GAMBERINI,G ET AL. (1989)
VP 6.31E-011	mm Hg	25	EST	NEELY,WB & BLAU,GE (1985)
DC	pKa			
HL 1.21E-012	atm m3/mol	25	EST	MEYLAN,WM & HOWARD,PH (1991)
OH 1.33E-011	cm3/molc sec	25	EST	MEYLAN,WM & HOWARD,PH (1993)

CAS #: 004831-62-3 — 2-METHYL-4-NITROQUINOLINE-1-OXIDE

Formula: $C_{10}H_{10}N_2O_3$

Mol Weight: 206.20

MP (deg C): | FP (deg C):

BP (deg C):

BP pressure (mm Hg):

Property/ Value	Units	Temp	Data Type	Reference
WS 1.45E+003	mg/L	25	EST	MEYLAN,WM ET AL. (1996)
logP 1.25			EXP	HANSCH,C & LEO,AJ (1985)
VP 1.06E-006	mm Hg	25	EST	NEELY,WB & BLAU,GE (1985)
DC	pKa			
HL 3.00E-014	atm m3/mol	25	EST	MEYLAN,WM & HOWARD,PH (1991)
OH 2.83E-012	cm3/molc sec	25	EST	MEYLAN,WM & HOWARD,PH (1993)

CAS #: 004832-17-1 — DECAHYDRO-2-NAPHTHALENONE

Formula: $C_{10}H_{16}O$

Mol Weight: 152.24

MP (deg C): | FP (deg C):

BP (deg C): 96

BP pressure (mm Hg): 2.50E+000

Property/ Value	Units	Temp	Data Type	Reference
WS 1.54E+003	mg/L	25	EST	MEYLAN,WM ET AL. (1996)
logP 1.97			EXP	HANSCH,C & LEO,AJ (1985)
VP 9.55E-002	mm Hg	25	EST	NEELY,WB & BLAU,GE (1985)
DC	pKa			
HL 7.00E-005	atm m3/mol	25	EST	MEYLAN,WM & HOWARD,PH (1991)
OH 2.69E-011	cm3/molc sec	25	EST	MEYLAN,WM & HOWARD,PH (1993)

CAS #: 004836-13-9				CYTIDINE, N-BENZOYL-2'-DEOXY-

Formula: $C_{16}H_{17}N_3O_5$

Mol Weight: 331.33

MP (deg C):   FP (deg C):

BP (deg C):

BP pressure (mm Hg):

Property/Value	Units	Temp	Data Type	Reference
WS 2.15E+003	mg/L	25	EST	MEYLAN,WM ET AL. (1996)
logP 0.67			EXP	SANGSTER,J (1994)
VP 9.48E-016	mm Hg	25	EST	NEELY,WB & BLAU,GE (1985)
DC	pKa			
HL 2.75E-021	atm m3/mol	25	EST	MEYLAN,WM & HOWARD,PH (1991)
OH 1.18E-010	cm3/molc sec	25	EST	MEYLAN,WM & HOWARD,PH (1993)

CAS #: 004837-88-1				1-MEO-2-ME-3-NO2-BENZENE

Formula: $C_8H_9NO_3$

Mol Weight: 167.17

MP (deg C): 54-56   FP (deg C):

BP (deg C):

BP pressure (mm Hg):

Property/Value	Units	Temp	Data Type	Reference
WS 1.33E+002	mg/L	25	EST	MEYLAN,WM ET AL. (1996)
logP 2.68			EXP	NAKAGAWA,Y ET AL. (1992)
VP 6.65E-003	mm Hg	25	EST	NEELY,WB & BLAU,GE (1985)
DC	pKa			
HL 1.39E-006	atm m3/mol	25	EST	MEYLAN,WM & HOWARD,PH (1991)
OH 3.27E-012	cm3/molc sec	25	EST	MEYLAN,WM & HOWARD,PH (1993)

CAS #: 004838-65-7				1-HEXYL-2-PYRROLIDONE

Formula: $C_{10}H_{19}NO$

Mol Weight: 169.27

MP (deg C):   FP (deg C):

BP (deg C):

BP pressure (mm Hg):

Property/Value	Units	Temp	Data Type	Reference
WS 3.73E+002	mg/L	25	EST	MEYLAN,WM ET AL. (1996)
logP 2.60			EXP	SASAKI,H ET AL. (1988)
VP 1.22E-003	mm Hg	25	EST	NEELY,WB & BLAU,GE (1985)
DC	pKa			
HL 1.30E-007	atm m3/mol	25	EST	MEYLAN,WM & HOWARD,PH (1991)
OH 3.33E-011	cm3/molc sec	25	EST	MEYLAN,WM & HOWARD,PH (1993)

CAS #: 004849-32-5				KASUGAMYCIN

Formula: $C_{14}H_{21}N_3O_3$

Mol Weight: 279.34

MP (deg C):   FP (deg C):

BP (deg C):

BP pressure (mm Hg):

Property/Value	Units	Temp	Data Type	Reference
WS 3.25E+002	mg/L		EXP	YALKOWSKY,SH & DANNENFELSER,RM (1992)
logP 1.66			EXP	HANSCH,C & LEO,AJ (1985)
VP 1.76E-007	mm Hg	25	EST	NEELY,WB & BLAU,GE (1985)
DC	pKa			
HL 1.35E-014	atm m3/mol	25	EST	MEYLAN,WM & HOWARD,PH (1991)
OH 3.94E-011	cm3/molc sec	25	EST	MEYLAN,WM & HOWARD,PH (1993)

CAS #: 004849-46-1				1,1-DIME-3-(3-HYDROXYPHENYL)UREA

Formula: $C_9H_{12}N_2O_2$

Mol Weight: 180.21

MP (deg C):   FP (deg C):

BP (deg C):

BP pressure (mm Hg):

Property/Value	Units	Temp	Data Type	Reference
WS 2.02E+004	mg/L	25	EST	MEYLAN,WM ET AL. (1996)
logP 0.51			EXP	HANSCH,C & LEO,AJ (1985)
VP 3.38E-006	mm Hg	25	EST	NEELY,WB & BLAU,GE (1985)
DC	pKa			
HL 1.01E-013	atm m3/mol	25	EST	MEYLAN,WM & HOWARD,PH (1991)
OH 2.02E-010	cm3/molc sec	25	EST	MEYLAN,WM & HOWARD,PH (1993)

CAS #: 004857-42-5				5-METHYLISOXAZOLE-3-CARBOXYLIC ACID

Formula: $C_5H_6NO_3$

Mol Weight: 128.11

MP (deg C):   FP (deg C):

BP (deg C):

BP pressure (mm Hg):

Property/Value	Units	Temp	Data Type	Reference
WS 2.48E+004	mg/L	25	EST	MEYLAN,WM ET AL. (1996)
logP 0.68			EXP	HANSCH,C & LEO,AJ (1985)
VP 1.46E-002	mm Hg	25	EST	NEELY,WB & BLAU,GE (1985)
DC	pKa			
HL 1.26E-009	atm m3/mol	25	EST	MEYLAN,WM & HOWARD,PH (1991)
OH 7.14E-012	cm3/molc sec	25	EST	MEYLAN,WM & HOWARD,PH (1993)

CAS #: 004859-06-7				2-(2,6-DIME PHENYLIMINO)IMIDAZOLINE

Formula: $C_{11}H_{15}N_3$

Mol Weight: 189.26

MP (deg C):   FP (deg C):

BP (deg C):

BP pressure (mm Hg):

Property/Value	Units	Temp	Data Type	Reference
WS 2.92E+004	mg/L	25	EST	MEYLAN,WM ET AL. (1996)
logP 1.45			EXP	HANSCH,C & LEO,AJ (1985)
VP 6.15E-005	mm Hg	25	EST	NEELY,WB & BLAU,GE (1985)
DC	pKa			
HL 3.34E-011	atm m3/mol	25	EST	MEYLAN,WM & HOWARD,PH (1991)
OH 1.58E-010	cm3/molc sec	25	EST	MEYLAN,WM & HOWARD,PH (1993)

CAS #: 004860-03-1				1-CHLOROHEXADECANE

Formula: $C_{16}H_{33}Cl$

Mol Weight: 260.89

MP (deg C): 17.9   FP (deg C):

BP (deg C): 322

BP pressure (mm Hg):

Property/Value	Units	Temp	Data Type	Reference
WS 1.24E-003	mg/L	25	EST	MEYLAN,WM ET AL. (1996)
logP 8.45			EST	MEYLAN,WM & HOWARD,PH (1995)
VP 8.26E-006	mm Hg	25	EXT	BOUBLIK,T ET AL. (1984)
DC	pKa			
HL 5.75E-001	atm m3/mol	25	EST	MEYLAN,WM & HOWARD,PH (1991)
OH 1.91E-011	cm3/molc sec	25	EST	MEYLAN,WM & HOWARD,PH (1993)

CAS #: 004871-13-0				2,4(1H,3H)-PYRIMIDINEDIONE, 5-FLUORO-1-(PHENYLME

Formula: $C_{11}H_9FN_2O_2$

Mol Weight: 220.20

MP (deg C):   **FP (deg C):**

BP (deg C):

BP pressure (mm Hg):

Property/ Value	Units	Temp	Data Type	Reference
WS 5.97E+003	mg/L	25	EST	MEYLAN,WM ET AL. (1996)
logP 0.89			EXP	SANGSTER,J (1994)
VP 4.05E-009	mm Hg	25	EST	NEELY,WB & BLAU,GE (1985)
DC	pKa			
HL 2.94E-011	atm m3/mol	25	EST	MEYLAN,WM & HOWARD,PH (1991)
OH 1.45E-011	cm3/molc sec	25	EST	MEYLAN,WM & HOWARD,PH (1993)

CAS #: 004888-33-9				2(1H)-QUINOLINONE, 3,4-DIHYDRO-4-PHENYL

Formula: $C_{15}H_{13}NO$

Mol Weight: 223.28

MP (deg C):   **FP (deg C):**

BP (deg C):

BP pressure (mm Hg):

Property/ Value	Units	Temp	Data Type	Reference
WS 1.37E+002	mg/L	25	EST	MEYLAN,WM ET AL. (1996)
logP 2.79			EXP	SANGSTER,J (1993)
VP 1.59E-007	mm Hg	25	EST	NEELY,WB & BLAU,GE (1985)
DC	pKa			
HL 2.42E-010	atm m3/mol	25	EST	MEYLAN,WM & HOWARD,PH (1991)
OH 2.80E-011	cm3/molc sec	25	EST	MEYLAN,WM & HOWARD,PH (1993)

CAS #: 004891-98-9				2,4-PYRIMIDINEDIAMINE, 5-[(2-BROMO-5-METHOXY-4-P

Formula: $C_{15}H_{19}BrN_4O_2$

Mol Weight: 367.25

MP (deg C):   **FP (deg C):**

BP (deg C):

BP pressure (mm Hg):

Property/ Value	Units	Temp	Data Type	Reference
WS 1.83E+001	mg/L	25	EST	MEYLAN,WM ET AL. (1996)
logP 2.84			EXP	SANGSTER,J (1993)
VP 1.19E-009	mm Hg	25	EST	NEELY,WB & BLAU,GE (1985)
DC	pKa			
HL 2.83E-013	atm m3/mol	25	EST	MEYLAN,WM & HOWARD,PH (1991)
OH 2.80E-011	cm3/molc sec	25	EST	MEYLAN,WM & HOWARD,PH (1993)

CAS #: 004894-75-1				CYCLOHEXANONE, 4-PHENYL-

Formula: $C_{12}H_{14}O$

Mol Weight: 174.24

MP (deg C): 78-81   **FP (deg C):**

BP (deg C):

BP pressure (mm Hg):

Property/ Value	Units	Temp	Data Type	Reference
WS 6.38E+002	mg/L	25	EST	MEYLAN,WM ET AL. (1996)
logP 2.30			EXP	SANGSTER,J (1993)
VP 2.33E-003	mm Hg	25	EST	NEELY,WB & BLAU,GE (1985)
DC	pKa			
HL 4.13E-006	atm m3/mol	25	EST	MEYLAN,WM & HOWARD,PH (1991)
OH 1.82E-011	cm3/molc sec	25	EST	MEYLAN,WM & HOWARD,PH (1993)

CAS #: 004897-25-0				5-CHLORO-1-METHYL-4-NITROIMIDAZOLE

Formula: $C_4H_4ClN_3O_2$

Mol Weight: 161.55

MP (deg C):   **FP (deg C):**

BP (deg C):

BP pressure (mm Hg):

Property/ Value	Units	Temp	Data Type	Reference
WS 1.52E+004	mg/L	25	EST	MEYLAN,WM ET AL. (1996)
logP 0.30			EXP	HANSCH,C & LEO,AJ (1985)
VP 2.83E-004	mm Hg	25	EST	NEELY,WB & BLAU,GE (1985)
DC	pKa			
HL 2.34E-007	atm m3/mol	25	EST	MEYLAN,WM & HOWARD,PH (1991)
OH 1.45E-012	cm3/molc sec	25	EST	MEYLAN,WM & HOWARD,PH (1993)

CAS #: 004897-31-8				4-CHLORO-1-METHYL-5-NITROIMIDAZOLE

Formula: $C_4H_4ClN_3O_2$

Mol Weight: 161.55

MP (deg C):   **FP (deg C):**

BP (deg C):

BP pressure (mm Hg):

Property/ Value	Units	Temp	Data Type	Reference
WS 7.06E+003	mg/L	25	EST	MEYLAN,WM ET AL. (1996)
logP 0.69			EXP	SUWINSKI,J ET AL. (1985)
VP 2.83E-004	mm Hg	25	EST	NEELY,WB & BLAU,GE (1985)
DC	pKa			
HL 2.34E-007	atm m3/mol	25	EST	MEYLAN,WM & HOWARD,PH (1991)
OH 1.45E-012	cm3/molc sec	25	EST	MEYLAN,WM & HOWARD,PH (1993)

CAS #: 004901-51-3				2,3,4,5-TETRACHLOROPHENOL

Formula: $C_6H_2Cl_4O$

Mol Weight: 231.89

MP (deg C): 116.5   **FP (deg C):**

BP (deg C):

BP pressure (mm Hg):

Property/ Value	Units	Temp	Data Type	Reference
WS 2.87E+001	mg/L	25	EST	MEYLAN,WM ET AL. (1996)
logP 4.21			EXP	HANSCH,C & LEO,AJ (1985)
VP 3.39E-004	mm Hg	25	EST	NEELY,WB & BLAU,GE (1985)
DC	pKa			
HL 1.69E-007	atm m3/mol	25	EST	MEYLAN,WM & HOWARD,PH (1991)
OH 1.55E-012	cm3/molc sec	25	EST	MEYLAN,WM & HOWARD,PH (1993)

CAS #: 004904-61-4				1,5,9-CYCLODODECATRIENE

Formula: $C_{12}H_{18}$

Mol Weight: 162.28

MP (deg C): -17   **FP (deg C):**

BP (deg C): 240

BP pressure (mm Hg):

Property/ Value	Units	Temp	Data Type	Reference
WS 3.90E-001	mg/L	25	EST	MEYLAN,WM ET AL. (1996)
logP 5.50			EXP	SANGSTER,J (1993)
VP 8.33E-002	mm Hg	25	EST	NEELY,WB & BLAU,GE (1985)
DC	pKa			
HL 4.99E-001	atm m3/mol	25	EST	MEYLAN,WM & HOWARD,PH (1991)
OH 1.76E-010	cm3/molc sec	25	EST	MEYLAN,WM & HOWARD,PH (1993)

004911-70-0 — 2,3-DIMETHYL-2-PENTANOL

CAS #:	004911-70-0
Formula:	$C_7H_{16}O$
Mol Weight:	116.20
MP (deg C):	
FP (deg C):	
BP (deg C):	
BP pressure (mm Hg):	

Property/Value	Units	Temp	Data Type	Reference
WS 1.56E+004	mg/L	25	EXP	SUZUKI,T (1991)
logP 2.13			EST	MEYLAN,WM & HOWARD,PH (1995)
VP 4.54E+000	mm Hg	25	EST	NEELY,WB & BLAU,GE (1985)
DC	pKa			
HL 2.34E-005	atm m3/mol	25	EST	MEYLAN,WM & HOWARD,PH (1991)
OH 1.17E-011	cm3/molc sec	25	EST	MEYLAN,WM & HOWARD,PH (1993)

004914-62-9 — 5-HYDROXY-OXACILLIN

CAS #:	004914-62-9
Formula:	$C_{19}H_{20}N_3O_6S$
Mol Weight:	418.45
MP (deg C):	
FP (deg C):	
BP (deg C):	
BP pressure (mm Hg):	

Property/Value	Units	Temp	Data Type	Reference
WS 6.93E+001	mg/L	25	EST	MEYLAN,WM ET AL. (1996)
logP 1.80			EXP	HANSCH,C ET AL. (1995)
VP 2.57E-016	mm Hg	25	EST	NEELY,WB & BLAU,GE (1985)
DC	pKa			
HL 1.61E-021	atm m3/mol	25	EST	MEYLAN,WM & HOWARD,PH (1991)
OH 1.57E-010	cm3/molc sec	25	EST	MEYLAN,WM & HOWARD,PH (1993)

004916-63-6 — 2,5-DIMETHYL-1,3,5-HEXATRIENE

CAS #:	004916-63-6
Formula:	C_8H_{12}
Mol Weight:	108.18
MP (deg C):	-9
FP (deg C):	
BP (deg C):	146
BP pressure (mm Hg):	

Property/Value	Units	Temp	Data Type	Reference
WS 1.51E+001	mg/L	25	EST	MEYLAN,WM ET AL. (1996)
logP 3.89			EST	MEYLAN,WM & HOWARD,PH (1995)
VP 2.76E+001	mm Hg	25	EST	NEELY,WB & BLAU,GE (1985)
DC	pKa			
HL 2.42E-001	atm m3/mol	25	EST	MEYLAN,WM & HOWARD,PH (1991)
OH 1.65E-010	cm3/molc sec	25	EST	MEYLAN,WM & HOWARD,PH (1993)

004917-89-9 — P-SEC-BUTYLPHENOXYACETIC ACID

CAS #:	004917-89-9
Formula:	$C_{12}H_{16}O_3$
Mol Weight:	208.26
MP (deg C):	
FP (deg C):	
BP (deg C):	
BP pressure (mm Hg):	

Property/Value	Units	Temp	Data Type	Reference
WS 2.13E+002	mg/L	25	EST	MEYLAN,WM ET AL. (1996)
logP 3.12			EXP	HANSCH,C & LEO,AJ (1985)
VP 5.83E-005	mm Hg	25	EST	NEELY,WB & BLAU,GE (1985)
DC	pKa			
HL 4.33E-008	atm m3/mol	25	EST	MEYLAN,WM & HOWARD,PH (1991)
OH 3.45E-011	cm3/molc sec	25	EST	MEYLAN,WM & HOWARD,PH (1993)

004919-33-9 — P-ETHOXYPHENYLACETIC ACID

CAS #:	004919-33-9
Formula:	$C_{10}H_{12}O_3$
Mol Weight:	180.21
MP (deg C):	87-90
FP (deg C):	
BP (deg C):	
BP pressure (mm Hg):	

Property/Value	Units	Temp	Data Type	Reference
WS 2.95E+003	mg/L	25	EST	MEYLAN,WM ET AL. (1996)
logP 1.95			EXP	HANSCH,C & LEO,AJ (1985)
VP 2.35E-004	mm Hg	25	EST	NEELY,WB & BLAU,GE (1985)
DC	pKa			
HL 3.47E-009	atm m3/mol	25	EST	MEYLAN,WM & HOWARD,PH (1991)
OH 2.96E-011	cm3/molc sec	25	EST	MEYLAN,WM & HOWARD,PH (1993)

004920-77-8 — 3-METHYL-2-NITROPHENOL

CAS #:	004920-77-8
Formula:	$C_7H_7NO_3$
Mol Weight:	153.14
MP (deg C):	35-39
FP (deg C):	
BP (deg C):	106-108
BP pressure (mm Hg):	9.50E+000

Property/Value	Units	Temp	Data Type	Reference
WS 3.51E+003	mg/L	20	EXP	SCHWARZENBACH,RP ET AL. (1988)
logP 2.29			EXP	SCHWARZENBACH,RP ET AL. (1988)
VP 3.72E-002	mm Hg	20	EXP	SCHWARZENBACH,RP ET AL. (1988)
DC	pKa			
HL 2.14E-006	atm m3/mol	20	EST	VP/WSOL
OH 1.11E-011	cm3/molc sec	25	EST	MEYLAN,WM & HOWARD,PH (1993)

004925-19-3 — BENZAMIDE,N(DIETAMET),2OME,3CL,5F

CAS #:	004925-19-3
Formula:	$C_{14}H_{20}ClFN_2O_2$
Mol Weight:	302.78
MP (deg C):	
FP (deg C):	
BP (deg C):	
BP pressure (mm Hg):	

Property/Value	Units	Temp	Data Type	Reference
WS 2.28E+001	mg/L	25	EST	MEYLAN,WM ET AL. (1996)
logP 3.18			EXP	HANSCH,C & LEO,AJ (1985)
VP 1.13E-007	mm Hg	25	EST	NEELY,WB & BLAU,GE (1985)
DC	pKa			
HL 2.89E-013	atm m3/mol	25	EST	MEYLAN,WM & HOWARD,PH (1991)
OH 1.06E-010	cm3/molc sec	25	EST	MEYLAN,WM & HOWARD,PH (1993)

004928-02-3 — 1,4-BENZDIAZEPIN-2-ONE-5-PH-7-NH2

CAS #:	004928-02-3
Formula:	$C_{15}H_{13}N_3O$
Mol Weight:	251.29
MP (deg C):	
FP (deg C):	
BP (deg C):	
BP pressure (mm Hg):	

Property/Value	Units	Temp	Data Type	Reference
WS 3.26E+003	mg/L	25	EST	MEYLAN,WM ET AL. (1996)
logP 1.00			EXP	HANSCH,C & LEO,AJ (1985)
VP 1.21E-009	mm Hg	25	EST	NEELY,WB & BLAU,GE (1985)
DC	pKa			
HL 8.48E-014	atm m3/mol	25	EST	MEYLAN,WM & HOWARD,PH (1991)
OH 1.29E-010	cm3/molc sec	25	EST	MEYLAN,WM & HOWARD,PH (1993)

CAS #: 004930-03-4				N,O-DIPHENYLCARBAMATE

Formula: $C_{13}H_{11}NO_2$

Mol Weight: 213.24

MP (deg C): **FP (deg C):**

BP (deg C):

BP pressure (mm Hg):

Property/Value	Units	Temp	Data Type	Reference
WS 5.68E+001	mg/L	25	EST	MEYLAN,WM ET AL. (1996)
logP 3.30			EXP	HANSCH,C & LEO,AJ (1985)
VP 8.56E-005	mm Hg	25	EST	NEELY,WB & BLAU,GE (1985)
DC	pKa			
HL 8.07E-009	atm m3/mol	25	EST	MEYLAN,WM & HOWARD,PH (1991)
OH 4.47E-011	cm3/molc sec	25	EST	MEYLAN,WM & HOWARD,PH (1993)

CAS #: 004932-53-0				BENZENESULFONAMIDE, N-[(PROPYLAMINO)CARBONYL]-

Formula: $C_{10}H_{14}N_2O_3S$

Mol Weight: 242.30

MP (deg C): **FP (deg C):**

BP (deg C):

BP pressure (mm Hg):

Property/Value	Units	Temp	Data Type	Reference
WS 1.54E+003	mg/L	25	EST	MEYLAN,WM ET AL. (1996)
logP 1.44			EXP	CLOUX,JL ET AL. (1988)
VP 1.59E-007	mm Hg	25	EST	NEELY,WB & BLAU,GE (1985)
DC	pKa			
HL 1.73E-010	atm m3/mol	25	EST	MEYLAN,WM & HOWARD,PH (1991)
OH 8.44E-012	cm3/molc sec	25	EST	MEYLAN,WM & HOWARD,PH (1993)

CAS #: 004940-39-0				4H-1-BENZOPYRAN-2-CARBOXYLIC ACID, 4-OXO-

Formula: $C_{10}H_6O_4$

Mol Weight: 190.16

MP (deg C): 260 dec **FP (deg C):**

BP (deg C):

BP pressure (mm Hg):

Property/Value	Units	Temp	Data Type	Reference
WS 4.94E+003	mg/L	25	EST	MEYLAN,WM ET AL. (1996)
logP 1.63			EXP	HANSCH,C ET AL. (1995)
VP 8.01E-006	mm Hg	25	EST	NEELY,WB & BLAU,GE (1985)
DC	pKa			
HL 2.67E-011	atm m3/mol	25	EST	MEYLAN,WM & HOWARD,PH (1991)
OH 3.08E-011	cm3/molc sec	25	EST	MEYLAN,WM & HOWARD,PH (1993)

CAS #: 004941-45-1				1,4-BENZODIAZEPIN-2-ONE,1,3-DIHYDRO-9-METHYL-7-N

Formula: $C_{16}H_{13}N_3O_3$

Mol Weight: 295.30

MP (deg C): **FP (deg C):**

BP (deg C):

BP pressure (mm Hg):

Property/Value	Units	Temp	Data Type	Reference
WS 4.39E+001	mg/L	25	EST	MEYLAN,WM ET AL. (1996)
logP 2.44			EXP	HANSCH,C ET AL. (1995)
VP 2.10E-010	mm Hg	25	EST	NEELY,WB & BLAU,GE (1985)
DC	pKa			
HL 1.04E-012	atm m3/mol	25	EST	MEYLAN,WM & HOWARD,PH (1991)
OH 5.64E-012	cm3/molc sec	25	EST	MEYLAN,WM & HOWARD,PH (1993)

CAS #: 004946-22-9				P-(DIMETHYLAMINO)BENZENETHIOL

Formula: $C_8H_{11}NS$

Mol Weight: 153.25

MP (deg C): **FP (deg C):**

BP (deg C):

BP pressure (mm Hg):

Property/Value	Units	Temp	Data Type	Reference
WS 2.63E+002	mg/L	25	EST	MEYLAN,WM ET AL. (1996)
logP 2.86			EST	MEYLAN,WM & HOWARD,PH (1995)
VP 6.20E-002	mm Hg	25	EST	NEELY,WB & BLAU,GE (1985)
DC	pKa			
HL 8.30E-006	atm m3/mol	25	EST	MEYLAN,WM & HOWARD,PH (1991)
OH 2.02E-010	cm3/molc sec	25	EST	MEYLAN,WM & HOWARD,PH (1993)

CAS #: 004949-13-7				2-FLUOROPYRAZINE

Formula: $C_4H_3FN_2$

Mol Weight: 98.08

MP (deg C): **FP (deg C):**

BP (deg C):

BP pressure (mm Hg):

Property/Value	Units	Temp	Data Type	Reference
WS 6.70E+004	mg/L	25	EST	MEYLAN,WM ET AL. (1996)
logP 0.29			EXP	YAMAGAMI,C ET AL. (1990A)
VP 1.64E+001	mm Hg	25	EST	NEELY,WB & BLAU,GE (1985)
DC	pKa			
HL 3.41E-006	atm m3/mol	25	EST	MEYLAN,WM & HOWARD,PH (1991)
OH 2.88E-013	cm3/molc sec	25	EST	MEYLAN,WM & HOWARD,PH (1993)

CAS #: 004959-16-4				1,4-BENZODIAZPN2-ONE,1-ME-5(PH)7NH2

Formula: $C_{16}H_{15}N_3O$

Mol Weight: 265.32

MP (deg C): **FP (deg C):**

BP (deg C):

BP pressure (mm Hg):

Property/Value	Units	Temp	Data Type	Reference
WS 1.10E+003	mg/L	25	EST	MEYLAN,WM ET AL. (1996)
logP 1.46			EXP	HANSCH,C & LEO,AJ (1985)
VP 5.28E-009	mm Hg	25	EST	NEELY,WB & BLAU,GE (1985)
DC	pKa			
HL 1.74E-012	atm m3/mol	25	EST	MEYLAN,WM & HOWARD,PH (1991)
OH 1.30E-010	cm3/molc sec	25	EST	MEYLAN,WM & HOWARD,PH (1993)

CAS #: 004964-76-5				7-METHOXYQUINOLINE

Formula: $C_{10}H_9NO$

Mol Weight: 159.19

MP (deg C): **FP (deg C):**

BP (deg C):

BP pressure (mm Hg):

Property/Value	Units	Temp	Data Type	Reference
WS 9.13E+002	mg/L	25	EST	MEYLAN,WM ET AL. (1996)
logP 2.37			EXP	HANSCH,C & LEO,AJ (1985)
VP 3.23E-003	mm Hg	25	EST	NEELY,WB & BLAU,GE (1985)
DC 5.03	pKa	20	EXP	PERRIN,DD (1965)
HL 4.07E-008	atm m3/mol	25	EST	MEYLAN,WM & HOWARD,PH (1991)
OH 1.60E-010	cm3/molc sec	25	EST	MEYLAN,WM & HOWARD,PH (1993)

779

004965-26-8 — 5-NITROBENZOTHIOPHENE

Formula: $C_8H_5NO_2S$

Mol Weight: 179.20

MP (deg C):
FP (deg C):
BP (deg C):
BP pressure (mm Hg):

Property/Value	Units	Temp	Data Type	Reference
WS 7.26E+001	mg/L	25	EST	MEYLAN,WM ET AL. (1996)
logP 2.92			EXP	HANSCH,C ET AL. (1995)
VP 1.38E-004	mm Hg	25	EST	NEELY,WB & BLAU,GE (1985)
DC	pKa			
HL 1.13E-006	atm m3/mol	25	EST	MEYLAN,WM & HOWARD,PH (1991)
OH 3.75E-012	cm3/molc sec	25	EST	MEYLAN,WM & HOWARD,PH (1993)

004965-30-4 — N-OH-1,4-NAPHTHOQUINONEMONIMINE

Formula: $C_{10}H_7NO_2$

Mol Weight: 173.17

MP (deg C):
FP (deg C):
BP (deg C):
BP pressure (mm Hg):

Property/Value	Units	Temp	Data Type	Reference
WS 6.84E+002	mg/L	25	EST	MEYLAN,WM ET AL. (1996)
logP 2.27			EXP	HANSCH,C & LEO,AJ (1985)
VP 1.86E-006	mm Hg	25	EST	NEELY,WB & BLAU,GE (1985)
DC 8.01	pKa		EXP	KORTUM,G ET AL (1961)
HL 1.79E-010	atm m3/mol	25	EST	MEYLAN,WM & HOWARD,PH (1991)
OH 1.52E-011	cm3/molc sec	25	EST	MEYLAN,WM & HOWARD,PH (1993)

004965-36-0 — 7-BROMOQUINOLINE

Formula: C_9H_6BrN

Mol Weight: 208.06

MP (deg C): 34
FP (deg C):
BP (deg C): 290
BP pressure (mm Hg):

Property/Value	Units	Temp	Data Type	Reference
WS 1.52E+002	mg/L	25	EST	MEYLAN,WM ET AL. (1996)
logP 2.92			EXP	HANSCH,C & LEO,AJ (1985)
VP 1.47E-003	mm Hg	25	EST	NEELY,WB & BLAU,GE (1985)
DC 3.87	pKa	25	EXP	PERRIN,DD (1965)
HL 2.74E-007	atm m3/mol	25	EST	MEYLAN,WM & HOWARD,PH (1991)
OH 9.06E-012	cm3/molc sec	25	EST	MEYLAN,WM & HOWARD,PH (1993)

004975-21-7 — FURAZAN, DIMETHYL-

Formula: $C_4H_6N_2O$

Mol Weight: 98.11

MP (deg C): -7
FP (deg C):
BP (deg C): 156
BP pressure (mm Hg):

Property/Value	Units	Temp	Data Type	Reference
WS 2.93E+004	mg/L	25	EST	MEYLAN,WM ET AL. (1996)
logP 0.71			EXP	CALVINO,R ET AL. (1992)
VP 9.87E+000	mm Hg	25	EST	NEELY,WB & BLAU,GE (1985)
DC	pKa			
HL 2.30E-004	atm m3/mol	25	EST	MEYLAN,WM & HOWARD,PH (1991)
OH 4.27E-012	cm3/molc sec	25	EST	MEYLAN,WM & HOWARD,PH (1993)

004980-73-8 — 1,4-BENZODIAZP2-ON,5(2NO2PH)7NO2

Formula: $C_{15}H_{10}N_4O_5$

Mol Weight: 326.27

MP (deg C):
FP (deg C):
BP (deg C):
BP pressure (mm Hg):

Property/Value	Units	Temp	Data Type	Reference
WS 1.48E+002	mg/L	25	EST	MEYLAN,WM ET AL. (1996)
logP 1.61			EXP	HANSCH,C & LEO,AJ (1985)
VP 8.90E-012	mm Hg	25	EST	NEELY,WB & BLAU,GE (1985)
DC	pKa			
HL 3.74E-015	atm m3/mol	25	EST	MEYLAN,WM & HOWARD,PH (1991)
OH 3.10E-012	cm3/molc sec	25	EST	MEYLAN,WM & HOWARD,PH (1993)

004988-64-1 — 6-MERCAPTOPURINE RIBOSIDE

Formula: $C_{10}H_{14}N_4O_4S$

Mol Weight: 286.31

MP (deg C):
FP (deg C):
BP (deg C):
BP pressure (mm Hg):

Property/Value	Units	Temp	Data Type	Reference
WS 2.41E+003	mg/L	25	EST	MEYLAN,WM ET AL. (1996)
logP -0.60			EXP	HANSCH,C & LEO,AJ (1985)
VP 9.83E-019	mm Hg	25	EST	NEELY,WB & BLAU,GE (1985)
DC	pKa			
HL 7.15E-020	atm m3/mol	25	EST	MEYLAN,WM & HOWARD,PH (1991)
OH 1.32E-010	cm3/molc sec	25	EST	MEYLAN,WM & HOWARD,PH (1993)

005001-33-2 — METANEPHRINE

Formula: $C_{10}H_{15}NO_3$

Mol Weight: 197.24

MP (deg C): 175 dec
FP (deg C):
BP (deg C):
BP pressure (mm Hg):

Property/Value	Units	Temp	Data Type	Reference
WS 1.00E+006	mg/L	25	EST	MEYLAN,WM ET AL. (1996)
logP -0.64			EXP	HANSCH,C & LEO,AJ (1985)
VP 2.16E-006	mm Hg	25	EST	NEELY,WB & BLAU,GE (1985)
DC	pKa			
HL 4.01E-016	atm m3/mol	25	EST	MEYLAN,WM & HOWARD,PH (1991)
OH 1.20E-010	cm3/molc sec	25	EST	MEYLAN,WM & HOWARD,PH (1993)

005003-48-5 — BENZOIC ACID, 2-(ACETYLOXY)-, 4-(ACETYLAMINO)PHE

Formula: $C_{17}H_{15}NO_5$

Mol Weight: 313.31

MP (deg C): 175-176
FP (deg C):
BP (deg C):
BP pressure (mm Hg):

Property/Value	Units	Temp	Data Type	Reference
WS 1.50E+002	mg/L	25	EST	MEYLAN,WM ET AL. (1996)
logP 2.15			EXP	NIELSEN,LS & BUNDGAARD,H (1989)
VP 7.50E-009	mm Hg	25	EST	NEELY,WB & BLAU,GE (1985)
DC	pKa			
HL 1.77E-013	atm m3/mol	25	EST	MEYLAN,WM & HOWARD,PH (1991)
OH 1.12E-011	cm3/molc sec	25	EST	MEYLAN,WM & HOWARD,PH (1993)

CAS #: 005004-45-5	1(2H)-PHTHALAZINONE, 4-PHENYL-

Formula: $C_{14}H_{10}N_2O$

Mol Weight: 222.25

MP (deg C):   FP (deg C):

BP (deg C):

BP pressure (mm Hg):

Property/Value	Units	Temp	Data Type	Reference
WS 2.55E+002	mg/L	25	EST	MEYLAN,WM ET AL. (1996)
logP 2.48			EXP	SANGSTER,J (1993)
VP 5.72E-009	mm Hg	25	EST	NEELY,WB & BLAU,GE (1985)
DC	pKa			
HL 1.75E-010	atm m3/mol	25	EST	MEYLAN,WM & HOWARD,PH (1991)
OH 1.37E-011	cm3/molc sec	25	EST	MEYLAN,WM & HOWARD,PH (1993)

CAS #: 005006-67-7	1-(2-HYDROXYET)-2-NITROIMIDAZOLE

Formula: $C_5H_9N_3O_3$

Mol Weight: 159.15

MP (deg C):   FP (deg C):

BP (deg C):

BP pressure (mm Hg):

Property/Value	Units	Temp	Data Type	Reference
WS 7.24E+004	mg/L	25	EST	MEYLAN,WM ET AL. (1996)
logP -0.47			EXP	HANSCH,C & LEO,AJ (1985)
VP 7.06E-007	mm Hg	25	EST	NEELY,WB & BLAU,GE (1985)
DC	pKa			
HL 1.53E-011	atm m3/mol	25	EST	MEYLAN,WM & HOWARD,PH (1991)
OH 9.81E-012	cm3/molc sec	25	EST	MEYLAN,WM & HOWARD,PH (1993)

CAS #: 005006-90-6	UREA, N-(3,4-DICHLOROPHENYL)-N'-(2-METHYLPROPYL)

Formula: $C_{11}H_{14}Cl_2N_2O$

Mol Weight: 261.15

MP (deg C):   FP (deg C):

BP (deg C):

BP pressure (mm Hg):

Property/Value	Units	Temp	Data Type	Reference
WS 3.06E+000	mg/L	25	EST	MEYLAN,WM ET AL. (1996)
logP 4.48			EXP	MITSUTAKE,KI ET AL. (1986)
VP 3.45E-006	mm Hg	25	EST	NEELY,WB & BLAU,GE (1985)
DC	pKa			
HL 5.68E-010	atm m3/mol	25	EST	MEYLAN,WM & HOWARD,PH (1991)
OH 1.72E-011	cm3/molc sec	25	EST	MEYLAN,WM & HOWARD,PH (1993)

CAS #: 005018-16-6	BENZENESULFONAMIDE, 4-AMINO-N-[6-METHOXY-5-(1-ME

Formula: $C_{14}H_{18}N_4O_4S$

Mol Weight: 338.39

MP (deg C):   FP (deg C):

BP (deg C):

BP pressure (mm Hg):

Property/Value	Units	Temp	Data Type	Reference
WS 9.60E+002	mg/L	25	EST	MEYLAN,WM ET AL. (1996)
logP 1.03			EXP	SANGSTER,J (1993)
VP 5.07E-010	mm Hg	25	EST	NEELY,WB & BLAU,GE (1985)
DC	pKa			
HL 2.51E-015	atm m3/mol	25	EST	MEYLAN,WM & HOWARD,PH (1991)
OH 3.96E-011	cm3/molc sec	25	EST	MEYLAN,WM & HOWARD,PH (1993)

CAS #: 005018-56-4	5MEO-6ETO-4-SULFANILAMIDOPYRIMIDIN

Formula: $C_{13}H_{16}N_4O_4S$

Mol Weight: 324.36

MP (deg C):   FP (deg C):

BP (deg C):

BP pressure (mm Hg):

Property/Value	Units	Temp	Data Type	Reference
WS 1.16E+003	mg/L	25	EST	MEYLAN,WM ET AL. (1996)
logP 1.18			EXP	HANSCH,C & LEO,AJ (1985)
VP 7.02E-010	mm Hg	25	EST	NEELY,WB & BLAU,GE (1985)
DC	pKa			
HL 1.89E-015	atm m3/mol	25	EST	MEYLAN,WM & HOWARD,PH (1991)
OH 3.36E-011	cm3/molc sec	25	EST	MEYLAN,WM & HOWARD,PH (1993)

CAS #: 005021-43-2	QUINOXALINE, 2-PHENYL-

Formula: $C_{14}H_{10}N_2$

Mol Weight: 206.25

MP (deg C):   FP (deg C):

BP (deg C):

BP pressure (mm Hg):

Property/Value	Units	Temp	Data Type	Reference
WS 4.34E+001	mg/L	25	EST	MEYLAN,WM ET AL. (1996)
logP 3.48			EXP	SANGSTER,J (1993)
VP 4.85E-006	mm Hg	25	EST	NEELY,WB & BLAU,GE (1985)
DC	pKa			
HL 2.19E-008	atm m3/mol	25	EST	MEYLAN,WM & HOWARD,PH (1991)
OH 6.86E-012	cm3/molc sec	25	EST	MEYLAN,WM & HOWARD,PH (1993)

CAS #: 005026-76-6	6-METHYL-1-HEPTENE

Formula: C_8H_{16}

Mol Weight: 112.22

MP (deg C): 113.2   FP (deg C):

BP (deg C):

BP pressure (mm Hg):

Property/Value	Units	Temp	Data Type	Reference
WS 1.06E+001	mg/L	25	EST	MEYLAN,WM ET AL. (1996)
logP 4.06			EST	MEYLAN,WM & HOWARD,PH (1995)
VP 2.52E+001	mm Hg	25	EXP	ZWOLINSKI,BJ & WILHOIT,RC (1971)
DC	pKa			
HL 6.32E-001	atm m3/mol	25	EST	MEYLAN,WM & HOWARD,PH (1991)
OH 3.30E-011	cm3/molc sec	25	EST	MEYLAN,WM & HOWARD,PH (1993)

CAS #: 005041-09-8	ISOBUTYLAMINE HYDROCHLORIDE

Formula: $C_4H_{11}ClN$

Mol Weight: 108.59

MP (deg C):   FP (deg C):

BP (deg C):

BP pressure (mm Hg):

Property/Value	Units	Temp	Data Type	Reference
WS 1.00E+006	mg/L	25	EXP	SEIDELL,A (1941)
logP -2.42			EST	MEYLAN,WM & HOWARD,PH (1995)
VP 9.87E-005	mm Hg	25	EST	NEELY,WB & BLAU,GE (1985)
DC	pKa			
HL 1.64E-012	atm m3/mol	25	EST	MEYLAN,WM & HOWARD,PH (1991)
OH 1.34E-011	cm3/molc sec	25	EST	MEYLAN,WM & HOWARD,PH (1993)

CAS #: 005048-08-8 — PYRIDINE, 4-(4-CHLOROPHENYL)-1,2,3,6-TETRAHYDRO-

Formula: $C_{12}H_{14}ClN$

Mol Weight: 207.71

MP (deg C):
FP (deg C):
BP (deg C):
BP pressure (mm Hg):

Property/Value	Units	Temp	Data Type	Reference
WS 5.94E+002	mg/L	25	EST	MEYLAN,WM ET AL. (1996)
logP 3.32			EXP	ALTOMARE,CA ET AL. (1992)
VP 1.09E-003	mm Hg	25	EST	NEELY,WB & BLAU,GE (1985)
DC	pKa			
HL 1.10E-006	atm m3/mol	25	EST	MEYLAN,WM & HOWARD,PH (1991)
OH 1.76E-010	cm3/molc sec	25	EST	MEYLAN,WM & HOWARD,PH (1993)

CAS #: 005049-61-6 — 2-AMINOPYRAZINE

Formula: $C_4H_5N_3$

Mol Weight: 95.10

MP (deg C):
FP (deg C):
BP (deg C):
BP pressure (mm Hg):

Property/Value	Units	Temp	Data Type	Reference
WS 1.39E+005	mg/L	25	EST	MEYLAN,WM ET AL. (1996)
logP -0.07			EXP	HANSCH,C & LEO,AJ (1985)
VP 2.46E-001	mm Hg	25	EST	NEELY,WB & BLAU,GE (1985)
DC 3.07	pKa	20	EXP	PERRIN,DD (1965)
HL 1.03E-009	atm m3/mol	25	EST	MEYLAN,WM & HOWARD,PH (1991)
OH 1.27E-011	cm3/molc sec	25	EST	MEYLAN,WM & HOWARD,PH (1993)

CAS #: 005053-43-0 — PYRIMIDINE, 2-METHYL-

Formula: $C_5H_6N_2$

Mol Weight: 94.12

MP (deg C): -4
FP (deg C):
BP (deg C): 138
BP pressure (mm Hg):

Property/Value	Units	Temp	Data Type	Reference
WS 1.34E+005	mg/L	25	EST	MEYLAN,WM ET AL. (1996)
logP -0.05			EXP	YAMAGAMI,C ET AL. (1990)
VP 4.63E+000	mm Hg	25	EST	NEELY,WB & BLAU,GE (1985)
DC	pKa			
HL 3.22E-006	atm m3/mol	25	EST	MEYLAN,WM & HOWARD,PH (1991)
OH 7.36E-013	cm3/molc sec	25	EST	MEYLAN,WM & HOWARD,PH (1993)

CAS #: 005057-96-5 — DIMETHYL TARTRATE (MESO)

Formula: $C_6H_{10}O_6$

Mol Weight: 178.14

MP (deg C): 57-60 dec
FP (deg C):
BP (deg C): 158
BP pressure (mm Hg): 2.00E-001

Property/Value	Units	Temp	Data Type	Reference
WS 4.57E+005	mg/L	25	EST	MEYLAN,WM ET AL. (1996)
logP -1.07			EST	MEYLAN,WM & HOWARD,PH (1995)
VP 1.38E-003	mm Hg	25	EST	NEELY,WB & BLAU,GE (1985)
DC	pKa			
HL 1.35E-005	atm m3/mol	25	EST	MEYLAN,WM & HOWARD,PH (1991)
OH 5.89E-012	cm3/molc sec	25	EST	MEYLAN,WM & HOWARD,PH (1993)

CAS #: 005075-92-3 — 2-PYRROLIDINONE, 1,5-DIMETHYL-

Formula: $C_6H_{11}NO$

Mol Weight: 113.16

MP (deg C):
FP (deg C):
BP (deg C): 217
BP pressure (mm Hg):

Property/Value	Units	Temp	Data Type	Reference
WS 1.22E+005	mg/L	25	EST	MEYLAN,WM ET AL. (1996)
logP -0.07			EXP	KIM,KH ET AL. (1993)
VP 1.47E-001	mm Hg	25	EST	NEELY,WB & BLAU,GE (1985)
DC	pKa			
HL 4.19E-008	atm m3/mol	25	EST	MEYLAN,WM & HOWARD,PH (1991)
OH 2.78E-011	cm3/molc sec	25	EST	MEYLAN,WM & HOWARD,PH (1993)

CAS #: 005102-79-4 — 2-DIPHACETYL-1,3INDANDIONE-1HYDRAZONE

Formula: $C_{23}H_{18}N_2O_2$

Mol Weight: 354.41

MP (deg C): 241-243
FP (deg C):
BP (deg C):
BP pressure (mm Hg):

Property/Value	Units	Temp	Data Type	Reference
WS 4.72E+000	mg/L	25	EST	MEYLAN,WM ET AL. (1996)
logP 3.62			EXP	HANSCH,C & LEO,AJ (1985)
VP 1.49E-010	mm Hg	25	EST	NEELY,WB & BLAU,GE (1985)
DC	pKa			
HL 2.62E-015	atm m3/mol	25	EST	MEYLAN,WM & HOWARD,PH (1991)
OH 1.57E-011	cm3/molc sec	25	EST	MEYLAN,WM & HOWARD,PH (1993)

CAS #: 005103-71-9 — ALPHA-CHLORDANE

Formula: $C_{10}H_6Cl_8$

Mol Weight: 409.78

MP (deg C):
FP (deg C):
BP (deg C):
BP pressure (mm Hg):

Property/Value	Units	Temp	Data Type	Reference
WS 5.60E-002	mg/L	25	EXP	MABEY,WR ET AL. (1981)
logP 6.60			EST	MEYLAN,WM & HOWARD,PH (1995)
VP 3.60E-005	mm Hg	25	EXP	HINCKLEY,DA ET AL. (1990)
DC	pKa			
HL 3.47E-004	atm m3/mol	25	EST	VP/WSOL
OH 5.29E-012	cm3/molc sec	25	EST	MEYLAN,WM & HOWARD,PH (1993)

CAS #: 005103-73-1 — CIS-NONACHLOR

Formula: $C_{10}H_5Cl_9$

Mol Weight: 444.23

MP (deg C):
FP (deg C):
BP (deg C):
BP pressure (mm Hg):

Property/Value	Units	Temp	Data Type	Reference
WS 1.04E-002	mg/L	25	EST	MEYLAN,WM ET AL. (1996)
logP 6.78			EST	MEYLAN,WM & HOWARD,PH (1995)
VP 1.25E-005	mm Hg	25	EXP	HINCKLEY,DA ET AL. (1990)
DC	pKa			
HL 2.48E-005	atm m3/mol	25	EST	MEYLAN,WM & HOWARD,PH (1991)
OH 4.95E-012	cm3/molc sec	25	EST	MEYLAN,WM & HOWARD,PH (1993)

CAS #: 005103-74-2 — CHLORDANE (TRANS)

Formula:	$C_{10}H_6Cl_8$			
Mol Weight:	409.78			
MP (deg C):		FP (deg C):		
BP (deg C):				
BP pressure (mm Hg):				

Property/Value	Units	Temp	Data Type	Reference
WS 5.60E-002	mg/L	25	EXP	MABEY,WR ET AL. (1981)
logP 6.60			EST	MEYLAN,WM & HOWARD,PH (1995)
VP 5.03E-005	mm Hg	25	EXP	HINCKLEY,DA ET AL. (1990)
DC	pKa			
HL 4.84E-004	atm m3/mol	25	EST	VP/WSOL
OH 5.29E-012	cm3/molc sec	25	EST	MEYLAN,WM & HOWARD,PH (1993)

CAS #: 005104-49-4 — FLURBIPROFEN

Formula:	$C_{15}H_{13}FO_2$			
Mol Weight:	244.27			
MP (deg C):	110-111	FP (deg C):		
BP (deg C):				
BP pressure (mm Hg):				

Property/Value	Units	Temp	Data Type	Reference
WS 8.00E+000	mg/L		EXP	YALKOWSKY,SH & DANNENFELSER,RM (1992)
logP 4.16			EXP	HANSCH,C & LEO,AJ (1985)
VP 1.96E-006	mm Hg	25	EST	NEELY,WB & BLAU,GE (1985)
DC	pKa			
HL 5.26E-009	atm m3/mol	25	EST	MEYLAN,WM & HOWARD,PH (1991)
OH 8.78E-012	cm3/molc sec	25	EST	MEYLAN,WM & HOWARD,PH (1993)

CAS #: 005108-54-3 — 124-BENZTHIADIAZIN-11SO2-3ME5BR7CL

Formula:	$C_8H_6BrClN_2O_2S$			
Mol Weight:	309.57			
MP (deg C):		FP (deg C):		
BP (deg C):				
BP pressure (mm Hg):				

Property/Value	Units	Temp	Data Type	Reference
WS 4.21E+002	mg/L	25	EST	MEYLAN,WM ET AL. (1996)
logP 1.65			EXP	HANSCH,C & LEO,AJ (1985)
VP 3.39E-009	mm Hg	25	EST	NEELY,WB & BLAU,GE (1985)
DC	pKa			
HL 2.73E-008	atm m3/mol	25	EST	MEYLAN,WM & HOWARD,PH (1991)
OH 3.70E-013	cm3/molc sec	25	EST	MEYLAN,WM & HOWARD,PH (1993)

CAS #: 005123-63-7 — 3-(DIETHYLAMINO)BENZENESULFONIC ACID, SODIUM SAL

Formula:	$C_{10}H_{14}NNaO_3S$			
Mol Weight:	251.28			
MP (deg C):		FP (deg C):		
BP (deg C):				
BP pressure (mm Hg):				

Property/Value	Units	Temp	Data Type	Reference
WS 5.41E+005	mg/L	25	EST	MEYLAN,WM ET AL. (1996)
logP -1.60			EXP	SANGSTER,J (1994)
VP 9.37E-013	mm Hg	25	EST	NEELY,WB & BLAU,GE (1985)
DC	pKa			
HL	atm m3/mol			
OH 4.90E-011	cm3/molc sec	25	EST	MEYLAN,WM & HOWARD,PH (1993)

CAS #: 005124-25-4 — C.I. DISPERSE YELLOW 42

Formula:	$C_{18}H_{15}N_3O_4S$			
Mol Weight:	369.40			
MP (deg C):		FP (deg C):		
BP (deg C):				
BP pressure (mm Hg):				

Property/Value	Units	Temp	Data Type	Reference
WS 2.00E-001	mg/L	25	EXP	BAUGHMAN,GL ET AL. (1993)
logP 4.60			EXP	BAUGHMAN,GL & WEBER,EJ (1991)
VP 2.11E-011	mm Hg	25	EST	NEELY,WB & BLAU,GE (1985)
DC	pKa			
HL 3.92E-012	atm m3/mol	25	EST	MEYLAN,WM & HOWARD,PH (1991)
OH 1.93E-010	cm3/molc sec	25	EST	MEYLAN,WM & HOWARD,PH (1993)

CAS #: 005131-24-8 — DITALIMFOS (LAPTRAN)

Formula:	$C_{12}H_{14}NO_4PS$			
Mol Weight:	299.29			
MP (deg C):		FP (deg C):		
BP (deg C):				
BP pressure (mm Hg):				

Property/Value	Units	Temp	Data Type	Reference
WS 1.33E+002	mg/L		EXP	YALKOWSKY,SH & DANNENFELSER,RM (1992)
logP 3.48			EXP	HANSCH,C & LEO,AJ (1985)
VP 2.38E-008	mm Hg	25	EST	NEELY,WB & BLAU,GE (1985)
DC	pKa			
HL 3.82E-009	atm m3/mol	25	EST	MEYLAN,WM & HOWARD,PH (1991)
OH 1.07E-010	cm3/molc sec	25	EST	MEYLAN,WM & HOWARD,PH (1993)

CAS #: 005131-60-2 — 1,3-BENZENEDIAMINE, 4-CHLORO-

Formula:	$C_6H_7ClN_2$			
Mol Weight:	142.59			
MP (deg C):		FP (deg C):		
BP (deg C):				
BP pressure (mm Hg):				

Property/Value	Units	Temp	Data Type	Reference
WS 1.54E+004	mg/L	25	EST	MEYLAN,WM ET AL. (1996)
logP 0.85			EXP	BRONAUGH,RL & CONGDON,ER (1984)
VP 2.06E-003	mm Hg	25	EST	NEELY,WB & BLAU,GE (1985)
DC	pKa			
HL 4.99E-010	atm m3/mol	25	EST	MEYLAN,WM & HOWARD,PH (1991)
OH 2.00E-010	cm3/molc sec	25	EST	MEYLAN,WM & HOWARD,PH (1993)

CAS #: 005145-99-3 — ETHYL ISOPROPYL SULFIDE

Formula:	$C_5H_{12}S$			
Mol Weight:	104.22			
MP (deg C):	-122.2	FP (deg C):		
BP (deg C):	107.5			
BP pressure (mm Hg):				

Property/Value	Units	Temp	Data Type	Reference
WS 1.19E+003	mg/L	25	EST	MEYLAN,WM ET AL. (1996)
logP 2.32			EST	MEYLAN,WM & HOWARD,PH (1995)
VP 3.36E+001	mm Hg	25	EST	NEELY,WB & BLAU,GE (1985)
DC	pKa			
HL 1.84E-003	atm m3/mol	25	EST	MEYLAN,WM & HOWARD,PH (1991)
OH 2.46E-011	cm3/molc sec	25	EST	MEYLAN,WM & HOWARD,PH (1993)

005152-83-0

CAS #: 005152-83-0 — 4-CINNOLINAMINE

Formula: $C_8H_7N_3$

Mol Weight: 145.17

MP (deg C): **FP (deg C):**

BP (deg C):

BP pressure (mm Hg):

Property/Value	Units	Temp	Data Type	Reference
WS 1.03E+004	mg/L	25	EST	MEYLAN,WM ET AL. (1996)
logP 1.04			EXP	HANSCH,C & LEO,AJ (1985)
VP 1.62E-005	mm Hg	25	EST	NEELY,WB & BLAU,GE (1985)
DC	pKa			
HL 9.90E-011	atm m3/mol	25	EST	MEYLAN,WM & HOWARD,PH (1991)
OH 1.10E-010	cm3/molc sec	25	EST	MEYLAN,WM & HOWARD,PH (1993)

005162-44-7

CAS #: 005162-44-7 — 4-BROMO-1-BUTENE

Formula: C_4H_7Br

Mol Weight: 135.01

MP (deg C): **FP (deg C):**

BP (deg C): 98.5

BP pressure (mm Hg):

Property/Value	Units	Temp	Data Type	Reference
WS 7.64E+002	mg/L	25	EXP	YALKOWSKY,SH & DANNENFELSER,RM (1992)
logP 2.53			EXP	HANSCH,C & LEO,AJ (1985)
VP 4.28E+001	mm Hg	25	EST	NEELY,WB & BLAU,GE (1985)
DC	pKa			
HL 1.48E-002	atm m3/mol	25	EST	MEYLAN,WM & HOWARD,PH (1991)
OH 2.71E-011	cm3/molc sec	25	EST	MEYLAN,WM & HOWARD,PH (1993)

005180-59-6

CAS #: 005180-59-6 — 2,6-BIS(1,1-DIMETHYLETHYL)-4-NITROBENZENAMINE

Formula: $C_{14}H_{22}N_2O_2$

Mol Weight: 250.34

MP (deg C): **FP (deg C):**

BP (deg C):

BP pressure (mm Hg):

Property/Value	Units	Temp	Data Type	Reference
WS 7.15E-001	mg/L	25	EST	MEYLAN,WM ET AL. (1996)
logP 5.29			EST	MEYLAN,WM & HOWARD,PH (1995)
VP 9.56E-006	mm Hg	25	EST	NEELY,WB & BLAU,GE (1985)
DC	pKa			
HL 5.01E-008	atm m3/mol	25	EST	MEYLAN,WM & HOWARD,PH (1991)
OH 2.36E-012	cm3/molc sec	25	EST	MEYLAN,WM & HOWARD,PH (1993)

005182-90-1

CAS #: 005182-90-1 — 2-QUINOXALINECARBOXAMIDE

Formula: $C_9H_7N_3O$

Mol Weight: 173.18

MP (deg C): **FP (deg C):**

BP (deg C):

BP pressure (mm Hg):

Property/Value	Units	Temp	Data Type	Reference
WS 8.65E+003	mg/L	25	EST	MEYLAN,WM ET AL. (1996)
logP 0.98			EXP	HANSCH,C & LEO,AJ (1985)
VP 1.17E-006	mm Hg	25	EST	NEELY,WB & BLAU,GE (1985)
DC	pKa			
HL 1.17E-013	atm m3/mol	25	EST	MEYLAN,WM & HOWARD,PH (1991)
OH 3.82E-012	cm3/molc sec	25	EST	MEYLAN,WM & HOWARD,PH (1993)

005183-78-8

CAS #: 005183-78-8 — N-METHYLBENZENESULFONAMIDE

Formula: $C_7H_9NO_2S$

Mol Weight: 171.22

MP (deg C): **FP (deg C):**

BP (deg C):

BP pressure (mm Hg):

Property/Value	Units	Temp	Data Type	Reference
WS 9.94E+003	mg/L	25	EST	MEYLAN,WM ET AL. (1996)
logP 0.92			EXP	HANSCH,C & LEO,AJ (1985)
VP 7.16E-004	mm Hg	25	EST	NEELY,WB & BLAU,GE (1985)
DC	pKa			
HL 9.27E-007	atm m3/mol	25	EST	MEYLAN,WM & HOWARD,PH (1991)
OH 1.68E-012	cm3/molc sec	25	EST	MEYLAN,WM & HOWARD,PH (1993)

005192-03-0

CAS #: 005192-03-0 — 5-AMINOINDOLE

Formula: $C_8H_8N_2$

Mol Weight: 132.17

MP (deg C): 131-133 **FP (deg C):**

BP (deg C):

BP pressure (mm Hg):

Property/Value	Units	Temp	Data Type	Reference
WS 2.28E+004	mg/L	25	EST	MEYLAN,WM ET AL. (1996)
logP 0.70			EXP	HANSCH,C & LEO,AJ (1985)
VP 3.02E-004	mm Hg	25	EST	NEELY,WB & BLAU,GE (1985)
DC	pKa			
HL 3.13E-010	atm m3/mol	25	EST	MEYLAN,WM & HOWARD,PH (1991)
OH 2.00E-010	cm3/molc sec	25	EST	MEYLAN,WM & HOWARD,PH (1993)

005194-50-3

CAS #: 005194-50-3 — 2,4-HEXADIENE (CIS,TRANS)

Formula: C_6H_{10}

Mol Weight: 82.15

MP (deg C): -96.1 **FP (deg C):**

BP (deg C): 83.5

BP pressure (mm Hg):

Property/Value	Units	Temp	Data Type	Reference
WS 1.01E+002	mg/L	25	EST	MEYLAN,WM ET AL. (1996)
logP 2.80			EXP	HANSCH,C ET AL. (1995)
VP 8.31E+001	mm Hg	25	EXP	DAUBERT,TE & DANNER,RP (1989)
DC	pKa			
HL 1.91E-001	atm m3/mol	25	EST	MEYLAN,WM & HOWARD,PH (1991)
OH 1.34E-010	cm3/molc sec	25	EXP	ATKINSON,R (1989)

005194-51-4

CAS #: 005194-51-4 — 2,4-HEXADIENE (TRANS,TRANS)

Formula: C_6H_{10}

Mol Weight: 82.15

MP (deg C): -44.9 **FP (deg C):**

BP (deg C): 82.2

BP pressure (mm Hg):

Property/Value	Units	Temp	Data Type	Reference
WS 1.01E+002	mg/L	25	EST	MEYLAN,WM ET AL. (1996)
logP 3.01			EXP	HANSCH,C ET AL. (1995)
VP 8.79E+001	mm Hg	25	EXP	DAUBERT,TE & DANNER,RP (1989)
DC	pKa			
HL 1.91E-001	atm m3/mol	25	EST	MEYLAN,WM & HOWARD,PH (1991)
OH 1.34E-010	cm3/molc sec	25	EXP	ATKINSON,R (1989)

CAS #:	005202-89-1	2-AMINO-5-CHLORO-METHYLBENZOATE

Formula: $C_8H_8ClNO_2$

Mol Weight: 185.61

MP (deg C): 66-68   FP (deg C):

BP (deg C): 168-170

BP pressure (mm Hg): 2.20E+001

Property/Value	Units	Temp	Data Type	Reference
WS 7.86E+001	mg/L	25	EST	MEYLAN,WM ET AL. (1996)
logP 3.30			EXP	HANSCH,C ET AL. (1995)
VP 1.15E-003	mm Hg	25	EST	NEELY,WB & BLAU,GE (1985)
DC	pKa			
HL 9.09E-009	atm m3/mol	25	EST	MEYLAN,WM & HOWARD,PH (1991)
OH 1.03E-011	cm3/molc sec	25	EST	MEYLAN,WM & HOWARD,PH (1993)

CAS #:	005213-49-0	2,4-DINITRO-1H-IMIDAZOLE

Formula: $C_3H_2N_4O_4$

Mol Weight: 158.07

MP (deg C):   FP (deg C):

BP (deg C):

BP pressure (mm Hg):

Property/Value	Units	Temp	Data Type	Reference
WS 1.92E+004	mg/L	25	EST	MEYLAN,WM ET AL. (1996)
logP 0.20			EXP	SUWINSKI,J ET AL. (1985)
VP 2.61E-006	mm Hg	25	EST	NEELY,WB & BLAU,GE (1985)
DC	pKa			
HL 5.85E-011	atm m3/mol	25	EST	MEYLAN,WM & HOWARD,PH (1991)
OH 3.93E-013	cm3/molc sec	25	EST	MEYLAN,WM & HOWARD,PH (1993)

CAS #:	005213-50-3	1-METHYL-2,4-DINITRO-1H-IMIDAZOLE

Formula: $C_4H_4N_4O_4$

Mol Weight: 172.10

MP (deg C):   FP (deg C):

BP (deg C):

BP pressure (mm Hg):

Property/Value	Units	Temp	Data Type	Reference
WS 3.42E+004	mg/L	25	EST	MEYLAN,WM ET AL. (1996)
logP -0.17			EXP	SUWINSKI,J ET AL. (1985)
VP 1.69E-005	mm Hg	25	EST	NEELY,WB & BLAU,GE (1985)
DC	pKa			
HL 1.25E-009	atm m3/mol	25	EST	MEYLAN,WM & HOWARD,PH (1991)
OH 5.29E-013	cm3/molc sec	25	EST	MEYLAN,WM & HOWARD,PH (1993)

CAS #:	005214-29-9	2-DIMETHYLAMINOPYRAZINE

Formula: $C_6H_9N_3$

Mol Weight: 123.16

MP (deg C):   FP (deg C):

BP (deg C):

BP pressure (mm Hg):

Property/Value	Units	Temp	Data Type	Reference
WS 1.57E+004	mg/L	25	EST	MEYLAN,WM ET AL. (1996)
logP 0.93			EXP	YAMAGAMI,C ET AL. (1990A)
VP 6.25E-001	mm Hg	25	EST	NEELY,WB & BLAU,GE (1985)
DC 3.24	pKa	25	EXP	PERRIN,DD (1965)
HL 4.64E-008	atm m3/mol	25	EST	MEYLAN,WM & HOWARD,PH (1991)
OH 4.62E-011	cm3/molc sec	25	EST	MEYLAN,WM & HOWARD,PH (1993)

CAS #:	005215-26-9	BENZENEACETAMIDE, N-(2-FLUOROPHENYL)-

Formula: $C_{14}H_{12}FNO$

Mol Weight: 229.26

MP (deg C):   FP (deg C):

BP (deg C):

BP pressure (mm Hg):

Property/Value	Units	Temp	Data Type	Reference
WS 1.78E+002	mg/L	25	EST	MEYLAN,WM ET AL. (1996)
logP 2.62			EXP	YAMAGAMI,C ET AL. (1984)
VP 1.13E-006	mm Hg	25	EST	NEELY,WB & BLAU,GE (1985)
DC	pKa			
HL 5.81E-010	atm m3/mol	25	EST	MEYLAN,WM & HOWARD,PH (1991)
OH 9.74E-012	cm3/molc sec	25	EST	MEYLAN,WM & HOWARD,PH (1993)

CAS #:	005215-27-0	ACETAMIDE, N-(3-FLUOROPHENYL)-2-PHENYL-

Formula: $C_{14}H_{12}FNO$

Mol Weight: 229.26

MP (deg C):   FP (deg C):

BP (deg C):

BP pressure (mm Hg):

Property/Value	Units	Temp	Data Type	Reference
WS 6.92E+001	mg/L	25	EST	MEYLAN,WM ET AL. (1996)
logP 3.10			EXP	YAMAGAMI,C ET AL. (1984)
VP 1.13E-006	mm Hg	25	EST	NEELY,WB & BLAU,GE (1985)
DC	pKa			
HL 5.81E-010	atm m3/mol	25	EST	MEYLAN,WM & HOWARD,PH (1991)
OH 2.11E-011	cm3/molc sec	25	EST	MEYLAN,WM & HOWARD,PH (1993)

CAS #:	005215-28-1	BENZENEACETAMIDE, N-(4-FLUOROPHENYL)-

Formula: $C_{14}H_{12}FNO$

Mol Weight: 229.26

MP (deg C):   FP (deg C):

BP (deg C):

BP pressure (mm Hg):

Property/Value	Units	Temp	Data Type	Reference
WS 1.11E+002	mg/L	25	EST	MEYLAN,WM ET AL. (1996)
logP 2.86			EXP	YAMAGAMI,C ET AL. (1984)
VP 1.13E-006	mm Hg	25	EST	NEELY,WB & BLAU,GE (1985)
DC	pKa			
HL 5.81E-010	atm m3/mol	25	EST	MEYLAN,WM & HOWARD,PH (1991)
OH 9.74E-012	cm3/molc sec	25	EST	MEYLAN,WM & HOWARD,PH (1993)

CAS #:	005216-25-1	4-CHLOROBENZOTRICHLORIDE

Formula: $C_7H_4Cl_4$

Mol Weight: 229.92

MP (deg C):   FP (deg C):

BP (deg C): 245

BP pressure (mm Hg):

Property/Value	Units	Temp	Data Type	Reference
WS 4.04E+000	mg/L	25	EST	MEYLAN,WM ET AL. (1996)
logP 4.54			EST	MEYLAN,WM & HOWARD,PH (1995)
VP 3.83E-002	mm Hg	25	EST	NEELY,WB & BLAU,GE (1985)
DC	pKa			
HL 1.93E-004	atm m3/mol	25	EST	MEYLAN,WM & HOWARD,PH (1991)
OH 2.40E-013	cm3/molc sec	25	EXP	ATKINSON,R (1989)

785

005221-42-1 — 4-ACETAMIDOPYRIDINE

Formula: $C_7H_8N_2O$

Mol Weight: 136.15

MP (deg C):
FP (deg C):
BP (deg C):
BP pressure (mm Hg):

Property/Value	Units	Temp	Data Type	Reference
WS 3.25E+004	mg/L	25	EST	MEYLAN,WM ET AL. (1996)
logP 0.50			EXP	HANSCH,C & LEO,AJ (1985)
VP 2.36E-004	mm Hg	25	EST	NEELY,WB & BLAU,GE (1985)
DC 5.87	pKa		EXP	PERRIN,DD (1965)
HL 8.08E-012	atm m3/mol	25	EST	MEYLAN,WM & HOWARD,PH (1991)
OH 2.46E-012	cm3/molc sec	25	EST	MEYLAN,WM & HOWARD,PH (1993)

005231-96-9 — 2-ACETAMIDOPYRIDINE

Formula: $C_7H_8N_2O$

Mol Weight: 136.15

MP (deg C):
FP (deg C):
BP (deg C):
BP pressure (mm Hg):

Property/Value	Units	Temp	Data Type	Reference
WS 2.62E+004	mg/L	25	EST	MEYLAN,WM ET AL. (1996)
logP 0.61			EXP	HANSCH,C & LEO,AJ (1985)
VP 2.36E-004	mm Hg	25	EST	NEELY,WB & BLAU,GE (1985)
DC 4.09	pKa		EXP	PERRIN,DD (1965)
HL 8.08E-012	atm m3/mol	25	EST	MEYLAN,WM & HOWARD,PH (1991)
OH 2.46E-012	cm3/molc sec	25	EST	MEYLAN,WM & HOWARD,PH (1993)

005234-26-4 — ACETOPHENONE, O-ACETYLAMINO

Formula: $C_{10}H_{11}NO_2$

Mol Weight: 177.20

MP (deg C):
FP (deg C):
BP (deg C):
BP pressure (mm Hg):

Property/Value	Units	Temp	Data Type	Reference
WS 2.98E+003	mg/L	25	EST	MEYLAN,WM ET AL. (1996)
logP 1.50			EXP	HANSCH,C & LEO,AJ (1985)
VP 1.15E-005	mm Hg	25	EST	NEELY,WB & BLAU,GE (1985)
DC	pKa			
HL 1.12E-011	atm m3/mol	25	EST	MEYLAN,WM & HOWARD,PH (1991)
OH 1.15E-011	cm3/molc sec	25	EST	MEYLAN,WM & HOWARD,PH (1993)

005234-68-4 — CARBOXIN

Formula: $C_{12}H_{13}NO_2S$

Mol Weight: 235.31

MP (deg C): 93-95
FP (deg C):
BP (deg C):
BP pressure (mm Hg):

Property/Value	Units	Temp	Data Type	Reference
WS 1.99E+002	mg/L	25	EXP	TOMLIN,C (1994)
logP 2.14			EXP	HANSCH,C & LEO,AJ (1985)
VP 1.80E-007	mm Hg	25	EXP	WAUCHOPE,RD ET AL. (1991A)
DC	pKa			
HL 2.80E-010	atm m3/mol	25	EST	VP/WSOL
OH 1.60E-010	cm3/molc sec	25	EST	MEYLAN,WM & HOWARD,PH (1993)

005250-39-5 — FLOXACILLIN

Formula: $C_{19}H_{18}ClFN_3O_5S$

Mol Weight: 454.89

MP (deg C):
FP (deg C):
BP (deg C):
BP pressure (mm Hg):

Property/Value	Units	Temp	Data Type	Reference
WS 8.82E+000	mg/L	25	EST	MEYLAN,WM ET AL. (1996)
logP 2.58			EXP	SANGSTER,J (1994)
VP 1.67E-015	mm Hg	25	EST	NEELY,WB & BLAU,GE (1985)
DC	pKa			
HL 2.20E-017	atm m3/mol	25	EST	MEYLAN,WM & HOWARD,PH (1991)
OH 9.64E-011	cm3/molc sec	25	EST	MEYLAN,WM & HOWARD,PH (1993)

005251-93-4 — BENZADOX

Formula: $C_9H_9NO_4$

Mol Weight: 195.18

MP (deg C): 140
FP (deg C):
BP (deg C):
BP pressure (mm Hg):

Property/Value	Units	Temp	Data Type	Reference
WS 7.87E+004	mg/L	25	EST	MEYLAN,WM ET AL. (1996)
logP -0.27			EXP	HANSCH,C & LEO,AJ (1985)
VP 2.99E-007	mm Hg	25	EST	NEELY,WB & BLAU,GE (1985)
DC	pKa			
HL 7.48E-014	atm m3/mol	25	EST	MEYLAN,WM & HOWARD,PH (1991)
OH 7.92E-012	cm3/molc sec	25	EST	MEYLAN,WM & HOWARD,PH (1993)

005259-86-9 — P-AMINOHIPPURIC ACID, ME ESTER

Formula: $C_{10}H_{12}N_2O_3$

Mol Weight: 208.22

MP (deg C):
FP (deg C):
BP (deg C):
BP pressure (mm Hg):

Property/Value	Units	Temp	Data Type	Reference
WS 3.06E+003	mg/L	25	EST	MEYLAN,WM ET AL. (1996)
logP -0.23			EXP	HANSCH,C & LEO,AJ (1985)
VP 4.23E-007	mm Hg	25	EST	NEELY,WB & BLAU,GE (1985)
DC	pKa			
HL 8.16E-014	atm m3/mol	25	EST	MEYLAN,WM & HOWARD,PH (1991)
OH 1.07E-010	cm3/molc sec	25	EST	MEYLAN,WM & HOWARD,PH (1993)

005259-88-1 — OXYCARBOXIN

Formula: $C_{12}H_{13}NO_4S$

Mol Weight: 267.31

MP (deg C):
FP (deg C):
BP (deg C):
BP pressure (mm Hg):

Property/Value	Units	Temp	Data Type	Reference
WS 1.00E+003	mg/L	25	EXP	WORTHING,CR & WALKER,SB (1987)
logP 0.74			EXP	HANSCH,C & LEO,AJ (1985)
VP 1.00E-005	mm Hg	25	EST	WAUCHOPE,RD ET AL. (1991A); <1E-5
DC	pKa			
HL 3.50E-009	atm m3/mol	25	EST	VP/WSOL
OH 2.28E-010	cm3/molc sec	25	EST	MEYLAN,WM & HOWARD,PH (1993)

CAS #:	005263-87-6	6-METHOXYQUINOLINE

Formula: $C_{10}H_9NO$

Mol Weight: 159.19

MP (deg C): 26.5 FP (deg C):

BP (deg C): 306

BP pressure (mm Hg):

Property/Value	Units	Temp	Data Type	Reference
WS 9.13E+002	mg/L	25	EST	MEYLAN,WM ET AL. (1996)
logP 2.20			EXP	HANSCH,C & LEO,AJ (1985)
VP 3.23E-003	mm Hg	25	EST	NEELY,WB & BLAU,GE (1985)
DC 5.03	pKa	20	EXP	PERRIN,DD (1965)
HL 4.07E-008	atm m3/mol	25	EST	MEYLAN,WM & HOWARD,PH (1991)
OH 1.60E-010	cm3/molc sec	25	EST	MEYLAN,WM & HOWARD,PH (1993)

CAS #:	005264-15-3	4-PYRIDINEBUTANOL

Formula: $C_9H_{13}NO$

Mol Weight: 151.21

MP (deg C): FP (deg C):

BP (deg C):

BP pressure (mm Hg):

Property/Value	Units	Temp	Data Type	Reference
WS 8.25E+005	mg/L	25	EST	MEYLAN,WM ET AL. (1996)
logP 0.90			EXP	HANSCH,C & LEO,AJ (1985)
VP 4.75E-004	mm Hg	25	EST	NEELY,WB & BLAU,GE (1985)
DC	pKa			
HL 6.66E-010	atm m3/mol	25	EST	MEYLAN,WM & HOWARD,PH (1991)
OH 9.06E-012	cm3/molc sec	25	EST	MEYLAN,WM & HOWARD,PH (1993)

CAS #:	005281-13-0	TROPITAL

Formula: $C_{24}H_{40}O_8$

Mol Weight: 456.58

MP (deg C): FP (deg C):

BP (deg C):

BP pressure (mm Hg):

Property/Value	Units	Temp	Data Type	Reference
WS 1.86E+000	mg/L	25	EST	MEYLAN,WM ET AL. (1996)
logP 3.35			EST	MEYLAN,WM & HOWARD,PH (1995)
VP 4.21E-010	mm Hg	25	EST	NEELY,WB & BLAU,GE (1985)
DC	pKa			
HL 1.52E-015	atm m3/mol	25	EST	MEYLAN,WM & HOWARD,PH (1991)
OH 1.35E-010	cm3/molc sec	25	EST	MEYLAN,WM & HOWARD,PH (1993)

CAS #:	005284-99-1	GLUCOPYRANOSIDE,CYCLOHEXYL

Formula: $C_{12}H_{22}O_6$

Mol Weight: 262.31

MP (deg C): FP (deg C):

BP (deg C):

BP pressure (mm Hg):

Property/Value	Units	Temp	Data Type	Reference
WS 6.97E+004	mg/L	25	EST	MEYLAN,WM ET AL. (1996)
logP -0.63			EXP	HANSCH,C & LEO,AJ (1985)
VP 1.65E-010	mm Hg	25	EST	NEELY,WB & BLAU,GE (1985)
DC	pKa			
HL 2.83E-014	atm m3/mol	25	EST	MEYLAN,WM & HOWARD,PH (1991)
OH 8.70E-011	cm3/molc sec	25	EST	MEYLAN,WM & HOWARD,PH (1993)

CAS #:	005285-87-0	PHENYLTHIOCYANATE

Formula: C_7H_5NS

Mol Weight: 135.19

MP (deg C): FP (deg C):

BP (deg C): 232.5

BP pressure (mm Hg):

Property/Value	Units	Temp	Data Type	Reference
WS 5.94E+002	mg/L	25	EST	MEYLAN,WM ET AL. (1996)
logP 2.54			EXP	HANSCH,C & LEO,AJ (1985)
VP 1.23E-002	mm Hg	25	EST	NEELY,WB & BLAU,GE (1985)
DC	pKa			
HL 8.76E-006	atm m3/mol	25	EST	MEYLAN,WM & HOWARD,PH (1991)
OH 9.02E-013	cm3/molc sec	25	EST	MEYLAN,WM & HOWARD,PH (1993)

CAS #:	005291-77-0	N-BENZYL-2-PYRROLIDINONE

Formula: $C_{11}H_{13}NO$

Mol Weight: 175.23

MP (deg C): FP (deg C):

BP (deg C):

BP pressure (mm Hg):

Property/Value	Units	Temp	Data Type	Reference
WS 4.97E+003	mg/L	25	EST	MEYLAN,WM ET AL. (1996)
logP 1.25			EXP	HANSCH,C & LEO,AJ (1985)
VP 1.22E-004	mm Hg	25	EST	NEELY,WB & BLAU,GE (1985)
DC	pKa			
HL 2.55E-009	atm m3/mol	25	EST	MEYLAN,WM & HOWARD,PH (1991)
OH 3.17E-011	cm3/molc sec	25	EST	MEYLAN,WM & HOWARD,PH (1993)

CAS #:	005292-21-7	CYCLOHEXYLACETIC ACID

Formula: $C_8H_{14}O_2$

Mol Weight: 142.20

MP (deg C): 33 FP (deg C):

BP (deg C): 245

BP pressure (mm Hg):

Property/Value	Units	Temp	Data Type	Reference
WS 2.88E+003	mg/L	25	EXP	NIYAZOV,AN ET AL. (1975)
logP 2.85			EST	MEYLAN,WM & HOWARD,PH (1995)
VP 9.40E-001	mm Hg	25	EXT	BOUBLIK,T ET AL. (1984)
DC 4.80	pKa		EXP	KORTUM,G ET AL (1961)
HL 1.32E-006	atm m3/mol	25	EST	MEYLAN,WM & HOWARD,PH (1991)
OH 1.21E-011	cm3/molc sec	25	EST	MEYLAN,WM & HOWARD,PH (1993)

CAS #:	005292-53-5	DIETHYLMALONATE, BENZAL

Formula: $C_{14}H_{16}O_4$

Mol Weight: 248.28

MP (deg C): 32 FP (deg C):

BP (deg C): 216

BP pressure (mm Hg): 3.00E+001

Property/Value	Units	Temp	Data Type	Reference
WS 5.13E+001	mg/L	25	EST	MEYLAN,WM ET AL. (1996)
logP 3.13			EXP	HANSCH,C & LEO,AJ (1985)
VP 1.01E-003	mm Hg	25	EST	NEELY,WB & BLAU,GE (1985)
DC	pKa			
HL 1.09E-008	atm m3/mol	25	EST	MEYLAN,WM & HOWARD,PH (1991)
OH 1.58E-011	cm3/molc sec	25	EST	MEYLAN,WM & HOWARD,PH (1993)

CAS #: 005296-64-0 — ALLYL PHENYL SULFIDE

Formula: $C_9H_{10}S$
Mol Weight: 150.24
MP (deg C):
FP (deg C):
BP (deg C):
BP pressure (mm Hg):

Property/Value	Units	Temp	Data Type	Reference
WS 7.61E+001	mg/L	25	EST	MEYLAN,WM ET AL. (1996)
logP 3.51			EXP	HANSCH,C ET AL. (1995)
VP 9.76E-002	mm Hg	25	EST	NEELY,WB & BLAU,GE (1985)
DC	pKa			
HL 2.06E-004	atm m3/mol	25	EST	MEYLAN,WM & HOWARD,PH (1991)
OH 4.62E-011	cm3/molc sec	25	EST	MEYLAN,WM & HOWARD,PH (1993)

CAS #: 005297-05-2 — CARD-20(22)-ENOLIDE DERIVATIVE

Formula: $C_{35}H_{54}O_{11}$
Mol Weight: 650.81
MP (deg C):
FP (deg C):
BP (deg C):
BP pressure (mm Hg):

Property/Value	Units	Temp	Data Type	Reference
WS 7.03E+000	mg/L	25	EST	MEYLAN,WM ET AL. (1996)
logP 1.20			EXP	SANGSTER,J (1993)
VP 1.78E-025	mm Hg	25	EST	NEELY,WB & BLAU,GE (1985)
DC	pKa			
HL 6.79E-022	atm m3/mol	25	EST	MEYLAN,WM & HOWARD,PH (1991)
OH 1.36E-010	cm3/molc sec	25	EST	MEYLAN,WM & HOWARD,PH (1993)

CAS #: 005297-10-9 — DIGOXIN (NON-OLEFINIC BOND)

Formula: $C_{41}H_{66}O_{14}$
Mol Weight: 782.97
MP (deg C):
FP (deg C):
BP (deg C):
BP pressure (mm Hg):

Property/Value	Units	Temp	Data Type	Reference
WS 6.73E-001	mg/L	25	EST	MEYLAN,WM ET AL. (1996)
logP 1.36			EXP	SANGSTER,J (1993)
VP 4.58E-030	mm Hg	25	EST	NEELY,WB & BLAU,GE (1985)
DC	pKa			
HL 8.42E-027	atm m3/mol	25	EST	MEYLAN,WM & HOWARD,PH (1991)
OH 1.44E-010	cm3/molc sec	25	EST	MEYLAN,WM & HOWARD,PH (1993)

CAS #: 005302-39-6 — N-(4-((4-HYDROXYPHENYL)AZO)PHENYL)ACETAMIDE

Formula: $C_{14}H_{13}N_3O_2$
Mol Weight: 255.28
MP (deg C):
FP (deg C):
BP (deg C):
BP pressure (mm Hg):

Property/Value	Units	Temp	Data Type	Reference
WS 1.43E+002	mg/L	25	EST	MEYLAN,WM ET AL. (1996)
logP 2.74			EST	MEYLAN,WM & HOWARD,PH (1995)
VP 1.07E-009	mm Hg	25	EST	NEELY,WB & BLAU,GE (1985)
DC	pKa			
HL 1.75E-015	atm m3/mol	25	EST	MEYLAN,WM & HOWARD,PH (1991)
OH 1.84E-011	cm3/molc sec	25	EST	MEYLAN,WM & HOWARD,PH (1993)

CAS #: 005305-00-0 — N-PHENYL-2,4-ISOOXAZOLIDIONE

Formula: $C_9H_7NO_3$
Mol Weight: 177.16
MP (deg C):
FP (deg C):
BP (deg C):
BP pressure (mm Hg):

Property/Value	Units	Temp	Data Type	Reference
WS 1.97E+004	mg/L	25	EST	MEYLAN,WM ET AL. (1996)
logP 0.54			EXP	LIPINSKI,CA ET AL. (1991)
VP 4.74E-006	mm Hg	25	EST	NEELY,WB & BLAU,GE (1985)
DC	pKa			
HL 2.49E-009	atm m3/mol	25	EST	MEYLAN,WM & HOWARD,PH (1991)
OH 1.28E-011	cm3/molc sec	25	EST	MEYLAN,WM & HOWARD,PH (1993)

CAS #: 005307-14-2 — 1,4-BENZENEDIAMINE, 2-NITRO-

Formula: $C_6H_7N_3O_2$
Mol Weight: 153.14
MP (deg C): 137-140
FP (deg C):
BP (deg C):
BP pressure (mm Hg):

Property/Value	Units	Temp	Data Type	Reference
WS 2.59E+004	mg/L	25	EST	MEYLAN,WM ET AL. (1996)
logP 0.53			EXP	BRONAUGH,RL & CONGDON,ER (1984)
VP 5.60E-005	mm Hg	25	EST	NEELY,WB & BLAU,GE (1985)
DC	pKa			
HL 5.81E-011	atm m3/mol	25	EST	MEYLAN,WM & HOWARD,PH (1991)
OH 2.20E-011	cm3/molc sec	25	EST	MEYLAN,WM & HOWARD,PH (1993)

CAS #: 005319-71-1 — 4-(3H)-QUINAZOLINONE,3-HYDROXY

Formula: $C_8H_6N_2O_2$
Mol Weight: 162.15
MP (deg C):
FP (deg C):
BP (deg C):
BP pressure (mm Hg):

Property/Value	Units	Temp	Data Type	Reference
WS 7.94E+003	mg/L	25	EST	MEYLAN,WM ET AL. (1996)
logP -0.45			EXP	HANSCH,C & LEO,AJ (1985)
VP 2.45E-007	mm Hg	25	EST	NEELY,WB & BLAU,GE (1985)
DC	pKa			
HL 7.32E-013	atm m3/mol	25	EST	MEYLAN,WM & HOWARD,PH (1991)
OH 2.41E-011	cm3/molc sec	25	EST	MEYLAN,WM & HOWARD,PH (1993)

CAS #: 005324-13-0 — 2,6-DIBROMO-4-CHLOROPHENOL

Formula: $C_6H_3Br_2ClO$
Mol Weight: 286.36
MP (deg C):
FP (deg C):
BP (deg C):
BP pressure (mm Hg):

Property/Value	Units	Temp	Data Type	Reference
WS 2.43E+001	mg/L	25	EST	MEYLAN,WM ET AL. (1996)
logP 3.94			EST	MEYLAN,WM & HOWARD,PH (1995)
VP 1.91E-004	mm Hg	25	EST	NEELY,WB & BLAU,GE (1985)
DC	pKa			
HL 6.59E-008	atm m3/mol	25	EST	MEYLAN,WM & HOWARD,PH (1991)
OH 5.14E-013	cm3/molc sec	25	EST	MEYLAN,WM & HOWARD,PH (1993)

CAS #: 005326-47-6				2-AMINO-5-IODOBENZOIC ACID

Formula: $C_7H_6INO_2$

Mol Weight: 263.04

MP (deg C): 219-221 de FP (deg C):

BP (deg C):

BP pressure (mm Hg):

Property/Value	Units	Temp	Data Type	Reference
WS 3.48E+002	mg/L	25	EST	MEYLAN,WM ET AL. (1996)
logP 2.06			EXP	HANSCH,C & LEO,AJ (1985)
VP 5.87E-006	mm Hg	25	EST	NEELY,WB & BLAU,GE (1985)
DC	pKa			
HL 8.88E-012	atm m3/mol	25	EST	MEYLAN,WM & HOWARD,PH (1991)
OH 1.37E-011	cm3/molc sec	25	EST	MEYLAN,WM & HOWARD,PH (1993)

CAS #: 005327-00-4				N-ET-BROMOACETAMIDE

Formula: C_4H_8BrNO

Mol Weight: 166.02

MP (deg C): FP (deg C):

BP (deg C):

BP pressure (mm Hg):

Property/Value	Units	Temp	Data Type	Reference
WS 3.29E+004	mg/L	25	EST	MEYLAN,WM ET AL. (1996)
logP 0.34			EXP	HANSCH,C & LEO,AJ (1985)
VP 3.95E-003	mm Hg	25	EST	NEELY,WB & BLAU,GE (1985)
DC				
HL 3.76E-009	atm m3/mol	25	EST	MEYLAN,WM & HOWARD,PH (1991)
OH 9.69E-012	cm3/molc sec	25	EST	MEYLAN,WM & HOWARD,PH (1993)

CAS #: 005330-38-1				BENZENEMETHANOL, 5-CHLORO-2-HYDROXY-

Formula: $C_7H_7ClO_2$

Mol Weight: 158.59

MP (deg C): FP (deg C):

BP (deg C):

BP pressure (mm Hg):

Property/Value	Units	Temp	Data Type	Reference
WS 5.04E+004	mg/L	25	EST	MEYLAN,WM ET AL. (1996)
logP 1.44			EXP	SANGSTER,J (1993)
VP 1.07E-004	mm Hg	25	EST	NEELY,WB & BLAU,GE (1985)
DC	pKa			
HL 1.68E-011	atm m3/mol	25	EST	MEYLAN,WM & HOWARD,PH (1991)
OH 1.54E-011	cm3/molc sec	25	EST	MEYLAN,WM & HOWARD,PH (1993)

CAS #: 005332-24-1				3-BROMOQUINOLINE

Formula: C_9H_6BrN

Mol Weight: 208.06

MP (deg C): 13.3 FP (deg C):

BP (deg C): 275

BP pressure (mm Hg):

Property/Value	Units	Temp	Data Type	Reference
WS 1.03E+002	mg/L	25	EST	MEYLAN,WM ET AL. (1996)
logP 3.03			EXP	HANSCH,C & LEO,AJ (1985)
VP 1.47E-003	mm Hg	25	EST	NEELY,WB & BLAU,GE (1985)
DC 2.69	pKa	25	EXP	PERRIN,DD (1965)
HL 2.74E-007	atm m3/mol	25	EST	MEYLAN,WM & HOWARD,PH (1991)
OH 9.06E-012	cm3/molc sec	25	EST	MEYLAN,WM & HOWARD,PH (1993)

CAS #: 005332-25-2				6-BROMOQUINOLINE

Formula: C_9H_6BrN

Mol Weight: 208.06

MP (deg C): 24 FP (deg C):

BP (deg C): 281

BP pressure (mm Hg):

Property/Value	Units	Temp	Data Type	Reference
WS 1.52E+002	mg/L	25	EST	MEYLAN,WM ET AL. (1996)
logP 2.83			EXP	HANSCH,C & LEO,AJ (1985)
VP 1.47E-003	mm Hg	25	EST	NEELY,WB & BLAU,GE (1985)
DC 3.87	pKa	25	EXP	PERRIN,DD (1965)
HL 2.74E-007	atm m3/mol	25	EST	MEYLAN,WM & HOWARD,PH (1991)
OH 9.06E-012	cm3/molc sec	25	EST	MEYLAN,WM & HOWARD,PH (1993)

CAS #: 005332-73-0				3-METHOXYPROPYLAMINE

Formula: $C_4H_{11}NO$

Mol Weight: 89.14

MP (deg C): FP (deg C):

BP (deg C): 117.5

BP pressure (mm Hg):

Property/Value	Units	Temp	Data Type	Reference
WS 1.00E+006	mg/L	25	EST	MEYLAN,WM ET AL. (1996)
logP -0.42			EST	MEYLAN,WM & HOWARD,PH (1995)
VP 2.17E+001	mm Hg	25	EST	NEELY,WB & BLAU,GE (1985)
DC	pKa			
HL 1.56E-007	atm m3/mol	25	EST	MEYLAN,WM & HOWARD,PH (1991)
OH 4.35E-011	cm3/molc sec	25	EST	MEYLAN,WM & HOWARD,PH (1993)

CAS #: 005334-39-4				3-ME-4-NO2 PYRAZOLE

Formula: $C_4H_5N_3O_2$

Mol Weight: 127.10

MP (deg C): FP (deg C):

BP (deg C):

BP pressure (mm Hg):

Property/Value	Units	Temp	Data Type	Reference
WS 5.17E+003	mg/L	25	EST	MEYLAN,WM ET AL. (1996)
logP 1.02			EXP	HANSCH,C ET AL. (1995)
VP 1.30E-003	mm Hg	25	EST	NEELY,WB & BLAU,GE (1985)
DC	pKa			
HL 1.61E-008	atm m3/mol	25	EST	MEYLAN,WM & HOWARD,PH (1991)
OH 3.99E-012	cm3/molc sec	25	EST	MEYLAN,WM & HOWARD,PH (1993)

CAS #: 005335-24-0				2,4-DICHLORO-6-PHENYLPHENOL

Formula: $C_{12}H_8Cl_2O$

Mol Weight: 239.10

MP (deg C): FP (deg C):

BP (deg C):

BP pressure (mm Hg):

Property/Value	Units	Temp	Data Type	Reference
WS 1.30E+001	mg/L	25	EST	MEYLAN,WM ET AL. (1996)
logP 4.57			EST	MEYLAN,WM & HOWARD,PH (1995)
VP 2.57E-006	mm Hg	25	EST	NEELY,WB & BLAU,GE (1985)
DC	pKa			
HL 2.36E-008	atm m3/mol	25	EST	MEYLAN,WM & HOWARD,PH (1991)
OH 3.07E-012	cm3/molc sec	25	EST	MEYLAN,WM & HOWARD,PH (1993)

CAS #: 005335-75-1				PYRIDINE, 4-BUTYL-

Formula: $C_9H_{13}N$
Mol Weight: 135.21
MP (deg C):   FP (deg C):
BP (deg C): 209
BP pressure (mm Hg):

Property/Value	Units	Temp	Data Type	Reference
WS 3.71E+003	mg/L	25	EST	MEYLAN,WM ET AL. (1996)
logP 3.13			EXP	SANGSTER,J (1993)
VP 3.57E-001	mm Hg	25	EST	NEELY,WB & BLAU,GE (1985)
DC	pKa			
HL 1.82E-005	atm m3/mol	25	EST	MEYLAN,WM & HOWARD,PH (1991)
OH 4.80E-012	cm3/molc sec	25	EST	MEYLAN,WM & HOWARD,PH (1993)

CAS #: 005337-72-4				2,6-DIMETHYLCYCLOHEXANOL

Formula: $C_8H_{16}O$
Mol Weight: 128.22
MP (deg C): 32.5   FP (deg C):
BP (deg C): 176
BP pressure (mm Hg):

Property/Value	Units	Temp	Data Type	Reference
WS 2.81E+003	mg/L	25	EST	MEYLAN,WM ET AL. (1996)
logP 2.47			EST	MEYLAN,WM & HOWARD,PH (1995)
VP 1.23E-001	mm Hg	25	EST	NEELY,WB & BLAU,GE (1985)
DC	pKa			
HL 8.63E-006	atm m3/mol	25	EST	MEYLAN,WM & HOWARD,PH (1991)
OH 2.09E-011	cm3/molc sec	25	EST	MEYLAN,WM & HOWARD,PH (1993)

CAS #: 005337-93-9				1-(4-METHYLPHENYL)-1-PROPANONE

Formula: $C_{10}H_{12}O$
Mol Weight: 148.21
MP (deg C): 7.2   FP (deg C):
BP (deg C): 236
BP pressure (mm Hg):

Property/Value	Units	Temp	Data Type	Reference
WS 3.73E+002	mg/L	25	EST	MEYLAN,WM ET AL. (1996)
logP 2.71			EST	MEYLAN,WM & HOWARD,PH (1995)
VP 9.37E-002	mm Hg	25	EXT	PERRY,RH & GREEN,D (1984)
DC	pKa			
HL 1.44E-005	atm m3/mol	25	EST	MEYLAN,WM & HOWARD,PH (1991)
OH 6.01E-012	cm3/molc sec	25	EST	MEYLAN,WM & HOWARD,PH (1993)

CAS #: 005341-95-7				2,3-BUTANEDIOL (MESO)

Formula: $C_4H_{10}O_2$
Mol Weight: 90.12
MP (deg C): 32-34   FP (deg C):
BP (deg C): 183-184
BP pressure (mm Hg):

Property/Value	Units	Temp	Data Type	Reference
WS 1.00E+006	mg/L	20	EXP	RIDDICK,JA ET AL. (1984)
logP -0.36			EST	MEYLAN,WM & HOWARD,PH (1995)
VP 3.80E-001	mm Hg	25	EXP	RIDDICK,JA ET AL. (1984)
DC	pKa			
HL 4.51E-008	atm m3/mol	25	EST	VP/WSOL
OH 1.73E-011	cm3/molc sec	25	EST	MEYLAN,WM & HOWARD,PH (1993)

CAS #: 005344-27-4				4-PYRIDINEETHANOL

Formula: C_7H_9NO
Mol Weight: 123.16
MP (deg C):   FP (deg C):
BP (deg C):
BP pressure (mm Hg):

Property/Value	Units	Temp	Data Type	Reference
WS 1.00E+006	mg/L	25	EST	MEYLAN,WM ET AL. (1996)
logP 0.10			EXP	HANSCH,C & LEO,AJ (1985)
VP 6.59E-003	mm Hg	25	EST	NEELY,WB & BLAU,GE (1985)
DC 5.60	pKa	25	EXP	PERRIN,DD (1972)
HL 3.78E-010	atm m3/mol	25	EST	MEYLAN,WM & HOWARD,PH (1991)
OH 6.23E-012	cm3/molc sec	25	EST	MEYLAN,WM & HOWARD,PH (1993)

CAS #: 005344-49-0				BENZOIC ACID, 2-CHLORO-6-NITRO-

Formula: $C_7H_4ClNO_4$
Mol Weight: 201.57
MP (deg C):   FP (deg C):
BP (deg C):
BP pressure (mm Hg):

Property/Value	Units	Temp	Data Type	Reference
WS 8.14E+002	mg/L	25	EST	MEYLAN,WM ET AL. (1996)
logP 1.56			EXP	SOTOMATSU,T ET AL. (1993)
VP 1.89E-005	mm Hg	25	EST	NEELY,WB & BLAU,GE (1985)
DC 1.34	pKa	25	EXP	KORTUM,G ET AL (1961)
HL 3.17E-010	atm m3/mol	25	EST	MEYLAN,WM & HOWARD,PH (1991)
OH 5.64E-013	cm3/molc sec	25	EST	MEYLAN,WM & HOWARD,PH (1993)

CAS #: 005344-82-1				1-(O-CHLOROPHENYL)-2-THIOUREA

Formula: $C_7H_7ClN_2S$
Mol Weight: 186.66
MP (deg C): 146   FP (deg C):
BP (deg C):
BP pressure (mm Hg):

Property/Value	Units	Temp	Data Type	Reference
WS 2.25E+004	mg/L	25	EST	MEYLAN,WM ET AL. (1996)
logP 1.60			EST	MEYLAN,WM & HOWARD,PH (1995)
VP 3.50E-004	mm Hg	25	EST	NEELY,WB & BLAU,GE (1985)
DC	pKa			
HL 6.46E-008	atm m3/mol	25	EST	MEYLAN,WM & HOWARD,PH (1991)
OH 3.35E-011	cm3/molc sec	25	EST	MEYLAN,WM & HOWARD,PH (1993)

CAS #: 005345-54-0				3-CHLORO-4-METHOXYANILINE

Formula: C_7H_8ClNO
Mol Weight: 157.60
MP (deg C): 50-55   FP (deg C):
BP (deg C):
BP pressure (mm Hg):

Property/Value	Units	Temp	Data Type	Reference
WS 1.85E+003	mg/L	25	EST	MEYLAN,WM ET AL. (1996)
logP 1.85			EXP	HANSCH,C & LEO,AJ (1985)
VP 1.32E-002	mm Hg	25	EST	NEELY,WB & BLAU,GE (1985)
DC	pKa			
HL 8.35E-008	atm m3/mol	25	EST	MEYLAN,WM & HOWARD,PH (1991)
OH 6.63E-011	cm3/molc sec	25	EST	MEYLAN,WM & HOWARD,PH (1993)

CAS #: 005346-38-3				2-PYRIDINETHIOCARBOXAMIDE

Formula: $C_6H_6N_2S$
Mol Weight: 138.19
MP (deg C):
FP (deg C):
BP (deg C):
BP pressure (mm Hg):

Property/ Value	Units	Temp	Data Type	Reference
WS 7.58E+004	mg/L	25	EST	MEYLAN,WM ET AL. (1996)
logP 1.24			EXP	HANSCH,C & LEO,AJ (1985)
VP 8.41E-003	mm Hg	25	EST	NEELY,WB & BLAU,GE (1985)
DC	pKa			
HL 5.13E-010	atm m3/mol	25	EST	MEYLAN,WM & HOWARD,PH (1991)
OH 2.18E-011	cm3/molc sec	25	EST	MEYLAN,WM & HOWARD,PH (1993)

CAS #: 005348-42-5				4,5-DICHLORO-O-PHENYLENEDIAMINE

Formula: $C_6H_6Cl_2N_2$
Mol Weight: 177.03
MP (deg C): 159-162
FP (deg C):
BP (deg C):
BP pressure (mm Hg):

Property/ Value	Units	Temp	Data Type	Reference
WS 1.12E+003	mg/L	25	EST	MEYLAN,WM ET AL. (1996)
logP 2.00			EXP	HANSCH,C ET AL. (1995)
VP 3.45E-004	mm Hg	25	EST	NEELY,WB & BLAU,GE (1985)
DC	pKa			
HL 3.69E-010	atm m3/mol	25	EST	MEYLAN,WM & HOWARD,PH (1991)
OH 3.62E-011	cm3/molc sec	25	EST	MEYLAN,WM & HOWARD,PH (1993)

CAS #: 005348-75-4				O-PHENYLPHENOXYACETIC ACID

Formula: $C_{14}H_{12}O_3$
Mol Weight: 228.25
MP (deg C):
FP (deg C):
BP (deg C):
BP pressure (mm Hg):

Property/ Value	Units	Temp	Data Type	Reference
WS 2.96E+002	mg/L	25	EST	MEYLAN,WM ET AL. (1996)
logP 2.83			EXP	HANSCH,C & LEO,AJ (1985)
VP 9.21E-007	mm Hg	25	EST	NEELY,WB & BLAU,GE (1985)
DC	pKa			
HL 1.29E-009	atm m3/mol	25	EST	MEYLAN,WM & HOWARD,PH (1991)
OH 2.35E-011	cm3/molc sec	25	EST	MEYLAN,WM & HOWARD,PH (1993)

CAS #: 005350-41-4				BENZYLTRIMETHYL AMMONIUM BROMIDE

Formula: $C_{10}H_{16}BrN$
Mol Weight: 230.15
MP (deg C): 230-232
FP (deg C):
BP (deg C):
BP pressure (mm Hg):

Property/ Value	Units	Temp	Data Type	Reference
WS 1.00E+006	mg/L	25	EST	MEYLAN,WM ET AL. (1996)
logP -2.38			EXP	HANSCH,C & LEO,AJ (1985)
VP 4.65E-008	mm Hg	25	EST	NEELY,WB & BLAU,GE (1985)
DC	pKa			
HL 2.68E-013	atm m3/mol	25	EST	MEYLAN,WM & HOWARD,PH (1991)
OH 1.73E-011	cm3/molc sec	25	EST	MEYLAN,WM & HOWARD,PH (1993)

CAS #: 005350-57-2				METHANONE, DIPHENYL-, HYDRAZONE

Formula: $C_{13}H_{12}N_2$
Mol Weight: 196.25
MP (deg C): 95-98
FP (deg C):
BP (deg C):
BP pressure (mm Hg):

Property/ Value	Units	Temp	Data Type	Reference
WS 1.89E+002	mg/L	25	EST	MEYLAN,WM ET AL. (1996)
logP 2.79			EXP	HANSCH,C ET AL. (1995)
VP 1.51E-004	mm Hg	25	EST	NEELY,WB & BLAU,GE (1985)
DC	pKa			
HL 6.02E-008	atm m3/mol	25	EST	MEYLAN,WM & HOWARD,PH (1991)
OH 8.59E-012	cm3/molc sec	25	EST	MEYLAN,WM & HOWARD,PH (1993)

CAS #: 005351-17-7				4-AMINOBENZOYLHYDRAZINE

Formula: $C_7H_9N_3O$
Mol Weight: 151.17
MP (deg C): 225-227
FP (deg C):
BP (deg C):
BP pressure (mm Hg):

Property/ Value	Units	Temp	Data Type	Reference
WS 1.61E+004	mg/L	25	EST	MEYLAN,WM ET AL. (1996)
logP -0.75			EXP	HANSCH,C & LEO,AJ (1985)
VP 3.14E-006	mm Hg	25	EST	NEELY,WB & BLAU,GE (1985)
DC	pKa			
HL 3.27E-015	atm m3/mol	25	EST	MEYLAN,WM & HOWARD,PH (1991)
OH 1.04E-010	cm3/molc sec	25	EST	MEYLAN,WM & HOWARD,PH (1993)

CAS #: 005351-23-5				BENZOYLHYDRAZINE, P-HYDROXY

Formula: $C_7H_8N_2O_2$
Mol Weight: 152.15
MP (deg C): 264-266
FP (deg C):
BP (deg C):
BP pressure (mm Hg):

Property/ Value	Units	Temp	Data Type	Reference
WS 2.65E+004	mg/L	25	EST	MEYLAN,WM ET AL. (1996)
logP -0.33			EXP	HANSCH,C & LEO,AJ (1985)
VP 1.24E-006	mm Hg	25	EST	NEELY,WB & BLAU,GE (1985)
DC	pKa			
HL 9.62E-016	atm m3/mol	25	EST	MEYLAN,WM & HOWARD,PH (1991)
OH 3.60E-011	cm3/molc sec	25	EST	MEYLAN,WM & HOWARD,PH (1993)

CAS #: 005352-63-6				CARD-20(22)-ENOLIDE, 3-[[(2,6-DIDEOXY-BETA-D-RIBO

Formula: $C_{29}H_{44}O_8$
Mol Weight: 520.67
MP (deg C):
FP (deg C):
BP (deg C):
BP pressure (mm Hg):

Property/ Value	Units	Temp	Data Type	Reference
WS 6.18E+001	mg/L	25	EST	MEYLAN,WM ET AL. (1996)
logP 1.09			EXP	SANGSTER,J (1993)
VP 4.06E-020	mm Hg	25	EST	NEELY,WB & BLAU,GE (1985)
DC	pKa			
HL 9.89E-017	atm m3/mol	25	EST	MEYLAN,WM & HOWARD,PH (1991)
OH 1.02E-010	cm3/molc sec	25	EST	MEYLAN,WM & HOWARD,PH (1993)

CAS #: 005382-47-8 — 6-QUINOLINECARBOXYLIC ACID HYDRAZIDE

Formula: $C_{10}H_9N_3O$

Mol Weight: 187.20

MP (deg C): | FP (deg C):

BP (deg C):

BP pressure (mm Hg):

Property/Value	Units	Temp	Data Type	Reference
WS 1.25E+003	mg/L	25	EST	MEYLAN,WM ET AL. (1996)
logP 0.35			EXP	HANSCH,C & LEO,AJ (1985)
VP 2.67E-007	mm Hg	25	EST	NEELY,WB & BLAU,GE (1985)
DC	pKa			
HL 1.18E-015	atm m3/mol	25	EST	MEYLAN,WM & HOWARD,PH (1991)
OH 1.86E-011	cm3/molc sec	25	EST	MEYLAN,WM & HOWARD,PH (1993)

CAS #: 005392-40-5 — CITRAL

Formula: $C_{10}H_{16}O$

Mol Weight: 152.24

MP (deg C): <-10 | FP (deg C):

BP (deg C): 226-228

BP pressure (mm Hg):

Property/Value	Units	Temp	Data Type	Reference
WS 1.34E+003	mg/L	37	EXP	YALKOWSKY,SH & DANNENFELSER,RM (1992)
logP 3.45			EST	MEYLAN,WM & HOWARD,PH (1995)
VP 9.13E-002	mm Hg	25	EST	NEELY,WB & BLAU,GE (1985)
DC	pKa			
HL 4.35E-005	atm m3/mol	25	EST	MEYLAN,WM & HOWARD,PH (1991)
OH 1.36E-010	cm3/molc sec	25	EST	MEYLAN,WM & HOWARD,PH (1993)

CAS #: 005394-83-2 — CIS-1,2,2-TRIMETHYL-1,3-CYCLOPENTANEDICARBOXYL*

Formula: $C_{10}H_{16}O_4$

Mol Weight: 200.24

MP (deg C): 186-188 | FP (deg C):

BP (deg C):

BP pressure (mm Hg):

Property/Value	Units	Temp	Data Type	Reference
WS 3.40E+003	mg/L	20	EXP	YALKOWSKY,SH & DANNENFELSER,RM (1992)
logP 1.78			EST	MEYLAN,WM & HOWARD,PH (1995)
VP 1.51E-005	mm Hg	25	EST	NEELY,WB & BLAU,GE (1985)
DC	pKa			
HL 1.31E-011	atm m3/mol	25	EST	MEYLAN,WM & HOWARD,PH (1991)
OH 5.56E-012	cm3/molc sec	25	EST	MEYLAN,WM & HOWARD,PH (1993)

CAS #: 005398-11-8 — 1(3H)-ISOBENZOFURANONE-3-PHENYL

Formula: $C_{14}H_{10}O_2$

Mol Weight: 210.23

MP (deg C): | FP (deg C):

BP (deg C):

BP pressure (mm Hg):

Property/Value	Units	Temp	Data Type	Reference
WS 2.16E+002	mg/L	25	EST	MEYLAN,WM ET AL. (1996)
logP 2.64			EXP	HANSCH,C & LEO,AJ (1985)
VP 1.01E-005	mm Hg	25	EST	NEELY,WB & BLAU,GE (1985)
DC	pKa			
HL 1.03E-006	atm m3/mol	25	EST	MEYLAN,WM & HOWARD,PH (1991)
OH 1.13E-011	cm3/molc sec	25	EST	MEYLAN,WM & HOWARD,PH (1993)

CAS #: 005401-94-5 — 1H-INDAZOLE, 5-NITRO-

Formula: $C_7H_{13}N_3O_2$

Mol Weight: 171.20

MP (deg C): | FP (deg C):

BP (deg C):

BP pressure (mm Hg):

Property/Value	Units	Temp	Data Type	Reference
WS 9.91E+002	mg/L	25	EST	MEYLAN,WM ET AL. (1996)
logP 1.68			EXP	SANGSTER,J (1993)
VP 1.28E-005	mm Hg	25	EST	NEELY,WB & BLAU,GE (1985)
DC	pKa			
HL 1.42E-009	atm m3/mol	25	EST	MEYLAN,WM & HOWARD,PH (1991)
OH 4.50E-012	cm3/molc sec	25	EST	MEYLAN,WM & HOWARD,PH (1993)

CAS #: 005405-53-8 — 9H-FLUORENE, 2,7-DINITRO-

Formula: $C_{13}H_8N_2O_4$

Mol Weight: 256.22

MP (deg C): 334 | FP (deg C):

BP (deg C):

BP pressure (mm Hg):

Property/Value	Units	Temp	Data Type	Reference
WS 9.52E-001	mg/L	25	EST	MEYLAN,WM ET AL. (1996)
logP 3.35			EXP	DEBNATH,AK & HANSCH,C (1992)
VP 1.88E-007	mm Hg	25	EST	NEELY,WB & BLAU,GE (1985)
DC	pKa			
HL 2.61E-009	atm m3/mol	25	EST	MEYLAN,WM & HOWARD,PH (1991)
OH 1.51E-012	cm3/molc sec	25	EST	MEYLAN,WM & HOWARD,PH (1993)

CAS #: 005406-33-7 — P-CHLOROBENZYL ACETATE

Formula: $C_9H_9ClO_2$

Mol Weight: 184.62

MP (deg C): | FP (deg C):

BP (deg C):

BP pressure (mm Hg):

Property/Value	Units	Temp	Data Type	Reference
WS 3.21E+002	mg/L	25	EST	MEYLAN,WM ET AL. (1996)
logP 2.59			EXP	HANSCH,C & LEO,AJ (1985)
VP 3.09E-002	mm Hg	25	EST	NEELY,WB & BLAU,GE (1985)
DC	pKa			
HL 1.05E-005	atm m3/mol	25	EST	MEYLAN,WM & HOWARD,PH (1991)
OH 3.27E-012	cm3/molc sec	25	EST	MEYLAN,WM & HOWARD,PH (1993)

CAS #: 005407-61-4 — B-PH-B-HYDROXY-N-ETHYLPYRROLIDINE

Formula: $C_{12}H_{17}NO$

Mol Weight: 191.28

MP (deg C): | FP (deg C):

BP (deg C):

BP pressure (mm Hg):

Property/Value	Units	Temp	Data Type	Reference
WS 3.27E+004	mg/L	25	EST	MEYLAN,WM ET AL. (1996)
logP 1.38			EXP	HANSCH,C & LEO,AJ (1985)
VP 2.78E-005	mm Hg	25	EST	NEELY,WB & BLAU,GE (1985)
DC	pKa			
HL 1.11E-010	atm m3/mol	25	EST	MEYLAN,WM & HOWARD,PH (1991)
OH 1.09E-010	cm3/molc sec	25	EST	MEYLAN,WM & HOWARD,PH (1993)

005408-86-6 — 2,3-DIBROMOBUTANE

Formula: $C_4H_8Br_2$
Mol Weight: 215.93
MP (deg C): -24
BP (deg C): 161
BP pressure (mm Hg):

Property/Value	Units	Temp	Data Type	Reference
WS 1.34E+002	mg/L	25	EST	MEYLAN,WM ET AL. (1996)
logP 2.85			EST	MEYLAN,WM & HOWARD,PH (1995)
VP 3.80E+000	mm Hg	25	EXP	PERRY,RH & GREEN,D (1984)
DC	pKa			
HL 2.29E-003	atm m3/mol	25	EST	MEYLAN,WM & HOWARD,PH (1991)
OH 6.25E-013	cm3/molc sec	25	EST	MEYLAN,WM & HOWARD,PH (1993)

005409-83-6 — 2,8-DICHLORODIBENZOFURAN

Formula: $C_{12}H_6Cl_2O$
Mol Weight: 237.09
MP (deg C):
BP (deg C):
BP pressure (mm Hg):

Property/Value	Units	Temp	Data Type	Reference
WS 1.45E-002	mg/L	25	EXP	DOUCETTE,WJ & ANDREN,AW (1988)
logP 5.44			EXP	DOUCETTE,WJ & ANDREN,AW (1987)
VP 2.54E-005	mm Hg	25	EST	NEELY,WB & BLAU,GE (1985)
DC	pKa			
HL 2.81E-005	atm m3/mol	25	EST	MEYLAN,WM & HOWARD,PH (1991)
OH 8.01E-013	cm3/molc sec	25	EST	MEYLAN,WM & HOWARD,PH (1993)

005415-44-1 — 1,3,7-TRIMETHYL URIC ACID

Formula: $C_8H_{10}N_4O_3$
Mol Weight: 210.19
MP (deg C):
BP (deg C):
BP pressure (mm Hg):

Property/Value	Units	Temp	Data Type	Reference
WS 3.93E+003	mg/L	25	EST	MEYLAN,WM ET AL. (1996)
logP -0.37			EXP	GASPARI,F & BONATI,M (1987)
VP 7.23E-009	mm Hg	25	EST	NEELY,WB & BLAU,GE (1985)
DC	pKa			
HL 1.27E-016	atm m3/mol	25	EST	MEYLAN,WM & HOWARD,PH (1991)
OH 1.40E-011	cm3/molc sec	25	EST	MEYLAN,WM & HOWARD,PH (1993)

005416-18-2 — 1,4-NAPHTHOQUINONE,2-ME-3-MEO

Formula: $C_{12}H_{10}O_3$
Mol Weight: 202.21
MP (deg C):
BP (deg C):
BP pressure (mm Hg):

Property/Value	Units	Temp	Data Type	Reference
WS 6.34E+002	mg/L	25	EST	MEYLAN,WM ET AL. (1996)
logP 2.14			EXP	HANSCH,C & LEO,AJ (1985)
VP 3.28E-005	mm Hg	25	EST	NEELY,WB & BLAU,GE (1985)
DC	pKa			
HL 2.03E-009	atm m3/mol	25	EST	MEYLAN,WM & HOWARD,PH (1991)
OH 1.92E-011	cm3/molc sec	25	EST	MEYLAN,WM & HOWARD,PH (1993)

005417-50-5 — 3-ACETYLAMINOQUINOLINE

Formula: $C_{11}H_{10}N_2O$
Mol Weight: 186.22
MP (deg C):
BP (deg C):
BP pressure (mm Hg):

Property/Value	Units	Temp	Data Type	Reference
WS 1.18E+003	mg/L	25	EST	MEYLAN,WM ET AL. (1996)
logP 1.92			EXP	HANSCH,C & LEO,AJ (1985)
VP 1.28E-006	mm Hg	25	EST	NEELY,WB & BLAU,GE (1985)
DC	pKa			
HL 7.88E-013	atm m3/mol	25	EST	MEYLAN,WM & HOWARD,PH (1991)
OH 9.18E-011	cm3/molc sec	25	EST	MEYLAN,WM & HOWARD,PH (1993)

005418-93-9 — 1H-BENZIMIDAZOL-2-AMINE, 5-CHLORO-

Formula: $C_7H_6ClN_3$
Mol Weight: 167.60
MP (deg C):
BP (deg C):
BP pressure (mm Hg):

Property/Value	Units	Temp	Data Type	Reference
WS 1.57E+003	mg/L	25	EST	MEYLAN,WM ET AL. (1996)
logP 1.88			EXP	HANSCH,C ET AL. (1995)
VP 6.30E-007	mm Hg	25	EST	NEELY,WB & BLAU,GE (1985)
DC	pKa			
HL 9.60E-011	atm m3/mol	25	EST	MEYLAN,WM & HOWARD,PH (1991)
OH 1.55E-010	cm3/molc sec	25	EST	MEYLAN,WM & HOWARD,PH (1993)

005419-96-5 — HYDRAZINECARBOTHIAMIDE, 2-(2-FURANYLMETHYLENE)-

Formula: $C_6H_7N_3OS$
Mol Weight: 169.21
MP (deg C):
BP (deg C):
BP pressure (mm Hg):

Property/Value	Units	Temp	Data Type	Reference
WS 2.47E+003	mg/L	25	EST	MEYLAN,WM ET AL. (1996)
logP 1.64			EXP	DE,AU ET AL. (1983)
VP 9.56E-004	mm Hg	25	EST	NEELY,WB & BLAU,GE (1985)
DC	pKa			
HL 3.40E-008	atm m3/mol	25	EST	MEYLAN,WM & HOWARD,PH (1991)
OH 1.82E-010	cm3/molc sec	25	EST	MEYLAN,WM & HOWARD,PH (1993)

005422-81-1 — N,N-PENTAMETHYLENE CINNAMAMIDE

Formula: $C_{14}H_{17}NO$
Mol Weight: 215.30
MP (deg C):
BP (deg C):
BP pressure (mm Hg):

Property/Value	Units	Temp	Data Type	Reference
WS 1.67E+002	mg/L	25	EST	MEYLAN,WM ET AL. (1996)
logP 2.74			EXP	HANSCH,C & LEO,AJ (1985)
VP 1.55E-006	mm Hg	25	EST	NEELY,WB & BLAU,GE (1985)
DC	pKa			
HL 7.94E-010	atm m3/mol	25	EST	MEYLAN,WM & HOWARD,PH (1991)
OH 4.12E-011	cm3/molc sec	25	EST	MEYLAN,WM & HOWARD,PH (1993)

005424-05-5 — 2-AMINOQUINOXALINE

Formula: $C_8H_7N_3$
Mol Weight: 145.17
MP (deg C):
FP (deg C):
BP (deg C):
BP pressure (mm Hg):

Property/Value	Units	Temp	Data Type	Reference
WS 4.17E+003	mg/L	25	EST	MEYLAN,WM ET AL. (1996)
logP 1.50			EXP	HANSCH,C & LEO,AJ (1985)
VP 2.50E-004	mm Hg	25	EST	NEELY,WB & BLAU,GE (1985)
DC	pKa			
HL 1.01E-010	atm m3/mol	25	EST	MEYLAN,WM & HOWARD,PH (1991)
OH 1.10E-010	cm3/molc sec	25	EST	MEYLAN,WM & HOWARD,PH (1993)

005424-19-1 — PHENYL-B-PYRIDYL KETONE

Formula: $C_{12}H_9NO$
Mol Weight: 183.21
MP (deg C): 36-40
FP (deg C):
BP (deg C): 307
BP pressure (mm Hg):

Property/Value	Units	Temp	Data Type	Reference
WS 1.32E+003	mg/L	25	EST	MEYLAN,WM ET AL. (1996)
logP 1.88			EXP	HANSCH,C & LEO,AJ (1985)
VP 3.67E-004	mm Hg	25	EST	NEELY,WB & BLAU,GE (1985)
DC	pKa			
HL 2.54E-009	atm m3/mol	25	EST	MEYLAN,WM & HOWARD,PH (1991)
OH 2.11E-012	cm3/molc sec	25	EST	MEYLAN,WM & HOWARD,PH (1993)

005428-54-6 — PHENOL, 2-METHYL-5-NITRO-

Formula: $C_7H_7NO_3$
Mol Weight: 153.14
MP (deg C):
FP (deg C):
BP (deg C):
BP pressure (mm Hg):

Property/Value	Units	Temp	Data Type	Reference
WS 1.00E+003	mg/L	25	EXP	CHEM INSPECT TEST INST (1992)
logP 2.47			EXP	CHEM INSPECT TEST INST (1992)
VP 6.32E-004	mm Hg	25	EST	NEELY,WB & BLAU,GE (1985)
DC 8.59	pKa	25	EXP	SERJEANT,EP & DEMPSEY,B (1979)
HL 2.44E-009	atm m3/mol	25	EST	MEYLAN,WM & HOWARD,PH (1991)
OH 3.85E-012	cm3/molc sec	25	EST	MEYLAN,WM & HOWARD,PH (1993)

005429-42-5 — N-(P-TOLYL)-3-N'-PIPERIDINOACETAMIDE

Formula: $C_{14}H_{20}N_2O$
Mol Weight: 232.33
MP (deg C):
FP (deg C):
BP (deg C):
BP pressure (mm Hg):

Property/Value	Units	Temp	Data Type	Reference
WS 8.77E+001	mg/L	25	EST	MEYLAN,WM ET AL. (1996)
logP 2.96			EXP	HANSCH,C & LEO,AJ (1985)
VP 4.23E-007	mm Hg	25	EST	NEELY,WB & BLAU,GE (1985)
DC	pKa			
HL 6.95E-011	atm m3/mol	25	EST	MEYLAN,WM & HOWARD,PH (1991)
OH 1.13E-010	cm3/molc sec	25	EST	MEYLAN,WM & HOWARD,PH (1993)

005432-28-0 — N-NITROSO-METHYLCYCLOHEXYLAMINE

Formula: $C_7H_{14}N_2O$
Mol Weight: 142.20
MP (deg C):
FP (deg C):
BP (deg C):
BP pressure (mm Hg):

Property/Value	Units	Temp	Data Type	Reference
WS 4.93E+003	mg/L	25	EST	MEYLAN,WM ET AL. (1996)
logP 1.43			EXP	HANSCH,C & LEO,AJ (1985)
VP 2.41E-002	mm Hg	25	EST	NEELY,WB & BLAU,GE (1985)
DC	pKa			
HL 3.74E-006	atm m3/mol	25	EST	MEYLAN,WM & HOWARD,PH (1991)
OH 3.56E-011	cm3/molc sec	25	EST	MEYLAN,WM & HOWARD,PH (1993)

005432-44-0 — PROPOXYETHYL TRIMETHYL AMMONIUM IODIDE

Formula: $C_8H_{20}INO$
Mol Weight: 273.16
MP (deg C):
FP (deg C):
BP (deg C):
BP pressure (mm Hg):

Property/Value	Units	Temp	Data Type	Reference
WS 8.44E+005	mg/L	25	EST	MEYLAN,WM ET AL. (1996)
logP -1.97			EXP	HANSCH,C & LEO,AJ (1985)
VP 9.86E-008	mm Hg	25	EST	NEELY,WB & BLAU,GE (1985)
DC	pKa			
HL 7.21E-014	atm m3/mol	25	EST	MEYLAN,WM & HOWARD,PH (1991)
OH 5.16E-011	cm3/molc sec	25	EST	MEYLAN,WM & HOWARD,PH (1993)

005435-82-5 — PYRIDO(12A)PYRIMIDIN-4-ON,3ETO-CO-7ME

Formula: $C_{12}H_{14}N_2O_3$
Mol Weight: 234.26
MP (deg C):
FP (deg C):
BP (deg C):
BP pressure (mm Hg):

Property/Value	Units	Temp	Data Type	Reference
WS 5.04E+003	mg/L	25	EST	MEYLAN,WM ET AL. (1996)
logP 0.90			EXP	HANSCH,C & LEO,AJ (1985)
VP 8.05E-006	mm Hg	25	EST	NEELY,WB & BLAU,GE (1985)
DC	pKa			
HL	atm m3/mol			
OH 9.15E-012	cm3/molc sec	25	EST	MEYLAN,WM & HOWARD,PH (1993)

005438-19-7 — 4-PROPOXYBENZOIC ACID

Formula: $C_{10}H_{12}O_3$
Mol Weight: 180.21
MP (deg C): 144-146
FP (deg C):
BP (deg C):
BP pressure (mm Hg):

Property/Value	Units	Temp	Data Type	Reference
WS 1.26E+002	mg/L	25	EST	MEYLAN,WM ET AL. (1996)
logP 3.09			EXP	DA,YZ ET AL. (1992)
VP 2.35E-004	mm Hg	25	EST	NEELY,WB & BLAU,GE (1985)
DC 4.78	pKa	20	EXP	KORTUM,G ET AL (1961)
HL 1.13E-008	atm m3/mol	25	EST	MEYLAN,WM & HOWARD,PH (1991)
OH 1.89E-011	cm3/molc sec	25	EST	MEYLAN,WM & HOWARD,PH (1993)

CAS #: 005441-52-1				3,5-DIMETHYLCYCLOHEXANOL

Formula: $C_8H_{16}O$

Mol Weight: 128.22

MP (deg C): 11.6   FP (deg C):

BP (deg C): 187

BP pressure (mm Hg):

Property/Value	Units	Temp	Data Type	Reference
WS 2.34E+003	mg/L	25	EST	MEYLAN,WM ET AL. (1996)
logP 2.47			EST	MEYLAN,WM & HOWARD,PH (1995)
VP 1.23E-001	mm Hg	25	EST	NEELY,WB & BLAU,GE (1985)
DC	pKa			
HL 8.63E-006	atm m3/mol	25	EST	MEYLAN,WM & HOWARD,PH (1991)
OH 2.09E-011	cm3/molc sec	25	EST	MEYLAN,WM & HOWARD,PH (1993)

CAS #: 005446-77-5				BENZOIC ACID, 4-(AMINOSULFONYL)-, ETHYL ESTER

Formula: $C_9H_{11}NO_4S$

Mol Weight: 229.26

MP (deg C):   FP (deg C):

BP (deg C):

BP pressure (mm Hg):

Property/Value	Units	Temp	Data Type	Reference
WS 3.08E+003	mg/L	25	EST	MEYLAN,WM ET AL. (1996)
logP 1.17			EXP	SANGSTER,J (1993)
VP 6.50E-006	mm Hg	25	EST	NEELY,WB & BLAU,GE (1985)
DC	pKa			
HL 3.61E-009	atm m3/mol	25	EST	MEYLAN,WM & HOWARD,PH (1991)
OH 1.80E-012	cm3/molc sec	25	EST	MEYLAN,WM & HOWARD,PH (1993)

CAS #: 005446-92-4				2-METHOXY-5-NITROPYRIDINE

Formula: $C_6H_6N_2O_3$

Mol Weight: 154.13

MP (deg C): 108-109   FP (deg C):

BP (deg C):

BP pressure (mm Hg):

Property/Value	Units	Temp	Data Type	Reference
WS 1.41E+003	mg/L	25	EST	MEYLAN,WM ET AL. (1996)
logP 1.55			EXP	HANSCH,C ET AL. (1995)
VP 9.47E-003	mm Hg	25	EST	NEELY,WB & BLAU,GE (1985)
DC	pKa			
HL 1.50E-008	atm m3/mol	25	EST	MEYLAN,WM & HOWARD,PH (1991)
OH 1.34E-012	cm3/molc sec	25	EST	MEYLAN,WM & HOWARD,PH (1993)

CAS #: 005447-87-0				MALONONITRILE, A-METHYLBENZAL

Formula: $C_{11}H_8N_2$

Mol Weight: 168.20

MP (deg C):   FP (deg C):

BP (deg C):

BP pressure (mm Hg):

Property/Value	Units	Temp	Data Type	Reference
WS 5.48E+002	mg/L	25	EST	MEYLAN,WM ET AL. (1996)
logP 2.10			EXP	HANSCH,C & LEO,AJ (1985)
VP 1.09E-004	mm Hg	25	EST	NEELY,WB & BLAU,GE (1985)
DC	pKa			
HL 2.15E-008	atm m3/mol	25	EST	MEYLAN,WM & HOWARD,PH (1991)
OH 4.79E-012	cm3/molc sec	25	EST	MEYLAN,WM & HOWARD,PH (1993)

CAS #: 005448-36-2				BENZAMIDE, 3-METHYL-N,N-BIS(1-METHYLETHYL)-

Formula: $C_{14}H_{21}NO$

Mol Weight: 219.33

MP (deg C):   FP (deg C):

BP (deg C):

BP pressure (mm Hg):

Property/Value	Units	Temp	Data Type	Reference
WS 9.51E+001	mg/L	25	EST	MEYLAN,WM ET AL. (1996)
logP 3.00			EXP	SURYANARAYANA,MVS ET AL. (1991)
VP 1.51E-001	mm Hg	30	EXP	SURYANARAYANA,MVS ET AL. (1991)
DC	pKa			
HL 3.66E-008	atm m3/mol	25	EST	MEYLAN,WM & HOWARD,PH (1991)
OH 3.39E-011	cm3/molc sec	25	EST	MEYLAN,WM & HOWARD,PH (1993)

CAS #: 005448-43-1				6-CHLOROQUINOXALINE

Formula: $C_8H_5ClN_2$

Mol Weight: 164.60

MP (deg C): 64   FP (deg C):

BP (deg C): 117

BP pressure (mm Hg): 1.00E+001

Property/Value	Units	Temp	Data Type	Reference
WS 2.13E+003	mg/L	25	EST	MEYLAN,WM ET AL. (1996)
logP 2.10			EXP	HANSCH,C & LEO,AJ (1985)
VP 2.46E-003	mm Hg	25	EST	NEELY,WB & BLAU,GE (1985)
DC	pKa			
HL 2.11E-007	atm m3/mol	25	EST	MEYLAN,WM & HOWARD,PH (1991)
OH 1.41E-012	cm3/molc sec	25	EST	MEYLAN,WM & HOWARD,PH (1993)

CAS #: 005450-54-4				3-PYRIDAZINECARBOXAMIDE

Formula: $C_5H_5N_3O$

Mol Weight: 123.12

MP (deg C):   FP (deg C):

BP (deg C):

BP pressure (mm Hg):

Property/Value	Units	Temp	Data Type	Reference
WS 2.01E+004	mg/L	25	EST	MEYLAN,WM ET AL. (1996)
logP -0.73			EXP	YAMAGAMI,C ET AL. (1990)
VP 1.49E-005	mm Hg	25	EST	NEELY,WB & BLAU,GE (1985)
DC 1.00	pKa		EXP	PERRIN,DD (1965)
HL 1.18E-012	atm m3/mol	25	EST	MEYLAN,WM & HOWARD,PH (1991)
OH 2.21E-012	cm3/molc sec	25	EST	MEYLAN,WM & HOWARD,PH (1993)

CAS #: 005451-39-8				2-PYRIDINEACETAMIDE

Formula: $C_7H_8N_2O$

Mol Weight: 136.15

MP (deg C):   FP (deg C):

BP (deg C):

BP pressure (mm Hg):

Property/Value	Units	Temp	Data Type	Reference
WS 1.00E+006	mg/L	25	EST	MEYLAN,WM ET AL. (1996)
logP -0.65			EXP	HANSCH,C & LEO,AJ (1985)
VP 1.57E-004	mm Hg	25	EST	NEELY,WB & BLAU,GE (1985)
DC	pKa			
HL 1.18E-012	atm m3/mol	25	EST	MEYLAN,WM & HOWARD,PH (1991)
OH 3.62E-012	cm3/molc sec	25	EST	MEYLAN,WM & HOWARD,PH (1993)

CAS #: 005451-83-2				ACETIC ACID, 3-METHOXYPHENYL ESTER

Formula: $C_9H_{10}O_3$

Mol Weight: 166.18

MP (deg C): | FP (deg C):

BP (deg C):

BP pressure (mm Hg):

Property/Value	Units	Temp	Data Type	Reference
WS 2.70E+003	mg/L	25	EST	MEYLAN,WM ET AL. (1996)
logP 1.61			EXP	SOTOMATSU,T ET AL. (1993)
VP 6.02E-002	mm Hg	25	EST	NEELY,WB & BLAU,GE (1985)
DC	pKa			
HL 3.83E-006	atm m3/mol	25	EST	MEYLAN,WM & HOWARD,PH (1991)
OH 2.31E-011	cm3/molc sec	25	EST	MEYLAN,WM & HOWARD,PH (1993)

CAS #: 005454-28-4				BUTYL HEPTANOATE

Formula: $C_{11}H_{22}O_2$

Mol Weight: 186.30

MP (deg C): -67.5 | FP (deg C):

BP (deg C): 226.2

BP pressure (mm Hg):

Property/Value	Units	Temp	Data Type	Reference
WS 1.09E+001	mg/L	25	EST	MEYLAN,WM ET AL. (1996)
logP 4.30			EST	MEYLAN,WM & HOWARD,PH (1995)
VP 7.95E-002	mm Hg	25	EST	NEELY,WB & BLAU,GE (1985)
DC	pKa			
HL 1.69E-003	atm m3/mol	25	EST	MEYLAN,WM & HOWARD,PH (1991)
OH 1.05E-011	cm3/molc sec	25	EST	MEYLAN,WM & HOWARD,PH (1993)

CAS #: 005454-79-5				3-METHYLCYCLOHEXANOL (CIS)

Formula: $C_7H_{14}O$

Mol Weight: 114.19

MP (deg C): -5.5 | FP (deg C):

BP (deg C): 168

BP pressure (mm Hg):

Property/Value	Units	Temp	Data Type	Reference
WS 5.99E+003	mg/L	25	EST	MEYLAN,WM ET AL. (1996)
logP 2.05			EST	MEYLAN,WM & HOWARD,PH (1995)
VP 5.31E-001	mm Hg	25	EXP	DAUBERT,TE & DANNER,RP (1989)
DC	pKa			
HL 6.50E-006	atm m3/mol	25	EST	MEYLAN,WM & HOWARD,PH (1991)
OH 1.92E-011	cm3/molc sec	25	EST	MEYLAN,WM & HOWARD,PH (1993)

CAS #: 005455-59-4				2-NITROBENZENESULFONAMIDE

Formula: $C_6H_6N_2O_4S$

Mol Weight: 202.19

MP (deg C): 190-192 | FP (deg C):

BP (deg C):

BP pressure (mm Hg):

Property/Value	Units	Temp	Data Type	Reference
WS 8.90E+003	mg/L	25	EST	MEYLAN,WM ET AL. (1996)
logP 0.34			EXP	HANSCH,C & LEO,AJ (1985)
VP 7.82E-006	mm Hg	25	EST	NEELY,WB & BLAU,GE (1985)
DC	pKa			
HL 1.67E-009	atm m3/mol	25	EST	MEYLAN,WM & HOWARD,PH (1991)
OH 5.21E-014	cm3/molc sec	25	EST	MEYLAN,WM & HOWARD,PH (1993)

CAS #: 005457-28-3				3-CYANOINDOLE

Formula: $C_9H_6N_2$

Mol Weight: 142.16

MP (deg C): 179-182 | FP (deg C):

BP (deg C):

BP pressure (mm Hg):

Property/Value	Units	Temp	Data Type	Reference
WS 4.65E+002	mg/L	25	EST	MEYLAN,WM ET AL. (1996)
logP 2.32			EXP	HANSCH,C & LEO,AJ (1985)
VP 2.32E-004	mm Hg	25	EST	NEELY,WB & BLAU,GE (1985)
DC	pKa			
HL 8.56E-009	atm m3/mol	25	EST	MEYLAN,WM & HOWARD,PH (1991)
OH 1.94E-011	cm3/molc sec	25	EST	MEYLAN,WM & HOWARD,PH (1993)

CAS #: 005458-59-3				ISOPROPYL OCTANOATE

Formula: $C_{11}H_{22}O_2$

Mol Weight: 186.30

MP (deg C): | FP (deg C):

BP (deg C): 93.8

BP pressure (mm Hg): 1.00E+001

Property/Value	Units	Temp	Data Type	Reference
WS 1.26E+001	mg/L	25	EST	MEYLAN,WM ET AL. (1996)
logP 4.23			EST	MEYLAN,WM & HOWARD,PH (1995)
VP 1.09E-001	mm Hg	25	EXT	BOUBLIK,T ET AL. (1984)
DC	pKa			
HL 1.69E-003	atm m3/mol	25	EST	MEYLAN,WM & HOWARD,PH (1991)
OH 1.08E-011	cm3/molc sec	25	EST	MEYLAN,WM & HOWARD,PH (1993)

CAS #: 005460-29-7				N-(3-BROMOPROPYL)PHTHALIMIDE

Formula: $C_{11}H_{10}BrNO_2$

Mol Weight: 268.12

MP (deg C): 74-76 | FP (deg C):

BP (deg C):

BP pressure (mm Hg):

Property/Value	Units	Temp	Data Type	Reference
WS 7.11E+001	mg/L	25	EST	MEYLAN,WM ET AL. (1996)
logP 2.83			EST	MEYLAN,WM & HOWARD,PH (1995)
VP 3.12E-008	mm Hg	25	EST	NEELY,WB & BLAU,GE (1985)
DC	pKa			
HL 4.56E-009	atm m3/mol	25	EST	MEYLAN,WM & HOWARD,PH (1991)
OH 2.13E-011	cm3/molc sec	25	EST	MEYLAN,WM & HOWARD,PH (1993)

CAS #: 005462-24-8				4-SULFAMYLBENZAMIDE,N-PROPYL

Formula: $C_{10}H_{14}N_2O_3S$

Mol Weight: 242.30

MP (deg C): | FP (deg C):

BP (deg C):

BP pressure (mm Hg):

Property/Value	Units	Temp	Data Type	Reference
WS 9.57E+003	mg/L	25	EST	MEYLAN,WM ET AL. (1996)
logP 0.51			EXP	HANSCH,C & LEO,AJ (1985)
VP 1.62E-008	mm Hg	25	EST	NEELY,WB & BLAU,GE (1985)
DC	pKa			
HL 6.72E-013	atm m3/mol	25	EST	MEYLAN,WM & HOWARD,PH (1991)
OH 1.19E-011	cm3/molc sec	25	EST	MEYLAN,WM & HOWARD,PH (1993)

CAS #: 005462-29-3 — C.I. VIOLET 2

Formula: $C_{18}H_{10}Cl_2O_2S_2$

Mol Weight: 393.31

MP (deg C):   FP (deg C):

BP (deg C):

BP pressure (mm Hg):

Property/Value	Units	Temp	Data Type	Reference
WS 2.99E-002	mg/L	25	EST	MEYLAN,WM ET AL. (1996)
logP 5.92			EST	MEYLAN,WM & HOWARD,PH (1995)
VP 3.00E-010	mm Hg	25	EXP	BAUGHMAN,GL & PERENICH,TA (1988)
DC	pKa			
HL 3.05E-013	atm m3/mol	25	EST	MEYLAN,WM & HOWARD,PH (1991)
OH 1.56E-011	cm3/molc sec	25	EST	MEYLAN,WM & HOWARD,PH (1993)

CAS #: 005465-00-9 — BENZENEACETAMIDE, N-ETHYL-

Formula: $C_{10}H_{13}NO$

Mol Weight: 163.22

MP (deg C):   FP (deg C):

BP (deg C):

BP pressure (mm Hg):

Property/Value	Units	Temp	Data Type	Reference
WS 9.26E+003	mg/L	25	EST	MEYLAN,WM ET AL. (1996)
logP 1.00			EXP	SURYANARAYANA,MVS ET AL. (1991)
VP 5.80E-002	mm Hg	30	EXP	SURYANARAYANA,MVS ET AL. (1991)
DC	pKa			
HL 2.63E-009	atm m3/mol	25	EST	MEYLAN,WM & HOWARD,PH (1991)
OH 1.50E-011	cm3/molc sec	25	EST	MEYLAN,WM & HOWARD,PH (1993)

CAS #: 005469-69-2 — 3-CL-6-PYRIDAZINAMINE

Formula: $C_4H_4ClN_3$

Mol Weight: 129.55

MP (deg C):   FP (deg C):

BP (deg C):

BP pressure (mm Hg):

Property/Value	Units	Temp	Data Type	Reference
WS 4.64E+004	mg/L	25	EST	MEYLAN,WM ET AL. (1996)
logP 0.35			EXP	HANSCH,C & LEO,AJ (1985)
VP 1.02E-003	mm Hg	25	EST	NEELY,WB & BLAU,GE (1985)
DC 3.85	pKa	20	EXP	PERRIN,DD (1972)
HL 4.74E-008	atm m3/mol	25	EST	MEYLAN,WM & HOWARD,PH (1991)
OH 3.71E-012	cm3/molc sec	25	EST	MEYLAN,WM & HOWARD,PH (1993)

CAS #: 005469-70-5 — 2-AMINOPYRIMIDINE

Formula: $C_4H_5N_3$

Mol Weight: 95.10

MP (deg C):   FP (deg C):

BP (deg C):

BP pressure (mm Hg):

Property/Value	Units	Temp	Data Type	Reference
WS 1.86E+005	mg/L	25	EST	MEYLAN,WM ET AL. (1996)
logP -0.22			EXP	HANSCH,C & LEO,AJ (1985)
VP 2.46E-001	mm Hg	25	EST	NEELY,WB & BLAU,GE (1985)
DC 3.45	pKa	20	EXP	PERRIN,DD (1965)
HL 3.66E-006	atm m3/mol	25	EST	MEYLAN,WM & HOWARD,PH (1991)
OH 1.27E-011	cm3/molc sec	25	EST	MEYLAN,WM & HOWARD,PH (1993)

CAS #: 005470-02-0 — 1-PROPYLPIPERIDINE

Formula: $C_8H_{17}N$

Mol Weight: 127.23

MP (deg C):   FP (deg C):

BP (deg C): 151.5

BP pressure (mm Hg):

Property/Value	Units	Temp	Data Type	Reference
WS 8.88E+003	mg/L	25	EST	MEYLAN,WM ET AL. (1996)
logP 2.38			EST	MEYLAN,WM & HOWARD,PH (1995)
VP 2.23E+000	mm Hg	25	EST	NEELY,WB & BLAU,GE (1985)
DC 10.41	pKa	27	EXP	PERRIN,DD (1965)
HL 6.64E-005	atm m3/mol	25	EST	MEYLAN,WM & HOWARD,PH (1991)
OH 1.04E-010	cm3/molc sec	25	EST	MEYLAN,WM & HOWARD,PH (1993)

CAS #: 005470-49-5 — 4-METHYLSULFONYLANILINE

Formula: $C_7H_9NO_2S$

Mol Weight: 171.22

MP (deg C):   FP (deg C):

BP (deg C):

BP pressure (mm Hg):

Property/Value	Units	Temp	Data Type	Reference
WS 7.68E+004	mg/L	25	EST	MEYLAN,WM ET AL. (1996)
logP -0.12			EXP	HANSCH,C & LEO,AJ (1985)
VP 1.25E-004	mm Hg	25	EST	NEELY,WB & BLAU,GE (1985)
DC 1.35	pKa	25	EXP	PERRIN,DD (1965)
HL 4.41E-010	atm m3/mol	25	EST	MEYLAN,WM & HOWARD,PH (1991)
OH 2.41E-011	cm3/molc sec	25	EST	MEYLAN,WM & HOWARD,PH (1993)

CAS #: 005470-65-5 — 3-BROMO-4-NITROPHENOL

Formula: $C_6H_4BrNO_3$

Mol Weight: 218.01

MP (deg C):   FP (deg C):

BP (deg C):

BP pressure (mm Hg):

Property/Value	Units	Temp	Data Type	Reference
WS 5.47E+002	mg/L	25	EST	MEYLAN,WM ET AL. (1996)
logP 2.80			EST	MEYLAN,WM & HOWARD,PH (1995)
VP 8.64E-005	mm Hg	25	EST	NEELY,WB & BLAU,GE (1985)
DC	pKa			
HL 8.81E-010	atm m3/mol	25	EST	MEYLAN,WM & HOWARD,PH (1991)
OH 2.76E-012	cm3/molc sec	25	EST	MEYLAN,WM & HOWARD,PH (1993)

CAS #: 005470-96-2 — 2-QUINOLINECARBOXALDEHYDE

Formula: $C_{10}H_7NO$

Mol Weight: 157.17

MP (deg C): 70-72   FP (deg C):

BP (deg C):

BP pressure (mm Hg):

Property/Value	Units	Temp	Data Type	Reference
WS 1.05E+003	mg/L	25	EST	MEYLAN,WM ET AL. (1996)
logP 2.14			EXP	HANSCH,C & LEO,AJ (1985)
VP 6.68E-004	mm Hg	25	EST	NEELY,WB & BLAU,GE (1985)
DC	pKa			
HL 1.72E-009	atm m3/mol	25	EST	MEYLAN,WM & HOWARD,PH (1991)
OH 2.42E-011	cm3/molc sec	25	EST	MEYLAN,WM & HOWARD,PH (1993)

CAS #: 005472-13-9 — PHENYL-O-TOLYLCARBINOL

Formula: $C_{14}H_{14}O$

Mol Weight: 198.27

MP (deg C): FP (deg C):

BP (deg C):

BP pressure (mm Hg):

Property/Value	Units	Temp	Data Type	Reference
WS 3.52E+002	mg/L	25	EST	MEYLAN,WM ET AL. (1996)
logP 3.06			EXP	HANSCH,C & LEO,AJ (1985)
VP 7.67E-006	mm Hg	25	EST	NEELY,WB & BLAU,GE (1985)
DC	pKa			
HL 1.94E-008	atm m3/mol	25	EST	MEYLAN,WM & HOWARD,PH (1991)
OH 1.78E-011	cm3/molc sec	25	EST	MEYLAN,WM & HOWARD,PH (1993)

CAS #: 005493-24-3 — 1H-ISOINDOLE-1,3(2H)-DIONE, 2-[(ACETYLOXY)METHYL]

Formula: $C_{11}H_9NO_4$

Mol Weight: 219.20

MP (deg C): FP (deg C):

BP (deg C):

BP pressure (mm Hg):

Property/Value	Units	Temp	Data Type	Reference
WS 3.62E+003	mg/L	25	EST	MEYLAN,WM ET AL. (1996)
logP 1.15			EXP	NISHIMURA,K ET AL. (1988)
VP 6.87E-008	mm Hg	25	EST	NEELY,WB & BLAU,GE (1985)
DC	pKa			
HL 1.07E-009	atm m3/mol	25	EST	MEYLAN,WM & HOWARD,PH (1991)
OH 2.20E-011	cm3/molc sec	25	EST	MEYLAN,WM & HOWARD,PH (1993)

CAS #: 005500-21-0 — CYCLOPROPANECARBONITRILE

Formula: C_4H_5N

Mol Weight: 67.09

MP (deg C): FP (deg C):

BP (deg C): 135.1

BP pressure (mm Hg):

Property/Value	Units	Temp	Data Type	Reference
WS 1.94E+004	mg/L	25	EST	MEYLAN,WM ET AL. (1996)
logP 0.68			EXP	HANSCH,C ET AL. (1995)
VP 1.30E+001	mm Hg	25	EST	NEELY,WB & BLAU,GE (1985)
DC	pKa			
HL 2.37E-005	atm m3/mol	25	EST	MEYLAN,WM & HOWARD,PH (1991)
OH 6.77E-014	cm3/molc sec	25	EST	MEYLAN,WM & HOWARD,PH (1993)

CAS #: 005510-99-6 — 2,6-DI-SEC-BUTYLPHENOL

Formula: $C_{14}H_{22}O$

Mol Weight: 206.33

MP (deg C): -10 FP (deg C):

BP (deg C): 255-260

BP pressure (mm Hg):

Property/Value	Units	Temp	Data Type	Reference
WS 1.80E+000	mg/L	25	EXP	CHEM INSPECT TEST INST (1992)
logP 4.36			EXP	HANSCH,C ET AL. (1995)
VP 3.12E-004	mm Hg	25	EST	NEELY,WB & BLAU,GE (1985)
DC	pKa			
HL 3.74E-006	atm m3/mol	25	EST	MEYLAN,WM & HOWARD,PH (1991)
OH 5.61E-011	cm3/molc sec	25	EST	MEYLAN,WM & HOWARD,PH (1993)

CAS #: 005511-98-8 — DIGOXIN ESTER DERIVATIVE

Formula: $C_{43}H_{66}O_{15}$

Mol Weight: 823.00

MP (deg C): FP (deg C):

BP (deg C):

BP pressure (mm Hg):

Property/Value	Units	Temp	Data Type	Reference
WS 1.09E-001	mg/L	25	EST	MEYLAN,WM ET AL. (1996)
logP 1.97			EXP	SANGSTER,J (1993)
VP 1.46E-030	mm Hg	25	EST	NEELY,WB & BLAU,GE (1985)
DC	pKa			
HL 3.03E-028	atm m3/mol	25	EST	MEYLAN,WM & HOWARD,PH (1991)
OH 1.67E-010	cm3/molc sec	25	EST	MEYLAN,WM & HOWARD,PH (1993)

CAS #: 005519-23-3 — BENZOIC ACID, 4-(DECYLOXY)-

Formula: $C_{17}H_{26}O_3$

Mol Weight: 278.39

MP (deg C): 94-133 FP (deg C):

BP (deg C):

BP pressure (mm Hg):

Property/Value	Units	Temp	Data Type	Reference
WS 6.96E-002	mg/L	25	EST	MEYLAN,WM ET AL. (1996)
logP 6.29			EXP	SANGSTER,J (1993)
VP 5.80E-007	mm Hg	25	EST	NEELY,WB & BLAU,GE (1985)
DC	pKa			
HL 8.21E-008	atm m3/mol	25	EST	MEYLAN,WM & HOWARD,PH (1991)
OH 2.93E-011	cm3/molc sec	25	EST	MEYLAN,WM & HOWARD,PH (1993)

CAS #: 005521-56-2 — PYRAZINAMINE, 6-METHYL-

Formula: $C_5H_7N_3$

Mol Weight: 109.13

MP (deg C): FP (deg C):

BP (deg C):

BP pressure (mm Hg):

Property/Value	Units	Temp	Data Type	Reference
WS 5.50E+004	mg/L	25	EST	MEYLAN,WM ET AL. (1996)
logP 0.35			EXP	YAMAGAMI,C ET AL. (1991)
VP 6.92E-002	mm Hg	25	EST	NEELY,WB & BLAU,GE (1985)
DC	pKa			
HL 1.14E-009	atm m3/mol	25	EST	MEYLAN,WM & HOWARD,PH (1991)
OH 3.33E-011	cm3/molc sec	25	EST	MEYLAN,WM & HOWARD,PH (1993)

CAS #: 005521-57-3 — PYRAZINECARBOXAMIDE, 5-METHYL-

Formula: $C_6H_7N_3O$

Mol Weight: 137.14

MP (deg C): FP (deg C):

BP (deg C):

BP pressure (mm Hg):

Property/Value	Units	Temp	Data Type	Reference
WS 6.89E+003	mg/L	25	EST	MEYLAN,WM ET AL. (1996)
logP -0.25			EXP	YAMAGAMI,C ET AL. (1991)
VP 8.78E-005	mm Hg	25	EST	NEELY,WB & BLAU,GE (1985)
DC	pKa			
HL 1.32E-012	atm m3/mol	25	EST	MEYLAN,WM & HOWARD,PH (1991)
OH 2.68E-012	cm3/molc sec	25	EST	MEYLAN,WM & HOWARD,PH (1993)

CAS #:	005521-62-0			PYRAZINECARBOXAMIDE, 6-METHYL-

Formula: $C_6H_7N_3O$

Mol Weight: 137.14

MP (deg C): | FP (deg C):

BP (deg C):

BP pressure (mm Hg):

Property/Value	Units	Temp	Data Type	Reference
WS 5.44E+003	mg/L	25	EST	MEYLAN,WM ET AL. (1996)
logP -0.13			EXP	YAMAGAMI,C ET AL. (1991)
VP 8.78E-005	mm Hg	25	EST	NEELY,WB & BLAU,GE (1985)
DC	pKa			
HL 1.32E-012	atm m3/mol	25	EST	MEYLAN,WM & HOWARD,PH (1991)
OH 2.39E-012	cm3/molc sec	25	EST	MEYLAN,WM & HOWARD,PH (1993)

CAS #:	005527-84-4			4H-NAPHTHO[2,3-B]PYRAN-2-CARBOXYLIC ACID, 6,7,8,

Formula: $C_{14}H_{12}O_4$

Mol Weight: 244.25

MP (deg C): | FP (deg C):

BP (deg C):

BP pressure (mm Hg):

Property/Value	Units	Temp	Data Type	Reference
WS 7.89E+001	mg/L	25	EST	MEYLAN,WM ET AL. (1996)
logP 3.40			EXP	HANSCH,C ET AL. (1995)
VP 1.43E-007	mm Hg	25	EST	NEELY,WB & BLAU,GE (1985)
DC	pKa			
HL 2.53E-011	atm m3/mol	25	EST	MEYLAN,WM & HOWARD,PH (1991)
OH 7.56E-011	cm3/molc sec	25	EST	MEYLAN,WM & HOWARD,PH (1993)

CAS #:	005532-46-7			5-ETO-6-MEO-4-SULFANILAMIDOPYRIMIDIN

Formula: $C_{13}H_{16}N_4O_4S$

Mol Weight: 324.36

MP (deg C): | FP (deg C):

BP (deg C):

BP pressure (mm Hg):

Property/Value	Units	Temp	Data Type	Reference
WS 1.16E+003	mg/L	25	EST	MEYLAN,WM ET AL. (1996)
logP 1.03			EXP	HANSCH,C & LEO,AJ (1985)
VP 7.02E-010	mm Hg	25	EST	NEELY,WB & BLAU,GE (1985)
DC	pKa			
HL 1.89E-015	atm m3/mol	25	EST	MEYLAN,WM & HOWARD,PH (1991)
OH 3.36E-011	cm3/molc sec	25	EST	MEYLAN,WM & HOWARD,PH (1993)

CAS #:	005532-90-1			PROPYL N-PHENYLCARBAMATE

Formula: $C_{10}H_{13}NO_2$

Mol Weight: 179.22

MP (deg C): | FP (deg C):

BP (deg C):

BP pressure (mm Hg):

Property/Value	Units	Temp	Data Type	Reference
WS 2.26E+002	mg/L	25	EST	MEYLAN,WM ET AL. (1996)
logP 2.80			EXP	BRIGGS,GG (1981)
VP 7.07E-003	mm Hg	25	EST	NEELY,WB & BLAU,GE (1985)
DC	pKa			
HL 3.85E-008	atm m3/mol	25	EST	MEYLAN,WM & HOWARD,PH (1991)
OH 4.58E-011	cm3/molc sec	25	EST	MEYLAN,WM & HOWARD,PH (1993)

CAS #:	005536-17-4			9H-PURIN-6-AMINE, 9-.BETA.-D-ARABINOFURANOSYL-

Formula: $C_{10}H_{13}N_5O_4$

Mol Weight: 267.25

MP (deg C): 257-257.5 | FP (deg C):

BP (deg C):

BP pressure (mm Hg):

Property/Value	Units	Temp	Data Type	Reference
WS 8.23E+003	mg/L	25	EST	MEYLAN,WM ET AL. (1996)
logP -1.11			EXP	HANSCH,C ET AL. (1995)
VP 7.00E-015	mm Hg	25	EST	NEELY,WB & BLAU,GE (1985)
DC	pKa			
HL 1.11E-022	atm m3/mol	25	EST	MEYLAN,WM & HOWARD,PH (1991)
OH 2.47E-010	cm3/molc sec	25	EST	MEYLAN,WM & HOWARD,PH (1993)

CAS #:	005541-67-3			5-METHYL-8-QUINOLINOL

Formula: $C_{10}H_9NO$

Mol Weight: 159.19

MP (deg C): | FP (deg C):

BP (deg C):

BP pressure (mm Hg):

Property/Value	Units	Temp	Data Type	Reference
WS 2.48E+003	mg/L	25	EST	MEYLAN,WM ET AL. (1996)
logP 2.37			EXP	HANSCH,C & LEO,AJ (1985)
VP 1.05E-004	mm Hg	25	EST	NEELY,WB & BLAU,GE (1985)
DC	pKa			
HL 7.90E-011	atm m3/mol	25	EST	MEYLAN,WM & HOWARD,PH (1991)
OH 8.86E-011	cm3/molc sec	25	EST	MEYLAN,WM & HOWARD,PH (1993)

CAS #:	005558-66-7			A,A-DIPHENYLPROPIONIC ACID

Formula: $C_{15}H_{14}O_2$

Mol Weight: 226.28

MP (deg C): 176 | FP (deg C):

BP (deg C):

BP pressure (mm Hg):

Property/Value	Units	Temp	Data Type	Reference
WS 3.99E+002	mg/L	25	EST	MEYLAN,WM ET AL. (1996)
logP 2.69			EXP	HANSCH,C & LEO,AJ (1985)
VP 5.97E-006	mm Hg	25	EST	NEELY,WB & BLAU,GE (1985)
DC	pKa			
HL 4.74E-009	atm m3/mol	25	EST	MEYLAN,WM & HOWARD,PH (1991)
OH 9.28E-012	cm3/molc sec	25	EST	MEYLAN,WM & HOWARD,PH (1993)

CAS #:	005566-34-7			GAMMA-CHLORDANE

Formula: $C_{10}H_6Cl_8$

Mol Weight: 409.78

MP (deg C): | FP (deg C):

BP (deg C):

BP pressure (mm Hg):

Property/Value	Units	Temp	Data Type	Reference
WS 1.43E-003	mg/L	25	EST	MEYLAN,WM ET AL. (1996)
logP 7.34				EST	MEYLAN,WM & HOWARD,PH (1995)
VP 2.52E-005	mm Hg	25	EST	NEELY,WB & BLAU,GE (1985)
DC	pKa			
HL 7.03E-005	atm m3/mol	25	EST	MEYLAN,WM & HOWARD,PH (1991)
OH 6.99E-012	cm3/molc sec	25	EST	MEYLAN,WM & HOWARD,PH (1993)

CAS #: 005571-62-0	1,4-BENZODIAZEPIN-2-ONE,1,3-DUHYDRO-7,9-DIMETHYL

Formula: C₁₇H₁₆N₂O

Mol Weight: 264.33

MP (deg C):   FP (deg C):

BP (deg C):

BP pressure (mm Hg):

Property/Value	Units	Temp	Data Type	Reference
WS 6.07E+001	mg/L	25	EST	MEYLAN,WM ET AL. (1996)
logP 2.94			EXP	HANSCH,C ET AL. (1995)
VP 4.70E-009	mm Hg	25	EST	NEELY,WB & BLAU,GE (1985)
DC	pKa			
HL 2.92E-010	atm m3/mol	25	EST	MEYLAN,WM & HOWARD,PH (1991)
OH 2.06E-011	cm3/molc sec	25	EST	MEYLAN,WM & HOWARD,PH (1993)

CAS #: 005571-63-1	2H-1,4-BENZODIAZEPIN-2-ONE, 1,3-DIHYDRO-7-METHYL

Formula: C₁₆H₁₄N₂O

Mol Weight: 250.30

MP (deg C):   FP (deg C):

BP (deg C):

BP pressure (mm Hg):

Property/Value	Units	Temp	Data Type	Reference
WS 1.36E+002	mg/L	25	EST	MEYLAN,WM ET AL. (1996)
logP 2.62			EXP	SANGSTER,J (1993)
VP 1.06E-008	mm Hg	25	EST	NEELY,WB & BLAU,GE (1985)
DC	pKa			
HL 2.65E-010	atm m3/mol	25	EST	MEYLAN,WM & HOWARD,PH (1991)
OH 2.34E-011	cm3/molc sec	25	EST	MEYLAN,WM & HOWARD,PH (1993)

CAS #: 005571-65-3	1,4-BENZODIAZP2-ON,1-ET-5-PH-7CL

Formula: C₁₇H₁₅ClN₂O

Mol Weight: 298.77

MP (deg C):   FP (deg C):

BP (deg C):

BP pressure (mm Hg):

Property/Value	Units	Temp	Data Type	Reference
WS 2.26E+001	mg/L	25	EST	MEYLAN,WM ET AL. (1996)
logP 3.21			EXP	HANSCH,C & LEO,AJ (1985)
VP 1.24E-008	mm Hg	25	EST	NEELY,WB & BLAU,GE (1985)
DC	pKa			
HL 4.84E-009	atm m3/mol	25	EST	MEYLAN,WM & HOWARD,PH (1991)
OH 1.33E-011	cm3/molc sec	25	EST	MEYLAN,WM & HOWARD,PH (1993)

CAS #: 005581-75-9	6-PHENYLCAPROIC ACID

Formula: C₁₂H₁₆O₂

Mol Weight: 192.26

MP (deg C): 23   FP (deg C):

BP (deg C): 206

BP pressure (mm Hg): 3.00E+001

Property/Value	Units	Temp	Data Type	Reference
WS 1.92E+002	mg/L	25	EST	MEYLAN,WM ET AL. (1996)
logP 3.27			EXP	HANSCH,C ET AL. (1995)
VP 9.74E-005	mm Hg	25	EST	NEELY,WB & BLAU,GE (1985)
DC	pKa			
HL 1.37E-007	atm m3/mol	25	EST	MEYLAN,WM & HOWARD,PH (1991)
OH 1.16E-011	cm3/molc sec	25	EST	MEYLAN,WM & HOWARD,PH (1993)

CAS #: 005585-14-8	FURAZAN, DIPHENYL-, 2-OXIDE

Formula: C₁₄H₁₀N₂O₂

Mol Weight: 238.25

MP (deg C):   FP (deg C):

BP (deg C):

BP pressure (mm Hg):

Property/Value	Units	Temp	Data Type	Reference
WS 2.14E+001	mg/L	25	EST	MEYLAN,WM ET AL. (1996)
logP 3.64			EXP	CALVINO,R ET AL. (1992)
VP 5.11E-008	mm Hg	25	EST	NEELY,WB & BLAU,GE (1985)
DC	pKa			
HL 1.12E-011	atm m3/mol	25	EST	MEYLAN,WM & HOWARD,PH (1991)
OH 1.08E-011	cm3/molc sec	25	EST	MEYLAN,WM & HOWARD,PH (1993)

CAS #: 005593-20-4	BETAMETHASONE 17,21-DIPROPIONATE

Formula: C₂₈H₃₇FO₇

Mol Weight: 504.60

MP (deg C): 231-234 de   FP (deg C):

BP (deg C):

BP pressure (mm Hg):

Property/Value	Units	Temp	Data Type	Reference
WS 7.24E-001	mg/L	25	EST	MEYLAN,WM ET AL. (1996)
logP 4.07			EXP	HANSCH,C & LEO,AJ (1985)
VP 9.01E-014	mm Hg	25	EST	NEELY,WB & BLAU,GE (1985)
DC	pKa			
HL 1.47E-015	atm m3/mol	25	EST	MEYLAN,WM & HOWARD,PH (1991)
OH 6.88E-011	cm3/molc sec	25	EST	MEYLAN,WM & HOWARD,PH (1993)

CAS #: 005594-16-1	ACETAMIDE, N-(6-METHYLPYRAZINYL)-

Formula: C₆H₈N₄O

Mol Weight: 152.16

MP (deg C):   FP (deg C):

BP (deg C):

BP pressure (mm Hg):

Property/Value	Units	Temp	Data Type	Reference
WS 3.52E+004	mg/L	25	EST	MEYLAN,WM ET AL. (1996)
logP 0.38			EXP	YAMAGAMI,C ET AL. (1991)
VP 8.78E-005	mm Hg	25	EST	NEELY,WB & BLAU,GE (1985)
DC	pKa			
HL 1.20E-013	atm m3/mol	25	EST	MEYLAN,WM & HOWARD,PH (1991)
OH 1.43E-011	cm3/molc sec	25	EST	MEYLAN,WM & HOWARD,PH (1993)

CAS #: 005598-13-0	CHLORPYRIFOS METHYL

Formula: C₇H₇Cl₃NO₃PS

Mol Weight: 322.54

MP (deg C): 43   FP (deg C):

BP (deg C):

BP pressure (mm Hg):

Property/Value	Units	Temp	Data Type	Reference
WS 4.76E+000	mg/L	20	EXP	CHIOU,CT ET AL. (1977)
logP 4.31			EXP	HANSCH,C ET AL. (1995)
VP 4.20E-005	mm Hg	25	EXP	TOMLIN,C (1994)
DC	pKa			
HL 3.75E-006	atm m3/mol	25	EST	VP/WSOL
OH 5.86E-011	cm3/molc sec	25	EST	MEYLAN,WM & HOWARD,PH (1993)

CAS #: 005602-96-0				CARBAMIC ACID, (4-METHYLPHENYL)-, METHYL ESTER
Formula: C9H11NO2				
Mol Weight: 165.19				

$C_9H_{11}NO_2$

Mol Weight: 165.19

MP (deg C): | FP (deg C):
BP (deg C):
BP pressure (mm Hg):

Property/Value	Units	Temp	Data Type	Reference
WS 7.17E+002	mg/L	25	EST	MEYLAN,WM ET AL. (1996)
logP 2.29			EXP	TAKAHASHI,J ET AL. (1988)
VP 1.87E-002	mm Hg	25	EST	NEELY,WB & BLAU,GE (1985)
DC	pKa			
HL 2.41E-008	atm m3/mol	25	EST	MEYLAN,WM & HOWARD,PH (1991)
OH 5.26E-011	cm3/molc sec	25	EST	MEYLAN,WM & HOWARD,PH (1993)

CAS #: 005617-41-4				HEPTYLCYCLOHEXANE

$C_{13}H_{26}$

Mol Weight: 182.35

MP (deg C): -30 | FP (deg C):
BP (deg C): 244
BP pressure (mm Hg):

Property/Value	Units	Temp	Data Type	Reference
WS 4.05E-002	mg/L	25	EST	MEYLAN,WM ET AL. (1996)
logP 6.54			EST	MEYLAN,WM & HOWARD,PH (1995)
VP 3.23E-002	mm Hg	25	EXT	ZWOLINSKI,BJ & WILHOIT,RC (1971)
DC	pKa			
HL 1.85E+000	atm m3/mol	25	EST	MEYLAN,WM & HOWARD,PH (1991)
OH 1.91E-011	cm3/molc sec	25	EST	MEYLAN,WM & HOWARD,PH (1993)

CAS #: 005621-02-3				2-PYRIMIDINAMINE, N,N-DIMETHYL-

$C_6H_9N_3$

Mol Weight: 123.16

MP (deg C): | FP (deg C):
BP (deg C):
BP pressure (mm Hg):

Property/Value	Units	Temp	Data Type	Reference
WS 1.19E+004	mg/L	25	EST	MEYLAN,WM ET AL. (1996)
logP 1.07			EXP	YAMAGAMI,C ET AL. (1990)
VP 6.25E-001	mm Hg	25	EST	NEELY,WB & BLAU,GE (1985)
DC	pKa			
HL 1.65E-004	atm m3/mol	25	EST	MEYLAN,WM & HOWARD,PH (1991)
OH 4.62E-011	cm3/molc sec	25	EST	MEYLAN,WM & HOWARD,PH (1993)

CAS #: 005632-47-3				N-NITROSOPIPERAZINE

$C_4H_9N_3O$

Mol Weight: 115.14

MP (deg C): | FP (deg C):
BP (deg C):
BP pressure (mm Hg):

Property/Value	Units	Temp	Data Type	Reference
WS 7.47E+005	mg/L	25	EST	MEYLAN,WM ET AL. (1996)
logP 0.18			EXP	HANSCH,C & LEO,AJ (1985)
VP 1.54E-002	mm Hg	25	EST	NEELY,WB & BLAU,GE (1985)
DC	pKa			
HL 2.72E-010	atm m3/mol	25	EST	MEYLAN,WM & HOWARD,PH (1991)
OH 1.06E-010	cm3/molc sec	25	EST	MEYLAN,WM & HOWARD,PH (1993)

CAS #: 005638-76-6				2-PYRIDINEETHANAMINE, N-METHYL-

$C_8H_{12}N_2$

Mol Weight: 136.20

MP (deg C): | FP (deg C):
BP (deg C): 113-114
BP pressure (mm Hg): 3.00E+001

Property/Value	Units	Temp	Data Type	Reference
WS 1.00E+006	mg/L	25	EST	MEYLAN,WM ET AL. (1996)
logP 0.68			EXP	YOUNG,RC ET AL. (1993)
VP 1.33E-001	mm Hg	25	EST	NEELY,WB & BLAU,GE (1985)
DC 10.13	pKa	10	EXP	PERRIN,DD (1965)
HL 2.33E-009	atm m3/mol	25	EST	MEYLAN,WM & HOWARD,PH (1991)
OH 7.70E-011	cm3/molc sec	25	EST	MEYLAN,WM & HOWARD,PH (1993)

CAS #: 005663-04-7				TRIPHENYLUREA

$C_{19}H_{16}N_2O$

Mol Weight: 288.35

MP (deg C): | FP (deg C):
BP (deg C):
BP pressure (mm Hg):

Property/Value	Units	Temp	Data Type	Reference
WS 2.27E+001	mg/L	25	EST	MEYLAN,WM ET AL. (1996)
logP 3.28			EXP	HANSCH,C & LEO,AJ (1985)
VP 5.24E-009	mm Hg	25	EST	NEELY,WB & BLAU,GE (1985)
DC	pKa			
HL 5.72E-010	atm m3/mol	25	EST	MEYLAN,WM & HOWARD,PH (1991)
OH 6.75E-011	cm3/molc sec	25	EST	MEYLAN,WM & HOWARD,PH (1993)

CAS #: 005663-74-1				N-BENZOYLSALICYLAMIDE

$C_{14}H_{11}NO_3$

Mol Weight: 241.25

MP (deg C): | FP (deg C):
BP (deg C):
BP pressure (mm Hg):

Property/Value	Units	Temp	Data Type	Reference
WS 6.66E+003	mg/L	25	EST	MEYLAN,WM ET AL. (1996)
logP 1.38			EXP	HANSCH,C & LEO,AJ (1985)
VP 7.18E-012	mm Hg	25	EST	NEELY,WB & BLAU,GE (1985)
DC	pKa			
HL 2.95E-010	atm m3/mol	25	EST	MEYLAN,WM & HOWARD,PH (1991)
OH 3.78E-011	cm3/molc sec	25	EST	MEYLAN,WM & HOWARD,PH (1993)

CAS #: 005666-21-7				ALLYLPROPYLAMINE

$C_6H_{13}N$

Mol Weight: 99.18

MP (deg C): | FP (deg C):
BP (deg C):
BP pressure (mm Hg): 0.00E+000

Property/Value	Units	Temp	Data Type	Reference
WS 8.76E+004	mg/L	25	EST	MEYLAN,WM ET AL. (1996)
logP 1.33			EXP	HANSCH,C & LEO,AJ (1985)
VP 1.94E+001	mm Hg	25	EST	NEELY,WB & BLAU,GE (1985)
DC	pKa			
HL 3.85E-005	atm m3/mol	25	EST	MEYLAN,WM & HOWARD,PH (1991)
OH 1.10E-010	cm3/molc sec	25	EST	MEYLAN,WM & HOWARD,PH (1993)

CAS #: 005673-07-4				1,3-DIMETHOXY-2-METHYLBENZENE

Formula: $C_9H_{12}O_2$
Mol Weight: 152.19
MP (deg C): 39-41
FP (deg C):
BP (deg C):
BP pressure (mm Hg):

Property/Value	Units	Temp	Data Type	Reference
WS 2.63E+002	mg/L	25	EST	MEYLAN,WM ET AL. (1996)
logP 2.87			EXP	NAKAGAWA,Y ET AL. (1992)
VP 1.96E-001	mm Hg	25	EST	NEELY,WB & BLAU,GE (1985)
DC	pKa			
HL 2.08E-005	atm m3/mol	25	EST	MEYLAN,WM & HOWARD,PH (1991)
OH 2.02E-010	cm3/molc sec	25	EST	MEYLAN,WM & HOWARD,PH (1993)

CAS #: 005681-57-2				3-BUTYLHYDROXYUREA

Formula: $C_5H_{12}N_2O_2$
Mol Weight: 132.16
MP (deg C):
FP (deg C):
BP (deg C):
BP pressure (mm Hg):

Property/Value	Units	Temp	Data Type	Reference
WS 2.35E+003	mg/L	25	EST	MEYLAN,WM ET AL. (1996)
logP 0.32			EXP	HANSCH,C & LEO,AJ (1985)
VP 2.77E-005	mm Hg	25	EST	NEELY,WB & BLAU,GE (1985)
DC	pKa			
HL 2.78E-010	atm m3/mol	25	EST	MEYLAN,WM & HOWARD,PH (1991)
OH 9.44E-012	cm3/molc sec	25	EST	MEYLAN,WM & HOWARD,PH (1993)

CAS #: 005683-33-0				2-DIMETHYLAMINOPYRIDINE

Formula: $C_7H_{10}N_2$
Mol Weight: 122.17
MP (deg C): 191
FP (deg C):
BP (deg C):
BP pressure (mm Hg):

Property/Value	Units	Temp	Data Type	Reference
WS 3.84E+003	mg/L	25	EST	MEYLAN,WM ET AL. (1996)
logP 1.65			EXP	YAMAGAMI,C ET AL. (1990A)
VP 1.00E+000	mm Hg	25	EST	NEELY,WB & BLAU,GE (1985)
DC	pKa			
HL 1.12E-007	atm m3/mol	25	EST	MEYLAN,WM & HOWARD,PH (1991)
OH 7.27E-011	cm3/molc sec	25	EST	MEYLAN,WM & HOWARD,PH (1993)

CAS #: 005692-23-9				N-(4-AMINOBUTYL)BENZAMIDE

Formula: $C_{11}H_{16}N_2O$
Mol Weight: 192.26
MP (deg C):
FP (deg C):
BP (deg C):
BP pressure (mm Hg):

Property/Value	Units	Temp	Data Type	Reference
WS 1.53E+004	mg/L	25	EST	MEYLAN,WM ET AL. (1996)
logP 0.58			EXP	HANSCH,C & LEO,AJ (1985)
VP 3.14E-006	mm Hg	25	EST	NEELY,WB & BLAU,GE (1985)
DC	pKa			
HL 1.17E-012	atm m3/mol	25	EST	MEYLAN,WM & HOWARD,PH (1991)
OH 4.65E-011	cm3/molc sec	25	EST	MEYLAN,WM & HOWARD,PH (1993)

CAS #: 005692-66-0				BIS(2,4,6-TRIMETHYLPHENYL)DIAZENE

Formula: $C_{18}H_{22}N_2$
Mol Weight: 266.39
MP (deg C):
FP (deg C):
BP (deg C):
BP pressure (mm Hg):

Property/Value	Units	Temp	Data Type	Reference
WS 3.43E-003	mg/L	25	EST	MEYLAN,WM ET AL. (1996)
logP 7.39			EST	MEYLAN,WM & HOWARD,PH (1995)
VP 5.28E-006	mm Hg	25	EST	NEELY,WB & BLAU,GE (1985)
DC	pKa			
HL 2.66E-005	atm m3/mol	25	EST	MEYLAN,WM & HOWARD,PH (1991)
OH 2.83E-011	cm3/molc sec	25	EST	MEYLAN,WM & HOWARD,PH (1993)

CAS #: 005710-11-2				3-ETHYLHYDROXYUREA

Formula: $C_3H_8N_2O_2$
Mol Weight: 104.11
MP (deg C):
FP (deg C):
BP (deg C):
BP pressure (mm Hg):

Property/Value	Units	Temp	Data Type	Reference
WS 2.48E+004	mg/L	25	EST	MEYLAN,WM ET AL. (1996)
logP -0.76			EXP	HANSCH,C & LEO,AJ (1985)
VP 2.61E-004	mm Hg	25	EST	NEELY,WB & BLAU,GE (1985)
DC	pKa			
HL 1.58E-010	atm m3/mol	25	EST	MEYLAN,WM & HOWARD,PH (1991)
OH 6.00E-012	cm3/molc sec	25	EST	MEYLAN,WM & HOWARD,PH (1993)

CAS #: 005710-12-3				3-PROPYLHYDROXYUREA

Formula: $C_4H_{10}N_2O_2$
Mol Weight: 118.14
MP (deg C):
FP (deg C):
BP (deg C):
BP pressure (mm Hg):

Property/Value	Units	Temp	Data Type	Reference
WS 7.70E+003	mg/L	25	EST	MEYLAN,WM ET AL. (1996)
logP -0.22			EXP	HANSCH,C & LEO,AJ (1985)
VP 8.93E-005	mm Hg	25	EST	NEELY,WB & BLAU,GE (1985)
DC	pKa			
HL 2.10E-010	atm m3/mol	25	EST	MEYLAN,WM & HOWARD,PH (1991)
OH 8.03E-012	cm3/molc sec	25	EST	MEYLAN,WM & HOWARD,PH (1993)

CAS #: 005715-02-6				2-CYANOPHENYL ACETATE

Formula: $C_9H_7NO_2$
Mol Weight: 161.16
MP (deg C):
FP (deg C):
BP (deg C):
BP pressure (mm Hg):

Property/Value	Units	Temp	Data Type	Reference
WS 2.69E+003	mg/L	25	EST	MEYLAN,WM ET AL. (1996)
logP 1.33			EXP	HANSCH,C & LEO,AJ (1985)
VP 5.21E-003	mm Hg	25	EST	NEELY,WB & BLAU,GE (1985)
DC	pKa			
HL 6.26E-007	atm m3/mol	25	EST	MEYLAN,WM & HOWARD,PH (1991)
OH 4.56E-013	cm3/molc sec	25	EST	MEYLAN,WM & HOWARD,PH (1993)

CAS #: 005722-68-9 — 3-T-BUTYL-4-NITROPHENOL

Formula: $C_{10}H_{13}NO_3$

Mol Weight: 195.22

MP (deg C): FP (deg C):

BP (deg C):

BP pressure (mm Hg):

Property/Value	Units	Temp	Data Type	Reference
WS 1.34E+002	mg/L	25	EST	MEYLAN,WM ET AL. (1996)
logP 3.65			EXP	HANSCH,C & LEO,AJ (1985)
VP 6.57E-005	mm Hg	25	EST	NEELY,WB & BLAU,GE (1985)
DC	pKa			
HL 5.71E-009	atm m3/mol	25	EST	MEYLAN,WM & HOWARD,PH (1991)
OH 9.82E-012	cm3/molc sec	25	EST	MEYLAN,WM & HOWARD,PH (1993)

CAS #: 005725-96-2 — METHANAMINE, N-HYDROXY-N-METHYL-

Formula: C_2H_7NO

Mol Weight: 61.08

MP (deg C): FP (deg C):

BP (deg C):

BP pressure (mm Hg):

Property/Value	Units	Temp	Data Type	Reference
WS 1.00E+006	mg/L	25	EST	MEYLAN,WM ET AL. (1996)
logP -1.15			EXP	HANSCH,C ET AL. (1995)
VP 2.45E-001	mm Hg	25	EST	NEELY,WB & BLAU,GE (1985)
DC 5.20	pKa		EXP	PERRIN,DD (1965)
HL 3.33E-008	atm m3/mol	25	EST	MEYLAN,WM & HOWARD,PH (1991)
OH 6.85E-011	cm3/molc sec	25	EST	MEYLAN,WM & HOWARD,PH (1993)

CAS #: 005728-21-2 — 1,2,5-THIADIAZOLE, 3,4-DIMETHYL-

Formula: $C_4H_6N_2S$

Mol Weight: 114.17

MP (deg C): FP (deg C):

BP (deg C):

BP pressure (mm Hg):

Property/Value	Units	Temp	Data Type	Reference
WS 8.50E+003	mg/L	25	EST	MEYLAN,WM ET AL. (1996)
logP 1.28			EXP	CALVINO,R ET AL. (1992)
VP 1.01E+000	mm Hg	25	EST	NEELY,WB & BLAU,GE (1985)
DC	pKa			
HL 2.41E-004	atm m3/mol	25	EST	MEYLAN,WM & HOWARD,PH (1991)
OH 9.72E-013	cm3/molc sec	25	EST	MEYLAN,WM & HOWARD,PH (1993)

CAS #: 005739-82-2 — 2-PROPENOIC ACID, 3-PHENOXY-, METHYL ESTER, (E)-

Formula: $C_{10}H_{10}O_3$

Mol Weight: 178.19

MP (deg C): FP (deg C):

BP (deg C):

BP pressure (mm Hg):

Property/Value	Units	Temp	Data Type	Reference
WS 2.89E+002	mg/L	25	EST	MEYLAN,WM ET AL. (1996)
logP 2.68			EXP	SANGSTER,J (1993)
VP 1.71E-002	mm Hg	25	EST	NEELY,WB & BLAU,GE (1985)
DC	pKa			
HL 3.18E-006	atm m3/mol	25	EST	MEYLAN,WM & HOWARD,PH (1991)
OH 4.15E-011	cm3/molc sec	25	EST	MEYLAN,WM & HOWARD,PH (1993)

CAS #: 005741-22-0 — 2-PROPANOL, 1-(2-METHOXYPHENOXY)-3-[(1-METHYLETH

Formula: $C_{13}H_{21}NO_3$

Mol Weight: 239.32

MP (deg C): 82-83 FP (deg C):

BP (deg C):

BP pressure (mm Hg):

Property/Value	Units	Temp	Data Type	Reference
WS 9.94E+003	mg/L	25	EST	MEYLAN,WM ET AL. (1996)
logP 1.69			EXP	RECANATINI,M (1992)
VP 2.02E-006	mm Hg	25	EST	NEELY,WB & BLAU,GE (1985)
DC	pKa			
HL 4.84E-013	atm m3/mol	25	EST	MEYLAN,WM & HOWARD,PH (1991)
OH 1.29E-010	cm3/molc sec	25	EST	MEYLAN,WM & HOWARD,PH (1993)

CAS #: 005765-44-6 — ISOXAZOLE, 5-METHYL-

Formula: C_4H_6NO

Mol Weight: 84.10

MP (deg C): FP (deg C):

BP (deg C): 122

BP pressure (mm Hg):

Property/Value	Units	Temp	Data Type	Reference
WS 5.33E+004	mg/L	25	EST	MEYLAN,WM ET AL. (1996)
logP 0.45			EXP	HANSCH,C ET AL. (1995)
VP 4.49E+001	mm Hg	25	EST	NEELY,WB & BLAU,GE (1985)
DC 2.30	pKa	25	EXP	PERRIN,DD (1965)
HL 6.26E-005	atm m3/mol	25	EST	MEYLAN,WM & HOWARD,PH (1991)
OH 2.39E-011	cm3/molc sec	25	EST	MEYLAN,WM & HOWARD,PH (1993)

CAS #: 005785-06-8 — BENZOYLHYDRAZINE, M-METHOXY

Formula: $C_8H_{10}N_2O_2$

Mol Weight: 166.18

MP (deg C): 93-95 FP (deg C):

BP (deg C):

BP pressure (mm Hg):

Property/Value	Units	Temp	Data Type	Reference
WS 2.92E+004	mg/L	25	EST	MEYLAN,WM ET AL. (1996)
logP 0.40			EXP	HANSCH,C & LEO,AJ (1985)
VP 1.43E-005	mm Hg	25	EST	NEELY,WB & BLAU,GE (1985)
DC	pKa			
HL 5.47E-013	atm m3/mol	25	EST	MEYLAN,WM & HOWARD,PH (1991)
OH 1.54E-011	cm3/molc sec	25	EST	MEYLAN,WM & HOWARD,PH (1993)

CAS #: 005786-21-0 — CLOZAPINE

Formula: $C_{18}H_{19}ClN_4$

Mol Weight: 326.83

MP (deg C): 183-184 FP (deg C):

BP (deg C):

BP pressure (mm Hg):

Property/Value	Units	Temp	Data Type	Reference
WS 1.18E+001	mg/L	25	EST	MEYLAN,WM ET AL. (1996)
logP 3.23			EXP	HANSCH,C ET AL. (1995)
VP 9.48E-009	mm Hg	25	EST	NEELY,WB & BLAU,GE (1985)
DC	pKa			
HL 9.29E-015	atm m3/mol	25	EST	MEYLAN,WM & HOWARD,PH (1991)
OH 3.76E-010	cm3/molc sec	25	EST	MEYLAN,WM & HOWARD,PH (1993)

CAS #: 005798-79-8	ALPHA-BROMOBENZENEACETONITRILE

Formula: C_8H_6BrN

Mol Weight: 196.05

MP (deg C): 29   FP (deg C):

BP (deg C): 133

BP pressure (mm Hg): 1.20E+001

Property/Value	Units	Temp	Data Type	Reference
WS 6.78E+002	mg/L	25	EST	MEYLAN,WM ET AL. (1996)
logP 1.83			EST	MEYLAN,WM & HOWARD,PH (1995)
VP 4.65E-003	mm Hg	25	EST	NEELY,WB & BLAU,GE (1985)
DC	pKa			
HL 2.84E-007	atm m3/mol	25	EST	MEYLAN,WM & HOWARD,PH (1991)
OH 2.11E-012	cm3/molc sec		EST	MEYLAN,WM & HOWARD,PH (1993)

CAS #: 005813-86-5	M-METHOXYBENZAMIDE

Formula: $C_8H_9NO_2$

Mol Weight: 151.17

MP (deg C): 134   FP (deg C):

BP (deg C):

BP pressure (mm Hg):

Property/Value	Units	Temp	Data Type	Reference
WS 1.41E+004	mg/L	25	EST	MEYLAN,WM ET AL. (1996)
logP 0.85			EXP	HANSCH,C & LEO,AJ (1985)
VP 1.11E-004	mm Hg	25	EST	NEELY,WB & BLAU,GE (1985)
DC	pKa			
HL 1.31E-010	atm m3/mol	25	EST	MEYLAN,WM & HOWARD,PH (1991)
OH 1.19E-011	cm3/molc sec	25	EST	MEYLAN,WM & HOWARD,PH (1993)

CAS #: 005813-89-8	THIOPHEN-2-CARBOXAMIDE

Formula: C_5H_7NOS

Mol Weight: 129.18

MP (deg C): 181-183   FP (deg C):

BP (deg C):

BP pressure (mm Hg):

Property/Value	Units	Temp	Data Type	Reference
WS 3.39E+004	mg/L	25	EST	MEYLAN,WM ET AL. (1996)
logP 0.52			EXP	HANSCH,C & LEO,AJ (1985)
VP 4.17E-004	mm Hg	25	EST	NEELY,WB & BLAU,GE (1985)
DC	pKa			
HL 1.20E-009	atm m3/mol	25	EST	MEYLAN,WM & HOWARD,PH (1991)
OH 1.07E-011	cm3/molc sec	25	EST	MEYLAN,WM & HOWARD,PH (1993)

CAS #: 005813-92-3	3,4-DIOXYMETHYLENECINNAMAMIDE

Formula: $C_{10}H_9NO_3$

Mol Weight: 191.19

MP (deg C):   FP (deg C):

BP (deg C):

BP pressure (mm Hg):

Property/Value	Units	Temp	Data Type	Reference
WS 3.09E+003	mg/L	25	EST	MEYLAN,WM ET AL. (1996)
logP 1.40			EXP	HANSCH,C & LEO,AJ (1985)
VP 2.88E-006	mm Hg	25	EST	NEELY,WB & BLAU,GE (1985)
DC	pKa			
HL 3.07E-013	atm m3/mol	25	EST	MEYLAN,WM & HOWARD,PH (1991)
OH 4.20E-011	cm3/molc sec	25	EST	MEYLAN,WM & HOWARD,PH (1993)

CAS #: 005814-05-1	BENZOYLHYDRAZINE, O-CHLORO

Formula: $C_7H_7ClN_2O$

Mol Weight: 170.60

MP (deg C): 118-120   FP (deg C):

BP (deg C):

BP pressure (mm Hg):

Property/Value	Units	Temp	Data Type	Reference
WS 2.27E+003	mg/L	25	EST	MEYLAN,WM ET AL. (1996)
logP 0.14			EXP	HANSCH,C & LEO,AJ (1985)
VP 2.02E-005	mm Hg	25	EST	NEELY,WB & BLAU,GE (1985)
DC	pKa			
HL 6.85E-012	atm m3/mol	25	EST	MEYLAN,WM & HOWARD,PH (1991)
OH 6.75E-012	cm3/molc sec	25	EST	MEYLAN,WM & HOWARD,PH (1993)

CAS #: 005818-06-4	BENZOYLHYDRAZINE, M-HYDROXY

Formula: $C_7H_8N_2O_2$

Mol Weight: 152.15

MP (deg C):   FP (deg C):

BP (deg C):

BP pressure (mm Hg):

Property/Value	Units	Temp	Data Type	Reference
WS 1.62E+004	mg/L	25	EST	MEYLAN,WM ET AL. (1996)
logP -0.08			EXP	HANSCH,C & LEO,AJ (1985)
VP 1.24E-006	mm Hg	25	EST	NEELY,WB & BLAU,GE (1985)
DC	pKa			
HL 9.62E-016	atm m3/mol	25	EST	MEYLAN,WM & HOWARD,PH (1991)
OH 1.97E-011	cm3/molc sec	25	EST	MEYLAN,WM & HOWARD,PH (1993)

CAS #: 005819-08-9	1,1'-SULFINYLBIS(2-CHLOROETHANE)

Formula: $C_4H_8Cl_2OS$

Mol Weight: 175.08

MP (deg C):   FP (deg C):

BP (deg C):

BP pressure (mm Hg):

Property/Value	Units	Temp	Data Type	Reference
WS 3.44E+004	mg/L	25	EST	MEYLAN,WM ET AL. (1996)
logP 0.27			EST	MEYLAN,WM & HOWARD,PH (1995)
VP 6.84E-003	mm Hg	25	EST	NEELY,WB & BLAU,GE (1985)
DC	pKa			
HL 1.08E-008	atm m3/mol	25	EST	MEYLAN,WM & HOWARD,PH (1991)
OH 6.61E-011	cm3/molc sec	25	EST	MEYLAN,WM & HOWARD,PH (1993)

CAS #: 005819-21-6	N-METHYL-4-NITROPHENYLCARBAMATE

Formula: $C_8H_8N_2O_4$

Mol Weight: 196.16

MP (deg C):   FP (deg C):

BP (deg C):

BP pressure (mm Hg):

Property/Value	Units	Temp	Data Type	Reference
WS 1.04E+003	mg/L	25	EST	MEYLAN,WM ET AL. (1996)
logP 1.47			EXP	HANSCH,C & LEO,AJ (1985)
VP 2.34E-004	mm Hg	25	EST	NEELY,WB & BLAU,GE (1985)
DC	pKa			
HL 1.27E-010	atm m3/mol	25	EST	MEYLAN,WM & HOWARD,PH (1991)
OH 6.31E-012	cm3/molc sec	25	EST	MEYLAN,WM & HOWARD,PH (1993)

| CAS #: | 005822-68-4 | PYRIDINIUM, 3-(ETHOXYCARBONYL)-1,2,5,6-TETRAHYDR |

Formula: $C_{10}H_{18}INO_2$

Mol Weight: 311.16

MP (deg C):   **FP (deg C):**

BP (deg C):

BP pressure (mm Hg):

Property/Value	Units	Temp	Data Type	Reference
WS 1.00E+006	mg/L	25	EST	MEYLAN,WM ET AL. (1996)
logP -2.35			EXP	SANGSTER,J (1993)
VP 2.91E-009	mm Hg	25	EST	NEELY,WB & BLAU,GE (1985)
DC	pKa			
HL 4.75E-015	atm m3/mol	25	EST	MEYLAN,WM & HOWARD,PH (1991)
OH 5.51E-011	cm3/molc sec	25	EST	MEYLAN,WM & HOWARD,PH (1993)

| CAS #: | 005822-97-9 | PHENYLMERCURY MONOETHANOL AMMONIUM ACETATE |

Formula: $C_{10}H_{15}HgNO_3$

Mol Weight: 397.83

MP (deg C):   **FP (deg C):**

BP (deg C):

BP pressure (mm Hg):

Property/Value	Units	Temp	Data Type	Reference
WS 1.16E+004	mg/L	25	EST	MEYLAN,WM ET AL. (1996)
logP -0.65			EST	MEYLAN,WM & HOWARD,PH (1995)
VP 1.14E-005	mm Hg	25	EST	NEELY,WB & BLAU,GE (1985)
DC	pKa			
HL	atm m3/mol			
OH 1.02E-010	cm3/molc sec	25	EST	MEYLAN,WM & HOWARD,PH (1993)

| CAS #: | 005826-91-5 | G-24622 |

Formula: $C_{12}H_{21}N_2O_3PS$

Mol Weight: 304.35

MP (deg C):   **FP (deg C):**

BP (deg C):

BP pressure (mm Hg):

Property/Value	Units	Temp	Data Type	Reference
WS 5.03E+000	mg/L	25	EST	MEYLAN,WM ET AL. (1996)
logP 3.94			EST	MEYLAN,WM & HOWARD,PH (1995)
VP 8.81E-006	mm Hg	25	EST	NEELY,WB & BLAU,GE (1985)
DC	pKa			
HL 8.73E-008	atm m3/mol	25	EST	MEYLAN,WM & HOWARD,PH (1991)
OH 9.70E-011	cm3/molc sec	25	EST	MEYLAN,WM & HOWARD,PH (1993)

| CAS #: | 005827-05-4 | APHIDAN |

Formula: $C_9H_{21}O_3PS_3$

Mol Weight: 304.43

MP (deg C):   **FP (deg C):**

BP (deg C):

BP pressure (mm Hg):

Property/Value	Units	Temp	Data Type	Reference
WS 1.50E+003	mg/L	15	EXP	YALKOWSKY,SH & DANNENFELSER,RM (1992)
logP 2.06			EST	MEYLAN,WM & HOWARD,PH (1995)
VP 4.16E-005	mm Hg	25	EST	NEELY,WB & BLAU,GE (1985)
DC	pKa			
HL 1.76E-010	atm m3/mol	25	EST	MEYLAN,WM & HOWARD,PH (1991)
OH 3.50E-010	cm3/molc sec	25	EST	MEYLAN,WM & HOWARD,PH (1993)

| CAS #: | 005830-33-1 | CYCLOHEXANEACETAMIDE, N,N-DIMETHYL- |

Formula: $C_{10}H_{19}NO$

Mol Weight: 169.27

MP (deg C):   **FP (deg C):**

BP (deg C):

BP pressure (mm Hg):

Property/Value	Units	Temp	Data Type	Reference
WS 8.36E+002	mg/L	25	EST	MEYLAN,WM ET AL. (1996)
logP 2.19			EXP	SURYANARAYANA,MVS ET AL. (1991)
VP 1.36E-001	mm Hg	30	EXP	SURYANARAYANA,MVS ET AL. (1991)
DC	pKa			
HL 1.30E-007	atm m3/mol	25	EST	MEYLAN,WM & HOWARD,PH (1991)
OH 3.55E-011	cm3/molc sec	25	EST	MEYLAN,WM & HOWARD,PH (1993)

| CAS #: | 005836-10-2 | ISOPROPYL-4,4'-DICHLOROBENZILATE |

Formula: $C_{17}H_{16}Cl_2O_3$

Mol Weight: 339.22

MP (deg C): 73   **FP (deg C):**

BP (deg C): 148-150

BP pressure (mm Hg): 5.00E-001

Property/Value	Units	Temp	Data Type	Reference
WS 1.00E+001	mg/L		EXP	YALKOWSKY,SH & DANNENFELSER,RM (1992)
logP 4.41			EST	MEYLAN,WM & HOWARD,PH (1995)
VP 1.80E-007	mm Hg	20	EXP	SPENCER,EY (1982)
DC	pKa			
HL 8.03E-009	atm m3/mol	20	EST	VP/WSOL
OH 6.87E-012	cm3/molc sec	25	EST	MEYLAN,WM & HOWARD,PH (1993)

| CAS #: | 005841-62-3 | 5-BENZYL-2,4-OXAZOLIDIONE |

Formula: $C_{10}H_9NO_3$

Mol Weight: 191.19

MP (deg C):   **FP (deg C):**

BP (deg C):

BP pressure (mm Hg):

Property/Value	Units	Temp	Data Type	Reference
WS 4.58E+003	mg/L	25	EST	MEYLAN,WM ET AL. (1996)
logP 1.20			EXP	LIPINSKI,CA ET AL. (1991)
VP 1.69E-008	mm Hg	25	EST	NEELY,WB & BLAU,GE (1985)
DC	pKa			
HL 5.12E-008	atm m3/mol	25	EST	MEYLAN,WM & HOWARD,PH (1991)
OH 1.63E-011	cm3/molc sec	25	EST	MEYLAN,WM & HOWARD,PH (1993)

| CAS #: | 005841-63-4 | 5-PHENYL-2,4-OXAZOLIDIONE |

Formula: $C_9H_7NO_3$

Mol Weight: 177.16

MP (deg C):   **FP (deg C):**

BP (deg C):

BP pressure (mm Hg):

Property/Value	Units	Temp	Data Type	Reference
WS 6.67E+003	mg/L	25	EST	MEYLAN,WM ET AL. (1996)
logP 1.09			EXP	LIPINSKI,CA ET AL. (1991)
VP 3.94E-008	mm Hg	25	EST	NEELY,WB & BLAU,GE (1985)
DC	pKa			
HL 3.86E-008	atm m3/mol	25	EST	MEYLAN,WM & HOWARD,PH (1991)
OH 1.20E-011	cm3/molc sec	25	EST	MEYLAN,WM & HOWARD,PH (1993)

CAS #: 005841-66-7	3-ME-5-PH OXAZOLIDINE-2,4-DIONE

Formula: $C_{10}H_9NO_3$

Mol Weight: 191.19

MP (deg C): | FP (deg C):

BP (deg C):

BP pressure (mm Hg):

Property/Value	Units	Temp	Data Type	Reference
WS 6.03E+003	mg/L	25	EST	MEYLAN,WM ET AL. (1996)
logP 1.06			EXP	HANSCH,C & LEO,AJ (1985)
VP 7.44E-007	mm Hg	25	EST	NEELY,WB & BLAU,GE (1985)
DC	pKa			
HL 8.47E-008	atm m3/mol	25	EST	MEYLAN,WM & HOWARD,PH (1991)
OH 2.20E-011	cm3/molc sec	25	EST	MEYLAN,WM & HOWARD,PH (1993)

CAS #: 005847-57-4	2,5-DICHLORO-4-NITROPHENOL

Formula: $C_6H_3Cl_2NO_3$

Mol Weight: 208.00

MP (deg C): | FP (deg C):

BP (deg C):

BP pressure (mm Hg):

Property/Value	Units	Temp	Data Type	Reference
WS 2.82E+002	mg/L	25	EST	MEYLAN,WM ET AL. (1996)
logP 3.20			EST	MEYLAN,WM & HOWARD,PH (1995)
VP 5.70E-005	mm Hg	25	EST	NEELY,WB & BLAU,GE (1985)
DC	pKa			
HL 1.21E-009	atm m3/mol	25	EST	MEYLAN,WM & HOWARD,PH (1991)
OH 9.95E-013	cm3/molc sec	25	EST	MEYLAN,WM & HOWARD,PH (1993)

CAS #: 005847-59-6	2-BROMO-4-NITROPHENOL

Formula: $C_6H_4BrNO_3$

Mol Weight: 218.01

MP (deg C): | FP (deg C):

BP (deg C):

BP pressure (mm Hg):

Property/Value	Units	Temp	Data Type	Reference
WS 2.20E+004	mg/L	100	EXP	YALKOWSKY,SH & DANNENFELSER,RM (1992)
logP 2.80			EST	MEYLAN,WM & HOWARD,PH (1995)
VP 8.64E-005	mm Hg	25	EST	NEELY,WB & BLAU,GE (1985)
DC	pKa			
HL 8.81E-010	atm m3/mol	25	EST	MEYLAN,WM & HOWARD,PH (1991)
OH 1.33E-012	cm3/molc sec	25	EST	MEYLAN,WM & HOWARD,PH (1993)

CAS #: 005850-16-8	C.I. ACID BROWN 14, DISODIUM SALT

Formula: $C_{26}H_{16}N_4Na_2O_8S_2$

Mol Weight: 622.55

MP (deg C): | FP (deg C):

BP (deg C):

BP pressure (mm Hg):

Property/Value	Units	Temp	Data Type	Reference
WS 4.99E-001	mg/L	25	EST	MEYLAN,WM ET AL. (1996)
logP 1.64			EST	MEYLAN,WM & HOWARD,PH (1995)
VP 3.36E-031	mm Hg	25	EST	NEELY,WB & BLAU,GE (1985)
DC	pKa			
HL	atm m3/mol			
OH 9.19E-011	cm3/molc sec	25	EST	MEYLAN,WM & HOWARD,PH (1993)

CAS #: 005850-86-2	C.I. ACID ORANGE 8, SODIUM SALT

Formula: $C_{17}H_{13}N_2NaO_4S$

Mol Weight: 364.36

MP (deg C): | FP (deg C):

BP (deg C):

BP pressure (mm Hg):

Property/Value	Units	Temp	Data Type	Reference
WS 8.05E+002	mg/L	25	EST	MEYLAN,WM ET AL. (1996)
logP 1.11			EST	MEYLAN,WM & HOWARD,PH (1995)
VP 7.68E-020	mm Hg	25	EST	NEELY,WB & BLAU,GE (1985)
DC	pKa			
HL	atm m3/mol			
OH 1.81E-011	cm3/molc sec	25	EST	MEYLAN,WM & HOWARD,PH (1993)

CAS #: 005853-29-2	CEPHAELINE DIHYDROCHLORIDE

Formula: $C_{28}H_{36}Cl_2N_2O_4$

Mol Weight: 535.52

MP (deg C): | FP (deg C):

BP (deg C):

BP pressure (mm Hg):

Property/Value	Units	Temp	Data Type	Reference
WS 2.65E+005	mg/L	18	EXP	STEPHEN,H & STEPHEN,T (1963)
logP 0.36			EST	MEYLAN,WM & HOWARD,PH (1995)
VP 2.58E-020	mm Hg	25	EST	NEELY,WB & BLAU,GE (1985)
DC	pKa			
HL 4.85E-026	atm m3/mol	25	EST	MEYLAN,WM & HOWARD,PH (1991)
OH 3.24E-010	cm3/molc sec	25	EST	MEYLAN,WM & HOWARD,PH (1993)

CAS #: 005858-39-9	C.I. ACID RED 4, MONOSODIUM SALT

Formula: $C_{17}H_{13}N_2NaO_5S$

Mol Weight: 380.36

MP (deg C): | FP (deg C):

BP (deg C):

BP pressure (mm Hg):

Property/Value	Units	Temp	Data Type	Reference
WS 1.61E+003	mg/L	25	EST	MEYLAN,WM ET AL. (1996)
logP 0.64			EST	MEYLAN,WM & HOWARD,PH (1995)
VP 2.87E-020	mm Hg	25	EST	NEELY,WB & BLAU,GE (1985)
DC	pKa			
HL	atm m3/mol			
OH 1.32E-011	cm3/molc sec	25	EST	MEYLAN,WM & HOWARD,PH (1993)

CAS #: 005867-45-8	3-ACETAMINOPYRIDINE

Formula: $C_7H_8N_2O$

Mol Weight: 136.15

MP (deg C): 133 | FP (deg C):

BP (deg C): 326.5

BP pressure (mm Hg):

Property/Value	Units	Temp	Data Type	Reference
WS 3.88E+004	mg/L	25	EST	MEYLAN,WM ET AL. (1996)
logP 0.41			EXP	HANSCH,C & LEO,AJ (1985)
VP 2.36E-004	mm Hg	25	EST	NEELY,WB & BLAU,GE (1985)
DC 4.36	pKa	24	EXP	PERRIN,DD (1965)
HL 8.08E-012	atm m3/mol	25	EST	MEYLAN,WM & HOWARD,PH (1991)
OH 2.46E-012	cm3/molc sec	25	EST	MEYLAN,WM & HOWARD,PH (1993)

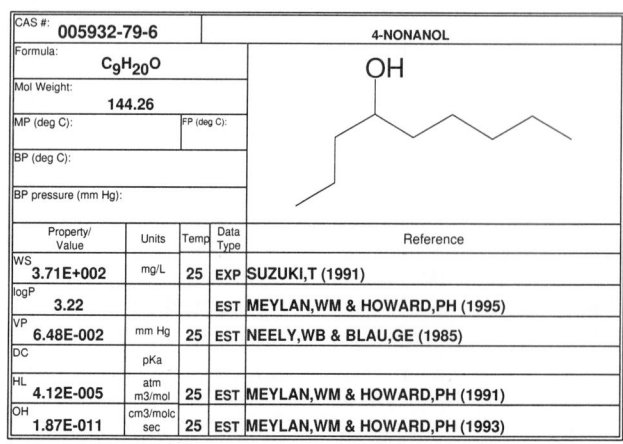

CAS #:	005878-19-3			1-METHOXY-2-PROPANONE
Formula:	$C_4H_8O_2$			
Mol Weight:	88.11			
MP (deg C):		FP (deg C):		
BP (deg C):	116			
BP pressure (mm Hg):				

	Property/Value	Units	Temp	Data Type	Reference
WS	3.37E+005	mg/L	25	EST	MEYLAN,WM ET AL. (1996)
logP	-0.50			EST	MEYLAN,WM & HOWARD,PH (1995)
VP	5.09E+001	mm Hg	25	EST	NEELY,WB & BLAU,GE (1985)
DC		pKa			
HL	4.11E-006	atm m3/mol	25	EST	MEYLAN,WM & HOWARD,PH (1991)
OH	6.77E-012	cm3/molc sec	25	EXP	ATKINSON,R (1989)

CAS #:	005884-48-0			PHENYLACETIC ACID,3-BR-4-PHMEO
Formula:	$C_{15}H_{13}BrO_3$			
Mol Weight:	321.17			
MP (deg C):		FP (deg C):		
BP (deg C):				
BP pressure (mm Hg):				

	Property/Value	Units	Temp	Data Type	Reference
WS	2.30E+001	mg/L	25	EST	MEYLAN,WM ET AL. (1996)
logP	3.51			EXP	HANSCH,C & LEO,AJ (1985)
VP	1.00E-007	mm Hg	25	EST	NEELY,WB & BLAU,GE (1985)
DC		pKa			
HL	8.41E-011	atm m3/mol	25	EST	MEYLAN,WM & HOWARD,PH (1991)
OH	1.81E-011	cm3/molc sec	25	EST	MEYLAN,WM & HOWARD,PH (1993)

CAS #:	005900-55-0			5-CHLORO-2-METHYLACETANILIDE
Formula:	$C_9H_{10}ClNO$			
Mol Weight:	183.64			
MP (deg C):		FP (deg C):		
BP (deg C):				
BP pressure (mm Hg):				

	Property/Value	Units	Temp	Data Type	Reference
WS	1.48E+003	mg/L	25	EST	MEYLAN,WM ET AL. (1996)
logP	1.82			EXP	NAKAGAWA,Y ET AL. (1992)
VP	3.66E-005	mm Hg	25	EST	NEELY,WB & BLAU,GE (1985)
DC		pKa			
HL	5.05E-009	atm m3/mol	25	EST	MEYLAN,WM & HOWARD,PH (1991)
OH	1.09E-011	cm3/molc sec	25	EST	MEYLAN,WM & HOWARD,PH (1993)

CAS #:	005902-51-2			TERBACIL
Formula:	$C_9H_{13}ClN_2O_2$			
Mol Weight:	216.67			
MP (deg C):	175-177	FP (deg C):		
BP (deg C):				
BP pressure (mm Hg):				

	Property/Value	Units	Temp	Data Type	Reference
WS	7.10E+002	mg/L	25	EXP	WORTHING,CR & WALKER,SB (1987)
logP	1.89			EXP	HANSCH,C & LEO,AJ (1985)
VP	2.97E-007	mm Hg	25	EST	HAMBURG,NE ET AL. (1989)
DC	9.00	pKa		EXP	WAUCHOPE,RD ET AL. (1991A)
HL	1.20E-010	atm m3/mol	25	EST	VP/WSOL
OH	3.63E-011	cm3/molc sec	25	EST	MEYLAN,WM & HOWARD,PH (1993)

CAS #:	005908-87-2			ETHYL DOCOSANOATE
Formula:	$C_{24}H_{48}O_2$			
Mol Weight:	368.65			
MP (deg C):	50	FP (deg C):		
BP (deg C):	240			
BP pressure (mm Hg):	1.00E+001			

	Property/Value	Units	Temp	Data Type	Reference
WS	3.57E-006	mg/L	25	EST	MEYLAN,WM ET AL. (1996)
logP	10.69			EST	MEYLAN,WM & HOWARD,PH (1995)
VP	5.42E-007	mm Hg	25	EST	NEELY,WB & BLAU,GE (1985)
DC		pKa			
HL	6.73E-002	atm m3/mol	25	EST	MEYLAN,WM & HOWARD,PH (1991)
OH	2.88E-011	cm3/molc sec	25	EST	MEYLAN,WM & HOWARD,PH (1993)

CAS #:	005910-89-4			PYRAZINE, 2,3-DIMETHYL-
Formula:	$C_6H_8N_2$			
Mol Weight:	108.14			
MP (deg C):		FP (deg C):		
BP (deg C):	156			
BP pressure (mm Hg):				

	Property/Value	Units	Temp	Data Type	Reference
WS	3.82E+004	mg/L	25	EST	MEYLAN,WM ET AL. (1996)
logP	0.54			EXP	YAMAGAMI,C ET AL. (1991)
VP	1.50E+000	mm Hg	25	EST	NEELY,WB & BLAU,GE (1985)
DC		pKa			
HL	3.55E-006	atm m3/mol	25	EST	MEYLAN,WM & HOWARD,PH (1991)
OH	1.01E-012	cm3/molc sec	25	EST	MEYLAN,WM & HOWARD,PH (1993)

CAS #:	005915-41-3			TERBUTHYLAZINE
Formula:	$C_9H_{15}ClN_4$			
Mol Weight:	214.70			
MP (deg C):	178	FP (deg C):		
BP (deg C):				
BP pressure (mm Hg):				

	Property/Value	Units	Temp	Data Type	Reference
WS	8.50E+000	mg/L	20	EXP	MARTIN,H & WORTHING,CR (1977)
logP	3.06			EXP	FINIZIO,A ET AL. (1991)
VP	1.12E-006	mm Hg	20	EXP	MARTIN,H & WORTHING,CR (1977)
DC		pKa			
HL	3.72E-008	atm m3/mol	20	EST	VP/WSOL
OH	9.47E-012	cm3/molc sec	25	EST	MEYLAN,WM & HOWARD,PH (1993)

CAS #:	005932-79-6			4-NONANOL
Formula:	$C_9H_{20}O$			
Mol Weight:	144.26			
MP (deg C):		FP (deg C):		
BP (deg C):				
BP pressure (mm Hg):				

	Property/Value	Units	Temp	Data Type	Reference
WS	3.71E+002	mg/L	25	EXP	SUZUKI,T (1991)
logP	3.22			EST	MEYLAN,WM & HOWARD,PH (1995)
VP	6.48E-002	mm Hg	25	EST	NEELY,WB & BLAU,GE (1985)
DC		pKa			
HL	4.12E-005	atm m3/mol	25	EST	MEYLAN,WM & HOWARD,PH (1991)
OH	1.87E-011	cm3/molc sec	25	EST	MEYLAN,WM & HOWARD,PH (1993)

CAS #: 005933-32-4				BENZOYLHYDRAZINE, P-BROMO
Formula: $C_7H_7BrN_2O$				
Mol Weight: 215.06				
MP (deg C): 165-167		FP (deg C):		
BP (deg C):				
BP pressure (mm Hg):				

Property/ Value	Units	Temp	Data Type	Reference
WS 2.95E+003	mg/L	25	EST	MEYLAN,WM ET AL. (1996)
logP 1.28			EXP	HANSCH,C & LEO,AJ (1985)
VP 7.82E-006	mm Hg	25	EST	NEELY,WB & BLAU,GE (1985)
DC	pKa			
HL 3.69E-012	atm m3/mol	25	EST	MEYLAN,WM & HOWARD,PH (1991)
OH 6.62E-012	cm3/molc sec	25	EST	MEYLAN,WM & HOWARD,PH (1993)

CAS #: 005944-41-2				2-T-BUTYLPYRIDINE
Formula: $C_9H_{13}N$				
Mol Weight: 135.21				
MP (deg C): -33		FP (deg C):		
BP (deg C): 170				
BP pressure (mm Hg):				

Property/ Value	Units	Temp	Data Type	Reference
WS 7.82E+002	mg/L	25	EST	MEYLAN,WM ET AL. (1996)
logP 2.40			EXP	HANSCH,C & LEO,AJ (1985)
VP 9.72E-001	mm Hg	25	EST	NEELY,WB & BLAU,GE (1985)
DC 5.76	pKa	25	EXP	PERRIN,DD (1965)
HL 1.82E-005	atm m3/mol	25	EST	MEYLAN,WM & HOWARD,PH (1991)
OH 1.32E-012	cm3/molc sec	25	EST	MEYLAN,WM & HOWARD,PH (1993)

CAS #: 005945-41-5				MEXICANIN-1
Formula: $C_{15}H_{18}O_4$				
Mol Weight: 262.31				
MP (deg C):		FP (deg C):		
BP (deg C):				
BP pressure (mm Hg):				

Property/ Value	Units	Temp	Data Type	Reference
WS 1.18E+004	mg/L	25	EST	MEYLAN,WM ET AL. (1996)
logP 0.36			EXP	HANSCH,C & LEO,AJ (1985)
VP 4.69E-009	mm Hg	25	EST	NEELY,WB & BLAU,GE (1985)
DC	pKa			
HL 1.38E-012	atm m3/mol	25	EST	MEYLAN,WM & HOWARD,PH (1991)
OH 1.04E-010	cm3/molc sec	25	EST	MEYLAN,WM & HOWARD,PH (1993)

CAS #: 005945-42-6				AROMATICIN
Formula: $C_{15}H_{18}O_3$				
Mol Weight: 246.31				
MP (deg C):		FP (deg C):		
BP (deg C):				
BP pressure (mm Hg):				

Property/ Value	Units	Temp	Data Type	Reference
WS 2.48E+003	mg/L	25	EST	MEYLAN,WM ET AL. (1996)
logP 1.17			EXP	HANSCH,C & LEO,AJ (1985)
VP 3.35E-006	mm Hg	25	EST	NEELY,WB & BLAU,GE (1985)
DC	pKa			
HL 3.78E-008	atm m3/mol	25	EST	MEYLAN,WM & HOWARD,PH (1991)
OH 8.81E-011	cm3/molc sec	25	EST	MEYLAN,WM & HOWARD,PH (1993)

CAS #: 005957-75-5				6H-DIBENZO[B,D]PYRAN-1-OL, 6A,7,10,10A-TETRAHYDR
Formula: $C_{21}H_{30}O_2$				
Mol Weight: 314.47				
MP (deg C):		FP (deg C):		
BP (deg C):				
BP pressure (mm Hg):				

Property/ Value	Units	Temp	Data Type	Reference
WS 1.80E-002	mg/L	25	EST	MEYLAN,WM ET AL. (1996)
logP 7.41			EXP	SANGSTER,J (1993)
VP 4.63E-008	mm Hg	25	EST	NEELY,WB & BLAU,GE (1985)
DC	pKa			
HL 2.44E-007	atm m3/mol	25	EST	MEYLAN,WM & HOWARD,PH (1991)
OH 3.09E-010	cm3/molc sec	25	EST	MEYLAN,WM & HOWARD,PH (1993)

CAS #: 005959-52-4				3-AMINO-2-NAPHTHOIC ACID
Formula: $C_{11}H_9NO_2$				
Mol Weight: 187.20				
MP (deg C): 216.5		FP (deg C):		
BP (deg C):				
BP pressure (mm Hg):				

Property/ Value	Units	Temp	Data Type	Reference
WS 3.45E+002	mg/L	25	EST	MEYLAN,WM ET AL. (1996)
logP 2.54			EST	MEYLAN,WM & HOWARD,PH (1995)
VP 1.48E-006	mm Hg	25	EST	NEELY,WB & BLAU,GE (1985)
DC	pKa			
HL 3.74E-012	atm m3/mol	25	EST	MEYLAN,WM & HOWARD,PH (1991)
OH 6.58E-011	cm3/molc sec	25	EST	MEYLAN,WM & HOWARD,PH (1993)

CAS #: 005963-49-5				1,2-DICHLORO-1,2-DIPHENYLETHANE
Formula: $C_{14}H_{12}Cl_2$				
Mol Weight: 251.16				
MP (deg C):		FP (deg C):		
BP (deg C):				
BP pressure (mm Hg):				

Property/ Value	Units	Temp	Data Type	Reference
WS 1.03E+000	mg/L	25	EST	MEYLAN,WM ET AL. (1996)
logP 5.10			EST	MEYLAN,WM & HOWARD,PH (1995)
VP 1.69E-004	mm Hg	25	EST	NEELY,WB & BLAU,GE (1985)
DC	pKa			
HL 7.91E-005	atm m3/mol	25	EST	MEYLAN,WM & HOWARD,PH (1991)
OH 4.55E-012	cm3/molc sec	25	EST	MEYLAN,WM & HOWARD,PH (1993)

CAS #: 005967-77-1				THEBAINE PICRATE
Formula: $C_{25}H_{24}N_4O_{10}$				
Mol Weight: 540.49				
MP (deg C):		FP (deg C):		
BP (deg C):				
BP pressure (mm Hg):				

Property/ Value	Units	Temp	Data Type	Reference
WS 3.42E-001	mg/L	25	EST	MEYLAN,WM ET AL. (1996)
logP 3.13			EST	MEYLAN,WM & HOWARD,PH (1995)
VP 4.56E-020	mm Hg	25	EST	NEELY,WB & BLAU,GE (1985)
DC	pKa			
HL 7.79E-027	atm m3/mol	25	EST	MEYLAN,WM & HOWARD,PH (1991)
OH 4.36E-010	cm3/molc sec	25	EST	MEYLAN,WM & HOWARD,PH (1993)

CAS #: 005974-93-6 — 2,4(1H,3H)-PYRIMIDINEDIONE, 1-[2,5-DIHYDRO-5-(HY

Formula: $C_9H_{10}N_2O_4$
Mol Weight: 210.19
MP (deg C): | FP (deg C):
BP (deg C):
BP pressure (mm Hg):

Property/Value	Units	Temp	Data Type	Reference
WS 1.56E+004	mg/L	25	EST	MEYLAN,WM ET AL. (1996)
logP -1.07			EXP	BALZARINI,JM ET AL. (1989)
VP 6.78E-012	mm Hg	25	EST	NEELY,WB & BLAU,GE (1985)
DC	pKa			
HL 1.48E-015	atm m3/mol	25	EST	MEYLAN,WM & HOWARD,PH (1991)
OH 1.20E-010	cm3/molc sec	25	EST	MEYLAN,WM & HOWARD,PH (1993)

CAS #: 005980-06-3 — 2-PYRROLIDINONE, 5-(3-PYRIDINYL)-

Formula: $C_9H_{10}N_2O$
Mol Weight: 162.19
MP (deg C): | FP (deg C):
BP (deg C):
BP pressure (mm Hg):

Property/Value	Units	Temp	Data Type	Reference
WS 1.00E+006	mg/L	25	EST	MEYLAN,WM ET AL. (1996)
logP -0.30			EXP	LI,NY & GORROD,JW (1992)
VP 1.66E-005	mm Hg	25	EST	NEELY,WB & BLAU,GE (1985)
DC	pKa			
HL 1.52E-012	atm m3/mol	25	EST	MEYLAN,WM & HOWARD,PH (1991)
OH 1.85E-011	cm3/molc sec	25	EST	MEYLAN,WM & HOWARD,PH (1993)

CAS #: 005983-08-4 — THYMIDINE, 2',3'-DIDEHYDRO-3'-DEOXY-4-THIO-

Formula: $C_{10}H_{12}N_2O_3S$
Mol Weight: 240.28
MP (deg C): | FP (deg C):
BP (deg C):
BP pressure (mm Hg):

Property/Value	Units	Temp	Data Type	Reference
WS 1.10E+003	mg/L	25	EST	MEYLAN,WM ET AL. (1996)
logP 0.09			EXP	PALOMINO,E ET AL. (1990)
VP 3.95E-011	mm Hg	25	EST	NEELY,WB & BLAU,GE (1985)
DC	pKa			
HL 2.74E-014	atm m3/mol	25	EST	MEYLAN,WM & HOWARD,PH (1991)
OH 1.65E-010	cm3/molc sec	25	EST	MEYLAN,WM & HOWARD,PH (1993)

CAS #: 005983-09-5 — URIDINE, 2',3'-DIDEOXY-

Formula: $C_9H_{12}N_2O_4$
Mol Weight: 212.21
MP (deg C): 127-129 | FP (deg C):
BP (deg C):
BP pressure (mm Hg):

Property/Value	Units	Temp	Data Type	Reference
WS 1.32E+004	mg/L	25	EST	MEYLAN,WM ET AL. (1996)
logP -1.00			EXP	SANGSTER,J (1993)
VP 7.69E-012	mm Hg	25	EST	NEELY,WB & BLAU,GE (1985)
DC	pKa			
HL 2.11E-016	atm m3/mol	25	EST	MEYLAN,WM & HOWARD,PH (1991)
OH 7.73E-011	cm3/molc sec	25	EST	MEYLAN,WM & HOWARD,PH (1993)

CAS #: 005989-27-5 — D-LIMONENE

Formula: $C_{10}H_{16}$
Mol Weight: 136.24
MP (deg C): -74.3 | FP (deg C):
BP (deg C): 176
BP pressure (mm Hg):

Property/Value	Units	Temp	Data Type	Reference
WS 1.38E+001	mg/L	25	EXP	MASSALDI,HA & KING,CJ (1973)
logP 4.83			EST	MEYLAN,WM & HOWARD,PH (1995)
VP 1.98E+000	mm Hg	25	EXP	YAWS,CL (1994B)
DC	pKa			
HL 2.57E-002	atm m3/mol	25	EST	VP/WSOL
OH 1.49E-010	cm3/molc sec	25	EXP	WINER,AM ET AL. (1976)

CAS #: 005989-54-8 — (-)-1-METHYL-4-(1-METHYLETHENYL)CYCLOHEXENE

Formula: $C_{10}H_{16}$
Mol Weight: 136.24
MP (deg C): | FP (deg C):
BP (deg C): 175-177
BP pressure (mm Hg):

Property/Value	Units	Temp	Data Type	Reference
WS 3.15E+000	mg/L	25	EST	MEYLAN,WM ET AL. (1996)
logP 4.83			EST	MEYLAN,WM & HOWARD,PH (1995)
VP 1.44E+000	mm Hg	25	EXP	NADAIS,MH & BERNARDO-GIL,MG (1993)
DC	pKa			
HL 3.80E-001	atm m3/mol	25	EST	MEYLAN,WM & HOWARD,PH (1991)
OH 1.45E-010	cm3/molc sec	25	EST	MEYLAN,WM & HOWARD,PH (1993)

CAS #: 006004-38-2 — TRICYCLO(5.2.1.0)DECANE

Formula: $C_{10}H_{16}$
Mol Weight: 136.24
MP (deg C): | FP (deg C):
BP (deg C):
BP pressure (mm Hg):

Property/Value	Units	Temp	Data Type	Reference
WS 1.10E+001	mg/L	25	EST	MEYLAN,WM ET AL. (1996)
logP 3.94			EST	MEYLAN,WM & HOWARD,PH (1995)
VP 1.98E+000	mm Hg	25	EST	NEELY,WB & BLAU,GE (1985)
DC	pKa			
HL 1.54E-001	atm m3/mol	25	EST	MEYLAN,WM & HOWARD,PH (1991)
OH 1.17E-011	cm3/molc sec	25	EST	MEYLAN,WM & HOWARD,PH (1993)

CAS #: 006004-44-0 — 1-PROPEN-1-ONE(METHYLKETENE)

Formula: C_3H_4O
Mol Weight: 56.06
MP (deg C): | FP (deg C):
BP (deg C):
BP pressure (mm Hg):

Property/Value	Units	Temp	Data Type	Reference
WS 1.70E+005	mg/L	25	EST	MEYLAN,WM ET AL. (1996)
logP -0.11			EST	MEYLAN,WM & HOWARD,PH (1995)
VP 2.67E+003	mm Hg	25	EST	NEELY,WB & BLAU,GE (1985)
DC	pKa			
HL	atm m3/mol			
OH 7.00E-011	cm3/molc sec	25	EXP	ATKINSON,R (1989)

CAS #: 006006-33-3				N-TRIDECYLCYCLOHEXANE

Formula: $C_{19}H_{38}$

Mol Weight: 266.51

MP (deg C): 18.5   FP (deg C):

BP (deg C): 346

BP pressure (mm Hg):

Property/ Value	Units	Temp	Data Type	Reference
WS 4.40E-005	mg/L	25	EST	MEYLAN,WM ET AL. (1996)
logP 9.49			EST	MEYLAN,WM & HOWARD,PH (1995)
VP 9.15E-006	mm Hg	25	EXT	ZWOLINSKI,BJ & WILHOIT,RC (1971)
DC	pKa			
HL 1.01E+001	atm m3/mol	25	EST	MEYLAN,WM & HOWARD,PH (1991)
OH 2.75E-011	cm3/molc sec	25	EST	MEYLAN,WM & HOWARD,PH (1993)

CAS #: 006006-95-7				N-PENTADECYLCYCLOHEXANE

Formula: $C_{21}H_{42}$

Mol Weight: 294.57

MP (deg C): 29   FP (deg C):

BP (deg C): 373

BP pressure (mm Hg):

Property/ Value	Units	Temp	Data Type	Reference
WS 4.40E-006	mg/L	25	EST	MEYLAN,WM ET AL. (1996)
logP 10.47			EST	MEYLAN,WM & HOWARD,PH (1995)
VP 3.46E-007	mm Hg	25	EXT	ZWOLINSKI,BJ & WILHOIT,RC (1971)
DC	pKa			
HL 1.79E+001	atm m3/mol	25	EST	MEYLAN,WM & HOWARD,PH (1991)
OH 3.04E-011	cm3/molc sec	25	EST	MEYLAN,WM & HOWARD,PH (1993)

CAS #: 006012-83-5				ETOXINOL

Formula: $C_{16}H_{16}Cl_2O_2$

Mol Weight: 311.21

MP (deg C):   FP (deg C):

BP (deg C):

BP pressure (mm Hg):

Property/ Value	Units	Temp	Data Type	Reference
WS 9.24E+000	mg/L	25	EST	MEYLAN,WM ET AL. (1996)
logP 4.18			EST	MEYLAN,WM & HOWARD,PH (1995)
VP 2.63E-008	mm Hg	25	EST	NEELY,WB & BLAU,GE (1985)
DC	pKa			
HL 1.99E-010	atm m3/mol	25	EST	MEYLAN,WM & HOWARD,PH (1991)
OH 3.13E-011	cm3/molc sec	25	EST	MEYLAN,WM & HOWARD,PH (1993)

CAS #: 006012-92-6				N-244

Formula: $C_{10}H_8ClNOS_2$

Mol Weight: 257.76

MP (deg C):   FP (deg C):

BP (deg C):

BP pressure (mm Hg):

Property/ Value	Units	Temp	Data Type	Reference
WS 7.86E+002	mg/L	25	EST	MEYLAN,WM ET AL. (1996)
logP 1.68			EST	MEYLAN,WM & HOWARD,PH (1995)
VP 3.55E-008	mm Hg	25	EST	NEELY,WB & BLAU,GE (1985)
DC	pKa			
HL 1.32E-007	atm m3/mol	25	EST	MEYLAN,WM & HOWARD,PH (1991)
OH 1.49E-011	cm3/molc sec	25	EST	MEYLAN,WM & HOWARD,PH (1993)

CAS #: 006021-23-4				3-PYRIDINEBUTANEAMINE

Formula: $C_9H_{14}N_2$

Mol Weight: 150.23

MP (deg C):   FP (deg C):

BP (deg C):

BP pressure (mm Hg):

Property/ Value	Units	Temp	Data Type	Reference
WS 1.00E+006	mg/L	25	EST	MEYLAN,WM ET AL. (1996)
logP 0.88			EXP	HANSCH,C & LEO,AJ (1985)
VP 1.44E-002	mm Hg	25	EST	NEELY,WB & BLAU,GE (1985)
DC	pKa			
HL 1.87E-009	atm m3/mol	25	EST	MEYLAN,WM & HOWARD,PH (1991)
OH 3.66E-011	cm3/molc sec	25	EST	MEYLAN,WM & HOWARD,PH (1993)

CAS #: 006032-29-7				2-PENTANOL

Formula: $C_5H_{12}O$

Mol Weight: 88.15

MP (deg C): -73   FP (deg C):

BP (deg C): 119.3

BP pressure (mm Hg):

Property/ Value	Units	Temp	Data Type	Reference
WS 4.46E+004	mg/L	25	EXP	YALKOWSKY,SH & DANNENFELSER,RM (1992)
logP 1.19			EXP	HANSCH,C ET AL. (1995)
VP 6.11E+000	mm Hg	25	EXP	YAWS,CL (1994A)
DC	pKa			
HL 1.48E-005	atm m3/mol	25	EXP	BUTLER,JAV ET AL. (1935)
OH 1.18E-011	cm3/molc sec	25	EXP	ATKINSON,R (1989)

CAS #: 006032-32-2				4-METHOXYPHENYL GLUCOPYRANOSIDE

Formula: $C_{13}H_{18}O_7$

Mol Weight: 286.28

MP (deg C): 158-160   FP (deg C):

BP (deg C):

BP pressure (mm Hg):

Property/ Value	Units	Temp	Data Type	Reference
WS 6.20E+004	mg/L	25	EST	MEYLAN,WM ET AL. (1996)
logP -0.73			EXP	HANSCH,C & LEO,AJ (1985)
VP 7.72E-012	mm Hg	25	EST	NEELY,WB & BLAU,GE (1985)
DC	pKa			
HL 6.80E-017	atm m3/mol	25	EST	MEYLAN,WM & HOWARD,PH (1991)
OH 8.15E-011	cm3/molc sec	25	EST	MEYLAN,WM & HOWARD,PH (1993)

CAS #: 006035-40-1				NARCOTINE

Formula: $C_{22}H_{23}NO_7$

Mol Weight: 413.43

MP (deg C): 176   FP (deg C):

BP (deg C):

BP pressure (mm Hg):

Property/ Value	Units	Temp	Data Type	Reference
WS 4.40E+001	mg/L	20	EXP	STEPHEN,H & STEPHEN,T (1963)
logP 1.97			EST	MEYLAN,WM & HOWARD,PH (1995)
VP 1.10E-011	mm Hg	25	EST	NEELY,WB & BLAU,GE (1985)
DC	pKa			
HL 7.92E-017	atm m3/mol	25	EST	MEYLAN,WM & HOWARD,PH (1991)
OH 2.96E-010	cm3/molc sec	25	EST	MEYLAN,WM & HOWARD,PH (1993)

CAS #: 006044-30-0 — GLUCOPYRANOSIDE,2-NAPHTHYL

Formula: $C_{16}H_{18}O_6$

Mol Weight: 306.32

MP (deg C):   FP (deg C):

BP (deg C):

BP pressure (mm Hg):

Property/Value	Units	Temp	Data Type	Reference
WS 2.53E+003	mg/L	25	EST	MEYLAN,WM ET AL. (1996)
logP 0.76			EXP	HANSCH,C & LEO,AJ (1985)
VP 9.54E-014	mm Hg	25	EST	NEELY,WB & BLAU,GE (1985)
DC	pKa			
HL 1.12E-016	atm m3/mol	25	EST	MEYLAN,WM & HOWARD,PH (1991)
OH 2.62E-010	cm3/molc sec	25	EST	MEYLAN,WM & HOWARD,PH (1993)

CAS #: 006054-48-4 — C.I..DISPERSE BLACK 1

Formula: $C_{16}H_{14}N_4$

Mol Weight: 262.32

MP (deg C):   FP (deg C):

BP (deg C):

BP pressure (mm Hg):

Property/Value	Units	Temp	Data Type	Reference
WS 7.90E-002	mg/L	25	EXP	BAUGHMAN,GL & PERENICH,TA (1988)
logP 3.45			EST	MEYLAN,WM & HOWARD,PH (1995)
VP 8.23E-009	mm Hg	25	EST	NEELY,WB & BLAU,GE (1985)
DC	pKa			
HL 1.79E-013	atm m3/mol	25	EST	MEYLAN,WM & HOWARD,PH (1991)
OH 9.08E-011	cm3/molc sec	25	EST	MEYLAN,WM & HOWARD,PH (1993)

CAS #: 006054-58-6 — C.I. DISPERSE DYE

Formula: $C_{18}H_{19}N_5O_3$

Mol Weight: 353.38

MP (deg C): 156   FP (deg C):

BP (deg C):

BP pressure (mm Hg):

Property/Value	Units	Temp	Data Type	Reference
WS 6.71E-002	mg/L	25	EXP	BAUGHMAN,GL & PERENICH,TA (1988)
logP 3.77			EST	MEYLAN,WM & HOWARD,PH (1995)
VP 1.94E-013	mm Hg	25	EST	NEELY,WB & BLAU,GE (1985)
DC	pKa			
HL 2.72E-017	atm m3/mol	25	EST	MEYLAN,WM & HOWARD,PH (1991)
OH 1.68E-010	cm3/molc sec	25	EST	MEYLAN,WM & HOWARD,PH (1993)

CAS #: 006061-06-9 — 1,1-DICHLORO-1,3-BUTADIENE

Formula: $C_4H_4Cl_2$

Mol Weight: 122.98

MP (deg C):   FP (deg C):

BP (deg C): 42-43

BP pressure (mm Hg): 9.00E+001

Property/Value	Units	Temp	Data Type	Reference
WS 3.38E+002	mg/L	25	EST	MEYLAN,WM ET AL. (1996)
logP 2.88			EST	MEYLAN,WM & HOWARD,PH (1995)
VP 5.25E+001	mm Hg	25	EST	NEELY,WB & BLAU,GE (1985)
DC	pKa			
HL 4.03E-002	atm m3/mol	25	EST	MEYLAN,WM & HOWARD,PH (1991)
OH 6.26E-012	cm3/molc sec	25	EST	MEYLAN,WM & HOWARD,PH (1993)

CAS #: 006064-63-7 — HEXANOIC ACID, 2-HYDROXY-

Formula: $C_6H_{12}O_3$

Mol Weight: 132.16

MP (deg C): 61   FP (deg C):

BP (deg C):

BP pressure (mm Hg):

Property/Value	Units	Temp	Data Type	Reference
WS 1.47E+005	mg/L	25	EST	MEYLAN,WM ET AL. (1996)
logP 0.81			EXP	SANGSTER,J (1993)
VP 3.14E-006	mm Hg	25	EXT	DAUBERT,TE & DANNER,RP (1991)
DC	pKa			
HL 2.65E-007	atm m3/mol	25	EST	MEYLAN,WM & HOWARD,PH (1991)
OH 1.11E-011	cm3/molc sec	25	EST	MEYLAN,WM & HOWARD,PH (1993)

CAS #: 006066-49-5 — 1(3H)-ISOBENZOFURANONE, 3-BUTYL-

Formula: $C_{12}H_{14}O_2$

Mol Weight: 190.24

MP (deg C):   FP (deg C):

BP (deg C):

BP pressure (mm Hg):

Property/Value	Units	Temp	Data Type	Reference
WS 1.99E+002	mg/L	25	EST	MEYLAN,WM ET AL. (1996)
logP 2.80			EXP	HANSCH,C & LEO,AJ (1985)
VP 1.96E-004	mm Hg	25	EST	NEELY,WB & BLAU,GE (1985)
DC	pKa			
HL 3.95E-005	atm m3/mol	25	EST	MEYLAN,WM & HOWARD,PH (1991)
OH 1.14E-011	cm3/molc sec	25	EST	MEYLAN,WM & HOWARD,PH (1993)

CAS #: 006083-47-2 — P-DIMETHYLAMINOBENZAMIDE

Formula: $C_9H_{12}N_2O$

Mol Weight: 164.21

MP (deg C):   FP (deg C):

BP (deg C):

BP pressure (mm Hg):

Property/Value	Units	Temp	Data Type	Reference
WS 6.96E+003	mg/L	25	EST	MEYLAN,WM ET AL. (1996)
logP 1.14			EXP	HANSCH,C & LEO,AJ (1985)
VP 4.44E-005	mm Hg	25	EST	NEELY,WB & BLAU,GE (1985)
DC	pKa			
HL 3.52E-011	atm m3/mol	25	EST	MEYLAN,WM & HOWARD,PH (1991)
OH 2.05E-010	cm3/molc sec	25	EST	MEYLAN,WM & HOWARD,PH (1993)

CAS #: 006091-44-7 — PIPERDINE, HYDROCHLOROIC ACID

Formula: $C_5H_{12}ClN$

Mol Weight: 121.61

MP (deg C): 245-248   FP (deg C):

BP (deg C):

BP pressure (mm Hg):

Property/Value	Units	Temp	Data Type	Reference
WS 1.00E+006	mg/L	25	EST	MEYLAN,WM ET AL. (1996)
logP -1.92			EST	MEYLAN,WM & HOWARD,PH (1995)
VP 1.48E-005	mm Hg	25	EST	NEELY,WB & BLAU,GE (1985)
DC	pKa			
HL 2.08E-011	atm m3/mol	25	EST	MEYLAN,WM & HOWARD,PH (1991)
OH 2.56E-011	cm3/molc sec	25	EST	MEYLAN,WM & HOWARD,PH (1993)

CAS #: 006091-81-2				2-METHYL-2,3-BENZODIAZIN-1-ONE

Formula: $C_9H_8N_2O$

Mol Weight: 160.18

MP (deg C): / FP (deg C):

BP (deg C):

BP pressure (mm Hg):

Property/Value	Units	Temp	Data Type	Reference
WS 2.06E+004	mg/L	25	EST	MEYLAN,WM ET AL. (1996)
logP 0.61			EXP	HANSCH,C & LEO,AJ (1985)
VP 1.24E-004	mm Hg	25	EST	NEELY,WB & BLAU,GE (1985)
DC	pKa			
HL 4.77E-009	atm m3/mol	25	EST	MEYLAN,WM & HOWARD,PH (1991)
OH 2.17E-011	cm3/molc sec	25	EST	MEYLAN,WM & HOWARD,PH (1993)

CAS #: 006092-24-6				2-MEO PHENYL GLUCOPYRANOSIDE

Formula: $C_{13}H_{18}O_7$

Mol Weight: 286.28

MP (deg C): / FP (deg C):

BP (deg C):

BP pressure (mm Hg):

Property/Value	Units	Temp	Data Type	Reference
WS 1.14E+005	mg/L	25	EST	MEYLAN,WM ET AL. (1996)
logP -1.04			EXP	HANSCH,C & LEO,AJ (1985)
VP 7.72E-012	mm Hg	25	EST	NEELY,WB & BLAU,GE (1985)
DC	pKa			
HL 6.80E-017	atm m3/mol	25	EST	MEYLAN,WM & HOWARD,PH (1991)
OH 8.15E-011	cm3/molc sec	25	EST	MEYLAN,WM & HOWARD,PH (1993)

CAS #: 006092-25-7				GLUCOPYRANOSIDE,3-METHYLPHENYL

Formula: $C_{13}H_{18}O_6$

Mol Weight: 270.28

MP (deg C): / FP (deg C):

BP (deg C):

BP pressure (mm Hg):

Property/Value	Units	Temp	Data Type	Reference
WS 2.70E+004	mg/L	25	EST	MEYLAN,WM ET AL. (1996)
logP -0.20			EXP	HANSCH,C & LEO,AJ (1985)
VP 2.48E-011	mm Hg	25	EST	NEELY,WB & BLAU,GE (1985)
DC	pKa			
HL 1.27E-015	atm m3/mol	25	EST	MEYLAN,WM & HOWARD,PH (1991)
OH 1.18E-010	cm3/molc sec	25	EST	MEYLAN,WM & HOWARD,PH (1993)

CAS #: 006094-02-6				2-METHYL-1-HEXENE

Formula: C_7H_{14}

Mol Weight: 98.19

MP (deg C): -102.8 / FP (deg C):

BP (deg C): 92

BP pressure (mm Hg):

Property/Value	Units	Temp	Data Type	Reference
WS 2.38E+001	mg/L	25	EST	MEYLAN,WM ET AL. (1996)
logP 3.70			EST	MEYLAN,WM & HOWARD,PH (1995)
VP 6.09E+001	mm Hg	25	EXP	YAWS,CL (1994A)
DC	pKa			
HL 5.62E-001	atm m3/mol	25	EST	MEYLAN,WM & HOWARD,PH (1991)
OH 5.54E-011	cm3/molc sec	25	EST	MEYLAN,WM & HOWARD,PH (1993)

CAS #: 006104-41-2				PYRIMIDINE, 4-METHOXY-

Formula: $C_5H_6N_2O$

Mol Weight: 110.12

MP (deg C): / FP (deg C):

BP (deg C):

BP pressure (mm Hg):

Property/Value	Units	Temp	Data Type	Reference
WS 3.76E+004	mg/L	25	EST	MEYLAN,WM ET AL. (1996)
logP 0.54			EXP	YAMAGAMI,C ET AL. (1990)
VP 1.60E+000	mm Hg	25	EST	NEELY,WB & BLAU,GE (1985)
DC 2.50	pKa	20	EXP	PERRIN,DD (1965)
HL 1.57E-006	atm m3/mol	25	EST	MEYLAN,WM & HOWARD,PH (1991)
OH 3.37E-012	cm3/molc sec	25	EST	MEYLAN,WM & HOWARD,PH (1993)

CAS #: 006108-74-3				BENZAMIDE,N-(3-AMINOPROPYL)

Formula: $C_{10}H_{14}N_2O$

Mol Weight: 178.24

MP (deg C): / FP (deg C):

BP (deg C):

BP pressure (mm Hg):

Property/Value	Units	Temp	Data Type	Reference
WS 1.53E+003	mg/L	25	EST	MEYLAN,WM ET AL. (1996)
logP 0.30			EXP	HANSCH,C & LEO,AJ (1985)
VP 7.22E-006	mm Hg	25	EST	NEELY,WB & BLAU,GE (1985)
DC	pKa			
HL 8.81E-013	atm m3/mol	25	EST	MEYLAN,WM & HOWARD,PH (1991)
OH 4.51E-011	cm3/molc sec	25	EST	MEYLAN,WM & HOWARD,PH (1993)

CAS #: 006117-80-2				2-BUTENE-1,4-DIOL (CIS)

Formula: $C_4H_8O_2$

Mol Weight: 88.11

MP (deg C): 4 / FP (deg C):

BP (deg C): 235

BP pressure (mm Hg):

Property/Value	Units	Temp	Data Type	Reference
WS 6.19E+005	mg/L	25	EST	MEYLAN,WM ET AL. (1996)
logP -0.43			EST	MEYLAN,WM & HOWARD,PH (1995)
VP 6.53E-003	mm Hg	25	EXP	DAUBERT,TE & DANNER,RP (1989)
DC	pKa			
HL 2.03E-007	atm m3/mol	25	EST	MEYLAN,WM & HOWARD,PH (1991)
OH 6.32E-011	cm3/molc sec	25	EST	MEYLAN,WM & HOWARD,PH (1993)

CAS #: 006125-24-2				B-NITROETHYL BENZENE

Formula: $C_8H_9NO_2$

Mol Weight: 151.17

MP (deg C): -23 / FP (deg C):

BP (deg C): 250

BP pressure (mm Hg):

Property/Value	Units	Temp	Data Type	Reference
WS 5.11E+002	mg/L	25	EST	MEYLAN,WM ET AL. (1996)
logP 2.08			EXP	HANSCH,C & LEO,AJ (1985)
VP 2.37E-002	mm Hg	25	EST	NEELY,WB & BLAU,GE (1985)
DC	pKa			
HL 4.51E-006	atm m3/mol	25	EST	MEYLAN,WM & HOWARD,PH (1991)
OH 5.11E-012	cm3/molc sec	25	EST	MEYLAN,WM & HOWARD,PH (1993)

CAS #: 006128-03-6	6,8-DIAZASPIRO[3.5]NONANE-5,7,9-TRIONE

Formula: $C_7H_8N_2O_3$

Mol Weight: 168.15

MP (deg C): | FP (deg C):

BP (deg C):

BP pressure (mm Hg):

Property/Value	Units	Temp	Data Type	Reference
WS 1.07E+005	mg/L	25	EST	MEYLAN,WM ET AL. (1996)
logP -0.27			EXP	PRANKERD,RJ & MCKEOWN,RH (1992)
VP 2.48E-010	mm Hg	25	EST	NEELY,WB & BLAU,GE (1985)
DC	pKa			
HL 1.20E-013	atm m3/mol	25	EST	MEYLAN,WM & HOWARD,PH (1991)
OH 4.90E-012	cm3/molc sec	25	EST	MEYLAN,WM & HOWARD,PH (1993)

CAS #: 006129-11-9	2ME2NO2-PROPANAL-N-MECARBAMOYL OXIME

Formula: $C_6H_{11}N_3O_4$

Mol Weight: 189.17

MP (deg C): | FP (deg C):

BP (deg C):

BP pressure (mm Hg):

Property/Value	Units	Temp	Data Type	Reference
WS 9.96E+003	mg/L	25	EST	MEYLAN,WM ET AL. (1996)
logP 0.36			EXP	HANSCH,C & LEO,AJ (1985)
VP 5.01E-003	mm Hg	25	EST	NEELY,WB & BLAU,GE (1985)
DC	pKa			
HL 1.91E-010	atm m3/mol	25	EST	MEYLAN,WM & HOWARD,PH (1991)
OH 6.50E-012	cm3/molc sec	25	EST	MEYLAN,WM & HOWARD,PH (1993)

CAS #: 006130-75-2	2,4,5-TRICHLOROANISOLE

Formula: $C_7H_5Cl_3O$

Mol Weight: 211.48

MP (deg C): 77.5 | FP (deg C):

BP (deg C): 254

BP pressure (mm Hg):

Property/Value	Units	Temp	Data Type	Reference
WS 1.97E+001	mg/L	25	EST	MEYLAN,WM ET AL. (1996)
logP 3.85			EXP	OPPERHUIZEN,A & VOORS,PI (1987)
VP 2.28E-002	mm Hg	25	EST	NEELY,WB & BLAU,GE (1985)
DC	pKa			
HL 1.30E-004	atm m3/mol	25	EST	MEYLAN,WM & HOWARD,PH (1991)
OH 2.12E-012	cm3/molc sec	25	EST	MEYLAN,WM & HOWARD,PH (1993)

CAS #: 006130-93-4	2,2,6,6-TETME-N-NITROSOPIPERDINE

Formula: $C_9H_{18}N_2O$

Mol Weight: 170.26

MP (deg C): | FP (deg C):

BP (deg C):

BP pressure (mm Hg):

Property/Value	Units	Temp	Data Type	Reference
WS 4.58E+002	mg/L	25	EST	MEYLAN,WM ET AL. (1996)
logP 2.49			EXP	HANSCH,C & LEO,AJ (1985)
VP 7.02E-003	mm Hg	25	EST	NEELY,WB & BLAU,GE (1985)
DC	pKa			
HL 6.59E-006	atm m3/mol	25	EST	MEYLAN,WM & HOWARD,PH (1991)
OH 4.91E-012	cm3/molc sec	25	EST	MEYLAN,WM & HOWARD,PH (1993)

CAS #: 006132-21-4	N-METHYL-3-NITROPHENYLCARBAMATE

Formula: $C_8H_8N_2O_4$

Mol Weight: 196.16

MP (deg C): | FP (deg C):

BP (deg C):

BP pressure (mm Hg):

Property/Value	Units	Temp	Data Type	Reference
WS 1.21E+003	mg/L	25	EST	MEYLAN,WM ET AL. (1996)
logP 1.39			EXP	HANSCH,C & LEO,AJ (1985)
VP 2.34E-004	mm Hg	25	EST	NEELY,WB & BLAU,GE (1985)
DC	pKa			
HL 1.27E-010	atm m3/mol	25	EST	MEYLAN,WM & HOWARD,PH (1991)
OH 6.24E-012	cm3/molc sec	25	EST	MEYLAN,WM & HOWARD,PH (1993)

CAS #: 006136-37-4	1-METHYLXANTHINE

Formula: $C_6H_6N_4O_2$

Mol Weight: 166.14

MP (deg C): | FP (deg C):

BP (deg C):

BP pressure (mm Hg):

Property/Value	Units	Temp	Data Type	Reference
WS 5.34E+003	mg/L	25	EST	MEYLAN,WM ET AL. (1996)
logP -0.27			EXP	GASPARI,F & BONATI,M (1987)
VP 1.17E-009	mm Hg	25	EST	NEELY,WB & BLAU,GE (1985)
DC 7.70	pKa	25	EXP	KORTUM,G ET AL (1961)
HL 8.19E-014	atm m3/mol	25	EST	MEYLAN,WM & HOWARD,PH (1991)
OH 5.53E-011	cm3/molc sec	25	EST	MEYLAN,WM & HOWARD,PH (1993)

CAS #: 006136-68-1	M-CYANOACETOPHENONE

Formula: C_9H_7NO

Mol Weight: 145.16

MP (deg C): 98-100 | FP (deg C):

BP (deg C):

BP pressure (mm Hg):

Property/Value	Units	Temp	Data Type	Reference
WS 4.42E+003	mg/L	25	EST	MEYLAN,WM ET AL. (1996)
logP 1.16			EXP	HANSCH,C & LEO,AJ (1985)
VP 6.42E-003	mm Hg	25	EST	NEELY,WB & BLAU,GE (1985)
DC	pKa			
HL 9.48E-008	atm m3/mol	25	EST	MEYLAN,WM & HOWARD,PH (1991)
OH 4.16E-013	cm3/molc sec	25	EST	MEYLAN,WM & HOWARD,PH (1993)

CAS #: 006138-53-0	2,4,6-TRIMETHYL-2,4,6-TRIPHENYLCYCLOTRISILOXANE

Formula: $C_{21}H_{24}O_3Si_3$

Mol Weight: 408.68

MP (deg C): | FP (deg C):

BP (deg C):

BP pressure (mm Hg):

Property/Value	Units	Temp	Data Type	Reference
WS 9.71E-005	mg/L	25	EST	MEYLAN,WM ET AL. (1996)
logP 8.12				MEYLAN,WM & HOWARD,PH (1995)
VP	mm Hg			
DC	pKa			
HL	atm m3/mol			
OH 6.30E-012	cm3/molc sec	25	EST	MEYLAN,WM & HOWARD,PH (1993)

CAS #:	006140-12-1			ACETOPHENONE,O-DICL ACETYLAMINO

Formula: $C_{10}H_9Cl_2NO_2$

Mol Weight: 246.09

MP (deg C):   FP (deg C):

BP (deg C):

BP pressure (mm Hg):

Property/Value	Units	Temp	Data Type	Reference
WS 1.36E+002	mg/L	25	EST	MEYLAN,WM ET AL. (1996)
logP 2.65			EXP	HANSCH,C & LEO,AJ (1985)
VP 6.33E-007	mm Hg	25	EST	NEELY,WB & BLAU,GE (1985)
DC	pKa			
HL 1.39E-012	atm m3/mol	25	EST	MEYLAN,WM & HOWARD,PH (1991)
OH 1.16E-011	cm3/molc sec	25	EST	MEYLAN,WM & HOWARD,PH (1993)

CAS #:	006141-18-0			2-METHYLTHIOQUINAZOLINE

Formula: $C_9H_8N_2S$

Mol Weight: 176.24

MP (deg C):   FP (deg C):

BP (deg C):

BP pressure (mm Hg):

Property/Value	Units	Temp	Data Type	Reference
WS 4.21E+002	mg/L	25	EST	MEYLAN,WM ET AL. (1996)
logP 2.50			EXP	HANSCH,C & LEO,AJ (1985)
VP 2.62E-004	mm Hg	25	EST	NEELY,WB & BLAU,GE (1985)
DC 1.60	pKa	20	EXP	PERRIN,DD (1965)
HL 3.23E-007	atm m3/mol	25	EST	MEYLAN,WM & HOWARD,PH (1991)
OH 1.40E-011	cm3/molc sec	25	EST	MEYLAN,WM & HOWARD,PH (1993)

CAS #:	006141-57-7			3-METHYLFUROIC ACID, ME ESTER

Formula: $C_7H_9O_3$

Mol Weight: 141.15

MP (deg C):   FP (deg C):

BP (deg C):

BP pressure (mm Hg):

Property/Value	Units	Temp	Data Type	Reference
WS 4.29E+003	mg/L	25	EST	MEYLAN,WM ET AL. (1996)
logP 1.51			EXP	HANSCH,C & LEO,AJ (1985)
VP 9.25E-001	mm Hg	25	EST	NEELY,WB & BLAU,GE (1985)
DC	pKa			
HL 3.82E-005	atm m3/mol	25	EST	MEYLAN,WM & HOWARD,PH (1991)
OH 3.43E-011	cm3/molc sec	25	EST	MEYLAN,WM & HOWARD,PH (1993)

CAS #:	006146-52-7			INDOLE, 5-NITRO

Formula: $C_8H_6N_2O_2$

Mol Weight: 162.15

MP (deg C): 140-142   FP (deg C):

BP (deg C):

BP pressure (mm Hg):

Property/Value	Units	Temp	Data Type	Reference
WS 1.88E+002	mg/L	25	EST	MEYLAN,WM ET AL. (1996)
logP 2.53			EXP	HANSCH,C & LEO,AJ (1985)
VP 8.99E-005	mm Hg	25	EST	NEELY,WB & BLAU,GE (1985)
DC	pKa			
HL 3.50E-009	atm m3/mol	25	EST	MEYLAN,WM & HOWARD,PH (1991)
OH 1.37E-011	cm3/molc sec	25	EST	MEYLAN,WM & HOWARD,PH (1993)

CAS #:	006149-23-1			1H-INDENE-1,3(2H)-DIONE, 2-(3-METHOXYPHENYL)-

Formula: $C_{16}H_{12}O_3$

Mol Weight: 252.27

MP (deg C):   FP (deg C):

BP (deg C):

BP pressure (mm Hg):

Property/Value	Units	Temp	Data Type	Reference
WS 7.98E+001	mg/L	25	EST	MEYLAN,WM ET AL. (1996)
logP 2.88			EXP	SANGSTER,J (1994)
VP 6.20E-007	mm Hg	25	EST	NEELY,WB & BLAU,GE (1985)
DC	pKa			
HL 2.83E-011	atm m3/mol	25	EST	MEYLAN,WM & HOWARD,PH (1991)
OH 5.35E-011	cm3/molc sec	25	EST	MEYLAN,WM & HOWARD,PH (1993)

CAS #:	006149-34-4			N-(4-NITROPHENYL)-1,4-BENZENDIAMINE

Formula: $C_{12}H_{11}N_3O_2$

Mol Weight: 229.24

MP (deg C):   FP (deg C):

BP (deg C):

BP pressure (mm Hg):

Property/Value	Units	Temp	Data Type	Reference
WS 3.91E+002	mg/L	25	EST	MEYLAN,WM ET AL. (1996)
logP 2.22			EST	MEYLAN,WM & HOWARD,PH (1995)
VP 6.89E-007	mm Hg	25	EST	NEELY,WB & BLAU,GE (1985)
DC 4.38	pKa	25	EXP	PERRIN,DD (1972)
HL 1.46E-012	atm m3/mol	25	EST	MEYLAN,WM & HOWARD,PH (1991)
OH 2.00E-010	cm3/molc sec	25	EST	MEYLAN,WM & HOWARD,PH (1993)

CAS #:	006153-44-2			4-PYRIMIDINECARBOXYLIC ACID, 1,2,3,6-TETRAHYDRO-

Formula: $C_6H_6N_2O_4$

Mol Weight: 170.13

MP (deg C):   FP (deg C):

BP (deg C):

BP pressure (mm Hg):

Property/Value	Units	Temp	Data Type	Reference
WS 1.78E+004	mg/L	25	EST	MEYLAN,WM ET AL. (1996)
logP 0.63			EXP	SANGSTER,J (1994)
VP 1.01E-007	mm Hg	25	EST	NEELY,WB & BLAU,GE (1985)
DC	pKa			
HL 2.59E-012	atm m3/mol	25	EST	MEYLAN,WM & HOWARD,PH (1991)
OH 8.60E-012	cm3/molc sec	25	EST	MEYLAN,WM & HOWARD,PH (1993)

CAS #:	006158-45-8			1-ISOPROPYLNAPHTHALENE

Formula: $C_{13}H_{14}$

Mol Weight: 170.26

MP (deg C): -16   FP (deg C):

BP (deg C): 268

BP pressure (mm Hg):

Property/Value	Units	Temp	Data Type	Reference
WS 6.89E+000	mg/L	25	EST	MEYLAN,WM ET AL. (1996)
logP 4.63			EST	MEYLAN,WM & HOWARD,PH (1995)
VP 6.06E-003	mm Hg	25	EXT	ZWOLINSKI,BJ & WILHOIT,RC (1971)
DC	pKa			
HL 1.02E-003	atm m3/mol	25	EST	MEYLAN,WM & HOWARD,PH (1991)
OH 5.35E-011	cm3/molc sec	25	EST	MEYLAN,WM & HOWARD,PH (1993)

CAS #: 006162-21-6	N',N'-DIMETHYLSULFANILAMIDE

Formula: $C_8H_{12}N_2O_2S$

Mol Weight: 200.26

MP (deg C): FP (deg C):

BP (deg C):

BP pressure (mm Hg):

Property/Value	Units	Temp	Data Type	Reference
WS 9.79E+003	mg/L	25	EST	MEYLAN,WM ET AL. (1996)
logP 0.76			EXP	HANSCH,C & LEO,AJ (1985)
VP 4.33E-005	mm Hg	25	EST	NEELY,WB & BLAU,GE (1985)
DC 1.54	pKa	24	EXP	PERRIN,DD (1965)
HL 6.71E-009	atm m3/mol	25	EST	MEYLAN,WM & HOWARD,PH (1991)
OH 8.16E-011	cm3/molc sec	25	EST	MEYLAN,WM & HOWARD,PH (1993)

CAS #: 006163-66-2	DI (TERT-BUTYL) ETHER

Formula: $C_8H_{18}O$

Mol Weight: 130.23

MP (deg C): FP (deg C):

BP (deg C): 107.2

BP pressure (mm Hg):

Property/Value	Units	Temp	Data Type	Reference
WS 3.79E+002	mg/L	25	EST	MEYLAN,WM ET AL. (1996)
logP 2.79			EST	MEYLAN,WM & HOWARD,PH (1995)
VP 3.26E+001	mm Hg	25	EXP	OHE,S (1976)
DC	pKa			
HL 4.72E-003	atm m3/mol	25	EST	MEYLAN,WM & HOWARD,PH (1991)
OH 2.86E-012	cm3/molc sec	25	EST	MEYLAN,WM & HOWARD,PH (1993)

CAS #: 006164-77-8	2,3-PYRAZINEDICARBOXYLIC ACID, DIMETHYL ESTER

Formula: $C_8H_8N_2O_4$

Mol Weight: 196.16

MP (deg C): FP (deg C):

BP (deg C):

BP pressure (mm Hg):

Property/Value	Units	Temp	Data Type	Reference
WS 4.76E+004	mg/L	25	EST	MEYLAN,WM ET AL. (1996)
logP -0.02			EXP	YAMAGAMI,C ET AL. (1991)
VP 8.61E-003	mm Hg	25	EST	NEELY,WB & BLAU,GE (1985)
DC	pKa			
HL 1.21E-010	atm m3/mol	25	EST	MEYLAN,WM & HOWARD,PH (1991)
OH 4.52E-013	cm3/molc sec	25	EST	MEYLAN,WM & HOWARD,PH (1993)

CAS #: 006164-79-0	2-METHYLPYRAZINE CARBOXYLATE

Formula: $C_6H_6N_2O_2$

Mol Weight: 138.13

MP (deg C): FP (deg C):

BP (deg C):

BP pressure (mm Hg):

Property/Value	Units	Temp	Data Type	Reference
WS 1.34E+005	mg/L	25	EST	MEYLAN,WM ET AL. (1996)
logP -0.23			EXP	YAMAGAMI,C ET AL. (1990A)
VP 1.38E-001	mm Hg	25	EST	NEELY,WB & BLAU,GE (1985)
DC	pKa			
HL 1.88E-008	atm m3/mol	25	EST	MEYLAN,WM & HOWARD,PH (1991)
OH 2.91E-013	cm3/molc sec	25	EST	MEYLAN,WM & HOWARD,PH (1993)

CAS #: 006164-98-3	CHLORDIMEFORM

Formula: $C_{10}H_{13}ClN_2$

Mol Weight: 196.68

MP (deg C): 35 FP (deg C):

BP (deg C): 156-157

BP pressure (mm Hg): 4.00E-001

Property/Value	Units	Temp	Data Type	Reference
WS 2.70E+002	mg/L	20	EXP	YALKOWSKY,SH & DANNENFELSER,RM (1992)
logP 2.89			EXP	HANSCH,C & LEO,AJ (1985)
VP 4.00E-007	mm Hg	25	EXP	WEBER,JB (1994)
DC	pKa			
HL 3.83E-010	atm m3/mol	25	EST	VP/WSOL
OH 9.06E-011	cm3/molc sec	25	EST	MEYLAN,WM & HOWARD,PH (1993)

CAS #: 006169-23-9	PHENOXYACETIC ACID, 3-UREA

Formula: $C_9H_{10}N_2O_4$

Mol Weight: 210.19

MP (deg C): FP (deg C):

BP (deg C):

BP pressure (mm Hg):

Property/Value	Units	Temp	Data Type	Reference
WS 1.14E+003	mg/L	25	EST	MEYLAN,WM ET AL. (1996)
logP 0.26			EXP	HANSCH,C & LEO,AJ (1985)
VP 2.23E-007	mm Hg	25	EST	NEELY,WB & BLAU,GE (1985)
DC	pKa			
HL 6.27E-016	atm m3/mol	25	EST	MEYLAN,WM & HOWARD,PH (1991)
OH 2.06E-010	cm3/molc sec	25	EST	MEYLAN,WM & HOWARD,PH (1993)

CAS #: 006174-95-4	ETHENE TETRACARBOXYLIC ACID,TETRAETHYL ESTER

Formula: $C_{14}H_{20}O_8$

Mol Weight: 316.31

MP (deg C): 58 FP (deg C):

BP (deg C): 203

BP pressure (mm Hg): 1.30E+001

Property/Value	Units	Temp	Data Type	Reference
WS 5.59E+001	mg/L	25	EST	MEYLAN,WM ET AL. (1996)
logP 2.63			EXP	HANSCH,C ET AL. (1995)
VP 1.32E-004	mm Hg	25	EST	NEELY,WB & BLAU,GE (1985)
DC	pKa			
HL 9.66E-013	atm m3/mol	25	EST	MEYLAN,WM & HOWARD,PH (1991)
OH 8.30E-012	cm3/molc sec	25	EST	MEYLAN,WM & HOWARD,PH (1993)

CAS #: 006175-49-1	METHYL DECYL KETONE

Formula: $C_{12}H_{24}O$

Mol Weight: 184.32

MP (deg C): 21 FP (deg C):

BP (deg C): 246.5

BP pressure (mm Hg):

Property/Value	Units	Temp	Data Type	Reference
WS 1.40E+001	mg/L	25	EST	MEYLAN,WM ET AL. (1996)
logP 4.18			EST	MEYLAN,WM & HOWARD,PH (1995)
VP 2.06E-002	mm Hg	25	EXT	PERRY,RH & GREEN,D (1984)
DC	pKa			
HL 6.35E-004	atm m3/mol	25	EST	MEYLAN,WM & HOWARD,PH (1991)
OH 1.52E-011	cm3/molc sec	25	EST	MEYLAN,WM & HOWARD,PH (1993)

CAS #: 006178-42-3				B-NITRO-4-OH-3-METHOXYSTYRENE

Formula: $C_9H_9NO_4$
Mol Weight: 195.18
MP (deg C): 168-170
FP (deg C):
BP (deg C):
BP pressure (mm Hg):

Property/Value	Units	Temp	Data Type	Reference
WS 4.68E+002	mg/L	25	EST	MEYLAN,WM ET AL. (1996)
logP 1.88			EXP	HANSCH,C & LEO,AJ (1985)
VP 1.83E-005	mm Hg	25	EST	NEELY,WB & BLAU,GE (1985)
DC	pKa			
HL 2.14E-011	atm m3/mol	25	EST	MEYLAN,WM & HOWARD,PH (1991)
OH 3.95E-011	cm3/molc sec	25	EST	MEYLAN,WM & HOWARD,PH (1993)

CAS #: 006187-24-2				3-NITRO-3-HEPTENE

Formula: $C_7H_{13}NO_2$
Mol Weight: 143.19
MP (deg C):
FP (deg C):
BP (deg C):
BP pressure (mm Hg):

Property/Value	Units	Temp	Data Type	Reference
WS 1.48E+002	mg/L	25	EST	MEYLAN,WM ET AL. (1996)
logP 2.75			EST	MEYLAN,WM & HOWARD,PH (1995)
VP 3.19E-001	mm Hg	25	EST	NEELY,WB & BLAU,GE (1985)
DC	pKa			
HL 4.46E-004	atm m3/mol	25	EST	MEYLAN,WM & HOWARD,PH (1991)
OH 2.11E-011	cm3/molc sec	25	EST	MEYLAN,WM & HOWARD,PH (1993)

CAS #: 006190-65-4				DESETHYLATRAZINE

Formula: $C_6H_{10}ClN_5$
Mol Weight: 187.63
MP (deg C):
FP (deg C):
BP (deg C):
BP pressure (mm Hg):

Property/Value	Units	Temp	Data Type	Reference
WS 3.20E+003	mg/L	22	EXP	MILLS,MS & THURMAN,EM (1994A)
logP 1.51			EXP	FINIZIO,A ET AL. (1991)
VP 1.49E-004	mm Hg	25	EST	NEELY,WB & BLAU,GE (1985)
DC	pKa			
HL 1.53E-009	atm m3/mol	25	EST	MEYLAN,WM & HOWARD,PH (1991)
OH 1.85E-011	cm3/molc sec	25	EST	MEYLAN,WM & HOWARD,PH (1993)

CAS #: 006231-18-1				PYRIDINE, 2,6-DIMETHOXY-

Formula: $C_7H_9NO_2$
Mol Weight: 139.16
MP (deg C):
FP (deg C):
BP (deg C): 179
BP pressure (mm Hg):

Property/Value	Units	Temp	Data Type	Reference
WS 9.17E+002	mg/L	25	EST	MEYLAN,WM ET AL. (1996)
logP 2.30			EXP	HANSCH,C ET AL. (1995)
VP 2.89E-001	mm Hg	25	EST	NEELY,WB & BLAU,GE (1985)
DC 1.60	pKa		EXP	PERRIN,DD (1972)
HL 2.05E-006	atm m3/mol	25	EST	MEYLAN,WM & HOWARD,PH (1991)
OH 4.67E-011	cm3/molc sec	25	EST	MEYLAN,WM & HOWARD,PH (1993)

CAS #: 006232-32-2				N-PHENYLANTHRANILIC ACID, SODIUM SALT

Formula: $C_{13}H_{10}NNaO_2$
Mol Weight: 235.22
MP (deg C):
FP (deg C):
BP (deg C):
BP pressure (mm Hg):

Property/Value	Units	Temp	Data Type	Reference
WS 2.74E+004	mg/L	25	EST	MEYLAN,WM ET AL. (1996)
logP 0.02			EXP	HANSCH,C & LEO,AJ (1985)
VP 1.72E-012	mm Hg	25	EST	NEELY,WB & BLAU,GE (1985)
DC	pKa			
HL	atm m3/mol			
OH 2.00E-010	cm3/molc sec	25	EST	MEYLAN,WM & HOWARD,PH (1993)

CAS #: 006232-56-0				C.I. DISPERSE ORANGE 5

Formula: $C_{15}H_{14}Cl_2N_4O_3$
Mol Weight: 369.21
MP (deg C): 127
FP (deg C):
BP (deg C):
BP pressure (mm Hg):

Property/Value	Units	Temp	Data Type	Reference
WS 1.59E-001	mg/L	25	EXP	BAUGHMAN,GL & PERENICH,TA (1988)
logP 5.00			EST	MEYLAN,WM & HOWARD,PH (1995)
VP 7.56E-012	mm Hg	25	EST	NEELY,WB & BLAU,GE (1985)
DC	pKa			
HL 2.46E-014	atm m3/mol	25	EST	MEYLAN,WM & HOWARD,PH (1991)
OH 7.42E-011	cm3/molc sec	25	EST	MEYLAN,WM & HOWARD,PH (1993)

CAS #: 006232-88-8				4-(BROMOMETHYL)BENZOIC ACID

Formula: $C_8H_7BrO_2$
Mol Weight: 215.05
MP (deg C): 228-232
FP (deg C):
BP (deg C):
BP pressure (mm Hg):

Property/Value	Units	Temp	Data Type	Reference
WS 1.59E+002	mg/L	25	EST	MEYLAN,WM ET AL. (1996)
logP 2.77			EST	MEYLAN,WM & HOWARD,PH (1995)
VP 1.79E-004	mm Hg	25	EST	NEELY,WB & BLAU,GE (1985)
DC	pKa			
HL 1.38E-008	atm m3/mol	25	EST	MEYLAN,WM & HOWARD,PH (1991)
OH 1.53E-012	cm3/molc sec	25	EST	MEYLAN,WM & HOWARD,PH (1993)

CAS #: 006232-91-3				2-AMINO-5-METHOXYBENZIMIDAZOLE

Formula: $C_8H_9N_3O$
Mol Weight: 163.18
MP (deg C):
FP (deg C):
BP (deg C):
BP pressure (mm Hg):

Property/Value	Units	Temp	Data Type	Reference
WS 2.16E+004	mg/L	25	EST	MEYLAN,WM ET AL. (1996)
logP 0.57			EXP	HANSCH,C ET AL. (1995)
VP 4.41E-007	mm Hg	25	EST	NEELY,WB & BLAU,GE (1985)
DC	pKa			
HL 7.66E-012	atm m3/mol	25	EST	MEYLAN,WM & HOWARD,PH (1991)
OH 2.01E-010	cm3/molc sec	25	EST	MEYLAN,WM & HOWARD,PH (1993)

CAS #: 006236-05-1				NIFUROXIME

Formula: $C_5H_5N_2O_4$

Mol Weight: 157.11

MP (deg C): 163-164 **FP (deg C):**

BP (deg C):

BP pressure (mm Hg):

Property/Value	Units	Temp	Data Type	Reference
WS 1.61E+004	mg/L	25	EST	MEYLAN,WM ET AL. (1996)
logP 0.30			EXP	HANSCH,C & LEO,AJ (1985)
VP 5.09E-005	mm Hg	25	EST	NEELY,WB & BLAU,GE (1985)
DC	pKa			
HL 1.41E-009	atm m3/mol	25	EST	MEYLAN,WM & HOWARD,PH (1991)
OH 1.39E-011	cm3/molc sec	25	EST	MEYLAN,WM & HOWARD,PH (1993)

CAS #: 006250-23-3				C.I. DISPERSE YELLOW 23

Formula: $C_{18}H_{14}N_4O$

Mol Weight: 302.34

MP (deg C): **FP (deg C):**

BP (deg C):

BP pressure (mm Hg):

Property/Value	Units	Temp	Data Type	Reference
WS 6.04E-005	mg/L	25	EXP	BAUGHMAN,GL & PERENICH,TA (1988)
logP 5.75			EST	MEYLAN,WM & HOWARD,PH (1995)
VP 2.41E-009	mm Hg	25	EST	NEELY,WB & BLAU,GE (1985)
DC	pKa			
HL 4.18E-012	atm m3/mol	25	EST	MEYLAN,WM & HOWARD,PH (1991)
OH 1.42E-011	cm3/molc sec	25	EST	MEYLAN,WM & HOWARD,PH (1993)

CAS #: 006253-10-7				C.I. DISPERSE ORANGE 13

Formula: $C_{22}H_{16}N_4O$

Mol Weight: 352.40

MP (deg C): 210.5 **FP (deg C):**

BP (deg C):

BP pressure (mm Hg):

Property/Value	Units	Temp	Data Type	Reference
WS 3.45E-001	mg/L	25	EXP	KUROIWA,S & OGASAWARA,S (1973)
logP 6.93			EST	MEYLAN,WM & HOWARD,PH (1995)
VP 9.20E-012	mm Hg	25	EST	NEELY,WB & BLAU,GE (1985)
DC	pKa			
HL 4.08E-013	atm m3/mol	25	EST	MEYLAN,WM & HOWARD,PH (1991)
OH 1.53E-011	cm3/molc sec	25	EST	MEYLAN,WM & HOWARD,PH (1993)

CAS #: 006265-73-2				N-HYDROXYETHYLNICOTINAMIDE

Formula: $C_8H_{10}N_2O_2$

Mol Weight: 166.18

MP (deg C): **FP (deg C):**

BP (deg C):

BP pressure (mm Hg):

Property/Value	Units	Temp	Data Type	Reference
WS 7.96E+004	mg/L	25	EST	MEYLAN,WM ET AL. (1996)
logP -0.11			EXP	HANSCH,C & LEO,AJ (1985)
VP 1.34E-007	mm Hg	25	EST	NEELY,WB & BLAU,GE (1985)
DC	pKa			
HL 3.09E-016	atm m3/mol	25	EST	MEYLAN,WM & HOWARD,PH (1991)
OH 1.47E-011	cm3/molc sec	25	EST	MEYLAN,WM & HOWARD,PH (1993)

CAS #: 006268-49-1				4-((4-(DIMETHYLAMINO)PHENYL)AZO)BENZOIC ACID

Formula: $C_{15}H_{15}N_3O_2$

Mol Weight: 269.31

MP (deg C): **FP (deg C):**

BP (deg C):

BP pressure (mm Hg):

Property/Value	Units	Temp	Data Type	Reference
WS 1.87E+000	mg/L	25	EST	MEYLAN,WM ET AL. (1996)
logP 4.17			EST	MEYLAN,WM & HOWARD,PH (1995)
VP 1.44E-007	mm Hg	25	EST	NEELY,WB & BLAU,GE (1985)
DC 2.35	pKa		EXP	PERRIN,DD (1965)
HL 4.71E-012	atm m3/mol	25	EST	MEYLAN,WM & HOWARD,PH (1991)
OH 1.50E-010	cm3/molc sec	25	EST	MEYLAN,WM & HOWARD,PH (1993)

CAS #: 006270-63-9				PYRAZINE-2-ONE

Formula: $C_4H_4N_2O$

Mol Weight: 96.09

MP (deg C): **FP (deg C):**

BP (deg C):

BP pressure (mm Hg):

Property/Value	Units	Temp	Data Type	Reference
WS 1.10E+005	mg/L	25	EST	MEYLAN,WM ET AL. (1996)
logP -1.49			EXP	HANSCH,C & LEO,AJ (1985)
VP 2.77E-003	mm Hg	25	EST	NEELY,WB & BLAU,GE (1985)
DC	pKa			
HL 3.92E-008	atm m3/mol	25	EST	MEYLAN,WM & HOWARD,PH (1991)
OH 2.73E-011	cm3/molc sec	25	EST	MEYLAN,WM & HOWARD,PH (1993)

CAS #: 006274-12-0				DIETHYLAMINE, HYDROBROMIDE

Formula: $C_4H_{12}BrN$

Mol Weight: 154.05

MP (deg C): 218-220 **FP (deg C):**

BP (deg C):

BP pressure (mm Hg):

Property/Value	Units	Temp	Data Type	Reference
WS 1.00E+006	mg/L	25	EXP	SEIDELL,A (1941)
logP -2.29			EST	MEYLAN,WM & HOWARD,PH (1995)
VP 1.34E-005	mm Hg	25	EST	NEELY,WB & BLAU,GE (1985)
DC	pKa			
HL 1.78E-012	atm m3/mol	25	EST	MEYLAN,WM & HOWARD,PH (1991)
OH 1.77E-011	cm3/molc sec	25	EST	MEYLAN,WM & HOWARD,PH (1993)

CAS #: 006280-88-2				2-NITRO-4-CHLOROBENZOIC ACID

Formula: $C_7H_4ClNO_4$

Mol Weight: 201.57

MP (deg C): 143-145 **FP (deg C):**

BP (deg C):

BP pressure (mm Hg):

Property/Value	Units	Temp	Data Type	Reference
WS 2.93E+002	mg/L	25	EST	MEYLAN,WM ET AL. (1996)
logP 2.08			EXP	HANSCH,C & LEO,AJ (1985)
VP 1.89E-005	mm Hg	25	EST	NEELY,WB & BLAU,GE (1985)
DC	pKa			
HL 3.17E-010	atm m3/mol	25	EST	MEYLAN,WM & HOWARD,PH (1991)
OH 5.64E-013	cm3/molc sec	25	EST	MEYLAN,WM & HOWARD,PH (1993)

CAS #: 006282-02-6	BENZAMIDE, N-(HYDROXYMETHYL)-

Formula: $C_8H_9NO_2$

Mol Weight: 151.17

MP (deg C):　　FP (deg C):

BP (deg C):

BP pressure (mm Hg):

Property/Value	Units	Temp	Data Type	Reference
WS 3.42E+004	mg/L	25	EST	MEYLAN,WM ET AL. (1996)
logP 0.40			EXP	SANGSTER,J (1994)
VP 6.25E-007	mm Hg	25	EST	NEELY,WB & BLAU,GE (1985)
DC	pKa			
HL 3.55E-012	atm m3/mol	25	EST	MEYLAN,WM & HOWARD,PH (1991)
OH 2.08E-011	cm3/molc sec	25	EST	MEYLAN,WM & HOWARD,PH (1993)

CAS #: 006285-68-3	2-AMINO-5-METHYLBENZIMIDAZOLE

Formula: $C_8H_9N_3$

Mol Weight: 147.18

MP (deg C):　　FP (deg C):

BP (deg C):

BP pressure (mm Hg):

Property/Value	Units	Temp	Data Type	Reference
WS 8.80E+003	mg/L	25	EST	MEYLAN,WM ET AL. (1996)
logP 1.11			EXP	HANSCH,C ET AL. (1995)
VP 1.14E-006	mm Hg	25	EST	NEELY,WB & BLAU,GE (1985)
DC	pKa			
HL 1.43E-010	atm m3/mol	25	EST	MEYLAN,WM & HOWARD,PH (1991)
OH 2.00E-010	cm3/molc sec	25	EST	MEYLAN,WM & HOWARD,PH (1993)

CAS #: 006287-38-3	3,4-DICHLOROBENZALDEHYDE

Formula: $C_7H_4Cl_2O$

Mol Weight: 175.02

MP (deg C): 44　　FP (deg C):

BP (deg C): 247.5

BP pressure (mm Hg):

Property/Value	Units	Temp	Data Type	Reference
WS 1.60E+002	mg/L	25	EST	MEYLAN,WM ET AL. (1996)
logP 3.00			EST	MEYLAN,WM & HOWARD,PH (1995)
VP 2.81E-002	mm Hg	25	EST	NEELY,WB & BLAU,GE (1985)
DC	pKa			
HL 7.37E-006	atm m3/mol	25	EST	MEYLAN,WM & HOWARD,PH (1991)
OH 1.71E-011	cm3/molc sec	25	EST	MEYLAN,WM & HOWARD,PH (1993)

CAS #: 006292-90-6	L-TYROSINE,BUTYL ESTER

Formula: $C_{13}H_{19}NO_3$

Mol Weight: 237.30

MP (deg C):　　FP (deg C):

BP (deg C):

BP pressure (mm Hg):

Property/Value	Units	Temp	Data Type	Reference
WS 3.12E+004	mg/L	25	EST	MEYLAN,WM ET AL. (1996)
logP 1.80			EXP	HUANG,CH ET AL. (1985)
VP 3.29E-006	mm Hg	25	EST	NEELY,WB & BLAU,GE (1985)
DC	pKa			
HL 9.39E-012	atm m3/mol	25	EST	MEYLAN,WM & HOWARD,PH (1991)
OH 8.39E-011	cm3/molc sec	25	EST	MEYLAN,WM & HOWARD,PH (1993)

CAS #: 006293-55-6	2-METHYL-6-BROMOBENZOQUINONE

Formula: $C_7H_5BrO_2$

Mol Weight: 201.03

MP (deg C):　　FP (deg C):

BP (deg C):

BP pressure (mm Hg):

Property/Value	Units	Temp	Data Type	Reference
WS 4.00E+003	mg/L	25	EST	MEYLAN,WM ET AL. (1996)
logP 1.21			EXP	HANSCH,C ET AL. (1995)
VP 1.95E-003	mm Hg	25	EST	NEELY,WB & BLAU,GE (1985)
DC	pKa			
HL 3.81E-010	atm m3/mol	25	EST	MEYLAN,WM & HOWARD,PH (1991)
OH 1.35E-011	cm3/molc sec	25	EST	MEYLAN,WM & HOWARD,PH (1993)

CAS #: 006293-56-7	3-PYRIDINEETHANOL

Formula: C_7H_9NO

Mol Weight: 123.16

MP (deg C):　　FP (deg C):

BP (deg C):

BP pressure (mm Hg):

Property/Value	Units	Temp	Data Type	Reference
WS 1.00E+006	mg/L	25	EST	MEYLAN,WM ET AL. (1996)
logP 0.12			EXP	HANSCH,C & LEO,AJ (1985)
VP 6.59E-003	mm Hg	25	EST	NEELY,WB & BLAU,GE (1985)
DC	pKa			
HL 3.78E-010	atm m3/mol	25	EST	MEYLAN,WM & HOWARD,PH (1991)
OH 6.23E-012	cm3/molc sec	25	EST	MEYLAN,WM & HOWARD,PH (1993)

CAS #: 006294-92-4	N-PROPYLQUINOLINIUM BROMIDE

Formula: $C_{12}H_{14}BrN$

Mol Weight: 252.16

MP (deg C):　　FP (deg C):

BP (deg C):

BP pressure (mm Hg):

Property/Value	Units	Temp	Data Type	Reference
WS 1.00E+006	mg/L	25	EST	MEYLAN,WM ET AL. (1996)
logP -2.52			EXP	HANSCH,C & LEO,AJ (1985)
VP 1.13E-004	mm Hg	25	EST	NEELY,WB & BLAU,GE (1985)
DC	pKa			
HL 2.59E-011	atm m3/mol	25	EST	MEYLAN,WM & HOWARD,PH (1991)
OH 2.69E-011	cm3/molc sec	25	EST	MEYLAN,WM & HOWARD,PH (1993)

CAS #: 006298-37-9	6-AMINOQUINOXALINE

Formula: $C_8H_7N_3$

Mol Weight: 145.17

MP (deg C): 159　　FP (deg C):

BP (deg C):

BP pressure (mm Hg):

Property/Value	Units	Temp	Data Type	Reference
WS 1.36E+004	mg/L	25	EST	MEYLAN,WM ET AL. (1996)
logP 0.90			EXP	HANSCH,C & LEO,AJ (1985)
VP 2.50E-004	mm Hg	25	EST	NEELY,WB & BLAU,GE (1985)
DC	pKa			
HL 1.01E-010	atm m3/mol	25	EST	MEYLAN,WM & HOWARD,PH (1991)
OH 1.10E-010	cm3/molc sec	25	EST	MEYLAN,WM & HOWARD,PH (1993)

CAS #: 006300-37-4 — C.I. DISPERSE YELLOW 7

Formula: $C_{19}H_{16}N_4O$

Mol Weight: 316.37

MP (deg C): | FP (deg C):

BP (deg C):

BP pressure (mm Hg):

	Property/Value	Units	Temp	Data Type	Reference
WS	5.81E-002	mg/L	25	EST	MEYLAN,WM ET AL. (1996)
logP	6.30			EST	MEYLAN,WM & HOWARD,PH (1995)
VP	1.03E-009	mm Hg	25	EST	NEELY,WB & BLAU,GE (1985)
DC		pKa			
HL	4.61E-012	atm m3/mol	25	EST	MEYLAN,WM & HOWARD,PH (1991)
OH	1.73E-011	cm3/molc sec	25	EST	MEYLAN,WM & HOWARD,PH (1993)

CAS #: 006304-27-4 — 2-PYRIDINEETHANAMINE, N,N-DIMETHYL-

Formula: $C_9H_{14}N_2$

Mol Weight: 150.23

MP (deg C): | FP (deg C):

BP (deg C):

BP pressure (mm Hg):

	Property/Value	Units	Temp	Data Type	Reference
WS	1.00E+006	mg/L	25	EST	MEYLAN,WM ET AL. (1996)
logP	1.12			EXP	YOUNG,RC ET AL. (1993)
VP	1.12E-001	mm Hg	25	EST	NEELY,WB & BLAU,GE (1985)
DC	8.75	pKa	25	EXP	PERRIN,DD (1965)
HL	5.12E-009	atm m3/mol	25	EST	MEYLAN,WM & HOWARD,PH (1991)
OH	8.13E-011	cm3/molc sec	25	EST	MEYLAN,WM & HOWARD,PH (1993)

CAS #: 006304-39-8 — OCTANOIC ACID HYDRAZIDE

Formula: $C_8H_{18}N_2O$

Mol Weight: 158.25

MP (deg C): 87-89 | FP (deg C):

BP (deg C):

BP pressure (mm Hg):

	Property/Value	Units	Temp	Data Type	Reference
WS	9.76E+003	mg/L	25	EST	MEYLAN,WM ET AL. (1996)
logP	1.00			EXP	HANSCH,C & LEO,AJ (1985)
VP	1.06E-004	mm Hg	25	EST	NEELY,WB & BLAU,GE (1985)
DC		pKa			
HL	2.56E-010	atm m3/mol	25	EST	MEYLAN,WM & HOWARD,PH (1991)
OH	1.64E-011	cm3/molc sec	25	EST	MEYLAN,WM & HOWARD,PH (1993)

CAS #: 006305-71-1 — 2,4-DIMETHYL-1-PENTANOL

Formula: $C_7H_{16}O$

Mol Weight: 116.20

MP (deg C): | FP (deg C):

BP (deg C):

BP pressure (mm Hg):

	Property/Value	Units	Temp	Data Type	Reference
WS	2.91E+003	mg/L	25	EXP	SUZUKI,T (1991)
logP	2.17			EST	MEYLAN,WM & HOWARD,PH (1995)
VP	1.04E+000	mm Hg	25	EST	NEELY,WB & BLAU,GE (1985)
DC		pKa			
HL	2.34E-005	atm m3/mol	25	EST	MEYLAN,WM & HOWARD,PH (1991)
OH	1.14E-011	cm3/molc sec	25	EST	MEYLAN,WM & HOWARD,PH (1993)

CAS #: 006306-24-7 — BENZAMIDE, 4-(AMINOSULFONYL)-

Formula: $C_7H_8N_2O_3S$

Mol Weight: 200.22

MP (deg C): | FP (deg C):

BP (deg C):

BP pressure (mm Hg):

	Property/Value	Units	Temp	Data Type	Reference
WS	1.09E+004	mg/L	25	EST	MEYLAN,WM ET AL. (1996)
logP	-0.83			EXP	CAROTTI,A ET AL. (1989)
VP	1.39E-007	mm Hg	25	EST	NEELY,WB & BLAU,GE (1985)
DC		pKa			
HL	1.74E-013	atm m3/mol	25	EST	MEYLAN,WM & HOWARD,PH (1991)
OH	2.38E-012	cm3/molc sec	25	EST	MEYLAN,WM & HOWARD,PH (1993)

CAS #: 006306-25-8 — BENZOIC ACID, 4-[(AMINOCARBONYL)AMINO]-

Formula: $C_8H_8N_2O_3$

Mol Weight: 180.16

MP (deg C): | FP (deg C):

BP (deg C):

BP pressure (mm Hg):

	Property/Value	Units	Temp	Data Type	Reference
WS	1.31E+004	mg/L	25	EST	MEYLAN,WM ET AL. (1996)
logP	0.73			EXP	SANGSTER,J (1994)
VP	1.16E-006	mm Hg	25	EST	NEELY,WB & BLAU,GE (1985)
DC		pKa			
HL	4.05E-015	atm m3/mol	25	EST	MEYLAN,WM & HOWARD,PH (1991)
OH	1.73E-011	cm3/molc sec	25	EST	MEYLAN,WM & HOWARD,PH (1993)

CAS #: 006310-41-4 — ACETAMIDE, N-[2-(METHYLTHIO)PHENYL]-

Formula: $C_9H_{11}NOS$

Mol Weight: 181.26

MP (deg C): | FP (deg C):

BP (deg C):

BP pressure (mm Hg):

	Property/Value	Units	Temp	Data Type	Reference
WS	2.68E+003	mg/L	25	EST	MEYLAN,WM ET AL. (1996)
logP	1.53			EXP	HANSCH,C ET AL. (1995)
VP	1.46E-005	mm Hg	25	EST	NEELY,WB & BLAU,GE (1985)
DC		pKa			
HL	1.80E-010	atm m3/mol	25	EST	MEYLAN,WM & HOWARD,PH (1991)
OH	8.79E-012	cm3/molc sec	25	EST	MEYLAN,WM & HOWARD,PH (1993)

CAS #: 006311-44-0 — 2,2',5,5'-TETRAMETHYLAZOBENZENE

Formula: $C_{16}H_{18}N_2$

Mol Weight: 238.34

MP (deg C): | FP (deg C):

BP (deg C):

BP pressure (mm Hg):

	Property/Value	Units	Temp	Data Type	Reference
WS	4.22E-002	mg/L	25	EST	MEYLAN,WM ET AL. (1996)
logP	6.30			EST	MEYLAN,WM & HOWARD,PH (1995)
VP	3.30E-005	mm Hg	25	EST	NEELY,WB & BLAU,GE (1985)
DC		pKa			
HL	2.18E-005	atm m3/mol	25	EST	MEYLAN,WM & HOWARD,PH (1991)
OH	5.49E-012	cm3/molc sec	25	EST	MEYLAN,WM & HOWARD,PH (1993)

CAS #: 006312-87-4				ACETAMIDE, N-(4-PHENOXYPHENYL)-

Formula: $C_{14}H_{13}NO_2$

Mol Weight: 227.27

MP (deg C): | FP (deg C):

BP (deg C):

BP pressure (mm Hg):

Property/ Value	Units	Temp	Data Type	Reference
WS 6.68E+001	mg/L	25	EST	MEYLAN,WM ET AL. (1996)
logP 3.13			EXP	HANSCH,C ET AL. (1995)
VP 3.74E-007	mm Hg	25	EST	NEELY,WB & BLAU,GE (1985)
DC	pKa			
HL 1.35E-010	atm m3/mol	25	EST	MEYLAN,WM & HOWARD,PH (1991)
OH 1.74E-011	cm3/molc sec	25	EST	MEYLAN,WM & HOWARD,PH (1993)

CAS #: 006317-89-1				M-ACETOXYACETANILIDE

Formula: $C_{10}H_{11}NO_3$

Mol Weight: 193.20

MP (deg C): | FP (deg C):

BP (deg C):

BP pressure (mm Hg):

Property/ Value	Units	Temp	Data Type	Reference
WS 6.25E+003	mg/L	25	EST	MEYLAN,WM ET AL. (1996)
logP 1.03			EXP	HANSCH,C & LEO,AJ (1985)
VP 9.60E-006	mm Hg	25	EST	NEELY,WB & BLAU,GE (1985)
DC	pKa			
HL 7.42E-011	atm m3/mol	25	EST	MEYLAN,WM & HOWARD,PH (1991)
OH 1.30E-011	cm3/molc sec	25	EST	MEYLAN,WM & HOWARD,PH (1993)

CAS #: 006318-51-0				4-(4-CHLOROBENZOYL)PYRIDINE

Formula: $C_{12}H_8ClNO$

Mol Weight: 217.66

MP (deg C): | FP (deg C):

BP (deg C):

BP pressure (mm Hg):

Property/ Value	Units	Temp	Data Type	Reference
WS 2.09E+002	mg/L	25	EST	MEYLAN,WM ET AL. (1996)
logP 2.61			EXP	HANSCH,C & LEO,AJ (1985)
VP 7.94E-005	mm Hg	25	EST	NEELY,WB & BLAU,GE (1985)
DC	pKa			
HL 1.88E-009	atm m3/mol	25	EST	MEYLAN,WM & HOWARD,PH (1991)
OH 1.59E-012	cm3/molc sec	25	EST	MEYLAN,WM & HOWARD,PH (1993)

CAS #: 006319-21-7				(2-METHOXYPHENYL)PHENYLDIAZENE

Formula: $C_{13}H_{12}N_2O$

Mol Weight: 212.25

MP (deg C): 41 | FP (deg C):

BP (deg C): 195-197

BP pressure (mm Hg): 1.40E+001

Property/ Value	Units	Temp	Data Type	Reference
WS 3.68E+000	mg/L	25	EST	MEYLAN,WM ET AL. (1996)
logP 4.19			EST	MEYLAN,WM & HOWARD,PH (1995)
VP 2.56E-004	mm Hg	25	EST	NEELY,WB & BLAU,GE (1985)
DC	pKa			
HL 8.70E-007	atm m3/mol	25	EST	MEYLAN,WM & HOWARD,PH (1991)
OH 1.01E-011	cm3/molc sec	25	EST	MEYLAN,WM & HOWARD,PH (1993)

CAS #: 006319-23-9				3,3'-DIMETHOXYAZOBENZENE

Formula: $C_{14}H_{14}N_2O_2$

Mol Weight: 242.28

MP (deg C): | FP (deg C):

BP (deg C):

BP pressure (mm Hg):

Property/ Value	Units	Temp	Data Type	Reference
WS 2.17E+000	mg/L	25	EST	MEYLAN,WM ET AL. (1996)
logP 4.27			EST	MEYLAN,WM & HOWARD,PH (1995)
VP 3.24E-005	mm Hg	25	EST	NEELY,WB & BLAU,GE (1985)
DC	pKa			
HL 5.15E-008	atm m3/mol	25	EST	MEYLAN,WM & HOWARD,PH (1991)
OH 7.63E-012	cm3/molc sec	25	EST	MEYLAN,WM & HOWARD,PH (1993)

CAS #: 006319-26-2				(2,6-DIMETHYLPHENYL)(2-METHYLPHENYL)DIAZENE

Formula: $C_{15}H_{16}N_2$

Mol Weight: 224.31

MP (deg C): | FP (deg C):

BP (deg C):

BP pressure (mm Hg):

Property/ Value	Units	Temp	Data Type	Reference
WS 1.48E-001	mg/L	25	EST	MEYLAN,WM ET AL. (1996)
logP 5.75			EST	MEYLAN,WM & HOWARD,PH (1995)
VP 9.09E-005	mm Hg	25	EST	NEELY,WB & BLAU,GE (1985)
DC	pKa			
HL 1.98E-005	atm m3/mol	25	EST	MEYLAN,WM & HOWARD,PH (1991)
OH 7.69E-012	cm3/molc sec	25	EST	MEYLAN,WM & HOWARD,PH (1993)

CAS #: 006321-94-4				P-CYANOFORMANILIDE

Formula: $C_8H_6N_2O$

Mol Weight: 146.15

MP (deg C): | FP (deg C):

BP (deg C):

BP pressure (mm Hg):

Property/ Value	Units	Temp	Data Type	Reference
WS 5.12E+003	mg/L	25	EST	MEYLAN,WM ET AL. (1996)
logP 1.08			EXP	HANSCH,C & LEO,AJ (1985)
VP 2.06E-005	mm Hg	25	EST	NEELY,WB & BLAU,GE (1985)
DC	pKa			
HL 8.16E-011	atm m3/mol	25	EST	MEYLAN,WM & HOWARD,PH (1991)
OH 7.53E-012	cm3/molc sec	25	EST	MEYLAN,WM & HOWARD,PH (1993)

CAS #: 006324-11-4				O-HYDROXYPHENOXYACETIC ACID

Formula: $C_8H_8O_4$

Mol Weight: 168.15

MP (deg C): | FP (deg C):

BP (deg C):

BP pressure (mm Hg):

Property/ Value	Units	Temp	Data Type	Reference
WS 2.93E+004	mg/L	25	EST	MEYLAN,WM ET AL. (1996)
logP 0.85			EXP	HANSCH,C & LEO,AJ (1985)
VP 1.85E-005	mm Hg	25	EST	NEELY,WB & BLAU,GE (1985)
DC 3.02	pKa	25	EXP	SERJEANT,EP & DEMPSEY,B (1979)
HL 1.75E-012	atm m3/mol	25	EST	MEYLAN,WM & HOWARD,PH (1991)
OH 3.38E-011	cm3/molc sec	25	EST	MEYLAN,WM & HOWARD,PH (1993)

CAS #: 006324-18-1				B-PHENYLETHYLPYRIDINIUM BROMIDE

Formula: $C_{13}H_{14}BrN$

Mol Weight: 264.17

MP (deg C): FP (deg C):

BP (deg C):

BP pressure (mm Hg):

Property/Value	Units	Temp	Data Type	Reference
WS 1.00E+006	mg/L	25	EST	MEYLAN,WM ET AL. (1996)
logP -2.35			EXP	HANSCH,C & LEO,AJ (1985)
VP 1.11E-004	mm Hg	25	EST	NEELY,WB & BLAU,GE (1985)
DC	pKa			
HL 1.61E-011	atm m3/mol	25	EST	MEYLAN,WM & HOWARD,PH (1991)
OH 1.75E-011	cm3/molc sec	25	EST	MEYLAN,WM & HOWARD,PH (1993)

CAS #: 006325-93-5				4-NITROBENZENESULFONAMIDE

Formula: $C_6H_6N_2O_4S$

Mol Weight: 202.19

MP (deg C): 178-180 FP (deg C):

BP (deg C):

BP pressure (mm Hg):

Property/Value	Units	Temp	Data Type	Reference
WS 4.94E+003	mg/L	25	EST	MEYLAN,WM ET AL. (1996)
logP 0.64			EXP	HANSCH,C & LEO,AJ (1985)
VP 7.82E-006	mm Hg	25	EST	NEELY,WB & BLAU,GE (1985)
DC	pKa			
HL 1.67E-009	atm m3/mol	25	EST	MEYLAN,WM & HOWARD,PH (1991)
OH 5.21E-014	cm3/molc sec	25	EST	MEYLAN,WM & HOWARD,PH (1993)

CAS #: 006326-19-8				CARBAMIC ACID, 2-PHENYLETHYL ESTER

Formula: $C_9H_{11}NO_2$

Mol Weight: 165.19

MP (deg C): FP (deg C):

BP (deg C):

BP pressure (mm Hg):

Property/Value	Units	Temp	Data Type	Reference
WS 3.39E+003	mg/L	25	EST	MEYLAN,WM ET AL. (1996)
logP 1.50			EXP	TANAKA,M ET AL. (1985)
VP 7.07E-003	mm Hg	25	EST	NEELY,WB & BLAU,GE (1985)
DC	pKa			
HL 4.24E-009	atm m3/mol	25	EST	MEYLAN,WM & HOWARD,PH (1991)
OH 9.83E-012	cm3/molc sec	25	EST	MEYLAN,WM & HOWARD,PH (1993)

CAS #: 006331-45-9				DIETHYLMALONATE,3-NO2 BENZAL

Formula: $C_{14}H_{15}NO_6$

Mol Weight: 293.28

MP (deg C): FP (deg C):

BP (deg C):

BP pressure (mm Hg):

Property/Value	Units	Temp	Data Type	Reference
WS 2.65E+001	mg/L	25	EST	MEYLAN,WM ET AL. (1996)
logP 2.71			EXP	HANSCH,C & LEO,AJ (1985)
VP 8.19E-006	mm Hg	25	EST	NEELY,WB & BLAU,GE (1985)
DC	pKa			
HL 4.31E-011	atm m3/mol	25	EST	MEYLAN,WM & HOWARD,PH (1991)
OH 1.42E-011	cm3/molc sec	25	EST	MEYLAN,WM & HOWARD,PH (1993)

CAS #: 006333-79-5				BENZENESULFONAMIDE, 4-CHLORO-N-METHYL

Formula: $C_7H_8ClNO_2S$

Mol Weight: 205.66

MP (deg C): FP (deg C):

BP (deg C):

BP pressure (mm Hg):

Property/Value	Units	Temp	Data Type	Reference
WS 9.38E+002	mg/L	25	EST	MEYLAN,WM ET AL. (1996)
logP 1.92			EXP	DUFFEL,MW ET AL. (1986)
VP 1.47E-004	mm Hg	25	EST	NEELY,WB & BLAU,GE (1985)
DC	pKa			
HL 6.87E-007	atm m3/mol	25	EST	MEYLAN,WM & HOWARD,PH (1991)
OH 1.56E-012	cm3/molc sec	25	EST	MEYLAN,WM & HOWARD,PH (1993)

CAS #: 006335-39-3				BENZENESULFONAMIDE, 4-(1-METHYLETHYL)-

Formula: $C_9H_{13}NO_2S$

Mol Weight: 199.27

MP (deg C): FP (deg C):

BP (deg C):

BP pressure (mm Hg):

Property/Value	Units	Temp	Data Type	Reference
WS 1.18E+003	mg/L	25	EST	MEYLAN,WM ET AL. (1996)
logP 1.84			EXP	SANGSTER,J (1993)
VP 8.66E-005	mm Hg	25	EST	NEELY,WB & BLAU,GE (1985)
DC	pKa			
HL 8.21E-007	atm m3/mol	25	EST	MEYLAN,WM & HOWARD,PH (1991)
OH 3.26E-012	cm3/molc sec	25	EST	MEYLAN,WM & HOWARD,PH (1993)

CAS #: 006337-56-0				2,6-DIME-4-HYDROXYACETANILIDE

Formula: $C_{10}H_{13}NO_2$

Mol Weight: 179.22

MP (deg C): FP (deg C):

BP (deg C):

BP pressure (mm Hg):

Property/Value	Units	Temp	Data Type	Reference
WS 3.02E+004	mg/L	25	EST	MEYLAN,WM ET AL. (1996)
logP 0.31			EXP	HANSCH,C & LEO,AJ (1985)
VP 1.05E-006	mm Hg	25	EST	NEELY,WB & BLAU,GE (1985)
DC	pKa			
HL 7.82E-013	atm m3/mol	25	EST	MEYLAN,WM & HOWARD,PH (1991)
OH 1.19E-010	cm3/molc sec	25	EST	MEYLAN,WM & HOWARD,PH (1993)

CAS #: 006339-04-4				PHENOXYACETIC ACID, M-ACETAMIDO-

Formula: $C_{10}H_{11}NO_4$

Mol Weight: 209.20

MP (deg C): FP (deg C):

BP (deg C):

BP pressure (mm Hg):

Property/Value	Units	Temp	Data Type	Reference
WS 1.53E+004	mg/L	25	EST	MEYLAN,WM ET AL. (1996)
logP 0.48			EXP	HANSCH,C & LEO,AJ (1985)
VP 1.29E-007	mm Hg	25	EST	NEELY,WB & BLAU,GE (1985)
DC	pKa			
HL 1.92E-014	atm m3/mol	25	EST	MEYLAN,WM & HOWARD,PH (1991)
OH 1.42E-010	cm3/molc sec	25	EST	MEYLAN,WM & HOWARD,PH (1993)

CAS #: 006344-63-4 — 9H-FLUOREN-1-AMINE

Formula: $C_{13}H_{11}N$

Mol Weight: 181.24

MP (deg C): 125-126

FP (deg C):

BP (deg C):

BP pressure (mm Hg):

Property/Value	Units	Temp	Data Type	Reference
WS 8.11E+000	mg/L	25	EST	MEYLAN,WM ET AL. (1996)
logP 3.18			EXP	DEBNATH,AK ET AL. (1992)
VP 2.79E-005	mm Hg	25	EST	NEELY,WB & BLAU,GE (1985)
DC 3.87	pKa	25	EXP	PERRIN,DD (1965)
HL 5.91E-008	atm m3/mol	25	EST	MEYLAN,WM & HOWARD,PH (1991)
OH 2.01E-010	cm3/molc sec	25	EST	MEYLAN,WM & HOWARD,PH (1993)

CAS #: 006358-09-4 — 2-AMINO-6-CHLORO-4-NITROPHENOL

Formula: $C_6H_5ClN_2O_3$

Mol Weight: 188.57

MP (deg C):

FP (deg C):

BP (deg C):

BP pressure (mm Hg):

Property/Value	Units	Temp	Data Type	Reference
WS 2.47E+003	mg/L	25	EST	MEYLAN,WM ET AL. (1996)
logP 1.53			EXP	HANSCH,C & LEO,AJ (1985)
VP 6.69E-006	mm Hg	25	EST	NEELY,WB & BLAU,GE (1985)
DC	pKa			
HL 5.79E-013	atm m3/mol	25	EST	MEYLAN,WM & HOWARD,PH (1991)
OH 4.70E-012	cm3/molc sec	25	EST	MEYLAN,WM & HOWARD,PH (1993)

CAS #: 006364-34-7 — 4-((4-AMINOPHENYL)AZO)1,3-BENZENEDIAMINE

Formula: $C_{12}H_{13}N_5$

Mol Weight: 227.27

MP (deg C):

FP (deg C):

BP (deg C):

BP pressure (mm Hg):

Property/Value	Units	Temp	Data Type	Reference
WS 1.08E+003	mg/L	25	EST	MEYLAN,WM ET AL. (1996)
logP 1.21			EST	MEYLAN,WM & HOWARD,PH (1995)
VP 5.69E-008	mm Hg	25	EST	NEELY,WB & BLAU,GE (1985)
DC	pKa			
HL 6.49E-016	atm m3/mol	25	EST	MEYLAN,WM & HOWARD,PH (1991)
OH 2.00E-010	cm3/molc sec	25	EST	MEYLAN,WM & HOWARD,PH (1993)

CAS #: 006367-14-2 — TYROSINE,N-ACETYL,N'-MEAM-AMIDE

Formula: $C_{12}H_{16}N_2O_3$

Mol Weight: 236.27

MP (deg C):

FP (deg C):

BP (deg C):

BP pressure (mm Hg):

Property/Value	Units	Temp	Data Type	Reference
WS 9.83E+003	mg/L	25	EST	MEYLAN,WM ET AL. (1996)
logP -0.32			EXP	HANSCH,C & LEO,AJ (1985)
VP 4.16E-011	mm Hg	25	EST	NEELY,WB & BLAU,GE (1985)
DC	pKa			
HL 2.89E-015	atm m3/mol	25	EST	MEYLAN,WM & HOWARD,PH (1991)
OH 1.31E-010	cm3/molc sec	25	EST	MEYLAN,WM & HOWARD,PH (1993)

CAS #: 006373-50-8 — BENZENAMINE, 4-CYCLOHEXYL-

Formula: $C_{12}H_{17}N$

Mol Weight: 175.28

MP (deg C): 53-56

FP (deg C):

BP (deg C): 166

BP pressure (mm Hg): 1.30E+001

Property/Value	Units	Temp	Data Type	Reference
WS 4.43E+001	mg/L	25	EST	MEYLAN,WM ET AL. (1996)
logP 3.65			EXP	HANSCH,C ET AL. (1995)
VP 7.95E-004	mm Hg	25	EST	NEELY,WB & BLAU,GE (1985)
DC	pKa			
HL 3.82E-006	atm m3/mol	25	EST	MEYLAN,WM & HOWARD,PH (1991)
OH 1.40E-010	cm3/molc sec	25	EST	MEYLAN,WM & HOWARD,PH (1993)

CAS #: 006373-73-5 — N-(2,4-DINITROPHENYL)-1,4-BENZENEDIAMINE

Formula: $C_{12}H_{10}N_4O_4$

Mol Weight: 274.24

MP (deg C): 187-190

FP (deg C):

BP (deg C):

BP pressure (mm Hg):

Property/Value	Units	Temp	Data Type	Reference
WS 3.15E+002	mg/L	25	EST	MEYLAN,WM ET AL. (1996)
logP 2.04			EST	MEYLAN,WM & HOWARD,PH (1995)
VP 6.00E-011	mm Hg	25	EXP	BAUGHMAN,GL & PERENICH,TA (1988)
DC	pKa			
HL 1.26E-013	atm m3/mol	25	EST	MEYLAN,WM & HOWARD,PH (1991)
OH 2.00E-010	cm3/molc sec	25	EST	MEYLAN,WM & HOWARD,PH (1993)

CAS #: 006382-06-5 — AMYL LACTATE

Formula: $C_8H_{16}O_3$

Mol Weight: 160.21

MP (deg C):

FP (deg C):

BP (deg C):

BP pressure (mm Hg):

Property/Value	Units	Temp	Data Type	Reference
WS 1.75E+004	mg/L	25	EST	MEYLAN,WM ET AL. (1996)
logP 1.29			EST	MEYLAN,WM & HOWARD,PH (1995)
VP 1.47E-002	mm Hg	25	EST	NEELY,WB & BLAU,GE (1985)
DC	pKa			
HL 1.13E-004	atm m3/mol	25	EST	MEYLAN,WM & HOWARD,PH (1991)
OH 1.13E-011	cm3/molc sec	25	EST	MEYLAN,WM & HOWARD,PH (1993)

CAS #: 006409-90-1 — DIRECT YELLOW 62, DISODIUM SALT

Formula: $C_{35}H_{27}N_5Na_2O_9S_2$

Mol Weight: 771.74

MP (deg C):

FP (deg C):

BP (deg C):

BP pressure (mm Hg):

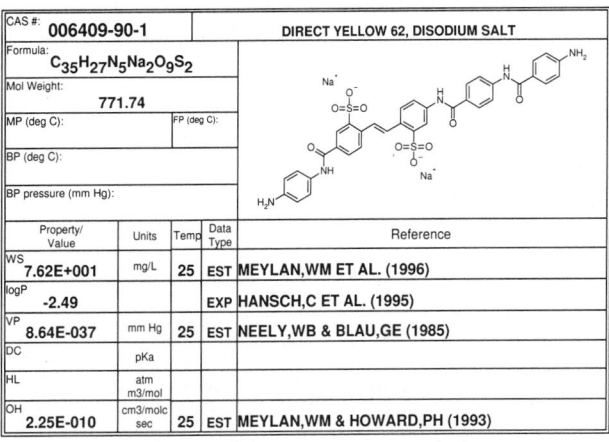

Property/Value	Units	Temp	Data Type	Reference
WS 7.62E+001	mg/L	25	EST	MEYLAN,WM ET AL. (1996)
logP -2.49			EXP	HANSCH,C ET AL. (1995)
VP 8.64E-037	mm Hg	25	EST	NEELY,WB & BLAU,GE (1985)
DC	pKa			
HL	atm m3/mol			
OH 2.25E-010	cm3/molc sec	25	EST	MEYLAN,WM & HOWARD,PH (1993)

CAS #: 006415-07-2	PHOSPHINE OXIDE, TRIS-(1-PYRROLIDINYL)-

Formula: $C_{12}H_{24}N_3OP$

Mol Weight: 257.32

MP (deg C): | FP (deg C):

BP (deg C):

BP pressure (mm Hg):

Property/Value	Units	Temp	Data Type	Reference
WS 8.07E+002	mg/L	25	EST	MEYLAN,WM ET AL. (1996)
logP 1.67			EXP	DEBORD,J & LABADIE,M (1985)
VP 2.72E-005	mm Hg	25	EST	NEELY,WB & BLAU,GE (1985)
DC	pKa			
HL 3.35E-012	atm m3/mol	25	EST	MEYLAN,WM & HOWARD,PH (1991)
OH 1.20E-010	cm3/molc sec	25	EST	MEYLAN,WM & HOWARD,PH (1993)

CAS #: 006422-86-2	BIS(2-ETHYLHEXYL) TEREPHTHALATE

Formula: $C_{24}H_{38}O_4$

Mol Weight: 390.57

MP (deg C): -48 | FP (deg C):

BP (deg C): 383

BP pressure (mm Hg):

Property/Value	Units	Temp	Data Type	Reference
WS 4.00E+000	mg/L	20	EXP	EASTMAN KODAK MSDS (1980)
logP 8.39			EST	MEYLAN,WM & HOWARD,PH (1995)
VP 2.14E-005	mm Hg	25	EST	NEELY,WB & BLAU,GE (1985)
DC	pKa			
HL 1.02E-005	atm m3/mol	25	EST	MEYLAN,WM & HOWARD,PH (1991)
OH 2.19E-011	cm3/molc sec	25	EST	MEYLAN,WM & HOWARD,PH (1993)

CAS #: 006424-85-7	ALIZARIN AZUROL A 2G

Formula: $C_{22}H_{16}N_3NaO_6S$

Mol Weight: 473.44

MP (deg C): | FP (deg C):

BP (deg C):

BP pressure (mm Hg):

Property/Value	Units	Temp	Data Type	Reference
WS 9.99E+000	mg/L	25	EST	MEYLAN,WM ET AL. (1996)
logP -0.46			EST	MEYLAN,WM & HOWARD,PH (1995)
VP 1.52E-024	mm Hg	25	EST	NEELY,WB & BLAU,GE (1985)
DC	pKa			
HL	atm m3/mol			
OH 9.72E-011	cm3/molc sec	25	EST	MEYLAN,WM & HOWARD,PH (1993)

CAS #: 006443-92-1	2-HEPTENE (CIS)

Formula: C_7H_{14}

Mol Weight: 98.19

MP (deg C): | FP (deg C):

BP (deg C): 98.4

BP pressure (mm Hg):

Property/Value	Units	Temp	Data Type	Reference
WS 3.11E+001	mg/L	25	EST	MEYLAN,WM ET AL. (1996)
logP 3.56			EST	MEYLAN,WM & HOWARD,PH (1995)
VP 4.85E+001	mm Hg	25	EXP	DAUBERT,TE & DANNER,RP (1989)
DC	pKa			
HL 5.62E-001	atm m3/mol	25	EST	MEYLAN,WM & HOWARD,PH (1991)
OH 6.04E-011	cm3/molc sec	25	EST	MEYLAN,WM & HOWARD,PH (1993)

CAS #: 006448-55-1	4-ME(5-NO2-2-FURFURILIDENE)THIAZOLE

Formula: $C_{10}H_9N_2O_3S$

Mol Weight: 237.26

MP (deg C): | FP (deg C):

BP (deg C):

BP pressure (mm Hg):

Property/Value	Units	Temp	Data Type	Reference
WS 1.89E+001	mg/L	25	EST	MEYLAN,WM ET AL. (1996)
logP 1.53			EXP	HANSCH,C & LEO,AJ (1985)
VP 6.57E-006	mm Hg	25	EST	NEELY,WB & BLAU,GE (1985)
DC	pKa			
HL 3.82E-010	atm m3/mol	25	EST	MEYLAN,WM & HOWARD,PH (1991)
OH 6.47E-011	cm3/molc sec	25	EST	MEYLAN,WM & HOWARD,PH (1993)

CAS #: 006451-55-4	124-BENZTHIADIZ-1-O2-3-ME-6-MEO

Formula: $C_9H_{10}N_2O_3S$

Mol Weight: 226.26

MP (deg C): | FP (deg C):

BP (deg C):

BP pressure (mm Hg):

Property/Value	Units	Temp	Data Type	Reference
WS 1.06E+004	mg/L	25	EST	MEYLAN,WM ET AL. (1996)
logP 0.56			EXP	HANSCH,C & LEO,AJ (1985)
VP 2.04E-008	mm Hg	25	EST	NEELY,WB & BLAU,GE (1985)
DC	pKa			
HL 5.47E-009	atm m3/mol	25	EST	MEYLAN,WM & HOWARD,PH (1991)
OH 3.89E-012	cm3/molc sec	25	EST	MEYLAN,WM & HOWARD,PH (1993)

CAS #: 006452-71-7	OXPRENALOL

Formula: $C_{15}H_{23}NO_3$

Mol Weight: 265.36

MP (deg C): | FP (deg C):

BP (deg C):

BP pressure (mm Hg):

Property/Value	Units	Temp	Data Type	Reference
WS 3.18E+003	mg/L	25	EST	MEYLAN,WM ET AL. (1996)
logP 2.10			EXP	HANSCH,C ET AL. (1995)
VP 3.12E-007	mm Hg	25	EST	NEELY,WB & BLAU,GE (1985)
DC	pKa			
HL 6.35E-013	atm m3/mol	25	EST	MEYLAN,WM & HOWARD,PH (1991)
OH 1.60E-010	cm3/molc sec	25	EST	MEYLAN,WM & HOWARD,PH (1993)

CAS #: 006457-92-7	2H-FURO[2,3-H]-1-BENZOPYRAN-2-ONE, 4-METHYL-

Formula: $C_{12}H_8O_3$

Mol Weight: 200.20

MP (deg C): | FP (deg C):

BP (deg C):

BP pressure (mm Hg):

Property/Value	Units	Temp	Data Type	Reference
WS 4.56E+002	mg/L	25	EST	MEYLAN,WM ET AL. (1996)
logP 2.32			EXP	SANGSTER,J (1993)
VP 1.06E-005	mm Hg	25	EST	NEELY,WB & BLAU,GE (1985)
DC	pKa			
HL 1.06E-006	atm m3/mol	25	EST	MEYLAN,WM & HOWARD,PH (1991)
OH 6.11E-011	cm3/molc sec	25	EST	MEYLAN,WM & HOWARD,PH (1993)

CAS #: 006476-18-2 — 3-PHENYL-2,4,5-PYRROLITRIONE

Formula: $C_{10}H_7NO_3$

Mol Weight: 189.17

MP (deg C): **FP (deg C):**

BP (deg C):

BP pressure (mm Hg):

Property/Value	Units	Temp	Data Type	Reference
WS 5.83E+002	mg/L	25	EST	MEYLAN,WM ET AL. (1996)
logP 2.26			EXP	LIPINSKI,CA ET AL. (1991)
VP 1.21E-009	mm Hg	25	EST	NEELY,WB & BLAU,GE (1985)
DC	pKa			
HL 6.13E-012	atm m3/mol	25	EST	MEYLAN,WM & HOWARD,PH (1991)
OH 1.10E-011	cm3/molc sec	25	EST	MEYLAN,WM & HOWARD,PH (1993)

CAS #: 006479-18-1 — N-METHYLQUINOXALINE-2-ONE

Formula: $C_9H_8N_2O$

Mol Weight: 160.18

MP (deg C): **FP (deg C):**

BP (deg C):

BP pressure (mm Hg):

Property/Value	Units	Temp	Data Type	Reference
WS 1.45E+004	mg/L	25	EST	MEYLAN,WM ET AL. (1996)
logP 0.79			EXP	HANSCH,C & LEO,AJ (1985)
VP 1.24E-004	mm Hg	25	EST	NEELY,WB & BLAU,GE (1985)
DC	pKa			
HL 1.90E-007	atm m3/mol	25	EST	MEYLAN,WM & HOWARD,PH (1991)
OH 1.56E-011	cm3/molc sec	25	EST	MEYLAN,WM & HOWARD,PH (1993)

CAS #: 006480-67-7 — 3-QUINOLINECARBOXAMIDE

Formula: $C_{10}H_8N_2O$

Mol Weight: 172.19

MP (deg C): **FP (deg C):**

BP (deg C):

BP pressure (mm Hg):

Property/Value	Units	Temp	Data Type	Reference
WS 7.47E+003	mg/L	25	EST	MEYLAN,WM ET AL. (1996)
logP 1.06			EXP	HANSCH,C & LEO,AJ (1985)
VP 2.09E-006	mm Hg	25	EST	NEELY,WB & BLAU,GE (1985)
DC	pKa			
HL 2.83E-013	atm m3/mol	25	EST	MEYLAN,WM & HOWARD,PH (1991)
OH 1.51E-011	cm3/molc sec	25	EST	MEYLAN,WM & HOWARD,PH (1993)

CAS #: 006485-40-1 — CARVONE

Formula: $C_{10}H_{14}O$

Mol Weight: 150.22

MP (deg C): **FP (deg C):**

BP (deg C): 227-230

BP pressure (mm Hg):

Property/Value	Units	Temp	Data Type	Reference
WS 1.31E+003	mg/L	25	EXP	SUZUKI,T (1991)
logP 3.07			EST	MEYLAN,WM & HOWARD,PH (1995)
VP 1.03E-001	mm Hg	25	EXT	PERRY,RH & GREEN,D (1984)
DC	pKa			
HL 7.73E-005	atm m3/mol	25	EST	MEYLAN,WM & HOWARD,PH (1991)
OH 1.41E-010	cm3/molc sec	25	EST	MEYLAN,WM & HOWARD,PH (1993)

CAS #: 006492-73-5 — C.I. VAT BLUE 3

Formula: $C_{16}H_9BrN_2O_2$

Mol Weight: 341.17

MP (deg C): **FP (deg C):**

BP (deg C):

BP pressure (mm Hg):

Property/Value	Units	Temp	Data Type	Reference
WS 2.67E+000	mg/L	25	EST	MEYLAN,WM ET AL. (1996)
logP 4.00			EST	MEYLAN,WM & HOWARD,PH (1995)
VP 3.00E-007	mm Hg	25	EXP	BAUGHMAN,GL & PERENICH,TA (1988)
DC	pKa			
HL 2.00E-014	atm m3/mol	25	EST	MEYLAN,WM & HOWARD,PH (1991)
OH 5.49E-011	cm3/molc sec	25	EST	MEYLAN,WM & HOWARD,PH (1993)

CAS #: 006497-78-5 — MORPHOLINE, 4-[2-(4-NITRO-1H-IMIDAZOL-1-YL)ETHYL

Formula: $C_9H_{16}N_4O_3$

Mol Weight: 228.25

MP (deg C): **FP (deg C):**

BP (deg C):

BP pressure (mm Hg):

Property/Value	Units	Temp	Data Type	Reference
WS 2.79E+005	mg/L	25	EST	MEYLAN,WM ET AL. (1996)
logP -0.38			EXP	SANGSTER,J (1993)
VP 8.82E-007	mm Hg	25	EST	NEELY,WB & BLAU,GE (1985)
DC	pKa			
HL 1.43E-012	atm m3/mol	25	EST	MEYLAN,WM & HOWARD,PH (1991)
OH 1.57E-010	cm3/molc sec	25	EST	MEYLAN,WM & HOWARD,PH (1993)

CAS #: 006506-37-2 — MORPHOLINE, 4-[2-(5-NITRO-1H-IMIDAZOL-1-YL)ETHYL

Formula: $C_9H_{14}N_4O_3$

Mol Weight: 226.24

MP (deg C): 110-111 **FP (deg C):**

BP (deg C):

BP pressure (mm Hg):

Property/Value	Units	Temp	Data Type	Reference
WS 1.15E+005	mg/L	25	EST	MEYLAN,WM ET AL. (1996)
logP 0.07			EXP	GUERRA,MC ET AL. (1981)
VP 8.82E-007	mm Hg	25	EST	NEELY,WB & BLAU,GE (1985)
DC	pKa			
HL 1.43E-012	atm m3/mol	25	EST	MEYLAN,WM & HOWARD,PH (1991)
OH 1.57E-010	cm3/molc sec	25	EST	MEYLAN,WM & HOWARD,PH (1993)

CAS #: 006513-23-1 — L-CYSTEINAMIDE, S,S'-BIS[N-ACETYL-

Formula: $C_{10}H_{18}N_4O_4S_2$

Mol Weight: 322.41

MP (deg C): **FP (deg C):**

BP (deg C):

BP pressure (mm Hg):

Property/Value	Units	Temp	Data Type	Reference
WS 1.26E+004	mg/L	25	EST	MEYLAN,WM ET AL. (1996)
logP -1.70			EXP	SANGSTER,J (1993)
VP 3.21E-017	mm Hg	25	EST	NEELY,WB & BLAU,GE (1985)
DC	pKa			
HL 3.46E-024	atm m3/mol	25	EST	MEYLAN,WM & HOWARD,PH (1991)
OH 3.68E-010	cm3/molc sec	25	EST	MEYLAN,WM & HOWARD,PH (1993)

CAS #: 006514-54-1	1-(N,N-DIETHYLAMINO-ACETYL-AMINO)INDANE

Formula: $C_{15}H_{22}N_2O$

Mol Weight: 246.36

MP (deg C): | FP (deg C):

BP (deg C):

BP pressure (mm Hg):

Property/ Value	Units	Temp	Data Type	Reference
WS 1.13E+002	mg/L	25	EST	MEYLAN,WM ET AL. (1996)
logP 2.74			EXP	HANSCH,C & LEO,AJ (1985)
VP 3.93E-007	mm Hg	25	EST	NEELY,WB & BLAU,GE (1985)
DC	pKa			
HL 2.23E-011	atm m3/mol	25	EST	MEYLAN,WM & HOWARD,PH (1991)
OH 1.11E-010	cm3/molc sec	25	EST	MEYLAN,WM & HOWARD,PH (1993)

CAS #: 006515-09-9	2,3,6-TRICHLOROPYRIDINE

Formula: $C_5H_2Cl_3N$

Mol Weight: 182.44

MP (deg C): | FP (deg C):

BP (deg C):

BP pressure (mm Hg):

Property/ Value	Units	Temp	Data Type	Reference
WS 2.31E+002	mg/L	25	EST	MEYLAN,WM ET AL. (1996)
logP 2.77			EXP	HANSCH,C & LEO,AJ (1985)
VP 1.13E-001	mm Hg	25	EST	NEELY,WB & BLAU,GE (1985)
DC	pKa			
HL 1.14E-002	atm m3/mol	25	EST	MEYLAN,WM & HOWARD,PH (1991)
OH 5.35E-014	cm3/molc sec	25	EST	MEYLAN,WM & HOWARD,PH (1993)

CAS #: 006515-38-4	3,5,6-TRICHLORO-2-PYRIDINOL

Formula: $C_5H_2Cl_3NO$

Mol Weight: 198.44

MP (deg C): | FP (deg C):

BP (deg C):

BP pressure (mm Hg):

Property/ Value	Units	Temp	Data Type	Reference
WS 8.08E+001	mg/L	25	EST	MEYLAN,WM ET AL. (1996)
logP 3.21			EXP	HANSCH,C & LEO,AJ (1985)
VP 1.03E-003	mm Hg	25	EST	NEELY,WB & BLAU,GE (1985)
DC	pKa			
HL 1.88E-008	atm m3/mol	25	EST	MEYLAN,WM & HOWARD,PH (1991)
OH 1.77E-013	cm3/molc sec	25	EST	MEYLAN,WM & HOWARD,PH (1993)

CAS #: 006526-67-6	BENZAMIDE, 2-CHLORO-N,N-DIMETHYL-

Formula: $C_9H_{10}ClNO$

Mol Weight: 183.64

MP (deg C): | FP (deg C):

BP (deg C):

BP pressure (mm Hg):

Property/ Value	Units	Temp	Data Type	Reference
WS 2.56E+003	mg/L	25	EST	MEYLAN,WM ET AL. (1996)
logP 1.54			EXP	SURYANARAYANA,MVS ET AL. (1991)
VP 7.60E-002	mm Hg	30	EXP	SURYANARAYANA,MVS ET AL. (1991)
DC	pKa			
HL 7.91E-009	atm m3/mol	25	EST	MEYLAN,WM & HOWARD,PH (1991)
OH 1.74E-011	cm3/molc sec	25	EST	MEYLAN,WM & HOWARD,PH (1993)

CAS #: 006526-72-3	1-NITRO-2-ISOPROPYLBENZENE

Formula: $C_9H_{11}NO_2$

Mol Weight: 165.19

MP (deg C): | FP (deg C):

BP (deg C): 103

BP pressure (mm Hg): 9.00E+000

Property/ Value	Units	Temp	Data Type	Reference
WS 3.71E+001	mg/L	25	EST	MEYLAN,WM ET AL. (1996)
logP 3.34			EXP	NAKAGAWA,Y ET AL. (1992)
VP 1.47E-002	mm Hg	25	EST	NEELY,WB & BLAU,GE (1985)
DC	pKa			
HL 4.13E-005	atm m3/mol	25	EST	MEYLAN,WM & HOWARD,PH (1991)
OH 2.85E-012	cm3/molc sec	25	EST	MEYLAN,WM & HOWARD,PH (1993)

CAS #: 006539-57-7	A-METHYLNORADRENALINE

Formula: $C_9H_{13}NO_3$

Mol Weight: 183.21

MP (deg C): | FP (deg C):

BP (deg C):

BP pressure (mm Hg):

Property/ Value	Units	Temp	Data Type	Reference
WS 1.00E+006	mg/L	25	EST	MEYLAN,WM ET AL. (1996)
logP -1.43			EXP	HANSCH,C & LEO,AJ (1985)
VP 6.26E-007	mm Hg	25	EST	NEELY,WB & BLAU,GE (1985)
DC	pKa			
HL 4.27E-019	atm m3/mol	25	EST	MEYLAN,WM & HOWARD,PH (1991)
OH 1.05E-010	cm3/molc sec	25	EST	MEYLAN,WM & HOWARD,PH (1993)

CAS #: 006547-53-1	PHENYLACETIC ACID,P-PHENYLMETHOXY

Formula: $C_{15}H_{14}O_3$

Mol Weight: 242.28

MP (deg C): | FP (deg C):

BP (deg C):

BP pressure (mm Hg):

Property/ Value	Units	Temp	Data Type	Reference
WS 3.08E+002	mg/L	25	EST	MEYLAN,WM ET AL. (1996)
logP 2.72			EXP	HANSCH,C & LEO,AJ (1985)
VP 1.01E-006	mm Hg	25	EST	NEELY,WB & BLAU,GE (1985)
DC	pKa			
HL 2.11E-010	atm m3/mol	25	EST	MEYLAN,WM & HOWARD,PH (1991)
OH 3.39E-011	cm3/molc sec	25	EST	MEYLAN,WM & HOWARD,PH (1993)

CAS #: 006549-64-0	1H-INDENE-1,3(2H)-DIONE, 2-(3,5-DIMETHYLPHENYL)-

Formula: $C_{17}H_{14}O_2$

Mol Weight: 250.30

MP (deg C): | FP (deg C):

BP (deg C):

BP pressure (mm Hg):

Property/ Value	Units	Temp	Data Type	Reference
WS 8.69E+000	mg/L	25	EST	MEYLAN,WM ET AL. (1996)
logP 4.02			EXP	SANGSTER,J (1993)
VP 6.34E-007	mm Hg	25	EST	NEELY,WB & BLAU,GE (1985)
DC	pKa			
HL 5.83E-010	atm m3/mol	25	EST	MEYLAN,WM & HOWARD,PH (1991)
OH 3.34E-011	cm3/molc sec	25	EST	MEYLAN,WM & HOWARD,PH (1993)

CAS #: 006556-12-3				GLUCURONIC ACID

Formula: $C_6H_{10}O_7$

Mol Weight: 194.14

MP (deg C): 165 FP (deg C):

BP (deg C):

BP pressure (mm Hg):

Property/Value	Units	Temp	Data Type	Reference
WS 1.00E+006	mg/L	25	EST	MEYLAN,WM ET AL. (1996)
logP -2.57			EXP	HANSCH,C & LEO,AJ (1985)
VP 7.39E-010	mm Hg	25	EST	NEELY,WB & BLAU,GE (1985)
DC	pKa			
HL 1.05E-017	atm m3/mol	25	EST	MEYLAN,WM & HOWARD,PH (1991)
OH 9.38E-011	cm3/molc sec	25	EST	MEYLAN,WM & HOWARD,PH (1993)

CAS #: 006560-83-4				2-IODOQUINOLINE

Formula: C_9H_6IN

Mol Weight: 255.06

MP (deg C): FP (deg C):

BP (deg C):

BP pressure (mm Hg):

Property/Value	Units	Temp	Data Type	Reference
WS 5.62E+001	mg/L	25	EST	MEYLAN,WM ET AL. (1996)
logP 3.04			EXP	HANSCH,C & LEO,AJ (1985)
VP 3.48E-004	mm Hg	25	EST	NEELY,WB & BLAU,GE (1985)
DC	pKa			
HL 1.60E-007	atm m3/mol	25	EST	MEYLAN,WM & HOWARD,PH (1991)
OH 9.49E-012	cm3/molc sec	25	EST	MEYLAN,WM & HOWARD,PH (1993)

CAS #: 006574-15-8				N-(P-NITROPHENYL)PIPERIDINE

Formula: $C_{11}H_{14}N_2O_2$

Mol Weight: 206.25

MP (deg C): FP (deg C):

BP (deg C):

BP pressure (mm Hg):

Property/Value	Units	Temp	Data Type	Reference
WS 2.72E+001	mg/L	25	EST	MEYLAN,WM ET AL. (1996)
logP 3.26			EXP	HANSCH,C & LEO,AJ (1985)
VP 1.03E-004	mm Hg	25	EST	NEELY,WB & BLAU,GE (1985)
DC	pKa			
HL 3.49E-007	atm m3/mol	25	EST	MEYLAN,WM & HOWARD,PH (1991)
OH 4.39E-011	cm3/molc sec	25	EST	MEYLAN,WM & HOWARD,PH (1993)

CAS #: 006577-34-0				1,4-OXATHIIN-3-CARBOXAMIDE, N-[1,1'-BIPHENYL]-2-

Formula: $C_{18}H_{17}NO_2S$

Mol Weight: 311.41

MP (deg C): FP (deg C):

BP (deg C):

BP pressure (mm Hg):

Property/Value	Units	Temp	Data Type	Reference
WS 7.13E+001	mg/L	25	EST	MEYLAN,WM ET AL. (1996)
logP 2.54			EXP	SANGSTER,J (1993)
VP 1.89E-010	mm Hg	25	EST	NEELY,WB & BLAU,GE (1985)
DC	pKa			
HL 5.73E-013	atm m3/mol	25	EST	MEYLAN,WM & HOWARD,PH (1991)
OH 1.18E-010	cm3/molc sec	25	EST	MEYLAN,WM & HOWARD,PH (1993)

CAS #: 006577-49-7				B-PHENYL-B-MEO-N-ET PYRROLIDINE

Formula: $C_{13}H_{19}NO$

Mol Weight: 205.30

MP (deg C): FP (deg C):

BP (deg C):

BP pressure (mm Hg):

Property/Value	Units	Temp	Data Type	Reference
WS 1.04E+004	mg/L	25	EST	MEYLAN,WM ET AL. (1996)
logP 1.88			EXP	HANSCH,C & LEO,AJ (1985)
VP 1.69E-003	mm Hg	25	EST	NEELY,WB & BLAU,GE (1985)
DC	pKa			
HL 3.56E-008	atm m3/mol	25	EST	MEYLAN,WM & HOWARD,PH (1991)
OH 1.16E-010	cm3/molc sec	25	EST	MEYLAN,WM & HOWARD,PH (1993)

CAS #: 006577-69-1				1,4-OXATHIIN-3-CARBOXYLIC ACID, 5,6-DIHYDRO-2-ME

Formula: $C_6H_8O_3S$

Mol Weight: 160.19

MP (deg C): FP (deg C):

BP (deg C):

BP pressure (mm Hg):

Property/Value	Units	Temp	Data Type	Reference
WS 1.57E+005	mg/L	25	EST	MEYLAN,WM ET AL. (1996)
logP 0.04			EXP	HANSCH,C & LEO,AJ (1985)
VP 8.15E-004	mm Hg	25	EST	NEELY,WB & BLAU,GE (1985)
DC	pKa			
HL 6.62E-010	atm m3/mol	25	EST	MEYLAN,WM & HOWARD,PH (1991)
OH 8.49E-011	cm3/molc sec	25	EST	MEYLAN,WM & HOWARD,PH (1993)

CAS #: 006587-21-9				2-CF3-5,6-DIBROMO-BENZIMIDAZOLE

Formula: $C_8H_3Br_2F_3N_2$

Mol Weight: 343.94

MP (deg C): FP (deg C):

BP (deg C):

BP pressure (mm Hg):

Property/Value	Units	Temp	Data Type	Reference
WS 1.93E+000	mg/L	25	EST	MEYLAN,WM ET AL. (1996)
logP 4.15			EXP	HANSCH,C & LEO,AJ (1985)
VP 4.86E-007	mm Hg	25	EST	NEELY,WB & BLAU,GE (1985)
DC	pKa			
HL 5.06E-007	atm m3/mol	25	EST	MEYLAN,WM & HOWARD,PH (1991)
OH 1.13E-012	cm3/molc sec	25	EST	MEYLAN,WM & HOWARD,PH (1993)

CAS #: 006600-40-4				NORVALINE

Formula: $C_5H_{11}NO_2$

Mol Weight: 117.15

MP (deg C): 307 FP (deg C):

BP (deg C):

BP pressure (mm Hg):

Property/Value	Units	Temp	Data Type	Reference
WS 5.54E+004	mg/L	25	EST	MEYLAN,WM ET AL. (1996)
logP -2.11			EXP	HANSCH,C ET AL. (1995)
VP 1.63E-008	mm Hg	25	EST	NEELY,WB & BLAU,GE (1985)
DC 2.34	pKa	13	EXP	KORTUM,G ET AL (1961)
HL 2.63E-009	atm m3/mol	25	EST	MEYLAN,WM & HOWARD,PH (1991)
OH 4.09E-011	cm3/molc sec	25	EST	MEYLAN,WM & HOWARD,PH (1993)

CAS #:	006602-28-4				3-HYDROXYPYRIDINE-N-OXIDE

Formula: $C_5H_7NO_2$

Mol Weight: 113.12

MP (deg C): 190-192 FP (deg C):

BP (deg C):

BP pressure (mm Hg):

Property/ Value	Units	Temp	Data Type	Reference
WS 3.58E+005	mg/L	25	EST	MEYLAN,WM ET AL. (1996)
logP -0.61			EXP	HANSCH,C ET AL. (1995)
VP 1.67E-003	mm Hg	25	EST	NEELY,WB & BLAU,GE (1985)
DC	pKa			
HL 7.34E-015	atm m3/mol	25	EST	MEYLAN,WM & HOWARD,PH (1991)
OH 6.46E-012	cm3/molc sec	25	EST	MEYLAN,WM & HOWARD,PH (1993)

CAS #:	006607-45-0				(1,2-DICHLOROETHENYL)BENZENE

Formula: $C_8H_6Cl_2$

Mol Weight: 173.04

MP (deg C): FP (deg C):

BP (deg C):

BP pressure (mm Hg):

Property/ Value	Units	Temp	Data Type	Reference
WS 3.77E+001	mg/L	25	EST	MEYLAN,WM ET AL. (1996)
logP 3.74				EST MEYLAN,WM & HOWARD,PH (1995)
VP 2.04E-001	mm Hg	25	EST	NEELY,WB & BLAU,GE (1985) .
DC	pKa			
HL 1.43E-003	atm m3/mol	25	EST	MEYLAN,WM & HOWARD,PH (1991)
OH 5.67E-012	cm3/molc sec	25	EST	MEYLAN,WM & HOWARD,PH (1993)

CAS #:	006609-40-1				5-CL-6-NO2-2-CF3-BENZIMIDAZOLE

Formula: $C_8H_3ClF_3N_3O_2$

Mol Weight: 265.58

MP (deg C): FP (deg C):

BP (deg C):

BP pressure (mm Hg):

Property/ Value	Units	Temp	Data Type	Reference
WS 1.43E+001	mg/L	25	EST	MEYLAN,WM ET AL. (1996)
logP 3.21			EXP	HANSCH,C & LEO,AJ (1985)
VP 1.51E-007	mm Hg	25	EST	NEELY,WB & BLAU,GE (1985)
DC	pKa			
HL 9.32E-009	atm m3/mol	25	EST	MEYLAN,WM & HOWARD,PH (1991)
OH 2.97E-013	cm3/molc sec	25	EST	MEYLAN,WM & HOWARD,PH (1993)

CAS #:	006625-74-7				TERT-VALERANILIDE

Formula: $C_{11}H_{15}NO$

Mol Weight: 177.25

MP (deg C): FP (deg C):

BP (deg C):

BP pressure (mm Hg):

Property/ Value	Units	Temp	Data Type	Reference
WS 1.13E+003	mg/L	25	EST	MEYLAN,WM ET AL. (1996)
logP 1.99			EXP	HANSCH,C & LEO,AJ (1985)
VP 7.21E-005	mm Hg	25	EST	NEELY,WB & BLAU,GE (1985)
DC	pKa			
HL 1.44E-008	atm m3/mol	25	EST	MEYLAN,WM & HOWARD,PH (1991)
OH 1.40E-011	cm3/molc sec	25	EST	MEYLAN,WM & HOWARD,PH (1993)

CAS #:	006626-22-8				BENZENAMINE, 4-[(4-BROMOPHENYL)SULFONYL]-

Formula: $C_{12}H_{10}BrNO_2S$

Mol Weight: 312.19

MP (deg C): FP (deg C):

BP (deg C):

BP pressure (mm Hg):

Property/ Value	Units	Temp	Data Type	Reference
WS 3.84E+001	mg/L	25	EST	MEYLAN,WM ET AL. (1996)
logP 2.85			EXP	ALTOMARE,C ET AL. (1991)
VP 5.66E-008	mm Hg	25	EST	NEELY,WB & BLAU,GE (1985)
DC	pKa			
HL 3.51E-011	atm m3/mol	25	EST	MEYLAN,WM & HOWARD,PH (1991)
OH 2.33E-011	cm3/molc sec	25	EST	MEYLAN,WM & HOWARD,PH (1993)

CAS #:	006626-84-2				DIMETHYLMALONATE, BENZAL

Formula: $C_{12}H_{12}O_4$

Mol Weight: 220.23

MP (deg C): FP (deg C):

BP (deg C):

BP pressure (mm Hg):

Property/ Value	Units	Temp	Data Type	Reference
WS 4.19E+002	mg/L	25	EST	MEYLAN,WM ET AL. (1996)
logP 2.24			EXP	HANSCH,C & LEO,AJ (1985)
VP 4.87E-003	mm Hg	25	EST	NEELY,WB & BLAU,GE (1985)
DC	pKa			
HL 6.20E-009	atm m3/mol	25	EST	MEYLAN,WM & HOWARD,PH (1991)
OH 1.29E-011	cm3/molc sec	25	EST	MEYLAN,WM & HOWARD,PH (1993)

CAS #:	006628-97-3				1,4-NAPHTHOQUINONE, 2-ANILINO

Formula: $C_{16}H_{11}NO_2$

Mol Weight: 249.27

MP (deg C): 193 FP (deg C):

BP (deg C):

BP pressure (mm Hg):

Property/ Value	Units	Temp	Data Type	Reference
WS 8.97E+001	mg/L	25	EST	MEYLAN,WM ET AL. (1996)
logP 2.84			EXP	HANSCH,C & LEO,AJ (1985)
VP 1.94E-007	mm Hg	25	EST	NEELY,WB & BLAU,GE (1985)
DC	pKa			
HL 4.66E-012	atm m3/mol	25	EST	MEYLAN,WM & HOWARD,PH (1991)
OH 4.98E-011	cm3/molc sec	25	EST	MEYLAN,WM & HOWARD,PH (1993)

CAS #:	006631-37-4				2-PYRIDINEAMINE, N-PHENYL

Formula: $C_{11}H_{10}N_2$

Mol Weight: 170.22

MP (deg C): FP (deg C):

BP (deg C):

BP pressure (mm Hg):

Property/ Value	Units	Temp	Data Type	Reference
WS 2.75E+002	mg/L	25	EST	MEYLAN,WM ET AL. (1996)
logP 2.75			EXP	HANSCH,C & LEO,AJ (1985)
VP 1.06E-003	mm Hg	25	EST	NEELY,WB & BLAU,GE (1985)
DC	pKa			
HL 1.37E-009	atm m3/mol	25	EST	MEYLAN,WM & HOWARD,PH (1991)
OH 1.74E-010	cm3/molc sec	25	EST	MEYLAN,WM & HOWARD,PH (1993)

006636-22-2 — 4-CH3COO-PHENYLALANINE

Formula: $C_{11}H_{13}NO_4$
Mol Weight: 223.23
MP (deg C):
FP (deg C):
BP (deg C):
BP pressure (mm Hg):

Property/ Value	Units	Temp	Data Type	Reference
WS 3.85E+003	mg/L	25	EST	MEYLAN,WM ET AL. (1996)
logP -1.33			EXP	HUANG,CH ET AL. (1985)
VP 6.93E-010	mm Hg	25	EST	NEELY,WB & BLAU,GE (1985)
DC	pKa			
HL 1.45E-012	atm m3/mol	25	EST	MEYLAN,WM & HOWARD,PH (1991)
OH 4.30E-011	cm3/molc sec	25	EST	MEYLAN,WM & HOWARD,PH (1993)

006638-60-4 — 9H-FLUOREN-2-AMINE, 7-BROMO-

Formula: $C_{13}H_{10}BrN$
Mol Weight: 260.14
MP (deg C):
FP (deg C):
BP (deg C):
BP pressure (mm Hg):

Property/ Value	Units	Temp	Data Type	Reference
WS 7.24E-001	mg/L	25	EXP	MEYLAN,WM ET AL. (1996)
logP 3.92			EXP	DEBNATH,AK & HANSCH,C (1992)
VP 2.77E-006	mm Hg	25	EST	NEELY,WB & BLAU,GE (1985)
DC	pKa			
HL 2.36E-008	atm m3/mol	25	EST	MEYLAN,WM & HOWARD,PH (1991)
OH 1.71E-010	cm3/molc sec	25	EST	MEYLAN,WM & HOWARD,PH (1993)

006639-06-1 — 1H-INDOLE-1-PROPANOIC ACID

Formula: $C_{11}H_{11}NO_2$
Mol Weight: 189.22
MP (deg C):
FP (deg C):
BP (deg C):
BP pressure (mm Hg):

Property/ Value	Units	Temp	Data Type	Reference
WS 1.63E+003	mg/L	25	EST	MEYLAN,WM ET AL. (1996)
logP 2.20			EXP	SANGSTER,J (1993)
VP 1.10E-005	mm Hg	25	EST	NEELY,WB & BLAU,GE (1985)
DC	pKa			
HL 1.87E-010	atm m3/mol	25	EST	MEYLAN,WM & HOWARD,PH (1991)
OH 1.13E-010	cm3/molc sec	25	EST	MEYLAN,WM & HOWARD,PH (1993)

006639-30-1 — 2,4,5-TRICHLOROTOLUENE

Formula: $C_7H_5Cl_3$
Mol Weight: 195.48
MP (deg C): 82.4
FP (deg C):
BP (deg C): 231
BP pressure (mm Hg):

Property/ Value	Units	Temp	Data Type	Reference
WS 6.97E+000	mg/L	25	EST	MEYLAN,WM ET AL. (1996)
logP 4.47			EST	MEYLAN,WM & HOWARD,PH (1995)
VP 7.32E-002	mm Hg	25	EST	NEELY,WB & BLAU,GE (1985)
DC	pKa			
HL 1.50E-003	atm m3/mol	20	EXP	OLIVER,BG (1985)
OH 5.01E-013	cm3/molc sec	25	EST	MEYLAN,WM & HOWARD,PH (1993)

006640-24-0 — PIPERAZINE, 1-(3-CHLOROPHENYL)-

Formula: $C_{10}H_{13}ClN_2$
Mol Weight: 196.68
MP (deg C):
FP (deg C):
BP (deg C):
BP pressure (mm Hg):

Property/ Value	Units	Temp	Data Type	Reference
WS 7.31E+003	mg/L	25	EST	MEYLAN,WM ET AL. (1996)
logP 2.11			EXP	CACCIA,S ET AL. (1985)
VP 4.79E-004	mm Hg	25	EST	NEELY,WB & BLAU,GE (1985)
DC	pKa			
HL 8.39E-009	atm m3/mol	25	EST	MEYLAN,WM & HOWARD,PH (1991)
OH 2.09E-010	cm3/molc sec	25	EST	MEYLAN,WM & HOWARD,PH (1993)

006640-27-3 — 2-METHYL-4-CHLOROPHENOL

Formula: C_7H_7ClO
Mol Weight: 142.59
MP (deg C):
FP (deg C):
BP (deg C): 195.5
BP pressure (mm Hg):

Property/ Value	Units	Temp	Data Type	Reference
WS 1.52E+003	mg/L	25	EST	MEYLAN,WM ET AL. (1996)
logP 2.70			EST	MEYLAN,WM & HOWARD,PH (1995)
VP 4.05E-002	mm Hg	25	EST	NEELY,WB & BLAU,GE (1985)
DC 8.74	pKa	20	EXP	SERJEANT,EP & DEMPSEY,B (1979)
HL 4.58E-007	atm m3/mol	25	EST	MEYLAN,WM & HOWARD,PH (1991)
OH 1.22E-011	cm3/molc sec	25	EST	MEYLAN,WM & HOWARD,PH (1993)

006641-02-7 — BENZENEMETHANOL, 3,5-DICHLORO-2-HYDROXY-

Formula: $C_7H_6Cl_2O_2$
Mol Weight: 193.03
MP (deg C):
FP (deg C):
BP (deg C):
BP pressure (mm Hg):

Property/ Value	Units	Temp	Data Type	Reference
WS 6.34E+003	mg/L	25	EST	MEYLAN,WM ET AL. (1996)
logP 2.30			EXP	SANGSTER,J (1993)
VP 1.43E-005	mm Hg	25	EST	NEELY,WB & BLAU,GE (1985)
DC	pKa			
HL 1.24E-011	atm m3/mol	25	EST	MEYLAN,WM & HOWARD,PH (1991)
OH 5.20E-012	cm3/molc sec	25	EST	MEYLAN,WM & HOWARD,PH (1993)

006642-30-4 — N,O-DIMETHYLCARBAMATE

Formula: $C_3H_7NO_2$
Mol Weight: 89.09
MP (deg C):
FP (deg C):
BP (deg C):
BP pressure (mm Hg):

Property/ Value	Units	Temp	Data Type	Reference
WS 1.41E+005	mg/L	25	EST	MEYLAN,WM ET AL. (1996)
logP -0.06			EXP	HANSCH,C & LEO,AJ (1985)
VP 3.88E+001	mm Hg	25	EST	NEELY,WB & BLAU,GE (1985)
DC	pKa			
HL 8.69E-008	atm m3/mol	25	EST	MEYLAN,WM & HOWARD,PH (1991)
OH 6.28E-012	cm3/molc sec	25	EST	MEYLAN,WM & HOWARD,PH (1993)

CAS #: 006647-92-3 — 4-ACETYLAMINOPYRAZOLE

Formula: $C_5H_7N_3O$

Mol Weight: 125.13

MP (deg C):　FP (deg C):

BP (deg C):

BP pressure (mm Hg):

Property/Value	Units	Temp	Data Type	Reference
WS 9.93E+003	mg/L	25	EST	MEYLAN,WM ET AL. (1996)
logP -0.38			EXP	HANSCH,C & LEO,AJ (1985)
VP 2.03E-005	mm Hg	25	EST	NEELY,WB & BLAU,GE (1985)
DC	pKa			
HL 4.23E-012	atm m3/mol	25	EST	MEYLAN,WM & HOWARD,PH (1991)
OH 2.00E-010	cm3/molc sec	25	EST	MEYLAN,WM & HOWARD,PH (1993)

CAS #: 006652-04-6 — 4-PHENYL-N-NITROSOPIPERIDINE

Formula: $C_{11}H_{14}N_2O$

Mol Weight: 190.25

MP (deg C):　FP (deg C):

BP (deg C):

BP pressure (mm Hg):

Property/Value	Units	Temp	Data Type	Reference
WS 3.01E+002	mg/L	25	EST	MEYLAN,WM ET AL. (1996)
logP 2.59			EXP	HANSCH,C & LEO,AJ (1985)
VP 1.52E-004	mm Hg	25	EST	NEELY,WB & BLAU,GE (1985)
DC	pKa			
HL 1.71E-007	atm m3/mol	25	EST	MEYLAN,WM & HOWARD,PH (1991)
OH 3.18E-011	cm3/molc sec	25	EST	MEYLAN,WM & HOWARD,PH (1993)

CAS #: 006657-05-2 — 2-CHLORO-4-PHENYLAZOPHENOL

Formula: $C_{12}H_9ClN_2O$

Mol Weight: 232.67

MP (deg C):　FP (deg C):

BP (deg C):

BP pressure (mm Hg):

Property/Value	Units	Temp	Data Type	Reference
WS 9.24E+000	mg/L	25	EST	MEYLAN,WM ET AL. (1996)
logP 4.28			EST	MEYLAN,WM & HOWARD,PH (1995)
VP 3.54E-006	mm Hg	25	EST	NEELY,WB & BLAU,GE (1985)
DC	pKa			
HL 1.13E-009	atm m3/mol	25	EST	MEYLAN,WM & HOWARD,PH (1991)
OH 4.77E-012	cm3/molc sec	25	EST	MEYLAN,WM & HOWARD,PH (1993)

CAS #: 006657-33-6 — PROPANENITRILE, 3- 4- (2-CHLORO-4-NITROPHENYL)A

Formula: $C_{17}H_{16}ClN_5O_3$

Mol Weight: 373.80

MP (deg C): 157　FP (deg C):

BP (deg C):

BP pressure (mm Hg):

Property/Value	Units	Temp	Data Type	Reference
WS 2.80E-001	mg/L	25	EXP	BAUGHMAN,GL ET AL. (1993)
logP 4.00			EXP	BAUGHMAN,GL & WEBER,EJ (1991)
VP 1.07E-013	mm Hg	25	EST	NEELY,WB & BLAU,GE (1985)
DC	pKa			
HL 1.83E-017	atm m3/mol	25	EST	MEYLAN,WM & HOWARD,PH (1991)
OH 7.42E-011	cm3/molc sec	25	EST	MEYLAN,WM & HOWARD,PH (1993)

CAS #: 006665-86-7 — 4H-1-BENZOPYRAN-4-ONE, 7-HYDROXY-2-PHENYL-

Formula: $C_{15}H_{10}O_3$

Mol Weight: 238.25

MP (deg C): 245-247　FP (deg C):

BP (deg C):

BP pressure (mm Hg):

Property/Value	Units	Temp	Data Type	Reference
WS 8.46E+001	mg/L	25	EST	MEYLAN,WM ET AL. (1996)
logP 3.62			EXP	HANSCH,C ET AL. (1995)
VP 1.25E-007	mm Hg	25	EST	NEELY,WB & BLAU,GE (1985)
DC	pKa			
HL 3.76E-012	atm m3/mol	25	EST	MEYLAN,WM & HOWARD,PH (1991)
OH 2.31E-010	cm3/molc sec	25	EST	MEYLAN,WM & HOWARD,PH (1993)

CAS #: 006673-35-4 — PRACTOLOL

Formula: $C_{14}H_{22}N_2O_3$

Mol Weight: 266.34

MP (deg C): 134-136　FP (deg C):

BP (deg C):

BP pressure (mm Hg):

Property/Value	Units	Temp	Data Type	Reference
WS 4.47E+003	mg/L	25	EST	MEYLAN,WM ET AL. (1996)
logP 0.79			EXP	HANSCH,C ET AL. (1995)
VP 4.29E-010	mm Hg	25	EST	NEELY,WB & BLAU,GE (1985)
DC	pKa			
HL 9.36E-018	atm m3/mol	25	EST	MEYLAN,WM & HOWARD,PH (1991)
OH 1.21E-010	cm3/molc sec	25	EST	MEYLAN,WM & HOWARD,PH (1993)

CAS #: 006673-38-7 — ACETAMIDE, N-[3-[2-HYDROXY-3-[(1-METHYLETHYL)AMI

Formula: $C_{14}H_{22}N_2O_3$

Mol Weight: 266.34

MP (deg C):　FP (deg C):

BP (deg C):

BP pressure (mm Hg):

Property/Value	Units	Temp	Data Type	Reference
WS 4.47E+003	mg/L	25	EST	MEYLAN,WM ET AL. (1996)
logP 0.74			EXP	SANGSTER,J (1993)
VP 4.29E-010	mm Hg	25	EST	NEELY,WB & BLAU,GE (1985)
DC	pKa			
HL 9.36E-018	atm m3/mol	25	EST	MEYLAN,WM & HOWARD,PH (1991)
OH 1.21E-010	cm3/molc sec	25	EST	MEYLAN,WM & HOWARD,PH (1993)

CAS #: 006676-90-0 — (2-METHYLPHENYL)PHENYLDIAZENE

Formula: $C_{13}H_{12}N_2$

Mol Weight: 196.25

MP (deg C):　FP (deg C):

BP (deg C):

BP pressure (mm Hg):

Property/Value	Units	Temp	Data Type	Reference
WS 1.78E+000	mg/L	25	EST	MEYLAN,WM ET AL. (1996)
logP 4.66			EST	MEYLAN,WM & HOWARD,PH (1995)
VP 9.42E-004	mm Hg	25	EST	NEELY,WB & BLAU,GE (1985)
DC	pKa			
HL 1.62E-005	atm m3/mol	25	EST	MEYLAN,WM & HOWARD,PH (1991)
OH 2.93E-012	cm3/molc sec	25	EST	MEYLAN,WM & HOWARD,PH (1993)

CAS #: 006677-98-1 — HYDROCORTISONE-21-PROPIONATE

Formula: $C_{24}H_{34}O_6$

Mol Weight: 418.53

MP (deg C): FP (deg C):

BP (deg C):

BP pressure (mm Hg):

Property/Value	Units	Temp	Data Type	Reference
WS 9.54E+000	mg/L	25	EST	MEYLAN,WM ET AL. (1996)
logP 2.80			EXP	HANSCH,C & LEO,AJ (1985)
VP 9.39E-014	mm Hg	25	EST	NEELY,WB & BLAU,GE (1985)
DC	pKa			
HL 8.28E-012	atm m3/mol	25	EST	MEYLAN,WM & HOWARD,PH (1991)
OH 1.10E-010	cm3/molc sec	25	EST	MEYLAN,WM & HOWARD,PH (1993)

CAS #: 006677-99-2 — PREGN-4-ENE-3,20-DIONE, 11,17-DIHYDROXY-21-(1-OX

Formula: $C_{25}H_{36}O_6$

Mol Weight: 432.56

MP (deg C): FP (deg C):

BP (deg C):

BP pressure (mm Hg):

Property/Value	Units	Temp	Data Type	Reference
WS 3.69E+000	mg/L	25	EST	MEYLAN,WM ET AL. (1996)
logP 2.91			EXP	SANGSTER,J (1993)
VP 3.43E-014	mm Hg	25	EST	NEELY,WB & BLAU,GE (1985)
DC	pKa			
HL 1.10E-011	atm m3/mol	25	EST	MEYLAN,WM & HOWARD,PH (1991)
OH 1.11E-010	cm3/molc sec	25	EST	MEYLAN,WM & HOWARD,PH (1993)

CAS #: 006678-00-8 — HYDROCORTISONE-21-VALERATE

Formula: $C_{26}H_{38}O_6$

Mol Weight: 446.59

MP (deg C): FP (deg C):

BP (deg C):

BP pressure (mm Hg):

Property/Value	Units	Temp	Data Type	Reference
WS 1.27E+000	mg/L	25	EST	MEYLAN,WM ET AL. (1996)
logP 3.62			EXP	HANSCH,C & LEO,AJ (1985)
VP 1.25E-014	mm Hg	25	EST	NEELY,WB & BLAU,GE (1985)
DC	pKa			
HL 1.46E-011	atm m3/mol	25	EST	MEYLAN,WM & HOWARD,PH (1991)
OH 1.13E-010	cm3/molc sec	25	EST	MEYLAN,WM & HOWARD,PH (1993)

CAS #: 006714-29-0 — 1H-IMIDAZO[1,2-B]PYRAZOLE, 2,3-DIHYDRO-

Formula: $C_5H_7N_3$

Mol Weight: 109.13

MP (deg C): FP (deg C):

BP (deg C):

BP pressure (mm Hg):

Property/Value	Units	Temp	Data Type	Reference
WS 1.79E+005	mg/L	25	EST	MEYLAN,WM ET AL. (1996)
logP -0.25			EXP	HANSCH,C ET AL. (1995)
VP 1.69E-001	mm Hg	25	EST	NEELY,WB & BLAU,GE (1985)
DC	pKa			
HL 2.69E-008	atm m3/mol	25	EST	MEYLAN,WM & HOWARD,PH (1991)
OH 2.09E-010	cm3/molc sec	25	EST	MEYLAN,WM & HOWARD,PH (1993)

CAS #: 006723-40-6 — 1H-INDENE-1,3(2H)-DIONE, 2-[4-(TRIFLUOROMETHYL)P

Formula: $C_{16}H_9F_3O_2$

Mol Weight: 290.24

MP (deg C): FP (deg C):

BP (deg C):

BP pressure (mm Hg):

Property/Value	Units	Temp	Data Type	Reference
WS 8.27E+000	mg/L	25	EST	MEYLAN,WM ET AL. (1996)
logP 3.78			EXP	SANGSTER,J (1994)
VP 2.26E-006	mm Hg	25	EST	NEELY,WB & BLAU,GE (1985)
DC	pKa			
HL 4.16E-009	atm m3/mol	25	EST	MEYLAN,WM & HOWARD,PH (1991)
OH 2.55E-012	cm3/molc sec	25	EST	MEYLAN,WM & HOWARD,PH (1993)

CAS #: 006728-26-3 — (E)-2-HEXENAL

Formula: $C_6H_{10}O$

Mol Weight: 98.15

MP (deg C): FP (deg C):

BP (deg C): 146.5

BP pressure (mm Hg):

Property/Value	Units	Temp	Data Type	Reference
WS 5.26E+003	mg/L	25	EST	MEYLAN,WM ET AL. (1996)
logP 1.58			EST	MEYLAN,WM & HOWARD,PH (1995)
VP 6.60E+000	mm Hg	25	EST	NEELY,WB & BLAU,GE (1985)
DC	pKa			
HL 9.88E-005	atm m3/mol	25	EST	MEYLAN,WM & HOWARD,PH (1991)
OH 3.85E-011	cm3/molc sec	25	EST	MEYLAN,WM & HOWARD,PH (1993)

CAS #: 006742-07-0 — 2(1H)-PYRIMIDINONE, 4-AMINO-1-(2,3,5-TRI-O-ACETY

Formula: $C_{15}H_{19}N_3O_8$

Mol Weight: 369.33

MP (deg C): FP (deg C):

BP (deg C):

BP pressure (mm Hg):

Property/Value	Units	Temp	Data Type	Reference
WS 1.12E+003	mg/L	25	EST	MEYLAN,WM ET AL. (1996)
logP -0.80			EXP	SANGSTER,J (1993)
VP 7.85E-009	mm Hg	25	EST	NEELY,WB & BLAU,GE (1985)
DC	pKa			
HL 4.32E-020	atm m3/mol	25	EST	MEYLAN,WM & HOWARD,PH (1991)
OH 1.28E-010	cm3/molc sec	25	EST	MEYLAN,WM & HOWARD,PH (1993)

CAS #: 006742-54-7 — 1-PHENYLUNDECANE

Formula: $C_{17}H_{28}$

Mol Weight: 232.41

MP (deg C): -5 FP (deg C):

BP (deg C): 316

BP pressure (mm Hg):

Property/Value	Units	Temp	Data Type	Reference
WS 3.30E-003	mg/L	25	EST	MEYLAN,WM ET AL. (1996)
logP 8.14			EXP	SHERBLOM,PM ET AL. (1992)
VP 2.42E-004	mm Hg	25	EXP	YAWS,CL (1994B)
DC	pKa			
HL 1.01E-001	atm m3/mol	25	EST	MEYLAN,WM & HOWARD,PH (1991)
OH 1.86E-011	cm3/molc sec	25	EST	MEYLAN,WM & HOWARD,PH (1993)

CAS #: 006754-13-8				HELENALIN
Formula: $C_{15}H_{18}O_4$				
Mol Weight: 262.31				
MP (deg C): 167-168		FP (deg C):		
BP (deg C):				
BP pressure (mm Hg):				

Property/ Value	Units	Temp	Data Type	Reference
WS 1.18E+004	mg/L	25	EST	MEYLAN,WM ET AL. (1996)
logP 0.87			EXP	HANSCH,C & LEO,AJ (1985)
VP 4.69E-009	mm Hg	25	EST	NEELY,WB & BLAU,GE (1985)
DC	pKa			
HL 1.38E-012	atm m3/mol	25	EST	MEYLAN,WM & HOWARD,PH (1991)
OH 1.04E-010	cm3/molc sec	25	EST	MEYLAN,WM & HOWARD,PH (1993)

CAS #: 006756-33-8				2(5-NO2-2-FURFURILIDENE)IMIDAZOLE
Formula: $C_9H_8N_3O_3$				
Mol Weight: 206.18				
MP (deg C):		FP (deg C):		
BP (deg C):				
BP pressure (mm Hg):				

Property/ Value	Units	Temp	Data Type	Reference
WS 1.25E+003	mg/L	25	EST	MEYLAN,WM ET AL. (1996)
logP 1.32			EXP	HANSCH,C & LEO,AJ (1985)
VP 2.06E-007	mm Hg	25	EST	NEELY,WB & BLAU,GE (1985)
DC	pKa			
HL 3.40E-010	atm m3/mol	25	EST	MEYLAN,WM & HOWARD,PH (1991)
OH 9.50E-011	cm3/molc sec	25	EST	MEYLAN,WM & HOWARD,PH (1993)

CAS #: 006765-39-5				1-HEPTADECENE
Formula: $C_{17}H_{34}$				
Mol Weight: 238.46				
MP (deg C): 11.5		FP (deg C):		
BP (deg C): 300				
BP pressure (mm Hg):				

Property/ Value	Units	Temp	Data Type	Reference
WS 3.94E-004	mg/L	25	EST	MEYLAN,WM ET AL. (1996)
logP 8.55			EST	MEYLAN,WM & HOWARD,PH (1995)
VP 4.54E-004	mm Hg	25	EXP	YAWS,CL (1994B)
DC	pKa			
HL 8.09E+000	atm m3/mol	25	EST	MEYLAN,WM & HOWARD,PH (1991)
OH 4.57E-011	cm3/molc sec	25	EST	MEYLAN,WM & HOWARD,PH (1993)

CAS #: 006768-20-3				DIETHYLMALONATE,2-CL BENZAL
Formula: $C_{14}H_{15}ClO_4$				
Mol Weight: 282.73				
MP (deg C):		FP (deg C):		
BP (deg C):				
BP pressure (mm Hg):				

Property/ Value	Units	Temp	Data Type	Reference
WS 3.16E+001	mg/L	25	EST	MEYLAN,WM ET AL. (1996)
logP 3.15			EXP	HANSCH,C & LEO,AJ (1985)
VP 2.92E-004	mm Hg	25	EST	NEELY,WB & BLAU,GE (1985)
DC	pKa			
HL 8.10E-009	atm m3/mol	25	EST	MEYLAN,WM & HOWARD,PH (1991)
OH 1.53E-011	cm3/molc sec	25	EST	MEYLAN,WM & HOWARD,PH (1993)

CAS #: 006768-21-4				DIETHYLMALONATE,3-CL BENZAL
Formula: $C_{14}H_{15}ClO_4$				
Mol Weight: 282.73				
MP (deg C):		FP (deg C):		
BP (deg C):				
BP pressure (mm Hg):				

Property/ Value	Units	Temp	Data Type	Reference
WS 1.62E+001	mg/L	25	EST	MEYLAN,WM ET AL. (1996)
logP 3.49			EXP	HANSCH,C & LEO,AJ (1985)
VP 2.92E-004	mm Hg	25	EST	NEELY,WB & BLAU,GE (1985)
DC	pKa			
HL 8.10E-009	atm m3/mol	25	EST	MEYLAN,WM & HOWARD,PH (1991)
OH 1.53E-011	cm3/molc sec	25	EST	MEYLAN,WM & HOWARD,PH (1993)

CAS #: 006768-22-5				DIETHYLMALONATE,2-MEO BENZAL
Formula: $C_{15}H_{18}O_5$				
Mol Weight: 278.31				
MP (deg C):		FP (deg C):		
BP (deg C):				
BP pressure (mm Hg):				

Property/ Value	Units	Temp	Data Type	Reference
WS 2.54E+001	mg/L	25	EST	MEYLAN,WM ET AL. (1996)
logP 3.29			EXP	HANSCH,C & LEO,AJ (1985)
VP 2.43E-004	mm Hg	25	EST	NEELY,WB & BLAU,GE (1985)
DC	pKa			
HL 6.47E-010	atm m3/mol	25	EST	MEYLAN,WM & HOWARD,PH (1991)
OH 3.50E-011	cm3/molc sec	25	EST	MEYLAN,WM & HOWARD,PH (1993)

CAS #: 006768-23-6				DIETHYLMALONATE,4-MEO BENZAL
Formula: $C_{15}H_{18}O_5$				
Mol Weight: 278.31				
MP (deg C):		FP (deg C):		
BP (deg C):				
BP pressure (mm Hg):				

Property/ Value	Units	Temp	Data Type	Reference
WS 2.59E+001	mg/L	25	EST	MEYLAN,WM ET AL. (1996)
logP 3.28			EXP	HANSCH,C & LEO,AJ (1985)
VP 2.43E-004	mm Hg	25	EST	NEELY,WB & BLAU,GE (1985)
DC	pKa			
HL 6.47E-010	atm m3/mol	25	EST	MEYLAN,WM & HOWARD,PH (1991)
OH 3.50E-011	cm3/molc sec	25	EST	MEYLAN,WM & HOWARD,PH (1993)

CAS #: 006770-38-3				BENZENE, 1,4-BIS(METHOXYMETHYL)-
Formula: $C_{10}H_{14}O_2$				
Mol Weight: 166.22				
MP (deg C):		FP (deg C):		
BP (deg C):				
BP pressure (mm Hg):				

Property/ Value	Units	Temp	Data Type	Reference
WS 5.17E+003	mg/L	25	EST	MEYLAN,WM ET AL. (1996)
logP 1.28			EXP	WANG,W ET AL. (1987)
VP 7.41E-002	mm Hg	25	EST	NEELY,WB & BLAU,GE (1985)
DC	pKa			
HL 8.99E-007	atm m3/mol	25	EST	MEYLAN,WM & HOWARD,PH (1991)
OH 1.90E-011	cm3/molc sec	25	EST	MEYLAN,WM & HOWARD,PH (1993)

CAS #: 006771-54-6				DIETHYLMALONATE,3-MEO BENZAL

Formula: $C_{15}H_{18}O_5$

Mol Weight: 278.31

MP (deg C): | FP (deg C):

BP (deg C):

BP pressure (mm Hg):

Property/Value	Units	Temp	Data Type	Reference
WS 3.15E+001	mg/L	25	EST	MEYLAN,WM ET AL. (1996)
logP 3.18			EXP	HANSCH,C & LEO,AJ (1985)
VP 2.43E-004	mm Hg	25	EST	NEELY,WB & BLAU,GE (1985)
DC	pKa			
HL 6.47E-010	atm m3/mol	25	EST	MEYLAN,WM & HOWARD,PH (1991)
OH 3.50E-011	cm3/molc sec	25	EST	MEYLAN,WM & HOWARD,PH (1993)

CAS #: 006777-05-5				N-BENZYL-2,4-IMIDAZOLIDIONE

Formula: $C_{10}H_{10}N_2O_2$

Mol Weight: 190.20

MP (deg C): | FP (deg C):

BP (deg C):

BP pressure (mm Hg):

Property/Value	Units	Temp	Data Type	Reference
WS 1.10E+004	mg/L	25	EST	MEYLAN,WM ET AL. (1996)
logP 0.76			EXP	LIPINSKI,CA ET AL. (1991)
VP 1.10E-008	mm Hg	25	EST	NEELY,WB & BLAU,GE (1985)
DC	pKa			
HL 2.78E-010	atm m3/mol	25	EST	MEYLAN,WM & HOWARD,PH (1991)
OH 1.30E-011	cm3/molc sec	25	EST	MEYLAN,WM & HOWARD,PH (1993)

CAS #: 006781-42-6				M-ACETYLACETOPHENONE

Formula: $C_{10}H_{10}O_2$

Mol Weight: 162.19

MP (deg C): 32 | FP (deg C):

BP (deg C): 152

BP pressure (mm Hg): 1.50E+001

Property/Value	Units	Temp	Data Type	Reference
WS 4.02E+003	mg/L	25	EST	MEYLAN,WM ET AL. (1996)
logP 1.43			EXP	HANSCH,C & LEO,AJ (1985)
VP 7.23E-003	mm Hg	25	EST	NEELY,WB & BLAU,GE (1985)
DC	pKa			
HL 1.79E-008	atm m3/mol	25	EST	MEYLAN,WM & HOWARD,PH (1991)
OH 1.82E-012	cm3/molc sec	25	EST	MEYLAN,WM & HOWARD,PH (1993)

CAS #: 006795-87-5				SEC-BUTYL METHYL ETHER

Formula: $C_5H_{12}O$

Mol Weight: 88.15

MP (deg C): | FP (deg C):

BP (deg C):

BP pressure (mm Hg):

Property/Value	Units	Temp	Data Type	Reference
WS 1.64E+004	mg/L	25	EXP	WAKITA,K ET AL. (1986)
logP 1.47			EST	MEYLAN,WM & HOWARD,PH (1995)
VP 2.08E+002	mm Hg	25	EXP	DAUBERT,TE & DANNER,RP (1989)
DC	pKa			
HL 1.47E-003	atm m3/mol	25	EST	VP/WSOL
OH 1.69E-011	cm3/molc sec	25	EST	MEYLAN,WM & HOWARD,PH (1993)

CAS #: 006815-52-7				N-BUTROYLCYCLOBUTANECARBOXAMIDE

Formula: $C_9H_{15}NO_2$

Mol Weight: 169.23

MP (deg C): | FP (deg C):

BP (deg C):

BP pressure (mm Hg):

Property/Value	Units	Temp	Data Type	Reference
WS 2.77E+004	mg/L	25	EST	MEYLAN,WM ET AL. (1996)
logP 0.41			EXP	HANSCH,C & LEO,AJ (1985)
VP 2.52E-007	mm Hg	25	EST	NEELY,WB & BLAU,GE (1985)
DC	pKa			
HL 1.04E-007	atm m3/mol	25	EST	MEYLAN,WM & HOWARD,PH (1991)
OH 1.37E-011	cm3/molc sec	25	EST	MEYLAN,WM & HOWARD,PH (1993)

CAS #: 006820-74-2				PERCHLOROISOBUTANE

Formula: C_4Cl_{10}

Mol Weight: 402.57

MP (deg C): | FP (deg C):

BP (deg C):

BP pressure (mm Hg):

Property/Value	Units	Temp	Data Type	Reference
WS 8.05E-003	mg/L	25	EST	MEYLAN,WM ET AL. (1996)
logP 6.52			EST	MEYLAN,WM & HOWARD,PH (1995)
VP 2.26E-003	mm Hg	25	EST	NEELY,WB & BLAU,GE (1985)
DC	pKa			
HL 1.13E-004	atm m3/mol	25	EST	MEYLAN,WM & HOWARD,PH (1991)
OH 0.00E+000	cm3/molc sec	25	EST	MEYLAN,WM & HOWARD,PH (1993)

CAS #: 006827-40-3				DIETHYLMALONATE,4-CL BENZAL

Formula: $C_{14}H_{15}ClO_4$

Mol Weight: 282.73

MP (deg C): | FP (deg C):

BP (deg C):

BP pressure (mm Hg):

Property/Value	Units	Temp	Data Type	Reference
WS 6.42E+000	mg/L	25	EST	MEYLAN,WM ET AL. (1996)
logP 3.96			EXP	HANSCH,C & LEO,AJ (1985)
VP 2.92E-004	mm Hg	25	EST	NEELY,WB & BLAU,GE (1985)
DC	pKa			
HL 8.10E-009	atm m3/mol	25	EST	MEYLAN,WM & HOWARD,PH (1991)
OH 1.53E-011	cm3/molc sec	25	EST	MEYLAN,WM & HOWARD,PH (1993)

CAS #: 006829-82-9				N1-(4-NO2-PHENYL)SULFANILAMIDE

Formula: $C_{12}H_{11}N_3O_4S$

Mol Weight: 293.30

MP (deg C): | FP (deg C):

BP (deg C):

BP pressure (mm Hg):

Property/Value	Units	Temp	Data Type	Reference
WS 8.14E+001	mg/L	25	EST	MEYLAN,WM ET AL. (1996)
logP 2.14			EXP	SANGSTER,J (1993)
VP 1.90E-009	mm Hg	25	EST	NEELY,WB & BLAU,GE (1985)
DC	pKa			
HL 3.25E-013	atm m3/mol	25	EST	MEYLAN,WM & HOWARD,PH (1991)
OH 2.83E-011	cm3/molc sec	25	EST	MEYLAN,WM & HOWARD,PH (1993)

833

006850-22-2 — 2-METHYL-6-NITROINDAZOLE

Formula: $C_8H_9N_3O_2$

Mol Weight: 179.18

MP (deg C):   FP (deg C):

BP (deg C):

BP pressure (mm Hg):

Property/Value	Units	Temp	Data Type	Reference
WS 7.27E+002	mg/L	25	EST	MEYLAN,WM ET AL. (1996)
logP 1.76			EXP	HANSCH,C ET AL. (1995)
VP 1.09E-004	mm Hg	25	EST	NEELY,WB & BLAU,GE (1985)
DC	pKa			
HL 3.03E-008	atm m3/mol	25	EST	MEYLAN,WM & HOWARD,PH (1991)
OH 4.64E-012	cm3/molc sec	25	EST	MEYLAN,WM & HOWARD,PH (1993)

006850-23-3 — 1-METHYL-6-NITROINDAZOLE

Formula: $C_8H_7N_3O_2$

Mol Weight: 177.16

MP (deg C):   FP (deg C):

BP (deg C):

BP pressure (mm Hg):

Property/Value	Units	Temp	Data Type	Reference
WS 5.00E+002	mg/L	25	EST	MEYLAN,WM ET AL. (1996)
logP 1.95			EXP	HANSCH,C ET AL. (1995)
VP 1.09E-004	mm Hg	25	EST	NEELY,WB & BLAU,GE (1985)
DC	pKa			
HL 3.03E-008	atm m3/mol	25	EST	MEYLAN,WM & HOWARD,PH (1991)
OH 4.64E-012	cm3/molc sec	25	EST	MEYLAN,WM & HOWARD,PH (1993)

006853-14-1 — 2-N,N-DIMEAMINO-1-PHENYLETHANOL

Formula: $C_{10}H_{15}NO$

Mol Weight: 165.24

MP (deg C):   FP (deg C):

BP (deg C):

BP pressure (mm Hg):

Property/Value	Units	Temp	Data Type	Reference
WS 3.19E+004	mg/L	25	EST	MEYLAN,WM ET AL. (1996)
logP 1.54			EXP	HANSCH,C & LEO,AJ (1985)
VP 1.23E-003	mm Hg	25	EST	NEELY,WB & BLAU,GE (1985)
DC	pKa			
HL 1.43E-010	atm m3/mol	25	EST	MEYLAN,WM & HOWARD,PH (1991)
OH 9.23E-011	cm3/molc sec	25	EST	MEYLAN,WM & HOWARD,PH (1993)

006863-58-7 — DI (SEC-BUTYL) ETHER

Formula: $C_8H_{18}O$

Mol Weight: 130.23

MP (deg C):   FP (deg C):

BP (deg C):

BP pressure (mm Hg):

Property/Value	Units	Temp	Data Type	Reference
WS 3.27E+002	mg/L	25	EST	MEYLAN,WM ET AL. (1996)
logP 2.87			EST	MEYLAN,WM & HOWARD,PH (1995)
VP 1.63E+001	mm Hg	25	EXP	YAWS,CL (1994B)
DC	pKa			
HL 4.72E-003	atm m3/mol	25	EST	MEYLAN,WM & HOWARD,PH (1991)
OH 3.21E-011	cm3/molc sec	25	EST	MEYLAN,WM & HOWARD,PH (1993)

006863-74-7 — 2-CHLORO-6-CYANO-PYRAZINE

Formula: $C_5H_2ClN_3$

Mol Weight: 139.54

MP (deg C):   FP (deg C):

BP (deg C):

BP pressure (mm Hg):

Property/Value	Units	Temp	Data Type	Reference
WS 9.66E+003	mg/L	25	EST	MEYLAN,WM ET AL. (1996)
logP 0.79			EXP	YAMAGAMI,C & TAKAO,N (1991)
VP 2.20E-002	mm Hg	25	EST	NEELY,WB & BLAU,GE (1985)
DC	pKa			
HL 1.32E-006	atm m3/mol	25	EST	MEYLAN,WM & HOWARD,PH (1991)
OH 2.12E-014	cm3/molc sec	25	EST	MEYLAN,WM & HOWARD,PH (1993)

006876-23-9 — 1,2-DIMETHYLCYCLOHEXANE (TRANS)

Formula: C_8H_{16}

Mol Weight: 112.22

MP (deg C): -90   FP (deg C):

BP (deg C): 123.5

BP pressure (mm Hg):

Property/Value	Units	Temp	Data Type	Reference
WS 5.20E+000	mg/L	20	EXP	YALKOWSKY,SH & DANNENFELSER,RM (1992)
logP 4.01			EST	MEYLAN,WM & HOWARD,PH (1995)
VP 1.94E+001	mm Hg	25	EXP	DAUBERT,TE & DANNER,RP (1989)
DC	pKa			
HL 5.43E-001	atm m3/mol	25	EST	VP/WSOL
OH 1.19E-011	cm3/molc sec	25	EST	MEYLAN,WM & HOWARD,PH (1993)

006876-65-9 — ACETAMIDE, N-PHENYL-N-(4-METHYLPHENYL)-

Formula: $C_{15}H_{15}NO$

Mol Weight: 225.29

MP (deg C):   FP (deg C):

BP (deg C):

BP pressure (mm Hg):

Property/Value	Units	Temp	Data Type	Reference
WS 8.02E+001	mg/L	25	EST	MEYLAN,WM ET AL. (1996)
logP 3.05			EXP	YAMAGAMI,C ET AL. (1984)
VP 8.18E-006	mm Hg	25	EST	NEELY,WB & BLAU,GE (1985)
DC	pKa			
HL 3.51E-008	atm m3/mol	25	EST	MEYLAN,WM & HOWARD,PH (1991)
OH 2.79E-011	cm3/molc sec	25	EST	MEYLAN,WM & HOWARD,PH (1993)

006893-02-3 — O-(4-HYDROXY-3-IODOPHENYL)-3,5-DIIODO-L-TYROSINE

Formula: $C_{15}H_{12}I_3NO_4$

Mol Weight: 650.98

MP (deg C): 236-237   FP (deg C):

BP (deg C):

BP pressure (mm Hg):

Property/Value	Units	Temp	Data Type	Reference
WS 3.96E+000	mg/L	37	EXP	YALKOWSKY,SH & DANNENFELSER,RM (1992)
logP 2.96			EST	MEYLAN,WM & HOWARD,PH (1995)
VP 3.90E-016	mm Hg	25	EST	NEELY,WB & BLAU,GE (1985)
DC	pKa			
HL 3.41E-018	atm m3/mol	25	EST	MEYLAN,WM & HOWARD,PH (1991)
OH 5.26E-011	cm3/molc sec	25	EST	MEYLAN,WM & HOWARD,PH (1993)

CAS #: 006898-86-8 — FURAZAN, METHYLPHENYL-, 2-OXIDE

Formula: $C_9H_8N_2O_2$

Mol Weight: 176.18

MP (deg C): | FP (deg C):

BP (deg C):

BP pressure (mm Hg):

Property/Value	Units	Temp	Data Type	Reference
WS 8.06E+002	mg/L	25	EST	MEYLAN,WM ET AL. (1996)
logP 2.17			EXP	CALVINO,R ET AL. (1992)
VP 2.73E-005	mm Hg	25	EST	NEELY,WB & BLAU,GE (1985)
DC	pKa			
HL 1.60E-010	atm m3/mol	25	EST	MEYLAN,WM & HOWARD,PH (1991)
OH 7.52E-012	cm3/molc sec	25	EST	MEYLAN,WM & HOWARD,PH (1993)

CAS #: 006898-87-9 — FURAZAN, METHYLPHENYL-, 5-OXIDE

Formula: $C_9H_8N_2O_2$

Mol Weight: 176.18

MP (deg C): | FP (deg C):

BP (deg C):

BP pressure (mm Hg):

Property/Value	Units	Temp	Data Type	Reference
WS 8.22E+002	mg/L	25	EST	MEYLAN,WM ET AL. (1996)
logP 2.16			EXP	CALVINO,R ET AL. (1992)
VP 2.73E-005	mm Hg	25	EST	NEELY,WB & BLAU,GE (1985)
DC	pKa			
HL 1.60E-010	atm m3/mol	25	EST	MEYLAN,WM & HOWARD,PH (1991)
OH 7.52E-012	cm3/molc sec	25	EST	MEYLAN,WM & HOWARD,PH (1993)

CAS #: 006905-47-1 — PYRAZINAMINE, 6-METHOXY-

Formula: $C_5H_7N_3O$

Mol Weight: 125.13

MP (deg C): | FP (deg C):

BP (deg C):

BP pressure (mm Hg):

Property/Value	Units	Temp	Data Type	Reference
WS 2.29E+004	mg/L	25	EST	MEYLAN,WM ET AL. (1996)
logP 0.73			EXP	YAMAGAMI,C ET AL. (1991)
VP 1.93E-002	mm Hg	25	EST	NEELY,WB & BLAU,GE (1985)
DC	pKa			
HL 5.56E-010	atm m3/mol	25	EST	MEYLAN,WM & HOWARD,PH (1991)
OH 1.41E-010	cm3/molc sec	25	EST	MEYLAN,WM & HOWARD,PH (1993)

CAS #: 006915-15-7 — MALIC ACID

Formula: $C_4H_6O_5$

Mol Weight: 134.09

MP (deg C): 131 | FP (deg C):

BP (deg C):

BP pressure (mm Hg):

Property/Value	Units	Temp	Data Type	Reference
WS 5.92E+005	mg/L	26	EXP	YALKOWSKY,SH & DANNENFELSER,RM (1992)
logP -1.26			EXP	HANSCH,C & LEO,AJ (1985)
VP 4.61E-006	mm Hg	25	EST	NEELY,WB & BLAU,GE (1985)
DC 3.40	pKa	25	EXP	LEUNG,HW & PAUSTENBACH,DJ (1990)
HL 8.40E-013	atm m3/mol	25	EST	MEYLAN,WM & HOWARD,PH (1991)
OH 7.76E-012	cm3/molc sec	25	EST	MEYLAN,WM & HOWARD,PH (1993)

CAS #: 006921-29-5 — TRIPROPYNYL-AMINE

Formula: C_9H_9N

Mol Weight: 131.18

MP (deg C): | FP (deg C):

BP (deg C): 79-85

BP pressure (mm Hg): 1.10E+001

Property/Value	Units	Temp	Data Type	Reference
WS 7.63E+004	mg/L	25	EST	MEYLAN,WM ET AL. (1996)
logP 1.27			EXP	HANSCH,C ET AL. (1995)
VP 7.17E-001	mm Hg	25	EST	NEELY,WB & BLAU,GE (1985)
DC 3.09	pKa		EXP	PERRIN,DD (1965)
HL 3.79E-007	atm m3/mol	25	EST	MEYLAN,WM & HOWARD,PH (1991)
OH 1.13E-010	cm3/molc sec	25	EST	MEYLAN,WM & HOWARD,PH (1993)

CAS #: 006923-22-4 — AZODRIN

Formula: $C_7H_{14}NO_5P$

Mol Weight: 223.17

MP (deg C): 55 | FP (deg C):

BP (deg C): 125

BP pressure (mm Hg): 5.00E-004

Property/Value	Units	Temp	Data Type	Reference
WS 1.00E+006	mg/L	20	EXP	SHIU,WY ET AL. (1990)
logP -1.31			EST	MEYLAN,WM & HOWARD,PH (1995)
VP 1.32E-005	mm Hg	25	EST	NEELY,WB & BLAU,GE (1985)
DC	pKa			
HL 5.45E-013	atm m3/mol	25	EST	MEYLAN,WM & HOWARD,PH (1991)
OH 4.22E-011	cm3/molc sec	25	EST	MEYLAN,WM & HOWARD,PH (1993)

CAS #: 006924-68-1 — 2-ETHYLPYRAZINE CARBOXYLATE

Formula: $C_7H_8N_2O_2$

Mol Weight: 152.15

MP (deg C): | FP (deg C):

BP (deg C):

BP pressure (mm Hg):

Property/Value	Units	Temp	Data Type	Reference
WS 4.28E+004	mg/L	25	EST	MEYLAN,WM ET AL. (1996)
logP 0.28			EXP	YAMAGAMI,C ET AL. (1990A)
VP 4.25E-002	mm Hg	25	EST	NEELY,WB & BLAU,GE (1985)
DC	pKa			
HL 2.50E-008	atm m3/mol	25	EST	MEYLAN,WM & HOWARD,PH (1991)
OH 1.74E-012	cm3/molc sec	25	EST	MEYLAN,WM & HOWARD,PH (1993)

CAS #: 006926-05-2 — PROPANOIC ACID, [(PHENYLMETHYL)SELENO]-

Formula: $C_{10}H_{12}O_2Se$

Mol Weight: 243.17

MP (deg C): | FP (deg C):

BP (deg C):

BP pressure (mm Hg):

Property/Value	Units	Temp	Data Type	Reference
WS 2.40E+003	mg/L	25	EST	MEYLAN,WM ET AL. (1996)
logP 1.67			EXP	SCHWARZ,K ET AL. (1972)
VP 1.70E-005	mm Hg	25	EST	NEELY,WB & BLAU,GE (1985)
DC	pKa			
HL 1.02E-009	atm m3/mol	25	EST	MEYLAN,WM & HOWARD,PH (1991)
OH 8.75E-011	cm3/molc sec	25	EST	MEYLAN,WM & HOWARD,PH (1993)

CAS #: 006926-58-5	4,4-DIMETHYLTHIOSEMICARBAZIDE

Formula: $C_3H_9N_3S$

Mol Weight: 119.19

MP (deg C): 153 dec | FP (deg C):

BP (deg C):

BP pressure (mm Hg):

Property/Value	Units	Temp	Data Type	Reference
WS 9.09E+004	mg/L	25	EST	MEYLAN,WM ET AL. (1996)
logP -0.30			EXP	HANSCH,C ET AL. (1995)
VP 2.63E-001	mm Hg	25	EST	NEELY,WB & BLAU,GE (1985)
DC	pKa			
HL 3.18E-009	atm m3/mol	25	EST	MEYLAN,WM & HOWARD,PH (1991)
OH 8.65E-011	cm3/molc sec	25	EST	MEYLAN,WM & HOWARD,PH (1993)

CAS #: 006932-05-4	N-ACETYLCYCLOBUTANECARBOXAMIDE

Formula: $C_7H_{11}NO_2$

Mol Weight: 141.17

MP (deg C): | FP (deg C):

BP (deg C):

BP pressure (mm Hg):

Property/Value	Units	Temp	Data Type	Reference
WS 1.82E+005	mg/L	25	EST	MEYLAN,WM ET AL. (1996)
logP -0.40			EXP	HANSCH,C & LEO,AJ (1985)
VP 1.30E-006	mm Hg	25	EST	NEELY,WB & BLAU,GE (1985)
DC	pKa			
HL 5.93E-008	atm m3/mol	25	EST	MEYLAN,WM & HOWARD,PH (1991)
OH 9.12E-012	cm3/molc sec	25	EST	MEYLAN,WM & HOWARD,PH (1993)

CAS #: 006933-10-4	3-METHYL-4-BROMOANILINE

Formula: C_7H_8BrN

Mol Weight: 186.06

MP (deg C): 81 | FP (deg C):

BP (deg C): 240

BP pressure (mm Hg):

Property/Value	Units	Temp	Data Type	Reference
WS 3.55E+002	mg/L	25	EST	MEYLAN,WM ET AL. (1996)
logP 2.53			EXP	HANSCH,C & LEO,AJ (1985)
VP 1.25E-002	mm Hg	25	EST	NEELY,WB & BLAU,GE (1985)
DC 4.05	pKa		EXP	PERRIN,DD (1972)
HL 8.37E-007	atm m3/mol	25	EST	MEYLAN,WM & HOWARD,PH (1991)
OH 8.07E-011	cm3/molc sec	25	EST	MEYLAN,WM & HOWARD,PH (1993)

CAS #: 006935-65-5	BENZAMIDE, N,N,3-TRIMETHYL-

Formula: $C_{10}H_{13}NO$

Mol Weight: 163.22

MP (deg C): | FP (deg C):

BP (deg C):

BP pressure (mm Hg):

Property/Value	Units	Temp	Data Type	Reference
WS 2.85E+003	mg/L	25	EST	MEYLAN,WM ET AL. (1996)
logP 1.60			EXP	SURYANARAYANA,MVS ET AL. (1991)
VP 1.10E-001	mm Hg	25	EXP	SURYANARAYANA,MVS ET AL. (1991)
DC	pKa			
HL 1.18E-008	atm m3/mol	25	EST	MEYLAN,WM & HOWARD,PH (1991)
OH 1.84E-011	cm3/molc sec	25	EST	MEYLAN,WM & HOWARD,PH (1993)

CAS #: 006936-40-9	BENZENE, 1,2,4,5-TETRACHLORO-3-METHOXY-

Formula: $C_7H_4Cl_4O$

Mol Weight: 245.92

MP (deg C): 87-90 | FP (deg C):

BP (deg C):

BP pressure (mm Hg):

Property/Value	Units	Temp	Data Type	Reference
WS 2.51E+000	mg/L	25	EST	MEYLAN,WM ET AL. (1996)
logP 4.68			EXP	OPPERHUIZEN,A & VOORS,PI (1987)
VP 3.19E-003	mm Hg	25	EST	NEELY,WB & BLAU,GE (1985)
DC	pKa			
HL 9.61E-005	atm m3/mol	25	EST	MEYLAN,WM & HOWARD,PH (1991)
OH 1.74E-012	cm3/molc sec	25	EST	MEYLAN,WM & HOWARD,PH (1993)

CAS #: 006937-52-6	BENZAMIDE, 4-METHYL-N,N-BIS(1-METHYLETHYL)-

Formula: $C_{14}H_{21}NO$

Mol Weight: 219.33

MP (deg C): | FP (deg C):

BP (deg C):

BP pressure (mm Hg):

Property/Value	Units	Temp	Data Type	Reference
WS 2.54E+002	mg/L	25	EST	MEYLAN,WM ET AL. (1996)
logP 2.50			EXP	SURYANARAYANA,MVS ET AL. (1991)
VP 1.59E-001	mm Hg	25	EXP	SURYANARAYANA,MVS ET AL. (1991)
DC	pKa			
HL 3.66E-008	atm m3/mol	25	EST	MEYLAN,WM & HOWARD,PH (1991)
OH 3.64E-011	cm3/molc sec	25	EST	MEYLAN,WM & HOWARD,PH (1993)

CAS #: 006938-06-3	NICOTINIC ACID, BUTYL ESTER

Formula: $C_{10}H_{13}NO_2$

Mol Weight: 179.22

MP (deg C): | FP (deg C):

BP (deg C): 122-123

BP pressure (mm Hg): 8.00E+000

Property/Value	Units	Temp	Data Type	Reference
WS 6.40E+002	mg/L	25	EST	MEYLAN,WM ET AL. (1996)
logP 2.27			EXP	HANSCH,C & LEO,AJ (1985)
VP 8.51E-003	mm Hg	25	EST	NEELY,WB & BLAU,GE (1985)
DC	pKa			
HL 1.06E-007	atm m3/mol	25	EST	MEYLAN,WM & HOWARD,PH (1991)
OH 4.69E-012	cm3/molc sec	25	EST	MEYLAN,WM & HOWARD,PH (1993)

CAS #: 006943-00-6	P-CHLORO-N-PHENYLSUCCINIMIDE

Formula: $C_{10}H_8ClNO_2$

Mol Weight: 209.63

MP (deg C): | FP (deg C):

BP (deg C):

BP pressure (mm Hg):

Property/Value	Units	Temp	Data Type	Reference
WS 5.91E+003	mg/L	25	EST	MEYLAN,WM ET AL. (1996)
logP 0.96			EXP	HANSCH,C & LEO,AJ (1985)
VP 2.25E-007	mm Hg	25	EST	NEELY,WB & BLAU,GE (1985)
DC	pKa			
HL 9.67E-008	atm m3/mol	25	EST	MEYLAN,WM & HOWARD,PH (1991)
OH 8.00E-012	cm3/molc sec	25	EST	MEYLAN,WM & HOWARD,PH (1993)

CAS #: 006947-77-9	5,7-DIAZASPIRO[2,5]OCTANE-4,6,8-TRIONE

Formula: $C_6H_6N_2O_3$

Mol Weight: 154.13

MP (deg C): FP (deg C):

BP (deg C):

BP pressure (mm Hg):

Property/Value	Units	Temp	Data Type	Reference
WS 2.06E+005	mg/L	25	EST	MEYLAN,WM ET AL. (1996)
logP -0.53			EXP	PRANKERD,RJ & MCKEOWN,RH (1992)
VP 6.09E-010	mm Hg	25	EST	NEELY,WB & BLAU,GE (1985)
DC	pKa			
HL 9.02E-014	atm m3/mol	25	EST	MEYLAN,WM & HOWARD,PH (1991)
OH 2.18E-012	cm3/molc sec	25	EST	MEYLAN,WM & HOWARD,PH (1993)

CAS #: 006952-59-6	3-BROMOBENZONITRILE

Formula: C_7H_4BrN

Mol Weight: 182.03

MP (deg C): 39.5 FP (deg C):

BP (deg C): 225

BP pressure (mm Hg):

Property/Value	Units	Temp	Data Type	Reference
WS 2.46E+002	mg/L	25	EST	MEYLAN,WM ET AL. (1996)
logP 2.43			EST	MEYLAN,WM & HOWARD,PH (1995)
VP 2.73E-002	mm Hg	25	EST	NEELY,WB & BLAU,GE (1985)
DC	pKa			
HL 2.07E-005	atm m3/mol	25	EST	MEYLAN,WM & HOWARD,PH (1991)
OH 1.61E-013	cm3/molc sec	25	EST	MEYLAN,WM & HOWARD,PH (1993)

CAS #: 006954-35-4	4-NITROIMIDAZOLE1-ME-5-SO3PH

Formula: $C_{10}H_{11}N_3O_5S$

Mol Weight: 285.28

MP (deg C): FP (deg C):

BP (deg C):

BP pressure (mm Hg):

Property/Value	Units	Temp	Data Type	Reference
WS 7.48E+002	mg/L	25	EST	MEYLAN,WM ET AL. (1996)
logP 1.08			EXP	HANSCH,C & LEO,AJ (1985)
VP 2.47E-009	mm Hg	25	EST	NEELY,WB & BLAU,GE (1985)
DC	pKa			
HL 1.75E-011	atm m3/mol	25	EST	MEYLAN,WM & HOWARD,PH (1991)
OH 2.26E-011	cm3/molc sec	25	EST	MEYLAN,WM & HOWARD,PH (1993)

CAS #: 006960-42-5	7-NITROINDOLE

Formula: $C_8H_6N_2O_2$

Mol Weight: 162.15

MP (deg C): FP (deg C):

BP (deg C):

BP pressure (mm Hg):

Property/Value	Units	Temp	Data Type	Reference
WS 1.55E+002	mg/L	25	EST	MEYLAN,WM ET AL. (1996)
logP 2.63			EXP	HANSCH,C ET AL. (1995)
VP 8.99E-005	mm Hg	25	EST	NEELY,WB & BLAU,GE (1985)
DC	pKa			
HL 3.50E-009	atm m3/mol	25	EST	MEYLAN,WM & HOWARD,PH (1991)
OH 1.37E-011	cm3/molc sec	25	EST	MEYLAN,WM & HOWARD,PH (1993)

CAS #: 006961-82-6	O-CHLOROBENZENESULFONAMIDE

Formula: $C_6H_6ClNO_2S$

Mol Weight: 191.64

MP (deg C): FP (deg C):

BP (deg C):

BP pressure (mm Hg):

Property/Value	Units	Temp	Data Type	Reference
WS 1.13E+004	mg/L	25	EST	MEYLAN,WM ET AL. (1996)
logP 0.74			EXP	HANSCH,C & LEO,AJ (1985)
VP 1.59E-004	mm Hg	25	EST	NEELY,WB & BLAU,GE (1985)
DC	pKa			
HL 3.13E-007	atm m3/mol	25	EST	MEYLAN,WM & HOWARD,PH (1991)
OH 2.93E-013	cm3/molc sec	25	EST	MEYLAN,WM & HOWARD,PH (1993)

CAS #: 006963-65-1	4-BROMO-5-NITRO-1H-IMIDAZOLE

Formula: $C_3H_2BrN_3O_2$

Mol Weight: 191.98

MP (deg C): FP (deg C):

BP (deg C):

BP pressure (mm Hg):

Property/Value	Units	Temp	Data Type	Reference
WS 3.68E+003	mg/L	25	EST	MEYLAN,WM ET AL. (1996)
logP 0.85			EXP	SUWINSKI,J ET AL. (1985)
VP 1.31E-005	mm Hg	25	EST	NEELY,WB & BLAU,GE (1985)
DC	pKa			
HL 5.90E-009	atm m3/mol	25	EST	MEYLAN,WM & HOWARD,PH (1991)
OH 1.29E-012	cm3/molc sec	25	EST	MEYLAN,WM & HOWARD,PH (1993)

CAS #: 006963-66-2	1H-IMIDAZOLE, 5-CHLORO-4-NITRO-

Formula: $C_3H_2ClN_3O_2$

Mol Weight: 147.52

MP (deg C): FP (deg C):

BP (deg C):

BP pressure (mm Hg):

Property/Value	Units	Temp	Data Type	Reference
WS 6.84E+003	mg/L	25	EST	MEYLAN,WM ET AL. (1996)
logP 0.78			EXP	SUWINSKI,J ET AL. (1985)
VP 3.52E-005	mm Hg	25	EST	NEELY,WB & BLAU,GE (1985)
DC	pKa			
HL 1.10E-008	atm m3/mol	25	EST	MEYLAN,WM & HOWARD,PH (1991)
OH 1.31E-012	cm3/molc sec	25	EST	MEYLAN,WM & HOWARD,PH (1993)

CAS #: 006964-19-8	4-CHLORO-3-PENTADECYLPHENOL

Formula: $C_{21}H_{35}ClO$

Mol Weight: 338.97

MP (deg C): FP (deg C):

BP (deg C):

BP pressure (mm Hg):

Property/Value	Units	Temp	Data Type	Reference
WS 1.81E-004	mg/L	25	EST	MEYLAN,WM ET AL. (1996)
logP 9.58			EST	MEYLAN,WM & HOWARD,PH (1995)
VP 3.54E-008	mm Hg	25	EST	NEELY,WB & BLAU,GE (1985)
DC	pKa			
HL 2.42E-005	atm m3/mol	25	EST	MEYLAN,WM & HOWARD,PH (1991)
OH 4.37E-011	cm3/molc sec	25	EST	MEYLAN,WM & HOWARD,PH (1993)

006966-38-7

CAS #: 006966-38-7

M-(BIS-METHYLSUFONYLAMIDO)BENZENE

Formula: $C_8H_{12}N_2O_4S_2$

Mol Weight: 264.32

MP (deg C):   **FP (deg C):**

BP (deg C):

BP pressure (mm Hg):

Property/Value	Units	Temp	Data Type	Reference
WS 8.40E+002	mg/L	25	EST	MEYLAN,WM ET AL. (1996)
logP 0.07			EXP	HANSCH,C & LEO,AJ (1985)
VP 8.88E-008	mm Hg	25	EST	NEELY,WB & BLAU,GE (1985)
DC	pKa			
HL 2.52E-010	atm m3/mol	25	EST	MEYLAN,WM & HOWARD,PH (1991)
OH 2.02E-010	cm3/molc sec	25	EST	MEYLAN,WM & HOWARD,PH (1993)

006969-90-0

CAS #: 006969-90-0

N,N-DIMETHYLPHENYLCARBAMATE

Formula: $C_9H_{11}NO_2$

Mol Weight: 165.19

MP (deg C):   **FP (deg C):**

BP (deg C):

BP pressure (mm Hg):

Property/Value	Units	Temp	Data Type	Reference
WS 3.01E+003	mg/L	25	EST	MEYLAN,WM ET AL. (1996)
logP 1.56			EXP	HANSCH,C & LEO,AJ (1985)
VP 6.41E-002	mm Hg	25	EST	NEELY,WB & BLAU,GE (1985)
DC	pKa			
HL 7.05E-008	atm m3/mol	25	EST	MEYLAN,WM & HOWARD,PH (1991)
OH 1.81E-011	cm3/molc sec	25	EST	MEYLAN,WM & HOWARD,PH (1993)

006971-38-6

CAS #: 006971-38-6

PHENOL, 2-METHYL-4-NITROSO-

Formula: $C_7H_7NO_2$

Mol Weight: 137.14

MP (deg C):   **FP (deg C):**

BP (deg C):

BP pressure (mm Hg):

Property/Value	Units	Temp	Data Type	Reference
WS 1.13E+004	mg/L	25	EST	MEYLAN,WM ET AL. (1996)
logP 1.71			EXP	SANGSTER,J (1993)
VP 3.91E-002	mm Hg	25	EST	NEELY,WB & BLAU,GE (1985)
DC	pKa			
HL 6.47E-009	atm m3/mol	25	EST	MEYLAN,WM & HOWARD,PH (1991)
OH 2.26E-011	cm3/molc sec	25	EST	MEYLAN,WM & HOWARD,PH (1993)

006972-05-0

CAS #: 006972-05-0

THIOUREA, N,N-DIMETHYL-

Formula: $C_3H_8N_2S$

Mol Weight: 104.17

MP (deg C):   **FP (deg C):**

BP (deg C):

BP pressure (mm Hg):

Property/Value	Units	Temp	Data Type	Reference
WS 1.32E+005	mg/L	25	EST	MEYLAN,WM ET AL. (1996)
logP -0.43			EXP	GOVERS,H ET AL. (1986)
VP 2.00E+000	mm Hg	25	EST	NEELY,WB & BLAU,GE (1985)
DC	pKa			
HL 7.61E-007	atm m3/mol	25	EST	MEYLAN,WM & HOWARD,PH (1991)
OH 8.95E-011	cm3/molc sec	25	EST	MEYLAN,WM & HOWARD,PH (1993)

006972-69-6

CAS #: 006972-69-6

N,N-DIMETHYLNICOTINAMIDE

Formula: $C_8H_{10}N_2O$

Mol Weight: 150.18

MP (deg C):   **FP (deg C):**

BP (deg C):

BP pressure (mm Hg):

Property/Value	Units	Temp	Data Type	Reference
WS 1.95E+005	mg/L	25	EST	MEYLAN,WM ET AL. (1996)
logP -0.48			EXP	HANSCH,C & LEO,AJ (1985)
VP 3.80E-003	mm Hg	25	EST	NEELY,WB & BLAU,GE (1985)
DC	pKa			
HL 1.40E-011	atm m3/mol	25	EST	MEYLAN,WM & HOWARD,PH (1991)
OH 1.65E-011	cm3/molc sec	25	EST	MEYLAN,WM & HOWARD,PH (1993)

006975-29-7

CAS #: 006975-29-7

2,4-DICHLOROACETANILIDE

Formula: $C_8H_7Cl_2NO$

Mol Weight: 204.06

MP (deg C):   **FP (deg C):**

BP (deg C):

BP pressure (mm Hg):

Property/Value	Units	Temp	Data Type	Reference
WS 5.74E+002	mg/L	25	EST	MEYLAN,WM ET AL. (1996)
logP 2.18			EXP	NAKAGAWA,Y ET AL. (1992)
VP 2.02E-005	mm Hg	25	EST	NEELY,WB & BLAU,GE (1985)
DC	pKa			
HL 3.39E-009	atm m3/mol	25	EST	MEYLAN,WM & HOWARD,PH (1991)
OH 1.16E-012	cm3/molc sec	25	EST	MEYLAN,WM & HOWARD,PH (1993)

006975-98-0

CAS #: 006975-98-0

2-METHYLDECANE

Formula: $C_{11}H_{24}$

Mol Weight: 156.31

MP (deg C): -48.9   **FP (deg C):**

BP (deg C): 189.3

BP pressure (mm Hg):

Property/Value	Units	Temp	Data Type	Reference
WS 2.97E-001	mg/L	25	EST	MEYLAN,WM ET AL. (1996)
logP 5.67			EST	MEYLAN,WM & HOWARD,PH (1995)
VP 6.02E-001	mm Hg	25	EXT	BOUBLIK,T ET AL. (1984)
DC	pKa			
HL 7.04E+000	atm m3/mol	25	EST	MEYLAN,WM & HOWARD,PH (1991)
OH 1.25E-011	cm3/molc sec	25	EST	MEYLAN,WM & HOWARD,PH (1993)

006981-01-7

CAS #: 006981-01-7

2,4-NH2PYRIMIDINE,5(2,4,5-MEO)BENZYL

Formula: $C_{14}H_{18}N_4O_3$

Mol Weight: 290.32

MP (deg C):   **FP (deg C):**

BP (deg C):

BP pressure (mm Hg):

Property/Value	Units	Temp	Data Type	Reference
WS 1.40E+003	mg/L	25	EST	MEYLAN,WM ET AL. (1996)
logP 1.17			EXP	HANSCH,C & LEO,AJ (1985)
VP 9.88E-009	mm Hg	25	EST	NEELY,WB & BLAU,GE (1985)
DC	pKa			
HL 2.39E-014	atm m3/mol	25	EST	MEYLAN,WM & HOWARD,PH (1991)
OH 2.03E-010	cm3/molc sec	25	EST	MEYLAN,WM & HOWARD,PH (1993)

CAS #: 006981-04-0 — 2,4-PYRIMIDINEDIAMINE, 5-[(2,3,4-TRIMETHOXYPHENY

Formula: $C_{14}H_{18}N_4O_3$
Mol Weight: 290.32
MP (deg C):
FP (deg C):
BP (deg C):
BP pressure (mm Hg):

Property/Value	Units	Temp	Data Type	Reference
WS 7.04E+002	mg/L	25	EST	MEYLAN,WM ET AL. (1996)
logP 1.52			EXP	HANSCH,C ET AL. (1995)
VP 9.88E-009	mm Hg	25	EST	NEELY,WB & BLAU,GE (1985)
DC	pKa			
HL 2.39E-014	atm m3/mol	25	EST	MEYLAN,WM & HOWARD,PH (1991)
OH 2.03E-010	cm3/molc sec	25	EST	MEYLAN,WM & HOWARD,PH (1993)

CAS #: 006981-11-9 — 2,4-DIAMINOPYRIMIDINE-5(245-CL)BENZYL

Formula: $C_{11}H_9Cl_3N_4$
Mol Weight: 303.58
MP (deg C):
FP (deg C):
BP (deg C):
BP pressure (mm Hg):

Property/Value	Units	Temp	Data Type	Reference
WS 6.79E+000	mg/L	25	EST	MEYLAN,WM ET AL. (1996)
logP 3.79			EXP	HANSCH,C & LEO,AJ (1985)
VP 2.77E-008	mm Hg	25	EST	NEELY,WB & BLAU,GE (1985)
DC	pKa			
HL 4.69E-011	atm m3/mol	25	EST	MEYLAN,WM & HOWARD,PH (1991)
OH 2.81E-012	cm3/molc sec	25	EST	MEYLAN,WM & HOWARD,PH (1993)

CAS #: 006981-18-6 — 2,4-NH2PYRIMIDIN,5(45-MEO-2-ME)BENZYL

Formula: $C_{14}H_{18}N_4O_2$
Mol Weight: 274.33
MP (deg C):
FP (deg C):
BP (deg C):
BP pressure (mm Hg):

Property/Value	Units	Temp	Data Type	Reference
WS 1.54E+003	mg/L	25	EST	MEYLAN,WM ET AL. (1996)
logP 1.23			EXP	HANSCH,C & LEO,AJ (1985)
VP 2.28E-008	mm Hg	25	EST	NEELY,WB & BLAU,GE (1985)
DC	pKa			
HL 4.45E-013	atm m3/mol	25	EST	MEYLAN,WM & HOWARD,PH (1991)
OH 6.34E-011	cm3/molc sec	25	EST	MEYLAN,WM & HOWARD,PH (1993)

CAS #: 006982-25-8 — DL-2,3-BUTANDIOL

Formula: $C_4H_{10}O_2$
Mol Weight: 90.12
MP (deg C):
FP (deg C):
BP (deg C):
BP pressure (mm Hg):

Property/Value	Units	Temp	Data Type	Reference
WS 7.59E+005	mg/L	25	EST	MEYLAN,WM ET AL. (1996)
logP -0.36			EST	MEYLAN,WM & HOWARD,PH (1995)
VP 1.82E-001	mm Hg	25	EXP	DAUBERT,TE & DANNER,RP (1989)
DC	pKa			
HL 2.30E-007	atm m3/mol	25	EST	MEYLAN,WM & HOWARD,PH (1991)
OH 1.73E-011	cm3/molc sec	25	EST	MEYLAN,WM & HOWARD,PH (1993)

CAS #: 006987-59-3 — SEARLE, SC-17127

Formula: $C_{24}H_{32}O_3$
Mol Weight: 368.52
MP (deg C):
FP (deg C):
BP (deg C):
BP pressure (mm Hg):

Property/Value	Units	Temp	Data Type	Reference
WS 9.76E+000	mg/L	25	EST	MEYLAN,WM ET AL. (1996)
logP 3.15			EXP	HANSCH,C & LEO,AJ (1985)
VP 1.96E-009	mm Hg	25	EST	NEELY,WB & BLAU,GE (1985)
DC	pKa			
HL 1.04E-007	atm m3/mol	25	EST	MEYLAN,WM & HOWARD,PH (1991)
OH 8.99E-011	cm3/molc sec	25	EST	MEYLAN,WM & HOWARD,PH (1993)

CAS #: 006988-21-2 — DIOXACARB

Formula: $C_{11}H_{13}NO_4$
Mol Weight: 223.23
MP (deg C):
FP (deg C):
BP (deg C):
BP pressure (mm Hg):

Property/Value	Units	Temp	Data Type	Reference
WS 6.00E+003	mg/L	20	EXP	YALKOWSKY,SH & DANNENFELSER,RM (1992)
logP 0.67			EXP	HANSCH,C & LEO,AJ (1985)
VP 3.00E-007	mm Hg	25	EXP	AUGUSTIJN-BECKERS,PWM ET AL. (1994)
DC	pKa			
HL 1.47E-011	atm m3/mol	25	EST	VP/WSOL
OH 2.63E-011	cm3/molc sec	25	EST	MEYLAN,WM & HOWARD,PH (1993)

CAS #: 006996-81-2 — Phosphonothioic acid, methyl-, diethyl ester

Formula: $C_5H_{13}O_2PS$
Mol Weight: 168.20
MP (deg C): -84.4
FP (deg C):
BP (deg C): 198.9
BP pressure (mm Hg):

Property/Value	Units	Temp	Data Type	Reference
WS 1.05E+003	mg/L	25	EST	MEYLAN,WM ET AL. (1996)
logP 2.08			EXP	KRIKORIAN,SE ET AL. (1987)
VP 6.55E-001	mm Hg	25	EST	NEELY,WB & BLAU,GE (1985)
DC	pKa			
HL 1.02E-003	atm m3/mol	25	EST	MEYLAN,WM & HOWARD,PH (1991)
OH 9.00E-011	cm3/molc sec	25	EST	MEYLAN,WM & HOWARD,PH (1993)

CAS #: 007005-72-3 — 4-CHLOROPHENYL PHENYL ETHER

Formula: $C_{12}H_9ClO$
Mol Weight: 204.66
MP (deg C):
FP (deg C):
BP (deg C): 284.5
BP pressure (mm Hg):

Property/Value	Units	Temp	Data Type	Reference
WS 3.30E+000	mg/L	25	EXP	BRANSON,DR (1977)
logP 4.08			EXP	SANGSTER,J (1994)
VP 2.70E-003	mm Hg	25	EXP	BRANSON,DR (1977)
DC	pKa			
HL 3.40E-002	atm m3/mol	25	EXP	BRANSON,DR (1977)
OH 1.25E-011	cm3/molc sec	25	EST	MEYLAN,WM & HOWARD,PH (1993)

CAS #: 007011-98-5	1(3H)-ISOBENZOFURANONE-3-BENZYL

Formula: $C_{15}H_{12}O_2$
Mol Weight: 224.26
MP (deg C):   FP (deg C):
BP (deg C):
BP pressure (mm Hg):

Property/Value	Units	Temp	Data Type	Reference
WS 7.07E+001	mg/L	25	EST	MEYLAN,WM ET AL. (1996)
logP 3.12			EXP	HANSCH,C & LEO,AJ (1985)
VP 4.29E-006	mm Hg	25	EST	NEELY,WB & BLAU,GE (1985)
DC	pKa			
HL 1.36E-006	atm m3/mol	25	EST	MEYLAN,WM & HOWARD,PH (1991)
OH 1.33E-011	cm3/molc sec	25	EST	MEYLAN,WM & HOWARD,PH (1993)

CAS #: 007012-37-5	2,4,4'-PCB

Formula: $C_{12}H_7Cl_3$
Mol Weight: 257.55
MP (deg C):   FP (deg C):
BP (deg C):
BP pressure (mm Hg):

Property/Value	Units	Temp	Data Type	Reference
WS 2.70E-001	mg/L	25	EXP	CHIOU,CT ET AL. (1983)
logP 5.62			EXP	HANSCH,C & LEO,AJ (1985)
VP 1.95E-004	mm Hg	25	EXP	BALLSCHMITER,K & WITTLINGER,R (1991)
DC	pKa			
HL 2.45E-004	atm m3/mol	25	EST	VP/WSOL
OH 1.19E-012	cm3/molc sec	25	EST	MEYLAN,WM & HOWARD,PH (1993)

CAS #: 007019-01-4	BENZENAMINE, 4-(PHENYLSULFONYL)-

Formula: $C_{12}H_{11}NO_2S$
Mol Weight: 233.29
MP (deg C):   FP (deg C):
BP (deg C):
BP pressure (mm Hg):

Property/Value	Units	Temp	Data Type	Reference
WS 9.17E+002	mg/L	25	EST	MEYLAN,WM ET AL. (1996)
logP 1.76			EXP	ALTOMARE,C ET AL. (1991)
VP 6.72E-007	mm Hg	25	EST	NEELY,WB & BLAU,GE (1985)
DC	pKa			
HL 8.81E-011	atm m3/mol	25	EST	MEYLAN,WM & HOWARD,PH (1991)
OH 2.34E-011	cm3/molc sec	25	EST	MEYLAN,WM & HOWARD,PH (1993)

CAS #: 007030-18-4	2-METHYL-4'-NITROAZOBENZENE

Formula: $C_{13}H_{11}N_3O_2$
Mol Weight: 241.25
MP (deg C):   FP (deg C):
BP (deg C):
BP pressure (mm Hg):

Property/Value	Units	Temp	Data Type	Reference
WS 4.72E-001	mg/L	25	EST	MEYLAN,WM ET AL. (1996)
logP 5.05			EST	MEYLAN,WM & HOWARD,PH (1995)
VP 5.35E-006	mm Hg	25	EST	NEELY,WB & BLAU,GE (1985)
DC	pKa			
HL 6.40E-008	atm m3/mol	25	EST	MEYLAN,WM & HOWARD,PH (1991)
OH 2.22E-012	cm3/molc sec	25	EST	MEYLAN,WM & HOWARD,PH (1993)

CAS #: 007040-58-6	BIS(1,2,2-TRIMETHYLPROPYL) METHYLPHOSPHONATE

Formula: $C_{13}H_{29}O_3P$
Mol Weight: 264.35
MP (deg C):   FP (deg C):
BP (deg C):
BP pressure (mm Hg):

Property/Value	Units	Temp	Data Type	Reference
WS 4.27E+000	mg/L	25	EST	MEYLAN,WM ET AL. (1996)
logP 3.96			EST	MEYLAN,WM & HOWARD,PH (1995)
VP 5.49E-003	mm Hg	25	EST	NEELY,WB & BLAU,GE (1985)
DC	pKa			
HL 2.12E-005	atm m3/mol	25	EST	MEYLAN,WM & HOWARD,PH (1991)
OH 9.93E-011	cm3/molc sec	25	EST	MEYLAN,WM & HOWARD,PH (1993)

CAS #: 007057-48-9	ADENOSINE, 2',3'-DIDEHYDRO-2',3'-DIDEOXY-

Formula: $C_{10}H_{11}N_5O_2$
Mol Weight: 233.23
MP (deg C):   FP (deg C):
BP (deg C):
BP pressure (mm Hg):

Property/Value	Units	Temp	Data Type	Reference
WS 3.83E+003	mg/L	25	EST	MEYLAN,WM ET AL. (1996)
logP -0.50			EXP	SANGSTER,J (1993)
VP 3.09E-011	mm Hg	25	EST	NEELY,WB & BLAU,GE (1985)
DC	pKa			
HL 5.81E-019	atm m3/mol	25	EST	MEYLAN,WM & HOWARD,PH (1991)
OH 2.82E-010	cm3/molc sec	25	EST	MEYLAN,WM & HOWARD,PH (1993)

CAS #: 007058-01-7	S-BUTYLCYCLOHEXANE

Formula: $C_{10}H_{20}$
Mol Weight: 140.27
MP (deg C):   FP (deg C):
BP (deg C): 179.3
BP pressure (mm Hg):

Property/Value	Units	Temp	Data Type	Reference
WS 1.32E+000	mg/L	25	EST	MEYLAN,WM ET AL. (1996)
logP 4.99			EST	MEYLAN,WM & HOWARD,PH (1995)
VP 2.35E+000	mm Hg	25	EXT	OHE,S (1976)
DC	pKa			
HL 7.92E-001	atm m3/mol	25	EST	MEYLAN,WM & HOWARD,PH (1991)
OH 1.51E-011	cm3/molc sec	25	EST	MEYLAN,WM & HOWARD,PH (1993)

CAS #: 007068-83-9	N-NITROSOMETHYLBUTYLAMINE

Formula: $C_5H_{12}N_2O$
Mol Weight: 116.16
MP (deg C):   FP (deg C):
BP (deg C):
BP pressure (mm Hg):

Property/Value	Units	Temp	Data Type	Reference
WS 1.24E+004	mg/L	25	EST	MEYLAN,WM ET AL. (1996)
logP 1.08			EXP	VERA,A ET AL. (1992)
VP 2.19E-001	mm Hg	25	EST	NEELY,WB & BLAU,GE (1985)
DC	pKa			
HL 4.81E-006	atm m3/mol	25	EST	MEYLAN,WM & HOWARD,PH (1991)
OH 1.47E-011	cm3/molc sec	25	EST	MEYLAN,WM & HOWARD,PH (1993)

CAS #: 007073-42-9				3-CHLORO-4-METHOXYACETANILIDE
Formula: $C_9H_{10}ClNO_2$				
Mol Weight: 199.64				
MP (deg C):		FP (deg C):		
BP (deg C):				
BP pressure (mm Hg):				

Property/ Value	Units	Temp	Data Type	Reference
WS 1.25E+003	mg/L	25	EST	MEYLAN,WM ET AL. (1996)
logP 1.81			EXP	NAKAGAWA,Y ET AL. (1992)
VP 1.42E-005	mm Hg	25	EST	NEELY,WB & BLAU,GE (1985)
DC	pKa			
HL 2.71E-010	atm m3/mol	25	EST	MEYLAN,WM & HOWARD,PH (1991)
OH 8.49E-012	cm3/molc sec	25	EST	MEYLAN,WM & HOWARD,PH (1993)

CAS #: 007085-19-0				MCPP
Formula: $C_{10}H_{11}ClO_3$				
Mol Weight: 214.65				
MP (deg C): 93-94		FP (deg C):		
BP (deg C):				
BP pressure (mm Hg):				

Property/ Value	Units	Temp	Data Type	Reference
WS 8.95E+002	mg/L	25	EXP	GUNTHER,FA ET AL. (1968)
logP 2.94			EST	MEYLAN,WM & HOWARD,PH (1995)
VP 8.36E-005	mm Hg	25	EST	NEELY,WB & BLAU,GE (1985)
DC	pKa			
HL 1.82E-008	atm m3/mol	25	EST	MEYLAN,WM & HOWARD,PH (1991)
OH 1.74E-011	cm3/molc sec	25	EST	MEYLAN,WM & HOWARD,PH (1993)

CAS #: 007090-25-7				N-ME-N-NITROSO-1-NAPHTHYLCARBAMAT
Formula: $C_{12}H_{10}N_2O_3$				
Mol Weight: 230.23				
MP (deg C):		FP (deg C):		
BP (deg C):				
BP pressure (mm Hg):				

Property/ Value	Units	Temp	Data Type	Reference
WS 1.12E+002	mg/L	25	EST	MEYLAN,WM ET AL. (1996)
logP 2.85			EXP	HANSCH,C & LEO,AJ (1985)
VP 1.99E-006	mm Hg	25	EST	NEELY,WB & BLAU,GE (1985)
DC	pKa			
HL 3.88E-010	atm m3/mol	25	EST	MEYLAN,WM & HOWARD,PH (1991)
OH 2.28E-011	cm3/molc sec	25	EST	MEYLAN,WM & HOWARD,PH (1993)

CAS #: 007090-41-7				BIPHENYLENE, 2,3,6,7-TETRACHLORO-
Formula: $C_{12}H_4Cl_4$				
Mol Weight: 289.98				
MP (deg C):		FP (deg C):		
BP (deg C):				
BP pressure (mm Hg):				

Property/ Value	Units	Temp	Data Type	Reference
WS 6.22E-003	mg/L	25	EST	MEYLAN,WM ET AL. (1996)
logP 6.14			EXP	BURKHARD,LP & KUEHL,DW (1986)
VP 2.37E-006	mm Hg	25	EST	NEELY,WB & BLAU,GE (1985)
DC	pKa			
HL 2.56E-005	atm m3/mol	25	EST	MEYLAN,WM & HOWARD,PH (1991)
OH 3.08E-013	cm3/molc sec	25	EST	MEYLAN,WM & HOWARD,PH (1993)

CAS #: 007093-67-6				GLYCINE, N-[N-[N-(N-GLYCYLGLYCYL)GLYCYL]GLYCYL]-
Formula: $C_{10}H_{17}N_5O_6$				
Mol Weight: 303.28				
MP (deg C):		FP (deg C):		
BP (deg C):				
BP pressure (mm Hg):				

Property/ Value	Units	Temp	Data Type	Reference
WS 3.86E+005	mg/L	25	EST	MEYLAN,WM ET AL. (1996)
logP -3.31			EXP	HANSCH,C ET AL. (1995)
VP 3.83E-018	mm Hg	25	EST	NEELY,WB & BLAU,GE (1985)
DC	pKa			
HL 3.49E-027	atm m3/mol	25	EST	MEYLAN,WM & HOWARD,PH (1991)
OH 6.15E-011	cm3/molc sec	25	EST	MEYLAN,WM & HOWARD,PH (1993)

CAS #: 007098-22-8				TETRATETRACONTANE
Formula: $C_{44}H_{90}$				
Mol Weight: 619.21				
MP (deg C): 87-88		FP (deg C):		
BP (deg C):				
BP pressure (mm Hg):				

Property/ Value	Units	Temp	Data Type	Reference
WS 6.28E-018	mg/L	25	EST	MEYLAN,WM ET AL. (1996)
logP 21.95			EST	MEYLAN,WM & HOWARD,PH (1995)
VP 2.48E-042	mm Hg	25	EXT	ZWOLINSKI,BJ & WILHOIT,RC (1971)
DC	pKa			
HL 8.10E+004	atm m3/mol	25	EST	MEYLAN,WM & HOWARD,PH (1991)
OH 5.92E-011	cm3/molc sec	25	EST	MEYLAN,WM & HOWARD,PH (1993)

CAS #: 007101-51-1				L-TYROSINE, 3-HYDROXY-, METHYL ESTER
Formula: $C_{10}H_{13}NO_4$				
Mol Weight: 211.22				
MP (deg C):		FP (deg C):		
BP (deg C):				
BP pressure (mm Hg):				

Property/ Value	Units	Temp	Data Type	Reference
WS 1.00E+006	mg/L	25	EST	MEYLAN,WM ET AL. (1996)
logP -0.21			EXP	FIX,JA ET AL. (1989)
VP 7.90E-007	mm Hg	25	EST	NEELY,WB & BLAU,GE (1985)
DC	pKa			
HL 4.18E-016	atm m3/mol	25	EST	MEYLAN,WM & HOWARD,PH (1991)
OH 9.61E-011	cm3/molc sec	25	EST	MEYLAN,WM & HOWARD,PH (1993)

CAS #: 007116-95-2				4-ISOPROPYLBIPHENYL
Formula: $C_{16}H_{18}$				
Mol Weight: 210.32				
MP (deg C):		FP (deg C):		
BP (deg C):				
BP pressure (mm Hg):				

Property/ Value	Units	Temp	Data Type	Reference
WS 9.00E-001	mg/L	25	EST	MEYLAN,WM ET AL. (1996)
logP 5.51			EXP	TSCATS
VP 8.18E-004	mm Hg	25	EST	NEELY,WB & BLAU,GE (1985)
DC	pKa			
HL 8.05E-004	atm m3/mol	25	EST	MEYLAN,WM & HOWARD,PH (1991)
OH 9.74E-012	cm3/molc sec	25	EST	MEYLAN,WM & HOWARD,PH (1993)

007120-01-6

CAS #: 007120-01-6

GUANIDINE, (4-METHYL-2-THIAZOLYL)-

Formula: $C_5H_8N_4S$

Mol Weight: 156.21

MP (deg C): **FP (deg C):**

BP (deg C):

BP pressure (mm Hg):

Property/Value	Units	Temp	Data Type	Reference
WS 6.33E+004	mg/L	25	EST	MEYLAN,WM ET AL. (1996)
logP 1.24			EXP	HANSCH,C ET AL. (1995)
VP 7.89E-004	mm Hg	25	EST	NEELY,WB & BLAU,GE (1985)
DC	pKa			
HL 1.01E-014	atm m3/mol	25	EST	MEYLAN,WM & HOWARD,PH (1991)
OH 4.61E-011	cm3/molc sec	25	EST	MEYLAN,WM & HOWARD,PH (1993)

007132-64-1

CAS #: 007132-64-1

METHYL PENTADECANOATE

Formula: $C_{16}H_{32}O_2$

Mol Weight: 256.43

MP (deg C): 18.5 **FP (deg C):**

BP (deg C): 153.5

BP pressure (mm Hg):

Property/Value	Units	Temp	Data Type	Reference
WS 3.69E-002	mg/L	25	EST	MEYLAN,WM ET AL. (1996)
logP 6.76			EST	MEYLAN,WM & HOWARD,PH (1995)
VP 1.63E-004	mm Hg	25	EXP	BACCANARI,DP ET AL (1968)
DC	pKa			
HL 6.97E-003	atm m3/mol	25	EST	MEYLAN,WM & HOWARD,PH (1991)
OH 1.79E-011	cm3/molc sec	25	EST	MEYLAN,WM & HOWARD,PH (1993)

007145-20-2

CAS #: 007145-20-2

2,3-DIMETHYL-2-HEXENE

Formula: C_8H_{16}

Mol Weight: 112.22

MP (deg C): -115.1 **FP (deg C):**

BP (deg C): 121.8

BP pressure (mm Hg):

Property/Value	Units	Temp	Data Type	Reference
WS 8.57E+000	mg/L	25	EST	MEYLAN,WM ET AL. (1996)
logP 4.17			EST	MEYLAN,WM & HOWARD,PH (1995)
VP 1.83E+001	mm Hg	25	EXP	ZWOLINSKI,BJ & WILHOIT,RC (1971)
DC	pKa			
HL 1.04E+000	atm m3/mol	25	EST	MEYLAN,WM & HOWARD,PH (1991)
OH 1.13E-010	cm3/molc sec	25	EST	MEYLAN,WM & HOWARD,PH (1993)

007145-60-0

CAS #: 007145-60-0

3-CL-6-PYRIDAZINAMINE,N,N-DIME

Formula: $C_6H_8ClN_3$

Mol Weight: 157.60

MP (deg C): **FP (deg C):**

BP (deg C):

BP pressure (mm Hg):

Property/Value	Units	Temp	Data Type	Reference
WS 1.02E+004	mg/L	25	EST	MEYLAN,WM ET AL. (1996)
logP 0.98			EXP	HANSCH,C & LEO,AJ (1985)
VP 2.04E-003	mm Hg	25	EST	NEELY,WB & BLAU,GE (1985)
DC	pKa			
HL 2.13E-006	atm m3/mol	25	EST	MEYLAN,WM & HOWARD,PH (1991)
OH 1.53E-011	cm3/molc sec	25	EST	MEYLAN,WM & HOWARD,PH (1993)

007146-68-1

CAS #: 007146-68-1

BENZENAMINE, 4-[(4-CHLOROPHENYL)SULFONYL]-

Formula: $C_{12}H_{10}ClNO_2S$

Mol Weight: 267.74

MP (deg C): **FP (deg C):**

BP (deg C):

BP pressure (mm Hg):

Property/Value	Units	Temp	Data Type	Reference
WS 1.20E+002	mg/L	25	EST	MEYLAN,WM ET AL. (1996)
logP 2.57			EXP	ALTOMARE,C ET AL. (1991)
VP 1.51E-007	mm Hg	25	EST	NEELY,WB & BLAU,GE (1985)
DC	pKa			
HL 6.53E-011	atm m3/mol	25	EST	MEYLAN,WM & HOWARD,PH (1991)
OH 2.33E-011	cm3/molc sec	25	EST	MEYLAN,WM & HOWARD,PH (1993)

007147-89-9

CAS #: 007147-89-9

4-CHLORO-5-METHYL-2-NITROPHENOL

Formula: $C_7H_6ClNO_3$

Mol Weight: 187.58

MP (deg C): **FP (deg C):**

BP (deg C):

BP pressure (mm Hg):

Property/Value	Units	Temp	Data Type	Reference
WS 7.64E+000	mg/L	20	EXP	SCHWARZENBACH,RP ET AL. (1988)
logP 2.93			EXP	SCHWARZENBACH,RP ET AL. (1988)
VP 8.53E-004	mm Hg	20	EXP	SCHWARZENBACH,RP ET AL. (1988)
DC	pKa			
HL 2.76E-005	atm m3/mol	20	EST	VP/WSOL
OH 3.45E-012	cm3/molc sec	25	EST	MEYLAN,WM & HOWARD,PH (1993)

007149-79-3

CAS #: 007149-79-3

3-CHLORO-4-METHYLACETANILIDE

Formula: $C_9H_{10}ClNO$

Mol Weight: 183.64

MP (deg C): **FP (deg C):**

BP (deg C):

BP pressure (mm Hg):

Property/Value	Units	Temp	Data Type	Reference
WS 2.67E+002	mg/L	25	EST	MEYLAN,WM ET AL. (1996)
logP 2.69			EXP	NAKAGAWA,Y ET AL. (1992)
VP 3.66E-005	mm Hg	25	EST	NEELY,WB & BLAU,GE (1985)
DC	pKa			
HL 5.05E-009	atm m3/mol	25	EST	MEYLAN,WM & HOWARD,PH (1991)
OH 1.09E-011	cm3/molc sec	25	EST	MEYLAN,WM & HOWARD,PH (1993)

007153-23-3

CAS #: 007153-23-3

6,7-DIMETHYLQUINOXALINE

Formula: $C_{10}H_{10}N_2$

Mol Weight: 158.20

MP (deg C): **FP (deg C):**

BP (deg C):

BP pressure (mm Hg):

Property/Value	Units	Temp	Data Type	Reference
WS 7.72E+002	mg/L	25	EST	MEYLAN,WM ET AL. (1996)
logP 2.29			EXP	HANSCH,C & LEO,AJ (1985)
VP 1.62E-003	mm Hg	25	EST	NEELY,WB & BLAU,GE (1985)
DC	pKa			
HL 3.47E-007	atm m3/mol	25	EST	MEYLAN,WM & HOWARD,PH (1991)
OH 6.67E-012	cm3/molc sec	25	EST	MEYLAN,WM & HOWARD,PH (1993)

CAS #: 007154-31-6 — 1,4-BENZENEDICARBOXAMIDE, N,N'-DIPHENYL-

Formula: $C_{20}H_{16}N_2O_2$

Mol Weight: 316.36

MP (deg C): | FP (deg C):

BP (deg C):

BP pressure (mm Hg):

Property/Value	Units	Temp	Data Type	Reference
WS 6.55E+000	mg/L	25	EST	MEYLAN,WM ET AL. (1996)
logP 3.72			EXP	HANSCH,C ET AL. (1995)
VP 1.75E-012	mm Hg	25	EST	NEELY,WB & BLAU,GE (1985)
DC	pKa			
HL 4.39E-013	atm m3/mol	25	EST	MEYLAN,WM & HOWARD,PH (1991)
OH 2.56E-011	cm3/molc sec	25	EST	MEYLAN,WM & HOWARD,PH (1993)

CAS #: 007154-73-6 — 1-PYRROLIDINEETHANAMINE

Formula: $C_6H_{14}N_2$

Mol Weight: 114.19

MP (deg C): | FP (deg C):

BP (deg C): 66-70

BP pressure (mm Hg): 2.30E+001

Property/Value	Units	Temp	Data Type	Reference
WS 9.16E+005	mg/L	25	EST	MEYLAN,WM ET AL. (1996)
logP 0.08			EXP	HANSCH,C ET AL. (1995)
VP 1.26E+000	mm Hg	25	EST	NEELY,WB & BLAU,GE (1985)
DC 9.74	pKa	30	EXP	PERRIN,DD (1965)
HL 3.88E-009	atm m3/mol	25	EST	MEYLAN,WM & HOWARD,PH (1991)
OH 1.28E-010	cm3/molc sec	25	EST	MEYLAN,WM & HOWARD,PH (1993)

CAS #: 007154-79-2 — 2,2,3,3-TETRAMETHYLPENTANE

Formula: C_9H_{20}

Mol Weight: 128.26

MP (deg C): -9.8 | FP (deg C):

BP (deg C): 140.2

BP pressure (mm Hg):

Property/Value	Units	Temp	Data Type	Reference
WS 3.61E+000	mg/L	25	EST	MEYLAN,WM ET AL. (1996)
logP 4.54			EST	MEYLAN,WM & HOWARD,PH (1995)
VP 9.45E+000	mm Hg	25	EXP	DAUBERT,TE & DANNER,RP (1989)
DC	pKa			
HL 4.00E+000	atm m3/mol	25	EST	MEYLAN,WM & HOWARD,PH (1991)
OH 2.15E-012	cm3/molc sec	25	EST	MEYLAN,WM & HOWARD,PH (1993)

CAS #: 007159-97-9 — 1,1-DIMETHYL-3-P-NITROPHENYLUREA

Formula: $C_9H_{11}N_3O_3$

Mol Weight: 209.21

MP (deg C): | FP (deg C):

BP (deg C):

BP pressure (mm Hg):

Property/Value	Units	Temp	Data Type	Reference
WS 2.01E+003	mg/L	25	EST	MEYLAN,WM ET AL. (1996)
logP 1.51			EXP	HANSCH,C & LEO,AJ (1985)
VP 2.53E-006	mm Hg	25	EST	NEELY,WB & BLAU,GE (1985)
DC	pKa			
HL 3.83E-012	atm m3/mol	25	EST	MEYLAN,WM & HOWARD,PH (1991)
OH 7.45E-012	cm3/molc sec	25	EST	MEYLAN,WM & HOWARD,PH (1993)

CAS #: 007159-98-0 — 1,1-DIMETHYL-3-M-NITROPHENYLUREA

Formula: $C_9H_{11}N_3O_3$

Mol Weight: 209.21

MP (deg C): | FP (deg C):

BP (deg C):

BP pressure (mm Hg):

Property/Value	Units	Temp	Data Type	Reference
WS 2.22E+003	mg/L	25	EST	MEYLAN,WM ET AL. (1996)
logP 1.46			EXP	HANSCH,C & LEO,AJ (1985)
VP 2.53E-006	mm Hg	25	EST	NEELY,WB & BLAU,GE (1985)
DC	pKa			
HL 3.83E-012	atm m3/mol	25	EST	MEYLAN,WM & HOWARD,PH (1991)
OH 5.84E-012	cm3/molc sec	25	EST	MEYLAN,WM & HOWARD,PH (1993)

CAS #: 007160-01-2 — 1,1-DIMETHYL-3-(P-TOLYL)UREA

Formula: $C_{10}H_{14}N_2O$

Mol Weight: 178.24

MP (deg C): | FP (deg C):

BP (deg C):

BP pressure (mm Hg):

Property/Value	Units	Temp	Data Type	Reference
WS 4.11E+003	mg/L	25	EST	MEYLAN,WM ET AL. (1996)
logP 1.33			EXP	HANSCH,C & LEO,AJ (1985)
VP 8.67E-005	mm Hg	25	EST	NEELY,WB & BLAU,GE (1985)
DC	pKa			
HL 1.07E-009	atm m3/mol	25	EST	MEYLAN,WM & HOWARD,PH (1991)
OH 5.45E-011	cm3/molc sec	25	EST	MEYLAN,WM & HOWARD,PH (1993)

CAS #: 007160-02-3 — N'-(4-METHOXYPHENYL)-N,N-DIMETHYLUREA

Formula: $C_{10}H_{14}N_2O_2$

Mol Weight: 194.24

MP (deg C): | FP (deg C):

BP (deg C):

BP pressure (mm Hg):

Property/Value	Units	Temp	Data Type	Reference
WS 9.15E+003	mg/L	25	EST	MEYLAN,WM ET AL. (1996)
logP 0.83			EXP	BRIGGS,GG (1981)
VP 3.28E-005	mm Hg	25	EST	NEELY,WB & BLAU,GE (1985)
DC	pKa			
HL 5.74E-011	atm m3/mol	25	EST	MEYLAN,WM & HOWARD,PH (1991)
OH 3.98E-011	cm3/molc sec	25	EST	MEYLAN,WM & HOWARD,PH (1993)

CAS #: 007160-22-7 — PROPANAMIDE, N-(3,4-DICHLOROPHENYL)-2,2-DIMETHYL

Formula: $C_{11}H_{13}Cl_2NO$

Mol Weight: 246.14

MP (deg C): | FP (deg C):

BP (deg C):

BP pressure (mm Hg):

Property/Value	Units	Temp	Data Type	Reference
WS 1.21E+001	mg/L	25	EST	MEYLAN,WM ET AL. (1996)
logP 3.88			EXP	HANSCH,C ET AL. (1995)
VP 4.29E-006	mm Hg	25	EST	NEELY,WB & BLAU,GE (1985)
DC	pKa			
HL 7.93E-009	atm m3/mol	25	EST	MEYLAN,WM & HOWARD,PH (1991)
OH 4.14E-012	cm3/molc sec	25	EST	MEYLAN,WM & HOWARD,PH (1993)

CAS #: 007160-25-0	BUTANAMIDE, N-(3,4-DICHLOROPHENYL)-2-METHYL-

Formula: $C_{11}H_{13}Cl_2NO$

Mol Weight: 246.14

MP (deg C): 　　　FP (deg C):

BP (deg C):

BP pressure (mm Hg):

Property/Value	Units	Temp	Data Type	Reference
WS 7.68E+000	mg/L	25	EST	MEYLAN,WM ET AL. (1996)
logP 4.11			EXP	HANSCH,C ET AL. (1995)
VP 3.45E-006	mm Hg	25	EST	NEELY,WB & BLAU,GE (1985)
DC	pKa			
HL 7.93E-009	atm m3/mol	25	EST	MEYLAN,WM & HOWARD,PH (1991)
OH 8.68E-012	cm3/molc sec	25	EST	MEYLAN,WM & HOWARD,PH (1993)

CAS #: 007194-84-5	N-HEPTATRIACONTANE

Formula: $C_{37}H_{76}$

Mol Weight: 521.02

MP (deg C): 　　　FP (deg C):

BP (deg C):

BP pressure (mm Hg):

Property/Value	Units	Temp	Data Type	Reference
WS 2.36E-014	mg/L	25	EST	MEYLAN,WM ET AL. (1996)
logP 18.51			EST	MEYLAN,WM & HOWARD,PH (1995)
VP 8.43E-025	mm Hg	25	EXT	ZWOLINSKI,BJ & WILHOIT,RC (1971)
DC	pKa			
HL 1.11E+004	atm m3/mol	25	EST	MEYLAN,WM & HOWARD,PH (1991)
OH 4.93E-011	cm3/molc sec	25	EST	MEYLAN,WM & HOWARD,PH (1993)

CAS #: 007194-85-6	N-OCTATRIACONTANE

Formula: $C_{38}H_{78}$

Mol Weight: 535.05

MP (deg C): 　　　FP (deg C):

BP (deg C):

BP pressure (mm Hg):

Property/Value	Units	Temp	Data Type	Reference
WS 7.30E-015	mg/L	25	EST	MEYLAN,WM ET AL. (1996)
logP 19.00			EST	MEYLAN,WM & HOWARD,PH (1995)
VP 1.22E-026	mm Hg	25	EXT	ZWOLINSKI,BJ & WILHOIT,RC (1971)
DC	pKa			
HL 1.48E+004	atm m3/mol	25	EST	MEYLAN,WM & HOWARD,PH (1991)
OH 5.07E-011	cm3/molc sec	25	EST	MEYLAN,WM & HOWARD,PH (1993)

CAS #: 007203-91-0	BENZOIC ACID, 4-(3,3-DIMETHYL-1-TRIAZENYL)-

Formula: $C_9H_{11}N_3O_2$

Mol Weight: 193.21

MP (deg C): 　　　FP (deg C):

BP (deg C):

BP pressure (mm Hg):

Property/Value	Units	Temp	Data Type	Reference
WS 4.85E+002	mg/L	25	EST	MEYLAN,WM ET AL. (1996)
logP 2.33			EXP	SANGSTER,J (1993)
VP 5.67E-005	mm Hg	25	EST	NEELY,WB & BLAU,GE (1985)
DC	pKa			
HL 1.08E-011	atm m3/mol	25	EST	MEYLAN,WM & HOWARD,PH (1991)
OH 3.26E-012	cm3/molc sec	25	EST	MEYLAN,WM & HOWARD,PH (1993)

CAS #: 007203-96-5	2-AZACYCLOHEPTANTHIONE

Formula: $C_6H_{11}NS$

Mol Weight: 129.23

MP (deg C): 107-110　　　FP (deg C):

BP (deg C):

BP pressure (mm Hg):

Property/Value	Units	Temp	Data Type	Reference
WS 2.12E+004	mg/L	25	EST	MEYLAN,WM ET AL. (1996)
logP 0.75			EXP	HANSCH,C & LEO,AJ (1985)
VP 3.59E-003	mm Hg	25	EST	NEELY,WB & BLAU,GE (1985)
DC	pKa			
HL 1.46E-005	atm m3/mol	25	EST	MEYLAN,WM & HOWARD,PH (1991)
OH 7.93E-011	cm3/molc sec	25	EST	MEYLAN,WM & HOWARD,PH (1993)

CAS #: 007205-98-3	CHLOROMETHYL PHENYL SULFONE

Formula: $C_7H_7ClO_2S$

Mol Weight: 190.65

MP (deg C): 53-55　　　FP (deg C):

BP (deg C):

BP pressure (mm Hg):

Property/Value	Units	Temp	Data Type	Reference
WS 4.26E+003	mg/L	25	EST	MEYLAN,WM ET AL. (1996)
logP 1.24			EXP	HANSCH,C ET AL. (1995)
VP 3.57E-004	mm Hg	25	EST	NEELY,WB & BLAU,GE (1985)
DC	pKa			
HL 4.39E-007	atm m3/mol	25	EST	MEYLAN,WM & HOWARD,PH (1991)
OH 3.19E-012	cm3/molc sec	25	EST	MEYLAN,WM & HOWARD,PH (1993)

CAS #: 007206-40-8	N-OCTADECYLPYRIDINIUM IODIDE

Formula: $C_{23}H_{42}IN$

Mol Weight: 459.50

MP (deg C): 　　　FP (deg C):

BP (deg C):

BP pressure (mm Hg):

Property/Value	Units	Temp	Data Type	Reference
WS 9.34E-001	mg/L	25	EST	MEYLAN,WM ET AL. (1996)
logP 3.68			EXP	HANSCH,C & LEO,AJ (1985)
VP 1.93E-008	mm Hg	25	EST	NEELY,WB & BLAU,GE (1985)
DC	pKa			
HL 1.86E-008	atm m3/mol	25	EST	MEYLAN,WM & HOWARD,PH (1991)
OH 3.40E-011	cm3/molc sec	25	EST	MEYLAN,WM & HOWARD,PH (1993)

CAS #: 007206-56-6	1-NITROAZULENE

Formula: $C_{10}H_7NO_2$

Mol Weight: 173.17

MP (deg C): 　　　FP (deg C):

BP (deg C):

BP pressure (mm Hg):

Property/Value	Units	Temp	Data Type	Reference
WS 1.49E+002	mg/L	25	EST	MEYLAN,WM ET AL. (1996)
logP 2.59			EXP	HANSCH,C & LEO,AJ (1985)
VP 1.45E-003	mm Hg	25	EST	NEELY,WB & BLAU,GE (1985)
DC	pKa			
HL 4.94E-005	atm m3/mol	25	EST	MEYLAN,WM & HOWARD,PH (1991)
OH 2.15E-010	cm3/molc sec	25	EST	MEYLAN,WM & HOWARD,PH (1993)

007206-57-7 — 1-ACETYLAZULENE

Formula: $C_{12}H_{10}O$
Mol Weight: 170.21
MP (deg C):
FP (deg C):
BP (deg C):
BP pressure (mm Hg):

Property/Value	Units	Temp	Data Type	Reference
WS 2.17E+002	mg/L	25	EST	MEYLAN,WM ET AL. (1996)
logP 2.87			EXP	HANSCH,C & LEO,AJ (1985)
VP 7.31E-003	mm Hg	25	EST	NEELY,WB & BLAU,GE (1985)
DC	pKa			
HL 1.66E-005	atm m3/mol	25	EST	MEYLAN,WM & HOWARD,PH (1991)
OH 2.63E-010	cm3/molc sec	25	EST	MEYLAN,WM & HOWARD,PH (1993)

007206-61-3 — 1-AZULENE CARBOXYALDEHYDE

Formula: $C_{11}H_8O$
Mol Weight: 156.19
MP (deg C):
FP (deg C):
BP (deg C):
BP pressure (mm Hg):

Property/Value	Units	Temp	Data Type	Reference
WS 3.96E+002	mg/L	25	EST	MEYLAN,WM ET AL. (1996)
logP 2.64			EXP	HANSCH,C & LEO,AJ (1985)
VP 7.92E-003	mm Hg	25	EST	NEELY,WB & BLAU,GE (1985)
DC	pKa			
HL 2.28E-005	atm m3/mol	25	EST	MEYLAN,WM & HOWARD,PH (1991)
OH 2.42E-010	cm3/molc sec	25	EST	MEYLAN,WM & HOWARD,PH (1993)

007207-62-7 — URIDINE, 2'-DEOXY-5-FLUORO-, 3',5'-DIPENTANOATE

Formula: $C_{19}H_{27}FN_2O_7$
Mol Weight: 414.43
MP (deg C):
FP (deg C):
BP (deg C):
BP pressure (mm Hg):

Property/Value	Units	Temp	Data Type	Reference
WS 1.18E+001	mg/L	25	EST	MEYLAN,WM ET AL. (1996)
logP 2.72			EXP	SANGSTER,J (1994)
VP 4.88E-013	mm Hg	25	EST	NEELY,WB & BLAU,GE (1985)
DC	pKa			
HL 3.43E-016	atm m3/mol	25	EST	MEYLAN,WM & HOWARD,PH (1991)
OH 8.18E-011	cm3/molc sec	25	EST	MEYLAN,WM & HOWARD,PH (1993)

007209-38-3 — 1,4-PIPERAZINEDIPROPANAMINE

Formula: $C_{10}H_{24}N_4$
Mol Weight: 200.33
MP (deg C): 15
FP (deg C):
BP (deg C): 150-152
BP pressure (mm Hg): 2.00E+000

Property/Value	Units	Temp	Data Type	Reference
WS 1.00E+006	mg/L	25	EST	MEYLAN,WM ET AL. (1996)
logP -1.43			EXP	HANSCH,C ET AL. (1995)
VP 1.63E-004	mm Hg	25	EST	NEELY,WB & BLAU,GE (1985)
DC	pKa			
HL 3.48E-016	atm m3/mol	25	EST	MEYLAN,WM & HOWARD,PH (1991)
OH 2.62E-010	cm3/molc sec	25	EST	MEYLAN,WM & HOWARD,PH (1993)

007225-96-9 — N-METHYL-3-ETHOXYPHENYLCARBAMATE

Formula: $C_{10}H_{13}NO_3$
Mol Weight: 195.22
MP (deg C):
FP (deg C):
BP (deg C):
BP pressure (mm Hg):

Property/Value	Units	Temp	Data Type	Reference
WS 1.48E+003	mg/L	25	EST	MEYLAN,WM ET AL. (1996)
logP 1.75			EXP	HANSCH,C & LEO,AJ (1985)
VP 1.79E-003	mm Hg	25	EST	NEELY,WB & BLAU,GE (1985)
DC	pKa			
HL 2.52E-009	atm m3/mol	25	EST	MEYLAN,WM & HOWARD,PH (1991)
OH 3.44E-011	cm3/molc sec	25	EST	MEYLAN,WM & HOWARD,PH (1993)

007227-91-0 — N,N-DIMETHYLTRIAZENOBENZENE

Formula: $C_8H_{11}N_3$
Mol Weight: 149.20
MP (deg C):
FP (deg C):
BP (deg C): 125-127
BP pressure (mm Hg): 1.90E+001

Property/Value	Units	Temp	Data Type	Reference
WS 4.70E+002	mg/L	25	EST	MEYLAN,WM ET AL. (1996)
logP 2.59			EXP	HANSCH,C & LEO,AJ (1985)
VP 1.22E-001	mm Hg	25	EST	NEELY,WB & BLAU,GE (1985)
DC	pKa			
HL 5.37E-007	atm m3/mol	25	EST	MEYLAN,WM & HOWARD,PH (1991)
OH 3.30E-012	cm3/molc sec	25	EST	MEYLAN,WM & HOWARD,PH (1993)

007234-29-9 — GLUCOPYRANOSIDE,2-IODOPHENYL

Formula: $C_{12}H_{15}IO_6$
Mol Weight: 382.15
MP (deg C):
FP (deg C):
BP (deg C):
BP pressure (mm Hg):

Property/Value	Units	Temp	Data Type	Reference
WS 2.32E+003	mg/L	25	EST	MEYLAN,WM ET AL. (1996)
logP 0.27			EXP	HANSCH,C & LEO,AJ (1985)
VP 7.37E-013	mm Hg	25	EST	NEELY,WB & BLAU,GE (1985)
DC	pKa			
HL 2.67E-016	atm m3/mol	25	EST	MEYLAN,WM & HOWARD,PH (1991)
OH 6.91E-011	cm3/molc sec	25	EST	MEYLAN,WM & HOWARD,PH (1993)

007234-31-3 — GLUCOPYRANOSIDE,2-METHYLPHENYL

Formula: $C_{13}H_{18}O_6$
Mol Weight: 270.28
MP (deg C):
FP (deg C):
BP (deg C):
BP pressure (mm Hg):

Property/Value	Units	Temp	Data Type	Reference
WS 2.49E+004	mg/L	25	EST	MEYLAN,WM ET AL. (1996)
logP -0.16			EXP	HANSCH,C & LEO,AJ (1985)
VP 2.48E-011	mm Hg	25	EST	NEELY,WB & BLAU,GE (1985)
DC	pKa			
HL 1.27E-015	atm m3/mol	25	EST	MEYLAN,WM & HOWARD,PH (1991)
OH 8.85E-011	cm3/molc sec	25	EST	MEYLAN,WM & HOWARD,PH (1993)

007236-57-9

THYMIDINE, 4-THIO-

Formula: $C_{10}H_{14}N_2O_4S$

Mol Weight: 258.30

MP (deg C): FP (deg C):

BP (deg C):

BP pressure (mm Hg):

Property/Value	Units	Temp	Data Type	Reference
WS 1.20E+003	mg/L	25	EST	MEYLAN,WM ET AL. (1996)
logP -0.07			EXP	PALOMINO,E ET AL. (1990)
VP 7.23E-013	mm Hg	25	EST	NEELY,WB & BLAU,GE (1985)
DC	pKa			
HL 1.43E-016	atm m3/mol	25	EST	MEYLAN,WM & HOWARD,PH (1991)
OH 1.30E-010	cm3/molc sec	25	EST	MEYLAN,WM & HOWARD,PH (1993)

007244-70-4

4-NITROPHENYLDIMETHYLCARBAMATE

Formula: $C_9H_{10}N_2O_4$

Mol Weight: 210.19

MP (deg C): FP (deg C):

BP (deg C):

BP pressure (mm Hg):

Property/Value	Units	Temp	Data Type	Reference
WS 8.27E+002	mg/L	25	EST	MEYLAN,WM ET AL. (1996)
logP 1.50			EXP	HANSCH,C & LEO,AJ (1985)
VP 2.25E-004	mm Hg	25	EST	NEELY,WB & BLAU,GE (1985)
DC	pKa			
HL 2.78E-010	atm m3/mol	25	EST	MEYLAN,WM & HOWARD,PH (1991)
OH 1.64E-011	cm3/molc sec	25	EST	MEYLAN,WM & HOWARD,PH (1993)

007244-78-2

P-NITROPHENYL BUTYL ETHER

Formula: $C_{10}H_{13}NO_3$

Mol Weight: 195.22

MP (deg C): FP (deg C):

BP (deg C):

BP pressure (mm Hg):

Property/Value	Units	Temp	Data Type	Reference
WS 1.93E+001	mg/L	25	EST	MEYLAN,WM ET AL. (1996)
logP 3.50			EXP	HANSCH,C & LEO,AJ (1985)
VP 7.41E-004	mm Hg	25	EST	NEELY,WB & BLAU,GE (1985)
DC	pKa			
HL 2.94E-006	atm m3/mol	25	EST	MEYLAN,WM & HOWARD,PH (1991)
OH 1.50E-011	cm3/molc sec	25	EST	MEYLAN,WM & HOWARD,PH (1993)

007247-89-4

2-METHYL-N-NITROSOPEPERIDINE

Formula: $C_6H_{12}N_2O$

Mol Weight: 128.18

MP (deg C): FP (deg C):

BP (deg C):

BP pressure (mm Hg):

Property/Value	Units	Temp	Data Type	Reference
WS 2.31E+004	mg/L	25	EST	MEYLAN,WM ET AL. (1996)
logP 0.71			EXP	HANSCH,C & LEO,AJ (1985)
VP 6.35E-002	mm Hg	25	EST	NEELY,WB & BLAU,GE (1985)
DC	pKa			
HL 2.82E-006	atm m3/mol	25	EST	MEYLAN,WM & HOWARD,PH (1991)
OH 3.73E-011	cm3/molc sec	25	EST	MEYLAN,WM & HOWARD,PH (1993)

007251-61-8

2-METHYLQUINOXALINE

Formula: $C_9H_8N_2$

Mol Weight: 144.18

MP (deg C): 180.5 FP (deg C):

BP (deg C): 244

BP pressure (mm Hg):

Property/Value	Units	Temp	Data Type	Reference
WS 3.39E+003	mg/L	25	EST	MEYLAN,WM ET AL. (1996)
logP 1.61			EXP	HANSCH,C & LEO,AJ (1985)
VP 5.10E-003	mm Hg	25	EST	NEELY,WB & BLAU,GE (1985)
DC 0.95	pKa	20	EXP	PERRIN,DD (1965)
HL 3.14E-007	atm m3/mol	25	EST	MEYLAN,WM & HOWARD,PH (1991)
OH 5.36E-012	cm3/molc sec	25	EST	MEYLAN,WM & HOWARD,PH (1993)

007261-97-4

DANTROLENE

Formula: $C_{14}H_{10}N_4O_5$

Mol Weight: 314.26

MP (deg C): 279-280 FP (deg C):

BP (deg C):

BP pressure (mm Hg):

Property/Value	Units	Temp	Data Type	Reference
WS 1.46E+002	mg/L	25	EST	MEYLAN,WM ET AL. (1996)
logP 1.70			EXP	JANSEN,ACA ET AL. (1991)
VP 2.31E-013	mm Hg	25	EST	NEELY,WB & BLAU,GE (1985)
DC	pKa			
HL 1.02E-013	atm m3/mol	25	EST	MEYLAN,WM & HOWARD,PH (1991)
OH 8.30E-011	cm3/molc sec	25	EST	MEYLAN,WM & HOWARD,PH (1993)

007265-01-2

GLUCOPYRANOSIDE,2-AMINOPHENYL

Formula: $C_{12}H_{17}NO_6$

Mol Weight: 271.27

MP (deg C): FP (deg C):

BP (deg C):

BP pressure (mm Hg):

Property/Value	Units	Temp	Data Type	Reference
WS 2.02E+005	mg/L	25	EST	MEYLAN,WM ET AL. (1996)
logP -1.23			EXP	HANSCH,C & LEO,AJ (1985)
VP 1.17E-012	mm Hg	25	EST	NEELY,WB & BLAU,GE (1985)
DC	pKa			
HL 4.06E-019	atm m3/mol	25	EST	MEYLAN,WM & HOWARD,PH (1991)
OH 1.55E-010	cm3/molc sec	25	EST	MEYLAN,WM & HOWARD,PH (1993)

007286-84-2

CHLORAMBEN, METHYL ESTER

Formula: $C_8H_7Cl_2NO_2$

Mol Weight: 220.06

MP (deg C): FP (deg C):

BP (deg C):

BP pressure (mm Hg):

Property/Value	Units	Temp	Data Type	Reference
WS 1.20E+002	mg/L	20	EXP	YALKOWSKY,SH & DANNENFELSER,RM (1992)
logP 2.60			EST	MEYLAN,WM & HOWARD,PH (1995)
VP 2.27E-004	mm Hg	25	EST	NEELY,WB & BLAU,GE (1985)
DC	pKa			
HL 6.74E-009	atm m3/mol	25	EST	MEYLAN,WM & HOWARD,PH (1991)
OH 5.11E-012	cm3/molc sec	25	EST	MEYLAN,WM & HOWARD,PH (1993)

PROMETRYNE

CAS #: 007287-19-6

Formula: $C_{10}H_{19}N_5S$

Mol Weight: 241.36

MP (deg C): 119

FP (deg C):

BP (deg C):

BP pressure (mm Hg):

Property/Value	Units	Temp	Data Type	Reference
WS 4.80E+001	mg/L	24	EXP	YALKOWSKY,SH & DANNENFELSER,RM (1992)
logP 3.51			EXP	HANSCH,C ET AL. (1995)
VP 2.00E-006	mm Hg	25	EXP	WEBER,JB (1994)
DC	pKa			
HL 1.32E-006	atm m3/mol	25	EST	VP/WSOL
OH 3.81E-011	cm3/molc sec	25	EST	MEYLAN,WM & HOWARD,PH (1993)

MONALIDE (POTABLAN)

CAS #: 007287-36-7

Formula: $C_{13}H_{18}ClNO$

Mol Weight: 239.75

MP (deg C): 87-88

FP (deg C):

BP (deg C):

BP pressure (mm Hg):

Property/Value	Units	Temp	Data Type	Reference
WS 2.28E+001	mg/L	23	EXP	SHIU,WY ET AL. (1990)
logP 3.83			EXP	MITSUTAKE,KI ET AL. (1986)
VP 2.92E-006	mm Hg	25	EST	NEELY,WB & BLAU,GE (1985)
DC	pKa			
HL 1.89E-008	atm m3/mol	25	EST	MEYLAN,WM & HOWARD,PH (1991)
OH 1.05E-011	cm3/molc sec	25	EST	MEYLAN,WM & HOWARD,PH (1993)

M-METHOXY-N,N-DIMETHYLBENZAMIDE

CAS #: 007290-99-5

Formula: $C_{10}H_{13}NO_2$

Mol Weight: 179.22

MP (deg C):

FP (deg C):

BP (deg C):

BP pressure (mm Hg):

Property/Value	Units	Temp	Data Type	Reference
WS 7.77E+003	mg/L	25	EST	MEYLAN,WM ET AL. (1996)
logP 1.00			EXP	HANSCH,C & LEO,AJ (1985)
VP 9.27E-004	mm Hg	25	EST	NEELY,WB & BLAU,GE (1985)
DC	pKa			
HL 6.31E-010	atm m3/mol	25	EST	MEYLAN,WM & HOWARD,PH (1991)
OH 2.60E-011	cm3/molc sec	25	EST	MEYLAN,WM & HOWARD,PH (1993)

P-METHOXY-N,N-DIMETHYLBENZAMIDE

CAS #: 007291-00-1

Formula: $C_{10}H_{13}NO_2$

Mol Weight: 179.22

MP (deg C):

FP (deg C):

BP (deg C):

BP pressure (mm Hg):

Property/Value	Units	Temp	Data Type	Reference
WS 8.41E+003	mg/L	25	EST	MEYLAN,WM ET AL. (1996)
logP 0.96			EXP	HANSCH,C & LEO,AJ (1985)
VP 9.27E-004	mm Hg	25	EST	NEELY,WB & BLAU,GE (1985)
DC	pKa			
HL 6.31E-010	atm m3/mol	25	EST	MEYLAN,WM & HOWARD,PH (1991)
OH 3.65E-011	cm3/molc sec	25	EST	MEYLAN,WM & HOWARD,PH (1993)

O-MEO N,N-DIMETHYLBENZAMIDE

CAS #: 007291-34-1

Formula: $C_{10}H_{13}NO_2$

Mol Weight: 179.22

MP (deg C):

FP (deg C):

BP (deg C):

BP pressure (mm Hg):

Property/Value	Units	Temp	Data Type	Reference
WS 1.38E+004	mg/L	25	EST	MEYLAN,WM ET AL. (1996)
logP 0.71			EXP	NAKAGAWA,Y ET AL. (1992)
VP 9.27E-004	mm Hg	25	EST	NEELY,WB & BLAU,GE (1985)
DC	pKa			
HL 6.31E-010	atm m3/mol	25	EST	MEYLAN,WM & HOWARD,PH (1991)
OH 3.65E-011	cm3/molc sec	25	EST	MEYLAN,WM & HOWARD,PH (1993)

PROPAPHOS

CAS #: 007292-16-2

Formula: $C_{13}H_{21}O_4PS$

Mol Weight: 304.35

MP (deg C):

FP (deg C):

BP (deg C):

BP pressure (mm Hg):

Property/Value	Units	Temp	Data Type	Reference
WS 1.25E+002	mg/L	25	EXP	YALKOWSKY,SH & DANNENFELSER,RM (1992)
logP 3.67			EXP	HANSCH,C ET AL. (1995)
VP 4.52E-006	mm Hg	25	EST	NEELY,WB & BLAU,GE (1985)
DC	pKa			
HL 8.34E-009	atm m3/mol	25	EST	MEYLAN,WM & HOWARD,PH (1991)
OH 6.00E-011	cm3/molc sec	25	EST	MEYLAN,WM & HOWARD,PH (1993)

2,2,3-TRIMETHYL-3-PENTANOL

CAS #: 007294-05-5

Formula: $C_8H_{18}O$

Mol Weight: 130.23

MP (deg C):

FP (deg C):

BP (deg C):

BP pressure (mm Hg):

Property/Value	Units	Temp	Data Type	Reference
WS 6.98E+003	mg/L	25	EXP	SUZUKI,T (1991)
logP 2.58			EST	MEYLAN,WM & HOWARD,PH (1995)
VP 2.14E+000	mm Hg	25	EST	NEELY,WB & BLAU,GE (1985)
DC	pKa			
HL 3.10E-005	atm m3/mol	25	EST	MEYLAN,WM & HOWARD,PH (1991)
OH 4.88E-012	cm3/molc sec	25	EST	MEYLAN,WM & HOWARD,PH (1993)

3-METHYOXYPYRIDINE

CAS #: 007295-76-3

Formula: C_6H_7NO

Mol Weight: 109.13

MP (deg C):

FP (deg C):

BP (deg C):

BP pressure (mm Hg):

Property/Value	Units	Temp	Data Type	Reference
WS 1.56E+004	mg/L	25	EST	MEYLAN,WM ET AL. (1996)
logP 0.99			EXP	HANSCH,C ET AL. (1995)
VP 2.58E+000	mm Hg	25	EST	NEELY,WB & BLAU,GE (1985)
DC 4.91	pKa	25	EXP	PERRIN,DD (1965)
HL 4.17E-007	atm m3/mol	25	EST	MEYLAN,WM & HOWARD,PH (1991)
OH 4.91E-012	cm3/molc sec	25	EST	MEYLAN,WM & HOWARD,PH (1993)

CAS #: 007295-91-2 — N-DECYLPYRIDINIUM IODIDE

Formula: $C_{15}H_{26}IN$

Mol Weight: 347.29

MP (deg C): FP (deg C):

BP (deg C):

BP pressure (mm Hg):

Property/ Value	Units	Temp	Data Type	Reference
WS 5.18E+003	mg/L	25	EST	MEYLAN,WM ET AL. (1996)
logP 0.11			EXP	HANSCH,C & LEO,AJ (1985)
VP 2.14E-005	mm Hg	25	EST	NEELY,WB & BLAU,GE (1985)
DC	pKa			
HL 1.93E-009	atm m3/mol	25	EST	MEYLAN,WM & HOWARD,PH (1991)
OH 2.27E-011	cm3/molc sec	25	EST	MEYLAN,WM & HOWARD,PH (1993)

CAS #: 007305-03-5 — 4-CHLOROPHENYLDIMETHYLCARBAMATE

Formula: $C_9H_{10}ClNO_2$

Mol Weight: 199.64

MP (deg C): FP (deg C):

BP (deg C):

BP pressure (mm Hg):

Property/ Value	Units	Temp	Data Type	Reference
WS 5.06E+002	mg/L	25	EST	MEYLAN,WM ET AL. (1996)
logP 2.27			EXP	HANSCH,C & LEO,AJ (1985)
VP 8.29E-003	mm Hg	25	EST	NEELY,WB & BLAU,GE (1985)
DC	pKa			
HL 5.23E-008	atm m3/mol	25	EST	MEYLAN,WM & HOWARD,PH (1991)
OH 1.71E-011	cm3/molc sec	25	EST	MEYLAN,WM & HOWARD,PH (1993)

CAS #: 007305-04-6 — 2-BROMOPHENYLDIMETHYLCARBAMATE

Formula: $C_9H_{10}BrNO_2$

Mol Weight: 244.09

MP (deg C): FP (deg C):

BP (deg C):

BP pressure (mm Hg):

Property/ Value	Units	Temp	Data Type	Reference
WS 3.58E+002	mg/L	25	EST	MEYLAN,WM ET AL. (1996)
logP 2.17			EXP	HANSCH,C & LEO,AJ (1985)
VP 2.58E-003	mm Hg	25	EST	NEELY,WB & BLAU,GE (1985)
DC	pKa			
HL 2.81E-008	atm m3/mol	25	EST	MEYLAN,WM & HOWARD,PH (1991)
OH 1.70E-011	cm3/molc sec	25	EST	MEYLAN,WM & HOWARD,PH (1993)

CAS #: 007305-06-8 — O-TOLYLDIMETHYLCARBAMATE

Formula: $C_{10}H_{13}NO_2$

Mol Weight: 179.22

MP (deg C): FP (deg C):

BP (deg C):

BP pressure (mm Hg):

Property/ Value	Units	Temp	Data Type	Reference
WS 1.43E+003	mg/L	25	EST	MEYLAN,WM ET AL. (1996)
logP 1.86			EXP	HANSCH,C & LEO,AJ (1985)
VP 1.86E-002	mm Hg	25	EST	NEELY,WB & BLAU,GE (1985)
DC	pKa			
HL 7.78E-008	atm m3/mol	25	EST	MEYLAN,WM & HOWARD,PH (1991)
OH 2.00E-011	cm3/molc sec	25	EST	MEYLAN,WM & HOWARD,PH (1993)

CAS #: 007305-07-9 — M-TOLYLDIMETHYLCARBAMATE

Formula: $C_{10}H_{13}NO_2$

Mol Weight: 179.22

MP (deg C): FP (deg C):

BP (deg C):

BP pressure (mm Hg):

Property/ Value	Units	Temp	Data Type	Reference
WS 9.86E+002	mg/L	25	EST	MEYLAN,WM ET AL. (1996)
logP 2.05			EXP	HANSCH,C & LEO,AJ (1985)
VP 1.86E-002	mm Hg	25	EST	NEELY,WB & BLAU,GE (1985)
DC	pKa			
HL 7.78E-008	atm m3/mol	25	EST	MEYLAN,WM & HOWARD,PH (1991)
OH 2.15E-011	cm3/molc sec	25	EST	MEYLAN,WM & HOWARD,PH (1993)

CAS #: 007305-08-0 — P-TOLYLDIMETHYLCARBAMATE

Formula: $C_{10}H_{13}NO_2$

Mol Weight: 179.22

MP (deg C): FP (deg C):

BP (deg C):

BP pressure (mm Hg):

Property/ Value	Units	Temp	Data Type	Reference
WS 1.03E+003	mg/L	25	EST	MEYLAN,WM ET AL. (1996)
logP 2.03			EXP	HANSCH,C & LEO,AJ (1985)
VP 1.86E-002	mm Hg	25	EST	NEELY,WB & BLAU,GE (1985)
DC	pKa			
HL 7.78E-008	atm m3/mol	25	EST	MEYLAN,WM & HOWARD,PH (1991)
OH 2.00E-011	cm3/molc sec	25	EST	MEYLAN,WM & HOWARD,PH (1993)

CAS #: 007305-09-1 — 3-METHOXYPHENYLDIMETHYLCARBAMATE

Formula: $C_{10}H_{13}NO_3$

Mol Weight: 195.22

MP (deg C): FP (deg C):

BP (deg C):

BP pressure (mm Hg):

Property/ Value	Units	Temp	Data Type	Reference
WS 1.99E+003	mg/L	25	EST	MEYLAN,WM ET AL. (1996)
logP 1.60			EXP	HANSCH,C & LEO,AJ (1985)
VP 5.26E-003	mm Hg	25	EST	NEELY,WB & BLAU,GE (1985)
DC	pKa			
HL 4.17E-009	atm m3/mol	25	EST	MEYLAN,WM & HOWARD,PH (1991)
OH 3.91E-011	cm3/molc sec	25	EST	MEYLAN,WM & HOWARD,PH (1993)

CAS #: 007305-10-4 — 4-METHOXYPHENYLDIMETHYLCARBAMATE

Formula: $C_{10}H_{13}NO_3$

Mol Weight: 195.22

MP (deg C): FP (deg C):

BP (deg C):

BP pressure (mm Hg):

Property/ Value	Units	Temp	Data Type	Reference
WS 2.28E+003	mg/L	25	EST	MEYLAN,WM ET AL. (1996)
logP 1.53			EXP	HANSCH,C & LEO,AJ (1985)
VP 5.26E-003	mm Hg	25	EST	NEELY,WB & BLAU,GE (1985)
DC	pKa			
HL 4.17E-009	atm m3/mol	25	EST	MEYLAN,WM & HOWARD,PH (1991)
OH 3.27E-011	cm3/molc sec	25	EST	MEYLAN,WM & HOWARD,PH (1993)

CAS #: 007307-04-2				5,5-DIMETHYL-2,4-HEXANEDIONE
Formula: $C_8H_{14}O_2$				
Mol Weight: 142.20				
MP (deg C):		FP (deg C):		
BP (deg C):				
BP pressure (mm Hg):				

Property/ Value	Units	Temp	Data Type	Reference
WS 3.07E+003	mg/L	25	EST	MEYLAN,WM ET AL. (1996)
logP 1.67			EXP	HANSCH,C & LEO,AJ (1985)
VP 1.11E+000	mm Hg	25	EST	NEELY,WB & BLAU,GE (1985)
DC	pKa			
HL 7.80E-008	atm m3/mol	25	EST	MEYLAN,WM & HOWARD,PH (1991)
OH 2.22E-012	cm3/molc sec	25	EST	MEYLAN,WM & HOWARD,PH (1993)

CAS #: 007319-00-8				TRANS-1,4-HEXADIENE
Formula: C_6H_{10}				
Mol Weight: 82.15				
MP (deg C): -138.7		FP (deg C):		
BP (deg C): 65				
BP pressure (mm Hg):				

Property/ Value	Units	Temp	Data Type	Reference
WS 1.17E+002	mg/L	25	EST	MEYLAN,WM ET AL. (1996)
logP 2.94			EST	MEYLAN,WM & HOWARD,PH (1995)
VP 1.07E+002	mm Hg	25	EST	NEELY,WB & BLAU,GE (1985)
DC	pKa			
HL 1.99E-001	atm m3/mol	25	EST	MEYLAN,WM & HOWARD,PH (1991)
OH 9.03E-011	cm3/molc sec	25	EXP	ATKINSON,R (1989)

CAS #: 007319-45-1				2,4-DIAMINO-5-BENZYLPYRIMIDINE
Formula: $C_{11}H_{12}N_4$				
Mol Weight: 200.25				
MP (deg C):		FP (deg C):		
BP (deg C):				
BP pressure (mm Hg):				

Property/ Value	Units	Temp	Data Type	Reference
WS 1.95E+003	mg/L	25	EST	MEYLAN,WM ET AL. (1996)
logP 1.58			EXP	HANSCH,C & LEO,AJ (1985)
VP 1.44E-006	mm Hg	25	EST	NEELY,WB & BLAU,GE (1985)
DC 7.27	pKa	20	EXP	PERRIN,DD (1972)
HL 1.15E-010	atm m3/mol	25	EST	MEYLAN,WM & HOWARD,PH (1991)
OH 7.31E-012	cm3/molc sec	25	EST	MEYLAN,WM & HOWARD,PH (1993)

CAS #: 007320-97-0				PENTANOIC ACID, 5-[(1,1'-BIPHENYL)-4-YL]-3-METHY
Formula: $C_{18}H_{18}O_3$				
Mol Weight: 282.34				
MP (deg C):		FP (deg C):		
BP (deg C):				
BP pressure (mm Hg):				

Property/ Value	Units	Temp	Data Type	Reference
WS 2.11E+001	mg/L	25	EST	MEYLAN,WM ET AL. (1996)
logP 3.82			EXP	SANGSTER,J (1993)
VP 2.56E-008	mm Hg	25	EST	NEELY,WB & BLAU,GE (1985)
DC	pKa			
HL 1.31E-011	atm m3/mol	25	EST	MEYLAN,WM & HOWARD,PH (1991)
OH 1.82E-011	cm3/molc sec	25	EST	MEYLAN,WM & HOWARD,PH (1993)

CAS #: 007321-55-3				DIMETHYLMALONONITRILE
Formula: $C_5H_6N_2$				
Mol Weight: 94.12				
MP (deg C): 31.5		FP (deg C):		
BP (deg C): 169.5				
BP pressure (mm Hg):				

Property/ Value	Units	Temp	Data Type	Reference
WS 4.23E+003	mg/L	25	EST	MEYLAN,WM ET AL. (1996)
logP 1.40			EST	MEYLAN,WM & HOWARD,PH (1995)
VP 1.45E+000	mm Hg	25	EXT	OHE,S (1976)
DC	pKa			
HL 2.23E-008	atm m3/mol	25	EST	MEYLAN,WM & HOWARD,PH (1991)
OH 3.35E-013	cm3/molc sec	25	EST	MEYLAN,WM & HOWARD,PH (1993)

CAS #: 007335-35-5				3-PHENYLHYDROXYUREA
Formula: $C_7H_8N_2O_2$				
Mol Weight: 152.15				
MP (deg C):		FP (deg C):		
BP (deg C):				
BP pressure (mm Hg):				

Property/ Value	Units	Temp	Data Type	Reference
WS 3.32E+004	mg/L	25	EST	MEYLAN,WM ET AL. (1996)
logP 0.41			EXP	HANSCH,C & LEO,AJ (1985)
VP 8.48E-007	mm Hg	25	EST	NEELY,WB & BLAU,GE (1985)
DC	pKa			
HL 2.99E-011	atm m3/mol	25	EST	MEYLAN,WM & HOWARD,PH (1991)
OH 4.37E-011	cm3/molc sec	25	EST	MEYLAN,WM & HOWARD,PH (1993)

CAS #: 007359-72-0				2,3,4-TRICHLOROTOLUENE
Formula: $C_7H_5Cl_3$				
Mol Weight: 195.48				
MP (deg C): 43.5		FP (deg C):		
BP (deg C): 244				
BP pressure (mm Hg):				

Property/ Value	Units	Temp	Data Type	Reference
WS 6.97E+000	mg/L	25	EST	MEYLAN,WM ET AL. (1996)
logP 4.47			EST	MEYLAN,WM & HOWARD,PH (1995)
VP 7.32E-002	mm Hg	25	EST	NEELY,WB & BLAU,GE (1985)
DC	pKa			
HL 2.42E-003	atm m3/mol	25	EST	MEYLAN,WM & HOWARD,PH (1991)
OH 4.81E-013	cm3/molc sec	25	EST	MEYLAN,WM & HOWARD,PH (1993)

CAS #: 007374-06-3				N-METHYL-2-NITROPHENYLCARBAMATE
Formula: $C_8H_8N_2O_4$				
Mol Weight: 196.16				
MP (deg C):		FP (deg C):		
BP (deg C):				
BP pressure (mm Hg):				

Property/ Value	Units	Temp	Data Type	Reference
WS 2.51E+003	mg/L	25	EST	MEYLAN,WM ET AL. (1996)
logP 1.02			EXP	HANSCH,C & LEO,AJ (1985)
VP 2.34E-004	mm Hg	25	EST	NEELY,WB & BLAU,GE (1985)
DC	pKa			
HL 1.27E-010	atm m3/mol	25	EST	MEYLAN,WM & HOWARD,PH (1991)
OH 6.31E-012	cm3/molc sec	25	EST	MEYLAN,WM & HOWARD,PH (1993)

007374-53-0

CAS #: 007374-53-0

Name: 4,6-BIS(ISOPROPYLAMINO)-S-TRIAZIN-2-OL

Formula: $C_9H_{17}N_5O$

Mol Weight: 211.27

MP (deg C):   **FP (deg C):**

BP (deg C):

BP pressure (mm Hg):

Property/Value	Units	Temp	Data Type	Reference
WS 3.26E+002	mg/L	26	EXP	SHIU,WY ET AL. (1990) @ pH=3
logP 2.51			EST	MEYLAN,WM & HOWARD,PH (1995)
VP 5.99E-006	mm Hg	25	EST	NEELY,WB & BLAU,GE (1985)
DC	pKa			
HL 8.34E-013	atm m3/mol	25	EST	MEYLAN,WM & HOWARD,PH (1991)
OH 3.72E-011	cm3/molc sec	25	EST	MEYLAN,WM & HOWARD,PH (1993)

007376-90-1

CAS #: 007376-90-1

Name: PHENYLALANINE-AMIDE,N-ACETYL

Formula: $C_{11}H_{14}N_2O_2$

Mol Weight: 206.25

MP (deg C):   **FP (deg C):**

BP (deg C):

BP pressure (mm Hg):

Property/Value	Units	Temp	Data Type	Reference
WS 2.15E+003	mg/L	25	EST	MEYLAN,WM ET AL. (1996)
logP -0.04			EXP	HANSCH,C & LEO,AJ (1985)
VP 4.42E-009	mm Hg	25	EST	NEELY,WB & BLAU,GE (1985)
DC	pKa			
HL 1.26E-011	atm m3/mol	25	EST	MEYLAN,WM & HOWARD,PH (1991)
OH 5.17E-011	cm3/molc sec	25	EST	MEYLAN,WM & HOWARD,PH (1993)

007388-31-0

CAS #: 007388-31-0

Name: METHOXYCHLOR-DDD

Formula: $C_{16}H_{16}Cl_2O_2$

Mol Weight: 311.21

MP (deg C):   **FP (deg C):**

BP (deg C):

BP pressure (mm Hg):

Property/Value	Units	Temp	Data Type	Reference
WS 1.61E+000	mg/L	25	EST	MEYLAN,WM ET AL. (1996)
logP 4.47			EXP	HANSCH,C & LEO,AJ (1985)
VP 3.43E-006	mm Hg	25	EST	NEELY,WB & BLAU,GE (1985)
DC	pKa			
HL 2.77E-007	atm m3/mol	25	EST	MEYLAN,WM & HOWARD,PH (1991)
OH 5.45E-011	cm3/molc sec	25	EST	MEYLAN,WM & HOWARD,PH (1993)

007388-32-1

CAS #: 007388-32-1

Name: 1,1-DICHLORO-2,2-BIS(P-ETHOXYPHENYL)ETHANE

Formula: $C_{18}H_{20}Cl_2O_2$

Mol Weight: 339.26

MP (deg C):   **FP (deg C):**

BP (deg C):

BP pressure (mm Hg):

Property/Value	Units	Temp	Data Type	Reference
WS 9.22E-002	mg/L	25	EST	MEYLAN,WM ET AL. (1996)
logP 5.73			EST	MEYLAN,WM & HOWARD,PH (1995)
VP 6.23E-007	mm Hg	25	EST	NEELY,WB & BLAU,GE (1985)
DC	pKa			
HL 4.88E-007	atm m3/mol	25	EST	MEYLAN,WM & HOWARD,PH (1991)
OH 6.51E-011	cm3/molc sec	25	EST	MEYLAN,WM & HOWARD,PH (1993)

007391-69-7

CAS #: 007391-69-7

Name: BARBITURIC ACID, 5-IPR

Formula: $C_7H_{10}N_2O_3$

Mol Weight: 170.17

MP (deg C):   **FP (deg C):**

BP (deg C):

BP pressure (mm Hg):

Property/Value	Units	Temp	Data Type	Reference
WS 6.39E+004	mg/L	25	EST	MEYLAN,WM ET AL. (1996)
logP -0.02			EXP	WONG,O & MCKEOWN,RH (1988)
VP 3.86E-010	mm Hg	25	EST	NEELY,WB & BLAU,GE (1985)
DC 4.94	pKa	25	EXP	KORTUM,G ET AL (1961)
HL 2.72E-013	atm m3/mol	25	EST	MEYLAN,WM & HOWARD,PH (1991)
OH 1.12E-011	cm3/molc sec	25	EST	MEYLAN,WM & HOWARD,PH (1993)

007400-08-0

CAS #: 007400-08-0

Name: CINNAMIC ACID, P-HYDROXY

Formula: $C_9H_8O_3$

Mol Weight: 164.16

MP (deg C): 211.5   **FP (deg C):**

BP (deg C):

BP pressure (mm Hg):

Property/Value	Units	Temp	Data Type	Reference
WS 1.83E+004	mg/L	25	EST	MEYLAN,WM ET AL. (1996)
logP 1.79			EXP	HANSCH,C & LEO,AJ (1985)
VP 1.61E-005	mm Hg	25	EST	NEELY,WB & BLAU,GE (1985)
DC 4.64	pKa	25	EXP	SERJEANT,EP & DEMPSEY,B (1979)
HL 1.35E-012	atm m3/mol	25	EST	MEYLAN,WM & HOWARD,PH (1991)
OH 5.17E-011	cm3/molc sec	25	EST	MEYLAN,WM & HOWARD,PH (1993)

007402-22-4

CAS #: 007402-22-4

Name: O-CHLORO-N-PHENYLSUCCINIMIDE

Formula: $C_{10}H_8ClNO_2$

Mol Weight: 209.63

MP (deg C):   **FP (deg C):**

BP (deg C):

BP pressure (mm Hg):

Property/Value	Units	Temp	Data Type	Reference
WS 1.30E+004	mg/L	25	EST	MEYLAN,WM ET AL. (1996)
logP 0.56			EXP	HANSCH,C & LEO,AJ (1985)
VP 2.25E-007	mm Hg	25	EST	NEELY,WB & BLAU,GE (1985)
DC	pKa			
HL 9.67E-008	atm m3/mol	25	EST	MEYLAN,WM & HOWARD,PH (1991)
OH 8.00E-012	cm3/molc sec	25	EST	MEYLAN,WM & HOWARD,PH (1993)

007403-41-0

CAS #: 007403-41-0

Name: BENZAMIDE, N-ETHYL-4-METHOXY-

Formula: $C_{10}H_{13}NO_2$

Mol Weight: 179.22

MP (deg C):   **FP (deg C):**

BP (deg C):

BP pressure (mm Hg):

Property/Value	Units	Temp	Data Type	Reference
WS 5.90E+003	mg/L	25	EST	MEYLAN,WM ET AL. (1996)
logP 1.14			EXP	SURYANARAYANA,MVS ET AL. (1991)
VP 6.20E-002	mm Hg	30	EXP	SURYANARAYANA,MVS ET AL. (1991)
DC	pKa			
HL 3.82E-010	atm m3/mol	25	EST	MEYLAN,WM & HOWARD,PH (1991)
OH 2.99E-011	cm3/molc sec	25	EST	MEYLAN,WM & HOWARD,PH (1993)

CAS #: 007409-30-5		P-NITROBENZYLAMINE		
Formula: $C_7H_8N_2O_2$				
Mol Weight: 152.15				
MP (deg C):	FP (deg C):			
BP (deg C):				
BP pressure (mm Hg):				

Property/Value	Units	Temp	Data Type	Reference
WS 3.83E+004	mg/L	25	EST	MEYLAN,WM ET AL. (1996)
logP 1.06			EXP	HANSCH,C & LEO,AJ (1985)
VP 2.87E-003	mm Hg	25	EST	NEELY,WB & BLAU,GE (1985)
DC 8.50	pKa	25	EXP	PERRIN,DD (1972)
HL 2.41E-009	atm m3/mol	25	EST	MEYLAN,WM & HOWARD,PH (1991)
OH 3.03E-011	cm3/molc sec	25	EST	MEYLAN,WM & HOWARD,PH (1993)

CAS #: 007411-23-6		1,2,4-TRIAZOLE,3,5-DIBROMO		
Formula: $C_2HBr_2N_3$				
Mol Weight: 226.87				
MP (deg C):	FP (deg C):			
BP (deg C):				
BP pressure (mm Hg):				

Property/Value	Units	Temp	Data Type	Reference
WS 3.87E+002	mg/L	25	EST	MEYLAN,WM ET AL. (1996)
logP 2.24			EXP	HANSCH,C & LEO,AJ (1985)
VP 9.52E-004	mm Hg	25	EST	NEELY,WB & BLAU,GE (1985)
DC	pKa			
HL 2.43E-007	atm m3/mol	25	EST	MEYLAN,WM & HOWARD,PH (1991)
OH 1.00E-013	cm3/molc sec	25	EST	MEYLAN,WM & HOWARD,PH (1993)

CAS #: 007411-79-2		BENZENESULFONAMIDE, 4-AMINO-N-(2,4-DIMETHYL-5-PY		
Formula: $C_{12}H_{14}N_4O_2S$				
Mol Weight: 278.33				
MP (deg C):	FP (deg C):			
BP (deg C):				
BP pressure (mm Hg):				

Property/Value	Units	Temp	Data Type	Reference
WS 5.13E+003	mg/L	25	EST	MEYLAN,WM ET AL. (1996)
logP 0.59			EXP	SANGSTER,J (1993)
VP 8.62E-009	mm Hg	25	EST	NEELY,WB & BLAU,GE (1985)
DC	pKa			
HL 5.43E-014	atm m3/mol	25	EST	MEYLAN,WM & HOWARD,PH (1991)
OH 3.08E-011	cm3/molc sec	25	EST	MEYLAN,WM & HOWARD,PH (1993)

CAS #: 007413-36-7		1(4-NO2 PH)-2-ISOPROPYLAMINO-ETHANOL		
Formula: $C_{11}H_{16}N_2O_3$				
Mol Weight: 224.26				
MP (deg C): 98	FP (deg C):			
BP (deg C):				
BP pressure (mm Hg):				

Property/Value	Units	Temp	Data Type	Reference
WS 1.09E+004	mg/L	25	EST	MEYLAN,WM ET AL. (1996)
logP 1.28			EXP	HANSCH,C & LEO,AJ (1985)
VP 7.22E-007	mm Hg	25	EST	NEELY,WB & BLAU,GE (1985)
DC	pKa			
HL 4.53E-013	atm m3/mol	25	EST	MEYLAN,WM & HOWARD,PH (1991)
OH 1.01E-010	cm3/molc sec	25	EST	MEYLAN,WM & HOWARD,PH (1993)

CAS #: 007420-37-3		M-OH BENZALDEHYDETHIOSEMICARBIZONE		
Formula: $C_8H_9N_3OS$				
Mol Weight: 195.24				
MP (deg C):	FP (deg C):			
BP (deg C):				
BP pressure (mm Hg):				

Property/Value	Units	Temp	Data Type	Reference
WS 7.42E+003	mg/L	25	EST	MEYLAN,WM ET AL. (1996)
logP 1.61			EXP	HANSCH,C & LEO,AJ (1985)
VP 2.93E-006	mm Hg	25	EST	NEELY,WB & BLAU,GE (1985)
DC	pKa			
HL 3.55E-012	atm m3/mol	25	EST	MEYLAN,WM & HOWARD,PH (1991)
OH 1.65E-010	cm3/molc sec	25	EST	MEYLAN,WM & HOWARD,PH (1993)

CAS #: 007421-93-4		ENDRIN ALDEHYDE		
Formula: $C_{12}H_8Cl_6O$				
Mol Weight: 380.91				
MP (deg C):	FP (deg C):			
BP (deg C):				
BP pressure (mm Hg):				

Property/Value	Units	Temp	Data Type	Reference
WS 2.40E-002	mg/L		EXP	SHIU,WY ET AL. (1990)
logP 4.80			EST	MEYLAN,WM & HOWARD,PH (1995)
VP 2.00E-007	mm Hg	25	EXP	MABEY,WR ET AL. (1981)
DC	pKa			
HL 4.18E-006	atm m3/mol	25	EST	VP/WSOL
OH 1.06E-010	cm3/molc sec	25	EST	MEYLAN,WM & HOWARD,PH (1993)

CAS #: 007423-55-4		N-MALEOYL-3-AMINOPROPIONIC ACID		
Formula: $C_7H_7NO_4$				
Mol Weight: 169.14				
MP (deg C): 103-106	FP (deg C):			
BP (deg C):				
BP pressure (mm Hg):				

Property/Value	Units	Temp	Data Type	Reference
WS 5.74E+004	mg/L	25	EST	MEYLAN,WM ET AL. (1996)
logP 0.04			EXP	HANSCH,C & LEO,AJ (1985)
VP 1.63E-007	mm Hg	25	EST	NEELY,WB & BLAU,GE (1985)
DC	pKa			
HL 1.37E-013	atm m3/mol	25	EST	MEYLAN,WM & HOWARD,PH (1991)
OH 2.33E-011	cm3/molc sec	25	EST	MEYLAN,WM & HOWARD,PH (1993)

CAS #: 007424-54-6		3,5-HEPTADIONE		
Formula: $C_7H_{12}O_2$				
Mol Weight: 128.17				
MP (deg C):	FP (deg C):			
BP (deg C):				
BP pressure (mm Hg):				

Property/Value	Units	Temp	Data Type	Reference
WS 1.03E+004	mg/L	25	EST	MEYLAN,WM ET AL. (1996)
logP 1.12			EXP	HANSCH,C & LEO,AJ (1985)
VP 1.11E+000	mm Hg	25	EST	NEELY,WB & BLAU,GE (1985)
DC	pKa			
HL 5.88E-008	atm m3/mol	25	EST	MEYLAN,WM & HOWARD,PH (1991)
OH 2.99E-012	cm3/molc sec	25	EST	MEYLAN,WM & HOWARD,PH (1993)

851

007433-42-3 — 1-ETHYLHYDROXYUREA

Formula: $C_3H_8N_2O_2$
Mol Weight: 104.11
MP (deg C):
FP (deg C):
BP (deg C):
BP pressure (mm Hg):

Property/Value	Units	Temp	Data Type	Reference
WS 6.78E+003	mg/L	25	EST	MEYLAN,WM ET AL. (1996)
logP -0.10			EXP	HANSCH,C & LEO,AJ (1985)
VP 2.28E-004	mm Hg	25	EST	NEELY,WB & BLAU,GE (1985)
DC	pKa			
HL 1.58E-010	atm m3/mol	25	EST	MEYLAN,WM & HOWARD,PH (1991)
OH 6.00E-012	cm3/molc sec	25	EST	MEYLAN,WM & HOWARD,PH (1993)

007433-43-4 — 1-METHYLHYDROXYUREA

Formula: $C_2H_6N_2O_2$
Mol Weight: 90.08
MP (deg C):
FP (deg C):
BP (deg C):
BP pressure (mm Hg):

Property/Value	Units	Temp	Data Type	Reference
WS 1.51E+004	mg/L	25	EST	MEYLAN,WM ET AL. (1996)
logP -0.46			EXP	HANSCH,C & LEO,AJ (1985)
VP 8.06E-004	mm Hg	25	EST	NEELY,WB & BLAU,GE (1985)
DC	pKa			
HL 1.19E-010	atm m3/mol	25	EST	MEYLAN,WM & HOWARD,PH (1991)
OH 2.56E-012	cm3/molc sec	25	EST	MEYLAN,WM & HOWARD,PH (1993)

007433-86-5 — 3-NITROQUINOLINE-1-OXIDE

Formula: $C_9H_8N_2O_3$
Mol Weight: 192.18
MP (deg C):
FP (deg C):
BP (deg C):
BP pressure (mm Hg):

Property/Value	Units	Temp	Data Type	Reference
WS 6.65E+003	mg/L	25	EST	MEYLAN,WM ET AL. (1996)
logP 0.56			EXP	HANSCH,C & LEO,AJ (1985)
VP 2.60E-006	mm Hg	25	EST	NEELY,WB & BLAU,GE (1985)
DC	pKa			
HL 2.72E-014	atm m3/mol	25	EST	MEYLAN,WM & HOWARD,PH (1991)
OH 1.80E-012	cm3/molc sec	25	EST	MEYLAN,WM & HOWARD,PH (1993)

007443-52-9 — 2-METHYLCYCLOHEXANOL (TRANS)

Formula: $C_7H_{14}O$
Mol Weight: 114.19
MP (deg C): -21
FP (deg C):
BP (deg C): 167.4
BP pressure (mm Hg):

Property/Value	Units	Temp	Data Type	Reference
WS 9.14E+003	mg/L	25	EST	MEYLAN,WM ET AL. (1996)
logP 1.82			EXP	FUNASAKI,N ET AL. (1986)
VP 1.20E+000	mm Hg	25	EXP	DAUBERT,TE & DANNER,RP (1989)
DC	pKa			
HL 6.50E-006	atm m3/mol	25	EST	MEYLAN,WM & HOWARD,PH (1991)
OH 1.92E-011	cm3/molc sec	25	EST	MEYLAN,WM & HOWARD,PH (1993)

007443-55-2 — 3-METHYLCYCLOHEXANOL (TRANS)

Formula: $C_7H_{14}O$
Mol Weight: 114.19
MP (deg C):
FP (deg C):
BP (deg C):
BP pressure (mm Hg):

Property/Value	Units	Temp	Data Type	Reference
WS 5.99E+003	mg/L	25	EST	MEYLAN,WM ET AL. (1996)
logP 2.05			EST	MEYLAN,WM & HOWARD,PH (1995)
VP 5.74E-001	mm Hg	25	EXP	DAUBERT,TE & DANNER,RP (1989)
DC	pKa			
HL 6.50E-006	atm m3/mol	25	EST	MEYLAN,WM & HOWARD,PH (1991)
OH 1.92E-011	cm3/molc sec	25	EST	MEYLAN,WM & HOWARD,PH (1993)

007443-70-1 — 2-METHYLCYCLOHEXANOL (CIS)

Formula: $C_7H_{14}O$
Mol Weight: 114.19
MP (deg C): 6-8
FP (deg C):
BP (deg C): 165
BP pressure (mm Hg):

Property/Value	Units	Temp	Data Type	Reference
WS 9.14E+003	mg/L	25	EST	MEYLAN,WM ET AL. (1996)
logP 1.84			EXP	FUNASAKI,N ET AL. (1986)
VP 1.45E+000	mm Hg	25	EXP	DAUBERT,TE & DANNER,RP (1989)
DC	pKa			
HL 6.50E-006	atm m3/mol	25	EST	MEYLAN,WM & HOWARD,PH (1991)
OH 1.92E-011	cm3/molc sec	25	EST	MEYLAN,WM & HOWARD,PH (1993)

007447-24-7 — 1-DECANAMINIUM, N,N,N-TRIMETHYL-, IODIDE

Formula: $C_{13}H_{30}IN$
Mol Weight: 327.30
MP (deg C):
FP (deg C):
BP (deg C):
BP pressure (mm Hg):

Property/Value	Units	Temp	Data Type	Reference
WS 1.16E+004	mg/L	25	EST	MEYLAN,WM ET AL. (1996)
logP -0.16			EXP	SANGSTER,J (1993)
VP 3.39E-009	mm Hg	25	EST	NEELY,WB & BLAU,GE (1985)
DC	pKa			
HL 3.37E-011	atm m3/mol	25	EST	MEYLAN,WM & HOWARD,PH (1991)
OH 2.57E-011	cm3/molc sec	25	EST	MEYLAN,WM & HOWARD,PH (1993)

007463-28-7 — P-ACETAMIDO-N,N-DIMETHYLANILINE

Formula: $C_{10}H_{14}N_2O$
Mol Weight: 178.24
MP (deg C):
FP (deg C):
BP (deg C):
BP pressure (mm Hg):

Property/Value	Units	Temp	Data Type	Reference
WS 7.86E+003	mg/L	25	EST	MEYLAN,WM ET AL. (1996)
logP 1.00			EXP	HANSCH,C ET AL. (1995)
VP 2.73E-005	mm Hg	25	EST	NEELY,WB & BLAU,GE (1985)
DC	pKa			
HL 9.81E-011	atm m3/mol	25	EST	MEYLAN,WM & HOWARD,PH (1991)
OH 1.96E-010	cm3/molc sec	25	EST	MEYLAN,WM & HOWARD,PH (1993)

007463-31-2 — M-ACETYLAMINOACETOPHENONE

Formula: $C_{10}H_{11}NO_2$

Mol Weight: 177.20

MP (deg C): FP (deg C):

BP (deg C):

BP pressure (mm Hg):

Property/Value	Units	Temp	Data Type	Reference
WS 7.07E+003	mg/L	25	EST	MEYLAN,WM ET AL. (1996)
logP 1.06			EXP	HANSCH,C & LEO,AJ (1985)
VP 1.15E-005	mm Hg	25	EST	NEELY,WB & BLAU,GE (1985)
DC	pKa			
HL 1.12E-011	atm m3/mol	25	EST	MEYLAN,WM & HOWARD,PH (1991)
OH 5.44E-012	cm3/molc sec	25	EST	MEYLAN,WM & HOWARD,PH (1993)

007464-68-8 — 1,5-DIMETHYL-4-NITRO-1H-IMIDAZOLE

Formula: $C_5H_7N_3O_2$

Mol Weight: 141.13

MP (deg C): FP (deg C):

BP (deg C):

BP pressure (mm Hg):

Property/Value	Units	Temp	Data Type	Reference
WS 5.00E+003	mg/L	25	EST	MEYLAN,WM ET AL. (1996)
logP 0.97			EST	MEYLAN,WM & HOWARD,PH (1995)
VP 5.58E-004	mm Hg	25	EST	NEELY,WB & BLAU,GE (1985)
DC	pKa			
HL 3.49E-007	atm m3/mol	25	EST	MEYLAN,WM & HOWARD,PH (1991)
OH 5.79E-012	cm3/molc sec	25	EST	MEYLAN,WM & HOWARD,PH (1993)

007466-54-8 — BENZOYLHYDRAZINE, O-METHOXY

Formula: $C_8H_{10}N_2O_2$

Mol Weight: 166.18

MP (deg C): FP (deg C):

BP (deg C):

BP pressure (mm Hg):

Property/Value	Units	Temp	Data Type	Reference
WS 1.92E+003	mg/L	25	EST	MEYLAN,WM ET AL. (1996)
logP 0.25			EXP	HANSCH,C & LEO,AJ (1985)
VP 1.43E-005	mm Hg	25	EST	NEELY,WB & BLAU,GE (1985)
DC	pKa			
HL 5.47E-013	atm m3/mol	25	EST	MEYLAN,WM & HOWARD,PH (1991)
OH 2.59E-011	cm3/molc sec	25	EST	MEYLAN,WM & HOWARD,PH (1993)

007481-88-1 — CYTIDINE, 2',3'-DIDEHYDRO-2',3'-DIDEOXY-

Formula: $C_9H_{11}N_3O_3$

Mol Weight: 209.21

MP (deg C): FP (deg C):

BP (deg C):

BP pressure (mm Hg):

Property/Value	Units	Temp	Data Type	Reference
WS 4.05E+004	mg/L	25	EST	MEYLAN,WM ET AL. (1996)
logP -1.55			EXP	SANGSTER,J (1993)
VP 5.15E-008	mm Hg	25	EST	NEELY,WB & BLAU,GE (1985)
DC	pKa			
HL 8.18E-016	atm m3/mol	25	EST	MEYLAN,WM & HOWARD,PH (1991)
OH 1.67E-010	cm3/molc sec	25	EST	MEYLAN,WM & HOWARD,PH (1993)

007481-89-2 — CYTIDINE, 2',3'-DIDEOXY-

Formula: $C_9H_{13}N_3O_3$

Mol Weight: 211.22

MP (deg C): 217-218 FP (deg C):

BP (deg C):

BP pressure (mm Hg):

Property/Value	Units	Temp	Data Type	Reference
WS 2.42E+004	mg/L	25	EST	MEYLAN,WM ET AL. (1996)
logP -1.30			EXP	SANGSTER,J (1993)
VP 5.83E-008	mm Hg	25	EST	NEELY,WB & BLAU,GE (1985)
DC	pKa			
HL 1.17E-016	atm m3/mol	25	EST	MEYLAN,WM & HOWARD,PH (1991)
OH 1.24E-010	cm3/molc sec	25	EST	MEYLAN,WM & HOWARD,PH (1993)

007495-11-6 — ACETAMIDE, N-(4-BROMOPHENYL)-2-PHENYL-

Formula: $C_{14}H_{12}BrNO$

Mol Weight: 290.17

MP (deg C): FP (deg C):

BP (deg C):

BP pressure (mm Hg):

Property/Value	Units	Temp	Data Type	Reference
WS 9.69E+000	mg/L	25	EST	MEYLAN,WM ET AL. (1996)
logP 3.70			EXP	YAMAGAMI,C ET AL. (1984)
VP 8.58E-008	mm Hg	25	EST	NEELY,WB & BLAU,GE (1985)
DC	pKa			
HL 1.98E-010	atm m3/mol	25	EST	MEYLAN,WM & HOWARD,PH (1991)
OH 9.10E-012	cm3/molc sec	25	EST	MEYLAN,WM & HOWARD,PH (1993)

007496-53-9 — BENZOIC ACID, 2-[(ACETYLAMINO)AMINO]-

Formula: $C_9H_{10}N_2O_3$

Mol Weight: 194.19

MP (deg C): FP (deg C):

BP (deg C):

BP pressure (mm Hg):

Property/Value	Units	Temp	Data Type	Reference
WS 7.23E+003	mg/L	25	EST	MEYLAN,WM ET AL. (1996)
logP 0.95			EXP	SANGSTER,J (1993)
VP 1.15E-007	mm Hg	25	EST	NEELY,WB & BLAU,GE (1985)
DC	pKa			
HL 2.55E-016	atm m3/mol	25	EST	MEYLAN,WM & HOWARD,PH (1991)
OH 2.48E-011	cm3/molc sec	25	EST	MEYLAN,WM & HOWARD,PH (1993)

007496-72-2 — 2-(4-BROMOPHENYL)INDOLIZINE

Formula: $C_{14}H_{12}BrN$

Mol Weight: 274.17

MP (deg C): FP (deg C):

BP (deg C):

BP pressure (mm Hg):

Property/Value	Units	Temp	Data Type	Reference
WS 4.10E-001	mg/L	25	EST	MEYLAN,WM ET AL. (1996)
logP 5.43			EXP	HANSCH,C & LEO,AJ (1985)
VP 4.51E-006	mm Hg	25	EST	NEELY,WB & BLAU,GE (1985)
DC	pKa			
HL 1.67E-009	atm m3/mol	25	EST	MEYLAN,WM & HOWARD,PH (1991)
OH 1.92E-010	cm3/molc sec	25	EST	MEYLAN,WM & HOWARD,PH (1993)

CAS #: 007496-73-3	2-(4-CHLOROPHENYL)INDOLIZINE

Formula: $C_{14}H_{12}ClN$

Mol Weight: 229.71

MP (deg C): 　FP (deg C):

BP (deg C):

BP pressure (mm Hg):

Property/Value	Units	Temp	Data Type	Reference
WS 1.14E+000	mg/L	25	EST	MEYLAN,WM ET AL. (1996)
logP 5.20			EXP	HANSCH,C & LEO,AJ (1985)
VP 1.05E-005	mm Hg	25	EST	NEELY,WB & BLAU,GE (1985)
DC	pKa			
HL 3.11E-009	atm m3/mol	25	EST	MEYLAN,WM & HOWARD,PH (1991)
OH 1.92E-010	cm3/molc sec	25	EST	MEYLAN,WM & HOWARD,PH (1993)

CAS #: 007496-81-3	2-(4-TOLYL)INDOLIZINE

Formula: $C_{15}H_{15}N$

Mol Weight: 209.29

MP (deg C): 　FP (deg C):

BP (deg C):

BP pressure (mm Hg):

Property/Value	Units	Temp	Data Type	Reference
WS 2.33E+000	mg/L	25	EST	MEYLAN,WM ET AL. (1996)
logP 4.96			EXP	HANSCH,C & LEO,AJ (1985)
VP 1.92E-005	mm Hg	25	EST	NEELY,WB & BLAU,GE (1985)
DC	pKa			
HL 4.63E-009	atm m3/mol	25	EST	MEYLAN,WM & HOWARD,PH (1991)
OH 1.95E-010	cm3/molc sec	25	EST	MEYLAN,WM & HOWARD,PH (1993)

CAS #: 007496-82-4	2-(4-METHOXYPHENYL)INDOLIZINE

Formula: $C_{15}H_{15}NO$

Mol Weight: 225.29

MP (deg C): 　FP (deg C):

BP (deg C):

BP pressure (mm Hg):

Property/Value	Units	Temp	Data Type	Reference
WS 8.56E+000	mg/L	25	EST	MEYLAN,WM ET AL. (1996)
logP 4.20			EXP	HANSCH,C & LEO,AJ (1985)
VP 7.37E-006	mm Hg	25	EST	NEELY,WB & BLAU,GE (1985)
DC	pKa			
HL 2.48E-010	atm m3/mol	25	EST	MEYLAN,WM & HOWARD,PH (1991)
OH 2.07E-010	cm3/molc sec	25	EST	MEYLAN,WM & HOWARD,PH (1993)

CAS #: 007500-53-0	1,2-BENZENEDIACETIC ACID

Formula: $C_{10}H_{10}O_4$

Mol Weight: 194.19

MP (deg C): 150-152 　FP (deg C):

BP (deg C):

BP pressure (mm Hg):

Property/Value	Units	Temp	Data Type	Reference
WS 6.74E+004	mg/L	25	EST	MEYLAN,WM ET AL. (1996)
logP 0.28			EST	MEYLAN,WM & HOWARD,PH (1995)
VP 2.25E-006	mm Hg	25	EST	NEELY,WB & BLAU,GE (1985)
DC	pKa			
HL 3.63E-013	atm m3/mol	25	EST	MEYLAN,WM & HOWARD,PH (1991)
OH 5.78E-012	cm3/molc sec	25	EST	MEYLAN,WM & HOWARD,PH (1993)

CAS #: 007515-80-2	2-PROPANAMINE, 2-METHYL-N-(1-METHYLETHYL)-

Formula: $C_7H_{17}N$

Mol Weight: 115.22

MP (deg C): 　FP (deg C):

BP (deg C): 98

BP pressure (mm Hg):

Property/Value	Units	Temp	Data Type	Reference
WS 4.95E+004	mg/L	25	EST	MEYLAN,WM ET AL. (1996)
logP 1.56			EXP	HANSCH,C ET AL. (1995)
VP 3.35E+001	mm Hg	25	EST	NEELY,WB & BLAU,GE (1985)
DC	pKa			
HL 6.86E-005	atm m3/mol	25	EST	MEYLAN,WM & HOWARD,PH (1991)
OH 8.19E-011	cm3/molc sec	25	EST	MEYLAN,WM & HOWARD,PH (1993)

CAS #: 007520-68-5	2,4-DIAMINOPYRIMIDINE-5(345-CL)BENZYL

Formula: $C_{11}H_9Cl_3N_4$

Mol Weight: 303.58

MP (deg C): 　FP (deg C):

BP (deg C):

BP pressure (mm Hg):

Property/Value	Units	Temp	Data Type	Reference
WS 9.11E+000	mg/L	25	EST	MEYLAN,WM ET AL. (1996)
logP 3.64			EXP	HANSCH,C & LEO,AJ (1985)
VP 2.77E-008	mm Hg	25	EST	NEELY,WB & BLAU,GE (1985)
DC	pKa			
HL 4.69E-011	atm m3/mol	25	EST	MEYLAN,WM & HOWARD,PH (1991)
OH 3.17E-012	cm3/molc sec	25	EST	MEYLAN,WM & HOWARD,PH (1993)

CAS #: 007525-62-4	M-ETHYLSTYRENE

Formula: $C_{10}H_{12}$

Mol Weight: 132.21

MP (deg C): -101 　FP (deg C):

BP (deg C): 190

BP pressure (mm Hg):

Property/Value	Units	Temp	Data Type	Reference
WS 2.79E+001	mg/L	25	EST	MEYLAN,WM ET AL. (1996)
logP 4.11			EXP	CHEM INSPECT TEST INST (1992)
VP 8.04E-001	mm Hg	25	EXT	OHE,S (1976)
DC	pKa			
HL 4.05E-003	atm m3/mol	25	EST	MEYLAN,WM & HOWARD,PH (1991)
OH 3.20E-011	cm3/molc sec	25	EST	MEYLAN,WM & HOWARD,PH (1993)

CAS #: 007556-99-2	7-CHLOROQUINAZOLINE

Formula: $C_8H_5ClN_2$

Mol Weight: 164.60

MP (deg C): 　FP (deg C):

BP (deg C):

BP pressure (mm Hg):

Property/Value	Units	Temp	Data Type	Reference
WS 2.13E+003	mg/L	25	EST	MEYLAN,WM ET AL. (1996)
logP 1.74			EXP	HANSCH,C & LEO,AJ (1985)
VP 2.46E-003	mm Hg	25	EST	NEELY,WB & BLAU,GE (1985)
DC	pKa			
HL 2.11E-007	atm m3/mol	25	EST	MEYLAN,WM & HOWARD,PH (1991)
OH 1.41E-012	cm3/molc sec	25	EST	MEYLAN,WM & HOWARD,PH (1993)

007559-36-6 — 4-METHYL-B-NITROSTYRENE

Formula: $C_9H_9NO_2$
Mol Weight: 163.18
MP (deg C):   **FP (deg C):**
BP (deg C):
BP pressure (mm Hg):

Property/ Value	Units	Temp	Data Type	Reference
WS 1.44E+002	mg/L	25	EST	MEYLAN,WM ET AL. (1996)
logP 2.66			EXP	HANSCH,C & LEO,AJ (1985)
VP 5.35E-003	mm Hg	25	EST	NEELY,WB & BLAU,GE (1985)
DC	pKa			
HL 3.83E-006	atm m3/mol	25	EST	MEYLAN,WM & HOWARD,PH (1991)
OH 1.63E-011	cm3/molc sec	25	EST	MEYLAN,WM & HOWARD,PH (1993)

007560-83-0 — CYCLOHEXANAMINE, N-CYCLOHEXYL-N-METHYL-

Formula: $C_{13}H_{25}N$
Mol Weight: 195.35
MP (deg C):   **FP (deg C):**
BP (deg C): 265
BP pressure (mm Hg):

Property/ Value	Units	Temp	Data Type	Reference
WS 3.19E+002	mg/L	25	EST	MEYLAN,WM ET AL. (1996)
logP 3.71			EXP	HANSCH,C ET AL. (1995)
VP 1.75E-002	mm Hg	25	EST	NEELY,WB & BLAU,GE (1985)
DC	pKa			
HL 1.21E-004	atm m3/mol	25	EST	MEYLAN,WM & HOWARD,PH (1991)
OH 1.36E-010	cm3/molc sec	25	EST	MEYLAN,WM & HOWARD,PH (1993)

007561-67-3 — 1H-INDENE-1,3(2H)-DIONE, 2-(4-ETHYLPHENYL)-

Formula: $C_{17}H_{14}O_2$
Mol Weight: 250.30
MP (deg C):   **FP (deg C):**
BP (deg C):
BP pressure (mm Hg):

Property/ Value	Units	Temp	Data Type	Reference
WS 1.06E+001	mg/L	25	EST	MEYLAN,WM ET AL. (1996)
logP 3.92			EXP	SANGSTER,J (1993)
VP 5.77E-007	mm Hg	25	EST	NEELY,WB & BLAU,GE (1985)
DC	pKa			
HL 7.02E-010	atm m3/mol	25	EST	MEYLAN,WM & HOWARD,PH (1991)
OH 8.55E-012	cm3/molc sec	25	EST	MEYLAN,WM & HOWARD,PH (1993)

007568-93-6 — PHENYLETHANOLAMINE

Formula: $C_8H_{11}NO$
Mol Weight: 137.18
MP (deg C): 56.5   **FP (deg C):**
BP (deg C): 160
BP pressure (mm Hg): 1.70E+001

Property/ Value	Units	Temp	Data Type	Reference
WS 4.58E+004	mg/L	20	EXP	YALKOWSKY,SH & DANNENFELSER,RM (1992)
logP -0.20			EST	MEYLAN,WM & HOWARD,PH (1995)
VP 1.32E-003	mm Hg	25	EST	NEELY,WB & BLAU,GE (1985)
DC	pKa			
HL 2.97E-011	atm m3/mol	25	EST	MEYLAN,WM & HOWARD,PH (1991)
OH 4.48E-011	cm3/molc sec	25	EST	MEYLAN,WM & HOWARD,PH (1993)

007576-65-0 — DISPERSE YELLOW 54

Formula: $C_{18}H_{11}NO_3$
Mol Weight: 289.29
MP (deg C):   **FP (deg C):**
BP (deg C):
BP pressure (mm Hg):

Property/ Value	Units	Temp	Data Type	Reference
WS 2.89E-003	mg/L	25	EXP	BAUGHMAN,GL & PERENICH,TA (1988)
logP 5.00			EXP	BAUGHMAN,GL & WEBER,EJ (1991)
VP 1.21E-010	mm Hg	25	EST	NEELY,WB & BLAU,GE (1985)
DC	pKa			
HL 6.37E-018	atm m3/mol	25	EST	MEYLAN,WM & HOWARD,PH (1991)
OH 8.52E-011	cm3/molc sec	25	EST	MEYLAN,WM & HOWARD,PH (1993)

007580-85-0 — 2-(T-BUTOXY)ETHANOL

Formula: $C_6H_{14}O_2$
Mol Weight: 118.18
MP (deg C):   **FP (deg C):**
BP (deg C):
BP pressure (mm Hg):

Property/ Value	Units	Temp	Data Type	Reference
WS 1.53E+005	mg/L	25	EST	MEYLAN,WM ET AL. (1996)
logP 0.39			EXP	BINTEIN,S ET AL. (1993)
VP 7.70E-001	mm Hg	25	EST	NEELY,WB & BLAU,GE (1985)
DC	pKa			
HL 9.79E-008	atm m3/mol	25	EST	MEYLAN,WM & HOWARD,PH (1991)
OH 1.26E-011	cm3/molc sec	25	EST	MEYLAN,WM & HOWARD,PH (1993)

007581-97-7 — 2,3-DICHLOROBUTANE

Formula: $C_4H_8Cl_2$
Mol Weight: 127.01
MP (deg C): -80   **FP (deg C):**
BP (deg C): 117-119
BP pressure (mm Hg):

Property/ Value	Units	Temp	Data Type	Reference
WS 5.62E+002	mg/L	20	EXP	YALKOWSKY,SH & DANNENFELSER,RM (1992)
logP 2.67			EST	MEYLAN,WM & HOWARD,PH (1995)
VP 2.44E+001	mm Hg	25	EXP	PERRY,RH & GREEN,D (1984)
DC	pKa			
HL 7.26E-003	atm m3/mol	25	EST	VP/WSOL
OH 6.29E-013	cm3/molc sec	25	EST	MEYLAN,WM & HOWARD,PH (1993)

007597-60-6 — 6-NH2-5-(N-FORMYLAMINO)-1,3-DIMETHYLURACIL

Formula: $C_7H_{10}N_4O_3$
Mol Weight: 198.18
MP (deg C):   **FP (deg C):**
BP (deg C):
BP pressure (mm Hg):

Property/ Value	Units	Temp	Data Type	Reference
WS 7.71E+003	mg/L	25	EST	MEYLAN,WM ET AL. (1996)
logP -0.64			EXP	GASPARI,F & BONATI,M (1987)
VP 3.83E-009	mm Hg	25	EST	NEELY,WB & BLAU,GE (1985)
DC	pKa			
HL 5.41E-016	atm m3/mol	25	EST	MEYLAN,WM & HOWARD,PH (1991)
OH 3.79E-011	cm3/molc sec	25	EST	MEYLAN,WM & HOWARD,PH (1993)

007597-97-9 — 2-(TERT-BUTYL)-4-ISOPROPYLPHENOL

CAS #: 007597-97-9

Formula: $C_{13}H_{20}O$

Mol Weight: 192.30

MP (deg C): FP (deg C):

BP (deg C):

BP pressure (mm Hg):

Property/Value	Units	Temp	Data Type	Reference
WS 1.24E+001	mg/L	25	EST	MEYLAN,WM ET AL. (1996)
logP 4.88			EST	MEYLAN,WM & HOWARD,PH (1995)
VP 1.45E-003	mm Hg	25	EST	NEELY,WB & BLAU,GE (1985)
DC	pKa			
HL 2.82E-006	atm m3/mol	25	EST	MEYLAN,WM & HOWARD,PH (1991)
OH 5.10E-011	cm3/molc sec	25	EST	MEYLAN,WM & HOWARD,PH (1993)

007598-61-0 — PHOSPHINIC ACID, (2,2-DIETHOXYETHYL)-, DIETHYL E

CAS #: 007598-61-0

Formula: $C_{10}H_{23}O_5P$

Mol Weight: 254.27

MP (deg C): FP (deg C):

BP (deg C): 146-149

BP pressure (mm Hg): 1.40E+001

Property/Value	Units	Temp	Data Type	Reference
WS 5.77E+003	mg/L	25	EST	MEYLAN,WM ET AL. (1996)
logP 0.69			EXP	HANSCH,C ET AL. (1995)
VP 8.33E-004	mm Hg	25	EST	NEELY,WB & BLAU,GE (1985)
DC	pKa			
HL 3.54E-009	atm m3/mol	25	EST	MEYLAN,WM & HOWARD,PH (1991)
OH 6.09E-011	cm3/molc sec	25	EST	MEYLAN,WM & HOWARD,PH (1993)

007606-79-3 — GLYCINE,N-ACETYL-N'-MEAMINO AMIDE

CAS #: 007606-79-3

Formula: $C_5H_{10}N_2O_2$

Mol Weight: 130.15

MP (deg C): FP (deg C):

BP (deg C):

BP pressure (mm Hg):

Property/Value	Units	Temp	Data Type	Reference
WS 9.67E+004	mg/L	25	EST	MEYLAN,WM ET AL. (1996)
logP -1.56			EXP	HANSCH,C & LEO,AJ (1985)
VP 1.65E-006	mm Hg	25	EST	NEELY,WB & BLAU,GE (1985)
DC	pKa			
HL 2.59E-010	atm m3/mol	25	EST	MEYLAN,WM & HOWARD,PH (1991)
OH 7.64E-011	cm3/molc sec	25	EST	MEYLAN,WM & HOWARD,PH (1993)

007612-96-6 — 4-ISOTHIOCYANOAZOBENZENE

CAS #: 007612-96-6

Formula: $C_{13}H_9N_3S$

Mol Weight: 239.30

MP (deg C): FP (deg C):

BP (deg C):

BP pressure (mm Hg):

Property/Value	Units	Temp	Data Type	Reference
WS 1.83E-001	mg/L	25	EST	MEYLAN,WM ET AL. (1996)
logP 5.55			EXP	HANSCH,C & LEO,AJ (1985)
VP 6.94E-006	mm Hg	25	EST	NEELY,WB & BLAU,GE (1985)
DC	pKa			
HL 2.13E-006	atm m3/mol	25	EST	MEYLAN,WM & HOWARD,PH (1991)
OH 1.26E-012	cm3/molc sec	25	EST	MEYLAN,WM & HOWARD,PH (1993)

007613-10-7 — 2-ISOTHIOCYANO-ANTHRACENE

CAS #: 007613-10-7

Formula: $C_{15}H_9NS$

Mol Weight: 235.31

MP (deg C): FP (deg C):

BP (deg C):

BP pressure (mm Hg):

Property/Value	Units	Temp	Data Type	Reference
WS 3.00E-002	mg/L	25	EST	MEYLAN,WM ET AL. (1996)
logP 5.70			EXP	HANSCH,C & LEO,AJ (1985)
VP 8.03E-007	mm Hg	25	EST	NEELY,WB & BLAU,GE (1985)
DC	pKa			
HL 7.44E-006	atm m3/mol	25	EST	MEYLAN,WM & HOWARD,PH (1991)
OH 4.39E-011	cm3/molc sec	25	EST	MEYLAN,WM & HOWARD,PH (1993)

007613-19-6 — 5-NITROQUINOLINE-1-OXIDE

CAS #: 007613-19-6

Formula: $C_9H_8N_2O_3$

Mol Weight: 192.18

MP (deg C): FP (deg C):

BP (deg C):

BP pressure (mm Hg):

Property/Value	Units	Temp	Data Type	Reference
WS 7.63E+003	mg/L	25	EST	MEYLAN,WM ET AL. (1996)
logP 0.49			EXP	HANSCH,C & LEO,AJ (1985)
VP 2.60E-006	mm Hg	25	EST	NEELY,WB & BLAU,GE (1985)
DC	pKa			
HL 2.72E-014	atm m3/mol	25	EST	MEYLAN,WM & HOWARD,PH (1991)
OH 1.80E-012	cm3/molc sec	25	EST	MEYLAN,WM & HOWARD,PH (1993)

007642-04-8 — 2-OCTENE (CIS)

CAS #: 007642-04-8

Formula: C_8H_{16}

Mol Weight: 112.22

MP (deg C): -100.2 FP (deg C):

BP (deg C): 125.6

BP pressure (mm Hg):

Property/Value	Units	Temp	Data Type	Reference
WS 1.07E+001	mg/L	25	EST	MEYLAN,WM ET AL. (1996)
logP 4.06			EST	MEYLAN,WM & HOWARD,PH (1995)
VP 1.40E+001	mm Hg	25	EST	NEELY,WB & BLAU,GE (1985)
DC	pKa			
HL 7.46E-001	atm m3/mol	25	EST	MEYLAN,WM & HOWARD,PH (1991)
OH 6.18E-011	cm3/molc sec	25	EST	MEYLAN,WM & HOWARD,PH (1993)

007642-10-6 — 3-HEPTENE (CIS)

CAS #: 007642-10-6

Formula: C_7H_{14}

Mol Weight: 98.19

MP (deg C): -136.6 FP (deg C):

BP (deg C): 95.8

BP pressure (mm Hg):

Property/Value	Units	Temp	Data Type	Reference
WS 3.11E+001	mg/L	25	EST	MEYLAN,WM ET AL. (1996)
logP 3.56			EST	MEYLAN,WM & HOWARD,PH (1995)
VP 5.31E+001	mm Hg	25	EXP	DAUBERT,TE & DANNER,RP (1989)
DC	pKa			
HL 5.62E-001	atm m3/mol	25	EST	MEYLAN,WM & HOWARD,PH (1991)
OH 6.00E-011	cm3/molc sec	25	EST	MEYLAN,WM & HOWARD,PH (1993)

007658-80-2 — BENZOYLHYDRAZINE, O-METHYL

Formula: $C_8H_{10}N_2O$
Mol Weight: 150.18
MP (deg C):
FP (deg C):
BP (deg C):
BP pressure (mm Hg):

Property/Value	Units	Temp	Data Type	Reference
WS 2.41E+003	mg/L	25	EST	MEYLAN,WM ET AL. (1996)
logP 0.22			EXP	HANSCH,C & LEO,AJ (1985)
VP 2.95E-005	mm Hg	25	EST	NEELY,WB & BLAU,GE (1985)
DC	pKa			
HL 1.02E-011	atm m3/mol	25	EST	MEYLAN,WM & HOWARD,PH (1991)
OH 1.03E-011	cm3/molc sec	25	EST	MEYLAN,WM & HOWARD,PH (1993)

007660-25-5 — BETA-FRUCTOPYRANOSE (D)

Formula: $C_6H_{12}O_6$
Mol Weight: 180.16
MP (deg C):
FP (deg C):
BP (deg C):
BP pressure (mm Hg):

Property/Value	Units	Temp	Data Type	Reference
WS 1.00E+006	mg/L	25	EST	MEYLAN,WM ET AL. (1996)
logP -1.55			EST	MEYLAN,WM & HOWARD,PH (1995)
VP 3.11E-007	mm Hg	25	EST	NEELY,WB & BLAU,GE (1985)
DC	pKa			
HL 1.17E-014	atm m3/mol	25	EST	MEYLAN,WM & HOWARD,PH (1991)
OH 5.82E-011	cm3/molc sec	25	EST	MEYLAN,WM & HOWARD,PH (1993)

007667-80-3 — N-HEXACONTANE

Formula: $C_{60}H_{122}$
Mol Weight: 843.64
MP (deg C):
FP (deg C):
BP (deg C):
BP pressure (mm Hg):

Property/Value	Units	Temp	Data Type	Reference
WS 3.87E-026	mg/L	25	EST	MEYLAN,WM ET AL. (1996)
logP 29.81			EST	MEYLAN,WM & HOWARD,PH (1995)
VP 5.11E-019	mm Hg	25	EST	NEELY,WB & BLAU,GE (1985)
DC	pKa			
HL 7.54E+006	atm m3/mol	25	EST	MEYLAN,WM & HOWARD,PH (1991)
OH 8.18E-011	cm3/molc sec	25	EST	MEYLAN,WM & HOWARD,PH (1993)

007673-07-6 — TRIBENZYLAMINE HYDROCHLORIDE

Formula: $C_{21}H_{22}ClN$
Mol Weight: 323.87
MP (deg C):
FP (deg C):
BP (deg C):
BP pressure (mm Hg):

Property/Value	Units	Temp	Data Type	Reference
WS 6.10E+003	mg/L	25	EXP	SEIDELL,A (1941)
logP 2.39			EST	MEYLAN,WM & HOWARD,PH (1995)
VP 2.13E-012	mm Hg	25	EST	NEELY,WB & BLAU,GE (1985)
DC	pKa			
HL 1.21E-015	atm m3/mol	25	EST	MEYLAN,WM & HOWARD,PH (1991)
OH 4.06E-011	cm3/molc sec	25	EST	MEYLAN,WM & HOWARD,PH (1993)

007681-15-4 — NICOTINIC ACID, PROPYL ESTER

Formula: $C_9H_{11}NO_2$
Mol Weight: 165.19
MP (deg C):
FP (deg C):
BP (deg C):
BP pressure (mm Hg):

Property/Value	Units	Temp	Data Type	Reference
WS 1.43E+003	mg/L	25	EST	MEYLAN,WM ET AL. (1996)
logP 1.94			EXP	HANSCH,C & LEO,AJ (1985)
VP 2.63E-002	mm Hg	25	EST	NEELY,WB & BLAU,GE (1985)
DC	pKa			
HL 8.01E-008	atm m3/mol	25	EST	MEYLAN,WM & HOWARD,PH (1991)
OH 3.27E-012	cm3/molc sec	25	EST	MEYLAN,WM & HOWARD,PH (1993)

007681-76-7 — RONIDAZOLE

Formula: $C_6H_{10}N_4O_4$
Mol Weight: 202.17
MP (deg C): 167-169
FP (deg C):
BP (deg C):
BP pressure (mm Hg):

Property/Value	Units	Temp	Data Type	Reference
WS 1.84E+003	mg/L	25	EST	MEYLAN,WM ET AL. (1996)
logP -0.38			EXP	SANGSTER,J (1994)
VP 2.45E-006	mm Hg	25	EST	NEELY,WB & BLAU,GE (1985)
DC	pKa			
HL 1.87E-013	atm m3/mol	25	EST	MEYLAN,WM & HOWARD,PH (1991)
OH 7.46E-012	cm3/molc sec	25	EST	MEYLAN,WM & HOWARD,PH (1993)

007682-32-8 — BENZIMIDAZOLE-4,5,6-TRIBR-2-CF3

Formula: $C_8H_2Br_3F_3N_2$
Mol Weight: 422.84
MP (deg C):
FP (deg C):
BP (deg C):
BP pressure (mm Hg):

Property/Value	Units	Temp	Data Type	Reference
WS 7.24E-001	mg/L	25	EST	MEYLAN,WM ET AL. (1996)
logP 4.08			EXP	HANSCH,C & LEO,AJ (1985)
VP 4.06E-008	mm Hg	25	EST	NEELY,WB & BLAU,GE (1985)
DC	pKa			
HL 2.02E-007	atm m3/mol	25	EST	MEYLAN,WM & HOWARD,PH (1991)
OH 4.81E-013	cm3/molc sec	25	EST	MEYLAN,WM & HOWARD,PH (1993)

007682-34-0 — BENZIMIDAZOLE-2,4,5,6,7-PENTACL

Formula: $C_7HCl_5N_2$
Mol Weight: 290.36
MP (deg C):
FP (deg C):
BP (deg C):
BP pressure (mm Hg):

Property/Value	Units	Temp	Data Type	Reference
WS 1.89E+000	mg/L	25	EST	MEYLAN,WM ET AL. (1996)
logP 4.53			EXP	HANSCH,C & LEO,AJ (1985)
VP 5.56E-008	mm Hg	25	EST	NEELY,WB & BLAU,GE (1985)
DC	pKa			
HL 8.19E-008	atm m3/mol	25	EST	MEYLAN,WM & HOWARD,PH (1991)
OH 6.88E-013	cm3/molc sec	25	EST	MEYLAN,WM & HOWARD,PH (1993)

CAS #: 007682-38-4				2,4,5-TRICHLOROIMIDAZOLE

Formula: $C_3HCl_3N_2$

Mol Weight: 171.41

MP (deg C): | FP (deg C):

BP (deg C):

BP pressure (mm Hg):

Property/ Value	Units	Temp	Data Type	Reference
WS 1.59E+003	mg/L	25	EST	MEYLAN,WM ET AL. (1996)
logP 1.85			EXP	HANSCH,C & LEO,AJ (1985)
VP 1.77E-004	mm Hg	25	EST	NEELY,WB & BLAU,GE (1985)
DC	pKa			
HL 1.53E-006	atm m3/mol	25	EST	MEYLAN,WM & HOWARD,PH (1991)
OH 3.60E-011	cm3/molc sec	25	EST	MEYLAN,WM & HOWARD,PH (1993)

CAS #: 007688-21-3				2-HEXENE (Z)

Formula: C_6H_{12}

Mol Weight: 84.16

MP (deg C): -141.1 | FP (deg C):

BP (deg C): 68.8

BP pressure (mm Hg):

Property/ Value	Units	Temp	Data Type	Reference
WS 8.85E+001	mg/L	25	EST	MEYLAN,WM ET AL. (1996)
logP 3.07			EST	MEYLAN,WM & HOWARD,PH (1995)
VP 1.50E+002	mm Hg	25	EXP	DAUBERT,TE & DANNER,RP (1989)
DC	pKa			
HL 4.23E-001	atm m3/mol	25	EST	MEYLAN,WM & HOWARD,PH (1991)
OH 5.90E-011	cm3/molc sec	25	EST	MEYLAN,WM & HOWARD,PH (1993)

CAS #: 007692-57-1				BENZIMIDAZOLE-5-CHLORO-2-MES

Formula: $C_8H_7ClN_2S$

Mol Weight: 198.68

MP (deg C): | FP (deg C):

BP (deg C):

BP pressure (mm Hg):

Property/ Value	Units	Temp	Data Type	Reference
WS 7.91E+001	mg/L	25	EST	MEYLAN,WM ET AL. (1996)
logP 3.22			EXP	HANSCH,C & LEO,AJ (1985)
VP 6.87E-007	mm Hg	25	EST	NEELY,WB & BLAU,GE (1985)
DC	pKa			
HL 7.92E-009	atm m3/mol	25	EST	MEYLAN,WM & HOWARD,PH (1991)
OH 3.35E-011	cm3/molc sec	25	EST	MEYLAN,WM & HOWARD,PH (1993)

CAS #: 007693-52-9				4-BROMO-2-NITROPHENOL

Formula: $C_6H_4BrNO_3$

Mol Weight: 218.01

MP (deg C): 92 | FP (deg C):

BP (deg C):

BP pressure (mm Hg):

Property/ Value	Units	Temp	Data Type	Reference
WS 1.44E+002	mg/L	25	EST	MEYLAN,WM ET AL. (1996)
logP 2.80			EST	MEYLAN,WM & HOWARD,PH (1995)
VP 8.64E-005	mm Hg	25	EST	NEELY,WB & BLAU,GE (1985)
DC	pKa			
HL 2.79E-006	atm m3/mol	25	EST	MEYLAN,WM & HOWARD,PH (1991)
OH 1.33E-012	cm3/molc sec	25	EST	MEYLAN,WM & HOWARD,PH (1993)

CAS #: 007695-63-8				2-OH-I-PROPYLAMINO PHENYL ETHER

Formula: $C_{12}H_{19}NO_2$

Mol Weight: 209.29

MP (deg C): | FP (deg C):

BP (deg C):

BP pressure (mm Hg):

Property/ Value	Units	Temp	Data Type	Reference
WS 1.36E+004	mg/L	25	EST	MEYLAN,WM ET AL. (1996)
logP 1.72			EXP	HANSCH,C ET AL. (1995)
VP 1.95E-005	mm Hg	25	EST	NEELY,WB & BLAU,GE (1985)
DC	pKa			
HL 8.18E-012	atm m3/mol	25	EST	MEYLAN,WM & HOWARD,PH (1991)
OH 1.31E-010	cm3/molc sec	25	EST	MEYLAN,WM & HOWARD,PH (1993)

CAS #: 007696-12-0				PHTHALTHRIN

Formula: $C_{19}H_{25}NO_4$

Mol Weight: 331.42

MP (deg C): 60-80 | FP (deg C):

BP (deg C): 185-190

BP pressure (mm Hg): 1.00E-001

Property/ Value	Units	Temp	Data Type	Reference
WS 4.60E+000	mg/L	30	EXP	WORTHING,CR & WALKER,SB (1989)
logP 4.73			EXP	HANSCH,C ET AL. (1995)
VP 3.50E-008	mm Hg	20	EXP	HARTLEY,D & KIDD,H (1983)
DC	pKa			
HL 3.32E-009	atm m3/mol	20	EST	VP/WSOL
OH 1.50E-010	cm3/molc sec	25	EST	MEYLAN,WM & HOWARD,PH (1993)

CAS #: 007700-17-6				CROTOXYPHOS

Formula: $C_{14}H_{19}O_6P$

Mol Weight: 314.28

MP (deg C): | FP (deg C):

BP (deg C): 135

BP pressure (mm Hg): 3.00E-002

Property/ Value	Units	Temp	Data Type	Reference
WS 1.00E+003	mg/L	20	EXP	BUDAVARI,S (1989)
logP 1.89			EST	MEYLAN,WM & HOWARD,PH (1995)
VP 1.40E-005	mm Hg	20	EXP	BUDAVARI,S ET AL. (1989)
DC	pKa			
HL 5.79E-009	atm m3/mol	20	EST	VP/WSOL
OH 9.28E-011	cm3/molc sec	25	EST	MEYLAN,WM & HOWARD,PH (1993)

CAS #: 007703-36-8				3-FURANCARBONITRILE, 5-[(DIMETHYLAMINO)METHYLENE

Formula: $C_{14}H_{12}N_2O_2$

Mol Weight: 240.26

MP (deg C): | FP (deg C):

BP (deg C):

BP pressure (mm Hg):

Property/ Value	Units	Temp	Data Type	Reference
WS 1.81E+004	mg/L	25	EST	MEYLAN,WM ET AL. (1996)
logP 1.07			EXP	SANGSTER,J (1994)
VP 1.86E-007	mm Hg	25	EST	NEELY,WB & BLAU,GE (1985)
DC	pKa			
HL 3.18E-011	atm m3/mol	25	EST	MEYLAN,WM & HOWARD,PH (1991)
OH 7.99E-011	cm3/molc sec	25	EST	MEYLAN,WM & HOWARD,PH (1993)

CAS #: 007705-14-8 — DIPENTENE (+-)

Formula: $C_{10}H_{16}$

Mol Weight: 136.24

MP (deg C): | FP (deg C):

BP (deg C):

BP pressure (mm Hg):

Property/Value	Units	Temp	Data Type	Reference
WS 3.15E+000	mg/L	25	EST	MEYLAN,WM ET AL. (1996)
logP 4.83			EST	MEYLAN,WM & HOWARD,PH (1995)
VP 2.00E+001	mm Hg	68	EXP	RIDDICK,JA ET AL. (1986)
DC	pKa			
HL 3.80E-001	atm m3/mol	25	EST	MEYLAN,WM & HOWARD,PH (1991)
OH 1.45E-010	cm3/molc sec	25	EST	MEYLAN,WM & HOWARD,PH (1993)

CAS #: 007722-15-8 — 3H-1,4-BENZODIAZEPIN-2-AMINE, 7-CHLORO-5-PHENYL-

Formula: $C_{15}H_{12}CIN_3O$

Mol Weight: 285.74

MP (deg C): | FP (deg C):

BP (deg C):

BP pressure (mm Hg):

Property/Value	Units	Temp	Data Type	Reference
WS 2.30E+003	mg/L	25	EST	MEYLAN,WM ET AL. (1996)
logP 2.13			EXP	SANGSTER,J (1993)
VP 1.23E-011	mm Hg	25	EST	NEELY,WB & BLAU,GE (1985)
DC	pKa			
HL 1.93E-017	atm m3/mol	25	EST	MEYLAN,WM & HOWARD,PH (1991)
OH 3.93E-011	cm3/molc sec	25	EST	MEYLAN,WM & HOWARD,PH (1993)

CAS #: 007722-19-2 — TRIMETHYL BUTYL AMMONIUM IODIDE

Formula: $C_7H_{18}IN$

Mol Weight: 243.13

MP (deg C): | FP (deg C):

BP (deg C):

BP pressure (mm Hg):

Property/Value	Units	Temp	Data Type	Reference
WS 1.00E+006	mg/L	25	EST	MEYLAN,WM ET AL. (1996)
logP -2.60			EXP	SANGSTER,J (1993)
VP 5.01E-007	mm Hg	25	EST	NEELY,WB & BLAU,GE (1985)
DC	pKa			
HL 6.16E-012	atm m3/mol	25	EST	MEYLAN,WM & HOWARD,PH (1991)
OH 1.72E-011	cm3/molc sec	25	EST	MEYLAN,WM & HOWARD,PH (1993)

CAS #: 007728-40-7 — ACETAMIDE, N-(2,6-DIMETHYLPHENYL)-2-(ETHYLAMINO)

Formula: $C_{12}H_{18}N_2O$

Mol Weight: 206.29

MP (deg C): | FP (deg C):

BP (deg C):

BP pressure (mm Hg):

Property/Value	Units	Temp	Data Type	Reference
WS 3.03E+003	mg/L	25	EST	MEYLAN,WM ET AL. (1996)
logP 1.32			EXP	SANGSTER,J (1993)
VP 2.58E-006	mm Hg	25	EST	NEELY,WB & BLAU,GE (1985)
DC	pKa			
HL 4.50E-011	atm m3/mol	25	EST	MEYLAN,WM & HOWARD,PH (1991)
OH 9.73E-011	cm3/molc sec	25	EST	MEYLAN,WM & HOWARD,PH (1993)

CAS #: 007729-78-4 — 4,6(1H,5H)-PYRIMIDINEDIONE, DIHYDRO-5,5-DIMETHYL

Formula: $C_6H_8N_2O_2S$

Mol Weight: 172.21

MP (deg C): | FP (deg C):

BP (deg C):

BP pressure (mm Hg):

Property/Value	Units	Temp	Data Type	Reference
WS 2.25E+004	mg/L	25	EST	MEYLAN,WM ET AL. (1996)
logP 0.50			EXP	WONG,O & MCKEOWN,RH (1988)
VP 7.81E-009	mm Hg	25	EST	NEELY,WB & BLAU,GE (1985)
DC	pKa			
HL 1.54E-009	atm m3/mol	25	EST	MEYLAN,WM & HOWARD,PH (1991)
OH 1.21E-011	cm3/molc sec	25	EST	MEYLAN,WM & HOWARD,PH (1993)

CAS #: 007731-28-4 — 4-METHYLCYCLOHEXANOL (CIS)

Formula: $C_7H_{14}O$

Mol Weight: 114.19

MP (deg C): -9.2 | FP (deg C):

BP (deg C): 173

BP pressure (mm Hg):

Property/Value	Units	Temp	Data Type	Reference
WS 1.01E+004	mg/L	25	EST	MEYLAN,WM ET AL. (1996)
logP 2.05			EST	MEYLAN,WM & HOWARD,PH (1995)
VP 3.28E-001	mm Hg	25	EXT	YAWS,CL (1994A)
DC	pKa			
HL 6.50E-006	atm m3/mol	25	EST	MEYLAN,WM & HOWARD,PH (1991)
OH 1.92E-011	cm3/molc sec	25	EST	MEYLAN,WM & HOWARD,PH (1993)

CAS #: 007731-29-5 — 4-METHYLCYCLOHEXANOL (TRANS)

Formula: $C_7H_{14}O$

Mol Weight: 114.19

MP (deg C): | FP (deg C):

BP (deg C): 174

BP pressure (mm Hg):

Property/Value	Units	Temp	Data Type	Reference
WS 1.01E+004	mg/L	25	EST	MEYLAN,WM ET AL. (1996)
logP 2.05			EST	MEYLAN,WM & HOWARD,PH (1995)
VP 3.89E-001	mm Hg	25	EXT	YAWS,CL (1994A)
DC	pKa			
HL 6.50E-006	atm m3/mol	25	EST	MEYLAN,WM & HOWARD,PH (1991)
OH 1.92E-011	cm3/molc sec	25	EST	MEYLAN,WM & HOWARD,PH (1993)

CAS #: 007738-99-0 — THIAZOLIDINE, 2-(P-CL PHENYL)

Formula: $C_9H_{10}CINS$

Mol Weight: 199.70

MP (deg C): | FP (deg C):

BP (deg C):

BP pressure (mm Hg):

Property/Value	Units	Temp	Data Type	Reference
WS 7.20E+003	mg/L	25	EST	MEYLAN,WM ET AL. (1996)
logP 2.10			EXP	HANSCH,C & LEO,AJ (1985)
VP 4.11E-004	mm Hg	25	EST	NEELY,WB & BLAU,GE (1985)
DC	pKa			
HL 2.19E-006	atm m3/mol	25	EST	MEYLAN,WM & HOWARD,PH (1991)
OH 1.95E-010	cm3/molc sec	25	EST	MEYLAN,WM & HOWARD,PH (1993)

CAS #: 007747-84-4 — 1-CHLOROPROPYNE

Formula: C_3H_3Cl

Mol Weight: 74.51

MP (deg C): FP (deg C):

BP (deg C):

BP pressure (mm Hg):

Property/ Value	Units	Temp	Data Type	Reference
WS 9.33E+003	mg/L	25	EST	MEYLAN,WM ET AL. (1996)
logP 1.35			EST	MEYLAN,WM & HOWARD,PH (1995)
VP 2.86E+002	mm Hg	25	EST	NEELY,WB & BLAU,GE (1985)
DC	pKa			
HL 1.48E-002	atm m3/mol	25	EST	MEYLAN,WM & HOWARD,PH (1991)
OH 5.81E-012	cm3/molc sec	25	EST	MEYLAN,WM & HOWARD,PH (1993)

CAS #: 007752-09-2 — 2H-1,2,4-BENZOTHIADIAZINE, 3-(1-ETHYLPROPYL)-, 1

Formula: $C_{12}H_{16}N_2O_2S$

Mol Weight: 252.34

MP (deg C): FP (deg C):

BP (deg C):

BP pressure (mm Hg):

Property/ Value	Units	Temp	Data Type	Reference
WS 5.70E+002	mg/L	25	EST	MEYLAN,WM ET AL. (1996)
logP 1.88			EXP	HANSCH,C ET AL. (1995)
VP 7.26E-009	mm Hg	25	EST	NEELY,WB & BLAU,GE (1985)
DC	pKa			
HL 2.87E-007	atm m3/mol	25	EST	MEYLAN,WM & HOWARD,PH (1991)
OH 6.70E-012	cm3/molc sec	25	EST	MEYLAN,WM & HOWARD,PH (1993)

CAS #: 007752-27-4 — BUTANAMIDE, N-[2-(AMINOSULFONYL)PHENYL]-2-ETHYL-

Formula: $C_{12}H_{18}N_2O_3S$

Mol Weight: 270.35

MP (deg C): FP (deg C):

BP (deg C):

BP pressure (mm Hg):

Property/ Value	Units	Temp	Data Type	Reference
WS 7.52E+002	mg/L	25	EST	MEYLAN,WM ET AL. (1996)
logP 1.62			EXP	MONZANI,A ET AL. (1985)
VP 5.24E-009	mm Hg	25	EST	NEELY,WB & BLAU,GE (1985)
DC	pKa			
HL 1.50E-012	atm m3/mol	25	EST	MEYLAN,WM & HOWARD,PH (1991)
OH 1.25E-011	cm3/molc sec	25	EST	MEYLAN,WM & HOWARD,PH (1993)

CAS #: 007756-44-7 — 5ET-6MEO-4SULFANILAMIDOPYRIMIDINE

Formula: $C_{13}H_{16}N_4O_3S$

Mol Weight: 308.36

MP (deg C): FP (deg C):

BP (deg C):

BP pressure (mm Hg):

Property/ Value	Units	Temp	Data Type	Reference
WS 1.54E+003	mg/L	25	EST	MEYLAN,WM ET AL. (1996)
logP 1.00			EXP	HANSCH,C & LEO,AJ (1985)
VP 1.64E-009	mm Hg	25	EST	NEELY,WB & BLAU,GE (1985)
DC	pKa			
HL 3.53E-014	atm m3/mol	25	EST	MEYLAN,WM & HOWARD,PH (1991)
OH 2.58E-011	cm3/molc sec	25	EST	MEYLAN,WM & HOWARD,PH (1993)

CAS #: 007761-45-7 — 2,4-NH2-5(3,4DICLPH)-6-MEPYRIMIDINE

Formula: $C_{11}H_{10}Cl_2N_4$

Mol Weight: 269.14

MP (deg C): FP (deg C):

BP (deg C):

BP pressure (mm Hg):

Property/ Value	Units	Temp	Data Type	Reference
WS 7.22E+001	mg/L	25	EST	MEYLAN,WM ET AL. (1996)
logP 2.82			EXP	HANSCH,C & LEO,AJ (1985)
VP 3.82E-008	mm Hg	25	EST	NEELY,WB & BLAU,GE (1985)
DC 7.15	pKa	20	EXP	PERRIN,DD (1972)
HL 6.02E-011	atm m3/mol	25	EST	MEYLAN,WM & HOWARD,PH (1991)
OH 2.00E-010	cm3/molc sec	25	EST	MEYLAN,WM & HOWARD,PH (1993)

CAS #: 007764-50-3 — 2-METHYL-5-(1-METHYLETHENYL)CYCLOHEXANONE

Formula: $C_{10}H_{16}O$

Mol Weight: 152.24

MP (deg C): FP (deg C):

BP (deg C): 87-88

BP pressure (mm Hg): 6.00E+000

Property/ Value	Units	Temp	Data Type	Reference
WS 1.02E+003	mg/L	15	EXP	STEPHEN,H & STEPHEN,T (1963)
logP 2.86			EST	MEYLAN,WM & HOWARD,PH (1995)
VP 2.19E-001	mm Hg	25	EST	NEELY,WB & BLAU,GE (1985)
DC	pKa			
HL 1.40E-004	atm m3/mol	25	EST	MEYLAN,WM & HOWARD,PH (1991)
OH 6.99E-011	cm3/molc sec	25	EST	MEYLAN,WM & HOWARD,PH (1993)

CAS #: 007778-06-5 — 1,3,2-DIAZAPHOSPHOLIDIN-2-AMINE, N,N,1,3-TETRAME

Formula: $C_6H_{16}N_3OP$

Mol Weight: 177.19

MP (deg C): FP (deg C):

BP (deg C):

BP pressure (mm Hg):

Property/ Value	Units	Temp	Data Type	Reference
WS 5.46E+004	mg/L	25	EST	MEYLAN,WM ET AL. (1996)
logP 0.02			EXP	DEBORD,J & LABADIE,M (1985)
VP 2.06E-002	mm Hg	25	EST	NEELY,WB & BLAU,GE (1985)
DC	pKa			
HL 3.14E-012	atm m3/mol	25	EST	MEYLAN,WM & HOWARD,PH (1991)
OH 1.05E-010	cm3/molc sec	25	EST	MEYLAN,WM & HOWARD,PH (1993)

CAS #: 007778-85-0 — 1,2-DIMETHOXYPROPANE

Formula: $C_5H_{12}O_2$

Mol Weight: 104.15

MP (deg C): FP (deg C):

BP (deg C): 96

BP pressure (mm Hg):

Property/ Value	Units	Temp	Data Type	Reference
WS 1.36E+005	mg/L	25	EST	MEYLAN,WM ET AL. (1996)
logP -0.09			EXP	HANSCH,C ET AL. (1995)
VP 8.46E+001	mm Hg	25	EST	NEELY,WB & BLAU,GE (1985)
DC	pKa			
HL 1.78E-005	atm m3/mol	25	EST	MEYLAN,WM & HOWARD,PH (1991)
OH 1.43E-011	cm3/molc sec	25	EXP	ATKINSON,R (1989)

ETHYL-3-HYDROXYBENZOATE

CAS #: 007781-98-8

Formula: $C_9H_{10}O_3$

Mol Weight: 166.18

MP (deg C): 74

FP (deg C):

BP (deg C):

BP pressure (mm Hg):

Property/ Value	Units	Temp	Data Type	Reference
WS 1.89E+003	mg/L	25	EST	MEYLAN,WM ET AL. (1996)
logP 2.47			EXP	SANGSTER,J (1993)
VP 1.73E-003	mm Hg	25	EST	NEELY,WB & BLAU,GE (1985)
DC	pKa			
HL 4.79E-009	atm m3/mol	25	EST	MEYLAN,WM & HOWARD,PH (1991)
OH 9.33E-012	cm3/molc sec	25	EST	MEYLAN,WM & HOWARD,PH (1993)

D-ALPHA-TERPINEOL

CAS #: 007785-53-7

Formula: $C_{10}H_{18}O$

Mol Weight: 154.25

MP (deg C):

FP (deg C):

BP (deg C):

BP pressure (mm Hg):

Property/ Value	Units	Temp	Data Type	Reference
WS 6.70E+002	mg/L	25	EST	MEYLAN,WM ET AL. (1996)
logP 3.33			EST	MEYLAN,WM & HOWARD,PH (1995)
VP 3.07E-002	mm Hg	25	EST	NEELY,WB & BLAU,GE (1985)
DC	pKa			
HL 1.58E-005	atm m3/mol	25	EST	MEYLAN,WM & HOWARD,PH (1991)
OH 1.03E-010	cm3/molc sec	25	EST	MEYLAN,WM & HOWARD,PH (1993)

MEVINPHOS

CAS #: 007786-34-7

Formula: $C_7H_{13}O_6P$

Mol Weight: 224.15

MP (deg C): 6.9-21

FP (deg C):

BP (deg C): 99-103

BP pressure (mm Hg): 3.00E-001

Property/ Value	Units	Temp	Data Type	Reference
WS 6.00E+005	mg/L		EXP	BENYON,KI ET AL. (1973)
logP 0.13			EXP	TOMLIN,C (1994)
VP 1.30E-004	mm Hg	25	EXP	WAUCHOPE,RD ET AL. (1991A)
DC	pKa			
HL 6.39E-011	atm m3/mol	25	EST	VP/WSOL
OH 8.52E-011	cm3/molc sec	25	EST	MEYLAN,WM & HOWARD,PH (1993)

1,1,1-TRICHLOROPROPANE

CAS #: 007789-89-1

Formula: $C_3H_5Cl_3$

Mol Weight: 147.43

MP (deg C):

FP (deg C):

BP (deg C): 108

BP pressure (mm Hg):

Property/ Value	Units	Temp	Data Type	Reference
WS 1.90E+003	mg/L	25	EXP	DILLING,WL (1977)
logP 3.17			EST	MEYLAN,WM & HOWARD,PH (1995)
VP 3.16E+001	mm Hg	25	EXP	PERRY,RH & GREEN,D (1984)
DC	pKa			
HL 3.23E-003	atm m3/mol	25	EST	VP/WSOL
OH 2.32E-013	cm3/molc sec	25	EST	MEYLAN,WM & HOWARD,PH (1993)

TERPINEOL

CAS #: 008000-41-7

Formula: $C_{10}H_{18}O$

Mol Weight: 154.25

MP (deg C):

FP (deg C):

BP (deg C):

BP pressure (mm Hg):

Property/ Value	Units	Temp	Data Type	Reference
WS 1.98E+003	mg/L	20	EXP	SEIDELL,A (1941)
logP 3.33			EST	MEYLAN,WM & HOWARD,PH (1995)
VP 3.07E-002	mm Hg	25	EST	NEELY,WB & BLAU,GE (1985)
DC	pKa			
HL 1.58E-005	atm m3/mol	25	EST	MEYLAN,WM & HOWARD,PH (1991)
OH 1.03E-010	cm3/molc sec	25	EST	MEYLAN,WM & HOWARD,PH (1993)

DEMETON

CAS #: 008000-97-3

Formula: $C_{16}H_{38}O_6P_2S_4$

Mol Weight: 516.68

MP (deg C):

FP (deg C):

BP (deg C):

BP pressure (mm Hg):

Property/ Value	Units	Temp	Data Type	Reference
WS 6.00E+001	mg/L	25	EXP	AUGUSTIJN-BECKERS,PWM ET AL. (1994)
logP				
VP 3.00E-004	mm Hg	20	EXP	AUGUSTIJN-BECKERS,PWM ET AL. (1994)
DC	pKa			
HL 3.40E-006	atm m3/mol	20	EST	VP/WSOL
OH	cm3/molc sec			

TOXAPHENE

CAS #: 008001-35-2

Formula:

Mol Weight:

MP (deg C):

FP (deg C):

BP (deg C):

BP pressure (mm Hg):

NO STRUCTURE DIAGRAM AVAILABLE

Property/ Value	Units	Temp	Data Type	Reference
WS 5.50E-001	mg/L	20	EXP	MURPHY,TJ ET AL. (1987)
logP 4.82			EST	LYMAN,WJ ET AL. (1982)
VP 6.69E-006	mm Hg	20	EXP	MURPHY,TJ ET AL. (1987)
DC	pKa			
HL 6.00E-006	atm m3/mol	20	EXP	MURPHY,TJ ET AL. (1987)
OH 2.50E-012	cm3/molc sec	25	EST	MEYLAN,WM & HOWARD,PH (1993)

SOLVENT YELLOW 33

CAS #: 008003-22-3

Formula: $C_{18}H_{11}NO_2$

Mol Weight: 273.29

MP (deg C):

FP (deg C):

BP (deg C):

BP pressure (mm Hg):

Property/ Value	Units	Temp	Data Type	Reference
WS 1.69E-001	mg/L	25	EXP	BAUGHMAN,GL & WEBER,EJ (1991)
logP 4.10			EXP	BAUGHMAN,GL & WEBER,EJ (1991)
VP 9.88E-009	mm Hg	25	EST	NEELY,WB & BLAU,GE (1985)
DC	pKa			
HL 6.12E-014	atm m3/mol	25	EST	MEYLAN,WM & HOWARD,PH (1991)
OH 3.58E-011	cm3/molc sec	25	EST	MEYLAN,WM & HOWARD,PH (1993)

CAS #: 008018-01-7 — MANCOZEB

Formula:	$C_8H_{12}MnN_4S_8Zn$
Mol Weight:	541.03

MP (deg C):		FP (deg C):	
BP (deg C):			
BP pressure (mm Hg):			

Property/Value	Units	Temp	Data Type	Reference
WS 6.00E+000	mg/L	25	EXP	WAUCHOPE,RD ET AL. (1991A)
logP				
VP 0.00E+000	mm Hg	25		WAUCHOPE,RD ET AL. (1991A)
DC	pKa			
HL	atm m3/mol			
OH	cm3/molc sec			

CAS #: 008032-32-4 — LIGROIN

Formula:	
Mol Weight:	

NO STRUCTURE DIAGRAM AVAILABLE

MP (deg C):		FP (deg C):	
BP (deg C):			
BP pressure (mm Hg):			

Property/Value	Units	Temp	Data Type	Reference
WS 3.40E+003	mg/L	22	EXP	STEPHEN,H & STEPHEN,T (1963)
logP				
VP 5.00E+001	mm Hg	20	EXP	USEPA (1989A)
DC	pKa			
HL	atm m3/mol			
OH	cm3/molc sec			

CAS #: 009006-42-2 — METIRAM

Formula:	$C_8H_{16}N_5S_8Zn$
Mol Weight:	504.13

MP (deg C): 140 dec		FP (deg C):	
BP (deg C):			
BP pressure (mm Hg):			

Property/Value	Units	Temp	Data Type	Reference
WS	mg/L			
logP 0.30			EXP	TOMLIN,C (1994) @ pH=7
VP	mm Hg		EXP	TOMLIN,C (1994); < 7.5E-8 @ 20 deg
DC	pKa			
HL	atm m3/mol			
OH	cm3/molc sec			

CAS #: 009011-05-6 — UREA, POLYMER WITH FORMALDEHYDE

Formula:	$C_2H_4N_2O_2$
Mol Weight:	88.07

MP (deg C):		FP (deg C):	
BP (deg C):			
BP pressure (mm Hg):			

Property/Value	Units	Temp	Data Type	Reference
WS 1.08E+005	mg/L	25	EST	MEYLAN,WM ET AL. (1996)
logP -1.45			EST	MEYLAN,WM & HOWARD,PH (1995)
VP 7.28E-005	mm Hg	25	EST	NEELY,WB & BLAU,GE (1985)
DC	pKa			
HL 1.48E-010	atm m3/mol	25	EST	MEYLAN,WM & HOWARD,PH (1991)
OH 2.00E-012	cm3/molc sec	25	EST	MEYLAN,WM & HOWARD,PH (1993)

CAS #: 010004-44-1 — 3-HYDROXY-5-METHYLISOXAZOLE

Formula:	$C_4H_5NO_2$
Mol Weight:	99.09

MP (deg C):		FP (deg C):	
BP (deg C):			
BP pressure (mm Hg):			

Property/Value	Units	Temp	Data Type	Reference
WS 8.50E+004	mg/L	25	EXP	YALKOWSKY,SH & DANNENFELSER,RM (1992)
logP 0.46			EXP	HANSCH,C & LEO,AJ (1985)
VP 1.24E+000	mm Hg	25	EST	NEELY,WB & BLAU,GE (1985)
DC	pKa			
HL 6.51E-009	atm m3/mol	25	EST	MEYLAN,WM & HOWARD,PH (1991)
OH 2.00E-010	cm3/molc sec	25	EST	MEYLAN,WM & HOWARD,PH (1993)

CAS #: 010031-96-6 — EUGENYL FORMATE

Formula:	$C_{11}H_{12}O_3$
Mol Weight:	192.22

MP (deg C):		FP (deg C):	
BP (deg C): 270			
BP pressure (mm Hg):			

Property/Value	Units	Temp	Data Type	Reference
WS 3.40E+002	mg/L	25	EST	MEYLAN,WM ET AL. (1996)
logP 2.52			EST	MEYLAN,WM & HOWARD,PH (1995)
VP 4.72E-003	mm Hg	25	EST	NEELY,WB & BLAU,GE (1985)
DC	pKa			
HL 7.61E-006	atm m3/mol	25	EST	MEYLAN,WM & HOWARD,PH (1991)
OH 7.43E-011	cm3/molc sec	25	EST	MEYLAN,WM & HOWARD,PH (1993)

CAS #: 010051-63-5 — N-ME-3-PROPIONYLPHENYLCARBAMATE

Formula:	$C_{11}H_{13}NO_3$
Mol Weight:	207.23

MP (deg C):		FP (deg C):	
BP (deg C):			
BP pressure (mm Hg):			

Property/Value	Units	Temp	Data Type	Reference
WS 2.15E+003	mg/L	25	EST	MEYLAN,WM ET AL. (1996)
logP 1.49			EXP	HANSCH,C & LEO,AJ (1985)
VP 1.68E-004	mm Hg	25	EST	NEELY,WB & BLAU,GE (1985)
DC	pKa			
HL 4.33E-011	atm m3/mol	25	EST	MEYLAN,WM & HOWARD,PH (1991)
OH 3.63E-011	cm3/molc sec	25	EST	MEYLAN,WM & HOWARD,PH (1993)

CAS #: 010054-21-4 — 1-LAURYL-4-CARBOXY-2-PYRROLIDONE

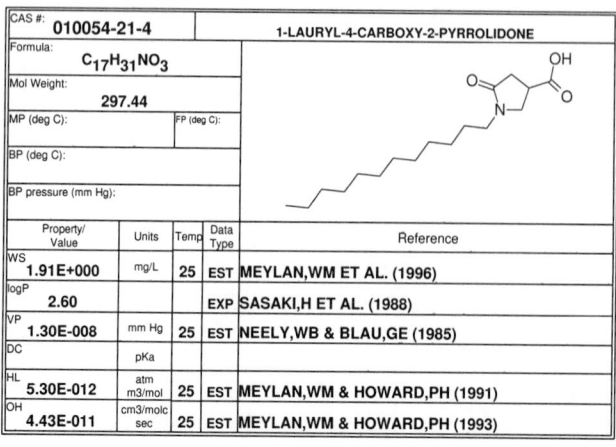

Formula:	$C_{17}H_{31}NO_3$
Mol Weight:	297.44

MP (deg C):		FP (deg C):	
BP (deg C):			
BP pressure (mm Hg):			

Property/Value	Units	Temp	Data Type	Reference
WS 1.91E+000	mg/L	25	EST	MEYLAN,WM ET AL. (1996)
logP 2.60			EXP	SASAKI,H ET AL. (1988)
VP 1.30E-008	mm Hg	25	EST	NEELY,WB & BLAU,GE (1985)
DC	pKa			
HL 5.30E-012	atm m3/mol	25	EST	MEYLAN,WM & HOWARD,PH (1991)
OH 4.43E-011	cm3/molc sec	25	EST	MEYLAN,WM & HOWARD,PH (1993)

010057-27-9 — HEXAFLUOROACETONE

Formula: C_3F_6O

Mol Weight: 166.02

MP (deg C): FP (deg C):

BP (deg C):

BP pressure (mm Hg):

Property/Value	Units	Temp	Data Type	Reference
WS 3.64E+003	mg/L	25	EST	MEYLAN,WM ET AL. (1996)
logP 1.46			EXP	HANSCH,C & LEO,AJ (1985)
VP 1.40E+003	mm Hg	25	EST	NEELY,WB & BLAU,GE (1985)
DC	pKa			
HL 3.07E-003	atm m3/mol	25	EST	MEYLAN,WM & HOWARD,PH (1991)
OH 0.00E+000	cm3/molc sec	25	EST	MEYLAN,WM & HOWARD,PH (1993)

010061-01-5 — 1,3-DICHLOROPROPENE, (Z)

Formula: $C_3H_4Cl_2$

Mol Weight: 110.97

MP (deg C): FP (deg C):

BP (deg C): 104.3

BP pressure (mm Hg):

Property/Value	Units	Temp	Data Type	Reference
WS 1.99E+003	mg/L	25	EST	MEYLAN,WM ET AL. (1996)
logP 2.06			EXP	TOMLIN,C (1994)
VP 2.67E+001	mm Hg	25	EST	NEELY,WB & BLAU,GE (1985)
DC	pKa			
HL 2.71E-003	atm m3/mol	20	EXP	LEISTRA,M (1970)
OH 8.41E-012	cm3/molc sec	25	EXP	ATKINSON,R (1989)

010061-02-6 — 1,3-DICHLOROPROPENE (TRANS)

Formula: $C_3H_4Cl_2$

Mol Weight: 110.97

MP (deg C): FP (deg C):

BP (deg C): 112

BP pressure (mm Hg):

Property/Value	Units	Temp	Data Type	Reference
WS 1.99E+003	mg/L	25	EST	MEYLAN,WM ET AL. (1996)
logP 2.03			EXP	TOMLIN,C (1994)
VP 2.67E+001	mm Hg	25	EST	NEELY,WB & BLAU,GE (1985)
DC	pKa			
HL 8.71E-004	atm m3/mol	20	EXP	LEISTRA,M (1970)
OH 9.36E-012	cm3/molc sec	25	EST	MEYLAN,WM & HOWARD,PH (1993)

010061-64-0 — L-CYSTEINAMIDE, N-ACETYL-

Formula: $C_5H_{10}N_2O_2S$

Mol Weight: 162.21

MP (deg C): FP (deg C):

BP (deg C):

BP pressure (mm Hg):

Property/Value	Units	Temp	Data Type	Reference
WS 5.79E+003	mg/L	25	EST	MEYLAN,WM ET AL. (1996)
logP -0.29			EXP	SANGSTER,J (1994)
VP 6.15E-007	mm Hg	25	EST	NEELY,WB & BLAU,GE (1985)
DC	pKa			
HL 5.56E-012	atm m3/mol	25	EST	MEYLAN,WM & HOWARD,PH (1991)
OH 7.59E-011	cm3/molc sec	25	EST	MEYLAN,WM & HOWARD,PH (1993)

010075-50-0 — 5-BROMOINDOLE

Formula: C_8H_6BrN

Mol Weight: 196.05

MP (deg C): 90-92 FP (deg C):

BP (deg C):

BP pressure (mm Hg):

Property/Value	Units	Temp	Data Type	Reference
WS 1.26E+002	mg/L	25	EST	MEYLAN,WM ET AL. (1996)
logP 3.00			EXP	HANSCH,C & LEO,AJ (1985)
VP 9.21E-004	mm Hg	25	EST	NEELY,WB & BLAU,GE (1985)
DC	pKa			
HL 3.53E-007	atm m3/mol	25	EST	MEYLAN,WM & HOWARD,PH (1991)
OH 6.92E-011	cm3/molc sec	25	EST	MEYLAN,WM & HOWARD,PH (1993)

010078-46-3 — 2-PROPEN-1-ONE, 3-(2,5-DIHYDRO-1H-PYRROL-1-YL)-1

Formula: $C_{16}H_{19}NO_4$

Mol Weight: 289.33

MP (deg C): FP (deg C):

BP (deg C):

BP pressure (mm Hg):

Property/Value	Units	Temp	Data Type	Reference
WS 4.44E+003	mg/L	25	EST	MEYLAN,WM ET AL. (1996)
logP 1.77			EXP	HANSCH,C ET AL. (1995)
VP 1.09E-006	mm Hg	25	EST	NEELY,WB & BLAU,GE (1985)
DC	pKa			
HL 7.59E-013	atm m3/mol	25	EST	MEYLAN,WM & HOWARD,PH (1991)
OH 2.37E-010	cm3/molc sec	25	EST	MEYLAN,WM & HOWARD,PH (1993)

010113-28-7 — O,O-DIMETHYL-O-PHENYLPHOSPHATE

Formula: $C_8H_{11}O_4P$

Mol Weight: 202.15

MP (deg C): FP (deg C):

BP (deg C):

BP pressure (mm Hg):

Property/Value	Units	Temp	Data Type	Reference
WS 3.87E+003	mg/L	25	EST	MEYLAN,WM ET AL. (1996)
logP 1.22			EXP	HANSCH,C & LEO,AJ (1985)
VP 2.45E-003	mm Hg	25	EST	NEELY,WB & BLAU,GE (1985)
DC	pKa			
HL 9.21E-008	atm m3/mol	25	EST	MEYLAN,WM & HOWARD,PH (1991)
OH 9.19E-012	cm3/molc sec	25	EST	MEYLAN,WM & HOWARD,PH (1993)

010118-89-5 — 2-NAPHTHACENECARBOXAMIDE DERIVATIVE

Formula: $C_{23}H_{27}N_3O_7$

Mol Weight: 457.49

MP (deg C): FP (deg C):

BP (deg C):

BP pressure (mm Hg):

Property/Value	Units	Temp	Data Type	Reference
WS 7.51E+001	mg/L	25	EST	MEYLAN,WM ET AL. (1996)
logP -0.07			EXP	SANGSTER,J (1993)
VP 3.71E-023	mm Hg	25	EST	NEELY,WB & BLAU,GE (1985)
DC	pKa			
HL 1.53E-024	atm m3/mol	25	EST	MEYLAN,WM & HOWARD,PH (1991)
OH 3.17E-010	cm3/molc sec	25	EST	MEYLAN,WM & HOWARD,PH (1993)

CAS #: 010118-90-8				MINOCYCLINE
Formula: $C_{23}H_{27}N_3O_7$				
Mol Weight: 457.49				
MP (deg C):		FP (deg C):		
BP (deg C):				
BP pressure (mm Hg):				

Property/ Value	Units	Temp	Data Type	Reference
WS 5.20E+004	mg/L	25	EXP	YALKOWSKY,SH & DANNENFELSER,RM (1992)
logP 0.05			EXP	SANGSTER,J (1993)
VP 3.71E-023	mm Hg	25	EST	NEELY,WB & BLAU,GE (1985)
DC	pKa			
HL 1.53E-024	atm m3/mol	25	EST	MEYLAN,WM & HOWARD,PH (1991)
OH 3.17E-010	cm3/molc sec	25	EST	MEYLAN,WM & HOWARD,PH (1993)

CAS #: 010124-65-9				LAURIC ACID, POTASSIUM SALT
Formula: $C_{12}H_{23}KO_2$				
Mol Weight: 238.42				
MP (deg C):		FP (deg C):		
BP (deg C):				
BP pressure (mm Hg):				

Property/ Value	Units	Temp	Data Type	Reference
WS 2.66E+003	mg/L	25	EST	MEYLAN,WM ET AL. (1996)
logP 1.19			EST	MEYLAN,WM & HOWARD,PH (1995)
VP 1.53E-010	mm Hg	25	EST	NEELY,WB & BLAU,GE (1985)
DC	pKa			
HL	atm m3/mol			
OH 1.30E-011	cm3/molc sec	25	EST	MEYLAN,WM & HOWARD,PH (1993)

CAS #: 010138-21-3				ETHYL VINYL DICHLOROSILANE
Formula: $C_4H_8Cl_2Si$				
Mol Weight: 155.10				
MP (deg C):		FP (deg C):		
BP (deg C):				
BP pressure (mm Hg):				

Property/ Value	Units	Temp	Data Type	Reference
WS 1.65E+002	mg/L	25	EST	MEYLAN,WM ET AL. (1996)
logP 3.09			EST	MEYLAN,WM & HOWARD,PH (1995)
VP 2.57E+000	mm Hg	25	EXT	BOUBLIK,T ET AL. (1984)
DC	pKa			
HL 2.47E-002	atm m3/mol	25	EST	MEYLAN,WM & HOWARD,PH (1991)
OH 2.75E-011	cm3/molc sec	25	EST	MEYLAN,WM & HOWARD,PH (1993)

CAS #: 010138-47-3				2-((1-ETHYLPENTYL)OXY)ETHANOL
Formula: $C_9H_{20}O_2$				
Mol Weight: 160.26				
MP (deg C):		FP (deg C):		
BP (deg C):				
BP pressure (mm Hg):				

Property/ Value	Units	Temp	Data Type	Reference
WS 4.62E+003	mg/L	25	EST	MEYLAN,WM ET AL. (1996)
logP 1.97			EST	MEYLAN,WM & HOWARD,PH (1995)
VP 9.75E-007	mm Hg	30	EXP	NASH,RG (1983B)
DC	pKa			
HL 2.29E-007	atm m3/mol	25	EST	MEYLAN,WM & HOWARD,PH (1991)
OH 3.45E-011	cm3/molc sec	25	EST	MEYLAN,WM & HOWARD,PH (1993)

CAS #: 010176-63-3				3-PYRIDINECARBONITRILE, 2-CHLORO-6-METHYL-5-PHEN
Formula: $C_{13}H_9ClN_2$				
Mol Weight: 228.68				
MP (deg C):		FP (deg C):		
BP (deg C):				
BP pressure (mm Hg):				

Property/ Value	Units	Temp	Data Type	Reference
WS 2.87E+001	mg/L	25	EST	MEYLAN,WM ET AL. (1996)
logP 3.24			EXP	SANGSTER,J (1993)
VP 4.49E-006	mm Hg	25	EST	NEELY,WB & BLAU,GE (1985)
DC	pKa			
HL 2.70E-007	atm m3/mol	25	EST	MEYLAN,WM & HOWARD,PH (1991)
OH 1.94E-012	cm3/molc sec	25	EST	MEYLAN,WM & HOWARD,PH (1993)

CAS #: 010190-66-6				C.I. DIRECT BROWN 191
Formula: $C_{12}H_{11}N_4NaO_3S$				
Mol Weight: 314.30				
MP (deg C):		FP (deg C):		
BP (deg C):				
BP pressure (mm Hg):				

Property/ Value	Units	Temp	Data Type	Reference
WS 9.54E+005	mg/L	25	EST	MEYLAN,WM ET AL. (1996)
logP -2.82			EST	MEYLAN,WM & HOWARD,PH (1995)
VP 2.50E-017	mm Hg	25	EST	NEELY,WB & BLAU,GE (1985)
DC	pKa			
HL	atm m3/mol			
OH 2.00E-010	cm3/molc sec	25	EST	MEYLAN,WM & HOWARD,PH (1993)

CAS #: 010206-21-0				CEPHACETRILE
Formula: $C_{13}H_{13}N_3O_6S$				
Mol Weight: 339.33				
MP (deg C):		FP (deg C):		
BP (deg C):				
BP pressure (mm Hg):				

Property/ Value	Units	Temp	Data Type	Reference
WS 4.63E+002	mg/L	25	EST	MEYLAN,WM ET AL. (1996)
logP -0.45			EXP	HANSCH,C ET AL. (1995)
VP 1.73E-013	mm Hg	25	EST	NEELY,WB & BLAU,GE (1985)
DC	pKa			
HL 1.65E-019	atm m3/mol	25	EST	MEYLAN,WM & HOWARD,PH (1991)
OH 1.10E-010	cm3/molc sec	25	EST	MEYLAN,WM & HOWARD,PH (1993)

CAS #: 010215-30-2				2-PROPOXY-1-PROPANOL
Formula: $C_6H_{14}O_2$				
Mol Weight: 118.18				
MP (deg C):		FP (deg C):		
BP (deg C):				
BP pressure (mm Hg):				

Property/ Value	Units	Temp	Data Type	Reference
WS 1.25E+005	mg/L	25	EST	MEYLAN,WM ET AL. (1996)
logP 0.49			EST	MEYLAN,WM & HOWARD,PH (1995)
VP 4.73E-001	mm Hg	25	EST	NEELY,WB & BLAU,GE (1985)
DC	pKa			
HL 9.79E-008	atm m3/mol	25	EST	MEYLAN,WM & HOWARD,PH (1991)
OH 2.93E-011	cm3/molc sec	25	EST	MEYLAN,WM & HOWARD,PH (1993)

CAS #: 010215-89-1 — SPIRO[AZULENO[4,5-B]FURAN-6(2H),2'-OXIRAN]-2-ONE

Formula: $C_{22}H_{26}O_8$

Mol Weight: 418.45

MP (deg C): FP (deg C):

BP (deg C):

BP pressure (mm Hg):

Property/Value	Units	Temp	Data Type	Reference
WS 1.71E+003	mg/L	25	EST	MEYLAN,WM ET AL. (1996)
logP 0.76			EXP	SANGSTER,J (1993)
VP 3.65E-012	mm Hg	25	EST	NEELY,WB & BLAU,GE (1985)
DC	pKa			
HL 7.23E-017	atm m3/mol	25	EST	MEYLAN,WM & HOWARD,PH (1991)
OH 1.48E-010	cm3/molc sec	25	EST	MEYLAN,WM & HOWARD,PH (1993)

CAS #: 010222-01-2 — 2,2-DIBROMO-2-CYANOACETAMIDE

Formula: $C_3H_2Br_2N_2O$

Mol Weight: 241.88

MP (deg C): FP (deg C):

BP (deg C):

BP pressure (mm Hg):

Property/Value	Units	Temp	Data Type	Reference
WS 1.97E+003	mg/L	25	EST	MEYLAN,WM ET AL. (1996)
logP 1.01			EST	MEYLAN,WM & HOWARD,PH (1995)
VP 1.52E-005	mm Hg	25	EST	NEELY,WB & BLAU,GE (1985)
DC	pKa			
HL 6.16E-014	atm m3/mol	25	EST	MEYLAN,WM & HOWARD,PH (1991)
OH 2.00E-012	cm3/molc sec	25	EST	MEYLAN,WM & HOWARD,PH (1993)

CAS #: 010233-13-3 — ISOPROPYL LAURATE

Formula: $C_{15}H_{30}O_2$

Mol Weight: 242.41

MP (deg C): FP (deg C):

BP (deg C): 196

BP pressure (mm Hg): 6.00E+001

Property/Value	Units	Temp	Data Type	Reference
WS 1.34E-001	mg/L	25	EST	MEYLAN,WM ET AL. (1996)
logP 6.19			EST	MEYLAN,WM & HOWARD,PH (1995)
VP 2.68E-003	mm Hg	25	EXT	BOUBLIK,T ET AL. (1984)
DC	pKa			
HL 5.25E-003	atm m3/mol	25	EST	MEYLAN,WM & HOWARD,PH (1991)
OH 1.69E-011	cm3/molc sec	25	EST	MEYLAN,WM & HOWARD,PH (1993)

CAS #: 010236-47-2 — NARINGINE

Formula: $C_{27}H_{32}O_{14}$

Mol Weight: 580.55

MP (deg C): ~83 FP (deg C):

BP (deg C):

BP pressure (mm Hg):

Property/Value	Units	Temp	Data Type	Reference
WS 1.95E+003	mg/L	25	EST	MEYLAN,WM ET AL. (1996)
logP -0.44			EXP	PERRISSOUD,D & TESTA,B (1986)
VP 4.91E-027	mm Hg	25	EST	NEELY,WB & BLAU,GE (1985)
DC	pKa			
HL 7.56E-032	atm m3/mol	25	EST	MEYLAN,WM & HOWARD,PH (1991)
OH 3.72E-010	cm3/molc sec	25	EST	MEYLAN,WM & HOWARD,PH (1993)

CAS #: 010245-51-9 — 1H-BENZ[DE]ISOQUINOLINE-1,3(2H)-DIONE, 2-[(ACETY

Formula: $C_{15}H_{11}NO_4$

Mol Weight: 269.26

MP (deg C): FP (deg C):

BP (deg C):

BP pressure (mm Hg):

Property/Value	Units	Temp	Data Type	Reference
WS 1.44E+001	mg/L	25	EST	MEYLAN,WM ET AL. (1996)
logP 2.34			EXP	NISHIMURA,K ET AL. (1988)
VP 4.85E-010	mm Hg	25	EST	NEELY,WB & BLAU,GE (1985)
DC	pKa			
HL 1.04E-010	atm m3/mol	25	EST	MEYLAN,WM & HOWARD,PH (1991)
OH 2.95E-011	cm3/molc sec	25	EST	MEYLAN,WM & HOWARD,PH (1993)

CAS #: 010247-71-9 — 5H-3,6-METHENOFURO[3,2-G]OXIRENO[D]OXACYCLOUNDEC

Formula: $C_{15}H_{16}O_6$

Mol Weight: 292.29

MP (deg C): FP (deg C):

BP (deg C):

BP pressure (mm Hg):

Property/Value	Units	Temp	Data Type	Reference
WS 1.55E+005	mg/L	25	EST	MEYLAN,WM ET AL. (1996)
logP -0.64			EXP	SANGSTER,J (1994)
VP 4.92E-011	mm Hg	25	EST	NEELY,WB & BLAU,GE (1985)
DC	pKa			
HL 7.10E-015	atm m3/mol	25	EST	MEYLAN,WM & HOWARD,PH (1991)
OH 6.07E-011	cm3/molc sec	25	EST	MEYLAN,WM & HOWARD,PH (1993)

CAS #: 010249-33-9 — 2-BUTENAMIDE, N-(3,4-DICHLOROPHENYL)-

Formula: $C_{10}H_9Cl_2NO$

Mol Weight: 230.10

MP (deg C): FP (deg C):

BP (deg C):

BP pressure (mm Hg):

Property/Value	Units	Temp	Data Type	Reference
WS 2.41E+001	mg/L	25	EST	MEYLAN,WM ET AL. (1996)
logP 3.63			EXP	HANSCH,C ET AL. (1995)
VP 3.09E-006	mm Hg	25	EST	NEELY,WB & BLAU,GE (1985)
DC	pKa			
HL 2.80E-009	atm m3/mol	25	EST	MEYLAN,WM & HOWARD,PH (1991)
OH 2.86E-011	cm3/molc sec	25	EST	MEYLAN,WM & HOWARD,PH (1993)

CAS #: 010264-18-3 — VALERANILIDE

Formula: $C_{11}H_{15}NO$

Mol Weight: 177.25

MP (deg C): FP (deg C):

BP (deg C):

BP pressure (mm Hg):

Property/Value	Units	Temp	Data Type	Reference
WS 3.35E+002	mg/L	25	EST	MEYLAN,WM ET AL. (1996)
logP 2.61			EXP	HANSCH,C & LEO,AJ (1985)
VP 3.28E-005	mm Hg	25	EST	NEELY,WB & BLAU,GE (1985)
DC	pKa			
HL 1.44E-008	atm m3/mol	25	EST	MEYLAN,WM & HOWARD,PH (1991)
OH 1.91E-011	cm3/molc sec	25	EST	MEYLAN,WM & HOWARD,PH (1993)

CAS #: 010265-92-6 — METHAMIDPHOS

Formula: $C_2H_8NO_2PS$

Mol Weight: 141.13

MP (deg C): 44.5

FP (deg C):

BP (deg C):

BP pressure (mm Hg):

Property/Value	Units	Temp	Data Type	Reference
WS 1.00E+006	mg/L	20	EXP	SHIU,WY ET AL. (1990)
logP -0.66			EXP	HANSCH,C & LEO,AJ (1985)
VP 3.00E-004	mm Hg	30	EXP	WORTHING,CR & WALKER,SB (1987)
DC	pKa			
HL 8.68E-010	atm m3/mol	25	EST	MEYLAN,WM & HOWARD,PH (1991)
OH 2.78E-011	cm3/molc sec	25	EST	ATKINSON,R (1988)

CAS #: 010268-78-7 — M-DIACETAMIDOBENZENE

Formula: $C_{10}H_{12}N_2O_2$

Mol Weight: 192.22

MP (deg C):

FP (deg C):

BP (deg C):

BP pressure (mm Hg):

Property/Value	Units	Temp	Data Type	Reference
WS 1.79E+004	mg/L	25	EST	MEYLAN,WM ET AL. (1996)
logP 0.50			EXP	HANSCH,C & LEO,AJ (1985)
VP 5.16E-008	mm Hg	25	EST	NEELY,WB & BLAU,GE (1985)
DC	pKa			
HL 1.12E-011	atm m3/mol	25	EST	MEYLAN,WM & HOWARD,PH (1991)
OH 7.93E-011	cm3/molc sec	25	EST	MEYLAN,WM & HOWARD,PH (1993)

CAS #: 010286-75-6 — N-(3,4-DICHLOROPHENYL)BENZAMIDE

Formula: $C_{13}H_9Cl_2NO$

Mol Weight: 266.13

MP (deg C):

FP (deg C):

BP (deg C):

BP pressure (mm Hg):

Property/Value	Units	Temp	Data Type	Reference
WS 3.23E+000	mg/L	25	EST	MEYLAN,WM ET AL. (1996)
logP 4.42			EXP	HANSCH,C ET AL. (1995)
VP 1.19E-007	mm Hg	25	EST	NEELY,WB & BLAU,GE (1985)
DC	pKa			
HL 6.71E-010	atm m3/mol	25	EST	MEYLAN,WM & HOWARD,PH (1991)
OH 4.33E-012	cm3/molc sec	25	EST	MEYLAN,WM & HOWARD,PH (1993)

CAS #: 010290-38-7 — 1,1-DIMETHYL-3-(3,5-DICLPH)UREA

Formula: $C_9H_{10}Cl_2N_2O$

Mol Weight: 233.10

MP (deg C):

FP (deg C):

BP (deg C):

BP pressure (mm Hg):

Property/Value	Units	Temp	Data Type	Reference
WS 7.00E+001	mg/L	25	EST	MEYLAN,WM ET AL. (1996)
logP 3.07			EXP	HANSCH,C & LEO,AJ (1985)
VP 1.04E-005	mm Hg	25	EST	NEELY,WB & BLAU,GE (1985)
DC	pKa			
HL 5.33E-010	atm m3/mol	25	EST	MEYLAN,WM & HOWARD,PH (1991)
OH 1.09E-011	cm3/molc sec	25	EST	MEYLAN,WM & HOWARD,PH (1993)

CAS #: 010291-06-2 — N-OCTYLPYRIDINIUM IODIDE

Formula: $C_{13}H_{22}IN$

Mol Weight: 319.23

MP (deg C):

FP (deg C):

BP (deg C):

BP pressure (mm Hg):

Property/Value	Units	Temp	Data Type	Reference
WS 4.94E+004	mg/L	25	EST	MEYLAN,WM ET AL. (1996)
logP -0.84			EXP	HANSCH,C & LEO,AJ (1985)
VP 1.20E-004	mm Hg	25	EST	NEELY,WB & BLAU,GE (1985)
DC	pKa			
HL 1.09E-009	atm m3/mol	25	EST	MEYLAN,WM & HOWARD,PH (1991)
OH 1.99E-011	cm3/molc sec	25	EST	MEYLAN,WM & HOWARD,PH (1993)

CAS #: 010311-84-9 — DIALIFOR

Formula: $C_{14}H_{17}ClNO_4PS_2$

Mol Weight: 393.85

MP (deg C): 67-69

FP (deg C):

BP (deg C):

BP pressure (mm Hg):

Property/Value	Units	Temp	Data Type	Reference
WS 1.80E-001	mg/L	25	EXP	FREED,VH ET AL. (1979A)
logP 4.69			EXP	HANSCH,C & LEO,AJ (1985)
VP 6.20E-008	mm Hg	25	EXP	FREED,VH ET AL. (1979A)
DC	pKa			
HL 1.78E-007	atm m3/mol	25	EST	VP/WSOL
OH 1.92E-010	cm3/molc sec	25	EST	MEYLAN,WM & HOWARD,PH (1993)

CAS #: 010315-98-7 — N-ISOBUTYLMORPHOLINE

Formula: $C_8H_{17}NO$

Mol Weight: 143.23

MP (deg C): < -10

FP (deg C):

BP (deg C): 181

BP pressure (mm Hg):

Property/Value	Units	Temp	Data Type	Reference
WS 2.80E+004	mg/L	25	EXP	CHEM INSPECT TEST INST (1992)
logP 1.17			EXP	CHEM INSPECT TEST INST (1992)
VP 1.13E+000	mm Hg	25	EST	NEELY,WB & BLAU,GE (1985)
DC	pKa			
HL 5.85E-007	atm m3/mol	25	EST	MEYLAN,WM & HOWARD,PH (1991)
OH 1.54E-010	cm3/molc sec	25	EST	MEYLAN,WM & HOWARD,PH (1993)

CAS #: 010316-00-4 — 4-BENZYLMORPHOLINE

Formula: $C_{11}H_{15}NO$

Mol Weight: 177.25

MP (deg C): 194

FP (deg C):

BP (deg C): 260.5

BP pressure (mm Hg):

Property/Value	Units	Temp	Data Type	Reference
WS 2.59E+004	mg/L	25	EST	MEYLAN,WM ET AL. (1996)
logP 1.58			EXP	HANSCH,C & LEO,AJ (1985)
VP 5.03E-003	mm Hg	25	EST	NEELY,WB & BLAU,GE (1985)
DC	pKa			
HL 2.02E-008	atm m3/mol	25	EST	MEYLAN,WM & HOWARD,PH (1991)
OH 1.54E-010	cm3/molc sec	25	EST	MEYLAN,WM & HOWARD,PH (1993)

CAS #: 010318-26-0				MITOLACTOL

Formula: $C_6H_{12}Br_2O_4$

Mol Weight: 307.98

MP (deg C): 187-188 de **FP (deg C):**

BP (deg C):

BP pressure (mm Hg):

Property/Value	Units	Temp	Data Type	Reference
WS 1.95E+004	mg/L	25	EST	MEYLAN,WM ET AL. (1996)
logP -0.29			EXP	HANSCH,C & LEO,AJ (1985)
VP 1.54E-008	mm Hg	25	EST	NEELY,WB & BLAU,GE (1985)
DC	pKa			
HL 7.21E-012	atm m3/mol	25	EST	MEYLAN,WM & HOWARD,PH (1991)
OH 2.94E-011	cm3/molc sec	25	EST	MEYLAN,WM & HOWARD,PH (1993)

CAS #: 010319-35-4				4-ISOTHIOCYANO-STILBENE

Formula: $C_{15}H_{11}NS$

Mol Weight: 237.33

MP (deg C): **FP (deg C):**

BP (deg C):

BP pressure (mm Hg):

Property/Value	Units	Temp	Data Type	Reference
WS 2.80E-001	mg/L	25	EST	MEYLAN,WM ET AL. (1996)
logP 5.85			EXP	HANSCH,C & LEO,AJ (1985)
VP 8.13E-006	mm Hg	25	EST	NEELY,WB & BLAU,GE (1985)
DC	pKa			
HL 1.79E-005	atm m3/mol	25	EST	MEYLAN,WM & HOWARD,PH (1991)
OH 6.02E-011	cm3/molc sec	25	EST	MEYLAN,WM & HOWARD,PH (1993)

CAS #: 010338-69-9				PYRIDINE, 1,2,3,6-TETRAHYDRO-4-PHENYL-

Formula: $C_{11}H_{13}N$

Mol Weight: 159.23

MP (deg C): **FP (deg C):**

BP (deg C):

BP pressure (mm Hg):

Property/Value	Units	Temp	Data Type	Reference
WS 1.40E+004	mg/L	25	EST	MEYLAN,WM ET AL. (1996)
logP 1.99			EXP	ALTOMARE,CA ET AL. (1992)
VP 3.97E-003	mm Hg	25	EST	NEELY,WB & BLAU,GE (1985)
DC	pKa			
HL 6.76E-007	atm m3/mol	25	EST	MEYLAN,WM & HOWARD,PH (1991)
OH 1.72E-010	cm3/molc sec	25	EST	MEYLAN,WM & HOWARD,PH (1993)

CAS #: 010342-59-3				4-PROPYLNITROBENZENE

Formula: $C_9H_{11}NO_2$

Mol Weight: 165.19

MP (deg C): **FP (deg C):**

BP (deg C):

BP pressure (mm Hg):

Property/Value	Units	Temp	Data Type	Reference
WS 2.99E+001	mg/L	25	EST	MEYLAN,WM ET AL. (1996)
logP 3.45			EXP	HANSCH,C ET AL. (1995)
VP 7.04E-003	mm Hg	25	EST	NEELY,WB & BLAU,GE (1985)
DC	pKa			
HL 4.13E-005	atm m3/mol	25	EST	MEYLAN,WM & HOWARD,PH (1991)
OH 3.07E-012	cm3/molc sec	25	EST	MEYLAN,WM & HOWARD,PH (1993)

CAS #: 010342-85-5				4'-PIPERIDINOACETOPHENONE

Formula: $C_{13}H_{17}NO$

Mol Weight: 203.29

MP (deg C): **FP (deg C):**

BP (deg C):

BP pressure (mm Hg):

Property/Value	Units	Temp	Data Type	Reference
WS 8.94E+001	mg/L	25	EST	MEYLAN,WM ET AL. (1996)
logP 3.13			EXP	HANSCH,C ET AL. (1995)
VP 2.90E-004	mm Hg	25	EST	NEELY,WB & BLAU,GE (1985)
DC	pKa			
HL 1.61E-007	atm m3/mol	25	EST	MEYLAN,WM & HOWARD,PH (1991)
OH 1.59E-010	cm3/molc sec	25	EST	MEYLAN,WM & HOWARD,PH (1993)

CAS #: 010345-79-6				BENZAMIDE, 2-CHLORO-N,N-DIETHYL-

Formula: $C_{11}H_{14}ClNO$

Mol Weight: 211.69

MP (deg C): **FP (deg C):**

BP (deg C):

BP pressure (mm Hg):

Property/Value	Units	Temp	Data Type	Reference
WS 6.13E+002	mg/L	25	EST	MEYLAN,WM ET AL. (1996)
logP 2.10			EXP	SURYANARAYANA,MVS ET AL. (1991)
VP 6.02E-001	mm Hg	30	EXP	SURYANARAYANA,MVS ET AL. (1991)
DC	pKa			
HL 1.39E-008	atm m3/mol	25	EST	MEYLAN,WM & HOWARD,PH (1991)
OH 2.42E-011	cm3/molc sec	25	EST	MEYLAN,WM & HOWARD,PH (1993)

CAS #: 010346-41-5				TRYPTOPHAN-AMIDE,N-ACETYL

Formula: $C_{13}H_{15}N_3O_2$

Mol Weight: 245.28

MP (deg C): **FP (deg C):**

BP (deg C):

BP pressure (mm Hg):

Property/Value	Units	Temp	Data Type	Reference
WS 1.10E+004	mg/L	25	EST	MEYLAN,WM ET AL. (1996)
logP 0.42			EXP	HANSCH,C & LEO,AJ (1985)
VP 1.23E-011	mm Hg	25	EST	NEELY,WB & BLAU,GE (1985)
DC	pKa			
HL 2.08E-015	atm m3/mol	25	EST	MEYLAN,WM & HOWARD,PH (1991)
OH 2.47E-010	cm3/molc sec	25	EST	MEYLAN,WM & HOWARD,PH (1993)

CAS #: 010349-09-4				FURAZAN, METHYLPHENYL-

Formula: $C_9H_8N_2O$

Mol Weight: 160.18

MP (deg C): **FP (deg C):**

BP (deg C):

BP pressure (mm Hg):

Property/Value	Units	Temp	Data Type	Reference
WS 4.19E+002	mg/L	25	EST	MEYLAN,WM ET AL. (1996)
logP 2.59			EXP	CALVINO,R ET AL. (1992)
VP 2.38E-003	mm Hg	25	EST	NEELY,WB & BLAU,GE (1985)
DC	pKa			
HL 1.60E-005	atm m3/mol	25	EST	MEYLAN,WM & HOWARD,PH (1991)
OH 7.52E-012	cm3/molc sec	25	EST	MEYLAN,WM & HOWARD,PH (1993)

CAS #: 010349-14-1 — FURAZANAMINE, 4-PHENYL-

Formula: $C_8H_7N_3O$

Mol Weight: 161.16

MP (deg C): FP (deg C):

BP (deg C):

BP pressure (mm Hg):

Property/Value	Units	Temp	Data Type	Reference
WS 1.92E+003	mg/L	25	EST	MEYLAN,WM ET AL. (1996)
logP 1.81			EXP	CALVINO,R ET AL. (1992)
VP 1.25E-004	mm Hg	25	EST	NEELY,WB & BLAU,GE (1985)
DC	pKa			
HL 5.13E-009	atm m3/mol	25	EST	MEYLAN,WM & HOWARD,PH (1991)
OH 7.39E-012	cm3/molc sec	25	EST	MEYLAN,WM & HOWARD,PH (1993)

CAS #: 010350-68-2 — 3(2-PYRIDYL)-5-ME-1,2,4-OXADIAZOLE

Formula: $C_8H_8N_3O$

Mol Weight: 162.17

MP (deg C): FP (deg C):

BP (deg C):

BP pressure (mm Hg):

Property/Value	Units	Temp	Data Type	Reference
WS 1.43E+004	mg/L	25	EST	MEYLAN,WM ET AL. (1996)
logP 0.79			EXP	HANSCH,C & LEO,AJ (1985)
VP 1.05E-003	mm Hg	25	EST	NEELY,WB & BLAU,GE (1985)
DC	pKa			
HL 2.60E-009	atm m3/mol	25	EST	MEYLAN,WM & HOWARD,PH (1991)
OH 4.78E-012	cm3/molc sec	25	EST	MEYLAN,WM & HOWARD,PH (1993)

CAS #: 010352-44-0 — ACETAMIDE, N-[4-(METHYLTHIO)PHENYL]

Formula: $C_9H_{11}NOS$

Mol Weight: 181.26

MP (deg C): FP (deg C):

BP (deg C):

BP pressure (mm Hg):

Property/Value	Units	Temp	Data Type	Reference
WS 1.00E+003	mg/L	25	EST	MEYLAN,WM ET AL. (1996)
logP 2.03			EXP	HANSCH,C ET AL. (1995)
VP 1.46E-005	mm Hg	25	EST	NEELY,WB & BLAU,GE (1985)
DC	pKa			
HL 1.80E-010	atm m3/mol	25	EST	MEYLAN,WM & HOWARD,PH (1991)
OH 8.79E-012	cm3/molc sec	25	EST	MEYLAN,WM & HOWARD,PH (1993)

CAS #: 010354-48-0 — N-BENZYL-2-FURAMIDE

Formula: $C_{12}H_{12}NO_2$

Mol Weight: 202.23

MP (deg C): FP (deg C):

BP (deg C):

BP pressure (mm Hg):

Property/Value	Units	Temp	Data Type	Reference
WS 1.23E+003	mg/L	25	EST	MEYLAN,WM ET AL. (1996)
logP 1.81			EXP	HANSCH,C & LEO,AJ (1985)
VP 3.15E-006	mm Hg	25	EST	NEELY,WB & BLAU,GE (1985)
DC	pKa			
HL 3.92E-010	atm m3/mol	25	EST	MEYLAN,WM & HOWARD,PH (1991)
OH 5.11E-011	cm3/molc sec	25	EST	MEYLAN,WM & HOWARD,PH (1993)

CAS #: 010354-55-9 — N-BUTYLNICOTINAMIDE

Formula: $C_{10}H_{14}N_2O$

Mol Weight: 178.24

MP (deg C): FP (deg C):

BP (deg C):

BP pressure (mm Hg):

Property/Value	Units	Temp	Data Type	Reference
WS 9.57E+003	mg/L	25	EST	MEYLAN,WM ET AL. (1996)
logP 0.90			EXP	HANSCH,C & LEO,AJ (1985)
VP 1.57E-005	mm Hg	25	EST	NEELY,WB & BLAU,GE (1985)
DC	pKa			
HL 1.49E-011	atm m3/mol	25	EST	MEYLAN,WM & HOWARD,PH (1991)
OH 1.33E-011	cm3/molc sec	25	EST	MEYLAN,WM & HOWARD,PH (1993)

CAS #: 010357-27-4 — _-D-MANNOPYRANOSIDE, 4-NITROPHENYL

Formula: $C_{12}H_{15}NO_8$

Mol Weight: 301.26

MP (deg C): FP (deg C):

BP (deg C):

BP pressure (mm Hg):

Property/Value	Units	Temp	Data Type	Reference
WS 7.01E+003	mg/L	25	EST	MEYLAN,WM ET AL. (1996)
logP -0.18			EXP	SANGSTER,J (1994)
VP 3.68E-013	mm Hg	25	EST	NEELY,WB & BLAU,GE (1985)
DC	pKa			
HL 4.54E-018	atm m3/mol	25	EST	MEYLAN,WM & HOWARD,PH (1991)
OH 6.47E-011	cm3/molc sec	25	EST	MEYLAN,WM & HOWARD,PH (1993)

CAS #: 010363-40-3 — O=P(NME)(OME)O-(2,4-DICLPHENYL)

Formula: $C_8H_{10}Cl_2NO_3P$

Mol Weight: 270.05

MP (deg C): FP (deg C):

BP (deg C):

BP pressure (mm Hg):

Property/Value	Units	Temp	Data Type	Reference
WS 1.60E+002	mg/L	25	EST	MEYLAN,WM ET AL. (1996)
logP 2.41			EXP	HANSCH,C & LEO,AJ (1985)
VP 5.63E-005	mm Hg	25	EST	NEELY,WB & BLAU,GE (1985)
DC	pKa			
HL 7.05E-010	atm m3/mol	25	EST	MEYLAN,WM & HOWARD,PH (1991)
OH 3.23E-011	cm3/molc sec	25	EST	MEYLAN,WM & HOWARD,PH (1993)

CAS #: 010369-88-7 — 2-PROPENOIC ACID, 3-PHENYL-, 2-(DIETHYLAMINO)ETH

Formula: $C_{15}H_{21}NO_2$

Mol Weight: 247.34

MP (deg C): FP (deg C):

BP (deg C):

BP pressure (mm Hg):

Property/Value	Units	Temp	Data Type	Reference
WS 2.06E+002	mg/L	25	EST	MEYLAN,WM ET AL. (1996)
logP 3.61			EXP	SANGSTER,J (1993)
VP 1.07E-004	mm Hg	25	EST	NEELY,WB & BLAU,GE (1985)
DC	pKa			
HL 4.80E-009	atm m3/mol	25	EST	MEYLAN,WM & HOWARD,PH (1991)
OH 1.18E-010	cm3/molc sec	25	EST	MEYLAN,WM & HOWARD,PH (1993)

CAS #: 010373-78-1 — DL-CAMPHORQUINONE

Formula: $C_{10}H_{14}O_2$
Mol Weight: 166.22
MP (deg C): 198-200
FP (deg C):
BP (deg C):
BP pressure (mm Hg):

Property/Value	Units	Temp	Data Type	Reference
WS 7.52E+003	mg/L	25	EST	MEYLAN,WM ET AL. (1996)
logP 1.09			EST	MEYLAN,WM & HOWARD,PH (1995)
VP 1.45E-002	mm Hg	25	EST	NEELY,WB & BLAU,GE (1985)
DC	pKa			
HL 2.10E-007	atm m3/mol	25	EST	MEYLAN,WM & HOWARD,PH (1991)
OH 8.01E-012	cm3/molc sec	25	EST	MEYLAN,WM & HOWARD,PH (1993)

CAS #: 010379-14-3 — TETRAZEPAM

Formula: $C_{16}H_{17}ClN_2O$
Mol Weight: 288.78
MP (deg C): 144
FP (deg C):
BP (deg C):
BP pressure (mm Hg):

Property/Value	Units	Temp	Data Type	Reference
WS 2.64E+001	mg/L	25	EST	MEYLAN,WM ET AL. (1996)
logP 3.20			EXP	HANSCH,C & LEO,AJ (1985)
VP 4.06E-008	mm Hg	25	EST	NEELY,WB & BLAU,GE (1985)
DC	pKa			
HL 1.13E-007	atm m3/mol	25	EST	MEYLAN,WM & HOWARD,PH (1991)
OH 9.76E-011	cm3/molc sec	25	EST	MEYLAN,WM & HOWARD,PH (1993)

CAS #: 010390-44-0 — PHENOXYMETHYLCEPHALOSPRIN

Formula: $C_{18}H_{18}N_2O_7S$
Mol Weight: 406.42
MP (deg C):
FP (deg C):
BP (deg C):
BP pressure (mm Hg):

Property/Value	Units	Temp	Data Type	Reference
WS 8.21E+001	mg/L	25	EST	MEYLAN,WM ET AL. (1996)
logP 0.26			EXP	HANSCH,C & LEO,AJ (1985)
VP 1.05E-014	mm Hg	25	EST	NEELY,WB & BLAU,GE (1985)
DC	pKa			
HL 1.22E-017	atm m3/mol	25	EST	MEYLAN,WM & HOWARD,PH (1991)
OH 1.36E-010	cm3/molc sec	25	EST	MEYLAN,WM & HOWARD,PH (1993)

CAS #: 010394-96-4 — MALONONITRILE, ALPHA-PHENYLBENZAL

Formula: $C_{16}H_{10}N_2$
Mol Weight: 230.27
MP (deg C):
FP (deg C):
BP (deg C):
BP pressure (mm Hg):

Property/Value	Units	Temp	Data Type	Reference
WS 1.53E+001	mg/L	25	EST	MEYLAN,WM ET AL. (1996)
logP 3.55			EXP	HANSCH,C & LEO,AJ (1985)
VP 4.25E-007	mm Hg	25	EST	NEELY,WB & BLAU,GE (1985)
DC	pKa			
HL 6.15E-010	atm m3/mol	25	EST	MEYLAN,WM & HOWARD,PH (1991)
OH 6.48E-012	cm3/molc sec	25	EST	MEYLAN,WM & HOWARD,PH (1993)

CAS #: 010397-59-8 — N,N-DIMETHYLPHENOXYACETAMIDE

Formula: $C_{10}H_{13}NO_2$
Mol Weight: 179.22
MP (deg C):
FP (deg C):
BP (deg C):
BP pressure (mm Hg):

Property/Value	Units	Temp	Data Type	Reference
WS 1.15E+004	mg/L	25	EST	MEYLAN,WM ET AL. (1996)
logP 0.80			EXP	HANSCH,C & LEO,AJ (1985)
VP 8.94E-004	mm Hg	25	EST	NEELY,WB & BLAU,GE (1985)
DC	pKa			
HL 1.65E-009	atm m3/mol	25	EST	MEYLAN,WM & HOWARD,PH (1991)
OH 4.19E-011	cm3/molc sec	25	EST	MEYLAN,WM & HOWARD,PH (1993)

CAS #: 010439-77-7 — N,2-DIMETHYL-4-NITROBENZENAMINE

Formula: $C_8H_{10}N_2O_2$
Mol Weight: 166.18
MP (deg C):
FP (deg C):
BP (deg C):
BP pressure (mm Hg):

Property/Value	Units	Temp	Data Type	Reference
WS 4.13E+002	mg/L	25	EST	MEYLAN,WM ET AL. (1996)
logP 2.57			EST	MEYLAN,WM & HOWARD,PH (1995)
VP 2.39E-003	mm Hg	25	EST	NEELY,WB & BLAU,GE (1985)
DC	pKa			
HL 1.82E-008	atm m3/mol	25	EST	MEYLAN,WM & HOWARD,PH (1991)
OH 7.94E-012	cm3/molc sec	25	EST	MEYLAN,WM & HOWARD,PH (1993)

CAS #: 010443-70-6 — MCPB-ETHYL

Formula: $C_{13}H_{17}ClO_3$
Mol Weight: 256.73
MP (deg C):
FP (deg C):
BP (deg C):
BP pressure (mm Hg):

Property/Value	Units	Temp	Data Type	Reference
WS 4.99E+000	mg/L	25	EST	MEYLAN,WM ET AL. (1996)
logP 4.26			EXP	TOMLIN,C (1994)
VP 1.27E-004	mm Hg	25	EST	NEELY,WB & BLAU,GE (1985)
DC	pKa			
HL 1.45E-006	atm m3/mol	25	EST	MEYLAN,WM & HOWARD,PH (1991)
OH 2.14E-011	cm3/molc sec	25	EST	MEYLAN,WM & HOWARD,PH (1993)

CAS #: 010448-09-6 — HEPTAMETHYLPHENYLCYCLOTETRASILOXANE

Formula: $C_{13}H_{26}O_4Si_4$
Mol Weight: 358.69
MP (deg C):
FP (deg C):
BP (deg C):
BP pressure (mm Hg):

Property/Value	Units	Temp	Data Type	Reference
WS 2.19E-003	mg/L	25	EST	MEYLAN,WM ET AL. (1996)
logP 6.30			EST	MEYLAN,WM & HOWARD,PH (1995)
VP	mm Hg			
DC	pKa			
HL	atm m3/mol			
OH 3.00E-012	cm3/molc sec	25	EST	MEYLAN,WM & HOWARD,PH (1993)

870

<table>
<tr><td colspan="2">CAS #: 010543-21-2</td><td>1,1-DI-(P-METHOXYPHENYL)ETHANE</td></tr>
</table>

CAS #: 010543-21-2 — 1,1-DI-(P-METHOXYPHENYL)ETHANE

Formula: $C_{16}H_{18}O_2$
Mol Weight: 242.32
MP (deg C): | FP (deg C):
BP (deg C):
BP pressure (mm Hg):

Property/Value	Units	Temp	Data Type	Reference
WS 3.60E+000	mg/L	25	EST	MEYLAN,WM ET AL. (1996)
logP 4.52			EXP	HANSCH,C & LEO,AJ (1985)
VP 7.09E-005	mm Hg	25	EST	NEELY,WB & BLAU,GE (1985)
DC	pKa			
HL 2.23E-006	atm m3/mol	25	EST	MEYLAN,WM & HOWARD,PH (1991)
OH 5.55E-011	cm3/molc sec	25	EST	MEYLAN,WM & HOWARD,PH (1993)

CAS #: 010548-10-4 — TERBUFOS SULFOXIDE

Formula: $C_9H_{21}O_3PS_3$
Mol Weight: 304.43
MP (deg C): | FP (deg C):
BP (deg C):
BP pressure (mm Hg):

Property/Value	Units	Temp	Data Type	Reference
WS 1.10E+003	mg/L	20	EXP	SHIU,WY ET AL. (1990)
logP 2.21			EXP	HANSCH,C & LEO,AJ (1985)
VP 3.42E-005	mm Hg	25	EST	NEELY,WB & BLAU,GE (1985)
DC	pKa			
HL 1.76E-010	atm m3/mol	25	EST	MEYLAN,WM & HOWARD,PH (1991)
OH 3.01E-010	cm3/molc sec	25	EST	MEYLAN,WM & HOWARD,PH (1993)

CAS #: 010552-94-0 — N-NITROSO-3-PYRROLINE

Formula: $C_4H_6N_2O$
Mol Weight: 98.11
MP (deg C): | FP (deg C):
BP (deg C):
BP pressure (mm Hg):

Property/Value	Units	Temp	Data Type	Reference
WS 7.10E+004	mg/L	25	EST	MEYLAN,WM ET AL. (1996)
logP 0.26			EXP	HANSCH,C & LEO,AJ (1985)
VP 2.95E-001	mm Hg	25	EST	NEELY,WB & BLAU,GE (1985)
DC	pKa			
HL 1.41E-006	atm m3/mol	25	EST	MEYLAN,WM & HOWARD,PH (1991)
OH 7.03E-011	cm3/molc sec	25	EST	MEYLAN,WM & HOWARD,PH (1993)

CAS #: 010574-37-5 — 2,3-DIMETHYL-2-PENTENE

Formula: C_7H_{14}
Mol Weight: 98.19
MP (deg C): | FP (deg C):
BP (deg C):
BP pressure (mm Hg):

Property/Value	Units	Temp	Data Type	Reference
WS 2.49E+001	mg/L	25	EST	MEYLAN,WM ET AL. (1996)
logP 3.68			EST	MEYLAN,WM & HOWARD,PH (1995)
VP 5.02E+001	mm Hg	25	EXP	DYKYJ,J & REPAS,M (1973)
DC	pKa			
HL 7.83E-001	atm m3/mol	25	EST	MEYLAN,WM & HOWARD,PH (1991)
OH 1.08E-010	cm3/molc sec	25	EXP	ATKINSON,R (1989)

CAS #: 010580-24-2 — UNDECANOIC ACID, BUTYL ESTER

Formula: $C_{15}H_{30}O_2$
Mol Weight: 242.41
MP (deg C): | FP (deg C):
BP (deg C):
BP pressure (mm Hg):

Property/Value	Units	Temp	Data Type	Reference
WS 1.16E-001	mg/L	25	EST	MEYLAN,WM ET AL. (1996)
logP 6.27			EST	MEYLAN,WM & HOWARD,PH (1995)
VP 1.88E-003	mm Hg	25	EST	NEELY,WB & BLAU,GE (1985)
DC	pKa			
HL 5.25E-003	atm m3/mol	25	EST	MEYLAN,WM & HOWARD,PH (1991)
OH 1.61E-011	cm3/molc sec	25	EST	MEYLAN,WM & HOWARD,PH (1993)

CAS #: 010586-17-1 — ISOPROPYL CYANOACRYLATE

Formula: $C_7H_9NO_2$
Mol Weight: 139.16
MP (deg C): | FP (deg C):
BP (deg C):
BP pressure (mm Hg):

Property/Value	Units	Temp	Data Type	Reference
WS 1.24E+003	mg/L	25	EST	MEYLAN,WM ET AL. (1996)
logP 1.83			EST	MEYLAN,WM & HOWARD,PH (1995)
VP 1.98E-001	mm Hg	25	EST	NEELY,WB & BLAU,GE (1985)
DC	pKa			
HL 3.63E-007	atm m3/mol	25	EST	MEYLAN,WM & HOWARD,PH (1991)
OH 6.32E-012	cm3/molc sec	25	EST	MEYLAN,WM & HOWARD,PH (1993)

CAS #: 010592-27-5 — 7-AZAINDOLE

Formula: $C_7H_6N_2$
Mol Weight: 118.14
MP (deg C): | FP (deg C):
BP (deg C):
BP pressure (mm Hg):

Property/Value	Units	Temp	Data Type	Reference
WS 2.85E+003	mg/L	25	EST	MEYLAN,WM ET AL. (1996)
logP 1.03			EST	MEYLAN,WM & HOWARD,PH (1995)
VP 8.58E-003	mm Hg	25	EST	NEELY,WB & BLAU,GE (1985)
DC	pKa			
HL 1.16E-009	atm m3/mol	25	EST	MEYLAN,WM & HOWARD,PH (1991)
OH 1.10E-010	cm3/molc sec	25	EST	MEYLAN,WM & HOWARD,PH (1993)

CAS #: 010595-95-6 — N-METHYL-N-NITROSOETHAMINE

Formula: $C_3H_8N_2O$
Mol Weight: 88.11
MP (deg C): | FP (deg C):
BP (deg C): 163
BP pressure (mm Hg): 7.47E+002

Property/Value	Units	Temp	Data Type	Reference
WS 3.00E+005	mg/L	20	EXP	IARC (1978)
logP 0.04			EXP	VERA,A ET AL. (1992)
VP 1.10E+000	mm Hg	20	EST	KLEIN,RG (1982)
DC	pKa			
HL 4.25E-007	atm m3/mol	20	EST	VP/WSOL
OH 1.00E-011	cm3/molc sec	25	EST	ATKINSON,R (1988)

CARBENDAZIM

CAS #: 010605-21-7
Formula: $C_9H_9N_3O_2$
Mol Weight: 191.19
MP (deg C): 300 dec
FP (deg C):
BP (deg C):
BP pressure (mm Hg):

Property/Value	Units	Temp	Data Type	Reference
WS 5.80E+000	mg/L	20	EXP	YALKOWSKY,SH & DANNENFELSER,RM (1992)
logP 1.52			EXP	HANSCH,C & LEO,AJ (1985)
VP 4.88E-010	mm Hg	20	EXP	AUGUSTIJN-BECKERS,PWM ET AL. (1994) @ pH=7
DC	pKa			
HL 2.12E-011	atm m3/mol	20	EST	VP/WSOL
OH 2.00E-010	cm3/molc sec	25	EST	MEYLAN,WM & HOWARD,PH (1993)

PLINOL

CAS #: 011039-70-6
Formula: $C_{10}H_{18}O$
Mol Weight: 154.25
MP (deg C):
FP (deg C):
BP (deg C):
BP pressure (mm Hg):

Property/Value	Units	Temp	Data Type	Reference
WS 8.32E+002	mg/L	25	EST	MEYLAN,WM ET AL. (1996)
logP 2.87			EXP	LI,J & PERDUE,EM (1995)
VP 6.22E-002	mm Hg	25	EST	NEELY,WB & BLAU,GE (1985)
DC	pKa			
HL 1.34E-005	atm m3/mol	25	EST	MEYLAN,WM & HOWARD,PH (1991)
OH 6.66E-011	cm3/molc sec	25	EST	MEYLAN,WM & HOWARD,PH (1993)

BENZOTHIOPHENE

CAS #: 011095-43-5
Formula: C_8H_6S
Mol Weight: 134.20
MP (deg C):
FP (deg C):
BP (deg C):
BP pressure (mm Hg):

Property/Value	Units	Temp	Data Type	Reference
WS 1.92E+002	mg/L	25	EST	MEYLAN,WM ET AL. (1996)
logP 2.99			EST	MEYLAN,WM & HOWARD,PH (1995)
VP 2.39E-001	mm Hg	25	EXT	YAWS,CL (1994B)
DC	pKa			
HL 2.86E-004	atm m3/mol	25	EST	MEYLAN,WM & HOWARD,PH (1991)
OH 3.00E-011	cm3/molc sec	25	EST	MEYLAN,WM & HOWARD,PH (1993)

AROCLOR 1260

CAS #: 011096-82-5
Formula: $C_{12}H_3Cl_7$
Mol Weight: 395.33
MP (deg C):
FP (deg C):
BP (deg C): 385-420
BP pressure (mm Hg):

Property/Value	Units	Temp	Data Type	Reference
WS 1.44E-002	mg/L	20	EXP	YALKOWSKY,SH & DANNENFELSER,RM (1992)
logP 6.80			EXP	HANSCH,C ET AL. (1995)
VP 4.05E-005	mm Hg	25	EXP	MABEY,WR ET AL. (1981)
DC	pKa			
HL 3.36E-004	atm m3/mol	25	EXP	BURKHARD,LP ET AL. (1985A)
OH	cm3/molc sec			

AROCLOR 1254

CAS #: 011097-69-1
Formula: $C_{12}H_5Cl_5$
Mol Weight: 326.44
MP (deg C):
FP (deg C):
BP (deg C): 365-390
BP pressure (mm Hg):

Property/Value	Units	Temp	Data Type	Reference
WS 4.30E-002	mg/L	20	EXP	YALKOWSKY,SH & DANNENFELSER,RM (1992)
logP 6.50			EXP	HANSCH,C ET AL. (1995)
VP 7.71E-005	mm Hg	25	EXP	MABEY,WR ET AL. (1981)
DC	pKa			
HL 2.83E-004	atm m3/mol	25	EXP	BURKHARD,LP ET AL. (1985A)
OH	cm3/molc sec			

AROCHLOR 1221

CAS #: 011104-28-2
Formula: $C_{12}H_9Cl$
Mol Weight: 188.66
MP (deg C):
FP (deg C):
BP (deg C): 275-320
BP pressure (mm Hg):

51% mono, 32% di, and 11% triphenyl

Property/Value	Units	Temp	Data Type	Reference
WS 1.50E+001	mg/L	25	EXP	MABEY,WR ET AL. (1981)
logP 4.70			EXP	HANSCH,C ET AL. (1995)
VP 6.70E-003	mm Hg	25	EXP	MABEY,WR ET AL. (1981)
DC	pKa			
HL 2.28E-004	atm m3/mol	25	EXP	BURKHARD,LP ET AL. (1985A)
OH	cm3/molc sec			

AROCLOR 1232

CAS #: 011141-16-5
Formula: $C_{12}H_9Cl$
Mol Weight: 188.66
MP (deg C): 290-325
FP (deg C):
BP (deg C):
BP pressure (mm Hg):

A mixture of isomers:
31% mono, 24% di, and 28% tri

Property/Value	Units	Temp	Data Type	Reference
WS 1.45E+000	mg/L	25	EXP	MABEY,WR ET AL. (1981)
logP 5.10			EXP	HANSCH,C ET AL. (1995)
VP 4.06E-003	mm Hg	25	EXP	MABEY,WR ET AL. (1981)
DC	pKa			
HL 6.95E-004	atm m3/mol	25	EST	VP/WSOL
OH	cm3/molc sec			

ETHEPHON

CAS #: 011672-87-0
Formula: $C_2H_6ClO_3P$
Mol Weight: 144.50
MP (deg C): 74-75
FP (deg C):
BP (deg C):
BP pressure (mm Hg):

Property/Value	Units	Temp	Data Type	Reference
WS 1.00E+006	mg/L		EXP	SHIU,WY ET AL. (1990)
logP 0.05			EST	MEYLAN,WM & HOWARD,PH (1995)
VP 3.78E-005	mm Hg	25	EST	NEELY,WB & BLAU,GE (1985)
DC	pKa			
HL 5.70E-012	atm m3/mol	25	EST	MEYLAN,WM & HOWARD,PH (1991)
OH 1.05E-012	cm3/molc sec	25	EST	MEYLAN,WM & HOWARD,PH (1993)

CAS #: 012122-67-7 — (ETHYLENEBIS(DITHIOCARBAMATO))ZINC

Formula: $C_4H_6N_2S_4Zn$
Mol Weight: 275.73
MP (deg C): 106.53
FP (deg C):
BP (deg C):
BP pressure (mm Hg):

Property/Value	Units	Temp	Data Type	Reference
WS 1.00E+001	mg/L	25	EXP	WORTHING,CR & WALKER,SB (1983)
logP				TOMLIN,C (1994) ; < 1.3
VP	mm Hg	20	EST	TOMLIN,C (1994) ; < 7.5E-8
DC	pKa			
HL	atm m3/mol			
OH	cm3/molc sec			

CAS #: 012239-34-8 — DISPERSE BLUE 79

Formula: $C_{24}H_{27}BrN_6O_{10}$
Mol Weight: 639.43
MP (deg C): 157
FP (deg C):
BP (deg C):
BP pressure (mm Hg):

Property/Value	Units	Temp	Data Type	Reference
WS 9.38E-004	mg/L	25	EXP	BAUGHMAN,GL & PERENICH,TA (1988)
logP 5.53			EST	MEYLAN,WM & HOWARD,PH (1995)
VP 7.20E-018	mm Hg	25	EST	NEELY,WB & BLAU,GE (1985)
DC	pKa			
HL 1.30E-024	atm m3/mol	25	EST	MEYLAN,WM & HOWARD,PH (1991)
OH 1.49E-010	cm3/molc sec	25	EST	MEYLAN,WM & HOWARD,PH (1993)

CAS #: 012427-38-2 — MANEB

Formula: $C_6H_{12}MnN_2S_4$
Mol Weight: 295.37
MP (deg C): 200 dec
FP (deg C):
BP (deg C):
BP pressure (mm Hg):

X NUMBER OF FRAGMENTS; X > 1

Property/Value	Units	Temp	Data Type	Reference
WS 6.00E+000	mg/L	25	EST	WAUCHOPE,RD ET AL. (1991A)
logP 0.62			EST	MEYLAN,WM & HOWARD,PH (1995)
VP 1.49E-005	mm Hg	25	EST	NEELY,WB & BLAU,GE (1985)
DC	pKa			
HL 5.64E-007	atm m3/mol	25	EST	MEYLAN,WM & HOWARD,PH (1991)
OH 2.04E-010	cm3/molc sec	25	EST	MEYLAN,WM & HOWARD,PH (1993)

CAS #: 012672-29-6 — AROCLOR 1248

Formula: $C_{24}H_{11}Cl_9$
Mol Weight: 618.43
MP (deg C):
FP (deg C):
BP (deg C): 340-375
BP pressure (mm Hg):

Property/Value	Units	Temp	Data Type	Reference
WS 1.00E-001	mg/L	20	EXP	YALKOWSKY,SH & DANNENFELSER,RM (1992)
logP 6.20			EXP	HANSCH,C ET AL. (1995)
VP 4.94E-004	mm Hg	25	EXP	CALLAHAN,MA ET AL (1979)
DC	pKa			
HL 4.40E-004	atm m3/mol	25	EXP	BURKHARD,LP ET AL. (1985A)
OH	cm3/molc sec			

CAS #: 012674-11-2 — AROCLOR 1016

Formula: $C_{24}H_{13}Cl_7$
Mol Weight: 549.54
MP (deg C):
FP (deg C):
BP (deg C):
BP pressure (mm Hg):

A mixture of various isomers, mostly 57% tri, 21% tetra, and 20% di.

Property/Value	Units	Temp	Data Type	Reference
WS 4.20E-001	mg/L	25	EXP	MABEY,WR ET AL. (1981)
logP 5.69			EST	MEYLAN,WM & HOWARD,PH (1995)
VP 4.00E-004	mm Hg	25	EXP	MABEY,WR ET AL. (1981)
DC	pKa			
HL 2.71E-004	atm m3/mol	25	EST	VP/WSOL
OH 1.19E-012	cm3/molc sec	25	EST	MEYLAN,WM & HOWARD,PH (1993)

CAS #: 012771-68-5 — ANCYMIDOL

Formula: $C_{15}H_{16}N_2O_2$
Mol Weight: 256.31
MP (deg C): 110-111
FP (deg C):
BP (deg C):
BP pressure (mm Hg):

Property/Value	Units	Temp	Data Type	Reference
WS 6.50E+002	mg/L	25	EXP	YALKOWSKY,SH & DANNENFELSER,RM (1992)
logP 1.91			EXP	TOMLIN,C (1994)
VP 2.00E-007	mm Hg	25	EXP	WAUCHOPE,RD ET AL. (1991A)
DC	pKa			
HL 1.04E-010	atm m3/mol	25	EXP	VP/WSOL
OH 2.76E-011	cm3/molc sec	25	EST	MEYLAN,WM & HOWARD,PH (1993)

CAS #: 012789-03-6 — CHLORDANE

Formula: $C_{10}H_6Cl_8$
Mol Weight: 409.78
MP (deg C):
FP (deg C):
BP (deg C):
BP pressure (mm Hg):

Property/Value	Units	Temp	Data Type	Reference
WS 1.30E-002	mg/L	25	EST	MEYLAN,WM ET AL. (1996)
logP 6.60			EST	MEYLAN,WM & HOWARD,PH (1995)
VP 9.98E-006	mm Hg	25	EXP	MACKAY,D ET AL. (1986)
DC	pKa			
HL 7.03E-005	atm m3/mol	25	EST	MEYLAN,WM & HOWARD,PH (1991)
OH 5.29E-012	cm3/molc sec	25	EST	MEYLAN,WM & HOWARD,PH (1993)

CAS #: 013004-52-9 — 2(4,4'-DIHYDROXYDIPHENLMETHYL)THIAZOLE

Formula: $C_{16}H_{15}NO_2S$
Mol Weight: 285.37
MP (deg C):
FP (deg C):
BP (deg C):
BP pressure (mm Hg):

Property/Value	Units	Temp	Data Type	Reference
WS 1.73E+002	mg/L	25	EST	MEYLAN,WM ET AL. (1996)
logP 2.96			EXP	HANSCH,C & LEO,AJ (1985)
VP 6.78E-010	mm Hg	25	EST	NEELY,WB & BLAU,GE (1985)
DC	pKa			
HL 2.98E-016	atm m3/mol	25	EST	MEYLAN,WM & HOWARD,PH (1991)
OH 8.58E-011	cm3/molc sec	25	EST	MEYLAN,WM & HOWARD,PH (1993)

CAS #: 013010-47-4

1-N=O-1-(2-CLET)-3-CYCLOHEXYLUREA

Formula: $C_9H_{16}ClN_3O_2$

Mol Weight: 233.70

MP (deg C): 88-90

FP (deg C):

BP (deg C):

BP pressure (mm Hg):

Property/Value	Units	Temp	Data Type	Reference
WS 1.11E+002	mg/L	25	EST	MEYLAN,WM ET AL. (1996)
logP 2.83			EXP	HANSCH,C & LEO,AJ (1985)
VP 1.01E-005	mm Hg	25	EST	NEELY,WB & BLAU,GE (1985)
DC	pKa			
HL 1.85E-010	atm m3/mol	25	EST	MEYLAN,WM & HOWARD,PH (1991)
OH 2.19E-011	cm3/molc sec	25	EST	MEYLAN,WM & HOWARD,PH (1993)

CAS #: 013017-11-3

PROPENOIC ACID, ESTER DERIVATIVE

Formula: $C_{19}H_{20}O_7$

Mol Weight: 360.37

MP (deg C):

FP (deg C):

BP (deg C):

BP pressure (mm Hg):

Property/Value	Units	Temp	Data Type	Reference
WS 1.38E+003	mg/L	25	EST	MEYLAN,WM ET AL. (1996)
logP 0.69			EXP	SANGSTER,J (1993)
VP 5.93E-010	mm Hg	25	EST	NEELY,WB & BLAU,GE (1985)
DC	pKa			
HL 3.82E-013	atm m3/mol	25	EST	MEYLAN,WM & HOWARD,PH (1991)
OH 7.87E-011	cm3/molc sec	25	EST	MEYLAN,WM & HOWARD,PH (1993)

CAS #: 013021-15-3

N,N,2,4,6-PENTAMETHYLANILINE

Formula: $C_{11}H_{17}N$

Mol Weight: 163.26

MP (deg C):

FP (deg C):

BP (deg C): 213-215

BP pressure (mm Hg):

Property/Value	Units	Temp	Data Type	Reference
WS 1.13E+001	mg/L	25	EST	MEYLAN,WM ET AL. (1996)
logP 4.41			EXP	HANSCH,C ET AL. (1995)
VP 7.19E-002	mm Hg	25	EST	NEELY,WB & BLAU,GE (1985)
DC	pKa			
HL 1.15E-004	atm m3/mol	25	EST	MEYLAN,WM & HOWARD,PH (1991)
OH 5.98E-011	cm3/molc sec	25	EST	MEYLAN,WM & HOWARD,PH (1993)

CAS #: 013029-08-8

2,2'-DICHLOROBIPHENYL

Formula: $C_{12}H_8Cl_2$

Mol Weight: 223.10

MP (deg C):

FP (deg C):

BP (deg C):

BP pressure (mm Hg):

Property/Value	Units	Temp	Data Type	Reference
WS 1.85E+000	mg/L	25	EXP	CHIOU,CT ET AL. (1983)
logP 4.97			EXP	HANSCH,C & LEO,AJ (1985)
VP 2.75E-003	mm Hg	25	EXP	BURKHARD,LP ET AL. (1985)
DC	pKa			
HL 4.36E-004	atm m3/mol	25	EST	VP/WSOL
OH 1.73E-012	cm3/molc sec	25	EST	MEYLAN,WM & HOWARD,PH (1993)

CAS #: 013031-39-5

3-CHLOROPHENYL ACETATE

Formula: $C_8H_7ClO_2$

Mol Weight: 170.60

MP (deg C):

FP (deg C):

BP (deg C):

BP pressure (mm Hg):

Property/Value	Units	Temp	Data Type	Reference
WS 6.38E+002	mg/L	25	EST	MEYLAN,WM ET AL. (1996)
logP 2.32			EXP	HANSCH,C & LEO,AJ (1985)
VP 9.07E-002	mm Hg	25	EST	NEELY,WB & BLAU,GE (1985)
DC	pKa			
HL 4.80E-005	atm m3/mol	25	EST	MEYLAN,WM & HOWARD,PH (1991)
OH 1.52E-012	cm3/molc sec	25	EST	MEYLAN,WM & HOWARD,PH (1993)

CAS #: 013031-43-1

P-ACETOXYACETOPHENONE

Formula: $C_{10}H_{10}O_3$

Mol Weight: 178.19

MP (deg C):

FP (deg C):

BP (deg C):

BP pressure (mm Hg):

Property/Value	Units	Temp	Data Type	Reference
WS 4.45E+003	mg/L	25	EST	MEYLAN,WM ET AL. (1996)
logP 1.29			EXP	HANSCH,C & LEO,AJ (1985)
VP 5.86E-003	mm Hg	25	EST	NEELY,WB & BLAU,GE (1985)
DC	pKa			
HL 1.18E-007	atm m3/mol	25	EST	MEYLAN,WM & HOWARD,PH (1991)
OH 2.04E-012	cm3/molc sec	25	EST	MEYLAN,WM & HOWARD,PH (1993)

CAS #: 013037-71-3

4-(2-BUTENYL)PHENOL

Formula: $C_{10}H_{12}O$

Mol Weight: 148.21

MP (deg C):

FP (deg C):

BP (deg C):

BP pressure (mm Hg):

Property/Value	Units	Temp	Data Type	Reference
WS 4.31E+002	mg/L	25	EST	MEYLAN,WM ET AL. (1996)
logP 3.32			EST	MEYLAN,WM & HOWARD,PH (1995)
VP 5.74E-003	mm Hg	25	EST	NEELY,WB & BLAU,GE (1985)
DC	pKa			
HL 1.27E-006	atm m3/mol	25	EST	MEYLAN,WM & HOWARD,PH (1991)
OH 9.82E-011	cm3/molc sec	25	EST	MEYLAN,WM & HOWARD,PH (1993)

CAS #: 013037-86-0

4-(HEPTYLOXY)PHENOL

Formula: $C_{13}H_{20}O_2$

Mol Weight: 208.30

MP (deg C): 60-63

FP (deg C):

BP (deg C):

BP pressure (mm Hg):

Property/Value	Units	Temp	Data Type	Reference
WS 2.00E+001	mg/L	25	EST	MEYLAN,WM ET AL. (1996)
logP 4.54			EST	MEYLAN,WM & HOWARD,PH (1995)
VP 8.63E-005	mm Hg	25	EST	NEELY,WB & BLAU,GE (1985)
DC	pKa			
HL 1.82E-007	atm m3/mol	25	EST	MEYLAN,WM & HOWARD,PH (1991)
OH 4.56E-011	cm3/molc sec	25	EST	MEYLAN,WM & HOWARD,PH (1993)

CAS #:	013074-09-4				3-ETHYL-DIMETHYLPARATHION
Formula:	$C_{10}H_{14}NO_5PS$				
Mol Weight:	291.26				
MP (deg C):		FP (deg C):			
BP (deg C):					
BP pressure (mm Hg):					

Property/Value	Units	Temp	Data Type	Reference
WS 3.60E+000	mg/L	25	EST	MEYLAN,WM ET AL. (1996)
logP 3.74			EXP	HANSCH,C & LEO,AJ (1985)
VP 9.73E-006	mm Hg	25	EST	NEELY,WB & BLAU,GE (1985)
DC	pKa			
HL 2.46E-007	atm m3/mol	25	EST	MEYLAN,WM & HOWARD,PH (1991)
OH 6.08E-011	cm3/molc sec	25	EST	MEYLAN,WM & HOWARD,PH (1993)

CAS #:	013074-11-8				3-ISOPROPYL-DIMETHYLPARA-OXON
Formula:	$C_{11}H_{16}NO_6P$				
Mol Weight:	289.23				
MP (deg C):		FP (deg C):			
BP (deg C):					
BP pressure (mm Hg):					

Property/Value	Units	Temp	Data Type	Reference
WS 3.69E+001	mg/L	25	EST	MEYLAN,WM ET AL. (1996)
logP 2.57			EXP	HANSCH,C & LEO,AJ (1985)
VP 7.16E-006	mm Hg	25	EST	NEELY,WB & BLAU,GE (1985)
DC	pKa			
HL 7.07E-010	atm m3/mol	25	EST	MEYLAN,WM & HOWARD,PH (1991)
OH 8.92E-012	cm3/molc sec	25	EST	MEYLAN,WM & HOWARD,PH (1993)

CAS #:	013087-49-5				3,7-DIMETHYL URIC ACID
Formula:	$C_7H_8N_4O_3$				
Mol Weight:	196.17				
MP (deg C):		FP (deg C):			
BP (deg C):					
BP pressure (mm Hg):					

Property/Value	Units	Temp	Data Type	Reference
WS 3.06E+004	mg/L	25	EST	MEYLAN,WM ET AL. (1996)
logP -1.33			EXP	GASPARI,F & BONATI,M (1987)
VP 1.77E-010	mm Hg	25	EST	NEELY,WB & BLAU,GE (1985)
DC	pKa			
HL 5.77E-017	atm m3/mol	25	EST	MEYLAN,WM & HOWARD,PH (1991)
OH 1.34E-011	cm3/molc sec	25	EST	MEYLAN,WM & HOWARD,PH (1993)

CAS #:	013089-48-0				CYTIDINE, N-BENZOYL-
Formula:	$C_{16}H_{17}N_3O_6$				
Mol Weight:	347.33				
MP (deg C):	230-234 de	FP (deg C):			
BP (deg C):					
BP pressure (mm Hg):					

Property/Value	Units	Temp	Data Type	Reference
WS 1.75E+002	mg/L	25	EST	MEYLAN,WM ET AL. (1996)
logP 0.30			EXP	HANSCH,C ET AL. (1995)
VP 8.99E-018	mm Hg	25	EST	NEELY,WB & BLAU,GE (1985)
DC	pKa			
HL 1.01E-022	atm m3/mol	25	EST	MEYLAN,WM & HOWARD,PH (1991)
OH 1.25E-010	cm3/molc sec	25	EST	MEYLAN,WM & HOWARD,PH (1993)

CAS #:	013093-12-4				HEXAMETHYL(SILAACENAPHTHENYL)CYCLOTETRASILOXAN*
Formula:	$C_{17}H_{26}O_4Si_4$				
Mol Weight:	406.74				
MP (deg C):		FP (deg C):			
BP (deg C):					
BP pressure (mm Hg):					

Property/Value	Units	Temp	Data Type	Reference
WS 9.53E-006	mg/L	25	EST	MEYLAN,WM ET AL. (1996)
logP 7.42			EST	MEYLAN,WM & HOWARD,PH (1995)
VP	mm Hg			
DC	pKa			
HL	atm m3/mol			
OH 3.93E-011	cm3/molc sec	25	EST	MEYLAN,WM & HOWARD,PH (1993)

CAS #:	013110-37-7				BENZOIC ACID, 4-AMINO-,PENTYL ESTER
Formula:	$C_{12}H_{17}NO_2$				
Mol Weight:	207.27				
MP (deg C):	52	FP (deg C):			
BP (deg C):					
BP pressure (mm Hg):					

Property/Value	Units	Temp	Data Type	Reference
WS 9.25E+001	mg/L	25	EXP	SUZUKI,T (1991)
logP 3.48			EXP	HANSCH,C ET AL. (1995)
VP 1.35E-004	mm Hg	25	EST	NEELY,WB & BLAU,GE (1985)
DC	pKa			
HL 3.81E-008	atm m3/mol	25	EST	MEYLAN,WM & HOWARD,PH (1991)
OH 4.08E-011	cm3/molc sec	25	EST	MEYLAN,WM & HOWARD,PH (1993)

CAS #:	013114-87-9				3-(TRIFLUOROMETHYL)PHENYLUREA
Formula:	$C_8H_7F_3N_2O$				
Mol Weight:	204.15				
MP (deg C):		FP (deg C):			
BP (deg C):					
BP pressure (mm Hg):					

Property/Value	Units	Temp	Data Type	Reference
WS 4.44E+002	mg/L	25	EST	MEYLAN,WM ET AL. (1996)
logP 2.31			EXP	BRIGGS,GG (1981)
VP 6.01E-004	mm Hg	25	EST	NEELY,WB & BLAU,GE (1985)
DC	pKa			
HL 1.75E-009	atm m3/mol	25	EST	MEYLAN,WM & HOWARD,PH (1991)
OH 6.46E-012	cm3/molc sec	25	EST	MEYLAN,WM & HOWARD,PH (1993)

CAS #:	013116-53-5				PROPANE, 1,2,2,3-TETRACHLORO-
Formula:	$C_3H_4Cl_4$				
Mol Weight:	181.88				
MP (deg C):		FP (deg C):			
BP (deg C):	165				
BP pressure (mm Hg):					

Property/Value	Units	Temp	Data Type	Reference
WS 4.80E+002	mg/L	25	EXP	CHEM INSPECT TEST INST (1992)
logP 2.72			EXP	CHEM INSPECT TEST INST (1992)
VP 3.05E+000	mm Hg	25	EXT	DYKYJ,J (1970)
DC	pKa			
HL 2.00E-003	atm m3/mol	25	EST	MEYLAN,WM & HOWARD,PH (1991)
OH 1.77E-013	cm3/molc sec	25	EST	MEYLAN,WM & HOWARD,PH (1993)

013129-23-2 — 3-FURANCARBOXYLIC ACID, METHYL ESTER

CAS #: 013129-23-2

Formula: $C_6H_6O_3$

Mol Weight: 126.11

MP (deg C): FP (deg C):

BP (deg C): 160

BP pressure (mm Hg):

Property/Value	Units	Temp	Data Type	Reference
WS 7.68E+003	mg/L	25	EST	MEYLAN,WM ET AL. (1996)
logP 1.28			EXP	YAMAGAMI,C ET AL. (1990)
VP 2.55E+000	mm Hg	25	EST	NEELY,WB & BLAU,GE (1985)
DC	pKa			
HL 3.46E-005	atm m3/mol	25	EST	MEYLAN,WM & HOWARD,PH (1991)
OH 1.32E-011	cm3/molc sec	25	EST	MEYLAN,WM & HOWARD,PH (1993)

013132-25-7 — PHENOL, P-(TRIMETHYL SILYL)-

CAS #: 013132-25-7

Formula: $C_9H_{14}OSi$

Mol Weight: 166.30

MP (deg C): FP (deg C):

BP (deg C):

BP pressure (mm Hg):

Property/Value	Units	Temp	Data Type	Reference
WS 1.28E+002	mg/L	25	EST	MEYLAN,WM ET AL. (1996)
logP 3.84			EXP	HANSCH,C & LEO,AJ (1985)
VP 2.87E-002	mm Hg	25	EST	NEELY,WB & BLAU,GE (1985)
DC	pKa			
HL 1.42E-006	atm m3/mol	25	EST	MEYLAN,WM & HOWARD,PH (1991)
OH 3.39E-011	cm3/molc sec	25	EST	MEYLAN,WM & HOWARD,PH (1993)

013138-51-7 — 1,2,3,3-TETRACHLOROBUTANE

CAS #: 013138-51-7

Formula: $C_4H_6Cl_4$

Mol Weight: 195.90

MP (deg C): FP (deg C):

BP (deg C): 90

BP pressure (mm Hg): 3.20E+001

Property/Value	Units	Temp	Data Type	Reference
WS 2.40E+001	mg/L	25	EST	MEYLAN,WM ET AL. (1996)
logP 3.84			EST	MEYLAN,WM & HOWARD,PH (1995)
VP 1.37E+000	mm Hg	25	EST	NEELY,WB & BLAU,GE (1985)
DC	pKa			
HL 2.65E-003	atm m3/mol	25	EST	MEYLAN,WM & HOWARD,PH (1991)
OH 2.28E-013	cm3/molc sec	25	EST	MEYLAN,WM & HOWARD,PH (1993)

013140-73-3 — ACETAMIDE, N-(3-BROMOPHENYL)-2-PHENYL-

CAS #: 013140-73-3

Formula: $C_{14}H_{11}BrO$

Mol Weight: 275.15

MP (deg C): FP (deg C):

BP (deg C):

BP pressure (mm Hg):

Property/Value	Units	Temp	Data Type	Reference
WS 1.03E+001	mg/L	25	EST	MEYLAN,WM ET AL. (1996)
logP 3.77			EXP	YAMAGAMI,C ET AL. (1984)
VP 2.29E-005	mm Hg	25	EST	NEELY,WB & BLAU,GE (1985)
DC	pKa			
HL 3.16E-007	atm m3/mol	25	EST	MEYLAN,WM & HOWARD,PH (1991)
OH 6.06E-012	cm3/molc sec	25	EST	MEYLAN,WM & HOWARD,PH (1993)

013140-76-6 — BENZENEACETAMIDE, N-(3-NITROPHENYL)-

CAS #: 013140-76-6

Formula: $C_{14}H_{12}N_2O_3$

Mol Weight: 256.26

MP (deg C): FP (deg C):

BP (deg C):

BP pressure (mm Hg):

Property/Value	Units	Temp	Data Type	Reference
WS 2.80E+001	mg/L	25	EST	MEYLAN,WM ET AL. (1996)
logP 2.93			EXP	YAMAGAMI,C ET AL. (1984)
VP 1.26E-008	mm Hg	25	EST	NEELY,WB & BLAU,GE (1985)
DC	pKa			
HL 1.97E-012	atm m3/mol	25	EST	MEYLAN,WM & HOWARD,PH (1991)
OH 6.63E-012	cm3/molc sec	25	EST	MEYLAN,WM & HOWARD,PH (1993)

013140-77-7 — BENZENEACETAMIDE, N-(4-NITROPHENYL)-

CAS #: 013140-77-7

Formula: $C_{14}H_{12}N_2O_3$

Mol Weight: 256.26

MP (deg C): FP (deg C):

BP (deg C):

BP pressure (mm Hg):

Property/Value	Units	Temp	Data Type	Reference
WS 1.71E+001	mg/L	25	EST	MEYLAN,WM ET AL. (1996)
logP 3.18			EXP	YAMAGAMI,C ET AL. (1984)
VP 1.26E-008	mm Hg	25	EST	NEELY,WB & BLAU,GE (1985)
DC	pKa			
HL 1.97E-012	atm m3/mol	25	EST	MEYLAN,WM & HOWARD,PH (1991)
OH 7.10E-012	cm3/molc sec	25	EST	MEYLAN,WM & HOWARD,PH (1993)

013140-86-8 — N-CYCLOPROPYL-N'-PHENYLUREA

CAS #: 013140-86-8

Formula: $C_{10}H_{12}N_2O$

Mol Weight: 176.22

MP (deg C): FP (deg C):

BP (deg C):

BP pressure (mm Hg):

Property/Value	Units	Temp	Data Type	Reference
WS 2.24E+003	mg/L	25	EST	MEYLAN,WM ET AL. (1996)
logP 1.65			EXP	HANSCH,C & LEO,AJ (1985)
VP 6.11E-005	mm Hg	25	EST	NEELY,WB & BLAU,GE (1985)
DC	pKa			
HL 3.43E-010	atm m3/mol	25	EST	MEYLAN,WM & HOWARD,PH (1991)
OH 4.40E-011	cm3/molc sec	25	EST	MEYLAN,WM & HOWARD,PH (1993)

013140-89-1 — N-CYCLOPENTYL-N'-PHENYLUREA

CAS #: 013140-89-1

Formula: $C_{12}H_{16}N_2O$

Mol Weight: 204.27

MP (deg C): FP (deg C):

BP (deg C):

BP pressure (mm Hg):

Property/Value	Units	Temp	Data Type	Reference
WS 2.27E+002	mg/L	25	EST	MEYLAN,WM ET AL. (1996)
logP 2.65			EXP	BRIGGS,GG (1981)
VP 1.02E-005	mm Hg	25	EST	NEELY,WB & BLAU,GE (1985)
DC	pKa			
HL 6.05E-010	atm m3/mol	25	EST	MEYLAN,WM & HOWARD,PH (1991)
OH 5.78E-011	cm3/molc sec	25	EST	MEYLAN,WM & HOWARD,PH (1993)

CAS #: 013147-25-6

(N-NITROSO-N-ETHYL)ETHANOLAMINE

Formula: $C_4H_{10}N_2O_2$

Mol Weight: 118.14

MP (deg C): FP (deg C):

BP (deg C):

BP pressure (mm Hg):

Property/Value	Units	Temp	Data Type	Reference
WS 3.82E+005	mg/L	25	EST	MEYLAN,WM ET AL. (1996)
logP -0.67			EST	MEYLAN,WM & HOWARD,PH (1995)
VP 9.06E-004	mm Hg	25	EST	NEELY,WB & BLAU,GE (1985)
DC	pKa			
HL 1.33E-010	atm m3/mol	25	EST	MEYLAN,WM & HOWARD,PH (1991)
OH 2.37E-011	cm3/molc sec	25	EST	MEYLAN,WM & HOWARD,PH (1993)

CAS #: 013153-62-3

9H-PURINE, 9-(á-D-ARABINOFURANOSYL)-6-(METHYLTHI

Formula: $C_{12}H_{15}N_3O_4S$

Mol Weight: 297.34

MP (deg C): FP (deg C):

BP (deg C):

BP pressure (mm Hg):

Property/Value	Units	Temp	Data Type	Reference
WS 1.07E+004	mg/L	25	EST	MEYLAN,WM ET AL. (1996)
logP 0.09			EXP	HANSCH,C ET AL. (1995)
VP 3.49E-013	mm Hg	25	EST	NEELY,WB & BLAU,GE (1985)
DC	pKa			
HL 8.63E-019	atm m3/mol	25	EST	MEYLAN,WM & HOWARD,PH (1991)
OH 2.48E-010	cm3/molc sec	25	EST	MEYLAN,WM & HOWARD,PH (1993)

CAS #: 013156-74-6

N,N-DIMETHYLCINNAMAMIDE

Formula: $C_{11}H_{13}NO$

Mol Weight: 175.23

MP (deg C): FP (deg C):

BP (deg C):

BP pressure (mm Hg):

Property/Value	Units	Temp	Data Type	Reference
WS 1.93E+003	mg/L	25	EST	MEYLAN,WM ET AL. (1996)
logP 1.73			EXP	HANSCH,C & LEO,AJ (1985)
VP 8.46E-004	mm Hg	25	EST	NEELY,WB & BLAU,GE (1985)
DC	pKa			
HL 1.27E-009	atm m3/mol	25	EST	MEYLAN,WM & HOWARD,PH (1991)
OH 3.77E-011	cm3/molc sec	25	EST	MEYLAN,WM & HOWARD,PH (1993)

CAS #: 013156-75-7

N,N-DIMETHYL-FURAN-2-CARBOXAMIDE

Formula: $C_7H_{10}NO_2$

Mol Weight: 140.16

MP (deg C): FP (deg C):

BP (deg C):

BP pressure (mm Hg):

Property/Value	Units	Temp	Data Type	Reference
WS 3.77E+004	mg/L	25	EST	MEYLAN,WM ET AL. (1996)
logP 0.41			EXP	HANSCH,C & LEO,AJ (1985)
VP 5.64E-002	mm Hg	25	EST	NEELY,WB & BLAU,GE (1985)
DC	pKa			
HL 1.07E-008	atm m3/mol	25	EST	MEYLAN,WM & HOWARD,PH (1991)
OH 5.30E-011	cm3/molc sec	25	EST	MEYLAN,WM & HOWARD,PH (1993)

CAS #: 013171-21-6

DIMECRON

Formula: $C_{10}H_{19}ClNO_5P$

Mol Weight: 299.69

MP (deg C): -45 FP (deg C):

BP (deg C): 162

BP pressure (mm Hg): 1.50E+000

Property/Value	Units	Temp	Data Type	Reference
WS 1.00E+006	mg/L	25	EXP	BUDAVARI,S (1989)
logP 0.79			EXP	TOMLIN,C (1994)
VP 2.50E-005	mm Hg	20	EXP	BUDAVARI,S (1989)
DC	pKa			
HL 1.52E-012	atm m3/mol	25	EST	MEYLAN,WM & HOWARD,PH (1991)
OH 4.51E-011	cm3/molc sec	25	EST	MEYLAN,WM & HOWARD,PH (1993)

CAS #: 013171-61-4

2,3,5,6-TETRAMETHYL-4-NITROBENZENAMINE

Formula: $C_{10}H_{14}N_2O_2$

Mol Weight: 194.24

MP (deg C): FP (deg C):

BP (deg C):

BP pressure (mm Hg):

Property/Value	Units	Temp	Data Type	Reference
WS 3.50E+001	mg/L	25	EST	MEYLAN,WM ET AL. (1996)
logP 3.66			EST	MEYLAN,WM & HOWARD,PH (1995)
VP 4.28E-005	mm Hg	25	EST	NEELY,WB & BLAU,GE (1985)
DC 2.36	pKa	25	EXP	PERRIN,DD (1965)
HL 1.11E-008	atm m3/mol	25	EST	MEYLAN,WM & HOWARD,PH (1991)
OH 1.38E-010	cm3/molc sec	25	EST	MEYLAN,WM & HOWARD,PH (1993)

CAS #: 013181-17-4

BROMOFENOXIM

Formula: $C_{13}H_7Br_2N_3O_6$

Mol Weight: 461.03

MP (deg C): FP (deg C):

BP (deg C):

BP pressure (mm Hg):

Property/Value	Units	Temp	Data Type	Reference
WS 1.00E-001	mg/L	20	EXP	SHIU,WY ET AL. (1990)
logP 3.30			EXP	HANSCH,C ET AL. (1995)
VP 8.67E-012	mm Hg	25	EST	NEELY,WB & BLAU,GE (1985)
DC	pKa			
HL 1.09E-014	atm m3/mol	25	EST	MEYLAN,WM & HOWARD,PH (1991)
OH 3.68E-012	cm3/molc sec	25	EST	MEYLAN,WM & HOWARD,PH (1993)

CAS #: 013182-82-6

METRONIDAZOLE ACETATE

Formula: $C_8H_{11}N_3O_4$

Mol Weight: 213.19

MP (deg C): FP (deg C):

BP (deg C):

BP pressure (mm Hg):

Property/Value	Units	Temp	Data Type	Reference
WS 8.44E+003	mg/L	25	EST	MEYLAN,WM ET AL. (1996)
logP 0.30			EXP	HANSCH,C & LEO,AJ (1985)
VP 5.54E-006	mm Hg	25	EST	NEELY,WB & BLAU,GE (1985)
DC	pKa			
HL 1.10E-009	atm m3/mol	25	EST	MEYLAN,WM & HOWARD,PH (1991)
OH 7.08E-012	cm3/molc sec	25	EST	MEYLAN,WM & HOWARD,PH (1993)

CAS #: 013194-48-4 — O-ETHYL S,S-DIPROPYL PHOSPHORODITHIOATE

Formula: $C_8H_{19}O_2PS_2$
Mol Weight: 242.34
MP (deg C):
FP (deg C):
BP (deg C): 86-91
BP pressure (mm Hg): 2.00E-001

Property/Value	Units	Temp	Data Type	Reference
WS 7.50E+002	mg/L	25	EXP	WAUCHOPE,RD ET AL. (1991)
logP 3.59			EXP	TOMLIN,C (1994)
VP 3.80E-004	mm Hg	25	EXP	WAUCHOPE,RD ET AL. (1991)
DC	pKa			
HL 1.62E-007	atm m3/mol	25	EST	VP/WSOL
OH 6.72E-011	cm3/molc sec	25	EST	MEYLAN,WM & HOWARD,PH (1993)

CAS #: 013205-48-6 — P-METHIOBENZOIC ACID

Formula: $C_8H_8O_2S$
Mol Weight: 168.22
MP (deg C): 193-196
FP (deg C):
BP (deg C):
BP pressure (mm Hg):

Property/Value	Units	Temp	Data Type	Reference
WS 2.87E+002	mg/L	25	EST	MEYLAN,WM ET AL. (1996)
logP 2.74			EXP	HANSCH,C & LEO,AJ (1985)
VP 2.52E-004	mm Hg	25	EST	NEELY,WB & BLAU,GE (1985)
DC	pKa			
HL 3.16E-009	atm m3/mol	25	EST	MEYLAN,WM & HOWARD,PH (1991)
OH 6.24E-012	cm3/molc sec	25	EST	MEYLAN,WM & HOWARD,PH (1993)

CAS #: 013206-67-2 — 1-NITROSO-1-(2-CLET)-3-PHENYLUREA

Formula: $C_9H_{10}ClN_3O_2$
Mol Weight: 227.65
MP (deg C):
FP (deg C):
BP (deg C):
BP pressure (mm Hg):

Property/Value	Units	Temp	Data Type	Reference
WS 2.34E+002	mg/L	25	EST	MEYLAN,WM ET AL. (1996)
logP 2.49			EXP	HANSCH,C & LEO,AJ (1985)
VP 4.90E-006	mm Hg	25	EST	NEELY,WB & BLAU,GE (1985)
DC	pKa			
HL 2.56E-011	atm m3/mol	25	EST	MEYLAN,WM & HOWARD,PH (1991)
OH 4.45E-011	cm3/molc sec	25	EST	MEYLAN,WM & HOWARD,PH (1993)

CAS #: 013207-50-6 — 4-METHYL-1-PHENYL-3-THIOSEMICARBAZIDE

Formula: $C_8H_{11}N_3S$
Mol Weight: 181.26
MP (deg C):
FP (deg C):
BP (deg C):
BP pressure (mm Hg):

Property/Value	Units	Temp	Data Type	Reference
WS 5.54E+004	mg/L	25	EST	MEYLAN,WM ET AL. (1996)
logP 1.17			EXP	HANSCH,C ET AL. (1995)
VP 2.40E-004	mm Hg	25	EST	NEELY,WB & BLAU,GE (1985)
DC	pKa			
HL 7.99E-010	atm m3/mol	25	EST	MEYLAN,WM & HOWARD,PH (1991)
OH 1.70E-010	cm3/molc sec	25	EST	MEYLAN,WM & HOWARD,PH (1993)

CAS #: 013207-66-4 — 8-QUINOLINOL, 5-AMINO-

Formula: $C_9H_8N_2O$
Mol Weight: 160.18
MP (deg C):
FP (deg C):
BP (deg C):
BP pressure (mm Hg):

Property/Value	Units	Temp	Data Type	Reference
WS 8.48E+004	mg/L	25	EST	MEYLAN,WM ET AL. (1996)
logP -0.11			EXP	SANGSTER,J (1993)
VP 7.36E-006	mm Hg	25	EST	NEELY,WB & BLAU,GE (1985)
DC 5.67	pKa	20	EXP	PERRIN,DD (1965)
HL 2.53E-014	atm m3/mol	25	EST	MEYLAN,WM & HOWARD,PH (1991)
OH 2.00E-010	cm3/molc sec	25	EST	MEYLAN,WM & HOWARD,PH (1993)

CAS #: 013214-66-9 — 4-PHENYLBUTYLAMINE

Formula: $C_{10}H_{15}N$
Mol Weight: 149.24
MP (deg C):
FP (deg C):
BP (deg C): 123-124
BP pressure (mm Hg): 1.70E+001

Property/Value	Units	Temp	Data Type	Reference
WS 6.95E+003	mg/L	25	EST	MEYLAN,WM ET AL. (1996)
logP 2.40			EXP	HANSCH,C & LEO,AJ (1985)
VP 3.48E-002	mm Hg	25	EST	NEELY,WB & BLAU,GE (1985)
DC 10.36	pKa	25	EXP	PERRIN,DD (1965)
HL 1.43E-006	atm m3/mol	25	EST	MEYLAN,WM & HOWARD,PH (1991)
OH 4.05E-011	cm3/molc sec	25	EST	MEYLAN,WM & HOWARD,PH (1993)

CAS #: 013229-35-1 — METHANESULFONAMIDE, N-METHYL-N-PHENYL-

Formula: $C_8H_{11}NO_2S$
Mol Weight: 185.25
MP (deg C):
FP (deg C):
BP (deg C):
BP pressure (mm Hg):

Property/Value	Units	Temp	Data Type	Reference
WS 6.98E+003	mg/L	25	EST	MEYLAN,WM ET AL. (1996)
logP 1.02			EXP	SANGSTER,J (1994)
VP 6.98E-004	mm Hg	25	EST	NEELY,WB & BLAU,GE (1985)
DC	pKa			
HL 2.39E-005	atm m3/mol	25	EST	MEYLAN,WM & HOWARD,PH (1991)
OH 1.47E-011	cm3/molc sec	25	EST	MEYLAN,WM & HOWARD,PH (1993)

CAS #: 013230-04-1 — 1,2-DIMETHYL-4-NITROIMIDAZOLE

Formula: $C_5H_7N_3O_2$
Mol Weight: 141.13
MP (deg C):
FP (deg C):
BP (deg C):
BP pressure (mm Hg):

Property/Value	Units	Temp	Data Type	Reference
WS 5.00E+003	mg/L	25	EST	MEYLAN,WM ET AL. (1996)
logP 0.97			EST	MEYLAN,WM & HOWARD,PH (1995)
VP 5.58E-004	mm Hg	25	EST	NEELY,WB & BLAU,GE (1985)
DC	pKa			
HL 3.49E-007	atm m3/mol	25	EST	MEYLAN,WM & HOWARD,PH (1991)
OH 4.13E-012	cm3/molc sec	25	EST	MEYLAN,WM & HOWARD,PH (1993)

CAS #: 013245-65-3 — 1,3-DICHLORO-2-CHLOROMETHYLPROPENE

Formula: $C_4H_5Cl_3$

Mol Weight: 159.44

MP (deg C): | FP (deg C):

BP (deg C):

BP pressure (mm Hg):

Property/Value	Units	Temp	Data Type	Reference
WS 1.58E+002	mg/L	25	EST	MEYLAN,WM ET AL. (1996)
logP 3.09			EST	MEYLAN,WM & HOWARD,PH (1995)
VP 8.19E-001	mm Hg	25	EST	NEELY,WB & BLAU,GE (1985)
DC	pKa			
HL 1.35E-002	atm m3/mol	25	EST	MEYLAN,WM & HOWARD,PH (1991)
OH 1.13E-011	cm3/molc sec	25	EST	MEYLAN,WM & HOWARD,PH (1993)

CAS #: 013256-06-9 — N-NITROSO-DIPHENYLAMINE

Formula: $C_{12}H_{10}N_2O$

Mol Weight: 198.23

MP (deg C): | FP (deg C):

BP (deg C):

BP pressure (mm Hg):

Property/Value	Units	Temp	Data Type	Reference
WS 2.59E+001	mg/L	25	EST	MEYLAN,WM ET AL. (1996)
logP 3.86			EXP	VERA,A ET AL. (1992)
VP 1.98E-003	mm Hg	25	EST	NEELY,WB & BLAU,GE (1985)
DC	pKa			
HL 1.98E-005	atm m3/mol	25	EST	MEYLAN,WM & HOWARD,PH (1991)
OH 2.97E-011	cm3/molc sec	25	EST	MEYLAN,WM & HOWARD,PH (1993)

CAS #: 013256-07-0 — N-NITROSO-METHYLAMYLAMINE

Formula: $C_6H_{14}N_2O$

Mol Weight: 130.19

MP (deg C): | FP (deg C):

BP (deg C):

BP pressure (mm Hg):

Property/Value	Units	Temp	Data Type	Reference
WS 6.77E+003	mg/L	25	EST	MEYLAN,WM ET AL. (1996)
logP 1.33			EST	MEYLAN,WM & HOWARD,PH (1995)
VP 8.74E-002	mm Hg	25	EST	NEELY,WB & BLAU,GE (1985)
DC	pKa			
HL 6.39E-006	atm m3/mol	25	EST	MEYLAN,WM & HOWARD,PH (1991)
OH 1.61E-011	cm3/molc sec	25	EST	MEYLAN,WM & HOWARD,PH (1993)

CAS #: 013256-13-8 — N-NITROSO-ETHYLVINYLAMINE

Formula: $C_4H_8N_2O$

Mol Weight: 100.12

MP (deg C): | FP (deg C):

BP (deg C):

BP pressure (mm Hg):

Property/Value	Units	Temp	Data Type	Reference
WS 7.78E+004	mg/L	25	EST	MEYLAN,WM ET AL. (1996)
logP 0.21			EST	MEYLAN,WM & HOWARD,PH (1995)
VP 6.03E-001	mm Hg	25	EST	NEELY,WB & BLAU,GE (1985)
DC	pKa			
HL 4.78E-006	atm m3/mol	25	EST	MEYLAN,WM & HOWARD,PH (1991)
OH 2.46E-011	cm3/molc sec	25	EST	MEYLAN,WM & HOWARD,PH (1993)

CAS #: 013256-22-9 — N-NITROSOSARCOSINE

Formula: $C_3H_6N_2O_3$

Mol Weight: 118.09

MP (deg C): 66-67 | FP (deg C):

BP (deg C):

BP pressure (mm Hg):

Property/Value	Units	Temp	Data Type	Reference
WS 1.00E+006	mg/L	25	EXP	IARC (1978)
logP -0.78			EST	MEYLAN,WM & HOWARD,PH (1995)
VP 2.13E-003	mm Hg	25	EST	NEELY,WB & BLAU,GE (1985)
DC	pKa			
HL 3.05E-010	atm m3/mol	25	EST	MEYLAN,WM & HOWARD,PH (1991)
OH 8.30E-012	cm3/molc sec	25	EST	MEYLAN,WM & HOWARD,PH (1993)

CAS #: 013258-63-4 — 4-PYRIDINE ETHANEAMINE

Formula: $C_7H_{10}N_2$

Mol Weight: 122.17

MP (deg C): | FP (deg C):

BP (deg C): 121

BP pressure (mm Hg): 1.00E+001

Property/Value	Units	Temp	Data Type	Reference
WS 1.00E+006	mg/L	25	EST	MEYLAN,WM ET AL. (1996)
logP -0.01			EXP	HANSCH,C & LEO,AJ (1985)
VP 1.48E-001	mm Hg	25	EST	NEELY,WB & BLAU,GE (1985)
DC	pKa			
HL 1.06E-009	atm m3/mol	25	EST	MEYLAN,WM & HOWARD,PH (1991)
OH 3.38E-011	cm3/molc sec	25	EST	MEYLAN,WM & HOWARD,PH (1993)

CAS #: 013263-99-5 — NSC #128,668

Formula: $C_{10}H_{14}N_6O_4$

Mol Weight: 282.26

MP (deg C): | FP (deg C):

BP (deg C):

BP pressure (mm Hg):

Property/Value	Units	Temp	Data Type	Reference
WS 1.24E+004	mg/L	25	EST	MEYLAN,WM ET AL. (1996)
logP -1.42			EXP	HANSCH,C ET AL. (1995)
VP 2.62E-015	mm Hg	25	EST	NEELY,WB & BLAU,GE (1985)
DC	pKa			
HL 2.36E-026	atm m3/mol	25	EST	MEYLAN,WM & HOWARD,PH (1991)
OH 2.47E-010	cm3/molc sec	25	EST	MEYLAN,WM & HOWARD,PH (1993)

CAS #: 013270-97-8 — 1,1,1,3,5,7,7,7-OCTAMETHYL-3,5-DIPHENYLTETRASIL*

Formula: $C_{20}H_{34}O_3Si_4$

Mol Weight: 434.84

MP (deg C): | FP (deg C):

BP (deg C):

BP pressure (mm Hg):

Property/Value	Units	Temp	Data Type	Reference
WS 3.58E-006	mg/L	25	EST	MEYLAN,WM ET AL. (1996)
logP 8.37			EST	MEYLAN,WM & HOWARD,PH (1995)
VP	mm Hg			
DC	pKa			
HL	atm m3/mol			
OH 5.10E-012	cm3/molc sec	25	EST	MEYLAN,WM & HOWARD,PH (1993)

CAS #: 013271-58-4 — 1,1,1,3,5,7,9,11,11,11-DECAMETHYL-3,5,7,9-TETRA*

Formula: $C_{34}H_{50}O_5Si_6$

Mol Weight: 707.29

MP (deg C): FP (deg C):

BP (deg C):

BP pressure (mm Hg):

Property/ Value	Units	Temp	Data Type	Reference
WS 8.15E-012	mg/L	25	EST	MEYLAN,WM ET AL. (1996)
logP 11.97			EST	MEYLAN,WM & HOWARD,PH (1995)
VP	mm Hg			
DC	pKa			
HL	atm m3/mol			
OH 9.30E-012	cm3/molc sec	25	EST	MEYLAN,WM & HOWARD,PH (1993)

CAS #: 013275-18-8 — 2,5-DICHLOROHEXANE

Formula: $C_6H_{12}Cl_2$

Mol Weight: 155.07

MP (deg C): FP (deg C):

BP (deg C):

BP pressure (mm Hg):

Property/ Value	Units	Temp	Data Type	Reference
WS 5.51E+001	mg/L	25	EST	MEYLAN,WM ET AL. (1996)
logP 3.65			EST	MEYLAN,WM & HOWARD,PH (1995)
VP 5.29E+000	mm Hg	25	EST	NEELY,WB & BLAU,GE (1985)
DC	pKa			
HL 3.77E-002	atm m3/mol	25	EST	MEYLAN,WM & HOWARD,PH (1991)
OH 2.74E-012	cm3/molc sec	25	EST	MEYLAN,WM & HOWARD,PH (1993)

CAS #: 013278-36-9 — BENZOIC ACID, 2-[(3-CHLOROPHENYL)AMINO]-

Formula: $C_{13}H_{10}ClNO_2$

Mol Weight: 247.68

MP (deg C): FP (deg C):

BP (deg C):

BP pressure (mm Hg):

Property/ Value	Units	Temp	Data Type	Reference
WS 5.62E-001	mg/L	25	EST	MEYLAN,WM ET AL. (1996)
logP 5.43			EXP	TERADA,H ET AL. (1987)
VP 6.45E-007	mm Hg	25	EST	NEELY,WB & BLAU,GE (1985)
DC	pKa			
HL 1.57E-011	atm m3/mol	25	EST	MEYLAN,WM & HOWARD,PH (1991)
OH 1.58E-010	cm3/molc sec	25	EST	MEYLAN,WM & HOWARD,PH (1993)

CAS #: 013290-48-7 — PIPERIDINE, 1-(3-METHYLBENZOYL)-

Formula: $C_{13}H_{17}NO$

Mol Weight: 203.29

MP (deg C): FP (deg C):

BP (deg C):

BP pressure (mm Hg):

Property/ Value	Units	Temp	Data Type	Reference
WS 2.53E+002	mg/L	25	EST	MEYLAN,WM ET AL. (1996)
logP 2.64			EST	MEYLAN,WM & HOWARD,PH (1995)
VP 7.83E-005	mm Hg	25	EST	NEELY,WB & BLAU,GE (1985)
DC	pKa			
HL 1.22E-008	atm m3/mol	25	EST	MEYLAN,WM & HOWARD,PH (1991)
OH 3.10E-011	cm3/molc sec	25	EST	MEYLAN,WM & HOWARD,PH (1993)

CAS #: 013294-71-8 — 2-BROMO-2-BUTENE (CIS)

Formula: C_4H_7Br

Mol Weight: 135.01

MP (deg C): FP (deg C):

BP (deg C): 82-90

BP pressure (mm Hg): 7.40E+002

Property/ Value	Units	Temp	Data Type	Reference
WS 6.64E+002	mg/L	25	EST	MEYLAN,WM ET AL. (1996)
logP 2.48			EST	MEYLAN,WM & HOWARD,PH (1995)
VP 5.69E+001	mm Hg	25	EXP	PERRY,RH & GREEN,D (1984)
DC	pKa			
HL 3.02E-002	atm m3/mol	25	EST	MEYLAN,WM & HOWARD,PH (1991)
OH 2.29E-011	cm3/molc sec	25	EST	MEYLAN,WM & HOWARD,PH (1993)

CAS #: 013308-82-2 — HYDRAZINECARBOXIMIDAMIDE, 2-[(4-METHOXYPHENYL)ME

Formula: $C_9H_{12}N_4O$

Mol Weight: 192.22

MP (deg C): FP (deg C):

BP (deg C):

BP pressure (mm Hg):

Property/ Value	Units	Temp	Data Type	Reference
WS 2.71E+004	mg/L	25	EST	MEYLAN,WM ET AL. (1996)
logP 0.29			EXP	SOMAN,G ET AL. (1986)
VP 4.08E-005	mm Hg	25	EST	NEELY,WB & BLAU,GE (1985)
DC	pKa			
HL 2.99E-013	atm m3/mol	25	EST	MEYLAN,WM & HOWARD,PH (1991)
OH 1.13E-010	cm3/molc sec	25	EST	MEYLAN,WM & HOWARD,PH (1993)

CAS #: 013308-88-8 — HYDRAZINECARBOXIMIDAMIDE, 2-[(4-CHLOROPHENYL)MET

Formula: $C_8H_9ClN_4$

Mol Weight: 196.64

MP (deg C): FP (deg C):

BP (deg C):

BP pressure (mm Hg):

Property/ Value	Units	Temp	Data Type	Reference
WS 6.89E+003	mg/L	25	EST	MEYLAN,WM ET AL. (1996)
logP 0.96			EXP	SOMAN,G ET AL. (1986)
VP 5.88E-005	mm Hg	25	EST	NEELY,WB & BLAU,GE (1985)
DC	pKa			
HL 3.74E-012	atm m3/mol	25	EST	MEYLAN,WM & HOWARD,PH (1991)
OH 8.76E-011	cm3/molc sec	25	EST	MEYLAN,WM & HOWARD,PH (1993)

CAS #: 013311-84-7 — PROPANAMIDE, 2-METHYL-N-[4-NITRO-3-(TRIFLUOROMET

Formula: $C_{11}H_{11}F_3N_2O_3$

Mol Weight: 276.22

MP (deg C): FP (deg C):

BP (deg C):

BP pressure (mm Hg):

Property/ Value	Units	Temp	Data Type	Reference
WS 9.45E+000	mg/L	25	EST	MEYLAN,WM ET AL. (1996)
logP 3.35			EXP	MORRIS,JJ ET AL. (1991)
VP 1.11E-006	mm Hg	25	EST	NEELY,WB & BLAU,GE (1985)
DC	pKa			
HL 3.73E-010	atm m3/mol	25	EST	MEYLAN,WM & HOWARD,PH (1991)
OH 2.75E-012	cm3/molc sec	25	EST	MEYLAN,WM & HOWARD,PH (1993)

CAS #: 013314-63-1				PYRIDINE, 1-ETHYL-1,2,3,6-TETRAHYDRO-4-PHENYL-

Formula: $C_{13}H_{17}N$

Mol Weight: 187.29

MP (deg C):   FP (deg C):

BP (deg C):

BP pressure (mm Hg):

Property/Value	Units	Temp	Data Type	Reference
WS 9.94E+002	mg/L	25	EST	MEYLAN,WM ET AL. (1996)
logP 3.18			EXP	ALTOMARE,CA ET AL. (1992)
VP 2.16E-003	mm Hg	25	EST	NEELY,WB & BLAU,GE (1985)
DC	pKa			
HL 1.97E-006	atm m3/mol	25	EST	MEYLAN,WM & HOWARD,PH (1991)
OH 1.84E-010	cm3/molc sec	25	EST	MEYLAN,WM & HOWARD,PH (1993)

CAS #: 013351-73-0				1-METHYLBENZOTRIAZOLE

Formula: $C_7H_7N_3$

Mol Weight: 133.15

MP (deg C): 64.5   FP (deg C):

BP (deg C): 270.5

BP pressure (mm Hg):

Property/Value	Units	Temp	Data Type	Reference
WS 9.68E+003	mg/L	25	EST	MEYLAN,WM ET AL. (1996)
logP 1.13			EXP	HANSCH,C & LEO,AJ (1985)
VP 1.04E-002	mm Hg	25	EST	NEELY,WB & BLAU,GE (1985)
DC	pKa			
HL 3.13E-006	atm m3/mol	25	EST	MEYLAN,WM & HOWARD,PH (1991)
OH 1.14E-012	cm3/molc sec	25	EST	MEYLAN,WM & HOWARD,PH (1993)

CAS #: 013358-11-7				N(2-ETHEXYL)-1-ISOPROPYL-4-METHYLBICYCLO[2.2.2]-

Formula: $C_{22}H_{35}NO_2$

Mol Weight: 345.53

MP (deg C):   FP (deg C):

BP (deg C):

BP pressure (mm Hg):

Property/Value	Units	Temp	Data Type	Reference
WS 5.36E-002	mg/L	25	EST	MEYLAN,WM ET AL. (1996)
logP 5.96			EXP	CHEM INSPECT TEST INST (1992)
VP 9.37E-010	mm Hg	25	EST	NEELY,WB & BLAU,GE (1985)
DC	pKa			
HL 1.18E-006	atm m3/mol	25	EST	MEYLAN,WM & HOWARD,PH (1991)
OH 1.01E-010	cm3/molc sec	25	EST	MEYLAN,WM & HOWARD,PH (1993)

CAS #: 013360-45-7				CHLORBROMURON

Formula: $C_9H_{10}BrClN_2O_2$

Mol Weight: 293.55

MP (deg C): 96   FP (deg C):

BP (deg C):

BP pressure (mm Hg):

Property/Value	Units	Temp	Data Type	Reference
WS 3.50E+001	mg/L	20	EXP	YALKOWSKY,SH & DANNENFELSER,RM (1992)
logP 3.09			EXP	HANSCH,C ET AL. (1995)
VP 3.97E-007	mm Hg	20	EXP	TOMLIN,C (1994)
DC	pKa			
HL 4.38E-009	atm m3/mol	20	EST	VP/WSOL
OH 1.02E-011	cm3/molc sec	25	EST	MEYLAN,WM & HOWARD,PH (1993)

CAS #: 013360-61-7				1-PENTADECENE

Formula: $C_{15}H_{30}$

Mol Weight: 210.41

MP (deg C): -2.8   FP (deg C):

BP (deg C): 268.2

BP pressure (mm Hg):

Property/Value	Units	Temp	Data Type	Reference
WS 3.84E-003	mg/L	25	EST	MEYLAN,WM ET AL. (1996)
logP 7.57			EST	MEYLAN,WM & HOWARD,PH (1995)
VP 4.54E-003	mm Hg	25	EXP	YAWS,CL (1994B)
DC	pKa			
HL 4.59E+000	atm m3/mol	25	EST	MEYLAN,WM & HOWARD,PH (1991)
OH 4.29E-011	cm3/molc sec	25	EST	MEYLAN,WM & HOWARD,PH (1993)

CAS #: 013373-32-5				1H-IMIDAZOLE, 2-(1-METHYLETHYL)-4-NITRO-

Formula: $C_6H_9N_3O_2$

Mol Weight: 155.16

MP (deg C):   FP (deg C):

BP (deg C):

BP pressure (mm Hg):

Property/Value	Units	Temp	Data Type	Reference
WS 5.00E+003	mg/L	25	EST	MEYLAN,WM ET AL. (1996)
logP 0.90			EXP	SANGSTER,J (1994)
VP 2.05E-005	mm Hg	25	EST	NEELY,WB & BLAU,GE (1985)
DC	pKa			
HL 2.88E-008	atm m3/mol	25	EST	MEYLAN,WM & HOWARD,PH (1991)
OH 6.06E-012	cm3/molc sec	25	EST	MEYLAN,WM & HOWARD,PH (1993)

CAS #: 013389-42-9				2-OCTENE (TRANS)

Formula: C_8H_{16}

Mol Weight: 112.22

MP (deg C): -87.7   FP (deg C):

BP (deg C): 125

BP pressure (mm Hg):

Property/Value	Units	Temp	Data Type	Reference
WS 1.07E+001	mg/L	25	EST	MEYLAN,WM ET AL. (1996)
logP 4.06			EST	MEYLAN,WM & HOWARD,PH (1995)
VP 1.64E+001	mm Hg	25	EXP	DAUBERT,TE & DANNER,RP (1989)
DC	pKa			
HL 7.46E-001	atm m3/mol	25	EST	MEYLAN,WM & HOWARD,PH (1991)
OH 6.18E-011	cm3/molc sec	25	EST	MEYLAN,WM & HOWARD,PH (1993)

CAS #: 013410-84-9				5-(5-NO2-2-FURFURYLIDENE-2-S-THIAZOLIDIN-4-ONE

Formula: $C_8H_5N_2O_4S_2$

Mol Weight: 257.27

MP (deg C):   FP (deg C):

BP (deg C):

BP pressure (mm Hg):

Property/Value	Units	Temp	Data Type	Reference
WS 2.08E+003	mg/L	25	EST	MEYLAN,WM ET AL. (1996)
logP 0.74			EXP	HANSCH,C & LEO,AJ (1985)
VP 7.05E-010	mm Hg	25	EST	NEELY,WB & BLAU,GE (1985)
DC	pKa			
HL 8.12E-012	atm m3/mol	25	EST	MEYLAN,WM & HOWARD,PH (1991)
OH 4.24E-011	cm3/molc sec	25	EST	MEYLAN,WM & HOWARD,PH (1993)

CAS #: 013411-16-0 — 2-PYRIDINEMETHANOL, 6-[2-(5-NITRO-2-FURANYL)ETHE

Formula: $C_{12}H_{10}N_2O_4$

Mol Weight: 246.22

MP (deg C): 170-171

FP (deg C):

BP (deg C):

BP pressure (mm Hg):

Property/Value	Units	Temp	Data Type	Reference
WS 6.84E+002	mg/L	25	EST	MEYLAN,WM ET AL. (1996)
logP 1.37			EXP	HANSCH,C ET AL. (1995)
VP 1.22E-008	mm Hg	25	EST	NEELY,WB & BLAU,GE (1985)
DC	pKa			
HL 2.57E-014	atm m3/mol	25	EST	MEYLAN,WM & HOWARD,PH (1991)
OH 6.54E-011	cm3/molc sec	25	EST	MEYLAN,WM & HOWARD,PH (1993)

CAS #: 013414-55-6 — 7-NITRO-2,2-DIMETHYL-2,3-DIHYDROBENZOFURAN

Formula: $C_{10}H_{11}NO_3$

Mol Weight: 193.20

MP (deg C):

FP (deg C):

BP (deg C):

BP pressure (mm Hg):

Property/Value	Units	Temp	Data Type	Reference
WS 1.20E+002	mg/L	25	EXP	FEDERAL REGISTER (1985)
logP 3.33			EXP	FEDERAL REGISTER (1985)
VP 6.87E-004	mm Hg	25	EST	NEELY,WB & BLAU,GE (1985)
DC	pKa			
HL 1.08E-006	atm m3/mol	25	EST	MEYLAN,WM & HOWARD,PH (1991)
OH 6.84E-012	cm3/molc sec	25	EST	MEYLAN,WM & HOWARD,PH (1993)

CAS #: 013419-31-3 — 4-(1-METHYLTRIDECYL)BENZENESULFONIC ACID, SODIU*

Formula: $C_{20}H_{33}NaO_3S$

Mol Weight: 376.54

MP (deg C):

FP (deg C):

BP (deg C):

BP pressure (mm Hg):

Property/Value	Units	Temp	Data Type	Reference
WS 1.97E+000	mg/L	25	EST	MEYLAN,WM ET AL. (1996)
logP 3.91			EST	MEYLAN,WM & HOWARD,PH (1995)
VP 6.80E-016	mm Hg	25	EST	NEELY,WB & BLAU,GE (1985)
DC	pKa			
HL	atm m3/mol			
OH 1.90E-011	cm3/molc sec	25	EST	MEYLAN,WM & HOWARD,PH (1993)

CAS #: 013425-39-3 — ETOFYLLINE NICOTINATE

Formula: $C_{15}H_{15}N_5O_4$

Mol Weight: 329.32

MP (deg C): 151-152

FP (deg C):

BP (deg C):

BP pressure (mm Hg):

Property/Value	Units	Temp	Data Type	Reference
WS 2.00E+004	mg/L	100	EXP	MERCK INDEX (1989)
logP 0.45			EST	MEYLAN,WM & HOWARD,PH (1995)
VP 1.84E-012	mm Hg	25	EST	NEELY,WB & BLAU,GE (1985)
DC	pKa			
HL 2.93E-017	atm m3/mol	25	EST	MEYLAN,WM & HOWARD,PH (1991)
OH 2.24E-011	cm3/molc sec	25	EST	MEYLAN,WM & HOWARD,PH (1993)

CAS #: 013425-98-4 — 1-PROPANOL, 3,3'-IMINOBIS-, DIMETHANESULFONATE (

Formula: $C_8H_{19}NO_6S_2$

Mol Weight: 289.37

MP (deg C): 113-118

FP (deg C):

BP (deg C):

BP pressure (mm Hg):

Property/Value	Units	Temp	Data Type	Reference
WS 1.00E+006	mg/L	25	EST	MEYLAN,WM ET AL. (1996)
logP -1.71			EXP	HANSCH,C ET AL. (1995)
VP 2.59E-007	mm Hg	25	EST	NEELY,WB & BLAU,GE (1985)
DC	pKa			
HL 1.55E-013	atm m3/mol	25	EST	MEYLAN,WM & HOWARD,PH (1991)
OH 8.91E-011	cm3/molc sec	25	EST	MEYLAN,WM & HOWARD,PH (1993)

CAS #: 013438-65-8 — 3-PYRIDINECARBOXAMIDE, 2-AMINO-

Formula: $C_6H_7N_3O$

Mol Weight: 137.14

MP (deg C):

FP (deg C):

BP (deg C):

BP pressure (mm Hg):

Property/Value	Units	Temp	Data Type	Reference
WS 1.53E+004	mg/L	25	EST	MEYLAN,WM ET AL. (1996)
logP 0.88			EXP	HANSCH,C ET AL. (1995)
VP 1.35E-005	mm Hg	25	EST	NEELY,WB & BLAU,GE (1985)
DC	pKa			
HL 1.02E-015	atm m3/mol	25	EST	MEYLAN,WM & HOWARD,PH (1991)
OH 2.06E-011	cm3/molc sec	25	EST	MEYLAN,WM & HOWARD,PH (1993)

CAS #: 013441-66-2 — L-THREO-ALPHA-D-GALACTO-OCTOPYRANOSIDE, DERIVATI

Formula: $C_{19}H_{35}ClN_2O_5S$

Mol Weight: 439.02

MP (deg C):

FP (deg C):

BP (deg C):

BP pressure (mm Hg):

Property/Value	Units	Temp	Data Type	Reference
WS 1.26E+001	mg/L	25	EST	MEYLAN,WM ET AL. (1996)
logP 2.51			EXP	HANSCH,C ET AL. (1995)
VP 1.83E-017	mm Hg	25	EST	NEELY,WB & BLAU,GE (1985)
DC	pKa			
HL 3.83E-022	atm m3/mol	25	EST	MEYLAN,WM & HOWARD,PH (1991)
OH 2.43E-010	cm3/molc sec	25	EST	MEYLAN,WM & HOWARD,PH (1993)

CAS #: 013450-66-3 — 3-PYRIDYLETHYL-2-N-PIPERIDINE

Formula: $C_{12}H_{18}N_2$

Mol Weight: 190.29

MP (deg C):

FP (deg C):

BP (deg C):

BP pressure (mm Hg):

Property/Value	Units	Temp	Data Type	Reference
WS 3.81E+005	mg/L	25	EST	MEYLAN,WM ET AL. (1996)
logP 1.66			EXP	HANSCH,C & LEO,AJ (1985)
VP 1.15E-003	mm Hg	25	EST	NEELY,WB & BLAU,GE (1985)
DC 8.81	pKa	25	EXP	PERRIN,DD (1965)
HL 5.28E-009	atm m3/mol	25	EST	MEYLAN,WM & HOWARD,PH (1991)
OH 1.04E-010	cm3/molc sec	25	EST	MEYLAN,WM & HOWARD,PH (1993)

CAS #: 013457-18-6				PYRAZOPHOS

Formula: $C_{14}H_{20}N_3O_5PS$

Mol Weight: 373.37

MP (deg C): 38-40   FP (deg C):

BP (deg C):

BP pressure (mm Hg):

Property/Value	Units	Temp	Data Type	Reference
WS 4.20E+000	mg/L	20	EXP	YALKOWSKY,SH & DANNENFELSER,RM (1992)
logP 3.80			EXP	TOMLIN,C (1994)
VP 9.80E-008	mm Hg	25	EST	NEELY,WB & BLAU,GE (1985)
DC	pKa			
HL 1.21E-012	atm m3/mol	25	EST	MEYLAN,WM & HOWARD,PH (1991)
OH 1.14E-010	cm3/molc sec	25	EST	MEYLAN,WM & HOWARD,PH (1993)

CAS #: 013460-15-6				124-BENZTHIADIAZIN-1-O2-3-ME-7-BR

Formula: $C_8H_7BrN_2O_2S$

Mol Weight: 275.13

MP (deg C):   FP (deg C):

BP (deg C):

BP pressure (mm Hg):

Property/Value	Units	Temp	Data Type	Reference
WS 1.16E+003	mg/L	25	EST	MEYLAN,WM ET AL. (1996)
logP 1.37			EXP	HANSCH,C & LEO,AJ (1985)
VP 1.25E-008	mm Hg	25	EST	NEELY,WB & BLAU,GE (1985)
DC	pKa			
HL 3.68E-008	atm m3/mol	25	EST	MEYLAN,WM & HOWARD,PH (1991)
OH 4.55E-013	cm3/molc sec	25	EST	MEYLAN,WM & HOWARD,PH (1993)

CAS #: 013460-16-7				124-Benzthiadiazin-1-O2-3-TBU-7-CL

Formula: $C_{11}H_{13}ClN_2O_2S$

Mol Weight: 272.76

MP (deg C):   FP (deg C):

BP (deg C):

BP pressure (mm Hg):

Property/Value	Units	Temp	Data Type	Reference
WS 1.57E+002	mg/L	25	EST	MEYLAN,WM ET AL. (1996)
logP 2.40			EXP	HANSCH,C & LEO,AJ (1985)
VP 5.72E-009	mm Hg	25	EST	NEELY,WB & BLAU,GE (1985)
DC	pKa			
HL 1.60E-007	atm m3/mol	25	EST	MEYLAN,WM & HOWARD,PH (1991)
OH 8.24E-013	cm3/molc sec	25	EST	MEYLAN,WM & HOWARD,PH (1993)

CAS #: 013460-17-8				124-BENZTHIDIAZIN-1-O2-3-CYPR-6-CL

Formula: $C_{10}H_9ClN_2O_2S$

Mol Weight: 256.71

MP (deg C):   FP (deg C):

BP (deg C):

BP pressure (mm Hg):

Property/Value	Units	Temp	Data Type	Reference
WS 4.42E+002	mg/L	25	EST	MEYLAN,WM ET AL. (1996)
logP 1.98			EXP	HANSCH,C & LEO,AJ (1985)
VP 4.99E-009	mm Hg	25	EST	NEELY,WB & BLAU,GE (1985)
DC	pKa			
HL 5.33E-008	atm m3/mol	25	EST	MEYLAN,WM & HOWARD,PH (1991)
OH 3.13E-013	cm3/molc sec	25	EST	MEYLAN,WM & HOWARD,PH (1993)

CAS #: 013466-78-9				3-CARENE (DELTA)

Formula: $C_{10}H_{16}$

Mol Weight: 136.24

MP (deg C):   FP (deg C):

BP (deg C): 168-169

BP pressure (mm Hg): 7.05E+002

Property/Value	Units	Temp	Data Type	Reference
WS 2.91E+000	mg/L	25	EST	MEYLAN,WM ET AL. (1996)
logP 4.61			EST	MEYLAN,WM & HOWARD,PH (1995)
VP 3.72E+000	mm Hg	25	EST	NEELY,WB & BLAU,GE (1985)
DC	pKa			
HL 1.07E-001	atm m3/mol	25	EST	MEYLAN,WM & HOWARD,PH (1991)
OH 8.78E-011	cm3/molc sec	25	EXP	ATKINSON,R (1989)

CAS #: 013475-81-5				2,2,3,3-TETRAMETHYLHEXANE

Formula: $C_{10}H_{22}$

Mol Weight: 142.29

MP (deg C): -54   FP (deg C):

BP (deg C): 160.3

BP pressure (mm Hg):

Property/Value	Units	Temp	Data Type	Reference
WS 1.20E+000	mg/L	25	EST	MEYLAN,WM ET AL. (1996)
logP 5.03			EST	MEYLAN,WM & HOWARD,PH (1995)
VP 3.89E+000	mm Hg	25	EXT	ZWOLINSKI,BJ & WILHOIT,RC (1971)
DC	pKa			
HL 5.30E+000	atm m3/mol	25	EST	MEYLAN,WM & HOWARD,PH (1991)
OH 3.57E-012	cm3/molc sec	25	EST	MEYLAN,WM & HOWARD,PH (1993)

CAS #: 013476-55-6				BENZOIC ACID, 4-AMINO-, HEXYL ESTER

Formula: $C_{13}H_{19}NO_2$

Mol Weight: 221.30

MP (deg C):   FP (deg C):

BP (deg C):

BP pressure (mm Hg):

Property/Value	Units	Temp	Data Type	Reference
WS 1.41E+001	mg/L	25	EST	MEYLAN,WM ET AL. (1996)
logP 3.96			EXP	LIANG,WQ & LIN,W (1992)
VP 5.52E-005	mm Hg	25	EST	NEELY,WB & BLAU,GE (1985)
DC	pKa			
HL 5.06E-008	atm m3/mol	25	EST	MEYLAN,WM & HOWARD,PH (1991)
OH 4.22E-011	cm3/molc sec	25	EST	MEYLAN,WM & HOWARD,PH (1993)

CAS #: 013481-25-9				2,3-PYRAZINEDICARBONITRILE

Formula: $C_6H_2N_4$

Mol Weight: 130.11

MP (deg C): 131-133   FP (deg C):

BP (deg C):

BP pressure (mm Hg):

Property/Value	Units	Temp	Data Type	Reference
WS 2.36E+004	mg/L	25	EST	MEYLAN,WM ET AL. (1996)
logP 0.38			EXP	YAMAGAMI,C ET AL. (1991)
VP 1.34E-003	mm Hg	25	EST	NEELY,WB & BLAU,GE (1985)
DC	pKa			
HL 2.72E-010	atm m3/mol	25	EST	MEYLAN,WM & HOWARD,PH (1991)
OH 5.32E-015	cm3/molc sec	25	EST	MEYLAN,WM & HOWARD,PH (1993)

013491-47-9 — ACETAMIDE, N-(1-beta-D-ARABINOFURANOSYL-1,2-DIHY

Formula: $C_{11}H_{15}N_3O_6$

Mol Weight: 285.26

MP (deg C): FP (deg C):

BP (deg C):

BP pressure (mm Hg):

Property/Value	Units	Temp	Data Type	Reference
WS 1.04E+004	mg/L	25	EST	MEYLAN,WM ET AL. (1996)
logP -1.35			EXP	SANGSTER,J (1994)
VP 6.52E-015	mm Hg	25	EST	NEELY,WB & BLAU,GE (1985)
DC	pKa			
HL 5.08E-022	atm m3/mol	25	EST	MEYLAN,WM & HOWARD,PH (1991)
OH 1.23E-010	cm3/molc sec	25	EST	MEYLAN,WM & HOWARD,PH (1993)

013509-38-1 — CARBANILIC ACID, THIO-, S-METHYL ESTER

Formula: C_8H_9NOS

Mol Weight: 167.23

MP (deg C): FP (deg C):

BP (deg C):

BP pressure (mm Hg):

Property/Value	Units	Temp	Data Type	Reference
WS 5.65E+002	mg/L	25	EST	MEYLAN,WM ET AL. (1996)
logP 2.40			EXP	HANSCH,C ET AL. (1995)
VP 3.63E-005	mm Hg	25	EST	NEELY,WB & BLAU,GE (1985)
DC	pKa			
HL 4.27E-007	atm m3/mol	25	EST	MEYLAN,WM & HOWARD,PH (1991)
OH 4.32E-011	cm3/molc sec	25	EST	MEYLAN,WM & HOWARD,PH (1993)

013515-66-7 — 2-METHYLPYRIDINIUM IODIDE,N-OCTYL

Formula: $C_{14}H_{24}IN$

Mol Weight: 333.26

MP (deg C): FP (deg C):

BP (deg C):

BP pressure (mm Hg):

Property/Value	Units	Temp	Data Type	Reference
WS 3.55E+004	mg/L	25	EST	MEYLAN,WM ET AL. (1996)
logP -0.77			EXP	HANSCH,C & LEO,AJ (1985)
VP 4.32E-005	mm Hg	25	EST	NEELY,WB & BLAU,GE (1985)
DC	pKa			
HL 1.21E-009	atm m3/mol	25	EST	MEYLAN,WM & HOWARD,PH (1991)
OH 2.06E-011	cm3/molc sec	25	EST	MEYLAN,WM & HOWARD,PH (1993)

013523-86-9 — PINDOLOL

Formula: $C_{14}H_{20}N_2O_2$

Mol Weight: 248.33

MP (deg C): 167-171 FP (deg C):

BP (deg C):

BP pressure (mm Hg):

Property/Value	Units	Temp	Data Type	Reference
WS 7.88E+003	mg/L	25	EST	MEYLAN,WM ET AL. (1996)
logP 1.75			EXP	SANGSTER,J (1994)
VP 1.23E-008	mm Hg	25	EST	NEELY,WB & BLAU,GE (1985)
DC	pKa			
HL 1.34E-015	atm m3/mol	25	EST	MEYLAN,WM & HOWARD,PH (1991)
OH 3.09E-010	cm3/molc sec	25	EST	MEYLAN,WM & HOWARD,PH (1993)

013532-26-8 — 2-DIETHYLAMINO-4-ETHYLAMINO-6-METHOXY-S-TRIAZI*

Formula: $C_{10}H_{19}N_5O$

Mol Weight: 225.30

MP (deg C): FP (deg C):

BP (deg C):

BP pressure (mm Hg):

Property/Value	Units	Temp	Data Type	Reference
WS 2.26E+002	mg/L	26	EXP	SHIU,WY ET AL. (1990)
logP 3.77			EST	MEYLAN,WM & HOWARD,PH (1995)
VP 4.59E-005	mm Hg	25	EST	NEELY,WB & BLAU,GE (1985)
DC	pKa			
HL 6.67E-008	atm m3/mol	25	EST	MEYLAN,WM & HOWARD,PH (1991)
OH 2.77E-011	cm3/molc sec	25	EST	MEYLAN,WM & HOWARD,PH (1993)

013534-98-0 — 4-AMINO-3-BROMOPYRIDINE

Formula: $C_5H_5BrN_2$

Mol Weight: 173.02

MP (deg C): FP (deg C):

BP (deg C):

BP pressure (mm Hg):

Property/Value	Units	Temp	Data Type	Reference
WS 1.29E+004	mg/L	25	EST	MEYLAN,WM ET AL. (1996)
logP 0.78			EST	MEYLAN,WM & HOWARD,PH (1995)
VP 2.01E-002	mm Hg	25	EST	NEELY,WB & BLAU,GE (1985)
DC 7.05	pKa	20	EXP	PERRIN,DD (1965)
HL 9.93E-010	atm m3/mol	25	EST	MEYLAN,WM & HOWARD,PH (1991)
OH 5.85E-012	cm3/molc sec	25	EST	MEYLAN,WM & HOWARD,PH (1993)

013538-14-2 — T-BU PHOSPHONIC ACID,ET,PNO2PH ESTR

Formula: $C_{12}H_{18}NO_5P$

Mol Weight: 287.25

MP (deg C): FP (deg C):

BP (deg C):

BP pressure (mm Hg):

Property/Value	Units	Temp	Data Type	Reference
WS 5.29E+001	mg/L	25	EST	MEYLAN,WM ET AL. (1996)
logP 2.40			EXP	HANSCH,C & LEO,AJ (1985)
VP 9.63E-006	mm Hg	25	EST	NEELY,WB & BLAU,GE (1985)
DC	pKa			
HL 5.66E-009	atm m3/mol	25	EST	MEYLAN,WM & HOWARD,PH (1991)
OH 2.05E-012	cm3/molc sec	25	EST	MEYLAN,WM & HOWARD,PH (1993)

013538-26-6 — N-ME-3,5-DICHLOROPHENYLCARBAMATE

Formula: $C_8H_7Cl_2NO_2$

Mol Weight: 220.06

MP (deg C): FP (deg C):

BP (deg C):

BP pressure (mm Hg):

Property/Value	Units	Temp	Data Type	Reference
WS 8.89E+001	mg/L	25	EST	MEYLAN,WM ET AL. (1996)
logP 3.03			EXP	HANSCH,C & LEO,AJ (1985)
VP 1.39E-003	mm Hg	25	EST	NEELY,WB & BLAU,GE (1985)
DC	pKa			
HL 1.76E-008	atm m3/mol	25	EST	MEYLAN,WM & HOWARD,PH (1991)
OH 7.05E-012	cm3/molc sec	25	EST	MEYLAN,WM & HOWARD,PH (1993)

CAS #: 013538-27-7 — N-METHYL-2-BROMOPHENYLCARBAMATE

Formula: $C_8H_8BrNO_2$
Mol Weight: 230.07
MP (deg C):
FP (deg C):
BP (deg C):
BP pressure (mm Hg):

Property/Value	Units	Temp	Data Type	Reference
WS 9.36E+002	mg/L	25	EST	MEYLAN,WM ET AL. (1996)
logP 1.77			EXP	HANSCH,C & LEO,AJ (1985)
VP 2.66E-003	mm Hg	25	EST	NEELY,WB & BLAU,GE (1985)
DC	pKa			
HL 1.28E-008	atm m3/mol	25	EST	MEYLAN,WM & HOWARD,PH (1991)
OH 6.96E-012	cm3/molc sec	25	EST	MEYLAN,WM & HOWARD,PH (1993)

CAS #: 013538-28-8 — N-METHYL-2-IODOPHENYLCARBAMATE

Formula: $C_8H_8INO_2$
Mol Weight: 277.06
MP (deg C):
FP (deg C):
BP (deg C):
BP pressure (mm Hg):

Property/Value	Units	Temp	Data Type	Reference
WS 3.67E+002	mg/L	25	EST	MEYLAN,WM ET AL. (1996)
logP 1.94			EXP	HANSCH,C & LEO,AJ (1985)
VP 5.32E-004	mm Hg	25	EST	NEELY,WB & BLAU,GE (1985)
DC	pKa			
HL 7.45E-009	atm m3/mol	25	EST	MEYLAN,WM & HOWARD,PH (1991)
OH 7.00E-012	cm3/molc sec	25	EST	MEYLAN,WM & HOWARD,PH (1993)

CAS #: 013538-50-6 — N-METHYL-4-BROMOPHENYLCARBAMATE

Formula: $C_8H_8BrNO_2$
Mol Weight: 230.07
MP (deg C):
FP (deg C):
BP (deg C):
BP pressure (mm Hg):

Property/Value	Units	Temp	Data Type	Reference
WS 4.26E+002	mg/L	25	EST	MEYLAN,WM ET AL. (1996)
logP 2.17			EXP	HANSCH,C & LEO,AJ (1985)
VP 2.66E-003	mm Hg	25	EST	NEELY,WB & BLAU,GE (1985)
DC	pKa			
HL 1.28E-008	atm m3/mol	25	EST	MEYLAN,WM & HOWARD,PH (1991)
OH 6.96E-012	cm3/molc sec	25	EST	MEYLAN,WM & HOWARD,PH (1993)

CAS #: 013538-51-7 — N-METHYL-4-IODOPHENYLCARBAMATE

Formula: $C_8H_8INO_2$
Mol Weight: 277.06
MP (deg C):
FP (deg C):
BP (deg C):
BP pressure (mm Hg):

Property/Value	Units	Temp	Data Type	Reference
WS 1.32E+002	mg/L	25	EST	MEYLAN,WM ET AL. (1996)
logP 2.46			EXP	HANSCH,C & LEO,AJ (1985)
VP 5.32E-004	mm Hg	25	EST	NEELY,WB & BLAU,GE (1985)
DC	pKa			
HL 7.45E-009	atm m3/mol	25	EST	MEYLAN,WM & HOWARD,PH (1991)
OH 7.00E-012	cm3/molc sec	25	EST	MEYLAN,WM & HOWARD,PH (1993)

CAS #: 013538-54-0 — N-METHYL-4-ETHOXYPHENYLCARBAMATE

Formula: $C_{10}H_{13}NO_3$
Mol Weight: 195.22
MP (deg C):
FP (deg C):
BP (deg C):
BP pressure (mm Hg):

Property/Value	Units	Temp	Data Type	Reference
WS 1.88E+003	mg/L	25	EST	MEYLAN,WM ET AL. (1996)
logP 1.63			EXP	HANSCH,C & LEO,AJ (1985)
VP 1.79E-003	mm Hg	25	EST	NEELY,WB & BLAU,GE (1985)
DC	pKa			
HL 2.52E-009	atm m3/mol	25	EST	MEYLAN,WM & HOWARD,PH (1991)
OH 2.80E-011	cm3/molc sec	25	EST	MEYLAN,WM & HOWARD,PH (1993)

CAS #: 013538-60-8 — N-METHYL-3-BROMOPHENYLCARBAMATE

Formula: $C_8H_8BrNO_2$
Mol Weight: 230.07
MP (deg C):
FP (deg C):
BP (deg C):
BP pressure (mm Hg):

Property/Value	Units	Temp	Data Type	Reference
WS 3.64E+002	mg/L	25	EST	MEYLAN,WM ET AL. (1996)
logP 2.25			EXP	HANSCH,C & LEO,AJ (1985)
VP 2.66E-003	mm Hg	25	EST	NEELY,WB & BLAU,GE (1985)
DC	pKa			
HL 1.28E-008	atm m3/mol	25	EST	MEYLAN,WM & HOWARD,PH (1991)
OH 7.32E-012	cm3/molc sec	25	EST	MEYLAN,WM & HOWARD,PH (1993)

CAS #: 013540-50-6 — PHENYL XYLYLMETHANE

Formula: $C_{15}H_{16}$
Mol Weight: 196.29
MP (deg C):
FP (deg C):
BP (deg C):
BP pressure (mm Hg):

Property/Value	Units	Temp	Data Type	Reference
WS 6.84E-001	mg/L	25	EST	MEYLAN,WM ET AL. (1996)
logP 5.65			EXP	CHEM INSPECT TEST INST (1992)
VP 8.63E-004	mm Hg	25	EST	NEELY,WB & BLAU,GE (1985)
DC	pKa			
HL 5.85E-004	atm m3/mol	25	EST	MEYLAN,WM & HOWARD,PH (1991)
OH 2.16E-011	cm3/molc sec	25	EST	MEYLAN,WM & HOWARD,PH (1993)

CAS #: 013547-07-4 — 2-CHLORO-4-(1-CHLOROVINYL)CYCLO-1-HEXENE

Formula: $C_8H_{10}Cl_2$
Mol Weight: 177.07
MP (deg C):
FP (deg C):
BP (deg C):
BP pressure (mm Hg):

Property/Value	Units	Temp	Data Type	Reference
WS 5.33E+000	mg/L	25	EST	MEYLAN,WM ET AL. (1996)
logP 4.72			EST	MEYLAN,WM & HOWARD,PH (1995)
VP 8.90E-001	mm Hg	25	EST	NEELY,WB & BLAU,GE (1985)
DC	pKa			
HL 8.00E-002	atm m3/mol	25	EST	MEYLAN,WM & HOWARD,PH (1991)
OH 3.57E-011	cm3/molc sec	25	EST	MEYLAN,WM & HOWARD,PH (1993)

CAS #: 013551-86-5 — 1(3-CL-2-OH PR)-2-NO2-IMIDAZOLE

Formula: $C_6H_{10}ClN_3O_3$

Mol Weight: 207.62

MP (deg C): FP (deg C):

BP (deg C):

BP pressure (mm Hg):

Property/Value	Units	Temp	Data Type	Reference
WS 1.24E+004	mg/L	25	EST	MEYLAN,WM ET AL. (1996)
logP 0.15			EXP	HANSCH,C & LEO,AJ (1985)
VP 6.14E-008	mm Hg	25	EST	NEELY,WB & BLAU,GE (1985)
DC	pKa			
HL 7.18E-012	atm m3/mol	25	EST	MEYLAN,WM & HOWARD,PH (1991)
OH 9.23E-012	cm3/molc sec	25	EST	MEYLAN,WM & HOWARD,PH (1993)

CAS #: 013551-87-6 — 1(2-OH-3-MEO-PR)-2-NITROIMIDAZOLE

Formula: $C_7H_{11}N_3O_4$

Mol Weight: 201.18

MP (deg C): FP (deg C):

BP (deg C):

BP pressure (mm Hg):

Property/Value	Units	Temp	Data Type	Reference
WS 3.64E+004	mg/L	25	EST	MEYLAN,WM ET AL. (1996)
logP -0.37			EXP	HANSCH,C & LEO,AJ (1985)
VP 1.08E-007	mm Hg	25	EST	NEELY,WB & BLAU,GE (1985)
DC	pKa			
HL 2.38E-013	atm m3/mol	25	EST	MEYLAN,WM & HOWARD,PH (1991)
OH 2.39E-011	cm3/molc sec	25	EST	MEYLAN,WM & HOWARD,PH (1993)

CAS #: 013551-89-8 — 1-(3-F-2-OHPROPYL)-2-NO2 IMIDAZOLE

Formula: $C_6H_{10}FN_3O_3$

Mol Weight: 191.16

MP (deg C): FP (deg C):

BP (deg C):

BP pressure (mm Hg):

Property/Value	Units	Temp	Data Type	Reference
WS 5.62E+004	mg/L	25	EST	MEYLAN,WM ET AL. (1996)
logP -0.52			EXP	HANSCH,C ET AL. (1995)
VP 6.18E-007	mm Hg	25	EST	NEELY,WB & BLAU,GE (1985)
DC	pKa			
HL 4.05E-011	atm m3/mol	25	EST	MEYLAN,WM & HOWARD,PH (1991)
OH 1.10E-011	cm3/molc sec	25	EST	MEYLAN,WM & HOWARD,PH (1993)

CAS #: 013551-91-2 — 1-(2-OH PROPYL)-2-NO2 IMIDAZOLE

Formula: $C_6H_{11}N_3O_3$

Mol Weight: 173.17

MP (deg C): FP (deg C):

BP (deg C):

BP pressure (mm Hg):

Property/Value	Units	Temp	Data Type	Reference
WS 3.39E+004	mg/L	25	EST	MEYLAN,WM ET AL. (1996)
logP -0.16			EXP	HANSCH,C & LEO,AJ (1985)
VP 6.77E-007	mm Hg	25	EST	NEELY,WB & BLAU,GE (1985)
DC	pKa			
HL 2.04E-011	atm m3/mol	25	EST	MEYLAN,WM & HOWARD,PH (1991)
OH 1.43E-011	cm3/molc sec	25	EST	MEYLAN,WM & HOWARD,PH (1993)

CAS #: 013551-92-3 — 3-(2-NITROIMIDAZOL-1-YL)-1,2-PROPANEDIOL

Formula: $C_6H_9N_3O_4$

Mol Weight: 187.16

MP (deg C): FP (deg C):

BP (deg C):

BP pressure (mm Hg):

Property/Value	Units	Temp	Data Type	Reference
WS 1.37E+005	mg/L	25	EST	MEYLAN,WM ET AL. (1996)
logP -0.96			EXP	HANSCH,C & LEO,AJ (1985)
VP 8.52E-009	mm Hg	25	EST	NEELY,WB & BLAU,GE (1985)
DC	pKa			
HL 7.45E-013	atm m3/mol	25	EST	MEYLAN,WM & HOWARD,PH (1991)
OH 2.02E-011	cm3/molc sec	25	EST	MEYLAN,WM & HOWARD,PH (1993)

CAS #: 013552-35-7 — 3-PYRIDYLMETHYL-2-PIPERIDINE

Formula: $C_{11}H_{16}N_2$

Mol Weight: 176.26

MP (deg C): FP (deg C):

BP (deg C):

BP pressure (mm Hg):

Property/Value	Units	Temp	Data Type	Reference
WS 6.23E+005	mg/L	25	EST	MEYLAN,WM ET AL. (1996)
logP 1.49			EXP	HANSCH,C & LEO,AJ (1985)
VP 1.05E-003	mm Hg	25	EST	NEELY,WB & BLAU,GE (1985)
DC	pKa			
HL 2.41E-009	atm m3/mol	25	EST	MEYLAN,WM & HOWARD,PH (1991)
OH 1.07E-010	cm3/molc sec	25	EST	MEYLAN,WM & HOWARD,PH (1993)

CAS #: 013552-46-0 — 1-PROPANONE, 1-(4-METHYLPHENYL)-3-(1-PIPERIDINYL

Formula: $C_{15}H_{21}NO$

Mol Weight: 231.34

MP (deg C): FP (deg C):

BP (deg C):

BP pressure (mm Hg):

Property/Value	Units	Temp	Data Type	Reference
WS 2.83E+002	mg/L	25	EST	MEYLAN,WM ET AL. (1996)
logP 3.55			EXP	SANGSTER,J (1994)
VP 4.55E-005	mm Hg	25	EST	NEELY,WB & BLAU,GE (1985)
DC	pKa			
HL 7.35E-009	atm m3/mol	25	EST	MEYLAN,WM & HOWARD,PH (1991)
OH 1.31E-010	cm3/molc sec	25	EST	MEYLAN,WM & HOWARD,PH (1993)

CAS #: 013575-92-3 — 1-NAPHTHALENOL, 2-AMINO-1,2,3,4-TETRAHYDRO-, CIS

Formula: $C_{10}H_{13}NO$

Mol Weight: 163.22

MP (deg C): FP (deg C):

BP (deg C):

BP pressure (mm Hg):

Property/Value	Units	Temp	Data Type	Reference
WS 1.37E+005	mg/L	25	EST	MEYLAN,WM ET AL. (1996)
logP 0.86			EXP	SANGSTER,J (1994)
VP 4.78E-005	mm Hg	25	EST	NEELY,WB & BLAU,GE (1985)
DC	pKa			
HL 1.92E-011	atm m3/mol	25	EST	MEYLAN,WM & HOWARD,PH (1991)
OH 6.52E-011	cm3/molc sec	25	EST	MEYLAN,WM & HOWARD,PH (1993)

013584-27-5

CAS #: 013584-27-5

1,4-OXATHIIN-3-CARBOXAMIDE, 5,6-DIHYDRO-2-METHYL

Formula: $C_6H_9NO_2S$

Mol Weight: 159.21

MP (deg C):

FP (deg C):

BP (deg C):

BP pressure (mm Hg):

Property/ Value	Units	Temp	Data Type	Reference
WS 1.06E+005	mg/L	25	EST	MEYLAN,WM ET AL. (1996)
logP -0.22			EXP	HANSCH,C & LEO,AJ (1985)
VP 6.76E-005	mm Hg	25	EST	NEELY,WB & BLAU,GE (1985)
DC	pKa			
HL 1.35E-011	atm m3/mol	25	EST	MEYLAN,WM & HOWARD,PH (1991)
OH 8.64E-011	cm3/molc sec	25	EST	MEYLAN,WM & HOWARD,PH (1993)

013593-03-8

CAS #: 013593-03-8

QUINALPHOS

Formula: $C_{12}H_{15}N_2O_3PS$

Mol Weight: 298.30

MP (deg C):

FP (deg C):

BP (deg C):

BP pressure (mm Hg):

Property/ Value	Units	Temp	Data Type	Reference
WS 2.20E+001	mg/L	24	EXP	YALKOWSKY,SH & DANNENFELSER,RM (1992)
logP 3.04			EST	MEYLAN,WM & HOWARD,PH (1995)
VP 2.47E-006	mm Hg	25	EST	NEELY,WB & BLAU,GE (1985)
DC	pKa			
HL 3.97E-009	atm m3/mol	25	EST	MEYLAN,WM & HOWARD,PH (1991)
OH 9.53E-011	cm3/molc sec	25	EST	MEYLAN,WM & HOWARD,PH (1993)

013602-12-5

CAS #: 013602-12-5

ISONICOTINIC ACID N-OXIDE

Formula: $C_6H_5NO_3$

Mol Weight: 139.11

MP (deg C): 270-271

FP (deg C):

BP (deg C):

BP pressure (mm Hg):

Property/ Value	Units	Temp	Data Type	Reference
WS 1.10E+004	mg/L		EXP	GOE,GL (1982)
logP -1.45			EST	MEYLAN,WM & HOWARD,PH (1995)
VP 5.82E-005	mm Hg	25	EST	NEELY,WB & BLAU,GE (1985)
DC -0.48	pKa	24	EXP	PERRIN,DD (1965)
HL 1.42E-015	atm m3/mol	25	EST	MEYLAN,WM & HOWARD,PH (1991)
OH 6.57E-013	cm3/molc sec	25	EST	MEYLAN,WM & HOWARD,PH (1993)

013603-07-1

CAS #: 013603-07-1

3-METHYL-N-NITROSOPIPERIDINE

Formula: $C_6H_{12}N_2O$

Mol Weight: 128.18

MP (deg C):

FP (deg C):

BP (deg C):

BP pressure (mm Hg):

Property/ Value	Units	Temp	Data Type	Reference
WS 1.33E+004	mg/L	25	EST	MEYLAN,WM ET AL. (1996)
logP 0.99			EXP	HANSCH,C & LEO,AJ (1985)
VP 6.35E-002	mm Hg	25	EST	NEELY,WB & BLAU,GE (1985)
DC	pKa			
HL 2.82E-006	atm m3/mol	25	EST	MEYLAN,WM & HOWARD,PH (1991)
OH 2.73E-011	cm3/molc sec	25	EST	MEYLAN,WM & HOWARD,PH (1993)

013609-67-1

CAS #: 013609-67-1

HYDROCORTISONE-17-BUTYRATE

Formula: $C_{25}H_{36}O_6$

Mol Weight: 432.56

MP (deg C): 217-220

FP (deg C):

BP (deg C):

BP pressure (mm Hg):

Property/ Value	Units	Temp	Data Type	Reference
WS 3.69E+000	mg/L	25	EST	MEYLAN,WM ET AL. (1996)
logP 3.18			EXP	HANSCH,C & LEO,AJ (1985)
VP 3.43E-014	mm Hg	25	EST	NEELY,WB & BLAU,GE (1985)
DC	pKa			
HL 1.10E-011	atm m3/mol	25	EST	MEYLAN,WM & HOWARD,PH (1991)
OH 1.11E-010	cm3/molc sec	25	EST	MEYLAN,WM & HOWARD,PH (1993)

013641-62-8

CAS #: 013641-62-8

7-PHENYL-4,6-DIYN-HEPT-2-EN-1-OL

Formula: $C_{13}H_{10}O$

Mol Weight: 182.22

MP (deg C):

FP (deg C):

BP (deg C):

BP pressure (mm Hg):

Property/ Value	Units	Temp	Data Type	Reference
WS 1.33E+003	mg/L	25	EST	MEYLAN,WM ET AL. (1996)
logP 2.48			EXP	MCLACHLAN,D ET AL. (1986)
VP 2.35E-006	mm Hg	25	EST	NEELY,WB & BLAU,GE (1985)
DC	pKa			
HL 4.33E-008	atm m3/mol	25	EST	MEYLAN,WM & HOWARD,PH (1991)
OH 1.43E-010	cm3/molc sec	25	EST	MEYLAN,WM & HOWARD,PH (1993)

013654-09-6

CAS #: 013654-09-6

DECABROMOBIPHENYL

Formula: $C_{12}Br_{10}$

Mol Weight: 943.22

MP (deg C):

FP (deg C):

BP (deg C):

BP pressure (mm Hg):

Property/ Value	Units	Temp	Data Type	Reference
WS 1.25E-011	mg/L	25	EST	MEYLAN,WM ET AL. (1996)
logP 12.66			EST	MEYLAN,WM & HOWARD,PH (1995)
VP 3.98E-010	mm Hg	25	EXT	BURKHARD,LP ET AL. (1984)
DC	pKa			
HL 4.17E-008	atm m3/mol	25	EST	MEYLAN,WM & HOWARD,PH (1991)
OH 1.26E-014	cm3/molc sec	25	EST	MEYLAN,WM & HOWARD,PH (1993)

013655-52-2

CAS #: 013655-52-2

ALPRENOLOL

Formula: $C_{15}H_{23}NO_2$

Mol Weight: 249.36

MP (deg C): 107-109

FP (deg C):

BP (deg C):

BP pressure (mm Hg):

Property/ Value	Units	Temp	Data Type	Reference
WS 5.47E+002	mg/L	25	EST	MEYLAN,WM ET AL. (1996)
logP 3.10			EXP	HANSCH,C & LEO,AJ (1985)
VP 7.97E-007	mm Hg	25	EST	NEELY,WB & BLAU,GE (1985)
DC	pKa			
HL 1.18E-011	atm m3/mol	25	EST	MEYLAN,WM & HOWARD,PH (1991)
OH 1.63E-010	cm3/molc sec	25	EST	MEYLAN,WM & HOWARD,PH (1993)

CAS #: 013671-00-6 — BENZOIC ACID, 2,6-DIFLUORO-, METHYL ESTER

Formula: $C_8H_6F_2O_2$

Mol Weight: 172.13

MP (deg C): FP (deg C):

BP (deg C):

BP pressure (mm Hg):

Property/Value	Units	Temp	Data Type	Reference
WS 1.27E+003	mg/L	25	EST	MEYLAN,WM ET AL. (1996)
logP 1.96			EXP	SOTOMATSU,T ET AL. (1993)
VP 7.01E-001	mm Hg	25	EST	NEELY,WB & BLAU,GE (1985)
DC	pKa			
HL 4.73E-005	atm m3/mol	25	EST	MEYLAN,WM & HOWARD,PH (1991)
OH 1.20E-012	cm3/molc sec	25	EST	MEYLAN,WM & HOWARD,PH (1993)

CAS #: 013674-84-5 — 2-PROPANOL, 1-CHLORO-, PHOSPHATE (3:1)

Formula: $C_9H_{18}Cl_3O_4P$

Mol Weight: 327.57

MP (deg C): -40   FP (deg C):

BP (deg C): >270

BP pressure (mm Hg):

Property/Value	Units	Temp	Data Type	Reference
WS 1.20E+003	mg/L	25	EXP	CHEM INSPECT TEST INST (1992)
logP 2.59			EXP	CHEM INSPECT TEST INST (1992)
VP 2.02E-005	mm Hg	25	EST	NEELY,WB & BLAU,GE (1985)
DC	pKa			
HL 5.96E-008	atm m3/mol	25	EST	MEYLAN,WM & HOWARD,PH (1991)
OH 4.48E-011	cm3/molc sec	25	EST	MEYLAN,WM & HOWARD,PH (1993)

CAS #: 013674-87-8 — TRIS(1,3-DICHLOROISOPROPYL) PHOSPHATE

Formula: $C_9H_{15}Cl_6O_4P$

Mol Weight: 430.91

MP (deg C): FP (deg C):

BP (deg C): 236-237

BP pressure (mm Hg): 5.00E+000

Property/Value	Units	Temp	Data Type	Reference
WS 7.00E+000	mg/L	24	EXP	YALKOWSKY,SH & DANNENFELSER,RM (1992)
logP 3.65			EXP	CHEM INSPECT TEST INST (1992)
VP 7.36E-008	mm Hg	25	EST	NEELY,WB & BLAU,GE (1985)
DC	pKa			
HL 2.61E-009	atm m3/mol	25	EST	MEYLAN,WM & HOWARD,PH (1991)
OH 1.81E-011	cm3/molc sec	25	EST	MEYLAN,WM & HOWARD,PH (1993)

CAS #: 013675-92-8 — 6-NITROQUINOLINE-1-OXIDE

Formula: $C_9H_8N_2O_3$

Mol Weight: 192.18

MP (deg C): FP (deg C):

BP (deg C):

BP pressure (mm Hg):

Property/Value	Units	Temp	Data Type	Reference
WS 1.85E+004	mg/L	25	EST	MEYLAN,WM ET AL. (1996)
logP 0.39			EXP	HANSCH,C & LEO,AJ (1985)
VP 2.60E-006	mm Hg	25	EST	NEELY,WB & BLAU,GE (1985)
DC	pKa			
HL 2.72E-014	atm m3/mol	25	EST	MEYLAN,WM & HOWARD,PH (1991)
OH 1.80E-012	cm3/molc sec	25	EST	MEYLAN,WM & HOWARD,PH (1993)

CAS #: 013678-98-3 — HEPT-1,3-DIYN-5-ENYLBENZENE

Formula: $C_{13}H_{10}$

Mol Weight: 166.22

MP (deg C): FP (deg C):

BP (deg C):

BP pressure (mm Hg):

Property/Value	Units	Temp	Data Type	Reference
WS 5.52E+000	mg/L	25	EST	MEYLAN,WM ET AL. (1996)
logP 4.76			EXP	MCLACHLAN,D ET AL. (1986)
VP 1.71E-003	mm Hg	25	EST	NEELY,WB & BLAU,GE (1985)
DC	pKa			
HL 1.18E-003	atm m3/mol	25	EST	MEYLAN,WM & HOWARD,PH (1991)
OH 1.39E-010	cm3/molc sec	25	EST	MEYLAN,WM & HOWARD,PH (1993)

CAS #: 013684-56-5 — DESMEDIPHAM

Formula: $C_{16}H_{16}N_2O_4$

Mol Weight: 300.32

MP (deg C): 120   FP (deg C):

BP (deg C):

BP pressure (mm Hg):

Property/Value	Units	Temp	Data Type	Reference
WS 7.00E+000	mg/L	20	EXP	SHIU,WY ET AL. (1990)
logP 3.39			EXP	TOMLIN,C (1994)
VP 3.00E-009	mm Hg	25	EXP	WAUCHOPE,RD ET AL. (1991A)
DC	pKa			
HL 1.69E-010	atm m3/mol	25	EST	VP/WSOL
OH 8.83E-011	cm3/molc sec	25	EST	MEYLAN,WM & HOWARD,PH (1993)

CAS #: 013684-63-4 — PHENMEDIPHAM

Formula: $C_{16}H_{16}N_2O_4$

Mol Weight: 300.32

MP (deg C): 143-144   FP (deg C):

BP (deg C):

BP pressure (mm Hg):

Property/Value	Units	Temp	Data Type	Reference
WS 4.70E+000	mg/L	25	EXP	WAUCHOPE,RD ET AL. (1991A)
logP 3.59			EXP	TOMLIN,C (1994)
VP 1.00E-011	mm Hg	25	EXP	WAUCHOPE,RD ET AL. (1991A)
DC	pKa			
HL 8.41E-013	atm m3/mol	25	EST	VP/WSOL
OH 1.91E-010	cm3/molc sec	25	EST	MEYLAM,WM & HOWARD,PH (1993)

CAS #: 013698-16-3 — ETHYL DICHLOROCARBAMATE

Formula: $C_3H_5Cl_2NO_2$

Mol Weight: 157.98

MP (deg C): FP (deg C):

BP (deg C): 55-56

BP pressure (mm Hg): 1.50E+001

Property/Value	Units	Temp	Data Type	Reference
WS 1.12E+005	mg/L	25	EST	MEYLAN,WM ET AL. (1996)
logP -0.24			EST	MEYLAN,WM & HOWARD,PH (1995)
VP 4.45E+000	mm Hg	25	EST	NEELY,WB & BLAU,GE (1985)
DC	pKa			
HL 1.94E-005	atm m3/mol	25	EST	MEYLAN,WM & HOWARD,PH (1991)
OH 7.35E-012	cm3/molc sec	25	EST	MEYLAN,WM & HOWARD,PH (1993)

889

CAS #: 013700-81-7				1,2-DIPHENYL TETRACHLOROETHANE
Formula: $C_{14}H_{10}Cl_4$				
Mol Weight: 320.05				
MP (deg C):		FP (deg C):		
BP (deg C):				
BP pressure (mm Hg):				

Property/Value	Units	Temp	Data Type	Reference
WS 1.10E-002	mg/L	25	EST	MEYLAN,WM ET AL. (1996)
logP 6.94			EST	MEYLAN,WM & HOWARD,PH (1995)
VP 2.70E-005	mm Hg	25	EST	NEELY,WB & BLAU,GE (1985)
DC	pKa			
HL 9.81E-006	atm m3/mol	25	EST	MEYLAN,WM & HOWARD,PH (1991)
OH 8.59E-012	cm3/molc sec	25	EST	MEYLAN,WM & HOWARD,PH (1993)

CAS #: 013707-23-8				PIPERIDINE, 1-(4-METHYLBENZOYL)-
Formula: $C_{13}H_{17}NO$				
Mol Weight: 203.29				
MP (deg C):		FP (deg C):		
BP (deg C):				
BP pressure (mm Hg):				

Property/Value	Units	Temp	Data Type	Reference
WS 1.71E+002	mg/L	25	EST	MEYLAN,WM ET AL. (1996)
logP 2.64			EST	MEYLAN,WM & HOWARD,PH (1995)
VP 7.83E-005	mm Hg	25	EST	NEELY,WB & BLAU,GE (1985)
DC	pKa			
HL 1.22E-008	atm m3/mol	25	EST	MEYLAN,WM & HOWARD,PH (1991)
OH 3.34E-011	cm3/molc sec	25	EST	MEYLAN,WM & HOWARD,PH (1993)

CAS #: 013708-12-8				5-METHYLQUINOXALINE
Formula: $C_9H_8N_2$				
Mol Weight: 144.18				
MP (deg C):		FP (deg C):		
BP (deg C): 120				
BP pressure (mm Hg): 1.50E+001				

Property/Value	Units	Temp	Data Type	Reference
WS 1.46E+003	mg/L	25	EST	MEYLAN,WM ET AL. (1996)
logP 2.04			EXP	HANSCH,C & LEO,AJ (1985)
VP 5.10E-003	mm Hg	25	EST	NEELY,WB & BLAU,GE (1985)
DC	pKa			
HL 3.14E-007	atm m3/mol	25	EST	MEYLAN,WM & HOWARD,PH (1991)
OH 5.36E-012	cm3/molc sec	25	EST	MEYLAN,WM & HOWARD,PH (1993)

CAS #: 013710-19-5				TOLFENAMIC ACID
Formula: $C_{14}H_{12}ClNO_2$				
Mol Weight: 261.71				
MP (deg C): 207-207.5		FP (deg C):		
BP (deg C):				
BP pressure (mm Hg):				

Property/Value	Units	Temp	Data Type	Reference
WS 7.82E-001	mg/L	25	EST	MEYLAN,WM ET AL. (1996)
logP 5.17			EXP	HANSCH,C ET AL. (1995)
VP 2.59E-007	mm Hg	25	EST	NEELY,WB & BLAU,GE (1985)
DC	pKa			
HL 1.73E-011	atm m3/mol	25	EST	MEYLAN,WM & HOWARD,PH (1991)
OH 1.81E-010	cm3/molc sec	25	EST	MEYLAN,WM & HOWARD,PH (1993)

CAS #: 013725-30-9				CARBAMIC ACID, (2-NITROPHENYL)-, METHYL ESTER
Formula: $C_8H_8N_2O_4$				
Mol Weight: 196.16				
MP (deg C):		FP (deg C):		
BP (deg C):				
BP pressure (mm Hg):				

Property/Value	Units	Temp	Data Type	Reference
WS 7.37E+002	mg/L	25	EST	MEYLAN,WM ET AL. (1996)
logP 2.10			EXP	TAKAHASHI,J ET AL. (1988)
VP 2.34E-004	mm Hg	25	EST	NEELY,WB & BLAU,GE (1985)
DC	pKa			
HL 1.88E-009	atm m3/mol	25	EST	MEYLAN,WM & HOWARD,PH (1991)
OH 5.55E-012	cm3/molc sec	25	EST	MEYLAN,WM & HOWARD,PH (1993)

CAS #: 013744-79-1				2,4-DINITRO-4'-(TRIFLUOROMETHYL)DIPHENYLAMINE
Formula: $C_{13}H_8F_3N_3O_4$				
Mol Weight: 327.22				
MP (deg C):		FP (deg C):		
BP (deg C):				
BP pressure (mm Hg):				

Property/Value	Units	Temp	Data Type	Reference
WS 1.30E+000	mg/L	25	EST	MEYLAN,WM ET AL. (1996)
logP 4.47			EST	MEYLAN,WM & HOWARD,PH (1995)
VP 2.25E-007	mm Hg	25	EST	NEELY,WB & BLAU,GE (1985)
DC	pKa			
HL 3.11E-009	atm m3/mol	25	EST	MEYLAN,WM & HOWARD,PH (1991)
OH 3.17E-011	cm3/molc sec	25	EST	MEYLAN,WM & HOWARD,PH (1993)

CAS #: 013749-94-5				1-(MES)-ACETALDEHYDE OXIME
Formula: C_3H_7NOS				
Mol Weight: 105.16				
MP (deg C):		FP (deg C):		
BP (deg C):				
BP pressure (mm Hg):				

Property/Value	Units	Temp	Data Type	Reference
WS 1.25E+004	mg/L	25	EST	MEYLAN,WM ET AL. (1996)
logP 1.12			EXP	TSCATS
VP 9.28E-002	mm Hg	25	EST	NEELY,WB & BLAU,GE (1985)
DC	pKa			
HL 6.27E-008	atm m3/mol	25	EST	MEYLAN,WM & HOWARD,PH (1991)
OH 2.93E-012	cm3/molc sec	25	EST	MEYLAN,WM & HOWARD,PH (1993)

CAS #: 013780-91-1				3-ISOPROPYL-4-HYDROXYACETANILIDE
Formula: $C_{11}H_{15}NO_2$				
Mol Weight: 193.25				
MP (deg C):		FP (deg C):		
BP (deg C):				
BP pressure (mm Hg):				

Property/Value	Units	Temp	Data Type	Reference
WS 1.64E+003	mg/L	25	EST	MEYLAN,WM ET AL. (1996)
logP 1.71			EXP	HANSCH,C & LEO,AJ (1985)
VP 8.20E-007	mm Hg	25	EST	NEELY,WB & BLAU,GE (1985)
DC	pKa			
HL 1.25E-012	atm m3/mol	25	EST	MEYLAN,WM & HOWARD,PH (1991)
OH 2.35E-011	cm3/molc sec	25	EST	MEYLAN,WM & HOWARD,PH (1993)

CAS #:	013781-53-8			3-CYANOMETHYLTHIOPHENE

Formula: C_6H_7NS

Mol Weight: 125.19

MP (deg C): | FP (deg C):

BP (deg C): 124-125

BP pressure (mm Hg): 1.60E+001

Property/Value	Units	Temp	Data Type	Reference
WS 4.45E+003	mg/L	25	EST	MEYLAN,WM ET AL. (1996)
logP 1.26			EXP	HANSCH,C & LEO,AJ (1985)
VP 3.38E-002	mm Hg	25	EST	NEELY,WB & BLAU,GE (1985)
DC	pKa			
HL 1.34E-006	atm m3/mol	25	EST	MEYLAN,WM & HOWARD,PH (1991)
OH 9.42E-012	cm3/molc sec	25	EST	MEYLAN,WM & HOWARD,PH (1993)

CAS #:	013781-66-3			3-ACETAMIDOTHIOPHENE

Formula: C_6H_9NOS

Mol Weight: 143.21

MP (deg C): | FP (deg C):

BP (deg C):

BP pressure (mm Hg):

Property/Value	Units	Temp	Data Type	Reference
WS 4.50E+004	mg/L	25	EST	MEYLAN,WM ET AL. (1996)
logP 0.31			EXP	HANSCH,C & LEO,AJ (1985)
VP 2.35E-004	mm Hg	25	EST	NEELY,WB & BLAU,GE (1985)
DC	pKa			
HL 3.35E-009	atm m3/mol	25	EST	MEYLAN,WM & HOWARD,PH (1991)
OH 6.08E-011	cm3/molc sec	25	EST	MEYLAN,WM & HOWARD,PH (1993)

CAS #:	013797-63-2			4-PYRIDINE IMIDAZOLE, 2-CF3

Formula: $C_7H_4F_3N_3$

Mol Weight: 187.13

MP (deg C): | FP (deg C):

BP (deg C):

BP pressure (mm Hg):

Property/Value	Units	Temp	Data Type	Reference
WS 8.00E+003	mg/L	25	EST	MEYLAN,WM ET AL. (1996)
logP 1.38			EXP	HANSCH,C & LEO,AJ (1985)
VP 2.78E-005	mm Hg	25	EST	NEELY,WB & BLAU,GE (1985)
DC	pKa			
HL 4.17E-009	atm m3/mol	25	EST	MEYLAN,WM & HOWARD,PH (1991)
OH 7.24E-012	cm3/molc sec	25	EST	MEYLAN,WM & HOWARD,PH (1993)

CAS #:	013838-16-9			ETHRANE [HALOETHER]

Formula: $C_3H_2ClF_5O$

Mol Weight: 184.49

MP (deg C): | FP (deg C):

BP (deg C): 56.5

BP pressure (mm Hg):

Property/Value	Units	Temp	Data Type	Reference
WS 5.62E+003	mg/L	37	EXP	YALKOWSKY,SH & DANNENFELSER,RM (1992)
logP 2.10			EXP	HANSCH,C & LEO,AJ (1985)
VP 5.91E+002	mm Hg	25	EST	NEELY,WB & BLAU,GE (1985)
DC	pKa			
HL 1.26E-002	atm m3/mol	25	EST	MEYLAN,WM & HOWARD,PH (1991)
OH 1.70E-014	cm3/molc sec	25	EXP	ATKINSON,R (1989)

CAS #:	013858-89-4			BENZOXAZOLE-2-AMINE,N,N-DIMETHYL

Formula: $C_9H_{10}N_2O$

Mol Weight: 162.19

MP (deg C): | FP (deg C):

BP (deg C):

BP pressure (mm Hg):

Property/Value	Units	Temp	Data Type	Reference
WS 1.08E+003	mg/L	25	EST	MEYLAN,WM ET AL. (1996)
logP 2.10			EXP	HANSCH,C & LEO,AJ (1985)
VP 6.85E-003	mm Hg	25	EST	NEELY,WB & BLAU,GE (1985)
DC	pKa			
HL 1.09E-008	atm m3/mol	25	EST	MEYLAN,WM & HOWARD,PH (1991)
OH 2.03E-010	cm3/molc sec	25	EST	MEYLAN,WM & HOWARD,PH (1993)

CAS #:	013871-68-6			P-AMINOPHENYL ACETATE

Formula: $C_8H_9NO_2$

Mol Weight: 151.17

MP (deg C): | FP (deg C):

BP (deg C):

BP pressure (mm Hg):

Property/Value	Units	Temp	Data Type	Reference
WS 3.10E+004	mg/L	25	EST	MEYLAN,WM ET AL. (1996)
logP 0.45			EXP	HANSCH,C ET AL. (1995)
VP 6.86E-003	mm Hg	25	EST	NEELY,WB & BLAU,GE (1985)
DC	pKa			
HL 2.29E-008	atm m3/mol	25	EST	MEYLAN,WM & HOWARD,PH (1991)
OH 7.92E-011	cm3/molc sec	25	EST	MEYLAN,WM & HOWARD,PH (1993)

CAS #:	013877-91-3			3,7-DIMETHYL-1,3,6-OCTATRIENE

Formula: $C_{10}H_{16}$

Mol Weight: 136.24

MP (deg C): | FP (deg C):

BP (deg C): 73

BP pressure (mm Hg): 2.10E+001

Property/Value	Units	Temp	Data Type	Reference
WS 2.01E+000	mg/L	25	EST	MEYLAN,WM ET AL. (1996)
logP 4.80			EST	MEYLAN,WM & HOWARD,PH (1995)
VP 2.68E+000	mm Hg	25	EST	NEELY,WB & BLAU,GE (1985)
DC	pKa			
HL 6.17E-001	atm m3/mol	25	EST	MEYLAN,WM & HOWARD,PH (1991)
OH 2.52E-010	cm3/molc sec	25	EXP	ATKINSON,R (1989)

CAS #:	013907-57-8			UREA, N,N''-1,4-CYCLOHEXANEDIYLBIS[N'-(2-CHLOROET

Formula: $C_{12}H_{20}Cl_2N_6O_4$

Mol Weight: 383.24

MP (deg C): | FP (deg C):

BP (deg C):

BP pressure (mm Hg):

Property/Value	Units	Temp	Data Type	Reference
WS 1.78E+001	mg/L	25	EST	MEYLAN,WM ET AL. (1996)
logP 2.74			EXP	HANSCH,C ET AL. (1995)
VP 3.86E-011	mm Hg	25	EST	NEELY,WB & BLAU,GE (1985)
DC	pKa			
HL 1.61E-022	atm m3/mol	25	EST	MEYLAN,WM & HOWARD,PH (1991)
OH 3.53E-011	cm3/molc sec	25	EST	MEYLAN,WM & HOWARD,PH (1993)

891

CAS #: 013907-78-3				UREA, N-(2-CHLOROETHYL)-N'-(4-CHLOROPHENYL)-N-NI
Formula: $C_9H_9Cl_2N_3O_2$				
Mol Weight: 262.10				
MP (deg C):		FP (deg C):		
BP (deg C):				
BP pressure (mm Hg):				

Property/ Value	Units	Temp	Data Type	Reference
WS 4.05E+001	mg/L	25	EST	MEYLAN,WM ET AL. (1996)
logP 3.16			EXP	HANSCH,C ET AL. (1995)
VP 1.12E-006	mm Hg	25	EST	NEELY,WB & BLAU,GE (1985)
DC	pKa			
HL 1.89E-011	atm m3/mol	25	EST	MEYLAN,WM & HOWARD,PH (1991)
OH 1.43E-011	cm3/molc sec	25	EST	MEYLAN,WM & HOWARD,PH (1993)

CAS #: 013908-92-4				3(2-CLET)-1-(2FET)-1-NITROSOUREA
Formula: $C_5H_9ClFN_3O_2$				
Mol Weight: 197.60				
MP (deg C):		FP (deg C):		
BP (deg C):				
BP pressure (mm Hg):				

Property/ Value	Units	Temp	Data Type	Reference
WS 6.81E+003	mg/L	25	EST	MEYLAN,WM ET AL. (1996)
logP 0.96			EXP	HANSCH,C & LEO,AJ (1985)
VP 4.25E-004	mm Hg	25	EST	NEELY,WB & BLAU,GE (1985)
DC	pKa			
HL 2.69E-010	atm m3/mol	25	EST	MEYLAN,WM & HOWARD,PH (1991)
OH 4.26E-012	cm3/molc sec	25	EST	MEYLAN,WM & HOWARD,PH (1993)

CAS #: 013908-98-0				CYCLOHEXANE,1,4-DI(1-NO-1ETF UREA)
Formula: $C_{12}H_{20}F_2N_6O_4$				
Mol Weight: 350.33				
MP (deg C):		FP (deg C):		
BP (deg C):				
BP pressure (mm Hg):				

Property/ Value	Units	Temp	Data Type	Reference
WS 4.42E+002	mg/L	25	EST	MEYLAN,WM ET AL. (1996)
logP 1.34			EXP	HANSCH,C & LEO,AJ (1985)
VP 2.71E-009	mm Hg	25	EST	NEELY,WB & BLAU,GE (1985)
DC	pKa			
HL 5.14E-021	atm m3/mol	25	EST	MEYLAN,WM & HOWARD,PH (1991)
OH 3.46E-011	cm3/molc sec	25	EST	MEYLAN,WM & HOWARD,PH (1993)

CAS #: 013909-02-9				1(2-CLET)-3(2-GLUTARIMIDYL)-1-NO-UREA
Formula: $C_8H_{11}ClN_4O_4$				
Mol Weight: 262.65				
MP (deg C):		FP (deg C):		
BP (deg C):				
BP pressure (mm Hg):				

Property/ Value	Units	Temp	Data Type	Reference
WS 9.71E+003	mg/L	25	EST	MEYLAN,WM ET AL. (1996)
logP 0.37			EXP	HANSCH,C & LEO,AJ (1985)
VP 6.73E-012	mm Hg	25	EST	NEELY,WB & BLAU,GE (1985)
DC	pKa			
HL 5.85E-019	atm m3/mol	25	EST	MEYLAN,WM & HOWARD,PH (1991)
OH 3.07E-011	cm3/molc sec	25	EST	MEYLAN,WM & HOWARD,PH (1993)

CAS #: 013909-03-0				3-CYPE-1-(2-CLET)-1-NITROSOUREA
Formula: $C_8H_{14}ClN_3O_2$				
Mol Weight: 219.67				
MP (deg C):		FP (deg C):		
BP (deg C):				
BP pressure (mm Hg):				

Property/ Value	Units	Temp	Data Type	Reference
WS 4.57E+002	mg/L	25	EST	MEYLAN,WM ET AL. (1996)
logP 2.20			EXP	HANSCH,C & LEO,AJ (1985)
VP 2.44E-005	mm Hg	25	EST	NEELY,WB & BLAU,GE (1985)
DC	pKa			
HL 1.39E-010	atm m3/mol	25	EST	MEYLAN,WM & HOWARD,PH (1991)
OH 1.70E-011	cm3/molc sec	25	EST	MEYLAN,WM & HOWARD,PH (1993)

CAS #: 013909-09-6				1-(2-CLET)-3(4-MECYHX)-1-N=O UREA
Formula: $C_{10}H_{18}ClN_3O_2$				
Mol Weight: 247.73				
MP (deg C):		FP (deg C):		
BP (deg C):				
BP pressure (mm Hg):				

Property/ Value	Units	Temp	Data Type	Reference
WS 3.70E+001	mg/L	25	EST	MEYLAN,WM ET AL. (1996)
logP 3.30			EXP	HANSCH,C & LEO,AJ (1985)
VP 5.61E-006	mm Hg	25	EST	NEELY,WB & BLAU,GE (1985)
DC	pKa			
HL 2.46E-010	atm m3/mol	25	EST	MEYLAN,WM & HOWARD,PH (1991)
OH 2.36E-011	cm3/molc sec	25	EST	MEYLAN,WM & HOWARD,PH (1993)

CAS #: 013909-11-0				1-(2-CLET)-3-(2-CLCYHX)1-N=O UREA
Formula: $C_9H_{15}Cl_2N_3O_2$				
Mol Weight: 268.14				
MP (deg C):		FP (deg C):		
BP (deg C):				
BP pressure (mm Hg):				

Property/ Value	Units	Temp	Data Type	Reference
WS 8.73E+001	mg/L	25	EST	MEYLAN,WM ET AL. (1996)
logP 2.73			EXP	HANSCH,C & LEO,AJ (1985)
VP 2.12E-006	mm Hg	25	EST	NEELY,WB & BLAU,GE (1985)
DC	pKa			
HL 6.52E-011	atm m3/mol	25	EST	MEYLAN,WM & HOWARD,PH (1991)
OH 1.11E-011	cm3/molc sec	25	EST	MEYLAN,WM & HOWARD,PH (1993)

CAS #: 013909-12-1				UREA, N'-(2-CHLOROCYCLOHEXYL)-N-(2-CHLOROETHYL)-
Formula: $C_9H_{15}Cl_2N_3O_2$				
Mol Weight: 268.14				
MP (deg C):		FP (deg C):		
BP (deg C):				
BP pressure (mm Hg):				

Property/ Value	Units	Temp	Data Type	Reference
WS 8.73E+001	mg/L	25	EST	MEYLAN,WM ET AL. (1996)
logP 2.73			EXP	SANGSTER,J (1994)
VP 2.12E-006	mm Hg	25	EST	NEELY,WB & BLAU,GE (1985)
DC	pKa			
HL 6.52E-011	atm m3/mol	25	EST	MEYLAN,WM & HOWARD,PH (1991)
OH 1.11E-011	cm3/molc sec	25	EST	MEYLAN,WM & HOWARD,PH (1993)

CAS #: 013909-13-2 — 3(2-NORBORNYL)-1-(2CLET)-1-NO UREA

Formula: $C_{10}H_{16}ClN_3O_2$

Mol Weight: 245.71

MP (deg C): FP (deg C):

BP (deg C):

BP pressure (mm Hg):

Property/Value	Units	Temp	Data Type	Reference
WS 7.12E+001	mg/L	25	EST	MEYLAN,WM ET AL. (1996)
logP 2.98			EXP	HANSCH,C & LEO,AJ (1985)
VP 5.00E-006	mm Hg	25	EST	NEELY,WB & BLAU,GE (1985)
DC	pKa			
HL 1.08E-010	atm m3/mol	25	EST	MEYLAN,WM & HOWARD,PH (1991)
OH 1.88E-011	cm3/molc sec	25	EST	MEYLAN,WM & HOWARD,PH (1993)

CAS #: 013909-14-3 — 1(2CLET)-1-(NO)-3-CYCLODODECYLUREA

Formula: $C_{15}H_{28}ClN_3O_2$

Mol Weight: 317.86

MP (deg C): FP (deg C):

BP (deg C):

BP pressure (mm Hg):

Property/Value	Units	Temp	Data Type	Reference
WS 1.47E-001	mg/L	25	EST	MEYLAN,WM ET AL. (1996)
logP 5.64			EXP	HANSCH,C & LEO,AJ (1985)
VP 4.60E-008	mm Hg	25	EST	NEELY,WB & BLAU,GE (1985)
DC				
HL 1.01E-009	atm m3/mol	25	EST	MEYLAN,WM & HOWARD,PH (1991)
OH 3.04E-011	cm3/molc sec	25	EST	MEYLAN,WM & HOWARD,PH (1993)

CAS #: 013909-21-2 — 1-(2-CLET)-3-(M-MEOPH)-1-N=O UREA

Formula: $C_{10}H_{12}ClN_3O_3$

Mol Weight: 257.68

MP (deg C): FP (deg C):

BP (deg C):

BP pressure (mm Hg):

Property/Value	Units	Temp	Data Type	Reference
WS 1.15E+002	mg/L	25	EST	MEYLAN,WM ET AL. (1996)
logP 2.66			EXP	HANSCH,C & LEO,AJ (1985)
VP 7.88E-007	mm Hg	25	EST	NEELY,WB & BLAU,GE (1985)
DC	pKa			
HL 1.51E-012	atm m3/mol	25	EST	MEYLAN,WM & HOWARD,PH (1991)
OH 2.03E-010	cm3/molc sec	25	EST	MEYLAN,WM & HOWARD,PH (1993)

CAS #: 013909-63-2 — BENZENESULFONAMIDE, 4-METHYL-N-[(PHENYLAMINO)CAR

Formula: $C_{14}H_{14}N_2O_3S$

Mol Weight: 290.34

MP (deg C): FP (deg C):

BP (deg C):

BP pressure (mm Hg):

Property/Value	Units	Temp	Data Type	Reference
WS 3.09E+001	mg/L	25	EST	MEYLAN,WM ET AL. (1996)
logP 3.11			EXP	SANGSTER,J (1993)
VP 1.97E-009	mm Hg	25	EST	NEELY,WB & BLAU,GE (1985)
DC	pKa			
HL 2.72E-011	atm m3/mol	25	EST	MEYLAN,WM & HOWARD,PH (1991)
OH 4.49E-011	cm3/molc sec	25	EST	MEYLAN,WM & HOWARD,PH (1993)

CAS #: 013909-64-3 — BENZENESULFONAMIDE, N-[(BUTYLAMINO)CARBONYL]-4-C

Formula: $C_{11}H_{15}ClN_2O_3S$

Mol Weight: 290.77

MP (deg C): FP (deg C):

BP (deg C):

BP pressure (mm Hg):

Property/Value	Units	Temp	Data Type	Reference
WS 5.53E+001	mg/L	25	EST	MEYLAN,WM ET AL. (1996)
logP 2.81			EXP	HANSCH,C ET AL. (1995)
VP 1.67E-008	mm Hg	25	EST	NEELY,WB & BLAU,GE (1985)
DC	pKa			
HL 1.70E-010	atm m3/mol	25	EST	MEYLAN,WM & HOWARD,PH (1991)
OH 9.73E-012	cm3/molc sec	25	EST	MEYLAN,WM & HOWARD,PH (1993)

CAS #: 013909-69-8 — UREA, N-METHYL-N'-[(4-METHYLPHENYL)SULFONYL]-

Formula: $C_9H_{12}N_2O_3S$

Mol Weight: 228.27

MP (deg C): FP (deg C):

BP (deg C):

BP pressure (mm Hg):

Property/Value	Units	Temp	Data Type	Reference
WS 4.80E+003	mg/L	25	EST	MEYLAN,WM ET AL. (1996)
logP 0.95			EXP	SANGSTER,J (1993)
VP 3.49E-007	mm Hg	25	EST	NEELY,WB & BLAU,GE (1985)
DC	pKa			
HL 1.08E-010	atm m3/mol	25	EST	MEYLAN,WM & HOWARD,PH (1991)
OH 3.78E-012	cm3/molc sec	25	EST	MEYLAN,WM & HOWARD,PH (1993)

CAS #: 013909-73-4 — 2,3,4-TRIMETHOXYACETOPHENONE

Formula: $C_{11}H_{14}O_4$

Mol Weight: 210.23

MP (deg C): 15.8 FP (deg C):

BP (deg C): 296

BP pressure (mm Hg):

Property/Value	Units	Temp	Data Type	Reference
WS 1.57E+003	mg/L	25	EST	MEYLAN,WM ET AL. (1996)
logP 1.63			EXP	HANSCH,C ET AL. (1995)
VP 6.61E-004	mm Hg	25	EST	NEELY,WB & BLAU,GE (1985)
DC	pKa			
HL 2.03E-009	atm m3/mol	25	EST	MEYLAN,WM & HOWARD,PH (1991)
OH 1.90E-010	cm3/molc sec	25	EST	MEYLAN,WM & HOWARD,PH (1993)

CAS #: 013917-17-4 — 1-NAPHTHALENOL, 2-AMINO-1,2,3,4-TETRAHYDRO-, TRA

Formula: $C_{10}H_{13}NO$

Mol Weight: 163.22

MP (deg C): FP (deg C):

BP (deg C):

BP pressure (mm Hg):

Property/Value	Units	Temp	Data Type	Reference
WS 1.37E+005	mg/L	25	EST	MEYLAN,WM ET AL. (1996)
logP 0.81			EXP	SANGSTER,J (1994)
VP 4.78E-005	mm Hg	25	EST	NEELY,WB & BLAU,GE (1985)
DC	pKa			
HL 1.92E-011	atm m3/mol	25	EST	MEYLAN,WM & HOWARD,PH (1991)
OH 6.52E-011	cm3/molc sec	25	EST	MEYLAN,WM & HOWARD,PH (1993)

CAS #: 013925-00-3				2-ETHYL PYRAZINE

Formula: $C_6H_8N_2$
Mol Weight: 108.14
MP (deg C): | **FP (deg C):**
BP (deg C): 152-153
BP pressure (mm Hg):

Property/Value	Units	Temp	Data Type	Reference
WS 2.84E+004	mg/L	25	EST	MEYLAN,WM ET AL. (1996)
logP 0.69			EXP	YAMAGAMI,C ET AL. (1990A)
VP 1.67E+000	mm Hg	25	EST	NEELY,WB & BLAU,GE (1985)
DC	pKa			
HL 2.45E-006	atm m3/mol	25	EXP	BUTTERY,RG ET AL. (1971)
OH 1.60E-012	cm3/molc sec	25	EST	MEYLAN,WM & HOWARD,PH (1993)

CAS #: 013932-40-6				2,4-NH2PYRIMIDINE,5-(3,4OCH2O)BENZYL

Formula: $C_{12}H_{12}N_4O_2$
Mol Weight: 244.26
MP (deg C): | **FP (deg C):**
BP (deg C):
BP pressure (mm Hg):

Property/Value	Units	Temp	Data Type	Reference
WS 1.31E+003	mg/L	25	EST	MEYLAN,WM ET AL. (1996)
logP 1.51			EXP	HANSCH,C & LEO,AJ (1985)
VP 5.42E-008	mm Hg	25	EST	NEELY,WB & BLAU,GE (1985)
DC	pKa			
HL 1.34E-013	atm m3/mol	25	EST	MEYLAN,WM & HOWARD,PH (1991)
OH 5.14E-011	cm3/molc sec	25	EST	MEYLAN,WM & HOWARD,PH (1993)

CAS #: 013935-94-9				1H-INDENE-1,3(2H)-DIONE, 2-[3-(1-METHYLETHYL)PHE

Formula: $C_{18}H_{16}O_2$
Mol Weight: 264.33
MP (deg C): | **FP (deg C):**
BP (deg C):
BP pressure (mm Hg):

Property/Value	Units	Temp	Data Type	Reference
WS 3.24E+000	mg/L	25	EST	MEYLAN,WM ET AL. (1996)
logP 4.43			EXP	SANGSTER,J (1994)
VP 4.34E-007	mm Hg	25	EST	NEELY,WB & BLAU,GE (1985)
DC	pKa			
HL 9.31E-010	atm m3/mol	25	EST	MEYLAN,WM & HOWARD,PH (1991)
OH 1.49E-011	cm3/molc sec	25	EST	MEYLAN,WM & HOWARD,PH (1993)

CAS #: 013935-95-0				1H-INDENE-1,3(2H)-DIONE, 2-[3-(1,1-DIMETHYLETHYL

Formula: $C_{19}H_{18}O_2$
Mol Weight: 278.35
MP (deg C): | **FP (deg C):**
BP (deg C):
BP pressure (mm Hg):

Property/Value	Units	Temp	Data Type	Reference
WS 1.11E+000	mg/L	25	EST	MEYLAN,WM ET AL. (1996)
logP 4.88			EXP	SANGSTER,J (1993)
VP 2.70E-007	mm Hg	25	EST	NEELY,WB & BLAU,GE (1985)
DC	pKa			
HL 1.24E-009	atm m3/mol	25	EST	MEYLAN,WM & HOWARD,PH (1991)
OH 1.23E-011	cm3/molc sec	25	EST	MEYLAN,WM & HOWARD,PH (1993)

CAS #: 013935-96-1				1H-INDENE-1,3(2H)-DIONE, 2-[4-(1,1-DIMETHYLETHYL

Formula: $C_{19}H_{18}O_2$
Mol Weight: 278.35
MP (deg C): | **FP (deg C):**
BP (deg C):
BP pressure (mm Hg):

Property/Value	Units	Temp	Data Type	Reference
WS 1.11E+000	mg/L	25	EST	MEYLAN,WM ET AL. (1996)
logP 4.88			EXP	SANGSTER,J (1993)
VP 2.70E-007	mm Hg	25	EST	NEELY,WB & BLAU,GE (1985)
DC	pKa			
HL 1.24E-009	atm m3/mol	25	EST	MEYLAN,WM & HOWARD,PH (1991)
OH 7.67E-012	cm3/molc sec	25	EST	MEYLAN,WM & HOWARD,PH (1993)

CAS #: 013936-01-1				1H-INDENE-1,3(2H)-DIONE, 2-[3,5-BIS(1,1-DIMETHYL

Formula: $C_{23}H_{26}O_2$
Mol Weight: 334.46
MP (deg C): | **FP (deg C):**
BP (deg C):
BP pressure (mm Hg):

Property/Value	Units	Temp	Data Type	Reference
WS 1.06E-002	mg/L	25	EST	MEYLAN,WM ET AL. (1996)
logP 6.86			EXP	SANGSTER,J (1993)
VP 1.79E-008	mm Hg	25	EST	NEELY,WB & BLAU,GE (1985)
DC	pKa			
HL 3.19E-009	atm m3/mol	25	EST	MEYLAN,WM & HOWARD,PH (1991)
OH 2.51E-011	cm3/molc sec	25	EST	MEYLAN,WM & HOWARD,PH (1993)

CAS #: 013941-09-8				N-METHYL-3-IODOPHENYLCARBAMATE

Formula: $C_8H_8INO_2$
Mol Weight: 277.06
MP (deg C): | **FP (deg C):**
BP (deg C):
BP pressure (mm Hg):

Property/Value	Units	Temp	Data Type	Reference
WS 1.17E+002	mg/L	25	EST	MEYLAN,WM ET AL. (1996)
logP 2.52			EXP	HANSCH,C & LEO,AJ (1985)
VP 5.32E-004	mm Hg	25	EST	NEELY,WB & BLAU,GE (1985)
DC	pKa			
HL 7.45E-009	atm m3/mol	25	EST	MEYLAN,WM & HOWARD,PH (1991)
OH 7.38E-012	cm3/molc sec	25	EST	MEYLAN,WM & HOWARD,PH (1993)

CAS #: 013952-84-6				SEC-BUTYLAMINE

Formula: $C_4H_{11}N$
Mol Weight: 73.14
MP (deg C): -104 | **FP (deg C):**
BP (deg C): 63
BP pressure (mm Hg):

Property/Value	Units	Temp	Data Type	Reference
WS 1.12E+005	mg/L	20	EXP	YALKOWSKY,SH & DANNENFELSER,RM (1992)
logP 0.76			EST	MEYLAN,WM & HOWARD,PH (1995)
VP 1.78E+002	mm Hg	25	EXP	DAUBERT,TE & DANNER,RP (1989)
DC 10.56	pKa		EXP	PERRIN,DD (1965)
HL 1.53E-004	atm m3/mol	25	EST	VP/WSOL
OH 4.47E-011	cm3/molc sec	25	EST	MEYLAN,WM & HOWARD,PH (1993)

CAS #: 013961-64-3	2(1H)-QUINAZOLINONE, 6-METHYL-4-PHENYL-

Formula: $C_{15}H_{12}N_2O$
Mol Weight: 236.28
MP (deg C):   FP (deg C):
BP (deg C):
BP pressure (mm Hg):

Property/Value	Units	Temp	Data Type	Reference
WS 2.94E+002	mg/L	25	EST	MEYLAN,WM ET AL. (1996)
logP 2.32			EXP	SANGSTER,J (1993)
VP 1.16E-007	mm Hg	25	EST	NEELY,WB & BLAU,GE (1985)
DC	pKa			
HL 4.21E-009	atm m3/mol	25	EST	MEYLAN,WM & HOWARD,PH (1991)
OH 6.72E-011	cm3/molc sec	25	EST	MEYLAN,WM & HOWARD,PH (1993)

CAS #: 013993-02-7	N-(O-TOLYL)-3-N'-PIPERIDINOACETAMIDE

Formula: $C_{14}H_{20}N_2O$
Mol Weight: 232.33
MP (deg C):   FP (deg C):
BP (deg C):
BP pressure (mm Hg):

Property/Value	Units	Temp	Data Type	Reference
WS 3.47E+002	mg/L	25	EST	MEYLAN,WM ET AL. (1996)
logP 2.26			EXP	HANSCH,C & LEO,AJ (1985)
VP 4.23E-007	mm Hg	25	EST	NEELY,WB & BLAU,GE (1985)
DC	pKa			
HL 6.95E-011	atm m3/mol	25	EST	MEYLAN,WM & HOWARD,PH (1991)
OH 1.13E-010	cm3/molc sec	25	EST	MEYLAN,WM & HOWARD,PH (1993)

CAS #: 014001-64-0	2-MES-4,6-DIMETHYLPYRIMIDINE

Formula: $C_7H_{10}N_2S$
Mol Weight: 154.24
MP (deg C):   FP (deg C):
BP (deg C):
BP pressure (mm Hg):

Property/Value	Units	Temp	Data Type	Reference
WS 1.88E+003	mg/L	25	EST	MEYLAN,WM ET AL. (1996)
logP 1.86			EXP	HANSCH,C & LEO,AJ (1985)
VP 1.99E-002	mm Hg	25	EST	NEELY,WB & BLAU,GE (1985)
DC 2.12	pKa	20	EXP	PERRIN,DD (1972)
HL 4.03E-006	atm m3/mol	25	EST	MEYLAN,WM & HOWARD,PH (1991)
OH 1.14E-011	cm3/molc sec	25	EST	MEYLAN,WM & HOWARD,PH (1993)

CAS #: 014003-66-8	4-METHYL-5-NITRO-1H-IMIDAZOLE

Formula: $C_4H_5N_3O_2$
Mol Weight: 127.10
MP (deg C):   FP (deg C):
BP (deg C):
BP pressure (mm Hg):

Property/Value	Units	Temp	Data Type	Reference
WS 2.45E+004	mg/L	25	EST	MEYLAN,WM ET AL. (1996)
logP 0.42			EST	MEYLAN,WM & HOWARD,PH (1995)
VP 6.48E-005	mm Hg	25	EST	NEELY,WB & BLAU,GE (1985)
DC	pKa			
HL 1.64E-008	atm m3/mol	25	EST	MEYLAN,WM & HOWARD,PH (1991)
OH 5.65E-012	cm3/molc sec	25	EST	MEYLAN,WM & HOWARD,PH (1993)

CAS #: 014005-50-6	2-AMINO-4-PHENYLQUINAZOLINE

Formula: $C_{14}H_{11}N_3$
Mol Weight: 221.26
MP (deg C):   FP (deg C):
BP (deg C):
BP pressure (mm Hg):

Property/Value	Units	Temp	Data Type	Reference
WS 1.03E+002	mg/L	25	EST	MEYLAN,WM ET AL. (1996)
logP 2.95			EXP	HANSCH,C & LEO,AJ (1985)
VP 1.94E-007	mm Hg	25	EST	NEELY,WB & BLAU,GE (1985)
DC	pKa			
HL 2.74E-008	atm m3/mol	25	EST	MEYLAN,WM & HOWARD,PH (1991)
OH 2.46E-011	cm3/molc sec	25	EST	MEYLAN,WM & HOWARD,PH (1993)

CAS #: 014008-60-7	BENZAMIDE, 2-HYDROXY-3-METHYL-

Formula: $C_8H_9NO_2$
Mol Weight: 151.17
MP (deg C):   FP (deg C):
BP (deg C):
BP pressure (mm Hg):

Property/Value	Units	Temp	Data Type	Reference
WS 4.41E+003	mg/L	25	EST	MEYLAN,WM ET AL. (1996)
logP 2.12			EXP	SANGSTER,J (1994)
VP 4.56E-006	mm Hg	25	EST	NEELY,WB & BLAU,GE (1985)
DC	pKa			
HL 3.20E-010	atm m3/mol	25	EST	MEYLAN,WM & HOWARD,PH (1991)
OH 3.95E-011	cm3/molc sec	25	EST	MEYLAN,WM & HOWARD,PH (1993)

CAS #: 014039-10-2	3(ADAMANTYL)-1(2CLET)-1-NO UREA

Formula: $C_{13}H_{20}ClN_3O_2$
Mol Weight: 285.78
MP (deg C):   FP (deg C):
BP (deg C):
BP pressure (mm Hg):

Property/Value	Units	Temp	Data Type	Reference
WS 1.35E+001	mg/L	25	EST	MEYLAN,WM ET AL. (1996)
logP 3.56			EXP	HANSCH,C & LEO,AJ (1985)
VP 4.80E-007	mm Hg	25	EST	NEELY,WB & BLAU,GE (1985)
DC	pKa			
HL 1.12E-010	atm m3/mol	25	EST	MEYLAN,WM & HOWARD,PH (1991)
OH 2.21E-011	cm3/molc sec	25	EST	MEYLAN,WM & HOWARD,PH (1993)

CAS #: 014047-09-7	3,3',4,4'-TETRACHLOROAZOBENZENE

Formula: $C_{12}H_6Cl_4N_2$
Mol Weight: 320.01
MP (deg C):   FP (deg C):
BP (deg C):
BP pressure (mm Hg):

Property/Value	Units	Temp	Data Type	Reference
WS 6.72E-003	mg/L	25	EST	MEYLAN,WM ET AL. (1996)
logP 6.69			EST	MEYLAN,WM & HOWARD,PH (1995)
VP 3.01E-006	mm Hg	25	EST	NEELY,WB & BLAU,GE (1985)
DC	pKa			
HL 4.43E-006	atm m3/mol	25	EST	MEYLAN,WM & HOWARD,PH (1991)
OH 3.17E-013	cm3/molc sec	25	EST	MEYLAN,WM & HOWARD,PH (1993)

CAS #:	014061-27-9			N-METHYL-3-TRIFLUOROMEPHCARBAMATE

Formula: $C_9H_8F_3NO_2$

Mol Weight: 219.16

MP (deg C): | FP (deg C):

BP (deg C):

BP pressure (mm Hg):

Property/Value	Units	Temp	Data Type	Reference
WS 3.29E+002	mg/L	25	EST	MEYLAN,WM ET AL. (1996)
logP 2.37			EXP	HANSCH,C & LEO,AJ (1985)
VP 3.69E-002	mm Hg	25	EST	NEELY,WB & BLAU,GE (1985)
DC	pKa			
HL 2.79E-007	atm m3/mol	25	EST	MEYLAN,WM & HOWARD,PH (1991)
OH 8.62E-013	cm3/molc sec	25	EST	MEYLAN,WM & HOWARD,PH (1993)

CAS #:	014062-34-1			3-AMINOBENZOYLHYDRAZINE

Formula: $C_7H_9N_3O$

Mol Weight: 151.17

MP (deg C): | FP (deg C):

BP (deg C):

BP pressure (mm Hg):

Property/Value	Units	Temp	Data Type	Reference
WS 1.99E+004	mg/L	25	EST	MEYLAN,WM ET AL. (1996)
logP -0.86			EXP	HANSCH,C & LEO,AJ (1985)
VP 3.14E-006	mm Hg	25	EST	NEELY,WB & BLAU,GE (1985)
DC	pKa			
HL 3.27E-015	atm m3/mol	25	EST	MEYLAN,WM & HOWARD,PH (1991)
OH 5.09E-011	cm3/molc sec	25	EST	MEYLAN,WM & HOWARD,PH (1993)

CAS #:	014062-61-4			3-MEO PHENYL GLUCOPYRANOSIDE

Formula: $C_{13}H_{18}O_7$

Mol Weight: 286.28

MP (deg C): | FP (deg C):

BP (deg C):

BP pressure (mm Hg):

Property/Value	Units	Temp	Data Type	Reference
WS 4.10E+004	mg/L	25	EST	MEYLAN,WM ET AL. (1996)
logP -0.52			EXP	HANSCH,C & LEO,AJ (1985)
VP 7.72E-012	mm Hg	25	EST	NEELY,WB & BLAU,GE (1985)
DC	pKa			
HL 6.80E-017	atm m3/mol	25	EST	MEYLAN,WM & HOWARD,PH (1991)
OH 2.63E-010	cm3/molc sec	25	EST	MEYLAN,WM & HOWARD,PH (1993)

CAS #:	014062-78-3			BENZAMIDE, N,N,4-TRIMETHYL-

Formula: $C_{10}H_{13}NO$

Mol Weight: 163.22

MP (deg C): 41 | FP (deg C):

BP (deg C): 156

BP pressure (mm Hg): 1.00E+001

Property/Value	Units	Temp	Data Type	Reference
WS 2.25E+003	mg/L	25	EST	MEYLAN,WM ET AL. (1996)
logP 1.72			EXP	SURYANARAYANA,MVS ET AL. (1991)
VP 1.10E-001	mm Hg	25	EXP	SURYANARAYANA,MVS ET AL. (1991)
DC	pKa			
HL 1.18E-008	atm m3/mol	25	EST	MEYLAN,WM & HOWARD,PH (1991)
OH 2.09E-011	cm3/molc sec	25	EST	MEYLAN,WM & HOWARD,PH (1993)

CAS #:	014073-00-8			3-METHYL-4-NITROQUINOLINE-1-OXIDE

Formula: $C_{10}H_{10}N_2O_3$

Mol Weight: 206.20

MP (deg C): | FP (deg C):

BP (deg C):

BP pressure (mm Hg):

Property/Value	Units	Temp	Data Type	Reference
WS 2.11E+003	mg/L	25	EST	MEYLAN,WM ET AL. (1996)
logP 1.06			EXP	HANSCH,C & LEO,AJ (1985)
VP 1.06E-006	mm Hg	25	EST	NEELY,WB & BLAU,GE (1985)
DC	pKa			
HL 3.00E-014	atm m3/mol	25	EST	MEYLAN,WM & HOWARD,PH (1991)
OH 2.83E-012	cm3/molc sec	25	EST	MEYLAN,WM & HOWARD,PH (1993)

CAS #:	014080-23-0			2-PYRIMIDINECARBONITRILE

Formula: $C_5H_3N_3$

Mol Weight: 105.10

MP (deg C): | FP (deg C):

BP (deg C):

BP pressure (mm Hg):

Property/Value	Units	Temp	Data Type	Reference
WS 1.13E+005	mg/L	25	EST	MEYLAN,WM ET AL. (1996)
logP -0.31			EXP	YAMAGAMI,C ET AL. (1990)
VP 1.68E-001	mm Hg	25	EST	NEELY,WB & BLAU,GE (1985)
DC	pKa			
HL 2.82E-008	atm m3/mol	25	EST	MEYLAN,WM & HOWARD,PH (1991)
OH 4.06E-014	cm3/molc sec	25	EST	MEYLAN,WM & HOWARD,PH (1993)

CAS #:	014080-32-1			PYRIMIDINE, 5-NITRO-

Formula: $C_4H_3N_3O_2$

Mol Weight: 125.09

MP (deg C): | FP (deg C):

BP (deg C):

BP pressure (mm Hg):

Property/Value	Units	Temp	Data Type	Reference
WS 4.32E+004	mg/L	25	EST	MEYLAN,WM ET AL. (1996)
logP -0.05			EXP	YAMAGAMI,C ET AL. (1994)
VP 4.89E-002	mm Hg	25	EST	NEELY,WB & BLAU,GE (1985)
DC 0.72	pKa	20	EXP	PERRIN,DD (1972)
HL 1.15E-008	atm m3/mol	25	EST	MEYLAN,WM & HOWARD,PH (1991)
OH 2.88E-014	cm3/molc sec	25	EST	MEYLAN,WM & HOWARD,PH (1993)

CAS #:	014080-56-9			THIENO[2,3-D]PYRIMIDINE, 4-AMINO-

Formula: $C_6H_5N_3S$

Mol Weight: 151.19

MP (deg C): | FP (deg C):

BP (deg C):

BP pressure (mm Hg):

Property/Value	Units	Temp	Data Type	Reference
WS 1.14E+004	mg/L	25	EST	MEYLAN,WM ET AL. (1996)
logP 0.96			EXP	HANSCH,C ET AL. (1995)
VP 1.43E-004	mm Hg	25	EST	NEELY,WB & BLAU,GE (1985)
DC	pKa			
HL 5.47E-011	atm m3/mol	25	EST	MEYLAN,WM & HOWARD,PH (1991)
OH 2.00E-010	cm3/molc sec	25	EST	MEYLAN,WM & HOWARD,PH (1993)

CAS #: 014088-71-2		PROCLONOL	

Formula: $C_{16}H_{14}Cl_2O$

Mol Weight: 293.20

MP (deg C): FP (deg C):

BP (deg C):

BP pressure (mm Hg):

Property/Value	Units	Temp	Data Type	Reference
WS 8.74E-001	mg/L	25	EST	MEYLAN,WM ET AL. (1996)
logP 5.50			EXP	HANSCH,C & LEO,AJ (1985)
VP 5.95E-008	mm Hg	25	EST	NEELY,WB & BLAU,GE (1985)
DC	pKa			
HL 9.94E-009	atm m3/mol	25	EST	MEYLAN,WM & HOWARD,PH (1991)
OH 3.71E-012	cm3/molc sec	25	EST	MEYLAN,WM & HOWARD,PH (1993)

CAS #: 014090-88-1		4,6-NONANEDIONE	

Formula: $C_9H_{16}O_2$

Mol Weight: 156.23

MP (deg C): FP (deg C):

BP (deg C):

BP pressure (mm Hg):

Property/Value	Units	Temp	Data Type	Reference
WS 8.87E+002	mg/L	25	EST	MEYLAN,WM ET AL. (1996)
logP 2.23			EXP	HANSCH,C & LEO,AJ (1985)
VP 1.69E-001	mm Hg	25	EST	NEELY,WB & BLAU,GE (1985)
DC	pKa			
HL 1.04E-007	atm m3/mol	25	EST	MEYLAN,WM & HOWARD,PH (1991)
OH 9.87E-012	cm3/molc sec	25	EST	MEYLAN,WM & HOWARD,PH (1993)

CAS #: 014094-43-0		5-METHYL-4-NITROQUINOLINE-1-OXIDE	

Formula: $C_{10}H_{10}N_2O_3$

Mol Weight: 206.20

MP (deg C): FP (deg C):

BP (deg C):

BP pressure (mm Hg):

Property/Value	Units	Temp	Data Type	Reference
WS 7.44E+002	mg/L	25	EST	MEYLAN,WM ET AL. (1996)
logP 1.36			EXP	HANSCH,C & LEO,AJ (1985)
VP 1.06E-006	mm Hg	25	EST	NEELY,WB & BLAU,GE (1985)
DC	pKa			
HL 3.00E-014	atm m3/mol	25	EST	MEYLAN,WM & HOWARD,PH (1991)
OH 2.83E-012	cm3/molc sec	25	EST	MEYLAN,WM & HOWARD,PH (1993)

CAS #: 014094-45-2		8-METHYL-4-NITROQUINOLINE-1-OXIDE	

Formula: $C_{10}H_{10}N_2O_3$

Mol Weight: 206.20

MP (deg C): FP (deg C):

BP (deg C):

BP pressure (mm Hg):

Property/Value	Units	Temp	Data Type	Reference
WS 7.44E+002	mg/L	25	EST	MEYLAN,WM ET AL. (1996)
logP 1.59			EXP	HANSCH,C & LEO,AJ (1985)
VP 1.06E-006	mm Hg	25	EST	NEELY,WB & BLAU,GE (1985)
DC	pKa			
HL 3.00E-014	atm m3/mol	25	EST	MEYLAN,WM & HOWARD,PH (1991)
OH 2.83E-012	cm3/molc sec	25	EST	MEYLAN,WM & HOWARD,PH (1993)

CAS #: 014098-20-5		BENZOBICYCLO(2.2.1)HEPTENE,9EX-NH2	

Formula: $C_{11}H_{13}N$

Mol Weight: 159.23

MP (deg C): FP (deg C):

BP (deg C):

BP pressure (mm Hg):

Property/Value	Units	Temp	Data Type	Reference
WS 1.18E+004	mg/L	25	EST	MEYLAN,WM ET AL. (1996)
logP 2.13			EXP	HANSCH,C & LEO,AJ (1985)
VP 8.71E-003	mm Hg	25	EST	NEELY,WB & BLAU,GE (1985)
DC	pKa			
HL 3.07E-007	atm m3/mol	25	EST	MEYLAN,WM & HOWARD,PH (1991)
OH 4.96E-011	cm3/molc sec	25	EST	MEYLAN,WM & HOWARD,PH (1993)

CAS #: 014098-24-9		B18C6-BENZO CROWN ETHER	

Formula: $C_{16}H_{24}O_6$

Mol Weight: 312.37

MP (deg C): FP (deg C):

BP (deg C):

BP pressure (mm Hg):

Property/Value	Units	Temp	Data Type	Reference
WS 3.32E+003	mg/L	25	EST	MEYLAN,WM ET AL. (1996)
logP 0.58			EXP	STOLWIJK,TB ET AL. (1989)
VP 3.17E-007	mm Hg	25	EST	NEELY,WB & BLAU,GE (1985)
DC	pKa			
HL 4.85E-013	atm m3/mol	25	EST	MEYLAN,WM & HOWARD,PH (1991)
OH 8.87E-011	cm3/molc sec	25	EST	MEYLAN,WM & HOWARD,PH (1993)

CAS #: 014098-44-3		BENZO-15-CROWN-5-ETHER	

Formula: $C_{14}H_{20}O_5$

Mol Weight: 268.31

MP (deg C): 78-80 FP (deg C):

BP (deg C):

BP pressure (mm Hg):

Property/Value	Units	Temp	Data Type	Reference
WS 3.12E+003	mg/L	25	EST	MEYLAN,WM ET AL. (1996)
logP 0.91			EXP	STOLWIJK,TB ET AL. (1989)
VP 5.66E-006	mm Hg	25	EST	NEELY,WB & BLAU,GE (1985)
DC	pKa			
HL 3.12E-011	atm m3/mol	25	EST	MEYLAN,WM & HOWARD,PH (1991)
OH 7.47E-011	cm3/molc sec	25	EST	MEYLAN,WM & HOWARD,PH (1993)

CAS #: 014100-52-8		3-CHLORO-4-NITROQUINOLINE-1-OXIDE	

Formula: $C_9H_7ClN_2O_3$

Mol Weight: 226.62

MP (deg C): FP (deg C):

BP (deg C):

BP pressure (mm Hg):

Property/Value	Units	Temp	Data Type	Reference
WS 9.69E+002	mg/L	25	EST	MEYLAN,WM ET AL. (1996)
logP 1.33			EXP	HANSCH,C & LEO,AJ (1985)
VP 5.92E-007	mm Hg	25	EST	NEELY,WB & BLAU,GE (1985)
DC	pKa			
HL 2.01E-014	atm m3/mol	25	EST	MEYLAN,WM & HOWARD,PH (1991)
OH 6.82E-013	cm3/molc sec	25	EST	MEYLAN,WM & HOWARD,PH (1993)

CAS #: 014112-00-6		1,2-DICHLORO-1,2-DIVINYLCYCLOBUTANE

Formula: $C_8H_{10}Cl_2$

Mol Weight: 177.07

MP (deg C): | FP (deg C):

BP (deg C):

BP pressure (mm Hg):

Property/Value	Units	Temp	Data Type	Reference
WS 1.56E+001	mg/L	25	EST	MEYLAN,WM ET AL. (1996)
logP 4.17			EST	MEYLAN,WM & HOWARD,PH (1995)
VP 2.14E+000	mm Hg	25	EST	NEELY,WB & BLAU,GE (1985)
DC	pKa			
HL 1.63E-002	atm m3/mol	25	EST	MEYLAN,WM & HOWARD,PH (1991)
OH 4.06E-011	cm3/molc sec	25	EST	MEYLAN,WM & HOWARD,PH (1993)

CAS #: 014129-82-9		1,3,3-TRICHLORO-2-(DICHLOROMETHYL)PROPENE

Formula: $C_4H_3Cl_5$

Mol Weight: 228.33

MP (deg C): | FP (deg C):

BP (deg C):

BP pressure (mm Hg):

Property/Value	Units	Temp	Data Type	Reference
WS 3.50E+001	mg/L	25	EST	MEYLAN,WM ET AL. (1996)
logP 3.45			EST	MEYLAN,WM & HOWARD,PH (1995)
VP 1.17E-001	mm Hg	25	EST	NEELY,WB & BLAU,GE (1985)
DC	pKa			
HL 1.68E-003	atm m3/mol	25	EST	MEYLAN,WM & HOWARD,PH (1991)
OH 1.11E-011	cm3/molc sec	25	EST	MEYLAN,WM & HOWARD,PH (1993)

CAS #: 014150-95-9		2-PYRIDINAMINE, 1-OXIDE

Formula: $C_5H_8N_2O$

Mol Weight: 112.13

MP (deg C): | FP (deg C):

BP (deg C):

BP pressure (mm Hg):

Property/Value	Units	Temp	Data Type	Reference
WS 6.01E+005	mg/L	25	EST	MEYLAN,WM ET AL. (1996)
logP -0.87			EXP	HANSCH,C ET AL. (1995)
VP 2.37E-003	mm Hg	25	EST	NEELY,WB & BLAU,GE (1985)
DC	pKa			
HL 2.49E-014	atm m3/mol	25	EST	MEYLAN,WM & HOWARD,PH (1991)
OH 2.04E-011	cm3/molc sec	25	EST	MEYLAN,WM & HOWARD,PH (1993)

CAS #: 014159-57-0		PHENYL-A-PYRIDYLCARBINOL

Formula: $C_{12}H_{11}NO$

Mol Weight: 185.23

MP (deg C): | FP (deg C):

BP (deg C):

BP pressure (mm Hg):

Property/Value	Units	Temp	Data Type	Reference
WS 1.69E+005	mg/L	25	EST	MEYLAN,WM ET AL. (1996)
logP 1.52			EXP	HANSCH,C & LEO,AJ (1985)
VP 1.01E-005	mm Hg	25	EST	NEELY,WB & BLAU,GE (1985)
DC	pKa			
HL 2.30E-011	atm m3/mol	25	EST	MEYLAN,WM & HOWARD,PH (1991)
OH 1.24E-011	cm3/molc sec	25	EST	MEYLAN,WM & HOWARD,PH (1993)

CAS #: 014167-59-0		N-TETRATRIACONTANE

Formula: $C_{34}H_{70}$

Mol Weight: 478.94

MP (deg C): 72.6 | FP (deg C):

BP (deg C): 285.4

BP pressure (mm Hg): 3.00E+000

Property/Value	Units	Temp	Data Type	Reference
WS 7.97E-013	mg/L	25	EST	MEYLAN,WM ET AL. (1996)
logP 17.04			EST	MEYLAN,WM & HOWARD,PH (1995)
VP 1.67E-020	mm Hg	25	EXT	ZWOLINSKI,BJ & WILHOIT,RC (1971)
DC	pKa			
HL 4.76E+003	atm m3/mol	25	EST	MEYLAN,WM & HOWARD,PH (1991)
OH 4.50E-011	cm3/molc sec	25	EST	MEYLAN,WM & HOWARD,PH (1993)

CAS #: 014168-01-5		BETA-DIHYDROHEPTACHLOR

Formula: $C_{10}H_7Cl_7$

Mol Weight: 375.34

MP (deg C): | FP (deg C):

BP (deg C):

BP pressure (mm Hg):

Property/Value	Units	Temp	Data Type	Reference
WS 1.43E-002	mg/L	25	EST	MEYLAN,WM ET AL. (1996)
logP 6.42			EST	MEYLAN,WM & HOWARD,PH (1995)
VP 5.06E-005	mm Hg	25	EST	NEELY,WB & BLAU,GE (1985)
DC	pKa			
HL 2.00E-004	atm m3/mol	25	EST	MEYLAN,WM & HOWARD,PH (1991)
OH 8.85E-012	cm3/molc sec	25	EST	MEYLAN,WM & HOWARD,PH (1993)

CAS #: 014174-09-5		[4,4]DB24C6-DIBENZO CROWN ETHER

Formula: $C_{24}H_{32}O_8$

Mol Weight: 448.52

MP (deg C): 103-105 | FP (deg C):

BP (deg C):

BP pressure (mm Hg):

Property/Value	Units	Temp	Data Type	Reference
WS 1.87E+000	mg/L	25	EST	MEYLAN,WM ET AL. (1996)
logP 2.11			EXP	STOLWIJK,TB ET AL. (1989)
VP 1.11E-011	mm Hg	25	EST	NEELY,WB & BLAU,GE (1985)
DC	pKa			
HL 1.65E-016	atm m3/mol	25	EST	MEYLAN,WM & HOWARD,PH (1991)
OH 1.21E-010	cm3/molc sec	25	EST	MEYLAN,WM & HOWARD,PH (1993)

CAS #: 014187-32-7		[3,3]DB18C6-DIBENZO CROWN ETHER

Formula: $C_{20}H_{24}O_6$

Mol Weight: 360.41

MP (deg C): 162-164 | FP (deg C):

BP (deg C):

BP pressure (mm Hg):

Property/Value	Units	Temp	Data Type	Reference
WS 7.30E+000	mg/L	25	EXP	YALKOWSKY,SH & DANNENFELSER,RM (1992)
logP 2.20			EXP	STOLWIJK,TB ET AL. (1989)
VP 3.97E-009	mm Hg	25	EST	NEELY,WB & BLAU,GE (1985)
DC	pKa			
HL 6.82E-013	atm m3/mol	25	EST	MEYLAN,WM & HOWARD,PH (1991)
OH 9.33E-011	cm3/molc sec	25	EST	MEYLAN,WM & HOWARD,PH (1993)

CAS #: 014191-95-8 — 4-HYDROXYBENZYLCYANIDE

Formula: C_8H_7NO

Mol Weight: 133.15

MP (deg C):

FP (deg C):

BP (deg C):

BP pressure (mm Hg):

Property/Value	Units	Temp	Data Type	Reference
WS 2.20E+004	mg/L	25	EST	MEYLAN,WM ET AL. (1996)
logP 1.08			EST	MEYLAN,WM & HOWARD,PH (1995)
VP 7.12E-004	mm Hg	25	EST	NEELY,WB & BLAU,GE (1985)
DC	pKa			
HL 2.57E-010	atm m3/mol	25	EST	MEYLAN,WM & HOWARD,PH (1991)
OH 2.00E-011	cm3/molc sec	25	EST	MEYLAN,WM & HOWARD,PH (1993)

CAS #: 014202-62-1 — 2,2-DIETHYL-1-PENTANOL

Formula: $C_9H_{20}O$

Mol Weight: 144.26

MP (deg C):

FP (deg C):

BP (deg C):

BP pressure (mm Hg):

Property/Value	Units	Temp	Data Type	Reference
WS 5.48E+002	mg/L	25	EXP	SHIU,WY ET AL. (1990)
logP 3.19			EST	MEYLAN,WM & HOWARD,PH (1995)
VP 7.08E-002	mm Hg	25	EST	NEELY,WB & BLAU,GE (1985)
DC	pKa			
HL 4.12E-005	atm m3/mol	25	EST	MEYLAN,WM & HOWARD,PH (1991)
OH 9.52E-012	cm3/molc sec	25	EST	MEYLAN,WM & HOWARD,PH (1993)

CAS #: 014206-62-3 — 2-NAPHTHALENECARBOXYLIC ACID,3-HYDROXY-,MONOSODI

Formula: $C_{11}H_7NaO_3$

Mol Weight: 210.17

MP (deg C):

FP (deg C):

BP (deg C):

BP pressure (mm Hg):

Property/Value	Units	Temp	Data Type	Reference
WS 1.06E+005	mg/L	25	EST	MEYLAN,WM ET AL. (1996)
logP 0.17			EXP	HANSCH,C ET AL. (1995)
VP 1.58E-013	mm Hg	25	EST	NEELY,WB & BLAU,GE (1985)
DC	pKa			
HL	atm m3/mol			
OH 4.61E-011	cm3/molc sec	25	EST	MEYLAN,WM & HOWARD,PH (1993)

CAS #: 014210-25-4 — 1-PHENYL-5-CHLOROTETRAZOLE

Formula: $C_7H_7ClN_4$

Mol Weight: 182.61

MP (deg C): 122-124

FP (deg C):

BP (deg C):

BP pressure (mm Hg):

Property/Value	Units	Temp	Data Type	Reference
WS 2.98E+003	mg/L	25	EST	MEYLAN,WM ET AL. (1996)
logP 1.48			EXP	HANSCH,C & LEO,AJ (1985)
VP 4.31E-004	mm Hg	25	EST	NEELY,WB & BLAU,GE (1985)
DC	pKa			
HL 1.65E-009	atm m3/mol	25	EST	MEYLAN,WM & HOWARD,PH (1991)
OH 2.57E-012	cm3/molc sec	25	EST	MEYLAN,WM & HOWARD,PH (1993)

CAS #: 014214-32-5 — DIFENOXURON

Formula: $C_{16}H_{18}N_2O_3$

Mol Weight: 286.33

MP (deg C):

FP (deg C):

BP (deg C):

BP pressure (mm Hg):

Property/Value	Units	Temp	Data Type	Reference
WS 2.00E+001	mg/L	20	EXP	YALKOWSKY,SH & DANNENFELSER,RM (1992)
logP 3.52			EST	MEYLAN,WM & HOWARD,PH (1995)
VP 2.89E-008	mm Hg	25	EST	NEELY,WB & BLAU,GE (1985)
DC	pKa			
HL 1.26E-012	atm m3/mol	25	EST	MEYLAN,WM & HOWARD,PH (1991)
OH 6.71E-011	cm3/molc sec	25	EST	MEYLAN,WM & HOWARD,PH (1993)

CAS #: 014251-76-4 — OCTYL-TRIMETHYL-AMMONIUM IODIDE

Formula: $C_{11}H_{26}IN$

Mol Weight: 299.24

MP (deg C):

FP (deg C):

BP (deg C):

BP pressure (mm Hg):

Property/Value	Units	Temp	Data Type	Reference
WS 1.02E+005	mg/L	25	EST	MEYLAN,WM ET AL. (1996)
logP -1.07			EXP	HANSCH,C & LEO,AJ (1985)
VP 1.93E-008	mm Hg	25	EST	NEELY,WB & BLAU,GE (1985)
DC	pKa			
HL 1.91E-011	atm m3/mol	25	EST	MEYLAN,WM & HOWARD,PH (1991)
OH 2.29E-011	cm3/molc sec	25	EST	MEYLAN,WM & HOWARD,PH (1993)

CAS #: 014255-72-2 — FENSULFOTHION SULFONE

Formula: $C_{11}H_{17}O_5PS_2$

Mol Weight: 324.36

MP (deg C):

FP (deg C):

BP (deg C):

BP pressure (mm Hg):

Property/Value	Units	Temp	Data Type	Reference
WS 8.54E+001	mg/L	20	EXP	BOWMAN,BT & SANS,WW (1983A)
logP 2.56			EXP	HANSCH,C & LEO,AJ (1985)
VP 1.33E-006	mm Hg	25	EST	NEELY,WB & BLAU,GE (1985)
DC	pKa			
HL 1.74E-008	atm m3/mol	25	EST	MEYLAN,WM & HOWARD,PH (1991)
OH 9.35E-011	cm3/molc sec	25	EST	MEYLAN,WM & HOWARD,PH (1993)

CAS #: 014294-09-8 — 1-PIPERIDINETHIOCARBOXAMIDE

Formula: $C_6H_{12}N_2S$

Mol Weight: 144.24

MP (deg C):

FP (deg C):

BP (deg C):

BP pressure (mm Hg):

Property/Value	Units	Temp	Data Type	Reference
WS 1.95E+005	mg/L	25	EST	MEYLAN,WM ET AL. (1996)
logP 0.73			EXP	HANSCH,C ET AL. (1995)
VP 2.18E-002	mm Hg	25	EST	NEELY,WB & BLAU,GE (1985)
DC	pKa			
HL 7.85E-007	atm m3/mol	25	EST	MEYLAN,WM & HOWARD,PH (1991)
OH 1.13E-010	cm3/molc sec	25	EST	MEYLAN,WM & HOWARD,PH (1993)

CAS #: 014309-40-1				
Formula: $C_{14}H_{21}NO_2$				BENZOIC ACID, 4-AMINO-, HEPTYL ESTER
Mol Weight: 235.33				
MP (deg C):		FP (deg C):		
BP (deg C):				
BP pressure (mm Hg):				

Property/ Value	Units	Temp	Data Type	Reference
WS 5.90E+000	mg/L	25	EXP	SUZUKI,T (1991)
logP 4.50			EXP	LIANG,WQ & LIN,W (1992)
VP 2.36E-005	mm Hg	25	EST	NEELY,WB & BLAU,GE (1985)
DC	pKa			
HL 6.71E-008	atm m3/mol	25	EST	MEYLAN,WM & HOWARD,PH (1991)
OH 4.36E-011	cm3/molc sec	25	EST	MEYLAN,WM & HOWARD,PH (1993)

CAS #: 014309-41-2				
Formula: $C_{15}H_{23}NO_2$				BENZOIC ACID, 4-AMINO-, OCTYL ESTER
Mol Weight: 249.36				
MP (deg C):		FP (deg C):		
BP (deg C):				
BP pressure (mm Hg):				

Property/ Value	Units	Temp	Data Type	Reference
WS 9.91E-001	mg/L	25	EXP	SUZUKI,T (1991)
logP 5.02			EXP	LIANG,WQ & LIN,W (1992)
VP 1.02E-005	mm Hg	25	EST	NEELY,WB & BLAU,GE (1985)
DC	pKa			
HL 8.91E-008	atm m3/mol	25	EST	MEYLAN,WM & HOWARD,PH (1991)
OH 4.50E-011	cm3/molc sec	25	EST	MEYLAN,WM & HOWARD,PH (1993)

CAS #: 014314-69-3				
Formula: $C_6H_3KN_2O_5$				PHENOL, 2,4-DINITRO-, POTASSIUM SALT
Mol Weight: 222.20				
MP (deg C):		FP (deg C):		
BP (deg C):				
BP pressure (mm Hg):				

Property/ Value	Units	Temp	Data Type	Reference
WS 2.94E+004	mg/L	25	EST	MEYLAN,WM ET AL. (1996)
logP -0.39			EXP	SANGSTER,J (1994)
VP 7.43E-012	mm Hg	25	EST	NEELY,WB & BLAU,GE (1985)
DC	pKa			
HL	atm m3/mol			
OH 3.36E-013	cm3/molc sec	25	EST	MEYLAN,WM & HOWARD,PH (1993)

CAS #: 014315-97-0				
Formula: $C_6H_{14}O_3$				1,1,3-TRIMETHOXYPROPANE
Mol Weight: 134.18				
MP (deg C):		FP (deg C):		
BP (deg C): 45-46				
BP pressure (mm Hg): 1.70E+001				

Property/ Value	Units	Temp	Data Type	Reference
WS 9.80E+004	mg/L	25	EST	MEYLAN,WM ET AL. (1996)
logP -0.05				MEYLAN,WM & HOWARD,PH (1995)
VP 1.14E+001	mm Hg	25	EST	NEELY,WB & BLAU,GE (1985)
DC	pKa			
HL 1.04E-006	atm m3/mol	25	EST	MEYLAN,WM & HOWARD,PH (1991)
OH 1.92E-011	cm3/molc sec	25	EXP	ATKINSON,R (1989)

CAS #: 014321-27-8				
Formula: $C_9H_{13}N$				BENZENEMETHANAMINE, N-ETHYL-
Mol Weight: 135.21				
MP (deg C):		FP (deg C):		
BP (deg C): 194				
BP pressure (mm Hg):				

Property/ Value	Units	Temp	Data Type	Reference
WS 2.49E+004	mg/L	25	EST	MEYLAN,WM ET AL. (1996)
logP 1.82			EXP	SANGSTER,J (1993)
VP 2.40E-001	mm Hg	25	EST	NEELY,WB & BLAU,GE (1985)
DC 9.64	pKa	25	EXP	PERRIN,DD (1965)
HL 1.78E-006	atm m3/mol	25	EST	MEYLAN,WM & HOWARD,PH (1991)
OH 8.54E-011	cm3/molc sec	25	EST	MEYLAN,WM & HOWARD,PH (1993)

CAS #: 014337-53-2				
Formula: $C_{15}H_{17}BrN_4O$				2(5BR-2-PYRIDYLAZO)-5(DIETAM)PHENOL
Mol Weight: 349.24				
MP (deg C): 157-158		FP (deg C):		
BP (deg C):				
BP pressure (mm Hg):				

Property/ Value	Units	Temp	Data Type	Reference
WS 5.65E-001	mg/L	25	EST	MEYLAN,WM ET AL. (1996)
logP 4.23			EXP	HANSCH,C & LEO,AJ (1985)
VP 7.63E-009	mm Hg	25	EST	NEELY,WB & BLAU,GE (1985)
DC	pKa			
HL 2.24E-014	atm m3/mol	25	EST	MEYLAN,WM & HOWARD,PH (1991)
OH 2.18E-010	cm3/molc sec	25	EST	MEYLAN,WM & HOWARD,PH (1993)

CAS #: 014339-54-9				
Formula: $C_9H_{13}N_3O$				2-PROPYLISONIAZID
Mol Weight: 179.22				
MP (deg C):		FP (deg C):		
BP (deg C):				
BP pressure (mm Hg):				

Property/ Value	Units	Temp	Data Type	Reference
WS 1.61E+004	mg/L	25	EST	MEYLAN,WM ET AL. (1996)
logP 0.63			EXP	HANSCH,C & LEO,AJ (1985)
VP 3.47E-006	mm Hg	25	EST	NEELY,WB & BLAU,GE (1985)
DC	pKa			
HL 2.35E-014	atm m3/mol	25	EST	MEYLAN,WM & HOWARD,PH (1991)
OH 8.38E-012	cm3/molc sec	25	EST	MEYLAN,WM & HOWARD,PH (1993)

CAS #: 014342-36-0				
Formula: $C_{12}H_{15}N$				2-EN-NH2 BENZOBICYCLO(222)OCTENE
Mol Weight: 173.26				
MP (deg C):		FP (deg C):		
BP (deg C):				
BP pressure (mm Hg):				

Property/ Value	Units	Temp	Data Type	Reference
WS 6.31E+003	mg/L	25	EST	MEYLAN,WM ET AL. (1996)
logP 2.29			EXP	HANSCH,C & LEO,AJ (1985)
VP 2.74E-003	mm Hg	25	EST	NEELY,WB & BLAU,GE (1985)
DC	pKa			
HL 4.08E-007	atm m3/mol	25	EST	MEYLAN,WM & HOWARD,PH (1991)
OH 6.40E-011	cm3/molc sec	25	EST	MEYLAN,WM & HOWARD,PH (1993)

CAS #: 014377-19-6 — BENZENEACETIC ACID, 4-BUTYL-

Formula: $C_{12}H_{16}O_2$

Mol Weight: 192.26

MP (deg C):   FP (deg C):

BP (deg C):

BP pressure (mm Hg):

Property/Value	Units	Temp	Data Type	Reference
WS 1.99E+002	mg/L	25	EST	MEYLAN,WM ET AL. (1996)
logP 3.25			EXP	SANGSTER,J (1993)
VP 8.81E-005	mm Hg	25	EST	NEELY,WB & BLAU,GE (1985)
DC	pKa			
HL 1.14E-007	atm m3/mol	25	EST	MEYLAN,WM & HOWARD,PH (1991)
OH 1.01E-011	cm3/molc sec	25	EST	MEYLAN,WM & HOWARD,PH (1993)

CAS #: 014377-21-0 — BENZENEACETIC ACID, 4-PENTYL-

Formula: $C_{13}H_{18}O_2$

Mol Weight: 206.29

MP (deg C):   FP (deg C):

BP (deg C):

BP pressure (mm Hg):

Property/Value	Units	Temp	Data Type	Reference
WS 6.33E+001	mg/L	25	EST	MEYLAN,WM ET AL. (1996)
logP 3.75			EXP	SANGSTER,J (1993)
VP 3.62E-005	mm Hg	25	EST	NEELY,WB & BLAU,GE (1985)
DC	pKa			
HL 1.52E-007	atm m3/mol	25	EST	MEYLAN,WM & HOWARD,PH (1991)
OH 1.15E-011	cm3/molc sec	25	EST	MEYLAN,WM & HOWARD,PH (1993)

CAS #: 014377-22-1 — BENZENEACETIC ACID, 4-HEXYL-

Formula: $C_{14}H_{20}O_2$

Mol Weight: 220.31

MP (deg C):   FP (deg C):

BP (deg C):

BP pressure (mm Hg):

Property/Value	Units	Temp	Data Type	Reference
WS 2.00E+001	mg/L	25	EST	MEYLAN,WM ET AL. (1996)
logP 4.25			EXP	SANGSTER,J (1993)
VP 1.84E-005	mm Hg	25	EST	NEELY,WB & BLAU,GE (1985)
DC	pKa			
HL 2.01E-007	atm m3/mol	25	EST	MEYLAN,WM & HOWARD,PH (1991)
OH 1.29E-011	cm3/molc sec	25	EST	MEYLAN,WM & HOWARD,PH (1993)

CAS #: 014387-10-1 — BENZENEACETIC ACID, 4-ETHYL-

Formula: $C_{10}H_{12}O_2$

Mol Weight: 164.21

MP (deg C):   FP (deg C):

BP (deg C):

BP pressure (mm Hg):

Property/Value	Units	Temp	Data Type	Reference
WS 1.95E+003	mg/L	25	EST	MEYLAN,WM ET AL. (1996)
logP 2.25			EXP	SANGSTER,J (1993)
VP 6.03E-004	mm Hg	25	EST	NEELY,WB & BLAU,GE (1985)
DC 4.37	pKa	25	EXP	KORTUM,G ET AL (1961)
HL 6.48E-008	atm m3/mol	25	EST	MEYLAN,WM & HOWARD,PH (1991)
OH 7.32E-012	cm3/molc sec	25	EST	MEYLAN,WM & HOWARD,PH (1993)

CAS #: 014402-22-3 — 2-METHYLPYRIDINIUM IODIDE, N-DODECYL

Formula: $C_{18}H_{32}IN$

Mol Weight: 389.37

MP (deg C):   FP (deg C):

BP (deg C):

BP pressure (mm Hg):

Property/Value	Units	Temp	Data Type	Reference
WS 2.66E+002	mg/L	25	EST	MEYLAN,WM ET AL. (1996)
logP 1.32			EXP	HANSCH,C & LEO,AJ (1985)
VP 1.38E-006	mm Hg	25	EST	NEELY,WB & BLAU,GE (1985)
DC	pKa			
HL 3.75E-009	atm m3/mol	25	EST	MEYLAN,WM & HOWARD,PH (1991)
OH 2.63E-011	cm3/molc sec	25	EST	MEYLAN,WM & HOWARD,PH (1993)

CAS #: 014402-23-4 — 2-METHYLPYRIDINIUM IODIDE, N-DECYL

Formula: $C_{16}H_{28}IN$

Mol Weight: 361.31

MP (deg C):   FP (deg C):

BP (deg C):

BP pressure (mm Hg):

Property/Value	Units	Temp	Data Type	Reference
WS 3.86E+003	mg/L	25	EST	MEYLAN,WM ET AL. (1996)
logP 0.16			EXP	HANSCH,C & LEO,AJ (1985)
VP 7.81E-006	mm Hg	25	EST	NEELY,WB & BLAU,GE (1985)
DC	pKa			
HL 2.13E-009	atm m3/mol	25	EST	MEYLAN,WM & HOWARD,PH (1991)
OH 2.35E-011	cm3/molc sec	25	EST	MEYLAN,WM & HOWARD,PH (1993)

CAS #: 014402-24-5 — 2-METHYLPYRIDINIUM IODIDE, N-HEXYL

Formula: $C_{12}H_{20}IN$

Mol Weight: 305.20

MP (deg C):   FP (deg C):

BP (deg C):

BP pressure (mm Hg):

Property/Value	Units	Temp	Data Type	Reference
WS 3.30E+005	mg/L	25	EST	MEYLAN,WM ET AL. (1996)
logP -1.71			EXP	HANSCH,C & LEO,AJ (1985)
VP 2.62E-004	mm Hg	25	EST	NEELY,WB & BLAU,GE (1985)
DC	pKa			
HL 6.84E-010	atm m3/mol	25	EST	MEYLAN,WM & HOWARD,PH (1991)
OH 1.78E-011	cm3/molc sec	25	EST	MEYLAN,WM & HOWARD,PH (1993)

CAS #: 014417-01-7 — N,N-DIMETHYLBENZENESULFONAMIDE

Formula: $C_8H_{11}NO_2S$

Mol Weight: 185.25

MP (deg C):   FP (deg C):

BP (deg C):

BP pressure (mm Hg):

Property/Value	Units	Temp	Data Type	Reference
WS 3.65E+003	mg/L	25	EST	MEYLAN,WM ET AL. (1996)
logP 1.35			EXP	HANSCH,C & LEO,AJ (1985)
VP 6.98E-004	mm Hg	25	EST	NEELY,WB & BLAU,GE (1985)
DC	pKa			
HL 2.04E-006	atm m3/mol	25	EST	MEYLAN,WM & HOWARD,PH (1991)
OH 2.95E-012	cm3/molc sec	25	EST	MEYLAN,WM & HOWARD,PH (1993)

CAS #:	014426-42-7			ETHYL 4-CHLOROPHENOXYACETATE

Formula: $C_{10}H_{11}ClO_3$

Mol Weight: 214.65

MP (deg C): FP (deg C):

BP (deg C):

BP pressure (mm Hg):

Property/ Value	Units	Temp	Data Type	Reference
WS 8.15E+002	mg/L	25	EXP	PARIS,DF ET AL. (1984)
logP 2.75			EST	MEYLAN,WM & HOWARD,PH (1995)
VP 2.95E-003	mm Hg		EST	NEELY,WB & BLAU,GE (1985)
DC	pKa			
HL 5.28E-006	atm m3/mol	25	EST	MEYLAN,WM & HOWARD,PH (1991)
OH 9.71E-012	cm3/molc sec	25	EST	MEYLAN,WM & HOWARD,PH (1993)

CAS #:	014433-36-4			P-ACETYL AMPHENICOL

Formula: $C_{13}H_{15}Cl_2NO_4$

Mol Weight: 320.17

MP (deg C): FP (deg C):

BP (deg C):

BP pressure (mm Hg):

Property/ Value	Units	Temp	Data Type	Reference
WS 3.05E+003	mg/L	25	EST	MEYLAN,WM ET AL. (1996)
logP 0.57			EXP	HANSCH,C & LEO,AJ (1985)
VP 7.04E-013	mm Hg	25	EST	NEELY,WB & BLAU,GE (1985)
DC	pKa			
HL 1.06E-018	atm m3/mol	25	EST	MEYLAN,WM & HOWARD,PH (1991)
OH 3.47E-011	cm3/molc sec	25	EST	MEYLAN,WM & HOWARD,PH (1993)

CAS #:	014437-17-3			CHLORFENPROP METHYL

Formula: $C_{10}H_{10}Cl_2O_2$

Mol Weight: 233.10

MP (deg C): FP (deg C):

BP (deg C):

BP pressure (mm Hg):

Property/ Value	Units	Temp	Data Type	Reference
WS 4.00E+001	mg/L	20	EXP	YALKOWSKY,SH & DANNENFELSER,RM (1992)
logP 3.40			EST	MEYLAN,WM & HOWARD,PH (1995)
VP 2.03E-003	mm Hg	25	EST	NEELY,WB & BLAU,GE (1985)
DC	pKa			
HL 4.90E-006	atm m3/mol	25	EST	MEYLAN,WM & HOWARD,PH (1991)
OH 2.81E-012	cm3/molc sec	25	EST	MEYLAN,WM & HOWARD,PH (1993)

CAS #:	014484-64-1			FERBAM

Formula: $C_9H_{18}FeN_3S_6$

Mol Weight: 416.49

MP (deg C): 180 dec FP (deg C):

BP (deg C):

BP pressure (mm Hg):

Property/ Value	Units	Temp	Data Type	Reference
WS 1.30E+002	mg/L	25	EXP	TOMLIN,C (1994)
logP 6.30			EXP	TOMLIN,C (1994)
VP	mm Hg			
DC	pKa			
HL	atm m3/mol			
OH	cm3/molc sec			

CAS #:	014496-34-5			FURANMETHAMINE, N,N-DIMETHYL

Formula: $C_7H_{12}NO$

Mol Weight: 126.18

MP (deg C): FP (deg C):

BP (deg C):

BP pressure (mm Hg):

Property/ Value	Units	Temp	Data Type	Reference
WS 9.42E+004	mg/L	25	EST	MEYLAN,WM ET AL. (1996)
logP 1.19			EXP	HANSCH,C & LEO,AJ (1985)
VP 3.41E+000	mm Hg	25	EST	NEELY,WB & BLAU,GE (1985)
DC	pKa			
HL 2.94E-006	atm m3/mol	25	EST	MEYLAN,WM & HOWARD,PH (1991)
OH 1.78E-010	cm3/molc sec	25	EST	MEYLAN,WM & HOWARD,PH (1993)

CAS #:	014499-87-7			2,2,3,3,-TETRACHLOROBUTANE

Formula: $C_4H_6Cl_4$

Mol Weight: 195.90

MP (deg C): FP (deg C):

BP (deg C):

BP pressure (mm Hg):

Property/ Value	Units	Temp	Data Type	Reference
WS 6.47E+000	mg/L	25	EST	MEYLAN,WM ET AL. (1996)
logP 4.51			EST	MEYLAN,WM & HOWARD,PH (1995)
VP 1.09E+001	mm Hg	25	EST	NEELY,WB & BLAU,GE (1985)
DC	pKa			
HL 2.65E-003	atm m3/mol	25	EST	MEYLAN,WM & HOWARD,PH (1991)
OH 6.80E-014	cm3/molc sec	25	EST	MEYLAN,WM & HOWARD,PH (1993)

CAS #:	014508-49-7			2-CHLOROPYRAZINE

Formula: $C_4H_3ClN_2$

Mol Weight: 114.53

MP (deg C): FP (deg C):

BP (deg C): 153-154

BP pressure (mm Hg):

Property/ Value	Units	Temp	Data Type	Reference
WS 2.65E+004	mg/L	25	EST	MEYLAN,WM ET AL. (1996)
logP 0.70			EXP	YAMAGAMI,C ET AL. (1990A)
VP 2.48E+000	mm Hg	25	EST	NEELY,WB & BLAU,GE (1985)
DC	pKa			
HL 1.36E-004	atm m3/mol	25	EST	MEYLAN,WM & HOWARD,PH (1991)
OH 1.62E-013	cm3/molc sec	25	EST	MEYLAN,WM & HOWARD,PH (1993)

CAS #:	014521-96-1			ETORPHINE

Formula: $C_{25}H_{33}NO_4$

Mol Weight: 411.55

MP (deg C): 214-217 FP (deg C):

BP (deg C):

BP pressure (mm Hg):

Property/ Value	Units	Temp	Data Type	Reference
WS 1.10E+002	mg/L	25	EST	MEYLAN,WM ET AL. (1996)
logP 2.79			EXP	SANGSTER,J (1993)
VP 3.26E-012	mm Hg	25	EST	NEELY,WB & BLAU,GE (1985)
DC	pKa			
HL 5.00E-018	atm m3/mol	25	EST	MEYLAN,WM & HOWARD,PH (1991)
OH 3.08E-010	cm3/molc sec	25	EST	MEYLAN,WM & HOWARD,PH (1993)

CAS #: 014529-53-4 — 2-ETHOXYPYRIDINE

Formula: C_7H_9NO

Mol Weight: 123.16

MP (deg C): **FP (deg C):**

BP (deg C):

BP pressure (mm Hg):

Property/Value	Units	Temp	Data Type	Reference
WS 2.78E+003	mg/L	25	EST	MEYLAN,WM ET AL. (1996)
logP 1.81			EXP	YAMAGAMI,C ET AL. (1990A)
VP 9.36E-001	mm Hg	25	EST	NEELY,WB & BLAU,GE (1985)
DC	pKa			
HL 5.05E-006	atm m3/mol	25	EST	MEYLAN,WM & HOWARD,PH (1991)
OH 1.03E-011	cm3/molc sec	25	EST	MEYLAN,WM & HOWARD,PH (1993)

CAS #: 014548-30-2 — 4-(4-METHYLBENZOYL)-PYRIDINE

Formula: $C_{13}H_{11}NO$

Mol Weight: 197.24

MP (deg C): **FP (deg C):**

BP (deg C):

BP pressure (mm Hg):

Property/Value	Units	Temp	Data Type	Reference
WS 3.25E+002	mg/L	25	EST	MEYLAN,WM ET AL. (1996)
logP 2.51			EXP	HANSCH,C & LEO,AJ (1985)
VP 1.28E-004	mm Hg	25	EST	NEELY,WB & BLAU,GE (1985)
DC	pKa			
HL 2.81E-009	atm m3/mol	25	EST	MEYLAN,WM & HOWARD,PH (1991)
OH 5.11E-012	cm3/molc sec	25	EST	MEYLAN,WM & HOWARD,PH (1993)

CAS #: 014548-46-0 — 4-BENZOYLPYRIDINE

Formula: $C_{12}H_9NO$

Mol Weight: 183.21

MP (deg C): 72 **FP (deg C):**

BP (deg C): 315

BP pressure (mm Hg):

Property/Value	Units	Temp	Data Type	Reference
WS 1.08E+003	mg/L	25	EST	MEYLAN,WM ET AL. (1996)
logP 1.98			EXP	HANSCH,C & LEO,AJ (1985)
VP 3.67E-004	mm Hg	25	EST	NEELY,WB & BLAU,GE (1985)
DC	pKa			
HL 2.54E-009	atm m3/mol	25	EST	MEYLAN,WM & HOWARD,PH (1991)
OH 2.11E-012	cm3/molc sec	25	EST	MEYLAN,WM & HOWARD,PH (1993)

CAS #: 014548-47-1 — 4-(4-METHOXYBENZOYL)PYRIDINE

Formula: $C_{13}H_{11}NO_2$

Mol Weight: 213.24

MP (deg C): **FP (deg C):**

BP (deg C):

BP pressure (mm Hg):

Property/Value	Units	Temp	Data Type	Reference
WS 8.24E+002	mg/L	25	EST	MEYLAN,WM ET AL. (1996)
logP 1.94			EXP	HANSCH,C & LEO,AJ (1985)
VP 5.48E-005	mm Hg	25	EST	NEELY,WB & BLAU,GE (1985)
DC	pKa			
HL 1.50E-010	atm m3/mol	25	EST	MEYLAN,WM & HOWARD,PH (1991)
OH 2.08E-011	cm3/molc sec	25	EST	MEYLAN,WM & HOWARD,PH (1993)

CAS #: 014556-46-8 — BUPRANOLOL

Formula: $C_{14}H_{22}ClNO_2$

Mol Weight: 271.79

MP (deg C): 220-222 **FP (deg C):**

BP (deg C):

BP pressure (mm Hg):

Property/Value	Units	Temp	Data Type	Reference
WS 7.39E+002	mg/L	25	EST	MEYLAN,WM ET AL. (1996)
logP 2.80			EXP	SANGSTER,J (1993)
VP 5.71E-007	mm Hg	25	EST	NEELY,WB & BLAU,GE (1985)
DC	pKa			
HL 8.88E-012	atm m3/mol	25	EST	MEYLAN,WM & HOWARD,PH (1991)
OH 1.08E-010	cm3/molc sec	25	EST	MEYLAN,WM & HOWARD,PH (1993)

CAS #: 014559-54-7 — 124-BENZTHIADIAZIN-1-O2-3-ME-6-CL

Formula: $C_8H_7ClN_2O_2S$

Mol Weight: 230.67

MP (deg C): **FP (deg C):**

BP (deg C):

BP pressure (mm Hg):

Property/Value	Units	Temp	Data Type	Reference
WS 2.80E+003	mg/L	25	EST	MEYLAN,WM ET AL. (1996)
logP 1.21			EXP	HANSCH,C & LEO,AJ (1985)
VP 2.88E-008	mm Hg	25	EST	NEELY,WB & BLAU,GE (1985)
DC	pKa			
HL 6.85E-008	atm m3/mol	25	EST	MEYLAN,WM & HOWARD,PH (1991)
OH 3.52E-013	cm3/molc sec	25	EST	MEYLAN,WM & HOWARD,PH (1993)

CAS #: 014559-55-8 — 124-BENZTHIADIAZIN-1-O2-3-ET-6-CL

Formula: $C_9H_9ClN_2O_2S$

Mol Weight: 244.70

MP (deg C): **FP (deg C):**

BP (deg C):

BP pressure (mm Hg):

Property/Value	Units	Temp	Data Type	Reference
WS 1.07E+003	mg/L	25	EST	MEYLAN,WM ET AL. (1996)
logP 1.61			EXP	HANSCH,C & LEO,AJ (1985)
VP 1.29E-008	mm Hg	25	EST	NEELY,WB & BLAU,GE (1985)
DC	pKa			
HL 9.10E-008	atm m3/mol	25	EST	MEYLAN,WM & HOWARD,PH (1991)
OH 1.50E-012	cm3/molc sec	25	EST	MEYLAN,WM & HOWARD,PH (1993)

CAS #: 014593-28-3 — 4-(TERT-BUTYL)-2-CHLORO-6-NITROPHENOL

Formula: $C_{10}H_{12}ClNO_3$

Mol Weight: 229.67

MP (deg C): **FP (deg C):**

BP (deg C):

BP pressure (mm Hg):

Property/Value	Units	Temp	Data Type	Reference
WS 4.73E+000	mg/L	25	EST	MEYLAN,WM ET AL. (1996)
logP 4.46			EST	MEYLAN,WM & HOWARD,PH (1995)
VP 1.43E-005	mm Hg	25	EST	NEELY,WB & BLAU,GE (1985)
DC	pKa			
HL 1.34E-005	atm m3/mol	25	EST	MEYLAN,WM & HOWARD,PH (1991)
OH 8.24E-013	cm3/molc sec	25	EST	MEYLAN,WM & HOWARD,PH (1993)

CAS #: 014609-74-6 — SODIUM P-NITROPHENOXIDE

Formula: $C_6H_4NNaO_3$

Mol Weight: 161.09

MP (deg C): FP (deg C):

BP (deg C):

BP pressure (mm Hg):

Property/Value	Units	Temp	Data Type	Reference
WS 3.62E+005	mg/L	25	EST	MEYLAN,WM ET AL. (1996)
logP -1.31			EXP	HANSCH,C & LEO,AJ (1985)
VP 4.00E-010	mm Hg	25	EST	NEELY,WB & BLAU,GE (1985)
DC	pKa			
HL	atm m3/mol			
OH 2.69E-012	cm3/molc sec	25	EST	MEYLAN,WM & HOWARD,PH (1993)

CAS #: 014667-55-1 — PYRAZINE, TRIMETHYL-

Formula: $C_7H_{10}N_2$

Mol Weight: 122.17

MP (deg C): FP (deg C):

BP (deg C): 171-172

BP pressure (mm Hg):

Property/Value	Units	Temp	Data Type	Reference
WS 1.52E+004	mg/L	25	EST	MEYLAN,WM ET AL. (1996)
logP 0.95			EXP	YAMAGAMI,C ET AL. (1991)
VP 6.08E-001	mm Hg	25	EST	NEELY,WB & BLAU,GE (1985)
DC 2.80	pKa	27	EXP	PERRIN,DD (1965)
HL 3.92E-006	atm m3/mol	25	EST	MEYLAN,WM & HOWARD,PH (1991)
OH 2.33E-012	cm3/molc sec	25	EST	MEYLAN,WM & HOWARD,PH (1993)

CAS #: 014676-37-0 — ANISOLE, O-DICHLOROACETYLAMINO-

Formula: $C_9H_9Cl_2NO_2$

Mol Weight: 234.08

MP (deg C): FP (deg C):

BP (deg C):

BP pressure (mm Hg):

Property/Value	Units	Temp	Data Type	Reference
WS 2.53E+002	mg/L	25	EST	MEYLAN,WM ET AL. (1996)
logP 2.41			EXP	HANSCH,C & LEO,AJ (1985)
VP 2.91E-006	mm Hg	25	EST	NEELY,WB & BLAU,GE (1985)
DC	pKa			
HL 4.53E-011	atm m3/mol	25	EST	MEYLAN,WM & HOWARD,PH (1991)
OH 1.23E-011	cm3/molc sec	25	EST	MEYLAN,WM & HOWARD,PH (1993)

CAS #: 014680-18-3 — N-(O-HYDROXYPHENYL)METHYLBENZAMIDE

Formula: $C_{14}H_{13}NO_2$

Mol Weight: 227.27

MP (deg C): FP (deg C):

BP (deg C):

BP pressure (mm Hg):

Property/Value	Units	Temp	Data Type	Reference
WS 7.95E+002	mg/L	25	EST	MEYLAN,WM ET AL. (1996)
logP 2.55			EXP	HANSCH,C & LEO,AJ (1985)
VP 9.97E-009	mm Hg	25	EST	NEELY,WB & BLAU,GE (1985)
DC	pKa			
HL 4.08E-014	atm m3/mol	25	EST	MEYLAN,WM & HOWARD,PH (1991)
OH 5.19E-011	cm3/molc sec	25	EST	MEYLAN,WM & HOWARD,PH (1993)

CAS #: 014686-13-6 — 2-HEPTENE (TRANS)

Formula: C_7H_{14}

Mol Weight: 98.19

MP (deg C): -109.5 FP (deg C):

BP (deg C): 98

BP pressure (mm Hg):

Property/Value	Units	Temp	Data Type	Reference
WS 1.45E+001	mg/L	25	EXP	HINE,J & MOOKERJEE,PK (1975)
logP 3.56			EST	MEYLAN,WM & HOWARD,PH (1995)
VP 4.89E+001	mm Hg	25	EXP	YAWS,CL (1994A)
DC	pKa			
HL 4.35E-001	atm m3/mol	25	EXP	VP/WSOL
OH 6.04E-011	cm3/molc sec	25	EST	MEYLAN,WM & HOWARD,PH (1993)

CAS #: 014686-14-7 — 3-HEPTENE (TRANS)

Formula: C_7H_{14}

Mol Weight: 98.19

MP (deg C): -136.6 FP (deg C):

BP (deg C): 95.7

BP pressure (mm Hg):

Property/Value	Units	Temp	Data Type	Reference
WS 3.11E+001	mg/L	25	EST	MEYLAN,WM ET AL. (1996)
logP 3.56			EST	MEYLAN,WM & HOWARD,PH (1995)
VP 5.24E+001	mm Hg	25	EXP	YAWS,CL (1994A)
DC	pKa			
HL 5.62E-001	atm m3/mol	25	EST	MEYLAN,WM & HOWARD,PH (1991)
OH 6.00E-011	cm3/molc sec	25	EST	MEYLAN,WM & HOWARD,PH (1993)

CAS #: 014691-88-4 — 1-PIPERIDINYLOXY, 4-AMINO-2,2,6,6-TETRAMETHYL-

Formula: $C_9H_{20}N_2O$

Mol Weight: 172.27

MP (deg C): FP (deg C):

BP (deg C):

BP pressure (mm Hg):

Property/Value	Units	Temp	Data Type	Reference
WS 2.03E+005	mg/L	25	EST	MEYLAN,WM ET AL. (1996)
logP 0.56			EXP	SANGSTER,J (1993)
VP 5.47E-006	mm Hg	25	EST	NEELY,WB & BLAU,GE (1985)
DC	pKa			
HL 1.10E-011	atm m3/mol	25	EST	MEYLAN,WM & HOWARD,PH (1991)
OH 1.18E-010	cm3/molc sec	25	EST	MEYLAN,WM & HOWARD,PH (1993)

CAS #: 014691-89-5 — 4-NHAC-2266-TETRAME-PIPERIDIN-N-OXIDE

Formula: $C_{11}H_{22}N_2O_2$

Mol Weight: 214.31

MP (deg C): 144-146 FP (deg C):

BP (deg C):

BP pressure (mm Hg):

Property/Value	Units	Temp	Data Type	Reference
WS 1.34E+004	mg/L	25	EST	MEYLAN,WM ET AL. (1996)
logP 0.52			EXP	HANSCH,C & LEO,AJ (1985)
VP 1.96E-009	mm Hg	25	EST	NEELY,WB & BLAU,GE (1985)
DC	pKa			
HL 2.74E-017	atm m3/mol	25	EST	MEYLAN,WM & HOWARD,PH (1991)
OH 4.13E-011	cm3/molc sec	25	EST	MEYLAN,WM & HOWARD,PH (1993)

CAS #: 014698-29-4				OXOLINIC ACID
Formula: $C_{13}H_{11}NO_5$				
Mol Weight: 261.24				
MP (deg C): 314-316 de		FP (deg C):		
BP (deg C):				
BP pressure (mm Hg):				

Property/ Value	Units	Temp	Data Type	Reference
WS 8.01E+003	mg/L	25	EST	MEYLAN,WM ET AL. (1996)
logP 0.94			EXP	SANGSTER,J (1994)
VP 4.75E-008	mm Hg	25	EST	NEELY,WB & BLAU,GE (1985)
DC	pKa			
HL 4.12E-016	atm m3/mol	25	EST	MEYLAN,WM & HOWARD,PH (1991)
OH 1.26E-010	cm3/molc sec	25	EST	MEYLAN,WM & HOWARD,PH (1993)

CAS #: 014737-91-8				2-PROPENOIC ACID, 3-(2-METHOXYPHENYL)-, (Z)-
Formula: $C_{10}H_{10}O_3$				
Mol Weight: 178.19				
MP (deg C): 94-96		FP (deg C):		
BP (deg C):				
BP pressure (mm Hg):				

Property/ Value	Units	Temp	Data Type	Reference
WS 3.97E+003	mg/L	25	EST	MEYLAN,WM ET AL. (1996)
logP 1.81			EXP	SANGSTER,J (1994)
VP 1.59E-004	mm Hg	25	EST	NEELY,WB & BLAU,GE (1985)
DC 4.46	pKa		EXP	KORTUM,G ET AL (1961)
HL 7.65E-010	atm m3/mol	25	EST	MEYLAN,WM & HOWARD,PH (1991)
OH 4.13E-011	cm3/molc sec	25	EST	MEYLAN,WM & HOWARD,PH (1993)

CAS #: 014753-13-0				7-METHYL-4-NITROQUINOLINE-1-OXIDE
Formula: $C_{10}H_{10}N_2O_3$				
Mol Weight: 206.20				
MP (deg C):		FP (deg C):		
BP (deg C):				
BP pressure (mm Hg):				

Property/ Value	Units	Temp	Data Type	Reference
WS 1.04E+003	mg/L	25	EST	MEYLAN,WM ET AL. (1996)
logP 1.42			EXP	HANSCH,C & LEO,AJ (1985)
VP 1.06E-006	mm Hg	25	EST	NEELY,WB & BLAU,GE (1985)
DC	pKa			
HL 3.00E-014	atm m3/mol	25	EST	MEYLAN,WM & HOWARD,PH (1991)
OH 2.83E-012	cm3/molc sec	25	EST	MEYLAN,WM & HOWARD,PH (1993)

CAS #: 014753-17-4				7-NITROQUINOLINE-1-OXIDE
Formula: $C_9H_8N_2O_3$				
Mol Weight: 192.18				
MP (deg C):		FP (deg C):		
BP (deg C):				
BP pressure (mm Hg):				

Property/ Value	Units	Temp	Data Type	Reference
WS 1.85E+004	mg/L	25	EST	MEYLAN,WM ET AL. (1996)
logP 0.36			EXP	HANSCH,C & LEO,AJ (1985)
VP 2.60E-006	mm Hg	25	EST	NEELY,WB & BLAU,GE (1985)
DC	pKa			
HL 2.72E-014	atm m3/mol	25	EST	MEYLAN,WM & HOWARD,PH (1991)
OH 1.80E-012	cm3/molc sec	25	EST	MEYLAN,WM & HOWARD,PH (1993)

CAS #: 014753-18-5				8-NITROQUINOLINE-1-OXIDE
Formula: $C_9H_8N_2O_3$				
Mol Weight: 192.18				
MP (deg C):		FP (deg C):		
BP (deg C):				
BP pressure (mm Hg):				

Property/ Value	Units	Temp	Data Type	Reference
WS 1.85E+004	mg/L	25	EST	MEYLAN,WM ET AL. (1996)
logP 0.04			EXP	HANSCH,C & LEO,AJ (1985)
VP 2.60E-006	mm Hg	25	EST	NEELY,WB & BLAU,GE (1985)
DC	pKa			
HL 2.72E-014	atm m3/mol	25	EST	MEYLAN,WM & HOWARD,PH (1991)
OH 1.80E-012	cm3/molc sec	25	EST	MEYLAN,WM & HOWARD,PH (1993)

CAS #: 014753-19-6				4,8-DINITROQUINOLINE-1-OXIDE
Formula: $C_9H_7N_3O_5$				
Mol Weight: 237.17				
MP (deg C):		FP (deg C):		
BP (deg C):				
BP pressure (mm Hg):				

Property/ Value	Units	Temp	Data Type	Reference
WS 1.80E+003	mg/L	25	EST	MEYLAN,WM ET AL. (1996)
logP 0.76			EXP	HANSCH,C & LEO,AJ (1985)
VP 4.65E-008	mm Hg	25	EST	NEELY,WB & BLAU,GE (1985)
DC	pKa			
HL 1.07E-016	atm m3/mol	25	EST	MEYLAN,WM & HOWARD,PH (1991)
OH 1.57E-013	cm3/molc sec	25	EST	MEYLAN,WM & HOWARD,PH (1993)

CAS #: 014757-80-3				N,N-DIMETHYL-FURAN-3-CARBOXAMIDE
Formula: $C_7H_{10}NO_2$				
Mol Weight: 140.16				
MP (deg C):		FP (deg C):		
BP (deg C):				
BP pressure (mm Hg):				

Property/ Value	Units	Temp	Data Type	Reference
WS 5.27E+004	mg/L	25	EST	MEYLAN,WM ET AL. (1996)
logP 0.24			EXP	HANSCH,C & LEO,AJ (1985)
VP 5.64E-002	mm Hg	25	EST	NEELY,WB & BLAU,GE (1985)
DC	pKa			
HL 1.07E-008	atm m3/mol	25	EST	MEYLAN,WM & HOWARD,PH (1991)
OH 5.30E-011	cm3/molc sec	25	EST	MEYLAN,WM & HOWARD,PH (1993)

CAS #: 014759-06-9				SULFORIDAZINE
Formula: $C_{20}H_{25}N_3O_2S_2$				
Mol Weight: 403.57				
MP (deg C): 121-123		FP (deg C):		
BP (deg C):				
BP pressure (mm Hg):				

Property/ Value	Units	Temp	Data Type	Reference
WS 3.65E-001	mg/L	25	EST	MEYLAN,WM ET AL. (1996)
logP 4.45			EXP	MANNHOLD,R ET AL. (1990)
VP 1.11E-011	mm Hg	25	EST	NEELY,WB & BLAU,GE (1985)
DC	pKa			
HL 4.02E-014	atm m3/mol	25	EST	MEYLAN,WM & HOWARD,PH (1991)
OH 2.31E-010	cm3/molc sec	25	EST	MEYLAN,WM & HOWARD,PH (1993)

CAS #:	014803-72-6				CARBAMIC ACID, (4-METHOXYPHENYL)-, METHYL ESTER
Formula:	$C_9H_{11}NO_3$				
Mol Weight:	181.19				
MP (deg C):		FP (deg C):			
BP (deg C):					
BP pressure (mm Hg):					

Property/Value	Units	Temp	Data Type	Reference
WS 1.81E+003	mg/L	25	EST	MEYLAN,WM ET AL. (1996)
logP 1.73			EXP	TAKAHASHI,J ET AL. (1988)
VP 5.26E-003	mm Hg	25	EST	NEELY,WB & BLAU,GE (1985)
DC	pKa			
HL 1.29E-009	atm m3/mol	25	EST	MEYLAN,WM & HOWARD,PH (1991)
OH 3.79E-011	cm3/molc sec	25	EST	MEYLAN,WM & HOWARD,PH (1993)

CAS #:	014803-73-7				CARBAMIC ACID, (2-METHOXYPHENYL)-, METHYL ESTER
Formula:	$C_9H_{11}NO_3$				
Mol Weight:	181.19				
MP (deg C):		FP (deg C):			
BP (deg C):					
BP pressure (mm Hg):					

Property/Value	Units	Temp	Data Type	Reference
WS 1.46E+003	mg/L	25	EST	MEYLAN,WM ET AL. (1996)
logP 1.84			EXP	TAKAHASHI,J ET AL. (1988)
VP 5.26E-003	mm Hg	25	EST	NEELY,WB & BLAU,GE (1985)
DC	pKa			
HL 1.29E-009	atm m3/mol	25	EST	MEYLAN,WM & HOWARD,PH (1991)
OH 3.79E-011	cm3/molc sec	25	EST	MEYLAN,WM & HOWARD,PH (1993)

CAS #:	014804-32-1				2-ETHYLANISOLE
Formula:	$C_9H_{12}O$				
Mol Weight:	136.20				
MP (deg C):		FP (deg C):			
BP (deg C):	187				
BP pressure (mm Hg):					

Property/Value	Units	Temp	Data Type	Reference
WS 1.67E+002	mg/L	25	EST	MEYLAN,WM ET AL. (1996)
logP 3.18			EXP	NAKAGAWA,Y ET AL. (1992)
VP 5.49E-001	mm Hg	25	EST	NEELY,WB & BLAU,GE (1985)
DC	pKa			
HL 4.67E-004	atm m3/mol	25	EST	MEYLAN,WM & HOWARD,PH (1991)
OH 2.81E-011	cm3/molc sec	25	EST	MEYLAN,WM & HOWARD,PH (1993)

CAS #:	014805-91-5				2-PYRIDINECARBOXAMIDE,NN-DIMETHYL
Formula:	$C_8H_{10}N_2O$				
Mol Weight:	150.18				
MP (deg C):		FP (deg C):			
BP (deg C):					
BP pressure (mm Hg):					

Property/Value	Units	Temp	Data Type	Reference
WS 1.84E+005	mg/L	25	EST	MEYLAN,WM ET AL. (1996)
logP -0.45			EXP	YAMAGAMI,C ET AL. (1990A)
VP 3.80E-003	mm Hg	25	EST	NEELY,WB & BLAU,GE (1985)
DC	pKa			
HL 1.40E-011	atm m3/mol	25	EST	MEYLAN,WM & HOWARD,PH (1991)
OH 1.65E-011	cm3/molc sec	25	EST	MEYLAN,WM & HOWARD,PH (1993)

CAS #:	014816-18-3				PHOXIM
Formula:	$C_{12}H_{15}N_2O_3PS$				
Mol Weight:	298.30				
MP (deg C):		FP (deg C):			
BP (deg C):	102				
BP pressure (mm Hg):	1.00E-002				

Property/Value	Units	Temp	Data Type	Reference
WS 4.10E+000	mg/L	20	EXP	YALKOWSKY,SH & DANNENFELSER,RM (1992)
logP 4.39			EXP	HANSCH,C & LEO,AJ (1985)
VP 8.50E-006	mm Hg	25	EST	NEELY,WB & BLAU,GE (1985)
DC	pKa			
HL 7.39E-005	atm m3/mol	25	EST	MEYLAN,WM & HOWARD,PH (1991)
OH 9.59E-011	cm3/molc sec	25	EST	MEYLAN,WM & HOWARD,PH (1993)

CAS #:	014838-15-4				PHENYLPROPANOLAMINE
Formula:	$C_9H_{13}NO$				
Mol Weight:	151.21				
MP (deg C):		FP (deg C):			
BP (deg C):					
BP pressure (mm Hg):					

Property/Value	Units	Temp	Data Type	Reference
WS 1.49E+005	mg/L	25	EST	MEYLAN,WM ET AL. (1996)
logP 0.67			EXP	SANGSTER,J (1994)
VP 8.67E-004	mm Hg	25	EST	NEELY,WB & BLAU,GE (1985)
DC 9.44	pKa	20	EXP	PERRIN,DD (1965)
HL 4.00E-011	atm m3/mol	25	EST	MEYLAM,WM & HOWARD,PH (1991)
OH 5.70E-011	cm3/molc sec	25	EST	ATKINSON,R (1988)

CAS #:	014850-23-8				4-OCTENE (TRANS)
Formula:	C_6H_{12}				
Mol Weight:	84.16				
MP (deg C):	-93.8	FP (deg C):			
BP (deg C):	122.3				
BP pressure (mm Hg):					

Property/Value	Units	Temp	Data Type	Reference
WS 1.07E+001	mg/L	25	EST	MEYLAN,WM ET AL. (1996)
logP 4.06			EST	MEYLAN,WM & HOWARD,PH (1995)
VP 1.78E+001	mm Hg	25	EXP	DAUBERT,TE & DANNER,RP (1989)
DC	pKa			
HL 7.46E-001	atm m3/mol	25	EST	MEYLAN,WM & HOWARD,PH (1991)
OH 6.13E-011	cm3/molc sec	25	EST	MEYLAN,WM & HOWARD,PH (1993)

CAS #:	014884-01-6				4-METHOXYPYRAZOLE
Formula:	$C_4H_6N_2O$				
Mol Weight:	98.11				
MP (deg C):		FP (deg C):			
BP (deg C):					
BP pressure (mm Hg):					

Property/Value	Units	Temp	Data Type	Reference
WS 5.72E+004	mg/L	25	EST	MEYLAN,WM ET AL. (1996)
logP 0.37			EXP	HANSCH,C & LEO,AJ (1985)
VP 1.38E-001	mm Hg	25	EST	NEELY,WB & BLAU,GE (1985)
DC	pKa			
HL 2.18E-007	atm m3/mol	25	EST	MEYLAN,WM & HOWARD,PH (1991)
OH 2.01E-010	cm3/molc sec	25	EST	MEYLAN,WM & HOWARD,PH (1993)

CAS #: 014884-03-8	4-ISOPROPOXYPYRAZOLE

Formula: $C_6H_{10}N_2O$

Mol Weight: 126.16

MP (deg C):   FP (deg C):

BP (deg C):

BP pressure (mm Hg):

Property/ Value	Units	Temp	Data Type	Reference
WS 1.03E+004	mg/L	25	EST	MEYLAN,WM ET AL. (1996)
logP 1.13			EXP	HANSCH,C & LEO,AJ (1985)
VP 2.52E-002	mm Hg	25	EST	NEELY,WB & BLAU,GE (1985)
DC	pKa			
HL 3.85E-007	atm m3/mol	25	EST	MEYLAN,WM & HOWARD,PH (1991)
OH 2.12E-010	cm3/molc sec	25	EST	MEYLAN,WM & HOWARD,PH (1993)

CAS #: 014885-29-1	1H-IMIDAZOLE, 1-METHYL-2-(1-METHYLETHYL)-5-NITRO

Formula: $C_7H_{11}N_3O_2$

Mol Weight: 169.18

MP (deg C): 60   FP (deg C):

BP (deg C):

BP pressure (mm Hg):

Property/ Value	Units	Temp	Data Type	Reference
WS 1.74E+003	mg/L	25	EST	MEYLAN,WM ET AL. (1996)
logP 1.36			EXP	KOSANOVIC,D ET AL. (1988)
VP 1.49E-004	mm Hg	25	EST	NEELY,WB & BLAU,GE (1985)
DC	pKa			
HL 6.15E-007	atm m3/mol	25	EST	MEYLAN,WM & HOWARD,PH (1991)
OH 6.20E-012	cm3/molc sec	25	EST	MEYLAN,WM & HOWARD,PH (1993)

CAS #: 014896-01-6	N7-PHENYL-MITOMYCIN C

Formula: $C_{21}H_{22}N_4O_5$

Mol Weight: 410.43

MP (deg C):   FP (deg C):

BP (deg C):

BP pressure (mm Hg):

Property/ Value	Units	Temp	Data Type	Reference
WS 2.09E+003	mg/L	25	EST	MEYLAN,WM ET AL. (1996)
logP 1.30			EXP	HANSCH,C & LEO,AJ (1985)
VP 3.13E-012	mm Hg	25	EST	NEELY,WB & BLAU,GE (1985)
DC	pKa			
HL 6.28E-025	atm m3/mol	25	EST	MEYLAN,WM & HOWARD,PH (1991)
OH 8.87E-011	cm3/molc sec	25	EST	MEYLAN,WM & HOWARD,PH (1993)

CAS #: 014906-59-3	4-CYANOPYRIDINE OXIDE

Formula: $C_6H_6N_2O$

Mol Weight: 122.13

MP (deg C): 223-225   FP (deg C):

BP (deg C):

BP pressure (mm Hg):

Property/ Value	Units	Temp	Data Type	Reference
WS 3.46E+005	mg/L	25	EST	MEYLAN,WM ET AL. (1996)
logP -0.94			EXP	HANSCH,C & LEO,AJ (1985)
VP 1.72E-003	mm Hg	25	EST	NEELY,WB & BLAU,GE (1985)
DC	pKa			
HL 6.81E-013	atm m3/mol	25	EST	MEYLAN,WM & HOWARD,PH (1991)
OH 6.53E-014	cm3/molc sec	25	EST	MEYLAN,WM & HOWARD,PH (1993)

CAS #: 014920-81-1	BENZOIC ACID, 2,6-DIMETHYL-, METHYL ESTER

Formula: $C_{10}H_{12}O_2$

Mol Weight: 164.21

MP (deg C):   FP (deg C):

BP (deg C):

BP pressure (mm Hg):

Property/ Value	Units	Temp	Data Type	Reference
WS 4.35E+002	mg/L	25	EST	MEYLAN,WM ET AL. (1996)
logP 2.55			EXP	SOTOMATSU,T ET AL. (1993)
VP 6.25E-002	mm Hg	25	EST	NEELY,WB & BLAU,GE (1985)
DC	pKa			
HL 4.23E-005	atm m3/mol	25	EST	MEYLAN,WM & HOWARD,PH (1991)
OH 4.76E-012	cm3/molc sec	25	EST	MEYLAN,WM & HOWARD,PH (1993)

CAS #: 014920-87-7	BENZOIC ACID, 2,6-DICHLORO-, METHYL ESTER

Formula: $C_8H_6Cl_2O_2$

Mol Weight: 205.04

MP (deg C):   FP (deg C):

BP (deg C):

BP pressure (mm Hg):

Property/ Value	Units	Temp	Data Type	Reference
WS 1.71E+002	mg/L	25	EST	MEYLAN,WM ET AL. (1996)
logP 2.79			EXP	SOTOMATSU,T ET AL. (1993)
VP 1.20E-002	mm Hg	25	EST	NEELY,WB & BLAU,GE (1985)
DC	pKa			
HL 1.91E-005	atm m3/mol	25	EST	MEYLAN,WM & HOWARD,PH (1991)
OH 5.28E-013	cm3/molc sec	25	EST	MEYLAN,WM & HOWARD,PH (1993)

CAS #: 014920-92-4	PENTAMETHYLPHENYLDISILOXANE

Formula: $C_{11}H_{20}OSi_2$

Mol Weight: 224.45

MP (deg C):   FP (deg C):

BP (deg C):

BP pressure (mm Hg):

Property/ Value	Units	Temp	Data Type	Reference
WS 4.26E-002	mg/L	25	EST	MEYLAN,WM ET AL. (1996)
logP 5.98			EST	MEYLAN,WM & HOWARD,PH (1995)
VP	mm Hg			
DC	pKa			
HL	atm m3/mol			
OH 2.70E-012	cm3/molc sec	25	EST	MEYLAN,WM & HOWARD,PH (1993)

CAS #: 014920-93-5	1,1,3-TRIMETHYL-1,3,3-TRIPHENYL-DISILOXANE

Formula: $C_{21}H_{24}OSi_2$

Mol Weight: 348.60

MP (deg C):   FP (deg C):

BP (deg C):

BP pressure (mm Hg):

Property/ Value	Units	Temp	Data Type	Reference
WS 6.90E-005	mg/L	25	EST	MEYLAN,WM ET AL. (1996)
logP 8.41			EST	MEYLAN,WM & HOWARD,PH (1995)
VP	mm Hg			
DC	pKa			
HL	atm m3/mol			
OH 6.30E-012	cm3/molc sec	25	EST	MEYLAN,WM & HOWARD,PH (1993)

CAS #: 014938-35-3	4-PENTYLPHENOL

Formula: $C_{11}H_{16}O$

Mol Weight: 164.25

MP (deg C): 23 FP (deg C):

BP (deg C): 250.5

BP pressure (mm Hg):

Property/Value	Units	Temp	Data Type	Reference
WS 8.48E+001	mg/L	25	EST	MEYLAN,WM ET AL. (1996)
logP 4.06			EXP	HANSCH,C & LEO,AJ (1985)
VP 2.29E-003	mm Hg	25	EST	NEELY,WB & BLAU,GE (1985)
DC	pKa			
HL 1.92E-006	atm m3/mol	25	EST	MEYLAN,WM & HOWARD,PH (1991)
OH 4.60E-011	cm3/molc sec	25	EST	MEYLAN,WM & HOWARD,PH (1993)

CAS #: 014949-00-9	2-NH2-1,3,4-THIADIAZOLE-SO2NH2

Formula: $C_2H_6N_4O_2S_2$

Mol Weight: 182.22

MP (deg C): FP (deg C):

BP (deg C):

BP pressure (mm Hg):

Property/Value	Units	Temp	Data Type	Reference
WS 3.23E+005	mg/L	25	EST	MEYLAN,WM ET AL. (1996)
logP -0.90			EXP	HANSCH,C & LEO,AJ (1985)
VP 8.16E-007	mm Hg	25	EST	NEELY,WB & BLAU,GE (1985)
DC	pKa			
HL 4.32E-014	atm m3/mol	25	EST	MEYLAN,WM & HOWARD,PH (1991)
OH 7.00E-013	cm3/molc sec	25	EST	MEYLAN,WM & HOWARD,PH (1993)

CAS #: 014959-84-3	4,8-DIHYDROXYQUINOLINE

Formula: $C_9H_7NO_2$

Mol Weight: 161.16

MP (deg C): FP (deg C):

BP (deg C):

BP pressure (mm Hg):

Property/Value	Units	Temp	Data Type	Reference
WS 7.60E+004	mg/L	25	EST	MEYLAN,WM ET AL. (1996)
logP 0.62			EXP	HANSCH,C & LEO,AJ (1985)
VP 4.02E-006	mm Hg	25	EST	NEELY,WB & BLAU,GE (1985)
DC	pKa			
HL 7.45E-015	atm m3/mol	25	EST	MEYLAN,WM & HOWARD,PH (1991)
OH 1.70E-010	cm3/molc sec	25	EST	MEYLAN,WM & HOWARD,PH (1993)

CAS #: 014959-86-5	7-DODECEN-1-OL ACETATE (Z)

Formula: $C_{14}H_{26}O_2$

Mol Weight: 226.36

MP (deg C): FP (deg C):

BP (deg C): 98-100

BP pressure (mm Hg): 5.00E-002

Property/Value	Units	Temp	Data Type	Reference
WS 5.68E-001	mg/L	25	EST	MEYLAN,WM ET AL. (1996)
logP 5.56			EST	MEYLAN,WM & HOWARD,PH (1995)
VP 4.10E-003	mm Hg	25	EST	NEELY,WB & BLAU,GE (1985)
DC	pKa			
HL 3.48E-003	atm m3/mol	25	EST	MEYLAN,WM & HOWARD,PH (1991)
OH 6.90E-011	cm3/molc sec	25	EST	MEYLAN,WM & HOWARD,PH (1993)

CAS #: 014983-92-7	CARBAMIC ACID, (2-METHYLPHENYL)-, METHYL ESTER

Formula: $C_9H_{11}NO_2$

Mol Weight: 165.19

MP (deg C): FP (deg C):

BP (deg C):

BP pressure (mm Hg):

Property/Value	Units	Temp	Data Type	Reference
WS 2.33E+003	mg/L	25	EST	MEYLAN,WM ET AL. (1996)
logP 1.69			EXP	TAKAHASHI,J ET AL. (1988)
VP 1.87E-002	mm Hg	25	EST	NEELY,WB & BLAU,GE (1985)
DC	pKa			
HL 2.41E-008	atm m3/mol	25	EST	MEYLAN,WM & HOWARD,PH (1991)
OH 5.26E-011	cm3/molc sec	25	EST	MEYLAN,WM & HOWARD,PH (1993)

CAS #: 014984-65-7	PHOSPHINIC ACID,BIS(DIMEAZ),ET EST

Formula: $C_{10}H_{21}N_2O_2P$

Mol Weight: 232.26

MP (deg C): FP (deg C):

BP (deg C):

BP pressure (mm Hg):

Property/Value	Units	Temp	Data Type	Reference
WS 3.98E+003	mg/L	25	EST	MEYLAN,WM ET AL. (1996)
logP 1.02			EXP	HANSCH,C & LEO,AJ (1985)
VP 7.79E-004	mm Hg	25	EST	NEELY,WB & BLAU,GE (1985)
DC	pKa			
HL 1.87E-010	atm m3/mol	25	EST	MEYLAN,WM & HOWARD,PH (1991)
OH 3.38E-011	cm3/molc sec	25	EST	MEYLAN,WM & HOWARD,PH (1993)

CAS #: 014985-34-3	URIDINE, 3-(PHENYLMETHYL)-

Formula: $C_{16}H_{18}N_2O_6$

Mol Weight: 334.33

MP (deg C): FP (deg C):

BP (deg C):

BP pressure (mm Hg):

Property/Value	Units	Temp	Data Type	Reference
WS 1.97E+002	mg/L	25	EST	MEYLAN,WM ET AL. (1996)
logP 0.33			EXP	YAMAMOTO,I ET AL. (1987)
VP 7.64E-017	mm Hg	25	EST	NEELY,WB & BLAU,GE (1985)
DC	pKa			
HL 5.00E-020	atm m3/mol	25	EST	MEYLAN,WM & HOWARD,PH (1991)
OH 1.00E-010	cm3/molc sec	25	EST	MEYLAN,WM & HOWARD,PH (1993)

CAS #: 015006-14-1	21,22-DIHYDROSTRYCHNINE

Formula: $C_{21}H_{24}N_2O_2$

Mol Weight: 336.44

MP (deg C): FP (deg C):

BP (deg C):

BP pressure (mm Hg):

Property/Value	Units	Temp	Data Type	Reference
WS 3.09E+002	mg/L	25	EST	MEYLAN,WM ET AL. (1996)
logP 1.62			EXP	HANSCH,C & LEO,AJ (1985)
VP 3.84E-009	mm Hg	25	EST	NEELY,WB & BLAU,GE (1985)
DC	pKa			
HL 7.23E-015	atm m3/mol	25	EST	MEYLAN,WM & HOWARD,PH (1991)
OH 1.12E-010	cm3/molc sec	25	EST	MEYLAN,WM & HOWARD,PH (1993)

CAS #: 015009-91-3 — 2-NITROPYRIDINE

Formula:	$C_5H_4N_2O_2$			
Mol Weight:	124.10			
MP (deg C):		FP (deg C):		
BP (deg C):				
BP pressure (mm Hg):				

Property/ Value	Units	Temp	Data Type	Reference
WS 1.54E+004	mg/L	25	EST	MEYLAN,WM ET AL. (1996)
logP 0.48			EXP	HANSCH,C & LEO,AJ (1985)
VP 1.15E-001	mm Hg	25	EST	NEELY,WB & BLAU,GE (1985)
DC	pKa			
HL 2.78E-008	atm m3/mol	25	EST	MEYLAN,WM & HOWARD,PH (1991)
OH 4.62E-014	cm3/molc sec	25	EST	MEYLAN,WM & HOWARD,PH (1993)

CAS #: 015011-28-6 — 4,5-DIMETHYL-8-QUINOLINOL

Formula:	$C_{11}H_{11}NO$			
Mol Weight:	173.22			
MP (deg C):		FP (deg C):		
BP (deg C):				
BP pressure (mm Hg):				

Property/ Value	Units	Temp	Data Type	Reference
WS 1.10E+003	mg/L	25	EST	MEYLAN,WM ET AL. (1996)
logP 2.71			EXP	HANSCH,C & LEO,AJ (1985)
VP 3.50E-005	mm Hg	25	EST	NEELY,WB & BLAU,GE (1985)
DC	pKa			
HL 8.72E-011	atm m3/mol	25	EST	MEYLAN,WM & HOWARD,PH (1991)
OH 1.58E-010	cm3/molc sec	25	EST	MEYLAN,WM & HOWARD,PH (1993)

CAS #: 015018-66-3 — 4-QUINAZOLINAMINE

Formula:	$C_8H_7N_3$			
Mol Weight:	145.17			
MP (deg C):		FP (deg C):		
BP (deg C):				
BP pressure (mm Hg):				

Property/ Value	Units	Temp	Data Type	Reference
WS 6.43E+003	mg/L	25	EST	MEYLAN,WM ET AL. (1996)
logP 1.28			EXP	HANSCH,C & LEO,AJ (1985)
VP 2.50E-004	mm Hg	25	EST	NEELY,WB & BLAU,GE (1985)
DC 5.70	pKa	20	EXP	PERRIN,DD (1965)
HL 1.01E-010	atm m3/mol	25	EST	MEYLAN,WM & HOWARD,PH (1991)
OH 1.10E-010	cm3/molc sec	25	EST	MEYLAN,WM & HOWARD,PH (1993)

CAS #: 015045-43-9 — 2,2,5,5-TETRAMETHYL-TETRAHYDROFURAN

Formula:	$C_8H_{16}O$			
Mol Weight:	128.22			
MP (deg C):		FP (deg C):		
BP (deg C):	112			
BP pressure (mm Hg):				

Property/ Value	Units	Temp	Data Type	Reference
WS 1.63E+003	mg/L	25	EST	MEYLAN,WM ET AL. (1996)
logP 2.06			EXP	HANSCH,C ET AL. (1995)
VP 1.44E+001	mm Hg	25	EST	NEELY,WB & BLAU,GE (1985)
DC	pKa			
HL 2.62E-004	atm m3/mol	25	EST	MEYLAN,WM & HOWARD,PH (1991)
OH 8.34E-012	cm3/molc sec	25	EST	MEYLAN,WM & HOWARD,PH (1993)

CAS #: 015062-75-6 — 5,7-DIME-2-OXO-1,3-BENZOXATHIOL

Formula:	$C_9H_8O_2S$			
Mol Weight:	180.23			
MP (deg C):		FP (deg C):		
BP (deg C):				
BP pressure (mm Hg):				

Property/ Value	Units	Temp	Data Type	Reference
WS 9.21E+001	mg/L	25	EST	MEYLAN,WM ET AL. (1996)
logP 3.25			EXP	HANSCH,C & LEO,AJ (1985)
VP 9.95E-005	mm Hg	25	EST	NEELY,WB & BLAU,GE (1985)
DC	pKa			
HL 9.02E-005	atm m3/mol	25	EST	MEYLAN,WM & HOWARD,PH (1991)
OH 8.67E-012	cm3/molc sec	25	EST	MEYLAN,WM & HOWARD,PH (1993)

CAS #: 015066-77-0 — TRIMETHYL HEXYL AMMONIUM IODIDE

Formula:	$C_9H_{22}IN$			
Mol Weight:	271.19			
MP (deg C):		FP (deg C):		
BP (deg C):				
BP pressure (mm Hg):				

Property/ Value	Units	Temp	Data Type	Reference
WS 6.71E+005	mg/L	25	EST	MEYLAN,WM ET AL. (1996)
logP -1.84			EXP	HANSCH,C & LEO,AJ (1985)
VP 1.08E-007	mm Hg	25	EST	NEELY,WB & BLAU,GE (1985)
DC	pKa			
HL 1.09E-011	atm m3/mol	25	EST	MEYLAN,WM & HOWARD,PH (1991)
OH 2.00E-011	cm3/molc sec	25	EST	MEYLAN,WM & HOWARD,PH (1993)

CAS #: 015104-03-7 — 4-METHYL-N-NITROSOPIPERIDINE

Formula:	$C_6H_{12}N_2O$			
Mol Weight:	128.18			
MP (deg C):		FP (deg C):		
BP (deg C):				
BP pressure (mm Hg):				

Property/ Value	Units	Temp	Data Type	Reference
WS 1.19E+004	mg/L	25	EST	MEYLAN,WM ET AL. (1996)
logP 1.05			EXP	HANSCH,C & LEO,AJ (1985)
VP 6.35E-002	mm Hg	25	EST	NEELY,WB & BLAU,GE (1985)
DC	pKa			
HL 2.82E-006	atm m3/mol	25	EST	MEYLAN,WM & HOWARD,PH (1991)
OH 2.73E-011	cm3/molc sec	25	EST	MEYLAN,WM & HOWARD,PH (1993)

CAS #: 015104-61-7 — 1,1,2,3,3-PENTACHLOROPROPANE

Formula:	$C_3H_3Cl_5$			
Mol Weight:	216.32			
MP (deg C):	-10	FP (deg C):		
BP (deg C):	204			
BP pressure (mm Hg):				

Property/ Value	Units	Temp	Data Type	Reference
WS 4.10E+002	mg/L	25	EXP	CHEM INSPECT TEST INST (1992)
logP 3.23			EXP	CHEM INSPECT TEST INST (1992)
VP 9.38E-001	mm Hg	25	EXT	DYKYJ,J (1970)
DC	pKa			
HL 7.04E-004	atm m3/mol	25	EST	MEYLAN,WM & HOWARD,PH (1991)
OH 2.97E-013	cm3/molc sec	25	EST	MEYLAN,WM & HOWARD,PH (1993)

CAS #: 015187-71-0				1,3,3-TRICHLOROBUTANE
Formula: $C_4H_7Cl_3$				
Mol Weight: 161.46				
MP (deg C):		FP (deg C):		
BP (deg C):				
BP pressure (mm Hg):				

Property/ Value	Units	Temp	Data Type	Reference
WS 5.03E+001	mg/L	25	EST	MEYLAN,WM ET AL. (1996)
logP 3.66			EST	MEYLAN,WM & HOWARD,PH (1995)
VP 6.52E+000	mm Hg	25	EST	NEELY,WB & BLAU,GE (1985)
DC	pKa			
HL 7.53E-003	atm m3/mol	25	EST	MEYLAN,WM & HOWARD,PH (1991)
OH 5.55E-013	cm3/molc sec	25	EST	MEYLAN,WM & HOWARD,PH (1993)

CAS #: 015210-60-3				1-AMINOADAMANTANE,3,5,7-TRIMETHYL
Formula: $C_{13}H_{23}N$				
Mol Weight: 193.33				
MP (deg C):		FP (deg C):		
BP (deg C):				
BP pressure (mm Hg):				

Property/ Value	Units	Temp	Data Type	Reference
WS 1.81E+002	mg/L	25	EST	MEYLAN,WM ET AL. (1996)
logP 4.01			EXP	HANSCH,C & LEO,AJ (1985)
VP 1.73E-002	mm Hg	25	EST	NEELY,WB & BLAU,GE (1985)
DC	pKa			
HL 1.95E-005	atm m3/mol	25	EST	MEYLAN,WM & HOWARD,PH (1991)
OH 3.00E-011	cm3/molc sec	25	EST	MEYLAN,WM & HOWARD,PH (1993)

CAS #: 015233-37-1				4,6-NH2 2,2-DIME-1(4-MEPH)S-TRIAZINE
Formula: $C_{12}H_{17}N_5$				
Mol Weight: 231.30				
MP (deg C):		FP (deg C):		
BP (deg C):				
BP pressure (mm Hg):				

Property/ Value	Units	Temp	Data Type	Reference
WS 5.63E+003	mg/L	25	EST	MEYLAN,WM ET AL. (1996)
logP 0.85			EXP	HANSCH,C & LEO,AJ (1985)
VP 2.34E-006	mm Hg	25	EST	NEELY,WB & BLAU,GE (1985)
DC	pKa			
HL 8.03E-013	atm m3/mol	25	EST	MEYLAN,WM & HOWARD,PH (1991)
OH 2.22E-010	cm3/molc sec	25	EST	MEYLAN,WM & HOWARD,PH (1993)

CAS #: 015258-73-8				BENZENEMETHANOL, 2,6-DICHLORO-
Formula: $C_7H_6Cl_2O$				
Mol Weight: 177.03				
MP (deg C): 98		FP (deg C):		
BP (deg C):				
BP pressure (mm Hg):				

Property/ Value	Units	Temp	Data Type	Reference
WS 1.40E+003	mg/L	25	EXP	CHEM TEST INSPEC INST (1992)
logP 2.02			EXP	MIYAKE,F ET AL. (1987)
VP 6.99E-004	mm Hg	25	EST	NEELY,WB & BLAU,GE (1985)
DC	pKa			
HL 1.19E-007	atm m3/mol	25	EST	MEYLAN,WM & HOWARD,PH (1991)
OH 4.58E-012	cm3/molc sec	25	EST	MEYLAN,WM & HOWARD,PH (1993)

CAS #: 015263-52-2				S,S'-(2-(DIMETHYLAMINO)-1,3-PROPANEDIYL) CARBAM*
Formula: $C_7H_{16}ClN_3O_2S_2$				
Mol Weight: 273.81				
MP (deg C):		FP (deg C):		
BP (deg C):				
BP pressure (mm Hg):				

Property/ Value	Units	Temp	Data Type	Reference
WS 8.91E+004	mg/L	25	EST	MEYLAN,WM ET AL. (1996)
logP -0.95			EST	MEYLAN,WM & HOWARD,PH (1995)
VP 7.05E-009	mm Hg	25	EST	NEELY,WB & BLAU,GE (1985)
DC	pKa			
HL 2.05E-013	atm m3/mol	25	EST	MEYLAN,WM & HOWARD,PH (1991)
OH 1.29E-010	cm3/molc sec	25	EST	MEYLAN,WM & HOWARD,PH (1993)

CAS #: 015267-77-3				ACETIC ACID, [3-METHYL-4-(METHYLSULFONYL)PHENOXY
Formula: $C_{10}H_{12}O_5S$				
Mol Weight: 244.27				
MP (deg C):		FP (deg C):		
BP (deg C):				
BP pressure (mm Hg):				

Property/ Value	Units	Temp	Data Type	Reference
WS 2.19E+004	mg/L	25	EST	MEYLAN,WM ET AL. (1996)
logP 0.54			EXP	SANGSTER,J (1994)
VP 4.07E-007	mm Hg	25	EST	NEELY,WB & BLAU,GE (1985)
DC	pKa			
HL 4.29E-012	atm m3/mol	25	EST	MEYLAN,WM & HOWARD,PH (1991)
OH 1.80E-011	cm3/molc sec	25	EST	MEYLAN,WM & HOWARD,PH (1993)

CAS #: 015299-99-7				N,N-DIETHYL-2-(1-NAPHTHALENYLOXY)PROPANAMIDE
Formula: $C_{17}H_{21}NO_2$				
Mol Weight: 271.36				
MP (deg C): 75		FP (deg C):		
BP (deg C):				
BP pressure (mm Hg):				

Property/ Value	Units	Temp	Data Type	Reference
WS 7.30E+001	mg/L	20	EXP	YALKOWSKY,SH & DANNENFELSER,RM (1992)
logP 3.36			EXP	HANSCH,C ET AL. (1995)
VP 4.00E-006	mm Hg	25	EXP	JURY,WA ET AL. (1983)
DC	pKa			
HL 1.96E-008	atm m3/mol	25	EST	VP/WSOL
OH 2.32E-010	cm3/molc sec	25	EST	MEYLAN,WM & HOWARD,PH (1993)

CAS #: 015301-48-1				BEZITRAMIDE
Formula: $C_{31}H_{32}N_4O_2$				
Mol Weight: 492.63				
MP (deg C): 145-149		FP (deg C):		
BP (deg C):				
BP pressure (mm Hg):				

Property/ Value	Units	Temp	Data Type	Reference
WS 9.17E-002	mg/L	25	EST	MEYLAN,WM ET AL. (1996)
logP 4.80			EXP	SANGSTER,J (1993)
VP 1.29E-014	mm Hg	25	EST	NEELY,WB & BLAU,GE (1985)
DC	pKa			
HL 2.22E-013	atm m3/mol	25	EST	MEYLAN,WM & HOWARD,PH (1991)
OH 1.26E-010	cm3/molc sec	25	EST	MEYLAN,WM & HOWARD,PH (1993)

CAS #: 015307-71-8				PHENYLACETIC ACID,2-(2',6'-DICHLORO-3-METHYL)ANI
Formula: $C_{15}H_{13}Cl_2NO_2$				
Mol Weight: 310.18				
MP (deg C):			FP (deg C):	
BP (deg C):				
BP pressure (mm Hg):				

Property/Value	Units	Temp	Data Type	Reference
WS 1.64E+000	mg/L	25	EST	MEYLAN,WM ET AL. (1996)
logP 4.93			EXP	HANSCH,C ET AL. (1995)
VP 2.59E-008	mm Hg	25	EST	NEELY,WB & BLAU,GE (1985)
DC	pKa			
HL 5.22E-012	atm m3/mol	25	EST	MEYLAN,WM & HOWARD,PH (1991)
OH 1.85E-010	cm3/molc sec	25	EST	MEYLAN,WM & HOWARD,PH (1993)

CAS #: 015307-79-6				BENZENEACETIC ACID, 2-[(2,6-DICHLOROPHENYL)AMINO
Formula: $C_{14}H_{10}Cl_2NNaO_2$				
Mol Weight: 318.14				
MP (deg C): 283-285			FP (deg C):	
BP (deg C):				
BP pressure (mm Hg):				

Property/Value	Units	Temp	Data Type	Reference
WS 2.43E+003	mg/L	25	EST	MEYLAN,WM ET AL. (1996)
logP 0.70			EXP	SANGSTER,J (1994)
VP 4.75E-014	mm Hg	25	EST	NEELY,WB & BLAU,GE (1985)
DC				
HL	atm m3/mol			
OH 1.64E-010	cm3/molc sec	25	EST	MEYLAN,WM & HOWARD,PH (1993)

CAS #: 015307-81-0				BENZENEACETIC ACID, 2-[(2,6-DICHLOROPHENYL)AMINO
Formula: $C_{14}H_{10}Cl_2KNO_2$				
Mol Weight: 334.25				
MP (deg C):			FP (deg C):	
BP (deg C):				
BP pressure (mm Hg):				

Property/Value	Units	Temp	Data Type	Reference
WS 2.15E+003	mg/L	25	EST	MEYLAN,WM ET AL. (1996)
logP 0.65			EXP	SANGSTER,J (1994)
VP 4.75E-014	mm Hg	25	EST	NEELY,WB & BLAU,GE (1985)
DC	pKa			
HL	atm m3/mol			
OH 1.64E-010	cm3/molc sec	25	EST	MEYLAN,WM & HOWARD,PH (1993)

CAS #: 015307-86-5				DICLOFENAC
Formula: $C_{14}H_{11}Cl_2NO_2$				
Mol Weight: 296.16				
MP (deg C):			FP (deg C):	
BP (deg C):				
BP pressure (mm Hg):				

Property/Value	Units	Temp	Data Type	Reference
WS 5.61E+000	mg/L	25	EST	MEYLAN,WM ET AL. (1996)
logP 4.40			EXP	SANGSTER,J (1993)
VP 6.14E-008	mm Hg	25	EST	NEELY,WB & BLAU,GE (1985)
DC	pKa			
HL 4.73E-012	atm m3/mol	25	EST	MEYLAN,WM & HOWARD,PH (1991)
OH 1.65E-010	cm3/molc sec	25	EST	MEYLAN,WM & HOWARD,PH (1993)

CAS #: 015327-44-3				2(2,3-DICLPHENYLIMINO)IMIDAZOLINE
Formula: $C_9H_9Cl_2N_3$				
Mol Weight: 230.10				
MP (deg C):			FP (deg C):	
BP (deg C):				
BP pressure (mm Hg):				

Property/Value	Units	Temp	Data Type	Reference
WS 5.84E+002	mg/L	25	EST	MEYLAN,WM ET AL. (1996)
logP 2.01			EXP	HANSCH,C & LEO,AJ (1985)
VP 2.76E-007	mm Hg	25	EST	NEELY,WB & BLAU,GE (1985)
DC	pKa			
HL 1.51E-011	atm m3/mol	25	EST	MEYLAN,WM & HOWARD,PH (1991)
OH 8.12E-011	cm3/molc sec	25	EST	MEYLAN,WM & HOWARD,PH (1993)

CAS #: 015327-50-1				2-IMIDAZOLIDINIMINE, N-(5-BROMO-2-CHLOROPHENYL)-
Formula: $C_9H_9BrClN_3$				
Mol Weight: 274.55				
MP (deg C):			FP (deg C):	
BP (deg C):				
BP pressure (mm Hg):				

Property/Value	Units	Temp	Data Type	Reference
WS 3.11E+003	mg/L	25	EST	MEYLAN,WM ET AL. (1996)
logP 2.05			EXP	SANGSTER,J (1993)
VP 8.80E-006	mm Hg	25	EST	NEELY,WB & BLAU,GE (1985)
DC	pKa			
HL 8.09E-012	atm m3/mol	25	EST	MEYLAN,WM & HOWARD,PH (1991)
OH 1.44E-010	cm3/molc sec	25	EST	MEYLAN,WM & HOWARD,PH (1993)

CAS #: 015379-32-5				BARBITURIC ACID,5ET-5-TBU
Formula: $C_{10}H_{16}N_2O_3$				
Mol Weight: 212.25				
MP (deg C):			FP (deg C):	
BP (deg C):				
BP pressure (mm Hg):				

Property/Value	Units	Temp	Data Type	Reference
WS 2.51E+003	mg/L	25	EST	MEYLAN,WM ET AL. (1996)
logP 1.38			EXP	WONG,O & MCKEOWN,RH (1988)
VP 6.94E-011	mm Hg	25	EST	NEELY,WB & BLAU,GE (1985)
DC	pKa			
HL 6.36E-013	atm m3/mol	25	EST	MEYLAN,WM & HOWARD,PH (1991)
OH 6.31E-012	cm3/molc sec	25	EST	MEYLAN,WM & HOWARD,PH (1993)

CAS #: 015386-80-8				ACETAMIDE, 2-CYANO-N-(3,4-DICHLOROPHENYL)-
Formula: $C_9H_6Cl_2N_2O$				
Mol Weight: 229.07				
MP (deg C):			FP (deg C):	
BP (deg C):				
BP pressure (mm Hg):				

Property/Value	Units	Temp	Data Type	Reference
WS 7.95E+001	mg/L	25	EST	MEYLAN,WM ET AL. (1996)
logP 2.72			EXP	HANSCH,C ET AL. (1995)
VP 3.31E-007	mm Hg	25	EST	NEELY,WB & BLAU,GE (1985)
DC	pKa			
HL 1.41E-012	atm m3/mol	25	EST	MEYLAN,WM & HOWARD,PH (1991)
OH 2.68E-012	cm3/molc sec	25	EST	MEYLAN,WM & HOWARD,PH (1993)

CAS #: 015386-99-9				M-CHLORO-N-PHENYLSUCCINIMIDE
Formula: $C_{10}H_8ClNO_2$				
Mol Weight: 209.63				
MP (deg C):		FP (deg C):		
BP (deg C):				
BP pressure (mm Hg):				

Property/Value	Units	Temp	Data Type	Reference
WS 5.68E+003	mg/L	25	EST	MEYLAN,WM ET AL. (1996)
logP 0.98			EXP	HANSCH,C & LEO,AJ (1985)
VP 2.25E-007	mm Hg	25	EST	NEELY,WB & BLAU,GE (1985)
DC	pKa			
HL 9.67E-008	atm m3/mol	25	EST	MEYLAN,WM & HOWARD,PH (1991)
OH 1.31E-011	cm3/molc sec	25	EST	MEYLAN,WM & HOWARD,PH (1993)

CAS #: 015402-79-6				2,4,6-TRIMETHOXYAMPHETAMINE
Formula: $C_{12}H_{19}NO_3$				
Mol Weight: 225.29				
MP (deg C):		FP (deg C):		
BP (deg C):				
BP pressure (mm Hg):				

Property/Value	Units	Temp	Data Type	Reference
WS 1.50E+004	mg/L	25	EST	MEYLAN,WM ET AL. (1996)
logP 1.57			EXP	HANSCH,C & LEO,AJ (1985)
VP 2.31E-004	mm Hg	25	EST	NEELY,WB & BLAU,GE (1985)
DC	pKa			
HL 2.23E-010	atm m3/mol	25	EST	MEYLAN,WM & HOWARD,PH (1991)
OH 2.47E-010	cm3/molc sec	25	EST	MEYLAN,WM & HOWARD,PH (1993)

CAS #: 015402-81-0				BENZENEETHANAMINE, 2,3-DIMETHOXY-à-METHYL-
Formula: $C_{11}H_{17}NO_2$				
Mol Weight: 195.26				
MP (deg C):		FP (deg C):		
BP (deg C):				
BP pressure (mm Hg):				

Property/Value	Units	Temp	Data Type	Reference
WS 2.52E+004	mg/L	25	EST	MEYLAN,WM ET AL. (1996)
logP 1.49			EXP	HANSCH,C ET AL. (1995)
VP 1.83E-003	mm Hg	25	EST	NEELY,WB & BLAU,GE (1985)
DC	pKa			
HL 3.77E-009	atm m3/mol	25	EST	MEYLAN,WM & HOWARD,PH (1991)
OH 9.24E-011	cm3/molc sec	25	EST	MEYLAN,WM & HOWARD,PH (1993)

CAS #: 015414-78-5				N-PHENYL-2,4-IMIDAZOLIDIONE
Formula: $C_9H_8N_2O_2$				
Mol Weight: 176.18				
MP (deg C):		FP (deg C):		
BP (deg C):				
BP pressure (mm Hg):				

Property/Value	Units	Temp	Data Type	Reference
WS 1.12E+004	mg/L	25	EST	MEYLAN,WM ET AL. (1996)
logP 0.83			EXP	LIPINSKI,CA ET AL. (1991)
VP 2.46E-008	mm Hg	25	EST	NEELY,WB & BLAU,GE (1985)
DC	pKa			
HL 8.08E-009	atm m3/mol	25	EST	MEYLAN,WM & HOWARD,PH (1991)
OH 1.57E-011	cm3/molc sec	25	EST	MEYLAN,WM & HOWARD,PH (1993)

CAS #: 015422-25-0				N-METHYLPHENOXYACETAMIDE
Formula: $C_9H_{11}NO_2$				
Mol Weight: 165.19				
MP (deg C):		FP (deg C):		
BP (deg C):				
BP pressure (mm Hg):				

Property/Value	Units	Temp	Data Type	Reference
WS 8.72E+003	mg/L	25	EST	MEYLAN,WM ET AL. (1996)
logP 1.02			EXP	HANSCH,C & LEO,AJ (1985)
VP 7.11E-005	mm Hg	25	EST	NEELY,WB & BLAU,GE (1985)
DC	pKa			
HL 7.52E-010	atm m3/mol	25	EST	MEYLAN,WM & HOWARD,PH (1991)
OH 3.18E-011	cm3/molc sec	25	EST	MEYLAN,WM & HOWARD,PH (1993)

CAS #: 015432-98-1				1H-INDENE-1,3(2H)-DIONE, 2-(3-METHYLPHENYL)-
Formula: $C_{16}H_{12}O_2$				
Mol Weight: 236.27				
MP (deg C):		FP (deg C):		
BP (deg C):				
BP pressure (mm Hg):				

Property/Value	Units	Temp	Data Type	Reference
WS 3.12E+001	mg/L	25	EST	MEYLAN,WM ET AL. (1996)
logP 3.46			EXP	SANGSTER,J (1993)
VP 1.35E-006	mm Hg	25	EST	NEELY,WB & BLAU,GE (1985)
DC	pKa			
HL 5.29E-010	atm m3/mol	25	EST	MEYLAN,WM & HOWARD,PH (1991)
OH 1.38E-011	cm3/molc sec	25	EST	MEYLAN,WM & HOWARD,PH (1993)

CAS #: 015438-85-4				2,4-BIS(DIETHYLAMINO)-6-METHOXY-S-TRIAZINE
Formula: $C_{12}H_{23}N_5O$				
Mol Weight: 253.35				
MP (deg C):		FP (deg C):		
BP (deg C):				
BP pressure (mm Hg):				

Property/Value	Units	Temp	Data Type	Reference
WS 1.77E+000	mg/L	25	EST	MEYLAN,WM ET AL. (1996)
logP 4.81			EST	MEYLAN,WM & HOWARD,PH (1995)
VP 1.85E-005	mm Hg	25	EST	NEELY,WB & BLAU,GE (1985)
DC	pKa			
HL 1.82E-006	atm m3/mol	25	EST	MEYLAN,WM & HOWARD,PH (1991)
OH 3.66E-011	cm3/molc sec	25	EST	MEYLAN,WM & HOWARD,PH (1993)

CAS #: 015446-39-6				N,N'-(AZODI-4,1-PHENYLENE)BISACETAMIDE
Formula: $C_{16}H_{16}N_4O_2$				
Mol Weight: 296.33				
MP (deg C):		FP (deg C):		
BP (deg C):				
BP pressure (mm Hg):				

Property/Value	Units	Temp	Data Type	Reference
WS 4.94E+001	mg/L	25	EST	MEYLAN,WM ET AL. (1996)
logP 2.32			EST	MEYLAN,WM & HOWARD,PH (1995)
VP 1.64E-011	mm Hg	25	EST	NEELY,WB & BLAU,GE (1985)
DC	pKa			
HL 3.06E-014	atm m3/mol	25	EST	MEYLAN,WM & HOWARD,PH (1991)
OH 1.00E-011	cm3/molc sec	25	EST	MEYLAN,WM & HOWARD,PH (1993)

CAS #: 015457-05-3	FLUORODIFEN

Formula: $C_{13}H_7F_3N_2O_5$

Mol Weight: 328.21

MP (deg C): 　　FP (deg C):

BP (deg C):

BP pressure (mm Hg):

Property/ Value	Units	Temp	Data Type	Reference
WS 2.00E+000	mg/L	20	EXP	YALKOWSKY,SH & DANNENFELSER,RM (1992)
logP 3.65			EXP	HANSCH,C & LEO,AJ (1985)
VP 5.07E-007	mm Hg	25	EST	NEELY,WB & BLAU,GE (1985)
DC	pKa			
HL 1.59E-008	atm m3/mol	25	EST	MEYLAN,WM & HOWARD,PH (1991)
OH 7.39E-013	cm3/molc sec	25	EST	MEYLAN,WM & HOWARD,PH (1993)

CAS #: 015468-86-7	1,3,5-TRIAZINE, 6-CHLORO-2,4-BIS[(2-PROPENYL)AMI

Formula: $C_9H_{12}ClN_5$

Mol Weight: 225.68

MP (deg C): 　　FP (deg C):

BP (deg C):

BP pressure (mm Hg):

Property/ Value	Units	Temp	Data Type	Reference
WS 1.44E+002	mg/L	25	EST	MEYLAN,WM ET AL. (1996)
logP 2.75			EXP	MITSUTAKE,KI ET AL. (1986)
VP 3.29E-005	mm Hg	25	EST	NEELY,WB & BLAU,GE (1985)
DC	pKa			
HL 3.30E-009	atm m3/mol	25	EST	MEYLAN,WM & HOWARD,PH (1991)
OH 7.01E-011	cm3/molc sec	25	EST	MEYLAN,WM & HOWARD,PH (1993)

CAS #: 015482-11-8	N-METHYL-2-PROPYLPHENYLCARBAMATE

Formula: $C_{11}H_{15}NO_2$

Mol Weight: 193.25

MP (deg C): 　　FP (deg C):

BP (deg C):

BP pressure (mm Hg):

Property/ Value	Units	Temp	Data Type	Reference
WS 4.22E+002	mg/L	25	EST	MEYLAN,WM ET AL. (1996)
logP 2.40			EXP	HANSCH,C & LEO,AJ (1985)
VP 2.11E-003	mm Hg	25	EST	NEELY,WB & BLAU,GE (1985)
DC	pKa			
HL 6.25E-008	atm m3/mol	25	EST	MEYLAN,WM & HOWARD,PH (1991)
OH 1.21E-011	cm3/molc sec	25	EST	MEYLAN,WM & HOWARD,PH (1993)

CAS #: 015506-53-3	1,3-CYCLOBUTANEDIONE

Formula: $C_4H_4O_2$

Mol Weight: 84.08

MP (deg C): 　　FP (deg C):

BP (deg C):

BP pressure (mm Hg):

Property/ Value	Units	Temp	Data Type	Reference
WS 2.64E+004	mg/L	25	EST	MEYLAN,WM ET AL. (1996)
logP 0.80			EST	MEYLAN,WM & HOWARD,PH (1995)
VP 2.22E+000	mm Hg	25	EST	NEELY,WB & BLAU,GE (1985)
DC	pKa			
HL 1.11E-008	atm m3/mol	25	EST	MEYLAN,WM & HOWARD,PH (1991)
OH 2.94E-013	cm3/molc sec	25	EST	MEYLAN,WM & HOWARD,PH (1993)

CAS #: 015532-75-9	PIPERAZINE, 1-[3-(TRIFLUOROMETHYL)PHENYL]-

Formula: $C_{11}H_{13}F_3N_2$

Mol Weight: 230.23

MP (deg C): 　　FP (deg C):

BP (deg C):

BP pressure (mm Hg):

Property/ Value	Units	Temp	Data Type	Reference
WS 2.60E+003	mg/L	25	EST	MEYLAN,WM ET AL. (1996)
logP 2.43			EXP	CACCIA,S ET AL. (1985)
VP 1.87E-003	mm Hg	25	EST	NEELY,WB & BLAU,GE (1985)
DC	pKa			
HL 9.84E-008	atm m3/mol	25	EST	MEYLAN,WM & HOWARD,PH (1991)
OH 6.49E-011	cm3/molc sec	25	EST	MEYLAN,WM & HOWARD,PH (1993)

CAS #: 015537-20-9	2-EX-NH2 BENZOBICYCLO(222)OCTENE

Formula: $C_{12}H_{15}N$

Mol Weight: 173.26

MP (deg C): 　　FP (deg C):

BP (deg C):

BP pressure (mm Hg):

Property/ Value	Units	Temp	Data Type	Reference
WS 6.31E+003	mg/L	25	EST	MEYLAN,WM ET AL. (1996)
logP 2.32			EXP	HANSCH,C & LEO,AJ (1985)
VP 2.74E-003	mm Hg	25	EST	NEELY,WB & BLAU,GE (1985)
DC	pKa			
HL 4.08E-007	atm m3/mol	25	EST	MEYLAN,WM & HOWARD,PH (1991)
OH 6.40E-011	cm3/molc sec	25	EST	MEYLAN,WM & HOWARD,PH (1993)

CAS #: 015545-48-9	CHLORTOLURON

Formula: $C_{10}H_{13}ClN_2O$

Mol Weight: 212.68

MP (deg C): 147-148　　FP (deg C):

BP (deg C):

BP pressure (mm Hg):

Property/ Value	Units	Temp	Data Type	Reference
WS 7.00E+001	mg/L	20	EXP	YALKOWSKY,SH & DANNENFELSER,RM (1992)
logP 2.41			EXP	HANSCH,C & LEO,AJ (1985)
VP 3.60E-008	mm Hg	20	EXP	KEARNEY,PC & KAUFMAN,DD (1975A)
DC	pKa			
HL 1.44E-010	atm m3/mol	20	EST	VP/WSOL
OH 3.90E-011	cm3/molc sec	25	EST	MEYLAN,WM & HOWARD,PH (1993)

CAS #: 015582-85-1	2-PYRIDINEBUTANAMIDE

Formula: $C_9H_{12}N_2O$

Mol Weight: 164.21

MP (deg C): 　　FP (deg C):

BP (deg C):

BP pressure (mm Hg):

Property/ Value	Units	Temp	Data Type	Reference
WS 1.00E+006	mg/L	25	EST	MEYLAN,WM ET AL. (1996)
logP -0.01			EXP	HANSCH,C & LEO,AJ (1985)
VP 3.11E-005	mm Hg	25	EST	NEELY,WB & BLAU,GE (1985)
DC	pKa			
HL 2.08E-012	atm m3/mol	25	EST	MEYLAN,WM & HOWARD,PH (1991)
OH 9.41E-012	cm3/molc sec	25	EST	MEYLAN,WM & HOWARD,PH (1993)

015583-16-1

CAS #: 015583-16-1

2-PYRIDINEPROPANEAMINE

Formula: $C_8H_{12}N_2$

Mol Weight: 136.20

MP (deg C): **FP (deg C):**

BP (deg C):

BP pressure (mm Hg):

Property/Value	Units	Temp	Data Type	Reference
WS 1.00E+006	mg/L	25	EST	MEYLAN,WM ET AL. (1996)
logP 0.49			EXP	HANSCH,C & LEO,AJ (1985)
VP 4.53E-002	mm Hg	25	EST	NEELY,WB & BLAU,GE (1985)
DC	pKa			
HL 1.41E-009	atm m3/mol	25	EST	MEYLAN,WM & HOWARD,PH (1991)
OH 3.52E-011	cm3/molc sec	25	EST	MEYLAN,WM & HOWARD,PH (1993)

015588-95-1

CAS #: 015588-95-1

2,5-DIMETHOXY-4-METHYLAMPHETAMINE

Formula: $C_{12}H_{19}NO_2$

Mol Weight: 209.29

MP (deg C): **FP (deg C):**

BP (deg C):

BP pressure (mm Hg):

Property/Value	Units	Temp	Data Type	Reference
WS 4.88E+003	mg/L	25	EST	MEYLAN,WM ET AL. (1996)
logP 2.24			EXP	HANSCH,C & LEO,AJ (1985)
VP 5.93E-004	mm Hg	25	EST	NEELY,WB & BLAU,GE (1985)
DC	pKa			
HL 4.16E-009	atm m3/mol	25	EST	MEYLAN,WM & HOWARD,PH (1991)
OH 1.05E-010	cm3/molc sec	25	EST	MEYLAN,WM & HOWARD,PH (1993)

015676-16-1

CAS #: 015676-16-1

SULPIRIDE

Formula: $C_{15}H_{23}N_3O_4S$

Mol Weight: 341.43

MP (deg C): 175-182 de **FP (deg C):**

BP (deg C):

BP pressure (mm Hg):

Property/Value	Units	Temp	Data Type	Reference
WS 2.28E+003	mg/L	25	EST	MEYLAN,WM ET AL. (1996)
logP 0.57			EXP	HOEGBERG,T ET AL. (1986)
VP 3.30E-011	mm Hg	25	EST	NEELY,WB & BLAU,GE (1985)
DC	pKa			
HL 1.53E-017	atm m3/mol	25	EST	MEYLAN,WM & HOWARD,PH (1991)
OH 1.23E-010	cm3/molc sec	25	EST	MEYLAN,WM & HOWARD,PH (1993)

015686-71-2

CAS #: 015686-71-2

CEPHALEXIN

Formula: $C_{16}H_{17}N_3O_4S$

Mol Weight: 347.40

MP (deg C): **FP (deg C):**

BP (deg C):

BP pressure (mm Hg):

Property/Value	Units	Temp	Data Type	Reference
WS 1.79E+003	mg/L	25	EST	MEYLAN,WM ET AL. (1996)
logP 0.65			EXP	HANSCH,C & LEO,AJ (1985)
VP 3.24E-015	mm Hg	25	EST	NEELY,WB & BLAU,GE (1985)
DC	pKa			
HL 2.77E-017	atm m3/mol	25	EST	MEYLAN,WM & HOWARD,PH (1991)
OH 1.48E-010	cm3/molc sec	25	EST	MEYLAN,WM & HOWARD,PH (1993)

015687-27-1

CAS #: 015687-27-1

IBUPROFEN

Formula: $C_{13}H_{18}O_2$

Mol Weight: 206.29

MP (deg C): 75-77 **FP (deg C):**

BP (deg C):

BP pressure (mm Hg):

Property/Value	Units	Temp	Data Type	Reference
WS 2.44E+003	mg/L	25	EST	YALKOWSKY,SH & DANNENFELSER,RM (1992)
logP 3.50			EXP	SANGSTER,J (1993)
VP 1.35E-003	mm Hg	25	EST	HL X WSOL
DC	pKa			
HL 1.50E-007	atm m3/mol	25	EST	MEYLAN,WM & HOWARD,PH (1991)
OH 1.16E-011	cm3/molc sec	25	EST	ATKINSON,R (1988)

015707-23-0

CAS #: 015707-23-0

PYRAZINE, 2-ETHYL-3-METHYL-

Formula: $C_7H_{10}N_2$

Mol Weight: 122.17

MP (deg C): **FP (deg C):**

BP (deg C): 57

BP pressure (mm Hg): 1.00E+001

Property/Value	Units	Temp	Data Type	Reference
WS 1.20E+004	mg/L	25	EST	MEYLAN,WM ET AL. (1996)
logP 1.07			EXP	YAMAGAMI,C ET AL. (1991)
VP 6.08E-001	mm Hg	25	EST	NEELY,WB & BLAU,GE (1985)
DC	pKa			
HL 4.72E-006	atm m3/mol	25	EST	MEYLAN,WM & HOWARD,PH (1991)
OH 1.97E-012	cm3/molc sec	25	EST	MEYLAN,WM & HOWARD,PH (1993)

015707-24-1

CAS #: 015707-24-1

PYRAZINE, 2,3-DIETHYL-

Formula: $C_8H_{12}N_2$

Mol Weight: 136.20

MP (deg C): **FP (deg C):**

BP (deg C): 180-182

BP pressure (mm Hg):

Property/Value	Units	Temp	Data Type	Reference
WS 4.46E+003	mg/L	25	EST	MEYLAN,WM ET AL. (1996)
logP 1.51			EXP	YAMAGAMI,C ET AL. (1991)
VP 1.92E-001	mm Hg	25	EST	NEELY,WB & BLAU,GE (1985)
DC	pKa			
HL 6.26E-006	atm m3/mol	25	EST	MEYLAN,WM & HOWARD,PH (1991)
OH 2.90E-012	cm3/molc sec	25	EST	MEYLAN,WM & HOWARD,PH (1993)

015725-14-1

CAS #: 015725-14-1

2-CHLOROBENZAL ACETYLACETONE

Formula: $C_{12}H_{11}ClO_2$

Mol Weight: 222.67

MP (deg C): **FP (deg C):**

BP (deg C):

BP pressure (mm Hg):

Property/Value	Units	Temp	Data Type	Reference
WS 4.49E+002	mg/L	25	EST	MEYLAN,WM ET AL. (1996)
logP 2.19			EXP	HANSCH,C & LEO,AJ (1985)
VP 1.85E-004	mm Hg	25	EST	NEELY,WB & BLAU,GE (1985)
DC	pKa			
HL 3.67E-010	atm m3/mol	25	EST	MEYLAN,WM & HOWARD,PH (1991)
OH 7.19E-011	cm3/molc sec	25	EST	MEYLAN,WM & HOWARD,PH (1993)

CAS #: 015725-15-2				3-CHLOROBENZAL ACETYLACETONE

Formula: $C_{12}H_{11}ClO_2$

Mol Weight: 222.67

MP (deg C):   FP (deg C):

BP (deg C):

BP pressure (mm Hg):

Property/Value	Units	Temp	Data Type	Reference
WS 3.91E+002	mg/L	25	EST	MEYLAN,WM ET AL. (1996)
logP 2.26			EXP	HANSCH,C & LEO,AJ (1985)
VP 1.85E-004	mm Hg	25	EST	NEELY,WB & BLAU,GE (1985)
DC	pKa			
HL 3.67E-010	atm m3/mol	25	EST	MEYLAN,WM & HOWARD,PH (1991)
OH 7.19E-011	cm3/molc sec	25	EST	MEYLAN,WM & HOWARD,PH (1993)

CAS #: 015725-16-3				2-METHOXYBENZAL ACETYLACETONE

Formula: $C_{13}H_{14}O_3$

Mol Weight: 218.25

MP (deg C):   FP (deg C):

BP (deg C):

BP pressure (mm Hg):

Property/Value	Units	Temp	Data Type	Reference
WS 1.17E+003	mg/L	25	EST	MEYLAN,WM ET AL. (1996)
logP 1.73			EXP	HANSCH,C & LEO,AJ (1985)
VP 1.26E-004	mm Hg	25	EST	NEELY,WB & BLAU,GE (1985)
DC	pKa			
HL 2.93E-011	atm m3/mol	25	EST	MEYLAN,WM & HOWARD,PH (1991)
OH 9.16E-011	cm3/molc sec	25	EST	MEYLAN,WM & HOWARD,PH (1993)

CAS #: 015725-17-4				4-METHOXYBENZAL ACETYLACETONE

Formula: $C_{13}H_{14}O_3$

Mol Weight: 218.25

MP (deg C):   FP (deg C):

BP (deg C):

BP pressure (mm Hg):

Property/Value	Units	Temp	Data Type	Reference
WS 1.17E+003	mg/L	25	EST	MEYLAN,WM ET AL. (1996)
logP 1.73			EXP	HANSCH,C & LEO,AJ (1985)
VP 1.26E-004	mm Hg	25	EST	NEELY,WB & BLAU,GE (1985)
DC	pKa			
HL 2.93E-011	atm m3/mol	25	EST	MEYLAN,WM & HOWARD,PH (1991)
OH 9.16E-011	cm3/molc sec	25	EST	MEYLAN,WM & HOWARD,PH (1993)

CAS #: 015725-18-5				ACETYLACETONE,2,4-DIMEO BENZAL

Formula: $C_{14}H_{16}O_4$

Mol Weight: 248.28

MP (deg C):   FP (deg C):

BP (deg C):

BP pressure (mm Hg):

Property/Value	Units	Temp	Data Type	Reference
WS 6.00E+002	mg/L	25	EST	MEYLAN,WM ET AL. (1996)
logP 1.88			EXP	HANSCH,C & LEO,AJ (1985)
VP 1.99E-005	mm Hg	25	EST	NEELY,WB & BLAU,GE (1985)
DC	pKa			
HL 1.73E-012	atm m3/mol	25	EST	MEYLAN,WM & HOWARD,PH (1991)
OH 2.72E-010	cm3/molc sec	25	EST	MEYLAN,WM & HOWARD,PH (1993)

CAS #: 015725-21-0				ETHYLACETOACETATE,3-FLUOROBENZAL

Formula: $C_{13}H_{13}FO_3$

Mol Weight: 236.25

MP (deg C):   FP (deg C):

BP (deg C):

BP pressure (mm Hg):

Property/Value	Units	Temp	Data Type	Reference
WS 2.23E+002	mg/L	25	EST	MEYLAN,WM ET AL. (1996)
logP 2.46			EXP	HANSCH,C & LEO,AJ (1985)
VP 3.90E-004	mm Hg	25	EST	NEELY,WB & BLAU,GE (1985)
DC	pKa			
HL 2.72E-009	atm m3/mol	25	EST	MEYLAN,WM & HOWARD,PH (1991)
OH 3.14E-011	cm3/molc sec	25	EST	MEYLAN,WM & HOWARD,PH (1993)

CAS #: 015725-22-1				ETHYLACETOACETATE,2-CHLOROBENZAL

Formula: $C_{13}H_{13}ClO_3$

Mol Weight: 252.70

MP (deg C):   FP (deg C):

BP (deg C):

BP pressure (mm Hg):

Property/Value	Units	Temp	Data Type	Reference
WS 8.58E+001	mg/L	25	EST	MEYLAN,WM ET AL. (1996)
logP 2.84			EXP	HANSCH,C & LEO,AJ (1985)
VP 6.36E-005	mm Hg	25	EST	NEELY,WB & BLAU,GE (1985)
DC	pKa			
HL 1.72E-009	atm m3/mol	25	EST	MEYLAN,WM & HOWARD,PH (1991)
OH 3.04E-011	cm3/molc sec	25	EST	MEYLAN,WM & HOWARD,PH (1993)

CAS #: 015725-23-2				ETHYLACETOACETATE,3-CHLOROBENZAL

Formula: $C_{13}H_{13}ClO_3$

Mol Weight: 252.70

MP (deg C):   FP (deg C):

BP (deg C):

BP pressure (mm Hg):

Property/Value	Units	Temp	Data Type	Reference
WS 6.65E+001	mg/L	25	EST	MEYLAN,WM ET AL. (1996)
logP 2.97			EXP	HANSCH,C & LEO,AJ (1985)
VP 6.36E-005	mm Hg	25	EST	NEELY,WB & BLAU,GE (1985)
DC	pKa			
HL 1.72E-009	atm m3/mol	25	EST	MEYLAN,WM & HOWARD,PH (1991)
OH 3.04E-011	cm3/molc sec	25	EST	MEYLAN,WM & HOWARD,PH (1993)

CAS #: 015725-24-3				ETHYLACETOACETATE,2-MEO BENZAL

Formula: $C_{14}H_{16}O_4$

Mol Weight: 248.28

MP (deg C):   FP (deg C):

BP (deg C):

BP pressure (mm Hg):

Property/Value	Units	Temp	Data Type	Reference
WS 2.20E+002	mg/L	25	EST	MEYLAN,WM ET AL. (1996)
logP 2.39			EXP	HANSCH,C & LEO,AJ (1985)
VP 4.44E-005	mm Hg	25	EST	NEELY,WB & BLAU,GE (1985)
DC	pKa			
HL 1.38E-010	atm m3/mol	25	EST	MEYLAN,WM & HOWARD,PH (1991)
OH 5.02E-011	cm3/molc sec	25	EST	MEYLAN,WM & HOWARD,PH (1993)

CAS #: 015725-25-4				ETHYLACETOACETATE,3-MEO BENZAL
Formula: C14H16O4				
Mol Weight: 248.28				
MP (deg C):		FP (deg C):		
BP (deg C):				
BP pressure (mm Hg):				

Property/Value	Units	Temp	Data Type	Reference
WS 2.43E+002	mg/L	25	EST	MEYLAN,WM ET AL. (1996)
logP 2.34			EXP	HANSCH,C & LEO,AJ (1985)
VP 4.44E-005	mm Hg	25	EST	NEELY,WB & BLAU,GE (1985)
DC	pKa			
HL 1.38E-010	atm m3/mol	25	EST	MEYLAN,WM & HOWARD,PH (1991)
OH 5.02E-011	cm3/molc sec	25	EST	MEYLAN,WM & HOWARD,PH (1993)

CAS #: 015725-26-5				ETHYLACETOACETATE,4-MEO BENZAL
Formula: C14H16O4				
Mol Weight: 248.28				
MP (deg C):		FP (deg C):		
BP (deg C):				
BP pressure (mm Hg):				

Property/Value	Units	Temp	Data Type	Reference
WS 1.92E+002	mg/L	25	EST	MEYLAN,WM ET AL. (1996)
logP 2.46			EXP	HANSCH,C & LEO,AJ (1985)
VP 4.44E-005	mm Hg	25	EST	NEELY,WB & BLAU,GE (1985)
DC	pKa			
HL 1.38E-010	atm m3/mol	25	EST	MEYLAN,WM & HOWARD,PH (1991)
OH 5.02E-011	cm3/molc sec	25	EST	MEYLAN,WM & HOWARD,PH (1993)

CAS #: 015725-27-6				ETHYLACETOACETATE, 3-ETO BENZAL
Formula: C15H18O4				
Mol Weight: 262.31				
MP (deg C):		FP (deg C):		
BP (deg C):				
BP pressure (mm Hg):				

Property/Value	Units	Temp	Data Type	Reference
WS 9.79E+001	mg/L	25	EST	MEYLAN,WM ET AL. (1996)
logP 2.71			EXP	HANSCH,C & LEO,AJ (1985)
VP 1.88E-005	mm Hg	25	EST	NEELY,WB & BLAU,GE (1985)
DC	pKa			
HL 1.83E-010	atm m3/mol	25	EST	MEYLAN,WM & HOWARD,PH (1991)
OH 5.55E-011	cm3/molc sec	25	EST	MEYLAN,WM & HOWARD,PH (1993)

CAS #: 015725-28-7				ETHYLACETOACETATE,3-PRO BENZAL
Formula: C16H20O4				
Mol Weight: 276.34				
MP (deg C):		FP (deg C):		
BP (deg C):				
BP pressure (mm Hg):				

Property/Value	Units	Temp	Data Type	Reference
WS 2.36E+001	mg/L	25	EST	MEYLAN,WM ET AL. (1996)
logP 3.34			EXP	HANSCH,C & LEO,AJ (1985)
VP 7.98E-006	mm Hg	25	EST	NEELY,WB & BLAU,GE (1985)
DC	pKa			
HL 2.43E-010	atm m3/mol	25	EST	MEYLAN,WM & HOWARD,PH (1991)
OH 5.98E-011	cm3/molc sec	25	EST	MEYLAN,WM & HOWARD,PH (1993)

CAS #: 015725-29-8				ETHYLACETOACETATE,2,4-DICLBENZAL
Formula: C13H12Cl2O3				
Mol Weight: 287.14				
MP (deg C):		FP (deg C):		
BP (deg C):				
BP pressure (mm Hg):				

Property/Value	Units	Temp	Data Type	Reference
WS 1.07E+001	mg/L	25	EST	MEYLAN,WM ET AL. (1996)
logP 3.67			EXP	HANSCH,C & LEO,AJ (1985)
VP 1.47E-005	mm Hg	25	EST	NEELY,WB & BLAU,GE (1985)
DC	pKa			
HL 1.28E-009	atm m3/mol	25	EST	MEYLAN,WM & HOWARD,PH (1991)
OH 3.00E-011	cm3/molc sec	25	EST	MEYLAN,WM & HOWARD,PH (1993)

CAS #: 015725-30-1				ETHYLACETOACETATE,2,6-DICLBENZAL
Formula: C13H12Cl2O3				
Mol Weight: 287.14				
MP (deg C):		FP (deg C):		
BP (deg C):				
BP pressure (mm Hg):				

Property/Value	Units	Temp	Data Type	Reference
WS 5.26E+001	mg/L	25	EST	MEYLAN,WM ET AL. (1996)
logP 2.86			EXP	HANSCH,C & LEO,AJ (1985)
VP 1.47E-005	mm Hg	25	EST	NEELY,WB & BLAU,GE (1985)
DC	pKa			
HL 1.28E-009	atm m3/mol	25	EST	MEYLAN,WM & HOWARD,PH (1991)
OH 3.00E-011	cm3/molc sec	25	EST	MEYLAN,WM & HOWARD,PH (1993)

CAS #: 015725-31-2				ETHYLACETOACETATE,3,4-DICLBENZAL
Formula: C13H12Cl2O3				
Mol Weight: 287.14				
MP (deg C):		FP (deg C):		
BP (deg C):				
BP pressure (mm Hg):				

Property/Value	Units	Temp	Data Type	Reference
WS 6.55E+000	mg/L	25	EST	MEYLAN,WM ET AL. (1996)
logP 3.92			EXP	HANSCH,C & LEO,AJ (1985)
VP 1.47E-005	mm Hg	25	EST	NEELY,WB & BLAU,GE (1985)
DC	pKa			
HL 1.28E-009	atm m3/mol	25	EST	MEYLAN,WM & HOWARD,PH (1991)
OH 2.95E-011	cm3/molc sec	25	EST	MEYLAN,WM & HOWARD,PH (1993)

CAS #: 015725-32-3				ETHYLACETOACETATE,2,4-DIMEO BENZAL
Formula: C15H18O5				
Mol Weight: 278.31				
MP (deg C):		FP (deg C):		
BP (deg C):				
BP pressure (mm Hg):				

Property/Value	Units	Temp	Data Type	Reference
WS 1.43E+002	mg/L	25	EST	MEYLAN,WM ET AL. (1996)
logP 2.41			EXP	HANSCH,C & LEO,AJ (1985)
VP 7.31E-006	mm Hg	25	EST	NEELY,WB & BLAU,GE (1985)
DC	pKa			
HL 8.14E-012	atm m3/mol	25	EST	MEYLAN,WM & HOWARD,PH (1991)
OH 2.31E-010	cm3/molc sec	25	EST	MEYLAN,WM & HOWARD,PH (1993)

CAS #: 015725-33-4 — DIETHYLMALONATE, 3-ME BENZAL

Formula: $C_{15}H_{18}O_4$				
Mol Weight: 262.31				
MP (deg C):		FP (deg C):		
BP (deg C):				
BP pressure (mm Hg):				

Property/Value	Units	Temp	Data Type	Reference
WS 1.88E+001	mg/L	25	EST	MEYLAN,WM ET AL. (1996)
logP 3.55			EXP	HANSCH,C & LEO,AJ (1985)
VP 4.92E-004	mm Hg	25	EST	NEELY,WB & BLAU,GE (1985)
DC	pKa			
HL 1.21E-008	atm m3/mol	25	EST	MEYLAN,WM & HOWARD,PH (1991)
OH 1.89E-011	cm3/molc sec	25	EST	MEYLAN,WM & HOWARD,PH (1993)

CAS #: 015725-37-8 — DIETHYLMALONATE, 3-ETHOXYBENZAL

Formula: $C_{16}H_{20}O_5$				
Mol Weight: 292.33				
MP (deg C):		FP (deg C):		
BP (deg C):				
BP pressure (mm Hg):				

Property/Value	Units	Temp	Data Type	Reference
WS 1.73E+001	mg/L	25	EST	MEYLAN,WM ET AL. (1996)
logP 3.39			EXP	HANSCH,C & LEO,AJ (1985)
VP 1.28E-004	mm Hg	25	EST	NEELY,WB & BLAU,GE (1985)
DC	pKa			
HL 8.59E-010	atm m3/mol	25	EST	MEYLAN,WM & HOWARD,PH (1991)
OH 4.04E-011	cm3/molc sec	25	EST	MEYLAN,WM & HOWARD,PH (1993)

CAS #: 015725-38-9 — DIETHYLMALONATE,3-PRO BENZAL

Formula: $C_{17}H_{22}O_5$				
Mol Weight: 306.36				
MP (deg C):		FP (deg C):		
BP (deg C):				
BP pressure (mm Hg):				

Property/Value	Units	Temp	Data Type	Reference
WS 4.32E+000	mg/L	25	EST	MEYLAN,WM ET AL. (1996)
logP 4.00			EXP	HANSCH,C & LEO,AJ (1985)
VP 6.78E-005	mm Hg	25	EST	NEELY,WB & BLAU,GE (1985)
DC	pKa			
HL 1.14E-009	atm m3/mol	25	EST	MEYLAN,WM & HOWARD,PH (1991)
OH 4.46E-011	cm3/molc sec	25	EST	MEYLAN,WM & HOWARD,PH (1993)

CAS #: 015725-39-0 — DIETHYLMALONATE,3-BUTOXYBENZAL

Formula: $C_{18}H_{24}O_5$				
Mol Weight: 320.39				
MP (deg C):		FP (deg C):		
BP (deg C):				
BP pressure (mm Hg):				

Property/Value	Units	Temp	Data Type	Reference
WS 3.37E+000	mg/L	25	EST	MEYLAN,WM ET AL. (1996)
logP 4.03			EXP	HANSCH,C & LEO,AJ (1985)
VP 3.16E-005	mm Hg	25	EST	NEELY,WB & BLAU,GE (1985)
DC	pKa			
HL 1.51E-009	atm m3/mol	25	EST	MEYLAN,WM & HOWARD,PH (1991)
OH 4.65E-011	cm3/molc sec	25	EST	MEYLAN,WM & HOWARD,PH (1993)

CAS #: 015795-19-4 — ACETYLACETONE, 4-BROMOBENZAL

Formula: $C_{12}H_{11}BrO_2$				
Mol Weight: 267.13				
MP (deg C):		FP (deg C):		
BP (deg C):				
BP pressure (mm Hg):				

Property/Value	Units	Temp	Data Type	Reference
WS 8.50E+001	mg/L	25	EST	MEYLAN,WM ET AL. (1996)
logP 2.75			EXP	HANSCH,C & LEO,AJ (1985)
VP 6.99E-005	mm Hg	25	EST	NEELY,WB & BLAU,GE (1985)
DC	pKa			
HL 1.97E-010	atm m3/mol	25	EST	MEYLAN,WM & HOWARD,PH (1991)
OH 7.17E-011	cm3/molc sec	25	EST	MEYLAN,WM & HOWARD,PH (1993)

CAS #: 015798-77-3 — 2H-FURO[2,3-H]-1-BENZOPYRAN-2-ONE, 8-METHYL-

Formula: $C_{12}H_8O_3$				
Mol Weight: 200.20				
MP (deg C):		FP (deg C):		
BP (deg C):				
BP pressure (mm Hg):				

Property/Value	Units	Temp	Data Type	Reference
WS 2.90E+002	mg/L	25	EST	MEYLAN,WM ET AL. (1996)
logP 2.55			EXP	SANGSTER,J (1993)
VP 9.99E-006	mm Hg	25	EST	NEELY,WB & BLAU,GE (1985)
DC	pKa			
HL 7.47E-007	atm m3/mol	25	EST	MEYLAN,WM & HOWARD,PH (1991)
OH 6.60E-011	cm3/molc sec	25	EST	MEYLAN,WM & HOWARD,PH (1993)

CAS #: 015804-19-0 — QUINOXALINE-1,4-DIHYDRO-2,3-DIONE

Formula: $C_8H_6N_2O_2$				
Mol Weight: 162.15				
MP (deg C): > 300		FP (deg C):		
BP (deg C):				
BP pressure (mm Hg):				

Property/Value	Units	Temp	Data Type	Reference
WS 2.21E+003	mg/L	25	EST	MEYLAN,WM ET AL. (1996)
logP 0.20			EXP	HANSCH,C & LEO,AJ (1985)
VP 5.49E-008	mm Hg	25	EST	NEELY,WB & BLAU,GE (1985)
DC	pKa			
HL 2.20E-011	atm m3/mol	25	EST	MEYLAN,WM & HOWARD,PH (1991)
OH 6.99E-011	cm3/molc sec	25	EST	MEYLAN,WM & HOWARD,PH (1993)

CAS #: 015804-61-2 — BENZALCYANOACETANILIDE

Formula: $C_{16}H_{12}N_2O$				
Mol Weight: 248.29				
MP (deg C):		FP (deg C):		
BP (deg C):				
BP pressure (mm Hg):				

Property/Value	Units	Temp	Data Type	Reference
WS 1.61E+001	mg/L	25	EST	MEYLAN,WM ET AL. (1996)
logP 3.41			EXP	HANSCH,C & LEO,AJ (1985)
VP 5.68E-009	mm Hg	25	EST	NEELY,WB & BLAU,GE (1985)
DC	pKa			
HL 3.25E-013	atm m3/mol	25	EST	MEYLAN,WM & HOWARD,PH (1991)
OH 4.94E-011	cm3/molc sec	25	EST	MEYLAN,WM & HOWARD,PH (1993)

CAS #: 015804-68-9				MALONAMIDE, 3-METHOXYBENZAL
Formula: $C_{11}H_{12}N_2O_3$				
Mol Weight: 220.23				
MP (deg C):			FP (deg C):	
BP (deg C):				
BP pressure (mm Hg):				

Property/ Value	Units	Temp	Data Type	Reference
WS 1.65E+003	mg/L	25	EST	MEYLAN,WM ET AL. (1996)
logP 0.01			EXP	HANSCH,C & LEO,AJ (1985)
VP 5.88E-009	mm Hg	25	EST	NEELY,WB & BLAU,GE (1985)
DC	pKa			
HL 2.37E-015	atm m3/mol	25	EST	MEYLAN,WM & HOWARD,PH (1991)
OH 3.57E-011	cm3/molc sec	25	EST	MEYLAN,WM & HOWARD,PH (1993)

CAS #: 015806-38-9				BUTANOIC ACID, 3-METHYL-, PHENYL ESTER
Formula: $C_{11}H_{14}O_2$				
Mol Weight: 178.23				
MP (deg C):			FP (deg C):	
BP (deg C):				
BP pressure (mm Hg):				

Property/ Value	Units	Temp	Data Type	Reference
WS 1.08E+002	mg/L	25	EST	MEYLAN,WM ET AL. (1996)
logP 3.18			EXP	YANG,HZ ET AL. (1987)
VP 4.28E-002	mm Hg	25	EST	NEELY,WB & BLAU,GE (1985)
DC	pKa			
HL 1.52E-004	atm m3/mol	25	EST	MEYLAN,WM & HOWARD,PH (1991)
OH 5.58E-012	cm3/molc sec	25	EST	MEYLAN,WM & HOWARD,PH (1993)

CAS #: 015818-09-4				4-METHYLBENZAL ACETYLACETONE
Formula: $C_{13}H_{14}O_2$				
Mol Weight: 202.26				
MP (deg C):			FP (deg C):	
BP (deg C):				
BP pressure (mm Hg):				

Property/ Value	Units	Temp	Data Type	Reference
WS 6.34E+002	mg/L	25	EST	MEYLAN,WM ET AL. (1996)
logP 2.14			EXP	HANSCH,C & LEO,AJ (1985)
VP 3.82E-004	mm Hg	25	EST	NEELY,WB & BLAU,GE (1985)
DC	pKa			
HL 5.47E-010	atm m3/mol	25	EST	MEYLAN,WM & HOWARD,PH (1991)
OH 7.55E-011	cm3/molc sec	25	EST	MEYLAN,WM & HOWARD,PH (1993)

CAS #: 015818-10-7				3-METHOXYBENZAL ACETYLACETONE
Formula: $C_{13}H_{14}O_3$				
Mol Weight: 218.25				
MP (deg C):			FP (deg C):	
BP (deg C):				
BP pressure (mm Hg):				

Property/ Value	Units	Temp	Data Type	Reference
WS 1.04E+003	mg/L	25	EST	MEYLAN,WM ET AL. (1996)
logP 1.79			EXP	HANSCH,C & LEO,AJ (1985)
VP 1.26E-004	mm Hg	25	EST	NEELY,WB & BLAU,GE (1985)
DC	pKa			
HL 2.93E-011	atm m3/mol	25	EST	MEYLAN,WM & HOWARD,PH (1991)
OH 9.16E-011	cm3/molc sec	25	EST	MEYLAN,WM & HOWARD,PH (1993)

CAS #: 015818-13-0				PROPANEDIOIC ACID, [(3,4-DIMETHOXYPHENYL)METHYLE
Formula: $C_{16}H_{20}O_6$				
Mol Weight: 308.33				
MP (deg C):			FP (deg C):	
BP (deg C):				
BP pressure (mm Hg):				

Property/ Value	Units	Temp	Data Type	Reference
WS 3.74E+001	mg/L	25	EST	MEYLAN,WM ET AL. (1996)
logP 2.89			EXP	HANSCH,C ET AL. (1995)
VP 4.83E-005	mm Hg	25	EST	NEELY,WB & BLAU,GE (1985)
DC	pKa			
HL 3.83E-011	atm m3/mol	25	EST	MEYLAN,WM & HOWARD,PH (1991)
OH 3.31E-011	cm3/molc sec	25	EST	MEYLAN,WM & HOWARD,PH (1993)

CAS #: 015845-96-2				PREGNA-1,4-DIENE-3,20-DIONE, 21-(2,2-DIMETHYL-1-
Formula: $C_{27}H_{36}F_2O_4$				
Mol Weight: 462.58				
MP (deg C):			FP (deg C):	
BP (deg C):				
BP pressure (mm Hg):				

Property/ Value	Units	Temp	Data Type	Reference
WS 2.03E+000	mg/L	25	EST	MEYLAN,WM ET AL. (1996)
logP 3.86			EXP	SANGSTER,J (1994)
VP 7.34E-013	mm Hg	25	EST	NEELY,WB & BLAU,GE (1985)
DC	pKa			
HL 9.21E-015	atm m3/mol	25	EST	MEYLAN,WM & HOWARD,PH (1991)
OH 7.10E-011	cm3/molc sec	25	EST	MEYLAN,WM & HOWARD,PH (1993)

CAS #: 015851-93-1				2-CHLORO-5-NITRO-B-NITROSTYRENE
Formula: $C_8H_5ClN_2O_4$				
Mol Weight: 228.59				
MP (deg C):			FP (deg C):	
BP (deg C):				
BP pressure (mm Hg):				

Property/ Value	Units	Temp	Data Type	Reference
WS 1.57E+002	mg/L	25	EST	MEYLAN,WM ET AL. (1996)
logP 2.23			EXP	HANSCH,C & LEO,AJ (1985)
VP 2.76E-005	mm Hg	25	EST	NEELY,WB & BLAU,GE (1985)
DC	pKa			
HL 1.02E-008	atm m3/mol	25	EST	MEYLAN,WM & HOWARD,PH (1991)
OH 1.16E-011	cm3/molc sec	25	EST	MEYLAN,WM & HOWARD,PH (1993)

CAS #: 015851-94-2				4-FLUOROBENZAL ACETYLACETONE
Formula: $C_{12}H_{11}FO_2$				
Mol Weight: 206.22				
MP (deg C):			FP (deg C):	
BP (deg C):				
BP pressure (mm Hg):				

Property/ Value	Units	Temp	Data Type	Reference
WS 1.59E+003	mg/L	25	EST	MEYLAN,WM ET AL. (1996)
logP 1.65			EXP	HANSCH,C & LEO,AJ (1985)
VP 1.26E-003	mm Hg	25	EST	NEELY,WB & BLAU,GE (1985)
DC	pKa			
HL 5.78E-010	atm m3/mol	25	EST	MEYLAN,WM & HOWARD,PH (1991)
OH 7.29E-011	cm3/molc sec	25	EST	MEYLAN,WM & HOWARD,PH (1993)

CAS #: 015861-24-2				INDOLE-5-CYANO-

Formula: $C_9H_6N_2$

Mol Weight: 142.16

MP (deg C): 106-108 FP (deg C):

BP (deg C):

BP pressure (mm Hg):

Property/Value	Units	Temp	Data Type	Reference
WS 4.13E+002	mg/L	25	EST	MEYLAN,WM ET AL. (1996)
logP 2.38			EXP	HANSCH,C ET AL. (1995)
VP 2.32E-004	mm Hg	25	EST	NEELY,WB & BLAU,GE (1985)
DC	pKa			
HL 8.56E-009	atm m3/mol	25	EST	MEYLAN,WM & HOWARD,PH (1991)
OH 1.94E-011	cm3/molc sec	25	EST	MEYLAN,WM & HOWARD,PH (1993)

CAS #: 015861-36-6				INDOLE-6-CYANO-

Formula: $C_9H_6N_2$

Mol Weight: 142.16

MP (deg C): FP (deg C):

BP (deg C):

BP pressure (mm Hg):

Property/Value	Units	Temp	Data Type	Reference
WS 4.13E+002	mg/L	25	EST	MEYLAN,WM ET AL. (1996)
logP 2.38			EXP	HANSCH,C ET AL. (1995)
VP 2.32E-004	mm Hg	25	EST	NEELY,WB & BLAU,GE (1985)
DC	pKa			
HL 8.56E-009	atm m3/mol	25	EST	MEYLAN,WM & HOWARD,PH (1991)
OH 1.94E-011	cm3/molc sec	25	EST	MEYLAN,WM & HOWARD,PH (1993)

CAS #: 015862-07-4				2,4,5-TRICHLOROBIPHENYL

Formula: $C_{12}H_7Cl_3$

Mol Weight: 257.55

MP (deg C): FP (deg C):

BP (deg C):

BP pressure (mm Hg):

Property/Value	Units	Temp	Data Type	Reference
WS 1.63E-001	mg/L	25	EXP	YALKOWSKY,SH & DANNENFELSER,RM (1992)
logP 5.81			EXP	HANSCH,C & LEO,AJ (1985)
VP 9.75E-004	mm Hg	25	EXP	SHIU,WY & MACKAY,D (1986)
DC	pKa			
HL 2.03E-003	atm m3/mol	25	EST	VP/WSOL
OH 1.26E-012	cm3/molc sec	25	EST	MEYLAN,WM & HOWARD,PH (1993)

CAS #: 015870-10-7				2-METHYL-1-HEPTENE

Formula: C_8H_{16}

Mol Weight: 112.22

MP (deg C): -90 FP (deg C):

BP (deg C): 119.3

BP pressure (mm Hg):

Property/Value	Units	Temp	Data Type	Reference
WS 8.20E+000	mg/L	25	EST	MEYLAN,WM ET AL. (1996)
logP 4.19			EST	MEYLAN,WM & HOWARD,PH (1995)
VP 2.02E+001	mm Hg	25	EXP	ZWOLINSKI,BJ & WILHOIT,RC (1971)
DC	pKa			
HL 7.46E-001	atm m3/mol	25	EST	MEYLAN,WM & HOWARD,PH (1991)
OH 5.68E-011	cm3/molc sec	25	EST	MEYLAN,WM & HOWARD,PH (1993)

CAS #: 015879-93-3				CHLORALOSE

Formula: $C_8H_{11}Cl_3O_6$

Mol Weight: 309.53

MP (deg C): 182-184 FP (deg C):

BP (deg C):

BP pressure (mm Hg):

Property/Value	Units	Temp	Data Type	Reference
WS 4.44E+003	mg/L	15	EXP	YALKOWSKY,SH & DANNENFELSER,RM (1992)
logP 1.02			EXP	SANGSTER,J (1994)
VP 2.81E-009	mm Hg	25	EST	NEELY,WB & BLAU,GE (1985)
DC	pKa			
HL 9.62E-017	atm m3/mol	25	EST	MEYLAN,WM & HOWARD,PH (1991)
OH 4.43E-011	cm3/molc sec	25	EST	MEYLAN,WM & HOWARD,PH (1993)

CAS #: 015888-02-5				MALONAMIDE, 3-METHYLBENZAL

Formula: $C_{11}H_{12}N_2O_2$

Mol Weight: 204.23

MP (deg C): FP (deg C):

BP (deg C):

BP pressure (mm Hg):

Property/Value	Units	Temp	Data Type	Reference
WS 1.90E+004	mg/L	25	EST	MEYLAN,WM ET AL. (1996)
logP 0.40			EXP	HANSCH,C & LEO,AJ (1985)
VP 1.36E-008	mm Hg	25	EST	NEELY,WB & BLAU,GE (1985)
DC	pKa			
HL 4.42E-014	atm m3/mol	25	EST	MEYLAN,WM & HOWARD,PH (1991)
OH 1.96E-011	cm3/molc sec	25	EST	MEYLAN,WM & HOWARD,PH (1993)

CAS #: 015890-40-1				1(TRANS), 2(CIS), 3-TRIMETHYLCYCLOPENTANE

Formula: C_8H_{16}

Mol Weight: 112.22

MP (deg C): -112 FP (deg C):

BP (deg C): 117.5

BP pressure (mm Hg):

Property/Value	Units	Temp	Data Type	Reference
WS 1.35E+001	mg/L	25	EST	MEYLAN,WM ET AL. (1996)
logP 3.94			EST	MEYLAN,WM & HOWARD,PH (1995)
VP 3.20E+001	mm Hg	25	EXP	ZWOLINSKI,BJ & WILHOIT,RC (1971)
DC	pKa			
HL 4.50E-001	atm m3/mol	25	EST	MEYLAN,WM & HOWARD,PH (1991)
OH 9.81E-012	cm3/molc sec	25	EST	MEYLAN,WM & HOWARD,PH (1993)

CAS #: 015893-46-6				PHENYLALANYLPHENYLALANAMIDE

Formula: $C_{18}H_{21}N_3O_2$

Mol Weight: 311.39

MP (deg C): FP (deg C):

BP (deg C):

BP pressure (mm Hg):

Property/Value	Units	Temp	Data Type	Reference
WS 1.14E+003	mg/L	25	EST	MEYLAN,WM ET AL. (1996)
logP 1.13			EXP	HANSCH,C & LEO,AJ (1985)
VP 6.16E-012	mm Hg	25	EST	NEELY,WB & BLAU,GE (1985)
DC	pKa			
HL 2.78E-015	atm m3/mol	25	EST	MEYLAN,WM & HOWARD,PH (1991)
OH 6.95E-011	cm3/molc sec	25	EST	MEYLAN,WM & HOWARD,PH (1993)

CAS #: 015907-82-1				CYCLOPENTANECARBOXAMIDE, N-(3,4-DICHLOROPHENYL)-

Formula: $C_{12}H_{13}Cl_2NO$

Mol Weight: 258.15

MP (deg C):   FP (deg C):

BP (deg C):

BP pressure (mm Hg):

Property/Value	Units	Temp	Data Type	Reference
WS 3.25E+000	mg/L	25	EST	MEYLAN,WM ET AL. (1996)
logP 4.47			EXP	HANSCH,C ET AL. (1995)
VP 6.17E-007	mm Hg	25	EST	NEELY,WB & BLAU,GE (1985)
DC	pKa			
HL 4.64E-009	atm m3/mol	25	EST	MEYLAN,WM & HOWARD,PH (1991)
OH 1.37E-011	cm3/molc sec	25	EST	MEYLAN,WM & HOWARD,PH (1993)

CAS #: 015907-85-4				CYCLOHEXANECARBOXAMIDE, N-(3,4-DICHLOROPHENYL)-

Formula: $C_{13}H_{15}Cl_2NO$

Mol Weight: 272.18

MP (deg C):   FP (deg C):

BP (deg C):

BP pressure (mm Hg):

Property/Value	Units	Temp	Data Type	Reference
WS 1.65E+000	mg/L	25	EST	MEYLAN,WM ET AL. (1996)
logP 4.72			EXP	HANSCH,C ET AL. (1995)
VP 2.50E-007	mm Hg	25	EST	NEELY,WB & BLAU,GE (1985)
DC	pKa			
HL 6.16E-009	atm m3/mol	25	EST	MEYLAN,WM & HOWARD,PH (1991)
OH 1.80E-011	cm3/molc sec	25	EST	MEYLAN,WM & HOWARD,PH (1993)

CAS #: 015910-22-2				2,5-DIMETHYL-3-HEXENE

Formula: C_8H_{16}

Mol Weight: 112.22

MP (deg C):   FP (deg C):

BP (deg C):

BP pressure (mm Hg):

Property/Value	Units	Temp	Data Type	Reference
WS 1.43E+001	mg/L	25	EST	MEYLAN,WM ET AL. (1996)
logP 3.91			EST	MEYLAN,WM & HOWARD,PH (1995)
VP 4.34E+001	mm Hg	25	EXP	ZWOLINSKI,BJ & WILHOIT,RC (1971)
DC	pKa			
HL 7.46E-001	atm m3/mol	25	EST	MEYLAN,WM & HOWARD,PH (1991)
OH 6.09E-011	cm3/molc sec	25	EST	MEYLAN,WM & HOWARD,PH (1993)

CAS #: 015913-35-6				3ME5(5NO2-2-FURFURIL)2-S-THIAZOL-4ONE

Formula: $C_9H_7N_2O_4S_2$

Mol Weight: 271.30

MP (deg C):   FP (deg C):

BP (deg C):

BP pressure (mm Hg):

Property/Value	Units	Temp	Data Type	Reference
WS 9.22E+002	mg/L	25	EST	MEYLAN,WM ET AL. (1996)
logP 1.06			EXP	HANSCH,C & LEO,AJ (1985)
VP 3.10E-009	mm Hg	25	EST	NEELY,WB & BLAU,GE (1985)
DC	pKa			
HL 1.78E-011	atm m3/mol	25	EST	MEYLAN,WM & HOWARD,PH (1991)
OH 5.24E-011	cm3/molc sec	25	EST	MEYLAN,WM & HOWARD,PH (1993)

CAS #: 015940-62-2				HYDRAZINECARBOXAMIDE, N-(3-METHYLPHENYL)-

Formula: $C_8H_{11}N_3O$

Mol Weight: 165.20

MP (deg C):   FP (deg C):

BP (deg C):

BP pressure (mm Hg):

Property/Value	Units	Temp	Data Type	Reference
WS 1.70E+004	mg/L	25	EST	MEYLAN,WM ET AL. (1996)
logP 0.68			EXP	KRAMER,CR & BECK,L (1981)
VP 2.95E-005	mm Hg	25	EST	NEELY,WB & BLAU,GE (1985)
DC	pKa			
HL 9.28E-013	atm m3/mol	25	EST	MEYLAN,WM & HOWARD,PH (1991)
OH 1.12E-010	cm3/molc sec	25	EST	MEYLAN,WM & HOWARD,PH (1993)

CAS #: 015940-63-3				HYDRAZINECARBOXAMIDE, N-(2-METHYLPHENYL)-

Formula: $C_8H_{11}N_3O$

Mol Weight: 165.20

MP (deg C):   FP (deg C):

BP (deg C):

BP pressure (mm Hg):

Property/Value	Units	Temp	Data Type	Reference
WS 3.19E+004	mg/L	25	EST	MEYLAN,WM ET AL. (1996)
logP 0.36			EXP	KRAMER,CR & BECK,L (1991)
VP 2.95E-005	mm Hg	25	EST	NEELY,WB & BLAU,GE (1985)
DC	pKa			
HL 9.28E-013	atm m3/mol	25	EST	MEYLAN,WM & HOWARD,PH (1991)
OH 5.34E-011	cm3/molc sec	25	EST	MEYLAN,WM & HOWARD,PH (1993)

CAS #: 015948-56-8				MALONAMIDE,3-FLUOROBENZAL

Formula: $C_{10}H_9FN_2O_2$

Mol Weight: 208.19

MP (deg C):   FP (deg C):

BP (deg C):

BP pressure (mm Hg):

Property/Value	Units	Temp	Data Type	Reference
WS 1.60E+003	mg/L	25	EST	MEYLAN,WM ET AL. (1996)
logP 0.10			EXP	HANSCH,C & LEO,AJ (1985)
VP 3.66E-008	mm Hg	25	EST	NEELY,WB & BLAU,GE (1985)
DC	pKa			
HL 4.67E-014	atm m3/mol	25	EST	MEYLAN,WM & HOWARD,PH (1991)
OH 1.69E-011	cm3/molc sec	25	EST	MEYLAN,WM & HOWARD,PH (1993)

CAS #: 015950-66-0				2,3,4-TRICHLOROPHENOL

Formula: $C_6H_3Cl_3O$

Mol Weight: 197.45

MP (deg C): 83.5   FP (deg C):

BP (deg C):

BP pressure (mm Hg):

Property/Value	Units	Temp	Data Type	Reference
WS 9.75E+001	mg/L	25	EST	MEYLAN,WM ET AL. (1996)
logP 3.80			EXP	HANSCH,C & LEO,AJ (1985)
VP 2.46E-003	mm Hg	25	EST	NEELY,WB & BLAU,GE (1985)
DC	pKa			
HL 2.28E-007	atm m3/mol	25	EST	MEYLAN,WM & HOWARD,PH (1991)
OH 2.14E-012	cm3/molc sec	25	EST	MEYLAN,WM & HOWARD,PH (1993)

CAS #: 015959-53-2 — PHENYLSULFAMIDE

Formula: $C_6H_8N_2O_2S$

Mol Weight: 172.21

MP (deg C): FP (deg C):

BP (deg C):

BP pressure (mm Hg):

Property/Value	Units	Temp	Data Type	Reference
WS 2.73E+004	mg/L	25	EST	MEYLAN,WM ET AL. (1996)
logP 0.40			EXP	HANSCH,C & LEO,AJ (1985)
VP 1.19E-004	mm Hg	25	EST	NEELY,WB & BLAU,GE (1985)
DC	pKa			
HL 1.76E-008	atm m3/mol	25	EST	MEYLAN,WM & HOWARD,PH (1991)
OH 4.27E-011	cm3/molc sec	25	EST	MEYLAN,WM & HOWARD,PH (1993)

CAS #: 015962-47-7 — ALANINE-AMIDE,N-ACETYL

Formula: $C_5H_{10}N_2O_2$

Mol Weight: 130.15

MP (deg C): FP (deg C):

BP (deg C):

BP pressure (mm Hg):

Property/Value	Units	Temp	Data Type	Reference
WS 8.93E+004	mg/L	25	EST	MEYLAN,WM ET AL. (1996)
logP -1.52			EXP	HANSCH,C & LEO,AJ (1985)
VP 1.44E-006	mm Hg	25	EST	NEELY,WB & BLAU,GE (1985)
DC	pKa			
HL 1.57E-010	atm m3/mol	25	EST	MEYLAN,WM & HOWARD,PH (1991)
OH 4.07E-011	cm3/molc sec	25	EST	MEYLAN,WM & HOWARD,PH (1993)

CAS #: 015968-05-5 — 2,2',6,6'-PCB

Formula: $C_{12}H_6Cl_4$

Mol Weight: 291.99

MP (deg C): FP (deg C):

BP (deg C):

BP pressure (mm Hg):

Property/Value	Units	Temp	Data Type	Reference
WS 1.19E-002	mg/L	25	EXP	YALKOWSKY,SH & DANNENFELSER,RM (1992)
logP 5.94			EXP	DEBRUIJN,J ET AL. (1989)
VP 8.45E-006	mm Hg	25	EST	NEELY,WB & BLAU,GE (1985)
DC	pKa			
HL 1.25E-004	atm m3/mol	25	EST	MEYLAN,WM & HOWARD,PH (1991)
OH 8.13E-013	cm3/molc sec	25	EST	MEYLAN,WM & HOWARD,PH (1993)

CAS #: 015972-60-8 — ALACHLOR

Formula: $C_{14}H_{20}ClNO_2$

Mol Weight: 269.77

MP (deg C): 39.5-41.5 FP (deg C):

BP (deg C): 100

BP pressure (mm Hg): 2.00E-002

Property/Value	Units	Temp	Data Type	Reference
WS 2.40E+002	mg/L	25	EXP	CHESTERS,G ET AL. (1989)
logP 3.52			EXP	HANSCH,C & LEO,AJ (1985)
VP 2.20E-005	mm Hg	25		CHESTERS,G ET AL. (1989)
DC	pKa			
HL 8.32E-009	atm m3/mol	23	EXP	FENDINGER,NJ & GLOTFELTY,DE (1988)
OH 1.85E-010	cm3/molc sec	25	EST	ATKINSON,R (1988)

CAS #: 015979-79-0 — 3-CHLORO-N-(4-CHLOROPHENYL)BENZENAMINE

Formula: $C_{12}H_9Cl_2N$

Mol Weight: 238.12

MP (deg C): FP (deg C):

BP (deg C):

BP pressure (mm Hg):

Property/Value	Units	Temp	Data Type	Reference
WS 3.37E+000	mg/L	25	EST	MEYLAN,WM ET AL. (1996)
logP 4.58			EST	MEYLAN,WM & HOWARD,PH (1995)
VP 7.80E-005	mm Hg	25	EST	NEELY,WB & BLAU,GE (1985)
DC	pKa			
HL 5.77E-007	atm m3/mol	25	EST	MEYLAN,WM & HOWARD,PH (1991)
OH 1.46E-010	cm3/molc sec	25	EST	MEYLAN,WM & HOWARD,PH (1993)

CAS #: 015979-81-4 — 4-METHYLSULFONYL-N-PHENYLBENZENAMINE

Formula: $C_{13}H_{13}NO_2S$

Mol Weight: 247.32

MP (deg C): FP (deg C):

BP (deg C):

BP pressure (mm Hg):

Property/Value	Units	Temp	Data Type	Reference
WS 6.30E+002	mg/L	25	EST	MEYLAN,WM ET AL. (1996)
logP 1.86			EST	MEYLAN,WM & HOWARD,PH (1995)
VP 1.19E-006	mm Hg	25	EST	NEELY,WB & BLAU,GE (1985)
DC	pKa			
HL 2.43E-010	atm m3/mol	25	EST	MEYLAN,WM & HOWARD,PH (1991)
OH 1.79E-010	cm3/molc sec	25	EST	MEYLAN,WM & HOWARD,PH (1993)

CAS #: 015979-82-5 — 3-METHYL-N-(4-NITROPHENYL)BENZENAMINE

Formula: $C_{13}H_{12}N_2O_2$

Mol Weight: 228.25

MP (deg C): FP (deg C):

BP (deg C):

BP pressure (mm Hg):

Property/Value	Units	Temp	Data Type	Reference
WS 7.53E+000	mg/L	25	EST	MEYLAN,WM ET AL. (1996)
logP 4.23			EST	MEYLAN,WM & HOWARD,PH (1995)
VP 6.67E-006	mm Hg	25	EST	NEELY,WB & BLAU,GE (1985)
DC	pKa			
HL 4.57E-009	atm m3/mol	25	EST	MEYLAN,WM & HOWARD,PH (1991)
OH 2.00E-010	cm3/molc sec	25	EST	MEYLAN,WM & HOWARD,PH (1993)

CAS #: 015979-85-8 — 3-CHLORO-N-(4-NITROPHENYL)BENZENAMINE

Formula: $C_{12}H_9ClN_2O_2$

Mol Weight: 248.67

MP (deg C): FP (deg C):

BP (deg C):

BP pressure (mm Hg):

Property/Value	Units	Temp	Data Type	Reference
WS 4.81E+000	mg/L	25	EST	MEYLAN,WM ET AL. (1996)
logP 4.33			EST	MEYLAN,WM & HOWARD,PH (1995)
VP 4.51E-006	mm Hg	25	EST	NEELY,WB & BLAU,GE (1985)
DC	pKa			
HL 3.07E-009	atm m3/mol	25	EST	MEYLAN,WM & HOWARD,PH (1991)
OH 1.21E-010	cm3/molc sec	25	EST	MEYLAN,WM & HOWARD,PH (1993)

CAS #: 015979-87-0	3-NITRO-N-(4-NITROPHENYL)BENZENAMINE

Formula: $C_{12}H_9N_3O_4$

Mol Weight: 259.22

MP (deg C): | FP (deg C):

BP (deg C):

BP pressure (mm Hg):

Property/Value	Units	Temp	Data Type	Reference
WS 6.86E+000	mg/L	25	EST	MEYLAN,WM ET AL. (1996)
logP 4.08			EST	MEYLAN,WM & HOWARD,PH (1995)
VP 3.28E-007	mm Hg	25	EST	NEELY,WB & BLAU,GE (1985)
DC	pKa			
HL 1.64E-011	atm m3/mol	25	EST	MEYLAN,WM & HOWARD,PH (1991)
OH 3.11E-011	cm3/molc sec	25	EST	MEYLAN,WM & HOWARD,PH (1993)

CAS #: 015981-93-8	O-TRIACETYL-N-ACETYL-ARA-C

Formula: $C_{17}H_{21}N_3O_9$

Mol Weight: 411.37

MP (deg C): | FP (deg C):

BP (deg C):

BP pressure (mm Hg):

Property/Value	Units	Temp	Data Type	Reference
WS 1.75E+002	mg/L	25	EST	MEYLAN,WM ET AL. (1996)
logP -0.16			EXP	HANSCH,C & LEO,AJ (1985)
VP 9.85E-012	mm Hg	25	EST	NEELY,WB & BLAU,GE (1985)
DC	pKa			
HL 1.40E-022	atm m3/mol	25	EST	MEYLAN,WM & HOWARD,PH (1991)
OH 1.12E-010	cm3/molc sec	25	EST	MEYLAN,WM & HOWARD,PH (1993)

CAS #: 015986-80-8	PYRAZINE, 2-METHYL-3-PROPYL-

Formula: $C_8H_{12}N_2$

Mol Weight: 136.20

MP (deg C): | FP (deg C):

BP (deg C): 189-190

BP pressure (mm Hg): 7.63E+002

Property/Value	Units	Temp	Data Type	Reference
WS 3.96E+003	mg/L	25	EST	MEYLAN,WM ET AL. (1996)
logP 1.57			EXP	YAMAGAMI,C ET AL. (1991)
VP 1.92E-001	mm Hg	25	EST	NEELY,WB & BLAU,GE (1985)
DC	pKa			
HL 6.26E-006	atm m3/mol	25	EST	MEYLAN,WM & HOWARD,PH (1991)
OH 3.33E-012	cm3/molc sec	25	EST	MEYLAN,WM & HOWARD,PH (1993)

CAS #: 015987-00-5	PYRAZINE, 2-BUTYL-3-METHYL-

Formula: $C_9H_{14}N_2$

Mol Weight: 150.23

MP (deg C): | FP (deg C):

BP (deg C):

BP pressure (mm Hg):

Property/Value	Units	Temp	Data Type	Reference
WS 1.22E+003	mg/L	25	EST	MEYLAN,WM ET AL. (1996)
logP 2.10			EXP	YAMAGAMI,C ET AL. (1991)
VP 5.86E-002	mm Hg	25	EST	NEELY,WB & BLAU,GE (1985)
DC	pKa			
HL 8.32E-006	atm m3/mol	25	EST	MEYLAN,WM & HOWARD,PH (1991)
OH 4.75E-012	cm3/molc sec	25	EST	MEYLAN,WM & HOWARD,PH (1993)

CAS #: 016018-79-4	3-NITROPHENYLAMIDINO UREA

Formula: $C_8H_{11}N_5O_3$

Mol Weight: 225.21

MP (deg C): | FP (deg C):

BP (deg C):

BP pressure (mm Hg):

Property/Value	Units	Temp	Data Type	Reference
WS 2.67E+003	mg/L	25	EST	MEYLAN,WM ET AL. (1996)
logP 1.28			EXP	HANSCH,C & LEO,AJ (1985)
VP 4.54E-008	mm Hg	25	EST	NEELY,WB & BLAU,GE (1985)
DC	pKa			
HL 5.38E-018	atm m3/mol	25	EST	MEYLAN,WM & HOWARD,PH (1991)
OH 2.57E-011	cm3/molc sec	25	EST	MEYLAN,WM & HOWARD,PH (1993)

CAS #: 016045-92-4	CHLOROSUCCINIC ACID

Formula: $C_4H_5ClO_4$

Mol Weight: 152.54

MP (deg C): | FP (deg C):

BP (deg C):

BP pressure (mm Hg):

Property/Value	Units	Temp	Data Type	Reference
WS 1.82E+005	mg/L	20	EXP	YALKOWSKY,SH & DANNENFELSER,RM (1992)
logP -0.57			EST	MEYLAN,WM & HOWARD,PH (1995)
VP 3.22E-004	mm Hg	25	EST	NEELY,WB & BLAU,GE (1985)
DC	pKa			
HL 1.90E-012	atm m3/mol	25	EST	MEYLAN,WM & HOWARD,PH (1991)
OH 1.97E-012	cm3/molc sec	25	EST	MEYLAN,WM & HOWARD,PH (1993)

CAS #: 016051-77-7	D-GLUCITOL, 1,4:3,6-DIANHYDRO-, 5-NITRATE

Formula: $C_6H_9NO_6$

Mol Weight: 191.14

MP (deg C): | FP (deg C):

BP (deg C):

BP pressure (mm Hg):

Property/Value	Units	Temp	Data Type	Reference
WS 1.07E+005	mg/L	25	EST	MEYLAN,WM ET AL. (1996)
logP -0.15			EXP	SANGSTER,J (1993)
VP 1.23E-004	mm Hg	25	EST	NEELY,WB & BLAU,GE (1985)
DC	pKa			
HL 5.69E-014	atm m3/mol	25	EST	MEYLAN,WM & HOWARD,PH (1991)
OH 2.58E-011	cm3/molc sec	25	EST	MEYLAN,WM & HOWARD,PH (1993)

CAS #: 016053-34-2	BENZENEACETAMIDE, N-(3-ETHOXYPHENYL)-

Formula: $C_{16}H_{17}NO_2$

Mol Weight: 255.32

MP (deg C): | FP (deg C):

BP (deg C):

BP pressure (mm Hg):

Property/Value	Units	Temp	Data Type	Reference
WS 3.85E+001	mg/L	25	EST	MEYLAN,WM ET AL. (1996)
logP 3.23			EXP	YAMAGAMI,C ET AL. (1984)
VP 6.79E-008	mm Hg	25	EST	NEELY,WB & BLAU,GE (1985)
DC	pKa			
HL 3.91E-011	atm m3/mol	25	EST	MEYLAN,WM & HOWARD,PH (1991)
OH 1.49E-010	cm3/molc sec	25	EST	MEYLAN,WM & HOWARD,PH (1993)

CAS #: 016063-69-7				2,4,6-TRICHLOROPYRIDINE

Formula: $C_5H_2Cl_3N$

Mol Weight: 182.44

MP (deg C): | FP (deg C):

BP (deg C):

BP pressure (mm Hg):

Property/Value	Units	Temp	Data Type	Reference
WS 2.76E+002	mg/L	25	EST	MEYLAN,WM ET AL. (1996)
logP 2.68			EXP	HANSCH,C & LEO,AJ (1985)
VP 1.13E-001	mm Hg	25	EST	NEELY,WB & BLAU,GE (1985)
DC	pKa			
HL 1.14E-002	atm m3/mol	25	EST	MEYLAN,WM & HOWARD,PH (1991)
OH 1.29E-013	cm3/molc sec	25	EST	MEYLAN,WM & HOWARD,PH (1993)

CAS #: 016063-70-0				2,3,5-TRICHLOROPYRIDINE

Formula: $C_5H_2Cl_3N$

Mol Weight: 182.44

MP (deg C): 48-50 | FP (deg C):

BP (deg C): 219

BP pressure (mm Hg):

Property/Value	Units	Temp	Data Type	Reference
WS 1.18E+002	mg/L	25	EST	MEYLAN,WM ET AL. (1996)
logP 3.11			EXP	HANSCH,C & LEO,AJ (1985)
VP 1.13E-001	mm Hg	25	EST	NEELY,WB & BLAU,GE (1985)
DC	pKa			
HL 1.81E-004	atm m3/mol	25	EST	MEYLAN,WM & HOWARD,PH (1991)
OH 5.35E-014	cm3/molc sec	25	EST	MEYLAN,WM & HOWARD,PH (1993)

CAS #: 016106-20-0				D-GLUCITOL, 1,4:3,6-DIANHYDRO-, 2-NITRATE

Formula: $C_6H_9NO_6$

Mol Weight: 191.14

MP (deg C): | FP (deg C):

BP (deg C):

BP pressure (mm Hg):

Property/Value	Units	Temp	Data Type	Reference
WS 1.07E+005	mg/L	25	EST	MEYLAN,WM ET AL. (1996)
logP -0.40			EXP	SANGSTER,J (1993)
VP 1.23E-004	mm Hg	25	EST	NEELY,WB & BLAU,GE (1985)
DC	pKa			
HL 5.69E-014	atm m3/mol	25	EST	MEYLAN,WM & HOWARD,PH (1991)
OH 2.58E-011	cm3/molc sec	25	EST	MEYLAN,WM & HOWARD,PH (1993)

CAS #: 016110-09-1				2,5-DICHLOROPYRIDINE

Formula: $C_5H_3Cl_2N$

Mol Weight: 147.99

MP (deg C): 59-62 | FP (deg C):

BP (deg C):

BP pressure (mm Hg):

Property/Value	Units	Temp	Data Type	Reference
WS 6.91E+002	mg/L	25	EST	MEYLAN,WM ET AL. (1996)
logP 2.40			EXP	HANSCH,C & LEO,AJ (1985)
VP 7.78E-001	mm Hg	25	EST	NEELY,WB & BLAU,GE (1985)
DC	pKa			
HL 2.44E-004	atm m3/mol	25	EST	MEYLAN,WM & HOWARD,PH (1991)
OH 7.60E-014	cm3/molc sec	25	EST	MEYLAN,WM & HOWARD,PH (1993)

CAS #: 016110-51-3				CROMOGYLYCIC ACID

Formula: $C_{23}H_{16}O_{11}$

Mol Weight: 468.38

MP (deg C): 241-242 de | FP (deg C):

BP (deg C):

BP pressure (mm Hg):

Property/Value	Units	Temp	Data Type	Reference
WS 2.10E+002	mg/L	25	EST	MEYLAN,WM ET AL. (1996)
logP 1.92			EXP	HANSCH,C ET AL. (1995)
VP 4.11E-020	mm Hg	25	EST	NEELY,WB & BLAU,GE (1985)
DC 1.10	pKa		EXP	SANGSTER,J (1993)
HL 2.19E-027	atm m3/mol	25	EST	MEYLAN,WM & HOWARD,PH (1991)
OH 2.41E-010	cm3/molc sec	25	EST	MEYLAN,WM & HOWARD,PH (1993)

CAS #: 016135-31-2				P-NITROFORMANILIDE

Formula: $C_7H_6N_2O_3$

Mol Weight: 166.14

MP (deg C): | FP (deg C):

BP (deg C):

BP pressure (mm Hg):

Property/Value	Units	Temp	Data Type	Reference
WS 3.85E+003	mg/L	25	EST	MEYLAN,WM ET AL. (1996)
logP 1.43			EXP	HANSCH,C & LEO,AJ (1985)
VP 9.83E-006	mm Hg	25	EST	NEELY,WB & BLAU,GE (1985)
DC	pKa			
HL 3.33E-011	atm m3/mol	25	EST	MEYLAN,WM & HOWARD,PH (1991)
OH 5.33E-012	cm3/molc sec	25	EST	MEYLAN,WM & HOWARD,PH (1993)

CAS #: 016136-52-0				INDOLE-4-CYANO-

Formula: $C_9H_6N_2$

Mol Weight: 142.16

MP (deg C): | FP (deg C):

BP (deg C):

BP pressure (mm Hg):

Property/Value	Units	Temp	Data Type	Reference
WS 3.53E+002	mg/L	25	EST	MEYLAN,WM ET AL. (1996)
logP 2.46			EXP	HANSCH,C ET AL. (1995)
VP 2.32E-004	mm Hg	25	EST	NEELY,WB & BLAU,GE (1985)
DC	pKa			
HL 8.56E-009	atm m3/mol	25	EST	MEYLAN,WM & HOWARD,PH (1991)
OH 1.94E-011	cm3/molc sec	25	EST	MEYLAN,WM & HOWARD,PH (1993)

CAS #: 016141-40-5				M-METHOXY-N-PHENYLSUCCINIMIDE

Formula: $C_{11}H_{11}NO_3$

Mol Weight: 205.22

MP (deg C): | FP (deg C):

BP (deg C):

BP pressure (mm Hg):

Property/Value	Units	Temp	Data Type	Reference
WS 2.89E+004	mg/L	25	EST	MEYLAN,WM ET AL. (1996)
logP 0.18			EXP	HANSCH,C & LEO,AJ (1985)
VP 1.58E-007	mm Hg	25	EST	NEELY,WB & BLAU,GE (1985)
DC	pKa			
HL 7.72E-009	atm m3/mol	25	EST	MEYLAN,WM & HOWARD,PH (1991)
OH 1.42E-010	cm3/molc sec	25	EST	MEYLAN,WM & HOWARD,PH (1993)

CAS #: 016156-59-5				PHENYLMETHANESULFONATE

Formula: $C_7H_8O_3S$

Mol Weight: 172.20

MP (deg C): | FP (deg C):

BP (deg C):

BP pressure (mm Hg):

Property/Value	Units	Temp	Data Type	Reference
WS 5.14E+003	mg/L	25	EST	MEYLAN,WM ET AL. (1996)
logP 1.25			EXP	HANSCH,C & LEO,AJ (1985)
VP 1.98E-003	mm Hg	25	EST	NEELY,WB & BLAU,GE (1985)
DC	pKa			
HL 1.49E-006	atm m3/mol	25	EST	MEYLAN,WM & HOWARD,PH (1991)
OH 2.16E-011	cm3/molc sec	25	EST	MEYLAN,WM & HOWARD,PH (1993)

CAS #: 016166-22-6				ACETAMIDE, N-4-PYRIMIDINYL-

Formula: $C_6H_7N_3O$

Mol Weight: 137.14

MP (deg C): | FP (deg C):

BP (deg C):

BP pressure (mm Hg):

Property/Value	Units	Temp	Data Type	Reference
WS 3.97E+003	mg/L	25	EST	MEYLAN,WM ET AL. (1996)
logP 0.03			EXP	YAMAGAMI,C ET AL. (1990)
VP 1.37E-004	mm Hg	25	EST	NEELY,WB & BLAU,GE (1985)
DC	pKa			
HL 3.34E-012	atm m3/mol	25	EST	MEYLAN,WM & HOWARD,PH (1991)
OH 1.57E-012	cm3/molc sec	25	EST	MEYLAN,WM & HOWARD,PH (1993)

CAS #: 016219-75-3				ETHYLIDENE NORBORNENE

Formula: C_9H_{12}

Mol Weight: 120.20

MP (deg C): | FP (deg C):

BP (deg C): 144-48

BP pressure (mm Hg):

Property/Value	Units	Temp	Data Type	Reference
WS 8.90E+000	mg/L	25	EXP	CHEM INSPECT TEST INST (1992)
logP 3.82			EXP	CHEM INSPECT TEST INST (1992)
VP 3.59E+000	mm Hg	25	EST	NEELY,WB & BLAU,GE (1985)
DC	pKa			
HL 1.26E-001	atm m3/mol	25	EST	MEYLAN,WM & HOWARD,PH (1991)
OH 1.48E-010	cm3/molc sec	25	EST	MEYLAN,WM & HOWARD,PH (1993)

CAS #: 016220-58-9				N-(3-CHLOROPHENYL)-2,4-DINITROBENZENAMINE

Formula: $C_{12}H_8ClN_3O_4$

Mol Weight: 293.67

MP (deg C): | FP (deg C):

BP (deg C):

BP pressure (mm Hg):

Property/Value	Units	Temp	Data Type	Reference
WS 3.83E+000	mg/L	25	EST	MEYLAN,WM ET AL. (1996)
logP 4.15				MEYLAN,WM & HOWARD,PH (1995)
VP 7.30E-008	mm Hg	25	EST	NEELY,WB & BLAU,GE (1985)
DC	pKa			
HL 2.65E-010	atm m3/mol	25	EST	MEYLAN,WM & HOWARD,PH (1991)
OH 1.05E-010	cm3/molc sec	25	EST	MEYLAN,WM & HOWARD,PH (1993)

CAS #: 016227-13-7				4-BENZYL-1,2,4-TRIAZOLE

Formula: $C_9H_{11}N_3$

Mol Weight: 161.21

MP (deg C): | FP (deg C):

BP (deg C):

BP pressure (mm Hg):

Property/Value	Units	Temp	Data Type	Reference
WS 1.61E+004	mg/L	25	EST	MEYLAN,WM ET AL. (1996)
logP 0.74			EXP	HANSCH,C ET AL. (1995)
VP 3.08E-003	mm Hg	25	EST	NEELY,WB & BLAU,GE (1985)
DC	pKa			
HL 2.63E-006	atm m3/mol	25	EST	MEYLAN,WM & HOWARD,PH (1991)
OH 5.88E-012	cm3/molc sec	25	EST	MEYLAN,WM & HOWARD,PH (1993)

CAS #: 016238-73-6				4,5-DINITROQUINOLINE-1-OXIDE

Formula: $C_9H_7N_3O_5$

Mol Weight: 237.17

MP (deg C): | FP (deg C):

BP (deg C):

BP pressure (mm Hg):

Property/Value	Units	Temp	Data Type	Reference
WS 1.80E+003	mg/L	25	EST	MEYLAN,WM ET AL. (1996)
logP 0.95			EXP	HANSCH,C & LEO,AJ (1985)
VP 4.65E-008	mm Hg	25	EST	NEELY,WB & BLAU,GE (1985)
DC	pKa			
HL 1.07E-016	atm m3/mol	25	EST	MEYLAN,WM & HOWARD,PH (1991)
OH 1.57E-013	cm3/molc sec	25	EST	MEYLAN,WM & HOWARD,PH (1993)

CAS #: 016268-62-5				PENTAMETHYLMELAMINE

Formula: $C_8H_{16}N_6$

Mol Weight: 196.26

MP (deg C): | FP (deg C):

BP (deg C):

BP pressure (mm Hg):

Property/Value	Units	Temp	Data Type	Reference
WS 1.25E+003	mg/L	25	EST	MEYLAN,WM ET AL. (1996)
logP 1.83			EXP	HANSCH,C & LEO,AJ (1985)
VP 3.11E-004	mm Hg	25	EST	NEELY,WB & BLAU,GE (1985)
DC	pKa			
HL 8.41E-010	atm m3/mol	25	EST	MEYLAN,WM & HOWARD,PH (1991)
OH 6.98E-012	cm3/molc sec	25	EST	MEYLAN,WM & HOWARD,PH (1993)

CAS #: 016268-75-0				N6,N6-DIET-TETRAME-MELAMINE

Formula: $C_{11}H_{22}N_6$

Mol Weight: 238.34

MP (deg C): | FP (deg C):

BP (deg C):

BP pressure (mm Hg):

Property/Value	Units	Temp	Data Type	Reference
WS 3.10E+001	mg/L	25	EST	MEYLAN,WM ET AL. (1996)
logP 3.45			EXP	HANSCH,C & LEO,AJ (1985)
VP 4.43E-005	mm Hg	25	EST	NEELY,WB & BLAU,GE (1985)
DC	pKa			
HL 3.04E-008	atm m3/mol	25	EST	MEYLAN,WM & HOWARD,PH (1991)
OH 2.34E-011	cm3/molc sec	25	EST	MEYLAN,WM & HOWARD,PH (1993)

CAS #: 016269-66-2				THIENO(2,3-D)-PYRMIDINE,4-CHLORO
Formula: C6H3ClN2S				
Mol Weight: 170.62				
MP (deg C):			**FP (deg C):**	
BP (deg C):				
BP pressure (mm Hg):				

Property/Value	Units	Temp	Data Type	Reference
WS 1.81E+003	mg/L	25	EST	MEYLAN,WM ET AL. (1996)
logP 1.79			EXP	HANSCH,C & LEO,AJ (1985)
VP 1.07E-003	mm Hg	25	EST	NEELY,WB & BLAU,GE (1985)
DC	pKa			
HL 7.24E-006	atm m3/mol	25	EST	MEYLAN,WM & HOWARD,PH (1991)
OH 2.11E-011	cm3/molc sec	25	EST	MEYLAN,WM & HOWARD,PH (1993)

CAS #: 016271-10-6				O,S-DIETHYLPHOSPHORAMIDOTHIOATE
Formula: C4H12NO2PS				
Mol Weight: 169.18				
MP (deg C):			**FP (deg C):**	
BP (deg C):				
BP pressure (mm Hg):				

Property/Value	Units	Temp	Data Type	Reference
WS 4.62E+004	mg/L	25	EST	MEYLAN,WM ET AL. (1996)
logP 0.15			EXP	HANSCH,C & LEO,AJ (1985)
VP 1.34E-002	mm Hg	25	EST	NEELY,WB & BLAU,GE (1985)
DC	pKa			
HL 1.53E-009	atm m3/mol	25	EST	MEYLAN,WM & HOWARD,PH (1991)
OH 6.61E-011	cm3/molc sec	25	EST	MEYLAN,WM & HOWARD,PH (1993)

CAS #: 016271-16-2				O,S-DIME-N-PR-PHOSPHORAMIDOTHIOAT
Formula: C5H14NO2PS				
Mol Weight: 183.21				
MP (deg C):			**FP (deg C):**	
BP (deg C):				
BP pressure (mm Hg):				

Property/Value	Units	Temp	Data Type	Reference
WS 1.48E+004	mg/L	25	EST	MEYLAN,WM ET AL. (1996)
logP 0.65			EXP	HANSCH,C & LEO,AJ (1985)
VP 1.32E-002	mm Hg	25	EST	NEELY,WB & BLAU,GE (1985)
DC	pKa			
HL 3.36E-009	atm m3/mol	25	EST	MEYLAN,WM & HOWARD,PH (1991)
OH 4.51E-011	cm3/molc sec	25	EST	MEYLAN,WM & HOWARD,PH (1993)

CAS #: 016308-92-2				2,4-DIMETHYLBENZYL ALCOHOL
Formula: C9H12O				
Mol Weight: 136.20				
MP (deg C): 22			**FP (deg C):**	
BP (deg C): 232				
BP pressure (mm Hg):				

Property/Value	Units	Temp	Data Type	Reference
WS 3.94E+003	mg/L	25	EST	MEYLAN,WM ET AL. (1996)
logP 2.17			EST	MEYLAN,WM & HOWARD,PH (1995)
VP 4.43E-003	mm Hg	25	EST	NEELY,WB & BLAU,GE (1985)
DC	pKa			
HL 2.65E-007	atm m3/mol	25	EST	MEYLAN,WM & HOWARD,PH (1991)
OH 1.99E-011	cm3/molc sec	25	EST	MEYLAN,WM & HOWARD,PH (1993)

CAS #: 016310-37-5				2-METHYLSULFONYLQUINOXALINE
Formula: C9H8N2O2S				
Mol Weight: 208.24				
MP (deg C):			**FP (deg C):**	
BP (deg C):				
BP pressure (mm Hg):				

Property/Value	Units	Temp	Data Type	Reference
WS 8.39E+003	mg/L	25	EST	MEYLAN,WM ET AL. (1996)
logP 0.79			EXP	HANSCH,C & LEO,AJ (1985)
VP 5.37E-006	mm Hg	25	EST	NEELY,WB & BLAU,GE (1985)
DC -1.66	pKa	20	EXP	PERRIN,DD (1972)
HL 2.57E-009	atm m3/mol	25	EST	MEYLAN,WM & HOWARD,PH (1991)
OH 1.49E-012	cm3/molc sec	25	EST	MEYLAN,WM & HOWARD,PH (1993)

CAS #: 016332-51-7				2-ETHOXYETHYL TRIMETHYL AMMONIUM IODIDE
Formula: C7H18INO				
Mol Weight: 259.13				
MP (deg C):			**FP (deg C):**	
BP (deg C):				
BP pressure (mm Hg):				

Property/Value	Units	Temp	Data Type	Reference
WS 1.00E+006	mg/L	25	EST	MEYLAN,WM ET AL. (1996)
logP -2.22			EXP	SANGSTER,J (1993)
VP 2.31E-007	mm Hg	25	EST	NEELY,WB & BLAU,GE (1985)
DC	pKa			
HL 5.43E-014	atm m3/mol	25	EST	MEYLAN,WM & HOWARD,PH (1991)
OH 4.74E-011	cm3/molc sec	25	EST	MEYLAN,WM & HOWARD,PH (1993)

CAS #: 016338-97-9				N-NITROSO-DIALLYLAMINE
Formula: C6H10N2O				
Mol Weight: 126.16				
MP (deg C):			**FP (deg C):**	
BP (deg C):				
BP pressure (mm Hg):				

Property/Value	Units	Temp	Data Type	Reference
WS 1.20E+004	mg/L	25	EST	MEYLAN,WM ET AL. (1996)
logP 1.05			EST	MEYLAN,WM & HOWARD,PH (1995)
VP 1.00E-001	mm Hg	25	EST	NEELY,WB & BLAU,GE (1985)
DC	pKa			
HL 3.55E-006	atm m3/mol	25	EST	MEYLAN,WM & HOWARD,PH (1991)
OH 7.00E-011	cm3/molc sec	25	EST	MEYLAN,WM & HOWARD,PH (1993)

CAS #: 016339-04-1				N-NITROSO-N-ETHYL-ISOPROPYLAMINE
Formula: C5H12N2O				
Mol Weight: 116.16				
MP (deg C):			**FP (deg C):**	
BP (deg C):				
BP pressure (mm Hg):				

Property/Value	Units	Temp	Data Type	Reference
WS 1.77E+004	mg/L	25	EST	MEYLAN,WM ET AL. (1996)
logP 0.90			EXP	HANSCH,C ET AL. (1995)
VP 3.86E-001	mm Hg	25	EST	NEELY,WB & BLAU,GE (1985)
DC	pKa			
HL 4.81E-006	atm m3/mol	25	EST	MEYLAN,WM & HOWARD,PH (1991)
OH 2.72E-011	cm3/molc sec	25	EST	MEYLAN,WM & HOWARD,PH (1993)

CAS #: 016339-07-4	4-METHYL-N-NITROSOPIPERAZINE
Formula: $C_5H_{11}N_3O$	
Mol Weight: 129.16	
MP (deg C):	FP (deg C):
BP (deg C):	
BP pressure (mm Hg):	

	Property/ Value	Units	Temp	Data Type	Reference
WS	6.37E+005	mg/L	25	EST	MEYLAN,WM ET AL. (1996)
logP	0.20			EXP	HANSCH,C & LEO,AJ (1985)
VP	2.48E-002	mm Hg	25	EST	NEELY,WB & BLAU,GE (1985)
DC	5.93	pKa		EXP	PERRIN,DD (1965)
HL	5.97E-010	atm m3/mol	25	EST	MEYLAN,WM & HOWARD,PH (1991)
OH	1.10E-010	cm3/molc sec	25	EST	MEYLAN,WM & HOWARD,PH (1993)

CAS #: 016339-18-7	N-NITROSO-BIS(CYANOMETHYL)AMINE
Formula: $C_4H_4N_4O$	
Mol Weight: 124.10	
MP (deg C):	FP (deg C):
BP (deg C):	
BP pressure (mm Hg):	

	Property/ Value	Units	Temp	Data Type	Reference
WS	4.87E+004	mg/L	25	EST	MEYLAN,WM ET AL. (1996)
logP	0.04			EXP	HANSCH,C & LEO,AJ (1985)
VP	1.98E-004	mm Hg	25	EST	NEELY,WB & BLAU,GE (1985)
DC		pKa			
HL	3.54E-013	atm m3/mol	25	EST	MEYLAN,WM & HOWARD,PH (1991)
OH	3.30E-012	cm3/molc sec	25	EST	MEYLAN,WM & HOWARD,PH (1993)

CAS #: 016347-95-8	4-METHOXYQUINAZOLINE
Formula: $C_9H_8N_2O$	
Mol Weight: 160.18	
MP (deg C):	FP (deg C):
BP (deg C):	
BP pressure (mm Hg):	

	Property/ Value	Units	Temp	Data Type	Reference
WS	1.42E+003	mg/L	25	EST	MEYLAN,WM ET AL. (1996)
logP	1.97			EXP	HANSCH,C & LEO,AJ (1985)
VP	1.57E-003	mm Hg	25	EST	NEELY,WB & BLAU,GE (1985)
DC	3.07	pKa	20	EXP	PERRIN,DD (1965)
HL	1.54E-007	atm m3/mol	25	EST	MEYLAN,WM & HOWARD,PH (1991)
OH	2.29E-011	cm3/molc sec	25	EST	MEYLAN,WM & HOWARD,PH (1993)

CAS #: 016350-99-5	N-BENZYL-N-FORMYLANILINE
Formula: $C_{14}H_{13}NO$	
Mol Weight: 211.27	
MP (deg C):	FP (deg C):
BP (deg C):	
BP pressure (mm Hg):	

	Property/ Value	Units	Temp	Data Type	Reference
WS	2.22E+002	mg/L	25	EST	MEYLAN,WM ET AL. (1996)
logP	2.62			EXP	HANSCH,C & LEO,AJ (1985)
VP	1.71E-005	mm Hg	25	EST	NEELY,WB & BLAU,GE (1985)
DC		pKa			
HL	1.40E-008	atm m3/mol	25	EST	MEYLAN,WM & HOWARD,PH (1991)
OH	2.59E-011	cm3/molc sec	25	EST	MEYLAN,WM & HOWARD,PH (1993)

CAS #: 016375-90-9	3-METHYL-4-HYDROXYACETANILIDE
Formula: $C_9H_{11}NO_2$	
Mol Weight: 165.19	
MP (deg C):	FP (deg C):
BP (deg C):	
BP pressure (mm Hg):	

	Property/ Value	Units	Temp	Data Type	Reference
WS	1.37E+004	mg/L	25	EST	MEYLAN,WM ET AL. (1996)
logP	0.79			EXP	HANSCH,C & LEO,AJ (1985)
VP	2.71E-006	mm Hg	25	EST	NEELY,WB & BLAU,GE (1985)
DC		pKa			
HL	7.09E-013	atm m3/mol	25	EST	MEYLAN,WM & HOWARD,PH (1991)
OH	2.27E-011	cm3/molc sec	25	EST	MEYLAN,WM & HOWARD,PH (1993)

CAS #: 016391-07-4	Phosphinic acid, methyl-, ethyl ester
Formula: $C_3H_9O_2P$	
Mol Weight: 108.08	
MP (deg C):	FP (deg C):
BP (deg C):	
BP pressure (mm Hg):	

	Property/ Value	Units	Temp	Data Type	Reference
WS	3.59E+005	mg/L	25	EST	MEYLAN,WM ET AL. (1996)
logP	-0.60			EXP	KRIKORIAN,SE ET AL. (1987)
VP	9.70E+000	mm Hg	25	EST	NEELY,WB & BLAU,GE (1985)
DC		pKa			
HL	2.18E-005	atm m3/mol	25	EST	MEYLAN,WM & HOWARD,PH (1991)
OH	1.76E-011	cm3/molc sec	25	EST	MEYLAN,WM & HOWARD,PH (1993)

CAS #: 016395-58-7	PROLIN-AMIDE, N-ACETYL
Formula: $C_7H_{12}N_2O_2$	
Mol Weight: 156.19	
MP (deg C):	FP (deg C):
BP (deg C):	
BP pressure (mm Hg):	

	Property/ Value	Units	Temp	Data Type	Reference
WS	4.86E+004	mg/L	25	EST	MEYLAN,WM ET AL. (1996)
logP	-1.34			EXP	HANSCH,C & LEO,AJ (1985)
VP	1.44E-007	mm Hg	25	EST	NEELY,WB & BLAU,GE (1985)
DC		pKa			
HL	2.01E-010	atm m3/mol	25	EST	MEYLAN,WM & HOWARD,PH (1991)
OH	9.52E-011	cm3/molc sec	25	EST	MEYLAN,WM & HOWARD,PH (1993)

CAS #: 016409-45-3	MENTHYL ACETATE
Formula: $C_{12}H_{22}O_2$	
Mol Weight: 198.31	
MP (deg C):	FP (deg C):
BP (deg C):	
BP pressure (mm Hg):	

	Property/ Value	Units	Temp	Data Type	Reference
WS	8.01E+000	mg/L	25	EST	MEYLAN,WM ET AL. (1996)
logP	4.39			EST	MEYLAN,WM & HOWARD,PH (1995)
VP	1.02E-001	mm Hg	25	EXP	PERRY,RH & GREEN,D (1984)
DC		pKa			
HL	9.90E-004	atm m3/mol	25	EST	MEYLAN,WM & HOWARD,PH (1991)
OH	1.85E-011	cm3/molc sec	25	EST	MEYLAN,WM & HOWARD,PH (1993)

CAS #: 016463-74-4 — PREGN-4-ENE-3,20-DIONE, 17-(ACETYLOXY)-11,21-DIH

Formula:	$C_{23}H_{32}O_6$
Mol Weight:	404.51
MP (deg C):	FP (deg C):
BP (deg C):	
BP pressure (mm Hg):	

Property/Value	Units	Temp	Data Type	Reference
WS 3.12E+001	mg/L	25	EST	MEYLAN,WM ET AL. (1996)
logP 2.30			EXP	SANGSTER,J (1993)
VP 1.16E-013	mm Hg	25	EST	NEELY,WB & BLAU,GE (1985)
DC	pKa			
HL 6.24E-012	atm m3/mol	25	EST	MEYLAN,WM & HOWARD,PH (1991)
OH 1.10E-010	cm3/molc sec	25	EST	MEYLAN,WM & HOWARD,PH (1993)

CAS #: 016479-50-8 — CARD-20(22)-ENOLIDE, 3-[[2,6-DIDEOXY-4-O-(2,6-DI

Formula:	$C_{35}H_{54}O_{10}$
Mol Weight:	634.81
MP (deg C):	FP (deg C):
BP (deg C):	
BP pressure (mm Hg):	

Property/Value	Units	Temp	Data Type	Reference
WS 1.38E+000	mg/L	25	EST	MEYLAN,WM ET AL. (1996)
logP 2.15			EXP	SANGSTER,J (1993)
VP 5.28E-024	mm Hg	25	EST	NEELY,WB & BLAU,GE (1985)
DC	pKa			
HL 1.86E-020	atm m3/mol	25	EST	MEYLAN,WM & HOWARD,PH (1991)
OH 1.31E-010	cm3/molc sec	25	EST	MEYLAN,WM & HOWARD,PH (1993)

CAS #: 016524-22-4 — BENZOIC ACID, 2-[(3-METHYLPHENYL)AMINO]-

Formula:	$C_{14}H_{13}NO_2$
Mol Weight:	227.27
MP (deg C):	FP (deg C):
BP (deg C):	
BP pressure (mm Hg):	

Property/Value	Units	Temp	Data Type	Reference
WS 2.14E+000	mg/L	25	EST	MEYLAN,WM ET AL. (1996)
logP 4.88			EXP	TERADA,H ET AL. (1974)
VP 1.15E-006	mm Hg	25	EST	NEELY,WB & BLAU,GE (1985)
DC	pKa			
HL 2.33E-011	atm m3/mol	25	EST	MEYLAN,WM & HOWARD,PH (1991)
OH 2.01E-010	cm3/molc sec	25	EST	MEYLAN,WM & HOWARD,PH (1993)

CAS #: 016533-50-9 — M-PROPOXYPHENOL

Formula:	$C_9H_{12}O_2$
Mol Weight:	152.19
MP (deg C):	FP (deg C):
BP (deg C):	
BP pressure (mm Hg):	

Property/Value	Units	Temp	Data Type	Reference
WS 1.73E+003	mg/L	25	EST	MEYLAN,WM ET AL. (1996)
logP 2.59			EXP	HANSCH,C & LEO,AJ (1985)
VP 7.65E-003	mm Hg	25	EST	NEELY,WB & BLAU,GE (1985)
DC	pKa			
HL 5.84E-008	atm m3/mol	25	EST	MEYLAN,WM & HOWARD,PH (1991)
OH 2.11E-010	cm3/molc sec	25	EST	MEYLAN,WM & HOWARD,PH (1993)

CAS #: 016543-55-8 — 3-(1-NITROSO-2-PYRROLIDINYL)PYRIDINE

Formula:	$C_9H_{11}N_3O$
Mol Weight:	177.21
MP (deg C): 47	FP (deg C):
BP (deg C): 154	
BP pressure (mm Hg): 2.00E-001	

Property/Value	Units	Temp	Data Type	Reference
WS 1.44E+001	mg/L	25	EST	MEYLAN,WM ET AL. (1996)
logP 0.32			EST	MEYLAN,WM & HOWARD,PH (1995)
VP 1.61E-004	mm Hg	25	EST	NEELY,WB & BLAU,GE (1985)
DC	pKa			
HL 1.69E-010	atm m3/mol	25	EST	MEYLAN,WM & HOWARD,PH (1991)
OH 3.06E-011	cm3/molc sec	25	EST	ATKINSON,R (1988)

CAS #: 016554-83-9 — BICYCLO(4.1.0)HEPT-3-ENE

Formula:	C_7H_{10}
Mol Weight:	94.16
MP (deg C):	FP (deg C):
BP (deg C):	
BP pressure (mm Hg):	

Property/Value	Units	Temp	Data Type	Reference
WS 6.63E+001	mg/L	25	EST	MEYLAN,WM ET AL. (1996)
logP 3.19			EST	MEYLAN,WM & HOWARD,PH (1995)
VP 3.05E-002	mm Hg	25	EXP	VARUSHCHENKO,RM ET AL. (1974)
DC	pKa			
HL 3.88E-002	atm m3/mol	25	EST	MEYLAN,WM & HOWARD,PH (1991)
OH 5.73E-011	cm3/molc sec	25	EST	MEYLAN,WM & HOWARD,PH (1993)

CAS #: 016563-41-0 — PROPANOIC ACID, 3-(1-NAPHTHALENYLOXY)-

Formula:	$C_{13}H_{12}O_3$
Mol Weight:	216.24
MP (deg C):	FP (deg C):
BP (deg C):	
BP pressure (mm Hg):	

Property/Value	Units	Temp	Data Type	Reference
WS 2.45E+002	mg/L	25	EST	MEYLAN,WM ET AL. (1996)
logP 3.00			EXP	CHAMBERLAIN,K ET AL. (1986)
VP 2.71E-006	mm Hg	25	EST	NEELY,WB & BLAU,GE (1985)
DC	pKa			
HL 3.07E-010	atm m3/mol	25	EST	MEYLAN,WM & HOWARD,PH (1991)
OH 2.10E-010	cm3/molc sec	25	EST	MEYLAN,WM & HOWARD,PH (1993)

CAS #: 016563-45-4 — BUTANOIC ACID, 4-(1-NAPHTHALENYLOXY)-

Formula:	$C_{14}H_{14}O_3$
Mol Weight:	230.27
MP (deg C):	FP (deg C):
BP (deg C):	
BP pressure (mm Hg):	

Property/Value	Units	Temp	Data Type	Reference
WS 7.73E+001	mg/L	25	EST	MEYLAN,WM ET AL. (1996)
logP 3.50			EXP	SANGSTER,J (1994)
VP 1.17E-006	mm Hg	25	EST	NEELY,WB & BLAU,GE (1985)
DC	pKa			
HL 4.08E-010	atm m3/mol	25	EST	MEYLAN,WM & HOWARD,PH (1991)
OH 2.12E-010	cm3/molc sec	25	EST	MEYLAN,WM & HOWARD,PH (1993)

016566-20-4 — 5-AMINOQUINOXALINE

CAS #:	016566-20-4		
Formula:	$C_8H_7N_3$		
Mol Weight:	145.17		
MP (deg C):		FP (deg C):	
BP (deg C):			
BP pressure (mm Hg):			

Property/Value	Units	Temp	Data Type	Reference
WS 6.30E+003	mg/L	25	EST	MEYLAN,WM ET AL. (1996)
logP 1.29			EXP	HANSCH,C & LEO,AJ (1985)
VP 2.50E-004	mm Hg	25	EST	NEELY,WB & BLAU,GE (1985)
DC	pKa			
HL 1.01E-010	atm m3/mol	25	EST	MEYLAN,WM & HOWARD,PH (1991)
OH 1.10E-010	cm3/molc sec	25	EST	MEYLAN,WM & HOWARD,PH (1993)

016583-98-5 — 2H-BENZOTRIAZOLE-2-ACETIC ACID, à,à-DIMETHYL-

CAS #:	016583-98-5		
Formula:	$C_{10}H_{11}N_3O_2$		
Mol Weight:	205.22		
MP (deg C):		FP (deg C):	
BP (deg C):			
BP pressure (mm Hg):			

Property/Value	Units	Temp	Data Type	Reference
WS 2.90E+004	mg/L	25	EST	MEYLAN,WM ET AL. (1996)
logP 0.64			EXP	SPARATORE,F ET AL. (1988)
VP 3.22E-006	mm Hg	25	EST	NEELY,WB & BLAU,GE (1985)
DC	pKa			
HL 4.10E-011	atm m3/mol	25	EST	MEYLAN,WM & HOWARD,PH (1991)
OH 1.85E-012	cm3/molc sec	25	EST	MEYLAN,WM & HOWARD,PH (1993)

016583-99-6 — 2-(2-CARBOXYET)BENZOTRIAZOLE

CAS #:	016583-99-6		
Formula:	$C_9H_{11}N_3O_2$		
Mol Weight:	193.21		
MP (deg C):		FP (deg C):	
BP (deg C):			
BP pressure (mm Hg):			

Property/Value	Units	Temp	Data Type	Reference
WS 1.28E+005	mg/L	25	EST	MEYLAN,WM ET AL. (1996)
logP -0.03			EXP	HANSCH,C & LEO,AJ (1985)
VP 3.40E-006	mm Hg	25	EST	NEELY,WB & BLAU,GE (1985)
DC	pKa			
HL 3.09E-011	atm m3/mol	25	EST	MEYLAN,WM & HOWARD,PH (1991)
OH 3.53E-012	cm3/molc sec	25	EST	MEYLAN,WM & HOWARD,PH (1993)

016584-00-2 — 2-METHYLBENZOTRIAZOLE

CAS #:	016584-00-2		
Formula:	$C_7H_9N_3$		
Mol Weight:	135.17		
MP (deg C):		FP (deg C):	
BP (deg C):			
BP pressure (mm Hg):			

Property/Value	Units	Temp	Data Type	Reference
WS 3.55E+003	mg/L	25	EST	MEYLAN,WM ET AL. (1996)
logP 1.64			EXP	HANSCH,C & LEO,AJ (1985)
VP 1.04E-002	mm Hg	25	EST	NEELY,WB & BLAU,GE (1985)
DC	pKa			
HL 3.13E-006	atm m3/mol	25	EST	MEYLAN,WM & HOWARD,PH (1991)
OH 1.14E-012	cm3/molc sec	25	EST	MEYLAN,WM & HOWARD,PH (1993)

016584-01-3 — 2-BUTYLBENZOTRIAZOLE

CAS #:	016584-01-3		
Formula:	$C_{10}H_{15}N_3$		
Mol Weight:	177.25		
MP (deg C):		FP (deg C):	
BP (deg C):			
BP pressure (mm Hg):			

Property/Value	Units	Temp	Data Type	Reference
WS 1.26E+002	mg/L	25	EST	MEYLAN,WM ET AL. (1996)
logP 3.12			EXP	HANSCH,C & LEO,AJ (1985)
VP 4.22E-004	mm Hg	25	EST	NEELY,WB & BLAU,GE (1985)
DC	pKa			
HL 7.32E-006	atm m3/mol	25	EST	MEYLAN,WM & HOWARD,PH (1991)
OH 4.88E-012	cm3/molc sec	25	EST	MEYLAN,WM & HOWARD,PH (1993)

016584-02-4 — 1-PROPYLBENZOTRIAZOLE

CAS #:	016584-02-4		
Formula:	$C_9H_{13}N_3$		
Mol Weight:	163.22		
MP (deg C):		FP (deg C):	
BP (deg C):			
BP pressure (mm Hg):			

Property/Value	Units	Temp	Data Type	Reference
WS 1.03E+003	mg/L	25	EST	MEYLAN,WM ET AL. (1996)
logP 2.13			EXP	HANSCH,C & LEO,AJ (1985)
VP 1.17E-003	mm Hg	25	EST	NEELY,WB & BLAU,GE (1985)
DC	pKa			
HL 5.51E-006	atm m3/mol	25	EST	MEYLAN,WM & HOWARD,PH (1991)
OH 3.46E-012	cm3/molc sec	25	EST	MEYLAN,WM & HOWARD,PH (1993)

016584-03-5 — 2-PROPYLBENZOTRIAZOLE

CAS #:	016584-03-5		
Formula:	$C_9H_{13}N_3$		
Mol Weight:	163.22		
MP (deg C):		FP (deg C):	
BP (deg C):			
BP pressure (mm Hg):			

Property/Value	Units	Temp	Data Type	Reference
WS 4.67E+002	mg/L	25	EST	MEYLAN,WM ET AL. (1996)
logP 2.53			EXP	HANSCH,C & LEO,AJ (1985)
VP 1.17E-003	mm Hg	25	EST	NEELY,WB & BLAU,GE (1985)
DC	pKa			
HL 5.51E-006	atm m3/mol	25	EST	MEYLAN,WM & HOWARD,PH (1991)
OH 3.46E-012	cm3/molc sec	25	EST	MEYLAN,WM & HOWARD,PH (1993)

016584-04-6 — 2-ETHYLBENZOTRIAZOLE

CAS #:	016584-04-6		
Formula:	$C_8H_{11}N_3$		
Mol Weight:	149.20		
MP (deg C):		FP (deg C):	
BP (deg C):			
BP pressure (mm Hg):			

Property/Value	Units	Temp	Data Type	Reference
WS 1.26E+003	mg/L	25	EST	MEYLAN,WM ET AL. (1996)
logP 2.10			EXP	HANSCH,C & LEO,AJ (1985)
VP 3.41E-003	mm Hg	25	EST	NEELY,WB & BLAU,GE (1985)
DC	pKa			
HL 4.15E-006	atm m3/mol	25	EST	MEYLAN,WM & HOWARD,PH (1991)
OH 2.10E-012	cm3/molc sec	25	EST	MEYLAN,WM & HOWARD,PH (1993)

CAS #: 016584-05-7				1-ETHYLBENZOTRIAZOLE

Formula: $C_8H_{11}N_3$

Mol Weight: 149.20

MP (deg C):   FP (deg C):

BP (deg C):

BP pressure (mm Hg):

Property/Value	Units	Temp	Data Type	Reference
WS 3.49E+003	mg/L	25	EST	MEYLAN,WM ET AL. (1996)
logP 1.58			EXP	HANSCH,C & LEO,AJ (1985)
VP 3.41E-003	mm Hg	25	EST	NEELY,WB & BLAU,GE (1985)
DC	pKa			
HL 4.15E-006	atm m3/mol	25	EST	MEYLAN,WM & HOWARD,PH (1991)
OH 2.10E-012	cm3/molc sec	25	EST	MEYLAN,WM & HOWARD,PH (1993)

CAS #: 016590-41-3				NALTREXONE

Formula: $C_{20}H_{29}NO_4$

Mol Weight: 347.46

MP (deg C): 168-170   FP (deg C):

BP (deg C):

BP pressure (mm Hg):

Property/Value	Units	Temp	Data Type	Reference
WS 1.63E+003	mg/L	25	EST	MEYLAN,WM ET AL. (1996)
logP 1.92			EXP	HANSCH,C & LEO,AJ (1985)
VP 1.30E-011	mm Hg	25	EST	NEELY,WB & BLAU,GE (1985)
DC	pKa			
HL 4.22E-019	atm m3/mol	25	EST	MEYLAN,WM & HOWARD,PH (1991)
OH 2.63E-010	cm3/molc sec	25	EST	MEYLAN,WM & HOWARD,PH (1993)

CAS #: 016593-50-3				3-PYRIDYLMETHYL TRIME AMMONIUM BROMIDE

Formula: $C_9H_{15}BrN_2$

Mol Weight: 231.14

MP (deg C):   FP (deg C):

BP (deg C):

BP pressure (mm Hg):

Property/Value	Units	Temp	Data Type	Reference
WS 1.00E+006	mg/L	25	EST	MEYLAN,WM ET AL. (1996)
logP -3.50			EXP	HANSCH,C & LEO,AJ (1985)
VP 2.54E-008	mm Hg	25	EST	NEELY,WB & BLAU,GE (1985)
DC	pKa			
HL 3.50E-016	atm m3/mol	25	EST	MEYLAN,WM & HOWARD,PH (1991)
OH 1.34E-011	cm3/molc sec	25	EST	MEYLAN,WM & HOWARD,PH (1993)

CAS #: 016605-91-7				2,3-DICHLOROBIPHENYL

Formula: $C_{12}H_8Cl_2$

Mol Weight: 223.10

MP (deg C):   FP (deg C):

BP (deg C):

BP pressure (mm Hg):

Property/Value	Units	Temp	Data Type	Reference
WS 1.71E+000	mg/L	25	EST	MEYLAN,WM ET AL. (1996)
logP 5.02			EXP	HANSCH,C & LEO,AJ (1985)
VP 1.91E-004	mm Hg	25	EST	NEELY,WB & BLAU,GE (1985)
DC	pKa			
HL 2.27E-004	atm m3/mol	25	EST	MEYLAN,WM & HOWARD,PH (1991)
OH 2.47E-012	cm3/molc sec	25	EST	MEYLAN,WM & HOWARD,PH (1993)

CAS #: 016606-02-3				2,4',5-TRICHLOROBIPHENYL

Formula: $C_{12}H_7Cl_3$

Mol Weight: 257.55

MP (deg C):   FP (deg C):

BP (deg C):

BP pressure (mm Hg):

Property/Value	Units	Temp	Data Type	Reference
WS 1.10E-001	mg/L	25	EXP	KILZER,L ET AL. (1979)
logP 5.69			EXP	HANSCH,C & LEO,AJ (1985)
VP 4.00E-004	mm Hg	25	EXP	KILZER,L ET AL. (1979)
DC	pKa			
HL 1.23E-003	atm m3/mol	25	EST	VP/WSOL
OH 1.09E-012	cm3/molc sec	25	EST	MEYLAN,WM & HOWARD,PH (1993)

CAS #: 016620-75-0				P,P'-DIMETHYLBENZOPHENONEIMINE

Formula: $C_{15}H_{15}N$

Mol Weight: 209.29

MP (deg C):   FP (deg C):

BP (deg C):

BP pressure (mm Hg):

Property/Value	Units	Temp	Data Type	Reference
WS 2.66E+001	mg/L	25	EST	MEYLAN,WM ET AL. (1996)
logP 3.71			EXP	HANSCH,C & LEO,AJ (1985)
VP 4.12E-005	mm Hg	25	EST	NEELY,WB & BLAU,GE (1985)
DC	pKa			
HL 1.75E-005	atm m3/mol	25	EST	MEYLAN,WM & HOWARD,PH (1991)
OH 1.25E-011	cm3/molc sec	25	EST	MEYLAN,WM & HOWARD,PH (1993)

CAS #: 016637-13-1				2,6-DIISOPROPYLACETANILIDE

Formula: $C_{14}H_{21}NO$

Mol Weight: 219.33

MP (deg C):   FP (deg C):

BP (deg C):

BP pressure (mm Hg):

Property/Value	Units	Temp	Data Type	Reference
WS 3.04E+002	mg/L	25	EST	MEYLAN,WM ET AL. (1996)
logP 2.41			EXP	NAKAGAWA,Y ET AL. (1992)
VP 6.98E-006	mm Hg	25	EST	NEELY,WB & BLAU,GE (1985)
DC	pKa			
HL 2.33E-008	atm m3/mol	25	EST	MEYLAN,WM & HOWARD,PH (1991)
OH 2.26E-011	cm3/molc sec	25	EST	MEYLAN,WM & HOWARD,PH (1993)

CAS #: 016645-06-0				N,N-DIMETHYLHYDROXYLAMINE

Formula: C_2H_7NO

Mol Weight: 61.08

MP (deg C): 107-109   FP (deg C):

BP (deg C):

BP pressure (mm Hg):

Property/Value	Units	Temp	Data Type	Reference
WS 1.00E+006	mg/L	25	EST	MEYLAN,WM ET AL. (1996)
logP -0.55			EST	MEYLAN,WM & HOWARD,PH (1995)
VP 2.12E+001	mm Hg	25	EXP	BOUBLIK,T ET AL. (1984)
DC 5.20	pKa	25	EXP	PERRIN,DD (1965)
HL 3.33E-008	atm m3/mol	25	EST	MEYLAN,WM & HOWARD,PH (1991)
OH 6.85E-011	cm3/molc sec	25	EST	MEYLAN,WM & HOWARD,PH (1993)

CAS #: 016650-10-5				TETRACHLOROETHYLENE OXIDE

Formula: C_2Cl_4O

Mol Weight: 181.83

MP (deg C): | FP (deg C):

BP (deg C):

BP pressure (mm Hg):

Property/ Value	Units	Temp	Data Type	Reference
WS 9.49E+003	mg/L	25	EST	MEYLAN,WM ET AL. (1996)
logP 0.88			EST	MEYLAN,WM & HOWARD,PH (1995)
VP 4.49E+000	mm Hg	25	EST	NEELY,WB & BLAU,GE (1985)
DC	pKa			
HL 2.91E-005	atm m3/mol	25	EST	MEYLAN,WM & HOWARD,PH (1991)
OH 0.00E+000	cm3/molc sec	25	EST	MEYLAN,WM & HOWARD,PH (1993)

CAS #: 016664-12-3				CARBAMIC ACID, (2-FLUOROPHENYL)-, METHYL ESTER

Formula: $C_8H_8FNO_2$

Mol Weight: 169.16

MP (deg C): | FP (deg C):

BP (deg C):

BP pressure (mm Hg):

Property/ Value	Units	Temp	Data Type	Reference
WS 2.37E+003	mg/L	25	EST	MEYLAN,WM ET AL. (1996)
logP 1.66			EXP	TAKAHASHI,J ET AL. (1988)
VP 7.45E-002	mm Hg	25	EST	NEELY,WB & BLAU,GE (1985)
DC	pKa			
HL 2.55E-008	atm m3/mol	25	EST	MEYLAN,WM & HOWARD,PH (1991)
OH 1.46E-011	cm3/molc sec	25	EST	MEYLAN,WM & HOWARD,PH (1993)

CAS #: 016665-89-7				2,6-DIETHYLACETANILIDE

Formula: $C_{12}H_{17}NO$

Mol Weight: 191.28

MP (deg C): | FP (deg C):

BP (deg C):

BP pressure (mm Hg):

Property/ Value	Units	Temp	Data Type	Reference
WS 1.27E+003	mg/L	25	EST	MEYLAN,WM ET AL. (1996)
logP 1.85			EXP	NAKAGAWA,Y ET AL. (1992)
VP 1.22E-005	mm Hg	25	EST	NEELY,WB & BLAU,GE (1985)
DC	pKa			
HL 1.32E-008	atm m3/mol	25	EST	MEYLAN,WM & HOWARD,PH (1991)
OH 2.07E-011	cm3/molc sec	25	EST	MEYLAN,WM & HOWARD,PH (1993)

CAS #: 016672-87-0				ETHEPHON

Formula: $C_2H_6ClO_3P$

Mol Weight: 144.50

MP (deg C): 74 | FP (deg C):

BP (deg C):

BP pressure (mm Hg):

Property/ Value	Units	Temp	Data Type	Reference
WS 7.30E+004	mg/L	25	EST	MEYLAN,WM ET AL. (1996)
logP 0.05			EST	MEYLAN,WM & HOWARD,PH (1995)
VP 3.78E-005	mm Hg	25	EST	NEELY,WB & BLAU,GE (1985)
DC	pKa			
HL 5.70E-012	atm m3/mol	25	EST	MEYLAN,WM & HOWARD,PH (1991)
OH 1.05E-012	cm3/molc sec	25	EST	MEYLAN,WM & HOWARD,PH (1993)

CAS #: 016681-63-3				2-BROMO-1-METHYL-4-NITRO-1H-IMIDAZOLE

Formula: $C_4H_4BrN_3O_2$

Mol Weight: 206.00

MP (deg C): | FP (deg C):

BP (deg C):

BP pressure (mm Hg):

Property/ Value	Units	Temp	Data Type	Reference
WS 8.68E+003	mg/L	25	EST	MEYLAN,WM ET AL. (1996)
logP 0.33			EXP	SUWINSKI,J ET AL. (1985)
VP 9.37E-005	mm Hg	25	EST	NEELY,WB & BLAU,GE (1985)
DC	pKa			
HL 1.26E-007	atm m3/mol	25	EST	MEYLAN,WM & HOWARD,PH (1991)
OH 1.04E-012	cm3/molc sec	25	EST	MEYLAN,WM & HOWARD,PH (1993)

CAS #: 016697-83-9				ACETAMIDE, N-METHYL-N-[(4-METHYLPHENYL)SULFONYL]

Formula: $C_{10}H_{13}NO_3S$

Mol Weight: 227.28

MP (deg C): | FP (deg C):

BP (deg C):

BP pressure (mm Hg):

Property/ Value	Units	Temp	Data Type	Reference
WS 1.05E+003	mg/L	25	EST	MEYLAN,WM ET AL. (1996)
logP 1.73			EXP	LARSEN,JD ET AL. (1988)
VP 3.27E-006	mm Hg	25	EST	NEELY,WB & BLAU,GE (1985)
DC	pKa			
HL 3.32E-009	atm m3/mol	25	EST	MEYLAN,WM & HOWARD,PH (1991)
OH 1.69E-011	cm3/molc sec	25	EST	MEYLAN,WM & HOWARD,PH (1993)

CAS #: 016725-53-4				9-TETRADECEN-1-OL ACETATE (Z)

Formula: $C_{16}H_{30}O_2$

Mol Weight: 254.42

MP (deg C): | FP (deg C):

BP (deg C):

BP pressure (mm Hg):

Property/ Value	Units	Temp	Data Type	Reference
WS 5.78E-002	mg/L	25	EST	MEYLAN,WM ET AL. (1996)
logP 6.54			EST	MEYLAN,WM & HOWARD,PH (1995)
VP 5.92E-004	mm Hg	25	EST	NEELY,WB & BLAU,GE (1985)
DC	pKa			
HL 6.13E-003	atm m3/mol	25	EST	MEYLAN,WM & HOWARD,PH (1991)
OH 7.18E-011	cm3/molc sec	25	EST	MEYLAN,WM & HOWARD,PH (1993)

CAS #: 016744-99-3				CARBAMIC ACID, (4-FLUOROPHENYL)-, METHYL ESTER

Formula: $C_8H_8FNO_2$

Mol Weight: 169.16

MP (deg C): | FP (deg C):

BP (deg C):

BP pressure (mm Hg):

Property/ Value	Units	Temp	Data Type	Reference
WS 1.34E+003	mg/L	25	EST	MEYLAN,WM ET AL. (1996)
logP 1.95			EXP	TAKAHASHI,J ET AL. (1988)
VP 7.45E-002	mm Hg	25	EST	NEELY,WB & BLAU,GE (1985)
DC	pKa			
HL 2.55E-008	atm m3/mol	25	EST	MEYLAN,WM & HOWARD,PH (1991)
OH 1.46E-011	cm3/molc sec	25	EST	MEYLAN,WM & HOWARD,PH (1993)

CAS #: 016747-25-4				2,2,3-TRIMETHYLHEXANE

Formula: C_9H_{20}

Mol Weight: 128.26

MP (deg C):   FP (deg C):

BP (deg C): 133.6

BP pressure (mm Hg):

Property/Value	Units	Temp	Data Type	Reference
WS 3.35E+000	mg/L	25	EST	MEYLAN,WM ET AL. (1996)
logP 4.58			EST	MEYLAN,WM & HOWARD,PH (1995)
VP 1.14E+001	mm Hg	25	EXP	BOUBLIK,T ET AL. (1984)
DC	pKa			
HL 4.00E+000	atm m3/mol	25	EST	MEYLAN,WM & HOWARD,PH (1991)
OH 6.33E-012	cm3/molc sec	25	EST	MEYLAN,WM & HOWARD,PH (1993)

CAS #: 016747-26-5				2,2,4-TRIMETHYLHEXANE

Formula: C_9H_{20}

Mol Weight: 128.26

MP (deg C): -120   FP (deg C):

BP (deg C): 126.5

BP pressure (mm Hg):

Property/Value	Units	Temp	Data Type	Reference
WS 3.35E+000	mg/L	25	EST	MEYLAN,WM ET AL. (1996)
logP 4.58			EST	MEYLAN,WM & HOWARD,PH (1995)
VP 1.59E+001	mm Hg	25	EXP	BOUBLIK,T ET AL. (1984)
DC	pKa			
HL 4.00E+000	atm m3/mol	25	EST	MEYLAN,WM & HOWARD,PH (1991)
OH 6.33E-012	cm3/molc sec	25	EST	MEYLAN,WM & HOWARD,PH (1993)

CAS #: 016747-28-7				2,3,3-TRIMETHYLHEXANE

Formula: C_9H_{20}

Mol Weight: 128.26

MP (deg C): -116.8   FP (deg C):

BP (deg C): 137.7

BP pressure (mm Hg):

Property/Value	Units	Temp	Data Type	Reference
WS 3.35E+000	mg/L	25	EST	MEYLAN,WM ET AL. (1996)
logP 4.58			EST	MEYLAN,WM & HOWARD,PH (1995)
VP 9.83E+000	mm Hg	25	EXP	ZWOLINSKI,BJ & WILHOIT,RC (1971)
DC	pKa			
HL 4.00E+000	atm m3/mol	25	EST	MEYLAN,WM & HOWARD,PH (1991)
OH 5.78E-012	cm3/molc sec	25	EST	MEYLAN,WM & HOWARD,PH (1993)

CAS #: 016747-30-1				2,4,4-TRIMETHYLHEXANE

Formula: C_9H_{20}

Mol Weight: 128.26

MP (deg C): -113.4   FP (deg C):

BP (deg C): 130.7

BP pressure (mm Hg):

Property/Value	Units	Temp	Data Type	Reference
WS 3.35E+000	mg/L	25	EST	MEYLAN,WM ET AL. (1996)
logP 4.58			EST	MEYLAN,WM & HOWARD,PH (1995)
VP 1.34E+001	mm Hg	25	EXP	ZWOLINSKI,BJ & WILHOIT,RC (1971)
DC	pKa			
HL 4.00E+000	atm m3/mol	25	EST	MEYLAN,WM & HOWARD,PH (1991)
OH 5.78E-012	cm3/molc sec	25	EST	MEYLAN,WM & HOWARD,PH (1993)

CAS #: 016747-31-2				3,3,4-TRIMETHYLHEXANE

Formula: C_9H_{20}

Mol Weight: 128.26

MP (deg C): -101.2   FP (deg C):

BP (deg C): 140.5

BP pressure (mm Hg):

Property/Value	Units	Temp	Data Type	Reference
WS 3.35E+000	mg/L	25	EST	MEYLAN,WM ET AL. (1996)
logP 4.58			EST	MEYLAN,WM & HOWARD,PH (1995)
VP 8.87E+000	mm Hg	25	EXT	ZWOLINSKI,BJ & WILHOIT,RC (1971)
DC	pKa			
HL 4.00E+000	atm m3/mol	25	EST	MEYLAN,WM & HOWARD,PH (1991)
OH 6.07E-012	cm3/molc sec	25	EST	MEYLAN,WM & HOWARD,PH (1993)

CAS #: 016747-32-3				2,2-DIMETHYL-3-ETHYLPENTANE

Formula: C_9H_{20}

Mol Weight: 128.26

MP (deg C): -99.3   FP (deg C):

BP (deg C): 138.8

BP pressure (mm Hg):

Property/Value	Units	Temp	Data Type	Reference
WS 3.35E+000	mg/L	25	EST	MEYLAN,WM ET AL. (1996)
logP 4.58			EST	MEYLAN,WM & HOWARD,PH (1995)
VP 1.13E+001	mm Hg	25	EXP	YAWS,CL (1994B)
DC	pKa			
HL 4.00E+000	atm m3/mol	25	EST	MEYLAN,WM & HOWARD,PH (1991)
OH 6.74E-012	cm3/molc sec	25	EST	MEYLAN,WM & HOWARD,PH (1993)

CAS #: 016747-38-9				2,3,3,4-TETRAMETHYLPENTANE

Formula: C_9H_{20}

Mol Weight: 128.26

MP (deg C): -102.1   FP (deg C):

BP (deg C): 141.5

BP pressure (mm Hg):

Property/Value	Units	Temp	Data Type	Reference
WS 3.87E+000	mg/L	25	EST	MEYLAN,WM ET AL. (1996)
logP 4.50			EST	MEYLAN,WM & HOWARD,PH (1995)
VP 8.83E+000	mm Hg	25	EXT	FORZIATI,AF ET AL. (1949)
DC	pKa			
HL 4.00E+000	atm m3/mol	25	EST	MEYLAN,WM & HOWARD,PH (1991)
OH 5.78E-012	cm3/molc sec	25	EST	MEYLAN,WM & HOWARD,PH (1993)

CAS #: 016747-42-5				2,2,4,5-TETRAMETHYLHEXANE

Formula: $C_{10}H_{22}$

Mol Weight: 142.29

MP (deg C):   FP (deg C):

BP (deg C): 147.9

BP pressure (mm Hg):

Property/Value	Units	Temp	Data Type	Reference
WS 1.29E+000	mg/L	25	EST	MEYLAN,WM ET AL. (1996)
logP 4.99			EST	MEYLAN,WM & HOWARD,PH (1995)
VP 6.02E+000	mm Hg	25	EXT	ZWOLINSKI,BJ & WILHOIT,RC (1971)
DC	pKa			
HL 5.30E+000	atm m3/mol	25	EST	MEYLAN,WM & HOWARD,PH (1991)
OH 7.74E-012	cm3/molc sec	25	EST	MEYLAN,WM & HOWARD,PH (1993)

931

016752-77-5 — METHOMYL

Formula: $C_5H_{10}N_2O_2S$
Mol Weight: 162.21
MP (deg C): 78

Property/Value	Units	Temp	Data Type	Reference
WS 5.80E+004	mg/L	25	EXP	YALKOWSKY,SH & DANNENFELSER,RM (1992)
logP 0.60			EXP	HANSCH,C & LEO,AJ (1985)
VP 5.00E-005	mm Hg	25	EXP	WAUCHOPE,RD ET AL. (1991A)
DC	pKa			
HL 1.84E-010	atm m3/mol	25	EST	VP/WSOL
OH 8.99E-012	cm3/molc sec	25	EST	MEYLAN,WM & HOWARD,PH (1993)

016758-34-2 — 1-PHENYLTHIO-B-GALACTOPYRANOSIDE

Formula: $C_{12}H_{16}O_5S$
Mol Weight: 272.32

Property/Value	Units	Temp	Data Type	Reference
WS 5.33E+004	mg/L	25	EST	MEYLAN,WM ET AL. (1996)
logP -0.56			EXP	HANSCH,C & LEO,AJ (1985)
VP 1.14E-011	mm Hg	25	EST	NEELY,WB & BLAU,GE (1985)
DC	pKa			
HL 5.67E-016	atm m3/mol	25	EST	MEYLAN,WM & HOWARD,PH (1991)
OH 1.79E-010	cm3/molc sec	25	EST	MEYLAN,WM & HOWARD,PH (1993)

016773-42-5 — 1(3CL2OHPR)-2-ME-5-NO2 IMIDAZOLE

Formula: $C_7H_{12}ClN_3O_3$
Mol Weight: 221.64
MP (deg C): 77-78

Property/Value	Units	Temp	Data Type	Reference
WS 4.33E+003	mg/L	25	EST	MEYLAN,WM ET AL. (1996)
logP 0.60			EXP	HANSCH,C & LEO,AJ (1985)
VP 2.17E-008	mm Hg	25	EST	NEELY,WB & BLAU,GE (1985)
DC	pKa			
HL 7.92E-012	atm m3/mol	25	EST	MEYLAN,WM & HOWARD,PH (1991)
OH 8.72E-012	cm3/molc sec	25	EST	MEYLAN,WM & HOWARD,PH (1993)

016803-92-2 — N1-(4-CHLOROPHENYL)SULFANILAMIDE

Formula: $C_{12}H_{11}ClN_2O_2S$
Mol Weight: 282.75

Property/Value	Units	Temp	Data Type	Reference
WS 7.49E+001	mg/L	25	EST	MEYLAN,WM ET AL. (1996)
logP 2.71			EXP	HANSCH,C & LEO,AJ (1985)
VP 2.57E-008	mm Hg	25	EST	NEELY,WB & BLAU,GE (1985)
DC	pKa			
HL 6.10E-011	atm m3/mol	25	EST	MEYLAN,WM & HOWARD,PH (1991)
OH 3.55E-011	cm3/molc sec	25	EST	MEYLAN,WM & HOWARD,PH (1993)

016803-95-5 — N1-(4-METHYLPHENYL)SULFANILAMIDE

Formula: $C_{13}H_{14}N_2O_2S$
Mol Weight: 262.33

Property/Value	Units	Temp	Data Type	Reference
WS 3.19E+002	mg/L	25	EST	MEYLAN,WM ET AL. (1996)
logP 2.11			EXP	HANSCH,C & LEO,AJ (1985)
VP 4.63E-008	mm Hg	25	EST	NEELY,WB & BLAU,GE (1985)
DC	pKa			
HL 9.09E-011	atm m3/mol	25	EST	MEYLAN,WM & HOWARD,PH (1991)
OH 7.54E-011	cm3/molc sec	25	EST	MEYLAN,WM & HOWARD,PH (1993)

016803-96-6 — N1-(2-METHYLPHENYL)SULFANILAMIDE

Formula: $C_{13}H_{14}N_2O_2S$
Mol Weight: 262.33

Property/Value	Units	Temp	Data Type	Reference
WS 3.73E+002	mg/L	25	EST	MEYLAN,WM ET AL. (1996)
logP 2.03			EXP	HANSCH,C & LEO,AJ (1985)
VP 4.63E-008	mm Hg	25	EST	NEELY,WB & BLAU,GE (1985)
DC	pKa			
HL 9.09E-011	atm m3/mol	25	EST	MEYLAN,WM & HOWARD,PH (1991)
OH 7.54E-011	cm3/molc sec	25	EST	MEYLAN,WM & HOWARD,PH (1993)

016805-99-5 — N1-(5-BR-2-PYRIDYL)SULFANILAMIDE

Formula: $C_{11}H_{10}BrN_3O_2S$
Mol Weight: 328.19

Property/Value	Units	Temp	Data Type	Reference
WS 9.09E+001	mg/L	25	EST	MEYLAN,WM ET AL. (1996)
logP 2.30			EXP	HANSCH,C & LEO,AJ (1985)
VP 7.46E-009	mm Hg	25	EST	NEELY,WB & BLAU,GE (1985)
DC	pKa			
HL 4.29E-014	atm m3/mol	25	EST	MEYLAN,WM & HOWARD,PH (1991)
OH 2.81E-011	cm3/molc sec	25	EST	MEYLAN,WM & HOWARD,PH (1993)

016822-80-3 — 2-IMIDAZOLIDINIMINE, N-(5-BROMO-2-METHYLPHENYL)-

Formula: $C_{10}H_{12}BrN_3$
Mol Weight: 254.14

Property/Value	Units	Temp	Data Type	Reference
WS 5.04E+003	mg/L	25	EST	MEYLAN,WM ET AL. (1996)
logP 1.94			EXP	SANGSTER,J (1993)
VP 1.55E-005	mm Hg	25	EST	NEELY,WB & BLAU,GE (1985)
DC	pKa			
HL 1.21E-011	atm m3/mol	25	EST	MEYLAN,WM & HOWARD,PH (1991)
OH 1.45E-010	cm3/molc sec	25	EST	MEYLAN,WM & HOWARD,PH (1993)

CAS #: 016867-53-1
PYRIDO(12A)PYRIMIDIN-4-ON,3ETO-CO-6ME

Formula: $C_{12}H_{12}N_2O_3$

Mol Weight: 232.24

MP (deg C):

FP (deg C):

BP (deg C):

BP pressure (mm Hg):

Property/Value	Units	Temp	Data Type	Reference
WS 4.48E+003	mg/L	25	EST	MEYLAN,WM ET AL. (1996)
logP 0.96			EXP	HANSCH,C & LEO,AJ (1985)
VP 1.93E-006	mm Hg	25	EST	NEELY,WB & BLAU,GE (1985)
DC	pKa			
HL 4.66E-012	atm m3/mol	25	EST	MEYLAN,WM & HOWARD,PH (1991)
OH 1.28E-010	cm3/molc sec	25	EST	MEYLAN,WM & HOWARD,PH (1993)

CAS #: 016891-79-5
P-METHYLAMINOBENZENESULFONAMIDE

Formula: $C_7H_{10}N_2O_2S$

Mol Weight: 186.23

MP (deg C):

FP (deg C):

BP (deg C):

BP pressure (mm Hg):

Property/Value	Units	Temp	Data Type	Reference
WS 4.39E+004	mg/L	25	EST	MEYLAN,WM ET AL. (1996)
logP 0.08			EXP	HANSCH,C & LEO,AJ (1985)
VP 4.45E-005	mm Hg	25	EST	NEELY,WB & BLAU,GE (1985)
DC 10.77	pKa		EXP	PERRIN,DD (1965)
HL 3.28E-010	atm m3/mol	25	EST	MEYLAN,WM & HOWARD,PH (1991)
OH 1.04E-011	cm3/molc sec	25	EST	MEYLAN,WM & HOWARD,PH (1993)

CAS #: 016926-87-7
2-PROPENOIC ACID, 2-METHYL-, 2-HYDROXY-3-PHENOXY

Formula: $C_{13}H_{16}O_4$

Mol Weight: 236.27

MP (deg C):

FP (deg C):

BP (deg C):

BP pressure (mm Hg):

Property/Value	Units	Temp	Data Type	Reference
WS 1.05E+003	mg/L	25	EST	MEYLAN,WM ET AL. (1996)
logP 2.27			EXP	SANGSTER,J (1993)
VP 4.12E-006	mm Hg	25	EST	NEELY,WB & BLAU,GE (1985)
DC	pKa			
HL 4.04E-011	atm m3/mol	25	EST	MEYLAN,WM & HOWARD,PH (1991)
OH 5.89E-011	cm3/molc sec	25	EST	MEYLAN,WM & HOWARD,PH (1993)

CAS #: 016935-34-5
2,4-IMIDAZOLIDINEDIONE, 5-(1-METHYLETHYL)-

Formula: $C_6H_{10}N_2O_2$

Mol Weight: 142.16

MP (deg C):

FP (deg C):

BP (deg C):

BP pressure (mm Hg):

Property/Value	Units	Temp	Data Type	Reference
WS 8.15E+003	mg/L	25	EST	MEYLAN,WM ET AL. (1996)
logP -0.36			EXP	SANGSTER,J (1994)
VP 1.36E-006	mm Hg	25	EST	NEELY,WB & BLAU,GE (1985)
DC	pKa			
HL 3.67E-009	atm m3/mol	25	EST	MEYLAN,WM & HOWARD,PH (1991)
OH 1.58E-011	cm3/molc sec	25	EST	MEYLAN,WM & HOWARD,PH (1993)

CAS #: 016947-63-0
2,6-DIMETHYL-4-NITROBENZENAMINE

Formula: $C_8H_{10}N_2O_2$

Mol Weight: 166.18

MP (deg C):

FP (deg C):

BP (deg C):

BP pressure (mm Hg):

Property/Value	Units	Temp	Data Type	Reference
WS 4.13E+002	mg/L	25	EST	MEYLAN,WM ET AL. (1996)
logP 2.57			EST	MEYLAN,WM & HOWARD,PH (1995)
VP 3.28E-004	mm Hg	25	EST	NEELY,WB & BLAU,GE (1985)
DC 0.98	pKa	22	EXP	PERRIN,DD (1965)
HL 9.15E-009	atm m3/mol	25	EST	MEYLAN,WM & HOWARD,PH (1991)
OH 2.17E-012	cm3/molc sec	25	EST	MEYLAN,WM & HOWARD,PH (1993)

CAS #: 016956-42-6
4-BENZYLSEMICARBAZIDE

Formula: $C_8H_{11}N_3O$

Mol Weight: 165.20

MP (deg C):

FP (deg C):

BP (deg C):

BP pressure (mm Hg):

Property/Value	Units	Temp	Data Type	Reference
WS 4.89E+003	mg/L	25	EST	MEYLAN,WM ET AL. (1996)
logP -0.22			EXP	HANSCH,C & LEO,AJ (1985)
VP 4.94E-005	mm Hg	25	EST	NEELY,WB & BLAU,GE (1985)
DC	pKa			
HL 2.70E-013	atm m3/mol	25	EST	MEYLAN,WM & HOWARD,PH (1991)
OH 7.85E-011	cm3/molc sec	25	EST	MEYLAN,WM & HOWARD,PH (1993)

CAS #: 016967-79-6
TRICHLOROOXIRANE

Formula: C_2HCl_3O

Mol Weight: 147.39

MP (deg C):

FP (deg C):

BP (deg C):

BP pressure (mm Hg):

Property/Value	Units	Temp	Data Type	Reference
WS 2.71E+004	mg/L	25	EST	MEYLAN,WM ET AL. (1996)
logP 0.54			EST	MEYLAN,WM & HOWARD,PH (1995)
VP 1.69E+001	mm Hg	25	EST	NEELY,WB & BLAU,GE (1985)
DC	pKa			
HL 5.25E-006	atm m3/mol	25	EST	MEYLAN,WM & HOWARD,PH (1991)
OH 1.11E-014	cm3/molc sec	25	EST	MEYLAN,WM & HOWARD,PH (1993)

CAS #: 016974-11-1
9-DODECEN-1-OL ACETATE (Z)

Formula: $C_{14}H_{26}O_2$

Mol Weight: 226.36

MP (deg C):

FP (deg C):

BP (deg C):

BP pressure (mm Hg):

Property/Value	Units	Temp	Data Type	Reference
WS 5.68E-001	mg/L	25	EST	MEYLAN,WM ET AL. (1996)
logP 5.56			EST	MEYLAN,WM & HOWARD,PH (1995)
VP 4.10E-003	mm Hg	25	EST	NEELY,WB & BLAU,GE (1985)
DC	pKa			
HL 3.48E-003	atm m3/mol	25	EST	MEYLAN,WM & HOWARD,PH (1991)
OH 6.91E-011	cm3/molc sec	25	EST	MEYLAN,WM & HOWARD,PH (1993)

016993-45-6 — 1,4-BENZENEDISULFONAMIDE

Formula: $C_6H_8N_2O_4S_2$
Mol Weight: 236.27
MP (deg C):
FP (deg C):
BP (deg C):
BP pressure (mm Hg):

Property/Value	Units	Temp	Data Type	Reference
WS 1.65E+004	mg/L	25	EST	MEYLAN,WM ET AL. (1996)
logP -0.96			EXP	HANSCH,C ET AL. (1995)
VP 1.26E-007	mm Hg	25	EST	NEELY,WB & BLAU,GE (1985)
DC	pKa			
HL 3.31E-011	atm m3/mol	25	EST	MEYLAN,WM & HOWARD,PH (1991)
OH 8.91E-014	cm3/molc sec	25	EST	MEYLAN,WM & HOWARD,PH (1993)

017004-62-5 — 8-HEXADECENOIC ACID

Formula: $C_{16}H_{30}O_2$
Mol Weight: 254.42
MP (deg C):
FP (deg C):
BP (deg C):
BP pressure (mm Hg):

Property/Value	Units	Temp	Data Type	Reference
WS 1.33E-001	mg/L	25	EST	MEYLAN,WM ET AL. (1996)
logP 6.58			EXP	SANGSTER,J (1994)
VP 8.41E-006	mm Hg	25	EST	NEELY,WB & BLAU,GE (1985)
DC	pKa			
HL 2.54E-005	atm m3/mol	25	EST	MEYLAN,WM & HOWARD,PH (1991)
OH 7.27E-011	cm3/molc sec	25	EST	MEYLAN,WM & HOWARD,PH (1993)

017012-22-5 — P-METHOXYCARBONYLACETANILIDE

Formula: $C_{10}H_{11}NO_3$
Mol Weight: 193.20
MP (deg C):
FP (deg C):
BP (deg C):
BP pressure (mm Hg):

Property/Value	Units	Temp	Data Type	Reference
WS 1.40E+003	mg/L	25	EST	MEYLAN,WM ET AL. (1996)
logP 1.79			EXP	HANSCH,C ET AL. (1995)
VP 9.60E-006	mm Hg	25	EST	NEELY,WB & BLAU,GE (1985)
DC	pKa			
HL 3.98E-011	atm m3/mol	25	EST	MEYLAN,WM & HOWARD,PH (1991)
OH 4.31E-012	cm3/molc sec	25	EST	MEYLAN,WM & HOWARD,PH (1993)

017013-41-1 — BARBITURIC ACID,5,5-DIBUTYL-

Formula: $C_{12}H_{20}N_2O_3$
Mol Weight: 240.30
MP (deg C):
FP (deg C):
BP (deg C):
BP pressure (mm Hg):

Property/Value	Units	Temp	Data Type	Reference
WS 1.32E+002	mg/L	25	EST	MEYLAN,WM ET AL. (1996)
logP 2.70			EXP	WONG,O & MCKEOWN,RH (1988)
VP 5.68E-012	mm Hg	25	EST	NEELY,WB & BLAU,GE (1985)
DC	pKa			
HL 1.12E-012	atm m3/mol	25	EST	MEYLAN,WM & HOWARD,PH (1991)
OH 1.64E-011	cm3/molc sec	25	EST	MEYLAN,WM & HOWARD,PH (1993)

017057-68-0 — D-ERYTHRO-ALPHA-D-GALACTO-OCTOPYRANOSIDE, METHYL

Formula: $C_{20}H_{38}N_2O_6S$
Mol Weight: 434.60
MP (deg C):
FP (deg C):
BP (deg C):
BP pressure (mm Hg):

Property/Value	Units	Temp	Data Type	Reference
WS 5.85E+001	mg/L	25	EST	MEYLAN,WM ET AL. (1996)
logP 1.76			EXP	HANSCH,C ET AL. (1995)
VP 1.59E-018	mm Hg	25	EST	NEELY,WB & BLAU,GE (1985)
DC	pKa			
HL 5.28E-023	atm m3/mol	25	EST	MEYLAN,WM & HOWARD,PH (1991)
OH 2.67E-010	cm3/molc sec	25	EST	MEYLAN,WM & HOWARD,PH (1993)

017066-68-1 — AZULENO[4,5-B]FURAN-2,9-DIONE, DECAHYDRO-4-HYDRO

Formula: $C_{15}H_{20}O_4$
Mol Weight: 264.32
MP (deg C):
FP (deg C):
BP (deg C):
BP pressure (mm Hg):

Property/Value	Units	Temp	Data Type	Reference
WS 4.05E+004	mg/L	25	EST	MEYLAN,WM ET AL. (1996)
logP 0.23			EXP	SANGSTER,J (1994)
VP 5.32E-009	mm Hg	25	EST	NEELY,WB & BLAU,GE (1985)
DC	pKa			
HL 2.95E-012	atm m3/mol	25	EST	MEYLAN,WM & HOWARD,PH (1991)
OH 5.03E-011	cm3/molc sec	25	EST	MEYLAN,WM & HOWARD,PH (1993)

017075-29-5 — 1-ACETYL-4-(4-ME PHENYL) SEMICARBAZIDE

Formula: $C_{10}H_{13}N_3O_2$
Mol Weight: 207.23
MP (deg C):
FP (deg C):
BP (deg C):
BP pressure (mm Hg):

Property/Value	Units	Temp	Data Type	Reference
WS 1.08E+004	mg/L	25	EST	MEYLAN,WM ET AL. (1996)
logP 0.67			EXP	STEIN,J ET AL. (1986)
VP 5.16E-008	mm Hg	25	EST	NEELY,WB & BLAU,GE (1985)
DC	pKa			
HL 3.01E-015	atm m3/mol	25	EST	MEYLAN,WM & HOWARD,PH (1991)
OH 5.90E-011	cm3/molc sec	25	EST	MEYLAN,WM & HOWARD,PH (1993)

017075-30-8 — 1-ACETYL-4-(4-OME PHENYL) SEMICARBAZIDE

Formula: $C_{10}H_{13}N_3O_3$
Mol Weight: 223.23
MP (deg C):
FP (deg C):
BP (deg C):
BP pressure (mm Hg):

Property/Value	Units	Temp	Data Type	Reference
WS 1.31E+003	mg/L	25	EST	MEYLAN,WM ET AL. (1996)
logP 0.11			EXP	STEIN,J ET AL. (1986)
VP 2.24E-008	mm Hg	25	EST	NEELY,WB & BLAU,GE (1985)
DC	pKa			
HL 1.61E-016	atm m3/mol	25	EST	MEYLAN,WM & HOWARD,PH (1991)
OH 4.43E-011	cm3/molc sec	25	EST	MEYLAN,WM & HOWARD,PH (1993)

CAS #: 017075-31-9

1-ACETYL-4-(4-CL PHENYL) SEMICARBAZIDE

Formula: $C_9H_{10}ClN_3O_2$

Mol Weight: 227.65

MP (deg C):　FP (deg C):

BP (deg C):

BP pressure (mm Hg):

Property/Value	Units	Temp	Data Type	Reference
WS 3.46E+003	mg/L	25	EST	MEYLAN,WM ET AL. (1996)
logP 1.12			EXP	STEIN,J ET AL. (1986)
VP 3.16E-008	mm Hg	25	EST	NEELY,WB & BLAU,GE (1985)
DC	pKa			
HL 2.02E-015	atm m3/mol	25	EST	MEYLAN,WM & HOWARD,PH (1991)
OH 1.91E-011	cm3/molc sec	25	EST	MEYLAN,WM & HOWARD,PH (1993)

CAS #: 017078-72-7

BENZENAMINE, 4-[(4-METHOXYPHENYL)SULFONYL]-

Formula: $C_{13}H_{13}NO_3S$

Mol Weight: 263.32

MP (deg C):　FP (deg C):

BP (deg C):

BP pressure (mm Hg):

Property/Value	Units	Temp	Data Type	Reference
WS 4.22E+002	mg/L	25	EST	MEYLAN,WM ET AL. (1996)
logP 1.96			EXP	ALTOMARE,C ET AL. (1991)
VP 1.05E-007	mm Hg	25	EST	NEELY,WB & BLAU,GE (1985)
DC	pKa			
HL 5.21E-012	atm m3/mol	25	EST	MEYLAN,WM & HOWARD,PH (1991)
OH 2.84E-011	cm3/molc sec	25	EST	MEYLAN,WM & HOWARD,PH (1993)

CAS #: 017080-02-3

FURETHRIN

Formula: $C_{21}H_{26}O_4$

Mol Weight: 342.44

MP (deg C):　FP (deg C):

BP (deg C):

BP pressure (mm Hg):

Property/Value	Units	Temp	Data Type	Reference
WS 8.50E-002	mg/L	25	EST	MEYLAN,WM ET AL. (1996)
logP 5.75			EST	MEYLAN,WM & HOWARD,PH (1995)
VP 1.52E-007	mm Hg	25	EST	NEELY,WB & BLAU,GE (1985)
DC	pKa			
HL 3.76E-008	atm m3/mol	25	EST	MEYLAN,WM & HOWARD,PH (1991)
OH 2.93E-010	cm3/molc sec	25	EST	MEYLAN,WM & HOWARD,PH (1993)

CAS #: 017109-49-8

EDIFENPHOS

Formula: $C_{14}H_{15}O_2PS_2$

Mol Weight: 310.38

MP (deg C):　FP (deg C):

BP (deg C): 154

BP pressure (mm Hg): 1.00E-002

Property/Value	Units	Temp	Data Type	Reference
WS 1.45E+002	mg/L		EXP	SHIU,WY ET AL. (1990)
logP 3.48			EXP	SAITO,H ET AL. (1993)
VP 2.70E-007	mm Hg	25	EXP	WATANABE,T ET AL. (1993)
DC	pKa			
HL 7.60E-010	atm m3/mol	25	EST	VP/WSOL
OH 4.45E-011	cm3/molc sec	25	EST	MEYLAN,WM & HOWARD,PH (1993)

CAS #: 017112-82-2

1-NAPHTHYLMETHYLISOTHIOCYANATE

Formula: $C_{12}H_9NS$

Mol Weight: 199.28

MP (deg C):　FP (deg C):

BP (deg C):

BP pressure (mm Hg):

Property/Value	Units	Temp	Data Type	Reference
WS 7.41E+000	mg/L	25	EST	MEYLAN,WM ET AL. (1996)
logP 4.42			EXP	HANSCH,C & LEO,AJ (1985)
VP 5.32E-005	mm Hg	25	EST	NEELY,WB & BLAU,GE (1985)
DC	pKa			
HL 2.45E-005	atm m3/mol	25	EST	MEYLAN,WM & HOWARD,PH (1991)
OH 5.46E-011	cm3/molc sec	25	EST	MEYLAN,WM & HOWARD,PH (1993)

CAS #: 017114-78-2

9-(1,1-DIMETHYLETHYL)-9H-FLUORENE

Formula: $C_{17}H_{18}$

Mol Weight: 222.33

MP (deg C):　FP (deg C):

BP (deg C):

BP pressure (mm Hg):

Property/Value	Units	Temp	Data Type	Reference
WS 5.09E-002	mg/L	25	EST	MEYLAN,WM ET AL. (1996)
logP 5.51			EST	MEYLAN,WM & HOWARD,PH (1995)
VP 7.84E-005	mm Hg	25	EST	NEELY,WB & BLAU,GE (1985)
DC	pKa			
HL 5.20E-004	atm m3/mol	25	EST	MEYLAN,WM & HOWARD,PH (1991)
OH 1.06E-011	cm3/molc sec	25	EST	MEYLAN,WM & HOWARD,PH (1993)

CAS #: 017165-86-5

9-(1,1'-BIPHENYL)-4-YL-9H-FLUORENE

Formula: $C_{25}H_{18}$

Mol Weight: 318.42

MP (deg C):　FP (deg C):

BP (deg C):

BP pressure (mm Hg):

Property/Value	Units	Temp	Data Type	Reference
WS 6.05E-004	mg/L	25	EST	MEYLAN,WM ET AL. (1996)
logP 7.13			EST	MEYLAN,WM & HOWARD,PH (1995)
VP 1.98E-009	mm Hg	25	EST	NEELY,WB & BLAU,GE (1985)
DC	pKa			
HL 1.04E-006	atm m3/mol	25	EST	MEYLAN,WM & HOWARD,PH (1991)
OH 1.72E-011	cm3/molc sec	25	EST	MEYLAN,WM & HOWARD,PH (1993)

CAS #: 017174-98-0

BENZAMIDE, 2-CYANO-

Formula: $C_8H_6N_2O$

Mol Weight: 146.15

MP (deg C):　FP (deg C):

BP (deg C):

BP pressure (mm Hg):

Property/Value	Units	Temp	Data Type	Reference
WS 2.10E+003	mg/L	25	EST	MEYLAN,WM ET AL. (1996)
logP 0.00			EXP	SOTOMATSU,T ET AL. (1993)
VP 1.69E-005	mm Hg	25	EST	NEELY,WB & BLAU,GE (1985)
DC	pKa			
HL 2.14E-011	atm m3/mol	25	EST	MEYLAN,WM & HOWARD,PH (1991)
OH 2.23E-012	cm3/molc sec	25	EST	MEYLAN,WM & HOWARD,PH (1993)

CAS #: 017180-93-7 — PYRIMIDINE, 4-CHLORO-

Formula: $C_4H_3ClN_2$
Mol Weight: 114.53
MP (deg C):
FP (deg C):
BP (deg C):
BP pressure (mm Hg):

Property/Value	Units	Temp	Data Type	Reference
WS 4.17E+004	mg/L	25	EST	MEYLAN,WM ET AL. (1996)
logP 0.47			EXP	YAMAGAMI,C ET AL. (1990)
VP 2.48E+000	mm Hg	25	EST	NEELY,WB & BLAU,GE (1985)
DC	pKa			
HL 1.36E-004	atm m3/mol	25	EST	MEYLAN,WM & HOWARD,PH (1991)
OH 1.62E-013	cm3/molc sec	25	EST	MEYLAN,WM & HOWARD,PH (1993)

CAS #: 017180-94-8 — PYRIMIDINE, 5-CHLORO-

Formula: $C_4H_3ClN_2$
Mol Weight: 114.53
MP (deg C):
FP (deg C):
BP (deg C):
BP pressure (mm Hg):

Property/Value	Units	Temp	Data Type	Reference
WS 4.17E+004	mg/L	25	EST	MEYLAN,WM ET AL. (1996)
logP 0.47			EXP	YAMAGAMI,C ET AL. (1990)
VP 2.48E+000	mm Hg	25	EST	NEELY,WB & BLAU,GE (1985)
DC	pKa			
HL 2.16E-006	atm m3/mol	25	EST	MEYLAN,WM & HOWARD,PH (1991)
OH 1.62E-013	cm3/molc sec	25	EST	MEYLAN,WM & HOWARD,PH (1993)

CAS #: 017186-60-6 — PH-ALANINE,N-ACETYL,N'-MEAM-AMIDE

Formula: $C_{12}H_{16}N_2O_2$
Mol Weight: 220.27
MP (deg C):
FP (deg C):
BP (deg C):
BP pressure (mm Hg):

Property/Value	Units	Temp	Data Type	Reference
WS 1.56E+004	mg/L	25	EST	MEYLAN,WM ET AL. (1996)
logP 0.40			EXP	HANSCH,C & LEO,AJ (1985)
VP 3.70E-009	mm Hg	25	EST	NEELY,WB & BLAU,GE (1985)
DC	pKa			
HL 2.78E-011	atm m3/mol	25	EST	MEYLAN,WM & HOWARD,PH (1991)
OH 9.50E-011	cm3/molc sec	25	EST	MEYLAN,WM & HOWARD,PH (1993)

CAS #: 017199-21-2 — 2-CHLORO-6-NITRO-4-(1,1,3,3-TETRAMETHYLBUTYL)-P*

Formula: $C_{14}H_{20}ClNO_3$
Mol Weight: 285.77
MP (deg C):
FP (deg C):
BP (deg C):
BP pressure (mm Hg):

Property/Value	Units	Temp	Data Type	Reference
WS 6.00E-002	mg/L	25	EST	MEYLAN,WM ET AL. (1996)
logP 6.32			EST	MEYLAN,WM & HOWARD,PH (1995)
VP 9.06E-007	mm Hg	25	EST	NEELY,WB & BLAU,GE (1985)
DC	pKa			
HL 4.16E-005	atm m3/mol	25	EST	MEYLAN,WM & HOWARD,PH (1991)
OH 2.57E-012	cm3/molc sec	25	EST	MEYLAN,WM & HOWARD,PH (1993)

CAS #: 017199-22-3 — 2-BROMO-6-NITRO-4-(1,1,3,3-TETRAMETHYLBUTYL)-PH*

Formula: $C_{14}H_{20}BrNO_3$
Mol Weight: 330.23
MP (deg C):
FP (deg C):
BP (deg C):
BP pressure (mm Hg):

Property/Value	Units	Temp	Data Type	Reference
WS 2.03E-002	mg/L	25	EST	MEYLAN,WM ET AL. (1996)
logP 6.56			EST	MEYLAN,WM & HOWARD,PH (1995)
VP 3.28E-007	mm Hg	25	EST	NEELY,WB & BLAU,GE (1985)
DC	pKa			
HL 2.23E-005	atm m3/mol	25	EST	MEYLAN,WM & HOWARD,PH (1991)
OH 2.55E-012	cm3/molc sec	25	EST	MEYLAN,WM & HOWARD,PH (1993)

CAS #: 017199-23-4 — 2-BROMO-4-(TERT-BUTYL)-6-NITROPHENOL

Formula: $C_{10}H_{12}BrNO_3$
Mol Weight: 274.12
MP (deg C):
FP (deg C):
BP (deg C):
BP pressure (mm Hg):

Property/Value	Units	Temp	Data Type	Reference
WS 1.65E+000	mg/L	25	EST	MEYLAN,WM ET AL. (1996)
logP 4.71			EST	MEYLAN,WM & HOWARD,PH (1995)
VP 5.06E-006	mm Hg	25	EST	NEELY,WB & BLAU,GE (1985)
DC	pKa			
HL 7.20E-006	atm m3/mol	25	EST	MEYLAN,WM & HOWARD,PH (1991)
OH 8.05E-013	cm3/molc sec	25	EST	MEYLAN,WM & HOWARD,PH (1993)

CAS #: 017199-24-5 — 2-CHLORO-4-(1,1,3,3-TETRAMETHYLBUTYL)PHENOL

Formula: $C_{14}H_{21}ClO$
Mol Weight: 240.78
MP (deg C):
FP (deg C):
BP (deg C):
BP pressure (mm Hg):

Property/Value	Units	Temp	Data Type	Reference
WS 8.89E-001	mg/L	25	EST	MEYLAN,WM ET AL. (1996)
logP 5.92			EST	MEYLAN,WM & HOWARD,PH (1995)
VP 1.17E-004	mm Hg	25	EST	NEELY,WB & BLAU,GE (1985)
DC	pKa			
HL 3.33E-006	atm m3/mol	25	EST	MEYLAN,WM & HOWARD,PH (1991)
OH 1.41E-011	cm3/molc sec	25	EST	MEYLAN,WM & HOWARD,PH (1993)

CAS #: 017199-55-2 — BENZENEETHANOL, beta-[2-(DIMETHYLAMINO)PROPYL]-

Formula: $C_{21}H_{29}NO$
Mol Weight: 311.47
MP (deg C):
FP (deg C):
BP (deg C):
BP pressure (mm Hg):

Property/Value	Units	Temp	Data Type	Reference
WS 2.42E+001	mg/L	25	EST	MEYLAN,WM ET AL. (1996)
logP 4.27			EXP	SANGSTER,J (1994)
VP 1.32E-008	mm Hg	25	EST	NEELY,WB & BLAU,GE (1985)
DC	pKa			
HL 4.76E-011	atm m3/mol	25	EST	MEYLAN,WM & HOWARD,PH (1991)
OH 1.13E-010	cm3/molc sec	25	EST	MEYLAN,WM & HOWARD,PH (1993)

CAS #: 017220-38-1	FURAZANDIAMINE

Formula: $C_2H_4N_4O$

Mol Weight: 100.08

MP (deg C):   FP (deg C):

BP (deg C):

BP pressure (mm Hg):

Property/ Value	Units	Temp	Data Type	Reference
WS 3.06E+005	mg/L	25	EST	MEYLAN,WM ET AL. (1996)
logP -0.49			EXP	CALVINO,R ET AL. (1992)
VP 1.57E-002	mm Hg	25	EST	NEELY,WB & BLAU,GE (1985)
DC	pKa			
HL 2.36E-011	atm m3/mol	25	EST	MEYLAN,WM & HOWARD,PH (1991)
OH 4.00E-012	cm3/molc sec	25	EST	MEYLAN,WM & HOWARD,PH (1993)

CAS #: 017227-47-3	4-QUINAZOLINOL

Formula: $C_8H_6N_2O$

Mol Weight: 146.15

MP (deg C):   FP (deg C):

BP (deg C):

BP pressure (mm Hg):

Property/ Value	Units	Temp	Data Type	Reference
WS 1.24E+004	mg/L	25	EST	MEYLAN,WM ET AL. (1996)
logP 0.94			EXP	HANSCH,C ET AL. (1995)
VP 1.43E-004	mm Hg	25	EST	NEELY,WB & BLAU,GE (1985)
DC 2.06	pKa	20	EXP	PERRIN,DD (1965)
HL 2.96E-011	atm m3/mol	25	EST	MEYLAN,WM & HOWARD,PH (1991)
OH 3.43E-011	cm3/molc sec	25	EST	MEYLAN,WM & HOWARD,PH (1993)

CAS #: 017258-31-0	3-PYRIDAZINAMINE,N,N-DIMETHYL

Formula: $C_6H_9N_3$

Mol Weight: 123.16

MP (deg C):   FP (deg C):

BP (deg C):

BP pressure (mm Hg):

Property/ Value	Units	Temp	Data Type	Reference
WS 5.52E+004	mg/L	25	EST	MEYLAN,WM ET AL. (1996)
logP 0.29			EXP	HANSCH,C & LEO,AJ (1985)
VP 1.43E-002	mm Hg	25	EST	NEELY,WB & BLAU,GE (1985)
DC	pKa			
HL 4.56E-008	atm m3/mol	25	EST	MEYLAN,WM & HOWARD,PH (1991)
OH 4.62E-011	cm3/molc sec	25	EST	MEYLAN,WM & HOWARD,PH (1993)

CAS #: 017258-38-7	4-PYRIDAZINAMINE,N,N-DIMETHYL

Formula: $C_6H_9N_3$

Mol Weight: 123.16

MP (deg C):   FP (deg C):

BP (deg C):

BP pressure (mm Hg):

Property/ Value	Units	Temp	Data Type	Reference
WS 1.17E+005	mg/L	25	EST	MEYLAN,WM ET AL. (1996)
logP -0.09			EXP	HANSCH,C & LEO,AJ (1985)
VP 1.43E-002	mm Hg	25	EST	NEELY,WB & BLAU,GE (1985)
DC	pKa			
HL 4.56E-008	atm m3/mol	25	EST	MEYLAN,WM & HOWARD,PH (1991)
OH 4.62E-011	cm3/molc sec	25	EST	MEYLAN,WM & HOWARD,PH (1993)

CAS #: 017260-71-8	M-CHLOROBENZENESULFONAMIDE

Formula: $C_6H_6ClNO_2S$

Mol Weight: 191.64

MP (deg C):   FP (deg C):

BP (deg C):

BP pressure (mm Hg):

Property/ Value	Units	Temp	Data Type	Reference
WS 3.82E+003	mg/L	25	EST	MEYLAN,WM ET AL. (1996)
logP 1.29			EXP	HANSCH,C & LEO,AJ (1985)
VP 1.59E-004	mm Hg	25	EST	NEELY,WB & BLAU,GE (1985)
DC	pKa			
HL 3.13E-007	atm m3/mol	25	EST	MEYLAN,WM & HOWARD,PH (1991)
OH 2.93E-013	cm3/molc sec	25	EST	MEYLAN,WM & HOWARD,PH (1993)

CAS #: 017264-58-3	P-BIPHENYLCARBOXYLIC ACID, NA SALT

Formula: $C_{13}H_9NaO_2$

Mol Weight: 220.21

MP (deg C):   FP (deg C):

BP (deg C):

BP pressure (mm Hg):

Property/ Value	Units	Temp	Data Type	Reference
WS 5.84E+004	mg/L	25	EST	MEYLAN,WM ET AL. (1996)
logP -0.27			EXP	HANSCH,C & LEO,AJ (1985)
VP 2.99E-012	mm Hg	25	EST	NEELY,WB & BLAU,GE (1985)
DC	pKa			
HL	atm m3/mol			
OH 6.48E-012	cm3/molc sec	25	EST	MEYLAN,WM & HOWARD,PH (1993)

CAS #: 017273-79-9	2-NAPHTHOIC ACID, SODIUM SALT

Formula: $C_{11}H_7NaO_2$

Mol Weight: 194.17

MP (deg C):   FP (deg C):

BP (deg C):

BP pressure (mm Hg):

Property/ Value	Units	Temp	Data Type	Reference
WS 3.84E+005	mg/L	25	EST	MEYLAN,WM ET AL. (1996)
logP -1.07			EXP	HANSCH,C & LEO,AJ (1985)
VP 2.15E-011	mm Hg	25	EST	NEELY,WB & BLAU,GE (1985)
DC	pKa			
HL	atm m3/mol			
OH 1.97E-011	cm3/molc sec	25	EST	MEYLAN,WM & HOWARD,PH (1993)

CAS #: 017278-57-8	MONOCHLOROAMPHENICOL

Formula: $C_{11}H_{13}ClN_2O_5$

Mol Weight: 288.69

MP (deg C):   FP (deg C):

BP (deg C):

BP pressure (mm Hg):

Property/ Value	Units	Temp	Data Type	Reference
WS 1.82E+003	mg/L	25	EST	MEYLAN,WM ET AL. (1996)
logP 0.59			EXP	HANSCH,C & LEO,AJ (1985)
VP 8.08E-013	mm Hg	25	EST	NEELY,WB & BLAU,GE (1985)
DC	pKa			
HL 6.49E-018	atm m3/mol	25	EST	MEYLAN,WM & HOWARD,PH (1991)
OH 3.10E-011	cm3/molc sec	25	EST	MEYLAN,WM & HOWARD,PH (1993)

CAS #: 017404-90-9 — 3-CHLOROCINNOLINE

Formula: $C_8H_5ClN_2$

Mol Weight: 164.60

MP (deg C): | FP (deg C):

BP (deg C):

BP pressure (mm Hg):

Property/Value	Units	Temp	Data Type	Reference
WS 2.97E+003	mg/L	25	EST	MEYLAN,WM ET AL. (1996)
logP 1.57			EXP	HANSCH,C & LEO,AJ (1985)
VP 9.91E-005	mm Hg	25	EST	NEELY,WB & BLAU,GE (1985)
DC	pKa			
HL 1.31E-005	atm m3/mol	25	EST	MEYLAN,WM & HOWARD,PH (1991)
OH 1.41E-012	cm3/molc sec	25	EST	MEYLAN,WM & HOWARD,PH (1993)

CAS #: 017413-90-0 — 2-(P-CLPHENOXY)-2-ET-PROPIONIC AC

Formula: $C_{11}H_{13}ClO_3$

Mol Weight: 228.68

MP (deg C): | FP (deg C):

BP (deg C):

BP pressure (mm Hg):

Property/Value	Units	Temp	Data Type	Reference
WS 2.56E+002	mg/L	25	EST	MEYLAN,WM ET AL. (1996)
logP 2.90			EXP	HANSCH,C & LEO,AJ (1985)
VP 4.61E-005	mm Hg	25	EST	NEELY,WB & BLAU,GE (1985)
DC	pKa			
HL 2.91E-008	atm m3/mol	25	EST	MEYLAN,WM & HOWARD,PH (1991)
OH 1.07E-011	cm3/molc sec	25	EST	MEYLAN,WM & HOWARD,PH (1993)

CAS #: 017418-58-5 — C.I. DISPERSE RED 60

Formula: $C_{20}H_{13}NO_4$

Mol Weight: 331.33

MP (deg C): | FP (deg C):

BP (deg C):

BP pressure (mm Hg):

Property/Value	Units	Temp	Data Type	Reference
WS 6.40E-004	mg/L	25	EXP	YEN,CPC ET AL. (1989)
logP 4.69			EST	MEYLAN,WM & HOWARD,PH (1995)
VP 2.10E-012	mm Hg	25	EST	NEELY,WB & BLAU,GE (1985)
DC	pKa			
HL 3.22E-015	atm m3/mol	25	EST	MEYLAN,WM & HOWARD,PH (1991)
OH 5.88E-011	cm3/molc sec	25	EST	MEYLAN,WM & HOWARD,PH (1993)

CAS #: 017422-56-9 — DIETHYLMALONATE,2-NO2 BENZAL

Formula: $C_{14}H_{15}NO_6$

Mol Weight: 293.28

MP (deg C): | FP (deg C):

BP (deg C):

BP pressure (mm Hg):

Property/Value	Units	Temp	Data Type	Reference
WS 3.71E+001	mg/L	25	EST	MEYLAN,WM ET AL. (1996)
logP 2.54			EXP	HANSCH,C & LEO,AJ (1985)
VP 8.19E-006	mm Hg	25	EST	NEELY,WB & BLAU,GE (1985)
DC	pKa			
HL 4.31E-011	atm m3/mol	25	EST	MEYLAN,WM & HOWARD,PH (1991)
OH 1.42E-011	cm3/molc sec	25	EST	MEYLAN,WM & HOWARD,PH (1993)

CAS #: 017433-31-7 — 1-ACETYL-2-PICOLINOYL HYDRAZINE

Formula: $C_8H_9N_3O_2$

Mol Weight: 179.18

MP (deg C): | FP (deg C):

BP (deg C):

BP pressure (mm Hg):

Property/Value	Units	Temp	Data Type	Reference
WS 8.35E+003	mg/L	25	EST	MEYLAN,WM ET AL. (1996)
logP -0.57			EXP	HANSCH,C & LEO,AJ (1985)
VP 7.89E-008	mm Hg	25	EST	NEELY,WB & BLAU,GE (1985)
DC	pKa			
HL 6.22E-014	atm m3/mol	25	EST	MEYLAN,WM & HOWARD,PH (1991)
OH 1.14E-011	cm3/molc sec	25	EST	MEYLAN,WM & HOWARD,PH (1993)

CAS #: 017433-92-0 — HYDRAZINECARBOXAMIDE, N-(3-NITROPHENYL)

Formula: $C_7H_8N_4O_3$

Mol Weight: 196.17

MP (deg C): | FP (deg C):

BP (deg C):

BP pressure (mm Hg):

Property/Value	Units	Temp	Data Type	Reference
WS 2.89E+003	mg/L	25	EST	MEYLAN,WM ET AL. (1996)
logP -0.13			EXP	KRAMER,CR & BECK,L (1981)
VP 1.08E-006	mm Hg	25	EST	NEELY,WB & BLAU,GE (1985)
DC	pKa			
HL 3.32E-015	atm m3/mol	25	EST	MEYLAN,WM & HOWARD,PH (1991)
OH 4.73E-012	cm3/molc sec	25	EST	MEYLAN,WM & HOWARD,PH (1993)

CAS #: 017433-93-1 — HYDRAZINECARBOXAMIDE, N-(4-NITROPHENYL)

Formula: $C_7H_8N_4O_3$

Mol Weight: 196.17

MP (deg C): | FP (deg C):

BP (deg C):

BP pressure (mm Hg):

Property/Value	Units	Temp	Data Type	Reference
WS 3.19E+003	mg/L	25	EST	MEYLAN,WM ET AL. (1996)
logP -0.18			EXP	KRAMER,CR & BECK,L (1981)
VP 1.08E-006	mm Hg	25	EST	NEELY,WB & BLAU,GE (1985)
DC	pKa			
HL 3.32E-015	atm m3/mol	25	EST	MEYLAN,WM & HOWARD,PH (1991)
OH 6.33E-012	cm3/molc sec	25	EST	MEYLAN,WM & HOWARD,PH (1993)

CAS #: 017433-94-2 — HYDRAZINECARBOXAMIDE, N-(2-NITROPHENYL)

Formula: $C_7H_8N_4O_3$

Mol Weight: 196.17

MP (deg C): | FP (deg C):

BP (deg C):

BP pressure (mm Hg):

Property/Value	Units	Temp	Data Type	Reference
WS 5.65E+003	mg/L	25	EST	MEYLAN,WM ET AL. (1996)
logP -0.47			EXP	KRAMER,CR & BECK,L (1981)
VP 1.08E-006	mm Hg	25	EST	NEELY,WB & BLAU,GE (1985)
DC	pKa			
HL 7.26E-014	atm m3/mol	25	EST	MEYLAN,WM & HOWARD,PH (1991)
OH 6.33E-012	cm3/molc sec	25	EST	MEYLAN,WM & HOWARD,PH (1993)

CAS #: 017455-13-9				1,4,7,10,13,16-HEXAOXACYCLOOCTADECANE

Formula: $C_{12}H_{26}O_6$

Mol Weight: 266.34

MP (deg C):   FP (deg C):

BP (deg C):

BP pressure (mm Hg):

Property/ Value	Units	Temp	Data Type	Reference
WS 7.49E+004	mg/L	25	EST	MEYLAN,WM ET AL. (1996)
logP -0.68			EXP	HANSCH,C ET AL. (1995)
VP 6.67E-005	mm Hg	25	EST	NEELY,WB & BLAU,GE (1985)
DC	pKa			
HL 3.45E-013	atm m3/mol	25	EST	MEYLAN,WM & HOWARD,PH (1991)
OH 8.41E-011	cm3/molc sec	25	EST	MEYLAN,WM & HOWARD,PH (1993)

CAS #: 017455-25-3				[5,5]DB30C10-DIBENZO CROWN ETHER

Formula: $C_{28}H_{40}O_{10}$

Mol Weight: 536.63

MP (deg C):   FP (deg C):

BP (deg C):

BP pressure (mm Hg):

Property/ Value	Units	Temp	Data Type	Reference
WS 9.37E-001	mg/L	25	EST	MEYLAN,WM ET AL. (1996)
logP 1.80			EXP	STOLWIJK,TB ET AL. (1989)
VP 6.32E-014	mm Hg	25	EST	NEELY,WB & BLAU,GE (1985)
DC	pKa			
HL 3.98E-020	atm m3/mol	25	EST	MEYLAN,WM & HOWARD,PH (1991)
OH 1.49E-010	cm3/molc sec	25	EST	MEYLAN,WM & HOWARD,PH (1993)

CAS #: 017505-12-3				1,2,5-SELENADIAZOLE, 3,4-DIMETHYL-

Formula: $C_4H_6N_2Se$

Mol Weight: 161.07

MP (deg C):   FP (deg C):

BP (deg C):

BP pressure (mm Hg):

Property/ Value	Units	Temp	Data Type	Reference
WS 1.18E+004	mg/L	25	EST	MEYLAN,WM ET AL. (1996)
logP 0.89			EXP	CALVINO,R ET AL. (1992)
VP 2.00E-004	mm Hg	25	EST	NEELY,WB & BLAU,GE (1985)
DC	pKa			
HL 2.41E-004	atm m3/mol	25	EST	MEYLAN,WM & HOWARD,PH (1991)
OH 9.72E-013	cm3/molc sec	25	EST	MEYLAN,WM & HOWARD,PH (1993)

CAS #: 017559-81-8				3-HEXENE-2,5-DIONE (CIS)

Formula: $C_6H_8O_2$

Mol Weight: 112.13

MP (deg C):   FP (deg C):

BP (deg C):

BP pressure (mm Hg):

Property/ Value	Units	Temp	Data Type	Reference
WS 3.46E+004	mg/L	25	EST	MEYLAN,WM ET AL. (1996)
logP 0.57			EST	MEYLAN,WM & HOWARD,PH (1995)
VP 2.17E+000	mm Hg	25	EST	NEELY,WB & BLAU,GE (1985)
DC	pKa			
HL 1.11E-008	atm m3/mol	25	EST	MEYLAN,WM & HOWARD,PH (1991)
OH 6.31E-011	cm3/molc sec	25	EXP	ATKINSON,R (1989)

CAS #: 017562-53-7				1,4-BENZDIAZEPIN-2-ONE-5-PH-7CN

Formula: $C_{16}H_{11}N_3O$

Mol Weight: 261.29

MP (deg C):   FP (deg C):

BP (deg C):

BP pressure (mm Hg):

Property/ Value	Units	Temp	Data Type	Reference
WS 3.10E+002	mg/L	25	EST	MEYLAN,WM ET AL. (1996)
logP 1.82			EXP	HANSCH,C & LEO,AJ (1985)
VP 8.96E-010	mm Hg	25	EST	NEELY,WB & BLAU,GE (1985)
DC	pKa			
HL 2.32E-012	atm m3/mol	25	EST	MEYLAN,WM & HOWARD,PH (1991)
OH 7.63E-012	cm3/molc sec	25	EST	MEYLAN,WM & HOWARD,PH (1993)

CAS #: 017570-26-2				2-PROPENOIC ACID, 3-(3-METHOXYPHENYL)-, (E)-

Formula: $C_{10}H_{10}O_3$

Mol Weight: 178.19

MP (deg C):   FP (deg C):

BP (deg C):

BP pressure (mm Hg):

Property/ Value	Units	Temp	Data Type	Reference
WS 1.32E+003	mg/L	25	EST	MEYLAN,WM ET AL. (1996)
logP 2.37			EXP	SANGSTER,J (1993)
VP 1.59E-004	mm Hg	25	EST	NEELY,WB & BLAU,GE (1985)
DC 4.37	pKa	25	EXP	KORTUM,G ET AL (1961)
HL 7.65E-010	atm m3/mol	25	EST	MEYLAN,WM & HOWARD,PH (1991)
OH 4.13E-011	cm3/molc sec	25	EST	MEYLAN,WM & HOWARD,PH (1993)

CAS #: 017576-53-3				3-NITROQUINOLINE

Formula: $C_9H_6N_2O_2$

Mol Weight: 174.16

MP (deg C):   FP (deg C):

BP (deg C):

BP pressure (mm Hg):

Property/ Value	Units	Temp	Data Type	Reference
WS 4.97E+002	mg/L	25	EST	MEYLAN,WM ET AL. (1996)
logP 1.97			EXP	HANSCH,C & LEO,AJ (1985)
VP 1.38E-004	mm Hg	25	EST	NEELY,WB & BLAU,GE (1985)
DC	pKa			
HL 2.72E-009	atm m3/mol	25	EST	MEYLAN,WM & HOWARD,PH (1991)
OH 1.80E-012	cm3/molc sec	25	EST	MEYLAN,WM & HOWARD,PH (1993)

CAS #: 017584-90-6				4H-1-BENZOPYRAN-4-ONE, 2,3-DIMETHYL-

Formula: $C_{11}H_{10}O_2$

Mol Weight: 174.20

MP (deg C):   FP (deg C):

BP (deg C):

BP pressure (mm Hg):

Property/ Value	Units	Temp	Data Type	Reference
WS 6.25E+002	mg/L	25	EST	MEYLAN,WM ET AL. (1996)
logP 2.31			EXP	HANSCH,C ET AL. (1995)
VP 1.04E-003	mm Hg	25	EST	NEELY,WB & BLAU,GE (1985)
DC	pKa			
HL 1.98E-006	atm m3/mol	25	EST	MEYLAN,WM & HOWARD,PH (1991)
OH 5.84E-011	cm3/molc sec	25	EST	MEYLAN,WM & HOWARD,PH (1993)

CAS #: 017587-22-3				1,1,1,2,2,3,3-HEPTAFLUORO-7,7-DIMETHYL-4,6-OCTA*
Formula: $C_{10}H_{11}F_7O_2$				
Mol Weight: 296.19				
MP (deg C): 38		FP (deg C):		
BP (deg C): 46-47				
BP pressure (mm Hg): 5.00E+000				

Property/Value	Units	Temp	Data Type	Reference
WS 7.91E+000	mg/L	25	EST	MEYLAN,WM ET AL. (1996)
logP 3.76			EST	MEYLAN,WM & HOWARD,PH (1995)
VP 1.50E+000	mm Hg	25	EST	NEELY,WB & BLAU,GE (1985)
DC	pKa			
HL 1.70E-005	atm m3/mol	25	EST	MEYLAN,WM & HOWARD,PH (1991)
OH 1.87E-012	cm3/molc sec	25	EST	MEYLAN,WM & HOWARD,PH (1993)

CAS #: 017590-87-3				(2,6-DIMETHYLPHENYL)PHENYLDIAZENE
Formula: $C_{14}H_{14}N_2$				
Mol Weight: 210.28				
MP (deg C):		FP (deg C):		
BP (deg C):				
BP pressure (mm Hg):				

Property/Value	Units	Temp	Data Type	Reference
WS 5.13E-001	mg/L	25	EST	MEYLAN,WM ET AL. (1996)
logP 5.21			EST	MEYLAN,WM & HOWARD,PH (1995)
VP 2.59E-004	mm Hg	25	EST	NEELY,WB & BLAU,GE (1985)
DC	pKa			
HL 1.79E-005	atm m3/mol	25	EST	MEYLAN,WM & HOWARD,PH (1991)
OH 6.31E-012	cm3/molc sec	25	EST	MEYLAN,WM & HOWARD,PH (1993)

CAS #: 017598-02-6				2,2-DIMETHYL-7-METHOXYBENZOPYRAN
Formula: $C_{12}H_{14}O_2$				
Mol Weight: 190.24				
MP (deg C):		FP (deg C):		
BP (deg C): 68				
BP pressure (mm Hg): 1.00E-001				

Property/Value	Units	Temp	Data Type	Reference
WS 6.75E+001	mg/L	25	EST	MEYLAN,WM ET AL. (1996)
logP 3.35			EXP	HANSCH,C ET AL. (1995)
VP 5.16E-003	mm Hg	25	EST	NEELY,WB & BLAU,GE (1985)
DC	pKa			
HL 8.87E-006	atm m3/mol	25	EST	MEYLAN,WM & HOWARD,PH (1991)
OH 2.58E-010	cm3/molc sec	25	EST	MEYLAN,WM & HOWARD,PH (1993)

CAS #: 017606-31-4				BENSULTAP
Formula: $C_{17}H_{21}NO_4S_4$				
Mol Weight: 431.62				
MP (deg C):		FP (deg C):		
BP (deg C):				
BP pressure (mm Hg):				

Property/Value	Units	Temp	Data Type	Reference
WS 7.50E-001	mg/L	25	EXP	SHIU,WY ET AL. (1990)
logP 3.36			EXP	HANSCH,C ET AL. (1995)
VP 2.83E-012	mm Hg	25	EST	NEELY,WB & BLAU,GE (1985)
DC	pKa			
HL 4.40E-014	atm m3/mol	25	EST	MEYLAN,WM & HOWARD,PH (1991)
OH 3.40E-010	cm3/molc sec	25	EST	MEYLAN,WM & HOWARD,PH (1993)

CAS #: 017618-77-8				3-METHYL-2-HEXENE
Formula: C_7H_{14}				
Mol Weight: 98.19				
MP (deg C):		FP (deg C):		
BP (deg C):				
BP pressure (mm Hg):				

Property/Value	Units	Temp	Data Type	Reference
WS 2.78E+001	mg/L	25	EST	MEYLAN,WM ET AL. (1996)
logP 3.62			EST	MEYLAN,WM & HOWARD,PH (1995)
VP 5.05E+001	mm Hg	25	EXP	DYKYJ,J & REPAS,M (1973)
DC	pKa			
HL 6.63E-001	atm m3/mol	25	EST	MEYLAN,WM & HOWARD,PH (1991)
OH 8.96E-011	cm3/molc sec	25	EST	MEYLAN,WM & HOWARD,PH (1993)

CAS #: 017629-01-5				4-PHENYLQUINAZOLINE
Formula: $C_{14}H_{10}N_2$				
Mol Weight: 206.25				
MP (deg C):		FP (deg C):		
BP (deg C):				
BP pressure (mm Hg):				

Property/Value	Units	Temp	Data Type	Reference
WS 1.97E+002	mg/L	25	EST	MEYLAN,WM ET AL. (1996)
logP 2.71			EXP	HANSCH,C & LEO,AJ (1985)
VP 4.85E-006	mm Hg	25	EST	NEELY,WB & BLAU,GE (1985)
DC	pKa			
HL 2.19E-008	atm m3/mol	25	EST	MEYLAN,WM & HOWARD,PH (1991)
OH 6.86E-012	cm3/molc sec	25	EST	MEYLAN,WM & HOWARD,PH (1993)

CAS #: 017629-04-8				QUINAZOLIN-2-ONE,1-ME-4-PHENYL
Formula: $C_{15}H_{12}N_2O$				
Mol Weight: 236.28				
MP (deg C):		FP (deg C):		
BP (deg C):				
BP pressure (mm Hg):				

Property/Value	Units	Temp	Data Type	Reference
WS 8.33E+002	mg/L	25	EST	MEYLAN,WM ET AL. (1996)
logP 1.79			EXP	HANSCH,C & LEO,AJ (1985)
VP 1.43E-006	mm Hg	25	EST	NEELY,WB & BLAU,GE (1985)
DC	pKa			
HL 7.82E-008	atm m3/mol	25	EST	MEYLAN,WM & HOWARD,PH (1991)
OH 1.97E-011	cm3/molc sec	25	EST	MEYLAN,WM & HOWARD,PH (1993)

CAS #: 017629-09-3				2-METHOXY-4-PHENYLQUINAZOLINE
Formula: $C_{15}H_{12}N_2O$				
Mol Weight: 236.28				
MP (deg C):		FP (deg C):		
BP (deg C):				
BP pressure (mm Hg):				

Property/Value	Units	Temp	Data Type	Reference
WS 2.89E+001	mg/L	25	EST	MEYLAN,WM ET AL. (1996)
logP 3.50			EXP	HANSCH,C & LEO,AJ (1985)
VP 9.07E-007	mm Hg	25	EST	NEELY,WB & BLAU,GE (1985)
DC	pKa			
HL 1.18E-008	atm m3/mol	25	EST	MEYLAN,WM & HOWARD,PH (1991)
OH 1.11E-011	cm3/molc sec	25	EST	MEYLAN,WM & HOWARD,PH (1993)

CAS #: 017647-69-7	FURAZAN, ETHYLMETHYL-

Formula: $C_5H_8N_2O$

Mol Weight: 112.13

MP (deg C):   FP (deg C):

BP (deg C):

BP pressure (mm Hg):

Property/Value	Units	Temp	Data Type	Reference
WS 1.05E+004	mg/L	25	EST	MEYLAN,WM ET AL. (1996)
logP 1.18			EXP	CALVINO,R ET AL. (1992)
VP 2.92E+000	mm Hg	25	EST	NEELY,WB & BLAU,GE (1985)
DC	pKa			
HL 3.06E-004	atm m3/mol	25	EST	MEYLAN,WM & HOWARD,PH (1991)
OH 5.24E-012	cm3/molc sec	25	EST	MEYLAN,WM & HOWARD,PH (1993)

CAS #: 017647-70-0	FURAZANAMINE, 4-METHYL-

Formula: $C_3H_5N_3O$

Mol Weight: 99.09

MP (deg C):   FP (deg C):

BP (deg C):

BP pressure (mm Hg):

Property/Value	Units	Temp	Data Type	Reference
WS 9.86E+004	mg/L	25	EST	MEYLAN,WM ET AL. (1996)
logP 0.09			EXP	CALVINO,R ET AL. (1992)
VP 4.98E-001	mm Hg	25	EST	NEELY,WB & BLAU,GE (1985)
DC	pKa			
HL 7.38E-008	atm m3/mol	25	EST	MEYLAN,WM & HOWARD,PH (1991)
OH 4.14E-012	cm3/molc sec	25	EST	MEYLAN,WM & HOWARD,PH (1993)

CAS #: 017650-76-9	2-CHLORO-DIMETHYL PARA-OXON

Formula: $C_8H_9ClNO_6P$

Mol Weight: 281.59

MP (deg C):   FP (deg C):

BP (deg C):

BP pressure (mm Hg):

Property/Value	Units	Temp	Data Type	Reference
WS 1.75E+002	mg/L	25	EST	MEYLAN,WM ET AL. (1996)
logP 1.83			EXP	HANSCH,C & LEO,AJ (1985)
VP 1.04E-005	mm Hg	25	EST	NEELY,WB & BLAU,GE (1985)
DC	pKa			
HL 2.69E-010	atm m3/mol	25	EST	MEYLAN,WM & HOWARD,PH (1991)
OH 5.71E-012	cm3/molc sec	25	EST	MEYLAN,WM & HOWARD,PH (1993)

CAS #: 017659-57-3	8-(3-NITROPHENYL)-ADENINE

Formula: $C_{11}H_8N_6O_2$

Mol Weight: 256.23

MP (deg C):   FP (deg C):

BP (deg C):

BP pressure (mm Hg):

Property/Value	Units	Temp	Data Type	Reference
WS 2.00E+002	mg/L	25	EST	MEYLAN,WM ET AL. (1996)
logP 1.93			EXP	HANSCH,C & LEO,AJ (1985)
VP 1.48E-011	mm Hg	25	EST	NEELY,WB & BLAU,GE (1985)
DC	pKa			
HL 2.13E-017	atm m3/mol	25	EST	MEYLAN,WM & HOWARD,PH (1991)
OH 2.00E-010	cm3/molc sec	25	EST	MEYLAN,WM & HOWARD,PH (1993)

CAS #: 017671-76-0	O-PROPYL-N-METHYLCARBAMATE

Formula: $C_5H_{11}NO_2$

Mol Weight: 117.15

MP (deg C):   FP (deg C):

BP (deg C):

BP pressure (mm Hg):

Property/Value	Units	Temp	Data Type	Reference
WS 1.59E+004	mg/L	25	EST	MEYLAN,WM ET AL. (1996)
logP 0.95			EXP	HANSCH,C & LEO,AJ (1985)
VP 4.94E+000	mm Hg	25	EST	NEELY,WB & BLAU,GE (1985)
DC	pKa			
HL 1.53E-007	atm m3/mol	25	EST	MEYLAN,WM & HOWARD,PH (1991)
OH 9.21E-012	cm3/molc sec	25	EST	MEYLAN,WM & HOWARD,PH (1993)

CAS #: 017696-95-6	4-METHYLSEMICARBAZIDE

Formula: $C_2H_7N_3O$

Mol Weight: 89.10

MP (deg C):   FP (deg C):

BP (deg C):

BP pressure (mm Hg):

Property/Value	Units	Temp	Data Type	Reference
WS 4.64E+005	mg/L	25	EST	MEYLAN,WM ET AL. (1996)
logP -2.20			EXP	HANSCH,C & LEO,AJ (1985)
VP 1.14E-001	mm Hg	25	EST	NEELY,WB & BLAU,GE (1985)
DC	pKa			
HL 3.35E-012	atm m3/mol	25	EST	MEYLAN,WM & HOWARD,PH (1991)
OH 6.63E-011	cm3/molc sec	25	EST	MEYLAN,WM & HOWARD,PH (1993)

CAS #: 017700-09-3	2,3,4-TRICHLORONITROBENZENE

Formula: $C_6H_2Cl_3NO_2$

Mol Weight: 226.45

MP (deg C): 55-56   FP (deg C):

BP (deg C):

BP pressure (mm Hg):

Property/Value	Units	Temp	Data Type	Reference
WS 2.60E+001	mg/L	20	EXP	YALKOWSKY,SH & DANNENFELSER,RM (1992)
logP 3.61			EXP	HANSCH,C & LEO,AJ (1985)
VP 7.92E-004	mm Hg	25	EST	NEELY,WB & BLAU,GE (1985)
DC	pKa			
HL 8.65E-006	atm m3/mol	25	EST	MEYLAN,WM & HOWARD,PH (1991)
OH 3.52E-014	cm3/molc sec	25	EST	MEYLAN,WM & HOWARD,PH (1993)

CAS #: 017700-54-8	2,6-DICHLOROACETANILIDE

Formula: $C_8H_7Cl_2NO$

Mol Weight: 204.06

MP (deg C):   FP (deg C):

BP (deg C):

BP pressure (mm Hg):

Property/Value	Units	Temp	Data Type	Reference
WS 3.11E+003	mg/L	25	EST	MEYLAN,WM ET AL. (1996)
logP 1.32			EXP	NAKAGAWA,Y ET AL. (1992)
VP 2.02E-005	mm Hg	25	EST	NEELY,WB & BLAU,GE (1985)
DC	pKa			
HL 3.39E-009	atm m3/mol	25	EST	MEYLAN,WM & HOWARD,PH (1991)
OH 1.16E-012	cm3/molc sec	25	EST	MEYLAN,WM & HOWARD,PH (1993)

CAS #: 017711-74-9	1,3,5-TRIAZINE-2,4-DIAMINE, 1,6-DIHYDRO-6,6-DIME

Formula: $C_{11}H_{14}N_6O_2$

Mol Weight: 262.27

MP (deg C):　FP (deg C):

BP (deg C):

BP pressure (mm Hg):

Property/Value	Units	Temp	Data Type	Reference
WS 7.31E+003	mg/L	25	EST	MEYLAN,WM ET AL. (1996)
logP 0.06			EXP	HANSCH,C ET AL. (1995)
VP 1.17E-007	mm Hg	25	EST	NEELY,WB & BLAU,GE (1985)
DC	pKa			
HL 2.87E-015	atm m3/mol	25	EST	MEYLAN,WM & HOWARD,PH (1991)
OH 5.51E-011	cm3/molc sec	25	EST	MEYLAN,WM & HOWARD,PH (1993)

CAS #: 017720-22-8	8-PHENYLADENINE

Formula: $C_{11}H_9N_5$

Mol Weight: 211.23

MP (deg C):　FP (deg C):

BP (deg C):

BP pressure (mm Hg):

Property/Value	Units	Temp	Data Type	Reference
WS 4.08E+002	mg/L	25	EST	MEYLAN,WM ET AL. (1996)
logP 2.31			EXP	HANSCH,C & LEO,AJ (1985)
VP 7.84E-010	mm Hg	25	EST	NEELY,WB & BLAU,GE (1985)
DC	pKa			
HL 5.39E-015	atm m3/mol	25	EST	MEYLAN,WM & HOWARD,PH (1991)
OH 2.03E-010	cm3/molc sec	25	EST	MEYLAN,WM & HOWARD,PH (1993)

CAS #: 017721-95-8	2,6-DIMETHYL-N-NITROSOPIPERIDINE

Formula: $C_7H_{14}N_2O$

Mol Weight: 142.20

MP (deg C):　FP (deg C):

BP (deg C):

BP pressure (mm Hg):

Property/Value	Units	Temp	Data Type	Reference
WS 5.65E+003	mg/L	25	EST	MEYLAN,WM ET AL. (1996)
logP 1.36			EXP	HANSCH,C & LEO,AJ (1985)
VP 3.37E-002	mm Hg	25	EST	NEELY,WB & BLAU,GE (1985)
DC	pKa			
HL 3.74E-006	atm m3/mol	25	EST	MEYLAN,WM & HOWARD,PH (1991)
OH 4.90E-011	cm3/molc sec	25	EST	MEYLAN,WM & HOWARD,PH (1993)

CAS #: 017751-47-2	3-PYRIDYLMETHYL-N-MORPHOLINE

Formula: $C_{10}H_{14}N_2O$

Mol Weight: 178.24

MP (deg C):　FP (deg C):

BP (deg C):

BP pressure (mm Hg):

Property/Value	Units	Temp	Data Type	Reference
WS 1.00E+006	mg/L	25	EST	MEYLAN,WM ET AL. (1996)
logP 0.04			EXP	HANSCH,C & LEO,AJ (1985)
VP 2.71E-003	mm Hg	25	EST	NEELY,WB & BLAU,GE (1985)
DC	pKa			
HL 2.64E-011	atm m3/mol	25	EST	MEYLAN,WM & HOWARD,PH (1991)
OH 1.50E-010	cm3/molc sec	25	EST	MEYLAN,WM & HOWARD,PH (1993)

CAS #: 017757-70-9	1,4-OXATHIIN-3-CARBOXAMIDE, 5,6-DIHYDRO-2-METHYL

Formula: $C_{12}H_{13}NO_3S$

Mol Weight: 251.31

MP (deg C):　FP (deg C):

BP (deg C):

BP pressure (mm Hg):

Property/Value	Units	Temp	Data Type	Reference
WS 5.88E+002	mg/L	25	EST	MEYLAN,WM ET AL. (1996)
logP 0.64			EXP	SANGSTER,J (1993)
VP 1.07E-008	mm Hg	25	EST	NEELY,WB & BLAU,GE (1985)
DC	pKa			
HL 4.71E-016	atm m3/mol	25	EST	MEYLAN,WM & HOWARD,PH (1991)
OH 1.85E-010	cm3/molc sec	25	EST	MEYLAN,WM & HOWARD,PH (1993)

CAS #: 017781-16-7	3-KETOCARBOFURANPHENOL

Formula: $C_{10}H_{10}O_3$

Mol Weight: 178.19

MP (deg C):　FP (deg C):

BP (deg C):

BP pressure (mm Hg):

Property/Value	Units	Temp	Data Type	Reference
WS 1.42E+003	mg/L	25	EST	MEYLAN,WM ET AL. (1996)
logP 1.87			EXP	HANSCH,C & LEO,AJ (1985)
VP 7.42E-005	mm Hg	25	EST	NEELY,WB & BLAU,GE (1985)
DC	pKa			
HL 2.50E-010	atm m3/mol	25	EST	MEYLAN,WM & HOWARD,PH (1991)
OH 1.47E-011	cm3/molc sec	25	EST	MEYLAN,WM & HOWARD,PH (1993)

CAS #: 017784-47-3	ACRIDINE, HYDROCHLORIDE

Formula: $C_{13}H_{10}ClN$

Mol Weight: 215.68

MP (deg C):　FP (deg C):

BP (deg C):

BP pressure (mm Hg):

Property/Value	Units	Temp	Data Type	Reference
WS 7.53E+003	mg/L	25	EST	MEYLAN,WM ET AL. (1996)
logP -0.50			EXP	HANSCH,C ET AL. (1995)
VP 1.71E-005	mm Hg	25	EST	NEELY,WB & BLAU,GE (1985)
DC	pKa			
HL 6.72E-014	atm m3/mol	25	EST	MEYLAN,WM & HOWARD,PH (1991)
OH 3.75E-011	cm3/molc sec	25	EST	MEYLAN,WM & HOWARD,PH (1993)

CAS #: 017795-32-3	ME-PHOSPHORAMIDATE,O-ME,O-4-MEOPH

Formula: $C_9H_{14}NO_4P$

Mol Weight: 231.19

MP (deg C):　FP (deg C):

BP (deg C):

BP pressure (mm Hg):

Property/Value	Units	Temp	Data Type	Reference
WS 3.73E+003	mg/L	25	EST	MEYLAN,WM ET AL. (1996)
logP 1.06			EXP	HANSCH,C & LEO,AJ (1985)
VP 1.37E-004	mm Hg	25	EST	NEELY,WB & BLAU,GE (1985)
DC	pKa			
HL 7.59E-011	atm m3/mol	25	EST	MEYLAN,WM & HOWARD,PH (1991)
OH 4.82E-011	cm3/molc sec	25	EST	MEYLAN,WM & HOWARD,PH (1993)

CAS #: 017804-35-2				BENOMYL
Formula: $C_{14}H_{18}N_4O_3$				
Mol Weight: 290.32				
MP (deg C):		FP (deg C):		
BP (deg C):				
BP pressure (mm Hg):				

Property/ Value	Units	Temp	Data Type	Reference
WS 3.80E+000	mg/L	20	EXP	YALKOWSKY,SH & DANNENFELSER,RM (1992)
logP 2.12			EXP	HANSCH,C ET AL. (1995)
VP 3.70E-009	mm Hg	25	EXP	WEBER,JB (1994)
DC	pKa			
HL 3.72E-010	atm m3/mol	25	EST	VP/WSOL
OH 3.10E-011	cm3/molc sec	25	EST	MEYLAN,WM & HOWARD,PH (1993)

CAS #: 017812-07-6				3-BENZOYLACRYLIC ACID (TRANS)
Formula: $C_{10}H_8O_3$				
Mol Weight: 176.17				
MP (deg C):		FP (deg C):		
BP (deg C):				
BP pressure (mm Hg):				

Property/ Value	Units	Temp	Data Type	Reference
WS 7.04E+003	mg/L	25	EST	MEYLAN,WM ET AL. (1996)
logP 1.53			EXP	HANSCH,C & LEO,AJ (1985)
VP 7.76E-005	mm Hg	25	EST	NEELY,WB & BLAU,GE (1985)
DC	pKa			
HL 2.42E-011	atm m3/mol	25	EST	MEYLAN,WM & HOWARD,PH (1991)
OH 9.21E-012	cm3/molc sec	25	EST	MEYLAN,WM & HOWARD,PH (1993)

CAS #: 017831-71-9				2-PROPENOIC ACID, OXYBIS(2,1-ETHANEDIYLOXY-2,1-E
Formula: $C_{14}H_{22}O_7$				
Mol Weight: 302.33				
MP (deg C):		FP (deg C):		
BP (deg C):				
BP pressure (mm Hg):				

Property/ Value	Units	Temp	Data Type	Reference
WS 9.99E+002	mg/L	25	EST	MEYLAN,WM ET AL. (1996)
logP 1.26			EXP	SANGSTER,J (1994)
VP 3.08E-004	mm Hg	25	EST	NEELY,WB & BLAU,GE (1985)
DC	pKa			
HL 5.78E-013	atm m3/mol	25	EST	MEYLAN,WM & HOWARD,PH (1991)
OH 6.41E-011	cm3/molc sec	25	EST	MEYLAN,WM & HOWARD,PH (1993)

CAS #: 017849-38-6				BENZENEMETHANOL, 2-CHLORO-
Formula: C_7H_7ClO				
Mol Weight: 142.59				
MP (deg C): 73		FP (deg C):		
BP (deg C): 230				
BP pressure (mm Hg):				

Property/ Value	Units	Temp	Data Type	Reference
WS 8.14E+003	mg/L	25	EST	MEYLAN,WM ET AL. (1996)
logP 1.77			EXP	MIYAKE,F ET AL. (1987)
VP 7.89E-003	mm Hg	25	EST	NEELY,WB & BLAU,GE (1985)
DC	pKa			
HL 1.61E-007	atm m3/mol	25	EST	MEYLAN,WM & HOWARD,PH (1991)
OH 5.08E-012	cm3/molc sec	25	EST	MEYLAN,WM & HOWARD,PH (1993)

CAS #: 017870-85-8				BENZOIC ACID, 4-CHLORO-2-[(3-CHLOROPHENYL)AMINO]
Formula: $C_{13}H_9Cl_2NO_2$				
Mol Weight: 282.13				
MP (deg C):		FP (deg C):		
BP (deg C):				
BP pressure (mm Hg):				

Property/ Value	Units	Temp	Data Type	Reference
WS 8.37E-001	mg/L	25	EST	MEYLAN,WM ET AL. (1996)
logP 5.00			EXP	SANGSTER,J (1993)
VP 1.45E-007	mm Hg	25	EST	NEELY,WB & BLAU,GE (1985)
DC	pKa			
HL 1.16E-011	atm m3/mol	25	EST	MEYLAN,WM & HOWARD,PH (1991)
OH 1.42E-010	cm3/molc sec	25	EST	MEYLAN,WM & HOWARD,PH (1993)

CAS #: 017883-59-9				2,2-DIME PROPIONIC ACID HYDRAZIDE
Formula: $C_5H_{12}N_2O$				
Mol Weight: 116.16				
MP (deg C):		FP (deg C):		
BP (deg C):				
BP pressure (mm Hg):				

Property/ Value	Units	Temp	Data Type	Reference
WS 1.01E+004	mg/L	25	EST	MEYLAN,WM ET AL. (1996)
logP -0.35			EXP	HANSCH,C & LEO,AJ (1985)
VP 5.87E-003	mm Hg	25	EST	NEELY,WB & BLAU,GE (1985)
DC	pKa			
HL 1.09E-010	atm m3/mol	25	EST	MEYLAN,WM & HOWARD,PH (1991)
OH 7.09E-012	cm3/molc sec	25	EST	MEYLAN,WM & HOWARD,PH (1993)

CAS #: 017902-23-7				TEGAFUR
Formula: $C_8H_9FN_2O_3$				
Mol Weight: 200.17				
MP (deg C): 171-173		FP (deg C):		
BP (deg C):				
BP pressure (mm Hg):				

Property/ Value	Units	Temp	Data Type	Reference
WS 3.64E+003	mg/L	25	EST	MEYLAN,WM ET AL. (1996)
logP -0.27			EXP	HANSCH,C & LEO,AJ (1985)
VP 3.38E-008	mm Hg	25	EST	NEELY,WB & BLAU,GE (1985)
DC	pKa			
HL 8.32E-012	atm m3/mol	25	EST	MEYLAN,WM & HOWARD,PH (1991)
OH 6.35E-011	cm3/molc sec	25	EST	MEYLAN,WM & HOWARD,PH (1993)

CAS #: 017906-09-1				1,1,1,3,3,5,7,7,7-NONAMETHYL-5-PHENYLTETRASILO*
Formula: $C_{15}H_{32}O_3Si_4$				
Mol Weight: 372.76				
MP (deg C):		FP (deg C):		
BP (deg C):				
BP pressure (mm Hg):				

Property/ Value	Units	Temp	Data Type	Reference
WS 9.50E-005	mg/L	25	EST	MEYLAN,WM ET AL. (1996)
logP 7.15			EST	MEYLAN,WM & HOWARD,PH (1995)
VP	mm Hg			
DC	pKa			
HL	atm m3/mol			
OH 3.30E-012	cm3/molc sec	25	EST	MEYLAN,WM & HOWARD,PH (1993)

CAS #: 017925-97-2 — CHLOROMETHYLCHLOR

Formula:	$C_{15}H_{12}Cl_4$			
Mol Weight:	334.07			
MP (deg C):		**FP (deg C):**		
BP (deg C):				
BP pressure (mm Hg):				

Property/ Value	Units	Temp	Data Type	Reference
WS 1.47E-002	mg/L	25	EST	MEYLAN,WM ET AL. (1996)
logP 6.70			EST	MEYLAN,WM & HOWARD,PH (1995)
VP 8.00E-006	mm Hg	25	EST	NEELY,WB & BLAU,GE (1985)
DC	pKa			
HL 2.28E-005	atm m3/mol	25	EST	MEYLAN,WM & HOWARD,PH (1991)
OH 8.05E-012	cm3/molc sec	25	EST	MEYLAN,WM & HOWARD,PH (1993)

CAS #: 017928-28-8 — METHYLTRIS(TRIMETHYLSILOXY)SILANE

Formula:	$C_{10}H_{30}O_3Si_4$			
Mol Weight:	310.69			
MP (deg C):		**FP (deg C):**		
BP (deg C):				
BP pressure (mm Hg):				

Property/ Value	Units	Temp	Data Type	Reference
WS 2.45E-003	mg/L	25	EST	MEYLAN,WM ET AL. (1996)
logP 5.93			EST	MEYLAN,WM & HOWARD,PH (1995)
VP 4.02E-002	mm Hg	25	EXT	FLANINGAM,OL (1986)
DC	pKa			
HL 6.01E-006	atm m3/mol	25	EST	MEYLAN,WM & HOWARD,PH (1991)
OH 1.50E-012	cm3/molc sec	25	EST	MEYLAN,WM & HOWARD,PH (1993)

CAS #: 017942-66-4 — 5-THIA-1-AZABICYCLO[4.2.0]OCT-2-ENE-2-CARBOXYLIC

Formula:	$C_{20}H_{18}N_2O_6S_2$			
Mol Weight:	446.50			
MP (deg C):		**FP (deg C):**		
BP (deg C):				
BP pressure (mm Hg):				

Property/ Value	Units	Temp	Data Type	Reference
WS 5.76E+001	mg/L	25	EST	MEYLAN,WM ET AL. (1996)
logP 1.68			EXP	SANGSTER,J (1994)
VP 8.35E-017	mm Hg	25	EST	NEELY,WB & BLAU,GE (1985)
DC	pKa			
HL 1.70E-018	atm m3/mol	25	EST	MEYLAN,WM & HOWARD,PH (1991)
OH 1.86E-010	cm3/molc sec	25	EST	MEYLAN,WM & HOWARD,PH (1993)

CAS #: 017945-79-8 — 2-PYRIDINEBUTANOL

Formula:	$C_9H_{13}NO$			
Mol Weight:	151.21			
MP (deg C):		**FP (deg C):**		
BP (deg C):				
BP pressure (mm Hg):				

Property/ Value	Units	Temp	Data Type	Reference
WS 8.92E+005	mg/L	25	EST	MEYLAN,WM ET AL. (1996)
logP 0.86			EXP	HANSCH,C & LEO,AJ (1985)
VP 4.75E-004	mm Hg	25	EST	NEELY,WB & BLAU,GE (1985)
DC	pKa			
HL 6.66E-010	atm m3/mol	25	EST	MEYLAN,WM & HOWARD,PH (1991)
OH 9.06E-012	cm3/molc sec	25	EST	MEYLAN,WM & HOWARD,PH (1993)

CAS #: 018005-40-8 — O-ETHYL METHYLPHOSPHONOTHIOATE

Formula:	$C_3H_9O_2PS$			
Mol Weight:	140.14			
MP (deg C):		**FP (deg C):**		
BP (deg C):				
BP pressure (mm Hg):				

Property/ Value	Units	Temp	Data Type	Reference
WS 8.34E+004	mg/L	25	EST	MEYLAN,WM ET AL. (1996)
logP 0.00			EST	MEYLAN,WM & HOWARD,PH (1995)
VP 3.18E-001	mm Hg	25	EST	NEELY,WB & BLAU,GE (1985)
DC	pKa			
HL 3.03E-006	atm m3/mol	25	EST	MEYLAN,WM & HOWARD,PH (1991)
OH 1.95E-011	cm3/molc sec	25	EST	MEYLAN,WM & HOWARD,PH (1993)

CAS #: 018035-92-2 — BENZOIC ACID, 3,4,5-TRIMETHOXY-, 7-CHLORO-2,3-DI

Formula:	$C_{25}H_{21}ClN_2O_6$			
Mol Weight:	480.91			
MP (deg C):		**FP (deg C):**		
BP (deg C):				
BP pressure (mm Hg):				

Property/ Value	Units	Temp	Data Type	Reference
WS 5.95E-001	mg/L	25	EST	MEYLAN,WM ET AL. (1996)
logP 3.75			EXP	SANGSTER,J (1993)
VP 7.37E-015	mm Hg	25	EST	NEELY,WB & BLAU,GE (1985)
DC	pKa			
HL 2.45E-015	atm m3/mol	25	EST	MEYLAN,WM & HOWARD,PH (1991)
OH 9.99E-011	cm3/molc sec	25	EST	MEYLAN,WM & HOWARD,PH (1993)

CAS #: 018046-21-4 — FENTIAZAC

Formula:	$C_{17}H_{14}ClNO_2S$			
Mol Weight:	331.82			
MP (deg C):	161-162	**FP (deg C):**		
BP (deg C):				
BP pressure (mm Hg):				

Property/ Value	Units	Temp	Data Type	Reference
WS 7.52E-001	mg/L	25	EST	MEYLAN,WM ET AL. (1996)
logP 5.19			EXP	HANSCH,C & LEO,AJ (1985)
VP 1.70E-010	mm Hg	25	EST	NEELY,WB & BLAU,GE (1985)
DC	pKa			
HL 1.37E-013	atm m3/mol	25	EST	MEYLAN,WM & HOWARD,PH (1991)
OH 7.00E-012	cm3/molc sec	25	EST	MEYLAN,WM & HOWARD,PH (1993)

CAS #: 018048-95-8 — DIMETHYLAMPHENICOL

Formula:	$C_{13}H_{18}N_2O_5$			
Mol Weight:	282.30			
MP (deg C):		**FP (deg C):**		
BP (deg C):				
BP pressure (mm Hg):				

Property/ Value	Units	Temp	Data Type	Reference
WS 1.34E+003	mg/L	25	EST	MEYLAN,WM ET AL. (1996)
logP 0.79			EXP	HANSCH,C & LEO,AJ (1985)
VP 2.88E-012	mm Hg	25	EST	NEELY,WB & BLAU,GE (1985)
DC	pKa			
HL 3.25E-017	atm m3/mol	25	EST	MEYLAN,WM & HOWARD,PH (1991)
OH 3.33E-011	cm3/molc sec	25	EST	MEYLAN,WM & HOWARD,PH (1993)

CAS #: 018063-03-1			2,6-DIFLUOROBENZAMIDE

Formula: $C_7H_5F_2NO$
Mol Weight: 157.12
MP (deg C): 145-148 FP (deg C):
BP (deg C):
BP pressure (mm Hg):

Property/ Value	Units	Temp	Data Type	Reference
WS 4.31E+004	mg/L	25	EST	MEYLAN,WM ET AL. (1996)
logP 0.25			EXP	NAKAGAWA,Y ET AL. (1992)
VP 1.26E-003	mm Hg	25	EST	NEELY,WB & BLAU,GE (1985)
DC	pKa			
HL 3.02E-009	atm m3/mol	25	EST	MEYLAN,WM & HOWARD,PH (1991)
OH 4.79E-012	cm3/molc sec	25	EST	MEYLAN,WM & HOWARD,PH (1993)

CAS #: 018077-53-7			1,1,1,3,5,5,5-HEPTAMETHYL-3-(2-(TRIMETHYLSILY*))

Formula: $C_{12}H_{34}O_2Si_4$
Mol Weight: 322.75
MP (deg C): FP (deg C):
BP (deg C):
BP pressure (mm Hg):

Property/ Value	Units	Temp	Data Type	Reference
WS 1.53E-004	mg/L	25	EST	MEYLAN,WM ET AL. (1996)
logP 7.72			EST	MEYLAN,WM & HOWARD,PH (1995)
VP	mm Hg			
DC	pKa			
HL	atm m3/mol			
OH 4.02E-012	cm3/molc sec	25	EST	MEYLAN,WM & HOWARD,PH (1993)

CAS #: 018109-43-8			BENZENEACETAMIDE, N-(3-CHLOROPHENYL)-

Formula: $C_{14}H_{12}ClNO$
Mol Weight: 245.71
MP (deg C): FP (deg C):
BP (deg C):
BP pressure (mm Hg):

Property/ Value	Units	Temp	Data Type	Reference
WS 2.06E+001	mg/L	25	EST	MEYLAN,WM ET AL. (1996)
logP 3.61			EXP	MITSUTAKE,KI ET AL. (1986)
VP 2.28E-007	mm Hg	25	EST	NEELY,WB & BLAU,GE (1985)
DC	pKa			
HL 3.69E-010	atm m3/mol	25	EST	MEYLAN,WM & HOWARD,PH (1991)
OH 1.43E-011	cm3/molc sec	25	EST	MEYLAN,WM & HOWARD,PH (1993)

CAS #: 018109-55-2			BENZENEACETAMIDE, N-[3-(DIETHYLAMINO)PROPYL]-N-(

Formula: $C_{23}H_{32}N_2O$
Mol Weight: 352.52
MP (deg C): FP (deg C):
BP (deg C):
BP pressure (mm Hg):

Property/ Value	Units	Temp	Data Type	Reference
WS 3.62E-001	mg/L	25	EST	MEYLAN,WM ET AL. (1996)
logP 4.94			EXP	SANGSTER,J (1993)
VP 4.19E-009	mm Hg	25	EST	NEELY,WB & BLAU,GE (1985)
DC	pKa			
HL 1.91E-011	atm m3/mol	25	EST	MEYLAN,WM & HOWARD,PH (1991)
OH 1.25E-010	cm3/molc sec	25	EST	MEYLAN,WM & HOWARD,PH (1993)

CAS #: 018138-04-0			PYRAZINE, 2,3-DIETHYL-5-METHYL-

Formula: $C_9H_{14}N_2$
Mol Weight: 150.23
MP (deg C): FP (deg C):
BP (deg C):
BP pressure (mm Hg):

Property/ Value	Units	Temp	Data Type	Reference
WS 1.64E+003	mg/L	25	EST	MEYLAN,WM ET AL. (1996)
logP 1.95			EXP	YAMAGAMI,C ET AL. (1991)
VP 5.49E-002	mm Hg	25	EST	NEELY,WB & BLAU,GE (1985)
DC	pKa			
HL 6.91E-006	atm m3/mol	25	EST	MEYLAN,WM & HOWARD,PH (1991)
OH 4.16E-012	cm3/molc sec	25	EST	MEYLAN,WM & HOWARD,PH (1993)

CAS #: 018143-31-2			SILANE, DIMETHYL PROPYL

Formula: $C_5H_{14}Si$
Mol Weight: 102.25
MP (deg C): FP (deg C):
BP (deg C):
BP pressure (mm Hg):

Property/ Value	Units	Temp	Data Type	Reference
WS 2.05E+002	mg/L	25	EST	MEYLAN,WM ET AL. (1996)
logP 3.22			EXP	HANSCH,C & LEO,AJ (1985)
VP 1.20E+002	mm Hg	25	EST	NEELY,WB & BLAU,GE (1985)
DC	pKa			
HL 2.83E-001	atm m3/mol	25	EST	MEYLAN,WM & HOWARD,PH (1991)
OH 2.88E-012	cm3/molc sec	25	EST	MEYLAN,WM & HOWARD,PH (1993)

CAS #: 018153-42-9			9-(3-METHYLPHENYL)-9H-FLUORENE

Formula: $C_{20}H_{16}$
Mol Weight: 256.35
MP (deg C): FP (deg C):
BP (deg C):
BP pressure (mm Hg):

Property/ Value	Units	Temp	Data Type	Reference
WS 1.51E-002	mg/L	25	EST	MEYLAN,WM ET AL. (1996)
logP 5.91			EST	MEYLAN,WM & HOWARD,PH (1995)
VP 9.47E-007	mm Hg	25	EST	NEELY,WB & BLAU,GE (1985)
DC	pKa			
HL 1.49E-005	atm m3/mol	25	EST	MEYLAN,WM & HOWARD,PH (1991)
OH 2.19E-011	cm3/molc sec	25	EST	MEYLAN,WM & HOWARD,PH (1993)

CAS #: 018153-43-0			9-(4-METHYLPHENYL)-9H-FLUORENE

Formula: $C_{20}H_{16}$
Mol Weight: 256.35
MP (deg C): FP (deg C):
BP (deg C):
BP pressure (mm Hg):

Property/ Value	Units	Temp	Data Type	Reference
WS 1.51E-002	mg/L	25	EST	MEYLAN,WM ET AL. (1996)
logP 5.91			EST	MEYLAN,WM & HOWARD,PH (1995)
VP 9.47E-007	mm Hg	25	EST	NEELY,WB & BLAU,GE (1985)
DC	pKa			
HL 1.49E-005	atm m3/mol	25	EST	MEYLAN,WM & HOWARD,PH (1991)
OH 1.60E-011	cm3/molc sec	25	EST	MEYLAN,WM & HOWARD,PH (1993)

CAS #: 018181-70-9 — IODOFENPHOS

Formula: $C_8H_8Cl_2IO_3PS$

Mol Weight: 413.00

MP (deg C): 72-73

FP (deg C):

BP (deg C):

BP pressure (mm Hg):

Property/Value	Units	Temp	Data Type	Reference
WS 1.00E-001	mg/L	20	EXP	YALKOWSKY,SH & DANNENFELSER,RM (1992)
logP 5.51			EXP	HANSCH,C ET AL. (1995)
VP 8.25E-007	mm Hg	20	EXP	METCALF,RL (1995)
DC	pKa			
HL 4.48E-006	atm m3/mol	20	EST	VP/WSOL
OH 5.88E-011	cm3/molc sec	25	EST	MEYLAN,WM & HOWARD,PH (1993)

CAS #: 018181-80-1 — ISOPROPYL 4,4'DIBROMOBENZILATE

Formula: $C_{17}H_{16}Br_2O_3$

Mol Weight: 428.13

MP (deg C): 77

FP (deg C):

BP (deg C):

BP pressure (mm Hg):

Property/Value	Units	Temp	Data Type	Reference
WS 5.00E+000	mg/L	20	EXP	YALKOWSKY,SH & DANNENFELSER,RM (1992)
logP 5.40			EXP	TOMLIN,C (1994)
VP 6.68E-010	mm Hg	25	EST	NEELY,WB & BLAU,GE (1985)
DC	pKa			
HL 4.99E-008	atm m3/mol	25	EST	MEYLAN,WM & HOWARD,PH (1991)
OH 6.52E-012	cm3/molc sec	25	EST	MEYLAN,WM & HOWARD,PH (1993)

CAS #: 018226-46-5 — 1,3,5-TRIS(CHLOROETHYL)BENZENE

Formula: $C_{12}H_{15}Cl_3$

Mol Weight: 265.61

MP (deg C):

FP (deg C):

BP (deg C):

BP pressure (mm Hg):

Property/Value	Units	Temp	Data Type	Reference
WS 1.88E-001	mg/L	25	EST	MEYLAN,WM ET AL. (1996)
logP 5.87			EST	MEYLAN,WM & HOWARD,PH (1995)
VP 3.11E-005	mm Hg	25	EST	NEELY,WB & BLAU,GE (1985)
DC	pKa			
HL 7.41E-004	atm m3/mol	25	EST	MEYLAN,WM & HOWARD,PH (1991)
OH 3.22E-011	cm3/molc sec	25	EST	MEYLAN,WM & HOWARD,PH (1993)

CAS #: 018259-05-7 — 2,3,4,5,6-PCB

Formula: $C_{12}H_5Cl_5$

Mol Weight: 326.44

MP (deg C):

FP (deg C):

BP (deg C):

BP pressure (mm Hg):

Property/Value	Units	Temp	Data Type	Reference
WS 4.01E-003	mg/L	25	EXP	YALKOWSKY,SH & DANNENFELSER,RM (1992)
logP 6.75			EXP	DEBRUIJN,J ET AL. (1989)
VP 2.22E-006	mm Hg	25	EST	NEELY,WB & BLAU,GE (1985)
DC	pKa			
HL 9.24E-005	atm m3/mol	25	EST	MEYLAN,WM & HOWARD,PH (1991)
OH 5.28E-013	cm3/molc sec	25	EST	MEYLAN,WM & HOWARD,PH (1993)

CAS #: 018281-04-4 — ETHYL NONADECANOATE

Formula: $C_{21}H_{42}O_2$

Mol Weight: 326.57

MP (deg C):

FP (deg C):

BP (deg C):

BP pressure (mm Hg):

Property/Value	Units	Temp	Data Type	Reference
WS 1.16E-004	mg/L	25	EST	MEYLAN,WM ET AL. (1996)
logP 9.21			EST	MEYLAN,WM & HOWARD,PH (1995)
VP 7.33E-006	mm Hg	25	EST	NEELY,WB & BLAU,GE (1985)
DC	pKa			
HL 2.88E-002	atm m3/mol	25	EST	MEYLAN,WM & HOWARD,PH (1991)
OH 2.45E-011	cm3/molc sec	25	EST	MEYLAN,WM & HOWARD,PH (1993)

CAS #: 018281-05-5 — ETHYL EICOSANOATE

Formula: $C_{22}H_{44}O_2$

Mol Weight: 340.59

MP (deg C): 50

FP (deg C):

BP (deg C): 295

BP pressure (mm Hg): 1.00E+002

Property/Value	Units	Temp	Data Type	Reference
WS 3.64E-005	mg/L	25	EST	MEYLAN,WM ET AL. (1996)
logP 9.70			EST	MEYLAN,WM & HOWARD,PH (1995)
VP 1.92E-008	mm Hg	25	EXT	OMAR,MM (1967)
DC	pKa			
HL 3.82E-002	atm m3/mol	25	EST	MEYLAN,WM & HOWARD,PH (1991)
OH 2.59E-011	cm3/molc sec	25	EST	MEYLAN,WM & HOWARD,PH (1993)

CAS #: 018281-07-7 — ETHYL TRICOSANOATE

Formula: $C_{25}H_{50}O_2$

Mol Weight: 382.68

MP (deg C):

FP (deg C):

BP (deg C):

BP pressure (mm Hg):

Property/Value	Units	Temp	Data Type	Reference
WS 1.12E-006	mg/L	25	EST	MEYLAN,WM ET AL. (1996)
logP 11.18			EST	MEYLAN,WM & HOWARD,PH (1995)
VP 2.25E-007	mm Hg	25	EST	NEELY,WB & BLAU,GE (1985)
DC	pKa			
HL 8.93E-002	atm m3/mol	25	EST	MEYLAN,WM & HOWARD,PH (1991)
OH 3.02E-011	cm3/molc sec	25	EST	MEYLAN,WM & HOWARD,PH (1993)

CAS #: 018315-50-9 — N-ME-3,4-DICHLOROPHENYLCARBAMATE

Formula: $C_8H_7Cl_2NO_2$

Mol Weight: 220.06

MP (deg C):

FP (deg C):

BP (deg C):

BP pressure (mm Hg):

Property/Value	Units	Temp	Data Type	Reference
WS 1.40E+002	mg/L	25	EST	MEYLAN,WM ET AL. (1996)
logP 2.80			EXP	HANSCH,C & LEO,AJ (1985)
VP 1.39E-003	mm Hg	25	EST	NEELY,WB & BLAU,GE (1985)
DC	pKa			
HL 1.76E-008	atm m3/mol	25	EST	MEYLAN,WM & HOWARD,PH (1991)
OH 6.47E-012	cm3/molc sec	25	EST	MEYLAN,WM & HOWARD,PH (1993)

CAS #: 018315-52-1 — N-ME-4-CYANOPHENYLCARBAMATE

Formula: $C_9H_8N_2O_2$

Mol Weight: 176.18

MP (deg C): FP (deg C):

BP (deg C):

BP pressure (mm Hg):

Property/ Value	Units	Temp	Data Type	Reference
WS 4.82E+003	mg/L	25	EST	MEYLAN,WM ET AL. (1996)
logP 0.95			EXP	HANSCH,C & LEO,AJ (1985)
VP 6.30E-004	mm Hg	25	EST	NEELY,WB & BLAU,GE (1985)
DC	pKa			
HL 3.10E-010	atm m3/mol	25	EST	MEYLAN,WM & HOWARD,PH (1991)
OH 6.41E-012	cm3/molc sec	25	EST	MEYLAN,WM & HOWARD,PH (1993)

CAS #: 018315-62-3 — N-ME-2,5-DICHLOROPHENYLCARBAMATE

Formula: $C_8H_7Cl_2NO_2$

Mol Weight: 220.06

MP (deg C): FP (deg C):

BP (deg C):

BP pressure (mm Hg):

Property/ Value	Units	Temp	Data Type	Reference
WS 2.84E+002	mg/L	25	EST	MEYLAN,WM ET AL. (1996)
logP 2.44			EXP	HANSCH,C & LEO,AJ (1985)
VP 1.39E-003	mm Hg	25	EST	NEELY,WB & BLAU,GE (1985)
DC	pKa			
HL 1.76E-008	atm m3/mol	25	EST	MEYLAN,WM & HOWARD,PH (1991)
OH 6.47E-012	cm3/molc sec	25	EST	MEYLAN,WM & HOWARD,PH (1993)

CAS #: 018323-44-9 — CLINDAMYCIN

Formula: $C_{18}H_{33}ClN_2O_5S$

Mol Weight: 424.99

MP (deg C): FP (deg C):

BP (deg C):

BP pressure (mm Hg):

Property/ Value	Units	Temp	Data Type	Reference
WS 3.06E+001	mg/L	25	EST	MEYLAN,WM ET AL. (1996)
logP 2.16			EXP	HANSCH,C ET AL. (1995)
VP 5.28E-017	mm Hg	25	EST	NEELY,WB & BLAU,GE (1985)
DC	pKa			
HL 2.89E-022	atm m3/mol	25	EST	MEYLAN,WM & HOWARD,PH (1991)
OH 2.36E-010	cm3/molc sec	25	EST	MEYLAN,WM & HOWARD,PH (1993)

CAS #: 018362-64-6 — 3,5-HEPTANEDIONE, 2,6-DIMETHYL-

Formula: $C_9H_{16}O_2$

Mol Weight: 156.23

MP (deg C): FP (deg C):

BP (deg C):

BP pressure (mm Hg):

Property/ Value	Units	Temp	Data Type	Reference
WS 9.05E+002	mg/L	25	EST	MEYLAN,WM ET AL. (1996)
logP 2.22			EXP	HANSCH,C & LEO,AJ (1985)
VP 5.23E-001	mm Hg	25	EST	NEELY,WB & BLAU,GE (1985)
DC	pKa			
HL 1.04E-007	atm m3/mol	25	EST	MEYLAN,WM & HOWARD,PH (1991)
OH 5.56E-012	cm3/molc sec	25	EST	MEYLAN,WM & HOWARD,PH (1993)

CAS #: 018371-12-5 — N-(4-((2,4-DIAMINOPHENYL)AZO)PHENYL)ACETAMIDE

Formula: $C_{14}H_{15}N_5O$

Mol Weight: 269.31

MP (deg C): FP (deg C):

BP (deg C):

BP pressure (mm Hg):

Property/ Value	Units	Temp	Data Type	Reference
WS 6.03E+002	mg/L	25	EST	MEYLAN,WM ET AL. (1996)
logP 1.23			EST	MEYLAN,WM & HOWARD,PH (1995)
VP 1.91E-010	mm Hg	25	EST	NEELY,WB & BLAU,GE (1985)
DC	pKa			
HL 2.10E-018	atm m3/mol	25	EST	MEYLAN,WM & HOWARD,PH (1991)
OH 2.00E-010	cm3/molc sec	25	EST	MEYLAN,WM & HOWARD,PH (1993)

CAS #: 018377-52-1 — O-NITRO-N-PHENYLSUCCINIMIDE

Formula: $C_{10}H_8N_2O_4$

Mol Weight: 220.19

MP (deg C): FP (deg C):

BP (deg C):

BP pressure (mm Hg):

Property/ Value	Units	Temp	Data Type	Reference
WS 6.89E+003	mg/L	25	EST	MEYLAN,WM ET AL. (1996)
logP 0.36			EXP	HANSCH,C & LEO,AJ (1985)
VP 1.72E-008	mm Hg	25	EST	NEELY,WB & BLAU,GE (1985)
DC	pKa			
HL 1.13E-008	atm m3/mol	25	EST	MEYLAN,WM & HOWARD,PH (1991)
OH 5.92E-012	cm3/molc sec	25	EST	MEYLAN,WM & HOWARD,PH (1993)

CAS #: 018381-45-8 — BUTANENITRILE, 4,4-DIETHOXY-

Formula: $C_8H_{15}NO_2$

Mol Weight: 157.21

MP (deg C): FP (deg C):

BP (deg C): 104-106

BP pressure (mm Hg): 1.00E+001

Property/ Value	Units	Temp	Data Type	Reference
WS 8.76E+003	mg/L	25	EST	MEYLAN,WM ET AL. (1996)
logP 0.75			EXP	HANSCH,C ET AL. (1995)
VP 6.37E-002	mm Hg	25	EST	NEELY,WB & BLAU,GE (1985)
DC	pKa			
HL 6.52E-008	atm m3/mol	25	EST	MEYLAN,WM & HOWARD,PH (1991)
OH 2.14E-011	cm3/molc sec	25	EST	MEYLAN,WM & HOWARD,PH (1993)

CAS #: 018404-43-8 — CARD-20(22)-ENOLIDE, 3-[(2,6-DIDEOXY-beta-D-RIBO

Formula: $C_{30}H_{46}O_8$

Mol Weight: 534.70

MP (deg C): FP (deg C):

BP (deg C):

BP pressure (mm Hg):

Property/ Value	Units	Temp	Data Type	Reference
WS 3.14E+000	mg/L	25	EST	MEYLAN,WM ET AL. (1996)
logP 2.50			EXP	SANGSTER,J (1993)
VP 1.24E-020	mm Hg	25	EST	NEELY,WB & BLAU,GE (1985)
DC	pKa			
HL 1.31E-016	atm m3/mol	25	EST	MEYLAN,WM & HOWARD,PH (1991)
OH 1.03E-010	cm3/molc sec	25	EST	MEYLAN,WM & HOWARD,PH (1993)

CAS #: 018407-16-4

1,1,1,3,3,5,5-HEPTAMETHYL-5-PHENYLTRISILOXANE

Formula: $C_{13}H_{26}O_2Si_3$

Mol Weight: 298.61

MP (deg C):

FP (deg C):

BP (deg C):

BP pressure (mm Hg):

Property/Value	Units	Temp	Data Type	Reference
WS 2.08E-003	mg/L	25	EST	MEYLAN,WM ET AL. (1996)
logP 6.56			EST	MEYLAN,WM & HOWARD,PH (1995)
VP	mm Hg			
DC	pKa			
HL	atm m3/mol			
OH 3.00E-012	cm3/molc sec	25	EST	MEYLAN,WM & HOWARD,PH (1993)

CAS #: 018429-70-4

BENZ(A)ANTHRACENE-4,5-DIMETHYLENE

Formula: $C_{20}H_{16}$

Mol Weight: 256.35

MP (deg C):

FP (deg C):

BP (deg C):

BP pressure (mm Hg):

Property/Value	Units	Temp	Data Type	Reference
WS 2.72E-003	mg/L	25	EXP	PEARLMAN,RS ET AL. (1984)
logP 6.62			EST	MEYLAN,WM & HOWARD,PH (1995)
VP 1.14E-007	mm Hg	25	EST	NEELY,WB & BLAU,GE (1985)
DC	pKa			
HL 6.10E-006	atm m3/mol	25	EST	MEYLAN,WM & HOWARD,PH (1991)
OH 1.60E-010	cm3/molc sec	25	EST	MEYLAN,WM & HOWARD,PH (1993)

CAS #: 018435-45-5

1-NONADECENE

Formula: $C_{19}H_{38}$

Mol Weight: 266.51

MP (deg C): 23

FP (deg C):

BP (deg C):

BP pressure (mm Hg):

Property/Value	Units	Temp	Data Type	Reference
WS 3.99E-005	mg/L	25	EST	MEYLAN,WM ET AL. (1996)
logP 9.54			EST	MEYLAN,WM & HOWARD,PH (1995)
VP 3.81E-005	mm Hg	25	EXP	YAWS,CL (1994B)
DC	pKa			
HL 1.43E+001	atm m3/mol	25	EST	MEYLAN,WM & HOWARD,PH (1991)
OH 4.85E-011	cm3/molc sec	25	EST	MEYLAN,WM & HOWARD,PH (1993)

CAS #: 018438-38-5

2-METHYLTHIOPYRIDINE

Formula: C_6H_7NS

Mol Weight: 125.19

MP (deg C):

FP (deg C):

BP (deg C):

BP pressure (mm Hg):

Property/Value	Units	Temp	Data Type	Reference
WS 3.33E+003	mg/L	25	EST	MEYLAN,WM ET AL. (1996)
logP 1.71			EXP	YAMAGAMI,C ET AL. (1990A)
VP 4.08E-001	mm Hg	25	EST	NEELY,WB & BLAU,GE (1985)
DC 3.59	pKa	20	EXP	PERRIN,DD (1965)
HL 8.00E-006	atm m3/mol	25	EST	MEYLAN,WM & HOWARD,PH (1991)
OH 3.45E-012	cm3/molc sec	25	EST	MEYLAN,WM & HOWARD,PH (1993)

CAS #: 018453-07-1

2-THIAZOLEETHANAMINE

Formula: $C_5H_8N_2S$

Mol Weight: 128.20

MP (deg C):

FP (deg C):

BP (deg C):

BP pressure (mm Hg):

Property/Value	Units	Temp	Data Type	Reference
WS 6.95E+005	mg/L	25	EST	MEYLAN,WM ET AL. (1996)
logP 0.16			EXP	YOUNG,RC ET AL. (1993)
VP 7.01E-002	mm Hg	25	EST	NEELY,WB & BLAU,GE (1985)
DC	pKa			
HL 5.77E-010	atm m3/mol	25	EST	MEYLAN,WM & HOWARD,PH (1991)
OH 3.63E-011	cm3/molc sec	25	EST	MEYLAN,WM & HOWARD,PH (1993)

CAS #: 018467-77-1

.ALPHA.-L-XYLO-2-HEXULOFURANOSONIC ACID, 2,3:4,6

Formula: $C_{12}H_{18}O_7$

Mol Weight: 274.27

MP (deg C): > 300

FP (deg C):

BP (deg C):

BP pressure (mm Hg):

Property/Value	Units	Temp	Data Type	Reference
WS 3.02E+003	mg/L	25	EST	MEYLAN,WM ET AL. (1996)
logP 1.35			EXP	TOMLIN,C (1994)
VP 2.82E-006	mm Hg	25	EST	NEELY,WB & BLAU,GE (1985)
DC	pKa			
HL 1.90E-016	atm m3/mol	25	EST	MEYLAN,WM & HOWARD,PH (1991)
OH 6.24E-011	cm3/molc sec	25	EST	MEYLAN,WM & HOWARD,PH (1993)

CAS #: 018489-25-3

1H-TETRAZOLE, 5-(PHENYLMETHYL)-

Formula: $C_8H_8N_4$

Mol Weight: 160.18

MP (deg C):

FP (deg C):

BP (deg C):

BP pressure (mm Hg):

Property/Value	Units	Temp	Data Type	Reference
WS 4.71E+003	mg/L	25	EST	MEYLAN,WM ET AL. (1996)
logP 1.36			EXP	HANSCH,C ET AL. (1995)
VP 3.55E-005	mm Hg	25	EST	NEELY,WB & BLAU,GE (1985)
DC	pKa			
HL 5.45E-008	atm m3/mol	25	EST	MEYLAN,WM & HOWARD,PH (1991)
OH 6.03E-012	cm3/molc sec	25	EST	MEYLAN,WM & HOWARD,PH (1993)

CAS #: 018495-00-6

N-PHENOXYACETYLMORPHOLINE

Formula: $C_{12}H_{15}NO_3$

Mol Weight: 221.26

MP (deg C):

FP (deg C):

BP (deg C):

BP pressure (mm Hg):

Property/Value	Units	Temp	Data Type	Reference
WS 7.91E+003	mg/L	25	EST	MEYLAN,WM ET AL. (1996)
logP 0.74			EXP	HANSCH,C & LEO,AJ (1985)
VP 2.32E-005	mm Hg	25	EST	NEELY,WB & BLAU,GE (1985)
DC	pKa			
HL 1.13E-011	atm m3/mol	25	EST	MEYLAN,WM & HOWARD,PH (1991)
OH 8.16E-011	cm3/molc sec	25	EST	MEYLAN,WM & HOWARD,PH (1993)

CAS #: 018495-30-2				1,1,2,3-TETRACHLOROPROPANE

Formula: $C_3H_4Cl_4$

Mol Weight: 181.88

MP (deg C): | FP (deg C):

BP (deg C):

BP pressure (mm Hg):

Property/Value	Units	Temp	Data Type	Reference
WS 2.75E+002	mg/L	25	EST	MEYLAN,WM ET AL. (1996)
logP 2.68			EST	MEYLAN,WM & HOWARD,PH (1995)
VP 8.37E-001	mm Hg	25	EST	NEELY,WB & BLAU,GE (1985)
DC	pKa			
HL 2.00E-003	atm m3/mol	25	EST	MEYLAN,WM & HOWARD,PH (1991)
OH 3.24E-013	cm3/molc sec	25	EST	MEYLAN,WM & HOWARD,PH (1993)

CAS #: 018514-84-6				CINNOLINE-4-ONE

Formula: $C_8H_6N_2O$

Mol Weight: 146.15

MP (deg C): | FP (deg C):

BP (deg C):

BP pressure (mm Hg):

Property/Value	Units	Temp	Data Type	Reference
WS 1.57E+004	mg/L	25	EST	MEYLAN,WM ET AL. (1996)
logP 0.82			EXP	HANSCH,C & LEO,AJ (1985)
VP 4.75E-005	mm Hg	25	EST	NEELY,WB & BLAU,GE (1985)
DC	pKa			
HL 6.14E-008	atm m3/mol	25	EST	MEYLAN,WM & HOWARD,PH (1991)
OH 4.03E-011	cm3/molc sec	25	EST	MEYLAN,WM & HOWARD,PH (1993)

CAS #: 018530-56-8				NOREA

Formula: $C_{13}H_{22}N_2O$

Mol Weight: 222.33

MP (deg C): 177 | FP (deg C):

BP (deg C):

BP pressure (mm Hg):

Property/Value	Units	Temp	Data Type	Reference
WS 1.50E+002	mg/L	25	EXP	YALKOWSKY,SH & DANNENFELSER,RM (1992)
logP 2.62			EST	MEYLAN,WM & HOWARD,PH (1995)
VP 1.71E-005	mm Hg	25	EST	NEELY,WB & BLAU,GE (1985)
DC	pKa			
HL 4.24E-009	atm m3/mol	25	EST	MEYLAN,WM & HOWARD,PH (1991)
OH 2.03E-011	cm3/molc sec	25	EST	MEYLAN,WM & HOWARD,PH (1993)

CAS #: 018584-93-5				CARBAMIC ACID, [3-(TRIFLUOROMETHYL)PHENYL]-, MET

Formula: $C_9H_8F_3NO_2$

Mol Weight: 219.16

MP (deg C): | FP (deg C):

BP (deg C):

BP pressure (mm Hg):

Property/Value	Units	Temp	Data Type	Reference
WS 1.59E+002	mg/L	25	EST	MEYLAN,WM ET AL. (1996)
logP 2.74			EXP	TAKAHASHI,J ET AL. (1988)
VP 3.69E-002	mm Hg	25	EST	NEELY,WB & BLAU,GE (1985)
DC	pKa			
HL 1.90E-007	atm m3/mol	25	EST	MEYLAN,WM & HOWARD,PH (1991)
OH 6.67E-012	cm3/molc sec	25	EST	MEYLAN,WM & HOWARD,PH (1993)

CAS #: 018585-38-1				1,2,3,4,5,6-HEXACHLOROHEXANE

Formula: $C_6H_8Cl_6$

Mol Weight: 292.85

MP (deg C): | FP (deg C):

BP (deg C):

BP pressure (mm Hg):

Property/Value	Units	Temp	Data Type	Reference
WS 1.87E+000	mg/L	25	EST	MEYLAN,WM ET AL. (1996)
logP 4.52			EST	MEYLAN,WM & HOWARD,PH (1995)
VP 1.98E-003	mm Hg	25	EST	NEELY,WB & BLAU,GE (1985)
DC	pKa			
HL 5.80E-004	atm m3/mol	25	EST	MEYLAN,WM & HOWARD,PH (1991)
OH 6.38E-013	cm3/molc sec	25	EST	MEYLAN,WM & HOWARD,PH (1993)

CAS #: 018588-57-3				2,4-PYRIMIDINEDIAMINE, 5-(3,4-DICHLOROPHENYL)-6-

Formula: $C_{12}H_{12}Cl_2N_4$

Mol Weight: 283.16

MP (deg C): | FP (deg C):

BP (deg C):

BP pressure (mm Hg):

Property/Value	Units	Temp	Data Type	Reference
WS 3.97E+001	mg/L	25	EST	MEYLAN,WM ET AL. (1996)
logP 3.03			EXP	SANGSTER,J (1993)
VP 1.71E-008	mm Hg	25	EST	NEELY,WB & BLAU,GE (1985)
DC 7.20	pKa	20	EXP	PERRIN,DD (1972)
HL 7.99E-011	atm m3/mol	25	EST	MEYLAN,WM & HOWARD,PH (1991)
OH 2.01E-010	cm3/molc sec	25	EST	MEYLAN,WM & HOWARD,PH (1993)

CAS #: 018591-79-2				ACETAMIDE, N-(4-HYDROXY-6-METHYL-3-PYRIDAZINYL)-

Formula: $C_7H_9N_3O_2$

Mol Weight: 167.17

MP (deg C): | FP (deg C):

BP (deg C):

BP pressure (mm Hg):

Property/Value	Units	Temp	Data Type	Reference
WS 1.79E+004	mg/L	25	EST	MEYLAN,WM ET AL. (1996)
logP -0.89			EXP	SANGSTER,J (1994)
VP 5.53E-008	mm Hg	25	EST	NEELY,WB & BLAU,GE (1985)
DC	pKa			
HL 3.77E-016	atm m3/mol	25	EST	MEYLAN,WM & HOWARD,PH (1991)
OH 5.75E-012	cm3/molc sec	25	EST	MEYLAN,WM & HOWARD,PH (1993)

CAS #: 018591-86-1				4-PYRIDAZINOL, 3-AMINO-6-METHYL-

Formula: $C_5H_7N_3O$

Mol Weight: 125.13

MP (deg C): | FP (deg C):

BP (deg C):

BP pressure (mm Hg):

Property/Value	Units	Temp	Data Type	Reference
WS 5.11E+005	mg/L	25	EST	MEYLAN,WM ET AL. (1996)
logP -0.85			EXP	SANGSTER,J (1994)
VP 1.99E-005	mm Hg	25	EST	NEELY,WB & BLAU,GE (1985)
DC	pKa			
HL 1.16E-013	atm m3/mol	25	EST	MEYLAN,WM & HOWARD,PH (1991)
OH 1.71E-011	cm3/molc sec	25	EST	MEYLAN,WM & HOWARD,PH (1993)

CAS #: 018597-53-0
PYRAZOL-3-ONE,2H1(4,6-ME2PYRIMID)5ME

Formula: $C_{10}H_{12}N_4O$

Mol Weight: 204.23

MP (deg C):   **FP (deg C):**

BP (deg C):

BP pressure (mm Hg):

Property/Value	Units	Temp	Data Type	Reference
WS 3.29E+003	mg/L	25	EST	MEYLAN,WM ET AL. (1996)
logP 0.72			EXP	HANSCH,C & LEO,AJ (1985)
VP 4.83E-007	mm Hg	25	EST	NEELY,WB & BLAU,GE (1985)
DC	pKa			
HL 7.09E-010	atm m3/mol	25	EST	MEYLAN,WM & HOWARD,PH (1991)
OH 3.41E-011	cm3/molc sec	25	EST	MEYLAN,WM & HOWARD,PH (1993)

CAS #: 018597-55-2
PYRAZOL-3-ONE,2(2ME-6MEO PYRIMID)5ME

Formula: $C_{10}H_{12}N_4O_2$

Mol Weight: 220.23

MP (deg C):   **FP (deg C):**

BP (deg C):

BP pressure (mm Hg):

Property/Value	Units	Temp	Data Type	Reference
WS 8.33E+003	mg/L	25	EST	MEYLAN,WM ET AL. (1996)
logP 0.72			EXP	HANSCH,C & LEO,AJ (1985)
VP 6.27E-007	mm Hg	25	EST	NEELY,WB & BLAU,GE (1985)
DC	pKa			
HL 9.77E-014	atm m3/mol	25	EST	MEYLAN,WM & HOWARD,PH (1991)
OH 1.25E-010	cm3/molc sec	25	EST	MEYLAN,WM & HOWARD,PH (1993)

CAS #: 018597-57-4
PYRAZOL-3-ONE,2(4,6-ME4PYRIMID)5ME

Formula: $C_{10}H_{12}N_4O$

Mol Weight: 204.23

MP (deg C):   **FP (deg C):**

BP (deg C):

BP pressure (mm Hg):

Property/Value	Units	Temp	Data Type	Reference
WS 7.23E+003	mg/L	25	EST	MEYLAN,WM ET AL. (1996)
logP 0.89			EXP	HANSCH,C & LEO,AJ (1985)
VP 1.42E-006	mm Hg	25	EST	NEELY,WB & BLAU,GE (1985)
DC	pKa			
HL 2.00E-013	atm m3/mol	25	EST	MEYLAN,WM & HOWARD,PH (1991)
OH 9.16E-011	cm3/molc sec	25	EST	MEYLAN,WM & HOWARD,PH (1993)

CAS #: 018604-02-9
2,2,4,4,6,8-HEXAMETHYL-6,8-DIPHENYLCYCLOTETRAS*

Formula: $C_{18}H_{28}O_4Si_4$

Mol Weight: 420.77

MP (deg C):   **FP (deg C):**

BP (deg C):

BP pressure (mm Hg):

Property/Value	Units	Temp	Data Type	Reference
WS 8.30E-005	mg/L	25	EST	MEYLAN,WM ET AL. (1996)
logP 7.52			EST	MEYLAN,WM & HOWARD,PH (1995)
VP	mm Hg			
DC	pKa			
HL	atm m3/mol			
OH 4.80E-012	cm3/molc sec	25	EST	MEYLAN,WM & HOWARD,PH (1993)

CAS #: 018608-30-5
PERCHLOROALLENE

Formula: C_3Cl_4

Mol Weight: 177.85

MP (deg C):   **FP (deg C):**

BP (deg C):

BP pressure (mm Hg):

Property/Value	Units	Temp	Data Type	Reference
WS 7.78E+001	mg/L	25	EST	MEYLAN,WM ET AL. (1996)
logP 3.35			EST	MEYLAN,WM & HOWARD,PH (1995)
VP 8.28E+000	mm Hg	25	EST	NEELY,WB & BLAU,GE (1985)
DC	pKa			
HL 1.65E-002	atm m3/mol	25	EST	MEYLAN,WM & HOWARD,PH (1991)
OH 2.14E-013	cm3/molc sec	25	EST	MEYLAN,WM & HOWARD,PH (1993)

CAS #: 018659-24-0
N-ME-3-ME-6-I-PR PHENYLCARBAMATE

Formula: $C_{12}H_{17}NO_2$

Mol Weight: 207.27

MP (deg C):   **FP (deg C):**

BP (deg C):

BP pressure (mm Hg):

Property/Value	Units	Temp	Data Type	Reference
WS 1.51E+002	mg/L	25	EST	MEYLAN,WM ET AL. (1996)
logP 2.84			EXP	HANSCH,C & LEO,AJ (1985)
VP 1.32E-003	mm Hg	25	EST	NEELY,WB & BLAU,GE (1985)
DC	pKa			
HL 6.89E-008	atm m3/mol	25	EST	MEYLAN,WM & HOWARD,PH (1991)
OH 1.48E-011	cm3/molc sec	25	EST	MEYLAN,WM & HOWARD,PH (1993)

CAS #: 018659-33-1
N-ME-3-ME-5-T-BU PHENYLCARBAMATE

Formula: $C_{13}H_{19}NO_2$

Mol Weight: 221.30

MP (deg C):   **FP (deg C):**

BP (deg C):

BP pressure (mm Hg):

Property/Value	Units	Temp	Data Type	Reference
WS 4.67E+001	mg/L	25	EST	MEYLAN,WM ET AL. (1996)
logP 3.35			EXP	HANSCH,C & LEO,AJ (1985)
VP 6.74E-004	mm Hg	25	EST	NEELY,WB & BLAU,GE (1985)
DC	pKa			
HL 9.15E-008	atm m3/mol	25	EST	MEYLAN,WM & HOWARD,PH (1991)
OH 1.83E-011	cm3/molc sec	25	EST	MEYLAN,WM & HOWARD,PH (1993)

CAS #: 018690-90-9
14NAPHTHOQUINONE,2ANILINO-3SULFONATE K SALT

Formula: $C_{16}H_{10}KNO_5S$

Mol Weight: 367.43

MP (deg C):   **FP (deg C):**

BP (deg C):

BP pressure (mm Hg):

Property/Value	Units	Temp	Data Type	Reference
WS 3.40E+004	mg/L	25	EST	MEYLAN,WM ET AL. (1996)
logP -0.99			EXP	HANSCH,C & LEO,AJ (1985)
VP 3.94E-018	mm Hg	25	EST	NEELY,WB & BLAU,GE (1985)
DC	pKa			
HL	atm m3/mol			
OH 5.15E-011	cm3/molc sec	25	EST	MEYLAN,WM & HOWARD,PH (1993)

CAS #: 018690-91-0 — 1,4-NAPHTHALENEDIONE, 2-(PHENYLAMINO)-3-(PHENYLS

Formula: $C_{22}H_{15}NO_4S$

Mol Weight: 389.43

MP (deg C):
FP (deg C):
BP (deg C):
BP pressure (mm Hg):

Property/Value	Units	Temp	Data Type	Reference
WS 4.62E+000	mg/L	25	EST	MEYLAN,WM ET AL. (1996)
logP 3.38			EXP	SANGSTER,J (1993)
VP 1.13E-012	mm Hg	25	EST	NEELY,WB & BLAU,GE (1985)
DC	pKa			
HL 2.64E-016	atm m3/mol	25	EST	MEYLAN,WM & HOWARD,PH (1991)
OH 5.19E-011	cm3/molc sec	25	EST	MEYLAN,WM & HOWARD,PH (1993)

CAS #: 018694-40-1 — PYRIMIDINE,2(5MEO-3ME-1PYRAZ)4MEO6ME

Formula: $C_{11}H_{16}N_4O_2$

Mol Weight: 236.28

MP (deg C): 90-92
FP (deg C):
BP (deg C):
BP pressure (mm Hg):

Property/Value	Units	Temp	Data Type	Reference
WS 2.42E+003	mg/L	25	EST	MEYLAN,WM ET AL. (1996)
logP 1.26			EXP	HANSCH,C & LEO,AJ (1985)
VP 6.09E-006	mm Hg	25	EST	NEELY,WB & BLAU,GE (1985)
DC	pKa			
HL 2.83E-013	atm m3/mol	25	EST	MEYLAN,WM & HOWARD,PH (1991)
OH 2.02E-010	cm3/molc sec	25	EST	MEYLAN,WM & HOWARD,PH (1993)

CAS #: 018694-41-2 — PYRIMIDINE,4(5MEO-3ME-1-PYRAZOL)

Formula: $C_9H_{12}N_4O$

Mol Weight: 192.22

MP (deg C):
FP (deg C):
BP (deg C):
BP pressure (mm Hg):

Property/Value	Units	Temp	Data Type	Reference
WS 1.84E+004	mg/L	25	EST	MEYLAN,WM ET AL. (1996)
logP 0.50			EXP	HANSCH,C & LEO,AJ (1985)
VP 1.02E-004	mm Hg	25	EST	NEELY,WB & BLAU,GE (1985)
DC	pKa			
HL 4.75E-013	atm m3/mol	25	EST	MEYLAN,WM & HOWARD,PH (1991)
OH 2.01E-010	cm3/molc sec	25	EST	MEYLAN,WM & HOWARD,PH (1993)

CAS #: 018694-42-3 — PYRIMIDINE,4(5MEO-3ME-1-PYRAZOL)6MEO

Formula: $C_{10}H_{14}N_4O_2$

Mol Weight: 222.25

MP (deg C):
FP (deg C):
BP (deg C):
BP pressure (mm Hg):

Property/Value	Units	Temp	Data Type	Reference
WS 1.69E+003	mg/L	25	EST	MEYLAN,WM ET AL. (1996)
logP 1.53			EXP	HANSCH,C & LEO,AJ (1985)
VP 1.49E-005	mm Hg	25	EST	NEELY,WB & BLAU,GE (1985)
DC	pKa			
HL 2.56E-013	atm m3/mol	25	EST	MEYLAN,WM & HOWARD,PH (1991)
OH 2.02E-010	cm3/molc sec	25	EST	MEYLAN,WM & HOWARD,PH (1993)

CAS #: 018694-43-4 — PYRIMIDINE,2(5MEO-3ME-1PYRAZ)4,6DIME

Formula: $C_{11}H_{16}N_4O$

Mol Weight: 220.28

MP (deg C):
FP (deg C):
BP (deg C):
BP pressure (mm Hg):

Property/Value	Units	Temp	Data Type	Reference
WS 8.37E+003	mg/L	25	EST	MEYLAN,WM ET AL. (1996)
logP 0.73			EXP	HANSCH,C & LEO,AJ (1985)
VP 1.53E-005	mm Hg	25	EST	NEELY,WB & BLAU,GE (1985)
DC	pKa			
HL 5.78E-013	atm m3/mol	25	EST	MEYLAN,WM & HOWARD,PH (1991)
OH 2.01E-010	cm3/molc sec	25	EST	MEYLAN,WM & HOWARD,PH (1993)

CAS #: 018695-02-8 — HELVETICOSOL

Formula: $C_{29}H_{44}O_9$

Mol Weight: 536.67

MP (deg C):
FP (deg C):
BP (deg C):
BP pressure (mm Hg):

Property/Value	Units	Temp	Data Type	Reference
WS 3.10E+002	mg/L	25	EST	MEYLAN,WM ET AL. (1996)
logP 0.15			EXP	HANSCH,C & LEO,AJ (1985)
VP 2.41E-021	mm Hg	25	EST	NEELY,WB & BLAU,GE (1985)
DC	pKa			
HL 3.07E-018	atm m3/mol	25	EST	MEYLAN,WM & HOWARD,PH (1991)
OH 9.74E-011	cm3/molc sec	25	EST	MEYLAN,WM & HOWARD,PH (1993)

CAS #: 018697-50-2 — PYRAZOL-3-ONE,1(4,6-ME2PYRIMID)5ME

Formula: $C_{10}H_{12}N_4O$

Mol Weight: 204.23

MP (deg C):
FP (deg C):
BP (deg C):
BP pressure (mm Hg):

Property/Value	Units	Temp	Data Type	Reference
WS 3.29E+003	mg/L	25	EST	MEYLAN,WM ET AL. (1996)
logP 1.29			EXP	HANSCH,C & LEO,AJ (1985)
VP 4.83E-007	mm Hg	25	EST	NEELY,WB & BLAU,GE (1985)
DC	pKa			
HL 7.09E-010	atm m3/mol	25	EST	MEYLAN,WM & HOWARD,PH (1991)
OH 3.41E-011	cm3/molc sec	25	EST	MEYLAN,WM & HOWARD,PH (1993)

CAS #: 018697-64-8 — 3-PYRAZOLIN-5-ONE, 3-METHYL-2-(4-PYRIMIDINYL)-

Formula: $C_8H_{10}N_4O$

Mol Weight: 178.20

MP (deg C):
FP (deg C):
BP (deg C):
BP pressure (mm Hg):

Property/Value	Units	Temp	Data Type	Reference
WS 3.66E+003	mg/L	25	EST	MEYLAN,WM ET AL. (1996)
logP 1.40			EXP	HANSCH,C & LEO,AJ (1985)
VP 9.36E-006	mm Hg	25	EST	NEELY,WB & BLAU,GE (1985)
DC	pKa			
HL 8.35E-016	atm m3/mol	25	EST	MEYLAN,WM & HOWARD,PH (1991)
OH 2.00E-010	cm3/molc sec	25	EST	MEYLAN,WM & HOWARD,PH (1993)

CAS #: 018698-96-9				BENZENEACETIC ACID, 2-IODO-

Formula: $C_8H_7IO_2$

Mol Weight: 262.05

MP (deg C): FP (deg C):

BP (deg C):

BP pressure (mm Hg):

Property/ Value	Units	Temp	Data Type	Reference
WS 5.15E+002	mg/L	25	EST	MEYLAN,WM ET AL. (1996)
logP 2.33			EXP	LAZNICEK,M & KVETINA,J (1988)
VP 7.89E-005	mm Hg	25	EST	NEELY,WB & BLAU,GE (1985)
DC 4.04	pKa	25	EXP	KORTUM,G ET AL (1961)
HL 1.02E-008	atm m3/mol	25	EST	MEYLAN,WM & HOWARD,PH (1991)
OH 2.55E-012	cm3/molc sec	25	EST	MEYLAN,WM & HOWARD,PH (1993)

CAS #: 018705-01-6				PHENOXYACETANILIDE

Formula: $C_{14}H_{13}NO_2$

Mol Weight: 227.27

MP (deg C): FP (deg C):

BP (deg C):

BP pressure (mm Hg):

Property/ Value	Units	Temp	Data Type	Reference
WS 1.47E+002	mg/L	25	EST	MEYLAN,WM ET AL. (1996)
logP 2.73			EXP	HANSCH,C & LEO,AJ (1985)
VP 3.99E-007	mm Hg	25	EST	NEELY,WB & BLAU,GE (1985)
DC	pKa			
HL 1.89E-010	atm m3/mol	25	EST	MEYLAN,WM & HOWARD,PH (1991)
OH 3.82E-011	cm3/molc sec	25	EST	MEYLAN,WM & HOWARD,PH (1993)

CAS #: 018705-45-8				4,4'-DIISOTHIOCYANATEBIPHENYL

Formula: $C_{14}H_8N_2S_2$

Mol Weight: 268.36

MP (deg C): FP (deg C):

BP (deg C):

BP pressure (mm Hg):

Property/ Value	Units	Temp	Data Type	Reference
WS 3.75E-001	mg/L	25	EST	MEYLAN,WM ET AL. (1996)
logP 5.50			EXP	HANSCH,C & LEO,AJ (1985)
VP 1.24E-007	mm Hg	25	EST	NEELY,WB & BLAU,GE (1985)
DC	pKa			
HL 8.69E-006	atm m3/mol	25	EST	MEYLAN,WM & HOWARD,PH (1991)
OH 7.43E-012	cm3/molc sec	25	EST	MEYLAN,WM & HOWARD,PH (1993)

CAS #: 018708-70-8				BENZENE, 1,3,5-TRICHLORO-2-NITRO-

Formula: $C_6H_2Cl_3NO_2$

Mol Weight: 226.45

MP (deg C): 72 FP (deg C):

BP (deg C):

BP pressure (mm Hg):

Property/ Value	Units	Temp	Data Type	Reference
WS 6.25E+000	mg/L	25	EXP	CHEM INSPECT TEST INST (1992)
logP 3.69			EXP	NIIMI,AJ ET AL. (1989)
VP 7.92E-004	mm Hg	25	EST	NEELY,WB & BLAU,GE (1985)
DC	pKa			
HL 8.65E-006	atm m3/mol	25	EST	MEYLAN,WM & HOWARD,PH (1991)
OH 8.48E-014	cm3/molc sec	25	EST	MEYLAN,WM & HOWARD,PH (1993)

CAS #: 018740-23-3				THIENO[2,3-D]PYRIMIDINE, 4-MORPHOLINYL-

Formula: $C_{10}H_{11}N_3OS$

Mol Weight: 221.28

MP (deg C): FP (deg C):

BP (deg C):

BP pressure (mm Hg):

Property/ Value	Units	Temp	Data Type	Reference
WS 1.32E+003	mg/L	25	EST	MEYLAN,WM ET AL. (1996)
logP 1.65			EXP	HANSCH,C ET AL. (1995)
VP 5.71E-006	mm Hg	25	EST	NEELY,WB & BLAU,GE (1985)
DC	pKa			
HL 1.69E-011	atm m3/mol	25	EST	MEYLAN,WM & HOWARD,PH (1991)
OH 2.75E-010	cm3/molc sec	25	EST	MEYLAN,WM & HOWARD,PH (1993)

CAS #: 018774-47-5				2-BENZOTHIOPHENAMINE,3-CYANO

Formula: $C_9H_6N_2S$

Mol Weight: 174.23

MP (deg C): FP (deg C):

BP (deg C):

BP pressure (mm Hg):

Property/ Value	Units	Temp	Data Type	Reference
WS 4.47E+002	mg/L	25	EST	MEYLAN,WM ET AL. (1996)
logP 2.17			EXP	HANSCH,C & LEO,AJ (1985)
VP 1.11E-005	mm Hg	25	EST	NEELY,WB & BLAU,GE (1985)
DC	pKa			
HL 9.76E-010	atm m3/mol	25	EST	MEYLAN,WM & HOWARD,PH (1991)
OH 4.34E-011	cm3/molc sec	25	EST	MEYLAN,WM & HOWARD,PH (1993)

CAS #: 018791-19-0				1,1,1,2,3,4,4,4-OCTACHLOROBUTANE

Formula: $C_4H_2Cl_8$

Mol Weight: 333.68

MP (deg C): FP (deg C):

BP (deg C):

BP pressure (mm Hg):

Property/ Value	Units	Temp	Data Type	Reference
WS 1.98E-001	mg/L	25	EST	MEYLAN,WM ET AL. (1996)
logP 5.38			EST	MEYLAN,WM & HOWARD,PH (1995)
VP 1.35E-002	mm Hg	25	EST	NEELY,WB & BLAU,GE (1985)
DC	pKa			
HL 4.09E-005	atm m3/mol	25	EST	MEYLAN,WM & HOWARD,PH (1991)
OH 3.66E-014	cm3/molc sec	25	EST	MEYLAN,WM & HOWARD,PH (1993)

CAS #: 018818-44-5				124-BENZTHIADIAZIN-1-O2-3-PHENYL

Formula: $C_{13}H_{10}N_2O_2S$

Mol Weight: 258.30

MP (deg C): FP (deg C):

BP (deg C):

BP pressure (mm Hg):

Property/ Value	Units	Temp	Data Type	Reference
WS 6.42E+002	mg/L	25	EST	MEYLAN,WM ET AL. (1996)
logP 1.78			EXP	HANSCH,C & LEO,AJ (1985)
VP 6.64E-010	mm Hg	25	EST	NEELY,WB & BLAU,GE (1985)
DC	pKa			
HL 5.62E-009	atm m3/mol	25	EST	MEYLAN,WM & HOWARD,PH (1991)
OH 4.75E-012	cm3/molc sec	25	EST	MEYLAN,WM & HOWARD,PH (1993)

CAS #: 018883-05-1				BENZOBICYCLO(222)OCTENE,2-EXO-MEAMINO

Formula: $C_{13}H_{17}N$

Mol Weight: 187.29

MP (deg C): | FP (deg C):

BP (deg C):

BP pressure (mm Hg):

Property/ Value	Units	Temp	Data Type	Reference
WS 3.17E+003	mg/L	25	EST	MEYLAN,WM ET AL. (1996)
logP 2.68			EXP	HANSCH,C & LEO,AJ (1985)
VP 2.23E-003	mm Hg	25	EST	NEELY,WB & BLAU,GE (1985)
DC	pKa			
HL 8.95E-007	atm m3/mol	25	EST	MEYLAN,WM & HOWARD,PH (1991)
OH 1.07E-010	cm3/molc sec	25	EST	MEYLAN,WM & HOWARD,PH (1993)

CAS #: 018883-06-2				BENZOBICYCLO(222)OCTENE,2ENDO-MEAMINO

Formula: $C_{13}H_{17}N$

Mol Weight: 187.29

MP (deg C): | FP (deg C):

BP (deg C):

BP pressure (mm Hg):

Property/ Value	Units	Temp	Data Type	Reference
WS 3.17E+003	mg/L	25	EST	MEYLAN,WM ET AL. (1996)
logP 2.59			EXP	HANSCH,C & LEO,AJ (1985)
VP 2.23E-003	mm Hg	25	EST	NEELY,WB & BLAU,GE (1985)
DC	pKa			
HL 8.95E-007	atm m3/mol	25	EST	MEYLAN,WM & HOWARD,PH (1991)
OH 1.07E-010	cm3/molc sec	25	EST	MEYLAN,WM & HOWARD,PH (1993)

CAS #: 018883-66-4				STREPTOZOCIN

Formula: $C_8H_{15}N_3O_7$

Mol Weight: 265.22

MP (deg C): 115 | FP (deg C):

BP (deg C):

BP pressure (mm Hg):

Property/ Value	Units	Temp	Data Type	Reference
WS 5.07E+003	mg/L	25	EST	MEYLAN,WM ET AL. (1996)
logP -1.45			EXP	HANSCH,C & LEO,AJ (1985)
VP 1.74E-012	mm Hg	25	EST	NEELY,WB & BLAU,GE (1985)
DC	pKa			
HL 7.85E-022	atm m3/mol	25	EST	MEYLAN,WM & HOWARD,PH (1991)
OH 1.08E-010	cm3/molc sec	25	EST	MEYLAN,WM & HOWARD,PH (1993)

CAS #: 018925-69-4				BENZENEACETAMIDE, N,N-DIMETHYL-

Formula: $C_{10}H_{13}NO$

Mol Weight: 163.22

MP (deg C): | FP (deg C):

BP (deg C):

BP pressure (mm Hg):

Property/ Value	Units	Temp	Data Type	Reference
WS 3.60E+003	mg/L	25	EST	MEYLAN,WM ET AL. (1996)
logP 1.48			EXP	SURYANARAYANA,MVS ET AL. (1991)
VP 2.00E-002	mm Hg	25	EXP	SURYANARAYANA,MVS ET AL. (1991)
DC	pKa			
HL 4.35E-009	atm m3/mol	25	EST	MEYLAN,WM & HOWARD,PH (1991)
OH 2.17E-011	cm3/molc sec	25	EST	MEYLAN,WM & HOWARD,PH (1993)

CAS #: 018936-17-9				2-METHYLBUTYRONITRILE

Formula: C_5H_9N

Mol Weight: 83.13

MP (deg C): | FP (deg C):

BP (deg C): 125

BP pressure (mm Hg):

Property/ Value	Units	Temp	Data Type	Reference
WS 8.89E+003	mg/L	25	EST	MEYLAN,WM ET AL. (1996)
logP 1.05			EXP	TANII,H & HASHIMOTO,K (1984)
VP 1.03E+001	mm Hg	25	EST	NEELY,WB & BLAU,GE (1985)
DC	pKa			
HL 7.15E-005	atm m3/mol	25	EST	MEYLAN,WM & HOWARD,PH (1991)
OH 1.94E-012	cm3/molc sec	25	EST	MEYLAN,WM & HOWARD,PH (1993)

CAS #: 018960-16-2				N-ISOPROPYLNICOTINAMIDE

Formula: $C_9H_{12}N_2O$

Mol Weight: 164.21

MP (deg C): | FP (deg C):

BP (deg C):

BP pressure (mm Hg):

Property/ Value	Units	Temp	Data Type	Reference
WS 2.05E+004	mg/L	25	EST	MEYLAN,WM ET AL. (1996)
logP 0.59			EXP	HANSCH,C & LEO,AJ (1985)
VP 6.66E-005	mm Hg	25	EST	NEELY,WB & BLAU,GE (1985)
DC	pKa			
HL 1.12E-011	atm m3/mol	25	EST	MEYLAN,WM & HOWARD,PH (1991)
OH 1.41E-011	cm3/molc sec	25	EST	MEYLAN,WM & HOWARD,PH (1993)

CAS #: 018979-50-5				PHENOL, 4-PROPOXY-

Formula: $C_9H_{12}O_2$

Mol Weight: 152.19

MP (deg C): 55-58 | FP (deg C):

BP (deg C):

BP pressure (mm Hg):

Property/ Value	Units	Temp	Data Type	Reference
WS 2.89E+003	mg/L	25	EST	MEYLAN,WM ET AL. (1996)
logP 2.33			EXP	HANSCH,C ET AL. (1995)
VP 7.65E-003	mm Hg	25	EST	NEELY,WB & BLAU,GE (1985)
DC	pKa			
HL 5.84E-008	atm m3/mol	25	EST	MEYLAN,WM & HOWARD,PH (1991)
OH 3.94E-011	cm3/molc sec	25	EST	MEYLAN,WM & HOWARD,PH (1993)

CAS #: 018979-53-8				PHENOL, 4-(PENTYLOXY)-

Formula: $C_{11}H_{16}O_2$

Mol Weight: 180.25

MP (deg C): | FP (deg C):

BP (deg C):

BP pressure (mm Hg):

Property/ Value	Units	Temp	Data Type	Reference
WS 2.14E+002	mg/L	25	EST	MEYLAN,WM ET AL. (1996)
logP 3.50			EXP	HANSCH,C ET AL. (1995)
VP 7.42E-004	mm Hg	25	EST	NEELY,WB & BLAU,GE (1985)
DC	pKa			
HL 1.03E-007	atm m3/mol	25	EST	MEYLAN,WM & HOWARD,PH (1991)
OH 4.27E-011	cm3/molc sec	25	EST	MEYLAN,WM & HOWARD,PH (1993)

018979-55-0 — 4-HEXYLOXYPHENOL

CAS #:	018979-55-0	4-HEXYLOXYPHENOL
Formula:	$C_{12}H_{18}O_2$	
Mol Weight:	194.28	
MP (deg C):	45-47	FP (deg C):
BP (deg C):		
BP pressure (mm Hg):		

Property/Value	Units	Temp	Data Type	Reference
WS 6.20E+001	mg/L	25	EST	MEYLAN,WM ET AL. (1996)
logP 4.05			EST	MEYLAN,WM & HOWARD,PH (1995)
VP 2.47E-004	mm Hg	25	EST	NEELY,WB & BLAU,GE (1985)
DC	pKa			
HL 1.37E-007	atm m3/mol	25	EST	MEYLAN,WM & HOWARD,PH (1991)
OH 4.41E-011	cm3/molc sec	25	EST	MEYLAN,WM & HOWARD,PH (1993)

018979-72-1 — M-BUTOXYPHENOL

CAS #:	018979-72-1	M-BUTOXYPHENOL
Formula:	$C_{10}H_{14}O_2$	
Mol Weight:	166.22	
MP (deg C):		FP (deg C):
BP (deg C):		
BP pressure (mm Hg):		

Property/Value	Units	Temp	Data Type	Reference
WS 6.30E+002	mg/L	25	EST	MEYLAN,WM ET AL. (1996)
logP 3.03			EXP	HANSCH,C & LEO,AJ (1985)
VP 2.33E-003	mm Hg	25	EST	NEELY,WB & BLAU,GE (1985)
DC				
HL 7.76E-008	atm m3/mol	25	EST	MEYLAN,WM & HOWARD,PH (1991)
OH 2.12E-010	cm3/molc sec	25	EST	MEYLAN,WM & HOWARD,PH (1993)

018979-73-2 — M-PENTOXYPHENOL

CAS #:	018979-73-2	M-PENTOXYPHENOL
Formula:	$C_{11}H_{16}O_2$	
Mol Weight:	180.25	
MP (deg C):		FP (deg C):
BP (deg C):		
BP pressure (mm Hg):		

Property/Value	Units	Temp	Data Type	Reference
WS 1.66E+002	mg/L	25	EST	MEYLAN,WM ET AL. (1996)
logP 3.63			EXP	HANSCH,C & LEO,AJ (1985)
VP 7.42E-004	mm Hg	25	EST	NEELY,WB & BLAU,GE (1985)
DC	pKa			
HL 1.03E-007	atm m3/mol	25	EST	MEYLAN,WM & HOWARD,PH (1991)
OH 2.14E-010	cm3/molc sec	25	EST	MEYLAN,WM & HOWARD,PH (1993)

018984-21-9 — 2,4-DICHLORO-B-NITROSTYRENE

CAS #:	018984-21-9	2,4-DICHLORO-B-NITROSTYRENE
Formula:	$C_8H_5Cl_2NO_2$	
Mol Weight:	218.04	
MP (deg C):		FP (deg C):
BP (deg C):		
BP pressure (mm Hg):		

Property/Value	Units	Temp	Data Type	Reference
WS 2.36E+001	mg/L	25	EST	MEYLAN,WM ET AL. (1996)
logP 3.26			EXP	HANSCH,C & LEO,AJ (1985)
VP 4.75E-004	mm Hg	25	EST	NEELY,WB & BLAU,GE (1985)
DC	pKa			
HL 1.91E-006	atm m3/mol	25	EST	MEYLAN,WM & HOWARD,PH (1991)
OH 1.23E-011	cm3/molc sec	25	EST	MEYLAN,WM & HOWARD,PH (1993)

018997-88-1 — á-D-GALACTOPYRANOSIDE, ETHYL

CAS #:	018997-88-1	á-D-GALACTOPYRANOSIDE, ETHYL
Formula:	$C_8H_{16}O_6$	
Mol Weight:	208.21	
MP (deg C):		FP (deg C):
BP (deg C):		
BP pressure (mm Hg):		

Property/Value	Units	Temp	Data Type	Reference
WS 1.00E+006	mg/L	25	EST	MEYLAN,WM ET AL. (1996)
logP -2.28			EXP	SANGSTER,J (1993)
VP 1.77E-008	mm Hg	25	EST	NEELY,WB & BLAU,GE (1985)
DC	pKa			
HL 2.07E-014	atm m3/mol	25	EST	MEYLAN,WM & HOWARD,PH (1991)
OH 6.82E-011	cm3/molc sec	25	EST	MEYLAN,WM & HOWARD,PH (1993)

019004-42-3 — 2-PYRIDYLMETHYL TRIME AMMONIUM BROMIDE

CAS #:	019004-42-3	2-PYRIDYLMETHYL TRIME AMMONIUM BROMIDE
Formula:	$C_9H_{15}BrN_2$	
Mol Weight:	231.14	
MP (deg C):		FP (deg C):
BP (deg C):		
BP pressure (mm Hg):		

Property/Value	Units	Temp	Data Type	Reference
WS 1.00E+006	mg/L	25	EST	MEYLAN,WM ET AL. (1996)
logP -3.40			EXP	HANSCH,C & LEO,AJ (1985)
VP 2.54E-008	mm Hg	25	EST	NEELY,WB & BLAU,GE (1985)
DC	pKa			
HL 3.50E-016	atm m3/mol	25	EST	MEYLAN,WM & HOWARD,PH (1991)
OH 1.34E-011	cm3/molc sec	25	EST	MEYLAN,WM & HOWARD,PH (1993)

019044-88-3 — ORYZALIN

CAS #:	019044-88-3	ORYZALIN
Formula:	$C_{12}H_{18}N_4O_6S$	
Mol Weight:	346.36	
MP (deg C):	141-142	FP (deg C):
BP (deg C):		
BP pressure (mm Hg):		

Property/Value	Units	Temp	Data Type	Reference
WS 2.50E+000	mg/L	25	EXP	WAUCHOPE,RD ET AL. (1991A)
logP 2.73			EST	MEYLAN,WM & HOWARD,PH (1995)
VP 1.00E-008	mm Hg	25	EST	WAUCHOPE,RD ET AL. (1991A)
DC 8.60	pKa	25	EXP	WAUCHOPE,RD ET AL. (1991A)
HL 1.91E-009	atm m3/mol	25	EST	MEYLAN,WM & HOWARD,PH (1991)
OH 1.04E-010	cm3/molc sec	25	EST	MEYLAN,WM & HOWARD,PH (1993)

019045-79-5 — DIPOTASSIUM MONOOCTYL ESTER

CAS #:	019045-79-5	DIPOTASSIUM MONOOCTYL ESTER
Formula:	$C_8H_{17}K_2O_4P$	
Mol Weight:	286.40	
MP (deg C):		FP (deg C):
BP (deg C):	289	
BP pressure (mm Hg):		

Property/Value	Units	Temp	Data Type	Reference
WS 7.00E+001	mg/L	25	EXP	CHEM INSPECT TEST INST (1992)
logP -2.31			EST	MEYLAN,WM & HOWARD,PH (1995)
VP 2.06E-008	mm Hg	25	EST	NEELY,WB & BLAU,GE (1985)
DC	pKa			
HL	atm m3/mol			
OH 2.98E-012	cm3/molc sec	25	EST	MEYLAN,WM & HOWARD,PH (1993)

CAS #: 019055-50-6				1H-INDENE-1,3(2H)-DIONE, 2-(3,5-DIMETHOXYPHENYL)
Formula: $C_{17}H_{14}O_4$				
Mol Weight: 282.30				
MP (deg C):		FP (deg C):		
BP (deg C):				
BP pressure (mm Hg):				

Property/ Value	Units	Temp	Data Type	Reference
WS 5.61E+001	mg/L	25	EST	MEYLAN,WM ET AL. (1996)
logP 2.86			EXP	SANGSTER,J (1993)
VP 9.71E-008	mm Hg	25	EST	NEELY,WB & BLAU,GE (1985)
DC	pKa			
HL 1.68E-012	atm m3/mol	25	EST	MEYLAN,WM & HOWARD,PH (1991)
OH 2.03E-010	cm3/molc sec	25	EST	MEYLAN,WM & HOWARD,PH (1993)

CAS #: 019055-70-0				1H-INDENE-1,3(2H)-DIONE, 2-[3-(TRIFLUOROMETHYL)P
Formula: $C_{16}H_9F_3O_2$				
Mol Weight: 290.24				
MP (deg C):		FP (deg C):		
BP (deg C):				
BP pressure (mm Hg):				

Property/ Value	Units	Temp	Data Type	Reference
WS 8.27E+000	mg/L	25	EST	MEYLAN,WM ET AL. (1996)
logP 3.78			EXP	SANGSTER,J (1993)
VP 2.26E-006	mm Hg	25	EST	NEELY,WB & BLAU,GE (1985)
DC	pKa			
HL 4.16E-009	atm m3/mol	25	EST	MEYLAN,WM & HOWARD,PH (1991)
OH 2.32E-012	cm3/molc sec	25	EST	MEYLAN,WM & HOWARD,PH (1993)

CAS #: 019064-18-7				BENZENEMETHANOL, 2,6-DIFLUORO-
Formula: $C_7H_6F_2O$				
Mol Weight: 144.12				
MP (deg C):		FP (deg C):		
BP (deg C):				
BP pressure (mm Hg):				

Property/ Value	Units	Temp	Data Type	Reference
WS 2.88E+004	mg/L	25	EST	MEYLAN,WM ET AL. (1996)
logP 1.12			EXP	EL TAYAR,N ET AL. (1991)
VP 9.10E-002	mm Hg	25	EST	NEELY,WB & BLAU,GE (1985)
DC	pKa			
HL 2.96E-007	atm m3/mol	25	EST	MEYLAN,WM & HOWARD,PH (1991)
OH 7.14E-012	cm3/molc sec	25	EST	MEYLAN,WM & HOWARD,PH (1993)

CAS #: 019064-65-4				PYRIDAZINE, 3-METHOXY-
Formula: $C_5H_6N_2O$				
Mol Weight: 110.12				
MP (deg C):		FP (deg C):		
BP (deg C):				
BP pressure (mm Hg):				

Property/ Value	Units	Temp	Data Type	Reference
WS 9.29E+004	mg/L	25	EST	MEYLAN,WM ET AL. (1996)
logP 0.08			EXP	YAMAGAMI,C ET AL. (1990)
VP 4.51E-002	mm Hg	25	EST	NEELY,WB & BLAU,GE (1985)
DC 2.48	pKa	20	EXP	PERRIN,DD (1965)
HL 1.55E-006	atm m3/mol	25	EST	MEYLAN,WM & HOWARD,PH (1991)
OH 3.37E-012	cm3/molc sec	25	EST	MEYLAN,WM & HOWARD,PH (1993)

CAS #: 019064-69-8				1-PHTHALAZINAMINE
Formula: $C_8H_7N_3$				
Mol Weight: 145.17				
MP (deg C):		FP (deg C):		
BP (deg C):				
BP pressure (mm Hg):				

Property/ Value	Units	Temp	Data Type	Reference
WS 1.59E+004	mg/L	25	EST	MEYLAN,WM ET AL. (1996)
logP 0.82			EXP	HANSCH,C & LEO,AJ (1985)
VP 1.62E-005	mm Hg	25	EST	NEELY,WB & BLAU,GE (1985)
DC	pKa			
HL 9.90E-011	atm m3/mol	25	EST	MEYLAN,WM & HOWARD,PH (1991)
OH 1.10E-010	cm3/molc sec	25	EST	MEYLAN,WM & HOWARD,PH (1993)

CAS #: 019067-63-1				4-PYRIDYLMETHYL TRIME AMMONIUM BROMIDE
Formula: $C_9H_{15}BrN_2$				
Mol Weight: 231.14				
MP (deg C):		FP (deg C):		
BP (deg C):				
BP pressure (mm Hg):				

Property/ Value	Units	Temp	Data Type	Reference
WS 1.00E+006	mg/L	25	EST	MEYLAN,WM ET AL. (1996)
logP -3.95			EXP	HANSCH,C & LEO,AJ (1985)
VP 2.54E-008	mm Hg	25	EST	NEELY,WB & BLAU,GE (1985)
DC	pKa			
HL 3.50E-016	atm m3/mol	25	EST	MEYLAN,WM & HOWARD,PH (1991)
OH 1.34E-011	cm3/molc sec	25	EST	MEYLAN,WM & HOWARD,PH (1993)

CAS #: 019095-79-5				N-CYCLOHEPTYL-N'-PHENYLUREA
Formula: $C_{14}H_{20}N_2O$				
Mol Weight: 232.33				
MP (deg C):		FP (deg C):		
BP (deg C):				
BP pressure (mm Hg):				

Property/ Value	Units	Temp	Data Type	Reference
WS 1.29E+001	mg/L	25	EST	MEYLAN,WM ET AL. (1996)
logP 3.93			EST	MEYLAN,WM & HOWARD,PH (1995)
VP 1.73E-006	mm Hg	25	EST	NEELY,WB & BLAU,GE (1985)
DC	pKa			
HL 1.07E-009	atm m3/mol	25	EST	MEYLAN,WM & HOWARD,PH (1991)
OH 6.42E-011	cm3/molc sec	25	EST	MEYLAN,WM & HOWARD,PH (1993)

CAS #: 019096-41-4				L-THREO-ALPHA-D-GALACTO-OCTOPYRANOSIDE
Formula: $C_{19}H_{35}ClN_2O_5S$				
Mol Weight: 439.02				
MP (deg C):		FP (deg C):		
BP (deg C):				
BP pressure (mm Hg):				

Property/ Value	Units	Temp	Data Type	Reference
WS 7.99E+000	mg/L	25	EST	MEYLAN,WM ET AL. (1996)
logP 2.74			EXP	HANSCH,C ET AL. (1995)
VP 3.93E-018	mm Hg	25	EST	NEELY,WB & BLAU,GE (1985)
DC	pKa			
HL 2.32E-022	atm m3/mol	25	EST	MEYLAN,WM & HOWARD,PH (1991)
OH 2.34E-010	cm3/molc sec	25	EST	MEYLAN,WM & HOWARD,PH (1993)

CAS #: 019109-66-1				TRIMETHYL PENTYL AMMONIUM IODIDE
Formula: $C_8H_{20}IN$				
Mol Weight: 257.16				
MP (deg C):		FP (deg C):		
BP (deg C):				
BP pressure (mm Hg):				

Property/Value	Units	Temp	Data Type	Reference
WS 1.00E+006	mg/L	25	EST	MEYLAN,WM ET AL. (1996)
logP -2.22			EXP	HANSCH,C & LEO,AJ (1985)
VP 2.52E-007	mm Hg	25	EST	NEELY,WB & BLAU,GE (1985)
DC	pKa			
HL 8.18E-012	atm m3/mol	25	EST	MEYLAN,WM & HOWARD,PH (1991)
OH 1.86E-011	cm3/molc sec	25	EST	MEYLAN,WM & HOWARD,PH (1993)

CAS #: 019111-74-1				2,6-DIMETHYLBENZONITRILE-N-OXIDE
Formula: C_9H_9NO				
Mol Weight: 147.18				
MP (deg C):		FP (deg C):		
BP (deg C):				
BP pressure (mm Hg):				

Property/Value	Units	Temp	Data Type	Reference
WS 3.57E+002	mg/L	25	EST	MEYLAN,WM ET AL. (1996)
logP 2.74			EXP	KIRCHNER,JJ ET AL. (1985)
VP 3.95E+000	mm Hg	25	EST	NEELY,WB & BLAU,GE (1985)
DC	pKa			
HL 6.35E-005	atm m3/mol	25	EST	MEYLAN,WM & HOWARD,PH (1991)
OH 9.77E-012	cm3/molc sec	25	EST	MEYLAN,WM & HOWARD,PH (1993)

CAS #: 019130-96-2				3,4,5-PIPERIDINETRIOL, 2-(HYDROXYMETHYL)-
Formula: $C_6H_{13}NO_4$				
Mol Weight: 163.17				
MP (deg C):		FP (deg C):		
BP (deg C):				
BP pressure (mm Hg):				

Property/Value	Units	Temp	Data Type	Reference
WS 1.00E+006	mg/L	25	EST	MEYLAN,WM ET AL. (1996)
logP -1.80			EXP	SANGSTER,J (1994)
VP 5.38E-008	mm Hg	25	EST	NEELY,WB & BLAU,GE (1985)
DC	pKa			
HL 4.07E-014	atm m3/mol	25	EST	MEYLAN,WM & HOWARD,PH (1991)
OH 1.36E-010	cm3/molc sec	25	EST	MEYLAN,WM & HOWARD,PH (1993)

CAS #: 019132-06-0				2,3-BUTANEDIOL (D)
Formula: $C_4H_{10}O_2$				
Mol Weight: 90.12				
MP (deg C):		FP (deg C):		
BP (deg C): 179-182				
BP pressure (mm Hg):				

Property/Value	Units	Temp	Data Type	Reference
WS 7.59E+005	mg/L	25	EST	MEYLAN,WM ET AL. (1996)
logP -0.36			EST	MEYLAN,WM & HOWARD,PH (1995)
VP 7.11E-001	mm Hg	25	EST	NEELY,WB & BLAU,GE (1985)
DC	pKa			
HL 2.30E-007	atm m3/mol	25	EST	MEYLAN,WM & HOWARD,PH (1991)
OH 1.73E-011	cm3/molc sec	25	EST	MEYLAN,WM & HOWARD,PH (1993)

CAS #: 019183-14-3				4,5-DINITROIMIDAZOLE
Formula: $C_3H_2N_4O_4$				
Mol Weight: 158.07				
MP (deg C):		FP (deg C):		
BP (deg C):				
BP pressure (mm Hg):				

Property/Value	Units	Temp	Data Type	Reference
WS 7.93E+003	mg/L	25	EST	MEYLAN,WM ET AL. (1996)
logP 0.65			EXP	SUWINSKI,J ET AL. (1985)
VP 2.61E-006	mm Hg	25	EST	NEELY,WB & BLAU,GE (1985)
DC	pKa			
HL 5.85E-011	atm m3/mol	25	EST	MEYLAN,WM & HOWARD,PH (1991)
OH 5.62E-013	cm3/molc sec	25	EST	MEYLAN,WM & HOWARD,PH (1993)

CAS #: 019183-15-4				1-METHYL-4,5-DINITROIMIDAZOLE
Formula: $C_4H_4N_4O_4$				
Mol Weight: 172.10				
MP (deg C):		FP (deg C):		
BP (deg C):				
BP pressure (mm Hg):				

Property/Value	Units	Temp	Data Type	Reference
WS 1.09E+004	mg/L	25	EST	MEYLAN,WM ET AL. (1996)
logP 0.41			EXP	SUWINSKI,J ET AL. (1985)
VP 1.69E-005	mm Hg	25	EST	NEELY,WB & BLAU,GE (1985)
DC	pKa			
HL 1.25E-009	atm m3/mol	25	EST	MEYLAN,WM & HOWARD,PH (1991)
OH 6.98E-013	cm3/molc sec	25	EST	MEYLAN,WM & HOWARD,PH (1993)

CAS #: 019183-16-5				2-METHYL-4,5-DINITROIMIDAZOLE
Formula: $C_4H_4N_4O_4$				
Mol Weight: 172.10				
MP (deg C):		FP (deg C):		
BP (deg C):				
BP pressure (mm Hg):				

Property/Value	Units	Temp	Data Type	Reference
WS 5.39E+003	mg/L	25	EST	MEYLAN,WM ET AL. (1996)
logP 0.77			EXP	SUWINSKI,J ET AL. (1985)
VP 1.25E-006	mm Hg	25	EST	NEELY,WB & BLAU,GE (1985)
DC	pKa			
HL 6.45E-011	atm m3/mol	25	EST	MEYLAN,WM & HOWARD,PH (1991)
OH 3.61E-011	cm3/molc sec	25	EST	MEYLAN,WM & HOWARD,PH (1993)

CAS #: 019183-17-6				1,2-DIMETHYL-4,5-DINITROIMIDAZOLE
Formula: $C_5H_6N_4O_4$				
Mol Weight: 186.13				
MP (deg C):		FP (deg C):		
BP (deg C):				
BP pressure (mm Hg):				

Property/Value	Units	Temp	Data Type	Reference
WS 7.99E+003	mg/L	25	EST	MEYLAN,WM ET AL. (1996)
logP 0.49			EXP	HANSCH,C ET AL. (1995)
VP 8.26E-006	mm Hg	25	EST	NEELY,WB & BLAU,GE (1985)
DC	pKa			
HL 1.38E-009	atm m3/mol	25	EST	MEYLAN,WM & HOWARD,PH (1991)
OH 3.63E-011	cm3/molc sec	25	EST	MEYLAN,WM & HOWARD,PH (1993)

019184-65-7 — MONOBROMONEOPENTYLTRIOL

CAS #:	019184-65-7
Formula:	$C_5H_{11}BrO_3$
Mol Weight:	199.05
MP (deg C):	75-77
FP (deg C):	
BP (deg C):	
BP pressure (mm Hg):	

Property/Value	Units	Temp	Data Type	Reference
WS 1.75E+005	mg/L	25	EST	MEYLAN,WM ET AL. (1996)
logP -0.70			EXP	HANSCH,C & LEO,AJ (1985)
VP 3.96E-006	mm Hg	25	EST	NEELY,WB & BLAU,GE (1985)
DC	pKa			
HL 1.29E-009	atm m3/mol	25	EST	MEYLAN,WM & HOWARD,PH (1991)
OH 1.28E-011	cm3/molc sec	25	EST	MEYLAN,WM & HOWARD,PH (1993)

019202-92-7 — TENULIN

CAS #:	019202-92-7
Formula:	$C_{17}H_{22}O_5$
Mol Weight:	306.36
MP (deg C):	
FP (deg C):	
BP (deg C):	
BP pressure (mm Hg):	

Property/Value	Units	Temp	Data Type	Reference
WS 2.02E+004	mg/L	25	EST	MEYLAN,WM ET AL. (1996)
logP 0.30			EXP	HANSCH,C ET AL. (1995)
VP 6.00E-010	mm Hg	25	EST	NEELY,WB & BLAU,GE (1985)
DC	pKa			
HL 2.54E-015	atm m3/mol	25	EST	MEYLAN,WM & HOWARD,PH (1991)
OH 9.25E-011	cm3/molc sec	25	EST	MEYLAN,WM & HOWARD,PH (1993)

019213-65-1 — 5-ETHYL-6-AZAURACIL

CAS #:	019213-65-1
Formula:	$C_5H_7N_3O_2$
Mol Weight:	141.13
MP (deg C):	
FP (deg C):	
BP (deg C):	
BP pressure (mm Hg):	

Property/Value	Units	Temp	Data Type	Reference
WS 2.63E+003	mg/L	25	EST	MEYLAN,WM ET AL. (1996)
logP 0.22			EXP	HANSCH,C & LEO,AJ (1985)
VP 3.78E-008	mm Hg	25	EST	NEELY,WB & BLAU,GE (1985)
DC	pKa			
HL 3.37E-009	atm m3/mol	25	EST	MEYLAN,WM & HOWARD,PH (1991)
OH 5.81E-012	cm3/molc sec	25	EST	MEYLAN,WM & HOWARD,PH (1993)

019216-39-8 — PROPANAMIDE, 2-(DIMETHYLAMINO)-N-(2,6-DIMETHYLPH

CAS #:	019216-39-8
Formula:	$C_{13}H_{20}N_2O$
Mol Weight:	220.32
MP (deg C):	
FP (deg C):	
BP (deg C):	
BP pressure (mm Hg):	

Property/Value	Units	Temp	Data Type	Reference
WS 1.69E+003	mg/L	25	EST	MEYLAN,WM ET AL. (1996)
logP 1.53			EXP	SANGSTER,J (1993)
VP 3.58E-006	mm Hg	25	EST	NEELY,WB & BLAU,GE (1985)
DC	pKa			
HL 9.87E-011	atm m3/mol	25	EST	MEYLAN,WM & HOWARD,PH (1991)
OH 1.02E-010	cm3/molc sec	25	EST	MEYLAN,WM & HOWARD,PH (1993)

019218-88-3 — BENZOIC ACID, 4-CHLORO-2-(PHENYLAMINO)-

CAS #:	019218-88-3
Formula:	$C_{13}H_{10}ClNO_2$
Mol Weight:	247.68
MP (deg C):	
FP (deg C):	
BP (deg C):	
BP pressure (mm Hg):	

Property/Value	Units	Temp	Data Type	Reference
WS 5.72E+000	mg/L	25	EST	MEYLAN,WM ET AL. (1996)
logP 4.25			EXP	SANGSTER,J (1993)
VP 6.45E-007	mm Hg	25	EST	NEELY,WB & BLAU,GE (1985)
DC	pKa			
HL 1.57E-011	atm m3/mol	25	EST	MEYLAN,WM & HOWARD,PH (1991)
OH 1.85E-010	cm3/molc sec	25	EST	MEYLAN,WM & HOWARD,PH (1993)

019218-94-1 — 1-IODOTETRADECANE

CAS #:	019218-94-1
Formula:	$C_{14}H_{29}I$
Mol Weight:	324.29
MP (deg C):	
FP (deg C):	
BP (deg C):	
BP pressure (mm Hg):	

Property/Value	Units	Temp	Data Type	Reference
WS 1.37E-003	mg/L	25	EST	MEYLAN,WM ET AL. (1996)
logP 7.98			EST	MEYLAN,WM & HOWARD,PH (1995)
VP 1.42E-004	mm Hg	25	EXT	LI,JCM & ROSSINI,FD (1953)
DC	pKa			
HL 2.19E-001	atm m3/mol	25	EST	MEYLAN,WM & HOWARD,PH (1991)
OH 1.66E-011	cm3/molc sec	25	EST	MEYLAN,WM & HOWARD,PH (1993)

019230-45-6 — ACETAMIDE, N-(3-IODOPHENYL)-

CAS #:	019230-45-6
Formula:	C_8H_8INO
Mol Weight:	261.06
MP (deg C):	
FP (deg C):	
BP (deg C):	
BP pressure (mm Hg):	

Property/Value	Units	Temp	Data Type	Reference
WS 1.14E+002	mg/L	25	EST	MEYLAN,WM ET AL. (1996)
logP 2.64			EXP	HANSCH,C ET AL. (1995)
VP 1.03E-005	mm Hg	25	EST	NEELY,WB & BLAU,GE (1985)
DC	pKa			
HL 1.43E-009	atm m3/mol	25	EST	MEYLAN,WM & HOWARD,PH (1991)
OH 8.29E-012	cm3/molc sec	25	EST	MEYLAN,WM & HOWARD,PH (1993)

019246-04-9 — ACETAMIDE, N-[2-(1-METHYLETHYL)PHENYL]-

CAS #:	019246-04-9
Formula:	$C_{11}H_{15}NO$
Mol Weight:	177.25
MP (deg C):	
FP (deg C):	
BP (deg C):	
BP pressure (mm Hg):	

Property/Value	Units	Temp	Data Type	Reference
WS 1.97E+003	mg/L	25	EST	MEYLAN,WM ET AL. (1996)
logP 1.71			EXP	HANSCH,C ET AL. (1995)
VP 5.37E-005	mm Hg	25	EST	NEELY,WB & BLAU,GE (1985)
DC	pKa			
HL 1.20E-008	atm m3/mol	25	EST	MEYLAN,WM & HOWARD,PH (1991)
OH 1.73E-011	cm3/molc sec	25	EST	MEYLAN,WM & HOWARD,PH (1993)

CAS #: 019271-90-0				O-T-BUTYLPHENOXYACETIC ACID

Formula: $C_{12}H_{16}O_3$

Mol Weight: 208.26

MP (deg C): | FP (deg C):

BP (deg C):

BP pressure (mm Hg):

Property/Value	Units	Temp	Data Type	Reference
WS 1.41E+002	mg/L	25	EST	MEYLAN,WM ET AL. (1996)
logP 3.33			EXP	HANSCH,C & LEO,AJ (1985)
VP 8.54E-005	mm Hg	25	EST	NEELY,WB & BLAU,GE (1985)
DC	pKa			
HL 4.33E-008	atm m3/mol	25	EST	MEYLAN,WM & HOWARD,PH (1991)
OH 3.11E-011	cm3/molc sec	25	EST	MEYLAN,WM & HOWARD,PH (1993)

CAS #: 019335-11-6				1H-INDAZOL-5-AMINE

Formula: $C_7H_7N_3$

Mol Weight: 133.15

MP (deg C): 175-178 | FP (deg C):

BP (deg C):

BP pressure (mm Hg):

Property/Value	Units	Temp	Data Type	Reference
WS 2.49E+004	mg/L	25	EST	MEYLAN,WM ET AL. (1996)
logP 0.65			EXP	SCHULTZ,TW & APPLEHANS,FM (1985)
VP 3.61E-005	mm Hg	25	EST	NEELY,WB & BLAU,GE (1985)
DC	pKa			
HL 1.27E-010	atm m3/mol	25	EST	MEYLAN,WM & HOWARD,PH (1991)
OH 2.00E-010	cm3/molc sec	25	EST	MEYLAN,WM & HOWARD,PH (1993)

CAS #: 019340-77-3				N-BENZYLGLYCOLAMIDE

Formula: $C_9H_{11}NO_2$

Mol Weight: 165.19

MP (deg C): | FP (deg C):

BP (deg C):

BP pressure (mm Hg):

Property/Value	Units	Temp	Data Type	Reference
WS 2.20E+004	mg/L	25	EST	MEYLAN,WM ET AL. (1996)
logP 0.55			EXP	HANSCH,C & LEO,AJ (1985)
VP 2.38E-007	mm Hg	25	EST	NEELY,WB & BLAU,GE (1985)
DC	pKa			
HL 3.09E-010	atm m3/mol	25	EST	MEYLAN,WM & HOWARD,PH (1991)
OH 1.68E-011	cm3/molc sec	25	EST	MEYLAN,WM & HOWARD,PH (1993)

CAS #: 019343-24-9				2-PROPANOL, 1-[(1-METHYLETHYL)AMINO]-3-(PHENYLME

Formula: $C_{13}H_{21}NO_2$

Mol Weight: 223.32

MP (deg C): | FP (deg C):

BP (deg C):

BP pressure (mm Hg):

Property/Value	Units	Temp	Data Type	Reference
WS 1.39E+004	mg/L	25	EST	MEYLAN,WM ET AL. (1996)
logP 1.62			EXP	MAULEON,D ET AL. (1988)
VP 6.51E-006	mm Hg	25	EST	NEELY,WB & BLAU,GE (1985)
DC	pKa			
HL 1.79E-012	atm m3/mol	25	EST	MEYLAN,WM & HOWARD,PH (1991)
OH 1.20E-010	cm3/molc sec	25	EST	MEYLAN,WM & HOWARD,PH (1993)

CAS #: 019353-97-0				2-METHOXY ISONIAZID

Formula: $C_7H_9N_3O_2$

Mol Weight: 167.17

MP (deg C): | FP (deg C):

BP (deg C):

BP pressure (mm Hg):

Property/Value	Units	Temp	Data Type	Reference
WS 3.78E+003	mg/L	25	EST	MEYLAN,WM ET AL. (1996)
logP -0.10			EXP	HANSCH,C & LEO,AJ (1985)
VP 7.33E-006	mm Hg	25	EST	NEELY,WB & BLAU,GE (1985)
DC	pKa			
HL 6.53E-015	atm m3/mol	25	EST	MEYLAN,WM & HOWARD,PH (1991)
OH 8.05E-012	cm3/molc sec	25	EST	MEYLAN,WM & HOWARD,PH (1993)

CAS #: 019353-98-1				2-DIETHYLAMINOISONIAZID

Formula: $C_{10}H_{16}N_4O$

Mol Weight: 208.27

MP (deg C): | FP (deg C):

BP (deg C):

BP pressure (mm Hg):

Property/Value	Units	Temp	Data Type	Reference
WS 4.39E+003	mg/L	25	EST	MEYLAN,WM ET AL. (1996)
logP 1.12			EXP	HANSCH,C & LEO,AJ (1985)
VP 6.55E-007	mm Hg	25	EST	NEELY,WB & BLAU,GE (1985)
DC	pKa			
HL 3.39E-016	atm m3/mol	25	EST	MEYLAN,WM & HOWARD,PH (1991)
OH 3.49E-011	cm3/molc sec	25	EST	MEYLAN,WM & HOWARD,PH (1993)

CAS #: 019365-01-6				2(1H)-PYRIDINONE, 3-HYDROXY-1-METHYL-

Formula: $C_6H_7NO_2$

Mol Weight: 125.13

MP (deg C): | FP (deg C):

BP (deg C):

BP pressure (mm Hg):

Property/Value	Units	Temp	Data Type	Reference
WS 2.11E+005	mg/L	25	EST	MEYLAN,WM ET AL. (1996)
logP -0.40			EXP	KONTOGHIORGHES,GJ (1988)
VP 3.40E-005	mm Hg	25	EST	NEELY,WB & BLAU,GE (1985)
DC	pKa			
HL 3.09E-008	atm m3/mol	25	EST	MEYLAN,WM & HOWARD,PH (1991)
OH 5.56E-011	cm3/molc sec	25	EST	MEYLAN,WM & HOWARD,PH (1993)

CAS #: 019367-61-4				BROMOTHIAZIDE

Formula: $C_7H_6BrN_3O_4S_2$

Mol Weight: 340.18

MP (deg C): | FP (deg C):

BP (deg C):

BP pressure (mm Hg):

Property/Value	Units	Temp	Data Type	Reference
WS 6.31E+002	mg/L	25	EST	MEYLAN,WM ET AL. (1996)
logP 0.00			EXP	SANGSTER,J (1994)
VP 8.79E-012	mm Hg	25	EST	NEELY,WB & BLAU,GE (1985)
DC	pKa			
HL 2.18E-012	atm m3/mol	25	EST	MEYLAN,WM & HOWARD,PH (1991)
OH 1.81E-011	cm3/molc sec	25	EST	MEYLAN,WM & HOWARD,PH (1993)

CAS #: 019411-75-7 — ACETYLACETONE,4-CHLOROBENZAL

Formula: $C_{12}H_{11}ClO_2$

Mol Weight: 222.67

MP (deg C): FP (deg C):

BP (deg C):

BP pressure (mm Hg):

Property/Value	Units	Temp	Data Type	Reference
WS 3.48E+002	mg/L	25	EST	MEYLAN,WM ET AL. (1996)
logP 2.32			EXP	HANSCH,C & LEO,AJ (1985)
VP 1.85E-004	mm Hg	25	EST	NEELY,WB & BLAU,GE (1985)
DC	pKa			
HL 3.67E-010	atm m3/mol	25	EST	MEYLAN,WM & HOWARD,PH (1991)
OH 7.19E-011	cm3/molc sec	25	EST	MEYLAN,WM & HOWARD,PH (1993)

CAS #: 019411-80-4 — ETHYLACETOACETATE,4-CHLOROBENZAL

Formula: $C_{13}H_{13}ClO_2$

Mol Weight: 236.70

MP (deg C): FP (deg C):

BP (deg C):

BP pressure (mm Hg):

Property/Value	Units	Temp	Data Type	Reference
WS 7.09E+001	mg/L	25	EST	MEYLAN,WM ET AL. (1996)
logP 3.04			EXP	HANSCH,C & LEO,AJ (1985)
VP 7.31E-005	mm Hg	25	EST	NEELY,WB & BLAU,GE (1985)
DC	pKa			
HL 4.87E-010	atm m3/mol	25	EST	MEYLAN,WM & HOWARD,PH (1991)
OH 7.30E-011	cm3/molc sec	25	EST	MEYLAN,WM & HOWARD,PH (1993)

CAS #: 019411-83-7 — MALONAMIDE,BENZAL

Formula: $C_{10}H_{10}N_2O_2$

Mol Weight: 190.20

MP (deg C): FP (deg C):

BP (deg C):

BP pressure (mm Hg):

Property/Value	Units	Temp	Data Type	Reference
WS 3.10E+003	mg/L	25	EST	MEYLAN,WM ET AL. (1996)
logP -0.13			EXP	HANSCH,C & LEO,AJ (1985)
VP 3.04E-008	mm Hg	25	EST	NEELY,WB & BLAU,GE (1985)
DC	pKa			
HL 4.00E-014	atm m3/mol	25	EST	MEYLAN,WM & HOWARD,PH (1991)
OH 1.65E-011	cm3/molc sec	25	EST	MEYLAN,WM & HOWARD,PH (1993)

CAS #: 019432-68-9 — THIOPHEN-2-ACETIC ACID, METHYL ESTER

Formula: $C_7H_{10}O_2S$

Mol Weight: 158.22

MP (deg C): FP (deg C):

BP (deg C):

BP pressure (mm Hg):

Property/Value	Units	Temp	Data Type	Reference
WS 3.66E+003	mg/L	25	EST	MEYLAN,WM ET AL. (1996)
logP 1.51			EXP	HANSCH,C & LEO,AJ (1985)
VP 8.41E-002	mm Hg	25	EST	NEELY,WB & BLAU,GE (1985)
DC	pKa			
HL 7.69E-006	atm m3/mol	25	EST	MEYLAN,WM & HOWARD,PH (1991)
OH 1.67E-011	cm3/molc sec	25	EST	MEYLAN,WM & HOWARD,PH (1993)

CAS #: 019434-42-5 — 4-AMINO-2-PHENYLPHENOL

Formula: $C_{12}H_{11}NO$

Mol Weight: 185.23

MP (deg C): FP (deg C):

BP (deg C):

BP pressure (mm Hg):

Property/Value	Units	Temp	Data Type	Reference
WS 9.99E+002	mg/L	25	EST	MEYLAN,WM ET AL. (1996)
logP 2.01			EST	MEYLAN,WM & HOWARD,PH (1995)
VP 1.67E-006	mm Hg	25	EST	NEELY,WB & BLAU,GE (1985)
DC	pKa			
HL 1.52E-011	atm m3/mol	25	EST	MEYLAN,WM & HOWARD,PH (1991)
OH 2.00E-010	cm3/molc sec	25	EST	MEYLAN,WM & HOWARD,PH (1993)

CAS #: 019438-10-9 — METHYL M-HYDROXYBENZOATE

Formula: $C_8H_8O_3$

Mol Weight: 152.15

MP (deg C): 73 FP (deg C):

BP (deg C): 281

BP pressure (mm Hg):

Property/Value	Units	Temp	Data Type	Reference
WS 6.86E+003	mg/L	25	EST	MEYLAN,WM ET AL. (1996)
logP 1.89			EXP	HANSCH,C & LEO,AJ (1985)
VP 5.63E-003	mm Hg	25	EST	NEELY,WB & BLAU,GE (1985)
DC	pKa			
HL 3.61E-009	atm m3/mol	25	EST	MEYLAN,WM & HOWARD,PH (1991)
OH 7.73E-012	cm3/molc sec	25	EST	MEYLAN,WM & HOWARD,PH (1993)

CAS #: 019477-12-4 — 124-BENZTHIADIAZINE-11-O2-7-CL

Formula: $C_7H_5ClN_2O_2S$

Mol Weight: 216.65

MP (deg C): FP (deg C):

BP (deg C):

BP pressure (mm Hg):

Property/Value	Units	Temp	Data Type	Reference
WS 4.37E+003	mg/L	25	EST	MEYLAN,WM ET AL. (1996)
logP 1.07			EXP	HANSCH,C & LEO,AJ (1985)
VP 6.77E-008	mm Hg	25	EST	NEELY,WB & BLAU,GE (1985)
DC	pKa			
HL 5.16E-008	atm m3/mol	25	EST	MEYLAN,WM & HOWARD,PH (1991)
OH 2.07E-011	cm3/molc sec	25	EST	MEYLAN,WM & HOWARD,PH (1993)

CAS #: 019477-31-7 — 124-BENZTHIADIAZINE-11-O2-6-CL

Formula: $C_7H_5ClN_2O_2S$

Mol Weight: 216.65

MP (deg C): FP (deg C):

BP (deg C):

BP pressure (mm Hg):

Property/Value	Units	Temp	Data Type	Reference
WS 4.92E+003	mg/L	25	EST	MEYLAN,WM ET AL. (1996)
logP 1.01			EXP	HANSCH,C & LEO,AJ (1985)
VP 6.77E-008	mm Hg	25	EST	NEELY,WB & BLAU,GE (1985)
DC	pKa			
HL 5.16E-008	atm m3/mol	25	EST	MEYLAN,WM & HOWARD,PH (1991)
OH 2.45E-011	cm3/molc sec	25	EST	MEYLAN,WM & HOWARD,PH (1993)

CAS #: 019480-39-8	(4-CHLORO-2-METHYLPHENOXY)ACETIC ACID, AMINE SA*

Formula: $C_9H_{12}ClNO_3$

Mol Weight: 217.65

MP (deg C): FP (deg C):

BP (deg C):

BP pressure (mm Hg):

Property/Value	Units	Temp	Data Type	Reference
WS 7.13E+004	mg/L	25	EST	MEYLAN,WM ET AL. (1996)
logP -0.36			EST	MEYLAN,WM & HOWARD,PH (1995)
VP 1.61E-008	mm Hg	25	EST	NEELY,WB & BLAU,GE (1985)
DC	pKa			
HL 6.59E-017	atm m3/mol	25	EST	MEYLAN,WM & HOWARD,PH (1991)
OH 9.60E-012	cm3/molc sec	25	EST	MEYLAN,WM & HOWARD,PH (1993)

CAS #: 019484-26-5	1-TRIDECANETHIOL

Formula: $C_{13}H_{28}S$

Mol Weight: 216.43

MP (deg C): FP (deg C):

BP (deg C):

BP pressure (mm Hg):

Property/Value	Units	Temp	Data Type	Reference
WS 7.25E-002	mg/L	25	EST	MEYLAN,WM ET AL. (1996)
logP 6.67			EST	MEYLAN,WM & HOWARD,PH (1995)
VP 6.93E-004	mm Hg	25	EXT	ZWOLINSKI,BJ & WILHOIT,RC (1971)
DC	pKa			
HL 7.83E-002	atm m3/mol	25	EST	MEYLAN,WM & HOWARD,PH (1991)
OH 5.69E-011	cm3/molc sec	25	EST	MEYLAN,WM & HOWARD,PH (1993)

CAS #: 019485-85-9	2-NAPHTHALENAMINE, 1,2,3,4-TETRAHYDRO-N-METHYL-

Formula: $C_{11}H_{15}N$

Mol Weight: 161.25

MP (deg C): FP (deg C):

BP (deg C):

BP pressure (mm Hg):

Property/Value	Units	Temp	Data Type	Reference
WS 6.38E+003	mg/L	25	EST	MEYLAN,WM ET AL. (1996)
logP 2.38			EXP	SANGSTER,J (1993)
VP 1.97E-002	mm Hg	25	EST	NEELY,WB & BLAU,GE (1985)
DC	pKa			
HL 1.15E-006	atm m3/mol	25	EST	MEYLAN,WM & HOWARD,PH (1991)
OH 1.01E-010	cm3/molc sec	25	EST	MEYLAN,WM & HOWARD,PH (1993)

CAS #: 019494-89-4	5-OHPICOLINALDEHYDE THIOSEMICARBAZONE

Formula: $C_7H_{10}N_4OS$

Mol Weight: 198.25

MP (deg C): FP (deg C):

BP (deg C):

BP pressure (mm Hg):

Property/Value	Units	Temp	Data Type	Reference
WS 1.21E+003	mg/L	25	EST	MEYLAN,WM ET AL. (1996)
logP 1.04			EXP	HANSCH,C & LEO,AJ (1985)
VP 1.47E-006	mm Hg	25	EST	NEELY,WB & BLAU,GE (1985)
DC	pKa			
HL 4.64E-015	atm m3/mol	25	EST	MEYLAN,WM & HOWARD,PH (1991)
OH 9.37E-011	cm3/molc sec	25	EST	MEYLAN,WM & HOWARD,PH (1993)

CAS #: 019495-05-7	B-NAPHTHYLMETHYLISOTHIOCYANATE

Formula: $C_{12}H_9NS$

Mol Weight: 199.28

MP (deg C): FP (deg C):

BP (deg C):

BP pressure (mm Hg):

Property/Value	Units	Temp	Data Type	Reference
WS 7.41E+000	mg/L	25	EST	MEYLAN,WM ET AL. (1996)
logP 4.42			EXP	HANSCH,C & LEO,AJ (1985)
VP 5.32E-005	mm Hg	25	EST	NEELY,WB & BLAU,GE (1985)
DC	pKa			
HL 2.45E-005	atm m3/mol	25	EST	MEYLAN,WM & HOWARD,PH (1991)
OH 5.46E-011	cm3/molc sec	25	EST	MEYLAN,WM & HOWARD,PH (1993)

CAS #: 019549-73-6	2,6-DIMETHYL-3-HEPTANOL

Formula: $C_9H_{20}O$

Mol Weight: 144.26

MP (deg C): FP (deg C):

BP (deg C): 175

BP pressure (mm Hg):

Property/Value	Units	Temp	Data Type	Reference
WS 6.14E+002	mg/L	25	EST	MEYLAN,WM ET AL. (1996)
logP 3.08			EST	MEYLAN,WM & HOWARD,PH (1995)
VP 2.79E-001	mm Hg	25	EST	NEELY,WB & BLAU,GE (1985)
DC	pKa			
HL 4.12E-005	atm m3/mol	25	EST	MEYLAN,WM & HOWARD,PH (1991)
OH 1.87E-011	cm3/molc sec	25	EST	MEYLAN,WM & HOWARD,PH (1993)

CAS #: 019549-79-2	3,5-DIMETHYL-4-HEPTANOL

Formula: $C_9H_{20}O$

Mol Weight: 144.26

MP (deg C): FP (deg C):

BP (deg C): 186

BP pressure (mm Hg):

Property/Value	Units	Temp	Data Type	Reference
WS 4.45E+002	mg/L	25	EXP	SUZUKI,T (1991)
logP 3.08			EST	MEYLAN,WM & HOWARD,PH (1995)
VP 2.79E-001	mm Hg	25	EST	NEELY,WB & BLAU,GE (1985)
DC	pKa			
HL 4.12E-005	atm m3/mol	25	EST	MEYLAN,WM & HOWARD,PH (1991)
OH 1.92E-011	cm3/molc sec	25	EST	MEYLAN,WM & HOWARD,PH (1993)

CAS #: 019561-31-0	4H-CYCLOHEPTA(DEF)PHENANTHRENE

Formula: $C_{17}H_{12}$

Mol Weight: 216.29

MP (deg C): FP (deg C):

BP (deg C):

BP pressure (mm Hg):

Property/Value	Units	Temp	Data Type	Reference
WS 8.82E-002	mg/L	25	EST	MEYLAN,WM ET AL. (1996)
logP 5.94			EST	MEYLAN,WM & HOWARD,PH (1995)
VP 1.30E-004	mm Hg	25	EST	NEELY,WB & BLAU,GE (1985)
DC	pKa			
HL	atm m3/mol			
OH 5.11E-010	cm3/molc sec	25	EST	MEYLAN,WM & HOWARD,PH (1993)

CAS #: 019590-04-6 — O=P(NME)(OME)O-(4-T-BU PHENYL)

Formula: $C_{12}H_{20}NO_3P$
Mol Weight: 257.27
MP (deg C): | FP (deg C):
BP (deg C):
BP pressure (mm Hg):

Property/Value	Units	Temp	Data Type	Reference
WS 1.05E+002	mg/L	25	EST	MEYLAN,WM ET AL. (1996)
logP 2.71			EXP	HANSCH,C & LEO,AJ (1985)
VP 6.62E-005	mm Hg	25	EST	NEELY,WB & BLAU,GE (1985)
DC	pKa			
HL 3.31E-009	atm m3/mol	25	EST	MEYLAN,WM & HOWARD,PH (1991)
OH 3.64E-011	cm3/molc sec	25	EST	MEYLAN,WM & HOWARD,PH (1993)

CAS #: 019590-05-7 — O=P(NC)(OC)O-3-T-BU PHENYL

Formula: $C_{12}H_{20}NO_3P$
Mol Weight: 257.27
MP (deg C): | FP (deg C):
BP (deg C):
BP pressure (mm Hg):

Property/Value	Units	Temp	Data Type	Reference
WS 1.18E+002	mg/L	25	EST	MEYLAN,WM ET AL. (1996)
logP 2.65			EXP	HANSCH,C & LEO,AJ (1985)
VP 6.62E-005	mm Hg	25	EST	NEELY,WB & BLAU,GE (1985)
DC	pKa			
HL 3.31E-009	atm m3/mol	25	EST	MEYLAN,WM & HOWARD,PH (1991)
OH 4.00E-011	cm3/molc sec	25	EST	MEYLAN,WM & HOWARD,PH (1993)

CAS #: 019591-17-4 — 2-IODOACETANILIDE

Formula: C_8H_8INO
Mol Weight: 261.06
MP (deg C): | FP (deg C):
BP (deg C):
BP pressure (mm Hg):

Property/Value	Units	Temp	Data Type	Reference
WS 8.49E+002	mg/L	25	EST	MEYLAN,WM ET AL. (1996)
logP 1.62			EXP	NAKAGAWA,Y ET AL. (1992)
VP 1.03E-005	mm Hg	25	EST	NEELY,WB & BLAU,GE (1985)
DC	pKa			
HL 1.43E-009	atm m3/mol	25	EST	MEYLAN,WM & HOWARD,PH (1991)
OH 4.20E-012	cm3/molc sec	25	EST	MEYLAN,WM & HOWARD,PH (1993)

CAS #: 019607-41-1 — PROPANEDIOIC ACID, 1,3-DITHIOLAN-2-YLIDENE-, DIE

Formula: $C_{10}H_{14}O_4S_2$
Mol Weight: 262.35
MP (deg C): | FP (deg C):
BP (deg C):
BP pressure (mm Hg):

Property/Value	Units	Temp	Data Type	Reference
WS 1.63E+002	mg/L	25	EST	MEYLAN,WM ET AL. (1996)
logP 2.45			EXP	SANGSTER,J (1993)
VP 2.66E-004	mm Hg	25	EST	NEELY,WB & BLAU,GE (1985)
DC	pKa			
HL 1.37E-010	atm m3/mol	25	EST	MEYLAN,WM & HOWARD,PH (1991)
OH 3.45E-011	cm3/molc sec	25	EST	MEYLAN,WM & HOWARD,PH (1993)

CAS #: 019608-64-1 — O=P(NC)(OC)O-2-CL PHENYL

Formula: $C_8H_{11}CINO_3P$
Mol Weight: 235.61
MP (deg C): | FP (deg C):
BP (deg C):
BP pressure (mm Hg):

Property/Value	Units	Temp	Data Type	Reference
WS 1.20E+003	mg/L	25	EST	MEYLAN,WM ET AL. (1996)
logP 1.61			EXP	HANSCH,C ET AL. (1995)
VP 1.91E-004	mm Hg	25	EST	NEELY,WB & BLAU,GE (1985)
DC	pKa			
HL 9.51E-010	atm m3/mol	25	EST	MEYLAN,WM & HOWARD,PH (1991)
OH 3.26E-011	cm3/molc sec	25	EST	MEYLAN,WM & HOWARD,PH (1993)

CAS #: 019666-30-9 — OXADIAZON

Formula: $C_{15}H_{18}Cl_2N_2O_3$
Mol Weight: 345.23
MP (deg C): 90 | FP (deg C):
BP (deg C):
BP pressure (mm Hg):

Property/Value	Units	Temp	Data Type	Reference
WS 7.00E-001	mg/L	24	EXP	YALKOWSKY,SH & DANNENFELSER,RM (1992)
logP 4.80			EXP	HANSCH,C ET AL. (1995)
VP 1.00E-005	mm Hg	25	EXP	WAUCHOPE,RD ET AL. (1991A)
DC	pKa			
HL 6.49E-006	atm m3/mol	25	EST	VP/WSOL
OH 2.43E-011	cm3/molc sec	25	EST	MEYLAN,WM & HOWARD,PH (1993)

CAS #: 019670-19-0 — O=P(NME)(OME)O-(4-CL-PHENYL)

Formula: $C_8H_{11}CINO_3P$
Mol Weight: 235.61
MP (deg C): | FP (deg C):
BP (deg C):
BP pressure (mm Hg):

Property/Value	Units	Temp	Data Type	Reference
WS 7.47E+002	mg/L	25	EST	MEYLAN,WM ET AL. (1996)
logP 1.85			EXP	HANSCH,C & LEO,AJ (1985)
VP 1.91E-004	mm Hg	25	EST	NEELY,WB & BLAU,GE (1985)
DC	pKa			
HL 9.51E-010	atm m3/mol	25	EST	MEYLAN,WM & HOWARD,PH (1991)
OH 3.26E-011	cm3/molc sec	25	EST	MEYLAN,WM & HOWARD,PH (1993)

CAS #: 019679-38-0 — 1,1'-(2,2,2-TRICHLOROETHYLIDENE)BIS(4-METHYLT*)

Formula: $C_{16}H_{15}Cl_3S_2$
Mol Weight: 377.78
MP (deg C): | FP (deg C):
BP (deg C):
BP pressure (mm Hg):

Property/Value	Units	Temp	Data Type	Reference
WS 7.85E-003	mg/L	25	EST	MEYLAN,WM ET AL. (1996)
logP 6.71			EST	MEYLAN,WM & HOWARD,PH (1995)
VP 1.28E-007	mm Hg	25	EST	NEELY,WB & BLAU,GE (1985)
DC	pKa			
HL 2.37E-008	atm m3/mol	25	EST	MEYLAN,WM & HOWARD,PH (1991)
OH 3.25E-011	cm3/molc sec	25	EST	MEYLAN,WM & HOWARD,PH (1993)

CAS #: 019700-21-1				GEOSMIN
Formula: $C_{12}H_{22}O$				
Mol Weight: 182.31				
MP (deg C):		FP (deg C):		
BP (deg C): 270				
BP pressure (mm Hg):				

Property/Value	Units	Temp	Data Type	Reference
WS 1.57E+002	mg/L	25	EST	MEYLAN,WM ET AL. (1996)
logP 3.57			EST	MEYLAN,WM & HOWARD,PH (1995)
VP 2.14E-003	mm Hg	25	EST	NEELY,WB & BLAU,GE (1985)
DC	pKa			
HL 1.18E-005	atm m3/mol	25	EST	MEYLAN,WM & HOWARD,PH (1991)
OH 2.24E-011	cm3/molc sec	25	EST	MEYLAN,WM & HOWARD,PH (1993)

CAS #: 019701-84-9				VALINE,N-ACETYL,N'-MEAM-AMIDE
Formula: $C_8H_{16}N_2O_2$				
Mol Weight: 172.23				
MP (deg C):		FP (deg C):		
BP (deg C):				
BP pressure (mm Hg):				

Property/Value	Units	Temp	Data Type	Reference
WS 5.85E+003	mg/L	25	EST	MEYLAN,WM ET AL. (1996)
logP -0.35			EXP	HANSCH,C & LEO,AJ (1985)
VP 4.04E-007	mm Hg	25	EST	NEELY,WB & BLAU,GE (1985)
DC	pKa			
HL 6.06E-010	atm m3/mol	25	EST	MEYLAN,WM & HOWARD,PH (1991)
OH 9.44E-011	cm3/molc sec	25	EST	MEYLAN,WM & HOWARD,PH (1993)

CAS #: 019723-86-5				PROPANEDIOIC ACID, 1,3-DITHIOLAN-2-YLIDENE,- DIM
Formula: $C_8H_{10}O_4S_2$				
Mol Weight: 234.29				
MP (deg C):		FP (deg C):		
BP (deg C):				
BP pressure (mm Hg):				

Property/Value	Units	Temp	Data Type	Reference
WS 1.48E+003	mg/L	25	EST	MEYLAN,WM ET AL. (1996)
logP 1.51			EXP	SANGSTER,J (1993)
VP 2.17E-003	mm Hg	25	EST	NEELY,WB & BLAU,GE (1985)
DC	pKa			
HL 7.79E-011	atm m3/mol	25	EST	MEYLAN,WM & HOWARD,PH (1991)
OH 3.16E-011	cm3/molc sec	25	EST	MEYLAN,WM & HOWARD,PH (1993)

CAS #: 019727-83-4				1H-INDOLE, 2,3-DIHYDRO-6-NITRO-
Formula: $C_8H_8N_2O_2$				
Mol Weight: 164.17				
MP (deg C): 67-69		FP (deg C):		
BP (deg C):				
BP pressure (mm Hg):				

Property/Value	Units	Temp	Data Type	Reference
WS 6.12E+002	mg/L	25	EST	MEYLAN,WM ET AL. (1996)
logP 1.92			EXP	DEBNATH,AK ET AL. (1991)
VP 4.75E-004	mm Hg	25	EST	NEELY,WB & BLAU,GE (1985)
DC	pKa			
HL 8.03E-009	atm m3/mol	25	EST	MEYLAN,WM & HOWARD,PH (1991)
OH 1.40E-011	cm3/molc sec	25	EST	MEYLAN,WM & HOWARD,PH (1993)

CAS #: 019730-12-2				N-I-PROPYL-3-PYRIDYLMETHYLAMINE
Formula: $C_9H_{14}N_2$				
Mol Weight: 150.23				
MP (deg C):		FP (deg C):		
BP (deg C):				
BP pressure (mm Hg):				

Property/Value	Units	Temp	Data Type	Reference
WS 1.00E+006	mg/L	25	EST	MEYLAN,WM ET AL. (1996)
logP 0.82			EXP	HANSCH,C & LEO,AJ (1985)
VP 7.75E-002	mm Hg	25	EST	NEELY,WB & BLAU,GE (1985)
DC	pKa			
HL 3.10E-009	atm m3/mol	25	EST	MEYLAN,WM & HOWARD,PH (1991)
OH 9.10E-011	cm3/molc sec	25	EST	MEYLAN,WM & HOWARD,PH (1993)

CAS #: 019730-13-3				N-N-PROPYL-3-PYRIDYLMETHYLAMINE
Formula: $C_9H_{14}N_2$				
Mol Weight: 150.23				
MP (deg C):		FP (deg C):		
BP (deg C):				
BP pressure (mm Hg):				

Property/Value	Units	Temp	Data Type	Reference
WS 1.00E+006	mg/L	25	EST	MEYLAN,WM ET AL. (1996)
logP 0.90			EXP	HANSCH,C & LEO,AJ (1985)
VP 4.11E-002	mm Hg	25	EST	NEELY,WB & BLAU,GE (1985)
DC	pKa			
HL 3.10E-009	atm m3/mol	25	EST	MEYLAN,WM & HOWARD,PH (1991)
OH 8.46E-011	cm3/molc sec	25	EST	MEYLAN,WM & HOWARD,PH (1993)

CAS #: 019730-15-5				N-ETHYL-2-(3-PYRIDYL)ETHYLAMINE
Formula: $C_9H_{14}N_2$				
Mol Weight: 150.23				
MP (deg C):		FP (deg C):		
BP (deg C):				
BP pressure (mm Hg):				

Property/Value	Units	Temp	Data Type	Reference
WS 1.00E+006	mg/L	25	EST	MEYLAN,WM ET AL. (1996)
logP 0.54			EXP	HANSCH,C & LEO,AJ (1985)
VP 4.11E-002	mm Hg	25	EST	NEELY,WB & BLAU,GE (1985)
DC	pKa			
HL 3.10E-009	atm m3/mol	25	EST	MEYLAN,WM & HOWARD,PH (1991)
OH 8.46E-011	cm3/molc sec	25	EST	MEYLAN,WM & HOWARD,PH (1993)

CAS #: 019745-07-4				PYRAZINE, 2,5-DICHLORO-
Formula: $C_4H_2Cl_2N_2$				
Mol Weight: 148.98				
MP (deg C):		FP (deg C):		
BP (deg C):				
BP pressure (mm Hg):				

Property/Value	Units	Temp	Data Type	Reference
WS 3.43E+003	mg/L	25	EST	MEYLAN,WM ET AL. (1996)
logP 1.58			EXP	YAMAGAMI,C ET AL. (1991)
VP 4.38E-001	mm Hg	25	EST	NEELY,WB & BLAU,GE (1985)
DC	pKa			
HL 6.38E-003	atm m3/mol	25	EST	MEYLAN,WM & HOWARD,PH (1991)
OH 4.72E-014	cm3/molc sec	25	EST	MEYLAN,WM & HOWARD,PH (1993)

CAS #: 019768-02-6 — FURAZAN, DIPHENYL-

Formula: $C_{14}H_{10}N_2O$

Mol Weight: 222.25

MP (deg C):   FP (deg C):

BP (deg C):

BP pressure (mm Hg):

Property/ Value	Units	Temp	Data Type	Reference
WS 1.80E+001	mg/L	25	EST	MEYLAN,WM ET AL. (1996)
logP 3.83			EXP	CALVINO,R ET AL. (1992)
VP 2.61E-006	mm Hg	25	EST	NEELY,WB & BLAU,GE (1985)
DC	pKa			
HL 1.12E-006	atm m3/mol	25	EST	MEYLAN,WM & HOWARD,PH (1991)
OH 1.08E-011	cm3/molc sec	25	EST	MEYLAN,WM & HOWARD,PH (1993)

CAS #: 019776-98-8 — BENZOXAZOLE, 2-METHYLAMINO-

Formula: $C_8H_8N_2O$

Mol Weight: 148.17

MP (deg C):   FP (deg C):

BP (deg C):

BP pressure (mm Hg):

Property/ Value	Units	Temp	Data Type	Reference
WS 2.16E+003	mg/L	25	EST	MEYLAN,WM ET AL. (1996)
logP 1.82			EXP	HANSCH,C & LEO,AJ (1985)
VP 7.25E-003	mm Hg	25	EST	NEELY,WB & BLAU,GE (1985)
DC	pKa			
HL 5.33E-010	atm m3/mol	25	EST	MEYLAN,WM & HOWARD,PH (1991)
OH 1.98E-010	cm3/molc sec	25	EST	MEYLAN,WM & HOWARD,PH (1993)

CAS #: 019789-60-7 — LYSIN-AMIDE, N-ACETYL

Formula: $C_8H_{17}N_3O_2$

Mol Weight: 187.24

MP (deg C):   FP (deg C):

BP (deg C):

BP pressure (mm Hg):

Property/ Value	Units	Temp	Data Type	Reference
WS 6.36E+005	mg/L	25	EST	MEYLAN,WM ET AL. (1996)
logP -2.82			EXP	HANSCH,C & LEO,AJ (1985)
VP 1.61E-008	mm Hg	25	EST	NEELY,WB & BLAU,GE (1985)
DC	pKa			
HL 3.77E-014	atm m3/mol	25	EST	MEYLAN,WM & HOWARD,PH (1991)
OH 8.22E-011	cm3/molc sec	25	EST	MEYLAN,WM & HOWARD,PH (1993)

CAS #: 019789-69-6 — 8-FLUORO-4-NITROQUINOLINE-1-OXIDE

Formula: $C_9H_7FN_2O_3$

Mol Weight: 210.17

MP (deg C):   FP (deg C):

BP (deg C):

BP pressure (mm Hg):

Property/ Value	Units	Temp	Data Type	Reference
WS 2.27E+003	mg/L	25	EST	MEYLAN,WM ET AL. (1996)
logP 1.00			EXP	HANSCH,C & LEO,AJ (1985)
VP 2.91E-006	mm Hg	25	EST	NEELY,WB & BLAU,GE (1985)
DC	pKa			
HL 3.17E-014	atm m3/mol	25	EST	MEYLAN,WM & HOWARD,PH (1991)
OH 9.78E-013	cm3/molc sec	25	EST	MEYLAN,WM & HOWARD,PH (1993)

CAS #: 019797-08-1 — N-ETHYL-EPSILON-CAPROLACTAM

Formula: $C_8H_{15}NO$

Mol Weight: 141.21

MP (deg C):   FP (deg C):

BP (deg C):

BP pressure (mm Hg):

Property/ Value	Units	Temp	Data Type	Reference
WS 2.31E+004	mg/L	25	EST	MEYLAN,WM ET AL. (1996)
logP 0.65			EXP	HANSCH,C & LEO,AJ (1985)
VP 7.96E-003	mm Hg	25	EST	NEELY,WB & BLAU,GE (1985)
DC	pKa			
HL 7.38E-008	atm m3/mol	25	EST	MEYLAN,WM & HOWARD,PH (1991)
OH 3.19E-011	cm3/molc sec	25	EST	MEYLAN,WM & HOWARD,PH (1993)

CAS #: 019829-29-9 — 4(1H)-PYRIDINETHIONE

Formula: C_5H_5NS

Mol Weight: 111.17

MP (deg C):   FP (deg C):

BP (deg C):

BP pressure (mm Hg):

Property/ Value	Units	Temp	Data Type	Reference
WS 7.41E+005	mg/L	25	EST	MEYLAN,WM ET AL. (1996)
logP 0.20			EXP	HANSCH,C ET AL. (1995)
VP 8.63E-003	mm Hg	25	EST	NEELY,WB & BLAU,GE (1985)
DC	pKa			
HL 2.67E-005	atm m3/mol	25	EST	MEYLAN,WM & HOWARD,PH (1991)
OH 1.31E-010	cm3/molc sec	25	EST	MEYLAN,WM & HOWARD,PH (1993)

CAS #: 019833-78-4 — DIETHYLAMINE, HYDROIODIDE

Formula: $C_4H_{12}IN$

Mol Weight: 201.05

MP (deg C):   FP (deg C):

BP (deg C):

BP pressure (mm Hg):

Property/ Value	Units	Temp	Data Type	Reference
WS 1.00E+006	mg/L	25	EXP	SEIDELL,A (1941)
logP -2.29			EST	MEYLAN,WM & HOWARD,PH (1995)
VP 4.45E-006	mm Hg	25	EST	NEELY,WB & BLAU,GE (1985)
DC	pKa			
HL 1.41E-012	atm m3/mol	25	EST	MEYLAN,WM & HOWARD,PH (1991)
OH 1.77E-011	cm3/molc sec	25	EST	MEYLAN,WM & HOWARD,PH (1993)

CAS #: 019837-74-2 — N1-(4-METHOXYPHENYL)SULFANILAMIDE

Formula: $C_{13}H_{14}N_2O_3S$

Mol Weight: 278.33

MP (deg C):   FP (deg C):

BP (deg C):

BP pressure (mm Hg):

Property/ Value	Units	Temp	Data Type	Reference
WS 8.41E+002	mg/L	25	EST	MEYLAN,WM ET AL. (1996)
logP 1.51			EXP	HANSCH,C & LEO,AJ (1985)
VP 1.79E-008	mm Hg	25	EST	NEELY,WB & BLAU,GE (1985)
DC	pKa			
HL 4.87E-012	atm m3/mol	25	EST	MEYLAN,WM & HOWARD,PH (1991)
OH 6.07E-011	cm3/molc sec	25	EST	MEYLAN,WM & HOWARD,PH (1993)

CAS #: 019837-84-4 — N1-(2-METHOXYPHENYL)SULFANILAMIDE

Formula: $C_{13}H_{14}N_2O_3S$

Mol Weight: 278.33

MP (deg C): FP (deg C):

BP (deg C):

BP pressure (mm Hg):

Property/Value	Units	Temp	Data Type	Reference
WS 7.62E+002	mg/L	25	EST	MEYLAN,WM ET AL. (1996)
logP 1.56			EXP	HANSCH,C & LEO,AJ (1985)
VP 1.79E-008	mm Hg	25	EST	NEELY,WB & BLAU,GE (1985)
DC	pKa			
HL 4.87E-012	atm m3/mol	25	EST	MEYLAN,WM & HOWARD,PH (1991)
OH 6.07E-011	cm3/molc sec	25	EST	MEYLAN,WM & HOWARD,PH (1993)

CAS #: 019837-85-5 — N1-(2-CHLOROPHENYL)SULFANILAMIDE

Formula: $C_{12}H_{11}ClN_2O_2S$

Mol Weight: 282.75

MP (deg C): FP (deg C):

BP (deg C):

BP pressure (mm Hg):

Property/Value	Units	Temp	Data Type	Reference
WS 3.03E+002	mg/L	25	EST	MEYLAN,WM ET AL. (1996)
logP 2.00			EXP	HANSCH,C & LEO,AJ (1985)
VP 2.57E-008	mm Hg	25	EST	NEELY,WB & BLAU,GE (1985)
DC	pKa			
HL 6.10E-011	atm m3/mol	25	EST	MEYLAN,WM & HOWARD,PH (1991)
OH 3.55E-011	cm3/molc sec	25	EST	MEYLAN,WM & HOWARD,PH (1993)

CAS #: 019837-88-8 — BENZENESULFONAMIDE, 4-AMINO-N-(2-NITROPHENYL)-

Formula: $C_{12}H_{11}N_3O_4S$

Mol Weight: 293.30

MP (deg C): FP (deg C):

BP (deg C):

BP pressure (mm Hg):

Property/Value	Units	Temp	Data Type	Reference
WS 8.63E+001	mg/L	25	EST	MEYLAN,WM ET AL. (1996)
logP 2.11			EXP	SANGSTER,J (1993)
VP 1.90E-009	mm Hg	25	EST	NEELY,WB & BLAU,GE (1985)
DC	pKa			
HL 7.11E-012	atm m3/mol	25	EST	MEYLAN,WM & HOWARD,PH (1991)
OH 2.83E-011	cm3/molc sec	25	EST	MEYLAN,WM & HOWARD,PH (1993)

CAS #: 019847-12-2 — 2-CYANOPYRAZINE

Formula: $C_5H_3N_3$

Mol Weight: 105.10

MP (deg C): FP (deg C):

BP (deg C): 87

BP pressure (mm Hg): 6.00E+000

Property/Value	Units	Temp	Data Type	Reference
WS 6.25E+004	mg/L	25	EST	MEYLAN,WM ET AL. (1996)
logP -0.01			EXP	YAMAGAMI,C ET AL. (1990A)
VP 1.68E-001	mm Hg	25	EST	NEELY,WB & BLAU,GE (1985)
DC	pKa			
HL 2.82E-008	atm m3/mol	25	EST	MEYLAN,WM & HOWARD,PH (1991)
OH 4.06E-014	cm3/molc sec	25	EST	MEYLAN,WM & HOWARD,PH (1993)

CAS #: 019852-25-6 — 5,3',4'-TRIHYDROXYFLAVONE

Formula: $C_{15}H_{10}O_5$

Mol Weight: 270.24

MP (deg C): FP (deg C):

BP (deg C):

BP pressure (mm Hg):

Property/Value	Units	Temp	Data Type	Reference
WS 1.03E+002	mg/L	25	EST	MEYLAN,WM ET AL. (1996)
logP 3.31			EXP	PERRISSOUD,D & TESTA,B (1986)
VP 1.01E-010	mm Hg	25	EST	NEELY,WB & BLAU,GE (1985)
DC	pKa			
HL 5.12E-017	atm m3/mol	25	EST	MEYLAN,WM & HOWARD,PH (1991)
OH 2.31E-010	cm3/molc sec	25	EST	MEYLAN,WM & HOWARD,PH (1993)

CAS #: 019860-27-6 — 5-PHENYL-2,4-PYRROLIDIONE

Formula: $C_{10}H_9NO_2$

Mol Weight: 175.19

MP (deg C): FP (deg C):

BP (deg C):

BP pressure (mm Hg):

Property/Value	Units	Temp	Data Type	Reference
WS 2.26E+004	mg/L	25	EST	MEYLAN,WM ET AL. (1996)
logP 0.48			EXP	LIPINSKI,CA ET AL. (1991)
VP 2.58E-006	mm Hg	25	EST	NEELY,WB & BLAU,GE (1985)
DC	pKa			
HL 8.84E-012	atm m3/mol	25	EST	MEYLAN,WM & HOWARD,PH (1991)
OH 1.53E-011	cm3/molc sec	25	EST	MEYLAN,WM & HOWARD,PH (1993)

CAS #: 019865-55-5 — ALPHA-(P-ME-PHENYL)-N-PHENYLNITRONE

Formula: $C_{14}H_{13}NO$

Mol Weight: 211.27

MP (deg C): FP (deg C):

BP (deg C):

BP pressure (mm Hg):

Property/Value	Units	Temp	Data Type	Reference
WS 3.35E+002	mg/L	25	EST	MEYLAN,WM ET AL. (1996)
logP 2.41			EXP	KIRCHNER,JJ ET AL. (1985)
VP 5.16E-009	mm Hg	25	EST	NEELY,WB & BLAU,GE (1985)
DC	pKa			
HL 8.72E-012	atm m3/mol	25	EST	MEYLAN,WM & HOWARD,PH (1991)
OH 8.62E-012	cm3/molc sec	25	EST	MEYLAN,WM & HOWARD,PH (1993)

CAS #: 019918-36-6 — 5-PYRIDINE IMADAZOLE, 2-CF3

Formula: $C_7H_4F_3N_3$

Mol Weight: 187.13

MP (deg C): FP (deg C):

BP (deg C):

BP pressure (mm Hg):

Property/Value	Units	Temp	Data Type	Reference
WS 8.00E+003	mg/L	25	EST	MEYLAN,WM ET AL. (1996)
logP 0.94			EXP	HANSCH,C & LEO,AJ (1985)
VP 2.78E-005	mm Hg	25	EST	NEELY,WB & BLAU,GE (1985)
DC	pKa			
HL 4.17E-009	atm m3/mol	25	EST	MEYLAN,WM & HOWARD,PH (1991)
OH 7.24E-012	cm3/molc sec	25	EST	MEYLAN,WM & HOWARD,PH (1993)

CAS #: 019934-51-1	TRICHLOROAMPHENICOL

Formula: $C_{11}H_{11}Cl_3N_2O_5$

Mol Weight: 357.58

MP (deg C):   FP (deg C):

BP (deg C):

BP pressure (mm Hg):

Property/Value	Units	Temp	Data Type	Reference
WS 4.72E+001	mg/L	25	EST	MEYLAN,WM ET AL. (1996)
logP 1.97			EXP	HANSCH,C & LEO,AJ (1985)
VP 2.73E-013	mm Hg	25	EST	NEELY,WB & BLAU,GE (1985)
DC	pKa			
HL 8.06E-019	atm m3/mol	25	EST	MEYLAN,WM & HOWARD,PH (1991)
OH 3.08E-011	cm3/molc sec	25	EST	MEYLAN,WM & HOWARD,PH (1993)

CAS #: 019937-59-8	METOXURON

Formula: $C_{10}H_{13}ClN_2O_2$

Mol Weight: 228.68

MP (deg C):   FP (deg C):

BP (deg C):

BP pressure (mm Hg):

Property/Value	Units	Temp	Data Type	Reference
WS 6.00E+002	mg/L	20	EXP	YALKOWSKY,SH & DANNENFELSER,RM (1992)
logP 1.64			EXP	HANSCH,C ET AL. (1995)
VP 7.35E-006	mm Hg	25	EST	NEELY,WB & BLAU,GE (1985)
DC	pKa			
HL 4.25E-011	atm m3/mol	25	EST	MEYLAN,WM & HOWARD,PH (1991)
OH 2.89E-011	cm3/molc sec	25	EST	MEYLAN,WM & HOWARD,PH (1993)

CAS #: 019961-27-4	ETHYL ISOPROPYLAMINE

Formula: $C_5H_{13}N$

Mol Weight: 87.17

MP (deg C):   FP (deg C):

BP (deg C): 69.6

BP pressure (mm Hg):

Property/Value	Units	Temp	Data Type	Reference
WS 2.07E+005	mg/L	25	EST	MEYLAN,WM ET AL. (1996)
logP 0.93			EXP	HANSCH,C & LEO,AJ (1985)
VP 9.40E+001	mm Hg	25	EST	NEELY,WB & BLAU,GE (1985)
DC	pKa			
HL 3.89E-005	atm m3/mol	25	EST	MEYLAN,WM & HOWARD,PH (1991)
OH 9.02E-011	cm3/molc sec	25	EST	MEYLAN,WM & HOWARD,PH (1993)

CAS #: 019982-08-2	1-AMINOADAMANTANE,3,5-DIMETHYL

Formula: $C_{12}H_{21}N$

Mol Weight: 179.31

MP (deg C): 258   FP (deg C):

BP (deg C):

BP pressure (mm Hg):

Property/Value	Units	Temp	Data Type	Reference
WS 8.94E+002	mg/L	25	EST	MEYLAN,WM ET AL. (1996)
logP 3.28			EXP	HANSCH,C & LEO,AJ (1985)
VP 3.56E-002	mm Hg	25	EST	NEELY,WB & BLAU,GE (1985)
DC	pKa			
HL 1.47E-005	atm m3/mol	25	EST	MEYLAN,WM & HOWARD,PH (1991)
OH 3.34E-011	cm3/molc sec	25	EST	MEYLAN,WM & HOWARD,PH (1993)

CAS #: 020020-02-4	1,2,3,4-TETRACHLORONAPHTHALENE

Formula: $C_{10}H_4Cl_4$

Mol Weight: 265.96

MP (deg C): 198   FP (deg C):

BP (deg C):

BP pressure (mm Hg):

Property/Value	Units	Temp	Data Type	Reference
WS 4.26E-003	mg/L	22	EXP	OPPERHUIZEN,A ET AL. (1985)
logP 5.75			EXP	HANSCH,C & LEO,AJ (1985)
VP 5.79E-005	mm Hg	25	EST	NEELY,WB & BLAU,GE (1985)
DC	pKa			
HL 1.59E+000	atm m3/mol	25	EST	MEYLAN,WM & HOWARD,PH (1991)
OH 8.90E-013	cm3/molc sec	25	EST	ATKINSON,R (1988)

CAS #: 020037-50-7	BENZAMIDE, N-[(4-AMINOPHENYL)SULFONYL]-2,5-DIMET

Formula: $C_{15}H_{16}N_2O_3S$

Mol Weight: 304.37

MP (deg C):   FP (deg C):

BP (deg C):

BP pressure (mm Hg):

Property/Value	Units	Temp	Data Type	Reference
WS 1.56E+002	mg/L	25	EST	MEYLAN,WM ET AL. (1996)
logP 2.19			EXP	SANGSTER,J (1994)
VP 4.20E-011	mm Hg	25	EST	NEELY,WB & BLAU,GE (1985)
DC	pKa			
HL 1.17E-013	atm m3/mol	25	EST	MEYLAN,WM & HOWARD,PH (1991)
OH 3.45E-011	cm3/molc sec	25	EST	MEYLAN,WM & HOWARD,PH (1993)

CAS #: 020039-91-2	2-BROMO-4-METHYL-6-NITROPHENOL

Formula: $C_7H_6BrNO_3$

Mol Weight: 232.04

MP (deg C):   FP (deg C):

BP (deg C):

BP pressure (mm Hg):

Property/Value	Units	Temp	Data Type	Reference
WS 4.12E+001	mg/L	25	EST	MEYLAN,WM ET AL. (1996)
logP 3.35			EST	MEYLAN,WM & HOWARD,PH (1995)
VP 3.60E-005	mm Hg	25	EST	NEELY,WB & BLAU,GE (1985)
DC	pKa			
HL 3.08E-006	atm m3/mol	25	EST	MEYLAN,WM & HOWARD,PH (1991)
OH 4.69E-013	cm3/molc sec	25	EST	MEYLAN,WM & HOWARD,PH (1993)

CAS #: 020046-63-3	124-BENZTHIADIAZN-1-O2-3-ME-7-CF3

Formula: $C_9H_7F_3N_2O_2S$

Mol Weight: 264.23

MP (deg C):   FP (deg C):

BP (deg C):

BP pressure (mm Hg):

Property/Value	Units	Temp	Data Type	Reference
WS 1.01E+003	mg/L	25	EST	MEYLAN,WM ET AL. (1996)
logP 1.51			EXP	HANSCH,C & LEO,AJ (1985)
VP 8.89E-008	mm Hg	25	EST	NEELY,WB & BLAU,GE (1985)
DC	pKa			
HL 8.04E-007	atm m3/mol	25	EST	MEYLAN,WM & HOWARD,PH (1991)
OH 2.36E-013	cm3/molc sec	25	EST	MEYLAN,WM & HOWARD,PH (1993)

CAS #: 020056-71-7 — PHENOL, 2-UNDECYL-

Formula: $C_{17}H_{28}O$

Mol Weight: 248.41

MP (deg C):   FP (deg C):

BP (deg C):

BP pressure (mm Hg):

Property/Value	Units	Temp	Data Type	Reference
WS 6.91E-002	mg/L	25	EST	MEYLAN,WM ET AL. (1996)
logP 7.17			EXP	ITOKAWA,H ET AL. (1989)
VP 5.62E-006	mm Hg	25	EST	NEELY,WB & BLAU,GE (1985)
DC	pKa			
HL 1.05E-005	atm m3/mol	25	EST	MEYLAN,WM & HOWARD,PH (1991)
OH 5.45E-011	cm3/molc sec	25	EST	MEYLAN,WM & HOWARD,PH (1993)

CAS #: 020056-72-8 — PHENOL, 3-UNDECYL-

Formula: $C_{17}H_{28}O$

Mol Weight: 248.41

MP (deg C):   FP (deg C):

BP (deg C):

BP pressure (mm Hg):

Property/Value	Units	Temp	Data Type	Reference
WS 1.00E-001	mg/L	25	EST	MEYLAN,WM ET AL. (1996)
logP 6.98			EXP	ITOKAWA,H ET AL. (1989)
VP 5.62E-006	mm Hg	25	EST	NEELY,WB & BLAU,GE (1985)
DC	pKa			
HL 1.05E-005	atm m3/mol	25	EST	MEYLAN,WM & HOWARD,PH (1991)
OH 9.67E-011	cm3/molc sec	25	EST	MEYLAN,WM & HOWARD,PH (1993)

CAS #: 020056-73-9 — PHENOL, 4-UNDECYL-

Formula: $C_{17}H_{28}O$

Mol Weight: 248.41

MP (deg C):   FP (deg C):

BP (deg C):

BP pressure (mm Hg):

Property/Value	Units	Temp	Data Type	Reference
WS 6.78E-002	mg/L	25	EST	MEYLAN,WM ET AL. (1996)
logP 7.18			EXP	ITOKAWA,H ET AL. (1989)
VP 5.62E-006	mm Hg	25	EST	NEELY,WB & BLAU,GE (1985)
DC	pKa			
HL 1.05E-005	atm m3/mol	25	EST	MEYLAN,WM & HOWARD,PH (1991)
OH 5.45E-011	cm3/molc sec	25	EST	MEYLAN,WM & HOWARD,PH (1993)

CAS #: 020056-92-2 — 7-DODECEN-1-OL (Z)

Formula: $C_{12}H_{24}O$

Mol Weight: 184.32

MP (deg C):   FP (deg C):

BP (deg C):

BP pressure (mm Hg):

Property/Value	Units	Temp	Data Type	Reference
WS 2.19E+001	mg/L	25	EST	MEYLAN,WM ET AL. (1996)
logP 4.55			EST	MEYLAN,WM & HOWARD,PH (1995)
VP 5.16E-004	mm Hg	25	EST	NEELY,WB & BLAU,GE (1985)
DC	pKa			
HL 8.47E-005	atm m3/mol	25	EST	MEYLAN,WM & HOWARD,PH (1991)
OH 7.12E-011	cm3/molc sec	25	EST	MEYLAN,WM & HOWARD,PH (1993)

CAS #: 020057-31-2 — A-(2,2,2-TRICHLOROETHYL)STYRENE

Formula: $C_{10}H_9Cl_3$

Mol Weight: 235.54

MP (deg C):   FP (deg C):

BP (deg C):

BP pressure (mm Hg):

Property/Value	Units	Temp	Data Type	Reference
WS 1.23E+001	mg/L	25	EXP	SHIU,WY ET AL. (1990)
logP 4.56			EXP	HANSCH,C & LEO,AJ (1985)
VP 1.72E-002	mm Hg	25	EST	NEELY,WB & BLAU,GE (1985)
DC	pKa			
HL 2.51E-004	atm m3/mol	25	EST	MEYLAN,WM & HOWARD,PH (1991)
OH 5.33E-011	cm3/molc sec	25	EST	MEYLAN,WM & HOWARD,PH (1993)

CAS #: 020071-53-8 — 2-BUTENOIC ACID, ESTER DERIVATIVE

Formula: $C_{20}H_{24}O_7$

Mol Weight: 376.41

MP (deg C):   FP (deg C):

BP (deg C):

BP pressure (mm Hg):

Property/Value	Units	Temp	Data Type	Reference
WS 1.54E+002	mg/L	25	EST	MEYLAN,WM ET AL. (1996)
logP 1.69			EXP	SANGSTER,J (1993)
VP 7.74E-013	mm Hg	25	EST	NEELY,WB & BLAU,GE (1985)
DC	pKa			
HL 9.41E-016	atm m3/mol	25	EST	MEYLAN,WM & HOWARD,PH (1991)
OH 1.16E-010	cm3/molc sec	25	EST	MEYLAN,WM & HOWARD,PH (1993)

CAS #: 020149-84-2 — ACETAMIDE, 2-CHLORO-N-(3,4-DICHLOROPHENYL)

Formula: $C_8H_6Cl_3NO$

Mol Weight: 238.50

MP (deg C):   FP (deg C):

BP (deg C):

BP pressure (mm Hg):

Property/Value	Units	Temp	Data Type	Reference
WS 4.24E+001	mg/L	25	EST	MEYLAN,WM ET AL. (1996)
logP 3.29			EXP	HANSCH,C ET AL. (1995)
VP 2.62E-006	mm Hg	25	EST	NEELY,WB & BLAU,GE (1985)
DC	pKa			
HL 1.19E-009	atm m3/mol	25	EST	MEYLAN,WM & HOWARD,PH (1991)
OH 2.82E-012	cm3/molc sec	25	EST	MEYLAN,WM & HOWARD,PH (1993)

CAS #: 020173-04-0 — N-METHYL-3-PYRIDYLMETHYLAMINE

Formula: $C_7H_{10}N_2$

Mol Weight: 122.17

MP (deg C):   FP (deg C):

BP (deg C):

BP pressure (mm Hg):

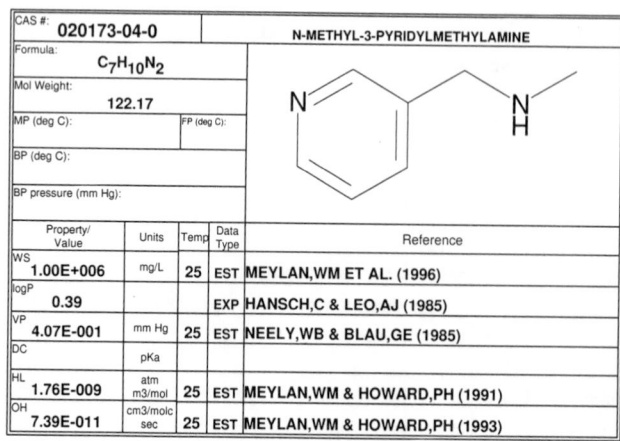

Property/Value	Units	Temp	Data Type	Reference
WS 1.00E+006	mg/L	25	EST	MEYLAN,WM ET AL. (1996)
logP 0.39			EXP	HANSCH,C & LEO,AJ (1985)
VP 4.07E-001	mm Hg	25	EST	NEELY,WB & BLAU,GE (1985)
DC	pKa			
HL 1.76E-009	atm m3/mol	25	EST	MEYLAN,WM & HOWARD,PH (1991)
OH 7.39E-011	cm3/molc sec	25	EST	MEYLAN,WM & HOWARD,PH (1993)

CAS #: 020173-12-0 — N-BUTYL-3-PYRIDYLMETHYLAMINE

Formula: $C_{10}H_{16}N_2$

Mol Weight: 164.25

MP (deg C):   FP (deg C):

BP (deg C):

BP pressure (mm Hg):

Property/ Value	Units	Temp	Data Type	Reference
WS 9.54E+005	mg/L	25	EST	MEYLAN,WM ET AL. (1996)
logP 1.34			EXP	HANSCH,C & LEO,AJ (1985)
VP 1.32E-002	mm Hg	25	EST	NEELY,WB & BLAU,GE (1985)
DC	pKa			
HL 4.11E-009	atm m3/mol	25	EST	MEYLAN,WM & HOWARD,PH (1991)
OH 8.60E-011	cm3/molc sec	25	EST	MEYLAN,WM & HOWARD,PH (1993)

CAS #: 020173-18-6 — N,N-DI-I-PR-3-PYRIDYLMETHYLAMINE

Formula: $C_{12}H_{20}N_2$

Mol Weight: 192.31

MP (deg C):   FP (deg C):

BP (deg C):

BP pressure (mm Hg):

Property/ Value	Units	Temp	Data Type	Reference
WS 1.12E+005	mg/L	25	EST	MEYLAN,WM ET AL. (1996)
logP 2.27			EXP	HANSCH,C & LEO,AJ (1985)
VP 1.67E-002	mm Hg	25	EST	NEELY,WB & BLAU,GE (1985)
DC	pKa			
HL 1.20E-008	atm m3/mol	25	EST	MEYLAN,WM & HOWARD,PH (1991)
OH 1.12E-010	cm3/molc sec	25	EST	MEYLAN,WM & HOWARD,PH (1993)

CAS #: 020173-24-4 — 3-PYRIDINE ETHANAMINE

Formula: $C_7H_{10}N_2$

Mol Weight: 122.17

MP (deg C):   FP (deg C):

BP (deg C):

BP pressure (mm Hg):

Property/ Value	Units	Temp	Data Type	Reference
WS 1.00E+006	mg/L	25	EST	MEYLAN,WM ET AL. (1996)
logP -0.11			EXP	HANSCH,C ET AL. (1995)
VP 1.48E-001	mm Hg	25	EST	NEELY,WB & BLAU,GE (1985)
DC	pKa			
HL 1.06E-009	atm m3/mol	25	EST	MEYLAN,WM & HOWARD,PH (1991)
OH 3.38E-011	cm3/molc sec	25	EST	MEYLAN,WM & HOWARD,PH (1993)

CAS #: 020173-26-6 — N,N-DIME-2-(3-PYRIDYL)ETHYLAMINE

Formula: $C_9H_{14}N_2$

Mol Weight: 150.23

MP (deg C):   FP (deg C):

BP (deg C):

BP pressure (mm Hg):

Property/ Value	Units	Temp	Data Type	Reference
WS 1.00E+006	mg/L	25	EST	MEYLAN,WM ET AL. (1996)
logP 0.82			EXP	HANSCH,C & LEO,AJ (1985)
VP 1.12E-001	mm Hg	25	EST	NEELY,WB & BLAU,GE (1985)
DC 8.86	pKa	25	EXP	PERRIN,DD (1965)
HL 5.12E-009	atm m3/mol	25	EST	MEYLAN,WM & HOWARD,PH (1991)
OH 8.13E-011	cm3/molc sec	25	EST	MEYLAN,WM & HOWARD,PH (1993)

CAS #: 020173-28-8 — 3-PYRIDYLETHYL-2-(N-PYRROLIDINE)

Formula: $C_{11}H_{16}N_2$

Mol Weight: 176.26

MP (deg C):   FP (deg C):

BP (deg C):

BP pressure (mm Hg):

Property/ Value	Units	Temp	Data Type	Reference
WS 8.37E+005	mg/L	25	EST	MEYLAN,WM ET AL. (1996)
logP 1.34			EXP	HANSCH,C & LEO,AJ (1985)
VP 4.17E-003	mm Hg	25	EST	NEELY,WB & BLAU,GE (1985)
DC 9.28	pKa	25	EXP	PERRIN,DD (1965)
HL 3.98E-009	atm m3/mol	25	EST	MEYLAN,WM & HOWARD,PH (1991)
OH 9.81E-011	cm3/molc sec	25	EST	MEYLAN,WM & HOWARD,PH (1993)

CAS #: 020173-34-6 — N,N-DIETHYL-3-PYRIDYLETHYLAMINE

Formula: $C_{11}H_{18}N_2$

Mol Weight: 178.28

MP (deg C):   FP (deg C):

BP (deg C):

BP pressure (mm Hg):

Property/ Value	Units	Temp	Data Type	Reference
WS 1.00E+006	mg/L	25	EST	MEYLAN,WM ET AL. (1996)
logP 1.23			EXP	HANSCH,C & LEO,AJ (1985)
VP 1.11E-002	mm Hg	25	EST	NEELY,WB & BLAU,GE (1985)
DC	pKa			
HL 9.03E-009	atm m3/mol	25	EST	MEYLAN,WM & HOWARD,PH (1991)
OH 9.65E-011	cm3/molc sec	25	EST	MEYLAN,WM & HOWARD,PH (1993)

CAS #: 020193-21-9 — PROPYL-BUTYLAMINE

Formula: $C_7H_{17}N$

Mol Weight: 115.22

MP (deg C):   FP (deg C):

BP (deg C):

BP pressure (mm Hg):

Property/ Value	Units	Temp	Data Type	Reference
WS 1.65E+004	mg/L	25	EST	MEYLAN,WM ET AL. (1996)
logP 2.12			EXP	HANSCH,C & LEO,AJ (1985)
VP 6.43E+000	mm Hg	25	EST	NEELY,WB & BLAU,GE (1985)
DC	pKa			
HL 6.86E-005	atm m3/mol	25	EST	MEYLAN,WM & HOWARD,PH (1991)
OH 8.84E-011	cm3/molc sec	25	EST	MEYLAN,WM & HOWARD,PH (1993)

CAS #: 020198-19-0 — 2-AMINOQUINAZOLINE-4-ONE

Formula: $C_8H_7N_3O$

Mol Weight: 161.16

MP (deg C):   FP (deg C):

BP (deg C):

BP pressure (mm Hg):

Property/ Value	Units	Temp	Data Type	Reference
WS 2.08E+004	mg/L	25	EST	MEYLAN,WM ET AL. (1996)
logP 0.60			EXP	HANSCH,C & LEO,AJ (1985)
VP 2.06E-008	mm Hg	25	EST	NEELY,WB & BLAU,GE (1985)
DC	pKa			
HL 3.76E-014	atm m3/mol	25	EST	MEYLAN,WM & HOWARD,PH (1991)
OH 2.76E-011	cm3/molc sec	25	EST	MEYLAN,WM & HOWARD,PH (1993)

CAS #: 020222-29-1				4-NITRO-N-(TRIPHENYLMETHYL)BENZENAMINE
Formula: $C_{25}H_{20}N_2O_2$				
Mol Weight: 380.45				
MP (deg C):		FP (deg C):		
BP (deg C):				
BP pressure (mm Hg):				

Property/ Value	Units	Temp	Data Type	Reference
WS 7.22E-003	mg/L	25	EST	MEYLAN,WM ET AL. (1996)
logP 6.73			EST	MEYLAN,WM & HOWARD,PH (1995)
VP 8.67E-011	mm Hg	25	EST	NEELY,WB & BLAU,GE (1985)
DC	pKa			
HL 8.68E-012	atm m3/mol	25	EST	MEYLAN,WM & HOWARD,PH (1991)
OH 1.82E-011	cm3/molc sec	25	EST	MEYLAN,WM & HOWARD,PH (1993)

CAS #: 020225-24-5				2-ETHYLPENTANOIC ACID
Formula: $C_7H_{14}O_2$				
Mol Weight: 130.19				
MP (deg C):		FP (deg C):		
BP (deg C):				
BP pressure (mm Hg):				

Property/ Value	Units	Temp	Data Type	Reference
WS 2.84E+003	mg/L	25	EST	MEYLAN,WM ET AL. (1996)
logP 2.23			EXP	SANGSTER,J (1994)
VP 1.48E-001	mm Hg	25	EST	NEELY,WB & BLAU,GE (1985)
DC 4.71	pKa	18	EXP	KORTUM,G ET AL (1961)
HL 2.26E-006	atm m3/mol	25	EST	MEYLAN,WM & HOWARD,PH (1991)
OH 6.77E-012	cm3/molc sec	25	EST	MEYLAN,WM & HOWARD,PH (1993)

CAS #: 020237-34-7				1,3-HEXADIENE (TRANS)
Formula: C_6H_{10}				
Mol Weight: 82.15				
MP (deg C): -102.4		FP (deg C):		
BP (deg C): 73.2				
BP pressure (mm Hg):				

Property/ Value	Units	Temp	Data Type	Reference
WS 7.53E+001	mg/L	25	EST	MEYLAN,WM ET AL. (1996)
logP 2.94			EST	MEYLAN,WM & HOWARD,PH (1995)
VP 1.07E+002	mm Hg	25	EST	NEELY,WB & BLAU,GE (1985)
DC	pKa			
HL 1.62E-001	atm m3/mol	25	EST	MEYLAN,WM & HOWARD,PH (1991)
OH 1.12E-010	cm3/molc sec	25	EXP	ATKINSON,R (1989)

CAS #: 020240-99-7				1-(O-ANISYL)-3,3-DIMETHYLTRIAZENE
Formula: $C_9H_{13}N_3O$				
Mol Weight: 179.22				
MP (deg C):		FP (deg C):		
BP (deg C):				
BP pressure (mm Hg):				

Property/ Value	Units	Temp	Data Type	Reference
WS 5.58E+002	mg/L	25	EST	MEYLAN,WM ET AL. (1996)
logP 2.34			EXP	HANSCH,C & LEO,AJ (1985)
VP 1.77E-002	mm Hg	25	EST	NEELY,WB & BLAU,GE (1985)
DC	pKa			
HL 3.17E-008	atm m3/mol	25	EST	MEYLAN,WM & HOWARD,PH (1991)
OH 1.19E-011	cm3/molc sec	25	EST	MEYLAN,WM & HOWARD,PH (1993)

CAS #: 020241-00-3				1-(O-CL PH)-3,3-DIMETHYLTRIAZENE
Formula: $C_8H_{10}ClN_3$				
Mol Weight: 183.64				
MP (deg C):		FP (deg C):		
BP (deg C):				
BP pressure (mm Hg):				

Property/ Value	Units	Temp	Data Type	Reference
WS 1.37E+002	mg/L	25	EST	MEYLAN,WM ET AL. (1996)
logP 3.03			EXP	HANSCH,C & LEO,AJ (1985)
VP 2.63E-002	mm Hg	25	EST	NEELY,WB & BLAU,GE (1985)
DC	pKa			
HL 3.98E-007	atm m3/mol	25	EST	MEYLAN,WM & HOWARD,PH (1991)
OH 3.07E-012	cm3/molc sec	25	EST	MEYLAN,WM & HOWARD,PH (1993)

CAS #: 020241-03-6				1-(3-TOLYL)-3,3-DIMETHYLTRIAZENE
Formula: $C_9H_{13}N_3$				
Mol Weight: 163.22				
MP (deg C):		FP (deg C):		
BP (deg C):				
BP pressure (mm Hg):				

Property/ Value	Units	Temp	Data Type	Reference
WS 2.44E+002	mg/L	25	EST	MEYLAN,WM ET AL. (1996)
logP 2.85			EXP	HANSCH,C & LEO,AJ (1985)
VP 4.68E-002	mm Hg	25	EST	NEELY,WB & BLAU,GE (1985)
DC	pKa			
HL 5.92E-007	atm m3/mol	25	EST	MEYLAN,WM & HOWARD,PH (1991)
OH 3.61E-012	cm3/molc sec	25	EST	MEYLAN,WM & HOWARD,PH (1993)

CAS #: 020241-05-8				1-(3-CL PH)-3,3-DIMETHYLTRIAZENE
Formula: $C_8H_{10}ClN_3$				
Mol Weight: 183.64				
MP (deg C):		FP (deg C):		
BP (deg C):				
BP pressure (mm Hg):				

Property/ Value	Units	Temp	Data Type	Reference
WS 1.37E+002	mg/L	25	EST	MEYLAN,WM ET AL. (1996)
logP 3.03			EXP	HANSCH,C & LEO,AJ (1985)
VP 2.63E-002	mm Hg	25	EST	NEELY,WB & BLAU,GE (1985)
DC	pKa			
HL 3.98E-007	atm m3/mol	25	EST	MEYLAN,WM & HOWARD,PH (1991)
OH 2.76E-012	cm3/molc sec	25	EST	MEYLAN,WM & HOWARD,PH (1993)

CAS #: 020241-06-9				1-(3-NO2 PH)-3,3-DIMETHYLTRIAZENE
Formula: $C_8H_{10}N_4O_2$				
Mol Weight: 194.19				
MP (deg C):		FP (deg C):		
BP (deg C):				
BP pressure (mm Hg):				

Property/ Value	Units	Temp	Data Type	Reference
WS 2.10E+002	mg/L	25	EST	MEYLAN,WM ET AL. (1996)
logP 2.75			EXP	HANSCH,C & LEO,AJ (1985)
VP 5.06E-004	mm Hg	25	EST	NEELY,WB & BLAU,GE (1985)
DC	pKa			
HL 2.12E-009	atm m3/mol	25	EST	MEYLAN,WM & HOWARD,PH (1991)
OH 2.63E-012	cm3/molc sec	25	EST	MEYLAN,WM & HOWARD,PH (1993)

973

CAS #: 020300-02-1 — THIOPHENE,2-THIOCARBOXAMIDE

Formula: $C_5H_7NS_2$

Mol Weight: 145.25

MP (deg C):   FP (deg C):

BP (deg C):

BP pressure (mm Hg):

Property/Value	Units	Temp	Data Type	Reference
WS 5.93E+004	mg/L	25	EST	MEYLAN,WM ET AL. (1996)
logP 1.34			EXP	HANSCH,C ET AL. (1995)
VP 8.32E-003	mm Hg	25	EST	NEELY,WB & BLAU,GE (1985)
DC	pKa			
HL 2.13E-007	atm m3/mol	25	EST	MEYLAN,WM & HOWARD,PH (1991)
OH 4.20E-011	cm3/molc sec	25	EST	MEYLAN,WM & HOWARD,PH (1993)

CAS #: 020308-67-2 — PIPERIDINE, 1-(2-ETHOXYBENZOYL)-

Formula: $C_{14}H_{19}NO_2$

Mol Weight: 233.31

MP (deg C):   FP (deg C):

BP (deg C):

BP pressure (mm Hg):

Property/Value	Units	Temp	Data Type	Reference
WS 1.76E+002	mg/L	25	EST	MEYLAN,WM ET AL. (1996)
logP 2.60			EXP	SURYANARAYANA,MVS ET AL. (1991)
VP 3.00E-002	mm Hg	25	EXP	SURYANARAYANA,MVS ET AL. (1991)
DC	pKa			
HL 8.65E-010	atm m3/mol	25	EST	MEYLAN,WM & HOWARD,PH (1991)
OH 5.44E-011	cm3/molc sec	25	EST	MEYLAN,WM & HOWARD,PH (1993)

CAS #: 020324-33-8 — TRIPROPYLENE GLYCOL METHYL ETHER

Formula: $C_{10}H_{22}O_4$

Mol Weight: 206.28

MP (deg C):   FP (deg C):

BP (deg C): 100

BP pressure (mm Hg): 2.00E+000

Property/Value	Units	Temp	Data Type	Reference
WS 1.96E+005	mg/L	25	EST	MEYLAN,WM ET AL. (1996)
logP -0.20			EST	MEYLAN,WM & HOWARD,PH (1995)
VP 2.00E-002	mm Hg	25	EXP	DOW CHEMICAL COMPANY (1990)
DC	pKa			
HL 2.36E-011	atm m3/mol	25	EST	MEYLAN,WM & HOWARD,PH (1991)
OH 6.00E-011	cm3/molc sec	25	EST	MEYLAN,WM & HOWARD,PH (1993)

CAS #: 020330-45-4 — ACETAMIDE, N-[4-(1,1-DIMETHYLETHYL)PHENYL]-

Formula: $C_{12}H_{17}NO$

Mol Weight: 191.28

MP (deg C):   FP (deg C):

BP (deg C):

BP pressure (mm Hg):

Property/Value	Units	Temp	Data Type	Reference
WS 1.20E+002	mg/L	25	EST	MEYLAN,WM ET AL. (1996)
logP 3.05			EXP	HANSCH,C ET AL. (1995)
VP 2.80E-005	mm Hg	25	EST	NEELY,WB & BLAU,GE (1985)
DC	pKa			
HL 1.59E-008	atm m3/mol	25	EST	MEYLAN,WM & HOWARD,PH (1991)
OH 1.55E-011	cm3/molc sec	25	EST	MEYLAN,WM & HOWARD,PH (1993)

CAS #: 020330-99-8 — ACETAMIDE, N-(4-PROPYLPHENYL)-

Formula: $C_{11}H_{15}NO$

Mol Weight: 177.25

MP (deg C):   FP (deg C):

BP (deg C):

BP pressure (mm Hg):

Property/Value	Units	Temp	Data Type	Reference
WS 2.55E+002	mg/L	25	EST	MEYLAN,WM ET AL. (1996)
logP 2.75			EXP	HANSCH,C ET AL. (1995)
VP 3.04E-005	mm Hg	25	EST	NEELY,WB & BLAU,GE (1985)
DC	pKa			
HL 1.20E-008	atm m3/mol	25	EST	MEYLAN,WM & HOWARD,PH (1991)
OH 1.77E-011	cm3/molc sec	25	EST	MEYLAN,WM & HOWARD,PH (1993)

CAS #: 020334-52-5 — 1-BUTEN-1-ONE(ETHYLKETENE)

Formula: C_4H_6O

Mol Weight: 70.09

MP (deg C):   FP (deg C):

BP (deg C):

BP pressure (mm Hg):

Property/Value	Units	Temp	Data Type	Reference
WS 6.40E+004	mg/L	25	EST	MEYLAN,WM ET AL. (1996)
logP 0.38			EST	MEYLAN,WM & HOWARD,PH (1995)
VP 1.03E+003	mm Hg	25	EST	NEELY,WB & BLAU,GE (1985)
DC	pKa			
HL	atm m3/mol			
OH 1.18E-010	cm3/molc sec	25	EXP	ATKINSON,R (1989)

CAS #: 020338-26-5 — 1,1,2,2,3,3,4,4-OCTACHLOROBUTANE

Formula: $C_4H_2Cl_8$

Mol Weight: 333.68

MP (deg C):   FP (deg C):

BP (deg C):

BP pressure (mm Hg):

Property/Value	Units	Temp	Data Type	Reference
WS 1.98E-001	mg/L	25	EST	MEYLAN,WM ET AL. (1996)
logP 5.38			EST	MEYLAN,WM & HOWARD,PH (1995)
VP 4.00E-003	mm Hg	25	EST	NEELY,WB & BLAU,GE (1985)
DC	pKa			
HL 4.09E-005	atm m3/mol	25	EST	MEYLAN,WM & HOWARD,PH (1991)
OH 1.40E-013	cm3/molc sec	25	EST	MEYLAN,WM & HOWARD,PH (1993)

CAS #: 020344-69-8 — 2,4-NH2PYRIMIDINE,5(3MEO-5-MEO)BENZYL

Formula: $C_{13}H_{16}N_4O_2$

Mol Weight: 260.30

MP (deg C):   FP (deg C):

BP (deg C):

BP pressure (mm Hg):

Property/Value	Units	Temp	Data Type	Reference
WS 9.46E+002	mg/L	25	EST	MEYLAN,WM ET AL. (1996)
logP 1.57			EXP	HANSCH,C & LEO,AJ (1985)
VP 5.09E-008	mm Hg	25	EST	NEELY,WB & BLAU,GE (1985)
DC	pKa			
HL 4.03E-013	atm m3/mol	25	EST	MEYLAN,WM & HOWARD,PH (1991)
OH 2.03E-010	cm3/molc sec	25	EST	MEYLAN,WM & HOWARD,PH (1993)

CAS #: 020354-26-1 — METHAZOLE

Formula: $C_9H_6Cl_2N_2O_3$

Mol Weight: 261.07

MP (deg C): 123

FP (deg C):

BP (deg C):

BP pressure (mm Hg):

Property/Value	Units	Temp	Data Type	Reference
WS 1.50E+000	mg/L	25	EXP	WAUCHOPE,RD ET AL. (1991A)
logP 3.22			EST	MEYLAN,WM & HOWARD,PH (1995)
VP 1.00E-006	mm Hg	25	EXP	WAUCHOPE,RD ET AL. (1991A)
DC	pKa			
HL 2.29E-007	atm m3/mol	25	EST	VP/WSOL
OH 2.47E-011	cm3/molc sec	25	EST	MEYLAN,WM & HOWARD,PH (1993)

CAS #: 020383-28-2 — BENZAMIDE, N,N-BIS(1-METHYLETHYL)-

Formula: $C_{13}H_{19}NO$

Mol Weight: 205.30

MP (deg C):

FP (deg C):

BP (deg C):

BP pressure (mm Hg):

Property/Value	Units	Temp	Data Type	Reference
WS 3.01E+002	mg/L	25	EST	MEYLAN,WM ET AL. (1996)
logP 2.50			EXP	SURYANARAYANA,MVS ET AL. (1991)
VP 1.16E-001	mm Hg	25	EXP	SURYANARAYANA,MVS ET AL. (1991)
DC	pKa			
HL 3.32E-008	atm m3/mol	25	EST	MEYLAN,WM & HOWARD,PH (1991)
OH 3.34E-011	cm3/molc sec	25	EST	MEYLAN,WM & HOWARD,PH (1993)

CAS #: 020404-02-8 — 2,3,6-TRICHLORO-4-NITROPHENOL

Formula: $C_6H_2Cl_3NO_3$

Mol Weight: 242.45

MP (deg C):

FP (deg C):

BP (deg C):

BP pressure (mm Hg):

Property/Value	Units	Temp	Data Type	Reference
WS 4.36E+001	mg/L	25	EST	MEYLAN,WM ET AL. (1996)
logP 3.93			EXP	HANSCH,C ET AL. (1995)
VP 1.04E-005	mm Hg	25	EST	NEELY,WB & BLAU,GE (1985)
DC	pKa			
HL 9.00E-010	atm m3/mol	25	EST	MEYLAN,WM & HOWARD,PH (1991)
OH 1.57E-013	cm3/molc sec	25	EST	MEYLAN,WM & HOWARD,PH (1993)

CAS #: 020417-83-8 — 3NO2-3ME-2BUTANON(N-ME-CARBMOY)OXIME

Formula: $C_7H_{13}N_3O_4$

Mol Weight: 203.20

MP (deg C):

FP (deg C):

BP (deg C):

BP pressure (mm Hg):

Property/Value	Units	Temp	Data Type	Reference
WS 4.69E+003	mg/L	25	EST	MEYLAN,WM ET AL. (1996)
logP 0.66			EXP	HANSCH,C & LEO,AJ (1985)
VP 2.13E-003	mm Hg	25	EST	NEELY,WB & BLAU,GE (1985)
DC	pKa			
HL 2.54E-010	atm m3/mol	25	EST	MEYLAN,WM & HOWARD,PH (1991)
OH 6.39E-012	cm3/molc sec	25	EST	MEYLAN,WM & HOWARD,PH (1993)

CAS #: 020434-64-4 — 124-BENZTHIADIAZN-1-O2-3-CYCLOPE

Formula: $C_{12}H_{14}N_2O_2S$

Mol Weight: 250.32

MP (deg C):

FP (deg C):

BP (deg C):

BP pressure (mm Hg):

Property/Value	Units	Temp	Data Type	Reference
WS 6.45E+002	mg/L	25	EST	MEYLAN,WM ET AL. (1996)
logP 1.83			EXP	HANSCH,C & LEO,AJ (1985)
VP 3.13E-009	mm Hg	25	EST	NEELY,WB & BLAU,GE (1985)
DC	pKa			
HL 1.27E-007	atm m3/mol	25	EST	MEYLAN,WM & HOWARD,PH (1991)
OH 7.87E-012	cm3/molc sec	25	EST	MEYLAN,WM & HOWARD,PH (1993)

CAS #: 020434-66-6 — 124-BENZTHIADIAZN-1-O2-3-BENZYL

Formula: $C_{14}H_{12}N_2O_2S$

Mol Weight: 272.33

MP (deg C):

FP (deg C):

BP (deg C):

BP pressure (mm Hg):

Property/Value	Units	Temp	Data Type	Reference
WS 4.31E+002	mg/L	25	EST	MEYLAN,WM ET AL. (1996)
logP 1.89			EXP	HANSCH,C & LEO,AJ (1985)
VP 2.92E-010	mm Hg	25	EST	NEELY,WB & BLAU,GE (1985)
DC	pKa			
HL 7.47E-009	atm m3/mol	25	EST	MEYLAN,WM & HOWARD,PH (1991)
OH 6.45E-012	cm3/molc sec	25	EST	MEYLAN,WM & HOWARD,PH (1993)

CAS #: 020451-62-1 — ETHYLACETOACETATE,3,4-DIMEO BENZAL

Formula: $C_{15}H_{18}O_5$

Mol Weight: 278.31

MP (deg C):

FP (deg C):

BP (deg C):

BP pressure (mm Hg):

Property/Value	Units	Temp	Data Type	Reference
WS 3.09E+002	mg/L	25	EST	MEYLAN,WM ET AL. (1996)
logP 2.02			EXP	HANSCH,C & LEO,AJ (1985)
VP 7.31E-006	mm Hg	25	EST	NEELY,WB & BLAU,GE (1985)
DC	pKa			
HL 8.14E-012	atm m3/mol	25	EST	MEYLAN,WM & HOWARD,PH (1991)
OH 4.83E-011	cm3/molc sec	25	EST	MEYLAN,WM & HOWARD,PH (1993)

CAS #: 020455-68-9 — DIBENZYLAMMONIUM CHLORIDE

Formula: $C_{14}H_{16}ClN$

Mol Weight: 233.74

MP (deg C):

FP (deg C):

BP (deg C):

BP pressure (mm Hg):

Property/Value	Units	Temp	Data Type	Reference
WS 2.12E+004	mg/L	25	EXP	STEPHEN,H & STEPHEN,T (1963)
logP 0.14			EST	MEYLAN,WM & HOWARD,PH (1995)
VP 1.86E-009	mm Hg	25	EST	NEELY,WB & BLAU,GE (1985)
DC	pKa			
HL 8.29E-015	atm m3/mol	25	EST	MEYLAN,WM & HOWARD,PH (1991)
OH 2.71E-011	cm3/molc sec	25	EST	MEYLAN,WM & HOWARD,PH (1993)

CAS #: 020501-52-4 — 2-BUTENOIC ACID, 2-METHYL-, 7-(ACETYLOXY)-6-(CHL

Formula: $C_{22}H_{27}ClO_8$

Mol Weight: 454.91

MP (deg C): | FP (deg C):

BP (deg C):

BP pressure (mm Hg):

Property/Value	Units	Temp	Data Type	Reference
WS 8.50E+001	mg/L	25	EST	MEYLAN,WM ET AL. (1996)
logP 1.42			EXP	SANGSTER,J (1993)
VP 1.75E-014	mm Hg	25	EST	NEELY,WB & BLAU,GE (1985)
DC	pKa			
HL 7.58E-016	atm m3/mol	25	EST	MEYLAN,WM & HOWARD,PH (1991)
OH 1.79E-010	cm3/molc sec	25	EST	MEYLAN,WM & HOWARD,PH (1993)

CAS #: 020562-02-1 — ALPHA-SOLANINE

Formula: $C_{45}H_{73}NO_{15}$

Mol Weight: 868.08

MP (deg C): 285 dec | FP (deg C):

BP (deg C):

BP pressure (mm Hg):

Property/Value	Units	Temp	Data Type	Reference
WS 1.38E+000	mg/L	25	EST	MEYLAN,WM ET AL. (1996)
logP 1.50			EST	MEYLAN,WM & HOWARD,PH (1995)
VP 1.67E-034	mm Hg	25	EST	NEELY,WB & BLAU,GE (1985)
DC	pKa			
HL 1.22E-031	atm m3/mol	25	EST	MEYLAN,WM & HOWARD,PH (1991)
OH 2.95E-010	cm3/molc sec	25	EST	MEYLAN,WM & HOWARD,PH (1993)

CAS #: 020587-61-5 — ETHANOL, 2- 2-(BENZOYLOXY)ETHOXY -

Formula: $C_{11}H_{14}O_4$

Mol Weight: 210.23

MP (deg C): | FP (deg C):

BP (deg C):

BP pressure (mm Hg):

Property/Value	Units	Temp	Data Type	Reference
WS 1.28E+004	mg/L	25	EST	MEYLAN,WM ET AL. (1996)
logP 1.16			EXP	HANSCH,C ET AL. (1995)
VP 7.29E-006	mm Hg	25	EST	NEELY,WB & BLAU,GE (1985)
DC	pKa			
HL 2.62E-011	atm m3/mol	25	EST	MEYLAN,WM & HOWARD,PH (1991)
OH 2.06E-011	cm3/molc sec	25	EST	MEYLAN,WM & HOWARD,PH (1993)

CAS #: 020605-19-0 — 4-BUTYLSEMICARBAZIDE

Formula: $C_5H_{13}N_3O$

Mol Weight: 131.18

MP (deg C): | FP (deg C):

BP (deg C):

BP pressure (mm Hg):

Property/Value	Units	Temp	Data Type	Reference
WS 1.45E+004	mg/L	25	EST	MEYLAN,WM ET AL. (1996)
logP -0.60			EXP	HANSCH,C & LEO,AJ (1985)
VP 3.63E-003	mm Hg	25	EST	NEELY,WB & BLAU,GE (1985)
DC	pKa			
HL 7.83E-012	atm m3/mol	25	EST	MEYLAN,WM & HOWARD,PH (1991)
OH 7.84E-011	cm3/molc sec	25	EST	MEYLAN,WM & HOWARD,PH (1993)

CAS #: 020634-92-8 — ETHYL DIPROPYLAMINE

Formula: $C_8H_{19}N$

Mol Weight: 129.25

MP (deg C): | FP (deg C):

BP (deg C): 138

BP pressure (mm Hg):

Property/Value	Units	Temp	Data Type	Reference
WS 4.85E+003	mg/L	25	EST	MEYLAN,WM ET AL. (1996)
logP 2.68			EXP	HANSCH,C & LEO,AJ (1985)
VP 5.78E+000	mm Hg	25	EST	NEELY,WB & BLAU,GE (1985)
DC	pKa			
HL 1.51E-004	atm m3/mol	25	EST	MEYLAN,WM & HOWARD,PH (1991)
OH 9.89E-011	cm3/molc sec	25	EST	MEYLAN,WM & HOWARD,PH (1993)

CAS #: 020651-71-2 — 4-BUTYLBENZOIC ACID

Formula: $C_{11}H_{14}O_2$

Mol Weight: 178.23

MP (deg C): 100-113 | FP (deg C):

BP (deg C):

BP pressure (mm Hg):

Property/Value	Units	Temp	Data Type	Reference
WS 2.29E+001	mg/L	25	EST	MEYLAN,WM ET AL. (1996)
logP 3.97			EXP	HANSCH,C ET AL. (1995)
VP 2.25E-004	mm Hg	25	EST	NEELY,WB & BLAU,GE (1985)
DC	pKa			
HL 2.80E-007	atm m3/mol	25	EST	MEYLAN,WM & HOWARD,PH (1991)
OH 6.19E-012	cm3/molc sec	25	EST	MEYLAN,WM & HOWARD,PH (1993)

CAS #: 020668-13-7 — CARBAMIC ACID, (2-CHLOROPHENYL)-, METHYL ESTER

Formula: $C_8H_8ClNO_2$

Mol Weight: 185.61

MP (deg C): | FP (deg C):

BP (deg C):

BP pressure (mm Hg):

Property/Value	Units	Temp	Data Type	Reference
WS 7.39E+002	mg/L	25	EST	MEYLAN,WM ET AL. (1996)
logP 2.16			EXP	TAKAHASHI,J ET AL. (1988)
VP 8.31E-003	mm Hg	25	EST	NEELY,WB & BLAU,GE (1985)
DC	pKa			
HL 1.62E-008	atm m3/mol	25	EST	MEYLAN,WM & HOWARD,PH (1991)
OH 1.27E-011	cm3/molc sec	25	EST	MEYLAN,WM & HOWARD,PH (1993)

CAS #: 020682-52-4 — FORMAMIDE, N-[2-(DIETHYLAMINO)ETHYL]-N-(2,6-DIME

Formula: $C_{15}H_{24}N_2O$

Mol Weight: 248.37

MP (deg C): | FP (deg C):

BP (deg C):

BP pressure (mm Hg):

Property/Value	Units	Temp	Data Type	Reference
WS 4.46E+002	mg/L	25	EST	MEYLAN,WM ET AL. (1996)
logP 3.21			EXP	SANGSTER,J (1993)
VP 8.76E-006	mm Hg	25	EST	NEELY,WB & BLAU,GE (1985)
DC	pKa			
HL 2.45E-010	atm m3/mol	25	EST	MEYLAN,WM & HOWARD,PH (1991)
OH 1.24E-010	cm3/molc sec	25	EST	MEYLAN,WM & HOWARD,PH (1993)

CAS #: 020691-84-3	3-((4-(DIMETHYLAMINO)PHENYL)AZO)BENZOIC ACID

Formula: $C_{15}H_{15}N_3O_2$

Mol Weight: 269.31

MP (deg C): | FP (deg C):

BP (deg C):

BP pressure (mm Hg):

Property/ Value	Units	Temp	Data Type	Reference
WS 1.87E+000	mg/L	25	EST	MEYLAN,WM ET AL. (1996)
logP 4.17			EST	MEYLAN,WM & HOWARD,PH (1995)
VP 1.44E-007	mm Hg	25	EST	NEELY,WB & BLAU,GE (1985)
DC 2.64	pKa		EXP	PERRIN,DD (1965)
HL 4.71E-012	atm m3/mol	25	EST	MEYLAN,WM & HOWARD,PH (1991)
OH 1.50E-010	cm3/molc sec	25	EST	MEYLAN,WM & HOWARD,PH (1993)

CAS #: 020725-03-5	FUSTIN

Formula: $C_{15}H_{12}O_6$

Mol Weight: 288.26

MP (deg C): 226-228 | FP (deg C):

BP (deg C):

BP pressure (mm Hg):

Property/ Value	Units	Temp	Data Type	Reference
WS 3.19E+004	mg/L	25	EST	MEYLAN,WM ET AL. (1996)
logP 0.87			EXP	PERRISSOUD,D & TESTA,B (1986)
VP 2.46E-012	mm Hg	25	EST	NEELY,WB & BLAU,GE (1985)
DC	pKa			
HL 3.63E-021	atm m3/mol	25	EST	MEYLAN,WM & HOWARD,PH (1991)
OH 2.53E-010	cm3/molc sec	25	EST	MEYLAN,WM & HOWARD,PH (1993)

CAS #: 020733-11-3	PYRIDAZINE, 4-METHOXY-

Formula: $C_5H_6N_2O$

Mol Weight: 110.12

MP (deg C): | FP (deg C):

BP (deg C):

BP pressure (mm Hg):

Property/ Value	Units	Temp	Data Type	Reference
WS 2.00E+005	mg/L	25	EST	MEYLAN,WM ET AL. (1996)
logP -0.31			EXP	YAMAGAMI,C ET AL. (1990)
VP 4.51E-002	mm Hg	25	EST	NEELY,WB & BLAU,GE (1985)
DC 3.66	pKa	20	EXP	PERRIN,DD (1965)
HL 1.70E-007	atm m3/mol	25	EST	MEYLAN,WM & HOWARD,PH (1991)
OH 3.37E-012	cm3/molc sec	25	EST	MEYLAN,WM & HOWARD,PH (1993)

CAS #: 020744-39-2	4-PYRIDAZINAMINE

Formula: $C_4H_5N_3$

Mol Weight: 95.10

MP (deg C): | FP (deg C):

BP (deg C):

BP pressure (mm Hg):

Property/ Value	Units	Temp	Data Type	Reference
WS 3.42E+005	mg/L	25	EST	MEYLAN,WM ET AL. (1996)
logP -0.53			EXP	YAMAGAMI,C ET AL. (1990)
VP 5.66E-003	mm Hg	25	EST	NEELY,WB & BLAU,GE (1985)
DC 6.65	pKa	20	EXP	PERRIN,DD (1965)
HL 1.01E-009	atm m3/mol	25	EST	MEYLAN,WM & HOWARD,PH (1991)
OH 1.27E-011	cm3/molc sec	25	EST	MEYLAN,WM & HOWARD,PH (1993)

CAS #: 020772-21-8	GLUCOPYRANOSIDE,3,5-DIMETHYLPHENYL

Formula: $C_{14}H_{20}O_6$

Mol Weight: 284.31

MP (deg C): | FP (deg C):

BP (deg C):

BP pressure (mm Hg):

Property/ Value	Units	Temp	Data Type	Reference
WS 9.08E+003	mg/L	25	EST	MEYLAN,WM ET AL. (1996)
logP 0.26			EXP	HANSCH,C & LEO,AJ (1985)
VP 7.97E-012	mm Hg	25	EST	NEELY,WB & BLAU,GE (1985)
DC	pKa			
HL 1.40E-015	atm m3/mol	25	EST	MEYLAN,WM & HOWARD,PH (1991)
OH 2.09E-010	cm3/molc sec	25	EST	MEYLAN,WM & HOWARD,PH (1993)

CAS #: 020772-23-0	GLUCOPYRANOSIDE,2-IPR-5-ME PHENYL

Formula: $C_{16}H_{24}O_6$

Mol Weight: 312.37

MP (deg C): | FP (deg C):

BP (deg C):

BP pressure (mm Hg):

Property/ Value	Units	Temp	Data Type	Reference
WS 1.27E+003	mg/L	25	EST	MEYLAN,WM ET AL. (1996)
logP 1.07			EXP	HANSCH,C & LEO,AJ (1985)
VP 1.85E-012	mm Hg	25	EST	NEELY,WB & BLAU,GE (1985)
DC	pKa			
HL 2.47E-015	atm m3/mol	25	EST	MEYLAN,WM & HOWARD,PH (1991)
OH 1.32E-010	cm3/molc sec	25	EST	MEYLAN,WM & HOWARD,PH (1993)

CAS #: 020772-25-2	3-CF3 PHENYL GLUCOPYRANOSIDE

Formula: $C_{13}H_{15}F_3O_6$

Mol Weight: 324.26

MP (deg C): | FP (deg C):

BP (deg C):

BP pressure (mm Hg):

Property/ Value	Units	Temp	Data Type	Reference
WS 3.37E+003	mg/L	25	EST	MEYLAN,WM ET AL. (1996)
logP 0.49			EXP	HANSCH,C & LEO,AJ (1985)
VP 4.93E-011	mm Hg	25	EST	NEELY,WB & BLAU,GE (1985)
DC	pKa			
HL 9.99E-015	atm m3/mol	25	EST	MEYLAN,WM & HOWARD,PH (1991)
OH 6.53E-011	cm3/molc sec	25	EST	MEYLAN,WM & HOWARD,PH (1993)

CAS #: 020782-57-4	1-ME-3-(3-CL-4-MEO PHENYL) UREA

Formula: $C_9H_{11}ClN_2O_2$

Mol Weight: 214.65

MP (deg C): | FP (deg C):

BP (deg C):

BP pressure (mm Hg):

Property/ Value	Units	Temp	Data Type	Reference
WS 1.15E+003	mg/L	25	EST	MEYLAN,WM ET AL. (1996)
logP 1.76			EXP	HANSCH,C & LEO,AJ (1985)
VP 1.42E-005	mm Hg	25	EST	NEELY,WB & BLAU,GE (1985)
DC	pKa			
HL 1.94E-011	atm m3/mol	25	EST	MEYLAN,WM & HOWARD,PH (1991)
OH 2.83E-011	cm3/molc sec	25	EST	MEYLAN,WM & HOWARD,PH (1993)

CAS #: 020818-25-1 — GLUCOPYRANOSIDE,4-AMINOPHENYL

Formula:	$C_{12}H_{17}NO_6$		
Mol Weight:	271.27		

MP (deg C):		FP (deg C):	
BP (deg C):			
BP pressure (mm Hg):			

Property/ Value	Units	Temp	Data Type	Reference
WS 1.00E+006	mg/L	25	EST	MEYLAN,WM ET AL. (1996)
logP -2.67			EXP	HANSCH,C & LEO,AJ (1985)
VP 1.17E-012	mm Hg	25	EST	NEELY,WB & BLAU,GE (1985)
DC	pKa			
HL 4.06E-019	atm m3/mol	25	EST	MEYLAN,WM & HOWARD,PH (1991)
OH 1.55E-010	cm3/molc sec	25	EST	MEYLAN,WM & HOWARD,PH (1993)

CAS #: 020830-75-5 — DIGOXIN

Formula:	$C_{41}H_{64}O_{14}$		
Mol Weight:	780.96		

MP (deg C): 248-250 de		FP (deg C):	
BP (deg C):			
BP pressure (mm Hg):			

Property/ Value	Units	Temp	Data Type	Reference
WS 6.48E+001	mg/L	25	EXP	YALKOWSKY,SH & DANNENFELSER,RM (1992)
logP 1.26			EXP	SANGSTER,J (1993)
VP 3.30E-030	mm Hg	25	EST	NEELY,WB & BLAU,GE (1985)
DC	pKa			
HL 4.66E-027	atm m3/mol	25	EST	MEYLAN,WM & HOWARD,PH (1991)
OH 1.70E-010	cm3/molc sec	25	EST	MEYLAN,WM & HOWARD,PH (1993)

CAS #: 020830-81-3 — DAUNOMYCIN

Formula:	$C_{27}H_{29}NO_{10}$		
Mol Weight:	527.53		

MP (deg C): 208-209		FP (deg C):	
BP (deg C):			
BP pressure (mm Hg):			

Property/ Value	Units	Temp	Data Type	Reference
WS 3.92E+001	mg/L	25	EST	MEYLAN,WM ET AL. (1996)
logP 1.83			EXP	SANGSTER,J (1993)
VP 3.46E-022	mm Hg	25	EST	NEELY,WB & BLAU,GE (1985)
DC	pKa			
HL 1.43E-025	atm m3/mol	25	EST	MEYLAN,WM & HOWARD,PH (1991)
OH 1.42E-010	cm3/molc sec	25	EST	MEYLAN,WM & HOWARD,PH (1993)

CAS #: 020838-34-0 — GLUCOPYRANOSIDE, 3-ETHYLPHENYL

Formula:	$C_{14}H_{20}O_6$		
Mol Weight:	284.31		

MP (deg C):		FP (deg C):	
BP (deg C):			
BP pressure (mm Hg):			

Property/ Value	Units	Temp	Data Type	Reference
WS 8.23E+003	mg/L	25	EST	MEYLAN,WM ET AL. (1996)
logP 0.31			EXP	HANSCH,C & LEO,AJ (1985)
VP 8.51E-012	mm Hg	25	EST	NEELY,WB & BLAU,GE (1985)
DC	pKa			
HL 1.68E-015	atm m3/mol	25	EST	MEYLAN,WM & HOWARD,PH (1991)
OH 1.17E-010	cm3/molc sec	25	EST	MEYLAN,WM & HOWARD,PH (1993)

CAS #: 020838-36-2 — GLUCOPYRANOSIDE, 3-T-BUTYL PHENYL

Formula:	$C_{16}H_{24}O_6$		
Mol Weight:	312.37		

MP (deg C):		FP (deg C):	
BP (deg C):			
BP pressure (mm Hg):			

Property/ Value	Units	Temp	Data Type	Reference
WS 1.43E+003	mg/L	25	EST	MEYLAN,WM ET AL. (1996)
logP 1.01			EXP	HANSCH,C & LEO,AJ (1985)
VP 2.70E-012	mm Hg	25	EST	NEELY,WB & BLAU,GE (1985)
DC	pKa			
HL 2.97E-015	atm m3/mol	25	EST	MEYLAN,WM & HOWARD,PH (1991)
OH 1.10E-010	cm3/molc sec	25	EST	MEYLAN,WM & HOWARD,PH (1993)

CAS #: 020838-40-8 — 4-IODOPHENYL GLUCOPYRANOSIDE

Formula:	$C_{12}H_{15}IO_6$		
Mol Weight:	382.15		

MP (deg C):		FP (deg C):	
BP (deg C):			
BP pressure (mm Hg):			

Property/ Value	Units	Temp	Data Type	Reference
WS 9.03E+002	mg/L	25	EST	MEYLAN,WM ET AL. (1996)
logP 0.75			EXP	HANSCH,C & LEO,AJ (1985)
VP 7.37E-013	mm Hg	25	EST	NEELY,WB & BLAU,GE (1985)
DC	pKa			
HL 2.67E-016	atm m3/mol	25	EST	MEYLAN,WM & HOWARD,PH (1991)
OH 6.91E-011	cm3/molc sec	25	EST	MEYLAN,WM & HOWARD,PH (1993)

CAS #: 020838-44-2 — 3-NITROPHENYL GLUCOPYRANOSIDE

Formula:	$C_{12}H_{15}NO_8$		
Mol Weight:	301.26		

MP (deg C):		FP (deg C):	
BP (deg C):			
BP pressure (mm Hg):			

Property/ Value	Units	Temp	Data Type	Reference
WS 1.34E+004	mg/L	25	EST	MEYLAN,WM ET AL. (1996)
logP -0.51			EXP	HANSCH,C & LEO,AJ (1985)
VP 3.68E-013	mm Hg	25	EST	NEELY,WB & BLAU,GE (1985)
DC	pKa			
HL 4.54E-018	atm m3/mol	25	EST	MEYLAN,WM & HOWARD,PH (1991)
OH 6.39E-011	cm3/molc sec	25	EST	MEYLAN,WM & HOWARD,PH (1993)

CAS #: 020863-54-1 — 3-ME,2-(N,N-DIMEAMINOME)FURAN

Formula:	$C_8H_{14}NO$		
Mol Weight:	140.21		

MP (deg C):		FP (deg C):	
BP (deg C):			
BP pressure (mm Hg):			

Property/ Value	Units	Temp	Data Type	Reference
WS 3.49E+004	mg/L	25	EST	MEYLAN,WM ET AL. (1996)
logP 1.63			EXP	HANSCH,C & LEO,AJ (1985)
VP 1.23E+000	mm Hg	25	EST	NEELY,WB & BLAU,GE (1985)
DC	pKa			
HL 3.25E-006	atm m3/mol	25	EST	MEYLAN,WM & HOWARD,PH (1991)
OH 2.06E-010	cm3/molc sec	25	EST	MEYLAN,WM & HOWARD,PH (1993)

CAS #: 020872-53-1	4-METHYLSULFONYLQUINOLINE-1-OXIDE

Formula: $C_{10}H_{11}NO_3S$

Mol Weight: 225.27

MP (deg C):   FP (deg C):

BP (deg C):

BP pressure (mm Hg):

Property/Value	Units	Temp	Data Type	Reference
WS 2.61E+004	mg/L	25	EST	MEYLAN,WM ET AL. (1996)
logP 0.12			EXP	HANSCH,C & LEO,AJ (1985)
VP 1.79E-007	mm Hg	25	EST	NEELY,WB & BLAU,GE (1985)
DC	pKa			
HL 1.59E-015	atm m3/mol	25	EST	MEYLAN,WM & HOWARD,PH (1991)
OH 4.14E-012	cm3/molc sec	25	EST	MEYLAN,WM & HOWARD,PH (1993)

CAS #: 020893-30-5	2-CYANOMETHYLTHIOPHENE

Formula: C_6H_5NS

Mol Weight: 123.18

MP (deg C):   FP (deg C):

BP (deg C): 115-120

BP pressure (mm Hg): 2.20E+001

Property/Value	Units	Temp	Data Type	Reference
WS 4.45E+003	mg/L	25	EST	MEYLAN,WM ET AL. (1996)
logP 1.26			EXP	HANSCH,C & LEO,AJ (1985)
VP 3.38E-002	mm Hg	25	EST	NEELY,WB & BLAU,GE (1985)
DC	pKa			
HL 1.34E-006	atm m3/mol	25	EST	MEYLAN,WM & HOWARD,PH (1991)
OH 9.42E-012	cm3/molc sec	25	EST	MEYLAN,WM & HOWARD,PH (1993)

CAS #: 020917-49-1	N-NITROSO-HEPTAMETHYLENEIMINE

Formula: $C_7H_{14}N_2O$

Mol Weight: 142.20

MP (deg C):   FP (deg C):

BP (deg C):

BP pressure (mm Hg):

Property/Value	Units	Temp	Data Type	Reference
WS 4.46E+003	mg/L	25	EST	MEYLAN,WM ET AL. (1996)
logP 1.48			EXP	HANSCH,C & LEO,AJ (1985)
VP 1.74E-002	mm Hg	25	EST	NEELY,WB & BLAU,GE (1985)
DC	pKa			
HL 3.74E-006	atm m3/mol	25	EST	MEYLAN,WM & HOWARD,PH (1991)
OH 2.84E-011	cm3/molc sec	25	EST	MEYLAN,WM & HOWARD,PH (1993)

CAS #: 020917-50-4	N-NITROSO-OCTAMETHYLENEIMINE

Formula: $C_8H_{16}N_2O$

Mol Weight: 156.23

MP (deg C):   FP (deg C):

BP (deg C):

BP pressure (mm Hg):

Property/Value	Units	Temp	Data Type	Reference
WS 1.29E+003	mg/L	25	EST	MEYLAN,WM ET AL. (1996)
logP 2.04			EXP	HANSCH,C & LEO,AJ (1985)
VP 7.30E-003	mm Hg	25	EST	NEELY,WB & BLAU,GE (1985)
DC	pKa			
HL 4.97E-006	atm m3/mol	25	EST	MEYLAN,WM & HOWARD,PH (1991)
OH 2.98E-011	cm3/molc sec	25	EST	MEYLAN,WM & HOWARD,PH (1993)

CAS #: 020940-42-5	N-(3-CHLOROPHENYL)-N'-METHYLUREA

Formula: $C_8H_9ClN_2O$

Mol Weight: 184.63

MP (deg C):   FP (deg C):

BP (deg C):

BP pressure (mm Hg):

Property/Value	Units	Temp	Data Type	Reference
WS 7.19E+002	mg/L	25	EST	MEYLAN,WM ET AL. (1996)
logP 2.18			EXP	BRIGGS,GG (1981)
VP 9.63E-005	mm Hg	25	EST	NEELY,WB & BLAU,GE (1985)
DC	pKa			
HL 3.28E-010	atm m3/mol	25	EST	MEYLAN,WM & HOWARD,PH (1991)
OH 3.16E-011	cm3/molc sec	25	EST	MEYLAN,WM & HOWARD,PH (1993)

CAS #: 020940-43-6	1,1-DIMETHYL-3-P-BR-PHENYLUREA

Formula: $C_9H_{11}BrN_2O$

Mol Weight: 243.11

MP (deg C):   FP (deg C):

BP (deg C):

BP pressure (mm Hg):

Property/Value	Units	Temp	Data Type	Reference
WS 3.48E+002	mg/L	25	EST	MEYLAN,WM ET AL. (1996)
logP 2.19			EXP	HANSCH,C & LEO,AJ (1985)
VP 1.74E-005	mm Hg	25	EST	NEELY,WB & BLAU,GE (1985)
DC	pKa			
HL 3.87E-010	atm m3/mol	25	EST	MEYLAN,WM & HOWARD,PH (1991)
OH 1.43E-011	cm3/molc sec	25	EST	MEYLAN,WM & HOWARD,PH (1993)

CAS #: 020944-88-1	2-(2-METHYLALLYL)PHENOL

Formula: $C_{10}H_{12}O$

Mol Weight: 148.21

MP (deg C):   FP (deg C):

BP (deg C):

BP pressure (mm Hg):

Property/Value	Units	Temp	Data Type	Reference
WS 3.30E+002	mg/L	25	EST	MEYLAN,WM ET AL. (1996)
logP 3.45			EST	MEYLAN,WM & HOWARD,PH (1995)
VP 1.32E-002	mm Hg	25	EST	NEELY,WB & BLAU,GE (1985)
DC	pKa			
HL 1.27E-006	atm m3/mol	25	EST	MEYLAN,WM & HOWARD,PH (1991)
OH 9.32E-011	cm3/molc sec	25	EST	MEYLAN,WM & HOWARD,PH (1993)

CAS #: 020980-22-7	PYRIMIDINE, 2-(1-PIPERAZINYL)-

Formula: $C_8H_{12}N_4$

Mol Weight: 164.21

MP (deg C):   FP (deg C):

BP (deg C):

BP pressure (mm Hg):

Property/Value	Units	Temp	Data Type	Reference
WS 2.54E+005	mg/L	25	EST	MEYLAN,WM ET AL. (1996)
logP 0.49			EXP	CACCIA,S ET AL. (1985)
VP 7.36E-004	mm Hg	25	EST	NEELY,WB & BLAU,GE (1985)
DC	pKa			
HL 2.18E-008	atm m3/mol	25	EST	MEYLAN,WM & HOWARD,PH (1991)
OH 1.23E-010	cm3/molc sec	25	EST	MEYLAN,WM & HOWARD,PH (1993)

CAS #: 020991-71-3					
Formula: $C_{43}H_{66}O_{15}$					CARD-20(22)-ENOLIDE DERIVATIVE
Mol Weight: 823.00					
MP (deg C):		FP (deg C):			
BP (deg C):					
BP pressure (mm Hg):					

Property/Value	Units	Temp	Data Type	Reference
WS 5.81E-002	mg/L	25	EST	MEYLAN,WM ET AL. (1996)
logP 2.29			EXP	SANGSTER,J (1993)
VP 1.46E-030	mm Hg	25	EST	NEELY,WB & BLAU,GE (1985)
DC	pKa			
HL 3.03E-028	atm m3/mol	25	EST	MEYLAN,WM & HOWARD,PH (1991)
OH 1.67E-010	cm3/molc sec	25	EST	MEYLAN,WM & HOWARD,PH (1993)

CAS #: 021003-78-1					
Formula: $C_{13}H_{11}NO_3$					3'-OH-N-PHENYL AMINOBENZOIC ACID
Mol Weight: 229.24					
MP (deg C):		FP (deg C):			
BP (deg C):					
BP pressure (mm Hg):					

Property/Value	Units	Temp	Data Type	Reference
WS 3.21E+001	mg/L	25	EST	MEYLAN,WM ET AL. (1996)
logP 3.49			EXP	HANSCH,C & LEO,AJ (1985)
VP 3.07E-008	mm Hg	25	EST	NEELY,WB & BLAU,GE (1985)
DC	pKa			
HL 2.20E-015	atm m3/mol	25	EST	MEYLAN,WM & HOWARD,PH (1991)
OH 2.01E-010	cm3/molc sec	25	EST	MEYLAN,WM & HOWARD,PH (1993)

CAS #: 021020-24-6					
Formula: C_4H_5Cl					3-CHLORO-1-BUTYNE
Mol Weight: 88.54					
MP (deg C):		FP (deg C):			
BP (deg C): 68.5					
BP pressure (mm Hg):					

Property/Value	Units	Temp	Data Type	Reference
WS 4.31E+003	mg/L	25	EST	MEYLAN,WM ET AL. (1996)
logP 1.71			EST	MEYLAN,WM & HOWARD,PH (1995)
VP 1.40E+002	mm Hg	25	EST	NEELY,WB & BLAU,GE (1985)
DC	pKa			
HL 7.51E-003	atm m3/mol	25	EST	MEYLAN,WM & HOWARD,PH (1991)
OH 7.79E-012	cm3/molc sec	25	EST	MEYLAN,WM & HOWARD,PH (1993)

CAS #: 021085-18-7					
Formula: $C_9H_{16}N_4OS$					3-SME-4-NH2-6-IPE-124-TRIAZIN-5-ONE
Mol Weight: 228.32					
MP (deg C):		FP (deg C):			
BP (deg C):					
BP pressure (mm Hg):					

Property/Value	Units	Temp	Data Type	Reference
WS 8.18E+002	mg/L	25	EST	MEYLAN,WM ET AL. (1996)
logP 1.85			EXP	HANSCH,C & LEO,AJ (1985)
VP 9.34E-007	mm Hg	25	EST	NEELY,WB & BLAU,GE (1985)
DC	pKa			
HL 2.40E-012	atm m3/mol	25	EST	MEYLAN,WM & HOWARD,PH (1991)
OH 2.64E-011	cm3/molc sec	25	EST	MEYLAN,WM & HOWARD,PH (1993)

CAS #: 021085-19-8					
Formula: $C_{11}H_{11}N_3OS$					3MES-4NH2-6CYHX-124-TRIAZIN-5-ONE
Mol Weight: 233.29					
MP (deg C):		FP (deg C):			
BP (deg C):					
BP pressure (mm Hg):					

Property/Value	Units	Temp	Data Type	Reference
WS 3.98E+002	mg/L	25	EST	MEYLAN,WM ET AL. (1996)
logP 2.14			EXP	HANSCH,C & LEO,AJ (1985)
VP 1.85E-007	mm Hg	25	EST	NEELY,WB & BLAU,GE (1985)
DC	pKa			
HL 1.40E-012	atm m3/mol	25	EST	MEYLAN,WM & HOWARD,PH (1991)
OH 3.63E-011	cm3/molc sec	25	EST	MEYLAN,WM & HOWARD,PH (1993)

CAS #: 021085-20-1					
Formula: $C_{10}H_{18}N_4OS$					3-SME-4-NH2-6-HX-124-TRIAZINE-5-ONE
Mol Weight: 242.35					
MP (deg C):		FP (deg C):			
BP (deg C):					
BP pressure (mm Hg):					

Property/Value	Units	Temp	Data Type	Reference
WS 1.34E+002	mg/L	25	EST	MEYLAN,WM ET AL. (1996)
logP 2.68			EXP	HANSCH,C & LEO,AJ (1985)
VP 2.64E-007	mm Hg	25	EST	NEELY,WB & BLAU,GE (1985)
DC	pKa			
HL 3.19E-012	atm m3/mol	25	EST	MEYLAN,WM & HOWARD,PH (1991)
OH 2.78E-011	cm3/molc sec	25	EST	MEYLAN,WM & HOWARD,PH (1993)

CAS #: 021085-65-4					
Formula: $C_{20}H_{38}N_2O_6S$					D-ERYTHRO-ALPHA-D-GALACTO-OCTOPYRANOSIDE, ETHYL
Mol Weight: 434.60					
MP (deg C):		FP (deg C):			
BP (deg C):					
BP pressure (mm Hg):					

Property/Value	Units	Temp	Data Type	Reference
WS 4.36E+001	mg/L	25	EST	MEYLAN,WM ET AL. (1996)
logP 1.91			EXP	HANSCH,C ET AL. (1995)
VP 1.59E-018	mm Hg	25	EST	NEELY,WB & BLAU,GE (1985)
DC	pKa			
HL 5.28E-023	atm m3/mol	25	EST	MEYLAN,WM & HOWARD,PH (1991)
OH 2.71E-010	cm3/molc sec	25	EST	MEYLAN,WM & HOWARD,PH (1993)

CAS #: 021087-57-0					
Formula: $C_7H_{13}N_5O$					6-IPR-4-NH2-3-MEAM-124-TRIAZ-5-ONE
Mol Weight: 183.21					
MP (deg C):		FP (deg C):			
BP (deg C):					
BP pressure (mm Hg):					

Property/Value	Units	Temp	Data Type	Reference
WS 1.44E+003	mg/L	25	EST	MEYLAN,WM ET AL. (1996)
logP 0.30			EXP	HANSCH,C & LEO,AJ (1985)
VP 9.62E-006	mm Hg	25	EST	NEELY,WB & BLAU,GE (1985)
DC	pKa			
HL 2.88E-014	atm m3/mol	25	EST	MEYLAN,WM & HOWARD,PH (1991)
OH 8.72E-011	cm3/molc sec	25	EST	MEYLAN,WM & HOWARD,PH (1993)

CAS #: 021087-58-1				124-TRIAZIN-5-ONE,4NH2-3MES-6PYRAN2YL

Formula: $C_9H_{14}N_4O_2S$

Mol Weight: 242.30

MP (deg C):   FP (deg C):

BP (deg C):

BP pressure (mm Hg):

Property/Value	Units	Temp	Data Type	Reference
WS 1.24E+004	mg/L	25	EST	MEYLAN,WM ET AL. (1996)
logP 0.38			EXP	HANSCH,C & LEO,AJ (1985)
VP 1.50E-007	mm Hg	25	EST	NEELY,WB & BLAU,GE (1985)
DC	pKa			
HL 1.17E-015	atm m3/mol	25	EST	MEYLAN,WM & HOWARD,PH (1991)
OH 8.84E-011	cm3/molc sec	25	EST	MEYLAN,WM & HOWARD,PH (1993)

CAS #: 021087-59-2				3-SME-4-NH2-6-ET-124-TRIAZIN-5-ONE

Formula: $C_6H_{10}N_4OS$

Mol Weight: 186.24

MP (deg C):   FP (deg C):

BP (deg C):

BP pressure (mm Hg):

Property/Value	Units	Temp	Data Type	Reference
WS 2.08E+004	mg/L	25	EST	MEYLAN,WM ET AL. (1996)
logP 0.46			EXP	HANSCH,C & LEO,AJ (1985)
VP 6.28E-006	mm Hg	25	EST	NEELY,WB & BLAU,GE (1985)
DC	pKa			
HL 1.03E-012	atm m3/mol	25	EST	MEYLAN,WM & HOWARD,PH (1991)
OH 2.16E-011	cm3/molc sec	25	EST	MEYLAN,WM & HOWARD,PH (1993)

CAS #: 021087-60-5				3-SME-4-NH2-6-PR-124-TRIAZIN-5-ONE

Formula: $C_7H_{12}N_4OS$

Mol Weight: 200.26

MP (deg C):   FP (deg C):

BP (deg C):

BP pressure (mm Hg):

Property/Value	Units	Temp	Data Type	Reference
WS 7.01E+003	mg/L	25	EST	MEYLAN,WM ET AL. (1996)
logP 0.93			EXP	HANSCH,C & LEO,AJ (1985)
VP 2.78E-006	mm Hg	25	EST	NEELY,WB & BLAU,GE (1985)
DC	pKa			
HL 1.36E-012	atm m3/mol	25	EST	MEYLAN,WM & HOWARD,PH (1991)
OH 2.36E-011	cm3/molc sec	25	EST	MEYLAN,WM & HOWARD,PH (1993)

CAS #: 021087-61-6				3-SME-4-NH2-6-IPR-124-TRIAZIN-5-ONE

Formula: $C_7H_{12}N_4OS$

Mol Weight: 200.26

MP (deg C):   FP (deg C):

BP (deg C):

BP pressure (mm Hg):

Property/Value	Units	Temp	Data Type	Reference
WS 5.99E+003	mg/L	25	EST	MEYLAN,WM ET AL. (1996)
logP 1.01			EXP	HANSCH,C & LEO,AJ (1985)
VP 4.71E-006	mm Hg	25	EST	NEELY,WB & BLAU,GE (1985)
DC	pKa			
HL 1.36E-012	atm m3/mol	25	EST	MEYLAN,WM & HOWARD,PH (1991)
OH 2.57E-011	cm3/molc sec	25	EST	MEYLAN,WM & HOWARD,PH (1993)

CAS #: 021087-62-7				3-SME-4-NH2-6-IBU-124-TRIAZIN-5-ONE

Formula: $C_8H_{14}N_4OS$

Mol Weight: 214.29

MP (deg C):   FP (deg C):

BP (deg C):

BP pressure (mm Hg):

Property/Value	Units	Temp	Data Type	Reference
WS 2.40E+003	mg/L	25	EST	MEYLAN,WM ET AL. (1996)
logP 1.39			EXP	HANSCH,C & LEO,AJ (1985)
VP 2.04E-006	mm Hg	25	EST	NEELY,WB & BLAU,GE (1985)
DC	pKa			
HL 1.81E-012	atm m3/mol	25	EST	MEYLAN,WM & HOWARD,PH (1991)
OH 2.50E-011	cm3/molc sec	25	EST	MEYLAN,WM & HOWARD,PH (1993)

CAS #: 021087-63-8				3-SME-4-NH2-6-PH-124-TRIAZIN-5-ONE

Formula: $C_{10}H_{10}N_4OS$

Mol Weight: 234.28

MP (deg C):   FP (deg C):

BP (deg C):

BP pressure (mm Hg):

Property/Value	Units	Temp	Data Type	Reference
WS 1.10E+003	mg/L	25	EST	MEYLAN,WM ET AL. (1996)
logP 1.66			EXP	HANSCH,C & LEO,AJ (1985)
VP 9.76E-008	mm Hg	25	EST	NEELY,WB & BLAU,GE (1985)
DC	pKa			
HL 4.70E-014	atm m3/mol	25	EST	MEYLAN,WM & HOWARD,PH (1991)
OH 2.21E-011	cm3/molc sec	25	EST	MEYLAN,WM & HOWARD,PH (1993)

CAS #: 021087-64-9				METRIBUZIN

Formula: $C_8H_{14}N_4OS$

Mol Weight: 214.29

MP (deg C): 126   FP (deg C):

BP (deg C):

BP pressure (mm Hg):

Property/Value	Units	Temp	Data Type	Reference
WS 1.22E+003	mg/L	20	EXP	WSSA (1989)
logP 1.70			EXP	HANSCH,C & LEO,AJ (1985)
VP 1.00E-005	mm Hg	20	EST	WSSA (1989)
DC	pKa			
HL 1.81E-012	atm m3/mol	20	EST	MEYLAN,WM & HOWARD,PH (1991)
OH 3.56E-011	cm3/molc sec	25	EST	ATKINSON,R (1988)

CAS #: 021117-51-1				1H-IMIDAZOLE-1-ETHANOL, 2-METHYL-5-NITRO-, NITRA

Formula: $C_6H_8N_4O_5$

Mol Weight: 216.15

MP (deg C):   FP (deg C):

BP (deg C):

BP pressure (mm Hg):

Property/Value	Units	Temp	Data Type	Reference
WS 2.99E+003	mg/L	25	EST	MEYLAN,WM ET AL. (1996)
logP 0.81			EXP	DE,AU ET AL. (1983)
VP 4.00E-006	mm Hg	25	EST	NEELY,WB & BLAU,GE (1985)
DC	pKa			
HL 1.25E-009	atm m3/mol	25	EST	MEYLAN,WM & HOWARD,PH (1991)
OH 4.58E-012	cm3/molc sec	25	EST	MEYLAN,WM & HOWARD,PH (1993)

CAS #: 021117-52-2	5-BROMO-1,2-DIMETHYL-4-NITROIMIDAZOLE

Formula: $C_5H_6BrN_3O_2$

Mol Weight: 220.03

MP (deg C):   FP (deg C):

BP (deg C):

BP pressure (mm Hg):

Property/Value	Units	Temp	Data Type	Reference
WS 3.61E+002	mg/L	25	EST	MEYLAN,WM ET AL. (1996)
logP 1.86			EST	MEYLAN,WM & HOWARD,PH (1995)
VP 4.35E-005	mm Hg	25	EST	NEELY,WB & BLAU,GE (1985)
DC	pKa			
HL 1.39E-007	atm m3/mol	25	EST	MEYLAN,WM & HOWARD,PH (1991)
OH 3.63E-011	cm3/molc sec	25	EST	MEYLAN,WM & HOWARD,PH (1993)

CAS #: 021149-88-2	BARBITURIC ACID,5-ET-5(3ME-2BUTENE)

Formula: $C_{11}H_{16}N_2O_3$

Mol Weight: 224.26

MP (deg C):   FP (deg C):

BP (deg C):

BP pressure (mm Hg):

Property/Value	Units	Temp	Data Type	Reference
WS 1.09E+003	mg/L	25	EST	MEYLAN,WM ET AL. (1996)
logP 1.73			EXP	PRANKERD,RJ & MCKEOWN,RH (1992)
VP 1.27E-011	mm Hg	25	EST	NEELY,WB & BLAU,GE (1985)
DC	pKa			
HL 8.76E-013	atm m3/mol	25	EST	MEYLAN,WM & HOWARD,PH (1991)
OH 9.66E-011	cm3/molc sec	25	EST	MEYLAN,WM & HOWARD,PH (1993)

CAS #: 021154-18-7	2-QUINOLINAMINE, N,N-DIMETHYL-

Formula: $C_{11}H_{12}N_2$

Mol Weight: 172.23

MP (deg C):   FP (deg C):

BP (deg C):

BP pressure (mm Hg):

Property/Value	Units	Temp	Data Type	Reference
WS 2.39E+002	mg/L	25	EST	MEYLAN,WM ET AL. (1996)
logP 2.81			EXP	SANGSTER,J (1993)
VP 1.11E-003	mm Hg	25	EST	NEELY,WB & BLAU,GE (1985)
DC	pKa			
HL 1.09E-008	atm m3/mol	25	EST	MEYLAN,WM & HOWARD,PH (1991)
OH 2.03E-010	cm3/molc sec	25	EST	MEYLAN,WM & HOWARD,PH (1993)

CAS #: 021156-62-7	PHENYLALANINE-N-ACETYL,METHYL ESTER

Formula: $C_{12}H_{15}NO_3$

Mol Weight: 221.26

MP (deg C):   FP (deg C):

BP (deg C):

BP pressure (mm Hg):

Property/Value	Units	Temp	Data Type	Reference
WS 5.55E+003	mg/L	25	EST	MEYLAN,WM ET AL. (1996)
logP 0.92			EXP	HANSCH,C ET AL. (1995)
VP 3.40E-006	mm Hg	25	EST	NEELY,WB & BLAU,GE (1985)
DC	pKa			
HL 1.25E-010	atm m3/mol	25	EST	MEYLAN,WM & HOWARD,PH (1991)
OH 1.91E-011	cm3/molc sec	25	EST	MEYLAN,WM & HOWARD,PH (1993)

CAS #: 021203-68-9	2-METHYL-5-NITROPYRIDINE

Formula: $C_6H_6N_2O_2$

Mol Weight: 138.13

MP (deg C):   FP (deg C):

BP (deg C):

BP pressure (mm Hg):

Property/Value	Units	Temp	Data Type	Reference
WS 4.96E+003	mg/L	25	EST	MEYLAN,WM ET AL. (1996)
logP 0.99			EXP	HANSCH,C & LEO,AJ (1985)
VP 3.31E-002	mm Hg	25	EST	NEELY,WB & BLAU,GE (1985)
DC	pKa			
HL 3.07E-008	atm m3/mol	25	EST	MEYLAN,WM & HOWARD,PH (1991)
OH 2.57E-013	cm3/molc sec	25	EST	MEYLAN,WM & HOWARD,PH (1993)

CAS #: 021236-52-2	PROPANAMIDE, 3-(DIETHYLAMINO)-N-(2,6-DIMETHYLPHE

Formula: $C_{15}H_{24}N_2O$

Mol Weight: 248.37

MP (deg C):   FP (deg C):

BP (deg C):

BP pressure (mm Hg):

Property/Value	Units	Temp	Data Type	Reference
WS 3.13E+002	mg/L	25	EST	MEYLAN,WM ET AL. (1996)
logP 2.21			EXP	SANGSTER,J (1993)
VP 4.63E-007	mm Hg	25	EST	NEELY,WB & BLAU,GE (1985)
DC	pKa			
HL 8.72E-012	atm m3/mol	25	EST	MEYLAN,WM & HOWARD,PH (1991)
OH 1.37E-010	cm3/molc sec	25	EST	MEYLAN,WM & HOWARD,PH (1993)

CAS #: 021236-53-3	BUTANAMIDE, 4-(DIETHYLAMINO)-N-(2,6-DIMETHYLPHEN

Formula: $C_{16}H_{26}N_2O$

Mol Weight: 262.40

MP (deg C):   FP (deg C):

BP (deg C):

BP pressure (mm Hg):

Property/Value	Units	Temp	Data Type	Reference
WS 3.51E+002	mg/L	25	EST	MEYLAN,WM ET AL. (1996)
logP 2.06			EXP	SANGSTER,J (1993)
VP 1.98E-007	mm Hg	25	EST	NEELY,WB & BLAU,GE (1985)
DC	pKa			
HL 1.16E-011	atm m3/mol	25	EST	MEYLAN,WM & HOWARD,PH (1991)
OH 1.19E-010	cm3/molc sec	25	EST	MEYLAN,WM & HOWARD,PH (1993)

CAS #: 021236-54-4	2(DIMEAMINO)-N-(2,6-DIMEPH)ACETAMIDE

Formula: $C_{12}H_{18}N_2O$

Mol Weight: 206.29

MP (deg C):   FP (deg C):

BP (deg C):

BP pressure (mm Hg):

Property/Value	Units	Temp	Data Type	Reference
WS 6.40E+003	mg/L	25	EST	MEYLAN,WM ET AL. (1996)
logP 0.94			EXP	HANSCH,C & LEO,AJ (1985)
VP 4.74E-006	mm Hg	25	EST	NEELY,WB & BLAU,GE (1985)
DC	pKa			
HL 7.44E-011	atm m3/mol	25	EST	MEYLAN,WM & HOWARD,PH (1991)
OH 9.40E-011	cm3/molc sec	25	EST	MEYLAN,WM & HOWARD,PH (1993)

CAS #: 021244-66-6 — IMIDAZOL-5-CONH2,4(33BIS(2OHET)1-NNN

Formula: $C_8H_{14}N_6O_3$

Mol Weight: 242.24

MP (deg C): FP (deg C):

BP (deg C):

BP pressure (mm Hg):

Property/Value	Units	Temp	Data Type	Reference
WS 1.09E+004	mg/L	25	EST	MEYLAN,WM ET AL. (1996)
logP -1.09			EXP	HANSCH,C & LEO,AJ (1985)
VP 8.88E-015	mm Hg	25	EST	NEELY,WB & BLAU,GE (1985)
DC	pKa			
HL 3.62E-022	atm m3/mol	25	EST	MEYLAN,WM & HOWARD,PH (1991)
OH 4.47E-011	cm3/molc sec	25	EST	MEYLAN,WM & HOWARD,PH (1993)

CAS #: 021253-58-7 — 2,4-NH2PYRIMIDINE,5(35MEO-4-OH)BENZYL

Formula: $C_{13}H_{16}N_4O_3$

Mol Weight: 276.30

MP (deg C): FP (deg C):

BP (deg C):

BP pressure (mm Hg):

Property/Value	Units	Temp	Data Type	Reference
WS 8.13E+003	mg/L	25	EST	MEYLAN,WM ET AL. (1996)
logP 0.37			EXP	HANSCH,C & LEO,AJ (1985)
VP 7.13E-010	mm Hg	25	EST	NEELY,WB & BLAU,GE (1985)
DC	pKa			
HL 4.20E-017	atm m3/mol	25	EST	MEYLAN,WM & HOWARD,PH (1991)
OH 2.03E-010	cm3/molc sec	25	EST	MEYLAN,WM & HOWARD,PH (1993)

CAS #: 021270-93-9 — 4-BUTYRAMID-2266-TEME-PIPERDIN-N-OXID

Formula: $C_{13}H_{26}N_2O_2$

Mol Weight: 242.36

MP (deg C): FP (deg C):

BP (deg C):

BP pressure (mm Hg):

Property/Value	Units	Temp	Data Type	Reference
WS 1.83E+003	mg/L	25	EST	MEYLAN,WM ET AL. (1996)
logP 1.35			EXP	HANSCH,C & LEO,AJ (1985)
VP 3.70E-010	mm Hg	25	EST	NEELY,WB & BLAU,GE (1985)
DC	pKa			
HL 6.27E-014	atm m3/mol	25	EST	MEYLAN,WM & HOWARD,PH (1991)
OH 9.17E-011	cm3/molc sec	25	EST	MEYLAN,WM & HOWARD,PH (1993)

CAS #: 021306-55-8 — IMIDODICARBONIMIDIC DIAMIDE, N-PENTYL

Formula: $C_7H_{17}N_5$

Mol Weight: 171.25

MP (deg C): FP (deg C):

BP (deg C):

BP pressure (mm Hg):

Property/Value	Units	Temp	Data Type	Reference
WS 4.51E+005	mg/L	25	EST	MEYLAN,WM ET AL. (1996)
logP -1.02			EXP	SANGSTER,J (1993)
VP 6.45E-005	mm Hg	25	EST	NEELY,WB & BLAU,GE (1985)
DC	pKa			
HL 1.08E-015	atm m3/mol	25	EST	MEYLAN,WM & HOWARD,PH (1991)
OH 1.62E-010	cm3/molc sec	25	EST	MEYLAN,WM & HOWARD,PH (1993)

CAS #: 021316-30-3 — 4,6-NH2 2,2-DIME-1(4-MEOPH)S-TRIAZINE

Formula: $C_{12}H_{17}N_5O$

Mol Weight: 247.30

MP (deg C): FP (deg C):

BP (deg C):

BP pressure (mm Hg):

Property/Value	Units	Temp	Data Type	Reference
WS 1.12E+004	mg/L	25	EST	MEYLAN,WM ET AL. (1996)
logP 0.40			EXP	HANSCH,C & LEO,AJ (1985)
VP 1.04E-006	mm Hg	25	EST	NEELY,WB & BLAU,GE (1985)
DC	pKa			
HL 4.31E-014	atm m3/mol	25	EST	MEYLAN,WM & HOWARD,PH (1991)
OH 1.70E-010	cm3/molc sec	25	EST	MEYLAN,WM & HOWARD,PH (1993)

CAS #: 021327-86-6 — BENZOIC ACID, 2-CHLORO-6-METHYL-

Formula: $C_8H_7ClO_2$

Mol Weight: 170.60

MP (deg C): FP (deg C):

BP (deg C):

BP pressure (mm Hg):

Property/Value	Units	Temp	Data Type	Reference
WS 6.51E+002	mg/L	25	EST	MEYLAN,WM ET AL. (1996)
logP 2.31			EXP	SOTOMATSU,T ET AL. (1993)
VP 8.86E-004	mm Hg	25	EST	NEELY,WB & BLAU,GE (1985)
DC 2.75	pKa	20	EXP	SERJEANT,EP & DEMPSEY,B (1979)
HL 8.87E-008	atm m3/mol	25	EST	MEYLAN,WM & HOWARD,PH (1991)
OH 1.98E-012	cm3/molc sec	25	EST	MEYLAN,WM & HOWARD,PH (1993)

CAS #: 021352-09-0 — 2-CHLORO-6-METHYLACETANILIDE

Formula: $C_9H_{10}ClNO$

Mol Weight: 183.64

MP (deg C): FP (deg C):

BP (deg C):

BP pressure (mm Hg):

Property/Value	Units	Temp	Data Type	Reference
WS 4.80E+003	mg/L	25	EST	MEYLAN,WM ET AL. (1996)
logP 1.22			EXP	NAKAGAWA,Y ET AL. (1992)
VP 3.66E-005	mm Hg	25	EST	NEELY,WB & BLAU,GE (1985)
DC	pKa			
HL 5.05E-009	atm m3/mol	25	EST	MEYLAN,WM & HOWARD,PH (1991)
OH 4.68E-012	cm3/molc sec	25	EST	MEYLAN,WM & HOWARD,PH (1993)

CAS #: 021352-21-6 — 2-ACETYLAMINOPYRAZINE

Formula: $C_6H_7N_3O$

Mol Weight: 137.14

MP (deg C): FP (deg C):

BP (deg C):

BP pressure (mm Hg):

Property/Value	Units	Temp	Data Type	Reference
WS 4.47E+003	mg/L	25	EST	MEYLAN,WM ET AL. (1996)
logP -0.03			EXP	YAMAGAMI,C ET AL. (1990A)
VP 1.37E-004	mm Hg	25	EST	NEELY,WB & BLAU,GE (1985)
DC	pKa			
HL 3.34E-012	atm m3/mol	25	EST	MEYLAN,WM & HOWARD,PH (1991)
OH 1.57E-012	cm3/molc sec	25	EST	MEYLAN,WM & HOWARD,PH (1993)

CAS #: 021392-61-0				PHENYLMETHYL MERCURY
Formula: C₇H₈Hg				
Mol Weight: 292.73				
MP (deg C):		FP (deg C):		
BP (deg C):				
BP pressure (mm Hg):				

Property/Value	Units	Temp	Data Type	Reference
WS 3.49E+002	mg/L	25	EST	MEYLAN,WM ET AL. (1996)
logP 1.86			EXP	HANSCH,C & LEO,AJ (1985)
VP 5.61E-002	mm Hg	25	EST	NEELY,WB & BLAU,GE (1985)
DC	pKa			
HL	atm m3/mol			
OH 2.09E-012	cm3/molc sec	25	EST	MEYLAN,WM & HOWARD,PH (1993)

CAS #: 021400-25-9				1,1,2-TRICHLOROPROPENE
Formula: C₃H₃Cl₃				
Mol Weight: 145.42				
MP (deg C):		FP (deg C):		
BP (deg C): 118				
BP pressure (mm Hg):				

Property/Value	Units	Temp	Data Type	Reference
WS 2.09E+002	mg/L	25	EST	MEYLAN,WM ET AL. (1996)
logP 3.02			EST	MEYLAN,WM & HOWARD,PH (1995)
VP 3.91E+001	mm Hg	25	EST	NEELY,WB & BLAU,GE (1985)
DC	pKa			
HL 3.60E-002	atm m3/mol	25	EST	MEYLAN,WM & HOWARD,PH (1991)
OH 1.15E-012	cm3/molc sec	25	EST	MEYLAN,WM & HOWARD,PH (1993)

CAS #: 021413-25-2				2,4-IMIDAZOLIDINEDIONE, 5-ETHYL-5-PHENYL-1-(PHEN
Formula: C₁₇H₁₆N₂O₄S				
Mol Weight: 344.39				
MP (deg C):		FP (deg C):		
BP (deg C):				
BP pressure (mm Hg):				

Property/Value	Units	Temp	Data Type	Reference
WS 5.52E+001	mg/L	25	EST	MEYLAN,WM ET AL. (1996)
logP 2.44			EXP	SANGSTER,J (1994)
VP 6.18E-014	mm Hg	25	EST	NEELY,WB & BLAU,GE (1985)
DC	pKa			
HL 2.73E-011	atm m3/mol	25	EST	MEYLAN,WM & HOWARD,PH (1991)
OH 1.05E-011	cm3/molc sec	25	EST	MEYLAN,WM & HOWARD,PH (1993)

CAS #: 021431-58-3				4-BROMO-1,2-DIMETHYL-5-NITROIMIDAZOLE
Formula: C₅H₆BrN₃O₂				
Mol Weight: 220.03				
MP (deg C):		FP (deg C):		
BP (deg C):				
BP pressure (mm Hg):				

Property/Value	Units	Temp	Data Type	Reference
WS 3.61E+002	mg/L	25	EST	MEYLAN,WM ET AL. (1996)
logP 1.86			EST	MEYLAN,WM & HOWARD,PH (1995)
VP 4.35E-005	mm Hg	25	EST	NEELY,WB & BLAU,GE (1985)
DC	pKa			
HL 1.39E-007	atm m3/mol	25	EST	MEYLAN,WM & HOWARD,PH (1991)
OH 3.63E-011	cm3/molc sec	25	EST	MEYLAN,WM & HOWARD,PH (1993)

CAS #: 021450-13-5				1-CHLORO-1-PENTENE
Formula: C₅H₉Cl				
Mol Weight: 104.58				
MP (deg C):		FP (deg C):		
BP (deg C):				
BP pressure (mm Hg):				

Property/Value	Units	Temp	Data Type	Reference
WS 3.00E+002	mg/L	25	EST	MEYLAN,WM ET AL. (1996)
logP 3.02			EST	MEYLAN,WM & HOWARD,PH (1995)
VP 6.10E+001	mm Hg	25	EST	NEELY,WB & BLAU,GE (1985)
DC	pKa			
HL 1.23E-001	atm m3/mol	25	EST	MEYLAN,WM & HOWARD,PH (1991)
OH 1.43E-011	cm3/molc sec	25	EST	MEYLAN,WM & HOWARD,PH (1993)

CAS #: 021466-00-2				3(33DIME-1-TRIAZENO)PYRAZOLE-4-CONH2
Formula: C₆H₁₀N₆O				
Mol Weight: 182.19				
MP (deg C):		FP (deg C):		
BP (deg C):				
BP pressure (mm Hg):				

Property/Value	Units	Temp	Data Type	Reference
WS 3.33E+003	mg/L	25	EST	MEYLAN,WM ET AL. (1996)
logP -0.12			EXP	HANSCH,C & LEO,AJ (1985)
VP 3.07E-007	mm Hg	25	EST	NEELY,WB & BLAU,GE (1985)
DC	pKa			
HL 1.51E-016	atm m3/mol	25	EST	MEYLAN,WM & HOWARD,PH (1991)
OH 1.05E-011	cm3/molc sec	25	EST	MEYLAN,WM & HOWARD,PH (1993)

CAS #: 021483-62-5				1,1,1,2,2,3,3,4,4-NONACHLOROBUTANE
Formula: C₄HCl₉				
Mol Weight: 368.13				
MP (deg C):		FP (deg C):		
BP (deg C):				
BP pressure (mm Hg):				

Property/Value	Units	Temp	Data Type	Reference
WS 2.01E-002	mg/L	25	EST	MEYLAN,WM ET AL. (1996)
logP 6.30			EST	MEYLAN,WM & HOWARD,PH (1995)
VP 2.84E-003	mm Hg	25	EST	NEELY,WB & BLAU,GE (1985)
DC	pKa			
HL 1.44E-005	atm m3/mol	25	EST	MEYLAN,WM & HOWARD,PH (1991)
OH 7.00E-014	cm3/molc sec	25	EST	MEYLAN,WM & HOWARD,PH (1993)

CAS #: 021507-14-2				ESTRA-1,3,5(10)-TRIENE-3,17-DIOL, 11-METHOXY-, (
Formula: C₁₉H₂₆O₃				
Mol Weight: 302.42				
MP (deg C):		FP (deg C):		
BP (deg C):				
BP pressure (mm Hg):				

Property/Value	Units	Temp	Data Type	Reference
WS 6.95E+002	mg/L	25	EST	MEYLAN,WM ET AL. (1996)
logP 2.72			EXP	SANGSTER,J (1993)
VP 2.14E-009	mm Hg	25	EST	NEELY,WB & BLAU,GE (1985)
DC	pKa			
HL 4.26E-013	atm m3/mol	25	EST	MEYLAN,WM & HOWARD,PH (1991)
OH 1.37E-010	cm3/molc sec	25	EST	MEYLAN,WM & HOWARD,PH (1993)

CAS #: 021704-31-4				1,2-BENZENEDIOL, 3-UNDECYL-
Formula: $C_{17}H_{28}O_2$				
Mol Weight: 264.41				
MP (deg C):		FP (deg C):		
BP (deg C):				
BP pressure (mm Hg):				

Property/ Value	Units	Temp	Data Type	Reference
WS 3.11E-001	mg/L	25	EST	MEYLAN,WM ET AL. (1996)
logP 6.30			EXP	ITOKAWA,H ET AL. (1989)
VP 8.66E-008	mm Hg	25	EST	NEELY,WB & BLAU,GE (1985)
DC	pKa			
HL 1.09E-009	atm m3/mol	25	EST	MEYLAN,WM & HOWARD,PH (1991)
OH 7.11E-011	cm3/molc sec	25	EST	MEYLAN,WM & HOWARD,PH (1993)

CAS #: 021724-58-3				3,4-DICHLOROPHENYLAMIDINOUREA
Formula: $C_8H_8Cl_2N_4O$				
Mol Weight: 247.09				
MP (deg C):		FP (deg C):		
BP (deg C):				
BP pressure (mm Hg):				

Property/ Value	Units	Temp	Data Type	Reference
WS 2.23E+002	mg/L	25	EST	MEYLAN,WM ET AL. (1996)
logP 2.39			EXP	HANSCH,C & LEO,AJ (1985)
VP 1.60E-007	mm Hg	25	EST	NEELY,WB & BLAU,GE (1985)
DC	pKa			
HL 7.49E-016	atm m3/mol	25	EST	MEYLAN,WM & HOWARD,PH (1991)
OH 3.08E-011	cm3/molc sec	25	EST	MEYLAN,WM & HOWARD,PH (1993)

CAS #: 021725-46-2				CYANAZINE
Formula: $C_9H_{13}ClN_6$				
Mol Weight: 240.70				
MP (deg C): 166.5-167		FP (deg C):		
BP (deg C):				
BP pressure (mm Hg):				

Property/ Value	Units	Temp	Data Type	Reference
WS 1.70E+002	mg/L	25	EXP	WAUCHOPE,RD ET AL. (1991A)
logP 2.22			EXP	HANSCH,C & LEO,AJ (1985)
VP 1.38E-007	mm Hg	25	EXP	GRAYSON,BT & FOSBRAEY,LA (1982)
DC 1.10	pKa		EXP	WEBER,JB (1972)
HL 2.57E-010	atm m3/mol	25	EST	VP/WSOL
OH 1.29E-010	cm3/molc sec	25	EST	ATKINSON,R (1988)

CAS #: 021727-39-9				CYCLOPENTANEMETHANAMINIUM, N,N,N-TRIMETHYL-, IOD
Formula: $C_9H_{20}IN$				
Mol Weight: 269.17				
MP (deg C):		FP (deg C):		
BP (deg C):				
BP pressure (mm Hg):				

Property/ Value	Units	Temp	Data Type	Reference
WS 9.43E+005	mg/L	25	EST	MEYLAN,WM ET AL. (1996)
logP -2.00			EXP	PRATESI,P ET AL. (1986)
VP 7.67E-008	mm Hg	25	EST	NEELY,WB & BLAU,GE (1985)
DC	pKa			
HL 4.79E-012	atm m3/mol	25	EST	MEYLAN,WM & HOWARD,PH (1991)
OH 2.19E-011	cm3/molc sec	25	EST	MEYLAN,WM & HOWARD,PH (1993)

CAS #: 021738-42-1				6-QUINOLINEMETHANOL, 1,2,3,4-TETRAHYDRO-2-[[(1-M
Formula: $C_{14}H_{21}N_3O_3$				
Mol Weight: 279.34				
MP (deg C): 147-149		FP (deg C):		
BP (deg C):				
BP pressure (mm Hg):				

Property/ Value	Units	Temp	Data Type	Reference
WS 8.20E+002	mg/L	25	EST	MEYLAN,WM ET AL. (1996)
logP 2.24			EXP	SANGSTER,J (1993)
VP 7.80E-010	mm Hg	25	EST	NEELY,WB & BLAU,GE (1985)
DC	pKa			
HL 2.27E-016	atm m3/mol	25	EST	MEYLAN,WM & HOWARD,PH (1991)
OH 1.31E-010	cm3/molc sec	25	EST	MEYLAN,WM & HOWARD,PH (1993)

CAS #: 021757-82-4				ACETIC ACID, 1-(3,4-DICHLOROPHENYL)-2,2,2-TRICHL
Formula: $C_{10}H_7Cl_5O_2$				
Mol Weight: 336.43				
MP (deg C):		FP (deg C):		
BP (deg C):				
BP pressure (mm Hg):				

Property/ Value	Units	Temp	Data Type	Reference
WS 5.00E+001	mg/L	20	EXP	SHIU,WY ET AL. (1990)
logP 4.92			EXP	HU,J & LENG,XF (1992)
VP 5.32E-005	mm Hg	25	EST	NEELY,WB & BLAU,GE (1985)
DC	pKa			
HL 4.51E-007	atm m3/mol	25	EST	MEYLAN,WM & HOWARD,PH (1991)
OH 1.26E-012	cm3/molc sec	25	EST	MEYLAN,WM & HOWARD,PH (1993)

CAS #: 021787-89-3				1(2-OH-3-ALLYLOXYPR)2-NO2IMIDAZOLE
Formula: $C_9H_{15}N_3O_4$				
Mol Weight: 229.24				
MP (deg C):		FP (deg C):		
BP (deg C):				
BP pressure (mm Hg):				

Property/ Value	Units	Temp	Data Type	Reference
WS 5.84E+003	mg/L	25	EST	MEYLAN,WM ET AL. (1996)
logP 0.40			EXP	HANSCH,C & LEO,AJ (1985)
VP 1.67E-008	mm Hg	25	EST	NEELY,WB & BLAU,GE (1985)
DC	pKa			
HL 3.13E-013	atm m3/mol	25	EST	MEYLAN,WM & HOWARD,PH (1991)
OH 5.51E-011	cm3/molc sec	25	EST	MEYLAN,WM & HOWARD,PH (1993)

CAS #: 021787-91-7				1(2-OH-6-CF3-ETOPR)-2-NO2-IMIDAZOLE
Formula: $C_8H_{12}F_3N_3O_4$				
Mol Weight: 271.20				
MP (deg C):		FP (deg C):		
BP (deg C):				
BP pressure (mm Hg):				

Property/ Value	Units	Temp	Data Type	Reference
WS 2.71E+003	mg/L	25	EST	MEYLAN,WM ET AL. (1996)
logP 0.52			EXP	HANSCH,C & LEO,AJ (1985)
VP 7.63E-008	mm Hg	25	EST	NEELY,WB & BLAU,GE (1985)
DC	pKa			
HL 2.49E-012	atm m3/mol	25	EST	MEYLAN,WM & HOWARD,PH (1991)
OH 1.73E-011	cm3/molc sec	25	EST	MEYLAN,WM & HOWARD,PH (1993)

CAS #:	021787-95-1				2-PROPANONE, 1-METHOXY-2-(2-NITRO-1H-IMIDAZOL-1-

Formula: $C_7H_9N_3O_4$

Mol Weight: 199.17

MP (deg C): FP (deg C):

BP (deg C):

BP pressure (mm Hg):

Property/Value	Units	Temp	Data Type	Reference
WS 3.00E+004	mg/L	25	EST	MEYLAN,WM ET AL. (1996)
logP -0.26			EXP	SANGSTER,J (1994)
VP 6.59E-006	mm Hg	25	EST	NEELY,WB & BLAU,GE (1985)
DC	pKa			
HL 1.76E-011	atm m3/mol	25	EST	MEYLAN,WM & HOWARD,PH (1991)
OH 1.03E-011	cm3/molc sec	25	EST	MEYLAN,WM & HOWARD,PH (1993)

CAS #:	021788-11-4				1-(3-PHO-2-OH-PROPYL)-2-NO2-IMIDAZOLE

Formula: $C_{12}H_{15}N_3O_4$

Mol Weight: 265.27

MP (deg C): FP (deg C):

BP (deg C):

BP pressure (mm Hg):

Property/Value	Units	Temp	Data Type	Reference
WS 4.88E+002	mg/L	25	EST	MEYLAN,WM ET AL. (1996)
logP 1.43			EXP	HANSCH,C & LEO,AJ (1985)
VP 2.13E-010	mm Hg	25	EST	NEELY,WB & BLAU,GE (1985)
DC	pKa			
HL 8.81E-014	atm m3/mol	25	EST	MEYLAN,WM & HOWARD,PH (1991)
OH 4.46E-011	cm3/molc sec	25	EST	MEYLAN,WM & HOWARD,PH (1993)

CAS #:	021822-34-4				24NH2-PYRIMIDINE,5(35MEO-4PHMEO)BENZYL

Formula: $C_{20}H_{22}N_4O_3$

Mol Weight: 366.42

MP (deg C): FP (deg C):

BP (deg C):

BP pressure (mm Hg):

Property/Value	Units	Temp	Data Type	Reference
WS 4.66E+001	mg/L	25	EST	MEYLAN,WM ET AL. (1996)
logP 2.46			EXP	HANSCH,C & LEO,AJ (1985)
VP 2.56E-011	mm Hg	25	EST	NEELY,WB & BLAU,GE (1985)
DC	pKa			
HL 1.93E-015	atm m3/mol	25	EST	MEYLAN,WM & HOWARD,PH (1991)
OH 2.08E-010	cm3/molc sec	25	EST	MEYLAN,WM & HOWARD,PH (1993)

CAS #:	021829-25-4				NIFEDIPINE

Formula: $C_{17}H_{18}N_2O_6$

Mol Weight: 346.34

MP (deg C): 172-174 FP (deg C):

BP (deg C):

BP pressure (mm Hg):

Property/Value	Units	Temp	Data Type	Reference
WS 5.63E+001	mg/L	25	EST	MEYLAN,WM ET AL. (1996)
logP 3.14			EXP	SANGSTER,J (1993)
VP 2.34E-008	mm Hg	25	EST	NEELY,WB & BLAU,GE (1985)
DC	pKa			
HL 7.31E-014	atm m3/mol	25	EST	MEYLAN,WM & HOWARD,PH (1991)
OH 1.12E-010	cm3/molc sec	25	EST	MEYLAN,WM & HOWARD,PH (1993)

CAS #:	021839-33-8				5-FLUOROURACIL-3-ETHOXYCARBONYL-

Formula: $C_7H_7FN_2O_4$

Mol Weight: 202.14

MP (deg C): FP (deg C):

BP (deg C):

BP pressure (mm Hg):

Property/Value	Units	Temp	Data Type	Reference
WS 2.92E+003	mg/L	25	EST	MEYLAN,WM ET AL. (1996)
logP -0.17			EXP	HANSCH,C ET AL. (1995)
VP 1.53E-008	mm Hg	25	EST	NEELY,WB & BLAU,GE (1985)
DC	pKa			
HL 2.53E-012	atm m3/mol	25	EST	MEYLAN,WM & HOWARD,PH (1991)
OH 7.49E-012	cm3/molc sec	25	EST	MEYLAN,WM & HOWARD,PH (1993)

CAS #:	021840-66-4				O,O-DIET-O-(3-CL-4-MESPH)PHOSPHOROTHIOATE

Formula: $C_{11}H_{16}ClO_3PS_2$

Mol Weight: 326.80

MP (deg C): FP (deg C):

BP (deg C):

BP pressure (mm Hg):

Property/Value	Units	Temp	Data Type	Reference
WS 4.07E-001	mg/L	25	EST	MEYLAN,WM ET AL. (1996)
logP 5.06			EXP	HANSCH,C & LEO,AJ (1985)
VP 1.59E-005	mm Hg	25	EST	NEELY,WB & BLAU,GE (1985)
DC	pKa			
HL 1.62E-006	atm m3/mol	25	EST	MEYLAN,WM & HOWARD,PH (1991)
OH 9.54E-011	cm3/molc sec	25	EST	MEYLAN,WM & HOWARD,PH (1993)

CAS #:	021846-07-1				9-(4-CHLOROPHENYL)-9H-FLUORENE

Formula: $C_{19}H_{13}Cl$

Mol Weight: 276.77

MP (deg C): FP (deg C):

BP (deg C):

BP pressure (mm Hg):

Property/Value	Units	Temp	Data Type	Reference
WS 9.55E-003	mg/L	25	EST	MEYLAN,WM ET AL. (1996)
logP 6.01			EST	MEYLAN,WM & HOWARD,PH (1995)
VP 6.05E-007	mm Hg	25	EST	NEELY,WB & BLAU,GE (1985)
DC	pKa			
HL 1.00E-005	atm m3/mol	25	EST	MEYLAN,WM & HOWARD,PH (1991)
OH 1.14E-011	cm3/molc sec	25	EST	MEYLAN,WM & HOWARD,PH (1993)

CAS #:	021846-08-2				9-(4-METHOXYPHENYL)-9H-FLUORENE

Formula: $C_{20}H_{16}O$

Mol Weight: 272.35

MP (deg C): FP (deg C):

BP (deg C):

BP pressure (mm Hg):

Property/Value	Units	Temp	Data Type	Reference
WS 3.06E-002	mg/L	25	EST	MEYLAN,WM ET AL. (1996)
logP 5.45			EST	MEYLAN,WM & HOWARD,PH (1995)
VP 4.24E-007	mm Hg	25	EST	NEELY,WB & BLAU,GE (1985)
DC	pKa			
HL 7.99E-007	atm m3/mol	25	EST	MEYLAN,WM & HOWARD,PH (1991)
OH 3.64E-011	cm3/molc sec	25	EST	MEYLAN,WM & HOWARD,PH (1993)

CAS #: 021862-63-5				CYCLOHEXANOL, 4-(1,1-DIMETHYLETHYL)-, TRANS-

Formula: $C_{10}H_{20}O$

Mol Weight: 156.27

MP (deg C):   FP (deg C):

BP (deg C):

BP pressure (mm Hg):

Property/Value	Units	Temp	Data Type	Reference
WS 5.29E+002	mg/L	25	EST	MEYLAN,WM ET AL. (1996)
logP 3.09			EXP	FUNASAKI,N ET AL. (1986)
VP 2.63E-002	mm Hg	25	EST	NEELY,WB & BLAU,GE (1985)
DC	pKa			
HL 1.52E-005	atm m3/mol	25	EST	MEYLAN,WM & HOWARD,PH (1991)
OH 2.02E-011	cm3/molc sec	25	EST	MEYLAN,WM & HOWARD,PH (1993)

CAS #: 021864-67-5				2,6-DIMETHOXYBENZAMIDE

Formula: $C_9H_{11}NO_3$

Mol Weight: 181.19

MP (deg C):   FP (deg C):

BP (deg C):

BP pressure (mm Hg):

Property/Value	Units	Temp	Data Type	Reference
WS 8.37E+004	mg/L	25	EST	MEYLAN,WM ET AL. (1996)
logP -0.22			EXP	NAKAGAWA,Y ET AL. (1992)
VP 1.59E-005	mm Hg	25	EST	NEELY,WB & BLAU,GE (1985)
DC	pKa			
HL 7.75E-012	atm m3/mol	25	EST	MEYLAN,WM & HOWARD,PH (1991)
OH 2.04E-010	cm3/molc sec	25	EST	MEYLAN,WM & HOWARD,PH (1993)

CAS #: 021888-98-2				DEXETIMIDE

Formula: $C_{23}H_{26}N_2O_2$

Mol Weight: 362.48

MP (deg C):   FP (deg C):

BP (deg C):

BP pressure (mm Hg):

Property/Value	Units	Temp	Data Type	Reference
WS 4.84E+000	mg/L	25	EST	MEYLAN,WM ET AL. (1996)
logP 3.55			EXP	HANSCH,C & LEO,AJ (1985)
VP 3.10E-014	mm Hg	25	EST	NEELY,WB & BLAU,GE (1985)
DC	pKa			
HL 1.49E-013	atm m3/mol	25	EST	MEYLAN,WM & HOWARD,PH (1991)
OH 1.40E-010	cm3/molc sec	25	EST	MEYLAN,WM & HOWARD,PH (1993)

CAS #: 021899-50-3				2-BUTENOIC ACID, ESTER DERIVATIVE

Formula: $C_{20}H_{22}O_7$

Mol Weight: 374.39

MP (deg C):   FP (deg C):

BP (deg C):

BP pressure (mm Hg):

Property/Value	Units	Temp	Data Type	Reference
WS 5.59E+002	mg/L	25	EST	MEYLAN,WM ET AL. (1996)
logP 1.05			EXP	SANGSTER,J (1993)
VP 1.82E-010	mm Hg	25	EST	NEELY,WB & BLAU,GE (1985)
DC	pKa			
HL 5.98E-013	atm m3/mol	25	EST	MEYLAN,WM & HOWARD,PH (1991)
OH 9.20E-011	cm3/molc sec	25	EST	MEYLAN,WM & HOWARD,PH (1993)

CAS #: 021905-82-8				3-NITROCINNOLINE

Formula: $C_8H_5N_3O_2$

Mol Weight: 175.15

MP (deg C):   FP (deg C):

BP (deg C):

BP pressure (mm Hg):

Property/Value	Units	Temp	Data Type	Reference
WS 5.64E+003	mg/L	25	EST	MEYLAN,WM ET AL. (1996)
logP 0.73			EXP	HANSCH,C & LEO,AJ (1985)
VP 5.80E-006	mm Hg	25	EST	NEELY,WB & BLAU,GE (1985)
DC	pKa			
HL 1.11E-009	atm m3/mol	25	EST	MEYLAN,WM & HOWARD,PH (1991)
OH 2.50E-013	cm3/molc sec	25	EST	MEYLAN,WM & HOWARD,PH (1993)

CAS #: 021914-07-8				BARBITURIC ACID,5,5-DIPHENYL-

Formula: $C_{16}H_{12}N_2O_3$

Mol Weight: 280.29

MP (deg C):   FP (deg C):

BP (deg C):

BP pressure (mm Hg):

Property/Value	Units	Temp	Data Type	Reference
WS 3.38E+002	mg/L	25	EST	MEYLAN,WM ET AL. (1996)
logP 1.96			EXP	PRANKERD,RJ & MCKEOWN,RH (1992)
VP 2.27E-014	mm Hg	25	EST	NEELY,WB & BLAU,GE (1985)
DC	pKa			
HL 7.57E-016	atm m3/mol	25	EST	MEYLAN,WM & HOWARD,PH (1991)
OH 1.06E-011	cm3/molc sec	25	EST	MEYLAN,WM & HOWARD,PH (1993)

CAS #: 021919-05-1				5-(1-AZIRIDINYL)-2,4-NO2BENZAMIDE

Formula: $C_9H_8N_4O_5$

Mol Weight: 252.19

MP (deg C):   FP (deg C):

BP (deg C):

BP pressure (mm Hg):

Property/Value	Units	Temp	Data Type	Reference
WS 4.77E+002	mg/L	25	EST	MEYLAN,WM ET AL. (1996)
logP -0.02			EXP	HANSCH,C ET AL. (1995)
VP 8.48E-009	mm Hg	25	EST	NEELY,WB & BLAU,GE (1985)
DC	pKa			
HL 5.29E-015	atm m3/mol	25	EST	MEYLAN,WM & HOWARD,PH (1991)
OH 3.39E-012	cm3/molc sec	25	EST	MEYLAN,WM & HOWARD,PH (1993)

CAS #: 021948-70-9				2-METHYLTHIOPYRAZINE

Formula: $C_5H_6N_2S$

Mol Weight: 126.18

MP (deg C):   FP (deg C):

BP (deg C):

BP pressure (mm Hg):

Property/Value	Units	Temp	Data Type	Reference
WS 9.53E+003	mg/L	25	EST	MEYLAN,WM ET AL. (1996)
logP 1.17			EXP	YAMAGAMI,C ET AL. (1990A)
VP 2.11E-001	mm Hg	25	EST	NEELY,WB & BLAU,GE (1985)
DC 0.48	pKa	20	EXP	PERRIN,DD (1965)
HL 3.31E-006	atm m3/mol	25	EST	MEYLAN,WM & HOWARD,PH (1991)
OH 2.54E-012	cm3/molc sec	25	EST	MEYLAN,WM & HOWARD,PH (1993)

CAS #: 021948-73-2 — QUINOXALINE, 2-(METHYLTHIO)-

Formula: $C_9H_8N_2S$

Mol Weight: 176.24

MP (deg C): | FP (deg C):

BP (deg C):

BP pressure (mm Hg):

Property/Value	Units	Temp	Data Type	Reference
WS 2.38E+002	mg/L	25	EST	MEYLAN,WM ET AL. (1996)
logP 2.79			EXP	SANGSTER,J (1993)
VP 2.62E-004	mm Hg	25	EST	NEELY,WB & BLAU,GE (1985)
DC 0.29	pKa	20	EXP	PERRIN,DD (1965)
HL 3.23E-007	atm m3/mol	25	EST	MEYLAN,WM & HOWARD,PH (1991)
OH 1.40E-011	cm3/molc sec	25	EST	MEYLAN,WM & HOWARD,PH (1993)

CAS #: 021949-06-4 — ISOPROPOXYETHYL TRIMETHYL AMMONIUM IODIDE

Formula: $C_8H_{20}INO$

Mol Weight: 273.16

MP (deg C): | FP (deg C):

BP (deg C):

BP pressure (mm Hg):

Property/Value	Units	Temp	Data Type	Reference
WS 5.16E+005	mg/L	25	EST	MEYLAN,WM ET AL. (1996)
logP -1.72			EXP	HANSCH,C & LEO,AJ (1985)
VP 1.72E-007	mm Hg	25	EST	NEELY,WB & BLAU,GE (1985)
DC	pKa			
HL 7.21E-014	atm m3/mol	25	EST	MEYLAN,WM & HOWARD,PH (1991)
OH 5.34E-011	cm3/molc sec	25	EST	MEYLAN,WM & HOWARD,PH (1993)

CAS #: 021949-11-1 — M-MEBENZYL-TRIMETHYL-AMMONIUM BR

Formula: $C_{11}H_{18}BrN$

Mol Weight: 244.18

MP (deg C): | FP (deg C):

BP (deg C):

BP pressure (mm Hg):

Property/Value	Units	Temp	Data Type	Reference
WS 1.00E+006	mg/L	25	EST	MEYLAN,WM ET AL. (1996)
logP -1.99			EXP	HANSCH,C & LEO,AJ (1985)
VP 1.84E-008	mm Hg	25	EST	NEELY,WB & BLAU,GE (1985)
DC	pKa			
HL 2.95E-013	atm m3/mol	25	EST	MEYLAN,WM & HOWARD,PH (1991)
OH 2.53E-011	cm3/molc sec	25	EST	MEYLAN,WM & HOWARD,PH (1993)

CAS #: 021959-57-9 — 2-(5-NO2-2-FURYLVINYL)5-NH2-3-OXOIMDAZOLINE

Formula: $C_8H_8N_4O_4$

Mol Weight: 224.18

MP (deg C): | FP (deg C):

BP (deg C):

BP pressure (mm Hg):

Property/Value	Units	Temp	Data Type	Reference
WS 1.06E+003	mg/L	25	EST	MEYLAN,WM ET AL. (1996)
logP 1.30			EXP	HANSCH,C & LEO,AJ (1985)
VP 2.88E-006	mm Hg	25	EST	NEELY,WB & BLAU,GE (1985)
DC	pKa			
HL 7.50E-013	atm m3/mol	25	EST	MEYLAN,WM & HOWARD,PH (1991)
OH 6.52E-011	cm3/molc sec	25	EST	MEYLAN,WM & HOWARD,PH (1993)

CAS #: 021988-05-6 — 3-CF3-4-NITROBENZENESULFONAMIDE

Formula: $C_7H_5F_3N_2O_4S$

Mol Weight: 270.19

MP (deg C): | FP (deg C):

BP (deg C):

BP pressure (mm Hg):

Property/Value	Units	Temp	Data Type	Reference
WS 2.47E+002	mg/L	25	EST	MEYLAN,WM ET AL. (1996)
logP 1.73			EXP	HANSCH,C & LEO,AJ (1985)
VP 5.40E-006	mm Hg	25	EST	NEELY,WB & BLAU,GE (1985)
DC	pKa			
HL 1.45E-008	atm m3/mol	25	EST	MEYLAN,WM & HOWARD,PH (1991)
OH 7.88E-015	cm3/molc sec	25	EST	MEYLAN,WM & HOWARD,PH (1993)

CAS #: 021998-12-9 — N-ME-4-CARBOMEO PHENYLCARBAMATE

Formula: $C_{10}H_{11}NO_4$

Mol Weight: 209.20

MP (deg C): | FP (deg C):

BP (deg C):

BP pressure (mm Hg):

Property/Value	Units	Temp	Data Type	Reference
WS 2.05E+003	mg/L	25	EST	MEYLAN,WM ET AL. (1996)
logP 1.50			EXP	HANSCH,C & LEO,AJ (1985)
VP 5.59E-003	mm Hg	25	EST	NEELY,WB & BLAU,GE (1985)
DC	pKa			
HL 2.07E-010	atm m3/mol	25	EST	MEYLAN,WM & HOWARD,PH (1991)
OH 6.92E-012	cm3/molc sec	25	EST	MEYLAN,WM & HOWARD,PH (1993)

CAS #: 022004-32-6 — 2,5-DIMETHOXY-4-ETHYLAMPHETAMINE

Formula: $C_{13}H_{21}NO_2$

Mol Weight: 223.32

MP (deg C): | FP (deg C):

BP (deg C):

BP pressure (mm Hg):

Property/Value	Units	Temp	Data Type	Reference
WS 1.34E+003	mg/L	25	EST	MEYLAN,WM ET AL. (1996)
logP 2.81			EXP	HANSCH,C & LEO,AJ (1985)
VP 2.22E-004	mm Hg	25	EST	NEELY,WB & BLAU,GE (1985)
DC	pKa			
HL 5.53E-009	atm m3/mol	25	EST	MEYLAN,WM & HOWARD,PH (1991)
OH 1.04E-010	cm3/molc sec	25	EST	MEYLAN,WM & HOWARD,PH (1993)

CAS #: 022010-25-9 — ACETAMIDE, N-(3,4-DICHLOROPHENYL)-2-(1-PIPERIDIN

Formula: $C_{13}H_{16}Cl_2N_2O$

Mol Weight: 287.19

MP (deg C): | FP (deg C):

BP (deg C):

BP pressure (mm Hg):

Property/Value	Units	Temp	Data Type	Reference
WS 1.11E+001	mg/L	25	EST	MEYLAN,WM ET AL. (1996)
logP 3.65			EXP	SANGSTER,J (1993)
VP 5.90E-008	mm Hg	25	EST	NEELY,WB & BLAU,GE (1985)
DC	pKa			
HL 3.46E-011	atm m3/mol	25	EST	MEYLAN,WM & HOWARD,PH (1991)
OH 1.01E-010	cm3/molc sec	25	EST	MEYLAN,WM & HOWARD,PH (1993)

989

CAS #: 022020-28-6 — 2-PROPENOIC ACID, 3-(3,4-DIHYDROXYPHENYL)-, BUTY

Formula: $C_{13}H_{16}O_4$

Mol Weight: 236.27

MP (deg C): | FP (deg C):
BP (deg C):
BP pressure (mm Hg):

Property/Value	Units	Temp	Data Type	Reference
WS 1.04E+002	mg/L	25	EST	MEYLAN,WM ET AL. (1996)
logP 3.53			EXP	NAITO,Y ET AL. (1991)
VP 3.20E-007	mm Hg	25	EST	NEELY,WB & BLAU,GE (1985)
DC	pKa			
HL 1.05E-013	atm m3/mol	25	EST	MEYLAN,WM & HOWARD,PH (1991)
OH 4.62E-011	cm3/molc sec	25	EST	MEYLAN,WM & HOWARD,PH (1993)

CAS #: 022026-39-7 — 1,3-BENZODIOLE-5-CARBOXYLIC ACID, HYDRAZIDE

Formula: $C_8H_8N_2O_3$

Mol Weight: 180.16

MP (deg C): | FP (deg C):
BP (deg C):
BP pressure (mm Hg):

Property/Value	Units	Temp	Data Type	Reference
WS 5.62E+003	mg/L	25	EST	MEYLAN,WM ET AL. (1996)
logP 1.16			EXP	SANGSTER,J (1994)
VP 2.24E-006	mm Hg	25	EST	NEELY,WB & BLAU,GE (1985)
DC	pKa			
HL 1.08E-014	atm m3/mol	25	EST	MEYLAN,WM & HOWARD,PH (1991)
OH 2.52E-011	cm3/molc sec	25	EST	MEYLAN,WM & HOWARD,PH (1993)

CAS #: 022037-28-1 — 3-BROMOFURAN

Formula: C_4H_4BrO

Mol Weight: 147.98

MP (deg C): | FP (deg C):
BP (deg C): 103
BP pressure (mm Hg):

Property/Value	Units	Temp	Data Type	Reference
WS 1.08E+003	mg/L	25	EST	MEYLAN,WM ET AL. (1996)
logP 2.18			EXP	HANSCH,C & LEO,AJ (1985)
VP 1.33E+001	mm Hg	25	EST	NEELY,WB & BLAU,GE (1985)
DC	pKa			
HL 2.14E-003	atm m3/mol	25	EST	MEYLAN,WM & HOWARD,PH (1991)
OH 2.55E-011	cm3/molc sec	25	EST	MEYLAN,WM & HOWARD,PH (1993)

CAS #: 022042-71-3 — P-FORMYLPHENOXYACETIC ACID

Formula: $C_9H_8O_4$

Mol Weight: 180.16

MP (deg C): | FP (deg C):
BP (deg C):
BP pressure (mm Hg):

Property/Value	Units	Temp	Data Type	Reference
WS 2.89E+004	mg/L	25	EST	MEYLAN,WM ET AL. (1996)
logP 0.79			EXP	HANSCH,C & LEO,AJ (1985)
VP 5.31E-005	mm Hg	25	EST	NEELY,WB & BLAU,GE (1985)
DC	pKa			
HL 4.18E-011	atm m3/mol	25	EST	MEYLAN,WM & HOWARD,PH (1991)
OH 2.88E-011	cm3/molc sec	25	EST	MEYLAN,WM & HOWARD,PH (1993)

CAS #: 022047-25-2 — ACETYLPYRAZINE

Formula: $C_6H_6N_2O$

Mol Weight: 122.13

MP (deg C): 76-78 | FP (deg C):
BP (deg C):
BP pressure (mm Hg):

Property/Value	Units	Temp	Data Type	Reference
WS 6.65E+004	mg/L	25	EST	MEYLAN,WM ET AL. (1996)
logP 0.20			EXP	YAMAGAMI,C ET AL. (1990A)
VP 1.88E-001	mm Hg	25	EST	NEELY,WB & BLAU,GE (1985)
DC	pKa			
HL 5.31E-009	atm m3/mol	25	EST	MEYLAN,WM & HOWARD,PH (1991)
OH 3.12E-013	cm3/molc sec	25	EST	MEYLAN,WM & HOWARD,PH (1993)

CAS #: 022071-15-4 — BENZENEACETIC ACID, 3-BENZOYL-.ALPHA.-METHYL-

Formula: $C_{16}H_{14}O_3$

Mol Weight: 254.29

MP (deg C): | FP (deg C):
BP (deg C):
BP pressure (mm Hg):

Property/Value	Units	Temp	Data Type	Reference
WS 5.10E+001	mg/L	20	EXP	YALKOWSKY,SH & DANNENFELSER,RM (1992)
logP 3.12			EXP	SANGSTER,J (1993)
VP 3.72E-007	mm Hg	25	EST	NEELY,WB & BLAU,GE (1985)
DC	pKa			
HL 2.12E-011	atm m3/mol	25	EST	MEYLAN,WM & HOWARD,PH (1991)
OH 6.06E-012	cm3/molc sec	25	EST	MEYLAN,WM & HOWARD,PH (1993)

CAS #: 022131-79-9 — 3-CL-4-ALLYLOXYPHENYL ACETIC ACID

Formula: $C_{11}H_{11}ClO_3$

Mol Weight: 226.66

MP (deg C): | FP (deg C):
BP (deg C):
BP pressure (mm Hg):

Property/Value	Units	Temp	Data Type	Reference
WS 6.00E+002	mg/L	25	EST	MEYLAN,WM ET AL. (1996)
logP 2.48			EXP	HANSCH,C & LEO,AJ (1985)
VP 2.12E-005	mm Hg	25	EST	NEELY,WB & BLAU,GE (1985)
DC	pKa			
HL 2.55E-009	atm m3/mol	25	EST	MEYLAN,WM & HOWARD,PH (1991)
OH 3.97E-011	cm3/molc sec	25	EST	MEYLAN,WM & HOWARD,PH (1993)

CAS #: 022138-39-2 — 1-PENTACHLOROCYCLOHEXANE

Formula: $C_6H_7Cl_5$

Mol Weight: 256.39

MP (deg C): | FP (deg C):
BP (deg C):
BP pressure (mm Hg):

Property/Value	Units	Temp	Data Type	Reference
WS 2.89E+001	mg/L	25	EST	MEYLAN,WM ET AL. (1996)
logP 3.53			EXP	HANSCH,C & LEO,AJ (1985)
VP 5.44E-003	mm Hg	25	EST	NEELY,WB & BLAU,GE (1985)
DC	pKa			
HL 7.26E-004	atm m3/mol	25	EST	MEYLAN,WM & HOWARD,PH (1991)
OH 1.06E-012	cm3/molc sec	25	EST	MEYLAN,WM & HOWARD,PH (1993)

CAS #: 022175-22-0				N-(3-CHLORO-4-METHYLPHENYL)-N'-METHYLUREA

Formula: $C_9H_{11}ClN_2O$

Mol Weight: 198.65

MP (deg C):   FP (deg C):

BP (deg C):

BP pressure (mm Hg):

Property/Value	Units	Temp	Data Type	Reference
WS 2.62E+002	mg/L	25	EST	MEYLAN,WM ET AL. (1996)
logP 2.61			EXP	BRIGGS,GG (1981)
VP 3.66E-005	mm Hg	25	EST	NEELY,WB & BLAU,GE (1985)
DC	pKa			
HL 3.61E-010	atm m3/mol	25	EST	MEYLAN,WM & HOWARD,PH (1991)
OH 3.85E-011	cm3/molc sec	25	EST	MEYLAN,WM & HOWARD,PH (1993)

CAS #: 022181-94-8				GLYCINE, N-[1-OXO-5-(1H-PURIN-6-YLTHIO)PENTYL]-,

Formula: $C_{14}H_{19}N_5O_3S$

Mol Weight: 337.40

MP (deg C):   FP (deg C):

BP (deg C):

BP pressure (mm Hg):

Property/Value	Units	Temp	Data Type	Reference
WS 1.23E+003	mg/L	25	EST	MEYLAN,WM ET AL. (1996)
logP 0.91			EXP	SANGSTER,J (1993)
VP 1.44E-013	mm Hg	25	EST	NEELY,WB & BLAU,GE (1985)
DC	pKa			
HL 1.11E-017	atm m3/mol	25	EST	MEYLAN,WM & HOWARD,PH (1991)
OH 2.26E-010	cm3/molc sec	25	EST	MEYLAN,WM & HOWARD,PH (1993)

CAS #: 022181-95-9				GLYCINE, N-[N-[1-OXO-5-(7H-PURIN-6-YLTHIO)PENTYL

Formula: $C_{16}H_{22}N_6O_4S$

Mol Weight: 394.46

MP (deg C):   FP (deg C):

BP (deg C):

BP pressure (mm Hg):

Property/Value	Units	Temp	Data Type	Reference
WS 1.24E+003	mg/L	25	EST	MEYLAN,WM ET AL. (1996)
logP 0.50			EXP	HANSCH,C ET AL. (1995)
VP 1.94E-017	mm Hg	25	EST	NEELY,WB & BLAU,GE (1985)
DC	pKa			
HL 1.17E-019	atm m3/mol	25	EST	MEYLAN,WM & HOWARD,PH (1991)
OH 2.34E-010	cm3/molc sec	25	EST	MEYLAN,WM & HOWARD,PH (1993)

CAS #: 022189-31-7				THIOTHIXENE

Formula: $C_{23}H_{29}N_3O_2S_2$

Mol Weight: 443.63

MP (deg C):   FP (deg C):

BP (deg C):

BP pressure (mm Hg):

Property/Value	Units	Temp	Data Type	Reference
WS 7.64E-001	mg/L	25	EST	MEYLAN,WM ET AL. (1996)
logP 3.78			EXP	HANSCH,C & LEO,AJ (1985)
VP 2.56E-012	mm Hg	25	EST	NEELY,WB & BLAU,GE (1985)
DC	pKa			
HL 3.70E-016	atm m3/mol	25	EST	MEYLAN,WM & HOWARD,PH (1991)
OH 2.92E-010	cm3/molc sec	25	EST	MEYLAN,WM & HOWARD,PH (1993)

CAS #: 022204-53-1				NAPROSYN

Formula: $C_{14}H_{14}O_3$

Mol Weight: 230.27

MP (deg C): 157-158   FP (deg C):

BP (deg C):

BP pressure (mm Hg):

Property/Value	Units	Temp	Data Type	Reference
WS 1.59E+001	mg/L	25	EXP	YALKOWSKY,SH & DANNENFELSER,RM (1992)
logP 3.18			EXP	HANSCH,C & LEO,AJ (1985)
VP 1.89E-006	mm Hg	25	EST	NEELY,WB & BLAU,GE (1985)
DC	pKa			
HL 3.39E-010	atm m3/mol	25	EST	MEYLAN,WM & HOWARD,PH (1991)
OH 1.16E-010	cm3/molc sec	25	EST	MEYLAN,WM & HOWARD,PH (1993)

CAS #: 022212-55-1				BENZOYLPROP ETHYL

Formula: $C_{18}H_{17}Cl_2NO_3$

Mol Weight: 366.25

MP (deg C):   FP (deg C):

BP (deg C):

BP pressure (mm Hg):

Property/Value	Units	Temp	Data Type	Reference
WS 2.00E+001	mg/L	25	EXP	YALKOWSKY,SH & DANNENFELSER,RM (1992)
logP 4.57			EST	MEYLAN,WM & HOWARD,PH (1995)
VP 1.03E-008	mm Hg	25	EST	NEELY,WB & BLAU,GE (1985)
DC	pKa			
HL 1.15E-009	atm m3/mol	25	EST	MEYLAN,WM & HOWARD,PH (1991)
OH 1.20E-011	cm3/molc sec	25	EST	MEYLAN,WM & HOWARD,PH (1993)

CAS #: 022224-92-6				FENAMIPHOS

Formula: $C_{13}H_{22}NO_3PS$

Mol Weight: 303.36

MP (deg C): 49.2   FP (deg C):

BP (deg C):

BP pressure (mm Hg):

Property/Value	Units	Temp	Data Type	Reference
WS 3.29E+002	mg/L	20	EXP	YALKOWSKY,SH & DANNENFELSER,RM (1992)
logP 3.23			EXP	HANSCH,C ET AL. (1995)
VP 1.00E-006	mm Hg	25	EXP	WAUCHOPE,RD ET AL. (1991A)
DC	pKa			
HL 1.21E-009	atm m3/mol	25	EST	VP/WSOL
OH 7.77E-011	cm3/molc sec	25	EST	MEYLAN,WM & HOWARD,PH (1993)

CAS #: 022248-79-9				TETRACHLORVINPHOS

Formula: $C_{10}H_9Cl_4O_4P$

Mol Weight: 365.97

MP (deg C): 97-98   FP (deg C):

BP (deg C):

BP pressure (mm Hg):

Property/Value	Units	Temp	Data Type	Reference
WS 1.10E+001	mg/L	20	EXP	YALKOWSKY,SH & DANNENFELSER,RM (1992)
logP 3.53			EXP	HANSCH,C ET AL. (1995)
VP 4.20E-008	mm Hg	20	EXP	TOMLIN,C (1994)
DC	pKa			
HL 1.80E-009	atm m3/mol	20	EST	VP/WSOL
OH 2.41E-011	cm3/molc sec	25	EST	MEYLAN,WM & HOWARD,PH (1993)

CAS #: 022259-30-9				FORMETANATE
Formula: $C_{11}H_{15}N_3O_2$				
Mol Weight: 221.26				
MP (deg C):		FP (deg C):		
BP (deg C):				
BP pressure (mm Hg):				

Property/ Value	Units	Temp	Data Type	Reference
WS 1.00E+003	mg/L		EXP	YALKOWSKY,SH & DANNENFELSER,RM (1992)
logP 0.88			EST	MEYLAN,WM & HOWARD,PH (1995)
VP 1.80E-004	mm Hg	25	EST	NEELY,WB & BLAU,GE (1985)
DC	pKa			
HL 3.60E-012	atm m3/mol	25	EST	MEYLAN,WM & HOWARD,PH (1991)
OH 9.42E-011	cm3/molc sec	25	EST	MEYLAN,WM & HOWARD,PH (1993)

CAS #: 022275-34-9				BARBITURIC ACID, 5-PHNEYL-
Formula: $C_{10}H_8N_2O_3$				
Mol Weight: 204.19				
MP (deg C):		FP (deg C):		
BP (deg C):				
BP pressure (mm Hg):				

Property/ Value	Units	Temp	Data Type	Reference
WS 5.82E+004	mg/L	25	EST	MEYLAN,WM ET AL. (1996)
logP -0.17			EXP	SANGSTER,J (1993)
VP 6.83E-012	mm Hg	25	EST	NEELY,WB & BLAU,GE (1985)
DC	pKa			
HL 9.37E-015	atm m3/mol	25	EST	MEYLAN,WM & HOWARD,PH (1991)
OH 7.72E-012	cm3/molc sec	25	EST	MEYLAN,WM & HOWARD,PH (1993)

CAS #: 022278-77-9				3-SH-4-NH2-6IPR-124TRIAZINE-5-ONE
Formula: $C_6H_{10}N_4OS$				
Mol Weight: 186.24				
MP (deg C):		FP (deg C):		
BP (deg C):				
BP pressure (mm Hg):				

Property/ Value	Units	Temp	Data Type	Reference
WS 4.66E+003	mg/L	25	EST	MEYLAN,WM ET AL. (1996)
logP 1.22			EXP	HANSCH,C & LEO,AJ (1985)
VP 6.73E-006	mm Hg	25	EST	NEELY,WB & BLAU,GE (1985)
DC	pKa			
HL 4.53E-012	atm m3/mol	25	EST	MEYLAN,WM & HOWARD,PH (1991)
OH 5.54E-011	cm3/molc sec	25	EST	MEYLAN,WM & HOWARD,PH (1993)

CAS #: 022305-44-8				3-METHYLBENZOTRIAZIN-4-ONE
Formula: $C_8H_7N_3O$				
Mol Weight: 161.16				
MP (deg C):		FP (deg C):		
BP (deg C):				
BP pressure (mm Hg):				

Property/ Value	Units	Temp	Data Type	Reference
WS 1.30E+004	mg/L	25	EST	MEYLAN,WM ET AL. (1996)
logP 0.84			EXP	HANSCH,C & LEO,AJ (1985)
VP 6.24E-005	mm Hg	25	EST	NEELY,WB & BLAU,GE (1985)
DC	pKa			
HL 7.12E-010	atm m3/mol	25	EST	MEYLAN,WM & HOWARD,PH (1991)
OH 1.59E-011	cm3/molc sec	25	EST	MEYLAN,WM & HOWARD,PH (1993)

CAS #: 022311-25-7				1,2,3,4-TETRABROMOBENZENE
Formula: $C_6H_2Br_4$				
Mol Weight: 393.72				
MP (deg C):		FP (deg C):		
BP (deg C):				
BP pressure (mm Hg):				

Property/ Value	Units	Temp	Data Type	Reference
WS 6.06E-002	mg/L	25	EST	MEYLAN,WM ET AL. (1996)
logP 5.55			EST	MEYLAN,WM & HOWARD,PH (1995)
VP 3.13E-004	mm Hg	25	EST	NEELY,WB & BLAU,GE (1985)
DC	pKa			
HL 1.36E-004	atm m3/mol	25	EST	MEYLAN,WM & HOWARD,PH (1991)
OH 6.35E-014	cm3/molc sec	25	EST	MEYLAN,WM & HOWARD,PH (1993)

CAS #: 022337-35-5				2-PYRIDYL-ME-(6NN-TETRA-ME)AMMONIUM BR
Formula: $C_{10}H_{17}BrN_2$				
Mol Weight: 245.16				
MP (deg C):		FP (deg C):		
BP (deg C):				
BP pressure (mm Hg):				

Property/ Value	Units	Temp	Data Type	Reference
WS 1.00E+006	mg/L	25	EST	MEYLAN,WM ET AL. (1996)
logP -2.90			EXP	HANSCH,C & LEO,AJ (1985)
VP 1.13E-008	mm Hg	25	EST	NEELY,WB & BLAU,GE (1985)
DC	pKa			
HL 3.87E-016	atm m3/mol	25	EST	MEYLAN,WM & HOWARD,PH (1991)
OH 1.38E-011	cm3/molc sec	25	EST	MEYLAN,WM & HOWARD,PH (1993)

CAS #: 022342-21-8				PIPERIDINE, 1-(2-CHLOROBENZOYL)-
Formula: $C_{12}H_{14}ClNO$				
Mol Weight: 223.70				
MP (deg C):		FP (deg C):		
BP (deg C):				
BP pressure (mm Hg):				

Property/ Value	Units	Temp	Data Type	Reference
WS 1.34E+002	mg/L	25	EST	MEYLAN,WM ET AL. (1996)
logP 2.73			EST	MEYLAN,WM & HOWARD,PH (1995)
VP 4.16E-005	mm Hg	25	EST	NEELY,WB & BLAU,GE (1985)
DC	pKa			
HL 8.16E-009	atm m3/mol	25	EST	MEYLAN,WM & HOWARD,PH (1991)
OH 2.99E-011	cm3/molc sec	25	EST	MEYLAN,WM & HOWARD,PH (1993)

CAS #: 022371-32-0				HEPTANOIC ACID HYDRAZIDE
Formula: $C_7H_{16}N_2O$				
Mol Weight: 144.22				
MP (deg C):		FP (deg C):		
BP (deg C):				
BP pressure (mm Hg):				

Property/ Value	Units	Temp	Data Type	Reference
WS 1.19E+004	mg/L	25	EST	MEYLAN,WM ET AL. (1996)
logP 0.97			EXP	HANSCH,C & LEO,AJ (1985)
VP 2.75E-004	mm Hg	25	EST	NEELY,WB & BLAU,GE (1985)
DC	pKa			
HL 1.93E-010	atm m3/mol	25	EST	MEYLAN,WM & HOWARD,PH (1991)
OH 1.50E-011	cm3/molc sec	25	EST	MEYLAN,WM & HOWARD,PH (1993)

CAS #: 022374-89-6 — BENZENEPROPANAMINE, à-METHYL-

Formula: $C_{10}H_{15}N$

Mol Weight: 149.24

MP (deg C): | FP (deg C):

BP (deg C): 221-222

BP pressure (mm Hg):

Property/ Value	Units	Temp	Data Type	Reference
WS 1.21E+004	mg/L	25	EST	MEYLAN,WM ET AL. (1996)
logP 2.12			EXP	KRIL,MB & FUNG,HL (1990)
VP 7.61E-002	mm Hg	25	EST	NEELY,WB & BLAU,GE (1985)
DC 9.79	pKa	25	EXP	PERRIN,DD (1965)
HL 1.43E-006	atm m3/mol	25	EST	MEYLAN,WM & HOWARD,PH (1991)
OH 5.08E-011	cm3/molc sec	25	EST	MEYLAN,WM & HOWARD,PH (1993)

CAS #: 022385-99-5 — 1-PROPANONE, 2-METHYL-1-PHENYL-3-(1-PIPERIDINYL)

Formula: $C_{15}H_{21}NO$

Mol Weight: 231.34

MP (deg C): | FP (deg C):

BP (deg C):

BP pressure (mm Hg):

Property/ Value	Units	Temp	Data Type	Reference
WS 3.07E+002	mg/L	25	EST	MEYLAN,WM ET AL. (1996)
logP 3.51			EXP	SANGSTER,J (1994)
VP 9.19E-005	mm Hg	25	EST	NEELY,WB & BLAU,GE (1985)
DC	pKa			
HL 8.84E-009	atm m3/mol	25	EST	MEYLAN,WM & HOWARD,PH (1991)
OH 1.30E-010	cm3/molc sec	25	EST	MEYLAN,WM & HOWARD,PH (1993)

CAS #: 022390-38-1 — 3-PYRIDAZINAMINE, 4-HYDROXY-N,N,6-TRIMETHYL-

Formula: $C_7H_{11}N_3O$

Mol Weight: 153.19

MP (deg C): | FP (deg C):

BP (deg C):

BP pressure (mm Hg):

Property/ Value	Units	Temp	Data Type	Reference
WS 4.58E+004	mg/L	25	EST	MEYLAN,WM ET AL. (1996)
logP 0.24			EXP	SANGSTER,J (1994)
VP 4.54E-005	mm Hg	25	EST	NEELY,WB & BLAU,GE (1985)
DC	pKa			
HL 5.24E-012	atm m3/mol	25	EST	MEYLAN,WM & HOWARD,PH (1991)
OH 1.96E-011	cm3/molc sec	25	EST	MEYLAN,WM & HOWARD,PH (1993)

CAS #: 022390-53-0 — 3,4-PYRIDAZINEDIOL, 6-METHYL-

Formula: $C_5H_6N_2O_2$

Mol Weight: 126.12

MP (deg C): | FP (deg C):

BP (deg C):

BP pressure (mm Hg):

Property/ Value	Units	Temp	Data Type	Reference
WS 1.72E+005	mg/L	25	EST	MEYLAN,WM ET AL. (1996)
logP -0.30			EXP	SANGSTER,J (1994)
VP 1.53E-005	mm Hg	25	EST	NEELY,WB & BLAU,GE (1985)
DC	pKa			
HL 3.43E-014	atm m3/mol	25	EST	MEYLAN,WM & HOWARD,PH (1991)
OH 7.48E-012	cm3/molc sec	25	EST	MEYLAN,WM & HOWARD,PH (1993)

CAS #: 022399-00-4 — DIETHYLMALONATE,4-NO2 BENZAL

Formula: $C_{14}H_{15}NO_6$

Mol Weight: 293.28

MP (deg C): | FP (deg C):

BP (deg C):

BP pressure (mm Hg):

Property/ Value	Units	Temp	Data Type	Reference
WS 1.41E+001	mg/L	25	EST	MEYLAN,WM ET AL. (1996)
logP 3.03			EXP	HANSCH,C & LEO,AJ (1985)
VP 8.19E-006	mm Hg	25	EST	NEELY,WB & BLAU,GE (1985)
DC	pKa			
HL 4.31E-011	atm m3/mol	25	EST	MEYLAN,WM & HOWARD,PH (1991)
OH 1.42E-011	cm3/molc sec	25	EST	MEYLAN,WM & HOWARD,PH (1993)

CAS #: 022399-01-5 — DIETHYLMALONATE,4-BR BENZAL

Formula: $C_{14}H_{15}BrO_4$

Mol Weight: 327.18

MP (deg C): | FP (deg C):

BP (deg C):

BP pressure (mm Hg):

Property/ Value	Units	Temp	Data Type	Reference
WS 3.13E+000	mg/L	25	EST	MEYLAN,WM ET AL. (1996)
logP 4.02			EXP	HANSCH,C & LEO,AJ (1985)
VP 1.64E-004	mm Hg	25	EST	NEELY,WB & BLAU,GE (1985)
DC	pKa			
HL 4.36E-009	atm m3/mol	25	EST	MEYLAN,WM & HOWARD,PH (1991)
OH 1.51E-011	cm3/molc sec	25	EST	MEYLAN,WM & HOWARD,PH (1993)

CAS #: 022433-76-7 — 6-ACETAMINOQUINOLINE

Formula: $C_{11}H_{10}N_2O$

Mol Weight: 186.22

MP (deg C): | FP (deg C):

BP (deg C):

BP pressure (mm Hg):

Property/ Value	Units	Temp	Data Type	Reference
WS 2.44E+003	mg/L	25	EST	MEYLAN,WM ET AL. (1996)
logP 1.55			EXP	HANSCH,C & LEO,AJ (1985)
VP 1.28E-006	mm Hg	25	EST	NEELY,WB & BLAU,GE (1985)
DC	pKa			
HL 7.88E-013	atm m3/mol	25	EST	MEYLAN,WM & HOWARD,PH (1991)
OH 9.18E-011	cm3/molc sec	25	EST	MEYLAN,WM & HOWARD,PH (1993)

CAS #: 022474-57-3 — 2,2,4,4,6,8,10-HEPTAMETHYL-6,8,10-TRIS(3,3,3-TR*

Formula: $C_{16}H_{33}F_9O_5Si_5$

Mol Weight: 616.85

MP (deg C): | FP (deg C):

BP (deg C):

BP pressure (mm Hg):

Property/ Value	Units	Temp	Data Type	Reference
WS 1.21E-007	mg/L	25	EST	MEYLAN,WM ET AL. (1996)
logP 9.90			EST	MEYLAN,WM & HOWARD,PH (1995)
VP	mm Hg			
DC	pKa			
HL	atm m3/mol			
OH 5.08E-012	cm3/molc sec	25	EST	MEYLAN,WM & HOWARD,PH (1993)

CAS #: 022494-42-4				DIFLUNISAL

Formula: $C_{13}H_8F_2O_3$

Mol Weight: 250.20

MP (deg C): 210-221   FP (deg C):

BP (deg C):

BP pressure (mm Hg):

Property/ Value	Units	Temp	Data Type	Reference
WS 1.45E+001	mg/L	25	EST	MEYLAN,WM ET AL. (1996)
logP 4.44			EXP	SANGSTER,J (1993)
VP 8.92E-008	mm Hg	25	EST	NEELY,WB & BLAU,GE (1985)
DC	pKa			
HL 1.49E-009	atm m3/mol	25	EST	MEYLAN,WM & HOWARD,PH (1991)
OH 8.05E-012	cm3/molc sec	25	EST	MEYLAN,WM & HOWARD,PH (1993)

CAS #: 022509-74-6				N-CARBETHOXYPHTHALIMIDE

Formula: $C_{11}H_9NO_4$

Mol Weight: 219.20

MP (deg C): 90-92   FP (deg C):

BP (deg C):

BP pressure (mm Hg):

Property/ Value	Units	Temp	Data Type	Reference
WS 3.58E+002	mg/L	25	EST	MEYLAN,WM ET AL. (1996)
logP 2.33			EST	MEYLAN,WM & HOWARD,PH (1995)
VP 2.32E-008	mm Hg	25	EST	NEELY,WB & BLAU,GE (1985)
DC	pKa			
HL 1.56E-010	atm m3/mol	25	EST	MEYLAN,WM & HOWARD,PH (1991)
OH 1.74E-011	cm3/molc sec	25	EST	MEYLAN,WM & HOWARD,PH (1993)

CAS #: 022526-29-0				4-(2-HYDROXYBENZOYL)-PYRIDINE

Formula: $C_{12}H_9NO_2$

Mol Weight: 199.21

MP (deg C):   FP (deg C):

BP (deg C):

BP pressure (mm Hg):

Property/ Value	Units	Temp	Data Type	Reference
WS 2.55E+003	mg/L	25	EST	MEYLAN,WM ET AL. (1996)
logP 2.13			EXP	HANSCH,C & LEO,AJ (1985)
VP 4.67E-006	mm Hg	25	EST	NEELY,WB & BLAU,GE (1985)
DC	pKa			
HL 3.33E-010	atm m3/mol	25	EST	MEYLAN,WM & HOWARD,PH (1991)
OH 3.09E-011	cm3/molc sec	25	EST	MEYLAN,WM & HOWARD,PH (1993)

CAS #: 022601-05-4				3-FURANMETHANAMINIUM, N,N,N-TRIMETHYL-, IODIDE

Formula: $C_8H_{14}INO$

Mol Weight: 267.11

MP (deg C):   FP (deg C):

BP (deg C):

BP pressure (mm Hg):

Property/ Value	Units	Temp	Data Type	Reference
WS 1.00E+006	mg/L	25	EST	MEYLAN,WM ET AL. (1996)
logP -2.40			EXP	SANGSTER,J (1994)
VP 5.65E-008	mm Hg	25	EST	NEELY,WB & BLAU,GE (1985)
DC	pKa			
HL 2.12E-013	atm m3/mol	25	EST	MEYLAN,WM & HOWARD,PH (1991)
OH 1.13E-010	cm3/molc sec	25	EST	MEYLAN,WM & HOWARD,PH (1993)

CAS #: 022608-53-3				O,S,S-TRIMETHYL PHOSPHORODITHIOATE

Formula: $C_3H_9O_2PS_2$

Mol Weight: 172.21

MP (deg C):   FP (deg C):

BP (deg C):

BP pressure (mm Hg):

Property/ Value	Units	Temp	Data Type	Reference
WS 1.56E+004	mg/L	25	EST	MEYLAN,WM ET AL. (1996)
logP 0.68			EST	MEYLAN,WM & HOWARD,PH (1995)
VP 3.54E-002	mm Hg	25	EST	NEELY,WB & BLAU,GE (1985)
DC	pKa			
HL 7.51E-008	atm m3/mol	25	EST	MEYLAN,WM & HOWARD,PH (1991)
OH 9.60E-012	cm3/molc sec	25	EXP	ATKINSON,R (1989)

CAS #: 022609-88-7				5-PHENYL-2,4-FURANDIONE

Formula: $C_{10}H_8O_3$

Mol Weight: 176.17

MP (deg C):   FP (deg C):

BP (deg C):

BP pressure (mm Hg):

Property/ Value	Units	Temp	Data Type	Reference
WS 3.19E+003	mg/L	25	EST	MEYLAN,WM ET AL. (1996)
logP 1.47			EXP	LIPINSKI,CA ET AL. (1991)
VP 6.73E-005	mm Hg	25	EST	NEELY,WB & BLAU,GE (1985)
DC	pKa			
HL 2.24E-008	atm m3/mol	25	EST	MEYLAN,WM & HOWARD,PH (1991)
OH 6.90E-012	cm3/molc sec	25	EST	MEYLAN,WM & HOWARD,PH (1993)

CAS #: 022627-00-5				O-METHYLBENZOPHENONEIMINE

Formula: $C_{14}H_{13}N$

Mol Weight: 195.27

MP (deg C):   FP (deg C):

BP (deg C):

BP pressure (mm Hg):

Property/ Value	Units	Temp	Data Type	Reference
WS 8.39E+001	mg/L	25	EST	MEYLAN,WM ET AL. (1996)
logP 3.21			EXP	HANSCH,C & LEO,AJ (1985)
VP 9.69E-005	mm Hg	25	EST	NEELY,WB & BLAU,GE (1985)
DC 7.21	pKa		EXP	PERRIN,DD (1965)
HL 1.59E-005	atm m3/mol	25	EST	MEYLAN,WM & HOWARD,PH (1991)
OH 1.05E-011	cm3/molc sec	25	EST	MEYLAN,WM & HOWARD,PH (1993)

CAS #: 022627-01-6				O,O'-DIMETHYLBENZOPHENONEIMINE

Formula: $C_{15}H_{15}N$

Mol Weight: 209.29

MP (deg C):   FP (deg C):

BP (deg C):

BP pressure (mm Hg):

Property/ Value	Units	Temp	Data Type	Reference
WS 5.84E+001	mg/L	25	EST	MEYLAN,WM ET AL. (1996)
logP 3.31			EXP	HANSCH,C & LEO,AJ (1985)
VP 4.12E-005	mm Hg	25	EST	NEELY,WB & BLAU,GE (1985)
DC	pKa			
HL 1.75E-005	atm m3/mol	25	EST	MEYLAN,WM & HOWARD,PH (1991)
OH 1.25E-011	cm3/molc sec	25	EST	MEYLAN,WM & HOWARD,PH (1993)

CAS #: 022627-02-7				2,6-DIMETHYLBENZOPHENONEIMINE

Formula: $C_{15}H_{15}N$

Mol Weight: 209.29

MP (deg C):   FP (deg C):

BP (deg C):

BP pressure (mm Hg):

Property/Value	Units	Temp	Data Type	Reference
WS 3.79E+001	mg/L	25	EST	MEYLAN,WM ET AL. (1996)
logP 3.53			EXP	HANSCH,C & LEO,AJ (1985)
VP 4.12E-005	mm Hg	25	EST	NEELY,WB & BLAU,GE (1985)
DC 7.71	pKa		EXP	PERRIN,DD (1965)
HL 1.75E-005	atm m3/mol	25	EST	MEYLAN,WM & HOWARD,PH (1991)
OH 2.05E-011	cm3/molc sec	25	EST	MEYLAN,WM & HOWARD,PH (1993)

CAS #: 022664-55-7				METIPRANOLOL

Formula: $C_{17}H_{27}NO_4$

Mol Weight: 309.41

MP (deg C): 105-107   FP (deg C):

BP (deg C):

BP pressure (mm Hg):

Property/Value	Units	Temp	Data Type	Reference
WS 5.89E+002	mg/L	25	EST	MEYLAN,WM ET AL. (1996)
logP 2.66			EXP	MANNHOLD,R ET AL. (1990)
VP 1.06E-008	mm Hg	25	EST	NEELY,WB & BLAU,GE (1985)
DC	pKa			
HL 1.32E-013	atm m3/mol	25	EST	MEYLAN,WM & HOWARD,PH (1991)
OH 2.42E-010	cm3/molc sec	25	EST	MEYLAN,WM & HOWARD,PH (1993)

CAS #: 022668-01-5				2(2NO2-IMIDZOL-1YL)N-2ETOH-ACETAMIDE

Formula: $C_7H_{10}N_4O_4$

Mol Weight: 214.18

MP (deg C):   FP (deg C):

BP (deg C):

BP pressure (mm Hg):

Property/Value	Units	Temp	Data Type	Reference
WS 1.03E+004	mg/L	25	EST	MEYLAN,WM ET AL. (1996)
logP -1.34			EXP	HANSCH,C & LEO,AJ (1985)
VP 3.36E-011	mm Hg	25	EST	NEELY,WB & BLAU,GE (1985)
DC	pKa			
HL 5.12E-018	atm m3/mol	25	EST	MEYLAN,WM & HOWARD,PH (1991)
OH 1.96E-011	cm3/molc sec	25	EST	MEYLAN,WM & HOWARD,PH (1993)

CAS #: 022680-31-5				124-BENZTHIADIAZN-1-O2-3-ME-8-CL

Formula: $C_8H_7ClN_2O_2S$

Mol Weight: 230.67

MP (deg C):   FP (deg C):

BP (deg C):

BP pressure (mm Hg):

Property/Value	Units	Temp	Data Type	Reference
WS 8.92E+003	mg/L	25	EST	MEYLAN,WM ET AL. (1996)
logP 0.62			EXP	HANSCH,C & LEO,AJ (1985)
VP 2.88E-008	mm Hg	25	EST	NEELY,WB & BLAU,GE (1985)
DC	pKa			
HL 6.85E-008	atm m3/mol	25	EST	MEYLAN,WM & HOWARD,PH (1991)
OH 3.52E-013	cm3/molc sec	25	EST	MEYLAN,WM & HOWARD,PH (1993)

CAS #: 022699-70-3				3-ETHYLACETOPHENONE

Formula: $C_{10}H_{12}O$

Mol Weight: 148.21

MP (deg C):   FP (deg C):

BP (deg C):

BP pressure (mm Hg):

Property/Value	Units	Temp	Data Type	Reference
WS 3.73E+002	mg/L	25	EST	MEYLAN,WM ET AL. (1996)
logP 2.71			EST	MEYLAN,WM & HOWARD,PH (1995)
VP 2.28E+000	mm Hg	25	EXP	BOUBLIK,T ET AL. (1986)
DC	pKa			
HL 1.44E-005	atm m3/mol	25	EST	MEYLAN,WM & HOWARD,PH (1991)
OH 3.37E-012	cm3/molc sec	25	EST	MEYLAN,WM & HOWARD,PH (1993)

CAS #: 022707-38-6				2-PYRROLIDINONE, 1-(1-OXOBUTYL)-

Formula: $C_8H_{13}NO_2$

Mol Weight: 155.20

MP (deg C):   FP (deg C):

BP (deg C):

BP pressure (mm Hg):

Property/Value	Units	Temp	Data Type	Reference
WS 7.07E+003	mg/L	25	EST	MEYLAN,WM ET AL. (1996)
logP 1.18			EXP	SANGSTER,J (1994)
VP 5.32E-005	mm Hg	25	EST	NEELY,WB & BLAU,GE (1985)
DC	pKa			
HL 1.30E-007	atm m3/mol	25	EST	MEYLAN,WM & HOWARD,PH (1991)
OH 2.77E-011	cm3/molc sec	25	EST	MEYLAN,WM & HOWARD,PH (1993)

CAS #: 022715-68-0				ALANINE,N-ACETYL,N'-MEAM-AMIDE

Formula: $C_6H_{12}N_2O_2$

Mol Weight: 144.17

MP (deg C):   FP (deg C):

BP (deg C):

BP pressure (mm Hg):

Property/Value	Units	Temp	Data Type	Reference
WS 4.25E+004	mg/L	25	EST	MEYLAN,WM ET AL. (1996)
logP -1.21			EXP	HANSCH,C & LEO,AJ (1985)
VP 1.22E-006	mm Hg	25	EST	NEELY,WB & BLAU,GE (1985)
DC	pKa			
HL 3.44E-010	atm m3/mol	25	EST	MEYLAN,WM & HOWARD,PH (1991)
OH 8.39E-011	cm3/molc sec	25	EST	MEYLAN,WM & HOWARD,PH (1993)

CAS #: 022718-51-0				4,4-DIMETHYLSEMICARBAZIDE

Formula: $C_3H_9N_3O$

Mol Weight: 103.12

MP (deg C):   FP (deg C):

BP (deg C):

BP pressure (mm Hg):

Property/Value	Units	Temp	Data Type	Reference
WS 1.23E+005	mg/L	25	EST	MEYLAN,WM ET AL. (1996)
logP -1.57			EXP	HANSCH,C & LEO,AJ (1985)
VP 1.06E-001	mm Hg	25	EST	NEELY,WB & BLAU,GE (1985)
DC	pKa			
HL 7.35E-012	atm m3/mol	25	EST	MEYLAN,WM & HOWARD,PH (1991)
OH 4.53E-012	cm3/molc sec	25	EST	MEYLAN,WM & HOWARD,PH (1993)

CAS #: 022726-00-7

M-BROMOBENZAMIDE

Formula: C_7H_6BrNO

Mol Weight: 200.04

MP (deg C): 155.3

FP (deg C):

BP (deg C):

BP pressure (mm Hg):

Property/Value	Units	Temp	Data Type	Reference
WS 1.71E+003	mg/L	25	EST	MEYLAN,WM ET AL. (1996)
logP 1.65			EXP	HANSCH,C & LEO,AJ (1985)
VP 5.64E-005	mm Hg	25	EST	NEELY,WB & BLAU,GE (1985)
DC	pKa			
HL 8.82E-010	atm m3/mol	25	EST	MEYLAN,WM & HOWARD,PH (1991)
OH 2.52E-012	cm3/molc sec	25	EST	MEYLAN,WM & HOWARD,PH (1993)

CAS #: 022731-28-8

CF3-SULFONANILIDE,P-BENZOYL

Formula: $C_{14}H_{10}F_3NO_3S$

Mol Weight: 329.30

MP (deg C):

FP (deg C):

BP (deg C):

BP pressure (mm Hg):

Property/Value	Units	Temp	Data Type	Reference
WS 1.83E+000	mg/L	25	EST	MEYLAN,WM ET AL. (1996)
logP 4.28			EXP	HANSCH,C & LEO,AJ (1985)
VP 1.69E-007	mm Hg	25	EST	NEELY,WB & BLAU,GE (1985)
DC	pKa			
HL 3.31E-009	atm m3/mol	25	EST	MEYLAN,WM & HOWARD,PH (1991)
OH 4.07E-011	cm3/molc sec	25	EST	MEYLAN,WM & HOWARD,PH (1993)

CAS #: 022736-85-2

DIFLUMIDONE

Formula: $C_{14}H_{11}F_2NO_3S$

Mol Weight: 311.31

MP (deg C):

FP (deg C):

BP (deg C):

BP pressure (mm Hg):

Property/Value	Units	Temp	Data Type	Reference
WS 3.81E+001	mg/L	25	EST	MEYLAN,WM ET AL. (1996)
logP 2.86			EXP	HANSCH,C & LEO,AJ (1985)
VP 1.46E-007	mm Hg	25	EST	NEELY,WB & BLAU,GE (1985)
DC	pKa			
HL 1.66E-009	atm m3/mol	25	EST	MEYLAN,WM & HOWARD,PH (1991)
OH 1.99E-011	cm3/molc sec	25	EST	MEYLAN,WM & HOWARD,PH (1993)

CAS #: 022752-98-3

N-(4-PYRIDYL)PYRIDINIUM CHLORIDE

Formula: $C_{10}H_9ClN_2$

Mol Weight: 192.65

MP (deg C):

FP (deg C):

BP (deg C):

BP pressure (mm Hg):

Property/Value	Units	Temp	Data Type	Reference
WS 1.00E+006	mg/L	25	EST	MEYLAN,WM ET AL. (1996)
logP -4.54			EST	MEYLAN,WM & HOWARD,PH (1995)
VP 6.26E-004	mm Hg	25	EST	NEELY,WB & BLAU,GE (1985)
DC	pKa			
HL 3.35E-017	atm m3/mol	25	EST	MEYLAN,WM & HOWARD,PH (1991)
OH 1.08E-011	cm3/molc sec	25	EST	MEYLAN,WM & HOWARD,PH (1993)

CAS #: 022754-37-6

4-PYRIMIDINECARBOXYLIC ACID, 1,2,3,6-TETRAHYDRO-

Formula: $C_9H_{12}N_2O_4$

Mol Weight: 212.21

MP (deg C):

FP (deg C):

BP (deg C):

BP pressure (mm Hg):

Property/Value	Units	Temp	Data Type	Reference
WS 4.71E+003	mg/L	25	EST	MEYLAN,WM ET AL. (1996)
logP 1.06			EXP	SANGSTER,J (1993)
VP 9.12E-009	mm Hg	25	EST	NEELY,WB & BLAU,GE (1985)
DC	pKa			
HL 6.06E-012	atm m3/mol	25	EST	MEYLAN,WM & HOWARD,PH (1991)
OH 1.30E-011	cm3/molc sec	25	EST	MEYLAN,WM & HOWARD,PH (1993)

CAS #: 022781-23-3

BENDIOCARB

Formula: $C_{11}H_{13}NO_4$

Mol Weight: 223.23

MP (deg C): 130

FP (deg C):

BP (deg C):

BP pressure (mm Hg):

Property/Value	Units	Temp	Data Type	Reference
WS 4.00E+001	mg/L	25	EXP	YALKOWSKY,SH & DANNENFELSER,RM (1992)
logP 1.70			EXP	HANSCH,C ET AL. (1995)
VP 5.00E-006	mm Hg	25	EXP	WRIGHT,CG ET AL. (1981)
DC 8.80	pKa	25	EXP	TOMLIN,C (1994)
HL 3.67E-008	atm m3/mol	25	EST	VP/WSOL
OH 2.78E-011	cm3/molc sec	25	EST	MEYLAN,WM & HOWARD,PH (1993)

CAS #: 022808-73-7

4-SULFAMYLBENZOIC ACID, ME ESTER

Formula: $C_8H_9NO_4S$

Mol Weight: 215.23

MP (deg C):

FP (deg C):

BP (deg C):

BP pressure (mm Hg):

Property/Value	Units	Temp	Data Type	Reference
WS 1.04E+004	mg/L	25	EST	MEYLAN,WM ET AL. (1996)
logP 0.64			EXP	HANSCH,C & LEO,AJ (1985)
VP 1.50E-005	mm Hg	25	EST	NEELY,WB & BLAU,GE (1985)
DC	pKa			
HL 2.72E-009	atm m3/mol	25	EST	MEYLAN,WM & HOWARD,PH (1991)
OH 3.52E-013	cm3/molc sec	25	EST	MEYLAN,WM & HOWARD,PH (1993)

CAS #: 022813-31-6

1-(CH2CO2ME)-2-NO2- IMIDAZOLE

Formula: $C_6H_9N_3O_4$

Mol Weight: 187.16

MP (deg C): 95-97

FP (deg C):

BP (deg C):

BP pressure (mm Hg):

Property/Value	Units	Temp	Data Type	Reference
WS 2.53E+004	mg/L	25	EST	MEYLAN,WM ET AL. (1996)
logP -0.09			EXP	HANSCH,C & LEO,AJ (1985)
VP 2.60E-005	mm Hg	25	EST	NEELY,WB & BLAU,GE (1985)
DC	pKa			
HL 7.53E-010	atm m3/mol	25	EST	MEYLAN,WM & HOWARD,PH (1991)
OH 5.41E-012	cm3/molc sec	25	EST	MEYLAN,WM & HOWARD,PH (1993)

CAS #: 022821-80-3				ACETAMIDE, N-[4-(METHYLSULFONYL)PHENYL]

Formula: $C_9H_{11}NO_3S$
Mol Weight: 213.26
MP (deg C):　　FP (deg C):
BP (deg C):
BP pressure (mm Hg):

Property/Value	Units	Temp	Data Type	Reference
WS 1.19E+004	mg/L	25	EST	MEYLAN,WM ET AL. (1996)
logP 0.58			EXP	HANSCH,C ET AL. (1995)
VP 3.49E-007	mm Hg	25	EST	NEELY,WB & BLAU,GE (1985)
DC	pKa			
HL 1.43E-012	atm m3/mol	25	EST	MEYLAN,WM & HOWARD,PH (1991)
OH 3.82E-012	cm3/molc sec	25	EST	MEYLAN,WM & HOWARD,PH (1993)

CAS #: 022839-47-0				ASPARTAME

Formula: $C_{14}H_{18}N_2O_5$
Mol Weight: 294.31
MP (deg C): 246.5　　FP (deg C):
BP (deg C):
BP pressure (mm Hg):

Property/Value	Units	Temp	Data Type	Reference
WS 1.00E+004	mg/L			MAZUR,R (1983)
logP -0.43			EST	MEYLAN,WM & HOWARD,PH (1995)
VP 4.54E-012	mm Hg	25	EST	NEELY,WB & BLAU,GE (1985)
DC	pKa			
HL 2.53E-018	atm m3/mol	25	EST	MEYLAN,WM & HOWARD,PH (1991)
OH 5.93E-011	cm3/molc sec	25	EST	MEYLAN,WM & HOWARD,PH (1993)

CAS #: 022900-79-4				3,5-DIMETHYL-4-HYDROXYACETANILIDE

Formula: $C_{10}H_{13}NO_2$
Mol Weight: 179.22
MP (deg C):　　FP (deg C):
BP (deg C):
BP pressure (mm Hg):

Property/Value	Units	Temp	Data Type	Reference
WS 6.26E+003	mg/L	25	EST	MEYLAN,WM ET AL. (1996)
logP 1.11			EXP	HANSCH,C & LEO,AJ (1985)
VP 1.05E-006	mm Hg	25	EST	NEELY,WB & BLAU,GE (1985)
DC	pKa			
HL 7.82E-013	atm m3/mol	25	EST	MEYLAN,WM & HOWARD,PH (1991)
OH 5.88E-011	cm3/molc sec	25	EST	MEYLAN,WM & HOWARD,PH (1993)

CAS #: 022903-73-7				2-(2-NO2-1-IMIDAZOLYL)ACETAMIDE

Formula: $C_5H_8N_4O_3$
Mol Weight: 172.14
MP (deg C):　　FP (deg C):
BP (deg C):
BP pressure (mm Hg):

Property/Value	Units	Temp	Data Type	Reference
WS 1.35E+004	mg/L	25	EST	MEYLAN,WM ET AL. (1996)
logP -1.22			EXP	HANSCH,C & LEO,AJ (1985)
VP 2.55E-007	mm Hg	25	EST	NEELY,WB & BLAU,GE (1985)
DC	pKa			
HL 4.80E-014	atm m3/mol	25	EST	MEYLAN,WM & HOWARD,PH (1991)
OH 7.20E-012	cm3/molc sec	25	EST	MEYLAN,WM & HOWARD,PH (1993)

CAS #: 022913-26-4				METHYL THIOPHEN-3-CARBOXYLATE

Formula: $C_6H_8O_2S$
Mol Weight: 144.19
MP (deg C):　　FP (deg C):
BP (deg C):
BP pressure (mm Hg):

Property/Value	Units	Temp	Data Type	Reference
WS 2.58E+003	mg/L	25	EST	MEYLAN,WM ET AL. (1996)
logP 1.76			EXP	HANSCH,C & LEO,AJ (1985)
VP 2.71E-001	mm Hg	25	EST	NEELY,WB & BLAU,GE (1985)
DC	pKa			
HL 1.89E-005	atm m3/mol	25	EST	MEYLAN,WM & HOWARD,PH (1991)
OH 3.28E-012	cm3/molc sec	25	EST	MEYLAN,WM & HOWARD,PH (1993)

CAS #: 022928-63-8				2-AZACYCLOOCTANTHIONE

Formula: $C_7H_{13}NS$
Mol Weight: 143.25
MP (deg C):　　FP (deg C):
BP (deg C):
BP pressure (mm Hg):

Property/Value	Units	Temp	Data Type	Reference
WS 1.14E+004	mg/L	25	EST	MEYLAN,WM ET AL. (1996)
logP 1.00			EXP	HANSCH,C & LEO,AJ (1985)
VP 1.19E-003	mm Hg	25	EST	NEELY,WB & BLAU,GE (1985)
DC	pKa			
HL 1.94E-005	atm m3/mol	25	EST	MEYLAN,WM & HOWARD,PH (1991)
OH 8.07E-011	cm3/molc sec	25	EST	MEYLAN,WM & HOWARD,PH (1993)

CAS #: 022936-75-0				DIMETHAMRTRYNE

Formula: $C_{11}H_{21}N_5S$
Mol Weight: 255.39
MP (deg C): 65　　FP (deg C):
BP (deg C):
BP pressure (mm Hg):

Property/Value	Units	Temp	Data Type	Reference
WS 5.00E+001	mg/L	20	EXP	SHIU,WY ET AL. (1990)
logP 3.90			EXP	HANSCH,C ET AL. (1995)
VP 5.84E-006	mm Hg	25	EST	NEELY,WB & BLAU,GE (1985)
DC	pKa			
HL 1.21E-008	atm m3/mol	25	EST	MEYLAN,WM & HOWARD,PH (1991)
OH 3.52E-011	cm3/molc sec	25	EST	MEYLAN,WM & HOWARD,PH (1993)

CAS #: 022936-85-2				1,3,5-TRIAZINE-2,4-DIAMINE, 6-CHLORO-N-CYCLOPROP

Formula: $C_8H_{12}ClN_5$
Mol Weight: 213.67
MP (deg C):　　FP (deg C):
BP (deg C):
BP pressure (mm Hg):

Property/Value	Units	Temp	Data Type	Reference
WS 5.21E+002	mg/L	25	EST	MEYLAN,WM ET AL. (1996)
logP 2.17			EXP	SANGSTER,J (1993)
VP 5.56E-005	mm Hg	25	EST	NEELY,WB & BLAU,GE (1985)
DC	pKa			
HL 1.97E-009	atm m3/mol	25	EST	MEYLAN,WM & HOWARD,PH (1991)
OH 9.57E-012	cm3/molc sec	25	EST	MEYLAN,WM & HOWARD,PH (1993)

CAS #: 023046-87-9				N-PROPIONYLCYCLOBUTANECARBOXAMIDE

Formula: $C_8H_{13}NO_2$

Mol Weight: 155.20

MP (deg C): FP (deg C):

BP (deg C):

BP pressure (mm Hg):

Property/ Value	Units	Temp	Data Type	Reference
WS 7.78E+004	mg/L	25	EST	MEYLAN,WM ET AL. (1996)
logP -0.04			EXP	HANSCH,C & LEO,AJ (1985)
VP 1.10E-004	mm Hg	25	EST	NEELY,WB & BLAU,GE (1985)
DC	pKa			
HL 3.10E-011	atm m3/mol	25	EST	MEYLAN,WM & HOWARD,PH (1991)
OH 4.56E-011	cm3/molc sec	25	EST	MEYLAN,WM & HOWARD,PH (1993)

CAS #: 023046-88-0				N-I-BUTYROYLCYCLOBUTANECARBOXAMIDE

Formula: $C_9H_{15}NO_2$

Mol Weight: 169.23

MP (deg C): FP (deg C):

BP (deg C):

BP pressure (mm Hg):

Property/ Value	Units	Temp	Data Type	Reference
WS 3.72E+004	mg/L	25	EST	MEYLAN,WM ET AL. (1996)
logP 0.26			EXP	HANSCH,C & LEO,AJ (1985)
VP 8.05E-005	mm Hg	25	EST	NEELY,WB & BLAU,GE (1985)
DC	pKa			
HL 4.11E-011	atm m3/mol	25	EST	MEYLAN,WM & HOWARD,PH (1991)
OH 6.18E-011	cm3/molc sec	25	EST	MEYLAN,WM & HOWARD,PH (1993)

CAS #: 023068-36-2				2,3-DICHLOROACETANILIDE

Formula: $C_8H_7Cl_2NO$

Mol Weight: 204.06

MP (deg C): FP (deg C):

BP (deg C):

BP pressure (mm Hg):

Property/ Value	Units	Temp	Data Type	Reference
WS 6.85E+002	mg/L	25	EST	MEYLAN,WM ET AL. (1996)
logP 2.09			EXP	NAKAGAWA,Y ET AL. (1992)
VP 2.02E-005	mm Hg	25	EST	NEELY,WB & BLAU,GE (1985)
DC	pKa			
HL 3.39E-009	atm m3/mol	25	EST	MEYLAN,WM & HOWARD,PH (1991)
OH 2.65E-012	cm3/molc sec	25	EST	MEYLAN,WM & HOWARD,PH (1993)

CAS #: 023088-24-6				6-CYANOQUINOXALINE

Formula: $C_9H_5N_3$

Mol Weight: 155.16

MP (deg C): FP (deg C):

BP (deg C):

BP pressure (mm Hg):

Property/ Value	Units	Temp	Data Type	Reference
WS 5.37E+003	mg/L	25	EST	MEYLAN,WM ET AL. (1996)
logP 1.01			EXP	HANSCH,C & LEO,AJ (1985)
VP 1.88E-004	mm Hg	25	EST	NEELY,WB & BLAU,GE (1985)
DC	pKa			
HL 2.75E-009	atm m3/mol	25	EST	MEYLAN,WM & HOWARD,PH (1991)
OH 3.53E-013	cm3/molc sec	25	EST	MEYLAN,WM & HOWARD,PH (1993)

CAS #: 023092-17-3				2H-1,4-BENZODIAZEPIN-2-ONE, 7-CHLORO-1,3-DIHYDRO

Formula: $C_{17}H_{12}ClF_3N_2O$

Mol Weight: 352.75

MP (deg C): 164-166 FP (deg C):

BP (deg C):

BP pressure (mm Hg):

Property/ Value	Units	Temp	Data Type	Reference
WS 2.43E+000	mg/L	25	EST	MEYLAN,WM ET AL. (1996)
logP 3.97			EXP	SANGSTER,J (1994)
VP 9.89E-009	mm Hg	25	EST	NEELY,WB & BLAU,GE (1985)
DC	pKa			
HL 1.12E-009	atm m3/mol	25	EST	MEYLAN,WM & HOWARD,PH (1991)
OH 5.77E-011	cm3/molc sec	25	EST	MEYLAN,WM & HOWARD,PH (1993)

CAS #: 023099-85-6				4-PH-1-PHTHALAZINAMINE,N,N-DIME

Formula: $C_{16}H_{15}N_3$

Mol Weight: 249.32

MP (deg C): FP (deg C):

BP (deg C):

BP pressure (mm Hg):

Property/ Value	Units	Temp	Data Type	Reference
WS 4.25E+001	mg/L	25	EST	MEYLAN,WM ET AL. (1996)
logP 3.22			EXP	HANSCH,C & LEO,AJ (1985)
VP 2.80E-008	mm Hg	25	EST	NEELY,WB & BLAU,GE (1985)
DC	pKa			
HL 3.42E-010	atm m3/mol	25	EST	MEYLAN,WM & HOWARD,PH (1991)
OH 4.52E-011	cm3/molc sec	25	EST	MEYLAN,WM & HOWARD,PH (1993)

CAS #: 023103-98-2				PIRIMICARB

Formula: $C_{11}H_{18}N_4O_2$

Mol Weight: 238.29

MP (deg C): 90.5 FP (deg C):

BP (deg C):

BP pressure (mm Hg):

Property/ Value	Units	Temp	Data Type	Reference
WS 2.70E+003	mg/L	25	EXP	YALKOWSKY,SH & DANNENFELSER,RM (1992)
logP 1.70			EXP	TOMLIN,C (1994)
VP 3.00E-005	mm Hg	30	EXP	AUGUSTIJN-BECKERS,PWM ET AL. (1994)
DC	pKa			
HL 3.48E-009	atm m3/mol	30	EST	VP/WSOL
OH 1.63E-010	cm3/molc sec	25	EST	MEYLAN,WM & HOWARD,PH (1993)

CAS #: 023135-22-0				OXAMYL

Formula: $C_7H_{13}N_3O_3S$

Mol Weight: 219.26

MP (deg C): 100-102 FP (deg C):

BP (deg C):

BP pressure (mm Hg):

Property/ Value	Units	Temp	Data Type	Reference
WS 2.80E+005	mg/L	25	EXP	YALKOWSKY,SH & DANNENFELSER,RM (1992)
logP -0.47			EXP	HANSCH,C ET AL. (1995)
VP 2.30E-004	mm Hg	25	EXP	WAUCHOPE,RD ET AL. (1991A)
DC	pKa			
HL 2.37E-010	atm m3/mol	25	EST	MEYLAN,WM & HOWARD,PH (1993)
OH 2.49E-011	cm3/molc sec	25	EST	MEYLAN,WM & HOWARD,PH (1993)

CAS #: 023138-95-6 — UREA, N-(3-ETHYLPHENYL)-N'-METHYL-

Formula: $C_{10}H_{14}N_2O$

Mol Weight: 178.24

MP (deg C):
FP (deg C):
BP (deg C):
BP pressure (mm Hg):

Property/Value	Units	Temp	Data Type	Reference
WS 9.04E+002	mg/L	25	EST	MEYLAN,WM ET AL. (1996)
logP 2.10			EXP	MITSUTAKE,KI ET AL. (1986)
VP 7.27E-005	mm Hg	25	EST	NEELY,WB & BLAU,GE (1985)
DC	pKa			
HL 6.48E-010	atm m3/mol	25	EST	MEYLAN,WM & HOWARD,PH (1991)
OH 5.46E-011	cm3/molc sec	25	EST	MEYLAN,WM & HOWARD,PH (1993)

CAS #: 023138-98-9 — UREA, N-(3-METHOXYPHENYL)-N'-METHYL-

Formula: $C_9H_{12}N_2O_2$

Mol Weight: 180.21

MP (deg C):
FP (deg C):
BP (deg C):
BP pressure (mm Hg):

Property/Value	Units	Temp	Data Type	Reference
WS 4.61E+003	mg/L	25	EST	MEYLAN,WM ET AL. (1996)
logP 1.26			EXP	MITSUTAKE,KI ET AL. (1986)
VP 6.59E-005	mm Hg	25	EST	NEELY,WB & BLAU,GE (1985)
DC	pKa			
HL 2.62E-011	atm m3/mol	25	EST	MEYLAN,WM & HOWARD,PH (1991)
OH 2.02E-010	cm3/molc sec	25	EST	MEYLAN,WM & HOWARD,PH (1993)

CAS #: 023141-75-5 — 124-BENZTHIADIAZIN,11-O2-7SULFAMYL

Formula: $C_7H_7N_3O_4S_2$

Mol Weight: 261.28

MP (deg C):
FP (deg C):
BP (deg C):
BP pressure (mm Hg):

Property/Value	Units	Temp	Data Type	Reference
WS 7.21E+003	mg/L	25	EST	MEYLAN,WM ET AL. (1996)
logP -0.70			EXP	HANSCH,C ET AL. (1995)
VP 7.98E-011	mm Hg	25	EST	NEELY,WB & BLAU,GE (1985)
DC	pKa			
HL 5.46E-012	atm m3/mol	25	EST	MEYLAN,WM & HOWARD,PH (1991)
OH 1.81E-011	cm3/molc sec	25	EST	MEYLAN,WM & HOWARD,PH (1993)

CAS #: 023141-80-2 — FLUETHIAZIDE

Formula: $C_9H_6F_5N_3O_4S_2$

Mol Weight: 379.29

MP (deg C):
FP (deg C):
BP (deg C):
BP pressure (mm Hg):

Property/Value	Units	Temp	Data Type	Reference
WS 4.64E+001	mg/L	25	EST	MEYLAN,WM ET AL. (1996)
logP 1.05			EXP	SANGSTER,J (1994)
VP 5.31E-011	mm Hg	25	EST	NEELY,WB & BLAU,GE (1985)
DC	pKa			
HL 2.49E-010	atm m3/mol	25	EST	MEYLAN,WM & HOWARD,PH (1991)
OH 2.23E-011	cm3/molc sec	25	EST	MEYLAN,WM & HOWARD,PH (1993)

CAS #: 023141-81-3 — 124-BENZTHIADIAZN,11-O2-7-NH2SO2-6NO2

Formula: $C_7H_6N_4O_6S_2$

Mol Weight: 306.28

MP (deg C):
FP (deg C):
BP (deg C):
BP pressure (mm Hg):

Property/Value	Units	Temp	Data Type	Reference
WS 3.49E+002	mg/L	25	EST	MEYLAN,WM ET AL. (1996)
logP 0.08			EXP	SANGSTER,J (1994)
VP 1.44E-012	mm Hg	25	EST	NEELY,WB & BLAU,GE (1985)
DC	pKa			
HL 2.15E-014	atm m3/mol	25	EST	MEYLAN,WM & HOWARD,PH (1991)
OH 1.81E-011	cm3/molc sec	25	EST	MEYLAN,WM & HOWARD,PH (1993)

CAS #: 023141-82-4 — HYDROTHIAZIDE

Formula: $C_7H_9N_3O_4S_2$

Mol Weight: 263.30

MP (deg C):
FP (deg C):
BP (deg C):
BP pressure (mm Hg):

Property/Value	Units	Temp	Data Type	Reference
WS 4.93E+003	mg/L	25	EST	MEYLAN,WM ET AL. (1996)
logP -0.52			EXP	HANSCH,C & LEO,AJ (1985)
VP 4.91E-009	mm Hg	25	EST	NEELY,WB & BLAU,GE (1985)
DC	pKa			
HL 5.92E-012	atm m3/mol	25	EST	MEYLAN,WM & HOWARD,PH (1991)
OH 8.27E-011	cm3/molc sec	25	EST	MEYLAN,WM & HOWARD,PH (1993)

CAS #: 023141-83-5 — HYDROBROMOTHIAZIDE

Formula: $C_7H_8BrN_3O_4S_2$

Mol Weight: 342.19

MP (deg C):
FP (deg C):
BP (deg C):
BP pressure (mm Hg):

Property/Value	Units	Temp	Data Type	Reference
WS 5.25E+002	mg/L	25	EST	MEYLAN,WM ET AL. (1996)
logP 0.08			EXP	HANSCH,C & LEO,AJ (1985)
VP 5.67E-010	mm Hg	25	EST	NEELY,WB & BLAU,GE (1985)
DC	pKa			
HL 2.36E-012	atm m3/mol	25	EST	MEYLAN,WM & HOWARD,PH (1991)
OH 8.20E-011	cm3/molc sec	25	EST	MEYLAN,WM & HOWARD,PH (1993)

CAS #: 023141-84-6 — 3-T-BUTYLHYDROCHLOROTHIAZIDE

Formula: $C_{11}H_{16}ClN_3O_4S_2$

Mol Weight: 353.85

MP (deg C):
FP (deg C):
BP (deg C):
BP pressure (mm Hg):

Property/Value	Units	Temp	Data Type	Reference
WS 2.57E+002	mg/L	25	EST	MEYLAN,WM ET AL. (1996)
logP 0.36			EXP	HANSCH,C & LEO,AJ (1985)
VP 1.42E-010	mm Hg	25	EST	NEELY,WB & BLAU,GE (1985)
DC	pKa			
HL 1.36E-011	atm m3/mol	25	EST	MEYLAN,WM & HOWARD,PH (1991)
OH 2.08E-010	cm3/molc sec	25	EST	MEYLAN,WM & HOWARD,PH (1993)

CAS #: 023141-86-8 — 3-CYCLOPENTYLMETHYLHYDROTHIAZIDE

Formula: $C_{13}H_{19}N_3O_4S_2$

Mol Weight: 345.44

MP (deg C):
FP (deg C):
BP (deg C):
BP pressure (mm Hg):

Property/Value	Units	Temp	Data Type	Reference
WS 5.12E+001	mg/L	25	EST	MEYLAN,WM ET AL. (1996)
logP 1.24			EXP	HANSCH,C & LEO,AJ (1985)
VP 3.36E-011	mm Hg	25	EST	NEELY,WB & BLAU,GE (1985)
DC	pKa			
HL 1.43E-011	atm m3/mol	25	EST	MEYLAN,WM & HOWARD,PH (1991)
OH 1.76E-010	cm3/molc sec	25	EST	MEYLAN,WM & HOWARD,PH (1993)

CAS #: 023141-87-9 — 3-DICHLOROMETHYLHYDROTHIAZIDE

Formula: $C_8H_9Cl_2N_3O_4S_2$

Mol Weight: 346.21

MP (deg C):
FP (deg C):
BP (deg C):
BP pressure (mm Hg):

Property/Value	Units	Temp	Data Type	Reference
WS 2.13E+002	mg/L	25	EST	MEYLAN,WM ET AL. (1996)
logP 0.51			EXP	SANGSTER,J (1994)
VP 1.37E-010	mm Hg	25	EST	NEELY,WB & BLAU,GE (1985)
DC	pKa			
HL 9.75E-013	atm m3/mol	25	EST	MEYLAN,WM & HOWARD,PH (1991)
OH 6.27E-011	cm3/molc sec	25	EST	MEYLAN,WM & HOWARD,PH (1993)

CAS #: 023141-88-0 — 6-NITROHYDROTHIAZIDE

Formula: $C_7H_8N_4O_6S_2$

Mol Weight: 308.29

MP (deg C):
FP (deg C):
BP (deg C):
BP pressure (mm Hg):

Property/Value	Units	Temp	Data Type	Reference
WS 2.68E+002	mg/L	25	EST	MEYLAN,WM ET AL. (1996)
logP 0.20			EXP	HANSCH,C & LEO,AJ (1985)
VP 9.60E-011	mm Hg	25	EST	NEELY,WB & BLAU,GE (1985)
DC	pKa			
HL 2.34E-014	atm m3/mol	25	EST	MEYLAN,WM & HOWARD,PH (1991)
OH 8.10E-011	cm3/molc sec	25	EST	MEYLAN,WM & HOWARD,PH (1993)

CAS #: 023184-66-9 — BUTACHLOR

Formula: $C_{17}H_{26}ClNO_2$

Mol Weight: 311.86

MP (deg C): < -5
FP (deg C):
BP (deg C): 156
BP pressure (mm Hg): 5.00E-001

Property/Value	Units	Temp	Data Type	Reference
WS 2.30E+001	mg/L	24	EXP	WORTHING,CR & WALKER,SB (1987)
logP 4.50			EXP	HANSCH,C ET AL. (1995)
VP 2.90E-006	mm Hg	25	EXP	WATANABE,T (1993)
DC	pKa			
HL 5.10E-008	atm m3/mol	25	EXP	WATANABE,T (1993)
OH 5.66E-011	cm3/molc sec	25	EST	MEYLAN,WM & HOWARD,PH (1993)

CAS #: 023186-94-9 — 2,4-IMIDAZOLIDINEDIONE, 5,5-BIS(PHENYLMETHYL)-

Formula: $C_{17}H_{16}N_2O_2$

Mol Weight: 280.33

MP (deg C):
FP (deg C):
BP (deg C):
BP pressure (mm Hg):

Property/Value	Units	Temp	Data Type	Reference
WS 1.27E+002	mg/L	25	EST	MEYLAN,WM ET AL. (1996)
logP 2.46			EXP	SANGSTER,J (1994)
VP 2.27E-011	mm Hg	25	EST	NEELY,WB & BLAU,GE (1985)
DC	pKa			
HL 1.80E-011	atm m3/mol	25	EST	MEYLAN,WM & HOWARD,PH (1991)
OH 1.90E-011	cm3/molc sec	25	EST	MEYLAN,WM & HOWARD,PH (1993)

CAS #: 023189-28-8 — PHENYLACETIC ACID,2-(2'-CL-6-ME)ANILINO

Formula: $C_{15}H_{14}ClNO_2$

Mol Weight: 275.74

MP (deg C):
FP (deg C):
BP (deg C):
BP pressure (mm Hg):

Property/Value	Units	Temp	Data Type	Reference
WS 7.80E+000	mg/L	25	EST	MEYLAN,WM ET AL. (1996)
logP 4.37			EXP	HANSCH,C ET AL. (1995)
VP 1.10E-007	mm Hg	25	EST	NEELY,WB & BLAU,GE (1985)
DC	pKa			
HL 7.05E-012	atm m3/mol	25	EST	MEYLAN,WM & HOWARD,PH (1991)
OH 2.01E-010	cm3/molc sec	25	EST	MEYLAN,WM & HOWARD,PH (1993)

CAS #: 023193-98-8 — 1,4BENZODIAZEPIN2ONE,1ME-5PH-7MES

Formula: $C_{17}H_{16}N_2OS$

Mol Weight: 296.39

MP (deg C):
FP (deg C):
BP (deg C):
BP pressure (mm Hg):

Property/Value	Units	Temp	Data Type	Reference
WS 3.90E+001	mg/L	25	EST	MEYLAN,WM ET AL. (1996)
logP 2.95			EXP	HANSCH,C & LEO,AJ (1985)
VP 4.50E-009	mm Hg	25	EST	NEELY,WB & BLAU,GE (1985)
DC	pKa			
HL 1.43E-010	atm m3/mol	25	EST	MEYLAN,WM & HOWARD,PH (1991)
OH 2.11E-011	cm3/molc sec	25	EST	MEYLAN,WM & HOWARD,PH (1993)

CAS #: 023214-92-8 — ADRIAMYCIN

Formula: $C_{27}H_{29}NO_{11}$

Mol Weight: 543.53

MP (deg C): 229-231
FP (deg C):
BP (deg C):
BP pressure (mm Hg):

Property/Value	Units	Temp	Data Type	Reference
WS 9.28E+001	mg/L	25	EST	MEYLAN,WM ET AL. (1996)
logP 1.27			EXP	HANSCH,C ET AL. (1995)
VP 8.99E-025	mm Hg	25	EST	NEELY,WB & BLAU,GE (1985)
DC	pKa			
HL 2.23E-023	atm m3/mol	25	EST	MEYLAN,WM & HOWARD,PH (1991)
OH 1.45E-010	cm3/molc sec	25	EST	MEYLAN,WM & HOWARD,PH (1993)

CAS #: 023222-62-0				2-PHENYL-4-I-PROPYLMORPHOLINE

Formula: $C_{13}H_{19}NO$

Mol Weight: 205.30

MP (deg C):   FP (deg C):

BP (deg C):

BP pressure (mm Hg):

Property/Value	Units	Temp	Data Type	Reference
WS 4.55E+003	mg/L	25	EST	MEYLAN,WM ET AL. (1996)
logP 2.30			EXP	HANSCH,C & LEO,AJ (1985)
VP 1.70E-003	mm Hg	25	EST	NEELY,WB & BLAU,GE (1985)
DC	pKa			
HL 3.56E-008	atm m3/mol	25	EST	MEYLAN,WM & HOWARD,PH (1991)
OH 1.52E-010	cm3/molc sec	25	EST	MEYLAN,WM & HOWARD,PH (1993)

CAS #: 023246-36-8				4-ISOTHIOCYANODIPHENYL AMINE

Formula: $C_{13}H_{10}N_2S$

Mol Weight: 226.30

MP (deg C):   FP (deg C):

BP (deg C):

BP pressure (mm Hg):

Property/Value	Units	Temp	Data Type	Reference
WS 1.93E+000	mg/L	25	EST	MEYLAN,WM ET AL. (1996)
logP 4.94			EXP	HANSCH,C & LEO,AJ (1985)
VP 9.96E-006	mm Hg	25	EST	NEELY,WB & BLAU,GE (1985)
DC	pKa			
HL 1.52E-007	atm m3/mol	25	EST	MEYLAN,WM & HOWARD,PH (1991)
OH 2.00E-010	cm3/molc sec	25	EST	MEYLAN,WM & HOWARD,PH (1993)

CAS #: 023251-68-5				2(5H)-FURANONE, 5-(4-HEXEN-2-YNYLIDENE)-

Formula: $C_{10}H_8O_2$

Mol Weight: 160.17

MP (deg C):   FP (deg C):

BP (deg C):

BP pressure (mm Hg):

Property/Value	Units	Temp	Data Type	Reference
WS 1.12E+003	mg/L	25	EST	MEYLAN,WM ET AL. (1996)
logP 2.09			EXP	MCLACHLAN,D ET AL. (1986)
VP 4.48E-004	mm Hg	25	EST	NEELY,WB & BLAU,GE (1985)
DC	pKa			
HL 1.21E-004	atm m3/mol	25	EST	MEYLAN,WM & HOWARD,PH (1991)
OH 1.79E-010	cm3/molc sec	25	EST	MEYLAN,WM & HOWARD,PH (1993)

CAS #: 023261-20-3				1,2:5,6-DIANHYDROGALACTITOL

Formula: $C_6H_{10}O_4$

Mol Weight: 146.14

MP (deg C):   FP (deg C):

BP (deg C):

BP pressure (mm Hg):

Property/Value	Units	Temp	Data Type	Reference
WS 9.97E+005	mg/L	25	EST	MEYLAN,WM ET AL. (1996)
logP -1.29			EXP	HANSCH,C & LEO,AJ (1985)
VP 1.01E-003	mm Hg	25	EST	NEELY,WB & BLAU,GE (1985)
DC	pKa			
HL 1.94E-012	atm m3/mol	25	EST	MEYLAN,WM & HOWARD,PH (1991)
OH 2.18E-011	cm3/molc sec	25	EST	MEYLAN,WM & HOWARD,PH (1993)

CAS #: 023264-57-5				2356-TETRAME-N,N'-DIN=O PIPERAZINE

Formula: $C_8H_{16}N_4O_2$

Mol Weight: 200.24

MP (deg C):   FP (deg C):

BP (deg C):

BP pressure (mm Hg):

Property/Value	Units	Temp	Data Type	Reference
WS 7.29E+003	mg/L	25	EST	MEYLAN,WM ET AL. (1996)
logP 0.91			EXP	HANSCH,C & LEO,AJ (1985)
VP 1.87E-005	mm Hg	25	EST	NEELY,WB & BLAU,GE (1985)
DC	pKa			
HL 1.05E-010	atm m3/mol	25	EST	MEYLAN,WM & HOWARD,PH (1991)
OH 8.94E-011	cm3/molc sec	25	EST	MEYLAN,WM & HOWARD,PH (1993)

CAS #: 023361-37-7				METHIONINE-AMIDE,N-ACETYL

Formula: $C_7H_{14}N_2O_2S$

Mol Weight: 190.27

MP (deg C):   FP (deg C):

BP (deg C):

BP pressure (mm Hg):

Property/Value	Units	Temp	Data Type	Reference
WS 7.81E+003	mg/L	25	EST	MEYLAN,WM ET AL. (1996)
logP -0.60			EXP	HANSCH,C & LEO,AJ (1985)
VP 3.03E-008	mm Hg	25	EST	NEELY,WB & BLAU,GE (1985)
DC	pKa			
HL 2.22E-012	atm m3/mol	25	EST	MEYLAN,WM & HOWARD,PH (1991)
OH 5.94E-011	cm3/molc sec	25	EST	MEYLAN,WM & HOWARD,PH (1993)

CAS #: 023361-38-8				L-SERINAMIDE, N-ACETYL-

Formula: $C_5H_{10}N_2O_3$

Mol Weight: 146.15

MP (deg C):   FP (deg C):

BP (deg C):

BP pressure (mm Hg):

Property/Value	Units	Temp	Data Type	Reference
WS 1.53E+005	mg/L	25	EST	MEYLAN,WM ET AL. (1996)
logP -1.87			EXP	SANGSTER,J (1994)
VP 1.15E-008	mm Hg	25	EST	NEELY,WB & BLAU,GE (1985)
DC	pKa			
HL 5.73E-015	atm m3/mol	25	EST	MEYLAN,WM & HOWARD,PH (1991)
OH 2.78E-011	cm3/molc sec	25	EST	MEYLAN,WM & HOWARD,PH (1993)

CAS #: 023374-45-0				GLYCINE, N-[N-[N-[1-OXO-5-(7H-PURIN-6-YLTHIO)PEN

Formula: $C_{18}H_{25}N_7O_5S$

Mol Weight: 451.51

MP (deg C):   FP (deg C):

BP (deg C):

BP pressure (mm Hg):

Property/Value	Units	Temp	Data Type	Reference
WS 4.73E+001	mg/L	25	EST	MEYLAN,WM ET AL. (1996)
logP 0.21			EXP	SANGSTER,J (1993)
VP 4.06E-021	mm Hg	25	EST	NEELY,WB & BLAU,GE (1985)
DC	pKa			
HL 7.77E-025	atm m3/mol	25	EST	MEYLAN,WM & HOWARD,PH (1991)
OH 2.42E-010	cm3/molc sec	25	EST	MEYLAN,WM & HOWARD,PH (1993)

CAS #: 023374-51-8 — á-D-ALANINE, N-[1-OXO-5-(1H-PURIN-6-YLTHIO)PENTY

Formula: $C_{15}H_{21}N_5O_3S$

Mol Weight: 351.43

MP (deg C):

FP (deg C):

BP (deg C):

BP pressure (mm Hg):

Property/Value	Units	Temp	Data Type	Reference
WS 3.12E+002	mg/L	25	EST	MEYLAN,WM ET AL. (1996)
logP 1.51			EXP	HANSCH,C ET AL. (1995)
VP 1.03E-013	mm Hg		EST	NEELY,WB & BLAU,GE (1985)
DC	pKa			
HL 1.47E-017	atm m3/mol	25	EST	MEYLAN,WM & HOWARD,PH (1991)
OH 2.29E-010	cm3/molc sec	25	EST	MEYLAN,WM & HOWARD,PH (1993)

CAS #: 023375-06-6 — P-MES CF3-METHANESULFONANILIDE

Formula: $C_8H_8F_3NO_2S_2$

Mol Weight: 271.28

MP (deg C):

FP (deg C):

BP (deg C):

BP pressure (mm Hg):

Property/Value	Units	Temp	Data Type	Reference
WS 1.15E+001	mg/L	25	EST	MEYLAN,WM ET AL. (1996)
logP 3.74			EXP	HANSCH,C & LEO,AJ (1985)
VP 3.64E-005	mm Hg	25	EST	NEELY,WB & BLAU,GE (1985)
DC	pKa			
HL 2.68E-007	atm m3/mol	25	EST	MEYLAN,WM & HOWARD,PH (1991)
OH 2.73E-011	cm3/molc sec	25	EST	MEYLAN,WM & HOWARD,PH (1993)

CAS #: 023375-08-8 — M-MESO2 CF3-METHANESULFONANILIDE

Formula: $C_8H_8F_3NO_4S_2$

Mol Weight: 303.28

MP (deg C):

FP (deg C):

BP (deg C):

BP pressure (mm Hg):

Property/Value	Units	Temp	Data Type	Reference
WS 3.09E+002	mg/L	25	EST	MEYLAN,WM ET AL. (1996)
logP 1.85			EXP	HANSCH,C & LEO,AJ (1985)
VP 8.73E-007	mm Hg		EST	NEELY,WB & BLAU,GE (1985)
DC	pKa			
HL 2.13E-009	atm m3/mol	25	EST	MEYLAN,WM & HOWARD,PH (1991)
OH 1.02E-011	cm3/molc sec	25	EST	MEYLAN,WM & HOWARD,PH (1993)

CAS #: 023375-10-2 — P-MESO2 CF3-METHANESULFONANILIDE

Formula: $C_8H_8F_3NO_4S_2$

Mol Weight: 303.28

MP (deg C):

FP (deg C):

BP (deg C):

BP pressure (mm Hg):

Property/Value	Units	Temp	Data Type	Reference
WS 2.35E+002	mg/L	25	EST	MEYLAN,WM ET AL. (1996)
logP 1.99			EXP	HANSCH,C & LEO,AJ (1985)
VP 8.73E-007	mm Hg	25	EST	NEELY,WB & BLAU,GE (1985)
DC	pKa			
HL 2.13E-009	atm m3/mol	25	EST	MEYLAN,WM & HOWARD,PH (1991)
OH 1.02E-011	cm3/molc sec	25	EST	MEYLAN,WM & HOWARD,PH (1993)

CAS #: 023375-11-3 — M-ACETYL CF3-METHANESULFONANILIDE

Formula: $C_9H_8F_3NO_3S$

Mol Weight: 267.23

MP (deg C):

FP (deg C):

BP (deg C):

BP pressure (mm Hg):

Property/Value	Units	Temp	Data Type	Reference
WS 8.01E+001	mg/L	25	EST	MEYLAN,WM ET AL. (1996)
logP 2.78			EXP	HANSCH,C & LEO,AJ (1985)
VP 2.88E-005	mm Hg	25	EST	NEELY,WB & BLAU,GE (1985)
DC	pKa			
HL 1.67E-008	atm m3/mol	25	EST	MEYLAN,WM & HOWARD,PH (1991)
OH 1.81E-011	cm3/molc sec	25	EST	MEYLAN,WM & HOWARD,PH (1993)

CAS #: 023375-12-4 — M-OH CF3-METHANESULFONANILIDE

Formula: $C_7H_6F_3NO_3S$

Mol Weight: 241.19

MP (deg C):

FP (deg C):

BP (deg C):

BP pressure (mm Hg):

Property/Value	Units	Temp	Data Type	Reference
WS 7.23E+002	mg/L	25	EST	MEYLAN,WM ET AL. (1996)
logP 2.51			EXP	HANSCH,C & LEO,AJ (1985)
VP 1.56E-005	mm Hg	25	EST	NEELY,WB & BLAU,GE (1985)
DC	pKa			
HL 9.55E-010	atm m3/mol	25	EST	MEYLAN,WM & HOWARD,PH (1991)
OH 2.00E-010	cm3/molc sec	25	EST	MEYLAN,WM & HOWARD,PH (1993)

CAS #: 023383-94-0 — P-ACETYL CF3-METHANESULFONANILIDE

Formula: $C_9H_8F_3NO_3S$

Mol Weight: 267.23

MP (deg C):

FP (deg C):

BP (deg C):

BP pressure (mm Hg):

Property/Value	Units	Temp	Data Type	Reference
WS 7.26E+001	mg/L	25	EST	MEYLAN,WM ET AL. (1996)
logP 2.83			EXP	HANSCH,C & LEO,AJ (1985)
VP 2.88E-005	mm Hg	25	EST	NEELY,WB & BLAU,GE (1985)
DC	pKa			
HL 1.67E-008	atm m3/mol	25	EST	MEYLAN,WM & HOWARD,PH (1991)
OH 3.90E-011	cm3/molc sec	25	EST	MEYLAN,WM & HOWARD,PH (1993)

CAS #: 023383-96-2 — 2,4-DICL CF3-METHANESULFONANILIDE

Formula: $C_7H_4Cl_2F_3NO_2S$

Mol Weight: 294.08

MP (deg C):

FP (deg C):

BP (deg C):

BP pressure (mm Hg):

Property/Value	Units	Temp	Data Type	Reference
WS 6.46E+000	mg/L	25	EST	MEYLAN,WM ET AL. (1996)
logP 3.88			EXP	HANSCH,C & LEO,AJ (1985)
VP 5.18E-005	mm Hg	25	EST	NEELY,WB & BLAU,GE (1985)
DC	pKa			
HL 5.04E-006	atm m3/mol	25	EST	MEYLAN,WM & HOWARD,PH (1991)
OH 3.64E-012	cm3/molc sec	25	EST	MEYLAN,WM & HOWARD,PH (1993)

CAS #: 023384-00-1 — P-FLUORO CF3-METHANESULFONANILIDE

Formula: $C_7H_5F_4NO_2S$
Mol Weight: 243.18
MP (deg C): · FP (deg C):
BP (deg C):
BP pressure (mm Hg):

Property/Value	Units	Temp	Data Type	Reference
WS 4.33E+001	mg/L	25	EST	MEYLAN,WM ET AL. (1996)
logP 3.25			EXP	HANSCH,C & LEO,AJ (1985)
VP 1.82E-003	mm Hg	25	EST	NEELY,WB & BLAU,GE (1985)
DC	pKa			
HL 1.07E-005	atm m3/mol	25	EST	MEYLAN,WM & HOWARD,PH (1991)
OH 1.44E-011	cm3/molc sec	25	EST	MEYLAN,WM & HOWARD,PH (1993)

CAS #: 023384-01-2 — M-FLUORO CF3-METHANESULFONANILIDE

Formula: $C_7H_5F_4NO_2S$
Mol Weight: 243.18
MP (deg C): · FP (deg C):
BP (deg C):
BP pressure (mm Hg):

Property/Value	Units	Temp	Data Type	Reference
WS 3.10E+001	mg/L	25	EST	MEYLAN,WM ET AL. (1996)
logP 3.42			EXP	HANSCH,C & LEO,AJ (1985)
VP 1.82E-003	mm Hg	25	EST	NEELY,WB & BLAU,GE (1985)
DC	pKa			
HL 1.07E-005	atm m3/mol	25	EST	MEYLAN,WM & HOWARD,PH (1991)
OH 5.34E-011	cm3/molc sec	25	EST	MEYLAN,WM & HOWARD,PH (1993)

CAS #: 023384-03-4 — M-CHLORO CF3-METHANESULFONANILIDE

Formula: $C_7H_5ClF_3NO_2S$
Mol Weight: 259.64
MP (deg C): · FP (deg C):
BP (deg C):
BP pressure (mm Hg):

Property/Value	Units	Temp	Data Type	Reference
WS 8.02E+000	mg/L	25	EST	MEYLAN,WM ET AL. (1996)
logP 4.00			EXP	HANSCH,C & LEO,AJ (1985)
VP 2.68E-004	mm Hg	25	EST	NEELY,WB & BLAU,GE (1985)
DC	pKa			
HL 6.80E-006	atm m3/mol	25	EST	MEYLAN,WM & HOWARD,PH (1991)
OH 3.00E-011	cm3/molc sec	25	EST	MEYLAN,WM & HOWARD,PH (1993)

CAS #: 023384-04-5 — P-CHLORO CF3-METHANESULFONANILIDE

Formula: $C_7H_5ClF_3NO_2S$
Mol Weight: 259.64
MP (deg C): · FP (deg C):
BP (deg C):
BP pressure (mm Hg):

Property/Value	Units	Temp	Data Type	Reference
WS 8.68E+000	mg/L	25	EST	MEYLAN,WM ET AL. (1996)
logP 3.96			EXP	HANSCH,C & LEO,AJ (1985)
VP 2.68E-004	mm Hg	25	EST	NEELY,WB & BLAU,GE (1985)
DC	pKa			
HL 6.80E-006	atm m3/mol	25	EST	MEYLAN,WM & HOWARD,PH (1991)
OH 1.25E-011	cm3/molc sec	25	EST	MEYLAN,WM & HOWARD,PH (1993)

CAS #: 023384-08-9 — M-BROMO CF3-METHANESULFONANILIDE

Formula: $C_7H_5BrF_3NO_2S$
Mol Weight: 304.09
MP (deg C): · FP (deg C):
BP (deg C):
BP pressure (mm Hg):

Property/Value	Units	Temp	Data Type	Reference
WS 6.35E+000	mg/L	25	EST	MEYLAN,WM ET AL. (1996)
logP 3.82			EXP	HANSCH,C & LEO,AJ (1985)
VP 9.05E-005	mm Hg	25	EST	NEELY,WB & BLAU,GE (1985)
DC	pKa			
HL 3.66E-006	atm m3/mol	25	EST	MEYLAN,WM & HOWARD,PH (1991)
OH 2.69E-011	cm3/molc sec	25	EST	MEYLAN,WM & HOWARD,PH (1993)

CAS #: 023384-10-3 — P-IODO CF3-METHANESULFONANILIDE

Formula: $C_7H_5F_3INO_2S$
Mol Weight: 351.09
MP (deg C): · FP (deg C):
BP (deg C):
BP pressure (mm Hg):

Property/Value	Units	Temp	Data Type	Reference
WS 1.09E+000	mg/L	25	EST	MEYLAN,WM ET AL. (1996)
logP 4.39			EXP	HANSCH,C & LEO,AJ (1985)
VP 2.56E-005	mm Hg	25	EST	NEELY,WB & BLAU,GE (1985)
DC	pKa			
HL 2.13E-006	atm m3/mol	25	EST	MEYLAN,WM & HOWARD,PH (1991)
OH 1.41E-011	cm3/molc sec	25	EST	MEYLAN,WM & HOWARD,PH (1993)

CAS #: 023384-11-4 — M-CF3 CF3-METHANESULFONANILIDE

Formula: $C_8H_5F_6NO_2S$
Mol Weight: 293.19
MP (deg C): · FP (deg C):
BP (deg C):
BP pressure (mm Hg):

Property/Value	Units	Temp	Data Type	Reference
WS 1.93E+000	mg/L	25	EST	MEYLAN,WM ET AL. (1996)
logP 4.50			EXP	HANSCH,C & LEO,AJ (1985)
VP 9.99E-004	mm Hg	25	EST	NEELY,WB & BLAU,GE (1985)
DC	pKa			
HL 7.98E-005	atm m3/mol	25	EST	MEYLAN,WM & HOWARD,PH (1991)
OH 6.46E-012	cm3/molc sec	25	EST	MEYLAN,WM & HOWARD,PH (1993)

CAS #: 023384-12-5 — P-CF3 CF3-METHANESULFONANILIDE

Formula: $C_8H_5F_6NO_2S$
Mol Weight: 293.19
MP (deg C): · FP (deg C):
BP (deg C):
BP pressure (mm Hg):

Property/Value	Units	Temp	Data Type	Reference
WS 2.05E+000	mg/L	25	EST	MEYLAN,WM ET AL. (1996)
logP 4.47			EXP	HANSCH,C & LEO,AJ (1985)
VP 9.99E-004	mm Hg	25	EST	NEELY,WB & BLAU,GE (1985)
DC	pKa			
HL 7.98E-005	atm m3/mol	25	EST	MEYLAN,WM & HOWARD,PH (1991)
OH 8.57E-012	cm3/molc sec	25	EST	MEYLAN,WM & HOWARD,PH (1993)

CAS #: 023384-22-7 — 2,4-DIF CF3-METHANESULFONANILIDE

Formula: $C_7H_4F_5NO_2S$

Mol Weight: 261.17

MP (deg C):　　FP (deg C):

BP (deg C):

BP pressure (mm Hg):

Property/Value	Units	Temp	Data Type	Reference
WS 4.99E+001	mg/L	25	EST	MEYLAN,WM ET AL. (1996)
logP 3.06			EXP	HANSCH,C & LEO,AJ (1985)
VP 2.12E-003	mm Hg	25	EST	NEELY,WB & BLAU,GE (1985)
DC	pKa			
HL 1.25E-005	atm m3/mol	25	EST	MEYLAN,WM & HOWARD,PH (1991)
OH 5.01E-012	cm3/molc sec	25	EST	MEYLAN,WM & HOWARD,PH (1993)

CAS #: 023384-33-0 — M-MEO CF3-METHANESULFONANILIDE

Formula: $C_8H_8F_3NO_3S$

Mol Weight: 255.22

MP (deg C):　　FP (deg C):

BP (deg C):

BP pressure (mm Hg):

Property/Value	Units	Temp	Data Type	Reference
WS 4.70E+001	mg/L	25	EST	MEYLAN,WM ET AL. (1996)
logP 3.13			EXP	HANSCH,C & LEO,AJ (1985)
VP 1.80E-004	mm Hg	25	EST	NEELY,WB & BLAU,GE (1985)
DC	pKa			
HL 5.43E-007	atm m3/mol	25	EST	MEYLAN,WM & HOWARD,PH (1991)
OH 2.01E-010	cm3/molc sec	25	EST	MEYLAN,WM & HOWARD,PH (1993)

CAS #: 023384-34-1 — P-MEO CF3-METHANESULFONANILIDE

Formula: $C_8H_8F_3NO_3S$

Mol Weight: 255.22

MP (deg C):　　FP (deg C):

BP (deg C):

BP pressure (mm Hg):

Property/Value	Units	Temp	Data Type	Reference
WS 6.07E+001	mg/L	25	EST	MEYLAN,WM ET AL. (1996)
logP 3.00			EXP	HANSCH,C & LEO,AJ (1985)
VP 1.80E-004	mm Hg	25	EST	NEELY,WB & BLAU,GE (1985)
DC	pKa			
HL 5.43E-007	atm m3/mol	25	EST	MEYLAN,WM & HOWARD,PH (1991)
OH 3.77E-011	cm3/molc sec	25	EST	MEYLAN,WM & HOWARD,PH (1993)

CAS #: 023401-43-6 — DL-VALINE, N-[1-OXO-5-(1H-PURIN-6-YLTHIO)PENTYL]

Formula: $C_{17}H_{25}N_5O_3S$

Mol Weight: 379.48

MP (deg C):　　FP (deg C):

BP (deg C):

BP pressure (mm Hg):

Property/Value	Units	Temp	Data Type	Reference
WS 9.97E+001	mg/L	25	EST	MEYLAN,WM ET AL. (1996)
logP 1.89			EXP	HANSCH,C ET AL. (1995)
VP 3.11E-014	mm Hg	25	EST	NEELY,WB & BLAU,GE (1985)
DC	pKa			
HL 2.59E-017	atm m3/mol	25	EST	MEYLAN,WM & HOWARD,PH (1991)
OH 2.33E-010	cm3/molc sec	25	EST	MEYLAN,WM & HOWARD,PH (1993)

CAS #: 023404-73-1 — L-ASPARTIC ACID, N-[1-OXO-5-(7H-PURIN-6-YLTHIO)P

Formula: $C_{18}H_{25}N_5O_5S$

Mol Weight: 423.49

MP (deg C):　　FP (deg C):

BP (deg C):

BP pressure (mm Hg):

Property/Value	Units	Temp	Data Type	Reference
WS 1.60E+002	mg/L	25	EST	MEYLAN,WM ET AL. (1996)
logP 1.33			EXP	SANGSTER,J (1993)
VP 3.90E-015	mm Hg	25	EST	NEELY,WB & BLAU,GE (1985)
DC	pKa			
HL 4.65E-020	atm m3/mol	25	EST	MEYLAN,WM & HOWARD,PH (1991)
OH 2.33E-010	cm3/molc sec	25	EST	MEYLAN,WM & HOWARD,PH (1993)

CAS #: 023409-17-8 — N-ME-2-ETHOXYPHENYLCARBAMATE

Formula: $C_{10}H_{13}NO_3$

Mol Weight: 195.22

MP (deg C):　　FP (deg C):

BP (deg C):

BP pressure (mm Hg):

Property/Value	Units	Temp	Data Type	Reference
WS 4.04E+003	mg/L	25	EST	MEYLAN,WM ET AL. (1996)
logP 1.24			EXP	HANSCH,C & LEO,AJ (1985)
VP 1.79E-003	mm Hg	25	EST	NEELY,WB & BLAU,GE (1985)
DC	pKa			
HL 2.52E-009	atm m3/mol	25	EST	MEYLAN,WM & HOWARD,PH (1991)
OH 2.80E-011	cm3/molc sec	25	EST	MEYLAN,WM & HOWARD,PH (1993)

CAS #: 023422-53-9 — FORMETANATE HYDROCHLORIDE

Formula: $C_{11}H_{16}ClN_3O_2$

Mol Weight: 257.72

MP (deg C):　　FP (deg C):

BP (deg C):

BP pressure (mm Hg):

Property/Value	Units	Temp	Data Type	Reference
WS 5.00E+005	mg/L	25	EXP	SHIU,WY ET AL. (1990)
logP -0.64			EST	MEYLAN,WM & HOWARD,PH (1995)
VP 6.80E-010	mm Hg		EST	LYMAN,WJ ET AL. (1990)
DC	pKa			
HL 2.27E-019	atm m3/mol	25	EST	MEYLAM,WM & HOWARD,PH (1991)
OH 7.52E-011	cm3/molc sec	25	EST	MEYLAN,WM & HOWARD,PH (1993)

CAS #: 023434-86-8 — 1H-INDENE-1,3(2H)-DIONE, 5-[3-OXO-3-(1-PIPERIDIN

Formula: $C_{15}H_{17}NO_3$

Mol Weight: 259.31

MP (deg C):　　FP (deg C):

BP (deg C):

BP pressure (mm Hg):

Property/Value	Units	Temp	Data Type	Reference
WS 8.20E+001	mg/L	25	EST	MEYLAN,WM ET AL. (1996)
logP 2.82			EXP	HANSCH,C ET AL. (1995)
VP 7.52E-007	mm Hg	25	EST	NEELY,WB & BLAU,GE (1985)
DC	pKa			
HL 1.53E-012	atm m3/mol	25	EST	MEYLAN,WM & HOWARD,PH (1991)
OH 6.59E-011	cm3/molc sec	25	EST	MEYLAN,WM & HOWARD,PH (1993)

CAS #: 023436-19-3				1-ISOBUTOXY-2-PROPANOL

Formula: $C_7H_{16}O_2$

Mol Weight: 132.20

MP (deg C): FP (deg C):

BP (deg C):

BP pressure (mm Hg):

Property/ Value	Units	Temp	Data Type	Reference
WS 4.87E+004	mg/L	25	EST	MEYLAN,WM ET AL. (1996)
logP 0.91			EST	MEYLAN,WM & HOWARD,PH (1995)
VP 1.30E+000	mm Hg	25	EXP	FLICK,EW (1991)
DC	pKa			
HL 1.30E-007	atm m3/mol	25	EST	MEYLAN,WM & HOWARD,PH (1991)
OH 2.98E-011	cm3/molc sec	25	EST	MEYLAN,WM & HOWARD,PH (1993)

CAS #: 023441-75-0				2(1H)-QUINAZOLINONE, 4-PHENYL-

Formula: $C_{14}H_{10}N_2O$

Mol Weight: 222.25

MP (deg C): FP (deg C):

BP (deg C):

BP pressure (mm Hg):

Property/ Value	Units	Temp	Data Type	Reference
WS 7.83E+002	mg/L	25	EST	MEYLAN,WM ET AL. (1996)
logP 1.91			EXP	SANGSTER,J (1993)
VP 2.91E-007	mm Hg	25	EST	NEELY,WB & BLAU,GE (1985)
DC	pKa			
HL 3.81E-009	atm m3/mol	25	EST	MEYLAN,WM & HOWARD,PH (1991)
OH 5.55E-011	cm3/molc sec	25	EST	MEYLAN,WM & HOWARD,PH (1993)

CAS #: 023443-10-9				4H-PYRIDO(1,2-A)PYRIMIDIN-4-ONE

Formula: $C_8H_8N_2O$

Mol Weight: 148.17

MP (deg C): FP (deg C):

BP (deg C):

BP pressure (mm Hg):

Property/ Value	Units	Temp	Data Type	Reference
WS 5.32E+004	mg/L	25	EST	MEYLAN,WM ET AL. (1996)
logP 0.20			EXP	HANSCH,C & LEO,AJ (1985)
VP 2.70E-003	mm Hg	25	EST	NEELY,WB & BLAU,GE (1985)
DC	pKa			
HL	atm m3/mol			
OH 7.14E-012	cm3/molc sec	25	EST	MEYLAN,WM & HOWARD,PH (1993)

CAS #: 023443-11-0				4H-PYRIDO(1,2-A)PYRIMIDIN-4-ONE,6-ME

Formula: $C_9H_{10}N_2O$

Mol Weight: 162.19

MP (deg C): FP (deg C):

BP (deg C):

BP pressure (mm Hg):

Property/ Value	Units	Temp	Data Type	Reference
WS 1.59E+004	mg/L	25	EST	MEYLAN,WM ET AL. (1996)
logP 0.74			EXP	HANSCH,C & LEO,AJ (1985)
VP 8.78E-004	mm Hg	25	EST	NEELY,WB & BLAU,GE (1985)
DC	pKa			
HL	atm m3/mol			
OH 7.87E-012	cm3/molc sec	25	EST	MEYLAN,WM & HOWARD,PH (1993)

CAS #: 023443-12-1				4H-PYRIDO[1,2-A]PYRIMIDIN-4-ONE, 9-METHYL-

Formula: $C_9H_8N_2O$

Mol Weight: 160.18

MP (deg C): FP (deg C):

BP (deg C):

BP pressure (mm Hg):

Property/ Value	Units	Temp	Data Type	Reference
WS 1.45E+004	mg/L	25	EST	MEYLAN,WM ET AL. (1996)
logP 0.79			EXP	SANGSTER,J (1994)
VP 1.60E-004	mm Hg	25	EST	NEELY,WB & BLAU,GE (1985)
DC	pKa			
HL 2.35E-009	atm m3/mol	25	EST	MEYLAN,WM & HOWARD,PH (1991)
OH 1.27E-010	cm3/molc sec	25	EST	MEYLAN,WM & HOWARD,PH (1993)

CAS #: 023443-20-1				4H-PYRIDO[1,2-A]PYRIMIDIN-4-ONE, 7-METHYL-

Formula: $C_9H_8N_2O$

Mol Weight: 160.18

MP (deg C): FP (deg C):

BP (deg C):

BP pressure (mm Hg):

Property/ Value	Units	Temp	Data Type	Reference
WS 2.27E+004	mg/L	25	EST	MEYLAN,WM ET AL. (1996)
logP 0.61			EXP	SANGSTER,J (1994)
VP 1.60E-004	mm Hg	25	EST	NEELY,WB & BLAU,GE (1985)
DC	pKa			
HL 2.35E-009	atm m3/mol	25	EST	MEYLAN,WM & HOWARD,PH (1991)
OH 1.27E-010	cm3/molc sec	25	EST	MEYLAN,WM & HOWARD,PH (1993)

CAS #: 023456-95-3				3,3-DIME-1-(4-CYANOPHENYL)TRIAZENE

Formula: $C_9H_{10}N_4$

Mol Weight: 174.21

MP (deg C): FP (deg C):

BP (deg C):

BP pressure (mm Hg):

Property/ Value	Units	Temp	Data Type	Reference
WS 2.90E+002	mg/L	25	EST	MEYLAN,WM ET AL. (1996)
logP 2.39			EXP	HANSCH,C & LEO,AJ (1985)
VP 1.72E-003	mm Hg	25	EST	NEELY,WB & BLAU,GE (1985)
DC	pKa			
HL 5.19E-009	atm m3/mol	25	EST	MEYLAN,WM & HOWARD,PH (1991)
OH 2.63E-012	cm3/molc sec	25	EST	MEYLAN,WM & HOWARD,PH (1993)

CAS #: 023478-26-4				BENZENEACETAMIDE, N-(3-HYDROXYPHENYL)-

Formula: $C_{14}H_{13}NO_2$

Mol Weight: 227.27

MP (deg C): FP (deg C):

BP (deg C):

BP pressure (mm Hg):

Property/ Value	Units	Temp	Data Type	Reference
WS 4.00E+002	mg/L	25	EST	MEYLAN,WM ET AL. (1996)
logP 2.22			EXP	YAMAGAMI,C ET AL. (1984)
VP 9.97E-009	mm Hg	25	EST	NEELY,WB & BLAU,GE (1985)
DC	pKa			
HL 5.18E-014	atm m3/mol	25	EST	MEYLAN,WM & HOWARD,PH (1991)
OH 2.01E-010	cm3/molc sec	25	EST	MEYLAN,WM & HOWARD,PH (1993)

CAS #: 023484-11-9				4-ACETAMIDOQUINOLINE-1-OXIDE

Formula: $C_{11}H_{12}N_2O_2$

Mol Weight: 204.23

MP (deg C):   FP (deg C):

BP (deg C):

BP pressure (mm Hg):

Property/Value	Units	Temp	Data Type	Reference
WS 5.29E+004	mg/L	25	EST	MEYLAN,WM ET AL. (1996)
logP -0.11			EXP	HANSCH,C & LEO,AJ (1985)
VP 2.66E-008	mm Hg	25	EST	NEELY,WB & BLAU,GE (1985)
DC	pKa			
HL 7.88E-018	atm m3/mol	25	EST	MEYLAN,WM & HOWARD,PH (1991)
OH 9.18E-011	cm3/molc sec	25	EST	MEYLAN,WM & HOWARD,PH (1993)

CAS #: 023505-41-1				PIRIMIPHOS ETHYL

Formula: $C_{13}H_{24}N_3O_3PS$

Mol Weight: 333.39

MP (deg C): 15   FP (deg C):

BP (deg C):

BP pressure (mm Hg):

Property/Value	Units	Temp	Data Type	Reference
WS 3.96E+000	mg/L	20	EXP	YALKOWSKY,SH & DANNENFELSER,RM (1992)
logP 4.85			EXP	HANSCH,C ET AL. (1995)
VP 2.90E-004	mm Hg	25	EXP	MERCK INDEX (1989)
DC	pKa			
HL 3.21E-005	atm m3/mol	25	EST	VP/WSOL
OH 1.93E-010	cm3/molc sec	25	EST	MEYLAN,WM & HOWARD,PH (1993)

CAS #: 023549-54-4				2-METHYLPROPANOYLUREA

Formula: $C_5H_{10}N_2O_2$

Mol Weight: 130.15

MP (deg C):   FP (deg C):

BP (deg C):

BP pressure (mm Hg):

Property/Value	Units	Temp	Data Type	Reference
WS 4.16E+003	mg/L	25	EST	MEYLAN,WM ET AL. (1996)
logP 0.04			EXP	HANSCH,C & LEO,AJ (1985)
VP 1.10E-005	mm Hg	25	EST	NEELY,WB & BLAU,GE (1985)
DC	pKa			
HL 1.90E-010	atm m3/mol	25	EST	MEYLAN,WM & HOWARD,PH (1991)
OH 4.52E-012	cm3/molc sec	25	EST	MEYLAN,WM & HOWARD,PH (1993)

CAS #: 023564-05-8				THIOPHANATE-METHYL

Formula: $C_{12}H_{14}N_4O_4S_2$

Mol Weight: 342.40

MP (deg C): 172 dec   FP (deg C):

BP (deg C):

BP pressure (mm Hg):

Property/Value	Units	Temp	Data Type	Reference
WS 4.39E+002	mg/L	25	EST	MEYLAN,WM ET AL. (1996)
logP 1.40			EXP	HANSCH,C ET AL. (1995)
VP 6.97E-009	mm Hg	25	EST	NEELY,WB & BLAU,GE (1985)
DC 7.28	pKa		EXP	TOMLIN,C (1994)
HL 2.94E-013	atm m3/mol	25	EST	MEYLAN,WM & HOWARD,PH (1991)
OH 8.13E-011	cm3/molc sec	25	EST	MEYLAN,WM & HOWARD,PH (1993)

CAS #: 023571-34-8				1-ME-2-NO2-5-IPR IMIDAZOLE

Formula: $C_7H_{13}N_3O_2$

Mol Weight: 171.20

MP (deg C):   FP (deg C):

BP (deg C):

BP pressure (mm Hg):

Property/Value	Units	Temp	Data Type	Reference
WS 2.96E+003	mg/L	25	EST	MEYLAN,WM ET AL. (1996)
logP 1.09			EXP	HANSCH,C & LEO,AJ (1985)
VP 1.49E-004	mm Hg	25	EST	NEELY,WB & BLAU,GE (1985)
DC	pKa			
HL 6.15E-007	atm m3/mol	25	EST	MEYLAN,WM & HOWARD,PH (1991)
OH 1.31E-011	cm3/molc sec	25	EST	MEYLAN,WM & HOWARD,PH (1993)

CAS #: 023571-38-2				1H-IMIDAZOLE-1-ETHANOL, 5-METHYL-2-NITRO-

Formula: $C_6H_9N_3O_3$

Mol Weight: 171.16

MP (deg C):   FP (deg C):

BP (deg C):

BP pressure (mm Hg):

Property/Value	Units	Temp	Data Type	Reference
WS 1.74E+004	mg/L	25	EST	MEYLAN,WM ET AL. (1996)
logP 0.18			EXP	SANGSTER,J (1993)
VP 3.05E-007	mm Hg	25	EST	NEELY,WB & BLAU,GE (1985)
DC	pKa			
HL 1.69E-011	atm m3/mol	25	EST	MEYLAN,WM & HOWARD,PH (1991)
OH 1.72E-011	cm3/molc sec	25	EST	MEYLAN,WM & HOWARD,PH (1993)

CAS #: 023597-82-2				N-HEXYL NICOTINOATE

Formula: $C_{12}H_{17}NO_2$

Mol Weight: 207.27

MP (deg C):   FP (deg C):

BP (deg C): 147

BP pressure (mm Hg): 2.00E+000

Property/Value	Units	Temp	Data Type	Reference
WS 4.04E+001	mg/L	25	EST	MEYLAN,WM ET AL. (1996)
logP 3.51			EXP	HOUK,J & GUY,RH (1988)
VP 9.95E-004	mm Hg	25	EST	NEELY,WB & BLAU,GE (1985)
DC	pKa			
HL 1.87E-007	atm m3/mol	25	EST	MEYLAN,WM & HOWARD,PH (1991)
OH 7.51E-012	cm3/molc sec	25	EST	MEYLAN,WM & HOWARD,PH (1993)

CAS #: 023611-75-8				2-CHLORO-6-CO2ME-PYRAZINE

Formula: $C_6H_5ClN_2O_2$

Mol Weight: 172.57

MP (deg C):   FP (deg C):

BP (deg C):

BP pressure (mm Hg):

Property/Value	Units	Temp	Data Type	Reference
WS 2.37E+004	mg/L	25	EST	MEYLAN,WM ET AL. (1996)
logP 0.47			EXP	YAMAGAMI,C & TAKAO,N (1991)
VP 1.84E-002	mm Hg	25	EST	NEELY,WB & BLAU,GE (1985)
DC	pKa			
HL 8.79E-007	atm m3/mol	25	EST	MEYLAN,WM & HOWARD,PH (1991)
OH 2.53E-013	cm3/molc sec	25	EST	MEYLAN,WM & HOWARD,PH (1993)

023676-09-7 — ETHYL 4-ETHOXYBENZOATE

Formula: $C_{11}H_{14}O_3$
Mol Weight: 194.23
MP (deg C):
FP (deg C):
BP (deg C): 275
BP pressure (mm Hg):

Property/Value	Units	Temp	Data Type	Reference
WS 3.94E+001	mg/L	25	EST	MEYLAN,WM ET AL. (1996)
logP 3.60			EXP	TSCATS
VP 6.45E-003	mm Hg	25	EST	NEELY,WB & BLAU,GE (1985)
DC	pKa			
HL 3.62E-006	atm m3/mol	25	EST	MEYLAN,WM & HOWARD,PH (1991)
OH 1.48E-011	cm3/molc sec	25	EST	MEYLAN,WM & HOWARD,PH (1993)

023690-13-3 — 2,4-DIMETHOXYAMPHETAMINE

Formula: $C_{11}H_{17}NO_2$
Mol Weight: 195.26
MP (deg C):
FP (deg C):
BP (deg C):
BP pressure (mm Hg):

Property/Value	Units	Temp	Data Type	Reference
WS 1.51E+004	mg/L	25	EST	MEYLAN,WM ET AL. (1996)
logP 1.75			EXP	HANSCH,C & LEO,AJ (1985)
VP 1.83E-003	mm Hg	25	EST	NEELY,WB & BLAU,GE (1985)
DC	pKa			
HL 3.77E-009	atm m3/mol	25	EST	MEYLAN,WM & HOWARD,PH (1991)
OH 2.46E-010	cm3/molc sec	25	EST	MEYLAN,WM & HOWARD,PH (1993)

023694-81-7 — 2-PROPANOL, 1-[(1-METHYLETHYL)AMINO]-3-[(2-METHY

Formula: $C_{15}H_{22}N_2O_2$
Mol Weight: 262.35
MP (deg C): 95-97
FP (deg C):
BP (deg C):
BP pressure (mm Hg):

Property/Value	Units	Temp	Data Type	Reference
WS 2.23E+003	mg/L	25	EST	MEYLAN,WM ET AL. (1996)
logP 2.30			EXP	RECANATINI,M (1992)
VP 4.96E-009	mm Hg	25	EST	NEELY,WB & BLAU,GE (1985)
DC	pKa			
HL 1.48E-015	atm m3/mol	25	EST	MEYLAN,WM & HOWARD,PH (1991)
OH 3.10E-010	cm3/molc sec	25	EST	MEYLAN,WM & HOWARD,PH (1993)

023702-21-8 — 1-ACETAMIDOAZULENE

Formula: $C_{12}H_{11}NO$
Mol Weight: 185.23
MP (deg C):
FP (deg C):
BP (deg C):
BP pressure (mm Hg):

Property/Value	Units	Temp	Data Type	Reference
WS 3.96E+002	mg/L	25	EST	MEYLAN,WM ET AL. (1996)
logP 2.48			EXP	HANSCH,C & LEO,AJ (1985)
VP 7.73E-006	mm Hg	25	EST	NEELY,WB & BLAU,GE (1985)
DC	pKa			
HL 2.74E-008	atm m3/mol	25	EST	MEYLAN,WM & HOWARD,PH (1991)
OH 2.48E-010	cm3/molc sec	25	EST	MEYLAN,WM & HOWARD,PH (1993)

023702-98-9 — A-(2-PIPERIDYL)PHENYLCARBINOL

Formula: $C_{12}H_{17}NO$
Mol Weight: 191.28
MP (deg C):
FP (deg C):
BP (deg C):
BP pressure (mm Hg):

Property/Value	Units	Temp	Data Type	Reference
WS 3.98E+004	mg/L	25	EST	MEYLAN,WM ET AL. (1996)
logP 1.28			EXP	HANSCH,C & LEO,AJ (1985)
VP 6.43E-006	mm Hg	25	EST	NEELY,WB & BLAU,GE (1985)
DC	pKa			
HL 6.72E-011	atm m3/mol	25	EST	MEYLAN,WM & HOWARD,PH (1991)
OH 1.18E-010	cm3/molc sec	25	EST	MEYLAN,WM & HOWARD,PH (1993)

023742-02-1 — A,A-DIET-PHENYLACETAMIDE,3,4-DIOXYME

Formula: $C_{13}H_{17}NO_3$
Mol Weight: 235.29
MP (deg C):
FP (deg C):
BP (deg C):
BP pressure (mm Hg):

Property/Value	Units	Temp	Data Type	Reference
WS 1.25E+003	mg/L	25	EST	MEYLAN,WM ET AL. (1996)
logP 1.59			EXP	HANSCH,C & LEO,AJ (1985)
VP 6.19E-007	mm Hg	25	EST	NEELY,WB & BLAU,GE (1985)
DC	pKa			
HL 3.26E-012	atm m3/mol	25	EST	MEYLAN,WM & HOWARD,PH (1991)
OH 5.33E-011	cm3/molc sec	25	EST	MEYLAN,WM & HOWARD,PH (1993)

023749-65-7 — 2,4,6-TRICHLOROTOLUENE

Formula: $C_7H_5Cl_3$
Mol Weight: 195.48
MP (deg C):
FP (deg C):
BP (deg C):
BP pressure (mm Hg):

Property/Value	Units	Temp	Data Type	Reference
WS 6.97E+000	mg/L	25	EST	MEYLAN,WM ET AL. (1996)
logP 4.47			EST	MEYLAN,WM & HOWARD,PH (1995)
VP 7.32E-002	mm Hg	25	EST	NEELY,WB & BLAU,GE (1985)
DC	pKa			
HL 2.42E-003	atm m3/mol	25	EST	MEYLAN,WM & HOWARD,PH (1991)
OH 9.68E-013	cm3/molc sec	25	EST	MEYLAN,WM & HOWARD,PH (1993)

023757-42-8 — 1H-IMIDAZOLE-4-AMINE, 2,5-DIHYDRO-2,2,5,5-TETRAK

Formula: $C_7H_3F_{12}N_3$
Mol Weight: 357.10
MP (deg C):
FP (deg C):
BP (deg C):
BP pressure (mm Hg):

Property/Value	Units	Temp	Data Type	Reference
WS 7.73E+000	mg/L	25	EST	MEYLAN,WM ET AL. (1996)
logP 3.35			EXP	HANSCH,C ET AL. (1995)
VP 9.84E-001	mm Hg	25	EST	NEELY,WB & BLAU,GE (1985)
DC	pKa			
HL 1.88E-004	atm m3/mol	25	EST	MEYLAN,WM & HOWARD,PH (1991)
OH 0.00E+000	cm3/molc sec	25	EST	MEYLAN,WM & HOWARD,PH (1993)

CAS #: 023795-32-6				N,N-DI-I-BU-3,4-(-OCH2O-)CINNAMAMIDE

Formula: $C_{18}H_{25}NO_3$

Mol Weight: 303.40

MP (deg C):　　FP (deg C):

BP (deg C):

BP pressure (mm Hg):

Property/Value	Units	Temp	Data Type	Reference
WS 2.17E+000	mg/L	25	EST	MEYLAN,WM ET AL. (1996)
logP 4.37			EXP	HANSCH,C & LEO,AJ (1985)
VP 3.30E-007	mm Hg	25	EST	NEELY,WB & BLAU,GE (1985)
DC	pKa			
HL 8.10E-012	atm m3/mol	25	EST	MEYLAN,WM & HOWARD,PH (1991)
OH 6.98E-011	cm3/molc sec	25	EST	MEYLAN,WM & HOWARD,PH (1993)

CAS #: 023813-24-3				PYRAZINECARBOXYLIC ACID, 6-METHOXY-, METHYL ESTE

Formula: $C_7H_8N_2O_3$

Mol Weight: 168.15

MP (deg C):　　FP (deg C):

BP (deg C):

BP pressure (mm Hg):

Property/Value	Units	Temp	Data Type	Reference
WS 1.62E+004	mg/L	25	EST	MEYLAN,WM ET AL. (1996)
logP 0.69			EXP	YAMAGAMI,C ET AL. (1991)
VP 1.14E-002	mm Hg	25	EST	NEELY,WB & BLAU,GE (1985)
DC	pKa			
HL 1.01E-008	atm m3/mol	25	EST	MEYLAN,WM & HOWARD,PH (1991)
OH 1.61E-012	cm3/molc sec	25	EST	MEYLAN,WM & HOWARD,PH (1993)

CAS #: 023815-28-3				BENZENESULFONAMIDE, 3,4-DICHLORO-

Formula: $C_6H_5Cl_2NO_2S$

Mol Weight: 226.08

MP (deg C):　　FP (deg C):

BP (deg C):

BP pressure (mm Hg):

Property/Value	Units	Temp	Data Type	Reference
WS 1.88E+003	mg/L	25	EST	MEYLAN,WM ET AL. (1996)
logP 1.44			EXP	SANGSTER,J (1993)
VP 3.18E-005	mm Hg	25	EST	NEELY,WB & BLAU,GE (1985)
DC	pKa			
HL 2.32E-007	atm m3/mol	25	EST	MEYLAN,WM & HOWARD,PH (1991)
OH 8.56E-014	cm3/molc sec	25	EST	MEYLAN,WM & HOWARD,PH (1993)

CAS #: 023830-88-8				2-(2,4-DICL PH IMINO)IMIDAZOLIDINE

Formula: $C_9H_9Cl_2N_3$

Mol Weight: 230.10

MP (deg C):　　FP (deg C):

BP (deg C):

BP pressure (mm Hg):

Property/Value	Units	Temp	Data Type	Reference
WS 6.44E+002	mg/L	25	EST	MEYLAN,WM ET AL. (1996)
logP 1.96			EXP	HANSCH,C & LEO,AJ (1985)
VP 2.76E-007	mm Hg	25	EST	NEELY,WB & BLAU,GE (1985)
DC	pKa			
HL 1.51E-011	atm m3/mol	25	EST	MEYLAN,WM & HOWARD,PH (1991)
OH 7.61E-011	cm3/molc sec	25	EST	MEYLAN,WM & HOWARD,PH (1993)

CAS #: 023885-56-5				P-PHENYL-AMPHENICOL

Formula: $C_{17}H_{17}Cl_2NO_3$

Mol Weight: 354.24

MP (deg C):　　FP (deg C):

BP (deg C):

BP pressure (mm Hg):

Property/Value	Units	Temp	Data Type	Reference
WS 2.11E+001	mg/L	25	EST	MEYLAN,WM ET AL. (1996)
logP 2.86			EXP	HANSCH,C & LEO,AJ (1985)
VP 7.56E-015	mm Hg	25	EST	NEELY,WB & BLAU,GE (1985)
DC	pKa			
HL 4.45E-017	atm m3/mol	25	EST	MEYLAN,WM & HOWARD,PH (1991)
OH 3.79E-011	cm3/molc sec	25	EST	MEYLAN,WM & HOWARD,PH (1993)

CAS #: 023885-59-8				P-BROMO-AMPHENICOL

Formula: $C_{11}H_{12}BrCl_2NO_3$

Mol Weight: 357.04

MP (deg C):　　FP (deg C):

BP (deg C):

BP pressure (mm Hg):

Property/Value	Units	Temp	Data Type	Reference
WS 1.42E+002	mg/L	25	EST	MEYLAN,WM ET AL. (1996)
logP 1.87			EXP	HANSCH,C & LEO,AJ (1985)
VP 2.58E-012	mm Hg	25	EST	NEELY,WB & BLAU,GE (1985)
DC	pKa			
HL 2.31E-016	atm m3/mol	25	EST	MEYLAN,WM & HOWARD,PH (1991)
OH 3.19E-011	cm3/molc sec	25	EST	MEYLAN,WM & HOWARD,PH (1993)

CAS #: 023885-61-2				P-CYANO-AMPHENICOL

Formula: $C_{12}H_{12}Cl_2N_2O_3$

Mol Weight: 303.15

MP (deg C):　　FP (deg C):

BP (deg C):

BP pressure (mm Hg):

Property/Value	Units	Temp	Data Type	Reference
WS 1.38E+003	mg/L	25	EST	MEYLAN,WM ET AL. (1996)
logP 0.78			EXP	HANSCH,C & LEO,AJ (1985)
VP 6.51E-013	mm Hg	25	EST	NEELY,WB & BLAU,GE (1985)
DC	pKa			
HL 5.60E-018	atm m3/mol	25	EST	MEYLAN,WM & HOWARD,PH (1991)
OH 3.12E-011	cm3/molc sec	25	EST	MEYLAN,WM & HOWARD,PH (1993)

CAS #: 023885-69-0				ETHYLAMPHENICOL

Formula: $C_{13}H_{18}N_2O_5$

Mol Weight: 282.30

MP (deg C):　　FP (deg C):

BP (deg C):

BP pressure (mm Hg):

Property/Value	Units	Temp	Data Type	Reference
WS 1.14E+003	mg/L	25	EST	MEYLAN,WM ET AL. (1996)
logP 0.87			EXP	HANSCH,C & LEO,AJ (1985)
VP 1.59E-012	mm Hg	25	EST	NEELY,WB & BLAU,GE (1985)
DC	pKa			
HL 3.25E-017	atm m3/mol	25	EST	MEYLAN,WM & HOWARD,PH (1991)
OH 3.54E-011	cm3/molc sec	25	EST	MEYLAN,WM & HOWARD,PH (1993)

CAS #: 023885-71-4 — CYANAMPHENICOL

Formula: $C_{12}H_{13}N_3O_5$
Mol Weight: 279.25
MP (deg C):
FP (deg C):
BP (deg C):
BP pressure (mm Hg):

Property/Value	Units	Temp	Data Type	Reference
WS 2.70E+002	mg/L	25	EST	MEYLAN,WM ET AL. (1996)
logP -0.22			EXP	HANSCH,C ET AL. (1995)
VP 7.83E-014	mm Hg	25	EST	NEELY,WB & BLAU,GE (1985)
DC	pKa			
HL 7.65E-021	atm m3/mol	25	EST	MEYLAN,WM & HOWARD,PH (1991)
OH 3.09E-011	cm3/molc sec	25	EST	MEYLAN,WM & HOWARD,PH (1993)

CAS #: 023893-07-4 — ERYTHROMYCIN 4''-ACETATE

Formula: $C_{39}H_{69}NO_{14}$
Mol Weight: 775.98
MP (deg C):
FP (deg C):
BP (deg C):
BP pressure (mm Hg):

Property/Value	Units	Temp	Data Type	Reference
WS 4.08E-001	mg/L	25	EST	MEYLAN,WM ET AL. (1996)
logP 2.85			EXP	SANGSTER,J (1993)
VP 1.02E-027	mm Hg	25	EST	NEELY,WB & BLAU,GE (1985)
DC	pKa			
HL 3.53E-030	atm m3/mol	25	EST	MEYLAN,WM & HOWARD,PH (1991)
OH 3.49E-010	cm3/molc sec	25	EST	MEYLAN,WM & HOWARD,PH (1993)

CAS #: 023898-86-4 — PYRAZOL-3-ON,2(26DIME-4PYRIMD)15DIME

Formula: $C_{11}H_{14}N_4O$
Mol Weight: 218.26
MP (deg C):
FP (deg C):
BP (deg C):
BP pressure (mm Hg):

Property/Value	Units	Temp	Data Type	Reference
WS 1.07E+003	mg/L	25	EST	MEYLAN,WM ET AL. (1996)
logP 0.24			EXP	HANSCH,C & LEO,AJ (1985)
VP 2.00E-006	mm Hg	25	EST	NEELY,WB & BLAU,GE (1985)
DC	pKa			
HL 4.39E-013	atm m3/mol	25	EST	MEYLAN,WM & HOWARD,PH (1991)
OH 2.99E-011	cm3/molc sec	25	EST	MEYLAN,WM & HOWARD,PH (1993)

CAS #: 023898-89-7 — PYRAZOL-3-ON,2(2ME6MEO-4PYRIMD)15DIME

Formula: $C_{11}H_{14}N_4O_2$
Mol Weight: 234.26
MP (deg C):
FP (deg C):
BP (deg C):
BP pressure (mm Hg):

Property/Value	Units	Temp	Data Type	Reference
WS 7.29E+003	mg/L	25	EST	MEYLAN,WM ET AL. (1996)
logP 0.70			EXP	HANSCH,C & LEO,AJ (1985)
VP 8.88E-007	mm Hg	25	EST	NEELY,WB & BLAU,GE (1985)
DC	pKa			
HL 2.15E-013	atm m3/mol	25	EST	MEYLAN,WM & HOWARD,PH (1991)
OH 6.28E-011	cm3/molc sec	25	EST	MEYLAN,WM & HOWARD,PH (1993)

CAS #: 023898-90-0 — PYRAZOL-3-ON,2(2MEO-6ME-4PYRIMD)5-ME

Formula: $C_{10}H_{12}N_4O_2$
Mol Weight: 220.23
MP (deg C):
FP (deg C):
BP (deg C):
BP pressure (mm Hg):

Property/Value	Units	Temp	Data Type	Reference
WS 4.53E+003	mg/L	25	EST	MEYLAN,WM ET AL. (1996)
logP 1.03			EXP	HANSCH,C & LEO,AJ (1985)
VP 6.27E-007	mm Hg	25	EST	NEELY,WB & BLAU,GE (1985)
DC	pKa			
HL 9.77E-014	atm m3/mol	25	EST	MEYLAN,WM & HOWARD,PH (1991)
OH 1.25E-010	cm3/molc sec	25	EST	MEYLAN,WM & HOWARD,PH (1993)

CAS #: 023898-92-2 — PYRAZOL-3-ON,1(46DIME-2PYRIMD)25DIME

Formula: $C_{11}H_{14}N_4O$
Mol Weight: 218.26
MP (deg C):
FP (deg C):
BP (deg C):
BP pressure (mm Hg):

Property/Value	Units	Temp	Data Type	Reference
WS 1.29E+004	mg/L	25	EST	MEYLAN,WM ET AL. (1996)
logP 0.51			EXP	HANSCH,C & LEO,AJ (1985)
VP 2.00E-006	mm Hg	25	EST	NEELY,WB & BLAU,GE (1985)
DC	pKa			
HL 1.56E-009	atm m3/mol	25	EST	MEYLAN,WM & HOWARD,PH (1991)
OH 4.42E-011	cm3/molc sec	25	EST	MEYLAN,WM & HOWARD,PH (1993)

CAS #: 023898-95-5 — PYRAZOL-3-ON,1(2ME6MEO-4PYRIMD)25DIME

Formula: $C_{11}H_{14}N_4O_2$
Mol Weight: 234.26
MP (deg C):
FP (deg C):
BP (deg C):
BP pressure (mm Hg):

Property/Value	Units	Temp	Data Type	Reference
WS 1.37E+004	mg/L	25	EST	MEYLAN,WM ET AL. (1996)
logP 0.38			EXP	HANSCH,C & LEO,AJ (1985)
VP 8.88E-007	mm Hg	25	EST	NEELY,WB & BLAU,GE (1985)
DC	pKa			
HL 2.15E-013	atm m3/mol	25	EST	MEYLAN,WM & HOWARD,PH (1991)
OH 7.71E-011	cm3/molc sec	25	EST	MEYLAN,WM & HOWARD,PH (1993)

CAS #: 023903-41-5 — PYRIMIDINE,4(3MEO-5ME-1PYRAZOL)26DIME

Formula: $C_{11}H_{16}N_4O$
Mol Weight: 220.28
MP (deg C):
FP (deg C):
BP (deg C):
BP pressure (mm Hg):

Property/Value	Units	Temp	Data Type	Reference
WS 2.12E+002	mg/L	25	EST	MEYLAN,WM ET AL. (1996)
logP 2.60			EXP	HANSCH,C & LEO,AJ (1985)
VP 1.53E-005	mm Hg	25	EST	NEELY,WB & BLAU,GE (1985)
DC	pKa			
HL 5.78E-013	atm m3/mol	25	EST	MEYLAN,WM & HOWARD,PH (1991)
OH 2.01E-010	cm3/molc sec	25	EST	MEYLAN,WM & HOWARD,PH (1993)

CAS #: 023903-42-6 — PYRIMIDINE,6(5MEO3ME-1PYRAZOL)2ME4MEO

Formula: $C_{11}H_{16}N_4O_2$

Mol Weight: 236.28

MP (deg C): | FP (deg C):

BP (deg C):

BP pressure (mm Hg):

Property/Value	Units	Temp	Data Type	Reference
WS 1.40E+003	mg/L	25	EST	MEYLAN,WM ET AL. (1996)
logP 1.54			EXP	HANSCH,C & LEO,AJ (1985)
VP 6.09E-006	mm Hg	25	EST	NEELY,WB & BLAU,GE (1985)
DC	pKa			
HL 2.83E-013	atm m3/mol	25	EST	MEYLAN,WM & HOWARD,PH (1991)
OH 2.02E-010	cm3/molc sec	25	EST	MEYLAN,WM & HOWARD,PH (1993)

CAS #: 023905-77-3 — PYRIMIDINE,2(3MEO-5ME-1PYRAZOL)46DIME

Formula: $C_{11}H_{16}N_4O$

Mol Weight: 220.28

MP (deg C): | FP (deg C):

BP (deg C):

BP pressure (mm Hg):

Property/Value	Units	Temp	Data Type	Reference
WS 9.81E+002	mg/L	25	EST	MEYLAN,WM ET AL. (1996)
logP 1.82			EXP	HANSCH,C & LEO,AJ (1985)
VP 1.53E-005	mm Hg	25	EST	NEELY,WB & BLAU,GE (1985)
DC	pKa			
HL 5.78E-013	atm m3/mol	25	EST	MEYLAN,WM & HOWARD,PH (1991)
OH 2.01E-010	cm3/molc sec	25	EST	MEYLAN,WM & HOWARD,PH (1993)

CAS #: 023905-85-3 — PYRAZOL-3-ON,1(46DIMEO-2PYRIMDIN)5ME

Formula: $C_{10}H_{12}N_4O_3$

Mol Weight: 236.23

MP (deg C): | FP (deg C):

BP (deg C):

BP pressure (mm Hg):

Property/Value	Units	Temp	Data Type	Reference
WS 2.88E+003	mg/L	25	EST	MEYLAN,WM ET AL. (1996)
logP 1.16			EXP	HANSCH,C & LEO,AJ (1985)
VP 9.29E-008	mm Hg	25	EST	NEELY,WB & BLAU,GE (1985)
DC	pKa			
HL 1.70E-010	atm m3/mol	25	EST	MEYLAN,WM & HOWARD,PH (1991)
OH 2.04E-010	cm3/molc sec	25	EST	MEYLAN,WM & HOWARD,PH (1993)

CAS #: 023905-98-8 — PYRIMIDINE,6(3ME5MEO-1PYRAZOL)24DIMEO

Formula: $C_{11}H_{16}N_4O_3$

Mol Weight: 252.28

MP (deg C): | FP (deg C):

BP (deg C):

BP pressure (mm Hg):

Property/Value	Units	Temp	Data Type	Reference
WS 6.08E+002	mg/L	25	EST	MEYLAN,WM ET AL. (1996)
logP 1.86			EXP	HANSCH,C & LEO,AJ (1985)
VP 2.70E-006	mm Hg	25	EST	NEELY,WB & BLAU,GE (1985)
DC	pKa			
HL 1.38E-013	atm m3/mol	25	EST	MEYLAN,WM & HOWARD,PH (1991)
OH 2.03E-010	cm3/molc sec	25	EST	MEYLAN,WM & HOWARD,PH (1993)

CAS #: 023906-03-8 — PYRAZOL-3-ON,2(6MEO-4-PYRIMDIN)15DIME

Formula: $C_{10}H_{12}N_4O_2$

Mol Weight: 220.23

MP (deg C): | FP (deg C):

BP (deg C):

BP pressure (mm Hg):

Property/Value	Units	Temp	Data Type	Reference
WS 9.32E+002	mg/L	25	EST	MEYLAN,WM ET AL. (1996)
logP 0.30			EXP	HANSCH,C & LEO,AJ (1985)
VP 1.94E-006	mm Hg	25	EST	NEELY,WB & BLAU,GE (1985)
DC	pKa			
HL 1.94E-013	atm m3/mol	25	EST	MEYLAN,WM & HOWARD,PH (1991)
OH 3.66E-011	cm3/molc sec	25	EST	MEYLAN,WM & HOWARD,PH (1993)

CAS #: 023917-23-9 — PYRAZOL-3-ON,2(6MEO-4-PYRIMDINYL)5ME

Formula: $C_9H_{10}N_4O_2$

Mol Weight: 206.21

MP (deg C): | FP (deg C):

BP (deg C):

BP pressure (mm Hg):

Property/Value	Units	Temp	Data Type	Reference
WS 5.80E+003	mg/L	25	EST	MEYLAN,WM ET AL. (1996)
logP 0.99			EXP	HANSCH,C & LEO,AJ (1985)
VP 1.37E-006	mm Hg	25	EST	NEELY,WB & BLAU,GE (1985)
DC	pKa			
HL 8.86E-014	atm m3/mol	25	EST	MEYLAN,WM & HOWARD,PH (1991)
OH 9.84E-011	cm3/molc sec	25	EST	MEYLAN,WM & HOWARD,PH (1993)

CAS #: 023917-24-0 — PYRIMIDINE,2-(3ME-5MEO-1PYRAZOL)4MEO

Formula: $C_{10}H_{14}N_4O_2$

Mol Weight: 222.25

MP (deg C): | FP (deg C):

BP (deg C):

BP pressure (mm Hg):

Property/Value	Units	Temp	Data Type	Reference
WS 4.71E+003	mg/L	25	EST	MEYLAN,WM ET AL. (1996)
logP 1.01			EXP	HANSCH,C & LEO,AJ (1985)
VP 1.49E-005	mm Hg	25	EST	NEELY,WB & BLAU,GE (1985)
DC	pKa			
HL 2.56E-013	atm m3/mol	25	EST	MEYLAN,WM & HOWARD,PH (1991)
OH 2.02E-010	cm3/molc sec	25	EST	MEYLAN,WM & HOWARD,PH (1993)

CAS #: 023950-58-5 — PRONAMIDE

Formula: $C_{12}H_{11}Cl_2NO$

Mol Weight: 256.13

MP (deg C): 155-156 | FP (deg C):

BP (deg C):

BP pressure (mm Hg):

Property/Value	Units	Temp	Data Type	Reference
WS 1.50E+001	mg/L	25	EXP	YALKOWSKY,SH & DANNENFELSER,RM (1992)
logP 3.43			EXP	ELLINGTON,JJ & STANCIL,FEJR (1988)
VP 8.50E-005	mm Hg	25	EXP	MERCK INDEX (1989)
DC	pKa			
HL 1.91E-006	atm m3/mol	25	EST	VP/WSOL
OH 1.32E-011	cm3/molc sec	25	EST	MEYLAN,WM & HOWARD,PH (1993)

CAS #: 023972-41-0	2-PHENYLMORPHOLINE

Formula: $C_{10}H_{13}NO$

Mol Weight: 163.22

MP (deg C): FP (deg C):

BP (deg C):

BP pressure (mm Hg):

Property/Value	Units	Temp	Data Type	Reference
WS 7.60E+004	mg/L	25	EST	MEYLAN,WM ET AL. (1996)
logP 1.11			EXP	HANSCH,C & LEO,AJ (1985)
VP 4.31E-003	mm Hg	25	EST	NEELY,WB & BLAU,GE (1985)
DC	pKa			
HL 9.20E-009	atm m3/mol	25	EST	MEYLAN,WM & HOWARD,PH (1991)
OH 1.30E-010	cm3/molc sec	25	EST	MEYLAN,WM & HOWARD,PH (1993)

CAS #: 023976-66-1	2-(2,4,6-TRIBROMOPHENOXY)ETHANOL

Formula: $C_8H_7Br_3O_2$

Mol Weight: 374.87

MP (deg C): FP (deg C):

BP (deg C):

BP pressure (mm Hg):

Property/Value	Units	Temp	Data Type	Reference
WS 1.70E+001	mg/L	25	EST	MEYLAN,WM ET AL. (1996)
logP 3.42			EXP	HANSCH,C & LEO,AJ (1985)
VP 2.90E-007	mm Hg	25	EST	NEELY,WB & BLAU,GE (1985)
DC	pKa			
HL 9.78E-010	atm m3/mol	25	EST	MEYLAN,WM & HOWARD,PH (1991)
OH 1.16E-011	cm3/molc sec	25	EST	MEYLAN,WM & HOWARD,PH (1993)

CAS #: 024017-47-8	TRIAZOPHOS

Formula: $C_{13}H_{17}N_2O_3PS$

Mol Weight: 312.33

MP (deg C): 2-5 FP (deg C):

BP (deg C):

BP pressure (mm Hg):

Property/Value	Units	Temp	Data Type	Reference
WS 2.47E+001	mg/L	20	EXP	YALKOWSKY,SH & DANNENFELSER,RM (1992)
logP 3.55			EXP	HANSCH,C & LEO,AJ (1985)
VP 2.90E-006	mm Hg	30	EXP	TOMLIN,C (1994)
DC	pKa			
HL 4.84E-008	atm m3/mol	30	EST	VP/WSOL
OH 9.52E-011	cm3/molc sec	25	EST	MEYLAN,WM & HOWARD,PH (1993)

CAS #: 024096-53-5	OHRIC

Formula: $C_{10}H_7Cl_2NO_2$

Mol Weight: 244.08

MP (deg C): FP (deg C):

BP (deg C):

BP pressure (mm Hg):

Property/Value	Units	Temp	Data Type	Reference
WS 1.63E+003	mg/L	25	EST	MEYLAN,WM ET AL. (1996)
logP 1.40			EXP	YANG,DJ ET AL. (1985)
VP 6.09E-008	mm Hg	25	EST	NEELY,WB & BLAU,GE (1985)
DC	pKa			
HL 7.16E-008	atm m3/mol	25	EST	MEYLAN,WM & HOWARD,PH (1991)
OH 1.05E-011	cm3/molc sec	25	EST	MEYLAN,WM & HOWARD,PH (1993)

CAS #: 024106-05-6	2-METHYL-4-BROMOACETANILIDE

Formula: $C_9H_{10}BrNO$

Mol Weight: 228.10

MP (deg C): FP (deg C):

BP (deg C):

BP pressure (mm Hg):

Property/Value	Units	Temp	Data Type	Reference
WS 5.75E+002	mg/L	25	EST	MEYLAN,WM ET AL. (1996)
logP 2.03			EXP	NAKAGAWA,Y ET AL. (1992)
VP 1.36E-005	mm Hg	25	EST	NEELY,WB & BLAU,GE (1985)
DC	pKa			
HL 2.71E-009	atm m3/mol	25	EST	MEYLAN,WM & HOWARD,PH (1991)
OH 4.60E-012	cm3/molc sec	25	EST	MEYLAN,WM & HOWARD,PH (1993)

CAS #: 024108-33-6	1,2,4-TRIAZINE, 3-METHYL-

Formula: $C_4H_5N_3$

Mol Weight: 95.10

MP (deg C): FP (deg C):

BP (deg C):

BP pressure (mm Hg):

Property/Value	Units	Temp	Data Type	Reference
WS 1.00E+006	mg/L	25	EST	MEYLAN,WM ET AL. (1996)
logP -1.31			EXP	HANSCH,C ET AL. (1995)
VP 2.41E-001	mm Hg	25	EST	NEELY,WB & BLAU,GE (1985)
DC	pKa			
HL 1.31E-006	atm m3/mol	25	EST	MEYLAN,WM & HOWARD,PH (1991)
OH 5.28E-013	cm3/molc sec	25	EST	MEYLAN,WM & HOWARD,PH (1993)

CAS #: 024141-52-4	TRANS-METHYL CHYRSANTHEMATE

Formula: $C_{11}H_{18}O_2$

Mol Weight: 182.26

MP (deg C): FP (deg C):

BP (deg C):

BP pressure (mm Hg):

Property/Value	Units	Temp	Data Type	Reference
WS 3.30E+001	mg/L	25	EST	MEYLAN,WM ET AL. (1996)
logP 3.76			EXP	HANSCH,C & LEO,AJ (1985)
VP 1.91E-001	mm Hg	25	EST	NEELY,WB & BLAU,GE (1985)
DC	pKa			
HL 7.74E-004	atm m3/mol	25	EST	MEYLAN,WM & HOWARD,PH (1991)
OH 8.78E-011	cm3/molc sec	25	EST	MEYLAN,WM & HOWARD,PH (1993)

CAS #: 024151-93-7	PIPEROPHOS

Formula: $C_{14}H_{28}NO_3PS_2$

Mol Weight: 353.49

MP (deg C): FP (deg C):

BP (deg C):

BP pressure (mm Hg):

Property/Value	Units	Temp	Data Type	Reference
WS 2.50E+001	mg/L	20	EXP	YALKOWSKY,SH & DANNENFELSER,RM (1992)
logP 4.04			EXP	SAITO,H ET AL. (1993)
VP 4.34E-007	mm Hg	25	EST	NEELY,WB & BLAU,GE (1985)
DC	pKa			
HL 1.97E-010	atm m3/mol	25	EST	MEYLAN,WM & HOWARD,PH (1991)
OH 1.51E-010	cm3/molc sec	25	EST	MEYLAN,WM & HOWARD,PH (1993)

CAS #: 024229-59-2 — OXAZOLIDINE, 4,4,5,5-TETRAMETHYL-2-[(5-NITRO-2-T

Formula: $C_{10}H_{14}N_4O_3S$

Mol Weight: 270.31

MP (deg C):   FP (deg C):

BP (deg C):

BP pressure (mm Hg):

Property/Value	Units	Temp	Data Type	Reference
WS 6.74E+002	mg/L	25	EST	MEYLAN,WM ET AL. (1996)
logP 2.40			EXP	SANGSTER,J (1993)
VP 9.41E-007	mm Hg	25	EST	NEELY,WB & BLAU,GE (1985)
DC	pKa			
HL 1.24E-014	atm m3/mol	25	EST	MEYLAN,WM & HOWARD,PH (1991)
OH 6.44E-011	cm3/molc sec	25	EST	MEYLAN,WM & HOWARD,PH (1993)

CAS #: 024240-60-6 — OXAZOLIDINE, 2-[(5-NITRO-2-THIAZOLYL)IMINO]-

Formula: $C_6H_6N_4O_3S$

Mol Weight: 214.20

MP (deg C):   FP (deg C):

BP (deg C):

BP pressure (mm Hg):

Property/Value	Units	Temp	Data Type	Reference
WS 5.00E+004	mg/L	25	EST	MEYLAN,WM ET AL. (1996)
logP 0.57			EXP	SANGSTER,J (1993)
VP 5.95E-006	mm Hg	25	EST	NEELY,WB & BLAU,GE (1985)
DC	pKa			
HL 3.99E-015	atm m3/mol	25	EST	MEYLAN,WM & HOWARD,PH (1991)
OH 9.31E-011	cm3/molc sec	25	EST	MEYLAN,WM & HOWARD,PH (1993)

CAS #: 024240-65-1 — 2-(3-HOET-1,3-OXAZOLIDINYLIDEN-2N)-5-NO2-THIAZOL

Formula: $C_8H_{12}N_4O_4S$

Mol Weight: 260.27

MP (deg C):   FP (deg C):

BP (deg C):

BP pressure (mm Hg):

Property/Value	Units	Temp	Data Type	Reference
WS 3.79E+004	mg/L	25	EST	MEYLAN,WM ET AL. (1996)
logP 0.43			EXP	HANSCH,C & LEO,AJ (1985)
VP 4.27E-009	mm Hg	25	EST	NEELY,WB & BLAU,GE (1985)
DC	pKa			
HL 4.25E-019	atm m3/mol	25	EST	MEYLAN,WM & HOWARD,PH (1991)
OH 1.11E-010	cm3/molc sec	25	EST	MEYLAN,WM & HOWARD,PH (1993)

CAS #: 024240-67-3 — 2(1-HOET-13THIAZOLIDINYLIDEN-2N)5NO2-THIAZOLE

Formula: $C_8H_{12}N_4O_3S_2$

Mol Weight: 276.34

MP (deg C):   FP (deg C):

BP (deg C):

BP pressure (mm Hg):

Property/Value	Units	Temp	Data Type	Reference
WS 5.78E+003	mg/L	25	EST	MEYLAN,WM ET AL. (1996)
logP 1.28			EXP	HANSCH,C & LEO,AJ (1985)
VP 1.07E-009	mm Hg	25	EST	NEELY,WB & BLAU,GE (1985)
DC	pKa			
HL 3.87E-019	atm m3/mol	25	EST	MEYLAN,WM & HOWARD,PH (1991)
OH 9.84E-011	cm3/molc sec	25	EST	MEYLAN,WM & HOWARD,PH (1993)

CAS #: 024240-69-5 — 2(1,3-IMIDAZOLIDINYLIDEN-2-AMINO)-5NO2-THIAZOLE

Formula: $C_6H_7N_5O_2S$

Mol Weight: 213.22

MP (deg C):   FP (deg C):

BP (deg C):

BP pressure (mm Hg):

Property/Value	Units	Temp	Data Type	Reference
WS 1.03E+004	mg/L	25	EST	MEYLAN,WM ET AL. (1996)
logP 1.38			EXP	HANSCH,C & LEO,AJ (1985)
VP 2.10E-006	mm Hg	25	EST	NEELY,WB & BLAU,GE (1985)
DC	pKa			
HL 7.69E-017	atm m3/mol	25	EST	MEYLAN,WM & HOWARD,PH (1991)
OH 1.43E-010	cm3/molc sec	25	EST	MEYLAN,WM & HOWARD,PH (1993)

CAS #: 024240-70-8 — 2(1-OHET-1,3-IMIDAZOLINYLIDEN-2-AMINO)5NO2-THIAZ

Formula: $C_8H_{11}N_5O_3S$

Mol Weight: 257.27

MP (deg C):   FP (deg C):

BP (deg C):

BP pressure (mm Hg):

Property/Value	Units	Temp	Data Type	Reference
WS 7.96E+003	mg/L	25	EST	MEYLAN,WM ET AL. (1996)
logP 1.23			EXP	HANSCH,C & LEO,AJ (1985)
VP 1.22E-009	mm Hg	25	EST	NEELY,WB & BLAU,GE (1985)
DC	pKa			
HL 8.20E-021	atm m3/mol	25	EST	MEYLAN,WM & HOWARD,PH (1991)
OH 1.61E-010	cm3/molc sec	25	EST	MEYLAN,WM & HOWARD,PH (1993)

CAS #: 024240-83-3 — 2(4-ME-1,3-OXAZOLIDINYLIDEN-2-AMINO)5NO2-THIAZOL

Formula: $C_7H_8N_4O_3S$

Mol Weight: 228.23

MP (deg C):   FP (deg C):

BP (deg C):

BP pressure (mm Hg):

Property/Value	Units	Temp	Data Type	Reference
WS 9.26E+003	mg/L	25	EST	MEYLAN,WM ET AL. (1996)
logP 1.34			EXP	HANSCH,C & LEO,AJ (1985)
VP 3.43E-006	mm Hg	25	EST	NEELY,WB & BLAU,GE (1985)
DC	pKa			
HL 5.29E-015	atm m3/mol	25	EST	MEYLAN,WM & HOWARD,PH (1991)
OH 8.35E-011	cm3/molc sec	25	EST	MEYLAN,WM & HOWARD,PH (1993)

CAS #: 024279-91-2 — 2,5-CYCLOHEXADIENE-1,4-DIONE, 2-[2-[(AMINOCARBON

Formula: $C_{15}H_{19}N_3O_5$

Mol Weight: 321.34

MP (deg C): 202 dec   FP (deg C):

BP (deg C):

BP pressure (mm Hg):

Property/Value	Units	Temp	Data Type	Reference
WS 6.16E+003	mg/L	25	EST	MEYLAN,WM ET AL. (1996)
logP -0.15			EXP	HANSCH,C ET AL. (1995)
VP 9.86E-009	mm Hg	25	EST	NEELY,WB & BLAU,GE (1985)
DC	pKa			
HL 2.08E-021	atm m3/mol	25	EST	MEYLAN,WM & HOWARD,PH (1991)
OH 4.84E-011	cm3/molc sec	25	EST	MEYLAN,WM & HOWARD,PH (1993)

CAS #: 024305-27-9	L-PROLINAMIDE, 5-OXO-L-PROLYL-L-HISTIDYL-

Formula: $C_{16}H_{22}N_6O_4$

Mol Weight: 362.39

MP (deg C):   FP (deg C):

BP (deg C):

BP pressure (mm Hg):

Property/Value	Units	Temp	Data Type	Reference
WS 3.22E+004	mg/L	25	EST	MEYLAN,WM ET AL. (1996)
logP -2.46			EXP	HANSCH,C ET AL. (1995)
VP 5.97E-019	mm Hg	25	EST	NEELY,WB & BLAU,GE (1985)
DC	pKa			
HL 3.89E-025	atm m3/mol	25	EST	MEYLAN,WM & HOWARD,PH (1991)
OH 1.54E-010	cm3/molc sec	25	EST	MEYLAN,WM & HOWARD,PH (1993)

CAS #: 024310-18-7	I-VALERIC ACID HYDRAZIDE

Formula: $C_5H_{12}N_2O$

Mol Weight: 116.16

MP (deg C):   FP (deg C):

BP (deg C):

BP pressure (mm Hg):

Property/Value	Units	Temp	Data Type	Reference
WS 9.34E+003	mg/L	25	EST	MEYLAN,WM ET AL. (1996)
logP -0.31			EXP	HANSCH,C & LEO,AJ (1985)
VP 4.30E-003	mm Hg	25	EST	NEELY,WB & BLAU,GE (1985)
DC	pKa			
HL 1.09E-010	atm m3/mol	25	EST	MEYLAN,WM & HOWARD,PH (1991)
OH 1.43E-011	cm3/molc sec	25	EST	MEYLAN,WM & HOWARD,PH (1993)

CAS #: 024316-19-6	CEPHALOTAXINE

Formula: $C_{18}H_{23}NO_4$

Mol Weight: 317.39

MP (deg C):   FP (deg C):

BP (deg C):

BP pressure (mm Hg):

Property/Value	Units	Temp	Data Type	Reference
WS 1.95E+004	mg/L	25	EST	MEYLAN,WM ET AL. (1996)
logP 0.84			EXP	HANSCH,C ET AL. (1995)
VP 6.74E-010	mm Hg	25	EST	NEELY,WB & BLAU,GE (1985)
DC	pKa			
HL 5.30E-015	atm m3/mol	25	EST	MEYLAN,WM & HOWARD,PH (1991)
OH 1.96E-010	cm3/molc sec	25	EST	MEYLAN,WM & HOWARD,PH (1993)

CAS #: 024324-17-2	9H-FLUORENE-9-METHANOL

Formula: $C_{14}H_{12}O$

Mol Weight: 196.25

MP (deg C): 105-107   FP (deg C):

BP (deg C):

BP pressure (mm Hg):

Property/Value	Units	Temp	Data Type	Reference
WS 3.91E+001	mg/L	25	EST	MEYLAN,WM ET AL. (1996)
logP 2.89			EXP	SANGSTER,J (1993)
VP 7.57E-007	mm Hg	25	EST	NEELY,WB & BLAU,GE (1985)
DC	pKa			
HL 8.12E-009	atm m3/mol	25	EST	MEYLAN,WM & HOWARD,PH (1991)
OH 1.42E-011	cm3/molc sec	25	EST	MEYLAN,WM & HOWARD,PH (1993)

CAS #: 024331-75-7	DIETHYLMALONATE,2-ME BENZAL

Formula: $C_{15}H_{18}O_4$

Mol Weight: 262.31

MP (deg C):   FP (deg C):

BP (deg C):

BP pressure (mm Hg):

Property/Value	Units	Temp	Data Type	Reference
WS 3.59E+001	mg/L	25	EST	MEYLAN,WM ET AL. (1996)
logP 3.22			EXP	HANSCH,C & LEO,AJ (1985)
VP 4.92E-004	mm Hg	25	EST	NEELY,WB & BLAU,GE (1985)
DC	pKa			
HL 1.21E-008	atm m3/mol	25	EST	MEYLAN,WM & HOWARD,PH (1991)
OH 1.89E-011	cm3/molc sec	25	EST	MEYLAN,WM & HOWARD,PH (1993)

CAS #: 024331-83-7	DIETHYLMALONATE,3-MEO 4-OH BENZAL

Formula: $C_{15}H_{18}O_6$

Mol Weight: 294.31

MP (deg C):   FP (deg C):

BP (deg C):

BP pressure (mm Hg):

Property/Value	Units	Temp	Data Type	Reference
WS 4.51E+001	mg/L	25	EST	MEYLAN,WM ET AL. (1996)
logP 2.89			EXP	HANSCH,C & LEO,AJ (1985)
VP 1.61E-006	mm Hg	25	EST	NEELY,WB & BLAU,GE (1985)
DC	pKa			
HL 6.73E-014	atm m3/mol	25	EST	MEYLAN,WM & HOWARD,PH (1991)
OH 4.20E-011	cm3/molc sec	25	EST	MEYLAN,WM & HOWARD,PH (1993)

CAS #: 024347-58-8	2,3-BUTANEDIOL (L)

Formula: $C_4H_{10}O_2$

Mol Weight: 90.12

MP (deg C):   FP (deg C):

BP (deg C): 77.3-77.4

BP pressure (mm Hg): 1.00E+001

Property/Value	Units	Temp	Data Type	Reference
WS 7.59E+005	mg/L	25	EST	MEYLAN,WM ET AL. (1996)
logP -0.36			EST	MEYLAN,WM & HOWARD,PH (1995)
VP 5.50E-001	mm Hg	25	EXP	RIDDICK,JA ET AL. (1986)
DC	pKa			
HL 2.30E-007	atm m3/mol	25	EST	MEYLAN,WM & HOWARD,PH (1991)
OH 1.73E-011	cm3/molc sec	25	EST	MEYLAN,WM & HOWARD,PH (1993)

CAS #: 024353-61-5	BENZOIC ACID, 2-[(AMINOMETHOXYPHOSPHINOTHIOYL)OX

Formula: $C_{11}H_{16}NO_4PS$

Mol Weight: 289.29

MP (deg C):   FP (deg C):

BP (deg C):

BP pressure (mm Hg):

Property/Value	Units	Temp	Data Type	Reference
WS 7.01E+001	mg/L	25	EST	MEYLAN,WM ET AL. (1996)
logP 2.70			EXP	HU,J & LENG,XF (1992)
VP 2.88E-005	mm Hg	25	EST	NEELY,WB & BLAU,GE (1985)
DC	pKa			
HL 3.07E-009	atm m3/mol	25	EST	MEYLAN,WM & HOWARD,PH (1991)
OH 2.41E-010	cm3/molc sec	25	EST	MEYLAN,WM & HOWARD,PH (1993)

024367-68-8 — O-(P-NITROBENZYL)BENZAMIDE

CAS #: 024367-68-8

Formula: $C_{14}H_{12}N_2O_3$

Mol Weight: 256.26

MP (deg C):

FP (deg C):

BP (deg C):

BP pressure (mm Hg):

Property/Value	Units	Temp	Data Type	Reference
WS 2.96E+002	mg/L	25	EST	MEYLAN,WM ET AL. (1996)
logP 1.73			EXP	HANSCH,C & LEO,AJ (1985)
VP 9.77E-009	mm Hg	25	EST	NEELY,WB & BLAU,GE (1985)
DC	pKa			
HL 7.78E-013	atm m3/mol	25	EST	MEYLAN,WM & HOWARD,PH (1991)
OH 7.96E-012	cm3/molc sec	25	EST	MEYLAN,WM & HOWARD,PH (1993)

024380-92-5 — METHYLANABASINE

CAS #: 024380-92-5

Formula: $C_{11}H_{16}N_2$

Mol Weight: 176.26

MP (deg C):

FP (deg C):

BP (deg C):

BP pressure (mm Hg):

Property/Value	Units	Temp	Data Type	Reference
WS 1.00E+006	mg/L	25	EST	MEYLAN,WM ET AL. (1996)
logP 0.96			EXP	HANSCH,C & LEO,AJ (1985)
VP 4.62E-003	mm Hg	25	EST	NEELY,WB & BLAU,GE (1985)
DC	pKa			
HL 3.98E-009	atm m3/mol	25	EST	MEYLAN,WM & HOWARD,PH (1991)
OH 1.05E-010	cm3/molc sec	25	EST	MEYLAN,WM & HOWARD,PH (1993)

024385-10-2 — 5,12-NAPHTHACENEDIONE, 7,8,9,10-TETRAHYDRO-6,8,1

CAS #: 024385-10-2

Formula: $C_{21}H_{18}O_9$

Mol Weight: 414.37

MP (deg C):

FP (deg C):

BP (deg C):

BP pressure (mm Hg):

Property/Value	Units	Temp	Data Type	Reference
WS 3.36E+000	mg/L	25	EST	MEYLAN,WM ET AL. (1996)
logP 2.74			EXP	HANSCH,C ET AL. (1995)
VP 9.26E-020	mm Hg	25	EST	NEELY,WB & BLAU,GE (1985)
DC	pKa			
HL 1.83E-015	atm m3/mol	25	EST	MEYLAN,WM & HOWARD,PH (1991)
OH 5.51E-011	cm3/molc sec	25	EST	MEYLAN,WM & HOWARD,PH (1993)

024397-14-6 — 2-HEXYLOXY-4-AMINOBENZOIC ACID

CAS #: 024397-14-6

Formula: $C_{13}H_{19}NO_3$

Mol Weight: 237.30

MP (deg C):

FP (deg C):

BP (deg C):

BP pressure (mm Hg):

Property/Value	Units	Temp	Data Type	Reference
WS 4.94E+001	mg/L	25	EST	MEYLAN,WM ET AL. (1996)
logP 3.22			EXP	HANSCH,C & LEO,AJ (1985)
VP 6.70E-007	mm Hg	25	EST	NEELY,WB & BLAU,GE (1985)
DC	pKa			
HL 9.35E-012	atm m3/mol	25	EST	MEYLAN,WM & HOWARD,PH (1991)
OH 2.16E-010	cm3/molc sec	25	EST	MEYLAN,WM & HOWARD,PH (1993)

024423-11-8 — PHENANTHRENE, 2-CHLORO-

CAS #: 024423-11-8

Formula: $C_{14}H_9Cl$

Mol Weight: 212.68

MP (deg C):

FP (deg C):

BP (deg C):

BP pressure (mm Hg):

Property/Value	Units	Temp	Data Type	Reference
WS 1.15E-001	mg/L	25	EST	MEYLAN,WM ET AL. (1996)
logP 5.16			EXP	VEITH,GD ET AL. (1979)
VP 2.56E-005	mm Hg	25	EST	NEELY,WB & BLAU,GE (1985)
DC	pKa			
HL 3.80E-005	atm m3/mol	25	EST	MEYLAN,WM & HOWARD,PH (1991)
OH 9.14E-012	cm3/molc sec	25	EST	MEYLAN,WM & HOWARD,PH (1993)

024448-94-0 — 2,4,6(1H,3H,5H)-PYRIMIDINETRIONE, 5,5-DIMETHYL-

CAS #: 024448-94-0

Formula: $C_6H_8N_2O_3$

Mol Weight: 156.14

MP (deg C):

FP (deg C):

BP (deg C):

BP pressure (mm Hg):

Property/Value	Units	Temp	Data Type	Reference
WS 1.69E+005	mg/L	25	EST	MEYLAN,WM ET AL. (1996)
logP -0.44			EXP	PRANKERD,RJ & MCKEOWN,RH (1992)
VP 8.24E-010	mm Hg	25	EST	NEELY,WB & BLAU,GE (1985)
DC	pKa			
HL 2.05E-013	atm m3/mol	25	EST	MEYLAN,WM & HOWARD,PH (1991)
OH 3.06E-012	cm3/molc sec	25	EST	MEYLAN,WM & HOWARD,PH (1993)

024454-46-4 — BENZAMIDE, 4-[(4-AMINOPHENYL)SULFONYL]-

CAS #: 024454-46-4

Formula: $C_{13}H_{12}N_2O_3S$

Mol Weight: 276.32

MP (deg C):

FP (deg C):

BP (deg C):

BP pressure (mm Hg):

Property/Value	Units	Temp	Data Type	Reference
WS 2.87E+003	mg/L	25	EST	MEYLAN,WM ET AL. (1996)
logP 0.90			EXP	ALTOMARE,C ET AL. (1991)
VP 1.70E-010	mm Hg	25	EST	NEELY,WB & BLAU,GE (1985)
DC	pKa			
HL 3.62E-017	atm m3/mol	25	EST	MEYLAN,WM & HOWARD,PH (1991)
OH 2.54E-011	cm3/molc sec	25	EST	MEYLAN,WM & HOWARD,PH (1993)

024478-72-6 — DIBENZOFURAN, 1,2,3,4-TETRACHLORO-

CAS #: 024478-72-6

Formula: $C_{12}H_4Cl_4O$

Mol Weight: 305.98

MP (deg C):

FP (deg C):

BP (deg C):

BP pressure (mm Hg):

Property/Value	Units	Temp	Data Type	Reference
WS 4.73E-003	mg/L	25	EST	MEYLAN,WM ET AL. (1996)
logP 6.17			EXP	SIJM,DTHM ET AL. (1989)
VP 1.53E-006	mm Hg	25	EST	NEELY,WB & BLAU,GE (1985)
DC	pKa			
HL 1.54E-005	atm m3/mol	25	EST	MEYLAN,WM & HOWARD,PH (1991)
OH 1.65E-013	cm3/molc sec	25	EST	MEYLAN,WM & HOWARD,PH (1993)

CAS #: 024493-78-5

OO-DIET-O(2-CL-4-MESPH)PHOSPHATE

Formula: $C_{11}H_{16}ClO_4PS$

Mol Weight: 310.74

MP (deg C):
FP (deg C):
BP (deg C):
BP pressure (mm Hg):

Property/Value	Units	Temp	Data Type	Reference
WS 2.64E+001	mg/L	25	EST	MEYLAN,WM ET AL. (1996)
logP 3.05			EXP	HANSCH,C & LEO,AJ (1985)
VP 5.90E-006	mm Hg	25	EST	NEELY,WB & BLAU,GE (1985)
DC	pKa			
HL 3.51E-009	atm m3/mol	25	EST	MEYLAN,WM & HOWARD,PH (1991)
OH 4.62E-011	cm3/molc sec	25	EST	MEYLAN,WM & HOWARD,PH (1993)

CAS #: 024535-11-3

PROPIONIC ACID HYDRAZIDE

Formula: $C_3H_8N_2O$

Mol Weight: 88.11

MP (deg C):
FP (deg C):
BP (deg C):
BP pressure (mm Hg):

Property/Value	Units	Temp	Data Type	Reference
WS 4.40E+004	mg/L	25	EST	MEYLAN,WM ET AL. (1996)
logP -1.00			EXP	HANSCH,C & LEO,AJ (1985)
VP 1.97E-002	mm Hg	25	EST	NEELY,WB & BLAU,GE (1985)
DC	pKa			
HL 6.20E-011	atm m3/mol	25	EST	MEYLAN,WM & HOWARD,PH (1991)
OH 6.73E-012	cm3/molc sec	25	EST	MEYLAN,WM & HOWARD,PH (1993)

CAS #: 024535-70-4

BENZENESULFONAMIDE, 4-CHLORO-N-[(ETHYLAMINO)CARB

Formula: $C_9H_{11}ClN_2O_3S$

Mol Weight: 262.72

MP (deg C):
FP (deg C):
BP (deg C):
BP pressure (mm Hg):

Property/Value	Units	Temp	Data Type	Reference
WS 5.95E+002	mg/L	25	EST	MEYLAN,WM ET AL. (1996)
logP 1.79			EXP	SANGSTER,J (1993)
VP 8.32E-008	mm Hg	25	EST	NEELY,WB & BLAU,GE (1985)
DC	pKa			
HL 9.64E-011	atm m3/mol	25	EST	MEYLAN,WM & HOWARD,PH (1991)
OH 6.29E-012	cm3/molc sec	25	EST	MEYLAN,WM & HOWARD,PH (1993)

CAS #: 024568-11-4

BROMOBENZENE,P-TRIFLUOROACETAMIDO

Formula: $C_8H_5BrF_3NO$

Mol Weight: 268.04

MP (deg C):
FP (deg C):
BP (deg C):
BP pressure (mm Hg):

Property/Value	Units	Temp	Data Type	Reference
WS 2.63E+001	mg/L	25	EST	MEYLAN,WM ET AL. (1996)
logP 3.34			EXP	HANSCH,C & LEO,AJ (1985)
VP 5.93E-005	mm Hg	25	EST	NEELY,WB & BLAU,GE (1985)
DC	pKa			
HL 1.94E-008	atm m3/mol	25	EST	MEYLAN,WM & HOWARD,PH (1991)
OH 3.56E-012	cm3/molc sec	25	EST	MEYLAN,WM & HOWARD,PH (1993)

CAS #: 024568-14-7

BENZOIC ACID, 4-[(TRIFLUOROACETYL)AMINO]-,ETHYL

Formula: $C_{11}H_{10}F_3NO_3$

Mol Weight: 261.20

MP (deg C):
FP (deg C):
BP (deg C):
BP pressure (mm Hg):

Property/Value	Units	Temp	Data Type	Reference
WS 2.45E+002	mg/L	25	EST	MEYLAN,WM ET AL. (1996)
logP 2.25			EXP	HANSCH,C ET AL. (1995)
VP 6.96E-006	mm Hg	25	EST	NEELY,WB & BLAU,GE (1985)
DC	pKa			
HL 4.16E-010	atm m3/mol	25	EST	MEYLAN,WM & HOWARD,PH (1991)
OH 5.68E-012	cm3/molc sec	25	EST	MEYLAN,WM & HOWARD,PH (1993)

CAS #: 024570-14-7

5(1-AZIRD)-2,4-NO2 BENZAMIDE,N-ET

Formula: $C_{11}H_{12}N_4O_5$

Mol Weight: 280.24

MP (deg C):
FP (deg C):
BP (deg C):
BP pressure (mm Hg):

Property/Value	Units	Temp	Data Type	Reference
WS 3.29E+003	mg/L	25	EST	MEYLAN,WM ET AL. (1996)
logP 0.44			EXP	SANGSTER,J (1993)
VP 4.86E-009	mm Hg	25	EST	NEELY,WB & BLAU,GE (1985)
DC	pKa			
HL 1.16E-014	atm m3/mol	25	EST	MEYLAN,WM & HOWARD,PH (1991)
OH 7.45E-012	cm3/molc sec	25	EST	MEYLAN,WM & HOWARD,PH (1993)

CAS #: 024570-16-9

5(1-AZIRD)-2,4-NO2 BENZAMIDE,N-CYPR

Formula: $C_{12}H_{12}N_4O_5$

Mol Weight: 292.25

MP (deg C):
FP (deg C):
BP (deg C):
BP pressure (mm Hg):

Property/Value	Units	Temp	Data Type	Reference
WS 1.30E+003	mg/L	25	EST	MEYLAN,WM ET AL. (1996)
logP 0.74			EXP	HANSCH,C & LEO,AJ (1985)
VP 8.27E-010	mm Hg	25	EST	NEELY,WB & BLAU,GE (1985)
DC	pKa			
HL 9.02E-015	atm m3/mol	25	EST	MEYLAN,WM & HOWARD,PH (1991)
OH 7.19E-012	cm3/molc sec	25	EST	MEYLAN,WM & HOWARD,PH (1993)

CAS #: 024632-47-1

NIFURPIPONE

Formula: $C_{12}H_{18}N_5O_4$

Mol Weight: 296.31

MP (deg C):
FP (deg C):
BP (deg C):
BP pressure (mm Hg):

Property/Value	Units	Temp	Data Type	Reference
WS 3.18E+002	mg/L	25	EST	MEYLAN,WM ET AL. (1996)
logP -0.10			EXP	HANSCH,C ET AL. (1995)
VP 3.64E-009	mm Hg	25	EST	NEELY,WB & BLAU,GE (1985)
DC	pKa			
HL 2.73E-017	atm m3/mol	25	EST	MEYLAN,WM & HOWARD,PH (1991)
OH 2.02E-010	cm3/molc sec	25	EST	MEYLAN,WM & HOWARD,PH (1993)

CAS #: 024683-00-9	2-ISOBUTYL-3-METHOXYPYRAZINE

Formula: $C_9H_{14}N_2O$

Mol Weight: 166.22

MP (deg C): | FP (deg C):

BP (deg C):

BP pressure (mm Hg):

Property/Value	Units	Temp	Data Type	Reference
WS 2.30E+002	mg/L	25	EST	MEYLAN,WM ET AL. (1996)
logP 2.86			EST	MEYLAN,WM & HOWARD,PH (1995)
VP 3.48E-002	mm Hg	25	EST	NEELY,WB & BLAU,GE (1985)
DC	pKa			
HL 4.89E-005	atm m3/mol	25	EXP	BUTTERY,RG ET AL. (1971)
OH 7.84E-012	cm3/molc sec	25	EST	MEYLAN,WM & HOWARD,PH (1993)

CAS #: 024691-76-7	PYRACARBOLID

Formula: $C_{13}H_{15}NO_2$

Mol Weight: 217.27

MP (deg C): 110-111 | FP (deg C):

BP (deg C):

BP pressure (mm Hg):

Property/Value	Units	Temp	Data Type	Reference
WS 6.00E+002	mg/L	40	EXP	YALKOWSKY,SH & DANNENFELSER,RM (1992)
logP 2.08			EST	MEYLAN,WM & HOWARD,PH (1995)
VP 1.11E-006	mm Hg	25	EST	NEELY,WB & BLAU,GE (1985)
DC	pKa			
HL 4.35E-010	atm m3/mol	25	EST	MEYLAN,WM & HOWARD,PH (1991)
OH 1.05E-010	cm3/molc sec	25	EST	MEYLAN,WM & HOWARD,PH (1993)

CAS #: 024786-13-8	FURAZAN, CHLOROPHENYL-

Formula: $C_8H_5ClN_2O$

Mol Weight: 180.59

MP (deg C): | FP (deg C):

BP (deg C):

BP pressure (mm Hg):

Property/Value	Units	Temp	Data Type	Reference
WS 1.01E+002	mg/L	25	EST	MEYLAN,WM ET AL. (1996)
logP 3.20			EXP	CALVINO,R ET AL. (1992)
VP 1.16E-003	mm Hg	25	EST	NEELY,WB & BLAU,GE (1985)
DC	pKa			
HL 1.08E-005	atm m3/mol	25	EST	MEYLAN,WM & HOWARD,PH (1991)
OH 6.50E-012	cm3/molc sec	25	EST	MEYLAN,WM & HOWARD,PH (1993)

CAS #: 024798-19-4	2,4-NH2PYRIMIDIN,5(34-MEO-5-ME)BENZYL

Formula: $C_{14}H_{18}N_4O_2$

Mol Weight: 274.33

MP (deg C): | FP (deg C):

BP (deg C):

BP pressure (mm Hg):

Property/Value	Units	Temp	Data Type	Reference
WS 7.14E+002	mg/L	25	EST	MEYLAN,WM ET AL. (1996)
logP 1.62			EXP	HANSCH,C & LEO,AJ (1985)
VP 2.28E-008	mm Hg	25	EST	NEELY,WB & BLAU,GE (1985)
DC	pKa			
HL 4.45E-013	atm m3/mol	25	EST	MEYLAN,WM & HOWARD,PH (1991)
OH 1.25E-010	cm3/molc sec	25	EST	MEYLAN,WM & HOWARD,PH (1993)

CAS #: 024809-26-5	D-PHENYLALANINAMIDE, N-ACETYL-D-PHENYLALANYL-

Formula: $C_{20}H_{23}N_3O_3$

Mol Weight: 353.42

MP (deg C): | FP (deg C):

BP (deg C):

BP pressure (mm Hg):

Property/Value	Units	Temp	Data Type	Reference
WS 5.69E+002	mg/L	25	EST	MEYLAN,WM ET AL. (1996)
logP 1.19			EXP	SANGSTER,J (1994)
VP 9.26E-015	mm Hg	25	EST	NEELY,WB & BLAU,GE (1985)
DC	pKa			
HL 9.01E-018	atm m3/mol	25	EST	MEYLAN,WM & HOWARD,PH (1991)
OH 4.48E-011	cm3/molc sec	25	EST	MEYLAN,WM & HOWARD,PH (1993)

CAS #: 024847-46-9	PROLINE,N-ACETYL N'-MEAMINO-AMIDE

Formula: $C_8H_{14}N_2O_2$

Mol Weight: 170.21

MP (deg C): | FP (deg C):

BP (deg C):

BP pressure (mm Hg):

Property/Value	Units	Temp	Data Type	Reference
WS 3.12E+004	mg/L	25	EST	MEYLAN,WM ET AL. (1996)
logP -1.19			EXP	HANSCH,C & LEO,AJ (1985)
VP 2.05E-007	mm Hg	25	EST	NEELY,WB & BLAU,GE (1985)
DC	pKa			
HL 4.42E-010	atm m3/mol	25	EST	MEYLAN,WM & HOWARD,PH (1991)
OH 9.94E-011	cm3/molc sec	25	EST	MEYLAN,WM & HOWARD,PH (1993)

CAS #: 024847-58-3	CARBAMIC ACID, 2-ETHYLBUTYL ESTER

Formula: $C_7H_{15}NO_2$

Mol Weight: 145.20

MP (deg C): | FP (deg C):

BP (deg C):

BP pressure (mm Hg):

Property/Value	Units	Temp	Data Type	Reference
WS 1.62E+003	mg/L	25	EST	MEYLAN,WM ET AL. (1996)
logP 1.98			EXP	SANGSTER,J (1994)
VP 5.41E-001	mm Hg	25	EST	NEELY,WB & BLAU,GE (1985)
DC	pKa			
HL 1.63E-007	atm m3/mol	25	EST	MEYLAN,WM & HOWARD,PH (1991)
OH 1.01E-011	cm3/molc sec	25	EST	MEYLAN,WM & HOWARD,PH (1993)

CAS #: 024849-83-0	2,4-PYRIMIDINEDIAMINE, 5-[(2-BROMO-4,5-DIMETHOXY

Formula: $C_{13}H_{15}BrN_4O_2$

Mol Weight: 339.20

MP (deg C): | FP (deg C):

BP (deg C):

BP pressure (mm Hg):

Property/Value	Units	Temp	Data Type	Reference
WS 2.86E+002	mg/L	25	EST	MEYLAN,WM ET AL. (1996)
logP 1.64			EXP	SANGSTER,J (1993)
VP 6.04E-009	mm Hg	25	EST	NEELY,WB & BLAU,GE (1985)
DC	pKa			
HL 1.61E-013	atm m3/mol	25	EST	MEYLAN,WM & HOWARD,PH (1991)
OH 1.84E-011	cm3/molc sec	25	EST	MEYLAN,WM & HOWARD,PH (1993)

CAS #: 024903-72-8				PYRIDAZINE, 4-ETHOXY-
Formula: $C_6H_8N_2O$				
Mol Weight: 124.14				
MP (deg C):		FP (deg C):		
BP (deg C):				
BP pressure (mm Hg):				

Property/Value	Units	Temp	Data Type	Reference
WS 6.54E+004	mg/L	25	EST	MEYLAN,WM ET AL. (1996)
logP 0.20			EXP	YAMAGAMI,C ET AL. (1990)
VP 1.44E-002	mm Hg	25	EST	NEELY,WB & BLAU,GE (1985)
DC	pKa			
HL 2.25E-007	atm m3/mol	25	EST	MEYLAN,WM & HOWARD,PH (1991)
OH 8.71E-012	cm3/molc sec	25	EST	MEYLAN,WM & HOWARD,PH (1993)

CAS #: 024910-63-2				3,4-DIMETHYL-2-PENTENE
Formula: C_7H_{14}				
Mol Weight: 98.19				
MP (deg C): -124.2		FP (deg C):		
BP (deg C): 89				
BP pressure (mm Hg):				

Property/Value	Units	Temp	Data Type	Reference
WS 3.22E+001	mg/L	25	EST	MEYLAN,WM ET AL. (1996)
logP 3.55			EST	MEYLAN,WM & HOWARD,PH (1995)
VP 6.76E+001	mm Hg	25	EXP	DYKYJ,J & REPAS,M (1973)
DC	pKa			
HL 6.63E-001	atm m3/mol	25	EST	MEYLAN,WM & HOWARD,PH (1991)
OH 8.94E-011	cm3/molc sec	25	EST	MEYLAN,WM & HOWARD,PH (1993)

CAS #: 024948-81-0				N-CHLORO-N-(1-METHYETHYL)-2-PROPANAMINE
Formula: $C_6H_{14}ClN$				
Mol Weight: 135.64				
MP (deg C):		FP (deg C):		
BP (deg C):				
BP pressure (mm Hg):				

Property/Value	Units	Temp	Data Type	Reference
WS 6.66E+003	mg/L	25	EST	MEYLAN,WM ET AL. (1996)
logP 1.31			EST	MEYLAN,WM & HOWARD,PH (1995)
VP 3.36E+001	mm Hg	25	EST	NEELY,WB & BLAU,GE (1985)
DC	pKa			
HL 9.92E-004	atm m3/mol	25	EST	MEYLAN,WM & HOWARD,PH (1991)
OH 3.68E-011	cm3/molc sec	25	EST	MEYLAN,WM & HOWARD,PH (1993)

CAS #: 024964-64-5				M-CYANOBENAZLDEHYDE
Formula: C_8H_5NO				
Mol Weight: 131.14				
MP (deg C): 76.5		FP (deg C):		
BP (deg C): 210				
BP pressure (mm Hg):				

Property/Value	Units	Temp	Data Type	Reference
WS 4.86E+003	mg/L	25	EST	MEYLAN,WM ET AL. (1996)
logP 1.18			EXP	HANSCH,C ET AL. (1995)
VP 1.19E-002	mm Hg	25	EST	NEELY,WB & BLAU,GE (1985)
DC	pKa			
HL 1.30E-007	atm m3/mol	25	EST	MEYLAN,WM & HOWARD,PH (1991)
OH 1.70E-011	cm3/molc sec	25	EST	MEYLAN,WM & HOWARD,PH (1993)

CAS #: 025006-32-0				LEPTOPHOS-O-ANALOG
Formula: $C_{13}H_{10}BrCl_2O_3P$				
Mol Weight: 396.01				
MP (deg C):		FP (deg C):		
BP (deg C):				
BP pressure (mm Hg):				

Property/Value	Units	Temp	Data Type	Reference
WS 3.98E-001	mg/L	25	EST	MEYLAN,WM ET AL. (1996)
logP 4.58			EXP	HANSCH,C & LEO,AJ (1985)
VP 5.51E-007	mm Hg	25	EST	NEELY,WB & BLAU,GE (1985)
DC	pKa			
HL 2.59E-009	atm m3/mol	25	EST	MEYLAN,WM & HOWARD,PH (1991)
OH 4.95E-012	cm3/molc sec	25	EST	MEYLAN,WM & HOWARD,PH (1993)

CAS #: 025007-30-1				2-I-PR-PHENYLDIMETHYLCARBAMATE
Formula: $C_{12}H_{17}NO_2$				
Mol Weight: 207.27				
MP (deg C):		FP (deg C):		
BP (deg C):				
BP pressure (mm Hg):				

Property/Value	Units	Temp	Data Type	Reference
WS 2.19E+002	mg/L	25	EST	MEYLAN,WM ET AL. (1996)
logP 2.65			EXP	HANSCH,C & LEO,AJ (1985)
VP 4.39E-003	mm Hg	25	EST	NEELY,WB & BLAU,GE (1985)
DC	pKa			
HL 1.37E-007	atm m3/mol	25	EST	MEYLAN,WM & HOWARD,PH (1991)
OH 2.18E-011	cm3/molc sec	25	EST	MEYLAN,WM & HOWARD,PH (1993)

CAS #: 025021-08-3				N-MALEOYLGLYCINE
Formula: $C_6H_5NO_4$				
Mol Weight: 155.11				
MP (deg C):		FP (deg C):		
BP (deg C):				
BP pressure (mm Hg):				

Property/Value	Units	Temp	Data Type	Reference
WS 9.48E+004	mg/L	25	EST	MEYLAN,WM ET AL. (1996)
logP -0.14			EXP	SANGSTER,J (1993)
VP 3.61E-007	mm Hg	25	EST	NEELY,WB & BLAU,GE (1985)
DC	pKa			
HL 2.06E-012	atm m3/mol	25	EST	MEYLAN,WM & HOWARD,PH (1991)
OH 2.06E-011	cm3/molc sec	25	EST	MEYLAN,WM & HOWARD,PH (1993)

CAS #: 025052-57-7				CARBANILIC ACID, 3,4-DICHLOROTHIO-, S-METHYL EST
Formula: $C_8H_7Cl_2NOS$				
Mol Weight: 236.12				
MP (deg C):		FP (deg C):		
BP (deg C):				
BP pressure (mm Hg):				

Property/Value	Units	Temp	Data Type	Reference
WS 6.88E+000	mg/L	25	EST	MEYLAN,WM ET AL. (1996)
logP 4.23			EXP	HANSCH,C ET AL. (1995)
VP 2.21E-006	mm Hg	25	EST	NEELY,WB & BLAU,GE (1985)
DC	pKa			
HL 2.34E-007	atm m3/mol	25	EST	MEYLAN,WM & HOWARD,PH (1991)
OH 9.32E-012	cm3/molc sec	25	EST	MEYLAN,WM & HOWARD,PH (1993)

CAS #: 025057-89-0				BENTAZONE
Formula: $C_{10}H_{12}N_2O_3S$				
Mol Weight: 240.28				
MP (deg C): 138		FP (deg C):		
BP (deg C):				
BP pressure (mm Hg):				

Property/Value	Units	Temp	Data Type	Reference
WS 5.00E+002	mg/L	20	EXP	YALKOWSKY,SH & DANNENFELSER,RM (1992)
logP 2.34			EXP	SAITO,H ET AL. (1993)
VP 3.45E-006	mm Hg	20	EXP	TOMLIN,C (1994)
DC	pKa			
HL 2.18E-009	atm m3/mol	20	EST	VP/WSOL
OH 6.22E-011	cm3/molc sec	25	EST	MEYLAN,WM & HOWARD,PH (1993)

CAS #: 025059-80-7				BENAZOLIN-ETHYL
Formula: $C_{11}H_{10}ClNO_3S$				
Mol Weight: 271.72				
MP (deg C):		FP (deg C):		
BP (deg C):				
BP pressure (mm Hg):				

Property/Value	Units	Temp	Data Type	Reference
WS 1.31E+002	mg/L	25	EST	MEYLAN,WM ET AL. (1996)
logP 2.50			EXP	TOMLIN,C (1994)
VP 3.81E-007	mm Hg	25	EST	NEELY,WB & BLAU,GE (1985)
DC	pKa			
HL 3.71E-007	atm m3/mol	25	EST	MEYLAN,WM & HOWARD,PH (1991)
OH 6.45E-012	cm3/molc sec	25	EST	MEYLAN,WM & HOWARD,PH (1993)

CAS #: 025079-96-3				N-((4-METHYLPHENYL)METHYL)ACETAMIDE
Formula: $C_{10}H_{13}NO$				
Mol Weight: 163.22				
MP (deg C):		FP (deg C):		
BP (deg C):				
BP pressure (mm Hg):				

Property/Value	Units	Temp	Data Type	Reference
WS 3.09E+003	mg/L	25	EST	MEYLAN,WM ET AL. (1996)
logP 1.56			EST	MEYLAN,WM & HOWARD,PH (1995)
VP 7.27E-005	mm Hg	25	EST	NEELY,WB & BLAU,GE (1985)
DC	pKa			
HL 2.19E-009	atm m3/mol	25	EST	MEYLAN,WM & HOWARD,PH (1991)
OH 1.58E-011	cm3/molc sec	25	EST	MEYLAN,WM & HOWARD,PH (1993)

CAS #: 025116-00-1				ACETAMIDE, N-(2-CYANOPHENYL)-
Formula: $C_9H_8N_2O$				
Mol Weight: 160.18				
MP (deg C):		FP (deg C):		
BP (deg C):				
BP pressure (mm Hg):				

Property/Value	Units	Temp	Data Type	Reference
WS 1.21E+004	mg/L	25	EST	MEYLAN,WM ET AL. (1996)
logP 0.57			EXP	HANSCH,C ET AL. (1995)
VP 1.06E-005	mm Hg	25	EST	NEELY,WB & BLAU,GE (1985)
DC	pKa			
HL 5.96E-011	atm m3/mol	25	EST	MEYLAN,WM & HOWARD,PH (1991)
OH 2.29E-012	cm3/molc sec	25	EST	MEYLAN,WM & HOWARD,PH (1993)

CAS #: 025122-41-2				CLOBETASOL
Formula: $C_{22}H_{28}ClFO_4$				
Mol Weight: 410.92				
MP (deg C): 195.5-197		FP (deg C):		
BP (deg C):				
BP pressure (mm Hg):				

Property/Value	Units	Temp	Data Type	Reference
WS 2.00E+001	mg/L	25	EST	MEYLAN,WM ET AL. (1996)
logP 2.48			EXP	HANSCH,C & LEO,AJ (1985)
VP 8.17E-013	mm Hg	25	EST	NEELY,WB & BLAU,GE (1985)
DC	pKa			
HL 1.62E-010	atm m3/mol	25	EST	MEYLAN,WM & HOWARD,PH (1991)
OH 6.64E-011	cm3/molc sec	25	EST	MEYLAN,WM & HOWARD,PH (1993)

CAS #: 025122-43-4				BROBETASONE-17-PROPIONATE
Formula: $C_{25}H_{30}BrFO_5$				
Mol Weight: 509.42				
MP (deg C):		FP (deg C):		
BP (deg C):				
BP pressure (mm Hg):				

Property/Value	Units	Temp	Data Type	Reference
WS 5.79E-001	mg/L	25	EST	MEYLAN,WM ET AL. (1996)
logP 3.55			EXP	HANSCH,C & LEO,AJ (1985)
VP 4.29E-011	mm Hg	25	EST	NEELY,WB & BLAU,GE (1985)
DC	pKa			
HL 7.91E-014	atm m3/mol	25	EST	MEYLAN,WM & HOWARD,PH (1991)
OH 6.03E-011	cm3/molc sec	25	EST	MEYLAN,WM & HOWARD,PH (1993)

CAS #: 025122-46-7				CLOBETASOL-17-PROPIONATE
Formula: $C_{25}H_{32}ClFO_5$				
Mol Weight: 466.98				
MP (deg C): 195.5-197		FP (deg C):		
BP (deg C):				
BP pressure (mm Hg):				

Property/Value	Units	Temp	Data Type	Reference
WS 3.86E+000	mg/L	25	EST	MEYLAN,WM ET AL. (1996)
logP 3.50			EXP	SANGSTER,J (1994)
VP 2.72E-013	mm Hg	25	EST	NEELY,WB & BLAU,GE (1985)
DC	pKa			
HL 2.32E-014	atm m3/mol	25	EST	MEYLAN,WM & HOWARD,PH (1991)
OH 6.71E-011	cm3/molc sec	25	EST	MEYLAN,WM & HOWARD,PH (1993)

CAS #: 025122-47-8				CLOBETASOL-17-BUTYRATE
Formula: $C_{26}H_{34}ClFO_5$				
Mol Weight: 481.01				
MP (deg C):		FP (deg C):		
BP (deg C):				
BP pressure (mm Hg):				

Property/Value	Units	Temp	Data Type	Reference
WS 2.43E+000	mg/L	25	EST	MEYLAN,WM ET AL. (1996)
logP 3.63			EXP	HANSCH,C & LEO,AJ (1985)
VP 1.03E-013	mm Hg	25	EST	NEELY,WB & BLAU,GE (1985)
DC	pKa			
HL 3.08E-014	atm m3/mol	25	EST	MEYLAN,WM & HOWARD,PH (1991)
OH 6.84E-011	cm3/molc sec	25	EST	MEYLAN,WM & HOWARD,PH (1993)

CAS #: 025122-49-0				CLOBETASOL-17-PROPIONATE-9-CL

Formula: $C_{25}H_{32}Cl_2O_5$

Mol Weight: 483.44

MP (deg C): | FP (deg C):

BP (deg C):

BP pressure (mm Hg):

Property/Value	Units	Temp	Data Type	Reference
WS 4.68E+000	mg/L	25	EST	MEYLAN,WM ET AL. (1996)
logP 3.28			EXP	HANSCH,C & LEO,AJ (1985)
VP 2.95E-014	mm Hg	25	EST	NEELY,WB & BLAU,GE (1985)
DC	pKa			
HL 4.10E-015	atm m3/mol	25	EST	MEYLAN,WM & HOWARD,PH (1991)
OH 6.38E-011	cm3/molc sec	25	EST	MEYLAN,WM & HOWARD,PH (1993)

CAS #: 025122-56-9				CLOBETASONE-17-PROPIONATE

Formula: $C_{25}H_{30}ClFO_5$

Mol Weight: 464.97

MP (deg C): | FP (deg C):

BP (deg C):

BP pressure (mm Hg):

Property/Value	Units	Temp	Data Type	Reference
WS 1.33E+000	mg/L	25	EST	MEYLAN,WM ET AL. (1996)
logP 3.46			EXP	HANSCH,C & LEO,AJ (1985)
VP 6.85E-011	mm Hg	25	EST	NEELY,WB & BLAU,GE (1985)
DC	pKa			
HL 2.42E-013	atm m3/mol	25	EST	MEYLAN,WM & HOWARD,PH (1991)
OH 6.04E-011	cm3/molc sec	25	EST	MEYLAN,WM & HOWARD,PH (1993)

CAS #: 025122-57-0				CLOBETASONE-17-BUTYRATE

Formula: $C_{26}H_{32}ClFO_5$

Mol Weight: 478.99

MP (deg C): 90-100 | FP (deg C):

BP (deg C):

BP pressure (mm Hg):

Property/Value	Units	Temp	Data Type	Reference
WS 6.00E-001	mg/L	25	EST	MEYLAN,WM ET AL. (1996)
logP 3.76			EXP	HANSCH,C & LEO,AJ (1985)
VP 2.98E-011	mm Hg	25	EST	NEELY,WB & BLAU,GE (1985)
DC	pKa			
HL 3.21E-013	atm m3/mol	25	EST	MEYLAN,WM & HOWARD,PH (1991)
OH 6.17E-011	cm3/molc sec	25	EST	MEYLAN,WM & HOWARD,PH (1993)

CAS #: 025126-19-6				H-(NO NO2) AMPHENICOL

Formula: $C_{11}H_{13}Cl_2NO_3$

Mol Weight: 278.14

MP (deg C): | FP (deg C):

BP (deg C):

BP pressure (mm Hg):

Property/Value	Units	Temp	Data Type	Reference
WS 2.59E+003	mg/L	25	EST	MEYLAN,WM ET AL. (1996)
logP 0.94			EXP	HANSCH,C & LEO,AJ (1985)
VP 4.15E-011	mm Hg	25	EST	NEELY,WB & BLAU,GE (1985)
DC	pKa			
HL 5.80E-016	atm m3/mol	25	EST	MEYLAN,WM & HOWARD,PH (1991)
OH 3.50E-011	cm3/molc sec	25	EST	MEYLAN,WM & HOWARD,PH (1993)

CAS #: 025141-58-6				O-I-PROPYLPHENOXYACETIC ACID

Formula: $C_{11}H_{14}O_3$

Mol Weight: 194.23

MP (deg C): | FP (deg C):

BP (deg C):

BP pressure (mm Hg):

Property/Value	Units	Temp	Data Type	Reference
WS 4.37E+002	mg/L	25	EST	MEYLAN,WM ET AL. (1996)
logP 2.84			EXP	HANSCH,C & LEO,AJ (1985)
VP 1.46E-004	mm Hg	25	EST	NEELY,WB & BLAU,GE (1985)
DC	pKa			
HL 3.26E-008	atm m3/mol	25	EST	MEYLAN,WM & HOWARD,PH (1991)
OH 3.29E-011	cm3/molc sec	25	EST	MEYLAN,WM & HOWARD,PH (1993)

CAS #: 025152-20-9				THYMIDINE, 5'-AMINO-5'-DEOXY-

Formula: $C_{10}H_{15}N_3O_4$

Mol Weight: 241.25

MP (deg C): | FP (deg C):

BP (deg C):

BP pressure (mm Hg):

Property/Value	Units	Temp	Data Type	Reference
WS 7.29E+004	mg/L	25	EST	MEYLAN,WM ET AL. (1996)
logP -2.05			EXP	LIN,TS (1984)
VP 4.46E-013	mm Hg	25	EST	NEELY,WB & BLAU,GE (1985)
DC	pKa			
HL 3.40E-020	atm m3/mol	25	EST	MEYLAN,WM & HOWARD,PH (1991)
OH 1.24E-010	cm3/molc sec	25	EST	MEYLAN,WM & HOWARD,PH (1993)

CAS #: 025155-15-1				P-I-PROPYLTOLUENE

Formula: $C_{10}H_{14}$

Mol Weight: 134.22

MP (deg C): -64 - -72 | FP (deg C):

BP (deg C): 175-178

BP pressure (mm Hg):

Property/Value	Units	Temp	Data Type	Reference
WS 2.34E+001	mg/L	25	EXP	SUNTIO,LR ET AL. (1988A)
logP 4.10			EXP	HANSCH,C & LEO,AJ (1985)
VP 1.64E+000	mm Hg	25	EXP	PERRY,RH & GREEN,D (1984)
DC	pKa			
HL 1.24E-002	atm m3/mol	25	EST	VP/WSOL
OH 8.54E-012	cm3/molc sec	25	EST	MEYLAN,WM & HOWARD,PH (1993)

CAS #: 025167-70-8				2,4,4-TRIMETHYLPENTENE

Formula: C_8H_{16}

Mol Weight: 112.22

MP (deg C): | FP (deg C):

BP (deg C): 101

BP pressure (mm Hg):

Property/Value	Units	Temp	Data Type	Reference
WS 4.04E+000	mg/L	25	EST	MEYLAN,WM ET AL. (1996)
logP 4.08			EST	MEYLAN,WM & HOWARD,PH (1995)
VP 7.10E+001	mm Hg	25	EST	NEELY,WB & BLAU,GE (1985)
DC	pKa			
HL 7.46E-001	atm m3/mol	25	EST	MEYLAN,WM & HOWARD,PH (1991)
OH 5.32E-011	cm3/molc sec	25	EST	MEYLAN,WM & HOWARD,PH (1993)

025168-15-4 — 2,4,5-T, ISOOCTYL ESTER

CAS #: 025168-15-4

Formula: $C_{16}H_{21}Cl_3O_3$

Mol Weight: 367.70

MP (deg C): FP (deg C):

BP (deg C):

BP pressure (mm Hg):

Property/Value	Units	Temp	Data Type	Reference
WS 3.02E-003	mg/L	25	EST	MEYLAN,WM ET AL. (1996)
logP 7.38			EST	MEYLAN,WM & HOWARD,PH (1995)
VP 1.05E-006	mm Hg	25	EXP	HAMILTON,DJ (1980)
DC	pKa			
HL 4.18E-005	atm m3/mol	25	EST	MEYLAN,WM & HOWARD,PH (1991)
OH 1.12E-011	cm3/molc sec	25	EST	MEYLAN,WM & HOWARD,PH (1993)

025168-26-7 — 2,4-D, ISOOCTYL ESTER

CAS #: 025168-26-7

Formula: $C_{16}H_{22}Cl_2O_3$

Mol Weight: 333.26

MP (deg C): FP (deg C):

BP (deg C):

BP pressure (mm Hg):

Property/Value	Units	Temp	Data Type	Reference
WS 1.73E-002	mg/L	25	EST	MEYLAN,WM ET AL. (1996)
logP 6.73			EST	MEYLAN,WM & HOWARD,PH (1995)
VP 4.53E-006	mm Hg	25	EXP	BOSCH,SJ (1983)
DC	pKa			
HL 5.65E-005	atm m3/mol	25	EST	MEYLAN,WM & HOWARD,PH (1991)
OH 1.19E-011	cm3/molc sec	25	EST	MEYLAN,WM & HOWARD,PH (1993)

025186-43-0 — N-(1-METHYLETHYL)-4-NITROBENZENAMINE

CAS #: 025186-43-0

Formula: $C_9H_{12}N_2O_2$

Mol Weight: 180.21

MP (deg C): FP (deg C):

BP (deg C):

BP pressure (mm Hg):

Property/Value	Units	Temp	Data Type	Reference
WS 1.74E+002	mg/L	25	EST	MEYLAN,WM ET AL. (1996)
logP 2.93			EST	MEYLAN,WM & HOWARD,PH (1995)
VP 1.77E-003	mm Hg	25	EST	NEELY,WB & BLAU,GE (1985)
DC	pKa			
HL 2.91E-008	atm m3/mol	25	EST	MEYLAN,WM & HOWARD,PH (1991)
OH 2.37E-011	cm3/molc sec	25	EST	MEYLAN,WM & HOWARD,PH (1993)

025186-47-4 — 3,5-DICHLOROTOLUENE

CAS #: 025186-47-4

Formula: $C_7H_6Cl_2$

Mol Weight: 161.03

MP (deg C): 26 FP (deg C):

BP (deg C): 201.5

BP pressure (mm Hg):

Property/Value	Units	Temp	Data Type	Reference
WS 3.63E+001	mg/L	25	EST	MEYLAN,WM ET AL. (1996)
logP 3.83			EST	MEYLAN,WM & HOWARD,PH (1995)
VP 4.57E-001	mm Hg	25	EST	NEELY,WB & BLAU,GE (1985)
DC	pKa			
HL 3.26E-003	atm m3/mol	25	EST	MEYLAN,WM & HOWARD,PH (1991)
OH 2.65E-012	cm3/molc sec	25	EST	MEYLAN,WM & HOWARD,PH (1993)

025204-89-1 — BUTANOIC ACID, 2-(4-CHLORO-2-METHYLPHENOXY)-

CAS #: 025204-89-1

Formula: $C_{11}H_{13}ClO_3$

Mol Weight: 228.68

MP (deg C): FP (deg C):

BP (deg C):

BP pressure (mm Hg):

Property/Value	Units	Temp	Data Type	Reference
WS 8.36E+001	mg/L	25	EST	MEYLAN,WM ET AL. (1996)
logP 3.47			EXP	ILCHMANN,A ET AL. (1993)
VP 3.47E-005	mm Hg	25	EST	NEELY,WB & BLAU,GE (1985)
DC	pKa			
HL 2.42E-008	atm m3/mol	25	EST	MEYLAN,WM & HOWARD,PH (1991)
OH 2.06E-011	cm3/molc sec	25	EST	MEYLAN,WM & HOWARD,PH (1993)

025209-50-1 — BENZENEACETIC ACID, 4-CYCLOPENTYL-

CAS #: 025209-50-1

Formula: $C_{13}H_{16}O_2$

Mol Weight: 204.27

MP (deg C): FP (deg C):

BP (deg C):

BP pressure (mm Hg):

Property/Value	Units	Temp	Data Type	Reference
WS 1.19E+002	mg/L	25	EST	MEYLAN,WM ET AL. (1996)
logP 3.44			EXP	SANGSTER,J (1993)
VP 3.02E-005	mm Hg	25	EST	NEELY,WB & BLAU,GE (1985)
DC	pKa			
HL 6.68E-008	atm m3/mol	25	EST	MEYLAN,WM & HOWARD,PH (1991)
OH 1.29E-011	cm3/molc sec	25	EST	MEYLAN,WM & HOWARD,PH (1993)

025216-70-0 — CARBAMIC ACID, (2-BROMOPHENYL)-, METHYL ESTER

CAS #: 025216-70-0

Formula: $C_8H_8BrNO_2$

Mol Weight: 230.07

MP (deg C): FP (deg C):

BP (deg C):

BP pressure (mm Hg):

Property/Value	Units	Temp	Data Type	Reference
WS 3.30E+002	mg/L	25	EST	MEYLAN,WM ET AL. (1996)
logP 2.30			EXP	TAKAHASHI,J ET AL. (1988)
VP 2.66E-003	mm Hg	25	EST	NEELY,WB & BLAU,GE (1985)
DC	pKa			
HL 8.70E-009	atm m3/mol	25	EST	MEYLAN,WM & HOWARD,PH (1991)
OH 1.24E-011	cm3/molc sec	25	EST	MEYLAN,WM & HOWARD,PH (1993)

025216-72-2 — CARBAMIC ACID, (3-BROMOPHENYL)-, METHYL ESTER

CAS #: 025216-72-2

Formula: $C_8H_8BrNO_2$

Mol Weight: 230.07

MP (deg C): FP (deg C):

BP (deg C):

BP pressure (mm Hg):

Property/Value	Units	Temp	Data Type	Reference
WS 1.36E+002	mg/L	25	EST	MEYLAN,WM ET AL. (1996)
logP 2.75			EXP	TAKAHASHI,J ET AL. (1988)
VP 2.66E-003	mm Hg	25	EST	NEELY,WB & BLAU,GE (1985)
DC	pKa			
HL 8.70E-009	atm m3/mol	25	EST	MEYLAN,WM & HOWARD,PH (1991)
OH 2.71E-011	cm3/molc sec	25	EST	MEYLAN,WM & HOWARD,PH (1993)

CAS #: 025251-56-3 — M-CL BENZYL TRIMETHYL AMMONIUM BROMIDE

Formula: $C_{10}H_{15}BrClN$

Mol Weight: 264.59

MP (deg C):
FP (deg C):
BP (deg C):
BP pressure (mm Hg):

Property/Value	Units	Temp	Data Type	Reference
WS 6.89E+005	mg/L	25	EST	MEYLAN,WM ET AL. (1996)
logP -1.81			EXP	HANSCH,C & LEO,AJ (1985)
VP 1.02E-008	mm Hg	25	EST	NEELY,WB & BLAU,GE (1985)
DC	pKa			
HL 1.98E-013	atm m3/mol	25	EST	MEYLAN,WM & HOWARD,PH (1991)
OH 1.59E-011	cm3/molc sec	25	EST	MEYLAN,WM & HOWARD,PH (1993)

CAS #: 025270-44-4 — BENZENESULFONAMIDE, 4-CHLORO-N-[(PHENYLAMINO)CAR

Formula: $C_{13}H_{11}ClN_2O_3S$

Mol Weight: 310.76

MP (deg C):
FP (deg C):
BP (deg C):
BP pressure (mm Hg):

Property/Value	Units	Temp	Data Type	Reference
WS 8.77E+000	mg/L	25	EST	MEYLAN,WM ET AL. (1996)
logP 3.61			EXP	HANSCH,C ET AL. (1995)
VP 1.19E-009	mm Hg	25	EST	NEELY,WB & BLAU,GE (1985)
DC	pKa			
HL 1.83E-011	atm m3/mol	25	EST	MEYLAN,WM & HOWARD,PH (1991)
OH 4.40E-011	cm3/molc sec	25	EST	MEYLAN,WM & HOWARD,PH (1993)

CAS #: 025277-05-8 — (3-CHLORO-4-METHOXYPHENYL)UREA

Formula: $C_8H_9ClN_2O_2$

Mol Weight: 200.63

MP (deg C):
FP (deg C):
BP (deg C):
BP pressure (mm Hg):

Property/Value	Units	Temp	Data Type	Reference
WS 2.94E+003	mg/L	25	EST	MEYLAN,WM ET AL. (1996)
logP 1.37			EXP	BRIGGS,GG (1981)
VP 2.27E-005	mm Hg	25	EST	NEELY,WB & BLAU,GE (1985)
DC	pKa			
HL 8.83E-012	atm m3/mol	25	EST	MEYLAN,WM & HOWARD,PH (1991)
OH 2.78E-011	cm3/molc sec	25	EST	MEYLAN,WM & HOWARD,PH (1993)

CAS #: 025283-63-0 — 3-QUINOLINECARBOXAMIDE,N,N-DIMETHYL

Formula: $C_{12}H_{12}N_2O$

Mol Weight: 200.24

MP (deg C):
FP (deg C):
BP (deg C):
BP pressure (mm Hg):

Property/Value	Units	Temp	Data Type	Reference
WS 6.23E+003	mg/L	25	EST	MEYLAN,WM ET AL. (1996)
logP 0.99			EXP	HANSCH,C & LEO,AJ (1985)
VP 1.20E-005	mm Hg	25	EST	NEELY,WB & BLAU,GE (1985)
DC	pKa			
HL 1.36E-012	atm m3/mol	25	EST	MEYLAN,WM & HOWARD,PH (1991)
OH 2.92E-011	cm3/molc sec	25	EST	MEYLAN,WM & HOWARD,PH (1993)

CAS #: 025311-71-1 — ISOFENPHOS

Formula: $C_{15}H_{24}NO_4PS$

Mol Weight: 345.40

MP (deg C): < -12
FP (deg C):
BP (deg C): 120
BP pressure (mm Hg): 1.00E-002

Property/Value	Units	Temp	Data Type	Reference
WS 2.21E+001	mg/L	20	EXP	YALKOWSKY,SH & DANNENFELSER,RM (1992)
logP 4.12			EXP	HANSCH,C & LEO,AJ (1985)
VP 3.00E-006	mm Hg	25	EXP	WAUCHOPE,RD ET AL. (1991A)
DC	pKa			
HL 6.17E-008	atm m3/mol	25	EST	VP/WSOL
OH 2.75E-010	cm3/molc sec	25	EST	MEYLAN,WM & HOWARD,PH (1993)

CAS #: 025314-91-4 — 3-ACETYL INDOLE

Formula: $C_{10}H_9NO$

Mol Weight: 159.19

MP (deg C):
FP (deg C):
BP (deg C):
BP pressure (mm Hg):

Property/Value	Units	Temp	Data Type	Reference
WS 1.20E+003	mg/L	25	EST	MEYLAN,WM ET AL. (1996)
logP 2.06			EXP	HANSCH,C & LEO,AJ (1985)
VP 2.56E-004	mm Hg	25	EST	NEELY,WB & BLAU,GE (1985)
DC	pKa			
HL 1.61E-009	atm m3/mol	25	EST	MEYLAN,WM & HOWARD,PH (1991)
OH 1.00E-010	cm3/molc sec	25	EST	MEYLAN,WM & HOWARD,PH (1993)

CAS #: 025327-89-3 — TETRABROMOBISPHENOL A BIS(ALLYL ETHER)

Formula: $C_{21}H_{20}Br_4O_2$

Mol Weight: 624.01

MP (deg C): 118-120
FP (deg C):
BP (deg C):
BP pressure (mm Hg):

Property/Value	Units	Temp	Data Type	Reference
WS 3.12E-007	mg/L	25	EST	MEYLAN,WM ET AL. (1996)
logP 10.02			EST	MEYLAN,WM & HOWARD,PH (1995)
VP 1.50E-010	mm Hg	25	EST	NEELY,WB & BLAU,GE (1985)
DC	pKa			
HL 1.29E-007	atm m3/mol	25	EST	MEYLAN,WM & HOWARD,PH (1991)
OH 6.73E-011	cm3/molc sec	25	EST	MEYLAN,WM & HOWARD,PH (1993)

CAS #: 025339-17-7 — ISODECANOL

Formula: $C_{10}H_{22}O$

Mol Weight: 158.29

MP (deg C):
FP (deg C):
BP (deg C):
BP pressure (mm Hg):

Property/Value	Units	Temp	Data Type	Reference
WS 1.52E+002	mg/L	25	EST	MEYLAN,WM ET AL. (1996)
logP 3.71			EST	MEYLAN,WM & HOWARD,PH (1995)
VP 2.07E-002	mm Hg	25	EXP	YAWS,CL (1994B)
DC	pKa			
HL 5.47E-005	atm m3/mol	25	EST	MEYLAN,WM & HOWARD,PH (1991)
OH 1.54E-011	cm3/molc sec	25	EST	MEYLAN,WM & HOWARD,PH (1993)

CAS #: 025366-23-8				THIAZAFLURON
Formula: $C_6H_9F_3N_4OS$				
Mol Weight: 242.22				
MP (deg C): 136-137		FP (deg C):		
BP (deg C):				
BP pressure (mm Hg):				

Property/Value	Units	Temp	Data Type	Reference
WS 2.10E+003	mg/L	20	EXP	SHIU,WY ET AL. (1990)
logP 1.85			EXP	SANGSTER,J (1994)
VP 3.66E-006	mm Hg	25	EST	NEELY,WB & BLAU,GE (1985)
DC	pKa			
HL 2.28E-011	atm m3/mol	25	EST	MEYLAN,WM & HOWARD,PH (1991)
OH 1.82E-012	cm3/molc sec	25	EST	MEYLAN,WM & HOWARD,PH (1993)

CAS #: 025371-75-9				O,O-DIPROPYL METHYLPHOSPHONOTHIOATE
Formula: $C_7H_{17}O_2PS$				
Mol Weight: 196.25				
MP (deg C):		FP (deg C):		
BP (deg C):				
BP pressure (mm Hg):				

Property/Value	Units	Temp	Data Type	Reference
WS 7.52E+001	mg/L	25	EST	MEYLAN,WM ET AL. (1996)
logP 3.26			EXP	KRIKORIAN,SE ET AL. (1987)
VP 9.12E-002	mm Hg	25	EST	NEELY,WB & BLAU,GE (1985)
DC	pKa			
HL 1.79E-003	atm m3/mol	25	EST	MEYLAN,WM & HOWARD,PH (1991)
OH 1.03E-010	cm3/molc sec	25	EST	MEYLAN,WM & HOWARD,PH (1993)

CAS #: 025379-20-8				2-PHENYLINDOLIZINE
Formula: $C_{14}H_{13}N$				
Mol Weight: 195.27				
MP (deg C):		FP (deg C):		
BP (deg C):				
BP pressure (mm Hg):				

Property/Value	Units	Temp	Data Type	Reference
WS 1.03E+001	mg/L	25	EST	MEYLAN,WM ET AL. (1996)
logP 4.29			EXP	HANSCH,C & LEO,AJ (1985)
VP 4.31E-005	mm Hg	25	EST	NEELY,WB & BLAU,GE (1985)
DC	pKa			
HL 4.19E-009	atm m3/mol	25	EST	MEYLAN,WM & HOWARD,PH (1991)
OH 1.94E-010	cm3/molc sec	25	EST	MEYLAN,WM & HOWARD,PH (1993)

CAS #: 025405-65-6				ERYTHROMYCIN, 4'-HYDROXY-
Formula: $C_{37}H_{67}NO_{14}$				
Mol Weight: 749.94				
MP (deg C):		FP (deg C):		
BP (deg C):				
BP pressure (mm Hg):				

Property/Value	Units	Temp	Data Type	Reference
WS 9.77E+000	mg/L	25	EST	MEYLAN,WM ET AL. (1996)
logP 1.44			EXP	HANSCH,C ET AL. (1995)
VP 1.11E-028	mm Hg	25	EST	NEELY,WB & BLAU,GE (1985)
DC	pKa			
HL 1.98E-030	atm m3/mol	25	EST	MEYLAN,WM & HOWARD,PH (1991)
OH 4.13E-010	cm3/molc sec	25	EST	MEYLAN,WM & HOWARD,PH (1993)

CAS #: 025413-64-3				N-NITROSOPROPYLBUTYLAMINE
Formula: $C_7H_{16}N_2O$				
Mol Weight: 144.22				
MP (deg C):		FP (deg C):		
BP (deg C):				
BP pressure (mm Hg):				

Property/Value	Units	Temp	Data Type	Reference
WS 1.32E+003	mg/L	25	EST	MEYLAN,WM ET AL. (1996)
logP 2.09			EXP	VERA,A ET AL. (1992)
VP 3.57E-002	mm Hg	25	EST	NEELY,WB & BLAU,GE (1985)
DC	pKa			
HL 8.48E-006	atm m3/mol	25	EST	MEYLAN,WM & HOWARD,PH (1991)
OH 2.54E-011	cm3/molc sec	25	EST	MEYLAN,WM & HOWARD,PH (1993)

CAS #: 025414-22-6				2-METHOXYFURAN
Formula: $C_5H_7O_2$				
Mol Weight: 99.11				
MP (deg C):		FP (deg C):		
BP (deg C): 110.5				
BP pressure (mm Hg):				

Property/Value	Units	Temp	Data Type	Reference
WS 6.98E+003	mg/L	25	EST	MEYLAN,WM ET AL. (1996)
logP 1.44			EXP	HANSCH,C & LEO,AJ (1985)
VP 2.49E+001	mm Hg	25	EST	NEELY,WB & BLAU,GE (1985)
DC	pKa			
HL 3.18E-004	atm m3/mol	25	EST	MEYLAN,WM & HOWARD,PH (1991)
OH 2.01E-010	cm3/molc sec	25	EST	MEYLAN,WM & HOWARD,PH (1993)

CAS #: 025423-22-7				ERYTHROMYCIN, 3'-AZIDO-3'-DE(DIMETHYLAMINO)-4'-H
Formula: $C_{35}H_{61}N_3O_{14}$				
Mol Weight: 747.89				
MP (deg C):		FP (deg C):		
BP (deg C):				
BP pressure (mm Hg):				

Property/Value	Units	Temp	Data Type	Reference
WS 1.84E-001	mg/L	25	EST	MEYLAN,WM ET AL. (1996)
logP 1.79			EXP	HANSCH,C ET AL. (1995)
VP 2.76E-035	mm Hg	25	EST	NEELY,WB & BLAU,GE (1985)
DC	pKa			
HL 8.78E-035	atm m3/mol	25	EST	MEYLAN,WM & HOWARD,PH (1991)
OH 3.20E-010	cm3/molc sec	25	EST	MEYLAN,WM & HOWARD,PH (1993)

CAS #: 025429-23-6				1,2-DIBROMOETHENE
Formula: $C_2H_2Br_2$				
Mol Weight: 185.86				
MP (deg C):		FP (deg C):		
BP (deg C):				
BP pressure (mm Hg):				

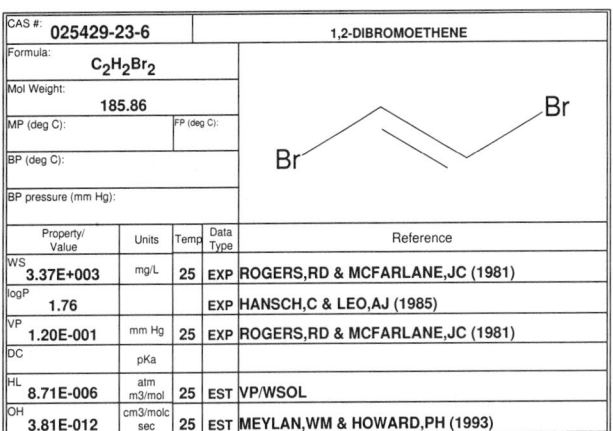

Property/Value	Units	Temp	Data Type	Reference
WS 3.37E+003	mg/L	25	EXP	ROGERS,RD & MCFARLANE,JC (1981)
logP 1.76			EXP	HANSCH,C & LEO,AJ (1985)
VP 1.20E-001	mm Hg	25	EXP	ROGERS,RD & MCFARLANE,JC (1981)
DC	pKa			
HL 8.71E-006	atm m3/mol	25	EST	VP/WSOL
OH 3.81E-012	cm3/molc sec	25	EST	MEYLAN,WM & HOWARD,PH (1993)

CAS #: 025444-83-1

THIOUREA, N-METHYL-N'-TRICYCLO[3.3.1.1,3,7]DEC-1

Formula: $C_{12}H_{20}N_2S$

Mol Weight: 224.37

MP (deg C):   FP (deg C):

BP (deg C):

BP pressure (mm Hg):

Property/ Value	Units	Temp	Data Type	Reference
WS 1.74E+003	mg/L	25	EST	MEYLAN,WM ET AL. (1996)
logP 2.67			EXP	HANSCH,C ET AL. (1995)
VP 1.08E-004	mm Hg	25	EST	NEELY,WB & BLAU,GE (1985)
DC	pKa			
HL 8.36E-007	atm m3/mol	25	EST	MEYLAN,WM & HOWARD,PH (1991)
OH 1.47E-010	cm3/molc sec	25	EST	MEYLAN,WM & HOWARD,PH (1993)

CAS #: 025445-77-6

4-PHENYL-1,2,3-THIADIAZOLE

Formula: $C_8H_8N_2S$

Mol Weight: 164.23

MP (deg C):   FP (deg C):

BP (deg C):

BP pressure (mm Hg):

Property/ Value	Units	Temp	Data Type	Reference
WS 5.74E+002	mg/L	25	EST	MEYLAN,WM ET AL. (1996)
logP 2.42			EXP	HANSCH,C ET AL. (1995)
VP 8.03E-005	mm Hg	25	EST	NEELY,WB & BLAU,GE (1985)
DC	pKa			
HL 1.34E-006	atm m3/mol	25	EST	MEYLAN,WM & HOWARD,PH (1991)
OH 4.60E-012	cm3/molc sec	25	EST	MEYLAN,WM & HOWARD,PH (1993)

CAS #: 025460-87-1

GLUTAMIC ACID-MONOAMIDE,N-ACETYL

Formula: $C_7H_{12}N_2O_4$

Mol Weight: 188.18

MP (deg C):   FP (deg C):

BP (deg C):

BP pressure (mm Hg):

Property/ Value	Units	Temp	Data Type	Reference
WS 3.16E+005	mg/L	25	EST	MEYLAN,WM ET AL. (1996)
logP -2.47			EXP	HANSCH,C & LEO,AJ (1985)
VP 1.47E-011	mm Hg	25	EST	NEELY,WB & BLAU,GE (1985)
DC	pKa			
HL 1.55E-015	atm m3/mol	25	EST	MEYLAN,WM & HOWARD,PH (1991)
OH 4.91E-011	cm3/molc sec	25	EST	MEYLAN,WM & HOWARD,PH (1993)

CAS #: 025523-14-2

3,3-DICHLOROPROPYNE

Formula: $C_3H_2Cl_2$

Mol Weight: 108.96

MP (deg C):   FP (deg C):

BP (deg C):

BP pressure (mm Hg):

Property/ Value	Units	Temp	Data Type	Reference
WS 6.00E+003	mg/L	25	EST	MEYLAN,WM ET AL. (1996)
logP 1.48			EST	MEYLAN,WM & HOWARD,PH (1995)
VP 4.36E+001	mm Hg	25	EST	NEELY,WB & BLAU,GE (1985)
DC	pKa			
HL 1.99E-003	atm m3/mol	25	EST	MEYLAN,WM & HOWARD,PH (1991)
OH 7.28E-012	cm3/molc sec	25	EST	MEYLAN,WM & HOWARD,PH (1993)

CAS #: 025526-93-6

THYMIDINE, 3'-DEOXY-3'-FLUORO-

Formula: $C_{10}H_{13}FN_2O_4$

Mol Weight: 244.22

MP (deg C): 176-178   FP (deg C):

BP (deg C):

BP pressure (mm Hg):

Property/ Value	Units	Temp	Data Type	Reference
WS 2.16E+003	mg/L	25	EST	MEYLAN,WM ET AL. (1996)
logP -0.28			EXP	BALZARINI,J ET AL. (1989)
VP 3.84E-012	mm Hg	25	EST	NEELY,WB & BLAU,GE (1985)
DC	pKa			
HL 6.58E-016	atm m3/mol	25	EST	MEYLAN,WM & HOWARD,PH (1991)
OH 3.33E-011	cm3/molc sec	25	EST	MEYLAN,WM & HOWARD,PH (1993)

CAS #: 025569-80-6

2,3'-PCB

Formula: $C_{12}H_8Cl_2$

Mol Weight: 223.10

MP (deg C):   FP (deg C):

BP (deg C):

BP pressure (mm Hg):

Property/ Value	Units	Temp	Data Type	Reference
WS 5.80E-001	mg/L	20	EXP	YALKOWSKY,SH & DANNENFELSER,RM (1992)
logP 5.02			EXP	HANSCH,C & LEO,AJ (1985)
VP 1.91E-004	mm Hg	25	EST	NEELY,WB & BLAU,GE (1985)
DC	pKa			
HL 2.27E-004	atm m3/mol	25	EST	MEYLAN,WM & HOWARD,PH (1991)
OH 2.62E-012	cm3/molc sec	25	EST	MEYLAN,WM & HOWARD,PH (1993)

CAS #: 025580-68-1

3ME-5-(5NO2-FURPROPENILIDENE)THIAZOLIDN-24-DIONE

Formula: $C_{11}H_9N_2O_5S$

Mol Weight: 281.27

MP (deg C):   FP (deg C):

BP (deg C):

BP pressure (mm Hg):

Property/ Value	Units	Temp	Data Type	Reference
WS 4.15E+002	mg/L	25	EST	MEYLAN,WM ET AL. (1996)
logP 1.40			EXP	HANSCH,C & LEO,AJ (1985)
VP 4.55E-010	mm Hg	25	EST	NEELY,WB & BLAU,GE (1985)
DC	pKa			
HL 1.14E-009	atm m3/mol	25	EST	MEYLAN,WM & HOWARD,PH (1991)
OH 8.37E-011	cm3/molc sec	25	EST	MEYLAN,WM & HOWARD,PH (1993)

CAS #: 025580-69-2

4-THIAZOLIDINONE, 2-IMINO-3-METHYL-5-[(5-NITRO-2

Formula: $C_9H_8N_3O_4S$

Mol Weight: 254.25

MP (deg C):   FP (deg C):

BP (deg C):

BP pressure (mm Hg):

Property/ Value	Units	Temp	Data Type	Reference
WS 5.45E+002	mg/L	25	EST	MEYLAN,WM ET AL. (1996)
logP 1.44			EXP	HANSCH,C & LEO,AJ (1985)
VP 4.13E-008	mm Hg	25	EST	NEELY,WB & BLAU,GE (1985)
DC	pKa			
HL 1.78E-015	atm m3/mol	25	EST	MEYLAN,WM & HOWARD,PH (1991)
OH 5.24E-011	cm3/molc sec	25	EST	MEYLAN,WM & HOWARD,PH (1993)

CAS #: 025602-39-5	FORMAMIDE, N-2-THIAZOLYL

Formula: $C_4H_6N_2OS$

Mol Weight: 130.17

MP (deg C):   FP (deg C):

BP (deg C):

BP pressure (mm Hg):

Property/Value	Units	Temp	Data Type	Reference
WS 3.23E+004	mg/L	25	EST	MEYLAN,WM ET AL. (1996)
logP 0.54			EXP	HANSCH,C & LEO,AJ (1985)
VP 3.04E-004	mm Hg	25	EST	NEELY,WB & BLAU,GE (1985)
DC	pKa			
HL 6.01E-012	atm m3/mol	25	EST	MEYLAN,WM & HOWARD,PH (1991)
OH 3.06E-011	cm3/molc sec	25	EST	MEYLAN,WM & HOWARD,PH (1993)

CAS #: 025603-06-9	2-MEAM-5-(5-NO2-2-FURFURYL)THIAZOLONE

Formula: $C_9H_8N_3O_4S$

Mol Weight: 254.25

MP (deg C):   FP (deg C):

BP (deg C):

BP pressure (mm Hg):

Property/Value	Units	Temp	Data Type	Reference
WS 9.45E+002	mg/L	25	EST	MEYLAN,WM ET AL. (1996)
logP 1.16			EXP	HANSCH,C & LEO,AJ (1985)
VP 1.04E-007	mm Hg	25	EST	NEELY,WB & BLAU,GE (1985)
DC	pKa			
HL 1.78E-015	atm m3/mol	25	EST	MEYLAN,WM & HOWARD,PH (1991)
OH 1.01E-010	cm3/molc sec	25	EST	MEYLAN,WM & HOWARD,PH (1993)

CAS #: 025603-07-0	2,4-THIAZOLIDINEDIONE, 3-METHYL-5-(5-NITROFURFUR

Formula: $C_9H_6N_2O_5S$

Mol Weight: 254.22

MP (deg C):   FP (deg C):

BP (deg C):

BP pressure (mm Hg):

Property/Value	Units	Temp	Data Type	Reference
WS 2.76E+002	mg/L	25	EST	MEYLAN,WM ET AL. (1996)
logP 1.78			EXP	HANSCH,C ET AL. (1995)
VP 3.00E-009	mm Hg	25	EST	NEELY,WB & BLAU,GE (1985)
DC	pKa			
HL 9.03E-010	atm m3/mol	25	EST	MEYLAN,WM & HOWARD,PH (1991)
OH 4.76E-011	cm3/molc sec	25	EST	MEYLAN,WM & HOWARD,PH (1993)

CAS #: 025603-08-1	4-THIAZOLIDINONE, 3-ETHYL-2-IMINO-5-[(5-NITRO-2-

Formula: $C_{10}H_{10}N_3O_4S$

Mol Weight: 268.27

MP (deg C):   FP (deg C):

BP (deg C):

BP pressure (mm Hg):

Property/Value	Units	Temp	Data Type	Reference
WS 4.55E+002	mg/L	25	EST	MEYLAN,WM ET AL. (1996)
logP 1.44			EXP	HANSCH,C & LEO,AJ (1985)
VP 1.85E-008	mm Hg	25	EST	NEELY,WB & BLAU,GE (1985)
DC	pKa			
HL 2.37E-015	atm m3/mol	25	EST	MEYLAN,WM & HOWARD,PH (1991)
OH 5.59E-011	cm3/molc sec	25	EST	MEYLAN,WM & HOWARD,PH (1993)

CAS #: 025603-09-2	4-THIAZOLIDINONE, 3-METHYL-2-(MEHTYLIMINO)-5-[(5

Formula: $C_{10}H_{10}N_3O_4S$

Mol Weight: 268.27

MP (deg C):   FP (deg C):

BP (deg C):

BP pressure (mm Hg):

Property/Value	Units	Temp	Data Type	Reference
WS 4.73E+002	mg/L	25	EST	MEYLAN,WM ET AL. (1996)
logP 1.42			EXP	HANSCH,C & LEO,AJ (1985)
VP 8.02E-008	mm Hg	25	EST	NEELY,WB & BLAU,GE (1985)
DC	pKa			
HL 3.92E-015	atm m3/mol	25	EST	MEYLAN,WM & HOWARD,PH (1991)
OH 5.26E-011	cm3/molc sec	25	EST	MEYLAN,WM & HOWARD,PH (1993)

CAS #: 025603-13-8	3ET-5-(5NO2-FURFURIL)THIAZOLIDIN-4-ON

Formula: $C_{10}H_9N_2O_5S$

Mol Weight: 269.26

MP (deg C):   FP (deg C):

BP (deg C):

BP pressure (mm Hg):

Property/Value	Units	Temp	Data Type	Reference
WS 8.43E+001	mg/L	25	EST	MEYLAN,WM ET AL. (1996)
logP 2.29			EXP	HANSCH,C & LEO,AJ (1985)
VP 1.33E-009	mm Hg	25	EST	NEELY,WB & BLAU,GE (1985)
DC	pKa			
HL 1.20E-009	atm m3/mol	25	EST	MEYLAN,WM & HOWARD,PH (1991)
OH 5.11E-011	cm3/molc sec	25	EST	MEYLAN,WM & HOWARD,PH (1993)

CAS #: 025603-14-9	3AL-5-(5NO2-FURFURIL)THIAZOLIDIN-2,4-ONE

Formula: $C_{11}H_9N_2O_5S$

Mol Weight: 281.27

MP (deg C):   FP (deg C):

BP (deg C):

BP pressure (mm Hg):

Property/Value	Units	Temp	Data Type	Reference
WS 8.27E+001	mg/L	25	EST	MEYLAN,WM ET AL. (1996)
logP 2.22			EXP	HANSCH,C & LEO,AJ (1985)
VP 6.22E-010	mm Hg	25	EST	NEELY,WB & BLAU,GE (1985)
DC	pKa			
HL 1.19E-009	atm m3/mol	25	EST	MEYLAN,WM & HOWARD,PH (1991)
OH 7.72E-011	cm3/molc sec	25	EST	MEYLAN,WM & HOWARD,PH (1993)

CAS #: 025612-07-1	N1-(4-CYANOPHENYL)SULFANILAMIDE

Formula: $C_{13}H_{11}N_3O_2S$

Mol Weight: 273.32

MP (deg C):   FP (deg C):

BP (deg C):

BP pressure (mm Hg):

Property/Value	Units	Temp	Data Type	Reference
WS 2.60E+002	mg/L	25	EST	MEYLAN,WM ET AL. (1996)
logP 1.83			EXP	HANSCH,C & LEO,AJ (1985)
VP 3.54E-009	mm Hg	25	EST	NEELY,WB & BLAU,GE (1985)
DC	pKa			
HL 7.95E-013	atm m3/mol	25	EST	MEYLAN,WM & HOWARD,PH (1991)
OH 3.05E-011	cm3/molc sec	25	EST	MEYLAN,WM & HOWARD,PH (1993)

CAS #: 025640-78-2 ISOPROPYLBIPHENYL

Formula: $C_{15}H_{16}$

Mol Weight: 196.29

MP (deg C): | FP (deg C):

BP (deg C):

BP pressure (mm Hg):

Property/Value	Units	Temp	Data Type	Reference
WS 6.00E-001	mg/L	25	EXP	ADDISON,RF ET AL. (1983)
logP 5.20			EXP	ADDISON,RF ET AL. (1983)
VP 5.00E-004	mm Hg	25	EXP	ADDISON,RF ET AL. (1983)
DC	pKa			
HL 2.15E-004	atm m3/mol	25	EST	VP/WSOL
OH 9.74E-012	cm3/molc sec	25	EST	MEYLAN,WM & HOWARD,PH (1993)

CAS #: 025653-16-1 TRI(3,5-XYLENYL) PHOSPHATE

Formula: $C_{24}H_{27}O_4P$

Mol Weight: 410.45

MP (deg C): | FP (deg C):

BP (deg C):

BP pressure (mm Hg):

Property/Value	Units	Temp	Data Type	Reference
WS 8.90E-001	mg/L	25	EXP	SAEGER,VW ET AL. (1979)
logP 7.98			EST	MEYLAN,WM & HOWARD,PH (1995)
VP 2.06E-008	mm Hg	25	EST	NEELY,WB & BLAU,GE (1985)
DC	pKa			
HL 7.19E-008	atm m3/mol	25	EST	MEYLAN,WM & HOWARD,PH (1991)
OH 7.47E-011	cm3/molc sec	25	EST	MEYLAN,WM & HOWARD,PH (1993)

CAS #: 025680-58-4 2-ETHYL-3-METHOXYPYRAZINE

Formula: $C_7H_{10}N_2O$

Mol Weight: 138.17

MP (deg C): | FP (deg C):

BP (deg C):

BP pressure (mm Hg):

Property/Value	Units	Temp	Data Type	Reference
WS 2.47E+003	mg/L	25	EST	MEYLAN,WM ET AL. (1996)
logP 1.80			EXP	YAMAGAMI,C ET AL. (1990)
VP 1.74E-001	mm Hg	25	EST	NEELY,WB & BLAU,GE (1985)
DC	pKa			
HL 1.47E-005	atm m3/mol	25	EXP	BUTTERY,RG ET AL. (1971)
OH 5.05E-012	cm3/molc sec	25	EST	MEYLAN,WM & HOWARD,PH (1993)

CAS #: 025694-31-9 2(N,N-DIME)-5(5-NO2-2-FURFURYL)THIAZOLONE

Formula: $C_{10}H_9N_3O_4S$

Mol Weight: 267.26

MP (deg C): | FP (deg C):

BP (deg C):

BP pressure (mm Hg):

Property/Value	Units	Temp	Data Type	Reference
WS 8.20E+002	mg/L	25	EST	MEYLAN,WM ET AL. (1996)
logP 1.14			EXP	HANSCH,C & LEO,AJ (1985)
VP 9.63E-008	mm Hg	25	EST	NEELY,WB & BLAU,GE (1985)
DC	pKa			
HL 3.92E-015	atm m3/mol	25	EST	MEYLAN,WM & HOWARD,PH (1991)
OH 1.05E-010	cm3/molc sec	25	EST	MEYLAN,WM & HOWARD,PH (1993)

CAS #: 025773-40-4 2-ISOPROPYL-3-METHOXYPYRAZINE

Formula: $C_8H_{12}N_2O$

Mol Weight: 152.20

MP (deg C): | FP (deg C):

BP (deg C):

BP pressure (mm Hg):

Property/Value	Units	Temp	Data Type	Reference
WS 6.99E+002	mg/L	25	EST	MEYLAN,WM ET AL. (1996)
logP 2.37			EST	MEYLAN,WM & HOWARD,PH (1995)
VP 1.13E-001	mm Hg	25	EST	NEELY,WB & BLAU,GE (1985)
DC	pKa			
HL 3.06E-006	atm m3/mol	25	EST	MEYLAN,WM & HOWARD,PH (1991)
OH 6.16E-012	cm3/molc sec	25	EST	MEYLAN,WM & HOWARD,PH (1993)

CAS #: 025784-00-3 O-(2-HYDROXYET)AMINOBENZOIC ACID

Formula: $C_9H_{11}NO_3$

Mol Weight: 181.19

MP (deg C): | FP (deg C):

BP (deg C):

BP pressure (mm Hg):

Property/Value	Units	Temp	Data Type	Reference
WS 3.82E+003	mg/L	25	EST	MEYLAN,WM ET AL. (1996)
logP 1.35			EXP	HANSCH,C & LEO,AJ (1985)
VP 2.94E-007	mm Hg	25	EST	NEELY,WB & BLAU,GE (1985)
DC	pKa			
HL 4.08E-015	atm m3/mol	25	EST	MEYLAN,WM & HOWARD,PH (1991)
OH 3.12E-011	cm3/molc sec	25	EST	MEYLAN,WM & HOWARD,PH (1993)

CAS #: 025790-55-0 4-CHLORO-1,2-BUTADIENE

Formula: C_4H_5Cl

Mol Weight: 88.54

MP (deg C): | FP (deg C):

BP (deg C): 88

BP pressure (mm Hg):

Property/Value	Units	Temp	Data Type	Reference
WS 1.32E+003	mg/L	25	EST	MEYLAN,WM ET AL. (1996)
logP 2.32			EST	MEYLAN,WM & HOWARD,PH (1995)
VP 6.12E+001	mm Hg	25	EST	NEELY,WB & BLAU,GE (1985)
DC	pKa			
HL 3.41E-002	atm m3/mol	25	EST	MEYLAN,WM & HOWARD,PH (1991)
OH 2.39E-011	cm3/molc sec	25	EST	MEYLAN,WM & HOWARD,PH (1993)

CAS #: 025800-28-6 BENZOIC ACID, 2-CHLORO-, PROPYL ESTER

Formula: $C_{10}H_{11}ClO_2$

Mol Weight: 198.65

MP (deg C): | FP (deg C):

BP (deg C):

BP pressure (mm Hg):

Property/Value	Units	Temp	Data Type	Reference
WS 5.03E+001	mg/L	25	EST	MEYLAN,WM ET AL. (1996)
logP 3.45			EXP	SANGSTER,J (1993)
VP 1.01E-002	mm Hg	25	EST	NEELY,WB & BLAU,GE (1985)
DC	pKa			
HL 4.53E-005	atm m3/mol	25	EST	MEYLAN,WM & HOWARD,PH (1991)
OH 3.60E-012	cm3/molc sec	25	EST	MEYLAN,WM & HOWARD,PH (1993)

CAS #: 025855-20-3 — QUINAZOLINE, 2-PHENYL-

Formula: $C_{14}H_{10}N_2$

Mol Weight: 206.25

MP (deg C):
FP (deg C):
BP (deg C):
BP pressure (mm Hg):

Property/Value	Units	Temp	Data Type	Reference
WS 4.88E+001	mg/L	25	EST	MEYLAN,WM ET AL. (1996)
logP 3.42			EXP	SANGSTER,J (1993)
VP 4.85E-006	mm Hg	25	EST	NEELY,WB & BLAU,GE (1985)
DC	pKa			
HL 2.19E-008	atm m3/mol	25	EST	MEYLAN,WM & HOWARD,PH (1991)
OH 6.86E-012	cm3/molc sec	25	EST	MEYLAN,WM & HOWARD,PH (1993)

CAS #: 025905-77-5 — MINAPRINE

Formula: $C_{17}H_{22}N_4O$

Mol Weight: 298.39

MP (deg C): 122
FP (deg C):
BP (deg C):
BP pressure (mm Hg):

Property/Value	Units	Temp	Data Type	Reference
WS 2.36E+003	mg/L	25	EST	MEYLAN,WM ET AL. (1996)
logP 2.03			EXP	HANSCH,C ET AL. (1995)
VP 1.78E-009	mm Hg	25	EST	NEELY,WB & BLAU,GE (1985)
DC	pKa			
HL 8.51E-016	atm m3/mol	25	EST	MEYLAN,WM & HOWARD,PH (1991)
OH 1.67E-010	cm3/molc sec	25	EST	MEYLAN,WM & HOWARD,PH (1993)

CAS #: 025933-30-6 — 5-CL-2'-ME SALICYLANILIDE

Formula: $C_{14}H_{12}ClNO_2$

Mol Weight: 261.71

MP (deg C):
FP (deg C):
BP (deg C):
BP pressure (mm Hg):

Property/Value	Units	Temp	Data Type	Reference
WS 1.96E+001	mg/L	25	EST	MEYLAN,WM ET AL. (1996)
logP 4.21			EXP	TERADA,H ET AL. (1988)
VP 2.50E-009	mm Hg	25	EST	NEELY,WB & BLAU,GE (1985)
DC	pKa			
HL 1.31E-010	atm m3/mol	25	EST	MEYLAN,WM & HOWARD,PH (1991)
OH 2.44E-011	cm3/molc sec	25	EST	MEYLAN,WM & HOWARD,PH (1993)

CAS #: 025953-19-9 — CEFAZOLIN

Formula: $C_{14}H_{14}N_8O_4S_3$

Mol Weight: 454.51

MP (deg C):
FP (deg C):
BP (deg C):
BP pressure (mm Hg):

Property/Value	Units	Temp	Data Type	Reference
WS 2.14E+002	mg/L	25	EST	MEYLAN,WM ET AL. (1996)
logP -0.58			EXP	HANSCH,C ET AL. (1995)
VP 1.50E-018	mm Hg	25	EST	NEELY,WB & BLAU,GE (1985)
DC	pKa			
HL 2.01E-023	atm m3/mol	25	EST	MEYLAN,WM & HOWARD,PH (1991)
OH 1.18E-010	cm3/molc sec	25	EST	MEYLAN,WM & HOWARD,PH (1993)

CAS #: 025954-13-6 — FOSAMINE AMMONIUM SALT

Formula: $C_3H_{11}N_2O_4P$

Mol Weight: 170.11

MP (deg C): 175
FP (deg C):
BP (deg C):
BP pressure (mm Hg):

Property/Value	Units	Temp	Data Type	Reference
WS 1.00E+006	mg/L	25	EXP	SHIU,WY ET AL. (1990)
logP -5.98			EST	MEYLAN,WM & HOWARD,PH (1995)
VP 2.72E-008	mm Hg	25	EST	NEELY,WB & BLAU,GE (1985)
DC	pKa			
HL 8.37E-023	atm m3/mol	25	EST	MEYLAN,WM & HOWARD,PH (1991)
OH 2.13E-011	cm3/molc sec	25	EST	MEYLAN,WM & HOWARD,PH (1993)

CAS #: 025963-47-7 — PHENOL, 4-(4-AMINOPHENYL)SULFONYL -

Formula: $C_{12}H_{11}NO_3S$

Mol Weight: 249.29

MP (deg C):
FP (deg C):
BP (deg C):
BP pressure (mm Hg):

Property/Value	Units	Temp	Data Type	Reference
WS 4.14E+003	mg/L	25	EST	MEYLAN,WM ET AL. (1996)
logP 1.57			EXP	ALTOMARE,C ET AL. (1991)
VP 6.72E-009	mm Hg	25	EST	NEELY,WB & BLAU,GE (1985)
DC	pKa			
HL 9.17E-015	atm m3/mol	25	EST	MEYLAN,WM & HOWARD,PH (1991)
OH 3.03E-011	cm3/molc sec	25	EST	MEYLAN,WM & HOWARD,PH (1993)

CAS #: 025973-55-1 — 2-(2H-BENZOTRIAZOL-2-YL)-4,6-DI-(TERT-PENTYL)PH*

Formula: $C_{22}H_{29}N_3O$

Mol Weight: 351.50

MP (deg C):
FP (deg C):
BP (deg C):
BP pressure (mm Hg):

Property/Value	Units	Temp	Data Type	Reference
WS 1.48E-002	mg/L	25	EST	MEYLAN,WM ET AL. (1996)
logP 7.25			EST	MEYLAN,WM & HOWARD,PH (1995)
VP 1.93E-010	mm Hg	25	EST	NEELY,WB & BLAU,GE (1985)
DC	pKa			
HL 6.52E-013	atm m3/mol	25	EST	MEYLAN,WM & HOWARD,PH (1991)
OH 1.58E-011	cm3/molc sec	25	EST	MEYLAN,WM & HOWARD,PH (1993)

CAS #: 025998-87-2 — 1-(4-BENZYLOXYPH)-3-MEO-3-ME UREA

Formula: $C_{16}H_{18}N_2O_3$

Mol Weight: 286.33

MP (deg C):
FP (deg C):
BP (deg C):
BP pressure (mm Hg):

Property/Value	Units	Temp	Data Type	Reference
WS 3.25E+001	mg/L	25	EST	MEYLAN,WM ET AL. (1996)
logP 3.11			EXP	HANSCH,C & LEO,AJ (1985)
VP 3.09E-008	mm Hg	25	EST	NEELY,WB & BLAU,GE (1985)
DC	pKa			
HL 1.00E-010	atm m3/mol	25	EST	MEYLAN,WM & HOWARD,PH (1991)
OH 4.90E-011	cm3/molc sec	25	EST	MEYLAN,WM & HOWARD,PH (1993)

CAS #: 026002-80-2	PHENOTHRIN

Formula: $C_{23}H_{26}O_3$

Mol Weight: 350.46

MP (deg C): FP (deg C):

BP (deg C):

BP pressure (mm Hg):

Property/Value	Units	Temp	Data Type	Reference
WS 2.00E+000	mg/L	30	EXP	YALKOWSKY,SH & DANNENFELSER,RM (1992)
logP 7.54			EST	MEYLAN,WM & HOWARD,PH (1995)
VP 1.43E-007	mm Hg	21	EXP	TOMLIN,C (1994)
DC	pKa			
HL 3.30E-008	atm m3/mol	21	EST	VP/WSOL
OH 1.23E-010	cm3/molc sec	25	EST	MEYLAN,WM & HOWARD,PH (1993)

CAS #: 026019-17-0	2-(N,N-DIME AMINOMETHYL)THIOPHENE

Formula: $C_7H_{13}NS$

Mol Weight: 143.25

MP (deg C): FP (deg C):

BP (deg C): 77-78

BP pressure (mm Hg): 2.20E+001

Property/Value	Units	Temp	Data Type	Reference
WS 3.16E+004	mg/L	25	EST	MEYLAN,WM ET AL. (1996)
logP 1.67			EXP	HANSCH,C & LEO,AJ (1985)
VP 3.59E-001	mm Hg	25	EST	NEELY,WB & BLAU,GE (1985)
DC	pKa			
HL 1.60E-006	atm m3/mol	25	EST	MEYLAN,WM & HOWARD,PH (1991)
OH 1.01E-010	cm3/molc sec	25	EST	MEYLAN,WM & HOWARD,PH (1993)

CAS #: 026019-18-1	2-CH2CH2NME2 THIOPHENE

Formula: $C_8H_{15}NS$

Mol Weight: 157.28

MP (deg C): FP (deg C):

BP (deg C):

BP pressure (mm Hg):

Property/Value	Units	Temp	Data Type	Reference
WS 1.46E+004	mg/L	25	EST	MEYLAN,WM ET AL. (1996)
logP 1.99			EXP	HANSCH,C & LEO,AJ (1985)
VP 1.12E-001	mm Hg	25	EST	NEELY,WB & BLAU,GE (1985)
DC	pKa			
HL 2.13E-006	atm m3/mol	25	EST	MEYLAN,WM & HOWARD,PH (1991)
OH 1.04E-010	cm3/molc sec	25	EST	MEYLAN,WM & HOWARD,PH (1993)

CAS #: 026060-30-0	BENZENECARBOTHIAMIDE, 4-(NITRO)-

Formula: $C_7H_6N_2O_2S$

Mol Weight: 182.20

MP (deg C): FP (deg C):

BP (deg C):

BP pressure (mm Hg):

Property/Value	Units	Temp	Data Type	Reference
WS 7.14E+003	mg/L	25	EST	MEYLAN,WM ET AL. (1996)
logP 1.75			EXP	SANGSTER,J (1994)
VP 7.50E-005	mm Hg	25	EST	NEELY,WB & BLAU,GE (1985)
DC	pKa			
HL 1.55E-009	atm m3/mol	25	EST	MEYLAN,WM & HOWARD,PH (1991)
OH 2.15E-011	cm3/molc sec	25	EST	MEYLAN,WM & HOWARD,PH (1993)

CAS #: 026074-74-8	CYCLOPROPANECARBOXYLIC ACID, 3-(3-METHOXY-2-METH

Formula: $C_{12}H_{18}O_4$

Mol Weight: 226.27

MP (deg C): FP (deg C):

BP (deg C):

BP pressure (mm Hg):

Property/Value	Units	Temp	Data Type	Reference
WS 1.67E+002	mg/L	25	EST	MEYLAN,WM ET AL. (1996)
logP 2.67			EXP	NISHIMURA,K ET AL. (1987)
VP 3.35E-002	mm Hg	25	EST	NEELY,WB & BLAU,GE (1985)
DC	pKa			
HL 7.40E-007	atm m3/mol	25	EST	MEYLAN,WM & HOWARD,PH (1991)
OH 3.14E-011	cm3/molc sec	25	EST	MEYLAN,WM & HOWARD,PH (1993)

CAS #: 026087-47-8	IBP (IPROBENFOS)

Formula: $C_{13}H_{21}O_3PS$

Mol Weight: 288.35

MP (deg C): FP (deg C):

BP (deg C):

BP pressure (mm Hg):

Property/Value	Units	Temp	Data Type	Reference
WS 4.00E+002	mg/L	25	EXP	SHIU,WY ET AL. (1990)
logP 3.34			EXP	SAITO,H ET AL. (1993)
VP 4.05E-005	mm Hg	25	EXP	WATANABE,T (1993)
DC	pKa			
HL 3.84E-008	atm m3/mol	25	EXP	VP/WSOL
OH 1.04E-010	cm3/molc sec	25	EST	MEYLAN,WM & HOWARD,PH (1993)

CAS #: 026114-12-5	BENZENEACETIC ACID, 4-PROPYL-

Formula: $C_{11}H_{14}O_2$

Mol Weight: 178.23

MP (deg C): FP (deg C):

BP (deg C):

BP pressure (mm Hg):

Property/Value	Units	Temp	Data Type	Reference
WS 3.82E+002	mg/L	25	EST	MEYLAN,WM ET AL. (1996)
logP 3.00			EXP	SANGSTER,J (1993)
VP 2.25E-004	mm Hg	25	EST	NEELY,WB & BLAU,GE (1985)
DC	pKa			
HL 8.60E-008	atm m3/mol	25	EST	MEYLAN,WM & HOWARD,PH (1991)
OH 8.68E-012	cm3/molc sec	25	EST	MEYLAN,WM & HOWARD,PH (1993)

CAS #: 026114-14-7	BENZENEACETIC ACID, 4-(1,1-DIMETHYLPROPYL)-

Formula: $C_{13}H_{18}O_2$

Mol Weight: 206.29

MP (deg C): FP (deg C):

BP (deg C):

BP pressure (mm Hg):

Property/Value	Units	Temp	Data Type	Reference
WS 1.08E+002	mg/L	25	EST	MEYLAN,WM ET AL. (1996)
logP 3.48			EXP	SANGSTER,J (1994)
VP 9.42E-005	mm Hg	25	EST	NEELY,WB & BLAU,GE (1985)
DC	pKa			
HL 1.52E-007	atm m3/mol	25	EST	MEYLAN,WM & HOWARD,PH (1991)
OH 7.30E-012	cm3/molc sec	25	EST	MEYLAN,WM & HOWARD,PH (1993)

CAS #: 026118-57-0				BENZENEACETIC ACID, 4-PROPOXY-
Formula: $C_{11}H_{14}O_3$				
Mol Weight: 194.23				
MP (deg C):		FP (deg C):		
BP (deg C):				
BP pressure (mm Hg):				

Property/Value	Units	Temp	Data Type	Reference
WS 1.24E+003	mg/L	25	EST	MEYLAN,WM ET AL. (1996)
logP 2.31			EXP	SANGSTER,J (1993)
VP 9.28E-005	mm Hg	25	EST	NEELY,WB & BLAU,GE (1985)
DC	pKa			
HL 4.61E-009	atm m3/mol	25	EST	MEYLAN,WM & HOWARD,PH (1991)
OH 3.38E-011	cm3/molc sec	25	EST	MEYLAN,WM & HOWARD,PH (1993)

CAS #: 026171-23-3				TOLMETIN
Formula: $C_{15}H_{15}NO_3$				
Mol Weight: 257.29				
MP (deg C): 155-157 de		FP (deg C):		
BP (deg C):				
BP pressure (mm Hg):				

Property/Value	Units	Temp	Data Type	Reference
WS 2.22E+002	mg/L	25	EST	MEYLAN,WM ET AL. (1996)
logP 2.79			EXP	SANGSTER,J (1993)
VP 1.34E-007	mm Hg	25	EST	NEELY,WB & BLAU,GE (1985)
DC	pKa			
HL 6.32E-013	atm m3/mol	25	EST	MEYLAN,WM & HOWARD,PH (1991)
OH 1.72E-010	cm3/molc sec	25	EST	MEYLAN,WM & HOWARD,PH (1993)

CAS #: 026225-79-6				ETHOFUMESATE
Formula: $C_{13}H_{18}O_5S$				
Mol Weight: 286.35				
MP (deg C): 71		FP (deg C):		
BP (deg C):				
BP pressure (mm Hg):				

Property/Value	Units	Temp	Data Type	Reference
WS 1.10E+002	mg/L	25	EXP	YALKOWSKY,SH & DANNENFELSER,RM (1992)
logP 2.70			EXP	TOMLIN,C (1994)
VP 4.90E-006	mm Hg	25	EXP	WAUCHOPE,RD ET AL. (1991A)
DC	pKa			
HL 1.68E-008	atm m3/mol	25	EST	VP/WSOL
OH 5.46E-011	cm3/molc sec	25	EST	MEYLAN,WM & HOWARD,PH (1993)

CAS #: 026232-35-9				6-CHLORO-BENZAZEPINE
Formula: $C_{10}H_{12}ClN$				
Mol Weight: 181.67				
MP (deg C):		FP (deg C):		
BP (deg C):				
BP pressure (mm Hg):				

Property/Value	Units	Temp	Data Type	Reference
WS 4.91E+003	mg/L	25	EST	MEYLAN,WM ET AL. (1996)
logP 2.40			EXP	HANSCH,C ET AL. (1995)
VP 2.25E-003	mm Hg	25	EST	NEELY,WB & BLAU,GE (1985)
DC	pKa			
HL 6.43E-007	atm m3/mol	25	EST	MEYLAN,WM & HOWARD,PH (1991)
OH 9.08E-011	cm3/molc sec	25	EST	MEYLAN,WM & HOWARD,PH (1993)

CAS #: 026245-56-7				N-CHLORO-2-PROPANAMINE
Formula: C_3H_8ClN				
Mol Weight: 93.56				
MP (deg C):		FP (deg C):		
BP (deg C):				
BP pressure (mm Hg):				

Property/Value	Units	Temp	Data Type	Reference
WS 8.41E+004	mg/L	25	EST	MEYLAN,WM ET AL. (1996)
logP 0.19			EST	MEYLAN,WM & HOWARD,PH (1995)
VP 1.57E+002	mm Hg	25	EST	NEELY,WB & BLAU,GE (1985)
DC	pKa			
HL 2.56E-004	atm m3/mol	25	EST	MEYLAN,WM & HOWARD,PH (1991)
OH 1.84E-011	cm3/molc sec	25	EST	MEYLAN,WM & HOWARD,PH (1993)

CAS #: 026258-70-8				1,1-BIS(4-ETHOXYPHENYL)-2-NITROPROPANE
Formula: $C_{19}H_{23}NO_4$				
Mol Weight: 329.40				
MP (deg C):		FP (deg C):		
BP (deg C):				
BP pressure (mm Hg):				

Property/Value	Units	Temp	Data Type	Reference
WS 2.46E-001	mg/L	25	EST	MEYLAN,WM ET AL. (1996)
logP 4.84			EST	MEYLAN,WM & HOWARD,PH (1995)
VP 1.36E-007	mm Hg	25	EST	NEELY,WB & BLAU,GE (1985)
DC	pKa			
HL 2.98E-009	atm m3/mol	25	EST	MEYLAN,WM & HOWARD,PH (1991)
OH 6.45E-011	cm3/molc sec	25	EST	MEYLAN,WM & HOWARD,PH (1993)

CAS #: 026259-45-0				SECBUMETON
Formula: $C_{10}H_{19}N_5O$				
Mol Weight: 225.30				
MP (deg C): 87		FP (deg C):		
BP (deg C):				
BP pressure (mm Hg):				

Property/Value	Units	Temp	Data Type	Reference
WS 6.20E+002	mg/L	20	EXP	YALKOWSKY,SH & DANNENFELSER,RM (1992)
logP 3.64			EST	MEYLAN,WM & HOWARD,PH (1995)
VP 7.30E-006	mm Hg	20	EXP	AUGUSTIJN-BECKERS,PWM ET AL. (1994)
DC	pKa			
HL 3.49E-009	atm m3/mol	20	EST	VP/WSOL
OH 3.37E-011	cm3/molc sec	25	EST	MEYLAN,WM & HOWARD,PH (1993)

CAS #: 026323-01-3				N-BENZYLQUINOLINIUM BROMIDE
Formula: $C_{16}H_{14}BrN$				
Mol Weight: 300.20				
MP (deg C):		FP (deg C):		
BP (deg C):				
BP pressure (mm Hg):				

Property/Value	Units	Temp	Data Type	Reference
WS 6.14E+004	mg/L	25	EST	MEYLAN,WM ET AL. (1996)
logP -0.82			EXP	HANSCH,C & LEO,AJ (1985)
VP 1.33E-006	mm Hg	25	EST	NEELY,WB & BLAU,GE (1985)
DC	pKa			
HL 1.19E-012	atm m3/mol	25	EST	MEYLAN,WM & HOWARD,PH (1991)
OH 3.02E-011	cm3/molc sec	25	EST	MEYLAN,WM & HOWARD,PH (1993)

CAS #: 026328-59-6 — 4-ISOTHIOCYANOBENZOPHENONE

Formula: $C_{14}H_9NOS$
Mol Weight: 239.30
MP (deg C):   **FP (deg C):**
BP (deg C):
BP pressure (mm Hg):

Property/Value	Units	Temp	Data Type	Reference
WS 1.84E+000	mg/L	25	EST	MEYLAN,WM ET AL. (1996)
logP 4.88			EXP	HANSCH,C & LEO,AJ (1985)
VP 4.12E-006	mm Hg	25	EST	NEELY,WB & BLAU,GE (1985)
DC	pKa			
HL 2.82E-007	atm m3/mol	25	EST	MEYLAN,WM & HOWARD,PH (1991)
OH 2.90E-012	cm3/molc sec	25	EST	MEYLAN,WM & HOWARD,PH (1993)

CAS #: 026399-36-0 — PROFLURALIN

Formula: $C_{14}H_{16}F_3N_3O_4$
Mol Weight: 347.30
MP (deg C): 33-36   **FP (deg C):**
BP (deg C):
BP pressure (mm Hg):

Property/Value	Units	Temp	Data Type	Reference
WS 1.00E-001	mg/L	20	EXP	YALKOWSKY,SH & DANNENFELSER,RM (1992)
logP 5.58			EXP	ELLGEHAUSEN,H ET AL (1981)
VP 6.90E-005	mm Hg	20	EXP	BESTE,CE (1983)
DC	pKa			
HL 3.15E-004	atm m3/mol	20	EST	VP/WSOL
OH 2.28E-011	cm3/molc sec	25	EST	MEYLAN,WM & HOWARD,PH (1993)

CAS #: 026444-49-5 — CRESYL DIPHENYL PHOSPHATE

Formula: $C_{19}H_{17}O_4P$
Mol Weight: 340.32
MP (deg C):   **FP (deg C):**
BP (deg C):
BP pressure (mm Hg):

Example - Attachment of methyl unspecified

Property/Value	Units	Temp	Data Type	Reference
WS 5.00E+001	mg/L	25	EXP	MIDWEST RESEARCH INSTITUTE (1977)
logP 4.51			EXP	SAEGER,VW ET AL. (1979)
VP 4.70E-006	mm Hg	25	EXP	BOETHLING,RS & COOPER,JC (1985)
DC	pKa			
HL 4.21E-008	atm m3/mol	25	EST	VP/WSOL
OH 7.14E-011	cm3/molc sec	25	EST	MEYLAN,WM & HOWARD,PH (1993)

CAS #: 026455-31-2 — P-NITROPHENYL-I-PROPYL ETHER

Formula: $C_9H_{11}NO_3$
Mol Weight: 181.19
MP (deg C):   **FP (deg C):**
BP (deg C):
BP pressure (mm Hg):

Property/Value	Units	Temp	Data Type	Reference
WS 8.99E+001	mg/L	25	EST	MEYLAN,WM ET AL. (1996)
logP 2.80			EXP	HANSCH,C & LEO,AJ (1985)
VP 4.29E-003	mm Hg	25	EST	NEELY,WB & BLAU,GE (1985)
DC	pKa			
HL 2.22E-006	atm m3/mol	25	EST	MEYLAN,WM & HOWARD,PH (1991)
OH 1.49E-011	cm3/molc sec	25	EST	MEYLAN,WM & HOWARD,PH (1993)

CAS #: 026489-01-0 — (+-)-CITRONELLOL

Formula: $C_{10}H_{20}O$
Mol Weight: 156.27
MP (deg C):   **FP (deg C):**
BP (deg C): 224
BP pressure (mm Hg):

Property/Value	Units	Temp	Data Type	Reference
WS 2.12E+002	mg/L	25	EST	MEYLAN,WM ET AL. (1996)
logP 3.56			EST	MEYLAN,WM & HOWARD,PH (1995)
VP 1.30E-002	mm Hg	25	EST	NEELY,WB & BLAU,GE (1985)
DC	pKa			
HL 5.68E-005	atm m3/mol	25	EST	MEYLAN,WM & HOWARD,PH (1991)
OH 9.84E-011	cm3/molc sec	25	EST	MEYLAN,WM & HOWARD,PH (1993)

CAS #: 026512-63-0 — O=P O,O-DIET-O-(2-ME-4-SMEPHENYL)

Formula: $C_{12}H_{19}O_4PS$
Mol Weight: 290.32
MP (deg C):   **FP (deg C):**
BP (deg C):
BP pressure (mm Hg):

Property/Value	Units	Temp	Data Type	Reference
WS 2.54E+001	mg/L	25	EST	MEYLAN,WM ET AL. (1996)
logP 3.21			EXP	HANSCH,C & LEO,AJ (1985)
VP 8.96E-006	mm Hg	25	EST	NEELY,WB & BLAU,GE (1985)
DC	pKa			
HL 5.22E-009	atm m3/mol	25	EST	MEYLAN,WM & HOWARD,PH (1991)
OH 6.39E-011	cm3/molc sec	25	EST	MEYLAN,WM & HOWARD,PH (1993)

CAS #: 026512-71-0 — O,O-DIET-O-(2-CL-4-MESPH)PHOSPHOROTHIOATE

Formula: $C_{11}H_{16}ClO_3PS_2$
Mol Weight: 326.80
MP (deg C):   **FP (deg C):**
BP (deg C):
BP pressure (mm Hg):

Property/Value	Units	Temp	Data Type	Reference
WS 1.49E+000	mg/L	25	EST	MEYLAN,WM ET AL. (1996)
logP 4.40			EXP	HANSCH,C & LEO,AJ (1985)
VP 1.59E-005	mm Hg	25	EST	NEELY,WB & BLAU,GE (1985)
DC	pKa			
HL 1.62E-006	atm m3/mol	25	EST	MEYLAN,WM & HOWARD,PH (1991)
OH 9.92E-011	cm3/molc sec	25	EST	MEYLAN,WM & HOWARD,PH (1993)

CAS #: 026530-20-1 — 3(2H)-ISOTHIAZOLONE, 2-OCTYL-

Formula: $C_{11}H_{19}NOS$
Mol Weight: 213.34
MP (deg C):   **FP (deg C):**
BP (deg C): 120
BP pressure (mm Hg): 1.00E-002

Property/Value	Units	Temp	Data Type	Reference
WS 3.02E+002	mg/L	25	EST	MEYLAN,WM ET AL. (1996)
logP 2.45			EXP	TOMLIN,C (1994)
VP 2.71E-005	mm Hg	25	EST	NEELY,WB & BLAU,GE (1985)
DC	pKa			
HL 3.60E-007	atm m3/mol	25	EST	MEYLAN,WM & HOWARD,PH (1991)
OH 3.94E-011	cm3/molc sec	25	EST	MEYLAN,WM & HOWARD,PH (1993)

CAS #: 026675-46-7				ISOFLURANE

Formula: $C_3H_2ClF_5O$

Mol Weight: 184.49

MP (deg C): 48.5 FP (deg C):

BP (deg C): 48.5

BP pressure (mm Hg):

Property/Value	Units	Temp	Data Type	Reference
WS 4.47E+003	mg/L	37	EXP	YALKOWSKY,SH & DANNENFELSER,RM (1992)
logP 2.06			EXP	HANSCH,C ET AL. (1995)
VP 5.91E+002	mm Hg	25	EST	NEELY,WB & BLAU,GE (1985)
DC	pKa			
HL 1.26E-002	atm m3/mol	25	EST	MEYLAN,WM & HOWARD,PH (1991)
OH 2.10E-014	cm3/molc sec	25	EXP	ATKINSON,R (1989)

CAS #: 026707-54-0				O=P(OET)(OET)O-2,6-CL-4-SMEPHENYL

Formula: $C_{11}H_{15}Cl_2O_4PS$

Mol Weight: 345.18

MP (deg C): FP (deg C):

BP (deg C):

BP pressure (mm Hg):

Property/Value	Units	Temp	Data Type	Reference
WS 4.41E+000	mg/L	25	EST	MEYLAN,WM ET AL. (1996)
logP 3.72			EXP	HANSCH,C & LEO,AJ (1985)
VP 1.95E-006	mm Hg	25	EST	NEELY,WB & BLAU,GE (1985)
DC	pKa			
HL 2.60E-009	atm m3/mol	25	EST	MEYLAN,WM & HOWARD,PH (1991)
OH 4.43E-011	cm3/molc sec	25	EST	MEYLAN,WM & HOWARD,PH (1993)

CAS #: 026722-85-0				CEPHALOTHIN ANALOG (3-CH2-S-CH3)

Formula: $C_{15}H_{18}N_2O_4S_3$

Mol Weight: 386.51

MP (deg C): FP (deg C):

BP (deg C):

BP pressure (mm Hg):

Property/Value	Units	Temp	Data Type	Reference
WS 5.24E+002	mg/L	25	EST	MEYLAN,WM ET AL. (1996)
logP 1.01			EXP	HANSCH,C & LEO,AJ (1985)
VP 2.34E-014	mm Hg	25	EST	NEELY,WB & BLAU,GE (1985)
DC	pKa			
HL 7.81E-017	atm m3/mol	25	EST	MEYLAN,WM & HOWARD,PH (1991)
OH 1.43E-010	cm3/molc sec	25	EST	MEYLAN,WM & HOWARD,PH (1993)

CAS #: 026761-40-0				DIISODECYL PHTHALATE

Formula: $C_{28}H_{46}O_4$

Mol Weight: 446.68

MP (deg C): FP (deg C):

BP (deg C):

BP pressure (mm Hg):

Property/Value	Units	Temp	Data Type	Reference
WS 5.00E+001	mg/L	25	EXP	PEAKALL,DB (1975)
logP 10.36			EST	MEYLAN,WM & HOWARD,PH (1995)
VP 1.00E-001	mm Hg	25	EXP	DORIGAN,J ET AL. (1976A)
DC	pKa			
HL 1.18E-003	atm m3/mol	25	EST	VP/WSOL
OH 2.62E-011	cm3/molc sec	25	EST	MEYLAN,WM & HOWARD,PH (1993)

CAS #: 026787-78-0				AMOXICILLIN

Formula: $C_{16}H_{19}N_3O_5S$

Mol Weight: 365.41

MP (deg C): FP (deg C):

BP (deg C):

BP pressure (mm Hg):

Property/Value	Units	Temp	Data Type	Reference
WS 3.43E+003	mg/L	25	EST	MEYLAN,WM ET AL. (1996)
logP 0.87			EXP	SANGSTER,J (1994)
VP 4.69E-017	mm Hg	25	EST	NEELY,WB & BLAU,GE (1985)
DC	pKa			
HL 2.49E-021	atm m3/mol	25	EST	MEYLAN,WM & HOWARD,PH (1991)
OH 1.61E-010	cm3/molc sec	25	EST	MEYLAN,WM & HOWARD,PH (1993)

CAS #: 026798-03-8				OO-DIET-O-(3-CL-4-MES PH)PHOSPHATE

Formula: $C_{11}H_{16}ClO_4PS$

Mol Weight: 310.74

MP (deg C): FP (deg C):

BP (deg C):

BP pressure (mm Hg):

Property/Value	Units	Temp	Data Type	Reference
WS 3.28E+001	mg/L	25	EST	MEYLAN,WM ET AL. (1996)
logP 2.94			EXP	HANSCH,C & LEO,AJ (1985)
VP 5.90E-006	mm Hg	25	EST	NEELY,WB & BLAU,GE (1985)
DC	pKa			
HL 3.51E-009	atm m3/mol	25	EST	MEYLAN,WM & HOWARD,PH (1991)
OH 4.24E-011	cm3/molc sec	25	EST	MEYLAN,WM & HOWARD,PH (1993)

CAS #: 026807-64-7				N1-(3-CL-2-PYRIDYL)SULFANILAMIDE

Formula: $C_{11}H_{10}ClN_3O_2S$

Mol Weight: 283.74

MP (deg C): FP (deg C):

BP (deg C):

BP pressure (mm Hg):

Property/Value	Units	Temp	Data Type	Reference
WS 3.04E+003	mg/L	25	EST	MEYLAN,WM ET AL. (1996)
logP 0.82			EXP	HANSCH,C & LEO,AJ (1985)
VP 1.72E-008	mm Hg	25	EST	NEELY,WB & BLAU,GE (1985)
DC	pKa			
HL 5.04E-012	atm m3/mol	25	EST	MEYLAN,WM & HOWARD,PH (1991)
OH 2.87E-011	cm3/molc sec	25	EST	MEYLAN,WM & HOWARD,PH (1993)

CAS #: 026819-07-8				BENZAMIDE, N-ETHYL-3-METHYL-

Formula: $C_{10}H_{13}NO$

Mol Weight: 163.22

MP (deg C): FP (deg C):

BP (deg C):

BP pressure (mm Hg):

Property/Value	Units	Temp	Data Type	Reference
WS 4.94E+003	mg/L	25	EST	MEYLAN,WM ET AL. (1996)
logP 1.32			EXP	SURYANARAYANA,MVS ET AL. (1991)
VP 1.30E-002	mm Hg	25	EXP	SURYANARAYANA,MVS ET AL. (1991)
DC	pKa			
HL 7.12E-009	atm m3/mol	25	EST	MEYLAN,WM & HOWARD,PH (1991)
OH 1.18E-011	cm3/molc sec	25	EST	MEYLAN,WM & HOWARD,PH (1993)

Formula: $C_{10}H_{13}NO$	
Mol Weight: 163.22	
MP (deg C):	FP (deg C):
BP (deg C):	
BP pressure (mm Hg):	

Property/Value	Units	Temp	Data Type	Reference
WS 4.39E+003	mg/L	25	EST	MEYLAN,WM ET AL. (1996)
logP 1.38			EXP	SURYANARAYANA,MVS ET AL. (1991)
VP 6.30E-002	mm Hg	25	EXP	SURYANARAYANA,MVS ET AL. (1991)
DC	pKa			
HL 7.12E-009	atm m3/mol	25	EST	MEYLAN,WM & HOWARD,PH (1991)
OH 1.43E-011	cm3/molc sec	25	EST	MEYLAN,WM & HOWARD,PH (1993)

| CAS #: 026839-75-8 | TIMOLOL |

Formula: $C_{13}H_{24}N_4O_3S$	
Mol Weight: 316.43	
MP (deg C): 201.5-203	FP (deg C):
BP (deg C):	
BP pressure (mm Hg):	

Property/Value	Units	Temp	Data Type	Reference
WS 2.74E+003	mg/L	25	EST	MEYLAN,WM ET AL. (1996)
logP 1.83			EXP	HANSCH,C ET AL. (1995)
VP 1.08E-009	mm Hg	25	EST	NEELY,WB & BLAU,GE (1985)
DC	pKa			
HL 4.35E-017	atm m3/mol	25	EST	MEYLAN,WM & HOWARD,PH (1991)
OH 1.67E-010	cm3/molc sec	25	EST	MEYLAN,WM & HOWARD,PH (1993)

| CAS #: 026863-17-2 | PYRIDINIUM, 1-METHYL-4-(4-METHYLPHENYL)-, IODIDE |

Formula: $C_{13}H_{14}IN$	
Mol Weight: 311.17	
MP (deg C):	FP (deg C):
BP (deg C):	
BP pressure (mm Hg):	

Property/Value	Units	Temp	Data Type	Reference
WS 4.43E+005	mg/L	25	EST	MEYLAN,WM ET AL. (1996)
logP -1.90			EXP	EL TAYAR,N ET AL. (1991)
VP 1.08E-005	mm Hg	25	EST	NEELY,WB & BLAU,GE (1985)
DC	pKa			
HL 1.27E-011	atm m3/mol	25	EST	MEYLAN,WM & HOWARD,PH (1991)
OH 1.41E-011	cm3/molc sec	25	EST	MEYLAN,WM & HOWARD,PH (1993)

| CAS #: 026952-21-6 | ISOOCTYL ALCOHOL |

Formula: $C_8H_{18}O$	
Mol Weight: 130.23	
MP (deg C): -117.2	FP (deg C):
BP (deg C):	
BP pressure (mm Hg):	

Property/Value	Units	Temp	Data Type	Reference
WS 6.40E+002	mg/L		EXP	NEELY,WB (1984)
logP 2.73			EST	MEYLAN,WM & HOWARD,PH (1995)
VP 1.95E+000	mm Hg	50	EXP	FLICK,EW (1991)
DC	pKa			
HL 3.10E-005	atm m3/mol	25	EST	MEYLAN,WM & HOWARD,PH (1991)
OH 1.25E-011	cm3/molc sec	25	EST	MEYLAN,WM & HOWARD,PH (1993)

| CAS #: 026970-95-6 | CEPHALOTHIN ANALOG |

Formula: $C_{17}H_{20}N_4O_4S_4$	
Mol Weight: 472.63	
MP (deg C):	FP (deg C):
BP (deg C):	
BP pressure (mm Hg):	

Property/Value	Units	Temp	Data Type	Reference
WS 3.04E+002	mg/L	25	EST	MEYLAN,WM ET AL. (1996)
logP 0.67			EXP	HANSCH,C & LEO,AJ (1985)
VP 1.65E-018	mm Hg	25	EST	NEELY,WB & BLAU,GE (1985)
DC	pKa			
HL 4.98E-021	atm m3/mol	25	EST	MEYLAN,WM & HOWARD,PH (1991)
OH 1.41E-010	cm3/molc sec	25	EST	MEYLAN,WM & HOWARD,PH (1993)

| CAS #: 026973-80-8 | CEPHALOSPORIC ACID,7-BR-ACAMINO |

Formula: $C_{12}H_{13}BrN_2O_6S$	
Mol Weight: 393.22	
MP (deg C):	FP (deg C):
BP (deg C):	
BP pressure (mm Hg):	

Property/Value	Units	Temp	Data Type	Reference
WS 6.54E+002	mg/L	25	EST	MEYLAN,WM ET AL. (1996)
logP -0.70			EXP	HANSCH,C & LEO,AJ (1985)
VP 7.79E-013	mm Hg	25	EST	NEELY,WB & BLAU,GE (1985)
DC	pKa			
HL 4.58E-017	atm m3/mol	25	EST	MEYLAN,WM & HOWARD,PH (1991)
OH 1.10E-010	cm3/molc sec	25	EST	MEYLAN,WM & HOWARD,PH (1993)

| CAS #: 027016-91-7 | 1H-1,4-BENZODIAZEPINE-1-CARBOXAMIDE, 2,3-DIHYDRO |

Formula: $C_{17}H_{14}N_4O_4$	
Mol Weight: 338.33	
MP (deg C):	FP (deg C):
BP (deg C):	
BP pressure (mm Hg):	

Property/Value	Units	Temp	Data Type	Reference
WS 9.32E+001	mg/L	25	EST	MEYLAN,WM ET AL. (1996)
logP 1.76			EXP	SANGSTER,J (1994)
VP 9.32E-013	mm Hg	25	EST	NEELY,WB & BLAU,GE (1985)
DC	pKa			
HL 1.87E-013	atm m3/mol	25	EST	MEYLAN,WM & HOWARD,PH (1991)
OH 8.42E-012	cm3/molc sec	25	EST	MEYLAN,WM & HOWARD,PH (1993)

| CAS #: 027052-09-1 | 2-PHENYL-1,1-DIOXO-THIAZOL-5-ONE |

Formula: $C_9H_9NO_3S$	
Mol Weight: 211.24	
MP (deg C):	FP (deg C):
BP (deg C):	
BP pressure (mm Hg):	

Property/Value	Units	Temp	Data Type	Reference
WS 7.06E+002	mg/L	25	EST	MEYLAN,WM ET AL. (1996)
logP 0.80			EXP	LIPINSKI,CA ET AL. (1991)
VP 1.45E-007	mm Hg	25	EST	NEELY,WB & BLAU,GE (1985)
DC	pKa			
HL 2.61E-011	atm m3/mol	25	EST	MEYLAN,WM & HOWARD,PH (1991)
OH 6.41E-011	cm3/molc sec	25	EST	MEYLAN,WM & HOWARD,PH (1993)

1033

CAS #:	027091-58-3				5(AZIRIDINYL)-2,4-DINO2-N-PH BENZAMIDE

Formula: $C_{15}H_{12}N_4O_5$

Mol Weight: 328.29

MP (deg C): FP (deg C):

BP (deg C):

BP pressure (mm Hg):

Property/Value	Units	Temp	Data Type	Reference
WS 4.25E+001	mg/L	25	EST	MEYLAN,WM ET AL. (1996)
logP 2.23			EXP	HANSCH,C & LEO,AJ (1985)
VP 2.83E-011	mm Hg	25	EST	NEELY,WB & BLAU,GE (1985)
DC	pKa			
HL 2.92E-015	atm m3/mol	25	EST	MEYLAN,WM & HOWARD,PH (1991)
OH 1.38E-011	cm3/molc sec	25	EST	MEYLAN,WM & HOWARD,PH (1993)

CAS #:	027138-31-4				DIPROPYLENE GLYCOL, DIBENZOATE

Formula: $C_{20}H_{22}O_5$

Mol Weight: 342.40

MP (deg C): FP (deg C):

BP (deg C):

BP pressure (mm Hg):

METHYL POSITIONS UNSPECIFIED

Property/Value	Units	Temp	Data Type	Reference
WS 1.50E+001	mg/L	25	EXP	BUTZ,RG ET AL. (1982A)
logP 3.88			EST	MEYLAN,WM & HOWARD,PH (1995)
VP 4.60E-007	mm Hg	25	EXP	BUTZ,RG ET AL. (1982A)
DC	pKa			
HL 1.38E-008	atm m3/mol	25	EST	VP/WSOL
OH 3.44E-011	cm3/molc sec	25	EST	MEYLAN,WM & HOWARD,PH (1993)

CAS #:	027147-69-9				BENZENAMINE, 2-[(4-AMINOPHENYL)SULFONYL]-

Formula: $C_{12}H_{12}N_2O_2S$

Mol Weight: 248.31

MP (deg C): FP (deg C):

BP (deg C):

BP pressure (mm Hg):

Property/Value	Units	Temp	Data Type	Reference
WS 1.13E+003	mg/L	25	EST	MEYLAN,WM ET AL. (1996)
logP 1.56			EXP	ALTOMARE,C ET AL. (1991)
VP 2.20E-008	mm Hg	25	EST	NEELY,WB & BLAU,GE (1985)
DC	pKa			
HL 3.11E-014	atm m3/mol	25	EST	MEYLAN,WM & HOWARD,PH (1991)
OH 4.60E-011	cm3/molc sec	25	EST	MEYLAN,WM & HOWARD,PH (1993)

CAS #:	027153-17-9				M-METHOXYFORMANILIDE

Formula: $C_8H_9NO_2$

Mol Weight: 151.17

MP (deg C): FP (deg C):

BP (deg C):

BP pressure (mm Hg):

Property/Value	Units	Temp	Data Type	Reference
WS 6.42E+003	mg/L	25	EST	MEYLAN,WM ET AL. (1996)
logP 1.25			EXP	HANSCH,C & LEO,AJ (1985)
VP 1.38E-004	mm Hg	25	EST	NEELY,WB & BLAU,GE (1985)
DC	pKa			
HL 5.00E-010	atm m3/mol	25	EST	MEYLAN,WM & HOWARD,PH (1991)
OH 2.01E-010	cm3/molc sec	25	EST	MEYLAN,WM & HOWARD,PH (1993)

CAS #:	027179-29-9				6H-DIBENZO[B,D]PYRAN, 6A,7,8,9,10,10A-HEXAHYDRO-

Formula: $C_{22}H_{32}O_2$

Mol Weight: 328.50

MP (deg C): FP (deg C):

BP (deg C):

BP pressure (mm Hg):

Property/Value	Units	Temp	Data Type	Reference
WS 2.59E-003	mg/L	25	EST	MEYLAN,WM ET AL. (1996)
logP 7.62			EXP	SANGSTER,J (1993)
VP 7.33E-007	mm Hg	25	EST	NEELY,WB & BLAU,GE (1985)
DC	pKa			
HL 1.18E-004	atm m3/mol	25	EST	MEYLAN,WM & HOWARD,PH (1991)
OH 2.75E-010	cm3/molc sec	25	EST	MEYLAN,WM & HOWARD,PH (1993)

CAS #:	027215-10-7				DIISOCTYL PHOSPHATE

Formula: $C_{16}H_{35}O_4P$

Mol Weight: 322.43

MP (deg C): FP (deg C):

BP (deg C):

BP pressure (mm Hg):

Property/Value	Units	Temp	Data Type	Reference
WS 5.93E-002	mg/L	25	EST	MEYLAN,WM ET AL. (1996)
logP 6.07			EST	MEYLAN,WM & HOWARD,PH (1995)
VP 4.65E-008	mm Hg	25	EST	NEELY,WB & BLAU,GE (1985)
DC	pKa			
HL 4.11E-008	atm m3/mol	25	EST	MEYLAN,WM & HOWARD,PH (1991)
OH 6.40E-011	cm3/molc sec	25	EST	MEYLAN,WM & HOWARD,PH (1993)

CAS #:	027221-03-0				5(1-AZIRD)-2,4-NO2 BENZAMIDE,NN-DIME

Formula: $C_{11}H_{12}N_4O_5$

Mol Weight: 280.24

MP (deg C): FP (deg C):

BP (deg C):

BP pressure (mm Hg):

Property/Value	Units	Temp	Data Type	Reference
WS 1.89E+003	mg/L	25	EST	MEYLAN,WM ET AL. (1996)
logP 0.63			EXP	HANSCH,C & LEO,AJ (1985)
VP 3.62E-008	mm Hg	25	EST	NEELY,WB & BLAU,GE (1985)
DC	pKa			
HL 2.55E-014	atm m3/mol	25	EST	MEYLAN,WM & HOWARD,PH (1991)
OH 1.75E-011	cm3/molc sec	25	EST	MEYLAN,WM & HOWARD,PH (1993)

CAS #:	027314-13-2				NORFLURAZON

Formula: $C_{12}H_9ClF_3N_3O$

Mol Weight: 303.67

MP (deg C): 174-180 FP (deg C):

BP (deg C):

BP pressure (mm Hg):

Property/Value	Units	Temp	Data Type	Reference
WS 2.80E+001	mg/L	25	EXP	WAUCHOPE,RD ET AL. (1991A)
logP 2.30			EXP	HANSCH,C ET AL. (1995)
VP 2.00E-008	mm Hg	25	EXP	WAUCHOPE,RD ET AL. (1991A)
DC	pKa			
HL 2.85E-010	atm m3/mol	25	EST	VP/WSOL
OH 7.18E-018	cm3/molc sec	25	EST	MEYLAN,WM & HOWARD,PH (1993)

1034

CAS #: 027472-90-8	4(5H)-THIAZOLONE, 5-[(5-NITRO-2-FURANYL)METHYLEN

Formula: $C_{11}H_{11}N_3O_4S$

Mol Weight: 281.29

MP (deg C): | FP (deg C):

BP (deg C):

BP pressure (mm Hg):

Property/Value	Units	Temp	Data Type	Reference
WS 5.60E+002	mg/L	25	EST	MEYLAN,WM ET AL. (1996)
logP 1.24			EXP	AKERBLOM,EB (1974)
VP 2.11E-008	mm Hg	25	EST	NEELY,WB & BLAU,GE (1985)
DC	pKa			
HL 3.14E-015	atm m3/mol	25	EST	MEYLAN,WM & HOWARD,PH (1991)
OH 1.12E-010	cm3/molc sec	25	EST	MEYLAN,WM & HOWARD,PH (1993)

CAS #: 027472-92-0	4(5H)-THIAZOLONE, 2-[(1-METHYLETHYL)AMINO]-5-[(5

Formula: $C_{11}H_{11}N_3O_4S$

Mol Weight: 281.29

MP (deg C): | FP (deg C):

BP (deg C):

BP pressure (mm Hg):

Property/Value	Units	Temp	Data Type	Reference
WS 5.08E+002	mg/L	25	EST	MEYLAN,WM ET AL. (1996)
logP 1.29			EXP	AKERBLOM,EB (1974)
VP 3.42E-008	mm Hg	25	EST	NEELY,WB & BLAU,GE (1985)
DC	pKa			
HL 3.14E-015	atm m3/mol	25	EST	MEYLAN,WM & HOWARD,PH (1991)
OH 1.18E-010	cm3/molc sec	25	EST	MEYLAN,WM & HOWARD,PH (1993)

CAS #: 027522-25-4	PYRIMIDINE, 5-ETHOXY-

Formula: $C_6H_8N_2O$

Mol Weight: 124.14

MP (deg C): | FP (deg C):

BP (deg C):

BP pressure (mm Hg):

Property/Value	Units	Temp	Data Type	Reference
WS 3.22E+004	mg/L	25	EST	MEYLAN,WM ET AL. (1996)
logP 0.56			EXP	YAMAGAMI,C ET AL. (1990)
VP 5.84E-001	mm Hg	25	EST	NEELY,WB & BLAU,GE (1985)
DC	pKa			
HL 2.29E-007	atm m3/mol	25	EST	MEYLAN,WM & HOWARD,PH (1991)
OH 8.71E-012	cm3/molc sec	25	EST	MEYLAN,WM & HOWARD,PH (1993)

CAS #: 027545-04-6	BENZAMIDE, O-ACETYLMETHYLAMINO

Formula: $C_{10}H_{12}N_2O_2$

Mol Weight: 192.22

MP (deg C): | FP (deg C):

BP (deg C):

BP pressure (mm Hg):

Property/Value	Units	Temp	Data Type	Reference
WS 1.44E+004	mg/L	25	EST	MEYLAN,WM ET AL. (1996)
logP 0.61			EXP	HANSCH,C & LEO,AJ (1985)
VP 1.74E-006	mm Hg	25	EST	NEELY,WB & BLAU,GE (1985)
DC	pKa			
HL 2.31E-014	atm m3/mol	25	EST	MEYLAN,WM & HOWARD,PH (1991)
OH 4.75E-011	cm3/molc sec	25	EST	MEYLAN,WM & HOWARD,PH (1993)

CAS #: 027550-11-4	2,4-THIAZOLIDINEDIONE, 3-(1-METHYLETHYL)-5-[(5-N

Formula: $C_{11}H_{10}N_2O_5S$

Mol Weight: 282.28

MP (deg C): | FP (deg C):

BP (deg C):

BP pressure (mm Hg):

Property/Value	Units	Temp	Data Type	Reference
WS 5.43E+001	mg/L	25	EST	MEYLAN,WM ET AL. (1996)
logP 2.42			EXP	SANGSTER,J (1993)
VP 9.60E-010	mm Hg	25	EST	NEELY,WB & BLAU,GE (1985)
DC	pKa			
HL 1.59E-009	atm m3/mol	25	EST	MEYLAN,WM & HOWARD,PH (1991)
OH 5.54E-011	cm3/molc sec	25	EST	MEYLAN,WM & HOWARD,PH (1993)

CAS #: 027554-26-3	DIISOOCTYL PHTHALATE

Formula: $C_{24}H_{38}O_4$

Mol Weight: 390.57

MP (deg C): <-50 | FP (deg C):

BP (deg C): 228-39

BP pressure (mm Hg):

Property/Value	Units	Temp	Data Type	Reference
WS 9.00E-002	mg/L	25	EXP	HOWARD,PH ET AL. (1985)
logP 8.39			EST	MEYLAN,WM & HOWARD,PH (1995)
VP 5.50E-006	mm Hg	25	EXP	HOWARD,PH ET AL. (1985)
DC	pKa			
HL 2.83E-004	atm m3/mol	25	EST	VP/WSOL
OH 2.06E-011	cm3/molc sec	25	EST	MEYLAN,WM & HOWARD,PH (1993)

CAS #: 027564-47-2	5(5NO2-2-FURIL)THIAZOLIDIN,24-DION

Formula: $C_8H_5N_2O_5S$

Mol Weight: 241.20

MP (deg C): | FP (deg C):

BP (deg C):

BP pressure (mm Hg):

Property/Value	Units	Temp	Data Type	Reference
WS 3.16E+003	mg/L	25	EST	MEYLAN,WM ET AL. (1996)
logP 0.63			EXP	HANSCH,C & LEO,AJ (1985)
VP 7.22E-011	mm Hg	25	EST	NEELY,WB & BLAU,GE (1985)
DC	pKa			
HL 4.11E-010	atm m3/mol	25	EST	MEYLAN,WM & HOWARD,PH (1991)
OH 4.71E-011	cm3/molc sec	25	EST	MEYLAN,WM & HOWARD,PH (1993)

CAS #: 027640-19-3	1,1'-PROPYLEN-BIS(1-NITROSOUREA)

Formula: $C_5H_{10}N_6O_4$

Mol Weight: 218.17

MP (deg C): | FP (deg C):

BP (deg C):

BP pressure (mm Hg):

Property/Value	Units	Temp	Data Type	Reference
WS 3.11E+003	mg/L	25	EST	MEYLAN,WM ET AL. (1996)
logP -0.30			EXP	HANSCH,C & LEO,AJ (1985)
VP 1.34E-006	mm Hg	25	EST	NEELY,WB & BLAU,GE (1985)
DC	pKa			
HL 4.69E-020	atm m3/mol	25	EST	MEYLAN,WM & HOWARD,PH (1991)
OH 4.54E-011	cm3/molc sec	25	EST	MEYLAN,WM & HOWARD,PH (1993)

CAS #: 027640-22-8	1,1'-HEXAMETHYLENE-BIS(1-NITROSOUREA)

Formula: $C_8H_{16}N_6O_4$

Mol Weight: 260.25

MP (deg C): FP (deg C):

BP (deg C):

BP pressure (mm Hg):

Property/ Value	Units	Temp	Data Type	Reference
WS 6.63E+003	mg/L	25	EST	MEYLAN,WM ET AL. (1996)
logP 0.58			EXP	HANSCH,C & LEO,AJ (1985)
VP 1.26E-007	mm Hg	25	EST	NEELY,WB & BLAU,GE (1985)
DC	pKa			
HL 1.10E-019	atm m3/mol	25	EST	MEYLAN,WM & HOWARD,PH (1991)
OH 4.97E-011	cm3/molc sec	25	EST	MEYLAN,WM & HOWARD,PH (1993)

CAS #: 027653-63-0	BARBITURIC ACID,5-ET-5-ME

Formula: $C_7H_{10}N_2O_3$

Mol Weight: 170.17

MP (deg C): FP (deg C):

BP (deg C):

BP pressure (mm Hg):

Property/ Value	Units	Temp	Data Type	Reference
WS 5.25E+004	mg/L	25	EST	MEYLAN,WM ET AL. (1996)
logP 0.08			EXP	PRANKERD,RJ & MCKEOWN,RH (1992)
VP 3.63E-010	mm Hg	25	EST	NEELY,WB & BLAU,GE (1985)
DC	pKa			
HL 2.72E-013	atm m3/mol	25	EST	MEYLAN,WM & HOWARD,PH (1991)
OH 6.34E-012	cm3/molc sec	25	EST	MEYLAN,WM & HOWARD,PH (1993)

CAS #: 027693-38-5	4-(4-NITROBENZOYL)-PYRIDINE

Formula: $C_{12}H_8N_2O_3$

Mol Weight: 228.21

MP (deg C): FP (deg C):

BP (deg C):

BP pressure (mm Hg):

Property/ Value	Units	Temp	Data Type	Reference
WS 3.98E+002	mg/L	25	EST	MEYLAN,WM ET AL. (1996)
logP 1.76			EXP	HANSCH,C & LEO,AJ (1985)
VP 4.13E-006	mm Hg	25	EST	NEELY,WB & BLAU,GE (1985)
DC	pKa			
HL 1.00E-011	atm m3/mol	25	EST	MEYLAN,WM & HOWARD,PH (1991)
OH 4.93E-013	cm3/molc sec	25	EST	MEYLAN,WM & HOWARD,PH (1993)

CAS #: 027693-70-5	N-PHENYLANTHRANILIC ACID,3'-NITRO

Formula: $C_{13}H_{10}N_2O_4$

Mol Weight: 258.24

MP (deg C): FP (deg C):

BP (deg C):

BP pressure (mm Hg):

Property/ Value	Units	Temp	Data Type	Reference
WS 2.66E+000	mg/L	25	EST	MEYLAN,WM ET AL. (1996)
logP 4.57			EXP	HANSCH,C & LEO,AJ (1985)
VP 3.63E-008	mm Hg	25	EST	NEELY,WB & BLAU,GE (1985)
DC	pKa			
HL 8.34E-014	atm m3/mol	25	EST	MEYLAN,WM & HOWARD,PH (1991)
OH 6.76E-011	cm3/molc sec	25	EST	MEYLAN,WM & HOWARD,PH (1993)

CAS #: 027693-73-8	BENZOIC ACID, 2-[(3-METHOXYPHENYL)AMINO]-

Formula: $C_{14}H_{13}NO_3$

Mol Weight: 243.26

MP (deg C): FP (deg C):

BP (deg C):

BP pressure (mm Hg):

Property/ Value	Units	Temp	Data Type	Reference
WS 5.38E+000	mg/L	25	EST	MEYLAN,WM ET AL. (1996)
logP 4.31			EXP	TERADA,H ET AL. (1987)
VP 4.52E-007	mm Hg	25	EST	NEELY,WB & BLAU,GE (1985)
DC	pKa			
HL 1.25E-012	atm m3/mol	25	EST	MEYLAN,WM & HOWARD,PH (1991)
OH 2.01E-010	cm3/molc sec	25	EST	MEYLAN,WM & HOWARD,PH (1993)

CAS #: 027696-28-2	BENZOIC ACID, 2-[(3-ACETYLPHENYL)AMINO]-

Formula: $C_{15}H_{13}NO_3$

Mol Weight: 255.28

MP (deg C): FP (deg C):

BP (deg C):

BP pressure (mm Hg):

Property/ Value	Units	Temp	Data Type	Reference
WS 7.39E+000	mg/L	25	EST	MEYLAN,WM ET AL. (1996)
logP 4.07			EXP	SANGSTER,J (1993)
VP 8.08E-008	mm Hg	25	EST	NEELY,WB & BLAU,GE (1985)
DC	pKa			
HL 3.85E-014	atm m3/mol	25	EST	MEYLAN,WM & HOWARD,PH (1991)
OH 1.17E-010	cm3/molc sec	25	EST	MEYLAN,WM & HOWARD,PH (1993)

CAS #: 027772-24-3	4-PYRIDAZINAMINE, 3-METHOXY-N,N-DIMETHYL-

Formula: $C_7H_{11}N_3O$

Mol Weight: 153.19

MP (deg C): FP (deg C):

BP (deg C):

BP pressure (mm Hg):

Property/ Value	Units	Temp	Data Type	Reference
WS 1.25E+004	mg/L	25	EST	MEYLAN,WM ET AL. (1996)
logP 0.90			EXP	HANSCH,C ET AL. (1995)
VP 1.31E-003	mm Hg	25	EST	NEELY,WB & BLAU,GE (1985)
DC	pKa			
HL 2.46E-008	atm m3/mol	25	EST	MEYLAN,WM & HOWARD,PH (1991)
OH 4.11E-011	cm3/molc sec	25	EST	MEYLAN,WM & HOWARD,PH (1993)

CAS #: 027810-94-2	24NH2-PYRIMIDINE,5(34MEO-5PHMEO)BENZYL

Formula: $C_{20}H_{22}N_4O_3$

Mol Weight: 366.42

MP (deg C): FP (deg C):

BP (deg C):

BP pressure (mm Hg):

Property/ Value	Units	Temp	Data Type	Reference
WS 4.66E+001	mg/L	25	EST	MEYLAN,WM ET AL. (1996)
logP 2.37			EXP	HANSCH,C & LEO,AJ (1985)
VP 2.56E-011	mm Hg	25	EST	NEELY,WB & BLAU,GE (1985)
DC	pKa			
HL 1.93E-015	atm m3/mol	25	EST	MEYLAN,WM & HOWARD,PH (1991)
OH 2.08E-010	cm3/molc sec	25	EST	MEYLAN,WM & HOWARD,PH (1993)

CAS #: 027816-82-6				BENZENEACETAMIDE, N-(3,4-DICHLOROPHENYL)-

Formula: $C_{14}H_{11}Cl_2NO$

Mol Weight: 280.16

MP (deg C): | FP (deg C):

BP (deg C):

BP pressure (mm Hg):

Property/ Value	Units	Temp	Data Type	Reference
WS 2.44E+000	mg/L	25	EST	MEYLAN,WM ET AL. (1996)
logP 4.47			EXP	HANSCH,C ET AL. (1995)
VP 5.07E-008	mm Hg	25	EST	NEELY,WB & BLAU,GE (1985)
DC	pKa			
HL 2.74E-010	atm m3/mol	25	EST	MEYLAN,WM & HOWARD,PH (1991)
OH 8.10E-012	cm3/molc sec	25	EST	MEYLAN,WM & HOWARD,PH (1993)

CAS #: 027843-08-9				W-PHENYLPROPIONALDEHYDE SEMICARBIZONE

Formula: $C_{10}H_{13}N_3O$

Mol Weight: 191.23

MP (deg C): | FP (deg C):

BP (deg C):

BP pressure (mm Hg):

Property/ Value	Units	Temp	Data Type	Reference
WS 5.06E+002	mg/L	25	EST	MEYLAN,WM ET AL. (1996)
logP 2.32			EXP	HANSCH,C & LEO,AJ (1985)
VP 1.75E-005	mm Hg	25	EST	NEELY,WB & BLAU,GE (1985)
DC	pKa			
HL 1.39E-010	atm m3/mol	25	EST	MEYLAN,WM & HOWARD,PH (1991)
OH 1.18E-011	cm3/molc sec	25	EST	MEYLAN,WM & HOWARD,PH (1993)

CAS #: 027876-24-0				P-HEXYLPYRIDINE

Formula: $C_{11}H_{17}N$

Mol Weight: 163.26

MP (deg C): | FP (deg C):

BP (deg C):

BP pressure (mm Hg):

Property/ Value	Units	Temp	Data Type	Reference
WS 2.54E+002	mg/L	25	EST	MEYLAN,WM ET AL. (1996)
logP 4.35			EXP	HANSCH,C & LEO,AJ (1985)
VP 4.22E-002	mm Hg	25	EST	NEELY,WB & BLAU,GE (1985)
DC	pKa			
HL 3.21E-005	atm m3/mol	25	EST	MEYLAN,WM & HOWARD,PH (1991)
OH 7.62E-012	cm3/molc sec	25	EST	MEYLAN,WM & HOWARD,PH (1993)

CAS #: 027910-26-5				CEPHALOSPORANIC ACID,7-MANDELAMIDO

Formula: $C_{18}H_{18}N_2O_7S$

Mol Weight: 406.42

MP (deg C): | FP (deg C):

BP (deg C):

BP pressure (mm Hg):

Property/ Value	Units	Temp	Data Type	Reference
WS 3.25E+002	mg/L	25	EST	MEYLAN,WM ET AL. (1996)
logP -0.44			EXP	HANSCH,C & LEO,AJ (1985)
VP 2.08E-018	mm Hg	25	EST	NEELY,WB & BLAU,GE (1985)
DC	pKa			
HL 5.00E-018	atm m3/mol	25	EST	MEYLAN,WM & HOWARD,PH (1991)
OH 1.20E-010	cm3/molc sec	25	EST	MEYLAN,WM & HOWARD,PH (1993)

CAS #: 027918-37-2				PHENYLSULFONYL (p-METHOXYBENZOYL) METHANE

Formula: $C_{15}H_{14}O_4S$

Mol Weight: 290.34

MP (deg C): | FP (deg C):

BP (deg C):

BP pressure (mm Hg):

Property/ Value	Units	Temp	Data Type	Reference
WS 1.92E+002	mg/L	25	EST	MEYLAN,WM ET AL. (1996)
logP 2.18			EXP	WANG,L ET AL. (1993)
VP 9.49E-008	mm Hg	25	EST	NEELY,WB & BLAU,GE (1985)
DC	pKa			
HL 9.83E-012	atm m3/mol	25	EST	MEYLAN,WM & HOWARD,PH (1991)
OH 2.63E-011	cm3/molc sec	25	EST	MEYLAN,WM & HOWARD,PH (1993)

CAS #: 027949-36-6				1,1-DIBROMO-2-CHLOROETHANE

Formula: $C_2H_3Br_2Cl$

Mol Weight: 222.32

MP (deg C): | FP (deg C):

BP (deg C):

BP pressure (mm Hg):

Property/ Value	Units	Temp	Data Type	Reference
WS 4.50E+002	mg/L	25	EST	MEYLAN,WM ET AL. (1996)
logP 2.19				MEYLAN,WM & HOWARD,PH (1995)
VP 1.15E+000	mm Hg	25	EST	NEELY,WB & BLAU,GE (1985)
DC	pKa			
HL 4.57E-004	atm m3/mol	25	EST	MEYLAN,WM & HOWARD,PH (1991)
OH 2.18E-013	cm3/molc sec	25	EST	MEYLAN,WM & HOWARD,PH (1993)

CAS #: 027955-87-9				1,1-BIS(P-ETHOXYPHENYL)-2,2-DIMETHYLPROPANE

Formula: $C_{21}H_{28}O_2$

Mol Weight: 312.46

MP (deg C): | FP (deg C):

BP (deg C):

BP pressure (mm Hg):

Property/ Value	Units	Temp	Data Type	Reference
WS 2.15E-002	mg/L	25	EST	MEYLAN,WM ET AL. (1996)
logP 6.66			EST	MEYLAN,WM & HOWARD,PH (1995)
VP 2.37E-006	mm Hg	25	EST	NEELY,WB & BLAU,GE (1985)
DC	pKa			
HL 9.19E-006	atm m3/mol	25	EST	MEYLAN,WM & HOWARD,PH (1991)
OH 6.70E-011	cm3/molc sec	25	EST	MEYLAN,WM & HOWARD,PH (1993)

CAS #: 028026-77-9				3-ETHYL-4-HYDROXYACETANILIDE

Formula: $C_{10}H_{13}NO_2$

Mol Weight: 179.22

MP (deg C): | FP (deg C):

BP (deg C):

BP pressure (mm Hg):

Property/ Value	Units	Temp	Data Type	Reference
WS 4.23E+003	mg/L	25	EST	MEYLAN,WM ET AL. (1996)
logP 1.31			EXP	HANSCH,C & LEO,AJ (1985)
VP 1.12E-006	mm Hg	25	EST	NEELY,WB & BLAU,GE (1985)
DC	pKa			
HL 9.41E-013	atm m3/mol	25	EST	MEYLAN,WM & HOWARD,PH (1991)
OH 2.26E-011	cm3/molc sec	25	EST	MEYLAN,WM & HOWARD,PH (1993)

028076-73-5 — 2,2',4,4'-TETRACHLORODIPHENYL ETHER

Formula: $C_{12}H_6Cl_4O$
Mol Weight: 307.99
MP (deg C):   **FP (deg C):**
BP (deg C):
BP pressure (mm Hg):

Property/Value	Units	Temp	Data Type	Reference
WS 1.26E-001	mg/L	25	EST	MEYLAN,WM ET AL. (1996)
logP 5.79			EST	MEYLAN,WM & HOWARD,PH (1995)
VP 9.89E-006	mm Hg	25	EST	NEELY,WB & BLAU,GE (1985)
DC	pKa			
HL 3.55E-005	atm m3/mol	25	EST	MEYLAN,WM & HOWARD,PH (1991)
OH 1.93E-012	cm3/molc sec	25	EST	MEYLAN,WM & HOWARD,PH (1993)

028099-10-7 — CARBAMIC ACID,O-ETHYL(TRIMEAMM)BR

Formula: $C_6H_{15}BrN_2O_2$
Mol Weight: 227.10
MP (deg C):   **FP (deg C):**
BP (deg C):
BP pressure (mm Hg):

Property/Value	Units	Temp	Data Type	Reference
WS 1.00E+006	mg/L	25	EST	MEYLAN,WM ET AL. (1996)
logP -3.90			EXP	HANSCH,C & LEO,AJ (1985)
VP 4.88E-008	mm Hg	25	EST	NEELY,WB & BLAU,GE (1985)
DC	pKa			
HL 2.37E-018	atm m3/mol	25	EST	MEYLAN,WM & HOWARD,PH (1991)
OH 1.83E-011	cm3/molc sec	25	EST	MEYLAN,WM & HOWARD,PH (1993)

028108-99-8 — ISOPROPYL PHENYL DIPHENYL PHOSPHATE

Formula: $C_{21}H_{21}O_4P$
Mol Weight: 368.37
MP (deg C):   **FP (deg C):**
BP (deg C): 220-230
BP pressure (mm Hg): 1.00E+000

MIXTURE OF ISOPROPYL ISOMERS

Property/Value	Units	Temp	Data Type	Reference
WS 2.20E+000	mg/L	25	EXP	SAEGER,VW ET AL. (1979)
logP 5.31			EXP	SAEGER,VW ET AL. (1979)
VP 3.52E-007	mm Hg	25	EST	HL X WSOL
DC	pKa			
HL 7.74E-008	atm m3/mol	25	EST	MEYLAN,WM & HOWARD,PH (1991)
OH 1.89E-011	cm3/molc sec	25	EST	ATKINSON,R (1988)

028125-87-3 — 2-IMIDAZOLIDINIMINE, N-(5-FLUORO-2-METHYLPHENYL)

Formula: $C_{10}H_{12}FN_3$
Mol Weight: 193.23
MP (deg C):   **FP (deg C):**
BP (deg C):
BP pressure (mm Hg):

Property/Value	Units	Temp	Data Type	Reference
WS 3.02E+004	mg/L	25	EST	MEYLAN,WM ET AL. (1996)
logP 1.41			EXP	SANGSTER,J (1993)
VP 1.92E-004	mm Hg	25	EST	NEELY,WB & BLAU,GE (1985)
DC	pKa			
HL 3.53E-011	atm m3/mol	25	EST	MEYLAN,WM & HOWARD,PH (1991)
OH 1.45E-010	cm3/molc sec	25	EST	MEYLAN,WM & HOWARD,PH (1993)

028165-52-8 — 2,5-DIBROMOPHENOL

Formula: $C_6H_4Br_2O$
Mol Weight: 251.92
MP (deg C):   **FP (deg C):**
BP (deg C):
BP pressure (mm Hg):

Property/Value	Units	Temp	Data Type	Reference
WS 1.35E+002	mg/L	25	EST	MEYLAN,WM ET AL. (1996)
logP 3.29			EST	MEYLAN,WM & HOWARD,PH (1995)
VP 1.34E-003	mm Hg	25	EST	NEELY,WB & BLAU,GE (1985)
DC	pKa			
HL 8.90E-008	atm m3/mol	25	EST	MEYLAN,WM & HOWARD,PH (1991)
OH 6.15E-012	cm3/molc sec	25	EST	MEYLAN,WM & HOWARD,PH (1993)

028165-71-1 — ACETIC ACID, 2,6-DICHLOROPHENYL ESTER

Formula: $C_8H_6Cl_2O_2$
Mol Weight: 205.04
MP (deg C):   **FP (deg C):**
BP (deg C):
BP pressure (mm Hg):

Property/Value	Units	Temp	Data Type	Reference
WS 1.27E+002	mg/L	25	EST	MEYLAN,WM ET AL. (1996)
logP 2.94			EXP	SOTOMATSU,T ET AL. (1993)
VP 1.20E-002	mm Hg	25	EST	NEELY,WB & BLAU,GE (1985)
DC	pKa			
HL 3.56E-005	atm m3/mol	25	EST	MEYLAN,WM & HOWARD,PH (1991)
OH 8.09E-013	cm3/molc sec	25	EST	MEYLAN,WM & HOWARD,PH (1993)

028165-91-5 — 1,2-BENZENEDIOL, 3-TRIDECYL-

Formula: $C_{19}H_{32}O_2$
Mol Weight: 292.47
MP (deg C):   **FP (deg C):**
BP (deg C):
BP pressure (mm Hg):

Property/Value	Units	Temp	Data Type	Reference
WS 1.24E-002	mg/L	25	EST	MEYLAN,WM ET AL. (1996)
logP 7.75			EXP	ITOKAWA,H ET AL. (1989)
VP 1.30E-008	mm Hg	25	EST	NEELY,WB & BLAU,GE (1985)
DC	pKa			
HL 1.93E-009	atm m3/mol	25	EST	MEYLAN,WM & HOWARD,PH (1991)
OH 7.39E-011	cm3/molc sec	25	EST	MEYLAN,WM & HOWARD,PH (1993)

028167-45-5 — O,S-DIPROPYLPHOSPHORAMIDOTHIOATE

Formula: $C_6H_{16}NO_2PS$
Mol Weight: 197.24
MP (deg C):   **FP (deg C):**
BP (deg C):
BP pressure (mm Hg):

Property/Value	Units	Temp	Data Type	Reference
WS 4.02E+003	mg/L	25	EST	MEYLAN,WM ET AL. (1996)
logP 1.23			EXP	HANSCH,C & LEO,AJ (1985)
VP 1.59E-003	mm Hg	25	EST	NEELY,WB & BLAU,GE (1985)
DC	pKa			
HL 2.70E-009	atm m3/mol	25	EST	MEYLAN,WM & HOWARD,PH (1991)
OH 7.72E-011	cm3/molc sec	25	EST	MEYLAN,WM & HOWARD,PH (1993)

CAS #: 028167-49-9				O,S-DIME-N-ME-PHOSPHORAMIDOTHIOAT

Formula: C₃H₁₀NO₂PS — $C_3H_{10}NO_2PS$

Mol Weight: 155.16

MP (deg C): | FP (deg C):

BP (deg C):

BP pressure (mm Hg):

Property/Value	Units	Temp	Data Type	Reference
WS 8.26E+004	mg/L	25	EST	MEYLAN,WM ET AL. (1996)
logP -0.07			EXP	HANSCH,C & LEO,AJ (1985)
VP 9.16E-002	mm Hg	25	EST	NEELY,WB & BLAU,GE (1985)
DC	pKa			
HL 1.91E-009	atm m3/mol	25	EST	MEYLAN,WM & HOWARD,PH (1991)
OH 3.43E-011	cm3/molc sec	25	EST	MEYLAN,WM & HOWARD,PH (1993)

CAS #: 028170-26-5				1-(M-F PHENYL)-3-MEO-3-ME UREA

Formula: $C_9H_{11}FN_2O_2$

Mol Weight: 198.20

MP (deg C): | FP (deg C):

BP (deg C):

BP pressure (mm Hg):

Property/Value	Units	Temp	Data Type	Reference
WS 1.64E+003	mg/L	25	EST	MEYLAN,WM ET AL. (1996)
logP 1.68			EXP	HANSCH,C & LEO,AJ (1985)
VP 9.49E-005	mm Hg	25	EST	NEELY,WB & BLAU,GE (1985)
DC	pKa			
HL 2.45E-008	atm m3/mol	25	EST	MEYLAN,WM & HOWARD,PH (1991)
OH 5.50E-011	cm3/molc sec	25	EST	MEYLAN,WM & HOWARD,PH (1993)

CAS #: 028170-54-9				N'-(3-METHOXYPHENYL)-N,N-DIMETHYLUREA

Formula: $C_{10}H_{14}N_2O_2$

Mol Weight: 194.24

MP (deg C): | FP (deg C):

BP (deg C):

BP pressure (mm Hg):

Property/Value	Units	Temp	Data Type	Reference
WS 1.56E+003	mg/L	25	EST	MEYLAN,WM ET AL. (1996)
logP 1.73			EXP	BRIGGS,GG (1981)
VP 3.28E-005	mm Hg	25	EST	NEELY,WB & BLAU,GE (1985)
DC	pKa			
HL 5.74E-011	atm m3/mol	25	EST	MEYLAN,WM & HOWARD,PH (1991)
OH 2.03E-010	cm3/molc sec	25	EST	MEYLAN,WM & HOWARD,PH (1993)

CAS #: 028177-48-2				PHENOL, 2,6-DIFLUORO-

Formula: $C_6H_4F_2O$

Mol Weight: 130.09

MP (deg C): 38-41 | FP (deg C):

BP (deg C): 59-61

BP pressure (mm Hg): 1.70E+001

Property/Value	Units	Temp	Data Type	Reference
WS 7.40E+003	mg/L	25	EST	MEYLAN,WM ET AL. (1996)
logP 1.96			EXP	SOTOMATSU,T ET AL. (1993)
VP 1.53E+000	mm Hg	25	EST	NEELY,WB & BLAU,GE (1985)
DC	pKa			
HL 7.63E-007	atm m3/mol	25	EST	MEYLAN,WM & HOWARD,PH (1991)
OH 3.94E-012	cm3/molc sec	25	EST	MEYLAN,WM & HOWARD,PH (1993)

CAS #: 028197-69-5				ACETAMIDE, N-[3-ACETYL-4-[2-HYDROXY-3-[(1-METHYL

Formula: $C_{16}H_{24}N_2O_4$

Mol Weight: 308.38

MP (deg C): | FP (deg C):

BP (deg C):

BP pressure (mm Hg):

Property/Value	Units	Temp	Data Type	Reference
WS 1.73E+003	mg/L	25	EST	MEYLAN,WM ET AL. (1996)
logP 0.94			EXP	RECANATINI,M (1992)
VP 9.20E-012	mm Hg	25	EST	NEELY,WB & BLAU,GE (1985)
DC	pKa			
HL 1.71E-020	atm m3/mol	25	EST	MEYLAN,WM & HOWARD,PH (1991)
OH 1.20E-010	cm3/molc sec	25	EST	MEYLAN,WM & HOWARD,PH (1993)

CAS #: 028238-55-3				CARBAMIC ACID, (4-ETHYLPHENYL)-, METHYL ESTER

Formula: $C_{10}H_{13}NO_2$

Mol Weight: 179.22

MP (deg C): | FP (deg C):

BP (deg C):

BP pressure (mm Hg):

Property/Value	Units	Temp	Data Type	Reference
WS 2.64E+002	mg/L	25	EST	MEYLAN,WM ET AL. (1996)
logP 2.72			EXP	TAKAHASHI,J ET AL. (1988)
VP 6.17E-003	mm Hg	25	EST	NEELY,WB & BLAU,GE (1985)
DC	pKa			
HL 3.20E-008	atm m3/mol	25	EST	MEYLAN,WM & HOWARD,PH (1991)
OH 5.33E-011	cm3/molc sec	25	EST	MEYLAN,WM & HOWARD,PH (1993)

CAS #: 028249-77-6				THIOBENCARB

Formula: $C_{12}H_{16}CINOS$

Mol Weight: 257.78

MP (deg C): 3.3 | FP (deg C):

BP (deg C): 126-129

BP pressure (mm Hg): 1.00E-002

Property/Value	Units	Temp	Data Type	Reference
WS 2.80E+001	mg/L	25	EXP	WAUCHOPE, RD ET AL. (1991A)
logP 3.40			EXP	HANSCH,C & LEO,AJ (1985)
VP 2.20E-005	mm Hg	25	EXP	WAUCHOPE, RD ET AL. (1991A)
DC	pKa			
HL 2.67E-007	atm m3/mol	25	EST	VP/WSOL
OH 2.25E-011	cm3/molc sec	25	EST	ATKINSON, R (1988)

CAS #: 028289-54-5				MPTP

Formula: $C_{12}H_{15}N$

Mol Weight: 173.26

MP (deg C): 37-40 | FP (deg C):

BP (deg C): 128-132

BP pressure (mm Hg): 1.20E+001

Property/Value	Units	Temp	Data Type	Reference
WS 2.93E+003	mg/L	25	EST	MEYLAN,WM ET AL. (1996)
logP 2.71			EXP	HANSCH,C ET AL. (1995)
VP 6.40E-003	mm Hg	25	EST	NEELY,WB & BLAU,GE (1985)
DC	pKa			
HL 1.48E-006	atm m3/mol	25	EST	MEYLAN,WM & HOWARD,PH (1991)
OH 1.77E-010	cm3/molc sec	25	EST	MEYLAN,WM & HOWARD,PH (1993)

CAS #: 028313-52-2				2-CL-5-CF3-C6H3NHN=C(CN)COOET
Formula: $C_{12}H_9ClF_3N_3O_2$				
Mol Weight: 319.67				
MP (deg C):		FP (deg C):		
BP (deg C):				
BP pressure (mm Hg):				

Property/Value	Units	Temp	Data Type	Reference
WS 1.49E+000	mg/L	25	EST	MEYLAN,WM ET AL. (1996)
logP 4.45			EXP	HANSCH,C & LEO,AJ (1985)
VP 3.00E-006	mm Hg	25	EST	NEELY,WB & BLAU,GE (1985)
DC	pKa			
HL 5.67E-006	atm m3/mol	25	EST	MEYLAN,WM & HOWARD,PH (1991)
OH 3.55E-012	cm3/molc sec	25	EST	MEYLAN,WM & HOWARD,PH (1993)

CAS #: 028313-57-7				4-CL-3-CF3-C6H3NHN=C(CN)COOET
Formula: $C_{12}H_9ClF_3N_3O_2$				
Mol Weight: 319.67				
MP (deg C):		FP (deg C):		
BP (deg C):				
BP pressure (mm Hg):				

Property/Value	Units	Temp	Data Type	Reference
WS 6.65E-001	mg/L	25	EST	MEYLAN,WM ET AL. (1996)
logP 4.86			EXP	HANSCH,C & LEO,AJ (1985)
VP 3.00E-006	mm Hg	25	EST	NEELY,WB & BLAU,GE (1985)
DC	pKa			
HL 5.67E-006	atm m3/mol	25	EST	MEYLAN,WM & HOWARD,PH (1991)
OH 3.55E-012	cm3/molc sec	25	EST	MEYLAN,WM & HOWARD,PH (1993)

CAS #: 028313-58-8				3,5-DICL-C6H3NHN=C(CN)CO-OME
Formula: $C_{10}H_7Cl_2N_3O_2$				
Mol Weight: 272.09				
MP (deg C):		FP (deg C):		
BP (deg C):				
BP pressure (mm Hg):				

Property/Value	Units	Temp	Data Type	Reference
WS 2.55E+000	mg/L	25	EST	MEYLAN,WM ET AL. (1996)
logP 4.50			EXP	HANSCH,C & LEO,AJ (1985)
VP 2.30E-006	mm Hg	25	EST	NEELY,WB & BLAU,GE (1985)
DC	pKa			
HL 3.64E-007	atm m3/mol	25	EST	MEYLAN,WM & HOWARD,PH (1991)
OH 2.13E-011	cm3/molc sec	25	EST	MEYLAN,WM & HOWARD,PH (1993)

CAS #: 028313-59-9				3,5-DICL-C6H3NHN=C(CN)COOET
Formula: $C_{11}H_9Cl_2N_3O_2$				
Mol Weight: 286.12				
MP (deg C):		FP (deg C):		
BP (deg C):				
BP pressure (mm Hg):				

Property/Value	Units	Temp	Data Type	Reference
WS 8.08E+000	mg/L	25	EST	MEYLAN,WM ET AL. (1996)
logP 3.82			EXP	HANSCH,C & LEO,AJ (1985)
VP 1.16E-006	mm Hg	25	EST	NEELY,WB & BLAU,GE (1985)
DC	pKa			
HL 4.83E-007	atm m3/mol	25	EST	MEYLAN,WM & HOWARD,PH (1991)
OH 2.28E-011	cm3/molc sec	25	EST	MEYLAN,WM & HOWARD,PH (1993)

CAS #: 028313-64-6				2,4-DINO2-C6H3NHN=C(CN)COOET
Formula: $C_{11}H_9N_5O_6$				
Mol Weight: 307.22				
MP (deg C):		FP (deg C):		
BP (deg C):				
BP pressure (mm Hg):				

Property/Value	Units	Temp	Data Type	Reference
WS 1.32E+000	mg/L	25	EST	MEYLAN,WM ET AL. (1996)
logP 4.14			EXP	HANSCH,C & LEO,AJ (1985)
VP 3.86E-009	mm Hg	25	EST	NEELY,WB & BLAU,GE (1985)
DC	pKa			
HL 3.00E-010	atm m3/mol	25	EST	MEYLAN,WM & HOWARD,PH (1991)
OH 2.33E-012	cm3/molc sec	25	EST	MEYLAN,WM & HOWARD,PH (1993)

CAS #: 028313-69-1				2,6-DICL-4-CF3-C6H2NHN=C(CN)COOME
Formula: $C_{11}H_6Cl_2F_3N_3O_2$				
Mol Weight: 340.09				
MP (deg C):		FP (deg C):		
BP (deg C):				
BP pressure (mm Hg):				

Property/Value	Units	Temp	Data Type	Reference
WS 1.34E+000	mg/L	25	EST	MEYLAN,WM ET AL. (1996)
logP 4.36			EXP	HANSCH,C & LEO,AJ (1985)
VP 1.80E-006	mm Hg	25	EST	NEELY,WB & BLAU,GE (1985)
DC	pKa			
HL 3.16E-006	atm m3/mol	25	EST	MEYLAN,WM & HOWARD,PH (1991)
OH 4.57E-013	cm3/molc sec	25	EST	MEYLAN,WM & HOWARD,PH (1993)

CAS #: 028313-74-8				3-CF3-C6H4NHN=C(CN)CO-OME
Formula: $C_{11}H_8F_3N_3O_2$				
Mol Weight: 271.20				
MP (deg C):		FP (deg C):		
BP (deg C):				
BP pressure (mm Hg):				

Property/Value	Units	Temp	Data Type	Reference
WS 1.06E+001	mg/L	25	EST	MEYLAN,WM ET AL. (1996)
logP 3.78			EXP	HANSCH,C & LEO,AJ (1985)
VP 2.81E-005	mm Hg	25	EST	NEELY,WB & BLAU,GE (1985)
DC	pKa			
HL 5.76E-006	atm m3/mol	25	EST	MEYLAN,WM & HOWARD,PH (1991)
OH 6.67E-012	cm3/molc sec	25	EST	MEYLAN,WM & HOWARD,PH (1993)

CAS #: 028313-76-0				4-CF3-C6H4NHN=C(CN)CO-OME
Formula: $C_{11}H_8F_3N_3O_2$				
Mol Weight: 271.20				
MP (deg C):		FP (deg C):		
BP (deg C):				
BP pressure (mm Hg):				

Property/Value	Units	Temp	Data Type	Reference
WS 1.04E+001	mg/L	25	EST	MEYLAN,WM ET AL. (1996)
logP 3.79			EXP	HANSCH,C & LEO,AJ (1985)
VP 2.81E-005	mm Hg	25	EST	NEELY,WB & BLAU,GE (1985)
DC	pKa			
HL 5.76E-006	atm m3/mol	25	EST	MEYLAN,WM & HOWARD,PH (1991)
OH 8.79E-012	cm3/molc sec	25	EST	MEYLAN,WM & HOWARD,PH (1993)

028313-77-1

CAS #: 028313-77-1

Formula: $C_{11}H_7ClF_3N_3O_2$

Mol Weight: 305.65

2-CL-5-CF3-C6H3NHN=C(CN)COOME

MP (deg C): | FP (deg C):

BP (deg C):

BP pressure (mm Hg):

Property/ Value	Units	Temp	Data Type	Reference
WS 1.91E+000	mg/L	25	EST	MEYLAN,WM ET AL. (1996)
logP 4.42			EXP	HANSCH,C & LEO,AJ (1985)
VP 6.99E-006	mm Hg	25	EST	NEELY,WB & BLAU,GE (1985)
DC	pKa			
HL 4.27E-006	atm m3/mol	25	EST	MEYLAN,WM & HOWARD,PH (1991)
OH 2.10E-012	cm3/molc sec	25	EST	MEYLAN,WM & HOWARD,PH (1993)

028313-79-3

CAS #: 028313-79-3

Formula: $C_{11}H_7F_3N_4O_4$

Mol Weight: 316.20

2-NO2-4-CF3-C6H3NHN=C(CN)COOME

MP (deg C): | FP (deg C):

BP (deg C):

BP pressure (mm Hg):

Property/ Value	Units	Temp	Data Type	Reference
WS 4.64E+000	mg/L	25	EST	MEYLAN,WM ET AL. (1996)
logP 3.44			EXP	HANSCH,C & LEO,AJ (1985)
VP 4.14E-007	mm Hg	25	EST	NEELY,WB & BLAU,GE (1985)
DC	pKa			
HL 4.97E-007	atm m3/mol	25	EST	MEYLAN,WM & HOWARD,PH (1991)
OH 1.29E-012	cm3/molc sec	25	EST	MEYLAN,WM & HOWARD,PH (1993)

028313-92-0

CAS #: 028313-92-0

Formula: $C_{11}H_8F_3N_3O_2S$

Mol Weight: 303.26

4-SCF3-C6H4NHN=C(CN)CO-OME

MP (deg C): | FP (deg C):

BP (deg C):

BP pressure (mm Hg):

Property/ Value	Units	Temp	Data Type	Reference
WS 2.50E+000	mg/L	25	EST	MEYLAN,WM ET AL. (1996)
logP 4.30			EXP	HANSCH,C & LEO,AJ (1985)
VP 2.91E-006	mm Hg	25	EST	NEELY,WB & BLAU,GE (1985)
DC	pKa			
HL 1.52E-007	atm m3/mol	25	EST	MEYLAN,WM & HOWARD,PH (1991)
OH 2.64E-011	cm3/molc sec	25	EST	MEYLAN,WM & HOWARD,PH (1993)

028317-56-8

CAS #: 028317-56-8

Formula: $C_{11}H_7ClF_3N_3O$

Mol Weight: 289.65

2-CL-5-CF3-C6H3NHN=C(CN)CO-ME

MP (deg C): | FP (deg C):

BP (deg C):

BP pressure (mm Hg):

Property/ Value	Units	Temp	Data Type	Reference
WS 6.47E-001	mg/L	25	EST	MEYLAN,WM ET AL. (1996)
logP 5.08			EXP	HANSCH,C & LEO,AJ (1985)
VP 9.73E-006	mm Hg	25	EST	NEELY,WB & BLAU,GE (1985)
DC	pKa			
HL 1.21E-006	atm m3/mol	25	EST	MEYLAN,WM & HOWARD,PH (1991)
OH 1.99E-012	cm3/molc sec	25	EST	MEYLAN,WM & HOWARD,PH (1993)

028317-58-0

CAS #: 028317-58-0

Formula: $C_{10}H_8ClN_3O$

Mol Weight: 221.65

BUTANENITRILE, 2-[(4-CHLOROPHENYL)HYDRAZONO]-3-O

MP (deg C): | FP (deg C):

BP (deg C):

BP pressure (mm Hg):

Property/ Value	Units	Temp	Data Type	Reference
WS 1.06E+001	mg/L	25	EST	MEYLAN,WM ET AL. (1996)
logP 4.10			EXP	HANSCH,C & LEO,AJ (1985)
VP 1.29E-005	mm Hg	25	EST	NEELY,WB & BLAU,GE (1985)
DC	pKa			
HL 1.39E-007	atm m3/mol	25	EST	MEYLAN,WM & HOWARD,PH (1991)
OH 1.26E-011	cm3/molc sec	25	EST	MEYLAN,WM & HOWARD,PH (1993)

028317-59-1

CAS #: 028317-59-1

Formula: $C_{10}H_8ClN_3O$

Mol Weight: 221.65

BUTANENITRILE, 2-[(2-CHLOROPHENYL)HYDRAZONO]-3-O

MP (deg C): | FP (deg C):

BP (deg C):

BP pressure (mm Hg):

Property/ Value	Units	Temp	Data Type	Reference
WS 1.29E+001	mg/L	25	EST	MEYLAN,WM ET AL. (1996)
logP 4.00			EXP	HANSCH,C & LEO,AJ (1985)
VP 1.29E-005	mm Hg	25	EST	NEELY,WB & BLAU,GE (1985)
DC	pKa			
HL 1.39E-007	atm m3/mol	25	EST	MEYLAN,WM & HOWARD,PH (1991)
OH 1.26E-011	cm3/molc sec	25	EST	MEYLAN,WM & HOWARD,PH (1993)

028317-60-4

CAS #: 028317-60-4

Formula: $C_{10}H_8ClN_3O$

Mol Weight: 221.65

BUTANENITRILE, 2-[(3-CHLOROPHENYL)HYDRAZONO]-3-O

MP (deg C): | FP (deg C):

BP (deg C):

BP pressure (mm Hg):

Property/ Value	Units	Temp	Data Type	Reference
WS 1.58E+001	mg/L	25	EST	MEYLAN,WM ET AL. (1996)
logP 3.90			EXP	HANSCH,C & LEO,AJ (1985)
VP 1.29E-005	mm Hg	25	EST	NEELY,WB & BLAU,GE (1985)
DC	pKa			
HL 1.39E-007	atm m3/mol	25	EST	MEYLAN,WM & HOWARD,PH (1991)
OH 3.01E-011	cm3/molc sec	25	EST	MEYLAN,WM & HOWARD,PH (1993)

028317-61-5

CAS #: 028317-61-5

Formula: $C_{10}H_7Cl_2N_3O$

Mol Weight: 256.09

3,4-DICL-C6H3NHN=C(CN)CO-ME

MP (deg C): | FP (deg C):

BP (deg C):

BP pressure (mm Hg):

Property/ Value	Units	Temp	Data Type	Reference
WS 2.79E+000	mg/L	25	EST	MEYLAN,WM ET AL. (1996)
logP 4.56			EXP	HANSCH,C & LEO,AJ (1985)
VP 3.25E-006	mm Hg	25	EST	NEELY,WB & BLAU,GE (1985)
DC	pKa			
HL 1.03E-007	atm m3/mol	25	EST	MEYLAN,WM & HOWARD,PH (1991)
OH 8.86E-012	cm3/molc sec	25	EST	MEYLAN,WM & HOWARD,PH (1993)

CAS #: 028317-62-6				3,5-DICL-C6H3NHN=C(CN)CO-ME

Formula: $C_{10}H_7Cl_2N_3O$
Mol Weight: 256.09
MP (deg C):   FP (deg C):
BP (deg C):
BP pressure (mm Hg):

Property/Value	Units	Temp	Data Type	Reference
WS 2.21E+000	mg/L	25	EST	MEYLAN,WM ET AL. (1996)
logP 4.68			EXP	HANSCH,C & LEO,AJ (1985)
VP 3.25E-006	mm Hg	25	EST	NEELY,WB & BLAU,GE (1985)
DC	pKa			
HL 1.03E-007	atm m3/mol	25	EST	MEYLAN,WM & HOWARD,PH (1991)
OH 2.12E-011	cm3/molc sec	25	EST	MEYLAN,WM & HOWARD,PH (1993)

CAS #: 028317-64-8				2-ME-4-CL-C6H3NHN=C(CN)CO-ME

Formula: $C_{11}H_{10}ClN_3O$
Mol Weight: 235.67
MP (deg C):   FP (deg C):
BP (deg C):
BP pressure (mm Hg):

Property/Value	Units	Temp	Data Type	Reference
WS 6.27E+000	mg/L	25	EST	MEYLAN,WM ET AL. (1996)
logP 4.28			EXP	HANSCH,C & LEO,AJ (1985)
VP 5.02E-006	mm Hg	25	EST	NEELY,WB & BLAU,GE (1985)
DC	pKa			
HL 1.53E-007	atm m3/mol	25	EST	MEYLAN,WM & HOWARD,PH (1991)
OH 1.55E-011	cm3/molc sec	25	EST	MEYLAN,WM & HOWARD,PH (1993)

CAS #: 028317-71-7				BUTANENITRILE, 2-[(2,4-DINITROPHENYL)HYDRAZONO]-

Formula: $C_{10}H_7N_5O_5$
Mol Weight: 277.20
MP (deg C):   FP (deg C):
BP (deg C):
BP pressure (mm Hg):

Property/Value	Units	Temp	Data Type	Reference
WS 8.46E+000	mg/L	25	EST	MEYLAN,WM ET AL. (1996)
logP 3.40			EXP	HANSCH,C & LEO,AJ (1985)
VP 1.13E-008	mm Hg	25	EST	NEELY,WB & BLAU,GE (1985)
DC	pKa			
HL 6.38E-011	atm m3/mol	25	EST	MEYLAN,WM & HOWARD,PH (1991)
OH 7.68E-013	cm3/molc sec	25	EST	MEYLAN,WM & HOWARD,PH (1993)

CAS #: 028317-78-4				4-SCF3-C6H4NHN=C(CN)CO-ME

Formula: $C_{11}H_8F_3N_3OS$
Mol Weight: 287.27
MP (deg C):   FP (deg C):
BP (deg C):
BP pressure (mm Hg):

Property/Value	Units	Temp	Data Type	Reference
WS 7.23E-001	mg/L	25	EST	MEYLAN,WM ET AL. (1996)
logP 5.04			EXP	HANSCH,C & LEO,AJ (1985)
VP 3.43E-006	mm Hg	25	EST	NEELY,WB & BLAU,GE (1985)
DC	pKa			
HL 4.30E-008	atm m3/mol	25	EST	MEYLAN,WM & HOWARD,PH (1991)
OH 2.63E-011	cm3/molc sec	25	EST	MEYLAN,WM & HOWARD,PH (1993)

CAS #: 028317-91-1				3-CL-C6H4NHN=C(CN)CO-T-BU

Formula: $C_{13}H_{14}ClN_3O$
Mol Weight: 263.73
MP (deg C):   FP (deg C):
BP (deg C):
BP pressure (mm Hg):

Property/Value	Units	Temp	Data Type	Reference
WS 2.25E+000	mg/L	25	EST	MEYLAN,WM ET AL. (1996)
logP 4.62			EXP	HANSCH,C & LEO,AJ (1985)
VP 2.12E-006	mm Hg	25	EST	NEELY,WB & BLAU,GE (1985)
DC	pKa			
HL 3.25E-007	atm m3/mol	25	EST	MEYLAN,WM & HOWARD,PH (1991)
OH 3.16E-011	cm3/molc sec	25	EST	MEYLAN,WM & HOWARD,PH (1993)

CAS #: 028322-78-3				2,4,5-TRICL-C6H2NHN=C(CN)CO-OET

Formula: $C_{11}H_8Cl_3N_3O_2$
Mol Weight: 320.56
MP (deg C):   FP (deg C):
BP (deg C):
BP pressure (mm Hg):

Property/Value	Units	Temp	Data Type	Reference
WS 3.30E-001	mg/L	25	EST	MEYLAN,WM ET AL. (1996)
logP 5.21			EXP	HANSCH,C & LEO,AJ (1985)
VP 2.52E-007	mm Hg	25	EST	NEELY,WB & BLAU,GE (1985)
DC	pKa			
HL 3.58E-007	atm m3/mol	25	EST	MEYLAN,WM & HOWARD,PH (1991)
OH 4.22E-012	cm3/molc sec	25	EST	MEYLAN,WM & HOWARD,PH (1993)

CAS #: 028343-28-4				2,4,5-TRICL-C6H2NHN=C(CN)CO-T-BU

Formula: $C_{13}H_{12}Cl_3N_3O$
Mol Weight: 332.62
MP (deg C):   FP (deg C):
BP (deg C):
BP pressure (mm Hg):

Property/Value	Units	Temp	Data Type	Reference
WS 7.80E-002	mg/L	25	EST	MEYLAN,WM ET AL. (1996)
logP 5.86			EXP	HANSCH,C & LEO,AJ (1985)
VP 1.42E-007	mm Hg	25	EST	NEELY,WB & BLAU,GE (1985)
DC	pKa			
HL 1.78E-007	atm m3/mol	25	EST	MEYLAN,WM & HOWARD,PH (1991)
OH 4.15E-012	cm3/molc sec	25	EST	MEYLAN,WM & HOWARD,PH (1993)

CAS #: 028354-19-0				BENZOIC ACID, O-CYANOMETHYLAMINO-

Formula: $C_9H_8N_2O_2$
Mol Weight: 176.18
MP (deg C):   FP (deg C):
BP (deg C):
BP pressure (mm Hg):

Property/Value	Units	Temp	Data Type	Reference
WS 9.06E+002	mg/L	25	EST	MEYLAN,WM ET AL. (1996)
logP 1.80			EXP	HANSCH,C & LEO,AJ (1985)
VP 6.42E-006	mm Hg	25	EST	NEELY,WB & BLAU,GE (1985)
DC	pKa			
HL 3.49E-014	atm m3/mol	25	EST	MEYLAN,WM & HOWARD,PH (1991)
OH 1.80E-011	cm3/molc sec	25	EST	MEYLAN,WM & HOWARD,PH (1993)

CAS #: 028354-25-8	ANISOLE, O-CYANOMETHYLAMINO-

Formula: $C_9H_{10}N_2O$

Mol Weight: 162.19

MP (deg C): FP (deg C):

BP (deg C):

BP pressure (mm Hg):

Property/Value	Units	Temp	Data Type	Reference
WS 2.82E+003	mg/L	25	EST	MEYLAN,WM ET AL. (1996)
logP 1.30			EXP	HANSCH,C & LEO,AJ (1985)
VP 7.90E-004	mm Hg	25	EST	NEELY,WB & BLAU,GE (1985)
DC	pKa			
HL 1.03E-010	atm m3/mol	25	EST	MEYLAN,WM & HOWARD,PH (1991)
OH 3.94E-011	cm3/molc sec	25	EST	MEYLAN,WM & HOWARD,PH (1993)

CAS #: 028356-58-3	4-ACETYLPYRIDINE

Formula: $C_7H_7NO_2$

Mol Weight: 137.14

MP (deg C): 141 dec FP (deg C):

BP (deg C):

BP pressure (mm Hg):

Property/Value	Units	Temp	Data Type	Reference
WS 1.00E+006	mg/L	25	EST	MEYLAN,WM ET AL. (1996)
logP 0.24			EST	MEYLAN,WM & HOWARD,PH (1995)
VP 2.64E-003	mm Hg	25	EST	NEELY,WB & BLAU,GE (1985)
DC	pKa			
HL 5.79E-011	atm m3/mol	25	EST	MEYLAN,WM & HOWARD,PH (1991)
OH 1.83E-012	cm3/molc sec	25	EST	MEYLAN,WM & HOWARD,PH (1993)

CAS #: 028384-50-1	2-CF3-C6H4NHN=C(CN)CO-OME

Formula: $C_{11}H_8F_3N_3O_2$

Mol Weight: 271.20

MP (deg C): FP (deg C):

BP (deg C):

BP pressure (mm Hg):

Property/Value	Units	Temp	Data Type	Reference
WS 1.25E+001	mg/L	25	EST	MEYLAN,WM ET AL. (1996)
logP 3.70			EXP	HANSCH,C & LEO,AJ (1985)
VP 2.81E-005	mm Hg	25	EST	NEELY,WB & BLAU,GE (1985)
DC	pKa			
HL 5.76E-006	atm m3/mol	25	EST	MEYLAN,WM & HOWARD,PH (1991)
OH 8.79E-012	cm3/molc sec	25	EST	MEYLAN,WM & HOWARD,PH (1993)

CAS #: 028466-26-4	4-AMINOPYRAZOLE

Formula: $C_3H_5N_3$

Mol Weight: 83.09

MP (deg C): 81 FP (deg C):

BP (deg C):

BP pressure (mm Hg):

Property/Value	Units	Temp	Data Type	Reference
WS 1.00E+006	mg/L	25	EST	MEYLAN,WM ET AL. (1996)
logP -1.09			EXP	HANSCH,C & LEO,AJ (1985)
VP 1.65E-002	mm Hg	25	EST	NEELY,WB & BLAU,GE (1985)
DC	pKa			
HL 1.30E-009	atm m3/mol	25	EST	MEYLAN,WM & HOWARD,PH (1991)
OH 2.00E-010	cm3/molc sec	25	EST	MEYLAN,WM & HOWARD,PH (1993)

CAS #: 028529-34-2	LEUCIN-AMIDE, N-ACETYL

Formula: $C_8H_{16}N_2O_2$

Mol Weight: 172.23

MP (deg C): FP (deg C):

BP (deg C):

BP pressure (mm Hg):

Property/Value	Units	Temp	Data Type	Reference
WS 3.80E+003	mg/L	25	EST	MEYLAN,WM ET AL. (1996)
logP -0.13			EXP	HANSCH,C & LEO,AJ (1985)
VP 2.18E-007	mm Hg	25	EST	NEELY,WB & BLAU,GE (1985)
DC	pKa			
HL 3.66E-010	atm m3/mol	25	EST	MEYLAN,WM & HOWARD,PH (1991)
OH 5.04E-011	cm3/molc sec	25	EST	MEYLAN,WM & HOWARD,PH (1993)

CAS #: 028553-12-0	DIISONONYL PHTHALATE

Formula: $C_{26}H_{42}O_4$

Mol Weight: 418.62

MP (deg C): FP (deg C):

BP (deg C):

BP pressure (mm Hg):

Property/Value	Units	Temp	Data Type	Reference
WS 2.00E-001	mg/L	20	EXP	HOWARD,PH ET AL. (1985)
logP 9.37			EST	MEYLAN,WM & HOWARD,PH (1995)
VP 5.40E-007	mm Hg	25	EXP	HOWARD,PH ET AL. (1985)
DC	pKa			
HL 1.49E-006	atm m3/mol	25	EST	VP/WSOL
OH 2.34E-011	cm3/molc sec	25	EST	MEYLAN,WM & HOWARD,PH (1993)

CAS #: 028610-09-5	GLYCINE, N-[1-OXO-5-(1H-PURIN-6-YLTHIO)PENTYL]-,

Formula: $C_{15}H_{21}N_5O_3S$

Mol Weight: 351.43

MP (deg C): FP (deg C):

BP (deg C):

BP pressure (mm Hg):

Property/Value	Units	Temp	Data Type	Reference
WS 4.11E+002	mg/L	25	EST	MEYLAN,WM ET AL. (1996)
logP 1.37			EXP	SANGSTER,J (1993)
VP 6.14E-014	mm Hg	25	EST	NEELY,WB & BLAU,GE (1985)
DC	pKa			
HL 1.47E-017	atm m3/mol	25	EST	MEYLAN,WM & HOWARD,PH (1991)
OH 2.27E-010	cm3/molc sec	25	EST	MEYLAN,WM & HOWARD,PH (1993)

CAS #: 028615-21-6	UREA, N-ETHYL-N'-BENZOYL-

Formula: $C_{10}H_{12}N_2O_2$

Mol Weight: 192.22

MP (deg C): FP (deg C):

BP (deg C):

BP pressure (mm Hg):

Property/Value	Units	Temp	Data Type	Reference
WS 1.10E+003	mg/L	25	EST	MEYLAN,WM ET AL. (1996)
logP 1.92			EXP	SOTOMATSU,T ET AL. (1987)
VP 5.16E-008	mm Hg	25	EST	NEELY,WB & BLAU,GE (1985)
DC	pKa			
HL 6.22E-011	atm m3/mol	25	EST	MEYLAN,WM & HOWARD,PH (1991)
OH 7.77E-012	cm3/molc sec	25	EST	MEYLAN,WM & HOWARD,PH (1993)

CAS #: 028616-93-5 — URIDINE, 5-BROMO-2',3'-DIDEOXY-

Formula: $C_9H_{11}BrN_2O_4$
Mol Weight: 291.11
MP (deg C): 179-182
FP (deg C):
BP (deg C):
BP pressure (mm Hg):

Property/Value	Units	Temp	Data Type	Reference
WS 4.75E+003	mg/L	25	EST	MEYLAN,WM ET AL. (1996)
logP -0.99			EXP	BALZARINI,J ET AL. (1989)
VP 3.66E-013	mm Hg	25	EST	NEELY,WB & BLAU,GE (1985)
DC	pKa			
HL 4.21E-017	atm m3/mol	25	EST	MEYLAN,WM & HOWARD,PH (1991)
OH 7.52E-011	cm3/molc sec	25	EST	MEYLAN,WM & HOWARD,PH (1993)

CAS #: 028648-86-4 — 4-PYRIMIDINECARBOXAMIDE

Formula: $C_5H_5N_3O$
Mol Weight: 123.12
MP (deg C):
FP (deg C):
BP (deg C):
BP pressure (mm Hg):

Property/Value	Units	Temp	Data Type	Reference
WS 1.82E+004	mg/L	25	EST	MEYLAN,WM ET AL. (1996)
logP -0.68			EXP	YAMAGAMI,C ET AL. (1990)
VP 2.44E-004	mm Hg	25	EST	NEELY,WB & BLAU,GE (1985)
DC	pKa			
HL 1.20E-012	atm m3/mol	25	EST	MEYLAN,WM & HOWARD,PH (1991)
OH 2.21E-012	cm3/molc sec	25	EST	MEYLAN,WM & HOWARD,PH (1993)

CAS #: 028772-56-7 — BROMADIOLONE

Formula: $C_{30}H_{23}BrO_4$
Mol Weight: 527.42
MP (deg C): 200-210
FP (deg C):
BP (deg C):
BP pressure (mm Hg):

Property/Value	Units	Temp	Data Type	Reference
WS 1.90E+001	mg/L	20	EXP	BUDAVARI,S ET AL. (1989)
logP 7.02			EST	MEYLAN,WM & HOWARD,PH (1995)
VP 1.50E-008	mm Hg	20	EXP	HARTLY,D & KIDD,H (1983)
DC 4.04	pKa	21	EXP	BUDAVARI,S ET AL. (1989)
HL 5.48E-010	atm m3/mol	20	EST	VP/WSOL
OH 6.14E-011	cm3/molc sec	25	EST	MEYLAN,WM & HOWARD,PH (1993)

CAS #: 028783-31-5 — 3-(B-NITROVINYL)THIOPHENE

Formula: $C_6H_7NO_2S$
Mol Weight: 157.19
MP (deg C):
FP (deg C):
BP (deg C):
BP pressure (mm Hg):

Property/Value	Units	Temp	Data Type	Reference
WS 6.46E+002	mg/L	25	EST	MEYLAN,WM ET AL. (1996)
logP 1.94			EXP	HANSCH,C & LEO,AJ (1985)
VP 8.27E-003	mm Hg	25	EST	NEELY,WB & BLAU,GE (1985)
DC	pKa			
HL 1.89E-006	atm m3/mol	25	EST	MEYLAN,WM & HOWARD,PH (1991)
OH 2.04E-011	cm3/molc sec	25	EST	MEYLAN,WM & HOWARD,PH (1993)

CAS #: 028795-24-6 — ME(3(2ME-5NO2-1-IMIDAZ)PROPYL)SULFONE

Formula: $C_8H_{15}N_3O_4S$
Mol Weight: 249.29
MP (deg C):
FP (deg C):
BP (deg C):
BP pressure (mm Hg):

Property/Value	Units	Temp	Data Type	Reference
WS 7.13E+004	mg/L	25	EST	MEYLAN,WM ET AL. (1996)
logP -1.00			EXP	HANSCH,C & LEO,AJ (1985)
VP 7.87E-008	mm Hg	25	EST	NEELY,WB & BLAU,GE (1985)
DC	pKa			
HL 5.21E-011	atm m3/mol	25	EST	MEYLAN,WM & HOWARD,PH (1991)
OH 1.66E-011	cm3/molc sec	25	EST	MEYLAN,WM & HOWARD,PH (1993)

CAS #: 028810-38-0 — TETRACHLOROCYCLOHEXENE

Formula: $C_6H_6Cl_4$
Mol Weight: 219.93
MP (deg C):
FP (deg C):
BP (deg C):
BP pressure (mm Hg):

Property/Value	Units	Temp	Data Type	Reference
WS 4.30E+001	mg/L	25	EST	MEYLAN,WM ET AL. (1996)
logP 3.74			EXP	HANSCH,C & LEO,AJ (1985)
VP 3.13E-002	mm Hg	25	EST	NEELY,WB & BLAU,GE (1985)
DC	pKa			
HL 1.81E-003	atm m3/mol	25	EST	MEYLAN,WM & HOWARD,PH (1991)
OH 3.33E-011	cm3/molc sec	25	EST	MEYLAN,WM & HOWARD,PH (1993)

CAS #: 028822-58-4 — 1H-PURINE-2,6-DIONE, 3,7-DIHYDRO-1-METHYL-3-(2-M

Formula: $C_{10}H_{14}N_4O_2$
Mol Weight: 222.25
MP (deg C): 200-201
FP (deg C):
BP (deg C):
BP pressure (mm Hg):

Property/Value	Units	Temp	Data Type	Reference
WS 3.49E+003	mg/L	25	EST	MEYLAN,WM ET AL. (1996)
logP 1.15			EXP	HANSCH,C ET AL. (1995)
VP 7.25E-010	mm Hg	25	EST	NEELY,WB & BLAU,GE (1985)
DC	pKa			
HL 3.92E-012	atm m3/mol	25	EST	MEYLAN,WM & HOWARD,PH (1991)
OH 2.62E-011	cm3/molc sec	25	EST	MEYLAN,WM & HOWARD,PH (1993)

CAS #: 028856-77-1 — 3-PHENYLTHIOPYRIDINE

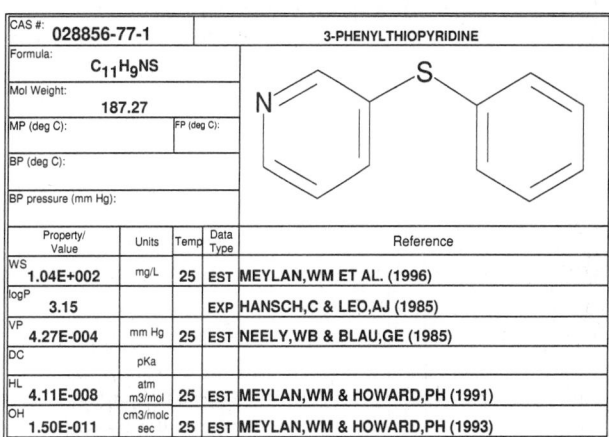

Formula: $C_{11}H_9NS$
Mol Weight: 187.27
MP (deg C):
FP (deg C):
BP (deg C):
BP pressure (mm Hg):

Property/Value	Units	Temp	Data Type	Reference
WS 1.04E+002	mg/L	25	EST	MEYLAN,WM ET AL. (1996)
logP 3.15			EXP	HANSCH,C & LEO,AJ (1985)
VP 4.27E-004	mm Hg	25	EST	NEELY,WB & BLAU,GE (1985)
DC	pKa			
HL 4.11E-008	atm m3/mol	25	EST	MEYLAN,WM & HOWARD,PH (1991)
OH 1.50E-011	cm3/molc sec	25	EST	MEYLAN,WM & HOWARD,PH (1993)

CAS #: 028879-93-8				21,22-DIHYDROBRUCINE
Formula: $C_{23}H_{28}N_2O_4$				
Mol Weight: 396.49				
MP (deg C):		FP (deg C):		
BP (deg C):				
BP pressure (mm Hg):				

Property/Value	Units	Temp	Data Type	Reference
WS 6.17E+002	mg/L	25	EST	MEYLAN,WM ET AL. (1996)
logP 0.84			EXP	HANSCH,C & LEO,AJ (1985)
VP 1.36E-010	mm Hg	25	EST	NEELY,WB & BLAU,GE (1985)
DC	pKa			
HL 2.53E-017	atm m3/mol	25	EST	MEYLAN,WM & HOWARD,PH (1991)
OH 2.41E-010	cm3/molc sec	25	EST	MEYLAN,WM & HOWARD,PH (1993)

CAS #: 028910-86-3				1,4-BENZDIAZEPIN-2-ONE-5(26F-PH)7CL
Formula: $C_{15}H_9ClF_2N_2O$				
Mol Weight: 306.70				
MP (deg C):		FP (deg C):		
BP (deg C):				
BP pressure (mm Hg):				

Property/Value	Units	Temp	Data Type	Reference
WS 5.77E+001	mg/L	25	EST	MEYLAN,WM ET AL. (1996)
logP 2.68			EXP	HANSCH,C & LEO,AJ (1985)
VP 9.35E-009	mm Hg	25	EST	NEELY,WB & BLAU,GE (1985)
DC	pKa			
HL 2.42E-010	atm m3/mol	25	EST	MEYLAN,WM & HOWARD,PH (1991)
OH 8.72E-012	cm3/molc sec	25	EST	MEYLAN,WM & HOWARD,PH (1993)

CAS #: 028915-24-4				1,3,4-OXADIAZOLE, 2-METHYL-5-[(PHENYLMETHYL)THIO]
Formula: $C_{10}H_{10}N_2OS$				
Mol Weight: 206.27				
MP (deg C):		FP (deg C):		
BP (deg C):				
BP pressure (mm Hg):				

Property/Value	Units	Temp	Data Type	Reference
WS 1.33E+003	mg/L	25	EST	MEYLAN,WM ET AL. (1996)
logP 1.74			EXP	HANSCH,C ET AL. (1995)
VP 2.30E-005	mm Hg	25	EST	NEELY,WB & BLAU,GE (1985)
DC	pKa			
HL 7.44E-009	atm m3/mol	25	EST	MEYLAN,WM & HOWARD,PH (1991)
OH 1.63E-011	cm3/molc sec	25	EST	MEYLAN,WM & HOWARD,PH (1993)

CAS #: 029074-04-2				4-T-BUTYLPHENYL GLUCOPYRANOSIDE
Formula: $C_{16}H_{24}O_6$				
Mol Weight: 312.37				
MP (deg C):		FP (deg C):		
BP (deg C):				
BP pressure (mm Hg):				

Property/Value	Units	Temp	Data Type	Reference
WS 1.02E+003	mg/L	25	EST	MEYLAN,WM ET AL. (1996)
logP 1.18			EXP	HANSCH,C & LEO,AJ (1985)
VP 2.70E-012	mm Hg	25	EST	NEELY,WB & BLAU,GE (1985)
DC	pKa			
HL 2.97E-015	atm m3/mol	25	EST	MEYLAN,WM & HOWARD,PH (1991)
OH 8.83E-011	cm3/molc sec	25	EST	MEYLAN,WM & HOWARD,PH (1993)

CAS #: 029082-74-4				OCTACHLOROSTYRENE
Formula: C_8Cl_8				
Mol Weight: 379.71				
MP (deg C):		FP (deg C):		
BP (deg C):				
BP pressure (mm Hg):				

Property/Value	Units	Temp	Data Type	Reference
WS 1.74E-003	mg/L	25	EST	MEYLAN,WM ET AL. (1996)
logP 7.46			EST	MEYLAN,WM & HOWARD,PH (1995)
VP 1.32E-005	mm Hg	25	EST	NEELY,WB & BLAU,GE (1985)
DC	pKa			
HL 2.30E-004	atm m3/mol	25	EST	MEYLAN,WM & HOWARD,PH (1991)
OH 1.07E-012	cm3/molc sec	25	EST	MEYLAN,WM & HOWARD,PH (1993)

CAS #: 029091-05-2				DINITRAMINE
Formula: $C_{11}H_{13}F_3N_4O_4$				
Mol Weight: 322.25				
MP (deg C): 98		FP (deg C):		
BP (deg C):				
BP pressure (mm Hg):				

Property/Value	Units	Temp	Data Type	Reference
WS 1.10E+000	mg/L	25	EXP	YALKOWSKY,SH & DANNENFELSER,RM (1992)
logP 3.96			EST	MEYLAN,WM & HOWARD,PH (1995)
VP 3.60E-006	mm Hg	25	EXP	AUGUSTIJN-BECKERS,PWM ET AL. (1994)
DC	pKa			
HL 1.39E-006	atm m3/mol	25	EST	VP/WSOL
OH 1.77E-011	cm3/molc sec	25	EST	MEYLAN,WM & HOWARD,PH (1993)

CAS #: 029094-61-9				GLIPIZIDE
Formula: $C_{21}H_{27}N_5O_4S$				
Mol Weight: 445.54				
MP (deg C):		FP (deg C):		
BP (deg C):				
BP pressure (mm Hg):				

Property/Value	Units	Temp	Data Type	Reference
WS 3.72E+001	mg/L	25	EST	MEYLAN,WM ET AL. (1996)
logP 1.91			EXP	HANSCH,C ET AL. (1995)
VP 2.73E-016	mm Hg	25	EST	NEELY,WB & BLAU,GE (1985)
DC	pKa			
HL 1.03E-020	atm m3/mol	25	EST	MEYLAN,WM & HOWARD,PH (1991)
OH 3.42E-011	cm3/molc sec	25	EST	MEYLAN,WM & HOWARD,PH (1993)

CAS #: 029104-30-1				BENZOXIMATE
Formula: $C_{18}H_{18}ClNO_5$				
Mol Weight: 363.80				
MP (deg C):		FP (deg C):		
BP (deg C):				
BP pressure (mm Hg):				

Property/Value	Units	Temp	Data Type	Reference
WS 3.33E+000	mg/L	25	EST	MEYLAN,WM ET AL. (1996)
logP 3.73			EXP	SAITO,H ET AL. (1993)
VP 4.40E-008	mm Hg	25	EST	NEELY,WB & BLAU,GE (1985)
DC	pKa			
HL 1.86E-010	atm m3/mol	25	EST	MEYLAN,WM & HOWARD,PH (1991)
OH 8.68E-011	cm3/molc sec	25	EST	MEYLAN,WM & HOWARD,PH (1993)

CAS #: 029121-29-7			BENZENEACETAMIDE, 3-[2-HYDROXY-3-[(1-METHYLETHYL			
Formula: $C_{14}H_{22}N_2O_3$					
Mol Weight: 266.34					
MP (deg C):		FP (deg C):			
BP (deg C):					
BP pressure (mm Hg):					

Property/Value	Units	Temp	Data Type	Reference	
WS 6.09E+002	mg/L	25	EST	MEYLAN,WM ET AL. (1996)	
logP 0.22			EXP	SANGSTER,J (1993)	
VP 2.92E-010	mm Hg	25	EST	NEELY,WB & BLAU,GE (1985)	
DC	pKa				
HL 1.37E-018	atm m3/mol	25	EST	MEYLAN,WM & HOWARD,PH (1991)	
OH 1.66E-010	cm3/molc sec	25	EST	MEYLAN,WM & HOWARD,PH (1993)	

CAS #: 029122-68-7			ATENOLOL			
Formula: $C_{14}H_{22}N_2O_3$					
Mol Weight: 266.34					
MP (deg C): 146-148		FP (deg C):			
BP (deg C):					
BP pressure (mm Hg):					

Property/Value	Units	Temp	Data Type	Reference	
WS 6.85E+002	mg/L	25	EST	MEYLAN,WM ET AL. (1996)	
logP 0.16			EXP	HANSCH,C & LEO,AJ (1985)	
VP 2.92E-010	mm Hg	25	EST	NEELY,WB & BLAU,GE (1985)	
DC	pKa				
HL 1.37E-018	atm m3/mol	25	EST	MEYLAN,WM & HOWARD,PH (1991)	
OH 1.38E-010	cm3/molc sec	25	EST	MEYLAN,WM & HOWARD,PH (1993)	

CAS #: 029152-10-1			SUDOXICAM			
Formula: $C_{13}H_{11}N_3O_4S_2$					
Mol Weight: 337.38					
MP (deg C):		FP (deg C):			
BP (deg C):					
BP pressure (mm Hg):					

Property/Value	Units	Temp	Data Type	Reference	
WS 2.61E+001	mg/L	25	EST	MEYLAN,WM ET AL. (1996)	
logP 1.64			EXP	HANSCH,C & LEO,AJ (1985)	
VP 5.41E-012	mm Hg	25	EST	NEELY,WB & BLAU,GE (1985)	
DC	pKa				
HL 1.46E-018	atm m3/mol	25	EST	MEYLAN,WM & HOWARD,PH (1991)	
OH 2.07E-011	cm3/molc sec	25	EST	MEYLAN,WM & HOWARD,PH (1993)	

CAS #: 029230-99-7			2-ACETYLPHENYLDIMETHYLCARBAMATE			
Formula: $C_{11}H_{13}NO_3$					
Mol Weight: 207.23					
MP (deg C):		FP (deg C):			
BP (deg C):					
BP pressure (mm Hg):					

Property/Value	Units	Temp	Data Type	Reference	
WS 6.45E+003	mg/L	25	EST	MEYLAN,WM ET AL. (1996)	
logP 0.93			EXP	HANSCH,C & LEO,AJ (1985)	
VP 6.80E-004	mm Hg	25	EST	NEELY,WB & BLAU,GE (1985)	
DC	pKa				
HL 1.28E-010	atm m3/mol	25	EST	MEYLAN,WM & HOWARD,PH (1991)	
OH 1.81E-011	cm3/molc sec	25	EST	MEYLAN,WM & HOWARD,PH (1993)	

CAS #: 029232-93-7			PIRIMIPHOS-METHYL			
Formula: $C_{11}H_{20}N_3O_3PS$					
Mol Weight: 305.34					
MP (deg C): 15		FP (deg C):			
BP (deg C):					
BP pressure (mm Hg):					

Property/Value	Units	Temp	Data Type	Reference	
WS 2.25E+001	mg/L	20	EXP	YALKOWSKY,SH & DANNENFELSER,RM (1992)	
logP 4.20			EXP	HANSCH,C & LEO,AJ (1985)	
VP 8.99E-006	mm Hg	25	EST	NEELY,WB & BLAU,GE (1985)	
DC	pKa				
HL 2.53E-006	atm m3/mol	25	EST	MEYLAN,WM & HOWARD,PH (1991)	
OH 1.60E-010	cm3/molc sec	25	EST	MEYLAN,WM & HOWARD,PH (1993)	

CAS #: 029247-87-8			4-PYRIDINECARBOXYLIC ACID, 2-IODO-, HYDRAZIDE			
Formula: $C_6H_6IN_3O$					
Mol Weight: 263.04					
MP (deg C):		FP (deg C):			
BP (deg C):					
BP pressure (mm Hg):					

Property/Value	Units	Temp	Data Type	Reference	
WS 7.20E+003	mg/L	25	EST	MEYLAN,WM ET AL. (1996)	
logP 0.52			EXP	SANGSTER,J (1993)	
VP 1.18E-006	mm Hg	25	EST	NEELY,WB & BLAU,GE (1985)	
DC	pKa				
HL 2.81E-015	atm m3/mol	25	EST	MEYLAN,WM & HOWARD,PH (1991)	
OH 5.61E-012	cm3/molc sec	25	EST	MEYLAN,WM & HOWARD,PH (1993)	

CAS #: 029268-77-7			2-METHOXY-6-METHYLAZOBENZENE			
Formula: $C_{14}H_{14}N_2O$					
Mol Weight: 226.28					
MP (deg C):		FP (deg C):			
BP (deg C):					
BP pressure (mm Hg):					

Property/Value	Units	Temp	Data Type	Reference	
WS 1.06E+000	mg/L	25	EST	MEYLAN,WM ET AL. (1996)	
logP 4.74			EST	MEYLAN,WM & HOWARD,PH (1995)	
VP 8.99E-005	mm Hg	25	EST	NEELY,WB & BLAU,GE (1985)	
DC	pKa				
HL 9.60E-007	atm m3/mol	25	EST	MEYLAN,WM & HOWARD,PH (1991)	
OH 2.40E-011	cm3/molc sec	25	EST	MEYLAN,WM & HOWARD,PH (1993)	

CAS #: 029268-78-8			2-METHOXY-2'-METHYLAZOBENZENE			
Formula: $C_{14}H_{14}N_2O$					
Mol Weight: 226.28					
MP (deg C):		FP (deg C):			
BP (deg C):					
BP pressure (mm Hg):					

Property/Value	Units	Temp	Data Type	Reference	
WS 1.06E+000	mg/L	25	EST	MEYLAN,WM ET AL. (1996)	
logP 4.74			EST	MEYLAN,WM & HOWARD,PH (1995)	
VP 8.99E-005	mm Hg	25	EST	NEELY,WB & BLAU,GE (1985)	
DC	pKa				
HL 9.60E-007	atm m3/mol	25	EST	MEYLAN,WM & HOWARD,PH (1991)	
OH 1.15E-011	cm3/molc sec	25	EST	MEYLAN,WM & HOWARD,PH (1993)	

CAS #: 029316-05-0				S-AMYLBENZENE

Formula: $C_{11}H_{16}$

Mol Weight: 148.25

MP (deg C): | FP (deg C):

BP (deg C):

BP pressure (mm Hg):

Property/Value	Units	Temp	Data Type	Reference
WS 1.27E+001	mg/L	25	EST	MEYLAN,WM ET AL. (1996)
logP 4.43			EST	MEYLAN,WM & HOWARD,PH (1995)
VP 6.16E-001	mm Hg	25	EXT	CHAO,J ET AL. (1983)
DC	pKa			
HL 1.85E-002	atm m3/mol	25	EST	MEYLAN,WM & HOWARD,PH (1991)
OH 9.91E-012	cm3/molc sec	25	EST	MEYLAN,WM & HOWARD,PH (1993)

CAS #: 029366-71-0				2'-NITROBENZOGUANAMINE

Formula: $C_9H_8N_6O_2$

Mol Weight: 232.20

MP (deg C): | FP (deg C):

BP (deg C):

BP pressure (mm Hg):

Property/Value	Units	Temp	Data Type	Reference
WS 4.60E+003	mg/L	25	EST	MEYLAN,WM ET AL. (1996)
logP 0.49			EXP	HANSCH,C & LEO,AJ (1985)
VP 1.65E-008	mm Hg	25	EST	NEELY,WB & BLAU,GE (1985)
DC	pKa			
HL 1.62E-013	atm m3/mol	25	EST	MEYLAN,WM & HOWARD,PH (1991)
OH 6.77E-013	cm3/molc sec	25	EST	MEYLAN,WM & HOWARD,PH (1993)

CAS #: 029366-72-1				3'-NITROBENZOGUANAMINE

Formula: $C_9H_8N_6O_2$

Mol Weight: 232.20

MP (deg C): | FP (deg C):

BP (deg C):

BP pressure (mm Hg):

Property/Value	Units	Temp	Data Type	Reference
WS 5.50E+002	mg/L	25	EST	MEYLAN,WM ET AL. (1996)
logP 1.57			EXP	HANSCH,C & LEO,AJ (1985)
VP 1.65E-008	mm Hg	25	EST	NEELY,WB & BLAU,GE (1985)
DC	pKa			
HL 1.62E-013	atm m3/mol	25	EST	MEYLAN,WM & HOWARD,PH (1991)
OH 5.49E-013	cm3/molc sec	25	EST	MEYLAN,WM & HOWARD,PH (1993)

CAS #: 029366-73-2				4'-NITROBENZOGUANAMINE

Formula: $C_9H_8N_6O_2$

Mol Weight: 232.20

MP (deg C): | FP (deg C):

BP (deg C):

BP pressure (mm Hg):

Property/Value	Units	Temp	Data Type	Reference
WS 5.19E+002	mg/L	25	EST	MEYLAN,WM ET AL. (1996)
logP 1.60			EXP	HANSCH,C & LEO,AJ (1985)
VP 1.65E-008	mm Hg	25	EST	NEELY,WB & BLAU,GE (1985)
DC	pKa			
HL 1.62E-013	atm m3/mol	25	EST	MEYLAN,WM & HOWARD,PH (1991)
OH 6.77E-013	cm3/molc sec	25	EST	MEYLAN,WM & HOWARD,PH (1993)

CAS #: 029366-77-6				2'-CHLOROBENZOGUANAMINE

Formula: $C_9H_8ClN_5$

Mol Weight: 221.65

MP (deg C): | FP (deg C):

BP (deg C):

BP pressure (mm Hg):

Property/Value	Units	Temp	Data Type	Reference
WS 3.25E+003	mg/L	25	EST	MEYLAN,WM ET AL. (1996)
logP 1.19			EXP	HANSCH,C & LEO,AJ (1985)
VP 2.11E-007	mm Hg	25	EST	NEELY,WB & BLAU,GE (1985)
DC	pKa			
HL 3.05E-011	atm m3/mol	25	EST	MEYLAN,WM & HOWARD,PH (1991)
OH 1.24E-012	cm3/molc sec	25	EST	MEYLAN,WM & HOWARD,PH (1993)

CAS #: 029418-21-1				2,4-DIMETHYLAZOBENZENE

Formula: $C_{14}H_{14}N_2$

Mol Weight: 210.28

MP (deg C): | FP (deg C):

BP (deg C):

BP pressure (mm Hg):

Property/Value	Units	Temp	Data Type	Reference
WS 5.13E-001	mg/L	25	EST	MEYLAN,WM ET AL. (1996)
logP 5.21			EST	MEYLAN,WM & HOWARD,PH (1995)
VP 2.59E-004	mm Hg	25	EST	NEELY,WB & BLAU,GE (1985)
DC	pKa			
HL 1.79E-005	atm m3/mol	25	EST	MEYLAN,WM & HOWARD,PH (1991)
OH 6.31E-012	cm3/molc sec	25	EST	MEYLAN,WM & HOWARD,PH (1993)

CAS #: 029418-22-2				O,P'-AZOTOLUENE

Formula: $C_{14}H_{14}N_2$

Mol Weight: 210.28

MP (deg C): | FP (deg C):

BP (deg C):

BP pressure (mm Hg):

Property/Value	Units	Temp	Data Type	Reference
WS 5.13E-001	mg/L	25	EST	MEYLAN,WM ET AL. (1996)
logP 5.21			EST	MEYLAN,WM & HOWARD,PH (1995)
VP 2.59E-004	mm Hg	25	EST	NEELY,WB & BLAU,GE (1985)
DC	pKa			
HL 1.79E-005	atm m3/mol	25	EST	MEYLAN,WM & HOWARD,PH (1991)
OH 4.31E-012	cm3/molc sec	25	EST	MEYLAN,WM & HOWARD,PH (1993)

CAS #: 029418-23-3				2,2',4-TRIMETHYLAZOBENZENE

Formula: $C_{15}H_{16}N_2$

Mol Weight: 224.31

MP (deg C): | FP (deg C):

BP (deg C):

BP pressure (mm Hg):

Property/Value	Units	Temp	Data Type	Reference
WS 1.48E-001	mg/L	25	EST	MEYLAN,WM ET AL. (1996)
logP 5.75			EST	MEYLAN,WM & HOWARD,PH (1995)
VP 9.09E-005	mm Hg	25	EST	NEELY,WB & BLAU,GE (1985)
DC	pKa			
HL 1.98E-005	atm m3/mol	25	EST	MEYLAN,WM & HOWARD,PH (1991)
OH 7.69E-012	cm3/molc sec	25	EST	MEYLAN,WM & HOWARD,PH (1993)

CAS #: 029418-24-4				2,4,4'-TRIMETHYLAZOBENZENE

Formula: $C_{15}H_{16}N_2$
Mol Weight: 224.31
MP (deg C): FP (deg C):
BP (deg C):
BP pressure (mm Hg):

Property/ Value	Units	Temp	Data Type	Reference
WS 1.48E-001	mg/L	25	EST	MEYLAN,WM ET AL. (1996)
logP 5.75			EST	MEYLAN,WM & HOWARD,PH (1995)
VP 9.09E-005	mm Hg	25	EST	NEELY,WB & BLAU,GE (1985)
DC	pKa			
HL 1.98E-005	atm m3/mol	25	EST	MEYLAN,WM & HOWARD,PH (1991)
OH 7.69E-012	cm3/molc sec	25	EST	MEYLAN,WM & HOWARD,PH (1993)

CAS #: 029418-25-5				BIS(2,4-DIMETHYLPHENYL)DIAZENE

Formula: $C_{16}H_{18}N_2$
Mol Weight: 238.34
MP (deg C): FP (deg C):
BP (deg C):
BP pressure (mm Hg):

Property/ Value	Units	Temp	Data Type	Reference
WS 4.22E-002	mg/L	25	EST	MEYLAN,WM ET AL. (1996)
logP 6.30			EST	MEYLAN,WM & HOWARD,PH (1995)
VP 3.30E-005	mm Hg	25	EST	NEELY,WB & BLAU,GE (1985)
DC	pKa			
HL 2.18E-005	atm m3/mol	25	EST	MEYLAN,WM & HOWARD,PH (1991)
OH 1.11E-011	cm3/molc sec	25	EST	MEYLAN,WM & HOWARD,PH (1993)

CAS #: 029418-26-6				2,4,6-TRIMETHYLAZOBENZENE

Formula: $C_{15}H_{16}N_2$
Mol Weight: 224.31
MP (deg C): FP (deg C):
BP (deg C):
BP pressure (mm Hg):

Property/ Value	Units	Temp	Data Type	Reference
WS 1.48E-001	mg/L	25	EST	MEYLAN,WM ET AL. (1996)
logP 5.75			EST	MEYLAN,WM & HOWARD,PH (1995)
VP 9.09E-005	mm Hg	25	EST	NEELY,WB & BLAU,GE (1985)
DC	pKa			
HL 1.98E-005	atm m3/mol	25	EST	MEYLAN,WM & HOWARD,PH (1991)
OH 1.49E-011	cm3/molc sec	25	EST	MEYLAN,WM & HOWARD,PH (1993)

CAS #: 029418-27-7				2,4',6-TRIMETHYLAZOBENZENE

Formula: $C_{15}H_{16}N_2$
Mol Weight: 224.31
MP (deg C): FP (deg C):
BP (deg C):
BP pressure (mm Hg):

Property/ Value	Units	Temp	Data Type	Reference
WS 1.48E-001	mg/L	25	EST	MEYLAN,WM ET AL. (1996)
logP 5.75			EST	MEYLAN,WM & HOWARD,PH (1995)
VP 9.09E-005	mm Hg	25	EST	NEELY,WB & BLAU,GE (1985)
DC	pKa			
HL 1.98E-005	atm m3/mol	25	EST	MEYLAN,WM & HOWARD,PH (1991)
OH 7.69E-012	cm3/molc sec	25	EST	MEYLAN,WM & HOWARD,PH (1993)

CAS #: 029418-29-9				2,4,4',6-TETRAMETHYLAZOBENZENE

Formula: $C_{16}H_{18}N_2$
Mol Weight: 238.34
MP (deg C): FP (deg C):
BP (deg C):
BP pressure (mm Hg):

Property/ Value	Units	Temp	Data Type	Reference
WS 4.22E-002	mg/L	25	EST	MEYLAN,WM ET AL. (1996)
logP 6.30			EST	MEYLAN,WM & HOWARD,PH (1995)
VP 3.30E-005	mm Hg	25	EST	NEELY,WB & BLAU,GE (1985)
DC	pKa			
HL 2.18E-005	atm m3/mol	25	EST	MEYLAN,WM & HOWARD,PH (1991)
OH 1.63E-011	cm3/molc sec	25	EST	MEYLAN,WM & HOWARD,PH (1993)

CAS #: 029418-30-2				2,2',4,6-TETRAMETHYLAZOBENZENE

Formula: $C_{16}H_{18}N_2$
Mol Weight: 238.34
MP (deg C): FP (deg C):
BP (deg C):
BP pressure (mm Hg):

Property/ Value	Units	Temp	Data Type	Reference
WS 4.22E-002	mg/L	25	EST	MEYLAN,WM ET AL. (1996)
logP 6.30			EST	MEYLAN,WM & HOWARD,PH (1995)
VP 3.30E-005	mm Hg	25	EST	NEELY,WB & BLAU,GE (1985)
DC	pKa			
HL 2.18E-005	atm m3/mol	25	EST	MEYLAN,WM & HOWARD,PH (1991)
OH 1.63E-011	cm3/molc sec	25	EST	MEYLAN,WM & HOWARD,PH (1993)

CAS #: 029418-31-3				BIS(2,6-DIMETHYLPHENYL)DIAZENE

Formula: $C_{16}H_{18}N_2$
Mol Weight: 238.34
MP (deg C): FP (deg C):
BP (deg C):
BP pressure (mm Hg):

Property/ Value	Units	Temp	Data Type	Reference
WS 4.22E-002	mg/L	25	EST	MEYLAN,WM ET AL. (1996)
logP 6.30			EST	MEYLAN,WM & HOWARD,PH (1995)
VP 3.30E-005	mm Hg	25	EST	NEELY,WB & BLAU,GE (1985)
DC	pKa			
HL 2.18E-005	atm m3/mol	25	EST	MEYLAN,WM & HOWARD,PH (1991)
OH 1.11E-011	cm3/molc sec	25	EST	MEYLAN,WM & HOWARD,PH (1993)

CAS #: 029418-34-6				2,2',3,3'-TETRAMETHYLAZOBENZENE

Formula: $C_{16}H_{18}N_2$
Mol Weight: 238.34
MP (deg C): FP (deg C):
BP (deg C):
BP pressure (mm Hg):

Property/ Value	Units	Temp	Data Type	Reference
WS 4.22E-002	mg/L	25	EST	MEYLAN,WM ET AL. (1996)
logP 6.30			EST	MEYLAN,WM & HOWARD,PH (1995)
VP 3.30E-005	mm Hg	25	EST	NEELY,WB & BLAU,GE (1985)
DC	pKa			
HL 2.18E-005	atm m3/mol	25	EST	MEYLAN,WM & HOWARD,PH (1991)
OH 5.49E-012	cm3/molc sec	25	EST	MEYLAN,WM & HOWARD,PH (1993)

CAS #: 029418-36-8	BIS(3,4-DIMETHYLPHENYL)DIAZENE

Formula:	$C_{16}H_{18}N_2$
Mol Weight:	238.34
MP (deg C):	FP (deg C):
BP (deg C):	
BP pressure (mm Hg):	

Property/Value	Units	Temp	Data Type	Reference
WS 4.22E-002	mg/L	25	EST	MEYLAN,WM ET AL. (1996)
logP 6.30			EST	MEYLAN,WM & HOWARD,PH (1995)
VP 3.30E-005	mm Hg	25	EST	NEELY,WB & BLAU,GE (1985)
DC	pKa			
HL 2.18E-005	atm m3/mol	25	EST	MEYLAN,WM & HOWARD,PH (1991)
OH 5.49E-012	cm3/molc sec	25	EST	MEYLAN,WM & HOWARD,PH (1993)

CAS #: 029418-37-9	O-(O-TOLYLAZO)PHENOL

Formula:	$C_{13}H_{12}N_2O$
Mol Weight:	212.25
MP (deg C):	FP (deg C):
BP (deg C):	
BP pressure (mm Hg):	

Property/Value	Units	Temp	Data Type	Reference
WS 3.63E+000	mg/L	25	EST	MEYLAN,WM ET AL. (1996)
logP 4.88			EST	MEYLAN,WM & HOWARD,PH (1995)
VP 6.72E-006	mm Hg	25	EST	NEELY,WB & BLAU,GE (1985)
DC	pKa			
HL 1.69E-009	atm m3/mol	25	EST	MEYLAN,WM & HOWARD,PH (1991)
OH 1.55E-011	cm3/molc sec	25	EST	MEYLAN,WM & HOWARD,PH (1993)

CAS #: 029418-38-0	2-(PHENYLAZO)-M-CRESOL

Formula:	$C_{13}H_{12}N_2O$
Mol Weight:	212.25
MP (deg C):	FP (deg C):
BP (deg C):	
BP pressure (mm Hg):	

Property/Value	Units	Temp	Data Type	Reference
WS 3.63E+000	mg/L	25	EST	MEYLAN,WM ET AL. (1996)
logP 4.88			EST	MEYLAN,WM & HOWARD,PH (1995)
VP 6.72E-006	mm Hg	25	EST	NEELY,WB & BLAU,GE (1985)
DC	pKa			
HL 1.69E-009	atm m3/mol	25	EST	MEYLAN,WM & HOWARD,PH (1991)
OH 3.55E-011	cm3/molc sec	25	EST	MEYLAN,WM & HOWARD,PH (1993)

CAS #: 029418-41-5	2-METHOXY-4'-METHYLAZOBENZENE

Formula:	$C_{14}H_{14}N_2O$
Mol Weight:	226.28
MP (deg C):	FP (deg C):
BP (deg C):	
BP pressure (mm Hg):	

Property/Value	Units	Temp	Data Type	Reference
WS 1.06E+000	mg/L	25	EST	MEYLAN,WM ET AL. (1996)
logP 4.74			EST	MEYLAN,WM & HOWARD,PH (1995)
VP 8.99E-005	mm Hg	25	EST	NEELY,WB & BLAU,GE (1985)
DC	pKa			
HL 9.60E-007	atm m3/mol	25	EST	MEYLAN,WM & HOWARD,PH (1991)
OH 1.15E-011	cm3/molc sec	25	EST	MEYLAN,WM & HOWARD,PH (1993)

CAS #: 029418-42-6	4-METHOXY-2-METHYLAZOBENZENE

Formula:	$C_{14}H_{14}N_2O$
Mol Weight:	226.28
MP (deg C):	FP (deg C):
BP (deg C):	
BP pressure (mm Hg):	

Property/Value	Units	Temp	Data Type	Reference
WS 1.06E+000	mg/L	25	EST	MEYLAN,WM ET AL. (1996)
logP 4.74			EST	MEYLAN,WM & HOWARD,PH (1995)
VP 8.99E-005	mm Hg	25	EST	NEELY,WB & BLAU,GE (1985)
DC	pKa			
HL 9.60E-007	atm m3/mol	25	EST	MEYLAN,WM & HOWARD,PH (1991)
OH 2.40E-011	cm3/molc sec	25	EST	MEYLAN,WM & HOWARD,PH (1993)

CAS #: 029418-43-7	(4-METHOXYPHENYL)(2-METHYLPHENYL)DIAZENE

Formula:	$C_{14}H_{14}N_2O$
Mol Weight:	226.28
MP (deg C):	FP (deg C):
BP (deg C):	
BP pressure (mm Hg):	

Property/Value	Units	Temp	Data Type	Reference
WS 1.06E+000	mg/L	25	EST	MEYLAN,WM ET AL. (1996)
logP 4.74			EST	MEYLAN,WM & HOWARD,PH (1995)
VP 8.99E-005	mm Hg	25	EST	NEELY,WB & BLAU,GE (1985)
DC	pKa			
HL 9.60E-007	atm m3/mol	25	EST	MEYLAN,WM & HOWARD,PH (1991)
OH 1.15E-011	cm3/molc sec	25	EST	MEYLAN,WM & HOWARD,PH (1993)

CAS #: 029418-44-8	(4-METHOXYPHENYL)(4-METHYLPHENYL)DIAZENE

Formula:	$C_{14}H_{14}N_2O$
Mol Weight:	226.28
MP (deg C):	FP (deg C):
BP (deg C):	
BP pressure (mm Hg):	

Property/Value	Units	Temp	Data Type	Reference
WS 1.06E+000	mg/L	25	EST	MEYLAN,WM ET AL. (1996)
logP 4.74			EST	MEYLAN,WM & HOWARD,PH (1995)
VP 8.99E-005	mm Hg	25	EST	NEELY,WB & BLAU,GE (1985)
DC	pKa			
HL 9.60E-007	atm m3/mol	25	EST	MEYLAN,WM & HOWARD,PH (1991)
OH 1.15E-011	cm3/molc sec	25	EST	MEYLAN,WM & HOWARD,PH (1993)

CAS #: 029418-45-9	4-METHOXY-O,O'-AZOTOLUENE

Formula:	$C_{15}H_{16}N_2O$
Mol Weight:	240.31
MP (deg C):	FP (deg C):
BP (deg C):	
BP pressure (mm Hg):	

Property/Value	Units	Temp	Data Type	Reference
WS 3.02E-001	mg/L	25	EST	MEYLAN,WM ET AL. (1996)
logP 5.29			EST	MEYLAN,WM & HOWARD,PH (1995)
VP 3.27E-005	mm Hg	25	EST	NEELY,WB & BLAU,GE (1985)
DC	pKa			
HL 1.06E-006	atm m3/mol	25	EST	MEYLAN,WM & HOWARD,PH (1991)
OH 2.54E-011	cm3/molc sec	25	EST	MEYLAN,WM & HOWARD,PH (1993)

CAS #: 029418-46-0				2,4-DIMETHOXYAZOBENZENE

Formula: $C_{14}H_{14}N_2O_2$

Mol Weight: 242.28

MP (deg C):   FP (deg C):

BP (deg C):

BP pressure (mm Hg):

Property/Value	Units	Temp	Data Type	Reference
WS 2.17E+000	mg/L	25	EST	MEYLAN,WM ET AL. (1996)
logP 4.27			EST	MEYLAN,WM & HOWARD,PH (1995)
VP 3.24E-005	mm Hg	25	EST	NEELY,WB & BLAU,GE (1985)
DC	pKa			
HL 5.15E-008	atm m3/mol	25	EST	MEYLAN,WM & HOWARD,PH (1991)
OH 9.64E-011	cm3/molc sec	25	EST	MEYLAN,WM & HOWARD,PH (1993)

CAS #: 029418-47-1				2,4'-DIMETHOXYAZOBENZENE

Formula: $C_{14}H_{14}N_2O_2$

Mol Weight: 242.28

MP (deg C):   FP (deg C):

BP (deg C):

BP pressure (mm Hg):

Property/Value	Units	Temp	Data Type	Reference
WS 2.17E+000	mg/L	25	EST	MEYLAN,WM ET AL. (1996)
logP 4.27			EST	MEYLAN,WM & HOWARD,PH (1995)
VP 3.24E-005	mm Hg	25	EST	NEELY,WB & BLAU,GE (1985)
DC	pKa			
HL 5.15E-008	atm m3/mol	25	EST	MEYLAN,WM & HOWARD,PH (1991)
OH 1.87E-011	cm3/molc sec	25	EST	MEYLAN,WM & HOWARD,PH (1993)

CAS #: 029418-48-2				2,6-DIMETHOXYAZOBENZENE

Formula: $C_{14}H_{14}N_2O_2$

Mol Weight: 242.28

MP (deg C):   FP (deg C):

BP (deg C):

BP pressure (mm Hg):

Property/Value	Units	Temp	Data Type	Reference
WS 2.17E+000	mg/L	25	EST	MEYLAN,WM ET AL. (1996)
logP 4.27			EST	MEYLAN,WM & HOWARD,PH (1995)
VP 3.24E-005	mm Hg	25	EST	NEELY,WB & BLAU,GE (1985)
DC	pKa			
HL 5.15E-008	atm m3/mol	25	EST	MEYLAN,WM & HOWARD,PH (1991)
OH 9.64E-011	cm3/molc sec	25	EST	MEYLAN,WM & HOWARD,PH (1993)

CAS #: 029418-49-3				2,4-DIMETHOXY-2'-METHYLAZOBENZENE

Formula: $C_{15}H_{16}N_2O_2$

Mol Weight: 256.31

MP (deg C):   FP (deg C):

BP (deg C):

BP pressure (mm Hg):

Property/Value	Units	Temp	Data Type	Reference
WS 6.17E-001	mg/L	25	EST	MEYLAN,WM ET AL. (1996)
logP 4.82			EST	MEYLAN,WM & HOWARD,PH (1995)
VP 1.19E-005	mm Hg	25	EST	NEELY,WB & BLAU,GE (1985)
DC	pKa			
HL 5.68E-008	atm m3/mol	25	EST	MEYLAN,WM & HOWARD,PH (1991)
OH 9.78E-011	cm3/molc sec	25	EST	MEYLAN,WM & HOWARD,PH (1993)

CAS #: 029418-50-6				2,4-DIMETHOXY-4'-METHYLAZOBENZENE

Formula: $C_{15}H_{16}N_2O_2$

Mol Weight: 256.31

MP (deg C):   FP (deg C):

BP (deg C):

BP pressure (mm Hg):

Property/Value	Units	Temp	Data Type	Reference
WS 6.17E-001	mg/L	25	EST	MEYLAN,WM ET AL. (1996)
logP 4.82			EST	MEYLAN,WM & HOWARD,PH (1995)
VP 1.19E-005	mm Hg	25	EST	NEELY,WB & BLAU,GE (1985)
DC	pKa			
HL 5.68E-008	atm m3/mol	25	EST	MEYLAN,WM & HOWARD,PH (1991)
OH 9.78E-011	cm3/molc sec	25	EST	MEYLAN,WM & HOWARD,PH (1993)

CAS #: 029418-51-7				2,6-DIMETHOXY-2'-METHYLAZOBENZENE

Formula: $C_{15}H_{16}N_2O_2$

Mol Weight: 256.31

MP (deg C):   FP (deg C):

BP (deg C):

BP pressure (mm Hg):

Property/Value	Units	Temp	Data Type	Reference
WS 6.17E-001	mg/L	25	EST	MEYLAN,WM ET AL. (1996)
logP 4.82			EST	MEYLAN,WM & HOWARD,PH (1995)
VP 1.19E-005	mm Hg	25	EST	NEELY,WB & BLAU,GE (1985)
DC	pKa			
HL 5.68E-008	atm m3/mol	25	EST	MEYLAN,WM & HOWARD,PH (1991)
OH 9.78E-011	cm3/molc sec	25	EST	MEYLAN,WM & HOWARD,PH (1993)

CAS #: 029418-52-8				2,6-DIMETHOXY-4'-METHYLAZOBENZENE

Formula: $C_{15}H_{16}N_2O_2$

Mol Weight: 256.31

MP (deg C):   FP (deg C):

BP (deg C):

BP pressure (mm Hg):

Property/Value	Units	Temp	Data Type	Reference
WS 6.17E-001	mg/L	25	EST	MEYLAN,WM ET AL. (1996)
logP 4.82			EST	MEYLAN,WM & HOWARD,PH (1995)
VP 1.19E-005	mm Hg	25	EST	NEELY,WB & BLAU,GE (1985)
DC	pKa			
HL 5.68E-008	atm m3/mol	25	EST	MEYLAN,WM & HOWARD,PH (1991)
OH 9.78E-011	cm3/molc sec	25	EST	MEYLAN,WM & HOWARD,PH (1993)

CAS #: 029418-53-9				4,4'-DIMETHOXY-2-METHYLAZOBENZENE

Formula: $C_{15}H_{16}N_2O_2$

Mol Weight: 256.31

MP (deg C):   FP (deg C):

BP (deg C):

BP pressure (mm Hg):

Property/Value	Units	Temp	Data Type	Reference
WS 6.17E-001	mg/L	25	EST	MEYLAN,WM ET AL. (1996)
logP 4.82			EST	MEYLAN,WM & HOWARD,PH (1995)
VP 1.19E-005	mm Hg	25	EST	NEELY,WB & BLAU,GE (1985)
DC	pKa			
HL 5.68E-008	atm m3/mol	25	EST	MEYLAN,WM & HOWARD,PH (1991)
OH 3.26E-011	cm3/molc sec	25	EST	MEYLAN,WM & HOWARD,PH (1993)

CAS #: 029418-54-0	4,4'-DIMETHOXY-O,O'-AZOTOLUENE

Formula: $C_{16}H_{18}N_2O_2$

Mol Weight: 270.33

MP (deg C):   FP (deg C):

BP (deg C):

BP pressure (mm Hg):

Property/Value	Units	Temp	Data Type	Reference
WS 1.75E-001	mg/L	25	EST	MEYLAN,WM ET AL. (1996)
logP 5.37			EST	MEYLAN,WM & HOWARD,PH (1995)
VP 5.10E-006	mm Hg	25	EST	NEELY,WB & BLAU,GE (1985)
DC	pKa			
HL 6.27E-008	atm m3/mol	25	EST	MEYLAN,WM & HOWARD,PH (1991)
OH 4.64E-011	cm3/molc sec	25	EST	MEYLAN,WM & HOWARD,PH (1993)

CAS #: 029418-55-1	BIS(2,4-DIMETHOXYPHENYL)DIAZENE

Formula: $C_{16}H_{18}N_2O_4$

Mol Weight: 302.33

MP (deg C):   FP (deg C):

BP (deg C):

BP pressure (mm Hg):

Property/Value	Units	Temp	Data Type	Reference
WS 7.18E-001	mg/L	25	EST	MEYLAN,WM ET AL. (1996)
logP 4.43			EST	MEYLAN,WM & HOWARD,PH (1995)
VP 7.29E-007	mm Hg	25	EST	NEELY,WB & BLAU,GE (1985)
DC	pKa			
HL 1.80E-010	atm m3/mol	25	EST	MEYLAN,WM & HOWARD,PH (1991)
OH 1.91E-010	cm3/molc sec	25	EST	MEYLAN,WM & HOWARD,PH (1993)

CAS #: 029418-57-3	2-METHYL-4-NITROAZOBENZENE

Formula: $C_{13}H_{11}N_3O_2$

Mol Weight: 241.25

MP (deg C):   FP (deg C):

BP (deg C):

BP pressure (mm Hg):

Property/Value	Units	Temp	Data Type	Reference
WS 4.72E-001	mg/L	25	EST	MEYLAN,WM ET AL. (1996)
logP 5.05			EST	MEYLAN,WM & HOWARD,PH (1995)
VP 5.35E-006	mm Hg	25	EST	NEELY,WB & BLAU,GE (1985)
DC	pKa			
HL 6.40E-008	atm m3/mol	25	EST	MEYLAN,WM & HOWARD,PH (1991)
OH 1.08E-012	cm3/molc sec	25	EST	MEYLAN,WM & HOWARD,PH (1993)

CAS #: 029418-58-4	(4-METHYLPHENYL)(4-NITROPHENYL)DIAZENE

Formula: $C_{13}H_{11}N_3O_2$

Mol Weight: 241.25

MP (deg C):   FP (deg C):

BP (deg C):

BP pressure (mm Hg):

Property/Value	Units	Temp	Data Type	Reference
WS 4.72E-001	mg/L	25	EST	MEYLAN,WM ET AL. (1996)
logP 5.05			EST	MEYLAN,WM & HOWARD,PH (1995)
VP 5.35E-006	mm Hg	25	EST	NEELY,WB & BLAU,GE (1985)
DC	pKa			
HL 6.40E-008	atm m3/mol	25	EST	MEYLAN,WM & HOWARD,PH (1991)
OH 2.22E-012	cm3/molc sec	25	EST	MEYLAN,WM & HOWARD,PH (1993)

CAS #: 029418-61-9	4'-METHOXY-2-METHYL-4-NITROAZOBENZENE

Formula: $C_{14}H_{13}N_3O_3$

Mol Weight: 271.28

MP (deg C):   FP (deg C):

BP (deg C):

BP pressure (mm Hg):

Property/Value	Units	Temp	Data Type	Reference
WS 2.74E-001	mg/L	25	EST	MEYLAN,WM ET AL. (1996)
logP 5.13			EST	MEYLAN,WM & HOWARD,PH (1995)
VP 8.31E-007	mm Hg	25	EST	NEELY,WB & BLAU,GE (1985)
DC	pKa			
HL 3.79E-009	atm m3/mol	25	EST	MEYLAN,WM & HOWARD,PH (1991)
OH 9.66E-012	cm3/molc sec	25	EST	MEYLAN,WM & HOWARD,PH (1993)

CAS #: 029418-67-5	BENZOYLHYDRAZINE, O-BROMO

Formula: $C_7H_7BrN_2O$

Mol Weight: 215.06

MP (deg C):   FP (deg C):

BP (deg C):

BP pressure (mm Hg):

Property/Value	Units	Temp	Data Type	Reference
WS 1.07E+003	mg/L	25	EST	MEYLAN,WM ET AL. (1996)
logP 0.26			EXP	HANSCH,C & LEO,AJ (1985)
VP 7.82E-006	mm Hg	25	EST	NEELY,WB & BLAU,GE (1985)
DC	pKa			
HL 3.69E-012	atm m3/mol	25	EST	MEYLAN,WM & HOWARD,PH (1991)
OH 6.62E-012	cm3/molc sec	25	EST	MEYLAN,WM & HOWARD,PH (1993)

CAS #: 029442-58-8	1,4-BENZDAZEPIN-2-ON-1MEOME-5PH-7NO2

Formula: $C_{17}H_{15}N_3O_4$

Mol Weight: 325.33

MP (deg C):   FP (deg C):

BP (deg C):

BP pressure (mm Hg):

Property/Value	Units	Temp	Data Type	Reference
WS 6.30E+001	mg/L	25	EST	MEYLAN,WM ET AL. (1996)
logP 2.05			EXP	HANSCH,C & LEO,AJ (1985)
VP 3.97E-010	mm Hg	25	EST	NEELY,WB & BLAU,GE (1985)
DC	pKa			
HL 4.53E-012	atm m3/mol	25	EST	MEYLAN,WM & HOWARD,PH (1991)
OH 3.10E-011	cm3/molc sec	25	EST	MEYLAN,WM & HOWARD,PH (1993)

CAS #: 029442-59-9	14-BENZODIAZP2-ON,1-MEOME-5(PH)-7CL

Formula: $C_{17}H_{15}ClN_2O_2$

Mol Weight: 314.77

MP (deg C):   FP (deg C):

BP (deg C):

BP pressure (mm Hg):

Property/Value	Units	Temp	Data Type	Reference
WS 3.63E+001	mg/L	25	EST	MEYLAN,WM ET AL. (1996)
logP 2.86			EXP	HANSCH,C & LEO,AJ (1985)
VP 5.37E-009	mm Hg	25	EST	NEELY,WB & BLAU,GE (1985)
DC	pKa			
HL 8.51E-010	atm m3/mol	25	EST	MEYLAN,WM & HOWARD,PH (1991)
OH 3.35E-011	cm3/molc sec	25	EST	MEYLAN,WM & HOWARD,PH (1993)

CAS #: 029442-60-2 — 14-BENZODIAZP2-ON,1-MEOME-5(2CLPH)7CL

Formula: $C_{17}H_{14}Cl_2N_2O_2$

Mol Weight: 349.22

MP (deg C): FP (deg C):

BP (deg C):

BP pressure (mm Hg):

Property/Value	Units	Temp	Data Type	Reference
WS 1.09E+001	mg/L	25	EST	MEYLAN,WM ET AL. (1996)
logP 3.23			EXP	HANSCH,C & LEO,AJ (1985)
VP 1.45E-009	mm Hg	25	EST	NEELY,WB & BLAU,GE (1985)
DC	pKa			
HL 6.31E-010	atm m3/mol	25	EST	MEYLAN,WM & HOWARD,PH (1991)
OH 3.09E-011	cm3/molc sec	25	EST	MEYLAN,WM & HOWARD,PH (1993)

CAS #: 029442-82-8 — 1,4-BENZDAZEPIN-2-ON-1-MESME-5-PH-7-CL

Formula: $C_{17}H_{15}ClN_2OS$

Mol Weight: 330.84

MP (deg C): FP (deg C):

BP (deg C):

BP pressure (mm Hg):

Property/Value	Units	Temp	Data Type	Reference
WS 1.18E+001	mg/L	25	EST	MEYLAN,WM ET AL. (1996)
logP 3.32			EXP	HANSCH,C & LEO,AJ (1985)
VP 1.21E-009	mm Hg	25	EST	NEELY,WB & BLAU,GE (1985)
DC	pKa			
HL 1.82E-008	atm m3/mol	25	EST	MEYLAN,WM & HOWARD,PH (1991)
OH 4.20E-011	cm3/molc sec	25	EST	MEYLAN,WM & HOWARD,PH (1993)

CAS #: 029446-15-9 — 2,3-DICHLORODIBENZO-P-DIOXIN

Formula: $C_{12}H_6Cl_2O_2$

Mol Weight: 253.09

MP (deg C): FP (deg C):

BP (deg C):

BP pressure (mm Hg):

Property/Value	Units	Temp	Data Type	Reference
WS 1.49E-002	mg/L	25	EXP	YALKOWSKY,SH & DANNENFELSER,RM (1992)
logP 5.63			EST	MEYLAN,WM & HOWARD,PH (1995)
VP 2.93E-006	mm Hg	25	EXP	SHIU,WY ET AL. (1988)
DC	pKa			
HL 6.55E-005	atm m3/mol	25	EST	VP/WSOL
OH 5.93E-012	cm3/molc sec	25	EST	MEYLAN,WM & HOWARD,PH (1993)

CAS #: 029460-92-2 — 2-ISOBUTYLPYRAZINE

Formula: $C_8H_{12}N_2$

Mol Weight: 136.20

MP (deg C): FP (deg C):

BP (deg C):

BP pressure (mm Hg):

Property/Value	Units	Temp	Data Type	Reference
WS 2.12E+003	mg/L	25	EST	MEYLAN,WM ET AL. (1996)
logP 1.89			EST	MEYLAN,WM & HOWARD,PH (1995)
VP 4.08E-001	mm Hg	25	EST	NEELY,WB & BLAU,GE (1985)
DC	pKa			
HL 7.54E-006	atm m3/mol	25	EST	MEYLAN,WM & HOWARD,PH (1991)
OH 4.44E-012	cm3/molc sec	25	EST	MEYLAN,WM & HOWARD,PH (1993)

CAS #: 029460-93-3 — ISOBUTYLPYRAZINE

Formula: $C_8H_{14}N_2$

Mol Weight: 138.21

MP (deg C): FP (deg C):

BP (deg C):

BP pressure (mm Hg):

Property/Value	Units	Temp	Data Type	Reference
WS 2.10E+003	mg/L	25	EST	MEYLAN,WM ET AL. (1996)
logP 1.89			EST	MEYLAN,WM & HOWARD,PH (1995)
VP 4.48E-001	mm Hg	25	EST	NEELY,WB & BLAU,GE (1985)
DC	pKa			
HL 4.89E-006	atm m3/mol	25	EXP	BUTTERY,RG ET AL. (1971)
OH 4.10E-012	cm3/molc sec	25	EST	MEYLAN,WM & HOWARD,PH (1993)

CAS #: 029526-42-9 — 8-DIMETHYLAMINOQUINOLINE

Formula: $C_{11}H_{12}N_2$

Mol Weight: 172.23

MP (deg C): FP (deg C):

BP (deg C):

BP pressure (mm Hg):

Property/Value	Units	Temp	Data Type	Reference
WS 2.80E+002	mg/L	25	EST	MEYLAN,WM ET AL. (1996)
logP 2.73			EXP	HANSCH,C & LEO,AJ (1985)
VP 1.11E-003	mm Hg	25	EST	NEELY,WB & BLAU,GE (1985)
DC	pKa			
HL 1.09E-008	atm m3/mol	25	EST	MEYLAN,WM & HOWARD,PH (1991)
OH 2.03E-010	cm3/molc sec	25	EST	MEYLAN,WM & HOWARD,PH (1993)

CAS #: 029604-73-7 — 3,6-PYRIDAZINAMINE-N,N'-TETRAME

Formula: $C_8H_{14}N_4$

Mol Weight: 166.23

MP (deg C): FP (deg C):

BP (deg C):

BP pressure (mm Hg):

Property/Value	Units	Temp	Data Type	Reference
WS 9.14E+003	mg/L	25	EST	MEYLAN,WM ET AL. (1996)
logP 0.99			EXP	HANSCH,C & LEO,AJ (1985)
VP 5.53E-004	mm Hg	25	EST	NEELY,WB & BLAU,GE (1985)
DC	pKa			
HL 7.25E-010	atm m3/mol	25	EST	MEYLAN,WM & HOWARD,PH (1991)
OH 7.65E-011	cm3/molc sec	25	EST	MEYLAN,WM & HOWARD,PH (1993)

CAS #: 029650-44-0 — 2-FLUOROPHENYL ACETATE

Formula: $C_8H_7FO_2$

Mol Weight: 154.14

MP (deg C): FP (deg C):

BP (deg C):

BP pressure (mm Hg):

Property/Value	Units	Temp	Data Type	Reference
WS 2.28E+003	mg/L	25	EST	MEYLAN,WM ET AL. (1996)
logP 1.76			EXP	HANSCH,C & LEO,AJ (1985)
VP 5.56E-001	mm Hg	25	EST	NEELY,WB & BLAU,GE (1985)
DC	pKa			
HL 7.56E-005	atm m3/mol	25	EST	MEYLAN,WM & HOWARD,PH (1991)
OH 1.89E-012	cm3/molc sec	25	EST	MEYLAN,WM & HOWARD,PH (1993)

CAS #: 029681-20-7				2-(2-(2-(TERT-BUTOXY)ETHOXY)ETHOXY)ETHANOL

Formula: $C_{10}H_{22}O_4$
Mol Weight: 206.28
MP (deg C):   FP (deg C):
BP (deg C):
BP pressure (mm Hg):

Property/Value	Units	Temp	Data Type	Reference
WS 1.47E+005	mg/L	25	EST	MEYLAN,WM ET AL. (1996)
logP -0.06			EST	MEYLAN,WM & HOWARD,PH (1995)
VP 2.87E-004	mm Hg	25	EST	NEELY,WB & BLAU,GE (1985)
DC	pKa			
HL 2.36E-011	atm m3/mol	25	EST	MEYLAN,WM & HOWARD,PH (1991)
OH 5.33E-011	cm3/molc sec	25	EST	MEYLAN,WM & HOWARD,PH (1993)

CAS #: 029682-39-1				1-BROMO-2-CHLORO-4-NITROBENZENE

Formula: $C_6H_3BrClNO_2$
Mol Weight: 236.45
MP (deg C):   FP (deg C):
BP (deg C):
BP pressure (mm Hg):

Property/Value	Units	Temp	Data Type	Reference
WS 1.92E+001	mg/L	25	EST	MEYLAN,WM ET AL. (1996)
logP 3.25			EXP	BRIGGS,GG (1981)
VP 1.50E-003	mm Hg	25	EST	NEELY,WB & BLAU,GE (1985)
DC	pKa			
HL 6.28E-006	atm m3/mol	25	EST	MEYLAN,WM & HOWARD,PH (1991)
OH 4.48E-014	cm3/molc sec	25	EST	MEYLAN,WM & HOWARD,PH (1993)

CAS #: 029701-43-7				GLYCINAMIDE, N-ACETYL-L-PHENYLALANYL-

Formula: $C_{13}H_{17}N_3O_3$
Mol Weight: 263.30
MP (deg C):   FP (deg C):
BP (deg C):
BP pressure (mm Hg):

Property/Value	Units	Temp	Data Type	Reference
WS 2.61E+003	mg/L	25	EST	MEYLAN,WM ET AL. (1996)
logP -0.50			EXP	HANSCH,C ET AL. (1995)
VP 6.20E-012	mm Hg	25	EST	NEELY,WB & BLAU,GE (1985)
DC	pKa			
HL 8.41E-017	atm m3/mol	25	EST	MEYLAN,WM & HOWARD,PH (1991)
OH 3.18E-011	cm3/molc sec	25	EST	MEYLAN,WM & HOWARD,PH (1993)

CAS #: 029726-60-1				ISOQUINOLINE, 1,2,3,4-TETRAHYDRO-3-METHYL-

Formula: $C_{10}H_{13}N$
Mol Weight: 147.22
MP (deg C):   FP (deg C):
BP (deg C):
BP pressure (mm Hg):

Property/Value	Units	Temp	Data Type	Reference
WS 1.79E+004	mg/L	25	EST	MEYLAN,WM ET AL. (1996)
logP 1.93			EXP	SANGSTER,J (1993)
VP 2.02E-002	mm Hg	25	EST	NEELY,WB & BLAU,GE (1985)
DC	pKa			
HL 8.67E-007	atm m3/mol	25	EST	MEYLAN,WM & HOWARD,PH (1991)
OH 1.01E-010	cm3/molc sec	25	EST	MEYLAN,WM & HOWARD,PH (1993)

CAS #: 029736-20-7				P-METHYLPHENYLHIPPURATE

Formula: $C_{16}H_{15}NO_3$
Mol Weight: 269.30
MP (deg C):   FP (deg C):
BP (deg C):
BP pressure (mm Hg):

Property/Value	Units	Temp	Data Type	Reference
WS 7.06E+001	mg/L	25	EST	MEYLAN,WM ET AL. (1996)
logP 2.83			EXP	HANSCH,C & LEO,AJ (1985)
VP 2.30E-008	mm Hg	25	EST	NEELY,WB & BLAU,GE (1985)
DC	pKa			
HL 9.42E-011	atm m3/mol	25	EST	MEYLAN,WM & HOWARD,PH (1991)
OH 1.40E-011	cm3/molc sec	25	EST	MEYLAN,WM & HOWARD,PH (1993)

CAS #: 029736-21-8				P-METHOXYPHENYLHIPPURATE

Formula: $C_{16}H_{15}NO_4$
Mol Weight: 285.30
MP (deg C):   FP (deg C):
BP (deg C):
BP pressure (mm Hg):

Property/Value	Units	Temp	Data Type	Reference
WS 1.69E+002	mg/L	25	EST	MEYLAN,WM ET AL. (1996)
logP 2.28			EXP	HANSCH,C & LEO,AJ (1985)
VP 8.85E-009	mm Hg	25	EST	NEELY,WB & BLAU,GE (1985)
DC	pKa			
HL 5.05E-012	atm m3/mol	25	EST	MEYLAN,WM & HOWARD,PH (1991)
OH 2.67E-011	cm3/molc sec	25	EST	MEYLAN,WM & HOWARD,PH (1993)

CAS #: 029736-22-9				P-FLUOROPHENYLHIPPURATE

Formula: $C_{15}H_{12}FNO_3$
Mol Weight: 273.27
MP (deg C):   FP (deg C):
BP (deg C):
BP pressure (mm Hg):

Property/Value	Units	Temp	Data Type	Reference
WS 1.39E+002	mg/L	25	EST	MEYLAN,WM ET AL. (1996)
logP 2.46			EXP	HANSCH,C & LEO,AJ (1985)
VP 6.53E-008	mm Hg	25	EST	NEELY,WB & BLAU,GE (1985)
DC	pKa			
HL 9.96E-011	atm m3/mol	25	EST	MEYLAN,WM & HOWARD,PH (1991)
OH 1.19E-011	cm3/molc sec	25	EST	MEYLAN,WM & HOWARD,PH (1993)

CAS #: 029761-21-5				ISODECYL DIPHENYL PHOSPHATE

Formula: $C_{22}H_{31}O_4P$
Mol Weight: 390.46
MP (deg C):   FP (deg C):
BP (deg C):
BP pressure (mm Hg):

Property/Value	Units	Temp	Data Type	Reference
WS 7.50E-001	mg/L	25	EXP	SAEGER,VW ET AL. (1979)
logP 5.44			EXP	SAEGER,VW ET AL. (1979)
VP 4.72E-008	mm Hg	25	EST	NEELY,WB & BLAU,GE (1985)
DC	pKa			
HL 4.36E-007	atm m3/mol	25	EST	MEYLAN,WM & HOWARD,PH (1991)
OH 4.38E-011	cm3/molc sec	25	EST	ATKINSON,R (1988)

029767-20-2 — TENIPOSIDE

Formula: $C_{32}H_{34}O_{13}S$

Mol Weight: 658.68

MP (deg C): 242-246

FP (deg C):

BP (deg C):

BP pressure (mm Hg):

Property/Value	Units	Temp	Data Type	Reference
WS 5.94E+000	mg/L	25	EST	MEYLAN,WM ET AL. (1996)
logP 1.24			EXP	HANSCH,C ET AL. (1995)
VP 6.80E-026	mm Hg	25	EST	NEELY,WB & BLAU,GE (1985)
DC	pKa			
HL 5.78E-032	atm m3/mol	25	EST	MEYLAN,WM & HOWARD,PH (1991)
OH 2.71E-010	cm3/molc sec	25	EST	MEYLAN,WM & HOWARD,PH (1993)

029812-79-1 — O-DECYLHYDROXYLAMINE

Formula: $C_{10}H_{23}NO$

Mol Weight: 173.30

MP (deg C):

FP (deg C):

BP (deg C):

BP pressure (mm Hg):

Property/Value	Units	Temp	Data Type	Reference
WS 2.20E+001	mg/L	25	EST	MEYLAN,WM ET AL. (1996)
logP 4.02			EST	MEYLAN,WM & HOWARD,PH (1995)
VP 4.52E-002	mm Hg	25	EST	NEELY,WB & BLAU,GE (1985)
DC	pKa			
HL 2.83E-005	atm m3/mol	25	EST	MEYLAN,WM & HOWARD,PH (1991)
OH 3.24E-011	cm3/molc sec	25	EST	MEYLAN,WM & HOWARD,PH (1993)

029849-15-8 — 2-BROMOISONIAZID

Formula: $C_6H_6BrN_3O$

Mol Weight: 216.04

MP (deg C):

FP (deg C):

BP (deg C):

BP pressure (mm Hg):

Property/Value	Units	Temp	Data Type	Reference
WS 1.08E+003	mg/L	25	EST	MEYLAN,WM ET AL. (1996)
logP 0.25			EXP	HANSCH,C & LEO,AJ (1985)
VP 4.04E-006	mm Hg	25	EST	NEELY,WB & BLAU,GE (1985)
DC	pKa			
HL 4.82E-015	atm m3/mol	25	EST	MEYLAN,WM & HOWARD,PH (1991)
OH 5.60E-012	cm3/molc sec	25	EST	MEYLAN,WM & HOWARD,PH (1993)

029854-52-2 — STRYCHNINE, 2-NITRO-

Formula: $C_{21}H_{23}N_3O_4$

Mol Weight: 381.44

MP (deg C):

FP (deg C):

BP (deg C):

BP pressure (mm Hg):

Property/Value	Units	Temp	Data Type	Reference
WS 7.05E+001	mg/L	25	EST	MEYLAN,WM ET AL. (1996)
logP 1.61			EXP	SANGSTER,J (1993)
VP 5.66E-011	mm Hg	25	EST	NEELY,WB & BLAU,GE (1985)
DC	pKa			
HL 2.35E-016	atm m3/mol	25	EST	MEYLAN,WM & HOWARD,PH (1991)
OH 1.79E-010	cm3/molc sec	25	EST	MEYLAN,WM & HOWARD,PH (1993)

029927-14-8 — 6-OH BENZOTHIAZOLE-2-SULFONAMIDE

Formula: $C_7H_6N_2O_3S_2$

Mol Weight: 230.27

MP (deg C):

FP (deg C):

BP (deg C):

BP pressure (mm Hg):

Property/Value	Units	Temp	Data Type	Reference
WS 1.25E+004	mg/L	25	EST	MEYLAN,WM ET AL. (1996)
logP 1.13			EXP	HANSCH,C & LEO,AJ (1985)
VP 1.87E-008	mm Hg	25	EST	NEELY,WB & BLAU,GE (1985)
DC	pKa			
HL 3.05E-015	atm m3/mol	25	EST	MEYLAN,WM & HOWARD,PH (1991)
OH 5.27E-012	cm3/molc sec	25	EST	MEYLAN,WM & HOWARD,PH (1993)

029945-54-8 — FURAZANAMINE, 4-PHENYL-, 2-OXIDE

Formula: $C_8H_7N_3O_2$

Mol Weight: 177.16

MP (deg C):

FP (deg C):

BP (deg C):

BP pressure (mm Hg):

Property/Value	Units	Temp	Data Type	Reference
WS 3.48E+003	mg/L	25	EST	MEYLAN,WM ET AL. (1996)
logP 1.42			EXP	CALVINO,R ET AL. (1992)
VP 2.39E-006	mm Hg	25	EST	NEELY,WB & BLAU,GE (1985)
DC	pKa			
HL 5.13E-014	atm m3/mol	25	EST	MEYLAN,WM & HOWARD,PH (1991)
OH 7.39E-012	cm3/molc sec	25	EST	MEYLAN,WM & HOWARD,PH (1993)

029964-84-9 — ISODECYL METHACRYLATE

Formula: $C_{14}H_{26}O_2$

Mol Weight: 226.36

MP (deg C):

FP (deg C):

BP (deg C): 126

BP pressure (mm Hg): 1.00E+001

Property/Value	Units	Temp	Data Type	Reference
WS 5.04E-001	mg/L	25	EST	MEYLAN,WM ET AL. (1996)
logP 5.62			EST	MEYLAN,WM & HOWARD,PH (1995)
VP 1.17E-002	mm Hg	25	EST	NEELY,WB & BLAU,GE (1985)
DC	pKa			
HL 1.86E-003	atm m3/mol	25	EST	MEYLAN,WM & HOWARD,PH (1991)
OH 3.12E-011	cm3/molc sec	25	EST	MEYLAN,WM & HOWARD,PH (1993)

029973-13-5 — CRONETON

Formula: $C_{11}H_{15}NO_2S$

Mol Weight: 225.31

MP (deg C):

FP (deg C):

BP (deg C):

BP pressure (mm Hg):

Property/Value	Units	Temp	Data Type	Reference
WS 1.82E+003	mg/L	20	EXP	YALKOWSKY,SH & DANNENFELSER,RM (1992)
logP 2.04			EST	MEYLAN,WM & HOWARD,PH (1995)
VP 1.19E-004	mm Hg	25	EST	NEELY,WB & BLAU,GE (1985)
DC	pKa			
HL 5.02E-010	atm m3/mol	25	EST	MEYLAN,WM & HOWARD,PH (1991)
OH 2.61E-011	cm3/molc sec	25	EST	MEYLAN,WM & HOWARD,PH (1993)

CAS #: 029973-91-9				PHENYLACETIC ACID, 3-MEO,4-PHMEO
Formula: $C_{16}H_{16}O_4$				
Mol Weight: 272.30				
MP (deg C):		FP (deg C):		
BP (deg C):				
BP pressure (mm Hg):				

Property/ Value	Units	Temp	Data Type	Reference
WS 2.70E+002	mg/L	25	EST	MEYLAN,WM ET AL. (1996)
logP 2.59			EXP	HANSCH,C & LEO,AJ (1985)
VP 1.87E-007	mm Hg	25	EST	NEELY,WB & BLAU,GE (1985)
DC	pKa			
HL 1.25E-011	atm m3/mol	25	EST	MEYLAN,WM & HOWARD,PH (1991)
OH 4.34E-011	cm3/molc sec	25	EST	MEYLAN,WM & HOWARD,PH (1993)

CAS #: 030022-13-0				P-AMINOPHENYLHIPPURATE
Formula: $C_{15}H_{14}N_2O_3$				
Mol Weight: 270.29				
MP (deg C):		FP (deg C):		
BP (deg C):				
BP pressure (mm Hg):				

Property/ Value	Units	Temp	Data Type	Reference
WS 2.50E+003	mg/L	25	EST	MEYLAN,WM ET AL. (1996)
logP 1.01			EXP	HANSCH,C & LEO,AJ (1985)
VP 2.24E-009	mm Hg	25	EST	NEELY,WB & BLAU,GE (1985)
DC	pKa			
HL 3.02E-014	atm m3/mol	25	EST	MEYLAN,WM & HOWARD,PH (1991)
OH 8.92E-011	cm3/molc sec	25	EST	MEYLAN,WM & HOWARD,PH (1993)

CAS #: 030031-64-2				1,2:3,4-DIEPOXYBUTANE (2S,3S)
Formula: $C_4H_6O_2$				
Mol Weight: 86.09				
MP (deg C): 4		FP (deg C):		
BP (deg C): 144				
BP pressure (mm Hg):				

Property/ Value	Units	Temp	Data Type	Reference
WS 1.00E+006	mg/L		EXP	DEAN,JA (1985)
logP -0.58			EST	MEYLAN,WM & HOWARD,PH (1995)
VP 5.68E+001	mm Hg	25	EST	NEELY,WB & BLAU,GE (1985)
DC	pKa			
HL 3.54E-008	atm m3/mol	25	EST	MEYLAN,WM & HOWARD,PH (1991)
OH 9.96E-013	cm3/molc sec	25	EST	MEYLAN,WM & HOWARD,PH (1993)

CAS #: 030059-86-0				FURAZANAMINE, 4-PHENYL-, 5-OXIDE
Formula: $C_8H_7N_3O_2$				
Mol Weight: 177.16				
MP (deg C):		FP (deg C):		
BP (deg C):				
BP pressure (mm Hg):				

Property/ Value	Units	Temp	Data Type	Reference
WS 3.16E+003	mg/L	25	EST	MEYLAN,WM ET AL. (1996)
logP 1.47			EXP	CALVINO,R ET AL. (1992)
VP 2.39E-006	mm Hg	25	EST	NEELY,WB & BLAU,GE (1985)
DC	pKa			
HL 5.13E-014	atm m3/mol	25	EST	MEYLAN,WM & HOWARD,PH (1991)
OH 7.39E-012	cm3/molc sec	25	EST	MEYLAN,WM & HOWARD,PH (1993)

CAS #: 030068-29-2				HYDRAZINECARBOXIMIDAMIDE, 2-[(4-NITROPHENYL)METH
Formula: $C_8H_9N_5O_2$				
Mol Weight: 207.19				
MP (deg C):		FP (deg C):		
BP (deg C):				
BP pressure (mm Hg):				

Property/ Value	Units	Temp	Data Type	Reference
WS 2.96E+003	mg/L	25	EST	MEYLAN,WM ET AL. (1996)
logP 0.87			EXP	SOMAN,G ET AL. (1986)
VP 3.13E-006	mm Hg	25	EST	NEELY,WB & BLAU,GE (1985)
DC	pKa			
HL 1.99E-014	atm m3/mol	25	EST	MEYLAN,WM & HOWARD,PH (1991)
OH 8.65E-011	cm3/molc sec	25	EST	MEYLAN,WM & HOWARD,PH (1993)

CAS #: 030085-34-8				3-(P-CHLOROPHENYL)HYDROXYUREA
Formula: $C_7H_7ClN_2O_2$				
Mol Weight: 186.60				
MP (deg C):		FP (deg C):		
BP (deg C):				
BP pressure (mm Hg):				

Property/ Value	Units	Temp	Data Type	Reference
WS 3.32E+003	mg/L	25	EST	MEYLAN,WM ET AL. (1996)
logP 1.39			EXP	HANSCH,C & LEO,AJ (1985)
VP 1.54E-007	mm Hg	25	EST	NEELY,WB & BLAU,GE (1985)
DC	pKa			
HL 2.22E-011	atm m3/mol	25	EST	MEYLAN,WM & HOWARD,PH (1991)
OH 1.35E-011	cm3/molc sec	25	EST	MEYLAN,WM & HOWARD,PH (1993)

CAS #: 030087-17-3				1-(3-OHPHENYL)-3-MEO-3-ME UREA
Formula: $C_9H_{12}N_2O_3$				
Mol Weight: 196.21				
MP (deg C):		FP (deg C):		
BP (deg C):				
BP pressure (mm Hg):				

Property/ Value	Units	Temp	Data Type	Reference
WS 8.94E+003	mg/L	25	EST	MEYLAN,WM ET AL. (1996)
logP 0.83			EXP	HANSCH,C & LEO,AJ (1985)
VP 1.28E-006	mm Hg	25	EST	NEELY,WB & BLAU,GE (1985)
DC	pKa			
HL 2.19E-012	atm m3/mol	25	EST	MEYLAN,WM & HOWARD,PH (1991)
OH 2.02E-010	cm3/molc sec	25	EST	MEYLAN,WM & HOWARD,PH (1993)

CAS #: 030101-52-1				3'-BROMOBENZOGUANAMINE
Formula: $C_9H_8BrN_5$				
Mol Weight: 266.11				
MP (deg C):		FP (deg C):		
BP (deg C):				
BP pressure (mm Hg):				

Property/ Value	Units	Temp	Data Type	Reference
WS 1.59E+002	mg/L	25	EST	MEYLAN,WM ET AL. (1996)
logP 2.44			EXP	HANSCH,C & LEO,AJ (1985)
VP 9.25E-008	mm Hg	25	EST	NEELY,WB & BLAU,GE (1985)
DC	pKa			
HL 1.64E-011	atm m3/mol	25	EST	MEYLAN,WM & HOWARD,PH (1991)
OH 2.39E-012	cm3/molc sec	25	EST	MEYLAN,WM & HOWARD,PH (1993)

CAS #: 030122-47-5 — 2(445ME-13OXAZOLID-2AM)5NO2THIAZOLE

Formula: $C_9H_{12}N_4O_3S$

Mol Weight: 256.28

MP (deg C): **FP (deg C):** **BP (deg C):** **BP pressure (mm Hg):**

Property/Value	Units	Temp	Data Type	Reference
WS 2.25E+003	mg/L	25	EST	MEYLAN,WM ET AL. (1996)
logP 1.88			EXP	HANSCH,C & LEO,AJ (1985)
VP 1.36E-006	mm Hg	25	EST	NEELY,WB & BLAU,GE (1985)
DC	pKa			
HL 9.33E-015	atm m3/mol	25	EST	MEYLAN,WM & HOWARD,PH (1991)
OH 1.15E-010	cm3/molc sec	25	EST	MEYLAN,WM & HOWARD,PH (1993)

CAS #: 030125-76-9 — N-BENZYL-2,4-PYRROLIDIONE

Formula: $C_{11}H_{11}NO_2$

Mol Weight: 189.22

MP (deg C): **FP (deg C):** **BP (deg C):** **BP pressure (mm Hg):**

Property/Value	Units	Temp	Data Type	Reference
WS 1.11E+004	mg/L	25	EST	MEYLAN,WM ET AL. (1996)
logP 0.76			EXP	LIPINSKI,CA ET AL. (1991)
VP 9.89E-006	mm Hg	25	EST	NEELY,WB & BLAU,GE (1985)
DC	pKa			
HL 1.94E-011	atm m3/mol	25	EST	MEYLAN,WM & HOWARD,PH (1991)
OH 2.64E-011	cm3/molc sec	25	EST	MEYLAN,WM & HOWARD,PH (1993)

CAS #: 030144-56-0 — 1,4-BENZDAZEPIN-2-ON-5PH-7-DIMEAMINO

Formula: $C_{17}H_{17}N_3O$

Mol Weight: 279.34

MP (deg C): **FP (deg C):** **BP (deg C):** **BP pressure (mm Hg):**

Property/Value	Units	Temp	Data Type	Reference
WS 1.36E+002	mg/L	25	EST	MEYLAN,WM ET AL. (1996)
logP 2.43			EXP	HANSCH,C & LEO,AJ (1985)
VP 2.13E-009	mm Hg	25	EST	NEELY,WB & BLAU,GE (1985)
DC	pKa			
HL 3.82E-012	atm m3/mol	25	EST	MEYLAN,WM & HOWARD,PH (1991)
OH 2.03E-010	cm3/molc sec	25	EST	MEYLAN,WM & HOWARD,PH (1993)

CAS #: 030144-75-3 — 1,4-DIAZEPIN-2-ONE,1,3-DIHYDRO-1-ME-7-DIMETHYLAM

Formula: $C_{18}H_{18}ClN_3O$

Mol Weight: 327.82

MP (deg C): **FP (deg C):** **BP (deg C):** **BP pressure (mm Hg):**

Property/Value	Units	Temp	Data Type	Reference
WS 4.00E+001	mg/L	25	EST	MEYLAN,WM ET AL. (1996)
logP 2.72			EXP	HANSCH,C ET AL. (1995)
VP 2.51E-009	mm Hg	25	EST	NEELY,WB & BLAU,GE (1985)
DC	pKa			
HL 5.79E-011	atm m3/mol	25	EST	MEYLAN,WM & HOWARD,PH (1991)
OH 2.04E-010	cm3/molc sec	25	EST	MEYLAN,WM & HOWARD,PH (1993)

CAS #: 030144-88-8 — 1,4-BENZODIAZPN2-ON,1-ME-5(26CLPH)7CL

Formula: $C_{16}H_{11}Cl_3N_2O$

Mol Weight: 353.64

MP (deg C): **FP (deg C):** **BP (deg C):** **BP pressure (mm Hg):**

Property/Value	Units	Temp	Data Type	Reference
WS 1.49E+001	mg/L	25	EST	MEYLAN,WM ET AL. (1996)
logP 3.04			EXP	HANSCH,C & LEO,AJ (1985)
VP 2.05E-009	mm Hg	25	EST	NEELY,WB & BLAU,GE (1985)
DC	pKa			
HL 2.00E-009	atm m3/mol	25	EST	MEYLAN,WM & HOWARD,PH (1991)
OH 6.76E-012	cm3/molc sec	25	EST	MEYLAN,WM & HOWARD,PH (1993)

CAS #: 030176-62-6 — 1-PHENYLADAMANTANE

Formula: $C_{16}H_{20}$

Mol Weight: 212.34

MP (deg C): **FP (deg C):** **BP (deg C):** **BP pressure (mm Hg):**

Property/Value	Units	Temp	Data Type	Reference
WS 6.88E-001	mg/L	25	EST	MEYLAN,WM ET AL. (1996)
logP 5.55			EXP	HANSCH,C ET AL. (1995)
VP 9.30E-004	mm Hg	25	EST	NEELY,WB & BLAU,GE (1985)
DC	pKa			
HL 6.53E-003	atm m3/mol	25	EST	MEYLAN,WM & HOWARD,PH (1991)
OH 2.36E-011	cm3/molc sec	25	EST	MEYLAN,WM & HOWARD,PH (1993)

CAS #: 030195-30-3 — 2H-1,4-BENZODIAZEPIN-2-ONE, 7-CHLORO-1,3-DIHYDRO

Formula: $C_{13}H_{10}ClN_3O$

Mol Weight: 259.70

MP (deg C): **FP (deg C):** **BP (deg C):** **BP pressure (mm Hg):**

Property/Value	Units	Temp	Data Type	Reference
WS 1.28E+002	mg/L	25	EST	MEYLAN,WM ET AL. (1996)
logP 2.59			EXP	HANSCH,C ET AL. (1995)
VP 2.80E-009	mm Hg	25	EST	NEELY,WB & BLAU,GE (1985)
DC	pKa			
HL 2.99E-013	atm m3/mol	25	EST	MEYLAN,WM & HOWARD,PH (1991)
OH 2.01E-010	cm3/molc sec	25	EST	MEYLAN,WM & HOWARD,PH (1993)

CAS #: 030273-97-3 — ETHANAMINIUM, 2-[(DIAZOACETYL)OXY]-N,N,N-TRIMETH

Formula: $C_7H_{14}BrN_3O_2$

Mol Weight: 252.11

MP (deg C): **FP (deg C):** **BP (deg C):** **BP pressure (mm Hg):**

Property/Value	Units	Temp	Data Type	Reference
WS 1.00E+006	mg/L	25	EST	MEYLAN,WM ET AL. (1996)
logP -3.52			EXP	SANGSTER,J (1993)
VP 1.03E-008	mm Hg	25	EST	NEELY,WB & BLAU,GE (1985)
DC	pKa			
HL 6.94E-014	atm m3/mol	25	EST	MEYLAN,WM & HOWARD,PH (1991)
OH 1.63E-011	cm3/molc sec	25	EST	MEYLAN,WM & HOWARD,PH (1993)

CAS #: 030345-85-8				N-BENZYL-2,4,5-IMIDAZOLITRIONE

Formula: $C_{10}H_8N_2O_3$

Mol Weight: 204.19

MP (deg C): | FP (deg C):

BP (deg C):

BP pressure (mm Hg):

Property/Value	Units	Temp	Data Type	Reference
WS 4.01E+003	mg/L	25	EST	MEYLAN,WM ET AL. (1996)
logP 1.19			EXP	LIPINSKI,CA ET AL. (1991)
VP 6.06E-011	mm Hg	25	EST	NEELY,WB & BLAU,GE (1985)
DC	pKa			
HL 1.22E-013	atm m3/mol	25	EST	MEYLAN,WM & HOWARD,PH (1991)
OH 1.07E-011	cm3/molc sec	25	EST	MEYLAN,WM & HOWARD,PH (1993)

CAS #: 030379-59-0				CARBAMIC ACID, METHYL-, PHENYLMETHYL ESTER

Formula: $C_9H_{11}NO_2$

Mol Weight: 165.19

MP (deg C): | FP (deg C):

BP (deg C):

BP pressure (mm Hg):

Property/Value	Units	Temp	Data Type	Reference
WS 2.38E+003	mg/L	25	EST	MEYLAN,WM ET AL. (1996)
logP 1.68			EXP	TANAKA,M ET AL. (1985)
VP 2.14E-002	mm Hg	25	EST	NEELY,WB & BLAU,GE (1985)
DC	pKa			
HL 7.01E-009	atm m3/mol	25	EST	MEYLAN,WM & HOWARD,PH (1991)
OH 1.24E-011	cm3/molc sec	25	EST	MEYLAN,WM & HOWARD,PH (1993)

CAS #: 030419-67-1				1,2:3,4-DIEPOXYBUTANE (2R,3R)

Formula: $C_4H_6O_2$

Mol Weight: 86.09

MP (deg C): 4 EST | FP (deg C):

BP (deg C): 144

BP pressure (mm Hg):

Property/Value	Units	Temp	Data Type	Reference
WS 1.00E+006	mg/L		EXP	DEAN,JA (1985)
logP -0.58			EST	MEYLAN,WM & HOWARD,PH (1995)
VP 5.68E+001	mm Hg	25	EST	NEELY,WB & BLAU,GE (1985)
DC	pKa			
HL 3.54E-008	atm m3/mol	25	EST	MEYLAN,WM & HOWARD,PH (1991)
OH 9.96E-013	cm3/molc sec	25	EST	MEYLAN,WM & HOWARD,PH (1993)

CAS #: 030453-31-7				ETHYL PROPYL DISULFIDE

Formula: $C_5H_{12}S_2$

Mol Weight: 136.28

MP (deg C): | FP (deg C):

BP (deg C):

BP pressure (mm Hg):

Property/Value	Units	Temp	Data Type	Reference
WS 1.20E+002	mg/L	25	EST	MEYLAN,WM ET AL. (1996)
logP 3.35			EST	MEYLAN,WM & HOWARD,PH (1995)
VP 9.62E-001	mm Hg	25	EST	NEELY,WB & BLAU,GE (1985)
DC	pKa			
HL 2.84E-003	atm m3/mol	25	EST	MEYLAN,WM & HOWARD,PH (1991)
OH 2.43E-010	cm3/molc sec	25	EST	MEYLAN,WM & HOWARD,PH (1993)

CAS #: 030508-25-9				2'-METHYLBENZOGUANAMINE

Formula: $C_{10}H_{11}N_5$

Mol Weight: 201.23

MP (deg C): | FP (deg C):

BP (deg C):

BP pressure (mm Hg):

Property/Value	Units	Temp	Data Type	Reference
WS 3.62E+003	mg/L	25	EST	MEYLAN,WM ET AL. (1996)
logP 1.26			EXP	HANSCH,C & LEO,AJ (1985)
VP 3.42E-007	mm Hg	25	EST	NEELY,WB & BLAU,GE (1985)
DC	pKa			
HL 4.54E-011	atm m3/mol	25	EST	MEYLAN,WM & HOWARD,PH (1991)
OH 4.58E-012	cm3/molc sec	25	EST	MEYLAN,WM & HOWARD,PH (1993)

CAS #: 030508-78-2				3'-TRIFLUOROMETHYLBENZOGUANAMINE

Formula: $C_{10}H_8F_3N_5$

Mol Weight: 255.20

MP (deg C): | FP (deg C):

BP (deg C):

BP pressure (mm Hg):

Property/Value	Units	Temp	Data Type	Reference
WS 1.16E+002	mg/L	25	EST	MEYLAN,WM ET AL. (1996)
logP 2.67			EXP	HANSCH,C & LEO,AJ (1985)
VP 5.97E-007	mm Hg	25	EST	NEELY,WB & BLAU,GE (1985)
DC	pKa			
HL 3.57E-010	atm m3/mol	25	EST	MEYLAN,WM & HOWARD,PH (1991)
OH 8.00E-013	cm3/molc sec	25	EST	MEYLAN,WM & HOWARD,PH (1993)

CAS #: 030516-87-1				THYMIDINE, 3'-AZIDO-3'-DEOXY-

Formula: $C_{11}H_{14}N_4O_4$

Mol Weight: 266.26

MP (deg C): 113-115 | FP (deg C):

BP (deg C):

BP pressure (mm Hg):

Property/Value	Units	Temp	Data Type	Reference
WS 8.52E+002	mg/L	25	EST	MEYLAN,WM ET AL. (1996)
logP 0.05			EXP	HANSCH,C ET AL. (1995)
VP 5.23E-020	mm Hg	25	EST	NEELY,WB & BLAU,GE (1985)
DC	pKa			
HL 2.01E-022	atm m3/mol	25	EST	MEYLAN,WM & HOWARD,PH (1991)
OH 9.30E-011	cm3/molc sec	25	EST	MEYLAN,WM & HOWARD,PH (1993)

CAS #: 030525-89-4				PARAFORMALDEHYDE

Formula: CH_2O

Mol Weight: 30.03

MP (deg C): 164 dec | FP (deg C):

BP (deg C):

BP pressure (mm Hg):

Property/Value	Units	Temp	Data Type	Reference
WS 2.00E+005	mg/L	18	EXP	YALKOWSKY,SH & DANNENFELSER,RM (1992)
logP -0.63			EST	MEYLAN,WM & HOWARD,PH (1995)
VP 1.06E+001	mm Hg	25	EXP	JORDAN,TE (1954)
DC 15.50	pKa	25	EXP	SERJEANT,EP & DEMPSEY,B (1979)
HL 2.12E-006	atm m3/mol	25	EST	VP/WSOL
OH 6.16E-013	cm3/molc sec	25	EST	MEYLAN,WM & HOWARD,PH (1993)

CAS #: 030530-43-9	3'-FLUOROBENZOGUANAMINE

Formula: $C_9H_8FN_5$

Mol Weight: 205.20

MP (deg C):
FP (deg C):
BP (deg C):
BP pressure (mm Hg):

Property/Value	Units	Temp	Data Type	Reference
WS 1.37E+003	mg/L	25	EST	MEYLAN,WM ET AL. (1996)
logP 1.73			EXP	HANSCH,C & LEO,AJ (1985)
VP 9.01E-007	mm Hg	25	EST	NEELY,WB & BLAU,GE (1985)
DC	pKa			
HL 4.80E-011	atm m3/mol	25	EST	MEYLAN,WM & HOWARD,PH (1991)
OH 4.50E-012	cm3/molc sec	25	EST	MEYLAN,WM & HOWARD,PH (1993)

CAS #: 030530-44-0	4'-FLUOROBENZOGUANAMINE

Formula: $C_9H_8FN_5$

Mol Weight: 205.20

MP (deg C):
FP (deg C):
BP (deg C):
BP pressure (mm Hg):

Property/Value	Units	Temp	Data Type	Reference
WS 1.74E+003	mg/L	25	EST	MEYLAN,WM ET AL. (1996)
logP 1.61			EXP	HANSCH,C & LEO,AJ (1985)
VP 9.01E-007	mm Hg	25	EST	NEELY,WB & BLAU,GE (1985)
DC	pKa			
HL 4.80E-011	atm m3/mol	25	EST	MEYLAN,WM & HOWARD,PH (1991)
OH 2.00E-012	cm3/molc sec	25	EST	MEYLAN,WM & HOWARD,PH (1993)

CAS #: 030530-48-4	2'-BROMOBENZOGUANAMINE

Formula: $C_9H_8BrN_5$

Mol Weight: 266.11

MP (deg C):
FP (deg C):
BP (deg C):
BP pressure (mm Hg):

Property/Value	Units	Temp	Data Type	Reference
WS 1.61E+003	mg/L	25	EST	MEYLAN,WM ET AL. (1996)
logP 1.26			EXP	HANSCH,C & LEO,AJ (1985)
VP 9.25E-008	mm Hg	25	EST	NEELY,WB & BLAU,GE (1985)
DC	pKa			
HL 1.64E-011	atm m3/mol	25	EST	MEYLAN,WM & HOWARD,PH (1991)
OH 1.22E-012	cm3/molc sec	25	EST	MEYLAN,WM & HOWARD,PH (1993)

CAS #: 030532-36-6	4-PYRIDINEPROPANEAMINE

Formula: $C_8H_{12}N_2$

Mol Weight: 136.20

MP (deg C):
FP (deg C):
BP (deg C):
BP pressure (mm Hg):

Property/Value	Units	Temp	Data Type	Reference
WS 1.00E+006	mg/L	25	EST	MEYLAN,WM ET AL. (1996)
logP 0.40			EXP	HANSCH,C & LEO,AJ (1985)
VP 4.53E-002	mm Hg	25	EST	NEELY,WB & BLAU,GE (1985)
DC	pKa			
HL 1.41E-009	atm m3/mol	25	EST	MEYLAN,WM & HOWARD,PH (1991)
OH 3.52E-011	cm3/molc sec	25	EST	MEYLAN,WM & HOWARD,PH (1993)

CAS #: 030544-61-7	CLANOBUTIN

Formula: $C_{18}H_{18}ClNO_4$

Mol Weight: 347.80

MP (deg C): 115-116
FP (deg C):
BP (deg C):
BP pressure (mm Hg):

Property/Value	Units	Temp	Data Type	Reference
WS 3.16E+001	mg/L	25	EST	MEYLAN,WM ET AL. (1996)
logP 2.70			EXP	HANSCH,C & LEO,AJ (1985)
VP 3.30E-010	mm Hg	25	EST	NEELY,WB & BLAU,GE (1985)
DC	pKa			
HL 1.44E-014	atm m3/mol	25	EST	MEYLAN,WM & HOWARD,PH (1991)
OH 2.08E-011	cm3/molc sec	25	EST	MEYLAN,WM & HOWARD,PH (1993)

CAS #: 030560-19-1	ACEPHATE

Formula: $C_4H_{10}NO_3PS$

Mol Weight: 183.17

MP (deg C): 82-89
FP (deg C):
BP (deg C):
BP pressure (mm Hg):

Property/Value	Units	Temp	Data Type	Reference
WS 8.18E+005	mg/L	25	EXP	WAUCHOPE,RD ET AL. (1991A)
logP -0.85			EXP	HANSCH,C & LEO,AJ (1985)
VP 1.70E-006	mm Hg	25	EXP	WAUCHOPE,RD ET AL. (1991A)
DC	pKa			
HL 5.01E-013	atm m3/mol	25	EST	VP/WSOL
OH 1.97E-011	cm3/molc sec	25	EST	MEYLAN,WM & HOWARD,PH (1993)

CAS #: 030630-07-0	8-METHYLTHIO CYCLIC AMP

Formula: $C_{11}H_{16}N_5O_6PS$

Mol Weight: 377.32

MP (deg C):
FP (deg C):
BP (deg C):
BP pressure (mm Hg):

Property/Value	Units	Temp	Data Type	Reference
WS 1.15E+004	mg/L	25	EST	MEYLAN,WM ET AL. (1996)
logP -2.03			EXP	KORTH,M & ENGELS,J (1987)
VP 6.06E-011	mm Hg	25	EST	NEELY,WB & BLAU,GE (1985)
DC	pKa			
HL 8.52E-030	atm m3/mol	25	EST	MEYLAN,WM & HOWARD,PH (1991)
OH 3.03E-010	cm3/molc sec	25	EST	MEYLAN,WM & HOWARD,PH (1993)

CAS #: 030652-21-2	4(1H)-PYRIDINONE, 3-HYDROXY-1-(2-HYDROXYETHYL)-2

Formula: $C_8H_{11}NO_3$

Mol Weight: 169.18

MP (deg C):
FP (deg C):
BP (deg C):
BP pressure (mm Hg):

Property/Value	Units	Temp	Data Type	Reference
WS 5.40E+005	mg/L	25	EST	MEYLAN,WM ET AL. (1996)
logP -1.10			EXP	DOBBIN,PS ET AL. (1993)
VP 3.80E-007	mm Hg	25	EST	NEELY,WB & BLAU,GE (1985)
DC	pKa			
HL 9.34E-010	atm m3/mol	25	EST	MEYLAN,WM & HOWARD,PH (1991)
OH 1.11E-010	cm3/molc sec	25	EST	MEYLAN,WM & HOWARD,PH (1993)

030673-36-0 — BUTYL DECANOATE

Formula: $C_{14}H_{28}O_2$
Mol Weight: 228.38
MP (deg C):
FP (deg C):
BP (deg C):
BP pressure (mm Hg):

Property/Value	Units	Temp	Data Type	Reference
WS 3.63E-001	mg/L	25	EST	MEYLAN,WM ET AL. (1996)
logP 5.78			EST	MEYLAN,WM & HOWARD,PH (1995)
VP 5.26E-003	mm Hg	25	EST	NEELY,WB & BLAU,GE (1985)
DC	pKa			
HL 3.96E-003	atm m3/mol	25	EST	MEYLAN,WM & HOWARD,PH (1991)
OH 1.47E-011	cm3/mol sec	25	EST	MEYLAN,WM & HOWARD,PH (1993)

030684-06-1 — BENZENEBUTANAMINE, HYDROCHLORIDE

Formula: $C_{10}H_{16}ClN$
Mol Weight: 185.70
MP (deg C):
FP (deg C):
BP (deg C):
BP pressure (mm Hg):

Property/Value	Units	Temp	Data Type	Reference
WS 2.05E+005	mg/L	25	EST	MEYLAN,WM ET AL. (1996)
logP -0.70			EXP	SANGSTER,J (1994)
VP 1.35E-007	mm Hg	25	EST	NEELY,WB & BLAU,GE (1985)
DC	pKa			
HL 1.33E-013	atm m3/mol	25	EST	MEYLAN,WM & HOWARD,PH (1991)
OH 1.95E-011	cm3/mol sec	25	EST	MEYLAN,WM & HOWARD,PH (1993)

030685-43-9 — CARD-20(22)-ENOLIDE DERIVATIVE

Formula: $C_{42}H_{66}O_{14}$
Mol Weight: 794.99
MP (deg C):
FP (deg C):
BP (deg C):
BP pressure (mm Hg):

Property/Value	Units	Temp	Data Type	Reference
WS 2.35E-001	mg/L	25	EST	MEYLAN,WM ET AL. (1996)
logP 1.80			EXP	HANSCH,C ET AL. (1995)
VP 1.27E-029	mm Hg	25	EST	NEELY,WB & BLAU,GE (1985)
DC	pKa			
HL 1.49E-027	atm m3/mol	25	EST	MEYLAN,WM & HOWARD,PH (1991)
OH 1.76E-010	cm3/mol sec	25	EST	MEYLAN,WM & HOWARD,PH (1993)

030717-57-8 — N,N-DIME-2-CARBOXAMIDE-THIOPHENE

Formula: $C_7H_{11}NOS$
Mol Weight: 157.24
MP (deg C):
FP (deg C):
BP (deg C):
BP pressure (mm Hg):

Property/Value	Units	Temp	Data Type	Reference
WS 1.65E+004	mg/L	25	EST	MEYLAN,WM ET AL. (1996)
logP 0.75			EXP	HANSCH,C & LEO,AJ (1985)
VP 3.78E-003	mm Hg	25	EST	NEELY,WB & BLAU,GE (1985)
DC	pKa			
HL 5.80E-009	atm m3/mol	25	EST	MEYLAN,WM & HOWARD,PH (1991)
OH 2.48E-011	cm3/mol sec	25	EST	MEYLAN,WM & HOWARD,PH (1993)

030746-58-8 — 1,2,3,4-TETRACHLORODIBENZO-P-DIOXIN

Formula: $C_{12}H_4Cl_4O_2$
Mol Weight: 321.98
MP (deg C):
FP (deg C):
BP (deg C):
BP pressure (mm Hg):

Property/Value	Units	Temp	Data Type	Reference
WS 4.70E-004	mg/L	25	EXP	YALKOWSKY,SH & DANNENFELSER,RM (1992)
logP 6.60			EXP	SHIU,WY ET AL. (1988)
VP 4.80E-008	mm Hg	25	EXP	SHIU,WY ET AL. (1988)
DC	pKa			
HL 4.33E-005	atm m3/mol	25	EST	VP/WSOL
OH 5.13E-012	cm3/mol sec	25	EST	MEYLAN,WM & HOWARD,PH (1993)

030777-18-5 — BENZO(A)FLUORENE

Formula: $C_{17}H_{12}$
Mol Weight: 216.29
MP (deg C):
FP (deg C):
BP (deg C):
BP pressure (mm Hg):

Property/Value	Units	Temp	Data Type	Reference
WS 4.54E-002	mg/L	25	EXP	PEARLMAN,RS ET AL. (1984)
logP 5.19			EST	MEYLAN,WM & HOWARD,PH (1995)
VP 4.68E-006	mm Hg	25	EST	NEELY,WB & BLAU,GE (1985)
DC	pKa			
HL 1.63E-005	atm m3/mol	25	EST	MEYLAN,WM & HOWARD,PH (1991)
OH 4.67E-011	cm3/mol sec	25	EST	MEYLAN,WM & HOWARD,PH (1993)

030777-19-6 — BENZO(B)FLUORENE

Formula: $C_{17}H_{12}$
Mol Weight: 216.29
MP (deg C):
FP (deg C):
BP (deg C):
BP pressure (mm Hg):

Property/Value	Units	Temp	Data Type	Reference
WS 2.01E-003	mg/L	25	EXP	PEARLMAN,RS ET AL. (1984)
logP 5.19			EST	MEYLAN,WM & HOWARD,PH (1995)
VP 4.68E-006	mm Hg	25	EST	NEELY,WB & BLAU,GE (1985)
DC	pKa			
HL 1.63E-005	atm m3/mol	25	EST	MEYLAN,WM & HOWARD,PH (1991)
OH 4.67E-011	cm3/mol sec	25	EST	MEYLAN,WM & HOWARD,PH (1993)

030802-37-0 — L-ALANINAMIDE, N-ACETYL-L-ALANYL-

Formula: $C_7H_{13}N_3O_3$
Mol Weight: 187.20
MP (deg C):
FP (deg C):
BP (deg C):
BP pressure (mm Hg):

Property/Value	Units	Temp	Data Type	Reference
WS 1.27E+005	mg/L	25	EST	MEYLAN,WM ET AL. (1996)
logP -2.00			EXP	AKAMATSU,M ET AL. (1990)
VP 4.19E-008	mm Hg	25	EST	NEELY,WB & BLAU,GE (1985)
DC	pKa			
HL 2.09E-015	atm m3/mol	25	EST	MEYLAN,WM & HOWARD,PH (1991)
OH 1.12E-011	cm3/mol sec	25	EST	MEYLAN,WM & HOWARD,PH (1993)

CAS #: 030961-42-3

N1(5-BR-3-ME-2-PYRIDYL)SULFANILAMIDE

Formula: $C_{12}H_{12}BrN_3O_2S$

Mol Weight: 342.22

MP (deg C): FP (deg C):

BP (deg C):

BP pressure (mm Hg):

Property/Value	Units	Temp	Data Type	Reference
WS 1.23E+002	mg/L	25	EST	MEYLAN,WM ET AL. (1996)
logP 2.05			EXP	HANSCH,C & LEO,AJ (1985)
VP 3.31E-009	mm Hg	25	EST	NEELY,WB & BLAU,GE (1985)
DC	pKa			
HL 4.74E-014	atm m3/mol	25	EST	MEYLAN,WM & HOWARD,PH (1991)
OH 3.65E-011	cm3/molc sec	25	EST	MEYLAN,WM & HOWARD,PH (1993)

CAS #: 030961-43-4

N1(3-BR-5-ME-2-PYRIDYL)SULFANILAMIDE

Formula: $C_{12}H_{12}BrN_3O_2S$

Mol Weight: 342.22

MP (deg C): FP (deg C):

BP (deg C):

BP pressure (mm Hg):

Property/Value	Units	Temp	Data Type	Reference
WS 2.04E+002	mg/L	25	EST	MEYLAN,WM ET AL. (1996)
logP 1.79			EXP	HANSCH,C & LEO,AJ (1985)
VP 3.31E-009	mm Hg	25	EST	NEELY,WB & BLAU,GE (1985)
DC	pKa			
HL 4.74E-014	atm m3/mol	25	EST	MEYLAN,WM & HOWARD,PH (1991)
OH 3.65E-011	cm3/molc sec	25	EST	MEYLAN,WM & HOWARD,PH (1993)

CAS #: 031017-40-0

1-PHENYLCYCLOHEXENE

Formula: $C_{12}H_{14}$

Mol Weight: 158.25

MP (deg C): FP (deg C):

BP (deg C):

BP pressure (mm Hg):

Property/Value	Units	Temp	Data Type	Reference
WS 9.43E+000	mg/L	25	EST	MEYLAN,WM ET AL. (1996)
logP 4.53			EXP	HANSCH,C ET AL. (1995)
VP 3.79E-002	mm Hg	25	EST	NEELY,WB & BLAU,GE (1985)
DC	pKa			
HL 5.27E-003	atm m3/mol	25	EST	MEYLAN,WM & HOWARD,PH (1991)
OH 9.39E-011	cm3/molc sec	25	EST	MEYLAN,WM & HOWARD,PH (1993)

CAS #: 031036-66-5

M-NITRO-N-PHENYLSUCCINIMIDE

Formula: $C_{10}H_8N_2O_4$

Mol Weight: 220.19

MP (deg C): FP (deg C):

BP (deg C):

BP pressure (mm Hg):

Property/Value	Units	Temp	Data Type	Reference
WS 6.63E+003	mg/L	25	EST	MEYLAN,WM ET AL. (1996)
logP 0.38			EXP	HANSCH,C & LEO,AJ (1985)
VP 1.72E-008	mm Hg	25	EST	NEELY,WB & BLAU,GE (1985)
DC	pKa			
HL 5.15E-010	atm m3/mol	25	EST	MEYLAN,WM & HOWARD,PH (1991)
OH 5.46E-012	cm3/molc sec	25	EST	MEYLAN,WM & HOWARD,PH (1993)

CAS #: 031052-76-3

2(1ME-13-IMIDAZOLINYLIDEN-2AM)5NO2-THIAZOLE

Formula: $C_7H_9N_5O_2S$

Mol Weight: 227.25

MP (deg C): FP (deg C):

BP (deg C):

BP pressure (mm Hg):

Property/Value	Units	Temp	Data Type	Reference
WS 3.72E+002	mg/L	25	EST	MEYLAN,WM ET AL. (1996)
logP 1.80			EXP	HANSCH,C & LEO,AJ (1985)
VP 3.83E-006	mm Hg	25	EST	NEELY,WB & BLAU,GE (1985)
DC	pKa			
HL 1.69E-016	atm m3/mol	25	EST	MEYLAN,WM & HOWARD,PH (1991)
OH 7.69E-011	cm3/molc sec	25	EST	MEYLAN,WM & HOWARD,PH (1993)

CAS #: 031052-77-4

2(1ET-13-IMIDAZOLINYLIDEN-2AM)5NO2-THIAZOLE

Formula: $C_8H_{11}N_5O_2S$

Mol Weight: 241.27

MP (deg C): FP (deg C):

BP (deg C):

BP pressure (mm Hg):

Property/Value	Units	Temp	Data Type	Reference
WS 1.22E+002	mg/L	25	EST	MEYLAN,WM ET AL. (1996)
logP 2.28			EXP	HANSCH,C & LEO,AJ (1985)
VP 1.76E-006	mm Hg	25	EST	NEELY,WB & BLAU,GE (1985)
DC	pKa			
HL 2.24E-016	atm m3/mol	25	EST	MEYLAN,WM & HOWARD,PH (1991)
OH 8.45E-011	cm3/molc sec	25	EST	MEYLAN,WM & HOWARD,PH (1993)

CAS #: 031052-78-5

2(14ME-13-IMIDAZOLN-2AM)5NO2-THIAZOLE

Formula: $C_8H_{11}N_5O_2S$

Mol Weight: 241.27

MP (deg C): FP (deg C):

BP (deg C):

BP pressure (mm Hg):

Property/Value	Units	Temp	Data Type	Reference
WS 2.33E+002	mg/L	25	EST	MEYLAN,WM ET AL. (1996)
logP 1.95			EXP	HANSCH,C & LEO,AJ (1985)
VP 2.20E-006	mm Hg	25	EST	NEELY,WB & BLAU,GE (1985)
DC	pKa			
HL 2.24E-016	atm m3/mol	25	EST	MEYLAN,WM & HOWARD,PH (1991)
OH 7.81E-011	cm3/molc sec	25	EST	MEYLAN,WM & HOWARD,PH (1993)

CAS #: 031052-79-6

2(13DIME-13-IMIDAZOLINYLIDEN-2AM)-5NO2-THIAZOLE

Formula: $C_8H_{11}N_5O_2S$

Mol Weight: 241.27

MP (deg C): FP (deg C):

BP (deg C):

BP pressure (mm Hg):

Property/Value	Units	Temp	Data Type	Reference
WS 1.69E+004	mg/L	25	EST	MEYLAN,WM ET AL. (1996)
logP 0.95			EXP	HANSCH,C & LEO,AJ (1985)
VP 4.18E-006	mm Hg	25	EST	NEELY,WB & BLAU,GE (1985)
DC	pKa			
HL 3.71E-016	atm m3/mol	25	EST	MEYLAN,WM & HOWARD,PH (1991)
OH 1.52E-010	cm3/molc sec	25	EST	MEYLAN,WM & HOWARD,PH (1993)

CAS #: 031052-84-3	2(15DIME-13-IMIDAZOLINYLIDEN-2AM)-5NO2-THIAZOLE

Formula: $C_8H_{11}N_5O_2S$

Mol Weight: 241.27

MP (deg C): | FP (deg C):

BP (deg C):

BP pressure (mm Hg):

Property/Value	Units	Temp	Data Type	Reference
WS 2.03E+002	mg/L	25	EST	MEYLAN,WM ET AL. (1996)
logP 2.02			EXP	HANSCH,C & LEO,AJ (1985)
VP 2.20E-006	mm Hg	25	EST	NEELY,WB & BLAU,GE (1985)
DC	pKa			
HL 2.24E-016	atm m3/mol	25	EST	MEYLAN,WM & HOWARD,PH (1991)
OH 8.63E-011	cm3/molc sec	25	EST	MEYLAN,WM & HOWARD,PH (1993)

CAS #: 031080-39-4	2-PROPYLHEPTANOIC ACID

Formula: $C_{10}H_{20}O_2$

Mol Weight: 172.27

MP (deg C): | FP (deg C):

BP (deg C):

BP pressure (mm Hg):

Property/Value	Units	Temp	Data Type	Reference
WS 2.76E+002	mg/L	25	EST	MEYLAN,WM ET AL. (1996)
logP 3.20			EXP	SANGSTER,J (1993)
VP 4.80E-003	mm Hg	25	EST	NEELY,WB & BLAU,GE (1985)
DC	pKa			
HL 5.28E-006	atm m3/mol	25	EST	MEYLAN,WM & HOWARD,PH (1991)
OH 1.10E-011	cm3/molc sec	25	EST	MEYLAN,WM & HOWARD,PH (1993)

CAS #: 031087-87-3	CARD-20(22)-ENOLIDE, 3-[(2,6-DIDEOXY-3-O-METHYL-

Formula: $C_{29}H_{44}O_7$

Mol Weight: 504.67

MP (deg C): | FP (deg C):

BP (deg C):

BP pressure (mm Hg):

Property/Value	Units	Temp	Data Type	Reference
WS 6.33E+000	mg/L	25	EST	MEYLAN,WM ET AL. (1996)
logP 2.37			EXP	SANGSTER,J (1994)
VP 6.90E-017	mm Hg	25	EST	NEELY,WB & BLAU,GE (1985)
DC	pKa			
HL 6.53E-016	atm m3/mol	25	EST	MEYLAN,WM & HOWARD,PH (1991)
OH 8.72E-011	cm3/molc sec	25	EST	MEYLAN,WM & HOWARD,PH (1993)

CAS #: 031088-06-9	2(1H)-PYRIMIDINONE, 4-AMINO-1-[5-O-(1-OXOHEXADEC

Formula: $C_{25}H_{43}N_3O_6$

Mol Weight: 481.64

MP (deg C): | FP (deg C):

BP (deg C):

BP pressure (mm Hg):

Property/Value	Units	Temp	Data Type	Reference
WS 4.75E-002	mg/L	25	EST	MEYLAN,WM ET AL. (1996)
logP 5.03			EXP	HANSCH,C ET AL. (1995)
VP 2.01E-017	mm Hg	25	EST	NEELY,WB & BLAU,GE (1985)
DC	pKa			
HL 5.39E-019	atm m3/mol	25	EST	MEYLAN,WM & HOWARD,PH (1991)
OH 1.56E-010	cm3/molc sec	25	EST	MEYLAN,WM & HOWARD,PH (1993)

CAS #: 031108-57-3	4-CYANOPYRAZOLE

Formula: $C_4H_3N_3$

Mol Weight: 93.09

MP (deg C): | FP (deg C):

BP (deg C):

BP pressure (mm Hg):

Property/Value	Units	Temp	Data Type	Reference
WS 4.14E+004	mg/L	25	EST	MEYLAN,WM ET AL. (1996)
logP 0.24			EXP	HANSCH,C & LEO,AJ (1985)
VP 1.19E-002	mm Hg	25	EST	NEELY,WB & BLAU,GE (1985)
DC	pKa			
HL 3.57E-008	atm m3/mol	25	EST	MEYLAN,WM & HOWARD,PH (1991)
OH 6.36E-012	cm3/molc sec	25	EST	MEYLAN,WM & HOWARD,PH (1993)

CAS #: 031112-62-6	METRIZAMIDE

Formula: $C_{18}H_{22}I_3N_3O_8$

Mol Weight: 789.10

MP (deg C): 222-224 de | FP (deg C):

BP (deg C):

BP pressure (mm Hg):

Property/Value	Units	Temp	Data Type	Reference
WS 1.79E+001	mg/L	25	EST	MEYLAN,WM ET AL. (1996)
logP -1.89			EXP	HANSCH,C ET AL. (1995)
VP 1.64E-028	mm Hg	25	EST	NEELY,WB & BLAU,GE (1985)
DC	pKa			
HL 5.63E-026	atm m3/mol	25	EST	MEYLAN,WM & HOWARD,PH (1991)
OH 9.78E-011	cm3/molc sec	25	EST	MEYLAN,WM & HOWARD,PH (1993)

CAS #: 031121-11-6	1-PROPANOL, 3-(PHENYLAMINO)-

Formula: $C_9H_{13}NO$

Mol Weight: 151.21

MP (deg C): | FP (deg C):

BP (deg C):

BP pressure (mm Hg):

Property/Value	Units	Temp	Data Type	Reference
WS 2.44E+004	mg/L	25	EST	MEYLAN,WM ET AL. (1996)
logP 0.57			EXP	KRIL,MB & FUNG,HL (1990)
VP 3.72E-004	mm Hg	25	EST	NEELY,WB & BLAU,GE (1985)
DC	pKa			
HL 2.69E-010	atm m3/mol	25	EST	MEYLAN,WM & HOWARD,PH (1991)
OH 5.89E-011	cm3/molc sec	25	EST	MEYLAN,WM & HOWARD,PH (1993)

CAS #: 031122-82-4	1,3-BENZENEDICARBOXAMIDE, TRIIODO DERIVATIVE

Formula: $C_{20}H_{28}I_3N_3O_8$

Mol Weight: 819.17

MP (deg C): | FP (deg C):

BP (deg C):

BP pressure (mm Hg):

Property/Value	Units	Temp	Data Type	Reference
WS 2.42E+001	mg/L	25	EST	MEYLAN,WM ET AL. (1996)
logP -2.28			EXP	SANGSTER,J (1993)
VP 1.43E-027	mm Hg	25	EST	NEELY,WB & BLAU,GE (1985)
DC	pKa			
HL 2.64E-027	atm m3/mol	25	EST	MEYLAN,WM & HOWARD,PH (1991)
OH 7.89E-011	cm3/molc sec	25	EST	MEYLAN,WM & HOWARD,PH (1993)

CAS #: 031122-84-6					1,3-BENZENEDICARBOXAMIDE, 5-(ACETYLAMINO)-N,N'-B

Formula: $C_{18}H_{24}I_3N_3O_7$

Mol Weight: 775.12

MP (deg C): FP (deg C):

BP (deg C):

BP pressure (mm Hg):

Property/ Value	Units	Temp	Data Type	Reference
WS 1.11E+002	mg/L	25	EST	MEYLAN,WM ET AL. (1996)
logP -2.71			EXP	SANGSTER,J (1993)
VP 1.25E-026	mm Hg	25	EST	NEELY,WB & BLAU,GE (1985)
DC	pKa			
HL 2.65E-027	atm m3/mol	25	EST	MEYLAN,WM & HOWARD,PH (1991)
OH 7.00E-011	cm3/mol sec	25	EST	MEYLAN,WM & HOWARD,PH (1993)

CAS #: 031208-76-1					5,7-DINITROBENZPYRAZOLE

Formula: $C_7H_4N_4O_4$

Mol Weight: 208.13

MP (deg C): FP (deg C):

BP (deg C):

BP pressure (mm Hg):

Property/ Value	Units	Temp	Data Type	Reference
WS 6.96E+002	mg/L	25	EST	MEYLAN,WM ET AL. (1996)
logP 1.60			EXP	HANSCH,C & LEO,AJ (1985)
VP 2.13E-007	mm Hg	25	EST	NEELY,WB & BLAU,GE (1985)
DC	pKa			
HL 5.61E-012	atm m3/mol	25	EST	MEYLAN,WM & HOWARD,PH (1991)
OH 3.93E-013	cm3/mol sec	25	EST	MEYLAN,WM & HOWARD,PH (1993)

CAS #: 031218-83-4					PROPETAMPHOS

Formula: $C_{10}H_{20}NO_4PS$

Mol Weight: 281.31

MP (deg C): FP (deg C):

BP (deg C): 88

BP pressure (mm Hg): 5.00E-003

Property/ Value	Units	Temp	Data Type	Reference
WS 1.10E+002	mg/L	24	EXP	YALKOWSKY,SH & DANNENFELSER,RM (1992)
logP 2.51			EST	MEYLAN,WM & HOWARD,PH (1995)
VP 3.94E-004	mm Hg	25	EST	NEELY,WB & BLAU,GE (1985)
DC	pKa			
HL 5.87E-008	atm m3/mol	25	EST	MEYLAN,WM & HOWARD,PH (1991)
OH 2.77E-010	cm3/mol sec	25	EST	MEYLAN,WM & HOWARD,PH (1993)

CAS #: 031357-16-1					ERYTHROMYCIN, 4"-FORMATE

Formula: $C_{38}H_{67}NO_{14}$

Mol Weight: 761.96

MP (deg C): FP (deg C):

BP (deg C):

BP pressure (mm Hg):

Property/ Value	Units	Temp	Data Type	Reference
WS 7.66E-001	mg/L	25	EST	MEYLAN,WM ET AL. (1996)
logP 2.64			EXP	SANGSTER,J (1993)
VP 2.45E-027	mm Hg	25	EST	NEELY,WB & BLAU,GE (1985)
DC	pKa			
HL 4.83E-030	atm m3/mol	25	EST	MEYLAN,WM & HOWARD,PH (1991)
OH 3.88E-010	cm3/mol sec	25	EST	MEYLAN,WM & HOWARD,PH (1993)

CAS #: 031357-18-3					ERYTHROMYCIN, 12-DEOXY-, 4"-ACETATE

Formula: $C_{39}H_{69}NO_{13}$

Mol Weight: 759.98

MP (deg C): FP (deg C):

BP (deg C):

BP pressure (mm Hg):

Property/ Value	Units	Temp	Data Type	Reference
WS 2.48E-001	mg/L	25	EST	MEYLAN,WM ET AL. (1996)
logP 3.23			EXP	SANGSTER,J (1993)
VP 1.05E-026	mm Hg	25	EST	NEELY,WB & BLAU,GE (1985)
DC	pKa			
HL 9.65E-029	atm m3/mol	25	EST	MEYLAN,WM & HOWARD,PH (1991)
OH 3.13E-010	cm3/mol sec	25	EST	MEYLAN,WM & HOWARD,PH (1993)

CAS #: 031357-41-2					ERYTHROMYCIN, 12-DEOXY-, 4"-FORMATE

Formula: $C_{38}H_{67}NO_{13}$

Mol Weight: 745.96

MP (deg C): FP (deg C):

BP (deg C):

BP pressure (mm Hg):

Property/ Value	Units	Temp	Data Type	Reference
WS 4.38E-001	mg/L	25	EST	MEYLAN,WM ET AL. (1996)
logP 3.05			EXP	SANGSTER,J (1993)
VP 2.51E-026	mm Hg	25	EST	NEELY,WB & BLAU,GE (1985)
DC	pKa			
HL 1.32E-028	atm m3/mol	25	EST	MEYLAN,WM & HOWARD,PH (1991)
OH 3.12E-010	cm3/mol sec	25	EST	MEYLAN,WM & HOWARD,PH (1993)

CAS #: 031357-44-5					ERYTHROMYCIN, 12-DEOXY-, 11-ACETATE

Formula: $C_{38}H_{66}NO_{13}$

Mol Weight: 744.95

MP (deg C): FP (deg C):

BP (deg C):

BP pressure (mm Hg):

Property/ Value	Units	Temp	Data Type	Reference
WS 2.07E-001	mg/L	25	EST	MEYLAN,WM ET AL. (1996)
logP 3.32			EXP	HANSCH,C ET AL. (1995)
VP 1.05E-026	mm Hg	25	EST	NEELY,WB & BLAU,GE (1985)
DC	pKa			
HL 9.65E-029	atm m3/mol	25	EST	MEYLAN,WM & HOWARD,PH (1991)
OH 3.11E-010	cm3/mol sec	25	EST	MEYLAN,WM & HOWARD,PH (1993)

CAS #: 031363-85-6					124-BENZTHIADIAZIN-1-O2-3-ME-5-CL

Formula: $C_8H_7ClN_2O_2S$

Mol Weight: 230.67

MP (deg C): FP (deg C):

BP (deg C):

BP pressure (mm Hg):

Property/ Value	Units	Temp	Data Type	Reference
WS 7.33E+003	mg/L	25	EST	MEYLAN,WM ET AL. (1996)
logP 0.72			EXP	HANSCH,C & LEO,AJ (1985)
VP 2.88E-008	mm Hg	25	EST	NEELY,WB & BLAU,GE (1985)
DC	pKa			
HL 6.85E-008	atm m3/mol	25	EST	MEYLAN,WM & HOWARD,PH (1991)
OH 4.89E-013	cm3/mol sec	25	EST	MEYLAN,WM & HOWARD,PH (1993)

CAS #: 031363-88-9 — 124-BENZTHIADIAZIN-1-O2-3-ME-6-ME

Formula: $C_9H_{10}N_2O_2S$
Mol Weight: 210.26
MP (deg C):　　FP (deg C):
BP (deg C):
BP pressure (mm Hg):

Property/Value	Units	Temp	Data Type	Reference
WS 9.04E+003	mg/L	25	EST	MEYLAN,WM ET AL. (1996)
logP 0.74			EXP	HANSCH,C & LEO,AJ (1985)
VP 5.17E-008	mm Hg	25	EST	NEELY,WB & BLAU,GE (1985)
DC	pKa			
HL 1.02E-007	atm m3/mol	25	EST	MEYLAN,WM & HOWARD,PH (1991)
OH 9.88E-013	cm3/molc sec	25	EST	MEYLAN,WM & HOWARD,PH (1993)

CAS #: 031363-89-0 — 124-BENZTHIADIAZIN-1-O2-3-ME-7-ME

Formula: $C_9H_{10}N_2O_2S$
Mol Weight: 210.26
MP (deg C):　　FP (deg C):
BP (deg C):
BP pressure (mm Hg):

Property/Value	Units	Temp	Data Type	Reference
WS 7.88E+003	mg/L	25	EST	MEYLAN,WM ET AL. (1996)
logP 0.81			EXP	HANSCH,C & LEO,AJ (1985)
VP 5.17E-008	mm Hg	25	EST	NEELY,WB & BLAU,GE (1985)
DC	pKa			
HL 1.02E-007	atm m3/mol	25	EST	MEYLAN,WM & HOWARD,PH (1991)
OH 1.50E-012	cm3/molc sec	25	EST	MEYLAN,WM & HOWARD,PH (1993)

CAS #: 031365-74-9 — 124-BENZTHIADIAZIN-1-O2-3-ME-7-F

Formula: $C_8H_7FN_2O_2S$
Mol Weight: 214.22
MP (deg C):　　FP (deg C):
BP (deg C):
BP pressure (mm Hg):

Property/Value	Units	Temp	Data Type	Reference
WS 1.09E+004	mg/L	25	EST	MEYLAN,WM ET AL. (1996)
logP 0.62			EXP	HANSCH,C & LEO,AJ (1985)
VP 1.46E-007	mm Hg	25	EST	NEELY,WB & BLAU,GE (1985)
DC	pKa			
HL 1.08E-007	atm m3/mol	25	EST	MEYLAN,WM & HOWARD,PH (1991)
OH 7.40E-013	cm3/molc sec	25	EST	MEYLAN,WM & HOWARD,PH (1993)

CAS #: 031365-75-0 — 124-BENZTHIADIAZIN-1-O2-3ME-6NO2-7CL

Formula: $C_8H_6ClN_3O_4S$
Mol Weight: 275.67
MP (deg C):　　FP (deg C):
BP (deg C):
BP pressure (mm Hg):

Property/Value	Units	Temp	Data Type	Reference
WS 4.24E+002	mg/L	25	EST	MEYLAN,WM ET AL. (1996)
logP 1.42			EXP	HANSCH,C & LEO,AJ (1985)
VP 5.83E-010	mm Hg	25	EST	NEELY,WB & BLAU,GE (1985)
DC	pKa			
HL 2.70E-010	atm m3/mol	25	EST	MEYLAN,WM & HOWARD,PH (1991)
OH 2.07E-013	cm3/molc sec	25	EST	MEYLAN,WM & HOWARD,PH (1993)

CAS #: 031365-88-5 — 124-BENZTHIADIAZIN-1-O2-3-ME-5-ME

Formula: $C_9H_{10}N_2O_2S$
Mol Weight: 210.26
MP (deg C):　　FP (deg C):
BP (deg C):
BP pressure (mm Hg):

Property/Value	Units	Temp	Data Type	Reference
WS 1.39E+004	mg/L	25	EST	MEYLAN,WM ET AL. (1996)
logP 0.52			EXP	HANSCH,C & LEO,AJ (1985)
VP 5.17E-008	mm Hg	25	EST	NEELY,WB & BLAU,GE (1985)
DC	pKa			
HL 1.02E-007	atm m3/mol	25	EST	MEYLAN,WM & HOWARD,PH (1991)
OH 1.50E-012	cm3/molc sec	25	EST	MEYLAN,WM & HOWARD,PH (1993)

CAS #: 031378-03-7 — PHENACYL p-TOLYL SULFONE

Formula: $C_{15}H_{14}O_3S$
Mol Weight: 274.34
MP (deg C):　　FP (deg C):
BP (deg C):
BP pressure (mm Hg):

Property/Value	Units	Temp	Data Type	Reference
WS 1.40E+002	mg/L	25	EST	MEYLAN,WM ET AL. (1996)
logP 2.45			EXP	WANG,L ET AL. (1993)
VP 2.43E-007	mm Hg	25	EST	NEELY,WB & BLAU,GE (1985)
DC	pKa			
HL 1.83E-010	atm m3/mol	25	EST	MEYLAN,WM & HOWARD,PH (1991)
OH 8.47E-012	cm3/molc sec	25	EST	MEYLAN,WM & HOWARD,PH (1993)

CAS #: 031401-45-3 — 4-PYRIMIDINAMINE, N,N-DIMETHYL-

Formula: $C_6H_9N_3$
Mol Weight: 123.16
MP (deg C):　　FP (deg C):
BP (deg C):
BP pressure (mm Hg):

Property/Value	Units	Temp	Data Type	Reference
WS 3.12E+004	mg/L	25	EST	MEYLAN,WM ET AL. (1996)
logP 0.58			EXP	YAMAGAMI,C ET AL. (1990)
VP 6.25E-001	mm Hg	25	EST	NEELY,WB & BLAU,GE (1985)
DC	pKa			
HL 4.64E-008	atm m3/mol	25	EST	MEYLAN,WM & HOWARD,PH (1991)
OH 4.62E-011	cm3/molc sec	25	EST	MEYLAN,WM & HOWARD,PH (1993)

CAS #: 031401-46-4 — 5-PYRIMIDINAMINE, N,N-DIMETHYL-

Formula: $C_6H_9N_3$
Mol Weight: 123.16
MP (deg C):　　FP (deg C):
BP (deg C):
BP pressure (mm Hg):

Property/Value	Units	Temp	Data Type	Reference
WS 3.95E+004	mg/L	25	EST	MEYLAN,WM ET AL. (1996)
logP 0.46			EXP	YAMAGAMI,C ET AL. (1990)
VP 6.25E-001	mm Hg	25	EST	NEELY,WB & BLAU,GE (1985)
DC	pKa			
HL 4.64E-008	atm m3/mol	25	EST	MEYLAN,WM & HOWARD,PH (1991)
OH 4.62E-011	cm3/molc sec	25	EST	MEYLAN,WM & HOWARD,PH (1993)

CAS #: 031423-92-4				1,4-DICHLORO-1-BUTENE
Formula: $C_4H_6Cl_2$				
Mol Weight: 125.00				
MP (deg C):		FP (deg C):		
BP (deg C):				
BP pressure (mm Hg):				

Property/Value	Units	Temp	Data Type	Reference
WS 4.06E+002	mg/L	25	EST	MEYLAN,WM ET AL. (1996)
logP 2.78			EST	MEYLAN,WM & HOWARD,PH (1995)
VP 9.53E+000	mm Hg	25	EST	NEELY,WB & BLAU,GE (1985)
DC	pKa			
HL 3.25E-002	atm m3/mol	25	EST	MEYLAN,WM & HOWARD,PH (1991)
OH 1.26E-011	cm3/molc sec	25	EST	MEYLAN,WM & HOWARD,PH (1993)

CAS #: 031431-39-7				MEBENDAZOLE
Formula: $C_{16}H_{13}N_3O_3$				
Mol Weight: 295.30				
MP (deg C): 288.5		FP (deg C):		
BP (deg C):				
BP pressure (mm Hg):				

Property/Value	Units	Temp	Data Type	Reference
WS 7.13E+001	mg/L	25	EXP	YALKOWSKY,SH & DANNENFELSER,RM (1992)
logP 2.83			EXP	SANGSTER,J (1994)
VP 5.33E-011	mm Hg	25	EST	NEELY,WB & BLAU,GE (1985)
DC	pKa			
HL 5.36E-016	atm m3/mol	25	EST	MEYLAN,WM & HOWARD,PH (1991)
OH 1.36E-010	cm3/molc sec	25	EST	MEYLAN,WM & HOWARD,PH (1993)

CAS #: 031458-33-0				PYRIMIDINE, 5-METHOXY-
Formula: $C_5H_6N_2O$				
Mol Weight: 110.12				
MP (deg C):		FP (deg C):		
BP (deg C):				
BP pressure (mm Hg):				

Property/Value	Units	Temp	Data Type	Reference
WS 9.47E+004	mg/L	25	EST	MEYLAN,WM ET AL. (1996)
logP 0.07			EXP	YAMAGAMI,C ET AL. (1990)
VP 1.60E+000	mm Hg	25	EST	NEELY,WB & BLAU,GE (1985)
DC	pKa			
HL 1.73E-007	atm m3/mol	25	EST	MEYLAN,WM & HOWARD,PH (1991)
OH 3.37E-012	cm3/molc sec	25	EST	MEYLAN,WM & HOWARD,PH (1993)

CAS #: 031478-45-2				CARBAMIC ACID, 2-(2-METHYL-5-NITRO-1H-IMIDAZOL-1
Formula: $C_7H_{10}N_4O_4$				
Mol Weight: 214.18				
MP (deg C):		FP (deg C):		
BP (deg C):				
BP pressure (mm Hg):				

Property/Value	Units	Temp	Data Type	Reference
WS 1.30E+003	mg/L	25	EST	MEYLAN,WM ET AL. (1996)
logP -0.29			EXP	SANGSTER,J (1994)
VP 1.06E-006	mm Hg	25	EST	NEELY,WB & BLAU,GE (1985)
DC	pKa			
HL 2.49E-013	atm m3/mol	25	EST	MEYLAN,WM & HOWARD,PH (1991)
OH 8.98E-012	cm3/molc sec	25	EST	MEYLAN,WM & HOWARD,PH (1993)

CAS #: 031508-00-6				2',3,4,4',5'-PCB
Formula: $C_{12}H_5Cl_5$				
Mol Weight: 326.44				
MP (deg C):		FP (deg C):		
BP (deg C):				
BP pressure (mm Hg):				

Property/Value	Units	Temp	Data Type	Reference
WS 1.34E-002	mg/L	20	EXP	YALKOWSKY,SH & DANNENFELSER,RM (1992)
logP 7.12			EXP	HANSCH,C & LEO,AJ (1985)
VP 8.97E-006	mm Hg	25	EXP	BIDLEMAN,TF (1984)
DC	pKa			
HL 2.88E-004	atm m3/mol	25	EST	VP/WSOL
OH 3.35E-013	cm3/molc sec	25	EST	MEYLAN,WM & HOWARD,PH (1993)

CAS #: 031542-48-0				1H-PURINE-2,6-DIONE, 3-BUTYL-3,9-DIHYDRO-1-METHY
Formula: $C_{10}H_{14}N_4O_2$				
Mol Weight: 222.25				
MP (deg C):		FP (deg C):		
BP (deg C):				
BP pressure (mm Hg):				

Property/Value	Units	Temp	Data Type	Reference
WS 2.65E+003	mg/L	25	EST	MEYLAN,WM ET AL. (1996)
logP 1.29			EXP	SANGSTER,J (1993)
VP 4.42E-010	mm Hg	25	EST	NEELY,WB & BLAU,GE (1985)
DC	pKa			
HL 3.92E-012	atm m3/mol	25	EST	MEYLAN,WM & HOWARD,PH (1991)
OH 2.62E-011	cm3/molc sec	25	EST	MEYLAN,WM & HOWARD,PH (1993)

CAS #: 031551-45-8				9H-FLUOREN-9-ONE, 2,7-DINITRO-
Formula: $C_{13}H_6N_2O_5$				
Mol Weight: 270.20				
MP (deg C): 292-295		FP (deg C):		
BP (deg C):				
BP pressure (mm Hg):				

Property/Value	Units	Temp	Data Type	Reference
WS 2.16E+000	mg/L	25	EST	MEYLAN,WM ET AL. (1996)
logP 2.84			EXP	DEBNATH,AK & HANSCH,C (1992)
VP 1.46E-008	mm Hg	25	EST	NEELY,WB & BLAU,GE (1985)
DC	pKa			
HL 1.05E-011	atm m3/mol	25	EST	MEYLAN,WM & HOWARD,PH (1991)
OH 5.69E-013	cm3/molc sec	25	EST	MEYLAN,WM & HOWARD,PH (1993)

CAS #: 031575-35-6				PYRIMIDINE, 2-FLUORO-
Formula: $C_4H_3FN_2$				
Mol Weight: 98.08				
MP (deg C):		FP (deg C):		
BP (deg C):				
BP pressure (mm Hg):				

Property/Value	Units	Temp	Data Type	Reference
WS 1.14E+005	mg/L	25	EST	MEYLAN,WM ET AL. (1996)
logP 0.02			EXP	YAMAGAMI,C ET AL. (1990)
VP 1.64E+001	mm Hg	25	EST	NEELY,WB & BLAU,GE (1985)
DC	pKa			
HL 3.41E-006	atm m3/mol	25	EST	MEYLAN,WM & HOWARD,PH (1991)
OH 2.88E-013	cm3/molc sec	25	EST	MEYLAN,WM & HOWARD,PH (1993)

2-NH2-5-BENZYL-1,3,4-OXADIAZOLE

Formula:	$C_9H_{10}N_3O$		
Mol Weight:	176.20		
MP (deg C):		FP (deg C):	
BP (deg C):			
BP pressure (mm Hg):			

Property/Value	Units	Temp	Data Type	Reference
WS 8.97E+003	mg/L	25	EST	MEYLAN,WM ET AL. (1996)
logP 0.95			EXP	HANSCH,C & LEO,AJ (1985)
VP 2.12E-005	mm Hg	25	EST	NEELY,WB & BLAU,GE (1985)
DC	pKa			
HL 9.01E-011	atm m3/mol	25	EST	MEYLAN,WM & HOWARD,PH (1991)
OH 9.78E-012	cm3/molc sec	25	EST	MEYLAN,WM & HOWARD,PH (1993)

CAS #: 031820-22-1

N-NITROSO-METHYL-NEOPENTYLAMINE

Formula:	$C_6H_{14}N_2O$		
Mol Weight:	130.19		
MP (deg C):		FP (deg C):	
BP (deg C):			
BP pressure (mm Hg):			

Property/Value	Units	Temp	Data Type	Reference
WS 6.58E+003	mg/L	25	EST	MEYLAN,WM ET AL. (1996)
logP 1.34			EXP	HANSCH,C & LEO,AJ (1985)
VP 2.19E-001	mm Hg	25	EST	NEELY,WB & BLAU,GE (1985)
DC	pKa			
HL 6.39E-006	atm m3/mol	25	EST	MEYLAN,WM & HOWARD,PH (1991)
OH 1.25E-011	cm3/molc sec	25	EST	MEYLAN,WM & HOWARD,PH (1993)

CAS #: 031822-03-4

BENZOYLHYDRAZINE, O-IODO

Formula:	$C_7H_7IN_2O$		
Mol Weight:	262.05		
MP (deg C):		FP (deg C):	
BP (deg C):			
BP pressure (mm Hg):			

Property/Value	Units	Temp	Data Type	Reference
WS 8.04E+003	mg/L	25	EST	MEYLAN,WM ET AL. (1996)
logP 0.47			EXP	HANSCH,C & LEO,AJ (1985)
VP 2.33E-006	mm Hg	25	EST	NEELY,WB & BLAU,GE (1985)
DC	pKa			
HL 2.14E-012	atm m3/mol	25	EST	MEYLAN,WM & HOWARD,PH (1991)
OH 6.67E-012	cm3/molc sec	25	EST	MEYLAN,WM & HOWARD,PH (1993)

CAS #: 031828-71-4

2-PROPANAMINE, 1-(2,6-DIMETHYLPHENOXY)-

Formula:	$C_{11}H_{17}NO$		
Mol Weight:	179.26		
MP (deg C): 203-205		FP (deg C):	
BP (deg C):			
BP pressure (mm Hg):			

Property/Value	Units	Temp	Data Type	Reference
WS 8.25E+003	mg/L	25	EST	MEYLAN,WM ET AL. (1996)
logP 2.15			EXP	MANNHOLD,R ET AL. (1990)
VP 6.01E-003	mm Hg	25	EST	NEELY,WB & BLAU,GE (1985)
DC	pKa			
HL 7.04E-008	atm m3/mol	25	EST	MEYLAN,WM & HOWARD,PH (1991)
OH 1.24E-010	cm3/molc sec	25	EST	MEYLAN,WM & HOWARD,PH (1993)

CAS #: 031835-53-7

2-METHYL-4-METHOXYQUINOLINE

Formula:	$C_{11}H_{11}NO$		
Mol Weight:	173.22		
MP (deg C):		FP (deg C):	
BP (deg C):			
BP pressure (mm Hg):			

Property/Value	Units	Temp	Data Type	Reference
WS 2.37E+002	mg/L	25	EST	MEYLAN,WM ET AL. (1996)
logP 2.81			EXP	HANSCH,C & LEO,AJ (1985)
VP 1.03E-003	mm Hg	25	EST	NEELY,WB & BLAU,GE (1985)
DC	pKa			
HL 4.49E-008	atm m3/mol	25	EST	MEYLAN,WM & HOWARD,PH (1991)
OH 8.05E-011	cm3/molc sec	25	EST	MEYLAN,WM & HOWARD,PH (1993)

CAS #: 031842-01-0

BENZENEACETIC ACID, 4-(1,4-DIHYDRO-1-OXO-2H-ISOI

Formula:	$C_{17}H_{15}NO_3$		
Mol Weight:	281.31		
MP (deg C): 213-214		FP (deg C):	
BP (deg C):			
BP pressure (mm Hg):			

Property/Value	Units	Temp	Data Type	Reference
WS 6.79E+001	mg/L	25	EST	MEYLAN,WM ET AL. (1996)
logP 2.77			EXP	SANGSTER,J (1993)
VP 2.41E-009	mm Hg	25	EST	NEELY,WB & BLAU,GE (1985)
DC	pKa			
HL 1.00E-013	atm m3/mol	25	EST	MEYLAN,WM & HOWARD,PH (1991)
OH 2.46E-011	cm3/molc sec	25	EST	MEYLAN,WM & HOWARD,PH (1993)

CAS #: 031844-98-1

1-BROMO-1-BUTENE

Formula:	C_4H_7Br		
Mol Weight:	135.01		
MP (deg C):		FP (deg C):	
BP (deg C):			
BP pressure (mm Hg):			

Property/Value	Units	Temp	Data Type	Reference
WS 7.42E+002	mg/L	25	EST	MEYLAN,WM ET AL. (1996)
logP 2.43			EST	MEYLAN,WM & HOWARD,PH (1995)
VP 7.68E+001	mm Hg	25	EXP	PERRY,RH & GREEN,D (1984)
DC	pKa			
HL 2.56E-002	atm m3/mol	25	EST	MEYLAN,WM & HOWARD,PH (1991)
OH 1.58E-011	cm3/molc sec	25	EST	MEYLAN,WM & HOWARD,PH (1993)

CAS #: 031857-31-5

1,3-DIMETHYL-2-CYANOGUANIDINE

Formula:	$C_4H_8N_4$		
Mol Weight:	112.14		
MP (deg C):		FP (deg C):	
BP (deg C):			
BP pressure (mm Hg):			

Property/Value	Units	Temp	Data Type	Reference
WS 2.35E+005	mg/L	25	EST	MEYLAN,WM ET AL. (1996)
logP -0.40			EXP	HANSCH,C & LEO,AJ (1985)
VP 1.09E-001	mm Hg	25	EST	NEELY,WB & BLAU,GE (1985)
DC	pKa			
HL 1.09E-009	atm m3/mol	25	EST	MEYLAN,WM & HOWARD,PH (1991)
OH 1.29E-010	cm3/molc sec	25	EST	MEYLAN,WM & HOWARD,PH (1993)

CAS #: 031883-05-3 — MORICIZINE

Formula: $C_{22}H_{25}N_3O_4S$

Mol Weight: 427.53

MP (deg C): 156-157

FP (deg C):

BP (deg C):

BP pressure (mm Hg):

Property/ Value	Units	Temp	Data Type	Reference
WS 4.57E-001	mg/L	25	EST	MEYLAN,WM ET AL. (1996)
logP 2.98			EXP	SANGSTER,J (1994)
VP 3.17E-012	mm Hg	25	EST	NEELY,WB & BLAU,GE (1985)
DC	pKa			
HL 3.23E-017	atm m3/mol	25	EST	MEYLAN,WM & HOWARD,PH (1991)
OH 3.52E-010	cm3/molc sec	25	EST	MEYLAN,WM & HOWARD,PH (1993)

CAS #: 031898-94-9 — 2-AZETIDINONE, 4-(2-NAPHTHALENYLOXY)-

Formula: $C_{13}H_{11}NO_2$

Mol Weight: 213.24

MP (deg C):

FP (deg C):

BP (deg C):

BP pressure (mm Hg):

Property/ Value	Units	Temp	Data Type	Reference
WS 4.14E+002	mg/L	25	EST	MEYLAN,WM ET AL. (1996)
logP 2.29			EXP	ARNOLDI,A ET AL. (1988)
VP 1.68E-007	mm Hg	25	EST	NEELY,WB & BLAU,GE (1985)
DC	pKa			
HL 9.13E-011	atm m3/mol	25	EST	MEYLAN,WM & HOWARD,PH (1991)
OH 2.59E-010	cm3/molc sec	25	EST	MEYLAN,WM & HOWARD,PH (1993)

CAS #: 031935-08-7 — 2,4-DIAMINO-5-(1-ADAMANTYL)-6-METHYLPYRIMIDINE

Formula: $C_{15}H_{22}N_4$

Mol Weight: 258.37

MP (deg C):

FP (deg C):

BP (deg C):

BP pressure (mm Hg):

Property/ Value	Units	Temp	Data Type	Reference
WS 1.18E+002	mg/L	25	EST	MEYLAN,WM ET AL. (1996)
logP 2.64			EXP	HANSCH,C ET AL. (1995)
VP 1.74E-007	mm Hg	25	EST	NEELY,WB & BLAU,GE (1985)
DC	pKa			
HL 1.73E-009	atm m3/mol	25	EST	MEYLAN,WM & HOWARD,PH (1991)
OH 2.19E-010	cm3/molc sec	25	EST	MEYLAN,WM & HOWARD,PH (1993)

CAS #: 031962-94-4 — DIGOXIN, ALPHA-METHYL

Formula: $C_{42}H_{66}O_{14}$

Mol Weight: 794.99

MP (deg C):

FP (deg C):

BP (deg C):

BP pressure (mm Hg):

Property/ Value	Units	Temp	Data Type	Reference
WS 2.60E-001	mg/L	25	EST	MEYLAN,WM ET AL. (1996)
logP 1.75			EXP	HANSCH,C ET AL. (1995)
VP 1.27E-029	mm Hg	25	EST	NEELY,WB & BLAU,GE (1985)
DC	pKa			
HL 1.49E-027	atm m3/mol	25	EST	MEYLAN,WM & HOWARD,PH (1991)
OH 1.76E-010	cm3/molc sec	25	EST	MEYLAN,WM & HOWARD,PH (1993)

CAS #: 031991-98-7 — 2,4(1H,3H)-PYRIMIDINEDIONE, 5-(BUTYLAMINO)-1,6-D

Formula: $C_{16}H_{21}N_3O_2$

Mol Weight: 287.36

MP (deg C):

FP (deg C):

BP (deg C):

BP pressure (mm Hg):

Property/ Value	Units	Temp	Data Type	Reference
WS 5.24E+002	mg/L	25	EST	MEYLAN,WM ET AL. (1996)
logP 1.69			EXP	HANSCH,C ET AL. (1995)
VP 1.58E-009	mm Hg	25	EST	NEELY,WB & BLAU,GE (1985)
DC	pKa			
HL 3.39E-011	atm m3/mol	25	EST	MEYLAN,WM & HOWARD,PH (1991)
OH 1.04E-010	cm3/molc sec	25	EST	MEYLAN,WM & HOWARD,PH (1993)

CAS #: 031991-99-8 — 2,4(1H,3H)-PYRIMIDINEDIONE, 5-(DIETHYLAMINO)-1,6

Formula: $C_{16}H_{21}N_3O_2$

Mol Weight: 287.36

MP (deg C):

FP (deg C):

BP (deg C):

BP pressure (mm Hg):

Property/ Value	Units	Temp	Data Type	Reference
WS 2.29E+002	mg/L	25	EST	MEYLAN,WM ET AL. (1996)
logP 2.11			EXP	HANSCH,C ET AL. (1995)
VP 3.29E-009	mm Hg	25	EST	NEELY,WB & BLAU,GE (1985)
DC	pKa			
HL 5.60E-011	atm m3/mol	25	EST	MEYLAN,WM & HOWARD,PH (1991)
OH 1.12E-010	cm3/molc sec	25	EST	MEYLAN,WM & HOWARD,PH (1993)

CAS #: 031992-01-5 — 2,4(1H,3H)-PYRIMIDINEDIONE, 5-(DIMETHYLAMINO)-6-

Formula: $C_{13}H_{15}N_3O_2$

Mol Weight: 245.28

MP (deg C):

FP (deg C):

BP (deg C):

BP pressure (mm Hg):

Property/ Value	Units	Temp	Data Type	Reference
WS 1.60E+003	mg/L	25	EST	MEYLAN,WM ET AL. (1996)
logP 1.40			EXP	HANSCH,C ET AL. (1995)
VP 4.14E-010	mm Hg	25	EST	NEELY,WB & BLAU,GE (1985)
DC	pKa			
HL 1.45E-011	atm m3/mol	25	EST	MEYLAN,WM & HOWARD,PH (1991)
OH 9.59E-011	cm3/molc sec	25	EST	MEYLAN,WM & HOWARD,PH (1993)

CAS #: 032092-14-1 — 4H-PYRIDO[1,2-A]PYRIMIDINE-3-CARBOXYLIC ACID, 6,

Formula: $C_{12}H_{16}N_2O_3$

Mol Weight: 236.27

MP (deg C):

FP (deg C):

BP (deg C):

BP pressure (mm Hg):

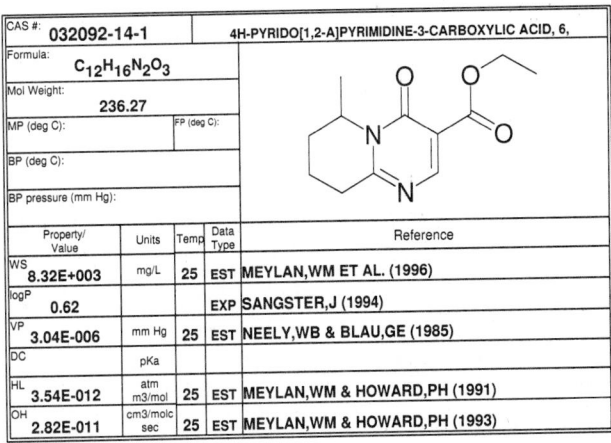

Property/ Value	Units	Temp	Data Type	Reference
WS 8.32E+003	mg/L	25	EST	MEYLAN,WM ET AL. (1996)
logP 0.62			EXP	SANGSTER,J (1994)
VP 3.04E-006	mm Hg	25	EST	NEELY,WB & BLAU,GE (1985)
DC	pKa			
HL 3.54E-012	atm m3/mol	25	EST	MEYLAN,WM & HOWARD,PH (1991)
OH 2.82E-011	cm3/molc sec	25	EST	MEYLAN,WM & HOWARD,PH (1993)

<table>
<tr><td colspan="2">

CAS #: 032092-18-5

Formula: $C_{11}H_{12}N_2O_3$

Mol Weight: 220.23

MP (deg C):   **FP (deg C):**

BP (deg C):

BP pressure (mm Hg):

PYRIDO(12A)PYRIMIDIN-4-ONE,3-ETO-CO

</td></tr>
</table>

Property/Value	Units	Temp	Data Type	Reference
WS 2.15E+004	mg/L	25	EST	MEYLAN,WM ET AL. (1996)
logP 0.25			EXP	HANSCH,C & LEO,AJ (1985)
VP 1.63E-005	mm Hg	25	EST	NEELY,WB & BLAU,GE (1985)
DC	pKa			
HL	atm m3/mol			
OH 8.42E-012	cm3/molc sec	25	EST	MEYLAN,WM & HOWARD,PH (1993)

CAS #: 032092-29-8

Formula: $C_9H_{12}N_2O$

Mol Weight: 164.21

MP (deg C):   **FP (deg C):**

BP (deg C):

BP pressure (mm Hg):

4H-PYRIDO[1,2-A]PYRIMIDIN-4-ONE, 6,7,8,9-TETRAHY

Property/Value	Units	Temp	Data Type	Reference
WS 1.78E+003	mg/L	25	EST	MEYLAN,WM ET AL. (1996)
logP 0.30			EXP	SANGSTER,J (1993)
VP 2.72E-004	mm Hg	25	EST	NEELY,WB & BLAU,GE (1985)
DC	pKa			
HL 1.78E-009	atm m3/mol	25	EST	MEYLAN,WM & HOWARD,PH (1991)
OH 2.70E-011	cm3/molc sec	25	EST	MEYLAN,WM & HOWARD,PH (1993)

CAS #: 032094-40-9

Formula: $C_8H_{10}N_2O_3S$

Mol Weight: 214.24

MP (deg C):   **FP (deg C):**

BP (deg C):

BP pressure (mm Hg):

BENZYLCARBONYL-NH-SO2-NH2

Property/Value	Units	Temp	Data Type	Reference
WS 9.83E+002	mg/L	25	EST	MEYLAN,WM ET AL. (1996)
logP 0.31			EXP	LIPINSKI,CA ET AL. (1991)
VP 8.06E-008	mm Hg	25	EST	NEELY,WB & BLAU,GE (1985)
DC	pKa			
HL 8.36E-012	atm m3/mol	25	EST	MEYLAN,WM & HOWARD,PH (1991)
OH 1.10E-011	cm3/molc sec	25	EST	MEYLAN,WM & HOWARD,PH (1993)

CAS #: 032111-28-7

Formula: $C_5H_7N_3$

Mol Weight: 109.13

MP (deg C):   **FP (deg C):**

BP (deg C):

BP pressure (mm Hg):

2-METHYLAMINOPYRAZINE

Property/Value	Units	Temp	Data Type	Reference
WS 3.64E+004	mg/L	25	EST	MEYLAN,WM ET AL. (1996)
logP 0.56			EXP	YAMAGAMI,C ET AL. (1990A)
VP 6.92E-001	mm Hg	25	EST	NEELY,WB & BLAU,GE (1985)
DC 3.39	pKa	25	EXP	PERRIN,DD (1965)
HL 2.26E-009	atm m3/mol	25	EST	MEYLAN,WM & HOWARD,PH (1991)
OH 6.30E-012	cm3/molc sec	25	EST	MEYLAN,WM & HOWARD,PH (1993)

CAS #: 032139-72-3

Formula: $C_6H_3Cl_3O_2$

Mol Weight: 213.45

MP (deg C):   **FP (deg C):**

BP (deg C):

BP pressure (mm Hg):

3,4,6-TRICHLOROCATECHOL

Property/Value	Units	Temp	Data Type	Reference
WS 1.19E+002	mg/L	25	EST	MEYLAN,WM ET AL. (1996)
logP 3.60			EXP	SANGSTER,J (1994)
VP 2.02E-005	mm Hg	25	EST	NEELY,WB & BLAU,GE (1985)
DC	pKa			
HL 2.37E-011	atm m3/mol	25	EST	MEYLAN,WM & HOWARD,PH (1991)
OH 3.59E-012	cm3/molc sec	25	EST	MEYLAN,WM & HOWARD,PH (1993)

CAS #: 032150-38-2

Formula: $C_{16}H_{21}N_3O_2$

Mol Weight: 287.36

MP (deg C):   **FP (deg C):**

BP (deg C):

BP pressure (mm Hg):

2,4(1H,3H)-PYRIMIDINEDIONE, 5-(DIETHYLAMINO)-3,6

Property/Value	Units	Temp	Data Type	Reference
WS 1.96E+002	mg/L	25	EST	MEYLAN,WM ET AL. (1996)
logP 2.19			EXP	HANSCH,C ET AL. (1995)
VP 3.29E-009	mm Hg	25	EST	NEELY,WB & BLAU,GE (1985)
DC	pKa			
HL 5.60E-011	atm m3/mol	25	EST	MEYLAN,WM & HOWARD,PH (1991)
OH 1.12E-010	cm3/molc sec	25	EST	MEYLAN,WM & HOWARD,PH (1993)

CAS #: 032150-60-0

Formula: $C_{16}H_{19}N_3O_2$

Mol Weight: 285.35

MP (deg C):   **FP (deg C):**

BP (deg C):

BP pressure (mm Hg):

2,4(1H,3H)-PYRIMIDINEDIONE, 5-(DIMETHYLAMINO)-6-

Property/Value	Units	Temp	Data Type	Reference
WS 3.04E+002	mg/L	25	EST	MEYLAN,WM ET AL. (1996)
logP 1.98			EXP	HANSCH,C ET AL. (1995)
VP 3.50E-009	mm Hg	25	EST	NEELY,WB & BLAU,GE (1985)
DC	pKa			
HL 4.17E-011	atm m3/mol	25	EST	MEYLAN,WM & HOWARD,PH (1991)
OH 1.26E-010	cm3/molc sec	25	EST	MEYLAN,WM & HOWARD,PH (1993)

CAS #: 032150-73-5

Formula: $C_{16}H_{19}N_3O_2$

Mol Weight: 285.35

MP (deg C):   **FP (deg C):**

BP (deg C):

BP pressure (mm Hg):

2,4(1H,3H)-PYRIMIDINEDIONE, 1,6-DIMETHYL-3-PHENY

Property/Value	Units	Temp	Data Type	Reference
WS 1.75E+002	mg/L	25	EST	MEYLAN,WM ET AL. (1996)
logP 2.26			EXP	HANSCH,C ET AL. (1995)
VP 1.66E-009	mm Hg	25	EST	NEELY,WB & BLAU,GE (1985)
DC	pKa			
HL 2.47E-011	atm m3/mol	25	EST	MEYLAN,WM & HOWARD,PH (1991)
OH 1.13E-010	cm3/molc sec	25	EST	MEYLAN,WM & HOWARD,PH (1993)

CAS #:	032150-74-6				2,4(1H,3H)-PYRIMIDINEDIONE, 1,6-DIMETHYL-5-(4-MO

Formula: $C_{16}H_{19}N_3O_3$

Mol Weight: 301.35

MP (deg C):   FP (deg C):

BP (deg C):

BP pressure (mm Hg):

Property/Value	Units	Temp	Data Type	Reference
WS 3.11E+003	mg/L	25	EST	MEYLAN,WM ET AL. (1996)
logP 0.69			EXP	HANSCH,C ET AL. (1995)
VP 5.45E-010	mm Hg	25	EST	NEELY,WB & BLAU,GE (1985)
DC	pKa			
HL 2.18E-013	atm m3/mol	25	EST	MEYLAN,WM & HOWARD,PH (1991)
OH 1.69E-010	cm3/molc sec	25	EST	MEYLAN,WM & HOWARD,PH (1993)

CAS #:	032150-76-8				2,4(1H,3H)-PYRIMIDINEDIONE, 5-(DIMETHYLAMINO)-6-

Formula: $C_{16}H_{19}N_3O_2$

Mol Weight: 285.35

MP (deg C):   FP (deg C):

BP (deg C):

BP pressure (mm Hg):

Property/Value	Units	Temp	Data Type	Reference
WS 8.80E+002	mg/L	25	EST	MEYLAN,WM ET AL. (1996)
logP 1.44			EXP	HANSCH,C ET AL. (1995)
VP 3.50E-009	mm Hg	25	EST	NEELY,WB & BLAU,GE (1985)
DC	pKa			
HL 4.17E-011	atm m3/mol	25	EST	MEYLAN,WM & HOWARD,PH (1991)
OH 1.26E-010	cm3/molc sec	25	EST	MEYLAN,WM & HOWARD,PH (1993)

CAS #:	032156-26-6				2,-5-DIMETHOXY-4-BR AMPHETAMINE

Formula: $C_{11}H_{16}BrNO_2$

Mol Weight: 274.16

MP (deg C):   FP (deg C):

BP (deg C):

BP pressure (mm Hg):

Property/Value	Units	Temp	Data Type	Reference
WS 1.10E+003	mg/L	25	EST	MEYLAN,WM ET AL. (1996)
logP 2.58			EXP	HANSCH,C & LEO,AJ (1985)
VP 1.16E-004	mm Hg	25	EST	NEELY,WB & BLAU,GE (1985)
DC	pKa			
HL 1.50E-009	atm m3/mol	25	EST	MEYLAN,WM & HOWARD,PH (1991)
OH 6.04E-011	cm3/molc sec	25	EST	MEYLAN,WM & HOWARD,PH (1993)

CAS #:	032222-42-7				PYRIDINIUM, 3-BROMO-1-METHYL-, IODIDE

Formula: C_6H_7BrIN

Mol Weight: 299.94

MP (deg C):   FP (deg C):

BP (deg C):

BP pressure (mm Hg):

Property/Value	Units	Temp	Data Type	Reference
WS 1.00E+006	mg/L	25	EST	MEYLAN,WM ET AL. (1996)
logP -2.91			EXP	SANGSTER,J (1993)
VP 5.76E-003	mm Hg	25	EST	NEELY,WB & BLAU,GE (1985)
DC	pKa			
HL 5.99E-011	atm m3/mol	25	EST	MEYLAN,WM & HOWARD,PH (1991)
OH 1.04E-011	cm3/molc sec	25	EST	MEYLAN,WM & HOWARD,PH (1993)

CAS #:	032288-17-8				DIETHYL DIMETHYLDIPHOSPHONATE

Formula: $C_6H_{16}O_5P_2$

Mol Weight: 230.14

MP (deg C):   FP (deg C):

BP (deg C):

BP pressure (mm Hg):

Property/Value	Units	Temp	Data Type	Reference
WS 8.12E+004	mg/L	25	EST	MEYLAN,WM ET AL. (1996)
logP -0.50			EST	MEYLAN,WM & HOWARD,PH (1995)
VP 1.41E-003	mm Hg	25	EST	NEELY,WB & BLAU,GE (1985)
DC	pKa			
HL 3.19E-009	atm m3/mol	25	EST	MEYLAN,WM & HOWARD,PH (1991)
OH 3.89E-011	cm3/molc sec	25	EST	MEYLAN,WM & HOWARD,PH (1993)

CAS #:	032319-90-7				1(2-FLET)-3(4-SULFAPYRANYL)-1-NO UREA

Formula: $C_8H_{14}FN_3O_2S$

Mol Weight: 235.28

MP (deg C):   FP (deg C):

BP (deg C):

BP pressure (mm Hg):

Property/Value	Units	Temp	Data Type	Reference
WS 1.71E+003	mg/L	25	EST	MEYLAN,WM ET AL. (1996)
logP 1.43			EXP	HANSCH,C & LEO,AJ (1985)
VP 1.58E-005	mm Hg	25	EST	NEELY,WB & BLAU,GE (1985)
DC	pKa			
HL 6.33E-012	atm m3/mol	25	EST	MEYLAN,WM & HOWARD,PH (1991)
OH 3.79E-011	cm3/molc sec	25	EST	MEYLAN,WM & HOWARD,PH (1993)

CAS #:	032324-41-7				BENZENESULFONAMIDE, N-[(ETHYLAMINO)CARBONYL]-

Formula: $C_9H_{12}N_2O_3S$

Mol Weight: 228.27

MP (deg C):   FP (deg C):

BP (deg C):

BP pressure (mm Hg):

Property/Value	Units	Temp	Data Type	Reference
WS 5.30E+003	mg/L	25	EST	MEYLAN,WM ET AL. (1996)
logP 0.90			EXP	CLOUX,JL ET AL. (1988)
VP 3.72E-007	mm Hg	25	EST	NEELY,WB & BLAU,GE (1985)
DC	pKa			
HL 1.30E-010	atm m3/mol	25	EST	MEYLAN,WM & HOWARD,PH (1991)
OH 6.41E-012	cm3/molc sec	25	EST	MEYLAN,WM & HOWARD,PH (1993)

CAS #:	032324-42-8				BENZENESULFONAMIDE, N-[(DIMETHYLAMINO)CARBONYL]-

Formula: $C_9H_{12}N_2O_3S$

Mol Weight: 228.27

MP (deg C):   FP (deg C):

BP (deg C):

BP pressure (mm Hg):

Property/Value	Units	Temp	Data Type	Reference
WS 8.78E+002	mg/L	25	EST	MEYLAN,WM ET AL. (1996)
logP 0.28			EXP	CLOUX,JL ET AL. (1988)
VP 4.36E-007	mm Hg	25	EST	NEELY,WB & BLAU,GE (1985)
DC	pKa			
HL 2.15E-010	atm m3/mol	25	EST	MEYLAN,WM & HOWARD,PH (1991)
OH 3.53E-012	cm3/molc sec	25	EST	MEYLAN,WM & HOWARD,PH (1993)

CAS #: 032345-29-2	O,O-DIET-O-PHENYLPHOSPHOROTHIOATE

Formula: $C_{10}H_{15}O_3PS$

Mol Weight: 246.27

MP (deg C):   FP (deg C):

BP (deg C):

BP pressure (mm Hg):

Property/Value	Units	Temp	Data Type	Reference
WS 2.75E+001	mg/L	25	EST	MEYLAN,WM ET AL. (1996)
logP 3.46			EXP	HANSCH,C & LEO,AJ (1985)
VP 9.74E-004	mm Hg	25	EST	NEELY,WB & BLAU,GE (1985)
DC	pKa			
HL 7.51E-005	atm m3/mol	25	EST	MEYLAN,WM & HOWARD,PH (1991)
OH 9.52E-011	cm3/molc sec	25	EST	MEYLAN,WM & HOWARD,PH (1993)

CAS #: 032377-09-6	9-PHENYL-9H-FLUORENE-4-CARBONITRILE

Formula: $C_{20}H_{13}N$

Mol Weight: 267.33

MP (deg C):   FP (deg C):

BP (deg C):

BP pressure (mm Hg):

Property/Value	Units	Temp	Data Type	Reference
WS 5.08E-002	mg/L	25	EST	MEYLAN,WM ET AL. (1996)
logP 4.91			EST	MEYLAN,WM & HOWARD,PH (1995)
VP 6.72E-008	mm Hg	25	EST	NEELY,WB & BLAU,GE (1985)
DC	pKa			
HL 1.31E-007	atm m3/mol	25	EST	MEYLAN,WM & HOWARD,PH (1991)
OH 9.90E-012	cm3/molc sec	25	EST	MEYLAN,WM & HOWARD,PH (1993)

CAS #: 032377-10-9	7-PHENYL-7H-BENZO(C)FLUORENE

Formula: $C_{23}H_{16}$

Mol Weight: 292.38

MP (deg C):   FP (deg C):

BP (deg C):

BP pressure (mm Hg):

Property/Value	Units	Temp	Data Type	Reference
WS 2.73E-003	mg/L	25	EST	MEYLAN,WM ET AL. (1996)
logP 6.54			EST	MEYLAN,WM & HOWARD,PH (1995)
VP 1.46E-008	mm Hg	25	EST	NEELY,WB & BLAU,GE (1985)
DC	pKa			
HL 1.32E-006	atm m3/mol	25	EST	MEYLAN,WM & HOWARD,PH (1991)
OH 5.09E-011	cm3/molc sec	25	EST	MEYLAN,WM & HOWARD,PH (1993)

CAS #: 032377-11-0	9-(3-CHLOROPHENYL)-9H-FLUORENE

Formula: $C_{19}H_{13}Cl$

Mol Weight: 276.77

MP (deg C):   FP (deg C):

BP (deg C):

BP pressure (mm Hg):

Property/Value	Units	Temp	Data Type	Reference
WS 9.55E-003	mg/L	25	EST	MEYLAN,WM ET AL. (1996)
logP 6.01			EST	MEYLAN,WM & HOWARD,PH (1995)
VP 6.05E-007	mm Hg	25	EST	NEELY,WB & BLAU,GE (1985)
DC	pKa			
HL 1.00E-005	atm m3/mol	25	EST	MEYLAN,WM & HOWARD,PH (1991)
OH 1.30E-011	cm3/molc sec	25	EST	MEYLAN,WM & HOWARD,PH (1993)

CAS #: 032377-12-1	9-(3-(TRIFLUOROMETHYL)PHENYL)-9H-FLUORENE

Formula: $C_{20}H_{13}F_3$

Mol Weight: 310.32

MP (deg C):   FP (deg C):

BP (deg C):

BP pressure (mm Hg):

Property/Value	Units	Temp	Data Type	Reference
WS 3.26E-003	mg/L	25	EST	MEYLAN,WM ET AL. (1996)
logP 6.33			EST	MEYLAN,WM & HOWARD,PH (1995)
VP 1.56E-006	mm Hg	25	EST	NEELY,WB & BLAU,GE (1985)
DC	pKa			
HL 7.46E-005	atm m3/mol	25	EST	MEYLAN,WM & HOWARD,PH (1991)
OH 1.04E-011	cm3/molc sec	25	EST	MEYLAN,WM & HOWARD,PH (1993)

CAS #: 032377-13-2	9-(3-METHOXYPHENYL)-9H-FLUORENE

Formula: $C_{20}H_{16}O$

Mol Weight: 272.35

MP (deg C):   FP (deg C):

BP (deg C):

BP pressure (mm Hg):

Property/Value	Units	Temp	Data Type	Reference
WS 3.06E-002	mg/L	25	EST	MEYLAN,WM ET AL. (1996)
logP 5.45			EST	MEYLAN,WM & HOWARD,PH (1995)
VP 4.24E-007	mm Hg	25	EST	NEELY,WB & BLAU,GE (1985)
DC	pKa			
HL 7.99E-007	atm m3/mol	25	EST	MEYLAN,WM & HOWARD,PH (1991)
OH 6.16E-011	cm3/molc sec	25	EST	MEYLAN,WM & HOWARD,PH (1993)

CAS #: 032377-15-4	4-(9H-FLUOREN-9-YL)-N,N-DIMETHYLBENZENAMINE

Formula: $C_{21}H_{19}N$

Mol Weight: 285.39

MP (deg C):   FP (deg C):

BP (deg C):

BP pressure (mm Hg):

Property/Value	Units	Temp	Data Type	Reference
WS 2.13E-002	mg/L	25	EST	MEYLAN,WM ET AL. (1996)
logP 5.54			EST	MEYLAN,WM & HOWARD,PH (1995)
VP 1.78E-007	mm Hg	25	EST	NEELY,WB & BLAU,GE (1985)
DC	pKa			
HL 2.15E-007	atm m3/mol	25	EST	MEYLAN,WM & HOWARD,PH (1991)
OH 2.04E-010	cm3/molc sec	25	EST	MEYLAN,WM & HOWARD,PH (1993)

CAS #: 032387-21-6	1H-INDOLE-3-CARBOXYLIC ACID, 1-METHYL-

Formula: $C_{10}H_9NO_2$

Mol Weight: 175.19

MP (deg C):   FP (deg C):

BP (deg C):

BP pressure (mm Hg):

Property/Value	Units	Temp	Data Type	Reference
WS 1.14E+003	mg/L	25	EST	MEYLAN,WM ET AL. (1996)
logP 2.00			EXP	HANSCH,C ET AL. (1995)
VP 2.35E-005	mm Hg	25	EST	NEELY,WB & BLAU,GE (1985)
DC	pKa			
HL 3.80E-010	atm m3/mol	25	EST	MEYLAN,WM & HOWARD,PH (1991)
OH 4.14E-011	cm3/molc sec	25	EST	MEYLAN,WM & HOWARD,PH (1993)

CAS #: 032407-99-1 — PHENYLMERCURY DIMETHYL DITHIOCARBAMATE

Formula: $C_9H_{12}HgNS_2$

Mol Weight: 398.92

MP (deg C): FP (deg C):

BP (deg C):

BP pressure (mm Hg):

Property/ Value	Units	Temp	Data Type	Reference
WS 6.00E+000	mg/L	20	EXP	YALKOWSKY,SH & DANNENFELSER,RM (1992)
logP 0.54			EST	MEYLAN,WM & HOWARD,PH (1995)
VP 2.32E-006	mm Hg	25	EST	NEELY,WB & BLAU,GE (1985)
DC	pKa			
HL	atm m3/mol			
OH 8.12E-011	cm3/molc sec	25	EST	MEYLAN,WM & HOWARD,PH (1993)

CAS #: 032451-61-9 — 8-UREIDO-QUINOLINE

Formula: $C_{10}H_9N_3O$

Mol Weight: 187.20

MP (deg C): FP (deg C):

BP (deg C):

BP pressure (mm Hg):

Property/ Value	Units	Temp	Data Type	Reference
WS 2.10E+003	mg/L	25	EST	MEYLAN,WM ET AL. (1996)
logP 1.62			EXP	HANSCH,C ET AL. (1995)
VP 2.09E-006	mm Hg	25	EST	NEELY,WB & BLAU,GE (1985)
DC	pKa			
HL 2.57E-014	atm m3/mol	25	EST	MEYLAN,WM & HOWARD,PH (1991)
OH 2.01E-010	cm3/molc sec	25	EST	MEYLAN,WM & HOWARD,PH (1993)

CAS #: 032483-15-1 — LEUCINE,N-ACETYL-N'-MEAMINO AMIDE

Formula: $C_9H_{18}N_2O_2$

Mol Weight: 186.26

MP (deg C): FP (deg C):

BP (deg C):

BP pressure (mm Hg):

Property/ Value	Units	Temp	Data Type	Reference
WS 1.91E+003	mg/L	25	EST	MEYLAN,WM ET AL. (1996)
logP 0.14			EXP	HANSCH,C & LEO,AJ (1985)
VP 1.83E-007	mm Hg	25	EST	NEELY,WB & BLAU,GE (1985)
DC	pKa			
HL 8.04E-010	atm m3/mol	25	EST	MEYLAN,WM & HOWARD,PH (1991)
OH 9.37E-011	cm3/molc sec	25	EST	MEYLAN,WM & HOWARD,PH (1993)

CAS #: 032483-16-2 — ISOLEUCINE,N-ACETYL-N'-MEAM AMIDE

Formula: $C_9H_{18}N_2O_2$

Mol Weight: 186.26

MP (deg C): FP (deg C):

BP (deg C):

BP pressure (mm Hg):

Property/ Value	Units	Temp	Data Type	Reference
WS 1.91E+003	mg/L	25	EST	MEYLAN,WM ET AL. (1996)
logP 0.14			EXP	HANSCH,C & LEO,AJ (1985)
VP 1.83E-007	mm Hg	25	EST	NEELY,WB & BLAU,GE (1985)
DC	pKa			
HL 8.04E-010	atm m3/mol	25	EST	MEYLAN,WM & HOWARD,PH (1991)
OH 9.73E-011	cm3/molc sec	25	EST	MEYLAN,WM & HOWARD,PH (1993)

CAS #: 032487-38-0 — 8-S-BENZYL CYCLIC AMP

Formula: $C_{17}H_{20}N_5O_6PS$

Mol Weight: 453.42

MP (deg C): FP (deg C):

BP (deg C):

BP pressure (mm Hg):

Property/ Value	Units	Temp	Data Type	Reference
WS 6.86E+002	mg/L	25	EST	MEYLAN,WM ET AL. (1996)
logP -1.15			EXP	KORTH,M & ENGELS,J (1987)
VP 6.06E-011	mm Hg	25	EST	NEELY,WB & BLAU,GE (1985)
DC	pKa			
HL 6.88E-031	atm m3/mol	25	EST	MEYLAN,WM & HOWARD,PH (1991)
OH 3.14E-010	cm3/molc sec	25	EST	MEYLAN,WM & HOWARD,PH (1993)

CAS #: 032527-55-2 — TIARAMIDE

Formula: $C_{15}H_{18}ClN_3O_3S$

Mol Weight: 355.85

MP (deg C): 159-161 FP (deg C):

BP (deg C):

BP pressure (mm Hg):

Property/ Value	Units	Temp	Data Type	Reference
WS 7.39E+002	mg/L	25	EST	MEYLAN,WM ET AL. (1996)
logP 1.04			EXP	HANSCH,C ET AL. (1995)
VP 7.42E-014	mm Hg	25	EST	NEELY,WB & BLAU,GE (1985)
DC	pKa			
HL 1.92E-015	atm m3/mol	25	EST	MEYLAN,WM & HOWARD,PH (1991)
OH 1.28E-010	cm3/molc sec	25	EST	MEYLAN,WM & HOWARD,PH (1993)

CAS #: 032557-55-4 — 2-BENZOYL-N-METHYLBENZAMIDE

Formula: $C_{15}H_{13}NO_2$

Mol Weight: 239.28

MP (deg C): FP (deg C):

BP (deg C):

BP pressure (mm Hg):

Property/ Value	Units	Temp	Data Type	Reference
WS 4.45E+002	mg/L	25	EST	MEYLAN,WM ET AL. (1996)
logP 2.09			EXP	HANSCH,C ET AL. (1995)
VP 6.67E-008	mm Hg	25	EST	NEELY,WB & BLAU,GE (1985)
DC	pKa			
HL 1.75E-012	atm m3/mol	25	EST	MEYLAN,WM & HOWARD,PH (1991)
OH 8.58E-012	cm3/molc sec	25	EST	MEYLAN,WM & HOWARD,PH (1993)

CAS #: 032562-52-0 — P(N-2CLET-N-ACETYL)AMINOPHACETIC ACID

Formula: $C_{12}H_{14}ClNO_3$

Mol Weight: 255.70

MP (deg C): FP (deg C):

BP (deg C):

BP pressure (mm Hg):

Property/ Value	Units	Temp	Data Type	Reference
WS 2.08E+003	mg/L	25	EST	MEYLAN,WM ET AL. (1996)
logP 1.20			EXP	HANSCH,C & LEO,AJ (1985)
VP 1.53E-007	mm Hg	25	EST	NEELY,WB & BLAU,GE (1985)
DC	pKa			
HL 4.85E-013	atm m3/mol	25	EST	MEYLAN,WM & HOWARD,PH (1991)
OH 1.59E-011	cm3/molc sec	25	EST	MEYLAN,WM & HOWARD,PH (1993)

CAS #: 032598-10-0				2,3',4,4'-TETRACHLOROBIPHENYL
Formula: $C_{12}H_6Cl_4$				
Mol Weight: 291.99				
MP (deg C):		FP (deg C):		
BP (deg C):				
BP pressure (mm Hg):				

Property/ Value	Units	Temp	Data Type	Reference
WS 3.68E-002	mg/L	20	EXP	YALKOWSKY,SH & DANNENFELSER,RM (1992)
logP 5.45			EXP	RAPAPORT,RA & EISENREICH,SJ (1984)
VP 4.62E-005	mm Hg	25	EXP	BIDLEMAN,TF (1984)
DC	pKa			
HL 4.81E-004	atm m3/mol	25	EST	VP/WSOL
OH 7.72E-013	cm3/molc sec	25	EST	MEYLAN,WM & HOWARD,PH (1993)

CAS #: 032598-11-1				2,3',4',5-PCB
Formula: $C_{12}H_6Cl_4$				
Mol Weight: 291.99				
MP (deg C):		FP (deg C):		
BP (deg C):				
BP pressure (mm Hg):				

Property/ Value	Units	Temp	Data Type	Reference
WS 4.10E-002	mg/L	25	EXP	OKOUCHI,S ET AL. (1992)
logP 6.23			EXP	HANSCH,C & LEO,AJ (1985)
VP 4.08E-005	mm Hg	25	EXP	BIDLEMAN,TF (1984)
DC	pKa			
HL 3.82E-004	atm m3/mol	25	EST	VP/WSOL
OH 7.30E-013	cm3/molc sec	25	EST	MEYLAN,WM & HOWARD,PH (1993)

CAS #: 032598-13-3				3,3',4,4'-PCB
Formula: $C_{12}H_6Cl_4$				
Mol Weight: 291.99				
MP (deg C):		FP (deg C):		
BP (deg C):				
BP pressure (mm Hg):				

Property/ Value	Units	Temp	Data Type	Reference
WS 5.69E-004	mg/L	25	EXP	YALKOWSKY,SH & DANNENFELSER,RM (1992)
logP 6.63			EXP	DEBRUIJN,J ET AL. (1989)
VP 1.64E-005	mm Hg	25	EXP	BIDLEMAN,TF (1984)
DC	pKa			
HL 1.11E-002	atm m3/mol	25	EST	VP/WSOL
OH 7.30E-013	cm3/molc sec	25	EST	MEYLAN,WM & HOWARD,PH (1993)

CAS #: 032598-14-4				2,3,3',4,4'-PENTACHLORO-1,1'-BIPHENYL
Formula: $C_{12}H_5Cl_5$				
Mol Weight: 326.44				
MP (deg C):		FP (deg C):		
BP (deg C):				
BP pressure (mm Hg):				

Property/ Value	Units	Temp	Data Type	Reference
WS 3.40E-003	mg/L	25	EXP	OZRETICH,RJ ET AL. (1995)
logP 6.98			EST	MEYLAN,WM & HOWARD,PH (1995)
VP 6.53E-006	mm Hg	25	EXP	BIDLEMAN,TF (1984)
DC	pKa			
HL 8.25E-004	atm m3/mol	25	EST	VP/WSOL
OH 3.35E-013	cm3/molc sec	25	EST	MEYLAN,WM & HOWARD,PH (1993)

CAS #: 032620-68-1				BENZOIC ACID, 2-(ACETYLOXY)-, (ACETYLOXY)METHYL
Formula: $C_{12}H_{12}O_6$				
Mol Weight: 252.23				
MP (deg C):		FP (deg C):		
BP (deg C):				
BP pressure (mm Hg):				

Property/ Value	Units	Temp	Data Type	Reference
WS 1.41E+003	mg/L	25	EST	MEYLAN,WM ET AL. (1996)
logP 1.42			EXP	NIELSEN,LS & BUNDGAARD,H (1989)
VP 7.91E-004	mm Hg	25	EST	NEELY,WB & BLAU,GE (1985)
DC	pKa			
HL 9.94E-010	atm m3/mol	25	EST	MEYLAN,WM & HOWARD,PH (1991)
OH 3.24E-012	cm3/molc sec	25	EST	MEYLAN,WM & HOWARD,PH (1993)

CAS #: 032620-72-7				BENZOIC ACID, 2-(ACETYLOXY)-, (2,5-DIOXO-1-PYRRO
Formula: $C_{14}H_{13}NO_6$				
Mol Weight: 291.26				
MP (deg C):		FP (deg C):		
BP (deg C):				
BP pressure (mm Hg):				

Property/ Value	Units	Temp	Data Type	Reference
WS 5.16E+003	mg/L	25	EST	MEYLAN,WM ET AL. (1996)
logP 0.50			EXP	NIELSEN,LS & BUNDGAARD,H (1989)
VP 3.31E-009	mm Hg	25	EST	NEELY,WB & BLAU,GE (1985)
DC	pKa			
HL 6.29E-012	atm m3/mol	25	EST	MEYLAN,WM & HOWARD,PH (1991)
OH 2.62E-011	cm3/molc sec	25	EST	MEYLAN,WM & HOWARD,PH (1993)

CAS #: 032654-59-4				4-ACETYLOXYAMINOQUINOLINE
Formula: $C_{11}H_{10}N_2O_2$				
Mol Weight: 202.21				
MP (deg C):		FP (deg C):		
BP (deg C):				
BP pressure (mm Hg):				

Property/ Value	Units	Temp	Data Type	Reference
WS 6.33E+003	mg/L	25	EST	MEYLAN,WM ET AL. (1996)
logP 0.97			EXP	HANSCH,C & LEO,AJ (1985)
VP 5.30E-005	mm Hg	25	EST	NEELY,WB & BLAU,GE (1985)
DC	pKa			
HL 3.17E-011	atm m3/mol	25	EST	MEYLAN,WM & HOWARD,PH (1991)
OH 2.00E-010	cm3/molc sec	25	EST	MEYLAN,WM & HOWARD,PH (1993)

CAS #: 032669-06-0				2-CHLOROETHYLBENZHYDRYL ETHER
Formula: $C_{15}H_{15}ClO$				
Mol Weight: 246.74				
MP (deg C):		FP (deg C):		
BP (deg C):				
BP pressure (mm Hg):				

Property/ Value	Units	Temp	Data Type	Reference
WS 4.48E+000	mg/L	25	EST	MEYLAN,WM ET AL. (1996)
logP 4.38			EXP	CHEM INSPECT TEST INST (1992)
VP 6.18E-005	mm Hg	25	EST	NEELY,WB & BLAU,GE (1985)
DC	pKa			
HL 2.63E-006	atm m3/mol	25	EST	MEYLAN,WM & HOWARD,PH (1991)
OH 1.45E-011	cm3/molc sec	25	EST	MEYLAN,WM & HOWARD,PH (1993)

1074

CAS #: 032690-93-0 — 2,4,4',5-PCB

Formula:	$C_{12}H_6Cl_4$
Mol Weight:	291.99
MP (deg C):	FP (deg C):
BP (deg C):	
BP pressure (mm Hg):	

Property/Value	Units	Temp	Data Type	Reference
WS 6.50E-003	mg/L	25	EXP	OZRETICH,RJ ET AL. (1995)
logP 6.67			EXP	HANSCH,C & LEO,AJ (1985)
VP 8.45E-006	mm Hg	25	EST	NEELY,WB & BLAU,GE (1985)
DC	pKa			
HL 1.25E-004	atm m3/mol	25	EST	MEYLAN,WM & HOWARD,PH (1991)
OH 5.77E-013	cm3/molc sec	25	EST	MEYLAN,WM & HOWARD,PH (1993)

CAS #: 032692-19-6 — 1H-INDOLE, 2,3-DIHYDRO-5-NITRO-

Formula:	$C_8H_8N_2O_2$
Mol Weight:	164.17
MP (deg C): 92-94	FP (deg C):
BP (deg C):	
BP pressure (mm Hg):	

Property/Value	Units	Temp	Data Type	Reference
WS 4.55E+002	mg/L	25	EST	MEYLAN,WM ET AL. (1996)
logP 2.07			EXP	DEBNATH,AK ET AL. (1991)
VP 4.75E-004	mm Hg	25	EST	NEELY,WB & BLAU,GE (1985)
DC	pKa			
HL 8.03E-009	atm m3/mol	25	EST	MEYLAN,WM & HOWARD,PH (1991)
OH 1.60E-011	cm3/molc sec	25	EST	MEYLAN,WM & HOWARD,PH (1993)

CAS #: 032694-76-1 — 1,1,1,2,3,3,4,4-OCTACHLOROBUTANE

Formula:	$C_4H_2Cl_8$
Mol Weight:	333.68
MP (deg C):	FP (deg C):
BP (deg C):	
BP pressure (mm Hg):	

Property/Value	Units	Temp	Data Type	Reference
WS 1.98E-001	mg/L	25	EST	MEYLAN,WM ET AL. (1996)
logP 5.38			EST	MEYLAN,WM & HOWARD,PH (1995)
VP 6.78E-003	mm Hg	25	EST	NEELY,WB & BLAU,GE (1985)
DC	pKa			
HL 4.09E-005	atm m3/mol	25	EST	MEYLAN,WM & HOWARD,PH (1991)
OH 8.28E-014	cm3/molc sec	25	EST	MEYLAN,WM & HOWARD,PH (1993)

CAS #: 032695-27-5 — BENZENAMINE, 4-[(4-AMINOPHENYL)SULFONYL]-N-HYDRO

Formula:	$C_{12}H_{12}N_2O_3S$
Mol Weight:	264.31
MP (deg C):	FP (deg C):
BP (deg C):	
BP pressure (mm Hg):	

Property/Value	Units	Temp	Data Type	Reference
WS 3.49E+003	mg/L	25	EST	MEYLAN,WM ET AL. (1996)
logP 0.88			EXP	ALTOMARE,C ET AL. (1991)
VP 2.43E-011	mm Hg	25	EST	NEELY,WB & BLAU,GE (1985)
DC	pKa			
HL 6.24E-017	atm m3/mol	25	EST	MEYLAN,WM & HOWARD,PH (1991)
OH 3.21E-011	cm3/molc sec	25	EST	MEYLAN,WM & HOWARD,PH (1993)

CAS #: 032723-67-4 — 3-METHYL-4-METHOXYBENZALDEHYDE

Formula:	$C_9H_{10}O_2$
Mol Weight:	150.18
MP (deg C):	FP (deg C):
BP (deg C): 80-85	
BP pressure (mm Hg): 1.00E+000	

Property/Value	Units	Temp	Data Type	Reference
WS 9.44E+002	mg/L	25	EST	MEYLAN,WM ET AL. (1996)
logP 2.23			EXP	HANSCH,C & LEO,AJ (1985)
VP 3.98E-002	mm Hg	25	EST	NEELY,WB & BLAU,GE (1985)
DC	pKa			
HL 8.77E-007	atm m3/mol	25	EST	MEYLAN,WM & HOWARD,PH (1991)
OH 2.65E-011	cm3/molc sec	25	EST	MEYLAN,WM & HOWARD,PH (1993)

CAS #: 032737-14-7 — PYRAZINE, 2-ETHOXY-3-METHYL-

Formula:	$C_7H_{10}N_2O$
Mol Weight:	138.17
MP (deg C):	FP (deg C):
BP (deg C):	
BP pressure (mm Hg):	

Property/Value	Units	Temp	Data Type	Reference
WS 2.38E+003	mg/L	25	EST	MEYLAN,WM ET AL. (1996)
logP 1.82			EXP	YAMAGAMI,C ET AL. (1990)
VP 1.74E-001	mm Hg	25	EST	NEELY,WB & BLAU,GE (1985)
DC	pKa			
HL 2.31E-006	atm m3/mol	25	EST	MEYLAN,WM & HOWARD,PH (1991)
OH 9.42E-012	cm3/molc sec	25	EST	MEYLAN,WM & HOWARD,PH (1993)

CAS #: 032743-27-4 — N,N-DIMETHYL 2-PYRAZINECARBOXAMIDE

Formula:	$C_7H_9N_3O$
Mol Weight:	151.17
MP (deg C):	FP (deg C):
BP (deg C):	
BP pressure (mm Hg):	

Property/Value	Units	Temp	Data Type	Reference
WS 1.77E+004	mg/L	25	EST	MEYLAN,WM ET AL. (1996)
logP -0.80			EXP	YAMAGAMI,C ET AL. (1990A)
VP 1.93E-003	mm Hg	25	EST	NEELY,WB & BLAU,GE (1985)
DC	pKa			
HL 5.78E-012	atm m3/mol	25	EST	MEYLAN,WM & HOWARD,PH (1991)
OH 1.63E-011	cm3/molc sec	25	EST	MEYLAN,WM & HOWARD,PH (1993)

CAS #: 032745-69-0 — 1,1-DIME-3-(P-T-BUTYLPHENYL)UREA

Formula:	$C_{13}H_{20}N_2O$
Mol Weight:	220.32
MP (deg C):	FP (deg C):
BP (deg C):	
BP pressure (mm Hg):	

Property/Value	Units	Temp	Data Type	Reference
WS 7.72E+001	mg/L	25	EST	MEYLAN,WM ET AL. (1996)
logP 3.10			EXP	HANSCH,C & LEO,AJ (1985)
VP 1.44E-005	mm Hg	25	EST	NEELY,WB & BLAU,GE (1985)
DC	pKa			
HL 2.51E-009	atm m3/mol	25	EST	MEYLAN,WM & HOWARD,PH (1991)
OH 5.38E-011	cm3/molc sec	25	EST	MEYLAN,WM & HOWARD,PH (1993)

CAS #: 032747-08-3 — GLYCOHYOCHOLIC ACID

Formula: $C_{26}H_{43}NO_6$			
Mol Weight: 465.64			
MP (deg C):		FP (deg C):	
BP (deg C):			
BP pressure (mm Hg):			

Property/ Value	Units	Temp	Data Type	Reference
WS 2.83E+001	mg/L	25	EST	MEYLAN,WM ET AL. (1996)
logP 1.90			EXP	RODA,A ET AL. (1990)
VP 6.48E-020	mm Hg	25	EST	NEELY,WB & BLAU,GE (1985)
DC	pKa			
HL 3.43E-018	atm m3/mol	25	EST	MEYLAN,WM & HOWARD,PH (1991)
OH 6.51E-011	cm3/molc sec	25	EST	MEYLAN,WM & HOWARD,PH (1993)

CAS #: 032774-16-6 — 3,3',4,4',5,5'-PCB

Formula: $C_{12}H_4Cl_6$			
Mol Weight: 360.88			
MP (deg C):		FP (deg C):	
BP (deg C):			
BP pressure (mm Hg):			

Property/ Value	Units	Temp	Data Type	Reference
WS 2.50E-003	mg/L	25	EST	MEYLAN,WM ET AL. (1996)
logP 7.41			EXP	DEBRUIJN,J ET AL. (1989)
VP 5.81E-007	mm Hg	25	EST	NEELY,WB & BLAU,GE (1985)
DC	pKa			
HL 6.85E-005	atm m3/mol	25	EST	MEYLAN,WM & HOWARD,PH (1991)
OH 3.04E-013	cm3/molc sec	25	EST	MEYLAN,WM & HOWARD,PH (1993)

CAS #: 032783-20-3 — BENZENE, [[(1-METHYLETHENYL)OXY]METHYL]-

Formula: $C_{10}H_{12}O$			
Mol Weight: 148.21			
MP (deg C):		FP (deg C):	
BP (deg C):			
BP pressure (mm Hg):			

Property/ Value	Units	Temp	Data Type	Reference
WS 6.76E+002	mg/L	25	EST	MEYLAN,WM ET AL. (1996)
logP 2.41			EXP	SANGSTER,J (1993)
VP 2.87E-001	mm Hg	25	EST	NEELY,WB & BLAU,GE (1985)
DC	pKa			
HL 7.25E-004	atm m3/mol	25	EST	MEYLAN,WM & HOWARD,PH (1991)
OH 7.75E-011	cm3/molc sec	25	EST	MEYLAN,WM & HOWARD,PH (1993)

CAS #: 032809-16-8 — PROCYMIDONE

Formula: $C_{13}H_{11}Cl_2NO_2$			
Mol Weight: 284.14			
MP (deg C): 166-166.5		FP (deg C):	
BP (deg C):			
BP pressure (mm Hg):			

Property/ Value	Units	Temp	Data Type	Reference
WS 4.50E+000	mg/L	25	EXP	TOMLIN,C (1994)
logP 3.08			EXP	SAITO,H ET AL. (1993)
VP 1.40E-004	mm Hg	25	EXP	AUGUSTIJN-BECKERS,PWM ET AL. (1984)
DC	pKa			
HL 1.16E-005	atm m3/mol	25	EST	VP/WSOL
OH 7.49E-012	cm3/molc sec	25	EST	MEYLAN,WM & HOWARD,PH (1993)

CAS #: 032857-62-8 — P-TRIFLUOROME PHENYLACETIC ACID

Formula: $C_9H_7F_3O_2$			
Mol Weight: 204.15			
MP (deg C): 83-85		FP (deg C):	
BP (deg C):			
BP pressure (mm Hg):			

Property/ Value	Units	Temp	Data Type	Reference
WS 8.36E+002	mg/L	25	EST	MEYLAN,WM ET AL. (1996)
logP 2.45			EXP	HANSCH,C & LEO,AJ (1985)
VP 3.27E-003	mm Hg	25	EST	NEELY,WB & BLAU,GE (1985)
DC	pKa			
HL 3.84E-007	atm m3/mol	25	EST	MEYLAN,WM & HOWARD,PH (1991)
OH 1.87E-012	cm3/molc sec	25	EST	MEYLAN,WM & HOWARD,PH (1993)

CAS #: 032857-63-9 — BENZENEACETIC ACID, 4-(1,1-DIMETHYLETHYL)-

Formula: $C_{12}H_{16}O_2$			
Mol Weight: 192.26			
MP (deg C):		FP (deg C):	
BP (deg C):			
BP pressure (mm Hg):			

Property/ Value	Units	Temp	Data Type	Reference
WS 3.39E+002	mg/L	25	EST	MEYLAN,WM ET AL. (1996)
logP 2.98			EXP	SANGSTER,J (1993)
VP 2.39E-004	mm Hg	25	EST	NEELY,WB & BLAU,GE (1985)
DC 4.42	pKa	25	EXP	KORTUM,G ET AL (1961)
HL 1.14E-007	atm m3/mol	25	EST	MEYLAN,WM & HOWARD,PH (1991)
OH 6.15E-012	cm3/molc sec	25	EST	MEYLAN,WM & HOWARD,PH (1993)

CAS #: 032861-85-1 — CHLOMETHOXYNIL

Formula: $C_{13}H_9Cl_2NO_4$			
Mol Weight: 314.13			
MP (deg C):		FP (deg C):	
BP (deg C):			
BP pressure (mm Hg):			

Property/ Value	Units	Temp	Data Type	Reference
WS 3.00E-001	mg/L	15	EXP	YALKOWSKY,SH & DANNENFELSER,RM (1992)
logP 4.40			EST	MEYLAN,WM & HOWARD,PH (1995)
VP 3.69E-007	mm Hg	25	EST	NEELY,WB & BLAU,GE (1985)
DC	pKa			
HL 1.51E-008	atm m3/mol	25	EST	MEYLAN,WM & HOWARD,PH (1991)
OH 8.58E-012	cm3/molc sec	25	EST	MEYLAN,WM & HOWARD,PH (1993)

CAS #: 032862-97-8 — M-BROMOCINNAMIC ACID

Formula: $C_9H_7BrO_2$			
Mol Weight: 227.06			
MP (deg C): 177-179		FP (deg C):	
BP (deg C):			
BP pressure (mm Hg):			

Property/ Value	Units	Temp	Data Type	Reference
WS 5.25E+002	mg/L	25	EXP	STEPHEN,H & STEPHEN,T (1963)
logP 2.96			EST	MEYLAN,WM & HOWARD,PH (1995)
VP 9.38E-005	mm Hg	25	EST	NEELY,WB & BLAU,GE (1985)
DC	pKa			
HL 5.15E-009	atm m3/mol	25	EST	MEYLAN,WM & HOWARD,PH (1991)
OH 2.14E-011	cm3/molc sec	25	EST	MEYLAN,WM & HOWARD,PH (1993)

CAS #: 032865-61-5		2-IODOPHENYLACETATE		

Formula: $C_8H_7IO_2$

Mol Weight: 262.05

MP (deg C):   **FP (deg C):**

BP (deg C):

BP pressure (mm Hg):

Property/Value	Units	Temp	Data Type	Reference
WS 1.35E+002	mg/L	25	EST	MEYLAN,WM ET AL. (1996)
logP 2.55			EXP	HANSCH,C & LEO,AJ (1985)
VP 5.32E-003	mm Hg	25	EST	NEELY,WB & BLAU,GE (1985)
DC	pKa			
HL 1.50E-005	atm m3/mol	25	EST	MEYLAN,WM & HOWARD,PH (1991)
OH 1.04E-012	cm3/molc sec	25	EST	MEYLAN,WM & HOWARD,PH (1993)

CAS #: 032924-66-6		CEPHALOTHIN ANALOG (S-TETRAZOLE)		

Formula: $C_{16}H_{20}N_6O_4S_3$

Mol Weight: 456.57

MP (deg C):   **FP (deg C):**

BP (deg C):

BP pressure (mm Hg):

Property/Value	Units	Temp	Data Type	Reference
WS 7.03E+001	mg/L	25	EST	MEYLAN,WM ET AL. (1996)
logP 0.00			EXP	HANSCH,C & LEO,AJ (1985)
VP 2.54E-017	mm Hg	25	EST	NEELY,WB & BLAU,GE (1985)
DC	pKa			
HL 3.78E-020	atm m3/mol	25	EST	MEYLAN,WM & HOWARD,PH (1991)
OH 1.41E-010	cm3/molc sec	25	EST	MEYLAN,WM & HOWARD,PH (1993)

CAS #: 032954-58-8		1-PENTANONE, 1-(3-FURANYL)-4-HYDROXY-		

Formula: $C_9H_{12}O_3$

Mol Weight: 168.19

MP (deg C):   **FP (deg C):**

BP (deg C):

BP pressure (mm Hg):

Property/Value	Units	Temp	Data Type	Reference
WS 3.89E+004	mg/L	25	EST	MEYLAN,WM ET AL. (1996)
logP 0.84			EXP	HANSCH,C ET AL. (1995)
VP 6.32E-004	mm Hg	25	EST	NEELY,WB & BLAU,GE (1985)
DC	pKa			
HL 8.37E-010	atm m3/mol	25	EST	MEYLAN,WM & HOWARD,PH (1991)
OH 5.09E-011	cm3/molc sec	25	EST	MEYLAN,WM & HOWARD,PH (1993)

CAS #: 033021-53-3		3-ACETYLQUINOLINE		

Formula: $C_{11}H_9NO$

Mol Weight: 171.20

MP (deg C):   **FP (deg C):**

BP (deg C):

BP pressure (mm Hg):

Property/Value	Units	Temp	Data Type	Reference
WS 1.91E+003	mg/L	25	EST	MEYLAN,WM ET AL. (1996)
logP 1.76			EXP	HANSCH,C & LEO,AJ (1985)
VP 3.93E-004	mm Hg	25	EST	NEELY,WB & BLAU,GE (1985)
DC	pKa			
HL 1.25E-009	atm m3/mol	25	EST	MEYLAN,WM & HOWARD,PH (1991)
OH 1.32E-011	cm3/molc sec	25	EST	MEYLAN,WM & HOWARD,PH (1993)

CAS #: 033021-93-1		1-NO-1-(2CLET)-3-(2(2BENZYL)PR)UREA		

Formula: $C_{13}H_{18}ClN_3O_2$

Mol Weight: 283.76

MP (deg C):   **FP (deg C):**

BP (deg C):

BP pressure (mm Hg):

Property/Value	Units	Temp	Data Type	Reference
WS 9.95E+000	mg/L	25	EST	MEYLAN,WM ET AL. (1996)
logP 3.73			EXP	HANSCH,C & LEO,AJ (1985)
VP 3.66E-007	mm Hg	25	EST	NEELY,WB & BLAU,GE (1985)
DC	pKa			
HL 1.92E-011	atm m3/mol	25	EST	MEYLAN,WM & HOWARD,PH (1991)
OH 9.14E-012	cm3/molc sec	25	EST	MEYLAN,WM & HOWARD,PH (1993)

CAS #: 033021-94-2		1-NO-1-(2CLET)-3(2CYHEXEN-1-YL)UREA		

Formula: $C_9H_{14}ClN_3O_2$

Mol Weight: 231.68

MP (deg C):   **FP (deg C):**

BP (deg C):

BP pressure (mm Hg):

Property/Value	Units	Temp	Data Type	Reference
WS 2.41E+002	mg/L	25	EST	MEYLAN,WM ET AL. (1996)
logP 2.45			EXP	HANSCH,C & LEO,AJ (1985)
VP 9.06E-006	mm Hg	25	EST	NEELY,WB & BLAU,GE (1985)
DC	pKa			
HL 1.63E-010	atm m3/mol	25	EST	MEYLAN,WM & HOWARD,PH (1991)
OH 7.30E-011	cm3/molc sec	25	EST	MEYLAN,WM & HOWARD,PH (1993)

CAS #: 033022-01-4		1-NO-1-(2CLET)-3(4SULFAPYRANYL)UREA		

Formula: $C_8H_{14}ClN_3O_2S$

Mol Weight: 251.74

MP (deg C):   **FP (deg C):**

BP (deg C):

BP pressure (mm Hg):

Property/Value	Units	Temp	Data Type	Reference
WS 3.95E+002	mg/L	25	EST	MEYLAN,WM ET AL. (1996)
logP 2.07			EXP	HANSCH,C & LEO,AJ (1985)
VP 1.84E-006	mm Hg	25	EST	NEELY,WB & BLAU,GE (1985)
DC	pKa			
HL 1.12E-012	atm m3/mol	25	EST	MEYLAN,WM & HOWARD,PH (1991)
OH 3.73E-011	cm3/molc sec	25	EST	MEYLAN,WM & HOWARD,PH (1993)

CAS #: 033022-02-5		1-NO-1-(2CLET)-3(4THIOPYRANSO2)UREA		

Formula: $C_8H_{14}ClN_3O_4S$

Mol Weight: 283.74

MP (deg C):   **FP (deg C):**

BP (deg C):

BP pressure (mm Hg):

Property/Value	Units	Temp	Data Type	Reference
WS 5.14E+002	mg/L	25	EST	MEYLAN,WM ET AL. (1996)
logP 0.19			EXP	HANSCH,C & LEO,AJ (1985)
VP 5.83E-008	mm Hg	25	EST	NEELY,WB & BLAU,GE (1985)
DC	pKa			
HL 8.90E-015	atm m3/mol	25	EST	MEYLAN,WM & HOWARD,PH (1991)
OH 3.56E-011	cm3/molc sec	25	EST	MEYLAN,WM & HOWARD,PH (1993)

CAS #: 033022-04-7				1-NO-1-(2CLET)-3(CYHX-35-DITHIAN)UREA
Formula: $C_7H_{12}ClN_3O_2S_2$				
Mol Weight: 269.77				
MP (deg C):		FP (deg C):		
BP (deg C):				
BP pressure (mm Hg):				

Property/Value	Units	Temp	Data Type	Reference
WS 3.07E+002	mg/L	25	EST	MEYLAN,WM ET AL. (1996)
logP 2.08			EXP	HANSCH,C & LEO,AJ (1985)
VP 4.18E-007	mm Hg	25	EST	NEELY,WB & BLAU,GE (1985)
DC	pKa			
HL 6.78E-015	atm m3/mol	25	EST	MEYLAN,WM & HOWARD,PH (1991)
OH 9.30E-011	cm3/molc sec	25	EST	MEYLAN,WM & HOWARD,PH (1993)

CAS #: 033022-05-8				1-NO-1-(2CLET)-3(M-DI-SO2PYRAN)UREA
Formula: $C_7H_{12}ClN_3O_6S_2$				
Mol Weight: 333.77				
MP (deg C):		FP (deg C):		
BP (deg C):				
BP pressure (mm Hg):				

Property/Value	Units	Temp	Data Type	Reference
WS 2.41E+003	mg/L	25	EST	MEYLAN,WM ET AL. (1996)
logP -0.94			EXP	HANSCH,C & LEO,AJ (1985)
VP 4.52E-010	mm Hg	25	EST	NEELY,WB & BLAU,GE (1985)
DC	pKa			
HL 4.28E-019	atm m3/mol	25	EST	MEYLAN,WM & HOWARD,PH (1991)
OH 8.96E-011	cm3/molc sec	25	EST	MEYLAN,WM & HOWARD,PH (1993)

CAS #: 033024-47-4				1-NO-1-(2FLET)-3(CYHX-35-DITHIAN)UREA
Formula: $C_7H_{12}FN_3O_2S_2$				
Mol Weight: 253.32				
MP (deg C):		FP (deg C):		
BP (deg C):				
BP pressure (mm Hg):				

Property/Value	Units	Temp	Data Type	Reference
WS 1.10E+003	mg/L	25	EST	MEYLAN,WM ET AL. (1996)
logP 1.54			EXP	HANSCH,C & LEO,AJ (1985)
VP 3.16E-006	mm Hg	25	EST	NEELY,WB & BLAU,GE (1985)
DC	pKa			
HL 3.83E-014	atm m3/mol	25	EST	MEYLAN,WM & HOWARD,PH (1991)
OH 9.36E-011	cm3/molc sec	25	EST	MEYLAN,WM & HOWARD,PH (1993)

CAS #: 033025-41-1				2,3,4,4'-PCB
Formula: $C_{12}H_6Cl_4$				
Mol Weight: 291.99				
MP (deg C):		FP (deg C):		
BP (deg C):				
BP pressure (mm Hg):				

Property/Value	Units	Temp	Data Type	Reference
WS 3.89E-002	mg/L	20	EXP	YALKOWSKY,SH & DANNENFELSER,RM (1992)
logP 5.84			EXP	HANSCH,C & LEO,AJ (1985)
VP 8.45E-006	mm Hg	25	EST	NEELY,WB & BLAU,GE (1985)
DC	pKa			
HL 1.25E-004	atm m3/mol	25	EST	MEYLAN,WM & HOWARD,PH (1991)
OH 5.77E-013	cm3/molc sec	25	EST	MEYLAN,WM & HOWARD,PH (1993)

CAS #: 033046-97-8				M-METHYL-TRIMETHYL ANILINIUM IODIDE
Formula: $C_{10}H_{16}IN$				
Mol Weight: 277.15				
MP (deg C):		FP (deg C):		
BP (deg C):				
BP pressure (mm Hg):				

Property/Value	Units	Temp	Data Type	Reference
WS 1.00E+006	mg/L	25	EST	MEYLAN,WM ET AL. (1996)
logP -2.31			EXP	HANSCH,C & LEO,AJ (1985)
VP 1.36E-008	mm Hg	25	EST	NEELY,WB & BLAU,GE (1985)
DC	pKa			
HL 7.30E-013	atm m3/mol	25	EST	MEYLAN,WM & HOWARD,PH (1991)
OH 5.02E-012	cm3/molc sec	25	EST	MEYLAN,WM & HOWARD,PH (1993)

CAS #: 033086-25-8				6H-DIBENZO[B,D]PYRAN-1-OL, 6A,7,8,10A-TETRAHYDRO
Formula: $C_{21}H_{30}O_2$				
Mol Weight: 314.47				
MP (deg C):		FP (deg C):		
BP (deg C):				
BP pressure (mm Hg):				

Property/Value	Units	Temp	Data Type	Reference
WS 4.28E-002	mg/L	25	EST	MEYLAN,WM ET AL. (1996)
logP 6.97			EXP	SANGSTER,J (1993)
VP 4.63E-008	mm Hg	25	EST	NEELY,WB & BLAU,GE (1985)
DC	pKa			
HL 2.44E-007	atm m3/mol	25	EST	MEYLAN,WM & HOWARD,PH (1991)
OH 3.09E-010	cm3/molc sec	25	EST	MEYLAN,WM & HOWARD,PH (1993)

CAS #: 033089-61-1				AMITRAZ
Formula: $C_{19}H_{23}N_3$				
Mol Weight: 293.42				
MP (deg C): 86		FP (deg C):		
BP (deg C):				
BP pressure (mm Hg):				

Property/Value	Units	Temp	Data Type	Reference
WS 1.00E+000	mg/L	25	EXP	SHIU,WY ET AL. (1990)
logP 5.50			EXP	HANSCH,C ET AL. (1995)
VP 2.00E-006	mm Hg	25	EXP	WAUCHOPE,RD ET AL. (1991A)
DC	pKa			
HL 7.72E-007	atm m3/mol	25	EST	VP/WSOL
OH 1.33E-010	cm3/molc sec	25	EST	MEYLAN,WM & HOWARD,PH (1993)

CAS #: 033089-74-6				METHANIMIDAMIDE, N-(2,4-DIMETHYLPHENYL)-N'-METHY
Formula: $C_{10}H_{14}N_2$				
Mol Weight: 162.24				
MP (deg C):		FP (deg C):		
BP (deg C):				
BP pressure (mm Hg):				

Property/Value	Units	Temp	Data Type	Reference
WS 2.19E+003	mg/L	25	EST	MEYLAN,WM ET AL. (1996)
logP 2.92			EXP	SANGSTER,J (1994)
VP 1.93E-002	mm Hg	25	EST	NEELY,WB & BLAU,GE (1985)
DC	pKa			
HL 3.35E-007	atm m3/mol	25	EST	MEYLAN,WM & HOWARD,PH (1991)
OH 9.72E-011	cm3/molc sec	25	EST	MEYLAN,WM & HOWARD,PH (1993)

033098-65-6 — 2-ETHYLACETANILIDE

Formula: $C_{10}H_{13}NO$

Mol Weight: 163.22

MP (deg C): FP (deg C):

BP (deg C):

BP pressure (mm Hg):

Property/Value	Units	Temp	Data Type	Reference
WS 4.94E+003	mg/L	25	EST	MEYLAN,WM ET AL. (1996)
logP 1.32			EXP	NAKAGAWA,Y ET AL. (1992)
VP 7.27E-005	mm Hg	25	EST	NEELY,WB & BLAU,GE (1985)
DC	pKa			
HL 9.04E-009	atm m3/mol	25	EST	MEYLAN,WM & HOWARD,PH (1991)
OH 1.63E-011	cm3/molc sec	25	EST	MEYLAN,WM & HOWARD,PH (1993)

033098-80-5 — 2,6-DIBROMOACETANILIDE

Formula: $C_8H_7Br_2NO$

Mol Weight: 292.97

MP (deg C): FP (deg C):

BP (deg C):

BP pressure (mm Hg):

Property/Value	Units	Temp	Data Type	Reference
WS 6.66E+002	mg/L	25	EST	MEYLAN,WM ET AL. (1996)
logP 1.53			EXP	NAKAGAWA,Y ET AL. (1992)
VP 3.61E-006	mm Hg	25	EST	NEELY,WB & BLAU,GE (1985)
DC	pKa			
HL 9.80E-010	atm m3/mol	25	EST	MEYLAN,WM & HOWARD,PH (1991)
OH 1.12E-012	cm3/molc sec	25	EST	MEYLAN,WM & HOWARD,PH (1993)

033100-27-5 — 1,4,7,10,13-PENTAOXACYCLOPENTADECANE

Formula: $C_{10}H_{20}O_5$

Mol Weight: 220.27

MP (deg C): FP (deg C):

BP (deg C): 100-135

BP pressure (mm Hg): 2.00E-001

Property/Value	Units	Temp	Data Type	Reference
WS 8.82E+004	mg/L	25	EST	MEYLAN,WM ET AL. (1996)
logP -0.48			EXP	HANSCH,C ET AL. (1995)
VP 1.44E-003	mm Hg	25	EST	NEELY,WB & BLAU,GE (1985)
DC	pKa			
HL 2.22E-011	atm m3/mol	25	EST	MEYLAN,WM & HOWARD,PH (1991)
OH 7.01E-011	cm3/molc sec	25	EST	MEYLAN,WM & HOWARD,PH (1993)

033101-81-4 — N-PHENYL-3,5-OXADIAZOLIDIONE

Formula: $C_8H_6N_2O_3$

Mol Weight: 178.15

MP (deg C): FP (deg C):

BP (deg C):

BP pressure (mm Hg):

Property/Value	Units	Temp	Data Type	Reference
WS 1.84E+003	mg/L	25	EST	MEYLAN,WM ET AL. (1996)
logP 1.74			EXP	LIPINSKI,CA ET AL. (1991)
VP 8.95E-008	mm Hg	25	EST	NEELY,WB & BLAU,GE (1985)
DC	pKa			
HL 2.42E-007	atm m3/mol	25	EST	MEYLAN,WM & HOWARD,PH (1991)
OH 1.34E-011	cm3/molc sec	25	EST	MEYLAN,WM & HOWARD,PH (1993)

033103-93-4 — 1-AMINOADAMATANE, 3-METHYL

Formula: $C_{11}H_{19}N$

Mol Weight: 165.28

MP (deg C): FP (deg C):

BP (deg C):

BP pressure (mm Hg):

Property/Value	Units	Temp	Data Type	Reference
WS 1.77E+003	mg/L	25	EST	MEYLAN,WM ET AL. (1996)
logP 3.01			EXP	HANSCH,C & LEO,AJ (1985)
VP 7.30E-002	mm Hg	25	EST	NEELY,WB & BLAU,GE (1985)
DC	pKa			
HL 1.10E-005	atm m3/mol	25	EST	MEYLAN,WM & HOWARD,PH (1991)
OH 3.69E-011	cm3/molc sec	25	EST	MEYLAN,WM & HOWARD,PH (1993)

033105-81-6 — T-BUTYLGLYCINE

Formula: $C_6H_{13}NO_2$

Mol Weight: 131.18

MP (deg C): > 300 FP (deg C):

BP (deg C):

BP pressure (mm Hg):

Property/Value	Units	Temp	Data Type	Reference
WS 2.52E+004	mg/L	25	EST	MEYLAN,WM ET AL. (1996)
logP -1.77			EXP	HANSCH,C & LEO,AJ (1985)
VP 1.92E-008	mm Hg	25	EST	NEELY,WB & BLAU,GE (1985)
DC	pKa			
HL 5.77E-009	atm m3/mol	25	EST	MEYLAN,WM & HOWARD,PH (1991)
OH 7.05E-011	cm3/molc sec	25	EST	MEYLAN,WM & HOWARD,PH (1993)

033125-97-2 — 1H-IMIDAZOLE-5-CARBOXYLIC ACID, 1-(1-PHENYLETHYL

Formula: $C_{14}H_{16}N_2O_2$

Mol Weight: 244.30

MP (deg C): 67 FP (deg C):

BP (deg C):

BP pressure (mm Hg):

Property/Value	Units	Temp	Data Type	Reference
WS 6.32E+001	mg/L	25	EST	MEYLAN,WM ET AL. (1996)
logP 3.05			EXP	HANSCH,C ET AL. (1995)
VP 1.51E-006	mm Hg	25	EST	NEELY,WB & BLAU,GE (1985)
DC	pKa			
HL 7.34E-008	atm m3/mol	25	EST	MEYLAN,WM & HOWARD,PH (1991)
OH 2.00E-011	cm3/molc sec	25	EST	MEYLAN,WM & HOWARD,PH (1993)

033130-54-0 — 6-NH2-5-(N-METHYLFORMYLAMINO)-1-METHYLURACIL

Formula: $C_7H_{10}N_4O_3$

Mol Weight: 198.18

MP (deg C): FP (deg C):

BP (deg C):

BP pressure (mm Hg):

Property/Value	Units	Temp	Data Type	Reference
WS 2.82E+004	mg/L	25	EST	MEYLAN,WM ET AL. (1996)
logP -1.30			EXP	GASPARI,F & BONATI,M (1987)
VP 7.23E-010	mm Hg	25	EST	NEELY,WB & BLAU,GE (1985)
DC	pKa			
HL 5.41E-016	atm m3/mol	25	EST	MEYLAN,WM & HOWARD,PH (1991)
OH 4.81E-011	cm3/molc sec	25	EST	MEYLAN,WM & HOWARD,PH (1993)

CAS #: 033130-55-1				6-NH2-5-(N-METHYLFORMYLAMINO)-1,3-DIMETHYLURACIL

Formula: $C_8H_{12}N_4O_3$

Mol Weight: 212.21

MP (deg C):   FP (deg C):

BP (deg C):

BP pressure (mm Hg):

Property/Value	Units	Temp	Data Type	Reference
WS 2.48E+004	mg/L	25	EST	MEYLAN,WM ET AL. (1996)
logP -1.32			EXP	GASPARI,F & BONATI,M (1987)
VP 2.87E-008	mm Hg	25	EST	NEELY,WB & BLAU,GE (1985)
DC	pKa			
HL 1.19E-015	atm m3/mol	25	EST	MEYLAN,WM & HOWARD,PH (1991)
OH 4.87E-011	cm3/molc sec	25	EST	MEYLAN,WM & HOWARD,PH (1993)

CAS #: 033146-45-1				2,6-DICHLOROBIPHENYL

Formula: $C_{12}H_8Cl_2$

Mol Weight: 223.10

MP (deg C):   FP (deg C):

BP (deg C):

BP pressure (mm Hg):

Property/Value	Units	Temp	Data Type	Reference
WS 2.41E+000	mg/L	25	EXP	YALKOWSKY,SH & DANNENFELSER,RM (1992)
logP 4.98			EXP	HANSCH,C & LEO,AJ (1985)
VP 1.91E-004	mm Hg	25	EST	NEELY,WB & BLAU,GE (1985)
DC	pKa			
HL 2.27E-004	atm m3/mol	25	EST	MEYLAN,WM & HOWARD,PH (1991)
OH 2.47E-012	cm3/molc sec	25	EST	MEYLAN,WM & HOWARD,PH (1993)

CAS #: 033213-65-9				BETA-ENDOSULFAN

Formula: $C_9H_6Cl_6O_3S$

Mol Weight: 406.93

MP (deg C): 108-110   FP (deg C):

BP (deg C):

BP pressure (mm Hg):

Property/Value	Units	Temp	Data Type	Reference
WS 4.50E-001	mg/L	20	EXP	BOWMAN,BT & SANS,WW (1983)
logP 3.83			EST	HANSCH,C & LEO,AJ (1985); FROM ALPHA
VP 1.00E-005	mm Hg	25	EXP	SUNTIO,LR ET AL. (1988)
DC	pKa			
HL 1.31E-005	atm m3/mol	25	EST	VP/WSOL
OH 7.59E-012	cm3/molc sec	25	EST	MEYLAN,WM & HOWARD,PH (1993)

CAS #: 033225-17-1				M-DIMETHYLAMINOBENZAMIDE

Formula: $C_9H_{12}N_2O$

Mol Weight: 164.21

MP (deg C):   FP (deg C):

BP (deg C):

BP pressure (mm Hg):

Property/Value	Units	Temp	Data Type	Reference
WS 1.01E+004	mg/L	25	EST	MEYLAN,WM ET AL. (1996)
logP 0.95			EXP	HANSCH,C & LEO,AJ (1985)
VP 4.44E-005	mm Hg	25	EST	NEELY,WB & BLAU,GE (1985)
DC	pKa			
HL 3.52E-011	atm m3/mol	25	EST	MEYLAN,WM & HOWARD,PH (1991)
OH 1.60E-010	cm3/molc sec	25	EST	MEYLAN,WM & HOWARD,PH (1993)

CAS #: 033228-44-3				BENZENAMINE, 4-PENTYL-

Formula: $C_{11}H_{17}N$

Mol Weight: 163.26

MP (deg C):   FP (deg C):

BP (deg C):

BP pressure (mm Hg):

Property/Value	Units	Temp	Data Type	Reference
WS 8.42E+001	mg/L	25	EST	MEYLAN,WM ET AL. (1996)
logP 3.39			EXP	HANSCH,C ET AL. (1995)
VP 3.68E-003	mm Hg	25	EST	NEELY,WB & BLAU,GE (1985)
DC	pKa			
HL 6.53E-006	atm m3/mol	25	EST	MEYLAN,WM & HOWARD,PH (1991)
OH 1.36E-010	cm3/molc sec	25	EST	MEYLAN,WM & HOWARD,PH (1993)

CAS #: 033245-39-5				FLUCHLORALIN

Formula: $C_{12}H_{13}ClF_3N_3O_4$

Mol Weight: 355.70

MP (deg C): 42-43   FP (deg C):

BP (deg C):

BP pressure (mm Hg):

Property/Value	Units	Temp	Data Type	Reference
WS 9.00E-001	mg/L	20	EXP	AUGUSTIJN-BECKERS,PWM ET AL. (1994)
logP 5.07			EST	MEYLAN,WM & HOWARD,PH (1995)
VP 2.10E-005	mm Hg	20	EXP	WORTHING,CR & WALKER,SB (1987)
DC	pKa			
HL 1.09E-005	atm m3/mol	20	EST	VP/WSOL
OH 7.73E-011	cm3/molc sec	25	EST	MEYLAM,WM & HOWARD,PH (1993)

CAS #: 033284-50-3				2,4-DICHLOROBIPHENYL

Formula: $C_{12}H_8Cl_2$

Mol Weight: 223.10

MP (deg C):   FP (deg C):

BP (deg C):

BP pressure (mm Hg):

Property/Value	Units	Temp	Data Type	Reference
WS 1.15E+000	mg/L	25	EXP	YALKOWSKY,SH & DANNENFELSER,RM (1992)
logP 5.16			EXP	SANGSTER,J (1993)
VP 1.38E-003	mm Hg	25	EXP	BIDLEMAN,TF (1984)
DC	pKa			
HL 3.52E-004	atm m3/mol	25	EST	VP/WSOL
OH 2.47E-012	cm3/molc sec	25	EST	MEYLAN,WM & HOWARD,PH (1993)

CAS #: 033284-52-5				3,3',5,5'-TETRACHLOROBIPHENYL

Formula: $C_{12}H_6Cl_4$

Mol Weight: 291.99

MP (deg C):   FP (deg C):

BP (deg C):

BP pressure (mm Hg):

Property/Value	Units	Temp	Data Type	Reference
WS 1.23E-003	mg/L	25	EXP	YALKOWSKY,SH & DANNENFELSER,RM (1992)
logP 6.34			EST	MEYLAN,WM & HOWARD,PH (1995)
VP 8.45E-006	mm Hg	25	EST	NEELY,WB & BLAU,GE (1985)
DC	pKa			
HL 1.25E-004	atm m3/mol	25	EST	MEYLAN,WM & HOWARD,PH (1991)
OH 1.76E-012	cm3/molc sec	25	EST	MEYLAN,WM & HOWARD,PH (1993)

CAS #: 033284-53-6				2,3,4,5-PCB
Formula: $C_{12}H_6Cl_4$				
Mol Weight: 291.99				
MP (deg C):		FP (deg C):		
BP (deg C):				
BP pressure (mm Hg):				

Property/ Value	Units	Temp	Data Type	Reference
WS 1.40E-002	mg/L	25	EXP	YALKOWSKY,SH & DANNENFELSER,RM (1992)
logP 6.41			EXP	HANSCH,C & LEO,AJ (1985)
VP 3.75E-005	mm Hg	25	EXP	HARNER,T & MACKAY,D (1995)
DC	pKa			
HL 1.03E-003	atm m3/mol	25	EST	VP/WSOL
OH 8.05E-013	cm3/molc sec	25	EST	MEYLAN,WM & HOWARD,PH (1993)

CAS #: 033286-22-5				DILTIAZEM
Formula: $C_{22}H_{26}N_2O_4S$				
Mol Weight: 414.53				
MP (deg C): 207.5-212		FP (deg C):		
BP (deg C):				
BP pressure (mm Hg):				

Property/ Value	Units	Temp	Data Type	Reference
WS 1.23E+001	mg/L	25	EST	MEYLAN,WM ET AL. (1996)
logP 2.70			EXP	HANSCH,C & LEO,AJ (1985)
VP 1.54E-011	mm Hg	25	EST	NEELY,WB & BLAU,GE (1985)
DC	pKa			
HL 8.61E-017	atm m3/mol	25	EST	MEYLAN,WM & HOWARD,PH (1991)
OH 1.80E-010	cm3/molc sec	25	EST	MEYLAN,WM & HOWARD,PH (1993)

CAS #: 033321-31-2				5-BENZYL-2,4-THIAZOLIDIONE
Formula: $C_{10}H_9NO_2S$				
Mol Weight: 207.25				
MP (deg C):		FP (deg C):		
BP (deg C):				
BP pressure (mm Hg):				

Property/ Value	Units	Temp	Data Type	Reference
WS 8.18E+002	mg/L	25	EST	MEYLAN,WM ET AL. (1996)
logP 1.98			EXP	LIPINSKI,CA ET AL. (1991)
VP 1.28E-009	mm Hg	25	EST	NEELY,WB & BLAU,GE (1985)
DC	pKa			
HL 1.42E-007	atm m3/mol	25	EST	MEYLAN,WM & HOWARD,PH (1991)
OH 2.63E-011	cm3/molc sec	25	EST	MEYLAN,WM & HOWARD,PH (1993)

CAS #: 033330-89-1				2-(3,3-DIME-1-TRIAZINO)BENZAMIDE
Formula: $C_9H_{12}N_4O$				
Mol Weight: 192.22				
MP (deg C):		FP (deg C):		
BP (deg C):				
BP pressure (mm Hg):				

Property/ Value	Units	Temp	Data Type	Reference
WS 1.60E+003	mg/L	25	EST	MEYLAN,WM ET AL. (1996)
logP 1.73			EXP	HANSCH,C & LEO,AJ (1985)
VP 6.00E-006	mm Hg	25	EST	NEELY,WB & BLAU,GE (1985)
DC	pKa			
HL 2.21E-013	atm m3/mol	25	EST	MEYLAN,WM & HOWARD,PH (1991)
OH 4.86E-012	cm3/molc sec	25	EST	MEYLAN,WM & HOWARD,PH (1993)

CAS #: 033330-91-5				4-(3,3-DIME-1-TRIAZINO)BENZAMIDE
Formula: $C_9H_{12}N_4O$				
Mol Weight: 192.22				
MP (deg C):		FP (deg C):		
BP (deg C):				
BP pressure (mm Hg):				

Property/ Value	Units	Temp	Data Type	Reference
WS 4.53E+003	mg/L	25	EST	MEYLAN,WM ET AL. (1996)
logP 1.20			EXP	HANSCH,C & LEO,AJ (1985)
VP 6.00E-006	mm Hg	25	EST	NEELY,WB & BLAU,GE (1985)
DC	pKa			
HL 2.21E-013	atm m3/mol	25	EST	MEYLAN,WM & HOWARD,PH (1991)
OH 4.86E-012	cm3/molc sec	25	EST	MEYLAN,WM & HOWARD,PH (1993)

CAS #: 033332-28-4				2-CHLORO-6-AMINO-PYRAZINE
Formula: $C_4H_4ClN_3$				
Mol Weight: 129.55				
MP (deg C):		FP (deg C):		
BP (deg C):				
BP pressure (mm Hg):				

Property/ Value	Units	Temp	Data Type	Reference
WS 1.43E+004	mg/L	25	EST	MEYLAN,WM ET AL. (1996)
logP 0.95			EXP	YAMAGAMI,C & TAKAO,N (1991)
VP 3.15E-002	mm Hg	25	EST	NEELY,WB & BLAU,GE (1985)
DC	pKa			
HL 4.82E-008	atm m3/mol	25	EST	MEYLAN,WM & HOWARD,PH (1991)
OH 8.93E-012	cm3/molc sec	25	EST	MEYLAN,WM & HOWARD,PH (1993)

CAS #: 033332-29-5				PYRAZINAMINE, 5-CHLORO-
Formula: $C_4H_4ClN_3$				
Mol Weight: 129.55				
MP (deg C):		FP (deg C):		
BP (deg C):				
BP pressure (mm Hg):				

Property/ Value	Units	Temp	Data Type	Reference
WS 2.47E+004	mg/L	25	EST	MEYLAN,WM ET AL. (1996)
logP 0.67			EXP	YAMAGAMI,C ET AL. (1990)
VP 3.15E-002	mm Hg	25	EST	NEELY,WB & BLAU,GE (1985)
DC	pKa			
HL 4.82E-008	atm m3/mol	25	EST	MEYLAN,WM & HOWARD,PH (1991)
OH 3.71E-012	cm3/molc sec	25	EST	MEYLAN,WM & HOWARD,PH (1993)

CAS #: 033332-30-8				2-CHLORO-6-METHOXY-PYRAZINE
Formula: $C_5H_5ClN_2O$				
Mol Weight: 144.56				
MP (deg C):		FP (deg C):		
BP (deg C):				
BP pressure (mm Hg):				

Property/ Value	Units	Temp	Data Type	Reference
WS 3.12E+003	mg/L	25	EST	MEYLAN,WM ET AL. (1996)
logP 1.65			EXP	YAMAGAMI,C & TAKAO,N (1991)
VP 2.62E-001	mm Hg	25	EST	NEELY,WB & BLAU,GE (1985)
DC	pKa			
HL 7.36E-005	atm m3/mol	25	EST	MEYLAN,WM & HOWARD,PH (1991)
OH 2.61E-012	cm3/molc sec	25	EST	MEYLAN,WM & HOWARD,PH (1993)

1081

033332-31-9 — PYRAZINE, 2-CHLORO-5-METHOXY-

Formula: $C_5H_5ClN_2O$

Mol Weight: 144.56

MP (deg C): — FP (deg C): —
BP (deg C): —
BP pressure (mm Hg): —

Property/Value	Units	Temp	Data Type	Reference
WS 4.03E+003	mg/L	25	EST	MEYLAN,WM ET AL. (1996)
logP 1.52			EXP	YAMAGAMI,C ET AL. (1991)
VP 2.62E-001	mm Hg	25	EST	NEELY,WB & BLAU,GE (1985)
DC	pKa			
HL 7.36E-005	atm m3/mol	25	EST	MEYLAN,WM & HOWARD,PH (1991)
OH 1.57E-012	cm3/molc sec	25	EST	MEYLAN,WM & HOWARD,PH (1993)

033332-53-5 — 1,2-BIS(2-MEO-ETO)BENZENE

Formula: $C_{12}H_{18}O_4$

Mol Weight: 226.27

MP (deg C): — FP (deg C): —
BP (deg C): —
BP pressure (mm Hg): —

Property/Value	Units	Temp	Data Type	Reference
WS 3.52E+003	mg/L	25	EST	MEYLAN,WM ET AL. (1996)
logP 1.12			EXP	HANSCH,C & LEO,AJ (1985)
VP 7.82E-004	mm Hg	25	EST	NEELY,WB & BLAU,GE (1985)
DC	pKa			
HL 4.55E-009	atm m3/mol	25	EST	MEYLAN,WM & HOWARD,PH (1991)
OH 4.83E-011	cm3/molc sec	25	EST	MEYLAN,WM & HOWARD,PH (1993)

033368-89-7 — 6-METHYLQUINOXALINE-1,4-DIOXIDE

Formula: $C_9H_{10}N_2O_2$

Mol Weight: 178.19

MP (deg C): — FP (deg C): —
BP (deg C): —
BP pressure (mm Hg): —

Property/Value	Units	Temp	Data Type	Reference
WS 1.14E+005	mg/L	25	EST	MEYLAN,WM ET AL. (1996)
logP -0.35			EXP	HANSCH,C & LEO,AJ (1985)
VP 8.75E-007	mm Hg	25	EST	NEELY,WB & BLAU,GE (1985)
DC	pKa			
HL 3.14E-017	atm m3/mol	25	EST	MEYLAN,WM & HOWARD,PH (1991)
OH 5.36E-012	cm3/molc sec	25	EST	MEYLAN,WM & HOWARD,PH (1993)

033376-25-9 — BARBITURIC ACID, 5-ET,5-PR

Formula: $C_9H_{14}N_2O_3$

Mol Weight: 198.22

MP (deg C): — FP (deg C): —
BP (deg C): —
BP pressure (mm Hg): —

Property/Value	Units	Temp	Data Type	Reference
WS 1.22E+004	mg/L	25	EST	MEYLAN,WM ET AL. (1996)
logP 0.66			EXP	HANSCH,C & LEO,AJ (1985)
VP 6.94E-011	mm Hg	25	EST	NEELY,WB & BLAU,GE (1985)
DC	pKa			
HL 4.79E-013	atm m3/mol	25	EST	MEYLAN,WM & HOWARD,PH (1991)
OH 1.16E-011	cm3/molc sec	25	EST	MEYLAN,WM & HOWARD,PH (1993)

033419-42-0 — ETOPSIDE

Formula: $C_{29}H_{32}O_{13}$

Mol Weight: 588.57

MP (deg C): 236-251 FP (deg C): —
BP (deg C): —
BP pressure (mm Hg): —

Property/Value	Units	Temp	Data Type	Reference
WS 5.87E+001	mg/L	25	EST	MEYLAN,WM ET AL. (1996)
logP 0.60			EXP	HANSCH,C ET AL. (1995)
VP 5.35E-023	mm Hg	25	EST	NEELY,WB & BLAU,GE (1985)
DC	pKa			
HL 1.75E-030	atm m3/mol	25	EST	MEYLAN,WM & HOWARD,PH (1991)
OH 2.71E-010	cm3/molc sec	25	EST	MEYLAN,WM & HOWARD,PH (1993)

033423-92-6 — 1,3,6,8-TETRACHLORODIBENZO-P-DIOXIN

Formula: $C_{12}H_4Cl_4O_2$

Mol Weight: 321.98

MP (deg C): — FP (deg C): —
BP (deg C): —
BP pressure (mm Hg): —

Property/Value	Units	Temp	Data Type	Reference
WS 3.20E-004	mg/L	20	EXP	YALKOWSKY,SH & DANNENFELSER,RM (1992)
logP 7.10			EXP	SHIU,WY ET AL. (1988)
VP 4.35E-007	mm Hg	25	EXP	SERVOS,MR ET AL. (1992)
DC	pKa			
HL 6.81E-004	atm m3/mol	20	EXP	WEBSTER,GRB ET AL (1985)
OH 6.97E-012	cm3/molc sec	25	EST	MEYLAN,WM & HOWARD,PH (1993)

033443-53-7 — 4-PHENYL-6-BROMOQUINAZOLIN-2-ONE

Formula: $C_{14}H_9BrN_2O$

Mol Weight: 301.15

MP (deg C): — FP (deg C): —
BP (deg C): —
BP pressure (mm Hg): —

Property/Value	Units	Temp	Data Type	Reference
WS 5.86E+001	mg/L	25	EST	MEYLAN,WM ET AL. (1996)
logP 2.71			EXP	HANSCH,C & LEO,AJ (1985)
VP 2.88E-008	mm Hg	25	EST	NEELY,WB & BLAU,GE (1985)
DC	pKa			
HL 1.52E-009	atm m3/mol	25	EST	MEYLAN,WM & HOWARD,PH (1991)
OH 1.90E-011	cm3/molc sec	25	EST	MEYLAN,WM & HOWARD,PH (1993)

033455-24-2 — 1,1,4,4-TETRACHLOROBUTANE

Formula: $C_4H_6Cl_4$

Mol Weight: 195.90

MP (deg C): — FP (deg C): —
BP (deg C): —
BP pressure (mm Hg): —

Property/Value	Units	Temp	Data Type	Reference
WS 8.92E+001	mg/L	25	EST	MEYLAN,WM ET AL. (1996)
logP 3.17			EST	MEYLAN,WM & HOWARD,PH (1995)
VP 5.31E-001	mm Hg	25	EST	NEELY,WB & BLAU,GE (1985)
DC	pKa			
HL 2.65E-003	atm m3/mol	25	EST	MEYLAN,WM & HOWARD,PH (1991)
OH 1.52E-012	cm3/molc sec	25	EST	MEYLAN,WM & HOWARD,PH (1993)

CAS #: 033484-38-7	4H-PYRIDO[1,2-A]PYRIMIDINE-3-CARBOXYLIC ACID, 6,

Formula: $C_{11}H_{14}N_2O_3$

Mol Weight: 222.25

MP (deg C): FP (deg C):

BP (deg C):

BP pressure (mm Hg):

Property/Value	Units	Temp	Data Type	Reference
WS 1.20E+003	mg/L	25	EST	MEYLAN,WM ET AL. (1996)
logP 0.16			EXP	SANGSTER,J (1994)
VP 7.00E-006	mm Hg	25	EST	NEELY,WB & BLAU,GE (1985)
DC	pKa			
HL 2.67E-012	atm m3/mol	25	EST	MEYLAN,WM & HOWARD,PH (1991)
OH 2.68E-011	cm3/molc sec	25	EST	MEYLAN,WM & HOWARD,PH (1993)

CAS #: 033484-45-6	4H-PYRIDO[1,2-A]PYRIMIDINE-3-CARBOXAMIDE, 6,7,8,

Formula: $C_{10}H_{13}N_3O_2$

Mol Weight: 207.23

MP (deg C): FP (deg C):

BP (deg C):

BP pressure (mm Hg):

Property/Value	Units	Temp	Data Type	Reference
WS 1.82E+003	mg/L	25	EST	MEYLAN,WM ET AL. (1996)
logP 0.04			EXP	SANGSTER,J (1994)
VP 6.78E-008	mm Hg	25	EST	NEELY,WB & BLAU,GE (1985)
DC	pKa			
HL 2.70E-013	atm m3/mol	25	EST	MEYLAN,WM & HOWARD,PH (1991)
OH 2.86E-011	cm3/molc sec	25	EST	MEYLAN,WM & HOWARD,PH (1993)

CAS #: 033489-27-9	3-I-PENTACHLOROCYCLOHEXANE

Formula: $C_6H_6Cl_5I$

Mol Weight: 382.28

MP (deg C): FP (deg C):

BP (deg C):

BP pressure (mm Hg):

Property/Value	Units	Temp	Data Type	Reference
WS 1.64E+000	mg/L	25	EST	MEYLAN,WM ET AL. (1996)
logP 4.05			EXP	HANSCH,C & LEO,AJ (1985)
VP 7.25E-005	mm Hg	25	EST	NEELY,WB & BLAU,GE (1985)
DC	pKa			
HL 5.42E-005	atm m3/mol	25	EST	MEYLAN,WM & HOWARD,PH (1991)
OH 7.01E-013	cm3/molc sec	25	EST	MEYLAN,WM & HOWARD,PH (1993)

CAS #: 033489-28-0	1-I-PENTACHLOROCYCLOHEXANE

Formula: $C_6H_6Cl_5I$

Mol Weight: 382.28

MP (deg C): FP (deg C):

BP (deg C):

BP pressure (mm Hg):

Property/Value	Units	Temp	Data Type	Reference
WS 1.64E+000	mg/L	25	EST	MEYLAN,WM ET AL. (1996)
logP 3.96			EXP	HANSCH,C & LEO,AJ (1985)
VP 7.25E-005	mm Hg	25	EST	NEELY,WB & BLAU,GE (1985)
DC	pKa			
HL 5.42E-005	atm m3/mol	25	EST	MEYLAN,WM & HOWARD,PH (1991)
OH 7.01E-013	cm3/molc sec	25	EST	MEYLAN,WM & HOWARD,PH (1993)

CAS #: 033533-53-8	MONO(ALPHA-METHYLBENZYL) PHTHALATE

Formula: $C_{16}H_{14}O_4$

Mol Weight: 270.29

MP (deg C): FP (deg C):

BP (deg C):

BP pressure (mm Hg):

Property/Value	Units	Temp	Data Type	Reference
WS 1.90E+001	mg/L	25	EST	MEYLAN,WM ET AL. (1996)
logP 3.49			EST	MEYLAN,WM & HOWARD,PH (1995)
VP 3.05E-007	mm Hg	25	EST	NEELY,WB & BLAU,GE (1985)
DC	pKa			
HL 7.49E-011	atm m3/mol	25	EST	MEYLAN,WM & HOWARD,PH (1991)
OH 8.90E-012	cm3/molc sec	25	EST	MEYLAN,WM & HOWARD,PH (1993)

CAS #: 033533-56-1	MONO(P-(TERT-BUTYL)ALPHA-METHYLBENZYL) PHTHALATE

Formula: $C_{20}H_{22}O_4$

Mol Weight: 326.40

MP (deg C): FP (deg C):

BP (deg C):

BP pressure (mm Hg):

Property/Value	Units	Temp	Data Type	Reference
WS 2.09E-001	mg/L	25	EST	MEYLAN,WM ET AL. (1996)
logP 5.40			EST	MEYLAN,WM & HOWARD,PH (1995)
VP 2.03E-008	mm Hg	25	EST	NEELY,WB & BLAU,GE (1985)
DC	pKa			
HL 1.93E-010	atm m3/mol	25	EST	MEYLAN,WM & HOWARD,PH (1991)
OH 1.03E-011	cm3/molc sec	25	EST	MEYLAN,WM & HOWARD,PH (1993)

CAS #: 033533-57-2	MONO(P-METHOXY-ALPHA-METHYLBENZYL) PHTHALATE

Formula: $C_{17}H_{16}O_5$

Mol Weight: 300.31

MP (deg C): FP (deg C):

BP (deg C):

BP pressure (mm Hg):

Property/Value	Units	Temp	Data Type	Reference
WS 1.09E+001	mg/L	25	EST	MEYLAN,WM ET AL. (1996)
logP 3.57			EST	MEYLAN,WM & HOWARD,PH (1995)
VP 4.74E-008	mm Hg	25	EST	NEELY,WB & BLAU,GE (1985)
DC	pKa			
HL 4.43E-012	atm m3/mol	25	EST	MEYLAN,WM & HOWARD,PH (1991)
OH 3.10E-011	cm3/molc sec	25	EST	MEYLAN,WM & HOWARD,PH (1993)

CAS #: 033576-92-0	O,O-DIMETHYL-O-PHENYLPHOSPHOROTHIOATE

Formula: $C_8H_{11}O_3PS$

Mol Weight: 218.21

MP (deg C): FP (deg C):

BP (deg C):

BP pressure (mm Hg):

Property/Value	Units	Temp	Data Type	Reference
WS 9.09E+001	mg/L	25	EST	MEYLAN,WM ET AL. (1996)
logP 3.03			EXP	DEBRUIJN,J & HERMENS,J (1991)
VP 4.67E-003	mm Hg	25	EST	NEELY,WB & BLAU,GE (1985)
DC	pKa			
HL 4.26E-005	atm m3/mol	25	EST	MEYLAN,WM & HOWARD,PH (1991)
OH 6.22E-011	cm3/molc sec	25	EST	MEYLAN,WM & HOWARD,PH (1993)

1083

CAS #: 033597-78-3	BENZENAMINE, 4-[(2,4,6-TRIMETHYLPHENYL)SULFONYL]

Formula: $C_{15}H_{17}NO_2S$

Mol Weight: 275.37

MP (deg C):   FP (deg C):

BP (deg C):

BP pressure (mm Hg):

Property/Value	Units	Temp	Data Type	Reference
WS 4.15E+001	mg/L	25	EST	MEYLAN,WM ET AL. (1996)
logP 3.06			EXP	ALTOMARE,C ET AL. (1991)
VP 4.28E-008	mm Hg	25	EST	NEELY,WB & BLAU,GE (1985)
DC	pKa			
HL 1.18E-010	atm m3/mol	25	EST	MEYLAN,WM & HOWARD,PH (1991)
OH 3.08E-011	cm3/molc sec	25	EST	MEYLAN,WM & HOWARD,PH (1993)

CAS #: 033602-03-8	2-PYRROLIDINONE, 1-(1-OXODECYL)-

Formula: $C_{14}H_{25}NO_2$

Mol Weight: 239.36

MP (deg C):   FP (deg C):

BP (deg C):

BP pressure (mm Hg):

Property/Value	Units	Temp	Data Type	Reference
WS 5.43E+000	mg/L	25	EST	MEYLAN,WM ET AL. (1996)
logP 4.33			EXP	HANSCH,C ET AL. (1995)
VP 4.07E-007	mm Hg	25	EST	NEELY,WB & BLAU,GE (1985)
DC	pKa			
HL 7.12E-007	atm m3/mol	25	EST	MEYLAN,WM & HOWARD,PH (1991)
OH 3.68E-011	cm3/molc sec	25	EST	MEYLAN,WM & HOWARD,PH (1993)

CAS #: 033607-91-9	6-CL-4-KETOCHROMAN-2-CARBOXLIC ACID

Formula: $C_{10}H_7ClO_4$

Mol Weight: 226.62

MP (deg C):   FP (deg C):

BP (deg C):

BP pressure (mm Hg):

Property/Value	Units	Temp	Data Type	Reference
WS 2.43E+003	mg/L	25	EST	MEYLAN,WM ET AL. (1996)
logP 1.77			EXP	HANSCH,C & LEO,AJ (1985)
VP 2.63E-006	mm Hg	25	EST	NEELY,WB & BLAU,GE (1985)
DC	pKa			
HL 9.98E-012	atm m3/mol	25	EST	MEYLAN,WM & HOWARD,PH (1991)
OH 4.17E-011	cm3/molc sec	25	EST	MEYLAN,WM & HOWARD,PH (1993)

CAS #: 033629-47-9	BUTRALIN

Formula: $C_{14}H_{21}N_3O_4$

Mol Weight: 295.34

MP (deg C): 60-61   FP (deg C):

BP (deg C):

BP pressure (mm Hg):

Property/Value	Units	Temp	Data Type	Reference
WS 1.00E+000	mg/L	25	EXP	SHIU,WY ET AL. (1990)
logP 5.15			EST	MEYLAN,WM & HOWARD,PH (1995)
VP 1.27E-005	mm Hg	25	EXP	WEBER,JB (1990)
DC	pKa			
HL 4.94E-006	atm m3/mol	25	EST	VP/WSOL
OH 2.42E-011	cm3/molc sec	25	EST	MEYLAN,WM & HOWARD,PH (1993)

CAS #: 033665-20-2	1,2,4-TRIAZIN-5(4H)-ONE, 4-AMINO-6-(1,1-DIMETHYL

Formula: $C_8H_{14}N_4O$

Mol Weight: 182.23

MP (deg C):   FP (deg C):

BP (deg C):

BP pressure (mm Hg):

Property/Value	Units	Temp	Data Type	Reference
WS 7.09E+003	mg/L	25	EST	MEYLAN,WM ET AL. (1996)
logP 1.03			EXP	SANGSTER,J (1994)
VP 2.86E-005	mm Hg	25	EST	NEELY,WB & BLAU,GE (1985)
DC	pKa			
HL 2.25E-010	atm m3/mol	25	EST	MEYLAN,WM & HOWARD,PH (1991)
OH 1.57E-011	cm3/molc sec	25	EST	MEYLAN,WM & HOWARD,PH (1993)

CAS #: 033665-71-3	3-N-BUAM-4NH2-6IPR-124-TRIAZIN-5-ONE

Formula: $C_{10}H_{19}N_5O$

Mol Weight: 225.30

MP (deg C):   FP (deg C):

BP (deg C):

BP pressure (mm Hg):

Property/Value	Units	Temp	Data Type	Reference
WS 9.74E+002	mg/L	25	EST	MEYLAN,WM ET AL. (1996)
logP 1.78			EXP	HANSCH,C & LEO,AJ (1985)
VP 9.31E-007	mm Hg	25	EST	NEELY,WB & BLAU,GE (1985)
DC	pKa			
HL 6.74E-014	atm m3/mol	25	EST	MEYLAN,WM & HOWARD,PH (1991)
OH 9.93E-011	cm3/molc sec	25	EST	MEYLAN,WM & HOWARD,PH (1993)

CAS #: 033671-46-4	2H-THIENO[2,3-E]-1,4-DIAZEPIN-2-ONE, 5-(2-CHLORO

Formula: $C_{16}H_{15}ClN_2OS$

Mol Weight: 318.83

MP (deg C): 118-120   FP (deg C):

BP (deg C):

BP pressure (mm Hg):

Property/Value	Units	Temp	Data Type	Reference
WS 1.83E+001	mg/L	25	EST	MEYLAN,WM ET AL. (1996)
logP 3.18			EXP	MARUYAMA,T ET AL. (1992)
VP 3.71E-009	mm Hg	25	EST	NEELY,WB & BLAU,GE (1985)
DC	pKa			
HL 2.90E-009	atm m3/mol	25	EST	MEYLAN,WM & HOWARD,PH (1991)
OH 3.13E-011	cm3/molc sec	25	EST	MEYLAN,WM & HOWARD,PH (1993)

CAS #: 033683-34-0	1-PIPERIDINYLOXY, 4-[[BIS(1-AZIRIDINYL)PHOSPHINO

Formula: $C_{13}H_{27}N_4OPS$

Mol Weight: 318.42

MP (deg C):   FP (deg C):

BP (deg C):

BP pressure (mm Hg):

Property/Value	Units	Temp	Data Type	Reference
WS 4.43E+002	mg/L	25	EST	MEYLAN,WM ET AL. (1996)
logP 1.57			EXP	SOSNOVSKY,G ET AL. (1986)
VP 2.06E-008	mm Hg	25	EST	NEELY,WB & BLAU,GE (1985)
DC	pKa			
HL 3.26E-019	atm m3/mol	25	EST	MEYLAN,WM & HOWARD,PH (1991)
OH 2.49E-010	cm3/molc sec	25	EST	MEYLAN,WM & HOWARD,PH (1993)

CAS #: 033693-04-8			TERBUMETON

Formula: $C_{10}H_{19}N_5O$

Mol Weight: 225.30

MP (deg C): | **FP (deg C):**

BP (deg C):

BP pressure (mm Hg):

Property/Value	Units	Temp	Data Type	Reference
WS 1.30E+002	mg/L	20	EXP	YALKOWSKY,SH & DANNENFELSER,RM (1992)
logP 3.10			EXP	FINIZIO,A ET AL. (1991)
VP 2.00E-006	mm Hg	25	EXP	KEARNEY,PC & KAUFMAN,DD (1975)
DC	pKa			
HL 4.56E-009	atm m3/mol	25	EST	VP/WSOL
OH 1.05E-011	cm3/molc sec	25	EST	MEYLAN,WM & HOWARD,PH (1993)

CAS #: 033715-62-7			3,4,5-TRICHLOROACETANILIDE

Formula: $C_8H_6Cl_3NO$

Mol Weight: 238.50

MP (deg C): | **FP (deg C):**

BP (deg C):

BP pressure (mm Hg):

Property/Value	Units	Temp	Data Type	Reference
WS 1.68E+001	mg/L	25	EST	MEYLAN,WM ET AL. (1996)
logP 3.76			EXP	NAKAGAWA,Y ET AL. (1992)
VP 5.63E-006	mm Hg	25	EST	NEELY,WB & BLAU,GE (1985)
DC	pKa			
HL 2.51E-009	atm m3/mol	25	EST	MEYLAN,WM & HOWARD,PH (1991)
OH 1.90E-012	cm3/molc sec	25	EST	MEYLAN,WM & HOWARD,PH (1993)

CAS #: 033719-74-3			3,5-DICHLOROANISOLE

Formula: $C_7H_6Cl_2O$

Mol Weight: 177.03

MP (deg C): 40-42 | **FP (deg C):**

BP (deg C):

BP pressure (mm Hg):

Property/Value	Units	Temp	Data Type	Reference
WS 3.24E+001	mg/L	25	EST	MEYLAN,WM ET AL. (1996)
logP 3.80			EXP	HANSCH,C & LEO,AJ (1985)
VP 1.64E-001	mm Hg	25	EST	NEELY,WB & BLAU,GE (1985)
DC	pKa			
HL 1.75E-004	atm m3/mol	25	EST	MEYLAN,WM & HOWARD,PH (1991)
OH 1.15E-011	cm3/molc sec	25	EST	MEYLAN,WM & HOWARD,PH (1993)

CAS #: 033721-54-9			ACETAMIDE, N-(2-METHOXY-5-NITROPHENYL)-

Formula: $C_9H_{10}N_2O_4$

Mol Weight: 210.19

MP (deg C): | **FP (deg C):**

BP (deg C):

BP pressure (mm Hg):

Property/Value	Units	Temp	Data Type	Reference
WS 6.79E+002	mg/L	25	EST	MEYLAN,WM ET AL. (1996)
logP 1.60			EXP	FURST,W & BECHER,M (1990)
VP 8.04E-007	mm Hg	25	EST	NEELY,WB & BLAU,GE (1985)
DC	pKa			
HL 1.44E-012	atm m3/mol	25	EST	MEYLAN,WM & HOWARD,PH (1991)
OH 2.34E-012	cm3/molc sec	25	EST	MEYLAN,WM & HOWARD,PH (1993)

CAS #: 033757-42-5			8-ACETYLAMINOQUINOLINE

Formula: $C_{11}H_{10}N_2O$

Mol Weight: 186.22

MP (deg C): | **FP (deg C):**

BP (deg C):

BP pressure (mm Hg):

Property/Value	Units	Temp	Data Type	Reference
WS 1.27E+004	mg/L	25	EST	MEYLAN,WM ET AL. (1996)
logP 1.91			EXP	HANSCH,C ET AL. (1995)
VP 1.28E-006	mm Hg	25	EST	NEELY,WB & BLAU,GE (1985)
DC	pKa			
HL 7.88E-013	atm m3/mol	25	EST	MEYLAN,WM & HOWARD,PH (1991)
OH 9.18E-011	cm3/molc sec	25	EST	MEYLAN,WM & HOWARD,PH (1993)

CAS #: 033784-05-3			SPIRONOLACTONE,7-BETA ANALOG

Formula: $C_{24}H_{32}O_4S$

Mol Weight: 416.58

MP (deg C): | **FP (deg C):**

BP (deg C):

BP pressure (mm Hg):

Property/Value	Units	Temp	Data Type	Reference
WS 2.84E+001	mg/L	25	EST	MEYLAN,WM ET AL. (1996)
logP 2.26			EXP	HANSCH,C & LEO,AJ (1985)
VP 3.24E-011	mm Hg	25	EST	NEELY,WB & BLAU,GE (1985)
DC	pKa			
HL 1.14E-010	atm m3/mol	25	EST	MEYLAN,WM & HOWARD,PH (1991)
OH 1.03E-010	cm3/molc sec	25	EST	MEYLAN,WM & HOWARD,PH (1993)

CAS #: 033802-91-4			N-(4-SO2F-BENZYL)-PYRIDINIUM BROMIDE

Formula: $C_{12}H_{11}BrFNO_2S$

Mol Weight: 332.19

MP (deg C): | **FP (deg C):**

BP (deg C):

BP pressure (mm Hg):

Property/Value	Units	Temp	Data Type	Reference
WS 1.35E+005	mg/L	25	EST	MEYLAN,WM ET AL. (1996)
logP -1.44			EXP	HANSCH,C & LEO,AJ (1985)
VP 4.98E-007	mm Hg	25	EST	NEELY,WB & BLAU,GE (1985)
DC	pKa			
HL 3.51E-014	atm m3/mol	25	EST	MEYLAN,WM & HOWARD,PH (1991)
OH 1.23E-011	cm3/molc sec	25	EST	MEYLAN,WM & HOWARD,PH (1993)

CAS #: 033803-13-3			4-ME-N-(4-SO2F-BENZYL)PYRIDINIUM BR

Formula: $C_{13}H_{13}BrFNO_2S$

Mol Weight: 346.22

MP (deg C): | **FP (deg C):**

BP (deg C):

BP pressure (mm Hg):

Property/Value	Units	Temp	Data Type	Reference
WS 5.57E+004	mg/L	25	EST	MEYLAN,WM ET AL. (1996)
logP -1.09			EXP	HANSCH,C & LEO,AJ (1985)
VP 2.00E-007	mm Hg	25	EST	NEELY,WB & BLAU,GE (1985)
DC	pKa			
HL 3.87E-014	atm m3/mol	25	EST	MEYLAN,WM & HOWARD,PH (1991)
OH 1.31E-011	cm3/molc sec	25	EST	MEYLAN,WM & HOWARD,PH (1993)

CAS #: 033809-77-7 — BENZAMIDE, O-ACETYLAMINO

Formula: $C_9H_{10}N_2O_2$

Mol Weight: 178.19

MP (deg C): FP (deg C):

BP (deg C):

BP pressure (mm Hg):

Property/Value	Units	Temp	Data Type	Reference
WS 1.47E+003	mg/L	25	EST	MEYLAN,WM ET AL. (1996)
logP 0.32			EXP	SANGSTER,J (1993)
VP 8.95E-008	mm Hg	25	EST	NEELY,WB & BLAU,GE (1985)
DC	pKa			
HL 4.02E-012	atm m3/mol	25	EST	MEYLAN,WM & HOWARD,PH (1991)
OH 1.34E-011	cm3/molc sec	25	EST	MEYLAN,WM & HOWARD,PH (1993)

CAS #: 033820-53-0 — ISOPROPALIN

Formula: $C_{15}H_{23}N_3O_4$

Mol Weight: 309.37

MP (deg C): FP (deg C):

BP (deg C):

BP pressure (mm Hg):

Property/Value	Units	Temp	Data Type	Reference
WS 1.10E-001	mg/L	25	EXP	PROBST,GW ET AL. (1975)
logP 5.80			EST	MEYLAN,WM & HOWARD,PH (1995)
VP 3.00E-005	mm Hg	25	EXP	WSSA (1983)
DC	pKa			
HL 1.11E-004	atm m3/mol	25	EST	VP/WSOL
OH 8.64E-011	cm3/molc sec	25	EST	MEYLAN,WM & HOWARD,PH (1993)

CAS #: 033854-15-8 — 2-BUTENOIC ACID, 2-METHYL-, 9-(ACETYLOXY)- ESTER

Formula: $C_{22}H_{28}O_7$

Mol Weight: 404.46

MP (deg C): FP (deg C):

BP (deg C):

BP pressure (mm Hg):

Property/Value	Units	Temp	Data Type	Reference
WS 9.50E+002	mg/L	25	EST	MEYLAN,WM ET AL. (1996)
logP 1.16			EXP	SANGSTER,J (1993)
VP 7.47E-012	mm Hg	25	EST	NEELY,WB & BLAU,GE (1985)
DC	pKa			
HL 1.39E-013	atm m3/mol	25	EST	MEYLAN,WM & HOWARD,PH (1991)
OH 2.42E-010	cm3/molc sec	25	EST	MEYLAN,WM & HOWARD,PH (1993)

CAS #: 033857-26-0 — 2,7-DICHLORODIBENZO-P-DIOXIN

Formula: $C_{12}H_6Cl_2O_2$

Mol Weight: 253.09

MP (deg C): FP (deg C):

BP (deg C):

BP pressure (mm Hg):

Property/Value	Units	Temp	Data Type	Reference
WS 3.75E-003	mg/L	25	EXP	YALKOWSKY,SH & DANNENFELSER,RM (1992)
logP 5.75			EXP	SHIU,WY ET AL. (1988)
VP 9.00E-007	mm Hg	25	EXP	SHIU,WY ET AL. (1988)
DC	pKa			
HL 7.99E-005	atm m3/mol	25	EST	VP/WSOL
OH 6.92E-012	cm3/molc sec	25	EST	MEYLAN,WM & HOWARD,PH (1993)

CAS #: 033857-28-2 — 2,3,7-TRICHLORODIBENZO-P-DIOXIN

Formula: $C_{12}H_5Cl_3O_2$

Mol Weight: 287.53

MP (deg C): FP (deg C):

BP (deg C):

BP pressure (mm Hg):

Property/Value	Units	Temp	Data Type	Reference
WS 4.93E-003	mg/L	25	EST	MEYLAN,WM ET AL. (1996)
logP 6.28			EST	MEYLAN,WM & HOWARD,PH (1995)
VP 6.36E-006	mm Hg	25	EST	NEELY,WB & BLAU,GE (1985)
DC	pKa			
HL 4.76E-006	atm m3/mol	25	EST	MEYLAN,WM & HOWARD,PH (1991)
OH 4.47E-012	cm3/molc sec	25	EST	MEYLAN,WM & HOWARD,PH (1993)

CAS #: 033862-44-1 — PHOSPHOROHYDRAZIDIC ACID, DIPHENYL ESTER

Formula: $C_{12}H_{13}N_2O_3P$

Mol Weight: 264.22

MP (deg C): FP (deg C):

BP (deg C):

BP pressure (mm Hg):

Property/Value	Units	Temp	Data Type	Reference
WS 1.01E+003	mg/L	25	EST	MEYLAN,WM ET AL. (1996)
logP 1.51			EXP	HANSCH,C ET AL. (1995)
VP 2.49E-006	mm Hg	25	EST	NEELY,WB & BLAU,GE (1985)
DC	pKa			
HL 9.03E-013	atm m3/mol	25	EST	MEYLAN,WM & HOWARD,PH (1991)
OH 3.47E-011	cm3/molc sec	25	EST	MEYLAN,WM & HOWARD,PH (1993)

CAS #: 033868-03-0 — 1,7-DIMETHYL URIC ACID

Formula: $C_7H_8N_4O_3$

Mol Weight: 196.17

MP (deg C): FP (deg C):

BP (deg C):

BP pressure (mm Hg):

Property/Value	Units	Temp	Data Type	Reference
WS 3.46E+003	mg/L	25	EST	MEYLAN,WM ET AL. (1996)
logP -0.22			EXP	GASPARI,F & BONATI,M (1987)
VP 1.66E-009	mm Hg	25	EST	NEELY,WB & BLAU,GE (1985)
DC	pKa			
HL 5.77E-017	atm m3/mol	25	EST	MEYLAN,WM & HOWARD,PH (1991)
OH 1.34E-011	cm3/molc sec	25	EST	MEYLAN,WM & HOWARD,PH (1993)

CAS #: 033870-85-8 — PYRAZINE, 2,6-DIETHOXY-

Formula: $C_8H_{12}N_2O_2$

Mol Weight: 168.20

MP (deg C): FP (deg C):

BP (deg C):

BP pressure (mm Hg):

Property/Value	Units	Temp	Data Type	Reference
WS 4.17E+002	mg/L	25	EST	MEYLAN,WM ET AL. (1996)
logP 2.55			EXP	YAMAGAMI,C ET AL. (1991)
VP 1.51E-002	mm Hg	25	EST	NEELY,WB & BLAU,GE (1985)
DC	pKa			
HL 1.50E-006	atm m3/mol	25	EST	MEYLAN,WM & HOWARD,PH (1991)
OH 4.03E-011	cm3/molc sec	25	EST	MEYLAN,WM & HOWARD,PH (1993)

033873-09-5 — 2,6-DIFLUOROPYRAZINE

Formula: $C_4H_2F_2N_2$

Mol Weight: 116.07

MP (deg C):　FP (deg C):

BP (deg C):

BP pressure (mm Hg):

Property/ Value	Units	Temp	Data Type	Reference
WS 2.42E+004	mg/L	25	EST	MEYLAN,WM ET AL. (1996)
logP 0.74			EXP	YAMAGAMI,C ET AL. (1990)
VP 2.08E+001	mm Hg	25	EST	NEELY,WB & BLAU,GE (1985)
DC	pKa			
HL 3.97E-006	atm m3/mol	25	EST	MEYLAN,WM & HOWARD,PH (1991)
OH 3.61E-013	cm3/molc sec	25	EST	MEYLAN,WM & HOWARD,PH (1993)

033873-10-8 — 2-CHLORO-6-FLUORO-PYRAZINE

Formula: $C_4H_2ClFN_2$

Mol Weight: 132.53

MP (deg C):　FP (deg C):

BP (deg C):

BP pressure (mm Hg):

Property/ Value	Units	Temp	Data Type	Reference
WS 9.18E+003	mg/L	25	EST	MEYLAN,WM ET AL. (1996)
logP 1.16			EXP	YAMAGAMI,C & TAKAO,N (1991)
VP 2.82E+000	mm Hg	25	EST	NEELY,WB & BLAU,GE (1985)
DC	pKa			
HL 1.59E-004	atm m3/mol	25	EST	MEYLAN,WM & HOWARD,PH (1991)
OH 2.03E-013	cm3/molc sec	25	EST	MEYLAN,WM & HOWARD,PH (1993)

033875-95-5 — TETRACHLOROCYCLOHEXENE

Formula: $C_6H_6Cl_4$

Mol Weight: 219.93

MP (deg C):　FP (deg C):

BP (deg C):

BP pressure (mm Hg):

Property/ Value	Units	Temp	Data Type	Reference
WS 4.30E+001	mg/L	25	EST	MEYLAN,WM ET AL. (1996)
logP 3.15			EXP	HANSCH,C & LEO,AJ (1985)
VP 3.13E-002	mm Hg	25	EST	NEELY,WB & BLAU,GE (1985)
DC	pKa			
HL 1.81E-003	atm m3/mol	25	EST	MEYLAN,WM & HOWARD,PH (1991)
OH 4.36E-011	cm3/molc sec	25	EST	MEYLAN,WM & HOWARD,PH (1993)

033979-03-2 — 2,2',4,4',6,6'-PCB

Formula: $C_{12}H_4Cl_6$

Mol Weight: 360.88

MP (deg C):　FP (deg C):

BP (deg C):

BP pressure (mm Hg):

Property/ Value	Units	Temp	Data Type	Reference
WS 2.27E-003	mg/L	25	EXP	YALKOWSKY,SH & DANNENFELSER,RM (1992)
logP 7.55			EXP	HANSCH,C & LEO,AJ (1985)
VP 1.20E-005	mm Hg	25	EXP	SHIU,WY & MACKAY,D (1986)
DC	pKa			
HL 2.51E-003	atm m3/mol	25	EST	VP/WSOL
OH 3.95E-013	cm3/molc sec	25	EST	MEYLAN,WM & HOWARD,PH (1993)

034010-15-6 — 11-TETRADECEN-1-OL (Z)

Formula: $C_{14}H_{28}O$

Mol Weight: 212.38

MP (deg C):　FP (deg C):

BP (deg C):

BP pressure (mm Hg):

Property/ Value	Units	Temp	Data Type	Reference
WS 2.28E+000	mg/L	25	EST	MEYLAN,WM ET AL. (1996)
logP 5.54			EST	MEYLAN,WM & HOWARD,PH (1995)
VP 5.98E-005	mm Hg	25	EST	NEELY,WB & BLAU,GE (1985)
DC	pKa			
HL 1.49E-004	atm m3/mol	25	EST	MEYLAN,WM & HOWARD,PH (1991)
OH 7.41E-011	cm3/molc sec	25	EST	MEYLAN,WM & HOWARD,PH (1993)

034014-18-1 — TEBUTHIURON

Formula: $C_9H_{16}N_4OS$

Mol Weight: 228.32

MP (deg C): 163 dec　FP (deg C):

BP (deg C):

BP pressure (mm Hg):

Property/ Value	Units	Temp	Data Type	Reference
WS 2.50E+003	mg/L	25	EXP	WORTHING,CR & WALKER,SB (1987)
logP 1.78			EST	MEYLAN,WM & HOWARD,PH (1995)
VP 2.00E-006	mm Hg	25	EXP	WORTHING,CR & WALKER,SB (1987)
DC 1.20	pKa			WEBER,JB & WHITACRE,DM (1982)
HL 2.40E-010	atm m3/mol	25	EST	VP/WSOL
OH 3.32E-012	cm3/molc sec	25	EST	MEYLAN,WM & HOWARD,PH (1993)

034017-16-8 — L-PHENYLALANINAMIDE, N-ACETYLGLYCYL-

Formula: $C_{13}H_{17}N_3O_3$

Mol Weight: 263.30

MP (deg C):　FP (deg C):

BP (deg C):

BP pressure (mm Hg):

Property/ Value	Units	Temp	Data Type	Reference
WS 2.94E+003	mg/L	25	EST	MEYLAN,WM ET AL. (1996)
logP -0.56			EXP	HANSCH,C ET AL. (1995)
VP 6.20E-012	mm Hg	25	EST	NEELY,WB & BLAU,GE (1985)
DC	pKa			
HL 8.41E-017	atm m3/mol	25	EST	MEYLAN,WM & HOWARD,PH (1991)
OH 3.18E-011	cm3/molc sec	25	EST	MEYLAN,WM & HOWARD,PH (1993)

034017-18-0 — L-VALINAMIDE, N-ACETYLGLYCYL-

Formula: $C_9H_{17}N_3O_3$

Mol Weight: 215.25

MP (deg C):　FP (deg C):

BP (deg C):

BP pressure (mm Hg):

Property/ Value	Units	Temp	Data Type	Reference
WS 2.44E+004	mg/L	25	EST	MEYLAN,WM ET AL. (1996)
logP -1.33			EXP	SANGSTER,J (1993)
VP 7.97E-010	mm Hg	25	EST	NEELY,WB & BLAU,GE (1985)
DC	pKa			
HL 1.84E-015	atm m3/mol	25	EST	MEYLAN,WM & HOWARD,PH (1991)
OH 3.12E-011	cm3/molc sec	25	EST	MEYLAN,WM & HOWARD,PH (1993)

CAS #: 034037-45-1	BENZOIC ACID, 4-[(4-AMINOPHENYL)SULFONYL]-

Formula: $C_{14}H_{13}NO_4S$

Mol Weight: 291.33

MP (deg C): FP (deg C):

BP (deg C):

BP pressure (mm Hg):

Property/Value	Units	Temp	Data Type	Reference
WS 2.09E+002	mg/L	25	EST	MEYLAN,WM ET AL. (1996)
logP 2.13			EXP	ALTOMARE,C ET AL. (1991)
VP 1.51E-008	mm Hg	25	EST	NEELY,WB & BLAU,GE (1985)
DC	pKa			
HL 5.68E-013	atm m3/mol	25	EST	MEYLAN,WM & HOWARD,PH (1991)
OH 2.34E-011	cm3/molc sec	25	EST	MEYLAN,WM & HOWARD,PH (1993)

CAS #: 034066-96-1	1,2-ETHANEDIAMINE, N-(2-AMINOETHYL)-N,N'-DIMETHY

Formula: $C_6H_{17}N_3$

Mol Weight: 131.22

MP (deg C): FP (deg C):

BP (deg C):

BP pressure (mm Hg):

Property/Value	Units	Temp	Data Type	Reference
WS 1.00E+006	mg/L	25	EST	MEYLAN,WM ET AL. (1996)
logP -0.92			EXP	HANSCH,C ET AL. (1995)
VP 3.96E-001	mm Hg	25	EST	NEELY,WB & BLAU,GE (1985)
DC	pKa			
HL 1.50E-012	atm m3/mol	25	EST	MEYLAN,WM & HOWARD,PH (1991)
OH 1.95E-010	cm3/molc sec	25	EST	MEYLAN,WM & HOWARD,PH (1993)

CAS #: 034084-50-9	14-BENZODIAZPN2-ON,1-ME-5(2-F PH)7NH2

Formula: $C_{16}H_{14}FN_3O$

Mol Weight: 283.31

MP (deg C): FP (deg C):

BP (deg C):

BP pressure (mm Hg):

Property/Value	Units	Temp	Data Type	Reference
WS 1.19E+003	mg/L	25	EST	MEYLAN,WM ET AL. (1996)
logP 1.30			EXP	HANSCH,C & LEO,AJ (1985)
VP 6.37E-009	mm Hg	25	EST	NEELY,WB & BLAU,GE (1985)
DC	pKa			
HL 2.03E-012	atm m3/mol	25	EST	MEYLAN,WM & HOWARD,PH (1991)
OH 1.28E-010	cm3/molc sec	25	EST	MEYLAN,WM & HOWARD,PH (1993)

CAS #: 034123-59-6	ISOPROTURON

Formula: $C_{12}H_{18}N_2O$

Mol Weight: 206.29

MP (deg C): 158 FP (deg C):

BP (deg C):

BP pressure (mm Hg):

Property/Value	Units	Temp	Data Type	Reference
WS 6.00E+001	mg/L	20	EXP	YALKOWSKY,SH & DANNENFELSER,RM (1992)
logP 2.87			EXP	HANSCH,C & LEO,AJ (1985)
VP 2.47E-008	mm Hg	25	EXP	TOMLIN,C (1994)
DC	pKa			
HL 1.12E-010	atm m3/mol	25	EST	VP/WSOL
OH 5.57E-011	cm3/molc sec	25	EST	MEYLAN,WM & HOWARD,PH (1993)

CAS #: 034175-79-6	2-BUTENOIC ACID, 2-[(ACETYLOXY)METHYL]- DERIVATI

Formula: $C_{22}H_{26}O_8$

Mol Weight: 418.45

MP (deg C): FP (deg C):

BP (deg C):

BP pressure (mm Hg):

Property/Value	Units	Temp	Data Type	Reference
WS 7.33E+002	mg/L	25	EST	MEYLAN,WM ET AL. (1996)
logP 1.19			EXP	SANGSTER,J (1993)
VP 2.16E-012	mm Hg	25	EST	NEELY,WB & BLAU,GE (1985)
DC	pKa			
HL 4.56E-016	atm m3/mol	25	EST	MEYLAN,WM & HOWARD,PH (1991)
OH 2.04E-010	cm3/molc sec	25	EST	MEYLAN,WM & HOWARD,PH (1993)

CAS #: 034190-59-5	ACETAMIDE, N-[[3-[2-HYDROXY-3-[(1-METHYLETHYL)AM

Formula: $C_{15}H_{24}N_2O_3$

Mol Weight: 280.37

MP (deg C): FP (deg C):

BP (deg C):

BP pressure (mm Hg):

Property/Value	Units	Temp	Data Type	Reference
WS 4.00E+002	mg/L	25	EST	MEYLAN,WM ET AL. (1996)
logP 0.34			EXP	SANGSTER,J (1993)
VP 1.57E-010	mm Hg	25	EST	NEELY,WB & BLAU,GE (1985)
DC	pKa			
HL 3.01E-018	atm m3/mol	25	EST	MEYLAN,WM & HOWARD,PH (1991)
OH 1.72E-010	cm3/molc sec	25	EST	MEYLAN,WM & HOWARD,PH (1993)

CAS #: 034197-05-2	METHYLETHOXYCHLOR

Formula: $C_{17}H_{17}Cl_3O$

Mol Weight: 343.68

MP (deg C): FP (deg C):

BP (deg C):

BP pressure (mm Hg):

Property/Value	Units	Temp	Data Type	Reference
WS 1.49E-002	mg/L	25	EST	MEYLAN,WM ET AL. (1996)
logP 6.62			EST	MEYLAN,WM & HOWARD,PH (1995)
VP 2.41E-006	mm Hg	25	EST	NEELY,WB & BLAU,GE (1985)
DC	pKa			
HL 2.41E-006	atm m3/mol	25	EST	MEYLAN,WM & HOWARD,PH (1991)
OH 3.84E-011	cm3/molc sec	25	EST	MEYLAN,WM & HOWARD,PH (1993)

CAS #: 034197-16-5	METHOXYMETHIOCHLOR

Formula: $C_{16}H_{15}Cl_3OS$

Mol Weight: 361.72

MP (deg C): FP (deg C):

BP (deg C):

BP pressure (mm Hg):

Property/Value	Units	Temp	Data Type	Reference
WS 2.74E-002	mg/L	25	EST	MEYLAN,WM ET AL. (1996)
logP 6.19			EST	MEYLAN,WM & HOWARD,PH (1995)
VP 5.80E-007	mm Hg	25	EST	NEELY,WB & BLAU,GE (1985)
DC	pKa			
HL 4.80E-008	atm m3/mol	25	EST	MEYLAN,WM & HOWARD,PH (1991)
OH 4.30E-011	cm3/molc sec	25	EST	MEYLAN,WM & HOWARD,PH (1993)

CAS #: 034197-26-7	1,1-BIS(P-MEO-PH)-2-NITROPROPANE

Formula: $C_{17}H_{19}NO_4$

Mol Weight: 301.35

MP (deg C):

FP (deg C):

BP (deg C):

BP pressure (mm Hg):

Property/Value	Units	Temp	Data Type	Reference
WS 2.43E+000	mg/L	25	EST	MEYLAN,WM ET AL. (1996)
logP 3.87			EXP	HANSCH,C & LEO,AJ (1985)
VP 6.29E-007	mm Hg	25	EST	NEELY,WB & BLAU,GE (1985)
DC	pKa			
HL 1.69E-009	atm m3/mol	25	EST	MEYLAN,WM & HOWARD,PH (1991)
OH 5.38E-011	cm3/molc sec	25	EST	MEYLAN,WM & HOWARD,PH (1993)

CAS #: 034202-69-2	HEXAFLUOROACETONE TRIHYDRATE

Formula: C_3F_6O

Mol Weight: 166.02

MP (deg C): 18-21

FP (deg C):

BP (deg C):

BP pressure (mm Hg):

Property/Value	Units	Temp	Data Type	Reference
WS 3.64E+003	mg/L	25	EST	MEYLAN,WM ET AL. (1996)
logP 0.60			EST	MEYLAN,WM & HOWARD,PH (1995)
VP 5.07E+003	mm Hg	25	EXP	YAWS,CL (1994)
DC	pKa			
HL 3.07E-003	atm m3/mol	25	EST	MEYLAN,WM & HOWARD,PH (1991)
OH 0.00E+000	cm3/molc sec	25	EST	MEYLAN,WM & HOWARD,PH (1993)

CAS #: 034205-21-5	DIMEFURON

Formula: $C_{15}H_{19}ClN_4O_3$

Mol Weight: 338.80

MP (deg C): 193

FP (deg C):

BP (deg C):

BP pressure (mm Hg):

Property/Value	Units	Temp	Data Type	Reference
WS 1.60E+001	mg/L	20	EXP	SHIU,WY ET AL. (1990)
logP 2.51			EXP	TOMLIN,C (1994)
VP 2.29E-010	mm Hg	25	EST	NEELY,WB & BLAU,GE (1985)
DC	pKa			
HL 7.51E-014	atm m3/mol	25	EST	MEYLAN,WM & HOWARD,PH (1991)
OH 1.83E-011	cm3/molc sec	25	EST	MEYLAN,WM & HOWARD,PH (1993)

CAS #: 034206-60-5	GRAYANOTOXANE-5,6,7,9,10,14,16-HEPTOL, 2,3-EPOXY

Formula: $C_{20}H_{32}O_8$

Mol Weight: 400.47

MP (deg C):

FP (deg C):

BP (deg C):

BP pressure (mm Hg):

Property/Value	Units	Temp	Data Type	Reference
WS 9.33E+003	mg/L	25	EST	MEYLAN,WM ET AL. (1996)
logP -0.57			EXP	HANSCH,C ET AL. (1995)
VP 1.60E-015	mm Hg	25	EST	NEELY,WB & BLAU,GE (1985)
DC	pKa			
HL 6.51E-017	atm m3/mol	25	EST	MEYLAN,WM & HOWARD,PH (1991)
OH 3.92E-011	cm3/molc sec	25	EST	MEYLAN,WM & HOWARD,PH (1993)

CAS #: 034213-80-4	3,5-DI-T-BU PHENYL GLUCOPYRANOSIDE

Formula: $C_{20}H_{32}O_6$

Mol Weight: 368.47

MP (deg C):

FP (deg C):

BP (deg C):

BP pressure (mm Hg):

Property/Value	Units	Temp	Data Type	Reference
WS 2.23E+001	mg/L	25	EST	MEYLAN,WM ET AL. (1996)
logP 2.73			EXP	HANSCH,C & LEO,AJ (1985)
VP 1.07E-013	mm Hg	25	EST	NEELY,WB & BLAU,GE (1985)
DC	pKa			
HL 7.67E-015	atm m3/mol	25	EST	MEYLAN,WM & HOWARD,PH (1991)
OH 1.67E-010	cm3/molc sec	25	EST	MEYLAN,WM & HOWARD,PH (1993)

CAS #: 034222-71-4	2-METHYL-B-NITROSTYRENE

Formula: $C_9H_9NO_2$

Mol Weight: 163.18

MP (deg C):

FP (deg C):

BP (deg C):

BP pressure (mm Hg):

Property/Value	Units	Temp	Data Type	Reference
WS 1.53E+002	mg/L	25	EST	MEYLAN,WM ET AL. (1996)
logP 2.63			EXP	HANSCH,C & LEO,AJ (1985)
VP 5.35E-003	mm Hg	25	EST	NEELY,WB & BLAU,GE (1985)
DC	pKa			
HL 3.83E-006	atm m3/mol	25	EST	MEYLAN,WM & HOWARD,PH (1991)
OH 1.63E-011	cm3/molc sec	25	EST	MEYLAN,WM & HOWARD,PH (1993)

CAS #: 034231-77-1	4-PYRIDAZINECARBOXYLIC ACID, METHYL ESTER

Formula: $C_6H_6N_2O_2$

Mol Weight: 138.13

MP (deg C):

FP (deg C):

BP (deg C):

BP pressure (mm Hg):

Property/Value	Units	Temp	Data Type	Reference
WS 1.45E+005	mg/L	25	EST	MEYLAN,WM ET AL. (1996)
logP -0.27			EXP	YAMAGAMI,C ET AL. (1990)
VP 3.61E-003	mm Hg	25	EST	NEELY,WB & BLAU,GE (1985)
DC	pKa			
HL 1.85E-008	atm m3/mol	25	EST	MEYLAN,WM & HOWARD,PH (1991)
OH 2.91E-013	cm3/molc sec	25	EST	MEYLAN,WM & HOWARD,PH (1993)

CAS #: 034251-46-2	BENZENEACETAMIDE, N,N-BIS(1-METHYLETHYL)-

Formula: $C_{14}H_{21}NO$

Mol Weight: 219.33

MP (deg C):

FP (deg C):

BP (deg C):

BP pressure (mm Hg):

Property/Value	Units	Temp	Data Type	Reference
WS 7.82E+001	mg/L	25	EST	MEYLAN,WM ET AL. (1996)
logP 3.10			EXP	SURYANARAYANA,MVS ET AL. (1991)
VP 1.40E-002	mm Hg	25	EXP	SURYANARAYANA,MVS ET AL. (1991)
DC	pKa			
HL 1.35E-008	atm m3/mol	25	EST	MEYLAN,WM & HOWARD,PH (1991)
OH 3.71E-011	cm3/molc sec	25	EST	MEYLAN,WM & HOWARD,PH (1993)

CAS #: 034253-01-5 — 5-PYRIMIDINECARBOXYLIC ACID, METHYL ESTER

Formula: $C_6H_6N_2O_2$
Mol Weight: 138.13
MP (deg C):　FP (deg C):
BP (deg C):
BP pressure (mm Hg):

Property/Value	Units	Temp	Data Type	Reference
WS 8.04E+004	mg/L	25	EST	MEYLAN,WM ET AL. (1996)
logP 0.03			EXP	YAMAGAMI,C ET AL. (1990)
VP 1.38E-001	mm Hg	25	EST	NEELY,WB & BLAU,GE (1985)
DC	pKa			
HL 1.88E-008	atm m3/mol	25	EST	MEYLAN,WM & HOWARD,PH (1991)
OH 2.91E-013	cm3/molc sec	25	EST	MEYLAN,WM & HOWARD,PH (1993)

CAS #: 034253-02-6 — 3-PYRIDAZINECARBOXYLIC ACID, METHYL ESTER

Formula: $C_6H_6N_2O_2$
Mol Weight: 138.13
MP (deg C):　FP (deg C):
BP (deg C):
BP pressure (mm Hg):

Property/Value	Units	Temp	Data Type	Reference
WS 1.99E+005	mg/L	25	EST	MEYLAN,WM ET AL. (1996)
logP -0.43			EXP	YAMAGAMI,C ET AL. (1990)
VP 3.61E-003	mm Hg	25	EST	NEELY,WB & BLAU,GE (1985)
DC	pKa			
HL 1.85E-008	atm m3/mol	25	EST	MEYLAN,WM & HOWARD,PH (1991)
OH 2.91E-013	cm3/molc sec	25	EST	MEYLAN,WM & HOWARD,PH (1993)

CAS #: 034253-03-7 — 2-PYRIMIDINECARBOXYLIC ACID, METHYL ESTER

Formula: $C_6H_6N_2O_2$
Mol Weight: 138.13
MP (deg C):　FP (deg C):
BP (deg C):
BP pressure (mm Hg):

Property/Value	Units	Temp	Data Type	Reference
WS 3.44E+005	mg/L	25	EST	MEYLAN,WM ET AL. (1996)
logP -0.71			EXP	YAMAGAMI,C ET AL. (1990)
VP 1.38E-001	mm Hg	25	EST	NEELY,WB & BLAU,GE (1985)
DC -0.68	pKa		EXP	PERRIN,DD (1965)
HL 1.88E-008	atm m3/mol	25	EST	MEYLAN,WM & HOWARD,PH (1991)
OH 2.91E-013	cm3/molc sec	25	EST	MEYLAN,WM & HOWARD,PH (1993)

CAS #: 034256-82-1 — ACETOCHLOR

Formula: $C_{14}H_{20}ClNO_2$
Mol Weight: 269.77
MP (deg C):　FP (deg C):
BP (deg C):
BP pressure (mm Hg):

Property/Value	Units	Temp	Data Type	Reference
WS 2.23E+002	mg/L	25	EXP	SHIU,WY ET AL. (1990)
logP 3.03			EXP	TOMLIN,C (1994)
VP 2.59E-006	mm Hg	25	EST	NEELY,WB & BLAU,GE (1985)
DC	pKa			
HL 2.23E-008	atm m3/mol	25	EST	MEYLAN,WM & HOWARD,PH (1991)
OH 4.96E-011	cm3/molc sec	25	EST	MEYLAN,WM & HOWARD,PH (1993)

CAS #: 034262-32-3 — BENZENAMINE, 3-[(4-AMINOPHENYL)SULFONYL]-

Formula: $C_{12}H_{12}N_2O_2S$
Mol Weight: 248.31
MP (deg C):　FP (deg C):
BP (deg C):
BP pressure (mm Hg):

Property/Value	Units	Temp	Data Type	Reference
WS 3.19E+003	mg/L	25	EST	MEYLAN,WM ET AL. (1996)
logP 1.03			EXP	ALTOMARE,C ET AL. (1991)
VP 2.20E-008	mm Hg	25	EST	NEELY,WB & BLAU,GE (1985)
DC	pKa			
HL 3.11E-014	atm m3/mol	25	EST	MEYLAN,WM & HOWARD,PH (1991)
OH 4.60E-011	cm3/molc sec	25	EST	MEYLAN,WM & HOWARD,PH (1993)

CAS #: 034378-65-9 — METHOTREXATE DIMETHYL ESTER

Formula: $C_{22}H_{26}N_8O_5$
Mol Weight: 482.50
MP (deg C):　FP (deg C):
BP (deg C):
BP pressure (mm Hg):

Property/Value	Units	Temp	Data Type	Reference
WS 3.20E+002	mg/L	25	EST	MEYLAN,WM ET AL. (1996)
logP 0.54			EXP	HANSCH,C ET AL. (1995)
VP 1.14E-016	mm Hg	25	EST	NEELY,WB & BLAU,GE (1985)
DC	pKa			
HL 1.58E-026	atm m3/mol	25	EST	MEYLAN,WM & HOWARD,PH (1991)
OH 3.16E-010	cm3/molc sec	25	EST	MEYLAN,WM & HOWARD,PH (1993)

CAS #: 034380-54-6 — P-SO2F PHENOXYACETIC ACID

Formula: $C_8H_7FO_5S$
Mol Weight: 234.20
MP (deg C):　FP (deg C):
BP (deg C):
BP pressure (mm Hg):

Property/Value	Units	Temp	Data Type	Reference
WS 2.00E+003	mg/L	25	EST	MEYLAN,WM ET AL. (1996)
logP 1.82			EXP	HANSCH,C & LEO,AJ (1985)
VP 2.16E-006	mm Hg	25	EST	NEELY,WB & BLAU,GE (1985)
DC	pKa			
HL 4.85E-011	atm m3/mol	25	EST	MEYLAN,WM & HOWARD,PH (1991)
OH 9.39E-012	cm3/molc sec	25	EST	MEYLAN,WM & HOWARD,PH (1993)

CAS #: 034385-93-8 — 1,4-BENZODIOXANE-2-CARBOXYLIC ACID

Formula: $C_9H_8O_4$
Mol Weight: 180.16
MP (deg C):　FP (deg C):
BP (deg C):
BP pressure (mm Hg):

Property/Value	Units	Temp	Data Type	Reference
WS 8.70E+003	mg/L	25	EST	MEYLAN,WM ET AL. (1996)
logP 1.40			EXP	HANSCH,C & LEO,AJ (1985)
VP 9.87E-005	mm Hg	25	EST	NEELY,WB & BLAU,GE (1985)
DC	pKa			
HL 4.38E-010	atm m3/mol	25	EST	MEYLAN,WM & HOWARD,PH (1991)
OH 3.70E-011	cm3/molc sec	25	EST	MEYLAN,WM & HOWARD,PH (1993)

CAS #: 034385-94-9 — 5-CL-2,3-DIHYDROBENZOFURAN-2-ACID

Formula: $C_9H_7ClO_3$
Mol Weight: 198.61
MP (deg C): FP (deg C):
BP (deg C):
BP pressure (mm Hg):

Property/Value	Units	Temp	Data Type	Reference
WS 1.74E+003	mg/L	25	EST	MEYLAN,WM ET AL. (1996)
logP 2.11			EXP	HANSCH,C & LEO,AJ (1985)
VP 7.11E-005	mm Hg	25	EST	NEELY,WB & BLAU,GE (1985)
DC	pKa			
HL 6.05E-009	atm m3/mol	25	EST	MEYLAN,WM & HOWARD,PH (1991)
OH 1.78E-011	cm3/molc sec	25	EST	MEYLAN,WM & HOWARD,PH (1993)

CAS #: 034392-61-5 — N1-(4-DIMEAMINOPH)SULFANILAMIDE

Formula: $C_{14}H_{17}N_3O_2S$
Mol Weight: 291.37
MP (deg C): FP (deg C):
BP (deg C):
BP pressure (mm Hg):

Property/Value	Units	Temp	Data Type	Reference
WS 6.29E+002	mg/L	25	EST	MEYLAN,WM ET AL. (1996)
logP 1.57			EXP	HANSCH,C & LEO,AJ (1985)
VP 8.37E-009	mm Hg	25	EST	NEELY,WB & BLAU,GE (1985)
DC	pKa			
HL 1.31E-012	atm m3/mol	25	EST	MEYLAN,WM & HOWARD,PH (1991)
OH 2.03E-010	cm3/molc sec	25	EST	MEYLAN,WM & HOWARD,PH (1993)

CAS #: 034392-79-5 — N1-(2-CL-3-PYRIDYL)SULFANILAMIDE

Formula: $C_{11}H_{10}ClN_3O_2S$
Mol Weight: 283.74
MP (deg C): FP (deg C):
BP (deg C):
BP pressure (mm Hg):

Property/Value	Units	Temp	Data Type	Reference
WS 6.82E+002	mg/L	25	EST	MEYLAN,WM ET AL. (1996)
logP 1.58			EXP	HANSCH,C & LEO,AJ (1985)
VP 1.72E-008	mm Hg	25	EST	NEELY,WB & BLAU,GE (1985)
DC	pKa			
HL 5.04E-012	atm m3/mol	25	EST	MEYLAN,WM & HOWARD,PH (1991)
OH 2.54E-011	cm3/molc sec	25	EST	MEYLAN,WM & HOWARD,PH (1993)

CAS #: 034392-82-0 — N1-(6-CL-3-PYRIDYL)SULFANILAMIDE

Formula: $C_{11}H_{10}ClN_3O_2S$
Mol Weight: 283.74
MP (deg C): FP (deg C):
BP (deg C):
BP pressure (mm Hg):

Property/Value	Units	Temp	Data Type	Reference
WS 1.86E+002	mg/L	25	EST	MEYLAN,WM ET AL. (1996)
logP 2.24			EXP	HANSCH,C & LEO,AJ (1985)
VP 1.72E-008	mm Hg	25	EST	NEELY,WB & BLAU,GE (1985)
DC	pKa			
HL 7.98E-014	atm m3/mol	25	EST	MEYLAN,WM & HOWARD,PH (1991)
OH 2.54E-011	cm3/molc sec	25	EST	MEYLAN,WM & HOWARD,PH (1993)

CAS #: 034569-18-1 — 2,4,6(1H,3H,5H)-PYRIMIDINETRIONE, 5-BUTYL-1-METH

Formula: $C_9H_{14}N_2O_3$
Mol Weight: 198.22
MP (deg C): FP (deg C):
BP (deg C):
BP pressure (mm Hg):

Property/Value	Units	Temp	Data Type	Reference
WS 5.14E+003	mg/L	25	EST	MEYLAN,WM ET AL. (1996)
logP 1.10			EXP	SANGSTER,J (1994)
VP 4.59E-010	mm Hg	25	EST	NEELY,WB & BLAU,GE (1985)
DC	pKa			
HL 7.92E-013	atm m3/mol	25	EST	MEYLAN,WM & HOWARD,PH (1991)
OH 1.11E-011	cm3/molc sec	25	EST	MEYLAN,WM & HOWARD,PH (1993)

CAS #: 034581-41-4 — 3-CHLORO-1,2-BUTADIENE

Formula: C_4H_5Cl
Mol Weight: 88.54
MP (deg C): FP (deg C):
BP (deg C):
BP pressure (mm Hg):

Property/Value	Units	Temp	Data Type	Reference
WS 8.27E+002	mg/L	25	EST	MEYLAN,WM ET AL. (1996)
logP 2.55			EST	MEYLAN,WM & HOWARD,PH (1995)
VP 2.84E+002	mm Hg	25	EST	NEELY,WB & BLAU,GE (1985)
DC	pKa			
HL 6.96E-002	atm m3/mol	25	EST	MEYLAN,WM & HOWARD,PH (1991)
OH 1.21E-011	cm3/molc sec	25	EST	MEYLAN,WM & HOWARD,PH (1993)

CAS #: 034593-75-4 — 4-(TERT-BUTYL)-2,6-DICHLOROPHENOL

Formula: $C_{10}H_{12}Cl_2O$
Mol Weight: 219.11
MP (deg C): FP (deg C):
BP (deg C):
BP pressure (mm Hg):

Property/Value	Units	Temp	Data Type	Reference
WS 1.25E+001	mg/L	25	EST	MEYLAN,WM ET AL. (1996)
logP 4.71			EST	MEYLAN,WM & HOWARD,PH (1995)
VP 4.94E-004	mm Hg	25	EST	NEELY,WB & BLAU,GE (1985)
DC	pKa			
HL 7.95E-007	atm m3/mol	25	EST	MEYLAN,WM & HOWARD,PH (1991)
OH 2.11E-012	cm3/molc sec	25	EST	MEYLAN,WM & HOWARD,PH (1993)

CAS #: 034622-58-7 — ORBENCARB

Formula: $C_{12}H_{16}ClNOS$
Mol Weight: 257.78
MP (deg C): FP (deg C):
BP (deg C):
BP pressure (mm Hg):

Property/Value	Units	Temp	Data Type	Reference
WS 2.52E+001	mg/L	25	EST	MEYLAN,WM ET AL. (1996)
logP 3.43			EXP	TOMLIN,C (1994)
VP 6.68E-006	mm Hg	25	EST	NEELY,WB & BLAU,GE (1985)
DC	pKa			
HL 3.93E-007	atm m3/mol	25	EST	MEYLAN,WM & HOWARD,PH (1991)
OH 2.54E-011	cm3/molc sec	25	EST	MEYLAN,WM & HOWARD,PH (1993)

034643-46-4 — PROTHIOPHOS

Formula: $C_{11}H_{15}Cl_2O_2PS_2$

Mol Weight: 345.25

MP (deg C):

FP (deg C):

BP (deg C): 126-128

BP pressure (mm Hg):

Property/Value	Units	Temp	Data Type	Reference
WS 7.00E-002	mg/L	20	EXP	TOMLIN,C (1994)
logP 5.67			EXP	TOMLIN,C (1994)
VP 9.40E-006	mm Hg	25	EST	NEELY,WB & BLAU,GE (1985)
DC	pKa			
HL 3.01E-005	atm m3/mol	25	EST	MEYLAN,WM & HOWARD,PH (1991)
OH 9.79E-011	cm3/molc sec	25	EST	MEYLAN,WM & HOWARD,PH (1993)

034645-84-6 — FENCLOFENAC

Formula: $C_{14}H_{10}Cl_2O_3$

Mol Weight: 297.14

MP (deg C): 109-115

FP (deg C):

BP (deg C):

BP pressure (mm Hg):

Property/Value	Units	Temp	Data Type	Reference
WS 2.52E+000	mg/L	25	EST	MEYLAN,WM ET AL. (1996)
logP 4.80			EXP	HANSCH,C & LEO,AJ (1985)
VP 1.40E-007	mm Hg	25	EST	NEELY,WB & BLAU,GE (1985)
DC	pKa			
HL 5.31E-010	atm m3/mol	25	EST	MEYLAN,WM & HOWARD,PH (1991)
OH 7.26E-012	cm3/molc sec	25	EST	MEYLAN,WM & HOWARD,PH (1993)

034661-75-1 — 2,4(1H,3H)-PYRIMIDINEDIONE, 6-[[3-[4-(2-METHOXYP

Formula: $C_{20}H_{29}N_5O_3$

Mol Weight: 387.49

MP (deg C): 156-158

FP (deg C):

BP (deg C):

BP pressure (mm Hg):

Property/Value	Units	Temp	Data Type	Reference
WS 1.57E+002	mg/L	25	EST	MEYLAN,WM ET AL. (1996)
logP 1.60			EXP	SANGSTER,J (1993)
VP 1.75E-012	mm Hg	25	EST	NEELY,WB & BLAU,GE (1985)
DC	pKa			
HL 2.20E-018	atm m3/mol	25	EST	MEYLAN,WM & HOWARD,PH (1991)
OH 3.36E-010	cm3/molc sec	25	EST	MEYLAN,WM & HOWARD,PH (1993)

034667-64-6 — PYRIDO(12A)PYRIMIDIN-4-ON,3ETO-CO-8ME

Formula: $C_{12}H_{14}N_2O_3$

Mol Weight: 234.26

MP (deg C):

FP (deg C):

BP (deg C):

BP pressure (mm Hg):

Property/Value	Units	Temp	Data Type	Reference
WS 5.35E+003	mg/L	25	EST	MEYLAN,WM ET AL. (1996)
logP 0.87			EXP	HANSCH,C & LEO,AJ (1985)
VP 8.05E-006	mm Hg	25	EST	NEELY,WB & BLAU,GE (1985)
DC	pKa			
HL	atm m3/mol			
OH 9.15E-012	cm3/molc sec	25	EST	MEYLAN,WM & HOWARD,PH (1993)

034691-02-6 — CEPHALOTHIN ANALOG (3-CH3)

Formula: $C_{14}H_{16}N_2O_4S_2$

Mol Weight: 340.42

MP (deg C):

FP (deg C):

BP (deg C):

BP pressure (mm Hg):

Property/Value	Units	Temp	Data Type	Reference
WS 2.19E+003	mg/L	25	EST	MEYLAN,WM ET AL. (1996)
logP 0.61			EXP	HANSCH,C & LEO,AJ (1985)
VP 6.32E-013	mm Hg	25	EST	NEELY,WB & BLAU,GE (1985)
DC	pKa			
HL 7.32E-015	atm m3/mol	25	EST	MEYLAN,WM & HOWARD,PH (1991)
OH 1.33E-010	cm3/molc sec	25	EST	MEYLAN,WM & HOWARD,PH (1993)

034761-82-5 — 3,5-DIMETHYL-4-NITROBENZENAMINE

Formula: $C_8H_{10}N_2O_2$

Mol Weight: 166.18

MP (deg C):

FP (deg C):

BP (deg C):

BP pressure (mm Hg):

Property/Value	Units	Temp	Data Type	Reference
WS 4.13E+002	mg/L	25	EST	MEYLAN,WM ET AL. (1996)
logP 2.57			EST	MEYLAN,WM & HOWARD,PH (1995)
VP 3.28E-004	mm Hg	25	EST	NEELY,WB & BLAU,GE (1985)
DC 2.54	pKa	21	EXP	PERRIN,DD (1965)
HL 9.15E-009	atm m3/mol	25	EST	MEYLAN,WM & HOWARD,PH (1991)
OH 9.20E-011	cm3/molc sec	25	EST	MEYLAN,WM & HOWARD,PH (1993)

034783-40-9 — HEPTENOPHOS

Formula: $C_9H_{12}ClO_4P$

Mol Weight: 250.62

MP (deg C):

FP (deg C):

BP (deg C):

BP pressure (mm Hg):

Property/Value	Units	Temp	Data Type	Reference
WS 2.20E+003	mg/L	20	EXP	SHIU,WY ET AL. (1990)
logP 1.41			EST	MEYLAN,WM & HOWARD,PH (1995)
VP 3.31E-004	mm Hg	25	EST	NEELY,WB & BLAU,GE (1985)
DC	pKa			
HL 1.17E-006	atm m3/mol	25	EST	MEYLAN,WM & HOWARD,PH (1991)
OH 8.69E-011	cm3/molc sec	25	EST	MEYLAN,WM & HOWARD,PH (1993)

034803-66-2 — PIPERAZINE, 1-(2-PYRIDINYL)-

Formula: $C_9H_{13}N_3$

Mol Weight: 163.22

MP (deg C):

FP (deg C):

BP (deg C): 120-122

BP pressure (mm Hg): 2.00E+000

Property/Value	Units	Temp	Data Type	Reference
WS 1.57E+005	mg/L	25	EST	MEYLAN,WM ET AL. (1996)
logP 0.74			EXP	CACCIA,S ET AL. (1985)
VP 1.30E-003	mm Hg	25	EST	NEELY,WB & BLAU,GE (1985)
DC	pKa			
HL 1.48E-011	atm m3/mol	25	EST	MEYLAN,WM & HOWARD,PH (1991)
OH 1.34E-010	cm3/molc sec	25	EST	MEYLAN,WM & HOWARD,PH (1993)

CAS #: 034816-55-2	19-NORPREGNA-1,3,5(10)-TRIEN-20-YNE-3,17-DIOL, 1

Formula: $C_{21}H_{26}O_3$

Mol Weight: 326.44

MP (deg C): 280 FP (deg C):

BP (deg C):

BP pressure (mm Hg):

Property/Value	Units	Temp	Data Type	Reference
WS 2.84E+002	mg/L	25	EST	MEYLAN,WM ET AL. (1996)
logP 3.01			EXP	SANGSTER,J (1993)
VP 5.21E-010	mm Hg	25	EST	NEELY,WB & BLAU,GE (1985)
DC	pKa			
HL 9.29E-014	atm m3/mol	25	EST	MEYLAN,WM & HOWARD,PH (1991)
OH 1.38E-010	cm3/molc sec	25	EST	MEYLAN,WM & HOWARD,PH (1993)

CAS #: 034819-62-0	4,4,4-TRICHLORO-1,2-BUTADIENE

Formula: $C_4H_3Cl_3$

Mol Weight: 157.43

MP (deg C): FP (deg C):

BP (deg C):

BP pressure (mm Hg):

Property/Value	Units	Temp	Data Type	Reference
WS 8.48E+001	mg/L	25	EST	MEYLAN,WM ET AL. (1996)
logP 3.42			EST	MEYLAN,WM & HOWARD,PH (1995)
VP 1.99E+001	mm Hg	25	EST	NEELY,WB & BLAU,GE (1985)
DC	pKa			
HL 4.23E-003	atm m3/mol	25	EST	MEYLAN,WM & HOWARD,PH (1991)
OH 2.36E-011	cm3/molc sec	25	EST	MEYLAN,WM & HOWARD,PH (1993)

CAS #: 034834-67-8	2-PYRROLIDINONE, 3-HYDROXY-1-METHYL-5-(3-PYRIDIN

Formula: $C_{10}H_{12}N_2O_2$

Mol Weight: 192.22

MP (deg C): FP (deg C):

BP (deg C):

BP pressure (mm Hg):

Property/Value	Units	Temp	Data Type	Reference
WS 1.00E+006	mg/L	25	EST	MEYLAN,WM ET AL. (1996)
logP -1.45			EXP	LI,NY & GORROD,JW (1992)
VP 2.16E-007	mm Hg	25	EST	NEELY,WB & BLAU,GE (1985)
DC	pKa			
HL 5.20E-013	atm m3/mol	25	EST	MEYLAN,WM & HOWARD,PH (1991)
OH 3.30E-011	cm3/molc sec	25	EST	MEYLAN,WM & HOWARD,PH (1993)

CAS #: 034835-05-7	P-NITROBENZYLSELENOPROPIONIC ACID

Formula: $C_{10}H_{11}NO_4Se$

Mol Weight: 288.16

MP (deg C): FP (deg C):

BP (deg C):

BP pressure (mm Hg):

Property/Value	Units	Temp	Data Type	Reference
WS 5.90E+002	mg/L	25	EST	MEYLAN,WM ET AL. (1996)
logP 1.63			EXP	HANSCH,C & LEO,AJ (1985)
VP 2.31E-007	mm Hg	25	EST	NEELY,WB & BLAU,GE (1985)
DC	pKa			
HL 4.02E-012	atm m3/mol	25	EST	MEYLAN,WM & HOWARD,PH (1991)
OH 8.32E-011	cm3/molc sec	25	EST	MEYLAN,WM & HOWARD,PH (1993)

CAS #: 034839-70-8	METIAMIDE

Formula: $C_9H_{16}N_4S_2$

Mol Weight: 244.38

MP (deg C): FP (deg C):

BP (deg C):

BP pressure (mm Hg):

Property/Value	Units	Temp	Data Type	Reference
WS 9.68E+004	mg/L	25	EST	MEYLAN,WM ET AL. (1996)
logP 0.50			EXP	HANSCH,C & LEO,AJ (1985)
VP 1.23E-008	mm Hg	25	EST	NEELY,WB & BLAU,GE (1985)
DC	pKa			
HL 6.70E-013	atm m3/mol	25	EST	MEYLAN,WM & HOWARD,PH (1991)
OH 2.10E-010	cm3/molc sec	25	EST	MEYLAN,WM & HOWARD,PH (1993)

CAS #: 034846-64-5	3-CYANOQUINOLINE

Formula: $C_{10}H_6N_2$

Mol Weight: 154.17

MP (deg C): 108-110 FP (deg C):

BP (deg C):

BP pressure (mm Hg):

Property/Value	Units	Temp	Data Type	Reference
WS 1.80E+003	mg/L	25	EST	MEYLAN,WM ET AL. (1996)
logP 1.57			EXP	HANSCH,C & LEO,AJ (1985)
VP 3.56E-004	mm Hg	25	EST	NEELY,WB & BLAU,GE (1985)
DC	pKa			
HL 6.65E-009	atm m3/mol	25	EST	MEYLAN,WM & HOWARD,PH (1991)
OH 2.54E-012	cm3/molc sec	25	EST	MEYLAN,WM & HOWARD,PH (1993)

CAS #: 034861-40-0	1-(4-I-PRPHENYL)-3-MEO-3-ME UREA

Formula: $C_{12}H_{18}N_2O_2$

Mol Weight: 222.29

MP (deg C): FP (deg C):

BP (deg C):

BP pressure (mm Hg):

Property/Value	Units	Temp	Data Type	Reference
WS 1.01E+002	mg/L	25	EST	MEYLAN,WM ET AL. (1996)
logP 2.95			EXP	HANSCH,C & LEO,AJ (1985)
VP 1.08E-005	mm Hg	25	EST	NEELY,WB & BLAU,GE (1985)
DC	pKa			
HL 4.09E-008	atm m3/mol	25	EST	MEYLAN,WM & HOWARD,PH (1991)
OH 5.52E-011	cm3/molc sec	25	EST	MEYLAN,WM & HOWARD,PH (1993)

CAS #: 034883-39-1	2,5-DICHLOROBIPHENYL

Formula: $C_{12}H_8Cl_2$

Mol Weight: 223.10

MP (deg C): FP (deg C):

BP (deg C):

BP pressure (mm Hg):

Property/Value	Units	Temp	Data Type	Reference
WS 1.12E+000	mg/L		EXP	YALKOWSKY,SH & DANNENFELSER,RM (1992)
logP 5.10			EXP	SANGSTER,J (1993)
VP 1.38E-003	mm Hg	25	EXP	BIDLEMAN,TF (1984)
DC	pKa			
HL 3.62E-004	atm m3/mol	25	EST	VP/WSOL
OH 2.47E-012	cm3/molc sec	25	EST	MEYLAN,WM & HOWARD,PH (1993)

CAS #: 034883-41-5 — 3,5-DICHLORO-1,1'-BIPHENYL

Formula: $C_{12}H_8Cl_2$

Mol Weight: 223.10

MP (deg C): FP (deg C):

BP (deg C):

BP pressure (mm Hg):

Property/Value	Units	Temp	Data Type	Reference
WS 7.95E-001	mg/L	25	EST	MEYLAN,WM ET AL. (1996)
logP 5.41			EXP	SANGSTER,J (1993)
VP 1.91E-004	mm Hg	25	EST	NEELY,WB & BLAU,GE (1985)
DC	pKa			
HL 2.27E-004	atm m3/mol	25	EST	MEYLAN,WM & HOWARD,PH (1991)
OH 3.45E-012	cm3/molc sec	25	EST	MEYLAN,WM & HOWARD,PH (1993)

CAS #: 034883-43-7 — 2,4'-DICHLOROBIPHENYL

Formula: $C_{12}H_8Cl_2$

Mol Weight: 223.10

MP (deg C): FP (deg C):

BP (deg C):

BP pressure (mm Hg):

Property/Value	Units	Temp	Data Type	Reference
WS 1.17E+000	mg/L	25	EXP	CHIOU,CT ET AL. (1983)
logP 5.09			EXP	SANGSTER,J (1993)
VP 2.09E-003	mm Hg	25	EXP	SHIU,WY & MACKAY,D (1986)
DC	pKa			
HL 5.24E-004	atm m3/mol	25	EST	VP/WSOL
OH 1.73E-012	cm3/molc sec	25	EST	MEYLAN,WM & HOWARD,PH (1993)

CAS #: 034915-68-9 — BUNITROLOL

Formula: $C_{14}H_{20}N_2O_2$

Mol Weight: 248.33

MP (deg C): 163-165 FP (deg C):

BP (deg C):

BP pressure (mm Hg):

Property/Value	Units	Temp	Data Type	Reference
WS 5.75E+003	mg/L	25	EST	MEYLAN,WM ET AL. (1996)
logP 1.60			EXP	MANNHOLD,R ET AL. (1990)
VP 1.39E-007	mm Hg	25	EST	NEELY,WB & BLAU,GE (1985)
DC	pKa			
HL 1.05E-013	atm m3/mol	25	EST	MEYLAN,WM & HOWARD,PH (1991)
OH 9.54E-011	cm3/molc sec	25	EST	MEYLAN,WM & HOWARD,PH (1993)

CAS #: 034918-50-8 — ACETAMIDE, 2-[[3-[2-HYDROXY-3-(1-METHYLETHYL)AMI

Formula: $C_{15}H_{24}N_2O_4$

Mol Weight: 296.37

MP (deg C): FP (deg C):

BP (deg C):

BP pressure (mm Hg):

Property/Value	Units	Temp	Data Type	Reference
WS 3.17E+002	mg/L	25	EST	MEYLAN,WM ET AL. (1996)
logP 0.35			EXP	SANGSTER,J (1993)
VP 5.71E-011	mm Hg	25	EST	NEELY,WB & BLAU,GE (1985)
DC	pKa			
HL 1.14E-018	atm m3/mol	25	EST	MEYLAN,WM & HOWARD,PH (1991)
OH 3.20E-010	cm3/molc sec	25	EST	MEYLAN,WM & HOWARD,PH (1993)

CAS #: 034919-98-7 — CETAMOLOL

Formula: $C_{16}H_{26}N_2O_4$

Mol Weight: 310.40

MP (deg C): 96-97 FP (deg C):

BP (deg C):

BP pressure (mm Hg):

Property/Value	Units	Temp	Data Type	Reference
WS 7.36E+002	mg/L	25	EST	MEYLAN,WM ET AL. (1996)
logP 1.36			EXP	RECANATINI,M (1992)
VP 3.21E-011	mm Hg	25	EST	NEELY,WB & BLAU,GE (1985)
DC	pKa			
HL 1.52E-018	atm m3/mol	25	EST	MEYLAN,WM & HOWARD,PH (1991)
OH 1.21E-010	cm3/molc sec	25	EST	MEYLAN,WM & HOWARD,PH (1993)

CAS #: 034932-78-0 — 1,4-BENZDIAZEPIN-2-ON-1ME-5(2FPH)-7I

Formula: $C_{16}H_{12}FIN_2O$

Mol Weight: 394.19

MP (deg C): FP (deg C):

BP (deg C):

BP pressure (mm Hg):

Property/Value	Units	Temp	Data Type	Reference
WS 6.93E+000	mg/L	25	EST	MEYLAN,WM ET AL. (1996)
logP 3.14			EXP	HANSCH,C & LEO,AJ (1985)
VP 4.08E-009	mm Hg	25	EST	NEELY,WB & BLAU,GE (1985)
DC	pKa			
HL 1.33E-009	atm m3/mol	25	EST	MEYLAN,WM & HOWARD,PH (1991)
OH 9.11E-012	cm3/molc sec	25	EST	MEYLAN,WM & HOWARD,PH (1993)

CAS #: 034933-71-6 — 4-PIPERIDINOL, 4-(4-CHLOROPHENYL)-1-[3-[(4-FLUOR

Formula: $C_{20}H_{24}FNO_2S$

Mol Weight: 361.48

MP (deg C): FP (deg C):

BP (deg C):

BP pressure (mm Hg):

Property/Value	Units	Temp	Data Type	Reference
WS 5.62E+001	mg/L	25	EST	MEYLAN,WM ET AL. (1996)
logP 4.17			EXP	SANGSTER,J (1993)
VP 1.40E-011	mm Hg	25	EST	NEELY,WB & BLAU,GE (1985)
DC	pKa			
HL 5.07E-017	atm m3/mol	25	EST	MEYLAN,WM & HOWARD,PH (1991)
OH 1.62E-010	cm3/molc sec	25	EST	MEYLAN,WM & HOWARD,PH (1993)

CAS #: 034939-39-4 — ACETAMIDE, 2-[4-[2-HYDROXY-3-[(1-METHYLETHYL)AMI

Formula: $C_{15}H_{24}N_2O_4$

Mol Weight: 296.37

MP (deg C): FP (deg C):

BP (deg C):

BP pressure (mm Hg):

Property/Value	Units	Temp	Data Type	Reference
WS 4.29E+003	mg/L	25	EST	MEYLAN,WM ET AL. (1996)
logP 0.56			EXP	SANGSTER,J (1993)
VP 5.71E-011	mm Hg	25	EST	NEELY,WB & BLAU,GE (1985)
DC	pKa			
HL 1.14E-018	atm m3/mol	25	EST	MEYLAN,WM & HOWARD,PH (1991)
OH 1.38E-010	cm3/molc sec	25	EST	MEYLAN,WM & HOWARD,PH (1993)

CAS #: 034970-69-9 — BURIAMIDE

Formula: $C_9H_{16}N_4S$

Mol Weight: 212.32

MP (deg C):   FP (deg C):

BP (deg C):

BP pressure (mm Hg):

Property/Value	Units	Temp	Data Type	Reference
WS 1.79E+005	mg/L	25	EST	MEYLAN,WM ET AL. (1996)
logP 0.39			EXP	HANSCH,C & LEO,AJ (1985)
VP 1.56E-007	mm Hg	25	EST	NEELY,WB & BLAU,GE (1985)
DC	pKa			
HL 1.00E-010	atm m3/mol	25	EST	MEYLAN,WM & HOWARD,PH (1991)
OH 2.31E-010	cm3/molc sec	25	EST	MEYLAN,WM & HOWARD,PH (1993)

CAS #: 034973-41-6 — 1,1,2,2,3,4,4-HEPTACHLOROBUTANE

Formula: $C_4H_3Cl_7$

Mol Weight: 299.24

MP (deg C):   FP (deg C):

BP (deg C): 137.5

BP pressure (mm Hg): 1.35E+001

Property/Value	Units	Temp	Data Type	Reference
WS 1.94E+000	mg/L	25	EST	MEYLAN,WM ET AL. (1996)
logP 4.46			EST	MEYLAN,WM & HOWARD,PH (1995)
VP 1.25E-002	mm Hg	25	EST	NEELY,WB & BLAU,GE (1985)
DC	pKa			
HL 1.16E-004	atm m3/mol	25	EST	MEYLAN,WM & HOWARD,PH (1991)
OH 2.37E-013	cm3/molc sec	25	EST	MEYLAN,WM & HOWARD,PH (1993)

CAS #: 034974-00-0 — 2-PYRIDINEBUTANEAMINE

Formula: $C_9H_{14}N_2$

Mol Weight: 150.23

MP (deg C):   FP (deg C):

BP (deg C):

BP pressure (mm Hg):

Property/Value	Units	Temp	Data Type	Reference
WS 1.00E+006	mg/L	25	EST	MEYLAN,WM ET AL. (1996)
logP 0.86			EXP	HANSCH,C & LEO,AJ (1985)
VP 1.44E-002	mm Hg	25	EST	NEELY,WB & BLAU,GE (1985)
DC	pKa			
HL 1.87E-009	atm m3/mol	25	EST	MEYLAN,WM & HOWARD,PH (1991)
OH 3.66E-011	cm3/molc sec	25	EST	MEYLAN,WM & HOWARD,PH (1993)

CAS #: 034993-08-3 — N-NITROSO-PHENMETRAZINE

Formula: $C_{11}H_{14}N_2O_2$

Mol Weight: 206.25

MP (deg C):   FP (deg C):

BP (deg C):

BP pressure (mm Hg):

Property/Value	Units	Temp	Data Type	Reference
WS 6.67E+002	mg/L	25	EST	MEYLAN,WM ET AL. (1996)
logP 2.09			EXP	HANSCH,C & LEO,AJ (1985)
VP 5.89E-005	mm Hg	25	EST	NEELY,WB & BLAU,GE (1985)
DC	pKa			
HL 1.51E-009	atm m3/mol	25	EST	MEYLAN,WM & HOWARD,PH (1991)
OH 7.90E-011	cm3/molc sec	25	EST	MEYLAN,WM & HOWARD,PH (1993)

CAS #: 035041-48-6 — 3-CARBAMOYLPYRIDINIUM IODIDE,N-OCTYL

Formula: $C_{14}H_{23}IN_2O$

Mol Weight: 362.26

MP (deg C):   FP (deg C):

BP (deg C):

BP pressure (mm Hg):

Property/Value	Units	Temp	Data Type	Reference
WS 2.07E+004	mg/L	25	EST	MEYLAN,WM ET AL. (1996)
logP -0.70			EXP	HANSCH,C & LEO,AJ (1985)
VP 1.67E-008	mm Hg	25	EST	NEELY,WB & BLAU,GE (1985)
DC	pKa			
HL 4.49E-016	atm m3/mol	25	EST	MEYLAN,WM & HOWARD,PH (1991)
OH 2.19E-011	cm3/molc sec	25	EST	MEYLAN,WM & HOWARD,PH (1993)

CAS #: 035041-49-7 — 3-CARBAMOYLPYRIDINIUM IODIDE, N-DECYL

Formula: $C_{16}H_{27}IN_2O$

Mol Weight: 390.31

MP (deg C):   FP (deg C):

BP (deg C):

BP pressure (mm Hg):

Property/Value	Units	Temp	Data Type	Reference
WS 6.22E+003	mg/L	25	EST	MEYLAN,WM ET AL. (1996)
logP -0.29			EXP	HANSCH,C & LEO,AJ (1985)
VP 3.01E-009	mm Hg	25	EST	NEELY,WB & BLAU,GE (1985)
DC	pKa			
HL 7.92E-016	atm m3/mol	25	EST	MEYLAN,WM & HOWARD,PH (1991)
OH 2.47E-011	cm3/molc sec	25	EST	MEYLAN,WM & HOWARD,PH (1993)

CAS #: 035045-02-4 — DEAMINOMETRIBUZIN

Formula: $C_8H_{13}N_3OS$

Mol Weight: 199.28

MP (deg C):   FP (deg C):

BP (deg C):

BP pressure (mm Hg):

Property/Value	Units	Temp	Data Type	Reference
WS 3.18E+002	mg/L	25	EST	MEYLAN,WM ET AL. (1996)
logP 2.51			EST	MEYLAN,WM & HOWARD,PH (1995)
VP 2.36E-008	mm Hg	25	EST	NEELY,WB & BLAU,GE (1985)
DC	pKa			
HL 4.33E-010	atm m3/mol	25	EST	MEYLAN,WM & HOWARD,PH (1991)
OH 8.76E-012	cm3/molc sec	25	EST	MEYLAN,WM & HOWARD,PH (1993)

CAS #: 035065-27-1 — 2,4,5,2',4',5'-PCB

Formula: $C_{12}H_4Cl_6$

Mol Weight: 360.88

MP (deg C):   FP (deg C):

BP (deg C):

BP pressure (mm Hg):

Property/Value	Units	Temp	Data Type	Reference
WS 9.50E-004	mg/L	24	EXP	CHIOU,CT ET AL. (1977)
logP 7.75			EXP	HANSCH,C & LEO,AJ (1985)
VP 3.43E-006	mm Hg	25	EXP	SHIU,WY & MACKAY,D (1986)
DC	pKa			
HL 1.71E-003	atm m3/mol	25	EST	VP/WSOL
OH 1.64E-013	cm3/molc sec	25	EST	MEYLAN,WM & HOWARD,PH (1993)

1095

CAS #: 035065-28-2				2,2',3,4,4',5'-HEXACHLOROBIPHENYL

Formula: $C_{12}H_4Cl_6$

Mol Weight: 360.88

MP (deg C): FP (deg C):

BP (deg C):

BP pressure (mm Hg):

Property/Value	Units	Temp	Data Type	Reference
WS 7.29E-003	mg/L	20	EXP	YALKOWSKY,SH & DANNENFELSER,RM (1992)
logP 7.44			EXP	RAPAPORT,RA & EISENREICH,SJ (1984)
VP 3.79E-006	mm Hg	25	EXP	SHIU,WY & MACKAY,D (1986)
DC	pKa			
HL 2.47E-004	atm m3/mol	25	EST	VP/WSOL
OH 1.64E-013	cm3/molc sec	25	EST	MEYLAN,WM & HOWARD,PH (1993)

CAS #: 035065-29-3				2,2',3,4,4',5,5'-HEPTACHLOROBIPHENYL

Formula: $C_{12}H_3Cl_7$

Mol Weight: 395.33

MP (deg C): FP (deg C):

BP (deg C):

BP pressure (mm Hg):

Property/Value	Units	Temp	Data Type	Reference
WS 3.85E-003	mg/L	20	EXP	YALKOWSKY,SH & DANNENFELSER,RM (1992)
logP 8.27			EST	MEYLAN,WM & HOWARD,PH (1995)
VP 9.77E-007	mm Hg	25	EXP	BIDLEMAN,TF (1984)
DC	pKa			
HL 1.32E-004	atm m3/mol	25	EST	VP/WSOL
OH 1.05E-013	cm3/molc sec	25	EST	MEYLAN,WM & HOWARD,PH (1993)

CAS #: 035065-30-6				2,2',3,3',4,4',5-HEPTACHLOROBIPHENYL

Formula: $C_{12}H_3Cl_7$

Mol Weight: 395.33

MP (deg C): FP (deg C):

BP (deg C):

BP pressure (mm Hg):

Property/Value	Units	Temp	Data Type	Reference
WS 3.47E-003	mg/L	20	EXP	YALKOWSKY,SH & DANNENFELSER,RM (1992)
logP 8.27			EST	MEYLAN,WM & HOWARD,PH (1995)
VP 6.28E-007	mm Hg	25	EXP	BIDLEMAN,TF (1984)
DC	pKa			
HL 9.41E-005	atm m3/mol	25	EST	VP/WSOL
OH 1.05E-013	cm3/molc sec	25	EST	MEYLAN,WM & HOWARD,PH (1993)

CAS #: 035096-55-0				3-CARBAMOYLPYRIDINIUM IODIDE,DODECYL

Formula: $C_{18}H_{31}IN_2O$

Mol Weight: 418.36

MP (deg C): FP (deg C):

BP (deg C):

BP pressure (mm Hg):

Property/Value	Units	Temp	Data Type	Reference
WS 7.24E+002	mg/L	25	EST	MEYLAN,WM ET AL. (1996)
logP 0.60			EXP	HANSCH,C & LEO,AJ (1985)
VP 5.87E-010	mm Hg	25	EST	NEELY,WB & BLAU,GE (1985)
DC	pKa			
HL 1.40E-015	atm m3/mol	25	EST	MEYLAN,WM & HOWARD,PH (1991)
OH 2.75E-011	cm3/molc sec	25	EST	MEYLAN,WM & HOWARD,PH (1993)

CAS #: 035103-34-5				N-((4-METHOXYPHENYL)METHYL)ACETAMIDE

Formula: $C_{10}H_{13}NO_2$

Mol Weight: 179.22

MP (deg C): FP (deg C):

BP (deg C):

BP pressure (mm Hg):

Property/Value	Units	Temp	Data Type	Reference
WS 6.48E+003	mg/L	25	EST	MEYLAN,WM ET AL. (1996)
logP 1.09			EST	MEYLAN,WM & HOWARD,PH (1995)
VP 2.76E-005	mm Hg	25	EST	NEELY,WB & BLAU,GE (1985)
DC	pKa			
HL 1.17E-010	atm m3/mol	25	EST	MEYLAN,WM & HOWARD,PH (1991)
OH 3.65E-011	cm3/molc sec	25	EST	MEYLAN,WM & HOWARD,PH (1993)

CAS #: 035129-56-7				3-CARBAMOYLPYRIDINIUM IODIDE,N-C14

Formula: $C_{20}H_{35}IN_2O$

Mol Weight: 446.42

MP (deg C): FP (deg C):

BP (deg C):

BP pressure (mm Hg):

Property/Value	Units	Temp	Data Type	Reference
WS 1.67E+002	mg/L	25	EST	MEYLAN,WM ET AL. (1996)
logP 1.14			EXP	HANSCH,C & LEO,AJ (1985)
VP 1.13E-010	mm Hg	25	EST	NEELY,WB & BLAU,GE (1985)
DC	pKa			
HL 2.46E-015	atm m3/mol	25	EST	MEYLAN,WM & HOWARD,PH (1991)
OH 3.03E-011	cm3/molc sec	25	EST	MEYLAN,WM & HOWARD,PH (1993)

CAS #: 035148-19-7				9-DODECENYL ACETATE (TRANS)

Formula: $C_{14}H_{26}O_2$

Mol Weight: 226.36

MP (deg C): FP (deg C):

BP (deg C):

BP pressure (mm Hg):

Property/Value	Units	Temp	Data Type	Reference
WS 5.68E-001	mg/L	25	EST	MEYLAN,WM ET AL. (1996)
logP 5.56			EST	MEYLAN,WM & HOWARD,PH (1995)
VP 4.10E-003	mm Hg	25	EST	NEELY,WB & BLAU,GE (1985)
DC	pKa			
HL 3.48E-003	atm m3/mol	25	EST	MEYLAN,WM & HOWARD,PH (1991)
OH 6.91E-011	cm3/molc sec	25	EST	MEYLAN,WM & HOWARD,PH (1993)

CAS #: 035153-15-2				9-TETRADECEN-1-OL (Z)

Formula: $C_{14}H_{28}O$

Mol Weight: 212.38

MP (deg C): FP (deg C):

BP (deg C):

BP pressure (mm Hg):

Property/Value	Units	Temp	Data Type	Reference
WS 2.28E+000	mg/L	25	EST	MEYLAN,WM ET AL. (1996)
logP 5.54			EST	MEYLAN,WM & HOWARD,PH (1995)
VP 5.98E-005	mm Hg	25	EST	NEELY,WB & BLAU,GE (1985)
DC	pKa			
HL 1.49E-004	atm m3/mol	25	EST	MEYLAN,WM & HOWARD,PH (1991)
OH 7.41E-011	cm3/molc sec	25	EST	MEYLAN,WM & HOWARD,PH (1993)

CAS #: 035187-28-1	7,11-DIMETHYLBENZ(A)ANTHRACENE

Formula: $C_{20}H_{16}$

Mol Weight: 256.35

MP (deg C): | FP (deg C):

BP (deg C):

BP pressure (mm Hg):

Property/Value	Units	Temp	Data Type	Reference
WS 3.79E-003	mg/L	25	EST	MEYLAN,WM ET AL. (1996)
logP 6.62			EST	MEYLAN,WM & HOWARD,PH (1995)
VP 1.14E-007	mm Hg	25	EST	NEELY,WB & BLAU,GE (1985)
DC	pKa			
HL 6.10E-006	atm m3/mol	25	EST	MEYLAN,WM & HOWARD,PH (1991)
OH 1.60E-010	cm3/molc sec	25	EST	MEYLAN,WM & HOWARD,PH (1993)

CAS #: 035203-88-4	BENZENESULFONAMIDE, 3-ACETYL-

Formula: $C_8H_9NO_3S$

Mol Weight: 199.23

MP (deg C): | FP (deg C):

BP (deg C):

BP pressure (mm Hg):

Property/Value	Units	Temp	Data Type	Reference
WS 2.60E+003	mg/L	25	EST	MEYLAN,WM ET AL. (1996)
logP 0.21			EXP	CAROTTI,A ET AL. (1989)
VP 1.79E-005	mm Hg	25	EST	NEELY,WB & BLAU,GE (1985)
DC	pKa			
HL 7.69E-010	atm m3/mol	25	EST	MEYLAN,WM & HOWARD,PH (1991)
OH 4.82E-013	cm3/molc sec	25	EST	MEYLAN,WM & HOWARD,PH (1993)

CAS #: 035203-91-9	8-SULFONAMIDOQUINOLINE

Formula: $C_9H_8N_2O_2S$

Mol Weight: 208.24

MP (deg C): | FP (deg C):

BP (deg C):

BP pressure (mm Hg):

Property/Value	Units	Temp	Data Type	Reference
WS 1.74E+004	mg/L	25	EST	MEYLAN,WM ET AL. (1996)
logP 0.42			EXP	HANSCH,C ET AL. (1995)
VP 2.02E-006	mm Hg	25	EST	NEELY,WB & BLAU,GE (1985)
DC	pKa			
HL 5.39E-011	atm m3/mol	25	EST	MEYLAN,WM & HOWARD,PH (1991)
OH 3.08E-012	cm3/molc sec	25	EST	MEYLAN,WM & HOWARD,PH (1993)

CAS #: 035204-11-6	N-(O-NO2PH)-3-N'-PIPERIDINOACETAMIDE

Formula: $C_{13}H_{17}N_3O_3$

Mol Weight: 263.30

MP (deg C): | FP (deg C):

BP (deg C):

BP pressure (mm Hg):

Property/Value	Units	Temp	Data Type	Reference
WS 3.64E+001	mg/L	25	EST	MEYLAN,WM ET AL. (1996)
logP 2.75			EXP	HANSCH,C & LEO,AJ (1985)
VP 1.66E-008	mm Hg	25	EST	NEELY,WB & BLAU,GE (1985)
DC	pKa			
HL 5.44E-012	atm m3/mol	25	EST	MEYLAN,WM & HOWARD,PH (1991)
OH 9.97E-011	cm3/molc sec	25	EST	MEYLAN,WM & HOWARD,PH (1993)

CAS #: 035207-08-0	BENZENESULFONAMIDE, N-(AMINOCARBONYL)-

Formula: $C_7H_8N_2O_3S$

Mol Weight: 200.22

MP (deg C): | FP (deg C):

BP (deg C):

BP pressure (mm Hg):

Property/Value	Units	Temp	Data Type	Reference
WS 1.23E+003	mg/L	25	EST	MEYLAN,WM ET AL. (1996)
logP 0.28			EXP	CLOUX,JL ET AL. (1988)
VP 1.41E-006	mm Hg	25	EST	NEELY,WB & BLAU,GE (1985)
DC	pKa			
HL 4.47E-011	atm m3/mol	25	EST	MEYLAN,WM & HOWARD,PH (1991)
OH 2.42E-012	cm3/molc sec	25	EST	MEYLAN,WM & HOWARD,PH (1993)

CAS #: 035231-36-8	PHENOL, 4-[2-(DIMETHYLAMINO)ETHOXY]-2-METHYL-5-(

Formula: $C_{14}H_{23}NO_2$

Mol Weight: 237.34

MP (deg C): | FP (deg C):

BP (deg C):

BP pressure (mm Hg):

Property/Value	Units	Temp	Data Type	Reference
WS 3.38E+003	mg/L	25	EST	MEYLAN,WM ET AL. (1996)
logP 2.93			EXP	DALLET ET AL. (1985)
VP 1.99E-005	mm Hg	25	EST	NEELY,WB & BLAU,GE (1985)
DC	pKa			
HL 4.69E-011	atm m3/mol	25	EST	MEYLAN,WM & HOWARD,PH (1991)
OH 1.92E-010	cm3/molc sec	25	EST	MEYLAN,WM & HOWARD,PH (1993)

CAS #: 035331-58-9	HEXAMETHYL(SILACYCLOHEXYL)CYCLOTETRASILOXANE

Formula: $C_{11}H_{28}O_4Si_4$

Mol Weight: 336.69

MP (deg C): | FP (deg C):

BP (deg C):

BP pressure (mm Hg):

Property/Value	Units	Temp	Data Type	Reference
WS 2.24E-003	mg/L	25	EST	MEYLAN,WM ET AL. (1996)
logP 6.45			EST	MEYLAN,WM & HOWARD,PH (1995)
VP	mm Hg			
DC	pKa			
HL	atm m3/mol			
OH 7.66E-012	cm3/molc sec	25	EST	MEYLAN,WM & HOWARD,PH (1993)

CAS #: 035367-38-5	DIFLUBENZURON

Formula: $C_{14}H_9ClF_2N_2O_2$

Mol Weight: 310.69

MP (deg C): 239 | FP (deg C):

BP (deg C):

BP pressure (mm Hg):

Property/Value	Units	Temp	Data Type	Reference
WS 3.00E-001	mg/L	24	EXP	CARRINGER,RD ET AL. (1975)
logP 3.88			EXP	SOTOMATSU,T ET AL. (1987)
VP 9.00E-010	mm Hg	25	EXP	WAUCHOPE,RD ET AL. (1991A)
DC	pKa			
HL 1.23E-009	atm m3/mol	25	EST	VP/WSOL
OH 1.40E-010	cm3/molc sec	25	EST	ATKINOSN,R (1988)

035367-40-9 — BENZAMIDE, 2,6-DIFLUORO-N-[[(4-FLUOROPHENYL)AMIN

CAS #: 035367-40-9
Formula: $C_{14}H_9F_3N_2O_2$
Mol Weight: 294.24
MP (deg C):
FP (deg C):
BP (deg C):
BP pressure (mm Hg):

Property/Value	Units	Temp	Data Type	Reference
WS 2.45E+001	mg/L	25	EST	MEYLAN,WM ET AL. (1996)
logP 3.20			EXP	SOTOMATSU,T ET AL. (1987)
VP 1.29E-009	mm Hg	25	EST	NEELY,WB & BLAU,GE (1985)
DC	pKa			
HL 1.87E-011	atm m3/mol	25	EST	MEYLAN,WM & HOWARD,PH (1991)
OH 1.82E-011	cm3/molc sec	25	EST	MEYLAN,WM & HOWARD,PH (1993)

035386-24-4 — PIPERAZINE, 1-(2-METHOXYPHENYL)-

CAS #: 035386-24-4
Formula: $C_{11}H_{16}N_2O$
Mol Weight: 192.26
MP (deg C): 37-40
FP (deg C):
BP (deg C): 130-133
BP pressure (mm Hg): 1.00E-001

Property/Value	Units	Temp	Data Type	Reference
WS 1.83E+004	mg/L	25	EST	MEYLAN,WM ET AL. (1996)
logP 1.67			EXP	CACCIA,S ET AL. (1985)
VP 3.15E-004	mm Hg	25	EST	NEELY,WB & BLAU,GE (1985)
DC	pKa			
HL 6.70E-010	atm m3/mol	25	EST	MEYLAN,WM & HOWARD,PH (1991)
OH 2.33E-010	cm3/molc sec	25	EST	MEYLAN,WM & HOWARD,PH (1993)

035400-43-2 — SULPROFOS

CAS #: 035400-43-2
Formula: $C_{12}H_{19}O_2PS_3$
Mol Weight: 322.45
MP (deg C):
FP (deg C):
BP (deg C): 155-158
BP pressure (mm Hg): 1.00E-001

Property/Value	Units	Temp	Data Type	Reference
WS 1.89E-001	mg/L	25	EST	MEYLAN,WM ET AL. (1996)
logP 5.48			EXP	TOMLIN,C (1994)
VP 9.87E-006	mm Hg	25	EST	NEELY,WB & BLAU,GE (1985)
DC	pKa			
HL 1.60E-006	atm m3/mol	25	EST	MEYLAN,WM & HOWARD,PH (1991)
OH 1.07E-010	cm3/molc sec	25	EST	MEYLAN,WM & HOWARD,PH (1993)

035407-50-2 — CARBAMIC ACID, (4-ETHOXYPHENYL)-, METHYL ESTER

CAS #: 035407-50-2
Formula: $C_{10}H_{13}NO_3$
Mol Weight: 195.22
MP (deg C):
FP (deg C):
BP (deg C):
BP pressure (mm Hg):

Property/Value	Units	Temp	Data Type	Reference
WS 6.75E+002	mg/L	25	EST	MEYLAN,WM ET AL. (1996)
logP 2.15			EXP	TAKAHASHI,J ET AL. (1988)
VP 1.79E-003	mm Hg	25	EST	NEELY,WB & BLAU,GE (1985)
DC	pKa			
HL 1.71E-009	atm m3/mol	25	EST	MEYLAN,WM & HOWARD,PH (1991)
OH 4.33E-011	cm3/molc sec	25	EST	MEYLAN,WM & HOWARD,PH (1993)

035413-85-5 — 2-PROPENOIC ACID, 2-METHYL-, 5A-ETHENYLDECAHYDRO

CAS #: 035413-85-5
Formula: $C_{19}H_{20}O_6$
Mol Weight: 344.37
MP (deg C):
FP (deg C):
BP (deg C):
BP pressure (mm Hg):

Property/Value	Units	Temp	Data Type	Reference
WS 3.12E+002	mg/L	25	EST	MEYLAN,WM ET AL. (1996)
logP 1.56			EXP	SANGSTER,J (1994)
VP 2.55E-009	mm Hg	25	EST	NEELY,WB & BLAU,GE (1985)
DC	pKa			
HL 1.96E-010	atm m3/mol	25	EST	MEYLAN,WM & HOWARD,PH (1991)
OH 9.52E-011	cm3/molc sec	25	EST	MEYLAN,WM & HOWARD,PH (1993)

035480-26-3 — M-METHOXYBENZYL ACETATE

CAS #: 035480-26-3
Formula: $C_{10}H_{12}O_3$
Mol Weight: 180.21
MP (deg C):
FP (deg C):
BP (deg C):
BP pressure (mm Hg):

Property/Value	Units	Temp	Data Type	Reference
WS 1.04E+003	mg/L	25	EST	MEYLAN,WM ET AL. (1996)
logP 2.02			EXP	HANSCH,C & LEO,AJ (1985)
VP 1.94E-002	mm Hg	25	EST	NEELY,WB & BLAU,GE (1985)
DC	pKa			
HL 8.37E-007	atm m3/mol	25	EST	MEYLAN,WM & HOWARD,PH (1991)
OH 5.59E-011	cm3/molc sec	25	EST	MEYLAN,WM & HOWARD,PH (1993)

035488-92-7 — P-NITRO-N-PHENYLSUCCINIMIDE

CAS #: 035488-92-7
Formula: $C_{10}H_8N_2O_4$
Mol Weight: 220.19
MP (deg C):
FP (deg C):
BP (deg C):
BP pressure (mm Hg):

Property/Value	Units	Temp	Data Type	Reference
WS 5.66E+003	mg/L	25	EST	MEYLAN,WM ET AL. (1996)
logP 0.46			EXP	HANSCH,C & LEO,AJ (1985)
VP 1.72E-008	mm Hg	25	EST	NEELY,WB & BLAU,GE (1985)
DC	pKa			
HL 5.15E-010	atm m3/mol	25	EST	MEYLAN,WM & HOWARD,PH (1991)
OH 5.92E-012	cm3/molc sec	25	EST	MEYLAN,WM & HOWARD,PH (1993)

035531-88-5 — CARBENICILLIN, INDANYL

CAS #: 035531-88-5
Formula: $C_{26}H_{26}N_2O_6S$
Mol Weight: 494.57
MP (deg C):
FP (deg C):
BP (deg C):
BP pressure (mm Hg):

Property/Value	Units	Temp	Data Type	Reference
WS 4.68E-001	mg/L	25	EST	MEYLAN,WM ET AL. (1996)
logP 3.77			EXP	HANSCH,C & LEO,AJ (1985)
VP 2.56E-017	mm Hg	25	EST	NEELY,WB & BLAU,GE (1985)
DC	pKa			
HL 7.30E-018	atm m3/mol	25	EST	MEYLAN,WM & HOWARD,PH (1991)
OH 2.47E-010	cm3/molc sec	25	EST	MEYLAN,WM & HOWARD,PH (1993)

035554-44-0 — IMAZALIL BASE

Formula: $C_{14}H_{16}Cl_2N_2O$

Mol Weight: 299.20
MP (deg C): 52.7
BP (deg C): > 340
BP pressure (mm Hg):

Property/Value	Units	Temp	Data Type	Reference
WS 1.80E+002	mg/L		EXP	TOMLIN,C (1994)
logP 3.82			EXP	TOMLIN,C (1994)
VP 7.00E-008	mm Hg	20	EXP	AUGUSTIJN-BECKERS,PWM ET AL. (1994)
DC	pKa			
HL 1.53E-010	atm m3/mol	20	EST	VP/WSOL
OH 8.49E-011	cm3/molc sec	25	EST	MEYLAN,WM & HOWARD,PH (1993)

035580-76-8 — HYDRAZINECARBOXAMIDE, N-(2-CHLOROPHENYL)

Formula: $C_7H_8ClN_3O$

Mol Weight: 185.61
MP (deg C):
BP (deg C):
BP pressure (mm Hg):

Property/Value	Units	Temp	Data Type	Reference
WS 2.13E+004	mg/L	25	EST	MEYLAN,WM ET AL. (1996)
logP 0.45			EXP	KRAMER,CR & BECK,L (1981)
VP 2.02E-005	mm Hg	25	EST	NEELY,WB & BLAU,GE (1985)
DC	pKa			
HL 6.23E-013	atm m3/mol	25	EST	MEYLAN,WM & HOWARD,PH (1991)
OH 1.35E-011	cm3/molc sec	25	EST	MEYLAN,WM & HOWARD,PH (1993)

035607-66-0 — CEFOXITIN

Formula: $C_{16}H_{17}N_3O_7S_2$

Mol Weight: 427.46
MP (deg C):
BP (deg C): 149-150 de
BP pressure (mm Hg):

Property/Value	Units	Temp	Data Type	Reference
WS 1.05E+002	mg/L	25	EST	MEYLAN,WM ET AL. (1996)
logP -0.02			EXP	SANGSTER,J (1993)
VP 1.09E-015	mm Hg	25	EST	NEELY,WB & BLAU,GE (1985)
DC	pKa			
HL 6.50E-021	atm m3/mol	25	EST	MEYLAN,WM & HOWARD,PH (1991)
OH 1.35E-010	cm3/molc sec	25	EST	MEYLAN,WM & HOWARD,PH (1993)

035681-63-1 — 1H-IMIDAZOLE, 5-IODO-1-METHYL-4-NITRO-

Formula: $C_4H_4IN_3O_2$

Mol Weight: 253.00
MP (deg C):
BP (deg C):
BP pressure (mm Hg):

Property/Value	Units	Temp	Data Type	Reference
WS 1.30E+003	mg/L	25	EST	MEYLAN,WM ET AL. (1996)
logP 1.00			EXP	GUPTA,RP ET AL. (1985)
VP 2.59E-005	mm Hg	25	EST	NEELY,WB & BLAU,GE (1985)
DC	pKa			
HL 7.33E-008	atm m3/mol	25	EST	MEYLAN,WM & HOWARD,PH (1991)
OH 1.62E-012	cm3/molc sec	25	EST	MEYLAN,WM & HOWARD,PH (1993)

035687-41-3 — 5-METHOXY-1-METHYL-3-NITRO-1H-IMIDAZOLE

Formula: $C_5H_7N_3O_3$

Mol Weight: 157.13
MP (deg C):
BP (deg C):
BP pressure (mm Hg):

Property/Value	Units	Temp	Data Type	Reference
WS 1.98E+004	mg/L	25	EST	MEYLAN,WM ET AL. (1996)
logP 0.96			EST	MEYLAN,WM & HOWARD,PH (1995)
VP 1.88E-004	mm Hg	25	EST	NEELY,WB & BLAU,GE (1985)
DC	pKa			
HL 1.87E-008	atm m3/mol	25	EST	MEYLAN,WM & HOWARD,PH (1991)
OH 4.86E-012	cm3/molc sec	25	EST	MEYLAN,WM & HOWARD,PH (1993)

035687-42-4 — 4-METHOXY-2-METHYL-5-NITRO-1H-IMIDAZOLE

Formula: $C_5H_7N_3O_3$

Mol Weight: 157.13
MP (deg C):
BP (deg C):
BP pressure (mm Hg):

Property/Value	Units	Temp	Data Type	Reference
WS 1.90E+004	mg/L	25	EST	MEYLAN,WM ET AL. (1996)
logP 0.96			EST	MEYLAN,WM & HOWARD,PH (1995)
VP 9.90E-006	mm Hg	25	EST	NEELY,WB & BLAU,GE (1985)
DC	pKa			
HL 9.67E-010	atm m3/mol	25	EST	MEYLAN,WM & HOWARD,PH (1991)
OH 3.70E-011	cm3/molc sec	25	EST	MEYLAN,WM & HOWARD,PH (1993)

035687-44-6 — 5-METHOXY-1,2-DIMETHYL-4-NITRO-1H-IMIDAZOLE

Formula: $C_6H_9N_3O_3$

Mol Weight: 171.16
MP (deg C):
BP (deg C):
BP pressure (mm Hg):

Property/Value	Units	Temp	Data Type	Reference
WS 1.28E+003	mg/L	25	EST	MEYLAN,WM ET AL. (1996)
logP 1.51			EST	MEYLAN,WM & HOWARD,PH (1995)
VP 6.80E-005	mm Hg	25	EST	NEELY,WB & BLAU,GE (1985)
DC	pKa			
HL 2.06E-008	atm m3/mol	25	EST	MEYLAN,WM & HOWARD,PH (1991)
OH 3.71E-011	cm3/molc sec	25	EST	MEYLAN,WM & HOWARD,PH (1993)

035693-92-6 — 2,4,6-TRICHLOROBIPHENYL

Formula: $C_{12}H_7Cl_3$

Mol Weight: 257.55
MP (deg C):
BP (deg C):
BP pressure (mm Hg):

Property/Value	Units	Temp	Data Type	Reference
WS 2.52E-001	mg/L	25	EXP	YALKOWSKY,SH & DANNENFELSER,RM (1992)
logP 5.47			EXP	HANSCH,C & LEO,AJ (1985)
VP 7.16E-004	mm Hg	25	EXP	BIDLEMAN,TF (1984)
DC	pKa			
HL 9.63E-004	atm m3/mol	25	EST	VP/WSOL
OH 1.54E-012	cm3/molc sec	25	EST	MEYLAN,WM & HOWARD,PH (1993)

CAS #: 035693-99-3				2,2',5,5'-PCB

Formula: $C_{12}H_6Cl_4$

Mol Weight: 291.99

MP (deg C): FP (deg C):

BP (deg C):

BP pressure (mm Hg):

Property/Value	Units	Temp	Data Type	Reference
WS 2.65E-002	mg/L	25	EXP	HAQUE,R & SCHMEDDING,D (1976)
logP 6.09			EXP	HANSCH,C & LEO,AJ (1985)
VP 8.45E-006	mm Hg	25	EST	NEELY,WB & BLAU,GE (1985)
DC	pKa			
HL 2.23E-004	atm m3/mol	25	EST	MEYLAN,WM & HOWARD,PH(1991)
OH 7.30E-013	cm3/molc sec	25	EST	MEYLAN,WM & HOWARD,PH (1993)

CAS #: 035694-06-5				2,2',3,4,4'-PCB

Formula: $C_{12}H_4Cl_6$

Mol Weight: 360.88

MP (deg C): FP (deg C):

BP (deg C):

BP pressure (mm Hg):

Property/Value	Units	Temp	Data Type	Reference
WS 8.40E-003	mg/L	20	EXP	YALKOWSKY,SH & DANNENFELSER,RM (1992)
logP 7.44			EXP	HANSCH,C & LEO,AJ (1985)
VP 5.81E-007	mm Hg	25	EST	NEELY,WB & BLAU,GE (1985)
DC	pKa			
HL 6.85E-005	atm m3/mol	25	EST	MEYLAN,WM & HOWARD,PH (1991)
OH 2.81E-013	cm3/molc sec	25	EST	MEYLAN,WM & HOWARD,PH (1993)

CAS #: 035694-08-7				2,2',3,3',4,4',5,5'-OCTACHLOROBIPHENYL

Formula: $C_{12}H_2Cl_8$

Mol Weight: 429.77

MP (deg C): FP (deg C):

BP (deg C):

BP pressure (mm Hg):

Property/Value	Units	Temp	Data Type	Reference
WS 2.72E-004	mg/L	25	EXP	YALKOWSKY,SH & DANNENFELSER,RM (1992)
logP 8.91			EST	MEYLAN,WM & HOWARD,PH (1995)
VP 2.87E-008	mm Hg	25	EST	NEELY,WB & BLAU,GE (1985)
DC	pKa			
HL 3.76E-005	atm m3/mol	25	EST	MEYLAN,WM & HOWARD,PH (1991)
OH 5.59E-014	cm3/molc sec	25	EST	MEYLAN,WM & HOWARD,PH (1993)

CAS #: 035704-19-9				ACETAMIDE, N-(4-CYANOPHENYL)-

Formula: $C_9H_8N_2O$

Mol Weight: 160.18

MP (deg C): 206-208 FP (deg C):

BP (deg C):

BP pressure (mm Hg):

Property/Value	Units	Temp	Data Type	Reference
WS 2.51E+003	mg/L	25	EST	MEYLAN,WM ET AL. (1996)
logP 1.37			EXP	HANSCH,C ET AL. (1995)
VP 1.06E-005	mm Hg	25	EST	NEELY,WB & BLAU,GE (1985)
DC	pKa			
HL 5.96E-011	atm m3/mol	25	EST	MEYLAN,WM & HOWARD,PH (1991)
OH 2.29E-012	cm3/molc sec	25	EST	MEYLAN,WM & HOWARD,PH (1993)

CAS #: 035763-43-0				N-(M-NO2PH)-3-N'-PIPERIDINOACETAMIDE

Formula: $C_{13}H_{17}N_3O_3$

Mol Weight: 263.30

MP (deg C): FP (deg C):

BP (deg C):

BP pressure (mm Hg):

Property/Value	Units	Temp	Data Type	Reference
WS 4.99E+001	mg/L	25	EST	MEYLAN,WM ET AL. (1996)
logP 2.59			EXP	HANSCH,C & LEO,AJ (1985)
VP 1.66E-008	mm Hg	25	EST	NEELY,WB & BLAU,GE (1985)
DC	pKa			
HL 2.49E-013	atm m3/mol	25	EST	MEYLAN,WM & HOWARD,PH (1991)
OH 9.92E-011	cm3/molc sec	25	EST	MEYLAN,WM & HOWARD,PH (1993)

CAS #: 035813-38-8				N-METHYL P-METHOXYACETANILIDE

Formula: $C_{10}H_{13}NO_2$

Mol Weight: 179.22

MP (deg C): FP (deg C):

BP (deg C):

BP pressure (mm Hg):

Property/Value	Units	Temp	Data Type	Reference
WS 1.35E+004	mg/L	25	EST	MEYLAN,WM ET AL. (1996)
logP 0.72			EXP	HANSCH,C & LEO,AJ (1985)
VP 9.27E-004	mm Hg	25	EST	NEELY,WB & BLAU,GE (1985)
DC	pKa			
HL 7.48E-009	atm m3/mol	25	EST	MEYLAN,WM & HOWARD,PH (1991)
OH 1.27E-011	cm3/molc sec	25	EST	MEYLAN,WM & HOWARD,PH (1993)

CAS #: 035822-46-9				1,2,3,4,6,7,8-HEPTACHLORODIBENZO-P-DIOXIN

Formula: $C_{12}HCl_7O_2$

Mol Weight: 425.31

MP (deg C): FP (deg C):

BP (deg C):

BP pressure (mm Hg):

Property/Value	Units	Temp	Data Type	Reference
WS 2.40E-006	mg/L	20	EXP	YALKOWSKY,SH & DANNENFELSER,RM (1992)
logP 8.20			EXP	SHIU,WY ET AL. (1988)
VP 7.50E-010	mm Hg	25	EXP	SHIU,WY ET AL. (1988)
DC	pKa			
HL 1.75E-004	atm m3/mol	25	EST	VP/WSOL
OH 9.18E-013	cm3/molc sec	25	EST	MEYLAN,WM & HOWARD,PH (1993)

CAS #: 035846-53-8				MAYTANSINE

Formula: $C_{34}H_{46}ClN_3O_{10}$

Mol Weight: 692.21

MP (deg C): 171-172 FP (deg C):

BP (deg C):

BP pressure (mm Hg):

Property/Value	Units	Temp	Data Type	Reference
WS 7.90E-001	mg/L	25	EST	MEYLAN,WM ET AL. (1996)
logP 1.99			EXP	HANSCH,C & LEO,AJ (1985)
VP 3.72E-025	mm Hg	25	EST	NEELY,WB & BLAU,GE (1985)
DC	pKa			
HL 4.23E-026	atm m3/mol	25	EST	MEYLAN,WM & HOWARD,PH (1991)
OH 4.13E-010	cm3/molc sec	25	EST	MEYLAN,WM & HOWARD,PH (1993)

035880-91-2 — 1,3-BENZENEDIAMINE, 4-[(4-AMINOPHENYL)SULFONYL]-

CAS #:	035880-91-2	1,3-BENZENEDIAMINE, 4-[(4-AMINOPHENYL)SULFONYL]-
Formula:	$C_{12}H_{13}N_3O_2S$	
Mol Weight:	263.32	
MP (deg C):		FP (deg C):
BP (deg C):		
BP pressure (mm Hg):		

Property/Value	Units	Temp	Data Type	Reference
WS 9.44E+003	mg/L	25	EST	MEYLAN,WM ET AL. (1996)
logP 0.38			EXP	ALTOMARE,C ET AL. (1991)
VP 1.10E-009	mm Hg	25	EST	NEELY,WB & BLAU,GE (1985)
DC	pKa			
HL 1.10E-017	atm m3/mol	25	EST	MEYLAN,WM & HOWARD,PH (1991)
OH 2.00E-010	cm3/molc sec	25	EST	MEYLAN,WM & HOWARD,PH (1993)

035889-00-0 — BENZENEACETIC ACID, 4-CYCLOHEXYL-

CAS #:	035889-00-0	BENZENEACETIC ACID, 4-CYCLOHEXYL-
Formula:	$C_{14}H_{18}O_2$	
Mol Weight:	218.30	
MP (deg C):		FP (deg C):
BP (deg C):		
BP pressure (mm Hg):		

Property/Value	Units	Temp	Data Type	Reference
WS 4.00E+001	mg/L	25	EST	MEYLAN,WM ET AL. (1996)
logP 3.91			EXP	SANGSTER,J (1994)
VP 1.25E-005	mm Hg	25	EST	NEELY,WB & BLAU,GE (1985)
DC	pKa			
HL 8.87E-008	atm m3/mol	25	EST	MEYLAN,WM & HOWARD,PH (1991)
OH 1.60E-011	cm3/molc sec	25	EST	MEYLAN,WM & HOWARD,PH (1993)

035975-00-9 — 5-QUINOLINAMINE, 6-NITRO-

CAS #:	035975-00-9	5-QUINOLINAMINE, 6-NITRO-
Formula:	$C_9H_7N_3O_2$	
Mol Weight:	189.18	
MP (deg C):	272-273 de	FP (deg C):
BP (deg C):		
BP pressure (mm Hg):		

Property/Value	Units	Temp	Data Type	Reference
WS 2.09E+003	mg/L	25	EST	MEYLAN,WM ET AL. (1996)
logP 1.61			EXP	SANGSTER,J (1993)
VP 5.24E-006	mm Hg	25	EST	NEELY,WB & BLAU,GE (1985)
DC	pKa			
HL 2.10E-011	atm m3/mol	25	EST	MEYLAN,WM & HOWARD,PH (1991)
OH 1.43E-011	cm3/molc sec	25	EST	MEYLAN,WM & HOWARD,PH (1993)

035990-32-0 — ACETAMIDE, N-(3,6-DIHYDRO-1(2H)-PYRIDINYL)-

CAS #:	035990-32-0	ACETAMIDE, N-(3,6-DIHYDRO-1(2H)-PYRIDINYL)-
Formula:	$C_7H_{12}N_2O$	
Mol Weight:	140.19	
MP (deg C):		FP (deg C):
BP (deg C):		
BP pressure (mm Hg):		

Property/Value	Units	Temp	Data Type	Reference
WS 3.01E+004	mg/L	25	EST	MEYLAN,WM ET AL. (1996)
logP 0.52			EXP	SANGSTER,J (1994)
VP 4.83E-004	mm Hg	25	EST	NEELY,WB & BLAU,GE (1985)
DC	pKa			
HL 2.04E-010	atm m3/mol	25	EST	MEYLAN,WM & HOWARD,PH (1991)
OH 8.25E-011	cm3/molc sec	25	EST	MEYLAN,WM & HOWARD,PH (1993)

036038-53-6 — 1,1,4,4-TETRACHLOROBUTA-1,3-DIENE

CAS #:	036038-53-6	1,1,4,4-TETRACHLOROBUTA-1,3-DIENE
Formula:	$C_4H_2Cl_4$	
Mol Weight:	191.87	
MP (deg C):		FP (deg C):
BP (deg C):		
BP pressure (mm Hg):		

Property/Value	Units	Temp	Data Type	Reference
WS 3.12E+001	mg/L	25	EST	MEYLAN,WM ET AL. (1996)
logP 3.73			EST	MEYLAN,WM & HOWARD,PH (1995)
VP 2.50E+000	mm Hg	25	EST	NEELY,WB & BLAU,GE (1985)
DC	pKa			
HL 2.09E-002	atm m3/mol	25	EST	MEYLAN,WM & HOWARD,PH (1991)
OH 5.06E-013	cm3/molc sec	25	EST	MEYLAN,WM & HOWARD,PH (1993)

036067-73-9 — 4H-OXAZOLO[4,5-D]AZEPIN-2-AMINE, 6-ETHYL-5,6,7,8

CAS #:	036067-73-9	4H-OXAZOLO[4,5-D]AZEPIN-2-AMINE, 6-ETHYL-5,6,7,8
Formula:	$C_{10}H_{17}N_3O$	
Mol Weight:	195.27	
MP (deg C):		FP (deg C):
BP (deg C):		
BP pressure (mm Hg):		

Property/Value	Units	Temp	Data Type	Reference
WS 2.66E+005	mg/L	25	EST	MEYLAN,WM ET AL. (1996)
logP 0.05			EXP	SANGSTER,J (1994)
VP 3.20E-004	mm Hg	25	EST	NEELY,WB & BLAU,GE (1985)
DC	pKa			
HL 1.17E-012	atm m3/mol	25	EST	MEYLAN,WM & HOWARD,PH (1991)
OH 1.08E-010	cm3/molc sec	25	EST	MEYLAN,WM & HOWARD,PH (1993)

036070-75-4 — PYRAZINECARBONITRILE, 5-CHLORO-

CAS #:	036070-75-4	PYRAZINECARBONITRILE, 5-CHLORO-
Formula:	$C_5H_2ClN_3$	
Mol Weight:	139.54	
MP (deg C):		FP (deg C):
BP (deg C):		
BP pressure (mm Hg):		

Property/Value	Units	Temp	Data Type	Reference
WS 7.48E+003	mg/L	25	EST	MEYLAN,WM ET AL. (1996)
logP 0.92			EXP	YAMAGAMI,C ET AL. (1991)
VP 2.20E-002	mm Hg	25	EST	NEELY,WB & BLAU,GE (1985)
DC	pKa			
HL 1.32E-006	atm m3/mol	25	EST	MEYLAN,WM & HOWARD,PH (1991)
OH 2.86E-014	cm3/molc sec	25	EST	MEYLAN,WM & HOWARD,PH (1993)

036070-79-8 — 2-CHLORO-6-CONH2-PYRAZINE

CAS #:	036070-79-8	2-CHLORO-6-CONH2-PYRAZINE
Formula:	$C_5H_4ClN_3O$	
Mol Weight:	157.56	
MP (deg C):		FP (deg C):
BP (deg C):		
BP pressure (mm Hg):		

Property/Value	Units	Temp	Data Type	Reference
WS 1.98E+003	mg/L	25	EST	MEYLAN,WM ET AL. (1996)
logP 0.28			EXP	YAMAGAMI,C & TAKAO,N (1991)
VP 4.71E-005	mm Hg	25	EST	NEELY,WB & BLAU,GE (1985)
DC	pKa			
HL 5.61E-011	atm m3/mol	25	EST	MEYLAN,WM & HOWARD,PH (1991)
OH 2.07E-012	cm3/molc sec	25	EST	MEYLAN,WM & HOWARD,PH (1993)

CAS #: 036070-86-7	PYRAZINECARBOXAMIDE, 6-METHOXY-

Formula: $C_6H_7N_3O_2$

Mol Weight: 153.14

MP (deg C): FP (deg C):

BP (deg C):

BP pressure (mm Hg):

Property/Value	Units	Temp	Data Type	Reference
WS 2.79E+003	mg/L	25	EST	MEYLAN,WM ET AL. (1996)
logP 0.13			EXP	YAMAGAMI,C ET AL. (1991)
VP 3.24E-005	mm Hg	25	EST	NEELY,WB & BLAU,GE (1985)
DC	pKa			
HL 6.47E-013	atm m3/mol	25	EST	MEYLAN,WM & HOWARD,PH (1991)
OH 3.90E-012	cm3/molc sec	25	EST	MEYLAN,WM & HOWARD,PH (1993)

CAS #: 036092-88-3	INDOLE-2-AMINE, N-METHYL

Formula: $C_9H_{10}N_2$

Mol Weight: 146.19

MP (deg C): FP (deg C):

BP (deg C):

BP pressure (mm Hg):

Property/Value	Units	Temp	Data Type	Reference
WS 2.68E+003	mg/L	25	EST	MEYLAN,WM ET AL. (1996)
logP 1.72			EXP	HANSCH,C & LEO,AJ (1985)
VP 7.25E-004	mm Hg	25	EST	NEELY,WB & BLAU,GE (1985)
DC	pKa			
HL 6.87E-010	atm m3/mol	25	EST	MEYLAN,WM & HOWARD,PH (1991)
OH 2.01E-010	cm3/molc sec	25	EST	MEYLAN,WM & HOWARD,PH (1993)

CAS #: 036093-54-6	14-BENZODIAZPN2-ON,5(2-F PH)7ACETYL

Formula: $C_{17}H_{13}FN_2O_2$

Mol Weight: 296.30

MP (deg C): FP (deg C):

BP (deg C):

BP pressure (mm Hg):

Property/Value	Units	Temp	Data Type	Reference
WS 4.13E+002	mg/L	25	EST	MEYLAN,WM ET AL. (1996)
logP 1.75			EXP	HANSCH,C & LEO,AJ (1985)
VP 1.15E-009	mm Hg	25	EST	NEELY,WB & BLAU,GE (1985)
DC	pKa			
HL 5.10E-013	atm m3/mol	25	EST	MEYLAN,WM & HOWARD,PH (1991)
OH 1.73E-011	cm3/molc sec	25	EST	MEYLAN,WM & HOWARD,PH (1993)

CAS #: 036137-88-9	5(33DIME-1-TRIAZENO)IMDAZL-4-COOET

Formula: $C_8H_{13}N_5O_2$

Mol Weight: 211.23

MP (deg C): FP (deg C):

BP (deg C):

BP pressure (mm Hg):

Property/Value	Units	Temp	Data Type	Reference
WS 1.61E+004	mg/L	25	EST	MEYLAN,WM ET AL. (1996)
logP 0.44			EXP	HANSCH,C & LEO,AJ (1985)
VP 1.39E-006	mm Hg	25	EST	NEELY,WB & BLAU,GE (1985)
DC	pKa			
HL 3.20E-012	atm m3/mol	25	EST	MEYLAN,WM & HOWARD,PH (1991)
OH 8.80E-012	cm3/molc sec	25	EST	MEYLAN,WM & HOWARD,PH (1993)

CAS #: 036164-42-8	7-ACETYLAMINOQUINOLINE

Formula: $C_{11}H_{10}N_2O$

Mol Weight: 186.22

MP (deg C): FP (deg C):

BP (deg C):

BP pressure (mm Hg):

Property/Value	Units	Temp	Data Type	Reference
WS 2.44E+003	mg/L	25	EST	MEYLAN,WM ET AL. (1996)
logP 1.55			EXP	HANSCH,C & LEO,AJ (1985)
VP 1.28E-006	mm Hg	25	EST	NEELY,WB & BLAU,GE (1985)
DC	pKa			
HL 7.88E-013	atm m3/mol	25	EST	MEYLAN,WM & HOWARD,PH (1991)
OH 9.18E-011	cm3/molc sec	25	EST	MEYLAN,WM & HOWARD,PH (1993)

CAS #: 036216-80-5	3-AMINO-BENZISOXAZOLE

Formula: $C_7H_7N_2O$

Mol Weight: 135.15

MP (deg C): FP (deg C):

BP (deg C):

BP pressure (mm Hg):

Property/Value	Units	Temp	Data Type	Reference
WS 6.73E+003	mg/L	25	EST	MEYLAN,WM ET AL. (1996)
logP 1.31			EXP	HANSCH,C & LEO,AJ (1985)
VP 2.86E-003	mm Hg	25	EST	NEELY,WB & BLAU,GE (1985)
DC	pKa			
HL 1.96E-009	atm m3/mol	25	EST	MEYLAN,WM & HOWARD,PH (1991)
OH 2.00E-010	cm3/molc sec	25	EST	MEYLAN,WM & HOWARD,PH (1993)

CAS #: 036318-56-6	2-IMIDAZOLIDINIMINE, N-(2-METHYLPHENYL)-

Formula: $C_{10}H_{13}N_3$

Mol Weight: 175.24

MP (deg C): FP (deg C):

BP (deg C):

BP pressure (mm Hg):

Property/Value	Units	Temp	Data Type	Reference
WS 4.77E+004	mg/L	25	EST	MEYLAN,WM ET AL. (1996)
logP 1.28			EXP	SANGSTER,J (1993)
VP 1.69E-004	mm Hg	25	EST	NEELY,WB & BLAU,GE (1985)
DC	pKa			
HL 3.03E-011	atm m3/mol	25	EST	MEYLAN,WM & HOWARD,PH (1991)
OH 1.49E-010	cm3/molc sec	25	EST	MEYLAN,WM & HOWARD,PH (1993)

CAS #: 036322-90-4	PIROXICAM

Formula: $C_{15}H_{13}N_3O_4S$

Mol Weight: 331.35

MP (deg C): 198-200 FP (deg C):

BP (deg C):

BP pressure (mm Hg):

Property/Value	Units	Temp	Data Type	Reference
WS 2.30E+001	mg/L	25	EXP	YALKOWSKY,SH & DANNENFELSER,RM (1992)
logP 0.26			EXP	SANGSTER,J (1994)
VP 4.67E-015	mm Hg	25	EST	NEELY,WB & BLAU,GE (1985)
DC	pKa			
HL 2.90E-019	atm m3/mol	25	EST	MEYLAN,WM & HOWARD,PH (1991)
OH 3.30E-011	cm3/molc sec	25	EST	MEYLAN,WM & HOWARD,PH (1993)

CAS #: 036330-85-5				FENBUFEN

Formula: $C_{16}H_{14}O_3$
Mol Weight: 254.29
MP (deg C): 185-187
FP (deg C):
BP (deg C):
BP pressure (mm Hg):

Property/Value	Units	Temp	Data Type	Reference
WS 2.21E+000	mg/L	25	EXP	YALKOWSKY,SH & DANNENFELSER,RM (1992)
logP 3.20			EXP	HANSCH,C ET AL. (1995)
VP 8.14E-008	mm Hg	25	EST	NEELY,WB & BLAU,GE (1985)
DC	pKa			
HL 7.44E-012	atm m3/mol	25	EST	MEYLAN,WM & HOWARD,PH (1991)
OH 1.06E-011	cm3/molc sec	25	EST	MEYLAN,WM & HOWARD,PH (1993)

CAS #: 036335-67-8				BUTAMIFOS

Formula: $C_{13}H_{21}N_2O_4PS$
Mol Weight: 332.36
MP (deg C):
FP (deg C):
BP (deg C):
BP pressure (mm Hg):

Property/Value	Units	Temp	Data Type	Reference
WS 5.10E+000	mg/L	20	EXP	SHIU,WY ET AL. (1990)
logP 4.62			EXP	TOMLIN,C (1994)
VP 1.38E-006	mm Hg	25	EST	NEELY,WB & BLAU,GE (1985)
DC	pKa			
HL 8.03E-009	atm m3/mol	25	EST	MEYLAN,WM & HOWARD,PH (1991)
OH 2.75E-010	cm3/molc sec	25	EST	MEYLAN,WM & HOWARD,PH (1993)

CAS #: 036371-18-3				PIPERAZINE-2-CARBOXANILIDE,2',6'-DIME

Formula: $C_{13}H_{19}N_3O$
Mol Weight: 233.32
MP (deg C):
FP (deg C):
BP (deg C):
BP pressure (mm Hg):

Property/Value	Units	Temp	Data Type	Reference
WS 1.38E+004	mg/L	25	EST	MEYLAN,WM ET AL. (1996)
logP 0.38			EXP	HANSCH,C & LEO,AJ (1985)
VP 3.82E-008	mm Hg	25	EST	NEELY,WB & BLAU,GE (1985)
DC	pKa			
HL 4.48E-015	atm m3/mol	25	EST	MEYLAN,WM & HOWARD,PH (1991)
OH 2.17E-010	cm3/molc sec	25	EST	MEYLAN,WM & HOWARD,PH (1993)

CAS #: 036380-48-0				BARBITURIC ACID,5-ET-5(2,3-DIMEBU)

Formula: $C_{12}H_{20}N_2O_3$
Mol Weight: 240.30
MP (deg C):
FP (deg C):
BP (deg C):
BP pressure (mm Hg):

Property/Value	Units	Temp	Data Type	Reference
WS 5.04E+002	mg/L	25	EST	MEYLAN,WM ET AL. (1996)
logP 2.02			EXP	HANSCH,C & LEO,AJ (1985)
VP 1.56E-011	mm Hg	25	EST	NEELY,WB & BLAU,GE (1985)
DC	pKa			
HL 1.12E-012	atm m3/mol	25	EST	MEYLAN,WM & HOWARD,PH (1991)
OH 1.61E-011	cm3/molc sec	25	EST	MEYLAN,WM & HOWARD,PH (1993)

CAS #: 036385-57-6				PIPERAZINE-2-CARBOXANILIDE

Formula: $C_{11}H_{15}N_3O$
Mol Weight: 205.26
MP (deg C):
FP (deg C):
BP (deg C):
BP pressure (mm Hg):

Property/Value	Units	Temp	Data Type	Reference
WS 1.60E+004	mg/L	25	EST	MEYLAN,WM ET AL. (1996)
logP 0.48			EXP	HANSCH,C & LEO,AJ (1985)
VP 1.88E-007	mm Hg	25	EST	NEELY,WB & BLAU,GE (1985)
DC	pKa			
HL 3.68E-015	atm m3/mol	25	EST	MEYLAN,WM & HOWARD,PH (1991)
OH 2.10E-010	cm3/molc sec	25	EST	MEYLAN,WM & HOWARD,PH (1993)

CAS #: 036385-59-8				PIPERAZINE-2-CARBOXANILIDE,2'-ME

Formula: $C_{12}H_{17}N_3O$
Mol Weight: 219.29
MP (deg C):
FP (deg C):
BP (deg C):
BP pressure (mm Hg):

Property/Value	Units	Temp	Data Type	Reference
WS 1.02E+003	mg/L	25	EST	MEYLAN,WM ET AL. (1996)
logP 0.26			EXP	HANSCH,C & LEO,AJ (1985)
VP 8.49E-008	mm Hg	25	EST	NEELY,WB & BLAU,GE (1985)
DC	pKa			
HL 4.06E-015	atm m3/mol	25	EST	MEYLAN,WM & HOWARD,PH (1991)
OH 2.13E-010	cm3/molc sec	25	EST	MEYLAN,WM & HOWARD,PH (1993)

CAS #: 036393-56-3				BENZENEMETHANOL, .ALPHA.-(1-AMINOETHYL)-, (R*,R*

Formula: $C_9H_{13}NO$
Mol Weight: 151.21
MP (deg C): 77.5-78
FP (deg C):
BP (deg C):
BP pressure (mm Hg):

Property/Value	Units	Temp	Data Type	Reference
WS 1.49E+005	mg/L	25	EST	MEYLAN,WM ET AL. (1996)
logP 0.83			EXP	SANGSTER,J (1994)
VP 8.67E-004	mm Hg	25	EST	NEELY,WB & BLAU,GE (1985)
DC 9.44	pKa	20	EXP	PERRIN,DD (1965)
HL 3.94E-011	atm m3/mol	25	EST	MEYLAN,WM & HOWARD,PH (1991)
OH 5.65E-011	cm3/molc sec	25	EST	MEYLAN,WM & HOWARD,PH (1993)

CAS #: 036405-75-1				N-PHENOXYACETYLPIPERIDINE

Formula: $C_{13}H_{17}NO_2$
Mol Weight: 219.29
MP (deg C):
FP (deg C):
BP (deg C):
BP pressure (mm Hg):

Property/Value	Units	Temp	Data Type	Reference
WS 9.88E+002	mg/L	25	EST	MEYLAN,WM ET AL. (1996)
logP 1.81			EXP	HANSCH,C & LEO,AJ (1985)
VP 3.23E-005	mm Hg	25	EST	NEELY,WB & BLAU,GE (1985)
DC	pKa			
HL 1.70E-009	atm m3/mol	25	EST	MEYLAN,WM & HOWARD,PH (1991)
OH 5.44E-011	cm3/molc sec	25	EST	MEYLAN,WM & HOWARD,PH (1993)

CAS #: 036417-16-0 — 2-OH NAPHTHOQUINONE,3(33DICLALLYL)

Formula: $C_{13}H_8Cl_2O_3$

Mol Weight: 283.11

MP (deg C):
FP (deg C):
BP (deg C):
BP pressure (mm Hg):

Property/ Value	Units	Temp	Data Type	Reference
WS 3.46E+001	mg/L	25	EST	MEYLAN,WM ET AL. (1996)
logP 3.10			EXP	HANSCH,C & LEO,AJ (1985)
VP 1.12E-009	mm Hg	25	EST	NEELY,WB & BLAU,GE (1985)
DC	pKa			
HL 2.59E-009	atm m3/mol	25	EST	MEYLAN,WM & HOWARD,PH (1991)
OH 1.91E-011	cm3/molc sec	25	EST	MEYLAN,WM & HOWARD,PH (1993)

CAS #: 036429-48-8 — 3-(2-PHENOXYETHOXY)PHENOL

Formula: $C_{14}H_{14}O_3$

Mol Weight: 230.27

MP (deg C):
FP (deg C):
BP (deg C):
BP pressure (mm Hg):

Property/ Value	Units	Temp	Data Type	Reference
WS 1.67E+002	mg/L	25	EST	MEYLAN,WM ET AL. (1996)
logP 3.33			EST	MEYLAN,WM & HOWARD,PH (1995)
VP 3.84E-006	mm Hg	25	EST	NEELY,WB & BLAU,GE (1985)
DC	pKa			
HL 1.91E-010	atm m3/mol	25	EST	MEYLAN,WM & HOWARD,PH (1991)
OH 2.14E-010	cm3/molc sec	25	EST	MEYLAN,WM & HOWARD,PH (1993)

CAS #: 036557-05-8 — 6H-DIBENZO[B,D]PYRAN-9-METHANOL, 6A,7,8,10A-TETR

Formula: $C_{21}H_{30}O_3$

Mol Weight: 330.47

MP (deg C):
FP (deg C):
BP (deg C):
BP pressure (mm Hg):

Property/ Value	Units	Temp	Data Type	Reference
WS 2.80E+000	mg/L	25	EST	MEYLAN,WM ET AL. (1996)
logP 5.33			EXP	SANGSTER,J (1993)
VP 1.10E-010	mm Hg	25	EST	NEELY,WB & BLAU,GE (1985)
DC	pKa			
HL 8.93E-012	atm m3/mol	25	EST	MEYLAN,WM & HOWARD,PH (1991)
OH 3.12E-010	cm3/molc sec	25	EST	MEYLAN,WM & HOWARD,PH (1993)

CAS #: 036592-59-3 — 2,3,6-TRIMETHYL-4-HYDROXYACETANILIDE

Formula: $C_{11}H_{15}NO_2$

Mol Weight: 193.25

MP (deg C):
FP (deg C):
BP (deg C):
BP pressure (mm Hg):

Property/ Value	Units	Temp	Data Type	Reference
WS 1.08E+004	mg/L	25	EST	MEYLAN,WM ET AL. (1996)
logP 0.75			EXP	HANSCH,C & LEO,AJ (1985)
VP 4.04E-007	mm Hg	25	EST	NEELY,WB & BLAU,GE (1985)
DC	pKa			
HL 8.63E-013	atm m3/mol	25	EST	MEYLAN,WM & HOWARD,PH (1991)
OH 1.46E-010	cm3/molc sec	25	EST	MEYLAN,WM & HOWARD,PH (1993)

CAS #: 036614-38-7 — ISOTHIOATE

Formula: $C_7H_{17}O_2PS_3$

Mol Weight: 260.38

MP (deg C):
FP (deg C):
BP (deg C):
BP pressure (mm Hg):

Property/ Value	Units	Temp	Data Type	Reference
WS 9.70E+001	mg/L	25	EXP	YALKOWSKY,SH & DANNENFELSER,RM (1992)
logP 3.29			EST	MEYLAN,WM & HOWARD,PH (1995)
VP 1.10E-003	mm Hg	25	EST	NEELY,WB & BLAU,GE (1985)
DC	pKa			
HL 1.58E-006	atm m3/mol	25	EST	MEYLAN,WM & HOWARD,PH (1991)
OH 1.08E-010	cm3/molc sec	25	EST	MEYLAN,WM & HOWARD,PH (1993)

CAS #: 036627-56-2 — 1,1-DIME-3-(3,5-DIME PH) UREA

Formula: $C_{11}H_{16}N_2O$

Mol Weight: 192.26

MP (deg C):
FP (deg C):
BP (deg C):
BP pressure (mm Hg):

Property/ Value	Units	Temp	Data Type	Reference
WS 1.14E+003	mg/L	25	EST	MEYLAN,WM ET AL. (1996)
logP 1.90			EXP	HANSCH,C & LEO,AJ (1985)
VP 3.35E-005	mm Hg	25	EST	NEELY,WB & BLAU,GE (1985)
DC	pKa			
HL 1.18E-009	atm m3/mol	25	EST	MEYLAN,WM & HOWARD,PH (1991)
OH 2.02E-010	cm3/molc sec	25	EST	MEYLAN,WM & HOWARD,PH (1993)

CAS #: 036635-03-7 — 1-BR-PENTACHLOROCYCLOHEXANE

Formula: $C_6H_6BrCl_5$

Mol Weight: 335.29

MP (deg C):
FP (deg C):
BP (deg C):
BP pressure (mm Hg):

Property/ Value	Units	Temp	Data Type	Reference
WS 4.24E+000	mg/L	25	EST	MEYLAN,WM ET AL. (1996)
logP 3.81			EXP	HANSCH,C & LEO,AJ (1985)
VP 2.20E-004	mm Hg	25	EST	NEELY,WB & BLAU,GE (1985)
DC	pKa			
HL 8.37E-005	atm m3/mol	25	EST	MEYLAN,WM & HOWARD,PH (1991)
OH 6.01E-013	cm3/molc sec	25	EST	MEYLAN,WM & HOWARD,PH (1993)

CAS #: 036637-18-0 — ETIDOCAINE (DURANEST)

Formula: $C_{17}H_{28}N_2O$

Mol Weight: 276.43

MP (deg C):
FP (deg C):
BP (deg C):
BP pressure (mm Hg):

Property/ Value	Units	Temp	Data Type	Reference
WS 1.19E+001	mg/L	25	EST	MEYLAN,WM ET AL. (1996)
logP 3.69			EXP	HANSCH,C ET AL. (1995)
VP 1.47E-007	mm Hg	25	EST	NEELY,WB & BLAU,GE (1985)
DC	pKa			
HL 3.07E-010	atm m3/mol	25	EST	MEYLAN,WM & HOWARD,PH (1991)
OH 1.26E-010	cm3/molc sec	25	EST	MEYLAN,WM & HOWARD,PH (1993)

CAS #: 036653-82-4				1-HEXADECANOL
Formula: C$_{16}$H$_{34}$O				
Mol Weight: 242.45				
MP (deg C): 49.3	FP (deg C):			
BP (deg C): 334				
BP pressure (mm Hg):				

Property/Value	Units	Temp	Data Type	Reference
WS 1.34E-002	mg/L	25	EXP	YALKOWSKY,SH & DANNENFELSER,RM (1992)
logP 6.73			EST	MEYLAN,WM & HOWARD,PH (1995)
VP 3.06E-006	mm Hg	30	EXP	LITTLEWOOD,R (1957)
DC 16.20	pKa	25	EST	BURKHARD,LP ET AL (1985B)
HL 7.28E-005	atm m3/mol	30	EST	VP/WSOL
OH 2.31E-011	cm3/molc sec	25	EST	MEYLAN,WM & HOWARD,PH (1993)

CAS #: 036664-18-3				MORPHOLINE, 4-[2-(2-NITRO-1H-IMIDAZOL-1-YL)ETHYL
Formula: C$_9$H$_{14}$N$_4$O$_3$				
Mol Weight: 226.24				
MP (deg C):	FP (deg C):			
BP (deg C):				
BP pressure (mm Hg):				

Property/Value	Units	Temp	Data Type	Reference
WS 1.61E+005	mg/L	25	EST	MEYLAN,WM ET AL. (1996)
logP -0.10			EXP	SANGSTER,J (1993)
VP 8.82E-007	mm Hg	25	EST	NEELY,WB & BLAU,GE (1985)
DC	pKa			
HL 1.43E-012	atm m3/mol	25	EST	MEYLAN,WM & HOWARD,PH (1991)
OH 1.57E-010	cm3/molc sec	25	EST	MEYLAN,WM & HOWARD,PH (1993)

CAS #: 036734-19-7				ROVRAL
Formula: C$_{13}$H$_{13}$Cl$_2$N$_3$O$_3$				
Mol Weight: 330.17				
MP (deg C): 136	FP (deg C):			
BP (deg C):				
BP pressure (mm Hg):				

Property/Value	Units	Temp	Data Type	Reference
WS 1.39E+001	mg/L	25	EXP	WAUCHOPE,RD ET AL. (1991A)
logP 3.00			EXP	TOMLIN,C (1994)
VP 1.00E-007	mm Hg	25	EST	WAUCHOPE,RD ET AL. (1991A)
DC	pKa			
HL 1.00E-013	atm m3/mol	25	EST	MEYLAN,WM & HOWARD,PH (1991)
OH 4.62E-011	cm3/molc sec	25	EST	MEYLAN,WM & HOWARD,PH (1993)

CAS #: 036735-22-5				2H-1,4-BENZODIAZEPINE-2-THIONE, 7-CHLORO-5-(2-FL
Formula: C$_{17}$H$_{11}$ClF$_4$N$_2$S				
Mol Weight: 386.80				
MP (deg C): 137.5-139	FP (deg C):			
BP (deg C):				
BP pressure (mm Hg):				

Property/Value	Units	Temp	Data Type	Reference
WS 1.34E+000	mg/L	25	EST	MEYLAN,WM ET AL. (1996)
logP 4.03			EXP	SANGSTER,J (1994)
VP 9.17E-009	mm Hg	25	EST	NEELY,WB & BLAU,GE (1985)
DC	pKa			
HL 1.28E-006	atm m3/mol	25	EST	MEYLAN,WM & HOWARD,PH (1991)
OH 5.60E-011	cm3/molc sec	25	EST	MEYLAN,WM & HOWARD,PH (1993)

CAS #: 036756-79-3				TIOCARBAZIL
Formula: C$_{16}$H$_{25}$NOS				
Mol Weight: 279.45				
MP (deg C):	FP (deg C):			
BP (deg C):				
BP pressure (mm Hg):				

Property/Value	Units	Temp	Data Type	Reference
WS 2.50E+000	mg/L	30	EXP	SHIU,WY ET AL. (1990)
logP 4.40			EXP	HANSCH,C ET AL. (1995)
VP 2.97E-006	mm Hg	25	EST	NEELY,WB & BLAU,GE (1985)
DC	pKa			
HL 1.65E-006	atm m3/mol	25	EST	MEYLAN,WM & HOWARD,PH (1991)
OH 4.31E-011	cm3/molc sec	25	EST	MEYLAN,WM & HOWARD,PH (1993)

CAS #: 036865-51-7				3,4,5-TRICL-C6H2NHN=C(CN)CO-OET
Formula: C$_{11}$H$_8$Cl$_3$N$_3$O$_2$				
Mol Weight: 320.56				
MP (deg C):	FP (deg C):			
BP (deg C):				
BP pressure (mm Hg):				

Property/Value	Units	Temp	Data Type	Reference
WS 4.71E-001	mg/L	25	EST	MEYLAN,WM ET AL. (1996)
logP 5.03			EXP	HANSCH,C & LEO,AJ (1985)
VP 2.52E-007	mm Hg	25	EST	NEELY,WB & BLAU,GE (1985)
DC	pKa			
HL 3.58E-007	atm m3/mol	25	EST	MEYLAN,WM & HOWARD,PH (1991)
OH 7.83E-012	cm3/molc sec	25	EST	MEYLAN,WM & HOWARD,PH (1993)

CAS #: 036865-53-9				4,5-(-OCF2CF2O-)C6H3NHN=C(CN)CO-OME
Formula: C$_{12}$H$_7$F$_4$N$_3$O$_4$				
Mol Weight: 333.20				
MP (deg C):	FP (deg C):			
BP (deg C):				
BP pressure (mm Hg):				

Property/Value	Units	Temp	Data Type	Reference
WS 5.53E-001	mg/L	25	EST	MEYLAN,WM ET AL. (1996)
logP 4.86			EXP	HANSCH,C & LEO,AJ (1985)
VP 1.24E-006	mm Hg	25	EST	NEELY,WB & BLAU,GE (1985)
DC	pKa			
HL 1.60E-008	atm m3/mol	25	EST	MEYLAN,WM & HOWARD,PH (1991)
OH 2.00E-010	cm3/molc sec	25	EST	MEYLAN,WM & HOWARD,PH (1993)

CAS #: 036865-54-0				3,5-DI-CF3-C6H3NHN=C(CN)COOME-CIS
Formula: C$_{12}$H$_7$F$_6$N$_3$O$_2$				
Mol Weight: 339.20				
MP (deg C):	FP (deg C):			
BP (deg C):				
BP pressure (mm Hg):				

Property/Value	Units	Temp	Data Type	Reference
WS 2.82E-001	mg/L	25	EST	MEYLAN,WM ET AL. (1996)
logP 5.02			EXP	HANSCH,C & LEO,AJ (1985)
VP 2.11E-005	mm Hg	25	EST	NEELY,WB & BLAU,GE (1985)
DC	pKa			
HL 5.01E-005	atm m3/mol	25	EST	MEYLAN,WM & HOWARD,PH (1991)
OH 1.19E-012	cm3/molc sec	25	EST	MEYLAN,WM & HOWARD,PH (1993)

CAS #: 036865-56-2 — 3,4,5-TRICL-C6H2NHN=C(CN)CO-OME

Formula: $C_{10}H_6Cl_3N_3O_2$
Mol Weight: 306.54
MP (deg C):   **FP (deg C):**
BP (deg C):
BP pressure (mm Hg):

Property/Value	Units	Temp	Data Type	Reference
WS 3.92E-001	mg/L	25	EST	MEYLAN,WM ET AL. (1996)
logP 5.22			EXP	HANSCH,C & LEO,AJ (1985)
VP 5.92E-007	mm Hg	25	EST	NEELY,WB & BLAU,GE (1985)
DC	pKa			
HL 2.70E-007	atm m3/mol	25	EST	MEYLAN,WM & HOWARD,PH (1991)
OH 6.38E-012	cm3/molc sec	25	EST	MEYLAN,WM & HOWARD,PH (1993)

CAS #: 036865-60-8 — 4,5-(-OCF2CF2O-)C6H3NHN=C(CN)CO-ME

Formula: $C_{12}H_7F_4N_3O_3$
Mol Weight: 317.20
MP (deg C):   **FP (deg C):**
BP (deg C):
BP pressure (mm Hg):

Property/Value	Units	Temp	Data Type	Reference
WS 2.84E-001	mg/L	25	EST	MEYLAN,WM ET AL. (1996)
logP 5.31			EXP	HANSCH,C & LEO,AJ (1985)
VP 1.51E-006	mm Hg	25	EST	NEELY,WB & BLAU,GE (1985)
DC	pKa			
HL 4.53E-009	atm m3/mol	25	EST	MEYLAN,WM & HOWARD,PH (1991)
OH 2.00E-010	cm3/molc sec	25	EST	MEYLAN,WM & HOWARD,PH (1993)

CAS #: 036865-72-2 — 2-CF3-4-CL-C6H3NHN=C(CN)COOME

Formula: $C_{11}H_7ClF_3N_3O_2$
Mol Weight: 305.65
MP (deg C):   **FP (deg C):**
BP (deg C):
BP pressure (mm Hg):

Property/Value	Units	Temp	Data Type	Reference
WS 1.19E+000	mg/L	25	EST	MEYLAN,WM ET AL. (1996)
logP 4.66			EXP	HANSCH,C & LEO,AJ (1985)
VP 6.99E-006	mm Hg	25	EST	NEELY,WB & BLAU,GE (1985)
DC	pKa			
HL 4.27E-006	atm m3/mol	25	EST	MEYLAN,WM & HOWARD,PH (1991)
OH 2.72E-012	cm3/molc sec	25	EST	MEYLAN,WM & HOWARD,PH (1993)

CAS #: 036865-77-7 — 3,4-DICL-C6H3NHN=C(CN)COOET

Formula: $C_{11}H_9Cl_2N_3O_2$
Mol Weight: 286.12
MP (deg C):   **FP (deg C):**
BP (deg C):
BP pressure (mm Hg):

Property/Value	Units	Temp	Data Type	Reference
WS 1.55E+000	mg/L	25	EST	MEYLAN,WM ET AL. (1996)
logP 4.66			EXP	HANSCH,C & LEO,AJ (1985)
VP 1.16E-006	mm Hg	25	EST	NEELY,WB & BLAU,GE (1985)
DC	pKa			
HL 4.83E-007	atm m3/mol	25	EST	MEYLAN,WM & HOWARD,PH (1991)
OH 1.04E-011	cm3/molc sec	25	EST	MEYLAN,WM & HOWARD,PH (1993)

CAS #: 036874-60-9 — 3-CHF2-4-CL-C6H3NHN=C(CN)CO-OME

Formula: $C_{11}H_8ClF_2N_3O_2$
Mol Weight: 287.66
MP (deg C):   **FP (deg C):**
BP (deg C):
BP pressure (mm Hg):

Property/Value	Units	Temp	Data Type	Reference
WS 8.23E+000	mg/L	25	EST	MEYLAN,WM ET AL. (1996)
logP 3.80			EXP	HANSCH,C & LEO,AJ (1985)
VP 6.49E-006	mm Hg	25	EST	NEELY,WB & BLAU,GE (1985)
DC	pKa			
HL 2.15E-006	atm m3/mol	25	EST	MEYLAN,WM & HOWARD,PH (1991)
OH 1.31E-011	cm3/molc sec	25	EST	MEYLAN,WM & HOWARD,PH (1993)

CAS #: 036874-67-6 — 3-CL-C6H4NHN=C(CN)COOET

Formula: $C_{11}H_{10}ClN_3O_2$
Mol Weight: 251.67
MP (deg C):   **FP (deg C):**
BP (deg C):
BP pressure (mm Hg):

Property/Value	Units	Temp	Data Type	Reference
WS 1.00E+001	mg/L	25	EST	MEYLAN,WM ET AL. (1996)
logP 3.94			EXP	HANSCH,C & LEO,AJ (1985)
VP 4.72E-006	mm Hg	25	EST	NEELY,WB & BLAU,GE (1985)
DC	pKa			
HL 6.52E-007	atm m3/mol	25	EST	MEYLAN,WM & HOWARD,PH (1991)
OH 3.17E-011	cm3/molc sec	25	EST	MEYLAN,WM & HOWARD,PH (1993)

CAS #: 036874-69-8 — ACETIC ACID, ((3-CHLOROPHENYL)HYDRAZONO)CYANO-,

Formula: $C_{10}H_8ClN_3O_2$
Mol Weight: 237.65
MP (deg C):   **FP (deg C):**
BP (deg C):
BP pressure (mm Hg):

Property/Value	Units	Temp	Data Type	Reference
WS 2.33E+001	mg/L	25	EST	MEYLAN,WM ET AL. (1996)
logP 3.60			EXP	HANSCH,C & LEO,AJ (1985)
VP 1.10E-005	mm Hg	25	EST	NEELY,WB & BLAU,GE (1985)
DC	pKa			
HL 4.91E-007	atm m3/mol	25	EST	MEYLAN,WM & HOWARD,PH (1991)
OH 3.02E-011	cm3/molc sec	25	EST	MEYLAN,WM & HOWARD,PH (1993)

CAS #: 036874-72-3 — 2,5-DICYANO-C6H3NHN=C(CN)COOME

Formula: $C_{12}H_7N_5O_2$
Mol Weight: 253.22
MP (deg C):   **FP (deg C):**
BP (deg C):
BP pressure (mm Hg):

Property/Value	Units	Temp	Data Type	Reference
WS 4.91E+001	mg/L	25	EST	MEYLAN,WM ET AL. (1996)
logP 2.81			EXP	HANSCH,C & LEO,AJ (1985)
VP 4.03E-008	mm Hg	25	EST	NEELY,WB & BLAU,GE (1985)
DC	pKa			
HL 6.19E-011	atm m3/mol	25	EST	MEYLAN,WM & HOWARD,PH (1991)
OH 1.20E-012	cm3/molc sec	25	EST	MEYLAN,WM & HOWARD,PH (1993)

CAS #: 036874-74-5				C6H5NHN=C(CN)CO-OME
Formula: $C_{10}H_9N_3O_2$				
Mol Weight: 203.20				
MP (deg C):		FP (deg C):		
BP (deg C):				
BP pressure (mm Hg):				

Property/ Value	Units	Temp	Data Type	Reference
WS 2.59E+002	mg/L	25	EST	MEYLAN,WM ET AL. (1996)
logP 2.59			EXP	HANSCH,C & LEO,AJ (1985)
VP 4.87E-005	mm Hg	25	EST	NEELY,WB & BLAU,GE (1985)
DC	pKa			
HL 6.63E-007	atm m3/mol	25	EST	MEYLAN,WM & HOWARD,PH (1991)
OH 4.29E-011	cm3/molc sec	25	EST	MEYLAN,WM & HOWARD,PH (1993)

CAS #: 036894-69-6				LABETALOL
Formula: $C_{19}H_{24}N_2O_3$				
Mol Weight: 328.41				
MP (deg C): 187-189		FP (deg C):		
BP (deg C):				
BP pressure (mm Hg):				

Property/ Value	Units	Temp	Data Type	Reference
WS 7.29E+001	mg/L	25	EST	MEYLAN,WM ET AL. (1996)
logP 3.09			EXP	HANSCH,C & LEO,AJ (1985)
VP 1.10E-013	mm Hg	25	EST	NEELY,WB & BLAU,GE (1985)
DC	pKa			
HL 6.63E-019	atm m3/mol	25	EST	MEYLAN,WM & HOWARD,PH (1991)
OH 1.51E-010	cm3/molc sec	25	EST	MEYLAN,WM & HOWARD,PH (1993)

CAS #: 036905-00-7				3,5-DI-CF3-C6H3NHN=C(CN)COOME-TRANS
Formula: $C_{12}H_7F_6N_3O_2$				
Mol Weight: 339.20				
MP (deg C):		FP (deg C):		
BP (deg C):				
BP pressure (mm Hg):				

Property/ Value	Units	Temp	Data Type	Reference
WS 2.82E-001	mg/L	25	EST	MEYLAN,WM ET AL. (1996)
logP 5.16			EXP	HANSCH,C & LEO,AJ (1985)
VP 2.11E-005	mm Hg	25	EST	NEELY,WB & BLAU,GE (1985)
DC	pKa			
HL 5.01E-005	atm m3/mol	25	EST	MEYLAN,WM & HOWARD,PH (1991)
OH 1.19E-012	cm3/molc sec	25	EST	MEYLAN,WM & HOWARD,PH (1993)

CAS #: 036905-04-1				ACETIC ACID, CYANO (2,4,5-TRICHLOROPHENYL) HYDRA
Formula: $C_{10}H_6Cl_3N_3O_2$				
Mol Weight: 306.54				
MP (deg C):		FP (deg C):		
BP (deg C):				
BP pressure (mm Hg):				

Property/ Value	Units	Temp	Data Type	Reference
WS 4.50E-001	mg/L	25	EST	MEYLAN,WM ET AL. (1996)
logP 5.15			EXP	HANSCH,C & LEO,AJ (1985)
VP 5.92E-007	mm Hg	25	EST	NEELY,WB & BLAU,GE (1985)
DC	pKa			
HL 2.70E-007	atm m3/mol	25	EST	MEYLAN,WM & HOWARD,PH (1991)
OH 2.78E-012	cm3/molc sec	25	EST	MEYLAN,WM & HOWARD,PH (1993)

CAS #: 036913-39-0				PYRIDINIUM, 1-METHYL-4-PHENYL-, IODIDE
Formula: $C_{12}H_{12}IN$				
Mol Weight: 297.14				
MP (deg C):		FP (deg C):		
BP (deg C):				
BP pressure (mm Hg):				

Property/ Value	Units	Temp	Data Type	Reference
WS 1.00E+006	mg/L	25	EST	MEYLAN,WM ET AL. (1996)
logP -2.28			EXP	EL TAYAR,N ET AL. (1991)
VP 2.94E-005	mm Hg	25	EST	NEELY,WB & BLAU,GE (1985)
DC	pKa			
HL 1.15E-011	atm m3/mol	25	EST	MEYLAN,WM & HOWARD,PH (1991)
OH 1.33E-011	cm3/molc sec	25	EST	MEYLAN,WM & HOWARD,PH (1993)

CAS #: 037077-84-2				2,3,3-TRICHLOROPROPENE
Formula: $C_3H_3Cl_3$				
Mol Weight: 145.42				
MP (deg C):		FP (deg C):		
BP (deg C):				
BP pressure (mm Hg):				

Property/ Value	Units	Temp	Data Type	Reference
WS 4.73E+002	mg/L	25	EST	MEYLAN,WM ET AL. (1996)
logP 2.61			EST	MEYLAN,WM & HOWARD,PH (1995)
VP 1.91E+001	mm Hg	25	EST	NEELY,WB & BLAU,GE (1985)
DC	pKa			
HL 8.63E-003	atm m3/mol	25	EST	MEYLAN,WM & HOWARD,PH (1991)
OH 8.48E-012	cm3/molc sec	25	EST	MEYLAN,WM & HOWARD,PH (1993)

CAS #: 037096-14-3				2-PYRROLIDINONE, 3-HYDROXY-1-METHYL-5-(3-PYRIDIN
Formula: $C_{10}H_{12}N_2O_2$				
Mol Weight: 192.22				
MP (deg C):		FP (deg C):		
BP (deg C):				
BP pressure (mm Hg):				

Property/ Value	Units	Temp	Data Type	Reference
WS 1.00E+006	mg/L	25	EST	MEYLAN,WM ET AL. (1996)
logP -1.48			EXP	LI,NY & GORROD,JW (1992)
VP 2.16E-007	mm Hg	25	EST	NEELY,WB & BLAU,GE (1985)
DC	pKa			
HL 5.20E-013	atm m3/mol	25	EST	MEYLAN,WM & HOWARD,PH (1991)
OH 3.30E-011	cm3/molc sec	25	EST	MEYLAN,WM & HOWARD,PH (1993)

CAS #: 037107-20-3				ACETAMIDE, N-[1-(4-CHLOROPHENYL)-2,5-DIOXO-3-PYR
Formula: $C_{12}H_9ClN_2O_3$				
Mol Weight: 264.67				
MP (deg C):		FP (deg C):		
BP (deg C):				
BP pressure (mm Hg):				

Property/ Value	Units	Temp	Data Type	Reference
WS 4.48E+003	mg/L	25	EST	MEYLAN,WM ET AL. (1996)
logP 0.75			EXP	HANSCH,C ET AL. (1995)
VP 4.92E-011	mm Hg	25	EST	NEELY,WB & BLAU,GE (1985)
DC	pKa			
HL 3.36E-013	atm m3/mol	25	EST	MEYLAN,WM & HOWARD,PH (1991)
OH 1.56E-011	cm3/molc sec	25	EST	MEYLAN,WM & HOWARD,PH (1993)

CAS #: 037148-00-8

124-BENZOTHIADIAZINE-1-O2-3-ME-6-BR

Formula: $C_8H_7BrN_2O_2S$

Mol Weight: 275.13

MP (deg C): FP (deg C):

BP (deg C):

BP pressure (mm Hg):

Property/Value	Units	Temp	Data Type	Reference
WS 1.16E+003	mg/L	25	EST	MEYLAN,WM ET AL. (1996)
logP 1.37			EXP	HANSCH,C & LEO,AJ (1985)
VP 1.25E-008	mm Hg	25	EST	NEELY,WB & BLAU,GE (1985)
DC	pKa			
HL 3.68E-008	atm m3/mol	25	EST	MEYLAN,WM & HOWARD,PH (1991)
OH 3.32E-013	cm3/molc sec	25	EST	MEYLAN,WM & HOWARD,PH (1993)

CAS #: 037148-02-0

124-BENZOTHIADIAZINE,7-I-3-ME,1,1-DIO

Formula: $C_8H_7IN_2O_2S$

Mol Weight: 322.13

MP (deg C): FP (deg C):

BP (deg C):

BP pressure (mm Hg):

Property/Value	Units	Temp	Data Type	Reference
WS 3.84E+002	mg/L	25	EST	MEYLAN,WM ET AL. (1996)
logP 1.61			EXP	HANSCH,C ET AL. (1995)
VP 3.50E-009	mm Hg	25	EST	NEELY,WB & BLAU,GE (1985)
DC	pKa			
HL 2.14E-008	atm m3/mol	25	EST	MEYLAN,WM & HOWARD,PH (1991)
OH 4.69E-013	cm3/molc sec	25	EST	MEYLAN,WM & HOWARD,PH (1993)

CAS #: 037148-03-1

124-BENZOTHIADIAZINE-1-O2-3-ME-6-ET

Formula: $C_{10}H_{12}N_2O_2S$

Mol Weight: 224.28

MP (deg C): FP (deg C):

BP (deg C):

BP pressure (mm Hg):

Property/Value	Units	Temp	Data Type	Reference
WS 2.80E+003	mg/L	25	EST	MEYLAN,WM ET AL. (1996)
logP 1.25			EXP	HANSCH,C & LEO,AJ (1985)
VP 2.18E-008	mm Hg	25	EST	NEELY,WB & BLAU,GE (1985)
DC	pKa			
HL 1.36E-007	atm m3/mol	25	EST	MEYLAN,WM & HOWARD,PH (1991)
OH 1.92E-012	cm3/molc sec	25	EST	MEYLAN,WM & HOWARD,PH (1993)

CAS #: 037148-07-5

124-BENZOTHIADIAZINE-1-O2-3-ME-6-NO2

Formula: $C_8H_7N_3O_4S$

Mol Weight: 241.23

MP (deg C): FP (deg C):

BP (deg C):

BP pressure (mm Hg):

Property/Value	Units	Temp	Data Type	Reference
WS 3.00E+003	mg/L	25	EST	MEYLAN,WM ET AL. (1996)
logP 0.65			EXP	HANSCH,C & LEO,AJ (1985)
VP 2.18E-009	mm Hg	25	EST	NEELY,WB & BLAU,GE (1985)
DC	pKa			
HL 3.65E-010	atm m3/mol	25	EST	MEYLAN,WM & HOWARD,PH (1991)
OH 2.24E-013	cm3/molc sec	25	EST	MEYLAN,WM & HOWARD,PH (1993)

CAS #: 037148-08-6

124-BENZTHIADIZ-1-O2-3-ME-6-NH2-7-CL

Formula: $C_8H_8ClN_3O_2S$

Mol Weight: 245.69

MP (deg C): FP (deg C):

BP (deg C):

BP pressure (mm Hg):

Property/Value	Units	Temp	Data Type	Reference
WS 7.24E+003	mg/L	25	EST	MEYLAN,WM ET AL. (1996)
logP 0.63			EXP	HANSCH,C & LEO,AJ (1985)
VP 1.47E-009	mm Hg	25	EST	NEELY,WB & BLAU,GE (1985)
DC	pKa			
HL 2.42E-011	atm m3/mol	25	EST	MEYLAN,WM & HOWARD,PH (1991)
OH 4.40E-012	cm3/molc sec	25	EST	MEYLAN,WM & HOWARD,PH (1993)

CAS #: 037148-09-7

124BENZOTHIADIAZN-11O2-3ME-6NHAC-7CL

Formula: $C_{10}H_{10}ClN_3O_3S$

Mol Weight: 287.73

MP (deg C): FP (deg C):

BP (deg C):

BP pressure (mm Hg):

Property/Value	Units	Temp	Data Type	Reference
WS 5.10E+003	mg/L	25	EST	MEYLAN,WM ET AL. (1996)
logP 0.53			EXP	HANSCH,C & LEO,AJ (1985)
VP 6.09E-012	mm Hg	25	EST	NEELY,WB & BLAU,GE (1985)
DC	pKa			
HL 7.85E-014	atm m3/mol	25	EST	MEYLAN,WM & HOWARD,PH (1991)
OH 1.72E-011	cm3/molc sec	25	EST	MEYLAN,WM & HOWARD,PH (1993)

CAS #: 037148-13-3

124-BENZTHIADIAZIN-11SO2-3ME5I7CL

Formula: $C_8H_6ClIN_2O_2S$

Mol Weight: 356.57

MP (deg C): FP (deg C):

BP (deg C):

BP pressure (mm Hg):

Property/Value	Units	Temp	Data Type	Reference
WS 1.61E+002	mg/L	25	EST	MEYLAN,WM ET AL. (1996)
logP 1.81			EXP	HANSCH,C & LEO,AJ (1985)
VP 9.39E-010	mm Hg	25	EST	NEELY,WB & BLAU,GE (1985)
DC	pKa			
HL 1.59E-008	atm m3/mol	25	EST	MEYLAN,WM & HOWARD,PH (1991)
OH 3.79E-013	cm3/molc sec	25	EST	MEYLAN,WM & HOWARD,PH (1993)

CAS #: 037148-19-9

124-BENZOTHIADIAZINE-1-O2-3-PR-6-CL

Formula: $C_{10}H_{11}ClN_2O_2S$

Mol Weight: 258.73

MP (deg C): FP (deg C):

BP (deg C):

BP pressure (mm Hg):

Property/Value	Units	Temp	Data Type	Reference
WS 3.83E+002	mg/L	25	EST	MEYLAN,WM ET AL. (1996)
logP 2.04			EXP	HANSCH,C & LEO,AJ (1985)
VP 5.72E-009	mm Hg	25	EST	NEELY,WB & BLAU,GE (1985)
DC	pKa			
HL 1.21E-007	atm m3/mol	25	EST	MEYLAN,WM & HOWARD,PH (1991)
OH 2.91E-012	cm3/molc sec	25	EST	MEYLAN,WM & HOWARD,PH (1993)

CAS #: 037148-20-2				124-BENZOTHIADIAZIN-1-O2-3-IPR-6-CL
Formula: $C_{10}H_{11}ClN_2O_2S$				
Mol Weight: 258.73				
MP (deg C):		FP (deg C):		
BP (deg C):				
BP pressure (mm Hg):				

Property/Value	Units	Temp	Data Type	Reference
WS 4.14E+002	mg/L	25	EST	MEYLAN,WM ET AL. (1996)
logP 2.00			EXP	HANSCH,C & LEO,AJ (1985)
VP 9.33E-009	mm Hg	25	EST	NEELY,WB & BLAU,GE (1985)
DC	pKa			
HL 1.21E-007	atm m3/mol	25	EST	MEYLAN,WM & HOWARD,PH (1991)
OH 2.91E-012	cm3/molc sec	25	EST	MEYLAN,WM & HOWARD,PH (1993)

CAS #: 037148-21-3				124-BENZOTHIADIAZIN-1-O2-3-IBU-6-CL
Formula: $C_{11}H_{13}ClN_2O_2S$				
Mol Weight: 272.76				
MP (deg C):		FP (deg C):		
BP (deg C):				
BP pressure (mm Hg):				

Property/Value	Units	Temp	Data Type	Reference
WS 1.74E+002	mg/L	25	EST	MEYLAN,WM ET AL. (1996)
logP 2.35			EXP	HANSCH,C & LEO,AJ (1985)
VP 4.15E-009	mm Hg	25	EST	NEELY,WB & BLAU,GE (1985)
DC	pKa			
HL 1.60E-007	atm m3/mol	25	EST	MEYLAN,WM & HOWARD,PH (1991)
OH 4.32E-012	cm3/molc sec	25	EST	MEYLAN,WM & HOWARD,PH (1993)

CAS #: 037148-22-4				124-BENZOTHIADIAZIN-1-O2-3-SBU-7-CL
Formula: $C_{11}H_{13}ClN_2O_2S$				
Mol Weight: 272.76				
MP (deg C):		FP (deg C):		
BP (deg C):				
BP pressure (mm Hg):				

Property/Value	Units	Temp	Data Type	Reference
WS 1.37E+002	mg/L	25	EST	MEYLAN,WM ET AL. (1996)
logP 2.47			EXP	HANSCH,C & LEO,AJ (1985)
VP 4.15E-009	mm Hg	25	EST	NEELY,WB & BLAU,GE (1985)
DC	pKa			
HL 1.60E-007	atm m3/mol	25	EST	MEYLAN,WM & HOWARD,PH (1991)
OH 4.74E-012	cm3/molc sec	25	EST	MEYLAN,WM & HOWARD,PH (1993)

CAS #: 037148-24-6				124-BENZTHIADIAZIN-1-O2-3-CYBU-7-CL
Formula: $C_{11}H_{11}ClN_2O_2S$				
Mol Weight: 270.74				
MP (deg C):		FP (deg C):		
BP (deg C):				
BP pressure (mm Hg):				

Property/Value	Units	Temp	Data Type	Reference
WS 2.21E+002	mg/L	25	EST	MEYLAN,WM ET AL. (1996)
logP 2.24			EXP	HANSCH,C & LEO,AJ (1985)
VP 2.05E-009	mm Hg	25	EST	NEELY,WB & BLAU,GE (1985)
DC	pKa			
HL 7.07E-008	atm m3/mol	25	EST	MEYLAN,WM & HOWARD,PH (1991)
OH 2.52E-012	cm3/molc sec	25	EST	MEYLAN,WM & HOWARD,PH (1993)

CAS #: 037157-54-3				124-BENZTHIADIAZN-1-O2-3-NEOPE-7-CL
Formula: $C_{12}H_{15}ClN_2O_2S$				
Mol Weight: 286.78				
MP (deg C):		FP (deg C):		
BP (deg C):				
BP pressure (mm Hg):				

Property/Value	Units	Temp	Data Type	Reference
WS 7.10E+001	mg/L	25	EST	MEYLAN,WM ET AL. (1996)
logP 2.71			EXP	HANSCH,C & LEO,AJ (1985)
VP 2.54E-009	mm Hg	25	EST	NEELY,WB & BLAU,GE (1985)
DC	pKa			
HL 2.13E-007	atm m3/mol	25	EST	MEYLAN,WM & HOWARD,PH (1991)
OH 2.24E-012	cm3/molc sec	25	EST	MEYLAN,WM & HOWARD,PH (1993)

CAS #: 037157-57-6				124-BENZTHIADIAZIN-1-O2-3-CLME-6-CL
Formula: $C_8H_6Cl_2N_2O_2S$				
Mol Weight: 265.12				
MP (deg C):		FP (deg C):		
BP (deg C):				
BP pressure (mm Hg):				

Property/Value	Units	Temp	Data Type	Reference
WS 7.16E+002	mg/L	25	EST	MEYLAN,WM ET AL. (1996)
logP 1.68			EXP	HANSCH,C & LEO,AJ (1985)
VP 3.30E-009	mm Hg	25	EST	NEELY,WB & BLAU,GE (1985)
DC	pKa			
HL 2.41E-008	atm m3/mol	25	EST	MEYLAN,WM & HOWARD,PH (1991)
OH 6.21E-013	cm3/molc sec	25	EST	MEYLAN,WM & HOWARD,PH (1993)

CAS #: 037157-59-8				124-BENZTHIADIAZN-1-O2-3-ALLYL-6-CL
Formula: $C_{10}H_9ClN_2O_2S$				
Mol Weight: 256.71				
MP (deg C):		FP (deg C):		
BP (deg C):				
BP pressure (mm Hg):				

Property/Value	Units	Temp	Data Type	Reference
WS 4.60E+002	mg/L	25	EST	MEYLAN,WM ET AL. (1996)
logP 1.96			EXP	HANSCH,C & LEO,AJ (1985)
VP 6.08E-009	mm Hg	25	EST	NEELY,WB & BLAU,GE (1985)
DC	pKa			
HL 9.00E-008	atm m3/mol	25	EST	MEYLAN,WM & HOWARD,PH (1991)
OH 2.76E-011	cm3/molc sec	25	EST	MEYLAN,WM & HOWARD,PH (1993)

CAS #: 037157-71-4				124-BENZTHIADIAZNE-1-O2-3-MEOME-6-CL
Formula: $C_9H_9ClN_2O_3S$				
Mol Weight: 260.70				
MP (deg C):		FP (deg C):		
BP (deg C):				
BP pressure (mm Hg):				

Property/Value	Units	Temp	Data Type	Reference
WS 2.28E+003	mg/L	25	EST	MEYLAN,WM ET AL. (1996)
logP 1.12			EXP	HANSCH,C & LEO,AJ (1985)
VP 5.55E-009	mm Hg	25	EST	NEELY,WB & BLAU,GE (1985)
DC	pKa			
HL 8.02E-010	atm m3/mol	25	EST	MEYLAN,WM & HOWARD,PH (1991)
OH 8.02E-012	cm3/molc sec	25	EST	MEYLAN,WM & HOWARD,PH (1993)

CAS #: 037157-79-2 — 124-BENZTHIADIAZIN-1-O2-3ME-5NO2-7CL

Formula: $C_8H_6ClN_3O_4S$

Mol Weight: 275.67

MP (deg C):   FP (deg C):

BP (deg C):

BP pressure (mm Hg):

Property/Value	Units	Temp	Data Type	Reference
WS 1.30E+003	mg/L	25	EST	MEYLAN,WM ET AL. (1996)
logP 0.85			EXP	HANSCH,C & LEO,AJ (1985)
VP 5.83E-010	mm Hg	25	EST	NEELY,WB & BLAU,GE (1985)
DC	pKa			
HL 2.70E-010	atm m3/mol	25	EST	MEYLAN,WM & HOWARD,PH (1991)
OH 1.95E-013	cm3/molc sec	25	EST	MEYLAN,WM & HOWARD,PH (1993)

CAS #: 037157-82-7 — 124-BENZTHIDIAZIN-1O2-3ME-7SO2N(DIME)

Formula: $C_{10}H_{13}N_3O_4S_2$

Mol Weight: 303.36

MP (deg C):   FP (deg C):

BP (deg C):

BP pressure (mm Hg):

Property/Value	Units	Temp	Data Type	Reference
WS 5.56E+002	mg/L	25	EST	MEYLAN,WM ET AL. (1996)
logP 0.32			EXP	HANSCH,C & LEO,AJ (1985)
VP 2.86E-011	mm Hg	25	EST	NEELY,WB & BLAU,GE (1985)
DC	pKa			
HL 3.49E-011	atm m3/mol	25	EST	MEYLAN,WM & HOWARD,PH (1991)
OH 2.79E-012	cm3/molc sec	25	EST	MEYLAN,WM & HOWARD,PH (1993)

CAS #: 037157-85-0 — 124-BENZTHIADIAZINE-1-O2-3-BU-7-CL

Formula: $C_{11}H_{13}ClN_2O_2S$

Mol Weight: 272.76

MP (deg C):   FP (deg C):

BP (deg C):

BP pressure (mm Hg):

Property/Value	Units	Temp	Data Type	Reference
WS 1.24E+002	mg/L	25	EST	MEYLAN,WM ET AL. (1996)
logP 2.52			EXP	HANSCH,C & LEO,AJ (1985)
VP 2.54E-009	mm Hg	25	EST	NEELY,WB & BLAU,GE (1985)
DC	pKa			
HL 1.60E-007	atm m3/mol	25	EST	MEYLAN,WM & HOWARD,PH (1991)
OH 4.46E-012	cm3/molc sec	25	EST	MEYLAN,WM & HOWARD,PH (1993)

CAS #: 037157-88-3 — 124-BENZTHIADIAZIN-1-O2-3-CYHEXYL

Formula: $C_{13}H_{16}N_2O_2S$

Mol Weight: 264.35

MP (deg C):   FP (deg C):

BP (deg C):

BP pressure (mm Hg):

Property/Value	Units	Temp	Data Type	Reference
WS 2.60E+002	mg/L	25	EST	MEYLAN,WM ET AL. (1996)
logP 2.20			EXP	HANSCH,C & LEO,AJ (1985)
VP 1.29E-009	mm Hg	25	EST	NEELY,WB & BLAU,GE (1985)
DC	pKa			
HL 1.68E-007	atm m3/mol	25	EST	MEYLAN,WM & HOWARD,PH (1991)
OH 1.98E-011	cm3/molc sec	25	EST	MEYLAN,WM & HOWARD,PH (1993)

CAS #: 037157-89-4 — 2H-1,2,4-BENZOTHIADIAZINE, 3-(1-CYCLOPENTEN-1-YL

Formula: $C_{12}H_{12}N_2O_2S$

Mol Weight: 248.31

MP (deg C):   FP (deg C):

BP (deg C):

BP pressure (mm Hg):

Property/Value	Units	Temp	Data Type	Reference
WS 8.71E+002	mg/L	25	EST	MEYLAN,WM ET AL. (1996)
logP 1.69			EXP	HANSCH,C ET AL. (1995)
VP 2.38E-009	mm Hg	25	EST	NEELY,WB & BLAU,GE (1985)
DC	pKa			
HL 1.32E-007	atm m3/mol	25	EST	MEYLAN,WM & HOWARD,PH (1991)
OH 9.03E-011	cm3/molc sec	25	EST	MEYLAN,WM & HOWARD,PH (1993)

CAS #: 037157-91-8 — 124-BENZTHIADIAZIN-1-O2-3CYHEXEN-3-YL

Formula: $C_{13}H_{14}N_2O_2S$

Mol Weight: 262.33

MP (deg C):   FP (deg C):

BP (deg C):

BP pressure (mm Hg):

Property/Value	Units	Temp	Data Type	Reference
WS 3.58E+002	mg/L	25	EST	MEYLAN,WM ET AL. (1996)
logP 2.05			EXP	HANSCH,C & LEO,AJ (1985)
VP 1.15E-009	mm Hg	25	EST	NEELY,WB & BLAU,GE (1985)
DC	pKa			
HL 1.48E-007	atm m3/mol	25	EST	MEYLAN,WM & HOWARD,PH (1991)
OH 6.42E-011	cm3/molc sec	25	EST	MEYLAN,WM & HOWARD,PH (1993)

CAS #: 037157-92-9 — 124-BENZTHIADIAZIN-1-O2-3(5NORBEN2YL)

Formula: $C_{14}H_{14}N_2O_2S$

Mol Weight: 274.34

MP (deg C):   FP (deg C):

BP (deg C):

BP pressure (mm Hg):

Property/Value	Units	Temp	Data Type	Reference
WS 3.01E+002	mg/L	25	EST	MEYLAN,WM ET AL. (1996)
logP 2.06			EXP	HANSCH,C & LEO,AJ (1985)
VP 6.52E-010	mm Hg	25	EST	NEELY,WB & BLAU,GE (1985)
DC	pKa			
HL 8.66E-008	atm m3/mol	25	EST	MEYLAN,WM & HOWARD,PH (1991)
OH 6.47E-011	cm3/molc sec	25	EST	MEYLAN,WM & HOWARD,PH (1993)

CAS #: 037157-95-2 — 124-BENZTHIADIAZIN-1-O2-3-(2-THIENYL)

Formula: $C_{11}H_8N_2O_2S_2$

Mol Weight: 264.33

MP (deg C):   FP (deg C):

BP (deg C):

BP pressure (mm Hg):

Property/Value	Units	Temp	Data Type	Reference
WS 7.23E+002	mg/L	25	EST	MEYLAN,WM ET AL. (1996)
logP 1.68			EXP	HANSCH,C & LEO,AJ (1985)
VP 4.43E-010	mm Hg	25	EST	NEELY,WB & BLAU,GE (1985)
DC	pKa			
HL 3.06E-009	atm m3/mol	25	EST	MEYLAN,WM & HOWARD,PH (1991)
OH 2.15E-011	cm3/molc sec	25	EST	MEYLAN,WM & HOWARD,PH (1993)

CAS #: 037157-96-3				124-BENZTHIADIAZIN-1-O2-3-(2-FURYL)
Formula: $C_{11}H_8N_2O_3S$				
Mol Weight: 248.26				
MP (deg C):		FP (deg C):		
BP (deg C):				
BP pressure (mm Hg):				

Property/Value	Units	Temp	Data Type	Reference
WS 3.01E+003	mg/L	25	EST	MEYLAN,WM ET AL. (1996)
logP 1.06			EXP	HANSCH,C & LEO,AJ (1985)
VP 2.74E-009	mm Hg	25	EST	NEELY,WB & BLAU,GE (1985)
DC	pKa			
HL 5.61E-009	atm m3/mol	25	EST	MEYLAN,WM & HOWARD,PH (1991)
OH 8.97E-011	cm3/molc sec	25	EST	MEYLAN,WM & HOWARD,PH (1993)

CAS #: 037157-97-4				124-BENZTHIADIAZN-1-O2-3-DICLME-6-CL
Formula: $C_8H_5Cl_3N_2O_2S$				
Mol Weight: 299.56				
MP (deg C):		FP (deg C):		
BP (deg C):				
BP pressure (mm Hg):				

Property/Value	Units	Temp	Data Type	Reference
WS 3.52E+002	mg/L	25	EST	MEYLAN,WM ET AL. (1996)
logP 1.81			EXP	HANSCH,C & LEO,AJ (1985)
VP 1.48E-009	mm Hg	25	EST	NEELY,WB & BLAU,GE (1985)
DC	pKa			
HL 8.50E-009	atm m3/mol	25	EST	MEYLAN,WM & HOWARD,PH (1991)
OH 5.29E-013	cm3/molc sec	25	EST	MEYLAN,WM & HOWARD,PH (1993)

CAS #: 037157-99-6				2H-1,2,4-BENZOTHIADIAZIN-3-AMINE,6-CHLORO-, 1,1-
Formula: $C_7H_6ClN_3O_2S$				
Mol Weight: 231.66				
MP (deg C):		FP (deg C):		
BP (deg C):				
BP pressure (mm Hg):				

Property/Value	Units	Temp	Data Type	Reference
WS 4.98E+003	mg/L	25	EST	MEYLAN,WM ET AL. (1996)
logP 0.91			EXP	HANSCH,C ET AL. (1995)
VP 7.59E-009	mm Hg	25	EST	NEELY,WB & BLAU,GE (1985)
DC	pKa			
HL 5.31E-012	atm m3/mol	25	EST	MEYLAN,WM & HOWARD,PH (1991)
OH 2.74E-011	cm3/molc sec	25	EST	MEYLAN,WM & HOWARD,PH (1993)

CAS #: 037158-00-2				124-BENZTHIADIAZNE-1-O2-3-ETSME-6-CL
Formula: $C_{10}H_{11}ClN_2O_2S_2$				
Mol Weight: 290.79				
MP (deg C):		FP (deg C):		
BP (deg C):				
BP pressure (mm Hg):				

Property/Value	Units	Temp	Data Type	Reference
WS 1.63E+002	mg/L	25	EST	MEYLAN,WM ET AL. (1996)
logP 2.26			EXP	HANSCH,C & LEO,AJ (1985)
VP 5.52E-010	mm Hg	25	EST	NEELY,WB & BLAU,GE (1985)
DC	pKa			
HL 9.71E-010	atm m3/mol	25	EST	MEYLAN,WM & HOWARD,PH (1991)
OH 1.83E-011	cm3/molc sec	25	EST	MEYLAN,WM & HOWARD,PH (1993)

CAS #: 037163-41-0				N(P-FPH)-3-N'-PIPERIDINOACETAMIDE
Formula: $C_{13}H_{17}FN_2O$				
Mol Weight: 236.29				
MP (deg C):		FP (deg C):		
BP (deg C):				
BP pressure (mm Hg):				

Property/Value	Units	Temp	Data Type	Reference
WS 1.39E+002	mg/L	25	EST	MEYLAN,WM ET AL. (1996)
logP 2.70			EXP	HANSCH,C & LEO,AJ (1985)
VP 1.18E-006	mm Hg	25	EST	NEELY,WB & BLAU,GE (1985)
DC	pKa			
HL 7.35E-011	atm m3/mol	25	EST	MEYLAN,WM & HOWARD,PH (1991)
OH 1.02E-010	cm3/molc sec	25	EST	MEYLAN,WM & HOWARD,PH (1993)

CAS #: 037174-09-7				2,2,2-TRIFLUOROET AMINE,N,N-DIET
Formula: $C_6H_{12}F_3N$				
Mol Weight: 155.16				
MP (deg C):		FP (deg C):		
BP (deg C):				
BP pressure (mm Hg):				

Property/Value	Units	Temp	Data Type	Reference
WS 1.82E+004	mg/L	25	EST	MEYLAN,WM ET AL. (1996)
logP 1.88			EXP	HANSCH,C & LEO,AJ (1985)
VP 9.40E+001	mm Hg	25	EST	NEELY,WB & BLAU,GE (1985)
DC	pKa			
HL 6.73E-004	atm m3/mol	25	EST	MEYLAN,WM & HOWARD,PH (1991)
OH 1.83E-011	cm3/molc sec	25	EST	MEYLAN,WM & HOWARD,PH (1993)

CAS #: 037178-37-3				L-TYROSINE, 3-HYDROXY-, ETHYL ESTER
Formula: $C_{11}H_{15}NO_4$				
Mol Weight: 225.25				
MP (deg C):		FP (deg C):		
BP (deg C):				
BP pressure (mm Hg):				

Property/Value	Units	Temp	Data Type	Reference
WS 7.21E+005	mg/L	25	EST	MEYLAN,WM ET AL. (1996)
logP 0.28			EXP	MARREL,C ET AL. (1985)
VP 3.11E-007	mm Hg	25	EST	NEELY,WB & BLAU,GE (1985)
DC	pKa			
HL 5.54E-016	atm m3/mol	25	EST	MEYLAN,WM & HOWARD,PH (1991)
OH 9.75E-011	cm3/molc sec	25	EST	MEYLAN,WM & HOWARD,PH (1993)

CAS #: 037183-26-9				2'-NO2 SALICYLANILIDE
Formula: $C_{13}H_{10}N_2O_4$				
Mol Weight: 258.24				
MP (deg C):		FP (deg C):		
BP (deg C):				
BP pressure (mm Hg):				

Property/Value	Units	Temp	Data Type	Reference
WS 1.17E+002	mg/L	25	EST	MEYLAN,WM ET AL. (1996)
logP 2.87			EXP	TERADA,H ET AL. (1988)
VP 3.75E-010	mm Hg	25	EST	NEELY,WB & BLAU,GE (1985)
DC	pKa			
HL 1.38E-011	atm m3/mol	25	EST	MEYLAN,WM & HOWARD,PH (1991)
OH 3.21E-011	cm3/molc sec	25	EST	MEYLAN,WM & HOWARD,PH (1993)

CAS #: 037183-28-1 — 2',4'-DICL SALICYLANILIDE

Formula: $C_{13}H_9Cl_2NO_2$

Mol Weight: 282.13

MP (deg C): | FP (deg C):

BP (deg C):

BP pressure (mm Hg):

Property/Value	Units	Temp	Data Type	Reference
WS 2.18E+001	mg/L	25	EST	MEYLAN,WM ET AL. (1996)
logP 5.49			EXP	TERADA,H ET AL. (1988)
VP 1.47E-009	mm Hg	25	EST	NEELY,WB & BLAU,GE (1985)
DC	pKa			
HL 8.79E-011	atm m3/mol	25	EST	MEYLAN,WM & HOWARD,PH (1991)
OH 3.16E-011	cm3/molc sec	25	EST	MEYLAN,WM & HOWARD,PH (1993)

CAS #: 037350-58-6 — METOPROLOL

Formula: $C_{15}H_{25}NO_3$

Mol Weight: 267.37

MP (deg C): | FP (deg C):

BP (deg C):

BP pressure (mm Hg):

Property/Value	Units	Temp	Data Type	Reference
WS 4.78E+003	mg/L	25	EST	MEYLAN,WM ET AL. (1996)
logP 1.88			EXP	HANSCH,C & LEO,AJ (1985)
VP 2.88E-007	mm Hg	25	EST	NEELY,WB & BLAU,GE (1985)
DC	pKa			
HL 1.40E-013	atm m3/mol	25	EST	MEYLAN,WM & HOWARD,PH (1991)
OH 1.47E-010	cm3/molc sec	25	EST	MEYLAN,WM & HOWARD,PH (1993)

CAS #: 037385-07-2 — 5-PYRIMIDINOL, HEXAHYDRO-1,3-DIMETHYL-2-[(5-NITR

Formula: $C_9H_{13}N_5O_3S$

Mol Weight: 271.30

MP (deg C): | FP (deg C):

BP (deg C):

BP pressure (mm Hg):

Property/Value	Units	Temp	Data Type	Reference
WS 5.44E+004	mg/L	25	EST	MEYLAN,WM ET AL. (1996)
logP 0.16			EXP	SANGSTER,J (1994)
VP 3.01E-009	mm Hg	25	EST	NEELY,WB & BLAU,GE (1985)
DC	pKa			
HL 1.80E-020	atm m3/mol	25	EST	MEYLAN,WM & HOWARD,PH (1991)
OH 1.66E-010	cm3/molc sec	25	EST	MEYLAN,WM & HOWARD,PH (1993)

CAS #: 037385-10-7 — 2(4ME-1,3-IMIDAZOLINYLIDEN-2AM)-5NO2-THIAZOLE

Formula: $C_7H_9N_5O_2S$

Mol Weight: 227.25

MP (deg C): | FP (deg C):

BP (deg C):

BP pressure (mm Hg):

Property/Value	Units	Temp	Data Type	Reference
WS 3.58E+002	mg/L	25	EST	MEYLAN,WM ET AL. (1996)
logP 1.82			EXP	HANSCH,C & LEO,AJ (1985)
VP 1.48E-008	mm Hg	25	EST	NEELY,WB & BLAU,GE (1985)
DC	pKa			
HL 1.02E-016	atm m3/mol	25	EST	MEYLAN,WM & HOWARD,PH (1991)
OH 7.38E-011	cm3/molc sec	25	EST	MEYLAN,WM & HOWARD,PH (1993)

CAS #: 037385-14-1 — 2(155TRIME-1,3-IMIDAZOLINYLIDEN-2AM)5NO2-THIAZOL

Formula: $C_9H_{13}N_5O_2S$

Mol Weight: 255.30

MP (deg C): | FP (deg C):

BP (deg C):

BP pressure (mm Hg):

Property/Value	Units	Temp	Data Type	Reference
WS 8.86E+001	mg/L	25	EST	MEYLAN,WM ET AL. (1996)
logP 2.35			EXP	HANSCH,C & LEO,AJ (1985)
VP 1.52E-006	mm Hg	25	EST	NEELY,WB & BLAU,GE (1985)
DC	pKa			
HL 2.98E-016	atm m3/mol	25	EST	MEYLAN,WM & HOWARD,PH (1991)
OH 6.87E-011	cm3/molc sec	25	EST	MEYLAN,WM & HOWARD,PH (1993)

CAS #: 037394-79-9 — 2(6H-5OH-PYRIMIDINYLIDEN-2AM)-5NO2-THIAZOLE

Formula: $C_7H_9N_5O_3S$

Mol Weight: 243.25

MP (deg C): | FP (deg C):

BP (deg C):

BP pressure (mm Hg):

Property/Value	Units	Temp	Data Type	Reference
WS 2.45E+003	mg/L	25	EST	MEYLAN,WM ET AL. (1996)
logP 0.74			EXP	HANSCH,C & LEO,AJ (1985)
VP 6.98E-012	mm Hg	25	EST	NEELY,WB & BLAU,GE (1985)
DC	pKa			
HL 3.73E-021	atm m3/mol	25	EST	MEYLAN,WM & HOWARD,PH (1991)
OH 8.54E-011	cm3/molc sec	25	EST	MEYLAN,WM & HOWARD,PH (1993)

CAS #: 037399-40-9 — 2'-NO2-4'-CL SALICYLANILIDE

Formula: $C_{13}H_9ClN_2O_4$

Mol Weight: 292.68

MP (deg C): | FP (deg C):

BP (deg C):

BP pressure (mm Hg):

Property/Value	Units	Temp	Data Type	Reference
WS 6.74E+000	mg/L	25	EST	MEYLAN,WM ET AL. (1996)
logP 4.09			EXP	TERADA,H ET AL. (1988)
VP 9.16E-011	mm Hg	25	EST	NEELY,WB & BLAU,GE (1985)
DC	pKa			
HL 1.02E-011	atm m3/mol	25	EST	MEYLAN,WM & HOWARD,PH (1991)
OH 3.10E-011	cm3/molc sec	25	EST	MEYLAN,WM & HOWARD,PH (1993)

CAS #: 037422-15-4 — 1(5NO2-2-THIAZOLYL)-2MES-2-IMIDAZOLINE

Formula: $C_7H_8N_4O_2S_2$

Mol Weight: 244.30

MP (deg C): | FP (deg C):

BP (deg C):

BP pressure (mm Hg):

Property/Value	Units	Temp	Data Type	Reference
WS 3.13E+002	mg/L	25	EST	MEYLAN,WM ET AL. (1996)
logP 1.78			EXP	HANSCH,C & LEO,AJ (1985)
VP 1.76E-006	mm Hg	25	EST	NEELY,WB & BLAU,GE (1985)
DC	pKa			
HL 7.45E-014	atm m3/mol	25	EST	MEYLAN,WM & HOWARD,PH (1991)
OH 1.24E-011	cm3/molc sec	25	EST	MEYLAN,WM & HOWARD,PH (1993)

CAS #: 037427-69-3 — 2(1-ACETOXYET-13THIAZOLIDINYLIDEN-2AM)5NO2-THIAZ

Formula: $C_{10}H_{12}N_4O_4S_2$

Mol Weight: 316.36

MP (deg C):
FP (deg C):
BP (deg C):
BP pressure (mm Hg):

Property/Value	Units	Temp	Data Type	Reference
WS 1.01E+003	mg/L	25	EST	MEYLAN,WM ET AL. (1996)
logP 1.88			EXP	HANSCH,C & LEO,AJ (1985)
VP 4.15E-008	mm Hg	25	EST	NEELY,WB & BLAU,GE (1985)
DC	pKa			
HL 2.52E-017	atm m3/mol	25	EST	MEYLAN,WM & HOWARD,PH (1991)
OH 9.62E-011	cm3/molc sec	25	EST	MEYLAN,WM & HOWARD,PH (1993)

CAS #: 037517-30-9 — ACEBUTOLOL

Formula: $C_{18}H_{28}N_2O_4$

Mol Weight: 336.43

MP (deg C): 119-123
FP (deg C):
BP (deg C):
BP pressure (mm Hg):

Property/Value	Units	Temp	Data Type	Reference
WS 2.59E+002	mg/L	25	EST	MEYLAN,WM ET AL. (1996)
logP 1.71			EXP	HANSCH,C ET AL. (1995)
VP 1.34E-012	mm Hg	25	EST	NEELY,WB & BLAU,GE (1985)
DC	pKa			
HL 3.01E-020	atm m3/mol	25	EST	MEYLAN,WM & HOWARD,PH (1991)
OH 1.25E-010	cm3/molc sec	25	EST	MEYLAN,WM & HOWARD,PH (1993)

CAS #: 037571-27-0 — BENZENESULFONIC ACID, 2-[2-[4-(2H-NAPHTHO[1,2-D]

Formula: $C_{24}H_{16}N_3NaO_3S$

Mol Weight: 449.47

MP (deg C):
FP (deg C):
BP (deg C):
BP pressure (mm Hg):

Property/Value	Units	Temp	Data Type	Reference
WS 3.38E+000	mg/L	25	EST	MEYLAN,WM ET AL. (1996)
logP 1.80			EXP	VEITH,GD ET AL. (1979)
VP 9.17E-022	mm Hg	25	EST	NEELY,WB & BLAU,GE (1985)
DC	pKa			
HL	atm m3/mol			
OH 6.09E-011	cm3/molc sec	25	EST	MEYLAN,WM & HOWARD,PH (1993)

CAS #: 037595-73-6 — P-ME CF3-METHANESULFONANILIDE

Formula: $C_8H_8F_3NO_2S$

Mol Weight: 239.22

MP (deg C):
FP (deg C):
BP (deg C):
BP pressure (mm Hg):

Property/Value	Units	Temp	Data Type	Reference
WS 2.29E+001	mg/L	25	EST	MEYLAN,WM ET AL. (1996)
logP 3.60			EXP	HANSCH,C & LEO,AJ (1985)
VP 5.25E-004	mm Hg	25	EST	NEELY,WB & BLAU,GE (1985)
DC	pKa			
HL 1.01E-005	atm m3/mol	25	EST	MEYLAN,WM & HOWARD,PH (1991)
OH 5.24E-011	cm3/molc sec	25	EST	MEYLAN,WM & HOWARD,PH (1993)

CAS #: 037619-24-2 — 1H-PYRROLE-2-CARBOXYLIC ACID, 1-METHYL-, METHYL

Formula: $C_7H_9NO_2$

Mol Weight: 139.16

MP (deg C):
FP (deg C):
BP (deg C):
BP pressure (mm Hg):

Property/Value	Units	Temp	Data Type	Reference
WS 2.45E+003	mg/L	25	EST	MEYLAN,WM ET AL. (1996)
logP 1.80			EXP	YAMAGAMI,C ET AL. (1994)
VP 3.63E-001	mm Hg	25	EST	NEELY,WB & BLAU,GE (1985)
DC	pKa			
HL 1.25E-006	atm m3/mol	25	EST	MEYLAN,WM & HOWARD,PH (1991)
OH 3.57E-011	cm3/molc sec	25	EST	MEYLAN,WM & HOWARD,PH (1993)

CAS #: 037640-71-4 — APRINDINE

Formula: $C_{22}H_{30}N_2$

Mol Weight: 322.50

MP (deg C): 120-121
FP (deg C):
BP (deg C):
BP pressure (mm Hg):

Property/Value	Units	Temp	Data Type	Reference
WS 6.52E+000	mg/L	25	EST	MEYLAN,WM ET AL. (1996)
logP 4.86			EXP	HANSCH,C & LEO,AJ (1985)
VP 2.49E-007	mm Hg	25	EST	NEELY,WB & BLAU,GE (1985)
DC	pKa			
HL 6.88E-009	atm m3/mol	25	EST	MEYLAN,WM & HOWARD,PH (1991)
OH 2.83E-010	cm3/molc sec	25	EST	MEYLAN,WM & HOWARD,PH (1993)

CAS #: 037680-65-2 — 2,2',5-TRICHLOROBIPHENYL

Formula: $C_{12}H_7Cl_3$

Mol Weight: 257.55

MP (deg C):
FP (deg C):
BP (deg C):
BP pressure (mm Hg):

Property/Value	Units	Temp	Data Type	Reference
WS 5.10E-001	mg/L	25	EXP	YALKOWSKY,SH & DANNENFELSER,RM (1992)
logP 5.55			EXP	HANSCH,C & LEO,AJ (1985)
VP 1.05E-003	mm Hg	25	EXP	SHIU,WY ET AL. (1988)
DC	pKa			
HL 6.98E-004	atm m3/mol	25	EST	VP/WSOL
OH 1.09E-012	cm3/molc sec	25	EST	MEYLAN,WM & HOWARD,PH (1993)

CAS #: 037680-66-3 — 2,2',4-TRICHLOROBIPHENYL

Formula: $C_{12}H_7Cl_3$

Mol Weight: 257.55

MP (deg C):
FP (deg C):
BP (deg C):
BP pressure (mm Hg):

Property/Value	Units	Temp	Data Type	Reference
WS 2.59E-001	mg/L	25	EST	MEYLAN,WM ET AL. (1996)
logP 5.76			EXP	HANSCH,C & LEO,AJ (1985)
VP 4.00E-005	mm Hg	25	EST	NEELY,WB & BLAU,GE (1985)
DC	pKa			
HL 1.68E-004	atm m3/mol	25	EST	MEYLAN,WM & HOWARD,PH (1991)
OH 1.19E-012	cm3/molc sec	25	EST	MEYLAN,WM & HOWARD,PH (1993)

CAS #: 037680-73-2				2,4,5,2',5'-PCB

Formula: $C_{12}H_5Cl_5$

Mol Weight: 326.44

MP (deg C):   FP (deg C):

BP (deg C):

BP pressure (mm Hg):

Property/Value	Units	Temp	Data Type	Reference
WS 1.00E-002	mg/L	24	EXP	CHIOU,CT ET AL. (1977)
logP 6.80			EXP	HANSCH,C & LEO,AJ (1985)
VP 2.52E-005	mm Hg	25	EXP	BIDLEMAN,TF (1984)
DC	pKa			
HL 1.08E-003	atm m3/mol	25	EST	VP/WSOL
OH 3.35E-013	cm3/molc sec	25	EST	MEYLAN,WM & HOWARD,PH (1993)

CAS #: 037761-96-9				7H-1,2,3-TRIAZOLO[4,5-D]PYRIMIDIN-7-ONE, 1,4-DIH

Formula: $C_{10}H_7N_5O$

Mol Weight: 213.20

MP (deg C):   FP (deg C):

BP (deg C):

BP pressure (mm Hg):

Property/Value	Units	Temp	Data Type	Reference
WS 3.08E+003	mg/L	25	EST	MEYLAN,WM ET AL. (1996)
logP 1.27			EXP	SANGSTER,J (1993)
VP 1.86E-011	mm Hg	25	EST	NEELY,WB & BLAU,GE (1985)
DC	pKa			
HL 8.23E-015	atm m3/mol	25	EST	MEYLAN,WM & HOWARD,PH (1991)
OH 9.90E-012	cm3/molc sec	25	EST	MEYLAN,WM & HOWARD,PH (1993)

CAS #: 037764-25-3				ACETAMIDE, 2,2-DICHLORO-N,N-DI-2-PROPENYL-

Formula: $C_8H_{11}Cl_2NO$

Mol Weight: 208.09

MP (deg C): 5.0-6.5   FP (deg C):

BP (deg C):

BP pressure (mm Hg):

Property/Value	Units	Temp	Data Type	Reference
WS 5.00E+003	mg/L	20	EXP	SHIU,WY ET AL. (1990)
logP 1.84			EXP	TOMLIN,C (1994)
VP 6.00E-003	mm Hg	25	EXP	AUGUSTIJN-BECKERS,PWM ET AL. (1994)
DC	pKa			
HL 3.29E-007	atm m3/mol		EST	VP/WSOL
OH 7.55E-011	cm3/molc sec	25	EST	MEYLAN,WM & HOWARD,PH (1993)

CAS #: 037793-01-4				á-D-GALACTOPYRANOSE, 2-DEOXY-2-[[(METHYLNITROSOA

Formula: $C_8H_{15}N_3O_7$

Mol Weight: 265.22

MP (deg C):   FP (deg C):

BP (deg C):

BP pressure (mm Hg):

Property/Value	Units	Temp	Data Type	Reference
WS 5.07E+003	mg/L	25	EST	MEYLAN,WM ET AL. (1996)
logP -0.89			EXP	SOSROVSKY,G & RAO,NUM (1991)
VP 1.74E-012	mm Hg	25	EST	NEELY,WB & BLAU,GE (1985)
DC	pKa			
HL 7.85E-022	atm m3/mol	25	EST	MEYLAN,WM & HOWARD,PH (1991)
OH 1.08E-010	cm3/molc sec	25	EST	MEYLAN,WM & HOWARD,PH (1993)

CAS #: 037793-02-5				á-D-MANNOPYRANOSE, 2-DEOXY-2-[[(METHYLNITROSOAMI

Formula: $C_8H_{15}N_3O_7$

Mol Weight: 265.22

MP (deg C):   FP (deg C):

BP (deg C):

BP pressure (mm Hg):

Property/Value	Units	Temp	Data Type	Reference
WS 5.07E+003	mg/L	25	EST	MEYLAN,WM ET AL. (1996)
logP -0.92			EXP	SOSROVSKY,G & RAO,NUM (1991)
VP 1.74E-012	mm Hg	25	EST	NEELY,WB & BLAU,GE (1985)
DC	pKa			
HL 7.85E-022	atm m3/mol	25	EST	MEYLAN,WM & HOWARD,PH (1991)
OH 1.08E-010	cm3/molc sec	25	EST	MEYLAN,WM & HOWARD,PH (1993)

CAS #: 037813-54-0				3,6-DIMETHYLTHIOPYRIDAZINE

Formula: $C_6H_8N_2S_2$

Mol Weight: 172.27

MP (deg C):   FP (deg C):

BP (deg C):

BP pressure (mm Hg):

Property/Value	Units	Temp	Data Type	Reference
WS 1.58E+003	mg/L	25	EST	MEYLAN,WM ET AL. (1996)
logP 1.85			EXP	HANSCH,C & LEO,AJ (1985)
VP 1.12E-004	mm Hg	25	EST	NEELY,WB & BLAU,GE (1985)
DC -6.00	pKa		EXP	PERRIN,DD (1972)
HL 3.69E-006	atm m3/mol	25	EST	MEYLAN,WM & HOWARD,PH (1991)
OH 3.03E-012	cm3/molc sec	25	EST	MEYLAN,WM & HOWARD,PH (1993)

CAS #: 037871-00-4				HEPTACHLORODIBENZO-P-DIOXIN

Formula: $C_{12}HCl_7O_2$

Mol Weight: 425.31

MP (deg C):   FP (deg C):

BP (deg C):

BP pressure (mm Hg):

Property/Value	Units	Temp	Data Type	Reference
WS 1.90E-003	mg/L	25	EXP	MIYATA,H ET AL. (1989A)
logP 7.78			EST	MCLACHLAN,MS ET AL. (1995)
VP	mm Hg			
DC	pKa			
HL 2.18E-005	atm m3/mol	25	EST	MEYLAN,WM & HOWARD,PH (1991)
OH	cm3/molc sec			

CAS #: 037873-43-1				HYDRAZINECARBOXIMIDAMIDE, 2-[(4-BROMOPHENYL)METH

Formula: $C_8H_9BrN_4$

Mol Weight: 241.10

MP (deg C):   FP (deg C):

BP (deg C):

BP pressure (mm Hg):

Property/Value	Units	Temp	Data Type	Reference
WS 2.06E+003	mg/L	25	EST	MEYLAN,WM ET AL. (1996)
logP 1.30			EXP	SOMAN,G ET AL. (1986)
VP 2.17E-005	mm Hg	25	EST	NEELY,WB & BLAU,GE (1985)
DC	pKa			
HL 2.01E-012	atm m3/mol	25	EST	MEYLAN,WM & HOWARD,PH (1991)
OH 8.74E-011	cm3/molc sec	25	EST	MEYLAN,WM & HOWARD,PH (1993)

CAS #: 037895-44-6	FURAZANCARBOXAMIDE, 4-METHYL-, 5-OXIDE
Formula: $C_4H_5N_3O_3$	
Mol Weight: 143.10	
MP (deg C):	FP (deg C):
BP (deg C):	
BP pressure (mm Hg):	

Property/Value	Units	Temp	Data Type	Reference
WS 3.75E+003	mg/L	25	EST	MEYLAN,WM ET AL. (1996)
logP 0.03			EXP	CALVINO,R ET AL. (1992)
VP 7.01E-006	mm Hg	25	EST	NEELY,WB & BLAU,GE (1985)
DC	pKa			
HL 8.58E-016	atm m3/mol	25	EST	MEYLAN,WM & HOWARD,PH (1991)
OH 6.14E-012	cm3/molc sec	25	EST	MEYLAN,WM & HOWARD,PH (1993)

CAS #: 037895-45-7	FURAZANCARBOXAMIDE, 4-METHYL-, 2-OXIDE
Formula: $C_4H_5N_3O_3$	
Mol Weight: 143.10	
MP (deg C):	FP (deg C):
BP (deg C):	
BP pressure (mm Hg):	

Property/Value	Units	Temp	Data Type	Reference
WS 4.00E+004	mg/L	25	EST	MEYLAN,WM ET AL. (1996)
logP 0.36			EXP	CALVINO,R ET AL. (1992)
VP 7.01E-006	mm Hg	25	EST	NEELY,WB & BLAU,GE (1985)
DC	pKa			
HL 8.58E-016	atm m3/mol	25	EST	MEYLAN,WM & HOWARD,PH (1991)
OH 6.14E-012	cm3/molc sec	25	EST	MEYLAN,WM & HOWARD,PH (1993)

CAS #: 037924-13-3	PERFLUIDONE
Formula: $C_{14}H_{12}F_3NO_4S_2$	
Mol Weight: 379.38	
MP (deg C): 143	FP (deg C):
BP (deg C):	
BP pressure (mm Hg):	

Property/Value	Units	Temp	Data Type	Reference
WS 6.00E+001	mg/L	22	EXP	YALKOWSKY,SH & DANNENFELSER,RM (1992)
logP 4.24			EST	MEYLAN,WM & HOWARD,PH (1995)
VP 9.00E-006	mm Hg	25	EXP	AUGUSTIJN-BECKERS,PWM ET AL. (1994)
DC	pKa			
HL 7.49E-008	atm m3/mol	25	EST	VP/WSOL
OH 1.17E-011	cm3/molc sec	25	EST	MEYLAN,WM & HOWARD,PH (1993)

CAS #: 037933-88-3	VALINE-AMIDE,N-ACETYL
Formula: $C_7H_{14}N_2O_2$	
Mol Weight: 158.20	
MP (deg C):	FP (deg C):
BP (deg C):	
BP pressure (mm Hg):	

Property/Value	Units	Temp	Data Type	Reference
WS 1.13E+004	mg/L	25	EST	MEYLAN,WM ET AL. (1996)
logP -0.61			EXP	HANSCH,C & LEO,AJ (1985)
VP 4.80E-007	mm Hg	25	EST	NEELY,WB & BLAU,GE (1985)
DC	pKa			
HL 2.76E-010	atm m3/mol	25	EST	MEYLAN,WM & HOWARD,PH (1991)
OH 5.11E-011	cm3/molc sec	25	EST	MEYLAN,WM & HOWARD,PH (1993)

CAS #: 037984-36-4	PHENYLACETIC ACID,2-(2'-ME-3'-CL)ANILINO
Formula: $C_{15}H_{14}ClNO_2$	
Mol Weight: 275.74	
MP (deg C):	FP (deg C):
BP (deg C):	
BP pressure (mm Hg):	

Property/Value	Units	Temp	Data Type	Reference
WS 4.59E+000	mg/L	25	EST	MEYLAN,WM ET AL. (1996)
logP 4.64			EXP	HANSCH,C ET AL. (1995)
VP 1.10E-007	mm Hg	25	EST	NEELY,WB & BLAU,GE (1985)
DC	pKa			
HL 7.05E-012	atm m3/mol	25	EST	MEYLAN,WM & HOWARD,PH (1991)
OH 2.01E-010	cm3/molc sec	25	EST	MEYLAN,WM & HOWARD,PH (1993)

CAS #: 037989-04-1	1-ISOQUINOLINAMINE, 3-(2-PYRIDINYL)-
Formula: $C_{14}H_{11}N_3$	
Mol Weight: 221.26	
MP (deg C):	FP (deg C):
BP (deg C):	
BP pressure (mm Hg):	

Property/Value	Units	Temp	Data Type	Reference
WS 1.55E+002	mg/L	25	EST	MEYLAN,WM ET AL. (1996)
logP 2.74			EXP	SANGSTER,J (1993)
VP 1.94E-007	mm Hg	25	EST	NEELY,WB & BLAU,GE (1985)
DC	pKa			
HL 2.44E-014	atm m3/mol	25	EST	MEYLAN,WM & HOWARD,PH (1991)
OH 1.53E-010	cm3/molc sec	25	EST	MEYLAN,WM & HOWARD,PH (1993)

CAS #: 038028-67-0	2-ETHOXYPYRAZINE
Formula: $C_6H_8N_2O$	
Mol Weight: 124.14	
MP (deg C):	FP (deg C):
BP (deg C):	
BP pressure (mm Hg):	

Property/Value	Units	Temp	Data Type	Reference
WS 7.82E+003	mg/L	25	EST	MEYLAN,WM ET AL. (1996)
logP 1.28			EXP	YAMAGAMI,C ET AL. (1990A)
VP 5.84E-001	mm Hg	25	EST	NEELY,WB & BLAU,GE (1985)
DC	pKa			
HL 2.09E-006	atm m3/mol	25	EST	MEYLAN,WM & HOWARD,PH (1991)
OH 8.71E-012	cm3/molc sec	25	EST	MEYLAN,WM & HOWARD,PH (1993)

CAS #: 038035-81-3	2(1H)-QUINOLINONE, 3-PHENYL-
Formula: $C_{15}H_{11}NO$	
Mol Weight: 221.26	
MP (deg C):	FP (deg C):
BP (deg C):	
BP pressure (mm Hg):	

Property/Value	Units	Temp	Data Type	Reference
WS 9.48E+001	mg/L	25	EST	MEYLAN,WM ET AL. (1996)
logP 2.99			EXP	SANGSTER,J (1993)
VP 1.11E-007	mm Hg	25	EST	NEELY,WB & BLAU,GE (1985)
DC	pKa			
HL 2.96E-011	atm m3/mol	25	EST	MEYLAN,WM & HOWARD,PH (1991)
OH 7.24E-011	cm3/molc sec	25	EST	MEYLAN,WM & HOWARD,PH (1993)

CAS #: 038064-90-3		1,3-DIMETHOXY-4-METHYLBENZENE

Formula: $C_9H_{12}O_2$

Mol Weight: 152.19

MP (deg C):   FP (deg C):

BP (deg C): 110-120

BP pressure (mm Hg): 3.00E+001

Property/Value	Units	Temp	Data Type	Reference
WS 3.26E+002	mg/L	25	EST	MEYLAN,WM ET AL. (1996)
logP 2.76			EXP	NAKAGAWA,Y ET AL. (1992)
VP 1.96E-001	mm Hg	25	EST	NEELY,WB & BLAU,GE (1985)
DC	pKa			
HL 2.08E-005	atm m3/mol	25	EST	MEYLAN,WM & HOWARD,PH (1991)
OH 2.02E-010	cm3/molc sec	25	EST	MEYLAN,WM & HOWARD,PH (1993)

CAS #: 038087-61-5		4-THIA-1-AZABICYCLO[3.2.0]HEPTANE DERIVATIVE

Formula: $C_{18}H_{21}N_3O_5$

Mol Weight: 359.39

MP (deg C):   FP (deg C):

BP (deg C):

BP pressure (mm Hg):

Property/Value	Units	Temp	Data Type	Reference
WS 9.26E+002	mg/L	25	EST	MEYLAN,WM ET AL. (1996)
logP 0.90			EXP	SANGSTER,J (1993)
VP 7.48E-015	mm Hg	25	EST	NEELY,WB & BLAU,GE (1985)
DC	pKa			
HL 2.06E-020	atm m3/mol	25	EST	MEYLAN,WM & HOWARD,PH (1991)
OH 4.33E-011	cm3/molc sec	25	EST	MEYLAN,WM & HOWARD,PH (1993)

CAS #: 038099-11-5		URIDINE, 5-(3,3-DIMETHYL-1-TRIAZENYL)-

Formula: $C_{11}H_{17}N_5O_6$

Mol Weight: 315.29

MP (deg C):   FP (deg C):

BP (deg C):

BP pressure (mm Hg):

Property/Value	Units	Temp	Data Type	Reference
WS 1.75E+004	mg/L	25	EST	MEYLAN,WM ET AL. (1996)
logP -1.82			EXP	HANSCH,C ET AL. (1995)
VP 9.86E-018	mm Hg	25	EST	NEELY,WB & BLAU,GE (1985)
DC	pKa			
HL 3.41E-022	atm m3/mol	25	EST	MEYLAN,WM & HOWARD,PH (1991)
OH 9.84E-011	cm3/molc sec	25	EST	MEYLAN,WM & HOWARD,PH (1993)

CAS #: 038103-61-6		BENZAMIDE, 4-[2-[[2-HYDROXY-3-(2-METHYLPHENOXY)P

Formula: $C_{19}H_{24}N_2O_4$

Mol Weight: 344.41

MP (deg C):   FP (deg C):

BP (deg C):

BP pressure (mm Hg):

Property/Value	Units	Temp	Data Type	Reference
WS 8.51E+001	mg/L	25	EST	MEYLAN,WM ET AL. (1996)
logP 2.22			EXP	SANGSTER,J (1993)
VP 1.51E-013	mm Hg	25	EST	NEELY,WB & BLAU,GE (1985)
DC	pKa			
HL 1.21E-020	atm m3/mol	25	EST	MEYLAN,WM & HOWARD,PH (1991)
OH 1.77E-010	cm3/molc sec	25	EST	MEYLAN,WM & HOWARD,PH (1993)

CAS #: 038178-38-0		1,6-DICHLORODIBENZO-P-DIOXIN

Formula: $C_{12}H_6Cl_2O_2$

Mol Weight: 253.09

MP (deg C):   FP (deg C):

BP (deg C):

BP pressure (mm Hg):

Property/Value	Units	Temp	Data Type	Reference
WS 2.74E-002	mg/L	25	EST	MEYLAN,WM ET AL. (1996)
logP 5.63			EST	MEYLAN,WM & HOWARD,PH (1995)
VP 2.84E-005	mm Hg	25	EST	NEELY,WB & BLAU,GE (1985)
DC	pKa			
HL 6.42E-006	atm m3/mol	25	EST	MEYLAN,WM & HOWARD,PH (1991)
OH 6.92E-012	cm3/molc sec	25	EST	MEYLAN,WM & HOWARD,PH (1993)

CAS #: 038178-99-3		1,2,4,5,7,8-HEXACHLORO-9H-XANTHENE

Formula: $C_{13}H_4Cl_6O$

Mol Weight: 388.89

MP (deg C):   FP (deg C):

BP (deg C):

BP pressure (mm Hg):

Property/Value	Units	Temp	Data Type	Reference
WS 2.91E-005	mg/L	25	EST	MEYLAN,WM ET AL. (1996)
logP 8.17			EST	MEYLAN,WM & HOWARD,PH (1995)
VP 1.19E-007	mm Hg	25	EST	NEELY,WB & BLAU,GE (1985)
DC	pKa			
HL 7.89E-006	atm m3/mol	25	EST	MEYLAN,WM & HOWARD,PH (1991)
OH 2.33E-012	cm3/molc sec	25	EST	MEYLAN,WM & HOWARD,PH (1993)

CAS #: 038191-26-3		HYDRAZINECARBOXAMIDE,N-(2-HYDROXYPHENYL)-

Formula: $C_7H_9N_3O_2$

Mol Weight: 167.17

MP (deg C):   FP (deg C):

BP (deg C):

BP pressure (mm Hg):

Property/Value	Units	Temp	Data Type	Reference
WS 1.62E+003	mg/L	25	EST	MEYLAN,WM ET AL. (1996)
logP 0.33			EXP	KRAMER,CR & BECK,L (1981)
VP 1.24E-006	mm Hg	25	EST	NEELY,WB & BLAU,GE (1985)
DC	pKa			
HL 8.75E-017	atm m3/mol	25	EST	MEYLAN,WM & HOWARD,PH (1991)
OH 5.57E-011	cm3/molc sec	25	EST	MEYLAN,WM & HOWARD,PH (1993)

CAS #: 038194-50-2		SULINDAC

Formula: $C_{20}H_{17}FO_3S$

Mol Weight: 356.42

MP (deg C): 182-185   FP (deg C):

BP (deg C):

BP pressure (mm Hg):

Property/Value	Units	Temp	Data Type	Reference
WS 1.69E+001	mg/L	25	EST	MEYLAN,WM ET AL. (1996)
logP 3.42			EXP	SANGSTER,J (1993)
VP 1.18E-010	mm Hg	25	EST	NEELY,WB & BLAU,GE (1985)
DC	pKa			
HL 1.25E-015	atm m3/mol	25	EST	MEYLAN,WM & HOWARD,PH (1991)
OH 3.50E-010	cm3/molc sec	25	EST	MEYLAN,WM & HOWARD,PH (1993)

CAS #: 038196-44-0				BETAMETHASONE 17,21-DIVALERATE

Formula: $C_{32}H_{45}FO_7$

Mol Weight: 560.71

MP (deg C):

FP (deg C):

BP (deg C):

BP pressure (mm Hg):

Property/ Value	Units	Temp	Data Type	Reference
WS 3.03E-002	mg/L	25	EST	MEYLAN,WM ET AL. (1996)
logP 5.26			EXP	HANSCH,C & LEO,AJ (1985)
VP 1.76E-015	mm Hg	25	EST	NEELY,WB & BLAU,GE (1985)
DC	pKa			
HL 4.57E-015	atm m3/mol	25	EST	MEYLAN,WM & HOWARD,PH (1991)
OH 7.43E-011	cm3/molc sec	25	EST	MEYLAN,WM & HOWARD,PH (1993)

CAS #: 038196-45-1				PREDISONE-17-VALERATE

Formula: $C_{26}H_{34}O_6$

Mol Weight: 442.56

MP (deg C):

FP (deg C):

BP (deg C):

BP pressure (mm Hg):

Property/ Value	Units	Temp	Data Type	Reference
WS 2.94E+000	mg/L	25	EST	MEYLAN,WM ET AL. (1996)
logP 3.82			EXP	HANSCH,C & LEO,AJ (1985)
VP 5.25E-014	mm Hg	25	EST	NEELY,WB & BLAU,GE (1985)
DC	pKa			
HL 7.15E-014	atm m3/mol	25	EST	MEYLAN,WM & HOWARD,PH (1991)
OH 6.81E-011	cm3/molc sec	25	EST	MEYLAN,WM & HOWARD,PH (1993)

CAS #: 038212-33-8				PIPERAZINE, 1-(4-CHLOROPHENYL)-

Formula: $C_{10}H_{13}ClN_2$

Mol Weight: 196.68

MP (deg C):

FP (deg C):

BP (deg C):

BP pressure (mm Hg):

Property/ Value	Units	Temp	Data Type	Reference
WS 8.90E+003	mg/L	25	EST	MEYLAN,WM ET AL. (1996)
logP 2.01			EXP	CACCIA,S ET AL. (1985)
VP 4.79E-004	mm Hg	25	EST	NEELY,WB & BLAU,GE (1985)
DC	pKa			
HL 8.39E-009	atm m3/mol	25	EST	MEYLAN,WM & HOWARD,PH (1991)
OH 1.49E-010	cm3/molc sec	25	EST	MEYLAN,WM & HOWARD,PH (1993)

CAS #: 038212-89-4				CARBENCILLIN, PHENYL

Formula: $C_{23}H_{22}N_2O_6S$

Mol Weight: 454.51

MP (deg C):

FP (deg C):

BP (deg C):

BP pressure (mm Hg):

Property/ Value	Units	Temp	Data Type	Reference
WS 2.91E+000	mg/L	25	EST	MEYLAN,WM ET AL. (1996)
logP 3.14			EXP	HANSCH,C & LEO,AJ (1985)
VP 5.27E-016	mm Hg	25	EST	NEELY,WB & BLAU,GE (1985)
DC	pKa			
HL 1.02E-017	atm m3/mol	25	EST	MEYLAN,WM & HOWARD,PH (1991)
OH 9.36E-011	cm3/molc sec	25	EST	MEYLAN,WM & HOWARD,PH (1993)

CAS #: 038260-54-7				ETRIMFOS

Formula: $C_{10}H_{17}N_2O_4PS$

Mol Weight: 292.30

MP (deg C): -3.35

FP (deg C):

BP (deg C):

BP pressure (mm Hg):

Property/ Value	Units	Temp	Data Type	Reference
WS 4.00E+001	mg/L	23	EXP	WORTHING,CR & WALKER,SB (1987)
logP 2.94			EST	MEYLAN,WM & HOWARD,PH (1995)
VP 6.45E-005	mm Hg	20	EXP	WORTHING,CR & WALKER,SB (1987)
DC	pKa			
HL 6.20E-007	atm m3/mol	20	EST	VP/WSOL
OH 7.96E-011	cm3/molc sec	25	EST	ATKINSON,R (1988)

CAS #: 038291-82-6				VALERIC ACID HYDRAZIDE

Formula: $C_5H_{12}N_2O$

Mol Weight: 116.16

MP (deg C):

FP (deg C):

BP (deg C):

BP pressure (mm Hg):

Property/ Value	Units	Temp	Data Type	Reference
WS 6.30E+003	mg/L	25	EST	MEYLAN,WM ET AL. (1996)
logP -0.11			EXP	HANSCH,C & LEO,AJ (1985)
VP 2.13E-003	mm Hg	25	EST	NEELY,WB & BLAU,GE (1985)
DC	pKa			
HL 1.09E-010	atm m3/mol	25	EST	MEYLAN,WM & HOWARD,PH (1991)
OH 1.22E-011	cm3/molc sec	25	EST	MEYLAN,WM & HOWARD,PH (1993)

CAS #: 038304-91-5				2,4-NH2-6PIPERIDINO-PYRIMIDINE-3-OXID

Formula: $C_9H_{17}N_5O$

Mol Weight: 211.27

MP (deg C): 248

FP (deg C):

BP (deg C):

BP pressure (mm Hg):

Property/ Value	Units	Temp	Data Type	Reference
WS 3.42E+003	mg/L	25	EST	MEYLAN,WM ET AL. (1996)
logP 1.24			EXP	HANSCH,C & LEO,AJ (1985)
VP 1.03E-007	mm Hg	25	EST	NEELY,WB & BLAU,GE (1985)
DC	pKa			
HL 2.12E-016	atm m3/mol	25	EST	MEYLAN,WM & HOWARD,PH (1991)
OH 2.26E-010	cm3/molc sec	25	EST	MEYLAN,WM & HOWARD,PH (1993)

CAS #: 038326-28-2				PYRIDO(12A)PYRIMIDIN-4-ONE,2ME,6ET

Formula: $C_{11}H_{14}N_2O$

Mol Weight: 190.25

MP (deg C):

FP (deg C):

BP (deg C):

BP pressure (mm Hg):

Property/ Value	Units	Temp	Data Type	Reference
WS 2.03E+003	mg/L	25	EST	MEYLAN,WM ET AL. (1996)
logP 1.63			EXP	HANSCH,C & LEO,AJ (1985)
VP 2.60E-005	mm Hg	25	EST	NEELY,WB & BLAU,GE (1985)
DC	pKa			
HL 4.88E-009	atm m3/mol	25	EST	MEYLAN,WM & HOWARD,PH (1991)
OH 1.39E-010	cm3/molc sec	25	EST	MEYLAN,WM & HOWARD,PH (1993)

038326-36-2 — 4H-PYRIDO[1,2-A]PYRIMIDINE-3-CARBOXYLIC ACID, 6,

CAS #:	038326-36-2		4H-PYRIDO[1,2-A]PYRIMIDINE-3-CARBOXYLIC ACID, 6,
Formula:	$C_{11}H_{14}N_2O_3$		
Mol Weight:	222.25		
MP (deg C):		FP (deg C):	
BP (deg C):			
BP pressure (mm Hg):			

Property/ Value	Units	Temp	Data Type	Reference
WS 1.22E+003	mg/L	25	EST	MEYLAN,WM ET AL. (1996)
logP 0.15			EXP	SANGSTER,J (1994)
VP 5.47E-006	mm Hg	25	EST	NEELY,WB & BLAU,GE (1985)
DC	pKa			
HL 2.67E-012	atm m3/mol	25	EST	MEYLAN,WM & HOWARD,PH (1991)
OH 2.30E-011	cm3/molc sec	25	EST	MEYLAN,WM & HOWARD,PH (1993)

038353-69-4 — BENZOIC ACID, 2-ACETYLOXY ET ESTER

CAS #:	038353-69-4		BENZOIC ACID, 2-ACETYLOXY ET ESTER
Formula:	$C_{11}H_{12}O_4$		
Mol Weight:	208.22		
MP (deg C):		FP (deg C):	
BP (deg C):			
BP pressure (mm Hg):			

Property/ Value	Units	Temp	Data Type	Reference
WS 1.04E+003	mg/L	25	EST	MEYLAN,WM ET AL. (1996)
logP 1.85			EXP	HANSCH,C & LEO,AJ (1985)
VP 1.15E-002	mm Hg	25	EST	NEELY,WB & BLAU,GE (1985)
DC	pKa			
HL 1.10E-007	atm m3/mol	25	EST	MEYLAN,WM & HOWARD,PH (1991)
OH 4.41E-012	cm3/molc sec	25	EST	MEYLAN,WM & HOWARD,PH (1993)

038363-40-5 — PENBUTOLOL

CAS #:	038363-40-5		PENBUTOLOL
Formula:	$C_{18}H_{29}NO_2$		
Mol Weight:	291.44		
MP (deg C):	68-72	FP (deg C):	
BP (deg C):			
BP pressure (mm Hg):			

Property/ Value	Units	Temp	Data Type	Reference
WS 7.00E+003	mg/L		EXP	YALKOWSKY,SH & DANNENFELSER,RM (1992)
logP 4.15			EXP	HANSCH,C & LEO,AJ (1985)
VP 4.46E-008	mm Hg	25	EST	NEELY,WB & BLAU,GE (1985)
DC	pKa			
HL 1.64E-011	atm m3/mol	25	EST	MEYLAN,WM & HOWARD,PH (1991)
OH 1.24E-010	cm3/molc sec	25	EST	MEYLAN,WM & HOWARD,PH (1993)

038367-19-0 — N(M-CLPH)-3-N'-PIPERIDINOACETAMIDE

CAS #:	038367-19-0		N(M-CLPH)-3-N'-PIPERIDINOACETAMIDE
Formula:	$C_{13}H_{17}ClN_2O$		
Mol Weight:	252.75		
MP (deg C):		FP (deg C):	
BP (deg C):			
BP pressure (mm Hg):			

Property/ Value	Units	Temp	Data Type	Reference
WS 2.59E+001	mg/L	25	EST	MEYLAN,WM ET AL. (1996)
logP 3.45			EXP	HANSCH,C & LEO,AJ (1985)
VP 2.37E-007	mm Hg	25	EST	NEELY,WB & BLAU,GE (1985)
DC	pKa			
HL 4.67E-011	atm m3/mol	25	EST	MEYLAN,WM & HOWARD,PH (1991)
OH 1.07E-010	cm3/molc sec	25	EST	MEYLAN,WM & HOWARD,PH (1993)

038367-20-3 — N-(M-TOLYL)-3-N'-PIPERIDINOACETAMIDE

CAS #:	038367-20-3		N-(M-TOLYL)-3-N'-PIPERIDINOACETAMIDE
Formula:	$C_{14}H_{20}N_2O$		
Mol Weight:	232.33		
MP (deg C):		FP (deg C):	
BP (deg C):			
BP pressure (mm Hg):			

Property/ Value	Units	Temp	Data Type	Reference
WS 8.60E+001	mg/L	25	EST	MEYLAN,WM ET AL. (1996)
logP 2.97			EXP	HANSCH,C & LEO,AJ (1985)
VP 4.23E-007	mm Hg	25	EST	NEELY,WB & BLAU,GE (1985)
DC	pKa			
HL 6.95E-011	atm m3/mol	25	EST	MEYLAN,WM & HOWARD,PH (1991)
OH 1.31E-010	cm3/molc sec	25	EST	MEYLAN,WM & HOWARD,PH (1993)

038367-22-5 — N(P-NO2PH)-3-N'-PIPERIDINOACETAMIDE

CAS #:	038367-22-5		N(P-NO2PH)-3-N'-PIPERIDINOACETAMIDE
Formula:	$C_{13}H_{17}N_3O_3$		
Mol Weight:	263.30		
MP (deg C):		FP (deg C):	
BP (deg C):			
BP pressure (mm Hg):			

Property/ Value	Units	Temp	Data Type	Reference
WS 3.86E+001	mg/L	25	EST	MEYLAN,WM ET AL. (1996)
logP 2.72			EXP	HANSCH,C & LEO,AJ (1985)
VP 1.66E-008	mm Hg	25	EST	NEELY,WB & BLAU,GE (1985)
DC	pKa			
HL 2.49E-013	atm m3/mol	25	EST	MEYLAN,WM & HOWARD,PH (1991)
OH 9.97E-011	cm3/molc sec	25	EST	MEYLAN,WM & HOWARD,PH (1993)

038367-23-6 — N(O-CLPH)-3-N'-PIPERIDINOACETAMIDE

CAS #:	038367-23-6		N(O-CLPH)-3-N'-PIPERIDINOACETAMIDE
Formula:	$C_{13}H_{17}ClN_2O$		
Mol Weight:	252.75		
MP (deg C):		FP (deg C):	
BP (deg C):			
BP pressure (mm Hg):			

Property/ Value	Units	Temp	Data Type	Reference
WS 1.00E+002	mg/L	25	EST	MEYLAN,WM ET AL. (1996)
logP 2.76			EXP	HANSCH,C & LEO,AJ (1985)
VP 2.37E-007	mm Hg	25	EST	NEELY,WB & BLAU,GE (1985)
DC	pKa			
HL 4.67E-011	atm m3/mol	25	EST	MEYLAN,WM & HOWARD,PH (1991)
OH 1.02E-010	cm3/molc sec	25	EST	MEYLAN,WM & HOWARD,PH (1993)

038379-99-6 — 2,2',3,5',6-PCB

CAS #:	038379-99-6		2,2',3,5',6-PCB
Formula:	$C_{12}H_5Cl_5$		
Mol Weight:	326.44		
MP (deg C):		FP (deg C):	
BP (deg C):			
BP pressure (mm Hg):			

Property/ Value	Units	Temp	Data Type	Reference
WS 5.41E-002	mg/L	20	EXP	YALKOWSKY,SH & DANNENFELSER,RM (1992)
logP 6.55			EXP	HANSCH,C & LEO,AJ (1985)
VP 2.22E-006	mm Hg	25	EST	NEELY,WB & BLAU,GE (1985)
DC	pKa			
HL 9.24E-005	atm m3/mol	25	EST	MEYLAN,WM & HOWARD,PH (1991)
OH 3.35E-013	cm3/molc sec	25	EST	MEYLAN,WM & HOWARD,PH (1993)

CAS #: 038380-01-7 2,2',4,4',5-PENTACHLOROBIPHENYL

Formula: $C_{12}H_5Cl_5$

Mol Weight: 326.44

MP (deg C): FP (deg C):

BP (deg C):

BP pressure (mm Hg):

Property/Value	Units	Temp	Data Type	Reference
WS 7.00E-003	mg/L	25	EXP	OZRETICH,RJ ET AL. (1995)
logP 7.21			EXP	RAPAPORT,RA & EISENREICH,SJ (1984)
VP 2.20E-005	mm Hg	25	EXP	BIDLEMAN,TF (1984)
DC	pKa			
HL 1.35E-003	atm m3/mol	25	EST	VP/WSOL
OH 4.00E-013	cm3/molc sec	25	EST	MEYLAN,WM & HOWARD,PH (1993)

CAS #: 038380-02-8 2,2',3,4,5'-PENTACHLOROBIPHENYL

Formula: $C_{12}H_5Cl_5$

Mol Weight: 326.44

MP (deg C): FP (deg C):

BP (deg C):

BP pressure (mm Hg):

Property/Value	Units	Temp	Data Type	Reference
WS 2.94E-002	mg/L	20	EXP	YALKOWSKY,SH & DANNENFELSER,RM (1992)
logP 6.37			EXP	RAPAPORT,RA & EISENREICH,SJ (1984)
VP 1.70E-005	mm Hg	25	EXP	BIDLEMAN,TF (1984)
DC	pKa			
HL 2.48E-004	atm m3/mol	25	EST	VP/WSOL
OH 3.35E-013	cm3/molc sec	25	EST	MEYLAN,WM & HOWARD,PH (1993)

CAS #: 038380-03-9 2,3,3',4',6-PENTACHLOROBIPHENYL

Formula: $C_{12}H_5Cl_5$

Mol Weight: 326.44

MP (deg C): FP (deg C):

BP (deg C):

BP pressure (mm Hg):

Property/Value	Units	Temp	Data Type	Reference
WS 5.40E-003	mg/L	25	EXP	OZRETICH,RJ ET AL. (1995)
logP 6.53			EXP	RAPAPORT,RA & EISENREICH,SJ (1984)
VP 2.22E-006	mm Hg	25	EST	NEELY,WB & BLAU,GE (1985)
DC	pKa			
HL 9.24E-005	atm m3/mol	25	EST	MEYLAN,WM & HOWARD,PH (1991)
OH 3.35E-013	cm3/molc sec	25	EST	MEYLAN,WM & HOWARD,PH (1993)

CAS #: 038380-04-0 2,2',3,4',5',6-PCB

Formula: $C_{12}H_4Cl_6$

Mol Weight: 360.88

MP (deg C): FP (deg C):

BP (deg C):

BP pressure (mm Hg):

Property/Value	Units	Temp	Data Type	Reference
WS 2.90E-003	mg/L	25	EXP	OZRETICH,RJ ET AL. (1995)
logP 7.28			EXP	HANSCH,C & LEO,AJ (1985)
VP 8.43E-006	mm Hg	25	EXP	BIDLEMAN,TF (1984)
DC	pKa			
HL 1.38E-003	atm m3/mol	25	EST	VP/WSOL
OH 1.64E-013	cm3/molc sec	25	EST	MEYLAN,WM & HOWARD,PH (1993)

CAS #: 038380-07-3 2,2',3,3',4,4'-PCB

Formula: $C_{12}H_4Cl_6$

Mol Weight: 360.88

MP (deg C): FP (deg C):

BP (deg C):

BP pressure (mm Hg):

Property/Value	Units	Temp	Data Type	Reference
WS 3.50E-004	mg/L	25	EXP	YALKOWSKY,SH & DANNENFELSER,RM (1992)
logP 7.31			EXP	HANSCH,C & LEO,AJ (1985)
VP 2.56E-006	mm Hg	25	EXP	BIDLEMAN,TF (1984)
DC	pKa			
HL 3.47E-003	atm m3/mol	25	EST	VP/WSOL
OH 1.64E-013	cm3/molc sec	25	EST	MEYLAN,WM & HOWARD,PH (1993)

CAS #: 038380-08-4 2,3,3',4,4',5-HEXACHLOROBIPHENYL

Formula: $C_{12}H_4Cl_6$

Mol Weight: 360.88

MP (deg C): FP (deg C):

BP (deg C):

BP pressure (mm Hg):

Property/Value	Units	Temp	Data Type	Reference
WS 5.33E-003	mg/L	20	EXP	YALKOWSKY,SH & DANNENFELSER,RM (1992)
logP 7.62			EST	MEYLAN,WM & HOWARD,PH (1995)
VP 1.61E-006	mm Hg	25	EXP	BIDLEMAN,TF (1984)
DC	pKa			
HL 1.43E-004	atm m3/mol	25	EST	VP/WSOL
OH 2.11E-013	cm3/molc sec	25	EST	MEYLAN,WM & HOWARD,PH (1993)

CAS #: 038397-06-7 6-NITRO-1-NAPHTHOL

Formula: $C_{10}H_7NO_3$

Mol Weight: 189.17

MP (deg C): FP (deg C):

BP (deg C):

BP pressure (mm Hg):

Property/Value	Units	Temp	Data Type	Reference
WS 5.56E+002	mg/L	25	EST	MEYLAN,WM ET AL. (1996)
logP 3.50			EXP	HANSCH,C & LEO,AJ (1985)
VP 4.24E-006	mm Hg	25	EST	NEELY,WB & BLAU,GE (1985)
DC	pKa			
HL 2.16E-010	atm m3/mol	25	EST	MEYLAN,WM & HOWARD,PH (1991)
OH 7.88E-012	cm3/molc sec	25	EST	MEYLAN,WM & HOWARD,PH (1993)

CAS #: 038407-85-1 HYDRAZINECARBOXIMIDAMIDE, 2-[[4-(DIMETHYLAMINO)P

Formula: $C_{10}H_{15}N_5$

Mol Weight: 205.26

MP (deg C): FP (deg C):

BP (deg C):

BP pressure (mm Hg):

Property/Value	Units	Temp	Data Type	Reference
WS 2.37E+004	mg/L	25	EST	MEYLAN,WM ET AL. (1996)
logP 0.28			EXP	SOMAN,G ET AL. (1986)
VP 1.74E-005	mm Hg	25	EST	NEELY,WB & BLAU,GE (1985)
DC	pKa			
HL 8.03E-014	atm m3/mol	25	EST	MEYLAN,WM & HOWARD,PH (1991)
OH 2.88E-010	cm3/molc sec	25	EST	MEYLAN,WM & HOWARD,PH (1993)

038411-22-2 — 2,2',3,3',6,6'-HEXACHLOROBIPHENYL

CAS #:	038411-22-2
Formula:	$C_{12}H_4Cl_6$
Mol Weight:	360.88
MP (deg C):	
FP (deg C):	
BP (deg C):	
BP pressure (mm Hg):	

Property/Value	Units	Temp	Data Type	Reference
WS 4.51E-003	mg/L	25	EXP	YALKOWSKY,SH & DANNENFELSER,RM (1992)
logP 6.51			EXP	RAPAPORT,RA & EISENREICH,SJ (1984)
VP 5.81E-007	mm Hg	25	EST	NEELY,WB & BLAU,GE (1985)
DC	pKa			
HL 6.85E-005	atm m3/mol	25	EST	MEYLAN,WM & HOWARD,PH (1991)
OH 1.64E-013	cm3/molc sec	25	EST	MEYLAN,WM & HOWARD,PH (1993)

038421-40-8 — 1,3,5,7-TETRAMETHYL-1,1,3,5,7,7-HEXAPHENYLTETR*

CAS #:	038421-40-8
Formula:	$C_{40}H_{42}O_3Si_4$
Mol Weight:	683.12
MP (deg C):	
FP (deg C):	
BP (deg C):	
BP pressure (mm Hg):	

Property/Value	Units	Temp	Data Type	Reference
WS 6.11E-012	mg/L	25	EST	MEYLAN,WM ET AL. (1996)
logP 13.23			EST	MEYLAN,WM & HOWARD,PH (1995)
VP	mm Hg			
DC	pKa			
HL	atm m3/mol			
OH 1.23E-011	cm3/molc sec	25	EST	MEYLAN,WM & HOWARD,PH (1993)

038421-90-8 — 5-DECEN-1-OL ACETATE (E)

CAS #:	038421-90-8
Formula:	$C_{12}H_{22}O_2$
Mol Weight:	198.31
MP (deg C):	
FP (deg C):	
BP (deg C):	85-105
BP pressure (mm Hg):	1.40E+001

Property/Value	Units	Temp	Data Type	Reference
WS 5.50E+000	mg/L	25	EST	MEYLAN,WM ET AL. (1996)
logP 4.58			EST	MEYLAN,WM & HOWARD,PH (1995)
VP 2.31E-002	mm Hg	25	EST	NEELY,WB & BLAU,GE (1985)
DC	pKa			
HL 1.98E-003	atm m3/mol	25	EST	MEYLAN,WM & HOWARD,PH (1991)
OH 6.62E-011	cm3/molc sec	25	EST	MEYLAN,WM & HOWARD,PH (1993)

038444-73-4 — 2,2',6-PCB

CAS #:	038444-73-4
Formula:	$C_{12}H_7Cl_3$
Mol Weight:	257.55
MP (deg C):	
FP (deg C):	
BP (deg C):	
BP pressure (mm Hg):	

Property/Value	Units	Temp	Data Type	Reference
WS 4.49E-001	mg/L	25	EST	MEYLAN,WM ET AL. (1996)
logP 5.48			EXP	HANSCH,C & LEO,AJ (1985)
VP 4.00E-005	mm Hg	25	EST	NEELY,WB & BLAU,GE (1985)
DC	pKa			
HL 1.68E-004	atm m3/mol	25	EST	MEYLAN,WM & HOWARD,PH (1991)
OH 1.19E-012	cm3/molc sec	25	EST	MEYLAN,WM & HOWARD,PH (1993)

038444-77-8 — 2,4',6-TRICHLOROBIPHENYL

CAS #:	038444-77-8
Formula:	$C_{12}H_7Cl_3$
Mol Weight:	257.55
MP (deg C):	
FP (deg C):	
BP (deg C):	
BP pressure (mm Hg):	

Property/Value	Units	Temp	Data Type	Reference
WS 2.64E-001	mg/L	25	EST	MEYLAN,WM ET AL. (1996)
logP 5.75			EXP	HANSCH,C & LEO,AJ (1985)
VP 4.00E-005	mm Hg	25	EST	NEELY,WB & BLAU,GE (1985)
DC	pKa			
HL 1.68E-004	atm m3/mol	25	EST	MEYLAN,WM & HOWARD,PH (1991)
OH 1.19E-012	cm3/molc sec	25	EST	MEYLAN,WM & HOWARD,PH (1993)

038444-78-9 — 2,2',3-PCB

CAS #:	038444-78-9
Formula:	$C_{12}H_7Cl_3$
Mol Weight:	257.55
MP (deg C):	
FP (deg C):	
BP (deg C):	
BP pressure (mm Hg):	

Property/Value	Units	Temp	Data Type	Reference
WS 2.93E-001	mg/L	20	EXP	YALKOWSKY,SH & DANNENFELSER,RM (1992)
logP 5.31			EXP	HANSCH,C & LEO,AJ (1985)
VP 1.14E-003	mm Hg		EXP	VALSARAJ,KT (1988)
DC	pKa			
HL 1.32E-003	atm m3/mol	20	EST	VP/WSOL
OH 1.09E-012	cm3/molc sec	25	EST	MEYLAN,WM & HOWARD,PH (1993)

038444-81-4 — 2,3',5-PCB

CAS #:	038444-81-4
Formula:	$C_{12}H_7Cl_3$
Mol Weight:	257.55
MP (deg C):	
FP (deg C):	
BP (deg C):	
BP pressure (mm Hg):	

Property/Value	Units	Temp	Data Type	Reference
WS 2.53E-001	mg/L	25	EXP	YALKOWSKY,SH & DANNENFELSER,RM (1992)
logP 5.76			EXP	HANSCH,C & LEO,AJ (1985)
VP 4.00E-005	mm Hg	25	EST	NEELY,WB & BLAU,GE (1985)
DC	pKa			
HL 1.68E-004	atm m3/mol	25	EST	MEYLAN,WM & HOWARD,PH (1991)
OH 1.76E-012	cm3/molc sec	25	EST	MEYLAN,WM & HOWARD,PH (1993)

038444-84-7 — 2,3,3'-PCB

CAS #:	038444-84-7
Formula:	$C_{12}H_7Cl_3$
Mol Weight:	257.55
MP (deg C):	
FP (deg C):	
BP (deg C):	
BP pressure (mm Hg):	

Property/Value	Units	Temp	Data Type	Reference
WS 3.76E-001	mg/L	25	EST	MEYLAN,WM ET AL. (1996)
logP 5.57			EXP	HANSCH,C & LEO,AJ (1985)
VP 4.00E-005	mm Hg	25	EST	NEELY,WB & BLAU,GE (1985)
DC	pKa			
HL 1.68E-004	atm m3/mol	25	EST	MEYLAN,WM & HOWARD,PH (1991)
OH 1.76E-012	cm3/molc sec	25	EST	MEYLAN,WM & HOWARD,PH (1993)

CAS #: 038444-85-8				2,3,4'-PCB

Formula: $C_{12}H_7Cl_3$

Mol Weight: 257.55

MP (deg C):   FP (deg C):

BP (deg C):

BP pressure (mm Hg):

Property/Value	Units	Temp	Data Type	Reference
WS 1.42E-001	mg/L	20	EXP	YALKOWSKY,SH & DANNENFELSER,RM (1992)
logP 5.42			EXP	HANSCH,C & LEO,AJ (1985)
VP 4.00E-005	mm Hg	25	EST	NEELY,WB & BLAU,GE (1985)
DC	pKa			
HL 1.68E-004	atm m3/mol	25	EST	MEYLAN,WM & HOWARD,PH (1991)
OH 1.09E-012	cm3/molc sec	25	EST	MEYLAN,WM & HOWARD,PH (1993)

CAS #: 038444-86-9				2',3,4-TRICHLOROBIPHENYL

Formula: $C_{12}H_7Cl_3$

Mol Weight: 257.55

MP (deg C):   FP (deg C):

BP (deg C):

BP pressure (mm Hg):

Property/Value	Units	Temp	Data Type	Reference
WS 1.33E-001	mg/L	20	EXP	YALKOWSKY,SH & DANNENFELSER,RM (1992)
logP 5.57			EXP	RAPAPORT,RA & EISENREICH,SJ (1984)
VP 1.03E-004	mm Hg	25	EXP	BANERJEE,S ET AL. (1990)
DC	pKa			
HL 2.63E-004	atm m3/mol	25	EST	VP/WSOL
OH 1.09E-012	cm3/molc sec	25	EST	MEYLAN,WM & HOWARD,PH (1993)

CAS #: 038444-87-0				3,3',5-TRICHLOROBIPHENYL

Formula: $C_{12}H_7Cl_3$

Mol Weight: 257.55

MP (deg C):   FP (deg C):

BP (deg C):

BP pressure (mm Hg):

Property/Value	Units	Temp	Data Type	Reference
WS 2.97E-001	mg/L	25	EST	MEYLAN,WM ET AL. (1996)
logP 4.15			EXP	RAPAPORT,RA & EISENREICH,SJ (1984)
VP 4.00E-005	mm Hg	25	EST	NEELY,WB & BLAU,GE (1985)
DC	pKa			
HL 1.68E-004	atm m3/mol	25	EST	MEYLAN,WM & HOWARD,PH (1991)
OH 2.49E-012	cm3/molc sec	25	EST	MEYLAN,WM & HOWARD,PH (1993)

CAS #: 038444-90-5				3,4,4'-TRICHLOROBIPHENYL

Formula: $C_{12}H_7Cl_3$

Mol Weight: 257.55

MP (deg C):   FP (deg C):

BP (deg C):

BP pressure (mm Hg):

Property/Value	Units	Temp	Data Type	Reference
WS 7.19E-002	mg/L	20	EXP	YALKOWSKY,SH & DANNENFELSER,RM (1992)
logP 4.94			EXP	RAPAPORT,RA & EISENREICH,SJ (1984)
VP 4.00E-005	mm Hg	25	EST	NEELY,WB & BLAU,GE (1985)
DC	pKa			
HL 1.68E-004	atm m3/mol	25	EST	MEYLAN,WM & HOWARD,PH (1991)
OH 1.09E-012	cm3/molc sec	25	EST	MEYLAN,WM & HOWARD,PH (1993)

CAS #: 038444-93-8				2,2',3,3'-TETRACHLOROBIPHENYL

Formula: $C_{12}H_6Cl_4$

Mol Weight: 291.99

MP (deg C):   FP (deg C):

BP (deg C):

BP pressure (mm Hg):

Property/Value	Units	Temp	Data Type	Reference
WS 1.56E-002	mg/L	25	EXP	YALKOWSKY,SH & DANNENFELSER,RM (1992)
logP 6.18			EXP	HANSCH,C & LEO,AJ (1985)
VP 7.35E-005	mm Hg	25	EXP	BIDLEMAN,TF (1984)
DC	pKa			
HL 1.81E-003	atm m3/mol	25	EST	VP/WSOL
OH 7.30E-013	cm3/molc sec	25	EST	MEYLAN,WM & HOWARD,PH (1993)

CAS #: 038488-14-1				ETHANONE, 1-(4-METHYLPHENYL)-2-(PHENYLSULFONYL)-

Formula: $C_{15}H_{14}O_3S$

Mol Weight: 274.34

MP (deg C):   FP (deg C):

BP (deg C):

BP pressure (mm Hg):

Property/Value	Units	Temp	Data Type	Reference
WS 8.54E+001	mg/L	25	EST	MEYLAN,WM ET AL. (1996)
logP 2.70			EXP	WANG,L ET AL. (1993)
VP 2.43E-007	mm Hg	25	EST	NEELY,WB & BLAU,GE (1985)
DC	pKa			
HL 1.83E-010	atm m3/mol	25	EST	MEYLAN,WM & HOWARD,PH (1991)
OH 1.07E-011	cm3/molc sec	25	EST	MEYLAN,WM & HOWARD,PH (1993)

CAS #: 038488-19-6				ETHANONE, 1-(4-CHLOROPHENYL)-2-(PHENYLSULFONYL)-

Formula: $C_{14}H_{11}ClO_3S$

Mol Weight: 294.76

MP (deg C):   FP (deg C):

BP (deg C):

BP pressure (mm Hg):

Property/Value	Units	Temp	Data Type	Reference
WS 6.64E+001	mg/L	25	EST	MEYLAN,WM ET AL. (1996)
logP 2.69			EXP	WANG,L ET AL. (1993)
VP 1.36E-007	mm Hg	25	EST	NEELY,WB & BLAU,GE (1985)
DC	pKa			
HL 1.23E-010	atm m3/mol	25	EST	MEYLAN,WM & HOWARD,PH (1991)
OH 7.13E-012	cm3/molc sec	25	EST	MEYLAN,WM & HOWARD,PH (1993)

CAS #: 038507-32-3				2,4(1H,3H)-PYRIMIDINEDIONE, 5-(DIMETHYLAMINO)-1,

Formula: $C_9H_{15}N_3O_2$

Mol Weight: 197.24

MP (deg C):   FP (deg C):

BP (deg C):

BP pressure (mm Hg):

Property/Value	Units	Temp	Data Type	Reference
WS 6.45E+003	mg/L	25	EST	MEYLAN,WM ET AL. (1996)
logP 0.99			EXP	HANSCH,C ET AL. (1995)
VP 2.96E-006	mm Hg	25	EST	NEELY,WB & BLAU,GE (1985)
DC	pKa			
HL 1.35E-011	atm m3/mol	25	EST	MEYLAN,WM & HOWARD,PH (1991)
OH 8.56E-011	cm3/molc sec	25	EST	MEYLAN,WM & HOWARD,PH (1993)

CAS #: 038557-71-0	2-CHLORO-6-METHYL-PYRAZINE

Formula: $C_5H_5ClN_2$

Mol Weight: 128.56

MP (deg C):　　FP (deg C):

BP (deg C):

BP pressure (mm Hg):

Property/Value	Units	Temp	Data Type	Reference
WS 1.21E+004	mg/L	25	EST	MEYLAN,WM ET AL. (1996)
logP 1.04			EXP	YAMAGAMI,C & TAKAO,N (1991)
VP 8.98E-001	mm Hg	25	EST	NEELY,WB & BLAU,GE (1985)
DC	pKa			
HL 1.51E-004	atm m3/mol	25	EST	MEYLAN,WM & HOWARD,PH (1991)
OH 5.58E-013	cm3/molc sec	25	EST	MEYLAN,WM & HOWARD,PH (1993)

CAS #: 038580-89-1	CARBAMIC ACID, DIMETHYL-, 1-METHYLETHYL ESTER

Formula: $C_6H_{13}NO_2$

Mol Weight: 131.18

MP (deg C):　　FP (deg C):

BP (deg C):

BP pressure (mm Hg):

Property/Value	Units	Temp	Data Type	Reference
WS 6.03E+003	mg/L	25	EST	MEYLAN,WM ET AL. (1996)
logP 1.38			EXP	TANAKA,M ET AL. (1985)
VP 8.27E+000	mm Hg	25	EST	NEELY,WB & BLAU,GE (1985)
DC	pKa			
HL 3.36E-007	atm m3/mol	25	EST	MEYLAN,WM & HOWARD,PH (1991)
OH 1.96E-011	cm3/molc sec	25	EST	MEYLAN,WM & HOWARD,PH (1993)

CAS #: 038603-23-5	THIABURIMAMIDE

Formula: $C_8H_{14}N_4S_2$

Mol Weight: 230.36

MP (deg C):　　FP (deg C):

BP (deg C):

BP pressure (mm Hg):

Property/Value	Units	Temp	Data Type	Reference
WS 1.10E+004	mg/L	25	EST	MEYLAN,WM ET AL. (1996)
logP 0.16			EXP	HANSCH,C & LEO,AJ (1985)
VP 3.11E-008	mm Hg	25	EST	NEELY,WB & BLAU,GE (1985)
DC	pKa			
HL 2.84E-010	atm m3/mol	25	EST	MEYLAN,WM & HOWARD,PH (1991)
OH 2.96E-010	cm3/molc sec	25	EST	MEYLAN,WM & HOWARD,PH (1993)

CAS #: 038603-64-4	1-ME-3(2(5ME IMIDAZOL4YL)MESET UREA

Formula: $C_9H_{16}N_4OS$

Mol Weight: 228.32

MP (deg C):　　FP (deg C):

BP (deg C):

BP pressure (mm Hg):

Property/Value	Units	Temp	Data Type	Reference
WS 1.71E+003	mg/L	25	EST	MEYLAN,WM ET AL. (1996)
logP -0.06			EXP	HANSCH,C & LEO,AJ (1985)
VP 4.09E-009	mm Hg	25	EST	NEELY,WB & BLAU,GE (1985)
DC	pKa			
HL 1.55E-015	atm m3/mol	25	EST	MEYLAN,WM & HOWARD,PH (1991)
OH 7.91E-011	cm3/molc sec	25	EST	MEYLAN,WM & HOWARD,PH (1993)

CAS #: 038604-70-5	1-ACETYL-1-METHYL-2-PHENYLHYDRAZINE

Formula: $C_9H_{12}N_2O$

Mol Weight: 164.21

MP (deg C):　　FP (deg C):

BP (deg C):

BP pressure (mm Hg):

Property/Value	Units	Temp	Data Type	Reference
WS 4.98E+003	mg/L	25	EST	MEYLAN,WM ET AL. (1996)
logP 1.31			EXP	HANSCH,C ET AL. (1995)
VP 1.03E-003	mm Hg	25	EST	NEELY,WB & BLAU,GE (1985)
DC	pKa			
HL 5.66E-011	atm m3/mol	25	EST	MEYLAN,WM & HOWARD,PH (1991)
OH 5.83E-011	cm3/molc sec	25	EST	MEYLAN,WM & HOWARD,PH (1993)

CAS #: 038695-58-8	THIOPHENE, 3-METHYLSULFONYL-

Formula: $C_5H_8O_2S_2$

Mol Weight: 164.25

MP (deg C):　　FP (deg C):

BP (deg C):

BP pressure (mm Hg):

Property/Value	Units	Temp	Data Type	Reference
WS 3.17E+004	mg/L	25	EST	MEYLAN,WM ET AL. (1996)
logP 0.38			EXP	HANSCH,C & LEO,AJ (1985)
VP 3.02E-003	mm Hg	25	EST	NEELY,WB & BLAU,GE (1985)
DC	pKa			
HL 6.78E-007	atm m3/mol	25	EST	MEYLAN,WM & HOWARD,PH (1991)
OH 3.10E-012	cm3/molc sec	25	EST	MEYLAN,WM & HOWARD,PH (1993)

CAS #: 038717-24-7	4-PYRIDAZINEAMINE,3-MEO,N,N-DIME

Formula: $C_7H_{11}N_3O$

Mol Weight: 153.19

MP (deg C):　　FP (deg C):

BP (deg C):

BP pressure (mm Hg):

Property/Value	Units	Temp	Data Type	Reference
WS 1.25E+004	mg/L	25	EST	MEYLAN,WM ET AL. (1996)
logP 0.90			EXP	HANSCH,C & LEO,AJ (1985)
VP 1.31E-003	mm Hg	25	EST	NEELY,WB & BLAU,GE (1985)
DC	pKa			
HL 2.46E-008	atm m3/mol	25	EST	MEYLAN,WM & HOWARD,PH (1991)
OH 4.11E-011	cm3/molc sec	25	EST	MEYLAN,WM & HOWARD,PH (1993)

CAS #: 038717-44-1	PYRIDAZINE-3,5-NH2,N,N'-TETRAME

Formula: $C_8H_{14}N_4$

Mol Weight: 166.23

MP (deg C):　　FP (deg C):

BP (deg C):

BP pressure (mm Hg):

Property/Value	Units	Temp	Data Type	Reference
WS 1.23E+004	mg/L	25	EST	MEYLAN,WM ET AL. (1996)
logP 0.84			EXP	HANSCH,C & LEO,AJ (1985)
VP 5.53E-004	mm Hg	25	EST	NEELY,WB & BLAU,GE (1985)
DC	pKa			
HL 7.25E-010	atm m3/mol	25	EST	MEYLAN,WM & HOWARD,PH (1991)
OH 2.05E-010	cm3/molc sec	25	EST	MEYLAN,WM & HOWARD,PH (1993)

CAS #: 038726-90-8				ACETAMIDE, 2-[(DIAZOACETYL)AMINO]-N-METHYL-
Formula: $C_5H_8N_4O_2$				
Mol Weight: 156.15				
MP (deg C):		FP (deg C):		
BP (deg C):				
BP pressure (mm Hg):				

Property/ Value	Units	Temp	Data Type	Reference
WS 7.14E+003	mg/L	25	EST	MEYLAN,WM ET AL. (1996)
logP -0.87			EXP	SANGSTER,J (1994)
VP 4.31E-007	mm Hg	25	EST	NEELY,WB & BLAU,GE (1985)
DC	pKa			
HL 1.72E-009	atm m3/mol	25	EST	MEYLAN,WM & HOWARD,PH (1991)
OH 1.45E-011	cm3/molc sec	25	EST	MEYLAN,WM & HOWARD,PH (1993)

CAS #: 038726-91-9				N-DIAZOACETYLGLYCINE-N'-ETHYLAMIDE
Formula: $C_6H_{10}N_4O_2$				
Mol Weight: 170.17				
MP (deg C):		FP (deg C):		
BP (deg C):				
BP pressure (mm Hg):				

Property/ Value	Units	Temp	Data Type	Reference
WS 1.02E+004	mg/L	25	EST	MEYLAN,WM ET AL. (1996)
logP -0.62			EXP	HANSCH,C & LEO,AJ (1985)
VP 2.54E-012	mm Hg	25	EST	NEELY,WB & BLAU,GE (1985)
DC	pKa			
HL 3.14E-015	atm m3/mol	25	EST	MEYLAN,WM & HOWARD,PH (1991)
OH 1.93E-011	cm3/molc sec	25	EST	MEYLAN,WM & HOWARD,PH (1993)

CAS #: 038727-55-8				ANTOR
Formula: $C_{16}H_{22}ClNO_3$				
Mol Weight: 311.81				
MP (deg C): 49-50		FP (deg C):		
BP (deg C):				
BP pressure (mm Hg):				

Property/ Value	Units	Temp	Data Type	Reference
WS 1.05E+002	mg/L	25	EXP	SHIU,WY ET AL. (1990)
logP 3.60			EXP	HANSCH,C ET AL. (1995)
VP 1.62E-007	mm Hg	25	EST	NEELY,WB & BLAU,GE (1985)
DC	pKa			
HL 6.03E-009	atm m3/mol	25	EST	MEYLAN,WM & HOWARD,PH (1991)
OH 2.54E-011	cm3/molc sec	25	EST	MEYLAN,WM & HOWARD,PH (1993)

CAS #: 038776-76-0				A-DIHYDROGRAYANOTOXIN
Formula: $C_{20}H_{34}O_5$				
Mol Weight: 354.49				
MP (deg C):		FP (deg C):		
BP (deg C):				
BP pressure (mm Hg):				

Property/ Value	Units	Temp	Data Type	Reference
WS 2.99E+002	mg/L	25	EST	MEYLAN,WM ET AL. (1996)
logP 1.51			EXP	HANSCH,C & LEO,AJ (1985)
VP 6.61E-013	mm Hg	25	EST	NEELY,WB & BLAU,GE (1985)
DC	pKa			
HL 3.96E-011	atm m3/mol	25	EST	MEYLAN,WM & HOWARD,PH (1991)
OH 6.01E-011	cm3/molc sec	25	EST	MEYLAN,WM & HOWARD,PH (1993)

CAS #: 038941-33-2				2-(246-TRIBR PH IMINO)IMIDAZOLIDINE
Formula: $C_9H_8Br_3N_3$				
Mol Weight: 397.91				
MP (deg C):		FP (deg C):		
BP (deg C):				
BP pressure (mm Hg):				

Property/ Value	Units	Temp	Data Type	Reference
WS 1.44E+002	mg/L	25	EST	MEYLAN,WM ET AL. (1996)
logP 2.75			EXP	HANSCH,C ET AL. (1995)
VP 3.79E-007	mm Hg	25	EST	NEELY,WB & BLAU,GE (1985)
DC	pKa			
HL 1.73E-012	atm m3/mol	25	EST	MEYLAN,WM & HOWARD,PH (1991)
OH 1.44E-010	cm3/molc sec	25	EST	MEYLAN,WM & HOWARD,PH (1993)

CAS #: 038954-41-5				1-PHENYL-2-ACETAMIDOCYCLOPROPANE
Formula: $C_{11}H_{13}NO$				
Mol Weight: 175.23				
MP (deg C):		FP (deg C):		
BP (deg C):				
BP pressure (mm Hg):				

Property/ Value	Units	Temp	Data Type	Reference
WS 3.49E+003	mg/L	25	EST	MEYLAN,WM ET AL. (1996)
logP 1.43			EXP	HANSCH,C & LEO,AJ (1985)
VP 3.32E-005	mm Hg	25	EST	NEELY,WB & BLAU,GE (1985)
DC	pKa			
HL 1.54E-009	atm m3/mol	25	EST	MEYLAN,WM & HOWARD,PH (1991)
OH 1.06E-011	cm3/molc sec	25	EST	MEYLAN,WM & HOWARD,PH (1993)

CAS #: 038964-22-6				2,8-DICHLORODIBENZO-P-DIOXIN
Formula: $C_{12}H_6Cl_2O_2$				
Mol Weight: 253.09				
MP (deg C):		FP (deg C):		
BP (deg C):				
BP pressure (mm Hg):				

Property/ Value	Units	Temp	Data Type	Reference
WS 1.67E-002	mg/L	25	EXP	YALKOWSKY,SH & DANNENFELSER,RM (1992)
logP 5.63			EST	MEYLAN,WM & HOWARD,PH (1995)
VP 1.05E-006	mm Hg	25	EXP	SHIU,WY ET AL. (1988)
DC	pKa			
HL 2.09E-005	atm m3/mol	25	EST	VP/WSOL
OH 6.92E-012	cm3/molc sec	25	EST	MEYLAN,WM & HOWARD,PH (1993)

CAS #: 039000-84-5				PROPANAMIDE, 2-(DIETHYLAMINO)-N-(2,6-DIMETHYLPHE
Formula: $C_{15}H_{24}N_2O$				
Mol Weight: 248.37				
MP (deg C):		FP (deg C):		
BP (deg C):				
BP pressure (mm Hg):				

Property/ Value	Units	Temp	Data Type	Reference
WS 1.13E+002	mg/L	25	EST	MEYLAN,WM ET AL. (1996)
logP 2.73			EXP	SANGSTER,J (1993)
VP 8.03E-007	mm Hg	25	EST	NEELY,WB & BLAU,GE (1985)
DC	pKa			
HL 1.74E-010	atm m3/mol	25	EST	MEYLAN,WM & HOWARD,PH (1991)
OH 1.17E-010	cm3/molc sec	25	EST	MEYLAN,WM & HOWARD,PH (1993)

1123

039001-02-0 — OCTACHLORODIBENZOFURAN

Formula: $C_{12}Cl_8O$
Mol Weight: 443.76
MP (deg C):
FP (deg C):
BP (deg C):
BP pressure (mm Hg):

Property/Value	Units	Temp	Data Type	Reference
WS 3.85E-006	mg/L	40	EXP	DOUCETTE,WJ & ANDREN,AW (1988)
logP 8.60			EXP	BURKHARD,LP & KUEHL,DW (1986)
VP 5.47E-009	mm Hg	25	EST	NEELY,WB & BLAU,GE (1985)
DC	pKa			
HL 4.66E-006	atm m3/mol	25	EST	MEYLAN,WM & HOWARD,PH (1991)
OH 6.94E-015	cm3/molc sec	25	EST	MEYLAN,WM & HOWARD,PH (1993)

039023-66-0 — 8-BROMO CYCLIC AMP

Formula: $C_{10}H_{13}BrN_5O_6P$
Mol Weight: 410.12
MP (deg C):
FP (deg C):
BP (deg C):
BP pressure (mm Hg):

Property/Value	Units	Temp	Data Type	Reference
WS 1.16E+004	mg/L	25	EST	MEYLAN,WM ET AL. (1996)
logP -2.27			EXP	KORTH,M & ENGELS,J (1987)
VP 6.06E-011	mm Hg	25	EST	NEELY,WB & BLAU,GE (1985)
DC	pKa			
HL 1.16E-028	atm m3/mol	25	EST	MEYLAN,WM & HOWARD,PH (1991)
OH 2.47E-010	cm3/molc sec	25	EST	MEYLAN,WM & HOWARD,PH (1993)

039070-08-1 — 1-ME-2-NO2-5-VINYL IMIDAZOLE

Formula: $C_6H_9N_3O_2$
Mol Weight: 155.16
MP (deg C):
FP (deg C):
BP (deg C):
BP pressure (mm Hg):

Property/Value	Units	Temp	Data Type	Reference
WS 8.51E+003	mg/L	25	EST	MEYLAN,WM ET AL. (1996)
logP 0.64			EXP	HANSCH,C & LEO,AJ (1985)
VP 2.24E-004	mm Hg	25	EST	NEELY,WB & BLAU,GE (1985)
DC	pKa			
HL 1.62E-007	atm m3/mol	25	EST	MEYLAN,WM & HOWARD,PH (1991)
OH 3.07E-011	cm3/molc sec	25	EST	MEYLAN,WM & HOWARD,PH (1993)

039070-09-2 — 1-ME-2-NO2-5-(1,2-DIOHET)IMIDAZOLE

Formula: $C_6H_{11}N_3O_4$
Mol Weight: 189.17
MP (deg C):
FP (deg C):
BP (deg C):
BP pressure (mm Hg):

Property/Value	Units	Temp	Data Type	Reference
WS 1.19E+005	mg/L	25	EST	MEYLAN,WM ET AL. (1996)
logP -0.89			EXP	HANSCH,C & LEO,AJ (1985)
VP 7.99E-009	mm Hg	25	EST	NEELY,WB & BLAU,GE (1985)
DC	pKa			
HL 6.19E-013	atm m3/mol	25	EST	MEYLAN,WM & HOWARD,PH (1991)
OH 2.35E-011	cm3/molc sec	25	EST	MEYLAN,WM & HOWARD,PH (1993)

039073-07-9 — 2,7-DIBROMODIBENZO-P-DIOXIN

Formula: $C_{12}H_6Br_2O_2$
Mol Weight: 342.00
MP (deg C):
FP (deg C):
BP (deg C):
BP pressure (mm Hg):

Property/Value	Units	Temp	Data Type	Reference
WS 3.18E-003	mg/L	25	EST	MEYLAN,WM ET AL. (1996)
logP 6.12			EST	MEYLAN,WM & HOWARD,PH (1995)
VP 4.87E-006	mm Hg	25	EST	NEELY,WB & BLAU,GE (1985)
DC	pKa			
HL 1.86E-006	atm m3/mol	25	EST	MEYLAN,WM & HOWARD,PH (1991)
OH 6.19E-012	cm3/molc sec	25	EST	MEYLAN,WM & HOWARD,PH (1993)

039076-02-3 — METHYL SEC-BUTYLCARBAMATE

Formula: $C_6H_{13}NO_2$
Mol Weight: 131.18
MP (deg C):
FP (deg C):
BP (deg C): 83
BP pressure (mm Hg): 1.60E+001

Property/Value	Units	Temp	Data Type	Reference
WS 6.33E+003	mg/L	25	EST	MEYLAN,WM ET AL. (1996)
logP 1.36			EST	MEYLAN,WM & HOWARD,PH (1995)
VP 3.29E+000	mm Hg	25	EST	NEELY,WB & BLAU,GE (1985)
DC	pKa			
HL 2.03E-007	atm m3/mol	25	EST	MEYLAN,WM & HOWARD,PH (1991)
OH 2.27E-011	cm3/molc sec	25	EST	MEYLAN,WM & HOWARD,PH (1993)

039076-18-1 — CARBAMIC ACID, (3-METHYLPHENYL)-, METHYL ESTER

Formula: $C_9H_{11}NO_2$
Mol Weight: 165.19
MP (deg C):
FP (deg C):
BP (deg C):
BP pressure (mm Hg):

Property/Value	Units	Temp	Data Type	Reference
WS 7.17E+002	mg/L	25	EST	MEYLAN,WM ET AL. (1996)
logP 2.29			EXP	TAKAHASHI,J ET AL. (1988)
VP 1.87E-002	mm Hg	25	EST	NEELY,WB & BLAU,GE (1985)
DC	pKa			
HL 2.41E-008	atm m3/mol	25	EST	MEYLAN,WM & HOWARD,PH (1991)
OH 1.12E-010	cm3/molc sec	25	EST	MEYLAN,WM & HOWARD,PH (1993)

039080-46-1 — PYRIDO(12A)PYRIMIDIN-4-ONE,3,6-DIME

Formula: $C_{10}H_{12}N_2O$
Mol Weight: 176.22
MP (deg C):
FP (deg C):
BP (deg C):
BP pressure (mm Hg):

Property/Value	Units	Temp	Data Type	Reference
WS 4.93E+003	mg/L	25	EST	MEYLAN,WM ET AL. (1996)
logP 1.26			EXP	HANSCH,C & LEO,AJ (1985)
VP 6.29E-005	mm Hg	25	EST	NEELY,WB & BLAU,GE (1985)
DC	pKa			
HL 3.68E-009	atm m3/mol	25	EST	MEYLAN,WM & HOWARD,PH (1991)
OH 1.39E-010	cm3/molc sec	25	EST	MEYLAN,WM & HOWARD,PH (1993)

CAS #: 039080-49-4		4H-PYRIDO[1,2-A]PYRIMIDIN-4-ONE, 6,7,8,9-TETRAHY	

Formula: $C_{10}H_{14}N_2O$
Mol Weight: 178.24
MP (deg C):   FP (deg C):
BP (deg C):
BP pressure (mm Hg):

Property/ Value	Units	Temp	Data Type	Reference
WS 1.50E+004	mg/L	25	EST	MEYLAN,WM ET AL. (1996)
logP 0.67			EXP	SANGSTER,J (1993)
VP 1.05E-004	mm Hg	25	EST	NEELY,WB & BLAU,GE (1985)
DC	pKa			
HL 2.79E-009	atm m3/mol	25	EST	MEYLAN,WM & HOWARD,PH (1991)
OH 3.86E-011	cm3/molc sec	25	EST	MEYLAN,WM & HOWARD,PH (1993)

CAS #: 039080-57-4		PYRIDO(12A)PYRIMIDIN-4-ONE,3-ME	

Formula: $C_9H_{10}N_2O$
Mol Weight: 162.19
MP (deg C):   FP (deg C):
BP (deg C):
BP pressure (mm Hg):

Property/ Value	Units	Temp	Data Type	Reference
WS 1.50E+004	mg/L	25	EST	MEYLAN,WM ET AL. (1996)
logP 0.77			EXP	HANSCH,C & LEO,AJ (1985)
VP 1.60E-004	mm Hg	25	EST	NEELY,WB & BLAU,GE (1985)
DC	pKa			
HL 2.35E-009	atm m3/mol	25	EST	MEYLAN,WM & HOWARD,PH (1991)
OH 1.10E-010	cm3/molc sec	25	EST	MEYLAN,WM & HOWARD,PH (1993)

CAS #: 039080-62-1		4H-PYRIDO[1,2-A]PYRIMIDINE-3-CARBOXYLIC ACID, 1,	

Formula: $C_{11}H_{16}N_2O_3$
Mol Weight: 224.26
MP (deg C):   FP (deg C):
BP (deg C):
BP pressure (mm Hg):

Property/ Value	Units	Temp	Data Type	Reference
WS 1.07E+004	mg/L	25	EST	MEYLAN,WM ET AL. (1996)
logP 0.57			EXP	SANGSTER,J (1994)
VP 3.08E-006	mm Hg	25	EST	NEELY,WB & BLAU,GE (1985)
DC	pKa			
HL 2.89E-012	atm m3/mol	25	EST	MEYLAN,WM & HOWARD,PH (1991)
OH 1.77E-010	cm3/molc sec	25	EST	MEYLAN,WM & HOWARD,PH (1993)

CAS #: 039083-26-6		1,2,3-TRICHLOROBUTA-1,3-DIENE(Z)	

Formula: $C_4H_3Cl_3$
Mol Weight: 157.43
MP (deg C):   FP (deg C):
BP (deg C):
BP pressure (mm Hg):

Property/ Value	Units	Temp	Data Type	Reference
WS 9.22E+001	mg/L	25	EST	MEYLAN,WM ET AL. (1996)
logP 3.38			EST	MEYLAN,WM & HOWARD,PH (1995)
VP 1.57E+001	mm Hg	25	EST	NEELY,WB & BLAU,GE (1985)
DC	pKa			
HL 2.90E-002	atm m3/mol	25	EST	MEYLAN,WM & HOWARD,PH (1991)
OH 1.76E-012	cm3/molc sec	25	EST	MEYLAN,WM & HOWARD,PH (1993)

CAS #: 039115-94-1		BENZOYLHYDRAZINE, M-IODO	

Formula: $C_7H_7IN_2O$
Mol Weight: 262.05
MP (deg C):   FP (deg C):
BP (deg C):
BP pressure (mm Hg):

Property/ Value	Units	Temp	Data Type	Reference
WS 1.00E+003	mg/L	25	EST	MEYLAN,WM ET AL. (1996)
logP 1.53			EXP	HANSCH,C & LEO,AJ (1985)
VP 2.33E-006	mm Hg	25	EST	NEELY,WB & BLAU,GE (1985)
DC	pKa			
HL 2.14E-012	atm m3/mol	25	EST	MEYLAN,WM & HOWARD,PH (1991)
OH 6.09E-012	cm3/molc sec	25	EST	MEYLAN,WM & HOWARD,PH (1993)

CAS #: 039115-95-2		BENZOYLHYDRAZINE, P-IODO	

Formula: $C_7H_7IN_2O$
Mol Weight: 262.05
MP (deg C):   FP (deg C):
BP (deg C):
BP pressure (mm Hg):

Property/ Value	Units	Temp	Data Type	Reference
WS 9.62E+002	mg/L	25	EST	MEYLAN,WM ET AL. (1996)
logP 1.55			EXP	HANSCH,C & LEO,AJ (1985)
VP 2.33E-006	mm Hg	25	EST	NEELY,WB & BLAU,GE (1985)
DC	pKa			
HL 2.14E-012	atm m3/mol	25	EST	MEYLAN,WM & HOWARD,PH (1991)
OH 6.67E-012	cm3/molc sec	25	EST	MEYLAN,WM & HOWARD,PH (1993)

CAS #: 039115-96-3		BENZOYLHYDRAZINE, M-BROMO	

Formula: $C_7H_7BrN_2O$
Mol Weight: 215.06
MP (deg C): 157-159   FP (deg C):
BP (deg C):
BP pressure (mm Hg):

Property/ Value	Units	Temp	Data Type	Reference
WS 3.07E+003	mg/L	25	EST	MEYLAN,WM ET AL. (1996)
logP 1.26			EXP	HANSCH,C & LEO,AJ (1985)
VP 7.82E-006	mm Hg	25	EST	NEELY,WB & BLAU,GE (1985)
DC	pKa			
HL 3.69E-012	atm m3/mol	25	EST	MEYLAN,WM & HOWARD,PH (1991)
OH 6.02E-012	cm3/molc sec	25	EST	MEYLAN,WM & HOWARD,PH (1993)

CAS #: 039123-39-2		4-PYRIDAZINECARBOXYLIC ACID, ETHYL ESTER	

Formula: $C_7H_8N_2O_2$
Mol Weight: 152.15
MP (deg C):   FP (deg C):
BP (deg C):
BP pressure (mm Hg):

Property/ Value	Units	Temp	Data Type	Reference
WS 4.72E+004	mg/L	25	EST	MEYLAN,WM ET AL. (1996)
logP 0.23			EXP	YAMAGAMI,C ET AL. (1990)
VP 1.26E-003	mm Hg	25	EST	NEELY,WB & BLAU,GE (1985)
DC	pKa			
HL 2.45E-008	atm m3/mol	25	EST	MEYLAN,WM & HOWARD,PH (1991)
OH 1.74E-012	cm3/molc sec	25	EST	MEYLAN,WM & HOWARD,PH (1993)

CAS #:	039170-83-7		CYCLOHEXANOL, 2,6-DIMETHYL-, (1à,2à,6á)-

Formula: $C_8H_{16}O$

Mol Weight: 128.22

MP (deg C): FP (deg C):

BP (deg C):

BP pressure (mm Hg):

Property/ Value	Units	Temp	Data Type	Reference
WS 2.81E+003	mg/L	25	EST	MEYLAN,WM ET AL. (1996)
logP 2.10			EXP	FUNASAKI,N ET AL. (1986)
VP 1.23E-001	mm Hg	25	EST	NEELY,WB & BLAU,GE (1985)
DC	pKa			
HL 8.63E-006	atm m3/mol	25	EST	MEYLAN,WM & HOWARD,PH (1991)
OH 2.09E-011	cm3/molc sec	25	EST	MEYLAN,WM & HOWARD,PH (1993)

CAS #:	039170-84-8		CYCLOHEXANOL, 2,6-DIMETHYL-, (1à,2à,6a)-

Formula: $C_8H_{16}O$

Mol Weight: 128.22

MP (deg C): FP (deg C):

BP (deg C):

BP pressure (mm Hg):

Property/ Value	Units	Temp	Data Type	Reference
WS 2.81E+003	mg/L	25	EST	MEYLAN,WM ET AL. (1996)
logP 2.37			EXP	FUNASAKI,N ET AL. (1986)
VP 1.23E-001	mm Hg	25	EST	NEELY,WB & BLAU,GE (1985)
DC	pKa			
HL 8.63E-006	atm m3/mol	25	EST	MEYLAN,WM & HOWARD,PH (1991)
OH 2.09E-011	cm3/molc sec	25	EST	MEYLAN,WM & HOWARD,PH (1993)

CAS #:	039190-66-4		PROPYL-ISOBUTYLAMINE

Formula: $C_7H_{17}N$

Mol Weight: 115.22

MP (deg C): FP (deg C):

BP (deg C):

BP pressure (mm Hg):

Property/ Value	Units	Temp	Data Type	Reference
WS 1.82E+004	mg/L	25	EST	MEYLAN,WM ET AL. (1996)
logP 2.07			EXP	HANSCH,C & LEO,AJ (1985)
VP 1.20E+001	mm Hg	25	EST	NEELY,WB & BLAU,GE (1985)
DC	pKa			
HL 6.86E-005	atm m3/mol	25	EST	MEYLAN,WM & HOWARD,PH (1991)
OH 8.84E-011	cm3/molc sec	25	EST	MEYLAN,WM & HOWARD,PH (1993)

CAS #:	039190-67-5		PROPYL SEC-BUTYLAMINE

Formula: $C_7H_{17}N$

Mol Weight: 115.22

MP (deg C): FP (deg C):

BP (deg C):

BP pressure (mm Hg):

Property/ Value	Units	Temp	Data Type	Reference
WS 2.49E+004	mg/L	25	EST	MEYLAN,WM ET AL. (1996)
logP 1.91			EXP	HANSCH,C & LEO,AJ (1985)
VP 1.20E+001	mm Hg	25	EST	NEELY,WB & BLAU,GE (1985)
DC	pKa			
HL 6.86E-005	atm m3/mol	25	EST	MEYLAN,WM & HOWARD,PH (1991)
OH 9.87E-011	cm3/molc sec	25	EST	MEYLAN,WM & HOWARD,PH (1993)

CAS #:	039201-78-0		3,5-DIMEO-4-PROPOXYPHENETHYLAMINE

Formula: $C_{13}H_{21}NO_3$

Mol Weight: 239.32

MP (deg C): FP (deg C):

BP (deg C):

BP pressure (mm Hg):

Property/ Value	Units	Temp	Data Type	Reference
WS 9.75E+003	mg/L	25	EST	MEYLAN,WM ET AL. (1996)
logP 1.70			EXP	HANSCH,C & LEO,AJ (1985)
VP 5.09E-005	mm Hg	25	EST	NEELY,WB & BLAU,GE (1985)
DC	pKa			
HL 2.96E-010	atm m3/mol	25	EST	MEYLAN,WM & HOWARD,PH (1991)
OH 2.45E-010	cm3/molc sec	25	EST	MEYLAN,WM & HOWARD,PH (1993)

CAS #:	039201-82-6		3,5-DIMEO-4-ETO-PHENETHYLAMINE

Formula: $C_{12}H_{19}NO_3$

Mol Weight: 225.29

MP (deg C): FP (deg C):

BP (deg C):

BP pressure (mm Hg):

Property/ Value	Units	Temp	Data Type	Reference
WS 3.70E+004	mg/L	25	EST	MEYLAN,WM ET AL. (1996)
logP 1.11			EXP	HANSCH,C & LEO,AJ (1985)
VP 1.25E-004	mm Hg	25	EST	NEELY,WB & BLAU,GE (1985)
DC	pKa			
HL 2.23E-010	atm m3/mol	25	EST	MEYLAN,WM & HOWARD,PH (1991)
OH 2.41E-010	cm3/molc sec	25	EST	MEYLAN,WM & HOWARD,PH (1993)

CAS #:	039209-88-6		2-METHOXYQUINOXALINE

Formula: $C_9H_8N_2O$

Mol Weight: 160.18

MP (deg C): FP (deg C):

BP (deg C):

BP pressure (mm Hg):

Property/ Value	Units	Temp	Data Type	Reference
WS 7.27E+002	mg/L	25	EST	MEYLAN,WM ET AL. (1996)
logP 2.31			EXP	HANSCH,C & LEO,AJ (1985)
VP 1.57E-003	mm Hg	25	EST	NEELY,WB & BLAU,GE (1985)
DC	pKa			
HL 1.54E-007	atm m3/mol	25	EST	MEYLAN,WM & HOWARD,PH (1991)
OH 2.29E-011	cm3/molc sec	25	EST	MEYLAN,WM & HOWARD,PH (1993)

CAS #:	039224-65-2		4,5-DICHLORO-2-NITROPHENOL

Formula: $C_6H_3Cl_2NO_3$

Mol Weight: 208.00

MP (deg C): FP (deg C):

BP (deg C):

BP pressure (mm Hg):

Property/ Value	Units	Temp	Data Type	Reference
WS 7.40E+001	mg/L	25	EST	MEYLAN,WM ET AL. (1996)
logP 3.20			EST	MEYLAN,WM & HOWARD,PH (1995)
VP 5.70E-005	mm Hg	25	EST	NEELY,WB & BLAU,GE (1985)
DC	pKa			
HL 3.84E-006	atm m3/mol	25	EST	MEYLAN,WM & HOWARD,PH (1991)
OH 9.95E-013	cm3/molc sec	25	EST	MEYLAN,WM & HOWARD,PH (1993)

1126

039227-28-6 — 1,2,3,4,7,8-HEXACHLORODIBENZO-P-DIOXIN

Formula: $C_{12}H_2Cl_6O_2$
Mol Weight: 390.87
MP (deg C):
FP (deg C):
BP (deg C):
BP pressure (mm Hg):

Property/Value	Units	Temp	Data Type	Reference
WS 4.40E-006	mg/L	20	EXP	YALKOWSKY,SH & DANNENFELSER,RM (1992)
logP 7.80			EXP	SHIU,WY ET AL. (1988)
VP 9.47E-008	mm Hg	25	EST	NEELY,WB & BLAU,GE (1985)
DC	pKa			
HL 1.94E-006	atm m3/mol	25	EST	MEYLAN,WM & HOWARD,PH (1991)
OH 1.22E-012	cm3/molc sec	25	EST	MEYLAN,WM & HOWARD,PH (1993)

039227-53-7 — 1-CHLORODIBENZO-P-DIOXIN

Formula: $C_{12}H_7ClO_2$
Mol Weight: 218.64
MP (deg C):
FP (deg C):
BP (deg C):
BP pressure (mm Hg):

Property/Value	Units	Temp	Data Type	Reference
WS 4.17E-001	mg/L	25	EXP	YALKOWSKY,SH & DANNENFELSER,RM (1992)
logP 4.75			EXP	SHIU,WY ET AL. (1988)
VP 9.00E-005	mm Hg	25	EXP	SHIU,WY ET AL. (1988)
DC	pKa			
HL 6.21E-005	atm m3/mol	25	EST	VP/WSOL
OH 8.38E-012	cm3/molc sec	25	EST	MEYLAN,WM & HOWARD,PH (1993)

039227-54-8 — 2-CHLORODIBENZO-P-DIOXIN

Formula: $C_{12}H_7ClO_2$
Mol Weight: 218.64
MP (deg C):
FP (deg C):
BP (deg C):
BP pressure (mm Hg):

Property/Value	Units	Temp	Data Type	Reference
WS 3.19E-001	mg/L	25	EXP	YALKOWSKY,SH & DANNENFELSER,RM (1992)
logP 5.00			EXP	SHIU,WY ET AL. (1988)
VP 1.28E-005	mm Hg	25	EXP	SHIU,WY ET AL. (1988)
DC	pKa			
HL 1.15E-005	atm m3/mol	25	EST	VP/WSOL
OH 8.38E-012	cm3/molc sec	25	EST	MEYLAN,WM & HOWARD,PH (1993)

039227-58-2 — 1,2,4-TRICHLORODIBENZO-P-DIOXIN

Formula: $C_{12}H_5Cl_3O_2$
Mol Weight: 287.53
MP (deg C):
FP (deg C):
BP (deg C):
BP pressure (mm Hg):

Property/Value	Units	Temp	Data Type	Reference
WS 8.41E-003	mg/L	25	EXP	YALKOWSKY,SH & DANNENFELSER,RM (1992)
logP 6.35			EXP	SHIU,WY ET AL. (1988)
VP 7.50E-007	mm Hg	25	EXP	SHIU,WY ET AL. (1988)
DC	pKa			
HL 3.37E-005	atm m3/mol	25	EST	VP/WSOL
OH 5.63E-012	cm3/molc sec	25	EST	MEYLAN,WM & HOWARD,PH (1993)

039227-61-7 — 1,2,3,4,7-PENTACHLORODIBENZO-P-DIOXIN

Formula: $C_{12}H_3Cl_5O_2$
Mol Weight: 356.42
MP (deg C):
FP (deg C):
BP (deg C):
BP pressure (mm Hg):

Property/Value	Units	Temp	Data Type	Reference
WS 1.20E-004	mg/L	20	EXP	YALKOWSKY,SH & DANNENFELSER,RM (1992)
logP 7.40			EXP	SHIU,WY ET AL. (1988)
VP 3.96E-007	mm Hg	25	EST	NEELY,WB & BLAU,GE (1985)
DC	pKa			
HL 2.61E-006	atm m3/mol	25	EST	MEYLAN,WM & HOWARD,PH (1991)
OH 3.67E-012	cm3/molc sec	25	EST	MEYLAN,WM & HOWARD,PH (1993)

039227-62-8 — 1,2,4,6,7,9-HEXACHLORODIBENZO-P-DIOXIN

Formula: $C_{12}H_2Cl_6O_2$
Mol Weight: 390.87
MP (deg C):
FP (deg C):
BP (deg C):
BP pressure (mm Hg):

Property/Value	Units	Temp	Data Type	Reference
WS 2.65E-005	mg/L	25	EST	MEYLAN,WM ET AL. (1996)
logP 8.21			EST	MEYLAN,WM & HOWARD,PH (1995)
VP 9.47E-008	mm Hg	25	EST	NEELY,WB & BLAU,GE (1985)
DC	pKa			
HL 1.94E-006	atm m3/mol	25	EST	MEYLAN,WM & HOWARD,PH (1991)
OH 1.42E-012	cm3/molc sec	25	EST	MEYLAN,WM & HOWARD,PH (1993)

039300-45-3 — DINOCAP

Formula: $C_{18}H_{24}N_2O_6$
Mol Weight: 364.40
MP (deg C):
FP (deg C):
BP (deg C): 138-140
BP pressure (mm Hg): 5.00E-002

Mixture of two isomers

Property/Value	Units	Temp	Data Type	Reference
WS 4.00E+000	mg/L	25	EXP	WAUCHOPE,RD ET AL. (1991A)
logP 5.98			EST	MEYLAN,WM & HOWARD,PH (1995)
VP 4.00E-008	mm Hg	25	EXP	WAUCHOPE,RD ET AL. (1991A)
DC 12.70	pKa		EXP	HEIMLICH,F & NOLTE,J (1993)
HL 4.79E-009	atm m3/mol	25	EST	VP/WSOL
OH 6.05E-011	cm3/molc sec	25	EST	MEYLAN,WM & HOWARD,PH (1993)

039485-83-1 — 2,2',4,4',6-PENTACHLOROBIPHENYL

Formula: $C_{12}H_5Cl_5$
Mol Weight: 326.44
MP (deg C):
FP (deg C):
BP (deg C):
BP pressure (mm Hg):

Property/Value	Units	Temp	Data Type	Reference
WS 9.39E-003	mg/L	25	EST	MEYLAN,WM ET AL. (1996)
logP 6.98			EST	MEYLAN,WM & HOWARD,PH (1995)
VP 2.22E-006	mm Hg	25	EST	NEELY,WB & BLAU,GE (1985)
DC	pKa			
HL 9.24E-005	atm m3/mol	25	EST	MEYLAN,WM & HOWARD,PH (1991)
OH 5.67E-013	cm3/molc sec	25	EST	MEYLAN,WM & HOWARD,PH (1993)

CAS #: 039495-15-3				2-METHYL-4-HYDROXYACETANILIDE
Formula: $C_9H_{11}NO_2$				
Mol Weight: 165.19				
MP (deg C):		FP (deg C):		
BP (deg C):				
BP pressure (mm Hg):				

Property/ Value	Units	Temp	Data Type	Reference
WS 4.64E+004	mg/L	25	EST	MEYLAN,WM ET AL. (1996)
logP 0.17			EXP	HANSCH,C & LEO,AJ (1985)
VP 2.71E-006	mm Hg	25	EST	NEELY,WB & BLAU,GE (1985)
DC	pKa			
HL 7.09E-013	atm m3/mol	25	EST	MEYLAN,WM & HOWARD,PH (1991)
OH 4.59E-011	cm3/molc sec	25	EST	MEYLAN,WM & HOWARD,PH (1993)

CAS #: 039512-50-0				PIPERAZINE, 1-(2-CHLOROPHENYL)-
Formula: $C_{10}H_{13}ClN_2$				
Mol Weight: 196.68				
MP (deg C):		FP (deg C):		
BP (deg C):				
BP pressure (mm Hg):				

Property/ Value	Units	Temp	Data Type	Reference
WS 7.03E+003	mg/L	25	EST	MEYLAN,WM ET AL. (1996)
logP 2.13			EXP	CACCIA,S ET AL. (1985)
VP 4.79E-004	mm Hg	25	EST	NEELY,WB & BLAU,GE (1985)
DC	pKa			
HL 8.39E-009	atm m3/mol	25	EST	MEYLAN,WM & HOWARD,PH (1991)
OH 1.49E-010	cm3/molc sec	25	EST	MEYLAN,WM & HOWARD,PH (1993)

CAS #: 039512-51-1				PIPERAZINE, 1-(2-METHYLPHENYL)-
Formula: $C_{11}H_{16}N_2$				
Mol Weight: 176.26				
MP (deg C):		FP (deg C):		
BP (deg C):				
BP pressure (mm Hg):				

Property/ Value	Units	Temp	Data Type	Reference
WS 8.70E+003	mg/L	25	EST	MEYLAN,WM ET AL. (1996)
logP 2.14			EXP	CACCIA,S ET AL. (1985)
VP 9.57E-004	mm Hg	25	EST	NEELY,WB & BLAU,GE (1985)
DC	pKa			
HL 1.25E-008	atm m3/mol	25	EST	MEYLAN,WM & HOWARD,PH (1991)
OH 2.86E-010	cm3/molc sec	25	EST	MEYLAN,WM & HOWARD,PH (1993)

CAS #: 039515-41-8				FENPROPATHRIN
Formula: $C_{22}H_{23}NO_3$				
Mol Weight: 349.43				
MP (deg C): 45-50		FP (deg C):		
BP (deg C):				
BP pressure (mm Hg):				

Property/ Value	Units	Temp	Data Type	Reference
WS 3.30E-001	mg/L	25	EXP	SHIU,WY ET AL. (1990)
logP 5.70			EXP	HANSCH,C ET AL. (1995)
VP 2.15E-006	mm Hg	25	EST	NEELY,WB & BLAU,GE (1985)
DC	pKa			
HL 2.05E-006	atm m3/mol	25	EST	MEYLAN,WM & HOWARD,PH (1991)
OH 1.79E-011	cm3/molc sec	25	EST	MEYLAN,WM & HOWARD,PH (1993)

CAS #: 039515-51-0				3-PHENOXYBENZALDEHYDE
Formula: $C_{13}H_{10}O_2$				
Mol Weight: 198.22				
MP (deg C):		FP (deg C):		
BP (deg C): 140				
BP pressure (mm Hg): 1.00E-001				

Property/ Value	Units	Temp	Data Type	Reference
WS 5.80E+001	mg/L	25	EST	MEYLAN,WM ET AL. (1996)
logP 3.38			EXP	HANSCH,C ET AL. (1995)
VP 1.64E-004	mm Hg	25	EST	NEELY,WB & BLAU,GE (1985)
DC	pKa			
HL 2.94E-007	atm m3/mol	25	EST	MEYLAN,WM & HOWARD,PH (1991)
OH 2.43E-011	cm3/molc sec	25	EST	MEYLAN,WM & HOWARD,PH (1993)

CAS #: 039552-01-7				ETHANONE, 1-[7-[2-HYDROXY-3-[(1-METHYLETHYL)AMIN
Formula: $C_{16}H_{21}NO_4$				
Mol Weight: 291.35				
MP (deg C): 115		FP (deg C):		
BP (deg C):				
BP pressure (mm Hg):				

Property/ Value	Units	Temp	Data Type	Reference
WS 2.64E+003	mg/L	25	EST	MEYLAN,WM ET AL. (1996)
logP 2.02			EXP	RECANATINI,M (1992)
VP 3.83E-009	mm Hg	25	EST	NEELY,WB & BLAU,GE (1985)
DC	pKa			
HL 1.45E-015	atm m3/mol	25	EST	MEYLAN,WM & HOWARD,PH (1991)
OH 1.81E-010	cm3/molc sec	25	EST	MEYLAN,WM & HOWARD,PH (1993)

CAS #: 039589-98-5				DIMALONE
Formula: $C_{11}H_{14}O_4$				
Mol Weight: 210.23				
MP (deg C): 38		FP (deg C):		
BP (deg C): 137				
BP pressure (mm Hg): 1.25E+001				

Property/ Value	Units	Temp	Data Type	Reference
WS 1.32E+004	mg/L	35	EXP	YALKOWSKY,SH & DANNENFELSER,RM (1992)
logP 1.78			EST	MEYLAN,WM & HOWARD,PH (1995)
VP 3.73E-002	mm Hg	25	EST	NEELY,WB & BLAU,GE (1985)
DC	pKa			
HL 3.91E-007	atm m3/mol	25	EST	MEYLAN,WM & HOWARD,PH (1991)
OH 6.43E-011	cm3/molc sec	25	EST	MEYLAN,WM & HOWARD,PH (1993)

CAS #: 039617-16-8				ACETAMIDE, N-[2-[3-[(1,1-DIMETHYLETHYL)AMINO]-2-
Formula: $C_{15}H_{24}N_2O_3$				
Mol Weight: 280.37				
MP (deg C):		FP (deg C):		
BP (deg C):				
BP pressure (mm Hg):				

Property/ Value	Units	Temp	Data Type	Reference
WS 4.16E+002	mg/L	25	EST	MEYLAN,WM ET AL. (1996)
logP 0.32			EXP	SANGSTER,J (1994)
VP 1.57E-010	mm Hg	25	EST	NEELY,WB & BLAU,GE (1985)
DC	pKa			
HL 3.01E-018	atm m3/mol	25	EST	MEYLAN,WM & HOWARD,PH (1991)
OH 1.45E-010	cm3/molc sec	25	EST	MEYLAN,WM & HOWARD,PH (1993)

039624-86-7 — DOWCO 275

Formula: $C_9H_{13}FNO_3PS$
Mol Weight: 265.25
MP (deg C): | FP (deg C):
BP (deg C):
BP pressure (mm Hg):

Property/Value	Units	Temp	Data Type	Reference
WS 1.96E+001	mg/L	25	EST	MEYLAN,WM ET AL. (1996)
logP 3.51			EXP	BRIGGS,GG (1981)
VP 8.18E-004	mm Hg	25	EST	NEELY,WB & BLAU,GE (1985)
DC	pKa			
HL 1.15E-007	atm m3/mol	25	EST	MEYLAN,WM & HOWARD,PH (1991)
OH 9.25E-011	cm3/molc sec	25	EST	MEYLAN,WM & HOWARD,PH (1993)

039638-51-2 — L-TYROSINE, 3-HYDROXY-, PROPYL ESTER

Formula: $C_{12}H_{17}NO_4$
Mol Weight: 239.27
MP (deg C): | FP (deg C):
BP (deg C):
BP pressure (mm Hg):

Property/Value	Units	Temp	Data Type	Reference
WS 2.36E+005	mg/L	25	EST	MEYLAN,WM ET AL. (1996)
logP 0.76			EXP	MARREL,C ET AL. (1985)
VP 1.21E-007	mm Hg	25	EST	NEELY,WB & BLAU,GE (1985)
DC	pKa			
HL 7.36E-016	atm m3/mol	25	EST	MEYLAN,WM & HOWARD,PH (1991)
OH 9.90E-011	cm3/molc sec	25	EST	MEYLAN,WM & HOWARD,PH (1993)

039640-62-5 — 4-PYRIDINEACETAMIDE

Formula: $C_7H_8N_2O$
Mol Weight: 136.15
MP (deg C): | FP (deg C):
BP (deg C):
BP pressure (mm Hg):

Property/Value	Units	Temp	Data Type	Reference
WS 1.00E+006	mg/L	25	EST	MEYLAN,WM ET AL. (1996)
logP -0.65			EXP	HANSCH,C & LEO,AJ (1985)
VP 1.57E-004	mm Hg	25	EST	NEELY,WB & BLAU,GE (1985)
DC	pKa			
HL 1.18E-012	atm m3/mol	25	EST	MEYLAN,WM & HOWARD,PH (1991)
OH 3.62E-012	cm3/molc sec	25	EST	MEYLAN,WM & HOWARD,PH (1993)

039700-44-2 — G-THIOBUTYROLACTONE

Formula: C_4H_6OS
Mol Weight: 102.16
MP (deg C): | FP (deg C):
BP (deg C):
BP pressure (mm Hg):

Property/Value	Units	Temp	Data Type	Reference
WS 3.54E+004	mg/L	25	EST	MEYLAN,WM ET AL. (1996)
logP 0.60			EXP	HANSCH,C & LEO,AJ (1985)
VP 2.17E-001	mm Hg	25	EST	NEELY,WB & BLAU,GE (1985)
DC	pKa			
HL 5.40E-005	atm m3/mol	25	EST	MEYLAN,WM & HOWARD,PH (1991)
OH 1.20E-011	cm3/molc sec	25	EST	MEYLAN,WM & HOWARD,PH (1993)

039765-80-5 — TRANS-NONACHLOR

Formula: $C_{10}H_5Cl_9$
Mol Weight: 444.23
MP (deg C): | FP (deg C):
BP (deg C):
BP pressure (mm Hg):

Property/Value	Units	Temp	Data Type	Reference
WS 1.04E-002	mg/L	25	EST	MEYLAN,WM ET AL. (1996)
logP 6.78			EST	MEYLAN,WM & HOWARD,PH (1995)
VP 2.78E-005	mm Hg	25	EXP	HINCKLEY,DA ET AL. (1990)
DC	pKa			
HL 2.48E-005	atm m3/mol	25	EST	MEYLAN,WM & HOWARD,PH (1991)
OH 4.95E-012	cm3/molc sec	25	EST	MEYLAN,WM & HOWARD,PH (1993)

039791-38-3 — PREGN-4-ENE-3,11,20-TRIONE, 17-(ACETYLOXY)-21-HY

Formula: $C_{23}H_{30}O_6$
Mol Weight: 402.49
MP (deg C): | FP (deg C):
BP (deg C):
BP pressure (mm Hg):

Property/Value	Units	Temp	Data Type	Reference
WS 7.73E+001	mg/L	25	EST	MEYLAN,WM ET AL. (1996)
logP 2.45			EXP	HANSCH,C ET AL. (1995)
VP 5.13E-013	mm Hg	25	EST	NEELY,WB & BLAU,GE (1985)
DC	pKa			
HL 6.51E-014	atm m3/mol	25	EST	MEYLAN,WM & HOWARD,PH (1991)
OH 1.07E-010	cm3/molc sec	25	EST	MEYLAN,WM & HOWARD,PH (1993)

039824-09-4 — 1-BUTYLGALACTOPYRANOSIDE

Formula: $C_{10}H_{20}O_6$
Mol Weight: 236.27
MP (deg C): | FP (deg C):
BP (deg C):
BP pressure (mm Hg):

Property/Value	Units	Temp	Data Type	Reference
WS 2.18E+005	mg/L	25	EST	MEYLAN,WM ET AL. (1996)
logP -1.04			EXP	HANSCH,C & LEO,AJ (1985)
VP 2.21E-009	mm Hg	25	EST	NEELY,WB & BLAU,GE (1985)
DC	pKa			
HL 3.65E-014	atm m3/mol	25	EST	MEYLAN,WM & HOWARD,PH (1991)
OH 7.44E-011	cm3/molc sec	25	EST	MEYLAN,WM & HOWARD,PH (1993)

039824-10-7 — B-PENTYLGALACTOPYRANOSIDE

Formula: $C_{11}H_{22}O_6$
Mol Weight: 250.29
MP (deg C): | FP (deg C):
BP (deg C):
BP pressure (mm Hg):

Property/Value	Units	Temp	Data Type	Reference
WS 5.38E+004	mg/L	25	EST	MEYLAN,WM ET AL. (1996)
logP -0.42			EXP	HANSCH,C & LEO,AJ (1985)
VP 7.73E-010	mm Hg	25	EST	NEELY,WB & BLAU,GE (1985)
DC	pKa			
HL 4.84E-014	atm m3/mol	25	EST	MEYLAN,WM & HOWARD,PH (1991)
OH 7.58E-011	cm3/molc sec	25	EST	MEYLAN,WM & HOWARD,PH (1993)

039824-11-8 — B-HEXYLGALACTOPYRANOSIDE

CAS #: 039824-11-8

Formula: $C_{12}H_{24}O_6$

Mol Weight: 264.32

MP (deg C): FP (deg C):

BP (deg C):

BP pressure (mm Hg):

Property/Value	Units	Temp	Data Type	Reference
WS 1.44E+004	mg/L	25	EST	MEYLAN,WM ET AL. (1996)
logP 0.16			EXP	HANSCH,C & LEO,AJ (1985)
VP 2.69E-010	mm Hg	25	EST	NEELY,WB & BLAU,GE (1985)
DC	pKa			
HL 6.43E-014	atm m3/mol	25	EST	MEYLAN,WM & HOWARD,PH (1991)
OH 7.72E-011	cm3/molc sec	25	EST	MEYLAN,WM & HOWARD,PH (1993)

039835-28-4 — BENZISOXAZOLE, 5-NITRO

CAS #: 039835-28-4

Formula: $C_7H_4N_2O_3$

Mol Weight: 164.12

MP (deg C): FP (deg C):

BP (deg C):

BP pressure (mm Hg):

Property/Value	Units	Temp	Data Type	Reference
WS 1.51E+003	mg/L	25	EST	MEYLAN,WM ET AL. (1996)
logP 1.46			EXP	HANSCH,C & LEO,AJ (1985)
VP 6.57E-004	mm Hg	25	EST	NEELY,WB & BLAU,GE (1985)
DC	pKa			
HL 2.18E-008	atm m3/mol	25	EST	MEYLAN,WM & HOWARD,PH (1991)
OH 1.12E-012	cm3/molc sec	25	EST	MEYLAN,WM & HOWARD,PH (1993)

039845-47-1 — 19-NORPREGNA-1,3,5(10)-TRIEN-20-YNE-3,17-DIOL, 1

CAS #: 039845-47-1

Formula: $C_{22}H_{28}O_2$

Mol Weight: 324.47

MP (deg C): FP (deg C):

BP (deg C):

BP pressure (mm Hg):

Property/Value	Units	Temp	Data Type	Reference
WS 2.40E+001	mg/L	25	EST	MEYLAN,WM ET AL. (1996)
logP 4.28			EXP	SANGSTER,J (1993)
VP 5.40E-010	mm Hg	25	EST	NEELY,WB & BLAU,GE (1985)
DC	pKa			
HL 1.40E-011	atm m3/mol	25	EST	MEYLAN,WM & HOWARD,PH (1991)
OH 1.27E-010	cm3/molc sec	25	EST	MEYLAN,WM & HOWARD,PH (1993)

039856-64-9 — 3-FURANEMETHANOL-5-BENZYL,ACETATE

CAS #: 039856-64-9

Formula: $C_{14}H_{15}O_3$

Mol Weight: 231.27

MP (deg C): FP (deg C):

BP (deg C):

BP pressure (mm Hg):

Property/Value	Units	Temp	Data Type	Reference
WS 5.19E+001	mg/L	25	EST	MEYLAN,WM ET AL. (1996)
logP 3.24			EXP	HANSCH,C & LEO,AJ (1985)
VP 1.38E-004	mm Hg	25	EST	NEELY,WB & BLAU,GE (1985)
DC	pKa			
HL 1.26E-006	atm m3/mol	25	EST	MEYLAN,WM & HOWARD,PH (1991)
OH 2.03E-010	cm3/molc sec	25	EST	MEYLAN,WM & HOWARD,PH (1993)

039923-25-6 — AZINPHOS-ETHYL, O-ANALOG

CAS #: 039923-25-6

Formula: $C_{12}H_{16}N_3O_4PS$

Mol Weight: 329.32

MP (deg C): FP (deg C):

BP (deg C):

BP pressure (mm Hg):

Property/Value	Units	Temp	Data Type	Reference
WS 3.34E+002	mg/L	25	EST	MEYLAN,WM ET AL. (1996)
logP 1.63			EXP	HANSCH,C & LEO,AJ (1985)
VP 2.06E-008	mm Hg	25	EST	NEELY,WB & BLAU,GE (1985)
DC	pKa			
HL 1.09E-012	atm m3/mol	25	EST	MEYLAN,WM & HOWARD,PH (1991)
OH 1.18E-010	cm3/molc sec	25	EST	MEYLAN,WM & HOWARD,PH (1993)

039928-74-0 — 1H-IMIDAZOLE-5-CARBOXALDEHYDE, 1-METHYL-2-NITRO-

CAS #: 039928-74-0

Formula: $C_5H_5N_3O_3$

Mol Weight: 155.11

MP (deg C): FP (deg C):

BP (deg C):

BP pressure (mm Hg):

Property/Value	Units	Temp	Data Type	Reference
WS 3.57E+004	mg/L	25	EST	MEYLAN,WM ET AL. (1996)
logP -0.10			EXP	HANSCH,C ET AL. (1995)
VP 4.61E-005	mm Hg	25	EST	NEELY,WB & BLAU,GE (1985)
DC	pKa			
HL 7.88E-010	atm m3/mol	25	EST	MEYLAN,WM & HOWARD,PH (1991)
OH 1.93E-011	cm3/molc sec	25	EST	MEYLAN,WM & HOWARD,PH (1993)

039935-49-4 — 3-DEAZAURIDINE

CAS #: 039935-49-4

Formula: $C_{10}H_{15}NO_6$

Mol Weight: 245.23

MP (deg C): FP (deg C):

BP (deg C):

BP pressure (mm Hg):

Property/Value	Units	Temp	Data Type	Reference
WS 1.36E+005	mg/L	25	EST	MEYLAN,WM ET AL. (1996)
logP -0.86			EXP	HANSCH,C ET AL. (1995)
VP 9.11E-013	mm Hg	25	EST	NEELY,WB & BLAU,GE (1985)
DC	pKa			
HL 8.45E-018	atm m3/mol	25	EST	MEYLAN,WM & HOWARD,PH (1991)
OH 1.33E-010	cm3/molc sec	25	EST	MEYLAN,WM & HOWARD,PH (1993)

039938-79-9 — 1,1-DIME-3-(3-AMINOPHENYL)UREA

CAS #: 039938-79-9

Formula: $C_9H_{13}N_3O$

Mol Weight: 179.22

MP (deg C): FP (deg C):

BP (deg C):

BP pressure (mm Hg):

Property/Value	Units	Temp	Data Type	Reference
WS 3.31E+003	mg/L	25	EST	MEYLAN,WM ET AL. (1996)
logP -0.10			EXP	HANSCH,C & LEO,AJ (1985)
VP 6.91E-006	mm Hg	25	EST	NEELY,WB & BLAU,GE (1985)
DC	pKa			
HL 3.43E-013	atm m3/mol	25	EST	MEYLAN,WM & HOWARD,PH (1991)
OH 2.02E-010	cm3/molc sec	25	EST	MEYLAN,WM & HOWARD,PH (1993)

CAS #:	039942-91-1	1H-IMIDAZOLE-4-CARBOXAMIDE, 5-[3-METHYL-3-(PHENY

Formula: $C_{12}H_{14}N_6O$

Mol Weight: 258.28

MP (deg C):   FP (deg C):

BP (deg C):

BP pressure (mm Hg):

Property/Value	Units	Temp	Data Type	Reference
WS 1.03E+003	mg/L	25	EST	MEYLAN,WM ET AL. (1996)
logP 1.54			EXP	HANSCH,C ET AL. (1995)
VP 5.44E-011	mm Hg	25	EST	NEELY,WB & BLAU,GE (1985)
DC	pKa			
HL 1.24E-017	atm m3/mol	25	EST	MEYLAN,WM & HOWARD,PH (1991)
OH 2.98E-011	cm3/molc sec	25	EST	MEYLAN,WM & HOWARD,PH (1993)

CAS #:	039960-99-1	P-AMIDO-AMPHENICOL

Formula: $C_{12}H_{14}Cl_2N_2O_4$

Mol Weight: 321.16

MP (deg C):   FP (deg C):

BP (deg C):

BP pressure (mm Hg):

Property/Value	Units	Temp	Data Type	Reference
WS 1.16E+003	mg/L	25	EST	MEYLAN,WM ET AL. (1996)
logP -0.48			EXP	HANSCH,C & LEO,AJ (1985)
VP 2.38E-015	mm Hg	25	EST	NEELY,WB & BLAU,GE (1985)
DC	pKa			
HL 3.78E-019	atm m3/mol	25	EST	MEYLAN,WM & HOWARD,PH (1991)
OH 3.66E-011	cm3/molc sec	25	EST	MEYLAN,WM & HOWARD,PH (1993)

CAS #:	039961-02-9	CYCLOPROPANECARBOXAMIDE, N-[4-[2-[(DICHLOROACETY

Formula: $C_{15}H_{18}Cl_2N_2O_4$

Mol Weight: 361.23

MP (deg C):   FP (deg C):

BP (deg C):

BP pressure (mm Hg):

Property/Value	Units	Temp	Data Type	Reference
WS 1.21E+003	mg/L	25	EST	MEYLAN,WM ET AL. (1996)
logP 0.75			EXP	HANSCH,C ET AL. (1995)
VP 1.28E-016	mm Hg	25	EST	NEELY,WB & BLAU,GE (1985)
DC	pKa			
HL 8.18E-019	atm m3/mol	25	EST	MEYLAN,WM & HOWARD,PH (1991)
OH 4.56E-011	cm3/molc sec	25	EST	MEYLAN,WM & HOWARD,PH (1993)

CAS #:	039961-07-4	P-UREA-AMPHENICOL

Formula: $C_{12}H_{15}Cl_2N_3O_4$

Mol Weight: 336.18

MP (deg C):   FP (deg C):

BP (deg C):

BP pressure (mm Hg):

Property/Value	Units	Temp	Data Type	Reference
WS 1.20E+003	mg/L	25	EST	MEYLAN,WM ET AL. (1996)
logP -0.60			EXP	HANSCH,C & LEO,AJ (1985)
VP 2.38E-015	mm Hg	25	EST	NEELY,WB & BLAU,GE (1985)
DC	pKa			
HL 2.17E-023	atm m3/mol	25	EST	MEYLAN,WM & HOWARD,PH (1991)
OH 8.27E-011	cm3/molc sec	25	EST	MEYLAN,WM & HOWARD,PH (1993)

CAS #:	039980-81-9	IMIDAZOLE-5-CONH2,4(3SBU-3MEN=NN)

Formula: $C_9H_{16}N_6O$

Mol Weight: 224.27

MP (deg C):   FP (deg C):

BP (deg C):

BP pressure (mm Hg):

Property/Value	Units	Temp	Data Type	Reference
WS 3.91E+003	mg/L	25	EST	MEYLAN,WM ET AL. (1996)
logP 1.08			EXP	HANSCH,C & LEO,AJ (1985)
VP 2.95E-009	mm Hg	25	EST	NEELY,WB & BLAU,GE (1985)
DC	pKa			
HL 3.60E-016	atm m3/mol	25	EST	MEYLAN,WM & HOWARD,PH (1991)
OH 3.99E-011	cm3/molc sec	25	EST	MEYLAN,WM & HOWARD,PH (1993)

CAS #:	040016-42-0	2-IMIDAZOLIDINETHIONE, 1-(5-NITRO-2-THIAZOLYL)-

Formula: $C_6H_6N_4O_2S_2$

Mol Weight: 230.27

MP (deg C):   FP (deg C):

BP (deg C):

BP pressure (mm Hg):

Property/Value	Units	Temp	Data Type	Reference
WS 4.45E+003	mg/L	25	EST	MEYLAN,WM ET AL. (1996)
logP 1.70			EXP	SANGSTER,J (1994)
VP 1.23E-007	mm Hg	25	EST	NEELY,WB & BLAU,GE (1985)
DC	pKa			
HL 4.85E-012	atm m3/mol	25	EST	MEYLAN,WM & HOWARD,PH (1991)
OH 8.03E-011	cm3/molc sec	25	EST	MEYLAN,WM & HOWARD,PH (1993)

CAS #:	040026-24-2	2H-1-BENZOPYRAN-2-CARBOXYLIC ACID, 6-CHLORO-2,3-

Formula: $C_{10}H_9ClO_3$

Mol Weight: 212.63

MP (deg C):   FP (deg C):

BP (deg C):

BP pressure (mm Hg):

Property/Value	Units	Temp	Data Type	Reference
WS 8.34E+002	mg/L	25	EST	MEYLAN,WM ET AL. (1996)
logP 2.40			EXP	SANGSTER,J (1993)
VP 2.81E-005	mm Hg	25	EST	NEELY,WB & BLAU,GE (1985)
DC	pKa			
HL 8.03E-009	atm m3/mol	25	EST	MEYLAN,WM & HOWARD,PH (1991)
OH 2.16E-011	cm3/molc sec	25	EST	MEYLAN,WM & HOWARD,PH (1993)

CAS #:	040027-72-3	MONOBROMOAMPHENICOL

Formula: $C_{11}H_{13}BrN_2O_5$

Mol Weight: 333.14

MP (deg C):   FP (deg C):

BP (deg C):

BP pressure (mm Hg):

Property/Value	Units	Temp	Data Type	Reference
WS 8.70E+002	mg/L	25	EST	MEYLAN,WM ET AL. (1996)
logP 0.66			EXP	HANSCH,C & LEO,AJ (1985)
VP 4.61E-013	mm Hg	25	EST	NEELY,WB & BLAU,GE (1985)
DC	pKa			
HL 2.12E-018	atm m3/mol	25	EST	MEYLAN,WM & HOWARD,PH (1991)
OH 3.10E-011	cm3/molc sec	25	EST	MEYLAN,WM & HOWARD,PH (1993)

040027-73-4 — MONOIODOAMPHENICOL

Formula: $C_{11}H_{13}IN_2O_5$

Mol Weight: 380.14

MP (deg C): FP (deg C):

BP (deg C):

BP pressure (mm Hg):

Property/Value	Units	Temp	Data Type	Reference
WS 2.18E+002	mg/L	25	EST	MEYLAN,WM ET AL. (1996)
logP 1.03			EXP	HANSCH,C & LEO,AJ (1985)
VP 1.07E-013	mm Hg	25	EST	NEELY,WB & BLAU,GE (1985)
DC	pKa			
HL 1.38E-018	atm m3/mol	25	EST	MEYLAN,WM & HOWARD,PH (1991)
OH 3.11E-011	cm3/molc sec	25	EST	MEYLAN,WM & HOWARD,PH (1993)

040065-00-7 — 2-(2-CF3-PH-IMINO)IMIDAZOLIDINE

Formula: $C_{10}H_{10}F_3N_3$

Mol Weight: 229.21

MP (deg C): FP (deg C):

BP (deg C):

BP pressure (mm Hg):

Property/Value	Units	Temp	Data Type	Reference
WS 1.88E+003	mg/L	25	EST	MEYLAN,WM ET AL. (1996)
logP 1.42			EXP	HANSCH,C & LEO,AJ (1985)
VP 3.12E-006	mm Hg	25	EST	NEELY,WB & BLAU,GE (1985)
DC	pKa			
HL 2.38E-010	atm m3/mol	25	EST	MEYLAN,WM & HOWARD,PH (1991)
OH 1.80E-011	cm3/molc sec	25	EST	MEYLAN,WM & HOWARD,PH (1993)

040065-09-6 — 2-(26-DICL-4-BRPHIMINO)IMIDAZOLIDINE

Formula: $C_9H_8BrCl_2N_3$

Mol Weight: 309.00

MP (deg C): FP (deg C):

BP (deg C):

BP pressure (mm Hg):

Property/Value	Units	Temp	Data Type	Reference
WS 5.93E+002	mg/L	25	EST	MEYLAN,WM ET AL. (1996)
logP 2.66			EXP	HANSCH,C & LEO,AJ (1985)
VP 2.03E-006	mm Hg	25	EST	NEELY,WB & BLAU,GE (1985)
DC	pKa			
HL 6.00E-012	atm m3/mol	25	EST	MEYLAN,WM & HOWARD,PH (1991)
OH 1.44E-010	cm3/molc sec	25	EST	MEYLAN,WM & HOWARD,PH (1993)

040089-90-5 — PYRIDINE, 4-HEPTYL-

Formula: $C_{12}H_{19}N$

Mol Weight: 177.29

MP (deg C): FP (deg C):

BP (deg C):

BP pressure (mm Hg):

Property/Value	Units	Temp	Data Type	Reference
WS 6.08E+001	mg/L	25	EST	MEYLAN,WM ET AL. (1996)
logP 5.00			EXP	SANGSTER,J (1993)
VP 1.36E-002	mm Hg	25	EST	NEELY,WB & BLAU,GE (1985)
DC	pKa			
HL 4.26E-005	atm m3/mol	25	EST	MEYLAN,WM & HOWARD,PH (1991)
OH 9.04E-012	cm3/molc sec	25	EST	MEYLAN,WM & HOWARD,PH (1993)

040120-74-9 — TRIS(1,3-DICHLOROPROPYL) PHOSPHATE

Formula: $C_9H_{15}Cl_6O_4P$

Mol Weight: 430.91

MP (deg C): 26.7 FP (deg C):

BP (deg C): 236-237

BP pressure (mm Hg): 5.00E+000

Property/Value	Units	Temp	Data Type	Reference
WS 1.00E+002	mg/L		EXP	MUIR,DCG (1984)
logP 3.65			EST	MEYLAN,WM & HOWARD,PH (1995)
VP 2.28E-007	mm Hg	25	EST	NEELY,WB & BLAU,GE (1985)
DC	pKa			
HL 2.61E-009	atm m3/mol	25	EST	MEYLAN,WM & HOWARD,PH (1991)
OH 5.74E-011	cm3/molc sec	25	EST	MEYLAN,WM & HOWARD,PH (1993)

040169-27-5 — 4,4,6,6,8,8-HEXAMETHYLSPIRO(CYCLOTETRASILOXANE*)

Formula: $C_{18}H_{26}O_5Si_4$

Mol Weight: 434.75

MP (deg C): FP (deg C):

BP (deg C):

BP pressure (mm Hg):

Property/Value	Units	Temp	Data Type	Reference
WS 1.32E-005	mg/L	25	EST	MEYLAN,WM ET AL. (1996)
logP 7.05			EST	MEYLAN,WM & HOWARD,PH (1995)
VP	mm Hg			
DC	pKa			
HL	atm m3/mol			
OH 1.96E-011	cm3/molc sec	25	EST	MEYLAN,WM & HOWARD,PH (1993)

040186-72-9 — 2,2',3,3',4,4',5,5',6-NONACHLOROBIPHENYL

Formula: $C_{12}HCl_9$

Mol Weight: 464.22

MP (deg C): FP (deg C):

BP (deg C):

BP pressure (mm Hg):

Property/Value	Units	Temp	Data Type	Reference
WS 2.50E-005	mg/L	25	EXP	YALKOWSKY,SH & DANNENFELSER,RM (1992)
logP 9.56			EST	MEYLAN,WM & HOWARD,PH (1995)
VP 7.60E-009	mm Hg	25	EST	NEELY,WB & BLAU,GE (1985)
DC	pKa			
HL 2.79E-005	atm m3/mol	25	EST	MEYLAN,WM & HOWARD,PH (1991)
OH 3.29E-014	cm3/molc sec	25	EST	MEYLAN,WM & HOWARD,PH (1993)

040246-14-8 — ALPHA-BETA-DIACETYLDIGOXIN

Formula: $C_{45}H_{68}O_{16}$

Mol Weight: 865.03

MP (deg C): FP (deg C):

BP (deg C):

BP pressure (mm Hg):

Property/Value	Units	Temp	Data Type	Reference
WS 2.20E-002	mg/L	25	EST	MEYLAN,WM ET AL. (1996)
logP 2.45			EXP	HANSCH,C ET AL. (1995)
VP 2.66E-030	mm Hg	25	EST	NEELY,WB & BLAU,GE (1985)
DC	pKa			
HL 1.97E-029	atm m3/mol	25	EST	MEYLAN,WM & HOWARD,PH (1991)
OH 1.63E-010	cm3/molc sec	25	EST	MEYLAN,WM & HOWARD,PH (1993)

CAS #: 040256-75-5				BENZAMIDE,N(DIETAMET),2MEO-5CL
Formula: $C_{14}H_{21}ClN_2O_2$				
Mol Weight: 284.79				
MP (deg C):		FP (deg C):		
BP (deg C):				
BP pressure (mm Hg):				

Property/ Value	Units	Temp	Data Type	Reference
WS 3.32E+001	mg/L	25	EST	MEYLAN,WM ET AL. (1996)
logP 3.11			EXP	HANSCH,C & LEO,AJ (1985)
VP 1.01E-007	mm Hg	25	EST	NEELY,WB & BLAU,GE (1985)
DC	pKa			
HL 2.47E-013	atm m3/mol	25	EST	MEYLAN,WM & HOWARD,PH (1991)
OH 1.11E-010	cm3/molc sec	25	EST	MEYLAN,WM & HOWARD,PH (1993)

CAS #: 040257-94-1				PROPANEDINITRILE, [(4-METHYLPHENYL)HYDRAZONO]-
Formula: $C_{10}H_8N_4$				
Mol Weight: 184.20				
MP (deg C):		FP (deg C):		
BP (deg C):				
BP pressure (mm Hg):				

Property/ Value	Units	Temp	Data Type	Reference
WS 9.72E+001	mg/L	25	EST	MEYLAN,WM ET AL. (1996)
logP 3.20			EXP	STURDIK,E ET AL. (1985)
VP 8.97E-006	mm Hg	25	EST	NEELY,WB & BLAU,GE (1985)
DC	pKa			
HL 6.39E-009	atm m3/mol	25	EST	MEYLAN,WM & HOWARD,PH (1991)
OH 5.24E-011	cm3/molc sec	25	EST	MEYLAN,WM & HOWARD,PH (1993)

CAS #: 040262-48-4				PYRAZINECARBONITRILE, 6-(DIMETHYLAMINO)-
Formula: $C_7H_8N_4$				
Mol Weight: 148.17				
MP (deg C):		FP (deg C):		
BP (deg C):				
BP pressure (mm Hg):				

Property/ Value	Units	Temp	Data Type	Reference
WS 4.64E+003	mg/L	25	EST	MEYLAN,WM ET AL. (1996)
logP 1.12			EXP	YAMAGAMI,C ET AL. (1991)
VP 4.47E-003	mm Hg	25	EST	NEELY,WB & BLAU,GE (1985)
DC	pKa			
HL 4.48E-010	atm m3/mol	25	EST	MEYLAN,WM & HOWARD,PH (1991)
OH 8.24E-012	cm3/molc sec	25	EST	MEYLAN,WM & HOWARD,PH (1993)

CAS #: 040262-54-2				PYRAZINECARBOXYLIC ACID, 6-(DIMETHYLAMINO)-, MET
Formula: $C_8H_{11}N_3O_2$				
Mol Weight: 181.20				
MP (deg C):		FP (deg C):		
BP (deg C):				
BP pressure (mm Hg):				

Property/ Value	Units	Temp	Data Type	Reference
WS 9.26E+003	mg/L	25	EST	MEYLAN,WM ET AL. (1996)
logP 0.90			EXP	YAMAGAMI,C ET AL. (1991)
VP 3.79E-003	mm Hg	25	EST	NEELY,WB & BLAU,GE (1985)
DC	pKa			
HL 2.99E-010	atm m3/mol	25	EST	MEYLAN,WM & HOWARD,PH (1991)
OH 1.24E-011	cm3/molc sec	25	EST	MEYLAN,WM & HOWARD,PH (1993)

CAS #: 040262-55-3				PYRAZINECARBOXAMIDE, 6-(DIMETHYLAMINO)
Formula: $C_7H_{10}N_4O$				
Mol Weight: 166.18				
MP (deg C):		FP (deg C):		
BP (deg C):				
BP pressure (mm Hg):				

Property/ Value	Units	Temp	Data Type	Reference
WS 3.04E+004	mg/L	25	EST	MEYLAN,WM ET AL. (1996)
logP 0.38			EXP	YAMAGAMI,C ET AL. (1991)
VP 1.37E-005	mm Hg	25	EST	NEELY,WB & BLAU,GE (1985)
DC	pKa			
HL 1.91E-014	atm m3/mol	25	EST	MEYLAN,WM & HOWARD,PH (1991)
OH 2.29E-011	cm3/molc sec	25	EST	MEYLAN,WM & HOWARD,PH (1993)

CAS #: 040274-67-7				XANTHONE-2-CARBOXYLIC ACID
Formula: $C_{14}H_8O_4$				
Mol Weight: 240.22				
MP (deg C):		FP (deg C):		
BP (deg C):				
BP pressure (mm Hg):				

Property/ Value	Units	Temp	Data Type	Reference
WS 4.50E+000	mg/L	25	EST	MEYLAN,WM ET AL. (1996)
logP 3.12			EXP	HANSCH,C & LEO,AJ (1985)
VP 9.76E-008	mm Hg	25	EST	NEELY,WB & BLAU,GE (1985)
DC	pKa			
HL 3.88E-012	atm m3/mol	25	EST	MEYLAN,WM & HOWARD,PH (1991)
OH 6.67E-012	cm3/molc sec	25	EST	MEYLAN,WM & HOWARD,PH (1993)

CAS #: 040279-26-3				2-METHYLTHIOQUINOLINE
Formula: $C_{10}H_9NS$				
Mol Weight: 175.25				
MP (deg C):		FP (deg C):		
BP (deg C):				
BP pressure (mm Hg):				

Property/ Value	Units	Temp	Data Type	Reference
WS 1.47E+002	mg/L	25	EST	MEYLAN,WM ET AL. (1996)
logP 3.04			EXP	HANSCH,C & LEO,AJ (1985)
VP 5.24E-004	mm Hg	25	EST	NEELY,WB & BLAU,GE (1985)
DC	pKa			
HL 7.81E-007	atm m3/mol	25	EST	MEYLAN,WM & HOWARD,PH (1991)
OH 9.39E-011	cm3/molc sec	25	EST	MEYLAN,WM & HOWARD,PH (1993)

CAS #: 040295-80-5				2-NAPHTHALENEMETHANOL, à-METHYL-, (\)-
Formula: $C_{12}H_{12}O$				
Mol Weight: 172.23				
MP (deg C):		FP (deg C):		
BP (deg C):				
BP pressure (mm Hg):				

Property/ Value	Units	Temp	Data Type	Reference
WS 9.99E+002	mg/L	25	EST	MEYLAN,WM ET AL. (1996)
logP 2.68			EXP	SANGSTER,J (1993)
VP 2.80E-005	mm Hg	25	EST	NEELY,WB & BLAU,GE (1985)
DC	pKa			
HL 2.82E-008	atm m3/mol	25	EST	MEYLAN,WM & HOWARD,PH (1991)
OH 5.83E-011	cm3/molc sec	25	EST	MEYLAN,WM & HOWARD,PH (1993)

CAS #: 040297-47-0	1-PIPERIDINEACETAMIDE, N-(3-METHOXYPHENYL)-

Formula: $C_{14}H_{20}N_2O_2$

Mol Weight: 248.33

MP (deg C):
FP (deg C):
BP (deg C):
BP pressure (mm Hg):

Property/Value	Units	Temp	Data Type	Reference
WS 1.17E+002	mg/L	25	EST	MEYLAN,WM ET AL. (1996)
logP 2.71			EXP	SANGSTER,J (1993)
VP 1.66E-007	mm Hg	25	EST	NEELY,WB & BLAU,GE (1985)
DC	pKa			
HL 3.73E-012	atm m3/mol	25	EST	MEYLAN,WM & HOWARD,PH (1991)
OH 2.36E-010	cm3/molc sec	25	EST	MEYLAN,WM & HOWARD,PH (1993)

CAS #: 040321-76-4	1,2,3,7,8-PENTACHLORODIBENZO-P-DIOXIN

Formula: $C_{12}H_3Cl_5O_2$

Mol Weight: 356.42

MP (deg C):
FP (deg C):
BP (deg C):
BP pressure (mm Hg):

Property/Value	Units	Temp	Data Type	Reference
WS 1.53E-004	mg/L	25	EST	MEYLAN,WM ET AL. (1996)
logP 7.56			EST	MEYLAN,WM & HOWARD,PH (1995)
VP 3.96E-007	mm Hg	25	EST	NEELY,WB & BLAU,GE (1985)
DC	pKa			
HL 2.61E-006	atm m3/mol	25	EST	MEYLAN,WM & HOWARD,PH (1991)
OH 1.72E-012	cm3/molc sec	25	EST	MEYLAN,WM & HOWARD,PH (1993)

CAS #: 040361-79-3	1-ME-2-NO2-5-MEOCARBONYL IMIDAZOLE

Formula: $C_6H_9N_3O_4$

Mol Weight: 187.16

MP (deg C):
FP (deg C):
BP (deg C):
BP pressure (mm Hg):

Property/Value	Units	Temp	Data Type	Reference
WS 7.77E+003	mg/L	25	EST	MEYLAN,WM ET AL. (1996)
logP 0.51			EXP	HANSCH,C & LEO,AJ (1985)
VP 2.41E-005	mm Hg	25	EST	NEELY,WB & BLAU,GE (1985)
DC	pKa			
HL 2.04E-009	atm m3/mol	25	EST	MEYLAN,WM & HOWARD,PH (1991)
OH 1.35E-012	cm3/molc sec	25	EST	MEYLAN,WM & HOWARD,PH (1993)

CAS #: 040427-75-6	beta-D-GALACTOPYRANOSIDE, OCTYL

Formula: $C_{14}H_{28}O_6$

Mol Weight: 292.38

MP (deg C):
FP (deg C):
BP (deg C):
BP pressure (mm Hg):

Property/Value	Units	Temp	Data Type	Reference
WS 7.70E+002	mg/L	25	EST	MEYLAN,WM ET AL. (1996)
logP 1.46			EXP	SANGSTER,J (1994)
VP 3.69E-011	mm Hg	25	EST	NEELY,WB & BLAU,GE (1985)
DC	pKa			
HL 1.13E-013	atm m3/mol	25	EST	MEYLAN,WM & HOWARD,PH (1991)
OH 8.00E-011	cm3/molc sec	25	EST	MEYLAN,WM & HOWARD,PH (1993)

CAS #: 040487-42-1	PENDIMETHALIN

Formula: $C_{13}H_{19}N_3O_4$

Mol Weight: 281.31

MP (deg C): 57
FP (deg C):
BP (deg C): 330
BP pressure (mm Hg):

Property/Value	Units	Temp	Data Type	Reference
WS 2.75E-001	mg/L	25	EXP	HUMBURG,NE ET AL. (1989)
logP 5.18			EXP	TOMLIN,C (1994)
VP 3.00E-005	mm Hg	25	EXP	HUMBURG,NE ET AL. (1989)
DC	pKa			
HL 8.56E-007	atm m3/mol	25	EXP	FENDINGER,NJ & GODFELTY,DE (1990)
OH 3.03E-011	cm3/molc sec	25	EST	MEYLAN,WM & HOWARD,PH (1993)

CAS #: 040533-52-6	1,3,5-TRIAZINE-2,4-DIAMINE, 6-CHLORO-N-(CYCLOPRO

Formula: $C_9H_{14}ClN_5$

Mol Weight: 227.70

MP (deg C):
FP (deg C):
BP (deg C):
BP pressure (mm Hg):

Property/Value	Units	Temp	Data Type	Reference
WS 1.15E+002	mg/L	25	EST	MEYLAN,WM ET AL. (1996)
logP 2.85			EXP	MITSUTAKE,KI ET AL. (1986)
VP 2.28E-005	mm Hg	25	EST	NEELY,WB & BLAU,GE (1985)
DC	pKa			
HL 2.62E-009	atm m3/mol	25	EST	MEYLAN,WM & HOWARD,PH (1991)
OH 1.98E-011	cm3/molc sec	25	EST	MEYLAN,WM & HOWARD,PH (1993)

CAS #: 040575-34-6	PHENOPYLATE

Formula: $C_{11}H_{11}Cl_2NO_2$

Mol Weight: 260.12

MP (deg C):
FP (deg C):
BP (deg C):
BP pressure (mm Hg):

Property/Value	Units	Temp	Data Type	Reference
WS 8.97E+000	mg/L	25	EST	MEYLAN,WM ET AL. (1996)
logP 3.94			EXP	NANDIHALLI,UB ET AL. (1993)
VP 7.31E-005	mm Hg	25	EST	NEELY,WB & BLAU,GE (1985)
DC	pKa			
HL 3.01E-008	atm m3/mol	25	EST	MEYLAN,WM & HOWARD,PH (1991)
OH 2.55E-011	cm3/molc sec	25	EST	MEYLAN,WM & HOWARD,PH (1993)

CAS #: 040596-69-8	ALTOSID

Formula: $C_{19}H_{34}O_3$

Mol Weight: 310.48

MP (deg C):
FP (deg C):
BP (deg C): 100
BP pressure (mm Hg): 5.00E-002

Property/Value	Units	Temp	Data Type	Reference
WS 2.00E+000	mg/L	25	EXP	YALKOWSKY,SH & DANNENFELSER,RM (1992)
logP 5.50			EXP	HANSCH,C ET AL. (1995)
VP 1.28E-004	mm Hg	25	EST	NEELY,WB & BLAU,GE (1985)
DC	pKa			
HL 5.71E-005	atm m3/mol	25	EST	MEYLAN,WM & HOWARD,PH (1991)
OH 8.30E-011	cm3/molc sec	25	EST	MEYLAN,WM & HOWARD,PH (1993)

CAS #: 040611-76-5 — 5-METHYLPYRROLE-3-CARBOXYLIC ACID,METHYL ESTER

Formula: $C_7H_9NO_2$

Mol Weight: 139.16

MP (deg C):   FP (deg C):

BP (deg C):

BP pressure (mm Hg):

Property/Value	Units	Temp	Data Type	Reference
WS 1.79E+003	mg/L	25	EST	MEYLAN,WM ET AL. (1996)
logP 1.96			EXP	HANSCH,C ET AL. (1995)
VP 4.48E-002	mm Hg	25	EST	NEELY,WB & BLAU,GE (1985)
DC	pKa			
HL 6.45E-008	atm m3/mol	25	EST	MEYLAN,WM & HOWARD,PH (1991)
OH 6.39E-011	cm3/molc sec	25	EST	MEYLAN,WM & HOWARD,PH (1993)

CAS #: 040647-30-1 — METHANAMINE, N-[(1-METHYL-2-NITRO-1H-IMIDAZOL-5-

Formula: $C_6H_{10}N_4O_3$

Mol Weight: 186.17

MP (deg C):   FP (deg C):

BP (deg C):

BP pressure (mm Hg):

Property/Value	Units	Temp	Data Type	Reference
WS 1.98E+004	mg/L	25	EST	MEYLAN,WM ET AL. (1996)
logP 0.04			EXP	HANSCH,C & LEO,AJ (1985)
VP 4.25E-008	mm Hg	25	EST	NEELY,WB & BLAU,GE (1985)
DC	pKa			
HL 2.78E-011	atm m3/mol	25	EST	MEYLAN,WM & HOWARD,PH (1991)
OH 1.49E-011	cm3/molc sec	25	EST	MEYLAN,WM & HOWARD,PH (1993)

CAS #: 040673-68-5 — ACETOPHENONE, O-(2-OHET)AMINO

Formula: $C_{10}H_{13}NO_2$

Mol Weight: 179.22

MP (deg C):   FP (deg C):

BP (deg C):

BP pressure (mm Hg):

Property/Value	Units	Temp	Data Type	Reference
WS 2.44E+003	mg/L	25	EST	MEYLAN,WM ET AL. (1996)
logP 1.59			EXP	HANSCH,C & LEO,AJ (1985)
VP 9.08E-006	mm Hg	25	EST	NEELY,WB & BLAU,GE (1985)
DC	pKa			
HL 3.69E-013	atm m3/mol	25	EST	MEYLAN,WM & HOWARD,PH (1991)
OH 5.38E-011	cm3/molc sec	25	EST	MEYLAN,WM & HOWARD,PH (1993)

CAS #: 040709-82-8 — ((2-CHLOROETHYL)SULFINYL)ETHENE

Formula: C_4H_7ClOS

Mol Weight: 138.62

MP (deg C):   FP (deg C):

BP (deg C):

BP pressure (mm Hg):

Property/Value	Units	Temp	Data Type	Reference
WS 1.08E+005	mg/L	25	EST	MEYLAN,WM ET AL. (1996)
logP -0.12			EST	MEYLAN,WM & HOWARD,PH (1995)
VP 1.04E-001	mm Hg	25	EST	NEELY,WB & BLAU,GE (1985)
DC	pKa			
HL 6.50E-008	atm m3/mol	25	EST	MEYLAN,WM & HOWARD,PH (1991)
OH 8.94E-011	cm3/molc sec	25	EST	MEYLAN,WM & HOWARD,PH (1993)

CAS #: 040725-71-1 — PHOSPHONIC DIAMIDE, N,N,N',N'-TETRAMETHYL-P-1-PY

Formula: $C_8H_{20}N_3OP$

Mol Weight: 205.24

MP (deg C):   FP (deg C):

BP (deg C):

BP pressure (mm Hg):

Property/Value	Units	Temp	Data Type	Reference
WS 8.20E+003	mg/L	25	EST	MEYLAN,WM ET AL. (1996)
logP 0.82			EXP	DEBORD,J & LABADIE,M (1985)
VP 2.74E-003	mm Hg	25	EST	NEELY,WB & BLAU,GE (1985)
DC	pKa			
HL 5.54E-012	atm m3/mol	25	EST	MEYLAN,WM & HOWARD,PH (1991)
OH 1.07E-010	cm3/molc sec	25	EST	MEYLAN,WM & HOWARD,PH (1993)

CAS #: 040805-79-6 — 5-PYRIMIDINECARBONITRILE

Formula: $C_5H_3N_3$

Mol Weight: 105.10

MP (deg C):   FP (deg C):

BP (deg C):

BP pressure (mm Hg):

Property/Value	Units	Temp	Data Type	Reference
WS 1.64E+005	mg/L	25	EST	MEYLAN,WM ET AL. (1996)
logP -0.50			EXP	YAMAGAMI,C ET AL. (1994)
VP 1.68E-001	mm Hg	25	EST	NEELY,WB & BLAU,GE (1985)
DC	pKa			
HL 2.82E-008	atm m3/mol	25	EST	MEYLAN,WM & HOWARD,PH (1991)
OH 4.06E-014	cm3/molc sec	25	EST	MEYLAN,WM & HOWARD,PH (1993)

CAS #: 040837-24-9 — 1,4-BENZDAZEPIN-2-ON-1-MEOME-5-PH-7AMINO

Formula: $C_{17}H_{17}N_3O_2$

Mol Weight: 295.34

MP (deg C):   FP (deg C):

BP (deg C):

BP pressure (mm Hg):

Property/Value	Units	Temp	Data Type	Reference
WS 9.75E+002	mg/L	25	EST	MEYLAN,WM ET AL. (1996)
logP 1.32			EXP	HANSCH,C & LEO,AJ (1985)
VP 1.00E-009	mm Hg	25	EST	NEELY,WB & BLAU,GE (1985)
DC	pKa			
HL 4.06E-013	atm m3/mol	25	EST	MEYLAN,WM & HOWARD,PH (1991)
OH 1.53E-010	cm3/molc sec	25	EST	MEYLAN,WM & HOWARD,PH (1993)

CAS #: 040837-34-1 — 14-BENZODIAZP2-ON,1-MEOME5(2FPH)-7AM

Formula: $C_{17}H_{16}FN_3O_2$

Mol Weight: 313.33

MP (deg C):   FP (deg C):

BP (deg C):

BP pressure (mm Hg):

Property/Value	Units	Temp	Data Type	Reference
WS 7.80E+002	mg/L	25	EST	MEYLAN,WM ET AL. (1996)
logP 1.31			EXP	HANSCH,C & LEO,AJ (1985)
VP 1.21E-009	mm Hg	25	EST	NEELY,WB & BLAU,GE (1985)
DC	pKa			
HL 4.74E-013	atm m3/mol	25	EST	MEYLAN,WM & HOWARD,PH (1991)
OH 1.52E-010	cm3/molc sec	25	EST	MEYLAN,WM & HOWARD,PH (1993)

CAS #: 040843-25-2				DICLOFOP

Formula: $C_{15}H_{12}Cl_2O_4$

Mol Weight: 327.17

MP (deg C): 39-41

FP (deg C):

BP (deg C):

BP pressure (mm Hg):

Property/Value	Units	Temp	Data Type	Reference
WS 3.00E+000	mg/L	22	EXP	SHIU,WY ET AL. (1990)
logP 4.58			EXP	TOMLIN,C (1994)
VP 4.03E-008	mm Hg	25	EST	NEELY,WB & BLAU,GE (1985)
DC	pKa			
HL 2.67E-010	atm m3/mol	25	EST	MEYLAN,WM & HOWARD,PH (1991)
OH 3.20E-011	cm3/molc sec	25	EST	MEYLAN,WM & HOWARD,PH (1993)

CAS #: 040852-07-1				4-PYRIDINE IMIDAZOLE,2-MES-6-CL

Formula: $C_7H_6ClN_3S$

Mol Weight: 199.66

MP (deg C):

FP (deg C):

BP (deg C):

BP pressure (mm Hg):

Property/Value	Units	Temp	Data Type	Reference
WS 4.00E+002	mg/L	25	EST	MEYLAN,WM ET AL. (1996)
logP 2.39			EXP	HANSCH,C & LEO,AJ (1985)
VP 3.88E-007	mm Hg	25	EST	NEELY,WB & BLAU,GE (1985)
DC	pKa			
HL 6.54E-010	atm m3/mol	25	EST	MEYLAN,WM & HOWARD,PH (1991)
OH 3.35E-011	cm3/molc sec	25	EST	MEYLAN,WM & HOWARD,PH (1993)

CAS #: 040912-73-0				BENZAMIDE, 5-BROMO-2-HYDROXY-3-METHYL

Formula: $C_8H_8BrNO_2$

Mol Weight: 230.07

MP (deg C):

FP (deg C):

BP (deg C):

BP pressure (mm Hg):

Property/Value	Units	Temp	Data Type	Reference
WS 5.08E+002	mg/L	25	EST	MEYLAN,WM ET AL. (1996)
logP 2.76			EXP	SANGSTER,J (1993)
VP 3.78E-007	mm Hg	25	EST	NEELY,WB & BLAU,GE (1985)
DC	pKa			
HL 1.28E-010	atm m3/mol	25	EST	MEYLAN,WM & HOWARD,PH (1991)
OH 3.21E-012	cm3/molc sec	25	EST	MEYLAN,WM & HOWARD,PH (1993)

CAS #: 040912-87-6				BENZAMIDE, 5-BROMO-2-HYDROXY-N,3-DIMETHYL-

Formula: $C_9H_{10}BrNO_2$

Mol Weight: 244.09

MP (deg C):

FP (deg C):

BP (deg C):

BP pressure (mm Hg):

Property/Value	Units	Temp	Data Type	Reference
WS 9.75E+001	mg/L	25	EST	MEYLAN,WM ET AL. (1996)
logP 3.51			EXP	SANGSTER,J (1993)
VP 2.12E-007	mm Hg	25	EST	NEELY,WB & BLAU,GE (1985)
DC	pKa			
HL 2.80E-010	atm m3/mol	25	EST	MEYLAN,WM & HOWARD,PH (1991)
OH 2.98E-011	cm3/molc sec	25	EST	MEYLAN,WM & HOWARD,PH (1993)

CAS #: 040912-88-7				BENZAMIDE, 5-BROMO-N-ETHYL-2-HYDROXY-3-METHYL-

Formula: $C_{10}H_{12}BrNO_2$

Mol Weight: 258.12

MP (deg C):

FP (deg C):

BP (deg C):

BP pressure (mm Hg):

Property/Value	Units	Temp	Data Type	Reference
WS 6.70E+001	mg/L	25	EST	MEYLAN,WM ET AL. (1996)
logP 3.61			EXP	SANGSTER,J (1993)
VP 9.14E-008	mm Hg	25	EST	NEELY,WB & BLAU,GE (1985)
DC	pKa			
HL 3.72E-010	atm m3/mol	25	EST	MEYLAN,WM & HOWARD,PH (1991)
OH 1.07E-011	cm3/molc sec	25	EST	MEYLAN,WM & HOWARD,PH (1993)

CAS #: 040912-89-8				BENZAMIDE, 5-BROMO-N-BUTYL-2-HYDROXY-3-METHYL-

Formula: $C_{12}H_{16}BrNO_2$

Mol Weight: 286.18

MP (deg C):

FP (deg C):

BP (deg C):

BP pressure (mm Hg):

Property/Value	Units	Temp	Data Type	Reference
WS 1.43E+001	mg/L	25	EST	MEYLAN,WM ET AL. (1996)
logP 4.21			EXP	SANGSTER,J (1993)
VP 1.67E-008	mm Hg	25	EST	NEELY,WB & BLAU,GE (1985)
DC	pKa			
HL 6.55E-010	atm m3/mol	25	EST	MEYLAN,WM & HOWARD,PH (1991)
OH 1.41E-011	cm3/molc sec	25	EST	MEYLAN,WM & HOWARD,PH (1993)

CAS #: 040919-33-3				URICYTIN

Formula: $C_8H_{11}N_3O_6$

Mol Weight: 245.19

MP (deg C):

FP (deg C):

BP (deg C):

BP pressure (mm Hg):

Property/Value	Units	Temp	Data Type	Reference
WS 5.42E+004	mg/L	25	EST	MEYLAN,WM ET AL. (1996)
logP -2.43			EXP	HANSCH,C ET AL. (1995)
VP 3.97E-016	mm Hg	25	EST	NEELY,WB & BLAU,GE (1985)
DC	pKa			
HL 7.82E-024	atm m3/mol	25	EST	MEYLAN,WM & HOWARD,PH (1991)
OH 1.04E-010	cm3/molc sec	25	EST	MEYLAN,WM & HOWARD,PH (1993)

CAS #: 040929-49-5				5-PYRIMIDINECARBOXAMIDE

Formula: $C_5H_5N_3O$

Mol Weight: 123.12

MP (deg C):

FP (deg C):

BP (deg C):

BP pressure (mm Hg):

Property/Value	Units	Temp	Data Type	Reference
WS 2.92E+004	mg/L	25	EST	MEYLAN,WM ET AL. (1996)
logP -0.92			EXP	YAMAGAMI,C ET AL. (1990)
VP 2.44E-004	mm Hg	25	EST	NEELY,WB & BLAU,GE (1985)
DC	pKa			
HL 1.20E-012	atm m3/mol	25	EST	MEYLAN,WM & HOWARD,PH (1991)
OH 2.21E-012	cm3/molc sec	25	EST	MEYLAN,WM & HOWARD,PH (1993)

CAS #: 040929-50-8				5-PYRIMIDINECARBOXYLIC ACID, ETHYL ESTER

Formula: $C_7H_8N_2O_2$

Mol Weight: 152.15

MP (deg C): **FP (deg C):**

BP (deg C):

BP pressure (mm Hg):

Property/ Value	Units	Temp	Data Type	Reference
WS 2.67E+004	mg/L	25	EST	MEYLAN,WM ET AL. (1996)
logP 0.52			EXP	YAMAGAMI,C ET AL. (1990)
VP 4.25E-002	mm Hg	25	EST	NEELY,WB & BLAU,GE (1985)
DC	pKa			
HL 2.50E-008	atm m3/mol	25	EST	MEYLAN,WM & HOWARD,PH (1991)
OH 1.74E-012	cm3/molc sec	25	EST	MEYLAN,WM & HOWARD,PH (1993)

CAS #: 040948-30-9				BENZAMIDE,N-(3-DIME AMINOPROPYL)

Formula: $C_{12}H_{18}N_2O$

Mol Weight: 206.29

MP (deg C): **FP (deg C):**

BP (deg C):

BP pressure (mm Hg):

Property/ Value	Units	Temp	Data Type	Reference
WS 3.34E+003	mg/L	25	EST	MEYLAN,WM ET AL. (1996)
logP 1.27			EXP	HANSCH,C & LEO,AJ (1985)
VP 5.55E-006	mm Hg	25	EST	NEELY,WB & BLAU,GE (1985)
DC	pKa			
HL 4.25E-012	atm m3/mol	25	EST	MEYLAN,WM & HOWARD,PH (1991)
OH 9.26E-011	cm3/molc sec	25	EST	MEYLAN,WM & HOWARD,PH (1993)

CAS #: 041020-65-9				MEXRENONE

Formula: $C_{24}H_{32}O_5$

Mol Weight: 400.52

MP (deg C): **FP (deg C):**

BP (deg C):

BP pressure (mm Hg):

Property/ Value	Units	Temp	Data Type	Reference
WS 6.70E+001	mg/L	25	EST	MEYLAN,WM ET AL. (1996)
logP 1.94			EXP	HANSCH,C & LEO,AJ (1985)
VP 2.71E-010	mm Hg	25	EST	NEELY,WB & BLAU,GE (1985)
DC	pKa			
HL 3.20E-010	atm m3/mol	25	EST	MEYLAN,WM & HOWARD,PH (1991)
OH 9.63E-011	cm3/molc sec	25	EST	MEYLAN,WM & HOWARD,PH (1993)

CAS #: 041020-77-3				ETHYL MEXRENONE

Formula: $C_{25}H_{34}O_5$

Mol Weight: 414.55

MP (deg C): **FP (deg C):**

BP (deg C):

BP pressure (mm Hg):

Property/ Value	Units	Temp	Data Type	Reference
WS 1.69E+001	mg/L	25	EST	MEYLAN,WM ET AL. (1996)
logP 2.54			EXP	HANSCH,C & LEO,AJ (1985)
VP 1.19E-010	mm Hg	25	EST	NEELY,WB & BLAU,GE (1985)
DC	pKa			
HL 4.25E-010	atm m3/mol	25	EST	MEYLAN,WM & HOWARD,PH (1991)
OH 9.77E-011	cm3/molc sec	25	EST	MEYLAN,WM & HOWARD,PH (1993)

CAS #: 041020-79-5				DICIRENONE

Formula: $C_{26}H_{36}O_5$

Mol Weight: 428.57

MP (deg C): **FP (deg C):**

BP (deg C):

BP pressure (mm Hg):

Property/ Value	Units	Temp	Data Type	Reference
WS 7.94E+000	mg/L	25	EST	MEYLAN,WM ET AL. (1996)
logP 2.82			EXP	HANSCH,C & LEO,AJ (1985)
VP 8.55E-011	mm Hg	25	EST	NEELY,WB & BLAU,GE (1985)
DC	pKa			
HL 5.64E-010	atm m3/mol	25	EST	MEYLAN,WM & HOWARD,PH (1991)
OH 9.95E-011	cm3/molc sec	25	EST	MEYLAN,WM & HOWARD,PH (1993)

CAS #: 041038-69-1				3-PYRIDINEPROPANEAMINE

Formula: $C_8H_{12}N_2$

Mol Weight: 136.20

MP (deg C): **FP (deg C):**

BP (deg C):

BP pressure (mm Hg):

Property/ Value	Units	Temp	Data Type	Reference
WS 1.00E+006	mg/L	25	EST	MEYLAN,WM ET AL. (1996)
logP 0.44			EXP	HANSCH,C & LEO,AJ (1985)
VP 4.53E-002	mm Hg	25	EST	NEELY,WB & BLAU,GE (1985)
DC	pKa			
HL 1.41E-009	atm m3/mol	25	EST	MEYLAN,WM & HOWARD,PH (1991)
OH 3.52E-011	cm3/molc sec	25	EST	MEYLAN,WM & HOWARD,PH (1993)

CAS #: 041078-01-7				1H-PURINE-2,6-DIONE, 3,9-DIHYDRO-3-ETHYL-

Formula: $C_7H_8N_4O_2$

Mol Weight: 180.17

MP (deg C): **FP (deg C):**

BP (deg C):

BP pressure (mm Hg):

Property/ Value	Units	Temp	Data Type	Reference
WS 3.34E+003	mg/L	25	EST	MEYLAN,WM ET AL. (1996)
logP -0.11			EXP	SANGSTER,J (1993)
VP 5.44E-011	mm Hg	25	EST	NEELY,WB & BLAU,GE (1985)
DC	pKa			
HL 1.01E-012	atm m3/mol	25	EST	MEYLAN,WM & HOWARD,PH (1991)
OH 2.22E-011	cm3/molc sec	25	EST	MEYLAN,WM & HOWARD,PH (1993)

CAS #: 041078-02-8				1H-PURINE-2,6-DIONE, 3,9-DIHYDRO-3-PROPYL-

Formula: $C_8H_{10}N_4O_2$

Mol Weight: 194.19

MP (deg C): 287-289 **FP (deg C):**

BP (deg C):

BP pressure (mm Hg):

Property/ Value	Units	Temp	Data Type	Reference
WS 1.20E+003	mg/L	25	EST	MEYLAN,WM ET AL. (1996)
logP 0.33			EXP	SANGSTER,J (1993)
VP 2.36E-011	mm Hg	25	EST	NEELY,WB & BLAU,GE (1985)
DC	pKa			
HL 1.35E-012	atm m3/mol	25	EST	MEYLAN,WM & HOWARD,PH (1991)
OH 2.42E-011	cm3/molc sec	25	EST	MEYLAN,WM & HOWARD,PH (1993)

CAS #: 041078-03-9				1H-PURINE-2,6-DIONE, 3,9-DIHYDRO-3-BUTYL-
Formula: $C_9H_{12}N_4O_2$				
Mol Weight: 208.22				
MP (deg C):			FP (deg C):	
BP (deg C):				
BP pressure (mm Hg):				

Property/Value	Units	Temp	Data Type	Reference
WS 7.61E+003	mg/L	25	EST	MEYLAN,WM ET AL. (1996)
logP 0.84			EXP	SANGSTER,J (1993)
VP 1.03E-011	mm Hg	25	EST	NEELY,WB & BLAU,GE (1985)
DC	pKa			
HL 1.79E-012	atm m3/mol	25	EST	MEYLAN,WM & HOWARD,PH (1991)
OH 2.56E-011	cm3/molc sec	25	EST	MEYLAN,WM & HOWARD,PH (1993)

CAS #: 041100-45-2				1-AMINOADAMANTANE,3-ETHYL
Formula: $C_{12}H_{21}N$				
Mol Weight: 179.31				
MP (deg C):			FP (deg C):	
BP (deg C):				
BP pressure (mm Hg):				

Property/Value	Units	Temp	Data Type	Reference
WS 7.06E+002	mg/L	25	EST	MEYLAN,WM ET AL. (1996)
logP 3.40			EXP	HANSCH,C & LEO,AJ (1985)
VP 2.27E-002	mm Hg	25	EST	NEELY,WB & BLAU,GE (1985)
DC	pKa			
HL 1.47E-005	atm m3/mol	25	EST	MEYLAN,WM & HOWARD,PH (1991)
OH 3.80E-011	cm3/molc sec	25	EST	MEYLAN,WM & HOWARD,PH (1993)

CAS #: 041107-56-6				URIDINE, 2',3'-DIDEOXY-3'-FLUORO-
Formula: $C_9H_{11}FN_2O_4$				
Mol Weight: 230.20				
MP (deg C): 184-188			FP (deg C):	
BP (deg C):				
BP pressure (mm Hg):				

Property/Value	Units	Temp	Data Type	Reference
WS 3.90E+003	mg/L	25	EST	MEYLAN,WM ET AL. (1996)
logP -0.49			EXP	BALZARINI,J ET AL. (1989)
VP 9.24E-012	mm Hg	25	EST	NEELY,WB & BLAU,GE (1985)
DC	pKa			
HL 4.20E-016	atm m3/mol	25	EST	MEYLAN,WM & HOWARD,PH (1991)
OH 2.17E-011	cm3/molc sec	25	EST	MEYLAN,WM & HOWARD,PH (1993)

CAS #: 041110-33-2				PYRAZINECARBOXYLIC ACID, 5-METHYL-, METHYL ESTER
Formula: $C_7H_8N_2O_2$				
Mol Weight: 152.15				
MP (deg C):			FP (deg C):	
BP (deg C):				
BP pressure (mm Hg):				

Property/Value	Units	Temp	Data Type	Reference
WS 5.31E+004	mg/L	25	EST	MEYLAN,WM ET AL. (1996)
logP 0.17			EXP	YAMAGAMI,C ET AL. (1991)
VP 3.97E-002	mm Hg	25	EST	NEELY,WB & BLAU,GE (1985)
DC	pKa			
HL 2.08E-008	atm m3/mol	25	EST	MEYLAN,WM & HOWARD,PH (1991)
OH 5.47E-013	cm3/molc sec	25	EST	MEYLAN,WM & HOWARD,PH (1993)

CAS #: 041110-34-3				PYRAZINECARBOXYLIC ACID, 5-METHYL-, ETHYL ESTER
Formula: $C_8H_{10}N_2O_2$				
Mol Weight: 166.18				
MP (deg C):			FP (deg C):	
BP (deg C):				
BP pressure (mm Hg):				

Property/Value	Units	Temp	Data Type	Reference
WS 1.93E+004	mg/L	25	EST	MEYLAN,WM ET AL. (1996)
logP 0.61			EXP	YAMAGAMI,C ET AL. (1991)
VP 1.26E-002	mm Hg	25	EST	NEELY,WB & BLAU,GE (1985)
DC	pKa			
HL 2.75E-008	atm m3/mol	25	EST	MEYLAN,WM & HOWARD,PH (1991)
OH 1.99E-012	cm3/molc sec	25	EST	MEYLAN,WM & HOWARD,PH (1993)

CAS #: 041110-38-7				PYRAZINECARBOXYLIC ACID, 6-METHYL-, METHYL ESTER
Formula: $C_7H_8N_2O_2$				
Mol Weight: 152.15				
MP (deg C):			FP (deg C):	
BP (deg C):				
BP pressure (mm Hg):				

Property/Value	Units	Temp	Data Type	Reference
WS 6.10E+004	mg/L	25	EST	MEYLAN,WM ET AL. (1996)
logP 0.10			EXP	YAMAGAMI,C ET AL. (1991)
VP 3.97E-002	mm Hg	25	EST	NEELY,WB & BLAU,GE (1985)
DC	pKa			
HL 2.08E-008	atm m3/mol	25	EST	MEYLAN,WM & HOWARD,PH (1991)
OH 4.86E-013	cm3/molc sec	25	EST	MEYLAN,WM & HOWARD,PH (1993)

CAS #: 041110-39-8				PYRAZINECARBOXYLIC ACID, 6-METHYL-, ETHYL ESTER
Formula: $C_8H_{10}N_2O_2$				
Mol Weight: 166.18				
MP (deg C):			FP (deg C):	
BP (deg C):				
BP pressure (mm Hg):				

Property/Value	Units	Temp	Data Type	Reference
WS 2.35E+004	mg/L	25	EST	MEYLAN,WM ET AL. (1996)
logP 0.51			EXP	YAMAGAMI,C ET AL. (1991)
VP 1.26E-002	mm Hg	25	EST	NEELY,WB & BLAU,GE (1985)
DC	pKa			
HL 2.75E-008	atm m3/mol	25	EST	MEYLAN,WM & HOWARD,PH (1991)
OH 1.93E-012	cm3/molc sec	25	EST	MEYLAN,WM & HOWARD,PH (1993)

CAS #: 041110-59-2				_-D-GLUCOPYRANOSIDE, METHYL 6-DEOXY-6-[[(METHYLN
Formula: $C_9H_{17}N_3O_7$				
Mol Weight: 279.25				
MP (deg C):			FP (deg C):	
BP (deg C):				
BP pressure (mm Hg):				

Property/Value	Units	Temp	Data Type	Reference
WS 1.74E+004	mg/L	25	EST	MEYLAN,WM ET AL. (1996)
logP -1.57			EXP	HANSCH,C ET AL. (1995)
VP 9.82E-012	mm Hg	25	EST	NEELY,WB & BLAU,GE (1985)
DC	pKa			
HL 1.26E-021	atm m3/mol	25	EST	MEYLAN,WM & HOWARD,PH (1991)
OH 6.50E-011	cm3/molc sec	25	EST	MEYLAN,WM & HOWARD,PH (1993)

CAS #: 041167-74-2 — P-BROMO-N-PHENYLSUCCINIMIDE

Formula: $C_{10}H_8BrNO_2$

Mol Weight: 254.08

MP (deg C):   FP (deg C):

BP (deg C):

BP pressure (mm Hg):

Property/ Value	Units	Temp	Data Type	Reference
WS 2.21E+003	mg/L	25	EST	MEYLAN,WM ET AL. (1996)
logP 1.18			EXP	HANSCH,C & LEO,AJ (1985)
VP 9.65E-008	mm Hg	25	EST	NEELY,WB & BLAU,GE (1985)
DC	pKa			
HL 5.20E-008	atm m3/mol	25	EST	MEYLAN,WM & HOWARD,PH (1991)
OH 7.93E-012	cm3/molc sec	25	EST	MEYLAN,WM & HOWARD,PH (1993)

CAS #: 041198-08-7 — PROFENOFOS

Formula: $C_{11}H_{15}BrClO_3PS$

Mol Weight: 373.64

MP (deg C):   FP (deg C):

BP (deg C): 110

BP pressure (mm Hg): 1.00E-003

Property/ Value	Units	Temp	Data Type	Reference
WS 2.00E+001	mg/L	20	EXP	YALKOWSKY,SH & DANNENFELSER,RM (1992)
logP 4.68			EXP	HANSCH,C & LEO,AJ (1985)
VP 9.00E-007	mm Hg	25	EXP	WAUCHOPE,RD ET AL. (1991A)
DC	pKa			
HL 2.21E-008	atm m3/mol	25	EST	VP/WSOL
OH 4.48E-011	cm3/molc sec	25	EST	MEYLAN,WM & HOWARD,PH (1993)

CAS #: 041205-21-4 — SPARTICIDE

Formula: $C_{10}H_4Cl_2FNO_2$

Mol Weight: 260.05

MP (deg C):   FP (deg C):

BP (deg C):

BP pressure (mm Hg):

Property/ Value	Units	Temp	Data Type	Reference
WS 2.26E+002	mg/L	25	EST	MEYLAN,WM ET AL. (1996)
logP 2.30			EXP	TOMLIN,C (1994)
VP 9.03E-008	mm Hg	25	EST	NEELY,WB & BLAU,GE (1985)
DC	pKa			
HL 1.97E-008	atm m3/mol	25	EST	MEYLAN,WM & HOWARD,PH (1991)
OH 4.79E-012	cm3/molc sec	25	EST	MEYLAN,WM & HOWARD,PH (1993)

CAS #: 041229-10-1 — 4-N-PIPERIDINYLQUINAZOLINE

Formula: $C_{13}H_{15}N_3$

Mol Weight: 213.28

MP (deg C):   FP (deg C):

BP (deg C):

BP pressure (mm Hg):

Property/ Value	Units	Temp	Data Type	Reference
WS 1.20E+002	mg/L	25	EST	MEYLAN,WM ET AL. (1996)
logP 2.92			EXP	HANSCH,C & LEO,AJ (1985)
VP 1.49E-005	mm Hg	25	EST	NEELY,WB & BLAU,GE (1985)
DC	pKa			
HL 4.67E-009	atm m3/mol	25	EST	MEYLAN,WM & HOWARD,PH (1991)
OH 1.76E-010	cm3/molc sec	25	EST	MEYLAN,WM & HOWARD,PH (1993)

CAS #: 041410-39-3 — 1-PHENYL-3-CYANOGUANIDINE

Formula: $C_8H_8N_4$

Mol Weight: 160.18

MP (deg C):   FP (deg C):

BP (deg C):

BP pressure (mm Hg):

Property/ Value	Units	Temp	Data Type	Reference
WS 8.67E+003	mg/L	25	EST	MEYLAN,WM ET AL. (1996)
logP 1.05			EXP	HANSCH,C & LEO,AJ (1985)
VP 5.29E-005	mm Hg	25	EST	NEELY,WB & BLAU,GE (1985)
DC	pKa			
HL 1.24E-010	atm m3/mol	25	EST	MEYLAN,WM & HOWARD,PH (1991)
OH 4.27E-011	cm3/molc sec	25	EST	MEYLAN,WM & HOWARD,PH (1993)

CAS #: 041451-75-6 — BRUCEANTIN

Formula: $C_{28}H_{36}O_{11}$

Mol Weight: 548.59

MP (deg C): 225-226   FP (deg C):

BP (deg C):

BP pressure (mm Hg):

Property/ Value	Units	Temp	Data Type	Reference
WS 1.33E+001	mg/L	25	EST	MEYLAN,WM ET AL. (1996)
logP 1.66			EXP	HANSCH,C & LEO,AJ (1985)
VP 1.16E-020	mm Hg	25	EST	NEELY,WB & BLAU,GE (1985)
DC	pKa			
HL 6.84E-017	atm m3/mol	25	EST	MEYLAN,WM & HOWARD,PH (1991)
OH 2.06E-010	cm3/molc sec	25	EST	MEYLAN,WM & HOWARD,PH (1993)

CAS #: 041464-39-5 — 2,2',3,5'-TETRACHLOROBIPHENYL

Formula: $C_{12}H_6Cl_4$

Mol Weight: 291.99

MP (deg C):   FP (deg C):

BP (deg C):

BP pressure (mm Hg):

Property/ Value	Units	Temp	Data Type	Reference
WS 1.00E-001	mg/L	20	EXP	YALKOWSKY,SH & DANNENFELSER,RM (1992)
logP 5.81			EXP	RAPAPORT,RA & EISENREICH,SJ (1984)
VP 8.45E-006	mm Hg	25	EST	NEELY,WB & BLAU,GE (1985)
DC	pKa			
HL 1.25E-004	atm m3/mol	25	EST	MEYLAN,WM & HOWARD,PH (1991)
OH 7.30E-013	cm3/molc sec	25	EST	MEYLAN,WM & HOWARD,PH (1993)

CAS #: 041464-40-8 — 2,2',4,5'-TETRACHLOROBIPHENYL

Formula: $C_{12}H_6Cl_4$

Mol Weight: 291.99

MP (deg C):   FP (deg C):

BP (deg C):

BP pressure (mm Hg):

Property/ Value	Units	Temp	Data Type	Reference
WS 7.81E-002	mg/L	20	EXP	YALKOWSKY,SH & DANNENFELSER,RM (1992)
logP 6.22			EXP	HANSCH,C & LEO,AJ (1985)
VP 8.48E-006	mm Hg	25	EXP	SHIU,WY & MACKAY,D (1986)
DC	pKa			
HL 4.17E-005	atm m3/mol	25	EST	VP/WSOL
OH 7.72E-013	cm3/molc sec	25	EST	MEYLAN,WM & HOWARD,PH (1993)

1139

CAS #: 041464-41-9				2,2',5,6'-TETRACHLOROBIPHENYL
Formula: $C_{12}H_6Cl_4$				
Mol Weight: 291.99				
MP (deg C):			FP (deg C):	
BP (deg C):				
BP pressure (mm Hg):				

Property/Value	Units	Temp	Data Type	Reference
WS 4.76E-002	mg/L	25	EXP	YALKOWSKY,SH & DANNENFELSER,RM (1992)
logP 6.34			EST	MEYLAN,WM & HOWARD,PH (1995)
VP 2.05E-004	mm Hg	25	EXP	BIDLEMAN,TF (1984)
DC	pKa			
HL 1.66E-003	atm m3/mol	25	EST	VP/WSOL
OH 7.72E-013	cm3/molc sec	25	EST	MEYLAN,WM & HOWARD,PH (1993)

CAS #: 041464-51-1				2,2',3',4,5-PENTACHLOROBIPHENYL
Formula: $C_{12}H_5Cl_5$				
Mol Weight: 326.44				
MP (deg C):			FP (deg C):	
BP (deg C):				
BP pressure (mm Hg):				

Property/Value	Units	Temp	Data Type	Reference
WS 2.84E-002	mg/L	20	EXP	YALKOWSKY,SH & DANNENFELSER,RM (1992)
logP 6.67			EXP	HANSCH,C & LEO,AJ (1985)
VP 2.22E-006	mm Hg	25	EST	NEELY,WB & BLAU,GE (1985)
DC	pKa			
HL 9.24E-005	atm m3/mol	25	EST	MEYLAN,WM & HOWARD,PH (1991)
OH 3.35E-013	cm3/molc sec	25	EST	MEYLAN,WM & HOWARD,PH (1993)

CAS #: 041468-25-1				PYRIDOXAL-5'-PHOSPHATE MONOHYDRATE
Formula: $C_8H_{12}NO_7P$				
Mol Weight: 265.16				
MP (deg C): 140-143			FP (deg C):	
BP (deg C):				
BP pressure (mm Hg):				

Property/Value	Units	Temp	Data Type	Reference
WS 1.19E+004	mg/L	25	EST	MEYLAN,WM ET AL. (1996)
logP 0.37			EST	MEYLAN,WM & HOWARD,PH (1995)
VP 3.28E-010	mm Hg	25	EST	NEELY,WB & BLAU,GE (1985)
DC	pKa			
HL 9.26E-020	atm m3/mol	25	EST	MEYLAN,WM & HOWARD,PH (1991)
OH 4.29E-011	cm3/molc sec	25	EST	MEYLAN,WM & HOWARD,PH (1993)

CAS #: 041483-43-6				BUPIRIMATE
Formula: $C_{13}H_{24}N_4O_3S$				
Mol Weight: 316.43				
MP (deg C): 50-51			FP (deg C):	
BP (deg C):				
BP pressure (mm Hg):				

Property/Value	Units	Temp	Data Type	Reference
WS 2.20E+001	mg/L	25	EXP	SHIU,WY ET AL. (1990)
logP 2.70			EXP	HANSCH,C ET AL. (1995)
VP 4.01E-008	mm Hg	25	EST	NEELY,WB & BLAU,GE (1985)
DC	pKa			
HL 1.37E-008	atm m3/mol	25	EST	MEYLAN,WM & HOWARD,PH (1991)
OH 1.92E-010	cm3/molc sec	25	EST	MEYLAN,WM & HOWARD,PH (1993)

CAS #: 041532-81-4				2-METHOXYETHOXYBENZENE
Formula: $C_9H_{12}O_2$				
Mol Weight: 152.19				
MP (deg C):			FP (deg C):	
BP (deg C):				
BP pressure (mm Hg):				

Property/Value	Units	Temp	Data Type	Reference
WS 2.47E+003	mg/L	25	EST	MEYLAN,WM ET AL. (1996)
logP 1.73			EXP	HANSCH,C & LEO,AJ (1985)
VP 1.96E-001	mm Hg	25	EST	NEELY,WB & BLAU,GE (1985)
DC	pKa			
HL 2.50E-005	atm m3/mol	25	EST	MEYLAN,WM & HOWARD,PH (1991)
OH 2.56E-011	cm3/molc sec	25	EST	MEYLAN,WM & HOWARD,PH (1993)

CAS #: 041536-80-5				2,6-PYRAZINEDIAMINE
Formula: $C_4H_6N_4$				
Mol Weight: 110.12				
MP (deg C):			FP (deg C):	
BP (deg C):				
BP pressure (mm Hg):				

Property/Value	Units	Temp	Data Type	Reference
WS 2.63E+005	mg/L	25	EST	MEYLAN,WM ET AL. (1996)
logP -0.45			EXP	YAMAGAMI,C ET AL. (1991)
VP 2.58E-003	mm Hg	25	EST	NEELY,WB & BLAU,GE (1985)
DC	pKa			
HL 3.64E-013	atm m3/mol	25	EST	MEYLAN,WM & HOWARD,PH (1991)
OH 2.00E-010	cm3/molc sec	25	EST	MEYLAN,WM & HOWARD,PH (1993)

CAS #: 041538-42-5				2,5-DIMEO-4-T-BU-AMPHETAMINE
Formula: $C_{15}H_{25}NO_2$				
Mol Weight: 251.37				
MP (deg C):			FP (deg C):	
BP (deg C):				
BP pressure (mm Hg):				

Property/Value	Units	Temp	Data Type	Reference
WS 1.08E+002	mg/L	25	EST	MEYLAN,WM ET AL. (1996)
logP 3.91			EXP	HANSCH,C & LEO,AJ (1985)
VP 9.27E-005	mm Hg	25	EST	NEELY,WB & BLAU,GE (1985)
DC	pKa			
HL 9.74E-009	atm m3/mol	25	EST	MEYLAN,WM & HOWARD,PH (1991)
OH 1.02E-010	cm3/molc sec	25	EST	MEYLAN,WM & HOWARD,PH (1993)

CAS #: 041541-11-1				DIS. A. 9
Formula: $C_{18}H_{22}N_4O_5$				
Mol Weight: 374.40				
MP (deg C):			FP (deg C):	
BP (deg C):				
BP pressure (mm Hg):				

Property/Value	Units	Temp	Data Type	Reference
WS 1.68E+000	mg/L	25	EXP	BAUGHMAN,GL & PERENICH,TA (1988)
logP 3.77			EST	MEYLAN,WM & HOWARD,PH (1995)
VP 2.67E-014	mm Hg	25	EST	NEELY,WB & BLAU,GE (1985)
DC	pKa			
HL 1.42E-016	atm m3/mol	25	EST	MEYLAN,WM & HOWARD,PH (1991)
OH 1.83E-010	cm3/molc sec	25	EST	MEYLAN,WM & HOWARD,PH (1993)

CAS #: 041541-13-3				DIS. A. 1

Formula: $C_{17}H_{19}N_5O_6$

Mol Weight: 389.37

MP (deg C): 190 FP (deg C):

BP (deg C):

BP pressure (mm Hg):

Property/Value	Units	Temp	Data Type	Reference
WS 3.89E-001	mg/L	25	EXP	BAUGHMAN,GL & PERENICH,TA (1988)
logP 3.51			EST	MEYLAN,WM & HOWARD,PH (1995)
VP 1.62E-015	mm Hg	25	EST	NEELY,WB & BLAU,GE (1985)
DC	pKa			
HL 9.47E-018	atm m3/mol	25	EST	MEYLAN,WM & HOWARD,PH (1991)
OH 1.81E-010	cm3/molc sec	25	EST	MEYLAN,WM & HOWARD,PH (1993)

CAS #: 041541-14-4				DIS. A. 13

Formula: $C_{17}H_{20}N_4O_5$

Mol Weight: 360.37

MP (deg C): FP (deg C):

BP (deg C):

BP pressure (mm Hg):

Property/Value	Units	Temp	Data Type	Reference
WS 7.20E+000	mg/L	25	EXP	BAUGHMAN,GL & PERENICH,TA (1988)
logP 3.22			EST	MEYLAN,WM & HOWARD,PH (1995)
VP 7.29E-014	mm Hg	25	EST	NEELY,WB & BLAU,GE (1985)
DC	pKa			
HL 1.29E-016	atm m3/mol	25	EST	MEYLAN,WM & HOWARD,PH (1991)
OH 8.93E-011	cm3/molc sec	25	EST	MEYLAN,WM & HOWARD,PH (1993)

CAS #: 041550-04-3				NNN-ME-N(3ME-CYPEME)AMMON I(Z)

Formula: $C_{10}H_{22}IN$

Mol Weight: 283.20

MP (deg C): FP (deg C):

BP (deg C):

BP pressure (mm Hg):

Property/Value	Units	Temp	Data Type	Reference
WS 2.93E+005	mg/L	25	EST	MEYLAN,WM ET AL. (1996)
logP -1.50			EXP	PRATESI,P ET AL. (1986)
VP 4.19E-008	mm Hg	25	EST	NEELY,WB & BLAU,GE (1985)
DC	pKa			
HL 6.36E-012	atm m3/mol	25	EST	MEYLAN,WM & HOWARD,PH (1991)
OH 2.33E-011	cm3/molc sec	25	EST	MEYLAN,WM & HOWARD,PH (1993)

CAS #: 041601-59-6				1,1,4-TRICHLORO-2-BUTENE

Formula: $C_4H_5Cl_3$

Mol Weight: 159.44

MP (deg C): FP (deg C):

BP (deg C):

BP pressure (mm Hg):

Property/Value	Units	Temp	Data Type	Reference
WS 2.91E+002	mg/L	25	EST	MEYLAN,WM ET AL. (1996)
logP 2.78			EST	MEYLAN,WM & HOWARD,PH (1995)
VP 1.05E+000	mm Hg	25	EST	NEELY,WB & BLAU,GE (1985)
DC	pKa			
HL 6.62E-003	atm m3/mol	25	EST	MEYLAN,WM & HOWARD,PH (1991)
OH 3.32E-011	cm3/molc sec	25	EST	MEYLAN,WM & HOWARD,PH (1993)

CAS #: 041643-81-6				ACETIC ACID, (1-NAPHTHALENYLOXY)-, ETHYL ESTER

Formula: $C_{14}H_{14}O_3$

Mol Weight: 230.27

MP (deg C): FP (deg C):

BP (deg C):

BP pressure (mm Hg):

Property/Value	Units	Temp	Data Type	Reference
WS 1.01E+002	mg/L	25	EST	MEYLAN,WM ET AL. (1996)
logP 2.90			EXP	CHAMBERLAIN,K ET AL. (1986)
VP 3.97E-005	mm Hg	25	EST	NEELY,WB & BLAU,GE (1985)
DC	pKa			
HL 6.96E-007	atm m3/mol	25	EST	MEYLAN,WM & HOWARD,PH (1991)
OH 2.06E-010	cm3/molc sec	25	EST	MEYLAN,WM & HOWARD,PH (1993)

CAS #: 041656-75-1				P-ACETYLFORMANILIDE

Formula: $C_9H_9NO_2$

Mol Weight: 163.18

MP (deg C): FP (deg C):

BP (deg C):

BP pressure (mm Hg):

Property/Value	Units	Temp	Data Type	Reference
WS 1.04E+004	mg/L	25	EST	MEYLAN,WM ET AL. (1996)
logP 0.94			EXP	HANSCH,C & LEO,AJ (1985)
VP 2.24E-005	mm Hg	25	EST	NEELY,WB & BLAU,GE (1985)
DC	pKa			
HL 1.54E-011	atm m3/mol	25	EST	MEYLAN,WM & HOWARD,PH (1991)
OH 3.90E-011	cm3/molc sec	25	EST	MEYLAN,WM & HOWARD,PH (1993)

CAS #: 041708-72-9				PROPANAMIDE, 2-AMINO-N-(2,6-DIMETHYLPHENYL)-

Formula: $C_{11}H_{16}N_2O$

Mol Weight: 192.26

MP (deg C): 246-266 FP (deg C):

BP (deg C):

BP pressure (mm Hg):

Property/Value	Units	Temp	Data Type	Reference
WS 1.07E+004	mg/L	25	EST	MEYLAN,WM ET AL. (1996)
logP 0.76			EXP	SANGSTER,J (1993)
VP 4.77E-006	mm Hg	25	EST	NEELY,WB & BLAU,GE (1985)
DC	pKa			
HL 2.05E-011	atm m3/mol	25	EST	MEYLAN,WM & HOWARD,PH (1991)
OH 5.40E-011	cm3/molc sec	25	EST	MEYLAN,WM & HOWARD,PH (1993)

CAS #: 041744-40-5				SULBENCILLIN

Formula: $C_{16}H_{18}N_2O_7S_2$

Mol Weight: 414.46

MP (deg C): 195-198 de FP (deg C):

BP (deg C):

BP pressure (mm Hg):

Property/Value	Units	Temp	Data Type	Reference
WS 6.94E+001	mg/L	25	EST	MEYLAN,WM ET AL. (1996)
logP 0.59			EXP	SANGSTER,J (1993)
VP 1.30E-019	mm Hg	25	EST	NEELY,WB & BLAU,GE (1985)
DC	pKa			
HL 1.99E-021	atm m3/mol	25	EST	MEYLAN,WM & HOWARD,PH (1991)
OH 9.21E-011	cm3/molc sec	25	EST	MEYLAN,WM & HOWARD,PH (1993)

CAS #: 041814-78-2				TRICYCLAZOLE

Formula: $C_9H_7N_3S$

Mol Weight: 189.24

MP (deg C): 187

FP (deg C):

BP (deg C):

BP pressure (mm Hg):

Property/Value	Units	Temp	Data Type	Reference
WS 1.60E+003	mg/L	25	EXP	YALKOWSKY,SH & DANNENFELSER,RM (1992)
logP 1.70			EXP	HANSCH,C ET AL. (1995)
VP 2.00E-007	mm Hg	25	EXP	AUGUSTIJN-BECKERS,PWM ET AL. (1994)
DC	pKa			
HL 3.11E-011	atm m3/mol	25	EST	VP/WSOL
OH 2.42E-011	cm3/molc sec	25	EST	MEYLAN,WM & HOWARD,PH (1993)

CAS #: 041851-50-7				5-CHLORO-1,3-CYCLOPENTADIENE

Formula: C_5H_5Cl

Mol Weight: 100.55

MP (deg C):

FP (deg C):

BP (deg C):

BP pressure (mm Hg):

Property/Value	Units	Temp	Data Type	Reference
WS 9.70E+002	mg/L	25	EST	MEYLAN,WM ET AL. (1996)
logP 2.43			EST	MEYLAN,WM & HOWARD,PH (1995)
VP 2.04E+001	mm Hg	25	EST	NEELY,WB & BLAU,GE (1985)
DC	pKa			
HL 2.24E-002	atm m3/mol	25	EST	MEYLAN,WM & HOWARD,PH (1991)
OH 1.43E-010	cm3/molc sec	25	EST	MEYLAN,WM & HOWARD,PH (1993)

CAS #: 041859-85-2				N-(P-HYDROXYPHENYL)METHYLBENZAMIDE

Formula: $C_{14}H_{13}NO_2$

Mol Weight: 227.27

MP (deg C):

FP (deg C):

BP (deg C):

BP pressure (mm Hg):

Property/Value	Units	Temp	Data Type	Reference
WS 3.03E+003	mg/L	25	EST	MEYLAN,WM ET AL. (1996)
logP 1.87			EXP	HANSCH,C & LEO,AJ (1985)
VP 9.97E-009	mm Hg	25	EST	NEELY,WB & BLAU,GE (1985)
DC	pKa			
HL 4.08E-014	atm m3/mol	25	EST	MEYLAN,WM & HOWARD,PH (1991)
OH 5.19E-011	cm3/molc sec	25	EST	MEYLAN,WM & HOWARD,PH (1993)

CAS #: 041941-66-6				8-(4-CL-PH)-S CYCLIC AMP

Formula: $C_{16}H_{17}ClN_5O_6PS$

Mol Weight: 473.83

MP (deg C):

FP (deg C):

BP (deg C):

BP pressure (mm Hg):

Property/Value	Units	Temp	Data Type	Reference
WS 3.65E+002	mg/L	25	EST	MEYLAN,WM ET AL. (1996)
logP -0.98			EXP	KORTH,M & ENGELS,J (1987)
VP 6.06E-011	mm Hg	25	EST	NEELY,WB & BLAU,GE (1985)
DC	pKa			
HL 1.26E-030	atm m3/mol	25	EST	MEYLAN,WM & HOWARD,PH (1991)
OH 3.05E-010	cm3/molc sec	25	EST	MEYLAN,WM & HOWARD,PH (1993)

CAS #: 041992-55-6				TETRACHLOROCYCLOHEXENE

Formula: $C_6H_6Cl_4$

Mol Weight: 219.93

MP (deg C):

FP (deg C):

BP (deg C):

BP pressure (mm Hg):

Property/Value	Units	Temp	Data Type	Reference
WS 4.30E+001	mg/L	25	EST	MEYLAN,WM ET AL. (1996)
logP 3.40			EXP	HANSCH,C & LEO,AJ (1985)
VP 3.13E-002	mm Hg	25	EST	NEELY,WB & BLAU,GE (1985)
DC	pKa			
HL 1.81E-003	atm m3/mol	25	EST	MEYLAN,WM & HOWARD,PH (1991)
OH 4.36E-011	cm3/molc sec	25	EST	MEYLAN,WM & HOWARD,PH (1993)

CAS #: 042013-74-1				1-PIPERIDINYLOXY, 4-[(2-HYDROXYETHYL)METHYLAMINO

Formula: $C_{12}H_{26}N_2O_2$

Mol Weight: 230.35

MP (deg C):

FP (deg C):

BP (deg C):

BP pressure (mm Hg):

Property/Value	Units	Temp	Data Type	Reference
WS 1.10E+005	mg/L	25	EST	MEYLAN,WM ET AL. (1996)
logP 0.53			EXP	SANGSTER,J (1993)
VP 1.84E-010	mm Hg	25	EST	NEELY,WB & BLAU,GE (1985)
DC	pKa			
HL 1.98E-018	atm m3/mol	25	EST	MEYLAN,WM & HOWARD,PH (1991)
OH 1.13E-010	cm3/molc sec	25	EST	MEYLAN,WM & HOWARD,PH (1993)

CAS #: 042087-80-9				METHYL 4-CHLORO-2-NITROBENZOATE

Formula: $C_8H_6ClNO_4$

Mol Weight: 215.59

MP (deg C): 43-45

FP (deg C):

BP (deg C):

BP pressure (mm Hg):

Property/Value	Units	Temp	Data Type	Reference
WS 1.64E+002	mg/L	25	EST	MEYLAN,WM ET AL. (1996)
logP 2.29			EST	MEYLAN,WM & HOWARD,PH (1995)
VP 3.15E-004	mm Hg	25	EST	NEELY,WB & BLAU,GE (1985)
DC	pKa			
HL 1.01E-007	atm m3/mol	25	EST	MEYLAN,WM & HOWARD,PH (1991)
OH 2.56E-013	cm3/molc sec	25	EST	MEYLAN,WM & HOWARD,PH (1993)

CAS #: 042103-65-1				2-PROPENOIC ACID, 3-PHENOXY-

Formula: $C_9H_8O_3$

Mol Weight: 164.16

MP (deg C):

FP (deg C):

BP (deg C):

BP pressure (mm Hg):

Property/Value	Units	Temp	Data Type	Reference
WS 1.87E+003	mg/L	25	EST	MEYLAN,WM ET AL. (1996)
logP 2.27			EXP	SANGSTER,J (1993)
VP 4.63E-004	mm Hg	25	EST	NEELY,WB & BLAU,GE (1985)
DC	pKa			
HL 9.92E-009	atm m3/mol	25	EST	MEYLAN,WM & HOWARD,PH (1991)
OH 4.18E-011	cm3/molc sec	25	EST	MEYLAN,WM & HOWARD,PH (1993)

CAS #: 042116-76-7		CARNIDAZOLE	

Formula: $C_8H_{12}N_4O_3S$

Mol Weight: 244.27

MP (deg C): 142.4 dec

FP (deg C):

BP (deg C):

BP pressure (mm Hg):

Property/Value	Units	Temp	Data Type	Reference
WS 1.08E+003	mg/L	25	EST	MEYLAN,WM ET AL. (1996)
logP 1.15			EXP	HANSCH,C & LEO,AJ (1985)
VP 1.48E-007	mm Hg	25	EST	NEELY,WB & BLAU,GE (1985)
DC	pKa			
HL 1.57E-011	atm m3/mol	25	EST	MEYLAN,WM & HOWARD,PH (1991)
OH 7.97E-011	cm3/molc sec	25	EST	MEYLAN,WM & HOWARD,PH (1993)

CAS #: 042126-84-1		1H-BENZO(CD)FLUOANTHENE	

Formula: $C_{19}H_{12}$

Mol Weight: 240.31

MP (deg C):

FP (deg C):

BP (deg C):

BP pressure (mm Hg):

Property/Value	Units	Temp	Data Type	Reference
WS 2.41E-002	mg/L	25	EST	MEYLAN,WM ET AL. (1996)
logP 5.78			EXP	HANSCH,C ET AL. (1995)
VP 1.85E-007	mm Hg	25	EST	NEELY,WB & BLAU,GE (1985)
DC	pKa			
HL 2.45E-006	atm m3/mol	25	EST	MEYLAN,WM & HOWARD,PH (1991)
OH 1.04E-010	cm3/molc sec	25	EST	MEYLAN,WM & HOWARD,PH (1993)

CAS #: 042135-22-8		9H-FLUOREN-9-ONE, 3-NITRO-	

Formula: $C_{13}H_7NO_3$

Mol Weight: 225.21

MP (deg C):

FP (deg C):

BP (deg C):

BP pressure (mm Hg):

Property/Value	Units	Temp	Data Type	Reference
WS 2.49E+000	mg/L	25	EST	MEYLAN,WM ET AL. (1996)
logP 3.06			EXP	DEBNATH,AK & HANSCH,C (1992)
VP 8.89E-007	mm Hg	25	EST	NEELY,WB & BLAU,GE (1985)
DC	pKa			
HL 2.67E-009	atm m3/mol	25	EST	MEYLAN,WM & HOWARD,PH (1991)
OH 2.54E-012	cm3/molc sec	25	EST	MEYLAN,WM & HOWARD,PH (1993)

CAS #: 042154-69-8		2-METHYL-3-HEXENE	

Formula: C_7H_{14}

Mol Weight: 98.19

MP (deg C):

FP (deg C):

BP (deg C):

BP pressure (mm Hg):

Property/Value	Units	Temp	Data Type	Reference
WS 3.59E+001	mg/L	25	EST	MEYLAN,WM ET AL. (1996)
logP 3.49			EST	MEYLAN,WM & HOWARD,PH (1995)
VP 7.61E+001	mm Hg	25	EXP	DYKYJ,J & REPAS,M (1973)
DC	pKa			
HL 5.62E-001	atm m3/mol	25	EST	MEYLAN,WM & HOWARD,PH (1991)
OH 5.98E-011	cm3/molc sec	25	EST	MEYLAN,WM & HOWARD,PH (1993)

CAS #: 042175-93-9		1-PYRROLIDINEACETAMIDE, N-(PHENYLMETHYL)-	

Formula: $C_{13}H_{18}N_2O$

Mol Weight: 218.30

MP (deg C):

FP (deg C):

BP (deg C):

BP pressure (mm Hg):

Property/Value	Units	Temp	Data Type	Reference
WS 1.27E+003	mg/L	25	EST	MEYLAN,WM ET AL. (1996)
logP 1.69			EXP	SANGSTER,J (1993)
VP 1.11E-006	mm Hg	25	EST	NEELY,WB & BLAU,GE (1985)
DC	pKa			
HL 1.52E-011	atm m3/mol	25	EST	MEYLAN,WM & HOWARD,PH (1991)
OH 1.06E-010	cm3/molc sec	25	EST	MEYLAN,WM & HOWARD,PH (1993)

CAS #: 042200-33-9		NADOLOL	

Formula: $C_{17}H_{27}NO_4$

Mol Weight: 309.41

MP (deg C): 124-136

FP (deg C):

BP (deg C):

BP pressure (mm Hg):

Property/Value	Units	Temp	Data Type	Reference
WS 2.24E+004	mg/L	25	EST	MEYLAN,WM ET AL. (1996)
logP 0.81			EXP	SANGSTER,J (1994)
VP 2.25E-011	mm Hg	25	EST	NEELY,WB & BLAU,GE (1985)
DC	pKa			
HL 1.37E-014	atm m3/mol	25	EST	MEYLAN,WM & HOWARD,PH (1991)
OH 1.80E-010	cm3/molc sec	25	EST	MEYLAN,WM & HOWARD,PH (1993)

CAS #: 042217-02-7		1-CHLOROEICOSANE	

Formula: $C_{20}H_{41}Cl$

Mol Weight: 317.00

MP (deg C):

FP (deg C):

BP (deg C):

BP pressure (mm Hg):

Property/Value	Units	Temp	Data Type	Reference
WS 1.24E-005	mg/L	25	EST	MEYLAN,WM ET AL. (1996)
logP 10.42			EST	MEYLAN,WM & HOWARD,PH (1995)
VP 2.40E-006	mm Hg	25	EXT	LI,JCM & ROSSINI,FD (1953)
DC	pKa			
HL 1.79E+000	atm m3/mol	25	EST	MEYLAN,WM & HOWARD,PH (1991)
OH 2.48E-011	cm3/molc sec	25	EST	MEYLAN,WM & HOWARD,PH (1993)

CAS #: 042270-37-1		PIPERAZINE, 1-(2-THIAZOLYL)-	

Formula: $C_7H_{11}N_3S$

Mol Weight: 169.25

MP (deg C):

FP (deg C):

BP (deg C):

BP pressure (mm Hg):

Property/Value	Units	Temp	Data Type	Reference
WS 1.90E+005	mg/L	25	EST	MEYLAN,WM ET AL. (1996)
logP 0.61			EXP	CACCIA,S ET AL. (1985)
VP 7.27E-004	mm Hg	25	EST	NEELY,WB & BLAU,GE (1985)
DC	pKa			
HL 8.05E-012	atm m3/mol	25	EST	MEYLAN,WM & HOWARD,PH (1991)
OH 2.11E-010	cm3/molc sec	25	EST	MEYLAN,WM & HOWARD,PH (1993)

CAS #: 042288-41-5 — ACETIC ACID, [4-(METHYLSULFONYL)PHENOXY]-

Formula: $C_9H_{10}O_5S$

Mol Weight: 230.24

MP (deg C):

FP (deg C):

BP (deg C):

BP pressure (mm Hg):

Property/Value	Units	Temp	Data Type	Reference
WS 6.70E+004	mg/L	25	EST	MEYLAN,WM ET AL. (1996)
logP 0.06			EXP	SANGSTER,J (1994)
VP 1.01E-006	mm Hg	25	EST	NEELY,WB & BLAU,GE (1985)
DC	pKa			
HL 3.88E-012	atm m3/mol	25	EST	MEYLAN,WM & HOWARD,PH (1991)
OH 1.05E-011	cm3/molc sec	25	EST	MEYLAN,WM & HOWARD,PH (1993)

CAS #: 042322-29-2 — QUINOLINE,8-ETHYLCARBONATE

Formula: $C_{12}H_{11}NO_3$

Mol Weight: 217.23

MP (deg C):

FP (deg C):

BP (deg C):

BP pressure (mm Hg):

Property/Value	Units	Temp	Data Type	Reference
WS 1.19E+003	mg/L	25	EST	MEYLAN,WM ET AL. (1996)
logP 1.73			EXP	HANSCH,C & LEO,AJ (1985)
VP 6.00E-005	mm Hg	25	EST	NEELY,WB & BLAU,GE (1985)
DC	pKa			
HL 3.89E-008	atm m3/mol	25	EST	MEYLAN,WM & HOWARD,PH (1991)
OH 1.65E-011	cm3/molc sec	25	EST	MEYLAN,WM & HOWARD,PH (1993)

CAS #: 042327-99-1 — 2-BUTANONE, 3-METHYL-4-(1-PIPERIDINYL)-

Formula: $C_{10}H_{19}NO$

Mol Weight: 169.27

MP (deg C):

FP (deg C):

BP (deg C):

BP pressure (mm Hg):

Property/Value	Units	Temp	Data Type	Reference
WS 5.10E+004	mg/L	25	EST	MEYLAN,WM ET AL. (1996)
logP 1.28			EXP	SANGSTER,J (1993)
VP 6.21E-002	mm Hg	25	EST	NEELY,WB & BLAU,GE (1985)
DC	pKa			
HL 4.47E-008	atm m3/mol	25	EST	MEYLAN,WM & HOWARD,PH (1991)
OH 1.28E-010	cm3/molc sec	25	EST	MEYLAN,WM & HOWARD,PH (1993)

CAS #: 042346-68-9 — 1-METHYL-4-CARBOXY-2-PYRROLIDONE

Formula: $C_6H_9NO_3$

Mol Weight: 143.14

MP (deg C):

FP (deg C):

BP (deg C):

BP pressure (mm Hg):

Property/Value	Units	Temp	Data Type	Reference
WS 1.00E+006	mg/L	25	EST	MEYLAN,WM ET AL. (1996)
logP -1.50			EXP	SASAKI,H ET AL. (1988)
VP 1.31E-004	mm Hg	25	EST	NEELY,WB & BLAU,GE (1985)
DC	pKa			
HL 2.35E-013	atm m3/mol	25	EST	MEYLAN,WM & HOWARD,PH (1991)
OH 2.61E-011	cm3/molc sec	25	EST	MEYLAN,WM & HOWARD,PH (1993)

CAS #: 042464-80-2 — ACETAMIDE, N-(5-QUINOLINYL)-

Formula: $C_{11}H_{10}N_2O$

Mol Weight: 186.22

MP (deg C):

FP (deg C):

BP (deg C):

BP pressure (mm Hg):

Property/Value	Units	Temp	Data Type	Reference
WS 1.27E+004	mg/L	25	EST	MEYLAN,WM ET AL. (1996)
logP 0.71			EXP	HANSCH,C & LEO,AJ (1985)
VP 1.28E-006	mm Hg	25	EST	NEELY,WB & BLAU,GE (1985)
DC	pKa			
HL 7.88E-013	atm m3/mol	25	EST	MEYLAN,WM & HOWARD,PH (1991)
OH 9.18E-011	cm3/molc sec	25	EST	MEYLAN,WM & HOWARD,PH (1993)

CAS #: 042465-53-2 — ETHANONE, 1-(2-AMINO-4-METHOXYPHENYL)

Formula: $C_9H_{11}NO_2$

Mol Weight: 165.19

MP (deg C):

FP (deg C):

BP (deg C):

BP pressure (mm Hg):

Property/Value	Units	Temp	Data Type	Reference
WS 2.03E+003	mg/L	25	EST	MEYLAN,WM ET AL. (1996)
logP 1.76			EXP	HANSCH,C ET AL. (1995)
VP 9.39E-004	mm Hg	25	EST	NEELY,WB & BLAU,GE (1985)
DC	pKa			
HL 2.05E-010	atm m3/mol	25	EST	MEYLAN,WM & HOWARD,PH (1991)
OH 2.01E-010	cm3/molc sec	25	EST	MEYLAN,WM & HOWARD,PH (1993)

CAS #: 042471-28-3 — NIMUSTINE

Formula: $C_9H_{13}ClN_6O_2$

Mol Weight: 272.70

MP (deg C):

FP (deg C):

BP (deg C):

BP pressure (mm Hg):

Property/Value	Units	Temp	Data Type	Reference
WS 8.19E+003	mg/L	25	EST	MEYLAN,WM ET AL. (1996)
logP 0.39			EXP	HANSCH,C ET AL. (1995)
VP 1.87E-008	mm Hg	25	EST	NEELY,WB & BLAU,GE (1985)
DC	pKa			
HL 1.74E-018	atm m3/mol	25	EST	MEYLAN,WM & HOWARD,PH (1991)
OH 7.93E-012	cm3/molc sec	25	EST	MEYLAN,WM & HOWARD,PH (1993)

CAS #: 042509-80-8 — ISAZOFOS

Formula: $C_9H_{17}ClN_3O_3PS$

Mol Weight: 313.74

MP (deg C):

FP (deg C):

BP (deg C): 170

BP pressure (mm Hg):

Property/Value	Units	Temp	Data Type	Reference
WS 1.50E+002	mg/L	20	EXP	YALKOWSKY,SH & DANNENFELSER,RM (1992)
logP 3.82			EXP	HANSCH,C & LEO,AJ (1985)
VP 8.70E-005	mm Hg	25	EXP	WEBER,JB (1994)
DC	pKa			
HL 2.39E-007	atm m3/mol	25	EST	VP/WSOL
OH 1.13E-010	cm3/molc sec	25	EST	MEYLAN,WM & HOWARD,PH (1993)

042548-73-2 — 1-(4-CARBAMOYLPHENYL)-3-METHYL-3-HYDROXYTRIAZENE

CAS #: 042548-73-2

Formula: $C_8H_{10}N_4O_2$

Mol Weight: 194.19

MP (deg C):
FP (deg C):
BP (deg C):
BP pressure (mm Hg):

Property/Value	Units	Temp	Data Type	Reference
WS 3.54E+003	mg/L	25	EST	MEYLAN,WM ET AL. (1996)
logP -0.22			EXP	HANSCH,C ET AL. (1995)
VP 2.89E-010	mm Hg	25	EST	NEELY,WB & BLAU,GE (1985)
DC	pKa			
HL 2.01E-016	atm m3/mol	25	EST	MEYLAN,WM & HOWARD,PH (1991)
OH 3.59E-012	cm3/molc sec	25	EST	MEYLAN,WM & HOWARD,PH (1993)

042558-93-0 — CIS-1-NO-1(CLET)-3(4COOH CYHEXYL)UREA

CAS #: 042558-93-0

Formula: $C_{10}H_{16}ClN_3O_4$

Mol Weight: 277.71

MP (deg C):
FP (deg C):
BP (deg C):
BP pressure (mm Hg):

Property/Value	Units	Temp	Data Type	Reference
WS 8.15E+002	mg/L	25	EST	MEYLAN,WM ET AL. (1996)
logP 1.68			EXP	HANSCH,C & LEO,AJ (1985)
VP 3.02E-008	mm Hg	25	EST	NEELY,WB & BLAU,GE (1985)
DC	pKa			
HL 1.38E-015	atm m3/mol	25	EST	MEYLAN,WM & HOWARD,PH (1991)
OH 2.32E-011	cm3/molc sec	25	EST	MEYLAN,WM & HOWARD,PH (1993)

042558-94-1 — TRAN-1NO-1(CLET)-3(4COOH CYHEXYL)UREA

CAS #: 042558-94-1

Formula: $C_{10}H_{16}ClN_3O_4$

Mol Weight: 277.71

MP (deg C):
FP (deg C):
BP (deg C):
BP pressure (mm Hg):

Property/Value	Units	Temp	Data Type	Reference
WS 8.15E+002	mg/L	25	EST	MEYLAN,WM ET AL. (1996)
logP 1.53			EXP	HANSCH,C & LEO,AJ (1985)
VP 3.02E-008	mm Hg	25	EST	NEELY,WB & BLAU,GE (1985)
DC	pKa			
HL 1.38E-015	atm m3/mol	25	EST	MEYLAN,WM & HOWARD,PH (1991)
OH 2.32E-011	cm3/molc sec	25	EST	MEYLAN,WM & HOWARD,PH (1993)

042576-02-3 — BIFENOX

CAS #: 042576-02-3

Formula: $C_{14}H_9Cl_2NO_5$

Mol Weight: 342.14

MP (deg C): 85
FP (deg C):
BP (deg C):
BP pressure (mm Hg):

Property/Value	Units	Temp	Data Type	Reference
WS 5.00E-001	mg/L		EXP	YALKOWSKY,SH & DANNENFELSER,RM (1992)
logP 4.48			EXP	HANSCH,C ET AL. (1995)
VP 1.10E-006	mm Hg	25	EXP	WEBER,JB (1994)
DC	pKa			
HL 9.90E-007	atm m3/mol	25	EST	VP/WSOL
OH 1.32E-012	cm3/molc sec	25	EST	MEYLAN,WM & HOWARD,PH (1993)

042583-67-5 — TRIFLUOROAMPHENICOL

CAS #: 042583-67-5

Formula: $C_{11}H_{11}F_3N_2O_5$

Mol Weight: 308.22

MP (deg C):
FP (deg C):
BP (deg C):
BP pressure (mm Hg):

Property/Value	Units	Temp	Data Type	Reference
WS 5.46E+002	mg/L	25	EST	MEYLAN,WM ET AL. (1996)
logP 1.07			EXP	HANSCH,C & LEO,AJ (1985)
VP 2.29E-011	mm Hg	25	EST	NEELY,WB & BLAU,GE (1985)
DC	pKa			
HL 1.45E-016	atm m3/mol	25	EST	MEYLAN,WM & HOWARD,PH (1991)
OH 2.53E-011	cm3/molc sec	25	EST	MEYLAN,WM & HOWARD,PH (1993)

042585-33-1 — 1-PIPERIDINYLOXY, 4-(METHYLAMINO)-2,2,6,6-TETRAM

CAS #: 042585-33-1

Formula: $C_{10}H_{22}N_2O$

Mol Weight: 186.30

MP (deg C):
FP (deg C):
BP (deg C):
BP pressure (mm Hg):

Property/Value	Units	Temp	Data Type	Reference
WS 1.01E+005	mg/L	25	EST	MEYLAN,WM ET AL. (1996)
logP 0.84			EXP	SANGSTER,J (1993)
VP 6.61E-007	mm Hg	25	EST	NEELY,WB & BLAU,GE (1985)
DC	pKa			
HL 1.86E-014	atm m3/mol	25	EST	MEYLAN,WM & HOWARD,PH (1991)
OH 9.51E-011	cm3/molc sec	25	EST	MEYLAN,WM & HOWARD,PH (1993)

042755-32-8 — 1,3,4-THIADIAZOLE, 2-METHYL-5-[(PHENYLMETHYL)THI

CAS #: 042755-32-8

Formula: $C_{10}H_{10}N_2S_2$

Mol Weight: 222.33

MP (deg C):
FP (deg C):
BP (deg C):
BP pressure (mm Hg):

Property/Value	Units	Temp	Data Type	Reference
WS 3.29E+002	mg/L	25	EST	MEYLAN,WM ET AL. (1996)
logP 2.35			EXP	HANSCH,C ET AL. (1995)
VP 3.67E-006	mm Hg	25	EST	NEELY,WB & BLAU,GE (1985)
DC	pKa			
HL 4.05E-009	atm m3/mol	25	EST	MEYLAN,WM & HOWARD,PH (1991)
OH 1.30E-011	cm3/molc sec	25	EST	MEYLAN,WM & HOWARD,PH (1993)

042839-04-3 — 4-PYRIMIDINECARBONITRILE

CAS #: 042839-04-3

Formula: $C_5H_3N_3$

Mol Weight: 105.10

MP (deg C):
FP (deg C):
BP (deg C):
BP pressure (mm Hg):

Property/Value	Units	Temp	Data Type	Reference
WS 7.18E+004	mg/L	25	EST	MEYLAN,WM ET AL. (1996)
logP -0.08			EXP	YAMAGAMI,C ET AL. (1990)
VP 1.68E-001	mm Hg	25	EST	NEELY,WB & BLAU,GE (1985)
DC	pKa			
HL 2.82E-008	atm m3/mol	25	EST	MEYLAN,WM & HOWARD,PH (1991)
OH 4.06E-014	cm3/molc sec	25	EST	MEYLAN,WM & HOWARD,PH (1993)

CAS #: 042839-08-7			2-PYRIMIDINECARBOXYLIC ACID, ETHYL ESTER

Formula: $C_7H_8N_2O_2$

Mol Weight: 152.15

MP (deg C): FP (deg C):

BP (deg C):

BP pressure (mm Hg):

Property/Value	Units	Temp	Data Type	Reference
WS 1.37E+005	mg/L	25	EST	MEYLAN,WM ET AL. (1996)
logP -0.31			EXP	YAMAGAMI,C ET AL. (1990)
VP 4.25E-002	mm Hg	25	EST	NEELY,WB & BLAU,GE (1985)
DC	pKa			
HL 2.50E-008	atm m3/mol	25	EST	MEYLAN,WM & HOWARD,PH (1991)
OH 1.74E-012	cm3/molc sec	25	EST	MEYLAN,WM & HOWARD,PH (1993)

CAS #: 042846-29-7			CYCLOHEXANOL, 2,6-DIMETHYL-, (1à,2á,6á)-

Formula: $C_8H_{16}O$

Mol Weight: 128.22

MP (deg C): FP (deg C):

BP (deg C):

BP pressure (mm Hg):

Property/Value	Units	Temp	Data Type	Reference
WS 2.81E+003	mg/L	25	EST	MEYLAN,WM ET AL. (1996)
logP 2.38			EXP	FUNASAKI,N ET AL. (1986)
VP 1.23E-001	mm Hg	25	EST	NEELY,WB & BLAU,GE (1985)
DC	pKa			
HL 8.63E-006	atm m3/mol	25	EST	MEYLAN,WM & HOWARD,PH (1991)
OH 2.09E-011	cm3/molc sec	25	EST	MEYLAN,WM & HOWARD,PH (1993)

CAS #: 042865-65-6			1,1-DIME-3-(M-ACETYLPHENYL)UREA

Formula: $C_{11}H_{14}N_2O_2$

Mol Weight: 206.25

MP (deg C): FP (deg C):

BP (deg C):

BP pressure (mm Hg):

Property/Value	Units	Temp	Data Type	Reference
WS 6.28E+003	mg/L	25	EST	MEYLAN,WM ET AL. (1996)
logP 0.95			EXP	HANSCH,C & LEO,AJ (1985)
VP 5.95E-006	mm Hg	25	EST	NEELY,WB & BLAU,GE (1985)
DC	pKa			
HL 1.77E-012	atm m3/mol	25	EST	MEYLAN,WM & HOWARD,PH (1991)
OH 2.02E-011	cm3/molc sec	25	EST	MEYLAN,WM & HOWARD,PH (1993)

CAS #: 042874-01-1			NITROFLUORFEN

Formula: $C_{13}H_7ClF_3NO_3$

Mol Weight: 317.65

MP (deg C): 68 FP (deg C):

BP (deg C):

BP pressure (mm Hg):

Property/Value	Units	Temp	Data Type	Reference
WS 5.23E-001	mg/L	25	EST	MEYLAN,WM ET AL. (1996)
logP 4.54			EXP	NANDIHALLI,UB ET AL. (1993)
VP 6.92E-006	mm Hg	25	EST	NEELY,WB & BLAU,GE (1985)
DC	pKa			
HL 2.99E-006	atm m3/mol	25	EST	MEYLAN,WM & HOWARD,PH (1991)
OH 9.04E-013	cm3/molc sec	25	EST	MEYLAN,WM & HOWARD,PH (1993)

CAS #: 042874-03-3			OXYFLUORFEN

Formula: $C_{15}H_{11}ClF_3NO_4$

Mol Weight: 361.71

MP (deg C): 65-84 FP (deg C):

BP (deg C):

BP pressure (mm Hg):

Property/Value	Units	Temp	Data Type	Reference
WS 1.00E-001	mg/L	25	EXP	SHIU,WY ET AL. (1990)
logP 4.73			EXP	NANDIHALLI,UB ET AL. (1993)
VP 4.82E-007	mm Hg	25	EST	NEELY,WB & BLAU,GE (1985)
DC	pKa			
HL 2.35E-007	atm m3/mol	25	EST	MEYLAN,WM & HOWARD,PH (1991)
OH 1.32E-011	cm3/molc sec	25	EST	MEYLAN,WM & HOWARD,PH (1993)

CAS #: 042975-18-8			4-(PHENYLAZO)BENZENESULFONIC ACID, SODIUM SALT

Formula: $C_{12}H_9N_2NaO_3S$

Mol Weight: 284.27

MP (deg C): FP (deg C):

BP (deg C):

BP pressure (mm Hg):

Property/Value	Units	Temp	Data Type	Reference
WS 2.89E+004	mg/L	25	EST	MEYLAN,WM ET AL. (1996)
logP -0.83			EST	MEYLAN,WM & HOWARD,PH (1995)
VP 1.48E-014	mm Hg	25	EST	NEELY,WB & BLAU,GE (1985)
DC	pKa			
HL	atm m3/mol			
OH 9.38E-013	cm3/molc sec	25	EST	MEYLAN,WM & HOWARD,PH (1993)

CAS #: 043013-70-3			2-PROPENOIC ACID, 3-PHENOXY-, ETHYL ESTER

Formula: $C_{11}H_{12}O_3$

Mol Weight: 192.22

MP (deg C): FP (deg C):

BP (deg C):

BP pressure (mm Hg):

Property/Value	Units	Temp	Data Type	Reference
WS 9.22E+001	mg/L	25	EST	MEYLAN,WM ET AL. (1996)
logP 3.18			EXP	SANGSTER,J (1993)
VP 5.80E-003	mm Hg	25	EST	NEELY,WB & BLAU,GE (1985)
DC	pKa			
HL 4.22E-006	atm m3/mol	25	EST	MEYLAN,WM & HOWARD,PH (1991)
OH 4.29E-011	cm3/molc sec	25	EST	MEYLAN,WM & HOWARD,PH (1993)

CAS #: 043032-67-3			METHYLSULFONYLANISOLE

Formula: $C_8H_{10}O_3S$

Mol Weight: 186.23

MP (deg C): FP (deg C):

BP (deg C):

BP pressure (mm Hg):

Property/Value	Units	Temp	Data Type	Reference
WS 1.15E+004	mg/L	25	EST	MEYLAN,WM ET AL. (1996)
logP 0.76			EXP	HANSCH,C ET AL. (1995)
VP 6.45E-004	mm Hg	25	EST	NEELY,WB & BLAU,GE (1985)
DC	pKa			
HL 7.38E-008	atm m3/mol	25	EST	MEYLAN,WM & HOWARD,PH (1991)
OH 6.49E-012	cm3/molc sec	25	EST	MEYLAN,WM & HOWARD,PH (1993)

CAS #:	043121-43-3			TRIADIMEFON

Formula: $C_{14}H_{16}ClN_3O_2$

Mol Weight: 293.76

MP (deg C): 82.3 FP (deg C):

BP (deg C):

BP pressure (mm Hg):

Property/Value	Units	Temp	Data Type	Reference
WS 7.15E+001	mg/L	25	EXP	WAUCHOPE,RD ET AL. (1991A)
logP 2.77			EXP	HANSCH,C ET AL. (1995)
VP 1.50E-008	mm Hg	25	EXP	WAUCHOPE,RD ET AL. (1991A)
DC	pKa			
HL 8.11E-011	atm m3/mol	25	EST	VP/WSOL
OH 2.70E-011	cm3/molc sec	25	EST	MEYLAN,WM & HOWARD,PH (1993)

CAS #:	043170-88-3			L-GLUTAMIC ACID, N-[4-[[(2,4-DIAMINO-6-PTERIDINY

Formula: $C_{24}H_{30}N_8O_5$

Mol Weight: 510.56

MP (deg C): FP (deg C):

BP (deg C):

BP pressure (mm Hg):

Property/Value	Units	Temp	Data Type	Reference
WS 8.92E+001	mg/L	25	EST	MEYLAN,WM ET AL. (1996)
logP 0.98			EXP	SANGSTER,J (1993)
VP 1.98E-017	mm Hg	25	EST	NEELY,WB & BLAU,GE (1985)
DC	pKa			
HL 2.78E-026	atm m3/mol	25	EST	MEYLAN,WM & HOWARD,PH (1991)
OH 3.19E-010	cm3/molc sec	25	EST	MEYLAN,WM & HOWARD,PH (1993)

CAS #:	043170-96-3			IMIDAZO[1,2-A]PYRIDINE-2-ETHANAMINE

Formula: $C_9H_{11}N_3$

Mol Weight: 161.21

MP (deg C): FP (deg C):

BP (deg C):

BP pressure (mm Hg):

Property/Value	Units	Temp	Data Type	Reference
WS 2.16E+005	mg/L	25	EST	MEYLAN,WM ET AL. (1996)
logP 0.59			EXP	SANGSTER,J (1994)
VP 3.01E-005	mm Hg	25	EST	NEELY,WB & BLAU,GE (1985)
DC	pKa			
HL 3.41E-012	atm m3/mol	25	EST	MEYLAN,WM & HOWARD,PH (1991)
OH 1.22E-010	cm3/molc sec	25	EST	MEYLAN,WM & HOWARD,PH (1993)

CAS #:	044652-67-7			1-HEXANAMINE, 2-METHYL-

Formula: $C_7H_{17}N$

Mol Weight: 115.22

MP (deg C): FP (deg C):

BP (deg C):

BP pressure (mm Hg):

Property/Value	Units	Temp	Data Type	Reference
WS 9.49E+003	mg/L	25	EST	MEYLAN,WM ET AL. (1996)
logP 2.40			EXP	SANGSTER,J (1993)
VP 5.35E+000	mm Hg	25	EST	NEELY,WB & BLAU,GE (1985)
DC 11.13	pKa	19	EXP	PERRIN,DD (1965)
HL 4.15E-005	atm m3/mol	25	EST	MEYLAN,WM & HOWARD,PH (1991)
OH 4.89E-011	cm3/molc sec	25	EST	MEYLAN,WM & HOWARD,PH (1993)

CAS #:	045803-84-7			4-VINYL-1,2-DICHLOROCYCLOHEXANE

Formula: $C_8H_{12}Cl_2$

Mol Weight: 179.09

MP (deg C): FP (deg C):

BP (deg C):

BP pressure (mm Hg):

Property/Value	Units	Temp	Data Type	Reference
WS 1.16E+001	mg/L	25	EST	MEYLAN,WM ET AL. (1996)
logP 4.31			EST	MEYLAN,WM & HOWARD,PH (1995)
VP 2.56E-001	mm Hg	25	EST	NEELY,WB & BLAU,GE (1985)
DC	pKa			
HL 2.18E-002	atm m3/mol	25	EST	MEYLAN,WM & HOWARD,PH (1991)
OH 3.21E-011	cm3/molc sec	25	EST	MEYLAN,WM & HOWARD,PH (1993)

CAS #:	045810-14-8			ACETAMIDE, N-5-PYRIMIDINYL-

Formula: $C_6H_7N_3O$

Mol Weight: 137.14

MP (deg C): FP (deg C):

BP (deg C):

BP pressure (mm Hg):

Property/Value	Units	Temp	Data Type	Reference
WS 6.50E+003	mg/L	25	EST	MEYLAN,WM ET AL. (1996)
logP -0.22			EXP	YAMAGAMI,C ET AL. (1990)
VP 1.37E-004	mm Hg	25	EST	NEELY,WB & BLAU,GE (1985)
DC	pKa			
HL 3.34E-012	atm m3/mol	25	EST	MEYLAN,WM & HOWARD,PH (1991)
OH 1.57E-012	cm3/molc sec	25	EST	MEYLAN,WM & HOWARD,PH (1993)

CAS #:	046061-25-0			4-T-BUTYL-N-NITROSOPIPERIDINE

Formula: $C_9H_{18}N_2O$

Mol Weight: 170.26

MP (deg C): FP (deg C):

BP (deg C):

BP pressure (mm Hg):

Property/Value	Units	Temp	Data Type	Reference
WS 4.15E+002	mg/L	25	EST	MEYLAN,WM ET AL. (1996)
logP 2.54			EXP	HANSCH,C & LEO,AJ (1985)
VP 8.05E-003	mm Hg	25	EST	NEELY,WB & BLAU,GE (1985)
DC	pKa			
HL 6.59E-006	atm .m3/mol	25	EST	MEYLAN,WM & HOWARD,PH (1991)
OH 2.83E-011	cm3/molc sec	25	EST	MEYLAN,WM & HOWARD,PH (1993)

CAS #:	046114-16-3			A-I-PROPYLBENZENE ETHANEAMINE

Formula: $C_{11}H_{17}N$

Mol Weight: 163.26

MP (deg C): FP (deg C):

BP (deg C):

BP pressure (mm Hg):

Property/Value	Units	Temp	Data Type	Reference
WS 4.14E+003	mg/L	25	EST	MEYLAN,WM ET AL. (1996)
logP 2.59			EXP	HANSCH,C & LEO,AJ (1985)
VP 5.21E-002	mm Hg	25	EST	NEELY,WB & BLAU,GE (1985)
DC	pKa			
HL 1.90E-006	atm m3/mol	25	EST	MEYLAN,WM & HOWARD,PH (1991)
OH 5.70E-011	cm3/molc sec	25	EST	MEYLAN,WM & HOWARD,PH (1993)

CAS #: 046122-12-7	N-HEXYLPYRIDINIUM IODIDE

Formula: $C_{11}H_{18}IN$

Mol Weight: 291.18

MP (deg C):   FP (deg C):

BP (deg C):

BP pressure (mm Hg):

Property/Value	Units	Temp	Data Type	Reference
WS 4.67E+005	mg/L	25	EST	MEYLAN,WM ET AL. (1996)
logP -1.79			EXP	HANSCH,C & LEO,AJ (1985)
VP 7.95E-004	mm Hg	25	EST	NEELY,WB & BLAU,GE (1985)
DC	pKa			
HL 6.20E-010	atm m3/mol	25	EST	MEYLAN,WM & HOWARD,PH (1991)
OH 1.71E-011	cm3/molc sec	25	EST	MEYLAN,WM & HOWARD,PH (1993)

CAS #: 046781-41-3	4,6-NH2-2,2-DIME-1(4IPH)S-TRIAZINE

Formula: $C_{11}H_{14}IN_5$

Mol Weight: 343.17

MP (deg C):   FP (deg C):

BP (deg C):

BP pressure (mm Hg):

Property/Value	Units	Temp	Data Type	Reference
WS 4.43E+002	mg/L	25	EST	MEYLAN,WM ET AL. (1996)
logP 1.39			EXP	HANSCH,C & LEO,AJ (1985)
VP 1.86E-007	mm Hg	25	EST	NEELY,WB & BLAU,GE (1985)
DC	pKa			
HL 1.69E-013	atm m3/mol	25	EST	MEYLAN,WM & HOWARD,PH (1991)
OH 9.07E-011	cm3/molc sec	25	EST	MEYLAN,WM & HOWARD,PH (1993)

CAS #: 046833-92-5	4-NO2-3-CHLOROPHENYLAMIDINOUREA

Formula: $C_8H_8ClN_5O_3$

Mol Weight: 257.64

MP (deg C):   FP (deg C):

BP (deg C):

BP pressure (mm Hg):

Property/Value	Units	Temp	Data Type	Reference
WS 6.35E+002	mg/L	25	EST	MEYLAN,WM ET AL. (1996)
logP 1.79			EXP	HANSCH,C & LEO,AJ (1985)
VP 1.24E-008	mm Hg	25	EST	NEELY,WB & BLAU,GE (1985)
DC	pKa			
HL 3.99E-018	atm m3/mol	25	EST	MEYLAN,WM & HOWARD,PH (1991)
OH 2.58E-011	cm3/molc sec	25	EST	MEYLAN,WM & HOWARD,PH (1993)

CAS #: 047071-11-4	4,6-NH2 2,2-DIME1(4-CF3)PH S-TRIAZENE

Formula: $C_{12}H_{14}F_3N_5$

Mol Weight: 285.27

MP (deg C):   FP (deg C):

BP (deg C):

BP pressure (mm Hg):

Property/Value	Units	Temp	Data Type	Reference
WS 1.36E+003	mg/L	25	EST	MEYLAN,WM ET AL. (1996)
logP 1.22			EXP	HANSCH,C & LEO,AJ (1985)
VP 4.05E-006	mm Hg	25	EST	NEELY,WB & BLAU,GE (1985)
DC	pKa			
HL 6.33E-012	atm m3/mol	25	EST	MEYLAN,WM & HOWARD,PH (1991)
OH 2.98E-011	cm3/molc sec	25	EST	MEYLAN,WM & HOWARD,PH (1993)

CAS #: 047082-97-3	2-PROPANOL, 1-[(1,1-DIMETHYLETHYL)AMINO]-3-[2-(2

Formula: $C_{16}H_{23}NO_3$

Mol Weight: 277.37

MP (deg C):   FP (deg C):

BP (deg C):

BP pressure (mm Hg):

Property/Value	Units	Temp	Data Type	Reference
WS 1.76E+003	mg/L	25	EST	MEYLAN,WM ET AL. (1996)
logP 2.32			EXP	RECANATINI,M (1992)
VP 8.51E-008	mm Hg	25	EST	NEELY,WB & BLAU,GE (1985)
DC	pKa			
HL 1.40E-013	atm m3/mol	25	EST	MEYLAN,WM & HOWARD,PH (1991)
OH 1.23E-010	cm3/molc sec	25	EST	MEYLAN,WM & HOWARD,PH (1993)

CAS #: 047141-42-4	LEVOBUNOLOL

Formula: $C_{17}H_{25}NO_3$

Mol Weight: 291.39

MP (deg C):   FP (deg C):

BP (deg C):

BP pressure (mm Hg):

Property/Value	Units	Temp	Data Type	Reference
WS 1.25E+003	mg/L	25	EST	MEYLAN,WM ET AL. (1996)
logP 2.40			EXP	HANSCH,C & LEO,AJ (1985)
VP 5.20E-009	mm Hg	25	EST	NEELY,WB & BLAU,GE (1985)
DC	pKa			
HL 1.28E-014	atm m3/mol	25	EST	MEYLAN,WM & HOWARD,PH (1991)
OH 1.09E-010	cm3/molc sec	25	EST	MEYLAN,WM & HOWARD,PH (1993)

CAS #: 049540-85-4	4-ISOTHIOCYANOPHENYL BENZOATE

Formula: $C_{14}H_9NO_2S$

Mol Weight: 255.30

MP (deg C):   FP (deg C):

BP (deg C):

BP pressure (mm Hg):

Property/Value	Units	Temp	Data Type	Reference
WS 1.45E+000	mg/L	25	EST	MEYLAN,WM ET AL. (1996)
logP 4.90			EXP	HANSCH,C & LEO,AJ (1985)
VP 2.95E-006	mm Hg	25	EST	NEELY,WB & BLAU,GE (1985)
DC	pKa			
HL 1.86E-006	atm m3/mol	25	EST	MEYLAN,WM & HOWARD,PH (1991)
OH 3.98E-012	cm3/molc sec	25	EST	MEYLAN,WM & HOWARD,PH (1993)

CAS #: 049547-83-3	2-(N,N-DIMEAMINOPROPYL)FURAN

Formula: $C_9H_{16}NO$

Mol Weight: 154.23

MP (deg C):   FP (deg C):

BP (deg C):

BP pressure (mm Hg):

Property/Value	Units	Temp	Data Type	Reference
WS 1.25E+004	mg/L	25	EST	MEYLAN,WM ET AL. (1996)
logP 2.08			EXP	HANSCH,C & LEO,AJ (1985)
VP 4.51E-001	mm Hg	25	EST	NEELY,WB & BLAU,GE (1985)
DC	pKa			
HL 5.19E-006	atm m3/mol	25	EST	MEYLAN,WM & HOWARD,PH (1991)
OH 1.82E-010	cm3/molc sec	25	EST	MEYLAN,WM & HOWARD,PH (1993)

CAS #: 049558-02-3				FURAZAN, METHYLNITRO-, 2-OXIDE
Formula: $C_3H_3N_3O_4$				
Mol Weight: 145.08				
MP (deg C):		FP (deg C):		
BP (deg C):				
BP pressure (mm Hg):				

Property/Value	Units	Temp	Data Type	Reference
WS 7.00E+003	mg/L	25	EST	MEYLAN,WM ET AL. (1996)
logP 0.78			EXP	CALVINO,R ET AL. (1992)
VP 5.56E-004	mm Hg	25	EST	NEELY,WB & BLAU,GE (1985)
DC	pKa			
HL 8.24E-012	atm m3/mol	25	EST	MEYLAN,WM & HOWARD,PH (1991)
OH 4.14E-012	cm3/molc sec	25	EST	MEYLAN,WM & HOWARD,PH (1993)

CAS #: 049561-47-9				2(5NO2-2-FURFURILIDENE)THIAZOLE
Formula: $C_9H_9N_2O_3S$				
Mol Weight: 225.25				
MP (deg C):		FP (deg C):		
BP (deg C):				
BP pressure (mm Hg):				

Property/Value	Units	Temp	Data Type	Reference
WS 2.84E+002	mg/L	25	EST	MEYLAN,WM ET AL. (1996)
logP 1.97			EXP	HANSCH,C & LEO,AJ (1985)
VP 1.61E-005	mm Hg	25	EST	NEELY,WB & BLAU,GE (1985)
DC	pKa			
HL 3.46E-010	atm m3/mol	25	EST	MEYLAN,WM & HOWARD,PH (1991)
OH 6.25E-011	cm3/molc sec	25	EST	MEYLAN,WM & HOWARD,PH (1993)

CAS #: 049606-44-2				2H-1,4-BENZODIAZEPIN-2-ONE, 7-CHLORO-5-(2-FLUORO
Formula: $C_{17}H_{11}ClF_4N_2O$				
Mol Weight: 370.74				
MP (deg C):		FP (deg C):		
BP (deg C):				
BP pressure (mm Hg):				

Property/Value	Units	Temp	Data Type	Reference
WS 6.26E+000	mg/L	25	EST	MEYLAN,WM ET AL. (1996)
logP 3.36			EXP	SANGSTER,J (1994)
VP 2.64E-008	mm Hg	25	EST	NEELY,WB & BLAU,GE (1985)
DC	pKa			
HL 4.45E-008	atm m3/mol	25	EST	MEYLAN,WM & HOWARD,PH (1991)
OH 8.25E-012	cm3/molc sec	25	EST	MEYLAN,WM & HOWARD,PH (1993)

CAS #: 049648-35-3				MONOCHLOROMETHYLAMPHENICOL
Formula: $C_{12}H_{15}ClN_2O_5$				
Mol Weight: 302.72				
MP (deg C):		FP (deg C):		
BP (deg C):				
BP pressure (mm Hg):				

Property/Value	Units	Temp	Data Type	Reference
WS 7.02E+002	mg/L	25	EST	MEYLAN,WM ET AL. (1996)
logP 0.98			EXP	HANSCH,C & LEO,AJ (1985)
VP 9.37E-013	mm Hg	25	EST	NEELY,WB & BLAU,GE (1985)
DC	pKa			
HL 8.62E-018	atm m3/mol	25	EST	MEYLAN,WM & HOWARD,PH (1991)
OH 3.17E-011	cm3/molc sec	25	EST	MEYLAN,WM & HOWARD,PH (1993)

CAS #: 049648-37-5				DIFLUOROAMPHENICOL
Formula: $C_{11}H_{12}F_2N_2O_5$				
Mol Weight: 290.23				
MP (deg C):		FP (deg C):		
BP (deg C):				
BP pressure (mm Hg):				

Property/Value	Units	Temp	Data Type	Reference
WS 2.50E+003	mg/L	25	EST	MEYLAN,WM ET AL. (1996)
logP 0.42			EXP	HANSCH,C & LEO,AJ (1985)
VP 1.71E-011	mm Hg	25	EST	NEELY,WB & BLAU,GE (1985)
DC	pKa			
HL 7.30E-017	atm m3/mol	25	EST	MEYLAN,WM & HOWARD,PH (1991)
OH 2.53E-011	cm3/molc sec	25	EST	MEYLAN,WM & HOWARD,PH (1993)

CAS #: 049648-38-6				MONOFLUOROAMPHENICOL
Formula: $C_{11}H_{13}FN_2O_5$				
Mol Weight: 272.24				
MP (deg C):		FP (deg C):		
BP (deg C):				
BP pressure (mm Hg):				

Property/Value	Units	Temp	Data Type	Reference
WS 2.64E+002	mg/L	25	EST	MEYLAN,WM ET AL. (1996)
logP 0.15			EXP	HANSCH,C & LEO,AJ (1985)
VP 1.04E-011	mm Hg	25	EST	NEELY,WB & BLAU,GE (1985)
DC	pKa			
HL 3.67E-017	atm m3/mol	25	EST	MEYLAN,WM & HOWARD,PH (1991)
OH 3.08E-011	cm3/molc sec	25	EST	MEYLAN,WM & HOWARD,PH (1993)

CAS #: 049648-42-2				TRIBROMOAMPHENICOL
Formula: $C_{11}H_{11}Br_3N_2O_5$				
Mol Weight: 490.93				
MP (deg C):		FP (deg C):		
BP (deg C):				
BP pressure (mm Hg):				

Property/Value	Units	Temp	Data Type	Reference
WS 4.68E+000	mg/L	25	EST	MEYLAN,WM ET AL. (1996)
logP 2.17			EXP	HANSCH,C & LEO,AJ (1985)
VP 1.83E-015	mm Hg	25	EST	NEELY,WB & BLAU,GE (1985)
DC	pKa			
HL 2.82E-020	atm m3/mol	25	EST	MEYLAN,WM & HOWARD,PH (1991)
OH 3.08E-011	cm3/molc sec	25	EST	MEYLAN,WM & HOWARD,PH (1993)

CAS #: 049648-47-7				PHENYLAMPHENICOL
Formula: $C_{17}H_{18}N_2O_5$				
Mol Weight: 330.34				
MP (deg C):		FP (deg C):		
BP (deg C):				
BP pressure (mm Hg):				

Property/Value	Units	Temp	Data Type	Reference
WS 1.67E+002	mg/L	25	EST	MEYLAN,WM ET AL. (1996)
logP 1.52			EXP	HANSCH,C & LEO,AJ (1985)
VP 8.18E-015	mm Hg	25	EST	NEELY,WB & BLAU,GE (1985)
DC	pKa			
HL 1.49E-018	atm m3/mol	25	EST	MEYLAN,WM & HOWARD,PH (1991)
OH 3.63E-011	cm3/molc sec	25	EST	MEYLAN,WM & HOWARD,PH (1993)

049648-48-8 — CYANO-PHENYL-AMPHENICOL

Formula: $C_{18}H_{17}N_3O_5$

Mol Weight: 355.35

MP (deg C): FP (deg C):

BP (deg C):

BP pressure (mm Hg):

Property/Value	Units	Temp	Data Type	Reference
WS 7.07E+001	mg/L	25	EST	MEYLAN,WM ET AL. (1996)
logP 1.47			EXP	HANSCH,C & LEO,AJ (1985)
VP 9.05E-017	mm Hg	25	EST	NEELY,WB & BLAU,GE (1985)
DC	pKa			
HL 6.17E-022	atm m3/mol	25	EST	MEYLAN,WM & HOWARD,PH (1991)
OH 3.57E-011	cm3/molc sec	25	EST	MEYLAN,WM & HOWARD,PH (1993)

049648-49-9 — DIETHYLAMPHENICOL

Formula: $C_{15}H_{22}N_2O_5$

Mol Weight: 310.35

MP (deg C): FP (deg C):

BP (deg C):

BP pressure (mm Hg):

Property/Value	Units	Temp	Data Type	Reference
WS 1.51E+002	mg/L	25	EST	MEYLAN,WM ET AL. (1996)
logP 1.71			EXP	HANSCH,C & LEO,AJ (1985)
VP 3.95E-013	mm Hg	25	EST	NEELY,WB & BLAU,GE (1985)
DC	pKa			
HL 5.73E-017	atm m3/mol	25	EST	MEYLAN,WM & HOWARD,PH (1991)
OH 4.06E-011	cm3/molc sec	25	EST	MEYLAN,WM & HOWARD,PH (1993)

049648-50-2 — TRIMETHYLAMPHENICOL

Formula: $C_{14}H_{20}N_2O_5$

Mol Weight: 296.33

MP (deg C): FP (deg C):

BP (deg C):

BP pressure (mm Hg):

Property/Value	Units	Temp	Data Type	Reference
WS 3.84E+002	mg/L	25	EST	MEYLAN,WM ET AL. (1996)
logP 1.33			EXP	HANSCH,C & LEO,AJ (1985)
VP 1.58E-012	mm Hg	25	EST	NEELY,WB & BLAU,GE (1985)
DC	pKa			
HL 4.31E-017	atm m3/mol	25	EST	MEYLAN,WM & HOWARD,PH (1991)
OH 3.24E-011	cm3/molc sec	25	EST	MEYLAN,WM & HOWARD,PH (1993)

049648-53-5 — P-IODOAMPHENICOL

Formula: $C_{11}H_{12}Cl_2INO_3$

Mol Weight: 404.03

MP (deg C): FP (deg C):

BP (deg C):

BP pressure (mm Hg):

Property/Value	Units	Temp	Data Type	Reference
WS 3.67E+001	mg/L	25	EST	MEYLAN,WM ET AL. (1996)
logP 2.22			EXP	HANSCH,C & LEO,AJ (1985)
VP 5.41E-013	mm Hg	25	EST	NEELY,WB & BLAU,GE (1985)
DC	pKa			
HL 1.34E-016	atm m3/mol	25	EST	MEYLAN,WM & HOWARD,PH (1991)
OH 3.19E-011	cm3/molc sec	25	EST	MEYLAN,WM & HOWARD,PH (1993)

049739-37-9 — FURAZAN, METHYL(PHENYLTHIO)-, 5-OXIDE

Formula: $C_9H_8N_2O_2S$

Mol Weight: 208.24

MP (deg C): FP (deg C):

BP (deg C):

BP pressure (mm Hg):

Property/Value	Units	Temp	Data Type	Reference
WS 1.81E+002	mg/L	25	EST	MEYLAN,WM ET AL. (1996)
logP 2.74			EXP	CALVINO,R ET AL. (1992)
VP 6.38E-006	mm Hg	25	EST	NEELY,WB & BLAU,GE (1985)
DC	pKa			
HL 1.22E-011	atm m3/mol	25	EST	MEYLAN,WM & HOWARD,PH (1991)
OH 1.67E-011	cm3/molc sec	25	EST	MEYLAN,WM & HOWARD,PH (1993)

049739-41-5 — FURAZAN, METHYL(PHENYLSULFONYL)-, 2-OXIDE

Formula: $C_9H_8N_2O_4S$

Mol Weight: 240.24

MP (deg C): FP (deg C):

BP (deg C):

BP pressure (mm Hg):

Property/Value	Units	Temp	Data Type	Reference
WS 4.94E+002	mg/L	25	EST	MEYLAN,WM ET AL. (1996)
logP 2.03			EXP	CALVINO,R ET AL. (1992)
VP 1.72E-007	mm Hg	25	EST	NEELY,WB & BLAU,GE (1985)
DC	pKa			
HL 9.66E-014	atm m3/mol	25	EST	MEYLAN,WM & HOWARD,PH (1991)
OH 4.55E-012	cm3/molc sec	25	EST	MEYLAN,WM & HOWARD,PH (1993)

049739-42-6 — FURAZAN, METHYL(PHENYLSULFONYL)-

Formula: $C_9H_8N_2O_3S$

Mol Weight: 224.24

MP (deg C): FP (deg C):

BP (deg C):

BP pressure (mm Hg):

Property/Value	Units	Temp	Data Type	Reference
WS 3.15E+002	mg/L	25	EST	MEYLAN,WM ET AL. (1996)
logP 2.36			EXP	CALVINO,R ET AL. (1992)
VP 7.58E-006	mm Hg	25	EST	NEELY,WB & BLAU,GE (1985)
DC	pKa			
HL 9.66E-009	atm m3/mol	25	EST	MEYLAN,WM & HOWARD,PH (1991)
OH 4.55E-012	cm3/molc sec	25	EST	MEYLAN,WM & HOWARD,PH (1993)

049739-43-7 — FURAZAN, METHYL(PHENYLSULFONYL)-, 5-OXIDE

Formula: $C_9H_8N_2O_4S$

Mol Weight: 240.24

MP (deg C): FP (deg C):

BP (deg C):

BP pressure (mm Hg):

Property/Value	Units	Temp	Data Type	Reference
WS 3.15E+002	mg/L	25	EST	MEYLAN,WM ET AL. (1996)
logP 2.26			EXP	CALVINO,R ET AL. (1992)
VP 1.72E-007	mm Hg	25	EST	NEELY,WB & BLAU,GE (1985)
DC	pKa			
HL 9.66E-014	atm m3/mol	25	EST	MEYLAN,WM & HOWARD,PH (1991)
OH 4.55E-012	cm3/molc sec	25	EST	MEYLAN,WM & HOWARD,PH (1993)

CAS #: 049845-48-9 — 2,4-NH2PYRIMIDIN,5(35-MEO-4-ME)BENZYL

Formula: $C_{14}H_{18}N_4O_2$

Mol Weight: 274.33

MP (deg C):　FP (deg C):

BP (deg C):

BP pressure (mm Hg):

Property/ Value	Units	Temp	Data Type	Reference
WS 2.57E+002	mg/L	25	EST	MEYLAN,WM ET AL. (1996)
logP 2.14			EXP	HANSCH,C & LEO,AJ (1985)
VP 2.28E-008	mm Hg	25	EST	NEELY,WB & BLAU,GE (1985)
DC	pKa			
HL 4.45E-013	atm m3/mol	25	EST	MEYLAN,WM & HOWARD,PH (1991)
OH 2.03E-010	cm3/molc sec	25	EST	MEYLAN,WM & HOWARD,PH (1993)

CAS #: 049871-96-7 — IMIDODICARBONIMIDIC DIAMIDE, N-(4-CHLOROPHENYL)-

Formula: $C_{11}H_{16}ClN_5$

Mol Weight: 253.74

MP (deg C):　FP (deg C):

BP (deg C):

BP pressure (mm Hg):

Property/ Value	Units	Temp	Data Type	Reference
WS 1.28E+002	mg/L	25	EST	MEYLAN,WM ET AL. (1996)
logP 2.63			EXP	WARNER,VD ET AL. (1976)
VP 4.26E-007	mm Hg	25	EST	NEELY,WB & BLAU,GE (1985)
DC	pKa			
HL 2.51E-016	atm m3/mol	25	EST	MEYLAN,WM & HOWARD,PH (1991)
OH 1.50E-010	cm3/molc sec	25	EST	MEYLAN,WM & HOWARD,PH (1993)

CAS #: 050285-70-6 — 1-NITROSOTRIETHYLUREA

Formula: $C_7H_{15}N_3O_2$

Mol Weight: 173.22

MP (deg C):　FP (deg C):

BP (deg C):

BP pressure (mm Hg):

Property/ Value	Units	Temp	Data Type	Reference
WS 2.87E+003	mg/L	25	EST	MEYLAN,WM ET AL. (1996)
logP 1.54			EXP	HANSCH,C & LEO,AJ (1985)
VP 6.46E-002	mm Hg	25	EST	NEELY,WB & BLAU,GE (1985)
DC	pKa			
HL 1.12E-009	atm m3/mol	25	EST	MEYLAN,WM & HOWARD,PH (1991)
OH 1.30E-011	cm3/molc sec	25	EST	MEYLAN,WM & HOWARD,PH (1993)

CAS #: 050285-71-7 — 1-NITROSO-1-ET-3,3-DIMETHYL-UREA

Formula: $C_5H_{11}N_3O_2$

Mol Weight: 145.16

MP (deg C):　FP (deg C):

BP (deg C):

BP pressure (mm Hg):

Property/ Value	Units	Temp	Data Type	Reference
WS 1.97E+004	mg/L	25	EST	MEYLAN,WM ET AL. (1996)
logP 0.71			EXP	HANSCH,C & LEO,AJ (1985)
VP 6.79E-001	mm Hg	25	EST	NEELY,WB & BLAU,GE (1985)
DC	pKa			
HL 6.34E-010	atm m3/mol	25	EST	MEYLAN,WM & HOWARD,PH (1991)
OH 6.11E-012	cm3/molc sec	25	EST	MEYLAN,WM & HOWARD,PH (1993)

CAS #: 050285-72-8 — 1-NITROSO-1-METHYL DIETHYLUREA

Formula: $C_6H_{13}N_3O_2$

Mol Weight: 159.19

MP (deg C):　FP (deg C):

BP (deg C):

BP pressure (mm Hg):

Property/ Value	Units	Temp	Data Type	Reference
WS 7.78E+003	mg/L	25	EST	MEYLAN,WM ET AL. (1996)
logP 1.11			EXP	HANSCH,C & LEO,AJ (1985)
VP 2.13E-001	mm Hg	25	EST	NEELY,WB & BLAU,GE (1985)
DC	pKa			
HL 8.42E-010	atm m3/mol	25	EST	MEYLAN,WM & HOWARD,PH (1991)
OH 9.55E-012	cm3/molc sec	25	EST	MEYLAN,WM & HOWARD,PH (1993)

CAS #: 050317-11-8 — 1,2-DIHYDROXY-2-BUTENE

Formula: $C_4H_8O_2$

Mol Weight: 88.11

MP (deg C):　FP (deg C):

BP (deg C):

BP pressure (mm Hg):

Property/ Value	Units	Temp	Data Type	Reference
WS 3.05E+005	mg/L	25	EST	MEYLAN,WM ET AL. (1996)
logP 0.15			EST	MEYLAN,WM & HOWARD,PH (1995)
VP 2.60E-002	mm Hg	25	EST	NEELY,WB & BLAU,GE (1985)
DC	pKa			
HL 1.61E-006	atm m3/mol	25	EST	MEYLAN,WM & HOWARD,PH (1991)
OH 9.06E-011	cm3/molc sec	25	EST	MEYLAN,WM & HOWARD,PH (1993)

CAS #: 050375-10-5 — 2,3,6-TRICHLOROANISOLE

Formula: $C_7H_5Cl_3O$

Mol Weight: 211.48

MP (deg C): 45　FP (deg C):

BP (deg C): 227

BP pressure (mm Hg):

Property/ Value	Units	Temp	Data Type	Reference
WS 2.97E+001	mg/L	25	EST	MEYLAN,WM ET AL. (1996)
logP 3.64			EXP	OPPERHUIZEN,A & VOORS,PI (1987)
VP 2.28E-002	mm Hg	25	EST	NEELY,WB & BLAU,GE (1985)
DC	pKa			
HL 2.88E-004	atm m3/mol	20	EXP	LALEZARY,S ET AL. (1984)
OH 2.22E-012	cm3/molc sec	25	EST	MEYLAN,WM & HOWARD,PH (1993)

CAS #: 050394-27-9 — PREGN-4ENE 11A-EPOXYLACTONE DERIV

Formula: $C_{22}H_{28}O_4$

Mol Weight: 356.47

MP (deg C):　FP (deg C):

BP (deg C):

BP pressure (mm Hg):

Property/ Value	Units	Temp	Data Type	Reference
WS 1.38E+002	mg/L	25	EST	MEYLAN,WM ET AL. (1996)
logP 1.89			EXP	HANSCH,C & LEO,AJ (1985)
VP 3.84E-009	mm Hg	25	EST	NEELY,WB & BLAU,GE (1985)
DC	pKa			
HL 1.65E-010	atm m3/mol	25	EST	MEYLAN,WM & HOWARD,PH (1991)
OH 9.93E-011	cm3/molc sec	25	EST	MEYLAN,WM & HOWARD,PH (1993)

1151

CAS #: 050405-18-0 — 4-PENTYLSEMICARBAZIDE

Formula: $C_6H_{15}N_3O$

Mol Weight: 145.21

MP (deg C): FP (deg C):

BP (deg C):

BP pressure (mm Hg):

Property/Value	Units	Temp	Data Type	Reference
WS 4.22E+003	mg/L	25	EST	MEYLAN,WM ET AL. (1996)
logP -0.04			EXP	HANSCH,C & LEO,AJ (1985)
VP 1.25E-003	mm Hg	25	EST	NEELY,WB & BLAU,GE (1985)
DC	pKa			
HL 1.04E-011	atm m3/mol	25	EST	MEYLAN,WM & HOWARD,PH (1991)
OH 7.98E-011	cm3/molc sec	25	EST	MEYLAN,WM & HOWARD,PH (1993)

CAS #: 050405-44-2 — 2H-PYRAN-2-ONE, 4-HYDROXY-3,5,6-TRIMETHYL-

Formula: $C_8H_{14}O_3$

Mol Weight: 158.20

MP (deg C): FP (deg C):

BP (deg C):

BP pressure (mm Hg):

Property/Value	Units	Temp	Data Type	Reference
WS 1.11E+005	mg/L	25	EST	MEYLAN,WM ET AL. (1996)
logP 0.36			EXP	SANGSTER,J (1994)
VP 2.30E-004	mm Hg	25	EST	NEELY,WB & BLAU,GE (1985)
DC	pKa			
HL 1.17E-008	atm m3/mol	25	EST	MEYLAN,WM & HOWARD,PH (1991)
OH 1.94E-011	cm3/molc sec	25	EST	MEYLAN,WM & HOWARD,PH (1993)

CAS #: 050463-48-4 — BENZENEPROPANOIC ACID, 4-(PHENYLMETHOXY)-

Formula: $C_{16}H_{16}O_3$

Mol Weight: 256.30

MP (deg C): FP (deg C):

BP (deg C):

BP pressure (mm Hg):

Property/Value	Units	Temp	Data Type	Reference
WS 8.08E+001	mg/L	25	EST	MEYLAN,WM ET AL. (1996)
logP 3.31			EXP	SANGSTER,J (1994)
VP 4.32E-007	mm Hg	25	EST	NEELY,WB & BLAU,GE (1985)
DC	pKa			
HL 2.80E-010	atm m3/mol	25	EST	MEYLAN,WM & HOWARD,PH (1991)
OH 3.93E-011	cm3/molc sec	25	EST	MEYLAN,WM & HOWARD,PH (1993)

CAS #: 050471-44-8 — VINCLOZOLIN

Formula: $C_{12}H_9Cl_2NO_3$

Mol Weight: 286.12

MP (deg C): 108 FP (deg C):

BP (deg C): 131

BP pressure (mm Hg): 5.00E-002

Property/Value	Units	Temp	Data Type	Reference
WS 1.00E+003	mg/L	20	EXP	YALKOWSKY,SH & DANNENFELSER,RM (1992)
logP 3.10			EXP	HANSCH,C ET AL. (1995)
VP 1.20E-007	mm Hg	20	EXP	AUGUSTIJN-BECKERS,PWM ET AL. (1994)
DC	pKa			
HL 4.52E-011	atm m3/mol	20	EST	VP/WSOL
OH 3.30E-011	cm3/molc sec	25	EST	MEYLAN,WM & HOWARD,PH (1993)

CAS #: 050512-35-1 — IPT (ISOPROTHIOLANE)

Formula: $C_{12}H_{18}O_4S_2$

Mol Weight: 290.40

MP (deg C): FP (deg C):

BP (deg C):

BP pressure (mm Hg):

Property/Value	Units	Temp	Data Type	Reference
WS 4.80E+001	mg/L	20	EXP	YALKOWSKY,SH & DANNENFELSER,RM (1992)
logP 2.88			EXP	SAITO,H ET AL. (1993)
VP 1.92E-004	mm Hg	25	EST	NEELY,WB & BLAU,GE (1985)
DC	pKa			
HL 2.42E-010	atm m3/mol	25	EST	MEYLAN,WM & HOWARD,PH (1991)
OH 3.81E-011	cm3/molc sec	25	EST	MEYLAN,WM & HOWARD,PH (1993)

CAS #: 050531-51-6 — 2(25-DIMEO PH IMINO)IMIDAZOLIDINE

Formula: $C_{11}H_{15}N_3O_2$

Mol Weight: 221.26

MP (deg C): FP (deg C):

BP (deg C):

BP pressure (mm Hg):

Property/Value	Units	Temp	Data Type	Reference
WS 5.03E+003	mg/L	25	EST	MEYLAN,WM ET AL. (1996)
logP 0.97			EXP	HANSCH,C & LEO,AJ (1985)
VP 1.35E-007	mm Hg	25	EST	NEELY,WB & BLAU,GE (1985)
DC	pKa			
HL 9.60E-014	atm m3/mol	25	EST	MEYLAN,WM & HOWARD,PH (1991)
OH 2.74E-010	cm3/molc sec	25	EST	MEYLAN,WM & HOWARD,PH (1993)

CAS #: 050550-65-7 — 2-PIPERIDINONE, 1-NITROSO-

Formula: $C_5H_8N_2O_2$

Mol Weight: 128.13

MP (deg C): FP (deg C):

BP (deg C):

BP pressure (mm Hg):

Property/Value	Units	Temp	Data Type	Reference
WS 3.61E+004	mg/L	25	EST	MEYLAN,WM ET AL. (1996)
logP -1.05			EXP	HANSCH,C ET AL. (1995)
VP 3.97E-002	mm Hg	25	EST	NEELY,WB & BLAU,GE (1985)
DC	pKa			
HL 2.36E-009	atm m3/mol	25	EST	MEYLAN,WM & HOWARD,PH (1991)
OH 1.15E-011	cm3/molc sec	25	EST	MEYLAN,WM & HOWARD,PH (1993)

CAS #: 050574-87-3 — MORPHOLINE, 4-[[4-(4,6-DIAMINO-2,2-DIMETHYL-1,3,

Formula: $C_{17}H_{24}N_6O_3$

Mol Weight: 360.42

MP (deg C): FP (deg C):

BP (deg C):

BP pressure (mm Hg):

Property/Value	Units	Temp	Data Type	Reference
WS 2.67E+003	mg/L	25	EST	MEYLAN,WM ET AL. (1996)
logP -1.18			EXP	HANSCH,C ET AL. (1995)
VP 8.05E-011	mm Hg	25	EST	NEELY,WB & BLAU,GE (1985)
DC	pKa			
HL 1.53E-021	atm m3/mol	25	EST	MEYLAN,WM & HOWARD,PH (1991)
OH 2.29E-010	cm3/molc sec	25	EST	MEYLAN,WM & HOWARD,PH (1993)

CAS #: 050585-37-0				2,3-DIBROMODIBENZO-P-DIOXIN

Formula: $C_{12}H_6Br_2O_2$
Mol Weight: 342.00
MP (deg C):   FP (deg C):
BP (deg C):
BP pressure (mm Hg):

Property/Value	Units	Temp	Data Type	Reference
WS 3.18E-003	mg/L	25	EST	MEYLAN,WM ET AL. (1996)
logP 6.12			EST	MEYLAN,WM & HOWARD,PH (1995)
VP 4.87E-006	mm Hg	25	EST	NEELY,WB & BLAU,GE (1985)
DC	pKa			
HL 1.86E-006	atm m3/mol	25	EST	MEYLAN,WM & HOWARD,PH (1991)
OH 5.81E-012	cm3/molc sec	25	EST	MEYLAN,WM & HOWARD,PH (1993)

CAS #: 050585-38-1				2,3-DIFLUORODIBENZO-P-DIOXIN

Formula: $C_{12}H_6F_2O_2$
Mol Weight: 220.18
MP (deg C):   FP (deg C):
BP (deg C):
BP pressure (mm Hg):

Property/Value	Units	Temp	Data Type	Reference
WS 2.38E-001	mg/L	25	EST	MEYLAN,WM ET AL. (1996)
logP 4.74			EST	MEYLAN,WM & HOWARD,PH (1995)
VP 8.39E-004	mm Hg	25	EST	NEELY,WB & BLAU,GE (1985)
DC	pKa			
HL 1.59E-005	atm m3/mol	25	EST	MEYLAN,WM & HOWARD,PH (1991)
OH 7.00E-012	cm3/molc sec	25	EST	MEYLAN,WM & HOWARD,PH (1993)

CAS #: 050585-39-2				1,3-DICHLORODIBENZO-P-DIOXIN

Formula: $C_{12}H_6Cl_2O_2$
Mol Weight: 253.09
MP (deg C):   FP (deg C):
BP (deg C):
BP pressure (mm Hg):

Property/Value	Units	Temp	Data Type	Reference
WS 2.74E-002	mg/L	25	EST	MEYLAN,WM ET AL. (1996)
logP 5.63			EST	MEYLAN,WM & HOWARD,PH (1995)
VP 2.84E-005	mm Hg	25	EST	NEELY,WB & BLAU,GE (1985)
DC	pKa			
HL 6.42E-006	atm m3/mol	25	EST	MEYLAN,WM & HOWARD,PH (1991)
OH 7.36E-012	cm3/molc sec	25	EST	MEYLAN,WM & HOWARD,PH (1993)

CAS #: 050585-40-5				2,3-DIBROMO-7,8-DICHLORODIBENZO-P-DIOXIN

Formula: $C_{12}H_4Br_2Cl_2O_2$
Mol Weight: 410.89
MP (deg C):   FP (deg C):
BP (deg C):
BP pressure (mm Hg):

Property/Value	Units	Temp	Data Type	Reference
WS 9.55E-005	mg/L	25	EST	MEYLAN,WM ET AL. (1996)
logP 7.41			EST	MEYLAN,WM & HOWARD,PH (1995)
VP 2.52E-007	mm Hg	25	EST	NEELY,WB & BLAU,GE (1985)
DC	pKa			
HL 1.02E-006	atm m3/mol	25	EST	MEYLAN,WM & HOWARD,PH (1991)
OH 1.90E-012	cm3/molc sec	25	EST	MEYLAN,WM & HOWARD,PH (1993)

CAS #: 050585-41-6				2,3,7,8-TETRABROMODIBENZO-P-DIOXIN

Formula: $C_{12}H_4Br_4O_2$
Mol Weight: 499.80
MP (deg C):   FP (deg C):
BP (deg C):
BP pressure (mm Hg):

Property/Value	Units	Temp	Data Type	Reference
WS 9.96E-006	mg/L	25	EST	MEYLAN,WM ET AL. (1996)
logP 7.90			EST	MEYLAN,WM & HOWARD,PH (1995)
VP 4.55E-008	mm Hg	25	EST	NEELY,WB & BLAU,GE (1985)
DC	pKa			
HL 2.95E-007	atm m3/mol	25	EST	MEYLAN,WM & HOWARD,PH (1991)
OH 1.78E-012	cm3/molc sec	25	EST	MEYLAN,WM & HOWARD,PH (1993)

CAS #: 050585-42-7				2,3-DICHLORO-7,8-DIFLUORODIBENZO-P-DIOXIN

Formula: $C_{12}H_4Cl_2F_2O_2$
Mol Weight: 289.07
MP (deg C):   FP (deg C):
BP (deg C):
BP pressure (mm Hg):

Property/Value	Units	Temp	Data Type	Reference
WS 7.79E-003	mg/L	25	EST	MEYLAN,WM ET AL. (1996)
logP 6.03			EST	MEYLAN,WM & HOWARD,PH (1995)
VP 3.47E-005	mm Hg	25	EST	NEELY,WB & BLAU,GE (1985)
DC	pKa			
HL 8.75E-006	atm m3/mol	25	EST	MEYLAN,WM & HOWARD,PH (1991)
OH 3.09E-012	cm3/molc sec	25	EST	MEYLAN,WM & HOWARD,PH (1993)

CAS #: 050585-43-8				2,3-DIBROMO-7,8-DIFLUORODIBENZO-P-DIOXIN

Formula: $C_{12}H_4Br_2F_2O_2$
Mol Weight: 377.98
MP (deg C):   FP (deg C):
BP (deg C):
BP pressure (mm Hg):

Property/Value	Units	Temp	Data Type	Reference
WS 8.74E-004	mg/L	25	EST	MEYLAN,WM ET AL. (1996)
logP 6.52			EST	MEYLAN,WM & HOWARD,PH (1995)
VP 6.10E-006	mm Hg	25	EST	NEELY,WB & BLAU,GE (1985)
DC	pKa			
HL 2.53E-006	atm m3/mol	25	EST	MEYLAN,WM & HOWARD,PH (1991)
OH 2.97E-012	cm3/molc sec	25	EST	MEYLAN,WM & HOWARD,PH (1993)

CAS #: 050585-46-1				1,3,7,8-TETRACHLORODIBENZO-P-DIOXIN

Formula: $C_{12}H_4Cl_4O_2$
Mol Weight: 321.98
MP (deg C):   FP (deg C):
BP (deg C):
BP pressure (mm Hg):

Property/Value	Units	Temp	Data Type	Reference
WS 6.12E-004	mg/L	25	EST	MEYLAN,WM ET AL. (1996)
logP 7.10			EXP	SHIU,WY ET AL. (1988)
VP 1.75E-006	mm Hg	25	EST	NEELY,WB & BLAU,GE (1985)
DC	pKa			
HL 3.53E-006	atm m3/mol	25	EST	MEYLAN,WM & HOWARD,PH (1991)
OH 3.45E-012	cm3/molc sec	25	EST	MEYLAN,WM & HOWARD,PH (1993)

CAS #: 050585-76-7	FLUOROMETHANESULFONANILIDE,P-CL

Formula: $C_7H_7ClFNO_2S$

Mol Weight: 223.65

MP (deg C):　　FP (deg C):

BP (deg C):

BP pressure (mm Hg):

Property/Value	Units	Temp	Data Type	Reference
WS 3.94E+002	mg/L	25	EST	MEYLAN,WM ET AL. (1996)
logP 2.25			EXP	HANSCH,C & LEO,AJ (1985)
VP 1.35E-004	mm Hg	25	EST	NEELY,WB & BLAU,GE (1985)
DC	pKa			
HL 1.72E-006	atm m3/mol	25	EST	MEYLAN,WM & HOWARD,PH (1991)
OH 1.31E-011	cm3/molc sec	25	EST	MEYLAN,WM & HOWARD,PH (1993)

CAS #: 050585-77-8	CF3-SULFONANILIDE, P-PHENYL

Formula: $C_{13}H_{10}F_3NO_2S$

Mol Weight: 301.29

MP (deg C):　　FP (deg C):

BP (deg C):

BP pressure (mm Hg):

Property/Value	Units	Temp	Data Type	Reference
WS 6.88E-001	mg/L	25	EST	MEYLAN,WM ET AL. (1996)
logP 4.97			EXP	HANSCH,C & LEO,AJ (1985)
VP 7.83E-007	mm Hg	25	EST	NEELY,WB & BLAU,GE (1985)
DC	pKa			
HL 7.05E-007	atm m3/mol	25	EST	MEYLAN,WM & HOWARD,PH (1991)
OH 3.39E-011	cm3/molc sec	25	EST	MEYLAN,WM & HOWARD,PH (1993)

CAS #: 050590-05-1	3,5-DIMETHYL-DIMETHYLPARATHION

Formula: $C_{10}H_{14}NO_5PS$

Mol Weight: 291.26

MP (deg C):　　FP (deg C):

BP (deg C):

BP pressure (mm Hg):

Property/Value	Units	Temp	Data Type	Reference
WS 2.68E+000	mg/L	25	EST	MEYLAN,WM ET AL. (1996)
logP 3.89			EXP	HANSCH,C & LEO,AJ (1985)
VP 9.74E-006	mm Hg	25	EST	NEELY,WB & BLAU,GE (1985)
DC	pKa			
HL 2.05E-007	atm m3/mol	25	EST	MEYLAN,WM & HOWARD,PH (1991)
OH 6.19E-011	cm3/molc sec	25	EST	MEYLAN,WM & HOWARD,PH (1993)

CAS #: 050590-06-2	3,5-DIMETHYL-DIMETHYLPARA-OXON

Formula: $C_{10}H_{14}NO_6P$

Mol Weight: 275.20

MP (deg C):　　FP (deg C):

BP (deg C):

BP pressure (mm Hg):

Property/Value	Units	Temp	Data Type	Reference
WS 4.44E+001	mg/L	25	EST	MEYLAN,WM ET AL. (1996)
logP 2.57			EXP	HANSCH,C & LEO,AJ (1985)
VP 9.39E-006	mm Hg	25	EST	NEELY,WB & BLAU,GE (1985)
DC	pKa			
HL 4.43E-010	atm m3/mol	25	EST	MEYLAN,WM & HOWARD,PH (1991)
OH 8.93E-012	cm3/molc sec	25	EST	MEYLAN,WM & HOWARD,PH (1993)

CAS #: 050594-66-6	BENZOIC ACID, 5- 2-CHLORO-4-(TRIFLUOROMETHYL)PHE

Formula: $C_{14}H_7ClF_3NO_5$

Mol Weight: 361.66

MP (deg C): 150　　FP (deg C):

BP (deg C):

BP pressure (mm Hg):

Property/Value	Units	Temp	Data Type	Reference
WS 1.20E+002	mg/L	25	EXP	TOMLIN,C (1994)
logP 3.70			EXP	NANDIHALLI,UB ET AL. (1992)
VP 1.53E-008	mm Hg	25	EST	NEELY,WB & BLAU,GE (1985)
DC	pKa			
HL 6.03E-011	atm m3/mol	25	EST	MEYLAN,WM & HOWARD,PH (1991)
OH 9.77E-013	cm3/molc sec	25	EST	MEYLAN,WM & HOWARD,PH (1993)

CAS #: 050594-67-7	ACIFLUORFEN-ME

Formula: $C_{15}H_9ClF_3NO_5$

Mol Weight: 375.69

MP (deg C):　　FP (deg C):

BP (deg C):

BP pressure (mm Hg):

Property/Value	Units	Temp	Data Type	Reference
WS 4.05E-001	mg/L	25	EST	MEYLAN,WM ET AL. (1996)
logP 4.26			EXP	NANDIHALLI,UB ET AL. (1993)
VP 7.50E-008	mm Hg	25	EXP	TOMLIN,C (1994)
DC	pKa			
HL 1.93E-008	atm m3/mol	25	EST	MEYLAN,WM & HOWARD,PH (1991)
OH 6.42E-013	cm3/molc sec	25	EST	MEYLAN,WM & HOWARD,PH (1993)

CAS #: 050615-69-5	beta-D-GALACTOPYRANOSIDE, HEPTYL 1-THIO-

Formula: $C_{13}H_{26}O_5S$

Mol Weight: 294.41

MP (deg C):　　FP (deg C):

BP (deg C):

BP pressure (mm Hg):

Property/Value	Units	Temp	Data Type	Reference
WS 6.93E+002	mg/L	25	EST	MEYLAN,WM ET AL. (1996)
logP 1.50			EXP	SANGSTER,J (1994)
VP 1.60E-011	mm Hg	25	EST	NEELY,WB & BLAU,GE (1985)
DC	pKa			
HL 1.55E-014	atm m3/mol	25	EST	MEYLAN,WM & HOWARD,PH (1991)
OH 1.84E-010	cm3/molc sec	25	EST	MEYLAN,WM & HOWARD,PH (1993)

CAS #: 050615-70-8	1(3-PHENPROPYL)-B-GALACTOPYRANOSIDE

Formula: $C_{15}H_{22}O_5S$

Mol Weight: 314.40

MP (deg C):　　FP (deg C):

BP (deg C):

BP pressure (mm Hg):

Property/Value	Units	Temp	Data Type	Reference
WS 4.79E+003	mg/L	25	EST	MEYLAN,WM ET AL. (1996)
logP 0.38			EXP	HANSCH,C & LEO,AJ (1985)
VP 4.51E-013	mm Hg	25	EST	NEELY,WB & BLAU,GE (1985)
DC	pKa			
HL 4.03E-016	atm m3/mol	25	EST	MEYLAN,WM & HOWARD,PH (1991)
OH 1.85E-010	cm3/molc sec	25	EST	MEYLAN,WM & HOWARD,PH (1993)

CAS #:	050618-71-8		BENZENESULFONAMIDE, N-[(DIETHYLAMINO)CARBONYL]-
Formula:	$C_{11}H_{16}N_2O_3S$		
Mol Weight:	256.33		
MP (deg C):		FP (deg C):	
BP (deg C):			
BP pressure (mm Hg):			

Property/ Value	Units	Temp	Data Type	Reference
WS 3.11E+003	mg/L	25	EST	MEYLAN,WM ET AL. (1996)
logP 0.99			EXP	CLOUX,JL ET AL. (1988)
VP 7.92E-008	mm Hg	25	EST	NEELY,WB & BLAU,GE (1985)
DC	pKa			
HL 3.79E-010	atm m3/mol	25	EST	MEYLAN,WM & HOWARD,PH (1991)
OH 1.04E-011	cm3/molc sec	25	EST	MEYLAN,WM & HOWARD,PH (1993)

CAS #:	050623-57-9		BUTYL NONANOATE
Formula:	$C_{13}H_{26}O_2$		
Mol Weight:	214.35		
MP (deg C):	-38	FP (deg C):	
BP (deg C):	122-124		
BP pressure (mm Hg):	2.00E+001		

Property/ Value	Units	Temp	Data Type	Reference
WS 1.13E+000	mg/L	25	EST	MEYLAN,WM ET AL. (1996)
logP 5.28			EST	MEYLAN,WM & HOWARD,PH (1995)
VP 2.14E-002	mm Hg	25	EXP	DAUBERT,TE & DANNER,RP (1989)
DC	pKa			
HL 2.98E-003	atm m3/mol	25	EST	MEYLAN,WM & HOWARD,PH (1991)
OH 1.33E-011	cm3/molc sec	25	EST	MEYLAN,WM & HOWARD,PH (1993)

CAS #:	050629-82-8		2-CHLOROFLUMETHASONE
Formula:	$C_{22}H_{27}ClF_2O_5$		
Mol Weight:	444.91		
MP (deg C):	220-222 de	FP (deg C):	
BP (deg C):			
BP pressure (mm Hg):			

Property/ Value	Units	Temp	Data Type	Reference
WS 1.01E+001	mg/L	25	EST	MEYLAN,WM ET AL. (1996)
logP 2.58			EXP	HANSCH,C & LEO,AJ (1985)
VP 2.50E-014	mm Hg	25	EST	NEELY,WB & BLAU,GE (1985)
DC	pKa			
HL 1.02E-007	atm m3/mol	25	EST	MEYLAN,WM & HOWARD,PH (1991)
OH 4.28E-011	cm3/molc sec	25	EST	MEYLAN,WM & HOWARD,PH (1993)

CAS #:	050700-61-3		N-METHYL-HYDROXYPYRID-4-ONE
Formula:	$C_6H_7NO_2$		
Mol Weight:	125.13		
MP (deg C):		FP (deg C):	
BP (deg C):			
BP pressure (mm Hg):			

Property/ Value	Units	Temp	Data Type	Reference
WS 3.80E+005	mg/L	25	EST	MEYLAN,WM ET AL. (1996)
logP -0.70			EXP	GYPARAKI,M ET AL. (1986)
VP 4.61E-004	mm Hg	25	EST	NEELY,WB & BLAU,GE (1985)
DC	pKa			
HL 1.23E-008	atm m3/mol	25	EST	MEYLAN,WM & HOWARD,PH (1991)
OH 9.24E-011	cm3/molc sec	25	EST	MEYLAN,WM & HOWARD,PH (1993)

CAS #:	050702-39-1		BENZENESULFONAMIDE, 3-IODO-
Formula:	$C_6H_6INO_2S$		
Mol Weight:	283.09		
MP (deg C):		FP (deg C):	
BP (deg C):			
BP pressure (mm Hg):			

Property/ Value	Units	Temp	Data Type	Reference
WS 6.88E+002	mg/L	25	EST	MEYLAN,WM ET AL. (1996)
logP 1.58			EXP	CAROTTI,A ET AL. (1989)
VP 1.61E-005	mm Hg	25	EST	NEELY,WB & BLAU,GE (1985)
DC	pKa			
HL 9.79E-008	atm m3/mol	25	EST	MEYLAN,WM & HOWARD,PH (1991)
OH 2.75E-013	cm3/molc sec	25	EST	MEYLAN,WM & HOWARD,PH (1993)

CAS #:	050728-06-8		O=P(OET)(OET-S-(2-THIOET)ET
Formula:	$C_8H_{19}O_2PS_2$		
Mol Weight:	242.34		
MP (deg C):		FP (deg C):	
BP (deg C):			
BP pressure (mm Hg):			

Property/ Value	Units	Temp	Data Type	Reference
WS 4.82E+002	mg/L	25	EST	MEYLAN,WM ET AL. (1996)
logP 2.03			EXP	HANSCH,C & LEO,AJ (1985)
VP 4.08E-004	mm Hg	25	EST	NEELY,WB & BLAU,GE (1985)
DC	pKa			
HL 2.27E-008	atm m3/mol	25	EST	MEYLAN,WM & HOWARD,PH (1991)
OH 6.21E-011	cm3/molc sec	25	EST	MEYLAN,WM & HOWARD,PH (1993)

CAS #:	050780-68-2		PROPANEDIOIC ACID, 1,3-DITHIOLAN-2-YLIDENE-, DIP
Formula:	$C_{12}H_{18}O_4S_2$		
Mol Weight:	290.40		
MP (deg C):		FP (deg C):	
BP (deg C):			
BP pressure (mm Hg):			

Property/ Value	Units	Temp	Data Type	Reference
WS 1.92E+001	mg/L	25	EST	MEYLAN,WM ET AL. (1996)
logP 3.35			EXP	SANGSTER,J (1993)
VP 4.55E-005	mm Hg	25	EST	NEELY,WB & BLAU,GE (1985)
DC	pKa			
HL 2.42E-010	atm m3/mol	25	EST	MEYLAN,WM & HOWARD,PH (1991)
OH 3.75E-011	cm3/molc sec	25	EST	MEYLAN,WM & HOWARD,PH (1993)

CAS #:	050780-71-7		PROPANEDIOIC ACID, 1,3-DITHIOLAN-2-YLIDENE-, BIS
Formula:	$C_{14}H_{22}O_4S_2$		
Mol Weight:	318.46		
MP (deg C):		FP (deg C):	
BP (deg C):			
BP pressure (mm Hg):			

Property/ Value	Units	Temp	Data Type	Reference
WS 1.30E+000	mg/L	25	EST	MEYLAN,WM ET AL. (1996)
logP 4.53			EXP	HANSCH,C ET AL. (1995)
VP 3.05E-005	mm Hg	25	EST	NEELY,WB & BLAU,GE (1985)
DC	pKa			
HL 4.27E-010	atm m3/mol	25	EST	MEYLAN,WM & HOWARD,PH (1991)
OH 3.63E-011	cm3/molc sec	25	EST	MEYLAN,WM & HOWARD,PH (1993)

CAS #: 050780-72-8			PROPANEDIOIC ACID, 1,3-DITHIOLAN-2-YLIDENE-, BIS
Formula: $C_{14}H_{22}O_4S_2$			
Mol Weight: 318.46			
MP (deg C):		FP (deg C):	
BP (deg C):			
BP pressure (mm Hg):			

Property/Value	Units	Temp	Data Type	Reference
WS 3.60E+000	mg/L	25	EST	MEYLAN,WM ET AL. (1996)
logP 4.01			EXP	SANGSTER,J (1994)
VP 3.29E-005	mm Hg	25	EST	NEELY,WB & BLAU,GE (1985)
DC	pKa			
HL 4.27E-010	atm m3/mol	25	EST	MEYLAN,WM & HOWARD,PH (1991)
OH 3.22E-011	cm3/molc sec	25	EST	MEYLAN,WM & HOWARD,PH (1993)

CAS #: 050780-76-2			PROPANEDIOIC ACID, 1,3-DITHIOLAN-2-YLIDENE-, ETH
Formula: $C_{11}H_{16}O_4S_2$			
Mol Weight: 276.38			
MP (deg C):		FP (deg C):	
BP (deg C):			
BP pressure (mm Hg):			

Property/Value	Units	Temp	Data Type	Reference
WS 5.61E+001	mg/L	25	EST	MEYLAN,WM ET AL. (1996)
logP 2.90			EXP	SANGSTER,J (1993)
VP 2.41E-004	mm Hg	25	EST	NEELY,WB & BLAU,GE (1985)
DC	pKa			
HL 1.82E-010	atm m3/mol	25	EST	MEYLAN,WM & HOWARD,PH (1991)
OH 3.63E-011	cm3/molc sec	25	EST	MEYLAN,WM & HOWARD,PH (1993)

CAS #: 050782-69-9			VX
Formula: $C_{11}H_{26}NO_2PS$			
Mol Weight: 267.37			
MP (deg C):		FP (deg C):	
BP (deg C): 298			
BP pressure (mm Hg):			

Property/Value	Units	Temp	Data Type	Reference
WS 3.00E+004	mg/L	25	EXP	CRABTREE,EV & SARVER,EW (1977)
logP 2.06			EST	MEYLAN,WM & HOWARD,PH (1995)
VP 7.00E-014	mm Hg	25	EXP	LEGGETT,DC (1987)
DC 9.12	pKa	25	EXP	DEMEK,MM ET AL. (1970)
HL 8.20E-009	atm m3/mol	25	EST	VP/WSOL
OH 1.49E-010	cm3/molc sec	25	EST	ATKINSON,R (1988)

CAS #: 050785-22-3			BENZOIC ACID, 2-(ACETYLOXY)-, 2-AMINO-2-OXOETHYL
Formula: $C_{11}H_{11}NO_5$			
Mol Weight: 237.21			
MP (deg C):		FP (deg C):	
BP (deg C):			
BP pressure (mm Hg):			

Property/Value	Units	Temp	Data Type	Reference
WS 1.80E+004	mg/L	25	EST	MEYLAN,WM ET AL. (1996)
logP 0.22			EXP	NIELSEN,LS & BUNDGAARD,H (1989)
VP 2.34E-006	mm Hg	25	EST	NEELY,WB & BLAU,GE (1985)
DC	pKa			
HL 4.49E-013	atm m3/mol	25	EST	MEYLAN,WM & HOWARD,PH (1991)
OH 3.87E-012	cm3/molc sec	25	EST	MEYLAN,WM & HOWARD,PH (1993)

CAS #: 050785-24-5			BENZOIC ACID, 2-(ACETYLOXY)-, 2-ETHOXY-2-OXOETHY
Formula: $C_{13}H_{14}O_6$			
Mol Weight: 266.25			
MP (deg C):		FP (deg C):	
BP (deg C):			
BP pressure (mm Hg):			

Property/Value	Units	Temp	Data Type	Reference
WS 7.19E+002	mg/L	25	EST	MEYLAN,WM ET AL. (1996)
logP 1.67			EXP	NIELSEN,LS & BUNDGAARD,H (1989)
VP 4.22E-004	mm Hg	25	EST	NEELY,WB & BLAU,GE (1985)
DC	pKa			
HL 9.34E-009	atm m3/mol	25	EST	MEYLAN,WM & HOWARD,PH (1991)
OH 3.52E-012	cm3/molc sec	25	EST	MEYLAN,WM & HOWARD,PH (1993)

CAS #: 050789-46-3			BENZAMIDE, N-METHYL-3-PHENOXY-
Formula: $C_{14}H_{13}NO_2$			
Mol Weight: 227.27			
MP (deg C):		FP (deg C):	
BP (deg C):			
BP pressure (mm Hg):			

Property/Value	Units	Temp	Data Type	Reference
WS 1.14E+002	mg/L	25	EST	MEYLAN,WM ET AL. (1996)
logP 2.86			EXP	SANGSTER,J (1994)
VP 3.74E-007	mm Hg	25	EST	NEELY,WB & BLAU,GE (1985)
DC	pKa			
HL 1.06E-010	atm m3/mol	25	EST	MEYLAN,WM & HOWARD,PH (1991)
OH 1.31E-011	cm3/molc sec	25	EST	MEYLAN,WM & HOWARD,PH (1993)

CAS #: 050816-31-4			N,N-DIETHYLUREA
Formula: $C_5H_{12}N_2O$			
Mol Weight: 116.16			
MP (deg C):		FP (deg C):	
BP (deg C):			
BP pressure (mm Hg):			

Property/Value	Units	Temp	Data Type	Reference
WS 4.14E+003	mg/L	25	EST	MEYLAN,WM ET AL. (1996)
logP 0.10			EST	MEYLAN,WM & HOWARD,PH (1995)
VP 3.79E-002	mm Hg	25	EST	NEELY,WB & BLAU,GE (1985)
DC	pKa			
HL 3.10E-009	atm m3/mol	25	EST	MEYLAN,WM & HOWARD,PH (1991)
OH 9.99E-012	cm3/molc sec	25	EST	MEYLAN,WM & HOWARD,PH (1993)

CAS #: 050832-71-8			3(5NO2-2-FURFURILIDENE)-124-TRIAZOLE
Formula: $C_8H_7N_4O_3$			
Mol Weight: 207.17			
MP (deg C):		FP (deg C):	
BP (deg C):			
BP pressure (mm Hg):			

Property/Value	Units	Temp	Data Type	Reference
WS 1.91E+003	mg/L	25	EST	MEYLAN,WM ET AL. (1996)
logP 1.10			EXP	HANSCH,C & LEO,AJ (1985)
VP 1.25E-006	mm Hg	25	EST	NEELY,WB & BLAU,GE (1985)
DC	pKa			
HL 1.38E-010	atm m3/mol	25	EST	MEYLAN,WM & HOWARD,PH (1991)
OH 6.13E-011	cm3/molc sec	25	EST	MEYLAN,WM & HOWARD,PH (1993)

050846-57-6 — 1-B-AMINOACETYLET-2ME-5-NO2IMIDAZOLE

Formula: $C_8H_{14}N_4O_3$

Mol Weight: 214.23

MP (deg C): FP (deg C):

BP (deg C):

BP pressure (mm Hg):

Property/Value	Units	Temp	Data Type	Reference
WS 1.90E+003	mg/L	25	EST	MEYLAN,WM ET AL. (1996)
logP -0.47			EXP	HANSCH,C & LEO,AJ (1985)
VP 2.99E-008	mm Hg	25	EST	NEELY,WB & BLAU,GE (1985)
DC	pKa			
HL 1.54E-013	atm m3/mol	25	EST	MEYLAN,WM & HOWARD,PH (1991)
OH 1.55E-011	cm3/molc sec	25	EST	MEYLAN,WM & HOWARD,PH (1993)

050868-74-1 — 2-METHYL-3-METHOXYACETANILIDE

Formula: $C_{10}H_{13}NO_2$

Mol Weight: 179.22

MP (deg C): FP (deg C):

BP (deg C):

BP pressure (mm Hg):

Property/Value	Units	Temp	Data Type	Reference
WS 6.02E+003	mg/L	25	EST	MEYLAN,WM ET AL. (1996)
logP 1.13			EXP	NAKAGAWA,Y ET AL. (1992)
VP 2.56E-005	mm Hg	25	EST	NEELY,WB & BLAU,GE (1985)
DC	pKa			
HL 4.03E-010	atm m3/mol	25	EST	MEYLAN,WM & HOWARD,PH (1991)
OH 1.69E-010	cm3/molc sec	25	EST	MEYLAN,WM & HOWARD,PH (1993)

050916-16-0 — BENZENEACETAMIDE, N-(3-METHYLPHENYL)-

Formula: $C_{15}H_{15}NO$

Mol Weight: 225.29

MP (deg C): FP (deg C):

BP (deg C):

BP pressure (mm Hg):

Property/Value	Units	Temp	Data Type	Reference
WS 6.72E+001	mg/L	25	EST	MEYLAN,WM ET AL. (1996)
logP 3.14			EXP	YAMAGAMI,C ET AL. (1984)
VP 4.08E-007	mm Hg	25	EST	NEELY,WB & BLAU,GE (1985)
DC	pKa			
HL 5.50E-010	atm m3/mol	25	EST	MEYLAN,WM & HOWARD,PH (1991)
OH 3.81E-011	cm3/molc sec	25	EST	MEYLAN,WM & HOWARD,PH (1993)

050916-19-3 — BENZENEACETAMIDE, N-(3-METHOXYPHENYL)-

Formula: $C_{15}H_{15}NO_2$

Mol Weight: 241.29

MP (deg C): FP (deg C):

BP (deg C):

BP pressure (mm Hg):

Property/Value	Units	Temp	Data Type	Reference
WS 9.54E+001	mg/L	25	EST	MEYLAN,WM ET AL. (1996)
logP 2.86			EXP	YAMAGAMI,C ET AL. (1984)
VP 1.60E-007	mm Hg	25	EST	NEELY,WB & BLAU,GE (1985)
DC	pKa			
HL 2.95E-011	atm m3/mol	25	EST	MEYLAN,WM & HOWARD,PH (1991)
OH 1.43E-010	cm3/molc sec	25	EST	MEYLAN,WM & HOWARD,PH (1993)

050916-21-7 — BENZENEACETAMIDE, N-(4-METHOXYPHENYL)-

Formula: $C_{15}H_{15}NO_2$

Mol Weight: 241.29

MP (deg C): FP (deg C):

BP (deg C):

BP pressure (mm Hg):

Property/Value	Units	Temp	Data Type	Reference
WS 1.98E+002	mg/L	25	EST	MEYLAN,WM ET AL. (1996)
logP 2.49			EXP	YAMAGAMI,C ET AL. (1984)
VP 1.60E-007	mm Hg	25	EST	NEELY,WB & BLAU,GE (1985)
DC	pKa			
HL 2.95E-011	atm m3/mol	25	EST	MEYLAN,WM & HOWARD,PH (1991)
OH 1.76E-011	cm3/molc sec	25	EST	MEYLAN,WM & HOWARD,PH (1993)

050917-19-6 — 3-MORPHOL-4-NH2-6-IPR-124-TRIAZINE-5-ONE

Formula: $C_{10}H_{17}N_5O_2$

Mol Weight: 239.28

MP (deg C): FP (deg C):

BP (deg C):

BP pressure (mm Hg):

Property/Value	Units	Temp	Data Type	Reference
WS 1.03E+004	mg/L	25	EST	MEYLAN,WM ET AL. (1996)
logP 0.49			EXP	HANSCH,C & LEO,AJ (1985)
VP 3.33E-007	mm Hg	25	EST	NEELY,WB & BLAU,GE (1985)
DC	pKa			
HL 4.34E-016	atm m3/mol	25	EST	MEYLAN,WM & HOWARD,PH (1991)
OH 1.64E-010	cm3/molc sec	25	EST	MEYLAN,WM & HOWARD,PH (1993)

050917-22-1 — 3-ETS-4-NH2-6-IPR-1,2,4-TRIAZIN-5-ON

Formula: $C_8H_{14}N_4OS$

Mol Weight: 214.29

MP (deg C): FP (deg C):

BP (deg C):

BP pressure (mm Hg):

Property/Value	Units	Temp	Data Type	Reference
WS 1.86E+003	mg/L	25	EST	MEYLAN,WM ET AL. (1996)
logP 1.52			EXP	HANSCH,C & LEO,AJ (1985)
VP 2.04E-006	mm Hg	25	EST	NEELY,WB & BLAU,GE (1985)
DC	pKa			
HL 1.81E-012	atm m3/mol	25	EST	MEYLAN,WM & HOWARD,PH (1991)
OH 3.21E-011	cm3/molc sec	25	EST	MEYLAN,WM & HOWARD,PH (1993)

050917-23-2 — 3-PRS-4-NH2-6-IPR-124-TRIAZINE-5-ONE

Formula: $C_9H_{16}N_4OS$

Mol Weight: 228.32

MP (deg C): FP (deg C):

BP (deg C):

BP pressure (mm Hg):

Property/Value	Units	Temp	Data Type	Reference
WS 4.81E+002	mg/L	25	EST	MEYLAN,WM ET AL. (1996)
logP 2.12			EXP	HANSCH,C & LEO,AJ (1985)
VP 9.34E-007	mm Hg	25	EST	NEELY,WB & BLAU,GE (1985)
DC	pKa			
HL 2.40E-012	atm m3/mol	25	EST	MEYLAN,WM & HOWARD,PH (1991)
OH 3.49E-011	cm3/molc sec	25	EST	MEYLAN,WM & HOWARD,PH (1993)

CAS #: 050917-24-3	3-IPRS-4-NH2-6-IPR-124-TRIAZINE-5-ONE

Formula: $C_9H_{16}N_4OS$

Mol Weight: 228.32

MP (deg C): FP (deg C):

BP (deg C):

BP pressure (mm Hg):

Property/Value	Units	Temp	Data Type	Reference
WS 5.41E+002	mg/L	25	EST	MEYLAN,WM ET AL. (1996)
logP 2.06			EXP	HANSCH,C & LEO,AJ (1985)
VP 1.52E-006	mm Hg	25	EST	NEELY,WB & BLAU,GE (1985)
DC	pKa			
HL 2.40E-012	atm m3/mol	25	EST	MEYLAN,WM & HOWARD,PH (1991)
OH 4.01E-011	cm3/molc sec	25	EST	MEYLAN,WM & HOWARD,PH (1993)

CAS #: 050917-25-4	3-BUS-4-NH2-6-IPR-124-TRIAZINE-5-ONE

Formula: $C_{10}H_{18}N_4OS$

Mol Weight: 242.35

MP (deg C): FP (deg C):

BP (deg C):

BP pressure (mm Hg):

Property/Value	Units	Temp	Data Type	Reference
WS 4.73E+001	mg/L	25	EST	MEYLAN,WM ET AL. (1996)
logP 3.21			EXP	SANGSTER,J (1994)
VP 4.25E-007	mm Hg	25	EST	NEELY,WB & BLAU,GE (1985)
DC	pKa			
HL 3.19E-012	atm m3/mol	25	EST	MEYLAN,WM & HOWARD,PH (1991)
OH 3.63E-011	cm3/molc sec	25	EST	MEYLAN,WM & HOWARD,PH (1993)

CAS #: 051008-91-4	ACETYLOLEANDRIN

Formula: $C_{34}H_{50}O_{10}$

Mol Weight: 618.77

MP (deg C): FP (deg C):

BP (deg C):

BP pressure (mm Hg):

Property/Value	Units	Temp	Data Type	Reference
WS 1.19E+000	mg/L	25	EST	MEYLAN,WM ET AL. (1996)
logP 2.95			EXP	HANSCH,C & LEO,AJ (1985)
VP 3.30E-018	mm Hg	25	EST	NEELY,WB & BLAU,GE (1985)
DC	pKa			
HL 1.34E-019	atm m3/mol	25	EST	MEYLAN,WM & HOWARD,PH (1991)
OH 1.01E-010	cm3/molc sec	25	EST	MEYLAN,WM & HOWARD,PH (1993)

CAS #: 051012-32-9	BENZAMIDE,N(DIETAMET),2-MEO,5-SO2ME

Formula: $C_{15}H_{24}N_2O_4S$

Mol Weight: 328.43

MP (deg C): 123-125 FP (deg C):

BP (deg C):

BP pressure (mm Hg):

Property/Value	Units	Temp	Data Type	Reference
WS 1.42E+003	mg/L	25	EST	MEYLAN,WM ET AL. (1996)
logP 0.90			EXP	HANSCH,C & LEO,AJ (1985)
VP 4.66E-010	mm Hg	25	EST	NEELY,WB & BLAU,GE (1985)
DC	pKa			
HL 7.73E-017	atm m3/mol	25	EST	MEYLAN,WM & HOWARD,PH (1991)
OH 1.11E-010	cm3/molc sec	25	EST	MEYLAN,WM & HOWARD,PH (1993)

CAS #: 051022-76-5	CARBAMOTHIOIC ACID, [2-(2-ETHYL-5-NITRO-1H-IMIDA

Formula: $C_9H_{14}N_4O_3S$

Mol Weight: 258.30

MP (deg C): FP (deg C):

BP (deg C):

BP pressure (mm Hg):

Property/Value	Units	Temp	Data Type	Reference
WS 4.91E+002	mg/L	25	EST	MEYLAN,WM ET AL. (1996)
logP 1.46			EXP	SANGSTER,J (1994)
VP 6.66E-008	mm Hg	25	EST	NEELY,WB & BLAU,GE (1985)
DC	pKa			
HL 2.08E-011	atm m3/mol	25	EST	MEYLAN,WM & HOWARD,PH (1991)
OH 8.06E-011	cm3/molc sec	25	EST	MEYLAN,WM & HOWARD,PH (1993)

CAS #: 051048-00-1	2H-1-BENZOPYRAN-2-CARBOXYLIC ACID, 3,4-DIHYDRO-4

Formula: $C_{10}H_8O_4$

Mol Weight: 192.17

MP (deg C): FP (deg C):

BP (deg C):

BP pressure (mm Hg):

Property/Value	Units	Temp	Data Type	Reference
WS 9.99E+003	mg/L	25	EST	MEYLAN,WM ET AL. (1996)
logP 1.26			EXP	HANSCH,C ET AL. (1995)
VP 9.45E-006	mm Hg	25	EST	NEELY,WB & BLAU,GE (1985)
DC	pKa			
HL 1.35E-011	atm m3/mol	25	EST	MEYLAN,WM & HOWARD,PH (1991)
OH 5.56E-011	cm3/molc sec	25	EST	MEYLAN,WM & HOWARD,PH (1993)

CAS #: 051055-31-3	N-PROPYLNICOTINAMIDE

Formula: $C_9H_{12}N_2O$

Mol Weight: 164.21

MP (deg C): FP (deg C):

BP (deg C):

BP pressure (mm Hg):

Property/Value	Units	Temp	Data Type	Reference
WS 1.79E+004	mg/L	25	EST	MEYLAN,WM ET AL. (1996)
logP 0.66			EXP	HANSCH,C & LEO,AJ (1985)
VP 3.70E-005	mm Hg	25	EST	NEELY,WB & BLAU,GE (1985)
DC	pKa			
HL 1.12E-011	atm m3/mol	25	EST	MEYLAN,WM & HOWARD,PH (1991)
OH 1.19E-011	cm3/molc sec	25	EST	MEYLAN,WM & HOWARD,PH (1993)

CAS #: 051074-95-4	P-ACETOXYBENZAMIDE

Formula: $C_9H_9NO_3$

Mol Weight: 179.18

MP (deg C): FP (deg C):

BP (deg C):

BP pressure (mm Hg):

Property/Value	Units	Temp	Data Type	Reference
WS 3.27E+004	mg/L	25	EST	MEYLAN,WM ET AL. (1996)
logP 0.27			EXP	HANSCH,C & LEO,AJ (1985)
VP 1.52E-005	mm Hg	25	EST	NEELY,WB & BLAU,GE (1985)
DC	pKa			
HL 2.66E-011	atm m3/mol	25	EST	MEYLAN,WM & HOWARD,PH (1991)
OH 3.93E-012	cm3/molc sec	25	EST	MEYLAN,WM & HOWARD,PH (1993)

CAS #: 051114-25-1 — ET-N-CL-ACETYL-N-(2CLPH)GLYCINATE

Formula: $C_{12}H_{13}Cl_2NO_3$

Mol Weight: 290.15

MP (deg C): | FP (deg C):

BP (deg C):

BP pressure (mm Hg):

Property/ Value	Units	Temp	Data Type	Reference
WS 1.52E+002	mg/L	25	EST	MEYLAN,WM ET AL. (1996)
logP 2.30			EXP	HANSCH,C & LEO,AJ (1985)
VP 1.24E-006	mm Hg	25	EST	NEELY,WB & BLAU,GE (1985)
DC	pKa			
HL 2.08E-009	atm m3/mol	25	EST	MEYLAN,WM & HOWARD,PH (1991)
OH 8.39E-012	cm3/molc sec	25	EST	MEYLAN,WM & HOWARD,PH (1993)

CAS #: 051114-26-2 — ET-N-CL-ACETYL-N-(PHENYL)GLYCINATE

Formula: $C_{12}H_{14}ClNO_3$

Mol Weight: 255.70

MP (deg C): | FP (deg C):

BP (deg C):

BP pressure (mm Hg):

Property/ Value	Units	Temp	Data Type	Reference
WS 9.46E+002	mg/L	25	EST	MEYLAN,WM ET AL. (1996)
logP 1.60			EXP	HANSCH,C & LEO,AJ (1985)
VP 4.92E-006	mm Hg	25	EST	NEELY,WB & BLAU,GE (1985)
DC	pKa			
HL 2.81E-009	atm m3/mol	25	EST	MEYLAN,WM & HOWARD,PH (1991)
OH 1.72E-011	cm3/molc sec	25	EST	MEYLAN,WM & HOWARD,PH (1993)

CAS #: 051170-56-0 — CARBAMIC ACID, DIETHYL-, PHENYLMETHYL ESTER

Formula: $C_{12}H_{17}NO_2$

Mol Weight: 207.27

MP (deg C): | FP (deg C):

BP (deg C):

BP pressure (mm Hg):

Property/ Value	Units	Temp	Data Type	Reference
WS 1.26E+002	mg/L	25	EST	MEYLAN,WM ET AL. (1996)
logP 2.93			EXP	TANAKA,M ET AL. (1985)
VP 2.43E-003	mm Hg	25	EST	NEELY,WB & BLAU,GE (1985)
DC	pKa			
HL 2.71E-008	atm m3/mol	25	EST	MEYLAN,WM & HOWARD,PH (1991)
OH 2.93E-011	cm3/molc sec	25	EST	MEYLAN,WM & HOWARD,PH (1993)

CAS #: 051207-31-9 — 2,3,7,8-TETRACHLORODIBENZOFURAN

Formula: $C_{12}H_4Cl_4O$

Mol Weight: 305.98

MP (deg C): | FP (deg C):

BP (deg C):

BP pressure (mm Hg):

Property/ Value	Units	Temp	Data Type	Reference
WS 4.19E-004	mg/L	23	EXP	FLETCHER,CL & MCKAY,WA (1993)
logP 6.53			EXP	SIJM,DTHM ET AL. (1989)
VP 1.53E-006	mm Hg	25	EST	NEELY,WB & BLAU,GE (1985)
DC	pKa			
HL 1.54E-005	atm m3/mol	25	EST	MEYLAN,WM & HOWARD,PH (1991)
OH 1.65E-013	cm3/molc sec	25	EST	MEYLAN,WM & HOWARD,PH (1993)

CAS #: 051218-45-2 — METOLACHLOR

Formula: $C_{15}H_{22}ClNO_2$

Mol Weight: 283.80

MP (deg C): | FP (deg C):

BP (deg C): 100

BP pressure (mm Hg): 0.00E+000

Property/ Value	Units	Temp	Data Type	Reference
WS 5.30E+002	mg/L	20	EXP	WAUCHOPE,RD ET AL. (1992)
logP 3.13			EXP	HANSCH,C & LEO,AJ (1985)
VP 3.14E-005	mm Hg	25	EXP	WAUCHOPE,RD ET AL. (1992)
DC	pKa			
HL 9.00E-009	atm m3/mol	20	EXP	CHESTERS,G ET AL. (1989)
OH 7.02E-011	cm3/molc sec	25	EST	ATKINSON,R (1988)

CAS #: 051218-49-6 — PRETILCHLOR

Formula: $C_{17}H_{26}ClNO_2$

Mol Weight: 311.86

MP (deg C): | FP (deg C):

BP (deg C): 135

BP pressure (mm Hg): 1.00E-003

Property/ Value	Units	Temp	Data Type	Reference
WS 5.00E+001	mg/L	20	EXP	SHIU,WY ET AL. (1990)
logP 4.08			EXP	TOMLIN,C (1994)
VP 2.03E-007	mm Hg	25	EST	NEELY,WB & BLAU,GE (1985)
DC	pKa			
HL 2.62E-009	atm m3/mol	25	EST	MEYLAN,WM & HOWARD,PH (1991)
OH 5.18E-011	cm3/molc sec	25	EST	MEYLAN,WM & HOWARD,PH (1993)

CAS #: 051234-28-7 — ORAFLEX (UNIPROFEN) (OPREN) (COXIGON)

Formula: $C_{16}H_{13}ClNO_3$

Mol Weight: 302.74

MP (deg C): | FP (deg C):

BP (deg C):

BP pressure (mm Hg):

Property/ Value	Units	Temp	Data Type	Reference
WS 5.20E+001	mg/L	25	EST	MEYLAN,WM ET AL. (1996)
logP 3.23			EXP	JACK,DB ET AL. (1988)
VP 5.26E-009	mm Hg	25	EST	NEELY,WB & BLAU,GE (1985)
DC	pKa			
HL 4.26E-013	atm m3/mol	25	EST	MEYLAN,WM & HOWARD,PH (1991)
OH 1.68E-011	cm3/molc sec	25	EST	MEYLAN,WM & HOWARD,PH (1993)

CAS #: 051235-04-2 — HEXAZINONE

Formula: $C_{12}H_{20}N_4O_2$

Mol Weight: 252.32

MP (deg C): 115-117 | FP (deg C):

BP (deg C):

BP pressure (mm Hg):

Property/ Value	Units	Temp	Data Type	Reference
WS 3.30E+004	mg/L	25	EXP	WORTHING,CR & WALKER,SB (1987)
logP 4.01			EST	MEYLAN,WM & HOWARD,PH (1995)
VP 2.00E-007	mm Hg	25	EST	WORTHING,CR & WALKER,SB (1987)
DC	pKa			
HL 4.11E-013	atm m3/mol	25	EST	MEYLAN,WM & HOWARD,PH (1991)
OH 9.02E-011	cm3/molc sec	25	EST	MEYLAN,WM & HOWARD,PH (1993)

CAS #: 051246-73-2	4-(2-METHOXYBENZOYL)-PYRIDINE

Formula: $C_{13}H_{11}NO_2$

Mol Weight: 213.24

MP (deg C): | FP (deg C):

BP (deg C):

BP pressure (mm Hg):

Property/Value	Units	Temp	Data Type	Reference
WS 9.64E+002	mg/L	25	EST	MEYLAN,WM ET AL. (1996)
logP 1.86			EXP	HANSCH,C & LEO,AJ (1985)
VP 5.48E-005	mm Hg	25	EST	NEELY,WB & BLAU,GE (1985)
DC	pKa			
HL 1.50E-010	atm m3/mol	25	EST	MEYLAN,WM & HOWARD,PH (1991)
OH 2.08E-011	cm3/molc sec	25	EST	MEYLAN,WM & HOWARD,PH (1993)

CAS #: 051246-76-5	4-(BENZOYL SULFONAMIDE)-PYRIDINE

Formula: $C_{12}H_{10}N_2O_3S$

Mol Weight: 262.29

MP (deg C): | FP (deg C):

BP (deg C):

BP pressure (mm Hg):

Property/Value	Units	Temp	Data Type	Reference
WS 5.97E+002	mg/L	25	EST	MEYLAN,WM ET AL. (1996)
logP 0.56			EXP	HANSCH,C & LEO,AJ (1985)
VP 5.95E-008	mm Hg	25	EST	NEELY,WB & BLAU,GE (1985)
DC	pKa			
HL 1.99E-013	atm m3/mol	25	EST	MEYLAN,WM & HOWARD,PH (1991)
OH 7.17E-013	cm3/molc sec	25	EST	MEYLAN,WM & HOWARD,PH (1993)

CAS #: 051246-77-6	4-(4-HYDROXYBENZOYL)-PYRIDINE

Formula: $C_{12}H_9NO_2$

Mol Weight: 199.21

MP (deg C): | FP (deg C):

BP (deg C):

BP pressure (mm Hg):

Property/Value	Units	Temp	Data Type	Reference
WS 1.14E+004	mg/L	25	EST	MEYLAN,WM ET AL. (1996)
logP 1.37			EXP	HANSCH,C & LEO,AJ (1985)
VP 4.67E-006	mm Hg	25	EST	NEELY,WB & BLAU,GE (1985)
DC	pKa			
HL 2.64E-013	atm m3/mol	25	EST	MEYLAN,WM & HOWARD,PH (1991)
OH 3.09E-011	cm3/molc sec	25	EST	MEYLAN,WM & HOWARD,PH (1993)

CAS #: 051246-79-8	CYTIDINE, 2',3'-DIDEOXY-3'-FLUORO-

Formula: $C_9H_{12}FN_3O_3$

Mol Weight: 229.21

MP (deg C): | FP (deg C):

BP (deg C):

BP pressure (mm Hg):

Property/Value	Units	Temp	Data Type	Reference
WS 8.66E+003	mg/L	25	EST	MEYLAN,WM ET AL. (1996)
logP -0.89			EXP	KERR,SG & KALMNA,TI (1992)
VP 7.07E-008	mm Hg	25	EST	NEELY,WB & BLAU,GE (1985)
DC	pKa			
HL 2.33E-016	atm m3/mol	25	EST	MEYLAN,WM & HOWARD,PH (1991)
OH 6.87E-011	cm3/molc sec	25	EST	MEYLAN,WM & HOWARD,PH (1993)

CAS #: 051264-00-7	THIOUREA, N-METHYL-N'-[4-(5-METHYL-1H-IMIDAZOL-4

Formula: $C_{10}H_{18}N_4S$

Mol Weight: 226.35

MP (deg C): | FP (deg C):

BP (deg C):

BP pressure (mm Hg):

Property/Value	Units	Temp	Data Type	Reference
WS 6.10E+004	mg/L	25	EST	MEYLAN,WM ET AL. (1996)
logP 0.85			EXP	SANGSTER,J (1994)
VP 6.20E-008	mm Hg	25	EST	NEELY,WB & BLAU,GE (1985)
DC	pKa			
HL 1.11E-010	atm m3/mol	25	EST	MEYLAN,WM & HOWARD,PH (1991)
OH 1.96E-010	cm3/molc sec	25	EST	MEYLAN,WM & HOWARD,PH (1993)

CAS #: 051264-74-5	PROPANOIC ACID, 2-(4-FORMYL-2-METHOXYPHENOXY)ETH

Formula: $C_{13}H_{16}O_5$

Mol Weight: 252.27

MP (deg C): | FP (deg C):

BP (deg C):

BP pressure (mm Hg):

Property/Value	Units	Temp	Data Type	Reference
WS 1.68E+003	mg/L	25	EST	MEYLAN,WM ET AL. (1996)
logP 1.33			EXP	HANSCH,C ET AL. (1995)
VP 2.38E-005	mm Hg	25	EST	NEELY,WB & BLAU,GE (1985)
DC	pKa			
HL 1.97E-010	atm m3/mol	25	EST	MEYLAN,WM & HOWARD,PH (1991)
OH 3.68E-011	cm3/molc sec	25	EST	MEYLAN,WM & HOWARD,PH (1993)

CAS #: 051307-86-9	1,4-BENZDAZEPIN-2-ONE-5-(2FPH)-7-ET

Formula: $C_{17}H_{15}FN_2O$

Mol Weight: 282.32

MP (deg C): | FP (deg C):

BP (deg C):

BP pressure (mm Hg):

Property/Value	Units	Temp	Data Type	Reference
WS 4.99E+001	mg/L	25	EST	MEYLAN,WM ET AL. (1996)
logP 2.92			EXP	HANSCH,C & LEO,AJ (1985)
VP 5.66E-009	mm Hg	25	EST	NEELY,WB & BLAU,GE (1985)
DC	pKa			
HL 4.10E-010	atm m3/mol	25	EST	MEYLAN,WM & HOWARD,PH (1991)
OH 2.29E-011	cm3/molc sec	25	EST	MEYLAN,WM & HOWARD,PH (1993)

CAS #: 051308-54-4	BUTHIOBATE

Formula: $C_{21}H_{28}N_2S_2$

Mol Weight: 372.60

MP (deg C): 31-33 | FP (deg C):

BP (deg C):

BP pressure (mm Hg):

Property/Value	Units	Temp	Data Type	Reference
WS 1.00E+000	mg/L	25	EXP	SHIU,WY ET AL. (1990)
logP 7.00			EXP	SANGSTER,J (1994)
VP 6.91E-009	mm Hg	25	EST	NEELY,WB & BLAU,GE (1985)
DC	pKa			
HL 8.86E-011	atm m3/mol	25	EST	MEYLAN,WM & HOWARD,PH (1991)
OH 2.91E-011	cm3/molc sec	25	EST	MEYLAN,WM & HOWARD,PH (1993)

CAS #: 051338-10-4				(4-(2,4-DICHLOROPHENOXY)PHENOXY)ACETIC ACID

Formula: $C_{14}H_{10}Cl_2O_4$

Mol Weight: 313.14

MP (deg C):   FP (deg C):

BP (deg C):

BP pressure (mm Hg):

Property/Value	Units	Temp	Data Type	Reference
WS 3.59E+000	mg/L	25	EST	MEYLAN,WM ET AL. (1996)
logP 4.51			EXP	SANGSTER,J (1994)
VP 5.43E-008	mm Hg	25	EST	NEELY,WB & BLAU,GE (1985)
DC	pKa			
HL 2.01E-010	atm m3/mol	25	EST	MEYLAN,WM & HOWARD,PH (1991)
OH 2.73E-011	cm3/molc sec	25	EST	MEYLAN,WM & HOWARD,PH (1993)

CAS #: 051338-27-3				DICLOFOP-METHYL

Formula: $C_{16}H_{14}Cl_2O_4$

Mol Weight: 341.19

MP (deg C): 39-41   FP (deg C):

BP (deg C): 175-177

BP pressure (mm Hg): 1.00E-001

Property/Value	Units	Temp	Data Type	Reference
WS 3.00E+000	mg/L	25	EXP	BUDAVARI,S (1989)
logP 4.62			EXP	KRAWCHUK,BP & WEBSTER,GRB (1987)
VP 2.55E-007	mm Hg	20	EXP	HARTLEY,D & KIDD,H (1987)
DC	pKa			
HL 3.82E-008	atm m3/mol	20	EST	VP/WSOL
OH 1.80E-011	cm3/molc sec	25	EST	MEYLAN,WM & HOWARD,PH (1993)

CAS #: 051344-12-8				ETHANAMINE, 2-(4-METHOXYPHENOXY)-N,N-DIMETHYL-

Formula: $C_{11}H_{17}NO_2$

Mol Weight: 195.26

MP (deg C):   FP (deg C):

BP (deg C):

BP pressure (mm Hg):

Property/Value	Units	Temp	Data Type	Reference
WS 4.20E+004	mg/L	25	EST	MEYLAN,WM ET AL. (1996)
logP 1.23			EXP	AL-SAADI,D ET AL. (1993)
VP 7.15E-003	mm Hg	25	EST	NEELY,WB & BLAU,GE (1985)
DC	pKa			
HL 1.24E-008	atm m3/mol	25	EST	MEYLAN,WM & HOWARD,PH (1991)
OH 1.25E-010	cm3/molc sec	25	EST	MEYLAN,WM & HOWARD,PH (1993)

CAS #: 051344-14-0				ETHANAMINE, N,N-DIMETHYL-2-(4-METHYLPHENOXY)-

Formula: $C_{11}H_{17}NO$

Mol Weight: 179.26

MP (deg C):   FP (deg C):

BP (deg C):

BP pressure (mm Hg):

Property/Value	Units	Temp	Data Type	Reference
WS 7.93E+003	mg/L	25	EST	MEYLAN,WM ET AL. (1996)
logP 2.17			EXP	AL-SAADI,D ET AL. (1993)
VP 2.54E-002	mm Hg	25	EST	NEELY,WB & BLAU,GE (1985)
DC	pKa			
HL 2.32E-007	atm m3/mol	25	EST	MEYLAN,WM & HOWARD,PH (1991)
OH 1.32E-010	cm3/molc sec	25	EST	MEYLAN,WM & HOWARD,PH (1993)

CAS #: 051422-54-9				ETHYL T-BUTYL ETHYL DIETHER

Formula: $C_8H_{18}O_2$

Mol Weight: 146.23

MP (deg C):   FP (deg C):

BP (deg C): 148

BP pressure (mm Hg):

Property/Value	Units	Temp	Data Type	Reference
WS 3.10E+003	mg/L	25	EST	MEYLAN,WM ET AL. (1996)
logP 1.65			EST	MEYLAN,WM & HOWARD,PH (1995)
VP 6.56E+000	mm Hg	25	EST	NEELY,WB & BLAU,GE (1985)
DC	pKa			
HL 4.16E-005	atm m3/mol	25	EST	MEYLAN,WM & HOWARD,PH (1991)
OH 2.16E-011	cm3/molc sec	25	EST	MEYLAN,WM & HOWARD,PH (1993)

CAS #: 051422-77-6				CARBAMIC ACID, (3-METHOXYPHENYL)-, METHYL ESTER

Formula: $C_9H_{11}NO_3$

Mol Weight: 181.19

MP (deg C):   FP (deg C):

BP (deg C):

BP pressure (mm Hg):

Property/Value	Units	Temp	Data Type	Reference
WS 1.25E+003	mg/L	25	EST	MEYLAN,WM ET AL. (1996)
logP 1.92			EXP	TAKAHASHI,J ET AL. (1988)
VP 5.26E-003	mm Hg	25	EST	NEELY,WB & BLAU,GE (1985)
DC	pKa			
HL 1.29E-009	atm m3/mol	25	EST	MEYLAN,WM & HOWARD,PH (1991)
OH 2.01E-010	cm3/molc sec	25	EST	MEYLAN,WM & HOWARD,PH (1993)

CAS #: 051450-97-6				DRAZOXOLON

Formula: $C_{10}H_8ClN_3O_2$

Mol Weight: 237.65

MP (deg C):   FP (deg C):

BP (deg C):

BP pressure (mm Hg):

Property/Value	Units	Temp	Data Type	Reference
WS 1.67E+002	mg/L	25	EST	MEYLAN,WM ET AL. (1996)
logP 2.60			EXP	HANSCH,C ET AL. (1995)
VP 1.54E-006	mm Hg	25	EST	NEELY,WB & BLAU,GE (1985)
DC	pKa			
HL 1.72E-007	atm m3/mol	25	EST	MEYLAN,WM & HOWARD,PH (1991)
OH 1.26E-011	cm3/molc sec	25	EST	MEYLAN,WM & HOWARD,PH (1993)

CAS #: 051460-47-0				THIOPHEN-3-CARBOXAMIDE

Formula: C_5H_7NOS

Mol Weight: 129.18

MP (deg C):   FP (deg C):

BP (deg C):

BP pressure (mm Hg):

Property/Value	Units	Temp	Data Type	Reference
WS 3.53E+004	mg/L	25	EST	MEYLAN,WM ET AL. (1996)
logP 0.50			EXP	HANSCH,C & LEO,AJ (1985)
VP 4.17E-004	mm Hg	25	EST	NEELY,WB & BLAU,GE (1985)
DC	pKa			
HL 1.20E-009	atm m3/mol	25	EST	MEYLAN,WM & HOWARD,PH (1991)
OH 1.07E-011	cm3/molc sec	25	EST	MEYLAN,WM & HOWARD,PH (1993)

1161

CAS #: 051543-29-4 — N1-METHYL-N1-(2-PYRIDYL)SULFANILAMIDE

Formula: $C_{12}H_{13}N_3O_2S$

Mol Weight: 263.32

MP (deg C): FP (deg C):

BP (deg C):

BP pressure (mm Hg):

Property/Value	Units	Temp	Data Type	Reference
WS 1.55E+003	mg/L	25	EST	MEYLAN,WM ET AL. (1996)
logP 1.30			EXP	HANSCH,C ET AL. (1995)
VP 2.81E-008	mm Hg	25	EST	NEELY,WB & BLAU,GE (1985)
DC	pKa			
HL 1.19E-013	atm m3/mol	25	EST	MEYLAN,WM & HOWARD,PH (1991)
OH 3.31E-011	cm3/molc sec	25	EST	MEYLAN,WM & HOWARD,PH (1993)

CAS #: 051575-56-5 — 1-HEXYLTHIO-B-GALACTOPYRANOSIDE

Formula: $C_{12}H_{24}O_5S$

Mol Weight: 280.39

MP (deg C): FP (deg C):

BP (deg C):

BP pressure (mm Hg):

Property/Value	Units	Temp	Data Type	Reference
WS 2.94E+003	mg/L	25	EST	MEYLAN,WM ET AL. (1996)
logP 0.86			EXP	HANSCH,C & LEO,AJ (1985)
VP 4.06E-011	mm Hg	25	EST	NEELY,WB & BLAU,GE (1985)
DC	pKa			
HL 1.17E-014	atm m3/mol	25	EST	MEYLAN,WM & HOWARD,PH (1991)
OH 1.83E-010	cm3/molc sec	25	EST	MEYLAN,WM & HOWARD,PH (1993)

CAS #: 051581-33-0 — 3-(N,N-DIMECARBAM)-2-NO2-PYRIDINE

Formula: $C_8H_9N_3O_4$

Mol Weight: 211.18

MP (deg C): FP (deg C):

BP (deg C):

BP pressure (mm Hg):

Property/Value	Units	Temp	Data Type	Reference
WS 5.40E+003	mg/L	25	EST	MEYLAN,WM ET AL. (1996)
logP 0.54			EXP	HANSCH,C & LEO,AJ (1985)
VP 1.03E-004	mm Hg	25	EST	NEELY,WB & BLAU,GE (1985)
DC	pKa			
HL 3.64E-013	atm m3/mol	25	EST	MEYLAN,WM & HOWARD,PH (1991)
OH 1.62E-011	cm3/molc sec	25	EST	MEYLAN,WM & HOWARD,PH (1993)

CAS #: 051581-34-1 — 3-(N,N-DIMECARBAM)-2-CL-PYRIDINE

Formula: $C_8H_9ClN_2O_2$

Mol Weight: 200.63

MP (deg C): FP (deg C):

BP (deg C):

BP pressure (mm Hg):

Property/Value	Units	Temp	Data Type	Reference
WS 5.62E+003	mg/L	25	EST	MEYLAN,WM ET AL. (1996)
logP 1.04			EXP	HANSCH,C & LEO,AJ (1985)
VP 3.58E-003	mm Hg	25	EST	NEELY,WB & BLAU,GE (1985)
DC	pKa			
HL 4.32E-009	atm m3/mol	25	EST	MEYLAN,WM & HOWARD,PH (1991)
OH 1.63E-011	cm3/molc sec	25	EST	MEYLAN,WM & HOWARD,PH (1993)

CAS #: 051581-35-2 — 3-(N,N-DIMECARBAM)-2-BR-PYRIDINE

Formula: $C_8H_9BrN_2O_2$

Mol Weight: 245.08

MP (deg C): FP (deg C):

BP (deg C):

BP pressure (mm Hg):

Property/Value	Units	Temp	Data Type	Reference
WS 2.68E+003	mg/L	25	EST	MEYLAN,WM ET AL. (1996)
logP 1.14				HANSCH,C & LEO,AJ (1985)
VP 1.25E-003	mm Hg	25	EST	NEELY,WB & BLAU,GE (1985)
DC	pKa			
HL 3.68E-011	atm m3/mol	25	EST	MEYLAN,WM & HOWARD,PH (1991)
OH 1.63E-011	cm3/molc sec	25	EST	MEYLAN,WM & HOWARD,PH (1993)

CAS #: 051581-36-3 — 3-(N,N-DIMECARBAM)-2-I-PYRIDINE

Formula: $C_8H_9IN_2O_2$

Mol Weight: 292.08

MP (deg C): FP (deg C):

BP (deg C):

BP pressure (mm Hg):

Property/Value	Units	Temp	Data Type	Reference
WS 1.15E+003	mg/L	25	EST	MEYLAN,WM ET AL. (1996)
logP 1.26			EXP	HANSCH,C & LEO,AJ (1985)
VP 2.89E-004	mm Hg	25	EST	NEELY,WB & BLAU,GE (1985)
DC	pKa			
HL 2.14E-011	atm m3/mol	25	EST	MEYLAN,WM & HOWARD,PH (1991)
OH 1.63E-011	cm3/molc sec	25	EST	MEYLAN,WM & HOWARD,PH (1993)

CAS #: 051630-58-1 — FENVALERATE

Formula: $C_{25}H_{22}ClNO_3$

Mol Weight: 419.91

MP (deg C): 59-60.2 FP (deg C):

BP (deg C): 300

BP pressure (mm Hg): 3.70E+001

Property/Value	Units	Temp	Data Type	Reference
WS 8.50E-002	mg/L		EXP	SHIU,WY ET AL. (1990)
logP 6.20			EXP	HANSCH,C & LEO,AJ (1985)
VP 2.80E-007	mm Hg	25	EXP	IARC (1991)
DC	pKa			
HL 1.19E-007	atm-m3/mol	25	EST	MEYLAN,WM & HOWARD,PH (1991)
OH 3.76E-011	cm3/molc sec	25	EST	ATKINSON,R (1988)

CAS #: 051688-32-5 — BENZENAMINE, 4-[(4-AMINOPHENYL)SULFONYL]-N,N-DIE

Formula: $C_{16}H_{20}N_2O_2S$

Mol Weight: 304.41

MP (deg C): FP (deg C):

BP (deg C):

BP pressure (mm Hg):

Property/Value	Units	Temp	Data Type	Reference
WS 3.17E+001	mg/L	25	EST	MEYLAN,WM ET AL. (1996)
logP 3.00			EXP	ALTOMARE,C ET AL. (1991)
VP 7.92E-009	mm Hg	25	EST	NEELY,WB & BLAU,GE (1985)
DC	pKa			
HL 2.47E-012	atm m3/mol	25	EST	MEYLAN,WM & HOWARD,PH (1991)
OH 7.21E-011	cm3/molc sec	25	EST	MEYLAN,WM & HOWARD,PH (1993)

CAS #: 051707-42-7		HYDRAZINECARBOXAMIDE, N-(3-CHLOROPHENYL)

Formula: $C_7H_8ClN_3O$

Mol Weight: 185.61

MP (deg C): | FP (deg C):

BP (deg C):

BP pressure (mm Hg):

Property/ Value	Units	Temp	Data Type	Reference
WS 1.01E+004	mg/L	25	EST	MEYLAN,WM ET AL. (1996)
logP 0.83			EXP	KRAMER,CR & BECK,L (1981)
VP 2.02E-005	mm Hg	25	EST	NEELY,WB & BLAU,GE (1985)
DC	pKa			
HL 6.23E-013	atm m3/mol	25	EST	MEYLAN,WM & HOWARD,PH (1991)
OH 3.10E-011	cm3/molc sec	25	EST	MEYLAN,WM & HOWARD,PH (1993)

CAS #: 051707-55-2		THIDIAZURON

Formula: $C_9H_8N_4OS$

Mol Weight: 220.25

MP (deg C): 213 | FP (deg C):

BP (deg C):

BP pressure (mm Hg):

Property/ Value	Units	Temp	Data Type	Reference
WS 2.00E+001	mg/L	23	EXP	SHIU,WY ET AL. (1990)
logP 2.10			EST	MEYLAN,WM & HOWARD,PH (1995)
VP 2.30E-011	mm Hg	25	EXP	WAUCHOPE,RD ET AL. (1991A)
DC	pKa			
HL 3.33E-013	atm m3/mol	25	EST	VP/WSOL
OH 5.80E-011	cm3/molc sec	25	EST	MEYLAN,WM & HOWARD,PH (1993)

CAS #: 051740-66-0		DIGOXIN,ALPHA-BETA-DIMETHYL-

Formula: $C_{43}H_{68}O_{14}$

Mol Weight: 809.01

MP (deg C): | FP (deg C):

BP (deg C):

BP pressure (mm Hg):

Property/ Value	Units	Temp	Data Type	Reference
WS 9.70E-002	mg/L	25	EST	MEYLAN,WM ET AL. (1996)
logP 2.14			EXP	HANSCH,C ET AL. (1995)
VP 4.86E-029	mm Hg	25	EST	NEELY,WB & BLAU,GE (1985)
DC	pKa			
HL 4.77E-028	atm m3/mol	25	EST	MEYLAN,WM & HOWARD,PH (1991)
OH 1.82E-010	cm3/molc sec	25	EST	MEYLAN,WM & HOWARD,PH (1993)

CAS #: 051740-69-3		DIGOXIN,12-ACETYL-BETA-METHYL-

Formula: $C_{44}H_{68}O_{15}$

Mol Weight: 837.02

MP (deg C): | FP (deg C):

BP (deg C):

BP pressure (mm Hg):

Property/ Value	Units	Temp	Data Type	Reference
WS 1.49E-002	mg/L	25	EST	MEYLAN,WM ET AL. (1996)
logP 2.87			EXP	HANSCH,C ET AL. (1995)
VP 5.64E-030	mm Hg	25	EST	NEELY,WB & BLAU,GE (1985)
DC	pKa			
HL 9.71E-029	atm m3/mol	25	EST	MEYLAN,WM & HOWARD,PH (1991)
OH 1.72E-010	cm3/molc sec	25	EST	MEYLAN,WM & HOWARD,PH (1993)

CAS #: 051753-57-2		1,4-BENZDAZEPIN-2-ONE-5-(2CLPH)-7BR

Formula: $C_{15}H_{10}BrClN_2O$

Mol Weight: 349.62

MP (deg C): | FP (deg C):

BP (deg C):

BP pressure (mm Hg):

Property/ Value	Units	Temp	Data Type	Reference
WS 9.47E+000	mg/L	25	EST	MEYLAN,WM ET AL. (1996)
logP 3.30			EXP	HANSCH,C & LEO,AJ (1985)
VP 7.45E-010	mm Hg	25	EST	NEELY,WB & BLAU,GE (1985)
DC	pKa			
HL 7.08E-011	atm m3/mol	25	EST	MEYLAN,WM & HOWARD,PH (1991)
OH 6.62E-012	cm3/molc sec	25	EST	MEYLAN,WM & HOWARD,PH (1993)

CAS #: 051754-90-6		PHOSPHORIC TRIAMIDE, N,N-DIETHYL-N',N',N'',N''-TET

Formula: $C_8H_{22}N_3OP$

Mol Weight: 207.26

MP (deg C): | FP (deg C):

BP (deg C):

BP pressure (mm Hg):

Property/ Value	Units	Temp	Data Type	Reference
WS 1.10E+004	mg/L	25	EST	MEYLAN,WM ET AL. (1996)
logP 0.66			EXP	DEBORD,J & LABADIE,M (1985)
VP 9.18E-003	mm Hg	25	EST	NEELY,WB & BLAU,GE (1985)
DC	pKa			
HL 1.26E-011	atm m3/mol	25	EST	MEYLAN,WM & HOWARD,PH (1991)
OH 1.05E-010	cm3/molc sec	25	EST	MEYLAN,WM & HOWARD,PH (1993)

CAS #: 051772-35-1		N-((1,1,3,3-TETRAMETHYLBUTYL)PHENYL)-1-NAPHTHA*

Formula: $C_{24}H_{29}N$

Mol Weight: 331.51

MP (deg C): | FP (deg C):

BP (deg C):

BP pressure (mm Hg):

Property/ Value	Units	Temp	Data Type	Reference
WS 7.49E-004	mg/L	25	EST	MEYLAN,WM ET AL. (1996)
logP 8.23			EST	MEYLAN,WM & HOWARD,PH (1995)
VP 1.09E-007	mm Hg	25	EST	NEELY,WB & BLAU,GE (1985)
DC	pKa			
HL 1.53E-005	atm m3/mol	25	EST	MEYLAN,WM & HOWARD,PH (1991)
OH 3.49E-010	cm3/molc sec	25	EST	MEYLAN,WM & HOWARD,PH (1993)

CAS #: 051795-30-3		PENTACHLOROCYCLOHEXENE

Formula: $C_6H_5Cl_5$

Mol Weight: 254.37

MP (deg C): | FP (deg C):

BP (deg C):

BP pressure (mm Hg):

Property/ Value	Units	Temp	Data Type	Reference
WS 1.15E+001	mg/L	25	EST	MEYLAN,WM ET AL. (1996)
logP 3.80			EXP	HANSCH,C & LEO,AJ (1985)
VP 5.99E-003	mm Hg	25	EST	NEELY,WB & BLAU,GE (1985)
DC	pKa			
HL 1.30E-003	atm m3/mol	25	EST	MEYLAN,WM & HOWARD,PH (1991)
OH 1.13E-011	cm3/molc sec	25	EST	MEYLAN,WM & HOWARD,PH (1993)

CAS #: 051803-78-2				METHANESULFONAMIDE, N-(4-NITRO-2-PHENOXYPHENYL)-
Formula: $C_{13}H_{12}N_2O_5S$				
Mol Weight: 308.31				
MP (deg C): 143-144.5		FP (deg C):		
BP (deg C):				
BP pressure (mm Hg):				

Property/ Value	Units	Temp	Data Type	Reference
WS 2.69E+001	mg/L	25	EST	MEYLAN,WM ET AL. (1996)
logP 2.60			EXP	SANGSTER,J (1993)
VP 7.14E-009	mm Hg	25	EST	NEELY,WB & BLAU,GE (1985)
DC	pKa			
HL 1.01E-010	atm m3/mol	25	EST	MEYLAN,WM & HOWARD,PH (1991)
OH 1.13E-011	cm3/molc sec	25	EST	MEYLAN,WM & HOWARD,PH (1993)

CAS #: 051908-16-8				2,2',3,4',5,5'-HEXACHLOROBIPHENYL
Formula: $C_{12}H_4Cl_6$				
Mol Weight: 360.88				
MP (deg C):		FP (deg C):		
BP (deg C):				
BP pressure (mm Hg):				

Property/ Value	Units	Temp	Data Type	Reference
WS 1.64E-003	mg/L	25	EST	MEYLAN,WM ET AL. (1996)
logP 7.62			EST	MEYLAN,WM & HOWARD,PH (1995)
VP 5.81E-007	mm Hg	25	EST	NEELY,WB & BLAU,GE (1985)
DC	pKa			
HL 6.85E-005	atm m3/mol	25	EST	MEYLAN,WM & HOWARD,PH (1991)
OH 2.34E-013	cm3/molc sec	25	EST	MEYLAN,WM & HOWARD,PH (1993)

CAS #: 051908-64-6				4-CHLORO-1-BUTYNE
Formula: C_4H_5Cl				
Mol Weight: 88.54				
MP (deg C):		FP (deg C):		
BP (deg C):				
BP pressure (mm Hg):				

Property/ Value	Units	Temp	Data Type	Reference
WS 3.73E+003	mg/L	25	EST	MEYLAN,WM ET AL. (1996)
logP 1.79			EST	MEYLAN,WM & HOWARD,PH (1995)
VP 4.28E+001	mm Hg	25	EST	NEELY,WB & BLAU,GE (1985)
DC	pKa			
HL 7.51E-003	atm m3/mol	25	EST	MEYLAN,WM & HOWARD,PH (1991)
OH 7.77E-012	cm3/molc sec	25	EST	MEYLAN,WM & HOWARD,PH (1993)

CAS #: 051939-71-0				CHROMAN-2-CARBOXYLIC ACID
Formula: $C_{10}H_{10}O_3$				
Mol Weight: 178.19				
MP (deg C):		FP (deg C):		
BP (deg C):				
BP pressure (mm Hg):				

Property/ Value	Units	Temp	Data Type	Reference
WS 3.33E+003	mg/L	25	EST	MEYLAN,WM ET AL. (1996)
logP 1.90			EXP	HANSCH,C & LEO,AJ (1985)
VP 1.36E-004	mm Hg	25	EST	NEELY,WB & BLAU,GE (1985)
DC	pKa			
HL 1.08E-008	atm m3/mol	25	EST	MEYLAN,WM & HOWARD,PH (1991)
OH 4.02E-011	cm3/molc sec	25	EST	MEYLAN,WM & HOWARD,PH (1993)

CAS #: 051940-44-4				PIPEMIDIC ACID
Formula: $C_{14}H_{17}N_5O_3$				
Mol Weight: 303.32				
MP (deg C): 253-255		FP (deg C):		
BP (deg C):				
BP pressure (mm Hg):				

Property/ Value	Units	Temp	Data Type	Reference
WS 3.22E+002	mg/L	25	EXP	YALKOWSKY,SH & DANNENFELSER,RM (1992)
logP -2.15			EXP	HANSCH,C ET AL. (1995)
VP 2.86E-013	mm Hg	25	EST	NEELY,WB & BLAU,GE (1985)
DC	pKa			
HL 1.43E-018	atm m3/mol	25	EST	MEYLAN,WM & HOWARD,PH (1991)
OH 1.22E-010	cm3/molc sec	25	EST	MEYLAN,WM & HOWARD,PH (1993)

CAS #: 051962-63-1				4-(1,2-DICHLOROETHYL)-1,2-DICHLOROCYCLOHEXANE
Formula: $C_8H_{12}Cl_4$				
Mol Weight: 250.00				
MP (deg C):		FP (deg C):		
BP (deg C):				
BP pressure (mm Hg):				

Property/ Value	Units	Temp	Data Type	Reference
WS 1.61E+000	mg/L	25	EST	MEYLAN,WM ET AL. (1996)
logP 4.88			EST	MEYLAN,WM & HOWARD,PH (1995)
VP 4.93E-003	mm Hg	25	EST	NEELY,WB & BLAU,GE (1985)
DC	pKa			
HL 3.63E-003	atm m3/mol	25	EST	MEYLAN,WM & HOWARD,PH (1991)
OH 4.40E-012	cm3/molc sec	25	EST	MEYLAN,WM & HOWARD,PH (1993)

CAS #: 051963-47-4				N-IPR-1-(PHENOXY)PROPIONIC HYDRAZIDE
Formula: $C_{11}H_{16}N_2O_2$				
Mol Weight: 208.26				
MP (deg C):		FP (deg C):		
BP (deg C):				
BP pressure (mm Hg):				

Property/ Value	Units	Temp	Data Type	Reference
WS 3.46E+003	mg/L	25	EST	MEYLAN,WM ET AL. (1996)
logP 1.39			EXP	HANSCH,C & LEO,AJ (1985)
VP 3.88E-006	mm Hg	25	EST	NEELY,WB & BLAU,GE (1985)
DC	pKa			
HL 5.54E-012	atm m3/mol	25	EST	MEYLAN,WM & HOWARD,PH (1991)
OH 1.13E-010	cm3/molc sec	25	EST	MEYLAN,WM & HOWARD,PH (1993)

CAS #: 051963-48-5				N'-IPR-(2-CLPHENOXY)ACETIC HYDRAZIDE
Formula: $C_{11}H_{15}ClN_2O_2$				
Mol Weight: 242.71				
MP (deg C):		FP (deg C):		
BP (deg C):				
BP pressure (mm Hg):				

Property/ Value	Units	Temp	Data Type	Reference
WS 7.24E+002	mg/L	25	EST	MEYLAN,WM ET AL. (1996)
logP 1.82			EXP	HANSCH,C & LEO,AJ (1985)
VP 1.06E-006	mm Hg	25	EST	NEELY,WB & BLAU,GE (1985)
DC	pKa			
HL 4.10E-012	atm m3/mol	25	EST	MEYLAN,WM & HOWARD,PH (1991)
OH 9.74E-011	cm3/molc sec	25	EST	MEYLAN,WM & HOWARD,PH (1993)

CAS #: 051963-49-6 — N'-IPR-(3-CLPHENOXY)ACETIC HYDRAZIDE

Formula: $C_{11}H_{15}ClN_2O_2$

Mol Weight: 242.71

MP (deg C): 　 FP (deg C):

BP (deg C):

BP pressure (mm Hg):

Property/Value	Units	Temp	Data Type	Reference
WS 6.07E+002	mg/L	25	EST	MEYLAN,WM ET AL. (1996)
logP 1.91			EXP	HANSCH,C & LEO,AJ (1985)
VP 1.06E-006	mm Hg	25	EST	NEELY,WB & BLAU,GE (1985)
DC	pKa			
HL 4.10E-012	atm m3/mol	25	EST	MEYLAN,WM & HOWARD,PH (1991)
OH 1.06E-010	cm3/molc sec	25	EST	MEYLAN,WM & HOWARD,PH (1993)

CAS #: 051963-50-9 — N-IPR1(2CL-PHENOXY)PROPIONIC HYDRAZID

Formula: $C_{12}H_{17}ClN_2O_2$

Mol Weight: 256.73

MP (deg C): 　 FP (deg C):

BP (deg C):

BP pressure (mm Hg):

Property/Value	Units	Temp	Data Type	Reference
WS 4.25E+002	mg/L	25	EST	MEYLAN,WM ET AL. (1996)
logP 2.00			EXP	HANSCH,C & LEO,AJ (1985)
VP 4.56E-007	mm Hg	25	EST	NEELY,WB & BLAU,GE (1985)
DC	pKa			
HL 7.70E-013	atm m3/mol	25	EST	MEYLAN,WM & HOWARD,PH (1991)
OH 1.18E-010	cm3/molc sec	25	EST	MEYLAN,WM & HOWARD,PH (1993)

CAS #: 051963-51-0 — N-IPR1(3CL-PHENOXY)PROPIONIC HYDRAZID

Formula: $C_{12}H_{17}ClN_2O_2$

Mol Weight: 256.73

MP (deg C): 　 FP (deg C):

BP (deg C):

BP pressure (mm Hg):

Property/Value	Units	Temp	Data Type	Reference
WS 2.98E+002	mg/L	25	EST	MEYLAN,WM ET AL. (1996)
logP 2.18			EXP	HANSCH,C & LEO,AJ (1985)
VP 4.56E-007	mm Hg	25	EST	NEELY,WB & BLAU,GE (1985)
DC	pKa			
HL 7.70E-013	atm m3/mol	25	EST	MEYLAN,WM & HOWARD,PH (1991)
OH 1.27E-010	cm3/molc sec	25	EST	MEYLAN,WM & HOWARD,PH (1993)

CAS #: 051963-52-1 — N-IPR1(4CL-PHENOXY)PROPIONIC HYDRAZID

Formula: $C_{12}H_{17}ClN_2O_2$

Mol Weight: 256.73

MP (deg C): 　 FP (deg C):

BP (deg C):

BP pressure (mm Hg):

Property/Value	Units	Temp	Data Type	Reference
WS 3.23E+002	mg/L	25	EST	MEYLAN,WM ET AL. (1996)
logP 2.14			EXP	HANSCH,C & LEO,AJ (1985)
VP 4.56E-007	mm Hg	25	EST	NEELY,WB & BLAU,GE (1985)
DC	pKa			
HL 7.70E-013	atm m3/mol	25	EST	MEYLAN,WM & HOWARD,PH (1991)
OH 1.18E-010	cm3/molc sec	25	EST	MEYLAN,WM & HOWARD,PH (1993)

CAS #: 051963-53-2 — N'-IPR-2ME-PHENOXYPROPIONIC HYDRAZIDE

Formula: $C_{13}H_{20}N_2O_2$

Mol Weight: 236.32

MP (deg C): 　 FP (deg C):

BP (deg C):

BP pressure (mm Hg):

Property/Value	Units	Temp	Data Type	Reference
WS 6.98E+002	mg/L	25	EST	MEYLAN,WM ET AL. (1996)
logP 1.88			EXP	HANSCH,C & LEO,AJ (1985)
VP 1.26E-006	mm Hg	25	EST	NEELY,WB & BLAU,GE (1985)
DC	pKa			
HL 1.38E-012	atm m3/mol	25	EST	MEYLAN,WM & HOWARD,PH (1991)
OH 1.36E-010	cm3/molc sec	25	EST	MEYLAN,WM & HOWARD,PH (1993)

CAS #: 051994-35-5 — PENTANOIC ACID, 5-[(1,1'-BIPHENYL)-4-YL]-5-OXO-

Formula: $C_{17}H_{16}O_3$

Mol Weight: 268.32

MP (deg C): 　 FP (deg C):

BP (deg C):

BP pressure (mm Hg):

Property/Value	Units	Temp	Data Type	Reference
WS 6.39E+001	mg/L	25	EST	MEYLAN,WM ET AL. (1996)
logP 3.35			EXP	KUCHAR,M ET AL. (1985)
VP 3.45E-008	mm Hg	25	EST	NEELY,WB & BLAU,GE (1985)
DC	pKa			
HL 9.87E-012	atm m3/mol	25	EST	MEYLAN,WM & HOWARD,PH (1991)
OH 1.32E-011	cm3/molc sec	25	EST	MEYLAN,WM & HOWARD,PH (1993)

CAS #: 052006-63-0 — 3-ETHYLTHIOPHENE

Formula: $C_6H_{10}S$

Mol Weight: 114.21

MP (deg C): 　 FP (deg C):

BP (deg C):

BP pressure (mm Hg):

Property/Value	Units	Temp	Data Type	Reference
WS 4.18E+002	mg/L	25	EST	MEYLAN,WM ET AL. (1996)
logP 2.82			EXP	HANSCH,C & LEO,AJ (1985)
VP 2.62E+000	mm Hg	25	EST	NEELY,WB & BLAU,GE (1985)
DC	pKa			
HL 4.29E-003	atm m3/mol	25	EST	MEYLAN,WM & HOWARD,PH (1991)
OH 2.48E-011	cm3/molc sec	25	EST	MEYLAN,WM & HOWARD,PH (1993)

CAS #: 052019-05-3 — ME-3-DEOXY-3(3ME3NOUREA)XYLOPYRANOSID

Formula: $C_8H_{15}N_3O_6$

Mol Weight: 249.23

MP (deg C): 　 FP (deg C):

BP (deg C):

BP pressure (mm Hg):

Property/Value	Units	Temp	Data Type	Reference
WS 9.79E+003	mg/L	25	EST	MEYLAN,WM ET AL. (1996)
logP -1.08			EXP	HANSCH,C & LEO,AJ (1985)
VP 1.49E-009	mm Hg	25	EST	NEELY,WB & BLAU,GE (1985)
DC	pKa			
HL 2.59E-020	atm m3/mol	25	EST	MEYLAN,WM & HOWARD,PH (1991)
OH 5.10E-011	cm3/molc sec	25	EST	MEYLAN,WM & HOWARD,PH (1993)

CAS #: 052019-10-0	_-D-ALTROPYRANOSIDE, METHYL 3-DEOXY-3-[[(METHYLN

Formula: C$_9$H$_{17}$N$_3$O$_7$	
Mol Weight: 279.25	
MP (deg C):	FP (deg C):
BP (deg C):	
BP pressure (mm Hg):	

Property/Value	Units	Temp	Data Type	Reference
WS 1.37E+004	mg/L	25	EST	MEYLAN,WM ET AL. (1996)
logP -0.82			EXP	HANSCH,C ET AL. (1995)
VP 7.08E-012	mm Hg	25	EST	NEELY,WB & BLAU,GE (1985)
DC	pKa			
HL 1.26E-021	atm m3/mol	25	EST	MEYLAN,WM & HOWARD,PH (1991)
OH 6.60E-011	cm3/molc sec	25	EST	MEYLAN,WM & HOWARD,PH (1993)

CAS #: 052019-12-2	à-D-GLUCOPYRANOSIDE, METHYL 3-DEOXY-3-[[(METHYLN

Formula: C$_9$H$_{17}$N$_3$O$_7$	
Mol Weight: 279.25	
MP (deg C):	FP (deg C):
BP (deg C):	
BP pressure (mm Hg):	

Property/Value	Units	Temp	Data Type	Reference
WS 1.37E+004	mg/L	25	EST	MEYLAN,WM ET AL. (1996)
logP -1.45			EXP	HANSCH,C ET AL. (1995)
VP 7.08E-012	mm Hg	25	EST	NEELY,WB & BLAU,GE (1985)
DC	pKa			
HL 1.26E-021	atm m3/mol	25	EST	MEYLAN,WM & HOWARD,PH (1991)
OH 6.60E-011	cm3/molc sec	25	EST	MEYLAN,WM & HOWARD,PH (1993)

CAS #: 052019-60-0	N'-IPR-PHENOXYACETIC ACID HYDRAZIDE

Formula: C$_{11}$H$_{16}$N$_2$O$_2$	
Mol Weight: 208.26	
MP (deg C):	FP (deg C):
BP (deg C):	
BP pressure (mm Hg):	

Property/Value	Units	Temp	Data Type	Reference
WS 3.46E+003	mg/L	25	EST	MEYLAN,WM ET AL. (1996)
logP 1.24			EXP	HANSCH,C & LEO,AJ (1985)
VP 3.88E-006	mm Hg	25	EST	NEELY,WB & BLAU,GE (1985)
DC	pKa			
HL 5.54E-012	atm m3/mol	25	EST	MEYLAN,WM & HOWARD,PH (1991)
OH 1.13E-010	cm3/molc sec	25	EST	MEYLAN,WM & HOWARD,PH (1993)

CAS #: 052049-26-0	1-NO-1(2CLET)-3(4-OH CYHEXYL)UREA

Formula: C$_9$H$_{16}$ClN$_3$O$_3$	
Mol Weight: 249.70	
MP (deg C):	FP (deg C):
BP (deg C):	
BP pressure (mm Hg):	

Property/Value	Units	Temp	Data Type	Reference
WS 3.32E+003	mg/L	25	EST	MEYLAN,WM ET AL. (1996)
logP 1.11			EXP	HANSCH,C & LEO,AJ (1985)
VP 1.64E-008	mm Hg	25	EST	NEELY,WB & BLAU,GE (1985)
DC	pKa			
HL 6.76E-015	atm m3/mol	25	EST	MEYLAN,WM & HOWARD,PH (1991)
OH 3.09E-011	cm3/molc sec	25	EST	MEYLAN,WM & HOWARD,PH (1993)

CAS #: 052067-44-4	O=P(OC)(SC)N-T-BU

Formula: C$_6$H$_{16}$NO$_2$PS	
Mol Weight: 197.24	
MP (deg C):	FP (deg C):
BP (deg C):	
BP pressure (mm Hg):	

Property/Value	Units	Temp	Data Type	Reference
WS 6.84E+003	mg/L	25	EST	MEYLAN,WM ET AL. (1996)
logP 0.96			EXP	HANSCH,C & LEO,AJ (1985)
VP 1.15E-002	mm Hg	25	EST	NEELY,WB & BLAU,GE (1985)
DC	pKa			
HL 4.46E-009	atm m3/mol	25	EST	MEYLAN,WM & HOWARD,PH (1991)
OH 3.36E-011	cm3/molc sec	25	EST	MEYLAN,WM & HOWARD,PH (1993)

CAS #: 052067-48-8	O=P(OC)(SC)N-ETHYL

Formula: C$_4$H$_{12}$NO$_2$PS	
Mol Weight: 169.18	
MP (deg C):	FP (deg C):
BP (deg C):	
BP pressure (mm Hg):	

Property/Value	Units	Temp	Data Type	Reference
WS 5.41E+004	mg/L	25	EST	MEYLAN,WM ET AL. (1996)
logP 0.07			EXP	HANSCH,C & LEO,AJ (1985)
VP 3.53E-002	mm Hg	25	EST	NEELY,WB & BLAU,GE (1985)
DC	pKa			
HL 2.53E-009	atm m3/mol	25	EST	MEYLAN,WM & HOWARD,PH (1991)
OH 4.19E-011	cm3/molc sec	25	EST	MEYLAN,WM & HOWARD,PH (1993)

CAS #: 052067-49-9	O=P(OC)(SC)N-N-BU

Formula: C$_6$H$_{16}$NO$_2$PS	
Mol Weight: 197.24	
MP (deg C):	FP (deg C):
BP (deg C):	
BP pressure (mm Hg):	

Property/Value	Units	Temp	Data Type	Reference
WS 7.12E+003	mg/L	25	EST	MEYLAN,WM ET AL. (1996)
logP 0.94			EXP	HANSCH,C & LEO,AJ (1985)
VP 4.54E-003	mm Hg	25	EST	NEELY,WB & BLAU,GE (1985)
DC	pKa			
HL 4.46E-009	atm m3/mol	25	EST	MEYLAN,WM & HOWARD,PH (1991)
OH 4.65E-011	cm3/molc sec	25	EST	MEYLAN,WM & HOWARD,PH (1993)

CAS #: 052067-50-2	O=P(OC)(SC)N-S-BU

Formula: C$_6$H$_{16}$NO$_2$PS	
Mol Weight: 197.24	
MP (deg C):	FP (deg C):
BP (deg C):	
BP pressure (mm Hg):	

Property/Value	Units	Temp	Data Type	Reference
WS 6.98E+003	mg/L	25	EST	MEYLAN,WM ET AL. (1996)
logP 0.95			EXP	HANSCH,C & LEO,AJ (1985)
VP 9.38E-003	mm Hg	25	EST	NEELY,WB & BLAU,GE (1985)
DC	pKa			
HL 4.46E-009	atm m3/mol	25	EST	MEYLAN,WM & HOWARD,PH (1991)
OH 5.68E-011	cm3/molc sec	25	EST	MEYLAN,WM & HOWARD,PH (1993)

CAS #: 052067-51-3

O=P(OC)(SC)N-I-BU

Formula: $C_6H_{16}NO_2PS$

Mol Weight: 197.24

MP (deg C): FP (deg C):

BP (deg C):

BP pressure (mm Hg):

Property/Value	Units	Temp	Data Type	Reference
WS 6.71E+003	mg/L	25	EST	MEYLAN,WM ET AL. (1996)
logP 0.97			EXP	HANSCH,C & LEO,AJ (1985)
VP 9.38E-003	mm Hg	25	EST	NEELY,WB & BLAU,GE (1985)
DC	pKa			
HL 4.46E-009	atm m3/mol	25	EST	MEYLAN,WM & HOWARD,PH (1991)
OH 4.65E-011	cm3/molc sec	25	EST	MEYLAN,WM & HOWARD,PH (1993)

CAS #: 052067-52-4

O,S-DIME-N-HEXYLPHOSPHOEAMIDITHIOATE

Formula: $C_8H_{20}NO_2PS$

Mol Weight: 225.29

MP (deg C): FP (deg C):

BP (deg C):

BP pressure (mm Hg):

Property/Value	Units	Temp	Data Type	Reference
WS 3.64E+003	mg/L	25	EST	MEYLAN,WM ET AL. (1996)
logP 1.11			EXP	HANSCH,C & LEO,AJ (1985)
VP 5.98E-004	mm Hg	25	EST	NEELY,WB & BLAU,GE (1985)
DC	pKa			
HL 7.86E-009	atm m3/mol	25	EST	MEYLAN,WM & HOWARD,PH (1991)
OH 4.93E-011	cm3/molc sec	25	EST	MEYLAN,WM & HOWARD,PH (1993)

CAS #: 052092-59-8

TETRAMETHOXYHYDROQUINONE

Formula: $C_{10}H_{14}O_6$

Mol Weight: 230.22

MP (deg C): FP (deg C):

BP (deg C):

BP pressure (mm Hg):

Property/Value	Units	Temp	Data Type	Reference
WS 1.41E+004	mg/L	25	EST	MEYLAN,WM ET AL. (1996)
logP 0.39			EXP	HANSCH,C ET AL. (1995)
VP 8.31E-007	mm Hg	25	EST	NEELY,WB & BLAU,GE (1985)
DC	pKa			
HL 7.14E-016	atm m3/mol	25	EST	MEYLAN,WM & HOWARD,PH (1991)
OH 2.04E-010	cm3/molc sec	25	EST	MEYLAN,WM & HOWARD,PH (1993)

CAS #: 052093-77-3

N'-ET-PHENOXYACETIC ACID HYDRAZIDE

Formula: $C_{10}H_{14}N_2O_2$

Mol Weight: 194.24

MP (deg C): FP (deg C):

BP (deg C):

BP pressure (mm Hg):

Property/Value	Units	Temp	Data Type	Reference
WS 6.55E+003	mg/L	25	EST	MEYLAN,WM ET AL. (1996)
logP 1.00			EXP	HANSCH,C & LEO,AJ (1985)
VP 5.14E-006	mm Hg	25	EST	NEELY,WB & BLAU,GE (1985)
DC	pKa			
HL 4.17E-012	atm m3/mol	25	EST	MEYLAN,WM & HOWARD,PH (1991)
OH 1.03E-010	cm3/molc sec	25	EST	MEYLAN,WM & HOWARD,PH (1993)

CAS #: 052093-78-4

N-BENZYL-PHENOXY ACETIC HYDRAZIDE

Formula: $C_{15}H_{16}N_2O_2$

Mol Weight: 256.31

MP (deg C): FP (deg C):

BP (deg C):

BP pressure (mm Hg):

Property/Value	Units	Temp	Data Type	Reference
WS 3.12E+002	mg/L	25	EST	MEYLAN,WM ET AL. (1996)
logP 2.16			EXP	HANSCH,C & LEO,AJ (1985)
VP 2.90E-008	mm Hg	25	EST	NEELY,WB & BLAU,GE (1985)
DC	pKa			
HL 2.54E-013	atm m3/mol	25	EST	MEYLAN,WM & HOWARD,PH (1991)
OH 1.08E-010	cm3/molc sec	25	EST	MEYLAN,WM & HOWARD,PH (1993)

CAS #: 052093-79-5

N'(1-ME-BENZYL)PHO-ACETIC ACID HYDRAZIDE

Formula: $C_{16}H_{18}N_2O_2$

Mol Weight: 270.33

MP (deg C): FP (deg C):

BP (deg C):

BP pressure (mm Hg):

Property/Value	Units	Temp	Data Type	Reference
WS 1.44E+002	mg/L	25	EST	MEYLAN,WM ET AL. (1996)
logP 2.46			EXP	HANSCH,C & LEO,AJ (1985)
VP 2.15E-008	mm Hg	25	EST	NEELY,WB & BLAU,GE (1985)
DC	pKa			
HL 3.37E-013	atm m3/mol	25	EST	MEYLAN,WM & HOWARD,PH (1991)
OH 1.17E-010	cm3/molc sec	25	EST	MEYLAN,WM & HOWARD,PH (1993)

CAS #: 052102-38-2

BENZENESULFONAMIDE, N-[(METHYLAMINO)CARBONYL]-

Formula: $C_8H_{10}N_2O_3S$

Mol Weight: 214.24

MP (deg C): FP (deg C):

BP (deg C):

BP pressure (mm Hg):

Property/Value	Units	Temp	Data Type	Reference
WS 1.55E+004	mg/L	25	EST	MEYLAN,WM ET AL. (1996)
logP 0.44			EXP	CLOUX,JL ET AL. (1988)
VP 8.67E-007	mm Hg	25	EST	NEELY,WB & BLAU,GE (1985)
DC	pKa			
HL 9.80E-011	atm m3/mol	25	EST	MEYLAN,WM & HOWARD,PH (1991)
OH 2.97E-012	cm3/molc sec	25	EST	MEYLAN,WM & HOWARD,PH (1993)

CAS #: 052102-41-7

BENZENESULFONAMIDE, N-[[(1,1-DIMETHYLETHYL)AMINO

Formula: $C_{11}H_{16}N_2O_3S$

Mol Weight: 256.33

MP (deg C): FP (deg C):

BP (deg C):

BP pressure (mm Hg):

Property/Value	Units	Temp	Data Type	Reference
WS 4.72E+002	mg/L	25	EST	MEYLAN,WM ET AL. (1996)
logP 1.95			EXP	CLOUX,JL ET AL. (1988)
VP 1.48E-007	mm Hg	25	EST	NEELY,WB & BLAU,GE (1985)
DC	pKa			
HL 2.29E-010	atm m3/mol	25	EST	MEYLAN,WM & HOWARD,PH (1991)
OH 2.92E-012	cm3/molc sec	25	EST	MEYLAN,WM & HOWARD,PH (1993)

CAS #: 052102-43-9			BENZENESULFONAMIDE, 4-CHLORO-N-[(METHYLAMINO)CAR

Formula: $C_8H_9ClN_2O_3S$

Mol Weight: 248.69

MP (deg C):   FP (deg C):

BP (deg C):

BP pressure (mm Hg):

Property/Value	Units	Temp	Data Type	Reference
WS 2.46E+003	mg/L	25	EST	MEYLAN,WM ET AL. (1996)
logP 1.16			EXP	SANGSTER,J (1993)
VP 1.95E-007	mm Hg	25	EST	NEELY,WB & BLAU,GE (1985)
DC	pKa			
HL 7.26E-011	atm m3/mol	25	EST	MEYLAN,WM & HOWARD,PH (1991)
OH 2.85E-012	cm3/molc sec	25	EST	MEYLAN,WM & HOWARD,PH (1993)

CAS #: 052109-40-7			8-SEBENZYL CYCLIC AMP

Formula: $C_{17}H_{20}N_5O_6PSe$

Mol Weight: 500.31

MP (deg C):   FP (deg C):

BP (deg C):

BP pressure (mm Hg):

Property/Value	Units	Temp	Data Type	Reference
WS 3.66E+002	mg/L	25	EST	MEYLAN,WM ET AL. (1996)
logP -1.18			EXP	KORTH,M & ENGELS,J (1987)
VP 6.06E-011	mm Hg	25	EST	NEELY,WB & BLAU,GE (1985)
DC	pKa			
HL 9.49E-031	atm m3/mol	25	EST	MEYLAN,WM & HOWARD,PH (1991)
OH 3.79E-010	cm3/molc sec	25	EST	MEYLAN,WM & HOWARD,PH (1993)

CAS #: 052113-78-7			CARBAMIC ACID, DIMETHYL-, 2-METHYLPROPYL ESTER

Formula: $C_7H_{15}NO_2$

Mol Weight: 145.20

MP (deg C):   FP (deg C):

BP (deg C):

BP pressure (mm Hg):

Property/Value	Units	Temp	Data Type	Reference
WS 1.56E+003	mg/L	25	EST	MEYLAN,WM ET AL. (1996)
logP 2.00			EXP	SANGSTER,J (1994)
VP 2.96E+000	mm Hg	25	EST	NEELY,WB & BLAU,GE (1985)
DC	pKa			
HL 4.46E-007	atm m3/mol	25	EST	MEYLAN,WM & HOWARD,PH (1991)
OH 2.07E-011	cm3/molc sec	25	EST	MEYLAN,WM & HOWARD,PH (1993)

CAS #: 052121-41-2			3-ETHYL-4-BROMOACETANILIDE

Formula: $C_{10}H_{12}BrNO$

Mol Weight: 242.12

MP (deg C):   FP (deg C):

BP (deg C):

BP pressure (mm Hg):

Property/Value	Units	Temp	Data Type	Reference
WS 3.75E+001	mg/L	25	EST	MEYLAN,WM ET AL. (1996)
logP 3.33			EXP	NAKAGAWA,Y ET AL. (1992)
VP 7.20E-006	mm Hg	25	EST	NEELY,WB & BLAU,GE (1985)
DC	pKa			
HL 3.60E-009	atm m3/mol	25	EST	MEYLAN,WM & HOWARD,PH (1991)
OH 1.00E-011	cm3/molc sec	25	EST	MEYLAN,WM & HOWARD,PH (1993)

CAS #: 052125-53-8			PROPYLENE GLYCOL MONOETHYL ETHER

Formula: $C_5H_{12}O_2$

Mol Weight: 104.15

MP (deg C):   FP (deg C):

BP (deg C):

BP pressure (mm Hg):

Property/Value	Units	Temp	Data Type	Reference
WS 3.66E+005	mg/L	25	EST	MEYLAN,WM ET AL. (1996)
logP 0.00			EST	MEYLAN,WM & HOWARD,PH (1995)
VP 2.42E+000	mm Hg	25	EST	NEELY,WB & BLAU,GE (1985)
DC	pKa			
HL 7.38E-008	atm m3/mol	25	EST	MEYLAN,WM & HOWARD,PH (1991)
OH 5.86E-011	cm3/molc sec	25	EST	MEYLAN,WM & HOWARD,PH (1993)

CAS #: 052128-35-5			24NH2-5ME-6(345MEOPHNME)QUINAZOLINE

Formula: $C_{19}H_{23}N_5O_3$

Mol Weight: 369.43

MP (deg C): 215-217   FP (deg C):

BP (deg C):

BP pressure (mm Hg):

Property/Value	Units	Temp	Data Type	Reference
WS 3.14E+001	mg/L	25	EST	MEYLAN,WM ET AL. (1996)
logP 2.55			EXP	HANSCH,C & LEO,AJ (1985)
VP 6.03E-012	mm Hg	25	EST	NEELY,WB & BLAU,GE (1985)
DC	pKa			
HL 1.81E-018	atm m3/mol	25	EST	MEYLAN,WM & HOWARD,PH (1991)
OH 4.11E-010	cm3/molc sec	25	EST	MEYLAN,WM & HOWARD,PH (1993)

CAS #: 052189-36-3			BENZOIC ACID, 3-(ACETYLAMINO)-, METHYL ESTER

Formula: $C_{10}H_{11}NO_3$

Mol Weight: 193.20

MP (deg C):   FP (deg C):

BP (deg C):

BP pressure (mm Hg):

Property/Value	Units	Temp	Data Type	Reference
WS 2.68E+003	mg/L	25	EST	MEYLAN,WM ET AL. (1996)
logP 1.46			EXP	HANSCH,C ET AL. (1995)
VP 9.60E-006	mm Hg	25	EST	NEELY,WB & BLAU,GE (1985)
DC	pKa			
HL 3.98E-011	atm m3/mol	25	EST	MEYLAN,WM & HOWARD,PH (1991)
OH 3.07E-012	cm3/molc sec	25	EST	MEYLAN,WM & HOWARD,PH (1993)

CAS #: 052190-69-9			1,3-BENZODIOXOLE-5-CARBOXYLIC ACID, 2-(AMINOTHIO

Formula: $C_9H_9N_3O_3S$

Mol Weight: 239.25

MP (deg C):   FP (deg C):

BP (deg C):

BP pressure (mm Hg):

Property/Value	Units	Temp	Data Type	Reference
WS 8.79E+002	mg/L	25	EST	MEYLAN,WM ET AL. (1996)
logP 0.21			EXP	SANGSTER,J (1994)
VP 1.41E-008	mm Hg	25	EST	NEELY,WB & BLAU,GE (1985)
DC	pKa			
HL 4.92E-016	atm m3/mol	25	EST	MEYLAN,WM & HOWARD,PH (1991)
OH 1.09E-010	cm3/molc sec	25	EST	MEYLAN,WM & HOWARD,PH (1993)

CAS #: 052222-71-6				4-PENTYLPYRAZOLE

Formula: $C_8H_{14}N_2$

Mol Weight: 138.21

MP (deg C):   FP (deg C):

BP (deg C):

BP pressure (mm Hg):

Property/Value	Units	Temp	Data Type	Reference
WS 2.53E+002	mg/L	25	EST	MEYLAN,WM ET AL. (1996)
logP 2.96			EXP	HANSCH,C & LEO,AJ (1985)
VP 4.48E-003	mm Hg	25	EST	NEELY,WB & BLAU,GE (1985)
DC	pKa			
HL 1.27E-005	atm m3/mol	25	EST	MEYLAN,WM & HOWARD,PH (1991)
OH 9.47E-011	cm3/molc sec	25	EST	MEYLAN,WM & HOWARD,PH (1993)

CAS #: 052236-30-3				6-(TERT-BUTYL)-1,2,4-TIRAZINE-3,5(2H,4H)-DIONE

Formula: $C_7H_{11}N_3O_2$

Mol Weight: 169.18

MP (deg C):   FP (deg C):

BP (deg C):

BP pressure (mm Hg):

Property/Value	Units	Temp	Data Type	Reference
WS 5.35E+003	mg/L	25	EST	MEYLAN,WM ET AL. (1996)
logP 1.25			EST	MEYLAN,WM & HOWARD,PH (1995)
VP 1.69E-008	mm Hg	25	EST	NEELY,WB & BLAU,GE (1985)
DC	pKa			
HL 5.94E-009	atm m3/mol	25	EST	MEYLAN,WM & HOWARD,PH (1991)
OH 2.50E-012	cm3/molc sec	25	EST	MEYLAN,WM & HOWARD,PH (1993)

CAS #: 052237-19-1				BENZENEPROPANOIC ACID, á-(AMINOMETHYL)-4-FLUORO-

Formula: $C_{10}H_{12}FNO_2$

Mol Weight: 197.21

MP (deg C):   FP (deg C):

BP (deg C):

BP pressure (mm Hg):

Property/Value	Units	Temp	Data Type	Reference
WS 8.95E+003	mg/L	25	EST	MEYLAN,WM ET AL. (1996)
logP -1.60			EXP	HANSCH,C ET AL. (1995)
VP 1.62E-008	mm Hg	25	EST	NEELY,WB & BLAU,GE (1985)
DC	pKa			
HL 9.36E-012	atm m3/mol	25	EST	MEYLAN,WM & HOWARD,PH (1991)
OH 3.89E-011	cm3/molc sec	25	EST	MEYLAN,WM & HOWARD,PH (1993)

CAS #: 052248-39-2				4-THIA-1-AZABICYCLO[3.2.0]HEPTANE-2-COOH DERIVAT

Formula: $C_{19}H_{17}Cl_2N_3O_6S$

Mol Weight: 486.33

MP (deg C):   FP (deg C):

BP (deg C):

BP pressure (mm Hg):

Property/Value	Units	Temp	Data Type	Reference
WS 5.59E+000	mg/L	25	EST	MEYLAN,WM ET AL. (1996)
logP 2.57			EXP	SANGSTER,J (1994)
VP 1.53E-017	mm Hg	25	EST	NEELY,WB & BLAU,GE (1985)
DC	pKa			
HL 8.82E-022	atm m3/mol	25	EST	MEYLAN,WM & HOWARD,PH (1991)
OH 1.54E-010	cm3/molc sec	25	EST	MEYLAN,WM & HOWARD,PH (1993)

CAS #: 052298-71-2				ETHANOL, 2-[[4,6-BIS(DIMETHYLAMINO)-1,3,5-TRIAZI

Formula: $C_{10}H_{20}N_6O$

Mol Weight: 240.31

MP (deg C):   FP (deg C):

BP (deg C):

BP pressure (mm Hg):

Property/Value	Units	Temp	Data Type	Reference
WS 5.67E+002	mg/L	25	EST	MEYLAN,WM ET AL. (1996)
logP 1.96			EXP	SANGSTER,J (1993)
VP 1.40E-007	mm Hg	25	EST	NEELY,WB & BLAU,GE (1985)
DC	pKa			
HL 8.36E-013	atm m3/mol	25	EST	MEYLAN,WM & HOWARD,PH (1991)
OH 2.18E-011	cm3/molc sec	25	EST	MEYLAN,WM & HOWARD,PH (1993)

CAS #: 052303-69-2				PROPANEDIOIC ACID, 1,3-DITHIOLAN-2-YLIDENE-, BIS

Formula: $C_{12}H_{18}O_5S_2$

Mol Weight: 306.40

MP (deg C):   FP (deg C):

BP (deg C):

BP pressure (mm Hg):

Property/Value	Units	Temp	Data Type	Reference
WS 4.85E+002	mg/L	25	EST	MEYLAN,WM ET AL. (1996)
logP 1.60			EXP	HANSCH,C ET AL. (1995)
VP 1.11E-005	mm Hg	25	EST	NEELY,WB & BLAU,GE (1985)
DC	pKa			
HL 1.53E-014	atm m3/mol	25	EST	MEYLAN,WM & HOWARD,PH (1991)
OH 9.64E-011	cm3/molc sec	25	EST	MEYLAN,WM & HOWARD,PH (1993)

CAS #: 052303-77-2				2-FURANMETHANAMINIUM, TETRAHYDRO-N,N,N-TRIMETHYL

Formula: $C_8H_{18}INO$

Mol Weight: 271.14

MP (deg C):   FP (deg C):

BP (deg C):

BP pressure (mm Hg):

Property/Value	Units	Temp	Data Type	Reference
WS 1.00E+006	mg/L	25	EST	MEYLAN,WM ET AL. (1996)
logP -2.30			EXP	PRATESI,P ET AL. (1984)
VP 5.71E-008	mm Hg	25	EST	NEELY,WB & BLAU,GE (1985)
DC	pKa			
HL 4.00E-015	atm m3/mol	25	EST	MEYLAN,WM & HOWARD,PH (1991)
OH 3.88E-011	cm3/molc sec	25	EST	MEYLAN,WM & HOWARD,PH (1993)

CAS #: 052315-07-8				CYPERMETHRIN

Formula: $C_{22}H_{19}Cl_2NO_3$

Mol Weight: 416.31

MP (deg C): 70   FP (deg C):

BP (deg C):

BP pressure (mm Hg):

Property/Value	Units	Temp	Data Type	Reference
WS 4.00E-003	mg/L	20	EXP	WAUCHOPE,RD ET AL. (1991A)
logP 6.00			EXP	SANGSTER,J (1993)
VP 1.40E-009	mm Hg	20	EXT	GRAYSON,BT ET AL. (1982)
DC	pKa			
HL 1.92E-007	atm m3/mol	20	EST	VP/WSOL
OH 3.70E-011	cm3/molc sec	25	EST	ATKINSON,R (1988)

CAS #: 052320-85-1	CLET-NITROSOUREIDOGLUTARIC ACID

Formula: $C_8H_{12}ClN_3O_6$

Mol Weight: 281.65

MP (deg C):   FP (deg C):

BP (deg C):

BP pressure (mm Hg):

Property/Value	Units	Temp	Data Type	Reference
WS 4.82E+003	mg/L	25	EST	MEYLAN,WM ET AL. (1996)
logP 0.60			EXP	HANSCH,C & LEO,AJ (1985)
VP 2.01E-009	mm Hg	25	EST	NEELY,WB & BLAU,GE (1985)
DC	pKa			
HL 1.98E-019	atm m3/mol	25	EST	MEYLAN,WM & HOWARD,PH (1991)
OH 1.35E-011	cm3/molc sec	25	EST	MEYLAN,WM & HOWARD,PH (1993)

CAS #: 052320-87-3	ALANINE, N-[[(2-CHLOROETHYL)NITROSOAMINO>CARBONY

Formula: $C_7H_{12}ClN_3O_4$

Mol Weight: 237.64

MP (deg C):   FP (deg C):

BP (deg C):

BP pressure (mm Hg):

Property/Value	Units	Temp	Data Type	Reference
WS 9.04E+002	mg/L	25	EST	MEYLAN,WM ET AL. (1996)
logP 1.74			EXP	HANSCH,C ET AL. (1995)
VP 8.19E-007	mm Hg	25	EST	NEELY,WB & BLAU,GE (1985)
DC	pKa			
HL 2.66E-014	atm m3/mol	25	EST	MEYLAN,WM & HOWARD,PH (1991)
OH 3.67E-012	cm3/molc sec	25	EST	MEYLAN,WM & HOWARD,PH (1993)

CAS #: 052320-88-4	CYCLOPENTANECARBOXYLIC ACID, 1-[[[(2-CHLOROETHYL

Formula: $C_9H_{14}ClN_3O_4$

Mol Weight: 263.68

MP (deg C):   FP (deg C):

BP (deg C):

BP pressure (mm Hg):

Property/Value	Units	Temp	Data Type	Reference
WS 3.81E+002	mg/L	25	EST	MEYLAN,WM ET AL. (1996)
logP 2.01			EXP	HANSCH,C ET AL. (1995)
VP 8.75E-008	mm Hg	25	EST	NEELY,WB & BLAU,GE (1985)
DC	pKa			
HL 2.07E-014	atm m3/mol	25	EST	MEYLAN,WM & HOWARD,PH (1991)
OH 7.86E-012	cm3/molc sec	25	EST	MEYLAN,WM & HOWARD,PH (1993)

CAS #: 052322-80-2	BENZENE, 1,2,4-TRICHLORO-5-PHENOXY-

Formula: $C_{12}H_7Cl_3O$

Mol Weight: 273.55

MP (deg C):   FP (deg C):

BP (deg C):

BP pressure (mm Hg):

Property/Value	Units	Temp	Data Type	Reference
WS 3.94E-001	mg/L	25	EST	MEYLAN,WM ET AL. (1996)
logP 5.44			EXP	OPPERHUIZEN,A & VOORS,PI (1987)
VP 3.82E-005	mm Hg	25	EST	NEELY,WB & BLAU,GE (1985)
DC	pKa			
HL 4.79E-005	atm m3/mol	25	EST	MEYLAN,WM & HOWARD,PH (1991)
OH 5.22E-012	cm3/molc sec	25	EST	MEYLAN,WM & HOWARD,PH (1993)

CAS #: 052329-50-7	L-PHENYLALANINAMIDE, N-ACETYL-L-TYROSYL-

Formula: $C_{20}H_{23}N_3O_4$

Mol Weight: 369.42

MP (deg C):   FP (deg C):

BP (deg C):

BP pressure (mm Hg):

Property/Value	Units	Temp	Data Type	Reference
WS 6.21E+003	mg/L	25	EST	MEYLAN,WM ET AL. (1996)
logP 0.54			EXP	HANSCH,C ET AL. (1995)
VP 3.65E-017	mm Hg	25	EST	NEELY,WB & BLAU,GE (1985)
DC	pKa			
HL 9.38E-022	atm m3/mol	25	EST	MEYLAN,WM & HOWARD,PH (1991)
OH 8.07E-011	cm3/molc sec	25	EST	MEYLAN,WM & HOWARD,PH (1993)

CAS #: 052386-94-4	GLYCINE, N-(3-IODOBENZOYL)-

Formula: $C_9H_8INO_3$

Mol Weight: 305.07

MP (deg C):   FP (deg C):

BP (deg C):

BP pressure (mm Hg):

Property/Value	Units	Temp	Data Type	Reference
WS 2.84E+002	mg/L	25	EST	MEYLAN,WM ET AL. (1996)
logP 1.88			EXP	LAZNICEK,M & KVETINA,J (1988)
VP 2.26E-008	mm Hg	25	EST	NEELY,WB & BLAU,GE (1985)
DC	pKa			
HL 1.67E-013	atm m3/mol	25	EST	MEYLAN,WM & HOWARD,PH (1991)
OH 9.48E-012	cm3/molc sec	25	EST	MEYLAN,WM & HOWARD,PH (1993)

CAS #: 052387-57-2	N-(ETHYLAMINOMETHYL)BENZAMIDE

Formula: $C_{10}H_{14}N_2O$

Mol Weight: 178.24

MP (deg C):   FP (deg C):

BP (deg C):

BP pressure (mm Hg):

Property/Value	Units	Temp	Data Type	Reference
WS 1.87E+004	mg/L	25	EST	MEYLAN,WM ET AL. (1996)
logP 0.56			EXP	HANSCH,C & LEO,AJ (1985)
VP 1.31E-005	mm Hg	25	EST	NEELY,WB & BLAU,GE (1985)
DC	pKa			
HL 4.61E-010	atm m3/mol	25	EST	MEYLAN,WM & HOWARD,PH (1991)
OH 1.15E-010	cm3/molc sec	25	EST	MEYLAN,WM & HOWARD,PH (1993)

CAS #: 052387-58-3	N-(ISO-BUTYLAMINOMETHYL)BENZAMIDE

Formula: $C_{12}H_{18}N_2O$

Mol Weight: 206.29

MP (deg C):   FP (deg C):

BP (deg C):

BP pressure (mm Hg):

Property/Value	Units	Temp	Data Type	Reference
WS 1.35E+003	mg/L	25	EST	MEYLAN,WM ET AL. (1996)
logP 1.73			EXP	HANSCH,C & LEO,AJ (1985)
VP 4.28E-006	mm Hg	25	EST	NEELY,WB & BLAU,GE (1985)
DC	pKa			
HL 8.12E-010	atm m3/mol	25	EST	MEYLAN,WM & HOWARD,PH (1991)
OH 1.19E-010	cm3/molc sec	25	EST	MEYLAN,WM & HOWARD,PH (1993)

CAS #: 052416-13-4 — 1-(O-THIOANISYL)-33-DIME TRIAZINE

Formula: $C_9H_{13}N_3S$
Mol Weight: 195.29
MP (deg C):
FP (deg C):
BP (deg C):
BP pressure (mm Hg):

Property/Value	Units	Temp	Data Type	Reference
WS 9.82E+001	mg/L	25	EST	MEYLAN,WM ET AL. (1996)
logP 3.13			EXP	HANSCH,C & LEO,AJ (1985)
VP 3.27E-003	mm Hg	25	EST	NEELY,WB & BLAU,GE (1985)
DC	pKa			
HL 1.56E-008	atm m3/mol	25	EST	MEYLAN,WM & HOWARD,PH (1991)
OH 8.57E-012	cm3/molc sec	25	EST	MEYLAN,WM & HOWARD,PH (1993)

CAS #: 052416-14-5 — 1-(3-THIOANISYL)-33-DIME TRIAZINE

Formula: $C_9H_{13}N_3S$
Mol Weight: 195.29
MP (deg C):
FP (deg C):
BP (deg C):
BP pressure (mm Hg):

Property/Value	Units	Temp	Data Type	Reference
WS 1.32E+002	mg/L	25	EST	MEYLAN,WM ET AL. (1996)
logP 2.98			EXP	HANSCH,C & LEO,AJ (1985)
VP 3.27E-003	mm Hg	25	EST	NEELY,WB & BLAU,GE (1985)
DC	pKa			
HL 1.56E-008	atm m3/mol	25	EST	MEYLAN,WM & HOWARD,PH (1991)
OH 5.34E-012	cm3/molc sec	25	EST	MEYLAN,WM & HOWARD,PH (1993)

CAS #: 052642-50-9 — O-T-HEXYMOPYRIMIDINE

Formula: $C_7H_{15}NO_2$
Mol Weight: 145.20
MP (deg C):
FP (deg C):
BP (deg C):
BP pressure (mm Hg):

Property/Value	Units	Temp	Data Type	Reference
WS 4.60E+003	mg/L	25	EST	MEYLAN,WM ET AL. (1996)
logP 1.45			EXP	HANSCH,C ET AL. (1995)
VP 8.06E-001	mm Hg	25	EST	NEELY,WB & BLAU,GE (1985)
DC	pKa			
HL 1.63E-007	atm m3/mol	25	EST	MEYLAN,WM & HOWARD,PH (1991)
OH 5.06E-012	cm3/molc sec	25	EST	MEYLAN,WM & HOWARD,PH (1993)

CAS #: 052643-52-4 — 1,4-BENZENEDIOL, 2,3-DIMETHOXY-

Formula: $C_8H_{10}O_4$
Mol Weight: 170.17
MP (deg C):
FP (deg C):
BP (deg C):
BP pressure (mm Hg):

Property/Value	Units	Temp	Data Type	Reference
WS 1.85E+004	mg/L	25	EST	MEYLAN,WM ET AL. (1996)
logP 0.61			EXP	SANGSTER,J (1994)
VP 5.52E-005	mm Hg	25	EST	NEELY,WB & BLAU,GE (1985)
DC	pKa			
HL 2.04E-013	atm m3/mol	25	EST	MEYLAN,WM & HOWARD,PH (1991)
OH 2.02E-010	cm3/molc sec	25	EST	MEYLAN,WM & HOWARD,PH (1993)

CAS #: 052645-53-1 — PERMETHRIN

Formula: $C_{21}H_{20}Cl_2O_3$
Mol Weight: 391.30
MP (deg C): 34-39
FP (deg C):
BP (deg C): 220
BP pressure (mm Hg): 5.00E-002

Property/Value	Units	Temp	Data Type	Reference
WS 2.00E-001	mg/L	20	EXP	HARTLEY,D & KIDD,H (1983)
logP 6.50			EXP	HANSCH,C & LEO,AJ (1985)
VP 9.75E-009	mm Hg	20	EXP	WORTHING,CR & WALKER,SB (1987)
DC	pKa			
HL 2.51E-008	atm m3/mol	20	EST	VP/WSOL
OH 3.90E-011	cm3/molc sec	25	EST	ATKINSON,R (1988)

CAS #: 052661-38-8 — 2-NH2-5-(5NO2-2-FURFURYL)THIAZOLONE

Formula: $C_8H_6N_3O_4S$
Mol Weight: 240.22
MP (deg C):
FP (deg C):
BP (deg C):
BP pressure (mm Hg):

Property/Value	Units	Temp	Data Type	Reference
WS 1.30E+003	mg/L	25	EST	MEYLAN,WM ET AL. (1996)
logP 1.09			EXP	HANSCH,C & LEO,AJ (1985)
VP 1.24E-007	mm Hg	25	EST	NEELY,WB & BLAU,GE (1985)
DC	pKa			
HL 8.13E-016	atm m3/mol	25	EST	MEYLAN,WM & HOWARD,PH (1991)
OH 5.79E-011	cm3/molc sec	25	EST	MEYLAN,WM & HOWARD,PH (1993)

CAS #: 052661-42-4 — ACETAMIDE, N-[5-[(5-NITRO-2-FURANYL)METHYLENE]-4

Formula: $C_{10}H_7N_3O_5S$
Mol Weight: 281.25
MP (deg C):
FP (deg C):
BP (deg C):
BP pressure (mm Hg):

Property/Value	Units	Temp	Data Type	Reference
WS 8.47E+002	mg/L	25	EST	MEYLAN,WM ET AL. (1996)
logP 1.03			EXP	SANGSTER,J (1994)
VP 2.54E-010	mm Hg	25	EST	NEELY,WB & BLAU,GE (1985)
DC	pKa			
HL 4.17E-015	atm m3/mol	25	EST	MEYLAN,WM & HOWARD,PH (1991)
OH 4.25E-011	cm3/molc sec	25	EST	MEYLAN,WM & HOWARD,PH (1993)

CAS #: 052661-43-5 — ACETAMIDE, N-METHYL-N-[5-[(5-NITRO-2-FURANYL)MET

Formula: $C_{11}H_9N_3O_5S$
Mol Weight: 295.28
MP (deg C):
FP (deg C):
BP (deg C):
BP pressure (mm Hg):

Property/Value	Units	Temp	Data Type	Reference
WS 6.76E+002	mg/L	25	EST	MEYLAN,WM ET AL. (1996)
logP 1.05			EXP	SANGSTER,J (1994)
VP 1.96E-009	mm Hg	25	EST	NEELY,WB & BLAU,GE (1985)
DC	pKa			
HL 9.17E-015	atm m3/mol	25	EST	MEYLAN,WM & HOWARD,PH (1991)
OH 5.25E-011	cm3/molc sec	25	EST	MEYLAN,WM & HOWARD,PH (1993)

CAS #:	052661-45-7	2-MES-5-(5NO2-2-FURFURILD)THIAZOLONE

Formula: $C_9H_7N_2O_4S_2$

Mol Weight: 271.30

MP (deg C): FP (deg C):

BP (deg C):

BP pressure (mm Hg):

Property/Value	Units	Temp	Data Type	Reference
WS 4.64E+002	mg/L	25	EST	MEYLAN,WM ET AL. (1996)
logP 1.41			EXP	HANSCH,C & LEO,AJ (1985)
VP 4.71E-008	mm Hg	25	EST	NEELY,WB & BLAU,GE (1985)
DC	pKa			
HL 8.43E-014	atm m3/mol	25	EST	MEYLAN,WM & HOWARD,PH (1991)
OH 3.96E-011	cm3/molc sec	25	EST	MEYLAN,WM & HOWARD,PH (1993)

CAS #:	052661-48-0	6(5NO2-2-FURYL)-13-THIAZIN-24-DIONE

Formula: $C_8H_5N_2O_5S$

Mol Weight: 241.20

MP (deg C): FP (deg C):

BP (deg C):

BP pressure (mm Hg):

Property/Value	Units	Temp	Data Type	Reference
WS 3.70E+003	mg/L	25	EST	MEYLAN,WM ET AL. (1996)
logP 0.55			EXP	HANSCH,C & LEO,AJ (1985)
VP 7.20E-011	mm Hg	25	EST	NEELY,WB & BLAU,GE (1985)
DC	pKa			
HL 4.11E-010	atm m3/mol	25	EST	MEYLAN,WM & HOWARD,PH (1991)
OH 4.71E-011	cm3/molc sec	25	EST	MEYLAN,WM & HOWARD,PH (1993)

CAS #:	052661-53-7	2MEAM-6(5NO2-2FURYL)-13-THIAZIN-4-ONE

Formula: $C_9H_8N_3O_4S$

Mol Weight: 254.25

MP (deg C): FP (deg C):

BP (deg C):

BP pressure (mm Hg):

Property/Value	Units	Temp	Data Type	Reference
WS 1.70E+003	mg/L	25	EST	MEYLAN,WM ET AL. (1996)
logP 0.86			EXP	HANSCH,C & LEO,AJ (1985)
VP 1.04E-007	mm Hg	25	EST	NEELY,WB & BLAU,GE (1985)
DC	pKa			
HL 1.78E-015	atm m3/mol	25	EST	MEYLAN,WM & HOWARD,PH (1991)
OH 1.01E-010	cm3/molc sec	25	EST	MEYLAN,WM & HOWARD,PH (1993)

CAS #:	052661-54-8	ACETAMIDE, N-[6-(5-NITRO-2-FURANYL)-4-OXO-4H-1,3

Formula: $C_{10}H_7N_3O_5S$

Mol Weight: 281.25

MP (deg C): FP (deg C):

BP (deg C):

BP pressure (mm Hg):

Property/Value	Units	Temp	Data Type	Reference
WS 8.81E+002	mg/L	25	EST	MEYLAN,WM ET AL. (1996)
logP 1.01			EXP	SANGSTER,J (1994)
VP 2.53E-010	mm Hg	25	EST	NEELY,WB & BLAU,GE (1985)
DC	pKa			
HL 4.17E-015	atm m3/mol	25	EST	MEYLAN,WM & HOWARD,PH (1991)
OH 4.25E-011	cm3/molc sec	25	EST	MEYLAN,WM & HOWARD,PH (1993)

CAS #:	052661-66-2	3-ME-2-ACETIMINO-5-(5NO2-2-FURFURILIDENE)THIAZOL

Formula: $C_{11}H_9N_3O_5S$

Mol Weight: 295.28

MP (deg C): FP (deg C):

BP (deg C):

BP pressure (mm Hg):

Property/Value	Units	Temp	Data Type	Reference
WS 4.94E+002	mg/L	25	EST	MEYLAN,WM ET AL. (1996)
logP 1.21			EXP	HANSCH,C & LEO,AJ (1985)
VP 1.96E-009	mm Hg	25	EST	NEELY,WB & BLAU,GE (1985)
DC	pKa			
HL 9.17E-015	atm m3/mol	25	EST	MEYLAN,WM & HOWARD,PH (1991)
OH 5.25E-011	cm3/molc sec	25	EST	MEYLAN,WM & HOWARD,PH (1993)

CAS #:	052661-68-4	ACETAMIDE, N-[5-[(5-NITRO-2-FURANYL)METHYLENE]-4

Formula: $C_{13}H_{13}N_3O_5S$

Mol Weight: 323.33

MP (deg C): FP (deg C):

BP (deg C):

BP pressure (mm Hg):

Property/Value	Units	Temp	Data Type	Reference
WS 3.65E+002	mg/L	25	EST	MEYLAN,WM ET AL. (1996)
logP 1.17			EXP	AKERBLOM,EB (1974)
VP 3.80E-010	mm Hg	25	EST	NEELY,WB & BLAU,GE (1985)
DC	pKa			
HL 1.62E-014	atm m3/mol	25	EST	MEYLAN,WM & HOWARD,PH (1991)
OH 5.80E-011	cm3/molc sec	25	EST	MEYLAN,WM & HOWARD,PH (1993)

CAS #:	052661-71-9	2,4-THIAZOLIDINEDIONE, 5-[(5-NITRO-2-FURANYL)MET

Formula: $C_{11}H_{10}N_2O_5S$

Mol Weight: 282.28

MP (deg C): FP (deg C):

BP (deg C):

BP pressure (mm Hg):

Property/Value	Units	Temp	Data Type	Reference
WS 5.65E+001	mg/L	25	EST	MEYLAN,WM ET AL. (1996)
logP 2.40			EXP	SANGSTER,J (1993)
VP 5.85E-010	mm Hg	25	EST	NEELY,WB & BLAU,GE (1985)
DC	pKa			
HL 1.59E-009	atm m3/mol	25	EST	MEYLAN,WM & HOWARD,PH (1991)
OH 5.31E-011	cm3/molc sec	25	EST	MEYLAN,WM & HOWARD,PH (1993)

CAS #:	052662-76-7	4-CYCLOHEXYLSEMICARBAZIDE

Formula: $C_7H_{15}N_3O$

Mol Weight: 157.22

MP (deg C): FP (deg C):

BP (deg C):

BP pressure (mm Hg):

Property/Value	Units	Temp	Data Type	Reference
WS 2.57E+003	mg/L	25	EST	MEYLAN,WM ET AL. (1996)
logP 0.15			EXP	HANSCH,C & LEO,AJ (1985)
VP 1.76E-004	mm Hg	25	EST	NEELY,WB & BLAU,GE (1985)
DC	pKa			
HL 6.08E-012	atm m3/mol	25	EST	MEYLAN,WM & HOWARD,PH (1991)
OH 2.11E-011	cm3/molc sec	25	EST	MEYLAN,WM & HOWARD,PH (1993)

052663-59-9 — 2,2',3,4-TETRACHLOROBIPHENYL

Formula: $C_{12}H_6Cl_4$

Mol Weight: 291.99

MP (deg C):   FP (deg C):

BP (deg C):

BP pressure (mm Hg):

Property/Value	Units	Temp	Data Type	Reference
WS 8.27E-002	mg/L	25	EST	MEYLAN,WM ET AL. (1996)
logP 6.11			EXP	HANSCH,C & LEO,AJ (1985)
VP 8.45E-006	mm Hg	25	EST	NEELY,WB & BLAU,GE (1985)
DC	pKa			
HL 1.25E-004	atm m3/mol	25	EST	MEYLAN,WM & HOWARD,PH (1991)
OH 5.77E-013	cm3/molc sec	25	EST	MEYLAN,WM & HOWARD,PH (1993)

052663-60-2 — 2,2',3,3',6-PENTACHLOROBIPHENYL

Formula: $C_{12}H_5Cl_5$

Mol Weight: 326.44

MP (deg C):   FP (deg C):

BP (deg C):

BP pressure (mm Hg):

Property/Value	Units	Temp	Data Type	Reference
WS 9.39E-003	mg/L	25	EST	MEYLAN,WM ET AL. (1996)
logP 6.04			EXP	RAPAPORT,RA & EISENREICH,SJ (1984)
VP 2.22E-006	mm Hg	25	EST	NEELY,WB & BLAU,GE (1985)
DC	pKa			
HL 9.24E-005	atm m3/mol	25	EST	MEYLAN,WM & HOWARD,PH (1991)
OH 3.35E-013	cm3/molc sec	25	EST	MEYLAN,WM & HOWARD,PH (1993)

052663-61-3 — 2,2',3,5,5'-PENTACHLOROBIPHENYL

Formula: $C_{12}H_5Cl_5$

Mol Weight: 326.44

MP (deg C):   FP (deg C):

BP (deg C):

BP pressure (mm Hg):

Property/Value	Units	Temp	Data Type	Reference
WS 1.36E-002	mg/L	25	EST	MEYLAN,WM ET AL. (1996)
logP 6.79			EXP	HANSCH,C & LEO,AJ (1985)
VP 2.22E-006	mm Hg	25	EST	NEELY,WB & BLAU,GE (1985)
DC	pKa			
HL 9.24E-005	atm m3/mol	25	EST	MEYLAN,WM & HOWARD,PH (1991)
OH 4.73E-013	cm3/molc sec	25	EST	MEYLAN,WM & HOWARD,PH (1993)

052663-62-4 — 2,2',3,4,5-PENTACHLOROBIPHENYL

Formula: $C_{12}H_5Cl_5$

Mol Weight: 326.44

MP (deg C):   FP (deg C):

BP (deg C):

BP pressure (mm Hg):

Property/Value	Units	Temp	Data Type	Reference
WS 2.91E-002	mg/L	20	EXP	YALKOWSKY,SH & DANNENFELSER,RM (1992)
logP 6.98			EST	MEYLAN,WM & HOWARD,PH (1995)
VP 2.22E-006	mm Hg	25	EST	NEELY,WB & BLAU,GE (1985)
DC	pKa			
HL 9.24E-005	atm m3/mol	25	EST	MEYLAN,WM & HOWARD,PH (1991)
OH 3.35E-013	cm3/molc sec	25	EST	MEYLAN,WM & HOWARD,PH (1993)

052663-63-5 — 2,2',3,5,5',6-HEXACHLOROBIPHENYL

Formula: $C_{12}H_4Cl_6$

Mol Weight: 360.88

MP (deg C):   FP (deg C):

BP (deg C):

BP pressure (mm Hg):

Property/Value	Units	Temp	Data Type	Reference
WS 1.36E-002	mg/L	20	EXP	YALKOWSKY,SH & DANNENFELSER,RM (1992)
logP 7.62			EST	MEYLAN,WM & HOWARD,PH (1995)
VP 2.29E-006	mm Hg	25	EXP	BIDLEMAN,TF (1984)
DC	pKa			
HL 8.03E-005	atm m3/mol	25	EST	VP/WSOL
OH 2.11E-013	cm3/molc sec	25	EST	MEYLAN,WM & HOWARD,PH (1993)

052663-64-6 — 2,2',3,3',5,6,6'-HEPTACHLOROBIPHENYL

Formula: $C_{12}H_3Cl_7$

Mol Weight: 395.33

MP (deg C):   FP (deg C):

BP (deg C):

BP pressure (mm Hg):

Property/Value	Units	Temp	Data Type	Reference
WS 2.84E-004	mg/L	25	EST	MEYLAN,WM ET AL. (1996)
logP 8.27			EST	MEYLAN,WM & HOWARD,PH (1995)
VP 1.30E-007	mm Hg	25	EST	NEELY,WB & BLAU,GE (1985)
DC	pKa			
HL 5.07E-005	atm m3/mol	25	EST	MEYLAN,WM & HOWARD,PH (1991)
OH 1.05E-013	cm3/molc sec	25	EST	MEYLAN,WM & HOWARD,PH (1993)

052663-66-8 — 2,2',3,3',4,5'-HEXACHLOROBIPHENYL

Formula: $C_{12}H_4Cl_6$

Mol Weight: 360.88

MP (deg C):   FP (deg C):

BP (deg C):

BP pressure (mm Hg):

Property/Value	Units	Temp	Data Type	Reference
WS 2.60E-003	mg/L	25	EST	MEYLAN,WM ET AL. (1996)
logP 7.39			EXP	HANSCH,C & LEO,AJ (1985)
VP 5.81E-007	mm Hg	25	EST	NEELY,WB & BLAU,GE (1985)
DC	pKa			
HL 6.85E-005	atm m3/mol	25	EST	MEYLAN,WM & HOWARD,PH (1991)
OH 2.34E-013	cm3/molc sec	25	EST	MEYLAN,WM & HOWARD,PH (1993)

052663-68-0 — 2,2',3,4',5,5',6-HEPTACHLOROBIPHENYL

Formula: $C_{12}H_3Cl_7$

Mol Weight: 395.33

MP (deg C):   FP (deg C):

BP (deg C):

BP pressure (mm Hg):

Property/Value	Units	Temp	Data Type	Reference
WS 4.51E-003	mg/L	20	EXP	YALKOWSKY,SH & DANNENFELSER,RM (1992)
logP 8.27			EST	MEYLAN,WM & HOWARD,PH (1995)
VP 1.30E-007	mm Hg	25	EST	NEELY,WB & BLAU,GE (1985)
DC	pKa			
HL 5.07E-005	atm m3/mol	25	EST	MEYLAN,WM & HOWARD,PH (1991)
OH 1.05E-013	cm3/molc sec	25	EST	MEYLAN,WM & HOWARD,PH (1993)

CAS #: 052663-71-5				2,2',3,3',4,4',6-HEPTACHLOROBIPHENYL
Formula: $C_{12}H_3Cl_7$				
Mol Weight: 395.33				
MP (deg C):			FP (deg C):	
BP (deg C):				
BP pressure (mm Hg):				

Property/Value	Units	Temp	Data Type	Reference
WS 2.17E-003	mg/L	25	EXP	YALKOWSKY,SH & DANNENFELSER,RM (1992)
logP 8.27			EST	MEYLAN,WM & HOWARD,PH (1995)
VP 1.40E-006	mm Hg	25	EXP	BIDLEMAN,TF (1994)
DC	pKa			
HL 3.36E-004	atm m3/mol	25	EST	VP/WSOL
OH 1.18E-013	cm3/molc sec	25	EST	MEYLAN,WM & HOWARD,PH (1993)

CAS #: 052663-72-6				2,3',4,4',5,5'-HEXACHLOROBIPHENYL
Formula: $C_{12}H_4Cl_6$				
Mol Weight: 360.88				
MP (deg C):			FP (deg C):	
BP (deg C):				
BP pressure (mm Hg):				

Property/Value	Units	Temp	Data Type	Reference
WS 1.64E-003	mg/L	25	EST	MEYLAN,WM ET AL. (1996)
logP 7.62			EST	MEYLAN,WM & HOWARD,PH (1995)
VP 5.81E-007	mm Hg	25	EST	NEELY,WB & BLAU,GE (1985)
DC	pKa			
HL 6.85E-005	atm m3/mol	25	EST	MEYLAN,WM & HOWARD,PH (1991)
OH 2.34E-013	cm3/molc sec	25	EST	MEYLAN,WM & HOWARD,PH (1993)

CAS #: 052663-77-1				2,3,4,5,6,2',3',5',6'-PCB
Formula: $C_{12}HCl_9$				
Mol Weight: 464.22				
MP (deg C):			FP (deg C):	
BP (deg C):				
BP pressure (mm Hg):				

Property/Value	Units	Temp	Data Type	Reference
WS 1.80E-005	mg/L	25	EXP	YALKOWSKY,SH & DANNENFELSER,RM (1992)
logP 8.16			EXP	HANSCH,C ET AL. (1995)
VP 7.60E-009	mm Hg	25	EST	NEELY,WB & BLAU,GE (1985)
DC	pKa			
HL 2.79E-005	atm m3/mol	25	EST	MEYLAN,WM & HOWARD,PH (1991)
OH 3.29E-014	cm3/molc sec	25	EST	MEYLAN,WM & HOWARD,PH (1993)

CAS #: 052704-70-8				2,2'3,3',5,6-HEXACHLOROBIPHENYL
Formula: $C_{12}H_4Cl_6$				
Mol Weight: 360.88				
MP (deg C):			FP (deg C):	
BP (deg C):				
BP pressure (mm Hg):				

Property/Value	Units	Temp	Data Type	Reference
WS 8.99E-004	mg/L	25	EXP	KUHNE,R ET AL. (1995)
logP 7.62			EST	MEYLAN,WM & HOWARD,PH (1995)
VP 1.10E-006	mm Hg	25	EXP	SHIU,WY & MACKAY,D (1986)
DC	pKa			
HL 5.81E-004	atm m3/mol	25	EST	VP/WSOL
OH 2.11E-013	cm3/molc sec	25	EST	MEYLAN,WM & HOWARD,PH (1993)

CAS #: 052712-04-6				2,2',3,4,5,5'-HEXACHLOROBIPHENYL
Formula: $C_{12}H_4Cl_6$				
Mol Weight: 360.88				
MP (deg C):			FP (deg C):	
BP (deg C):				
BP pressure (mm Hg):				

Property/Value	Units	Temp	Data Type	Reference
WS 1.64E-003	mg/L	25	EST	MEYLAN,WM ET AL. (1996)
logP 7.19			EXP	HANSCH,C ET AL. (1995)
VP 5.81E-007	mm Hg	25	EST	NEELY,WB & BLAU,GE (1985)
DC	pKa			
HL 6.85E-005	atm m3/mol	25	EST	MEYLAN,WM & HOWARD,PH (1991)
OH 2.11E-013	cm3/molc sec	25	EST	MEYLAN,WM & HOWARD,PH (1993)

CAS #: 052712-05-7				2,2',3,4,5,5',6-HEPTACHLORO-1,1'-BIPHENYL
Formula: $C_{12}H_3Cl_7$				
Mol Weight: 395.33				
MP (deg C):			FP (deg C):	
BP (deg C):				
BP pressure (mm Hg):				

Property/Value	Units	Temp	Data Type	Reference
WS 5.46E-003	mg/L	20	EXP	YALKOWSKY,SH & DANNENFELSER,RM (1992)
logP 8.27			EST	MEYLAN,WM & HOWARD,PH (1995)
VP 1.30E-007	mm Hg	25	EST	NEELY,WB & BLAU,GE (1985)
DC	pKa			
HL 5.07E-005	atm m3/mol	25	EST	MEYLAN,WM & HOWARD,PH (1991)
OH 1.24E-013	cm3/molc sec	25	EST	MEYLAN,WM & HOWARD,PH (1993)

CAS #: 052744-13-5				2,2',3,3',5,6'-HEXACHLOROBIPHENYL
Formula: $C_{12}H_4Cl_6$				
Mol Weight: 360.88				
MP (deg C):			FP (deg C):	
BP (deg C):				
BP pressure (mm Hg):				

Property/Value	Units	Temp	Data Type	Reference
WS 4.17E-003	mg/L	25	EST	MEYLAN,WM ET AL. (1996)
logP 7.15			EXP	HANSCH,C & LEO,AJ (1985)
VP 5.81E-007	mm Hg	25	EST	NEELY,WB & BLAU,GE (1985)
DC	pKa			
HL 6.85E-005	atm m3/mol	25	EST	MEYLAN,WM & HOWARD,PH (1991)
OH 2.34E-013	cm3/molc sec	25	EST	MEYLAN,WM & HOWARD,PH (1993)

CAS #: 052748-69-3				STRYCHNINE HYDRIODIDE
Formula: $C_{21}H_{23}IN_2O_2$				
Mol Weight: 462.33				
MP (deg C):			FP (deg C):	
BP (deg C):				
BP pressure (mm Hg):				

Property/Value	Units	Temp	Data Type	Reference
WS 2.87E+003	mg/L	25	EXP	STEPHEN,H & STEPHEN,T (1963)
logP -1.67			EST	MEYLAN,WM & HOWARD,PH (1995)
VP 1.38E-015	mm Hg	25	EST	NEELY,WB & BLAU,GE (1985)
DC	pKa			
HL 3.76E-020	atm m3/mol	25	EST	MEYLAN,WM & HOWARD,PH (1991)
OH 1.86E-010	cm3/molc sec	25	EST	MEYLAN,WM & HOWARD,PH (1993)

CAS #: 052756-22-6 — FLAMPROP-ISOPROPYL

Formula: $C_{19}H_{19}ClFNO_3$

Mol Weight: 363.82

MP (deg C):

FP (deg C):

BP (deg C):

BP pressure (mm Hg):

Property/Value	Units	Temp	Data Type	Reference
WS 1.80E+001	mg/L	20	EXP	YALKOWSKY,SH & DANNENFELSER,RM (1992)
logP 4.24			EST	MEYLAN,WM & HOWARD,PH (1995)
VP 3.91E-008	mm Hg	25	EST	NEELY,WB & BLAU,GE (1985)
DC	pKa			
HL 2.41E-009	atm m3/mol	25	EST	MEYLAN,WM & HOWARD,PH (1991)
OH 1.42E-011	cm3/molc sec	25	EST	MEYLAN,WM & HOWARD,PH (1993)

CAS #: 052756-25-9 — FLAMPROP-METHYL

Formula: $C_{17}H_{15}ClFNO_3$

Mol Weight: 335.77

MP (deg C): 84-86

FP (deg C):

BP (deg C):

BP pressure (mm Hg):

Property/Value	Units	Temp	Data Type	Reference
WS 3.50E+001	mg/L	20	EXP	SHIU,WY ET AL. (1990)
logP 3.63			EST	MEYLAN,WM & HOWARD,PH (1995)
VP 1.24E-007	mm Hg	25	EST	NEELY,WB & BLAU,GE (1985)
DC	pKa			
HL 1.37E-009	atm m3/mol	25	EST	MEYLAN,WM & HOWARD,PH (1991)
OH 1.10E-011	cm3/molc sec	25	EST	MEYLAN,WM & HOWARD,PH (1993)

CAS #: 052780-43-5 — PHENOL, 4-TRIDECYL-

Formula: $C_{19}H_{32}O$

Mol Weight: 276.47

MP (deg C):

FP (deg C):

BP (deg C):

BP pressure (mm Hg):

Property/Value	Units	Temp	Data Type	Reference
WS 6.34E-003	mg/L	25	EST	MEYLAN,WM ET AL. (1996)
logP 8.20			EXP	SANGSTER,J (1993)
VP 9.37E-007	mm Hg	25	EST	NEELY,WB & BLAU,GE (1985)
DC	pKa			
HL 1.85E-005	atm m3/mol	25	EST	MEYLAN,WM & HOWARD,PH (1991)
OH 5.73E-011	cm3/molc sec	25	EST	MEYLAN,WM & HOWARD,PH (1993)

CAS #: 052783-43-4 — NONADECANOL

Formula: $C_{19}H_{40}O$

Mol Weight: 284.53

MP (deg C):

FP (deg C):

BP (deg C):

BP pressure (mm Hg):

Property/Value	Units	Temp	Data Type	Reference
WS 4.78E-003	mg/L	25	EST	MEYLAN,WM ET AL. (1996)
logP 8.21			EST	MEYLAN,WM & HOWARD,PH (1995)
VP 3.65E-007	mm Hg	25	EST	NEELY,WB & BLAU,GE (1985)
DC	pKa			
HL 7.00E-004	atm m3/mol	25	EST	MEYLAN,WM & HOWARD,PH (1991)
OH 2.81E-011	cm3/molc sec	25	EST	MEYLAN,WM & HOWARD,PH (1993)

CAS #: 052806-53-8 — PROPANAMIDE, 2-HYDROXY-2-METHYL-N-[4-NITRO-3-(TR

Formula: $C_{11}H_{11}F_3N_2O_4$

Mol Weight: 292.22

MP (deg C):

FP (deg C):

BP (deg C):

BP pressure (mm Hg):

Property/Value	Units	Temp	Data Type	Reference
WS 2.75E+001	mg/L	25	EST	MEYLAN,WM ET AL. (1996)
logP 2.70			EXP	MORRIS,JJ ET AL. (1991)
VP 2.73E-009	mm Hg	25	EST	NEELY,WB & BLAU,GE (1985)
DC	pKa			
HL 5.82E-011	atm m3/mol	25	EST	MEYLAN,WM & HOWARD,PH (1991)
OH 1.44E-012	cm3/molc sec	25	EST	MEYLAN,WM & HOWARD,PH (1993)

CAS #: 052819-97-3 — N-DIAZOACETYLGLYCINE-N'-N-PR AMIDE

Formula: $C_7H_{12}N_4O_2$

Mol Weight: 184.20

MP (deg C):

FP (deg C):

BP (deg C):

BP pressure (mm Hg):

Property/Value	Units	Temp	Data Type	Reference
WS 4.12E+003	mg/L	25	EST	MEYLAN,WM ET AL. (1996)
logP -0.24			EXP	HANSCH,C & LEO,AJ (1985)
VP 2.24E-013	mm Hg	25	EST	NEELY,WB & BLAU,GE (1985)
DC	pKa			
HL 4.17E-015	atm m3/mol	25	EST	MEYLAN,WM & HOWARD,PH (1991)
OH 8.85E-011	cm3/molc sec	25	EST	MEYLAN,WM & HOWARD,PH (1993)

CAS #: 052819-98-4 — N-DIAZOACETYLGLYCINE-N'-I-PR AMIDE

Formula: $C_7H_{12}N_4O_2$

Mol Weight: 184.20

MP (deg C):

FP (deg C):

BP (deg C):

BP pressure (mm Hg):

Property/Value	Units	Temp	Data Type	Reference
WS 5.76E+003	mg/L	25	EST	MEYLAN,WM ET AL. (1996)
logP -0.41			EXP	HANSCH,C & LEO,AJ (1985)
VP 3.73E-013	mm Hg	25	EST	NEELY,WB & BLAU,GE (1985)
DC	pKa			
HL 4.17E-015	atm m3/mol	25	EST	MEYLAN,WM & HOWARD,PH (1991)
OH 9.48E-011	cm3/molc sec	25	EST	MEYLAN,WM & HOWARD,PH (1993)

CAS #: 052829-30-8 — 14-BENZDIAZPN-2-ON,1(23DIOHPR)5(2FPH)7CL

Formula: $C_{18}H_{16}ClFN_2O_3$

Mol Weight: 362.79

MP (deg C):

FP (deg C):

BP (deg C):

BP pressure (mm Hg):

Property/Value	Units	Temp	Data Type	Reference
WS 2.66E+002	mg/L	25	EST	MEYLAN,WM ET AL. (1996)
logP 1.51			EXP	HANSCH,C & LEO,AJ (1985)
VP 4.18E-014	mm Hg	25	EST	NEELY,WB & BLAU,GE (1985)
DC	pKa			
HL 1.00E-014	atm m3/mol	25	EST	MEYLAN,WM & HOWARD,PH (1991)
OH 2.73E-011	cm3/molc sec	25	EST	MEYLAN,WM & HOWARD,PH (1993)

CAS #:	052838-38-7			2-AMINO-5-BU-1,3,4-OXADIAZOLE

Formula: $C_6H_{12}N_3O$

Mol Weight: 142.18

MP (deg C):

FP (deg C):

BP (deg C):

BP pressure (mm Hg):

Property/Value	Units	Temp	Data Type	Reference
WS 2.31E+004	mg/L	25	EST	MEYLAN,WM ET AL. (1996)
logP 0.65			EXP	HANSCH,C & LEO,AJ (1985)
VP 1.30E-003	mm Hg	25	EST	NEELY,WB & BLAU,GE (1985)
DC	pKa			
HL 2.61E-009	atm m3/mol	25	EST	MEYLAN,WM & HOWARD,PH (1991)
OH 7.88E-012	cm3/molc sec	25	EST	MEYLAN,WM & HOWARD,PH (1993)

CAS #:	052838-39-8			2-AMINO-5-ME-1,3,4-OXADIAZOLE

Formula: $C_3H_6N_3O$

Mol Weight: 100.10

MP (deg C):

FP (deg C):

BP (deg C):

BP pressure (mm Hg):

Property/Value	Units	Temp	Data Type	Reference
WS 7.18E+005	mg/L	25	EST	MEYLAN,WM ET AL. (1996)
logP -0.92			EXP	HANSCH,C & LEO,AJ (1985)
VP 3.67E-002	mm Hg	25	EST	NEELY,WB & BLAU,GE (1985)
DC	pKa			
HL 1.12E-009	atm m3/mol	25	EST	MEYLAN,WM & HOWARD,PH (1991)
OH 4.14E-012	cm3/molc sec	25	EST	MEYLAN,WM & HOWARD,PH (1993)

CAS #:	052845-72-4			1-CHLOROAMITRIPTYLINE

Formula: $C_{20}H_{22}ClN$

Mol Weight: 311.86

MP (deg C):

FP (deg C):

BP (deg C):

BP pressure (mm Hg):

Property/Value	Units	Temp	Data Type	Reference
WS 1.51E-001	mg/L	25	EST	MEYLAN,WM ET AL. (1996)
logP 5.55			EXP	HANSCH,C & LEO,AJ (1985)
VP 4.73E-007	mm Hg	25	EST	NEELY,WB & BLAU,GE (1985)
DC	pKa			
HL 5.08E-008	atm m3/mol	25	EST	MEYLAN,WM & HOWARD,PH (1991)
OH 1.76E-010	cm3/molc sec	25	EST	MEYLAN,WM & HOWARD,PH (1993)

CAS #:	052888-80-9			PROSULFOCARB

Formula: $C_{14}H_{21}NOS$

Mol Weight: 251.39

MP (deg C):

FP (deg C):

BP (deg C):

BP pressure (mm Hg):

Property/Value	Units	Temp	Data Type	Reference
WS 2.48E+000	mg/L	25	EST	MEYLAN,WM ET AL. (1996)
logP 4.65			EXP	TOMLIN,C (1994)
VP 5.88E-006	mm Hg	25	EST	NEELY,WB & BLAU,GE (1985)
DC	pKa			
HL 9.36E-007	atm m3/mol	25	EST	MEYLAN,WM & HOWARD,PH (1991)
OH 3.26E-011	cm3/molc sec	25	EST	MEYLAN,WM & HOWARD,PH (1993)

CAS #:	052918-63-5			DELTAMETHRIN

Formula: $C_{22}H_{19}Br_2NO_3$

Mol Weight: 505.22

MP (deg C): 98-101

FP (deg C):

BP (deg C):

BP pressure (mm Hg):

Property/Value	Units	Temp	Data Type	Reference
WS 2.00E-003	mg/L	20	EXP	WORTHING,CR & WALKER,SB (1987)
logP 6.20			EXP	HANSCH,C ET AL. (1995)
VP 1.50E-008	mm Hg	20	EST	WORTHING,CR & WALKER,SB (1987)
DC	pKa			
HL 5.00E-006	atm m3/mol	25	EST	VP/WSOL
OH 3.94E-011	cm3/molc sec	25	EST	ATKINSON,R (1988)

CAS #:	052980-49-1			BENZAMIDE, N,N-DIMETHYL-2-[[(4-METHYLPHENYL)SULF

Formula: $C_{16}H_{17}NO_4S$

Mol Weight: 319.38

MP (deg C):

FP (deg C):

BP (deg C):

BP pressure (mm Hg):

Property/Value	Units	Temp	Data Type	Reference
WS 1.81E+002	mg/L	25	EST	MEYLAN,WM ET AL. (1996)
logP 2.01			EXP	HANSCH,C ET AL. (1995)
VP 2.84E-009	mm Hg	25	EST	NEELY,WB & BLAU,GE (1985)
DC	pKa			
HL 6.51E-013	atm m3/mol	25	EST	MEYLAN,WM & HOWARD,PH (1991)
OH 3.69E-011	cm3/molc sec	25	EST	MEYLAN,WM & HOWARD,PH (1993)

CAS #:	052986-70-6			6-METHOXYISOQUINOLINE

Formula: $C_{10}H_9NO$

Mol Weight: 159.19

MP (deg C):

FP (deg C):

BP (deg C):

BP pressure (mm Hg):

Property/Value	Units	Temp	Data Type	Reference
WS 8.77E+002	mg/L	25	EST	MEYLAN,WM ET AL. (1996)
logP 2.22			EXP	HANSCH,C & LEO,AJ (1985)
VP 3.23E-003	mm Hg	25	EST	NEELY,WB & BLAU,GE (1985)
DC	pKa			
HL 4.07E-008	atm m3/mol	25	EST	MEYLAN,WM & HOWARD,PH (1991)
OH 1.60E-010	cm3/molc sec	25	EST	MEYLAN,WM & HOWARD,PH (1993)

CAS #:	053179-12-7			2H-BENZIMIDAZOL-2-ONE, 1-[1-[4,4-BIS(4-FLUOROPHE

Formula: $C_{28}H_{28}ClF_2N_3O$

Mol Weight: 496.00

MP (deg C):

FP (deg C):

BP (deg C):

BP pressure (mm Hg):

Property/Value	Units	Temp	Data Type	Reference
WS 6.56E-004	mg/L	25	EST	MEYLAN,WM ET AL. (1996)
logP 7.10			EXP	SANGSTER,J (1994)
VP 1.59E-013	mm Hg	25	EST	NEELY,WB & BLAU,GE (1985)
DC	pKa			
HL 8.13E-014	atm m3/mol	25	EST	MEYLAN,WM & HOWARD,PH (1991)
OH 1.32E-010	cm3/molc sec	25	EST	MEYLAN,WM & HOWARD,PH (1993)

CAS #: 053207-52-6				1,3,3-TRIME-2,3-DIHYDRO-4-QUINOLONE
Formula: $C_{12}H_{15}NO$				
Mol Weight: 189.26				
MP (deg C):		FP (deg C):		
BP (deg C):				
BP pressure (mm Hg):				

Property/Value	Units	Temp	Data Type	Reference
WS 2.36E+002	mg/L	25	EST	MEYLAN,WM ET AL. (1996)
logP 2.72			EXP	HANSCH,C & LEO,AJ (1985)
VP 7.33E-004	mm Hg	25	EST	NEELY,WB & BLAU,GE (1985)
DC	pKa			
HL 1.21E-007	atm m3/mol	25	EST	MEYLAN,WM & HOWARD,PH (1991)
OH 2.08E-010	cm3/molc sec	25	EST	MEYLAN,WM & HOWARD,PH (1993)

CAS #: 053207-61-7				1(5-NO2-2-FURYL)-2-IMIDAZOLIDINONE
Formula: $C_7H_8N_3O_4$				
Mol Weight: 198.16				
MP (deg C):		FP (deg C):		
BP (deg C):				
BP pressure (mm Hg):				

Property/Value	Units	Temp	Data Type	Reference
WS 5.52E+002	mg/L	25	EST	MEYLAN,WM ET AL. (1996)
logP 0.25			EXP	HANSCH,C & LEO,AJ (1985)
VP 3.34E-006	mm Hg	25	EST	NEELY,WB & BLAU,GE (1985)
DC	pKa			
HL 2.47E-013	atm m3/mol	25	EST	MEYLAN,WM & HOWARD,PH (1991)
OH 2.56E-011	cm3/molc sec	25	EST	MEYLAN,WM & HOWARD,PH (1993)

CAS #: 053207-62-8				1(5-NO2-2-FURYL)-2-IMIDAZOLIDINTHIONE
Formula: $C_7H_8N_3O_3S$				
Mol Weight: 214.22				
MP (deg C):		FP (deg C):		
BP (deg C):				
BP pressure (mm Hg):				

Property/Value	Units	Temp	Data Type	Reference
WS 2.26E+004	mg/L	25	EST	MEYLAN,WM ET AL. (1996)
logP 0.98			EXP	HANSCH,C & LEO,AJ (1985)
VP 4.20E-007	mm Hg	25	EST	NEELY,WB & BLAU,GE (1985)
DC	pKa			
HL 1.07E-010	atm m3/mol	25	EST	MEYLAN,WM & HOWARD,PH (1991)
OH 1.64E-010	cm3/molc sec	25	EST	MEYLAN,WM & HOWARD,PH (1993)

CAS #: 053207-64-0				N-IPR-3(5NO2-2-THIAZOLYL)ACRYLAMIDE
Formula: $C_9H_{13}N_3O_3S$				
Mol Weight: 243.29				
MP (deg C):		FP (deg C):		
BP (deg C):				
BP pressure (mm Hg):				

Property/Value	Units	Temp	Data Type	Reference
WS 3.38E+002	mg/L	25	EST	MEYLAN,WM ET AL. (1996)
logP 1.76			EXP	HANSCH,C & LEO,AJ (1985)
VP 1.00E-007	mm Hg	25	EST	NEELY,WB & BLAU,GE (1985)
DC	pKa			
HL 2.87E-015	atm m3/mol	25	EST	MEYLAN,WM & HOWARD,PH (1991)
OH 3.36E-011	cm3/molc sec	25	EST	MEYLAN,WM & HOWARD,PH (1993)

CAS #: 053207-66-2				1(5-NO2-2-FURYL)-2-MES IMIDAZOLINE
Formula: $C_8H_{10}N_3O_3S$				
Mol Weight: 228.25				
MP (deg C):		FP (deg C):		
BP (deg C):				
BP pressure (mm Hg):				

Property/Value	Units	Temp	Data Type	Reference
WS 1.24E+003	mg/L	25	EST	MEYLAN,WM ET AL. (1996)
logP 1.19			EXP	HANSCH,C & LEO,AJ (1985)
VP 7.32E-006	mm Hg	25	EST	NEELY,WB & BLAU,GE (1985)
DC	pKa			
HL 2.42E-007	atm m3/mol	25	EST	MEYLAN,WM & HOWARD,PH (1991)
OH 2.10E-010	cm3/molc sec	25	EST	MEYLAN,WM & HOWARD,PH (1993)

CAS #: 053242-40-3				3,4-DIMEPYRIDINIUMIODIDE,N-OCTYL
Formula: $C_{15}H_{26}IN$				
Mol Weight: 347.29				
MP (deg C):		FP (deg C):		
BP (deg C):				
BP pressure (mm Hg):				

Property/Value	Units	Temp	Data Type	Reference
WS 1.44E+004	mg/L	25	EST	MEYLAN,WM ET AL. (1996)
logP -0.41			EXP	HANSCH,C & LEO,AJ (1985)
VP 1.58E-005	mm Hg	25	EST	NEELY,WB & BLAU,GE (1985)
DC	pKa			
HL 1.33E-009	atm m3/mol	25	EST	MEYLAN,WM & HOWARD,PH (1991)
OH 2.10E-011	cm3/molc sec	25	EST	MEYLAN,WM & HOWARD,PH (1993)

CAS #: 053285-95-3				IMIDODICARBONIC DIAMIDE, N-(2-FLUOROPHENYL)-N',2
Formula: $C_{10}H_{12}FN_3O_2$				
Mol Weight: 225.22				
MP (deg C):		FP (deg C):		
BP (deg C):				
BP pressure (mm Hg):				

Property/Value	Units	Temp	Data Type	Reference
WS 5.00E+002	mg/L	25	EST	MEYLAN,WM ET AL. (1996)
logP 2.12			EXP	CAMILLERI,P ET AL. (1989)
VP 6.68E-008	mm Hg	25	EST	NEELY,WB & BLAU,GE (1985)
DC	pKa			
HL 6.29E-013	atm m3/mol	25	EST	MEYLAN,WM & HOWARD,PH (1991)
OH 1.75E-011	cm3/molc sec	25	EST	MEYLAN,WM & HOWARD,PH (1993)

CAS #: 053294-05-6				6-CHLOROINDOLE
Formula: C_8H_6ClN				
Mol Weight: 151.60				
MP (deg C):		FP (deg C):		
BP (deg C):				
BP pressure (mm Hg):				

Property/Value	Units	Temp	Data Type	Reference
WS 1.25E+002	mg/L	25	EST	MEYLAN,WM ET AL. (1996)
logP 3.25			EXP	HANSCH,C ET AL. (1995)
VP 2.73E-003	mm Hg	25	EST	NEELY,WB & BLAU,GE (1985)
DC	pKa			
HL 6.56E-007	atm m3/mol	25	EST	MEYLAN,WM & HOWARD,PH (1991)
OH 7.74E-011	cm3/molc sec	25	EST	MEYLAN,WM & HOWARD,PH (1993)

CAS #: 053309-89-0				1-BENZYLPROPYLAMINE

Formula: $C_{10}H_{15}N$

Mol Weight: 149.24

MP (deg C): FP (deg C):

BP (deg C):

BP pressure (mm Hg):

Property/Value	Units	Temp	Data Type	Reference
WS 8.80E+003	mg/L	25	EST	MEYLAN,WM ET AL. (1996)
logP 2.28			EXP	HANSCH,C & LEO,AJ (1985)
VP 7.61E-002	mm Hg	25	EST	NEELY,WB & BLAU,GE (1985)
DC	pKa			
HL 1.43E-006	atm m3/mol	25	EST	MEYLAN,WM & HOWARD,PH (1991)
OH 5.56E-011	cm3/molc sec	25	EST	MEYLAN,WM & HOWARD,PH (1993)

CAS #: 053316-92-0				N(M-BRPH)-3-N'-PIPERIDINOACETAMIDE

Formula: $C_{13}H_{17}BrN_2O$

Mol Weight: 297.20

MP (deg C): FP (deg C):

BP (deg C):

BP pressure (mm Hg):

Property/Value	Units	Temp	Data Type	Reference
WS 1.03E+001	mg/L	25	EST	MEYLAN,WM ET AL. (1996)
logP 3.62			EXP	HANSCH,C & LEO,AJ (1985)
VP 9.35E-008	mm Hg	25	EST	NEELY,WB & BLAU,GE (1985)
DC	pKa			
HL 2.51E-011	atm m3/mol	25	EST	MEYLAN,WM & HOWARD,PH (1991)
OH 1.06E-010	cm3/molc sec	25	EST	MEYLAN,WM & HOWARD,PH (1993)

CAS #: 053393-06-9				THIOUREA, N-HEXYL-N'-METHYL-

Formula: $C_8H_{18}N_2S$

Mol Weight: 174.31

MP (deg C): FP (deg C):

BP (deg C):

BP pressure (mm Hg):

Property/Value	Units	Temp	Data Type	Reference
WS 6.36E+003	mg/L	25	EST	MEYLAN,WM ET AL. (1996)
logP 2.31			EXP	LEAHY,DE ET AL. (1989)
VP 7.02E-003	mm Hg	25	EST	NEELY,WB & BLAU,GE (1985)
DC	pKa			
HL 3.14E-006	atm m3/mol	25	EST	MEYLAN,WM & HOWARD,PH (1991)
OH 1.44E-010	cm3/molc sec	25	EST	MEYLAN,WM & HOWARD,PH (1993)

CAS #: 053394-58-4				G-PHENYLPROPYLPYRIDINIUM BROMIDE

Formula: $C_{14}H_{16}BrN$

Mol Weight: 278.19

MP (deg C): FP (deg C):

BP (deg C):

BP pressure (mm Hg):

Property/Value	Units	Temp	Data Type	Reference
WS 6.36E+005	mg/L	25	EST	MEYLAN,WM ET AL. (1996)
logP -1.86			EXP	HANSCH,C & LEO,AJ (1985)
VP 4.52E-005	mm Hg	25	EST	NEELY,WB & BLAU,GE (1985)
DC	pKa			
HL 2.14E-011	atm m3/mol	25	EST	MEYLAN,WM & HOWARD,PH (1991)
OH 1.89E-011	cm3/molc sec	25	EST	MEYLAN,WM & HOWARD,PH (1993)

CAS #: 053409-75-9				2-PYRIMIDINAMINE, 4-[2-(1-METHYL-5-NITRO-1H-IMID

Formula: $C_{10}H_9N_5O_2$

Mol Weight: 231.22

MP (deg C): FP (deg C):

BP (deg C):

BP pressure (mm Hg):

Property/Value	Units	Temp	Data Type	Reference
WS 2.30E+003	mg/L	25	EST	MEYLAN,WM ET AL. (1996)
logP 0.85			EXP	SANGSTER,J (1993)
VP 1.26E-007	mm Hg	25	EST	NEELY,WB & BLAU,GE (1985)
DC	pKa			
HL 3.93E-012	atm m3/mol	25	EST	MEYLAN,WM & HOWARD,PH (1991)
OH 5.97E-011	cm3/molc sec	25	EST	MEYLAN,WM & HOWARD,PH (1993)

CAS #: 053429-15-5				BENZENEPENTANAMINE, HYDROCHLORIDE

Formula: $C_{11}H_{18}ClN$

Mol Weight: 199.73

MP (deg C): FP (deg C):

BP (deg C):

BP pressure (mm Hg):

Property/Value	Units	Temp	Data Type	Reference
WS 4.75E+004	mg/L	25	EST	MEYLAN,WM ET AL. (1996)
logP -0.04			EXP	SANGSTER,J (1994)
VP 5.75E-008	mm Hg	25	EST	NEELY,WB & BLAU,GE (1985)
DC	pKa			
HL 1.76E-013	atm m3/mol	25	EST	MEYLAN,WM & HOWARD,PH (1991)
OH 2.09E-011	cm3/molc sec	25	EST	MEYLAN,WM & HOWARD,PH (1993)

CAS #: 053439-65-9				PHOSPHINIC AMIDE, N,N-DIMETHYL-P,P-BIS(1-PYRROLI

Formula: $C_{10}H_{22}N_3OP$

Mol Weight: 231.28

MP (deg C): FP (deg C):

BP (deg C):

BP pressure (mm Hg):

Property/Value	Units	Temp	Data Type	Reference
WS 7.43E+002	mg/L	25	EST	MEYLAN,WM ET AL. (1996)
logP 1.88			EXP	DEBORD,J & LABADIE,M (1985)
VP 2.27E-004	mm Hg	25	EST	NEELY,WB & BLAU,GE (1985)
DC	pKa			
HL 4.31E-012	atm m3/mol	25	EST	MEYLAN,WM & HOWARD,PH (1991)
OH 1.16E-010	cm3/molc sec	25	EST	MEYLAN,WM & HOWARD,PH (1993)

CAS #: 053446-96-1				9,10-PHENANTHRENEDIOL, 1,2,3,4,4A,9,10,10A-OCTAH

Formula: $C_{14}H_{18}O_2$

Mol Weight: 218.30

MP (deg C): FP (deg C):

BP (deg C):

BP pressure (mm Hg):

Property/Value	Units	Temp	Data Type	Reference
WS 3.67E+002	mg/L	25	EST	MEYLAN,WM ET AL. (1996)
logP 2.47			EXP	SANGSTER,J (1993)
VP 1.84E-007	mm Hg	25	EST	NEELY,WB & BLAU,GE (1985)
DC	pKa			
HL 9.34E-009	atm m3/mol	25	EST	MEYLAN,WM & HOWARD,PH (1991)
OH 3.67E-011	cm3/molc sec	25	EST	MEYLAN,WM & HOWARD,PH (1993)

1180

CAS #: 053555-63-8				1,2,3,5-TETRACHLORONAPHTHALENE

Formula: $C_{10}H_4Cl_4$

Mol Weight: 265.96

MP (deg C):   FP (deg C):

BP (deg C):

BP pressure (mm Hg):

Property/Value	Units	Temp	Data Type	Reference
WS 2.28E-001	mg/L	25	EST	MEYLAN,WM ET AL. (1996)
logP 5.77			EXP	OPPERHUIZEN,A (1987)
VP 5.79E-005	mm Hg	25	EST	NEELY,WB & BLAU,GE (1985)
DC	pKa			
HL 1.59E-004	atm m3/mol	25	EST	MEYLAN,WM & HOWARD,PH (1991)
OH 9.11E-013	cm3/molc sec	25	EST	MEYLAN,WM & HOWARD,PH (1993)

CAS #: 053555-64-9				1,3,5,7-TETRACHLORONAPHTHALENE

Formula: $C_{10}H_4Cl_4$

Mol Weight: 265.96

MP (deg C):   FP (deg C):

BP (deg C):

BP pressure (mm Hg):

Property/Value	Units	Temp	Data Type	Reference
WS 9.96E-002	mg/L	25	EST	MEYLAN,WM ET AL. (1996)
logP 6.19			EXP	OPPERHUIZEN,A (1987)
VP 5.79E-005	mm Hg	25	EST	NEELY,WB & BLAU,GE (1985)
DC	pKa			
HL 1.59E-004	atm m3/mol	25	EST	MEYLAN,WM & HOWARD,PH (1991)
OH 9.11E-013	cm3/molc sec	25	EST	MEYLAN,WM & HOWARD,PH (1993)

CAS #: 053583-79-2				SULTOPRIDE

Formula: $C_{17}H_{26}N_2O_4S$

Mol Weight: 354.47

MP (deg C): 181-182   FP (deg C):

BP (deg C):

BP pressure (mm Hg):

Property/Value	Units	Temp	Data Type	Reference
WS 7.24E+002	mg/L	25	EST	MEYLAN,WM ET AL. (1996)
logP 1.06			EXP	MANNHOLD,R ET AL. (1990)
VP 5.67E-011	mm Hg	25	EST	NEELY,WB & BLAU,GE (1985)
DC	pKa			
HL 6.00E-017	atm m3/mol	25	EST	MEYLAN,WM & HOWARD,PH (1991)
OH 1.30E-010	cm3/molc sec	25	EST	MEYLAN,WM & HOWARD,PH (1993)

CAS #: 053596-19-3				2-PROPENOIC ACID, 3-(2,4-DICHLOROPHENOXY)-, (E)-

Formula: $C_9H_6Cl_2O_3$

Mol Weight: 233.05

MP (deg C):   FP (deg C):

BP (deg C):

BP pressure (mm Hg):

Property/Value	Units	Temp	Data Type	Reference
WS 8.07E+001	mg/L	25	EST	MEYLAN,WM ET AL. (1996)
logP 3.46			EXP	SANGSTER,J (1993)
VP 2.08E-005	mm Hg	25	EST	NEELY,WB & BLAU,GE (1985)
DC	pKa			
HL 5.45E-009	atm m3/mol	25	EST	MEYLAN,WM & HOWARD,PH (1991)
OH 2.21E-011	cm3/molc sec	25	EST	MEYLAN,WM & HOWARD,PH (1993)

CAS #: 053634-34-7				1-CHLORO-1,3-DIMETHYL-1,3,3-TRIPHENYLDISILOXANE

Formula: $C_{20}H_{21}ClOSi_2$

Mol Weight: 369.01

MP (deg C):   FP (deg C):

BP (deg C):

BP pressure (mm Hg):

Property/Value	Units	Temp	Data Type	Reference
WS 8.27E-005	mg/L	25	EST	MEYLAN,WM ET AL. (1996)
logP 8.17			EST	MEYLAN,WM & HOWARD,PH (1995)
VP	mm Hg			
DC	pKa			
HL	atm m3/mol			
OH 6.15E-012	cm3/molc sec	25	EST	MEYLAN,WM & HOWARD,PH (1993)

CAS #: 053643-53-1				TUBERIN

Formula: $C_{10}H_{11}NO_2$

Mol Weight: 177.20

MP (deg C): 132-133   FP (deg C):

BP (deg C):

BP pressure (mm Hg):

Property/Value	Units	Temp	Data Type	Reference
WS 1.25E+003	mg/L	25	EST	MEYLAN,WM ET AL. (1996)
logP 1.94			EXP	HANSCH,C & LEO,AJ (1985)
VP 1.89E-005	mm Hg	25	EST	NEELY,WB & BLAU,GE (1985)
DC	pKa			
HL 1.56E-010	atm m3/mol	25	EST	MEYLAN,WM & HOWARD,PH (1991)
OH 6.04E-011	cm3/molc sec	25	EST	MEYLAN,WM & HOWARD,PH (1993)

CAS #: 053660-20-1				BENZENEETHANAMINE, N-METHYL-à-(PHENYLMETHYL)-

Formula: $C_{16}H_{19}N$

Mol Weight: 225.34

MP (deg C):   FP (deg C):

BP (deg C):

BP pressure (mm Hg):

Property/Value	Units	Temp	Data Type	Reference
WS 1.87E+002	mg/L	25	EST	MEYLAN,WM ET AL. (1996)
logP 3.80			EXP	SANGSTER,J (1993)
VP 1.01E-004	mm Hg	25	EST	NEELY,WB & BLAU,GE (1985)
DC	pKa			
HL 1.91E-007	atm m3/mol	25	EST	MEYLAN,WM & HOWARD,PH (1991)
OH 1.04E-010	cm3/molc sec	25	EST	MEYLAN,WM & HOWARD,PH (1993)

CAS #: 053670-99-8				BENZENEACETAMIDE, N-[2-[(3-PHENOXY-2-HYDROXYPROP

Formula: $C_{19}H_{24}N_2O_3$

Mol Weight: 328.41

MP (deg C):   FP (deg C):

BP (deg C):

BP pressure (mm Hg):

Property/Value	Units	Temp	Data Type	Reference
WS 2.42E+003	mg/L	25	EST	MEYLAN,WM ET AL. (1996)
logP 0.63			EXP	MAULEON,D ET AL. (1988)
VP 5.61E-013	mm Hg	25	EST	NEELY,WB & BLAU,GE (1985)
DC	pKa			
HL 1.66E-019	atm m3/mol	25	EST	MEYLAN,WM & HOWARD,PH (1991)
OH 1.39E-010	cm3/molc sec	25	EST	MEYLAN,WM & HOWARD,PH (1993)

CAS #: 053673-10-2				
Formula: $C_{10}H_{14}N_2O_2$			O-MEO BENZAMIDE, N-(2-AMINOETHYL)	
Mol Weight: 194.24				
MP (deg C):		FP (deg C):		
BP (deg C):				
BP pressure (mm Hg):				

Property/Value	Units	Temp	Data Type	Reference
WS 2.29E+003	mg/L	25	EST	MEYLAN,WM ET AL. (1996)
logP 0.00			EXP	HANSCH,C & LEO,AJ (1985)
VP 2.71E-006	mm Hg	25	EST	NEELY,WB & BLAU,GE (1985)
DC	pKa			
HL 3.93E-014	atm m3/mol	25	EST	MEYLAN,WM & HOWARD,PH (1991)
OH 6.23E-011	cm3/molc sec	25	EST	MEYLAN,WM & HOWARD,PH (1993)

CAS #: 053742-07-7				
Formula: $C_{12}HCl_9$			NONACHLORO-1,1'-BIPHENYL	
Mol Weight: 464.22				
MP (deg C):		FP (deg C):		
BP (deg C):				
BP pressure (mm Hg):				

Property/Value	Units	Temp	Data Type	Reference
WS 1.10E-004	mg/L	182	EXP	OPPERHUIZEN,A (1987)
logP 9.56			EST	MEYLAN,WM & HOWARD,PH (1995)
VP 7.60E-009	mm Hg	25	EST	NEELY,WB & BLAU,GE (1985)
DC	pKa			
HL 2.79E-005	atm m3/mol	25	EST	MEYLAN,WM & HOWARD,PH (1991)
OH 3.29E-014	cm3/molc sec	25	EST	MEYLAN,WM & HOWARD,PH (1993)

CAS #: 053766-80-6				
Formula: $C_{10}H_{11}N_5O_3$			GUANOSINE, 2',3'-DIDEHYDRO-2',3'-DIDEOXY-	
Mol Weight: 249.23				
MP (deg C):		FP (deg C):		
BP (deg C):				
BP pressure (mm Hg):				

Property/Value	Units	Temp	Data Type	Reference
WS 1.26E+004	mg/L	25	EST	MEYLAN,WM ET AL. (1996)
logP -1.21			EXP	BALZARINI,J ET AL. (1989)
VP 6.77E-016	mm Hg	25	EST	NEELY,WB & BLAU,GE (1985)
DC	pKa			
HL 2.17E-022	atm m3/mol	25	EST	MEYLAN,WM & HOWARD,PH (1991)
OH 1.44E-010	cm3/molc sec	25	EST	MEYLAN,WM & HOWARD,PH (1993)

CAS #: 053772-82-0				
Formula: $C_{23}H_{25}F_3N_2OS$			1-PIPERAZINEETHANOL, 4-[3-[2-(TRIFLUOROMETHYL)-9	
Mol Weight: 434.53				
MP (deg C):		FP (deg C):		
BP (deg C):				
BP pressure (mm Hg):				

Property/Value	Units	Temp	Data Type	Reference
WS 3.46E-001	mg/L	25	EST	MEYLAN,WM ET AL. (1996)
logP 4.25			EXP	SANGSTER,J (1993)
VP 1.05E-012	mm Hg	25	EST	NEELY,WB & BLAU,GE (1985)
DC	pKa			
HL 4.13E-016	atm m3/mol	25	EST	MEYLAN,WM & HOWARD,PH (1991)
OH 1.71E-010	cm3/molc sec	25	EST	MEYLAN,WM & HOWARD,PH (1993)

CAS #: 053772-85-3				
Formula: $C_{23}H_{25}F_3N_2OS$			1-PIPERAZINEETHANOL, 4-[3-[2-(TRIFLUOROMETHYL)-9	
Mol Weight: 434.53				
MP (deg C):		FP (deg C):		
BP (deg C):				
BP pressure (mm Hg):				

Property/Value	Units	Temp	Data Type	Reference
WS 3.46E-001	mg/L	25	EST	MEYLAN,WM ET AL. (1996)
logP 4.25			EXP	SANGSTER,J (1993)
VP 1.05E-012	mm Hg	25	EST	NEELY,WB & BLAU,GE (1985)
DC	pKa			
HL 4.13E-016	atm m3/mol	25	EST	MEYLAN,WM & HOWARD,PH (1991)
OH 1.71E-010	cm3/molc sec	25	EST	MEYLAN,WM & HOWARD,PH (1993)

CAS #: 053780-34-0				
Formula: $C_{11}H_{13}F_3N_2O_3S$			MEFLUIDIDE	
Mol Weight: 310.30				
MP (deg C): 184		FP (deg C):		
BP (deg C):				
BP pressure (mm Hg):				

Property/Value	Units	Temp	Data Type	Reference
WS 1.80E+002	mg/L	23	EXP	YALKOWSKY,SH & DANNENFELSER,RM (1992)
logP 2.72			EST	MEYLAN,WM & HOWARD,PH (1995)
VP 2.40E-008	mm Hg	25	EST	NEELY,WB & BLAU,GE (1985)
DC	pKa			
HL 1.28E-011	atm m3/mol	25	EST	MEYLAN,WM & HOWARD,PH (1991)
OH 2.00E-010	cm3/molc sec	25	EST	MEYLAN,WM & HOWARD,PH (1993)

CAS #: 053808-87-0				
Formula: $C_{16}H_{22}N_4O_4$			24NH2PYRIMIDINE,5(35-MEO-4-OETOME)BENZYL	
Mol Weight: 334.38				
MP (deg C): 153-156		FP (deg C):		
BP (deg C):				
BP pressure (mm Hg):				

Property/Value	Units	Temp	Data Type	Reference
WS 2.56E+003	mg/L	25	EST	MEYLAN,WM ET AL. (1996)
logP 0.56			EXP	HANSCH,C & LEO,AJ (1985)
VP 8.33E-010	mm Hg	25	EST	NEELY,WB & BLAU,GE (1985)
DC	pKa			
HL 3.71E-016	atm m3/mol	25	EST	MEYLAN,WM & HOWARD,PH (1991)
OH 2.17E-010	cm3/molc sec	25	EST	MEYLAN,WM & HOWARD,PH (1993)

CAS #: 053861-64-6				
Formula: $C_6H_7Cl_5O$			1-HYDROXYPENTACHLOROCYCLOHEXANE	
Mol Weight: 272.39				
MP (deg C):		FP (deg C):		
BP (deg C):				
BP pressure (mm Hg):				

Property/Value	Units	Temp	Data Type	Reference
WS 3.88E+002	mg/L	25	EST	MEYLAN,WM ET AL. (1996)
logP 2.54			EXP	HANSCH,C & LEO,AJ (1985)
VP 1.02E-005	mm Hg	25	EST	NEELY,WB & BLAU,GE (1985)
DC	pKa			
HL 2.66E-008	atm m3/mol	25	EST	MEYLAN,WM & HOWARD,PH (1991)
OH 1.96E-012	cm3/molc sec	25	EST	MEYLAN,WM & HOWARD,PH (1993)

CAS #: 053916-74-8				3-(B-NITROVINYL)FURAN

Formula: $C_6H_6NO_3$
Mol Weight: 140.12
MP (deg C): **FP (deg C):**
BP (deg C):
BP pressure (mm Hg):

Property/Value	Units	Temp	Data Type	Reference
WS 2.15E+003	mg/L	25	EST	MEYLAN,WM ET AL. (1996)
logP 1.41			EXP	HANSCH,C & LEO,AJ (1985)
VP 1.11E-001	mm Hg	25	EST	NEELY,WB & BLAU,GE (1985)
DC	pKa			
HL 3.47E-006	atm m3/mol	25	EST	MEYLAN,WM & HOWARD,PH (1991)
OH 4.95E-011	cm3/molc sec	25	EST	MEYLAN,WM & HOWARD,PH (1993)

CAS #: 053943-59-2				BARBITURIC ACID,5-ME-5-IPR

Formula: $C_8H_{12}N_2O_3$
Mol Weight: 184.20
MP (deg C): **FP (deg C):**
BP (deg C):
BP pressure (mm Hg):

Property/Value	Units	Temp	Data Type	Reference
WS 1.82E+004	mg/L	25	EST	MEYLAN,WM ET AL. (1996)
logP 0.54			EXP	WONG,O & MCKEOWN,RH (1988)
VP 2.61E-010	mm Hg	25	EST	NEELY,WB & BLAU,GE (1985)
DC	pKa			
HL 3.61E-013	atm m3/mol	25	EST	MEYLAN,WM & HOWARD,PH (1991)
OH 1.04E-011	cm3/molc sec	25	EST	MEYLAN,WM & HOWARD,PH (1993)

CAS #: 053951-84-1				METHYL3-QUINOLINECARBOXYLATE

Formula: $C_{11}H_9NO_2$
Mol Weight: 187.20
MP (deg C): **FP (deg C):**
BP (deg C):
BP pressure (mm Hg):

Property/Value	Units	Temp	Data Type	Reference
WS 6.08E+002	mg/L	25	EST	MEYLAN,WM ET AL. (1996)
logP 2.25			EXP	HANSCH,C & LEO,AJ (1985)
VP 3.19E-004	mm Hg	25	EST	NEELY,WB & BLAU,GE (1985)
DC	pKa			
HL 4.44E-009	atm m3/mol	25	EST	MEYLAN,WM & HOWARD,PH (1991)
OH 4.84E-012	cm3/molc sec	25	EST	MEYLAN,WM & HOWARD,PH (1993)

CAS #: 053978-04-4				1,2,3-TRICHLOROBUTA-1,3-DIENE(E)

Formula: $C_4H_3Cl_3$
Mol Weight: 157.43
MP (deg C): **FP (deg C):**
BP (deg C):
BP pressure (mm Hg):

Property/Value	Units	Temp	Data Type	Reference
WS 9.22E+001	mg/L	25	EST	MEYLAN,WM ET AL. (1996)
logP 3.38			EST	MEYLAN,WM & HOWARD,PH (1995)
VP 1.57E+001	mm Hg	25	EST	NEELY,WB & BLAU,GE (1985)
DC	pKa			
HL 2.90E-002	atm m3/mol	25	EST	MEYLAN,WM & HOWARD,PH (1991)
OH 1.76E-012	cm3/molc sec	25	EST	MEYLAN,WM & HOWARD,PH (1993)

CAS #: 054010-81-0				1-ETHOXY-4-(1(4-METHYLPHENYL)-2-NITROPROPYL)-B*-

Formula: $C_{18}H_{21}NO_3$
Mol Weight: 299.37
MP (deg C): **FP (deg C):**
BP (deg C):
BP pressure (mm Hg):

Property/Value	Units	Temp	Data Type	Reference
WS 3.88E-001	mg/L	25	EST	MEYLAN,WM ET AL. (1996)
logP 4.82			EST	MEYLAN,WM & HOWARD,PH (1995)
VP 6.96E-007	mm Hg	25	EST	NEELY,WB & BLAU,GE (1985)
DC	pKa			
HL 4.19E-008	atm m3/mol	25	EST	MEYLAN,WM & HOWARD,PH (1991)
OH 3.87E-011	cm3/molc sec	25	EST	MEYLAN,WM & HOWARD,PH (1993)

CAS #: 054013-07-9				PYRAZINAMINE, 5-METHOXY-

Formula: $C_5H_7N_3O$
Mol Weight: 125.13
MP (deg C): **FP (deg C):**
BP (deg C):
BP pressure (mm Hg):

Property/Value	Units	Temp	Data Type	Reference
WS 2.78E+004	mg/L	25	EST	MEYLAN,WM ET AL. (1996)
logP 0.63			EXP	YAMAGAMI,C ET AL. (1991)
VP 1.93E-002	mm Hg	25	EST	NEELY,WB & BLAU,GE (1985)
DC	pKa			
HL 5.56E-010	atm m3/mol	25	EST	MEYLAN,WM & HOWARD,PH (1991)
OH 1.18E-011	cm3/molc sec	25	EST	MEYLAN,WM & HOWARD,PH (1993)

CAS #: 054015-45-1				PYRAZINECARBOXAMIDE, 6-ETHOXY-

Formula: $C_7H_9N_3O_2$
Mol Weight: 167.17
MP (deg C): **FP (deg C):**
BP (deg C):
BP pressure (mm Hg):

Property/Value	Units	Temp	Data Type	Reference
WS 1.91E+004	mg/L	25	EST	MEYLAN,WM ET AL. (1996)
logP 0.61			EXP	YAMAGAMI,C ET AL. (1991)
VP 1.37E-005	mm Hg	25	EST	NEELY,WB & BLAU,GE (1985)
DC	pKa			
HL 8.59E-013	atm m3/mol	25	EST	MEYLAN,WM & HOWARD,PH (1991)
OH 9.24E-012	cm3/molc sec	25	EST	MEYLAN,WM & HOWARD,PH (1993)

CAS #: 054063-32-0				CLOBETASONE

Formula: $C_{22}H_{26}ClFO_4$
Mol Weight: 408.90
MP (deg C): **FP (deg C):**
BP (deg C):
BP pressure (mm Hg):

Property/Value	Units	Temp	Data Type	Reference
WS 5.15E+001	mg/L	25	EST	MEYLAN,WM ET AL. (1996)
logP 2.61			EXP	HANSCH,C & LEO,AJ (1985)
VP 3.40E-012	mm Hg	25	EST	NEELY,WB & BLAU,GE (1985)
DC	pKa			
HL 1.69E-012	atm m3/mol	25	EST	MEYLAN,WM & HOWARD,PH (1991)
OH 5.97E-011	cm3/molc sec	25	EST	MEYLAN,WM & HOWARD,PH (1993)

CAS #: 054083-24-8				PENTACHLOROCYCLOHEXENE
Formula: $C_6H_5Cl_5$				
Mol Weight: 254.37				
MP (deg C):		FP (deg C):		
BP (deg C):				
BP pressure (mm Hg):				

Property/ Value	Units	Temp	Data Type	Reference
WS 1.15E+001	mg/L	25	EST	MEYLAN,WM ET AL. (1996)
logP 3.60			EXP	HANSCH,C & LEO,AJ (1985)
VP 5.99E-003	mm Hg	25	EST	NEELY,WB & BLAU,GE (1985)
DC	pKa			
HL 1.30E-003	atm m3/mol	25	EST	MEYLAN,WM & HOWARD,PH (1991)
OH 1.13E-011	cm3/molc sec	25	EST	MEYLAN,WM & HOWARD,PH (1993)

CAS #: 054083-25-9				PENTACHLOROCYCLOHEXENE
Formula: $C_6H_5Cl_5$				
Mol Weight: 254.37				
MP (deg C):		FP (deg C):		
BP (deg C):				
BP pressure (mm Hg):				

Property/ Value	Units	Temp	Data Type	Reference
WS 1.15E+001	mg/L	25	EST	MEYLAN,WM ET AL. (1996)
logP 3.85			EXP	HANSCH,C & LEO,AJ (1985)
VP 5.99E-003	mm Hg	25	EST	NEELY,WB & BLAU,GE (1985)
DC	pKa			
HL 1.30E-003	atm m3/mol	25	EST	MEYLAN,WM & HOWARD,PH (1991)
OH 1.13E-011	cm3/molc sec	25	EST	MEYLAN,WM & HOWARD,PH (1993)

CAS #: 054108-51-9				NAPHTHALENE, 1-(METHYLSULFONYL)-
Formula: $C_{11}H_{10}O_2S$				
Mol Weight: 206.27				
MP (deg C):		FP (deg C):		
BP (deg C):				
BP pressure (mm Hg):				

Property/ Value	Units	Temp	Data Type	Reference
WS 1.33E+003	mg/L	25	EST	MEYLAN,WM ET AL. (1996)
logP 1.74			EXP	ALCORN,CJ ET AL. (1993)
VP 1.60E-005	mm Hg	25	EST	NEELY,WB & BLAU,GE (1985)
DC	pKa			
HL 1.22E-007	atm m3/mol	25	EST	MEYLAN,WM & HOWARD,PH (1991)
OH 5.68E-012	cm3/molc sec	25	EST	MEYLAN,WM & HOWARD,PH (1993)

CAS #: 054135-80-7				1,2,3-TRICHLORO-4-METHOXYBENZENE
Formula: $C_7H_5Cl_3O$				
Mol Weight: 211.48				
MP (deg C): 69-70		FP (deg C):		
BP (deg C):				
BP pressure (mm Hg):				

Property/ Value	Units	Temp	Data Type	Reference
WS 2.44E+001	mg/L	25	EST	MEYLAN,WM ET AL. (1996)
logP 3.74			EXP	OPPERHUIZEN,A & VOORS,PI (1987)
VP 2.28E-002	mm Hg	25	EST	NEELY,WB & BLAU,GE (1985)
DC	pKa			
HL 1.30E-004	atm m3/mol	25	EST	MEYLAN,WM & HOWARD,PH (1991)
OH 2.12E-012	cm3/molc sec	25	EST	MEYLAN,WM & HOWARD,PH (1993)

CAS #: 054135-81-8				2,3,5-TRICHLOROANISOLE
Formula: $C_7H_5Cl_3O$				
Mol Weight: 211.48				
MP (deg C):		FP (deg C):		
BP (deg C):				
BP pressure (mm Hg):				

Property/ Value	Units	Temp	Data Type	Reference
WS 1.68E+001	mg/L	25	EST	MEYLAN,WM ET AL. (1996)
logP 3.93			EXP	OPPERHUIZEN,A & VOORS,PI (1987)
VP 2.28E-002	mm Hg	25	EST	NEELY,WB & BLAU,GE (1985)
DC	pKa			
HL 1.30E-004	atm m3/mol	25	EST	MEYLAN,WM & HOWARD,PH (1991)
OH 3.94E-012	cm3/molc sec	25	EST	MEYLAN,WM & HOWARD,PH (1993)

CAS #: 054135-82-9				BENZENE, 1,2,3-TRICHLORO-5-METHOXY-
Formula: $C_7H_5Cl_3O$				
Mol Weight: 211.48				
MP (deg C):		FP (deg C):		
BP (deg C):				
BP pressure (mm Hg):				

Property/ Value	Units	Temp	Data Type	Reference
WS 9.50E+000	mg/L	25	EST	MEYLAN,WM ET AL. (1996)
logP 4.22			EXP	OPPERHUIZEN,A & VOORS,PI (1987)
VP 2.28E-002	mm Hg	25	EST	NEELY,WB & BLAU,GE (1985)
DC	pKa			
HL 1.30E-004	atm m3/mol	25	EST	MEYLAN,WM & HOWARD,PH (1991)
OH 3.94E-012	cm3/molc sec	25	EST	MEYLAN,WM & HOWARD,PH (1993)

CAS #: 054138-85-1				UREA, N-(2-CHLOROETHYL)-N-NITROSO-N'-(2,3,4-TRI-
Formula: $C_{13}H_{18}ClN_3O_9$				
Mol Weight: 395.76				
MP (deg C):		FP (deg C):		
BP (deg C):				
BP pressure (mm Hg):				

Property/ Value	Units	Temp	Data Type	Reference
WS 3.53E+002	mg/L	25	EST	MEYLAN,WM ET AL. (1996)
logP 1.13			EXP	HANSCH,C ET AL. (1995)
VP 3.38E-009	mm Hg	25	EST	NEELY,WB & BLAU,GE (1985)
DC	pKa			
HL 3.14E-020	atm m3/mol	25	EST	MEYLAN,WM & HOWARD,PH (1991)
OH 8.35E-011	cm3/molc sec	25	EST	MEYLAN,WM & HOWARD,PH (1993)

CAS #: 054143-55-4				FLECANIDE
Formula: $C_{17}H_{20}F_6N_2O_3$				
Mol Weight: 414.35				
MP (deg C): 145-147		FP (deg C):		
BP (deg C):				
BP pressure (mm Hg):				

Property/ Value	Units	Temp	Data Type	Reference
WS 1.48E+000	mg/L	25	EST	MEYLAN,WM ET AL. (1996)
logP 3.78			EXP	MANNHOLD,R ET AL. (1990)
VP 7.29E-009	mm Hg	25	EST	NEELY,WB & BLAU,GE (1985)
DC	pKa			
HL 5.75E-013	atm m3/mol	25	EST	MEYLAN,WM & HOWARD,PH (1991)
OH 6.47E-011	cm3/molc sec	25	EST	MEYLAN,WM & HOWARD,PH (1993)

CAS #: 054266-28-3		N-ME-4-PROPIONYLPHENYLCARBAMATE

Formula: $C_{11}H_{13}NO_3$

Mol Weight: 207.23

MP (deg C):   FP (deg C):

BP (deg C):

BP pressure (mm Hg):

Property/Value	Units	Temp	Data Type	Reference
WS 1.91E+003	mg/L	25	EST	MEYLAN,WM ET AL. (1996)
logP 1.55			EXP	HANSCH,C & LEO,AJ (1985)
VP 2.61E-004	mm Hg	25	EST	NEELY,WB & BLAU,GE (1985)
DC	pKa			
HL 7.76E-011	atm m3/mol	25	EST	MEYLAN,WM & HOWARD,PH (1991)
OH 9.12E-012	cm3/molc sec	25	EST	MEYLAN,WM & HOWARD,PH (1993)

CAS #: 054266-44-3		1,3-OXATHIOLAN-4-ONE, O-[(METHYLAMINO)CARBONYL]O

Formula: $C_5H_8N_2O_3S$

Mol Weight: 176.20

MP (deg C):   FP (deg C):

BP (deg C):

BP pressure (mm Hg):

Property/Value	Units	Temp	Data Type	Reference
WS 2.50E+003	mg/L	25	EST	MEYLAN,WM ET AL. (1996)
logP 0.06			EXP	KURTZ,AP & DURDEN,JA (1987)
VP 6.94E-003	mm Hg	25	EST	NEELY,WB & BLAU,GE (1985)
DC	pKa			
HL 7.86E-012	atm m3/mol	25	EST	MEYLAN,WM & HOWARD,PH (1991)
OH 4.89E-011	cm3/molc sec	25	EST	MEYLAN,WM & HOWARD,PH (1993)

CAS #: 054266-46-5		1,3-OXATHIOLAN-4-ONE, 5-METHYL-, O-[(METHYLAMINO

Formula: $C_6H_{10}N_2O_3S$

Mol Weight: 190.22

MP (deg C):   FP (deg C):

BP (deg C):

BP pressure (mm Hg):

Property/Value	Units	Temp	Data Type	Reference
WS 1.98E+004	mg/L	25	EST	MEYLAN,WM ET AL. (1996)
logP 0.46			EXP	KURTZ,AP & DURDEN,JA (1987)
VP 3.22E-003	mm Hg	25	EST	NEELY,WB & BLAU,GE (1985)
DC	pKa			
HL 1.04E-011	atm m3/mol	25	EST	MEYLAN,WM & HOWARD,PH (1991)
OH 5.51E-011	cm3/molc sec	25	EST	MEYLAN,WM & HOWARD,PH (1993)

CAS #: 054266-55-6		1,4-OXATHIAN-3-ONE, O-[(METHYLAMINO)CARBONYL]OXI

Formula: $C_6H_{10}N_2O_3S$

Mol Weight: 190.22

MP (deg C):   FP (deg C):

BP (deg C):

BP pressure (mm Hg):

Property/Value	Units	Temp	Data Type	Reference
WS 2.40E+003	mg/L	25	EST	MEYLAN,WM ET AL. (1996)
logP 0.00			EXP	KURTZ,AP & DURDEN,JA (1987)
VP 2.20E-003	mm Hg	25	EST	NEELY,WB & BLAU,GE (1985)
DC	pKa			
HL 1.04E-011	atm m3/mol	25	EST	MEYLAN,WM & HOWARD,PH (1991)
OH 4.73E-011	cm3/molc sec	25	EST	MEYLAN,WM & HOWARD,PH (1993)

CAS #: 054266-80-7		1,3-OXATHIOLAN-4-ONE, 5-PROPYL-, O-(METHYLAMINO

Formula: $C_8H_{14}N_2O_3S$

Mol Weight: 218.28

MP (deg C):   FP (deg C):

BP (deg C):

BP pressure (mm Hg):

Property/Value	Units	Temp	Data Type	Reference
WS 1.77E+003	mg/L	25	EST	MEYLAN,WM ET AL. (1996)
logP 1.52			EXP	KURTZ,AP & DURDEN,JA (1987)
VP 4.08E-004	mm Hg	25	EST	NEELY,WB & BLAU,GE (1985)
DC	pKa			
HL 1.84E-011	atm m3/mol	25	EST	MEYLAN,WM & HOWARD,PH (1991)
OH 6.04E-011	cm3/molc sec	25	EST	MEYLAN,WM & HOWARD,PH (1993)

CAS #: 054266-83-0		1,3-OXATHIOLAN-4-ONE, 5-(1-METHYLETHYL)-, O-[(ME

Formula: $C_8H_{14}N_2O_3S$

Mol Weight: 218.28

MP (deg C):   FP (deg C):

BP (deg C):

BP pressure (mm Hg):

Property/Value	Units	Temp	Data Type	Reference
WS 1.92E+003	mg/L	25	EST	MEYLAN,WM ET AL. (1996)
logP 1.48			EXP	KURTZ,AP & DURDEN,JA (1987)
VP 7.96E-004	mm Hg	25	EST	NEELY,WB & BLAU,GE (1985)
DC	pKa			
HL 1.84E-011	atm m3/mol	25	EST	MEYLAN,WM & HOWARD,PH (1991)
OH 6.04E-011	cm3/molc sec	25	EST	MEYLAN,WM & HOWARD,PH (1993)

CAS #: 054266-84-1		1,3-OXATHIOLAN-4-ONE, 5-(2-METHOXYETHYL)-, O-[(M

Formula: $C_8H_{14}N_2O_4S$

Mol Weight: 234.28

MP (deg C):   FP (deg C):

BP (deg C):

BP pressure (mm Hg):

Property/Value	Units	Temp	Data Type	Reference
WS 1.06E+004	mg/L	25	EST	MEYLAN,WM ET AL. (1996)
logP 0.51			EXP	KURTZ,AP & DURDEN,JA (1987)
VP 1.38E-004	mm Hg	25	EST	NEELY,WB & BLAU,GE (1985)
DC	pKa			
HL 1.62E-013	atm m3/mol	25	EST	MEYLAN,WM & HOWARD,PH (1991)
OH 6.95E-011	cm3/molc sec	25	EST	MEYLAN,WM & HOWARD,PH (1993)

CAS #: 054267-24-2		O=P(NME)(OME)-O-(4-NITROPHENYL)

Formula: $C_8H_{11}N_2O_5P$

Mol Weight: 246.16

MP (deg C):   FP (deg C):

BP (deg C):

BP pressure (mm Hg):

Property/Value	Units	Temp	Data Type	Reference
WS 8.33E+002	mg/L	25	EST	MEYLAN,WM ET AL. (1996)
logP 1.27			EXP	HANSCH,C & LEO,AJ (1985)
VP 1.82E-005	mm Hg	25	EST	NEELY,WB & BLAU,GE (1985)
DC	pKa			
HL 5.06E-012	atm m3/mol	25	EST	MEYLAN,WM & HOWARD,PH (1991)
OH 3.20E-011	cm3/molc sec	25	EST	MEYLAN,WM & HOWARD,PH (1993)

CAS #: 054335-82-9				N-ME-3-FORMYLPHENYLCARBAMATE

Formula: C9H9NO3

Mol Weight: 179.18

MP (deg C): | FP (deg C):

BP (deg C):

BP pressure (mm Hg):

Property/Value	Units	Temp	Data Type	Reference
WS 9.10E+003	mg/L	25	EST	MEYLAN,WM ET AL. (1996)
logP 0.92			EXP	HANSCH,C & LEO,AJ (1985)
VP 1.03E-003	mm Hg	25	EST	NEELY,WB & BLAU,GE (1985)
DC	pKa			
HL 8.01E-011	atm m3/mol	25	EST	MEYLAN,WM & HOWARD,PH (1991)
OH 2.40E-011	cm3/molc sec	25	EST	MEYLAN,WM & HOWARD,PH (1993)

CAS #: 054335-83-0				N-ME-4-FORMYLPHENYLCARBAMATE

Formula: C9H9NO3

Mol Weight: 179.18

MP (deg C): | FP (deg C):

BP (deg C):

BP pressure (mm Hg):

Property/Value	Units	Temp	Data Type	Reference
WS 7.93E+003	mg/L	25	EST	MEYLAN,WM ET AL. (1996)
logP 0.99			EXP	HANSCH,C & LEO,AJ (1985)
VP 1.03E-003	mm Hg	25	EST	NEELY,WB & BLAU,GE (1985)
DC	pKa			
HL 8.01E-011	atm m3/mol	25	EST	MEYLAN,WM & HOWARD,PH (1991)
OH 2.37E-011	cm3/molc sec	25	EST	MEYLAN,WM & HOWARD,PH (1993)

CAS #: 054340-62-4				BUFURALOL

Formula: C16H23NO2

Mol Weight: 261.37

MP (deg C): | FP (deg C):

BP (deg C):

BP pressure (mm Hg):

Property/Value	Units	Temp	Data Type	Reference
WS 2.14E+002	mg/L	25	EST	MEYLAN,WM ET AL. (1996)
logP 3.50			EXP	HANSCH,C & LEO,AJ (1985)
VP 9.96E-008	mm Hg	25	EST	NEELY,WB & BLAU,GE (1985)
DC	pKa			
HL 2.18E-011	atm m3/mol	25	EST	MEYLAN,WM & HOWARD,PH (1991)
OH 1.93E-010	cm3/molc sec	25	EST	MEYLAN,WM & HOWARD,PH (1993)

CAS #: 054387-29-0				PIPERAZINE, 1-METHYL-4-[(1-METHYL-5-NITRO-1H-IMI

Formula: C10H17N5O2

Mol Weight: 239.28

MP (deg C): | FP (deg C):

BP (deg C):

BP pressure (mm Hg):

Property/Value	Units	Temp	Data Type	Reference
WS 2.11E+005	mg/L	25	EST	MEYLAN,WM ET AL. (1996)
logP -0.32			EXP	SANGSTER,J (1994)
VP 3.89E-007	mm Hg	25	EST	NEELY,WB & BLAU,GE (1985)
DC	pKa			
HL 5.02E-014	atm m3/mol	25	EST	MEYLAN,WM & HOWARD,PH (1991)
OH 1.89E-010	cm3/molc sec	25	EST	MEYLAN,WM & HOWARD,PH (1993)

CAS #: 054394-78-4				SEC-VALERANILIDE

Formula: C11H15NO

Mol Weight: 177.25

MP (deg C): | FP (deg C):

BP (deg C):

BP pressure (mm Hg):

Property/Value	Units	Temp	Data Type	Reference
WS 3.70E+002	mg/L	25	EST	MEYLAN,WM ET AL. (1996)
logP 2.56			EXP	HANSCH,C & LEO,AJ (1985)
VP 5.80E-005	mm Hg	25	EST	NEELY,WB & BLAU,GE (1985)
DC	pKa			
HL 1.44E-008	atm m3/mol	25	EST	MEYLAN,WM & HOWARD,PH (1991)
OH 1.85E-011	cm3/molc sec	25	EST	MEYLAN,WM & HOWARD,PH (1993)

CAS #: 054446-78-5				1-(2-BUTOXYETHOXY)ETHANOL

Formula: C8H18O3

Mol Weight: 162.23

MP (deg C): | FP (deg C):

BP (deg C):

BP pressure (mm Hg):

Property/Value	Units	Temp	Data Type	Reference
WS 1.41E+005	mg/L	25	EST	MEYLAN,WM ET AL. (1996)
logP 0.22			EST	MEYLAN,WM & HOWARD,PH (1995)
VP 1.81E-002	mm Hg	25	EST	NEELY,WB & BLAU,GE (1985)
DC	pKa			
HL 1.52E-009	atm m3/mol	25	EST	MEYLAN,WM & HOWARD,PH (1991)
OH 6.81E-011	cm3/molc sec	25	EST	MEYLAN,WM & HOWARD,PH (1993)

CAS #: 054466-88-5				INDOL-3-YL-THIOACETICACID

Formula: C10H9NOS

Mol Weight: 191.25

MP (deg C): | FP (deg C):

BP (deg C):

BP pressure (mm Hg):

Property/Value	Units	Temp	Data Type	Reference
WS 3.01E+003	mg/L	25	EST	MEYLAN,WM ET AL. (1996)
logP 1.78			EXP	HANSCH,C & LEO,AJ (1985)
VP 5.60E-007	mm Hg	25	EST	NEELY,WB & BLAU,GE (1985)
DC	pKa			
HL 1.92E-013	atm m3/mol	25	EST	MEYLAN,WM & HOWARD,PH (1991)
OH 2.06E-010	cm3/molc sec	25	EST	MEYLAN,WM & HOWARD,PH (1993)

CAS #: 054504-62-0				4H-PYRIDO[1,2-A]PYRIMIDIN-4-ONE, 6,7,8,9-TETRAHY

Formula: C10H14N2O

Mol Weight: 178.24

MP (deg C): | FP (deg C):

BP (deg C):

BP pressure (mm Hg):

Property/Value	Units	Temp	Data Type	Reference
WS 1.12E+004	mg/L	25	EST	MEYLAN,WM ET AL. (1996)
logP 0.82			EXP	SANGSTER,J (1993)
VP 1.05E-004	mm Hg	25	EST	NEELY,WB & BLAU,GE (1985)
DC	pKa			
HL 2.79E-009	atm m3/mol	25	EST	MEYLAN,WM & HOWARD,PH (1991)
OH 3.86E-011	cm3/molc sec	25	EST	MEYLAN,WM & HOWARD,PH (1993)

CAS #: 054548-50-4				CHLORO-M-CRESOL

Formula: C_7H_7ClO

Mol Weight: 142.59

MP (deg C): FP (deg C):

BP (deg C):

BP pressure (mm Hg):

Property/ Value	Units	Temp	Data Type	Reference
WS 1.52E+003	mg/L	25	EST	MEYLAN,WM ET AL. (1996)
logP 2.70			EST	MEYLAN,WM & HOWARD,PH (1995)
VP 4.05E-002	mm Hg	25	EST	NEELY,WB & BLAU,GE (1985)
DC	pKa			
HL 4.58E-007	atm m3/mol	25	EST	MEYLAN,WM & HOWARD,PH (1991)
OH 6.15E-011	cm3/molc sec	25	EST	MEYLAN,WM & HOWARD,PH (1993)

CAS #: 054593-83-8				CHLORETHOXYFOS

Formula: $C_6H_{11}Cl_4O_3PS$

Mol Weight: 336.00

MP (deg C): FP (deg C):

BP (deg C):

BP pressure (mm Hg):

Property/ Value	Units	Temp	Data Type	Reference
WS 9.05E-001	mg/L	25	EST	MEYLAN,WM ET AL. (1996)
logP 4.59			EXP	TOMLIN,C (1994)
VP 9.44E-004	mm Hg	25	EST	NEELY,WB & BLAU,GE (1985)
DC	pKa			
HL 4.15E-006	atm m3/mol	25	EST	MEYLAN,WM & HOWARD,PH (1991)
OH 9.27E-011	cm3/molc sec	25	EST	MEYLAN,WM & HOWARD,PH (1993)

CAS #: 054606-23-4				4H-PYRIDO[1,2-A]PYRIMIDINE-3-CARBOXAMIDE, 6,7,8,

Formula: $C_9H_{11}N_3O_2$

Mol Weight: 193.21

MP (deg C): FP (deg C):

BP (deg C):

BP pressure (mm Hg):

Property/ Value	Units	Temp	Data Type	Reference
WS 4.90E+003	mg/L	25	EST	MEYLAN,WM ET AL. (1996)
logP -0.38			EXP	SANGSTER,J (1994)
VP 1.19E-007	mm Hg	25	EST	NEELY,WB & BLAU,GE (1985)
DC	pKa			
HL 2.03E-013	atm m3/mol	25	EST	MEYLAN,WM & HOWARD,PH (1991)
OH 2.33E-011	cm3/molc sec	25	EST	MEYLAN,WM & HOWARD,PH (1993)

CAS #: 054606-24-5				4H-PYRIDO[1,2-A]PYRIMIDIN-4-ONE, 6,7,8,9-TETRAHY

Formula: $C_9H_{12}N_2O$

Mol Weight: 164.21

MP (deg C): FP (deg C):

BP (deg C):

BP pressure (mm Hg):

Property/ Value	Units	Temp	Data Type	Reference
WS 2.53E+003	mg/L	25	EST	MEYLAN,WM ET AL. (1996)
logP 0.12			EXP	SANGSTER,J (1993)
VP 2.04E-004	mm Hg	25	EST	NEELY,WB & BLAU,GE (1985)
DC	pKa			
HL 2.10E-009	atm m3/mol	25	EST	MEYLAN,WM & HOWARD,PH (1991)
OH 3.33E-011	cm3/molc sec	25	EST	MEYLAN,WM & HOWARD,PH (1993)

CAS #: 054680-35-2				1-ETHYL-1-NITROSO-3-PHENYLUREA

Formula: $C_9H_{11}N_3O_2$

Mol Weight: 193.21

MP (deg C): FP (deg C):

BP (deg C):

BP pressure (mm Hg):

Property/ Value	Units	Temp	Data Type	Reference
WS 5.68E+002	mg/L	25	EST	MEYLAN,WM ET AL. (1996)
logP 2.25			EXP	HANSCH,C & LEO,AJ (1985)
VP 4.63E-005	mm Hg	25	EST	NEELY,WB & BLAU,GE (1985)
DC	pKa			
HL 7.26E-011	atm m3/mol	25	EST	MEYLAN,WM & HOWARD,PH (1991)
OH 4.67E-011	cm3/molc sec	25	EST	MEYLAN,WM & HOWARD,PH (1993)

CAS #: 054704-34-6				CYCLOHEXANEETHANAMINE, à-METHYL-

Formula: $C_9H_{19}N$

Mol Weight: 141.26

MP (deg C): FP (deg C):

BP (deg C):

BP pressure (mm Hg):

Property/ Value	Units	Temp	Data Type	Reference
WS 2.50E+003	mg/L	25	EST	MEYLAN,WM ET AL. (1996)
logP 2.96			EXP	SANGSTER,J (1993)
VP 4.48E-001	mm Hg	25	EST	NEELY,WB & BLAU,GE (1985)
DC	pKa			
HL 3.22E-005	atm m3/mol	25	EST	MEYLAN,WM & HOWARD,PH (1991)
OH 5.54E-011	cm3/molc sec	25	EST	MEYLAN,WM & HOWARD,PH (1993)

CAS #: 054705-16-7				BENZAMIDE, 2-(METHYLTHIO)-

Formula: C_8H_9NOS

Mol Weight: 167.23

MP (deg C): FP (deg C):

BP (deg C):

BP pressure (mm Hg):

Property/ Value	Units	Temp	Data Type	Reference
WS 1.57E+004	mg/L	25	EST	MEYLAN,WM ET AL. (1996)
logP 0.71			EXP	SOTOMATSU,T ET AL. (1993)
VP 2.32E-005	mm Hg	25	EST	NEELY,WB & BLAU,GE (1985)
DC	pKa			
HL 6.45E-011	atm m3/mol	25	EST	MEYLAN,WM & HOWARD,PH (1991)
OH 1.45E-011	cm3/molc sec	25	EST	MEYLAN,WM & HOWARD,PH (1993)

CAS #: 054749-90-5				CHLOROZOTOCIN

Formula: $C_9H_{16}ClN_3O_4$

Mol Weight: 265.70

MP (deg C): 147-148 de FP (deg C):

BP (deg C):

BP pressure (mm Hg):

Property/ Value	Units	Temp	Data Type	Reference
WS 1.83E+003	mg/L	25	EST	MEYLAN,WM ET AL. (1996)
logP -1.02			EXP	HANSCH,C & LEO,AJ (1985)
VP 3.98E-014	mm Hg	25	EST	NEELY,WB & BLAU,GE (1985)
DC	pKa			
HL 3.67E-022	atm m3/mol	25	EST	MEYLAN,WM & HOWARD,PH (1991)
OH 1.09E-010	cm3/molc sec	25	EST	MEYLAN,WM & HOWARD,PH (1993)

054797-20-5 — PHENYL AZOXYCYANIDE

CAS #: 054797-20-5

Formula: $C_7H_5N_3O$

Mol Weight: 147.14

MP (deg C):　FP (deg C):

BP (deg C):

BP pressure (mm Hg):

Property/Value	Units	Temp	Data Type	Reference
WS 7.30E+002	mg/L	25	EST	MEYLAN,WM ET AL. (1996)
logP 1.87			EXP	CALVINO,R R ET AL (1991)
VP 7.08E-008	mm Hg	25	EST	NEELY,WB & BLAU,GE (1985)
DC	pKa			
HL 2.06E-011	atm m3/mol	25	EST	MEYLAN,WM & HOWARD,PH (1991)
OH 4.17E-013	cm3/molc sec	25	EST	MEYLAN,WM & HOWARD,PH (1993)

054797-21-6 — P-METHYLPHENYL AZOXYCYANIDE

CAS #: 054797-21-6

Formula: $C_8H_7N_3O$

Mol Weight: 161.16

MP (deg C):　FP (deg C):

BP (deg C):

BP pressure (mm Hg):

Property/Value	Units	Temp	Data Type	Reference
WS 2.28E+002	mg/L	25	EST	MEYLAN,WM ET AL. (1996)
logP 2.39			EXP	CALVINO,R R ET AL (1991)
VP 2.81E-008	mm Hg	25	EST	NEELY,WB & BLAU,GE (1985)
DC	pKa			
HL 2.27E-011	atm m3/mol	25	EST	MEYLAN,WM & HOWARD,PH (1991)
OH 1.22E-012	cm3/molc sec	25	EST	MEYLAN,WM & HOWARD,PH (1993)

054797-22-7 — P-CHLOROPHENYL AZOXYCYANIDE

CAS #: 054797-22-7

Formula: $C_7H_4ClN_3O$

Mol Weight: 181.58

MP (deg C):　FP (deg C):

BP (deg C):

BP pressure (mm Hg):

Property/Value	Units	Temp	Data Type	Reference
WS 2.05E+002	mg/L	25	EST	MEYLAN,WM ET AL. (1996)
logP 2.33			EXP	CALVINO,R R ET AL (1991)
VP 1.72E-008	mm Hg	25	EST	NEELY,WB & BLAU,GE (1985)
DC	pKa			
HL 1.53E-011	atm m3/mol	25	EST	MEYLAN,WM & HOWARD,PH (1991)
OH 2.93E-013	cm3/molc sec	25	EST	MEYLAN,WM & HOWARD,PH (1993)

054831-57-1 — HISTIDINE-AMIDE,N-ACETYL

CAS #: 054831-57-1

Formula: $C_8H_{12}N_4O_2$

Mol Weight: 196.21

MP (deg C):　FP (deg C):

BP (deg C):

BP pressure (mm Hg):

Property/Value	Units	Temp	Data Type	Reference
WS 6.34E+004	mg/L	25	EST	MEYLAN,WM ET AL. (1996)
logP -1.70			EXP	HANSCH,C & LEO,AJ (1985)
VP 1.97E-011	mm Hg	25	EST	NEELY,WB & BLAU,GE (1985)
DC	pKa			
HL 8.81E-015	atm m3/mol	25	EST	MEYLAN,WM & HOWARD,PH (1991)
OH 1.36E-010	cm3/molc sec	25	EST	MEYLAN,WM & HOWARD,PH (1993)

054850-02-1 — 3,5-DICL-4'-F SALICYLANILIDE

CAS #: 054850-02-1

Formula: $C_{13}H_8Cl_2FNO_2$

Mol Weight: 300.12

MP (deg C):　FP (deg C):

BP (deg C):

BP pressure (mm Hg):

Property/Value	Units	Temp	Data Type	Reference
WS 4.01E+000	mg/L	25	EST	MEYLAN,WM ET AL. (1996)
logP 4.76			EXP	TERADA,H ET AL. (1988)
VP 1.80E-009	mm Hg	25	EST	NEELY,WB & BLAU,GE (1985)
DC	pKa			
HL 1.03E-010	atm m3/mol	25	EST	MEYLAN,WM & HOWARD,PH (1991)
OH 4.61E-012	cm3/molc sec	25	EST	MEYLAN,WM & HOWARD,PH (1993)

054855-60-6 — CIMETIDINE,PYRIMIDIN-2-ONE ANALOG

CAS #: 054855-60-6

Formula: $C_{11}H_{15}N_5OS$

Mol Weight: 265.34

MP (deg C):　FP (deg C):

BP (deg C):

BP pressure (mm Hg):

Property/Value	Units	Temp	Data Type	Reference
WS 2.09E+003	mg/L	25	EST	MEYLAN,WM ET AL. (1996)
logP -0.40			EXP	HANSCH,C ET AL. (1995)
VP 5.21E-011	mm Hg	25	EST	NEELY,WB & BLAU,GE (1985)
DC	pKa			
HL 9.29E-017	atm m3/mol	25	EST	MEYLAN,WM & HOWARD,PH (1991)
OH 1.80E-010	cm3/molc sec	25	EST	MEYLAN,WM & HOWARD,PH (1993)

054855-69-5 — CIMETIDINE,2(PYRIDIN-4-ONE) ANALOG

CAS #: 054855-69-5

Formula: $C_{12}H_{16}N_4OS$

Mol Weight: 264.35

MP (deg C):　FP (deg C):

BP (deg C):

BP pressure (mm Hg):

Property/Value	Units	Temp	Data Type	Reference
WS 4.16E+003	mg/L	25	EST	MEYLAN,WM ET AL. (1996)
logP 0.79			EXP	HANSCH,C ET AL. (1995)
VP 1.99E-011	mm Hg	25	EST	NEELY,WB & BLAU,GE (1985)
DC	pKa			
HL 1.49E-015	atm m3/mol	25	EST	MEYLAN,WM & HOWARD,PH (1991)
OH 1.75E-010	cm3/molc sec	25	EST	MEYLAN,WM & HOWARD,PH (1993)

054855-72-0 — CIMETIDINE,4-THIOPYRIMIDIN-2-YL-DIOXIDE ANALOG

CAS #: 054855-72-0

Formula: $C_{10}H_{17}N_5O_2S_2$

Mol Weight: 303.41

MP (deg C):　FP (deg C):

BP (deg C):

BP pressure (mm Hg):

Property/Value	Units	Temp	Data Type	Reference
WS 2.46E+003	mg/L	25	EST	MEYLAN,WM ET AL. (1996)
logP -0.74			EXP	HANSCH,C ET AL. (1995)
VP 1.26E-013	mm Hg	25	EST	NEELY,WB & BLAU,GE (1985)
DC	pKa			
HL 2.68E-017	atm m3/mol	25	EST	MEYLAN,WM & HOWARD,PH (1991)
OH 2.28E-010	cm3/molc sec	25	EST	MEYLAN,WM & HOWARD,PH (1993)

CAS #: 054855-84-4				CIMETIDINE,PYRIMID-4-ONE-2-YL ANALOG

Formula: $C_{11}H_{15}N_5OS$
Mol Weight: 265.34
MP (deg C): | FP (deg C):
BP (deg C):
BP pressure (mm Hg):

Property/ Value	Units	Temp	Data Type	Reference
WS 1.87E+004	mg/L	25	EST	MEYLAN,WM ET AL. (1996)
logP 0.02			EXP	HANSCH,C ET AL. (1995)
VP 3.83E-013	mm Hg	25	EST	NEELY,WB & BLAU,GE (1985)
DC	pKa			
HL 1.18E-019	atm m3/mol	25	EST	MEYLAN,WM & HOWARD,PH (1991)
OH 2.15E-010	cm3/molc sec	25	EST	MEYLAN,WM & HOWARD,PH (1993)

CAS #: 054870-28-9				MEGLITINIDE

Formula: $C_{17}H_{16}ClNO_4$
Mol Weight: 333.77
MP (deg C): | FP (deg C):
BP (deg C):
BP pressure (mm Hg):

Property/ Value	Units	Temp	Data Type	Reference
WS 9.68E+000	mg/L	25	EST	MEYLAN,WM ET AL. (1996)
logP 3.40			EXP	HANSCH,C ET AL. (1995)
VP 4.20E-011	mm Hg	25	EST	NEELY,WB & BLAU,GE (1985)
DC	pKa			
HL 4.60E-016	atm m3/mol	25	EST	MEYLAN,WM & HOWARD,PH (1991)
OH 2.02E-011	cm3/molc sec	25	EST	MEYLAN,WM & HOWARD,PH (1993)

CAS #: 054922-60-0				N'-IPR(3-TOLYLOXY)ACETIC AC HYDRAZID

Formula: $C_{12}H_{18}N_2O_2$
Mol Weight: 222.29
MP (deg C): | FP (deg C):
BP (deg C):
BP pressure (mm Hg):

Property/ Value	Units	Temp	Data Type	Reference
WS 1.03E+003	mg/L	25	EST	MEYLAN,WM ET AL. (1996)
logP 1.77			EXP	HANSCH,C & LEO,AJ (1985)
VP 1.88E-006	mm Hg	25	EST	NEELY,WB & BLAU,GE (1985)
DC	pKa			
HL 6.11E-012	atm m3/mol	25	EST	MEYLAN,WM & HOWARD,PH (1991)
OH 1.47E-010	cm3/molc sec	25	EST	MEYLAN,WM & HOWARD,PH (1993)

CAS #: 054922-61-1				N'-IPR(4-TOLYLOXY)ACETIC AC HYDRAZID

Formula: $C_{12}H_{18}N_2O_2$
Mol Weight: 222.29
MP (deg C): | FP (deg C):
BP (deg C):
BP pressure (mm Hg):

Property/ Value	Units	Temp	Data Type	Reference
WS 9.53E+002	mg/L	25	EST	MEYLAN,WM ET AL. (1996)
logP 1.81			EXP	HANSCH,C & LEO,AJ (1985)
VP 1.88E-006	mm Hg	25	EST	NEELY,WB & BLAU,GE (1985)
DC	pKa			
HL 6.11E-012	atm m3/mol	25	EST	MEYLAN,WM & HOWARD,PH (1991)
OH 1.18E-010	cm3/molc sec	25	EST	MEYLAN,WM & HOWARD,PH (1993)

CAS #: 054922-62-2				N-IPR-1(4-MEO PHENOXY)ACETIC HYDRAZID

Formula: $C_{12}H_{18}N_2O_3$
Mol Weight: 238.29
MP (deg C): | FP (deg C):
BP (deg C):
BP pressure (mm Hg):

Property/ Value	Units	Temp	Data Type	Reference
WS 2.54E+003	mg/L	25	EST	MEYLAN,WM ET AL. (1996)
logP 1.21			EXP	HANSCH,C & LEO,AJ (1985)
VP 7.45E-007	mm Hg	25	EST	NEELY,WB & BLAU,GE (1985)
DC	pKa			
HL 3.28E-013	atm m3/mol	25	EST	MEYLAN,WM & HOWARD,PH (1991)
OH 1.11E-010	cm3/molc sec	25	EST	MEYLAN,WM & HOWARD,PH (1993)

CAS #: 054922-63-3				N'-IPR-2-ME-2(4CL-PHO)PROPIONICACIDHYDRAZIDE

Formula: $C_{13}H_{19}ClN_2O_2$
Mol Weight: 270.76
MP (deg C): | FP (deg C):
BP (deg C):
BP pressure (mm Hg):

Property/ Value	Units	Temp	Data Type	Reference
WS 1.18E+002	mg/L	25	EST	MEYLAN,WM ET AL. (1996)
logP 2.56			EXP	HANSCH,C & LEO,AJ (1985)
VP 4.25E-007	mm Hg	25	EST	NEELY,WB & BLAU,GE (1985)
DC	pKa			
HL 7.23E-012	atm m3/mol	25	EST	MEYLAN,WM & HOWARD,PH (1991)
OH 9.42E-011	cm3/molc sec	25	EST	MEYLAN,WM & HOWARD,PH (1993)

CAS #: 054922-65-5				N'-ET(4-CLPHENOXY)ACETIC A HYDRAZIDE

Formula: $C_{10}H_{13}ClN_2O_2$
Mol Weight: 228.68
MP (deg C): | FP (deg C):
BP (deg C):
BP pressure (mm Hg):

Property/ Value	Units	Temp	Data Type	Reference
WS 1.03E+003	mg/L	25	EST	MEYLAN,WM ET AL. (1996)
logP 1.73			EXP	HANSCH,C & LEO,AJ (1985)
VP 1.42E-006	mm Hg	25	EST	NEELY,WB & BLAU,GE (1985)
DC	pKa			
HL 3.09E-012	atm m3/mol	25	EST	MEYLAN,WM & HOWARD,PH (1991)
OH 8.79E-011	cm3/molc sec	25	EST	MEYLAN,WM & HOWARD,PH (1993)

CAS #: 054922-68-8				N'-BENZYL-(4-CLPHO)ACETICACIDHYDRAZIDE

Formula: $C_{15}H_{15}ClN_2O_2$
Mol Weight: 290.75
MP (deg C): | FP (deg C):
BP (deg C):
BP pressure (mm Hg):

Property/ Value	Units	Temp	Data Type	Reference
WS 4.46E+001	mg/L	25	EST	MEYLAN,WM ET AL. (1996)
logP 2.92			EXP	HANSCH,C & LEO,AJ (1985)
VP 7.01E-009	mm Hg	25	EST	NEELY,WB & BLAU,GE (1985)
DC	pKa			
HL 1.88E-013	atm m3/mol	25	EST	MEYLAN,WM & HOWARD,PH (1991)
OH 9.26E-011	cm3/molc sec	25	EST	MEYLAN,WM & HOWARD,PH (1993)

CAS #: 055040-03-4 — PIPERIDINE, 1-(4-METHYL-1-PHENYLCYCLOHEXYL)-, TR

Formula: $C_{18}H_{27}N$
Mol Weight: 257.42
MP (deg C):
FP (deg C):
BP (deg C):
BP pressure (mm Hg):

Property/Value	Units	Temp	Data Type	Reference
WS 3.21E+000	mg/L	25	EST	MEYLAN,WM ET AL. (1996)
logP 5.66			EXP	SANGSTER,J (1993)
VP 2.75E-005	mm Hg	25	EST	NEELY,WB & BLAU,GE (1985)
DC	pKa			
HL 7.34E-006	atm m3/mol	25	EST	MEYLAN,WM & HOWARD,PH (1991)
OH 1.05E-010	cm3/molc sec	25	EST	MEYLAN,WM & HOWARD,PH (1993)

CAS #: 055049-18-8 — PIPERIDINOACETIC ACID

Formula: $C_7H_{13}NO_2$
Mol Weight: 143.19
MP (deg C):
FP (deg C):
BP (deg C):
BP pressure (mm Hg):

Property/Value	Units	Temp	Data Type	Reference
WS 5.45E+003	mg/L	25	EST	MEYLAN,WM ET AL. (1996)
logP -1.05			EXP	HANSCH,C & LEO,AJ (1985)
VP 1.95E-009	mm Hg	25	EST	NEELY,WB & BLAU,GE (1985)
DC	pKa			
HL 5.59E-009	atm m3/mol	25	EST	MEYLAN,WM & HOWARD,PH (1991)
OH 9.86E-011	cm3/molc sec	25	EST	MEYLAN,WM & HOWARD,PH (1993)

CAS #: 055079-83-9 — 2,4,6,8-NONATETRAENOIC ACID, 9-(4-METHOXY-2,3,6-

Formula: $C_{21}H_{26}O_3$
Mol Weight: 326.44
MP (deg C): 228-230
FP (deg C):
BP (deg C):
BP pressure (mm Hg):

Property/Value	Units	Temp	Data Type	Reference
WS 7.29E-002	mg/L	25	EST	MEYLAN,WM ET AL. (1996)
logP 6.40			EXP	SANGSTER,J (1993)
VP 1.34E-008	mm Hg	25	EST	NEELY,WB & BLAU,GE (1985)
DC	pKa			
HL 5.08E-009	atm m3/mol	25	EST	MEYLAN,WM & HOWARD,PH (1991)
OH 3.89E-010	cm3/molc sec	25	EST	MEYLAN,WM & HOWARD,PH (1993)

CAS #: 055096-26-9 — NALMEFENE

Formula: $C_{21}H_{25}NO_3$
Mol Weight: 339.44
MP (deg C): 188-190
FP (deg C):
BP (deg C):
BP pressure (mm Hg):

Property/Value	Units	Temp	Data Type	Reference
WS 1.40E+002	mg/L	25	EST	NEELY,WB & BLAU,GE (1985)
logP 2.66			EST	MEYLAN,WM & HOWARD,PH (1995)
VP 5.70E-014	mm Hg	25	EST	HL X WSOL
DC	pKa			
HL 1.82E-016	atm m3/mol	25	EST	MEYLAN,WM & HOWARD,PH (1991)
OH	cm3/molc sec			

CAS #: 055121-29-4 — ACETAMIDE, N-[4-[(DICYANOMETHYLENE)HYDRAZONO]PHE

Formula: $C_{11}H_9N_5O$
Mol Weight: 227.23
MP (deg C):
FP (deg C):
BP (deg C):
BP pressure (mm Hg):

Property/Value	Units	Temp	Data Type	Reference
WS 1.73E+002	mg/L	25	EST	MEYLAN,WM ET AL. (1996)
logP 1.83			EXP	STURDIK,E ET AL. (1985)
VP 7.24E-009	mm Hg	25	EST	NEELY,WB & BLAU,GE (1985)
DC	pKa			
HL 1.15E-017	atm m3/mol	25	EST	MEYLAN,WM & HOWARD,PH (1991)
OH 5.09E-012	cm3/molc sec	25	EST	MEYLAN,WM & HOWARD,PH (1993)

CAS #: 055147-64-3 — 4'-DEHYDROOLEANDRIN

Formula: $C_{32}H_{46}O_9$
Mol Weight: 574.72
MP (deg C):
FP (deg C):
BP (deg C):
BP pressure (mm Hg):

Property/Value	Units	Temp	Data Type	Reference
WS 9.17E+000	mg/L	25	EST	MEYLAN,WM ET AL. (1996)
logP 1.65			EXP	HANSCH,C & LEO,AJ (1985)
VP 3.15E-019	mm Hg	25	EST	NEELY,WB & BLAU,GE (1985)
DC	pKa			
HL 1.81E-018	atm m3/mol	25	EST	MEYLAN,WM & HOWARD,PH (1991)
OH 1.57E-010	cm3/molc sec	25	EST	MEYLAN,WM & HOWARD,PH (1993)

CAS #: 055179-31-2 — BITERTANOL

Formula: $C_{20}H_{25}N_3O_2$
Mol Weight: 339.44
MP (deg C): 125-129
FP (deg C):
BP (deg C):
BP pressure (mm Hg):

Property/Value	Units	Temp	Data Type	Reference
WS 6.69E+000	mg/L	25	EST	MEYLAN,WM ET AL. (1996)
logP 4.16			EXP	HANSCH,C ET AL. (1995)
VP 3.07E-011	mm Hg	25	EST	NEELY,WB & BLAU,GE (1985)
DC	pKa			
HL 1.63E-012	atm m3/mol	25	EST	MEYLAN,WM & HOWARD,PH (1991)
OH 4.43E-011	cm3/molc sec	25	EST	MEYLAN,WM & HOWARD,PH (1993)

CAS #: 055205-89-5 — 3,5-DIETHYL-4-HYDROXYACETANILIDE

Formula: $C_{12}H_{17}NO_2$
Mol Weight: 207.27
MP (deg C):
FP (deg C):
BP (deg C):
BP pressure (mm Hg):

Property/Value	Units	Temp	Data Type	Reference
WS 1.02E+003	mg/L	25	EST	MEYLAN,WM ET AL. (1996)
logP 1.87			EXP	HANSCH,C & LEO,AJ (1985)
VP 1.76E-007	mm Hg	25	EST	NEELY,WB & BLAU,GE (1985)
DC	pKa			
HL 1.38E-012	atm m3/mol	25	EST	MEYLAN,WM & HOWARD,PH (1991)
OH 5.52E-011	cm3/molc sec	25	EST	MEYLAN,WM & HOWARD,PH (1993)

CAS #: 055215-17-3			2,2',3,4,6-PENTACHLOROBIPHENYL

Formula: $C_{12}H_5Cl_5$

Mol Weight: 326.44

MP (deg C): FP (deg C):

BP (deg C):

BP pressure (mm Hg):

Property/Value	Units	Temp	Data Type	Reference
WS 1.20E-002	mg/L	25	EXP	YALKOWSKY,SH & DANNENFELSER,RM (1992)
logP 6.98			EST	MEYLAN,WM & HOWARD,PH (1995)
VP 2.22E-006	mm Hg	25	EST	NEELY,WB & BLAU,GE (1985)
DC	pKa			
HL 9.24E-005	atm m3/mol	25	EST	MEYLAN,WM & HOWARD,PH (1991)
OH 4.18E-013	cm3/molc sec	25	EST	MEYLAN,WM & HOWARD,PH (1993)

CAS #: 055215-18-4			2,2',3,3',4,5-HEXACHLOROBIPHENYL

Formula: $C_{12}H_4Cl_6$

Mol Weight: 360.88

MP (deg C): FP (deg C):

BP (deg C):

BP pressure (mm Hg):

Property/Value	Units	Temp	Data Type	Reference
WS 5.81E-003	mg/L	25	EXP	YALKOWSKY,SH & DANNENFELSER,RM (1992)
logP 7.32			EXP	RAPAPORT,RA & EISENREICH,SJ (1984)
VP 5.81E-007	mm Hg	25	EST	NEELY,WB & BLAU,GE (1985)
DC	pKa			
HL 6.85E-005	atm m3/mol	25	EST	MEYLAN,WM & HOWARD,PH (1991)
OH 2.11E-013	cm3/molc sec	25	EST	MEYLAN,WM & HOWARD,PH (1993)

CAS #: 055219-65-3			TRIADIMENOL

Formula: $C_{14}H_{20}ClN_3O_2$

Mol Weight: 297.79

MP (deg C): 121-127 FP (deg C):

BP (deg C):

BP pressure (mm Hg):

Property/Value	Units	Temp	Data Type	Reference
WS 9.50E+001	mg/L	20	EXP	SHIU,WY ET AL. (1990)
logP 2.90			EXP	BAKER,EA ET AL. (1992)
VP 3.10E-010	mm Hg	20	EXP	AUGUSTIJN-BECKERS,PWM ET AL. (1994)
DC	pKa			
HL 1.28E-012	atm m3/mol	20	EST	VP/WSOL
OH 3.18E-011	cm3/molc sec	25	EST	MEYLAN,WM & HOWARD,PH (1993)

CAS #: 055236-14-1			BENZAMIDE,N-(DIETAMET),2-MEO,5-F

Formula: $C_{14}H_{21}FN_2O_2$

Mol Weight: 268.33

MP (deg C): FP (deg C):

BP (deg C):

BP pressure (mm Hg):

Property/Value	Units	Temp	Data Type	Reference
WS 1.08E+002	mg/L	25	EST	MEYLAN,WM ET AL. (1996)
logP 2.62			EXP	HANSCH,C & LEO,AJ (1985)
VP 5.07E-007	mm Hg	25	EST	NEELY,WB & BLAU,GE (1985)
DC	pKa			
HL 3.89E-013	atm m3/mol	25	EST	MEYLAN,WM & HOWARD,PH (1991)
OH 1.12E-010	cm3/molc sec	25	EST	MEYLAN,WM & HOWARD,PH (1993)

CAS #: 055265-51-5			2-F-3-BR-TETRACHLOROCYCLOHEXANE

Formula: $C_6H_6BrCl_4F$

Mol Weight: 318.83

MP (deg C): FP (deg C):

BP (deg C):

BP pressure (mm Hg):

Property/Value	Units	Temp	Data Type	Reference
WS 1.50E+001	mg/L	25	EST	MEYLAN,WM ET AL. (1996)
logP 3.28			EXP	HANSCH,C & LEO,AJ (1985)
VP 1.76E-003	mm Hg	25	EST	NEELY,WB & BLAU,GE (1985)
DC	pKa			
HL 4.73E-004	atm m3/mol	25	EST	MEYLAN,WM & HOWARD,PH (1991)
OH 4.40E-013	cm3/molc sec	25	EST	MEYLAN,WM & HOWARD,PH (1993)

CAS #: 055265-69-5			2,4-DIBR-TETRACHLOROCYCLOHEXANE

Formula: $C_6H_6Br_2Cl_4$

Mol Weight: 379.74

MP (deg C): FP (deg C):

BP (deg C):

BP pressure (mm Hg):

Property/Value	Units	Temp	Data Type	Reference
WS 1.60E+000	mg/L	25	EST	MEYLAN,WM ET AL. (1996)
logP 3.98			EXP	HANSCH,C & LEO,AJ (1985)
VP 6.49E-005	mm Hg	25	EST	NEELY,WB & BLAU,GE (1985)
DC	pKa			
HL 2.74E-005	atm m3/mol	25	EST	MEYLAN,WM & HOWARD,PH (1991)
OH 6.36E-013	cm3/molc sec	25	EST	MEYLAN,WM & HOWARD,PH (1993)

CAS #: 055268-75-2			CEFUROXIM

Formula: $C_{16}H_{16}N_4O_8S$

Mol Weight: 424.39

MP (deg C): FP (deg C):

BP (deg C):

BP pressure (mm Hg):

Property/Value	Units	Temp	Data Type	Reference
WS 1.45E+002	mg/L	25	EST	MEYLAN,WM ET AL. (1996)
logP -0.16			EXP	SANGSTER,J (1993)
VP 1.72E-015	mm Hg	25	EST	NEELY,WB & BLAU,GE (1985)
DC	pKa			
HL 2.77E-021	atm m3/mol	25	EST	MEYLAN,WM & HOWARD,PH (1991)
OH 2.01E-010	cm3/molc sec	25	EST	MEYLAN,WM & HOWARD,PH (1993)

CAS #: 055283-68-6			BENZENAMINE, N-ETHYL-N-(2-METHYL-2-PROPENYL)-2,6

Formula: $C_{13}H_{14}F_3N_3O_4$

Mol Weight: 333.27

MP (deg C): 57 FP (deg C):

BP (deg C):

BP pressure (mm Hg):

Property/Value	Units	Temp	Data Type	Reference
WS 2.00E-001	mg/L	25	EXP	SHIU,WY ET AL. (1990) @ pH=7
logP 5.11			EXP	TOMLIN,C (1994)
VP 8.20E-007	mm Hg	25	EXP	WEBER,JB (1994)
DC	pKa			
HL 1.80E-006	atm m3/mol	25	EST	VP/WSOL
OH 6.91E-011	cm3/molc sec	25	EST	MEYLAN,WM & HOWARD,PH (1993)

CAS #: 055290-64-7 — DIMETHIPIN

Formula: $C_6H_{10}O_4S_2$
Mol Weight: 210.27
MP (deg C): 162-167
FP (deg C):
BP (deg C):
BP pressure (mm Hg):

Property/Value	Units	Temp	Data Type	Reference
WS 3.00E+003	mg/L	25	EXP	SHIU,WY ET AL. (1990)
logP -0.17			EXP	TOMLIN,C (1994)
VP 2.20E-005	mm Hg	25	EST	NEELY,WB & BLAU,GE (1985)
DC	pKa			
HL 5.38E-009	atm m3/mol	25	EST	MEYLAN,WM & HOWARD,PH (1991)
OH 1.28E-010	cm3/molc sec	25	EST	MEYLAN,WM & HOWARD,PH (1993)

CAS #: 055298-45-8 — 1,2-DIBR-TETRACHLOROCYCLOHEXANE

Formula: $C_6H_6Br_2Cl_4$
Mol Weight: 379.74
MP (deg C):
FP (deg C):
BP (deg C):
BP pressure (mm Hg):

Property/Value	Units	Temp	Data Type	Reference
WS 1.60E+000	mg/L	25	EST	MEYLAN,WM ET AL. (1996)
logP 3.88			EXP	HANSCH,C & LEO,AJ (1985)
VP 6.49E-005	mm Hg	25	EST	NEELY,WB & BLAU,GE (1985)
DC	pKa			
HL 2.74E-005	atm m3/mol	25	EST	MEYLAN,WM & HOWARD,PH (1991)
OH 6.15E-013	cm3/molc sec	25	EST	MEYLAN,WM & HOWARD,PH (1993)

CAS #: 055312-69-1 — 2,2',3,4,5-PENTACHLOROBIPHENYL

Formula: $C_{12}H_5Cl_5$
Mol Weight: 326.44
MP (deg C):
FP (deg C):
BP (deg C):
BP pressure (mm Hg):

Property/Value	Units	Temp	Data Type	Reference
WS 9.80E-003	mg/L	25	EXP	YALKOWSKY,SH & DANNENFELSER,RM (1992)
logP 6.98			EST	MEYLAN,WM & HOWARD,PH (1995)
VP 6.98E-005	mm Hg	25	EXP	SHIU,WY & MACKAY,D (1986)
DC	pKa			
HL 3.06E-003	atm m3/mol	25	EST	VP/WSOL
OH 3.99E-013	cm3/molc sec	25	EST	MEYLAN,WM & HOWARD,PH (1993)

CAS #: 055321-98-7 — 2,6-DIMETHYLBENZAMIDE

Formula: $C_9H_{11}NO$
Mol Weight: 149.19
MP (deg C):
FP (deg C):
BP (deg C):
BP pressure (mm Hg):

Property/Value	Units	Temp	Data Type	Reference
WS 1.30E+004	mg/L	25	EST	MEYLAN,WM ET AL. (1996)
logP 0.90			EXP	NAKAGAWA,Y ET AL. (1992)
VP 1.14E-004	mm Hg	25	EST	NEELY,WB & BLAU,GE (1985)
DC	pKa			
HL 2.70E-009	atm m3/mol	25	EST	MEYLAN,WM & HOWARD,PH (1991)
OH 1.44E-011	cm3/molc sec	25	EST	MEYLAN,WM & HOWARD,PH (1993)

CAS #: 055332-89-3 — 2,3-DIBR-TETRACHLOROCYCLOHEXANE

Formula: $C_6H_6Br_2Cl_4$
Mol Weight: 379.74
MP (deg C):
FP (deg C):
BP (deg C):
BP pressure (mm Hg):

Property/Value	Units	Temp	Data Type	Reference
WS 1.60E+000	mg/L	25	EST	MEYLAN,WM ET AL. (1996)
logP 3.99			EXP	HANSCH,C & LEO,AJ (1985)
VP 6.49E-005	mm Hg	25	EST	NEELY,WB & BLAU,GE (1985)
DC	pKa			
HL 2.74E-005	atm m3/mol	25	EST	MEYLAN,WM & HOWARD,PH (1991)
OH 6.15E-013	cm3/molc sec	25	EST	MEYLAN,WM & HOWARD,PH (1993)

CAS #: 055335-06-3 — TRICLOPYR

Formula: $C_7H_4Cl_3NO_3$
Mol Weight: 256.47
MP (deg C): 148-150
FP (deg C):
BP (deg C):
BP pressure (mm Hg):

Property/Value	Units	Temp	Data Type	Reference
WS 4.40E+002	mg/L	25	EXP	SHIU,WY ET AL. (1990)
logP 2.53			EST	MEYLAN,WM & HOWARD,PH (1995)
VP 1.26E-006	mm Hg	25	EXP	WEBER,JB (1994)
DC	pKa			
HL 9.66E-010	atm m3/mol	25	EST	VP/WSOL
OH 4.84E-012	cm3/molc sec	25	EST	MEYLAN,WM & HOWARD,PH (1993)

CAS #: 055379-70-9 — 2-S-BUTYLPHENYLDIMETHYLCARBAMATE

Formula: $C_{13}H_{19}NO_2$
Mol Weight: 221.30
MP (deg C):
FP (deg C):
BP (deg C):
BP pressure (mm Hg):

Property/Value	Units	Temp	Data Type	Reference
WS 5.05E+001	mg/L	25	EST	MEYLAN,WM ET AL. (1996)
logP 3.31			EXP	HANSCH,C & LEO,AJ (1985)
VP 1.54E-003	mm Hg	25	EST	NEELY,WB & BLAU,GE (1985)
DC	pKa			
HL 1.82E-007	atm m3/mol	25	EST	MEYLAN,WM & HOWARD,PH (1991)
OH 2.34E-011	cm3/molc sec	25	EST	MEYLAN,WM & HOWARD,PH (1993)

CAS #: 055379-71-0 — RH-1908 O-PH N-BUTYLENE CARBAMATE

Formula: $C_{11}H_{13}NO_2$
Mol Weight: 191.23
MP (deg C):
FP (deg C):
BP (deg C):
BP pressure (mm Hg):

Property/Value	Units	Temp	Data Type	Reference
WS 2.26E+002	mg/L	25	EST	MEYLAN,WM ET AL. (1996)
logP 2.73			EXP	NANDIHALLI,UB ET AL. (1993)
VP 2.05E-003	mm Hg	25	EST	NEELY,WB & BLAU,GE (1985)
DC	pKa			
HL 5.48E-008	atm m3/mol	25	EST	MEYLAN,WM & HOWARD,PH (1991)
OH 2.68E-011	cm3/molc sec	25	EST	MEYLAN,WM & HOWARD,PH (1993)

CAS #: 055380-34-2			2,6-DIME-N,N'-DINITROSOPIPERAZINE

Formula: C6H12N4O2

Mol Weight: 172.19

MP (deg C): | FP (deg C):

BP (deg C):

BP pressure (mm Hg):

Property/ Value	Units	Temp	Data Type	Reference
WS 5.13E+004	mg/L	25	EST	MEYLAN,WM ET AL. (1996)
logP 0.08			EXP	HANSCH,C & LEO,AJ (1985)
VP 5.90E-005	mm Hg	25	EST	NEELY,WB & BLAU,GE (1985)
DC	pKa			
HL 5.93E-011	atm m3/mol	25	EST	MEYLAN,WM & HOWARD,PH (1991)
OH 6.61E-011	cm3/molc sec	25	EST	MEYLAN,WM & HOWARD,PH (1993)

CAS #: 055390-39-1			4-THIA-1-AZABICYCLO[3.2.0]HEPTANE-2-CARBOXYLIC A

Formula: C19H18ClN3O6S

Mol Weight: 451.89

MP (deg C): | FP (deg C):

BP (deg C):

BP pressure (mm Hg):

Property/ Value	Units	Temp	Data Type	Reference
WS 2.84E+001	mg/L	25	EST	MEYLAN,WM ET AL. (1996)
logP 2.00			EXP	SANGSTER,J (1994)
VP 6.30E-017	mm Hg	25	EST	NEELY,WB & BLAU,GE (1985)
DC	pKa			
HL 1.19E-021	atm m3/mol	25	EST	MEYLAN,WM & HOWARD,PH (1991)
OH 1.54E-010	cm3/molc sec	25	EST	MEYLAN,WM & HOWARD,PH (1993)

CAS #: 055469-64-2			4(33-DIMETRIAZENO)BENZENESULFONAMIDE

Formula: C8H12N4O2S

Mol Weight: 228.27

MP (deg C): | FP (deg C):

BP (deg C):

BP pressure (mm Hg):

Property/ Value	Units	Temp	Data Type	Reference
WS 3.87E+003	mg/L	25	EST	MEYLAN,WM ET AL. (1996)
logP 1.06			EXP	HANSCH,C & LEO,AJ (1985)
VP 5.89E-006	mm Hg	25	EST	NEELY,WB & BLAU,GE (1985)
DC	pKa			
HL 4.21E-011	atm m3/mol	25	EST	MEYLAN,WM & HOWARD,PH (1991)
OH 2.69E-012	cm3/molc sec	25	EST	MEYLAN,WM & HOWARD,PH (1993)

CAS #: 055505-26-5			8-MEHTYL-1-NONANOL

Formula: C10H22O

Mol Weight: 158.29

MP (deg C): | FP (deg C):

BP (deg C): 108

BP pressure (mm Hg): 1.00E+001

Property/ Value	Units	Temp	Data Type	Reference
WS 1.52E+002	mg/L	25	EST	MEYLAN,WM ET AL. (1996)
logP 3.71			EST	MEYLAN,WM & HOWARD,PH (1995)
VP 2.07E-002	mm Hg	25	EXP	DAUBERT,TE & DANNER,RP (1989)
DC	pKa			
HL 5.47E-005	atm m3/mol	25	EST	MEYLAN,WM & HOWARD,PH (1991)
OH 1.54E-011	cm3/molc sec	25	EST	MEYLAN,WM & HOWARD,PH (1993)

CAS #: 055511-98-3			BUTHIDAZOLE

Formula: C10H16N4O2S

Mol Weight: 256.33

MP (deg C): | FP (deg C):

BP (deg C):

BP pressure (mm Hg):

Property/ Value	Units	Temp	Data Type	Reference
WS 3.40E+003	mg/L	25	EXP	YALKOWSKY,SH & DANNENFELSER,RM (1992)
logP -0.38			EST	MEYLAN,WM & HOWARD,PH (1995)
VP 9.64E-010	mm Hg	25	EST	NEELY,WB & BLAU,GE (1985)
DC	pKa			
HL 4.78E-015	atm m3/mol	25	EST	MEYLAN,WM & HOWARD,PH (1991)
OH 3.41E-011	cm3/molc sec	25	EST	MEYLAN,WM & HOWARD,PH (1993)

CAS #: 055556-85-9			3-OH-N-NITROSOPIPERIDINE

Formula: C5H10N2O2

Mol Weight: 130.15

MP (deg C): | FP (deg C):

BP (deg C):

BP pressure (mm Hg):

Property/ Value	Units	Temp	Data Type	Reference
WS 2.31E+005	mg/L	25	EST	MEYLAN,WM ET AL. (1996)
logP -0.47			EXP	HANSCH,C & LEO,AJ (1985)
VP 2.25E-004	mm Hg	25	EST	NEELY,WB & BLAU,GE (1985)
DC	pKa			
HL 7.76E-011	atm m3/mol	25	EST	MEYLAN,WM & HOWARD,PH (1991)
OH 3.46E-011	cm3/molc sec	25	EST	MEYLAN,WM & HOWARD,PH (1993)

CAS #: 055556-86-0			2,5-DIMETHYL-N-NITROSOPYRROLIDINE

Formula: C6H12N2O

Mol Weight: 128.18

MP (deg C): | FP (deg C):

BP (deg C):

BP pressure (mm Hg):

Property/ Value	Units	Temp	Data Type	Reference
WS 1.72E+004	mg/L	25	EST	MEYLAN,WM ET AL. (1996)
logP 0.86			EXP	HANSCH,C & LEO,AJ (1985)
VP 8.97E-002	mm Hg	25	EST	NEELY,WB & BLAU,GE (1985)
DC	pKa			
HL 2.82E-006	atm m3/mol	25	EST	MEYLAN,WM & HOWARD,PH (1991)
OH 3.81E-011	cm3/molc sec	25	EST	MEYLAN,WM & HOWARD,PH (1993)

CAS #: 055556-88-2			2,5-DIME-N,N'-DINITROSOPIPERAZINE

Formula: C6H12N4O2

Mol Weight: 172.19

MP (deg C): | FP (deg C):

BP (deg C):

BP pressure (mm Hg):

Property/ Value	Units	Temp	Data Type	Reference
WS 4.47E+004	mg/L	25	EST	MEYLAN,WM ET AL. (1996)
logP 0.15			EXP	HANSCH,C & LEO,AJ (1985)
VP 5.90E-005	mm Hg	25	EST	NEELY,WB & BLAU,GE (1985)
DC	pKa			
HL 5.93E-011	atm m3/mol	25	EST	MEYLAN,WM & HOWARD,PH (1991)
OH 6.61E-011	cm3/molc sec	25	EST	MEYLAN,WM & HOWARD,PH (1993)

055556-91-7 — N-NITROSO-4-PIPERIDONE

CAS #:	055556-91-7
Formula:	$C_5H_8N_2O_2$
Mol Weight:	128.13
MP (deg C):	
FP (deg C):	
BP (deg C):	
BP pressure (mm Hg):	

Property/Value	Units	Temp	Data Type	Reference
WS 2.36E+005	mg/L	25	EST	MEYLAN,WM ET AL. (1996)
logP -0.47			EXP	HANSCH,C & LEO,AJ (1985)
VP 5.55E-003	mm Hg	25	EST	NEELY,WB & BLAU,GE (1985)
DC	pKa			
HL 8.10E-010	atm m3/mol	25	EST	MEYLAN,WM & HOWARD,PH (1991)
OH 6.95E-011	cm3/molc sec	25	EST	MEYLAN,WM & HOWARD,PH (1993)

055556-92-8 — PYRIDINE, 1,2,3,6-TETRAHYDRO-1-NITROSO-

CAS #:	055556-92-8
Formula:	$C_5H_8N_2O$
Mol Weight:	112.13
MP (deg C):	
FP (deg C):	
BP (deg C):	
BP pressure (mm Hg):	

Property/Value	Units	Temp	Data Type	Reference
WS 1.05E+004	mg/L	25	EST	MEYLAN,WM ET AL. (1996)
logP 1.18			EXP	SANGSTER,J (1993)
VP 1.07E-001	mm Hg	25	EST	NEELY,WB & BLAU,GE (1985)
DC	pKa			
HL 1.87E-006	atm m3/mol	25	EST	MEYLAN,WM & HOWARD,PH (1991)
OH 7.69E-011	cm3/molc sec	25	EST	MEYLAN,WM & HOWARD,PH (1993)

055556-93-9 — 4-OH-N-NITROSOPIPERIDINE

CAS #:	055556-93-9
Formula:	$C_5H_{10}N_2O_2$
Mol Weight:	130.15
MP (deg C):	
FP (deg C):	
BP (deg C):	
BP pressure (mm Hg):	

Property/Value	Units	Temp	Data Type	Reference
WS 5.29E+005	mg/L	25	EST	MEYLAN,WM ET AL. (1996)
logP -0.89			EXP	HANSCH,C & LEO,AJ (1985)
VP 2.25E-004	mm Hg	25	EST	NEELY,WB & BLAU,GE (1985)
DC	pKa			
HL 7.76E-011	atm m3/mol	25	EST	MEYLAN,WM & HOWARD,PH (1991)
OH 3.46E-011	cm3/molc sec	25	EST	MEYLAN,WM & HOWARD,PH (1993)

055556-94-0 — 2-ME-N,N'-DINITROSOPIPERAZINE

CAS #:	055556-94-0
Formula:	$C_5H_{10}N_4O_2$
Mol Weight:	158.16
MP (deg C):	
FP (deg C):	
BP (deg C):	
BP pressure (mm Hg):	

Property/Value	Units	Temp	Data Type	Reference
WS 1.21E+005	mg/L	25	EST	MEYLAN,WM ET AL. (1996)
logP -0.28			EXP	HANSCH,C & LEO,AJ (1985)
VP 1.05E-004	mm Hg	25	EST	NEELY,WB & BLAU,GE (1985)
DC	pKa			
HL 4.47E-011	atm m3/mol	25	EST	MEYLAN,WM & HOWARD,PH (1991)
OH 5.44E-011	cm3/molc sec	25	EST	MEYLAN,WM & HOWARD,PH (1993)

055557-00-1 — 1H-1,4-DIAZEPINE, HEXAHYDRO-1,4-DINITROSO-

CAS #:	055557-00-1
Formula:	$C_5H_{10}N_4O_2$
Mol Weight:	158.16
MP (deg C):	
FP (deg C):	
BP (deg C):	
BP pressure (mm Hg):	

Property/Value	Units	Temp	Data Type	Reference
WS 1.90E+005	mg/L	25	EST	MEYLAN,WM ET AL. (1996)
logP -0.51			EXP	HANSCH,C & LEO,AJ (1985)
VP 8.00E-005	mm Hg	25	EST	NEELY,WB & BLAU,GE (1985)
DC	pKa			
HL 4.47E-011	atm m3/mol	25	EST	MEYLAN,WM & HOWARD,PH (1991)
OH 4.41E-011	cm3/molc sec	25	EST	MEYLAN,WM & HOWARD,PH (1993)

055565-48-5 — BENZYLOXYCARBONYL-PHALANINYLVALINE

CAS #:	055565-48-5
Formula:	$C_{22}H_{26}N_2O_5$
Mol Weight:	398.46
MP (deg C):	
FP (deg C):	
BP (deg C):	
BP pressure (mm Hg):	

Property/Value	Units	Temp	Data Type	Reference
WS 1.09E+001	mg/L	25	EST	MEYLAN,WM ET AL. (1996)
logP 2.88			EXP	HANSCH,C & LEO,AJ (1985)
VP 8.35E-013	mm Hg	25	EST	NEELY,WB & BLAU,GE (1985)
DC	pKa			
HL 2.75E-015	atm m3/mol	25	EST	MEYLAN,WM & HOWARD,PH (1991)
OH 4.37E-011	cm3/molc sec	25	EST	MEYLAN,WM & HOWARD,PH (1993)

055653-13-9 — PROPANEDINITRILE, [(2-NITROPHENYL)HYDRAZONO]-

CAS #:	055653-13-9
Formula:	$C_9H_5N_5O_2$
Mol Weight:	215.17
MP (deg C):	
FP (deg C):	
BP (deg C):	
BP pressure (mm Hg):	

Property/Value	Units	Temp	Data Type	Reference
WS 4.96E+001	mg/L	25	EST	MEYLAN,WM ET AL. (1996)
logP 2.90			EXP	STURDIK,E ET AL. (1985)
VP 3.13E-007	mm Hg	25	EST	NEELY,WB & BLAU,GE (1985)
DC	pKa			
HL 5.00E-010	atm m3/mol	25	EST	MEYLAN,WM & HOWARD,PH (1991)
OH 5.33E-012	cm3/molc sec	25	EST	MEYLAN,WM & HOWARD,PH (1993)

055653-16-2 — PROPANEDINITRILE, [(4-ACETYLPHENYL)HYDRAZONO]-

CAS #:	055653-16-2
Formula:	$C_{11}H_8N_4O$
Mol Weight:	212.21
MP (deg C):	
FP (deg C):	
BP (deg C):	
BP pressure (mm Hg):	

Property/Value	Units	Temp	Data Type	Reference
WS 6.09E+002	mg/L	25	EST	MEYLAN,WM ET AL. (1996)
logP 2.10			EXP	STURDIK,E ET AL. (1985)
VP 6.73E-007	mm Hg	25	EST	NEELY,WB & BLAU,GE (1985)
DC	pKa			
HL 1.05E-011	atm m3/mol	25	EST	MEYLAN,WM & HOWARD,PH (1991)
OH 3.90E-011	cm3/molc sec	25	EST	MEYLAN,WM & HOWARD,PH (1993)

055687-40-6 — 24-NH2PYRIMIDN,5(35MEO-4-CO2ME)BENZYL

CAS #:	055687-40-6	24-NH2PYRIMIDN,5(35MEO-4-CO2ME)BENZYL
Formula:	$C_{15}H_{18}N_4O_4$	
Mol Weight:	318.34	
MP (deg C):		FP (deg C):
BP (deg C):		
BP pressure (mm Hg):		

Property/Value	Units	Temp	Data Type	Reference
WS 1.48E+003	mg/L	25	EST	MEYLAN,WM ET AL. (1996)
logP 0.95			EXP	HANSCH,C & LEO,AJ (1985)
VP 1.63E-009	mm Hg	25	EST	NEELY,WB & BLAU,GE (1985)
DC	pKa			
HL 2.60E-015	atm m3/mol	25	EST	MEYLAN,WM & HOWARD,PH (1991)
OH 1.94E-010	cm3/molc sec	25	EST	MEYLAN,WM & HOWARD,PH (1993)

055687-46-2 — 24NH2PYRIMIDN,5(35MEO-4CO2IPR)BENZYL

CAS #:	055687-46-2	24NH2PYRIMIDN,5(35MEO-4CO2IPR)BENZYL
Formula:	$C_{17}H_{22}N_4O_4$	
Mol Weight:	346.39	
MP (deg C):		FP (deg C):
BP (deg C):		
BP pressure (mm Hg):		

Property/Value	Units	Temp	Data Type	Reference
WS 2.21E+002	mg/L	25	EST	MEYLAN,WM ET AL. (1996)
logP 1.72			EXP	HANSCH,C & LEO,AJ (1985)
VP 5.19E-010	mm Hg	25	EST	NEELY,WB & BLAU,GE (1985)
DC	pKa			
HL 4.58E-015	atm m3/mol	25	EST	MEYLAN,WM & HOWARD,PH (1991)
OH 2.06E-010	cm3/molc sec	25	EST	MEYLAN,WM & HOWARD,PH (1993)

055687-48-4 — 24-NH2PYRIMIDN,5(35-MEO-4CO2ET)BENZYL

CAS #:	055687-48-4	24-NH2PYRIMIDN,5(35-MEO-4CO2ET)BENZYL
Formula:	$C_{16}H_{20}N_4O_4$	
Mol Weight:	332.36	
MP (deg C):		FP (deg C):
BP (deg C):		
BP pressure (mm Hg):		

Property/Value	Units	Temp	Data Type	Reference
WS 6.77E+002	mg/L	25	EST	MEYLAN,WM ET AL. (1996)
logP 1.25			EXP	HANSCH,C & LEO,AJ (1985)
VP 7.20E-010	mm Hg	25	EST	NEELY,WB & BLAU,GE (1985)
DC	pKa			
HL 3.45E-015	atm m3/mol	25	EST	MEYLAN,WM & HOWARD,PH (1991)
OH 1.96E-010	cm3/molc sec	25	EST	MEYLAN,WM & HOWARD,PH (1993)

055687-49-5 — 24-NH2PYRIMIDN,5(35-MEO-4-CH2OH)BENZYL

CAS #:	055687-49-5	24-NH2PYRIMIDN,5(35-MEO-4-CH2OH)BENZYL
Formula:	$C_{14}H_{18}N_4O_3$	
Mol Weight:	290.32	
MP (deg C):		FP (deg C):
BP (deg C):		
BP pressure (mm Hg):		

Property/Value	Units	Temp	Data Type	Reference
WS 6.12E+003	mg/L	25	EST	MEYLAN,WM ET AL. (1996)
logP 0.42			EXP	HANSCH,C & LEO,AJ (1985)
VP 9.39E-012	mm Hg	25	EST	NEELY,WB & BLAU,GE (1985)
DC	pKa			
HL 1.63E-017	atm m3/mol	25	EST	MEYLAN,WM & HOWARD,PH (1991)
OH 2.06E-010	cm3/molc sec	25	EST	MEYLAN,WM & HOWARD,PH (1993)

055687-53-1 — 24NH2PYRIMIDINE,5(35-ETO-4CO2ET)BENZYL

CAS #:	055687-53-1	24NH2PYRIMIDINE,5(35-ETO-4CO2ET)BENZYL
Formula:	$C_{18}H_{24}N_4O_4$	
Mol Weight:	360.42	
MP (deg C):		FP (deg C):
BP (deg C):		
BP pressure (mm Hg):		

Property/Value	Units	Temp	Data Type	Reference
WS 6.05E+001	mg/L	25	EST	MEYLAN,WM ET AL. (1996)
logP 2.28			EXP	HANSCH,C & LEO,AJ (1985)
VP 1.39E-010	mm Hg	25	EST	NEELY,WB & BLAU,GE (1985)
DC	pKa			
HL 6.08E-015	atm m3/mol	25	EST	MEYLAN,WM & HOWARD,PH (1991)
OH 2.07E-010	cm3/molc sec	25	EST	MEYLAN,WM & HOWARD,PH (1993)

055687-56-4 — 24NH2PYRIMD,5(35ETO4COSO2ME)BENZYL

CAS #:	055687-56-4	24NH2PYRIMD,5(35ETO4COSO2ME)BENZYL
Formula:	$C_{18}H_{26}N_4O_3$	
Mol Weight:	346.43	
MP (deg C):		FP (deg C):
BP (deg C):		
BP pressure (mm Hg):		

Property/Value	Units	Temp	Data Type	Reference
WS 5.16E+001	mg/L	25	EST	MEYLAN,WM ET AL. (1996)
logP 2.46			EXP	HANSCH,C & LEO,AJ (1985)
VP 1.12E-012	mm Hg	25	EST	NEELY,WB & BLAU,GE (1985)
DC	pKa			
HL 5.05E-017	atm m3/mol	25	EST	MEYLAN,WM & HOWARD,PH (1991)
OH 2.14E-010	cm3/molc sec	25	EST	MEYLAN,WM & HOWARD,PH (1993)

055687-57-5 — 24-NH2PYRIMIDIN,5(35MEO-4CONH2)BENZYL

CAS #:	055687-57-5	24-NH2PYRIMIDIN,5(35MEO-4CONH2)BENZYL
Formula:	$C_{14}H_{17}N_5O_3$	
Mol Weight:	303.32	
MP (deg C):		FP (deg C):
BP (deg C):		
BP pressure (mm Hg):		

Property/Value	Units	Temp	Data Type	Reference
WS 3.06E+003	mg/L	25	EST	MEYLAN,WM ET AL. (1996)
logP -0.85			EXP	HANSCH,C & LEO,AJ (1985)
VP 1.97E-011	mm Hg	25	EST	NEELY,WB & BLAU,GE (1985)
DC	pKa			
HL 1.66E-019	atm m3/mol	25	EST	MEYLAN,WM & HOWARD,PH (1991)
OH 2.05E-010	cm3/molc sec	25	EST	MEYLAN,WM & HOWARD,PH (1993)

055687-58-6 — 2,4-NH2PYRIMIDIN,5(3,5MEO-4CòN)BENZYL

CAS #:	055687-58-6	2,4-NH2PYRIMIDIN,5(3,5MEO-4CòN)BENZYL
Formula:	$C_{14}H_{15}N_5O_2$	
Mol Weight:	285.31	
MP (deg C):		FP (deg C):
BP (deg C):		
BP pressure (mm Hg):		

Property/Value	Units	Temp	Data Type	Reference
WS 5.28E+002	mg/L	25	EST	MEYLAN,WM ET AL. (1996)
logP 1.39			EXP	HANSCH,C & LEO,AJ (1985)
VP 1.96E-009	mm Hg	25	EST	NEELY,WB & BLAU,GE (1985)
DC	pKa			
HL 3.90E-015	atm m3/mol	25	EST	MEYLAN,WM & HOWARD,PH (1991)
OH 1.08E-010	cm3/molc sec	25	EST	MEYLAN,WM & HOWARD,PH (1993)

CAS #: 055687-72-4				24NH2PYRIMD,5(35ETO4COSO2ME)BENZYL
Formula: $C_{18}H_{24}N_4O_5S$				
Mol Weight: 408.48				
MP (deg C):		FP (deg C):		
BP (deg C):				
BP pressure (mm Hg):				

Property/ Value	Units	Temp	Data Type	Reference
WS 5.74E+002	mg/L	25	EST	MEYLAN,WM ET AL. (1996)
logP 0.79			EXP	HANSCH,C & LEO,AJ (1985)
VP 4.72E-013	mm Hg	25	EST	NEELY,WB & BLAU,GE (1985)
DC	pKa			
HL 1.10E-019	atm m3/mol	25	EST	MEYLAN,WM & HOWARD,PH (1991)
OH 2.20E-010	cm3/molc sec	25	EST	MEYLAN,WM & HOWARD,PH (1993)

CAS #: 055687-73-5				24NH2PYRIMIDN,5(35ETO-4-ACETYL)BENZYL
Formula: $C_{17}H_{22}N_4O_3$				
Mol Weight: 330.39				
MP (deg C):		FP (deg C):		
BP (deg C):				
BP pressure (mm Hg):				

Property/ Value	Units	Temp	Data Type	Reference
WS 1.47E+002	mg/L	25	EST	MEYLAN,WM ET AL. (1996)
logP 2.04			EXP	HANSCH,C & LEO,AJ (1985)
VP 4.06E-010	mm Hg	25	EST	NEELY,WB & BLAU,GE (1985)
DC	pKa			
HL 1.29E-015	atm m3/mol	25	EST	MEYLAN,WM & HOWARD,PH (1991)
OH 2.13E-010	cm3/molc sec	25	EST	MEYLAN,WM & HOWARD,PH (1993)

CAS #: 055687-74-6				24NH2PYRIMIDN,5(35MEO-4COCH2SO2ME)BENZYL
Formula: $C_{16}H_{20}N_4O_5S$				
Mol Weight: 380.43				
MP (deg C):		FP (deg C):		
BP (deg C):				
BP pressure (mm Hg):				

Property/ Value	Units	Temp	Data Type	Reference
WS 7.44E+003	mg/L	25	EST	MEYLAN,WM ET AL. (1996)
logP -0.31			EXP	HANSCH,C & LEO,AJ (1985)
VP 2.56E-012	mm Hg	25	EST	NEELY,WB & BLAU,GE (1985)
DC	pKa			
HL 6.22E-020	atm m3/mol	25	EST	MEYLAN,WM & HOWARD,PH (1991)
OH 2.09E-010	cm3/molc sec	25	EST	MEYLAN,WM & HOWARD,PH (1993)

CAS #: 055687-76-8				2,4-NH2PYRIMIDN,5(3,5MEO-4-AC)BENZYL
Formula: $C_{15}H_{18}N_4O_3$				
Mol Weight: 302.34				
MP (deg C):		FP (deg C):		
BP (deg C):				
BP pressure (mm Hg):				

Property/ Value	Units	Temp	Data Type	Reference
WS 1.37E+003	mg/L	25	EST	MEYLAN,WM ET AL. (1996)
logP 1.10			EXP	HANSCH,C & LEO,AJ (1985)
VP 2.09E-009	mm Hg	25	EST	NEELY,WB & BLAU,GE (1985)
DC	pKa			
HL 7.34E-016	atm m3/mol	25	EST	MEYLAN,WM & HOWARD,PH (1991)
OH 2.03E-010	cm3/molc sec	25	EST	MEYLAN,WM & HOWARD,PH (1993)

CAS #: 055687-86-0				24NH2PYRIMIDIN,5(35-MEO-4-MEO-CH2)BENZYL
Formula: $C_{15}H_{20}N_4O_3$				
Mol Weight: 304.35				
MP (deg C):		FP (deg C):		
BP (deg C):				
BP pressure (mm Hg):				

Property/ Value	Units	Temp	Data Type	Reference
WS 1.12E+003	mg/L	25	EST	MEYLAN,WM ET AL. (1996)
logP 1.19			EXP	HANSCH,C & LEO,AJ (1985)
VP 4.39E-009	mm Hg	25	EST	NEELY,WB & BLAU,GE (1985)
DC	pKa			
HL 5.21E-015	atm m3/mol	25	EST	MEYLAN,WM & HOWARD,PH (1991)
OH 2.09E-010	cm3/molc sec	25	EST	MEYLAN,WM & HOWARD,PH (1993)

CAS #: 055687-91-7				24NH2PYRIMIDINE,5(35-MEO-4-COHME2)BENZYL
Formula: $C_{16}H_{22}N_4O_3$				
Mol Weight: 318.38				
MP (deg C):		FP (deg C):		
BP (deg C):				
BP pressure (mm Hg):				

Property/ Value	Units	Temp	Data Type	Reference
WS 5.11E+002	mg/L	25	EST	MEYLAN,WM ET AL. (1996)
logP 1.49			EXP	HANSCH,C & LEO,AJ (1985)
VP 7.68E-012	mm Hg	25	EST	NEELY,WB & BLAU,GE (1985)
DC	pKa			
HL 2.87E-017	atm m3/mol	25	EST	MEYLAN,WM & HOWARD,PH (1991)
OH 2.04E-010	cm3/molc sec	25	EST	MEYLAN,WM & HOWARD,PH (1993)

CAS #: 055692-91-6				DIGITOXIN, BETA-METHYL
Formula: $C_{42}H_{68}O_{13}$				
Mol Weight: 781.00				
MP (deg C):		FP (deg C):		
BP (deg C):				
BP pressure (mm Hg):				

Property/ Value	Units	Temp	Data Type	Reference
WS 2.22E-002	mg/L	25	EST	MEYLAN,WM ET AL. (1996)
logP 3.11			EXP	HANSCH,C ET AL. (1995)
VP 3.65E-028	mm Hg	25	EST	NEELY,WB & BLAU,GE (1985)
DC	pKa			
HL 7.37E-026	atm m3/mol	25	EST	MEYLAN,WM & HOWARD,PH (1991)
OH 1.43E-010	cm3/molc sec	25	EST	MEYLAN,WM & HOWARD,PH (1993)

CAS #: 055702-45-9				2,3,6-TRICHLOROBIPHENYL
Formula: $C_{12}H_7Cl_3$				
Mol Weight: 257.55				
MP (deg C):		FP (deg C):		
BP (deg C):				
BP pressure (mm Hg):				

Property/ Value	Units	Temp	Data Type	Reference
WS 1.31E-001	mg/L	25	EXP	KUHNE,R ET AL. (1995)
logP 5.67			EXP	HANSCH,C & LEO,AJ (1985)
VP 4.00E-005	mm Hg	25	EST	NEELY,WB & BLAU,GE (1985)
DC	pKa			
HL 1.68E-004	atm m3/mol	25	EST	MEYLAN,WM & HOWARD,PH (1991)
OH 1.26E-012	cm3/molc sec	25	EST	MEYLAN,WM & HOWARD,PH (1993)

055702-46-0 — 2,3,4-TRICHLORO-1,1'-BIPHENYL

Formula: $C_{12}H_7Cl_3$

Mol Weight: 257.55

MP (deg C): FP (deg C):

BP (deg C):

BP pressure (mm Hg):

Property/Value	Units	Temp	Data Type	Reference
WS 2.97E-001	mg/L	25	EST	MEYLAN,WM ET AL. (1996)
logP 5.69			EST	MEYLAN,WM & HOWARD,PH (1995)
VP 4.00E-005	mm Hg	25	EST	NEELY,WB & BLAU,GE (1985)
DC	pKa			
HL 1.68E-004	atm m3/mol	25	EST	MEYLAN,WM & HOWARD,PH (1991)
OH 1.26E-012	cm3/molc sec	25	EST	MEYLAN,WM & HOWARD,PH (1993)

055720-37-1 — 1,3,7-TRICHLORONAPHTHALENE

Formula: $C_{10}H_5Cl_3$

Mol Weight: 231.51

MP (deg C): FP (deg C):

BP (deg C):

BP pressure (mm Hg):

Property/Value	Units	Temp	Data Type	Reference
WS 8.06E-001	mg/L	25	EST	MEYLAN,WM ET AL. (1996)
logP 5.35			EXP	OPPERHUIZEN,A (1987)
VP 3.14E-004	mm Hg	25	EST	NEELY,WB & BLAU,GE (1985)
DC	pKa			
HL 2.14E-004	atm m3/mol	25	EST	MEYLAN,WM & HOWARD,PH (1991)
OH 2.01E-012	cm3/molc sec	25	EST	MEYLAN,WM & HOWARD,PH (1993)

055720-40-6 — 2,3,6-TRICHLORONAPHTHALENE

Formula: $C_{10}H_5Cl_3$

Mol Weight: 231.51

MP (deg C): FP (deg C):

BP (deg C):

BP pressure (mm Hg):

Property/Value	Units	Temp	Data Type	Reference
WS 1.27E+000	mg/L	25	EST	MEYLAN,WM ET AL. (1996)
logP 5.12			EXP	OPPERHUIZEN,A (1987)
VP 3.14E-004	mm Hg	25	EST	NEELY,WB & BLAU,GE (1985)
DC	pKa			
HL 2.14E-004	atm m3/mol	25	EST	MEYLAN,WM & HOWARD,PH (1991)
OH 2.01E-012	cm3/molc sec	25	EST	MEYLAN,WM & HOWARD,PH (1993)

055720-43-9 — 1,4,6,7-TETRACHLORONAPHTHALENE

Formula: $C_{10}H_4Cl_4$

Mol Weight: 265.96

MP (deg C): FP (deg C):

BP (deg C):

BP pressure (mm Hg):

Property/Value	Units	Temp	Data Type	Reference
WS 2.10E-001	mg/L	25	EST	MEYLAN,WM ET AL. (1996)
logP 5.81			EXP	OPPERHUIZEN,A (1987)
VP 5.79E-005	mm Hg	25	EST	NEELY,WB & BLAU,GE (1985)
DC	pKa			
HL 1.59E-004	atm m3/mol	25	EST	MEYLAN,WM & HOWARD,PH (1991)
OH 9.11E-013	cm3/molc sec	25	EST	MEYLAN,WM & HOWARD,PH (1993)

055720-47-3 — L-TYROSINE, 3-HYDROXY-, PHENYLMETHYL ESTER

Formula: $C_{16}H_{17}NO_4$

Mol Weight: 287.32

MP (deg C): FP (deg C):

BP (deg C):

BP pressure (mm Hg):

Property/Value	Units	Temp	Data Type	Reference
WS 2.95E+004	mg/L	25	EST	MEYLAN,WM ET AL. (1996)
logP 1.50			EXP	MARREL,C ET AL. (1985)
VP 8.96E-010	mm Hg	25	EST	NEELY,WB & BLAU,GE (1985)
DC	pKa			
HL 3.37E-017	atm m3/mol	25	EST	MEYLAN,WM & HOWARD,PH (1991)
OH 1.02E-010	cm3/molc sec	25	EST	MEYLAN,WM & HOWARD,PH (1993)

055750-53-3 — N-MALEOYL-6-AMINOHEXANOIC ACID

Formula: $C_{10}H_{13}NO_4$

Mol Weight: 211.22

MP (deg C): FP (deg C):

BP (deg C):

BP pressure (mm Hg):

Property/Value	Units	Temp	Data Type	Reference
WS 3.99E+003	mg/L	25	EST	MEYLAN,WM ET AL. (1996)
logP 1.15			EXP	HANSCH,C & LEO,AJ (1985)
VP 1.48E-008	mm Hg	25	EST	NEELY,WB & BLAU,GE (1985)
DC	pKa			
HL 3.21E-013	atm m3/mol	25	EST	MEYLAN,WM & HOWARD,PH (1991)
OH 2.76E-011	cm3/molc sec	25	EST	MEYLAN,WM & HOWARD,PH (1993)

055752-22-2 — 1,2,4-OXADIAZOLE,5-METHYL-3-(P-TOLYL)

Formula: $C_{10}H_{11}N_2O$

Mol Weight: 175.21

MP (deg C): FP (deg C):

BP (deg C):

BP pressure (mm Hg):

Property/Value	Units	Temp	Data Type	Reference
WS 2.69E+002	mg/L	25	EST	MEYLAN,WM ET AL. (1996)
logP 2.74			EXP	HANSCH,C ET AL. (1995)
VP 7.75E-004	mm Hg	25	EST	NEELY,WB & BLAU,GE (1985)
DC	pKa			
HL 2.19E-006	atm m3/mol	25	EST	MEYLAN,WM & HOWARD,PH (1991)
OH 8.42E-012	cm3/molc sec	25	EST	MEYLAN,WM & HOWARD,PH (1993)

055755-16-3 — BENZENEETHANAMINE, 2-METHYL-

Formula: $C_9H_{13}N$

Mol Weight: 135.21

MP (deg C): FP (deg C):

BP (deg C):

BP pressure (mm Hg):

Property/Value	Units	Temp	Data Type	Reference
WS 3.84E+004	mg/L	25	EST	MEYLAN,WM ET AL. (1996)
logP 1.60			EXP	KRIL,MB & FUNG,HL (1990)
VP 9.74E-002	mm Hg	25	EST	NEELY,WB & BLAU,GE (1985)
DC	pKa			
HL 8.96E-007	atm m3/mol	25	EST	MEYLAN,WM & HOWARD,PH (1991)
OH 3.92E-011	cm3/molc sec	25	EST	MEYLAN,WM & HOWARD,PH (1993)

055782-76-8 — 6-NH2-5-(N-METHYLFORMYLAMINO)-3-METHYLURACIL

Formula: $C_7H_{10}N_4O_3$
Mol Weight: 198.18
MP (deg C):
FP (deg C):
BP (deg C):
BP pressure (mm Hg):

Property/Value	Units	Temp	Data Type	Reference
WS 1.19E+005	mg/L	25	EST	MEYLAN,WM ET AL. (1996)
logP -2.03			EXP	GASPARI,F & BONATI,M (1987)
VP 6.64E-009	mm Hg	25	EST	NEELY,WB & BLAU,GE (1985)
DC	pKa			
HL 5.41E-016	atm m3/mol	25	EST	MEYLAN,WM & HOWARD,PH (1991)
OH 4.81E-011	cm3/molc sec	25	EST	MEYLAN,WM & HOWARD,PH (1993)

055790-22-2 — GLYCINE, N-(4-IODOBENZOYL)-

Formula: $C_9H_8INO_3$
Mol Weight: 305.07
MP (deg C):
FP (deg C):
BP (deg C):
BP pressure (mm Hg):

Property/Value	Units	Temp	Data Type	Reference
WS 2.53E+002	mg/L	25	EST	MEYLAN,WM ET AL. (1996)
logP 1.94			EXP	LAZNICEK,M & KVETINA,J (1988)
VP 2.26E-008	mm Hg	25	EST	NEELY,WB & BLAU,GE (1985)
DC	pKa			
HL 1.67E-013	atm m3/mol	25	EST	MEYLAN,WM & HOWARD,PH (1991)
OH 1.01E-011	cm3/molc sec	25	EST	MEYLAN,WM & HOWARD,PH (1993)

055814-41-0 — MEPRONIL

Formula: $C_{17}H_{19}NO_2$
Mol Weight: 269.35
MP (deg C):
FP (deg C):
BP (deg C):
BP pressure (mm Hg):

Property/Value	Units	Temp	Data Type	Reference
WS 1.38E+001	mg/L	25	EST	MEYLAN,WM ET AL. (1996)
logP 3.66			EXP	HANSCH,C ET AL. (1995)
VP 4.72E-008	mm Hg	25	EST	NEELY,WB & BLAU,GE (1985)
DC	pKa			
HL 1.41E-010	atm m3/mol	25	EST	MEYLAN,WM & HOWARD,PH (1991)
OH 1.54E-010	cm3/molc sec	25	EST	MEYLAN,WM & HOWARD,PH (1993)

055836-71-0 — P-ETHOXYBENZAMIDE

Formula: $C_9H_{11}NO_2$
Mol Weight: 165.19
MP (deg C): 208-210
FP (deg C):
BP (deg C):
BP pressure (mm Hg):

Property/Value	Units	Temp	Data Type	Reference
WS 5.03E+003	mg/L	25	EST	MEYLAN,WM ET AL. (1996)
logP 1.30			EXP	NAKAGAWA,Y ET AL. (1992)
VP 4.48E-005	mm Hg	25	EST	NEELY,WB & BLAU,GE (1985)
DC	pKa			
HL 1.74E-010	atm m3/mol	25	EST	MEYLAN,WM & HOWARD,PH (1991)
OH 2.78E-011	cm3/molc sec	25	EST	MEYLAN,WM & HOWARD,PH (1993)

055861-78-4 — ISOURON

Formula: $C_{10}H_{18}N_3O_2$
Mol Weight: 212.27
MP (deg C): 119-120
FP (deg C):
BP (deg C):
BP pressure (mm Hg):

Property/Value	Units	Temp	Data Type	Reference
WS 7.90E+002	mg/L	25	EXP	SHIU,WY ET AL. (1990)
logP 1.98			EXP	TOMLIN,C (1994)
VP 3.19E-005	mm Hg	25	EST	NEELY,WB & BLAU,GE (1985)
DC	pKa			
HL 2.64E-011	atm m3/mol	25	EST	MEYLAN,WM & HOWARD,PH (1991)
OH 2.03E-010	cm3/molc sec	25	EST	MEYLAN,WM & HOWARD,PH (1993)

055863-26-8 — DIBENZO[B,E]THIEPIN-2-ACETIC ACID, 6,11-DIHYDRO-

Formula: $C_{16}H_{12}O_3S$
Mol Weight: 284.34
MP (deg C):
FP (deg C):
BP (deg C):
BP pressure (mm Hg):

Property/Value	Units	Temp	Data Type	Reference
WS 8.48E+000	mg/L	25	EST	MEYLAN,WM ET AL. (1996)
logP 2.97			EXP	SANGSTER,J (1993)
VP 5.38E-009	mm Hg	25	EST	NEELY,WB & BLAU,GE (1985)
DC	pKa			
HL 1.70E-013	atm m3/mol	25	EST	MEYLAN,WM & HOWARD,PH (1991)
OH 2.47E-011	cm3/molc sec	25	EST	MEYLAN,WM & HOWARD,PH (1993)

055884-23-6 — 1,1-ETHENEDIAMINE, N-METHYL-N'-[2-[[(5-METHYL-1H

Formula: $C_{10}H_{17}N_5O_2S$
Mol Weight: 271.34
MP (deg C):
FP (deg C):
BP (deg C):
BP pressure (mm Hg):

Property/Value	Units	Temp	Data Type	Reference
WS 1.64E+005	mg/L	25	EST	MEYLAN,WM ET AL. (1996)
logP -0.40			EXP	SANGSTER,J (1994)
VP 1.44E-009	mm Hg	25	EST	NEELY,WB & BLAU,GE (1985)
DC	pKa			
HL 4.81E-015	atm m3/mol	25	EST	MEYLAN,WM & HOWARD,PH (1991)
OH 2.16E-010	cm3/molc sec	25	EST	MEYLAN,WM & HOWARD,PH (1993)

055903-47-4 — NNN-ME-N(3OH4MECYPEME)AMMON I

Formula: $C_{10}H_{22}INO$
Mol Weight: 299.20
MP (deg C):
FP (deg C):
BP (deg C):
BP pressure (mm Hg):

Property/Value	Units	Temp	Data Type	Reference
WS 1.00E+006	mg/L	25	EST	MEYLAN,WM ET AL. (1996)
logP -2.60			EXP	PRATESI,P ET AL. (1986)
VP 2.42E-011	mm Hg	25	EST	NEELY,WB & BLAU,GE (1985)
DC	pKa			
HL 2.32E-016	atm m3/mol	25	EST	MEYLAN,WM & HOWARD,PH (1991)
OH 3.05E-011	cm3/molc sec	25	EST	MEYLAN,WM & HOWARD,PH (1993)

CAS #: 055921-54-5				24-NH2-6-ETAM PYRIMIDINE-3-OXIDE

Formula: $C_6H_{13}N_5O$
Mol Weight: 171.20
MP (deg C): FP (deg C):
BP (deg C):
BP pressure (mm Hg):

Property/Value	Units	Temp	Data Type	Reference
WS 6.71E+004	mg/L	25	EST	MEYLAN,WM ET AL. (1996)
logP -0.04			EXP	HANSCH,C & LEO,AJ (1985)
VP 1.13E-006	mm Hg	25	EST	NEELY,WB & BLAU,GE (1985)
DC	pKa			
HL 1.33E-017	atm m3/mol	25	EST	MEYLAN,WM & HOWARD,PH (1991)
OH 2.09E-010	cm3/molc sec	25	EST	MEYLAN,WM & HOWARD,PH (1993)

CAS #: 055921-55-6				2,4-DINH2-6-BUAM-PYRIMIDINE-3-OXIDE

Formula: $C_8H_{17}N_5O$
Mol Weight: 199.26
MP (deg C): FP (deg C):
BP (deg C):
BP pressure (mm Hg):

Property/Value	Units	Temp	Data Type	Reference
WS 6.08E+003	mg/L	25	EST	MEYLAN,WM ET AL. (1996)
logP 1.02			EXP	HANSCH,C & LEO,AJ (1985)
VP 2.34E-007	mm Hg	25	EST	NEELY,WB & BLAU,GE (1985)
DC	pKa			
HL 2.35E-017	atm m3/mol	25	EST	MEYLAN,WM & HOWARD,PH (1991)
OH 2.13E-010	cm3/molc sec	25	EST	MEYLAN,WM & HOWARD,PH (1993)

CAS #: 055921-56-7				24-NH2-6-DECYLAMINOPYRIMIDINE-3-OXIDE

Formula: $C_{14}H_{29}N_5O$
Mol Weight: 283.42
MP (deg C): FP (deg C):
BP (deg C):
BP pressure (mm Hg):

Property/Value	Units	Temp	Data Type	Reference
WS 5.16E+000	mg/L	25	EST	MEYLAN,WM ET AL. (1996)
logP 4.08			EXP	HANSCH,C & LEO,AJ (1985)
VP 1.88E-009	mm Hg	25	EST	NEELY,WB & BLAU,GE (1985)
DC	pKa			
HL 1.29E-016	atm m3/mol	25	EST	MEYLAN,WM & HOWARD,PH (1991)
OH 2.22E-010	cm3/molc sec	25	EST	MEYLAN,WM & HOWARD,PH (1993)

CAS #: 055921-57-8				24NH2-6-CYHXAM-PYRIMIDINE-3-OXIDE

Formula: $C_{10}H_{19}N_5O$
Mol Weight: 225.30
MP (deg C): FP (deg C):
BP (deg C):
BP pressure (mm Hg):

Property/Value	Units	Temp	Data Type	Reference
WS 1.37E+003	mg/L	25	EST	MEYLAN,WM ET AL. (1996)
logP 1.62			EXP	HANSCH,C & LEO,AJ (1985)
VP 3.34E-008	mm Hg	25	EST	NEELY,WB & BLAU,GE (1985)
DC	pKa			
HL 1.82E-017	atm m3/mol	25	EST	MEYLAN,WM & HOWARD,PH (1991)
OH 2.34E-010	cm3/molc sec	25	EST	MEYLAN,WM & HOWARD,PH (1993)

CAS #: 055921-58-9				24-NH2-6-DIMEAM PYRIMIDINE-3-OXIDE

Formula: $C_6H_{13}N_5O$
Mol Weight: 171.20
MP (deg C): FP (deg C):
BP (deg C):
BP pressure (mm Hg):

Property/Value	Units	Temp	Data Type	Reference
WS 4.53E+004	mg/L	25	EST	MEYLAN,WM ET AL. (1996)
logP 0.16			EXP	HANSCH,C & LEO,AJ (1985)
VP 2.29E-006	mm Hg	25	EST	NEELY,WB & BLAU,GE (1985)
DC	pKa			
HL 2.06E-016	atm m3/mol	25	EST	MEYLAN,WM & HOWARD,PH (1991)
OH 2.03E-010	cm3/molc sec	25	EST	MEYLAN,WM & HOWARD,PH (1993)

CAS #: 055921-60-3				24NH2-6DIPROPYLAMINO-PYRIMIDINE-3-OXIDE

Formula: $C_{10}H_{21}N_5O$
Mol Weight: 227.31
MP (deg C): FP (deg C):
BP (deg C):
BP pressure (mm Hg):

Property/Value	Units	Temp	Data Type	Reference
WS 4.61E+002	mg/L	25	EST	MEYLAN,WM ET AL. (1996) .
logP 2.16			EXP	HANSCH,C & LEO,AJ (1985)
VP 9.76E-008	mm Hg	25	EST	NEELY,WB & BLAU,GE (1985)
DC	pKa			
HL 6.39E-016	atm m3/mol	25	EST	MEYLAN,WM & HOWARD,PH (1991)
OH 2.24E-010	cm3/molc sec	25	EST	MEYLAN,WM & HOWARD,PH (1993)

CAS #: 055921-61-4				24NH2-6DIALLYLAMINO-PYRIMIDINE-3-OXIDE

Formula: $C_{10}H_{17}N_5O$
Mol Weight: 223.28
MP (deg C): FP (deg C):
BP (deg C):
BP pressure (mm Hg):

Property/Value	Units	Temp	Data Type	Reference
WS 1.46E+003	mg/L	25	EST	MEYLAN,WM ET AL. (1996)
logP 1.60			EXP	HANSCH,C & LEO,AJ (1985)
VP 1.10E-007	mm Hg	25	EST	NEELY,WB & BLAU,GE (1985)
DC	pKa			
HL 3.55E-016	atm m3/mol	25	EST	MEYLAN,WM & HOWARD,PH (1991)
OH 2.70E-010	cm3/molc sec	25	EST	MEYLAN,WM & HOWARD,PH (1993)

CAS #: 055921-62-5				24NH2-6DIBUTYLAMINO-PYRIMIDINE-3-OXIDE

Formula: $C_{12}H_{25}N_5O$
Mol Weight: 255.37
MP (deg C): FP (deg C):
BP (deg C):
BP pressure (mm Hg):

Property/Value	Units	Temp	Data Type	Reference
WS 4.28E+001	mg/L	25	EST	MEYLAN,WM ET AL. (1996)
logP 3.19			EXP	HANSCH,C & LEO,AJ (1985)
VP 1.97E-008	mm Hg	25	EST	NEELY,WB & BLAU,GE (1985)
DC	pKa			
HL 1.13E-015	atm m3/mol	25	EST	MEYLAN,WM & HOWARD,PH (1991)
OH 2.27E-010	cm3/molc sec	25	EST	MEYLAN,WM & HOWARD,PH (1993)

CAS #: 055921-63-6	24NH2-6-DICYCLOHXAMPYRIMIDIN-3-OXIDE

Formula: $C_{16}H_{29}N_5O$

Mol Weight: 307.44

MP (deg C): FP (deg C):

BP (deg C):

BP pressure (mm Hg):

Property/Value	Units	Temp	Data Type	Reference
WS 5.44E+000	mg/L	25	EST	MEYLAN,WM ET AL. (1996)
logP 3.89			EXP	HANSCH,C & LEO,AJ (1985)
VP 3.67E-010	mm Hg	25	EST	NEELY,WB & BLAU,GE (1985)
DC	pKa			
HL 6.80E-016	atm m3/mol	25	EST	MEYLAN,WM & HOWARD,PH (1991)
OH 2.69E-010	cm3/molc sec	25	EST	MEYLAN,WM & HOWARD,PH (1993)

CAS #: 055921-64-7	2,4-PYRIMIDINEDIAMINE, 6-(4-MORPHOLINYL)-, 3-OXI

Formula: $C_8H_{15}N_5O_2$

Mol Weight: 213.24

MP (deg C): FP (deg C):

BP (deg C):

BP pressure (mm Hg):

Property/Value	Units	Temp	Data Type	Reference
WS 1.89E+004	mg/L	25	EST	MEYLAN,WM ET AL. (1996)
logP 0.36			EXP	HANSCH,C & LEO,AJ (1985)
VP 8.31E-008	mm Hg	25	EST	NEELY,WB & BLAU,GE (1985)
DC	pKa			
HL 1.41E-018	atm m3/mol	25	EST	MEYLAN,WM & HOWARD,PH (1991)
OH 2.75E-010	cm3/molc sec	25	EST	MEYLAN,WM & HOWARD,PH (1993)

CAS #: 055921-65-8	24-NH2-6-PYRROLIDIN-PYRIMIDINE-3-OXID

Formula: $C_8H_{15}N_5O$

Mol Weight: 197.24

MP (deg C): FP (deg C):

BP (deg C):

BP pressure (mm Hg):

Property/Value	Units	Temp	Data Type	Reference
WS 8.87E+003	mg/L	25	EST	MEYLAN,WM ET AL. (1996)
logP 0.84			EXP	HANSCH,C & LEO,AJ (1985)
VP 2.45E-007	mm Hg	25	EST	NEELY,WB & BLAU,GE (1985)
DC	pKa			
HL 1.60E-016	atm m3/mol	25	EST	MEYLAN,WM & HOWARD,PH (1991)
OH 2.19E-010	cm3/molc sec	25	EST	MEYLAN,WM & HOWARD,PH (1993)

CAS #: 055936-40-8	2-NITRO-4-PHENYLAZOPHENOL

Formula: $C_{12}H_9N_3O_3$

Mol Weight: 243.22

MP (deg C): FP (deg C):

BP (deg C):

BP pressure (mm Hg):

Property/Value	Units	Temp	Data Type	Reference
WS 1.32E+001	mg/L	25	EST	MEYLAN,WM ET AL. (1996)
logP 4.03			EST	MEYLAN,WM & HOWARD,PH (1995)
VP 1.76E-007	mm Hg	25	EST	NEELY,WB & BLAU,GE (1985)
DC	pKa			
HL 6.04E-012	atm m3/mol	25	EST	MEYLAN,WM & HOWARD,PH (1991)
OH 1.34E-011	cm3/molc sec	25	EST	MEYLAN,WM & HOWARD,PH (1993)

CAS #: 055973-02-9	2,4,6-PYRIMIDINETRIAMINE, N -METHYL-1-OXIDE

Formula: $C_5H_{11}N_5O$

Mol Weight: 157.18

MP (deg C): FP (deg C):

BP (deg C):

BP pressure (mm Hg):

Property/Value	Units	Temp	Data Type	Reference
WS 1.89E+005	mg/L	25	EST	MEYLAN,WM ET AL. (1996)
logP -0.49			EXP	HANSCH,C & LEO,AJ (1985)
VP 2.48E-006	mm Hg	25	EST	NEELY,WB & BLAU,GE (1985)
DC	pKa			
HL 1.00E-017	atm m3/mol	25	EST	MEYLAN,WM & HOWARD,PH (1991)
OH 2.01E-010	cm3/molc sec	25	EST	MEYLAN,WM & HOWARD,PH (1993)

CAS #: 055985-32-5	3,5-PYRIDINEDICARBOXYLIC ACID, DERIVATIVE

Formula: $C_{26}H_{29}N_3O_6$

Mol Weight: 479.54

MP (deg C): 136-138 FP (deg C):

BP (deg C):

BP pressure (mm Hg):

Property/Value	Units	Temp	Data Type	Reference
WS 2.20E+000	mg/L	25	EST	MEYLAN,WM ET AL. (1996)
logP 3.82			EXP	SANGSTER,J (1993)
VP 1.79E-012	mm Hg	25	EST	NEELY,WB & BLAU,GE (1985)
DC	pKa			
HL 3.88E-018	atm m3/mol	25	EST	MEYLAN,WM & HOWARD,PH (1991)
OH 2.05E-010	cm3/molc sec	25	EST	MEYLAN,WM & HOWARD,PH (1993)

CAS #: 055994-21-3	1H-INDENE-1,3(2H)-DIONE, 2-[4-(1-METHYLETHYL)PHE

Formula: $C_{18}H_{16}O_2$

Mol Weight: 264.33

MP (deg C): FP (deg C):

BP (deg C):

BP pressure (mm Hg):

Property/Value	Units	Temp	Data Type	Reference
WS 3.24E+000	mg/L	25	EST	MEYLAN,WM ET AL. (1996)
logP 4.43			EXP	SANGSTER,J (1994)
VP 4.34E-007	mm Hg	25	EST	NEELY,WB & BLAU,GE (1985)
DC	pKa			
HL 9.31E-010	atm m3/mol	25	EST	MEYLAN,WM & HOWARD,PH (1991)
OH 9.46E-012	cm3/molc sec	25	EST	MEYLAN,WM & HOWARD,PH (1993)

CAS #: 055994-22-4	1H-INDENE-1,3(2H)-DIONE, 2-(4-OCTYLPHENYL)-

Formula: $C_{23}H_{26}O_2$

Mol Weight: 334.46

MP (deg C): FP (deg C):

BP (deg C):

BP pressure (mm Hg):

Property/Value	Units	Temp	Data Type	Reference
WS 9.27E-003	mg/L	25	EST	MEYLAN,WM ET AL. (1996)
logP 6.93			EXP	SANGSTER,J (1993)
VP 3.91E-009	mm Hg	25	EST	NEELY,WB & BLAU,GE (1985)
DC	pKa			
HL 3.84E-009	atm m3/mol	25	EST	MEYLAN,WM & HOWARD,PH (1991)
OH 1.70E-011	cm3/molc sec	25	EST	MEYLAN,WM & HOWARD,PH (1993)

CAS #: 055994-23-5	1H-INDENE-1,3(2H)-DIONE, 2-(3,5-DIETHYLPHENYL)-

Formula: $C_{19}H_{18}O_2$
Mol Weight: 278.35
MP (deg C): FP (deg C):
BP (deg C):
BP pressure (mm Hg):

Property/Value	Units	Temp	Data Type	Reference
WS 9.90E-001	mg/L	25	EST	MEYLAN,WM ET AL. (1996)
logP 4.94			EXP	SANGSTER,J (1993)
VP 1.16E-007	mm Hg	25	EST	NEELY,WB & BLAU,GE (1985)
DC	pKa			
HL 1.03E-009	atm m3/mol	25	EST	MEYLAN,WM & HOWARD,PH (1991)
OH 3.24E-011	cm3/molc sec	25	EST	MEYLAN,WM & HOWARD,PH (1993)

CAS #: 055994-24-6	1H-INDENE-1,3(2H)-DIONE, 2-[3,5-BIS(1-METHYLETHY

Formula: $C_{21}H_{22}O_2$
Mol Weight: 306.41
MP (deg C): FP (deg C):
BP (deg C):
BP pressure (mm Hg):

Property/Value	Units	Temp	Data Type	Reference
WS 9.16E-002	mg/L	25	EST	MEYLAN,WM ET AL. (1996)
logP 5.96			EXP	SANGSTER,J (1993)
VP 6.39E-008	mm Hg	25	EST	NEELY,WB & BLAU,GE (1985)
DC	pKa			
HL 1.81E-009	atm m3/mol	25	EST	MEYLAN,WM & HOWARD,PH (1991)
OH 3.22E-011	cm3/molc sec	25	EST	MEYLAN,WM & HOWARD,PH (1993)

CAS #: 056010-86-7	1H-IMIDAZOLE-4-CARBOXAMIDE, 5-(3-METHYL-3-OCTYL-

Formula: $C_{13}H_{24}N_6O$
Mol Weight: 280.38
MP (deg C): FP (deg C):
BP (deg C):
BP pressure (mm Hg):

Property/Value	Units	Temp	Data Type	Reference
WS 2.84E+001	mg/L	25	EST	MEYLAN,WM ET AL. (1996)
logP 3.22			EXP	SANGSTER,J (1994)
VP 6.71E-011	mm Hg	25	EST	NEELY,WB & BLAU,GE (1985)
DC	pKa			
HL 1.12E-015	atm m3/mol	25	EST	MEYLAN,WM & HOWARD,PH (1991)
OH 3.53E-011	cm3/molc sec	25	EST	MEYLAN,WM & HOWARD,PH (1993)

CAS #: 056030-54-7	SUFENTANIL

Formula: $C_{22}H_{30}N_2O_2S$
Mol Weight: 386.56
MP (deg C): 97 FP (deg C):
BP (deg C):
BP pressure (mm Hg):

Property/Value	Units	Temp	Data Type	Reference
WS 7.60E+001	mg/L	25	EXP	ROY,SD & FLYNN,GL (1988)
logP 3.95			EXP	SANGSTER,J (1993)
VP 6.10E-013	mm Hg	25	EST	HL X WSOL
DC	pKa			
HL 4.11E-015	atm m3/mol	25	EST	MEYLAN,WM & HOWARD,PH (1991)
OH 1.66E-010	cm3/molc sec	25	EST	ATKINSON,R (1988)

CAS #: 056046-08-3	3-METHOXYPENTACHLOROCYCLOHEXANE

Formula: $C_7H_9Cl_5O$
Mol Weight: 286.41
MP (deg C): FP (deg C):
BP (deg C):
BP pressure (mm Hg):

Property/Value	Units	Temp	Data Type	Reference
WS 3.06E+001	mg/L	25	EST	MEYLAN,WM ET AL. (1996)
logP 3.51			EXP	HANSCH,C & LEO,AJ (1985)
VP 8.16E-004	mm Hg	25	EST	NEELY,WB & BLAU,GE (1985)
DC	pKa			
HL 8.50E-006	atm m3/mol	25	EST	MEYLAN,WM & HOWARD,PH (1991)
OH 2.61E-012	cm3/molc sec	25	EST	MEYLAN,WM & HOWARD,PH (1993)

CAS #: 056046-09-4	3-METHIOPENTACHLOROCYCLOHEXANE

Formula: $C_7H_9Cl_5S$
Mol Weight: 302.48
MP (deg C): FP (deg C):
BP (deg C):
BP pressure (mm Hg):

Property/Value	Units	Temp	Data Type	Reference
WS 7.45E+000	mg/L	25	EST	MEYLAN,WM ET AL. (1996)
logP 3.85			EXP	HANSCH,C & LEO,AJ (1985)
VP 1.58E-004	mm Hg	25	EST	NEELY,WB & BLAU,GE (1985)
DC	pKa			
HL 7.75E-006	atm m3/mol	25	EST	MEYLAN,WM & HOWARD,PH (1991)
OH 5.66E-012	cm3/molc sec	25	EST	MEYLAN,WM & HOWARD,PH (1993)

CAS #: 056046-62-9	N-(2-ETHYL(3-METHYL-4-NITROPHENYL)AMINO)ETHYL)-M

Formula: $C_{12}H_{19}N_3O_3S$
Mol Weight: 285.37
MP (deg C): FP (deg C):
BP (deg C):
BP pressure (mm Hg):

Property/Value	Units	Temp	Data Type	Reference
WS 2.26E+002	mg/L	25	EST	MEYLAN,WM ET AL. (1996)
logP 2.13			EST	MEYLAN,WM & HOWARD,PH (1995)
VP 7.00E-007	mm Hg	25	EST	NEELY,WB & BLAU,GE (1985)
DC	pKa			
HL 1.00E-010	atm m3/mol	25	EST	MEYLAN,WM & HOWARD,PH (1991)
OH 2.30E-010	cm3/molc sec	25	EST	MEYLAN,WM & HOWARD,PH (1993)

CAS #: 056058-99-2	2,4(1H,3H)-PYRIMIDINEDIONE, 5-FLUORO-1-(1-OXOPRO

Formula: $C_7H_7FN_2O_3$
Mol Weight: 186.14
MP (deg C): FP (deg C):
BP (deg C):
BP pressure (mm Hg):

Property/Value	Units	Temp	Data Type	Reference
WS 1.73E+003	mg/L	25	EST	MEYLAN,WM ET AL. (1996)
logP 0.19			EXP	SANGSTER,J (1994)
VP 4.85E-008	mm Hg	25	EST	NEELY,WB & BLAU,GE (1985)
DC	pKa			
HL 7.14E-013	atm m3/mol	25	EST	MEYLAN,WM & HOWARD,PH (1991)
OH 7.06E-012	cm3/molc sec	25	EST	MEYLAN,WM & HOWARD,PH (1993)

CAS #:	056066-19-4			2,4-NH2PYRIMIDIN,5(3,5MEO-4NH2)BENZYL
Formula:	$C_{13}H_{17}N_5O_2$			
Mol Weight:	275.31			
MP (deg C):		FP (deg C):		
BP (deg C):				
BP pressure (mm Hg):				

Property/Value	Units	Temp	Data Type	Reference
WS 3.33E+003	mg/L	25	EST	MEYLAN,WM ET AL. (1996)
logP 0.83			EXP	HANSCH,C & LEO,AJ (1985)
VP 2.63E-009	mm Hg	25	EST	NEELY,WB & BLAU,GE (1985)
DC	pKa			
HL 1.42E-016	atm m3/mol	25	EST	MEYLAN,WM & HOWARD,PH (1991)
OH 2.03E-010	cm3/molc sec	25	EST	MEYLAN,WM & HOWARD,PH (1993)

CAS #:	056066-43-4			24NH2PYRIMIDINE,5(3,5-ETO-4-NH2)BENZYL
Formula:	$C_{15}H_{21}N_5O_2$			
Mol Weight:	303.37			
MP (deg C):		FP (deg C):		
BP (deg C):				
BP pressure (mm Hg):				

Property/Value	Units	Temp	Data Type	Reference
WS 4.23E+002	mg/L	25	EST	MEYLAN,WM ET AL. (1996)
logP 1.69			EXP	HANSCH,C & LEO,AJ (1985)
VP 5.12E-010	mm Hg	25	EST	NEELY,WB & BLAU,GE (1985)
DC	pKa			
HL 2.51E-016	atm m3/mol	25	EST	MEYLAN,WM & HOWARD,PH (1991)
OH 2.13E-010	cm3/molc sec	25	EST	MEYLAN,WM & HOWARD,PH (1993)

CAS #:	056066-63-8			2,4-PYRIMIDINEDIAMINE, 5-[[4-(DIMETHYLAMINO)-3,5
Formula:	$C_{15}H_{21}N_5O_2$			
Mol Weight:	303.37			
MP (deg C):		FP (deg C):		
BP (deg C):				
BP pressure (mm Hg):				

Property/Value	Units	Temp	Data Type	Reference
WS 6.02E+002	mg/L	25	EST	MEYLAN,WM ET AL. (1996)
logP 1.51			EXP	HANSCH,C ET AL. (1995)
VP 4.64E-009	mm Hg	25	EST	NEELY,WB & BLAU,GE (1985)
DC	pKa			
HL 6.41E-015	atm m3/mol	25	EST	MEYLAN,WM & HOWARD,PH (1991)
OH 2.05E-010	cm3/molc sec	25	EST	MEYLAN,WM & HOWARD,PH (1993)

CAS #:	056070-15-6			PHOSPHOROTHIOIC ACID, S-[[(1,1-DIMETHYLETHYL)SUL
Formula:	$C_9H_{21}O_5PS_2$			
Mol Weight:	304.37			
MP (deg C):		FP (deg C):		
BP (deg C):				
BP pressure (mm Hg):				

Property/Value	Units	Temp	Data Type	Reference
WS 5.09E+001	mg/L	25	EST	MEYLAN,WM ET AL. (1996)
logP 2.76			EXP	SANGSTER,J (1994)
VP 5.21E-006	mm Hg	25	EST	NEELY,WB & BLAU,GE (1985)
DC	pKa			
HL 4.78E-011	atm m3/mol	25	EST	MEYLAN,WM & HOWARD,PH (1991)
OH 1.88E-010	cm3/molc sec	25	EST	MEYLAN,WM & HOWARD,PH (1993)

CAS #:	056070-16-7			TERBUFOS SULFONE
Formula:	$C_9H_{21}O_4PS_3$			
Mol Weight:	320.43			
MP (deg C):		FP (deg C):		
BP (deg C):				
BP pressure (mm Hg):				

Property/Value	Units	Temp	Data Type	Reference
WS 4.08E+002	mg/L	19	EXP	SHIU,WY ET AL. (1990)
logP 2.48			EXP	HANSCH,C & LEO,AJ (1985)
VP 7.88E-006	mm Hg	25	EST	NEELY,WB & BLAU,GE (1985)
DC	pKa			
HL 2.21E-008	atm m3/mol	25	EST	MEYLAN,WM & HOWARD,PH (1991)
OH 2.41E-010	cm3/molc sec	25	EST	MEYLAN,WM & HOWARD,PH (1993)

CAS #:	056086-55-6			2-F-PENTACHLOROCYCLOHEXANE
Formula:	$C_6H_6Cl_5F$			
Mol Weight:	274.38			
MP (deg C):		FP (deg C):		
BP (deg C):				
BP pressure (mm Hg):				

Property/Value	Units	Temp	Data Type	Reference
WS 3.26E+001	mg/L	25	EST	MEYLAN,WM ET AL. (1996)
logP 3.19			EXP	HANSCH,C & LEO,AJ (1985)
VP 6.78E-003	mm Hg	25	EST	NEELY,WB & BLAU,GE (1985)
DC	pKa			
HL 1.44E-003	atm m3/mol	25	EST	MEYLAN,WM & HOWARD,PH (1991)
OH 4.22E-013	cm3/molc sec	25	EST	MEYLAN,WM & HOWARD,PH (1993)

CAS #:	056108-12-4			4-TERT-BUTYLBENZAMIDE
Formula:	$C_{11}H_{15}NO$			
Mol Weight:	177.25			
MP (deg C):		FP (deg C):		
BP (deg C):				
BP pressure (mm Hg):				

Property/Value	Units	Temp	Data Type	Reference
WS 4.08E+002	mg/L	25	EST	MEYLAN,WM ET AL. (1996)
logP 2.51			EXP	NAKAGAWA,Y ET AL. (1992)
VP 4.54E-005	mm Hg	25	EST	NEELY,WB & BLAU,GE (1985)
DC	pKa			
HL 5.72E-009	atm m3/mol	25	EST	MEYLAN,WM & HOWARD,PH (1991)
OH 6.42E-012	cm3/molc sec	25	EST	MEYLAN,WM & HOWARD,PH (1993)

CAS #:	056165-57-2			PHOSPHOROTHIOIC ACID, S-[[(1,1-DIMETHYLETHYL)SUL
Formula:	$C_9H_{21}O_4PS_2$			
Mol Weight:	288.37			
MP (deg C):		FP (deg C):		
BP (deg C):				
BP pressure (mm Hg):				

Property/Value	Units	Temp	Data Type	Reference
WS 7.67E+001	mg/L	25	EST	MEYLAN,WM ET AL. (1996)
logP 2.66			EXP	SANGSTER,J (1994)
VP 8.43E-006	mm Hg	25	EST	NEELY,WB & BLAU,GE (1985)
DC	pKa			
HL 3.80E-013	atm m3/mol	25	EST	MEYLAN,WM & HOWARD,PH (1991)
OH 2.48E-010	cm3/molc sec	25	EST	MEYLAN,WM & HOWARD,PH (1993)

CAS #: 056177-33-4	2-ISOPROPYLBENZAMIDE

Formula: $C_{10}H_{13}NO$

Mol Weight: 163.22

MP (deg C): | FP (deg C):

BP (deg C):

BP pressure (mm Hg):

Property/ Value	Units	Temp	Data Type	Reference
WS 4.30E+003	mg/L	25	EST	MEYLAN,WM ET AL. (1996)
logP 1.39			EXP	NAKAGAWA,Y ET AL. (1992)
VP 8.93E-005	mm Hg	25	EST	NEELY,WB & BLAU,GE (1985)
DC	pKa			
HL 4.31E-009	atm m3/mol	25	EST	MEYLAN,WM & HOWARD,PH (1991)
OH 8.49E-012	cm3/molc sec	25	EST	MEYLAN,WM & HOWARD,PH (1993)

CAS #: 056183-32-5	24NH2-5(3-BR-4-MEO-BENZYL)PYRIMIDINE

Formula: $C_{12}H_{13}BrN_4O$

Mol Weight: 309.17

MP (deg C): | FP (deg C):

BP (deg C):

BP pressure (mm Hg):

Property/ Value	Units	Temp	Data Type	Reference
WS 2.69E+002	mg/L	25	EST	MEYLAN,WM ET AL. (1996)
logP 1.88			EXP	HANSCH,C & LEO,AJ (1985)
VP 3.13E-008	mm Hg	25	EST	NEELY,WB & BLAU,GE (1985)
DC	pKa			
HL 2.72E-012	atm m3/mol	25	EST	MEYLAN,WM & HOWARD,PH (1991)
OH 1.08E-011	cm3/molc sec	25	EST	MEYLAN,WM & HOWARD,PH (1993)

CAS #: 056189-70-9	4-CL-3-ME-PHENYLAMIDINO UREA

Formula: $C_9H_{11}ClN_4O$

Mol Weight: 226.67

MP (deg C): | FP (deg C):

BP (deg C):

BP pressure (mm Hg):

Property/ Value	Units	Temp	Data Type	Reference
WS 4.28E+002	mg/L	25	EST	MEYLAN,WM ET AL. (1996)
logP 2.19			EXP	HANSCH,C & LEO,AJ (1985)
VP 2.59E-007	mm Hg	25	EST	NEELY,WB & BLAU,GE (1985)
DC	pKa			
HL 1.12E-015	atm m3/mol	25	EST	MEYLAN,WM & HOWARD,PH (1991)
OH 5.47E-011	cm3/molc sec	25	EST	MEYLAN,WM & HOWARD,PH (1993)

CAS #: 056209-30-4	7,9-DIAZASPIRO[4.5]DECANE-6,8,10-TRIONE

Formula: $C_8H_{10}N_2O_3$

Mol Weight: 182.18

MP (deg C): | FP (deg C):

BP (deg C):

BP pressure (mm Hg):

Property/ Value	Units	Temp	Data Type	Reference
WS 3.35E+004	mg/L	25	EST	MEYLAN,WM ET AL. (1996)
logP 0.24			EXP	PRANKERD,RJ & MCKEOWN,RH (1992)
VP 1.01E-010	mm Hg	25	EST	NEELY,WB & BLAU,GE (1985)
DC	pKa			
HL 1.59E-013	atm m3/mol	25	EST	MEYLAN,WM & HOWARD,PH (1991)
OH 1.14E-011	cm3/molc sec	25	EST	MEYLAN,WM & HOWARD,PH (1993)

CAS #: 056222-10-7	N-((4-NITROPHENYL)METHYL)ACETAMIDE

Formula: $C_9H_{10}N_2O_3$

Mol Weight: 194.19

MP (deg C): | FP (deg C):

BP (deg C):

BP pressure (mm Hg):

Property/ Value	Units	Temp	Data Type	Reference
WS 3.19E+003	mg/L	25	EST	MEYLAN,WM ET AL. (1996)
logP 1.46			EST	MEYLAN,WM & HOWARD,PH (1995)
VP 4.62E-005	mm Hg	25	EST	NEELY,WB & BLAU,GE (1985)
DC	pKa			
HL 4.99E-010	atm m3/mol	25	EST	MEYLAN,WM & HOWARD,PH (1991)
OH 2.21E-012	cm3/molc sec	25	EST	MEYLAN,WM & HOWARD,PH (1993)

CAS #: 056239-24-8	TR-1-NO-1(2CLET)-3(4-OH CYHEXYL)UREA

Formula: $C_9H_{16}ClN_3O_3$

Mol Weight: 249.70

MP (deg C): | FP (deg C):

BP (deg C):

BP pressure (mm Hg):

Property/ Value	Units	Temp	Data Type	Reference
WS 3.32E+003	mg/L	25	EST	MEYLAN,WM ET AL. (1996)
logP 1.00			EXP	HANSCH,C & LEO,AJ (1985)
VP 1.64E-008	mm Hg	25	EST	NEELY,WB & BLAU,GE (1985)
DC	pKa			
HL 6.76E-015	atm m3/mol	25	EST	MEYLAN,WM & HOWARD,PH (1991)
OH 3.09E-011	cm3/molc sec	25	EST	MEYLAN,WM & HOWARD,PH (1993)

CAS #: 056245-60-4	GALACTOPYRANOSIDE, ETHYL 1-THIO-

Formula: $C_8H_{16}O_5S$

Mol Weight: 224.28

MP (deg C): | FP (deg C):

BP (deg C):

BP pressure (mm Hg):

Property/ Value	Units	Temp	Data Type	Reference
WS 6.36E+005	mg/L	25	EST	MEYLAN,WM ET AL. (1996)
logP -1.51			EXP	SANGSTER,J (1993)
VP 2.79E-009	mm Hg	25	EST	NEELY,WB & BLAU,GE (1985)
DC	pKa			
HL 3.76E-015	atm m3/mol	25	EST	MEYLAN,WM & HOWARD,PH (1991)
OH 1.76E-010	cm3/molc sec	25	EST	MEYLAN,WM & HOWARD,PH (1993)

CAS #: 056265-21-5	1,1'-(2-METHYLPROPYLIDENE)BIS(4-ETHOXYBENZENE)

Formula: $C_{20}H_{26}O_2$

Mol Weight: 298.43

MP (deg C): | FP (deg C):

BP (deg C):

BP pressure (mm Hg):

Property/ Value	Units	Temp	Data Type	Reference
WS 6.33E-002	mg/L	25	EST	MEYLAN,WM ET AL. (1996)
logP 6.20			EST	MEYLAN,WM & HOWARD,PH (1995)
VP 4.18E-006	mm Hg	25	EST	NEELY,WB & BLAU,GE (1985)
DC	pKa			
HL 6.93E-006	atm m3/mol	25	EST	MEYLAN,WM & HOWARD,PH (1991)
OH 6.92E-011	cm3/molc sec	25	EST	MEYLAN,WM & HOWARD,PH (1993)

CAS #: 056265-22-6 — 1,1-BIS(P-ETHOXYPHENYL)-2-CHLOROPROPANE

Formula: $C_{19}H_{23}ClO_2$

Mol Weight: 318.85

MP (deg C): FP (deg C):

BP (deg C):

BP pressure (mm Hg):

Property/ Value	Units	Temp	Data Type	Reference
WS 7.65E-002	mg/L	25	EST	MEYLAN,WM ET AL. (1996)
logP 5.97			EST	MEYLAN,WM & HOWARD,PH (1995)
VP 1.72E-006	mm Hg	25	EST	NEELY,WB & BLAU,GE (1985)
DC	pKa			
HL 1.84E-006	atm m3/mol	25	EST	MEYLAN,WM & HOWARD,PH (1991)
OH 6.58E-011	cm3/molc sec	25	EST	MEYLAN,WM & HOWARD,PH (1993)

CAS #: 056265-23-7 — 1,1-BIS(P-ETHOXYPHENYL)-2,2-DICHLOROPROPANE

Formula: $C_{19}H_{22}Cl_2O_2$

Mol Weight: 353.29

MP (deg C): FP (deg C):

BP (deg C):

BP pressure (mm Hg):

Property/ Value	Units	Temp	Data Type	Reference
WS 7.78E-003	mg/L	25	EST	MEYLAN,WM ET AL. (1996)
logP 6.89			EST	MEYLAN,WM & HOWARD,PH (1995)
VP 8.44E-007	mm Hg	25	EST	NEELY,WB & BLAU,GE (1985)
DC	pKa			
HL 6.47E-007	atm m3/mol	25	EST	MEYLAN,WM & HOWARD,PH (1991)
OH 6.46E-011	cm3/molc sec	25	EST	MEYLAN,WM & HOWARD,PH (1993)

CAS #: 056265-24-8 — 1,1'-(2-CHLORO-2-METHYLPROPYLIDENE)BIS-4-ETHOX*

Formula: $C_{20}H_{25}ClO_2$

Mol Weight: 332.87

MP (deg C): FP (deg C):

BP (deg C):

BP pressure (mm Hg):

Property/ Value	Units	Temp	Data Type	Reference
WS 2.59E-002	mg/L	25	EST	MEYLAN,WM ET AL. (1996)
logP 6.42			EST	MEYLAN,WM & HOWARD,PH (1995)
VP 1.30E-006	mm Hg	25	EST	NEELY,WB & BLAU,GE (1985)
DC	pKa			
HL 2.44E-006	atm m3/mol	25	EST	MEYLAN,WM & HOWARD,PH (1991)
OH 6.63E-011	cm3/molc sec	25	EST	MEYLAN,WM & HOWARD,PH (1993)

CAS #: 056265-26-0 — 1-ETHOXY-4-(2-METHYL-1-(4-METHYLPHENYL)PROPYL)*

Formula: $C_{19}H_{24}O$

Mol Weight: 268.40

MP (deg C): FP (deg C):

BP (deg C):

BP pressure (mm Hg):

Property/ Value	Units	Temp	Data Type	Reference
WS 9.88E-002	mg/L	25	EST	MEYLAN,WM ET AL. (1996)
logP 6.18			EST	MEYLAN,WM & HOWARD,PH (1995)
VP 2.73E-005	mm Hg	25	EST	NEELY,WB & BLAU,GE (1985)
DC	pKa			
HL 9.73E-005	atm m3/mol	25	EST	MEYLAN,WM & HOWARD,PH (1991)
OH 4.34E-011	cm3/molc sec	25	EST	MEYLAN,WM & HOWARD,PH (1993)

CAS #: 056265-27-1 — 1-(2-CHLORO-1-(4-ETHOXYPHENYL)PROPYL)-4-METHBE*

Formula: $C_{18}H_{21}ClO$

Mol Weight: 288.82

MP (deg C): FP (deg C):

BP (deg C):

BP pressure (mm Hg):

Property/ Value	Units	Temp	Data Type	Reference
WS 1.20E-001	mg/L	25	EST	MEYLAN,WM ET AL. (1996)
logP 5.94			EST	MEYLAN,WM & HOWARD,PH (1995)
VP 9.67E-006	mm Hg	25	EST	NEELY,WB & BLAU,GE (1985)
DC	pKa			
HL 2.58E-005	atm m3/mol	25	EST	MEYLAN,WM & HOWARD,PH (1991)
OH 4.00E-011	cm3/molc sec	25	EST	MEYLAN,WM & HOWARD,PH (1993)

CAS #: 056323-43-4 — 1-NO-1(CLET)-3(2-OH CYHEXYL)UREA

Formula: $C_9H_{16}ClN_3O_3$

Mol Weight: 249.70

MP (deg C): FP (deg C):

BP (deg C):

BP pressure (mm Hg):

Property/ Value	Units	Temp	Data Type	Reference
WS 1.70E+003	mg/L	25	EST	MEYLAN,WM ET AL. (1996)
logP 1.75			EXP	HANSCH,C & LEO,AJ (1985)
VP 1.64E-008	mm Hg	25	EST	NEELY,WB & BLAU,GE (1985)
DC	pKa			
HL 6.76E-015	atm m3/mol	25	EST	MEYLAN,WM & HOWARD,PH (1991)
OH 3.09E-011	cm3/molc sec	25	EST	MEYLAN,WM & HOWARD,PH (1993)

CAS #: 056343-48-7 — PYRAZINAMINE, N,N,6-TRIMETHYL-

Formula: $C_7H_{11}N_3$

Mol Weight: 137.19

MP (deg C): FP (deg C):

BP (deg C):

BP pressure (mm Hg):

Property/ Value	Units	Temp	Data Type	Reference
WS 3.93E+003	mg/L	25	EST	MEYLAN,WM ET AL. (1996)
logP 1.57			EXP	YAMAGAMI,C ET AL. (1991)
VP 1.78E-001	mm Hg	25	EST	NEELY,WB & BLAU,GE (1985)
DC	pKa			
HL 5.12E-008	atm m3/mol	25	EST	MEYLAN,WM & HOWARD,PH (1991)
OH 1.17E-010	cm3/molc sec	25	EST	MEYLAN,WM & HOWARD,PH (1993)

CAS #: 056344-90-2 — BARBITURIC ACID,5-PR-5-ET-1-ME

Formula: $C_{10}H_{16}N_2O_3$

Mol Weight: 212.25

MP (deg C): FP (deg C):

BP (deg C):

BP pressure (mm Hg):

Property/ Value	Units	Temp	Data Type	Reference
WS 1.47E+003	mg/L	25	EST	MEYLAN,WM ET AL. (1996)
logP 1.65			EXP	HANSCH,C & LEO,AJ (1985)
VP 3.11E-010	mm Hg	25	EST	NEELY,WB & BLAU,GE (1985)
DC	pKa			
HL 1.05E-012	atm m3/mol	25	EST	MEYLAN,WM & HOWARD,PH (1991)
OH 1.22E-011	cm3/molc sec	25	EST	MEYLAN,WM & HOWARD,PH (1993)

CAS #: 056356-13-9				2-TOLYL 4-AMINOSALICYLATE

Formula: $C_{14}H_{13}NO_3$

Mol Weight: 243.26

MP (deg C):   FP (deg C):

BP (deg C):

BP pressure (mm Hg):

Property/ Value	Units	Temp	Data Type	Reference
WS 5.37E+001	mg/L	25	EST	MEYLAN,WM ET AL. (1996)
logP 3.14			EXP	HANSCH,C & LEO,AJ (1985)
VP 1.02E-007	mm Hg	25	EST	NEELY,WB & BLAU,GE (1985)
DC	pKa			
HL 6.56E-010	atm m3/mol	25	EST	MEYLAN,WM & HOWARD,PH (1991)
OH 2.00E-010	cm3/molc sec	25	EST	MEYLAN,WM & HOWARD,PH (1993)

CAS #: 056356-14-0				3-TOLYL 4-AMINOSALICYLATE

Formula: $C_{14}H_{13}NO_3$

Mol Weight: 243.26

MP (deg C):   FP (deg C):

BP (deg C):

BP pressure (mm Hg):

Property/ Value	Units	Temp	Data Type	Reference
WS 2.01E+001	mg/L	25	EST	MEYLAN,WM ET AL. (1996)
logP 3.64			EXP	HANSCH,C & LEO,AJ (1985)
VP 1.02E-007	mm Hg	25	EST	NEELY,WB & BLAU,GE (1985)
DC	pKa			
HL 6.56E-010	atm m3/mol	25	EST	MEYLAN,WM & HOWARD,PH (1991)
OH 2.00E-010	cm3/molc sec	25	EST	MEYLAN,WM & HOWARD,PH (1993)

CAS #: 056356-15-1				4-TOLYL 4-AMINOSALICYLATE

Formula: $C_{14}H_{13}NO_3$

Mol Weight: 243.26

MP (deg C):   FP (deg C):

BP (deg C):

BP pressure (mm Hg):

Property/ Value	Units	Temp	Data Type	Reference
WS 3.35E+001	mg/L	25	EST	MEYLAN,WM ET AL. (1996)
logP 3.38			EXP	HANSCH,C & LEO,AJ (1985)
VP 1.02E-007	mm Hg	25	EST	NEELY,WB & BLAU,GE (1985)
DC	pKa			
HL 6.56E-010	atm m3/mol	25	EST	MEYLAN,WM & HOWARD,PH (1991)
OH 2.00E-010	cm3/molc sec	25	EST	MEYLAN,WM & HOWARD,PH (1993)

CAS #: 056356-16-2				BENZOIC ACID, 4-AMINO-2-HYDROXY-, 2,6-DIMETHYLPH

Formula: $C_{15}H_{15}NO_3$

Mol Weight: 257.29

MP (deg C):   FP (deg C):

BP (deg C):

BP pressure (mm Hg):

Property/ Value	Units	Temp	Data Type	Reference
WS 1.05E+001	mg/L	25	EST	MEYLAN,WM ET AL. (1996)
logP 3.88			EXP	SANGSTER,J (1993)
VP 3.88E-008	mm Hg	25	EST	NEELY,WB & BLAU,GE (1985)
DC	pKa			
HL 7.23E-010	atm m3/mol	25	EST	MEYLAN,WM & HOWARD,PH (1991)
OH 2.00E-010	cm3/molc sec	25	EST	MEYLAN,WM & HOWARD,PH (1993)

CAS #: 056356-17-3				2-MEO PHENYL 4-AMINOSALICYLATE

Formula: $C_{14}H_{13}NO_4$

Mol Weight: 259.26

MP (deg C):   FP (deg C):

BP (deg C):

BP pressure (mm Hg):

Property/ Value	Units	Temp	Data Type	Reference
WS 7.29E+001	mg/L	25	EST	MEYLAN,WM ET AL. (1996)
logP 2.88			EXP	HANSCH,C & LEO,AJ (1985)
VP 3.79E-008	mm Hg	25	EST	NEELY,WB & BLAU,GE (1985)
DC	pKa			
HL 3.51E-011	atm m3/mol	25	EST	MEYLAN,WM & HOWARD,PH (1991)
OH 2.01E-010	cm3/molc sec	25	EST	MEYLAN,WM & HOWARD,PH (1993)

CAS #: 056356-18-4				3-MEO PHENYL 4-AMINOSALICYLATE

Formula: $C_{14}H_{13}NO_4$

Mol Weight: 259.26

MP (deg C):   FP (deg C):

BP (deg C):

BP pressure (mm Hg):

Property/ Value	Units	Temp	Data Type	Reference
WS 3.52E+001	mg/L	25	EST	MEYLAN,WM ET AL. (1996)
logP 3.25			EXP	HANSCH,C & LEO,AJ (1985)
VP 3.79E-008	mm Hg	25	EST	NEELY,WB & BLAU,GE (1985)
DC	pKa			
HL 3.51E-011	atm m3/mol	25	EST	MEYLAN,WM & HOWARD,PH (1991)
OH 2.01E-010	cm3/molc sec	25	EST	MEYLAN,WM & HOWARD,PH (1993)

CAS #: 056356-19-5				4-AMINOSALICYLIC ACID,4-IPR ESTER

Formula: $C_{16}H_{17}NO_3$

Mol Weight: 271.32

MP (deg C):   FP (deg C):

BP (deg C):

BP pressure (mm Hg):

Property/ Value	Units	Temp	Data Type	Reference
WS 5.44E+000	mg/L	25	EST	MEYLAN,WM ET AL. (1996)
logP 4.12			EXP	HANSCH,C & LEO,AJ (1985)
VP 3.01E-008	mm Hg	25	EST	NEELY,WB & BLAU,GE (1985)
DC	pKa			
HL 1.16E-009	atm m3/mol	25	EST	MEYLAN,WM & HOWARD,PH (1991)
OH 2.02E-010	cm3/molc sec	25	EST	MEYLAN,WM & HOWARD,PH (1993)

CAS #: 056356-20-8				4-AMINOSALICYLIC ACID,2-IPR ESTER

Formula: $C_{16}H_{17}NO_3$

Mol Weight: 271.32

MP (deg C):   FP (deg C):

BP (deg C):

BP pressure (mm Hg):

Property/ Value	Units	Temp	Data Type	Reference
WS 3.97E+000	mg/L	25	EST	MEYLAN,WM ET AL. (1996)
logP 4.28			EXP	HANSCH,C & LEO,AJ (1985)
VP 3.01E-008	mm Hg	25	EST	NEELY,WB & BLAU,GE (1985)
DC	pKa			
HL 1.16E-009	atm m3/mol	25	EST	MEYLAN,WM & HOWARD,PH (1991)
OH 2.02E-010	cm3/molc sec	25	EST	MEYLAN,WM & HOWARD,PH (1993)

056356-31-1 — 2-BROMOPHENYL 4-AMINOSALICYLATE

Formula: $C_{13}H_{10}BrNO_3$
Mol Weight: 308.14
MP (deg C):
FP (deg C):
BP (deg C):
BP pressure (mm Hg):

Property/Value	Units	Temp	Data Type	Reference
WS 7.04E+000	mg/L	25	EST	MEYLAN,WM ET AL. (1996)
logP 3.74			EXP	HANSCH,C & LEO,AJ (1985)
VP 1.97E-008	mm Hg	25	EST	NEELY,WB & BLAU,GE (1985)
DC	pKa			
HL 2.37E-010	atm m3/mol	25	EST	MEYLAN,WM & HOWARD,PH (1991)
OH 2.00E-010	cm3/molc sec	25	EST	MEYLAN,WM & HOWARD,PH (1993)

056356-32-2 — 3-BROMOPHENYL 4-AMINOSALICYLATE

Formula: $C_{13}H_{10}BrNO_3$
Mol Weight: 308.14
MP (deg C):
FP (deg C):
BP (deg C):
BP pressure (mm Hg):

Property/Value	Units	Temp	Data Type	Reference
WS 5.78E+000	mg/L	25	EST	MEYLAN,WM ET AL. (1996)
logP 3.84			EXP	HANSCH,C & LEO,AJ (1985)
VP 1.97E-008	mm Hg	25	EST	NEELY,WB & BLAU,GE (1985)
DC	pKa			
HL 2.37E-010	atm m3/mol	25	EST	MEYLAN,WM & HOWARD,PH (1991)
OH 2.00E-010	cm3/molc sec	25	EST	MEYLAN,WM & HOWARD,PH (1993)

056356-33-3 — 4-BROMOPHENYL 4-AMINOSALICYLATE

Formula: $C_{13}H_{10}BrNO_3$
Mol Weight: 308.14
MP (deg C):
FP (deg C):
BP (deg C):
BP pressure (mm Hg):

Property/Value	Units	Temp	Data Type	Reference
WS 1.22E+001	mg/L	25	EST	MEYLAN,WM ET AL. (1996)
logP 3.46			EXP	HANSCH,C & LEO,AJ (1985)
VP 1.97E-008	mm Hg	25	EST	NEELY,WB & BLAU,GE (1985)
DC	pKa			
HL 2.37E-010	atm m3/mol	25	EST	MEYLAN,WM & HOWARD,PH (1991)
OH 2.00E-010	cm3/molc sec	25	EST	MEYLAN,WM & HOWARD,PH (1993)

056356-34-4 — 3-CF3 PHENYL 4-AMINOSALICYLATE

Formula: $C_{14}H_{10}F_3NO_3$
Mol Weight: 297.24
MP (deg C):
FP (deg C):
BP (deg C):
BP pressure (mm Hg):

Property/Value	Units	Temp	Data Type	Reference
WS 1.05E+001	mg/L	25	EST	MEYLAN,WM ET AL. (1996)
logP 3.61			EXP	HANSCH,C & LEO,AJ (1985)
VP 1.81E-007	mm Hg	25	EST	NEELY,WB & BLAU,GE (1985)
DC	pKa			
HL 5.16E-009	atm m3/mol	25	EST	MEYLAN,WM & HOWARD,PH (1991)
OH 2.00E-010	cm3/molc sec	25	EST	MEYLAN,WM & HOWARD,PH (1993)

056400-11-4 — 3(CF3-MEO)PENTACHLOROCYCLOHEXANE

Formula: $C_8H_8Cl_5F_3O$
Mol Weight: 354.41
MP (deg C):
FP (deg C):
BP (deg C):
BP pressure (mm Hg):

Property/Value	Units	Temp	Data Type	Reference
WS 1.67E+000	mg/L	25	EST	MEYLAN,WM ET AL. (1996)
logP 4.15			EXP	HANSCH,C & LEO,AJ (1985)
VP 5.32E-004	mm Hg	25	EST	NEELY,WB & BLAU,GE (1985)
DC	pKa			
HL 8.89E-005	atm m3/mol	25	EST	MEYLAN,WM & HOWARD,PH (1991)
OH 1.19E-012	cm3/molc sec	25	EST	MEYLAN,WM & HOWARD,PH (1993)

056400-12-5 — 2,3-DIMEO TETRACHLOROCYCLOHEXANE

Formula: $C_8H_{12}Cl_4O_2$
Mol Weight: 282.00
MP (deg C):
FP (deg C):
BP (deg C):
BP pressure (mm Hg):

Property/Value	Units	Temp	Data Type	Reference
WS 1.04E+002	mg/L	25	EST	MEYLAN,WM ET AL. (1996)
logP 2.82			EXP	HANSCH,C & LEO,AJ (1985)
VP 8.50E-004	mm Hg	25	EST	NEELY,WB & BLAU,GE (1985)
DC	pKa			
HL 2.82E-007	atm m3/mol	25	EST	MEYLAN,WM & HOWARD,PH (1991)
OH 1.16E-011	cm3/molc sec	25	EST	MEYLAN,WM & HOWARD,PH (1993)

056400-36-3 — 3,6-DIMEO TETRACHLOROCYCLOHEXANE

Formula: $C_8H_{12}Cl_4O_2$
Mol Weight: 282.00
MP (deg C):
FP (deg C):
BP (deg C):
BP pressure (mm Hg):

Property/Value	Units	Temp	Data Type	Reference
WS 1.04E+002	mg/L	25	EST	MEYLAN,WM ET AL. (1996)
logP 3.15			EXP	HANSCH,C & LEO,AJ (1985)
VP 8.50E-004	mm Hg	25	EST	NEELY,WB & BLAU,GE (1985)
DC	pKa			
HL 2.82E-007	atm m3/mol	25	EST	MEYLAN,WM & HOWARD,PH (1991)
OH 4.65E-012	cm3/molc sec	25	EST	MEYLAN,WM & HOWARD,PH (1993)

056400-43-2 — 1-HOME PENTACHLOROCYCLOHEXANE

Formula: $C_7H_9Cl_5O$
Mol Weight: 286.41
MP (deg C):
FP (deg C):
BP (deg C):
BP pressure (mm Hg):

Property/Value	Units	Temp	Data Type	Reference
WS 1.47E+002	mg/L	25	EST	MEYLAN,WM ET AL. (1996)
logP 2.94			EXP	HANSCH,C & LEO,AJ (1985)
VP 2.67E-006	mm Hg	25	EST	NEELY,WB & BLAU,GE (1985)
DC	pKa			
HL 3.53E-008	atm m3/mol	25	EST	MEYLAN,WM & HOWARD,PH (1991)
OH 5.41E-012	cm3/molc sec	25	EST	MEYLAN,WM & HOWARD,PH (1993)

CAS #: 056406-50-9 — M-NITROBENZAMIDINE HYDROCHLORIDE

Formula: $C_7H_7N_3O_2$

Mol Weight: 165.15

MP (deg C): 199-201

FP (deg C):

BP (deg C):

BP pressure (mm Hg):

Property/Value	Units	Temp	Data Type	Reference
WS 6.41E+003	mg/L	25	EST	MEYLAN,WM ET AL. (1996)
logP 0.72			EXP	HANSCH,C ET AL. (1995)
VP 1.11E-004	mm Hg	25	EST	NEELY,WB & BLAU,GE (1985)
DC	pKa			
HL 7.24E-011	atm m3/mol	25	EST	MEYLAN,WM & HOWARD,PH (1991)
OH 2.14E-011	cm3/molc sec	25	EST	MEYLAN,WM & HOWARD,PH (1993)

CAS #: 056421-31-9 — 2-METHOXYPENTACHLOROCYCLOHEXANE

Formula: $C_7H_9Cl_5O$

Mol Weight: 286.41

MP (deg C):

FP (deg C):

BP (deg C):

BP pressure (mm Hg):

Property/Value	Units	Temp	Data Type	Reference
WS 3.06E+001	mg/L	25	EST	MEYLAN,WM ET AL. (1996)
logP 3.14			EXP	HANSCH,C & LEO,AJ (1985)
VP 8.16E-004	mm Hg	25	EST	NEELY,WB & BLAU,GE (1985)
DC	pKa			
HL 8.50E-006	atm m3/mol	25	EST	MEYLAN,WM & HOWARD,PH (1991)
OH 2.61E-012	cm3/molc sec	25	EST	MEYLAN,WM & HOWARD,PH (1993)

CAS #: 056421-32-0 — 2-ETHOXYPENTACHLOROCYCLOHEXANE

Formula: $C_8H_{11}Cl_5O$

Mol Weight: 300.44

MP (deg C):

FP (deg C):

BP (deg C):

BP pressure (mm Hg):

Property/Value	Units	Temp	Data Type	Reference
WS 4.97E+000	mg/L	25	EST	MEYLAN,WM ET AL. (1996)
logP 3.69			EXP	HANSCH,C & LEO,AJ (1985)
VP 3.09E-004	mm Hg	25	EST	NEELY,WB & BLAU,GE (1985)
DC	pKa			
HL 1.13E-005	atm m3/mol	25	EST	MEYLAN,WM & HOWARD,PH (1991)
OH 3.88E-012	cm3/molc sec	25	EST	MEYLAN,WM & HOWARD,PH (1993)

CAS #: 056421-35-3 — 3-ETHOXYPENTACHLOROCYCLOHEXANE

Formula: $C_8H_{11}Cl_5O$

Mol Weight: 300.44

MP (deg C):

FP (deg C):

BP (deg C):

BP pressure (mm Hg):

Property/Value	Units	Temp	Data Type	Reference
WS 4.97E+000	mg/L	25	EST	MEYLAN,WM ET AL. (1996)
logP 3.97			EXP	HANSCH,C & LEO,AJ (1985)
VP 3.09E-004	mm Hg	25	EST	NEELY,WB & BLAU,GE (1985)
DC	pKa			
HL 1.13E-005	atm m3/mol	25	EST	MEYLAN,WM & HOWARD,PH (1991)
OH 3.88E-012	cm3/molc sec	25	EST	MEYLAN,WM & HOWARD,PH (1993)

CAS #: 056421-36-4 — 1(CF3-MEO)PENTACHLOROCYCLOHEXANE

Formula: $C_8H_8Cl_5F_3O$

Mol Weight: 354.41

MP (deg C):

FP (deg C):

BP (deg C):

BP pressure (mm Hg):

Property/Value	Units	Temp	Data Type	Reference
WS 1.99E+000	mg/L	25	EST	MEYLAN,WM ET AL. (1996)
logP 4.06			EXP	HANSCH,C & LEO,AJ (1985)
VP 6.39E-004	mm Hg	25	EST	NEELY,WB & BLAU,GE (1985)
DC	pKa			
HL 8.89E-005	atm m3/mol	25	EST	MEYLAN,WM & HOWARD,PH (1991)
OH 2.25E-012	cm3/molc sec	25	EST	MEYLAN,WM & HOWARD,PH (1993)

CAS #: 056421-44-4 — 1-METHYLPENTACHLOROCYCLOHEXANE

Formula: $C_7H_9Cl_5$

Mol Weight: 270.41

MP (deg C):

FP (deg C):

BP (deg C):

BP pressure (mm Hg):

Property/Value	Units	Temp	Data Type	Reference
WS 6.45E+000	mg/L	25	EST	MEYLAN,WM ET AL. (1996)
logP 4.04			EXP	HANSCH,C & LEO,AJ (1985)
VP 2.62E-003	mm Hg	25	EST	NEELY,WB & BLAU,GE (1985)
DC	pKa			
HL 9.64E-004	atm m3/mol	25	EST	MEYLAN,WM & HOWARD,PH (1991)
OH 1.36E-012	cm3/molc sec	25	EST	MEYLAN,WM & HOWARD,PH (1993)

CAS #: 056421-45-5 — 1,4-DIMETHYLTETRACHLOROCYCLOHEXANE

Formula: $C_8H_{12}Cl_4$

Mol Weight: 250.00

MP (deg C):

FP (deg C):

BP (deg C):

BP pressure (mm Hg):

Property/Value	Units	Temp	Data Type	Reference
WS 4.13E+000	mg/L	25	EST	MEYLAN,WM ET AL. (1996)
logP 4.40			EXP	HANSCH,C & LEO,AJ (1985)
VP 9.16E-003	mm Hg	25	EST	NEELY,WB & BLAU,GE (1985)
DC	pKa			
HL 3.63E-003	atm m3/mol	25	EST	MEYLAN,WM & HOWARD,PH (1991)
OH 2.14E-012	cm3/molc sec	25	EST	MEYLAN,WM & HOWARD,PH (1993)

CAS #: 056423-63-3 — 2-BROMOPYRAZINE

Formula: $C_4H_3BrN_2$

Mol Weight: 158.99

MP (deg C):

FP (deg C):

BP (deg C):

BP pressure (mm Hg):

Property/Value	Units	Temp	Data Type	Reference
WS 1.11E+004	mg/L	25	EST	MEYLAN,WM ET AL. (1996)
logP 0.93			EXP	YAMAGAMI,C ET AL. (1990A)
VP 8.67E-001	mm Hg	25	EST	NEELY,WB & BLAU,GE (1985)
DC	pKa			
HL 1.16E-006	atm m3/mol	25	EST	MEYLAN,WM & HOWARD,PH (1991)
OH 1.45E-013	cm3/molc sec	25	EST	MEYLAN,WM & HOWARD,PH (1993)

056425-91-3 — FLURPRIMIDOL

CAS #:	056425-91-3		FLURPRIMIDOL	
Formula:	$C_{15}H_{15}F_3N_2O_2$			
Mol Weight:	312.29			
MP (deg C):	94-96	FP (deg C):		
BP (deg C):				
BP pressure (mm Hg):				

Property/Value	Units	Temp	Data Type	Reference
WS 4.73E+001	mg/L	25	EST	MEYLAN,WM ET AL. (1996)
logP 3.34			EXP	TOMLIN,C (1994)
VP 2.51E-007	mm Hg	25	EST	NEELY,WB & BLAU,GE (1985)
DC	pKa			
HL 1.04E-011	atm m3/mol	25	EST	MEYLAN,WM & HOWARD,PH (1991)
OH 3.42E-011	cm3/molc sec	25	EST	MEYLAN,WM & HOWARD,PH (1993)

056462-00-1 — N-METHYL-2,6-DICHLOROANILINE

CAS #:	056462-00-1		N-METHYL-2,6-DICHLOROANILINE	
Formula:	$C_7H_7Cl_2N$			
Mol Weight:	176.05			
MP (deg C):		FP (deg C):		
BP (deg C):				
BP pressure (mm Hg):				

Property/Value	Units	Temp	Data Type	Reference
WS 1.88E+002	mg/L	25	EST	MEYLAN,WM ET AL. (1996)
logP 2.91			EXP	HANSCH,C ET AL. (1995)
VP 5.01E-002	mm Hg	25	EST	NEELY,WB & BLAU,GE (1985)
DC	pKa			
HL 2.30E-006	atm m3/mol	25	EST	MEYLAN,WM & HOWARD,PH (1991)
OH 4.90E-012	cm3/molc sec	25	EST	MEYLAN,WM & HOWARD,PH (1993)

056507-37-0 — DIKETOMETRIBUZIN

CAS #:	056507-37-0		DIKETOMETRIBUZIN	
Formula:	$C_7H_{12}N_4O_2$			
Mol Weight:	184.20			
MP (deg C):		FP (deg C):		
BP (deg C):				
BP pressure (mm Hg):				

Property/Value	Units	Temp	Data Type	Reference
WS 1.65E+003	mg/L	25	EST	MEYLAN,WM ET AL. (1996)
logP 0.23			EST	MEYLAN,WM & HOWARD,PH (1995)
VP 1.93E-008	mm Hg	25	EST	NEELY,WB & BLAU,GE (1985)
DC	pKa			
HL 2.48E-011	atm m3/mol	25	EST	MEYLAN,WM & HOWARD,PH (1991)
OH 1.65E-011	cm3/molc sec	25	EST	MEYLAN,WM & HOWARD,PH (1993)

056514-55-7 — 2-(2,5-DICL PH IMINO)IMIDAZOLIDINE

CAS #:	056514-55-7		2-(2,5-DICL PH IMINO)IMIDAZOLIDINE	
Formula:	$C_9H_9Cl_2N_3$			
Mol Weight:	230.10			
MP (deg C):		FP (deg C):		
BP (deg C):				
BP pressure (mm Hg):				

Property/Value	Units	Temp	Data Type	Reference
WS 5.61E+003	mg/L	25	EST	MEYLAN,WM ET AL. (1996)
logP 2.04			EXP	HANSCH,C & LEO,AJ (1985)
VP 2.28E-005	mm Hg	25	EST	NEELY,WB & BLAU,GE (1985)
DC	pKa			
HL 1.51E-011	atm m3/mol	25	EST	MEYLAN,WM & HOWARD,PH (1991)
OH 1.44E-010	cm3/molc sec	25	EST	MEYLAN,WM & HOWARD,PH (1993)

056518-41-3 — 24-NH2PYRIMIDINE,5(3,5MEO-4-BR)BENZYL

CAS #:	056518-41-3		24-NH2PYRIMIDINE,5(3,5MEO-4-BR)BENZYL	
Formula:	$C_{13}H_{15}BrN_4O_2$			
Mol Weight:	339.20			
MP (deg C):	225-228	FP (deg C):		
BP (deg C):				
BP pressure (mm Hg):				

Property/Value	Units	Temp	Data Type	Reference
WS 1.38E+002	mg/L	25	EST	MEYLAN,WM ET AL. (1996)
logP 2.01			EXP	HANSCH,C & LEO,AJ (1985)
VP 6.04E-009	mm Hg	25	EST	NEELY,WB & BLAU,GE (1985)
DC	pKa			
HL 1.61E-013	atm m3/mol	25	EST	MEYLAN,WM & HOWARD,PH (1991)
OH 1.73E-010	cm3/molc sec	25	EST	MEYLAN,WM & HOWARD,PH (1993)

056518-45-7 — 24-NH2PYRIMIDINE,5(3,5MEO-4-CL)BENZYL

CAS #:	056518-45-7		24-NH2PYRIMIDINE,5(3,5MEO-4-CL)BENZYL	
Formula:	$C_{13}H_{15}ClN_4O_2$			
Mol Weight:	294.74			
MP (deg C):		FP (deg C):		
BP (deg C):				
BP pressure (mm Hg):				

Property/Value	Units	Temp	Data Type	Reference
WS 3.33E+002	mg/L	25	EST	MEYLAN,WM ET AL. (1996)
logP 1.87			EXP	HANSCH,C & LEO,AJ (1985)
VP 1.39E-008	mm Hg	25	EST	NEELY,WB & BLAU,GE (1985)
DC	pKa			
HL 2.99E-013	atm m3/mol	25	EST	MEYLAN,WM & HOWARD,PH (1991)
OH 1.76E-010	cm3/molc sec	25	EST	MEYLAN,WM & HOWARD,PH (1993)

056518-58-2 — 2,4-NH2PYRIMIDINE,5(3,5MEO-4-I)BENZYL

CAS #:	056518-58-2		2,4-NH2PYRIMIDINE,5(3,5MEO-4-I)BENZYL	
Formula:	$C_{13}H_{15}IN_4O_2$			
Mol Weight:	386.19			
MP (deg C):		FP (deg C):		
BP (deg C):				
BP pressure (mm Hg):				

Property/Value	Units	Temp	Data Type	Reference
WS 3.32E+001	mg/L	25	EST	MEYLAN,WM ET AL. (1996)
logP 2.40			EXP	HANSCH,C & LEO,AJ (1985)
VP 1.68E-009	mm Hg	25	EST	NEELY,WB & BLAU,GE (1985)
DC	pKa			
HL 9.34E-014	atm m3/mol	25	EST	MEYLAN,WM & HOWARD,PH (1991)
OH 1.99E-010	cm3/molc sec	25	EST	MEYLAN,WM & HOWARD,PH (1993)

056563-17-8 — 1(2H)-PYRIMIDINECARBOXAMIDE, 5-FLUORO-3,4-DICHLO

CAS #:	056563-17-8		1(2H)-PYRIMIDINECARBOXAMIDE, 5-FLUORO-3,4-DICHLO	
Formula:	$C_{11}H_8FN_3O_3$			
Mol Weight:	249.20			
MP (deg C):		FP (deg C):		
BP (deg C):				
BP pressure (mm Hg):				

Property/Value	Units	Temp	Data Type	Reference
WS 1.25E+003	mg/L	25	EST	MEYLAN,WM ET AL. (1996)
logP 1.50			EXP	BUUR,A & BUNDGAARD,H (1985)
VP 4.25E-012	mm Hg	25	EST	NEELY,WB & BLAU,GE (1985)
DC	pKa			
HL 5.57E-016	atm m3/mol	25	EST	MEYLAN,WM & HOWARD,PH (1991)
OH 4.85E-011	cm3/molc sec	25	EST	MEYLAN,WM & HOWARD,PH (1993)

CAS #: 056563-18-9 — 1(2H)-PYRIMIDINECARBOXAMIDE, 5-FLUORO-3,4-DIHYDR

Formula: $C_6H_6FN_3O_3$

Mol Weight: 187.13

MP (deg C):　FP (deg C):

BP (deg C):

BP pressure (mm Hg):

Property/Value	Units	Temp	Data Type	Reference
WS 7.53E+004	mg/L	25	EST	MEYLAN,WM ET AL. (1996)
logP -0.20			EXP	SANGSTER,J (1993)
VP 7.65E-010	mm Hg	25	EST	NEELY,WB & BLAU,GE (1985)
DC	pKa			
HL 2.22E-015	atm m3/mol	25	EST	MEYLAN,WM & HOWARD,PH (1991)
OH 7.39E-012	cm3/molc sec	25	EST	MEYLAN,WM & HOWARD,PH (1993)

CAS #: 056610-79-8 — 1H-TETRAZOLE, 1-METHYL-5-[(PHENYLMETHYL)THIO]-

Formula: $C_9H_{10}N_4S$

Mol Weight: 206.27

MP (deg C):　FP (deg C):

BP (deg C):

BP pressure (mm Hg):

Property/Value	Units	Temp	Data Type	Reference
WS 1.65E+003	mg/L	25	EST	MEYLAN,WM ET AL. (1996)
logP 1.63			EXP	HANSCH,C ET AL. (1995)
VP 3.61E-005	mm Hg	25	EST	NEELY,WB & BLAU,GE (1985)
DC	pKa			
HL 3.07E-008	atm m3/mol	25	EST	MEYLAN,WM & HOWARD,PH (1991)
OH 1.29E-011	cm3/molc sec	25	EST	MEYLAN,WM & HOWARD,PH (1993)

CAS #: 056656-96-3 — PROPANEDINITRILE, [(5-BROMO-2-FURANYL)METHYLENE]

Formula: $C_8H_3BrN_2O$

Mol Weight: 223.03

MP (deg C):　FP (deg C):

BP (deg C):

BP pressure (mm Hg):

Property/Value	Units	Temp	Data Type	Reference
WS 2.58E+002	mg/L	25	EST	MEYLAN,WM ET AL. (1996)
logP 2.16			EXP	BALAZ,S ET AL. (1985)
VP 4.61E-005	mm Hg	25	EST	NEELY,WB & BLAU,GE (1985)
DC	pKa			
HL 5.46E-009	atm m3/mol	25	EST	MEYLAN,WM & HOWARD,PH (1991)
OH 2.62E-011	cm3/molc sec	25	EST	MEYLAN,WM & HOWARD,PH (1993)

CAS #: 056656-97-4 — 2-PROPENOIC ACID, 3-(5-BROMO-2-FURANYL)-2-CYANO-

Formula: $C_9H_6BrNO_3$

Mol Weight: 256.06

MP (deg C):　FP (deg C):

BP (deg C):

BP pressure (mm Hg):

Property/Value	Units	Temp	Data Type	Reference
WS 1.04E+002	mg/L	25	EST	MEYLAN,WM ET AL. (1996)
logP 2.41			EXP	BALAZ,S ET AL. (1985)
VP 9.24E-005	mm Hg	25	EST	NEELY,WB & BLAU,GE (1985)
DC	pKa			
HL 3.67E-009	atm m3/mol	25	EST	MEYLAN,WM & HOWARD,PH (1991)
OH 2.91E-011	cm3/molc sec	25	EST	MEYLAN,WM & HOWARD,PH (1993)

CAS #: 056682-87-2 — 1,3,5-TRIS(DICHLOROMETHYL)BENZENE

Formula: $C_9H_6Cl_6$

Mol Weight: 326.87

MP (deg C):　FP (deg C):

BP (deg C):

BP pressure (mm Hg):

Property/Value	Units	Temp	Data Type	Reference
WS 5.17E-001	mg/L	25	EST	MEYLAN,WM ET AL. (1996)
logP 4.94			EST	MEYLAN,WM & HOWARD,PH (1995)
VP 2.82E-005	mm Hg	25	EST	NEELY,WB & BLAU,GE (1985)
DC	pKa			
HL 1.38E-005	atm m3/mol	25	EST	MEYLAN,WM & HOWARD,PH (1991)
OH 2.98E-012	cm3/molc sec	25	EST	MEYLAN,WM & HOWARD,PH (1993)

CAS #: 056683-54-6 — 11-HEXADECEN-1-OL (Z)

Formula: $C_{16}H_{32}O$

Mol Weight: 240.43

MP (deg C):　FP (deg C):

BP (deg C):

BP pressure (mm Hg):

Property/Value	Units	Temp	Data Type	Reference
WS 2.34E-001	mg/L	25	EST	MEYLAN,WM ET AL. (1996)
logP 6.52			EST	MEYLAN,WM & HOWARD,PH (1995)
VP 5.75E-006	mm Hg	25	EST	NEELY,WB & BLAU,GE (1985)
DC	pKa			
HL 2.63E-004	atm m3/mol	25	EST	MEYLAN,WM & HOWARD,PH (1991)
OH 7.69E-011	cm3/molc sec	25	EST	MEYLAN,WM & HOWARD,PH (1993)

CAS #: 056689-45-3 — JOSAMYCIN

Formula: $C_{42}H_{69}NO_{15}$

Mol Weight: 828.02

MP (deg C):　FP (deg C):

BP (deg C):

BP pressure (mm Hg):

Property/Value	Units	Temp	Data Type	Reference
WS 4.50E-001	mg/L	25	EST	MEYLAN,WM ET AL. (1996)
logP 2.39			EXP	HANSCH,C ET AL. (1995)
VP 1.41E-028	mm Hg	25	EST	NEELY,WB & BLAU,GE (1985)
DC	pKa			
HL 3.49E-031	atm m3/mol	25	EST	MEYLAN,WM & HOWARD,PH (1991)
OH 4.30E-010	cm3/molc sec	25	EST	MEYLAN,WM & HOWARD,PH (1993)

CAS #: 056711-06-9 — ISO-LEUCIN-AMIDE, N-ACETYL

Formula: $C_8H_{16}N_2O_2$

Mol Weight: 172.23

MP (deg C):　FP (deg C):

BP (deg C):

BP pressure (mm Hg):

Property/Value	Units	Temp	Data Type	Reference
WS 3.12E+003	mg/L	25	EST	MEYLAN,WM ET AL. (1996)
logP -0.03			EXP	HANSCH,C & LEO,AJ (1985)
VP 2.18E-007	mm Hg	25	EST	NEELY,WB & BLAU,GE (1985)
DC	pKa			
HL 3.66E-010	atm m3/mol	25	EST	MEYLAN,WM & HOWARD,PH (1991)
OH 5.40E-011	cm3/molc sec	25	EST	MEYLAN,WM & HOWARD,PH (1993)

CAS #:	056875-82-2			2H-1-BENZOPYRAN-2-ONE, 3-(ACETYLOXY)-1-[2-(ACETY

Formula: $C_{21}H_{19}ClN_2O_5$

Mol Weight: 414.85

MP (deg C):   FP (deg C):

BP (deg C):

BP pressure (mm Hg):

Property/Value	Units	Temp	Data Type	Reference
WS 1.13E+001	mg/L	25	EST	MEYLAN,WM ET AL. (1996)
logP 2.74			EXP	SANGSTER,J (1993)
VP 4.82E-011	mm Hg	25	EST	NEELY,WB & BLAU,GE (1985)
DC	pKa			
HL 3.87E-012	atm m3/mol	25	EST	MEYLAN,WM & HOWARD,PH (1991)
OH 1.77E-011	cm3/molc sec	25	EST	MEYLAN,WM & HOWARD,PH (1993)

CAS #:	056875-83-3			2H-1,4-BENZODIAZEPIN-2-ONE, 3-(ACETYLOXY)-1-[3-(

Formula: $C_{22}H_{21}ClN_2O_5$

Mol Weight: 428.88

MP (deg C):   FP (deg C):

BP (deg C):

BP pressure (mm Hg):

Property/Value	Units	Temp	Data Type	Reference
WS 4.74E+000	mg/L	25	EST	MEYLAN,WM ET AL. (1996)
logP 3.08			EXP	SANGSTER,J (1993)
VP 2.10E-011	mm Hg	25	EST	NEELY,WB & BLAU,GE (1985)
DC	pKa			
HL 5.14E-012	atm m3/mol	25	EST	MEYLAN,WM & HOWARD,PH (1991)
OH 1.91E-011	cm3/molc sec	25	EST	MEYLAN,WM & HOWARD,PH (1993)

CAS #:	056892-91-2			CYCLOPROPANECARBOXYLIC ACID, 2-(PHENYLETHYNYL)-,

Formula: $C_{12}H_9NaO_2$

Mol Weight: 208.19

MP (deg C):   FP (deg C):

BP (deg C):

BP pressure (mm Hg):

Property/Value	Units	Temp	Data Type	Reference
WS 1.32E+005	mg/L	25	EST	MEYLAN,WM ET AL. (1996)
logP -0.61			EXP	HANSCH,C ET AL. (1995)
VP 3.28E-011	mm Hg	25	EST	NEELY,WB & BLAU,GE (1985)
DC	pKa			
HL	atm m3/mol			
OH 2.82E-011	cm3/molc sec	25	EST	MEYLAN,WM & HOWARD,PH (1993)

CAS #:	056933-31-4			2-METHOXYTETRACHLOROCYCLOHEXANE

Formula: $C_7H_{10}Cl_4O$

Mol Weight: 251.97

MP (deg C):   FP (deg C):

BP (deg C):

BP pressure (mm Hg):

Property/Value	Units	Temp	Data Type	Reference
WS 6.45E+001	mg/L	25	EST	MEYLAN,WM ET AL. (1996)
logP 2.99			EXP	HANSCH,C & LEO,AJ (1985)
VP 5.72E-003	mm Hg	25	EST	NEELY,WB & BLAU,GE (1985)
DC	pKa			
HL 2.41E-005	atm m3/mol	25	EST	MEYLAN,WM & HOWARD,PH (1991)
OH 6.63E-012	cm3/molc sec	25	EST	MEYLAN,WM & HOWARD,PH (1993)

CAS #:	056933-60-9			BETAMETHASONE-21-BUTYRATE

Formula: $C_{26}H_{35}FO_6$

Mol Weight: 462.56

MP (deg C):   FP (deg C):

BP (deg C):

BP pressure (mm Hg):

Property/Value	Units	Temp	Data Type	Reference
WS 1.15E+000	mg/L	25	EST	MEYLAN,WM ET AL. (1996)
logP 3.55			EXP	HANSCH,C & LEO,AJ (1985)
VP 2.27E-014	mm Hg	25	EST	NEELY,WB & BLAU,GE (1985)
DC	pKa			
HL 1.36E-011	atm m3/mol	25	EST	MEYLAN,WM & HOWARD,PH (1991)
OH 6.94E-011	cm3/molc sec	25	EST	MEYLAN,WM & HOWARD,PH (1993)

CAS #:	056961-20-7			3,4,5-TRICHLOROCATECHOL

Formula: $C_6H_3Cl_3O_2$

Mol Weight: 213.45

MP (deg C):   FP (deg C):

BP (deg C):

BP pressure (mm Hg):

Property/Value	Units	Temp	Data Type	Reference
WS 9.62E+001	mg/L	25	EST	MEYLAN,WM ET AL. (1996)
logP 3.71			EXP	SANGSTER,J (1994)
VP 2.02E-005	mm Hg	25	EST	NEELY,WB & BLAU,GE (1985)
DC	pKa			
HL 2.37E-011	atm m3/mol	25	EST	MEYLAN,WM & HOWARD,PH (1991)
OH 3.59E-012	cm3/molc sec	25	EST	MEYLAN,WM & HOWARD,PH (1993)

CAS #:	056980-93-9			UREA, N'-[3-ACETYL-4-[3-[(1,1-DIMETHYLETHYL)AMIN

Formula: $C_{20}H_{33}N_3O_4$

Mol Weight: 379.50

MP (deg C): 110-112   FP (deg C):

BP (deg C):

BP pressure (mm Hg):

Property/Value	Units	Temp	Data Type	Reference
WS 9.39E+001	mg/L	25	EST	MEYLAN,WM ET AL. (1996)
logP 1.92			EXP	SANGSTER,J (1994)
VP 3.11E-013	mm Hg	25	EST	NEELY,WB & BLAU,GE (1985)
DC	pKa			
HL 6.27E-021	atm m3/mol	25	EST	MEYLAN,WM & HOWARD,PH (1991)
OH 1.33E-010	cm3/molc sec	25	EST	MEYLAN,WM & HOWARD,PH (1993)

CAS #:	056986-36-8			N-NITROSO-ACETOXYMETHYL-N-BUTYLAMINE

Formula: $C_7H_{14}N_2O_3$

Mol Weight: 174.20

MP (deg C):   FP (deg C):

BP (deg C):

BP pressure (mm Hg):

Property/Value	Units	Temp	Data Type	Reference
WS 2.93E+003	mg/L	25	EST	MEYLAN,WM ET AL. (1996)
logP 1.52			EST	MEYLAN,WM & HOWARD,PH (1995)
VP 3.39E-003	mm Hg	25	EST	NEELY,WB & BLAU,GE (1985)
DC	pKa			
HL 2.29E-007	atm m3/mol	25	EST	MEYLAN,WM & HOWARD,PH (1991)
OH 2.74E-011	cm3/molc sec	25	EST	MEYLAN,WM & HOWARD,PH (1993)

CAS #: 056994-23-1				3-PENTACHLOROCYCLOHEXANE
Formula: C₆H₇Cl₅				

CAS #:	056994-23-1		3-PENTACHLOROCYCLOHEXANE
Formula:	$C_6H_7Cl_5$		
Mol Weight:	256.39		
MP (deg C):		FP (deg C):	
BP (deg C):			
BP pressure (mm Hg):			

Property/Value	Units	Temp	Data Type	Reference
WS 2.89E+001	mg/L	25	EST	MEYLAN,WM ET AL. (1996)
logP 3.37			EXP	HANSCH,C & LEO,AJ (1985)
VP 5.44E-003	mm Hg	25	EST	NEELY,WB & BLAU,GE (1985)
DC	pKa			
HL 7.26E-004	atm m3/mol	25	EST	MEYLAN,WM & HOWARD,PH (1991)
OH 1.06E-012	cm3/molc sec	25	EST	MEYLAN,WM & HOWARD,PH (1993)

CAS #:	056994-25-3		TRICHLOROCYCLOHEXENE (345)
Formula:	$C_6H_7Cl_3$		
Mol Weight:	185.48		
MP (deg C):		FP (deg C):	
BP (deg C):			
BP pressure (mm Hg):			

Property/Value	Units	Temp	Data Type	Reference
WS 1.94E+002	mg/L	25	EST	MEYLAN,WM ET AL. (1996)
logP 2.84			EXP	HANSCH,C & LEO,AJ (1985)
VP 1.84E-001	mm Hg	25	EST	NEELY,WB & BLAU,GE (1985)
DC	pKa			
HL 5.15E-003	atm m3/mol	25	EST	MEYLAN,WM & HOWARD,PH (1991)
OH 4.39E-011	cm3/molc sec	25	EST	MEYLAN,WM & HOWARD,PH (1993)

CAS #:	057018-04-9		TOLCLOFOS-METHYL
Formula:	$C_9H_{11}Cl_2O_3PS$		
Mol Weight:	301.13		
MP (deg C):		FP (deg C):	
BP (deg C):			
BP pressure (mm Hg):			

Property/Value	Units	Temp	Data Type	Reference
WS 1.54E+000	mg/L	25	EST	MEYLAN,WM ET AL. (1996)
logP 4.56			EXP	TOMLIN,C (1994)
VP 4.30E-004	mm Hg	25	EXP	AUGUSTIJN-BECKERS,PWM ET AL. (1994)
DC	pKa			
HL 2.58E-005	atm m3/mol	25	EST	MEYLAN,WM & HOWARD,PH (1991)
OH 6.06E-011	cm3/molc sec	25	EST	MEYLAN,WM & HOWARD,PH (1993)

CAS #:	057025-76-0		MC-15608 [2,4'-DICL-4-CF3-3'-CO2ME-DIPH ETHER]
Formula:	$C_{15}H_9Cl_2F_3O_3$		
Mol Weight:	365.14		
MP (deg C):		FP (deg C):	
BP (deg C):			
BP pressure (mm Hg):			

Property/Value	Units	Temp	Data Type	Reference
WS 2.49E-001	mg/L	25	EST	MEYLAN,WM ET AL. (1996)
logP 5.04			EXP	NANDIHALLI,UB ET AL. (1993)
VP 3.15E-006	mm Hg	25	EST	NEELY,WB & BLAU,GE (1985)
DC	pKa			
HL 3.62E-006	atm m3/mol	25	EST	MEYLAN,WM & HOWARD,PH (1991)
OH 9.47E-013	cm3/molc sec	25	EST	MEYLAN,WM & HOWARD,PH (1993)

CAS #:	057057-83-7		3,4,5-TRICHLORO-2-METHOXYPHENOL
Formula:	$C_7H_5Cl_3O_2$		
Mol Weight:	227.48		
MP (deg C):		FP (deg C):	
BP (deg C):			
BP pressure (mm Hg):			

Property/Value	Units	Temp	Data Type	Reference
WS 1.89E+001	mg/L	25	EST	MEYLAN,WM ET AL. (1996)
logP 3.77			EXP	HANSCH,C & LEO,AJ (1985)
VP 4.80E-003	mm Hg	25	EXP	BIDLEMAN,TF & RENBERG,L (1985)
DC	pKa			
HL 1.35E-008	atm m3/mol	25	EST	MEYLAN,WM & HOWARD,PH (1991)
OH 5.13E-012	cm3/molc sec	25	EST	MEYLAN,WM & HOWARD,PH (1993)

CAS #:	057058-33-0		N-((4-CHLOROPHENYL)METHYL)ACETAMIDE
Formula:	$C_9H_{10}ClNO$		
Mol Weight:	183.64		
MP (deg C):		FP (deg C):	
BP (deg C):			
BP pressure (mm Hg):			

Property/Value	Units	Temp	Data Type	Reference
WS 2.04E+003	mg/L	25	EST	MEYLAN,WM ET AL. (1996)
logP 1.66			EST	MEYLAN,WM & HOWARD,PH (1995)
VP 3.95E-005	mm Hg	25	EST	NEELY,WB & BLAU,GE (1985)
DC	pKa			
HL 1.47E-009	atm m3/mol	25	EST	MEYLAN,WM & HOWARD,PH (1991)
OH 1.11E-011	cm3/molc sec	25	EST	MEYLAN,WM & HOWARD,PH (1993)

CAS #:	057078-98-5		N-MALEOYL-4-AMINOBUTYRIC ACID
Formula:	$C_8H_9NO_4$		
Mol Weight:	183.17		
MP (deg C):		FP (deg C):	
BP (deg C):			
BP pressure (mm Hg):			

Property/Value	Units	Temp	Data Type	Reference
WS 2.42E+004	mg/L	25	EST	MEYLAN,WM ET AL. (1996)
logP 0.40			EXP	HANSCH,C & LEO,AJ (1985)
VP 7.37E-008	mm Hg	25	EST	NEELY,WB & BLAU,GE (1985)
DC	pKa			
HL 1.82E-013	atm m3/mol	25	EST	MEYLAN,WM & HOWARD,PH (1991)
OH 2.48E-011	cm3/molc sec	25	EST	MEYLAN,WM & HOWARD,PH (1993)

CAS #:	057078-99-6		N-MALEOYL-5-AMINOPENTANOIC ACID
Formula:	$C_9H_{11}NO_4$		
Mol Weight:	197.19		
MP (deg C):		FP (deg C):	
BP (deg C):			
BP pressure (mm Hg):			

Property/Value	Units	Temp	Data Type	Reference
WS 1.39E+004	mg/L	25	EST	MEYLAN,WM ET AL. (1996)
logP 0.60			EXP	HANSCH,C & LEO,AJ (1985)
VP 3.31E-008	mm Hg	25	EST	NEELY,WB & BLAU,GE (1985)
DC	pKa			
HL 2.42E-013	atm m3/mol	25	EST	MEYLAN,WM & HOWARD,PH (1991)
OH 2.62E-011	cm3/molc sec	25	EST	MEYLAN,WM & HOWARD,PH (1993)

CAS #: 057079-00-2				N-MALEOYL-8-AMINOOCTANOIC ACID
Formula: $C_{12}H_{17}NO_4$				
Mol Weight: 239.27				
MP (deg C):		FP (deg C):		
BP (deg C):				
BP pressure (mm Hg):				

Property/Value	Units	Temp	Data Type	Reference
WS 5.41E+002	mg/L	25	EST	MEYLAN,WM ET AL. (1996)
logP 1.99			EXP	HANSCH,C & LEO,AJ (1985)
VP 2.93E-009	mm Hg	25	EST	NEELY,WB & BLAU,GE (1985)
DC	pKa			
HL 5.66E-013	atm m3/mol	25	EST	MEYLAN,WM & HOWARD,PH (1991)
OH 3.04E-011	cm3/molc sec	25	EST	MEYLAN,WM & HOWARD,PH (1993)

CAS #: 057079-05-7				N-MALEOYL-5-NH2PROPIONIC ACID,ET EST
Formula: $C_9H_{11}NO_4$				
Mol Weight: 197.19				
MP (deg C):		FP (deg C):		
BP (deg C):				
BP pressure (mm Hg):				

Property/Value	Units	Temp	Data Type	Reference
WS 2.02E+004	mg/L	25	EST	MEYLAN,WM ET AL. (1996)
logP 0.41			EXP	HANSCH,C & LEO,AJ (1985)
VP 8.76E-007	mm Hg	25	EST	NEELY,WB & BLAU,GE (1985)
DC	pKa			
HL 5.83E-011	atm m3/mol	25	EST	MEYLAN,WM & HOWARD,PH (1991)
OH 2.45E-011	cm3/molc sec	25	EST	MEYLAN,WM & HOWARD,PH (1993)

CAS #: 057079-07-9				N-MALEOYL-6-NH2HEXANOIC ACID,ET ESTER
Formula: $C_{12}H_{17}NO_4$				
Mol Weight: 239.27				
MP (deg C):		FP (deg C):		
BP (deg C):				
BP pressure (mm Hg):				

Property/Value	Units	Temp	Data Type	Reference
WS 1.42E+003	mg/L	25	EST	MEYLAN,WM ET AL. (1996)
logP 1.50			EXP	HANSCH,C & LEO,AJ (1985)
VP 6.82E-008	mm Hg	25	EST	NEELY,WB & BLAU,GE (1985)
DC	pKa			
HL 1.37E-010	atm m3/mol	25	EST	MEYLAN,WM & HOWARD,PH (1991)
OH 2.87E-011	cm3/molc sec	25	EST	MEYLAN,WM & HOWARD,PH (1993)

CAS #: 057101-48-1				2(2,6-DICL-4-OHPH IMINO)IMIDAZOLIDIN
Formula: $C_9H_9Cl_2N_3O$				
Mol Weight: 246.10				
MP (deg C):		FP (deg C):		
BP (deg C):				
BP pressure (mm Hg):				

Property/Value	Units	Temp	Data Type	Reference
WS 4.85E+004	mg/L	25	EST	MEYLAN,WM ET AL. (1996)
logP 1.52			EXP	HANSCH,C & LEO,AJ (1985)
VP 3.44E-007	mm Hg	25	EST	NEELY,WB & BLAU,GE (1985)
DC	pKa			
HL 1.57E-015	atm m3/mol	25	EST	MEYLAN,WM & HOWARD,PH (1991)
OH 1.61E-010	cm3/molc sec	25	EST	MEYLAN,WM & HOWARD,PH (1993)

CAS #: 057117-31-4				2,3,4,7,8-PENTACHLORODIBENZOFURAN
Formula: $C_{12}H_3Cl_5O$				
Mol Weight: 340.42				
MP (deg C):		FP (deg C):		
BP (deg C):				
BP pressure (mm Hg):				

Property/Value	Units	Temp	Data Type	Reference
WS 2.35E-004	mg/L	23	EXP	FRIESEN,KJ ET AL. (1990B)
logP 6.92			EXP	SIJM,DTHM ET AL. (1989)
VP 3.46E-007	mm Hg	25	EST	NEELY,WB & BLAU,GE (1985)
DC	pKa			
HL 1.14E-005	atm m3/mol	25	EST	MEYLAN,WM & HOWARD,PH (1991)
OH 7.46E-014	cm3/molc sec	25	EST	MEYLAN,WM & HOWARD,PH (1993)

CAS #: 057117-37-0				DIBENZOFURAN, 2,3,6,8-TETRACHLORO-
Formula: $C_{12}H_4Cl_4O$				
Mol Weight: 305.98				
MP (deg C):		FP (deg C):		
BP (deg C):				
BP pressure (mm Hg):				

Property/Value	Units	Temp	Data Type	Reference
WS 1.57E-003	mg/L	25	EST	MEYLAN,WM ET AL. (1996)
logP 6.73			EXP	SIJM,DTHM ET AL. (1989)
VP 1.53E-006	mm Hg	25	EST	NEELY,WB & BLAU,GE (1985)
DC	pKa			
HL 1.54E-005	atm m3/mol	25	EST	MEYLAN,WM & HOWARD,PH (1991)
OH 1.65E-013	cm3/molc sec	25	EST	MEYLAN,WM & HOWARD,PH (1993)

CAS #: 057117-38-1				DIBENZOFURAN, 2,4,6,7-TETRACHLORO-
Formula: $C_{12}H_4Cl_4O$				
Mol Weight: 305.98				
MP (deg C):		FP (deg C):		
BP (deg C):				
BP pressure (mm Hg):				

Property/Value	Units	Temp	Data Type	Reference
WS 4.04E-003	mg/L	25	EST	MEYLAN,WM ET AL. (1996)
logP 6.25			EXP	SIJM,DTHM ET AL. (1989)
VP 1.53E-006	mm Hg	25	EST	NEELY,WB & BLAU,GE (1985)
DC	pKa			
HL 1.54E-005	atm m3/mol	25	EST	MEYLAN,WM & HOWARD,PH (1991)
OH 1.65E-013	cm3/molc sec	25	EST	MEYLAN,WM & HOWARD,PH (1993)

CAS #: 057117-39-2				DIBENZOFURAN, 2,3,6,7-TETRACHLORO-
Formula: $C_{12}H_4Cl_4O$				
Mol Weight: 305.98				
MP (deg C):		FP (deg C):		
BP (deg C):				
BP pressure (mm Hg):				

Property/Value	Units	Temp	Data Type	Reference
WS 4.21E-003	mg/L	25	EST	MEYLAN,WM ET AL. (1996)
logP 6.31			EXP	SIJM,DTHM ET AL. (1989)
VP 1.53E-006	mm Hg	25	EST	NEELY,WB & BLAU,GE (1985)
DC	pKa			
HL 1.54E-005	atm m3/mol	25	EST	MEYLAN,WM & HOWARD,PH (1991)
OH 1.65E-013	cm3/molc sec	25	EST	MEYLAN,WM & HOWARD,PH (1993)

CAS #: 057117-41-6				DIBENZOFURAN, 1,2,3,7,8-PENTACHLORO-

Formula: $C_{12}H_3Cl_5O$

Mol Weight: 340.42

MP (deg C): FP (deg C):

BP (deg C):

BP pressure (mm Hg):

Property/Value	Units	Temp	Data Type	Reference
WS 8.73E-004	mg/L	25	EST	MEYLAN,WM ET AL. (1996)
logP 6.79			EXP	SIJM,DTHM ET AL. (1989)
VP 3.46E-007	mm Hg	25	EST	NEELY,WB & BLAU,GE (1985)
DC	pKa			
HL 1.14E-005	atm m3/mol	25	EST	MEYLAN,WM & HOWARD,PH (1991)
OH 7.46E-014	cm3/molc sec	25	EST	MEYLAN,WM & HOWARD,PH (1993)

CAS #: 057117-42-7				DIBENZOFURAN, 1,2,3,6,7-PENTACHLORO-

Formula: $C_{12}H_3Cl_5O$

Mol Weight: 340.42

MP (deg C): FP (deg C):

BP (deg C):

BP pressure (mm Hg):

Property/Value	Units	Temp	Data Type	Reference
WS 2.48E-003	mg/L	25	EST	MEYLAN,WM ET AL. (1996)
logP 6.26			EXP	HANSCH,C ET AL. (1995)
VP 3.46E-007	mm Hg	25	EST	NEELY,WB & BLAU,GE (1985)
DC	pKa			
HL 1.14E-005	atm m3/mol	25	EST	MEYLAN,WM & HOWARD,PH (1991)
OH 7.46E-014	cm3/molc sec	25	EST	MEYLAN,WM & HOWARD,PH (1993)

CAS #: 057237-97-5				1H-BENZIMIDAZOLE, 2-[(2-PYRIDINYLMETHYL)SULFINYL

Formula: $C_{13}H_{11}N_3OS$

Mol Weight: 257.32

MP (deg C): FP (deg C):

BP (deg C):

BP pressure (mm Hg):

Property/Value	Units	Temp	Data Type	Reference
WS 3.14E+004	mg/L	25	EST	MEYLAN,WM ET AL. (1996)
logP 1.33			EXP	HANSCH,C ET AL. (1995)
VP 1.48E-010	mm Hg	25	EST	NEELY,WB & BLAU,GE (1985)
DC	pKa			
HL 7.12E-017	atm m3/mol	25	EST	MEYLAN,WM & HOWARD,PH (1991)
OH 7.59E-011	cm3/molc sec	25	EST	MEYLAN,WM & HOWARD,PH (1993)

CAS #: 057315-37-4				1,4-DIMETHOXYPHTHALAZINE

Formula: $C_{10}H_{10}N_2O_2$

Mol Weight: 190.20

MP (deg C): FP (deg C):

BP (deg C):

BP pressure (mm Hg):

Property/Value	Units	Temp	Data Type	Reference
WS 6.23E+002	mg/L	25	EST	MEYLAN,WM ET AL. (1996)
logP 2.22			EXP	HANSCH,C & LEO,AJ (1985)
VP 1.22E-005	mm Hg	25	EST	NEELY,WB & BLAU,GE (1985)
DC	pKa			
HL 8.15E-008	atm m3/mol	25	EST	MEYLAN,WM & HOWARD,PH (1991)
OH 2.07E-011	cm3/molc sec	25	EST	MEYLAN,WM & HOWARD,PH (1993)

CAS #: 057334-34-6				4-METHYL-5,8-DIHYDROXYQUINOLINE

Formula: $C_{10}H_9NO_2$

Mol Weight: 175.19

MP (deg C): FP (deg C):

BP (deg C):

BP pressure (mm Hg):

Property/Value	Units	Temp	Data Type	Reference
WS 9.69E+003	mg/L	25	EST	MEYLAN,WM ET AL. (1996)
logP 1.59			EXP	HANSCH,C & LEO,AJ (1985)
VP 1.45E-006	mm Hg	25	EST	NEELY,WB & BLAU,GE (1985)
DC	pKa			
HL 8.23E-015	atm m3/mol	25	EST	MEYLAN,WM & HOWARD,PH (1991)
OH 2.00E-010	cm3/molc sec	25	EST	MEYLAN,WM & HOWARD,PH (1993)

CAS #: 057334-35-7				5-METHOXY-8-QUINOLINOL

Formula: $C_{10}H_9NO_2$

Mol Weight: 175.19

MP (deg C): FP (deg C):

BP (deg C):

BP pressure (mm Hg):

Property/Value	Units	Temp	Data Type	Reference
WS 3.85E+003	mg/L	25	EST	MEYLAN,WM ET AL. (1996)
logP 2.06			EXP	HANSCH,C & LEO,AJ (1985)
VP 3.41E-005	mm Hg	25	EST	NEELY,WB & BLAU,GE (1985)
DC	pKa			
HL 4.24E-012	atm m3/mol	25	EST	MEYLAN,WM & HOWARD,PH (1991)
OH 1.54E-010	cm3/molc sec	25	EST	MEYLAN,WM & HOWARD,PH (1993)

CAS #: 057334-36-8				4-CHLORO-8-QUINOLINOL

Formula: C_9H_6ClNO

Mol Weight: 179.61

MP (deg C): FP (deg C):

BP (deg C):

BP pressure (mm Hg):

Property/Value	Units	Temp	Data Type	Reference
WS 1.10E+003	mg/L	25	EST	MEYLAN,WM ET AL. (1996)
logP 2.67			EXP	HANSCH,C & LEO,AJ (1985)
VP 5.20E-005	mm Hg	25	EST	NEELY,WB & BLAU,GE (1985)
DC	pKa			
HL 5.31E-011	atm m3/mol	25	EST	MEYLAN,WM & HOWARD,PH (1991)
OH 2.25E-011	cm3/molc sec	25	EST	MEYLAN,WM & HOWARD,PH (1993)

CAS #: 057334-38-0				4-ME-5-MEO-8-QUINOLINOL

Formula: $C_{11}H_{11}NO_2$

Mol Weight: 189.22

MP (deg C): FP (deg C):

BP (deg C):

BP pressure (mm Hg):

Property/Value	Units	Temp	Data Type	Reference
WS 8.45E+002	mg/L	25	EST	MEYLAN,WM ET AL. (1996)
logP 2.75			EXP	HANSCH,C & LEO,AJ (1985)
VP 1.20E-005	mm Hg	25	EST	NEELY,WB & BLAU,GE (1985)
DC	pKa			
HL 4.68E-012	atm m3/mol	25	EST	MEYLAN,WM & HOWARD,PH (1991)
OH 2.01E-010	cm3/molc sec	25	EST	MEYLAN,WM & HOWARD,PH (1993)

CAS #: 057369-32-1				PYROQUILON

Formula: $C_{11}H_{11}NO$

Mol Weight: 173.22

MP (deg C): 112 FP (deg C):

BP (deg C):

BP pressure (mm Hg):

Property/ Value	Units	Temp	Data Type	Reference
WS 4.00E+003	mg/L	20	EXP	SHIU,WY ET AL. (1990)
logP 1.57			EXP	TOMLIN,C (1994)
VP 7.96E-005	mm Hg	25	EST	NEELY,WB & BLAU,GE (1985)
DC	pKa			
HL 2.99E-008	atm m3/mol	25	EST	MEYLAN,WM & HOWARD,PH (1991)
OH 2.67E-011	cm3/molc sec	25	EST	MEYLAN,WM & HOWARD,PH (1993)

CAS #: 057381-26-7				2',5'-DICHLOROBENZOGUANAMINE

Formula: $C_9H_7Cl_2N_5$

Mol Weight: 256.10

MP (deg C): 268-269 FP (deg C):

BP (deg C):

BP pressure (mm Hg):

Property/ Value	Units	Temp	Data Type	Reference
WS 4.55E+002	mg/L	25	EST	MEYLAN,WM ET AL. (1996)
logP 1.97			EXP	HANSCH,C & LEO,AJ (1985)
VP 5.84E-008	mm Hg	25	EST	NEELY,WB & BLAU,GE (1985)
DC	pKa			
HL 2.26E-011	atm m3/mol	25	EST	MEYLAN,WM & HOWARD,PH (1991)
OH 8.65E-013	cm3/molc sec	25	EST	MEYLAN,WM & HOWARD,PH (1993)

CAS #: 057381-35-8				5'-CL-2'-FLUOROBENZOGUANAMINE

Formula: $C_9H_7ClFN_5$

Mol Weight: 239.64

MP (deg C): FP (deg C):

BP (deg C):

BP pressure (mm Hg):

Property/ Value	Units	Temp	Data Type	Reference
WS 7.68E+002	mg/L	25	EST	MEYLAN,WM ET AL. (1996)
logP 1.81			EXP	HANSCH,C & LEO,AJ (1985)
VP 2.53E-007	mm Hg	25	EST	NEELY,WB & BLAU,GE (1985)
DC	pKa			
HL 3.55E-011	atm m3/mol	25	EST	MEYLAN,WM & HOWARD,PH (1991)
OH 9.74E-013	cm3/molc sec	25	EST	MEYLAN,WM & HOWARD,PH (1993)

CAS #: 057381-38-1				2'-BR-5'-CHLOROBENZOGUANAMINE

Formula: $C_9H_7BrClN_5$

Mol Weight: 300.55

MP (deg C): FP (deg C):

BP (deg C):

BP pressure (mm Hg):

Property/ Value	Units	Temp	Data Type	Reference
WS 3.21E+002	mg/L	25	EST	MEYLAN,WM ET AL. (1996)
logP 1.85			EXP	HANSCH,C & LEO,AJ (1985)
VP 2.55E-008	mm Hg	25	EST	NEELY,WB & BLAU,GE (1985)
DC	pKa			
HL 1.21E-011	atm m3/mol	25	EST	MEYLAN,WM & HOWARD,PH (1991)
OH 8.53E-013	cm3/molc sec	25	EST	MEYLAN,WM & HOWARD,PH (1993)

CAS #: 057381-40-5				2'-BR-5'-FLUOROBENZOGUANAMINE

Formula: $C_9H_7BrFN_5$

Mol Weight: 284.10

MP (deg C): FP (deg C):

BP (deg C):

BP pressure (mm Hg):

Property/ Value	Units	Temp	Data Type	Reference
WS 8.77E+002	mg/L	25	EST	MEYLAN,WM ET AL. (1996)
logP 1.45			EXP	HANSCH,C & LEO,AJ (1985)
VP 1.11E-007	mm Hg	25	EST	NEELY,WB & BLAU,GE (1985)
DC	pKa			
HL 1.91E-011	atm m3/mol	25	EST	MEYLAN,WM & HOWARD,PH (1991)
OH 1.39E-012	cm3/molc sec	25	EST	MEYLAN,WM & HOWARD,PH (1993)

CAS #: 057381-42-7				2',5'-DIBROMOBENZOGUANAMINE

Formula: $C_9H_7Br_2N_5$

Mol Weight: 345.01

MP (deg C): FP (deg C):

BP (deg C):

BP pressure (mm Hg):

Property/ Value	Units	Temp	Data Type	Reference
WS 1.09E+002	mg/L	25	EST	MEYLAN,WM ET AL. (1996)
logP 2.09			EXP	HANSCH,C & LEO,AJ (1985)
VP 1.11E-008	mm Hg	25	EST	NEELY,WB & BLAU,GE (1985)
DC	pKa			
HL 6.52E-012	atm m3/mol	25	EST	MEYLAN,WM & HOWARD,PH (1991)
OH 7.81E-013	cm3/molc sec	25	EST	MEYLAN,WM & HOWARD,PH (1993)

CAS #: 057381-45-0				2'-CL-5'-BROMOBENZOGUANAMINE

Formula: $C_9H_7BrClN_5$

Mol Weight: 300.55

MP (deg C): FP (deg C):

BP (deg C):

BP pressure (mm Hg):

Property/ Value	Units	Temp	Data Type	Reference
WS 2.21E+002	mg/L	25	EST	MEYLAN,WM ET AL. (1996)
logP 2.04			EXP	HANSCH,C & LEO,AJ (1985)
VP 2.55E-008	mm Hg	25	EST	NEELY,WB & BLAU,GE (1985)
DC	pKa			
HL 1.21E-011	atm m3/mol	25	EST	MEYLAN,WM & HOWARD,PH (1991)
OH 7.92E-013	cm3/molc sec	25	EST	MEYLAN,WM & HOWARD,PH (1993)

CAS #: 057381-46-1				2',4'-DICHLOROBENZOGUANAMINE

Formula: $C_9H_7Cl_2N_5$

Mol Weight: 256.10

MP (deg C): FP (deg C):

BP (deg C):

BP pressure (mm Hg):

Property/ Value	Units	Temp	Data Type	Reference
WS 3.74E+002	mg/L	25	EST	MEYLAN,WM ET AL. (1996)
logP 2.07			EXP	HANSCH,C & LEO,AJ (1985)
VP 5.84E-008	mm Hg	25	EST	NEELY,WB & BLAU,GE (1985)
DC	pKa			
HL 2.26E-011	atm m3/mol	25	EST	MEYLAN,WM & HOWARD,PH (1991)
OH 8.59E-013	cm3/molc sec	25	EST	MEYLAN,WM & HOWARD,PH (1993)

057381-50-7 — 2'-BR-4'-CHLOROBENZOGUANAMINE

Formula: $C_9H_7BrClN_5$

Mol Weight: 300.55

MP (deg C):   FP (deg C):

BP (deg C):

BP pressure (mm Hg):

Property/Value	Units	Temp	Data Type	Reference
WS 2.00E+002	mg/L	25	EST	MEYLAN,WM ET AL. (1996)
logP 2.09			EXP	HANSCH,C & LEO,AJ (1985)
VP 2.55E-008	mm Hg	25	EST	NEELY,WB & BLAU,GE (1985)
DC	pKa			
HL 1.21E-011	atm m3/mol	25	EST	MEYLAN,WM & HOWARD,PH (1991)
OH 7.87E-013	cm3/molc sec	25	EST	MEYLAN,WM & HOWARD,PH (1993)

057381-54-1 — 2',6'-DICHLOROBENZOGUANAMINE

Formula: $C_9H_7Cl_2N_5$

Mol Weight: 256.10

MP (deg C):   FP (deg C):

BP (deg C):

BP pressure (mm Hg):

Property/Value	Units	Temp	Data Type	Reference
WS 1.45E+003	mg/L	25	EST	MEYLAN,WM ET AL. (1996)
logP 1.38			EXP	HANSCH,C & LEO,AJ (1985)
VP 5.84E-008	mm Hg	25	EST	NEELY,WB & BLAU,GE (1985)
DC	pKa			
HL 2.26E-011	atm m3/mol	25	EST	MEYLAN,WM & HOWARD,PH (1991)
OH 8.59E-013	cm3/molc sec	25	EST	MEYLAN,WM & HOWARD,PH (1993)

057381-57-4 — 2'-CL-5'-FLUOROBENZOGUANAMINE

Formula: $C_9H_7ClFN_5$

Mol Weight: 239.64

MP (deg C):   FP (deg C):

BP (deg C):

BP pressure (mm Hg):

Property/Value	Units	Temp	Data Type	Reference
WS 1.79E+003	mg/L	25	EST	MEYLAN,WM ET AL. (1996)
logP 1.38			EXP	HANSCH,C & LEO,AJ (1985)
VP 2.53E-007	mm Hg	25	EST	NEELY,WB & BLAU,GE (1985)
DC	pKa			
HL 3.55E-011	atm m3/mol	25	EST	MEYLAN,WM & HOWARD,PH (1991)
OH 1.41E-012	cm3/molc sec	25	EST	MEYLAN,WM & HOWARD,PH (1993)

057381-60-9 — 5'-BR-2'-FLUOROBENZOGUANAMINE

Formula: $C_9H_7BrFN_5$

Mol Weight: 284.10

MP (deg C):   FP (deg C):

BP (deg C):

BP pressure (mm Hg):

Property/Value	Units	Temp	Data Type	Reference
WS 3.35E+002	mg/L	25	EST	MEYLAN,WM ET AL. (1996)
logP 1.94			EXP	HANSCH,C & LEO,AJ (1985)
VP 1.11E-007	mm Hg	25	EST	NEELY,WB & BLAU,GE (1985)
DC	pKa			
HL 1.91E-011	atm m3/mol	25	EST	MEYLAN,WM & HOWARD,PH (1991)
OH 8.89E-013	cm3/molc sec	25	EST	MEYLAN,WM & HOWARD,PH (1993)

057383-80-9 — 2,3-DIBROMOPHENOL

Formula: $C_6H_4Br_2O$

Mol Weight: 251.92

MP (deg C):   FP (deg C):

BP (deg C):

BP pressure (mm Hg):

Property/Value	Units	Temp	Data Type	Reference
WS 1.35E+002	mg/L	25	EST	MEYLAN,WM ET AL. (1996)
logP 3.29				MEYLAN,WM & HOWARD,PH (1995)
VP 1.34E-003	mm Hg	25	EST	NEELY,WB & BLAU,GE (1985)
DC	pKa			
HL 8.90E-008	atm m3/mol	25	EST	MEYLAN,WM & HOWARD,PH (1991)
OH 6.15E-012	cm3/molc sec	25	EST	MEYLAN,WM & HOWARD,PH (1993)

057421-72-4 — 4-PROPYLSEMICARBAZIDE

Formula: $C_4H_{11}N_3O$

Mol Weight: 117.15

MP (deg C):   FP (deg C):

BP (deg C):

BP pressure (mm Hg):

Property/Value	Units	Temp	Data Type	Reference
WS 4.38E+004	mg/L	25	EST	MEYLAN,WM ET AL. (1996)
logP -1.10			EXP	HANSCH,C & LEO,AJ (1985)
VP 1.10E-002	mm Hg	25	EST	NEELY,WB & BLAU,GE (1985)
DC	pKa			
HL 5.90E-012	atm m3/mol	25	EST	MEYLAN,WM & HOWARD,PH (1991)
OH 7.70E-011	cm3/molc sec	25	EST	MEYLAN,WM & HOWARD,PH (1993)

057421-73-5 — 4-ALLYLSEMICARBAZIDE

Formula: $C_4H_9N_3O$

Mol Weight: 115.14

MP (deg C):   FP (deg C):

BP (deg C):

BP pressure (mm Hg):

Property/Value	Units	Temp	Data Type	Reference
WS 1.08E+005	mg/L	25	EST	MEYLAN,WM ET AL. (1996)
logP -1.55			EXP	HANSCH,C & LEO,AJ (1985)
VP 1.21E-002	mm Hg	25	EST	NEELY,WB & BLAU,GE (1985)
DC	pKa			
HL 4.40E-012	atm m3/mol	25	EST	MEYLAN,WM & HOWARD,PH (1991)
OH 1.00E-010	cm3/molc sec	25	EST	MEYLAN,WM & HOWARD,PH (1993)

057440-16-1 — ACETAMIDE, N-[2-(BENZOYLOXY)ETHYL]-N-METHYL-

Formula: $C_{12}H_{15}NO_3$

Mol Weight: 221.26

MP (deg C):   FP (deg C):

BP (deg C):

BP pressure (mm Hg):

Property/Value	Units	Temp	Data Type	Reference
WS 2.96E+003	mg/L	25	EST	MEYLAN,WM ET AL. (1996)
logP 1.24			EXP	NIELSEN,LS & BUNDGAARD,H (1988)
VP 4.70E-005	mm Hg	25	EST	NEELY,WB & BLAU,GE (1985)
DC	pKa			
HL 3.37E-011	atm m3/mol	25	EST	MEYLAN,WM & HOWARD,PH (1991)
OH 2.28E-011	cm3/molc sec	25	EST	MEYLAN,WM & HOWARD,PH (1993)

CAS #: 057453-98-2				BENZOIC ACID, 2-METHOXYETHYL ESTER
Formula: $C_{10}H_{12}O_3$				
Mol Weight: 180.21				
MP (deg C):		FP (deg C):		
BP (deg C):				
BP pressure (mm Hg):				

Property/Value	Units	Temp	Data Type	Reference
WS 1.66E+003	mg/L	25	EST	MEYLAN,WM ET AL. (1996)
logP 1.78			EXP	HANSCH,C & LEO,AJ (1985)
VP 2.20E-002	mm Hg	25	EST	NEELY,WB & BLAU,GE (1985)
DC	pKa			
HL 5.39E-007	atm m3/mol	25	EST	MEYLAN,WM & HOWARD,PH (1991)
OH 1.03E-011	cm3/molc sec	25	EST	MEYLAN,WM & HOWARD,PH (1993)

CAS #: 057524-89-7				HYDROCORTISONE-17-VALERATE
Formula: $C_{26}H_{38}O_6$				
Mol Weight: 446.59				
MP (deg C): 217-220		FP (deg C):		
BP (deg C):				
BP pressure (mm Hg):				

Property/Value	Units	Temp	Data Type	Reference
WS 9.08E-001	mg/L	25	EST	MEYLAN,WM ET AL. (1996)
logP 3.79			EXP	HANSCH,C & LEO,AJ (1985)
VP 5.62E-015	mm Hg	25	EST	NEELY,WB & BLAU,GE (1985)
DC	pKa			
HL 1.46E-011	atm m3/mol	25	EST	MEYLAN,WM & HOWARD,PH (1991)
OH 1.14E-010	cm3/molc sec	25	EST	MEYLAN,WM & HOWARD,PH (1993)

CAS #: 057531-37-0				2-CHLORO-4-NITRO-1H-IMIDAZOLE
Formula: $C_3H_2ClN_3O_2$				
Mol Weight: 147.52				
MP (deg C):		FP (deg C):		
BP (deg C):				
BP pressure (mm Hg):				

Property/Value	Units	Temp	Data Type	Reference
WS 8.66E+003	mg/L	25	EST	MEYLAN,WM ET AL. (1996)
logP 0.66			EXP	SUWINSKI,J ET AL. (1985)
VP 3.52E-005	mm Hg	25	EST	NEELY,WB & BLAU,GE (1985)
DC	pKa			
HL 1.10E-008	atm m3/mol	25	EST	MEYLAN,WM & HOWARD,PH (1991)
OH 9.18E-013	cm3/molc sec	25	EST	MEYLAN,WM & HOWARD,PH (1993)

CAS #: 057531-38-1				4-CHLORO-5-NITRO-1H-IMIDAZOLE
Formula: $C_3H_2ClN_3O_2$				
Mol Weight: 147.52				
MP (deg C):		FP (deg C):		
BP (deg C):				
BP pressure (mm Hg):				

Property/Value	Units	Temp	Data Type	Reference
WS 6.84E+003	mg/L	25	EST	MEYLAN,WM ET AL. (1996)
logP 0.52			EST	MEYLAN,WM & HOWARD,PH (1995)
VP 3.52E-005	mm Hg	25	EST	NEELY,WB & BLAU,GE (1985)
DC	pKa			
HL 1.10E-008	atm m3/mol	25	EST	MEYLAN,WM & HOWARD,PH (1991)
OH 1.31E-012	cm3/molc sec	25	EST	MEYLAN,WM & HOWARD,PH (1993)

CAS #: 057541-72-7				3,4-DICHLORO-N-NITROSOPIPERIDINE
Formula: $C_5H_8Cl_2N_2O$				
Mol Weight: 183.04				
MP (deg C):		FP (deg C):		
BP (deg C):				
BP pressure (mm Hg):				

Property/Value	Units	Temp	Data Type	Reference
WS 6.88E+003	mg/L	25	EST	MEYLAN,WM ET AL. (1996)
logP 1.04			EXP	HANSCH,C & LEO,AJ (1985)
VP 3.12E-003	mm Hg	25	EST	NEELY,WB & BLAU,GE (1985)
DC	pKa			
HL 2.63E-007	atm m3/mol	25	EST	MEYLAN,WM & HOWARD,PH (1991)
OH 1.49E-011	cm3/molc sec	25	EST	MEYLAN,WM & HOWARD,PH (1993)

CAS #: 057541-73-8				3,4-DIBROMO-N-NITROSOPIPERIDINE
Formula: $C_5H_8Br_2N_2O$				
Mol Weight: 271.95				
MP (deg C):		FP (deg C):		
BP (deg C):				
BP pressure (mm Hg):				

Property/Value	Units	Temp	Data Type	Reference
WS 1.59E+003	mg/L	25	EST	MEYLAN,WM ET AL. (1996)
logP 1.23			EXP	HANSCH,C & LEO,AJ (1985)
VP 2.59E-004	mm Hg	25	EST	NEELY,WB & BLAU,GE (1985)
DC	pKa			
HL 2.82E-008	atm m3/mol	25	EST	MEYLAN,WM & HOWARD,PH (1991)
OH 1.58E-011	cm3/molc sec	25	EST	MEYLAN,WM & HOWARD,PH (1993)

CAS #: 057553-97-6				OHET-(3,4,5-TRIMEO)PHENYL SULFONE
Formula: $C_{11}H_{16}O_6S$				
Mol Weight: 276.31				
MP (deg C):		FP (deg C):		
BP (deg C):				
BP pressure (mm Hg):				

Property/Value	Units	Temp	Data Type	Reference
WS 1.18E+004	mg/L	25	EST	MEYLAN,WM ET AL. (1996)
logP 0.18			EXP	HANSCH,C & LEO,AJ (1985)
VP 5.63E-009	mm Hg	25	EST	NEELY,WB & BLAU,GE (1985)
DC	pKa			
HL 1.25E-014	atm m3/mol	25	EST	MEYLAN,WM & HOWARD,PH (1991)
OH 5.95E-011	cm3/molc sec	25	EST	MEYLAN,WM & HOWARD,PH (1993)

CAS #: 057553-99-8				ALLYL-(3,4,5-MEO)PHENYL SULFONE
Formula: $C_{12}H_{16}O_5S$				
Mol Weight: 272.32				
MP (deg C):		FP (deg C):		
BP (deg C):				
BP pressure (mm Hg):				

Property/Value	Units	Temp	Data Type	Reference
WS 1.27E+003	mg/L	25	EST	MEYLAN,WM ET AL. (1996)
logP 1.34			EXP	HANSCH,C & LEO,AJ (1985)
VP 2.76E-006	mm Hg	25	EST	NEELY,WB & BLAU,GE (1985)
DC	pKa			
HL 3.39E-010	atm m3/mol	25	EST	MEYLAN,WM & HOWARD,PH (1991)
OH 7.99E-011	cm3/molc sec	25	EST	MEYLAN,WM & HOWARD,PH (1993)

CAS #: 057554-04-8	(N-PIPERIDINYL)ET-(345-MEO)PH SULFONE

Formula: $C_{16}H_{25}NO_5S$

Mol Weight: 343.45

MP (deg C):   FP (deg C):

BP (deg C):

BP pressure (mm Hg):

Property/Value	Units	Temp	Data Type	Reference
WS 2.97E+003	mg/L	25	EST	MEYLAN,WM ET AL. (1996)
logP 1.60			EXP	HANSCH,C & LEO,AJ (1985)
VP 1.53E-008	mm Hg	25	EST	NEELY,WB & BLAU,GE (1985)
DC	pKa			
HL 1.75E-013	atm m3/mol	25	EST	MEYLAN,WM & HOWARD,PH (1991)
OH 1.58E-010	cm3/molc sec	25	EST	MEYLAN,WM & HOWARD,PH (1993)

CAS #: 057554-05-9	(N-MORPHOLINYL)ET-(345-MEO)PH SULFONE

Formula: $C_{15}H_{23}NO_6S$

Mol Weight: 345.42

MP (deg C):   FP (deg C):

BP (deg C):

BP pressure (mm Hg):

Property/Value	Units	Temp	Data Type	Reference
WS 5.32E+003	mg/L	25	EST	MEYLAN,WM ET AL. (1996)
logP 1.29			EXP	HANSCH,C & LEO,AJ (1985)
VP 1.23E-008	mm Hg	25	EST	NEELY,WB & BLAU,GE (1985)
DC	pKa			
HL 1.16E-015	atm m3/mol	25	EST	MEYLAN,WM & HOWARD,PH (1991)
OH 2.07E-010	cm3/molc sec	25	EST	MEYLAN,WM & HOWARD,PH (1993)

CAS #: 057562-99-9	BARBITURIC ACID,5(1MEBU)-5-ET-1-ME

Formula: $C_{12}H_{20}N_2O_3$

Mol Weight: 240.30

MP (deg C):   FP (deg C):

BP (deg C):

BP pressure (mm Hg):

Property/Value	Units	Temp	Data Type	Reference
WS 2.16E+002	mg/L	25	EST	MEYLAN,WM ET AL. (1996)
logP 2.45			EXP	HANSCH,C & LEO,AJ (1985)
VP 9.81E-011	mm Hg	25	EST	NEELY,WB & BLAU,GE (1985)
DC	pKa			
HL 1.85E-012	atm m3/mol	25	EST	MEYLAN,WM & HOWARD,PH (1991)
OH 1.86E-011	cm3/molc sec	25	EST	MEYLAN,WM & HOWARD,PH (1993)

CAS #: 057646-30-7	N(26DIMEPH)-N-(2FURANOYL)ALANINE,ME

Formula: $C_{17}H_{19}NO_4$

Mol Weight: 301.35

MP (deg C): 70   FP (deg C):

BP (deg C):

BP pressure (mm Hg):

Property/Value	Units	Temp	Data Type	Reference
WS 2.30E+002	mg/L	20	EXP	SHIU,WY ET AL. (1990)
logP 2.61			EXP	HANSCH,C & LEO,AJ (1985)
VP 1.74E-007	mm Hg	25	EST	NEELY,WB & BLAU,GE (1985)
DC	pKa			
HL 9.62E-011	atm m3/mol	25	EST	MEYLAN,WM & HOWARD,PH (1991)
OH 6.16E-011	cm3/molc sec	25	EST	MEYLAN,WM & HOWARD,PH (1993)

CAS #: 057653-85-7	1,2,3,6,7,8-HEXACHLORODIBENZO-P-DIOXIN

Formula: $C_{12}H_2Cl_6O_2$

Mol Weight: 390.87

MP (deg C):   FP (deg C):

BP (deg C):

BP pressure (mm Hg):

Property/Value	Units	Temp	Data Type	Reference
WS 2.65E-005	mg/L	25	EST	MEYLAN,WM ET AL. (1996)
logP 8.21			EST	MEYLAN,WM & HOWARD,PH (1995)
VP 9.47E-008	mm Hg	25	EST	NEELY,WB & BLAU,GE (1985)
DC	pKa			
HL 1.94E-006	atm m3/mol	25	EST	MEYLAN,WM & HOWARD,PH (1991)
OH 1.42E-012	cm3/molc sec	25	EST	MEYLAN,WM & HOWARD,PH (1993)

CAS #: 057700-94-4	PIPERIDINE, 1-(4-METHOXYBENZOYL)-

Formula: $C_{13}H_{17}NO_2$

Mol Weight: 219.29

MP (deg C):   FP (deg C):

BP (deg C):

BP pressure (mm Hg):

Property/Value	Units	Temp	Data Type	Reference
WS 3.10E+002	mg/L	25	EST	MEYLAN,WM ET AL. (1996)
logP 2.17			EST	MEYLAN,WM & HOWARD,PH (1995)
VP 2.92E-005	mm Hg	25	EST	NEELY,WB & BLAU,GE (1985)
DC	pKa			
HL 6.51E-010	atm m3/mol	25	EST	MEYLAN,WM & HOWARD,PH (1991)
OH 4.91E-011	cm3/molc sec	25	EST	MEYLAN,WM & HOWARD,PH (1993)

CAS #: 057721-93-4	1,2-BIS(2-(2-MEO-ETO)-ETO)BENZENE

Formula: $C_{16}H_{26}O_6$

Mol Weight: 314.38

MP (deg C):   FP (deg C):

BP (deg C):

BP pressure (mm Hg):

Property/Value	Units	Temp	Data Type	Reference
WS 2.27E+003	mg/L	25	EST	MEYLAN,WM ET AL. (1996)
logP 0.76			EXP	STOLWIJK,TB ET AL. (1989)
VP 3.27E-006	mm Hg	25	EST	NEELY,WB & BLAU,GE (1985)
DC	pKa			
HL 1.10E-012	atm m3/mol	25	EST	MEYLAN,WM & HOWARD,PH (1991)
OH 7.63E-011	cm3/molc sec	25	EST	MEYLAN,WM & HOWARD,PH (1993)

CAS #: 057721-95-6	1,2-BIS(2-(2-MEO-ETO)-ETO-ETO)BENZENE

Formula: $C_{20}H_{34}O_8$

Mol Weight: 402.49

MP (deg C):   FP (deg C):

BP (deg C):

BP pressure (mm Hg):

Property/Value	Units	Temp	Data Type	Reference
WS 1.40E+003	mg/L	25	EST	MEYLAN,WM ET AL. (1996)
logP 0.38			EXP	STOLWIJK,TB ET AL. (1989)
VP 2.09E-008	mm Hg	25	EST	NEELY,WB & BLAU,GE (1985)
DC	pKa			
HL 2.66E-016	atm m3/mol	25	EST	MEYLAN,WM & HOWARD,PH (1991)
OH 1.04E-010	cm3/molc sec	25	EST	MEYLAN,WM & HOWARD,PH (1993)

CAS #: 057722-15-3				HEXACHLOROCYCLOHEXENE

Formula: $C_6H_4Cl_6$

Mol Weight: 288.82

MP (deg C): | FP (deg C):

BP (deg C):

BP pressure (mm Hg):

Property/ Value	Units	Temp	Data Type	Reference
WS 2.75E+000	mg/L	25	EST	MEYLAN,WM ET AL. (1996)
logP 4.31			EXP	HANSCH,C & LEO,AJ (1985)
VP 1.01E-003	mm Hg	25	EST	NEELY,WB & BLAU,GE (1985)
DC	pKa			
HL 9.38E-004	atm m3/mol	25	EST	MEYLAN,WM & HOWARD,PH (1991)
OH 3.52E-012	cm3/molc sec	25	EST	MEYLAN,WM & HOWARD,PH (1993)

CAS #: 057722-16-4				HEXACHLOROCYCLOHEXENE

Formula: $C_6H_4Cl_6$

Mol Weight: 288.82

MP (deg C): | FP (deg C):

BP (deg C):

BP pressure (mm Hg):

Property/ Value	Units	Temp	Data Type	Reference
WS 2.75E+000	mg/L	25	EST	MEYLAN,WM ET AL. (1996)
logP 4.34			EXP	HANSCH,C & LEO,AJ (1985)
VP 1.01E-003	mm Hg	25	EST	NEELY,WB & BLAU,GE (1985)
DC	pKa			
HL 9.38E-004	atm m3/mol	25	EST	MEYLAN,WM & HOWARD,PH (1991)
OH 3.52E-012	cm3/molc sec	25	EST	MEYLAN,WM & HOWARD,PH (1993)

CAS #: 057722-17-5				HEXACHLOROCYCLOHEXENE

Formula: $C_6H_4Cl_6$

Mol Weight: 288.82

MP (deg C): | FP (deg C):

BP (deg C):

BP pressure (mm Hg):

Property/ Value	Units	Temp	Data Type	Reference
WS 2.75E+000	mg/L	25	EST	MEYLAN,WM ET AL. (1996)
logP 4.35			EXP	HANSCH,C & LEO,AJ (1985)
VP 1.01E-003	mm Hg	25	EST	NEELY,WB & BLAU,GE (1985)
DC	pKa			
HL 9.38E-004	atm m3/mol	25	EST	MEYLAN,WM & HOWARD,PH (1991)
OH 3.52E-012	cm3/molc sec	25	EST	MEYLAN,WM & HOWARD,PH (1993)

CAS #: 057773-19-0				PYRIDO(12A)PYRIMIDINE-4-ONE,3ET,6ME

Formula: $C_{11}H_{14}N_2O$

Mol Weight: 190.25

MP (deg C): | FP (deg C):

BP (deg C):

BP pressure (mm Hg):

Property/ Value	Units	Temp	Data Type	Reference
WS 1.49E+003	mg/L	25	EST	MEYLAN,WM ET AL. (1996)
logP 1.79			EXP	HANSCH,C & LEO,AJ (1985)
VP 2.60E-005	mm Hg	25	EST	NEELY,WB & BLAU,GE (1985)
DC	pKa			
HL 4.88E-009	atm m3/mol	25	EST	MEYLAN,WM & HOWARD,PH (1991)
OH 1.39E-010	cm3/molc sec	25	EST	MEYLAN,WM & HOWARD,PH (1993)

CAS #: 057801-81-7				BROTIZOLAM

Formula: $C_{15}H_{12}BrClN_4S$

Mol Weight: 395.72

MP (deg C): 212-214 | FP (deg C):

BP (deg C):

BP pressure (mm Hg):

Property/ Value	Units	Temp	Data Type	Reference
WS 1.08E+000	mg/L	25	EST	MEYLAN,WM ET AL. (1996)
logP 2.79			EXP	HANSCH,C ET AL. (1995)
VP 1.29E-009	mm Hg	25	EST	NEELY,WB & BLAU,GE (1985)
DC	pKa			
HL 2.11E-012	atm m3/mol	25	EST	MEYLAN,WM & HOWARD,PH (1991)
OH 1.23E-011	cm3/molc sec	25	EST	MEYLAN,WM & HOWARD,PH (1993)

CAS #: 057808-36-3				1,1,4-TRICHLOROBUT-2-ENE(E)

Formula: $C_4H_5Cl_3$

Mol Weight: 159.44

MP (deg C): | FP (deg C):

BP (deg C):

BP pressure (mm Hg):

Property/ Value	Units	Temp	Data Type	Reference
WS 2.91E+002	mg/L	25	EST	MEYLAN,WM ET AL. (1996)
logP 2.78			EST	MEYLAN,WM & HOWARD,PH (1995)
VP 1.05E+000	mm Hg	25	EST	NEELY,WB & BLAU,GE (1985)
DC	pKa			
HL 6.62E-003	atm m3/mol	25	EST	MEYLAN,WM & HOWARD,PH (1991)
OH 3.32E-011	cm3/molc sec	25	EST	MEYLAN,WM & HOWARD,PH (1993)

CAS #: 057808-66-9				DOMPERIDONE

Formula: $C_{22}H_{24}ClN_5O_2$

Mol Weight: 425.92

MP (deg C): | FP (deg C):

BP (deg C):

BP pressure (mm Hg):

Property/ Value	Units	Temp	Data Type	Reference
WS 9.86E-001	mg/L	25	EST	MEYLAN,WM ET AL. (1996)
logP 3.90			EXP	EL TAYER,N ET AL. (1985)
VP 1.15E-014	mm Hg	25	EST	NEELY,WB & BLAU,GE (1985)
DC	pKa			
HL 6.11E-022	atm m3/mol	25	EST	MEYLAN,WM & HOWARD,PH (1991)
OH 1.44E-010	cm3/molc sec	25	EST	MEYLAN,WM & HOWARD,PH (1993)

CAS #: 057837-19-1				METALAXYL

Formula: $C_{15}H_{21}NO_4$

Mol Weight: 279.34

MP (deg C): 71-72 | FP (deg C):

BP (deg C):

BP pressure (mm Hg):

Property/ Value	Units	Temp	Data Type	Reference
WS 7.00E+003	mg/L	20	EXP	YALKOWSKY,SH & DANNENFELSER,RM (1992)
logP 1.65			EXP	HANSCH,C & LEO,AJ (1985)
VP 5.62E-006	mm Hg	25	EXP	WAUCHOPE,RD ET AL. (1991A)
DC	pKa			
HL 2.95E-009	atm m3/mol	25	EST	VP/WSOL
OH 3.03E-011	cm3/molc sec	25	EST	MEYLAN,WM & HOWARD,PH (1993)

CAS #: 057856-12-9	BIS(2-DIISOPROPYLAMINO)ETHYL METHYLPHOSPHONITE

Formula: $C_{17}H_{39}N_2O_2P$

Mol Weight: 334.49

MP (deg C): 77.5 FP (deg C):

BP (deg C): 338

BP pressure (mm Hg):

Property/ Value	Units	Temp	Data Type	Reference
WS 1.46E+001	mg/L	25	EST	MEYLAN,WM ET AL. (1996)
logP 4.37			EST	MEYLAN,WM & HOWARD,PH (1995)
VP 8.10E-005	mm Hg	25	EST	NEELY,WB & BLAU,GE (1985)
DC	pKa			
HL 2.25E-010	atm m3/mol	25	EST	MEYLAN,WM & HOWARD,PH (1991)
OH 3.13E-010	cm3/molc sec	25	EST	MEYLAN,WM & HOWARD,PH (1993)

CAS #: 057875-61-3	1,2-DINITROANTHRAQUINONE

Formula: $C_{14}H_6N_2O_6$

Mol Weight: 298.21

MP (deg C): FP (deg C):

BP (deg C):

BP pressure (mm Hg):

Property/ Value	Units	Temp	Data Type	Reference
WS 1.13E+000	mg/L	25	EST	MEYLAN,WM ET AL. (1996)
logP 2.98			EST	MEYLAN,WM & HOWARD,PH (1995)
VP 7.00E-012	mm Hg	25	EXP	BAUGHMAN,GL & PERENICH,TA (1988)
DC	pKa			
HL 4.95E-014	atm m3/mol	25	EST	MEYLAN,WM & HOWARD,PH (1991)
OH 7.57E-013	cm3/molc sec	25	EST	MEYLAN,WM & HOWARD,PH (1993)

CAS #: 057930-20-8	4-I-PROPYLSEMICARBAZIDE

Formula: $C_4H_{11}N_3O$

Mol Weight: 117.15

MP (deg C): FP (deg C):

BP (deg C):

BP pressure (mm Hg):

Property/ Value	Units	Temp	Data Type	Reference
WS 6.49E+004	mg/L	25	EST	MEYLAN,WM ET AL. (1996)
logP -1.30			EXP	HANSCH,C & LEO,AJ (1985)
VP 2.30E-002	mm Hg	25	EST	NEELY,WB & BLAU,GE (1985)
DC	pKa			
HL 5.90E-012	atm m3/mol	25	EST	MEYLAN,WM & HOWARD,PH (1991)
OH 8.34E-011	cm3/molc sec	25	EST	MEYLAN,WM & HOWARD,PH (1993)

CAS #: 057998-68-2	DIAZIQUONE

Formula: $C_{16}H_{20}N_4O_6$

Mol Weight: 364.36

MP (deg C): 230 dec FP (deg C):

BP (deg C):

BP pressure (mm Hg):

Property/ Value	Units	Temp	Data Type	Reference
WS 2.63E+003	mg/L	25	EST	MEYLAN,WM ET AL. (1996)
logP -0.02			EXP	HANSCH,C ET AL. (1995)
VP 1.33E-009	mm Hg	25	EST	NEELY,WB & BLAU,GE (1985)
DC	pKa			
HL 6.86E-019	atm m3/mol	25	EST	MEYLAN,WM & HOWARD,PH (1991)
OH 3.69E-011	cm3/molc sec	25	EST	MEYLAN,WM & HOWARD,PH (1993)

CAS #: 058011-68-0	PYRAZOLATE

Formula: $C_{19}H_{18}Cl_2N_2O_4S$

Mol Weight: 441.34

MP (deg C): FP (deg C):

BP (deg C):

BP pressure (mm Hg):

Property/ Value	Units	Temp	Data Type	Reference
WS 8.04E-002	mg/L	25	EST	MEYLAN,WM ET AL. (1996)
logP 3.90			EXP	SAITO,H ET AL. (1993)
VP 6.86E-012	mm Hg	25	EST	NEELY,WB & BLAU,GE (1985)
DC	pKa			
HL 1.05E-012	atm m3/mol	25	EST	MEYLAN,WM & HOWARD,PH (1991)
OH 3.84E-011	cm3/molc sec	25	EST	MEYLAN,WM & HOWARD,PH (1993)

CAS #: 058171-47-4	ETHANAMINE, N-ETHYL-2,2,2-TRIFLUORO-

Formula: $C_4H_8F_3N$

Mol Weight: 127.11

MP (deg C): FP (deg C):

BP (deg C):

BP pressure (mm Hg):

Property/ Value	Units	Temp	Data Type	Reference
WS 2.24E+005	mg/L	25	EST	MEYLAN,WM ET AL. (1996)
logP 0.74			EXP	HANSCH,C ET AL. (1995)
VP 2.90E+002	mm Hg	25	EST	NEELY,WB & BLAU,GE (1985)
DC	pKa			
HL 2.31E-004	atm m3/mol	25	EST	MEYLAN,WM & HOWARD,PH (1991)
OH 9.47E-012	cm3/molc sec	25	EST	MEYLAN,WM & HOWARD,PH (1993)

CAS #: 058175-07-8	N,N-DIME-QUINOXALINE-2,3-DIONE

Formula: $C_{10}H_{10}N_2O_2$

Mol Weight: 190.20

MP (deg C): FP (deg C):

BP (deg C):

BP pressure (mm Hg):

Property/ Value	Units	Temp	Data Type	Reference
WS 2.05E+003	mg/L	25	EST	MEYLAN,WM ET AL. (1996)
logP 0.08			EXP	HANSCH,C & LEO,AJ (1985)
VP 1.22E-006	mm Hg	25	EST	NEELY,WB & BLAU,GE (1985)
DC	pKa			
HL 9.24E-009	atm m3/mol	25	EST	MEYLAN,WM & HOWARD,PH (1991)
OH 7.61E-012	cm3/molc sec	25	EST	MEYLAN,WM & HOWARD,PH (1993)

CAS #: 058182-63-1	ITANOXONE

Formula: $C_{17}H_{13}ClO_3$

Mol Weight: 300.74

MP (deg C): FP (deg C):

BP (deg C):

BP pressure (mm Hg):

Property/ Value	Units	Temp	Data Type	Reference
WS 4.41E+001	mg/L	25	EST	MEYLAN,WM ET AL. (1996)
logP 3.32			EXP	HANSCH,C & LEO,AJ (1985)
VP 1.08E-008	mm Hg	25	EST	NEELY,WB & BLAU,GE (1985)
DC	pKa			
HL 3.43E-012	atm m3/mol	25	EST	MEYLAN,WM & HOWARD,PH (1991)
OH 2.25E-011	cm3/molc sec	25	EST	MEYLAN,WM & HOWARD,PH (1993)

CAS #: 058202-83-8 — P-N-ACETYLAMINOBENZAMIDE

Formula: $C_9H_{10}N_2O_2$

Mol Weight: 178.19

MP (deg C): FP (deg C):

BP (deg C):

BP pressure (mm Hg):

Property/Value	Units	Temp	Data Type	Reference
WS 2.70E+003	mg/L	25	EST	MEYLAN,WM ET AL. (1996)
logP 0.01			EXP	NAKAGAWA,Y ET AL. (1992)
VP 8.95E-008	mm Hg	25	EST	NEELY,WB & BLAU,GE (1985)
DC	pKa			
HL 4.02E-012	atm m3/mol	25	EST	MEYLAN,WM & HOWARD,PH (1991)
OH 1.34E-011	cm3/molc sec	25	EST	MEYLAN,WM & HOWARD,PH (1993)

CAS #: 058202-84-9 — ACETAMIDE, N-(3-CYANOPHENYL)-

Formula: $C_9H_8N_2O$

Mol Weight: 160.18

MP (deg C): FP (deg C):

BP (deg C):

BP pressure (mm Hg):

Property/Value	Units	Temp	Data Type	Reference
WS 3.51E+003	mg/L	25	EST	MEYLAN,WM ET AL. (1996)
logP 1.20			EXP	HANSCH,C ET AL. (1995)
VP 1.06E-005	mm Hg	25	EST	NEELY,WB & BLAU,GE (1985)
DC	pKa			
HL 5.96E-011	atm m3/mol	25	EST	MEYLAN,WM & HOWARD,PH (1991)
OH 1.73E-012	cm3/molc sec	25	EST	MEYLAN,WM & HOWARD,PH (1993)

CAS #: 058247-23-7 — 4-METHOXYPHENYLAMIDINO UREA

Formula: $C_9H_{12}N_4O_2$

Mol Weight: 208.22

MP (deg C): FP (deg C):

BP (deg C):

BP pressure (mm Hg):

Property/Value	Units	Temp	Data Type	Reference
WS 1.32E+004	mg/L	25	EST	MEYLAN,WM ET AL. (1996)
logP 0.56			EXP	HANSCH,C & LEO,AJ (1985)
VP 4.18E-007	mm Hg	25	EST	NEELY,WB & BLAU,GE (1985)
DC	pKa			
HL 8.07E-017	atm m3/mol	25	EST	MEYLAN,WM & HOWARD,PH (1991)
OH 5.97E-011	cm3/molc sec	25	EST	MEYLAN,WM & HOWARD,PH (1993)

CAS #: 058247-24-8 — 4-CHLOROPHENYLAMIDINO UREA

Formula: $C_8H_9ClN_4O$

Mol Weight: 212.64

MP (deg C): FP (deg C):

BP (deg C):

BP pressure (mm Hg):

Property/Value	Units	Temp	Data Type	Reference
WS 1.14E+003	mg/L	25	EST	MEYLAN,WM ET AL. (1996)
logP 1.78			EXP	HANSCH,C & LEO,AJ (1985)
VP 5.96E-007	mm Hg	25	EST	NEELY,WB & BLAU,GE (1985)
DC	pKa			
HL 1.01E-015	atm m3/mol	25	EST	MEYLAN,WM & HOWARD,PH (1991)
OH 3.45E-011	cm3/molc sec	25	EST	MEYLAN,WM & HOWARD,PH (1993)

CAS #: 058247-25-9 — 4-CYANO-3-MEO-PHENYLAMIDINO UREA

Formula: $C_{10}H_{11}N_5O_2$

Mol Weight: 233.23

MP (deg C): FP (deg C):

BP (deg C):

BP pressure (mm Hg):

Property/Value	Units	Temp	Data Type	Reference
WS 1.28E+003	mg/L	25	EST	MEYLAN,WM ET AL. (1996)
logP 1.28			EXP	HANSCH,C & LEO,AJ (1985)
VP 1.63E-008	mm Hg	25	EST	NEELY,WB & BLAU,GE (1985)
DC	pKa			
HL 7.80E-019	atm m3/mol	25	EST	MEYLAN,WM & HOWARD,PH (1991)
OH 1.06E-010	cm3/molc sec	25	EST	MEYLAN,WM & HOWARD,PH (1993)

CAS #: 058283-42-4 — PIPERIDINE, 1-(3-METHYL-1-PHENYLCYCLOHEXYL)-, Cl

Formula: $C_{18}H_{27}N$

Mol Weight: 257.42

MP (deg C): FP (deg C):

BP (deg C):

BP pressure (mm Hg):

Property/Value	Units	Temp	Data Type	Reference
WS 2.17E+000	mg/L	25	EST	MEYLAN,WM ET AL. (1996)
logP 5.86			EXP	SANGSTER,J (1993)
VP 2.75E-005	mm Hg	25	EST	NEELY,WB & BLAU,GE (1985)
DC	pKa			
HL 7.34E-006	atm m3/mol	25	EST	MEYLAN,WM & HOWARD,PH (1991)
OH 1.05E-010	cm3/molc sec	25	EST	MEYLAN,WM & HOWARD,PH (1993)

CAS #: 058326-97-9 — 1,3-OXATHIOLANE-5-METHANAMINIUM, N,N,N-TRIMETHYL

Formula: $C_7H_{16}INOS$

Mol Weight: 289.18

MP (deg C): FP (deg C):

BP (deg C):

BP pressure (mm Hg):

Property/Value	Units	Temp	Data Type	Reference
WS 1.00E+006	mg/L	25	EST	MEYLAN,WM ET AL. (1996)
logP -2.70			EXP	SANGSTER,J (1994)
VP 1.03E-008	mm Hg	25	EST	NEELY,WB & BLAU,GE (1985)
DC	pKa			
HL 1.93E-016	atm m3/mol	25	EST	MEYLAN,WM & HOWARD,PH (1991)
OH 7.32E-011	cm3/molc sec	25	EST	MEYLAN,WM & HOWARD,PH (1993)

CAS #: 058333-58-7 — 5-THIA-1-AZABICYCLO[4.2.0]OCT-2-ENE-2-CARBOXYLIC

Formula: $C_{24}H_{28}N_6O_7S_2$

Mol Weight: 576.65

MP (deg C): FP (deg C):

BP (deg C):

BP pressure (mm Hg):

Property/Value	Units	Temp	Data Type	Reference
WS 4.30E+001	mg/L	25	EST	MEYLAN,WM ET AL. (1996)
logP 0.85			EXP	SANGSTER,J (1993)
VP 1.31E-022	mm Hg	25	EST	NEELY,WB & BLAU,GE (1985)
DC	pKa			
HL 1.93E-020	atm m3/mol	25	EST	MEYLAN,WM & HOWARD,PH (1991)
OH 1.29E-010	cm3/molc sec	25	EST	MEYLAN,WM & HOWARD,PH (1993)

CAS #: 058349-01-2				4-BROMO-2-CHLORO-6-NITROPHENOL

Formula: $C_6H_3BrClNO_3$

Mol Weight: 252.46

MP (deg C): **FP (deg C):**

BP (deg C):

BP pressure (mm Hg):

Property/ Value	Units	Temp	Data Type	Reference
WS 2.63E+001	mg/L	25	EST	MEYLAN,WM ET AL. (1996)
logP 3.44			EST	MEYLAN,WM & HOWARD,PH (1995)
VP 1.86E-005	mm Hg	25	EST	NEELY,WB & BLAU,GE (1985)
DC	pKa			
HL 2.06E-006	atm m3/mol	25	EST	MEYLAN,WM & HOWARD,PH (1991)
OH 1.92E-013	cm3/molc sec	25	EST	MEYLAN,WM & HOWARD,PH (1993)

CAS #: 058414-52-1				METHYL THIOPHEN-3-ACETATE

Formula: $C_7H_{10}O_2S$

Mol Weight: 158.22

MP (deg C): **FP (deg C):**

BP (deg C):

BP pressure (mm Hg):

Property/ Value	Units	Temp	Data Type	Reference
WS 4.03E+003	mg/L	25	EST	MEYLAN,WM ET AL. (1996)
logP 1.46			EXP	HANSCH,C & LEO,AJ (1985)
VP 8.41E-002	mm Hg	25	EST	NEELY,WB & BLAU,GE (1985)
DC	pKa			
HL 7.69E-006	atm m3/mol	25	EST	MEYLAN,WM & HOWARD,PH (1991)
OH 1.67E-011	cm3/molc sec	25	EST	MEYLAN,WM & HOWARD,PH (1993)

CAS #: 058471-47-9				1(2H)-PYRIMIDINECARBOXAMIDE, N-ETHYL-5-FLUORO-3,

Formula: $C_7H_8FN_3O_3$

Mol Weight: 201.16

MP (deg C): **FP (deg C):**

BP (deg C):

BP pressure (mm Hg):

Property/ Value	Units	Temp	Data Type	Reference
WS 2.17E+004	mg/L	25	EST	MEYLAN,WM ET AL. (1996)
logP 0.35			EXP	SANGSTER,J (1994)
VP 6.32E-009	mm Hg	25	EST	NEELY,WB & BLAU,GE (1985)
DC	pKa			
HL 2.94E-015	atm m3/mol	25	EST	MEYLAN,WM & HOWARD,PH (1991)
OH 1.08E-011	cm3/molc sec	25	EST	MEYLAN,WM & HOWARD,PH (1993)

CAS #: 058479-86-0				N(P-BRPH)3-N'-PIPERIDINOACETAMIDE

Formula: $C_{13}H_{17}BrN_2O$

Mol Weight: 297.20

MP (deg C): **FP (deg C):**

BP (deg C):

BP pressure (mm Hg):

Property/ Value	Units	Temp	Data Type	Reference
WS 1.14E+001	mg/L	25	EST	MEYLAN,WM ET AL. (1996)
logP 3.57			EXP	HANSCH,C & LEO,AJ (1985)
VP 9.35E-008	mm Hg	25	EST	NEELY,WB & BLAU,GE (1985)
DC	pKa			
HL 2.51E-011	atm m3/mol	25	EST	MEYLAN,WM & HOWARD,PH (1991)
OH 1.02E-010	cm3/molc sec	25	EST	MEYLAN,WM & HOWARD,PH (1993)

CAS #: 058479-87-1				N-(P-IPH)3-N'-PIPERIDINOACETAMIDE

Formula: $C_{13}H_{17}IN_2O$

Mol Weight: 344.20

MP (deg C): **FP (deg C):**

BP (deg C):

BP pressure (mm Hg):

Property/ Value	Units	Temp	Data Type	Reference
WS 1.22E+001	mg/L	25	EST	MEYLAN,WM ET AL. (1996)
logP 3.21			EXP	HANSCH,C & LEO,AJ (1985)
VP 2.66E-008	mm Hg	25	EST	NEELY,WB & BLAU,GE (1985)
DC	pKa			
HL 1.46E-011	atm m3/mol	25	EST	MEYLAN,WM & HOWARD,PH (1991)
OH 1.02E-010	cm3/molc sec	25	EST	MEYLAN,WM & HOWARD,PH (1993)

CAS #: 058479-88-2				N-(O-FPH)3-N'-PIPERIDINOACETAMIDE

Formula: $C_{13}H_{17}FN_2O$

Mol Weight: 236.29

MP (deg C): **FP (deg C):**

BP (deg C):

BP pressure (mm Hg):

Property/ Value	Units	Temp	Data Type	Reference
WS 4.02E+002	mg/L	25	EST	MEYLAN,WM ET AL. (1996)
logP 2.16			EXP	HANSCH,C & LEO,AJ (1985)
VP 1.18E-006	mm Hg	25	EST	NEELY,WB & BLAU,GE (1985)
DC	pKa			
HL 7.35E-011	atm m3/mol	25	EST	MEYLAN,WM & HOWARD,PH (1991)
OH 1.02E-010	cm3/molc sec	25	EST	MEYLAN,WM & HOWARD,PH (1993)

CAS #: 058479-89-3				N(O-BRPH)3-N'-PIPERIDINOACETAMIDE

Formula: $C_{13}H_{17}BrN_2O$

Mol Weight: 297.20

MP (deg C): **FP (deg C):**

BP (deg C):

BP pressure (mm Hg):

Property/ Value	Units	Temp	Data Type	Reference
WS 3.93E+001	mg/L	25	EST	MEYLAN,WM ET AL. (1996)
logP 2.94			EXP	HANSCH,C & LEO,AJ (1985)
VP 9.35E-008	mm Hg	25	EST	NEELY,WB & BLAU,GE (1985)
DC	pKa			
HL 2.51E-011	atm m3/mol	25	EST	MEYLAN,WM & HOWARD,PH (1991)
OH 1.02E-010	cm3/molc sec	25	EST	MEYLAN,WM & HOWARD,PH (1993)

CAS #: 058479-90-6				N-(M-FPH)3-N'-PIPERIDINOACETAMIDE

Formula: $C_{13}H_{17}FN_2O$

Mol Weight: 236.29

MP (deg C): **FP (deg C):**

BP (deg C):

BP pressure (mm Hg):

Property/ Value	Units	Temp	Data Type	Reference
WS 1.04E+002	mg/L	25	EST	MEYLAN,WM ET AL. (1996)
logP 2.85			EXP	HANSCH,C & LEO,AJ (1985)
VP 1.18E-006	mm Hg	25	EST	NEELY,WB & BLAU,GE (1985)
DC	pKa			
HL 7.35E-011	atm m3/mol	25	EST	MEYLAN,WM & HOWARD,PH (1991)
OH 1.14E-010	cm3/molc sec	25	EST	MEYLAN,WM & HOWARD,PH (1993)

CAS #: 058479-91-7 — N-(M-IPH)3-N'-PIPERIDINOACETAMIDE

Formula: $C_{13}H_{17}IN_2O$

Mol Weight: 344.20

MP (deg C): — FP (deg C): —
BP (deg C): —
BP pressure (mm Hg): —

Property/Value	Units	Temp	Data Type	Reference
WS 1.17E+001	mg/L	25	EST	MEYLAN,WM ET AL. (1996)
logP 3.23			EXP	HANSCH,C & LEO,AJ (1985)
VP 2.66E-008	mm Hg	25	EST	NEELY,WB & BLAU,GE (1985)
DC	pKa			
HL 1.46E-011	atm m3/mol	25	EST	MEYLAN,WM & HOWARD,PH (1991)
OH 1.06E-010	cm3/molc sec	25	EST	MEYLAN,WM & HOWARD,PH (1993)

CAS #: 058479-93-9 — 1-PIPERIDINEACETAMIDE, N-(4-METHOXYPHENYL)-

Formula: $C_{14}H_{20}N_2O_2$

Mol Weight: 248.33

MP (deg C): — FP (deg C): —
BP (deg C): —
BP pressure (mm Hg): —

Property/Value	Units	Temp	Data Type	Reference
WS 1.70E+002	mg/L	25	EST	MEYLAN,WM ET AL. (1996)
logP 2.52			EXP	SANGSTER,J (1993)
VP 1.66E-007	mm Hg	25	EST	NEELY,WB & BLAU,GE (1985)
DC	pKa			
HL 3.73E-012	atm m3/mol	25	EST	MEYLAN,WM & HOWARD,PH (1991)
OH 1.10E-010	cm3/molc sec	25	EST	MEYLAN,WM & HOWARD,PH (1993)

CAS #: 058479-94-0 — 1-PIPERIDINEACETAMIDE, N-(4-ETHOXYPHENYL)-

Formula: $C_{15}H_{22}N_2O_2$

Mol Weight: 262.35

MP (deg C): — FP (deg C): —
BP (deg C): —
BP pressure (mm Hg): —

Property/Value	Units	Temp	Data Type	Reference
WS 7.88E+001	mg/L	25	EST	MEYLAN,WM ET AL. (1996)
logP 2.82			EXP	SANGSTER,J (1993)
VP 7.03E-008	mm Hg	25	EST	NEELY,WB & BLAU,GE (1985)
DC	pKa			
HL 4.95E-012	atm m3/mol	25	EST	MEYLAN,WM & HOWARD,PH (1991)
OH 1.16E-010	cm3/molc sec	25	EST	MEYLAN,WM & HOWARD,PH (1993)

CAS #: 058480-99-2 — 2-I-BUTYLISONIAZID

Formula: $C_{10}H_{15}N_3O$

Mol Weight: 193.25

MP (deg C): — FP (deg C): —
BP (deg C): —
BP pressure (mm Hg): —

Property/Value	Units	Temp	Data Type	Reference
WS 6.37E+003	mg/L	25	EST	MEYLAN,WM ET AL. (1996)
logP 1.02			EXP	HANSCH,C & LEO,AJ (1985)
VP 2.60E-006	mm Hg	25	EST	NEELY,WB & BLAU,GE (1985)
DC	pKa			
HL 3.13E-014	atm m3/mol	25	EST	MEYLAN,WM & HOWARD,PH (1991)
OH 9.78E-012	cm3/molc sec	25	EST	MEYLAN,WM & HOWARD,PH (1993)

CAS #: 058481-00-8 — 2-ETHOXYISONIAZID

Formula: $C_8H_{11}N_3O_2$

Mol Weight: 181.20

MP (deg C): — FP (deg C): —
BP (deg C): —
BP pressure (mm Hg): —

Property/Value	Units	Temp	Data Type	Reference
WS 2.11E+004	mg/L	25	EST	MEYLAN,WM ET AL. (1996)
logP 0.48			EXP	HANSCH,C & LEO,AJ (1985)
VP 3.19E-006	mm Hg	25	EST	NEELY,WB & BLAU,GE (1985)
DC	pKa			
HL 8.67E-015	atm m3/mol	25	EST	MEYLAN,WM & HOWARD,PH (1991)
OH 1.34E-011	cm3/molc sec	25	EST	MEYLAN,WM & HOWARD,PH (1993)

CAS #: 058481-01-9 — 2-AMINOISONIAZID

Formula: $C_6H_8N_4O$

Mol Weight: 152.16

MP (deg C): — FP (deg C): —
BP (deg C): —
BP pressure (mm Hg): —

Property/Value	Units	Temp	Data Type	Reference
WS 2.50E+004	mg/L	25	EST	MEYLAN,WM ET AL. (1996)
logP -0.98			EXP	HANSCH,C & LEO,AJ (1985)
VP 1.82E-006	mm Hg	25	EST	NEELY,WB & BLAU,GE (1985)
DC	pKa			
HL 4.28E-018	atm m3/mol	25	EST	MEYLAN,WM & HOWARD,PH (1991)
OH 1.41E-011	cm3/molc sec	25	EST	MEYLAN,WM & HOWARD,PH (1993)

CAS #: 058481-02-0 — 2-ACETAMIDOISONIAZID

Formula: $C_8H_{10}N_4O_2$

Mol Weight: 194.19

MP (deg C): — FP (deg C): —
BP (deg C): —
BP pressure (mm Hg): —

Property/Value	Units	Temp	Data Type	Reference
WS 2.68E+004	mg/L	25	EST	MEYLAN,WM ET AL. (1996)
logP -1.25			EXP	HANSCH,C & LEO,AJ (1985)
VP 9.37E-009	mm Hg	25	EST	NEELY,WB & BLAU,GE (1985)
DC	pKa			
HL 2.20E-017	atm m3/mol	25	EST	MEYLAN,WM & HOWARD,PH (1991)
OH 6.60E-012	cm3/molc sec	25	EST	MEYLAN,WM & HOWARD,PH (1993)

CAS #: 058481-03-1 — 2-ACETAMIDOMETHYLISONIAZID

Formula: $C_9H_{12}N_4O_2$

Mol Weight: 208.22

MP (deg C): — FP (deg C): —
BP (deg C): —
BP pressure (mm Hg): —

Property/Value	Units	Temp	Data Type	Reference
WS 4.35E+004	mg/L	25	EST	MEYLAN,WM ET AL. (1996)
logP -1.58			EXP	HANSCH,C & LEO,AJ (1985)
VP 4.17E-009	mm Hg	25	EST	NEELY,WB & BLAU,GE (1985)
DC	pKa			
HL 7.06E-018	atm m3/mol	25	EST	MEYLAN,WM & HOWARD,PH (1991)
OH 1.53E-011	cm3/molc sec	25	EST	MEYLAN,WM & HOWARD,PH (1993)

CAS #: 058481-04-2 — 2-CHLOROISONIAZID

Formula: $C_6H_6ClN_3O$
Mol Weight: 171.59
MP (deg C):　FP (deg C):
BP (deg C):
BP pressure (mm Hg):

Property/Value	Units	Temp	Data Type	Reference
WS 2.38E+003	mg/L	25	EST	MEYLAN,WM ET AL. (1996)
logP 0.11			EXP	HANSCH,C & LEO,AJ (1985)
VP 1.04E-005	mm Hg	25	EST	NEELY,WB & BLAU,GE (1985)
DC	pKa			
HL 5.66E-013	atm m3/mol	25	EST	MEYLAN,WM & HOWARD,PH (1991)
OH 5.61E-012	cm3/molc sec	25	EST	MEYLAN,WM & HOWARD,PH (1993)

CAS #: 058481-05-3 — 2-NITROISONIAZID

Formula: $C_6H_6N_4O_3$
Mol Weight: 182.14
MP (deg C):　FP (deg C):
BP (deg C):
BP pressure (mm Hg):

Property/Value	Units	Temp	Data Type	Reference
WS 4.78E+003	mg/L	25	EST	MEYLAN,WM ET AL. (1996)
logP -0.76			EXP	HANSCH,C & LEO,AJ (1985)
VP 7.51E-007	mm Hg	25	EST	NEELY,WB & BLAU,GE (1985)
DC	pKa			
HL 4.78E-017	atm m3/mol	25	EST	MEYLAN,WM & HOWARD,PH (1991)
OH 5.54E-012	cm3/molc sec	25	EST	MEYLAN,WM & HOWARD,PH (1993)

CAS #: 058481-06-4 — 2-PHENYLISONIAZID

Formula: $C_{12}H_{11}N_3O$
Mol Weight: 213.24
MP (deg C):　FP (deg C):
BP (deg C):
BP pressure (mm Hg):

Property/Value	Units	Temp	Data Type	Reference
WS 2.58E+003	mg/L	25	EST	MEYLAN,WM ET AL. (1996)
logP 1.36			EXP	HANSCH,C & LEO,AJ (1985)
VP 4.08E-008	mm Hg	25	EST	NEELY,WB & BLAU,GE (1985)
DC	pKa			
HL 9.30E-016	atm m3/mol	25	EST	MEYLAN,WM & HOWARD,PH (1991)
OH 9.16E-012	cm3/molc sec	25	EST	MEYLAN,WM & HOWARD,PH (1993)

CAS #: 058481-07-5 — 2-BENZYLISONIAZID

Formula: $C_{13}H_{13}N_3O$
Mol Weight: 227.27
MP (deg C):　FP (deg C):
BP (deg C):
BP pressure (mm Hg):

Property/Value	Units	Temp	Data Type	Reference
WS 4.32E+003	mg/L	25	EST	MEYLAN,WM ET AL. (1996)
logP 1.01			EXP	HANSCH,C & LEO,AJ (1985)
VP 4.84E-008	mm Hg	25	EST	NEELY,WB & BLAU,GE (1985)
DC	pKa			
HL 1.08E-015	atm m3/mol	25	EST	MEYLAN,WM & HOWARD,PH (1991)
OH 1.17E-011	cm3/molc sec	25	EST	MEYLAN,WM & HOWARD,PH (1993)

CAS #: 058484-09-6 — 1-NO-1-(2-CLET)-3(2,6-OHCYHEX)UREA

Formula: $C_9H_{16}ClN_3O_4$
Mol Weight: 265.70
MP (deg C):　FP (deg C):
BP (deg C):
BP pressure (mm Hg):

Property/Value	Units	Temp	Data Type	Reference
WS 6.91E+002	mg/L	25	EST	MEYLAN,WM ET AL. (1996)
logP 0.16			EXP	HANSCH,C & LEO,AJ (1985)
VP 3.00E-010	mm Hg	25	EST	NEELY,WB & BLAU,GE (1985)
DC	pKa			
HL 2.47E-016	atm m3/mol	25	EST	MEYLAN,WM & HOWARD,PH (1991)
OH 3.99E-011	cm3/molc sec	25	EST	MEYLAN,WM & HOWARD,PH (1993)

CAS #: 058484-10-9 — UREA, N-(2-CHLOROETHYL)-N-NITROSO-N'-(2,3,4,6-TE

Formula: $C_{17}H_{24}ClN_3O_{11}$
Mol Weight: 481.85
MP (deg C):　FP (deg C):
BP (deg C):
BP pressure (mm Hg):

Property/Value	Units	Temp	Data Type	Reference
WS 1.21E+002	mg/L	25	EST	MEYLAN,WM ET AL. (1996)
logP 1.04			EXP	HANSCH,C ET AL. (1995)
VP 2.10E-011	mm Hg	25	EST	NEELY,WB & BLAU,GE (1985)
DC	pKa			
HL 1.32E-022	atm m3/mol	25	EST	MEYLAN,WM & HOWARD,PH (1991)
OH 9.67E-011	cm3/molc sec	25	EST	MEYLAN,WM & HOWARD,PH (1993)

CAS #: 058494-43-2 — TR-1-NO-1-(2-CLET)-3(2-OHCYHEX)UREA

Formula: $C_9H_{16}ClN_3O_3$
Mol Weight: 249.70
MP (deg C):　FP (deg C):
BP (deg C):
BP pressure (mm Hg):

Property/Value	Units	Temp	Data Type	Reference
WS 1.70E+003	mg/L	25	EST	MEYLAN,WM ET AL. (1996)
logP 1.34			EXP	HANSCH,C & LEO,AJ (1985)
VP 1.64E-008	mm Hg	25	EST	NEELY,WB & BLAU,GE (1985)
DC	pKa			
HL 6.76E-015	atm m3/mol	25	EST	MEYLAN,WM & HOWARD,PH (1991)
OH 3.09E-011	cm3/molc sec	25	EST	MEYLAN,WM & HOWARD,PH (1993)

CAS #: 058528-60-2 — DIS. A. 10

Formula: $C_{17}H_{18}Cl_2N_4O_4$
Mol Weight: 413.26
MP (deg C):　FP (deg C):
BP (deg C):
BP pressure (mm Hg):

Property/Value	Units	Temp	Data Type	Reference
WS 4.54E-001	mg/L	25	EXP	BAUGHMAN,GL & PERENICH,TA (1988)
logP 4.98			EST	MEYLAN,WM & HOWARD,PH (1995)
VP 8.05E-015	mm Hg	25	EST	NEELY,WB & BLAU,GE (1985)
DC	pKa			
HL 1.32E-015	atm m3/mol	25	EST	MEYLAN,WM & HOWARD,PH (1991)
OH 1.81E-010	cm3/molc sec	25	EST	MEYLAN,WM & HOWARD,PH (1993)

1225

CAS #: 058742-04-4	BENZOBICYCLO(2,2,1)HEPTENE,2-EN-AMINO

Formula: $C_{11}H_{13}N$

Mol Weight: 159.23

MP (deg C):   FP (deg C):

BP (deg C):

BP pressure (mm Hg):

Property/Value	Units	Temp	Data Type	Reference
WS 1.15E+004	mg/L	25	EST	MEYLAN,WM ET AL. (1996)
logP 2.00			EXP	HANSCH,C & LEO,AJ (1985)
VP 8.71E-003	mm Hg	25	EST	NEELY,WB & BLAU,GE (1985)
DC	pKa			
HL 3.07E-007	atm m3/mol	25	EST	MEYLAN,WM & HOWARD,PH (1991)
OH 4.96E-011	cm3/molc sec	25	EST	MEYLAN,WM & HOWARD,PH (1993)

CAS #: 058742-05-5	1,4-METHANONAPHTHALEN-2-AMINE, 1,2,3,4-TETRAHYDR

Formula: $C_{12}H_{15}N$

Mol Weight: 173.26

MP (deg C):   FP (deg C):

BP (deg C):

BP pressure (mm Hg):

Property/Value	Units	Temp	Data Type	Reference
WS 4.70E+003	mg/L	25	EST	MEYLAN,WM ET AL. (1996)
logP 2.32			EXP	SANGSTER,J (1993)
VP 6.95E-003	mm Hg	25	EST	NEELY,WB & BLAU,GE (1985)
DC				
HL 6.74E-007	atm m3/mol	25	EST	MEYLAN,WM & HOWARD,PH (1991)
OH 9.29E-011	cm3/molc sec	25	EST	MEYLAN,WM & HOWARD,PH (1993)

CAS #: 058802-08-7	1,2,4,7,8-PENTACHLORODIBENZO-P-DIOXIN

Formula: $C_{12}H_3Cl_5O_2$

Mol Weight: 356.42

MP (deg C):   FP (deg C):

BP (deg C):

BP pressure (mm Hg):

Property/Value	Units	Temp	Data Type	Reference
WS 1.53E-004	mg/L	25	EST	MEYLAN,WM ET AL. (1996)
logP 7.56			EST	MEYLAN,WM & HOWARD,PH (1995)
VP 3.96E-007	mm Hg	25	EST	NEELY,WB & BLAU,GE (1985)
DC	pKa			
HL 2.61E-006	atm m3/mol	25	EST	MEYLAN,WM & HOWARD,PH (1991)
OH 1.72E-012	cm3/molc sec	25	EST	MEYLAN,WM & HOWARD,PH (1993)

CAS #: 058802-20-3	DIBENZOFURAN, 1,2,7,8-TETRACHLORO-

Formula: $C_{12}H_4Cl_4O$

Mol Weight: 305.98

MP (deg C):   FP (deg C):

BP (deg C):

BP pressure (mm Hg):

Property/Value	Units	Temp	Data Type	Reference
WS 4.21E-003	mg/L	25	EST	MEYLAN,WM ET AL. (1996)
logP 6.23			EXP	SIJM,DTHM ET AL. (1989)
VP 1.53E-006	mm Hg	25	EST	NEELY,WB & BLAU,GE (1985)
DC	pKa			
HL 1.54E-005	atm m3/mol	25	EST	MEYLAN,WM & HOWARD,PH (1991)
OH 1.65E-013	cm3/molc sec	25	EST	MEYLAN,WM & HOWARD,PH (1993)

CAS #: 058814-67-8	ISOQUINOLINE, 3-(2-PYRIDINYL)-

Formula: $C_{14}H_{10}N_2$

Mol Weight: 206.25

MP (deg C):   FP (deg C):

BP (deg C):

BP pressure (mm Hg):

Property/Value	Units	Temp	Data Type	Reference
WS 1.01E+002	mg/L	25	EST	MEYLAN,WM ET AL. (1996)
logP 3.05			EXP	SANGSTER,J (1993)
VP 4.85E-006	mm Hg	25	EST	NEELY,WB & BLAU,GE (1985)
DC	pKa			
HL 6.92E-011	atm m3/mol	25	EST	MEYLAN,WM & HOWARD,PH (1991)
OH 2.57E-011	cm3/molc sec	25	EST	MEYLAN,WM & HOWARD,PH (1993)

CAS #: 058845-59-3	1NO1(2CLET)3(TRIACETYLXYLOSYL)UREA

Formula: $C_{14}H_{20}ClN_3O_9$

Mol Weight: 409.78

MP (deg C):   FP (deg C):

BP (deg C):

BP pressure (mm Hg):

Property/Value	Units	Temp	Data Type	Reference
WS 3.38E+002	mg/L	25	EST	MEYLAN,WM ET AL. (1996)
logP 1.05			EXP	HANSCH,C & LEO,AJ (1985)
VP 1.41E-009	mm Hg	25	EST	NEELY,WB & BLAU,GE (1985)
DC	pKa			
HL 4.16E-020	atm m3/mol	25	EST	MEYLAN,WM & HOWARD,PH (1991)
OH 8.39E-011	cm3/molc sec	25	EST	MEYLAN,WM & HOWARD,PH (1993)

CAS #: 058889-08-0	PROPANENITRILE, 3-(1-NAPHTHALENYLOXY)-

Formula: $C_{13}H_{11}NO$

Mol Weight: 197.24

MP (deg C):   FP (deg C):

BP (deg C):

BP pressure (mm Hg):

Property/Value	Units	Temp	Data Type	Reference
WS 9.98E+001	mg/L	25	EST	MEYLAN,WM ET AL. (1996)
logP 2.80			EXP	CHAMBERLAIN,K ET AL. (1986)
VP 2.00E-005	mm Hg	25	EST	NEELY,WB & BLAU,GE (1985)
DC	pKa			
HL 1.71E-008	atm m3/mol	25	EST	MEYLAN,WM & HOWARD,PH (1991)
OH 2.01E-010	cm3/molc sec	25	EST	MEYLAN,WM & HOWARD,PH (1993)

CAS #: 058907-81-6	1,4-NAPHTHOQUINONE,2-ACETAMIDO,3-ANILINO

Formula: $C_{18}H_{14}N_2O_3$

Mol Weight: 306.32

MP (deg C):   FP (deg C):

BP (deg C):

BP pressure (mm Hg):

Property/Value	Units	Temp	Data Type	Reference
WS 3.76E+002	mg/L	25	EST	MEYLAN,WM ET AL. (1996)
logP 1.73			EXP	HANSCH,C & LEO,AJ (1985)
VP 3.36E-011	mm Hg	25	EST	NEELY,WB & BLAU,GE (1985)
DC	pKa			
HL 6.48E-017	atm m3/mol	25	EST	MEYLAN,WM & HOWARD,PH (1991)
OH 5.39E-011	cm3/molc sec	25	EST	MEYLAN,WM & HOWARD,PH (1993)

CAS #: 058910-96-6	9H-CARBAZOLE, 1,3,6,8-TETRACHLORO-

Formula: $C_{12}H_5Cl_4N$

Mol Weight: 304.99

MP (deg C):   FP (deg C):

BP (deg C):

BP pressure (mm Hg):

Property/Value	Units	Temp	Data Type	Reference
WS 1.10E-002	mg/L	25	EST	MEYLAN,WM ET AL. (1996)
logP 5.75			EXP	BURKHARD,LP & KUEHL,DW (1986)
VP 1.37E-007	mm Hg	25	EST	NEELY,WB & BLAU,GE (1985)
DC	pKa			
HL 2.61E-008	atm m3/mol	25	EST	MEYLAN,WM & HOWARD,PH (1991)
OH 5.48E-012	cm3/molc sec	25	EST	MEYLAN,WM & HOWARD,PH (1993)

CAS #: 058942-03-3	2,4-IMIDAZOLIDINEDIONE, 5-(2-METHYLPROPYL)-

Formula: $C_7H_{12}N_2O_2$

Mol Weight: 156.19

MP (deg C):   FP (deg C):

BP (deg C):

BP pressure (mm Hg):

Property/Value	Units	Temp	Data Type	Reference
WS 1.75E+003	mg/L	25	EST	MEYLAN,WM ET AL. (1996)
logP 0.35			EXP	SANGSTER,J (1994)
VP 6.19E-007	mm Hg	25	EST	NEELY,WB & BLAU,GE (1985)
DC	pKa			
HL 4.87E-009	atm m3/mol	25	EST	MEYLAN,WM & HOWARD,PH (1991)
OH 1.51E-011	cm3/molc sec	25	EST	MEYLAN,WM & HOWARD,PH (1993)

CAS #: 058944-11-9	NNN-ME-N(1ME-3-CYPENTENE)AMMON I

Formula: $C_{10}H_{20}IN$

Mol Weight: 281.18

MP (deg C):   FP (deg C):

BP (deg C):

BP pressure (mm Hg):

Property/Value	Units	Temp	Data Type	Reference
WS 6.62E+005	mg/L	25	EST	MEYLAN,WM ET AL. (1996)
logP -1.90			EXP	PRATESI,P ET AL. (1986)
VP 2.92E-008	mm Hg	25	EST	NEELY,WB & BLAU,GE (1985)
DC	pKa			
HL 6.60E-012	atm m3/mol	25	EST	MEYLAN,WM & HOWARD,PH (1991)
OH 1.06E-010	cm3/molc sec	25	EST	MEYLAN,WM & HOWARD,PH (1993)

CAS #: 059040-30-1	NAFAZATROM

Formula: $C_{16}H_{16}N_2O_2$

Mol Weight: 268.32

MP (deg C):   FP (deg C):

BP (deg C):

BP pressure (mm Hg):

Property/Value	Units	Temp	Data Type	Reference
WS 8.54E+001	mg/L	25	EST	MEYLAN,WM ET AL. (1996)
logP 2.74			EXP	HANSCH,C ET AL. (1995)
VP 2.21E-008	mm Hg	25	EST	NEELY,WB & BLAU,GE (1985)
DC	pKa			
HL 2.16E-011	atm m3/mol	25	EST	MEYLAN,WM & HOWARD,PH (1991)
OH 2.36E-010	cm3/molc sec	25	EST	MEYLAN,WM & HOWARD,PH (1993)

CAS #: 059082-52-9	4-PYRIDINEPENTANEAMINE

Formula: $C_{10}H_{16}N_2$

Mol Weight: 164.25

MP (deg C):   FP (deg C):

BP (deg C):

BP pressure (mm Hg):

Property/Value	Units	Temp	Data Type	Reference
WS 8.48E+005	mg/L	25	EST	MEYLAN,WM ET AL. (1996)
logP 1.40			EXP	HANSCH,C & LEO,AJ (1985)
VP 4.71E-003	mm Hg	25	EST	NEELY,WB & BLAU,GE (1985)
DC	pKa			
HL 2.49E-009	atm m3/mol	25	EST	MEYLAN,WM & HOWARD,PH (1991)
OH 3.80E-011	cm3/molc sec	25	EST	MEYLAN,WM & HOWARD,PH (1993)

CAS #: 059082-57-4	2-PYRIDINEPENTANEAMINE

Formula: $C_{10}H_{16}N_2$

Mol Weight: 164.25

MP (deg C):   FP (deg C):

BP (deg C):

BP pressure (mm Hg):

Property/Value	Units	Temp	Data Type	Reference
WS 9.92E+005	mg/L	25	EST	MEYLAN,WM ET AL. (1996)
logP 1.32			EXP	HANSCH,C & LEO,AJ (1985)
VP 4.71E-003	mm Hg	25	EST	NEELY,WB & BLAU,GE (1985)
DC	pKa			
HL 2.49E-009	atm m3/mol	25	EST	MEYLAN,WM & HOWARD,PH (1991)
OH 3.80E-011	cm3/molc sec	25	EST	MEYLAN,WM & HOWARD,PH (1993)

CAS #: 059094-49-4	1,4-NAPHTHOQUINONE,2-BUTYLTHIO

Formula: $C_{14}H_{14}O_2S$

Mol Weight: 246.33

MP (deg C):   FP (deg C):

BP (deg C):

BP pressure (mm Hg):

Property/Value	Units	Temp	Data Type	Reference
WS 3.84E+001	mg/L	25	EST	MEYLAN,WM ET AL. (1996)
logP 3.29			EXP	HANSCH,C & LEO,AJ (1985)
VP 1.42E-006	mm Hg	25	EST	NEELY,WB & BLAU,GE (1985)
DC	pKa			
HL 1.65E-010	atm m3/mol	25	EST	MEYLAN,WM & HOWARD,PH (1991)
OH 2.48E-011	cm3/molc sec	25	EST	MEYLAN,WM & HOWARD,PH (1993)

CAS #: 059094-50-7	1,4-NAPHTHOQUINONE,2-BUTYLAMINO-

Formula: $C_{14}H_{15}NO_2$

Mol Weight: 229.28

MP (deg C):   FP (deg C):

BP (deg C):

BP pressure (mm Hg):

Property/Value	Units	Temp	Data Type	Reference
WS 6.91E+002	mg/L	25	EST	MEYLAN,WM ET AL. (1996)
logP 3.11			EXP	HANSCH,C & LEO,AJ (1985)
VP 2.73E-006	mm Hg	25	EST	NEELY,WB & BLAU,GE (1985)
DC	pKa			
HL 4.34E-011	atm m3/mol	25	EST	MEYLAN,WM & HOWARD,PH (1991)
OH 8.35E-011	cm3/molc sec	25	EST	MEYLAN,WM & HOWARD,PH (1993)

CAS #: 059122-46-2				MISOPROSTOL

Formula: $C_{22}H_{38}O_5$

Mol Weight: 382.55

MP (deg C): | FP (deg C):

BP (deg C):

BP pressure (mm Hg):

Property/ Value	Units	Temp	Data Type	Reference
WS 1.86E-001	mg/L	25	EST	MEYLAN,WM ET AL. (1996)
logP 4.96			EST	MEYLAN,WM & HOWARD,PH (1995)
VP	mm Hg			
DC	pKa			
HL 1.00E-011	atm m3/mol	25	EST	MEYLAN,WM & HOWARD,PH (1991)
OH 1.15E-010	cm3/molc sec	25	EST	MEYLAN,WM & HOWARD,PH (1993)

CAS #: 059170-23-9				BEVANTOLOL

Formula: $C_{20}H_{27}NO_4$

Mol Weight: 345.44

MP (deg C): 137-138 | FP (deg C):

BP (deg C):

BP pressure (mm Hg):

Property/ Value	Units	Temp	Data Type	Reference
WS 1.84E+002	mg/L	25	EST	MEYLAN,WM ET AL. (1996)
logP 3.00			EXP	HANSCH,C & LEO,AJ (1985)
VP 1.10E-010	mm Hg	25	EST	NEELY,WB & BLAU,GE (1985)
DC	pKa			
HL 1.92E-015	atm m3/mol	25	EST	MEYLAN,WM & HOWARD,PH (1991)
OH 2.07E-010	cm3/molc sec	25	EST	MEYLAN,WM & HOWARD,PH (1993)

CAS #: 059177-47-8				4-BROMO-1-METHYL-5-NITRO-1H-IMIDAZOLE

Formula: $C_4H_4BrN_3O_2$

Mol Weight: 206.00

MP (deg C): | FP (deg C):

BP (deg C):

BP pressure (mm Hg):

Property/ Value	Units	Temp	Data Type	Reference
WS 3.65E+003	mg/L	25	EST	MEYLAN,WM ET AL. (1996)
logP 0.77			EXP	SUWINSKI,J ET AL. (1985)
VP 9.37E-005	mm Hg	25	EST	NEELY,WB & BLAU,GE (1985)
DC	pKa			
HL 1.26E-007	atm m3/mol	25	EST	MEYLAN,WM & HOWARD,PH (1991)
OH 1.43E-012	cm3/molc sec	25	EST	MEYLAN,WM & HOWARD,PH (1993)

CAS #: 059277-86-0				ETHANOL, 2-[(2,6-DIAMINO-9H-PURIN-9-YL)METHOXY]-

Formula: $C_8H_{12}N_6O_2$

Mol Weight: 224.22

MP (deg C): | FP (deg C):

BP (deg C):

BP pressure (mm Hg):

Property/ Value	Units	Temp	Data Type	Reference
WS 1.29E+004	mg/L	25	EST	MEYLAN,WM ET AL. (1996)
logP -1.06			EXP	HANSCH,C ET AL. (1995)
VP 9.42E-012	mm Hg	25	EST	NEELY,WB & BLAU,GE (1985)
DC	pKa			
HL 1.07E-018	atm m3/mol	25	EST	MEYLAN,WM & HOWARD,PH (1991)
OH 2.17E-010	cm3/molc sec	25	EST	MEYLAN,WM & HOWARD,PH (1993)

CAS #: 059277-89-3				ACYCLOVIR

Formula: $C_8H_{13}N_5O_3$

Mol Weight: 227.22

MP (deg C): | FP (deg C):

BP (deg C):

BP pressure (mm Hg):

Property/ Value	Units	Temp	Data Type	Reference
WS 3.40E+004	mg/L	25	EST	MEYLAN,WM ET AL. (1996)
logP -1.56			EXP	KRISTL,A ET AL. (1993)
VP 7.47E-015	mm Hg	25	EST	NEELY,WB & BLAU,GE (1985)
DC	pKa			
HL 3.18E-022	atm m3/mol	25	EST	MEYLAN,WM & HOWARD,PH (1991)
OH 7.94E-011	cm3/molc sec	25	EST	MEYLAN,WM & HOWARD,PH (1993)

CAS #: 059280-70-5				2-FLUORO-4-CHLOROACETANILIDE

Formula: C_8H_7ClFNO

Mol Weight: 187.60

MP (deg C): 47-50 | FP (deg C):

BP (deg C):

BP pressure (mm Hg):

Property/ Value	Units	Temp	Data Type	Reference
WS 1.18E+003	mg/L	25	EST	MEYLAN,WM ET AL. (1996)
logP 1.91			EXP	NAKAGAWA,Y ET AL. (1992)
VP 1.08E-004	mm Hg	25	EST	NEELY,WB & BLAU,GE (1985)
DC	pKa			
HL 5.34E-009	atm m3/mol	25	EST	MEYLAN,WM & HOWARD,PH (1991)
OH 1.33E-012	cm3/molc sec	25	EST	MEYLAN,WM & HOWARD,PH (1993)

CAS #: 059303-10-5				PYRAZINE, 2-CHLORO-5-METHYL-

Formula: $C_5H_5ClN_2$

Mol Weight: 128.56

MP (deg C): | FP (deg C):

BP (deg C):

BP pressure (mm Hg):

Property/ Value	Units	Temp	Data Type	Reference
WS 1.11E+004	mg/L	25	EST	MEYLAN,WM ET AL. (1996)
logP 1.08			EXP	YAMAGAMI,C ET AL. (1991)
VP 8.98E-001	mm Hg	25	EST	NEELY,WB & BLAU,GE (1985)
DC	pKa			
HL 1.51E-004	atm m3/mol	25	EST	MEYLAN,WM & HOWARD,PH (1991)
OH 3.34E-013	cm3/molc sec	25	EST	MEYLAN,WM & HOWARD,PH (1993)

CAS #: 059338-93-1				1H-BENZOTRIAZOLE-5-CARBOXAMIDE, 6-METHOXY-N-[[1-

Formula: $C_{16}H_{21}N_5O_2$

Mol Weight: 315.38

MP (deg C): 139 | FP (deg C):

BP (deg C):

BP pressure (mm Hg):

Property/ Value	Units	Temp	Data Type	Reference
WS 2.95E+002	mg/L	25	EST	MEYLAN,WM ET AL. (1996)
logP 1.79			EXP	MANNHOLD,R ET AL. (1990)
VP 1.17E-011	mm Hg	25	EST	NEELY,WB & BLAU,GE (1985)
DC	pKa			
HL 5.26E-018	atm m3/mol	25	EST	MEYLAN,WM & HOWARD,PH (1991)
OH 1.47E-010	cm3/molc sec	25	EST	MEYLAN,WM & HOWARD,PH (1993)

CAS #: 059359-46-5				BUTANAMIDE, 2-AMINO-N-(2,6-DIMETHYLPHENYL)-

Formula: $C_{12}H_{18}N_2O$

Mol Weight: 206.29

MP (deg C): FP (deg C):

BP (deg C):

BP pressure (mm Hg):

Property/ Value	Units	Temp	Data Type	Reference
WS 7.79E+003	mg/L	25	EST	MEYLAN,WM ET AL. (1996)
logP 0.84			EXP	SANGSTER,J (1993)
VP 2.07E-006	mm Hg	25	EST	NEELY,WB & BLAU,GE (1985)
DC	pKa			
HL 1.36E-012	atm m3/mol	25	EST	MEYLAN,WM & HOWARD,PH (1991)
OH 1.11E-010	cm3/molc sec	25	EST	MEYLAN,WM & HOWARD,PH (1993)

CAS #: 059365-60-5				1-CHLORO-2-(1,2-ETHANEDIOL)BENZENE

Formula: $C_8H_9ClO_2$

Mol Weight: 172.61

MP (deg C): FP (deg C):

BP (deg C):

BP pressure (mm Hg):

Property/ Value	Units	Temp	Data Type	Reference
WS 5.99E+003	mg/L	25	EST	MEYLAN,WM ET AL. (1996)
logP 1.17			EXP	CHEM INSPECT TEST INST (1992)
VP 2.85E-005	mm Hg	25	EST	NEELY,WB & BLAU,GE (1985)
DC	pKa			
HL 7.82E-009	atm m3/mol	25	EST	MEYLAN,WM & HOWARD,PH (1991)
OH 1.43E-011	cm3/molc sec	25	EST	MEYLAN,WM & HOWARD,PH (1993)

CAS #: 059414-37-8				CIS-1-NO-1(2-CLET)3(AC OXYCYHEX)UREA

Formula: $C_{11}H_{18}ClN_3O_4$

Mol Weight: 291.74

MP (deg C): FP (deg C):

BP (deg C):

BP pressure (mm Hg):

Property/ Value	Units	Temp	Data Type	Reference
WS 2.79E+002	mg/L	25	EST	MEYLAN,WM ET AL. (1996)
logP 1.98			EXP	HANSCH,C & LEO,AJ (1985)
VP 3.34E-007	mm Hg	25	EST	NEELY,WB & BLAU,GE (1985)
DC	pKa			
HL 4.40E-013	atm m3/mol	25	EST	MEYLAN,WM & HOWARD,PH (1991)
OH 2.53E-011	cm3/molc sec	25	EST	MEYLAN,WM & HOWARD,PH (1993)

CAS #: 059414-38-9				1-NO-1-(2-CLET)-3(ACETYLOXYCYHEX)UREA

Formula: $C_{11}H_{18}ClN_3O_4$

Mol Weight: 291.74

MP (deg C): FP (deg C):

BP (deg C):

BP pressure (mm Hg):

Property/ Value	Units	Temp	Data Type	Reference
WS 3.08E+002	mg/L	25	EST	MEYLAN,WM ET AL. (1996)
logP 1.93			EXP	HANSCH,C & LEO,AJ (1985)
VP 3.34E-007	mm Hg	25	EST	NEELY,WB & BLAU,GE (1985)
DC	pKa			
HL 4.40E-013	atm m3/mol	25	EST	MEYLAN,WM & HOWARD,PH (1991)
OH 2.53E-011	cm3/molc sec	25	EST	MEYLAN,WM & HOWARD,PH (1993)

CAS #: 059445-83-9				1-(THIOPHEN-2-YL)ACETALDEHYDE OXIME

Formula: C_6H_9NOS

Mol Weight: 143.21

MP (deg C): FP (deg C):

BP (deg C):

BP pressure (mm Hg):

Property/ Value	Units	Temp	Data Type	Reference
WS 2.36E+003	mg/L	25	EST	MEYLAN,WM ET AL. (1996)
logP 1.81			EXP	HANSCH,C & LEO,AJ (1985)
VP 8.07E-004	mm Hg	25	EST	NEELY,WB & BLAU,GE (1985)
DC	pKa			
HL 2.58E-007	atm m3/mol	25	EST	MEYLAN,WM & HOWARD,PH (1991)
OH 2.12E-011	cm3/molc sec	25	EST	MEYLAN,WM & HOWARD,PH (1993)

CAS #: 059465-40-6				2-(245-TRICL PH IMINO)IMIDAZOLIDINE

Formula: $C_9H_8Cl_3N_3$

Mol Weight: 264.54

MP (deg C): FP (deg C):

BP (deg C):

BP pressure (mm Hg):

Property/ Value	Units	Temp	Data Type	Reference
WS 3.10E+002	mg/L	25	EST	MEYLAN,WM ET AL. (1996)
logP 2.11			EXP	HANSCH,C & LEO,AJ (1985)
VP 6.13E-008	mm Hg	25	EST	NEELY,WB & BLAU,GE (1985)
DC	pKa			
HL 1.12E-011	atm m3/mol	25	EST	MEYLAN,WM & HOWARD,PH (1991)
OH 7.50E-011	cm3/molc sec	25	EST	MEYLAN,WM & HOWARD,PH (1993)

CAS #: 059465-51-9				2-(246-TRICL PH IMINO)IMIDAZOLIDINE

Formula: $C_9H_8Cl_3N_3$

Mol Weight: 264.54

MP (deg C): FP (deg C):

BP (deg C):

BP pressure (mm Hg):

Property/ Value	Units	Temp	Data Type	Reference
WS 2.70E+002	mg/L	25	EST	MEYLAN,WM ET AL. (1996)
logP 2.18			EXP	HANSCH,C & LEO,AJ (1985)
VP 6.13E-008	mm Hg	25	EST	NEELY,WB & BLAU,GE (1985)
DC	pKa			
HL 1.12E-011	atm m3/mol	25	EST	MEYLAN,WM & HOWARD,PH (1991)
OH 7.36E-011	cm3/molc sec	25	EST	MEYLAN,WM & HOWARD,PH (1993)

CAS #: 059477-92-8				1-PHENYL-3,3-DIMETHYLTRIAZINE OXIDE

Formula: $C_8H_{11}N_3O$

Mol Weight: 165.20

MP (deg C): FP (deg C):

BP (deg C):

BP pressure (mm Hg):

Property/ Value	Units	Temp	Data Type	Reference
WS 1.88E+003	mg/L	25	EST	MEYLAN,WM ET AL. (1996)
logP 1.80			EXP	HANSCH,C & LEO,AJ (1985)
VP 4.00E-007	mm Hg	25	EST	NEELY,WB & BLAU,GE (1985)
DC	pKa			
HL 4.31E-014	atm m3/mol	25	EST	MEYLAN,WM & HOWARD,PH (1991)
OH 2.95E-012	cm3/molc sec	25	EST	MEYLAN,WM & HOWARD,PH (1993)

CAS #: 059481-28-6				2,4-DIAMINO-5-(3-MEOBENZYL)PYRIMIDINE
Formula: $C_{12}H_{14}N_4O$				
Mol Weight: 230.27				
MP (deg C):		FP (deg C):		
BP (deg C):				
BP pressure (mm Hg):				

Property/ Value	Units	Temp	Data Type	Reference
WS 1.47E+003	mg/L	25	EST	MEYLAN,WM ET AL. (1996)
logP 1.54			EXP	HANSCH,C & LEO,AJ (1985)
VP 2.58E-007	mm Hg	25	EST	NEELY,WB & BLAU,GE (1985)
DC	pKa			
HL 6.82E-012	atm m3/mol	25	EST	MEYLAN,WM & HOWARD,PH (1991)
OH 5.67E-011	cm3/molc sec	25	EST	MEYLAN,WM & HOWARD,PH (1993)

CAS #: 059521-86-7				CEPHALOTHIN ANALOG(7-DICLPH-S-ACAM
Formula: $C_{18}H_{16}Cl_2N_2O_6S_2$				
Mol Weight: 491.37				
MP (deg C):		FP (deg C):		
BP (deg C):				
BP pressure (mm Hg):				

Property/ Value	Units	Temp	Data Type	Reference
WS 3.11E+001	mg/L	25	EST	MEYLAN,WM ET AL. (1996)
logP 1.66			EXP	HANSCH,C & LEO,AJ (1985)
VP 1.31E-016	mm Hg	25	EST	NEELY,WB & BLAU,GE (1985)
DC	pKa			
HL 4.65E-019	atm m3/mol	25	EST	MEYLAN,WM & HOWARD,PH (1991)
OH 1.18E-010	cm3/molc sec	25	EST	MEYLAN,WM & HOWARD,PH (1993)

CAS #: 059660-23-0				GUANIDINE, N-METHYL-N'-[2-[[(5-METHYL-1H-IMIDAZO
Formula: $C_9H_{16}N_6O_2S$				
Mol Weight: 272.33				
MP (deg C):		FP (deg C):		
BP (deg C):				
BP pressure (mm Hg):				

Property/ Value	Units	Temp	Data Type	Reference
WS 1.24E+004	mg/L	25	EST	MEYLAN,WM ET AL. (1996)
logP 0.18			EXP	HANSCH,C ET AL. (1995)
VP 1.29E-009	mm Hg	25	EST	NEELY,WB & BLAU,GE (1985)
DC	pKa			
HL 1.90E-017	atm m3/mol	25	EST	MEYLAN,WM & HOWARD,PH (1991)
OH 2.10E-010	cm3/molc sec	25	EST	MEYLAN,WM & HOWARD,PH (1993)

CAS #: 059669-26-0				THIODICARB
Formula: $C_{10}H_{18}N_4O_4S_3$				
Mol Weight: 354.47				
MP (deg C): 173-174		FP (deg C):		
BP (deg C):				
BP pressure (mm Hg):				

Property/ Value	Units	Temp	Data Type	Reference
WS 3.50E+001	mg/L	25	EXP	SHIU,WY ET AL. (1990)
logP 1.70			EXP	HANSCH,C & LEO,AJ (1985)
VP 7.00E-005	mm Hg	25	EXP	WEBER,JB (1994)
DC	pKa			
HL 9.33E-007	atm m3/mol	25	EST	VP/WSOL
OH 3.70E-011	cm3/molc sec	25	EST	MEYLAN,WM & HOWARD,PH (1993)

CAS #: 059700-57-1				PENTACHLORO(2,2,3,3-TETRAFLUOROPROPOXY)CYCLOTRI*
Formula: $C_3H_3Cl_5F_4N_3OP_3$				
Mol Weight: 443.26				
MP (deg C):		FP (deg C):		
BP (deg C):				
BP pressure (mm Hg):				

Property/ Value	Units	Temp	Data Type	Reference
WS 2.29E-001	mg/L	25	EST	MEYLAN,WM ET AL. (1996)
logP 4.52			EST	MEYLAN,WM & HOWARD,PH (1995)
VP 1.90E-005	mm Hg	25	EST	NEELY,WB & BLAU,GE (1985)
DC	pKa			
HL 7.25E-008	atm m3/mol	25	EST	MEYLAN,WM & HOWARD,PH (1991)
OH 3.45E-013	cm3/molc sec	25	EST	MEYLAN,WM & HOWARD,PH (1993)

CAS #: 059708-19-9				1-(4-CARBAMOYLPHENYL)-3-METHYL-3-ETHYLTRIAZENE
Formula: $C_{10}H_{14}N_4O$				
Mol Weight: 206.25				
MP (deg C):		FP (deg C):		
BP (deg C):				
BP pressure (mm Hg):				

Property/ Value	Units	Temp	Data Type	Reference
WS 1.46E+003	mg/L	25	EST	MEYLAN,WM ET AL. (1996)
logP 1.69			EXP	HANSCH,C ET AL. (1995)
VP 2.59E-006	mm Hg	25	EST	NEELY,WB & BLAU,GE (1985)
DC	pKa			
HL 2.93E-013	atm m3/mol	25	EST	MEYLAN,WM & HOWARD,PH (1991)
OH 1.24E-011	cm3/molc sec	25	EST	MEYLAN,WM & HOWARD,PH (1993)

CAS #: 059708-20-2				1-(4-CARBAMOYLPHENYL)-3-METHYL-3-HYDROXYETHYLTRI
Formula: $C_{10}H_{14}N_4O_2$				
Mol Weight: 222.25				
MP (deg C):		FP (deg C):		
BP (deg C):				
BP pressure (mm Hg):				

Property/ Value	Units	Temp	Data Type	Reference
WS 9.84E+002	mg/L	25	EST	MEYLAN,WM ET AL. (1996)
logP 0.26			EXP	HANSCH,C ET AL. (1995)
VP 2.13E-009	mm Hg	25	EST	NEELY,WB & BLAU,GE (1985)
DC	pKa			
HL 1.07E-017	atm m3/mol	25	EST	MEYLAN,WM & HOWARD,PH (1991)
OH 1.84E-011	cm3/molc sec	25	EST	MEYLAN,WM & HOWARD,PH (1993)

CAS #: 059708-21-3				4-(3-ME-3-BU-1-TRIAZENO)BENZAMIDE
Formula: $C_{12}H_{18}N_4O$				
Mol Weight: 234.30				
MP (deg C):		FP (deg C):		
BP (deg C):				
BP pressure (mm Hg):				

Property/ Value	Units	Temp	Data Type	Reference
WS 2.29E+002	mg/L	25	EST	MEYLAN,WM ET AL. (1996)
logP 2.46			EXP	SANGSTER,J (1994)
VP 5.70E-007	mm Hg	25	EST	NEELY,WB & BLAU,GE (1985)
DC	pKa			
HL 5.16E-013	atm m3/mol	25	EST	MEYLAN,WM & HOWARD,PH (1991)
OH 1.70E-011	cm3/molc sec	25	EST	MEYLAN,WM & HOWARD,PH (1993)

059708-22-4 — 1-(4-CARBAMOYLPHENYL)-3-METHYL-3-PENTYLTRIAZENE

CAS #: 059708-22-4

Formula: $C_{13}H_{20}N_4O$

Mol Weight: 248.33

MP (deg C):   **FP (deg C):**

BP (deg C):

BP pressure (mm Hg):

Property/Value	Units	Temp	Data Type	Reference
WS 4.65E+001	mg/L	25	EST	MEYLAN,WM ET AL. (1996)
logP 3.18			EXP	HANSCH,C ET AL. (1995)
VP 2.44E-007	mm Hg	25	EST	NEELY,WB & BLAU,GE (1985)
DC	pKa			
HL 6.85E-013	atm m3/mol	25	EST	MEYLAN,WM & HOWARD,PH (1991)
OH 1.84E-011	cm3/molc sec	25	EST	MEYLAN,WM & HOWARD,PH (1993)

059708-23-5 — 1-(4-CARBAMOYLPHENYL)-3-I-PROPYL-3-METHYLTRIAZEN

CAS #: 059708-23-5

Formula: $C_{11}H_{16}N_4O$

Mol Weight: 220.28

MP (deg C):   **FP (deg C):**

BP (deg C):

BP pressure (mm Hg):

Property/Value	Units	Temp	Data Type	Reference
WS 6.46E+002	mg/L	25	EST	MEYLAN,WM ET AL. (1996)
logP 2.02			EXP	HANSCH,C ET AL. (1995)
VP 1.95E-006	mm Hg	25	EST	NEELY,WB & BLAU,GE (1985)
DC	pKa			
HL 3.89E-013	atm m3/mol	25	EST	MEYLAN,WM & HOWARD,PH (1991)
OH 2.20E-011	cm3/molc sec	25	EST	MEYLAN,WM & HOWARD,PH (1993)

059708-24-6 — P-(3-BENZYL-3ME-1-TRIAZENO)BENZAMIDE

CAS #: 059708-24-6

Formula: $C_{15}H_{16}N_4O$

Mol Weight: 268.32

MP (deg C):   **FP (deg C):**

BP (deg C):

BP pressure (mm Hg):

Property/Value	Units	Temp	Data Type	Reference
WS 8.21E+001	mg/L	25	EST	MEYLAN,WM ET AL. (1996)
logP 2.76			EXP	HANSCH,C & LEO,AJ (1985)
VP 1.40E-008	mm Hg	25	EST	NEELY,WB & BLAU,GE (1985)
DC	pKa			
HL 1.78E-014	atm m3/mol	25	EST	MEYLAN,WM & HOWARD,PH (1991)
OH 1.71E-011	cm3/molc sec	25	EST	MEYLAN,WM & HOWARD,PH (1993)

059708-25-7 — 4-(3-ME-3-T-BU-1-TRIAZENO)BENZAMIDE

CAS #: 059708-25-7

Formula: $C_{12}H_{18}N_4O$

Mol Weight: 234.30

MP (deg C):   **FP (deg C):**

BP (deg C):

BP pressure (mm Hg):

Property/Value	Units	Temp	Data Type	Reference
WS 1.70E+002	mg/L	25	EST	MEYLAN,WM ET AL. (1996)
logP 2.61			EXP	HANSCH,C & LEO,AJ (1985)
VP 1.23E-006	mm Hg	25	EST	NEELY,WB & BLAU,GE (1985)
DC	pKa			
HL 5.16E-013	atm m3/mol	25	EST	MEYLAN,WM & HOWARD,PH (1991)
OH 4.09E-012	cm3/molc sec	25	EST	MEYLAN,WM & HOWARD,PH (1993)

059729-31-6 — LORCAINIDE

CAS #: 059729-31-6

Formula: $C_{22}H_{27}ClN_2O$

Mol Weight: 370.93

MP (deg C): 263   **FP (deg C):**

BP (deg C):

BP pressure (mm Hg):

Property/Value	Units	Temp	Data Type	Reference
WS 3.34E-001	mg/L	25	EST	MEYLAN,WM ET AL. (1996)
logP 4.85			EXP	MANNHOLD,R ET AL. (1990)
VP 2.01E-009	mm Hg	25	EST	NEELY,WB & BLAU,GE (1985)
DC	pKa			
HL 6.82E-012	atm m3/mol	25	EST	MEYLAN,WM & HOWARD,PH (1991)
OH 1.30E-010	cm3/molc sec	25	EST	MEYLAN,WM & HOWARD,PH (1993)

059729-37-2 — 1H-IMIDAZOLE, 1-METHYL-2-[[4-(METHYLTHIO)PHENOXY

CAS #: 059729-37-2

Formula: $C_{12}H_{13}N_3O_3S$

Mol Weight: 279.32

MP (deg C):   **FP (deg C):**

BP (deg C):

BP pressure (mm Hg):

Property/Value	Units	Temp	Data Type	Reference
WS 4.83E+001	mg/L	25	EST	MEYLAN,WM ET AL. (1996)
logP 2.50			EXP	SANGSTER,J (1994)
VP 1.49E-008	mm Hg	25	EST	NEELY,WB & BLAU,GE (1985)
DC	pKa			
HL 4.40E-011	atm m3/mol	25	EST	MEYLAN,WM & HOWARD,PH (1991)
OH 2.39E-011	cm3/molc sec	25	EST	MEYLAN,WM & HOWARD,PH (1993)

059756-60-4 — FLURIDONE

CAS #: 059756-60-4

Formula: $C_{19}H_{14}F_3NO$

Mol Weight: 329.32

MP (deg C): 151-154   **FP (deg C):**

BP (deg C):

BP pressure (mm Hg):

Property/Value	Units	Temp	Data Type	Reference
WS 1.20E+001	mg/L	25	EXP	SHIU,WY ET AL. (1990) @ pH=7
logP 3.16			EXP	HANSCH,C ET AL. (1995)
VP 9.75E-003	mm Hg		EXP	WORTHING,CR & WALKER,SB (1983)
DC	pKa			
HL 8.10E-004	atm m3/mol	25	EST	VP/WSOL
OH 1.59E-010	cm3/molc sec	25	EST	MEYLAN,WM & HOWARD,PH (1993)

059772-33-7 — 2-(2,6-DIFL PH IMINO)IMIDAZOLIDINE

CAS #: 059772-33-7

Formula: $C_9H_9F_2N_3$

Mol Weight: 197.19

MP (deg C):   **FP (deg C):**

BP (deg C):

BP pressure (mm Hg):

Property/Value	Units	Temp	Data Type	Reference
WS 7.26E+003	mg/L	25	EST	MEYLAN,WM ET AL. (1996)
logP 0.93			EXP	HANSCH,C & LEO,AJ (1985)
VP 5.63E-006	mm Hg	25	EST	NEELY,WB & BLAU,GE (1985)
DC	pKa			
HL 3.73E-011	atm m3/mol	25	EST	MEYLAN,WM & HOWARD,PH (1991)
OH 7.75E-011	cm3/molc sec	25	EST	MEYLAN,WM & HOWARD,PH (1993)

CAS #: 059772-37-1	2-(2-CL-6-F PH IMINO)IMIDAZOLIDINE

Formula: $C_9H_9ClFN_3$

Mol Weight: 213.64

MP (deg C):　FP (deg C):

BP (deg C):

BP pressure (mm Hg):

Property/ Value	Units	Temp	Data Type	Reference
WS 2.19E+004	mg/L	25	EST	MEYLAN,WM ET AL. (1996)
logP 1.45			EXP	HANSCH,C & LEO,AJ (1985)
VP 9.94E-005	mm Hg	25	EST	NEELY,WB & BLAU,GE (1985)
DC	pKa			
HL 2.37E-011	atm m3/mol	25	EST	MEYLAN,WM & HOWARD,PH (1991)
OH 1.45E-010	cm3/molc sec	25	EST	MEYLAN,WM & HOWARD,PH (1993)

CAS #: 059777-58-1	PROPYL 4-SULFAMYLBENZOATE

Formula: $C_{10}H_{13}NO_4S$

Mol Weight: 243.28

MP (deg C):　FP (deg C):

BP (deg C):

BP pressure (mm Hg):

Property/ Value	Units	Temp	Data Type	Reference
WS 8.25E+002	mg/L	25	EST	MEYLAN,WM ET AL. (1996)
logP 1.75			EXP	HANSCH,C & LEO,AJ (1985)
VP 2.81E-006	mm Hg	25	EST	NEELY,WB & BLAU,GE (1985)
DC				
HL 4.79E-009	atm m3/mol	25	EST	MEYLAN,WM & HOWARD,PH (1991)
OH 3.29E-012	cm3/molc sec	25	EST	MEYLAN,WM & HOWARD,PH (1993)

CAS #: 059777-59-2	BUTYL 4-SULFAMYLBENZOATE

Formula: $C_{11}H_{15}NO_4S$

Mol Weight: 257.31

MP (deg C):　FP (deg C):

BP (deg C):

BP pressure (mm Hg):

Property/ Value	Units	Temp	Data Type	Reference
WS 2.16E+002	mg/L	25	EST	MEYLAN,WM ET AL. (1996)
logP 2.34			EXP	HANSCH,C & LEO,AJ (1985)
VP 1.45E-006	mm Hg	25	EST	NEELY,WB & BLAU,GE (1985)
DC	pKa			
HL 6.36E-009	atm m3/mol	25	EST	MEYLAN,WM & HOWARD,PH (1991)
OH 4.70E-012	cm3/molc sec	25	EST	MEYLAN,WM & HOWARD,PH (1993)

CAS #: 059777-60-5	PENTYL 4-SULFAMYLBENZOATE

Formula: $C_{12}H_{17}NO_4S$

Mol Weight: 271.34

MP (deg C):　FP (deg C):

BP (deg C):

BP pressure (mm Hg):

Property/ Value	Units	Temp	Data Type	Reference
WS 8.71E+001	mg/L	25	EST	MEYLAN,WM ET AL. (1996)
logP 2.71			EXP	HANSCH,C & LEO,AJ (1985)
VP 6.26E-007	mm Hg	25	EST	NEELY,WB & BLAU,GE (1985)
DC	pKa			
HL 8.45E-009	atm m3/mol	25	EST	MEYLAN,WM & HOWARD,PH (1991)
OH 6.12E-012	cm3/molc sec	25	EST	MEYLAN,WM & HOWARD,PH (1993)

CAS #: 059777-62-7	N-ETHYL 4-SULFAMYLBENZAMIDE

Formula: $C_9H_{12}N_2O_3S$

Mol Weight: 228.27

MP (deg C):　FP (deg C):

BP (deg C):

BP pressure (mm Hg):

Property/ Value	Units	Temp	Data Type	Reference
WS 1.44E+003	mg/L	25	EST	MEYLAN,WM ET AL. (1996)
logP 0.03			EXP	HANSCH,C & LEO,AJ (1985)
VP 3.62E-008	mm Hg	25	EST	NEELY,WB & BLAU,GE (1985)
DC	pKa			
HL 5.06E-013	atm m3/mol	25	EST	MEYLAN,WM & HOWARD,PH (1991)
OH 9.88E-012	cm3/molc sec	25	EST	MEYLAN,WM & HOWARD,PH (1993)

CAS #: 059777-63-8	N-BUTYL 4-SULFAMYLBENZAMIDE

Formula: $C_{11}H_{16}N_2O_3S$

Mol Weight: 256.33

MP (deg C):　FP (deg C):

BP (deg C):

BP pressure (mm Hg):

Property/ Value	Units	Temp	Data Type	Reference
WS 2.77E+003	mg/L	25	EST	MEYLAN,WM ET AL. (1996)
logP 1.05			EXP	HANSCH,C & LEO,AJ (1985)
VP 7.22E-009	mm Hg	25	EST	NEELY,WB & BLAU,GE (1985)
DC	pKa			
HL 8.91E-013	atm m3/mol	25	EST	MEYLAN,WM & HOWARD,PH (1991)
OH 1.33E-011	cm3/molc sec	25	EST	MEYLAN,WM & HOWARD,PH (1993)

CAS #: 059777-64-9	N-PENTYL 4-SULFAMYLBENZAMIDE

Formula: $C_{12}H_{18}N_2O_3S$

Mol Weight: 270.35

MP (deg C):　FP (deg C):

BP (deg C):

BP pressure (mm Hg):

Property/ Value	Units	Temp	Data Type	Reference
WS 9.34E+002	mg/L	25	EST	MEYLAN,WM ET AL. (1996)
logP 1.51			EXP	HANSCH,C & LEO,AJ (1985)
VP 3.21E-009	mm Hg	25	EST	NEELY,WB & BLAU,GE (1985)
DC	pKa			
HL 1.18E-012	atm m3/mol	25	EST	MEYLAN,WM & HOWARD,PH (1991)
OH 1.47E-011	cm3/molc sec	25	EST	MEYLAN,WM & HOWARD,PH (1993)

CAS #: 059777-65-0	4-SULFAMYLBENZAMIDE,N-HEXYL

Formula: $C_{13}H_{20}N_2O_3S$

Mol Weight: 284.38

MP (deg C):　FP (deg C):

BP (deg C):

BP pressure (mm Hg):

Property/ Value	Units	Temp	Data Type	Reference
WS 2.69E+002	mg/L	25	EST	MEYLAN,WM ET AL. (1996)
logP 2.05			EXP	HANSCH,C & LEO,AJ (1985)
VP 1.42E-009	mm Hg	25	EST	NEELY,WB & BLAU,GE (1985)
DC	pKa			
HL 1.57E-012	atm m3/mol	25	EST	MEYLAN,WM & HOWARD,PH (1991)
OH 1.61E-011	cm3/molc sec	25	EST	MEYLAN,WM & HOWARD,PH (1993)

CAS #: 059857-86-2 — 1-METHYL-4-METHOXYCARBONYL-2-PYRROLIDONE

Formula:	$C_7H_{11}NO_3$	
Mol Weight:	157.17	
MP (deg C):		FP (deg C):
BP (deg C):		
BP pressure (mm Hg):		

Property/Value	Units	Temp	Data Type	Reference
WS 3.02E+005	mg/L	25	EST	MEYLAN,WM ET AL. (1996)
logP -0.74			EXP	SASAKI,H ET AL. (1988)
VP 3.21E-003	mm Hg	25	EST	NEELY,WB & BLAU,GE (1985)
DC	pKa			
HL 7.51E-011	atm m3/mol	25	EST	MEYLAN,WM & HOWARD,PH (1991)
OH 2.57E-011	cm3/molc sec	25	EST	MEYLAN,WM & HOWARD,PH (1993)

CAS #: 059863-59-1 — 3,4-DICHLORO-N-NITROSOPYRROLIDINE

Formula:	$C_4H_6Cl_2N_2O$	
Mol Weight:	169.01	
MP (deg C):		FP (deg C):
BP (deg C):		
BP pressure (mm Hg):		

Property/Value	Units	Temp	Data Type	Reference
WS 1.22E+004	mg/L	25	EST	MEYLAN,WM ET AL. (1996)
logP 0.83			EXP	HANSCH,C & LEO,AJ (1985)
VP 8.91E-003	mm Hg	25	EST	NEELY,WB & BLAU,GE (1985)
DC	pKa			
HL 1.98E-007	atm m3/mol	25	EST	MEYLAN,WM & HOWARD,PH (1991)
OH 5.53E-012	cm3/molc sec	25	EST	MEYLAN,WM & HOWARD,PH (1993)

CAS #: 059886-40-7 — 2,5-CYCLOHEXADIENE-1,4-DIONE, 2,5-BIS(1-AZIRIDIN

Formula:	$C_{22}H_{34}N_6O_4$	
Mol Weight:	446.55	
MP (deg C):		FP (deg C):
BP (deg C):		
BP pressure (mm Hg):		

Property/Value	Units	Temp	Data Type	Reference
WS 7.25E+004	mg/L	25	EST	MEYLAN,WM ET AL. (1996)
logP -0.77			EXP	SANGSTER,J (1993)
VP 9.13E-018	mm Hg	25	EST	NEELY,WB & BLAU,GE (1985)
DC	pKa			
HL 8.61E-030	atm m3/mol	25	EST	MEYLAN,WM & HOWARD,PH (1991)
OH 5.42E-010	cm3/molc sec	25	EST	MEYLAN,WM & HOWARD,PH (1993)

CAS #: 059886-46-3 — 2,5-CYCLOHEXADIENE-1,4-DIONE, 2,5-BIS(1-AZIRIDIN

Formula:	$C_{16}H_{24}N_4O_4$	
Mol Weight:	336.39	
MP (deg C):		FP (deg C):
BP (deg C):		
BP pressure (mm Hg):		

Property/Value	Units	Temp	Data Type	Reference
WS 1.58E+005	mg/L	25	EST	MEYLAN,WM ET AL. (1996)
logP -0.37			EXP	HANSCH,C ET AL. (1995)
VP 2.97E-013	mm Hg	25	EST	NEELY,WB & BLAU,GE (1985)
DC	pKa			
HL 1.02E-022	atm m3/mol	25	EST	MEYLAN,WM & HOWARD,PH (1991)
OH 1.87E-010	cm3/molc sec	25	EST	MEYLAN,WM & HOWARD,PH (1993)

CAS #: 059886-55-4 — 2,5-CYCLOHEXADIENE-1,4-DIONE, 2,5-BIS(1-AZIRIDIN

Formula:	$C_{16}H_{24}N_4O_6$	
Mol Weight:	368.39	
MP (deg C):		FP (deg C):
BP (deg C):		
BP pressure (mm Hg):		

Property/Value	Units	Temp	Data Type	Reference
WS 1.00E+006	mg/L	25	EST	MEYLAN,WM ET AL. (1996)
logP -1.97			EXP	HANSCH,C ET AL. (1995)
VP 3.93E-017	mm Hg	25	EST	NEELY,WB & BLAU,GE (1985)
DC	pKa			
HL 5.00E-026	atm m3/mol	25	EST	MEYLAN,WM & HOWARD,PH (1991)
OH 1.99E-010	cm3/molc sec	25	EST	MEYLAN,WM & HOWARD,PH (1993)

CAS #: 059887-89-7 — 4H-1-BENZOPYRAN-4-ONE, 7-HYDROXY-

Formula:	$C_9H_6O_3$	
Mol Weight:	162.15	
MP (deg C):		FP (deg C):
BP (deg C):		
BP pressure (mm Hg):		

Property/Value	Units	Temp	Data Type	Reference
WS 1.14E+004	mg/L	25	EST	MEYLAN,WM ET AL. (1996)
logP 1.58			EXP	HANSCH,C ET AL. (1995)
VP 8.74E-005	mm Hg	25	EST	NEELY,WB & BLAU,GE (1985)
DC	pKa			
HL 8.39E-011	atm m3/mol	25	EST	MEYLAN,WM & HOWARD,PH (1991)
OH 2.11E-010	cm3/molc sec	25	EST	MEYLAN,WM & HOWARD,PH (1993)

CAS #: 059897-94-8 — TR-CYPR-CO2ME-22-DIME-3(2DICLETHENYL)

Formula:	$C_9H_{12}Cl_2O_2$	
Mol Weight:	223.10	
MP (deg C):		FP (deg C):
BP (deg C):		
BP pressure (mm Hg):		

Property/Value	Units	Temp	Data Type	Reference
WS 2.63E+001	mg/L	25	EST	MEYLAN,WM ET AL. (1996)
logP 3.63			EXP	HANSCH,C & LEO,AJ (1985)
VP 6.44E-002	mm Hg	25	EST	NEELY,WB & BLAU,GE (1985)
DC	pKa			
HL 1.63E-004	atm m3/mol	25	EST	MEYLAN,WM & HOWARD,PH (1991)
OH 4.49E-012	cm3/molc sec	25	EST	MEYLAN,WM & HOWARD,PH (1993)

CAS #: 059906-37-5 — N,N-DIME THIOPHENE-3-CARBOXAMIDE

Formula:	$C_7H_{11}NOS$	
Mol Weight:	157.24	
MP (deg C):		FP (deg C):
BP (deg C):		
BP pressure (mm Hg):		

Property/Value	Units	Temp	Data Type	Reference
WS 2.44E+004	mg/L	25	EST	MEYLAN,WM ET AL. (1996)
logP 0.55			EXP	HANSCH,C & LEO,AJ (1985)
VP 3.78E-003	mm Hg	25	EST	NEELY,WB & BLAU,GE (1985)
DC	pKa			
HL 5.80E-009	atm m3/mol	25	EST	MEYLAN,WM & HOWARD,PH (1991)
OH 2.48E-011	cm3/molc sec	25	EST	MEYLAN,WM & HOWARD,PH (1993)

CAS #: 059907-22-1				2-METHYL-6-NITROACETANILIDE

Formula: $C_9H_{10}N_2O_3$

Mol Weight: 194.19

MP (deg C): FP (deg C):

BP (deg C):

BP pressure (mm Hg):

Property/Value	Units	Temp	Data Type	Reference
WS 5.75E+003	mg/L	25	EST	MEYLAN,WM ET AL. (1996)
logP 0.61			EXP	NAKAGAWA,Y ET AL. (1992)
VP 2.03E-006	mm Hg	25	EST	NEELY,WB & BLAU,GE (1985)
DC	pKa			
HL 5.88E-010	atm m3/mol	25	EST	MEYLAN,WM & HOWARD,PH (1991)
OH 2.14E-012	cm3/molc sec	25	EST	MEYLAN,WM & HOWARD,PH (1993)

CAS #: 059917-58-7				N-(DIMETHYLAMINOMETHYL)BENZAMIDE

Formula: $C_{10}H_{14}N_2O$

Mol Weight: 178.24

MP (deg C): FP (deg C):

BP (deg C):

BP pressure (mm Hg):

Property/Value	Units	Temp	Data Type	Reference
WS 1.63E+004	mg/L	25	EST	MEYLAN,WM ET AL. (1996)
logP 0.63			EXP	HANSCH,C & LEO,AJ (1985)
VP 2.95E-005	mm Hg	25	EST	NEELY,WB & BLAU,GE (1985)
DC	pKa			
HL 7.62E-010	atm m3/mol	25	EST	MEYLAN,WM & HOWARD,PH (1991)
OH 1.11E-010	cm3/molc sec	25	EST	MEYLAN,WM & HOWARD,PH (1993)

CAS #: 059937-28-9				MALOTILATE

Formula: $C_{12}H_{16}O_4S_2$

Mol Weight: 288.39

MP (deg C): 60.5 FP (deg C):

BP (deg C):

BP pressure (mm Hg):

Property/Value	Units	Temp	Data Type	Reference
WS 9.00E+000	mg/L	25	EST	MEYLAN,WM ET AL. (1996)
logP 3.75			EXP	SANGSTER,J (1994)
VP 3.51E-002	mm Hg	25	EST	NEELY,WB & BLAU,GE (1985)
DC	pKa			
HL 4.34E-007	atm m3/mol	25	EST	MEYLAN,WM & HOWARD,PH (1991)
OH 5.06E-011	cm3/molc sec	25	EST	MEYLAN,WM & HOWARD,PH (1993)

CAS #: 060003-46-5				BENZENEACETIC ACID, 4-(HEXYLOXY)-

Formula: $C_{14}H_{20}O_3$

Mol Weight: 236.31

MP (deg C): FP (deg C):

BP (deg C):

BP pressure (mm Hg):

Property/Value	Units	Temp	Data Type	Reference
WS 2.90E+001	mg/L	25	EST	MEYLAN,WM ET AL. (1996)
logP 3.96			EXP	SANGSTER,J (1994)
VP 7.27E-006	mm Hg	25	EST	NEELY,WB & BLAU,GE (1985)
DC	pKa			
HL 1.08E-008	atm m3/mol	25	EST	MEYLAN,WM & HOWARD,PH (1991)
OH 3.86E-011	cm3/molc sec	25	EST	MEYLAN,WM & HOWARD,PH (1993)

CAS #: 060010-81-3				1,2,4-TRIAZOLE,3,5-DIACETAMIDO-

Formula: $C_6H_9N_5O_2$

Mol Weight: 183.17

MP (deg C): FP (deg C):

BP (deg C):

BP pressure (mm Hg):

Property/Value	Units	Temp	Data Type	Reference
WS 1.93E+004	mg/L	25	EST	MEYLAN,WM ET AL. (1996)
logP -1.02			EXP	HANSCH,C & LEO,AJ (1985)
VP 2.13E-009	mm Hg	25	EST	NEELY,WB & BLAU,GE (1985)
DC	pKa			
HL 3.18E-015	atm m3/mol	25	EST	MEYLAN,WM & HOWARD,PH (1991)
OH 3.04E-013	cm3/molc sec	25	EST	MEYLAN,WM & HOWARD,PH (1993)

CAS #: 060035-83-8				N-(TRIOHMEMETHANE)-N'-PHENYL UREA

Formula: $C_{11}H_{16}N_2O_4$

Mol Weight: 240.26

MP (deg C): FP (deg C):

BP (deg C):

BP pressure (mm Hg):

Property/Value	Units	Temp	Data Type	Reference
WS 1.15E+004	mg/L	25	EST	MEYLAN,WM ET AL. (1996)
logP 0.43			EXP	HANSCH,C & LEO,AJ (1985)
VP 6.87E-012	mm Hg	25	EST	NEELY,WB & BLAU,GE (1985)
DC	pKa			
HL 5.06E-017	atm m3/mol	25	EST	MEYLAN,WM & HOWARD,PH (1991)
OH 5.61E-011	cm3/molc sec	25	EST	MEYLAN,WM & HOWARD,PH (1993)

CAS #: 060044-33-9				2-CHLORO-4-NONYLPHENOL

Formula: $C_{15}H_{23}ClO$

Mol Weight: 254.80

MP (deg C): FP (deg C):

BP (deg C):

BP pressure (mm Hg):

Property/Value	Units	Temp	Data Type	Reference
WS 1.83E-001	mg/L	25	EST	MEYLAN,WM ET AL. (1996)
logP 6.63			EST	MEYLAN,WM & HOWARD,PH (1995)
VP 6.54E-006	mm Hg	25	EST	NEELY,WB & BLAU,GE (1985)
DC	pKa			
HL 4.42E-006	atm m3/mol	25	EST	MEYLAN,WM & HOWARD,PH (1991)
OH 2.29E-011	cm3/molc sec	25	EST	MEYLAN,WM & HOWARD,PH (1993)

CAS #: 060050-38-6				M-CF3-N-PHENYLSUCCINIMIDE

Formula: $C_{11}H_8F_3NO_2$

Mol Weight: 243.19

MP (deg C): FP (deg C):

BP (deg C):

BP pressure (mm Hg):

Property/Value	Units	Temp	Data Type	Reference
WS 2.17E+003	mg/L	25	EST	MEYLAN,WM ET AL. (1996)
logP 1.26			EXP	HANSCH,C & LEO,AJ (1985)
VP 6.87E-007	mm Hg	25	EST	NEELY,WB & BLAU,GE (1985)
DC	pKa			
HL 1.13E-006	atm m3/mol	25	EST	MEYLAN,WM & HOWARD,PH (1991)
OH 6.25E-012	cm3/molc sec	25	EST	MEYLAN,WM & HOWARD,PH (1993)

CAS #: 060067-82-5				1,2,3-TRIMEO TRICHLOROCYCLOHEXANE
Formula: $C_9H_{15}Cl_3O_3$				
Mol Weight: 277.58				
MP (deg C):		FP (deg C):		
BP (deg C):				
BP pressure (mm Hg):				

Property/Value	Units	Temp	Data Type	Reference
WS 4.18E+002	mg/L	25	EST	MEYLAN,WM ET AL. (1996)
logP 1.87			EXP	HANSCH,C & LEO,AJ (1985)
VP 8.86E-004	mm Hg	25	EST	NEELY,WB & BLAU,GE (1985)
DC	pKa			
HL 9.39E-009	atm m3/mol	25	EST	MEYLAN,WM & HOWARD,PH (1991)
OH 3.02E-011	cm3/molc sec	25	EST	MEYLAN,WM & HOWARD,PH (1993)

CAS #: 060067-85-8				2-MEO 3-MES TETRACHLOROCYCLOHEXANE
Formula: $C_8H_{12}Cl_4OS$				
Mol Weight: 298.06				
MP (deg C):		FP (deg C):		
BP (deg C):				
BP pressure (mm Hg):				

Property/Value	Units	Temp	Data Type	Reference
WS 1.92E+001	mg/L	25	EST	MEYLAN,WM ET AL. (1996)
logP 3.30			EXP	HANSCH,C & LEO,AJ (1985)
VP 1.63E-004	mm Hg	25	EST	NEELY,WB & BLAU,GE (1985)
DC	pKa			
HL 2.57E-007	atm m3/mol	25	EST	MEYLAN,WM & HOWARD,PH (1991)
OH 1.57E-011	cm3/molc sec	25	EST	MEYLAN,WM & HOWARD,PH (1993)

CAS #: 060084-10-8				4-THIAZOLECARBOXAMIDE, 2-(beta-D-RIBOFURANOSYL)-
Formula: $C_9H_{12}N_2O_5S$				
Mol Weight: 260.27				
MP (deg C):		FP (deg C):		
BP (deg C):				
BP pressure (mm Hg):				

Property/Value	Units	Temp	Data Type	Reference
WS 3.05E+004	mg/L	25	EST	MEYLAN,WM ET AL. (1996)
logP -1.73			EXP	SANGSTER,J (1994)
VP 2.19E-013	mm Hg	25	EST	NEELY,WB & BLAU,GE (1985)
DC	pKa			
HL 1.29E-022	atm m3/mol	25	EST	MEYLAN,WM & HOWARD,PH (1991)
OH 4.96E-011	cm3/molc sec	25	EST	MEYLAN,WM & HOWARD,PH (1993)

CAS #: 060129-67-1				DIS. A. 14
Formula: $C_{16}H_{17}N_5O_6$				
Mol Weight: 375.34				
MP (deg C):		FP (deg C):		
BP (deg C):				
BP pressure (mm Hg):				

Property/Value	Units	Temp	Data Type	Reference
WS 2.25E+000	mg/L	25	EXP	BAUGHMAN,GL & PERENICH,TA (1988)
logP 2.96			EST	MEYLAN,WM & HOWARD,PH (1995)
VP 4.49E-015	mm Hg	25	EST	NEELY,WB & BLAU,GE (1985)
DC	pKa			
HL 8.58E-018	atm m3/mol	25	EST	MEYLAN,WM & HOWARD,PH (1991)
OH 8.78E-011	cm3/molc sec	25	EST	MEYLAN,WM & HOWARD,PH (1993)

CAS #: 060132-40-3				1,2-DIMEO TETRACHLOROCYCLOHEXANE
Formula: $C_8H_{12}Cl_4O_2$				
Mol Weight: 282.00				
MP (deg C):		FP (deg C):		
BP (deg C):				
BP pressure (mm Hg):				

Property/Value	Units	Temp	Data Type	Reference
WS 1.04E+002	mg/L	25	EST	MEYLAN,WM ET AL. (1996)
logP 2.55			EXP	HANSCH,C & LEO,AJ (1985)
VP 8.50E-004	mm Hg	25	EST	NEELY,WB & BLAU,GE (1985)
DC	pKa			
HL 2.82E-007	atm m3/mol	25	EST	MEYLAN,WM & HOWARD,PH (1991)
OH 1.16E-011	cm3/molc sec	25	EST	MEYLAN,WM & HOWARD,PH (1993)

CAS #: 060132-42-5				1-METHIOPENTACHLOROCYCLOHEXANE
Formula: $C_7H_9Cl_5S$				
Mol Weight: 302.48				
MP (deg C):		FP (deg C):		
BP (deg C):				
BP pressure (mm Hg):				

Property/Value	Units	Temp	Data Type	Reference
WS 7.45E+000	mg/L	25	EST	MEYLAN,WM ET AL. (1996)
logP 3.75			EXP	HANSCH,C & LEO,AJ (1985)
VP 1.58E-004	mm Hg	25	EST	NEELY,WB & BLAU,GE (1985)
DC	pKa			
HL 7.75E-006	atm m3/mol	25	EST	MEYLAN,WM & HOWARD,PH (1991)
OH 5.66E-012	cm3/molc sec	25	EST	MEYLAN,WM & HOWARD,PH (1993)

CAS #: 060141-98-2				DIAZOACETYLGLYCINE-N'-IBU AMIDE
Formula: $C_8H_{14}N_4O_2$				
Mol Weight: 198.23				
MP (deg C):		FP (deg C):		
BP (deg C):				
BP pressure (mm Hg):				

Property/Value	Units	Temp	Data Type	Reference
WS 2.16E+004	mg/L	25	EST	MEYLAN,WM ET AL. (1996)
logP 0.37			EXP	HANSCH,C & LEO,AJ (1985)
VP 9.54E-014	mm Hg	25	EST	NEELY,WB & BLAU,GE (1985)
DC	pKa			
HL 5.53E-015	atm m3/mol	25	EST	MEYLAN,WM & HOWARD,PH (1991)
OH 8.99E-011	cm3/molc sec	25	EST	MEYLAN,WM & HOWARD,PH (1993)

CAS #: 060141-99-3				N-DIAZOACETYLGLYCINE-N'-HEXYLAMIDE
Formula: $C_{10}H_{18}N_4O_2$				
Mol Weight: 226.28				
MP (deg C):		FP (deg C):		
BP (deg C):				
BP pressure (mm Hg):				

Property/Value	Units	Temp	Data Type	Reference
WS 1.64E+003	mg/L	25	EST	MEYLAN,WM ET AL. (1996)
logP 1.51			EXP	HANSCH,C & LEO,AJ (1985)
VP 8.53E-014	mm Hg	25	EST	NEELY,WB & BLAU,GE (1985)
DC	pKa			
HL 9.75E-015	atm m3/mol	25	EST	MEYLAN,WM & HOWARD,PH (1991)
OH 2.56E-011	cm3/molc sec	25	EST	MEYLAN,WM & HOWARD,PH (1993)

CAS #: 060142-00-9				N-(DIAZOACETYLAMINOACETYL)MORPHOLINE

Formula: $C_8H_{12}N_4O_3$

Mol Weight: 212.21

MP (deg C):   FP (deg C):

BP (deg C):

BP pressure (mm Hg):

Property/Value	Units	Temp	Data Type	Reference
WS 4.76E+003	mg/L	25	EST	MEYLAN,WM ET AL. (1996)
logP -0.48			EXP	HANSCH,C & LEO,AJ (1985)
VP 3.13E-014	mm Hg	25	EST	NEELY,WB & BLAU,GE (1985)
DC	pKa			
HL 3.56E-017	atm m3/mol	25	EST	MEYLAN,WM & HOWARD,PH (1991)
OH 1.54E-010	cm3/molc sec	25	EST	MEYLAN,WM & HOWARD,PH (1993)

CAS #: 060142-49-6				DIAZENECARBOXAMIDE, PHENYL-, 2-OXIDE

Formula: $C_7H_7N_3O_2$

Mol Weight: 165.15

MP (deg C):   FP (deg C):

BP (deg C):

BP pressure (mm Hg):

Property/Value	Units	Temp	Data Type	Reference
WS 7.66E+003	mg/L	25	EST	MEYLAN,WM ET AL. (1996)
logP 0.58			EXP	HANSCH,C ET AL. (1995)
VP 4.39E-009	mm Hg	25	EST	NEELY,WB & BLAU,GE (1985)
DC	pKa			
HL 2.26E-016	atm m3/mol	25	EST	MEYLAN,WM & HOWARD,PH (1991)
OH 1.42E-012	cm3/molc sec	25	EST	MEYLAN,WM & HOWARD,PH (1993)

CAS #: 060142-50-9				P-NITROPHENYL AZOXYCYANIDE

Formula: $C_7H_4N_4O_3$

Mol Weight: 192.13

MP (deg C):   FP (deg C):

BP (deg C):

BP pressure (mm Hg):

Property/Value	Units	Temp	Data Type	Reference
WS 3.71E+002	mg/L	25	EST	MEYLAN,WM ET AL. (1996)
logP 1.51			EXP	CALVINO,R R ET AL. (1991)
VP 1.29E-009	mm Hg	25	EST	NEELY,WB & BLAU,GE (1985)
DC	pKa			
HL 8.12E-014	atm m3/mol	25	EST	MEYLAN,WM & HOWARD,PH (1991)
OH 5.21E-014	cm3/molc sec	25	EST	MEYLAN,WM & HOWARD,PH (1993)

CAS #: 060142-96-3				CYCLOHEXANEACETIC ACID, 1-(AMINOMETHYL)-

Formula: $C_9H_{17}NO_2$

Mol Weight: 171.24

MP (deg C):   FP (deg C):

BP (deg C):

BP pressure (mm Hg):

Property/Value	Units	Temp	Data Type	Reference
WS 4.49E+003	mg/L	25	EST	MEYLAN,WM ET AL. (1996)
logP -1.10			EXP	SANGSTER,J (1994)
VP 2.94E-010	mm Hg	25	EST	NEELY,WB & BLAU,GE (1985)
DC	pKa			
HL 1.81E-010	atm m3/mol	25	EST	MEYLAN,WM & HOWARD,PH (1991)
OH 4.01E-011	cm3/molc sec	25	EST	MEYLAN,WM & HOWARD,PH (1993)

CAS #: 060145-76-8				ET-N-CL-ACETYL-N-(2ETPH)GLYCINATE

Formula: $C_{14}H_{18}ClNO_3$

Mol Weight: 283.76

MP (deg C):   FP (deg C):

BP (deg C):

BP pressure (mm Hg):

Property/Value	Units	Temp	Data Type	Reference
WS 9.74E+001	mg/L	25	EST	MEYLAN,WM ET AL. (1996)
logP 2.57			EXP	HANSCH,C & LEO,AJ (1985)
VP 9.64E-007	mm Hg	25	EST	NEELY,WB & BLAU,GE (1985)
DC	pKa			
HL 4.12E-009	atm m3/mol	25	EST	MEYLAN,WM & HOWARD,PH (1991)
OH 2.10E-011	cm3/molc sec	25	EST	MEYLAN,WM & HOWARD,PH (1993)

CAS #: 060145-77-9				GLYCINE, N-(CHLOROACETYL)-N-[1,1'-BIPHENYL]-2-YL

Formula: $C_{18}H_{18}ClNO_3$

Mol Weight: 331.80

MP (deg C):   FP (deg C):

BP (deg C):

BP pressure (mm Hg):

Property/Value	Units	Temp	Data Type	Reference
WS 1.06E+001	mg/L	25	EST	MEYLAN,WM ET AL. (1996)
logP 3.37			EXP	SANGSTER,J (1993)
VP 4.24E-009	mm Hg	25	EST	NEELY,WB & BLAU,GE (1985)
DC	pKa			
HL 2.16E-010	atm m3/mol	25	EST	MEYLAN,WM & HOWARD,PH (1991)
OH 1.70E-011	cm3/molc sec	25	EST	MEYLAN,WM & HOWARD,PH (1993)

CAS #: 060145-78-0				ET-N-CHLOROACETYL-N-(2MEOPH)GLYCINATE

Formula: $C_{13}H_{16}ClNO_4$

Mol Weight: 285.73

MP (deg C):   FP (deg C):

BP (deg C):

BP pressure (mm Hg):

Property/Value	Units	Temp	Data Type	Reference
WS 4.49E+002	mg/L	25	EST	MEYLAN,WM ET AL. (1996)
logP 1.78			EXP	HANSCH,C & LEO,AJ (1985)
VP 8.73E-007	mm Hg	25	EST	NEELY,WB & BLAU,GE (1985)
DC	pKa			
HL 1.66E-010	atm m3/mol	25	EST	MEYLAN,WM & HOWARD,PH (1991)
OH 1.68E-011	cm3/molc sec	25	EST	MEYLAN,WM & HOWARD,PH (1993)

CAS #: 060145-79-1				GLYCINE, N-(CHLOROACETYL)-N-[2-(1-METHYLETHOXY)P

Formula: $C_{15}H_{20}ClNO_4$

Mol Weight: 313.78

MP (deg C):   FP (deg C):

BP (deg C):

BP pressure (mm Hg):

Property/Value	Units	Temp	Data Type	Reference
WS 4.57E+001	mg/L	25	EST	MEYLAN,WM ET AL. (1996)
logP 2.75			EXP	SANGSTER,J (1994)
VP 2.80E-007	mm Hg	25	EST	NEELY,WB & BLAU,GE (1985)
DC	pKa			
HL 2.93E-010	atm m3/mol	25	EST	MEYLAN,WM & HOWARD,PH (1991)
OH 2.82E-011	cm3/molc sec	25	EST	MEYLAN,WM & HOWARD,PH (1993)

CAS #: 060165-07-3 — 3-ISOPROPYLHYDROXYUREA

Formula: $C_4H_{10}N_2O_2$
Mol Weight: 118.14
MP (deg C):
FP (deg C):
BP (deg C):
BP pressure (mm Hg):

Property/Value	Units	Temp	Data Type	Reference
WS 3.37E+003	mg/L	25	EST	MEYLAN,WM ET AL. (1996)
logP 0.20			EXP	HANSCH,C & LEO,AJ (1985)
VP 1.70E-004	mm Hg	25	EST	NEELY,WB & BLAU,GE (1985)
DC	pKa			
HL 2.10E-010	atm m3/mol	25	EST	MEYLAN,WM & HOWARD,PH (1991)
OH 1.03E-011	cm3/molc sec	25	EST	MEYLAN,WM & HOWARD,PH (1993)

CAS #: 060168-88-9 — FENARIMOL

Formula: $C_{17}H_{12}Cl_2N_2O$
Mol Weight: 331.20
MP (deg C): 118
FP (deg C):
BP (deg C):
BP pressure (mm Hg):

Property/Value	Units	Temp	Data Type	Reference
WS 1.37E+001	mg/L	25	EXP	YALKOWSKY,SH & DANNENFELSER,RM (1992)
logP 3.60			EXP	HANSCH,C ET AL. (1995)
VP 2.20E-007	mm Hg	25	EXP	WAUCHOPE,RD ET AL. (1991A)
DC	pKa			
HL 7.00E-009	atm m3/mol	25	EST	VP/WSOL
OH 3.94E-012	cm3/molc sec	25	EST	MEYLAN,WM & HOWARD,PH (1993)

CAS #: 060174-20-1 — 1-(2-OH-3-MEOPR)-2-ME-5-NO2IMIDAZOLE

Formula: $C_8H_{15}N_3O_4$
Mol Weight: 217.23
MP (deg C):
FP (deg C):
BP (deg C):
BP pressure (mm Hg):

Property/Value	Units	Temp	Data Type	Reference
WS 1.37E+004	mg/L	25	EST	MEYLAN,WM ET AL. (1996)
logP 0.04			EXP	HANSCH,C & LEO,AJ (1985)
VP 3.84E-008	mm Hg	25	EST	NEELY,WB & BLAU,GE (1985)
DC	pKa			
HL 2.63E-013	atm m3/mol	25	EST	MEYLAN,WM & HOWARD,PH (1991)
OH 2.34E-011	cm3/molc sec	25	EST	MEYLAN,WM & HOWARD,PH (1993)

CAS #: 060207-31-0 — AZACONAZOL

Formula: $C_{12}H_{13}Cl_2N_3O_2$
Mol Weight: 302.16
MP (deg C):
FP (deg C):
BP (deg C):
BP pressure (mm Hg):

Property/Value	Units	Temp	Data Type	Reference
WS 1.28E+002	mg/L	25	EST	MEYLAN,WM ET AL. (1996)
logP 2.32			EXP	HANSCH,C ET AL. (1995)
VP 9.04E-007	mm Hg	25	EST	NEELY,WB & BLAU,GE (1985)
DC	pKa			
HL 5.81E-010	atm m3/mol	25	EST	MEYLAN,WM & HOWARD,PH (1991)
OH 1.57E-011	cm3/molc sec	25	EST	MEYLAN,WM & HOWARD,PH (1993)

CAS #: 060207-90-1 — PROPICONAZOLE

Formula: $C_{15}H_{19}Cl_2N_3O_2$
Mol Weight: 344.24
MP (deg C):
FP (deg C):
BP (deg C): 180
BP pressure (mm Hg): 1.00E-001

Property/Value	Units	Temp	Data Type	Reference
WS 1.10E+002	mg/L	20	EXP	SHIU,WY ET AL. (1990)
logP 3.50			EXP	HANSCH,C ET AL. (1995)
VP 1.00E-006	mm Hg	25	EXP	WEBER,JB (1994)
DC	pKa			
HL 4.12E-009	atm m3/mol	25	EST	VP/WSOL
OH 2.72E-011	cm3/molc sec	25	EST	MEYLAN,WM & HOWARD,PH (1993)

CAS #: 060207-93-4 — 1-[[2-(2,4-DICHLOROPHENYL)-4-ETHYL-1,3-DIOXOLAN-

Formula: $C_{14}H_{17}Cl_2N_3O_2$
Mol Weight: 330.22
MP (deg C): 75-93
FP (deg C):
BP (deg C):
BP pressure (mm Hg):

Property/Value	Units	Temp	Data Type	Reference
WS 8.00E+001	mg/L	20	EXP	SHIU,WY ET AL. (1990)
logP 3.10			EXP	HANSCH,C ET AL. (1995)
VP 2.34E-007	mm Hg	25	EST	NEELY,WB & BLAU,GE (1985)
DC	pKa			
HL 1.02E-009	atm m3/mol	25	EST	MEYLAN,WM & HOWARD,PH (1991)
OH 2.58E-011	cm3/molc sec	25	EST	MEYLAN,WM & HOWARD,PH (1993)

CAS #: 060211-57-6 — BENZENEMETHANOL, 3,5-DICHLORO-

Formula: $C_7H_6Cl_2O$
Mol Weight: 177.03
MP (deg C): 79-82
FP (deg C):
BP (deg C):
BP pressure (mm Hg):

Property/Value	Units	Temp	Data Type	Reference
WS 6.15E+002	mg/L	25	EST	MEYLAN,WM ET AL. (1996)
logP 2.90			EXP	MIYAKE,F ET AL. (1987)
VP 6.99E-004	mm Hg	25	EST	NEELY,WB & BLAU,GE (1985)
DC	pKa			
HL 1.19E-007	atm m3/mol	25	EST	MEYLAN,WM & HOWARD,PH (1991)
OH 5.81E-012	cm3/molc sec	25	EST	MEYLAN,WM & HOWARD,PH (1993)

CAS #: 060221-92-3 — N1-P-CHLOROPHENYL-N5-METHYLBIGUANIDE

Formula: $C_9H_{12}ClN_5$
Mol Weight: 225.68
MP (deg C):
FP (deg C):
BP (deg C):
BP pressure (mm Hg):

Property/Value	Units	Temp	Data Type	Reference
WS 1.44E+003	mg/L	25	EST	MEYLAN,WM ET AL. (1996)
logP 1.58			EXP	HANSCH,C & LEO,AJ (1985)
VP 2.30E-006	mm Hg	25	EST	NEELY,WB & BLAU,GE (1985)
DC	pKa			
HL 1.42E-016	atm m3/mol	25	EST	MEYLAN,WM & HOWARD,PH (1991)
OH 1.40E-010	cm3/molc sec	25	EST	MEYLAN,WM & HOWARD,PH (1993)

CAS #: 060221-93-4 — N1-P-CHLOROPHENYL-N5-ETHYLBIGUANIDE

Formula: $C_{10}H_{14}ClN_5$

Mol Weight: 239.71

MP (deg C): FP (deg C):

BP (deg C):

BP pressure (mm Hg):

Property/Value	Units	Temp	Data Type	Reference
WS 4.01E+002	mg/L	25	EST	MEYLAN,WM ET AL. (1996)
logP 2.14			EXP	HANSCH,C & LEO,AJ (1985)
VP 9.91E-007	mm Hg	25	EST	NEELY,WB & BLAU,GE (1985)
DC	pKa			
HL 1.89E-016	atm m3/mol	25	EST	MEYLAN,WM & HOWARD,PH (1991)
OH 1.47E-010	cm3/molc sec	25	EST	MEYLAN,WM & HOWARD,PH (1993)

CAS #: 060331-52-4 — 4,6-DIIDORESORCYL-1,3-DIGLUCOSIDE

Formula: $C_{18}H_{24}I_2O_{12}$

Mol Weight: 686.19

MP (deg C): FP (deg C):

BP (deg C):

BP pressure (mm Hg):

Property/Value	Units	Temp	Data Type	Reference
WS 7.09E+001	mg/L	25	EST	MEYLAN,WM ET AL. (1996)
logP -0.25			EXP	HANSCH,C & LEO,AJ (1985)
VP 6.54E-025	mm Hg	25	EST	NEELY,WB & BLAU,GE (1985)
DC	pKa			
HL 1.05E-025	atm m3/mol	25	EST	MEYLAN,WM & HOWARD,PH (1991)
OH 1.50E-010	cm3/molc sec	25	EST	MEYLAN,WM & HOWARD,PH (1993)

CAS #: 060404-04-8 — 3-FURANCARBOXYLIC ACID, 4,5-DIHYDRO-5-[(4-METHOX

Formula: $C_{16}H_{16}O_5$

Mol Weight: 288.30

MP (deg C): FP (deg C):

BP (deg C):

BP pressure (mm Hg):

Property/Value	Units	Temp	Data Type	Reference
WS 4.26E+001	mg/L	25	EST	MEYLAN,WM ET AL. (1996)
logP 2.96			EXP	SANGSTER,J (1993)
VP 5.31E-007	mm Hg	25	EST	NEELY,WB & BLAU,GE (1985)
DC	pKa			
HL 1.97E-010	atm m3/mol	25	EST	MEYLAN,WM & HOWARD,PH (1991)
OH 6.67E-011	cm3/molc sec	25	EST	MEYLAN,WM & HOWARD,PH (1993)

CAS #: 060406-79-3 — 2,4-DIIDORESORCYL-1,3-DIGLUCOSIDE

Formula: $C_{18}H_{24}I_2O_{12}$

Mol Weight: 686.19

MP (deg C): FP (deg C):

BP (deg C):

BP pressure (mm Hg):

Property/Value	Units	Temp	Data Type	Reference
WS 5.70E+002	mg/L	25	EST	MEYLAN,WM ET AL. (1996)
logP -1.31			EXP	HANSCH,C & LEO,AJ (1985)
VP 6.54E-025	mm Hg	25	EST	NEELY,WB & BLAU,GE (1985)
DC	pKa			
HL 1.05E-025	atm m3/mol	25	EST	MEYLAN,WM & HOWARD,PH (1991)
OH 1.50E-010	cm3/molc sec	25	EST	MEYLAN,WM & HOWARD,PH (1993)

CAS #: 060423-87-2 — 1-ACETOXYPENTACHLOROCYCLOHEXANE

Formula: $C_8H_9Cl_5O_2$

Mol Weight: 314.42

MP (deg C): FP (deg C):

BP (deg C):

BP pressure (mm Hg):

Property/Value	Units	Temp	Data Type	Reference
WS 1.26E+001	mg/L	25	EST	MEYLAN,WM ET AL. (1996)
logP 3.40			EXP	HANSCH,C & LEO,AJ (1985)
VP 9.36E-005	mm Hg	25	EST	NEELY,WB & BLAU,GE (1985)
DC	pKa			
HL 1.73E-006	atm m3/mol	25	EST	MEYLAN,WM & HOWARD,PH (1991)
OH 1.44E-012	cm3/molc sec	25	EST	MEYLAN,WM & HOWARD,PH (1993)

CAS #: 060577-35-7 — 3-OXAZOLIDINYLOXY, 2,2,5,5-TETRAMETHYL-

Formula: $C_7H_{15}NO_2$

Mol Weight: 145.20

MP (deg C): FP (deg C):

BP (deg C):

BP pressure (mm Hg):

Property/Value	Units	Temp	Data Type	Reference
WS 3.13E+004	mg/L	25	EST	MEYLAN,WM ET AL. (1996)
logP 0.48			EXP	SANGSTER,J (1994)
VP 1.09E-005	mm Hg	25	EST	NEELY,WB & BLAU,GE (1985)
DC	pKa			
HL 8.20E-012	atm m3/mol	25	EST	MEYLAN,WM & HOWARD,PH (1991)
OH 6.89E-012	cm3/molc sec	25	EST	MEYLAN,WM & HOWARD,PH (1993)

CAS #: 060597-20-8 — TRIMETHYL HYDRAZINE, MONOHYDROCHLORIDE

Formula: $C_3H_{11}ClN_2$

Mol Weight: 110.59

MP (deg C): FP (deg C):

BP (deg C):

BP pressure (mm Hg):

Property/Value	Units	Temp	Data Type	Reference
WS 2.00E+001	mg/L		EXP	LINDGAARD-JORGENSEN,P & JACOBSEN,BN (1986)
logP -2.98			EST	MEYLAN,WM & HOWARD,PH (1995)
VP 2.93E-005	mm Hg	25	EST	NEELY,WB & BLAU,GE (1985)
DC	pKa			
HL 9.63E-015	atm m3/mol	25	EST	MEYLAN,WM & HOWARD,PH (1991)
OH 6.68E-011	cm3/molc sec	25	EST	MEYLAN,WM & HOWARD,PH (1993)

CAS #: 060617-40-5 — 1,1,1,3,3,5,7,7,9,11,11,13,15,15,15-PENTADECAM*

Formula: $C_{33}H_{60}O_7Si_8$

Mol Weight: 793.53

MP (deg C): FP (deg C):

BP (deg C):

BP pressure (mm Hg):

Property/Value	Units	Temp	Data Type	Reference
WS 3.81E-013	mg/L	25	EST	MEYLAN,WM ET AL. (1996)
logP 11.93			EST	MEYLAN,WM & HOWARD,PH (1995)
VP	mm Hg			
DC	pKa			
HL	atm m3/mol			
OH 8.09E-012	cm3/molc sec	25	EST	MEYLAN,WM & HOWARD,PH (1993)

CAS #: 060628-92-4	1-ME-2-NO2-5-(1-OH-1-MEET)IMIDAZOLE

Formula: $C_7H_{13}N_3O_3$

Mol Weight: 187.20

MP (deg C):   FP (deg C):

BP (deg C):

BP pressure (mm Hg):

Property/Value	Units	Temp	Data Type	Reference
WS 9.08E+003	mg/L	25	EST	MEYLAN,WM ET AL. (1996)
logP 0.43			EXP	HANSCH,C & LEO,AJ (1985)
VP 5.81E-007	mm Hg	25	EST	NEELY,WB & BLAU,GE (1985)
DC	pKa			
HL 2.25E-011	atm m3/mol	25	EST	MEYLAN,WM & HOWARD,PH (1991)
OH 1.12E-011	cm3/molc sec	25	EST	MEYLAN,WM & HOWARD,PH (1993)

CAS #: 060693-35-8	M-FLUORO-N-PHENYLSUCCINIMIDE

Formula: $C_{10}H_8FNO_2$

Mol Weight: 193.18

MP (deg C):   FP (deg C):

BP (deg C):

BP pressure (mm Hg):

Property/Value	Units	Temp	Data Type	Reference
WS 2.24E+004	mg/L	25	EST	MEYLAN,WM ET AL. (1996)
logP 0.38			EXP	HANSCH,C & LEO,AJ (1985)
VP 1.12E-006	mm Hg	25	EST	NEELY,WB & BLAU,GE (1985)
DC	pKa			
HL 1.52E-007	atm m3/mol	25	EST	MEYLAN,WM & HOWARD,PH (1991)
OH 1.99E-011	cm3/molc sec	25	EST	MEYLAN,WM & HOWARD,PH (1993)

CAS #: 060693-37-0	P-FLUORO-N-PHENYLSUCCINIMIDE

Formula: $C_{10}H_8FNO_2$

Mol Weight: 193.18

MP (deg C):   FP (deg C):

BP (deg C):

BP pressure (mm Hg):

Property/Value	Units	Temp	Data Type	Reference
WS 2.68E+004	mg/L	25	EST	MEYLAN,WM ET AL. (1996)
logP 0.29			EXP	HANSCH,C & LEO,AJ (1985)
VP 1.12E-006	mm Hg	25	EST	NEELY,WB & BLAU,GE (1985)
DC	pKa			
HL 1.52E-007	atm m3/mol	25	EST	MEYLAN,WM & HOWARD,PH (1991)
OH 8.56E-012	cm3/molc sec	25	EST	MEYLAN,WM & HOWARD,PH (1993)

CAS #: 060708-27-2	2-FURANCARBOXAMIDE, N-PROPYL-

Formula: $C_8H_{11}NO_2$

Mol Weight: 153.18

MP (deg C):   FP (deg C):

BP (deg C):

BP pressure (mm Hg):

Property/Value	Units	Temp	Data Type	Reference
WS 8.45E+003	mg/L	25	EST	MEYLAN,WM ET AL. (1996)
logP 1.10			EXP	SANGSTER,J (1994)
VP 3.43E-004	mm Hg	25	EST	NEELY,WB & BLAU,GE (1985)
DC	pKa			
HL 8.55E-009	atm m3/mol	25	EST	MEYLAN,WM & HOWARD,PH (1991)
OH 4.84E-011	cm3/molc sec	25	EST	MEYLAN,WM & HOWARD,PH (1993)

CAS #: 060736-83-6	PHENYLACETIC ACID,3-CL-4-PHMEO

Formula: $C_{15}H_{13}ClO_3$

Mol Weight: 276.72

MP (deg C):   FP (deg C):

BP (deg C):

BP pressure (mm Hg):

Property/Value	Units	Temp	Data Type	Reference
WS 4.89E+001	mg/L	25	EST	MEYLAN,WM ET AL. (1996)
logP 3.43			EXP	HANSCH,C & LEO,AJ (1985)
VP 2.67E-007	mm Hg	25	EST	NEELY,WB & BLAU,GE (1985)
DC	pKa			
HL 1.56E-010	atm m3/mol	25	EST	MEYLAN,WM & HOWARD,PH (1991)
OH 1.82E-011	cm3/molc sec	25	EST	MEYLAN,WM & HOWARD,PH (1993)

CAS #: 060753-14-2	3-PYRIDINEBUTANOL

Formula: $C_9H_{13}NO$

Mol Weight: 151.21

MP (deg C):   FP (deg C):

BP (deg C):

BP pressure (mm Hg):

Property/Value	Units	Temp	Data Type	Reference
WS 7.93E+005	mg/L	25	EST	MEYLAN,WM ET AL. (1996)
logP 0.92			EXP	HANSCH,C & LEO,AJ (1985)
VP 4.75E-004	mm Hg	25	EST	NEELY,WB & BLAU,GE (1985)
DC	pKa			
HL 6.66E-010	atm m3/mol	25	EST	MEYLAN,WM & HOWARD,PH (1991)
OH 9.06E-012	cm3/molc sec	25	EST	MEYLAN,WM & HOWARD,PH (1993)

CAS #: 060784-43-2	1,1'-TETRAMETHYLENE-BIS-CNU

Formula: $C_{10}H_{18}Cl_2N_6O_4$

Mol Weight: 357.20

MP (deg C):   FP (deg C):

BP (deg C):

BP pressure (mm Hg):

Property/Value	Units	Temp	Data Type	Reference
WS 9.55E+002	mg/L	25	EST	MEYLAN,WM ET AL. (1996)
logP 0.90			EXP	HANSCH,C & LEO,AJ (1985)
VP 2.30E-010	mm Hg	25	EST	NEELY,WB & BLAU,GE (1985)
DC	pKa			
HL 2.08E-022	atm m3/mol	25	EST	MEYLAN,WM & HOWARD,PH (1991)
OH 1.79E-011	cm3/molc sec	25	EST	MEYLAN,WM & HOWARD,PH (1993)

CAS #: 060784-46-5	HYDROXYETHYL-CLET-NO UREA

Formula: $C_5H_{10}ClN_3O_3$

Mol Weight: 195.61

MP (deg C):   FP (deg C):

BP (deg C):

BP pressure (mm Hg):

Property/Value	Units	Temp	Data Type	Reference
WS 1.25E+003	mg/L	25	EST	MEYLAN,WM ET AL. (1996)
logP 0.30			EXP	HANSCH,C & LEO,AJ (1985)
VP 6.85E-007	mm Hg	25	EST	NEELY,WB & BLAU,GE (1985)
DC	pKa			
HL 4.94E-015	atm m3/mol	25	EST	MEYLAN,WM & HOWARD,PH (1991)
OH 1.17E-011	cm3/molc sec	25	EST	MEYLAN,WM & HOWARD,PH (1993)

060784-70-5 — BARBITURIC ACID,5-ET,5-HEPTYL

Formula: $C_{13}H_{22}N_2O_3$
Mol Weight: 254.33
MP (deg C):
FP (deg C):
BP (deg C):
BP pressure (mm Hg):

Property/Value	Units	Temp	Data Type	Reference
WS 7.32E+001	mg/L	25	EST	MEYLAN,WM ET AL. (1996)
logP 2.91			EXP	HANSCH,C & LEO,AJ (1985)
VP 2.45E-012	mm Hg	25	EST	NEELY,WB & BLAU,GE (1985)
DC	pKa			
HL 1.49E-012	atm m3/mol	25	EST	MEYLAN,WM & HOWARD,PH (1991)
OH 1.73E-011	cm3/molc sec	25	EST	MEYLAN,WM & HOWARD,PH (1993)

060828-33-3 — THREONIN-AMIDE, N-ACETYL

Formula: $C_6H_{12}N_2O_3$
Mol Weight: 160.17
MP (deg C):
FP (deg C):
BP (deg C):
BP pressure (mm Hg):

Property/Value	Units	Temp	Data Type	Reference
WS 7.33E+004	mg/L	25	EST	MEYLAN,WM ET AL. (1996)
logP -1.57			EXP	HANSCH,C & LEO,AJ (1985)
VP 1.22E-009	mm Hg	25	EST	NEELY,WB & BLAU,GE (1985)
DC	pKa			
HL 7.60E-015	atm m3/mol	25	EST	MEYLAN,WM & HOWARD,PH (1991)
OH 7.00E-011	cm3/molc sec	25	EST	MEYLAN,WM & HOWARD,PH (1993)

060835-75-8 — BENZO-18-CROWN-6-ETHER,M-CARBOXY-

Formula: $C_{17}H_{24}O_8$
Mol Weight: 356.38
MP (deg C):
FP (deg C):
BP (deg C):
BP pressure (mm Hg):

Property/Value	Units	Temp	Data Type	Reference
WS 9.83E+003	mg/L	25	EST	MEYLAN,WM ET AL. (1996)
logP -0.28			EXP	HANSCH,C ET AL. (1995)
VP 4.71E-010	mm Hg	25	EST	NEELY,WB & BLAU,GE (1985)
DC	pKa			
HL 9.76E-018	atm m3/mol	25	EST	MEYLAN,WM & HOWARD,PH (1991)
OH 7.75E-011	cm3/molc sec	25	EST	MEYLAN,WM & HOWARD,PH (1993)

060845-51-4 — 2-CHLOROMETHYL-3,3-DICHLOROPROPENE

Formula: $C_4H_5Cl_3$
Mol Weight: 159.44
MP (deg C):
FP (deg C):
BP (deg C):
BP pressure (mm Hg):

Property/Value	Units	Temp	Data Type	Reference
WS 9.88E+001	mg/L	25	EST	MEYLAN,WM ET AL. (1996)
logP 3.33			EST	MEYLAN,WM & HOWARD,PH (1995)
VP 3.58E+000	mm Hg	25	EST	NEELY,WB & BLAU,GE (1985)
DC	pKa			
HL 2.76E-002	atm m3/mol	25	EST	MEYLAN,WM & HOWARD,PH (1991)
OH 4.18E-012	cm3/molc sec	25	EST	MEYLAN,WM & HOWARD,PH (1993)

060903-21-1 — 4-THIA-1-AZABICYCLO[3.2.0]HEPTANE-2-CARBOXYLIC A

Formula: $C_{21}H_{27}N_3O_6S$
Mol Weight: 449.53
MP (deg C):
FP (deg C):
BP (deg C):
BP pressure (mm Hg):

Property/Value	Units	Temp	Data Type	Reference
WS 1.55E+002	mg/L	25	EST	MEYLAN,WM ET AL. (1996)
logP 0.30			EXP	SANGSTER,J (1993)
VP 2.09E-015	mm Hg	25	EST	NEELY,WB & BLAU,GE (1985)
DC	pKa			
HL 8.32E-018	atm m3/mol	25	EST	MEYLAN,WM & HOWARD,PH (1991)
OH 2.12E-010	cm3/molc sec	25	EST	MEYLAN,WM & HOWARD,PH (1993)

060908-29-4 — 1(2H)-PYRIMIDINECARBOXAMIDE, 5-FLUORO-3,4-DIHYDR

Formula: $C_7H_8FN_3O_3$
Mol Weight: 201.16
MP (deg C):
FP (deg C):
BP (deg C):
BP pressure (mm Hg):

Property/Value	Units	Temp	Data Type	Reference
WS 8.94E+004	mg/L	25	EST	MEYLAN,WM ET AL. (1996)
logP -0.37			EXP	SANGSTER,J (1994)
VP 5.83E-009	mm Hg	25	EST	NEELY,WB & BLAU,GE (1985)
DC	pKa			
HL 4.87E-015	atm m3/mol	25	EST	MEYLAN,WM & HOWARD,PH (1991)
OH 7.95E-012	cm3/molc sec	25	EST	MEYLAN,WM & HOWARD,PH (1993)

060980-82-7 — THIAZOLIDINE, 2-(M-CHLOROPHENYL)

Formula: $C_9H_{10}ClNS$
Mol Weight: 199.70
MP (deg C):
FP (deg C):
BP (deg C):
BP pressure (mm Hg):

Property/Value	Units	Temp	Data Type	Reference
WS 8.10E+003	mg/L	25	EST	MEYLAN,WM ET AL. (1996)
logP 2.04			EXP	HANSCH,C & LEO,AJ (1985)
VP 4.11E-004	mm Hg	25	EST	NEELY,WB & BLAU,GE (1985)
DC	pKa			
HL 2.19E-006	atm m3/mol	25	EST	MEYLAN,WM & HOWARD,PH (1991)
OH 1.96E-010	cm3/molc sec	25	EST	MEYLAN,WM & HOWARD,PH (1993)

061019-78-1 — FENAPANIL

Formula: $C_{16}H_{21}N_3$
Mol Weight: 255.37
MP (deg C): 160-162
FP (deg C):
BP (deg C):
BP pressure (mm Hg):

Property/Value	Units	Temp	Data Type	Reference
WS 3.80E+001	mg/L	25	EST	MEYLAN,WM ET AL. (1996)
logP 2.94			EXP	HANSCH,C ET AL. (1995)
VP 9.30E-008	mm Hg	25	EST	NEELY,WB & BLAU,GE (1985)
DC	pKa			
HL 1.11E-008	atm m3/mol	25	EST	MEYLAN,WM & HOWARD,PH (1991)
OH 4.56E-011	cm3/molc sec	25	EST	MEYLAN,WM & HOWARD,PH (1993)

CAS #: 061137-49-3				UREA, N'-[4-[(ACETYLOXY)METHYL]CYCLOHEXYL]-N-(2-

Formula: $C_{12}H_{20}ClN_3O_4$

Mol Weight: 305.76

MP (deg C):   FP (deg C):

BP (deg C):

BP pressure (mm Hg):

Property/Value	Units	Temp	Data Type	Reference
WS 1.39E+002	mg/L	25	EST	MEYLAN,WM ET AL. (1996)
logP 2.24			EXP	HANSCH,C ET AL. (1995)
VP 1.42E-007	mm Hg	25	EST	NEELY,WB & BLAU,GE (1985)
DC	pKa			
HL 5.85E-013	atm m3/mol	25	EST	MEYLAN,WM & HOWARD,PH (1991)
OH 2.60E-011	cm3/molc sec	25	EST	MEYLAN,WM & HOWARD,PH (1993)

CAS #: 061151-66-4				2-(M-CL-PHENYL)-4-IPR-MORPHOLINE

Formula: $C_{13}H_{18}ClNO$

Mol Weight: 239.75

MP (deg C):   FP (deg C):

BP (deg C):

BP pressure (mm Hg):

Property/Value	Units	Temp	Data Type	Reference
WS 4.42E+002	mg/L	25	EST	MEYLAN,WM ET AL. (1996)
logP 3.27			EXP	HANSCH,C & LEO,AJ (1985)
VP 3.10E-004	mm Hg	25	EST	NEELY,WB & BLAU,GE (1985)
DC	pKa			
HL 2.64E-008	atm m3/mol	25	EST	MEYLAN,WM & HOWARD,PH (1991)
OH 1.50E-010	cm3/molc sec	25	EST	MEYLAN,WM & HOWARD,PH (1993)

CAS #: 061192-50-5				2-PYRROLIDINONE, 5-HYDROXY-1-METHYL-5-(3-PYRIDIN

Formula: $C_{10}H_{12}N_2O_2$

Mol Weight: 192.22

MP (deg C):   FP (deg C):

BP (deg C):

BP pressure (mm Hg):

Property/Value	Units	Temp	Data Type	Reference
WS 1.47E+005	mg/L	25	EST	MEYLAN,WM ET AL. (1996)
logP -0.57			EXP	LI,NY & GORROD,JW (1992)
VP 3.55E-007	mm Hg	25	EST	NEELY,WB & BLAU,GE (1985)
DC	pKa			
HL 2.43E-015	atm m3/mol	25	EST	MEYLAN,WM & HOWARD,PH (1991)
OH 2.83E-011	cm3/molc sec	25	EST	MEYLAN,WM & HOWARD,PH (1993)

CAS #: 061213-25-0				FLUROCHLORIDONE

Formula: $C_{12}H_{10}Cl_2F_3NO$

Mol Weight: 312.12

MP (deg C): 42-47   FP (deg C):

BP (deg C):

BP pressure (mm Hg):

Property/Value	Units	Temp	Data Type	Reference
WS 2.80E+001	mg/L	20	EXP	SHIU,WY ET AL. (1990)
logP 3.36			EXP	TOMLIN,C (1994)
VP 3.30E-006	mm Hg	25	EXP	HEATH,J ET AL. (1992)
DC	pKa			
HL 4.84E-008	atm m3/mol	25	EST	VP/WSOL
OH 8.76E-012	cm3/molc sec	25	EST	MEYLAN,WM & HOWARD,PH (1993)

CAS #: 061224-32-6				P-AMINOBENZYLALCOHOL

Formula: C_7H_9NO

Mol Weight: 123.16

MP (deg C):   FP (deg C):

BP (deg C):

BP pressure (mm Hg):

Property/Value	Units	Temp	Data Type	Reference
WS 1.51E+005	mg/L	25	EST	MEYLAN,WM ET AL. (1996)
logP -0.22			EXP	HANSCH,C & LEO,AJ (1985)
VP 3.71E-004	mm Hg	25	EST	NEELY,WB & BLAU,GE (1985)
DC	pKa			
HL 7.68E-011	atm m3/mol	25	EST	MEYLAN,WM & HOWARD,PH (1991)
OH 1.35E-010	cm3/molc sec	25	EST	MEYLAN,WM & HOWARD,PH (1993)

CAS #: 061251-77-2				3-BENZOYL-5-FLUOROURACIL

Formula: $C_{11}H_7FN_2O_3$

Mol Weight: 234.19

MP (deg C):   FP (deg C):

BP (deg C):

BP pressure (mm Hg):

Property/Value	Units	Temp	Data Type	Reference
WS 5.99E+003	mg/L	25	EST	MEYLAN,WM ET AL. (1996)
logP 0.80			EXP	HANSCH,C ET AL. (1995)
VP 6.86E-010	mm Hg	25	EST	NEELY,WB & BLAU,GE (1985)
DC	pKa			
HL 1.06E-013	atm m3/mol	25	EST	MEYLAN,WM & HOWARD,PH (1991)
OH 7.61E-012	cm3/molc sec	25	EST	MEYLAN,WM & HOWARD,PH (1993)

CAS #: 061260-08-0				[2,7]DB27C9-DIBENZO CROWN ETHER

Formula: $C_{26}H_{36}O_9$

Mol Weight: 492.57

MP (deg C):   FP (deg C):

BP (deg C):

BP pressure (mm Hg):

Property/Value	Units	Temp	Data Type	Reference
WS 3.24E+001	mg/L	25	EST	MEYLAN,WM ET AL. (1996)
logP 1.63			EXP	STOLWIJK,TB ET AL. (1989)
VP 7.18E-013	mm Hg	25	EST	NEELY,WB & BLAU,GE (1985)
DC	pKa			
HL 2.56E-018	atm m3/mol	25	EST	MEYLAN,WM & HOWARD,PH (1991)
OH 1.35E-010	cm3/molc sec	25	EST	MEYLAN,WM & HOWARD,PH (1993)

CAS #: 061345-66-2				A-CHLORO-I-VALERYLUREA

Formula: $C_6H_{11}ClN_2O_2$

Mol Weight: 178.62

MP (deg C):   FP (deg C):

BP (deg C):

BP pressure (mm Hg):

Property/Value	Units	Temp	Data Type	Reference
WS 7.83E+003	mg/L	25	EST	MEYLAN,WM ET AL. (1996)
logP 1.00			EXP	HANSCH,C & LEO,AJ (1985)
VP 1.35E-006	mm Hg	25	EST	NEELY,WB & BLAU,GE (1985)
DC	pKa			
HL 8.89E-011	atm m3/mol	25	EST	MEYLAN,WM & HOWARD,PH (1991)
OH 9.03E-012	cm3/molc sec	25	EST	MEYLAN,WM & HOWARD,PH (1993)

061346-84-7 — BARBITURIC ACID,1-ME-5-ALLYL-5-ET

Formula: $C_{10}H_{14}N_2O_3$

Mol Weight: 210.23

MP (deg C): FP (deg C):

BP (deg C):

BP pressure (mm Hg):

Property/Value	Units	Temp	Data Type	Reference
WS 2.73E+003	mg/L	25	EST	MEYLAN,WM ET AL. (1996)
logP 1.35			EXP	HANSCH,C & LEO,AJ (1985)
VP 3.31E-010	mm Hg	25	EST	NEELY,WB & BLAU,GE (1985)
DC	pKa			
HL 7.83E-013	atm m3/mol	25	EST	MEYLAN,WM & HOWARD,PH (1991)
OH 3.63E-011	cm3/molc sec	25	EST	MEYLAN,WM & HOWARD,PH (1993)

061346-87-0 — BARBITURIC ACID,1-ME-5-ET-5(PHET)

Formula: $C_{15}H_{18}N_2O_3$

Mol Weight: 274.32

MP (deg C): FP (deg C):

BP (deg C):

BP pressure (mm Hg):

Property/Value	Units	Temp	Data Type	Reference
WS 3.59E+001	mg/L	25	EST	MEYLAN,WM ET AL. (1996)
logP 3.14			EXP	HANSCH,C & LEO,AJ (1985)
VP 1.69E-012	mm Hg	25	EST	NEELY,WB & BLAU,GE (1985)
DC	pKa			
HL 6.39E-014	atm m3/mol	25	EST	MEYLAN,WM & HOWARD,PH (1991)
OH 1.68E-011	cm3/molc sec	25	EST	MEYLAN,WM & HOWARD,PH (1993)

061367-70-2 — 3,5-DIMEO-4-I-PROPOXYPHENETHYLAMINE

Formula: $C_{13}H_{21}NO_3$

Mol Weight: 239.32

MP (deg C): FP (deg C):

BP (deg C):

BP pressure (mm Hg):

Property/Value	Units	Temp	Data Type	Reference
WS 1.39E+004	mg/L	25	EST	MEYLAN,WM ET AL. (1996)
logP 1.52			EXP	HANSCH,C & LEO,AJ (1985)
VP 9.13E-005	mm Hg	25	EST	NEELY,WB & BLAU,GE (1985)
DC	pKa			
HL 2.96E-010	atm m3/mol	25	EST	MEYLAN,WM & HOWARD,PH (1991)
OH 2.47E-010	cm3/molc sec	25	EST	MEYLAN,WM & HOWARD,PH (1993)

061367-72-4 — 3,5-DIMEO-4-BROMOPHENETHYLAMINE

Formula: $C_{10}H_{14}BrNO_2$

Mol Weight: 260.14

MP (deg C): FP (deg C):

BP (deg C):

BP pressure (mm Hg):

Property/Value	Units	Temp	Data Type	Reference
WS 3.91E+003	mg/L	25	EST	MEYLAN,WM ET AL. (1996)
logP 2.03			EXP	HANSCH,C & LEO,AJ (1985)
VP 1.60E-004	mm Hg	25	EST	NEELY,WB & BLAU,GE (1985)
DC	pKa			
HL 1.13E-009	atm m3/mol	25	EST	MEYLAN,WM & HOWARD,PH (1991)
OH 2.03E-010	cm3/molc sec	25	EST	MEYLAN,WM & HOWARD,PH (1993)

061380-40-3 — LOFENTANIL

Formula: $C_{25}H_{32}N_2O_3$

Mol Weight: 408.55

MP (deg C): 177 FP (deg C):

BP (deg C):

BP pressure (mm Hg):

Property/Value	Units	Temp	Data Type	Reference
WS 6.75E-001	mg/L	25	EST	MEYLAN,WM ET AL. (1996)
logP 4.22			EXP	SANGSTER,J (1993)
VP 1.13E-010	mm Hg	25	EST	NEELY,WB & BLAU,GE (1985)
DC	pKa			
HL 5.80E-013	atm m3/mol	25	EST	MEYLAN,WM & HOWARD,PH (1991)
OH 1.22E-010	cm3/molc sec	25	EST	MEYLAN,WM & HOWARD,PH (1993)

061405-48-9 — RH-1460 DIPH THIOETHER

Formula: $C_{15}H_9ClF_3NO_4S$

Mol Weight: 391.76

MP (deg C): FP (deg C):

BP (deg C):

BP pressure (mm Hg):

Property/Value	Units	Temp	Data Type	Reference
WS 1.21E-001	mg/L	25	EST	MEYLAN,WM ET AL. (1996)
logP 4.76			EXP	NANDIHALLI,UB ET AL. (1993)
VP 3.61E-008	mm Hg	25	EST	NEELY,WB & BLAU,GE (1985)
DC	pKa			
HL 5.14E-009	atm m3/mol	25	EST	MEYLAN,WM & HOWARD,PH (1991)
OH 1.30E-012	cm3/molc sec	25	EST	MEYLAN,WM & HOWARD,PH (1993)

061421-89-4 — A-CHLORO-A-ETHYLBUTANOYLUREA

Formula: $C_7H_{13}ClN_2O_2$

Mol Weight: 192.65

MP (deg C): FP (deg C):

BP (deg C):

BP pressure (mm Hg):

Property/Value	Units	Temp	Data Type	Reference
WS 3.04E+003	mg/L	25	EST	MEYLAN,WM ET AL. (1996)
logP 1.40			EXP	HANSCH,C & LEO,AJ (1985)
VP 7.37E-007	mm Hg	25	EST	NEELY,WB & BLAU,GE (1985)
DC	pKa			
HL 1.18E-010	atm m3/mol	25	EST	MEYLAN,WM & HOWARD,PH (1991)
OH 4.63E-012	cm3/molc sec	25	EST	MEYLAN,WM & HOWARD,PH (1993)

061422-45-5 — CARMOFUR

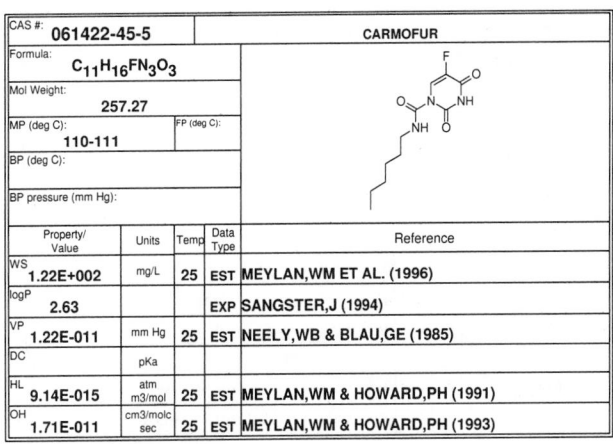

Formula: $C_{11}H_{16}FN_3O_3$

Mol Weight: 257.27

MP (deg C): 110-111 FP (deg C):

BP (deg C):

BP pressure (mm Hg):

Property/Value	Units	Temp	Data Type	Reference
WS 1.22E+002	mg/L	25	EST	MEYLAN,WM ET AL. (1996)
logP 2.63			EXP	SANGSTER,J (1994)
VP 1.22E-011	mm Hg	25	EST	NEELY,WB & BLAU,GE (1985)
DC	pKa			
HL 9.14E-015	atm m3/mol	25	EST	MEYLAN,WM & HOWARD,PH (1991)
OH 1.71E-011	cm3/molc sec	25	EST	MEYLAN,WM & HOWARD,PH (1993)

1244

CAS #: 061544-27-2	2,4-NH2PYRIMIDINE,5(3,5-NHME)BENZYL

Formula: $C_{13}H_{18}N_6$

Mol Weight: 258.33

MP (deg C): FP (deg C):

BP (deg C):

BP pressure (mm Hg):

Property/ Value	Units	Temp	Data Type	Reference
WS 8.95E+003	mg/L	25	EST	MEYLAN,WM ET AL. (1996)
logP 0.44			EXP	HANSCH,C & LEO,AJ (1985)
VP 1.34E-008	mm Hg	25	EST	NEELY,WB & BLAU,GE (1985)
DC	pKa			
HL 6.94E-017	atm m3/mol	25	EST	MEYLAN,WM & HOWARD,PH (1991)
OH 2.03E-010	cm3/molc sec	25	EST	MEYLAN,WM & HOWARD,PH (1993)

CAS #: 061544-28-3	24NH2PYRIMDN,5(3NHME-4ME-5NMECOME)BENZYL

Formula: $C_{16}H_{22}N_6O$

Mol Weight: 314.39

MP (deg C): FP (deg C):

BP (deg C):

BP pressure (mm Hg):

Property/ Value	Units	Temp	Data Type	Reference
WS 3.57E+003	mg/L	25	EST	MEYLAN,WM ET AL. (1996)
logP 0.53			EXP	HANSCH,C & LEO,AJ (1985)
VP 1.05E-010	mm Hg	25	EST	NEELY,WB & BLAU,GE (1985)
DC	pKa			
HL 2.32E-018	atm m3/mol	25	EST	MEYLAN,WM & HOWARD,PH (1991)
OH 2.03E-010	cm3/molc sec	25	EST	MEYLAN,WM & HOWARD,PH (1993)

CAS #: 061554-12-9	14-BENZDIAZPN-2-ON,1(23DIOHPR)5PH7NO2

Formula: $C_{18}H_{17}N_3O_5$

Mol Weight: 355.35

MP (deg C): FP (deg C):

BP (deg C):

BP pressure (mm Hg):

Property/ Value	Units	Temp	Data Type	Reference
WS 3.55E+002	mg/L	25	EST	MEYLAN,WM ET AL. (1996)
logP 0.96			EXP	HANSCH,C & LEO,AJ (1985)
VP 1.31E-015	mm Hg	25	EST	NEELY,WB & BLAU,GE (1985)
DC	pKa			
HL 4.57E-017	atm m3/mol	25	EST	MEYLAN,WM & HOWARD,PH (1991)
OH 2.61E-011	cm3/molc sec	25	EST	MEYLAN,WM & HOWARD,PH (1993)

CAS #: 061566-10-7	AMBAZONE [SEMICARBAZONE]

Formula: $C_8H_{11}N_7O$

Mol Weight: 221.22

MP (deg C): FP (deg C):

BP (deg C):

BP pressure (mm Hg):

Property/ Value	Units	Temp	Data Type	Reference
WS 1.58E+003	mg/L	25	EST	MEYLAN,WM ET AL. (1996)
logP 1.56			EXP	KRAMARCZYK,K (1987)
VP 5.89E-008	mm Hg	25	EST	NEELY,WB & BLAU,GE (1985)
DC	pKa			
HL 1.17E-018	atm m3/mol	25	EST	MEYLAN,WM & HOWARD,PH (1991)
OH 1.99E-010	cm3/molc sec	25	EST	MEYLAN,WM & HOWARD,PH (1993)

CAS #: 061566-63-0	2-PENTOXY-4-AMINOBENZOIC ACID

Formula: $C_{12}H_{17}NO_3$

Mol Weight: 223.27

MP (deg C): FP (deg C):

BP (deg C):

BP pressure (mm Hg):

Property/ Value	Units	Temp	Data Type	Reference
WS 3.07E+002	mg/L	25	EST	MEYLAN,WM ET AL. (1996)
logP 2.38			EXP	HANSCH,C & LEO,AJ (1985)
VP 1.56E-006	mm Hg	25	EST	NEELY,WB & BLAU,GE (1985)
DC	pKa			
HL 7.04E-012	atm m3/mol	25	EST	MEYLAN,WM & HOWARD,PH (1991)
OH 2.14E-010	cm3/molc sec	25	EST	MEYLAN,WM & HOWARD,PH (1993)

CAS #: 061566-64-1	2-I-PENTOXY-4-AMINOBENZOIC ACID

Formula: $C_{12}H_{17}NO_3$

Mol Weight: 223.27

MP (deg C): FP (deg C):

BP (deg C):

BP pressure (mm Hg):

Property/ Value	Units	Temp	Data Type	Reference
WS 3.59E+002	mg/L	25	EST	MEYLAN,WM ET AL. (1996)
logP 2.30			EXP	HANSCH,C & LEO,AJ (1985)
VP 2.69E-006	mm Hg	25	EST	NEELY,WB & BLAU,GE (1985)
DC	pKa			
HL 7.04E-012	atm m3/mol	25	EST	MEYLAN,WM & HOWARD,PH (1991)
OH 2.14E-010	cm3/molc sec	25	EST	MEYLAN,WM & HOWARD,PH (1993)

CAS #: 061573-39-5	24-NH2PYRIMIDN,5(3,5-DIMEAMINO)BENZYL

Formula: $C_{15}H_{22}N_6$

Mol Weight: 286.38

MP (deg C): FP (deg C):

BP (deg C):

BP pressure (mm Hg):

Property/ Value	Units	Temp	Data Type	Reference
WS 6.33E+002	mg/L	25	EST	MEYLAN,WM ET AL. (1996)
logP 1.60			EXP	HANSCH,C & LEO,AJ (1985)
VP 1.14E-008	mm Hg	25	EST	NEELY,WB & BLAU,GE (1985)
DC	pKa			
HL 2.91E-014	atm m3/mol	25	EST	MEYLAN,WM & HOWARD,PH (1991)
OH 2.06E-010	cm3/molc sec	25	EST	MEYLAN,WM & HOWARD,PH (1993)

CAS #: 061573-40-8	24NH2PYRIMDN,5(3MEAM4MEO5DIMEAM)BNZYL

Formula: $C_{15}H_{22}N_6O$

Mol Weight: 302.38

MP (deg C): FP (deg C):

BP (deg C):

BP pressure (mm Hg):

Property/ Value	Units	Temp	Data Type	Reference
WS 5.53E+002	mg/L	25	EST	MEYLAN,WM ET AL. (1996)
logP 1.56			EXP	HANSCH,C & LEO,AJ (1985)
VP 2.36E-009	mm Hg	25	EST	NEELY,WB & BLAU,GE (1985)
DC	pKa			
HL 8.41E-017	atm m3/mol	25	EST	MEYLAN,WM & HOWARD,PH (1991)
OH 2.06E-010	cm3/molc sec	25	EST	MEYLAN,WM & HOWARD,PH (1993)

CAS #: 061597-27-1	6H-DIBENZO[B,D]PYRAN-1-OL, 3-(1,1-DIMETHYLHEPTYL

Formula: $C_{25}H_{38}O_2$

Mol Weight: 370.58

MP (deg C): FP (deg C):

BP (deg C):

BP pressure (mm Hg):

Property/Value	Units	Temp	Data Type	Reference
WS 5.51E-005	mg/L	25	EST	MEYLAN,WM ET AL. (1996)
logP 9.96			EXP	SANGSTER,J (1993)
VP 2.81E-009	mm Hg	25	EST	NEELY,WB & BLAU,GE (1985)
DC	pKa			
HL 7.59E-007	atm m3/mol	25	EST	MEYLAN,WM & HOWARD,PH (1991)
OH 3.11E-010	cm3/molc sec	25	EST	MEYLAN,WM & HOWARD,PH (1993)

CAS #: 061601-23-8	1,4-NAPHTHALENEDIOL, 2,3-DIMETHOXY-, DIACETATE

Formula: $C_{16}H_{16}O_6$

Mol Weight: 304.30

MP (deg C): FP (deg C):

BP (deg C):

BP pressure (mm Hg):

Property/Value	Units	Temp	Data Type	Reference
WS 1.05E+002	mg/L	25	EST	MEYLAN,WM ET AL. (1996)
logP 2.39			EXP	JONES,GH ET AL. (1986)
VP 2.64E-006	mm Hg	25	EST	NEELY,WB & BLAU,GE (1985)
DC	pKa			
HL 2.66E-010	atm m3/mol	25	EST	MEYLAN,WM & HOWARD,PH (1991)
OH 1.58E-010	cm3/molc sec	25	EST	MEYLAN,WM & HOWARD,PH (1993)

CAS #: 061638-07-1	2,5-DIMETHOXY-4-MES AMPHETAMINE

Formula: $C_{12}H_{19}NO_2S$

Mol Weight: 241.35

MP (deg C): FP (deg C):

BP (deg C):

BP pressure (mm Hg):

Property/Value	Units	Temp	Data Type	Reference
WS 3.77E+003	mg/L	25	EST	MEYLAN,WM ET AL. (1996)
logP 2.17			EXP	HANSCH,C & LEO,AJ (1985)
VP 3.94E-005	mm Hg	25	EST	NEELY,WB & BLAU,GE (1985)
DC	pKa			
HL 1.10E-010	atm m3/mol	25	EST	MEYLAN,WM & HOWARD,PH (1991)
OH 1.93E-010	cm3/molc sec	25	EST	MEYLAN,WM & HOWARD,PH (1993)

CAS #: 061638-09-3	2,5-DIMETHOXY-4-MES-PHENETHYLAMINE

Formula: $C_{11}H_{17}NO_2S$

Mol Weight: 227.33

MP (deg C): FP (deg C):

BP (deg C):

BP pressure (mm Hg):

Property/Value	Units	Temp	Data Type	Reference
WS 9.12E+003	mg/L	25	EST	MEYLAN,WM ET AL. (1996)
logP 1.81			EXP	HANSCH,C & LEO,AJ (1985)
VP 6.23E-005	mm Hg	25	EST	NEELY,WB & BLAU,GE (1985)
DC	pKa			
HL 8.28E-011	atm m3/mol	25	EST	MEYLAN,WM & HOWARD,PH (1991)
OH 1.82E-010	cm3/molc sec	25	EST	MEYLAN,WM & HOWARD,PH (1993)

CAS #: 061655-72-9	2-CHLORO-6-(N,N-DIMETHYLAMINO)-PYRAZINE

Formula: $C_6H_8ClN_3$

Mol Weight: 157.60

MP (deg C): FP (deg C):

BP (deg C):

BP pressure (mm Hg):

Property/Value	Units	Temp	Data Type	Reference
WS 1.52E+003	mg/L	25	EST	MEYLAN,WM ET AL. (1996)
logP 1.95			EXP	YAMAGAMI,C & TAKAO,N (1991)
VP 8.03E-002	mm Hg	25	EST	NEELY,WB & BLAU,GE (1985)
DC	pKa			
HL 2.17E-006	atm m3/mol	25	EST	MEYLAN,WM & HOWARD,PH (1991)
OH 3.32E-011	cm3/molc sec	25	EST	MEYLAN,WM & HOWARD,PH (1993)

CAS #: 061706-06-7	1,1-DIME-3-(4-BENZOYLPHENYL)UREA

Formula: $C_{16}H_{16}N_2O_2$

Mol Weight: 268.32

MP (deg C): FP (deg C):

BP (deg C):

BP pressure (mm Hg):

Property/Value	Units	Temp	Data Type	Reference
WS 1.15E+002	mg/L	25	EST	MEYLAN,WM ET AL. (1996)
logP 2.59			EXP	HANSCH,C & LEO,AJ (1985)
VP 3.31E-008	mm Hg	25	EST	NEELY,WB & BLAU,GE (1985)
DC	pKa			
HL 3.50E-013	atm m3/mol	25	EST	MEYLAN,WM & HOWARD,PH (1991)
OH 4.28E-011	cm3/molc sec	25	EST	MEYLAN,WM & HOWARD,PH (1993)

CAS #: 061820-94-8	ETHANONE,1-(4-CLPH)-2-[(4MEPH)SULFONYL]

Formula: $C_{15}H_{13}ClO_3S$

Mol Weight: 308.79

MP (deg C): FP (deg C):

BP (deg C):

BP pressure (mm Hg):

Property/Value	Units	Temp	Data Type	Reference
WS 1.28E+001	mg/L	25	EST	MEYLAN,WM ET AL. (1996)
logP 3.43			EXP	WANG,L ET AL. (1993)
VP 5.41E-008	mm Hg	25	EST	NEELY,WB & BLAU,GE (1985)
DC	pKa			
HL 1.36E-010	atm m3/mol	25	EST	MEYLAN,WM & HOWARD,PH (1991)
OH 7.94E-012	cm3/molc sec	25	EST	MEYLAN,WM & HOWARD,PH (1993)

CAS #: 061820-95-9	ETHANONE,1-(4-MEPH)-2-[(4MEPH)SULFONYL]

Formula: $C_{16}H_{16}O_3S$

Mol Weight: 288.37

MP (deg C): FP (deg C):

BP (deg C):

BP pressure (mm Hg):

Property/Value	Units	Temp	Data Type	Reference
WS 2.18E+001	mg/L	25	EST	MEYLAN,WM ET AL. (1996)
logP 3.30			EXP	WANG,L ET AL. (1993)
VP 9.72E-008	mm Hg	25	EST	NEELY,WB & BLAU,GE (1985)
DC	pKa			
HL 2.02E-010	atm m3/mol	25	EST	MEYLAN,WM & HOWARD,PH (1991)
OH 1.15E-011	cm3/molc sec	25	EST	MEYLAN,WM & HOWARD,PH (1993)

062100-54-3 — PROPANAMIDE, 2-HYDROXY-2-METHYL-N-(4-NITROPHENYL

Formula: $C_{10}H_{12}N_2O_4$
Mol Weight: 224.22
MP (deg C): **FP (deg C):**
BP (deg C):
BP pressure (mm Hg):

Property/Value	Units	Temp	Data Type	Reference
WS 5.62E+002	mg/L	25	EST	MEYLAN,WM ET AL. (1996)
logP 1.61			EXP	MORRIS,JJ ET AL. (1991)
VP 3.59E-009	mm Hg	25	EST	NEELY,WB & BLAU,GE (1985)
DC	pKa			
HL 6.69E-012	atm m3/mol	25	EST	MEYLAN,WM & HOWARD,PH (1991)
OH 2.75E-012	cm3/molc sec	25	EST	MEYLAN,WM & HOWARD,PH (1993)

062100-83-8 — ACETAMIDE, N-(2,3-DIMETHYL-8-NITRO-4-OXO-4H-1-BE

Formula: $C_{13}H_{12}N_2O_5$
Mol Weight: 276.25
MP (deg C): **FP (deg C):**
BP (deg C):
BP pressure (mm Hg):

Property/Value	Units	Temp	Data Type	Reference
WS 2.89E+002	mg/L	25	EST	MEYLAN,WM ET AL. (1996)
logP 1.61			EXP	HANSCH,C ET AL. (1995)
VP 1.87E-009	mm Hg	25	EST	NEELY,WB & BLAU,GE (1985)
DC	pKa			
HL 1.96E-013	atm m3/mol	25	EST	MEYLAN,WM & HOWARD,PH (1991)
OH 5.45E-011	cm3/molc sec	25	EST	MEYLAN,WM & HOWARD,PH (1993)

062163-09-1 — 5-CHLOROQUINOXALINE

Formula: $C_8H_5ClN_2$
Mol Weight: 164.60
MP (deg C): **FP (deg C):**
BP (deg C):
BP pressure (mm Hg):

Property/Value	Units	Temp	Data Type	Reference
WS 2.09E+003	mg/L	25	EST	MEYLAN,WM ET AL. (1996)
logP 1.75			EXP	HANSCH,C & LEO,AJ (1985)
VP 2.46E-003	mm Hg	25	EST	NEELY,WB & BLAU,GE (1985)
DC	pKa			
HL 2.11E-007	atm m3/mol	25	EST	MEYLAN,WM & HOWARD,PH (1991)
OH 1.41E-012	cm3/molc sec	25	EST	MEYLAN,WM & HOWARD,PH (1993)

062174-83-8 — 4-PYRIDINEBUTANEAMINE

Formula: $C_9H_{14}N_2$
Mol Weight: 150.23
MP (deg C): **FP (deg C):**
BP (deg C):
BP pressure (mm Hg):

Property/Value	Units	Temp	Data Type	Reference
WS 1.00E+006	mg/L	25	EST	MEYLAN,WM ET AL. (1996)
logP 0.86			EXP	HANSCH,C & LEO,AJ (1985)
VP 1.44E-002	mm Hg	25	EST	NEELY,WB & BLAU,GE (1985)
DC	pKa			
HL 1.87E-009	atm m3/mol	25	EST	MEYLAN,WM & HOWARD,PH (1991)
OH 3.66E-011	cm3/molc sec	25	EST	MEYLAN,WM & HOWARD,PH (1993)

062178-32-9 — B-PROPYLGALACTOPYRANOSIDE

Formula: $C_9H_{18}O_6$
Mol Weight: 222.24
MP (deg C): **FP (deg C):**
BP (deg C):
BP pressure (mm Hg):

Property/Value	Units	Temp	Data Type	Reference
WS 8.76E+005	mg/L	25	EST	MEYLAN,WM ET AL. (1996)
logP -1.66			EXP	HANSCH,C & LEO,AJ (1985)
VP 6.27E-009	mm Hg	25	EST	NEELY,WB & BLAU,GE (1985)
DC	pKa			
HL 2.75E-014	atm m3/mol	25	EST	MEYLAN,WM & HOWARD,PH (1991)
OH 7.25E-011	cm3/molc sec	25	EST	MEYLAN,WM & HOWARD,PH (1993)

062251-73-4 — N'-IPR(2-TOLYLOXY)ACETIC AC HYDRAZID

Formula: $C_{12}H_{18}N_2O_2$
Mol Weight: 222.29
MP (deg C): **FP (deg C):**
BP (deg C):
BP pressure (mm Hg):

Property/Value	Units	Temp	Data Type	Reference
WS 9.16E+002	mg/L	25	EST	MEYLAN,WM ET AL. (1996)
logP 1.83			EXP	HANSCH,C & LEO,AJ (1985)
VP 1.88E-006	mm Hg	25	EST	NEELY,WB & BLAU,GE (1985)
DC	pKa			
HL 6.11E-012	atm m3/mol	25	EST	MEYLAN,WM & HOWARD,PH (1991)
OH 1.18E-010	cm3/molc sec	25	EST	MEYLAN,WM & HOWARD,PH (1993)

062251-74-5 — N-IPR-1(2-MEO PHENOXY)ACETIC HYDRAZID

Formula: $C_{12}H_{18}N_2O_3$
Mol Weight: 238.29
MP (deg C): **FP (deg C):**
BP (deg C):
BP pressure (mm Hg):

Property/Value	Units	Temp	Data Type	Reference
WS 3.09E+003	mg/L	25	EST	MEYLAN,WM ET AL. (1996)
logP 1.11			EXP	HANSCH,C & LEO,AJ (1985)
VP 7.45E-007	mm Hg	25	EST	NEELY,WB & BLAU,GE (1985)
DC	pKa			
HL 3.28E-013	atm m3/mol	25	EST	MEYLAN,WM & HOWARD,PH (1991)
OH 1.11E-010	cm3/molc sec	25	EST	MEYLAN,WM & HOWARD,PH (1993)

062251-75-6 — N-IPR-1(3-MEO PHENOXY)ACETIC HYDRAZID

Formula: $C_{12}H_{18}N_2O_3$
Mol Weight: 238.29
MP (deg C): **FP (deg C):**
BP (deg C):
BP pressure (mm Hg):

Property/Value	Units	Temp	Data Type	Reference
WS 2.01E+003	mg/L	25	EST	MEYLAN,WM ET AL. (1996)
logP 1.33			EXP	HANSCH,C & LEO,AJ (1985)
VP 7.45E-007	mm Hg	25	EST	NEELY,WB & BLAU,GE (1985)
DC	pKa			
HL 3.28E-013	atm m3/mol	25	EST	MEYLAN,WM & HOWARD,PH (1991)
OH 2.92E-010	cm3/molc sec	25	EST	MEYLAN,WM & HOWARD,PH (1993)

CAS #: 062308-10-5	3-METHYL-4-((4-NITROPHENYL)AZO)BENZENAMINE

Formula: $C_{13}H_{12}N_4O_2$

Mol Weight: 256.27

MP (deg C): FP (deg C):

BP (deg C):

BP pressure (mm Hg):

Property/Value	Units	Temp	Data Type	Reference
WS 2.37E+000	mg/L	25	EST	MEYLAN,WM ET AL. (1996)
logP 4.14			EST	MEYLAN,WM & HOWARD,PH (1995)
VP 7.13E-010	mm Hg	25	EXP	SHIMIZU,T ET AL. (1987)
DC	pKa			
HL 2.26E-011	atm m3/mol	25	EST	MEYLAN,WM & HOWARD,PH (1991)
OH 1.12E-010	cm3/molc sec	25	EST	MEYLAN,WM & HOWARD,PH (1993)

CAS #: 062433-26-5	METHANONE, (4-CHLOROPHENYL)(5-FLUORO-2-HYDROXYPH

Formula: $C_{13}H_8ClFO_2$

Mol Weight: 250.66

MP (deg C): FP (deg C):

BP (deg C):

BP pressure (mm Hg):

Property/Value	Units	Temp	Data Type	Reference
WS 3.10E+001	mg/L	25	EST	MEYLAN,WM ET AL. (1996)
logP 4.05			EXP	FARRAJ,NF ET AL. (1988)
VP 2.29E-006	mm Hg	25	EST	NEELY,WB & BLAU,GE (1985)
DC	pKa			
HL 2.20E-007	atm m3/mol	25	EST	MEYLAN,WM & HOWARD,PH (1991)
OH 1.16E-011	cm3/molc sec	25	EST	MEYLAN,WM & HOWARD,PH (1993)

CAS #: 062470-53-5	1,3,7,9-TETRACHLORODIBENZO-P-DIOXIN

Formula: $C_{12}H_4Cl_4O_2$

Mol Weight: 321.98

MP (deg C): FP (deg C):

BP (deg C):

BP pressure (mm Hg):

Property/Value	Units	Temp	Data Type	Reference
WS 6.12E-004	mg/L	25	EST	MEYLAN,WM ET AL. (1996)
logP 6.92			EST	MEYLAN,WM & HOWARD,PH (1995)
VP 1.75E-006	mm Hg	25	EST	NEELY,WB & BLAU,GE (1985)
DC	pKa			
HL 3.53E-006	atm m3/mol	25	EST	MEYLAN,WM & HOWARD,PH (1991)
OH 4.87E-012	cm3/molc sec	25	EST	MEYLAN,WM & HOWARD,PH (1993)

CAS #: 062484-16-6	QUINAZOLINE-2,4-DIONE,6-ME

Formula: $C_9H_8N_2O_2$

Mol Weight: 176.18

MP (deg C): FP (deg C):

BP (deg C):

BP pressure (mm Hg):

Property/Value	Units	Temp	Data Type	Reference
WS 1.95E+004	mg/L	25	EST	MEYLAN,WM ET AL. (1996)
logP 0.55			EXP	HANSCH,C & LEO,AJ (1985)
VP 2.46E-008	mm Hg	25	EST	NEELY,WB & BLAU,GE (1985)
DC	pKa			
HL 5.90E-011	atm m3/mol	25	EST	MEYLAN,WM & HOWARD,PH (1991)
OH 4.88E-011	cm3/molc sec	25	EST	MEYLAN,WM & HOWARD,PH (1993)

CAS #: 062567-44-6	PYRIDAZINE, 3-ETHOXY-

Formula: $C_6H_8N_2O$

Mol Weight: 124.14

MP (deg C): FP (deg C):

BP (deg C):

BP pressure (mm Hg):

Property/Value	Units	Temp	Data Type	Reference
WS 2.81E+004	mg/L	25	EST	MEYLAN,WM ET AL. (1996)
logP 0.63			EXP	YAMAGAMI,C ET AL. (1990)
VP 1.44E-002	mm Hg	25	EST	NEELY,WB & BLAU,GE (1985)
DC	pKa			
HL 2.06E-006	atm m3/mol	25	EST	MEYLAN,WM & HOWARD,PH (1991)
OH 8.71E-012	cm3/molc sec	25	EST	MEYLAN,WM & HOWARD,PH (1993)

CAS #: 062570-20-1	DIS. A. 12

Formula: $C_{16}H_{17}N_5O_5$

Mol Weight: 359.34

MP (deg C): FP (deg C):

BP (deg C):

BP pressure (mm Hg):

Property/Value	Units	Temp	Data Type	Reference
WS 7.18E-001	mg/L	25	EXP	BAUGHMAN,GL & PERENICH,TA (1988)
logP 4.02			EST	MEYLAN,WM & HOWARD,PH (1995)
VP 6.31E-013	mm Hg	25	EST	NEELY,WB & BLAU,GE (1985)
DC	pKa			
HL 2.35E-016	atm m3/mol	25	EST	MEYLAN,WM & HOWARD,PH (1991)
OH 8.18E-011	cm3/molc sec	25	EST	MEYLAN,WM & HOWARD,PH (1993)

CAS #: 062571-86-2	CAPTOPRIL

Formula: $C_9H_{15}NO_3S$

Mol Weight: 217.29

MP (deg C): 106 FP (deg C):

BP (deg C):

BP pressure (mm Hg):

Property/Value	Units	Temp	Data Type	Reference
WS 1.82E+004	mg/L	25	EST	MEYLAN,WM ET AL. (1996)
logP 0.34			EXP	RANADIVE,SA ET AL. (1992)
VP 3.64E-006	mm Hg	25	EST	NEELY,WB & BLAU,GE (1985)
DC	pKa			
HL 3.89E-013	atm m3/mol	25	EST	MEYLAN,WM & HOWARD,PH (1991)
OH 9.07E-011	cm3/molc sec	25	EST	MEYLAN,WM & HOWARD,PH (1993)

CAS #: 062580-80-7	1-(2,3-DIOHPR)-2-ME-5-NO2IMIDAZOLE

Formula: $C_7H_{13}N_3O_4$

Mol Weight: 203.20

MP (deg C): FP (deg C):

BP (deg C):

BP pressure (mm Hg):

Property/Value	Units	Temp	Data Type	Reference
WS 5.61E+004	mg/L	25	EST	MEYLAN,WM ET AL. (1996)
logP -0.59			EXP	HANSCH,C & LEO,AJ (1985)
VP 2.91E-009	mm Hg	25	EST	NEELY,WB & BLAU,GE (1985)
DC	pKa			
HL 8.22E-013	atm m3/mol	25	EST	MEYLAN,WM & HOWARD,PH (1991)
OH 1.97E-011	cm3/molc sec	25	EST	MEYLAN,WM & HOWARD,PH (1993)

062624-26-4 — BENZOBICYCLO(2,2,1)HEPTENE,2-EX-AMINO

Formula: $C_{11}H_{13}N$
Mol Weight: 159.23
MP (deg C):
FP (deg C):
BP (deg C):
BP pressure (mm Hg):

Property/Value	Units	Temp	Data Type	Reference
WS 1.15E+004	mg/L	25	EST	MEYLAN,WM ET AL. (1996)
logP 2.09			EXP	HANSCH,C & LEO,AJ (1985)
VP 8.71E-003	mm Hg	25	EST	NEELY,WB & BLAU,GE (1985)
DC	pKa			
HL 3.07E-007	atm m3/mol	25	EST	MEYLAN,WM & HOWARD,PH (1991)
OH 4.96E-011	cm3/molc sec	25	EST	MEYLAN,WM & HOWARD,PH (1993)

062624-27-5 — BENZOBICYCLO(2,2,1)HEPTENE,2-EX-MEAM

Formula: $C_{12}H_{15}N$
Mol Weight: 173.26
MP (deg C):
FP (deg C):
BP (deg C):
BP pressure (mm Hg):

Property/Value	Units	Temp	Data Type	Reference
WS 4.70E+003	mg/L	25	EST	MEYLAN,WM ET AL. (1996)
logP 2.41			EXP	HANSCH,C & LEO,AJ (1985)
VP 6.95E-003	mm Hg	25	EST	NEELY,WB & BLAU,GE (1985)
DC	pKa			
HL 6.74E-007	atm m3/mol	25	EST	MEYLAN,WM & HOWARD,PH (1991)
OH 9.29E-011	cm3/molc sec	25	EST	MEYLAN,WM & HOWARD,PH (1993)

062666-20-0 — PROGABIDE

Formula: $C_{17}H_{16}ClFN_2O_2$
Mol Weight: 334.78
MP (deg C): 133-135
FP (deg C):
BP (deg C):
BP pressure (mm Hg):

Property/Value	Units	Temp	Data Type	Reference
WS 7.08E+001	mg/L	25	EST	MEYLAN,WM ET AL. (1996)
logP 3.06			EXP	FARRAJ,NF ET AL. (1988)
VP 6.13E-011	mm Hg	25	EST	NEELY,WB & BLAU,GE (1985)
DC	pKa			
HL 7.61E-016	atm m3/mol	25	EST	MEYLAN,WM & HOWARD,PH (1991)
OH 2.38E-011	cm3/molc sec	25	EST	MEYLAN,WM & HOWARD,PH (1993)

062774-57-6 — HYDRAZINECARBOXAMIDE, N-(4-METHYLPHENYL)-

Formula: $C_8H_{11}N_3O$
Mol Weight: 165.20
MP (deg C):
FP (deg C):
BP (deg C):
BP pressure (mm Hg):

Property/Value	Units	Temp	Data Type	Reference
WS 1.67E+004	mg/L	25	EST	MEYLAN,WM ET AL. (1996)
logP 0.69			EXP	KRAMER,CR & BECK,L (1981)
VP 2.95E-005	mm Hg	25	EST	NEELY,WB & BLAU,GE (1985)
DC	pKa			
HL 9.28E-013	atm m3/mol	25	EST	MEYLAN,WM & HOWARD,PH (1991)
OH 5.34E-011	cm3/molc sec	25	EST	MEYLAN,WM & HOWARD,PH (1993)

062774-58-7 — HYDRAZINECARBOXAMIDE, N-(2-METHOXYPHENYL)-

Formula: $C_8H_{11}N_3O_2$
Mol Weight: 181.20
MP (deg C):
FP (deg C):
BP (deg C):
BP pressure (mm Hg):

Property/Value	Units	Temp	Data Type	Reference
WS 3.57E+003	mg/L	25	EST	MEYLAN,WM ET AL. (1996)
logP -0.15			EXP	KRAMER,CR & BECK,L (1981)
VP 1.43E-005	mm Hg	25	EST	NEELY,WB & BLAU,GE (1985)
DC	pKa			
HL 4.98E-014	atm m3/mol	25	EST	MEYLAN,WM & HOWARD,PH (1991)
OH 3.87E-011	cm3/molc sec	25	EST	MEYLAN,WM & HOWARD,PH (1993)

062774-59-8 — HYDRAZINECARBOXAMIDE, N-(4-METHOXYPHENYL)-

Formula: $C_8H_{11}N_3O_2$
Mol Weight: 181.20
MP (deg C):
FP (deg C):
BP (deg C):
BP pressure (mm Hg):

Property/Value	Units	Temp	Data Type	Reference
WS 1.91E+003	mg/L	25	EST	MEYLAN,WM ET AL. (1996)
logP 0.17			EXP	KRAMER,CR & BECK,L (1981)
VP 1.43E-005	mm Hg	25	EST	NEELY,WB & BLAU,GE (1985)
DC	pKa			
HL 4.98E-014	atm m3/mol	25	EST	MEYLAN,WM & HOWARD,PH (1991)
OH 3.87E-011	cm3/molc sec	25	EST	MEYLAN,WM & HOWARD,PH (1993)

062783-49-7 — BENZENEMETHANAMINE, N,3-DIMETHYL-N-NITROSO-

Formula: $C_9H_{12}N_2O$
Mol Weight: 164.21
MP (deg C):
FP (deg C):
BP (deg C):
BP pressure (mm Hg):

Property/Value	Units	Temp	Data Type	Reference
WS 1.39E+003	mg/L	25	EST	MEYLAN,WM ET AL. (1996)
logP 1.96			EXP	SINGER,GM ET AL. (1986)
VP 1.18E-003	mm Hg	25	EST	NEELY,WB & BLAU,GE (1985)
DC	pKa			
HL 1.83E-007	atm m3/mol	25	EST	MEYLAN,WM & HOWARD,PH (1991)
OH 2.27E-011	cm3/molc sec	25	EST	MEYLAN,WM & HOWARD,PH (1993)

062783-50-0 — BENZENEMETHANAMINE, N,4-DIMETHYL-N-NITROSO-

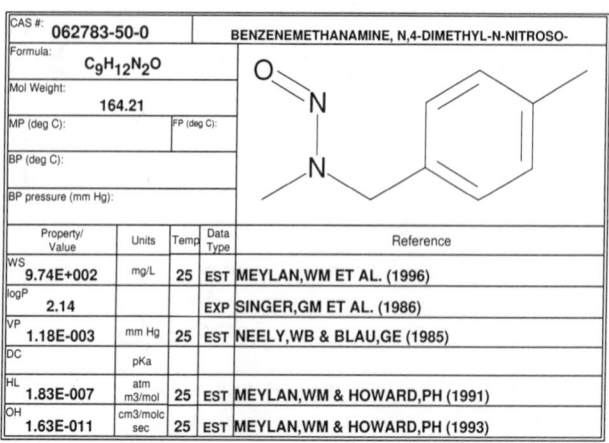

Formula: $C_9H_{12}N_2O$
Mol Weight: 164.21
MP (deg C):
FP (deg C):
BP (deg C):
BP pressure (mm Hg):

Property/Value	Units	Temp	Data Type	Reference
WS 9.74E+002	mg/L	25	EST	MEYLAN,WM ET AL. (1996)
logP 2.14			EXP	SINGER,GM ET AL. (1986)
VP 1.18E-003	mm Hg	25	EST	NEELY,WB & BLAU,GE (1985)
DC	pKa			
HL 1.83E-007	atm m3/mol	25	EST	MEYLAN,WM & HOWARD,PH (1991)
OH 1.63E-011	cm3/molc sec	25	EST	MEYLAN,WM & HOWARD,PH (1993)

CAS #: 062796-65-0				2,2',4,6-TETRACHLOROBIPHENYL
Formula: $C_{12}H_6Cl_4$				
Mol Weight: 291.99				
MP (deg C):		FP (deg C):		
BP (deg C):				
BP pressure (mm Hg):				

Property/Value	Units	Temp	Data Type	Reference
WS 5.32E-002	mg/L	25	EST	MEYLAN,WM ET AL. (1996)
logP 6.34			EST	MEYLAN,WM & HOWARD,PH (1995)
VP 8.45E-006	mm Hg	25	EST	NEELY,WB & BLAU,GE (1985)
DC	pKa			
HL 1.25E-004	atm m3/mol	25	EST	MEYLAN,WM & HOWARD,PH (1991)
OH 8.27E-013	cm3/molc sec	25	EST	MEYLAN,WM & HOWARD,PH (1993)

CAS #: 062825-07-4				M-CHLOROPHENYL AZOXYCYANIDE
Formula: $C_7H_4ClN_3O$				
Mol Weight: 181.58				
MP (deg C):		FP (deg C):		
BP (deg C):				
BP pressure (mm Hg):				

Property/Value	Units	Temp	Data Type	Reference
WS 1.79E+002	mg/L	25	EST	MEYLAN,WM ET AL. (1996)
logP 2.40			EXP	CALVINO,R R ET AL (1991)
VP 1.72E-008	mm Hg	25	EST	NEELY,WB & BLAU,GE (1985)
DC	pKa			
HL 1.53E-011	atm m3/mol	25	EST	MEYLAN,WM & HOWARD,PH (1991)
OH 2.93E-013	cm3/molc sec	25	EST	MEYLAN,WM & HOWARD,PH (1993)

CAS #: 062825-09-6				M-BROMOPHENYL AZOXYCYANIDE
Formula: $C_7H_4BrN_3O$				
Mol Weight: 226.04				
MP (deg C):		FP (deg C):		
BP (deg C):				
BP pressure (mm Hg):				

Property/Value	Units	Temp	Data Type	Reference
WS 7.70E+001	mg/L	25	EST	MEYLAN,WM ET AL. (1996)
logP 2.56			EXP	CALVINO,R R ET AL (1991)
VP 7.46E-009	mm Hg	25	EST	NEELY,WB & BLAU,GE (1985)
DC	pKa			
HL 8.20E-012	atm m3/mol	25	EST	MEYLAN,WM & HOWARD,PH (1991)
OH 2.62E-013	cm3/molc sec	25	EST	MEYLAN,WM & HOWARD,PH (1993)

CAS #: 062825-10-9				P-BROMOPHENYL AZOXYCYANIDE
Formula: $C_7H_4BrN_3O$				
Mol Weight: 226.04				
MP (deg C):		FP (deg C):		
BP (deg C):				
BP pressure (mm Hg):				

Property/Value	Units	Temp	Data Type	Reference
WS 8.66E+001	mg/L	25	EST	MEYLAN,WM ET AL. (1996)
logP 2.50			EXP	CALVINO,R R ET AL (1991)
VP 7.46E-009	mm Hg	25	EST	NEELY,WB & BLAU,GE (1985)
DC	pKa			
HL 8.20E-012	atm m3/mol	25	EST	MEYLAN,WM & HOWARD,PH (1991)
OH 2.62E-013	cm3/molc sec	25	EST	MEYLAN,WM & HOWARD,PH (1993)

CAS #: 062825-12-1				M-NITROPHENYL AZOXYCYANIDE
Formula: $C_7H_4N_4O_3$				
Mol Weight: 192.13				
MP (deg C):		FP (deg C):		
BP (deg C):				
BP pressure (mm Hg):				

Property/Value	Units	Temp	Data Type	Reference
WS 4.17E+002	mg/L	25	EST	MEYLAN,WM ET AL. (1996)
logP 1.45			EXP	CALVINO,R R ET AL (1991)
VP 1.29E-009	mm Hg	25	EST	NEELY,WB & BLAU,GE (1985)
DC	pKa			
HL 8.12E-014	atm m3/mol	25	EST	MEYLAN,WM & HOWARD,PH (1991)
OH 5.21E-014	cm3/molc sec	25	EST	MEYLAN,WM & HOWARD,PH (1993)

CAS #: 062825-14-3				M-METHOXYPHENYL AZOXYCYANIDE
Formula: $C_8H_7N_3O_2$				
Mol Weight: 177.16				
MP (deg C):		FP (deg C):		
BP (deg C):				
BP pressure (mm Hg):				

Property/Value	Units	Temp	Data Type	Reference
WS 2.57E+002	mg/L	25	EST	MEYLAN,WM ET AL. (1996)
logP 2.24			EXP	CALVINO,R R ET AL (1991)
VP 1.22E-008	mm Hg	25	EST	NEELY,WB & BLAU,GE (1985)
DC	pKa			
HL 1.22E-012	atm m3/mol	25	EST	MEYLAN,WM & HOWARD,PH (1991)
OH 5.43E-012	cm3/molc sec	25	EST	MEYLAN,WM & HOWARD,PH (1993)

CAS #: 062825-15-4				P-METHOXY PHENYL AZOXYCYANIDE
Formula: $C_8H_7N_3O_2$				
Mol Weight: 177.16				
MP (deg C):		FP (deg C):		
BP (deg C):				
BP pressure (mm Hg):				

Property/Value	Units	Temp	Data Type	Reference
WS 3.07E+002	mg/L	25	EST	MEYLAN,WM ET AL. (1996)
logP 2.15			EXP	CALVINO,R R ET AL (1991)
VP 1.22E-008	mm Hg	25	EST	NEELY,WB & BLAU,GE (1985)
DC	pKa			
HL 1.22E-012	atm m3/mol	25	EST	MEYLAN,WM & HOWARD,PH (1991)
OH 5.43E-012	cm3/molc sec	25	EST	MEYLAN,WM & HOWARD,PH (1993)

CAS #: 062825-16-5				P-(N,N-DIMETHYL)PHENYL AZOXYCYANIDE
Formula: $C_9H_{10}N_4O$				
Mol Weight: 190.21				
MP (deg C):		FP (deg C):		
BP (deg C):				
BP pressure (mm Hg):				

Property/Value	Units	Temp	Data Type	Reference
WS 1.97E+002	mg/L	25	EST	MEYLAN,WM ET AL. (1996)
logP 2.30			EXP	CALVINO,R R ET AL (1991)
VP 5.73E-009	mm Hg	25	EST	NEELY,WB & BLAU,GE (1985)
DC	pKa			
HL 3.27E-013	atm m3/mol	25	EST	MEYLAN,WM & HOWARD,PH (1991)
OH 8.16E-011	cm3/molc sec	25	EST	MEYLAN,WM & HOWARD,PH (1993)

CAS #: 062850-32-2				FENOTHIOCARB

Formula: $C_{13}H_{19}NO_2S$

Mol Weight: 253.37

MP (deg C): 40-41

FP (deg C):

BP (deg C):

BP pressure (mm Hg):

Property/Value	Units	Temp	Data Type	Reference
WS 3.00E+001	mg/L	20	EXP	SHIU,WY ET AL. (1990)
logP 3.28			EXP	TOMLIN,C (1994)
VP 5.22E-006	mm Hg	25	EST	NEELY,WB & BLAU,GE (1985)
DC	pKa			
HL 3.78E-008	atm m3/mol	25	EST	MEYLAN,WM & HOWARD,PH (1991)
OH 5.17E-011	cm3/molc sec	25	EST	MEYLAN,WM & HOWARD,PH (1993)

CAS #: 062893-19-0				CEFOPERAZONE

Formula: $C_{25}H_{29}N_9O_8S_2$

Mol Weight: 647.69

MP (deg C): 167-171

FP (deg C):

BP (deg C):

BP pressure (mm Hg):

Property/Value	Units	Temp	Data Type	Reference
WS 6.42E+001	mg/L	25	EST	MEYLAN,WM ET AL. (1996)
logP -0.74			EXP	HANSCH,C ET AL. (1995)
VP 9.61E-030	mm Hg	25	EST	NEELY,WB & BLAU,GE (1985)
DC	pKa			
HL 1.63E-039	atm m3/mol	25	EST	MEYLAN,WM & HOWARD,PH (1991)
OH 2.78E-010	cm3/molc sec	25	EST	MEYLAN,WM & HOWARD,PH (1993)

CAS #: 062937-22-8				8-FORMYLAMIDOQUINOLINE

Formula: $C_{10}H_8N_2O$

Mol Weight: 172.19

MP (deg C):

FP (deg C):

BP (deg C):

BP pressure (mm Hg):

Property/Value	Units	Temp	Data Type	Reference
WS 1.74E+003	mg/L	25	EST	MEYLAN,WM ET AL. (1996)
logP 1.80			EXP	HANSCH,C & LEO,AJ (1985)
VP 2.54E-006	mm Hg	25	EST	NEELY,WB & BLAU,GE (1985)
DC	pKa			
HL 1.08E-012	atm m3/mol	25	EST	MEYLAN,WM & HOWARD,PH (1991)
OH 2.00E-010	cm3/molc sec	25	EST	MEYLAN,WM & HOWARD,PH (1993)

CAS #: 062959-39-1				4-((4-METHYLPHENYL)AZO)BENZENESULFONIC ACID,SOD*

Formula: $C_{13}H_{11}N_2NaO_3S$

Mol Weight: 298.30

MP (deg C):

FP (deg C):

BP (deg C):

BP pressure (mm Hg):

Property/Value	Units	Temp	Data Type	Reference
WS 8.16E+003	mg/L	25	EST	MEYLAN,WM ET AL. (1996)
logP -0.29			EST	MEYLAN,WM & HOWARD,PH (1995)
VP 6.24E-015	mm Hg	25	EST	NEELY,WB & BLAU,GE (1985)
DC	pKa			
HL	atm m3/mol			
OH 2.32E-012	cm3/molc sec	25	EST	MEYLAN,WM & HOWARD,PH (1993)

CAS #: 062959-40-4				4-((2,4-DIMETHYLPHENYL)AZO)BENZENESULFONIC ACID*

Formula: $C_{14}H_{13}N_2NaO_3S$

Mol Weight: 312.33

MP (deg C):

FP (deg C):

BP (deg C):

BP pressure (mm Hg):

Property/Value	Units	Temp	Data Type	Reference
WS 2.30E+003	mg/L	25	EST	MEYLAN,WM ET AL. (1996)
logP 0.26			EST	MEYLAN,WM & HOWARD,PH (1995)
VP 2.63E-015	mm Hg	25	EST	NEELY,WB & BLAU,GE (1985)
DC	pKa			
HL	atm m3/mol			
OH 5.70E-012	cm3/molc sec	25	EST	MEYLAN,WM & HOWARD,PH (1993)

CAS #: 062959-41-5				4-((2,4-DICHLOROPHENYL)AZO)BENZENESULFONIC ACID*

Formula: $C_{12}H_7Cl_2N_2NaO_3S$

Mol Weight: 353.16

MP (deg C):

FP (deg C):

BP (deg C):

BP pressure (mm Hg):

Property/Value	Units	Temp	Data Type	Reference
WS 8.96E+002	mg/L	25	EST	MEYLAN,WM ET AL. (1996)
logP 0.46			EST	MEYLAN,WM & HOWARD,PH (1995)
VP 9.11E-016	mm Hg	25	EST	NEELY,WB & BLAU,GE (1985)
DC	pKa			
HL	atm m3/mol			
OH 5.48E-013	cm3/molc sec	25	EST	MEYLAN,WM & HOWARD,PH (1993)

CAS #: 062973-76-6				AZANIDAZOLE

Formula: $C_{10}H_{10}N_6O_2$

Mol Weight: 246.23

MP (deg C): 232-235

FP (deg C):

BP (deg C):

BP pressure (mm Hg):

Property/Value	Units	Temp	Data Type	Reference
WS 1.90E+003	mg/L	25	EST	MEYLAN,WM ET AL. (1996)
logP 0.85			EXP	SANGSTER,J (1993)
VP 6.60E-009	mm Hg	25	EST	NEELY,WB & BLAU,GE (1985)
DC	pKa			
HL 4.93E-012	atm m3/mol	25	EST	MEYLAN,WM & HOWARD,PH (1991)
OH 7.14E-011	cm3/molc sec	25	EST	MEYLAN,WM & HOWARD,PH (1993)

CAS #: 062981-74-2				1-CHLORO-1-BUTYNE

Formula: C_4H_5Cl

Mol Weight: 88.54

MP (deg C):

FP (deg C):

BP (deg C):

BP pressure (mm Hg):

Property/Value	Units	Temp	Data Type	Reference
WS 3.34E+003	mg/L	25	EST	MEYLAN,WM ET AL. (1996)
logP 1.84			EST	MEYLAN,WM & HOWARD,PH (1995)
VP 1.08E+002	mm Hg	25	EST	NEELY,WB & BLAU,GE (1985)
DC	pKa			
HL 1.96E-002	atm m3/mol	25	EST	MEYLAN,WM & HOWARD,PH (1991)
OH 6.77E-012	cm3/molc sec	25	EST	MEYLAN,WM & HOWARD,PH (1993)

CAS #: 063020-20-2	2-ACETAMIDOCARBAZOLE

Formula: C_{14}H_{12}N_2O

Mol Weight: 224.26

MP (deg C):   FP (deg C):

BP (deg C):

BP pressure (mm Hg):

Property/ Value	Units	Temp	Data Type	Reference
WS 6.68E+000	mg/L	25	EST	MEYLAN,WM ET AL. (1996)
logP 3.02			EXP	HANSCH,C & LEO,AJ (1985)
VP 4.78E-009	mm Hg	25	EST	NEELY,WB & BLAU,GE (1985)
DC	pKa			
HL 9.91E-014	atm m3/mol	25	EST	MEYLAN,WM & HOWARD,PH (1991)
OH 2.00E-010	cm3/molc sec	25	EST	MEYLAN,WM & HOWARD,PH (1993)

CAS #: 063075-06-9	N-PENTYL N-PHENYLCARBAMATE

Formula: C_{12}H_{17}NO_2

Mol Weight: 207.27

MP (deg C):   FP (deg C):

BP (deg C):

BP pressure (mm Hg):

Property/ Value	Units	Temp	Data Type	Reference
WS 2.28E+001	mg/L	25	EST	MEYLAN,WM ET AL. (1996)
logP 3.80			EXP	HANSCH,C & LEO,AJ (1985)
VP 8.64E-004	mm Hg	25	EST	NEELY,WB & BLAU,GE (1985)
DC				
HL 6.78E-008	atm m3/mol	25	EST	MEYLAN,WM & HOWARD,PH (1991)
OH 4.86E-011	cm3/molc sec	25	EST	MEYLAN,WM & HOWARD,PH (1993)

CAS #: 063144-76-3	B27C9-BENZO CROWN ETHER

Formula: C_{21}H_{34}O_9

Mol Weight: 430.50

MP (deg C):   FP (deg C):

BP (deg C):

BP pressure (mm Hg):

Property/ Value	Units	Temp	Data Type	Reference
WS 1.03E+003	mg/L	25	EST	MEYLAN,WM ET AL. (1996)
logP 0.23			EXP	STOLWIJK,TB ET AL. (1989)
VP 7.74E-011	mm Hg	25	EST	NEELY,WB & BLAU,GE (1985)
DC	pKa			
HL 1.82E-018	atm m3/mol	25	EST	MEYLAN,WM & HOWARD,PH (1991)
OH 1.31E-010	cm3/molc sec	25	EST	MEYLAN,WM & HOWARD,PH (1993)

CAS #: 063151-11-1	2,2-DICHLOROPROPANOL

Formula: C_3H_6Cl_2O

Mol Weight: 128.99

MP (deg C):   FP (deg C):

BP (deg C):

BP pressure (mm Hg):

Property/ Value	Units	Temp	Data Type	Reference
WS 1.73E+004	mg/L	25	EST	MEYLAN,WM ET AL. (1996)
logP 1.45			EST	MEYLAN,WM & HOWARD,PH (1995)
VP 5.97E+000	mm Hg	25	EST	NEELY,WB & BLAU,GE (1985)
DC	pKa			
HL 5.89E-007	atm m3/mol	25	EST	MEYLAN,WM & HOWARD,PH (1991)
OH 5.41E-013	cm3/molc sec	25	EST	MEYLAN,WM & HOWARD,PH (1993)

CAS #: 063224-18-0	BENZAMIDE,N-(2-DIMETHYLAMINOETHYL)

Formula: C_{11}H_{16}N_2O

Mol Weight: 192.26

MP (deg C):   FP (deg C):

BP (deg C):

BP pressure (mm Hg):

Property/ Value	Units	Temp	Data Type	Reference
WS 5.51E+003	mg/L	25	EST	MEYLAN,WM ET AL. (1996)
logP 1.10			EXP	HANSCH,C & LEO,AJ (1985)
VP 1.28E-005	mm Hg	25	EST	NEELY,WB & BLAU,GE (1985)
DC	pKa			
HL 3.20E-012	atm m3/mol	25	EST	MEYLAN,WM & HOWARD,PH (1991)
OH 9.12E-011	cm3/molc sec	25	EST	MEYLAN,WM & HOWARD,PH (1993)

CAS #: 063253-20-3	ACETIC ACID, 2,2,2-TRICHLORO-1-(4-METHOXYPHENYL)

Formula: C_{11}H_{11}Cl_3O_3

Mol Weight: 297.57

MP (deg C):   FP (deg C):

BP (deg C):

BP pressure (mm Hg):

Property/ Value	Units	Temp	Data Type	Reference
WS 8.44E+000	mg/L	25	EST	MEYLAN,WM ET AL. (1996)
logP 3.72			EXP	HU,J & LENG,XF (1992)
VP 1.93E-004	mm Hg	25	EST	NEELY,WB & BLAU,GE (1985)
DC	pKa			
HL 4.86E-008	atm m3/mol	25	EST	MEYLAN,WM & HOWARD,PH (1991)
OH 2.70E-011	cm3/molc sec	25	EST	MEYLAN,WM & HOWARD,PH (1993)

CAS #: 063284-71-9	5-PYRIMIDINEMETHANOL, .ALPHA.-(2-CHLOROPHENYL)-.

Formula: C_{17}H_{12}ClFN_2O

Mol Weight: 314.75

MP (deg C): 126-27   FP (deg C):

BP (deg C):

BP pressure (mm Hg):

Property/ Value	Units	Temp	Data Type	Reference
WS 2.60E+001	mg/L	25	EXP	SHIU,WY ET AL. (1990)
logP 3.18			EXP	TOMLIN,C (1994)
VP 1.82E-009	mm Hg	25	EST	NEELY,WB & BLAU,GE (1985)
DC	pKa			
HL 6.64E-013	atm m3/mol	25	EST	MEYLAN,WM & HOWARD,PH (1991)
OH 5.22E-012	cm3/molc sec	25	EST	MEYLAN,WM & HOWARD,PH (1993)

CAS #: 063346-71-4	2(5-CL-2,4-MEO PH IMINO)IMIDAZOLINE

Formula: C_{11}H_{14}ClN_3O_2

Mol Weight: 255.71

MP (deg C):   FP (deg C):

BP (deg C):

BP pressure (mm Hg):

Property/ Value	Units	Temp	Data Type	Reference
WS 1.49E+004	mg/L	25	EST	MEYLAN,WM ET AL. (1996)
logP 1.38			EXP	HANSCH,C & LEO,AJ (1985)
VP 2.62E-006	mm Hg	25	EST	NEELY,WB & BLAU,GE (1985)
DC	pKa			
HL 7.11E-014	atm m3/mol	25	EST	MEYLAN,WM & HOWARD,PH (1991)
OH 2.21E-010	cm3/molc sec	25	EST	MEYLAN,WM & HOWARD,PH (1993)

CAS #: 063346-72-5 — 2-(2-MEO-4-ME PH IMINO)IMIDAZOLIDINE

Formula: $C_{11}H_{15}N_3O$

Mol Weight: 205.26

MP (deg C): FP (deg C):

BP (deg C):

BP pressure (mm Hg):

Property/Value	Units	Temp	Data Type	Reference
WS 1.67E+004	mg/L	25	EST	MEYLAN,WM ET AL. (1996)
logP 1.64			EXP	HANSCH,C & LEO,AJ (1985)
VP 2.84E-005	mm Hg	25	EST	NEELY,WB & BLAU,GE (1985)
DC	pKa			
HL 1.79E-012	atm m3/mol	25	EST	MEYLAN,WM & HOWARD,PH (1991)
OH 2.06E-010	cm3/molc sec	25	EST	MEYLAN,WM & HOWARD,PH (1993)

CAS #: 063346-73-6 — 2-(4-BR PHENYLIMINO)IMIDAZOLIDINE

Formula: $C_9H_{10}BrN_3$

Mol Weight: 240.11

MP (deg C): FP (deg C):

BP (deg C):

BP pressure (mm Hg):

Property/Value	Units	Temp	Data Type	Reference
WS 8.59E+002	mg/L	25	EST	MEYLAN,WM ET AL. (1996)
logP 1.75			EXP	HANSCH,C & LEO,AJ (1985)
VP 4.64E-007	mm Hg	25	EST	NEELY,WB & BLAU,GE (1985)
DC	pKa			
HL 1.09E-011	atm m3/mol	25	EST	MEYLAN,WM & HOWARD,PH (1991)
OH 8.47E-011	cm3/molc sec	25	EST	MEYLAN,WM & HOWARD,PH (1993)

CAS #: 063346-74-7 — 2-IMIDAZOLIDINIMINE, N-[2,6-BIS(1-METHYLETHYL)PH

Formula: $C_{15}H_{23}N_3$

Mol Weight: 245.37

MP (deg C): FP (deg C):

BP (deg C):

BP pressure (mm Hg):

Property/Value	Units	Temp	Data Type	Reference
WS 6.48E+002	mg/L	25	EST	MEYLAN,WM ET AL. (1996)
logP 3.04			EXP	SANGSTER,J (1993)
VP 6.27E-006	mm Hg	25	EST	NEELY,WB & BLAU,GE (1985)
DC	pKa			
HL 1.04E-010	atm m3/mol	25	EST	MEYLAN,WM & HOWARD,PH (1991)
OH 1.60E-010	cm3/molc sec	25	EST	MEYLAN,WM & HOWARD,PH (1993)

CAS #: 063407-51-2 — 1-PROPYLTHIO-B-GALACTOPYRANSIDE

Formula: $C_9H_{18}O_5S$

Mol Weight: 238.30

MP (deg C): FP (deg C):

BP (deg C):

BP pressure (mm Hg):

Property/Value	Units	Temp	Data Type	Reference
WS 2.79E+005	mg/L	25	EST	MEYLAN,WM ET AL. (1996)
logP -1.18			EXP	HANSCH,C & LEO,AJ (1985)
VP 9.79E-010	mm Hg	25	EST	NEELY,WB & BLAU,GE (1985)
DC	pKa			
HL 5.00E-015	atm m3/mol	25	EST	MEYLAN,WM & HOWARD,PH (1991)
OH 1.79E-010	cm3/molc sec	25	EST	MEYLAN,WM & HOWARD,PH (1993)

CAS #: 063407-52-3 — 1-BUTYLTHIO-B-GALACTOPYRANOSIDE

Formula: $C_{10}H_{20}O_5S$

Mol Weight: 252.33

MP (deg C): FP (deg C):

BP (deg C):

BP pressure (mm Hg):

Property/Value	Units	Temp	Data Type	Reference
WS 6.64E+004	mg/L	25	EST	MEYLAN,WM ET AL. (1996)
logP -0.54			EXP	HANSCH,C & LEO,AJ (1985)
VP 3.41E-010	mm Hg	25	EST	NEELY,WB & BLAU,GE (1985)
DC	pKa			
HL 6.63E-015	atm m3/mol	25	EST	MEYLAN,WM & HOWARD,PH (1991)
OH 1.80E-010	cm3/molc sec	25	EST	MEYLAN,WM & HOWARD,PH (1993)

CAS #: 063407-53-4 — 1-BENZYLTHIO-B-GALACTOPYRANOSIDE

Formula: $C_{13}H_{18}O_5S$

Mol Weight: 286.35

MP (deg C): FP (deg C):

BP (deg C):

BP pressure (mm Hg):

Property/Value	Units	Temp	Data Type	Reference
WS 3.71E+004	mg/L	25	EST	MEYLAN,WM ET AL. (1996)
logP -0.47			EXP	HANSCH,C & LEO,AJ (1985)
VP 3.91E-012	mm Hg	25	EST	NEELY,WB & BLAU,GE (1985)
DC	pKa			
HL 2.29E-016	atm m3/mol	25	EST	MEYLAN,WM & HOWARD,PH (1991)
OH 1.81E-010	cm3/molc sec	25	EST	MEYLAN,WM & HOWARD,PH (1993)

CAS #: 063407-54-5 — 1(2-PHENETS)-B-GALACTOPYRANOSIDE

Formula: $C_{14}H_{20}O_5S$

Mol Weight: 300.38

MP (deg C): FP (deg C):

BP (deg C):

BP pressure (mm Hg):

Property/Value	Units	Temp	Data Type	Reference
WS 1.49E+004	mg/L	25	EST	MEYLAN,WM ET AL. (1996)
logP -0.10			EXP	HANSCH,C & LEO,AJ (1985)
VP 1.33E-012	mm Hg	25	EST	NEELY,WB & BLAU,GE (1985)
DC	pKa			
HL 3.04E-016	atm m3/mol	25	EST	MEYLAN,WM & HOWARD,PH (1991)
OH 1.83E-010	cm3/molc sec	25	EST	MEYLAN,WM & HOWARD,PH (1993)

CAS #: 063472-04-8 — BUTANOIC ACID, 4-[(1,1'-BIPHENYL)-4-YL]-2-METHYL

Formula: $C_{17}H_{16}O_3$

Mol Weight: 268.32

MP (deg C): FP (deg C):

BP (deg C):

BP pressure (mm Hg):

Property/Value	Units	Temp	Data Type	Reference
WS 5.79E+001	mg/L	25	EST	MEYLAN,WM ET AL. (1996)
logP 3.40			EXP	KUCHAR,M ET AL. (1985)
VP 6.04E-008	mm Hg	25	EST	NEELY,WB & BLAU,GE (1985)
DC	pKa			
HL 9.87E-012	atm m3/mol	25	EST	MEYLAN,WM & HOWARD,PH (1991)
OH 1.37E-011	cm3/molc sec	25	EST	MEYLAN,WM & HOWARD,PH (1993)

063474-05-5

CAS #: 063474-05-5
2-PROPANOL, 1-[(1,1-DIMETHYLETHYL)AMINO]-3-(PHEN

Formula: $C_{14}H_{23}NO_2$
Mol Weight: 237.34
MP (deg C):
FP (deg C):
BP (deg C):
BP pressure (mm Hg):

Property/Value	Units	Temp	Data Type	Reference
WS 9.99E+003	mg/L	25	EST	MEYLAN,WM ET AL. (1996)
logP 1.70			EXP	MAULEON,D ET AL. (1988)
VP 3.15E-006	mm Hg	25	EST	NEELY,WB & BLAU,GE (1985)
DC	pKa			
HL 2.37E-012	atm m3/mol	25	EST	MEYLAN,WM & HOWARD,PH (1991)
OH 1.02E-010	cm3/molc sec	25	EST	MEYLAN,WM & HOWARD,PH (1993)

063491-77-0

CAS #: 063491-77-0
3-T-BUTYLHYDROXYUREA

Formula: $C_5H_{12}N_2O_2$
Mol Weight: 132.16
MP (deg C):
FP (deg C):
BP (deg C):
BP pressure (mm Hg):

Property/Value	Units	Temp	Data Type	Reference
WS 3.37E+004	mg/L	25	EST	MEYLAN,WM ET AL. (1996)
logP 0.50			EXP	HANSCH,C & LEO,AJ (1985)
VP 8.24E-005	mm Hg	25	EST	NEELY,WB & BLAU,GE (1985)
DC	pKa			
HL 2.78E-010	atm m3/mol	25	EST	MEYLAN,WM & HOWARD,PH (1991)
OH 2.50E-012	cm3/molc sec	25	EST	MEYLAN,WM & HOWARD,PH (1993)

063574-83-4

CAS #: 063574-83-4
14-BENZDIAZEPIN-2-ONE,5-(2-BRPH)7-CL

Formula: $C_{15}H_{10}BrClN_2O$
Mol Weight: 349.62
MP (deg C):
FP (deg C):
BP (deg C):
BP pressure (mm Hg):

Property/Value	Units	Temp	Data Type	Reference
WS 1.22E+001	mg/L	25	EST	MEYLAN,WM ET AL. (1996)
logP 3.17			EXP	HANSCH,C & LEO,AJ (1985)
VP 7.45E-010	mm Hg	25	EST	NEELY,WB & BLAU,GE (1985)
DC	pKa			
HL 7.08E-011	atm m3/mol	25	EST	MEYLAN,WM & HOWARD,PH (1991)
OH 6.52E-012	cm3/molc sec	25	EST	MEYLAN,WM & HOWARD,PH (1993)

063634-21-9

CAS #: 063634-21-9
2-CHLORO-1-METHYL-4-NITRO-1H-IMIDAZOLE

Formula: $C_4H_4ClN_3O_2$
Mol Weight: 161.55
MP (deg C):
FP (deg C):
BP (deg C):
BP pressure (mm Hg):

Property/Value	Units	Temp	Data Type	Reference
WS 1.89E+004	mg/L	25	EST	MEYLAN,WM ET AL. (1996)
logP 0.19			EXP	SUWINSKI,J ET AL. (1985)
VP 2.83E-004	mm Hg	25	EST	NEELY,WB & BLAU,GE (1985)
DC	pKa			
HL 2.34E-007	atm m3/mol	25	EST	MEYLAN,WM & HOWARD,PH (1991)
OH 1.05E-012	cm3/molc sec	25	EST	MEYLAN,WM & HOWARD,PH (1993)

063655-40-3

CAS #: 063655-40-3
2-AMINO-5-CYANOBENZIMIDAZOLE

Formula: $C_8H_6N_4$
Mol Weight: 158.16
MP (deg C):
FP (deg C):
BP (deg C):
BP pressure (mm Hg):

Property/Value	Units	Temp	Data Type	Reference
WS 4.11E+003	mg/L	25	EST	MEYLAN,WM ET AL. (1996)
logP 1.13			EXP	HANSCH,C ET AL. (1995)
VP 8.53E-008	mm Hg	25	EST	NEELY,WB & BLAU,GE (1985)
DC	pKa			
HL 1.25E-012	atm m3/mol	25	EST	MEYLAN,WM & HOWARD,PH (1991)
OH 5.21E-011	cm3/molc sec	25	EST	MEYLAN,WM & HOWARD,PH (1993)

063659-18-7

CAS #: 063659-18-7
2-PROPANOL, 1-[4-[2-(CYCLOPROPYLMETHOXY)ETHYL]PH

Formula: $C_{18}H_{29}NO_3$
Mol Weight: 307.44
MP (deg C): 70-72
FP (deg C):
BP (deg C):
BP pressure (mm Hg):

Property/Value	Units	Temp	Data Type	Reference
WS 4.51E+002	mg/L	25	EST	MEYLAN,WM ET AL. (1996)
logP 2.81			EXP	RECANATINI,M (1992)
VP 1.33E-008	mm Hg	25	EST	NEELY,WB & BLAU,GE (1985)
DC	pKa			
HL 1.45E-013	atm m3/mol	25	EST	MEYLAN,WM & HOWARD,PH (1991)
OH 1.53E-010	cm3/molc sec	25	EST	MEYLAN,WM & HOWARD,PH (1993)

063662-67-9

CAS #: 063662-67-9
4-CHLORO-2-METHYL-5-NITRO-1H-IMIDAZOLE

Formula: $C_4H_4ClN_3O_2$
Mol Weight: 161.55
MP (deg C):
FP (deg C):
BP (deg C):
BP pressure (mm Hg):

Property/Value	Units	Temp	Data Type	Reference
WS 3.41E+003	mg/L	25	EST	MEYLAN,WM ET AL. (1996)
logP 1.07			EST	MEYLAN,WM & HOWARD,PH (1995)
VP 1.40E-005	mm Hg	25	EST	NEELY,WB & BLAU,GE (1985)
DC	pKa			
HL 1.21E-008	atm m3/mol	25	EST	MEYLAN,WM & HOWARD,PH (1991)
OH 3.61E-011	cm3/molc sec	25	EST	MEYLAN,WM & HOWARD,PH (1993)

063677-95-2

CAS #: 063677-95-2
PENTOSTATIN

Formula: $C_{11}H_{18}N_4O_4$
Mol Weight: 270.29
MP (deg C):
FP (deg C):
BP (deg C):
BP pressure (mm Hg):

Property/Value	Units	Temp	Data Type	Reference
WS 1.00E+006	mg/L	25	EST	MEYLAN,WM ET AL. (1996)
logP -2.09			EXP	HANSCH,C ET AL. (1995)
VP 2.52E-016	mm Hg	25	EST	NEELY,WB & BLAU,GE (1985)
DC	pKa			
HL 1.48E-019	atm m3/mol	25	EST	MEYLAN,WM & HOWARD,PH (1991)
OH 1.80E-010	cm3/molc sec	25	EST	MEYLAN,WM & HOWARD,PH (1993)

CAS #: 063736-04-9 — PYRIDO(12A)PYRIMIDINE-4-ONE,2ET,6ME

Formula: $C_{11}H_{14}N_2O$
Mol Weight: 190.25
MP (deg C): FP (deg C):
BP (deg C):
BP pressure (mm Hg):

Property/Value	Units	Temp	Data Type	Reference
WS 2.20E+003	mg/L	25	EST	MEYLAN,WM ET AL. (1996)
logP 1.59			EXP	HANSCH,C & LEO,AJ (1985)
VP 1.23E-004	mm Hg	25	EST	NEELY,WB & BLAU,GE (1985)
DC	pKa			
HL	atm m3/mol			
OH 2.05E-011	cm3/molc sec	25	EST	MEYLAN,WM & HOWARD,PH (1993)

CAS #: 063837-33-2 — C1CCCCC1OC2CCC(OCC3OC(CC)OC3)CC2

Formula: $C_{18}H_{20}O_4$
Mol Weight: 300.36
MP (deg C): FP (deg C):
BP (deg C):
BP pressure (mm Hg):

Property/Value	Units	Temp	Data Type	Reference
WS 2.36E+000	mg/L	25	EST	MEYLAN,WM ET AL. (1996)
logP 4.35			EXP	TOMLIN,C (1994)
VP 7.60E-007	mm Hg	25	EST	NEELY,WB & BLAU,GE (1985)
DC	pKa			
HL 4.94E-009	atm m3/mol	25	EST	MEYLAN,WM & HOWARD,PH (1991)
OH 6.17E-011	cm3/molc sec	25	EST	MEYLAN,WM & HOWARD,PH (1993)

CAS #: 063893-52-7 — THIENO(23D)PYRIMIDINE-4-DEMEAMINO

Formula: $C_8H_9N_3S$
Mol Weight: 179.25
MP (deg C): FP (deg C):
BP (deg C):
BP pressure (mm Hg):

Property/Value	Units	Temp	Data Type	Reference
WS 1.55E+003	mg/L	25	EST	MEYLAN,WM ET AL. (1996)
logP 1.82			EXP	HANSCH,C & LEO,AJ (1985)
VP 2.52E-004	mm Hg	25	EST	NEELY,WB & BLAU,GE (1985)
DC	pKa			
HL 2.46E-009	atm m3/mol	25	EST	MEYLAN,WM & HOWARD,PH (1991)
OH 2.03E-010	cm3/molc sec	25	EST	MEYLAN,WM & HOWARD,PH (1993)

CAS #: 063936-04-9 — 1-S-BUTYLBENZOTRIAZOLE

Formula: $C_{10}H_{13}N_3$
Mol Weight: 175.24
MP (deg C): FP (deg C):
BP (deg C):
BP pressure (mm Hg):

Property/Value	Units	Temp	Data Type	Reference
WS 6.18E+002	mg/L	25	EST	MEYLAN,WM ET AL. (1996)
logP 2.31			EXP	HANSCH,C & LEO,AJ (1985)
VP 8.16E-004	mm Hg	25	EST	NEELY,WB & BLAU,GE (1985)
DC	pKa			
HL 7.32E-006	atm m3/mol	25	EST	MEYLAN,WM & HOWARD,PH (1991)
OH 4.87E-012	cm3/molc sec	25	EST	MEYLAN,WM & HOWARD,PH (1993)

CAS #: 063951-01-9 — A-PROPYL BENZENEETHANEAMINE

Formula: $C_{11}H_{17}N$
Mol Weight: 163.26
MP (deg C): FP (deg C):
BP (deg C):
BP pressure (mm Hg):

Property/Value	Units	Temp	Data Type	Reference
WS 2.79E+003	mg/L	25	EST	MEYLAN,WM ET AL. (1996)
logP 2.79			EXP	HANSCH,C & LEO,AJ (1985)
VP 2.41E-002	mm Hg	25	EST	NEELY,WB & BLAU,GE (1985)
DC	pKa			
HL 1.90E-006	atm m3/mol	25	EST	MEYLAN,WM & HOWARD,PH (1991)
OH 5.70E-011	cm3/molc sec	25	EST	MEYLAN,WM & HOWARD,PH (1993)

CAS #: 064002-57-9 — 1(1H)-ISOBENZOFURANONE-3-I-PROPYL

Formula: $C_{11}H_{12}O_2$
Mol Weight: 176.22
MP (deg C): FP (deg C):
BP (deg C):
BP pressure (mm Hg):

Property/Value	Units	Temp	Data Type	Reference
WS 9.24E+002	mg/L	25	EST	MEYLAN,WM ET AL. (1996)
logP 2.10			EXP	HANSCH,C & LEO,AJ (1985)
VP 1.17E-003	mm Hg	25	EST	NEELY,WB & BLAU,GE (1985)
DC	pKa			
HL 2.98E-005	atm m3/mol	25	EST	MEYLAN,WM & HOWARD,PH (1991)
OH 9.99E-012	cm3/molc sec	25	EST	MEYLAN,WM & HOWARD,PH (1993)

CAS #: 064028-99-5 — N-(N"-ME-ETHOXYCARBAMYLTHIO)METHOMYL

Formula: $C_9H_{17}N_3O_4S_2$
Mol Weight: 295.38
MP (deg C): FP (deg C):
BP (deg C):
BP pressure (mm Hg):

Property/Value	Units	Temp	Data Type	Reference
WS 3.12E+002	mg/L	25	EST	MEYLAN,WM ET AL. (1996)
logP 1.90			EXP	HANSCH,C & LEO,AJ (1985)
VP 1.73E-004	mm Hg	25	EST	NEELY,WB & BLAU,GE (1985)
DC	pKa			
HL 5.58E-009	atm m3/mol	25	EST	MEYLAN,WM & HOWARD,PH (1991)
OH 3.57E-011	cm3/molc sec	25	EST	MEYLAN,WM & HOWARD,PH (1993)

CAS #: 064029-01-2 — N(N"-ME-TBUO CARBAMYLTHIO)METHOMYL

Formula: $C_{11}H_{21}N_3O_4S_2$
Mol Weight: 323.44
MP (deg C): FP (deg C):
BP (deg C):
BP pressure (mm Hg):

Property/Value	Units	Temp	Data Type	Reference
WS 3.63E+001	mg/L	25	EST	MEYLAN,WM ET AL. (1996)
logP 2.80			EXP	HANSCH,C & LEO,AJ (1985)
VP 8.59E-005	mm Hg	25	EST	NEELY,WB & BLAU,GE (1985)
DC	pKa			
HL 9.83E-009	atm m3/mol	25	EST	MEYLAN,WM & HOWARD,PH (1991)
OH 3.45E-011	cm3/molc sec	25	EST	MEYLAN,WM & HOWARD,PH (1993)

CAS #: 064055-10-3				ETHANIMIDOTHIOIC ACID, N-[[[[(METHOXYCARBONYLMET
Formula: $C_8H_{15}N_3O_4S_2$				
Mol Weight: 281.35				
MP (deg C):		FP (deg C):		
BP (deg C):				
BP pressure (mm Hg):				

Property/Value	Units	Temp	Data Type	Reference
WS 1.00E+003	mg/L	25	EST	MEYLAN,WM ET AL. (1996)
logP 1.40			EXP	SANGSTER,J (1993)
VP 3.03E-004	mm Hg	25	EST	NEELY,WB & BLAU,GE (1985)
DC	pKa			
HL 4.20E-009	atm m3/mol	25	EST	MEYLAN,WM & HOWARD,PH (1991)
OH 3.43E-011	cm3/molc sec	25	EST	MEYLAN,WM & HOWARD,PH (1993)

CAS #: 064098-82-4				1(2H)-PYRIMIDINECARBOXAMIDE, N-BUTYL-5-FLUORO-3,
Formula: $C_9H_{12}FN_3O_3$				
Mol Weight: 229.21				
MP (deg C):		FP (deg C):		
BP (deg C):				
BP pressure (mm Hg):				

Property/Value	Units	Temp	Data Type	Reference
WS 1.81E+003	mg/L	25	EST	MEYLAN,WM ET AL. (1996)
logP 1.44			EXP	SANGSTER,J (1993)
VP 6.44E-011	mm Hg	25	EST	NEELY,WB & BLAU,GE (1985)
DC	pKa			
HL 5.19E-015	atm m3/mol	25	EST	MEYLAN,WM & HOWARD,PH (1991)
OH 1.43E-011	cm3/molc sec	25	EST	MEYLAN,WM & HOWARD,PH (1993)

CAS #: 064098-85-7				1-PYRIMIDINECARBOXAMIDE, 5-FLUORO-1,3-DIHYDRO-N-
Formula: $C_{13}H_{20}FN_3O_3$				
Mol Weight: 285.32				
MP (deg C):		FP (deg C):		
BP (deg C):				
BP pressure (mm Hg):				

Property/Value	Units	Temp	Data Type	Reference
WS 6.98E+000	mg/L	25	EST	MEYLAN,WM ET AL. (1996)
logP 3.90			EXP	SASAKI,H ET AL. (1990)
VP 2.27E-012	mm Hg	25	EST	NEELY,WB & BLAU,GE (1985)
DC	pKa			
HL 1.61E-014	atm m3/mol	25	EST	MEYLAN,WM & HOWARD,PH (1991)
OH 1.99E-011	cm3/molc sec	25	EST	MEYLAN,WM & HOWARD,PH (1993)

CAS #: 064118-84-9				PHENYLACETIC ACID,2-(2',6'-DICHLORO-4'-HYDROXY)A
Formula: $C_{14}H_{11}Cl_2NO_3$				
Mol Weight: 312.15				
MP (deg C):		FP (deg C):		
BP (deg C):				
BP pressure (mm Hg):				

Property/Value	Units	Temp	Data Type	Reference
WS 1.79E+001	mg/L	25	EST	MEYLAN,WM ET AL. (1996)
logP 3.70			EXP	HANSCH,C ET AL. (1995)
VP 8.17E-010	mm Hg	25	EST	NEELY,WB & BLAU,GE (1985)
DC	pKa			
HL 4.93E-016	atm m3/mol	25	EST	MEYLAN,WM & HOWARD,PH (1991)
OH 1.80E-010	cm3/molc sec	25	EST	MEYLAN,WM & HOWARD,PH (1993)

CAS #: 064124-14-7				1,3,5-TRIAZINE-2,4,6-TRIAMINE, N-HYDROXY-N,N',N'
Formula: $C_8H_{16}N_6O$				
Mol Weight: 212.26				
MP (deg C):		FP (deg C):		
BP (deg C):				
BP pressure (mm Hg):				

Property/Value	Units	Temp	Data Type	Reference
WS 1.60E+003	mg/L	25	EST	MEYLAN,WM ET AL. (1996)
logP 1.61			EXP	SANGSTER,J (1993)
VP 2.24E-008	mm Hg	25	EST	NEELY,WB & BLAU,GE (1985)
DC	pKa			
HL 1.57E-011	atm m3/mol	25	EST	MEYLAN,WM & HOWARD,PH (1991)
OH 6.53E-012	cm3/molc sec	25	EST	MEYLAN,WM & HOWARD,PH (1993)

CAS #: 064124-20-5				N2,N4,N6-TRIME-N,N,N-TRIME-MELAMINE
Formula: $C_{12}H_{24}N_6$				
Mol Weight: 252.37				
MP (deg C):		FP (deg C):		
BP (deg C):				
BP pressure (mm Hg):				

Property/Value	Units	Temp	Data Type	Reference
WS 1.07E+001	mg/L	25	EST	MEYLAN,WM ET AL. (1996)
logP 3.90			EXP	HANSCH,C & LEO,AJ (1985)
VP 1.86E-005	mm Hg	25	EST	NEELY,WB & BLAU,GE (1985)
DC	pKa			
HL 4.03E-008	atm m3/mol	25	EST	MEYLAN,WM & HOWARD,PH (1991)
OH 3.10E-011	cm3/molc sec	25	EST	MEYLAN,WM & HOWARD,PH (1993)

CAS #: 064230-41-7				1H-PYRROLE-2-CARBOXAMIDE, 1-METHYL-
Formula: $C_6H_8N_2O$				
Mol Weight: 124.14				
MP (deg C):		FP (deg C):		
BP (deg C):				
BP pressure (mm Hg):				

Property/Value	Units	Temp	Data Type	Reference
WS 4.00E+004	mg/L	25	EST	MEYLAN,WM ET AL. (1996)
logP 0.45			EXP	YAMAGAMI,C ET AL. (1994)
VP 5.88E-004	mm Hg	25	EST	NEELY,WB & BLAU,GE (1985)
DC	pKa			
HL 7.96E-011	atm m3/mol	25	EST	MEYLAN,WM & HOWARD,PH (1991)
OH 1.02E-010	cm3/molc sec	25	EST	MEYLAN,WM & HOWARD,PH (1993)

CAS #: 064249-01-0				ANILOFOS
Formula: $C_{13}H_{19}ClNO_3PS_2$				
Mol Weight: 367.86				
MP (deg C):		FP (deg C):		
BP (deg C):				
BP pressure (mm Hg):				

Property/Value	Units	Temp	Data Type	Reference
WS 1.36E+001	mg/L	20	EXP	SHIU,WY ET AL. (1990)
logP 3.81			EXP	TOMLIN,C (1994)
VP 2.33E-007	mm Hg	25	EST	NEELY,WB & BLAU,GE (1985)
DC	pKa			
HL 1.42E-010	atm m3/mol	25	EST	MEYLAN,WM & HOWARD,PH (1991)
OH 8.49E-011	cm3/molc sec	25	EST	MEYLAN,WM & HOWARD,PH (1993)

CAS #: 064272-13-5	PREGNA-1,4-DIENE-3,20-DIONE, 2,9-DICHLORO-21-(2,

Formula: $C_{27}H_{35}Cl_2FO_6$

Mol Weight: 545.48

MP (deg C): FP (deg C):

BP (deg C):

BP pressure (mm Hg):

Property/ Value	Units	Temp	Data Type	Reference
WS 3.33E-002	mg/L	25	EST	MEYLAN,WM ET AL. (1996)
logP 4.73			EXP	SANGSTER,J (1993)
VP 1.45E-015	mm Hg	25	EST	NEELY,WB & BLAU,GE (1985)
DC	pKa			
HL 4.58E-012	atm m3/mol	25	EST	MEYLAN,WM & HOWARD,PH (1991)
OH 5.39E-011	cm3/molc sec	25	EST	MEYLAN,WM & HOWARD,PH (1993)

CAS #: 064272-18-0	2-CHLOROFLUMETHASONE-17-PROPIONATE

Formula: $C_{25}H_{31}ClF_2O_6$

Mol Weight: 500.97

MP (deg C): FP (deg C):

BP (deg C):

BP pressure (mm Hg):

Property/ Value	Units	Temp	Data Type	Reference
WS 1.97E+000	mg/L	25	EST	MEYLAN,WM ET AL. (1996)
logP 2.99			EXP	HANSCH,C & LEO,AJ (1985)
VP 8.15E-015	mm Hg	25	EST	NEELY,WB & BLAU,GE (1985)
DC	pKa			
HL 1.47E-011	atm m3/mol	25	EST	MEYLAN,WM & HOWARD,PH (1991)
OH 4.35E-011	cm3/molc sec	25	EST	MEYLAN,WM & HOWARD,PH (1993)

CAS #: 064312-78-3	BENZYL-T-CHYRSANTHEMATE

Formula: $C_{17}H_{22}O_2$

Mol Weight: 258.36

MP (deg C): FP (deg C):

BP (deg C):

BP pressure (mm Hg):

Property/ Value	Units	Temp	Data Type	Reference
WS 4.35E-001	mg/L	25	EST	MEYLAN,WM ET AL. (1996)
logP 5.49			EXP	HANSCH,C & LEO,AJ (1985)
VP 1.10E-004	mm Hg	25	EST	NEELY,WB & BLAU,GE (1985)
DC	pKa			
HL 6.25E-005	atm m3/mol	25	EST	MEYLAN,WM & HOWARD,PH (1991)
OH 9.39E-011	cm3/molc sec	25	EST	MEYLAN,WM & HOWARD,PH (1993)

CAS #: 064360-42-5	PHENYLACETIC ACID,3-ME-4-PHMEO

Formula: $C_{16}H_{16}O_3$

Mol Weight: 256.30

MP (deg C): FP (deg C):

BP (deg C):

BP pressure (mm Hg):

Property/ Value	Units	Temp	Data Type	Reference
WS 7.18E+001	mg/L	25	EST	MEYLAN,WM ET AL. (1996)
logP 3.37			EXP	HANSCH,C & LEO,AJ (1985)
VP 4.77E-007	mm Hg	25	EST	NEELY,WB & BLAU,GE (1985)
DC	pKa			
HL 2.33E-010	atm m3/mol	25	EST	MEYLAN,WM & HOWARD,PH (1991)
OH 3.91E-011	cm3/molc sec	25	EST	MEYLAN,WM & HOWARD,PH (1993)

CAS #: 064365-27-1	ARGININ-AMIDE, N-ACETYL

Formula: $C_8H_{17}N_5O_2$

Mol Weight: 215.26

MP (deg C): FP (deg C):

BP (deg C):

BP pressure (mm Hg):

Property/ Value	Units	Temp	Data Type	Reference
WS 4.76E+005	mg/L	25	EST	MEYLAN,WM ET AL. (1996)
logP -2.84			EXP	HANSCH,C & LEO,AJ (1985)
VP 3.04E-010	mm Hg	25	EST	NEELY,WB & BLAU,GE (1985)
DC	pKa			
HL 1.92E-019	atm m3/mol	25	EST	MEYLAN,WM & HOWARD,PH (1991)
OH 1.44E-010	cm3/molc sec	25	EST	MEYLAN,WM & HOWARD,PH (1993)

CAS #: 064399-29-7	4H-PYRIDO[1,2-A]PYRIMIDINE-3-CARBOXAMIDE, 1,6,7,

Formula: $C_{11}H_{15}N_3O_2$

Mol Weight: 221.26

MP (deg C): FP (deg C):

BP (deg C):

BP pressure (mm Hg):

Property/ Value	Units	Temp	Data Type	Reference
WS 1.31E+003	mg/L	25	EST	MEYLAN,WM ET AL. (1996)
logP 0.12			EXP	SANGSTER,J (1994)
VP 4.44E-008	mm Hg	25	EST	NEELY,WB & BLAU,GE (1985)
DC	pKa			
HL 2.09E-012	atm m3/mol	25	EST	MEYLAN,WM & HOWARD,PH (1991)
OH 1.25E-010	cm3/molc sec	25	EST	MEYLAN,WM & HOWARD,PH (1993)

CAS #: 064405-35-2	4H-PYRIDO[1,2-A]PYRIMIDINE-3-CARBOXYLIC ACID, 6,

Formula: $C_{12}H_{16}N_2O_3$

Mol Weight: 236.27

MP (deg C): FP (deg C):

BP (deg C):

BP pressure (mm Hg):

Property/ Value	Units	Temp	Data Type	Reference
WS 6.97E+003	mg/L	25	EST	MEYLAN,WM ET AL. (1996)
logP 0.71			EXP	SANGSTER,J (1994)
VP 3.04E-006	mm Hg	25	EST	NEELY,WB & BLAU,GE (1985)
DC	pKa			
HL 3.54E-012	atm m3/mol	25	EST	MEYLAN,WM & HOWARD,PH (1991)
OH 2.47E-011	cm3/molc sec	25	EST	MEYLAN,WM & HOWARD,PH (1993)

CAS #: 064431-77-2	2-NITRO-6-METHOXYBENZAMIDE

Formula: $C_8H_8N_2O_4$

Mol Weight: 196.16

MP (deg C): FP (deg C):

BP (deg C):

BP pressure (mm Hg):

Property/ Value	Units	Temp	Data Type	Reference
WS 1.52E+003	mg/L	25	EST	MEYLAN,WM ET AL. (1996)
logP -0.26			EXP	NAKAGAWA,Y ET AL. (1992)
VP 1.37E-006	mm Hg	25	EST	NEELY,WB & BLAU,GE (1985)
DC	pKa			
HL 5.17E-013	atm m3/mol	25	EST	MEYLAN,WM & HOWARD,PH (1991)
OH 4.54E-012	cm3/molc sec	25	EST	MEYLAN,WM & HOWARD,PH (1993)

CAS #: 064462-06-2			M-METHYLPHENYL AZOXYCYANIDE
Formula: $C_8H_7N_3O$			
Mol Weight: 161.16			
MP (deg C):		FP (deg C):	
BP (deg C):			
BP pressure (mm Hg):			

Property/Value	Units	Temp	Data Type	Reference
WS 1.87E+002	mg/L	25	EST	MEYLAN,WM ET AL. (1996)
logP 2.49			EXP	CALVINO,R R ET AL (1991)
VP 2.81E-008	mm Hg	25	EST	NEELY,WB & BLAU,GE (1985)
DC	pKa			
HL 2.27E-011	atm m3/mol	25	EST	MEYLAN,WM & HOWARD,PH (1991)
OH 1.22E-012	cm3/molc sec	25	EST	MEYLAN,WM & HOWARD,PH (1993)

CAS #: 064480-66-6			GLYCOURSODEOXYCHOLIC ACID
Formula: $C_{26}H_{43}NO_5$			
Mol Weight: 449.64			
MP (deg C):		FP (deg C):	
BP (deg C):			
BP pressure (mm Hg):			

Property/Value	Units	Temp	Data Type	Reference
WS 2.82E+001	mg/L	25	EST	MEYLAN,WM ET AL. (1996)
logP 2.02			EXP	RODA,A ET AL. (1990)
VP 8.36E-018	mm Hg	25	EST	NEELY,WB & BLAU,GE (1985)
DC	pKa			
HL 9.38E-017	atm m3/mol	25	EST	MEYLAN,WM & HOWARD,PH (1991)
OH 5.79E-011	cm3/molc sec	25	EST	MEYLAN,WM & HOWARD,PH (1993)

CAS #: 064532-97-4			4-NONYLPHENYL DIPHENYL PHOSPATE
Formula: $C_{27}H_{33}O_4P$			
Mol Weight: 452.54			
MP (deg C):		FP (deg C):	
BP (deg C): 471			
BP pressure (mm Hg):			

Property/Value	Units	Temp	Data Type	Reference
WS 7.70E-001	mg/L	25	EXP	MAYER,FL ET AL. (1981)
logP 4.93			EXP	MAYER,FL ET AL. (1981)
VP 1.90E-008	mm Hg	25	EXP	BOETHLING,RS & COOPER,JC (1985)
DC	pKa			
HL 1.40E-008	atm m3/mol	25	EXP	BOETHLING,RS & COOPER,JC (1985)
OH 2.26E-011	cm3/molc sec	25	EST	MEYLAN,WM & HOWARD,PH (1993)

CAS #: 064544-07-6			CEFUROXIME AXETIL
Formula: $C_{20}H_{22}N_4O_{10}S$			
Mol Weight: 510.48			
MP (deg C):		FP (deg C):	
BP (deg C):			
BP pressure (mm Hg):			

Property/Value	Units	Temp	Data Type	Reference
WS 1.07E+002	mg/L	25	EST	MEYLAN,WM ET AL. (1996)
logP 0.89			EXP	HANSCH,C ET AL. (1995)
VP 9.95E-016	mm Hg	25	EST	NEELY,WB & BLAU,GE (1985)
DC	pKa			
HL 2.81E-021	atm m3/mol	25	EST	MEYLAN,WM & HOWARD,PH (1991)
OH 2.19E-010	cm3/molc sec	25	EST	MEYLAN,WM & HOWARD,PH (1993)

CAS #: 064560-17-4			DIBENZOFURAN, 1,3,7,9-TETRACHLORO-
Formula: $C_{12}H_4Cl_4O$			
Mol Weight: 305.98			
MP (deg C):		FP (deg C):	
BP (deg C):			
BP pressure (mm Hg):			

Property/Value	Units	Temp	Data Type	Reference
WS 3.19E-003	mg/L	25	EST	MEYLAN,WM ET AL. (1996)
logP 6.34			EXP	SIJM,DTHM ET AL. (1989)
VP 1.53E-006	mm Hg	25	EST	NEELY,WB & BLAU,GE (1985)
DC	pKa			
HL 1.54E-005	atm m3/mol	25	EST	MEYLAN,WM & HOWARD,PH (1991)
OH 1.65E-013	cm3/molc sec	25	EST	MEYLAN,WM & HOWARD,PH (1993)

CAS #: 064638-08-0			2,5-DIMEO-4-PROPYLAMPHETAMINE
Formula: $C_{14}H_{23}NO_2$			
Mol Weight: 237.34			
MP (deg C):		FP (deg C):	
BP (deg C):			
BP pressure (mm Hg):			

Property/Value	Units	Temp	Data Type	Reference
WS 3.75E+002	mg/L	25	EST	MEYLAN,WM ET AL. (1996)
logP 3.37			EXP	HANSCH,C & LEO,AJ (1985)
VP 8.68E-005	mm Hg	25	EST	NEELY,WB & BLAU,GE (1985)
DC	pKa			
HL 7.34E-009	atm m3/mol	25	EST	MEYLAN,WM & HOWARD,PH (1991)
OH 1.05E-010	cm3/molc sec	25	EST	MEYLAN,WM & HOWARD,PH (1993)

CAS #: 064638-09-1			BENZENEETHANAMINE, 4-BUTYL-2,5-DIMETHOXY-à-METHY
Formula: $C_{15}H_{25}NO_2$			
Mol Weight: 251.37			
MP (deg C):		FP (deg C):	
BP (deg C):			
BP pressure (mm Hg):			

Property/Value	Units	Temp	Data Type	Reference
WS 9.08E+001	mg/L	25	EST	MEYLAN,WM ET AL. (1996)
logP 4.00			EXP	SANGSTER,J (1993)
VP 3.57E-005	mm Hg	25	EST	NEELY,WB & BLAU,GE (1985)
DC	pKa			
HL 9.74E-009	atm m3/mol	25	EST	MEYLAN,WM & HOWARD,PH (1991)
OH 1.06E-010	cm3/molc sec	25	EST	MEYLAN,WM & HOWARD,PH (1993)

CAS #: 064638-10-4			2,5-DIMETHOXY-4-PENTYLAMPHETAMINE
Formula: $C_{16}H_{27}NO_2$			
Mol Weight: 265.40			
MP (deg C):		FP (deg C):	
BP (deg C):			
BP pressure (mm Hg):			

Property/Value	Units	Temp	Data Type	Reference
WS 3.26E+001	mg/L	25	EST	MEYLAN,WM ET AL. (1996)
logP 4.43			EXP	HANSCH,C & LEO,AJ (1985)
VP 1.81E-005	mm Hg	25	EST	NEELY,WB & BLAU,GE (1985)
DC	pKa			
HL 1.29E-008	atm m3/mol	25	EST	MEYLAN,WM & HOWARD,PH (1991)
OH 1.08E-010	cm3/molc sec	25	EST	MEYLAN,WM & HOWARD,PH (1993)

CAS #: 064649-43-0 — ACETAMIDE, 2-(BENZOYLOXY)-

Formula: $C_9H_9NO_3$
Mol Weight: 179.18
MP (deg C): FP (deg C):
BP (deg C):
BP pressure (mm Hg):

Property/Value	Units	Temp	Data Type	Reference
WS 1.43E+004	mg/L	25	EST	MEYLAN,WM ET AL. (1996)
logP 0.69			EXP	NIELSEN,LS & BUNDGAARD,H (1988)
VP 1.65E-005	mm Hg	25	EST	NEELY,WB & BLAU,GE (1985)
DC	pKa			
HL 3.73E-011	atm m3/mol	25	EST	MEYLAN,WM & HOWARD,PH (1991)
OH 3.75E-012	cm3/molc sec	25	EST	MEYLAN,WM & HOWARD,PH (1993)

CAS #: 064649-57-6 — ACETAMIDE, 2-(BENZOYLOXY)-N-ETHYL-

Formula: $C_{11}H_{13}NO_3$
Mol Weight: 207.23
MP (deg C): FP (deg C):
BP (deg C):
BP pressure (mm Hg):

Property/Value	Units	Temp	Data Type	Reference
WS 3.24E+003	mg/L	25	EST	MEYLAN,WM ET AL. (1996)
logP 1.28			EXP	NIELSEN,LS & BUNDGAARD,H (1988)
VP 4.50E-006	mm Hg	25	EST	NEELY,WB & BLAU,GE (1985)
DC	pKa			
HL 1.09E-010	atm m3/mol	25	EST	MEYLAN,WM & HOWARD,PH (1991)
OH 1.12E-011	cm3/molc sec	25	EST	MEYLAN,WM & HOWARD,PH (1993)

CAS #: 064649-63-4 — BENZOIC ACID, 2-(DIETHYLAMINO)-2-OXOETHYL ESTER

Formula: $C_{13}H_{17}NO_3$
Mol Weight: 235.29
MP (deg C): FP (deg C):
BP (deg C):
BP pressure (mm Hg):

Property/Value	Units	Temp	Data Type	Reference
WS 4.96E+002	mg/L	25	EST	MEYLAN,WM ET AL. (1996)
logP 2.06			EXP	NIELSEN,LS & BUNDGAARD,H (1988)
VP 2.01E-005	mm Hg	25	EST	NEELY,WB & BLAU,GE (1985)
DC	pKa			
HL 3.17E-010	atm m3/mol	25	EST	MEYLAN,WM & HOWARD,PH (1991)
OH 2.47E-011	cm3/molc sec	25	EST	MEYLAN,WM & HOWARD,PH (1993)

CAS #: 064663-59-8 — 2-BENZOFURANCARBOXAMIDE, N-METHYL-

Formula: $C_{10}H_9NO_2$
Mol Weight: 175.19
MP (deg C): FP (deg C):
BP (deg C):
BP pressure (mm Hg):

Property/Value	Units	Temp	Data Type	Reference
WS 1.53E+003	mg/L	25	EST	MEYLAN,WM ET AL. (1996)
logP 1.85			EXP	YAMAGAMI,C ET AL. (1994)
VP 7.75E-006	mm Hg	25	EST	NEELY,WB & BLAU,GE (1985)
DC	pKa			
HL 4.74E-010	atm m3/mol	25	EST	MEYLAN,WM & HOWARD,PH (1991)
OH 4.01E-011	cm3/molc sec	25	EST	MEYLAN,WM & HOWARD,PH (1993)

CAS #: 064673-04-7 — BENZALDEHYDE, 4-(2-HYDROXYETHOXY)-3-METHOXY-

Formula: $C_{10}H_{12}O_4$
Mol Weight: 196.20
MP (deg C): FP (deg C):
BP (deg C):
BP pressure (mm Hg):

Property/Value	Units	Temp	Data Type	Reference
WS 4.73E+004	mg/L	25	EST	MEYLAN,WM ET AL. (1996)
logP 0.58			EXP	HANSCH,C ET AL. (1995)
VP 4.33E-006	mm Hg	25	EST	NEELY,WB & BLAU,GE (1985)
DC	pKa			
HL 2.28E-012	atm m3/mol	25	EST	MEYLAN,WM & HOWARD,PH (1991)
OH 3.83E-011	cm3/molc sec	25	EST	MEYLAN,WM & HOWARD,PH (1993)

CAS #: 064674-86-8 — ACETIC ACID, [4-OXO-5-(1-PIPERIDINYL)-2-THIAZOLI

Formula: $C_{12}H_{18}N_2O_3S$
Mol Weight: 270.35
MP (deg C): FP (deg C):
BP (deg C):
BP pressure (mm Hg):

Property/Value	Units	Temp	Data Type	Reference
WS 1.83E+002	mg/L	25	EST	MEYLAN,WM ET AL. (1996)
logP 2.34			EXP	SANGSTER,J (1994)
VP 3.34E-008	mm Hg	25	EST	NEELY,WB & BLAU,GE (1985)
DC	pKa			
HL 2.27E-009	atm m3/mol	25	EST	MEYLAN,WM & HOWARD,PH (1991)
OH 1.98E-010	cm3/molc sec	25	EST	MEYLAN,WM & HOWARD,PH (1993)

CAS #: 064693-23-8 — N-BROMOBENZOQUINONEMONIMINE

Formula: C_6H_4BrNO
Mol Weight: 186.01
MP (deg C): FP (deg C):
BP (deg C):
BP pressure (mm Hg):

Property/Value	Units	Temp	Data Type	Reference
WS 5.69E+003	mg/L	25	EST	MEYLAN,WM ET AL. (1996)
logP 1.12			EXP	HANSCH,C & LEO,AJ (1985)
VP 2.26E-002	mm Hg	25	EST	NEELY,WB & BLAU,GE (1985)
DC	pKa			
HL 5.15E-006	atm m3/mol	25	EST	MEYLAN,WM & HOWARD,PH (1991)
OH 2.26E-011	cm3/molc sec	25	EST	MEYLAN,WM & HOWARD,PH (1993)

CAS #: 064694-85-5 — ACETAMIDE, N-(3,4-DICHLOROPHENYL)-2,2,2-TRIFLUOR

Formula: $C_8H_4Cl_2F_3NO$
Mol Weight: 258.03
MP (deg C): FP (deg C):
BP (deg C):
BP pressure (mm Hg):

Property/Value	Units	Temp	Data Type	Reference
WS 9.97E+000	mg/L	25	EST	MEYLAN,WM ET AL. (1996)
logP 3.90			EXP	HANSCH,C ET AL. (1995)
VP 3.42E-005	mm Hg	25	EST	NEELY,WB & BLAU,GE (1985)
DC	pKa			
HL 2.67E-008	atm m3/mol	25	EST	MEYLAN,WM & HOWARD,PH (1991)
OH 2.55E-012	cm3/molc sec	25	EST	MEYLAN,WM & HOWARD,PH (1993)

CAS #: 064700-56-7 — TRICLOPYR ESTER

Formula: $C_{13}H_{16}Cl_3NO_4$

Mol Weight: 356.64

MP (deg C): FP (deg C):

BP (deg C):

BP pressure (mm Hg):

Property/Value	Units	Temp	Data Type	Reference
WS 1.00E+006	mg/L		EXP	SHIU,WY ET AL. (1990)
logP 4.01			EST	MEYLAN,WM & HOWARD,PH (1995)
VP 7.92E-007	mm Hg	25	EST	NEELY,WB & BLAU,GE (1985)
DC	pKa			
HL 5.98E-008	atm m3/mol	25	EST	MEYLAN,WM & HOWARD,PH (1991)
OH 2.30E-011	cm3/molc sec	25	EST	MEYLAN,WM & HOWARD,PH (1993)

CAS #: 064778-75-2 — 3,5-DIMEO-4-BUO-PHENETHYLAMINE

Formula: $C_{14}H_{23}NO_3$

Mol Weight: 253.34

MP (deg C): FP (deg C):

BP (deg C):

BP pressure (mm Hg):

Property/Value	Units	Temp	Data Type	Reference
WS 2.41E+003	mg/L	25	EST	MEYLAN,WM ET AL. (1996)
logP 2.32			EXP	HANSCH,C & LEO,AJ (1985)
VP 2.17E-005	mm Hg	25	EST	NEELY,WB & BLAU,GE (1985)
DC	pKa			
HL 3.93E-010	atm m3/mol	25	EST	MEYLAN,WM & HOWARD,PH (1991)
OH 2.47E-010	cm3/molc sec	25	EST	MEYLAN,WM & HOWARD,PH (1993)

CAS #: 064778-79-6 — 3,5-DIMETHOXY-4-BR AMPHETAMINE

Formula: $C_{11}H_{16}BrNO_2$

Mol Weight: 274.16

MP (deg C): FP (deg C):

BP (deg C):

BP pressure (mm Hg):

Property/Value	Units	Temp	Data Type	Reference
WS 1.45E+003	mg/L	25	EST	MEYLAN,WM ET AL. (1996)
logP 2.44			EXP	HANSCH,C & LEO,AJ (1985)
VP 1.16E-004	mm Hg	25	EST	NEELY,WB & BLAU,GE (1985)
DC	pKa			
HL 1.50E-009	atm m3/mol	25	EST	MEYLAN,WM & HOWARD,PH (1991)
OH 2.15E-010	cm3/molc sec	25	EST	MEYLAN,WM & HOWARD,PH (1993)

CAS #: 064810-90-8 — BARBITURIC ACID,5-ET-5-OCTYL

Formula: $C_{14}H_{24}N_2O_3$

Mol Weight: 268.36

MP (deg C): FP (deg C):

BP (deg C):

BP pressure (mm Hg):

Property/Value	Units	Temp	Data Type	Reference
WS 1.10E+001	mg/L	25	EST	MEYLAN,WM ET AL. (1996)
logP 3.78			EXP	YIH,TD & VAN ROSSUM,JM (1977)
VP 1.05E-012	mm Hg	25	EST	NEELY,WB & BLAU,GE (1985)
DC	pKa			
HL 1.97E-012	atm m3/mol	25	EST	MEYLAN,WM & HOWARD,PH (1991)
OH 1.87E-011	cm3/molc sec	25	EST	MEYLAN,WM & HOWARD,PH (1993)

CAS #: 064810-91-9 — BARBITURIC ACID,5-ET-5-NONYL

Formula: $C_{15}H_{26}N_2O_3$

Mol Weight: 282.39

MP (deg C): FP (deg C):

BP (deg C):

BP pressure (mm Hg):

Property/Value	Units	Temp	Data Type	Reference
WS 5.96E+000	mg/L	25	EST	MEYLAN,WM ET AL. (1996)
logP 4.00			EXP	YIH,TD & VAN ROSSUM,JM (1977)
VP 4.51E-013	mm Hg	25	EST	NEELY,WB & BLAU,GE (1985)
DC	pKa			
HL 2.62E-012	atm m3/mol	25	EST	MEYLAN,WM & HOWARD,PH (1991)
OH 2.01E-011	cm3/molc sec	25	EST	MEYLAN,WM & HOWARD,PH (1993)

CAS #: 064835-48-9 — ACETAMIDE, N-(4-BROMO-3,5-DIMETHYLPHENYL)-

Formula: $C_{10}H_{12}BrNO$

Mol Weight: 242.12

MP (deg C): FP (deg C):

BP (deg C):

BP pressure (mm Hg):

Property/Value	Units	Temp	Data Type	Reference
WS 4.13E+001	mg/L	25	EST	MEYLAN,WM ET AL. (1996)
logP 3.28			EXP	NAKAGAWA,Y ET AL. (1992)
VP 6.76E-006	mm Hg	25	EST	NEELY,WB & BLAU,GE (1985)
DC	pKa			
HL 2.99E-009	atm m3/mol	25	EST	MEYLAN,WM & HOWARD,PH (1991)
OH 2.46E-011	cm3/molc sec	25	EST	MEYLAN,WM & HOWARD,PH (1993)

CAS #: 064889-77-6 — BARBITURIC ACID,N-ME-5ALLYL,5-BR-ALLYL

Formula: $C_{11}H_{13}BrN_2O_3$

Mol Weight: 301.14

MP (deg C): FP (deg C):

BP (deg C):

BP pressure (mm Hg):

Property/Value	Units	Temp	Data Type	Reference
WS 3.65E+002	mg/L	25	EST	MEYLAN,WM ET AL. (1996)
logP 1.78			EXP	HANSCH,C & LEO,AJ (1985)
VP 7.54E-012	mm Hg	25	EST	NEELY,WB & BLAU,GE (1985)
DC	pKa			
HL 1.54E-013	atm m3/mol	25	EST	MEYLAN,WM & HOWARD,PH (1991)
OH 5.08E-011	cm3/molc sec	25	EST	MEYLAN,WM & HOWARD,PH (1993)

CAS #: 064902-72-3 — CHLORSULFURON (PH 7)

Formula: $C_{12}H_{12}ClN_5O_4S$

Mol Weight: 357.78

MP (deg C): 174-178 FP (deg C):

BP (deg C):

BP pressure (mm Hg):

Property/Value	Units	Temp	Data Type	Reference
WS 2.80E+004	mg/L	25	EXP	BEYER,EMJR ET AL. (1988); PH 7
logP 2.00			EXP	RIBO,JM (1988)
VP 4.60E-006	mm Hg	25	EXP	WSSA (1989)
DC 3.60	pKa			WAUCHOPE,RD ET AL. (1992)
HL 7.73E-011	atm m3/mol	25	EST	VP/WSOL
OH 2.52E-011	cm3/molc sec	25	EST	ATKINSON,R (1988)

064980-40-1

CAS #: 064980-40-1

2-PROPANOL, 1-[(1,1-DIMETHYLETHYL)AMINO]-3-PHENO

Formula: $C_{13}H_{21}NO_2$

Mol Weight: 223.32

MP (deg C): FP (deg C):

BP (deg C):

BP pressure (mm Hg):

Property/Value	Units	Temp	Data Type	Reference
WS 1.76E+004	mg/L	25	EST	MEYLAN,WM ET AL. (1996)
logP 1.50			EXP	MAULEON,D ET AL. (1988)
VP 7.71E-006	mm Hg	25	EST	NEELY,WB & BLAU,GE (1985)
DC	pKa			
HL 1.09E-011	atm m3/mol	25	EST	MEYLAN,WM & HOWARD,PH (1991)
OH 1.13E-010	cm3/molc sec	25	EST	MEYLAN,WM & HOWARD,PH (1993)

065016-34-4

CAS #: 065016-34-4

O-MEO BENZAMIDE,N-(DIETAMINOET)

Formula: $C_{14}H_{22}N_2O_2$

Mol Weight: 250.34

MP (deg C): FP (deg C):

BP (deg C):

BP pressure (mm Hg):

Property/Value	Units	Temp	Data Type	Reference
WS 4.80E+002	mg/L	25	EST	MEYLAN,WM ET AL. (1996)
logP 1.98			EXP	HANSCH,C & LEO,AJ (1985)
VP 4.52E-007	mm Hg	25	EST	NEELY,WB & BLAU,GE (1985)
DC	pKa			
HL 3.34E-013	atm m3/mol	25	EST	MEYLAN,WM & HOWARD,PH (1991)
OH 1.25E-010	cm3/molc sec	25	EST	MEYLAN,WM & HOWARD,PH (1993)

065016-46-8

CAS #: 065016-46-8

BENZAMIDE,N(DIETAMET),2MEO-4NH2-5C∂N

Formula: $C_{15}H_{22}N_4O_2$

Mol Weight: 290.37

MP (deg C): FP (deg C):

BP (deg C):

BP pressure (mm Hg):

Property/Value	Units	Temp	Data Type	Reference
WS 1.67E+002	mg/L	25	EST	MEYLAN,WM ET AL. (1996)
logP 1.94			EXP	HANSCH,C & LEO,AJ (1985)
VP 6.38E-010	mm Hg	25	EST	NEELY,WB & BLAU,GE (1985)
DC	pKa			
HL 1.14E-018	atm m3/mol	25	EST	MEYLAN,WM & HOWARD,PH (1991)
OH 2.97E-010	cm3/molc sec	25	EST	MEYLAN,WM & HOWARD,PH (1993)

065052-47-3

CAS #: 065052-47-3

3'-(CF3-THIO)BENZOGUANAMINE

Formula: $C_{10}H_8F_3N_5S$

Mol Weight: 287.27

MP (deg C): FP (deg C):

BP (deg C):

BP pressure (mm Hg):

Property/Value	Units	Temp	Data Type	Reference
WS 2.69E+001	mg/L	25	EST	MEYLAN,WM ET AL. (1996)
logP 3.20			EXP	HANSCH,C & LEO,AJ (1985)
VP 6.14E-008	mm Hg	25	EST	NEELY,WB & BLAU,GE (1985)
DC	pKa			
HL 9.43E-012	atm m3/mol	25	EST	MEYLAN,WM & HOWARD,PH (1991)
OH 2.21E-011	cm3/molc sec	25	EST	MEYLAN,WM & HOWARD,PH (1993)

065052-49-5

CAS #: 065052-49-5

3'-METHYLTHIOBENZOGUANAMINE

Formula: $C_{10}H_{11}N_5S$

Mol Weight: 233.30

MP (deg C): FP (deg C):

BP (deg C):

BP pressure (mm Hg):

Property/Value	Units	Temp	Data Type	Reference
WS 4.18E+002	mg/L	25	EST	MEYLAN,WM ET AL. (1996)
logP 2.16			EXP	HANSCH,C & LEO,AJ (1985)
VP 3.49E-008	mm Hg	25	EST	NEELY,WB & BLAU,GE (1985)
DC	pKa			
HL 1.20E-012	atm m3/mol	25	EST	MEYLAN,WM & HOWARD,PH (1991)
OH 2.32E-011	cm3/molc sec	25	EST	MEYLAN,WM & HOWARD,PH (1993)

065052-50-8

CAS #: 065052-50-8

3'-METHYLSULFONYLBENZOGUANAMINE

Formula: $C_{10}H_{11}N_5O_2S$

Mol Weight: 265.30

MP (deg C): FP (deg C):

BP (deg C):

BP pressure (mm Hg):

Property/Value	Units	Temp	Data Type	Reference
WS 1.10E+004	mg/L	25	EST	MEYLAN,WM ET AL. (1996)
logP 0.29			EXP	HANSCH,C & LEO,AJ (1985)
VP 1.14E-009	mm Hg	25	EST	NEELY,WB & BLAU,GE (1985)
DC	pKa			
HL 9.52E-015	atm m3/mol	25	EST	MEYLAN,WM & HOWARD,PH (1991)
OH 2.07E-012	cm3/molc sec	25	EST	MEYLAN,WM & HOWARD,PH (1993)

065052-53-1

CAS #: 065052-53-1

2'-CL-5'-CF3-BENZOGUANAMINE

Formula: $C_{10}H_7ClF_3N_5$ ·

Mol Weight: 289.65

MP (deg C): FP (deg C):

BP (deg C):

BP pressure (mm Hg):

Property/Value	Units	Temp	Data Type	Reference
WS 1.62E+002	mg/L	25	EST	MEYLAN,WM ET AL. (1996)
logP 2.27			EXP	HANSCH,C & LEO,AJ (1985)
VP 1.67E-007	mm Hg	25	EST	NEELY,WB & BLAU,GE (1985)
DC	pKa			
HL 2.65E-010	atm m3/mol	25	EST	MEYLAN,WM & HOWARD,PH (1991)
OH 4.52E-013	cm3/molc sec	25	EST	MEYLAN,WM & HOWARD,PH (1993)

065052-55-3

CAS #: 065052-55-3

2'-MEO-5'-BR BENZOGUANAMINE

Formula: $C_{10}H_{10}BrN_5O$

Mol Weight: 296.13

MP (deg C): FP (deg C):

BP (deg C):

BP pressure (mm Hg):

Property/Value	Units	Temp	Data Type	Reference
WS 6.38E+002	mg/L	25	EST	MEYLAN,WM ET AL. (1996)
logP 1.53			EXP	HANSCH,C & LEO,AJ (1985)
VP 1.81E-008	mm Hg	25	EST	NEELY,WB & BLAU,GE (1985)
DC	pKa			
HL 9.69E-013	atm m3/mol	25	EST	MEYLAN,WM & HOWARD,PH (1991)
OH 5.49E-012	cm3/molc sec	25	EST	MEYLAN,WM & HOWARD,PH (1993)

CAS #: 065101-39-5				TEPA ADAMANTYL-UREA DERIVATIVE
Formula: $C_{15}H_{25}N_4O_2P$				
Mol Weight: 324.37				
MP (deg C):		FP (deg C):		
BP (deg C):				
BP pressure (mm Hg):				

Property/ Value	Units	Temp	Data Type	Reference
WS 2.51E+002	mg/L	25	EST	MEYLAN,WM ET AL. (1996)
logP 1.81			EXP	SOSNOVSKY,G ET AL. (1986)
VP 7.58E-008	mm Hg	25	EST	NEELY,WB & BLAU,GE (1985)
DC	pKa			
HL 7.33E-017	atm m3/mol	25	EST	MEYLAN,WM & HOWARD,PH (1991)
OH 2.52E-011	cm3/molc sec	25	EST	MEYLAN,WM & HOWARD,PH (1993)

CAS #: 065118-48-1				L-VALINAMIDE, N-ACETYL-L-PHENYLALANYL-
Formula: $C_{16}H_{23}N_3O_3$				
Mol Weight: 305.38				
MP (deg C):		FP (deg C):		
BP (deg C):				
BP pressure (mm Hg):				

Property/ Value	Units	Temp	Data Type	Reference
WS 4.90E+003	mg/L	25	EST	MEYLAN,WM ET AL. (1996)
logP 0.43			EXP	HANSCH,C ET AL. (1995)
VP 1.37E-012	mm Hg	25	EST	NEELY,WB & BLAU,GE (1985)
DC	pKa			
HL 1.97E-016	atm m3/mol	25	EST	MEYLAN,WM & HOWARD,PH (1991)
OH 4.42E-011	cm3/molc sec	25	EST	MEYLAN,WM & HOWARD,PH (1993)

CAS #: 065118-51-6				L-LEUCINAMIDE, N-ACETYL-L-PHENYLALANYLGLYCYL-
Formula: $C_{19}H_{28}N_4O_4$				
Mol Weight: 376.46				
MP (deg C):		FP (deg C):		
BP (deg C):				
BP pressure (mm Hg):				

Property/ Value	Units	Temp	Data Type	Reference
WS 1.31E+003	mg/L	25	EST	MEYLAN,WM ET AL. (1996)
logP 0.60			EXP	HANSCH,C ET AL. (1995)
VP 8.24E-017	mm Hg	25	EST	NEELY,WB & BLAU,GE (1985)
DC	pKa			
HL 1.74E-021	atm m3/mol	25	EST	MEYLAN,WM & HOWARD,PH (1991)
OH 5.18E-011	cm3/molc sec	25	EST	MEYLAN,WM & HOWARD,PH (1993)

CAS #: 065122-21-6				1,4-DICHLORO-4-VINYLCYCLO-1-HEXENE
Formula: $C_8H_{10}Cl_2$				
Mol Weight: 177.07				
MP (deg C):		FP (deg C):		
BP (deg C):				
BP pressure (mm Hg):				

Property/ Value	Units	Temp	Data Type	Reference
WS 9.16E+000	mg/L	25	EST	MEYLAN,WM ET AL. (1996)
logP 4.44			EST	MEYLAN,WM & HOWARD,PH (1995)
VP 7.53E-001	mm Hg	25	EST	NEELY,WB & BLAU,GE (1985)
DC	pKa			
HL 3.92E-002	atm m3/mol	25	EST	MEYLAN,WM & HOWARD,PH (1991)
OH 4.15E-011	cm3/molc sec	25	EST	MEYLAN,WM & HOWARD,PH (1993)

CAS #: 065125-87-3				DIS. A. 8
Formula: $C_{17}H_{18}ClN_5O_6$				
Mol Weight: 423.82				
MP (deg C):		FP (deg C):		
BP (deg C):				
BP pressure (mm Hg):				

Property/ Value	Units	Temp	Data Type	Reference
WS 2.12E-001	mg/L	25	EXP	BAUGHMAN,GL & PERENICH,TA (1988)
logP 4.15			EST	MEYLAN,WM & HOWARD,PH (1995)
VP 3.13E-016	mm Hg	25	EST	NEELY,WB & BLAU,GE (1985)
DC	pKa			
HL 7.01E-018	atm m3/mol	25	EST	MEYLAN,WM & HOWARD,PH (1991)
OH 1.81E-010	cm3/molc sec	25	EST	MEYLAN,WM & HOWARD,PH (1993)

CAS #: 065243-35-8				CEFMENOXIME,PIVALOYL-OME ESTER
Formula: $C_{22}H_{27}N_9O_7S_3$				
Mol Weight: 625.71				
MP (deg C):		FP (deg C):		
BP (deg C):				
BP pressure (mm Hg):				

Property/ Value	Units	Temp	Data Type	Reference
WS 3.56E+000	mg/L	25	EST	MEYLAN,WM ET AL. (1996)
logP 1.74			EXP	HANSCH,C ET AL. (1995)
VP 1.05E-020	mm Hg	25	EST	NEELY,WB & BLAU,GE (1985)
DC	pKa			
HL 1.20E-026	atm m3/mol	25	EST	MEYLAN,WM & HOWARD,PH (1991)
OH 1.24E-010	cm3/molc sec	25	EST	MEYLAN,WM & HOWARD,PH (1993)

CAS #: 065243-53-0				5-THIA-1-AZABICYCLO[4.2.0]OCT-2-ENE-2-CARBOXYLIC
Formula: $C_{22}H_{27}N_5O_9S_2$				
Mol Weight: 569.62				
MP (deg C):		FP (deg C):		
BP (deg C):				
BP pressure (mm Hg):				

Property/ Value	Units	Temp	Data Type	Reference
WS 1.47E+000	mg/L	25	EST	MEYLAN,WM ET AL. (1996)
logP 2.62			EXP	HANSCH,C ET AL. (1995)
VP 3.93E-018	mm Hg	25	EST	NEELY,WB & BLAU,GE (1985)
DC	pKa			
HL 5.52E-024	atm m3/mol	25	EST	MEYLAN,WM & HOWARD,PH (1991)
OH 1.20E-010	cm3/molc sec	25	EST	MEYLAN,WM & HOWARD,PH (1993)

CAS #: 065247-10-1				14-BENZDIAZEPIN-2-ONE,5-(2-BRPH)7-BR
Formula: $C_{15}H_{10}Br_2N_2O$				
Mol Weight: 394.08				
MP (deg C):		FP (deg C):		
BP (deg C):				
BP pressure (mm Hg):				

Property/ Value	Units	Temp	Data Type	Reference
WS 5.07E+000	mg/L	25	EST	MEYLAN,WM ET AL. (1996)
logP 3.30			EXP	HANSCH,C & LEO,AJ (1985)
VP 3.18E-010	mm Hg	25	EST	NEELY,WB & BLAU,GE (1985)
DC	pKa			
HL 3.81E-011	atm m3/mol	25	EST	MEYLAN,WM & HOWARD,PH (1991)
OH 6.44E-012	cm3/molc sec	25	EST	MEYLAN,WM & HOWARD,PH (1993)

CAS #: 065247-11-2 — 14-BENZDIAZEPIN-2-ONE,5-(3-BRPH)7-BR

Formula: $C_{15}H_{10}Br_2N_2O$

Mol Weight: 394.08

MP (deg C): FP (deg C):

BP (deg C):

BP pressure (mm Hg):

Property/Value	Units	Temp	Data Type	Reference
WS 1.62E+000	mg/L	25	EST	MEYLAN,WM ET AL. (1996)
logP 3.88			EXP	HANSCH,C & LEO,AJ (1985)
VP 3.18E-010	mm Hg	25	EST	NEELY,WB & BLAU,GE (1985)
DC	pKa			
HL 3.81E-011	atm m3/mol	25	EST	MEYLAN,WM & HOWARD,PH (1991)
OH 7.67E-012	cm3/molc sec	25	EST	MEYLAN,WM & HOWARD,PH (1993)

CAS #: 065247-12-3 — 14-BENZDIAZEPIN-2-ONE,5-(4-BRPH)7-BR

Formula: $C_{15}H_{10}Br_2N_2O$

Mol Weight: 394.08

MP (deg C): FP (deg C):

BP (deg C):

BP pressure (mm Hg):

Property/Value	Units	Temp	Data Type	Reference
WS 1.18E+000	mg/L	25	EST	MEYLAN,WM ET AL. (1996)
logP 4.04			EXP	HANSCH,C & LEO,AJ (1985)
VP 3.18E-010	mm Hg	25	EST	NEELY,WB & BLAU,GE (1985)
DC	pKa			
HL 3.81E-011	atm m3/mol	25	EST	MEYLAN,WM & HOWARD,PH (1991)
OH 6.44E-012	cm3/molc sec	25	EST	MEYLAN,WM & HOWARD,PH (1993)

CAS #: 065247-13-4 — 14-BENZDIAZEPIN-2-ONE,5-(3-CLPH)7-BR

Formula: $C_{15}H_{10}BrClN_2O$

Mol Weight: 349.62

MP (deg C): FP (deg C):

BP (deg C):

BP pressure (mm Hg):

Property/Value	Units	Temp	Data Type	Reference
WS 4.95E+000	mg/L	25	EST	MEYLAN,WM ET AL. (1996)
logP 3.63			EXP	HANSCH,C & LEO,AJ (1985)
VP 7.45E-010	mm Hg	25	EST	NEELY,WB & BLAU,GE (1985)
DC	pKa			
HL 7.08E-011	atm m3/mol	25	EST	MEYLAN,WM & HOWARD,PH (1991)
OH 7.99E-012	cm3/molc sec	25	EST	MEYLAN,WM & HOWARD,PH (1993)

CAS #: 065247-14-5 — 14-BENZDIAZEPIN-2-ONE,5-(4-CLPH)7-BR

Formula: $C_{15}H_{10}BrClN_2O$

Mol Weight: 349.62

MP (deg C): FP (deg C):

BP (deg C):

BP pressure (mm Hg):

Property/Value	Units	Temp	Data Type	Reference
WS 3.68E+000	mg/L	25	EST	MEYLAN,WM ET AL. (1996)
logP 3.78			EXP	HANSCH,C & LEO,AJ (1985)
VP 7.45E-010	mm Hg	25	EST	NEELY,WB & BLAU,GE (1985)
DC	pKa			
HL 7.08E-011	atm m3/mol	25	EST	MEYLAN,WM & HOWARD,PH (1991)
OH 6.62E-012	cm3/molc sec	25	EST	MEYLAN,WM & HOWARD,PH (1993)

CAS #: 065261-13-4 — BENZAMIDE, N,N-DIMETHYL-3-PHENOXY-

Formula: $C_{15}H_{15}NO_2$

Mol Weight: 241.29

MP (deg C): FP (deg C):

BP (deg C):

BP pressure (mm Hg):

Property/Value	Units	Temp	Data Type	Reference
WS 1.69E+002	mg/L	25	EST	MEYLAN,WM ET AL. (1996)
logP 2.57			EXP	SANGSTER,J (1994)
VP 3.70E-006	mm Hg	25	EST	NEELY,WB & BLAU,GE (1985)
DC	pKa			
HL 2.33E-010	atm m3/mol	25	EST	MEYLAN,WM & HOWARD,PH (1991)
OH 2.31E-011	cm3/molc sec	25	EST	MEYLAN,WM & HOWARD,PH (1993)

CAS #: 065261-51-0 — BENZAMIDE, N-(1-METHYLETHYL)-3-(2,5-DIMETHYLPHEN

Formula: $C_{18}H_{21}NO_2$

Mol Weight: 283.37

MP (deg C): FP (deg C):

BP (deg C):

BP pressure (mm Hg):

Property/Value	Units	Temp	Data Type	Reference
WS 2.96E+000	mg/L	25	EST	MEYLAN,WM ET AL. (1996)
logP 4.35			EXP	SANGSTER,J (1994)
VP 1.86E-008	mm Hg	25	EST	NEELY,WB & BLAU,GE (1985)
DC	pKa			
HL 2.28E-010	atm m3/mol	25	EST	MEYLAN,WM & HOWARD,PH (1991)
OH 3.19E-011	cm3/molc sec	25	EST	MEYLAN,WM & HOWARD,PH (1993)

CAS #: 065261-99-6 — BENZAMIDE, N-(1,1-DIMETHYLETHYL)-3-(2,5-DIMETHYL

Formula: $C_{19}H_{23}NO_2$

Mol Weight: 297.40

MP (deg C): FP (deg C):

BP (deg C):

BP pressure (mm Hg):

Property/Value	Units	Temp	Data Type	Reference
WS 1.05E+000	mg/L	25	EST	MEYLAN,WM ET AL. (1996)
logP 4.78			EXP	SANGSTER,J (1994)
VP 1.04E-008	mm Hg	25	EST	NEELY,WB & BLAU,GE (1985)
DC	pKa			
HL 3.03E-010	atm m3/mol	25	EST	MEYLAN,WM & HOWARD,PH (1991)
OH 2.41E-011	cm3/molc sec	25	EST	MEYLAN,WM & HOWARD,PH (1993)

CAS #: 065262-21-7 — BENZAMIDE, 3-(2,5-DIMETHYLPHENOXY)-N-METHYL-

Formula: $C_{16}H_{17}NO_2$

Mol Weight: 255.32

MP (deg C): FP (deg C):

BP (deg C):

BP pressure (mm Hg):

Property/Value	Units	Temp	Data Type	Reference
WS 1.36E+001	mg/L	25	EST	MEYLAN,WM ET AL. (1996)
logP 3.76			EXP	SANGSTER,J (1994)
VP 5.95E-008	mm Hg	25	EST	NEELY,WB & BLAU,GE (1985)
DC	pKa			
HL 1.29E-010	atm m3/mol	25	EST	MEYLAN,WM & HOWARD,PH (1991)
OH 2.41E-011	cm3/molc sec	25	EST	MEYLAN,WM & HOWARD,PH (1993)

CAS #:	065277-42-1		PIPERAZINE, 1-ACETYL-4-[4-[[2-(2,4-DICHLOROPHENY

Formula: $C_{26}H_{28}Cl_2N_4O_4$

Mol Weight: 531.44

MP (deg C): 146 | **FP (deg C):**

BP (deg C):

BP pressure (mm Hg):

Property/Value	Units	Temp	Data Type	Reference
WS 8.66E-002	mg/L	25	EST	MEYLAN,WM ET AL. (1996)
logP 4.35			EXP	SANGSTER,J (1993)
VP 1.17E-015	mm Hg	25	EST	NEELY,WB & BLAU,GE (1985)
DC	pKa			
HL 5.59E-020	atm m3/mol	25	EST	MEYLAN,WM & HOWARD,PH (1991)
OH 2.34E-010	cm3/molc sec	25	EST	MEYLAN,WM & HOWARD,PH (1993)

CAS #:	065283-97-8		2-AMINO-5-I-PR-1,3,4-OXADIAZOLE

Formula: $C_5H_{10}N_3O$

Mol Weight: 128.15

MP (deg C): | **FP (deg C):**

BP (deg C):

BP pressure (mm Hg):

Property/Value	Units	Temp	Data Type	Reference
WS 6.37E+004	mg/L	25	EST	MEYLAN,WM ET AL. (1996)
logP 0.20			EXP	HANSCH,C & LEO,AJ (1985)
VP 7.75E-003	mm Hg	25	EST	NEELY,WB & BLAU,GE (1985)
DC	pKa			
HL 1.97E-009	atm m3/mol	25	EST	MEYLAN,WM & HOWARD,PH (1991)
OH 6.27E-012	cm3/molc sec	25	EST	MEYLAN,WM & HOWARD,PH (1993)

CAS #:	065300-53-0		N-(2-(P-NO2-PHO)ETHYL)MORPHOLINE

Formula: $C_{12}H_{16}N_2O_4$

Mol Weight: 252.27

MP (deg C): | **FP (deg C):**

BP (deg C):

BP pressure (mm Hg):

Property/Value	Units	Temp	Data Type	Reference
WS 4.62E+003	mg/L	25	EST	MEYLAN,WM ET AL. (1996)
logP 1.54			EXP	HANSCH,C & LEO,AJ (1985)
VP 6.39E-006	mm Hg	25	EST	NEELY,WB & BLAU,GE (1985)
DC	pKa			
HL 5.67E-012	atm m3/mol	25	EST	MEYLAN,WM & HOWARD,PH (1991)
OH 1.81E-010	cm3/molc sec	25	EST	MEYLAN,WM & HOWARD,PH (1993)

CAS #:	065329-13-7		BARBITURIC ACID,5-(1-CYCLOHEPTEN-1-YL)-5-ETHYL-1

Formula: $C_{14}H_{20}N_2O_3$

Mol Weight: 264.33

MP (deg C): | **FP (deg C):**

BP (deg C):

BP pressure (mm Hg):

Property/Value	Units	Temp	Data Type	Reference
WS 7.68E+001	mg/L	25	EST	MEYLAN,WM ET AL. (1996)
logP 2.82			EXP	YIH,TD & VAN ROSSUM,JM (1977)
VP 5.41E-012	mm Hg	25	EST	NEELY,WB & BLAU,GE (1985)
DC	pKa			
HL 1.49E-012	atm m3/mol	25	EST	MEYLAN,WM & HOWARD,PH (1991)
OH 9.98E-011	cm3/molc sec	25	EST	MEYLAN,WM & HOWARD,PH (1993)

CAS #:	065332-44-7		N,N'-(DITHIODI-2,1-ETHANEDIYL)BIS(N-(1-METHYL*))

Formula: $C_{16}H_{36}N_2S_2$

Mol Weight: 320.61

MP (deg C): | **FP (deg C):**

BP (deg C):

BP pressure (mm Hg):

Property/Value	Units	Temp	Data Type	Reference
WS 6.16E+000	mg/L	25	EST	MEYLAN,WM ET AL. (1996)
logP 4.90			EST	MEYLAN,WM & HOWARD,PH (1995)
VP 4.11E-005	mm Hg	25	EST	NEELY,WB & BLAU,GE (1985)
DC	pKa			
HL 5.07E-009	atm m3/mol	25	EST	MEYLAN,WM & HOWARD,PH (1991)
OH 4.70E-010	cm3/molc sec	25	EST	MEYLAN,WM & HOWARD,PH (1993)

CAS #:	065356-76-5		L-LEUCINAMIDE, N-ACETYL-L-TYROSYL-

Formula: $C_{17}H_{25}N_3O_4$

Mol Weight: 335.41

MP (deg C): | **FP (deg C):**

BP (deg C):

BP pressure (mm Hg):

Property/Value	Units	Temp	Data Type	Reference
WS 7.53E+002	mg/L	25	EST	MEYLAN,WM ET AL. (1996)
logP 0.32			EXP	HANSCH,C ET AL. (1995)
VP 3.22E-015	mm Hg	25	EST	NEELY,WB & BLAU,GE (1985)
DC	pKa			
HL 2.72E-020	atm m3/mol	25	EST	MEYLAN,WM & HOWARD,PH (1991)
OH 7.94E-011	cm3/molc sec	25	EST	MEYLAN,WM & HOWARD,PH (1993)

CAS #:	065445-59-2		3,5-DIME-N-NITROSOPIPERIDINE

Formula: $C_7H_{14}N_2O$

Mol Weight: 142.20

MP (deg C): | **FP (deg C):**

BP (deg C):

BP pressure (mm Hg):

Property/Value	Units	Temp	Data Type	Reference
WS 4.05E+003	mg/L	25	EST	MEYLAN,WM ET AL. (1996)
logP 1.53			EXP	HANSCH,C & LEO,AJ (1985)
VP 3.37E-002	mm Hg	25	EST	NEELY,WB & BLAU,GE (1985)
DC	pKa			
HL 3.74E-006	atm m3/mol	25	EST	MEYLAN,WM & HOWARD,PH (1991)
OH 2.90E-011	cm3/molc sec	25	EST	MEYLAN,WM & HOWARD,PH (1993)

CAS #:	065446-98-2		1-PIPERIDINEACETAMIDE, N-(4-ETHYLPHENYL)-

Formula: $C_{15}H_{22}N_2O$

Mol Weight: 246.36

MP (deg C): | **FP (deg C):**

BP (deg C):

BP pressure (mm Hg):

Property/Value	Units	Temp	Data Type	Reference
WS 2.44E+001	mg/L	25	EST	MEYLAN,WM ET AL. (1996)
logP 3.52			EXP	SANGSTER,J (1993)
VP 1.81E-007	mm Hg	25	EST	NEELY,WB & BLAU,GE (1985)
DC	pKa			
HL 9.23E-011	atm m3/mol	25	EST	MEYLAN,WM & HOWARD,PH (1991)
OH 1.14E-010	cm3/molc sec	25	EST	MEYLAN,WM & HOWARD,PH (1993)

CAS #: 065448-74-0		CARBAMIC ACID, (3-CYANOPHENYL)-, METHYL ESTER	
Formula: $C_9H_8N_2O_2$			
Mol Weight: 176.18			
MP (deg C):		FP (deg C):	
BP (deg C):			
BP pressure (mm Hg):			

Property/Value	Units	Temp	Data Type	Reference
WS 9.61E+002	mg/L	25	EST	MEYLAN,WM ET AL. (1996)
logP 1.77			EXP	TAKAHASHI,J ET AL. (1988)
VP 6.30E-004	mm Hg	25	EST	NEELY,WB & BLAU,GE (1985)
DC	pKa			
HL 2.11E-010	atm m3/mol	25	EST	MEYLAN,WM & HOWARD,PH (1991)
OH 5.80E-012	cm3/molc sec	25	EST	MEYLAN,WM & HOWARD,PH (1993)

CAS #: 065466-50-4		PREGNA-1,4-DIENE-3,20-DIONE, 2-CHLORO-6,9-DIFLUO	
Formula: $C_{24}H_{29}ClF_2O_6$			
Mol Weight: 486.94			
MP (deg C):		FP (deg C):	
BP (deg C):			
BP pressure (mm Hg):			

Property/Value	Units	Temp	Data Type	Reference
WS 2.43E+000	mg/L	25	EST	MEYLAN,WM ET AL. (1996)
logP 2.99			EXP	HANSCH,C ET AL. (1995)
VP 2.41E-014	mm Hg	25	EST	NEELY,WB & BLAU,GE (1985)
DC	pKa			
HL 1.41E-012	atm m3/mol	25	EST	MEYLAN,WM & HOWARD,PH (1991)
OH 8.15E-011	cm3/molc sec	25	EST	MEYLAN,WM & HOWARD,PH (1993)

CAS #: 065481-90-5		3-CARBAMOYLPYRIDINIUM IODIDE, N-HEXYL	
Formula: $C_{12}H_{19}IN_2O$			
Mol Weight: 334.20			
MP (deg C):		FP (deg C):	
BP (deg C):			
BP pressure (mm Hg):			

Property/Value	Units	Temp	Data Type	Reference
WS 9.18E+004	mg/L	25	EST	MEYLAN,WM ET AL. (1996)
logP -1.26			EXP	HANSCH,C & LEO,AJ (1985)
VP 9.33E-008	mm Hg	25	EST	NEELY,WB & BLAU,GE (1985)
DC	pKa			
HL 2.55E-016	atm m3/mol	25	EST	MEYLAN,WM & HOWARD,PH (1991)
OH 1.90E-011	cm3/molc sec	25	EST	MEYLAN,WM & HOWARD,PH (1993)

CAS #: 065510-45-4		2,2',3,4,4'-PENTACHLOROBIPHENYL	
Formula: $C_{12}H_5Cl_5$			
Mol Weight: 326.44			
MP (deg C):		FP (deg C):	
BP (deg C):			
BP pressure (mm Hg):			

Property/Value	Units	Temp	Data Type	Reference
WS 1.94E-002	mg/L	25	EST	MEYLAN,WM ET AL. (1996)
logP 6.61			EXP	HANSCH,C & LEO,AJ (1985)
VP 2.22E-006	mm Hg	25	EST	NEELY,WB & BLAU,GE (1985)
DC	pKa			
HL 9.24E-005	atm m3/mol	25	EST	MEYLAN,WM & HOWARD,PH (1991)
OH 4.00E-013	cm3/molc sec	25	EST	MEYLAN,WM & HOWARD,PH (1993)

CAS #: 065514-04-7		FURAZAN, CHLOROMETHYL-, 2-OXIDE	
Formula: $C_3H_3ClN_2O_2$			
Mol Weight: 134.52			
MP (deg C):		FP (deg C):	
BP (deg C):			
BP pressure (mm Hg):			

Property/Value	Units	Temp	Data Type	Reference
WS 1.50E+004	mg/L	25	EST	MEYLAN,WM ET AL. (1996)
logP 0.90			EXP	CALVINO,R ET AL. (1992)
VP 2.23E-002	mm Hg	25	EST	NEELY,WB & BLAU,GE (1985)
DC	pKa			
HL 1.55E-009	atm m3/mol	25	EST	MEYLAN,WM & HOWARD,PH (1991)
OH 4.14E-012	cm3/molc sec	25	EST	MEYLAN,WM & HOWARD,PH (1993)

CAS #: 065542-14-5		BENZOIC ACID, 4-[3-[(4-CHLOROPHENYL)METHYL]-3-ME	
Formula: $C_{15}H_{14}ClN_3O_2$			
Mol Weight: 303.75			
MP (deg C):		FP (deg C):	
BP (deg C):			
BP pressure (mm Hg):			

Property/Value	Units	Temp	Data Type	Reference
WS 8.42E-001	mg/L	25	EST	MEYLAN,WM ET AL. (1996)
logP 4.85			EXP	SANGSTER,J (1993)
VP 3.40E-008	mm Hg	25	EST	NEELY,WB & BLAU,GE (1985)
DC	pKa			
HL 6.46E-013	atm m3/mol	25	EST	MEYLAN,WM & HOWARD,PH (1991)
OH 1.24E-011	cm3/molc sec	25	EST	MEYLAN,WM & HOWARD,PH (1993)

CAS #: 065542-15-6		P(3-PNO2BENZYL-3ME-1TRIAZN)BENZOIC AC	
Formula: $C_{15}H_{14}N_4O_4$			
Mol Weight: 314.30			
MP (deg C):		FP (deg C):	
BP (deg C):			
BP pressure (mm Hg):			

Property/Value	Units	Temp	Data Type	Reference
WS 1.49E+000	mg/L	25	EST	MEYLAN,WM ET AL. (1996)
logP 4.03			EXP	HANSCH,C & LEO,AJ (1985)
VP 1.84E-009	mm Hg	25	EST	NEELY,WB & BLAU,GE (1985)
DC	pKa			
HL 3.44E-015	atm m3/mol	25	EST	MEYLAN,WM & HOWARD,PH (1991)
OH 1.13E-011	cm3/molc sec	25	EST	MEYLAN,WM & HOWARD,PH (1993)

CAS #: 065542-16-7		P-(3-(P-CYANOBENZYL)-3-METHYL-1-TRIAZENO)-	
Formula: $C_{16}H_{14}N_4O_2$			
Mol Weight: 294.32			
MP (deg C):		FP (deg C):	
BP (deg C):			
BP pressure (mm Hg):			

Property/Value	Units	Temp	Data Type	Reference
WS 6.07E+000	mg/L	25	EST	MEYLAN,WM ET AL. (1996)
logP 3.60			EXP	HANSCH,C & LEO,AJ (1985)
VP 3.76E-009	mm Hg	25	EST	NEELY,WB & BLAU,GE (1985)
DC	pKa			
HL 8.42E-015	atm m3/mol	25	EST	MEYLAN,WM & HOWARD,PH (1991)
OH 1.15E-011	cm3/molc sec	25	EST	MEYLAN,WM & HOWARD,PH (1993)

CAS #: 065542-17-8				P(3(P-MEBENZYL)3ME-1TRIAZENO)BENZOICACID
Formula: C₁₆H₁₇N₃O₂				

Formula: $C_{16}H_{17}N_3O_2$

Mol Weight: 283.33

MP (deg C): FP (deg C):

BP (deg C):

BP pressure (mm Hg):

Property/ Value	Units	Temp	Data Type	Reference
WS 1.35E+000	mg/L	25	EST	MEYLAN,WM ET AL. (1996)
logP 4.75			EXP	HANSCH,C & LEO,AJ (1985)
VP 6.12E-008	mm Hg	25	EST	NEELY,WB & BLAU,GE (1985)
DC			pKa	
HL 9.62E-013	atm m3/mol	25	EST	MEYLAN,WM & HOWARD,PH (1991)
OH 1.70E-011	cm3/molc sec	25	EST	MEYLAN,WM & HOWARD,PH (1993)

CAS #: 065542-18-9				P(3(PMEBENZYL)3MEO-1TRIAZENO)BENZOICACID

Formula: $C_{16}H_{17}N_3O_3$

Mol Weight: 299.33

MP (deg C): FP (deg C):

BP (deg C):

BP pressure (mm Hg):

Property/ Value	Units	Temp	Data Type	Reference
WS 2.04E+000	mg/L	25	EST	MEYLAN,WM ET AL. (1996)
logP 4.43			EXP	HANSCH,C & LEO,AJ (1985)
VP 2.37E-008	mm Hg	25	EST	NEELY,WB & BLAU,GE (1985)
DC			pKa	
HL 5.16E-014	atm m3/mol	25	EST	MEYLAN,WM & HOWARD,PH (1991)
OH 3.77E-011	cm3/molc sec	25	EST	MEYLAN,WM & HOWARD,PH (1993)

CAS #: 065542-19-0				BENZOIC ACID, 4-[3-[[4-(ACETYLAMINO)PHENYL]METHY

Formula: $C_{17}H_{18}N_4O_3$

Mol Weight: 326.36

MP (deg C): FP (deg C):

BP (deg C):

BP pressure (mm Hg):

Property/ Value	Units	Temp	Data Type	Reference
WS 1.18E+001	mg/L	25	EST	MEYLAN,WM ET AL. (1996)
logP 3.35			EXP	HANSCH,C ET AL. (1995)
VP 1.97E-011	mm Hg	25	EST	NEELY,WB & BLAU,GE (1985)
DC			pKa	
HL 9.99E-019	atm m3/mol	25	EST	MEYLAN,WM & HOWARD,PH (1991)
OH 2.60E-011	cm3/molc sec	25	EST	MEYLAN,WM & HOWARD,PH (1993)

CAS #: 065542-21-4				P(3-BENZYL-3ME-1TRIAZENO)BENZONITRILE

Formula: $C_{15}H_{14}N_4$

Mol Weight: 250.31

MP (deg C): FP (deg C):

BP (deg C):

BP pressure (mm Hg):

Property/ Value	Units	Temp	Data Type	Reference
WS 7.00E+000	mg/L	25	EST	MEYLAN,WM ET AL. (1996)
logP 3.82			EXP	HANSCH,C & LEO,AJ (1985)
VP 2.08E-006	mm Hg	25	EST	NEELY,WB & BLAU,GE (1985)
DC			pKa	
HL 4.19E-010	atm m3/mol	25	EST	MEYLAN,WM & HOWARD,PH (1991)
OH 1.49E-011	cm3/molc sec	25	EST	MEYLAN,WM & HOWARD,PH (1993)

CAS #: 065570-20-9				ANISOLE, O-BUTYLAMINO-

Formula: $C_{11}H_{17}NO$

Mol Weight: 179.26

MP (deg C): FP (deg C):

BP (deg C):

BP pressure (mm Hg):

Property/ Value	Units	Temp	Data Type	Reference
WS 2.04E+002	mg/L	25	EST	MEYLAN,WM ET AL. (1996)
logP 2.85			EXP	HANSCH,C & LEO,AJ (1985)
VP 8.39E-003	mm Hg	25	EST	NEELY,WB & BLAU,GE (1985)
DC			pKa	
HL 5.79E-007	atm m3/mol	25	EST	MEYLAN,WM & HOWARD,PH (1991)
OH 5.11E-011	cm3/molc sec	25	EST	MEYLAN,WM & HOWARD,PH (1993)

CAS #: 065570-49-2				1-PIPERIDINEACETAMIDE, N-(3-ETHOXYPHENYL)-

Formula: $C_{15}H_{22}N_2O_2$

Mol Weight: 262.35

MP (deg C): FP (deg C):

BP (deg C):

BP pressure (mm Hg):

Property/ Value	Units	Temp	Data Type	Reference
WS 3.96E+001	mg/L	25	EST	MEYLAN,WM ET AL. (1996)
logP 3.17			EXP	SANGSTER,J (1993)
VP 7.03E-008	mm Hg	25	EST	NEELY,WB & BLAU,GE (1985)
DC			pKa	
HL 4.95E-012	atm m3/mol	25	EST	MEYLAN,WM & HOWARD,PH (1991)
OH 2.41E-010	cm3/molc sec	25	EST	MEYLAN,WM & HOWARD,PH (1993)

CAS #: 065587-38-4				BENZOIC ACID, 4-[3-METHYL-3-(PHENYLMETHYL)-1-TRI

Formula: $C_{15}H_{15}N_3O_2$

Mol Weight: 269.31

MP (deg C): FP (deg C):

BP (deg C):

BP pressure (mm Hg):

Property/ Value	Units	Temp	Data Type	Reference
WS 6.54E+000	mg/L	25	EST	MEYLAN,WM ET AL. (1996)
logP 4.04			EXP	HANSCH,C ET AL. (1995)
VP 1.32E-007	mm Hg	25	EST	NEELY,WB & BLAU,GE (1985)
DC			pKa	
HL 8.72E-013	atm m3/mol	25	EST	MEYLAN,WM & HOWARD,PH (1991)
OH 1.55E-011	cm3/molc sec	25	EST	MEYLAN,WM & HOWARD,PH (1993)

CAS #: 065601-40-3				2,2,4,4-TETRACHLORO-2,2,4,4,6,6-HEXAHYDRO-6,6-B*

Formula: $C_6H_6Cl_4F_8N_3O_2P_3$

Mol Weight: 538.85

MP (deg C): FP (deg C):

BP (deg C):

BP pressure (mm Hg):

Property/ Value	Units	Temp	Data Type	Reference
WS 7.30E-003	mg/L	25	EST	MEYLAN,WM ET AL. (1996)
logP 5.55			EST	MEYLAN,WM & HOWARD,PH (1995)
VP 7.61E-006	mm Hg	25	EST	NEELY,WB & BLAU,GE (1985)
DC			pKa	
HL 6.05E-007	atm m3/mol	25	EST	MEYLAN,WM & HOWARD,PH (1991)
OH 6.90E-013	cm3/molc sec	25	EST	MEYLAN,WM & HOWARD,PH (1993)

CAS #: 065601-41-4				2,2,4-TRICHLORO-2,2,4,4,6,6-HEXAHYDRO-4,6,6-TRI*
Formula: $C_9H_9Cl_3F_{12}N_3O_3P_3$				
Mol Weight: 634.45				
MP (deg C):		FP (deg C):		
BP (deg C):				
BP pressure (mm Hg):				

Property/Value	Units	Temp	Data Type	Reference
WS 2.26E-004	mg/L	25	EST	MEYLAN,WM ET AL. (1996)
logP 6.59			EST	MEYLAN,WM & HOWARD,PH (1995)
VP 2.51E-006	mm Hg	25	EST	NEELY,WB & BLAU,GE (1985)
DC	pKa			
HL 5.05E-006	atm m3/mol	25	EST	MEYLAN,WM & HOWARD,PH (1991)
OH 1.03E-012	cm3/molc sec	25	EST	MEYLAN,WM & HOWARD,PH (1993)

CAS #: 065601-42-5				2,2-DICHLORO-2,2,4,4,6,6-HEXAHYDRO-4,4,6,6-TETR*
Formula: $C_{12}H_{12}Cl_2F_{16}N_3O_4P_3$				
Mol Weight: 730.05				
MP (deg C):		FP (deg C):		
BP (deg C):				
BP pressure (mm Hg):				

Property/Value	Units	Temp	Data Type	Reference
WS 6.82E-006	mg/L	25	EST	MEYLAN,WM ET AL. (1996)
logP 7.62			EST	MEYLAN,WM & HOWARD,PH (1995)
VP 8.17E-007	mm Hg	25	EST	NEELY,WB & BLAU,GE (1985)
DC	pKa			
HL 4.22E-005	atm m3/mol	25	EST	MEYLAN,WM & HOWARD,PH (1991)
OH 1.38E-012	cm3/molc sec	25	EST	MEYLAN,WM & HOWARD,PH (1993)

CAS #: 065611-47-4				6,6,6-TRIFLUORO-1-HEXANOL
Formula: $C_6H_{11}F_3O$				
Mol Weight: 156.15				
MP (deg C):		FP (deg C):		
BP (deg C):				
BP pressure (mm Hg):				

Property/Value	Units	Temp	Data Type	Reference
WS 9.17E+003	mg/L	25	EST	MEYLAN,WM ET AL. (1996)
logP 1.64			EXP	MULLER,N (1986)
VP 1.93E+000	mm Hg	25	EST	NEELY,WB & BLAU,GE (1985)
DC	pKa			
HL 8.75E-005	atm m3/mol	25	EST	MEYLAN,WM & HOWARD,PH (1991)
OH 8.48E-012	cm3/molc sec	25	EST	MEYLAN,WM & HOWARD,PH (1993)

CAS #: 065731-84-2				BETA-CYPERMETHRIN ISOMER
Formula: $C_{22}H_{19}Cl_2NO_3$				
Mol Weight: 416.31				
MP (deg C):		FP (deg C):		
BP (deg C):				
BP pressure (mm Hg):				

Property/Value	Units	Temp	Data Type	Reference
WS 8.80E-003	mg/L	25	EST	MEYLAN,WM ET AL. (1996)
logP 6.05			EXP	SANGSTER,J (1993)
VP 1.36E-007	mm Hg	25	EST	NEELY,WB & BLAU,GE (1985)
DC	pKa			
HL 7.89E-007	atm m3/mol	25	EST	MEYLAN,WM & HOWARD,PH (1991)
OH 2.15E-011	cm3/molc sec	25	EST	MEYLAN,WM & HOWARD,PH (1993)

CAS #: 065732-07-2				BETA-CYPERMETHRIN ISOMER
Formula: $C_{22}H_{19}Cl_2NO_3$				
Mol Weight: 416.31				
MP (deg C):		FP (deg C):		
BP (deg C):				
BP pressure (mm Hg):				

Property/Value	Units	Temp	Data Type	Reference
WS 8.80E-003	mg/L	25	EST	MEYLAN,WM ET AL. (1996)
logP 6.06			EXP	SANGSTER,J (1993)
VP 1.36E-007	mm Hg	25	EST	NEELY,WB & BLAU,GE (1985)
DC	pKa			
HL 7.89E-007	atm m3/mol	25	EST	MEYLAN,WM & HOWARD,PH (1991)
OH 2.15E-011	cm3/molc sec	25	EST	MEYLAN,WM & HOWARD,PH (1993)

CAS #: 065754-05-4				4H-PYRIDO[1,2-A]PYRIMIDIN-4-ONE, 6,7,8,9-TETRAHY
Formula: $C_8H_{10}N_2O$				
Mol Weight: 150.18				
MP (deg C):		FP (deg C):		
BP (deg C):				
BP pressure (mm Hg):				

Property/Value	Units	Temp	Data Type	Reference
WS 4.89E+003	mg/L	25	EST	MEYLAN,WM ET AL. (1996)
logP -0.14			EXP	SANGSTER,J (1994)
VP 5.48E-004	mm Hg	25	EST	NEELY,WB & BLAU,GE (1985)
DC	pKa			
HL 1.34E-009	atm m3/mol	25	EST	MEYLAN,WM & HOWARD,PH (1991)
OH 2.17E-011	cm3/molc sec	25	EST	MEYLAN,WM & HOWARD,PH (1993)

CAS #: 065873-69-0				2,4NH2PYRIMIDIN,5(26MEOPYRIDIN3YL)CH2
Formula: $C_{12}H_{15}N_5O_2$				
Mol Weight: 261.29				
MP (deg C):		FP (deg C):		
BP (deg C):				
BP pressure (mm Hg):				

Property/Value	Units	Temp	Data Type	Reference
WS 3.93E+002	mg/L	25	EST	MEYLAN,WM ET AL. (1996)
logP 2.01			EXP	HANSCH,C & LEO,AJ (1985)
VP 3.50E-008	mm Hg	25	EST	NEELY,WB & BLAU,GE (1985)
DC	pKa			
HL 4.39E-014	atm m3/mol	25	EST	MEYLAN,WM & HOWARD,PH (1991)
OH 5.90E-011	cm3/molc sec	25	EST	MEYLAN,WM & HOWARD,PH (1993)

CAS #: 065886-71-7				1,3,5-TRIAZIN-2(1H)-ONE, 4-AMINO-1-(beta-D-ARABI
Formula: $C_8H_{12}N_4O_5$				
Mol Weight: 244.21				
MP (deg C): 220 dec		FP (deg C):		
BP (deg C):				
BP pressure (mm Hg):				

Property/Value	Units	Temp	Data Type	Reference
WS 8.89E+004	mg/L	25	EST	MEYLAN,WM ET AL. (1996)
logP -2.17			EXP	SANGSTER,J (1994)
VP 8.66E-012	mm Hg	25	EST	NEELY,WB & BLAU,GE (1985)
DC	pKa			
HL 3.76E-021	atm m3/mol	25	EST	MEYLAN,WM & HOWARD,PH (1991)
OH 1.11E-010	cm3/molc sec	25	EST	MEYLAN,WM & HOWARD,PH (1993)

CAS #: 065902-59-2	2-BROMO-4-NITRO-1H-IMIDAZOLE

Formula: $C_3H_2BrN_3O_2$

Mol Weight: 191.97

MP (deg C): | FP (deg C):

BP (deg C):

BP pressure (mm Hg):

Property/Value	Units	Temp	Data Type	Reference
WS 4.85E+003	mg/L	25	EST	MEYLAN,WM ET AL. (1996)
logP 0.71			EXP	SUWINSKI,J ET AL. (1985)
VP 1.31E-005	mm Hg	25	EST	NEELY,WB & BLAU,GE (1985)
DC	pKa			
HL 5.90E-009	atm m3/mol	25	EST	MEYLAN,WM & HOWARD,PH (1991)
OH 9.02E-013	cm3/molc sec	25	EST	MEYLAN,WM & HOWARD,PH (1993)

CAS #: 065907-26-8	N(THIO-N,O-DIME CARBAMYL)CARBOFURAN

Formula: $C_{15}H_{20}N_2O_5S$

Mol Weight: 340.40

MP (deg C): | FP (deg C):

BP (deg C):

BP pressure (mm Hg):

Property/Value	Units	Temp	Data Type	Reference
WS 1.94E+001	mg/L	25	EST	MEYLAN,WM ET AL. (1996)
logP 3.00			EXP	HANSCH,C & LEO,AJ (1985)
VP 1.09E-006	mm Hg	25	EST	NEELY,WB & BLAU,GE (1985)
DC	pKa			
HL 3.39E-009	atm m3/mol	25	EST	MEYLAN,WM & HOWARD,PH (1991)
OH 5.41E-011	cm3/molc sec	25	EST	MEYLAN,WM & HOWARD,PH (1993)

CAS #: 065907-27-9	CARBOFURAN,N-(N'ME-O-ET-CARBAMYLTHIO)

Formula: $C_{16}H_{22}N_2O_5S$

Mol Weight: 354.43

MP (deg C): | FP (deg C):

BP (deg C):

BP pressure (mm Hg):

Property/Value	Units	Temp	Data Type	Reference
WS 5.98E+000	mg/L	25	EST	MEYLAN,WM ET AL. (1996)
logP 3.50			EXP	HANSCH,C & LEO,AJ (1985)
VP 4.62E-007	mm Hg	25	EST	NEELY,WB & BLAU,GE (1985)
DC	pKa			
HL 4.49E-009	atm m3/mol	25	EST	MEYLAN,WM & HOWARD,PH (1991)
OH 5.56E-011	cm3/molc sec	25	EST	MEYLAN,WM & HOWARD,PH (1993)

CAS #: 065907-30-4	N-THIO-(N'-ME-O-BU-CARBAMYL)CARBOFURAN

Formula: $C_{18}H_{26}N_2O_5S$

Mol Weight: 382.48

MP (deg C): | FP (deg C):

BP (deg C):

BP pressure (mm Hg):

Property/Value	Units	Temp	Data Type	Reference
WS 1.00E+001	mg/L	20	EXP	SHIU,WY ET AL. (1990)
logP 4.70			EXP	HANSCH,C & LEO,AJ (1985)
VP 8.17E-008	mm Hg	25	EST	NEELY,WB & BLAU,GE (1985)
DC	pKa			
HL 7.92E-009	atm m3/mol	25	EST	MEYLAN,WM & HOWARD,PH (1991)
OH 5.85E-011	cm3/molc sec	25	EST	MEYLAN,WM & HOWARD,PH (1993)

CAS #: 065907-54-2	6-OXA-3-THIA-2,4,7,10-TETRAAZAUNDEC-7-ENOIC ACID

Formula: $C_{10}H_{18}N_4O_5S_2$

Mol Weight: 338.41

MP (deg C): | FP (deg C):

BP (deg C):

BP pressure (mm Hg):

Property/Value	Units	Temp	Data Type	Reference
WS 4.09E+002	mg/L	25	EST	MEYLAN,WM ET AL. (1996)
logP -0.07			EXP	SANGSTER,J (1993)
VP 4.58E-007	mm Hg	25	EST	NEELY,WB & BLAU,GE (1985)
DC	pKa			
HL 4.62E-014	atm m3/mol	25	EST	MEYLAN,WM & HOWARD,PH (1991)
OH 5.02E-011	cm3/molc sec	25	EST	MEYLAN,WM & HOWARD,PH (1993)

CAS #: 065936-23-4	2-IMIDAZOLIDINIMINE, N-(4-BROMO-2,6-DIMETHYLPHEN

Formula: $C_{11}H_{14}BrN_3$

Mol Weight: 268.16

MP (deg C): | FP (deg C):

BP (deg C):

BP pressure (mm Hg):

Property/Value	Units	Temp	Data Type	Reference
WS 7.75E+002	mg/L	25	EST	MEYLAN,WM ET AL. (1996)
logP 2.80			EXP	SANGSTER,J (1993)
VP 6.35E-006	mm Hg	25	EST	NEELY,WB & BLAU,GE (1985)
DC	pKa			
HL 1.33E-011	atm m3/mol	25	EST	MEYLAN,WM & HOWARD,PH (1991)
OH 1.53E-010	cm3/molc sec	25	EST	MEYLAN,WM & HOWARD,PH (1993)

CAS #: 065936-24-5	2(2,6-DICL-4MEOPHIMINO)IMIDAZOLIDINE

Formula: $C_{10}H_{11}Cl_2N_3O$

Mol Weight: 260.12

MP (deg C): | FP (deg C):

BP (deg C):

BP pressure (mm Hg):

Property/Value	Units	Temp	Data Type	Reference
WS 8.94E+002	mg/L	25	EST	MEYLAN,WM ET AL. (1996)
logP 1.60			EXP	HANSCH,C & LEO,AJ (1985)
VP 4.27E-008	mm Hg	25	EST	NEELY,WB & BLAU,GE (1985)
DC	pKa			
HL 8.91E-013	atm m3/mol	25	EST	MEYLAN,WM & HOWARD,PH (1991)
OH 9.07E-011	cm3/molc sec	25	EST	MEYLAN,WM & HOWARD,PH (1993)

CAS #: 065936-25-6	2-IMIDAZOLIDINIMINE, N-(2-BROMO-6-CHLOROPHENYL)-

Formula: $C_9H_9BrClN_3$

Mol Weight: 274.55

MP (deg C): | FP (deg C):

BP (deg C):

BP pressure (mm Hg):

Property/Value	Units	Temp	Data Type	Reference
WS 5.84E+003	mg/L	25	EST	MEYLAN,WM ET AL. (1996)
logP 1.73			EXP	SANGSTER,J (1993)
VP 8.80E-006	mm Hg	25	EST	NEELY,WB & BLAU,GE (1985)
DC	pKa			
HL 8.09E-012	atm m3/mol	25	EST	MEYLAN,WM & HOWARD,PH (1991)
OH 1.44E-010	cm3/molc sec	25	EST	MEYLAN,WM & HOWARD,PH (1993)

1269

CAS #: 065936-26-7	2-(26-DICL-4-NO2-PHIMINO)IMIDAZOLIDN
Formula: $C_9H_8Cl_2N_4O_2$	
Mol Weight: 275.10	
MP (deg C):	FP (deg C):
BP (deg C):	
BP pressure (mm Hg):	

Property/Value	Units	Temp	Data Type	Reference
WS 2.80E+003	mg/L	25	EST	MEYLAN,WM ET AL. (1996)
logP 2.10			EXP	HANSCH,C & LEO,AJ (1985)
VP 3.55E-007	mm Hg	25	EST	NEELY,WB & BLAU,GE (1985)
DC	pKa			
HL 5.94E-014	atm m3/mol	25	EST	MEYLAN,WM & HOWARD,PH (1991)
OH 1.43E-010	cm3/molc sec	25	EST	MEYLAN,WM & HOWARD,PH (1993)

CAS #: 065980-97-4	HYDROCORTISONE-17-PROPIONATE
Formula: $C_{24}H_{34}O_6$	
Mol Weight: 418.53	
MP (deg C):	FP (deg C):
BP (deg C):	
BP pressure (mm Hg):	

Property/Value	Units	Temp	Data Type	Reference
WS 1.16E+001	mg/L	25	EST	MEYLAN,WM ET AL. (1996)
logP 2.70			EXP	HANSCH,C & LEO,AJ (1985)
VP 4.24E-014	mm Hg	25	EST	NEELY,WB & BLAU,GE (1985)
DC	pKa			
HL 8.28E-012	atm m3/mol	25	EST	MEYLAN,WM & HOWARD,PH (1991)
OH 1.11E-010	cm3/molc sec	25	EST	MEYLAN,WM & HOWARD,PH (1993)

CAS #: 065988-94-5	1-(2-OH-3-ETOPR)-2-NO2 IMIDAZOLE
Formula: $C_8H_{15}N_3O_4$	
Mol Weight: 217.23	
MP (deg C):	FP (deg C):
BP (deg C):	
BP pressure (mm Hg):	

Property/Value	Units	Temp	Data Type	Reference
WS 1.37E+004	mg/L	25	EST	MEYLAN,WM ET AL. (1996)
logP 0.04			EXP	HANSCH,C & LEO,AJ (1985)
VP 4.09E-008	mm Hg	25	EST	NEELY,WB & BLAU,GE (1985)
DC	pKa			
HL 3.17E-013	atm m3/mol	25	EST	MEYLAN,WM & HOWARD,PH (1991)
OH 2.92E-011	cm3/molc sec	25	EST	MEYLAN,WM & HOWARD,PH (1993)

CAS #: 066017-91-2	N-NITROSO-ACETOXYMETHYL-N-PROPYLAMINE
Formula: $C_6H_{12}N_2O_3$	
Mol Weight: 160.17	
MP (deg C):	FP (deg C):
BP (deg C):	
BP pressure (mm Hg):	

Property/Value	Units	Temp	Data Type	Reference
WS 8.95E+003	mg/L	25	EST	MEYLAN,WM ET AL. (1996)
logP 1.03			EST	MEYLAN,WM & HOWARD,PH (1995)
VP 9.26E-003	mm Hg	25	EST	NEELY,WB & BLAU,GE (1985)
DC	pKa			
HL 1.72E-007	atm m3/mol	25	EST	MEYLAN,WM & HOWARD,PH (1991)
OH 2.59E-011	cm3/molc sec	25	EST	MEYLAN,WM & HOWARD,PH (1993)

CAS #: 066063-05-6	PENCYCURON
Formula: $C_{19}H_{21}ClN_2O$	
Mol Weight: 328.85	
MP (deg C):	FP (deg C):
BP (deg C):	
BP pressure (mm Hg):	

Property/Value	Units	Temp	Data Type	Reference
WS 4.00E-001	mg/L	20	EXP	SHIU,WY ET AL. (1990)
logP 4.82			EXP	HANSCH,C ET AL. (1995)
VP 2.54E-009	mm Hg	25	EST	NEELY,WB & BLAU,GE (1985)
DC	pKa			
HL 7.95E-011	atm m3/mol	25	EST	MEYLAN,WM & HOWARD,PH (1991)
OH 6.33E-011	cm3/molc sec	25	EST	MEYLAN,WM & HOWARD,PH (1993)

CAS #: 066073-53-8	2-CHLORO-6-BROMOBENZAMIDE
Formula: $C_7H_5BrClNO$	
Mol Weight: 234.49	
MP (deg C):	FP (deg C):
BP (deg C):	
BP pressure (mm Hg):	

Property/Value	Units	Temp	Data Type	Reference
WS 4.03E+003	mg/L	25	EST	MEYLAN,WM ET AL. (1996)
logP 1.00			EXP	NAKAGAWA,Y ET AL. (1992)
VP 1.51E-005	mm Hg	25	EST	NEELY,WB & BLAU,GE (1985)
DC	pKa			
HL 6.54E-010	atm m3/mol	25	EST	MEYLAN,WM & HOWARD,PH (1991)
OH 2.79E-012	cm3/molc sec	25	EST	MEYLAN,WM & HOWARD,PH (1993)

CAS #: 066073-54-9	2-FLUORO-6-CHLOROBENZAMIDE
Formula: C_7H_5ClFNO	
Mol Weight: 173.58	
MP (deg C):	FP (deg C):
BP (deg C):	
BP pressure (mm Hg):	

Property/Value	Units	Temp	Data Type	Reference
WS 2.26E+004	mg/L	25	EST	MEYLAN,WM ET AL. (1996)
logP 0.49			EXP	NAKAGAWA,Y ET AL. (1992)
VP 1.87E-004	mm Hg	25	EST	NEELY,WB & BLAU,GE (1985)
DC	pKa			
HL 1.91E-009	atm m3/mol	25	EST	MEYLAN,WM & HOWARD,PH (1991)
OH 3.57E-012	cm3/molc sec	25	EST	MEYLAN,WM & HOWARD,PH (1993)

CAS #: 066079-85-4	2(2-PYRIDYL)-5-ME-1,3,4-OXADIAZOLE
Formula: $C_8H_8N_3O$	
Mol Weight: 162.17	
MP (deg C):	FP (deg C):
BP (deg C):	
BP pressure (mm Hg):	

Property/Value	Units	Temp	Data Type	Reference
WS 2.74E+004	mg/L	25	EST	MEYLAN,WM ET AL. (1996)
logP 0.46			EXP	HANSCH,C & LEO,AJ (1985)
VP 1.26E-004	mm Hg	25	EST	NEELY,WB & BLAU,GE (1985)
DC	pKa			
HL 3.18E-010	atm m3/mol	25	EST	MEYLAN,WM & HOWARD,PH (1991)
OH 4.78E-012	cm3/molc sec	25	EST	MEYLAN,WM & HOWARD,PH (1993)

CAS #: 066108-93-8				M-BENZAMIDE,NNBIS(23DIOHPR),5N....
Formula: $C_{18}H_{24}I_3N_3O_8$				
Mol Weight: 791.12				
MP (deg C):			FP (deg C):	
BP (deg C):				
BP pressure (mm Hg):				

Property/ Value	Units	Temp	Data Type	Reference
WS 5.42E+001	mg/L	25	EST	MEYLAN,WM ET AL. (1996)
logP -2.47			EXP	HANSCH,C & LEO,AJ (1985)
VP 1.15E-029	mm Hg	25	EST	NEELY,WB & BLAU,GE (1985)
DC	pKa			
HL 5.48E-028	atm m3/mol	25	EST	MEYLAN,WM & HOWARD,PH (1991)
OH 5.88E-011	cm3/molc sec	25	EST	MEYLAN,WM & HOWARD,PH (1993)

CAS #: 066108-95-0				IOHEXOL
Formula: $C_{19}H_{26}I_3N_3O_9$				
Mol Weight: 821.15				
MP (deg C): 174-180			FP (deg C):	
BP (deg C):				
BP pressure (mm Hg):				

Property/ Value	Units	Temp	Data Type	Reference
WS 1.07E+002	mg/L	25	EST	MEYLAN,WM ET AL. (1996)
logP -3.05			EXP	HANSCH,C & LEO,AJ (1985)
VP 2.77E-031	mm Hg	25	EST	NEELY,WB & BLAU,GE (1985)
DC	pKa			
HL 2.66E-029	atm m3/mol	25	EST	MEYLAN,WM & HOWARD,PH (1991)
OH 6.92E-011	cm3/molc sec	25	EST	MEYLAN,WM & HOWARD,PH (1993)

CAS #: 066121-41-3				DIS. A. 18
Formula: $C_{14}H_6Cl_2N_2O_4$				
Mol Weight: 337.12				
MP (deg C):			FP (deg C):	
BP (deg C):				
BP pressure (mm Hg):				

Property/ Value	Units	Temp	Data Type	Reference
WS 2.04E-002	mg/L	25	EST	MEYLAN,WM ET AL. (1996)
logP 5.21			EST	MEYLAN,WM & HOWARD,PH (1995)
VP 4.00E-013	mm Hg	25	EXP	BAUGHMAN,GL & PERENICH,TA (1988)
DC	pKa			
HL 2.43E-015	atm m3/mol	25	EST	MEYLAN,WM & HOWARD,PH (1991)
OH 5.32E-012	cm3/molc sec	25	EST	MEYLAN,WM & HOWARD,PH (1993)

CAS #: 066203-00-7				UREA, N-[4,7-DIMETHOXY-6-[2-(1-PYRROLIDINYL)ETHO
Formula: $C_{18}H_{25}N_3O_5$				
Mol Weight: 363.42				
MP (deg C):			FP (deg C):	
BP (deg C):				
BP pressure (mm Hg):				

Property/ Value	Units	Temp	Data Type	Reference
WS 3.41E+002	mg/L	25	EST	MEYLAN,WM ET AL. (1996)
logP 1.38			EXP	MANNHOLD,R ET AL. (1990)
VP 1.73E-010	mm Hg	25	EST	NEELY,WB & BLAU,GE (1985)
DC	pKa			
HL 4.56E-018	atm m3/mol	25	EST	MEYLAN,WM & HOWARD,PH (1991)
OH 3.26E-010	cm3/molc sec	25	EST	MEYLAN,WM & HOWARD,PH (1993)

CAS #: 066224-65-5				9H-PURIN-6-AMINE
Formula: $C_5H_5N_5$				
Mol Weight: 135.13				
MP (deg C):			FP (deg C):	
BP (deg C):				
BP pressure (mm Hg):				

Property/ Value	Units	Temp	Data Type	Reference
WS 4.74E+003	mg/L	25	EST	MEYLAN,WM ET AL. (1996)
logP -0.05			EXP	LAM,SP ET AL. (1989)
VP 8.38E-007	mm Hg	25	EST	NEELY,WB & BLAU,GE (1985)
DC 4.15	pKa	25	EXP	KORTUM,G ET AL (1961)
HL 7.02E-014	atm m3/mol	25	EST	MEYLAN,WM & HOWARD,PH (1991)
OH 2.00E-010	cm3/molc sec	25	EST	MEYLAN,WM & HOWARD,PH (1993)

CAS #: 066230-04-4				ESFENVALERATE
Formula: $C_{25}H_{22}ClNO_3$				
Mol Weight: 419.91				
MP (deg C): 59-60.2			FP (deg C):	
BP (deg C):				
BP pressure (mm Hg):				

Property/ Value	Units	Temp	Data Type	Reference
WS 1.00E+000	mg/L	20	EST	WORTHING,CR & WALKER,SB (1987)
logP 6.76			EST	MEYLAN,WM & HOWARD,PH (1995)
VP 5.00E-007	mm Hg	25	EXP	IARC (1991)
DC	pKa			
HL 1.19E-007	atm m3/mol	25	EST	MEYLAN,WM & HOWARD,PH (1991)
OH 3.76E-011	cm3/molc sec	25	EST	ATKINSON,R (1988)

CAS #: 066287-52-3				BENZOIC ACID, 3-NITRO-2-[[3-(TRIFLUOROMETHYL)PHE
Formula: $C_{14}H_7F_3NNaO_4$				
Mol Weight: 333.20				
MP (deg C):			FP (deg C):	
BP (deg C):				
BP pressure (mm Hg):				

Property/ Value	Units	Temp	Data Type	Reference
WS 1.02E+003	mg/L	25	EST	MEYLAN,WM ET AL. (1996)
logP 0.58			EXP	SANGSTER,J (1993)
VP 3.95E-014	mm Hg	25	EST	NEELY,WB & BLAU,GE (1985)
DC	pKa			
HL	atm m3/mol			
OH 7.64E-013	cm3/molc sec	25	EST	MEYLAN,WM & HOWARD,PH (1993)

CAS #: 066287-53-4				BENZOIC ACID, 4-CHLORO-2-[[3-(TRIFLUOROMETHYL)PH
Formula: $C_{14}H_8ClF_3NNaO_2$				
Mol Weight: 337.66				
MP (deg C):			FP (deg C):	
BP (deg C):				
BP pressure (mm Hg):				

Property/ Value	Units	Temp	Data Type	Reference
WS 8.14E+001	mg/L	25	EST	MEYLAN,WM ET AL. (1996)
logP 2.29			EXP	SANGSTER,J (1993)
VP 3.42E-013	mm Hg	25	EST	NEELY,WB & BLAU,GE (1985)
DC	pKa			
HL	atm m3/mol			
OH 1.16E-010	cm3/molc sec	25	EST	MEYLAN,WM & HOWARD,PH (1993)

CAS #: 066308-18-7	CARBAMIC ACID, (4,5-DIHYDRO-1-METHYL-5-PHENYL-1H

Formula: $C_{12}H_{15}N_3O_2$
Mol Weight: 233.27
MP (deg C):   FP (deg C):
BP (deg C):
BP pressure (mm Hg):

Property/Value	Units	Temp	Data Type	Reference
WS 1.33E+003	mg/L	25	EST	MEYLAN,WM ET AL. (1996)
logP 1.57			EXP	SANGSTER,J (1994)
VP 2.19E-005	mm Hg	25	EST	NEELY,WB & BLAU,GE (1985)
DC	pKa			
HL 1.01E-013	atm m3/mol	25	EST	MEYLAN,WM & HOWARD,PH (1991)
OH 9.63E-011	cm3/molc sec	25	EST	MEYLAN,WM & HOWARD,PH (1993)

CAS #: 066308-26-7	CARBAMIC ACID, [5-(2-FLUOROPHENYL)-4,5-DIHYDRO-1

Formula: $C_{12}H_{14}FN_3O_2$
Mol Weight: 251.26
MP (deg C):   FP (deg C):
BP (deg C):
BP pressure (mm Hg):

Property/Value	Units	Temp	Data Type	Reference
WS 1.15E+003	mg/L	25	EST	MEYLAN,WM ET AL. (1996)
logP 1.53			EXP	SANGSTER,J (1994)
VP 2.99E-005	mm Hg	25	EST	NEELY,WB & BLAU,GE (1985)
DC	pKa			
HL 1.18E-013	atm m3/mol	25	EST	MEYLAN,WM & HOWARD,PH (1991)
OH 9.46E-011	cm3/molc sec	25	EST	MEYLAN,WM & HOWARD,PH (1993)

CAS #: 066308-27-8	CARBAMIC ACID, [5-(2-CHLOROPHENYL)-4,5-DIHYDRO-1

Formula: $C_{12}H_{14}ClN_3O_2$
Mol Weight: 267.72
MP (deg C):   FP (deg C):
BP (deg C):
BP pressure (mm Hg):

Property/Value	Units	Temp	Data Type	Reference
WS 3.15E+002	mg/L	25	EST	MEYLAN,WM ET AL. (1996)
logP 2.08			EXP	SANGSTER,J (1994)
VP 6.19E-006	mm Hg	25	EST	NEELY,WB & BLAU,GE (1985)
DC	pKa			
HL 7.49E-014	atm m3/mol	25	EST	MEYLAN,WM & HOWARD,PH (1991)
OH 9.33E-011	cm3/molc sec	25	EST	MEYLAN,WM & HOWARD,PH (1993)

CAS #: 066308-28-9	CARBAMIC ACID, [5-(4-CHLOROPHENYL)-4,5-DIHYDRO-1

Formula: $C_{12}H_{14}ClN_3O_2$
Mol Weight: 267.72
MP (deg C):   FP (deg C):
BP (deg C):
BP pressure (mm Hg):

Property/Value	Units	Temp	Data Type	Reference
WS 2.64E+002	mg/L	25	EST	MEYLAN,WM ET AL. (1996)
logP 2.17			EXP	SANGSTER,J (1994)
VP 6.19E-006	mm Hg	25	EST	NEELY,WB & BLAU,GE (1985)
DC	pKa			
HL 7.49E-014	atm m3/mol	25	EST	MEYLAN,WM & HOWARD,PH (1991)
OH 9.33E-011	cm3/molc sec	25	EST	MEYLAN,WM & HOWARD,PH (1993)

CAS #: 066308-35-8	CARBAMIC ACID, [5-(2,3-DIMETHYLPHENYL)-4,5-DIHYD

Formula: $C_{15}H_{21}N_3O_2$
Mol Weight: 275.35
MP (deg C):   FP (deg C):
BP (deg C):
BP pressure (mm Hg):

Property/Value	Units	Temp	Data Type	Reference
WS 1.09E+002	mg/L	25	EST	MEYLAN,WM ET AL. (1996)
logP 2.57			EXP	SANGSTER,J (1994)
VP 1.93E-006	mm Hg	25	EST	NEELY,WB & BLAU,GE (1985)
DC	pKa			
HL 1.63E-013	atm m3/mol	25	EST	MEYLAN,WM & HOWARD,PH (1991)
OH 1.08E-010	cm3/molc sec	25	EST	MEYLAN,WM & HOWARD,PH (1993)

CAS #: 066323-44-2	ADENOSINE, 3'-AZIDO-2',3'-DIDEOXY-

Formula: $C_{10}H_{12}N_8O_2$
Mol Weight: 276.26
MP (deg C):   FP (deg C):
BP (deg C):
BP pressure (mm Hg):

Property/Value	Units	Temp	Data Type	Reference
WS 1.53E+002	mg/L	25	EST	MEYLAN,WM ET AL. (1996)
logP 0.35			EXP	BALZARINI,J ET AL. (1989)
VP 5.82E-019	mm Hg	25	EST	NEELY,WB & BLAU,GE (1985)
DC	pKa			
HL 1.83E-027	atm m3/mol	25	EST	MEYLAN,WM & HOWARD,PH (1991)
OH 2.34E-010	cm3/molc sec	25	EST	MEYLAN,WM & HOWARD,PH (1993)

CAS #: 066346-83-6	3-CL-6-PYRIDAZINECARBOXAMIDE

Formula: $C_5H_4ClN_3O$
Mol Weight: 157.56
MP (deg C):   FP (deg C):
BP (deg C):
BP pressure (mm Hg):

Property/Value	Units	Temp	Data Type	Reference
WS 3.44E+003	mg/L	25	EST	MEYLAN,WM ET AL. (1996)
logP 0.00			EXP	HANSCH,C & LEO,AJ (1985)
VP 3.46E-006	mm Hg	25	EST	NEELY,WB & BLAU,GE (1985)
DC	pKa			
HL 5.51E-011	atm m3/mol	25	EST	MEYLAN,WM & HOWARD,PH (1991)
OH 2.15E-012	cm3/molc sec	25	EST	MEYLAN,WM & HOWARD,PH (1993)

CAS #: 066357-35-5	ZANTAC (RANITIDINE HYDROCHLORIDE)

Formula: $C_{13}H_{22}N_4O_3S$
Mol Weight: 314.41
MP (deg C): 69-70   FP (deg C):
BP (deg C):
BP pressure (mm Hg):

Property/Value	Units	Temp	Data Type	Reference
WS 2.47E+004	mg/L	25	EST	MEYLAN,WM ET AL. (1996)
logP 0.27			EXP	SANGSTER,J (1993)
VP 1.20E-007	mm Hg	25	EST	NEELY,WB & BLAU,GE (1985)
DC	pKa			
HL 3.42E-015	atm m3/mol	25	EST	MEYLAN,WM & HOWARD,PH (1991)
OH 3.62E-010	cm3/molc sec	25	EST	MEYLAN,WM & HOWARD,PH (1993)

CAS #: 066381-49-5 — N-(N''-ME-HXOCARBAMYLTHIO)METHOMYL

Formula: $C_{13}H_{25}N_3O_4S_2$

Mol Weight: 351.49

MP (deg C): FP (deg C):

BP (deg C):

BP pressure (mm Hg):

Property/Value	Units	Temp	Data Type	Reference
WS 1.57E+000	mg/L	25	EST	MEYLAN,WM ET AL. (1996)
logP 4.20			EXP	HANSCH,C & LEO,AJ (1985)
VP 8.19E-006	mm Hg	25	EST	NEELY,WB & BLAU,GE (1985)
DC	pKa			
HL 1.73E-008	atm m3/mol	25	EST	MEYLAN,WM & HOWARD,PH (1991)
OH 4.14E-011	cm3/molc sec	25	EST	MEYLAN,WM & HOWARD,PH (1993)

CAS #: 066381-51-9 — N-(N''-BU-BUO-CARBAMYLTHIO)METHOMYL

Formula: $C_{14}H_{27}N_3O_4S_2$

Mol Weight: 365.52

MP (deg C): FP (deg C):

BP (deg C):

BP pressure (mm Hg):

Property/Value	Units	Temp	Data Type	Reference
WS 1.29E+000	mg/L	25	EST	MEYLAN,WM ET AL. (1996)
logP 4.20			EXP	HANSCH,C & LEO,AJ (1985)
VP 3.39E-006	mm Hg	25	EST	NEELY,WB & BLAU,GE (1985)
DC	pKa			
HL 2.30E-008	atm m3/mol	25	EST	MEYLAN,WM & HOWARD,PH (1991)
OH 4.55E-011	cm3/molc sec	25	EST	MEYLAN,WM & HOWARD,PH (1993)

CAS #: 066381-55-3 — N(N''-BU-MEO CARBAMYLTHIO)METHOMYL

Formula: $C_{11}H_{21}N_3O_4S_2$

Mol Weight: 323.44

MP (deg C): FP (deg C):

BP (deg C):

BP pressure (mm Hg):

Property/Value	Units	Temp	Data Type	Reference
WS 6.55E+001	mg/L	25	EST	MEYLAN,WM ET AL. (1996)
logP 2.50			EXP	HANSCH,C & LEO,AJ (1985)
VP 4.63E-005	mm Hg	25	EST	NEELY,WB & BLAU,GE (1985)
DC	pKa			
HL 9.83E-009	atm m3/mol	25	EST	MEYLAN,WM & HOWARD,PH (1991)
OH 4.11E-011	cm3/molc sec	25	EST	MEYLAN,WM & HOWARD,PH (1993)

CAS #: 066382-00-1 — 2-S-BUTYLBENZOTRIAZOLE

Formula: $C_{10}H_{15}N_3$

Mol Weight: 177.25

MP (deg C): FP (deg C):

BP (deg C):

BP pressure (mm Hg):

Property/Value	Units	Temp	Data Type	Reference
WS 1.76E+002	mg/L	25	EST	MEYLAN,WM ET AL. (1996)
logP 2.95			EXP	HANSCH,C & LEO,AJ (1985)
VP 8.16E-004	mm Hg	25	EST	NEELY,WB & BLAU,GE (1985)
DC	pKa			
HL 7.32E-006	atm m3/mol	25	EST	MEYLAN,WM & HOWARD,PH (1991)
OH 4.87E-012	cm3/molc sec	25	EST	MEYLAN,WM & HOWARD,PH (1993)

CAS #: 066395-17-3 — á-D-GLUCOPYRANOSE, 2-DEOXY-2-[[(METHYLNITROSOAMI

Formula: $C_8H_{15}N_3O_7$

Mol Weight: 265.22

MP (deg C): FP (deg C):

BP (deg C):

BP pressure (mm Hg):

Property/Value	Units	Temp	Data Type	Reference
WS 5.07E+003	mg/L	25	EST	MEYLAN,WM ET AL. (1996)
logP -0.85			EXP	SOSROVSKY,G & RAO,NUM (1991)
VP 1.74E-012	mm Hg	25	EST	NEELY,WB & BLAU,GE (1985)
DC	pKa			
HL 7.85E-022	atm m3/mol	25	EST	MEYLAN,WM & HOWARD,PH (1991)
OH 1.08E-010	cm3/molc sec	25	EST	MEYLAN,WM & HOWARD,PH (1993)

CAS #: 066432-03-9 — N-FORMYL-STYRYLAMINE

Formula: C_9H_9NO

Mol Weight: 147.18

MP (deg C): FP (deg C):

BP (deg C):

BP pressure (mm Hg):

Property/Value	Units	Temp	Data Type	Reference
WS 1.69E+003	mg/L	25	EST	MEYLAN,WM ET AL. (1996)
logP 1.95			EXP	HANSCH,C & LEO,AJ (1985)
VP 1.30E-004	mm Hg	25	EST	NEELY,WB & BLAU,GE (1985)
DC	pKa			
HL 2.64E-009	atm m3/mol	25	EST	MEYLAN,WM & HOWARD,PH (1991)
OH 4.12E-011	cm3/molc sec	25	EST	MEYLAN,WM & HOWARD,PH (1993)

CAS #: 066432-08-4 — N-FORMYL-P-ACETOXYSTYRLAMINE

Formula: $C_{11}H_{11}NO_3$

Mol Weight: 205.22

MP (deg C): FP (deg C):

BP (deg C):

BP pressure (mm Hg):

Property/Value	Units	Temp	Data Type	Reference
WS 3.07E+003	mg/L	25	EST	MEYLAN,WM ET AL. (1996)
logP 1.32			EXP	HANSCH,C & LEO,AJ (1985)
VP 2.85E-006	mm Hg	25	EST	NEELY,WB & BLAU,GE (1985)
DC	pKa			
HL 3.17E-011	atm m3/mol	25	EST	MEYLAN,WM & HOWARD,PH (1991)
OH 4.13E-011	cm3/molc sec	25	EST	MEYLAN,WM & HOWARD,PH (1993)

CAS #: 066451-06-7 — 2-PROPANOL, 1-(2-BICYCLO[2.2.1]HEPT-2-YLPHENOXY)

Formula: $C_{19}H_{29}NO_2$

Mol Weight: 303.45

MP (deg C): FP (deg C):

BP (deg C):

BP pressure (mm Hg):

Property/Value	Units	Temp	Data Type	Reference
WS 3.28E+001	mg/L	25	EST	MEYLAN,WM ET AL. (1996)
logP 4.17			EXP	RECANATINI,M (1992)
VP 1.42E-008	mm Hg	25	EST	NEELY,WB & BLAU,GE (1985)
DC	pKa			
HL 9.60E-012	atm m3/mol	25	EST	MEYLAN,WM & HOWARD,PH (1991)
OH 1.45E-010	cm3/molc sec	25	EST	MEYLAN,WM & HOWARD,PH (1993)

066469-80-5 — N(N''-ME-CYPR CARBAMYLTHIO)METHOMYL

Formula: $C_{10}H_{17}N_3O_3S_2$
Mol Weight: 291.39
MP (deg C):   **FP (deg C):**
BP (deg C):
BP pressure (mm Hg):

Property/Value	Units	Temp	Data Type	Reference
WS 4.87E+002	mg/L	25	EST	MEYLAN,WM ET AL. (1996)
logP 1.70			EXP	HANSCH,C & LEO,AJ (1985)
VP 8.10E-007	mm Hg	25	EST	NEELY,WB & BLAU,GE (1985)
DC	pKa			
HL 9.23E-010	atm m3/mol	25	EST	MEYLAN,WM & HOWARD,PH (1991)
OH 3.43E-011	cm3/molc sec	25	EST	MEYLAN,WM & HOWARD,PH (1993)

066473-27-6 — NNN-TRIME-N-(5CLME-2-FURANME) I

Formula: $C_9H_{15}ClINO$
Mol Weight: 315.58
MP (deg C):   **FP (deg C):**
BP (deg C):
BP pressure (mm Hg):

Property/Value	Units	Temp	Data Type	Reference
WS 5.08E+005	mg/L	25	EST	MEYLAN,WM ET AL. (1996)
logP -2.00			EXP	PRATESI,P ET AL. (1986)
VP 2.36E-009	mm Hg	25	EST	NEELY,WB & BLAU,GE (1985)
DC	pKa			
HL 8.25E-014	atm m3/mol	25	EST	MEYLAN,WM & HOWARD,PH (1991)
OH 7.82E-011	cm3/molc sec	25	EST	MEYLAN,WM & HOWARD,PH (1993)

066480-64-6 — 1H-BENZOTRIAZOLE, 1-CYCLOPENTYL-

Formula: $C_{11}H_{13}N_3$
Mol Weight: 187.25
MP (deg C):   **FP (deg C):**
BP (deg C):
BP pressure (mm Hg):

Property/Value	Units	Temp	Data Type	Reference
WS 2.99E+002	mg/L	25	EST	MEYLAN,WM ET AL. (1996)
logP 2.61			EXP	SPARATORE,F ET AL. (1988)
VP 1.09E-004	mm Hg	25	EST	NEELY,WB & BLAU,GE (1985)
DC	pKa			
HL 4.28E-006	atm m3/mol	25	EST	MEYLAN,WM & HOWARD,PH (1991)
OH 7.87E-012	cm3/molc sec	25	EST	MEYLAN,WM & HOWARD,PH (1993)

066480-65-7 — 2H-BENZOTRIAZOLE, 2-CYCLOHEXYL-

Formula: $C_{12}H_{15}N_3$
Mol Weight: 201.27
MP (deg C):   **FP (deg C):**
BP (deg C):
BP pressure (mm Hg):

Property/Value	Units	Temp	Data Type	Reference
WS 3.42E+001	mg/L	25	EST	MEYLAN,WM ET AL. (1996)
logP 3.63			EXP	SPARATORE,F ET AL. (1988)
VP 4.17E-005	mm Hg	25	EST	NEELY,WB & BLAU,GE (1985)
DC	pKa			
HL 5.69E-006	atm m3/mol	25	EST	MEYLAN,WM & HOWARD,PH (1991)
OH 1.10E-011	cm3/molc sec	25	EST	MEYLAN,WM & HOWARD,PH (1993)

066516-96-9 — N-FORMYL-P-CYANOSTYRYLAMINE

Formula: $C_{10}H_8N_2O$
Mol Weight: 172.19
MP (deg C):   **FP (deg C):**
BP (deg C):
BP pressure (mm Hg):

Property/Value	Units	Temp	Data Type	Reference
WS 1.35E+003	mg/L	25	EST	MEYLAN,WM ET AL. (1996)
logP 1.62			EXP	HANSCH,C & LEO,AJ (1985)
VP 3.83E-006	mm Hg	25	EST	NEELY,WB & BLAU,GE (1985)
DC	pKa			
HL 2.55E-011	atm m3/mol	25	EST	MEYLAN,WM & HOWARD,PH (1991)
OH 3.97E-011	cm3/molc sec	25	EST	MEYLAN,WM & HOWARD,PH (1993)

066521-48-0 — 4-(3,3-DIME-1-TRIAZINO)PHENYLUREA

Formula: $C_9H_{13}N_5O$
Mol Weight: 207.24
MP (deg C):   **FP (deg C):**
BP (deg C):
BP pressure (mm Hg):

Property/Value	Units	Temp	Data Type	Reference
WS 3.44E+003	mg/L	25	EST	MEYLAN,WM ET AL. (1996)
logP 1.25			EXP	HANSCH,C & LEO,AJ (1985)
VP 6.00E-006	mm Hg	25	EST	NEELY,WB & BLAU,GE (1985)
DC	pKa			
HL 2.01E-014	atm m3/mol	25	EST	MEYLAN,WM & HOWARD,PH (1991)
OH 2.04E-011	cm3/molc sec	25	EST	MEYLAN,WM & HOWARD,PH (1993)

066521-49-1 — 1-(4-CARBAMOYLPHENYL)-3-METHYL-3-OCTYLTRIAZENE

Formula: $C_{16}H_{26}N_4O$
Mol Weight: 290.41
MP (deg C):   **FP (deg C):**
BP (deg C):
BP pressure (mm Hg):

Property/Value	Units	Temp	Data Type	Reference
WS 1.35E+000	mg/L	25	EST	MEYLAN,WM ET AL. (1996)
logP 4.70			EXP	HANSCH,C ET AL. (1995)
VP 1.86E-008	mm Hg	25	EST	NEELY,WB & BLAU,GE (1985)
DC	pKa			
HL 1.60E-012	atm m3/mol	25	EST	MEYLAN,WM & HOWARD,PH (1991)
OH 2.27E-011	cm3/molc sec	25	EST	MEYLAN,WM & HOWARD,PH (1993)

066536-69-4 — 2H-BENZOTRIAZOLE, 2-CYCLOPENTYL-

Formula: $C_{11}H_{13}N_3$
Mol Weight: 187.25
MP (deg C):   **FP (deg C):**
BP (deg C):
BP pressure (mm Hg):

Property/Value	Units	Temp	Data Type	Reference
WS 1.29E+002	mg/L	25	EST	MEYLAN,WM ET AL. (1996)
logP 3.04			EXP	SPARATORE,F ET AL. (1988)
VP 1.09E-004	mm Hg	25	EST	NEELY,WB & BLAU,GE (1985)
DC	pKa			
HL 4.28E-006	atm m3/mol	25	EST	MEYLAN,WM & HOWARD,PH (1991)
OH 7.87E-012	cm3/molc sec	25	EST	MEYLAN,WM & HOWARD,PH (1993)

CAS #: 066536-70-7	1H-BENZOTRIAZOLE, 1-CYCLOHEXYL-

Formula: $C_{12}H_{15}N_3$
Mol Weight: 201.27
MP (deg C): FP (deg C):
BP (deg C):
BP pressure (mm Hg):

Property/Value	Units	Temp	Data Type	Reference
WS 9.52E+001	mg/L	25	EST	MEYLAN,WM ET AL. (1996)
logP 3.11			EXP	SPARATORE,F ET AL. (1988)
VP 4.17E-005	mm Hg	25	EST	NEELY,WB & BLAU,GE (1985)
DC	pKa			
HL 5.69E-006	atm m3/mol	25	EST	MEYLAN,WM & HOWARD,PH (1991)
OH 1.10E-011	cm3/molc sec	25	EST	MEYLAN,WM & HOWARD,PH (1993)

CAS #: 066635-84-5	1H-PYRROLIZINE-1-CARBOXYLIC ACID, 2,3-DIHYDRO-5-

Formula: $C_{16}H_{15}NO_3$
Mol Weight: 269.30
MP (deg C): FP (deg C):
BP (deg C):
BP pressure (mm Hg):

Property/Value	Units	Temp	Data Type	Reference
WS 1.53E+002	mg/L	25	EST	MEYLAN,WM ET AL. (1996)
logP 2.90			EXP	SANGSTER,J (1993)
VP 4.80E-008	mm Hg	25	EST	NEELY,WB & BLAU,GE (1985)
DC	pKa			
HL 3.70E-013	atm m3/mol	25	EST	MEYLAN,WM & HOWARD,PH (1991)
OH 2.04E-010	cm3/molc sec	25	EST	MEYLAN,WM & HOWARD,PH (1993)

CAS #: 066635-90-3	1H-PYRROLIZINE-1-CARBOXYLIC ACID, 5-(4-FLUOROBEN

Formula: $C_{15}H_{12}FNO_3$
Mol Weight: 273.27
MP (deg C): FP (deg C):
BP (deg C):
BP pressure (mm Hg):

Property/Value	Units	Temp	Data Type	Reference
WS 1.70E+002	mg/L	25	EST	MEYLAN,WM ET AL. (1996)
logP 2.82			EXP	SANGSTER,J (1993)
VP 1.28E-007	mm Hg	25	EST	NEELY,WB & BLAU,GE (1985)
DC	pKa			
HL 3.91E-013	atm m3/mol	25	EST	MEYLAN,WM & HOWARD,PH (1991)
OH 2.04E-010	cm3/molc sec	25	EST	MEYLAN,WM & HOWARD,PH (1993)

CAS #: 066644-81-3	VERALIPRIDE

Formula: $C_{17}H_{25}N_3O_5S$
Mol Weight: 383.47
MP (deg C): FP (deg C):
BP (deg C):
BP pressure (mm Hg):

Property/Value	Units	Temp	Data Type	Reference
WS 2.15E+002	mg/L	25	EST	MEYLAN,WM ET AL. (1996)
logP 1.47			EXP	MANNHOLD,R ET AL. (1990)
VP 2.76E-012	mm Hg	25	EST	NEELY,WB & BLAU,GE (1985)
DC	pKa			
HL 8.96E-019	atm m3/mol	25	EST	MEYLAN,WM & HOWARD,PH (1991)
OH 1.47E-010	cm3/molc sec	25	EST	MEYLAN,WM & HOWARD,PH (1993)

CAS #: 066675-75-0	PROPANAMIDE, 3-AMINO-N-(2,6-DIMETHYLPHENYL)-

Formula: $C_{11}H_{16}N_2O$
Mol Weight: 192.26
MP (deg C): FP (deg C):
BP (deg C):
BP pressure (mm Hg):

Property/Value	Units	Temp	Data Type	Reference
WS 1.42E+004	mg/L	25	EST	MEYLAN,WM ET AL. (1996)
logP 0.62			EXP	SANGSTER,J (1993)
VP 2.77E-006	mm Hg	25	EST	NEELY,WB & BLAU,GE (1985)
DC	pKa			
HL 1.03E-012	atm m3/mol	25	EST	MEYLAN,WM & HOWARD,PH (1991)
OH 7.47E-011	cm3/molc sec	25	EST	MEYLAN,WM & HOWARD,PH (1993)

CAS #: 066722-44-9	2-PROPANOL, 1-[(1-METHYLETHYL)AMINO]-3-[4-[[2-(1

Formula: $C_{18}H_{31}NO_4$
Mol Weight: 325.45
MP (deg C): 100 FP (deg C):
BP (deg C):
BP pressure (mm Hg):

Property/Value	Units	Temp	Data Type	Reference
WS 2.24E+003	mg/L	25	EST	MEYLAN,WM ET AL. (1996)
logP 1.87			EXP	RECANATINI,M (1992)
VP 9.54E-009	mm Hg	25	EST	NEELY,WB & BLAU,GE (1985)
DC	pKa			
HL 2.89E-015	atm m3/mol	25	EST	MEYLAN,WM & HOWARD,PH (1991)
OH 1.68E-010	cm3/molc sec	25	EST	MEYLAN,WM & HOWARD,PH (1993)

CAS #: 066843-01-4	2,4,6(1H,3H,5H)-PYRIMIDINETRIONE, 5-METHYL-5-(3-

Formula: $C_{10}H_{14}N_2O_3$
Mol Weight: 210.23
MP (deg C): FP (deg C):
BP (deg C):
BP pressure (mm Hg):

Property/Value	Units	Temp	Data Type	Reference
WS 3.88E+003	mg/L	25	EST	MEYLAN,WM ET AL. (1996)
logP 1.17			EXP	PRANKERD,RJ & MCKEOWN,RH (1992)
VP 2.94E-011	mm Hg	25	EST	NEELY,WB & BLAU,GE (1985)
DC	pKa			
HL 6.60E-013	atm m3/mol	25	EST	MEYLAN,WM & HOWARD,PH (1991)
OH 9.33E-011	cm3/molc sec	25	EST	MEYLAN,WM & HOWARD,PH (1993)

CAS #: 066895-70-3	M-METHYLPHENYLHIPPURATE

Formula: $C_{16}H_{15}NO_3$
Mol Weight: 269.30
MP (deg C): FP (deg C):
BP (deg C):
BP pressure (mm Hg):

Property/Value	Units	Temp	Data Type	Reference
WS 8.77E+001	mg/L	25	EST	MEYLAN,WM ET AL. (1996)
logP 2.72			EXP	HANSCH,C & LEO,AJ (1985)
VP 2.30E-008	mm Hg	25	EST	NEELY,WB & BLAU,GE (1985)
DC	pKa			
HL 9.42E-011	atm m3/mol	25	EST	MEYLAN,WM & HOWARD,PH (1991)
OH 1.55E-011	cm3/molc sec	25	EST	MEYLAN,WM & HOWARD,PH (1993)

CAS #: 066974-58-1	1-(O-I PH)-3,3-DIMETHYLTRIAZENE

Formula: $C_8H_{10}IN_3$

Mol Weight: 275.09

MP (deg C): 　 FP (deg C):

BP (deg C):

BP pressure (mm Hg):

Property/Value	Units	Temp	Data Type	Reference
WS 1.75E+001	mg/L	25	EST	MEYLAN,WM ET AL. (1996)
logP 3.50			EXP	HANSCH,C & LEO,AJ (1985)
VP 1.78E-003	mm Hg	25	EST	NEELY,WB & BLAU,GE (1985)
DC	pKa			
HL 1.24E-007	atm m3/mol	25	EST	MEYLAN,WM & HOWARD,PH (1991)
OH 3.04E-012	cm3/molc sec	25	EST	MEYLAN,WM & HOWARD,PH (1993)

CAS #: 066974-61-6	1(O-HYDROXYMEPH)-3,3-DIME TRIAZENE

Formula: $C_9H_{13}N_3O$

Mol Weight: 179.22

MP (deg C): 　 FP (deg C):

BP (deg C):

BP pressure (mm Hg):

Property/Value	Units	Temp	Data Type	Reference
WS 1.85E+003	mg/L	25	EST	MEYLAN,WM ET AL. (1996)
logP 1.73			EXP	HANSCH,C & LEO,AJ (1985)
VP 6.55E-005	mm Hg	25	EST	NEELY,WB & BLAU,GE (1985)
DC	pKa			
HL 2.17E-011	atm m3/mol	25	EST	MEYLAN,WM & HOWARD,PH (1991)
OH 7.86E-012	cm3/molc sec	25	EST	MEYLAN,WM & HOWARD,PH (1993)

CAS #: 066974-73-0	BENZAMIDE, 4-[3-METHYL-3-(2-PROPENYL)-1-TRIAZENY

Formula: $C_{11}H_{14}N_4O$

Mol Weight: 218.26

MP (deg C): 　 FP (deg C):

BP (deg C):

BP pressure (mm Hg):

Property/Value	Units	Temp	Data Type	Reference
WS 5.77E+002	mg/L	25	EST	MEYLAN,WM ET AL. (1996)
logP 2.09			EXP	HANSCH,C ET AL. (1995)
VP 1.19E-006	mm Hg	25	EST	NEELY,WB & BLAU,GE (1985)
DC	pKa			
HL 2.90E-013	atm m3/mol	25	EST	MEYLAN,WM & HOWARD,PH (1991)
OH 3.86E-011	cm3/molc sec	25	EST	MEYLAN,WM & HOWARD,PH (1993)

CAS #: 066974-76-3	1-(4-CARBAMOYLPHENYL)-3-METHYL-3-METHOXYTRIAZENE

Formula: $C_9H_{12}N_4O_2$

Mol Weight: 208.22

MP (deg C): 　 FP (deg C):

BP (deg C):

BP pressure (mm Hg):

Property/Value	Units	Temp	Data Type	Reference
WS 3.82E+003	mg/L	25	EST	MEYLAN,WM ET AL. (1996)
logP 1.19			EXP	HANSCH,C ET AL. (1995)
VP 2.35E-006	mm Hg	25	EST	NEELY,WB & BLAU,GE (1985)
DC	pKa			
HL 6.44E-014	atm m3/mol	25	EST	MEYLAN,WM & HOWARD,PH (1991)
OH 3.61E-012	cm3/molc sec	25	EST	MEYLAN,WM & HOWARD,PH (1993)

CAS #: 066974-78-5	4-MEO-3-(33-DIME TRIAZENO)BENZAMIDE

Formula: $C_{10}H_{14}N_4O_2$

Mol Weight: 222.25

MP (deg C): 　 FP (deg C):

BP (deg C):

BP pressure (mm Hg):

Property/Value	Units	Temp	Data Type	Reference
WS 1.41E+004	mg/L	25	EST	MEYLAN,WM ET AL. (1996)
logP 0.44			EXP	HANSCH,C & LEO,AJ (1985)
VP 1.14E-006	mm Hg	25	EST	NEELY,WB & BLAU,GE (1985)
DC	pKa			
HL 1.30E-014	atm m3/mol	25	EST	MEYLAN,WM & HOWARD,PH (1991)
OH 1.31E-011	cm3/molc sec	25	EST	MEYLAN,WM & HOWARD,PH (1993)

CAS #: 066974-82-1	1-(3,4-DICYANOPH)-33-DIME TRIAZENE

Formula: $C_{10}H_9N_5$

Mol Weight: 199.22

MP (deg C): 　 FP (deg C):

BP (deg C):

BP pressure (mm Hg):

Property/Value	Units	Temp	Data Type	Reference
WS 3.30E+002	mg/L	25	EST	MEYLAN,WM ET AL. (1996)
logP 2.18			EXP	HANSCH,C & LEO,AJ (1985)
VP 2.86E-005	mm Hg	25	EST	NEELY,WB & BLAU,GE (1985)
DC	pKa			
HL 5.01E-011	atm m3/mol	25	EST	MEYLAN,WM & HOWARD,PH (1991)
OH 2.55E-012	cm3/molc sec	25	EST	MEYLAN,WM & HOWARD,PH (1993)

CAS #: 066974-88-7	N-ME-2-(33-DIME-1-TRIAZINO)BENZAMIDE

Formula: $C_{10}H_{14}N_4O$

Mol Weight: 206.25

MP (deg C): 　 FP (deg C):

BP (deg C):

BP pressure (mm Hg):

Property/Value	Units	Temp	Data Type	Reference
WS 1.11E+003	mg/L	25	EST	MEYLAN,WM ET AL. (1996)
logP 1.83			EXP	HANSCH,C & LEO,AJ (1985)
VP 3.77E-006	mm Hg	25	EST	NEELY,WB & BLAU,GE (1985)
DC	pKa			
HL 4.84E-013	atm m3/mol	25	EST	MEYLAN,WM & HOWARD,PH (1991)
OH 8.91E-012	cm3/molc sec	25	EST	MEYLAN,WM & HOWARD,PH (1993)

CAS #: 066974-90-1	BENZAMIDE, N-(CYANOMETHYL)-2-(3,3-DIMETHYL-1-THI

Formula: $C_{11}H_{13}N_5O$

Mol Weight: 231.26

MP (deg C): 　 FP (deg C):

BP (deg C):

BP pressure (mm Hg):

Property/Value	Units	Temp	Data Type	Reference
WS 4.82E+002	mg/L	25	EST	MEYLAN,WM ET AL. (1996)
logP 1.79			EXP	HANSCH,C ET AL. (1995)
VP 5.65E-008	mm Hg	25	EST	NEELY,WB & BLAU,GE (1985)
DC	pKa			
HL 2.01E-016	atm m3/mol	25	EST	MEYLAN,WM & HOWARD,PH (1991)
OH 3.58E-012	cm3/molc sec	25	EST	MEYLAN,WM & HOWARD,PH (1993)

CAS #: 066974-91-2	BENZAMIDE, N-(2-AMINO-2-OXOETHYL)-2-(3,3-DIMETHY

Formula: $C_{11}H_{15}N_5O_2$

Mol Weight: 249.27

MP (deg C):   FP (deg C):

BP (deg C):

BP pressure (mm Hg):

Property/ Value	Units	Temp	Data Type	Reference
WS 2.35E+003	mg/L	25	EST	MEYLAN,WM ET AL. (1996)
logP 1.18			EXP	HANSCH,C & LEO,AJ (1985)
VP 1.29E-009	mm Hg	25	EST	NEELY,WB & BLAU,GE (1985)
DC	pKa			
HL 4.73E-015	atm m3/mol	25	EST	MEYLAN,WM & HOWARD,PH (1991)
OH 1.40E-011	cm3/molc sec	25	EST	MEYLAN,WM & HOWARD,PH (1993)

CAS #: 066974-92-3	N-CYANO-2(33DIME-1-TRIAZEN)BENZAMIDE

Formula: $C_{10}H_{11}N_5O$

Mol Weight: 217.23

MP (deg C):   FP (deg C):

BP (deg C):

BP pressure (mm Hg):

Property/ Value	Units	Temp	Data Type	Reference
WS 7.38E+003	mg/L	25	EST	MEYLAN,WM ET AL. (1996)
logP 0.80			EXP	HANSCH,C & LEO,AJ (1985)
VP 1.33E-007	mm Hg	25	EST	NEELY,WB & BLAU,GE (1985)
DC	pKa			
HL 2.12E-012	atm m3/mol	25	EST	MEYLAN,WM & HOWARD,PH (1991)
OH 8.36E-012	cm3/molc sec	25	EST	MEYLAN,WM & HOWARD,PH (1993)

CAS #: 066975-11-9	P-(3-ME-3-ACETYL-1-TRIAZENO)TOLUENE

Formula: $C_{10}H_{13}N_3O$

Mol Weight: 191.23

MP (deg C):   FP (deg C):

BP (deg C):

BP pressure (mm Hg):

Property/ Value	Units	Temp	Data Type	Reference
WS 8.45E+001	mg/L	25	EST	MEYLAN,WM ET AL. (1996)
logP 3.23			EXP	HANSCH,C & LEO,AJ (1985)
VP 2.64E-004	mm Hg	25	EST	NEELY,WB & BLAU,GE (1985)
DC	pKa			
HL 8.74E-010	atm m3/mol	25	EST	MEYLAN,WM & HOWARD,PH (1991)
OH 3.81E-012	cm3/molc sec	25	EST	MEYLAN,WM & HOWARD,PH (1993)

CAS #: 066975-13-1	BENZOIC ACID HYDRAZD,O(33DIME)TRIAZIN

Formula: $C_9H_{13}N_5O$

Mol Weight: 207.24

MP (deg C):   FP (deg C):

BP (deg C):

BP pressure (mm Hg):

Property/ Value	Units	Temp	Data Type	Reference
WS 3.24E+003	mg/L	25	EST	MEYLAN,WM ET AL. (1996)
logP 1.28			EXP	HANSCH,C & LEO,AJ (1985)
VP 8.29E-007	mm Hg	25	EST	NEELY,WB & BLAU,GE (1985)
DC	pKa			
HL 9.21E-016	atm m3/mol	25	EST	MEYLAN,WM & HOWARD,PH (1991)
OH 8.36E-012	cm3/molc sec	25	EST	MEYLAN,WM & HOWARD,PH (1993)

CAS #: 066975-19-7	1H-IMIDAZOLE-4-CARBOXAMIDE, 5-[3-METHYL-3-(2-PR

Formula: $C_8H_{10}N_6O$

Mol Weight: 206.21

MP (deg C):   FP (deg C):

BP (deg C):

BP pressure (mm Hg):

Property/ Value	Units	Temp	Data Type	Reference
WS 1.04E+003	mg/L	25	EST	MEYLAN,WM ET AL. (1996)
logP 0.33			EXP	HANSCH,C & LEO,AJ (1985)
VP 3.06E-009	mm Hg	25	EST	NEELY,WB & BLAU,GE (1985)
DC	pKa			
HL 3.35E-017	atm m3/mol	25	EST	MEYLAN,WM & HOWARD,PH (1991)
OH 3.20E-011	cm3/molc sec	25	EST	MEYLAN,WM & HOWARD,PH (1993)

CAS #: 066999-80-2	BENZAMIDE,N-(3-DIETHYLAMINOPROPYL)

Formula: $C_{14}H_{22}N_2O$

Mol Weight: 234.34

MP (deg C):   FP (deg C):

BP (deg C):

BP pressure (mm Hg):

Property/ Value	Units	Temp	Data Type	Reference
WS 1.34E+003	mg/L	25	EST	MEYLAN,WM ET AL. (1996)
logP 1.56			EXP	HANSCH,C & LEO,AJ (1985)
VP 1.03E-006	mm Hg	25	EST	NEELY,WB & BLAU,GE (1985)
DC	pKa			
HL 7.49E-012	atm m3/mol	25	EST	MEYLAN,WM & HOWARD,PH (1991)
OH 1.08E-010	cm3/molc sec	25	EST	MEYLAN,WM & HOWARD,PH (1993)

CAS #: 066999-97-1	1-PHENYLOXYCARBONYL-5-FLUOROURACIL

Formula: $C_{11}H_7FN_2O_4$

Mol Weight: 250.19

MP (deg C):   FP (deg C):

BP (deg C):

BP pressure (mm Hg):

Property/ Value	Units	Temp	Data Type	Reference
WS 6.70E+003	mg/L	25	EST	MEYLAN,WM ET AL. (1996)
logP 0.64			EXP	HANSCH,C ET AL. (1995)
VP 2.11E-010	mm Hg	25	EST	NEELY,WB & BLAU,GE (1985)
DC	pKa			
HL 7.04E-013	atm m3/mol	25	EST	MEYLAN,WM & HOWARD,PH (1991)
OH 7.84E-012	cm3/molc sec	25	EST	MEYLAN,WM & HOWARD,PH (1993)

CAS #: 066999-98-2	URACIL,1-BENZYLOXYCARBONYL-5-FLUORO-

Formula: $C_{12}H_9FN_2O_4$

Mol Weight: 264.21

MP (deg C):   FP (deg C):

BP (deg C):

BP pressure (mm Hg):

Property/ Value	Units	Temp	Data Type	Reference
WS 1.94E+003	mg/L	25	EST	MEYLAN,WM ET AL. (1996)
logP 1.18			EXP	HANSCH,C ET AL. (1995)
VP 9.21E-011	mm Hg	25	EST	NEELY,WB & BLAU,GE (1985)
DC	pKa			
HL 1.54E-013	atm m3/mol	25	EST	MEYLAN,WM & HOWARD,PH (1991)
OH 1.22E-011	cm3/molc sec	25	EST	MEYLAN,WM & HOWARD,PH (1993)

CAS #: 067023-19-2 — BENZENEACETAMIDE, N-(2,6-DIMETHYLPHENYL)-N-[3-(1

Formula: $C_{24}H_{32}N_2O$

Mol Weight: 364.54

MP (deg C): FP (deg C):

BP (deg C):

BP pressure (mm Hg):

Property/ Value	Units	Temp	Data Type	Reference
WS 1.57E-001	mg/L	25	EST	MEYLAN,WM ET AL. (1996)
logP 5.28			EXP	SANGSTER,J (1993)
VP 6.64E-010	mm Hg	25	EST	NEELY,WB & BLAU,GE (1985)
DC	pKa			
HL 1.12E-011	atm m3/mol	25	EST	MEYLAN,WM & HOWARD,PH (1991)
OH 1.33E-010	cm3/molc sec	25	EST	MEYLAN,WM & HOWARD,PH (1993)

CAS #: 067027-56-9 — 14-BENZDIAZPIN-2-ONE,3ME5-(2CLPH)7NO2

Formula: $C_{16}H_{12}ClN_3O_3$

Mol Weight: 329.75

MP (deg C): FP (deg C):

BP (deg C):

BP pressure (mm Hg):

Property/ Value	Units	Temp	Data Type	Reference
WS 1.59E+001	mg/L	25	EST	MEYLAN,WM ET AL. (1996)
logP 2.72			EXP	HANSCH,C & LEO,AJ (1985)
VP 7.02E-011	mm Hg	25	EST	NEELY,WB & BLAU,GE (1985)
DC	pKa			
HL 9.31E-013	atm m3/mol	25	EST	MEYLAN,WM & HOWARD,PH (1991)
OH 5.49E-012	cm3/molc sec	25	EST	MEYLAN,WM & HOWARD,PH (1993)

CAS #: 067028-18-6 — 1,2,3,7-TETRACHLORODIBENZODIOXIN

Formula: $C_{12}H_4Cl_4O_2$

Mol Weight: 321.98

MP (deg C): FP (deg C):

BP (deg C):

BP pressure (mm Hg):

Property/ Value	Units	Temp	Data Type	Reference
WS 4.30E-004	mg/L	20	EXP	YALKOWSKY,SH & DANNENFELSER,RM (1992)
logP 6.90			EXP	SHIU,WY ET AL. (1988)
VP 7.50E-009	mm Hg	25	EXP	SHIU,WY ET AL. (1988)
DC	pKa			
HL 7.60E-006	atm m3/mol	25	EXP	SHIU,WY ET AL. (1988)
OH 4.17E-012	cm3/molc sec	25	EST	MEYLAN,WM & HOWARD,PH (1993)

CAS #: 067246-30-4 — 1-S-PENTABENZOTRIAZOLE

Formula: $C_{11}H_{15}N_3$

Mol Weight: 189.26

MP (deg C): FP (deg C):

BP (deg C):

BP pressure (mm Hg):

Property/ Value	Units	Temp	Data Type	Reference
WS 1.50E+002	mg/L	25	EST	MEYLAN,WM ET AL. (1996)
logP 2.95			EXP	HANSCH,C & LEO,AJ (1985)
VP 2.99E-004	mm Hg	25	EST	NEELY,WB & BLAU,GE (1985)
DC	pKa			
HL 9.72E-006	atm m3/mol	25	EST	MEYLAN,WM & HOWARD,PH (1991)
OH 6.28E-012	cm3/molc sec	25	EST	MEYLAN,WM & HOWARD,PH (1993)

CAS #: 067309-36-8 — A-BUTYL BENZENEETHANEAMINE

Formula: $C_{12}H_{19}N$

Mol Weight: 177.29

MP (deg C): FP (deg C):

BP (deg C):

BP pressure (mm Hg):

Property/ Value	Units	Temp	Data Type	Reference
WS 7.81E+002	mg/L	25	EST	MEYLAN,WM ET AL. (1996)
logP 3.36			EXP	HANSCH,C & LEO,AJ (1985)
VP 7.96E-003	mm Hg	25	EST	NEELY,WB & BLAU,GE (1985)
DC	pKa			
HL 2.52E-006	atm m3/mol	25	EST	MEYLAN,WM & HOWARD,PH (1991)
OH 5.84E-011	cm3/molc sec	25	EST	MEYLAN,WM & HOWARD,PH (1993)

CAS #: 067309-37-9 — A-T-BUTYL BENZENEETHANEAMINE

Formula: $C_{12}H_{19}N$

Mol Weight: 177.29

MP (deg C): FP (deg C):

BP (deg C):

BP pressure (mm Hg):

Property/ Value	Units	Temp	Data Type	Reference
WS 1.89E+003	mg/L	25	EST	MEYLAN,WM ET AL. (1996)
logP 2.91			EXP	HANSCH,C & LEO,AJ (1985)
VP 2.09E-002	mm Hg	25	EST	NEELY,WB & BLAU,GE (1985)
DC	pKa			
HL 2.52E-006	atm m3/mol	25	EST	MEYLAN,WM & HOWARD,PH (1991)
OH 5.48E-011	cm3/molc sec	25	EST	MEYLAN,WM & HOWARD,PH (1993)

CAS #: 067309-38-0 — A-I-BUTYL BENZENEETHANEAMINE

Formula: $C_{12}H_{19}N$

Mol Weight: 177.29

MP (deg C): FP (deg C):

BP (deg C):

BP pressure (mm Hg):

Property/ Value	Units	Temp	Data Type	Reference
WS 8.96E+002	mg/L	25	EST	MEYLAN,WM ET AL. (1996)
logP 3.29			EXP	HANSCH,C & LEO,AJ (1985)
VP 1.69E-002	mm Hg	25	EST	NEELY,WB & BLAU,GE (1985)
DC	pKa			
HL 2.52E-006	atm m3/mol	25	EST	MEYLAN,WM & HOWARD,PH (1991)
OH 5.84E-011	cm3/molc sec	25	EST	MEYLAN,WM & HOWARD,PH (1993)

CAS #: 067410-39-3 — 3,5-DIOXA-6-AZA-4-PHOSPAOCT-6-ENE-8-NITRILE, 4-A

Formula: $C_{10}H_{12}N_3O_3P$

Mol Weight: 253.20

MP (deg C): FP (deg C):

BP (deg C):

BP pressure (mm Hg):

Property/ Value	Units	Temp	Data Type	Reference
WS 5.51E+003	mg/L	25	EST	MEYLAN,WM ET AL. (1996)
logP 0.72			EXP	SANGSTER,J (1993)
VP 2.38E-006	mm Hg	25	EST	NEELY,WB & BLAU,GE (1985)
DC	pKa			
HL 7.64E-010	atm m3/mol	25	EST	MEYLAN,WM & HOWARD,PH (1991)
OH 5.11E-011	cm3/molc sec	25	EST	MEYLAN,WM & HOWARD,PH (1993)

CAS #: 067410-45-1			3,5-DIOXA-6-AZA-4-PHOSPHAOCT-6-ENE-8-NITRILE, 4-

Formula: $C_{12}H_{16}N_3O_3P$

Mol Weight: 281.25

MP (deg C):   FP (deg C):

BP (deg C):

BP pressure (mm Hg):

Property/Value	Units	Temp	Data Type	Reference
WS 2.44E+001	mg/L	25	EST	MEYLAN,WM ET AL. (1996)
logP 3.29			EXP	HANSCH,C ET AL. (1995)
VP 1.91E-006	mm Hg	25	EST	NEELY,WB & BLAU,GE (1985)
DC	pKa			
HL 3.68E-009	atm m3/mol	25	EST	MEYLAN,WM & HOWARD,PH (1991)
OH 5.36E-011	cm3/molc sec	25	EST	MEYLAN,WM & HOWARD,PH (1993)

CAS #: 067410-46-2			3,5-DIOXA-6-AZA-4-PHOSPHAOCT-6-ENE-8-NITRILE, 7-

Formula: $C_{11}H_{13}ClN_3O_3P$

Mol Weight: 301.67

MP (deg C):   FP (deg C):

BP (deg C):

BP pressure (mm Hg):

Property/Value	Units	Temp	Data Type	Reference
WS 8.80E+001	mg/L	25	EST	MEYLAN,WM ET AL. (1996)
logP 2.50			EXP	SANGSTER,J (1993)
VP 7.51E-007	mm Hg	25	EST	NEELY,WB & BLAU,GE (1985)
DC	pKa			
HL 1.24E-009	atm m3/mol	25	EST	MEYLAN,WM & HOWARD,PH (1991)
OH 4.97E-011	cm3/molc sec	25	EST	MEYLAN,WM & HOWARD,PH (1993)

CAS #: 067410-48-4			3,5-DIOXA-6-AZA-4-PHOSPHAOCT-6-ENE-8-NITRILE, 4-

Formula: $C_{12}H_{15}N_4O_5P$

Mol Weight: 326.25

MP (deg C):   FP (deg C):

BP (deg C):

BP pressure (mm Hg):

Property/Value	Units	Temp	Data Type	Reference
WS 1.10E+002	mg/L	25	EST	MEYLAN,WM ET AL. (1996)
logP 1.76			EXP	SANGSTER,J (1994)
VP 4.21E-008	mm Hg	25	EST	NEELY,WB & BLAU,GE (1985)
DC	pKa			
HL 8.78E-012	atm m3/mol	25	EST	MEYLAN,WM & HOWARD,PH (1991)
OH 5.62E-011	cm3/molc sec	25	EST	MEYLAN,WM & HOWARD,PH (1993)

CAS #: 067410-49-5			3,5-DIOXA-6-AZA-4-PHOSPHAOCT-6-ENE-8-NITRILE, 4-

Formula: $C_{13}H_{18}N_3O_4P$

Mol Weight: 311.28

MP (deg C):   FP (deg C):

BP (deg C):

BP pressure (mm Hg):

Property/Value	Units	Temp	Data Type	Reference
WS 9.59E+001	mg/L	25	EST	MEYLAN,WM ET AL. (1996)
logP 2.39			EXP	SANGSTER,J (1994)
VP 2.83E-007	mm Hg	25	EST	NEELY,WB & BLAU,GE (1985)
DC	pKa			
HL 1.32E-010	atm m3/mol	25	EST	MEYLAN,WM & HOWARD,PH (1991)
OH 8.23E-011	cm3/molc sec	25	EST	MEYLAN,WM & HOWARD,PH (1993)

CAS #: 067446-07-5			5-DECEN-1-OL ACETATE (Z)

Formula: $C_{12}H_{22}O_2$

Mol Weight: 198.31

MP (deg C):   FP (deg C):

BP (deg C):

BP pressure (mm Hg):

Property/Value	Units	Temp	Data Type	Reference
WS 5.50E+000	mg/L	25	EST	MEYLAN,WM ET AL. (1996)
logP 4.58			EST	MEYLAN,WM & HOWARD,PH (1995)
VP 2.31E-002	mm Hg	25	EST	NEELY,WB & BLAU,GE (1985)
DC	pKa			
HL 1.98E-003	atm m3/mol	25	EST	MEYLAN,WM & HOWARD,PH (1991)
OH 6.61E-011	cm3/molc sec	25	EST	MEYLAN,WM & HOWARD,PH (1993)

CAS #: 067452-27-1			4-DECENYL ACETATE (Z)

Formula: $C_{12}H_{22}O_2$

Mol Weight: 198.31

MP (deg C):   FP (deg C):

BP (deg C):

BP pressure (mm Hg):

Property/Value	Units	Temp	Data Type	Reference
WS 5.50E+000	mg/L	25	EST	MEYLAN,WM ET AL. (1996)
logP 4.58			EST	MEYLAN,WM & HOWARD,PH (1995)
VP 2.31E-002	mm Hg	25	EST	NEELY,WB & BLAU,GE (1985)
DC	pKa			
HL 1.98E-003	atm m3/mol	25	EST	MEYLAN,WM & HOWARD,PH (1991)
OH 6.61E-011	cm3/molc sec	25	EST	MEYLAN,WM & HOWARD,PH (1993)

CAS #: 067460-68-8			2,5-DIMEO-4-NO2-AMPHETAMINE

Formula: $C_{11}H_{16}N_2O_4$

Mol Weight: 240.26

MP (deg C):   FP (deg C):

BP (deg C):

BP pressure (mm Hg):

Property/Value	Units	Temp	Data Type	Reference
WS 3.63E+003	mg/L	25	EST	MEYLAN,WM ET AL. (1996)
logP 1.74			EXP	HANSCH,C & LEO,AJ (1985)
VP 1.52E-005	mm Hg	25	EST	NEELY,WB & BLAU,GE (1985)
DC	pKa			
HL 1.49E-011	atm m3/mol	25	EST	MEYLAN,WM & HOWARD,PH (1991)
OH 5.19E-011	cm3/molc sec	25	EST	MEYLAN,WM & HOWARD,PH (1993)

CAS #: 067481-22-5			DIBENZOFURAN, 2,3,4,6,8-PENTACHLORO-

Formula: $C_{12}H_3Cl_5O$

Mol Weight: 340.42

MP (deg C):   FP (deg C):

BP (deg C):

BP pressure (mm Hg):

Property/Value	Units	Temp	Data Type	Reference
WS 1.29E-003	mg/L	25	EST	MEYLAN,WM ET AL. (1996)
logP 6.59			EXP	SIJM,DTHM ET AL. (1989)
VP 3.46E-007	mm Hg	25	EST	NEELY,WB & BLAU,GE (1985)
DC	pKa			
HL 1.14E-005	atm m3/mol	25	EST	MEYLAN,WM & HOWARD,PH (1991)
OH 7.46E-014	cm3/molc sec	25	EST	MEYLAN,WM & HOWARD,PH (1993)

CAS #: 067517-48-0	DIBENZOFURAN, 1,2,3,4,8-PENTACHLORO-
Formula: $C_{12}H_3Cl_5O$	
Mol Weight: 340.42	
MP (deg C):	FP (deg C):
BP (deg C):	
BP pressure (mm Hg):	

Property/ Value	Units	Temp	Data Type	Reference
WS 8.73E-004	mg/L	25	EST	MEYLAN,WM ET AL. (1996)
logP 6.79			EXP	SIJM,DTHM ET AL. (1989)
VP 3.46E-007	mm Hg	25	EST	NEELY,WB & BLAU,GE (1985)
DC	pKa			
HL 1.14E-005	atm m3/mol	25	EST	MEYLAN,WM & HOWARD,PH (1991)
OH 7.46E-014	cm3/molc sec	25	EST	MEYLAN,WM & HOWARD,PH (1993)

CAS #: 067523-85-7	1-(4-NITRO),2-DIBENZOYL-1-TERT-BUTYLHYDRAZINE
Formula: $C_{18}H_{19}N_3O_4$	
Mol Weight: 341.37	
MP (deg C):	FP (deg C):
BP (deg C):	
BP pressure (mm Hg):	

Property/ Value	Units	Temp	Data Type	Reference
WS 1.61E+001	mg/L	25	EST	MEYLAN,WM ET AL. (1996)
logP 2.63			EXP	OIKAWA,N ET AL. (1994)
VP 2.07E-011	mm Hg	25	EST	NEELY,WB & BLAU,GE (1985)
DC	pKa			
HL 1.91E-013	atm m3/mol	25	EST	MEYLAN,WM & HOWARD,PH (1991)
OH 2.29E-011	cm3/molc sec	25	EST	MEYLAN,WM & HOWARD,PH (1993)

CAS #: 067546-51-4	1,6-DICHLOROHEXA-1,5-DIENE
Formula: $C_6H_8Cl_2$	
Mol Weight: 151.04	
MP (deg C):	FP (deg C):
BP (deg C):	
BP pressure (mm Hg):	

Property/ Value	Units	Temp	Data Type	Reference
WS 4.90E+001	mg/L	25	EST	MEYLAN,WM ET AL. (1996)
logP 3.73			EST	MEYLAN,WM & HOWARD,PH (1995)
VP 3.88E+000	mm Hg	25	EST	NEELY,WB & BLAU,GE (1985)
DC	pKa			
HL 8.72E-002	atm m3/mol	25	EST	MEYLAN,WM & HOWARD,PH (1991)
OH 2.60E-011	cm3/molc sec	25	EST	MEYLAN,WM & HOWARD,PH (1993)

CAS #: 067562-39-4	DIBENZOFURAN, 1,2,3,4,6,7,8-HEPTACHLORO-
Formula: $C_{12}HCl_7O$	
Mol Weight: 409.31	
MP (deg C):	FP (deg C):
BP (deg C):	
BP pressure (mm Hg):	

Property/ Value	Units	Temp	Data Type	Reference
WS 1.35E-006	mg/L	25	EXP	FLETCHER,CL & MACKAY,WA (1993)
logP 7.92			EXP	SIJM,DTHM ET AL. (1989)
VP 2.01E-008	mm Hg	25	EST	NEELY,WB & BLAU,GE (1985)
DC	pKa			
HL 6.28E-006	atm m3/mol	25	EST	MEYLAN,WM & HOWARD,PH (1991)
OH 1.53E-014	cm3/molc sec	25	EST	MEYLAN,WM & HOWARD,PH (1993)

CAS #: 067747-09-5	PROCHLORAZ
Formula: $C_{15}H_{18}Cl_3N_3O_2$	
Mol Weight: 378.69	
MP (deg C): 46.5-49.3	FP (deg C):
BP (deg C): 208-210	
BP pressure (mm Hg): 2.00E-001	

Property/ Value	Units	Temp	Data Type	Reference
WS 3.44E+001	mg/L	25	EXP	TOMLIN,C (1994)
logP 4.10			EXP	BAKER,EA ET AL. (1992)
VP 1.13E-006	mm Hg	25	EXP	TOMLIN,C (1994)
DC	pKa			
HL 1.64E-008	atm m3/mol	25	EST	VP/WSOL
OH 7.80E-011	cm3/molc sec	25	EST	MEYLAN,WM & HOWARD,PH (1993)

CAS #: 067774-11-2	1H-IMIDAZOLE, 2-NITRO-1-(2,3,4,6-TETRA-O-ACETYL-
Formula: $C_{17}H_{21}N_3O_{11}$	
Mol Weight: 443.37	
MP (deg C):	FP (deg C):
BP (deg C):	
BP pressure (mm Hg):	

Property/ Value	Units	Temp	Data Type	Reference
WS 2.40E+002	mg/L	25	EST	MEYLAN,WM ET AL. (1996)
logP 0.52			EXP	HANSCH,C ET AL. (1995)
VP 2.14E-011	mm Hg	25	EST	NEELY,WB & BLAU,GE (1985)
DC	pKa			
HL 2.05E-020	atm m3/mol	25	EST	MEYLAN,WM & HOWARD,PH (1991)
OH 5.33E-011	cm3/molc sec	25	EST	MEYLAN,WM & HOWARD,PH (1993)

CAS #: 067774-31-6	N-NITROSO-3,5-DIME PIPERAZINE
Formula: $C_6H_{13}N_3O$	
Mol Weight: 143.19	
MP (deg C):	FP (deg C):
BP (deg C):	
BP pressure (mm Hg):	

Property/ Value	Units	Temp	Data Type	Reference
WS 1.97E+005	mg/L	25	EST	MEYLAN,WM ET AL. (1996)
logP 0.73			EXP	HANSCH,C & LEO,AJ (1985)
VP 3.55E-003	mm Hg	25	EST	NEELY,WB & BLAU,GE (1985)
DC	pKa			
HL 4.79E-010	atm m3/mol	25	EST	MEYLAN,WM & HOWARD,PH (1991)
OH 1.29E-010	cm3/molc sec	25	EST	MEYLAN,WM & HOWARD,PH (1993)

CAS #: 067832-97-7	2-ETHYLBENZAMIDE
Formula: $C_9H_{11}NO$	
Mol Weight: 149.19	
MP (deg C):	FP (deg C):
BP (deg C):	
BP pressure (mm Hg):	

Property/ Value	Units	Temp	Data Type	Reference
WS 7.66E+003	mg/L	25	EST	MEYLAN,WM ET AL. (1996)
logP 1.17			EXP	NAKAGAWA,Y ET AL. (1992)
VP 1.23E-004	mm Hg	25	EST	NEELY,WB & BLAU,GE (1985)
DC	pKa			
HL 3.24E-009	atm m3/mol	25	EST	MEYLAN,WM & HOWARD,PH (1991)
OH 7.52E-012	cm3/molc sec	25	EST	MEYLAN,WM & HOWARD,PH (1993)

CAS #: 067845-28-7				2-METHYL-6-PROPOXYPYRAZINE
Formula: $C_8H_{12}N_2O$				
Mol Weight: 152.20				
MP (deg C):		FP (deg C):		
BP (deg C):				
BP pressure (mm Hg):				

Property/Value	Units	Temp	Data Type	Reference
WS 8.22E+002	mg/L	25	EST	MEYLAN,WM ET AL. (1996)
logP 2.29			EXP	HANSCH,C & LEO,AJ (1985)
VP 5.30E-002	mm Hg	25	EST	NEELY,WB & BLAU,GE (1985)
DC	pKa			
HL 3.06E-006	atm m3/mol	25	EST	MEYLAN,WM & HOWARD,PH (1991)
OH 1.72E-011	cm3/molc sec	25	EST	MEYLAN,WM & HOWARD,PH (1993)

CAS #: 067856-65-9				N-NITROSO-BIS(2-MEOET)AMINE
Formula: $C_6H_{14}N_2O_3$				
Mol Weight: 162.19				
MP (deg C):		FP (deg C):		
BP (deg C):				
BP pressure (mm Hg):				

Property/Value	Units	Temp	Data Type	Reference
WS 1.41E+005	mg/L	25	EST	MEYLAN,WM ET AL. (1996)
logP -0.38			EXP	HANSCH,C & LEO,AJ (1985)
VP 1.06E-002	mm Hg	25	EST	NEELY,WB & BLAU,GE (1985)
DC	pKa			
HL 4.97E-010	atm m3/mol	25	EST	MEYLAN,WM & HOWARD,PH (1991)
OH 7.65E-011	cm3/molc sec	25	EST	MEYLAN,WM & HOWARD,PH (1993)

CAS #: 067856-66-0				N-NITROSO-BIS(2-ETOET)AMINE
Formula: $C_8H_{18}N_2O_3$				
Mol Weight: 190.24				
MP (deg C):		FP (deg C):		
BP (deg C):				
BP pressure (mm Hg):				

Property/Value	Units	Temp	Data Type	Reference
WS 1.04E+004	mg/L	25	EST	MEYLAN,WM ET AL. (1996)
logP 0.79			EXP	HANSCH,C & LEO,AJ (1985)
VP 1.46E-003	mm Hg	25	EST	NEELY,WB & BLAU,GE (1985)
DC	pKa			
HL 8.76E-010	atm m3/mol	25	EST	MEYLAN,WM & HOWARD,PH (1991)
OH 8.72E-011	cm3/molc sec	25	EST	MEYLAN,WM & HOWARD,PH (1993)

CAS #: 067888-96-4				2,2',4,5,5'-PENTABROMOBIPHENYL
Formula: $C_{12}H_5Br_5$				
Mol Weight: 548.72				
MP (deg C):		FP (deg C):		
BP (deg C):				
BP pressure (mm Hg):				

Property/Value	Units	Temp	Data Type	Reference
WS 1.03E-004	mg/L	25	EXP	DOUCETTE,WJ & ANDREN,AW (1988)
logP 8.21			EST	MEYLAN,WM & HOWARD,PH (1995)
VP 2.12E-008	mm Hg	25	EST	NEELY,WB & BLAU,GE (1985)
DC	pKa			
HL 4.15E-006	atm m3/mol	25	EST	MEYLAN,WM & HOWARD,PH (1991)
OH 2.83E-013	cm3/molc sec	25	EST	MEYLAN,WM & HOWARD,PH (1993)

CAS #: 067950-78-1				B21C7-BENZO CROWN ETHER
Formula: $C_{18}H_{28}O_7$				
Mol Weight: 356.42				
MP (deg C):		FP (deg C):		
BP (deg C):				
BP pressure (mm Hg):				

Property/Value	Units	Temp	Data Type	Reference
WS 1.85E+003	mg/L	25	EST	MEYLAN,WM ET AL. (1996)
logP 0.57			EXP	STOLWIJK,TB ET AL. (1989)
VP 1.45E-008	mm Hg	25	EST	NEELY,WB & BLAU,GE (1985)
DC	pKa			
HL 7.53E-015	atm m3/mol	25	EST	MEYLAN,WM & HOWARD,PH (1991)
OH 1.03E-010	cm3/molc sec	25	EST	MEYLAN,WM & HOWARD,PH (1993)

CAS #: 067952-57-2				6-METHYL-2-HEPTANOL ACETATE
Formula: $C_{10}H_{20}O_2$				
Mol Weight: 172.27				
MP (deg C):		FP (deg C):		
BP (deg C):				
BP pressure (mm Hg):				

Property/Value	Units	Temp	Data Type	Reference
WS 4.46E+001	mg/L	25	EST	MEYLAN,WM ET AL. (1996)
logP 3.66			EST	MEYLAN,WM & HOWARD,PH (1995)
VP 6.99E-001	mm Hg	25	EST	NEELY,WB & BLAU,GE (1985)
DC	pKa			
HL 1.27E-003	atm m3/mol	25	EST	MEYLAN,WM & HOWARD,PH (1991)
OH 1.10E-011	cm3/molc sec	25	EST	MEYLAN,WM & HOWARD,PH (1993)

CAS #: 068019-78-3				4-METHOXY-5-NITRO-1H-IMIDAZOLE
Formula: $C_4H_5N_3O_3$				
Mol Weight: 143.10				
MP (deg C):		FP (deg C):		
BP (deg C):				
BP pressure (mm Hg):				

Property/Value	Units	Temp	Data Type	Reference
WS 2.56E+004	mg/L	25	EST	MEYLAN,WM ET AL. (1996)
logP 0.41			EST	MEYLAN,WM & HOWARD,PH (1995)
VP 2.47E-005	mm Hg	25	EST	NEELY,WB & BLAU,GE (1985)
DC	pKa			
HL 8.77E-010	atm m3/mol	25	EST	MEYLAN,WM & HOWARD,PH (1991)
OH 4.72E-012	cm3/molc sec	25	EST	MEYLAN,WM & HOWARD,PH (1993)

CAS #: 068085-85-8				CYHALOTHRIN
Formula: $C_{23}H_{19}ClF_3NO_3$				
Mol Weight: 449.86				
MP (deg C):		FP (deg C):		
BP (deg C): 187-190				
BP pressure (mm Hg): 2.00E+000				

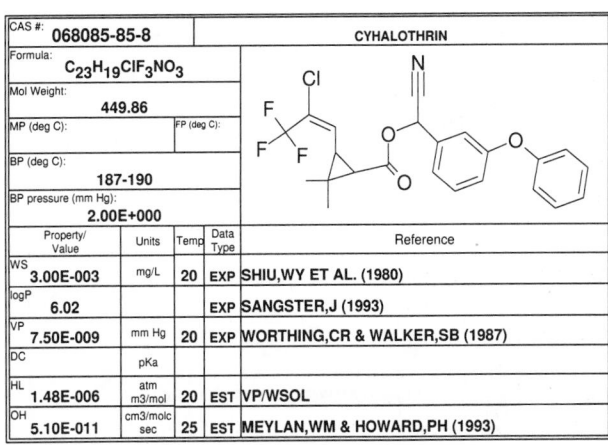

Property/Value	Units	Temp	Data Type	Reference
WS 3.00E-003	mg/L	20	EXP	SHIU,WY ET AL. (1980)
logP 6.02			EXP	SANGSTER,J (1993)
VP 7.50E-009	mm Hg	20	EXP	WORTHING,CR & WALKER,SB (1987)
DC	pKa			
HL 1.48E-006	atm m3/mol	20	EST	VP/WSOL
OH 5.10E-011	cm3/molc sec	25	EST	MEYLAN,WM & HOWARD,PH (1993)

CAS #: 068358-79-2 — 3(ME-MEO)1(2(4MEPH)ETO) PH UREA

Formula: $C_{18}H_{22}N_2O_3$

Mol Weight: 314.39

MP (deg C):

FP (deg C):

BP (deg C):

BP pressure (mm Hg):

Property/Value	Units	Temp	Data Type	Reference
WS 5.64E+000	mg/L	25	EST	MEYLAN,WM ET AL. (1996)
logP 3.81			EXP	HANSCH,C & LEO,AJ (1985)
VP 5.10E-009	mm Hg	25	EST	NEELY,WB & BLAU,GE (1985)
DC	pKa			
HL 1.47E-010	atm m3/mol	25	EST	MEYLAN,WM & HOWARD,PH (1991)
OH 5.51E-011	cm3/molc sec	25	EST	MEYLAN,WM & HOWARD,PH (1993)

CAS #: 068359-37-5 — CYFLUTHRIN

Formula: $C_{22}H_{18}Cl_2FNO_3$

Mol Weight: 434.30

MP (deg C): 60

FP (deg C):

BP (deg C):

BP pressure (mm Hg):

Property/Value	Units	Temp	Data Type	Reference
WS 2.00E+000	mg/L	20	EXP	SHIU,WY ET AL. (1990)
logP 5.94			EXP	TOMLIN,C (1994)
VP 2.03E-009	mm Hg	25	EXP	TOMLIN,C (1994)
DC	pKa			
HL 5.80E-010	atm m3/mol	25	EST	VP/WSOL
OH 1.37E-011	cm3/molc sec	25	EST	MEYLAN,WM & HOWARD,PH (1993)

CAS #: 068505-69-1 — BENFURESATE

Formula: $C_{12}H_{16}O_4S$

Mol Weight: 256.32

MP (deg C):

FP (deg C):

BP (deg C):

BP pressure (mm Hg):

Property/Value	Units	Temp	Data Type	Reference
WS 1.91E+002	mg/L	25	EST	MEYLAN,WM ET AL. (1996)
logP 2.41			EXP	TOMLIN,C (1994)
VP 5.85E-006	mm Hg	25	EST	NEELY,WB & BLAU,GE (1985)
DC	pKa			
HL 5.43E-008	atm m3/mol	25	EST	MEYLAN,WM & HOWARD,PH (1991)
OH 1.23E-011	cm3/molc sec	25	EST	MEYLAN,WM & HOWARD,PH (1993)

CAS #: 068648-87-3 — ALKYLBENZENE (C10-C15)

Formula: $C_{18}H_{30}$

Mol Weight: 246.44

REPRESENTATIVE COMPOUND

MP (deg C):

FP (deg C):

BP (deg C):

BP pressure (mm Hg):

Property/Value	Units	Temp	Data Type	Reference
WS 1.02E-003	mg/L	25	EST	MEYLAN,WM ET AL. (1996)
logP 7.94			EST	MEYLAN,WM & HOWARD,PH (1995)
VP 2.45E-004	mm Hg	25	EST	NEELY,WB & BLAU,GE (1985)
DC	pKa			
HL 1.34E-001	atm m3/mol	25	EST	MEYLAN,WM & HOWARD,PH (1991)
OH 2.00E-011	cm3/molc sec	25	EST	MEYLAN,WM & HOWARD,PH (1993)

CAS #: 068659-48-3 — ACETAMIDE, N-ACETYL-2-(BENZOYLOXY)-

Formula: $C_{11}H_{11}NO_4$

Mol Weight: 221.21

MP (deg C):

FP (deg C):

BP (deg C):

BP pressure (mm Hg):

Property/Value	Units	Temp	Data Type	Reference
WS 2.96E+003	mg/L	25	EST	MEYLAN,WM ET AL. (1996)
logP 1.24			EXP	NIELSEN,LS & BUNDGAARD,H (1988)
VP 3.73E-009	mm Hg	25	EST	NEELY,WB & BLAU,GE (1985)
DC	pKa			
HL 1.92E-010	atm m3/mol	25	EST	MEYLAN,WM & HOWARD,PH (1991)
OH 7.35E-012	cm3/molc sec	25	EST	MEYLAN,WM & HOWARD,PH (1993)

CAS #: 068661-74-5 — 5-THIA-1-AZABICYCLO[4.2.0]OCT-2-ENE-2-CARBOXYLIC

Formula: $C_{22}H_{26}N_2O_8S_2$

Mol Weight: 510.59

MP (deg C):

FP (deg C):

BP (deg C):

BP pressure (mm Hg):

Property/Value	Units	Temp	Data Type	Reference
WS 2.79E+001	mg/L	25	EST	MEYLAN,WM ET AL. (1996)
logP 1.57			EXP	SANGSTER,J (1993)
VP 3.08E-015	mm Hg	25	EST	NEELY,WB & BLAU,GE (1985)
DC	pKa			
HL 3.11E-017	atm m3/mol	25	EST	MEYLAN,WM & HOWARD,PH (1991)
OH 1.37E-010	cm3/molc sec	25	EST	MEYLAN,WM & HOWARD,PH (1993)

CAS #: 068672-91-3 — N-THIO-(N'-BU-O-ME-CARBAMYL)CARBOFURAN

Formula: $C_{18}H_{26}N_2O_5S$

Mol Weight: 382.48

MP (deg C):

FP (deg C):

BP (deg C):

BP pressure (mm Hg):

Property/Value	Units	Temp	Data Type	Reference
WS 4.63E-001	mg/L	25	EST	MEYLAN,WM ET AL. (1996)
logP 4.60			EXP	HANSCH,C & LEO,AJ (1985)
VP 8.17E-008	mm Hg	25	EST	NEELY,WB & BLAU,GE (1985)
DC	pKa			
HL 7.92E-009	atm m3/mol	25	EST	MEYLAN,WM & HOWARD,PH (1991)
OH 6.10E-011	cm3/molc sec	25	EST	MEYLAN,WM & HOWARD,PH (1993)

CAS #: 068694-11-1 — TRIFLUMIZOLE

Formula: $C_{15}H_{17}ClF_3N_3O$

Mol Weight: 347.77

MP (deg C): 63.5

FP (deg C):

BP (deg C):

BP pressure (mm Hg):

Property/Value	Units	Temp	Data Type	Reference
WS 1.25E+004	mg/L	20	EXP	SHIU,WY ET AL. (1990)
logP 1.40			EXP	HANSCH,C ET AL. (1995)
VP 1.10E-008	mm Hg	25	EXP	AUGUSTIJN-BECKERS,PWM ET AL. (1994)
DC	pKa			
HL 4.03E-013	atm m3/mol	25	EST	VP/WSOL
OH 5.37E-011	cm3/molc sec	25	EST	MEYLAN,WM & HOWARD,PH (1993)

CAS #: 068709-03-5 — CARBAMIC ACID, (1,2,3,6-TETRAHYDRO-1-PYRIDINYL)-

Formula: $C_8H_{14}N_2O_2$
Mol Weight: 170.21
MP (deg C):
FP (deg C):
BP (deg C):
BP pressure (mm Hg):

Property/Value	Units	Temp	Data Type	Reference
WS 7.19E+003	mg/L	25	EST	MEYLAN,WM ET AL. (1996)
logP 1.09			EXP	SANGSTER,J (1993)
VP 1.97E-002	mm Hg	25	EST	NEELY,WB & BLAU,GE (1985)
DC	pKa			
HL 9.60E-010	atm m3/mol	25	EST	MEYLAN,WM & HOWARD,PH (1991)
OH 8.41E-011	cm3/molc sec	25	EST	MEYLAN,WM & HOWARD,PH (1993)

CAS #: 068724-15-2 — 3-FURANMETHANAMINIUM, N,N,N,5-TETRAMETHYL-, IODI

Formula: $C_9H_{16}INO$
Mol Weight: 281.14
MP (deg C):
FP (deg C):
BP (deg C):
BP pressure (mm Hg):

Property/Value	Units	Temp	Data Type	Reference
WS 8.06E+005	mg/L	25	EST	MEYLAN,WM ET AL. (1996)
logP -2.00			EXP	PRATESI,P ET AL. (1986)
VP 2.23E-008	mm Hg	25	EST	NEELY,WB & BLAU,GE (1985)
DC	pKa			
HL 2.34E-013	atm m3/mol	25	EST	MEYLAN,WM & HOWARD,PH (1991)
OH 2.13E-010	cm3/molc sec	25	EST	MEYLAN,WM & HOWARD,PH (1993)

CAS #: 068724-25-4 — 3-FURANMETHANAMINIUM, TETRAHYDRO-N,N,N,5-TETRAME

Formula: $C_9H_{20}INO$
Mol Weight: 285.17
MP (deg C):
FP (deg C):
BP (deg C):
BP pressure (mm Hg):

Property/Value	Units	Temp	Data Type	Reference
WS 1.00E+006	mg/L	25	EST	MEYLAN,WM ET AL. (1996)
logP -2.15			EXP	PRATESI,P ET AL. (1986)
VP 3.11E-008	mm Hg	25	EST	NEELY,WB & BLAU,GE (1985)
DC	pKa			
HL 5.32E-015	atm m3/mol	25	EST	MEYLAN,WM & HOWARD,PH (1991)
OH 4.12E-011	cm3/molc sec	25	EST	MEYLAN,WM & HOWARD,PH (1993)

CAS #: 068760-70-3 — 6-DECENYL ACETATE (Z)

Formula: $C_{12}H_{22}O_2$
Mol Weight: 198.31
MP (deg C):
FP (deg C):
BP (deg C):
BP pressure (mm Hg):

Property/Value	Units	Temp	Data Type	Reference
WS 5.50E+000	mg/L	25	EST	MEYLAN,WM ET AL. (1996)
logP 4.58			EST	MEYLAN,WM & HOWARD,PH (1995)
VP 2.31E-002	mm Hg	25	EST	NEELY,WB & BLAU,GE (1985)
DC	pKa			
HL 1.98E-003	atm m3/mol	25	EST	MEYLAN,WM & HOWARD,PH (1991)
OH 6.61E-011	cm3/molc sec	25	EST	MEYLAN,WM & HOWARD,PH (1993)

CAS #: 068776-62-5 — 4-PYRIDAZINECARBONITRILE

Formula: $C_5H_3N_3$
Mol Weight: 105.10
MP (deg C):
FP (deg C):
BP (deg C):
BP pressure (mm Hg):

Property/Value	Units	Temp	Data Type	Reference
WS 2.12E+005	mg/L	25	EST	MEYLAN,WM ET AL. (1996)
logP -0.63			EXP	YAMAGAMI,C ET AL. (1990)
VP 4.14E-003	mm Hg	25	EST	NEELY,WB & BLAU,GE (1985)
DC	pKa			
HL 2.77E-008	atm m3/mol	25	EST	MEYLAN,WM & HOWARD,PH (1991)
OH 4.06E-014	cm3/molc sec	25	EST	MEYLAN,WM & HOWARD,PH (1993)

CAS #: 068840-05-1 — CARBOFURAN,N-(N'-ME,MECARBAMIDOTHIO)

Formula: $C_{15}H_{20}N_2O_4S$
Mol Weight: 324.40
MP (deg C):
FP (deg C):
BP (deg C):
BP pressure (mm Hg):

Property/Value	Units	Temp	Data Type	Reference
WS 4.36E+001	mg/L	25	EST	MEYLAN,WM ET AL. (1996)
logP 2.70			EXP	HANSCH,C & LEO,AJ (1985)
VP 4.83E-008	mm Hg	25	EST	NEELY,WB & BLAU,GE (1985)
DC	pKa			
HL 9.57E-010	atm m3/mol	25	EST	MEYLAN,WM & HOWARD,PH (1991)
OH 5.40E-011	cm3/molc sec	25	EST	MEYLAN,WM & HOWARD,PH (1993)

CAS #: 068877-63-4 — AZO DYE N1

Formula: $C_{21}H_{20}BrN_7O_6$
Mol Weight: 546.34
MP (deg C):
FP (deg C):
BP (deg C):
BP pressure (mm Hg):

Property/Value	Units	Temp	Data Type	Reference
WS 6.90E-004	mg/L	25	EXP	BAUGHMAN,GL ET AL. (1993)
logP 5.40			EXP	BAUGHMAN,GL & WEBER,EJ (1991)
VP 4.68E-017	mm Hg	25	EST	NEELY,WB & BLAU,GE (1985)
DC	pKa			
HL 7.11E-023	atm m3/mol	25	EST	MEYLAN,WM & HOWARD,PH (1991)
OH 2.37E-010	cm3/molc sec	25	EST	MEYLAN,WM & HOWARD,PH (1993)

CAS #: 068885-48-3 — 2,4-NH2PYRIMIDINE,5(3MEO-4-MES)BENZYL

Formula: $C_{13}H_{16}N_4OS$
Mol Weight: 276.36
MP (deg C):
FP (deg C):
BP (deg C):
BP pressure (mm Hg):

Property/Value	Units	Temp	Data Type	Reference
WS 3.85E+002	mg/L	25	EST	MEYLAN,WM ET AL. (1996)
logP 1.92			EXP	HANSCH,C & LEO,AJ (1985)
VP 1.17E-008	mm Hg	25	EST	NEELY,WB & BLAU,GE (1985)
DC	pKa			
HL 1.99E-013	atm m3/mol	25	EST	MEYLAN,WM & HOWARD,PH (1991)
OH 3.72E-011	cm3/molc sec	25	EST	MEYLAN,WM & HOWARD,PH (1993)

CAS #: 068902-57-8 — 24-NH2PYRIMIDIN,5(35-MEO-4-MES)BENZYL

Formula:	$C_{14}H_{18}N_4O_2S$
Mol Weight:	306.39
MP (deg C):	FP (deg C):
BP (deg C):	
BP pressure (mm Hg):	

Property/Value	Units	Temp	Data Type	Reference
WS 5.45E+002	mg/L	25	EST	MEYLAN,WM ET AL. (1996)
logP 1.54			EXP	HANSCH,C & LEO,AJ (1985)
VP 2.24E-009	mm Hg	25	EST	NEELY,WB & BLAU,GE (1985)
DC	pKa			
HL 1.18E-014	atm m3/mol	25	EST	MEYLAN,WM & HOWARD,PH (1991)
OH 2.04E-010	cm3/molc sec	25	EST	MEYLAN,WM & HOWARD,PH (1993)

CAS #: 068960-60-1 — 2-AZETIDINONE, 4-[(4-CHLOROPHENYL)THIO]-

Formula:	C_9H_8ClNOS
Mol Weight:	213.69
MP (deg C):	FP (deg C):
BP (deg C):	
BP pressure (mm Hg):	

Property/Value	Units	Temp	Data Type	Reference
WS 3.96E+002	mg/L	25	EST	MEYLAN,WM ET AL. (1996)
logP 2.31			EXP	ARNOLDI,A ET AL. (1988)
VP 1.66E-006	mm Hg	25	EST	NEELY,WB & BLAU,GE (1985)
DC	pKa			
HL 8.00E-009	atm m3/mol	25	EST	MEYLAN,WM & HOWARD,PH (1991)
OH 7.72E-011	cm3/molc sec	25	EST	MEYLAN,WM & HOWARD,PH (1993)

CAS #: 069014-14-8 — GUANIDINE, N-[2-[[[2-[(AMINOIMINOMETHYL)AMINO]-4

Formula:	$C_{10}H_{16}N_8S_2$
Mol Weight:	312.42
MP (deg C):	FP (deg C):
BP (deg C):	
BP pressure (mm Hg):	

Property/Value	Units	Temp	Data Type	Reference
WS 2.73E+003	mg/L	25	EST	MEYLAN,WM ET AL. (1996)
logP 0.68			EXP	HANSCH,C ET AL. (1995)
VP 5.31E-010	mm Hg	25	EST	NEELY,WB & BLAU,GE (1985)
DC	pKa			
HL 2.11E-024	atm m3/mol	25	EST	MEYLAN,WM & HOWARD,PH (1991)
OH 2.02E-010	cm3/molc sec	25	EST	MEYLAN,WM & HOWARD,PH (1993)

CAS #: 069049-73-6 — NEDOCROMIL

Formula:	$C_{19}H_{17}NO_7$
Mol Weight:	371.35
MP (deg C): 298-300 de	FP (deg C):
BP (deg C):	
BP pressure (mm Hg):	

Property/Value	Units	Temp	Data Type	Reference
WS 1.45E+002	mg/L	25	EST	MEYLAN,WM ET AL. (1996)
logP 2.22			EXP	SANGSTER,J (1994)
VP 5.40E-013	mm Hg	25	EST	NEELY,WB & BLAU,GE (1985)
DC	pKa			
HL 6.83E-020	atm m3/mol	25	EST	MEYLAN,WM & HOWARD,PH (1991)
OH 2.98E-011	cm3/molc sec	25	EST	MEYLAN,WM & HOWARD,PH (1993)

CAS #: 069148-74-9 — 2-IPRO PHENYLDIMETHYLCARBAMATE

Formula:	$C_{12}H_{17}NO_3$
Mol Weight:	223.27
MP (deg C):	FP (deg C):
BP (deg C):	
BP pressure (mm Hg):	

Property/Value	Units	Temp	Data Type	Reference
WS 5.76E+002	mg/L	25	EST	MEYLAN,WM ET AL. (1996)
logP 2.06			EXP	HANSCH,C & LEO,AJ (1985)
VP 1.31E-003	mm Hg	25	EST	NEELY,WB & BLAU,GE (1985)
DC	pKa			
HL 7.35E-009	atm m3/mol	25	EST	MEYLAN,WM & HOWARD,PH (1991)
OH 4.41E-011	cm3/molc sec	25	EST	MEYLAN,WM & HOWARD,PH (1993)

CAS #: 069163-87-7 — 3-PH PROPIONALDEHYDE OXIME,ME ETHER

Formula:	$C_{10}H_{13}NO$
Mol Weight:	163.22
MP (deg C):	FP (deg C):
BP (deg C):	
BP pressure (mm Hg):	

Property/Value	Units	Temp	Data Type	Reference
WS 9.84E+002	mg/L	25	EST	MEYLAN,WM ET AL. (1996)
logP 2.14			EXP	HANSCH,C & LEO,AJ (1985)
VP 5.01E-002	mm Hg	25	EST	NEELY,WB & BLAU,GE (1985)
DC	pKa			
HL 2.02E-004	atm m3/mol	25	EST	MEYLAN,WM & HOWARD,PH (1991)
OH 9.82E-012	cm3/molc sec	25	EST	MEYLAN,WM & HOWARD,PH (1993)

CAS #: 069194-89-4 — HYDRAZINECARBOXAMIDE, N-(4-CHLOROPHENYL)

Formula:	$C_7H_8ClN_3O$
Mol Weight:	185.61
MP (deg C):	FP (deg C):
BP (deg C):	
BP pressure (mm Hg):	

Property/Value	Units	Temp	Data Type	Reference
WS 9.34E+003	mg/L	25	EST	MEYLAN,WM ET AL. (1996)
logP 0.87			EXP	KRAMER,CR & BECK,L (1981)
VP 2.02E-005	mm Hg	25	EST	NEELY,WB & BLAU,GE (1985)
DC	pKa			
HL 6.23E-013	atm m3/mol	25	EST	MEYLAN,WM & HOWARD,PH (1991)
OH 1.35E-011	cm3/molc sec	25	EST	MEYLAN,WM & HOWARD,PH (1993)

CAS #: 069194-91-8 — 24-NH2PYRIMIDINE,5(35MEO-4-PRENYL)BENZYL

Formula:	$C_{16}H_{20}N_4O_2$
Mol Weight:	300.36
MP (deg C):	FP (deg C):
BP (deg C):	
BP pressure (mm Hg):	

Property/Value	Units	Temp	Data Type	Reference
WS 2.17E+002	mg/L	25	EST	MEYLAN,WM ET AL. (1996)
logP 2.05			EXP	HANSCH,C & LEO,AJ (1985)
VP 3.53E-009	mm Hg	25	EST	NEELY,WB & BLAU,GE (1985)
DC	pKa			
HL 3.24E-013	atm m3/mol	25	EST	MEYLAN,WM & HOWARD,PH (1991)
OH 2.59E-010	cm3/molc sec	25	EST	MEYLAN,WM & HOWARD,PH (1993)

CAS #: 069218-26-4				1-PENTABENZOTRIAZOLE

Formula: $C_{11}H_{15}N_3$
Mol Weight: 189.26
MP (deg C):   FP (deg C):
BP (deg C):
BP pressure (mm Hg):

Property/Value	Units	Temp	Data Type	Reference
WS 8.82E+001	mg/L	25	EST	MEYLAN,WM ET AL. (1996)
logP 3.22			EXP	HANSCH,C & LEO,AJ (1985)
VP 1.59E-004	mm Hg	25	EST	NEELY,WB & BLAU,GE (1985)
DC	pKa			
HL 9.72E-006	atm m3/mol	25	EST	MEYLAN,WM & HOWARD,PH (1991)
OH 6.29E-012	cm3/molc sec	25	EST	MEYLAN,WM & HOWARD,PH (1993)

CAS #: 069218-27-5				2-PENTABENZOTRIAZOLE

Formula: $C_{11}H_{17}N_3$
Mol Weight: 191.28
MP (deg C):   FP (deg C):
BP (deg C):
BP pressure (mm Hg):

Property/Value	Units	Temp	Data Type	Reference
WS 3.94E+001	mg/L	25	EST	MEYLAN,WM ET AL. (1996)
logP 3.63			EXP	HANSCH,C & LEO,AJ (1985)
VP 1.59E-004	mm Hg	25	EST	NEELY,WB & BLAU,GE (1985)
DC	pKa			
HL 9.72E-006	atm m3/mol	25	EST	MEYLAN,WM & HOWARD,PH (1991)
OH 6.29E-012	cm3/molc sec	25	EST	MEYLAN,WM & HOWARD,PH (1993)

CAS #: 069218-28-6				2-HEXYLBENZOTRIAZOLE

Formula: $C_{12}H_{19}N_3$
Mol Weight: 205.31
MP (deg C):   FP (deg C):
BP (deg C):
BP pressure (mm Hg):

Property/Value	Units	Temp	Data Type	Reference
WS 1.01E+001	mg/L	25	EST	MEYLAN,WM ET AL. (1996)
logP 4.24			EXP	HANSCH,C & LEO,AJ (1985)
VP 6.29E-005	mm Hg	25	EST	NEELY,WB & BLAU,GE (1985)
DC	pKa			
HL 1.29E-005	atm m3/mol	25	EST	MEYLAN,WM & HOWARD,PH (1991)
OH 7.70E-012	cm3/molc sec	25	EST	MEYLAN,WM & HOWARD,PH (1993)

CAS #: 069218-29-7				1-I-PROPYLBENZOTRIAZOLE

Formula: $C_9H_{11}N_3$
Mol Weight: 161.21
MP (deg C):   FP (deg C):
BP (deg C):
BP pressure (mm Hg):

Property/Value	Units	Temp	Data Type	Reference
WS 1.38E+003	mg/L	25	EST	MEYLAN,WM ET AL. (1996)
logP 1.98			EXP	HANSCH,C & LEO,AJ (1985)
VP 2.33E-003	mm Hg	25	EST	NEELY,WB & BLAU,GE (1985)
DC	pKa			
HL 5.51E-006	atm m3/mol	25	EST	MEYLAN,WM & HOWARD,PH (1991)
OH 3.27E-012	cm3/molc sec	25	EST	MEYLAN,WM & HOWARD,PH (1993)

CAS #: 069218-30-0				1-S-HEXYLBENZOTRIAZOLE

Formula: $C_{12}H_{17}N_3$
Mol Weight: 203.29
MP (deg C):   FP (deg C):
BP (deg C):
BP pressure (mm Hg):

Property/Value	Units	Temp	Data Type	Reference
WS 4.86E+001	mg/L	25	EST	MEYLAN,WM ET AL. (1996)
logP 3.44			EXP	HANSCH,C & LEO,AJ (1985)
VP 1.15E-004	mm Hg	25	EST	NEELY,WB & BLAU,GE (1985)
DC	pKa			
HL 1.29E-005	atm m3/mol	25	EST	MEYLAN,WM & HOWARD,PH (1991)
OH 7.70E-012	cm3/molc sec	25	EST	MEYLAN,WM & HOWARD,PH (1993)

CAS #: 069218-31-1				2-I-PROPYLBENZOTRIAZOLE

Formula: $C_9H_{13}N_3$
Mol Weight: 163.22
MP (deg C):   FP (deg C):
BP (deg C):
BP pressure (mm Hg):

Property/Value	Units	Temp	Data Type	Reference
WS 4.67E+002	mg/L	25	EST	MEYLAN,WM ET AL. (1996)
logP 2.53			EXP	HANSCH,C & LEO,AJ (1985)
VP 2.33E-003	mm Hg	25	EST	NEELY,WB & BLAU,GE (1985)
DC	pKa			
HL 5.51E-006	atm m3/mol	25	EST	MEYLAN,WM & HOWARD,PH (1991)
OH 3.27E-012	cm3/molc sec	25	EST	MEYLAN,WM & HOWARD,PH (1993)

CAS #: 069218-32-2				2-S-PENTABENZOTRIAZOLE

Formula: $C_{11}H_{17}N_3$
Mol Weight: 191.28
MP (deg C):   FP (deg C):
BP (deg C):
BP pressure (mm Hg):

Property/Value	Units	Temp	Data Type	Reference
WS 5.95E+001	mg/L	25	EST	MEYLAN,WM ET AL. (1996)
logP 3.42			EXP	HANSCH,C & LEO,AJ (1985)
VP 2.99E-004	mm Hg	25	EST	NEELY,WB & BLAU,GE (1985)
DC	pKa			
HL 9.72E-006	atm m3/mol	25	EST	MEYLAN,WM & HOWARD,PH (1991)
OH 6.28E-012	cm3/molc sec	25	EST	MEYLAN,WM & HOWARD,PH (1993)

CAS #: 069218-33-3				2-S-HEXYLBENZOTRIAZOLE

Formula: $C_{12}H_{19}N_3$
Mol Weight: 205.31
MP (deg C):   FP (deg C):
BP (deg C):
BP pressure (mm Hg):

Property/Value	Units	Temp	Data Type	Reference
WS 2.49E+001	mg/L	25	EST	MEYLAN,WM ET AL. (1996)
logP 3.78			EXP	HANSCH,C & LEO,AJ (1985)
VP 1.15E-004	mm Hg	25	EST	NEELY,WB & BLAU,GE (1985)
DC	pKa			
HL 1.29E-005	atm m3/mol	25	EST	MEYLAN,WM & HOWARD,PH (1991)
OH 7.70E-012	cm3/molc sec	25	EST	MEYLAN,WM & HOWARD,PH (1993)

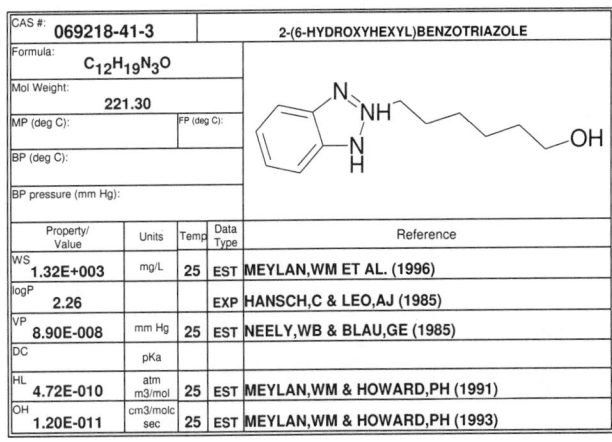

CAS #: 069218-34-4 — 1-(3-HYDROXYPROPYL)BENZOTRIAZOLE

Formula: $C_9H_{11}N_3O$

Mol Weight: 177.21

MP (deg C): | FP (deg C):

BP (deg C):

BP pressure (mm Hg):

Property/Value	Units	Temp	Data Type	Reference
WS 3.07E+004	mg/L	25	EST	MEYLAN,WM ET AL. (1996)
logP 0.91			EXP	HANSCH,C & LEO,AJ (1985)
VP 1.33E-006	mm Hg	25	EST	NEELY,WB & BLAU,GE (1985)
DC	pKa			
HL 2.02E-010	atm m3/mol	25	EST	MEYLAN,WM & HOWARD,PH (1991)
OH 7.72E-012	cm3/molc sec	25	EST	MEYLAN,WM & HOWARD,PH (1993)

CAS #: 069218-35-5 — 1-(4-HYDROXYBUTYL)BENZOTRIAZOLE

Formula: $C_{10}H_{13}N_3O$

Mol Weight: 191.23

MP (deg C): | FP (deg C):

BP (deg C):

BP pressure (mm Hg):

Property/Value	Units	Temp	Data Type	Reference
WS 2.07E+004	mg/L	25	EST	MEYLAN,WM ET AL. (1996)
logP 1.03			EXP	HANSCH,C & LEO,AJ (1985)
VP 6.10E-007	mm Hg	25	EST	NEELY,WB & BLAU,GE (1985)
DC	pKa			
HL 2.68E-010	atm m3/mol	25	EST	MEYLAN,WM & HOWARD,PH (1991)
OH 9.14E-012	cm3/molc sec	25	EST	MEYLAN,WM & HOWARD,PH (1993)

CAS #: 069218-36-6 — 1-(5-HYDROXYPENTYL)BENZOTRIAZOLE

Formula: $C_{11}H_{15}N_3O$

Mol Weight: 205.26

MP (deg C): | FP (deg C):

BP (deg C):

BP pressure (mm Hg):

Property/Value	Units	Temp	Data Type	Reference
WS 6.83E+003	mg/L	25	EST	MEYLAN,WM ET AL. (1996)
logP 1.51			EXP	HANSCH,C & LEO,AJ (1985)
VP 2.34E-007	mm Hg	25	EST	NEELY,WB & BLAU,GE (1985)
DC	pKa			
HL 3.55E-010	atm m3/mol	25	EST	MEYLAN,WM & HOWARD,PH (1991)
OH 1.05E-011	cm3/molc sec	25	EST	MEYLAN,WM & HOWARD,PH (1993)

CAS #: 069218-37-7 — 1-(6-HYDROXYHEXYL)BENZOTRIAZOLE

Formula: $C_{12}H_{17}N_3O$

Mol Weight: 219.29

MP (deg C): | FP (deg C):

BP (deg C):

BP pressure (mm Hg):

Property/Value	Units	Temp	Data Type	Reference
WS 3.89E+003	mg/L	25	EST	MEYLAN,WM ET AL. (1996)
logP 1.71			EXP	HANSCH,C & LEO,AJ (1985)
VP 8.90E-008	mm Hg	25	EST	NEELY,WB & BLAU,GE (1985)
DC	pKa			
HL 4.72E-010	atm m3/mol	25	EST	MEYLAN,WM & HOWARD,PH (1991)
OH 1.20E-011	cm3/molc sec	25	EST	MEYLAN,WM & HOWARD,PH (1993)

CAS #: 069218-38-8 — 2-(3-HYDROXYPROPYL)BENZOTRIAZOLE

Formula: $C_9H_{13}N_3O$

Mol Weight: 179.22

MP (deg C): | FP (deg C):

BP (deg C):

BP pressure (mm Hg):

Property/Value	Units	Temp	Data Type	Reference
WS 1.17E+004	mg/L	25	EST	MEYLAN,WM ET AL. (1996)
logP 1.40			EXP	HANSCH,C & LEO,AJ (1985)
VP 1.33E-006	mm Hg	25	EST	NEELY,WB & BLAU,GE (1985)
DC	pKa			
HL 2.02E-010	atm m3/mol	25	EST	MEYLAN,WM & HOWARD,PH (1991)
OH 7.72E-012	cm3/molc sec	25	EST	MEYLAN,WM & HOWARD,PH (1993)

CAS #: 069218-39-9 — 2-(4-HYDROXYBUTYL)BENZOTRIAZOLE

Formula: $C_{10}H_{15}N_3O$

Mol Weight: 193.25

MP (deg C): | FP (deg C):

BP (deg C):

BP pressure (mm Hg):

Property/Value	Units	Temp	Data Type	Reference
WS 7.16E+003	mg/L	25	EST	MEYLAN,WM ET AL. (1996)
logP 1.57			EXP	HANSCH,C & LEO,AJ (1985)
VP 6.10E-007	mm Hg	25	EST	NEELY,WB & BLAU,GE (1985)
DC	pKa			
HL 2.68E-010	atm m3/mol	25	EST	MEYLAN,WM & HOWARD,PH (1991)
OH 9.14E-012	cm3/molc sec	25	EST	MEYLAN,WM & HOWARD,PH (1993)

CAS #: 069218-40-2 — 2-(5-HYDROXYPENTYL)BENZOTRIAZOLE

Formula: $C_{11}H_{17}N_3O$

Mol Weight: 207.28

MP (deg C): | FP (deg C):

BP (deg C):

BP pressure (mm Hg):

Property/Value	Units	Temp	Data Type	Reference
WS 2.18E+003	mg/L	25	EST	MEYLAN,WM ET AL. (1996)
logP 2.09			EXP	SANGSTER,J (1994)
VP 2.34E-007	mm Hg	25	EST	NEELY,WB & BLAU,GE (1985)
DC	pKa			
HL 3.55E-010	atm m3/mol	25	EST	MEYLAN,WM & HOWARD,PH (1991)
OH 1.05E-011	cm3/molc sec	25	EST	MEYLAN,WM & HOWARD,PH (1993)

CAS #: 069218-41-3 — 2-(6-HYDROXYHEXYL)BENZOTRIAZOLE

Formula: $C_{12}H_{19}N_3O$

Mol Weight: 221.30

MP (deg C): | FP (deg C):

BP (deg C):

BP pressure (mm Hg):

Property/Value	Units	Temp	Data Type	Reference
WS 1.32E+003	mg/L	25	EST	MEYLAN,WM ET AL. (1996)
logP 2.26			EXP	HANSCH,C & LEO,AJ (1985)
VP 8.90E-008	mm Hg	25	EST	NEELY,WB & BLAU,GE (1985)
DC	pKa			
HL 4.72E-010	atm m3/mol	25	EST	MEYLAN,WM & HOWARD,PH (1991)
OH 1.20E-011	cm3/molc sec	25	EST	MEYLAN,WM & HOWARD,PH (1993)

CAS #:	069218-42-4				1H-BENZOTRIAZOLE-1-PENTANOIC ACID

Formula: $C_{11}H_{13}N_3O_2$

Mol Weight: 219.25

MP (deg C): | FP (deg C):

BP (deg C):

BP pressure (mm Hg):

Property/ Value	Units	Temp	Data Type	Reference
WS 3.36E+004	mg/L	25	EST	MEYLAN,WM ET AL. (1996)
logP 0.48			EXP	SPARATORE,F ET AL. (1978)
VP 6.73E-007	mm Hg	25	EST	NEELY,WB & BLAU,GE (1985)
DC	pKa			
HL 5.44E-011	atm m3/mol	25	EST	MEYLAN,WM & HOWARD,PH (1991)
OH 6.36E-012	cm3/molc sec	25	EST	MEYLAN,WM & HOWARD,PH (1993)

CAS #:	069218-43-5				1-(5-CARBOXYPENTYL)BENZOTRIAZOLE

Formula: $C_{12}H_{15}N_3O_2$

Mol Weight: 233.27

MP (deg C): | FP (deg C):

BP (deg C):

BP pressure (mm Hg):

Property/ Value	Units	Temp	Data Type	Reference
WS 1.34E+004	mg/L	25	EST	MEYLAN,WM ET AL. (1996)
logP 0.86			EXP	HANSCH,C & LEO,AJ (1985)
VP 3.06E-007	mm Hg	25	EST	NEELY,WB & BLAU,GE (1985)
DC	pKa			
HL 7.23E-011	atm m3/mol	25	EST	MEYLAN,WM & HOWARD,PH (1991)
OH 7.77E-012	cm3/molc sec	25	EST	MEYLAN,WM & HOWARD,PH (1993)

CAS #:	069218-44-6				2H-BENZOTRIAZOLE-2-PENTANOIC ACID

Formula: $C_{11}H_{13}N_3O_2$

Mol Weight: 219.25

MP (deg C): | FP (deg C):

BP (deg C):

BP pressure (mm Hg):

Property/ Value	Units	Temp	Data Type	Reference
WS 1.36E+004	mg/L	25	EST	MEYLAN,WM ET AL. (1996)
logP 0.94			EXP	SPARATORE,F ET AL. (1978)
VP 6.73E-007	mm Hg	25	EST	NEELY,WB & BLAU,GE (1985)
DC	pKa			
HL 5.44E-011	atm m3/mol	25	EST	MEYLAN,WM & HOWARD,PH (1991)
OH 6.36E-012	cm3/molc sec	25	EST	MEYLAN,WM & HOWARD,PH (1993)

CAS #:	069218-45-7				2-(5-CARBOXYPENTYL)BENZOTRIAZOLE

Formula: $C_{12}H_{17}N_3O_2$

Mol Weight: 235.29

MP (deg C): | FP (deg C):

BP (deg C):

BP pressure (mm Hg):

Property/ Value	Units	Temp	Data Type	Reference
WS 3.73E+003	mg/L	25	EST	MEYLAN,WM ET AL. (1996)
logP 1.51			EXP	HANSCH,C & LEO,AJ (1985)
VP 3.06E-007	mm Hg	25	EST	NEELY,WB & BLAU,GE (1985)
DC	pKa			
HL 7.23E-011	atm m3/mol	25	EST	MEYLAN,WM & HOWARD,PH (1991)
OH 7.77E-012	cm3/molc sec	25	EST	MEYLAN,WM & HOWARD,PH (1993)

CAS #:	069218-46-8				1H-BENZOTRIAZOLE-1-ACETIC ACID, ETHYL ESTER

Formula: $C_{10}H_{11}N_3O_2$

Mol Weight: 205.22

MP (deg C): | FP (deg C):

BP (deg C):

BP pressure (mm Hg):

Property/ Value	Units	Temp	Data Type	Reference
WS 3.19E+003	mg/L	25	EST	MEYLAN,WM ET AL. (1996)
logP 1.30			EXP	SANGSTER,J (1993)
VP 4.51E-005	mm Hg	25	EST	NEELY,WB & BLAU,GE (1985)
DC	pKa			
HL 9.89E-009	atm m3/mol	25	EST	MEYLAN,WM & HOWARD,PH (1991)
OH 3.35E-012	cm3/molc sec	25	EST	MEYLAN,WM & HOWARD,PH (1993)

CAS #:	069218-47-9				1H-BENZOTRIAZOLE-1-PROPANOIC ACID, ETHYL ESTER

Formula: $C_{11}H_{13}N_3O_2$

Mol Weight: 219.25

MP (deg C): | FP (deg C):

BP (deg C):

BP pressure (mm Hg):

Property/ Value	Units	Temp	Data Type	Reference
WS 1.49E+003	mg/L	25	EST	MEYLAN,WM ET AL. (1996)
logP 1.60			EXP	SPARATORE,F ET AL. (1978)
VP 1.90E-005	mm Hg	25	EST	NEELY,WB & BLAU,GE (1985)
DC	pKa			
HL 1.31E-008	atm m3/mol	25	EST	MEYLAN,WM & HOWARD,PH (1991)
OH 4.66E-012	cm3/molc sec	25	EST	MEYLAN,WM & HOWARD,PH (1993)

CAS #:	069218-48-0				1-(3-CARBETHOXYPROPYL)BENZOTRIAZOLE

Formula: $C_{12}H_{15}N_3O_2$

Mol Weight: 233.27

MP (deg C): | FP (deg C):

BP (deg C):

BP pressure (mm Hg):

Property/ Value	Units	Temp	Data Type	Reference
WS 7.54E+002	mg/L	25	EST	MEYLAN,WM ET AL. (1996)
logP 1.86			EXP	HANSCH,C & LEO,AJ (1985)
VP 1.01E-005	mm Hg	25	EST	NEELY,WB & BLAU,GE (1985)
DC	pKa			
HL 1.74E-008	atm m3/mol	25	EST	MEYLAN,WM & HOWARD,PH (1991)
OH 6.26E-012	cm3/molc sec	25	EST	MEYLAN,WM & HOWARD,PH (1993)

CAS #:	069218-49-1				1-(4-CARBETHOXYBUTYL)BENZOTRIAZOLE

Formula: $C_{13}H_{17}N_3O_2$

Mol Weight: 247.30

MP (deg C): | FP (deg C):

BP (deg C):

BP pressure (mm Hg):

Property/ Value	Units	Temp	Data Type	Reference
WS 3.24E+002	mg/L	25	EST	MEYLAN,WM ET AL. (1996)
logP 2.20			EXP	HANSCH,C & LEO,AJ (1985)
VP 4.39E-006	mm Hg	25	EST	NEELY,WB & BLAU,GE (1985)
DC	pKa			
HL 2.31E-008	atm m3/mol	25	EST	MEYLAN,WM & HOWARD,PH (1991)
OH 7.49E-012	cm3/molc sec	25	EST	MEYLAN,WM & HOWARD,PH (1993)

CAS #: 069218-50-4				1(5-CARBETHOXYPENTYL)BENZOTRIAZOLE
Formula: $C_{14}H_{19}N_3O_2$				
Mol Weight: 261.33				
MP (deg C):		FP (deg C):		
BP (deg C):				
BP pressure (mm Hg):				

Property/ Value	Units	Temp	Data Type	Reference
WS 1.05E+002	mg/L	25	EST	MEYLAN,WM ET AL. (1996)
logP 2.68			EXP	HANSCH,C & LEO,AJ (1985)
VP 1.90E-006	mm Hg	25	EST	NEELY,WB & BLAU,GE (1985)
DC	pKa			
HL 3.07E-008	atm m3/mol	25	EST	MEYLAN,WM & HOWARD,PH (1991)
OH 8.90E-012	cm3/molc sec	25	EST	MEYLAN,WM & HOWARD,PH (1993)

CAS #: 069218-51-5				2H-BENZOTRIAZOLE-2-ACETIC ACID, ETHYL ESTER
Formula: $C_{10}H_{11}N_3O_2$				
Mol Weight: 205.22				
MP (deg C):		FP (deg C):		
BP (deg C):				
BP pressure (mm Hg):				

Property/ Value	Units	Temp	Data Type	Reference
WS 1.08E+003	mg/L	25	EST	MEYLAN,WM ET AL. (1996)
logP 1.85			EXP	SANGSTER,J (1993)
VP 4.51E-005	mm Hg	25	EST	NEELY,WB & BLAU,GE (1985)
DC	pKa			
HL 9.89E-009	atm m3/mol	25	EST	MEYLAN,WM & HOWARD,PH (1991)
OH 3.35E-012	cm3/molc sec	25	EST	MEYLAN,WM & HOWARD,PH (1993)

CAS #: 069218-52-6				2H-BENZOTRIAZOLE-2-PROPANOIC ACID, ETHYL ESTER
Formula: $C_{11}H_{13}N_3O_2$				
Mol Weight: 219.25				
MP (deg C):		FP (deg C):		
BP (deg C):				
BP pressure (mm Hg):				

Property/ Value	Units	Temp	Data Type	Reference
WS 5.38E+002	mg/L	25	EST	MEYLAN,WM ET AL. (1996)
logP 2.12			EXP	SPARATORE,F ET AL. (1978)
VP 1.90E-005	mm Hg	25	EST	NEELY,WB & BLAU,GE (1985)
DC	pKa			
HL 1.31E-008	atm m3/mol	25	EST	MEYLAN,WM & HOWARD,PH (1991)
OH 4.66E-012	cm3/molc sec	25	EST	MEYLAN,WM & HOWARD,PH (1993)

CAS #: 069218-53-7				2-(3-CARBETHOXYPROPYL)BENZOTRIAZOLE
Formula: $C_{12}H_{17}N_3O_2$				
Mol Weight: 235.29				
MP (deg C):		FP (deg C):		
BP (deg C):				
BP pressure (mm Hg):				

Property/ Value	Units	Temp	Data Type	Reference
WS 2.56E+002	mg/L	25	EST	MEYLAN,WM ET AL. (1996)
logP 2.41			EXP	HANSCH,C & LEO,AJ (1985)
VP 1.01E-005	mm Hg	25	EST	NEELY,WB & BLAU,GE (1985)
DC	pKa			
HL 1.74E-008	atm m3/mol	25	EST	MEYLAN,WM & HOWARD,PH (1991)
OH 6.26E-012	cm3/molc sec	25	EST	MEYLAN,WM & HOWARD,PH (1993)

CAS #: 069218-54-8				2-(4-CARBETHOXYBUTYL)BENZOTRIAZOLE
Formula: $C_{13}H_{19}N_3O_2$				
Mol Weight: 249.32				
MP (deg C):		FP (deg C):		
BP (deg C):				
BP pressure (mm Hg):				

Property/ Value	Units	Temp	Data Type	Reference
WS 9.02E+001	mg/L	25	EST	MEYLAN,WM ET AL. (1996)
logP 2.85			EXP	HANSCH,C & LEO,AJ (1985)
VP 4.39E-006	mm Hg	25	EST	NEELY,WB & BLAU,GE (1985)
DC	pKa			
HL 2.31E-008	atm m3/mol	25	EST	MEYLAN,WM & HOWARD,PH (1991)
OH 7.49E-012	cm3/molc sec	25	EST	MEYLAN,WM & HOWARD,PH (1993)

CAS #: 069218-55-9				2(5-CARBETHOXYPENTYL)BENZOTRIAZOLE
Formula: $C_{14}H_{21}N_3O_2$				
Mol Weight: 263.34				
MP (deg C):		FP (deg C):		
BP (deg C):				
BP pressure (mm Hg):				

Property/ Value	Units	Temp	Data Type	Reference
WS 4.01E+001	mg/L	25	EST	MEYLAN,WM ET AL. (1996)
logP 3.17			EXP	HANSCH,C & LEO,AJ (1985)
VP 1.90E-006	mm Hg	25	EST	NEELY,WB & BLAU,GE (1985)
DC	pKa			
HL 3.07E-008	atm m3/mol	25	EST	MEYLAN,WM & HOWARD,PH (1991)
OH 8.90E-012	cm3/molc sec	25	EST	MEYLAN,WM & HOWARD,PH (1993)

CAS #: 069218-56-0				1-CARBAMYLMETHYLBENZOTRIAZOLE
Formula: $C_8H_8N_4O$				
Mol Weight: 176.18				
MP (deg C):		FP (deg C):		
BP (deg C):				
BP pressure (mm Hg):				

Property/ Value	Units	Temp	Data Type	Reference
WS 3.43E+003	mg/L	25	EST	MEYLAN,WM ET AL. (1996)
logP -0.10			EXP	HANSCH,C & LEO,AJ (1985)
VP 9.97E-007	mm Hg	25	EST	NEELY,WB & BLAU,GE (1985)
DC	pKa			
HL 4.75E-013	atm m3/mol	25	EST	MEYLAN,WM & HOWARD,PH (1991)
OH 3.70E-012	cm3/molc sec	25	EST	MEYLAN,WM & HOWARD,PH (1993)

CAS #: 069218-57-1				1-(2-CARBAMYLETHYL)BENZOTRIAZOLE
Formula: $C_9H_{10}N_4O$				
Mol Weight: 190.21				
MP (deg C):		FP (deg C):		
BP (deg C):				
BP pressure (mm Hg):				

Property/ Value	Units	Temp	Data Type	Reference
WS 1.41E+003	mg/L	25	EST	MEYLAN,WM ET AL. (1996)
logP 0.27			EXP	HANSCH,C & LEO,AJ (1985)
VP 4.54E-007	mm Hg	25	EST	NEELY,WB & BLAU,GE (1985)
DC	pKa			
HL 6.31E-013	atm m3/mol	25	EST	MEYLAN,WM & HOWARD,PH (1991)
OH 7.50E-012	cm3/molc sec	25	EST	MEYLAN,WM & HOWARD,PH (1993)

CAS #: 069218-58-2 — 1-(3-CARBAMYLPROPYL)BENZOTRIAZOLE

Formula: $C_{10}H_{12}N_4O$

Mol Weight: 204.23

MP (deg C): FP (deg C):

BP (deg C):

BP pressure (mm Hg):

Property/Value	Units	Temp	Data Type	Reference
WS 2.01E+004	mg/L	25	EST	MEYLAN,WM ET AL. (1996)
logP 0.37			EXP	HANSCH,C & LEO,AJ (1985)
VP 2.06E-007	mm Hg	25	EST	NEELY,WB & BLAU,GE (1985)
DC	pKa			
HL 8.37E-013	atm m3/mol	25	EST	MEYLAN,WM & HOWARD,PH (1991)
OH 9.49E-012	cm3/molc sec	25	EST	MEYLAN,WM & HOWARD,PH (1993)

CAS #: 069218-59-3 — 1-(4-CARBAMOYLBUTYL)BENZOTRIAZOLE

Formula: $C_{11}H_{14}N_4O$

Mol Weight: 218.26

MP (deg C): FP (deg C):

BP (deg C):

BP pressure (mm Hg):

Property/Value	Units	Temp	Data Type	Reference
WS 1.12E+004	mg/L	25	EST	MEYLAN,WM ET AL. (1996)
logP 0.58			EXP	HANSCH,C & LEO,AJ (1985)
VP 9.29E-008	mm Hg	25	EST	NEELY,WB & BLAU,GE (1985)
DC	pKa			
HL 1.11E-012	atm m3/mol	25	EST	MEYLAN,WM & HOWARD,PH (1991)
OH 1.09E-011	cm3/molc sec	25	EST	MEYLAN,WM & HOWARD,PH (1993)

CAS #: 069218-60-6 — 1-(5-CARBAMYLPENTYL)BENZOTRIAZOLE

Formula: $C_{12}H_{16}N_4O$

Mol Weight: 232.29

MP (deg C): FP (deg C):

BP (deg C):

BP pressure (mm Hg):

Property/Value	Units	Temp	Data Type	Reference
WS 6.26E+003	mg/L	25	EST	MEYLAN,WM ET AL. (1996)
logP 0.79			EXP	HANSCH,C & LEO,AJ (1985)
VP 4.18E-008	mm Hg	25	EST	NEELY,WB & BLAU,GE (1985)
DC	pKa			
HL 1.48E-012	atm m3/mol	25	EST	MEYLAN,WM & HOWARD,PH (1991)
OH 1.23E-011	cm3/molc sec	25	EST	MEYLAN,WM & HOWARD,PH (1993)

CAS #: 069218-61-7 — 2-CARBAMYLMETHYLBENZOTRIAZOLE

Formula: $C_8H_{10}N_4O$

Mol Weight: 178.20

MP (deg C): FP (deg C):

BP (deg C):

BP pressure (mm Hg):

Property/Value	Units	Temp	Data Type	Reference
WS 1.62E+003	mg/L	25	EST	MEYLAN,WM ET AL. (1996)
logP 0.28			EXP	HANSCH,C & LEO,AJ (1985)
VP 9.97E-007	mm Hg	25	EST	NEELY,WB & BLAU,GE (1985)
DC	pKa			
HL 4.75E-013	atm m3/mol	25	EST	MEYLAN,WM & HOWARD,PH (1991)
OH 3.70E-012	cm3/molc sec	25	EST	MEYLAN,WM & HOWARD,PH (1993)

CAS #: 069218-62-8 — 2-(2-CARBAMYLETHYL)BENZOTRIAZOLE

Formula: $C_9H_{12}N_4O$

Mol Weight: 192.22

MP (deg C): FP (deg C):

BP (deg C):

BP pressure (mm Hg):

Property/Value	Units	Temp	Data Type	Reference
WS 1.34E+004	mg/L	25	EST	MEYLAN,WM ET AL. (1996)
logP 0.66			EXP	HANSCH,C & LEO,AJ (1985)
VP 4.54E-007	mm Hg	25	EST	NEELY,WB & BLAU,GE (1985)
DC	pKa			
HL 6.31E-013	atm m3/mol	25	EST	MEYLAN,WM & HOWARD,PH (1991)
OH 7.50E-012	cm3/molc sec	25	EST	MEYLAN,WM & HOWARD,PH (1993)

CAS #: 069218-63-9 — 2-(3-CARBAMYLPROPYL)BENZOTRIAZOLE

Formula: $C_{10}H_{14}N_4O$

Mol Weight: 206.25

MP (deg C): FP (deg C):

BP (deg C):

BP pressure (mm Hg):

Property/Value	Units	Temp	Data Type	Reference
WS 8.98E+003	mg/L	25	EST	MEYLAN,WM ET AL. (1996)
logP 0.78			EXP	HANSCH,C & LEO,AJ (1985)
VP 2.06E-007	mm Hg	25	EST	NEELY,WB & BLAU,GE (1985)
DC	pKa			
HL 8.37E-013	atm m3/mol	25	EST	MEYLAN,WM & HOWARD,PH (1991)
OH 9.49E-012	cm3/molc sec	25	EST	MEYLAN,WM & HOWARD,PH (1993)

CAS #: 069218-64-0 — 2-(4-CARBAMOYLBUTYL)BENZOTRIAZOLE

Formula: $C_{11}H_{16}N_4O$

Mol Weight: 220.28

MP (deg C): FP (deg C):

BP (deg C):

BP pressure (mm Hg):

Property/Value	Units	Temp	Data Type	Reference
WS 4.83E+003	mg/L	25	EST	MEYLAN,WM ET AL. (1996)
logP 1.01			EXP	HANSCH,C & LEO,AJ (1985)
VP 9.29E-008	mm Hg	25	EST	NEELY,WB & BLAU,GE (1985)
DC	pKa			
HL 1.11E-012	atm m3/mol	25	EST	MEYLAN,WM & HOWARD,PH (1991)
OH 1.09E-011	cm3/molc sec	25	EST	MEYLAN,WM & HOWARD,PH (1993)

CAS #: 069218-65-1 — 2-(5-CARBAMYLPENTYL)BENZOTRIAZOLE

Formula: $C_{12}H_{18}N_4O$

Mol Weight: 234.30

MP (deg C): FP (deg C):

BP (deg C):

BP pressure (mm Hg):

Property/Value	Units	Temp	Data Type	Reference
WS 1.85E+003	mg/L	25	EST	MEYLAN,WM ET AL. (1996)
logP 1.41			EXP	HANSCH,C & LEO,AJ (1985)
VP 4.18E-008	mm Hg	25	EST	NEELY,WB & BLAU,GE (1985)
DC	pKa			
HL 1.48E-012	atm m3/mol	25	EST	MEYLAN,WM & HOWARD,PH (1991)
OH 1.23E-011	cm3/molc sec	25	EST	MEYLAN,WM & HOWARD,PH (1993)

CAS #: 069380-11-6	1-(1-IMIDAZOLYL)ADAMANTANE

Formula: $C_{13}H_{18}N_2$
Mol Weight: 202.30
MP (deg C): | FP (deg C):
BP (deg C):
BP pressure (mm Hg):

Property/ Value	Units	Temp	Data Type	Reference
WS 1.42E+002	mg/L	25	EST	MEYLAN,WM ET AL. (1996)
logP 2.90			EXP	HANSCH,C ET AL. (1995)
VP 4.36E-005	mm Hg	25	EST	NEELY,WB & BLAU,GE (1985)
DC	pKa			
HL 8.79E-005	atm m3/mol	25	EST	MEYLAN,WM & HOWARD,PH (1991)
OH 5.53E-011	cm3/molc sec	25	EST	MEYLAN,WM & HOWARD,PH (1993)

CAS #: 069397-85-9	1H-INDOLE-3-PROPANAMIDE, N-METHYL-

Formula: $C_{12}H_{14}N_2O$
Mol Weight: 202.26
MP (deg C): | FP (deg C):
BP (deg C):
BP pressure (mm Hg):

Property/ Value	Units	Temp	Data Type	Reference
WS 3.79E+003	mg/L	25	EST	MEYLAN,WM ET AL. (1996)
logP 1.23			EXP	RADZICKA,A & WOLFENDEN,R (1988)
VP 1.59E-007	mm Hg	25	EST	NEELY,WB & BLAU,GE (1985)
DC	pKa			
HL 4.33E-013	atm m3/mol	25	EST	MEYLAN,WM & HOWARD,PH (1991)
OH 2.11E-010	cm3/molc sec	25	EST	MEYLAN,WM & HOWARD,PH (1993)

CAS #: 069409-94-5	FLUVALINATE

Formula: $C_{26}H_{22}ClF_3N_2O_3$
Mol Weight: 502.93
MP (deg C): | FP (deg C):
BP (deg C):
BP pressure (mm Hg):

Property/ Value	Units	Temp	Data Type	Reference
WS 5.00E-003	mg/L	25	EXP	WAUCHOPE,RD ET AL. (1991A)
logP 6.81			EST	MEYLAN,WM & HOWARD,PH (1995)
VP 1.00E-007	mm Hg	25	EST	WAUCHOPE,RD ET AL (1991A)
DC	pKa			
HL 1.45E-008	atm m3/mol	25	EST	MEYLAN,WM & HOWARD,PH (1991)
OH 9.85E-011	cm3/molc sec	25	EST	ATKINSON,R (1988)

CAS #: 069472-19-1	DIS. A. 6

Formula: $C_{19}H_{21}N_5O_2$
Mol Weight: 351.41
MP (deg C): 118 | FP (deg C):
BP (deg C):
BP pressure (mm Hg):

Property/ Value	Units	Temp	Data Type	Reference
WS 7.02E-003	mg/L	25	EXP	BAUGHMAN,GL & PERENICH,TA (1988)
logP 5.04			EST	MEYLAN,WM & HOWARD,PH (1995)
VP 1.84E-010	mm Hg	25	EST	NEELY,WB & BLAU,GE (1985)
DC	pKa			
HL 1.86E-014	atm m3/mol	25	EST	MEYLAN,WM & HOWARD,PH (1991)
OH 1.79E-011	cm3/molc sec	25	EST	MEYLAN,WM & HOWARD,PH (1993)

CAS #: 069477-71-0	2,5-DIME-4-HYDROXYACETANILIDE

Formula: $C_{10}H_{13}NO_2$
Mol Weight: 179.22
MP (deg C): | FP (deg C):
BP (deg C):
BP pressure (mm Hg):

Property/ Value	Units	Temp	Data Type	Reference
WS 1.71E+004	mg/L	25	EST	MEYLAN,WM ET AL. (1996)
logP 0.60			EXP	NAKAGAWA,Y ET AL. (1992)
VP 1.05E-006	mm Hg	25	EST	NEELY,WB & BLAU,GE (1985)
DC	pKa			
HL 7.82E-013	atm m3/mol	25	EST	MEYLAN,WM & HOWARD,PH (1991)
OH 5.63E-011	cm3/molc sec	25	EST	MEYLAN,WM & HOWARD,PH (1993)

CAS #: 069477-72-1	2,3,5-TRIME-4-HYDROXYACETANILIDE

Formula: $C_{11}H_{15}NO_2$
Mol Weight: 193.25
MP (deg C): | FP (deg C):
BP (deg C):
BP pressure (mm Hg):

Property/ Value	Units	Temp	Data Type	Reference
WS 9.44E+003	mg/L	25	EST	MEYLAN,WM ET AL. (1996)
logP 0.82			EXP	HANSCH,C & LEO,AJ (1985)
VP 4.04E-007	mm Hg	25	EST	NEELY,WB & BLAU,GE (1985)
DC	pKa			
HL 8.63E-013	atm m3/mol	25	EST	MEYLAN,WM & HOWARD,PH (1991)
OH 7.21E-011	cm3/molc sec	25	EST	MEYLAN,WM & HOWARD,PH (1993)

CAS #: 069519-61-5	BARBITURIC ACID,N-ME,5-S-BU,5-BROMALLYL

Formula: $C_{12}H_{17}BrN_2O_3$
Mol Weight: 317.19
MP (deg C): | FP (deg C):
BP (deg C):
BP pressure (mm Hg):

Property/ Value	Units	Temp	Data Type	Reference
WS 8.18E+001	mg/L	25	EST	MEYLAN,WM ET AL. (1996)
logP 2.43			EXP	HANSCH,C & LEO,AJ (1985)
VP 8.80E-012	mm Hg	25	EST	NEELY,WB & BLAU,GE (1985)
DC	pKa			
HL 2.75E-013	atm m3/mol	25	EST	MEYLAN,WM & HOWARD,PH (1991)
OH 3.04E-011	cm3/molc sec	25	EST	MEYLAN,WM & HOWARD,PH (1993)

CAS #: 069527-40-8	PROPANEDINITRILE, [(5-IODO-2-FURANYL)METHYLENE]-

Formula: $C_8H_3IN_2O$
Mol Weight: 270.03
MP (deg C): | FP (deg C):
BP (deg C):
BP pressure (mm Hg):

Property/ Value	Units	Temp	Data Type	Reference
WS 8.51E+001	mg/L	25	EST	MEYLAN,WM ET AL. (1996)
logP 2.42			EXP	BALAZ,S ET AL. (1985)
VP 1.43E-005	mm Hg	25	EST	NEELY,WB & BLAU,GE (1985)
DC	pKa			
HL 3.17E-009	atm m3/mol	25	EST	MEYLAN,WM & HOWARD,PH (1991)
OH 2.73E-011	cm3/molc sec	25	EST	MEYLAN,WM & HOWARD,PH (1993)

1291

CAS #: 069527-41-9	2-PROPENOIC ACID, 2-CYANO-3-(5-IODO-2-FURANYL)-,

Formula: $C_9H_6INO_3$

Mol Weight: 303.06

MP (deg C): | FP (deg C):

BP (deg C):

BP pressure (mm Hg):

Property/Value	Units	Temp	Data Type	Reference
WS 2.36E+001	mg/L	25	EST	MEYLAN,WM ET AL. (1996)
logP 2.85			EXP	BALAZ,S ET AL. (1985)
VP 2.76E-005	mm Hg	25	EST	NEELY,WB & BLAU,GE (1985)
DC	pKa			
HL 2.13E-009	atm m3/mol	25	EST	MEYLAN,WM & HOWARD,PH (1991)
OH 3.02E-011	cm3/molc sec	25	EST	MEYLAN,WM & HOWARD,PH (1993)

CAS #: 069567-10-8	3,4,5-PIPERIDINETRIOL, 2-(HYDROXYMETHYL)-1-METHY

Formula: $C_7H_{15}NO_5$

Mol Weight: 193.20

MP (deg C): | FP (deg C):

BP (deg C):

BP pressure (mm Hg):

Property/Value	Units	Temp	Data Type	Reference
WS 1.00E+006	mg/L	25	EST	MEYLAN,WM ET AL. (1996)
logP -2.40			EXP	SANGSTER,J (1994)
VP 3.89E-009	mm Hg	25	EST	NEELY,WB & BLAU,GE (1985)
DC	pKa			
HL 6.52E-014	atm m3/mol	25	EST	MEYLAN,WM & HOWARD,PH (1991)
OH 2.08E-010	cm3/molc sec	25	EST	MEYLAN,WM & HOWARD,PH (1993)

CAS #: 069645-07-4	1,6-DICHLOROHEXA-1,3,5-TRIENE

Formula: $C_6H_6Cl_2$

Mol Weight: 149.02

MP (deg C): | FP (deg C):

BP (deg C):

BP pressure (mm Hg):

Property/Value	Units	Temp	Data Type	Reference
WS 7.63E+001	mg/L	25	EST	MEYLAN,WM ET AL. (1996)
logP 3.52			EST	MEYLAN,WM & HOWARD,PH (1995)
VP 2.88E+000	mm Hg	25	EST	NEELY,WB & BLAU,GE (1985)
DC	pKa			
HL 5.09E-002	atm m3/mol	25	EST	MEYLAN,WM & HOWARD,PH (1991)
OH 3.54E-011	cm3/molc sec	25	EST	MEYLAN,WM & HOWARD,PH (1993)

CAS #: 069655-05-6	INOSINE, 2',3'-DIDEOXY-

Formula: $C_{10}H_{12}N_4O_3$

Mol Weight: 236.23

MP (deg C): 160-163 | FP (deg C):

BP (deg C):

BP pressure (mm Hg):

Property/Value	Units	Temp	Data Type	Reference
WS 1.58E+004	mg/L	25	EST	MEYLAN,WM ET AL. (1996)
logP -1.24			EXP	SANGSTER,J (1993)
VP 6.59E-011	mm Hg	25	EST	NEELY,WB & BLAU,GE (1985)
DC	pKa			
HL 2.45E-020	atm m3/mol	25	EST	MEYLAN,WM & HOWARD,PH (1991)
OH 2.33E-010	cm3/molc sec	25	EST	MEYLAN,WM & HOWARD,PH (1993)

CAS #: 069684-88-4	CYCLOPROPANAMINE, 2-PHENYL-, CIS-(+)-

Formula: $C_9H_{11}N$

Mol Weight: 133.19

MP (deg C): | FP (deg C):

BP (deg C):

BP pressure (mm Hg):

Property/Value	Units	Temp	Data Type	Reference
WS 4.86E+004	mg/L	25	EST	MEYLAN,WM ET AL. (1996)
logP 1.49			EXP	SANGSTER,J (1993)
VP 1.09E-001	mm Hg	25	EST	NEELY,WB & BLAU,GE (1985)
DC	pKa			
HL 4.75E-007	atm m3/mol	25	EST	MEYLAN,WM & HOWARD,PH (1991)
OH 2.63E-011	cm3/molc sec	25	EST	MEYLAN,WM & HOWARD,PH (1993)

CAS #: 069712-34-1	CEPHALOSPORIN DERIVATIVE

Formula: $C_{18}H_{20}N_6O_7S_4$

Mol Weight: 560.65

MP (deg C): | FP (deg C):

BP (deg C):

BP pressure (mm Hg):

Property/Value	Units	Temp	Data Type	Reference
WS 1.56E+000	mg/L	25	EST	MEYLAN,WM ET AL. (1996)
logP 2.66			EXP	MATSUDA,H ET AL. (1980)
VP 3.91E-021	mm Hg	25	EST	NEELY,WB & BLAU,GE (1985)
DC	pKa			
HL 1.19E-028	atm m3/mol	25	EST	MEYLAN,WM & HOWARD,PH (1991)
OH 1.88E-010	cm3/molc sec	25	EST	MEYLAN,WM & HOWARD,PH (1993)

CAS #: 069741-89-5	2-AMINO-5-PROPYL-1,3,4-OXADIAZOLE

Formula: $C_5H_{10}N_3O$

Mol Weight: 128.15

MP (deg C): | FP (deg C):

BP (deg C):

BP pressure (mm Hg):

Property/Value	Units	Temp	Data Type	Reference
WS 6.00E+004	mg/L	25	EST	MEYLAN,WM ET AL. (1996)
logP 0.23			EXP	HANSCH,C & LEO,AJ (1985)
VP 3.80E-003	mm Hg	25	EST	NEELY,WB & BLAU,GE (1985)
DC	pKa			
HL 1.97E-009	atm m3/mol	25	EST	MEYLAN,WM & HOWARD,PH (1991)
OH 6.46E-012	cm3/molc sec	25	EST	MEYLAN,WM & HOWARD,PH (1993)

CAS #: 069741-90-8	2-AMINO-5-I-BU-1,3,4-OXADIAZOLE

Formula: $C_6H_{12}N_3O$

Mol Weight: 142.18

MP (deg C): | FP (deg C):

BP (deg C):

BP pressure (mm Hg):

Property/Value	Units	Temp	Data Type	Reference
WS 3.77E+004	mg/L	25	EST	MEYLAN,WM ET AL. (1996)
logP 0.40			EXP	HANSCH,C & LEO,AJ (1985)
VP 2.59E-003	mm Hg	25	EST	NEELY,WB & BLAU,GE (1985)
DC	pKa			
HL 2.61E-009	atm m3/mol	25	EST	MEYLAN,WM & HOWARD,PH (1991)
OH 7.87E-012	cm3/molc sec	25	EST	MEYLAN,WM & HOWARD,PH (1993)

069741-91-9 — 2-AMINO-5-SEC-BUTYL-1,3,4-OXADIAZOLE

Formula:	$C_6H_{12}N_3O$
Mol Weight:	142.18
MP (deg C):	FP (deg C):
BP (deg C):	
BP pressure (mm Hg):	

Property/Value	Units	Temp	Data Type	Reference
WS 2.45E+004	mg/L	25	EST	MEYLAN,WM ET AL. (1996)
logP 0.62			EXP	HANSCH,C & LEO,AJ (1985)
VP 2.59E-003	mm Hg	25	EST	NEELY,WB & BLAU,GE (1985)
DC	pKa			
HL 2.61E-009	atm m3/mol	25	EST	MEYLAN,WM & HOWARD,PH (1991)
OH 7.87E-012	cm3/molc sec	25	EST	MEYLAN,WM & HOWARD,PH (1993)

069741-92-0 — 2-AMINO-5-T-BU-1,3,4-OXADIAZOLE

Formula:	$C_6H_{12}N_3O$
Mol Weight:	142.18
MP (deg C):	FP (deg C):
BP (deg C):	
BP pressure (mm Hg):	

Property/Value	Units	Temp	Data Type	Reference
WS 3.22E+004	mg/L	25	EST	MEYLAN,WM ET AL. (1996)
logP 0.48			EXP	HANSCH,C & LEO,AJ (1985)
VP 3.50E-003	mm Hg	25	EST	NEELY,WB & BLAU,GE (1985)
DC	pKa			
HL 2.61E-009	atm m3/mol	25	EST	MEYLAN,WM & HOWARD,PH (1991)
OH 4.50E-012	cm3/molc sec	25	EST	MEYLAN,WM & HOWARD,PH (1993)

069741-93-1 — 2-AMINO-5-AMYL-1,3,4-OXADIAZOLE

Formula:	$C_7H_{14}N_3O$
Mol Weight:	156.21
MP (deg C):	FP (deg C):
BP (deg C):	
BP pressure (mm Hg):	

Property/Value	Units	Temp	Data Type	Reference
WS 8.27E+003	mg/L	25	EST	MEYLAN,WM ET AL. (1996)
logP 1.10			EXP	HANSCH,C & LEO,AJ (1985)
VP 4.64E-004	mm Hg	25	EST	NEELY,WB & BLAU,GE (1985)
DC	pKa			
HL 3.47E-009	atm m3/mol	25	EST	MEYLAN,WM & HOWARD,PH (1991)
OH 9.29E-012	cm3/molc sec	25	EST	MEYLAN,WM & HOWARD,PH (1993)

069741-94-2 — 2-NH2-5-HEXYL-1,3,4-OXADIAZOLE

Formula:	$C_8H_{16}N_3O$
Mol Weight:	170.24
MP (deg C):	FP (deg C):
BP (deg C):	
BP pressure (mm Hg):	

Property/Value	Units	Temp	Data Type	Reference
WS 3.73E+003	mg/L	25	EST	MEYLAN,WM ET AL. (1996)
logP 1.43			EXP	HANSCH,C & LEO,AJ (1985)
VP 1.74E-004	mm Hg	25	EST	NEELY,WB & BLAU,GE (1985)
DC	pKa			
HL 4.60E-009	atm m3/mol	25	EST	MEYLAN,WM & HOWARD,PH (1991)
OH 1.07E-011	cm3/molc sec	25	EST	MEYLAN,WM & HOWARD,PH (1993)

069806-40-2 — PROPANOIC ACID, 2-[4-[[3-CHLORO-5-(TRIFLUOROMETH

Formula:	$C_{16}H_{13}ClF_3NO_4$
Mol Weight:	375.73
MP (deg C): 35-37	FP (deg C):
BP (deg C):	
BP pressure (mm Hg):	

Property/Value	Units	Temp	Data Type	Reference
WS 9.30E+000	mg/L	25	EXP	SHIU,WY ET AL. (1990)
logP 3.52			EXP	SANGSTER,J (1994)
VP 6.45E-007	mm Hg	25	EXP	AUGUSTIJN-BECKERS,PWM ET AL. (1994)
DC	pKa			
HL 3.43E-008	atm m3/mol	25	EST	VP/WSOL
OH 3.07E-011	cm3/molc sec	25	EST	MEYLAN,WM & HOWARD,PH (1993)

069808-71-5 — 1H-INDOLE-2-CARBOXAMIDE, N-METHYL-

Formula:	$C_{10}H_{10}N_2O$
Mol Weight:	174.20
MP (deg C):	FP (deg C):
BP (deg C):	
BP pressure (mm Hg):	

Property/Value	Units	Temp	Data Type	Reference
WS 1.40E+003	mg/L	25	EST	MEYLAN,WM ET AL. (1996)
logP 1.90			EXP	YAMAGAMI,C ET AL. (1994)
VP 8.66E-007	mm Hg	25	EST	NEELY,WB & BLAU,GE (1985)
DC	pKa			
HL 7.99E-013	atm m3/mol	25	EST	MEYLAN,WM & HOWARD,PH (1991)
OH 1.06E-010	cm3/molc sec	25	EST	MEYLAN,WM & HOWARD,PH (1993)

069808-72-6 — 1H-INDOLE-2-CARBOXAMIDE, N-ETHYL-

Formula:	$C_{11}H_{12}N_2O$
Mol Weight:	188.23
MP (deg C):	FP (deg C):
BP (deg C):	
BP pressure (mm Hg):	

Property/Value	Units	Temp	Data Type	Reference
WS 5.24E+002	mg/L	25	EST	MEYLAN,WM ET AL. (1996)
logP 2.32			EXP	YAMAGAMI,C ET AL. (1994)
VP 3.72E-007	mm Hg	25	EST	NEELY,WB & BLAU,GE (1985)
DC	pKa			
HL 1.06E-012	atm m3/mol	25	EST	MEYLAN,WM & HOWARD,PH (1991)
OH 1.10E-010	cm3/molc sec	25	EST	MEYLAN,WM & HOWARD,PH (1993)

069810-98-6 — CARBAMIC ACID, [4,5-DIHYDRO-4-[4-(1-METHYLETHYL)

Formula:	$C_{14}H_{19}N_3O_2$
Mol Weight:	261.33
MP (deg C):	FP (deg C):
BP (deg C):	
BP pressure (mm Hg):	

Property/Value	Units	Temp	Data Type	Reference
WS 1.23E+002	mg/L	25	EST	MEYLAN,WM ET AL. (1996)
logP 2.60			EXP	SANGSTER,J (1994)
VP 2.17E-008	mm Hg	25	EST	NEELY,WB & BLAU,GE (1985)
DC	pKa			
HL 8.95E-014	atm m3/mol	25	EST	MEYLAN,WM & HOWARD,PH (1991)
OH 9.52E-011	cm3/molc sec	25	EST	MEYLAN,WM & HOWARD,PH (1993)

CAS #:	069981-38-0		PHOSPHONIC DIAMIDE, N,N,N',N'-TETRAETHYL-P-(1-PY

Formula: $C_{12}H_{28}N_3OP$

Mol Weight: 261.35

MP (deg C): FP (deg C):

BP (deg C):

BP pressure (mm Hg):

Property/ Value	Units	Temp	Data Type	Reference
WS 1.86E+002	mg/L	25	EST	MEYLAN,WM ET AL. (1996)
logP 2.39			EXP	DEBORD,J & LABADIE,M (1985)
VP 9.19E-005	mm Hg	25	EST	NEELY,WB & BLAU,GE (1985)
DC	pKa			
HL 1.72E-011	atm m3/mol	25	EST	MEYLAN,WM & HOWARD,PH (1991)
OH 1.37E-010	cm3/molc sec	25	EST	MEYLAN,WM & HOWARD,PH (1993)

CAS #:	070026-41-4		CYCLOPENTA[D]PYRIDO[1,2-A]PYRIMIDIN-10(1H)-ONE,

Formula: $C_{11}H_{10}N_2O$

Mol Weight: 186.22

MP (deg C): FP (deg C):

BP (deg C):

BP pressure (mm Hg):

Property/ Value	Units	Temp	Data Type	Reference
WS 4.94E+003	mg/L	25	EST	MEYLAN,WM ET AL. (1996)
logP 1.19			EXP	SANGSTER,J (1994)
VP 1.97E-005	mm Hg	25	EST	NEELY,WB & BLAU,GE (1985)
DC	pKa			
HL 2.15E-009	atm m3/mol	25	EST	MEYLAN,WM & HOWARD,PH (1991)
OH 1.17E-010	cm3/molc sec	25	EST	MEYLAN,WM & HOWARD,PH (1993)

CAS #:	070026-51-6		CYCLOPENTA[D]PYRIDO[1,2-A]PYRIMIDIN-10(1H)-ONE,

Formula: $C_{12}H_{12}N_2O$

Mol Weight: 200.24

MP (deg C): FP (deg C):

BP (deg C):

BP pressure (mm Hg):

Property/ Value	Units	Temp	Data Type	Reference
WS 1.77E+003	mg/L	25	EST	MEYLAN,WM ET AL. (1996)
logP 1.63			EXP	SANGSTER,J (1994)
VP 8.64E-006	mm Hg	25	EST	NEELY,WB & BLAU,GE (1985)
DC	pKa			
HL 3.37E-009	atm m3/mol	25	EST	MEYLAN,WM & HOWARD,PH (1991)
OH 1.46E-010	cm3/molc sec	25	EST	MEYLAN,WM & HOWARD,PH (1993)

CAS #:	070026-79-8		CYCLOPENTA[D]PYRIDO[1,2-A]PYRIMIDIN-10(1H)-ONE,

Formula: $C_{12}H_{16}N_2O$

Mol Weight: 204.27

MP (deg C): FP (deg C):

BP (deg C):

BP pressure (mm Hg):

Property/ Value	Units	Temp	Data Type	Reference
WS 9.58E+003	mg/L	25	EST	MEYLAN,WM ET AL. (1996)
logP 0.72			EXP	SANGSTER,J (1994)
VP 2.44E-005	mm Hg	25	EST	NEELY,WB & BLAU,GE (1985)
DC	pKa			
HL 1.93E-009	atm m3/mol	25	EST	MEYLAN,WM & HOWARD,PH (1991)
OH 3.92E-011	cm3/molc sec	25	EST	MEYLAN,WM & HOWARD,PH (1993)

CAS #:	070090-28-7		2-PROPOXYPYRAZINE

Formula: $C_7H_{10}N_2O$

Mol Weight: 138.17

MP (deg C): FP (deg C):

BP (deg C):

BP pressure (mm Hg):

Property/ Value	Units	Temp	Data Type	Reference
WS 2.29E+003	mg/L	25	EST	MEYLAN,WM ET AL. (1996)
logP 1.84			EXP	YAMAGAMI,C ET AL. (1990A)
VP 1.62E-001	mm Hg	25	EST	NEELY,WB & BLAU,GE (1985)
DC	pKa			
HL 2.77E-006	atm m3/mol	25	EST	MEYLAN,WM & HOWARD,PH (1991)
OH 1.30E-011	cm3/molc sec	25	EST	MEYLAN,WM & HOWARD,PH (1993)

CAS #:	070116-00-6		CINCHONINE OXALATE

Formula: $C_{20}H_{24}N_2O_3$

Mol Weight: 340.43

MP (deg C): FP (deg C):

BP (deg C):

BP pressure (mm Hg):

Property/ Value	Units	Temp	Data Type	Reference
WS 9.60E+003	mg/L	20	EXP	SEIDELL,A (1941)
logP 4.13			EST	MEYLAN,WM & HOWARD,PH (1995)
VP 4.86E-016	mm Hg	25	EST	NEELY,WB & BLAU,GE (1985)
DC	pKa			
HL 6.88E-023	atm m3/mol	25	EST	MEYLAN,WM & HOWARD,PH (1991)
OH 1.69E-010	cm3/molc sec	25	EST	MEYLAN,WM & HOWARD,PH (1993)

CAS #:	070124-77-5		FLUCYTHRINATE

Formula: $C_{26}H_{23}F_2NO_4$

Mol Weight: 451.47

MP (deg C): FP (deg C):

BP (deg C): 108

BP pressure (mm Hg): 3.50E-001

Property/ Value	Units	Temp	Data Type	Reference
WS 6.00E-002	mg/L	25	EXP	WAUCHOPE,RD ET AL. (1991A)
logP 6.20			EXP	HANSCH,C & LEO,AJ (1985)
VP 8.70E-009	mm Hg	25	EXP	WAUCHOPE,RD ET AL. (1991A)
DC	pKa			
HL 8.61E-008	atm m3/mol	25	EST	VP/WSOL
OH 6.36E-011	cm3/molc sec	25	EST	ATKINSON,R (1988)

CAS #:	070132-50-2		PIMONIDAZOLE

Formula: $C_{11}H_{18}N_4O_3$

Mol Weight: 254.29

MP (deg C): FP (deg C):

BP (deg C):

BP pressure (mm Hg):

Property/ Value	Units	Temp	Data Type	Reference
WS 1.30E+004	mg/L	25	EST	MEYLAN,WM ET AL. (1996)
logP 1.00			EXP	HANSCH,C ET AL. (1995)
VP 7.70E-010	mm Hg	25	EST	NEELY,WB & BLAU,GE (1985)
DC	pKa			
HL 1.04E-014	atm m3/mol	25	EST	MEYLAN,WM & HOWARD,PH (1991)
OH 1.18E-010	cm3/molc sec	25	EST	MEYLAN,WM & HOWARD,PH (1993)

CAS #: 070135-16-9	2-FURANMETHANAMINIUM, TETRAHYDRO-N,N,N,5-TETRAME

Formula: $C_9H_{20}INO$

Mol Weight: 285.17

MP (deg C): | FP (deg C):

BP (deg C):

BP pressure (mm Hg):

Property/Value	Units	Temp	Data Type	Reference
WS 5.16E+005	mg/L	25	EST	MEYLAN,WM ET AL. (1996)
logP -1.80			EXP	PRATESI,P ET AL. (1986)
VP 3.11E-008	mm Hg	25	EST	NEELY,WB & BLAU,GE (1985)
DC	pKa			
HL 5.32E-015	atm m3/mol	25	EST	MEYLAN,WM & HOWARD,PH (1991)
OH 4.29E-011	cm3/molc sec	25	EST	MEYLAN,WM & HOWARD,PH (1993)

CAS #: 070172-32-6	PHENYLACETIC ACID,2-(2',3'-DICHLORO)ANILINO

Formula: $C_{14}H_{11}Cl_2NO_2$

Mol Weight: 296.16

MP (deg C): | FP (deg C):

BP (deg C):

BP pressure (mm Hg):

Property/Value	Units	Temp	Data Type	Reference
WS 2.27E+000	mg/L	25	EST	MEYLAN,WM ET AL. (1996)
logP 4.86			EXP	HANSCH,C ET AL. (1995)
VP 6.14E-008	mm Hg	25	EST	NEELY,WB & BLAU,GE (1985)
DC	pKa			
HL 4.73E-012	atm m3/mol	25	EST	MEYLAN,WM & HOWARD,PH (1991)
OH 1.82E-010	cm3/molc sec	25	EST	MEYLAN,WM & HOWARD,PH (1993)

CAS #: 070172-33-7	2-ANILINOPHENYLACETIC ACID

Formula: $C_{14}H_{13}NO_2$

Mol Weight: 227.27

MP (deg C): | FP (deg C):

BP (deg C):

BP pressure (mm Hg):

Property/Value	Units	Temp	Data Type	Reference
WS 8.18E+001	mg/L	25	EST	MEYLAN,WM ET AL. (1996)
logP 3.49			EXP	HANSCH,C ET AL. (1995)
VP 1.22E-006	mm Hg	25	EST	NEELY,WB & BLAU,GE (1985)
DC	pKa			
HL 8.62E-012	atm m3/mol	25	EST	MEYLAN,WM & HOWARD,PH (1991)
OH 2.01E-010	cm3/molc sec	25	EST	MEYLAN,WM & HOWARD,PH (1993)

CAS #: 070195-11-8	2-METHYLBUTYRIC ACID HYDRAZIDE

Formula: $C_5H_{12}N_2O$

Mol Weight: 116.16

MP (deg C): | FP (deg C):

BP (deg C):

BP pressure (mm Hg):

Property/Value	Units	Temp	Data Type	Reference
WS 9.71E+003	mg/L	25	EST	MEYLAN,WM ET AL. (1996)
logP -0.33			EXP	HANSCH,C & LEO,AJ (1985)
VP 4.30E-003	mm Hg	25	EST	NEELY,WB & BLAU,GE (1985)
DC	pKa			
HL 1.09E-010	atm m3/mol	25	EST	MEYLAN,WM & HOWARD,PH (1991)
OH 1.16E-011	cm3/molc sec	25	EST	MEYLAN,WM & HOWARD,PH (1993)

CAS #: 070290-53-8	O-METHYL-N-PHENYLSUCCINIMIDE

Formula: $C_{11}H_{11}NO_2$

Mol Weight: 189.22

MP (deg C): | FP (deg C):

BP (deg C):

BP pressure (mm Hg):

Property/Value	Units	Temp	Data Type	Reference
WS 2.80E+004	mg/L	25	EST	MEYLAN,WM ET AL. (1996)
logP 0.29			EXP	HANSCH,C & LEO,AJ (1985)
VP 4.03E-007	mm Hg	25	EST	NEELY,WB & BLAU,GE (1985)
DC	pKa			
HL 1.44E-007	atm m3/mol	25	EST	MEYLAN,WM & HOWARD,PH (1991)
OH 1.97E-011	cm3/molc sec	25	EST	MEYLAN,WM & HOWARD,PH (1993)

CAS #: 070362-47-9	2,2'4,5-TETRACHLOROBIPHENYL

Formula: $C_{12}H_6Cl_4$

Mol Weight: 291.99

MP (deg C): | FP (deg C):

BP (deg C):

BP pressure (mm Hg):

Property/Value	Units	Temp	Data Type	Reference
WS 1.64E-002	mg/L	25	EXP	SHIU,WY & MACKAY,D (1986)
logP 6.34			EST	MEYLAN,WM & HOWARD,PH (1995)
VP 8.45E-006	mm Hg	25	EST	NEELY,WB & BLAU,GE (1985)
DC	pKa			
HL 1.25E-004	atm m3/mol	25	EST	MEYLAN,WM & HOWARD,PH (1991)
OH 5.77E-013	cm3/molc sec	25	EST	MEYLAN,WM & HOWARD,PH (1993)

CAS #: 070381-44-1	4H-PYRIDO[1,2-A]PYRIMIDIN-4-ONE, 3-ETHYL-2,6-DIM

Formula: $C_{12}H_{14}N_2O$

Mol Weight: 202.26

MP (deg C): | FP (deg C):

BP (deg C):

BP pressure (mm Hg):

Property/Value	Units	Temp	Data Type	Reference
WS 1.08E+003	mg/L	25	EST	MEYLAN,WM ET AL. (1996)
logP 1.86			EXP	SANGSTER,J (1994)
VP 1.38E-005	mm Hg	25	EST	NEELY,WB & BLAU,GE (1985)
DC	pKa			
HL 7.65E-009	atm m3/mol	25	EST	MEYLAN,WM & HOWARD,PH (1991)
OH 1.44E-010	cm3/molc sec	25	EST	MEYLAN,WM & HOWARD,PH (1993)

CAS #: 070381-46-3	4H-PYRIDO[1,2-A]PYRIMIDIN-4-ONE, 3-ETHYL-2,6-DIM

Formula: $C_{12}H_{18}N_2O$

Mol Weight: 206.29

MP (deg C): | FP (deg C):

BP (deg C):

BP pressure (mm Hg):

Property/Value	Units	Temp	Data Type	Reference
WS 2.44E+003	mg/L	25	EST	MEYLAN,WM ET AL. (1996)
logP 1.43			EXP	SANGSTER,J (1994)
VP 1.77E-005	mm Hg	25	EST	NEELY,WB & BLAU,GE (1985)
DC	pKa			
HL 5.80E-009	atm m3/mol	25	EST	MEYLAN,WM & HOWARD,PH (1991)
OH 4.45E-011	cm3/molc sec	25	EST	MEYLAN,WM & HOWARD,PH (1993)

CAS #: 070381-53-2				4H-PYRIDO[1,2-A]PYRIMIDIN-4-ONE, 2-METHYL-3-PROP

Formula: $C_{12}H_{14}N_2O$

Mol Weight: 202.26

MP (deg C):   FP (deg C):

BP (deg C):

BP pressure (mm Hg):

Property/Value	Units	Temp	Data Type	Reference
WS 1.36E+003	mg/L	25	EST	MEYLAN,WM ET AL. (1996)
logP 1.75			EXP	SANGSTER,J (1994)
VP 5.28E-005	mm Hg	25	EST	NEELY,WB & BLAU,GE (1985)
DC	pKa			
HL	atm m3/mol			
OH 2.61E-011	cm3/molc sec	25	EST	MEYLAN,WM & HOWARD,PH (1993)

CAS #: 070381-71-4				4H-PYRIDO[1,2-A]PYRIMIDIN-4-ONE, 2-ETHYL-3,6-DIM

Formula: $C_{12}H_{14}N_2O$

Mol Weight: 202.26

MP (deg C):   FP (deg C):

BP (deg C):

BP pressure (mm Hg):

Property/Value	Units	Temp	Data Type	Reference
WS 1.08E+003	mg/L	25	EST	MEYLAN,WM ET AL. (1996)
logP 1.87			EXP	SANGSTER,J (1994)
VP 1.38E-005	mm Hg	25	EST	NEELY,WB & BLAU,GE (1985)
DC	pKa			
HL 7.65E-009	atm m3/mol	25	EST	MEYLAN,WM & HOWARD,PH (1991)
OH 1.44E-010	cm3/molc sec	25	EST	MEYLAN,WM & HOWARD,PH (1993)

CAS #: 070458-92-3				PEFLOXACIN

Formula: $C_{17}H_{20}FN_3O_3$

Mol Weight: 333.37

MP (deg C): 270-272 de   FP (deg C):

BP (deg C):

BP pressure (mm Hg):

Property/Value	Units	Temp	Data Type	Reference
WS 1.14E+004	mg/L	25	EST	MEYLAN,WM ET AL. (1996)
logP 0.27			EXP	HANSCH,C ET AL. (1995)
VP 9.39E-013	mm Hg	25	EST	NEELY,WB & BLAU,GE (1985)
DC	pKa			
HL 1.91E-018	atm m3/mol	25	EST	MEYLAN,WM & HOWARD,PH (1991)
OH 3.26E-010	cm3/molc sec	25	EST	MEYLAN,WM & HOWARD,PH (1993)

CAS #: 070458-96-7				NORFLOXACIN

Formula: $C_{16}H_{18}FN_3O_3$

Mol Weight: 319.34

MP (deg C): 227-228   FP (deg C):

BP (deg C):

BP pressure (mm Hg):

Property/Value	Units	Temp	Data Type	Reference
WS 1.78E+005	mg/L	25	EST	MEYLAN,WM ET AL. (1996)
logP -1.03			EXP	HANSCH,C ET AL. (1995)
VP 6.77E-013	mm Hg	25	EST	NEELY,WB & BLAU,GE (1985)
DC	pKa			
HL 8.70E-019	atm m3/mol	25	EST	MEYLAN,WM & HOWARD,PH (1991)
OH 3.21E-010	cm3/molc sec	25	EST	MEYLAN,WM & HOWARD,PH (1993)

CAS #: 070477-14-4				1-(3-T-BUPH)-3-MEO-3-METHYL UREA

Formula: $C_{13}H_{20}N_2O_2$

Mol Weight: 236.32

MP (deg C):   FP (deg C):

BP (deg C):

BP pressure (mm Hg):

Property/Value	Units	Temp	Data Type	Reference
WS 5.10E+001	mg/L	25	EST	MEYLAN,WM ET AL. (1996)
logP 3.21			EXP	HANSCH,C & LEO,AJ (1985)
VP 5.67E-006	mm Hg	25	EST	NEELY,WB & BLAU,GE (1985)
DC	pKa			
HL 5.43E-008	atm m3/mol	25	EST	MEYLAN,WM & HOWARD,PH (1991)
OH 9.61E-011	cm3/molc sec	25	EST	MEYLAN,WM & HOWARD,PH (1993)

CAS #: 070528-90-4				AZO DYE N2

Formula: $C_{15}H_{12}ClN_5O_4$

Mol Weight: 361.75

MP (deg C):   FP (deg C):

BP (deg C):

BP pressure (mm Hg):

Property/Value	Units	Temp	Data Type	Reference
WS 7.40E-002	mg/L	25	EXP	BAUGHMAN,GL ET AL. (1993)
logP 3.40			EXP	BAUGHMAN,GL & WEBER,EJ (1991)
VP 4.06E-015	mm Hg	25	EST	NEELY,WB & BLAU,GE (1985)
DC	pKa			
HL 3.25E-018	atm m3/mol	25	EST	MEYLAN,WM & HOWARD,PH (1991)
OH 2.64E-011	cm3/molc sec	25	EST	MEYLAN,WM & HOWARD,PH (1993)

CAS #: 070579-34-9				4,6-NH2 2,2-DIME-1-(3-ACPH)S-TRIAZENE

Formula: $C_{13}H_{17}N_5O$

Mol Weight: 259.31

MP (deg C):   FP (deg C):

BP (deg C):

BP pressure (mm Hg):

Property/Value	Units	Temp	Data Type	Reference
WS 3.24E+004	mg/L	25	EST	MEYLAN,WM ET AL. (1996)
logP -0.22			EXP	HANSCH,C & LEO,AJ (1985)
VP 2.29E-007	mm Hg	25	EST	NEELY,WB & BLAU,GE (1985)
DC	pKa			
HL 1.33E-015	atm m3/mol	25	EST	MEYLAN,WM & HOWARD,PH (1991)
OH 1.04E-010	cm3/molc sec	25	EST	MEYLAN,WM & HOWARD,PH (1993)

CAS #: 070579-43-0				46NH2-22ME-1(3URIDOPHO)MEPH S-TRIAZINE

Formula: $C_{19}H_{23}N_7O_2$

Mol Weight: 381.44

MP (deg C):   FP (deg C):

BP (deg C):

BP pressure (mm Hg):

Property/Value	Units	Temp	Data Type	Reference
WS 6.79E+002	mg/L	25	EST	MEYLAN,WM ET AL. (1996)
logP 0.90			EXP	HANSCH,C & LEO,AJ (1985)
VP 1.09E-012	mm Hg	25	EST	NEELY,WB & BLAU,GE (1985)
DC	pKa			
HL 1.91E-022	atm m3/mol	25	EST	MEYLAN,WM & HOWARD,PH (1991)
OH 2.48E-010	cm3/molc sec	25	EST	MEYLAN,WM & HOWARD,PH (1993)

CAS #: 070606-63-2

4,6-NH2 2,2-DIME-1-(3-BUO PH)S-TRIAZINE

Formula: $C_{15}H_{23}N_5O$

Mol Weight: 289.38

MP (deg C): FP (deg C):

BP (deg C):

BP pressure (mm Hg):

Property/Value	Units	Temp	Data Type	Reference
WS 2.61E+002	mg/L	25	EST	MEYLAN,WM ET AL. (1996)
logP 2.03			EXP	HANSCH,C & LEO,AJ (1985)
VP 9.68E-008	mm Hg	25	EST	NEELY,WB & BLAU,GE (1985)
DC	pKa			
HL 1.01E-013	atm m3/mol	25	EST	MEYLAN,WM & HOWARD,PH (1991)
OH 2.55E-010	cm3/molc sec	25	EST	MEYLAN,WM & HOWARD,PH (1993)

CAS #: 070648-16-7

DIBENZOFURAN, 1,3,4,7-TETRACHLORO-

Formula: $C_{12}H_4Cl_4O$

Mol Weight: 305.98

MP (deg C): FP (deg C):

BP (deg C):

BP pressure (mm Hg):

Property/Value	Units	Temp	Data Type	Reference
WS 4.21E-003	mg/L	25	EST	MEYLAN,WM ET AL. (1996)
logP 6.23			EXP	SIJM,DTHM ET AL. (1989)
VP 1.53E-006	mm Hg	25	EST	NEELY,WB & BLAU,GE (1985)
DC	pKa			
HL 1.54E-005	atm m3/mol	25	EST	MEYLAN,WM & HOWARD,PH (1991)
OH 1.65E-013	cm3/molc sec	25	EST	MEYLAN,WM & HOWARD,PH (1993)

CAS #: 070648-24-7

DIBENZOFURAN, 1,2,4,6,9-PENTACHLORO-

Formula: $C_{12}H_3Cl_5O$

Mol Weight: 340.42

MP (deg C): FP (deg C):

BP (deg C):

BP pressure (mm Hg):

Property/Value	Units	Temp	Data Type	Reference
WS 1.29E-003	mg/L	25	EST	MEYLAN,WM ET AL. (1996)
logP 6.59			EXP	SIJM,DTHM ET AL. (1989)
VP 3.46E-007	mm Hg	25	EST	NEELY,WB & BLAU,GE (1985)
DC	pKa			
HL 1.14E-005	atm m3/mol	25	EST	MEYLAN,WM & HOWARD,PH (1991)
OH 7.46E-014	cm3/molc sec	25	EST	MEYLAN,WM & HOWARD,PH (1993)

CAS #: 070657-63-5

BENZAMIDE, 2-CHLORO-N,N-BIS(1-METHYLETHYL)-

Formula: $C_{13}H_{18}ClNO$

Mol Weight: 239.75

MP (deg C): FP (deg C):

BP (deg C):

BP pressure (mm Hg):

Property/Value	Units	Temp	Data Type	Reference
WS 7.39E+001	mg/L	25	EST	MEYLAN,WM ET AL. (1996)
logP 3.00			EXP	SURYANARAYANA,MVS ET AL. (1991)
VP 7.73E+000	mm Hg	30	EXP	SURYANARAYANA,MVS ET AL. (1991)
DC	pKa			
HL 2.46E-008	atm m3/mol	25	EST	MEYLAN,WM & HOWARD,PH (1991)
OH 3.28E-011	cm3/molc sec	25	EST	MEYLAN,WM & HOWARD,PH (1993)

CAS #: 070732-30-8

BENZENEMETHANOL, _-[(CYCLOHEXYLAMINO)METHYL]-, H

Formula: $C_{14}H_{20}ClNO$

Mol Weight: 253.77

MP (deg C): FP (deg C):

BP (deg C):

BP pressure (mm Hg):

Property/Value	Units	Temp	Data Type	Reference
WS 2.73E+004	mg/L	25	EST	MEYLAN,WM ET AL. (1996)
logP -0.11			EXP	HANSCH,C ET AL. (1995)
VP 2.54E-012	mm Hg	25	EST	NEELY,WB & BLAU,GE (1985)
DC	pKa			
HL 9.06E-018	atm m3/mol	25	EST	MEYLAN,WM & HOWARD,PH (1991)
OH 5.82E-011	cm3/molc sec	25	EST	MEYLAN,WM & HOWARD,PH (1993)

CAS #: 070785-03-4

BENZENEPROPANAMINE, 4-(1,1-DIMETHYLETHYL)-N,N-DI

Formula: $C_{21}H_{29}N$

Mol Weight: 295.47

MP (deg C): FP (deg C):

BP (deg C):

BP pressure (mm Hg):

Property/Value	Units	Temp	Data Type	Reference
WS 5.86E+000	mg/L	25	EST	MEYLAN,WM ET AL. (1996)
logP 5.10			EXP	SANGSTER,J (1993)
VP 5.71E-006	mm Hg	25	EST	NEELY,WB & BLAU,GE (1985)
DC	pKa			
HL 1.08E-006	atm m3/mol	25	EST	MEYLAN,WM & HOWARD,PH (1991)
OH 9.37E-011	cm3/molc sec	25	EST	MEYLAN,WM & HOWARD,PH (1993)

CAS #: 070785-06-7

3(4-T-BU BENZHYDRYLPROPYL)AMINE,NN-DIME

Formula: $C_{22}H_{31}N$

Mol Weight: 309.50

MP (deg C): FP (deg C):

BP (deg C):

BP pressure (mm Hg):

Property/Value	Units	Temp	Data Type	Reference
WS 1.43E+000	mg/L	25	EST	MEYLAN,WM ET AL. (1996)
logP 5.72			EXP	HANSCH,C & LEO,AJ (1985)
VP 2.86E-006	mm Hg	25	EST	NEELY,WB & BLAU,GE (1985)
DC	pKa			
HL 1.44E-006	atm m3/mol	25	EST	MEYLAN,WM & HOWARD,PH (1991)
OH 9.51E-011	cm3/molc sec	25	EST	MEYLAN,WM & HOWARD,PH (1993)

CAS #: 070785-07-8

N-PH-N(4-TBUBNZYL)-N(NNDIME)ET AMINE

Formula: $C_{21}H_{30}N_2$

Mol Weight: 310.49

MP (deg C): FP (deg C):

BP (deg C):

BP pressure (mm Hg):

Property/Value	Units	Temp	Data Type	Reference
WS 2.99E+000	mg/L	25	EST	MEYLAN,WM ET AL. (1996)
logP 5.34			EXP	HANSCH,C & LEO,AJ (1985)
VP 1.37E-006	mm Hg	25	EST	NEELY,WB & BLAU,GE (1985)
DC	pKa			
HL 1.18E-008	atm m3/mol	25	EST	MEYLAN,WM & HOWARD,PH (1991)
OH 2.51E-010	cm3/molc sec	25	EST	MEYLAN,WM & HOWARD,PH (1993)

CAS #: 070859-35-7			1,1-DIME-3-(3-PHENYLETOPH) UREA

Formula: $C_{17}H_{20}N_2O_2$

Mol Weight: 284.36

MP (deg C):   FP (deg C):

BP (deg C):

BP pressure (mm Hg):

Property/ Value	Units	Temp	Data Type	Reference
WS 1.93E+001	mg/L	25	EST	MEYLAN,WM ET AL. (1996)
logP 3.39			EXP	HANSCH,C & LEO,AJ (1985)
VP 3.37E-008	mm Hg	25	EST	NEELY,WB & BLAU,GE (1985)
DC	pKa			
HL 6.15E-012	atm m3/mol	25	EST	MEYLAN,WM & HOWARD,PH (1991)
OH 2.12E-010	cm3/molc sec	25	EST	MEYLAN,WM & HOWARD,PH (1993)

CAS #: 070866-07-8			beta-D-GLUCOPYRANOSE, 2-[[[(2-CHLOROETHYL)NITROS

Formula: $C_9H_{16}ClN_3O_7$

Mol Weight: 313.70

MP (deg C):   FP (deg C):

BP (deg C):

BP pressure (mm Hg):

Property/ Value	Units	Temp	Data Type	Reference
WS 1.83E+003	mg/L	25	EST	MEYLAN,WM ET AL. (1996)
logP -0.66			EXP	SANGSTER,J (1994)
VP 3.98E-014	mm Hg	25	EST	NEELY,WB & BLAU,GE (1985)
DC	pKa			
HL 3.67E-022	atm m3/mol	25	EST	MEYLAN,WM & HOWARD,PH (1991)
OH 1.09E-010	cm3/molc sec	25	EST	MEYLAN,WM & HOWARD,PH (1993)

CAS #: 070872-82-1			DIBENZOFURAN, 1,2,6,7,9-PENTACHLORO-

Formula: $C_{12}H_3Cl_5O$

Mol Weight: 340.42

MP (deg C):   FP (deg C):

BP (deg C):

BP pressure (mm Hg):

Property/ Value	Units	Temp	Data Type	Reference
WS 1.51E-003	mg/L	25	EST	MEYLAN,WM ET AL. (1996)
logP 6.51			EXP	SIJM,DTHM ET AL. (1989)
VP 3.46E-007	mm Hg	25	EST	NEELY,WB & BLAU,GE (1985)
DC	pKa			
HL 1.14E-005	atm m3/mol	25	EST	MEYLAN,WM & HOWARD,PH (1991)
OH 7.46E-014	cm3/molc sec	25	EST	MEYLAN,WM & HOWARD,PH (1993)

CAS #: 070873-69-7			ADAMANTYLALANINE

Formula: $C_{13}H_{21}NO_2$

Mol Weight: 223.32

MP (deg C):   FP (deg C):

BP (deg C):

BP pressure (mm Hg):

Property/ Value	Units	Temp	Data Type	Reference
WS 1.21E+002	mg/L	25	EST	MEYLAN,WM ET AL. (1996)
logP 0.43			EXP	HANSCH,C ET AL. (1995)
VP 2.81E-011	mm Hg	25	EST	NEELY,WB & BLAU,GE (1985)
DC	pKa			
HL 2.17E-009	atm m3/mol	25	EST	MEYLAN,WM & HOWARD,PH (1991)
OH 5.89E-011	cm3/molc sec	25	EST	MEYLAN,WM & HOWARD,PH (1993)

CAS #: 070934-02-0			PYRROLO[2,1-B]QUINAZOLIN-9(1H)-ONE, 2,3-DIHYDRO-

Formula: $C_{12}H_{12}N_2O$

Mol Weight: 200.24

MP (deg C):   FP (deg C):

BP (deg C):

BP pressure (mm Hg):

Property/ Value	Units	Temp	Data Type	Reference
WS 1.91E+003	mg/L	25	EST	MEYLAN,WM ET AL. (1996)
logP 1.59			EXP	SANGSTER,J (1994)
VP 6.44E-006	mm Hg	25	EST	NEELY,WB & BLAU,GE (1985)
DC	pKa			
HL 6.88E-010	atm m3/mol	25	EST	MEYLAN,WM & HOWARD,PH (1991)
OH 1.51E-011	cm3/molc sec	25	EST	MEYLAN,WM & HOWARD,PH (1993)

CAS #: 070959-58-9			[3,4'-BIPYRIDINE]-5-CARBONITRILE,6-CHLORO-

Formula: $C_{11}H_6ClN_3$

Mol Weight: 215.64

MP (deg C):   FP (deg C):

BP (deg C):

BP pressure (mm Hg):

Property/ Value	Units	Temp	Data Type	Reference
WS 6.44E+002	mg/L	25	EST	MEYLAN,WM ET AL. (1996)
logP 1.74			EXP	HAGEN,V ET AL. (1989)
VP 6.55E-006	mm Hg	25	EST	NEELY,WB & BLAU,GE (1985)
DC	pKa			
HL 3.20E-010	atm m3/mol	25	EST	MEYLAN,WM & HOWARD,PH (1991)
OH 3.62E-013	cm3/molc sec	25	EST	MEYLAN,WM & HOWARD,PH (1993)

CAS #: 070959-61-4			[3,4'-BIPYRIDINE]-5-CARBONITRILE,6-METHOXY-

Formula: $C_{12}H_9N_3O$

Mol Weight: 211.23

MP (deg C):   FP (deg C):

BP (deg C):

BP pressure (mm Hg):

Property/ Value	Units	Temp	Data Type	Reference
WS 3.92E+002	mg/L	25	EST	MEYLAN,WM ET AL. (1996)
logP 2.02			EXP	SANGSTER,J (1994)
VP 4.62E-006	mm Hg	25	EST	NEELY,WB & BLAU,GE (1985)
DC	pKa			
HL 3.69E-012	atm m3/mol	25	EST	MEYLAN,WM & HOWARD,PH (1991)
OH 1.38E-012	cm3/molc sec	25	EST	MEYLAN,WM & HOWARD,PH (1993)

CAS #: 070965-23-0			1-TRIAZOLEACETIC ACID-3-NITRO,METHYL ESTER

Formula: $C_5H_8N_4O_4$

Mol Weight: 188.14

MP (deg C):   FP (deg C):

BP (deg C):

BP pressure (mm Hg):

Property/ Value	Units	Temp	Data Type	Reference
WS 4.42E+004	mg/L	25	EST	MEYLAN,WM ET AL. (1996)
logP -0.38			EXP	HANSCH,C ET AL. (1995)
VP 3.25E-004	mm Hg	25	EST	NEELY,WB & BLAU,GE (1985)
DC	pKa			
HL 3.06E-010	atm m3/mol	25	EST	MEYLAN,WM & HOWARD,PH (1991)
OH 9.21E-013	cm3/molc sec	25	EST	MEYLAN,WM & HOWARD,PH (1993)

CAS #: 070999-47-2

4H-PYRIDO[1,2-A]PYRIMIDINE-3-CARBOXYLIC ACID, 6,

Formula: $C_{12}H_{16}N_2O_3$

Mol Weight: 236.27

MP (deg C): **FP (deg C):**

BP (deg C):

BP pressure (mm Hg):

Property/Value	Units	Temp	Data Type	Reference
WS 7.69E+003	mg/L	25	EST	MEYLAN,WM ET AL. (1996)
logP 0.66			EXP	SANGSTER,J (1994)
VP 3.04E-006	mm Hg	25	EST	NEELY,WB & BLAU,GE (1985)
DC	pKa			
HL 3.54E-012	atm m3/mol	25	EST	MEYLAN,WM & HOWARD,PH (1991)
OH 2.47E-011	cm3/molc sec	25	EST	MEYLAN,WM & HOWARD,PH (1993)

CAS #: 071006-72-9

1-(3-IPRO-2-OH-PR)-2-NO2 IMIDAZOLE

Formula: $C_9H_{17}N_3O_4$

Mol Weight: 231.25

MP (deg C): **FP (deg C):**

BP (deg C):

BP pressure (mm Hg):

Property/Value	Units	Temp	Data Type	Reference
WS 4.59E+003	mg/L	25	EST	MEYLAN,WM ET AL. (1996)
logP 0.51			EXP	HANSCH,C & LEO,AJ (1985)
VP 2.90E-008	mm Hg	25	EST	NEELY,WB & BLAU,GE (1985)
DC	pKa			
HL 4.20E-013	atm m3/mol	25	EST	MEYLAN,WM & HOWARD,PH (1991)
OH 3.52E-011	cm3/molc sec	25	EST	MEYLAN,WM & HOWARD,PH (1993)

CAS #: 071006-73-0

1-(2-MEO ETHYL)-2-NO2 IMIDAZOLE

Formula: $C_6H_{11}N_3O_3$

Mol Weight: 173.17

MP (deg C): **FP (deg C):**

BP (deg C):

BP pressure (mm Hg):

Property/Value	Units	Temp	Data Type	Reference
WS 1.95E+004	mg/L	25	EST	MEYLAN,WM ET AL. (1996)
logP 0.12			EXP	HANSCH,C & LEO,AJ (1985)
VP 7.92E-005	mm Hg	25	EST	NEELY,WB & BLAU,GE (1985)
DC	pKa			
HL 4.91E-009	atm m3/mol	25	EST	MEYLAN,WM & HOWARD,PH (1991)
OH 1.56E-011	cm3/molc sec	25	EST	MEYLAN,WM & HOWARD,PH (1993)

CAS #: 071006-74-1

1-(3-PHENOXYPROPYL)-2-NO2 -IMIDAZOLE

Formula: $C_{12}H_{15}N_3O_3$

Mol Weight: 249.27

MP (deg C): **FP (deg C):**

BP (deg C):

BP pressure (mm Hg):

Property/Value	Units	Temp	Data Type	Reference
WS 9.26E+001	mg/L	25	EST	MEYLAN,WM ET AL. (1996)
logP 2.38			EXP	HANSCH,C & LEO,AJ (1985)
VP 1.92E-007	mm Hg	25	EST	NEELY,WB & BLAU,GE (1985)
DC	pKa			
HL 2.41E-009	atm m3/mol	25	EST	MEYLAN,WM & HOWARD,PH (1991)
OH 3.82E-011	cm3/molc sec	25	EST	MEYLAN,WM & HOWARD,PH (1993)

CAS #: 071006-75-2

1-(ACETYLMETHYL)-2-NO2 IMIDAZOLE

Formula: $C_6H_9N_3O_3$

Mol Weight: 171.16

MP (deg C): **FP (deg C):**

BP (deg C):

BP pressure (mm Hg):

Property/Value	Units	Temp	Data Type	Reference
WS 4.94E+004	mg/L	25	EST	MEYLAN,WM ET AL. (1996)
logP -0.34			EXP	HANSCH,C & LEO,AJ (1985)
VP 3.15E-005	mm Hg	25	EST	NEELY,WB & BLAU,GE (1985)
DC	pKa			
HL 2.13E-010	atm m3/mol	25	EST	MEYLAN,WM & HOWARD,PH (1991)
OH 5.30E-012	cm3/molc sec	25	EST	MEYLAN,WM & HOWARD,PH (1993)

CAS #: 071006-76-3

1-ETOET-2-NO2 IMIDAZOLE

Formula: $C_7H_{13}N_3O_3$

Mol Weight: 187.20

MP (deg C): **FP (deg C):**

BP (deg C):

BP pressure (mm Hg):

Property/Value	Units	Temp	Data Type	Reference
WS 7.46E+003	mg/L	25	EST	MEYLAN,WM ET AL. (1996)
logP 0.53			EXP	HANSCH,C & LEO,AJ (1985)
VP 3.26E-005	mm Hg	25	EST	NEELY,WB & BLAU,GE (1985)
DC	pKa			
HL 6.52E-009	atm m3/mol	25	EST	MEYLAN,WM & HOWARD,PH (1991)
OH 2.09E-011	cm3/molc sec	25	EST	MEYLAN,WM & HOWARD,PH (1993)

CAS #: 071006-77-4

1-(2-(PHENOXY)ETHYL)2-NO2IMIDAZOLE

Formula: $C_{11}H_{13}N_3O_3$

Mol Weight: 235.24

MP (deg C): **FP (deg C):**

BP (deg C):

BP pressure (mm Hg):

Property/Value	Units	Temp	Data Type	Reference
WS 2.90E+002	mg/L	25	EST	MEYLAN,WM ET AL. (1996)
logP 1.89			EXP	HANSCH,C & LEO,AJ (1985)
VP 4.48E-007	mm Hg	25	EST	NEELY,WB & BLAU,GE (1985)
DC	pKa			
HL 1.82E-009	atm m3/mol	25	EST	MEYLAN,WM & HOWARD,PH (1991)
OH 3.63E-011	cm3/molc sec	25	EST	MEYLAN,WM & HOWARD,PH (1993)

CAS #: 071006-78-5

1-(2-OH-3-MEOPR)-2-ME-4-NO2IMIDAZOLE

Formula: $C_8H_{15}N_3O_4$

Mol Weight: 217.23

MP (deg C): **FP (deg C):**

BP (deg C):

BP pressure (mm Hg):

Property/Value	Units	Temp	Data Type	Reference
WS 3.02E+004	mg/L	25	EST	MEYLAN,WM ET AL. (1996)
logP -0.36			EXP	HANSCH,C & LEO,AJ (1985)
VP 3.84E-008	mm Hg	25	EST	NEELY,WB & BLAU,GE (1985)
DC	pKa			
HL 2.63E-013	atm m3/mol	25	EST	MEYLAN,WM & HOWARD,PH (1991)
OH 2.34E-011	cm3/molc sec	25	EST	MEYLAN,WM & HOWARD,PH (1993)

CAS #: 071006-79-6				ME(2(2-NO2-1-IMIDAZOLYL)ET)SULFONE
Formula: $C_6H_{11}N_3O_4S$				
Mol Weight: 221.24				
MP (deg C):		FP (deg C):		
BP (deg C):				
BP pressure (mm Hg):				

Property/Value	Units	Temp	Data Type	Reference
WS 1.83E+005	mg/L	25	EST	MEYLAN,WM ET AL. (1996)
logP -1.30			EXP	HANSCH,C & LEO,AJ (1985)
VP 4.18E-007	mm Hg	25	EST	NEELY,WB & BLAU,GE (1985)
DC	pKa			
HL 3.56E-011	atm m3/mol	25	EST	MEYLAN,WM & HOWARD,PH (1991)
OH 1.57E-011	cm3/molc sec	25	EST	MEYLAN,WM & HOWARD,PH (1993)

CAS #: 071006-80-9				PHENYL(2-NO2-1-IMIDAZOLYL)ET)SULFONE
Formula: $C_{11}H_{13}N_3O_4S$				
Mol Weight: 283.31				
MP (deg C):		FP (deg C):		
BP (deg C):				
BP pressure (mm Hg):				

Property/Value	Units	Temp	Data Type	Reference
WS 2.31E+003	mg/L	25	EST	MEYLAN,WM ET AL. (1996)
logP 0.52			EXP	HANSCH,C & LEO,AJ (1985)
VP 2.55E-009	mm Hg	25	EST	NEELY,WB & BLAU,GE (1985)
DC	pKa			
HL 7.11E-012	atm m3/mol	25	EST	MEYLAN,WM & HOWARD,PH (1991)
OH 1.50E-011	cm3/molc sec	25	EST	MEYLAN,WM & HOWARD,PH (1993)

CAS #: 071048-87-8				1,9-PHENANTHRIDINEDIOL, 5,6,6A,7,8,9,10,10A-OCTA
Formula: $C_{27}H_{35}NO_4$				
Mol Weight: 437.58				
MP (deg C):		FP (deg C):		
BP (deg C):				
BP pressure (mm Hg):				

Property/Value	Units	Temp	Data Type	Reference
WS 2.24E-002	mg/L	25	EST	MEYLAN,WM ET AL. (1996)
logP 5.70			EXP	SANGSTER,J (1993)
VP 4.25E-014	mm Hg	25	EST	NEELY,WB & BLAU,GE (1985)
DC	pKa			
HL 3.20E-014	atm m3/mol	25	EST	MEYLAN,WM & HOWARD,PH (1991)
OH 2.62E-010	cm3/molc sec	25	EST	MEYLAN,WM & HOWARD,PH (1993)

CAS #: 071059-53-5				4-CL BENZALDEHYDE,O-((MEAM)CO)OXIME
Formula: $C_9H_9ClN_2O_2$				
Mol Weight: 212.64				
MP (deg C):		FP (deg C):		
BP (deg C):				
BP pressure (mm Hg):				

Property/Value	Units	Temp	Data Type	Reference
WS 4.34E+002	mg/L	25	EST	MEYLAN,WM ET AL. (1996)
logP 2.27			EXP	HANSCH,C & LEO,AJ (1985)
VP 7.66E-004	mm Hg	25	EST	NEELY,WB & BLAU,GE (1985)
DC	pKa			
HL 8.54E-009	atm m3/mol	25	EST	MEYLAN,WM & HOWARD,PH (1991)
OH 9.65E-012	cm3/molc sec	25	EST	MEYLAN,WM & HOWARD,PH (1993)

CAS #: 071125-73-0				24NH2-PYRIMIDINE,5(3,5-MEO-4-ETS)BENZYL
Formula: $C_{15}H_{20}N_4O_2S$				
Mol Weight: 320.42				
MP (deg C):		FP (deg C):		
BP (deg C):				
BP pressure (mm Hg):				

Property/Value	Units	Temp	Data Type	Reference
WS 2.76E+002	mg/L	25	EST	MEYLAN,WM ET AL. (1996)
logP 1.79			EXP	HANSCH,C & LEO,AJ (1985)
VP 9.90E-010	mm Hg	25	EST	NEELY,WB & BLAU,GE (1985)
DC	pKa			
HL 1.56E-014	atm m3/mol	25	EST	MEYLAN,WM & HOWARD,PH (1991)
OH 2.10E-010	cm3/molc sec	25	EST	MEYLAN,WM & HOWARD,PH (1993)

CAS #: 071125-75-2				24NH2PYRIMIDIN,5(35MEO-4(1PYRRYL))BENZYL
Formula: $C_{17}H_{21}N_5O_2$				
Mol Weight: 327.39				
MP (deg C):		FP (deg C):		
BP (deg C):				
BP pressure (mm Hg):				

Property/Value	Units	Temp	Data Type	Reference
WS 2.63E+002	mg/L	25	EST	MEYLAN,WM ET AL. (1996)
logP 1.78			EXP	HANSCH,C & LEO,AJ (1985)
VP 1.63E-010	mm Hg	25	EST	NEELY,WB & BLAU,GE (1985)
DC	pKa			
HL 2.47E-018	atm m3/mol	25	EST	MEYLAN,WM & HOWARD,PH (1991)
OH 2.03E-010	cm3/molc sec	25	EST	MEYLAN,WM & HOWARD,PH (1993)

CAS #: 071195-58-9				ALFENTANIL
Formula: $C_{21}H_{32}N_6O_3$				
Mol Weight: 416.53				
MP (deg C): 140.8		FP (deg C):		
BP (deg C):				
BP pressure (mm Hg):				

Property/Value	Units	Temp	Data Type	Reference
WS 3.46E+001	mg/L	25	EST	MEYLAN,WM ET AL. (1996)
logP 2.16			EXP	HANSCH,C ET AL. (1995)
VP 4.52E-012	mm Hg	25	EST	NEELY,WB & BLAU,GE (1985)
DC	pKa			
HL 9.73E-020	atm m3/mol	25	EST	MEYLAN,WM & HOWARD,PH (1991)
OH 1.42E-010	cm3/molc sec	25	EST	MEYLAN,WM & HOWARD,PH (1993)

CAS #: 071203-60-6				ETHYL-(3,4,5-TRIMEO)PHENYL SULFONE
Formula: $C_{11}H_{16}O_5S$				
Mol Weight: 260.31				
MP (deg C):		FP (deg C):		
BP (deg C):				
BP pressure (mm Hg):				

Property/Value	Units	Temp	Data Type	Reference
WS 2.63E+003	mg/L	25	EST	MEYLAN,WM ET AL. (1996)
logP 1.05			EXP	HANSCH,C & LEO,AJ (1985)
VP 5.95E-006	mm Hg	25	EST	NEELY,WB & BLAU,GE (1985)
DC	pKa			
HL 3.43E-010	atm m3/mol	25	EST	MEYLAN,WM & HOWARD,PH (1991)
OH 5.38E-011	cm3/molc sec	25	EST	MEYLAN,WM & HOWARD,PH (1993)

1301

CAS #: 071203-61-7	I-PENTYL-(345-MEO)PHENYL-SULFONE

Formula: $C_{14}H_{22}O_5S$

Mol Weight: 302.39

MP (deg C):   FP (deg C):

BP (deg C):

BP pressure (mm Hg):

Property/Value	Units	Temp	Data Type	Reference
WS 1.84E+002	mg/L	25	EST	MEYLAN,WM ET AL. (1996)
logP 2.12			EXP	HANSCH,C & LEO,AJ (1985)
VP 8.32E-007	mm Hg	25	EST	NEELY,WB & BLAU,GE (1985)
DC	pKa			
HL 8.02E-010	atm m3/mol	25	EST	MEYLAN,WM & HOWARD,PH (1991)
OH 5.94E-011	cm3/molc sec	25	EST	MEYLAN,WM & HOWARD,PH (1993)

CAS #: 071203-62-8	PROPYL-(3,4,5-MEO)PHENYL-SULFONE

Formula: $C_{12}H_{18}O_5S$

Mol Weight: 274.34

MP (deg C):   FP (deg C):

BP (deg C):

BP pressure (mm Hg):

Property/Value	Units	Temp	Data Type	Reference
WS 8.03E+002	mg/L	25	EST	MEYLAN,WM ET AL. (1996)
logP 1.56			EXP	HANSCH,C & LEO,AJ (1985)
VP 2.57E-006	mm Hg	25	EST	NEELY,WB & BLAU,GE (1985)
DC	pKa			
HL 4.55E-010	atm m3/mol	25	EST	MEYLAN,WM & HOWARD,PH (1991)
OH 5.66E-011	cm3/molc sec	25	EST	MEYLAN,WM & HOWARD,PH (1993)

CAS #: 071203-63-9	2-OH-2-PH-ET-(3,4,5-MEO)PHENYLSUFONE

Formula: $C_{17}H_{20}O_6S$

Mol Weight: 352.41

MP (deg C):   FP (deg C):

BP (deg C):

BP pressure (mm Hg):

Property/Value	Units	Temp	Data Type	Reference
WS 2.34E+002	mg/L	25	EST	MEYLAN,WM ET AL. (1996)
logP 1.65			EXP	HANSCH,C & LEO,AJ (1985)
VP 1.01E-011	mm Hg	25	EST	NEELY,WB & BLAU,GE (1985)
DC	pKa			
HL 1.01E-015	atm m3/mol	25	EST	MEYLAN,WM & HOWARD,PH (1991)
OH 6.84E-011	cm3/molc sec	25	EST	MEYLAN,WM & HOWARD,PH (1993)

CAS #: 071203-64-0	(N-BU-ME)AMET-(345-MEO)PHENYL-SULFONE

Formula: $C_{16}H_{27}NO_5S$

Mol Weight: 345.46

MP (deg C):   FP (deg C):

BP (deg C):

BP pressure (mm Hg):

Property/Value	Units	Temp	Data Type	Reference
WS 1.19E+003	mg/L	25	EST	MEYLAN,WM ET AL. (1996)
logP 2.05			EXP	HANSCH,C & LEO,AJ (1985)
VP 3.99E-008	mm Hg	25	EST	NEELY,WB & BLAU,GE (1985)
DC	pKa			
HL 3.98E-013	atm m3/mol	25	EST	MEYLAN,WM & HOWARD,PH (1991)
OH 1.47E-010	cm3/molc sec	25	EST	MEYLAN,WM & HOWARD,PH (1993)

CAS #: 071203-66-2	N-DIBUAMET 3,4,5-TRIMEO PH SULFONE

Formula: $C_{19}H_{33}NO_5S$

Mol Weight: 387.54

MP (deg C):   FP (deg C):

BP (deg C):

BP pressure (mm Hg):

Property/Value	Units	Temp	Data Type	Reference
WS 3.46E+001	mg/L	25	EST	MEYLAN,WM ET AL. (1996)
logP 3.55			EXP	HANSCH,C & LEO,AJ (1985)
VP 2.95E-009	mm Hg	25	EST	NEELY,WB & BLAU,GE (1985)
DC	pKa			
HL 9.31E-013	atm m3/mol	25	EST	MEYLAN,WM & HOWARD,PH (1991)
OH 1.59E-010	cm3/molc sec	25	EST	MEYLAN,WM & HOWARD,PH (1993)

CAS #: 071225-60-0	BENZAMIDE,N(DIETAMET),2-MEO,5-BR

Formula: $C_{14}H_{21}BrN_2O_2$

Mol Weight: 329.24

MP (deg C):   FP (deg C):

BP (deg C):

BP pressure (mm Hg):

Property/Value	Units	Temp	Data Type	Reference
WS 1.53E+001	mg/L	25	EST	MEYLAN,WM ET AL. (1996)
logP 3.20			EXP	HANSCH,C & LEO,AJ (1985)
VP 3.86E-008	mm Hg	25	EST	NEELY,WB & BLAU,GE (1985)
DC	pKa			
HL 1.33E-013	atm m3/mol	25	EST	MEYLAN,WM & HOWARD,PH (1991)
OH 1.11E-010	cm3/molc sec	25	EST	MEYLAN,WM & HOWARD,PH (1993)

CAS #: 071225-61-1	BENZAMIDE,N(DIETAMINOET),3,5-DICL

Formula: $C_{13}H_{18}Cl_2N_2O$

Mol Weight: 289.21

MP (deg C):   FP (deg C):

BP (deg C):

BP pressure (mm Hg):

Property/Value	Units	Temp	Data Type	Reference
WS 8.56E+000	mg/L	25	EST	MEYLAN,WM ET AL. (1996)
logP 3.77			EXP	HANSCH,C & LEO,AJ (1985)
VP 1.45E-007	mm Hg	25	EST	NEELY,WB & BLAU,GE (1985)
DC	pKa			
HL 3.10E-012	atm m3/mol	25	EST	MEYLAN,WM & HOWARD,PH (1991)
OH 1.05E-010	cm3/molc sec	25	EST	MEYLAN,WM & HOWARD,PH (1993)

CAS #: 071323-93-8	M-NITROBENZYL TRIMETHYL AMMONIUM BROMIDE

Formula: $C_{10}H_{15}BrN_2O_2$

Mol Weight: 275.15

MP (deg C):   FP (deg C):

BP (deg C):

BP pressure (mm Hg):

Property/Value	Units	Temp	Data Type	Reference
WS 1.00E+006	mg/L	25	EST	MEYLAN,WM ET AL. (1996)
logP -2.66			EXP	HANSCH,C & LEO,AJ (1985)
VP 7.55E-010	mm Hg	25	EST	NEELY,WB & BLAU,GE (1985)
DC	pKa			
HL 1.06E-015	atm m3/mol	25	EST	MEYLAN,WM & HOWARD,PH (1991)
OH 1.29E-011	cm3/molc sec	25	EST	MEYLAN,WM & HOWARD,PH (1993)

CAS #: 071323-94-9				M-METHOXYBENZYL TRIMETHYL AMMONIUM BROMIDE
Formula: $C_{11}H_{18}BrNO$				
Mol Weight: 260.18				
MP (deg C):		FP (deg C):		
BP (deg C):				
BP pressure (mm Hg):				

Property/Value	Units	Temp	Data Type	Reference
WS 8.71E+005	mg/L	25	EST	MEYLAN,WM ET AL. (1996)
logP -1.90			EXP	HANSCH,C & LEO,AJ (1985)
VP 7.15E-009	mm Hg	25	EST	NEELY,WB & BLAU,GE (1985)
DC	pKa			
HL 1.58E-014	atm m3/mol	25	EST	MEYLAN,WM & HOWARD,PH (1991)
OH 6.67E-011	cm3/molc sec	25	EST	MEYLAN,WM & HOWARD,PH (1993)

CAS #: 071323-95-0				M-CARBAMYLBENZYL TRIMETHYL AMMONIUM BROMIDE
Formula: $C_{11}H_{17}BrN_2O$				
Mol Weight: 273.17				
MP (deg C):		FP (deg C):		
BP (deg C):				
BP pressure (mm Hg):				

Property/Value	Units	Temp	Data Type	Reference
WS 2.11E+005	mg/L	25	EST	MEYLAN,WM ET AL. (1996)
logP -2.80			EXP	HANSCH,C & LEO,AJ (1985)
VP 1.41E-011	mm Hg	25	EST	NEELY,WB & BLAU,GE (1985)
DC	pKa			
HL 1.10E-019	atm m3/mol	25	EST	MEYLAN,WM & HOWARD,PH (1991)
OH 1.66E-011	cm3/molc sec	25	EST	MEYLAN,WM & HOWARD,PH (1993)

CAS #: 071323-96-1				M-AMINOBENZYL TRIMETHYL AMMONIUM BROMIDE
Formula: $C_{10}H_{17}BrN_2$				
Mol Weight: 245.16				
MP (deg C):		FP (deg C):		
BP (deg C):				
BP pressure (mm Hg):				

Property/Value	Units	Temp	Data Type	Reference
WS 1.00E+006	mg/L	25	EST	MEYLAN,WM ET AL. (1996)
logP -2.50			EXP	HANSCH,C & LEO,AJ (1985)
VP 1.90E-009	mm Hg	25	EST	NEELY,WB & BLAU,GE (1985)
DC	pKa			
HL 9.46E-017	atm m3/mol	25	EST	MEYLAN,WM & HOWARD,PH (1991)
OH 2.12E-010	cm3/molc sec	25	EST	MEYLAN,WM & HOWARD,PH (1993)

CAS #: 071323-97-2				NNN-ME-N(3(CYANOPHENOL)METHYL) AMMONIUM BROMIDE
Formula: $C_{11}H_{15}BrN_2$				
Mol Weight: 255.16				
MP (deg C):		FP (deg C):		
BP (deg C):				
BP pressure (mm Hg):				

Property/Value	Units	Temp	Data Type	Reference
WS 1.00E+006	mg/L	25	EST	MEYLAN,WM ET AL. (1996)
logP -3.05			EXP	HANSCH,C ET AL. (1995)
VP 1.41E-009	mm Hg	25	EST	NEELY,WB & BLAU,GE (1985)
DC	pKa			
HL 2.59E-015	atm m3/mol	25	EST	MEYLAN,WM & HOWARD,PH (1991)
OH 1.31E-011	cm3/molc sec	25	EST	MEYLAN,WM & HOWARD,PH (1993)

CAS #: 071323-98-3				M-HYDROXYBENZYL TRIMETHYL AMMONIUM BROMIDE
Formula: $C_{10}H_{16}BrNO$				
Mol Weight: 246.15				
MP (deg C):		FP (deg C):		
BP (deg C):				
BP pressure (mm Hg):				

Property/Value	Units	Temp	Data Type	Reference
WS 1.00E+006	mg/L	25	EST	MEYLAN,WM ET AL. (1996)
logP -2.40			EXP	HANSCH,C & LEO,AJ (1985)
VP 5.05E-010	mm Hg	25	EST	NEELY,WB & BLAU,GE (1985)
DC	pKa			
HL 2.79E-017	atm m3/mol	25	EST	MEYLAN,WM & HOWARD,PH (1991)
OH 9.54E-011	cm3/molc sec	25	EST	MEYLAN,WM & HOWARD,PH (1993)

CAS #: 071323-99-4				M-BROMOBENZYL TRIMETHYL AMMONIUM BROMIDE
Formula: $C_{10}H_{15}Br_2N$				
Mol Weight: 309.05				
MP (deg C):		FP (deg C):		
BP (deg C):				
BP pressure (mm Hg):				

Property/Value	Units	Temp	Data Type	Reference
WS 3.82E+005	mg/L	25	EST	MEYLAN,WM ET AL. (1996)
logP -1.81			EXP	HANSCH,C & LEO,AJ (1985)
VP 4.37E-009	mm Hg	25	EST	NEELY,WB & BLAU,GE (1985)
DC	pKa			
HL 1.07E-013	atm m3/mol	25	EST	MEYLAN,WM & HOWARD,PH (1991)
OH 1.55E-011	cm3/molc sec	25	EST	MEYLAN,WM & HOWARD,PH (1993)

CAS #: 071324-00-0				M-FLUOROBENZYL TRIMETHYL AMMONIUM BROMIDE
Formula: $C_{10}H_{15}BrFN$				
Mol Weight: 248.14				
MP (deg C):		FP (deg C):		
BP (deg C):				
BP pressure (mm Hg):				

Property/Value	Units	Temp	Data Type	Reference
WS 1.00E+006	mg/L	25	EST	MEYLAN,WM ET AL. (1996)
logP -2.20			EXP	HANSCH,C & LEO,AJ (1985)
VP 5.22E-008	mm Hg	25	EST	NEELY,WB & BLAU,GE (1985)
DC	pKa			
HL 3.12E-013	atm m3/mol	25	EST	MEYLAN,WM & HOWARD,PH (1991)
OH 1.85E-011	cm3/molc sec	25	EST	MEYLAN,WM & HOWARD,PH (1993)

CAS #: 071324-01-1				NNN-ME-N(3-CF3PHME) AMMONIUM BROMIDE
Formula: $C_{11}H_{15}BrF_3N$				
Mol Weight: 298.15				
MP (deg C):		FP (deg C):		
BP (deg C):				
BP pressure (mm Hg):				

Property/Value	Units	Temp	Data Type	Reference
WS 1.79E+005	mg/L	25	EST	MEYLAN,WM ET AL. (1996)
logP -1.35			EXP	HANSCH,C ET AL. (1995)
VP 3.17E-008	mm Hg	25	EST	NEELY,WB & BLAU,GE (1985)
DC	pKa			
HL 2.33E-012	atm m3/mol	25	EST	MEYLAN,WM & HOWARD,PH (1991)
OH 1.32E-011	cm3/molc sec	25	EST	MEYLAN,WM & HOWARD,PH (1993)

CAS #: 071548-53-3	1-HEXYL-4-METHOXYCARBONYL-2-PYRROLIDONE
Formula: $C_{12}H_{21}NO_3$	
Mol Weight: 227.31	
MP (deg C):	FP (deg C):
BP (deg C):	
BP pressure (mm Hg):	

Property/ Value	Units	Temp	Data Type	Reference
WS 3.42E+002	mg/L	25	EST	MEYLAN,WM ET AL. (1996)
logP 2.30			EXP	SASAKI,H ET AL. (1988)
VP 2.60E-005	mm Hg	25	EST	NEELY,WB & BLAU,GE (1985)
DC	pKa			
HL 3.10E-010	atm m3/mol	25	EST	MEYLAN,WM & HOWARD,PH (1991)
OH 3.54E-011	cm3/molc sec	25	EST	MEYLAN,WM & HOWARD,PH (1993)

CAS #: 071626-11-4	BENALAXYL
Formula: $C_{20}H_{23}NO_3$	
Mol Weight: 325.41	
MP (deg C): 78-80	FP (deg C):
BP (deg C):	
BP pressure (mm Hg):	

Property/ Value	Units	Temp	Data Type	Reference
WS 3.70E+001	mg/L	25	EXP	SHIU,WY ET AL. (1990)
logP 3.40			EXP	HANSCH,C ET AL. (1995)
VP 1.00E-005	mm Hg	25	EXP	AUGUSTIJN-BECKERS,PWM ET AL. (1994)
DC	pKa			
HL 1.16E-007	atm m3/mol	25	EST	VP/WSOL
OH 3.07E-011	cm3/molc sec	25	EST	MEYLAN,WM & HOWARD,PH (1993)

CAS #: 071675-85-9	AMISULPRIDE
Formula: $C_{17}H_{27}N_3O_4S$	
Mol Weight: 369.49	
MP (deg C): 126-127	FP (deg C):
BP (deg C):	
BP pressure (mm Hg):	

Property/ Value	Units	Temp	Data Type	Reference
WS 5.43E+002	mg/L	25	EST	MEYLAN,WM ET AL. (1996)
logP 1.10			EXP	MANNHOLD,R ET AL. (1990)
VP 2.63E-012	mm Hg	25	EST	NEELY,WB & BLAU,GE (1985)
DC	pKa			
HL 2.12E-020	atm m3/mol	25	EST	MEYLAN,WM & HOWARD,PH (1991)
OH 3.26E-010	cm3/molc sec	25	EST	MEYLAN,WM & HOWARD,PH (1993)

CAS #: 071879-58-8	4-PYRIDINEBUTANAMIDE
Formula: $C_9H_{12}N_2O$	
Mol Weight: 164.21	
MP (deg C):	FP (deg C):
BP (deg C):	
BP pressure (mm Hg):	

Property/ Value	Units	Temp	Data Type	Reference
WS 1.00E+006	mg/L	25	EST	MEYLAN,WM ET AL. (1996)
logP -0.09			EXP	HANSCH,C & LEO,AJ (1985)
VP 3.11E-005	mm Hg	25	EST	NEELY,WB & BLAU,GE (1985)
DC	pKa			
HL 2.08E-012	atm m3/mol	25	EST	MEYLAN,WM & HOWARD,PH (1991)
OH 9.41E-012	cm3/molc sec	25	EST	MEYLAN,WM & HOWARD,PH (1993)

CAS #: 071972-66-2	4H-1-BENZOPYRAN-4-ONE, 3-METHYL-2-PHENYL-
Formula: $C_{16}H_{12}O_2$	
Mol Weight: 236.27	
MP (deg C):	FP (deg C):
BP (deg C):	
BP pressure (mm Hg):	

Property/ Value	Units	Temp	Data Type	Reference
WS 1.26E+001	mg/L	25	EST	MEYLAN,WM ET AL. (1996)
logP 3.92			EXP	HANSCH,C ET AL. (1995)
VP 4.08E-006	mm Hg	25	EST	NEELY,WB & BLAU,GE (1985)
DC	pKa			
HL 5.66E-008	atm m3/mol	25	EST	MEYLAN,WM & HOWARD,PH (1991)
OH 6.01E-011	cm3/molc sec	25	EST	MEYLAN,WM & HOWARD,PH (1993)

CAS #: 071990-00-6	2,6-METHANO-3-BENZAZOCIN-8-OL, 6-ETHYL-1,2,3,4,5
Formula: $C_{20}H_{29}NO_2$	
Mol Weight: 315.46	
MP (deg C):	FP (deg C):
BP (deg C):	
BP pressure (mm Hg):	

Property/ Value	Units	Temp	Data Type	Reference
WS 2.11E+002	mg/L	25	EST	MEYLAN,WM ET AL. (1996)
logP 3.82			EXP	SANGSTER,J (1994)
VP 1.03E-009	mm Hg	25	EST	NEELY,WB & BLAU,GE (1985)
DC	pKa			
HL 1.80E-014	atm m3/mol	25	EST	MEYLAN,WM & HOWARD,PH (1991)
OH 3.76E-010	cm3/molc sec	25	EST	MEYLAN,WM & HOWARD,PH (1993)

CAS #: 071998-72-6	DIBENZOFURAN, 1,3,6,8-TETRACHLORO-
Formula: $C_{12}H_4Cl_4O$	
Mol Weight: 305.98	
MP (deg C):	FP (deg C):
BP (deg C):	
BP pressure (mm Hg):	

Property/ Value	Units	Temp	Data Type	Reference
WS 3.19E-003	mg/L	25	EST	MEYLAN,WM ET AL. (1996)
logP 6.37			EXP	SIJM,DTHM ET AL. (1989)
VP 1.53E-006	mm Hg	25	EST	NEELY,WB & BLAU,GE (1985)
DC	pKa			
HL 1.54E-005	atm m3/mol	25	EST	MEYLAN,WM & HOWARD,PH (1991)
OH 1.65E-013	cm3/molc sec	25	EST	MEYLAN,WM & HOWARD,PH (1993)

CAS #: 072111-57-0	2,3-PYRAZINEDICARBONITRILE, 5-CHLORO-
Formula: C_6HClN_4	
Mol Weight: 164.55	
MP (deg C):	FP (deg C):
BP (deg C):	
BP pressure (mm Hg):	

Property/ Value	Units	Temp	Data Type	Reference
WS 5.47E+003	mg/L	25	EST	MEYLAN,WM ET AL. (1996)
logP 0.95			EXP	YAMAGAMI,C ET AL. (1990)
VP 2.24E-004	mm Hg	25	EST	NEELY,WB & BLAU,GE (1985)
DC	pKa			
HL 1.27E-008	atm m3/mol	25	EST	MEYLAN,WM & HOWARD,PH (1991)
OH 3.74E-015	cm3/molc sec	25	EST	MEYLAN,WM & HOWARD,PH (1993)

CAS #: 072111-58-1				2,3-PYRAZINEDICARBONITRILE, 5-CHLORO-6-ETHYL-

Formula: $C_8H_5ClN_4$

Mol Weight: 192.61

MP (deg C):   FP (deg C):

BP (deg C):

BP pressure (mm Hg):

Property/Value	Units	Temp	Data Type	Reference
WS 7.67E+002	mg/L	25	EST	MEYLAN,WM ET AL. (1996)
logP 1.79			EXP	YAMAGAMI,C ET AL. (1991)
VP 4.04E-005	mm Hg	25	EST	NEELY,WB & BLAU,GE (1985)
DC	pKa			
HL 1.87E-008	atm m3/mol	25	EST	MEYLAN,WM & HOWARD,PH (1991)
OH 1.11E-012	cm3/molc sec	25	EST	MEYLAN,WM & HOWARD,PH (1993)

CAS #: 072111-60-5				2,3-PYRAZINEDICARBONITRILE, 5-BUTYL-6-CHLORO-

Formula: $C_{10}H_9ClN_4$

Mol Weight: 220.66

MP (deg C):   FP (deg C):

BP (deg C):

BP pressure (mm Hg):

Property/Value	Units	Temp	Data Type	Reference
WS 6.31E+001	mg/L	25	EST	MEYLAN,WM ET AL. (1996)
logP 2.89			EXP	YAMAGAMI,C ET AL. (1991)
VP 7.50E-006	mm Hg	25	EST	NEELY,WB & BLAU,GE (1985)
DC	pKa			
HL 3.29E-008	atm m3/mol	25	EST	MEYLAN,WM & HOWARD,PH (1991)
OH 3.88E-012	cm3/molc sec	25	EST	MEYLAN,WM & HOWARD,PH (1993)

CAS #: 072112-24-4				2,3-PYRAZINEDICARBONITRILE, 5-BUTYL-6-METHYL-

Formula: $C_{11}H_{12}N_4$

Mol Weight: 200.25

MP (deg C):   FP (deg C):

BP (deg C):

BP pressure (mm Hg):

Property/Value	Units	Temp	Data Type	Reference
WS 2.57E+002	mg/L	25	EST	MEYLAN,WM ET AL. (1996)
logP 2.30			EXP	YAMAGAMI,C ET AL. (1990)
VP 1.32E-005	mm Hg	25	EST	NEELY,WB & BLAU,GE (1985)
DC	pKa			
HL 7.76E-010	atm m3/mol	25	EST	MEYLAN,WM & HOWARD,PH (1991)
OH 4.03E-012	cm3/molc sec	25	EST	MEYLAN,WM & HOWARD,PH (1993)

CAS #: 072115-11-8				PROPANAMIDE, 3,3,3-TRIFLUORO-2-HYDROXY-2-METHYL-

Formula: $C_{11}H_8F_6N_2O_4$

Mol Weight: 346.19

MP (deg C):   FP (deg C):

BP (deg C):

BP pressure (mm Hg):

Property/Value	Units	Temp	Data Type	Reference
WS 4.83E+000	mg/L	25	EST	MEYLAN,WM ET AL. (1996)
logP 3.21			EXP	MORRIS,JJ ET AL. (1991)
VP 5.26E-009	mm Hg	25	EST	NEELY,WB & BLAU,GE (1985)
DC	pKa			
HL 4.58E-010	atm m3/mol	25	EST	MEYLAN,WM & HOWARD,PH (1991)
OH 9.05E-013	cm3/molc sec	25	EST	MEYLAN,WM & HOWARD,PH (1993)

CAS #: 072116-69-9				BENZENEACETAMIDE, N-(3-ACETYLPHENYL)-

Formula: $C_{16}H_{15}NO_2$

Mol Weight: 253.30

MP (deg C):   FP (deg C):

BP (deg C):

BP pressure (mm Hg):

Property/Value	Units	Temp	Data Type	Reference
WS 1.31E+002	mg/L	25	EST	MEYLAN,WM ET AL. (1996)
logP 2.62			EXP	YAMAGAMI,C ET AL. (1984)
VP 2.82E-008	mm Hg	25	EST	NEELY,WB & BLAU,GE (1985)
DC	pKa			
HL 9.07E-013	atm m3/mol	25	EST	MEYLAN,WM & HOWARD,PH (1991)
OH 1.09E-011	cm3/molc sec	25	EST	MEYLAN,WM & HOWARD,PH (1993)

CAS #: 072126-54-6				2,3-PYRAZINEDICARBONITRILE, 5-CHLORO-6-METHYL-

Formula: $C_7H_3ClN_4$

Mol Weight: 178.58

MP (deg C):   FP (deg C):

BP (deg C):

BP pressure (mm Hg):

Property/Value	Units	Temp	Data Type	Reference
WS 2.41E+003	mg/L	25	EST	MEYLAN,WM ET AL. (1996)
logP 1.29			EXP	YAMAGAMI,C ET AL. (1991)
VP 9.99E-005	mm Hg	25	EST	NEELY,WB & BLAU,GE (1985)
DC	pKa			
HL 1.41E-008	atm m3/mol	25	EST	MEYLAN,WM & HOWARD,PH (1991)
OH 1.41E-013	cm3/molc sec	25	EST	MEYLAN,WM & HOWARD,PH (1993)

CAS #: 072165-35-6				PIPERIDINE, 1-(4,4-DIMETHYL-1-PHENYLCYCLOHEXYL)-

Formula: $C_{19}H_{29}N$

Mol Weight: 271.45

MP (deg C):   FP (deg C):

BP (deg C):

BP pressure (mm Hg):

Property/Value	Units	Temp	Data Type	Reference
WS 6.38E-001	mg/L	25	EST	MEYLAN,WM ET AL. (1996)
logP 6.39			EXP	SANGSTER,J (1993)
VP 1.18E-005	mm Hg	25	EST	NEELY,WB & BLAU,GE (1985)
DC	pKa			
HL 9.74E-006	atm m3/mol	25	EST	MEYLAN,WM & HOWARD,PH (1991)
OH 1.06E-010	cm3/molc sec	25	EST	MEYLAN,WM & HOWARD,PH (1993)

CAS #: 072178-02-0				FOMESAFEN

Formula: $C_{15}H_{10}ClF_3N_2O_6S$

Mol Weight: 438.77

MP (deg C): 220-21   FP (deg C):

BP (deg C):

BP pressure (mm Hg):

Property/Value	Units	Temp	Data Type	Reference
WS 5.00E+001	mg/L	20	EXP	SHIU,WY ET AL. (1990)
logP 2.90			EXP	TOMLIN,C (1994)
VP 3.21E-012	mm Hg	25	EST	NEELY,WB & BLAU,GE (1985)
DC	pKa			
HL 7.53E-013	atm m3/mol	25	EST	MEYLAN,WM & HOWARD,PH (1991)
OH 1.61E-012	cm3/molc sec	25	EST	MEYLAN,WM & HOWARD,PH (1993)

CAS #: 072189-64-1 — HYDRAZINECARBOXIMIDAMIDE, 2-[1-(2-THIENYL)ETHYLI

Formula: $C_7H_{10}N_4S$
Mol Weight: 182.25
MP (deg C):
FP (deg C):
BP (deg C):
BP pressure (mm Hg):

Property/Value	Units	Temp	Data Type	Reference
WS 6.81E+003	mg/L	25	EST	MEYLAN,WM ET AL. (1996)
logP 1.05			EXP	JIRA,T ET AL. (1985)
VP 7.52E-005	mm Hg	25	EST	NEELY,WB & BLAU,GE (1985)
DC	pKa			
HL 3.64E-012	atm m3/mol	25	EST	MEYLAN,WM & HOWARD,PH (1991)
OH 1.05E-010	cm3/molc sec	25	EST	MEYLAN,WM & HOWARD,PH (1993)

CAS #: 072216-45-6 — B24C8-BENZO CROWN ETHER

Formula: $C_{20}H_{32}O_8$
Mol Weight: 400.47
MP (deg C):
FP (deg C):
BP (deg C):
BP pressure (mm Hg):

Property/Value	Units	Temp	Data Type	Reference
WS 1.26E+003	mg/L	25	EST	MEYLAN,WM ET AL. (1996)
logP 0.45			EXP	STOLWIJK,TB ET AL. (1989)
VP 1.20E-009	mm Hg	25	EST	NEELY,WB & BLAU,GE (1985)
DC	pKa			
HL 1.17E-016	atm m3/mol	25	EST	MEYLAN,WM & HOWARD,PH (1991)
OH 1.17E-010	cm3/molc sec	25	EST	MEYLAN,WM & HOWARD,PH (1993)

CAS #: 072242-01-4 — PIPERIDINE, 1-(3,3,5-TRIMETHYL-1-PHENYLCYCLOHEXY

Formula: $C_{20}H_{31}N$
Mol Weight: 285.48
MP (deg C):
FP (deg C):
BP (deg C):
BP pressure (mm Hg):

Property/Value	Units	Temp	Data Type	Reference
WS 5.73E-001	mg/L	25	EST	MEYLAN,WM ET AL. (1996)
logP 6.35			EXP	SANGSTER,J (1993)
VP 9.92E-006	mm Hg	25	EST	NEELY,WB & BLAU,GE (1985)
DC	pKa			
HL 1.29E-005	atm m3/mol	25	EST	MEYLAN,WM & HOWARD,PH (1991)
OH 1.04E-010	cm3/molc sec	25	EST	MEYLAN,WM & HOWARD,PH (1993)

CAS #: 072299-07-1 — PIPERIDINE, 1-(CYCLOHEXYLACETYL)-

Formula: $C_{13}H_{23}NO$
Mol Weight: 209.33
MP (deg C):
FP (deg C):
BP (deg C):
BP pressure (mm Hg):

Property/Value	Units	Temp	Data Type	Reference
WS 7.25E+001	mg/L	25	EST	MEYLAN,WM ET AL. (1996)
logP 3.20			EXP	SURYANARAYANA,MVS ET AL. (1991)
VP 3.15E-001	mm Hg	30	EXP	SURYANARAYANA,MVS ET AL. (1991)
DC	pKa			
HL 1.34E-007	atm m3/mol	25	EST	MEYLAN,WM & HOWARD,PH (1991)
OH 4.80E-011	cm3/molc sec	25	EST	MEYLAN,WM & HOWARD,PH (1993)

CAS #: 072301-79-2 — 1H-BENZIMIDAZOL-2-AMINE, 6-[(HYDROXYIMINO)PHENYL

Formula: $C_{17}H_{18}N_4O_3S$
Mol Weight: 358.42
MP (deg C): 198-199
FP (deg C):
BP (deg C):
BP pressure (mm Hg):

Property/Value	Units	Temp	Data Type	Reference
WS 1.24E+001	mg/L	25	EST	MEYLAN,WM ET AL. (1996)
logP 3.10			EXP	HANSCH,C ET AL. (1995)
VP 1.99E-016	mm Hg	25	EST	NEELY,WB & BLAU,GE (1985)
DC	pKa			
HL 1.74E-018	atm m3/mol	25	EST	MEYLAN,WM & HOWARD,PH (1991)
OH 2.20E-010	cm3/molc sec	25	EST	MEYLAN,WM & HOWARD,PH (1993)

CAS #: 072336-12-0 — 9((DIETAMINO)ACETYLAMINO) FLUORENE

Formula: $C_{19}H_{22}N_2O$
Mol Weight: 294.40
MP (deg C):
FP (deg C):
BP (deg C):
BP pressure (mm Hg):

Property/Value	Units	Temp	Data Type	Reference
WS 8.00E-001	mg/L	25	EST	MEYLAN,WM ET AL. (1996)
logP 3.64			EXP	HANSCH,C & LEO,AJ (1985)
VP 1.83E-009	mm Hg	25	EST	NEELY,WB & BLAU,GE (1985)
DC	pKa			
HL 9.73E-013	atm m3/mol	25	EST	MEYLAN,WM & HOWARD,PH (1991)
OH 1.10E-010	cm3/molc sec	25	EST	MEYLAN,WM & HOWARD,PH (1993)

CAS #: 072336-16-4 — ACETAMIDE, 2-(DIETHYLAMINO)-N-(PHENYLMETHYL)-

Formula: $C_{13}H_{20}N_2O$
Mol Weight: 220.32
MP (deg C):
FP (deg C):
BP (deg C):
BP pressure (mm Hg):

Property/Value	Units	Temp	Data Type	Reference
WS 5.10E+002	mg/L	25	EST	MEYLAN,WM ET AL. (1996)
logP 2.14			EXP	SANGSTER,J (1993)
VP 2.40E-006	mm Hg	25	EST	NEELY,WB & BLAU,GE (1985)
DC	pKa			
HL 3.46E-011	atm m3/mol	25	EST	MEYLAN,WM & HOWARD,PH (1991)
OH 1.04E-010	cm3/molc sec	25	EST	MEYLAN,WM & HOWARD,PH (1993)

CAS #: 072336-17-5 — ACETAMIDE, 2-(DIPROPYLAMINO)-N-(PHENYLMETHYL)-

Formula: $C_{15}H_{24}N_2O$
Mol Weight: 248.37
MP (deg C):
FP (deg C):
BP (deg C):
BP pressure (mm Hg):

Property/Value	Units	Temp	Data Type	Reference
WS 5.33E+001	mg/L	25	EST	MEYLAN,WM ET AL. (1996)
logP 3.11			EXP	SANGSTER,J (1993)
VP 5.26E-007	mm Hg	25	EST	NEELY,WB & BLAU,GE (1985)
DC	pKa			
HL 6.09E-011	atm m3/mol	25	EST	MEYLAN,WM & HOWARD,PH (1991)
OH 1.11E-010	cm3/molc sec	25	EST	MEYLAN,WM & HOWARD,PH (1993)

CAS #: 072336-19-7				1-PIPERIDINEACETAMIDE, N-(PHENYLMETHYL)-

Formula: $C_{14}H_{20}N_2O$

Mol Weight: 232.33

MP (deg C): FP (deg C):

BP (deg C):

BP pressure (mm Hg):

Property/Value	Units	Temp	Data Type	Reference
WS 3.41E+002	mg/L	25	EST	MEYLAN,WM ET AL. (1996)
logP 2.27			EXP	SANGSTER,J (1993)
VP 4.51E-007	mm Hg	25	EST	NEELY,WB & BLAU,GE (1985)
DC	pKa			
HL 2.02E-011	atm m3/mol	25	EST	MEYLAN,WM & HOWARD,PH (1991)
OH 1.12E-010	cm3/molc sec	25	EST	MEYLAN,WM & HOWARD,PH (1993)

CAS #: 072336-20-0				4-MORPHOLINEACETAMIDE, N-(PHENYLMETHYL)-

Formula: $C_{13}H_{18}N_2O_2$

Mol Weight: 234.30

MP (deg C): FP (deg C):

BP (deg C):

BP pressure (mm Hg):

Property/Value	Units	Temp	Data Type	Reference
WS 7.73E+003	mg/L	25	EST	MEYLAN,WM ET AL. (1996)
logP 0.67			EXP	SANGSTER,J (1993)
VP 3.37E-007	mm Hg	25	EST	NEELY,WB & BLAU,GE (1985)
DC	pKa			
HL 1.34E-013	atm m3/mol	25	EST	MEYLAN,WM & HOWARD,PH (1991)
OH 1.62E-010	cm3/molc sec	25	EST	MEYLAN,WM & HOWARD,PH (1993)

CAS #: 072336-22-2				ACETAMIDE, 2-(DIPROPYLAMINO)-N-(9H-FLUOREN-9-YL)

Formula: $C_{21}H_{26}N_2O$

Mol Weight: 322.45

MP (deg C): FP (deg C):

BP (deg C):

BP pressure (mm Hg):

Property/Value	Units	Temp	Data Type	Reference
WS 8.13E-002	mg/L	25	EST	MEYLAN,WM ET AL. (1996)
logP 4.61			EXP	SANGSTER,J (1993)
VP 3.55E-010	mm Hg	25	EST	NEELY,WB & BLAU,GE (1985)
DC	pKa			
HL 1.71E-012	atm m3/mol	25	EST	MEYLAN,WM & HOWARD,PH (1991)
OH 1.17E-010	cm3/molc sec	25	EST	MEYLAN,WM & HOWARD,PH (1993)

CAS #: 072391-46-9				CHLOZOLINATE

Formula: $C_{13}H_{11}Cl_2NO_5$

Mol Weight: 332.14

MP (deg C): FP (deg C):

BP (deg C):

BP pressure (mm Hg):

Property/Value	Units	Temp	Data Type	Reference
WS 1.01E+002	mg/L	25	EST	MEYLAN,WM ET AL. (1996)
logP 2.22			EST	MEYLAN,WM & HOWARD,PH (1995)
VP 9.50E-008	mm Hg	25	EXP	AUGUSTIJN-BECKERS,PWM ET AL. (1994)
DC	pKa			
HL 5.67E-009	atm m3/mol	25	EST	MEYLAN,WM & HOWARD,PH (1991)
OH 8.34E-012	cm3/molc sec	25	EST	MEYLAN,WM & HOWARD,PH (1993)

CAS #: 072424-02-3				PHENOL, 3-TRIDECYL-

Formula: $C_{19}H_{32}O$

Mol Weight: 276.47

MP (deg C): FP (deg C):

BP (deg C):

BP pressure (mm Hg):

Property/Value	Units	Temp	Data Type	Reference
WS 1.80E-003	mg/L	25	EST	MEYLAN,WM ET AL. (1996)
logP 8.84			EXP	SANGSTER,J (1993)
VP 9.37E-007	mm Hg	25	EST	NEELY,WB & BLAU,GE (1985)
DC	pKa			
HL 1.85E-005	atm m3/mol	25	EST	MEYLAN,WM & HOWARD,PH (1991)
OH 9.95E-011	cm3/molc sec	25	EST	MEYLAN,WM & HOWARD,PH (1993)

CAS #: 072430-63-8				14-BENZDIAZPIN-2-ONE,1ME5-(4MEOPH)7CL

Formula: $C_{17}H_{15}ClN_2O_2$

Mol Weight: 314.77

MP (deg C): FP (deg C):

BP (deg C):

BP pressure (mm Hg):

Property/Value	Units	Temp	Data Type	Reference
WS 2.05E+001	mg/L	25	EST	MEYLAN,WM ET AL. (1996)
logP 3.15			EXP	HANSCH,C & LEO,AJ (1985)
VP 5.37E-009	mm Hg	25	EST	NEELY,WB & BLAU,GE (1985)
DC	pKa			
HL 2.16E-010	atm m3/mol	25	EST	MEYLAN,WM & HOWARD,PH (1991)
OH 3.22E-011	cm3/molc sec	25	EST	MEYLAN,WM & HOWARD,PH (1993)

CAS #: 072459-58-6				TRIAZOXIDE

Formula: $C_{10}H_{10}ClN_5O$

Mol Weight: 251.68

MP (deg C): FP (deg C):

BP (deg C):

BP pressure (mm Hg):

Property/Value	Units	Temp	Data Type	Reference
WS 4.41E+002	mg/L	25	EST	MEYLAN,WM ET AL. (1996)
logP 2.04			EXP	TOMLIN,C (1994)
VP 1.85E-009	mm Hg	25	EST	NEELY,WB & BLAU,GE (1985)
DC	pKa			
HL 2.18E-018	atm m3/mol	25	EST	MEYLAN,WM & HOWARD,PH (1991)
OH 3.65E-011	cm3/molc sec	25	EST	MEYLAN,WM & HOWARD,PH (1993)

CAS #: 072490-01-8				FENOXYCARB

Formula: $C_{17}H_{19}NO_4$

Mol Weight: 301.35

MP (deg C): 53-54 FP (deg C):

BP (deg C):

BP pressure (mm Hg):

Property/Value	Units	Temp	Data Type	Reference
WS 6.00E+000	mg/L	20	EXP	SHIU,WY ET AL. (1990)
logP 4.30			EXP	HANSCH,C ET AL. (1995)
VP 6.50E-009	mm Hg	25	EXP	TOMLIN,C (1994)
DC	pKa			
HL 4.30E-010	atm m3/mol	25	EST	VP/WSOL
OH 5.40E-011	cm3/molc sec	25	EST	MEYLAN,WM & HOWARD,PH (1993)

CAS #: 072570-99-1 — ACETAMIDE, N-(7-NITRO-9H-FLUOREN-2-YL)-

Formula: $C_{15}H_{12}N_2O_3$

Mol Weight: 268.27

MP (deg C): 261-264

FP (deg C):

BP (deg C):

BP pressure (mm Hg):

Property/ Value	Units	Temp	Data Type	Reference
WS 1.38E+000	mg/L	25	EST	MEYLAN,WM ET AL. (1996)
logP 3.08			EXP	DEBNATH,AK ET AL. (1991)
VP 1.67E-009	mm Hg	25	EST	NEELY,WB & BLAU,GE (1985)
DC	pKa			
HL 7.56E-013	atm m3/mol	25	EST	MEYLAN,WM & HOWARD,PH (1991)
OH 2.08E-011	cm3/molc sec	25	EST	MEYLAN,WM & HOWARD,PH (1993)

CAS #: 072586-68-6 — 1,3-DIME-1-NITROSO-3-PHENYLUREA

Formula: $C_9H_{11}N_3O_2$

Mol Weight: 193.21

MP (deg C):

FP (deg C):

BP (deg C):

BP pressure (mm Hg):

Property/ Value	Units	Temp	Data Type	Reference
WS 2.12E+003	mg/L	25	EST	MEYLAN,WM ET AL. (1996)
logP 1.58			EXP	HANSCH,C & LEO,AJ (1985)
VP 1.60E-003	mm Hg	25	EST	NEELY,WB & BLAU,GE (1985)
DC	pKa			
HL 1.12E-009	atm m3/mol	25	EST	MEYLAN,WM & HOWARD,PH (1991)
OH 1.35E-011	cm3/molc sec	25	EST	MEYLAN,WM & HOWARD,PH (1993)

CAS #: 072593-17-0 — BENZAMIDE, N-[2-(3,4,6,7,12,12A-HEXAHYDROPYRAZIN

Formula: $C_{23}H_{25}N_5O_3$

Mol Weight: 419.49

MP (deg C):

FP (deg C):

BP (deg C):

BP pressure (mm Hg):

Property/ Value	Units	Temp	Data Type	Reference
WS 2.74E+001	mg/L	25	EST	MEYLAN,WM ET AL. (1996)
logP 1.80			EXP	SANGSTER,J (1993)
VP 4.03E-014	mm Hg	25	EST	NEELY,WB & BLAU,GE (1985)
DC	pKa			
HL 8.02E-023	atm m3/mol	25	EST	MEYLAN,WM & HOWARD,PH (1991)
OH 2.51E-010	cm3/molc sec	25	EST	MEYLAN,WM & HOWARD,PH (1993)

CAS #: 072593-21-6 — BENZAMIDE, N-[2-(3,4,6,7,12,12A-HEXAHYDROPYRAZIN

Formula: $C_{23}H_{26}N_4O$

Mol Weight: 374.49

MP (deg C):

FP (deg C):

BP (deg C):

BP pressure (mm Hg):

Property/ Value	Units	Temp	Data Type	Reference
WS 2.21E+002	mg/L	25	EST	MEYLAN,WM ET AL. (1996)
logP 1.52			EXP	SANGSTER,J (1993)
VP 2.38E-012	mm Hg	25	EST	NEELY,WB & BLAU,GE (1985)
DC	pKa			
HL 2.03E-020	atm m3/mol	25	EST	MEYLAN,WM & HOWARD,PH (1991)
OH 2.52E-010	cm3/molc sec	25	EST	MEYLAN,WM & HOWARD,PH (1993)

CAS #: 072597-35-4 — BENZOBICYCLO(2,2,1)HEPTENE,9-EN-AMINO

Formula: $C_{11}H_{13}N$

Mol Weight: 159.23

MP (deg C):

FP (deg C):

BP (deg C):

BP pressure (mm Hg):

Property/ Value	Units	Temp	Data Type	Reference
WS 1.18E+004	mg/L	25	EST	MEYLAN,WM ET AL. (1996)
logP 2.08			EXP	HANSCH,C & LEO,AJ (1985)
VP 8.71E-003	mm Hg	25	EST	NEELY,WB & BLAU,GE (1985)
DC	pKa			
HL 3.07E-007	atm m3/mol	25	EST	MEYLAN,WM & HOWARD,PH (1991)
OH 4.96E-011	cm3/molc sec	25	EST	MEYLAN,WM & HOWARD,PH (1993)

CAS #: 072601-44-6 — O-FLUORO-N-PHENYLSUCCINIMIDE

Formula: $C_{10}H_8FNO_2$

Mol Weight: 193.18

MP (deg C):

FP (deg C):

BP (deg C):

BP pressure (mm Hg):

Property/ Value	Units	Temp	Data Type	Reference
WS 3.07E+004	mg/L	25	EST	MEYLAN,WM ET AL. (1996)
logP 0.22			EXP	HANSCH,C & LEO,AJ (1985)
VP 1.12E-006	mm Hg	25	EST	NEELY,WB & BLAU,GE (1985)
DC	pKa			
HL 1.52E-007	atm m3/mol	25	EST	MEYLAN,WM & HOWARD,PH (1991)
OH 8.56E-012	cm3/molc sec	25	EST	MEYLAN,WM & HOWARD,PH (1993)

CAS #: 072601-45-7 — O-IODO-N-PHENYLSUCCINIMIDE

Formula: $C_{10}H_8INO_2$

Mol Weight: 301.09

MP (deg C):

FP (deg C):

BP (deg C):

BP pressure (mm Hg):

Property/ Value	Units	Temp	Data Type	Reference
WS 2.51E+003	mg/L	25	EST	MEYLAN,WM ET AL. (1996)
logP 0.80			EXP	HANSCH,C & LEO,AJ (1985)
VP 2.74E-008	mm Hg	25	EST	NEELY,WB & BLAU,GE (1985)
DC	pKa			
HL 3.02E-008	atm m3/mol	25	EST	MEYLAN,WM & HOWARD,PH (1991)
OH 8.47E-012	cm3/molc sec	25	EST	MEYLAN,WM & HOWARD,PH (1993)

CAS #: 072601-46-8 — M-IODO-N-PHENYLSUCCINIMIDE

Formula: $C_{10}H_8INO_2$

Mol Weight: 301.09

MP (deg C):

FP (deg C):

BP (deg C):

BP pressure (mm Hg):

Property/ Value	Units	Temp	Data Type	Reference
WS 7.87E+002	mg/L	25	EST	MEYLAN,WM ET AL. (1996)
logP 1.39			EXP	HANSCH,C & LEO,AJ (1985)
VP 2.74E-008	mm Hg	25	EST	NEELY,WB & BLAU,GE (1985)
DC	pKa			
HL 3.02E-008	atm m3/mol	25	EST	MEYLAN,WM & HOWARD,PH (1991)
OH 1.26E-011	cm3/molc sec	25	EST	MEYLAN,WM & HOWARD,PH (1993)

CAS #: 072601-47-9 — M-ETHYL-N-PHENYLSUCCINIMIDE

Formula: $C_{12}H_{13}NO_2$
Mol Weight: 203.24
MP (deg C):
FP (deg C):
BP (deg C):
BP pressure (mm Hg):

Property/Value	Units	Temp	Data Type	Reference
WS 5.45E+003	mg/L	25	EST	MEYLAN,WM ET AL. (1996)
logP 1.04			EXP	HANSCH,C & LEO,AJ (1985)
VP 1.72E-007	mm Hg	25	EST	NEELY,WB & BLAU,GE (1985)
DC	pKa			
HL 1.91E-007	atm m3/mol	25	EST	MEYLAN,WM & HOWARD,PH (1991)
OH 3.63E-011	cm3/molc sec	25	EST	MEYLAN,WM & HOWARD,PH (1993)

CAS #: 072601-52-6 — M-METHOXYCARBONYL-N-PHENYLSUCCINIMIDE

Formula: $C_{12}H_{11}NO_4$
Mol Weight: 233.23
MP (deg C):
FP (deg C):
BP (deg C):
BP pressure (mm Hg):

Property/Value	Units	Temp	Data Type	Reference
WS 1.44E+004	mg/L	25	EST	MEYLAN,WM ET AL. (1996)
logP 0.36			EXP	HANSCH,C & LEO,AJ (1985)
VP 2.66E-008	mm Hg	25	EST	NEELY,WB & BLAU,GE (1985)
DC	pKa			
HL 8.40E-010	atm m3/mol	25	EST	MEYLAN,WM & HOWARD,PH (1991)
OH 7.33E-012	cm3/molc sec	25	EST	MEYLAN,WM & HOWARD,PH (1993)

CAS #: 072601-54-8 — P-ETHYL-N-PHENYLSUCCINIMIDE

Formula: $C_{12}H_{13}NO_2$
Mol Weight: 203.24
MP (deg C):
FP (deg C):
BP (deg C):
BP pressure (mm Hg):

Property/Value	Units	Temp	Data Type	Reference
WS 4.84E+003	mg/L	25	EST	MEYLAN,WM ET AL. (1996)
logP 1.10			EXP	HANSCH,C & LEO,AJ (1985)
VP 1.72E-007	mm Hg	25	EST	NEELY,WB & BLAU,GE (1985)
DC	pKa			
HL 1.91E-007	atm m3/mol	25	EST	MEYLAN,WM & HOWARD,PH (1991)
OH 2.06E-011	cm3/molc sec	25	EST	MEYLAN,WM & HOWARD,PH (1993)

CAS #: 072615-27-1 — BENZENEACETIC ACID, à-(1-METHYLETHYL)-, METHYL E

Formula: $C_{12}H_{16}O_2$
Mol Weight: 192.26
MP (deg C):
FP (deg C):
BP (deg C):
BP pressure (mm Hg):

Property/Value	Units	Temp	Data Type	Reference
WS 6.86E+001	mg/L	25	EST	MEYLAN,WM ET AL. (1996)
logP 3.33			EXP	YANG,HZ ET AL. (1987)
VP 2.94E-002	mm Hg	25	EST	NEELY,WB & BLAU,GE (1985)
DC	pKa			
HL 3.31E-005	atm m3/mol	25	EST	MEYLAN,WM & HOWARD,PH (1991)
OH 9.33E-012	cm3/molc sec	25	EST	MEYLAN,WM & HOWARD,PH (1993)

CAS #: 072699-09-3 — 3,5-DIBR-2'-NO2-4'-CL SALICYLANILIDE

Formula: $C_{13}H_7Br_2ClN_2O_4$
Mol Weight: 450.47
MP (deg C):
FP (deg C):
BP (deg C):
BP pressure (mm Hg):

Property/Value	Units	Temp	Data Type	Reference
WS 1.12E-001	mg/L	25	EST	MEYLAN,WM ET AL. (1996)
logP 5.05			EXP	TERADA,H ET AL. (1988)
VP 8.38E-013	mm Hg	25	EST	NEELY,WB & BLAU,GE (1985)
DC	pKa			
HL 1.62E-012	atm m3/mol	25	EST	MEYLAN,WM & HOWARD,PH (1991)
OH 8.17E-013	cm3/molc sec	25	EST	MEYLAN,WM & HOWARD,PH (1993)

CAS #: 072732-50-4 — MOXISYLYTE,N-DESMETHYL,DESACETYL

Formula: $C_{13}H_{21}NO_2$
Mol Weight: 223.32
MP (deg C):
FP (deg C):
BP (deg C):
BP pressure (mm Hg):

Property/Value	Units	Temp	Data Type	Reference
WS 1.26E+004	mg/L	25	EST	MEYLAN,WM ET AL. (1996)
logP 2.35			EXP	HANSCH,C ET AL. (1995)
VP 2.06E-005	mm Hg	25	EST	NEELY,WB & BLAU,GE (1985)
DC	pKa			
HL 2.13E-011	atm m3/mol	25	EST	MEYLAN,WM & HOWARD,PH (1991)
OH 1.88E-010	cm3/molc sec	25	EST	MEYLAN,WM & HOWARD,PH (1993)

CAS #: 072732-56-0 — PIRITREXIM

Formula: $C_{17}H_{19}N_5O_2$
Mol Weight: 325.37
MP (deg C):
FP (deg C):
BP (deg C):
BP pressure (mm Hg):

Property/Value	Units	Temp	Data Type	Reference
WS 1.32E+002	mg/L	25	EST	MEYLAN,WM ET AL. (1996)
logP 2.13			EXP	HANSCH,C ET AL. (1995)
VP 1.06E-010	mm Hg	25	EST	NEELY,WB & BLAU,GE (1985)
DC	pKa			
HL 5.69E-017	atm m3/mol	25	EST	MEYLAN,WM & HOWARD,PH (1991)
OH 2.49E-010	cm3/molc sec	25	EST	MEYLAN,WM & HOWARD,PH (1993)

CAS #: 072755-13-6 — CARBAMIC ACID, (3-FLUOROPHENYL)-, METHYL ESTER

Formula: $C_8H_8FNO_2$
Mol Weight: 169.16
MP (deg C):
FP (deg C):
BP (deg C):
BP pressure (mm Hg):

Property/Value	Units	Temp	Data Type	Reference
WS 8.04E+002	mg/L	25	EST	MEYLAN,WM ET AL. (1996)
logP 2.21			EXP	TAKAHASHI,J ET AL. (1993)
VP 7.45E-002	mm Hg	25	EST	NEELY,WB & BLAU,GE (1985)
DC	pKa			
HL 2.55E-008	atm m3/mol	25	EST	MEYLAN,WM & HOWARD,PH (1991)
OH 5.37E-011	cm3/molc sec	25	EST	MEYLAN,WM & HOWARD,PH (1993)

072775-79-2 — 2'-ETHYLBENZOGUANAMINE

Formula: $C_{11}H_{13}N_5$

Mol Weight: 215.26

MP (deg C): FP (deg C):

BP (deg C):

BP pressure (mm Hg):

Property/Value	Units	Temp	Data Type	Reference
WS 1.42E+003	mg/L	25	EST	MEYLAN,WM ET AL. (1996)
logP 1.65			EXP	HANSCH,C & LEO,AJ (1985)
VP 1.55E-007	mm Hg	25	EST	NEELY,WB & BLAU,GE (1985)
DC	pKa			
HL 6.02E-011	atm m3/mol	25	EST	MEYLAN,WM & HOWARD,PH (1991)
OH 5.52E-012	cm3/molc sec	25	EST	MEYLAN,WM & HOWARD,PH (1993)

072775-80-5 — 2'-METHOXYBENZOGUANAMINE

Formula: $C_{10}H_{11}N_5O$

Mol Weight: 217.23

MP (deg C): FP (deg C):

BP (deg C):

BP pressure (mm Hg):

Property/Value	Units	Temp	Data Type	Reference
WS 1.14E+004	mg/L	25	EST	MEYLAN,WM ET AL. (1996)
logP 0.58			EXP	HANSCH,C & LEO,AJ (1985)
VP 1.50E-007	mm Hg	25	EST	NEELY,WB & BLAU,GE (1985)
DC	pKa			
HL 2.43E-012	atm m3/mol	25	EST	MEYLAN,WM & HOWARD,PH (1991)
OH 1.65E-011	cm3/molc sec	25	EST	MEYLAN,WM & HOWARD,PH (1993)

072775-81-6 — 2'-ETHOXYLBENZOGUANAMINE

Formula: $C_{11}H_{13}N_5O$

Mol Weight: 231.26

MP (deg C): FP (deg C):

BP (deg C):

BP pressure (mm Hg):

Property/Value	Units	Temp	Data Type	Reference
WS 3.95E+003	mg/L	25	EST	MEYLAN,WM ET AL. (1996)
logP 1.03			EXP	HANSCH,C & LEO,AJ (1985)
VP 6.76E-008	mm Hg	25	EST	NEELY,WB & BLAU,GE (1985)
DC	pKa			
HL 3.23E-012	atm m3/mol	25	EST	MEYLAN,WM & HOWARD,PH (1991)
OH 2.18E-011	cm3/molc sec	25	EST	MEYLAN,WM & HOWARD,PH (1993)

072775-85-0 — 2,4(1H,3H)-PYRIMIDINEDIONE, 1,6-DIMETHYL-5-[(2-M

Formula: $C_{16}H_{21}N_3O_2$

Mol Weight: 287.36

MP (deg C): FP (deg C):

BP (deg C):

BP pressure (mm Hg):

Property/Value	Units	Temp	Data Type	Reference
WS 7.92E+002	mg/L	25	EST	MEYLAN,WM ET AL. (1996)
logP 1.48			EXP	HANSCH,C ET AL. (1995)
VP 2.59E-009	mm Hg	25	EST	NEELY,WB & BLAU,GE (1985)
DC	pKa			
HL 3.39E-011	atm m3/mol	25	EST	MEYLAN,WM & HOWARD,PH (1991)
OH 1.04E-010	cm3/molc sec	25	EST	MEYLAN,WM & HOWARD,PH (1993)

072781-91-0 — 2'-IODOBENZOGUANAMINE

Formula: $C_9H_8IN_5$

Mol Weight: 313.10

MP (deg C): FP (deg C):

BP (deg C):

BP pressure (mm Hg):

Property/Value	Units	Temp	Data Type	Reference
WS 6.43E+002	mg/L	25	EST	MEYLAN,WM ET AL. (1996)
logP 1.41			EXP	HANSCH,C & LEO,AJ (1985)
VP 2.63E-008	mm Hg	25	EST	NEELY,WB & BLAU,GE (1985)
DC	pKa			
HL 9.53E-012	atm m3/mol	25	EST	MEYLAN,WM & HOWARD,PH (1991)
OH 1.37E-012	cm3/molc sec	25	EST	MEYLAN,WM & HOWARD,PH (1993)

072828-64-9 — AZO DYE N5

Formula: $C_{23}H_{24}N_6O_4$

Mol Weight: 448.49

MP (deg C): FP (deg C):

BP (deg C):

BP pressure (mm Hg):

Property/Value	Units	Temp	Data Type	Reference
WS 5.90E-004	mg/L		EXP	YEN,C ET AL. (1989)
logP 5.50			EXP	BAUGHMAN,GL & WEBER,EJ (1991)
VP 3.70E-013	mm Hg	25	EST	NEELY,WB & BLAU,GE (1985)
DC	pKa			
HL 7.03E-016	atm m3/mol	25	EST	MEYLAN,WM & HOWARD,PH (1991)
OH 1.78E-010	cm3/molc sec	25	EST	MEYLAN,WM & HOWARD,PH (1993)

072836-46-5 — P-MEO-CINNAMIC ACID, DIET-AMINO-ET ESTER

Formula: $C_{16}H_{23}NO_3$

Mol Weight: 277.37

MP (deg C): FP (deg C):

BP (deg C):

BP pressure (mm Hg):

Property/Value	Units	Temp	Data Type	Reference
WS 1.57E+002	mg/L	25	EST	MEYLAN,WM ET AL. (1996)
logP 3.55			EXP	HANSCH,C & LEO,AJ (1985)
VP 1.39E-005	mm Hg	25	EST	NEELY,WB & BLAU,GE (1985)
DC	pKa			
HL 2.84E-010	atm m3/mol	25	EST	MEYLAN,WM & HOWARD,PH (1991)
OH 1.37E-010	cm3/molc sec	25	EST	MEYLAN,WM & HOWARD,PH (1993)

072836-47-6 — 3,4,5-TRIMEO-CINNAMIC ACID, DIETAMET ESTER

Formula: $C_{18}H_{27}NO_5$

Mol Weight: 337.42

MP (deg C): FP (deg C):

BP (deg C):

BP pressure (mm Hg):

Property/Value	Units	Temp	Data Type	Reference
WS 2.82E+002	mg/L	25	EST	MEYLAN,WM ET AL. (1996)
logP 2.84			EXP	HANSCH,C & LEO,AJ (1985)
VP 4.17E-007	mm Hg	25	EST	NEELY,WB & BLAU,GE (1985)
DC	pKa			
HL 9.94E-013	atm m3/mol	25	EST	MEYLAN,WM & HOWARD,PH (1991)
OH 3.11E-010	cm3/molc sec	25	EST	MEYLAN,WM & HOWARD,PH (1993)

CAS #: 072836-48-7 — P-F CINNAMIC ACID,DIETHYLAMINOETHYLESTER

Formula: $C_{15}H_{20}FNO_2$

Mol Weight: 265.33

MP (deg C):   FP (deg C):

BP (deg C):

BP pressure (mm Hg):

Property/Value	Units	Temp	Data Type	Reference
WS 1.95E+002	mg/L	25	EST	MEYLAN,WM ET AL. (1996)
logP 3.52			EXP	HANSCH,C & LEO,AJ (1985)
VP 1.10E-004	mm Hg	25	EST	NEELY,WB & BLAU,GE (1985)
DC	pKa			
HL 5.61E-009	atm m3/mol	25	EST	MEYLAN,WM & HOWARD,PH (1991)
OH 1.18E-010	cm3/molc sec	25	EST	MEYLAN,WM & HOWARD,PH (1993)

CAS #: 072962-38-0 — 3-FURANCARBOXAMIDE, N-(AMINOCARBONYL)-TETRAHYDRO

Formula: $C_{10}H_{14}N_2O_4$

Mol Weight: 226.23

MP (deg C):   FP (deg C):

BP (deg C):

BP pressure (mm Hg):

Property/Value	Units	Temp	Data Type	Reference
WS 1.01E+003	mg/L	25	EST	MEYLAN,WM ET AL. (1996)
logP 0.34			EXP	WITTEKIND,HH ET AL. (1988)
VP 1.40E-009	mm Hg	25	EST	NEELY,WB & BLAU,GE (1985)
DC	pKa			
HL 3.48E-013	atm m3/mol	25	EST	MEYLAN,WM & HOWARD,PH (1991)
OH 3.87E-011	cm3/molc sec	25	EST	MEYLAN,WM & HOWARD,PH (1993)

CAS #: 072962-39-1 — 3-FURANCARBOXAMIDE, N-(AMINOCARBONYL)-TETRAHYDRO

Formula: $C_{10}H_{14}N_2O_4$

Mol Weight: 226.23

MP (deg C):   FP (deg C):

BP (deg C):

BP pressure (mm Hg):

Property/Value	Units	Temp	Data Type	Reference
WS 1.01E+003	mg/L	25	EST	MEYLAN,WM ET AL. (1996)
logP 0.22			EXP	WITTEKIND,HH ET AL. (1988)
VP 1.40E-009	mm Hg	25	EST	NEELY,WB & BLAU,GE (1985)
DC	pKa			
HL 3.48E-013	atm m3/mol	25	EST	MEYLAN,WM & HOWARD,PH (1991)
OH 3.87E-011	cm3/molc sec	25	EST	MEYLAN,WM & HOWARD,PH (1993)

CAS #: 073013-68-0 — 6-ACETYLQUINOLINE

Formula: $C_{11}H_9NO$

Mol Weight: 171.20

MP (deg C):   FP (deg C):

BP (deg C):

BP pressure (mm Hg):

Property/Value	Units	Temp	Data Type	Reference
WS 2.72E+003	mg/L	25	EST	MEYLAN,WM ET AL. (1996)
logP 1.58			EXP	HANSCH,C & LEO,AJ (1985)
VP 3.93E-004	mm Hg	25	EST	NEELY,WB & BLAU,GE (1985)
DC	pKa			
HL 1.25E-009	atm m3/mol	25	EST	MEYLAN,WM & HOWARD,PH (1991)
OH 1.32E-011	cm3/molc sec	25	EST	MEYLAN,WM & HOWARD,PH (1993)

CAS #: 073046-15-8 — 2,4-PYRIMIDINEDIAMINE, 5-[(3-ETHOXY-4-METHOXYPHE

Formula: $C_{14}H_{18}N_4O_2$

Mol Weight: 274.33

MP (deg C):   FP (deg C):

BP (deg C):

BP pressure (mm Hg):

Property/Value	Units	Temp	Data Type	Reference
WS 1.24E+003	mg/L	25	EST	MEYLAN,WM ET AL. (1996)
logP 1.20			EXP	SANGSTER,J (1993)
VP 2.28E-008	mm Hg	25	EST	NEELY,WB & BLAU,GE (1985)
DC	pKa			
HL 5.35E-013	atm m3/mol	25	EST	MEYLAN,WM & HOWARD,PH (1991)
OH 5.57E-011	cm3/molc sec	25	EST	MEYLAN,WM & HOWARD,PH (1993)

CAS #: 073090-06-9 — 1-ISOBENZOFURANONE-3-SPIROPENTYL

Formula: $C_{12}H_{12}O_2$

Mol Weight: 188.23

MP (deg C):   FP (deg C):

BP (deg C):

BP pressure (mm Hg):

Property/Value	Units	Temp	Data Type	Reference
WS 3.27E+002	mg/L	25	EST	MEYLAN,WM ET AL. (1996)
logP 2.56			EXP	HANSCH,C & LEO,AJ (1985)
VP 1.17E-004	mm Hg	25	EST	NEELY,WB & BLAU,GE (1985)
DC	pKa			
HL 1.74E-005	atm m3/mol	25	EST	MEYLAN,WM & HOWARD,PH (1991)
OH 7.53E-012	cm3/molc sec	25	EST	MEYLAN,WM & HOWARD,PH (1993)

CAS #: 073090-70-7 — 24NH2PYRIMIDINE,5(35ETO4(1PYRRYL))BENZYL

Formula: $C_{19}H_{25}N_5O_2$

Mol Weight: 355.44

MP (deg C):   FP (deg C):

BP (deg C):

BP pressure (mm Hg):

Property/Value	Units	Temp	Data Type	Reference
WS 2.01E+001	mg/L	25	EST	MEYLAN,WM ET AL. (1996)
logP 2.89			EXP	HANSCH,C & LEO,AJ (1985)
VP 3.11E-011	mm Hg	25	EST	NEELY,WB & BLAU,GE (1985)
DC	pKa			
HL 4.35E-018	atm m3/mol	25	EST	MEYLAN,WM & HOWARD,PH (1991)
OH 2.13E-010	cm3/molc sec	25	EST	MEYLAN,WM & HOWARD,PH (1993)

CAS #: 073159-84-9 — 9-OXABENZOBICYCLO(221)HEPTENE,2EXNH2

Formula: $C_{10}H_{11}NO$

Mol Weight: 161.21

MP (deg C):   FP (deg C):

BP (deg C):

BP pressure (mm Hg):

Property/Value	Units	Temp	Data Type	Reference
WS 3.14E+005	mg/L	25	EST	MEYLAN,WM ET AL. (1996)
logP 0.75			EXP	HANSCH,C & LEO,AJ (1985)
VP 5.89E-003	mm Hg	25	EST	NEELY,WB & BLAU,GE (1985)
DC	pKa			
HL 2.57E-010	atm m3/mol	25	EST	MEYLAN,WM & HOWARD,PH (1991)
OH 6.36E-011	cm3/molc sec	25	EST	MEYLAN,WM & HOWARD,PH (1993)

CAS #: 073207-98-4	S-(2-(BIS(1-METHYLETHYL)AMINO)ETHYL) METHYL PHO*

Formula: $C_9H_{22}NO_2PS$

Mol Weight: 239.32

MP (deg C):   FP (deg C):

BP (deg C):

BP pressure (mm Hg):

Property/Value	Units	Temp	Data Type	Reference
WS 1.40E+004	mg/L	25	EST	MEYLAN,WM ET AL. (1996)
logP 1.52			EST	MEYLAN,WM & HOWARD,PH (1995)
VP 5.14E-006	mm Hg	25	EST	NEELY,WB & BLAU,GE (1985)
DC	pKa			
HL 4.38E-012	atm m3/mol	25	EST	MEYLAN,WM & HOWARD,PH (1991)
OH 1.37E-010	cm3/molc sec	25	EST	MEYLAN,WM & HOWARD,PH (1993)

CAS #: 073208-84-1	9-OXABENZOBICYCLO(221)HEPTENE,2ENNH2

Formula: $C_{10}H_{11}NO$

Mol Weight: 161.21

MP (deg C):   FP (deg C):

BP (deg C):

BP pressure (mm Hg):

Property/Value	Units	Temp	Data Type	Reference
WS 3.14E+005	mg/L	25	EST	MEYLAN,WM ET AL. (1996)
logP 0.40			EXP	HANSCH,C & LEO,AJ (1985)
VP 5.89E-003	mm Hg	25	EST	NEELY,WB & BLAU,GE (1985)
DC	pKa			
HL 2.57E-010	atm m3/mol	25	EST	MEYLAN,WM & HOWARD,PH (1991)
OH 6.36E-011	cm3/molc sec	25	EST	MEYLAN,WM & HOWARD,PH (1993)

CAS #: 073250-68-7	MEFENACET

Formula: $C_{16}H_{14}N_2O_2S$

Mol Weight: 298.37

MP (deg C):   FP (deg C):

BP (deg C):

BP pressure (mm Hg):

Property/Value	Units	Temp	Data Type	Reference
WS 4.00E+000	mg/L	20	EXP	SHIU,WY ET AL. (1990)
logP 3.23			EXP	TOMLIN,C (1994)
VP 6.47E-009	mm Hg	25	EST	NEELY,WB & BLAU,GE (1985)
DC	pKa			
HL 2.69E-013	atm m3/mol	25	EST	MEYLAN,WM & HOWARD,PH (1991)
OH 9.44E-011	cm3/molc sec	25	EST	MEYLAN,WM & HOWARD,PH (1993)

CAS #: 073258-87-4	P-BUTOXYBENZAMIDE

Formula: $C_{11}H_{15}NO_2$

Mol Weight: 193.25

MP (deg C):   FP (deg C):

BP (deg C):

BP pressure (mm Hg):

Property/Value	Units	Temp	Data Type	Reference
WS 3.61E+002	mg/L	25	EST	MEYLAN,WM ET AL. (1996)
logP 2.48			EXP	HANSCH,C & LEO,AJ (1985)
VP 8.26E-006	mm Hg	25	EST	NEELY,WB & BLAU,GE (1985)
DC	pKa			
HL 3.06E-010	atm m3/mol	25	EST	MEYLAN,WM & HOWARD,PH (1991)
OH 3.39E-011	cm3/molc sec	25	EST	MEYLAN,WM & HOWARD,PH (1993)

CAS #: 073264-18-3	2,4-DINH2PYRIMIDINE,5(4MEO-3ME)BENZYL

Formula: $C_{13}H_{16}N_4O$

Mol Weight: 244.30

MP (deg C):   FP (deg C):

BP (deg C):

BP pressure (mm Hg):

Property/Value	Units	Temp	Data Type	Reference
WS 5.95E+002	mg/L	25	EST	MEYLAN,WM ET AL. (1996)
logP 1.91			EXP	HANSCH,C & LEO,AJ (1985)
VP 1.17E-007	mm Hg	25	EST	NEELY,WB & BLAU,GE (1985)
DC	pKa			
HL 7.52E-012	atm m3/mol	25	EST	MEYLAN,WM & HOWARD,PH (1991)
OH 3.55E-011	cm3/molc sec	25	EST	MEYLAN,WM & HOWARD,PH (1993)

CAS #: 073264-19-4	2,4-DINH2PYRIMIDINE,5(3MEO-4ME)BENZYL

Formula: $C_{13}H_{16}N_4O$

Mol Weight: 244.30

MP (deg C):   FP (deg C):

BP (deg C):

BP pressure (mm Hg):

Property/Value	Units	Temp	Data Type	Reference
WS 3.78E+002	mg/L	25	EST	MEYLAN,WM ET AL. (1996)
logP 2.14			EXP	HANSCH,C & LEO,AJ (1985)
VP 1.17E-007	mm Hg	25	EST	NEELY,WB & BLAU,GE (1985)
DC	pKa			
HL 7.52E-012	atm m3/mol	25	EST	MEYLAN,WM & HOWARD,PH (1991)
OH 6.89E-011	cm3/molc sec	25	EST	MEYLAN,WM & HOWARD,PH (1993)

CAS #: 073264-20-7	2,4-PYRIMIDINEDIAMINE, 5-[[4-METHOXY-3-(1-METHYL

Formula: $C_{15}H_{20}N_4O$

Mol Weight: 272.35

MP (deg C):   FP (deg C):

BP (deg C):

BP pressure (mm Hg):

Property/Value	Units	Temp	Data Type	Reference
WS 7.34E+001	mg/L	25	EST	MEYLAN,WM ET AL. (1996)
logP 2.79			EXP	SANGSTER,J (1993)
VP 3.82E-008	mm Hg	25	EST	NEELY,WB & BLAU,GE (1985)
DC	pKa			
HL 1.33E-011	atm m3/mol	25	EST	MEYLAN,WM & HOWARD,PH (1991)
OH 3.71E-011	cm3/molc sec	25	EST	MEYLAN,WM & HOWARD,PH (1993)

CAS #: 073264-21-8	2,4-PYRIMIDINEDIAMINE, 5-[[3-METHOXY-4-(1-METHYL

Formula: $C_{15}H_{20}N_4O$

Mol Weight: 272.35

MP (deg C):   FP (deg C):

BP (deg C):

BP pressure (mm Hg):

Property/Value	Units	Temp	Data Type	Reference
WS 3.69E+001	mg/L	25	EST	MEYLAN,WM ET AL. (1996)
logP 3.14			EXP	SANGSTER,J (1993)
VP 3.82E-008	mm Hg	25	EST	NEELY,WB & BLAU,GE (1985)
DC	pKa			
HL 1.33E-011	atm m3/mol	25	EST	MEYLAN,WM & HOWARD,PH (1991)
OH 6.99E-011	cm3/molc sec	25	EST	MEYLAN,WM & HOWARD,PH (1993)

CAS #: 073264-25-2

24-NH2-5(4-CL-3-MEO-BENZYL)PYRIMIDINE

Formula: $C_{12}H_{13}ClN_4O$

Mol Weight: 264.72

MP (deg C):　　FP (deg C):

BP (deg C):

BP pressure (mm Hg):

Property/ Value	Units	Temp	Data Type	Reference
WS 4.95E+002	mg/L	25	EST	MEYLAN,WM ET AL. (1996)
logP 1.87			EXP	HANSCH,C & LEO,AJ (1985)
VP 7.16E-008	mm Hg	25	EST	NEELY,WB & BLAU,GE (1985)
DC	pKa			
HL 5.05E-012	atm m3/mol	25	EST	MEYLAN,WM & HOWARD,PH (1991)
OH 1.89E-011	cm3/molc sec	25	EST	MEYLAN,WM & HOWARD,PH (1993)

CAS #: 073275-70-4

24-NH2-5(3-CL-4-MEO-BENZYL)PYRIMIDINE

Formula: $C_{12}H_{13}ClN_4O$

Mol Weight: 264.72

MP (deg C):　　FP (deg C):

BP (deg C):

BP pressure (mm Hg):

Property/ Value	Units	Temp	Data Type	Reference
WS 7.63E+002	mg/L	25	EST	MEYLAN,WM ET AL. (1996)
logP 1.65			EXP	HANSCH,C & LEO,AJ (1985)
VP 7.16E-008	mm Hg	25	EST	NEELY,WB & BLAU,GE (1985)
DC	pKa			
HL 5.05E-012	atm m3/mol	25	EST	MEYLAN,WM & HOWARD,PH (1991)
OH 1.09E-011	cm3/molc sec	25	EST	MEYLAN,WM & HOWARD,PH (1993)

CAS #: 073334-07-3

IOPROMIDE

Formula: $C_{18}H_{24}I_3N_3O_8$

Mol Weight: 791.12

MP (deg C):　　FP (deg C):

BP (deg C):

BP pressure (mm Hg):

Property/ Value	Units	Temp	Data Type	Reference
WS 2.38E+001	mg/L	25	EST	MEYLAN,WM ET AL. (1996)
logP -2.05			EXP	HANSCH,C ET AL. (1995)
VP 1.59E-028	mm Hg	25	EST	NEELY,WB & BLAU,GE (1985)
DC	pKa			
HL 1.00E-028	atm m3/mol	25	EST	MEYLAN,WM & HOWARD,PH (1991)
OH 6.50E-011	cm3/molc sec	25	EST	MEYLAN,WM & HOWARD,PH (1993)

CAS #: 073356-40-8

2,4-NH2PYRIMIDINE,5(4-OH-3MEO)BENZYL

Formula: $C_{12}H_{14}N_4O_2$

Mol Weight: 246.27

MP (deg C):　　FP (deg C):

BP (deg C):

BP pressure (mm Hg):

Property/ Value	Units	Temp	Data Type	Reference
WS 7.05E+003	mg/L	25	EST	MEYLAN,WM ET AL. (1996)
logP 0.64			EXP	HANSCH,C & LEO,AJ (1985)
VP 4.15E-009	mm Hg	25	EST	NEELY,WB & BLAU,GE (1985)
DC	pKa			
HL 7.09E-016	atm m3/mol	25	EST	MEYLAN,WM & HOWARD,PH (1991)
OH 4.02E-011	cm3/molc sec	25	EST	MEYLAN,WM & HOWARD,PH (1993)

CAS #: 073384-60-8

1H-IMIDAZO[4,5-B]PYRIDINE, 2-[2-METHOXY-4-(METHY

Formula: $C_{14}H_{13}N_3O_2S$

Mol Weight: 287.34

MP (deg C): 203-205　　FP (deg C):

BP (deg C):

BP pressure (mm Hg):

Property/ Value	Units	Temp	Data Type	Reference
WS 1.43E+003	mg/L	25	EST	MEYLAN,WM ET AL. (1996)
logP 1.18			EXP	SANGSTER,J (1993)
VP 9.87E-012	mm Hg	25	EST	NEELY,WB & BLAU,GE (1985)
DC	pKa			
HL 4.01E-018	atm m3/mol	25	EST	MEYLAN,WM & HOWARD,PH (1991)
OH 1.28E-010	cm3/molc sec	25	EST	MEYLAN,WM & HOWARD,PH (1993)

CAS #: 073403-31-3

2-PROPENOIC ACID, 2-CYANO-3-(5-METHYL-2-FURANYL)

Formula: $C_{10}H_9NO_3$

Mol Weight: 191.19

MP (deg C):　　FP (deg C):

BP (deg C):

BP pressure (mm Hg):

Property/ Value	Units	Temp	Data Type	Reference
WS 3.35E+002	mg/L	25	EST	MEYLAN,WM ET AL. (1996)
logP 2.22			EXP	BALAZ,S ET AL. (1985)
VP 5.09E-004	mm Hg	25	EST	NEELY,WB & BLAU,GE (1985)
DC	pKa			
HL 1.02E-008	atm m3/mol	25	EST	MEYLAN,WM & HOWARD,PH (1991)
OH 1.05E-010	cm3/molc sec	25	EST	MEYLAN,WM & HOWARD,PH (1993)

CAS #: 073459-03-7

2H-FURO(2,3-H)-1-BENZOPYRAN-2-ONE, 5-METHYL-

Formula: $C_{12}H_8O_3$

Mol Weight: 200.20

MP (deg C):　　FP (deg C):

BP (deg C):

BP pressure (mm Hg):

Property/ Value	Units	Temp	Data Type	Reference
WS 3.82E+002	mg/L	25	EST	MEYLAN,WM ET AL. (1996)
logP 2.41			EXP	SANGSTER,J (1993)
VP 9.99E-006	mm Hg	25	EST	NEELY,WB & BLAU,GE (1985)
DC	pKa			
HL 7.47E-007	atm m3/mol	25	EST	MEYLAN,WM & HOWARD,PH (1991)
OH 6.60E-011	cm3/molc sec	25	EST	MEYLAN,WM & HOWARD,PH (1993)

CAS #: 073469-85-9

HYDRAZINECARBOXAMIDE, N-(3-METHOXYPHENYL)-

Formula: $C_8H_{11}N_3O_2$

Mol Weight: 181.20

MP (deg C):　　FP (deg C):

BP (deg C):

BP pressure (mm Hg):

Property/ Value	Units	Temp	Data Type	Reference
WS 1.91E+003	mg/L	25	EST	MEYLAN,WM ET AL. (1996)
logP 0.17			EXP	KRAMER,CR & BECK,L (1981)
VP 1.43E-005	mm Hg	25	EST	NEELY,WB & BLAU,GE (1985)
DC	pKa			
HL 4.98E-014	atm m3/mol	25	EST	MEYLAN,WM & HOWARD,PH (1991)
OH 2.02E-010	cm3/molc sec	25	EST	MEYLAN,WM & HOWARD,PH (1993)

CAS #: 073469-87-1 — HYDRAZINECARBOXAMIDE, N-(2,5-DICHLOROPHENYL)-

Formula: $C_7H_7Cl_2N_3O$

Mol Weight: 220.06

MP (deg C): FP (deg C):
BP (deg C):
BP pressure (mm Hg):

Property/Value	Units	Temp	Data Type	Reference
WS 4.20E+003	mg/L	25	EST	MEYLAN,WM ET AL. (1996)
logP 1.07			EXP	KRAMER,CR & BECK,L (1981)
VP 4.70E-006	mm Hg	25	EST	NEELY,WB & BLAU,GE (1985)
DC	pKa			
HL 4.62E-013	atm m3/mol	25	EST	MEYLAN,WM & HOWARD,PH (1991)
OH 9.76E-012	cm3/molc sec	25	EST	MEYLAN,WM & HOWARD,PH (1993)

CAS #: 073469-88-2 — HYDRAZINECARBOXAMIDE, N-(2-BROMOPHENYL)

Formula: $C_7H_8BrN_3O$

Mol Weight: 230.07

MP (deg C): FP (deg C):
BP (deg C):
BP pressure (mm Hg):

Property/Value	Units	Temp	Data Type	Reference
WS 8.31E+003	mg/L	25	EST	MEYLAN,WM ET AL. (1996)
logP 0.66			EXP	KRAMER,CR & BECK,L (1981)
VP 7.82E-006	mm Hg	25	EST	NEELY,WB & BLAU,GE (1985)
DC	pKa			
HL 3.35E-013	atm m3/mol	25	EST	MEYLAN,WM & HOWARD,PH (1991)
OH 1.32E-011	cm3/molc sec	25	EST	MEYLAN,WM & HOWARD,PH (1993)

CAS #: 073469-89-3 — HYDRAZINECARBOXAMIDE, N-(3-BROMOPHENYL)

Formula: $C_8H_{10}BrN_3O$

Mol Weight: 244.10

MP (deg C): FP (deg C):
BP (deg C):
BP pressure (mm Hg):

Property/Value	Units	Temp	Data Type	Reference
WS 3.86E+003	mg/L	25	EST	MEYLAN,WM ET AL. (1996)
logP 0.96			EXP	KRAMER,CR & BECK,L (1981)
VP 6.17E-007	mm Hg	25	EST	NEELY,WB & BLAU,GE (1985)
DC	pKa			
HL 2.11E-014	atm m3/mol	25	EST	MEYLAN,WM & HOWARD,PH (1991)
OH 3.89E-011	cm3/molc sec	25	EST	MEYLAN,WM & HOWARD,PH (1993)

CAS #: 073469-90-6 — HYDRAZINECARBOXAMIDE, N-(4-IODOPHENYL)

Formula: $C_7H_8IN_3O$

Mol Weight: 277.07

MP (deg C): FP (deg C):
BP (deg C):
BP pressure (mm Hg):

Property/Value	Units	Temp	Data Type	Reference
WS 1.45E+003	mg/L	25	EST	MEYLAN,WM ET AL. (1996)
logP 1.24			EXP	KRAMER,CR & BECK,L (1981)
VP 2.33E-006	mm Hg	25	EST	NEELY,WB & BLAU,GE (1985)
DC	pKa			
HL 1.95E-013	atm m3/mol	25	EST	MEYLAN,WM & HOWARD,PH (1991)
OH 1.51E-011	cm3/molc sec	25	EST	MEYLAN,WM & HOWARD,PH (1993)

CAS #: 073469-91-7 — HYDRAZINECARBOXAMIDE, N-(3-HYDROXYPHENYL)-

Formula: $C_7H_9N_3O_2$

Mol Weight: 167.17

MP (deg C): FP (deg C):
BP (deg C):
BP pressure (mm Hg):

Property/Value	Units	Temp	Data Type	Reference
WS 8.30E+003	mg/L	25	EST	MEYLAN,WM ET AL. (1996)
logP -0.50			EXP	KRAMER,CR & BECK,L (1981)
VP 1.24E-006	mm Hg	25	EST	NEELY,WB & BLAU,GE (1985)
DC	pKa			
HL 8.75E-017	atm m3/mol	25	EST	MEYLAN,WM & HOWARD,PH (1991)
OH 2.01E-010	cm3/molc sec	25	EST	MEYLAN,WM & HOWARD,PH (1993)

CAS #: 073469-92-8 — HYDRAZINECARBOXAMIDE, N-(4-HYDROXYPHENYL)-

Formula: $C_7H_9N_3O_2$

Mol Weight: 167.17

MP (deg C): FP (deg C):
BP (deg C):
BP pressure (mm Hg):

Property/Value	Units	Temp	Data Type	Reference
WS 9.16E+003	mg/L	25	EST	MEYLAN,WM ET AL. (1996)
logP -0.55			EXP	KRAMER,CR & BECK,L (1981)
VP 1.24E-006	mm Hg	25	EST	NEELY,WB & BLAU,GE (1985)
DC	pKa			
HL 8.75E-017	atm m3/mol	25	EST	MEYLAN,WM & HOWARD,PH (1991)
OH 5.57E-011	cm3/molc sec	25	EST	MEYLAN,WM & HOWARD,PH (1993)

CAS #: 073469-93-9 — BENZOIC ACID, 2-[(HYDRAZINECARBONYL)AMINO]-

Formula: $C_8H_9N_3O_3$

Mol Weight: 195.18

MP (deg C): FP (deg C):
BP (deg C):
BP pressure (mm Hg):

Property/Value	Units	Temp	Data Type	Reference
WS 5.94E+003	mg/L	25	EST	MEYLAN,WM ET AL. (1996)
logP -0.49			EXP	KRAMER,CR & BECK,L (1981)
VP 1.76E-007	mm Hg	25	EST	NEELY,WB & BLAU,GE (1985)
DC	pKa			
HL 1.69E-017	atm m3/mol	25	EST	MEYLAN,WM & HOWARD,PH (1991)
OH 1.73E-011	cm3/molc sec	25	EST	MEYLAN,WM & HOWARD,PH (1993)

CAS #: 073469-94-0 — BENZOIC ACID, 3-[(HYDRAZINOCARBONYL)AMINO]-

Formula: $C_8H_9N_3O_3$

Mol Weight: 195.18

MP (deg C): FP (deg C):
BP (deg C):
BP pressure (mm Hg):

Property/Value	Units	Temp	Data Type	Reference
WS 3.11E+003	mg/L	25	EST	MEYLAN,WM ET AL. (1996)
logP -0.16			EXP	KRAMER,CR & BECK,L (1981)
VP 1.76E-007	mm Hg	25	EST	NEELY,WB & BLAU,GE (1985)
DC	pKa			
HL 1.69E-017	atm m3/mol	25	EST	MEYLAN,WM & HOWARD,PH (1991)
OH 1.32E-011	cm3/molc sec	25	EST	MEYLAN,WM & HOWARD,PH (1993)

CAS #: 073469-95-1				BENZOIC ACID, 4-[(HYDRAZINOCARBONYL)AMINO]-
Formula: $C_8H_9N_3O_3$				
Mol Weight: 195.18				
MP (deg C):		FP (deg C):		
BP (deg C):				
BP pressure (mm Hg):				

Property/ Value	Units	Temp	Data Type	Reference
WS 3.43E+003	mg/L	25	EST	MEYLAN,WM ET AL. (1996)
logP -0.21			EXP	KRAMER,CR & BECK,L (1981)
VP 1.76E-007	mm Hg	25	EST	NEELY,WB & BLAU,GE (1985)
DC	pKa			
HL 1.69E-017	atm m3/mol	25	EST	MEYLAN,WM & HOWARD,PH (1991)
OH 1.73E-011	cm3/molc sec	25	EST	MEYLAN,WM & HOWARD,PH (1993)

CAS #: 073498-01-8				N1-(5-IPR-2-PYRIMIDYL)SULFANILIDE
Formula: $C_{13}H_{16}N_4O_2S$				
Mol Weight: 292.36				
MP (deg C):		FP (deg C):		
BP (deg C):				
BP pressure (mm Hg):				

Property/ Value	Units	Temp	Data Type	Reference
WS 1.21E+003	mg/L	25	EST	MEYLAN,WM ET AL. (1996)
logP 1.23			EXP	HANSCH,C & LEO,AJ (1985)
VP 6.24E-009	mm Hg	25	EST	NEELY,WB & BLAU,GE (1985)
DC	pKa			
HL 3.08E-010	atm m3/mol	25	EST	MEYLAN,WM & HOWARD,PH (1991)
OH 2.62E-011	cm3/molc sec	25	EST	MEYLAN,WM & HOWARD,PH (1993)

CAS #: 073573-88-3				MEVASTATIN
Formula: $C_{23}H_{36}O_5$				
Mol Weight: 392.54				
MP (deg C): 152		FP (deg C):		
BP (deg C):				
BP pressure (mm Hg):				

Property/ Value	Units	Temp	Data Type	Reference
WS 4.80E+000	mg/L	25	EST	MEYLAN,WM ET AL. (1996)
logP 3.95			EXP	SANGSTER,J (1994)
VP 2.19E-012	mm Hg	25	EST	NEELY,WB & BLAU,GE (1985)
DC	pKa			
HL 1.60E-010	atm m3/mol	25	EST	MEYLAN,WM & HOWARD,PH (1991)
OH 2.30E-010	cm3/molc sec	25	EST	MEYLAN,WM & HOWARD,PH (1993)

CAS #: 073576-30-4				2,4-NH2PYRIMIDIN,5(35-ME-4-MEO)BENZYL
Formula: $C_{14}H_{18}N_4O$				
Mol Weight: 258.33				
MP (deg C):		FP (deg C):		
BP (deg C):				
BP pressure (mm Hg):				

Property/ Value	Units	Temp	Data Type	Reference
WS 2.87E+002	mg/L	25	EST	MEYLAN,WM ET AL. (1996)
logP 2.19			EXP	HANSCH,C & LEO,AJ (1985)
VP 5.25E-008	mm Hg	25	EST	NEELY,WB & BLAU,GE (1985)
DC	pKa			
HL 8.30E-012	atm m3/mol	25	EST	MEYLAN,WM & HOWARD,PH (1991)
OH 3.21E-011	cm3/molc sec	25	EST	MEYLAN,WM & HOWARD,PH (1993)

CAS #: 073590-58-6				OMEPRAZOLE
Formula: $C_{17}H_{19}N_3O_3S$				
Mol Weight: 345.42				
MP (deg C): 156		FP (deg C):		
BP (deg C):				
BP pressure (mm Hg):				

Property/ Value	Units	Temp	Data Type	Reference
WS 8.23E+001	mg/L	25	EST	MEYLAN,WM ET AL. (1996)
logP 2.23			EXP	SANGSTER,J (1994)
VP 9.16E-013	mm Hg	25	EST	NEELY,WB & BLAU,GE (1985)
DC	pKa			
HL 3.04E-019	atm m3/mol	25	EST	MEYLAN,WM & HOWARD,PH (1991)
OH 9.56E-011	cm3/molc sec	25	EST	MEYLAN,WM & HOWARD,PH (1993)

CAS #: 073602-65-0				(1-METHYLNONYL)BENZENESULFONIC ACID, SODIUM SALT
Formula: $C_{16}H_{25}NaO_3S$				
Mol Weight: 320.43				
MP (deg C):		FP (deg C):		
BP (deg C):				
BP pressure (mm Hg):				

Property/ Value	Units	Temp	Data Type	Reference
WS 2.05E+002	mg/L	25	EST	MEYLAN,WM ET AL. (1996)
logP 1.94			EST	MEYLAN,WM & HOWARD,PH (1995)
VP 2.16E-014	mm Hg	25	EST	NEELY,WB & BLAU,GE (1985)
DC	pKa			
HL	atm m3/mol			
OH 1.33E-011	cm3/molc sec	25	EST	MEYLAN,WM & HOWARD,PH (1993)

CAS #: 073602-67-2				(1-BUTYLHEXYL)BENZENESULFONIC ACID, SODIUM SALT
Formula: $C_{16}H_{25}NaO_3S$				
Mol Weight: 320.43				
MP (deg C):		FP (deg C):		
BP (deg C):				
BP pressure (mm Hg):				

Property/ Value	Units	Temp	Data Type	Reference
WS 2.05E+002	mg/L	25	EST	MEYLAN,WM ET AL. (1996)
logP 1.94			EST	MEYLAN,WM & HOWARD,PH (1995)
VP 2.16E-014	mm Hg	25	EST	NEELY,WB & BLAU,GE (1985)
DC	pKa			
HL	atm m3/mol			
OH 1.36E-011	cm3/molc sec	25	EST	MEYLAN,WM & HOWARD,PH (1993)

CAS #: 073632-76-5				BUTANOIC ACID, 2-[[(AMINOCARBONYL)AMINO]CARBONYL
Formula: $C_{10}H_{18}N_2O_4$				
Mol Weight: 230.27				
MP (deg C):		FP (deg C):		
BP (deg C):				
BP pressure (mm Hg):				

Property/ Value	Units	Temp	Data Type	Reference
WS 1.75E+003	mg/L	25	EST	MEYLAN,WM ET AL. (1996)
logP 1.45			EXP	SANGSTER,J (1993)
VP 4.01E-008	mm Hg	25	EST	NEELY,WB & BLAU,GE (1985)
DC	pKa			
HL 1.06E-012	atm m3/mol	25	EST	MEYLAN,WM & HOWARD,PH (1991)
OH 1.13E-011	cm3/molc sec	25	EST	MEYLAN,WM & HOWARD,PH (1993)

CAS #: 073632-77-6 — BUTANOIC ACID, 2-[[(AMINOCARBONYL)AMINO]CARBONYL

Formula: $C_{11}H_{20}N_2O_4$

Mol Weight: 244.29

MP (deg C):
FP (deg C):
BP (deg C):
BP pressure (mm Hg):

Property/Value	Units	Temp	Data Type	Reference
WS 5.08E+002	mg/L	25	EST	MEYLAN,WM ET AL. (1996)
logP 1.99			EXP	SANGSTER,J (1994)
VP 2.91E-008	mm Hg	25	EST	NEELY,WB & BLAU,GE (1985)
DC	pKa			
HL 1.41E-012	atm m3/mol	25	EST	MEYLAN,WM & HOWARD,PH (1991)
OH 1.31E-011	cm3/molc sec	25	EST	MEYLAN,WM & HOWARD,PH (1993)

CAS #: 073632-78-7 — BUTANOIC ACID, 2-[[(AMINOCARBONYL)AMINO]CARBONYL

Formula: $C_{15}H_{20}N_2O_4$

Mol Weight: 292.34

MP (deg C):
FP (deg C):
BP (deg C):
BP pressure (mm Hg):

Property/Value	Units	Temp	Data Type	Reference
WS 6.22E+001	mg/L	25	EST	MEYLAN,WM ET AL. (1996)
logP 2.74			EXP	SANGSTER,J (1993)
VP 2.48E-010	mm Hg	25	EST	NEELY,WB & BLAU,GE (1985)
DC	pKa			
HL 6.44E-014	atm m3/mol	25	EST	MEYLAN,WM & HOWARD,PH (1991)
OH 1.60E-011	cm3/molc sec	25	EST	MEYLAN,WM & HOWARD,PH (1993)

CAS #: 073632-79-8 — BUTANOIC ACID, 2-[[(AMINOCARBONYL)AMINO]CARBONYL

Formula: $C_{10}H_{18}N_2O_5$

Mol Weight: 246.27

MP (deg C):
FP (deg C):
BP (deg C):
BP pressure (mm Hg):

Property/Value	Units	Temp	Data Type	Reference
WS 3.47E+003	mg/L	25	EST	MEYLAN,WM ET AL. (1996)
logP 1.00			EXP	SANGSTER,J (1994)
VP 1.74E-008	mm Hg	25	EST	NEELY,WB & BLAU,GE (1985)
DC	pKa			
HL 9.34E-015	atm m3/mol	25	EST	MEYLAN,WM & HOWARD,PH (1991)
OH 1.96E-011	cm3/molc sec	25	EST	MEYLAN,WM & HOWARD,PH (1993)

CAS #: 073632-80-1 — BUTANOIC ACID, 2-[[(AMINOCARBONYL)AMINO]CARBONYL

Formula: $C_{11}H_{18}N_2O_6$

Mol Weight: 274.28

MP (deg C):
FP (deg C):
BP (deg C):
BP pressure (mm Hg):

Property/Value	Units	Temp	Data Type	Reference
WS 2.47E+003	mg/L	25	EST	MEYLAN,WM ET AL. (1996)
logP 0.99			EXP	SANGSTER,J (1994)
VP 9.50E-009	mm Hg	25	EST	NEELY,WB & BLAU,GE (1985)
DC	pKa			
HL 1.34E-014	atm m3/mol	25	EST	MEYLAN,WM & HOWARD,PH (1991)
OH 1.09E-011	cm3/molc sec	25	EST	MEYLAN,WM & HOWARD,PH (1993)

CAS #: 073632-81-2 — BENZENEACETIC ACID, à-[[(AMINOCARBONYL)AMINO]CAR

Formula: $C_{13}H_{16}N_2O_4$

Mol Weight: 264.28

MP (deg C):
FP (deg C):
BP (deg C):
BP pressure (mm Hg):

Property/Value	Units	Temp	Data Type	Reference
WS 8.81E+002	mg/L	25	EST	MEYLAN,WM ET AL. (1996)
logP 1.58			EXP	SANGSTER,J (1993)
VP 1.28E-009	mm Hg	25	EST	NEELY,WB & BLAU,GE (1985)
DC	pKa			
HL 3.65E-014	atm m3/mol	25	EST	MEYLAN,WM & HOWARD,PH (1991)
OH 1.03E-011	cm3/molc sec	25	EST	MEYLAN,WM & HOWARD,PH (1993)

CAS #: 073632-82-3 — 4-PENTENOIC ACID, 2-[[(AMINOCARBONYL)AMINO]CARBO

Formula: $C_{11}H_{16}N_2O_4$

Mol Weight: 240.26

MP (deg C):
FP (deg C):
BP (deg C):
BP pressure (mm Hg):

Property/Value	Units	Temp	Data Type	Reference
WS 1.81E+003	mg/L	25	EST	MEYLAN,WM ET AL. (1996)
logP 1.37			EXP	SANGSTER,J (1994)
VP 2.02E-008	mm Hg	25	EST	NEELY,WB & BLAU,GE (1985)
DC	pKa			
HL 7.80E-013	atm m3/mol	25	EST	MEYLAN,WM & HOWARD,PH (1991)
OH 6.21E-011	cm3/molc sec	25	EST	MEYLAN,WM & HOWARD,PH (1993)

CAS #: 073632-84-5 — PROPANAMIDE, 2-(1-CYCLOHEXENYL)-N-[(METHYLAMINO)

Formula: $C_{13}H_{20}N_2O_4$

Mol Weight: 268.32

MP (deg C):
FP (deg C):
BP (deg C):
BP pressure (mm Hg):

Property/Value	Units	Temp	Data Type	Reference
WS 5.98E+002	mg/L	25	EST	MEYLAN,WM ET AL. (1996)
logP 1.75			EXP	SANGSTER,J (1994)
VP 2.87E-009	mm Hg	25	EST	NEELY,WB & BLAU,GE (1985)
DC	pKa			
HL 1.59E-012	atm m3/mol	25	EST	MEYLAN,WM & HOWARD,PH (1991)
OH 7.30E-011	cm3/molc sec	25	EST	MEYLAN,WM & HOWARD,PH (1993)

CAS #: 073632-86-7 — 1-CYCLOHEXENE-1-ACETIC ACID, _-METHYL-_-[[[(METH

Formula: $C_{14}H_{22}N_2O_5$

Mol Weight: 298.34

MP (deg C):
FP (deg C):
BP (deg C):
BP pressure (mm Hg):

Property/Value	Units	Temp	Data Type	Reference
WS 3.64E+002	mg/L	25	EST	MEYLAN,WM ET AL. (1996)
logP 1.80			EXP	SANGSTER,J (1994)
VP 4.58E-010	mm Hg	25	EST	NEELY,WB & BLAU,GE (1985)
DC	pKa			
HL 2.20E-014	atm m3/mol	25	EST	MEYLAN,WM & HOWARD,PH (1991)
OH 1.05E-010	cm3/molc sec	25	EST	MEYLAN,WM & HOWARD,PH (1993)

073655-51-3 — 2-IODO-4-AMINOBENZOIC ACID

CAS #:	073655-51-3			
Formula:	$C_7H_6INO_2$			
Mol Weight:	263.04			
MP (deg C):		FP (deg C):		
BP (deg C):				
BP pressure (mm Hg):				

Property/Value	Units	Temp	Data Type	Reference
WS 7.80E+002	mg/L	25	EST	MEYLAN,WM ET AL. (1996)
logP 1.65			EXP	HANSCH,C & LEO,AJ (1985)
VP 5.87E-006	mm Hg	25	EST	NEELY,WB & BLAU,GE (1985)
DC	pKa			
HL 8.88E-012	atm m3/mol	25	EST	MEYLAN,WM & HOWARD,PH (1991)
OH 2.68E-011	cm3/molc sec	25	EST	MEYLAN,WM & HOWARD,PH (1993)

073930-96-8 — BUTANOIC ACID, 2-[[(AMINOTHIOXOMETHYL)AMINO]CARB

CAS #:	073930-96-8			
Formula:	$C_9H_{16}N_2O_3S$			
Mol Weight:	232.30			
MP (deg C):		FP (deg C):		
BP (deg C):				
BP pressure (mm Hg):				

Property/Value	Units	Temp	Data Type	Reference
WS 8.25E+002	mg/L	25	EST	MEYLAN,WM ET AL. (1996)
logP 1.82			EXP	SANGSTER,J (1994)
VP 7.23E-007	mm Hg	25	EST	NEELY,WB & BLAU,GE (1985)
DC	pKa			
HL 3.78E-012	atm m3/mol	25	EST	MEYLAN,WM & HOWARD,PH (1991)
OH 3.43E-011	cm3/molc sec	25	EST	MEYLAN,WM & HOWARD,PH (1993)

073931-96-1 — DENZIMOL

CAS #:	073931-96-1			
Formula:	$C_{19}H_{20}N_2O$			
Mol Weight:	292.38			
MP (deg C):		FP (deg C):		
BP (deg C):				
BP pressure (mm Hg):				

Property/Value	Units	Temp	Data Type	Reference
WS 1.70E+001	mg/L	25	EST	MEYLAN,WM ET AL. (1996)
logP 3.40			EXP	HANSCH,C ET AL. (1995)
VP 2.99E-011	mm Hg	25	EST	NEELY,WB & BLAU,GE (1985)
DC	pKa			
HL 3.71E-011	atm m3/mol	25	EST	MEYLAN,WM & HOWARD,PH (1991)
OH 5.86E-011	cm3/molc sec	25	EST	MEYLAN,WM & HOWARD,PH (1993)

073986-52-4 — 2,6-DICHLORO-4-OCTYLPHENOL

CAS #:	073986-52-4			
Formula:	$C_{14}H_{20}Cl_2O$			
Mol Weight:	275.22			
MP (deg C):		FP (deg C):		
BP (deg C):				
BP pressure (mm Hg):				

Property/Value	Units	Temp	Data Type	Reference
WS 1.04E-001	mg/L	25	EST	MEYLAN,WM ET AL. (1996)
logP 6.79				MEYLAN,WM & HOWARD,PH (1995)
VP 3.85E-006	mm Hg	25	EST	NEELY,WB & BLAU,GE (1985)
DC	pKa			
HL 2.47E-006	atm m3/mol	25	EST	MEYLAN,WM & HOWARD,PH (1991)
OH 1.13E-011	cm3/molc sec	25	EST	MEYLAN,WM & HOWARD,PH (1993)

073987-38-9 — 6-QUINOLINECARBOXYLIC ACID,ET ESTER

CAS #:	073987-38-9			
Formula:	$C_{12}H_{11}NO_2$			
Mol Weight:	201.23			
MP (deg C):		FP (deg C):		
BP (deg C):				
BP pressure (mm Hg):				

Property/Value	Units	Temp	Data Type	Reference
WS 2.40E+002	mg/L	25	EST	MEYLAN,WM ET AL. (1996)
logP 2.64			EXP	HANSCH,C & LEO,AJ (1985)
VP 1.23E-004	mm Hg	25	EST	NEELY,WB & BLAU,GE (1985)
DC	pKa			
HL 5.89E-009	atm m3/mol	25	EST	MEYLAN,WM & HOWARD,PH (1991)
OH 6.32E-012	cm3/molc sec	25	EST	MEYLAN,WM & HOWARD,PH (1993)

074011-58-8 — ENOXACIN

CAS #:	074011-58-8			
Formula:	$C_{15}H_{17}FN_4O_3$			
Mol Weight:	320.33			
MP (deg C):	220-224	FP (deg C):		
BP (deg C):				
BP pressure (mm Hg):				

Property/Value	Units	Temp	Data Type	Reference
WS 3.43E+004	mg/L	25	EST	MEYLAN,WM ET AL. (1996)
logP -0.20			EXP	SANGSTER,J (1994) (ION-CORRECT
VP 4.79E-013	mm Hg	25	EST	NEELY,WB & BLAU,GE (1985)
DC	pKa			
HL 1.14E-021	atm m3/mol	25	EST	MEYLAN,WM & HOWARD,PH (1991)
OH 1.22E-010	cm3/molc sec	25	EST	MEYLAN,WM & HOWARD,PH (1993)

074014-51-0 — ROKITAMYCIN

CAS #:	074014-51-0			
Formula:	$C_{42}H_{69}NO_{15}$			
Mol Weight:	828.02			
MP (deg C):	116	FP (deg C):		
BP (deg C):				
BP pressure (mm Hg):				

Property/Value	Units	Temp	Data Type	Reference
WS 1.16E-001	mg/L	25	EST	MEYLAN,WM ET AL. (1996)
logP 3.08			EXP	HANSCH,C ET AL. (1994)
VP 5.02E-029	mm Hg	25	EST	NEELY,WB & BLAU,GE (1985)
DC	pKa			
HL 3.49E-031	atm m3/mol	25	EST	MEYLAN,WM & HOWARD,PH (1991)
OH 4.21E-010	cm3/molc sec	25	EST	MEYLAN,WM & HOWARD,PH (1993)

074054-79-8 — 3-BUTENAMIDE, N-(2,4-DICHLOROPHENYL)-2-METHYL-

CAS #:	074054-79-8			
Formula:	$C_{11}H_{11}Cl_2NO$			
Mol Weight:	244.12			
MP (deg C):		FP (deg C):		
BP (deg C):				
BP pressure (mm Hg):				

Property/Value	Units	Temp	Data Type	Reference
WS 2.67E+001	mg/L	25	EST	MEYLAN,WM ET AL. (1996)
logP 3.49			EXP	MITSUTAKE,KI ET AL. (1986)
VP 3.06E-006	mm Hg	25	EST	NEELY,WB & BLAU,GE (1985)
DC	pKa			
HL 5.91E-009	atm m3/mol	25	EST	MEYLAN,WM & HOWARD,PH (1991)
OH 2.93E-011	cm3/molc sec	25	EST	MEYLAN,WM & HOWARD,PH (1993)

CAS #: 074070-46-5 — ACLONIFEN

Formula: $C_{12}H_9ClN_2O_3$

Mol Weight: 264.67

MP (deg C): FP (deg C):

BP (deg C):

BP pressure (mm Hg):

Property/Value	Units	Temp	Data Type	Reference
WS 2.50E+000	mg/L	20	EXP	SHIU,WY ET AL. (1990)
logP 4.04			EXP	NANDIHALLI,UB ET AL. (1993)
VP 3.50E-007	mm Hg	25	EST	NEELY,WB & BLAU,GE (1985)
DC	pKa			
HL 2.66E-009	atm m3/mol	25	EST	MEYLAN,WM & HOWARD,PH (1991)
OH 1.48E-011	cm3/molc sec	25	EST	MEYLAN,WM & HOWARD,PH (1993)

CAS #: 074099-07-3 — HYDRAZINECARBOXAMIDE, N-(4-ETHOXYPHENYL)-

Formula: $C_9H_{13}N_3O_2$

Mol Weight: 195.22

MP (deg C): FP (deg C):

BP (deg C):

BP pressure (mm Hg):

Property/Value	Units	Temp	Data Type	Reference
WS 1.63E+004	mg/L	25	EST	MEYLAN,WM ET AL. (1996)
logP 0.53			EXP	KRAMER,CR & BECK,L (1981)
VP 6.23E-006	mm Hg	25	EST	NEELY,WB & BLAU,GE (1985)
DC	pKa			
HL 6.61E-014	atm m3/mol	25	EST	MEYLAN,WM & HOWARD,PH (1991)
OH 4.41E-011	cm3/molc sec	25	EST	MEYLAN,WM & HOWARD,PH (1993)

CAS #: 074099-08-4 — HYDRAZINECARBOXAMIDE, N-(1-NAPHTHALENYL)-

Formula: $C_{11}H_{11}N_3O$

Mol Weight: 201.23

MP (deg C): FP (deg C):

BP (deg C):

BP pressure (mm Hg):

Property/Value	Units	Temp	Data Type	Reference
WS 7.35E+003	mg/L	25	EST	MEYLAN,WM ET AL. (1996)
logP 0.90			EXP	KRAMER,CR & BECK,L (1981)
VP 4.77E-007	mm Hg	25	EST	NEELY,WB & BLAU,GE (1985)
DC	pKa			
HL 8.21E-014	atm m3/mol	25	EST	MEYLAN,WM & HOWARD,PH (1991)
OH 2.01E-010	cm3/molc sec	25	EST	MEYLAN,WM & HOWARD,PH (1993)

CAS #: 074109-81-2 — 1(3-BENZYLOXYPHENYL)-3-(ME-MEO) UREA

Formula: $C_{16}H_{18}N_2O_3$

Mol Weight: 286.33

MP (deg C): FP (deg C):

BP (deg C):

BP pressure (mm Hg):

Property/Value	Units	Temp	Data Type	Reference
WS 3.25E+001	mg/L	25	EST	MEYLAN,WM ET AL. (1996)
logP 3.11			EXP	HANSCH,C & LEO,AJ (1985)
VP 3.09E-008	mm Hg	25	EST	NEELY,WB & BLAU,GE (1985)
DC	pKa			
HL 1.00E-010	atm m3/mol	25	EST	MEYLAN,WM & HOWARD,PH (1991)
OH 2.07E-010	cm3/molc sec	25	EST	MEYLAN,WM & HOWARD,PH (1993)

CAS #: 074115-24-5 — CLOFENTEZINE

Formula: $C_{14}H_8Cl_2N_4$

Mol Weight: 303.15

MP (deg C): 182 FP (deg C):

BP (deg C):

BP pressure (mm Hg):

Property/Value	Units	Temp	Data Type	Reference
WS 2.65E+001	mg/L	25	EST	MEYLAN,WM ET AL. (1996)
logP 3.10			EXP	TOMLIN,C (1994)
VP 9.75E-010	mm Hg	25	EXP	TOMLIN,C (1994)
DC	pKa			
HL 1.56E-009	atm m3/mol	25	EST	MEYLAN,WM & HOWARD,PH (1991)
OH 2.10E-012	cm3/molc sec	25	EST	MEYLAN,WM & HOWARD,PH (1993)

CAS #: 074134-16-0 — ACETAMIDE, N-(9-OXO-9H-THIOXANTHEN-3-YL)-, S,S-D

Formula: $C_{15}H_{11}NO_4S$

Mol Weight: 301.32

MP (deg C): FP (deg C):

BP (deg C):

BP pressure (mm Hg):

Property/Value	Units	Temp	Data Type	Reference
WS 9.22E+000	mg/L	25	EST	MEYLAN,WM ET AL. (1996)
logP 2.35			EXP	HANSCH,C ET AL. (1995)
VP 6.76E-011	mm Hg	25	EST	NEELY,WB & BLAU,GE (1985)
DC	pKa			
HL 4.67E-016	atm m3/mol	25	EST	MEYLAN,WM & HOWARD,PH (1991)
OH 2.90E-012	cm3/molc sec	25	EST	MEYLAN,WM & HOWARD,PH (1993)

CAS #: 074134-17-1 — ACETAMIDE, N-(7-METHYL-9-OXO-9H-THIOXANTHEN-3-YL

Formula: $C_{16}H_{13}NO_4S$

Mol Weight: 315.35

MP (deg C): FP (deg C):

BP (deg C):

BP pressure (mm Hg):

Property/Value	Units	Temp	Data Type	Reference
WS 3.40E+000	mg/L	25	EST	MEYLAN,WM ET AL. (1996)
logP 2.76			EXP	SANGSTER,J (1994)
VP 2.94E-011	mm Hg	25	EST	NEELY,WB & BLAU,GE (1985)
DC	pKa			
HL 5.16E-016	atm m3/mol	25	EST	MEYLAN,WM & HOWARD,PH (1991)
OH 3.12E-012	cm3/molc sec	25	EST	MEYLAN,WM & HOWARD,PH (1993)

CAS #: 074134-18-2 — ACETAMIDE, N-(7-ETHYL-9-OXO-9H-THIOXANTHEN-3-YL)

Formula: $C_{17}H_{15}NO_4S$

Mol Weight: 329.38

MP (deg C): FP (deg C):

BP (deg C):

BP pressure (mm Hg):

Property/Value	Units	Temp	Data Type	Reference
WS 7.38E-001	mg/L	25	EST	MEYLAN,WM ET AL. (1996)
logP 3.44			EXP	HANSCH,C ET AL. (1995)
VP 1.28E-011	mm Hg	25	EST	NEELY,WB & BLAU,GE (1985)
DC	pKa			
HL 6.85E-016	atm m3/mol	25	EST	MEYLAN,WM & HOWARD,PH (1991)
OH 4.09E-012	cm3/molc sec	25	EST	MEYLAN,WM & HOWARD,PH (1993)

CAS #: 074134-19-3 — ACETAMIDE, N-(9-OXO-7-PROPYL-9H-THIOXANTHEN-3-YL

Formula: $C_{18}H_{17}NO_4S$

Mol Weight: 343.40

MP (deg C): | FP (deg C):

BP (deg C):

BP pressure (mm Hg):

Property/Value	Units	Temp	Data Type	Reference
WS 5.30E-001	mg/L	25	EST	MEYLAN,WM ET AL. (1996)
logP 3.51			EXP	HANSCH,C ET AL. (1995)
VP 5.52E-012	mm Hg	25	EST	NEELY,WB & BLAU,GE (1985)
DC	pKa			
HL 9.09E-016	atm m3/mol	25	EST	MEYLAN,WM & HOWARD,PH (1991)
OH 5.45E-012	cm3/molc sec	25	EST	MEYLAN,WM & HOWARD,PH (1993)

CAS #: 074141-73-4 — 1H-IMIDAZOLE, 1-[2-(METHYLSULFINYL)ETHYL]-2-NITR

Formula: $C_6H_9N_3O_3S$

Mol Weight: 203.22

MP (deg C): | FP (deg C):

BP (deg C):

BP pressure (mm Hg):

Property/Value	Units	Temp	Data Type	Reference
WS 1.89E+005	mg/L	25	EST	MEYLAN,WM ET AL. (1996)
logP -1.22			EXP	SANGSTER,J (1993)
VP 9.43E-007	mm Hg	25	EST	NEELY,WB & BLAU,GE (1985)
DC	pKa			
HL 2.82E-013	atm m3/mol	25	EST	MEYLAN,WM & HOWARD,PH (1991)
OH 7.57E-011	cm3/molc sec	25	EST	MEYLAN,WM & HOWARD,PH (1993)

CAS #: 074141-74-5 — 1H-IMIDAZOLE-1-ACETAMIDE, N,N-BIS(2-HYDROXYETHYL

Formula: $C_9H_{14}N_4O_5$

Mol Weight: 258.24

MP (deg C): | FP (deg C):

BP (deg C):

BP pressure (mm Hg):

Property/Value	Units	Temp	Data Type	Reference
WS 1.07E+004	mg/L	25	EST	MEYLAN,WM ET AL. (1996)
logP -1.64			EXP	SANGSTER,J (1993)
VP 1.28E-012	mm Hg	25	EST	NEELY,WB & BLAU,GE (1985)
DC	pKa			
HL 5.45E-019	atm m3/mol	25	EST	MEYLAN,WM & HOWARD,PH (1991)
OH 3.79E-011	cm3/molc sec	25	EST	MEYLAN,WM & HOWARD,PH (1993)

CAS #: 074141-75-6 — 1H-IMIDAZOLE-1-ACETAMIDE, N-(2,3-DIHYDROXYPROPYL

Formula: $C_8H_{12}N_4O_5$

Mol Weight: 244.21

MP (deg C): | FP (deg C):

BP (deg C):

BP pressure (mm Hg):

Property/Value	Units	Temp	Data Type	Reference
WS 1.93E+004	mg/L	25	EST	MEYLAN,WM ET AL. (1996)
logP -1.85			EXP	SANGSTER,J (1993)
VP 2.69E-013	mm Hg	25	EST	NEELY,WB & BLAU,GE (1985)
DC	pKa			
HL 2.48E-019	atm m3/mol	25	EST	MEYLAN,WM & HOWARD,PH (1991)
OH 3.00E-011	cm3/molc sec	25	EST	MEYLAN,WM & HOWARD,PH (1993)

CAS #: 074188-69-5 — 1H-PYRROL-2-AMINE, N-[2-[[(5-METHYL-1H-IMIDAZOL-

Formula: $C_{11}H_{15}N_5O_2S$

Mol Weight: 281.34

MP (deg C): | FP (deg C):

BP (deg C):

BP pressure (mm Hg):

Property/Value	Units	Temp	Data Type	Reference
WS 8.30E+002	mg/L	25	EST	MEYLAN,WM ET AL. (1996)
logP 1.04			EXP	SANGSTER,J (1994)
VP 8.98E-011	mm Hg	25	EST	NEELY,WB & BLAU,GE (1985)
DC	pKa			
HL 2.44E-017	atm m3/mol	25	EST	MEYLAN,WM & HOWARD,PH (1991)
OH 1.05E-010	cm3/molc sec	25	EST	MEYLAN,WM & HOWARD,PH (1993)

CAS #: 074222-97-2 — SULFOMETURON (pH 5-7)

Formula: $C_{15}H_{16}N_4O_5S$

Mol Weight: 364.38

MP (deg C): 203-205 | FP (deg C):

BP (deg C):

BP pressure (mm Hg):

Property/Value	Units	Temp	Data Type	Reference
WS 1.00E+001	mg/L	25	EXP	SHIU,WY ET AL. (1990)
logP 1.20			EXP	HARVEY,J JR ET AL. (1985)
VP 1.18E-011	mm Hg	25	EST	NEELY,WB & BLAU,GE (1985)
DC	pKa			
HL 3.72E-013	atm m3/mol	25	EST	MEYLAN,WM & HOWARD,PH (1991)
OH 3.59E-011	cm3/molc sec	25	EST	MEYLAN,WM & HOWARD,PH (1993)

CAS #: 074223-56-6 — SULFOMETURON

Formula: $C_{14}H_{14}N_4O_5S$

Mol Weight: 350.36

MP (deg C): | FP (deg C):

BP (deg C):

BP pressure (mm Hg):

Property/Value	Units	Temp	Data Type	Reference
WS 1.00E+001	mg/L		EXP	NEARY,DG ET AL. (1993) @ pH=5
logP -1.13			EST	MEYLAN,WM & HOWARD,PH (1995)
VP 1.07E-012	mm Hg	25	EST	NEELY,WB & BLAU,GE (1985)
DC	pKa			
HL 1.16E-015	atm m3/mol	25	EST	MEYLAN,WM & HOWARD,PH (1991)
OH 3.62E-011	cm3/molc sec	25	EST	MEYLAN,WM & HOWARD,PH (1993)

CAS #: 074223-64-6 — METSULFURON-METHYL

Formula: $C_{14}H_{15}N_5O_6S$

Mol Weight: 381.37

MP (deg C): 158 | FP (deg C):

BP (deg C):

BP pressure (mm Hg):

Property/Value	Units	Temp	Data Type	Reference
WS 9.50E+003	mg/L	25	EXP	BEYER,EMJR ET AL. (1988); PH 7
logP 2.20			EXP	SANGSTER,J (1994)
VP 2.50E-012	mm Hg	25		BEYER,EMJR ET AL. (1988)
DC 3.30	pKa		EXP	BEYER,EMJR ET AL. (1988)
HL 1.32E-016	atm m3/mol		EST	VP/WSOL
OH 2.52E-011	cm3/molc sec	25	EST	ATKINSON,R ET AL. (1988)

1320

CAS #: 074255-46-2 — 4-T-BUTYLSEMICARBAZIDE

Formula: $C_5H_{13}N_3O$

Mol Weight: 131.18

MP (deg C):   **FP (deg C):**

BP (deg C):

BP pressure (mm Hg):

Property/Value	Units	Temp	Data Type	Reference
WS 2.19E+004	mg/L	25	EST	MEYLAN,WM ET AL. (1996)
logP -0.81			EXP	HANSCH,C & LEO,AJ (1985)
VP 5.87E-003	mm Hg	25	EST	NEELY,WB & BLAU,GE (1985)
DC	pKa			
HL 7.83E-012	atm m3/mol	25	EST	MEYLAN,WM & HOWARD,PH (1991)
OH 2.50E-012	cm3/molc sec	25	EST	MEYLAN,WM & HOWARD,PH (1993)

CAS #: 074339-86-9 — ACETAMIDE, 2-CHLORO-N-[2-OXO-1-(PHENYLAMINO)ETHY

Formula: $C_{16}H_{15}ClN_2O_2$

Mol Weight: 302.76

MP (deg C):   **FP (deg C):**

BP (deg C):

BP pressure (mm Hg):

Property/Value	Units	Temp	Data Type	Reference
WS 1.24E+002	mg/L	25	EST	MEYLAN,WM ET AL. (1996)
logP 2.32			EXP	HANSCH,C ET AL. (1995)
VP 2.76E-010	mm Hg	25	EST	NEELY,WB & BLAU,GE (1985)
DC	pKa			
HL 1.18E-010	atm m3/mol	25	EST	MEYLAN,WM & HOWARD,PH (1991)
OH 2.80E-011	cm3/molc sec	25	EST	MEYLAN,WM & HOWARD,PH (1993)

CAS #: 074339-88-1 — ACETAMIDE, 2-(ACETYLPHENYLAMINO)-N-PHENYL-

Formula: $C_{16}H_{16}N_2O_2$

Mol Weight: 268.32

MP (deg C):   **FP (deg C):**

BP (deg C):

BP pressure (mm Hg):

Property/Value	Units	Temp	Data Type	Reference
WS 3.59E+002	mg/L	25	EST	MEYLAN,WM ET AL. (1996)
logP 2.01			EXP	HANSCH,C ET AL. (1995)
VP 2.47E-009	mm Hg	25	EST	NEELY,WB & BLAU,GE (1985)
DC	pKa			
HL 3.35E-010	atm m3/mol	25	EST	MEYLAN,WM & HOWARD,PH (1991)
OH 2.78E-011	cm3/molc sec	25	EST	MEYLAN,WM & HOWARD,PH (1993)

CAS #: 074339-89-2 — PROPANAMIDE, 2-CHLORO-N-[2-OXO-2-(PHENYLAMINO)ET

Formula: $C_{17}H_{17}ClN_2O_2$

Mol Weight: 316.79

MP (deg C):   **FP (deg C):**

BP (deg C):

BP pressure (mm Hg):

Property/Value	Units	Temp	Data Type	Reference
WS 2.58E+001	mg/L	25	EST	MEYLAN,WM ET AL. (1996)
logP 3.02			EXP	HANSCH,C ET AL. (1995)
VP 3.12E-010	mm Hg	25	EST	NEELY,WB & BLAU,GE (1985)
DC	pKa			
HL 1.57E-010	atm m3/mol	25	EST	MEYLAN,WM & HOWARD,PH (1991)
OH 2.87E-011	cm3/molc sec	25	EST	MEYLAN,WM & HOWARD,PH (1993)

CAS #: 074350-90-6 — ACETAMIDE, 2,2-DICHLORO-N-[2-OXO-2-(PHENYLAMINO)

Formula: $C_{16}H_{14}Cl_2N_2O_2$

Mol Weight: 337.21

MP (deg C):   **FP (deg C):**

BP (deg C):

BP pressure (mm Hg):

Property/Value	Units	Temp	Data Type	Reference
WS 2.19E+001	mg/L	25	EST	MEYLAN,WM ET AL. (1996)
logP 2.96			EXP	HANSCH,C ET AL. (1995)
VP 1.22E-010	mm Hg	25	EST	NEELY,WB & BLAU,GE (1985)
DC	pKa			
HL 4.16E-011	atm m3/mol	25	EST	MEYLAN,WM & HOWARD,PH (1991)
OH 2.79E-011	cm3/molc sec	25	EST	MEYLAN,WM & HOWARD,PH (1993)

CAS #: 074385-72-1 — N-(N''-DI-ME-N'-ME-URYLTHIO)METHOMYL

Formula: $C_9H_{18}N_4O_3S_2$

Mol Weight: 294.40

MP (deg C):   **FP (deg C):**

BP (deg C):

BP pressure (mm Hg):

Property/Value	Units	Temp	Data Type	Reference
WS 1.25E+003	mg/L	25	EST	MEYLAN,WM ET AL. (1996)
logP 1.20			EXP	HANSCH,C & LEO,AJ (1985)
VP 2.67E-006	mm Hg	25	EST	NEELY,WB & BLAU,GE (1985)
DC	pKa			
HL 1.18E-013	atm m3/mol	25	EST	MEYLAN,WM & HOWARD,PH (1991)
OH 2.22E-011	cm3/molc sec	25	EST	MEYLAN,WM & HOWARD,PH (1993)

CAS #: 074385-86-7 — ETHANIMIDOTHIOIC ACID, N-[(4-CYCLOPROPYL-2,6-DIM

Formula: $C_{11}H_{20}N_4O_3S_2$

Mol Weight: 320.44

MP (deg C):   **FP (deg C):**

BP (deg C):

BP pressure (mm Hg):

Property/Value	Units	Temp	Data Type	Reference
WS 1.82E+002	mg/L	25	EST	MEYLAN,WM ET AL. (1996)
logP 2.00			EXP	SANGSTER,J (1993)
VP 4.23E-007	mm Hg	25	EST	NEELY,WB & BLAU,GE (1985)
DC	pKa			
HL 9.16E-014	atm m3/mol	25	EST	MEYLAN,WM & HOWARD,PH (1991)
OH 2.19E-011	cm3/molc sec	25	EST	MEYLAN,WM & HOWARD,PH (1993)

CAS #: 074385-89-0 — N-(N''-DI-IPR-N'-ME-URYLTHIO)METHOMYL

Formula: $C_{13}H_{26}N_4O_3S_2$

Mol Weight: 350.51

MP (deg C):   **FP (deg C):**

BP (deg C):

BP pressure (mm Hg):

Property/Value	Units	Temp	Data Type	Reference
WS 3.70E+001	mg/L	25	EST	MEYLAN,WM ET AL. (1996)
logP 2.60			EXP	HANSCH,C & LEO,AJ (1985)
VP 1.33E-008	mm Hg	25	EST	NEELY,WB & BLAU,GE (1985)
DC	pKa			
HL 2.21E-013	atm m3/mol	25	EST	MEYLAN,WM & HOWARD,PH (1991)
OH 3.82E-011	cm3/molc sec	25	EST	MEYLAN,WM & HOWARD,PH (1993)

<table>
<tr><td colspan="2">

CAS #: 074399-87-4 — N-(N''-DI-ME-N'-HEX-URYLTHIO)METHOMYL

</td></tr>
</table>

Formula: $C_{14}H_{28}N_4O_3S_2$	
Mol Weight: 364.53	
MP (deg C):	**FP (deg C):**
BP (deg C):	
BP pressure (mm Hg):	

Property/Value	Units	Temp	Data Type	Reference
WS 7.69E+000	mg/L	25	EST	MEYLAN,WM ET AL. (1996)
logP 3.30			EXP	HANSCH,C & LEO,AJ (1985)
VP 4.15E-008	mm Hg	25	EST	NEELY,WB & BLAU,GE (1985)
DC	pKa			
HL 4.86E-013	atm m3/mol	25	EST	MEYLAN,WM & HOWARD,PH (1991)
OH 3.19E-011	cm3/molc sec	25	EST	MEYLAN,WM & HOWARD,PH (1993)

CAS #: 074407-83-3 — 2-PROPENOIC ACID, 3-(2-CHLOROPHENOXY)-, ETHYL ES

Formula: $C_{11}H_{11}ClO_3$	
Mol Weight: 226.66	
MP (deg C):	**FP (deg C):**
BP (deg C):	
BP pressure (mm Hg):	

Property/Value	Units	Temp	Data Type	Reference
WS 2.95E+001	mg/L	25	EST	MEYLAN,WM ET AL. (1996)
logP 3.55			EXP	SANGSTER,J (1993)
VP 8.39E-004	mm Hg	25	EST	NEELY,WB & BLAU,GE (1985)
DC	pKa			
HL 3.12E-006	atm m3/mol	25	EST	MEYLAN,WM & HOWARD,PH (1991)
OH 2.77E-011	cm3/molc sec	25	EST	MEYLAN,WM & HOWARD,PH (1993)

CAS #: 074430-97-0 — 1,9-PHENANTHRIDINEDIOL, 5,6,6A,7,8,9,10,10A-OCTA

Formula: $C_{27}H_{35}NO_4$	
Mol Weight: 437.58	
MP (deg C):	**FP (deg C):**
BP (deg C):	
BP pressure (mm Hg):	

Property/Value	Units	Temp	Data Type	Reference
WS 2.24E-002	mg/L	25	EST	MEYLAN,WM ET AL. (1996)
logP 5.74			EXP	SANGSTER,J (1993)
VP 4.25E-014	mm Hg	25	EST	NEELY,WB & BLAU,GE (1985)
DC	pKa			
HL 3.20E-014	atm m3/mol	25	EST	MEYLAN,WM & HOWARD,PH (1991)
OH 2.62E-010	cm3/molc sec	25	EST	MEYLAN,WM & HOWARD,PH (1993)

CAS #: 074433-34-4 — BENZAMIDE, 4-[(DIETHYLAMINO)METHYL]-N-(2,6-DIMET

Formula: $C_{20}H_{26}N_2O$	
Mol Weight: 310.44	
MP (deg C):	**FP (deg C):**
BP (deg C):	
BP pressure (mm Hg):	

Property/Value	Units	Temp	Data Type	Reference
WS 1.12E+001	mg/L	25	EST	MEYLAN,WM ET AL. (1996)
logP 3.49			EXP	HANSCH,C ET AL. (1995)
VP 2.16E-009	mm Hg	25	EST	NEELY,WB & BLAU,GE (1985)
DC	pKa			
HL 1.44E-012	atm m3/mol	25	EST	MEYLAN,WM & HOWARD,PH (1991)
OH 1.16E-010	cm3/molc sec	25	EST	MEYLAN,WM & HOWARD,PH (1993)

CAS #: 074440-81-6 — HEXYLPYRIDINIUM BROMIDE

Formula: $C_{11}H_{18}BrN$	
Mol Weight: 244.18	
MP (deg C):	**FP (deg C):**
BP (deg C):	
BP pressure (mm Hg):	

Property/Value	Units	Temp	Data Type	Reference
WS 1.00E+006	mg/L	25	EST	MEYLAN,WM ET AL. (1996)
logP -2.03			EXP	HANSCH,C & LEO,AJ (1985)
VP 3.20E-003	mm Hg	25	EST	NEELY,WB & BLAU,GE (1985)
DC	pKa			
HL 6.20E-010	atm m3/mol	25	EST	MEYLAN,WM & HOWARD,PH (1991)
OH 1.71E-011	cm3/molc sec	25	EST	MEYLAN,WM & HOWARD,PH (1993)

CAS #: 074472-46-1 — 2,3,3',5,5',6-HEXACHLOROBIPHENYL

Formula: $C_{12}H_4Cl_6$	
Mol Weight: 360.88'	
MP (deg C):	**FP (deg C):**
BP (deg C):	
BP pressure (mm Hg):	

Property/Value	Units	Temp	Data Type	Reference
WS 2.71E-003	mg/L	25	EST	MEYLAN,WM ET AL. (1996)
logP 7.37			EXP	HANSCH,C & LEO,AJ (1985)
VP 5.81E-007	mm Hg	25	EST	NEELY,WB & BLAU,GE (1985)
DC	pKa			
HL 6.85E-005	atm m3/mol	25	EST	MEYLAN,WM & HOWARD,PH (1991)
OH 4.03E-013	cm3/molc sec	25	EST	MEYLAN,WM & HOWARD,PH (1993)

CAS #: 074517-78-5 — 9H-FLUORENE-9-CARBOXAMIDE, 9-[3-[(1-METHYLETHYL)

Formula: $C_{20}H_{24}N_2O$	
Mol Weight: 308.43	
MP (deg C): 94-95	**FP (deg C):**
BP (deg C):	
BP pressure (mm Hg):	

Property/Value	Units	Temp	Data Type	Reference
WS 1.88E+000	mg/L	25	EST	MEYLAN,WM ET AL. (1996)
logP 3.11			EXP	MANNHOLD,R ET AL. (1990)
VP 7.58E-010	mm Hg	25	EST	NEELY,WB & BLAU,GE (1985)
DC	pKa			
HL 2.37E-014	atm m3/mol	25	EST	MEYLAN,WM & HOWARD,PH (1991)
OH 1.08E-010	cm3/molc sec	25	EST	MEYLAN,WM & HOWARD,PH (1993)

CAS #: 074550-86-0 — 1H-IMIDAZOLE-1-ETHANOL, 2-[2-(1,3-BENZODIOXOL-5-

Formula: $C_{14}H_{13}N_3O_5$	
Mol Weight: 303.28	
MP (deg C):	**FP (deg C):**
BP (deg C):	
BP pressure (mm Hg):	

Property/Value	Units	Temp	Data Type	Reference
WS 8.84E+001	mg/L	25	EST	MEYLAN,WM ET AL. (1996)
logP 2.03			EXP	SANGSTER,J (1993)
VP 1.47E-012	mm Hg	25	EST	NEELY,WB & BLAU,GE (1985)
DC	pKa			
HL 4.10E-016	atm m3/mol	25	EST	MEYLAN,WM & HOWARD,PH (1991)
OH 8.49E-011	cm3/molc sec	25	EST	MEYLAN,WM & HOWARD,PH (1993)

CAS #: 074550-87-1	1H-IMIDAZOLE, 2-CYCLOPROPYL-1-METHYL-5-NITRO-

Formula: $C_7H_9N_3O_2$

Mol Weight: 167.17

MP (deg C): FP (deg C):

BP (deg C):

BP pressure (mm Hg):

Property/Value	Units	Temp	Data Type	Reference
WS 3.62E+003	mg/L	25	EST	MEYLAN,WM ET AL. (1996)
logP 1.00			EXP	SANGSTER,J (1994)
VP 6.30E-005	mm Hg	25	EST	NEELY,WB & BLAU,GE (1985)
DC	pKa			
HL 2.71E-007	atm m3/mol	25	EST	MEYLAN,WM & HOWARD,PH (1991)
OH 4.04E-012	cm3/molc sec	25	EST	MEYLAN,WM & HOWARD,PH (1993)

CAS #: 074550-92-8	1H-IMIDAZOLE-1-ETHANOL, 2-METHYL-_-[(2-METHYL-5-

Formula: $C_{11}H_{14}N_6O_5$

Mol Weight: 310.27

MP (deg C): FP (deg C):

BP (deg C):

BP pressure (mm Hg):

Property/Value	Units	Temp	Data Type	Reference
WS 1.16E+002	mg/L	25	EST	MEYLAN,WM ET AL. (1996)
logP 0.31			EXP	SANGSTER,J (1994)
VP 1.17E-012	mm Hg	25	EST	NEELY,WB & BLAU,GE (1985)
DC	pKa			
HL 1.06E-016	atm m3/mol	25	EST	MEYLAN,WM & HOWARD,PH (1991)
OH 2.07E-011	cm3/molc sec	25	EST	MEYLAN,WM & HOWARD,PH (1993)

CAS #: 074550-94-0	1H-IMIDAZOLE-1-ETHANOL, à-[(DIETHYLAMINO)METHYL]

Formula: $C_{11}H_{20}N_4O_3$

Mol Weight: 256.31

MP (deg C): FP (deg C):

BP (deg C):

BP pressure (mm Hg):

Property/Value	Units	Temp	Data Type	Reference
WS 1.06E+005	mg/L	25	EST	MEYLAN,WM ET AL. (1996)
logP -0.08			EXP	CANTELLI-FORTI,G ET AL. (1986)
VP 2.03E-009	mm Hg	25	EST	NEELY,WB & BLAU,GE (1985)
DC	pKa			
HL 1.96E-014	atm m3/mol	25	EST	MEYLAN,WM & HOWARD,PH (1991)
OH 1.10E-010	cm3/molc sec	25	EST	MEYLAN,WM & HOWARD,PH (1993)

CAS #: 074571-56-5	PIPERAZINE, 1-METHYL-4-[2-(5-NITRO-1H-IMIDAZOL-1

Formula: $C_{10}H_{17}N_5O_2$

Mol Weight: 239.28

MP (deg C): FP (deg C):

BP (deg C):

BP pressure (mm Hg):

Property/Value	Units	Temp	Data Type	Reference
WS 1.67E+005	mg/L	25	EST	MEYLAN,WM ET AL. (1996)
logP -0.20			EXP	CANTELLI-FORTI,G ET AL. (1986)
VP 3.88E-007	mm Hg	25	EST	NEELY,WB & BLAU,GE (1985)
DC	pKa			
HL 6.04E-014	atm m3/mol	25	EST	MEYLAN,WM & HOWARD,PH (1991)
OH 1.92E-010	cm3/molc sec	25	EST	MEYLAN,WM & HOWARD,PH (1993)

CAS #: 074602-43-0	BENZAMIDE, N-(2,6-DIMETHYLPHENYL)-3-(1-PYRROLIDI

Formula: $C_{20}H_{24}N_2O$

Mol Weight: 308.43

MP (deg C): FP (deg C):

BP (deg C):

BP pressure (mm Hg):

Property/Value	Units	Temp	Data Type	Reference
WS 2.47E+001	mg/L	25	EST	MEYLAN,WM ET AL. (1996)
logP 3.10			EXP	HANSCH,C ET AL. (1995)
VP 1.09E-009	mm Hg	25	EST	NEELY,WB & BLAU,GE (1985)
DC	pKa			
HL 6.33E-013	atm m3/mol	25	EST	MEYLAN,WM & HOWARD,PH (1991)
OH 1.15E-010	cm3/molc sec	25	EST	MEYLAN,WM & HOWARD,PH (1993)

CAS #: 074602-46-3	BENZAMIDE, N-(2,6-DIMETHYLPHENYL)-4-(1-PIPERIDIN

Formula: $C_{21}H_{26}N_2O$

Mol Weight: 322.45

MP (deg C): FP (deg C):

BP (deg C):

BP pressure (mm Hg):

Property/Value	Units	Temp	Data Type	Reference
WS 4.58E+000	mg/L	25	EST	MEYLAN,WM ET AL. (1996)
logP 3.86			EXP	HANSCH,C ET AL. (1995)
VP 4.44E-010	mm Hg	25	EST	NEELY,WB & BLAU,GE (1985)
DC	pKa			
HL 8.40E-013	atm m3/mol	25	EST	MEYLAN,WM & HOWARD,PH (1991)
OH 1.24E-010	cm3/molc sec	25	EST	MEYLAN,WM & HOWARD,PH (1993)

CAS #: 074602-48-5	BENZAMIDE, N-(2,6-DIMETHYLPHENYL)-3-(4-MORPHOLIN

Formula: $C_{20}H_{24}N_2O_2$

Mol Weight: 324.43

MP (deg C): FP (deg C):

BP (deg C):

BP pressure (mm Hg):

Property/Value	Units	Temp	Data Type	Reference
WS 1.31E+002	mg/L	25	EST	MEYLAN,WM ET AL. (1996)
logP 2.14			EXP	HANSCH,C ET AL. (1995)
VP 3.55E-010	mm Hg	25	EST	NEELY,WB & BLAU,GE (1985)
DC	pKa			
HL 5.58E-015	atm m3/mol	25	EST	MEYLAN,WM & HOWARD,PH (1991)
OH 1.71E-010	cm3/molc sec	25	EST	MEYLAN,WM & HOWARD,PH (1993)

CAS #: 074604-76-5	N[4(14BENZDIOXAN6YL)2THIAZOLYL]OXAMIC ACID

Formula: $C_{13}H_{12}N_2O_5S$

Mol Weight: 308.31

MP (deg C): FP (deg C):

BP (deg C):

BP pressure (mm Hg):

Property/Value	Units	Temp	Data Type	Reference
WS 3.27E+002	mg/L	25	EST	MEYLAN,WM ET AL. (1996)
logP 1.80			EXP	PANDIT,NK ET AL. (1989)
VP 1.09E-011	mm Hg	25	EST	NEELY,WB & BLAU,GE (1985)
DC	pKa			
HL 2.29E-020	atm m3/mol	25	EST	MEYLAN,WM & HOWARD,PH (1991)
OH 5.08E-011	cm3/molc sec	25	EST	MEYLAN,WM & HOWARD,PH (1993)

CAS #: 074633-67-3				THIOUREA, N-METHYL-N'-[2-[[(1-METHYL-1H-IMIDAZOL-

Formula: $C_9H_{16}N_4S_2$

Mol Weight: 244.38

MP (deg C):   FP (deg C):

BP (deg C):

BP pressure (mm Hg):

Property/Value	Units	Temp	Data Type	Reference
WS 6.76E+003	mg/L	25	EST	MEYLAN,WM ET AL. (1996)
logP 0.32			EXP	SANGSTER,J (1993)
VP 2.16E-007	mm Hg	25	EST	NEELY,WB & BLAU,GE (1985)
DC	pKa			
HL 1.29E-011	atm m3/mol	25	EST	MEYLAN,WM & HOWARD,PH (1991)
OH 2.45E-010	cm3/molc sec	25	EST	MEYLAN,WM & HOWARD,PH (1993)

CAS #: 074639-30-8				2-ME PYRIDINIUM IODIDE,N-TETRADECYL

Formula: $C_{20}H_{36}IN$

Mol Weight: 417.42

MP (deg C):   FP (deg C):

BP (deg C):

BP pressure (mm Hg):

Property/Value	Units	Temp	Data Type	Reference
WS 2.75E+001	mg/L	25	EST	MEYLAN,WM ET AL. (1996)
logP 2.27			EXP	HANSCH,C & LEO,AJ (1985)
VP 2.38E-007	mm Hg	25	EST	NEELY,WB & BLAU,GE (1985)
DC	pKa			
HL 6.60E-009	atm m3/mol	25	EST	MEYLAN,WM & HOWARD,PH (1991)
OH 2.91E-011	cm3/molc sec	25	EST	MEYLAN,WM & HOWARD,PH (1993)

CAS #: 074729-54-7				à-D-GLUCOPYRANOSE, 2-[[[(2-CHLOROETHYL)NITROSOAM

Formula: $C_{15}H_{22}ClN_3O_9$

Mol Weight: 423.81

MP (deg C):   FP (deg C):

BP (deg C):

BP pressure (mm Hg):

Property/Value	Units	Temp	Data Type	Reference
WS 1.62E+002	mg/L	25	EST	MEYLAN,WM ET AL. (1996)
logP 1.32			EXP	HANSCH,C ET AL. (1995)
VP 7.57E-010	mm Hg	25	EST	NEELY,WB & BLAU,GE (1985)
DC	pKa			
HL 2.77E-021	atm m3/mol	25	EST	MEYLAN,WM & HOWARD,PH (1991)
OH 6.26E-011	cm3/molc sec	25	EST	MEYLAN,WM & HOWARD,PH (1993)

CAS #: 074731-64-9				2,4-NH2PYRIMD,5(35-MEO-4CONME2)BENZYL

Formula: $C_{16}H_{21}N_5O_3$

Mol Weight: 331.38

MP (deg C):   FP (deg C):

BP (deg C):

BP pressure (mm Hg):

Property/Value	Units	Temp	Data Type	Reference
WS 5.93E+002	mg/L	25	EST	MEYLAN,WM ET AL. (1996)
logP -0.21			EXP	HANSCH,C & LEO,AJ (1985)
VP 8.81E-011	mm Hg	25	EST	NEELY,WB & BLAU,GE (1985)
DC	pKa			
HL 7.99E-019	atm m3/mol	25	EST	MEYLAN,WM & HOWARD,PH (1991)
OH 2.19E-010	cm3/molc sec	25	EST	MEYLAN,WM & HOWARD,PH (1993)

CAS #: 074731-66-1				2,4-NH2PYRIMD,5(35-MEO-4-IPR)BENZYL

Formula: $C_{16}H_{22}N_4O_2$

Mol Weight: 302.38

MP (deg C):   FP (deg C):

BP (deg C):

BP pressure (mm Hg):

Property/Value	Units	Temp	Data Type	Reference
WS 1.58E+001	mg/L	25	EST	MEYLAN,WM ET AL. (1996)
logP 3.37			EXP	HANSCH,C & LEO,AJ (1985)
VP 7.39E-009	mm Hg	25	EST	NEELY,WB & BLAU,GE (1985)
DC	pKa			
HL 7.84E-013	atm m3/mol	25	EST	MEYLAN,WM & HOWARD,PH (1991)
OH 2.05E-010	cm3/molc sec	25	EST	MEYLAN,WM & HOWARD,PH (1993)

CAS #: 074738-17-3				FENPICLONIL

Formula: $C_{11}H_6Cl_2N_2$

Mol Weight: 237.09

MP (deg C):   FP (deg C):

BP (deg C):

BP pressure (mm Hg):

Property/Value	Units	Temp	Data Type	Reference
WS 7.65E+000	mg/L	25	EST	MEYLAN,WM ET AL. (1996)
logP 3.86			EXP	TOMLIN,C (1994)
VP 1.71E-006	mm Hg	25	EST	NEELY,WB & BLAU,GE (1985)
DC	pKa			
HL 3.70E-009	atm m3/mol	25	EST	MEYLAN,WM & HOWARD,PH (1991)
OH 1.82E-011	cm3/molc sec	25	EST	MEYLAN,WM & HOWARD,PH (1993)

CAS #: 074746-42-2				ACETAMIDE, N-[4-[(DIMETHYLAMINO)SULFONYL]PHENYL]

Formula: $C_{11}H_{16}N_2O_3S$

Mol Weight: 256.33

MP (deg C):   FP (deg C):

BP (deg C):

BP pressure (mm Hg):

Property/Value	Units	Temp	Data Type	Reference
WS 6.80E+002	mg/L	25	EST	MEYLAN,WM ET AL. (1996)
logP 0.23			EXP	HANSCH,C ET AL. (1995)
VP 6.61E-007	mm Hg	25	EST	NEELY,WB & BLAU,GE (1985)
DC	pKa			
HL 4.78E-011	atm m3/mol	25	EST	MEYLAN,WM & HOWARD,PH (1991)
OH 5.84E-012	cm3/molc sec	25	EST	MEYLAN,WM & HOWARD,PH (1993)

CAS #: 074746-43-3				NN-DIME-4(N-ME-N(PROPIONYL)BZSO2N

Formula: $C_{12}H_{18}N_2O_3S$

Mol Weight: 270.35

MP (deg C):   FP (deg C):

BP (deg C):

BP pressure (mm Hg):

Property/Value	Units	Temp	Data Type	Reference
WS 4.50E+003	mg/L	25	EST	MEYLAN,WM ET AL. (1996)
logP 0.71			EXP	HANSCH,C & LEO,AJ (1985)
VP 2.83E-007	mm Hg	25	EST	NEELY,WB & BLAU,GE (1985)
DC	pKa			
HL 6.34E-011	atm m3/mol	25	EST	MEYLAN,WM & HOWARD,PH (1991)
OH 6.97E-012	cm3/molc sec	25	EST	MEYLAN,WM & HOWARD,PH (1993)

074746-44-4 — HEPTANAMIDE, N-[4-(DIMETHYLAMINO)SULFONYL]PHENYL

Formula: $C_{16}H_{26}N_2O_3S$
Mol Weight: 326.46
MP (deg C): **FP (deg C):**
BP (deg C):
BP pressure (mm Hg):

Property/Value	Units	Temp	Data Type	Reference
WS 5.06E+001	mg/L	25	EST	MEYLAN,WM ET AL. (1996)
logP 2.61			EXP	SANGSTER,J (1994)
VP 9.09E-009	mm Hg	25	EST	NEELY,WB & BLAU,GE (1985)
DC	pKa			
HL 1.97E-010	atm m3/mol	25	EST	MEYLAN,WM & HOWARD,PH (1991)
OH 1.52E-011	cm3/molc sec	25	EST	MEYLAN,WM & HOWARD,PH (1993)

074751-37-4 — BETA-D-GLUCOPYRANOSIDE, METHYL 3-[[[(2-CHLOROETH

Formula: $C_{16}H_{24}ClN_3O_{10}$
Mol Weight: 453.84
MP (deg C): **FP (deg C):**
BP (deg C):
BP pressure (mm Hg):

Property/Value	Units	Temp	Data Type	Reference
WS 3.77E+002	mg/L	25	EST	MEYLAN,WM ET AL. (1996)
logP 0.67			EXP	HANSCH,C ET AL. (1995)
VP 1.33E-010	mm Hg	25	EST	NEELY,WB & BLAU,GE (1985)
DC	pKa			
HL 1.63E-022	atm m3/mol	25	EST	MEYLAN,WM & HOWARD,PH (1991)
OH 5.38E-011	cm3/molc sec	25	EST	MEYLAN,WM & HOWARD,PH (1993)

074754-55-5 — 6-NITRATOHEXANOIC ACID

Formula: $C_6H_{11}NO_5$
Mol Weight: 177.16
MP (deg C): **FP (deg C):**
BP (deg C):
BP pressure (mm Hg):

Property/Value	Units	Temp	Data Type	Reference
WS 2.90E+003	mg/L	25	EST	MEYLAN,WM ET AL. (1996)
logP 1.98			EST	MEYLAN,WM & HOWARD,PH (1995)
VP 1.50E-005	mm Hg	30	EXP	HEISLER,SL & FRIEDLANDER,SK (1977)
DC	pKa			
HL 4.58E-009	atm m3/mol	25	EST	MEYLAN,WM & HOWARD,PH (1991)
OH 4.84E-012	cm3/molc sec	25	EST	MEYLAN,WM & HOWARD,PH (1993)

074772-77-3 — CIGLITAZONE

Formula: $C_{17}H_{21}NO_3S$
Mol Weight: 319.43
MP (deg C): **FP (deg C):**
BP (deg C):
BP pressure (mm Hg):

Property/Value	Units	Temp	Data Type	Reference
WS 3.62E+000	mg/L	25	EST	MEYLAN,WM ET AL. (1996)
logP 4.00			EXP	LIPINSKI,CA ET AL. (1991)
VP 1.71E-012	mm Hg	25	EST	NEELY,WB & BLAU,GE (1985)
DC	pKa			
HL 2.02E-008	atm m3/mol	25	EST	MEYLAN,WM & HOWARD,PH (1991)
OH 8.05E-011	cm3/molc sec	25	EST	MEYLAN,WM & HOWARD,PH (1993)

074782-23-3 — OXABETRINIL

Formula: $C_{12}H_{12}N_2O_3$
Mol Weight: 232.24
MP (deg C): 77.7 **FP (deg C):**
BP (deg C):
BP pressure (mm Hg):

Property/Value	Units	Temp	Data Type	Reference
WS 2.00E+001	mg/L	20	EXP	SHIU,WY ET AL. (1990)
logP 2.76			EXP	TOMLIN,C (1994)
VP 1.12E-005	mm Hg	25	EST	NEELY,WB & BLAU,GE (1985)
DC	pKa			
HL 1.26E-007	atm m3/mol	25	EST	MEYLAN,WM & HOWARD,PH (1991)
OH 2.27E-011	cm3/molc sec	25	EST	MEYLAN,WM & HOWARD,PH (1993)

074819-88-8 — NNN-ME-N(1ME2CYPENTENE)AMMON I

Formula: $C_{10}H_{20}IN$
Mol Weight: 281.18
MP (deg C): **FP (deg C):**
BP (deg C):
BP pressure (mm Hg):

Property/Value	Units	Temp	Data Type	Reference
WS 6.00E+005	mg/L	25	EST	MEYLAN,WM ET AL. (1996)
logP -1.85			EXP	PRATESI,P ET AL. (1986)
VP 2.92E-008	mm Hg	25	EST	NEELY,WB & BLAU,GE (1985)
DC	pKa			
HL 6.60E-012	atm m3/mol	25	EST	MEYLAN,WM & HOWARD,PH (1991)
OH 1.06E-010	cm3/molc sec	25	EST	MEYLAN,WM & HOWARD,PH (1993)

074974-49-5 — P-DIETHYLAMINOBENZYL ALCOHOL

Formula: $C_{11}H_{17}NO$
Mol Weight: 179.26
MP (deg C): **FP (deg C):**
BP (deg C):
BP pressure (mm Hg):

Property/Value	Units	Temp	Data Type	Reference
WS 6.15E+002	mg/L	25	EST	MEYLAN,WM ET AL. (1996)
logP 2.29			EXP	HANSCH,C & LEO,AJ (1985)
VP 9.42E-005	mm Hg	25	EST	NEELY,WB & BLAU,GE (1985)
DC	pKa			
HL 6.09E-009	atm m3/mol	25	EST	MEYLAN,WM & HOWARD,PH (1991)
OH 2.00E-010	cm3/molc sec	25	EST	MEYLAN,WM & HOWARD,PH (1993)

075032-34-7 — BENZENE, 1-[3-(METHYLTHIO)PROPOXY]-4-NITRO-

Formula: $C_{10}H_{13}NO_3S$
Mol Weight: 227.28
MP (deg C): **FP (deg C):**
BP (deg C):
BP pressure (mm Hg):

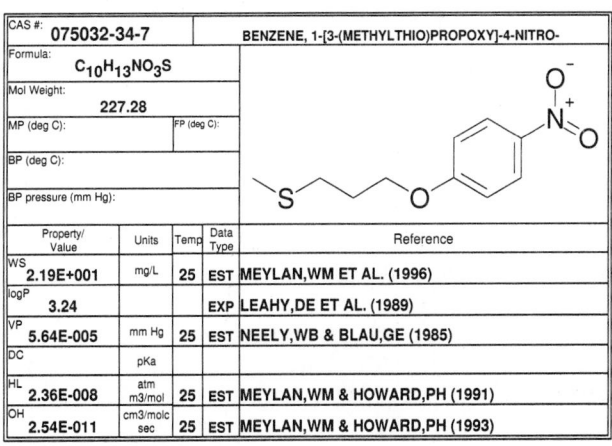

Property/Value	Units	Temp	Data Type	Reference
WS 2.19E+001	mg/L	25	EST	MEYLAN,WM ET AL. (1996)
logP 3.24			EXP	LEAHY,DE ET AL. (1989)
VP 5.64E-005	mm Hg	25	EST	NEELY,WB & BLAU,GE (1985)
DC	pKa			
HL 2.36E-008	atm m3/mol	25	EST	MEYLAN,WM & HOWARD,PH (1991)
OH 2.54E-011	cm3/molc sec	25	EST	MEYLAN,WM & HOWARD,PH (1993)

CAS #: 075221-43-1				PHENYLACETIC ACID,3-MEO-4-CYHXMEO

Formula: $C_{16}H_{22}O_4$

Mol Weight: 278.35

MP (deg C): FP (deg C):

BP (deg C):

BP pressure (mm Hg):

Property/ Value	Units	Temp	Data Type	Reference
WS 5.60E+001	mg/L	25	EST	MEYLAN,WM ET AL. (1996)
logP 3.35			EXP	HANSCH,C & LEO,AJ (1985)
VP 3.90E-007	mm Hg	25	EST	NEELY,WB & BLAU,GE (1985)
DC	pKa			
HL 3.73E-010	atm m3/mol	25	EST	MEYLAN,WM & HOWARD,PH (1991)
OH 5.72E-011	cm3/molc sec	25	EST	MEYLAN,WM & HOWARD,PH (1993)

CAS #: 075330-75-5				LOVASTATIN

Formula: $C_{24}H_{38}O_5$

Mol Weight: 406.57

MP (deg C): FP (deg C):

BP (deg C):

BP pressure (mm Hg):

Property/ Value	Units	Temp	Data Type	Reference
WS 2.14E+000	mg/L	25	EST	MEYLAN,WM ET AL. (1996)
logP 4.26			EXP	HANSCH,C ET AL. (1995)
VP 1.34E-012	mm Hg	25	EST	NEELY,WB & BLAU,GE (1985)
DC	pKa			
HL 2.12E-010	atm m3/mol	25	EST	MEYLAN,WM & HOWARD,PH (1991)
OH 2.31E-010	cm3/molc sec	25	EST	MEYLAN,WM & HOWARD,PH (1993)

CAS #: 075333-79-8				BENZENAMINE, 4-[(2,4-DINITROPHENYL)SULFONYL]-

Formula: $C_{12}H_9N_3O_6S$

Mol Weight: 323.29

MP (deg C): FP (deg C):

BP (deg C):

BP pressure (mm Hg):

Property/ Value	Units	Temp	Data Type	Reference
WS 9.79E+001	mg/L	25	EST	MEYLAN,WM ET AL. (1996)
logP 1.84			EXP	ALTOMARE,C ET AL. (1991)
VP 1.71E-010	mm Hg	25	EST	NEELY,WB & BLAU,GE (1985)
DC	pKa			
HL 1.37E-015	atm m3/mol	25	EST	MEYLAN,WM & HOWARD,PH (1991)
OH 2.30E-011	cm3/molc sec	25	EST	MEYLAN,WM & HOWARD,PH (1993)

CAS #: 075338-41-9				8-FLUORO-PEFLOXACIN

Formula: $C_{17}H_{19}F_2N_3O_3$

Mol Weight: 351.36

MP (deg C): FP (deg C):

BP (deg C):

BP pressure (mm Hg):

Property/ Value	Units	Temp	Data Type	Reference
WS 8.37E+003	mg/L	25	EST	MEYLAN,WM ET AL. (1996)
logP 0.30			EXP	SANGSTER,J (1994)
VP 1.11E-012	mm Hg	25	EST	NEELY,WB & BLAU,GE (1985)
DC	pKa			
HL 2.23E-018	atm m3/mol	25	EST	MEYLAN,WM & HOWARD,PH (1991)
OH 1.27E-010	cm3/molc sec	25	EST	MEYLAN,WM & HOWARD,PH (1993)

CAS #: 075349-23-4				1-METHYL-4-BENZOYLPIPERAZINE

Formula: $C_{12}H_{16}N_2O$

Mol Weight: 204.27

MP (deg C): FP (deg C):

BP (deg C):

BP pressure (mm Hg):

Property/ Value	Units	Temp	Data Type	Reference
WS 1.02E+003	mg/L	25	EST	MEYLAN,WM ET AL. (1996)
logP 0.35			EXP	HANSCH,C & LEO,AJ (1985)
VP 5.66E-005	mm Hg	25	EST	NEELY,WB & BLAU,GE (1985)
DC 6.78	pKa		EXP	PERRIN,DD (1965)
HL 3.10E-012	atm m3/mol	25	EST	MEYLAN,WM & HOWARD,PH (1991)
OH 1.15E-010	cm3/molc sec	25	EST	MEYLAN,WM & HOWARD,PH (1993)

CAS #: 075389-00-3				FURAZAN, BIS(PHENYLMETHYL)-, 2-OXIDE

Formula: $C_{16}H_{14}N_2O_2$

Mol Weight: 266.30

MP (deg C): FP (deg C):

BP (deg C):

BP pressure (mm Hg):

Property/ Value	Units	Temp	Data Type	Reference
WS 1.46E+001	mg/L	25	EST	MEYLAN,WM ET AL. (1996)
logP 3.65			EXP	CALVINO,R ET AL. (1992)
VP 6.77E-008	mm Hg	25	EST	NEELY,WB & BLAU,GE (1985)
DC	pKa			
HL 1.50E-011	atm m3/mol	25	EST	MEYLAN,WM & HOWARD,PH (1991)
OH 1.56E-011	cm3/molc sec	25	EST	MEYLAN,WM & HOWARD,PH (1993)

CAS #: 075410-15-0				2,4(1H,3H)-PYRIMIDINEDIONE, 3-ACETYL-5-FLUORO-

Formula: $C_6H_5FN_2O_3$

Mol Weight: 172.12

MP (deg C): FP (deg C):

BP (deg C):

BP pressure (mm Hg):

Property/ Value	Units	Temp	Data Type	Reference
WS 5.74E+003	mg/L	25	EST	MEYLAN,WM ET AL. (1996)
logP -0.34			EXP	SANGSTER,J (1994)
VP 1.08E-007	mm Hg	25	EST	NEELY,WB & BLAU,GE (1985)
DC	pKa			
HL 5.38E-013	atm m3/mol	25	EST	MEYLAN,WM & HOWARD,PH (1991)
OH 5.93E-012	cm3/molc sec	25	EST	MEYLAN,WM & HOWARD,PH (1993)

CAS #: 075455-41-3				4-HYDROXY-3-CHLORO-1-BUTENE

Formula: C_4H_7ClO

Mol Weight: 106.55

MP (deg C): FP (deg C):

BP (deg C): 66-67

BP pressure (mm Hg): 3.00E+001

Property/ Value	Units	Temp	Data Type	Reference
WS 6.34E+004	mg/L	25	EST	MEYLAN,WM ET AL. (1996)
logP 0.89			EST	MEYLAN,WM & HOWARD,PH (1995)
VP 1.59E+000	mm Hg	25	EST	NEELY,WB & BLAU,GE (1985)
DC	pKa			
HL 1.65E-006	atm m3/mol	25	EST	MEYLAN,WM & HOWARD,PH (1991)
OH 2.22E-011	cm3/molc sec	25	EST	MEYLAN,WM & HOWARD,PH (1993)

CAS #:	075523-35-2	2-PYRIDINEMETHANAMINIUM, 6-FLUORO-N,N,N-TRIMETHY

Formula: $C_9H_{14}BrFN_2$

Mol Weight: 249.13

MP (deg C): FP (deg C):

BP (deg C):

BP pressure (mm Hg):

Property/Value	Units	Temp	Data Type	Reference
WS 1.00E+006	mg/L	25	EST	MEYLAN,WM ET AL. (1996)
logP -3.20			EXP	SANGSTER,J (1993)
VP 3.05E-008	mm Hg	25	EST	NEELY,WB & BLAU,GE (1985)
DC	pKa			
HL 4.09E-016	atm m3/mol	25	EST	MEYLAN,WM & HOWARD,PH (1991)
OH 1.36E-011	cm3/molc sec	25	EST	MEYLAN,WM & HOWARD,PH (1993)

CAS #:	075523-36-3	2-PYRIDINEMETHANAMINIUM, 6-CHLORO-N,N,N-TRIMETHY

Formula: $C_9H_{14}BrClN_2$

Mol Weight: 265.58

MP (deg C): FP (deg C):

BP (deg C):

BP pressure (mm Hg):

Property/Value	Units	Temp	Data Type	Reference
WS 1.00E+006	mg/L	25	EST	MEYLAN,WM ET AL. (1996)
logP -2.60			EXP	SANGSTER,J (1993)
VP 6.92E-009	mm Hg	25	EST	NEELY,WB & BLAU,GE (1985)
DC	pKa			
HL 1.64E-014	atm m3/mol	25	EST	MEYLAN,WM & HOWARD,PH (1991)
OH 1.31E-011	cm3/molc sec	25	EST	MEYLAN,WM & HOWARD,PH (1993)

CAS #:	075523-37-4	2-PYRIDINEMETHANAMINIUM, 6-BROMO-N,N,N-TRIMETHYL

Formula: $C_9H_{14}Br_2N_2$

Mol Weight: 310.03

MP (deg C): FP (deg C):

BP (deg C):

BP pressure (mm Hg):

Property/Value	Units	Temp	Data Type	Reference
WS 1.00E+006	mg/L	25	EST	MEYLAN,WM ET AL. (1996)
logP -2.49			EXP	SANGSTER,J (1993)
VP 2.99E-009	mm Hg	25	EST	NEELY,WB & BLAU,GE (1985)
DC	pKa			
HL 1.40E-016	atm m3/mol	25	EST	MEYLAN,WM & HOWARD,PH (1991)
OH 1.31E-011	cm3/molc sec	25	EST	MEYLAN,WM & HOWARD,PH (1993)

CAS #:	075523-38-5	4-PYRIDYL-ME(6NN-TETRA-ME)AMMONIUM BR

Formula: $C_{10}H_{17}BrN_2$

Mol Weight: 245.16

MP (deg C): FP (deg C):

BP (deg C):

BP pressure (mm Hg):

Property/Value	Units	Temp	Data Type	Reference
WS 1.00E+006	mg/L	25	EST	MEYLAN,WM ET AL. (1996)
logP -3.20			EXP	HANSCH,C & LEO,AJ (1985)
VP 1.13E-008	mm Hg	25	EST	NEELY,WB & BLAU,GE (1985)
DC	pKa			
HL 3.87E-016	atm m3/mol	25	EST	MEYLAN,WM & HOWARD,PH (1991)
OH 1.38E-011	cm3/molc sec	25	EST	MEYLAN,WM & HOWARD,PH (1993)

CAS #:	075523-39-6	2-FURANMETHANAMINIUM, 5-BROMO-N,N,N-TRIMETHYL-,

Formula: $C_8H_{13}ClINO$

Mol Weight: 301.56

MP (deg C): FP (deg C):

BP (deg C):

BP pressure (mm Hg):

Property/Value	Units	Temp	Data Type	Reference
WS 3.54E+005	mg/L	25	EST	MEYLAN,WM ET AL. (1996)
logP -1.80			EXP	SANGSTER,J (1994)
VP 1.26E-008	mm Hg	25	EST	NEELY,WB & BLAU,GE (1985)
DC	pKa			
HL 1.57E-013	atm m3/mol	25	EST	MEYLAN,WM & HOWARD,PH (1991)
OH 4.72E-011	cm3/molc sec	25	EST	MEYLAN,WM & HOWARD,PH (1993)

CAS #:	075523-40-9	2-FURANMETHANAMINIUM, 5-CHLORO-N,N,N-TRIMETHYL-,

Formula: $C_8H_{13}ClINO$

Mol Weight: 301.56

MP (deg C): FP (deg C):

BP (deg C):

BP pressure (mm Hg):

Property/Value	Units	Temp	Data Type	Reference
WS 3.54E+005	mg/L	25	EST	MEYLAN,WM ET AL. (1996)
logP -1.72			EXP	SANGSTER,J (1994)
VP 1.26E-008	mm Hg	25	EST	NEELY,WB & BLAU,GE (1985)
DC	pKa			
HL 1.57E-013	atm m3/mol	25	EST	MEYLAN,WM & HOWARD,PH (1991)
OH 4.72E-011	cm3/molc sec	25	EST	MEYLAN,WM & HOWARD,PH (1993)

CAS #:	075523-41-0	2-FURANMETHANAMINIUM, 5-ETHYL-N,N,N-TRIMETHYL-,

Formula: $C_{10}H_{18}INO$

Mol Weight: 295.17

MP (deg C): FP (deg C):

BP (deg C):

BP pressure (mm Hg):

Property/Value	Units	Temp	Data Type	Reference
WS 1.98E+005	mg/L	25	EST	MEYLAN,WM ET AL. (1996)
logP -1.38			EXP	PRATESI,P ET AL. (1986)
VP 9.39E-009	mm Hg	25	EST	NEELY,WB & BLAU,GE (1985)
DC	pKa			
HL 3.11E-013	atm m3/mol	25	EST	MEYLAN,WM & HOWARD,PH (1991)
OH 1.36E-010	cm3/molc sec	25	EST	MEYLAN,WM & HOWARD,PH (1993)

CAS #:	075584-51-9	2H-BENZOTRIAZOLE-2-ACETIC ACID, à-METHYL-, ETHYL

Formula: $C_{11}H_{13}N_3O_2$

Mol Weight: 219.25

MP (deg C): FP (deg C):

BP (deg C):

BP pressure (mm Hg):

Property/Value	Units	Temp	Data Type	Reference
WS 3.22E+002	mg/L	25	EST	MEYLAN,WM ET AL. (1996)
logP 2.38			EXP	SPARATORE,F ET AL. (1988)
VP 3.34E-005	mm Hg	25	EST	NEELY,WB & BLAU,GE (1985)
DC	pKa			
HL 1.31E-008	atm m3/mol	25	EST	MEYLAN,WM & HOWARD,PH (1991)
OH 4.26E-012	cm3/molc sec	25	EST	MEYLAN,WM & HOWARD,PH (1993)

CAS #: 075584-60-0	1H-BENZOTRIAZOLE-1-ACETIC ACID, à-METHYL-, ETHYL

Formula: $C_{11}H_{13}N_3O_2$

Mol Weight: 219.25

MP (deg C):   FP (deg C):

BP (deg C):

BP pressure (mm Hg):

Property/ Value	Units	Temp	Data Type	Reference
WS 8.79E+002	mg/L	25	EST	MEYLAN,WM ET AL. (1996)
logP 1.87			EXP	SPARATORE,F ET AL. (1988)
VP 3.34E-005	mm Hg	25	EST	NEELY,WB & BLAU,GE (1985)
DC	pKa			
HL 1.31E-008	atm m3/mol	25	EST	MEYLAN,WM & HOWARD,PH (1991)
OH 4.26E-012	cm3/molc sec	25	EST	MEYLAN,WM & HOWARD,PH (1993)

CAS #: 075680-92-1	3-CL-6-ET PYRIDAZINECARBOXYLATE

Formula: $C_7H_7ClN_2O_2$

Mol Weight: 186.60

MP (deg C):   FP (deg C):

BP (deg C):

BP pressure (mm Hg):

Property/ Value	Units	Temp	Data Type	Reference
WS 4.04E+004	mg/L	25	EST	MEYLAN,WM ET AL. (1996)
logP 0.12			EXP	HANSCH,C & LEO,AJ (1985)
VP 2.44E-004	mm Hg	25	EST	NEELY,WB & BLAU,GE (1985)
DC	pKa			
HL 1.15E-006	atm m3/mol	25	EST	MEYLAN,WM & HOWARD,PH (1991)
OH 1.71E-012	cm3/molc sec	25	EST	MEYLAN,WM & HOWARD,PH (1993)

CAS #: 075736-33-3	DICLOBUTRAZOL ISOMERS

Formula: $C_{15}H_{21}Cl_2N_3O$

Mol Weight: 330.26

MP (deg C): 147-149   FP (deg C):

BP (deg C):

BP pressure (mm Hg):

Property/ Value	Units	Temp	Data Type	Reference
WS 9.00E+000	mg/L	25	EXP	SHIU,WY ET AL. (1990)
logP 3.80			EXP	BAKER,EA ET AL. (1992)
VP 4.51E-009	mm Hg	25	EST	NEELY,WB & BLAU,GE (1985)
DC	pKa			
HL 2.89E-010	atm m3/mol	25	EST	MEYLAN,WM & HOWARD,PH (1991)
OH 1.63E-011	cm3/molc sec	25	EST	MEYLAN,WM & HOWARD,PH (1993)

CAS #: 075773-74-9	2-AMINO-4-NITROSOPHENOL

Formula: $C_6H_6N_2O_2$

Mol Weight: 138.13

MP (deg C):   FP (deg C):

BP (deg C):

BP pressure (mm Hg):

Property/ Value	Units	Temp	Data Type	Reference
WS 8.04E+004	mg/L	25	EST	MEYLAN,WM ET AL. (1996)
logP 0.03			EXP	HANSCH,C & LEO,AJ (1985)
VP 1.15E-003	mm Hg	25	EST	NEELY,WB & BLAU,GE (1985)
DC	pKa			
HL 2.07E-012	atm m3/mol	25	EST	MEYLAN,WM & HOWARD,PH (1991)
OH 2.00E-010	cm3/molc sec	25	EST	MEYLAN,WM & HOWARD,PH (1993)

CAS #: 075808-93-4	THIAZOLIDINE,2-(M-NITROPHENYL)

Formula: $C_9H_{10}N_2O_2S$

Mol Weight: 210.26

MP (deg C):   FP (deg C):

BP (deg C):

BP pressure (mm Hg):

Property/ Value	Units	Temp	Data Type	Reference
WS 7.05E+003	mg/L	25	EST	MEYLAN,WM ET AL. (1996)
logP 1.59			EXP	HANSCH,C & LEO,AJ (1985)
VP 1.77E-005	mm Hg	25	EST	NEELY,WB & BLAU,GE (1985)
DC	pKa			
HL 1.17E-008	atm m3/mol	25	EST	MEYLAN,WM & HOWARD,PH (1991)
OH 1.93E-010	cm3/molc sec	25	EST	MEYLAN,WM & HOWARD,PH (1993)

CAS #: 075841-82-6	MOPIDRALAZINE

Formula: $C_{14}H_{19}N_5O$

Mol Weight: 273.34

MP (deg C):   FP (deg C):

BP (deg C):

BP pressure (mm Hg):

Property/ Value	Units	Temp	Data Type	Reference
WS 3.86E+002	mg/L	25	EST	MEYLAN,WM ET AL. (1996)
logP 1.94			EXP	HANSCH,C ET AL. (1995)
VP 2.06E-008	mm Hg	25	EST	NEELY,WB & BLAU,GE (1985)
DC	pKa			
HL 4.80E-016	atm m3/mol	25	EST	MEYLAN,WM & HOWARD,PH (1991)
OH 2.75E-010	cm3/molc sec	25	EST	MEYLAN,WM & HOWARD,PH (1993)

CAS #: 075841-91-7	3-PYRIDAZINAMINE, N-(2,5-DIMETHYL-1H-PYRROL-1-YL

Formula: $C_{15}H_{22}N_6$

Mol Weight: 286.38

MP (deg C):   FP (deg C):

BP (deg C):

BP pressure (mm Hg):

Property/ Value	Units	Temp	Data Type	Reference
WS 3.31E+003	mg/L	25	EST	MEYLAN,WM ET AL. (1996)
logP 1.94			EXP	SANGSTER,J (1993)
VP 9.96E-009	mm Hg	25	EST	NEELY,WB & BLAU,GE (1985)
DC	pKa			
HL 2.04E-017	atm m3/mol	25	EST	MEYLAN,WM & HOWARD,PH (1991)
OH 3.10E-010	cm3/molc sec	25	EST	MEYLAN,WM & HOWARD,PH (1993)

CAS #: 075841-95-1	2-PROPANOL, 1-[[6-[(2,5-DIMETHYL-1H-PYRROL-1-YL)

Formula: $C_{15}H_{23}N_5O_2$

Mol Weight: 305.38

MP (deg C):   FP (deg C):

BP (deg C):

BP pressure (mm Hg):

Property/ Value	Units	Temp	Data Type	Reference
WS 8.86E+002	mg/L	25	EST	MEYLAN,WM ET AL. (1996)
logP 1.30			EXP	SANGSTER,J (1993)
VP 5.63E-013	mm Hg	25	EST	NEELY,WB & BLAU,GE (1985)
DC	pKa			
HL 2.19E-019	atm m3/mol	25	EST	MEYLAN,WM & HOWARD,PH (1991)
OH 2.34E-010	cm3/molc sec	25	EST	MEYLAN,WM & HOWARD,PH (1993)

CAS #: 075841-97-3 — 2-PROPANOL, 1,1'-[[6-[(2,5-DIMETHYL-1H-PYRROL-1-

Formula: $C_{16}H_{25}N_5O_2$

Mol Weight: 319.41

MP (deg C): FP (deg C):

BP (deg C):

BP pressure (mm Hg):

Property/Value	Units	Temp	Data Type	Reference
WS 4.48E+002	mg/L	25	EST	MEYLAN,WM ET AL. (1996)
logP 1.55			EXP	HANSCH,C ET AL. (1995)
VP 5.23E-013	mm Hg	25	EST	NEELY,WB & BLAU,GE (1985)
DC	pKa			
HL 2.91E-019	atm m3/mol	25	EST	MEYLAN,WM & HOWARD,PH (1991)
OH 2.39E-010	cm3/molc sec	25	EST	MEYLAN,WM & HOWARD,PH (1993)

CAS #: 075842-02-3 — 3,6-PYRIDAZINEDIAMINE, N'-(2,5-DIMETHYL-1H-PYRRO

Formula: $C_{16}H_{25}N_5O_2$

Mol Weight: 319.41

MP (deg C): FP (deg C):

BP (deg C):

BP pressure (mm Hg):

Property/Value	Units	Temp	Data Type	Reference
WS 1.64E+002	mg/L	25	EST	MEYLAN,WM ET AL. (1996)
logP 2.06			EXP	HANSCH,C ET AL. (1995)
VP 4.51E-009	mm Hg	25	EST	NEELY,WB & BLAU,GE (1985)
DC	pKa			
HL 1.69E-017	atm m3/mol	25	EST	MEYLAN,WM & HOWARD,PH (1991)
OH 2.77E-010	cm3/molc sec	25	EST	MEYLAN,WM & HOWARD,PH (1993)

CAS #: 075842-06-7 — 3,6-PYRIDAZINEDIAMINE, N'-(2,5-DIMETHYL-1H-PYRRO

Formula: $C_{12}H_{17}N_5$

Mol Weight: 231.30

MP (deg C): FP (deg C):

BP (deg C):

BP pressure (mm Hg):

Property/Value	Units	Temp	Data Type	Reference
WS 4.12E+002	mg/L	25	EST	MEYLAN,WM ET AL. (1996)
logP 2.18			EXP	SANGSTER,J (1993)
VP 6.95E-007	mm Hg	25	EST	NEELY,WB & BLAU,GE (1985)
DC	pKa			
HL 7.01E-014	atm m3/mol	25	EST	MEYLAN,WM & HOWARD,PH (1991)
OH 2.03E-010	cm3/molc sec	25	EST	MEYLAN,WM & HOWARD,PH (1993)

CAS #: 075851-96-6 — NN-DIME4(N-ME-N(2MESOPROPIONYL)BZSO2N

Formula: $C_{13}H_{20}N_2O_4S_2$

Mol Weight: 332.44

MP (deg C): FP (deg C):

BP (deg C):

BP pressure (mm Hg):

Property/Value	Units	Temp	Data Type	Reference
WS 1.29E+002	mg/L	25	EST	MEYLAN,WM ET AL. (1996)
logP -0.67			EXP	HANSCH,C & LEO,AJ (1985)
VP 8.65E-010	mm Hg	25	EST	NEELY,WB & BLAU,GE (1985)
DC	pKa			
HL 4.27E-017	atm m3/mol	25	EST	MEYLAN,WM & HOWARD,PH (1991)
OH 7.87E-011	cm3/molc sec	25	EST	MEYLAN,WM & HOWARD,PH (1993)

CAS #: 075851-97-7 — BUTANEDIAMIDE, N-[4-[(DIMETHYLAMINO)SULFONYL]PHE

Formula: $C_{13}H_{19}N_3O_4S$

Mol Weight: 313.38

MP (deg C): FP (deg C):

BP (deg C):

BP pressure (mm Hg):

Property/Value	Units	Temp	Data Type	Reference
WS 1.51E+003	mg/L	25	EST	MEYLAN,WM ET AL. (1996)
logP -0.56			EXP	HANSCH,C ET AL. (1995)
VP 8.41E-011	mm Hg	25	EST	NEELY,WB & BLAU,GE (1985)
DC	pKa			
HL 1.53E-014	atm m3/mol	25	EST	MEYLAN,WM & HOWARD,PH (1991)
OH 1.32E-011	cm3/molc sec	25	EST	MEYLAN,WM & HOWARD,PH (1993)

CAS #: 075851-98-8 — PROPANAMIDE, 3-(ACETYLAMINO)-N-[4-[(DIMETHYLAMIN

Formula: $C_{14}H_{21}N_3O_4S$

Mol Weight: 327.41

MP (deg C): FP (deg C):

BP (deg C):

BP pressure (mm Hg):

Property/Value	Units	Temp	Data Type	Reference
WS 8.25E+002	mg/L	25	EST	MEYLAN,WM ET AL. (1996)
logP -0.35			EXP	HANSCH,C ET AL. (1995)
VP 4.75E-011	mm Hg	25	EST	NEELY,WB & BLAU,GE (1985)
DC	pKa			
HL 3.35E-014	atm m3/mol	25	EST	MEYLAN,WM & HOWARD,PH (1991)
OH 2.71E-011	cm3/molc sec	25	EST	MEYLAN,WM & HOWARD,PH (1993)

CAS #: 075851-99-9 — PROPANAMIDE, N-[4-[(DIMETHYLAMINO)SULFONYL]PHENY

Formula: $C_{12}H_{18}N_2O_4S$

Mol Weight: 286.35

MP (deg C): FP (deg C):

BP (deg C):

BP pressure (mm Hg):

Property/Value	Units	Temp	Data Type	Reference
WS 1.20E+003	mg/L	25	EST	MEYLAN,WM ET AL. (1996)
logP -0.26			EXP	HANSCH,C ET AL. (1995)
VP 1.29E-010	mm Hg	25	EST	NEELY,WB & BLAU,GE (1985)
DC	pKa			
HL 2.32E-015	atm m3/mol	25	EST	MEYLAN,WM & HOWARD,PH (1991)
OH 1.95E-011	cm3/molc sec	25	EST	MEYLAN,WM & HOWARD,PH (1993)

CAS #: 075852-01-6 — PROPANAMIDE, 3-CYANO-N-[4-[(DIMETHYLAMINO)SULFON

Formula: $C_{13}H_{17}N_3O_3S$

Mol Weight: 295.36

MP (deg C): FP (deg C):

BP (deg C):

BP pressure (mm Hg):

Property/Value	Units	Temp	Data Type	Reference
WS 4.86E+002	mg/L	25	EST	MEYLAN,WM ET AL. (1996)
logP -0.17			EXP	HANSCH,C ET AL. (1995)
VP 3.65E-009	mm Hg	25	EST	NEELY,WB & BLAU,GE (1985)
DC	pKa			
HL 2.63E-014	atm m3/mol	25	EST	MEYLAN,WM & HOWARD,PH (1991)
OH 6.52E-012	cm3/molc sec	25	EST	MEYLAN,WM & HOWARD,PH (1993)

CAS #: 075852-03-8	PROPANAMIDE, 3-(ACETYLOXY)-N-[4-[(DIMETHYLAMINO)

Formula: $C_{14}H_{20}N_2O_5S$

Mol Weight: 328.39

MP (deg C):   FP (deg C):

BP (deg C):

BP pressure (mm Hg):

Property/ Value	Units	Temp	Data Type	Reference
WS 2.45E+002	mg/L	25	EST	MEYLAN,WM ET AL. (1996)
logP 0.26			EXP	HANSCH,C ET AL. (1995)
VP 6.75E-009	mm Hg	25	EST	NEELY,WB & BLAU,GE (1985)
DC	pKa			
HL 1.51E-013	atm m3/mol	25	EST	MEYLAN,WM & HOWARD,PH (1991)
OH 1.25E-011	cm3/molc sec	25	EST	MEYLAN,WM & HOWARD,PH (1993)

CAS #: 075852-04-9	NN-DIME-4(N-ME-N(2NO2PROPIONYL)BZSO2N

Formula: $C_{12}H_{17}N_3O_5S$

Mol Weight: 315.35

MP (deg C):   FP (deg C):

BP (deg C):

BP pressure (mm Hg):

Property/ Value	Units	Temp	Data Type	Reference
WS 1.10E+002	mg/L	25	EST	MEYLAN,WM ET AL. (1996)
logP 0.30			EXP	HANSCH,C & LEO,AJ (1985)
VP 7.08E-009	mm Hg	25	EST	NEELY,WB & BLAU,GE (1985)
DC	pKa			
HL 3.63E-014	atm m3/mol	25	EST	MEYLAN,WM & HOWARD,PH (1991)
OH 6.05E-012	cm3/molc sec	25	EST	MEYLAN,WM & HOWARD,PH (1993)

CAS #: 075852-05-0	NN-DIME-4(N-ME-N(2MEOPROPIONYL)BZSO2N

Formula: $C_{13}H_{20}N_2O_4S$

Mol Weight: 300.38

MP (deg C):   FP (deg C):

BP (deg C):

BP pressure (mm Hg):

Property/ Value	Units	Temp	Data Type	Reference
WS 3.13E+002	mg/L	25	EST	MEYLAN,WM ET AL. (1996)
logP 0.33			EXP	HANSCH,C & LEO,AJ (1985)
VP 8.21E-008	mm Hg	25	EST	NEELY,WB & BLAU,GE (1985)
DC	pKa			
HL 5.26E-012	atm m3/mol	25	EST	MEYLAN,WM & HOWARD,PH (1991)
OH 1.60E-011	cm3/molc sec	25	EST	MEYLAN,WM & HOWARD,PH (1993)

CAS #: 075852-08-3	PROPANAMIDE, N-[4-[(DIMETHYLAMINO)SULFONYL]PHENY

Formula: $C_{13}H_{17}N_3O_3S_2$

Mol Weight: 327.43

MP (deg C):   FP (deg C):

BP (deg C):

BP pressure (mm Hg):

Property/ Value	Units	Temp	Data Type	Reference
WS 1.62E+003	mg/L	25	EST	MEYLAN,WM ET AL. (1996)
logP 0.84			EXP	HANSCH,C ET AL. (1995)
VP 3.50E-010	mm Hg	25	EST	NEELY,WB & BLAU,GE (1985)
DC	pKa			
HL 3.78E-014	atm m3/mol	25	EST	MEYLAN,WM & HOWARD,PH (1991)
OH 3.50E-011	cm3/molc sec	25	EST	MEYLAN,WM & HOWARD,PH (1993)

CAS #: 075852-09-4	BUTANOIC ACID, 4-[[4-[(DIMETHYLAMINO)SULFONYL]PH

Formula: $C_{15}H_{22}N_2O_5S$

Mol Weight: 342.42

MP (deg C):   FP (deg C):

BP (deg C):

BP pressure (mm Hg):

Property/ Value	Units	Temp	Data Type	Reference
WS 1.32E+003	mg/L	25	EST	MEYLAN,WM ET AL. (1996)
logP 0.84			EXP	HANSCH,C ET AL. (1995)
VP 3.00E-009	mm Hg	25	EST	NEELY,WB & BLAU,GE (1985)
DC	pKa			
HL 2.00E-013	atm m3/mol	25	EST	MEYLAN,WM & HOWARD,PH (1991)
OH 1.10E-011	cm3/molc sec	25	EST	MEYLAN,WM & HOWARD,PH (1993)

CAS #: 075852-11-8	NN-DIME-4(N-ME-N(2CLPROPIONYL)BZSO2N

Formula: $C_{12}H_{17}ClN_2O_3S$

Mol Weight: 304.80

MP (deg C):   FP (deg C):

BP (deg C):

BP pressure (mm Hg):

Property/ Value	Units	Temp	Data Type	Reference
WS 1.68E+003	mg/L	25	EST	MEYLAN,WM ET AL. (1996)
logP 0.98			EXP	HANSCH,C & LEO,AJ (1985)
VP 7.77E-008	mm Hg	25	EST	NEELY,WB & BLAU,GE (1985)
DC	pKa			
HL 2.23E-011	atm m3/mol	25	EST	MEYLAN,WM & HOWARD,PH (1991)
OH 6.72E-012	cm3/molc sec	25	EST	MEYLAN,WM & HOWARD,PH (1993)

CAS #: 075852-12-9	NN-DIME-4(N-ME-N(2MESPROPIONYL)BZSO2N

Formula: $C_{13}H_{20}N_2O_3S_2$

Mol Weight: 316.44

MP (deg C):   FP (deg C):

BP (deg C):

BP pressure (mm Hg):

Property/ Value	Units	Temp	Data Type	Reference
WS 1.40E+003	mg/L	25	EST	MEYLAN,WM ET AL. (1996)
logP 0.99			EXP	HANSCH,C & LEO,AJ (1985)
VP 1.78E-008	mm Hg	25	EST	NEELY,WB & BLAU,GE (1985)
DC	pKa			
HL 6.77E-013	atm m3/mol	25	EST	MEYLAN,WM & HOWARD,PH (1991)
OH 2.04E-011	cm3/molc sec	25	EST	MEYLAN,WM & HOWARD,PH (1993)

CAS #: 075852-13-0	NN-DIME-4(N-ME-N(2BRPROPIONYL)BZSO2N

Formula: $C_{12}H_{17}BrN_2O_3S$

Mol Weight: 349.25

MP (deg C):   FP (deg C):

BP (deg C):

BP pressure (mm Hg):

Property/ Value	Units	Temp	Data Type	Reference
WS 6.79E+002	mg/L	25	EST	MEYLAN,WM ET AL. (1996)
logP 1.13			EXP	HANSCH,C & LEO,AJ (1985)
VP 2.73E-008	mm Hg	25	EST	NEELY,WB & BLAU,GE (1985)
DC	pKa			
HL 7.31E-012	atm m3/mol	25	EST	MEYLAN,WM & HOWARD,PH (1991)
OH 6.57E-012	cm3/molc sec	25	EST	MEYLAN,WM & HOWARD,PH (1993)

CAS #:	075852-14-1	NN-DIME-4(N-ME-N-BUTROYL)BZ SULFONAMIDE

Formula: $C_{13}H_{20}N_2O_3S$

Mol Weight: 284.38

MP (deg C):   FP (deg C):

BP (deg C):

BP pressure (mm Hg):

Property/Value	Units	Temp	Data Type	Reference
WS 1.46E+003	mg/L	25	EST	MEYLAN,WM ET AL. (1996)
logP 1.19			EXP	HANSCH,C & LEO,AJ (1985)
VP 1.21E-007	mm Hg	25	EST	NEELY,WB & BLAU,GE (1985)
DC	pKa			
HL 8.42E-011	atm m3/mol	25	EST	MEYLAN,WM & HOWARD,PH (1991)
OH 1.04E-011	cm3/molc sec	25	EST	MEYLAN,WM & HOWARD,PH (1993)

CAS #:	075852-15-2	PROPANAMIDE, N-[4-(DIMETHYLAMINO)SULFONYL]PHENYL

Formula: $C_{15}H_{24}N_2O_4S$

Mol Weight: 328.43

MP (deg C):   FP (deg C):

BP (deg C):

BP pressure (mm Hg):

Property/Value	Units	Temp	Data Type	Reference
WS 7.01E+002	mg/L	25	EST	MEYLAN,WM ET AL. (1996)
logP 1.26			EXP	HANSCH,C ET AL. (1995)
VP 8.31E-009	mm Hg	25	EST	NEELY,WB & BLAU,GE (1985)
DC				
HL 1.31E-012	atm m3/mol	25	EST	MEYLAN,WM & HOWARD,PH (1991)
OH 4.09E-011	cm3/molc sec	25	EST	MEYLAN,WM & HOWARD,PH (1993)

CAS #:	075852-16-3	NN-DIME-4(N-ME-N-(2I PROPIONYL)BZSO2N

Formula: $C_{12}H_{17}IN_2O_3S$

Mol Weight: 396.25

MP (deg C):   FP (deg C):

BP (deg C):

BP pressure (mm Hg):

Property/Value	Units	Temp	Data Type	Reference
WS 2.10E+002	mg/L	25	EST	MEYLAN,WM ET AL. (1996)
logP 1.39			EXP	HANSCH,C & LEO,AJ (1985)
VP 8.48E-009	mm Hg	25	EST	NEELY,WB & BLAU,GE (1985)
DC	pKa			
HL 4.73E-012	atm m3/mol	25	EST	MEYLAN,WM & HOWARD,PH (1991)
OH 6.93E-012	cm3/molc sec	25	EST	MEYLAN,WM & HOWARD,PH (1993)

CAS #:	075852-17-4	PENTANAMIDE, N-[[4-(DIMETHYLAMINO)SULFONYL]PHENY

Formula: $C_{14}H_{22}N_2O_3S$

Mol Weight: 298.41

MP (deg C):   FP (deg C):

BP (deg C):

BP pressure (mm Hg):

Property/Value	Units	Temp	Data Type	Reference
WS 5.09E+002	mg/L	25	EST	MEYLAN,WM ET AL. (1996)
logP 1.63			EXP	HANSCH,C ET AL. (1995)
VP 5.12E-008	mm Hg	25	EST	NEELY,WB & BLAU,GE (1985)
DC	pKa			
HL 1.12E-010	atm m3/mol	25	EST	MEYLAN,WM & HOWARD,PH (1991)
OH 1.24E-011	cm3/molc sec	25	EST	MEYLAN,WM & HOWARD,PH (1993)

CAS #:	075852-19-6	HEXANAMIDE, N-[4-(DIMETHYLAMINO)SULFONYL]PHENYL]

Formula: $C_{15}H_{24}N_2O_3S$

Mol Weight: 312.43

MP (deg C):   FP (deg C):

BP (deg C):

BP pressure (mm Hg):

Property/Value	Units	Temp	Data Type	Reference
WS 1.49E+002	mg/L	25	EST	MEYLAN,WM ET AL. (1996)
logP 2.16			EXP	HANSCH,C ET AL. (1995)
VP 2.16E-008	mm Hg	25	EST	NEELY,WB & BLAU,GE (1985)
DC	pKa			
HL 1.48E-010	atm m3/mol	25	EST	MEYLAN,WM & HOWARD,PH (1991)
OH 1.38E-011	cm3/molc sec	25	EST	MEYLAN,WM & HOWARD,PH (1993)

CAS #:	075881-16-2	2-ME-N-NITROSOMORPHOLINE

Formula: $C_5H_{10}N_2O_2$

Mol Weight: 130.15

MP (deg C):   FP (deg C):

BP (deg C):

BP pressure (mm Hg):

Property/Value	Units	Temp	Data Type	Reference
WS 9.74E+004	mg/L	25	EST	MEYLAN,WM ET AL. (1996)
logP -0.03			EXP	HANSCH,C & LEO,AJ (1985)
VP 4.98E-002	mm Hg	25	EST	NEELY,WB & BLAU,GE (1985)
DC	pKa			
HL 1.87E-008	atm m3/mol	25	EST	MEYLAN,WM & HOWARD,PH (1991)
OH 1.08E-010	cm3/molc sec	25	EST	MEYLAN,WM & HOWARD,PH (1993)

CAS #:	075883-07-7	BETAMETHASONE-21-PROPIONATE

Formula: $C_{25}H_{33}FO_6$

Mol Weight: 448.54

MP (deg C):   FP (deg C):

BP (deg C):

BP pressure (mm Hg):

Property/Value	Units	Temp	Data Type	Reference
WS 3.71E+000	mg/L	25	EST	MEYLAN,WM ET AL. (1996)
logP 3.06			EXP	HANSCH,C & LEO,AJ (1985)
VP 6.21E-014	mm Hg	25	EST	NEELY,WB & BLAU,GE (1985)
DC	pKa			
HL 1.03E-011	atm m3/mol	25	EST	MEYLAN,WM & HOWARD,PH (1991)
OH 6.81E-011	cm3/molc sec	25	EST	MEYLAN,WM & HOWARD,PH (1993)

CAS #:	075887-63-7	CARBONIC ACID, DECAHYDRO-3,6,9-TRIMETHYL-3,12-EP

Formula: $C_{22}H_{28}O_7$

Mol Weight: 404.46

MP (deg C):   FP (deg C):

BP (deg C):

BP pressure (mm Hg):

Property/Value	Units	Temp	Data Type	Reference
WS 4.92E-001	mg/L	25	EST	MEYLAN,WM ET AL. (1996)
logP 4.41			EXP	SANGSTER,J (1993)
VP 5.65E-009	mm Hg	25	EST	NEELY,WB & BLAU,GE (1985)
DC	pKa			
HL 5.01E-010	atm m3/mol	25	EST	MEYLAN,WM & HOWARD,PH (1991)
OH 5.62E-011	cm3/molc sec	25	EST	MEYLAN,WM & HOWARD,PH (1993)

CAS #: 075889-44-0	24-NH2PYRIMIDINE,5(2ME-45OCH2O)BENZYL

Formula: $C_{13}H_{14}N_4O_2$

Mol Weight: 258.28

MP (deg C):   FP (deg C):

BP (deg C):

BP pressure (mm Hg):

Property/ Value	Units	Temp	Data Type	Reference
WS 5.17E+002	mg/L	25	EST	MEYLAN,WM ET AL. (1996)
logP 1.89			EXP	HANSCH,C & LEO,AJ (1985)
VP 2.43E-008	mm Hg	25	EST	NEELY,WB & BLAU,GE (1985)
DC	pKa			
HL 1.48E-013	atm m3/mol	25	EST	MEYLAN,WM & HOWARD,PH (1991)
OH 6.45E-011	cm3/molc sec	25	EST	MEYLAN,WM & HOWARD,PH (1993)

CAS #: 075889-62-2	PHOSPHONIC ACID, [[4-(2-BENZOTHIAZOLYL)PHENYL]ME

Formula: $C_{18}H_{20}NO_3PS$

Mol Weight: 361.40

MP (deg C):   FP (deg C):

BP (deg C):

BP pressure (mm Hg):

Property/ Value	Units	Temp	Data Type	Reference
WS 5.42E+000	mg/L	25	EST	MEYLAN,WM ET AL. (1996)
logP 3.50			EXP	SANGSTER,J (1993)
VP 2.06E-008	mm Hg	25	EST	NEELY,WB & BLAU,GE (1985)
DC	pKa			
HL 9.47E-013	atm m3/mol	25	EST	MEYLAN,WM & HOWARD,PH (1991)
OH 5.59E-011	cm3/molc sec	25	EST	MEYLAN,WM & HOWARD,PH (1993)

CAS #: 075912-69-5	1-(2-(P-NO2PHENOXY)ETHYL)IMIDAZOLE

Formula: $C_{11}H_{13}N_3O_3$

Mol Weight: 235.24

MP (deg C):   FP (deg C):

BP (deg C):

BP pressure (mm Hg):

Property/ Value	Units	Temp	Data Type	Reference
WS 4.29E+002	mg/L	25	EST	MEYLAN,WM ET AL. (1996)
logP 1.69			EXP	HANSCH,C & LEO,AJ (1985)
VP 4.48E-007	mm Hg	25	EST	NEELY,WB & BLAU,GE (1985)
DC	pKa			
HL 1.82E-009	atm m3/mol	25	EST	MEYLAN,WM & HOWARD,PH (1991)
OH 4.90E-011	cm3/molc sec	25	EST	MEYLAN,WM & HOWARD,PH (1993)

CAS #: 075922-48-4	1,3-DIETHYL-8-PHENYLXANTHINE

Formula: $C_{15}H_{16}N_4O_2$

Mol Weight: 284.32

MP (deg C):   FP (deg C):

BP (deg C):

BP pressure (mm Hg):

Property/ Value	Units	Temp	Data Type	Reference
WS 4.76E+001	mg/L	25	EST	MEYLAN,WM ET AL. (1996)
logP 2.93			EXP	HANSCH,C ET AL. (1995)
VP 8.70E-013	mm Hg	25	EST	NEELY,WB & BLAU,GE (1985)
DC	pKa			
HL 2.27E-013	atm m3/mol	25	EST	MEYLAN,WM & HOWARD,PH (1991)
OH 4.84E-011	cm3/molc sec	25	EST	MEYLAN,WM & HOWARD,PH (1993)

CAS #: 075994-55-7	CEPHALOTHIN ANALOG(7PHOACETYLAMINO

Formula: $C_{18}H_{18}N_2O_6S$

Mol Weight: 390.42

MP (deg C):   FP (deg C):

BP (deg C):

BP pressure (mm Hg):

Property/ Value	Units	Temp	Data Type	Reference
WS 1.16E+002	mg/L	25	EST	MEYLAN,WM ET AL. (1996)
logP 0.20			EXP	HANSCH,C & LEO,AJ (1985)
VP 2.82E-014	mm Hg	25	EST	NEELY,WB & BLAU,GE (1985)
DC	pKa			
HL 1.93E-016	atm m3/mol	25	EST	MEYLAN,WM & HOWARD,PH (1991)
OH 1.80E-010	cm3/molc sec	25	EST	MEYLAN,WM & HOWARD,PH (1993)

CAS #: 076016-68-7	FURAZAN, NITROPHENYL-

Formula: $C_8H_5N_3O_3$

Mol Weight: 191.15

MP (deg C):   FP (deg C):

BP (deg C):

BP pressure (mm Hg):

Property/ Value	Units	Temp	Data Type	Reference
WS 1.12E+002	mg/L	25	EST	MEYLAN,WM ET AL. (1996)
logP 2.63			EXP	CALVINO,R ET AL. (1992)
VP 4.25E-005	mm Hg	25	EST	NEELY,WB & BLAU,GE (1985)
DC	pKa			
HL 5.73E-008	atm m3/mol	25	EST	MEYLAN,WM & HOWARD,PH (1991)
OH 6.50E-012	cm3/molc sec	25	EST	MEYLAN,WM & HOWARD,PH (1993)

CAS #: 076252-06-7	QUINOLINE, 1,2,3,4-TETRAHYDRO-8-[2-HYDROXY-3-[(1

Formula: $C_{21}H_{27}N_3O_3$

Mol Weight: 369.47

MP (deg C):   FP (deg C):

BP (deg C):

BP pressure (mm Hg):

Property/ Value	Units	Temp	Data Type	Reference
WS 1.91E+002	mg/L	25	EST	MEYLAN,WM ET AL. (1996)
logP 1.63			EXP	MANNHOLD,R ET AL. (1990)
VP 6.59E-013	mm Hg	25	EST	NEELY,WB & BLAU,GE (1985)
DC	pKa			
HL 3.21E-020	atm m3/mol	25	EST	MEYLAN,WM & HOWARD,PH (1991)
OH 1.45E-010	cm3/molc sec	25	EST	MEYLAN,WM & HOWARD,PH (1993)

CAS #: 076343-98-1	O-SEC-BUTYLPHENOXYACETIC ACID

Formula: $C_{12}H_{16}O_3$

Mol Weight: 208.26

MP (deg C):   FP (deg C):

BP (deg C):

BP pressure (mm Hg):

Property/ Value	Units	Temp	Data Type	Reference
WS 1.44E+002	mg/L	25	EST	MEYLAN,WM ET AL. (1996)
logP 3.32			EXP	HANSCH,C & LEO,AJ (1985)
VP 5.83E-005	mm Hg	25	EST	NEELY,WB & BLAU,GE (1985)
DC	pKa			
HL 4.33E-008	atm m3/mol	25	EST	MEYLAN,WM & HOWARD,PH (1991)
OH 3.45E-011	cm3/molc sec	25	EST	MEYLAN,WM & HOWARD,PH (1993)

CAS #: 076350-28-2	M-BENZAMIDE,NN'BIS(23DIOHPR)5ACE..

Formula: $C_{17}H_{22}I_3N_3O_7$

Mol Weight: 761.09

MP (deg C):   FP (deg C):

BP (deg C):

BP pressure (mm Hg):

Property/Value	Units	Temp	Data Type	Reference
WS 3.85E+001	mg/L	25	EST	MEYLAN,WM ET AL. (1996)
logP -2.06			EXP	HANSCH,C & LEO,AJ (1985)
VP 1.14E-027	mm Hg	25	EST	NEELY,WB & BLAU,GE (1985)
DC	pKa			
HL 1.13E-026	atm m3/mol	25	EST	MEYLAN,WM & HOWARD,PH (1991)
OH 5.05E-011	cm3/molc sec	25	EST	MEYLAN,WM & HOWARD,PH (1993)

CAS #: 076420-72-9	L-PROLINE, 1-[N-(1-CARBOXY-3-PHENYLPROPYL)-L-ALA

Formula: $C_{18}H_{24}N_2O_5$

Mol Weight: 348.40

MP (deg C): 148-151   FP (deg C):

BP (deg C):

BP pressure (mm Hg):

Property/Value	Units	Temp	Data Type	Reference
WS 8.94E+000	mg/L	25	EST	MEYLAN,WM ET AL. (1996)
logP -0.62			EXP	SANGSTER,J (1993)
VP 2.70E-013	mm Hg	25	EST	NEELY,WB & BLAU,GE (1985)
DC	pKa			
HL 7.85E-019	atm m3/mol	25	EST	MEYLAN,WM & HOWARD,PH (1991)
OH 1.29E-010	cm3/molc sec	25	EST	MEYLAN,WM & HOWARD,PH (1993)

CAS #: 076432-30-9	BENZOIC ACID, 2-(ACETYLOXY)-, (METHYLTHIO)METHYL

Formula: $C_{11}H_{12}O_4S$

Mol Weight: 240.28

MP (deg C):   FP (deg C):

BP (deg C):

BP pressure (mm Hg):

Property/Value	Units	Temp	Data Type	Reference
WS 5.90E+002	mg/L	25	EST	MEYLAN,WM ET AL. (1996)
logP 1.94			EXP	NIELSEN,LS & BUNDGAARD,H (1989)
VP 1.15E-003	mm Hg	25	EST	NEELY,WB & BLAU,GE (1985)
DC	pKa			
HL 4.46E-009	atm m3/mol	25	EST	MEYLAN,WM & HOWARD,PH (1991)
OH 1.52E-011	cm3/molc sec	25	EST	MEYLAN,WM & HOWARD,PH (1993)

CAS #: 076432-33-2	BENZOIC ACID, 2-(ACETYLOXY)-, (METHYLSULFINYL)ME

Formula: $C_{11}H_{12}O_5S$

Mol Weight: 256.28

MP (deg C):   FP (deg C):

BP (deg C):

BP pressure (mm Hg):

Property/Value	Units	Temp	Data Type	Reference
WS 1.76E+004	mg/L	25	EST	MEYLAN,WM ET AL. (1996)
logP 0.11			EXP	NIELSEN,LS & BUNDGAARD,H (1989)
VP 7.08E-005	mm Hg	25	EST	NEELY,WB & BLAU,GE (1985)
DC	pKa			
HL 2.81E-013	atm m3/mol	25	EST	MEYLAN,WM & HOWARD,PH (1991)
OH 7.35E-011	cm3/molc sec	25	EST	MEYLAN,WM & HOWARD,PH (1993)

CAS #: 076432-35-4	BENZOIC ACID, 2-(ACETYLOXY)-, (METHYLSULFONYL)ME

Formula: $C_{11}H_{12}O_6S$

Mol Weight: 272.28

MP (deg C):   FP (deg C):

BP (deg C):

BP pressure (mm Hg):

Property/Value	Units	Temp	Data Type	Reference
WS 5.78E+003	mg/L	25	EST	MEYLAN,WM ET AL. (1996)
logP 0.57			EXP	NIELSEN,LS & BUNDGAARD,H (1989)
VP 1.55E-005	mm Hg	25	EST	NEELY,WB & BLAU,GE (1985)
DC	pKa			
HL 3.54E-011	atm m3/mol	25	EST	MEYLAN,WM & HOWARD,PH (1991)
OH 1.35E-011	cm3/molc sec	25	EST	MEYLAN,WM & HOWARD,PH (1993)

CAS #: 076529-47-0	1H-IMIDAZOLE, 4-IODO-1-METHYL-5-NITRO-

Formula: $C_4H_4IN_3O_2$

Mol Weight: 253.00

MP (deg C):   FP (deg C):

BP (deg C):

BP pressure (mm Hg):

Property/Value	Units	Temp	Data Type	Reference
WS 7.34E+002	mg/L	25	EST	MEYLAN,WM ET AL. (1996)
logP 1.29			EXP	GUPTA,RP ET AL. (1985)
VP 2.59E-005	mm Hg	25	EST	NEELY,WB & BLAU,GE (1985)
DC	pKa			
HL 7.33E-008	atm m3/mol	25	EST	MEYLAN,WM & HOWARD,PH (1991)
OH 1.62E-012	cm3/molc sec	25	EST	MEYLAN,WM & HOWARD,PH (1993)

CAS #: 076529-48-1	1H-IMIDAZOLE, 4-IODO-5-NITRO-

Formula: $C_3H_2IN_3O_2$

Mol Weight: 238.97

MP (deg C):   FP (deg C):

BP (deg C):

BP pressure (mm Hg):

Property/Value	Units	Temp	Data Type	Reference
WS 6.39E+003	mg/L	25	EST	MEYLAN,WM ET AL. (1996)
logP 0.28			EXP	GUPTA,RP ET AL. (1985)
VP 3.99E-006	mm Hg	25	EST	NEELY,WB & BLAU,GE (1985)
DC	pKa			
HL 3.43E-009	atm m3/mol	25	EST	MEYLAN,WM & HOWARD,PH (1991)
OH 1.49E-012	cm3/molc sec	25	EST	MEYLAN,WM & HOWARD,PH (1993)

CAS #: 076542-53-5	PROPANEDINITRILE, [[5-(METHYLTHIO)-2-FURANYL]MET

Formula: $C_9H_6N_2OS$

Mol Weight: 190.22

MP (deg C):   FP (deg C):

BP (deg C):

BP pressure (mm Hg):

Property/Value	Units	Temp	Data Type	Reference
WS 2.15E+002	mg/L	25	EST	MEYLAN,WM ET AL. (1996)
logP 2.45			EXP	BALAZ,S ET AL. (1985)
VP 2.06E-005	mm Hg	25	EST	NEELY,WB & BLAU,GE (1985)
DC	pKa			
HL 3.99E-010	atm m3/mol	25	EST	MEYLAN,WM & HOWARD,PH (1991)
OH 2.03E-010	cm3/molc sec	25	EST	MEYLAN,WM & HOWARD,PH (1993)

CAS #: 076542-54-6 — 2-PROPENOIC ACID, 2-CYANO-3-[5-(METHYLTHIO)-2-FU

Formula: $C_{10}H_9NO_3S$

Mol Weight: 223.25

MP (deg C): FP (deg C):

BP (deg C):

BP pressure (mm Hg):

Property/ Value	Units	Temp	Data Type	Reference
WS 1.06E+002	mg/L	25	EST	MEYLAN,WM ET AL. (1996)
logP 2.61			EXP	SANGSTER,J (1994)
VP 3.46E-005	mm Hg	25	EST	NEELY,WB & BLAU,GE (1985)
DC	pKa			
HL 2.68E-010	atm m3/mol	25	EST	MEYLAN,WM & HOWARD,PH (1991)
OH 2.06E-010	cm3/molc sec	25	EST	MEYLAN,WM & HOWARD,PH (1993)

CAS #: 076547-98-3 — L-PROLINE, 1-[N -(1-CARBOXY-3-PHENYLPROPYL)-L-LY

Formula: $C_{21}H_{31}N_3O_5$

Mol Weight: 405.50

MP (deg C): FP (deg C):

BP (deg C):

BP pressure (mm Hg):

Property/ Value	Units	Temp	Data Type	Reference
WS 1.30E+001	mg/L	25	EST	MEYLAN,WM ET AL. (1996)
logP -1.22			EXP	SANGSTER,J (1993)
VP 3.95E-015	mm Hg	25	EST	NEELY,WB & BLAU,GE (1985)
DC	pKa			
HL 1.89E-022	atm m3/mol	25	EST	MEYLAN,WM & HOWARD,PH (1991)
OH 1.71E-010	cm3/molc sec	25	EST	MEYLAN,WM & HOWARD,PH (1993)

CAS #: 076578-14-8 — QUIZALOFOP-ETHYL

Formula: $C_{19}H_{17}ClN_2O_4$

Mol Weight: 372.81

MP (deg C): 91.7-92.1 FP (deg C):

BP (deg C):

BP pressure (mm Hg):

Property/ Value	Units	Temp	Data Type	Reference
WS 3.00E-001	mg/L	20	EXP	SHIU,WY ET AL. (1990)
logP 4.28			EXP	TOMLIN,C (1994)
VP 6.49E-012	mm Hg	25	EXP	TOMLIN,C (1994)
DC	pKa			
HL 1.06E-011	atm m3/mol	25	EST	VP/WSOL
OH 3.35E-011	cm3/molc sec	25	EST	MEYLAN,WM & HOWARD,PH (1993)

CAS #: 076674-21-0 — FLUTRIAFOL (PP450)

Formula: $C_{16}H_{13}F_2N_3O$

Mol Weight: 301.30

MP (deg C): 130 FP (deg C):

BP (deg C):

BP pressure (mm Hg):

Property/ Value	Units	Temp	Data Type	Reference
WS 1.30E+002	mg/L	20	EXP	SHIU,WY ET AL. (1990) @ pH=7
logP 2.29			EXP	PATIL,SG ET AL. (1988)
VP 8.37E-009	mm Hg	25	EST	NEELY,WB & BLAU,GE (1985)
DC	pKa			
HL 1.40E-011	atm m3/mol	25	EST	MEYLAN,WM & HOWARD,PH (1991)
OH 9.65E-012	cm3/molc sec	25	EST	MEYLAN,WM & HOWARD,PH (1993)

CAS #: 076738-62-0 — PACLOBUTRAZOL

Formula: $C_{15}H_{20}ClN_3O$

Mol Weight: 293.80

MP (deg C): 165-166 FP (deg C):

BP (deg C):

BP pressure (mm Hg):

Property/ Value	Units	Temp	Data Type	Reference
WS 3.50E+001	mg/L		EXP	SHIU,WY ET AL. (1990)
logP 3.20			EXP	BAKER,EA ET AL. (1992)
VP 7.50E-009	mm Hg	20	EXP	TOMLIN,C (1994)
DC	pKa			
HL 8.28E-011	atm m3/mol	20	EST	VP/WSOL
OH 1.68E-011	cm3/molc sec	25	EST	MEYLAN,WM & HOWARD,PH (1993)

CAS #: 076824-35-6 — FAMOTIDINE

Formula: $C_8H_{15}N_7O_2S_3$

Mol Weight: 337.45

MP (deg C): 163-164 FP (deg C):

BP (deg C):

BP pressure (mm Hg):

Property/ Value	Units	Temp	Data Type	Reference
WS 1.27E+003	mg/L	25	EST	MEYLAN,WM ET AL. (1996)
logP -0.64			EXP	ISLAM,MS & NARURKAR,MM (1993)
VP 6.02E-011	mm Hg	25	EST	NEELY,WB & BLAU,GE (1985)
DC	pKa			
HL 5.44E-024	atm m3/mol	25	EST	MEYLAN,WM & HOWARD,PH (1991)
OH 8.62E-011	cm3/molc sec	25	EST	MEYLAN,WM & HOWARD,PH (1993)

CAS #: 077067-91-5 — 2'-ME-4'-CL SALICYLANILIDE

Formula: $C_{14}H_{12}ClNO_2$

Mol Weight: 261.71

MP (deg C): FP (deg C):

BP (deg C):

BP pressure (mm Hg):

Property/ Value	Units	Temp	Data Type	Reference
WS 3.44E+001	mg/L	25	EST	MEYLAN,WM ET AL. (1996)
logP 5.18			EXP	TERADA,H ET AL. (1988)
VP 2.50E-009	mm Hg	25	EST	NEELY,WB & BLAU,GE (1985)
DC	pKa			
HL 1.31E-010	atm m3/mol	25	EST	MEYLAN,WM & HOWARD,PH (1991)
OH 3.51E-011	cm3/molc sec	25	EST	MEYLAN,WM & HOWARD,PH (1993)

CAS #: 077068-02-1 — 5-F-2'-ME-4'-CL SALICYANILIDE

Formula: $C_{14}H_{11}ClFNO_2$

Mol Weight: 279.70

MP (deg C): FP (deg C):

BP (deg C):

BP pressure (mm Hg):

Property/ Value	Units	Temp	Data Type	Reference
WS 4.08E+000	mg/L	25	EST	MEYLAN,WM ET AL. (1996)
logP 4.89			EXP	TERADA,H ET AL. (1988)
VP 3.05E-009	mm Hg	25	EST	NEELY,WB & BLAU,GE (1985)
DC	pKa			
HL 1.53E-010	atm m3/mol	25	EST	MEYLAN,WM & HOWARD,PH (1991)
OH 1.50E-011	cm3/molc sec	25	EST	MEYLAN,WM & HOWARD,PH (1993)

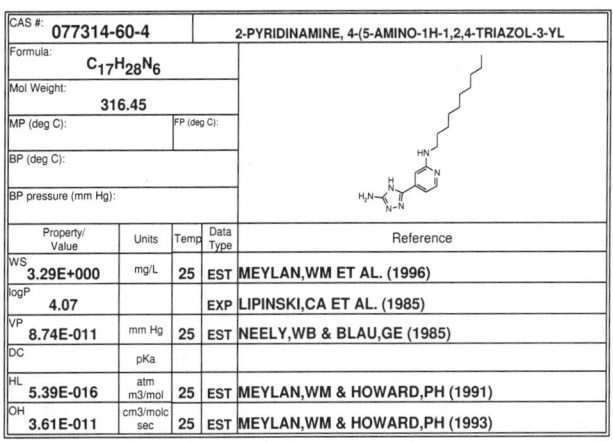

CAS #: 077314-61-5	2-PYRIDINAMINE, 4-(5-AMINO-1H-1,2,4-TRIAZOL-3-YL

Formula: $C_{10}H_{11}N_5$

Mol Weight: 201.23

MP (deg C): FP (deg C):

BP (deg C):

BP pressure (mm Hg):

Property/Value	Units	Temp	Data Type	Reference
WS 3.62E+003	mg/L	25	EST	MEYLAN,WM ET AL. (1996)
logP 1.26			EXP	LIPINSKI,CA ET AL. (1985)
VP 5.65E-007	mm Hg	25	EST	NEELY,WB & BLAU,GE (1985)
DC	pKa			
HL 1.56E-013	atm m3/mol	25	EST	MEYLAN,WM & HOWARD,PH (1991)
OH 4.92E-011	cm3/molc sec	25	EST	MEYLAN,WM & HOWARD,PH (1993)

CAS #: 077314-62-6	ETHANOL, 2-[[4-(5-AMINO-1H-1,2,4-TRIAZOL-3-YL)-2

Formula: $C_9H_{12}N_6O$

Mol Weight: 220.24

MP (deg C): FP (deg C):

BP (deg C):

BP pressure (mm Hg):

Property/Value	Units	Temp	Data Type	Reference
WS 1.05E+003	mg/L	25	EST	MEYLAN,WM ET AL. (1996)
logP 0.24			EXP	LIPINSKI,CA ET AL. (1985)
VP 2.92E-011	mm Hg	25	EST	NEELY,WB & BLAU,GE (1985)
DC	pKa			
HL 2.04E-021	atm m3/mol	25	EST	MEYLAN,WM & HOWARD,PH (1991)
OH 2.90E-011	cm3/molc sec	25	EST	MEYLAN,WM & HOWARD,PH (1993)

CAS #: 077314-64-8	2-PYRIDINAMINE, 4-(5-AMINO-1H-1,2,4-TRIAZOL-3-YL

Formula: $C_{10}H_{14}N_6O$

Mol Weight: 234.26

MP (deg C): FP (deg C):

BP (deg C):

BP pressure (mm Hg):

Property/Value	Units	Temp	Data Type	Reference
WS 4.55E+003	mg/L	25	EST	MEYLAN,WM ET AL. (1996)
logP 0.94			EXP	LIPINSKI,CA ET AL. (1985)
VP 1.15E-008	mm Hg	25	EST	NEELY,WB & BLAU,GE (1985)
DC	pKa			
HL 6.54E-019	atm m3/mol	25	EST	MEYLAN,WM & HOWARD,PH (1991)
OH 5.24E-011	cm3/molc sec	25	EST	MEYLAN,WM & HOWARD,PH (1993)

CAS #: 077314-69-3	2-PYRIDINAMINE, 4-(5-AMINO-1H-1,2,4-TRIAZOL-3-YL

Formula: $C_{15}H_{16}N_6$

Mol Weight: 280.33

MP (deg C): FP (deg C):

BP (deg C):

BP pressure (mm Hg):

Property/Value	Units	Temp	Data Type	Reference
WS 9.05E+001	mg/L	25	EST	MEYLAN,WM ET AL. (1996)
logP 2.63			EXP	LIPINSKI,CA ET AL. (1985)
VP 1.62E-010	mm Hg	25	EST	NEELY,WB & BLAU,GE (1985)
DC	pKa			
HL 4.51E-018	atm m3/mol	25	EST	MEYLAN,WM & HOWARD,PH (1991)
OH 3.08E-011	cm3/molc sec	25	EST	MEYLAN,WM & HOWARD,PH (1993)

CAS #: 077314-73-9	2-PYRIDINAMINE, 4-(5-AMINO-1H-1,2,4-TRIAZOL-3-YL

Formula: $C_{15}H_{16}N_6$

Mol Weight: 280.33

MP (deg C): FP (deg C):

BP (deg C):

BP pressure (mm Hg):

Property/Value	Units	Temp	Data Type	Reference
WS 8.05E+001	mg/L	25	EST	MEYLAN,WM ET AL. (1996)
logP 2.69			EXP	LIPINSKI,CA ET AL. (1985)
VP 1.62E-010	mm Hg	25	EST	NEELY,WB & BLAU,GE (1985)
DC	pKa			
HL 3.75E-018	atm m3/mol	25	EST	MEYLAN,WM & HOWARD,PH (1991)
OH 2.92E-011	cm3/molc sec	25	EST	MEYLAN,WM & HOWARD,PH (1993)

CAS #: 077314-75-1	2-PYRIDINAMINE, 4-(5-AMINO-1H-1,2,4-TRIAZOL-3-YL

Formula: $C_7H_8N_6$

Mol Weight: 176.18

MP (deg C): FP (deg C):

BP (deg C):

BP pressure (mm Hg):

Property/Value	Units	Temp	Data Type	Reference
WS 2.27E+003	mg/L	25	EST	MEYLAN,WM ET AL. (1996)
logP 0.11			EXP	LIPINSKI,CA ET AL. (1985)
VP 6.94E-008	mm Hg	25	EST	NEELY,WB & BLAU,GE (1985)
DC	pKa			
HL 1.92E-017	atm m3/mol	25	EST	MEYLAN,WM & HOWARD,PH (1991)
OH 3.56E-011	cm3/molc sec	25	EST	MEYLAN,WM & HOWARD,PH (1993)

CAS #: 077314-76-2	2-PYRIDINAMINE, 4-(5-AMINO-1H-1,2,4-TRIAZOL-3-YL

Formula: $C_8H_{10}N_6$

Mol Weight: 190.21

MP (deg C): FP (deg C):

BP (deg C):

BP pressure (mm Hg):

Property/Value	Units	Temp	Data Type	Reference
WS 1.26E+004	mg/L	25	EST	MEYLAN,WM ET AL. (1996)
logP 0.69			EXP	LIPINSKI,CA ET AL. (1985)
VP 1.31E-007	mm Hg	25	EST	NEELY,WB & BLAU,GE (1985)
DC	pKa			
HL 4.21E-017	atm m3/mol	25	EST	MEYLAN,WM & HOWARD,PH (1991)
OH 1.54E-011	cm3/molc sec	25	EST	MEYLAN,WM & HOWARD,PH (1993)

CAS #: 077314-77-3	2-PYRIDINAMINE, 4-(5-AMINO-1H-1,2,4-TRIAZOL-3-YL

Formula: $C_9H_{12}N_6$

Mol Weight: 204.24

MP (deg C): FP (deg C):

BP (deg C):

BP pressure (mm Hg):

Property/Value	Units	Temp	Data Type	Reference
WS 4.51E+003	mg/L	25	EST	MEYLAN,WM ET AL. (1996)
logP 1.13			EXP	LIPINSKI,CA ET AL. (1985)
VP 5.91E-008	mm Hg	25	EST	NEELY,WB & BLAU,GE (1985)
DC	pKa			
HL 5.59E-017	atm m3/mol	25	EST	MEYLAN,WM & HOWARD,PH (1991)
OH 2.30E-011	cm3/molc sec	25	EST	MEYLAN,WM & HOWARD,PH (1993)

CAS #: 077314-79-5 — 2-PYRIDINAMINE, 4-(5-AMINO-1H-1,2,4-TRIAZOL-3-YL

Formula: $C_{11}H_{16}N_6$

Mol Weight: 232.29

MP (deg C):
FP (deg C):
BP (deg C):
BP pressure (mm Hg):

Property/ Value	Units	Temp	Data Type	Reference
WS 4.31E+002	mg/L	25	EST	MEYLAN,WM ET AL. (1996)
logP 2.15			EXP	LIPINSKI,CA ET AL. (1985)
VP 1.19E-008	mm Hg	25	EST	NEELY,WB & BLAU,GE (1985)
DC	pKa			
HL 9.85E-017	atm m3/mol	25	EST	MEYLAN,WM & HOWARD,PH (1991)
OH 2.76E-011	cm3/molc sec	25	EST	MEYLAN,WM & HOWARD,PH (1993)

CAS #: 077317-64-7 — 3-OH-5-BR-6-ME-PYRIMIDINE-2,4-DIONE

Formula: $C_5H_5BrN_2O_3$

Mol Weight: 221.02

MP (deg C):
FP (deg C):
BP (deg C):
BP pressure (mm Hg):

Property/ Value	Units	Temp	Data Type	Reference
WS 1.51E+004	mg/L	25	EST	MEYLAN,WM ET AL. (1996)
logP -1.12			EXP	HANSCH,C & LEO,AJ (1985)
VP 4.34E-010	mm Hg	25	EST	NEELY,WB & BLAU,GE (1985)
DC	pKa			
HL 3.67E-013	atm m3/mol	25	EST	MEYLAN,WM & HOWARD,PH (1991)
OH 8.14E-012	cm3/molc sec	25	EST	MEYLAN,WM & HOWARD,PH (1993)

CAS #: 077317-65-8 — 3-MEO-5-BR-6-ME-PYRIMIDIN-2,4-DIONE

Formula: $C_6H_7BrN_2O_3$

Mol Weight: 235.04

MP (deg C):
FP (deg C):
BP (deg C):
BP pressure (mm Hg):

Property/ Value	Units	Temp	Data Type	Reference
WS 2.53E+003	mg/L	25	EST	MEYLAN,WM ET AL. (1996)
logP -0.30			EXP	HANSCH,C & LEO,AJ (1985)
VP 2.03E-007	mm Hg	25	EST	NEELY,WB & BLAU,GE (1985)
DC	pKa			
HL 1.17E-010	atm m3/mol	25	EST	MEYLAN,WM & HOWARD,PH (1991)
OH 8.17E-012	cm3/molc sec	25	EST	MEYLAN,WM & HOWARD,PH (1993)

CAS #: 077317-66-9 — 3-ETO-5-BR-6-ME-PYRIMIDIN-2,4-DIONE

Formula: $C_7H_9BrN_2O_3$

Mol Weight: 249.07

MP (deg C):
FP (deg C):
BP (deg C):
BP pressure (mm Hg):

Property/ Value	Units	Temp	Data Type	Reference
WS 9.44E+002	mg/L	25	EST	MEYLAN,WM ET AL. (1996)
logP 0.11			EXP	HANSCH,C & LEO,AJ (1985)
VP 9.17E-008	mm Hg	25	EST	NEELY,WB & BLAU,GE (1985)
DC	pKa			
HL 1.56E-010	atm m3/mol	25	EST	MEYLAN,WM & HOWARD,PH (1991)
OH 8.48E-012	cm3/molc sec	25	EST	MEYLAN,WM & HOWARD,PH (1993)

CAS #: 077317-67-0 — 3-CLETO-5-BR-6-ME-PYRIMID-2,4-DIONE

Formula: $C_7H_8BrClN_2O_3$

Mol Weight: 283.52

MP (deg C):
FP (deg C):
BP (deg C):
BP pressure (mm Hg):

Property/ Value	Units	Temp	Data Type	Reference
WS 4.15E+002	mg/L	25	EST	MEYLAN,WM ET AL. (1996)
logP 0.30			EXP	HANSCH,C & LEO,AJ (1985)
VP 1.07E-008	mm Hg	25	EST	NEELY,WB & BLAU,GE (1985)
DC	pKa			
HL 5.49E-011	atm m3/mol	25	EST	MEYLAN,WM & HOWARD,PH (1991)
OH 8.64E-012	cm3/molc sec	25	EST	MEYLAN,WM & HOWARD,PH (1993)

CAS #: 077317-68-1 — 3-PRO-5-BR-6-ME-PYRIMIDIN-2,4-DIONE

Formula: $C_8H_{11}BrN_2O_3$

Mol Weight: 263.10

MP (deg C):
FP (deg C):
BP (deg C):
BP pressure (mm Hg):

Property/ Value	Units	Temp	Data Type	Reference
WS 4.49E+003	mg/L	25	EST	MEYLAN,WM ET AL. (1996)
logP 0.76			EXP	HANSCH,C & LEO,AJ (1985)
VP 4.12E-008	mm Hg	25	EST	NEELY,WB & BLAU,GE (1985)
DC	pKa			
HL 2.07E-010	atm m3/mol	25	EST	MEYLAN,WM & HOWARD,PH (1991)
OH 9.66E-012	cm3/molc sec	25	EST	MEYLAN,WM & HOWARD,PH (1993)

CAS #: 077317-69-2 — 3-IPRO-5-BR-6-ME-PYRIMIDI-2,4-DIONE

Formula: $C_8H_{11}BrN_2O_3$

Mol Weight: 263.10

MP (deg C):
FP (deg C):
BP (deg C):
BP pressure (mm Hg):

Property/ Value	Units	Temp	Data Type	Reference
WS 8.25E+003	mg/L	25	EST	MEYLAN,WM ET AL. (1996)
logP 0.45			EXP	HANSCH,C & LEO,AJ (1985)
VP 6.68E-008	mm Hg	25	EST	NEELY,WB & BLAU,GE (1985)
DC	pKa			
HL 2.07E-010	atm m3/mol	25	EST	MEYLAN,WM & HOWARD,PH (1991)
OH 8.83E-012	cm3/molc sec	25	EST	MEYLAN,WM & HOWARD,PH (1993)

CAS #: 077317-70-5 — 3-ALLYLO-5-BR-6-ME-PYRIMID-2,4-DION

Formula: $C_8H_9BrN_2O_3$

Mol Weight: 261.08

MP (deg C):
FP (deg C):
BP (deg C):
BP pressure (mm Hg):

Property/ Value	Units	Temp	Data Type	Reference
WS 5.36E+002	mg/L	25	EST	MEYLAN,WM ET AL. (1996)
logP 0.32			EXP	HANSCH,C & LEO,AJ (1985)
VP 4.38E-008	mm Hg	25	EST	NEELY,WB & BLAU,GE (1985)
DC	pKa			
HL 1.54E-010	atm m3/mol	25	EST	MEYLAN,WM & HOWARD,PH (1991)
OH 3.46E-011	cm3/molc sec	25	EST	MEYLAN,WM & HOWARD,PH (1993)

CAS #: 077317-71-6 — 3-C≡CCO-5-BR-6-ME-PYRIMID-2,4-DIONE

Formula: $C_8H_7BrN_2O_3$

Mol Weight: 259.07

MP (deg C): | FP (deg C):

BP (deg C):

BP pressure (mm Hg):

Property/ Value	Units	Temp	Data Type	Reference
WS 1.26E+003	mg/L	25	EST	MEYLAN,WM ET AL. (1996)
logP -0.10			EXP	HANSCH,C & LEO,AJ (1985)
VP 3.11E-008	mm Hg	25	EST	NEELY,WB & BLAU,GE (1985)
DC	pKa			
HL 2.56E-011	atm m3/mol	25	EST	MEYLAN,WM & HOWARD,PH (1991)
OH 1.53E-011	cm3/molc sec	25	EST	MEYLAN,WM & HOWARD,PH (1993)

CAS #: 077317-72-7 — 3-BUO-5-BR-6-ME-PYRIMIDIN-2,4-DIONE

Formula: $C_9H_{13}BrN_2O_3$

Mol Weight: 277.12

MP (deg C): | FP (deg C):

BP (deg C):

BP pressure (mm Hg):

Property/ Value	Units	Temp	Data Type	Reference
WS 1.45E+003	mg/L	25	EST	MEYLAN,WM ET AL. (1996)
logP 1.24			EXP	HANSCH,C & LEO,AJ (1985)
VP 1.85E-008	mm Hg	25	EST	NEELY,WB & BLAU,GE (1985)
DC	pKa			
HL 2.75E-010	atm m3/mol	25	EST	MEYLAN,WM & HOWARD,PH (1991)
OH 1.11E-011	cm3/molc sec	25	EST	MEYLAN,WM & HOWARD,PH (1993)

CAS #: 077317-73-8 — 3-SBUO-5-BR-6-ME-PYRIMIDIN-2,4-DION

Formula: $C_9H_{13}BrN_2O_3$

Mol Weight: 277.12

MP (deg C): | FP (deg C):

BP (deg C):

BP pressure (mm Hg):

Property/ Value	Units	Temp	Data Type	Reference
WS 1.88E+003	mg/L	25	EST	MEYLAN,WM ET AL. (1996)
logP 1.11			EXP	HANSCH,C & LEO,AJ (1985)
VP 3.00E-008	mm Hg	25	EST	NEELY,WB & BLAU,GE (1985)
DC	pKa			
HL 2.75E-010	atm m3/mol	25	EST	MEYLAN,WM & HOWARD,PH (1991)
OH 1.01E-011	cm3/molc sec	25	EST	MEYLAN,WM & HOWARD,PH (1993)

CAS #: 077317-74-9 — 3-IBUO-5-BR-6-ME-PYRIMIDIN-2,4-DION

Formula: $C_9H_{13}BrN_2O_3$

Mol Weight: 277.12

MP (deg C): | FP (deg C):

BP (deg C):

BP pressure (mm Hg):

Property/ Value	Units	Temp	Data Type	Reference
WS 1.48E+003	mg/L	25	EST	MEYLAN,WM ET AL. (1996)
logP 1.23			EXP	HANSCH,C & LEO,AJ (1985)
VP 3.00E-008	mm Hg	25	EST	NEELY,WB & BLAU,GE (1985)
DC	pKa			
HL 2.75E-010	atm m3/mol	25	EST	MEYLAN,WM & HOWARD,PH (1991)
OH 1.11E-011	cm3/molc sec	25	EST	MEYLAN,WM & HOWARD,PH (1993)

CAS #: 077317-75-0 — 3-PEO-5-BR-6-ME-PYRIMIDIN-2,4-DIONE

Formula: $C_{10}H_{15}BrN_2O_3$

Mol Weight: 291.15

MP (deg C): | FP (deg C):

BP (deg C):

BP pressure (mm Hg):

Property/ Value	Units	Temp	Data Type	Reference
WS 4.98E+002	mg/L	25	EST	MEYLAN,WM ET AL. (1996)
logP 1.69			EXP	HANSCH,C & LEO,AJ (1985)
VP 8.24E-009	mm Hg	25	EST	NEELY,WB & BLAU,GE (1985)
DC	pKa			
HL 3.65E-010	atm m3/mol	25	EST	MEYLAN,WM & HOWARD,PH (1991)
OH 1.25E-011	cm3/molc sec	25	EST	MEYLAN,WM & HOWARD,PH (1993)

CAS #: 077317-76-1 — 3-HXO-5-BR-6-ME-PYRIMIDIN-2,4-DIONE

Formula: $C_{11}H_{17}BrN_2O_3$

Mol Weight: 305.18

MP (deg C): | FP (deg C):

BP (deg C):

BP pressure (mm Hg):

Property/ Value	Units	Temp	Data Type	Reference
WS 8.73E+001	mg/L	25	EST	MEYLAN,WM ET AL. (1996)
logP 2.48			EXP	HANSCH,C & LEO,AJ (1985)
VP 3.66E-009	mm Hg	25	EST	NEELY,WB & BLAU,GE (1985)
DC	pKa			
HL 4.84E-010	atm m3/mol	25	EST	MEYLAN,WM & HOWARD,PH (1991)
OH 1.39E-011	cm3/molc sec	25	EST	MEYLAN,WM & HOWARD,PH (1993)

CAS #: 077317-78-3 — 3-OCO-5-BR-6-ME-PYRIMIDIN-2,4-DIONE

Formula: $C_{13}H_{21}BrN_2O_3$

Mol Weight: 333.23

MP (deg C): | FP (deg C):

BP (deg C):

BP pressure (mm Hg):

Property/ Value	Units	Temp	Data Type	Reference
WS 9.95E+000	mg/L	25	EST	MEYLAN,WM ET AL. (1996)
logP 3.39			EXP	HANSCH,C & LEO,AJ (1985)
VP 7.15E-010	mm Hg	25	EST	NEELY,WB & BLAU,GE (1985)
DC	pKa			
HL 8.53E-010	atm m3/mol	25	EST	MEYLAN,WM & HOWARD,PH (1991)
OH 1.67E-011	cm3/molc sec	25	EST	MEYLAN,WM & HOWARD,PH (1993)

CAS #: 077317-79-4 — 3-DEO-5-BR-6-ME-PYRIMIDIN-2,4-DIONE

Formula: $C_{15}H_{25}BrN_2O_3$

Mol Weight: 361.29

MP (deg C): | FP (deg C):

BP (deg C):

BP pressure (mm Hg):

Property/ Value	Units	Temp	Data Type	Reference
WS 5.66E-001	mg/L	25	EST	MEYLAN,WM ET AL. (1996)
logP 4.65			EXP	HANSCH,C & LEO,AJ (1985)
VP 1.38E-010	mm Hg	25	EST	NEELY,WB & BLAU,GE (1985)
DC	pKa			
HL 1.50E-009	atm m3/mol	25	EST	MEYLAN,WM & HOWARD,PH (1991)
OH 1.96E-011	cm3/molc sec	25	EST	MEYLAN,WM & HOWARD,PH (1993)

CAS #: 077317-85-2			PYRIMIDINE-2,4-DIONE,3-PRO-6-ME
Formula: $C_8H_{12}N_2O_3$			
Mol Weight: 184.20			
MP (deg C):		FP (deg C):	
BP (deg C):			
BP pressure (mm Hg):			

Property/Value	Units	Temp	Data Type	Reference
WS 1.43E+003	mg/L	25	EST	MEYLAN,WM ET AL. (1996)
logP 0.30			EXP	HANSCH,C & LEO,AJ (1985)
VP 5.09E-007	mm Hg	25	EST	NEELY,WB & BLAU,GE (1985)
DC	pKa			
HL 1.04E-009	atm m3/mol	25	EST	MEYLAN,WM & HOWARD,PH (1991)
OH 2.19E-011	cm3/molc sec	25	EST	MEYLAN,WM & HOWARD,PH (1993)

CAS #: 077317-86-3			3-PRO-5-SCN-6-ME-PYRIMIDIN-2,4-DION
Formula: $C_9H_{11}N_3O_3S$			
Mol Weight: 241.27			
MP (deg C):		FP (deg C):	
BP (deg C):			
BP pressure (mm Hg):			

Property/Value	Units	Temp	Data Type	Reference
WS 2.00E+003	mg/L	25	EST	MEYLAN,WM ET AL. (1996)
logP -0.22			EXP	HANSCH,C & LEO,AJ (1985)
VP 9.63E-010	mm Hg	25	EST	NEELY,WB & BLAU,GE (1985)
DC	pKa			
HL 2.07E-012	atm m3/mol	25	EST	MEYLAN,WM & HOWARD,PH (1991)
OH 2.68E-011	cm3/molc sec	25	EST	MEYLAN,WM & HOWARD,PH (1993)

CAS #: 077317-87-4			3-PRO-5-BR-6-PR-PYRIMIDIN-2,4-DIONE
Formula: $C_{10}H_{15}BrN_2O_3$			
Mol Weight: 291.15			
MP (deg C):		FP (deg C):	
BP (deg C):			
BP pressure (mm Hg):			

Property/Value	Units	Temp	Data Type	Reference
WS 4.60E+002	mg/L	25	EST	MEYLAN,WM ET AL. (1996)
logP 1.73			EXP	HANSCH,C & LEO,AJ (1985)
VP 8.24E-009	mm Hg	25	EST	NEELY,WB & BLAU,GE (1985)
DC	pKa			
HL 3.65E-010	atm m3/mol	25	EST	MEYLAN,WM & HOWARD,PH (1991)
OH 1.20E-011	cm3/molc sec	25	EST	MEYLAN,WM & HOWARD,PH (1993)

CAS #: 077317-88-5			3-PRO-5-BR-1,6-ME-PYRIMIDIN-2,4-DION
Formula: $C_9H_{13}BrN_2O_3$			
Mol Weight: 277.12			
MP (deg C):		FP (deg C):	
BP (deg C):			
BP pressure (mm Hg):			

Property/Value	Units	Temp	Data Type	Reference
WS 4.04E+003	mg/L	25	EST	MEYLAN,WM ET AL. (1996)
logP 0.72			EXP	HANSCH,C & LEO,AJ (1985)
VP 1.75E-007	mm Hg	25	EST	NEELY,WB & BLAU,GE (1985)
DC	pKa			
HL 4.54E-010	atm m3/mol	25	EST	MEYLAN,WM & HOWARD,PH (1991)
OH 1.02E-011	cm3/molc sec	25	EST	MEYLAN,WM & HOWARD,PH (1993)

CAS #: 077453-35-1			24NH2PYRIMIDIN,5(3ETO4COHME2-5MEO)BENZYL
Formula: $C_{17}H_{24}N_4O_3$			
Mol Weight: 332.41			
MP (deg C):		FP (deg C):	
BP (deg C):			
BP pressure (mm Hg):			

Property/Value	Units	Temp	Data Type	Reference
WS 1.25E+002	mg/L	25	EST	MEYLAN,WM ET AL. (1996)
logP 2.11			EXP	HANSCH,C & LEO,AJ (1985)
VP 2.94E-012	mm Hg	25	EST	NEELY,WB & BLAU,GE (1985)
DC	pKa			
HL 3.81E-017	atm m3/mol	25	EST	MEYLAN,WM & HOWARD,PH (1991)
OH 2.09E-010	cm3/molc sec	25	EST	MEYLAN,WM & HOWARD,PH (1993)

CAS #: 077456-53-2			ISOMAZOLE,PYRIMIDYL ANALOG
Formula: $C_{13}H_{12}N_4O_2S$			
Mol Weight: 288.33			
MP (deg C):		FP (deg C):	
BP (deg C):			
BP pressure (mm Hg):			

Property/Value	Units	Temp	Data Type	Reference
WS 5.06E+003	mg/L	25	EST	MEYLAN,WM ET AL. (1996)
logP 0.53			EXP	HANSCH,C ET AL. (1995)
VP 6.67E-012	mm Hg	25	EST	NEELY,WB & BLAU,GE (1985)
DC	pKa			
HL 1.66E-018	atm m3/mol	25	EST	MEYLAN,WM & HOWARD,PH (1991)
OH 1.28E-010	cm3/molc sec	25	EST	MEYLAN,WM & HOWARD,PH (1993)

CAS #: 077470-84-9			HEPTANAMIDE, 7-(DIETHYLAMINO)-N-(2,6-DIMETHYLPHE
Formula: $C_{19}H_{32}N_2O$			
Mol Weight: 304.48			
MP (deg C):		FP (deg C):	
BP (deg C):			
BP pressure (mm Hg):			

Property/Value	Units	Temp	Data Type	Reference
WS 1.23E+001	mg/L	25	EST	MEYLAN,WM ET AL. (1996)
logP 3.48			EXP	SANGSTER,J (1993)
VP 1.50E-008	mm Hg	25	EST	NEELY,WB & BLAU,GE (1985)
DC	pKa			
HL 2.71E-011	atm m3/mol	25	EST	MEYLAN,WM & HOWARD,PH (1991)
OH 1.23E-010	cm3/molc sec	25	EST	MEYLAN,WM & HOWARD,PH (1993)

CAS #: 077470-88-3			HEXANAMIDE, 5-AMINO-N-(2,6-DIMETHYLPHENYL)-
Formula: $C_{14}H_{22}N_2O$			
Mol Weight: 234.34			
MP (deg C):		FP (deg C):	
BP (deg C):			
BP pressure (mm Hg):			

Property/Value	Units	Temp	Data Type	Reference
WS 1.80E+003	mg/L	25	EST	MEYLAN,WM ET AL. (1996)
logP 1.41			EXP	SANGSTER,J (1993)
VP 3.83E-007	mm Hg	25	EST	NEELY,WB & BLAU,GE (1985)
DC	pKa			
HL 2.40E-012	atm m3/mol	25	EST	MEYLAN,WM & HOWARD,PH (1991)
OH 6.90E-011	cm3/molc sec	25	EST	MEYLAN,WM & HOWARD,PH (1993)

077470-99-6 — HEXANAMIDE, 5-(DIETHYLAMINO)-N-(2,6-DIMETHYLPHEN

CAS #:	077470-99-6
Formula:	$C_{18}H_{30}N_2O$
Mol Weight:	290.45
MP (deg C):	
FP (deg C):	
BP (deg C):	
BP pressure (mm Hg):	

Property/Value	Units	Temp	Data Type	Reference
WS 7.76E+001	mg/L	25	EST	MEYLAN,WM ET AL. (1996)
logP 2.64			EXP	SANGSTER,J (1993)
VP 6.25E-008	mm Hg	25	EST	NEELY,WB & BLAU,GE (1985)
DC	pKa			
HL 2.04E-011	atm m3/mol	25	EST	MEYLAN,WM & HOWARD,PH (1991)
OH 1.32E-010	cm3/molc sec	25	EST	MEYLAN,WM & HOWARD,PH (1993)

077471-00-2 — BUTANAMIDE, 3-(DIETHYLAMINO)-N-(2,6-DIMETHYLPHEN

CAS #:	077471-00-2
Formula:	$C_{16}H_{26}N_2O$
Mol Weight:	262.40
MP (deg C):	
FP (deg C):	
BP (deg C):	
BP pressure (mm Hg):	

Property/Value	Units	Temp	Data Type	Reference
WS 1.15E+002	mg/L	25	EST	MEYLAN,WM ET AL. (1996)
logP 2.63			EXP	SANGSTER,J (1993)
VP 3.45E-007	mm Hg	25	EST	NEELY,WB & BLAU,GE (1985)
DC	pKa			
HL 1.16E-011	atm m3/mol	25	EST	MEYLAN,WM & HOWARD,PH (1991)
OH 1.74E-010	cm3/molc sec	25	EST	MEYLAN,WM & HOWARD,PH (1993)

077501-60-1 — BENZOFLUORFEN

CAS #:	077501-60-1
Formula:	$C_{16}H_9ClF_3NO_7$
Mol Weight:	419.70
MP (deg C):	
FP (deg C):	
BP (deg C):	
BP pressure (mm Hg):	

Property/Value	Units	Temp	Data Type	Reference
WS 5.27E-001	mg/L	25	EST	MEYLAN,WM ET AL. (1996)
logP 4.27			EXP	NANDIHALLI,UB ET AL. (1993)
VP 4.79E-010	mm Hg	25	EST	NEELY,WB & BLAU,GE (1985)
DC	pKa			
HL 1.02E-012	atm m3/mol	25	EST	MEYLAN,WM & HOWARD,PH (1991)
OH 2.07E-012	cm3/molc sec	25	EST	MEYLAN,WM & HOWARD,PH (1993)

077501-63-4 — LACTOFEN

CAS #:	077501-63-4
Formula:	$C_{19}H_{15}ClF_3NO_7$
Mol Weight:	461.78
MP (deg C):	
FP (deg C):	
BP (deg C):	
BP pressure (mm Hg):	

Property/Value	Units	Temp	Data Type	Reference
WS 1.00E-001	mg/L		EXP	REDDY,KN ET AL (1994)
logP 4.81			EXP	NANDIHALLI,UB ET AL. (1993)
VP 7.69E-009	mm Hg	25	EST	NEELY,WB & BLAU,GE (1985)
DC	pKa			
HL 5.73E-010	atm m3/mol	25	EST	MEYLAN,WM & HOWARD,PH (1991)
OH 4.67E-012	cm3/molc sec	25	EST	MEYLAN,WM & HOWARD,PH (1993)

077580-78-0 — FURAZAN, METHYLPROPYL-

CAS #:	077580-78-0
Formula:	$C_6H_{10}N_2O$
Mol Weight:	126.16
MP (deg C):	
FP (deg C):	
BP (deg C):	
BP pressure (mm Hg):	

Property/Value	Units	Temp	Data Type	Reference
WS 3.30E+003	mg/L	25	EST	MEYLAN,WM ET AL. (1996)
logP 1.71			EXP	CALVINO,R ET AL. (1992)
VP 1.28E+000	mm Hg	25	EST	NEELY,WB & BLAU,GE (1985)
DC	pKa			
HL 4.06E-004	atm m3/mol	25	EST	MEYLAN,WM & HOWARD,PH (1991)
OH 6.60E-012	cm3/molc sec	25	EST	MEYLAN,WM & HOWARD,PH (1993)

077666-53-6 — FURAZAN, METHYLNITRO

CAS #:	077666-53-6
Formula:	$C_3H_3N_3O_3$
Mol Weight:	129.08
MP (deg C):	
FP (deg C):	
BP (deg C):	
BP pressure (mm Hg):	

Property/Value	Units	Temp	Data Type	Reference
WS 5.61E+003	mg/L	25	EST	MEYLAN,WM ET AL. (1996)
logP 0.97			EXP	CALVINO,R ET AL. (1992)
VP 9.56E-002	mm Hg	25	EST	NEELY,WB & BLAU,GE (1985)
DC	pKa			
HL 8.24E-007	atm m3/mol	25	EST	MEYLAN,WM & HOWARD,PH (1991)
OH 4.14E-012	cm3/molc sec	25	EST	MEYLAN,WM & HOWARD,PH (1993)

077753-24-3 — 1,1,2,3,4-PENTACHLOROBUTANE

CAS #:	077753-24-3
Formula:	$C_4H_5Cl_5$
Mol Weight:	230.35
MP (deg C):	
FP (deg C):	
BP (deg C):	95.5
BP pressure (mm Hg):	1.10E+001

Property/Value	Units	Temp	Data Type	Reference
WS 4.13E+001	mg/L	25	EST	MEYLAN,WM ET AL. (1996)
logP 3.36			EST	MEYLAN,WM & HOWARD,PH (1995)
VP 6.92E-002	mm Hg	25	EST	NEELY,WB & BLAU,GE (1985)
DC	pKa			
HL 9.35E-004	atm m3/mol	25	EST	MEYLAN,WM & HOWARD,PH (1991)
OH 4.20E-013	cm3/molc sec	25	EST	MEYLAN,WM & HOWARD,PH (1993)

077812-86-3 — N,N-DIME-4-MEAM BENZENESULFONAMIDE

CAS #:	077812-86-3
Formula:	$C_9H_{14}N_2O_2S$
Mol Weight:	214.29
MP (deg C):	
FP (deg C):	
BP (deg C):	
BP pressure (mm Hg):	

Property/Value	Units	Temp	Data Type	Reference
WS 2.22E+003	mg/L	25	EST	MEYLAN,WM ET AL. (1996)
logP 1.43			EXP	HANSCH,C & LEO,AJ (1985)
VP 4.09E-005	mm Hg	25	EST	NEELY,WB & BLAU,GE (1985)
DC	pKa			
HL 1.58E-009	atm m3/mol	25	EST	MEYLAN,WM & HOWARD,PH (1991)
OH 1.29E-011	cm3/molc sec	25	EST	MEYLAN,WM & HOWARD,PH (1993)

077812-87-4 — NN-DIME-4-(N-ME-N-FORMYL)BENZENE-SO2N

Formula: $C_{10}H_{14}N_2O_3S$

Mol Weight: 242.30

MP (deg C): FP (deg C):

BP (deg C):

BP pressure (mm Hg):

Property/Value	Units	Temp	Data Type	Reference
WS 1.19E+004	mg/L	25	EST	MEYLAN,WM ET AL. (1996)
logP 0.40			EXP	HANSCH,C & LEO,AJ (1985)
VP 1.31E-006	mm Hg	25	EST	NEELY,WB & BLAU,GE (1985)
DC	pKa			
HL 6.54E-011	atm m3/mol	25	EST	MEYLAN,WM & HOWARD,PH (1991)
OH 6.45E-012	cm3/molc sec	25	EST	MEYLAN,WM & HOWARD,PH (1993)

077812-88-5 — 4-PENTYNAMIDE, N-[4-[(DIMETHYLAMINO)SULFONYL]PHE

Formula: $C_{14}H_{18}N_2O_3S$

Mol Weight: 294.38

MP (deg C): FP (deg C):

BP (deg C):

BP pressure (mm Hg):

Property/Value	Units	Temp	Data Type	Reference
WS 2.35E+003	mg/L	25	EST	MEYLAN,WM ET AL. (1996)
logP 0.88			EXP	HANSCH,C ET AL. (1995)
VP 3.13E-008	mm Hg	25	EST	NEELY,WB & BLAU,GE (1985)
DC	pKa			
HL 1.38E-011	atm m3/mol	25	EST	MEYLAN,WM & HOWARD,PH (1991)
OH 1.72E-011	cm3/molc sec	25	EST	MEYLAN,WM & HOWARD,PH (1993)

077963-50-9 — B30C10-BENZO CROWN ETHER

Formula: $C_{24}H_{40}O_{10}$

Mol Weight: 488.58

MP (deg C): FP (deg C):

BP (deg C):

BP pressure (mm Hg):

Property/Value	Units	Temp	Data Type	Reference
WS 7.99E+002	mg/L	25	EST	MEYLAN,WM ET AL. (1996)
logP 0.03			EXP	STOLWIJK,TB ET AL. (1989)
VP 4.78E-012	mm Hg	25	EST	NEELY,WB & BLAU,GE (1985)
DC	pKa			
HL 2.83E-020	atm m3/mol	25	EST	MEYLAN,WM & HOWARD,PH (1991)
OH 1.45E-010	cm3/molc sec	25	EST	MEYLAN,WM & HOWARD,PH (1993)

078025-68-0 — 24NH2-PYRIMIDINE,5(3,5-MEO-4-ETO)BENZYL

Formula: $C_{15}H_{20}N_4O_3$

Mol Weight: 304.35

MP (deg C): FP (deg C):

BP (deg C):

BP pressure (mm Hg):

Property/Value	Units	Temp	Data Type	Reference
WS 8.31E+002	mg/L	25	EST	MEYLAN,WM ET AL. (1996)
logP 1.34			EXP	HANSCH,C & LEO,AJ (1985)
VP 4.39E-009	mm Hg	25	EST	NEELY,WB & BLAU,GE (1985)
DC	pKa			
HL 3.17E-014	atm m3/mol	25	EST	MEYLAN,WM & HOWARD,PH (1991)
OH 2.09E-010	cm3/molc sec	25	EST	MEYLAN,WM & HOWARD,PH (1993)

078026-01-4 — 24NH2-PYRIMIDINE,5(3,5-MEO-4-ET)BENZYL

Formula: $C_{15}H_{20}N_4O_2$

Mol Weight: 288.35

MP (deg C): FP (deg C):

BP (deg C):

BP pressure (mm Hg):

Property/Value	Units	Temp	Data Type	Reference
WS 6.82E+001	mg/L	25	EST	MEYLAN,WM ET AL. (1996)
logP 2.72			EXP	HANSCH,C & LEO,AJ (1985)
VP 1.02E-008	mm Hg	25	EST	NEELY,WB & BLAU,GE (1985)
DC	pKa			
HL 5.91E-013	atm m3/mol	25	EST	MEYLAN,WM & HOWARD,PH (1991)
OH 2.04E-010	cm3/molc sec	25	EST	MEYLAN,WM & HOWARD,PH (1993)

078090-11-6 — 1H-BENZIMIDAZOLE-5-CARBOXYLIC ACID, 6-METHYL-2-[

Formula: $C_{17}H_{17}N_3O_3S$

Mol Weight: 343.41

MP (deg C): FP (deg C):

BP (deg C):

BP pressure (mm Hg):

Property/Value	Units	Temp	Data Type	Reference
WS 1.21E+003	mg/L	25	EST	MEYLAN,WM ET AL. (1996)
logP 2.40			EXP	HANSCH,C ET AL. (1995)
VP 7.88E-013	mm Hg	25	EST	NEELY,WB & BLAU,GE (1985)
DC	pKa			
HL 5.59E-019	atm m3/mol	25	EST	MEYLAN,WM & HOWARD,PH (1991)
OH 7.26E-011	cm3/molc sec	25	EST	MEYLAN,WM & HOWARD,PH (1993)

078142-97-9 — N7-(3-PYRIDYL)-MITOMYCIN C

Formula: $C_{20}H_{21}N_5O_5$

Mol Weight: 411.42

MP (deg C): FP (deg C):

BP (deg C):

BP pressure (mm Hg):

Property/Value	Units	Temp	Data Type	Reference
WS 9.53E+003	mg/L	25	EST	MEYLAN,WM ET AL. (1996)
logP 0.52			EXP	HANSCH,C & LEO,AJ (1985)
VP 2.11E-012	mm Hg	25	EST	NEELY,WB & BLAU,GE (1985)
DC	pKa			
HL 8.22E-028	atm m3/mol	25	EST	MEYLAN,WM & HOWARD,PH (1991)
OH 5.41E-011	cm3/molc sec	25	EST	MEYLAN,WM & HOWARD,PH (1993)

078182-93-1 — (DES-IPR) N,N-DIETAMINOET-CLOFAZIMINE

Formula: $C_{30}H_{29}Cl_2N_5$

Mol Weight: 530.51

MP (deg C): FP (deg C):

BP (deg C):

BP pressure (mm Hg):

Property/Value	Units	Temp	Data Type	Reference
WS 1.03E-001	mg/L	25	EST	MEYLAN,WM ET AL. (1996)
logP 4.27			EXP	QUIGLEY,JM ET AL. (1990)
VP 2.95E-014	mm Hg	25	EST	NEELY,WB & BLAU,GE (1985)
DC	pKa			
HL 2.37E-014	atm m3/mol	25	EST	MEYLAN,WM & HOWARD,PH (1991)
OH 2.30E-010	cm3/molc sec	25	EST	MEYLAN,WM & HOWARD,PH (1993)

CAS #: 078191-89-6				2,4-DIME PYRIDINIUM I,N-HEXADECYL

Formula: $C_{23}H_{42}IN$

Mol Weight: 459.50

MP (deg C):   FP (deg C):

BP (deg C):

BP pressure (mm Hg):

Property/Value	Units	Temp	Data Type	Reference
WS 6.06E-001	mg/L	25	EST	MEYLAN,WM ET AL. (1996)
logP 3.90			EXP	HANSCH,C & LEO,AJ (1985)
VP 1.84E-008	mm Hg	25	EST	NEELY,WB & BLAU,GE (1985)
DC	pKa			
HL 1.28E-008	atm m3/mol	25	EST	MEYLAN,WM & HOWARD,PH (1991)
OH 3.36E-011	cm3/molc sec	25	EST	MEYLAN,WM & HOWARD,PH (1993)

CAS #: 078233-72-4				L-LEUCINAMIDE, N-ACETYL-L-ALANYL-

Formula: $C_{11}H_{21}N_3O_3$

Mol Weight: 243.31

MP (deg C):   FP (deg C):

BP (deg C):

BP pressure (mm Hg):

Property/Value	Units	Temp	Data Type	Reference
WS 3.65E+003	mg/L	25	EST	MEYLAN,WM ET AL. (1996)
logP -0.54			EXP	HANSCH,C ET AL. (1995)
VP 2.53E-010	mm Hg	25	EST	NEELY,WB & BLAU,GE (1985)
DC	pKa			
HL 3.24E-015	atm m3/mol	25	EST	MEYLAN,WM & HOWARD,PH (1991)
OH 3.41E-011	cm3/molc sec	25	EST	MEYLAN,WM & HOWARD,PH (1993)

CAS #: 078233-99-5				24-NH2PYRIMIDINE,5(4MEO-3PHMEO)BENZYL

Formula: $C_{19}H_{20}N_4O_2$

Mol Weight: 336.40

MP (deg C):   FP (deg C):

BP (deg C):

BP pressure (mm Hg):

Property/Value	Units	Temp	Data Type	Reference
WS 4.00E+001	mg/L	25	EST	MEYLAN,WM ET AL. (1996)
logP 2.44			EXP	HANSCH,C & LEO,AJ (1985)
VP 1.39E-010	mm Hg	25	EST	NEELY,WB & BLAU,GE (1985)
DC	pKa			
HL 3.25E-014	atm m3/mol	25	EST	MEYLAN,WM & HOWARD,PH (1991)
OH 6.01E-011	cm3/molc sec	25	EST	MEYLAN,WM & HOWARD,PH (1993)

CAS #: 078266-06-5				BR-N(246-MEPH)CARBAMOYLME IMINODIACETIC ACID

Formula: $C_{15}H_{19}BrN_2O_5$

Mol Weight: 387.23

MP (deg C): 198-200   FP (deg C):

BP (deg C):

BP pressure (mm Hg):

Property/Value	Units	Temp	Data Type	Reference
WS 1.23E+000	mg/L	25	EST	MEYLAN,WM ET AL. (1996)
logP 0.11			EXP	GODOY,N ET AL. (1989)
VP 2.53E-014	mm Hg	25	EST	NEELY,WB & BLAU,GE (1985)
DC	pKa			
HL 7.20E-019	atm m3/mol	25	EST	MEYLAN,WM & HOWARD,PH (1991)
OH 9.22E-011	cm3/molc sec	25	EST	MEYLAN,WM & HOWARD,PH (1993)

CAS #: 078326-57-5				1-I-PROPYL-1-NITROSO-3-PHENYLUREA

Formula: $C_{10}H_{13}N_3O_2$

Mol Weight: 207.23

MP (deg C):   FP (deg C):

BP (deg C):

BP pressure (mm Hg):

Property/Value	Units	Temp	Data Type	Reference
WS 2.94E+002	mg/L	25	EST	MEYLAN,WM ET AL. (1996)
logP 2.50			EXP	HANSCH,C & LEO,AJ (1985)
VP 2.78E-005	mm Hg	25	EST	NEELY,WB & BLAU,GE (1985)
DC	pKa			
HL 9.64E-011	atm m3/mol	25	EST	MEYLAN,WM & HOWARD,PH (1991)
OH 5.09E-011	cm3/molc sec	25	EST	MEYLAN,WM & HOWARD,PH (1993)

CAS #: 078327-27-2				N7-(2-THIAZOLYL)-MITOMYCIN C

Formula: $C_{18}H_{19}N_5O_5S$

Mol Weight: 417.45

MP (deg C):   FP (deg C):

BP (deg C):

BP pressure (mm Hg):

Property/Value	Units	Temp	Data Type	Reference
WS 7.18E+003	mg/L	25	EST	MEYLAN,WM ET AL. (1996)
logP 0.62			EXP	HANSCH,C & LEO,AJ (1985)
VP 1.39E-012	mm Hg	25	EST	NEELY,WB & BLAU,GE (1985)
DC	pKa			
HL 4.47E-028	atm m3/mol	25	EST	MEYLAN,WM & HOWARD,PH (1991)
OH 7.66E-011	cm3/molc sec	25	EST	MEYLAN,WM & HOWARD,PH (1993)

CAS #: 078508-43-7				N'-(4-BROMO-3,5-DIMETHYLPHENYL)-N,N-DIMETHYLU*

Formula: $C_{11}H_{15}BrN_2O$

Mol Weight: 271.16

MP (deg C):   FP (deg C):

BP (deg C):

BP pressure (mm Hg):

Property/Value	Units	Temp	Data Type	Reference
WS 5.77E+001	mg/L	25	EST	MEYLAN,WM ET AL. (1996)
logP 2.92			EXP	HANSCH,C & LEO,AJ (1985)
VP 3.43E-006	mm Hg	25	EST	NEELY,WB & BLAU,GE (1985)
DC	pKa			
HL 4.71E-010	atm m3/mol	25	EST	MEYLAN,WM & HOWARD,PH (1991)
OH 8.57E-011	cm3/molc sec	25	EST	MEYLAN,WM & HOWARD,PH (1993)

CAS #: 078508-44-8				(4-PHENOXYPHENYL)UREA

Formula: $C_{13}H_{12}N_2O_2$

Mol Weight: 228.25

MP (deg C):   FP (deg C):

BP (deg C):

BP pressure (mm Hg):

Property/Value	Units	Temp	Data Type	Reference
WS 1.26E+002	mg/L	25	EST	MEYLAN,WM ET AL. (1996)
logP 2.80			EXP	HANSCH,C & LEO,AJ (1985)
VP 6.11E-007	mm Hg	25	EST	NEELY,WB & BLAU,GE (1985)
DC	pKa			
HL 4.40E-012	atm m3/mol	25	EST	MEYLAN,WM & HOWARD,PH (1991)
OH 4.86E-011	cm3/molc sec	25	EST	MEYLAN,WM & HOWARD,PH (1993)

CAS #: 078508-45-9				(4-FLUORO-3-METHYLPHENYL)UREA
Formula: C8H9FN2O				
Mol Weight: 168.17				
MP (deg C):		FP (deg C):		
BP (deg C):				
BP pressure (mm Hg):				

Property/ Value	Units	Temp	Data Type	Reference
WS 2.75E+003	mg/L	25	EST	MEYLAN,WM ET AL. (1996)
logP 1.59			EXP	BRIGGS,GG (1981)
VP 3.65E-004	mm Hg	25	EST	NEELY,WB & BLAU,GE (1985)
DC	pKa			
HL 2.59E-010	atm m3/mol	25	EST	MEYLAN,WM & HOWARD,PH (1991)
OH 3.87E-011	cm3/molc sec	25	EST	MEYLAN,WM & HOWARD,PH (1993)

CAS #: 078508-46-0				(3-METHYL-4-BROMOPHENYL)UREA
Formula: C8H9BrN2O				
Mol Weight: 229.08				
MP (deg C):		FP (deg C):		
BP (deg C):				
BP pressure (mm Hg):				

Property/ Value	Units	Temp	Data Type	Reference
WS 2.30E+002	mg/L	25	EST	MEYLAN,WM ET AL. (1996)
logP 2.49			EXP	HANSCH,C & LEO,AJ (1985)
VP 2.68E-005	mm Hg	25	EST	NEELY,WB & BLAU,GE (1985)
DC	pKa			
HL 8.85E-011	atm m3/mol	25	EST	MEYLAN,WM & HOWARD,PH (1991)
OH 3.31E-011	cm3/molc sec	25	EST	MEYLAN,WM & HOWARD,PH (1993)

CAS #: 078513-74-3				1,9-PHENANTHRIDINEDIOL, 5,6,6A,7,8,9,10,10A-OCTA
Formula: C28H37NO4				
Mol Weight: 451.61				
MP (deg C):		FP (deg C):		
BP (deg C):				
BP pressure (mm Hg):				

Property/ Value	Units	Temp	Data Type	Reference
WS 1.90E-002	mg/L	25	EST	MEYLAN,WM ET AL. (1996)
logP 5.72			EXP	SANGSTER,J (1993)
VP 6.56E-014	mm Hg	25	EST	NEELY,WB & BLAU,GE (1985)
DC	pKa			
HL 6.56E-013	atm m3/mol	25	EST	MEYLAN,WM & HOWARD,PH (1991)
OH 2.64E-010	cm3/molc sec	25	EST	MEYLAN,WM & HOWARD,PH (1993)

CAS #: 078593-33-6				CINNOLINE, 3-BROMO-
Formula: C8H5BrN2				
Mol Weight: 209.05				
MP (deg C):		FP (deg C):		
BP (deg C):				
BP pressure (mm Hg):				

Property/ Value	Units	Temp	Data Type	Reference
WS 1.34E+003	mg/L	25	EST	MEYLAN,WM ET AL. (1996)
logP 1.72			EXP	SANGSTER,J (1993)
VP 4.13E-005	mm Hg	25	EST	NEELY,WB & BLAU,GE (1985)
DC	pKa			
HL 1.12E-007	atm m3/mol	25	EST	MEYLAN,WM & HOWARD,PH (1991)
OH 1.26E-012	cm3/molc sec	25	EST	MEYLAN,WM & HOWARD,PH (1993)

CAS #: 078617-12-6				á-D-GALACTOPYRANOSIDE, HEPTYL
Formula: C13H26O6				
Mol Weight: 278.35				
MP (deg C):		FP (deg C):		
BP (deg C):				
BP pressure (mm Hg):				

Property/ Value	Units	Temp	Data Type	Reference
WS 3.90E+003	mg/L	25	EST	MEYLAN,WM ET AL. (1996)
logP 0.73			EXP	SANGSTER,J (1993)
VP 9.30E-011	mm Hg	25	EST	NEELY,WB & BLAU,GE (1985)
DC	pKa			
HL 8.53E-014	atm m3/mol	25	EST	MEYLAN,WM & HOWARD,PH (1991)
OH 7.86E-011	cm3/molc sec	25	EST	MEYLAN,WM & HOWARD,PH (1993)

CAS #: 078755-81-4				FLUMAZENIL
Formula: C15H14FN3O3				
Mol Weight: 303.30				
MP (deg C): 201-203		FP (deg C):		
BP (deg C):				
BP pressure (mm Hg):				

Property/ Value	Units	Temp	Data Type	Reference
WS 1.28E+002	mg/L	25	EST	MEYLAN,WM ET AL. (1996)
logP 1.00			EXP	MFG DATA SHEET
VP 8.30E-010	mm Hg	25	EST	NEELY,WB & BLAU,GE (1985)
DC	pKa			
HL 9.89E-017	atm m3/mol	25	EST	MEYLAN,WM & HOWARD,PH (1991)
OH 3.63E-011	cm3/molc sec	25	EST	MEYLAN,WM & HOWARD,PH (1993)

CAS #: 078798-46-6				O-MEO BENZAMIDE, N-(2-DIMEAMINOET)
Formula: C12H18N2O2				
Mol Weight: 222.29				
MP (deg C):		FP (deg C):		
BP (deg C):				
BP pressure (mm Hg):				

Property/ Value	Units	Temp	Data Type	Reference
WS 3.56E+003	mg/L	25	EST	MEYLAN,WM ET AL. (1996)
logP 1.14			EXP	HANSCH,C & LEO,AJ (1985)
VP 2.43E-006	mm Hg	25	EST	NEELY,WB & BLAU,GE (1985)
DC	pKa			
HL 1.89E-013	atm m3/mol	25	EST	MEYLAN,WM & HOWARD,PH (1991)
OH 1.10E-010	cm3/molc sec	25	EST	MEYLAN,WM & HOWARD,PH (1993)

CAS #: 078834-86-3				2-IMIDAZOLIDINIMINE, N-(3-BROMO-2-CHLOROPHENYL)-
Formula: C9H9BrClN3				
Mol Weight: 274.55				
MP (deg C):		FP (deg C):		
BP (deg C):				
BP pressure (mm Hg):				

Property/ Value	Units	Temp	Data Type	Reference
WS 3.11E+003	mg/L	25	EST	MEYLAN,WM ET AL. (1996)
logP 2.05			EXP	SANGSTER,J (1993)
VP 8.80E-006	mm Hg	25	EST	NEELY,WB & BLAU,GE (1985)
DC	pKa			
HL 8.09E-012	atm m3/mol	25	EST	MEYLAN,WM & HOWARD,PH (1991)
OH 1.44E-010	cm3/molc sec	25	EST	MEYLAN,WM & HOWARD,PH (1993)

CAS #: 078834-88-5 — 2-IMIDAZOLIDINIMINE, N-(2,3-DIBROMOPHENYL)-

Formula: $C_9H_9Br_2N_3$

Mol Weight: 319.01

MP (deg C): | FP (deg C):

BP (deg C):

BP pressure (mm Hg):

Property/Value	Units	Temp	Data Type	Reference
WS 1.38E+003	mg/L	25	EST	MEYLAN,WM ET AL. (1996)
logP 2.16			EXP	SANGSTER,J (1993)
VP 3.39E-006	mm Hg	25	EST	NEELY,WB & BLAU,GE (1985)
DC	pKa			
HL 4.35E-012	atm m3/mol	25	EST	MEYLAN,WM & HOWARD,PH (1991)
OH 1.43E-010	cm3/molc sec	25	EST	MEYLAN,WM & HOWARD,PH (1993)

CAS #: 078879-25-1 — N(2,4-NO2-6-CF3 PH)-2(3,5CL)PYRIDINAMINE

Formula: $C_{12}H_5Cl_2F_3N_4O_4$

Mol Weight: 397.10

MP (deg C): | FP (deg C):

BP (deg C):

BP pressure (mm Hg):

Property/Value	Units	Temp	Data Type	Reference
WS 3.47E+000	mg/L	25	EST	MEYLAN,WM ET AL. (1996)
logP 3.47			EXP	BRANDT,U ET AL. (1992)
VP 9.40E-009	mm Hg	25	EST	NEELY,WB & BLAU,GE (1985)
DC	pKa			
HL 2.23E-012	atm m3/mol	25	EST	MEYLAN,WM & HOWARD,PH (1991)
OH 3.04E-013	cm3/molc sec	25	EST	MEYLAN,WM & HOWARD,PH (1993)

CAS #: 078879-38-6 — N(24-NO2-6CF3 PH)-2(3CL-5CF3)PYRIDINAMIN

Formula: $C_{13}H_6ClF_5N_4O_4$

Mol Weight: 412.66

MP (deg C): | FP (deg C):

BP (deg C):

BP pressure (mm Hg):

Property/Value	Units	Temp	Data Type	Reference
WS 2.52E+000	mg/L	25	EST	MEYLAN,WM ET AL. (1996)
logP 3.52			EXP	BRANDT,U ET AL. (1992)
VP 2.14E-008	mm Hg	25	EST	NEELY,WB & BLAU,GE (1985)
DC	pKa			
HL 1.32E-011	atm m3/mol	25	EST	MEYLAN,WM & HOWARD,PH (1991)
OH 1.07E-013	cm3/molc sec	25	EST	MEYLAN,WM & HOWARD,PH (1993)

CAS #: 078944-67-9 — BENZOIC ACID,O-ACETYLMETHYLAMINO

Formula: $C_{10}H_{11}NO_3$

Mol Weight: 193.20

MP (deg C): | FP (deg C):

BP (deg C):

BP pressure (mm Hg):

Property/Value	Units	Temp	Data Type	Reference
WS 1.25E+003	mg/L	25	EST	MEYLAN,WM ET AL. (1996)
logP 1.85			EXP	HANSCH,C & LEO,AJ (1985)
VP 1.47E-005	mm Hg	25	EST	NEELY,WB & BLAU,GE (1985)
DC	pKa			
HL 1.13E-012	atm m3/mol	25	EST	MEYLAN,WM & HOWARD,PH (1991)
OH 2.29E-011	cm3/molc sec	25	EST	MEYLAN,WM & HOWARD,PH (1993)

CAS #: 078982-40-8 — 2H-FURO[2,3-H]-1-BENZOPYRAN-2-ONE, 9-METHYL-

Formula: $C_{12}H_8O_3$

Mol Weight: 200.20

MP (deg C): | FP (deg C):

BP (deg C):

BP pressure (mm Hg):

Property/Value	Units	Temp	Data Type	Reference
WS 4.05E+002	mg/L	25	EST	MEYLAN,WM ET AL. (1996)
logP 2.38			EXP	SANGSTER,J (1993)
VP 9.99E-006	mm Hg	25	EST	NEELY,WB & BLAU,GE (1985)
DC	pKa			
HL 7.47E-007	atm m3/mol	25	EST	MEYLAN,WM & HOWARD,PH (1991)
OH 6.60E-011	cm3/molc sec	25	EST	MEYLAN,WM & HOWARD,PH (1993)

CAS #: 079127-80-3 — FENOXYCARB

Formula: $C_{17}H_{19}NO_4$

Mol Weight: 301.35

MP (deg C): 53-54 | FP (deg C):

BP (deg C):

BP pressure (mm Hg):

Property/Value	Units	Temp	Data Type	Reference
WS 6.00E+000	mg/L	20	EXP	WORTHING,CR & WALKER,SB (1987)
logP 3.69			EST	LYMAN,WJ ET AL. (1990)
VP 1.28E-005	mm Hg	25	EXP	WORTHING,CR & WALKER,SB (1987)
DC	pKa			
HL 8.42E-007	atm m3/mol	25	EST	VP/WSOL
OH 1.32E-010	cm3/molc sec	25	EST	ATKINSON,R (1987)

CAS #: 079277-27-3 — HARMONY

Formula: $C_{12}H_{15}N_5O_6S_2$

Mol Weight: 389.41

MP (deg C): | FP (deg C):

BP (deg C):

BP pressure (mm Hg):

Property/Value	Units	Temp	Data Type	Reference
WS 1.71E+002	mg/L	25	EST	MEYLAN,WM ET AL. (1996)
logP 1.56			EXP	HAY,JV (1990)
VP 2.20E-012	mm Hg	25	EST	NEELY,WB & BLAU,GE (1985)
DC	pKa			
HL 4.08E-014	atm m3/mol	25	EST	MEYLAN,WM & HOWARD,PH (1991)
OH 3.10E-012	cm3/molc sec	25	EST	MEYLAN,WM & HOWARD,PH (1993)

CAS #: 079353-74-5 — 4-I-BUTYLSEMICARBAZIDE

Formula: $C_5H_{13}N_3O$

Mol Weight: 131.18

MP (deg C): | FP (deg C):

BP (deg C):

BP pressure (mm Hg):

Property/Value	Units	Temp	Data Type	Reference
WS 2.32E+004	mg/L	25	EST	MEYLAN,WM ET AL. (1996)
logP -0.84			EXP	HANSCH,C & LEO,AJ (1985)
VP 4.30E-003	mm Hg	25	EST	NEELY,WB & BLAU,GE (1985)
DC	pKa			
HL 7.83E-012	atm m3/mol	25	EST	MEYLAN,WM & HOWARD,PH (1991)
OH 9.43E-012	cm3/molc sec	25	EST	MEYLAN,WM & HOWARD,PH (1993)

CAS #: 079353-76-7				4-HEXYLSEMICARBAZIDE

Formula: $C_7H_{17}N_3O$
Mol Weight: 159.23
MP (deg C): **FP (deg C):**
BP (deg C):
BP pressure (mm Hg):

Property/ Value	Units	Temp	Data Type	Reference
WS 2.79E+004	mg/L	25	EST	MEYLAN,WM ET AL. (1996)
logP 0.46			EXP	HANSCH,C & LEO,AJ (1985)
VP 2.75E-004	mm Hg	25	EST	NEELY,WB & BLAU,GE (1985)
DC	pKa			
HL 1.38E-011	atm m3/mol	25	EST	MEYLAN,WM & HOWARD,PH (1991)
OH 1.23E-011	cm3/molc sec	25	EST	MEYLAN,WM & HOWARD,PH (1993)

CAS #: 079402-07-6				3,5-DIBR-2',4'-DIFLUORO SALICYLANILIDE

Formula: $C_{13}H_7Br_2F_2NO_2$
Mol Weight: 407.02
MP (deg C): **FP (deg C):**
BP (deg C):
BP pressure (mm Hg):

Property/ Value	Units	Temp	Data Type	Reference
WS 5.77E-001	mg/L	25	EST	MEYLAN,WM ET AL. (1996)
logP 4.99			EXP	TERADA,H ET AL. (1988)
VP 3.64E-010	mm Hg	25	EST	NEELY,WB & BLAU,GE (1985)
DC	pKa			
HL 3.46E-011	atm m3/mol	25	EST	MEYLAN,WM & HOWARD,PH (1991)
OH 1.96E-012	cm3/molc sec	25	EST	MEYLAN,WM & HOWARD,PH (1993)

CAS #: 079458-54-1				1,1,1,4,4,4-HEXACHLOROBUTANE

Formula: $C_4H_4Cl_6$
Mol Weight: 264.79
MP (deg C): **FP (deg C):**
BP (deg C):
BP pressure (mm Hg):

Property/ Value	Units	Temp	Data Type	Reference
WS 1.02E+000	mg/L	25	EST	MEYLAN,WM ET AL. (1996)
logP 5.02			EST	MEYLAN,WM & HOWARD,PH (1995)
VP 4.31E-001	mm Hg	25	EST	NEELY,WB & BLAU,GE (1985)
DC	pKa			
HL 3.29E-004	atm m3/mol	25	EST	MEYLAN,WM & HOWARD,PH (1991)
OH 1.59E-013	cm3/molc sec	25	EST	MEYLAN,WM & HOWARD,PH (1993)

CAS #: 079467-22-4				BENZENEMETHANOL, 2-[[2-(AMINOMETHYL)PHENYL]THIO]

Formula: $C_{14}H_{15}NOS$
Mol Weight: 245.35
MP (deg C): 96-98 **FP (deg C):**
BP (deg C):
BP pressure (mm Hg):

Property/ Value	Units	Temp	Data Type	Reference
WS 2.33E+003	mg/L	25	EST	MEYLAN,WM ET AL. (1996)
logP 2.39			EXP	SANGSTER,J (1994)
VP 9.48E-009	mm Hg	25	EST	NEELY,WB & BLAU,GE (1985)
DC	pKa			
HL 1.44E-013	atm m3/mol	25	EST	MEYLAN,WM & HOWARD,PH (1991)
OH 6.37E-011	cm3/molc sec	25	EST	MEYLAN,WM & HOWARD,PH (1993)

CAS #: 079508-78-4				46-NH2 22-DIME-1(3-PHOMEPH)S-TRIAZINE

Formula: $C_{18}H_{21}N_5O$
Mol Weight: 323.40
MP (deg C): **FP (deg C):**
BP (deg C):
BP pressure (mm Hg):

Property/ Value	Units	Temp	Data Type	Reference
WS 2.31E+002	mg/L	25	EST	MEYLAN,WM ET AL. (1996)
logP 1.86			EXP	HANSCH,C & LEO,AJ (1985)
VP 3.15E-009	mm Hg	25	EST	NEELY,WB & BLAU,GE (1985)
DC	pKa			
HL 3.48E-015	atm m3/mol	25	EST	MEYLAN,WM & HOWARD,PH (1991)
OH 2.48E-010	cm3/molc sec	25	EST	MEYLAN,WM & HOWARD,PH (1993)

CAS #: 079508-86-4				ACETAMIDE, N-[3-[[3-(4,6-DIAMINO-2,2-DIMETHYL-1,

Formula: $C_{20}H_{24}N_6O_2$
Mol Weight: 380.45
MP (deg C): **FP (deg C):**
BP (deg C):
BP pressure (mm Hg):

Property/ Value	Units	Temp	Data Type	Reference
WS 4.21E+002	mg/L	25	EST	MEYLAN,WM ET AL. (1996)
logP 1.15			EXP	HANSCH,C ET AL. (1995)
VP 6.07E-013	mm Hg	25	EST	NEELY,WB & BLAU,GE (1985)
DC	pKa			
HL 3.98E-021	atm m3/mol	25	EST	MEYLAN,WM & HOWARD,PH (1991)
OH 2.48E-010	cm3/molc sec	25	EST	MEYLAN,WM & HOWARD,PH (1993)

CAS #: 079510-48-8				METSULFURON-METHYL (PH5-7)

Formula: $C_{13}H_{13}N_5O_6S$
Mol Weight: 367.34
MP (deg C): **FP (deg C):**
BP (deg C):
BP pressure (mm Hg):

Property/ Value	Units	Temp	Data Type	Reference
WS 1.72E+002	mg/L	25	EST	MEYLAN,WM ET AL. (1996)
logP 1.70			EXP	HAY,JV (1990)
VP 2.99E-013	mm Hg	25	EST	NEELY,WB & BLAU,GE (1985)
DC	pKa			
HL 2.35E-016	atm m3/mol	25	EST	MEYLAN,WM & HOWARD,PH (1991)
OH 2.90E-012	cm3/molc sec	25	EST	MEYLAN,WM & HOWARD,PH (1993)

CAS #: 079538-32-2				TETRAFLUTHRIN

Formula: $C_{17}H_{14}ClF_7O_2$
Mol Weight: 418.74
MP (deg C): 44.6 **FP (deg C):**
BP (deg C):
BP pressure (mm Hg):

Property/ Value	Units	Temp	Data Type	Reference
WS 2.00E-002	mg/L	20	EXP	SHIU,WY ET AL. (1990)
logP 6.50			EXP	TOMLIN,C (1994)
VP 6.00E-005	mm Hg	20	EXP	BEWICK,DW ET AL. (1986)
DC	pKa			
HL 1.65E-003	atm m3/mol	20	EST	VP/WSOL
OH 1.70E-011	cm3/molc sec	25	EST	MEYLAN,WM & HOWARD,PH (1993)

CAS #: 079606-43-2 — BENZAMIDE, 4-METHOXY-N,N-BIS(1-METHYLETHYL)-

Formula: $C_{14}H_{21}NO_2$

Mol Weight: 235.33

MP (deg C): | FP (deg C):

BP (deg C):

BP pressure (mm Hg):

Property/Value	Units	Temp	Data Type	Reference
WS 1.16E+002	mg/L	25	EST	MEYLAN,WM ET AL. (1996)
logP 2.80			EXP	SURYANARAYANA,MVS ET AL. (1991)
VP 1.55E-001	mm Hg	25	EXP	SURYANARAYANA,MVS ET AL. (1991)
DC	pKa			
HL 1.96E-009	atm m3/mol	25	EST	MEYLAN,WM & HOWARD,PH (1991)
OH 5.20E-011	cm3/molc sec	25	EST	MEYLAN,WM & HOWARD,PH (1993)

CAS #: 079622-59-6 — FLUAZINAM

Formula: $C_{13}H_4Cl_2F_6N_4O_4$

Mol Weight: 465.10

MP (deg C): | FP (deg C):

BP (deg C):

BP pressure (mm Hg):

Property/Value	Units	Temp	Data Type	Reference
WS 1.93E+000	mg/L	25	EST	MEYLAN,WM ET AL. (1996)
logP 3.27			EXP	BRANDT,U ET AL. (1992)
VP 7.41E-009	mm Hg	25	EST	NEELY,WB & BLAU,GE (1985)
DC	pKa			
HL 5.23E-009	atm m3/mol	25	EST	MEYLAN,WM & HOWARD,PH (1991)
OH 6.56E-014	cm3/molc sec	25	EST	MEYLAN,WM & HOWARD,PH (1993)

CAS #: 079660-72-3 — FLEROXACIN

Formula: $C_{17}H_{18}F_3N_3O_3$

Mol Weight: 369.35

MP (deg C): 269-271 de | FP (deg C):

BP (deg C):

BP pressure (mm Hg):

Property/Value	Units	Temp	Data Type	Reference
WS 7.32E+003	mg/L	25	EST	MEYLAN,WM ET AL. (1996)
logP 0.24			EXP	ROSS,DL ET AL. (1992)
VP 1.04E-012	mm Hg	25	EST	NEELY,WB & BLAU,GE (1985)
DC	pKa			
HL 4.44E-018	atm m3/mol	25	EST	MEYLAN,WM & HOWARD,PH (1991)
OH 1.23E-010	cm3/molc sec	25	EST	MEYLAN,WM & HOWARD,PH (1993)

CAS #: 079661-41-9 — ETHANAMINIUM, 2-[(CYCLOPROPYLCARBONYL)OXY]-N,N,N

Formula: $C_9H_{18}INO_2$

Mol Weight: 299.15

MP (deg C): | FP (deg C):

BP (deg C):

BP pressure (mm Hg):

Property/Value	Units	Temp	Data Type	Reference
WS 1.00E+006	mg/L	25	EST	MEYLAN,WM ET AL. (1996)
logP -2.70			EXP	SANGSTER,J (1994)
VP 1.12E-008	mm Hg	25	EST	NEELY,WB & BLAU,GE (1985)
DC	pKa			
HL 6.47E-015	atm m3/mol	25	EST	MEYLAN,WM & HOWARD,PH (1991)
OH 1.64E-011	cm3/molc sec	25	EST	MEYLAN,WM & HOWARD,PH (1993)

CAS #: 079661-42-0 — ETHANAMINIUM, 2-(FORMYLOXY)-N,N,N-TRIMETHYL-, IO

Formula: $C_6H_{14}INO_2$

Mol Weight: 259.09

MP (deg C): | FP (deg C):

BP (deg C):

BP pressure (mm Hg):

Property/Value	Units	Temp	Data Type	Reference
WS 1.00E+006	mg/L	25	EST	MEYLAN,WM ET AL. (1996)
logP -3.83			EXP	SANGSTER,J (1994)
VP 1.59E-007	mm Hg	25	EST	NEELY,WB & BLAU,GE (1985)
DC	pKa			
HL 1.14E-014	atm m3/mol	25	EST	MEYLAN,WM & HOWARD,PH (1991)
OH 1.63E-011	cm3/molc sec	25	EST	MEYLAN,WM & HOWARD,PH (1993)

CAS #: 079794-75-5 — LORATADINE

Formula: $C_{22}H_{23}ClN_2O_2$

Mol Weight: 382.89

MP (deg C): 134-136 | FP (deg C):

BP (deg C):

BP pressure (mm Hg):

Property/Value	Units	Temp	Data Type	Reference
WS 1.10E-002	mg/L	25	EST	MEYLAN,WM ET AL. (1996)
logP 5.20			EXP	SANGSTER,J (1994)
VP 1.56E-009	mm Hg	25	EST	NEELY,WB & BLAU,GE (1985)
DC	pKa			
HL 3.19E-013	atm m3/mol	25	EST	MEYLAN,WM & HOWARD,PH (1991)
OH 1.45E-010	cm3/molc sec	25	EST	MEYLAN,WM & HOWARD,PH (1993)

CAS #: 079867-78-0 — 5,12-NAPHTHACENEDIONE PYRANOSYL DERIVATIVE

Formula: $C_{31}H_{35}NO_{11}$

Mol Weight: 597.62

MP (deg C): | FP (deg C):

BP (deg C):

BP pressure (mm Hg):

Property/Value	Units	Temp	Data Type	Reference
WS 5.33E+000	mg/L	25	EST	MEYLAN,WM ET AL. (1996)
logP 2.31			EXP	SANGSTER,J (1993)
VP 2.71E-024	mm Hg	25	EST	NEELY,WB & BLAU,GE (1985)
DC	pKa			
HL 4.71E-027	atm m3/mol	25	EST	MEYLAN,WM & HOWARD,PH (1991)
OH 2.62E-010	cm3/molc sec	25	EST	MEYLAN,WM & HOWARD,PH (1993)

CAS #: 079902-63-9 — SIMVASTATIN

Formula: $C_{25}H_{40}O_5$

Mol Weight: 420.59

MP (deg C): 135-138 | FP (deg C):

BP (deg C):

BP pressure (mm Hg):

Property/Value	Units	Temp	Data Type	Reference
WS 7.65E-001	mg/L	25	EST	MEYLAN,WM ET AL. (1996)
logP 4.68			EXP	HANSCH,C ET AL. (1995)
VP 6.41E-013	mm Hg	25	EST	NEELY,WB & BLAU,GE (1985)
DC	pKa			
HL 2.81E-010	atm m3/mol	25	EST	MEYLAN,WM & HOWARD,PH (1991)
OH 2.30E-010	cm3/molc sec	25	EST	MEYLAN,WM & HOWARD,PH (1993)

079951-58-9 — MORPHOLINO-DAUNORUBICIN,13-DIHYDRO

CAS #:	079951-58-9		MORPHOLINO-DAUNORUBICIN,13-DIHYDRO
Formula:	$C_{31}H_{37}NO_{11}$		
Mol Weight:	599.64		
MP (deg C):		FP (deg C):	
BP (deg C):			
BP pressure (mm Hg):			

Property/Value	Units	Temp	Data Type	Reference
WS 8.97E+000	mg/L	25	EST	MEYLAN,WM ET AL. (1996)
logP 2.03			EXP	HANSCH,C ET AL. (1995)
VP 1.61E-025	mm Hg	25	EST	NEELY,WB & BLAU,GE (1985)
DC	pKa			
HL 1.06E-028	atm m3/mol	25	EST	MEYLAN,WM & HOWARD,PH (1991)
OH 2.88E-010	cm3/molc sec	25	EST	MEYLAN,WM & HOWARD,PH (1993)

080033-99-4 — 3,5-DICL-2',4'-DIFLUORO SALICYLANILIDE

CAS #:	080033-99-4		3,5-DICL-2',4'-DIFLUORO SALICYLANILIDE
Formula:	$C_{13}H_7Cl_2F_2NO_2$		
Mol Weight:	318.11		
MP (deg C):		FP (deg C):	
BP (deg C):			
BP pressure (mm Hg):			

Property/Value	Units	Temp	Data Type	Reference
WS 4.75E+000	mg/L	25	EST	MEYLAN,WM ET AL. (1996)
logP 4.55			EXP	TERADA,H ET AL. (1988)
VP 2.19E-009	mm Hg	25	EST	NEELY,WB & BLAU,GE (1985)
DC	pKa			
HL 1.20E-010	atm m3/mol	25	EST	MEYLAN,WM & HOWARD,PH (1991)
OH 2.02E-012	cm3/molc sec	25	EST	MEYLAN,WM & HOWARD,PH (1993)

080121-60-4 — 1-AMINOADAMANTANE,3,5-DIETHYL

CAS #:	080121-60-4		1-AMINOADAMANTANE,3,5-DIETHYL
Formula:	$C_{14}H_{25}N$		
Mol Weight:	207.36		
MP (deg C):		FP (deg C):	
BP (deg C):			
BP pressure (mm Hg):			

Property/Value	Units	Temp	Data Type	Reference
WS 7.72E+001	mg/L	25	EST	MEYLAN,WM ET AL. (1996)
logP 4.36			EXP	HANSCH,C & LEO,AJ (1985)
VP 3.67E-003	mm Hg	25	EST	NEELY,WB & BLAU,GE (1985)
DC	pKa			
HL 2.58E-005	atm m3/mol	25	EST	MEYLAN,WM & HOWARD,PH (1991)
OH 3.57E-011	cm3/molc sec	25	EST	MEYLAN,WM & HOWARD,PH (1993)

080121-61-5 — 1-AMINOADAMANTANE,3-PROPYL

CAS #:	080121-61-5		1-AMINOADAMANTANE,3-PROPYL
Formula:	$C_{13}H_{23}N$		
Mol Weight:	193.33		
MP (deg C):		FP (deg C):	
BP (deg C):			
BP pressure (mm Hg):			

Property/Value	Units	Temp	Data Type	Reference
WS 2.96E+002	mg/L	25	EST	MEYLAN,WM ET AL. (1996)
logP 3.76			EXP	HANSCH,C & LEO,AJ (1985)
VP 7.34E-003	mm Hg	25	EST	NEELY,WB & BLAU,GE (1985)
DC	pKa			
HL 1.95E-005	atm m3/mol	25	EST	MEYLAN,WM & HOWARD,PH (1991)
OH 3.94E-011	cm3/molc sec	25	EST	MEYLAN,WM & HOWARD,PH (1993)

080125-14-0 — REMOXIPRIDE

CAS #:	080125-14-0		REMOXIPRIDE
Formula:	$C_{16}H_{23}BrN_2O_3$		
Mol Weight:	371.28		
MP (deg C):		FP (deg C):	
BP (deg C):			
BP pressure (mm Hg):			

Property/Value	Units	Temp	Data Type	Reference
WS 7.40E+001	mg/L	25	EST	MEYLAN,WM ET AL. (1996)
logP 2.10			EXP	HOEGBERG,T ET AL. (1986)
VP 2.12E-009	mm Hg	25	EST	NEELY,WB & BLAU,GE (1985)
DC	pKa			
HL 4.60E-015	atm m3/mol	25	EST	MEYLAN,WM & HOWARD,PH (1991)
OH 1.81E-010	cm3/molc sec	25	EST	MEYLAN,WM & HOWARD,PH (1993)

080171-69-3 — CARBAMIC ACID, [2-(ETHOXYMETHYL)PHENYL]-, 2-(1-P

CAS #:	080171-69-3		CARBAMIC ACID, [2-(ETHOXYMETHYL)PHENYL]-, 2-(1-P
Formula:	$C_{17}H_{26}N_2O_3$		
Mol Weight:	306.41		
MP (deg C):		FP (deg C):	
BP (deg C):			
BP pressure (mm Hg):			

Property/Value	Units	Temp	Data Type	Reference
WS 5.46E+002	mg/L	25	EST	MEYLAN,WM ET AL. (1996)
logP 2.72			EXP	HANSCH,C ET AL. (1995)
VP 5.48E-007	mm Hg	25	EST	NEELY,WB & BLAU,GE (1985)
DC	pKa			
HL 2.54E-013	atm m3/mol	25	EST	MEYLAN,WM & HOWARD,PH (1991)
OH 1.68E-010	cm3/molc sec	25	EST	MEYLAN,WM & HOWARD,PH (1993)

080267-16-9 — 2,4-PYRIMIDINEDIAMINE, 5-[(3,4-DIETHOXYPHENYL)ME

CAS #:	080267-16-9		2,4-PYRIMIDINEDIAMINE, 5-[(3,4-DIETHOXYPHENYL)ME
Formula:	$C_{15}H_{20}N_4O_2$		
Mol Weight:	288.35		
MP (deg C):		FP (deg C):	
BP (deg C):			
BP pressure (mm Hg):			

Property/Value	Units	Temp	Data Type	Reference
WS 4.87E+002	mg/L	25	EST	MEYLAN,WM ET AL. (1996)
logP 1.72			EXP	SANGSTER,J (1993)
VP 1.02E-008	mm Hg	25	EST	NEELY,WB & BLAU,GE (1985)
DC	pKa			
HL 7.11E-013	atm m3/mol	25	EST	MEYLAN,WM & HOWARD,PH (1991)
OH 6.10E-011	cm3/molc sec	25	EST	MEYLAN,WM & HOWARD,PH (1993)

080267-17-0 — 2,4-PYRIMIDINEDIAMINE, 5-[(3-METHOXY-4-PROPOXYPH

CAS #:	080267-17-0		2,4-PYRIMIDINEDIAMINE, 5-[(3-METHOXY-4-PROPOXYPH
Formula:	$C_{15}H_{20}N_4O_2$		
Mol Weight:	288.35		
MP (deg C):		FP (deg C):	
BP (deg C):			
BP pressure (mm Hg):			

Property/Value	Units	Temp	Data Type	Reference
WS 3.56E+002	mg/L	25	EST	MEYLAN,WM ET AL. (1996)
logP 2.02			EXP	SANGSTER,J (1993)
VP 1.02E-008	mm Hg	25	EST	NEELY,WB & BLAU,GE (1985)
DC	pKa			
HL 7.11E-013	atm m3/mol	25	EST	MEYLAN,WM & HOWARD,PH (1991)
OH 6.00E-011	cm3/molc sec	25	EST	MEYLAN,WM & HOWARD,PH (1993)

CAS #: 080353-09-9 — PYRIDINE, 1,2,3,6-TETRAHYDRO-1-METHYL-4-(2-THIEN

Formula: $C_{10}H_{13}NS$

Mol Weight: 179.29

MP (deg C):   FP (deg C):

BP (deg C):

BP pressure (mm Hg):

Property/Value	Units	Temp	Data Type	Reference
WS 4.06E+003	mg/L	25	EST	MEYLAN,WM ET AL. (1996)
logP 2.51			EXP	SANGSTER,J (1993)
VP 3.40E-003	mm Hg	25	EST	NEELY,WB & BLAU,GE (1985)
DC	pKa			
HL 8.06E-007	atm m3/mol	25	EST	MEYLAN,WM & HOWARD,PH (1991)
OH 1.84E-010	cm3/molc sec	25	EST	MEYLAN,WM & HOWARD,PH (1993)

CAS #: 080407-58-5 — 46-NH2-(2-OCH2-CON)BENZYLPYRIMIDINE

Formula: $C_{13}H_{15}N_5O_2$

Mol Weight: 273.30

MP (deg C):   FP (deg C):

BP (deg C):

BP pressure (mm Hg):

Property/Value	Units	Temp	Data Type	Reference
WS 1.67E+003	mg/L	25	EST	MEYLAN,WM ET AL. (1996)
logP -0.34			EXP	HANSCH,C & LEO,AJ (1985)
VP 1.07E-010	mm Hg	25	EST	NEELY,WB & BLAU,GE (1985)
DC	pKa			
HL 2.07E-021	atm m3/mol	25	EST	MEYLAN,WM & HOWARD,PH (1991)
OH 2.07E-010	cm3/molc sec	25	EST	MEYLAN,WM & HOWARD,PH (1993)

CAS #: 080412-20-0 — INDOL-3-YL-2-THIOPROPIONIC ACID

Formula: $C_{11}H_{11}NO_2S$

Mol Weight: 221.28

MP (deg C):   FP (deg C):

BP (deg C):

BP pressure (mm Hg):

Property/Value	Units	Temp	Data Type	Reference
WS 1.23E+003	mg/L	25	EST	MEYLAN,WM ET AL. (1996)
logP 2.15			EXP	HANSCH,C & LEO,AJ (1985)
VP 2.40E-007	mm Hg	25	EST	NEELY,WB & BLAU,GE (1985)
DC	pKa			
HL 2.55E-013	atm m3/mol	25	EST	MEYLAN,WM & HOWARD,PH (1991)
OH 2.10E-010	cm3/molc sec	25	EST	MEYLAN,WM & HOWARD,PH (1993)

CAS #: 080431-02-3 — 3-PYRIDINEACETAMIDE, N-(3,6-DIHYDRO-1(2H)-PYRIDI

Formula: $C_{12}H_{15}N_3O$

Mol Weight: 217.27

MP (deg C):   FP (deg C):

BP (deg C):

BP pressure (mm Hg):

Property/Value	Units	Temp	Data Type	Reference
WS 3.15E+004	mg/L	25	EST	MEYLAN,WM ET AL. (1996)
logP 0.05			EXP	SANGSTER,J (1994)
VP 4.78E-007	mm Hg	25	EST	NEELY,WB & BLAU,GE (1985)
DC	pKa			
HL 2.16E-014	atm m3/mol	25	EST	MEYLAN,WM & HOWARD,PH (1991)
OH 8.40E-011	cm3/molc sec	25	EST	MEYLAN,WM & HOWARD,PH (1993)

CAS #: 080431-09-0 — 1234-H4-PYRIDINE-1-T-BU-CARBONYLAMINO

Formula: $C_{10}H_{18}N_2O$

Mol Weight: 182.27

MP (deg C):   FP (deg C):

BP (deg C):

BP pressure (mm Hg):

Property/Value	Units	Temp	Data Type	Reference
WS 7.08E+003	mg/L	25	EST	MEYLAN,WM ET AL. (1996)
logP 1.03			EXP	HANSCH,C & LEO,AJ (1985)
VP 6.50E-005	mm Hg	25	EST	NEELY,WB & BLAU,GE (1985)
DC	pKa			
HL 4.78E-010	atm m3/mol	25	EST	MEYLAN,WM & HOWARD,PH (1991)
OH 8.40E-011	cm3/molc sec	25	EST	MEYLAN,WM & HOWARD,PH (1993)

CAS #: 080431-10-3 — CYCLOPROPANECARBOXAMIDE, N-(3,6-DIHYDRO-1(2H)-PY

Formula: $C_9H_{14}N_2O$

Mol Weight: 166.22

MP (deg C):   FP (deg C):

BP (deg C):

BP pressure (mm Hg):

Property/Value	Units	Temp	Data Type	Reference
WS 7.08E+003	mg/L	25	EST	MEYLAN,WM ET AL. (1996)
logP 1.12			EXP	SANGSTER,J (1994)
VP 5.49E-005	mm Hg	25	EST	NEELY,WB & BLAU,GE (1985)
DC	pKa			
HL 1.59E-010	atm m3/mol	25	EST	MEYLAN,WM & HOWARD,PH (1991)
OH 6.51E-011	cm3/molc sec	25	EST	MEYLAN,WM & HOWARD,PH (1993)

CAS #: 080431-14-7 — 2-FURANCARBOXAMIDE, N-(3,6-DIHYDRO-1(2H)-PYRIDIN

Formula: $C_{10}H_{12}N_2O_2$

Mol Weight: 192.22

MP (deg C):   FP (deg C):

BP (deg C):

BP pressure (mm Hg):

Property/Value	Units	Temp	Data Type	Reference
WS 9.74E+003	mg/L	25	EST	MEYLAN,WM ET AL. (1996)
logP 0.81			EXP	SANGSTER,J (1993)
VP 8.13E-006	mm Hg	25	EST	NEELY,WB & BLAU,GE (1985)
DC	pKa			
HL 4.04E-011	atm m3/mol	25	EST	MEYLAN,WM & HOWARD,PH (1991)
OH 1.19E-010	cm3/molc sec	25	EST	MEYLAN,WM & HOWARD,PH (1993)

CAS #: 080431-15-8 — 2-THIOPHENECARBOXAMIDE, N-(3,6-DIHYDRO-1(2H)-PYR

Formula: $C_{10}H_{12}N_2OS$

Mol Weight: 208.28

MP (deg C):   FP (deg C):

BP (deg C):

BP pressure (mm Hg):

Property/Value	Units	Temp	Data Type	Reference
WS 9.84E+002	mg/L	25	EST	MEYLAN,WM ET AL. (1996)
logP 1.88			EXP	SANGSTER,J (1993)
VP 1.11E-006	mm Hg	25	EST	NEELY,WB & BLAU,GE (1985)
DC	pKa			
HL 2.20E-011	atm m3/mol	25	EST	MEYLAN,WM & HOWARD,PH (1991)
OH 9.11E-011	cm3/molc sec	25	EST	MEYLAN,WM & HOWARD,PH (1993)

CAS #: 080488-98-8				2-(3-BROMOPHENYL)INDOLIZINE

Formula: $C_{14}H_{12}BrN$

Mol Weight: 274.17

MP (deg C):   FP (deg C):

BP (deg C):

BP pressure (mm Hg):

Property/ Value	Units	Temp	Data Type	Reference
WS 5.19E-001	mg/L	25	EST	MEYLAN,WM ET AL. (1996)
logP 5.31			EXP	HANSCH,C & LEO,AJ (1985)
VP 4.51E-006	mm Hg	25	EST	NEELY,WB & BLAU,GE (1985)
DC	pKa			
HL 1.67E-009	atm m3/mol	25	EST	MEYLAN,WM & HOWARD,PH (1991)
OH 1.93E-010	cm3/molc sec	25	EST	MEYLAN,WM & HOWARD,PH (1993)

CAS #: 080563-87-7				BENZOIC ACID, 2-CHLORO-6-NITRO-, METHYL ESTER

Formula: $C_8H_6ClNO_4$

Mol Weight: 215.59

MP (deg C):   FP (deg C):

BP (deg C):

BP pressure (mm Hg):

Property/ Value	Units	Temp	Data Type	Reference
WS 1.92E+002	mg/L	25	EST	MEYLAN,WM ET AL. (1996)
logP 2.21			EXP	SOTOMATSU,T ET AL. (1993)
VP 3.15E-004	mm Hg	25	EST	NEELY,WB & BLAU,GE (1985)
DC	pKa			
HL 1.01E-007	atm m3/mol	25	EST	MEYLAN,WM & HOWARD,PH (1991)
OH 2.56E-013	cm3/molc sec	25	EST	MEYLAN,WM & HOWARD,PH (1993)

CAS #: 080616-56-4				ACETAMIDE, N,N'-1H-1,2,4-TRIAZOLE-3,5-DIYLBIS[2,

Formula: $C_6H_3F_6N_5O_2$

Mol Weight: 291.11

MP (deg C):   FP (deg C):

BP (deg C):

BP pressure (mm Hg):

Property/ Value	Units	Temp	Data Type	Reference
WS 4.09E+003	mg/L	25	EST	MEYLAN,WM ET AL. (1996)
logP 0.62			EXP	HANSCH,C & LEO,AJ (1985)
VP 6.71E-009	mm Hg	25	EST	NEELY,WB & BLAU,GE (1985)
DC	pKa			
HL 1.97E-013	atm m3/mol	25	EST	MEYLAN,WM & HOWARD,PH (1991)
OH 1.00E-013	cm3/molc sec	25	EST	MEYLAN,WM & HOWARD,PH (1993)

CAS #: 080616-57-5				PROPANAMIDE, N,N'-1H-1,2,4-TRIAZOLE-3,5-DIYLBIS-

Formula: $C_8H_{13}N_5O_2$

Mol Weight: 211.23

MP (deg C):   FP (deg C):

BP (deg C):

BP pressure (mm Hg):

Property/ Value	Units	Temp	Data Type	Reference
WS 1.63E+003	mg/L	25	EST	MEYLAN,WM ET AL. (1996)
logP 0.07			EXP	HANSCH,C & LEO,AJ (1985)
VP 4.15E-010	mm Hg	25	EST	NEELY,WB & BLAU,GE (1985)
DC	pKa			
HL 5.60E-015	atm m3/mol	25	EST	MEYLAN,WM & HOWARD,PH (1991)
OH 2.56E-012	cm3/molc sec	25	EST	MEYLAN,WM & HOWARD,PH (1993)

CAS #: 080616-58-6				1,2,4-TRIAZOLE-3,5-BUTYRAMIDO-

Formula: $C_{10}H_{17}N_5O_2$

Mol Weight: 239.28

MP (deg C):   FP (deg C):

BP (deg C):

BP pressure (mm Hg):

Property/ Value	Units	Temp	Data Type	Reference
WS 5.73E+003	mg/L	25	EST	MEYLAN,WM ET AL. (1996)
logP 0.79			EXP	HANSCH,C & LEO,AJ (1985)
VP 7.95E-011	mm Hg	25	EST	NEELY,WB & BLAU,GE (1985)
DC	pKa			
HL 9.87E-015	atm m3/mol	25	EST	MEYLAN,WM & HOWARD,PH (1991)
OH 9.44E-012	cm3/molc sec	25	EST	MEYLAN,WM & HOWARD,PH (1993)

CAS #: 080616-59-7				BUTANAMIDE, N-(5-AMINO-1H-1,2,4-TRIAZOL-3-YL)-2,

Formula: $C_6H_4F_7N_5O$

Mol Weight: 295.12

MP (deg C):   FP (deg C):

BP (deg C):

BP pressure (mm Hg):

Property/ Value	Units	Temp	Data Type	Reference
WS 6.34E+002	mg/L	25	EST	MEYLAN,WM ET AL. (1996)
logP 1.54			EXP	ALHAIDER,AA ET AL. (1982)
VP 5.56E-007	mm Hg	25	EST	NEELY,WB & BLAU,GE (1985)
DC	pKa			
HL 1.35E-013	atm m3/mol	25	EST	MEYLAN,WM & HOWARD,PH (1991)
OH 1.00E-013	cm3/molc sec	25	EST	MEYLAN,WM & HOWARD,PH (1993)

CAS #: 080617-74-9				2-PROPANOL, 1-[(1-METHYLETHYL)AMINO]-3-(3-PHENYL

Formula: $C_{15}H_{25}NO_2$

Mol Weight: 251.37

MP (deg C):   FP (deg C):

BP (deg C):

BP pressure (mm Hg):

Property/ Value	Units	Temp	Data Type	Reference
WS 1.15E+004	mg/L	25	EST	MEYLAN,WM ET AL. (1996)
logP 1.54			EXP	MAULEON,D ET AL. (1988)
VP 8.52E-007	mm Hg	25	EST	NEELY,WB & BLAU,GE (1985)
DC	pKa			
HL 3.15E-012	atm m3/mol	25	EST	MEYLAN,WM & HOWARD,PH (1991)
OH 1.27E-010	cm3/molc sec	25	EST	MEYLAN,WM & HOWARD,PH (1993)

CAS #: 080617-76-1				2-PROPANOL, 1-[(1,1-DIMETHYLETHYL)AMINO]-3-(3-PH

Formula: $C_{16}H_{27}NO_2$

Mol Weight: 265.40

MP (deg C):   FP (deg C):

BP (deg C):

BP pressure (mm Hg):

Property/ Value	Units	Temp	Data Type	Reference
WS 2.72E+003	mg/L	25	EST	MEYLAN,WM ET AL. (1996)
logP 2.18			EXP	MAULEON,D ET AL. (1988)
VP 4.43E-007	mm Hg	25	EST	NEELY,WB & BLAU,GE (1985)
DC	pKa			
HL 4.18E-012	atm m3/mol	25	EST	MEYLAN,WM & HOWARD,PH (1991)
OH 1.09E-010	cm3/molc sec	25	EST	MEYLAN,WM & HOWARD,PH (1993)

CAS #: 080617-78-3 — 2-PROPANOL, 1-[(1,1-DIMETHYLETHYL)AMINO]-3-[(3-P

Formula: $C_{16}H_{25}NO_2$

Mol Weight: 263.38

MP (deg C): FP (deg C):

BP (deg C):

BP pressure (mm Hg):

Property/ Value	Units	Temp	Data Type	Reference
WS 2.25E+003	mg/L	25	EST	MEYLAN,WM ET AL. (1996)
logP 2.29			EXP	MAULEON,D ET AL. (1988)
VP 3.52E-007	mm Hg	25	EST	NEELY,WB & BLAU,GE (1985)
DC	pKa			
HL 1.73E-012	atm m3/mol	25	EST	MEYLAN,WM & HOWARD,PH (1991)
OH 1.56E-010	cm3/molc sec	25	EST	MEYLAN,WM & HOWARD,PH (1993)

CAS #: 080617-86-3 — 2-PROPANOL, 1-[(1,1-DIMETHYLETHYL)AMINO]-3-[(1,2

Formula: $C_{17}H_{27}NO_2$

Mol Weight: 277.41

MP (deg C): FP (deg C):

BP (deg C):

BP pressure (mm Hg):

Property/ Value	Units	Temp	Data Type	Reference
WS 1.57E+002	mg/L	25	EST	MEYLAN,WM ET AL. (1996)
logP 3.55			EXP	MAULEON,D ET AL. (1988)
VP 1.26E-007	mm Hg	25	EST	NEELY,WB & BLAU,GE (1985)
DC	pKa			
HL 2.03E-012	atm m3/mol	25	EST	MEYLAN,WM & HOWARD,PH (1991)
OH 1.16E-010	cm3/molc sec	25	EST	MEYLAN,WM & HOWARD,PH (1993)

CAS #: 080715-99-7 — 4H-PYRIDO[1,2-A]PYRIMIDIN-3-CARBOXYLIC ACID, 6,7

Formula: $C_{10}H_{12}N_2O_3$

Mol Weight: 208.22

MP (deg C): FP (deg C):

BP (deg C):

BP pressure (mm Hg):

Property/ Value	Units	Temp	Data Type	Reference
WS 3.18E+003	mg/L	25	EST	MEYLAN,WM ET AL. (1996)
logP -0.25			EXP	SANGSTER,J (1994)
VP 1.25E-005	mm Hg	25	EST	NEELY,WB & BLAU,GE (1985)
DC	pKa			
HL 2.01E-012	atm m3/mol	25	EST	MEYLAN,WM & HOWARD,PH (1991)
OH 2.16E-011	cm3/molc sec	25	EST	MEYLAN,WM & HOWARD,PH (1993)

CAS #: 080790-68-7 — 5,12-NAPHTHACENEDIONE PYRANOSYL DERIVATIVE

Formula: $C_{31}H_{35}NO_{12}$

Mol Weight: 613.62

MP (deg C): FP (deg C):

BP (deg C):

BP pressure (mm Hg):

Property/ Value	Units	Temp	Data Type	Reference
WS 1.31E+001	mg/L	25	EST	MEYLAN,WM ET AL. (1996)
logP 1.73			EXP	SANGSTER,J (1993)
VP 1.03E-026	mm Hg	25	EST	NEELY,WB & BLAU,GE (1985)
DC	pKa			
HL 7.35E-025	atm m3/mol	25	EST	MEYLAN,WM & HOWARD,PH (1991)
OH 2.65E-010	cm3/molc sec	25	EST	MEYLAN,WM & HOWARD,PH (1993)

CAS #: 080844-07-1 — ETOFENPROX

Formula: $C_{25}H_{28}O_3$

Mol Weight: 376.50

MP (deg C): FP (deg C):

BP (deg C):

BP pressure (mm Hg):

Property/ Value	Units	Temp	Data Type	Reference
WS 4.08E-003	mg/L	25	EST	MEYLAN,WM ET AL. (1996)
logP 7.05			EXP	TOMLIN,C (1994)
VP 6.79E-009	mm Hg	25	EST	NEELY,WB & BLAU,GE (1985)
DC	pKa			
HL 2.26E-008	atm m3/mol	25	EST	MEYLAN,WM & HOWARD,PH (1991)
OH 6.22E-011	cm3/molc sec	25	EST	MEYLAN,WM & HOWARD,PH (1993)

CAS #: 080866-90-6 — 2-ETHYL-2-PHENYL-MALONAMIDE

Formula: $C_{11}H_{14}N_2O_2$

Mol Weight: 206.25

MP (deg C): 120 dec FP (deg C):

BP (deg C):

BP pressure (mm Hg):

Property/ Value	Units	Temp	Data Type	Reference
WS 1.54E+003	mg/L	25	EST	MEYLAN,WM ET AL. (1996)
logP 0.13			EXP	HANSCH,C ET AL. (1995)
VP 3.13E-008	mm Hg	25	EST	NEELY,WB & BLAU,GE (1985)
DC	pKa			
HL 3.83E-013	atm m3/mol	25	EST	MEYLAN,WM & HOWARD,PH (1991)
OH 1.21E-011	cm3/molc sec	25	EST	MEYLAN,WM & HOWARD,PH (1993)

CAS #: 080916-23-0 — 2-PROPENOIC ACID, 2-METHYL-, 2-[(1-NAPHTHALENYLO

Formula: $C_{18}H_{20}O_4$

Mol Weight: 300.36

MP (deg C): FP (deg C):

BP (deg C):

BP pressure (mm Hg):

Property/ Value	Units	Temp	Data Type	Reference
WS 2.39E+001	mg/L	25	EST	MEYLAN,WM ET AL. (1996)
logP 3.77			EXP	SANGSTER,J (1993)
VP 6.05E-009	mm Hg	25	EST	NEELY,WB & BLAU,GE (1985)
DC	pKa			
HL 5.24E-012	atm m3/mol	25	EST	MEYLAN,WM & HOWARD,PH (1991)
OH 2.46E-010	cm3/molc sec	25	EST	MEYLAN,WM & HOWARD,PH (1993)

CAS #: 080929-58-4 — 5-THIA-1-AZABICYCLO[4.2.0]OCT-2-ENE-2-CARBOXYLIC

Formula: $C_{23}H_{27}N_3O_8S_2$

Mol Weight: 537.61

MP (deg C): FP (deg C):

BP (deg C):

BP pressure (mm Hg):

Property/ Value	Units	Temp	Data Type	Reference
WS 1.66E+001	mg/L	25	EST	MEYLAN,WM ET AL. (1996)
logP 1.63			EXP	SANGSTER,J (1994)
VP 2.61E-016	mm Hg	25	EST	NEELY,WB & BLAU,GE (1985)
DC	pKa			
HL 1.98E-021	atm m3/mol	25	EST	MEYLAN,WM & HOWARD,PH (1991)
OH 1.20E-010	cm3/molc sec	25	EST	MEYLAN,WM & HOWARD,PH (1993)

CAS #: 081020-79-3				PROPANEDINITRILE, [(5-METHYL-2-FURANYL)METHYLENE
Formula: $C_9H_6N_2O$				
Mol Weight: 158.16				
MP (deg C):		FP (deg C):		
BP (deg C):				
BP pressure (mm Hg):				

Property/ Value	Units	Temp	Data Type	Reference
WS 7.00E+002	mg/L	25	EST	MEYLAN,WM ET AL. (1996)
logP 2.03			EXP	BALAZ,S ET AL. (1985)
VP 2.35E-004	mm Hg	25	EST	NEELY,WB & BLAU,GE (1985)
DC	pKa			
HL 1.51E-008	atm m3/mol	25	EST	MEYLAN,WM & HOWARD,PH (1991)
OH 1.02E-010	cm3/molc sec	25	EST	MEYLAN,WM & HOWARD,PH (1993)

CAS #: 081060-10-8				2-IMIDAZOLIDINIMINE, N-(2-CHLORO-3-FLUOROPHENYL)
Formula: $C_9H_9ClFN_3$				
Mol Weight: 213.64				
MP (deg C):		FP (deg C):		
BP (deg C):				
BP pressure (mm Hg):				

Property/ Value	Units	Temp	Data Type	Reference
WS 1.91E+004	mg/L	25	EST	MEYLAN,WM ET AL. (1996)
logP 1.52			EXP	SANGSTER,J (1993)
VP 9.94E-005	mm Hg	25	EST	NEELY,WB & BLAU,GE (1985)
DC	pKa			
HL 2.37E-011	atm m3/mol	25	EST	MEYLAN,WM & HOWARD,PH (1991)
OH 1.44E-010	cm3/molc sec	25	EST	MEYLAN,WM & HOWARD,PH (1993)

CAS #: 081060-11-9				2-IMIDAZOLIDINIMINE, N-(2-CHLORO-5-FLUOROPHENYL)
Formula: $C_9H_9ClFN_3$				
Mol Weight: 213.64				
MP (deg C):		FP (deg C):		
BP (deg C):				
BP pressure (mm Hg):				

Property/ Value	Units	Temp	Data Type	Reference
WS 1.91E+004	mg/L	25	EST	MEYLAN,WM ET AL. (1996)
logP 1.52			EXP	SANGSTER,J (1993)
VP 9.94E-005	mm Hg	25	EST	NEELY,WB & BLAU,GE (1985)
DC	pKa			
HL 2.37E-011	atm m3/mol	25	EST	MEYLAN,WM & HOWARD,PH (1991)
OH 1.44E-010	cm3/molc sec	25	EST	MEYLAN,WM & HOWARD,PH (1993)

CAS #: 081098-60-4				CISAPRIDE
Formula: $C_{23}H_{29}ClFN_3O_4$				
Mol Weight: 465.96				
MP (deg C):		FP (deg C):		
BP (deg C):				
BP pressure (mm Hg):				

Property/ Value	Units	Temp	Data Type	Reference
WS 2.71E+000	mg/L	25	EST	MEYLAN,WM ET AL. (1996)
logP 3.09			EST	MEYLAN,WM & HOWARD,PH (1995)
VP	mm Hg			
DC	pKa			
HL 5.44E-024	atm m3/mol	25	EST	MEYLAN,WM & HOWARD,PH (1991)
OH 3.47E-014	cm3/molc sec	25	EST	MEYLAN,WM & HOWARD,PH (1993)

CAS #: 081310-40-9				BENZENEACETONITRILE, à-(1-METHYLETHYL)-4-NITRO-
Formula: $C_{11}H_{12}N_2O_2$				
Mol Weight: 204.23				
MP (deg C):		FP (deg C):		
BP (deg C):				
BP pressure (mm Hg):				

Property/ Value	Units	Temp	Data Type	Reference
WS 3.74E+001	mg/L	25	EST	MEYLAN,WM ET AL. (1996)
logP 2.80			EXP	WANG,W ET AL. (1987)
VP 5.56E-005	mm Hg	25	EST	NEELY,WB & BLAU,GE (1985)
DC	pKa			
HL 2.28E-008	atm m3/mol	25	EST	MEYLAN,WM & HOWARD,PH (1991)
OH 3.75E-012	cm3/molc sec	25	EST	MEYLAN,WM & HOWARD,PH (1993)

CAS #: 081334-34-1				IMAZAPYR
Formula: $C_{13}H_{15}N_3O_3$				
Mol Weight: 261.28				
MP (deg C): 167-73		FP (deg C):		
BP (deg C):				
BP pressure (mm Hg):				

Property/ Value	Units	Temp	Data Type	Reference
WS 1.13E+004	mg/L	25	EXP	TOMLIN,C (1994)
logP 1.30			EXP	TOMLIN,C (1994)
VP 1.79E-011	mm Hg	25	EST	NEELY,WB & BLAU,GE (1985)
DC	pKa			
HL 7.08E-017	atm m3/mol	25	EST	MEYLAN,WM & HOWARD,PH (1991)
OH 1.48E-011	cm3/molc sec	25	EST	MEYLAN,WM & HOWARD,PH (1993)

CAS #: 081437-10-7				4-ETHOXYPYRAZOLE
Formula: $C_5H_8N_2O$				
Mol Weight: 112.13				
MP (deg C):		FP (deg C):		
BP (deg C):				
BP pressure (mm Hg):				

Property/ Value	Units	Temp	Data Type	Reference
WS 2.22E+004	mg/L	25	EST	MEYLAN,WM ET AL. (1996)
logP 0.80			EXP	HANSCH,C & LEO,AJ (1985)
VP 4.27E-002	mm Hg	25	EST	NEELY,WB & BLAU,GE (1985)
DC	pKa			
HL 2.90E-007	atm m3/mol	25	EST	MEYLAN,WM & HOWARD,PH (1991)
OH 2.06E-010	cm3/molc sec	25	EST	MEYLAN,WM & HOWARD,PH (1993)

CAS #: 081496-81-3				3,12-EPOXY-12H-PYRANO[4,3-J]-1,2-BENZODIOXEPIN-1
Formula: $C_{15}H_{26}O_5$				
Mol Weight: 286.37				
MP (deg C):		FP (deg C):		
BP (deg C):				
BP pressure (mm Hg):				

Property/ Value	Units	Temp	Data Type	Reference
WS 6.60E+002	mg/L	25	EST	MEYLAN,WM ET AL. (1996)
logP 2.19			EXP	SANGSTER,J (1994)
VP 1.99E-007	mm Hg	25	EST	NEELY,WB & BLAU,GE (1985)
DC	pKa			
HL 5.89E-012	atm m3/mol	25	EST	MEYLAN,WM & HOWARD,PH (1991)
OH 8.20E-011	cm3/molc sec	25	EST	MEYLAN,WM & HOWARD,PH (1993)

CAS #: 081592-05-4 — P-CYANO PHENYLHIPPURATE

Formula: $C_{16}H_{12}N_2O_3$

Mol Weight: 280.29

MP (deg C): | FP (deg C):

BP (deg C):

BP pressure (mm Hg):

Property/Value	Units	Temp	Data Type	Reference
WS 1.37E+002	mg/L	25	EST	MEYLAN,WM ET AL. (1996)
logP 2.11			EXP	HANSCH,C & LEO,AJ (1985)
VP 1.67E-009	mm Hg	25	EST	NEELY,WB & BLAU,GE (1985)
DC	pKa			
HL 8.25E-013	atm m3/mol	25	EST	MEYLAN,WM & HOWARD,PH (1991)
OH 1.05E-011	cm3/molc sec	25	EST	MEYLAN,WM & HOWARD,PH (1993)

CAS #: 081592-09-8 — M-ACETAMIDOPHENYLHIPPURATE

Formula: $C_{17}H_{16}N_2O_4$

Mol Weight: 312.33

MP (deg C): | FP (deg C):

BP (deg C):

BP pressure (mm Hg):

Property/Value	Units	Temp	Data Type	Reference
WS 3.67E+002	mg/L	25	EST	MEYLAN,WM ET AL. (1996)
logP 1.70			EXP	HANSCH,C & LEO,AJ (1985)
VP 9.38E-012	mm Hg	25	EST	NEELY,WB & BLAU,GE (1985)
DC	pKa			
HL 1.55E-013	atm m3/mol	25	EST	MEYLAN,WM & HOWARD,PH (1991)
OH 2.30E-011	cm3/molc sec	25	EST	MEYLAN,WM & HOWARD,PH (1993)

CAS #: 081592-11-2 — M-CONH2 PHENYLHIPPURATE

Formula: $C_{16}H_{14}N_2O_4$

Mol Weight: 298.30

MP (deg C): | FP (deg C):

BP (deg C):

BP pressure (mm Hg):

Property/Value	Units	Temp	Data Type	Reference
WS 1.19E+003	mg/L	25	EST	MEYLAN,WM ET AL. (1996)
logP 1.20			EXP	HANSCH,C & LEO,AJ (1985)
VP 1.67E-011	mm Hg	25	EST	NEELY,WB & BLAU,GE (1985)
DC	pKa			
HL 5.56E-014	atm m3/mol	25	EST	MEYLAN,WM & HOWARD,PH (1991)
OH 1.34E-011	cm3/molc sec	25	EST	MEYLAN,WM & HOWARD,PH (1993)

CAS #: 081592-12-3 — M-CYANO PHENYLHIPPURATE

Formula: $C_{16}H_{12}N_2O_3$

Mol Weight: 280.29

MP (deg C): | FP (deg C):

BP (deg C):

BP pressure (mm Hg):

Property/Value	Units	Temp	Data Type	Reference
WS 1.37E+002	mg/L	25	EST	MEYLAN,WM ET AL. (1996)
logP 2.11			EXP	HANSCH,C & LEO,AJ (1985)
VP 1.67E-009	mm Hg	25	EST	NEELY,WB & BLAU,GE (1985)
DC	pKa			
HL 8.25E-013	atm m3/mol	25	EST	MEYLAN,WM & HOWARD,PH (1991)
OH 1.04E-011	cm3/molc sec	25	EST	MEYLAN,WM & HOWARD,PH (1993)

CAS #: 081592-15-6 — 3,5-DIMETHYLPHENYLHIPPURATE

Formula: $C_{17}H_{17}NO_3$

Mol Weight: 283.33

MP (deg C): | FP (deg C):

BP (deg C):

BP pressure (mm Hg):

Property/Value	Units	Temp	Data Type	Reference
WS 1.91E+001	mg/L	25	EST	MEYLAN,WM ET AL. (1996)
logP 3.40			EXP	HANSCH,C & LEO,AJ (1985)
VP 9.06E-009	mm Hg	25	EST	NEELY,WB & BLAU,GE (1985)
DC	pKa			
HL 1.04E-010	atm m3/mol	25	EST	MEYLAN,WM & HOWARD,PH (1991)
OH 2.41E-011	cm3/molc sec	25	EST	MEYLAN,WM & HOWARD,PH (1993)

CAS #: 081592-16-7 — 3,5-DIMETHOXYPHENYLHIPPURATE

Formula: $C_{17}H_{17}NO_5$

Mol Weight: 315.33

MP (deg C): | FP (deg C):

BP (deg C):

BP pressure (mm Hg):

Property/Value	Units	Temp	Data Type	Reference
WS 5.03E+001	mg/L	25	EST	MEYLAN,WM ET AL. (1996)
logP 2.69			EXP	HANSCH,C & LEO,AJ (1985)
VP 1.61E-009	mm Hg	25	EST	NEELY,WB & BLAU,GE (1985)
DC	pKa			
HL 2.99E-013	atm m3/mol	25	EST	MEYLAN,WM & HOWARD,PH (1991)
OH 2.10E-010	cm3/molc sec	25	EST	MEYLAN,WM & HOWARD,PH (1993)

CAS #: 081634-99-3 — 3-DECENYL ACETATE (Z)

Formula: $C_{12}H_{22}O_2$

Mol Weight: 198.31

MP (deg C): | FP (deg C):

BP (deg C):

BP pressure (mm Hg):

Property/Value	Units	Temp	Data Type	Reference
WS 5.50E+000	mg/L	25	EST	MEYLAN,WM ET AL. (1996)
logP 4.58			EST	MEYLAN,WM & HOWARD,PH (1995)
VP 2.31E-002	mm Hg	25	EST	NEELY,WB & BLAU,GE (1985)
DC	pKa			
HL 1.98E-003	atm m3/mol	25	EST	MEYLAN,WM & HOWARD,PH (1991)
OH 6.61E-011	cm3/molc sec	25	EST	MEYLAN,WM & HOWARD,PH (1993)

CAS #: 081777-89-1 — CLOMAZONE

Formula: $C_{12}H_{14}ClNO_2$

Mol Weight: 239.70

MP (deg C): | FP (deg C):

BP (deg C):

BP pressure (mm Hg):

Property/Value	Units	Temp	Data Type	Reference
WS 1.10E+003	mg/L	25	EXP	SHIU,WY ET AL. (1990)
logP 2.50			EXP	TOMLIN,C (1994)
VP 1.44E-004	mm Hg	25	EXP	TOMLIN,C (1994)
DC	pKa			
HL 4.13E-008	atm m3/mol	25	EST	VP/WSOL
OH 2.21E-011	cm3/molc sec	25	EST	MEYLAN,WM & HOWARD,PH (1993)

1352

CAS #: 081792-70-3	BUTANAMIDE, 2-ETHYL-N-[[3-NITRO-5-(TRIFLUOROMETH

Formula: $C_{13}H_{15}F_3N_2O_5S$

Mol Weight: 368.33

MP (deg C): | FP (deg C):

BP (deg C):

BP pressure (mm Hg):

Property/Value	Units	Temp	Data Type	Reference
WS 1.75E+000	mg/L	25	EST	MEYLAN,WM ET AL. (1996)
logP 3.57			EXP	MONZANI,A ET AL. (1985)
VP 7.15E-010	mm Hg	25	EST	NEELY,WB & BLAU,GE (1985)
DC	pKa			
HL 1.46E-010	atm m3/mol	25	EST	MEYLAN,WM & HOWARD,PH (1991)
OH 9.83E-012	cm3/molc sec	25	EST	MEYLAN,WM & HOWARD,PH (1993)

CAS #: 081792-73-6	BENZENESULFONAMIDE, 2-AMINO-N-(2-METHYL-1-OXOPRO

Formula: $C_{10}H_{14}N_2O_3S$

Mol Weight: 242.30

MP (deg C): | FP (deg C):

BP (deg C):

BP pressure (mm Hg):

Property/Value	Units	Temp	Data Type	Reference
WS 2.61E+003	mg/L	25	EST	MEYLAN,WM ET AL. (1996)
logP 1.17			EXP	MONZANI,A ET AL. (1985)
VP 1.16E-008	mm Hg	25	EST	NEELY,WB & BLAU,GE (1985)
DC	pKa			
HL 8.52E-013	atm m3/mol	25	EST	MEYLAN,WM & HOWARD,PH (1991)
OH 3.10E-011	cm3/molc sec	25	EST	MEYLAN,WM & HOWARD,PH (1993)

CAS #: 081792-74-7	PROPANAMIDE, N-[(2-AMINO-4-METHYLPHENYL)SULFONYL

Formula: $C_{11}H_{16}N_2O_3S$

Mol Weight: 256.33

MP (deg C): | FP (deg C):

BP (deg C):

BP pressure (mm Hg):

Property/Value	Units	Temp	Data Type	Reference
WS 8.51E+002	mg/L	25	EST	MEYLAN,WM ET AL. (1996)
logP 1.65			EXP	MONZANI,A ET AL. (1985)
VP 5.16E-009	mm Hg	25	EST	NEELY,WB & BLAU,GE (1985)
DC	pKa			
HL 9.41E-013	atm m3/mol	25	EST	MEYLAN,WM & HOWARD,PH (1991)
OH 6.82E-011	cm3/molc sec	25	EST	MEYLAN,WM & HOWARD,PH (1993)

CAS #: 081792-76-9	BENZENESULFONAMIDE, 2-AMINO-4-METHOXY-N-(2-METHY

Formula: $C_{11}H_{16}N_2O_4S$

Mol Weight: 272.33

MP (deg C): | FP (deg C):

BP (deg C):

BP pressure (mm Hg):

Property/Value	Units	Temp	Data Type	Reference
WS 7.93E+002	mg/L	25	EST	MEYLAN,WM ET AL. (1996)
logP 1.58			EXP	MONZANI,A ET AL. (1985)
VP 2.22E-009	mm Hg	25	EST	NEELY,WB & BLAU,GE (1985)
DC	pKa			
HL 5.04E-014	atm m3/mol	25	EST	MEYLAN,WM & HOWARD,PH (1991)
OH 2.09E-010	cm3/molc sec	25	EST	MEYLAN,WM & HOWARD,PH (1993)

CAS #: 081792-78-1	BUTANAMIDE, N-[(2-AMINOPHENYL)SULFONYL]-2-ETHYL-

Formula: $C_{12}H_{18}N_2O_3S$

Mol Weight: 270.35

MP (deg C): | FP (deg C):

BP (deg C):

BP pressure (mm Hg):

Property/Value	Units	Temp	Data Type	Reference
WS 7.23E+002	mg/L	25	EST	MEYLAN,WM ET AL. (1996)
logP 1.64			EXP	MONZANI,A ET AL. (1985)
VP 2.29E-009	mm Hg	25	EST	NEELY,WB & BLAU,GE (1985)
DC	pKa			
HL 1.50E-012	atm m3/mol	25	EST	MEYLAN,WM & HOWARD,PH (1991)
OH 3.83E-011	cm3/molc sec	25	EST	MEYLAN,WM & HOWARD,PH (1993)

CAS #: 081792-79-2	BUTANAMIDE, N-[(2-AMINO-4-METHYLPHENYL)SULFONYL]

Formula: $C_{13}H_{20}N_2O_3S$

Mol Weight: 284.38

MP (deg C): | FP (deg C):

BP (deg C):

BP pressure (mm Hg):

Property/Value	Units	Temp	Data Type	Reference
WS 2.08E+002	mg/L	25	EST	MEYLAN,WM ET AL. (1996)
logP 2.18			EXP	MONZANI,A ET AL. (1985)
VP 1.01E-009	mm Hg	25	EST	NEELY,WB & BLAU,GE (1985)
DC	pKa			
HL 1.66E-012	atm m3/mol	25	EST	MEYLAN,WM & HOWARD,PH (1991)
OH 7.55E-011	cm3/molc sec	25	EST	MEYLAN,WM & HOWARD,PH (1993)

CAS #: 081792-81-6	BUTANAMIDE, N-[(2-AMINO-4-METHOXYPHENYL)SULFONYL

Formula: $C_{13}H_{20}N_2O_4S$

Mol Weight: 300.38

MP (deg C): | FP (deg C):

BP (deg C):

BP pressure (mm Hg):

Property/Value	Units	Temp	Data Type	Reference
WS 2.26E+002	mg/L	25	EST	MEYLAN,WM ET AL. (1996)
logP 2.03			EXP	MONZANI,A ET AL. (1985)
VP 4.31E-010	mm Hg	25	EST	NEELY,WB & BLAU,GE (1985)
DC	pKa			
HL 8.89E-014	atm m3/mol	25	EST	MEYLAN,WM & HOWARD,PH (1991)
OH 2.16E-010	cm3/molc sec	25	EST	MEYLAN,WM & HOWARD,PH (1993)

CAS #: 081792-82-7	BUTANAMIDE, N-[(2-AMINO-4-CHLOROPHENYL)SULFONYL]

Formula: $C_{12}H_{17}ClN_2O_3S$

Mol Weight: 304.80

MP (deg C): | FP (deg C):

BP (deg C):

BP pressure (mm Hg):

Property/Value	Units	Temp	Data Type	Reference
WS 7.95E+001	mg/L	25	EST	MEYLAN,WM ET AL. (1996)
logP 2.53			EXP	MONZANI,A ET AL. (1985)
VP 6.13E-010	mm Hg	25	EST	NEELY,WB & BLAU,GE (1985)
DC	pKa			
HL 1.11E-012	atm m3/mol	25	EST	MEYLAN,WM & HOWARD,PH (1991)
OH 3.15E-011	cm3/molc sec	25	EST	MEYLAN,WM & HOWARD,PH (1993)

CAS #: 081792-98-5	BENZENESULFONAMIDE, 2-NITRO-N-(2-METHYL-1-OXOPRO

Formula: $C_{10}H_{12}N_2O_5S$

Mol Weight: 272.28

MP (deg C): | FP (deg C):

BP (deg C):

BP pressure (mm Hg):

Property/Value	Units	Temp	Data Type	Reference
WS 2.45E+002	mg/L	25	EST	MEYLAN,WM ET AL. (1996)
logP 1.72			EXP	MONZANI,A ET AL. (1985)
VP 4.65E-009	mm Hg	25	EST	NEELY,WB & BLAU,GE (1985)
DC	pKa			
HL 9.52E-012	atm m3/mol	25	EST	MEYLAN,WM & HOWARD,PH (1991)
OH 8.07E-012	cm3/molc sec	25	EST	MEYLAN,WM & HOWARD,PH (1993)

CAS #: 081792-99-6	PROPANAMIDE, 2-METHYL-N-[(4-METHYL-2-NITROPHENYL

Formula: $C_{11}H_{14}N_2O_5S$

Mol Weight: 286.31

MP (deg C): | FP (deg C):

BP (deg C):

BP pressure (mm Hg):

Property/Value	Units	Temp	Data Type	Reference
WS 8.42E+001	mg/L	25	EST	MEYLAN,WM ET AL. (1996)
logP 2.17			EXP	MONZANI,A ET AL. (1985)
VP 2.06E-009	mm Hg	25	EST	NEELY,WB & BLAU,GE (1985)
DC	pKa			
HL 1.05E-011	atm m3/mol	25	EST	MEYLAN,WM & HOWARD,PH (1991)
OH 8.25E-012	cm3/molc sec	25	EST	MEYLAN,WM & HOWARD,PH (1993)

CAS #: 081793-03-5	BUTANAMIDE, 2-ETHYL-N-[(2-NITROPHENYL)SULFONYL-

Formula: $C_{12}H_{16}N_2O_5S$

Mol Weight: 300.34

MP (deg C): | FP (deg C):

BP (deg C):

BP pressure (mm Hg):

Property/Value	Units	Temp	Data Type	Reference
WS 6.98E+001	mg/L	25	EST	MEYLAN,WM ET AL. (1996)
logP 2.17			EXP	MONZANI,A ET AL. (1985)
VP 9.09E-010	mm Hg	25	EST	NEELY,WB & BLAU,GE (1985)
DC	pKa			
HL 1.68E-011	atm m3/mol	25	EST	MEYLAN,WM & HOWARD,PH (1991)
OH 1.54E-011	cm3/molc sec	25	EST	MEYLAN,WM & HOWARD,PH (1993)

CAS #: 081793-04-6	BUTANAMIDE, 2-ETHYL-N-[(4-METHYL-2-NITROPHENYL)S

Formula: $C_{13}H_{18}N_2O_5S$

Mol Weight: 314.36

MP (deg C): | FP (deg C):

BP (deg C):

BP pressure (mm Hg):

Property/Value	Units	Temp	Data Type	Reference
WS 2.16E+001	mg/L	25	EST	MEYLAN,WM ET AL. (1996)
logP 2.67			EXP	MONZANI,A ET AL. (1985)
VP 4.00E-010	mm Hg	25	EST	NEELY,WB & BLAU,GE (1985)
DC	pKa			
HL 1.85E-011	atm m3/mol	25	EST	MEYLAN,WM & HOWARD,PH (1991)
OH 1.56E-011	cm3/molc sec	25	EST	MEYLAN,WM & HOWARD,PH (1993)

CAS #: 081835-01-0	8-S-T-BUTYL CYCLIC AMP

Formula: $C_{14}H_{22}N_5O_6PS$

Mol Weight: 419.40

MP (deg C): | FP (deg C):

BP (deg C):

BP pressure (mm Hg):

Property/Value	Units	Temp	Data Type	Reference
WS 9.22E+002	mg/L	25	EST	MEYLAN,WM ET AL. (1996)
logP -1.05			EXP	KORTH,M & ENGELS,J (1987)
VP 6.06E-011	mm Hg	25	EST	NEELY,WB & BLAU,GE (1985)
DC	pKa			
HL 1.99E-029	atm m3/mol	25	EST	MEYLAN,WM & HOWARD,PH (1991)
OH 3.02E-010	cm3/molc sec	25	EST	MEYLAN,WM & HOWARD,PH (1993)

CAS #: 081865-11-4	PROPANEDINITRILE, [[4-(2-CHLOROETHYL)PHENYL]HYDR

Formula: $C_{11}H_9ClN_4$

Mol Weight: 232.67

MP (deg C): | FP (deg C):

BP (deg C):

BP pressure (mm Hg):

Property/Value	Units	Temp	Data Type	Reference
WS 2.58E+001	mg/L	25	EST	MEYLAN,WM ET AL. (1996)
logP 3.58			EXP	STURDIK,E ET AL. (1985)
VP 4.58E-007	mm Hg	25	EST	NEELY,WB & BLAU,GE (1985)
DC	pKa			
HL 2.99E-009	atm m3/mol	25	EST	MEYLAN,WM & HOWARD,PH (1991)
OH 5.27E-011	cm3/molc sec	25	EST	MEYLAN,WM & HOWARD,PH (1993)

CAS #: 081872-10-8	L-PROLINE, 1-[3-(BENZOYLTHIO)-2-METHYL-1-OXOPROP

Formula: $C_{22}H_{23}NO_4S_2$

Mol Weight: 429.56

MP (deg C): | FP (deg C):

BP (deg C):

BP pressure (mm Hg):

Property/Value	Units	Temp	Data Type	Reference
WS 3.50E-001	mg/L	25	EST	MEYLAN,WM ET AL. (1996)
logP 4.40			EXP	RANADIVE,SA ET AL. (1992)
VP 2.90E-013	mm Hg	25	EST	NEELY,WB & BLAU,GE (1985)
DC	pKa			
HL 3.92E-018	atm m3/mol	25	EST	MEYLAN,WM & HOWARD,PH (1991)
OH 7.72E-011	cm3/molc sec	25	EST	MEYLAN,WM & HOWARD,PH (1993)

CAS #: 081892-67-3	1H-IMIDAZOLE-1-ACETAMIDE, N-(2-HYDROXYETHYL)-à-M

Formula: $C_8H_{12}N_4O_4$

Mol Weight: 228.21

MP (deg C): | FP (deg C):

BP (deg C):

BP pressure (mm Hg):

Property/Value	Units	Temp	Data Type	Reference
WS 6.08E+003	mg/L	25	EST	MEYLAN,WM ET AL. (1996)
logP -1.16			EXP	SANGSTER,J (1993)
VP 2.30E-011	mm Hg	25	EST	NEELY,WB & BLAU,GE (1985)
DC	pKa			
HL 6.79E-018	atm m3/mol	25	EST	MEYLAN,WM & HOWARD,PH (1991)
OH 2.09E-011	cm3/molc sec	25	EST	MEYLAN,WM & HOWARD,PH (1993)

CAS #: 081892-68-4				1H-IMIDAZOLE-1-ACETAMIDE, N-(2,3-DIHYDROXYPROPYL

Formula: $C_9H_{14}N_4O_5$

Mol Weight: 258.24

MP (deg C): FP (deg C):

BP (deg C):

BP pressure (mm Hg):

Property/ Value	Units	Temp	Data Type	Reference
WS 1.07E+004	mg/L	25	EST	MEYLAN,WM ET AL. (1996)
logP -1.64			EXP	SANGSTER,J (1993)
VP 1.81E-013	mm Hg	25	EST	NEELY,WB & BLAU,GE (1985)
DC	pKa			
HL 3.30E-019	atm m3/mol	25	EST	MEYLAN,WM & HOWARD,PH (1991)
OH 3.13E-011	cm3/molc sec	25	EST	MEYLAN,WM & HOWARD,PH (1993)

CAS #: 081892-69-5				1H-IMIDAZOLE-1-ACETAMIDE, N-[2-HYDROXY-1,1-BIS(H

Formula: $C_9H_{14}N_4O_6$

Mol Weight: 274.24

MP (deg C): FP (deg C):

BP (deg C):

BP pressure (mm Hg):

Property/ Value	Units	Temp	Data Type	Reference
WS 1.12E+004	mg/L	25	EST	MEYLAN,WM ET AL. (1996)
logP -1.77			EXP	SANGSTER,J (1993)
VP 6.52E-016	mm Hg	25	EST	NEELY,WB & BLAU,GE (1985)
DC	pKa			
HL 1.21E-020	atm m3/mol	25	EST	MEYLAN,WM & HOWARD,PH (1991)
OH 2.32E-011	cm3/molc sec	25	EST	MEYLAN,WM & HOWARD,PH (1993)

CAS #: 082110-72-3				FENOXAPROP-ETHYL

Formula: $C_{18}H_{16}ClNO_5$

Mol Weight: 361.78

MP (deg C): 84-85 FP (deg C):

BP (deg C): 200

BP pressure (mm Hg): 7.50E-001

Property/ Value	Units	Temp	Data Type	Reference
WS 3.12E-001	mg/L	25	EST	MEYLAN,WM ET AL. (1996)
logP 4.95				MEYLAN,WM & HOWARD,PH (1995)
VP 1.20E-008	mm Hg	25	EST	NEELY,WB & BLAU,GE (1985)
DC	pKa			
HL 1.96E-011	atm m3/mol	25	EST	MEYLAN,WM & HOWARD,PH (1991)
OH 3.86E-011	cm3/molc sec	25	EST	MEYLAN,WM & HOWARD,PH (1993)

CAS #: 082249-23-8				2-PROPANOL, 1-BUTOXY-3-(2-NITRO-1H-IMIDAZOL-1-YL

Formula: $C_{10}H_{17}N_3O_4$

Mol Weight: 243.26

MP (deg C): FP (deg C):

BP (deg C):

BP pressure (mm Hg):

Property/ Value	Units	Temp	Data Type	Reference
WS 1.18E+003	mg/L	25	EST	MEYLAN,WM ET AL. (1996)
logP 1.11			EXP	SANGSTER,J (1993)
VP 5.77E-009	mm Hg	25	EST	NEELY,WB & BLAU,GE (1985)
DC	pKa			
HL 5.58E-013	atm m3/mol	25	EST	MEYLAN,WM & HOWARD,PH (1991)
OH 3.54E-011	cm3/molc sec	25	EST	MEYLAN,WM & HOWARD,PH (1993)

CAS #: 082410-32-0				6H-PURIN-6-ONE, 2-AMINO-1,9-DIHYDRO-9-[[2-HYDROX

Formula: $C_9H_{13}N_5O_4$

Mol Weight: 255.24

MP (deg C): 250 dec FP (deg C):

BP (deg C):

BP pressure (mm Hg):

Property/ Value	Units	Temp	Data Type	Reference
WS 2.83E+004	mg/L	25	EST	MEYLAN,WM ET AL. (1996)
logP -1.66			EXP	SANGSTER,J (1993)
VP 2.86E-017	mm Hg	25	EST	NEELY,WB & BLAU,GE (1985)
DC	pKa			
HL 1.54E-023	atm m3/mol	25	EST	MEYLAN,WM & HOWARD,PH (1991)
OH 9.44E-011	cm3/molc sec	25	EST	MEYLAN,WM & HOWARD,PH (1993)

CAS #: 082412-82-6				BENZENEACETIC ACID, 4-BUTYL-à-HYDROXY-à-METHYL-

Formula: $C_{13}H_{18}O_3$

Mol Weight: 222.29

MP (deg C): FP (deg C):

BP (deg C):

BP pressure (mm Hg):

Property/ Value	Units	Temp	Data Type	Reference
WS 1.21E+003	mg/L	25	EST	MEYLAN,WM ET AL. (1996)
logP 2.75			EXP	KUCHAR,M ET AL. (1985)
VP 4.22E-007	mm Hg	25	EST	NEELY,WB & BLAU,GE (1985)
DC	pKa			
HL 2.36E-008	atm m3/mol	25	EST	MEYLAN,WM & HOWARD,PH (1991)
OH 1.09E-011	cm3/molc sec	25	EST	MEYLAN,WM & HOWARD,PH (1993)

CAS #: 082419-36-1				OFLOXACIN

Formula: $C_{18}H_{20}FN_3O_4$

Mol Weight: 361.38

MP (deg C): 250-257 de FP (deg C):

BP (deg C):

BP pressure (mm Hg):

Property/ Value	Units	Temp	Data Type	Reference
WS 2.83E+004	mg/L	25	EST	MEYLAN,WM ET AL. (1996)
logP -0.39			EXP	HANSCH,C ET AL. (1995)
VP 1.55E-013	mm Hg	25	EST	NEELY,WB & BLAU,GE (1985)
DC	pKa			
HL 4.98E-020	atm m3/mol	25	EST	MEYLAN,WM & HOWARD,PH (1991)
OH 1.97E-010	cm3/molc sec	25	EST	MEYLAN,WM & HOWARD,PH (1993)

CAS #: 082488-02-6				M-NITROBENZYL-TR-CHRYSANTHEMATE

Formula: $C_{17}H_{21}NO_4$

Mol Weight: 303.36

MP (deg C): FP (deg C):

BP (deg C):

BP pressure (mm Hg):

Property/ Value	Units	Temp	Data Type	Reference
WS 1.73E-001	mg/L	25	EST	MEYLAN,WM ET AL. (1996)
logP 5.20			EXP	HANSCH,C & LEO,AJ (1985)
VP 1.12E-006	mm Hg	25	EST	NEELY,WB & BLAU,GE (1985)
DC	pKa			
HL 2.47E-007	atm m3/mol	25	EST	MEYLAN,WM & HOWARD,PH (1991)
OH 8.95E-011	cm3/molc sec	25	EST	MEYLAN,WM & HOWARD,PH (1993)

CAS #: 082530-96-9 — HYDRAZINECARBOXIMIDAMIDE, 2-[(4-ETHOXYPHENYL)MET

Formula: $C_{10}H_{14}N_4O$

Mol Weight: 206.25

MP (deg C): **FP (deg C):**

BP (deg C):

BP pressure (mm Hg):

Property/ Value	Units	Temp	Data Type	Reference
WS 2.54E+004	mg/L	25	EST	MEYLAN,WM ET AL. (1996)
logP 0.24			EXP	SANGSTER,J (1994)
VP 1.76E-005	mm Hg	25	EST	NEELY,WB & BLAU,GE (1985)
DC	pKa			
HL 3.97E-013	atm m3/mol	25	EST	MEYLAN,WM & HOWARD,PH (1991)
OH 1.18E-010	cm3/molc sec	25	EST	MEYLAN,WM & HOWARD,PH (1993)

CAS #: 082535-32-8 — P-CYANOBENZYL-TR-CHRYSANTHEMATE

Formula: $C_{18}H_{21}NO_2$

Mol Weight: 283.37

MP (deg C): **FP (deg C):**

BP (deg C):

BP pressure (mm Hg):

Property/ Value	Units	Temp	Data Type	Reference
WS 5.34E-001	mg/L	25	EST	MEYLAN,WM ET AL. (1996)
logP 4.91			EXP	HANSCH,C & LEO,AJ (1985)
VP 2.66E-006	mm Hg	25	EST	NEELY,WB & BLAU,GE (1985)
DC	pKa			
HL 6.04E-007	atm m3/mol	25	EST	MEYLAN,WM & HOWARD,PH (1991)
OH 9.00E-011	cm3/molc sec	25	EST	MEYLAN,WM & HOWARD,PH (1993)

CAS #: 082560-54-1 — BENFURACARB

Formula: $C_{20}H_{30}N_2O_5S$

Mol Weight: 410.54

MP (deg C): **FP (deg C):**

BP (deg C):

BP pressure (mm Hg):

Property/ Value	Units	Temp	Data Type	Reference
WS 8.00E+000	mg/L	20	EXP	SHIU,WY ET AL. (1990)
logP 4.30			EXP	TOMLIN,C (1994)
VP 2.59E-008	mm Hg	25	EST	NEELY,WB & BLAU,GE (1985)
DC	pKa			
HL 3.02E-012	atm m3/mol	25	EST	MEYLAN,WM & HOWARD,PH (1991)
OH 6.99E-011	cm3/molc sec	25	EST	MEYLAN,WM & HOWARD,PH (1993)

CAS #: 082568-68-1 — BENZAMIDE, 2-[(4-NITROPHENYL)METHOXY]-

Formula: $C_{14}H_{12}N_2O_4$

Mol Weight: 272.26

MP (deg C): **FP (deg C):**

BP (deg C):

BP pressure (mm Hg):

Property/ Value	Units	Temp	Data Type	Reference
WS 2.41E+002	mg/L	25	EST	MEYLAN,WM ET AL. (1996)
logP 1.73			EXP	SANGSTER,J (1993)
VP 4.21E-009	mm Hg	25	EST	NEELY,WB & BLAU,GE (1985)
DC	pKa			
HL 4.17E-014	atm m3/mol	25	EST	MEYLAN,WM & HOWARD,PH (1991)
OH 2.79E-011	cm3/molc sec	25	EST	MEYLAN,WM & HOWARD,PH (1993)

CAS #: 082568-69-2 — BENZAMIDE, 2-[(4-METHOXYPHENYL)METHOXY]-

Formula: $C_{15}H_{15}NO_3$

Mol Weight: 257.29

MP (deg C): **FP (deg C):**

BP (deg C):

BP pressure (mm Hg):

Property/ Value	Units	Temp	Data Type	Reference
WS 8.74E+002	mg/L	25	EST	MEYLAN,WM ET AL. (1996)
logP 1.63			EXP	SANGSTER,J (1993)
VP 4.05E-008	mm Hg	25	EST	NEELY,WB & BLAU,GE (1985)
DC	pKa			
HL 6.26E-013	atm m3/mol	25	EST	MEYLAN,WM & HOWARD,PH (1991)
OH 5.43E-011	cm3/molc sec	25	EST	MEYLAN,WM & HOWARD,PH (1993)

CAS #: 082609-03-8 — ACETIC ACID, (1-NAPHTHALENYLOXY)-, SODIUM SALT

Formula: $C_{12}H_9NaO_3$

Mol Weight: 224.19

MP (deg C): **FP (deg C):**

BP (deg C):

BP pressure (mm Hg):

Property/ Value	Units	Temp	Data Type	Reference
WS 6.24E+005	mg/L	25	EST	MEYLAN,WM ET AL. (1996)
logP -1.50			EXP	CHAMBERLAIN,K ET AL. (1986)
VP 3.90E-012	mm Hg	25	EST	NEELY,WB & BLAU,GE (1985)
DC	pKa			
HL	atm m3/mol			
OH 2.04E-010	cm3/molc sec	25	EST	MEYLAN,WM & HOWARD,PH (1993)

CAS #: 082682-68-6 — 2-(3-METHOXYPHENYL)INDOLIZINE

Formula: $C_{15}H_{15}NO$

Mol Weight: 225.29

MP (deg C): **FP (deg C):**

BP (deg C):

BP pressure (mm Hg):

Property/ Value	Units	Temp	Data Type	Reference
WS 9.64E+000	mg/L	25	EST	MEYLAN,WM ET AL. (1996)
logP 4.14			EXP	HANSCH,C & LEO,AJ (1985)
VP 7.37E-006	mm Hg	25	EST	NEELY,WB & BLAU,GE (1985)
DC	pKa			
HL 2.48E-010	atm m3/mol	25	EST	MEYLAN,WM & HOWARD,PH (1991)
OH 2.29E-010	cm3/molc sec	25	EST	MEYLAN,WM & HOWARD,PH (1993)

CAS #: 082682-69-7 — 2-(3-TOLYL)INDOLIZINE

Formula: $C_{15}H_{15}N$

Mol Weight: 209.29

MP (deg C): **FP (deg C):**

BP (deg C):

BP pressure (mm Hg):

Property/ Value	Units	Temp	Data Type	Reference
WS 2.78E+000	mg/L	25	EST	MEYLAN,WM ET AL. (1996)
logP 4.87			EXP	HANSCH,C & LEO,AJ (1985)
VP 1.92E-005	mm Hg	25	EST	NEELY,WB & BLAU,GE (1985)
DC	pKa			
HL 4.63E-009	atm m3/mol	25	EST	MEYLAN,WM & HOWARD,PH (1991)
OH 2.00E-010	cm3/molc sec	25	EST	MEYLAN,WM & HOWARD,PH (1993)

CAS #: 082682-70-0				2-(3-CHLOROPHENYL)INDOLIZINE
Formula: $C_{14}H_{12}ClN$				
Mol Weight: 229.71				
MP (deg C):		FP (deg C):		
BP (deg C):				
BP pressure (mm Hg):				

Property/Value	Units	Temp	Data Type	Reference
WS 1.28E+000	mg/L	25	EST	MEYLAN,WM ET AL. (1996)
logP 5.14			EXP	HANSCH,C & LEO,AJ (1985)
VP 1.05E-005	mm Hg	25	EST	NEELY,WB & BLAU,GE (1985)
DC	pKa			
HL 3.11E-009	atm m3/mol	25	EST	MEYLAN,WM & HOWARD,PH (1991)
OH 1.93E-010	cm3/molc sec	25	EST	MEYLAN,WM & HOWARD,PH (1993)

CAS #: 082700-02-5				CYCLOOCTANEETHANAMINE, à-METHYL-
Formula: $C_{11}H_{23}N$				
Mol Weight: 169.31				
MP (deg C):		FP (deg C):		
BP (deg C):				
BP pressure (mm Hg):				

Property/Value	Units	Temp	Data Type	Reference
WS 4.37E+002	mg/L	25	EST	MEYLAN,WM ET AL. (1996)
logP 3.70			EXP	SANGSTER,J (1993)
VP 5.31E-002	mm Hg	25	EST	NEELY,WB & BLAU,GE (1985)
DC	pKa			
HL 5.68E-005	atm m3/mol	25	EST	MEYLAN,WM & HOWARD,PH (1991)
OH 5.83E-011	cm3/molc sec	25	EST	MEYLAN,WM & HOWARD,PH (1993)

CAS #: 082774-00-3				N(N''-ME-(4ME-PH)CARBAMYLTHIO)METHOMYL
Formula: $C_{14}H_{19}N_3O_3S_2$				
Mol Weight: 341.45				
MP (deg C):		FP (deg C):		
BP (deg C):				
BP pressure (mm Hg):				

Property/Value	Units	Temp	Data Type	Reference
WS 5.11E+001	mg/L	25	EST	MEYLAN,WM ET AL. (1996)
logP 2.50			EXP	HANSCH,C & LEO,AJ (1985)
VP 1.29E-008	mm Hg	25	EST	NEELY,WB & BLAU,GE (1985)
DC	pKa			
HL 2.59E-010	atm m3/mol	25	EST	MEYLAN,WM & HOWARD,PH (1991)
OH 3.88E-011	cm3/molc sec	25	EST	MEYLAN,WM & HOWARD,PH (1993)

CAS #: 082774-01-4				N(N''-ME(PHENYL)CARBAMYLTHIO)METHOMYL
Formula: $C_{13}H_{17}N_3O_3S_2$				
Mol Weight: 327.43				
MP (deg C):		FP (deg C):		
BP (deg C):				
BP pressure (mm Hg):				

Property/Value	Units	Temp	Data Type	Reference
WS 1.12E+002	mg/L	25	EST	MEYLAN,WM ET AL. (1996)
logP 2.20			EXP	HANSCH,C & LEO,AJ (1985)
VP 2.92E-008	mm Hg	25	EST	NEELY,WB & BLAU,GE (1985)
DC	pKa			
HL 2.35E-010	atm m3/mol	25	EST	MEYLAN,WM & HOWARD,PH (1991)
OH 3.58E-011	cm3/molc sec	25	EST	MEYLAN,WM & HOWARD,PH (1993)

CAS #: 082774-06-9				N(N''-ME(4-CLPH)CARBAMYLTHIO)METHOMYL
Formula: $C_{13}H_{16}ClN_3O_3S_2$				
Mol Weight: 361.87				
MP (deg C):		FP (deg C):		
BP (deg C):				
BP pressure (mm Hg):				

Property/Value	Units	Temp	Data Type	Reference
WS 3.16E+001	mg/L	25	EST	MEYLAN,WM ET AL. (1996)
logP 2.60			EXP	HANSCH,C & LEO,AJ (1985)
VP 7.14E-009	mm Hg	25	EST	NEELY,WB & BLAU,GE (1985)
DC	pKa			
HL 1.74E-010	atm m3/mol	25	EST	MEYLAN,WM & HOWARD,PH (1991)
OH 3.53E-011	cm3/molc sec	25	EST	MEYLAN,WM & HOWARD,PH (1993)

CAS #: 082774-07-0				N(N''-ME(2-CLPH)CARBAMYLTHIO)METHOMYL
Formula: $C_{13}H_{16}ClN_3O_3S_2$				
Mol Weight: 361.87				
MP (deg C):		FP (deg C):		
BP (deg C):				
BP pressure (mm Hg):				

Property/Value	Units	Temp	Data Type	Reference
WS 3.85E+001	mg/L	25	EST	MEYLAN,WM ET AL. (1996)
logP 2.50			EXP	HANSCH,C & LEO,AJ (1985)
VP 7.14E-009	mm Hg	25	EST	NEELY,WB & BLAU,GE (1985)
DC	pKa			
HL 1.74E-010	atm m3/mol	25	EST	MEYLAN,WM & HOWARD,PH (1991)
OH 3.53E-011	cm3/molc sec	25	EST	MEYLAN,WM & HOWARD,PH (1993)

CAS #: 082774-14-9				N(N''-ME(3-CLPH)CARBAMYLTHIO)METHOMYL
Formula: $C_{13}H_{16}ClN_3O_3S_2$				
Mol Weight: 361.87				
MP (deg C):		FP (deg C):		
BP (deg C):				
BP pressure (mm Hg):				

Property/Value	Units	Temp	Data Type	Reference
WS 2.60E+001	mg/L	25	EST	MEYLAN,WM ET AL. (1996)
logP 2.70			EXP	HANSCH,C & LEO,AJ (1985)
VP 7.14E-009	mm Hg	25	EST	NEELY,WB & BLAU,GE (1985)
DC	pKa			
HL 1.74E-010	atm m3/mol	25	EST	MEYLAN,WM & HOWARD,PH (1991)
OH 3.46E-011	cm3/molc sec	25	EST	MEYLAN,WM & HOWARD,PH (1993)

CAS #: 082774-15-0				N(N''-ME(3-CF3PH)CARBAMYLTHIO)METHOMYL
Formula: $C_{14}H_{16}F_3N_3O_3S_2$				
Mol Weight: 395.43				
MP (deg C):		FP (deg C):		
BP (deg C):				
BP pressure (mm Hg):				

Property/Value	Units	Temp	Data Type	Reference
WS 2.40E+001	mg/L	25	EST	MEYLAN,WM ET AL. (1996)
logP 2.50			EXP	HANSCH,C & LEO,AJ (1985)
VP 2.23E-008	mm Hg	25	EST	NEELY,WB & BLAU,GE (1985)
DC	pKa			
HL 2.04E-009	atm m3/mol	25	EST	MEYLAN,WM & HOWARD,PH (1991)
OH 4.40E-012	cm3/molc sec	25	EST	MEYLAN,WM & HOWARD,PH (1993)

082774-16-1 — N(N"ME(24DICLPH)CARBAMYLTHIO)METHOMYL

Formula: $C_{13}H_{15}Cl_2N_3O_3S_2$

Mol Weight: 396.32

MP (deg C):
FP (deg C):
BP (deg C):
BP pressure (mm Hg):

Property/Value	Units	Temp	Data Type	Reference
WS 5.97E+000	mg/L	25	EST	MEYLAN,WM ET AL. (1996)
logP 3.20			EXP	HANSCH,C & LEO,AJ (1985)
VP 1.53E-009	mm Hg	25	EST	NEELY,WB & BLAU,GE (1985)
DC	pKa			
HL 1.29E-010	atm m3/mol	25	EST	MEYLAN,WM & HOWARD,PH (1991)
OH 3.49E-011	cm3/molc sec	25	EST	MEYLAN,WM & HOWARD,PH (1993)

082774-17-2 — N(N"-IPR(4ME-PH)CARBAMYLTHIO)METHOMYL

Formula: $C_{16}H_{23}N_3O_3S_2$

Mol Weight: 369.51

MP (deg C):
FP (deg C):
BP (deg C):
BP pressure (mm Hg):

Property/Value	Units	Temp	Data Type	Reference
WS 5.89E+000	mg/L	25	EST	MEYLAN,WM ET AL. (1996)
logP 3.40			EXP	HANSCH,C & LEO,AJ (1985)
VP 4.01E-009	mm Hg	25	EST	NEELY,WB & BLAU,GE (1985)
DC	pKa			
HL 4.57E-010	atm m3/mol	25	EST	MEYLAN,WM & HOWARD,PH (1991)
OH 4.66E-011	cm3/molc sec	25	EST	MEYLAN,WM & HOWARD,PH (1993)

082774-18-3 — N(N"-BU-(PHENYL)CARBAMYLTHIO)METHOMYL

Formula: $C_{16}H_{23}N_3O_3S_2$

Mol Weight: 369.51

MP (deg C):
FP (deg C):
BP (deg C):
BP pressure (mm Hg):

Property/Value	Units	Temp	Data Type	Reference
WS 5.89E+000	mg/L	25	EST	MEYLAN,WM ET AL. (1996)
logP 3.40			EXP	HANSCH,C & LEO,AJ (1985)
VP 2.42E-009	mm Hg	25	EST	NEELY,WB & BLAU,GE (1985)
DC	pKa			
HL 5.50E-010	atm m3/mol	25	EST	MEYLAN,WM & HOWARD,PH (1991)
OH 4.27E-011	cm3/molc sec	25	EST	MEYLAN,WM & HOWARD,PH (1993)

082775-80-2 — FURAZAN, CHLOROPHENYL-, 5-OXIDE

Formula: $C_8H_5ClN_2O_2$

Mol Weight: 196.59

MP (deg C):
FP (deg C):
BP (deg C):
BP pressure (mm Hg):

Property/Value	Units	Temp	Data Type	Reference
WS 1.78E+002	mg/L	25	EST	MEYLAN,WM ET AL. (1996)
logP 2.82			EXP	CALVINO,R ET AL. (1992)
VP 1.54E-005	mm Hg	25	EST	NEELY,WB & BLAU,GE (1985)
DC	pKa			
HL 1.08E-010	atm m3/mol	25	EST	MEYLAN,WM & HOWARD,PH (1991)
OH 6.50E-012	cm3/molc sec	25	EST	MEYLAN,WM & HOWARD,PH (1993)

082775-81-3 — FURAZAN, CHLOROPHENYL-, 2-OXIDE

Formula: $C_8H_5ClN_2O_2$

Mol Weight: 196.59

MP (deg C):
FP (deg C):
BP (deg C):
BP pressure (mm Hg):

Property/Value	Units	Temp	Data Type	Reference
WS 2.25E+002	mg/L	25	EST	MEYLAN,WM ET AL. (1996)
logP 2.70			EXP	CALVINO,R ET AL. (1992)
VP 1.54E-005	mm Hg	25	EST	NEELY,WB & BLAU,GE (1985)
DC	pKa			
HL 1.08E-010	atm m3/mol	25	EST	MEYLAN,WM & HOWARD,PH (1991)
OH 6.50E-012	cm3/molc sec	25	EST	MEYLAN,WM & HOWARD,PH (1993)

082780-90-3 — 2-IMIDAZOLIDINIMINE, N-(2,3,6-TRICHLOROPHENYL)-

Formula: $C_9H_8Cl_3N_3$

Mol Weight: 264.54

MP (deg C):
FP (deg C):
BP (deg C):
BP pressure (mm Hg):

Property/Value	Units	Temp	Data Type	Reference
WS 1.58E+003	mg/L	25	EST	MEYLAN,WM ET AL. (1996)
logP 2.46			EXP	DE JONGE,A ET AL. (1984)
VP 5.29E-006	mm Hg	25	EST	NEELY,WB & BLAU,GE (1985)
DC	pKa			
HL 1.12E-011	atm m3/mol	25	EST	MEYLAN,WM & HOWARD,PH (1991)
OH 1.43E-010	cm3/molc sec	25	EST	MEYLAN,WM & HOWARD,PH (1993)

082784-14-3 — N(N"ME(26DICLPH)CARBAMYLTHIO)METHOMYL

Formula: $C_{13}H_{15}Cl_2N_3O_3S_2$

Mol Weight: 396.32

MP (deg C):
FP (deg C):
BP (deg C):
BP pressure (mm Hg):

Property/Value	Units	Temp	Data Type	Reference
WS 1.94E+001	mg/L	25	EST	MEYLAN,WM ET AL. (1996)
logP 2.60			EXP	HANSCH,C & LEO,AJ (1985)
VP 1.53E-009	mm Hg	25	EST	NEELY,WB & BLAU,GE (1985)
DC	pKa			
HL 1.29E-010	atm m3/mol	25	EST	MEYLAN,WM & HOWARD,PH (1991)
OH 3.49E-011	cm3/molc sec	25	EST	MEYLAN,WM & HOWARD,PH (1993)

082801-84-1 — 2(2,3-DICL-6MEPH IMINO)IMIDAZOLIDINE

Formula: $C_{10}H_{11}Cl_2N_3$

Mol Weight: 244.13

MP (deg C):
FP (deg C):
BP (deg C):
BP pressure (mm Hg):

Property/Value	Units	Temp	Data Type	Reference
WS 2.56E+003	mg/L	25	EST	MEYLAN,WM ET AL. (1996)
logP 2.35			EXP	HANSCH,C & LEO,AJ (1985)
VP 9.32E-006	mm Hg	25	EST	NEELY,WB & BLAU,GE (1985)
DC	pKa			
HL 1.66E-011	atm m3/mol	25	EST	MEYLAN,WM & HOWARD,PH (1991)
OH 1.44E-010	cm3/molc sec	25	EST	MEYLAN,WM & HOWARD,PH (1993)

CAS #: 083002-04-4 — PHENOL, 5-(1,1-DIMETHYLHEPTYL)-2-[5-HYDROXY-2-(3

Formula: $C_{24}H_{40}O_3$

Mol Weight: 376.58

MP (deg C): | FP (deg C):

BP (deg C):

BP pressure (mm Hg):

Property/Value	Units	Temp	Data Type	Reference
WS 9.45E-002	mg/L	25	EST	MEYLAN,WM ET AL. (1996)
logP 6.13			EXP	SANGSTER,J (1993)
VP 5.98E-013	mm Hg	25	EST	NEELY,WB & BLAU,GE (1985)
DC	pKa			
HL 3.75E-011	atm m3/mol	25	EST	MEYLAN,WM & HOWARD,PH (1991)
OH 1.24E-010	cm3/molc sec	25	EST	MEYLAN,WM & HOWARD,PH (1993)

CAS #: 083029-57-6 — PROPANAMIDE, N-[2-[(2-HYDROXY-3-PHENOXYPROPYL)AM

Formula: $C_{15}H_{24}N_2O_3$

Mol Weight: 280.37

Property/Value	Units	Temp	Data Type	Reference
WS 1.70E+003	mg/L	25	EST	MEYLAN,WM ET AL. (1996)
logP 1.14			EXP	MAULEON,D ET AL. (1988)
VP 1.68E-010	mm Hg	25	EST	NEELY,WB & BLAU,GE (1985)
DC	pKa			
HL 3.62E-018	atm m3/mol	25	EST	MEYLAN,WM & HOWARD,PH (1991)
OH 1.36E-010	cm3/molc sec	25	EST	MEYLAN,WM & HOWARD,PH (1993)

CAS #: 083040-20-4 — 1-PYRROLIDINYLOXY, 3,4-DICARBOXY-2,2,5,5-TETRAME

Formula: $C_{10}H_{17}NO_5$

Mol Weight: 231.25

Property/Value	Units	Temp	Data Type	Reference
WS 3.05E+004	mg/L	25	EST	MEYLAN,WM ET AL. (1996)
logP 0.46			EXP	FUCHS,J ET AL. (1990)
VP 2.14E-010	mm Hg	25	EST	NEELY,WB & BLAU,GE (1985)
DC	pKa			
HL 3.42E-021	atm m3/mol	25	EST	MEYLAN,WM & HOWARD,PH (1991)
OH 5.23E-012	cm3/molc sec	25	EST	MEYLAN,WM & HOWARD,PH (1993)

CAS #: 083055-99-6 — BENSULFURON-METHYL (PH7)

Formula: $C_{16}H_{18}N_4O_7S$

Mol Weight: 410.41

MP (deg C): 187

Property/Value	Units	Temp	Data Type	Reference
WS 3.00E+000	mg/L	25	EXP	HAY,JV (1990) @ pH=5
logP 1.80			EXP	HAY,JV (1990)
VP 2.10E-014	mm Hg	25	EXP	TOMLIN,C (1994)
DC	pKa			
HL 3.78E-015	atm m3/mol	25	EST	VP/WSOL
OH 2.10E-010	cm3/molc sec	25	EST	MEYLAN,WM & HOWARD,PH (1993)

CAS #: 083073-36-3 — 1H-PYRROLIZINE-1-CARBOXYLIC ACID, 5-(4-CHLOROBEN

Formula: $C_{16}H_{14}ClNO_3$

Mol Weight: 303.75

Property/Value	Units	Temp	Data Type	Reference
WS 2.44E+001	mg/L	25	EST	MEYLAN,WM ET AL. (1996)
logP 3.60			EXP	SANGSTER,J (1994)
VP 1.31E-008	mm Hg	25	EST	NEELY,WB & BLAU,GE (1985)
DC	pKa			
HL 2.74E-013	atm m3/mol	25	EST	MEYLAN,WM & HOWARD,PH (1991)
OH 2.04E-010	cm3/molc sec	25	EST	MEYLAN,WM & HOWARD,PH (1993)

CAS #: 083073-37-4 — 1H-PYRROLIZINE-1-CARBOXYLIC ACID, 6-ETHYL-5-(4-F

Formula: $C_{17}H_{16}FNO_3$

Mol Weight: 301.32

Property/Value	Units	Temp	Data Type	Reference
WS 2.38E+001	mg/L	25	EST	MEYLAN,WM ET AL. (1996)
logP 3.63			EXP	SANGSTER,J (1994)
VP 2.59E-008	mm Hg	25	EST	NEELY,WB & BLAU,GE (1985)
DC	pKa			
HL 5.73E-013	atm m3/mol	25	EST	MEYLAN,WM & HOWARD,PH (1991)
OH 2.05E-010	cm3/molc sec	25	EST	MEYLAN,WM & HOWARD,PH (1993)

CAS #: 083107-46-4 — 1-(BD-GLUCOPYRANOSYL)-2-NO2IMIDAZOLE

Formula: $C_9H_{13}N_3O_7$

Mol Weight: 275.22

Property/Value	Units	Temp	Data Type	Reference
WS 1.09E+005	mg/L	25	EST	MEYLAN,WM ET AL. (1996)
logP -1.40			EXP	HANSCH,C & LEO,AJ (1985)
VP 4.06E-014	mm Hg	25	EST	NEELY,WB & BLAU,GE (1985)
DC	pKa			
HL 1.14E-018	atm m3/mol	25	EST	MEYLAN,WM & HOWARD,PH (1991)
OH 7.24E-011	cm3/molc sec	25	EST	MEYLAN,WM & HOWARD,PH (1993)

CAS #: 083107-47-5 — 1(BD-GLUCOTHIOPYRANOSYL)2NO2IMIDAZOLE

Formula: $C_9H_{13}N_3O_6S$

Mol Weight: 291.28

Property/Value	Units	Temp	Data Type	Reference
WS 2.06E+004	mg/L	25	EST	MEYLAN,WM ET AL. (1996)
logP -0.66			EXP	HANSCH,C & LEO,AJ (1985)
VP 8.69E-015	mm Hg	25	EST	NEELY,WB & BLAU,GE (1985)
DC	pKa			
HL 8.26E-018	atm m3/mol	25	EST	MEYLAN,WM & HOWARD,PH (1991)
OH 8.31E-011	cm3/molc sec	25	EST	MEYLAN,WM & HOWARD,PH (1993)

CAS #:	083107-50-0			D-GLYCEROL-A-D-GALACTO DERIVATIVE

Formula:	$C_{23}H_{30}N_4O_{14}$		
Mol Weight:	586.51		
MP (deg C):		FP (deg C):	
BP (deg C):			
BP pressure (mm Hg):			

Property/Value	Units	Temp	Data Type	Reference
WS 3.58E+001	mg/L	25	EST	MEYLAN,WM ET AL. (1996)
logP 0.41			EXP	HANSCH,C & LEO,AJ (1985)
VP 5.15E-017	mm Hg	25	EST	NEELY,WB & BLAU,GE (1985)
DC	pKa			
HL 2.03E-028	atm m3/mol	25	EST	MEYLAN,WM & HOWARD,PH (1991)
OH 6.06E-011	cm3/molc sec	25	EST	MEYLAN,WM & HOWARD,PH (1993)

CAS #:	083116-90-9			1H-IMIDAZOLE, 2-NITRO-1-(2,3,4,6-TETRA-O-ACETYL-

Formula:	$C_{17}H_{21}N_3O_{10}S$		
Mol Weight:	459.44		
MP (deg C):		FP (deg C):	
BP (deg C):			
BP pressure (mm Hg):			

Property/Value	Units	Temp	Data Type	Reference
WS 1.53E+002	mg/L	25	EST	MEYLAN,WM ET AL. (1996)
logP 0.63			EXP	SANGSTER,J (1993)
VP 6.16E-012	mm Hg	25	EST	NEELY,WB & BLAU,GE (1985)
DC	pKa			
HL 1.48E-019	atm m3/mol	25	EST	MEYLAN,WM & HOWARD,PH (1991)
OH 6.40E-011	cm3/molc sec	25	EST	MEYLAN,WM & HOWARD,PH (1993)

CAS #:	083118-48-3			BENZOBICYCLO(221)HEPTENE,2EN-NH2-6CF3

Formula:	$C_{12}H_{12}F_3N$		
Mol Weight:	227.23		
MP (deg C):		FP (deg C):	
BP (deg C):			
BP pressure (mm Hg):			

Property/Value	Units	Temp	Data Type	Reference
WS 1.18E+003	mg/L	25	EST	MEYLAN,WM ET AL. (1996)
logP 2.91			EXP	HANSCH,C & LEO,AJ (1985)
VP 6.14E-003	mm Hg	25	EST	NEELY,WB & BLAU,GE (1985)
DC	pKa			
HL 2.67E-006	atm m3/mol	25	EST	MEYLAN,WM & HOWARD,PH (1991)
OH 2.42E-011	cm3/molc sec	25	EST	MEYLAN,WM & HOWARD,PH (1993)

CAS #:	083118-50-7			BENZOBICYCLO(221)HEPTENE,2EX-NH2-6CF3

Formula:	$C_{12}H_{12}F_3N$		
Mol Weight:	227.23		
MP (deg C):		FP (deg C):	
BP (deg C):			
BP pressure (mm Hg):			

Property/Value	Units	Temp	Data Type	Reference
WS 1.18E+003	mg/L	25	EST	MEYLAN,WM ET AL. (1996)
logP 3.21			EXP	HANSCH,C & LEO,AJ (1985)
VP 6.14E-003	mm Hg	25	EST	NEELY,WB & BLAU,GE (1985)
DC	pKa			
HL 2.67E-006	atm m3/mol	25	EST	MEYLAN,WM & HOWARD,PH (1991)
OH 2.42E-011	cm3/molc sec	25	EST	MEYLAN,WM & HOWARD,PH (1993)

CAS #:	083118-51-8			BENZOBICYCLO(221)HEPTENE,2EX-NH2-7CF3

Formula:	$C_{12}H_{12}F_3N$		
Mol Weight:	227.23		
MP (deg C):		FP (deg C):	
BP (deg C):			
BP pressure (mm Hg):			

Property/Value	Units	Temp	Data Type	Reference
WS 1.18E+003	mg/L	25	EST	MEYLAN,WM ET AL. (1996)
logP 3.19			EXP	HANSCH,C & LEO,AJ (1985)
VP 6.14E-003	mm Hg	25	EST	NEELY,WB & BLAU,GE (1985)
DC	pKa			
HL 2.67E-006	atm m3/mol	25	EST	MEYLAN,WM & HOWARD,PH (1991)
OH 2.42E-011	cm3/molc sec	25	EST	MEYLAN,WM & HOWARD,PH (1993)

CAS #:	083121-18-0			TEFLUBENZURON

Formula:	$C_{14}H_6Cl_2F_4N_2O_2$		
Mol Weight:	381.12		
MP (deg C):	222.5	FP (deg C):	
BP (deg C):			
BP pressure (mm Hg):			

Property/Value	Units	Temp	Data Type	Reference
WS 2.00E-002	mg/L	20	EXP	SHIU,WY ET AL. (1990)
logP 4.56			EXP	SAITO,H ET AL. (1993)
VP 1.10E-010	mm Hg	25	EST	NEELY,WB & BLAU,GE (1985)
DC	pKa			
HL 1.20E-011	atm m3/mol	25	EST	MEYLAN,WM & HOWARD,PH (1991)
OH 6.19E-012	cm3/molc sec	25	EST	MEYLAN,WM & HOWARD,PH (1993)

CAS #:	083123-89-1			TR-CROTYL-CHRYSANTHEMATE

Formula:	$C_{14}H_{22}O_2$		
Mol Weight:	222.33		
MP (deg C):		FP (deg C):	
BP (deg C):			
BP pressure (mm Hg):			

Property/Value	Units	Temp	Data Type	Reference
WS 1.51E+000	mg/L	25	EST	MEYLAN,WM ET AL. (1996)
logP 5.09			EXP	HANSCH,C & LEO,AJ (1985)
VP 7.01E-003	mm Hg	25	EST	NEELY,WB & BLAU,GE (1985)
DC	pKa			
HL 1.59E-003	atm m3/mol	25	EST	MEYLAN,WM & HOWARD,PH (1991)
OH 1.46E-010	cm3/molc sec	25	EST	MEYLAN,WM & HOWARD,PH (1993)

CAS #:	083128-50-1			BARBITURIC ACID, N-ME,5-ETHYL

Formula:	$C_7H_{10}N_2O_3$		
Mol Weight:	170.17		
MP (deg C):		FP (deg C):	
BP (deg C):			
BP pressure (mm Hg):			

Property/Value	Units	Temp	Data Type	Reference
WS 4.57E+004	mg/L	25	EST	MEYLAN,WM ET AL. (1996)
logP 0.15			EXP	HANSCH,C & LEO,AJ (1985)
VP 2.36E-009	mm Hg	25	EST	NEELY,WB & BLAU,GE (1985)
DC	pKa			
HL 4.49E-013	atm m3/mol	25	EST	MEYLAN,WM & HOWARD,PH (1991)
OH 7.71E-012	cm3/molc sec	25	EST	MEYLAN,WM & HOWARD,PH (1993)

083130-01-2

CAS #: 083130-01-2 — ALANYCARB

Formula: $C_{17}H_{25}N_3O_4S_2$

Mol Weight: 399.53

MP (deg C): FP (deg C):

BP (deg C):

BP pressure (mm Hg):

Property/Value	Units	Temp	Data Type	Reference
WS 3.63E+000	mg/L	25	EST	MEYLAN,WM ET AL. (1996)
logP 3.43			EXP	TOMLIN,C (1994)
VP 5.20E-008	mm Hg	25	EST	NEELY,WB & BLAU,GE (1985)
DC	pKa			
HL 1.72E-013	atm m3/mol	25	EST	MEYLAN,WM & HOWARD,PH (1991)
OH 4.52E-011	cm3/molc sec	25	EST	MEYLAN,WM & HOWARD,PH (1993)

083157-96-4

CAS #: 083157-96-4 — 2,4-PYRIMIDINEDIAMINE, 5-[(2,6-DICHLORO-3,5-DIME

Formula: $C_{13}H_{14}Cl_2N_4O_2$

Mol Weight: 329.19

MP (deg C): FP (deg C):

BP (deg C):

BP pressure (mm Hg):

Property/Value	Units	Temp	Data Type	Reference
WS 5.49E+001	mg/L	25	EST	MEYLAN,WM ET AL. (1996)
logP 2.55			EXP	SANGSTER,J (1993)
VP 3.78E-009	mm Hg	25	EST	NEELY,WB & BLAU,GE (1985)
DC	pKa			
HL 2.21E-013	atm m3/mol	25	EST	MEYLAN,WM & HOWARD,PH (1991)
OH 5.44E-011	cm3/molc sec	25	EST	MEYLAN,WM & HOWARD,PH (1993)

083157-97-5

CAS #: 083157-97-5 — 2,4-PYRIMIDINEDIAMINE, 5-[(6-BROMO-1,3-BENZODIOX

Formula: $C_{12}H_{11}BrN_4O_2$

Mol Weight: 323.16

MP (deg C): FP (deg C):

BP (deg C):

BP pressure (mm Hg):

Property/Value	Units	Temp	Data Type	Reference
WS 6.84E+001	mg/L	25	EST	MEYLAN,WM ET AL. (1996)
logP 2.48			EXP	SANGSTER,J (1993)
VP 6.43E-009	mm Hg	25	EST	NEELY,WB & BLAU,GE (1985)
DC	pKa			
HL 5.33E-014	atm m3/mol	25	EST	MEYLAN,WM & HOWARD,PH (1991)
OH 1.95E-011	cm3/molc sec	25	EST	MEYLAN,WM & HOWARD,PH (1993)

083157-99-7

CAS #: 083157-99-7 — 2,4NH2PYRIMIDIN,5(26MEOPYRIDIN4YL)CH2

Formula: $C_{12}H_{15}N_5O_2$

Mol Weight: 261.29

MP (deg C): FP (deg C):

BP (deg C):

BP pressure (mm Hg):

Property/Value	Units	Temp	Data Type	Reference
WS 8.14E+002	mg/L	25	EST	MEYLAN,WM ET AL. (1996)
logP 1.64			EXP	HANSCH,C & LEO,AJ (1985)
VP 3.50E-008	mm Hg	25	EST	NEELY,WB & BLAU,GE (1985)
DC	pKa			
HL 4.39E-014	atm m3/mol	25	EST	MEYLAN,WM & HOWARD,PH (1991)
OH 1.16E-010	cm3/molc sec	25	EST	MEYLAN,WM & HOWARD,PH (1993)

083158-00-3

CAS #: 083158-00-3 — 2,4-NH2PYRIMIDINE,5(35-OH-4MEO)BENZYL

Formula: $C_{12}H_{14}N_4O_3$

Mol Weight: 262.27

MP (deg C): FP (deg C):

BP (deg C):

BP pressure (mm Hg):

Property/Value	Units	Temp	Data Type	Reference
WS 2.19E+004	mg/L	25	EST	MEYLAN,WM ET AL. (1996)
logP -0.04			EXP	HANSCH,C & LEO,AJ (1985)
VP 5.69E-011	mm Hg	25	EST	NEELY,WB & BLAU,GE (1985)
DC	pKa			
HL 7.38E-020	atm m3/mol	25	EST	MEYLAN,WM & HOWARD,PH (1991)
OH 2.02E-010	cm3/molc sec	25	EST	MEYLAN,WM & HOWARD,PH (1993)

083158-01-4

CAS #: 083158-01-4 — 24NH2-PYRIMIDIN,5(3MEO-45PHCO)BENZYL

Formula: $C_{26}H_{26}N_4O_3$

Mol Weight: 442.52

MP (deg C): FP (deg C):

BP (deg C):

BP pressure (mm Hg):

Property/Value	Units	Temp	Data Type	Reference
WS 7.91E-001	mg/L	25	EST	MEYLAN,WM ET AL. (1996)
logP 4.09			EXP	HANSCH,C & LEO,AJ (1985)
VP 5.58E-014	mm Hg	25	EST	NEELY,WB & BLAU,GE (1985)
DC	pKa			
HL 1.55E-016	atm m3/mol	25	EST	MEYLAN,WM & HOWARD,PH (1991)
OH 2.13E-010	cm3/molc sec	25	EST	MEYLAN,WM & HOWARD,PH (1993)

083158-02-5

CAS #: 083158-02-5 — 24NH2-PYRIMIDIN,5(4MEO-35PHCO)BENZYL

Formula: $C_{26}H_{26}N_4O_3$

Mol Weight: 442.52

MP (deg C): FP (deg C):

BP (deg C):

BP pressure (mm Hg):

Property/Value	Units	Temp	Data Type	Reference
WS 7.91E-001	mg/L	25	EST	MEYLAN,WM ET AL. (1996)
logP 3.89			EXP	HANSCH,C & LEO,AJ (1985)
VP 5.58E-014	mm Hg	25	EST	NEELY,WB & BLAU,GE (1985)
DC	pKa			
HL 1.55E-016	atm m3/mol	25	EST	MEYLAN,WM & HOWARD,PH (1991)
OH 2.13E-010	cm3/molc sec	25	EST	MEYLAN,WM & HOWARD,PH (1993)

083158-03-6

CAS #: 083158-03-6 — 2,4-NH2PYRIMD,5(2-ME-4MEO-5BUO)BENZYL

Formula: $C_{17}H_{24}N_4O_2$

Mol Weight: 316.41

MP (deg C): FP (deg C):

BP (deg C):

BP pressure (mm Hg):

Property/Value	Units	Temp	Data Type	Reference
WS 4.16E+001	mg/L	25	EST	MEYLAN,WM ET AL. (1996)
logP 2.78			EXP	HANSCH,C & LEO,AJ (1985)
VP 2.01E-009	mm Hg	25	EST	NEELY,WB & BLAU,GE (1985)
DC	pKa			
HL 1.04E-012	atm m3/mol	25	EST	MEYLAN,WM & HOWARD,PH (1991)
OH 7.49E-011	cm3/molc sec	25	EST	MEYLAN,WM & HOWARD,PH (1993)

083158-04-7

CAS #: 083158-04-7

Title: 2,4-NH2PYRIMIDINE,5(35-CL-4MEO)BENZYL

Formula: $C_{12}H_{12}Cl_2N_4O$

Mol Weight: 299.16

MP (deg C):
FP (deg C):
BP (deg C):
BP pressure (mm Hg):

Property/Value	Units	Temp	Data Type	Reference
WS 5.25E+001	mg/L	25	EST	MEYLAN,WM ET AL. (1996)
logP 2.78			EXP	HANSCH,C & LEO,AJ (1985)
VP 1.97E-008	mm Hg	25	EST	NEELY,WB & BLAU,GE (1985)
DC	pKa			
HL 3.74E-012	atm m3/mol	25	EST	MEYLAN,WM & HOWARD,PH (1991)
OH 5.37E-012	cm3/molc sec	25	EST	MEYLAN,WM & HOWARD,PH (1993)

083158-05-8

CAS #: 083158-05-8

Title: 24NH2-PYRIMD,5(2ME-4MEO-5PHMEO)BENZYL

Formula: $C_{20}H_{22}N_4O_2$

Mol Weight: 350.42

MP (deg C):
FP (deg C):
BP (deg C):
BP pressure (mm Hg):

Property/Value	Units	Temp	Data Type	Reference
WS 3.11E+001	mg/L	25	EST	MEYLAN,WM ET AL. (1996)
logP 2.69			EXP	HANSCH,C & LEO,AJ (1985)
VP 6.07E-011	mm Hg	25	EST	NEELY,WB & BLAU,GE (1985)
DC	pKa			
HL 3.59E-014	atm m3/mol	25	EST	MEYLAN,WM & HOWARD,PH (1991)
OH 7.31E-011	cm3/molc sec	25	EST	MEYLAN,WM & HOWARD,PH (1993)

083158-06-9

CAS #: 083158-06-9

Title: 24-NH2PYRIMIDINE,5(3MEO-4PHMEO)BENZYL

Formula: $C_{19}H_{20}N_4O_2$

Mol Weight: 336.40

MP (deg C):
FP (deg C):
BP (deg C):
BP pressure (mm Hg):

Property/Value	Units	Temp	Data Type	Reference
WS 4.00E+001	mg/L	25	EST	MEYLAN,WM ET AL. (1996)
logP 2.66			EXP	HANSCH,C & LEO,AJ (1985)
VP 1.39E-010	mm Hg	25	EST	NEELY,WB & BLAU,GE (1985)
DC	pKa			
HL 3.25E-014	atm m3/mol	25	EST	MEYLAN,WM & HOWARD,PH (1991)
OH 6.01E-011	cm3/molc sec	25	EST	MEYLAN,WM & HOWARD,PH (1993)

083164-33-4

CAS #: 083164-33-4

Title: 3-PYRIDINECARBOXAMIDE, N-(2,4-DIFLUOROPHENYL)-2-

Formula: $C_{19}H_{11}F_5N_2O_2$

Mol Weight: 394.30

MP (deg C): 161-162
FP (deg C):
BP (deg C):
BP pressure (mm Hg):

Property/Value	Units	Temp	Data Type	Reference
WS 5.00E-002	mg/L		EXP	SHIU,WY ET AL. (1990)
logP 4.90			EXP	TOMLIN,C (1994)
VP 1.30E-009	mm Hg	25	EST	NEELY,WB & BLAU,GE (1985)
DC	pKa			
HL 3.77E-012	atm m3/mol	25	EST	MEYLAN,WM & HOWARD,PH (1991)
OH 3.20E-012	cm3/molc sec	25	EST	MEYLAN,WM & HOWARD,PH (1993)

083166-73-8

CAS #: 083166-73-8

Title: 2,4-NH2PYRIMIDINE,5(3,4MEO-5-I)BENZYL

Formula: $C_{13}H_{15}IN_4O_2$

Mol Weight: 386.19

MP (deg C):
FP (deg C):
BP (deg C):
BP pressure (mm Hg):

Property/Value	Units	Temp	Data Type	Reference
WS 3.82E+001	mg/L	25	EST	MEYLAN,WM ET AL. (1996)
logP 2.33			EXP	HANSCH,C & LEO,AJ (1985)
VP 1.68E-009	mm Hg	25	EST	NEELY,WB & BLAU,GE (1985)
DC	pKa			
HL 9.34E-014	atm m3/mol	25	EST	MEYLAN,WM & HOWARD,PH (1991)
OH 3.46E-011	cm3/molc sec	25	EST	MEYLAN,WM & HOWARD,PH (1993)

083166-74-9

CAS #: 083166-74-9

Title: 24-NH2PYRIMIDINE,5(3,5MEO-2-CL)BENZYL

Formula: $C_{13}H_{15}ClN_4O_2$

Mol Weight: 294.74

MP (deg C):
FP (deg C):
BP (deg C):
BP pressure (mm Hg):

Property/Value	Units	Temp	Data Type	Reference
WS 2.00E+002	mg/L	25	EST	MEYLAN,WM ET AL. (1996)
logP 2.13			EXP	HANSCH,C & LEO,AJ (1985)
VP 1.39E-008	mm Hg	25	EST	NEELY,WB & BLAU,GE (1985)
DC	pKa			
HL 2.99E-013	atm m3/mol	25	EST	MEYLAN,WM & HOWARD,PH (1991)
OH 1.76E-010	cm3/molc sec	25	EST	MEYLAN,WM & HOWARD,PH (1993)

083166-75-0

CAS #: 083166-75-0

Title: 24NH2PYRIMIDIN,5(3-MEO-4-C(OH)ME2)BENZYL

Formula: $C_{15}H_{20}N_4O_2$

Mol Weight: 288.35

MP (deg C):
FP (deg C):
BP (deg C):
BP pressure (mm Hg):

Property/Value	Units	Temp	Data Type	Reference
WS 9.51E+002	mg/L	25	EST	MEYLAN,WM ET AL. (1996)
logP 1.38			EXP	HANSCH,C & LEO,AJ (1985)
VP 5.37E-011	mm Hg	25	EST	NEELY,WB & BLAU,GE (1985)
DC	pKa			
HL 4.85E-016	atm m3/mol	25	EST	MEYLAN,WM & HOWARD,PH (1991)
OH 6.86E-011	cm3/molc sec	25	EST	MEYLAN,WM & HOWARD,PH (1993)

083166-76-1

CAS #: 083166-76-1

Title: 2,4-NH2PYRIMIDINE,5(3-OH-4MEO)BENZYL

Formula: $C_{12}H_{14}N_4O_2$

Mol Weight: 246.27

MP (deg C):
FP (deg C):
BP (deg C):
BP pressure (mm Hg):

Property/Value	Units	Temp	Data Type	Reference
WS 5.79E+003	mg/L	25	EST	MEYLAN,WM ET AL. (1996)
logP 0.74			EXP	HANSCH,C & LEO,AJ (1985)
VP 4.15E-009	mm Hg	25	EST	NEELY,WB & BLAU,GE (1985)
DC	pKa			
HL 7.09E-016	atm m3/mol	25	EST	MEYLAN,WM & HOWARD,PH (1991)
OH 7.51E-011	cm3/molc sec	25	EST	MEYLAN,WM & HOWARD,PH (1993)

CAS #: 083166-77-2				24NH2-PYRIMIDIN,5(3,5-MEO-4OCH2CN)BENZYL
Formula: $C_{15}H_{17}N_5O_3$				
Mol Weight: 315.33				
MP (deg C):		FP (deg C):		
BP (deg C):				
BP pressure (mm Hg):				

Property/Value	Units	Temp	Data Type	Reference
WS 1.32E+003	mg/L	25	EST	MEYLAN,WM ET AL. (1996)
logP 0.72			EXP	HANSCH,C & LEO,AJ (1985)
VP 1.64E-010	mm Hg	25	EST	NEELY,WB & BLAU,GE (1985)
DC	pKa			
HL 6.54E-014	atm m3/mol	25	EST	MEYLAN,WM & HOWARD,PH (1991)
OH 2.04E-010	cm3/molc sec	25	EST	MEYLAN,WM & HOWARD,PH (1993)

CAS #: 083166-78-3				ACETIC ACID, [4-[(2,4-DIAMINO-5-PYRIMIDINYL)METH
Formula: $C_{16}H_{20}N_4O_5$				
Mol Weight: 348.36				
MP (deg C):		FP (deg C):		
BP (deg C):				
BP pressure (mm Hg):				

Property/Value	Units	Temp	Data Type	Reference
WS 2.15E+003	mg/L	25	EST	MEYLAN,WM ET AL. (1996)
logP 0.55			EXP	HANSCH,C & LEO,AJ (1985)
VP 3.06E-010	mm Hg	25	EST	NEELY,WB & BLAU,GE (1985)
DC	pKa			
HL 4.02E-016	atm m3/mol	25	EST	MEYLAN,WM & HOWARD,PH (1991)
OH 2.07E-010	cm3/molc sec	25	EST	MEYLAN,WM & HOWARD,PH (1993)

CAS #: 083200-90-2				1H-IMIDAZOLE-1-ETHANOL, 4-BROMO-à-(METHOXYMETHYL
Formula: $C_7H_{10}BrN_3O_4$				
Mol Weight: 280.08				
MP (deg C):		FP (deg C):		
BP (deg C):				
BP pressure (mm Hg):				

Property/Value	Units	Temp	Data Type	Reference
WS 2.64E+003	mg/L	25	EST	MEYLAN,WM ET AL. (1996)
logP 0.46			EXP	SANGSTER,J (1993)
VP 6.56E-009	mm Hg	25	EST	NEELY,WB & BLAU,GE (1985)
DC	pKa			
HL 9.50E-014	atm m3/mol	25	EST	MEYLAN,WM & HOWARD,PH (1991)
OH 2.22E-011	cm3/molc sec	25	EST	MEYLAN,WM & HOWARD,PH (1993)

CAS #: 083283-07-2				NNN-ME-N(1ME-1-CYPENTENE)AMMON I
Formula: $C_{10}H_{20}IN$				
Mol Weight: 281.18				
MP (deg C):		FP (deg C):		
BP (deg C):				
BP pressure (mm Hg):				

Property/Value	Units	Temp	Data Type	Reference
WS 7.30E+005	mg/L	25	EST	MEYLAN,WM ET AL. (1996)
logP -1.95			EXP	PRATESI,P ET AL. (1986)
VP 2.92E-008	mm Hg	25	EST	NEELY,WB & BLAU,GE (1985)
DC	pKa			
HL 6.60E-012	atm m3/mol	25	EST	MEYLAN,WM & HOWARD,PH (1991)
OH 1.04E-010	cm3/molc sec	25	EST	MEYLAN,WM & HOWARD,PH (1993)

CAS #: 083293-76-9				5(33-DIME-1-TRIAZENO)ISOPHTHALAMIDE
Formula: $C_{10}H_{13}N_5O_2$				
Mol Weight: 235.25				
MP (deg C):		FP (deg C):		
BP (deg C):				
BP pressure (mm Hg):				

Property/Value	Units	Temp	Data Type	Reference
WS 9.06E+002	mg/L	25	EST	MEYLAN,WM ET AL. (1996)
logP 0.22			EXP	HANSCH,C & LEO,AJ (1985)
VP 1.76E-009	mm Hg	25	EST	NEELY,WB & BLAU,GE (1985)
DC	pKa			
HL 1.44E-016	atm m3/mol	25	EST	MEYLAN,WM & HOWARD,PH (1991)
OH 6.58E-012	cm3/molc sec	25	EST	MEYLAN,WM & HOWARD,PH (1993)

CAS #: 083293-82-7				1,2,2,3,3-PENTACHLOROBUTANE
Formula: $C_4H_5Cl_5$				
Mol Weight: 230.35				
MP (deg C):		FP (deg C):		
BP (deg C):				
BP pressure (mm Hg):				

Property/Value	Units	Temp	Data Type	Reference
WS 2.59E+000	mg/L	25	EST	MEYLAN,WM ET AL. (1996)
logP 4.76			EST	MEYLAN,WM & HOWARD,PH (1995)
VP 7.10E-001	mm Hg	25	EST	NEELY,WB & BLAU,GE (1985)
DC	pKa			
HL 9.35E-004	atm m3/mol	25	EST	MEYLAN,WM & HOWARD,PH (1991)
OH 1.23E-013	cm3/molc sec	25	EST	MEYLAN,WM & HOWARD,PH (1993)

CAS #: 083593-63-9				2-N,N-DIBUAMINO-1-PHENYLETHANOL
Formula: $C_{16}H_{27}NO$				
Mol Weight: 249.40				
MP (deg C):		FP (deg C):		
BP (deg C):				
BP pressure (mm Hg):				

Property/Value	Units	Temp	Data Type	Reference
WS 4.86E+002	mg/L	25	EST	MEYLAN,WM ET AL. (1996)
logP 3.16			EXP	HANSCH,C & LEO,AJ (1985)
VP 1.37E-006	mm Hg	25	EST	NEELY,WB & BLAU,GE (1985)
DC	pKa			
HL 7.83E-010	atm m3/mol	25	EST	MEYLAN,WM & HOWARD,PH (1991)
OH 1.17E-010	cm3/molc sec	25	EST	MEYLAN,WM & HOWARD,PH (1993)

CAS #: 083657-24-3				DINICONAZOLE
Formula: $C_{15}H_{19}Cl_2N_3O$				
Mol Weight: 328.24				
MP (deg C): 148-149		FP (deg C):		
BP (deg C):				
BP pressure (mm Hg):				

Property/Value	Units	Temp	Data Type	Reference
WS 4.00E+001	mg/L	25	EXP	WEBER,JB & SWAIN,LR (1993)
logP 4.30			EXP	TOMLIN,C (1994)
VP 3.68E-005	mm Hg	25	EXP	TOMLIN,C (1994)
DC	pKa			
HL 3.97E-007	atm m3/mol	25	EST	VP/WSOL
OH 9.69E-011	cm3/molc sec	25	EST	MEYLAN,WM & HOWARD,PH (1993)

CAS #: 083682-28-4				2,2,3,4,5,5-HEXACHLOROHEXANE
Formula: $C_6H_8Cl_6$				
Mol Weight: 292.85				
MP (deg C):		FP (deg C):		
BP (deg C):				
BP pressure (mm Hg):				

Property/Value	Units	Temp	Data Type	Reference
WS 1.36E-001	mg/L	25	EST	MEYLAN,WM ET AL. (1996)
logP 5.85			EST	MEYLAN,WM & HOWARD,PH (1995)
VP 5.79E-002	mm Hg	25	EST	NEELY,WB & BLAU,GE (1985)
DC	pKa			
HL 5.80E-004	atm m3/mol	25	EST	MEYLAN,WM & HOWARD,PH (1991)
OH 2.01E-013	cm3/molc sec	25	EST	MEYLAN,WM & HOWARD,PH (1993)

CAS #: 083682-29-5				1,2,2,5,5,6-HEXACHLOROHEXANE
Formula: $C_6H_8Cl_6$				
Mol Weight: 292.85				
MP (deg C):		FP (deg C):		
BP (deg C):				
BP pressure (mm Hg):				

Property/Value	Units	Temp	Data Type	Reference
WS 1.02E-001	mg/L	25	EST	MEYLAN,WM ET AL. (1996)
logP 6.00			EST	MEYLAN,WM & HOWARD,PH (1995)
VP 4.21E-003	mm Hg	25	EST	NEELY,WB & BLAU,GE (1985)
DC	pKa			
HL 5.80E-004	atm m3/mol	25	EST	MEYLAN,WM & HOWARD,PH (1991)
OH 7.52E-013	cm3/molc sec	25	EST	MEYLAN,WM & HOWARD,PH (1993)

CAS #: 083682-30-8				1,1,4,4-TETRACHLORO-2,3-DICHLOROMETHYL-2-BUTENE
Formula: $C_6H_6Cl_6$				
Mol Weight: 290.83				
MP (deg C):		FP (deg C):		
BP (deg C):				
BP pressure (mm Hg):				

Property/Value	Units	Temp	Data Type	Reference
WS 1.76E+000	mg/L	25	EST	MEYLAN,WM ET AL. (1996)
logP 4.56			EST	MEYLAN,WM & HOWARD,PH (1995)
VP 9.47E-004	mm Hg	25	EST	NEELY,WB & BLAU,GE (1985)
DC	pKa			
HL 7.11E-004	atm m3/mol	25	EST	MEYLAN,WM & HOWARD,PH (1991)
OH 3.80E-011	cm3/molc sec	25	EST	MEYLAN,WM & HOWARD,PH (1993)

CAS #: 083682-31-9				2,2,3,4,5,5-HEXACHLORO-3-HEXENE
Formula: $C_6H_6Cl_6$				
Mol Weight: 290.83				
MP (deg C):		FP (deg C):		
BP (deg C):				
BP pressure (mm Hg):				

Property/Value	Units	Temp	Data Type	Reference
WS 6.25E-002	mg/L	25	EST	MEYLAN,WM ET AL. (1996)
logP 6.26			EST	MEYLAN,WM & HOWARD,PH (1995)
VP 8.16E-002	mm Hg	25	EST	NEELY,WB & BLAU,GE (1985)
DC	pKa			
HL 2.13E-003	atm m3/mol	25	EST	MEYLAN,WM & HOWARD,PH (1991)
OH 2.87E-012	cm3/molc sec	25	EST	MEYLAN,WM & HOWARD,PH (1993)

CAS #: 083682-32-0				1,3-DICHLOROALLENE
Formula: $C_3H_2Cl_2$				
Mol Weight: 108.96				
MP (deg C):		FP (deg C):		
BP (deg C):				
BP pressure (mm Hg):				

Property/Value	Units	Temp	Data Type	Reference
WS 1.05E+003	mg/L	25	EST	MEYLAN,WM ET AL. (1996)
logP 2.36			EST	MEYLAN,WM & HOWARD,PH (1995)
VP 9.22E+001	mm Hg	25	EST	NEELY,WB & BLAU,GE (1985)
DC	pKa			
HL 3.19E-002	atm m3/mol	25	EST	MEYLAN,WM & HOWARD,PH (1991)
OH 2.51E-012	cm3/molc sec	25	EST	MEYLAN,WM & HOWARD,PH (1993)

CAS #: 083682-33-1				3,4-DICHLOROHEXA-1,5-DIENE
Formula: $C_6H_8Cl_2$				
Mol Weight: 151.04				
MP (deg C):		FP (deg C):		
BP (deg C):				
BP pressure (mm Hg):				

Property/Value	Units	Temp	Data Type	Reference
WS 9.82E+001	mg/L	25	EST	MEYLAN,WM ET AL. (1996)
logP 3.38			EST	MEYLAN,WM & HOWARD,PH (1995)
VP 6.16E+000	mm Hg	25	EST	NEELY,WB & BLAU,GE (1985)
DC	pKa			
HL 2.09E-002	atm m3/mol	25	EST	MEYLAN,WM & HOWARD,PH (1991)
OH 4.05E-011	cm3/molc sec	25	EST	MEYLAN,WM & HOWARD,PH (1993)

CAS #: 083682-34-2				PERCHLOROHEXANE
Formula: C_6Cl_{14}				
Mol Weight: 568.41				
MP (deg C):		FP (deg C):		
BP (deg C):				
BP pressure (mm Hg):				

Property/Value	Units	Temp	Data Type	Reference
WS 3.36E-007	mg/L	25	EST	MEYLAN,WM ET AL. (1996)
logP 10.41			EST	MEYLAN,WM & HOWARD,PH (1995)
VP 5.02E-006	mm Hg	25	EST	NEELY,WB & BLAU,GE (1985)
DC	pKa			
HL 3.08E-006	atm m3/mol	25	EST	MEYLAN,WM & HOWARD,PH (1991)
OH 0.00E+000	cm3/molc sec	25	EST	MEYLAN,WM & HOWARD,PH (1993)

CAS #: 083682-35-3				3-CHLOROHEXA-1,2,5-TRIENE
Formula: C_6H_7Cl				
Mol Weight: 114.58				
MP (deg C):		FP (deg C):		
BP (deg C):				
BP pressure (mm Hg):				

Property/Value	Units	Temp	Data Type	Reference
WS 1.31E+002	mg/L	25	EST	MEYLAN,WM ET AL. (1996)
logP 3.40			EST	MEYLAN,WM & HOWARD,PH (1995)
VP 4.33E+001	mm Hg	25	EST	NEELY,WB & BLAU,GE (1985)
DC	pKa			
HL 9.14E-002	atm m3/mol	25	EST	MEYLAN,WM & HOWARD,PH (1991)
OH 3.92E-011	cm3/molc sec	25	EST	MEYLAN,WM & HOWARD,PH (1993)

CAS #: 083682-36-4 — 1-CHLOROHEPT-1-YNE-5-ENE

Formula: C_7H_9Cl

Mol Weight: 128.60

MP (deg C):   FP (deg C):

BP (deg C):

BP pressure (mm Hg):

Property/Value	Units	Temp	Data Type	Reference
WS 2.09E+002	mg/L	25	EST	MEYLAN,WM ET AL. (1996)
logP 3.10			EST	MEYLAN,WM & HOWARD,PH (1995)
VP 4.21E+000	mm Hg	25	EST	NEELY,WB & BLAU,GE (1985)
DC	pKa			
HL 4.04E-002	atm m3/mol	25	EST	MEYLAN,WM & HOWARD,PH (1991)
OH 6.45E-011	cm3/molc sec	25	EST	MEYLAN,WM & HOWARD,PH (1993)

CAS #: 083682-37-5 — 3,4-DICHLOROHEXA-1,3,5-TRIENE

Formula: $C_6H_6Cl_2$

Mol Weight: 149.02

MP (deg C):   FP (deg C):

BP (deg C):

BP pressure (mm Hg):

Property/Value	Units	Temp	Data Type	Reference
WS 4.49E+001	mg/L	25	EST	MEYLAN,WM ET AL. (1996)
logP 3.78			EST	MEYLAN,WM & HOWARD,PH (1995)
VP 1.04E+001	mm Hg	25	EST	NEELY,WB & BLAU,GE (1985)
DC	pKa			
HL 5.09E-002	atm m3/mol	25	EST	MEYLAN,WM & HOWARD,PH (1991)
OH 2.95E-011	cm3/molc sec	25	EST	MEYLAN,WM & HOWARD,PH (1993)

CAS #: 083682-38-6 — 2-TRICHLOROMETHYL-3,3,3-TRICHLOROPROPENE

Formula: $C_4H_2Cl_6$

Mol Weight: 262.78

MP (deg C):   FP (deg C):

BP (deg C):

BP pressure (mm Hg):

Property/Value	Units	Temp	Data Type	Reference
WS 1.22E+000	mg/L	25	EST	MEYLAN,WM ET AL. (1996)
logP 4.94			EST	MEYLAN,WM & HOWARD,PH (1995)
VP 6.11E-001	mm Hg	25	EST	NEELY,WB & BLAU,GE (1985)
DC	pKa			
HL 2.90E-004	atm m3/mol	25	EST	MEYLAN,WM & HOWARD,PH (1991)
OH 2.97E-011	cm3/molc sec	25	EST	MEYLAN,WM & HOWARD,PH (1993)

CAS #: 083682-39-7 — 1-CHLORO-2-TRICHLOROMETHYL-3,3,3-TRICHLOROPROP*

Formula: C_4HCl_7

Mol Weight: 297.22

MP (deg C):   FP (deg C):

BP (deg C):

BP pressure (mm Hg):

Property/Value	Units	Temp	Data Type	Reference
WS 3.84E-001	mg/L	25	EST	MEYLAN,WM ET AL. (1996)
logP 5.29			EST	MEYLAN,WM & HOWARD,PH (1995)
VP 6.89E-002	mm Hg	25	EST	NEELY,WB & BLAU,GE (1985)
DC	pKa			
HL 2.08E-004	atm m3/mol	25	EST	MEYLAN,WM & HOWARD,PH (1991)
OH 1.05E-011	cm3/molc sec	25	EST	MEYLAN,WM & HOWARD,PH (1993)

CAS #: 083682-40-0 — 1,1-DICHLORO-2,3-BIS(TRICHLOROMETHYL)PROPENE

Formula: $C_5H_2Cl_8$

Mol Weight: 345.70

MP (deg C):   FP (deg C):

BP (deg C):

BP pressure (mm Hg):

Property/Value	Units	Temp	Data Type	Reference
WS 2.86E-002	mg/L	25	EST	MEYLAN,WM ET AL. (1996)
logP 6.28			EST	MEYLAN,WM & HOWARD,PH (1995)
VP 6.22E-003	mm Hg	25	EST	NEELY,WB & BLAU,GE (1985)
DC	pKa			
HL 1.99E-004	atm m3/mol	25	EST	MEYLAN,WM & HOWARD,PH (1991)
OH 3.75E-012	cm3/molc sec	25	EST	MEYLAN,WM & HOWARD,PH (1993)

CAS #: 083682-41-1 — 4,4-DICHLORO-1,2-BUTADIENE

Formula: $C_4H_4Cl_2$

Mol Weight: 122.98

MP (deg C):   FP (deg C):

BP (deg C):

BP pressure (mm Hg):

Property/Value	Units	Temp	Data Type	Reference
WS 7.22E+002	mg/L	25	EST	MEYLAN,WM ET AL. (1996)
logP 2.50			EST	MEYLAN,WM & HOWARD,PH (1995)
VP 2.22E+001	mm Hg	25	EST	NEELY,WB & BLAU,GE (1985)
DC	pKa			
HL 1.20E-002	atm m3/mol	25	EST	MEYLAN,WM & HOWARD,PH (1991)
OH 2.38E-011	cm3/molc sec	25	EST	MEYLAN,WM & HOWARD,PH (1993)

CAS #: 083682-42-2 — 4,4-DICHLORO-1-BUTYNE

Formula: $C_4H_4Cl_2$

Mol Weight: 122.98

MP (deg C):   FP (deg C):

BP (deg C):

BP pressure (mm Hg):

Property/Value	Units	Temp	Data Type	Reference
WS 2.04E+003	mg/L	25	EST	MEYLAN,WM ET AL. (1996)
logP 1.97			EST	MEYLAN,WM & HOWARD,PH (1995)
VP 1.55E+001	mm Hg	25	EST	NEELY,WB & BLAU,GE (1985)
DC	pKa			
HL 2.65E-003	atm m3/mol	25	EST	MEYLAN,WM & HOWARD,PH (1991)
OH 7.68E-012	cm3/molc sec	25	EST	MEYLAN,WM & HOWARD,PH (1993)

CAS #: 083682-43-3 — 4,4,4-TRICHLORO-1-BUTYNE

Formula: $C_4H_3Cl_3$

Mol Weight: 157.43

MP (deg C):   FP (deg C):

BP (deg C):

BP pressure (mm Hg):

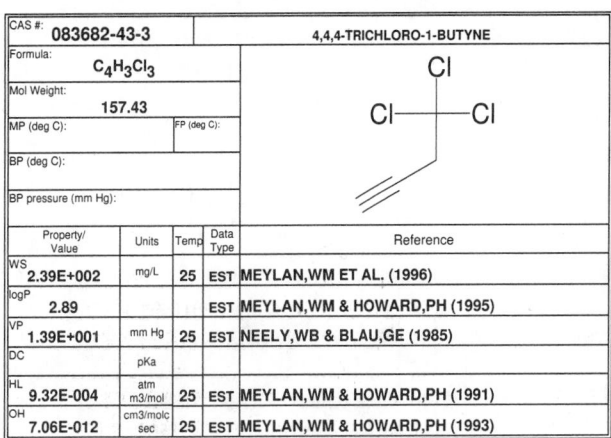

Property/Value	Units	Temp	Data Type	Reference
WS 2.39E+002	mg/L	25	EST	MEYLAN,WM ET AL. (1996)
logP 2.89			EST	MEYLAN,WM & HOWARD,PH (1995)
VP 1.39E+001	mm Hg	25	EST	NEELY,WB & BLAU,GE (1985)
DC	pKa			
HL 9.32E-004	atm m3/mol	25	EST	MEYLAN,WM & HOWARD,PH (1991)
OH 7.06E-012	cm3/molc sec	25	EST	MEYLAN,WM & HOWARD,PH (1993)

CAS #: 083682-44-4				1,4-DICHLORO-1,2-BUTADIENE

Formula: $C_4H_4Cl_2$
Mol Weight: 122.98
MP (deg C):　FP (deg C):
BP (deg C):
BP pressure (mm Hg):

Property/Value	Units	Temp	Data Type	Reference
WS 5.10E+002	mg/L	25	EST	MEYLAN,WM ET AL. (1996)
logP 2.67			EST	MEYLAN,WM & HOWARD,PH (1995)
VP 8.78E+000	mm Hg	25	EST	NEELY,WB & BLAU,GE (1985)
DC	pKa			
HL 2.45E-002	atm m3/mol	25	EST	MEYLAN,WM & HOWARD,PH (1991)
OH 9.45E-012	cm3/molc sec	25	EST	MEYLAN,WM & HOWARD,PH (1993)

CAS #: 083682-45-5				1,4-DICHLORO-1-BUTYNE

Formula: $C_4H_4Cl_2$
Mol Weight: 122.98
MP (deg C):　FP (deg C):
BP (deg C):
BP pressure (mm Hg):

Property/Value	Units	Temp	Data Type	Reference
WS 1.58E+003	mg/L	25	EST	MEYLAN,WM ET AL. (1996)
logP 2.10			EST	MEYLAN,WM & HOWARD,PH (1995)
VP 6.21E+000	mm Hg	25	EST	NEELY,WB & BLAU,GE (1985)
DC	pKa			
HL 6.92E-003	atm m3/mol	25	EST	MEYLAN,WM & HOWARD,PH (1991)
OH 6.44E-012	cm3/molc sec	25	EST	MEYLAN,WM & HOWARD,PH (1993)

CAS #: 083682-46-6				1,1,4-TRICHLOROBUTADIENE

Formula: $C_4H_3Cl_3$
Mol Weight: 157.43
MP (deg C):　FP (deg C):
BP (deg C):
BP pressure (mm Hg):

Property/Value	Units	Temp	Data Type	Reference
WS 1.20E+002	mg/L	25	EST	MEYLAN,WM ET AL. (1996)
logP 3.24			EST	MEYLAN,WM & HOWARD,PH (1995)
VP 8.27E+000	mm Hg	25	EST	NEELY,WB & BLAU,GE (1985)
DC	pKa			
HL 2.90E-002	atm m3/mol	25	EST	MEYLAN,WM & HOWARD,PH (1991)
OH 1.76E-012	cm3/molc sec	25	EST	MEYLAN,WM & HOWARD,PH (1993)

CAS #: 083682-47-7				1-HEXACHLOROBUTYNE

Formula: C_4Cl_6
Mol Weight: 260.76
MP (deg C):　FP (deg C):
BP (deg C):
BP pressure (mm Hg):

Property/Value	Units	Temp	Data Type	Reference
WS 4.37E+000	mg/L	25	EST	MEYLAN,WM ET AL. (1996)
logP 4.30			EST	MEYLAN,WM & HOWARD,PH (1995)
VP 1.41E-001	mm Hg	25	EST	NEELY,WB & BLAU,GE (1985)
DC	pKa			
HL 1.06E-004	atm m3/mol	25	EST	MEYLAN,WM & HOWARD,PH (1991)
OH 5.67E-012	cm3/molc sec	25	EST	MEYLAN,WM & HOWARD,PH (1993)

CAS #: 083682-48-8				1,6-DICHLORO-2,5-DIMETHYL-1,5-HEXADIENE

Formula: $C_8H_{12}Cl_2$
Mol Weight: 179.09
MP (deg C):　FP (deg C):
BP (deg C):
BP pressure (mm Hg):

Property/Value	Units	Temp	Data Type	Reference
WS 4.22E+000	mg/L	25	EST	MEYLAN,WM ET AL. (1996)
logP 4.82			EST	MEYLAN,WM & HOWARD,PH (1995)
VP 9.54E-001	mm Hg	25	EST	NEELY,WB & BLAU,GE (1985)
DC	pKa			
HL 2.14E-001	atm m3/mol	25	EST	MEYLAN,WM & HOWARD,PH (1991)
OH 3.91E-011	cm3/molc sec	25	EST	MEYLAN,WM & HOWARD,PH (1993)

CAS #: 083682-49-9				3,4-DICHLORO-2,5-DIMETHYL-1,5-HEXADIENE

Formula: $C_8H_{12}Cl_2$
Mol Weight: 179.09
MP (deg C):　FP (deg C):
BP (deg C):
BP pressure (mm Hg):

Property/Value	Units	Temp	Data Type	Reference
WS 8.46E+000	mg/L	25	EST	MEYLAN,WM ET AL. (1996)
logP 4.47			EST	MEYLAN,WM & HOWARD,PH (1995)
VP 1.40E+000	mm Hg	25	EST	NEELY,WB & BLAU,GE (1985)
DC	pKa			
HL 5.14E-002	atm m3/mol	25	EST	MEYLAN,WM & HOWARD,PH (1991)
OH 7.89E-011	cm3/molc sec	25	EST	MEYLAN,WM & HOWARD,PH (1993)

CAS #: 083682-50-2				2,5-DIMETHYL-3,3,4,4-TETRACHLORO-1,5-HEXADIENE

Formula: $C_8H_{10}Cl_4$
Mol Weight: 247.98
MP (deg C):　FP (deg C):
BP (deg C):
BP pressure (mm Hg):

Property/Value	Units	Temp	Data Type	Reference
WS 9.86E-002	mg/L	25	EST	MEYLAN,WM ET AL. (1996)
logP 6.31			EST	MEYLAN,WM & HOWARD,PH (1995)
VP 3.80E-001	mm Hg	25	EST	NEELY,WB & BLAU,GE (1985)
DC	pKa			
HL 6.37E-003	atm m3/mol	25	EST	MEYLAN,WM & HOWARD,PH (1991)
OH 7.84E-011	cm3/molc sec	25	EST	MEYLAN,WM & HOWARD,PH (1993)

CAS #: 083682-51-3				2,5-BIS(CHLOROMETHYL)-1,5-HEXADIENE

Formula: $C_8H_{12}Cl_2$
Mol Weight: 179.09
MP (deg C):　FP (deg C):
BP (deg C):
BP pressure (mm Hg):

Property/Value	Units	Temp	Data Type	Reference
WS 6.33E+000	mg/L	25	EST	MEYLAN,WM ET AL. (1996)
logP 4.62			EST	MEYLAN,WM & HOWARD,PH (1995)
VP 1.41E-001	mm Hg	25	EST	NEELY,WB & BLAU,GE (1985)
DC	pKa			
HL 5.14E-002	atm m3/mol	25	EST	MEYLAN,WM & HOWARD,PH (1991)
OH 8.11E-011	cm3/molc sec	25	EST	MEYLAN,WM & HOWARD,PH (1993)

CAS #: **083682-52-4**				2,5-BIS(CHLOROMETHYL)-3,4-DICHLORO-1,5-HEXADIE*

Formula: C$_8$H$_{10}$Cl$_4$

Mol Weight: 247.98

MP (deg C): | FP (deg C):

BP (deg C):

BP pressure (mm Hg):

Property/ Value	Units	Temp	Data Type	Reference
WS 1.36E+000	mg/L	25	EST	MEYLAN,WM ET AL. (1996)
logP 4.98			EST	MEYLAN,WM & HOWARD,PH (1995)
VP 8.02E-003	mm Hg	25	EST	NEELY,WB & BLAU,GE (1985)
DC	pKa			
HL 6.37E-003	atm m3/mol	25	EST	MEYLAN,WM & HOWARD,PH (1991)
OH 6.06E-011	cm3/molc sec	25	EST	MEYLAN,WM & HOWARD,PH (1993)

CAS #: **083682-53-5**				2,5-BIS(CHLOROMETHYL)-3,3,4,4-TETRACHLORO-1,5-*

Formula: C$_8$H$_8$Cl$_6$

Mol Weight: 316.87

MP (deg C): | FP (deg C):

BP (deg C):

BP pressure (mm Hg):

Property/ Value	Units	Temp	Data Type	Reference
WS 1.46E-002	mg/L	25	EST	MEYLAN,WM ET AL. (1996)
logP 6.82			EST	MEYLAN,WM & HOWARD,PH (1995)
VP 1.39E-003	mm Hg	25	EST	NEELY,WB & BLAU,GE (1985)
DC	pKa			
HL 7.91E-004	atm m3/mol	25	EST	MEYLAN,WM & HOWARD,PH (1991)
OH 6.01E-011	cm3/molc sec	25	EST	MEYLAN,WM & HOWARD,PH (1993)

CAS #: **083682-54-6**				2,5-DIMETHYL-1,6-DICHLORO-1,3,5-HEXATRIENE

Formula: C$_8$H$_{10}$Cl$_2$

Mol Weight: 177.07

MP (deg C): | FP (deg C):

BP (deg C):

BP pressure (mm Hg):

Property/ Value	Units	Temp	Data Type	Reference
WS 6.58E+000	mg/L	25	EST	MEYLAN,WM ET AL. (1996)
logP 4.61			EST	MEYLAN,WM & HOWARD,PH (1995)
VP 7.11E-001	mm Hg	25	EST	NEELY,WB & BLAU,GE (1985)
DC	pKa			
HL 1.25E-001	atm m3/mol	25	EST	MEYLAN,WM & HOWARD,PH (1991)
OH 4.97E-011	cm3/molc sec	25	EST	MEYLAN,WM & HOWARD,PH (1993)

CAS #: **083682-55-7**				2,5-BIS(CHLOROMETHYL)-3,4-DICHLORO-1,3,5-HEXAT*

Formula: C$_8$H$_8$Cl$_4$

Mol Weight: 245.96

MP (deg C): | FP (deg C):

BP (deg C):

BP pressure (mm Hg):

Property/ Value	Units	Temp	Data Type	Reference
WS 6.24E-001	mg/L	25	EST	MEYLAN,WM ET AL. (1996)
logP 5.39			EST	MEYLAN,WM & HOWARD,PH (1995)
VP 1.05E-002	mm Hg	25	EST	NEELY,WB & BLAU,GE (1985)
DC	pKa			
HL 1.55E-002	atm m3/mol	25	EST	MEYLAN,WM & HOWARD,PH (1991)
OH 4.13E-011	cm3/molc sec	25	EST	MEYLAN,WM & HOWARD,PH (1993)

CAS #: **083682-56-8**				2,5-BIS(CHLOROMETHYL)-1,3,5-HEXATRIENE

Formula: C$_8$H$_{10}$Cl$_2$

Mol Weight: 177.07

MP (deg C): | FP (deg C):

BP (deg C):

BP pressure (mm Hg):

Property/ Value	Units	Temp	Data Type	Reference
WS 9.89E+000	mg/L	25	EST	MEYLAN,WM ET AL. (1996)
logP 4.40			EST	MEYLAN,WM & HOWARD,PH (1995)
VP 1.05E-001	mm Hg	25	EST	NEELY,WB & BLAU,GE (1985)
DC	pKa			
HL 3.00E-002	atm m3/mol	25	EST	MEYLAN,WM & HOWARD,PH (1991)
OH 1.26E-010	cm3/molc sec	25	EST	MEYLAN,WM & HOWARD,PH (1993)

CAS #: **083682-57-9**				2,5-BIS(DICHLOROMETHYL)-3,4-DICHLORO-1,3,5-HEX*

Formula: C$_8$H$_6$Cl$_6$

Mol Weight: 314.86

MP (deg C): | FP (deg C):

BP (deg C):

BP pressure (mm Hg):

Property/ Value	Units	Temp	Data Type	Reference
WS 1.24E-001	mg/L	25	EST	MEYLAN,WM ET AL. (1996)
logP 5.75			EST	MEYLAN,WM & HOWARD,PH (1995)
VP 2.06E-003	mm Hg	25	EST	NEELY,WB & BLAU,GE (1985)
DC	pKa			
HL 1.93E-003	atm m3/mol	25	EST	MEYLAN,WM & HOWARD,PH (1991)
OH 4.12E-011	cm3/molc sec	25	EST	MEYLAN,WM & HOWARD,PH (1993)

CAS #: **083682-58-0**				2,5-BIS(TRICHLOROMETHYL)-3,4-DICHLORO-1,3,5-HE*

Formula: C$_8$H$_4$Cl$_8$

Mol Weight: 383.75

MP (deg C): | FP (deg C):

BP (deg C):

BP pressure (mm Hg):

Property/ Value	Units	Temp	Data Type	Reference
WS 1.27E-003	mg/L	25	EST	MEYLAN,WM ET AL. (1996)
logP 7.59			EST	MEYLAN,WM & HOWARD,PH (1995)
VP 6.11E-004	mm Hg	25	EST	NEELY,WB & BLAU,GE (1985)
DC	pKa			
HL 2.39E-004	atm m3/mol	25	EST	MEYLAN,WM & HOWARD,PH (1991)
OH 4.06E-011	cm3/molc sec	25	EST	MEYLAN,WM & HOWARD,PH (1993)

CAS #: **083682-59-1**				1,3,5-TRIS(2,2-DICHLOROETHYL)BENZENE

Formula: C$_{12}$H$_{12}$Cl$_6$

Mol Weight: 368.95

MP (deg C): | FP (deg C):

BP (deg C):

BP pressure (mm Hg):

Property/ Value	Units	Temp	Data Type	Reference
WS 1.59E-002	mg/L	25	EST	MEYLAN,WM ET AL. (1996)
logP 6.41			EST	MEYLAN,WM & HOWARD,PH (1995)
VP 2.14E-006	mm Hg	25	EST	NEELY,WB & BLAU,GE (1985)
DC	pKa			
HL 3.24E-005	atm m3/mol	25	EST	MEYLAN,WM & HOWARD,PH (1991)
OH 3.20E-011	cm3/molc sec	25	EST	MEYLAN,WM & HOWARD,PH (1993)

083682-60-4 — 1,3,5-TRIS(2,2,2-TRICHLOROETHYL)BENZENE

Formula: $C_{12}H_9Cl_9$

Mol Weight: 472.28

MP (deg C):   FP (deg C):

BP (deg C):

BP pressure (mm Hg):

Property/Value	Units	Temp	Data Type	Reference
WS 1.57E-005	mg/L	25	EST	MEYLAN,WM ET AL. (1996)
logP 9.17			EST	MEYLAN,WM & HOWARD,PH (1995)
VP 1.01E-006	mm Hg	25	EST	NEELY,WB & BLAU,GE (1985)
DC	pKa			
HL 1.41E-006	atm m3/mol	25	EST	MEYLAN,WM & HOWARD,PH (1991)
OH 3.01E-011	cm3/molc sec	25	EST	MEYLAN,WM & HOWARD,PH (1993)

083682-61-5 — 1,2-DICHLORO-1,5-OCTADIENE

Formula: $C_9H_{14}Cl_2$

Mol Weight: 193.12

MP (deg C):   FP (deg C):

BP (deg C):

BP pressure (mm Hg):

Property/Value	Units	Temp	Data Type	Reference
WS 1.53E+000	mg/L	25	EST	MEYLAN,WM ET AL. (1996)
logP 5.26			EST	MEYLAN,WM & HOWARD,PH (1995)
VP 2.86E-001	mm Hg	25	EST	NEELY,WB & BLAU,GE (1985)
DC	pKa			
HL 2.41E-001	atm m3/mol	25	EST	MEYLAN,WM & HOWARD,PH (1991)
OH 6.50E-011	cm3/molc sec	25	EST	MEYLAN,WM & HOWARD,PH (1993)

083682-62-6 — 1,2-BIS(1,2-DICHLOROETHYL)CYCLOBUTANE

Formula: $C_8H_{12}Cl_4$

Mol Weight: 250.00

MP (deg C):   FP (deg C):

BP (deg C):

BP pressure (mm Hg):

Property/Value	Units	Temp	Data Type	Reference
WS 1.61E+000	mg/L	25	EST	MEYLAN,WM ET AL. (1996)
logP 4.88			EST	MEYLAN,WM & HOWARD,PH (1995)
VP 4.34E-003	mm Hg	25	EST	NEELY,WB & BLAU,GE (1985)
DC	pKa			
HL 3.63E-003	atm m3/mol	25	EST	MEYLAN,WM & HOWARD,PH (1991)
OH 2.29E-012	cm3/molc sec	25	EST	MEYLAN,WM & HOWARD,PH (1993)

083682-63-7 — 1,4-DICHLORO-4-(1-CHLOROETHENYL)CYCLO-1-HEXENE

Formula: $C_8H_9Cl_3$

Mol Weight: 211.52

MP (deg C):   FP (deg C):

BP (deg C):

BP pressure (mm Hg):

Property/Value	Units	Temp	Data Type	Reference
WS 2.33E+000	mg/L	25	EST	MEYLAN,WM ET AL. (1996)
logP 4.93			EST	MEYLAN,WM & HOWARD,PH (1995)
VP 2.22E-001	mm Hg	25	EST	NEELY,WB & BLAU,GE (1985)
DC	pKa			
HL 2.82E-002	atm m3/mol	25	EST	MEYLAN,WM & HOWARD,PH (1991)
OH 2.97E-011	cm3/molc sec	25	EST	MEYLAN,WM & HOWARD,PH (1993)

083682-64-8 — 1,2,4-TRICHLORO-4-(1,1,2-TRICHLOROETHYL)CYCLOH*

Formula: $C_8H_{10}Cl_6$

Mol Weight: 318.89

MP (deg C):   FP (deg C):

BP (deg C):

BP pressure (mm Hg):

Property/Value	Units	Temp	Data Type	Reference
WS 6.91E-002	mg/L	25	EST	MEYLAN,WM ET AL. (1996)
logP 6.02			EST	MEYLAN,WM & HOWARD,PH (1995)
VP 3.70E-004	mm Hg	25	EST	NEELY,WB & BLAU,GE (1985)
DC	pKa			
HL 4.51E-004	atm m3/mol	25	EST	MEYLAN,WM & HOWARD,PH (1991)
OH 2.64E-012	cm3/molc sec	25	EST	MEYLAN,WM & HOWARD,PH (1993)

083682-65-9 — 1,5-DICHLOROCYCLO-1,4-OCTADIENE

Formula: $C_8H_{10}Cl_2$

Mol Weight: 177.07

MP (deg C):   FP (deg C):

BP (deg C):

BP pressure (mm Hg):

Property/Value	Units	Temp	Data Type	Reference
WS 5.38E+000	mg/L	25	EST	MEYLAN,WM ET AL. (1996)
logP 4.71			EST	MEYLAN,WM & HOWARD,PH (1995)
VP 3.91E-001	mm Hg	25	EST	NEELY,WB & BLAU,GE (1985)
DC	pKa			
HL 9.45E-002	atm m3/mol	25	EST	MEYLAN,WM & HOWARD,PH (1991)
OH 4.11E-011	cm3/molc sec	25	EST	MEYLAN,WM & HOWARD,PH (1993)

083682-66-0 — BIS(2-CHLORO-3-BUTENYL) ETHER

Formula: $C_8H_{12}Cl_2O$

Mol Weight: 195.09

MP (deg C):   FP (deg C):

BP (deg C):

BP pressure (mm Hg):

Property/Value	Units	Temp	Data Type	Reference
WS 1.04E+002	mg/L	25	EST	MEYLAN,WM ET AL. (1996)
logP 3.10			EST	MEYLAN,WM & HOWARD,PH (1995)
VP 2.85E-001	mm Hg	25	EST	NEELY,WB & BLAU,GE (1985)
DC	pKa			
HL 3.25E-004	atm m3/mol	25	EST	MEYLAN,WM & HOWARD,PH (1991)
OH 4.58E-011	cm3/molc sec	25	EST	MEYLAN,WM & HOWARD,PH (1993)

083682-67-1 — 1,1'-OXYBIS-3-BUTEN-2-OL

Formula: $C_8H_{14}O_3$

Mol Weight: 158.20

MP (deg C):   FP (deg C):

BP (deg C):

BP pressure (mm Hg):

Property/Value	Units	Temp	Data Type	Reference
WS 6.07E+004	mg/L	25	EST	MEYLAN,WM ET AL. (1996)
logP 0.07			EST	MEYLAN,WM & HOWARD,PH (1995)
VP 1.65E-003	mm Hg	25	EST	NEELY,WB & BLAU,GE (1985)
DC	pKa			
HL 3.50E-009	atm m3/mol	25	EST	MEYLAN,WM & HOWARD,PH (1991)
OH 8.36E-011	cm3/molc sec	25	EST	MEYLAN,WM & HOWARD,PH (1993)

CAS #: 083682-68-2		2,2'-OXYBIS-3-BUTEN-1-OL		
Formula: $C_8H_{14}O_3$				
Mol Weight: 158.20				
MP (deg C):	FP (deg C):			
BP (deg C):				
BP pressure (mm Hg):				

Property/Value	Units	Temp	Data Type	Reference
WS 6.07E+004	mg/L	25	EST	MEYLAN,WM ET AL. (1996)
logP 0.07			EST	MEYLAN,WM & HOWARD,PH (1995)
VP 7.54E-004	mm Hg	25	EST	NEELY,WB & BLAU,GE (1985)
DC	pKa			
HL 3.50E-009	atm m3/mol	25	EST	MEYLAN,WM & HOWARD,PH (1991)
OH 9.00E-011	cm3/molc sec	25	EST	MEYLAN,WM & HOWARD,PH (1993)

CAS #: 083682-69-3		1,1,2,2,3,3-HEXACHLOROBUTANE		
Formula: $C_4H_4Cl_6$				
Mol Weight: 264.79				
MP (deg C):	FP (deg C):			
BP (deg C):				
BP pressure (mm Hg):				

Property/Value	Units	Temp	Data Type	Reference
WS 1.17E+000	mg/L	25	EST	MEYLAN,WM ET AL. (1996)
logP 4.94			EST	MEYLAN,WM & HOWARD,PH (1995)
VP 2.66E-001	mm Hg	25	EST	NEELY,WB & BLAU,GE (1985)
DC	pKa			
HL 3.29E-004	atm m3/mol	25	EST	MEYLAN,WM & HOWARD,PH (1991)
OH 1.04E-013	cm3/molc sec	25	EST	MEYLAN,WM & HOWARD,PH (1993)

CAS #: 083682-70-6		1,1,1,2,2,3,3-HEPTACHLOROBUTANE		
Formula: $C_4H_3Cl_7$				
Mol Weight: 299.24				
MP (deg C):	FP (deg C):			
BP (deg C):				
BP pressure (mm Hg):				

Property/Value	Units	Temp	Data Type	Reference
WS 1.22E-001	mg/L	25	EST	MEYLAN,WM ET AL. (1996)
logP 5.86			EST	MEYLAN,WM & HOWARD,PH (1995)
VP 1.94E-001	mm Hg	25	EST	NEELY,WB & BLAU,GE (1985)
DC	pKa			
HL 1.16E-004	atm m3/mol	25	EST	MEYLAN,WM & HOWARD,PH (1991)
OH 3.40E-014	cm3/molc sec	25	EST	MEYLAN,WM & HOWARD,PH (1993)

CAS #: 083682-71-7		2,3,3,4,4,4-HEXACHLOROBUT-1-ENE		
Formula: $C_4H_2Cl_6$				
Mol Weight: 262.78				
MP (deg C):	FP (deg C):			
BP (deg C):				
BP pressure (mm Hg):				

Property/Value	Units	Temp	Data Type	Reference
WS 8.53E-001	mg/L	25	EST	MEYLAN,WM ET AL. (1996)
logP 5.12			EST	MEYLAN,WM & HOWARD,PH (1995)
VP 6.68E-001	mm Hg	25	EST	NEELY,WB & BLAU,GE (1985)
DC	pKa			
HL 5.01E-004	atm m3/mol	25	EST	MEYLAN,WM & HOWARD,PH (1991)
OH 8.20E-012	cm3/molc sec	25	EST	MEYLAN,WM & HOWARD,PH (1993)

CAS #: 083682-72-8		3,3-DICHLOROPROPANOL		
Formula: $C_3H_6Cl_2O$				
Mol Weight: 128.99				
MP (deg C):	FP (deg C):			
BP (deg C): 82-83				
BP pressure (mm Hg): 2.00E+001				

Property/Value	Units	Temp	Data Type	Reference
WS 6.42E+004	mg/L	25	EST	MEYLAN,WM ET AL. (1996)
logP 0.78			EST	MEYLAN,WM & HOWARD,PH (1995)
VP 3.59E-001	mm Hg	25	EST	NEELY,WB & BLAU,GE (1985)
DC	pKa			
HL 5.89E-007	atm m3/mol	25	EST	MEYLAN,WM & HOWARD,PH (1991)
OH 4.92E-012	cm3/molc sec	25	EST	MEYLAN,WM & HOWARD,PH (1993)

CAS #: 083704-27-2		DIBENZOFURAN, 1,3,4,6-TETRACHLORO-		
Formula: $C_{12}H_4Cl_4O$				
Mol Weight: 305.98				
MP (deg C):	FP (deg C):			
BP (deg C):				
BP pressure (mm Hg):				

Property/Value	Units	Temp	Data Type	Reference
WS 3.59E-003	mg/L	25	EST	MEYLAN,WM ET AL. (1996)
logP 6.31			EXP	SIJM,DTHM ET AL. (1989)
VP 1.53E-006	mm Hg	25	EST	NEELY,WB & BLAU,GE (1985)
DC	pKa			
HL 1.54E-005	atm m3/mol	25	EST	MEYLAN,WM & HOWARD,PH (1991)
OH 1.65E-013	cm3/molc sec	25	EST	MEYLAN,WM & HOWARD,PH (1993)

CAS #: 083704-30-7		DIBENZOFURAN, 2,3,4,6-TETRACHLORO-		
Formula: $C_{12}H_4Cl_4O$				
Mol Weight: 305.98				
MP (deg C):	FP (deg C):			
BP (deg C):				
BP pressure (mm Hg):				

Property/Value	Units	Temp	Data Type	Reference
WS 5.32E-003	mg/L	25	EST	MEYLAN,WM ET AL. (1996)
logP 6.11			EXP	SIJM,DTHM ET AL. (1989)
VP 1.53E-006	mm Hg	25	EST	NEELY,WB & BLAU,GE (1985)
DC	pKa			
HL 1.54E-005	atm m3/mol	25	EST	MEYLAN,WM & HOWARD,PH (1991)
OH 1.65E-013	cm3/molc sec	25	EST	MEYLAN,WM & HOWARD,PH (1993)

CAS #: 083801-60-9		1-(3-BROMOPHENOXY)SILATRANE		
Formula: $C_{12}H_{16}BrNO_4Si$				
Mol Weight: 346.26				
MP (deg C):	FP (deg C):			
BP (deg C):				
BP pressure (mm Hg):				

Property/Value	Units	Temp	Data Type	Reference
WS 3.99E+004	mg/L	25	EST	MEYLAN,WM ET AL. (1996)
logP 0.26			EXP	LUKASIAK,J (1984)
VP 3.34E-006	mm Hg	25	EST	NEELY,WB & BLAU,GE (1985)
DC	pKa			
HL 5.73E-014	atm m3/mol	25	EST	MEYLAN,WM & HOWARD,PH (1991)
OH 1.32E-010	cm3/molc sec	25	EST	MEYLAN,WM & HOWARD,PH (1993)

CAS #: 083870-87-5				BENZENESULFONYL-L-GLUTAMINE

Formula: $C_{11}H_{14}N_2O_5S$

Mol Weight: 286.31

MP (deg C): | FP (deg C):

BP (deg C):

BP pressure (mm Hg):

Property/Value	Units	Temp	Data Type	Reference
WS 3.77E+002	mg/L	25	EST	MEYLAN,WM ET AL. (1996)
logP 0.33			EXP	JHA,T ET AL. (1986)
VP 1.17E-010	mm Hg	25	EST	NEELY,WB & BLAU,GE (1985)
DC	pKa			
HL 3.68E-017	atm m3/mol	25	EST	MEYLAN,WM & HOWARD,PH (1991)
OH 2.49E-011	cm3/molc sec	25	EST	MEYLAN,WM & HOWARD,PH (1993)

CAS #: 083870-89-7				BENZENESULFONYL-L-(N'-ET)GLUTAMINE

Formula: $C_{13}H_{18}N_2O_5S$

Mol Weight: 314.36

MP (deg C): | FP (deg C):

BP (deg C):

BP pressure (mm Hg):

Property/Value	Units	Temp	Data Type	Reference
WS 3.69E+002	mg/L	25	EST	MEYLAN,WM ET AL. (1996)
logP 0.15			EXP	JHA,T ET AL. (1986)
VP 2.89E-011	mm Hg	25	EST	NEELY,WB & BLAU,GE (1985)
DC	pKa			
HL 1.07E-016	atm m3/mol	25	EST	MEYLAN,WM & HOWARD,PH (1991)
OH 3.24E-011	cm3/molc sec	25	EST	MEYLAN,WM & HOWARD,PH (1993)

CAS #: 083870-90-0				BENZENESULFONYL-L-(N'-PR)GLUTAMINE

Formula: $C_{14}H_{20}N_2O_5S$

Mol Weight: 328.39

MP (deg C): | FP (deg C):

BP (deg C):

BP pressure (mm Hg):

Property/Value	Units	Temp	Data Type	Reference
WS 4.01E+002	mg/L	25	EST	MEYLAN,WM ET AL. (1996)
logP 0.01			EXP	JHA,T ET AL. (1986)
VP 1.25E-011	mm Hg	25	EST	NEELY,WB & BLAU,GE (1985)
DC	pKa			
HL 1.42E-016	atm m3/mol	25	EST	MEYLAN,WM & HOWARD,PH (1991)
OH 3.44E-011	cm3/molc sec	25	EST	MEYLAN,WM & HOWARD,PH (1993)

CAS #: 083870-92-2				BENZENESULFONYL-L-(N'-PENTYL)GLUTAMINE

Formula: $C_{16}H_{24}N_2O_5S$

Mol Weight: 356.44

MP (deg C): | FP (deg C):

BP (deg C):

BP pressure (mm Hg):

Property/Value	Units	Temp	Data Type	Reference
WS 3.37E+002	mg/L	25	EST	MEYLAN,WM ET AL. (1996)
logP 0.03			EXP	JHA,T ET AL. (1986)
VP 2.34E-012	mm Hg	25	EST	NEELY,WB & BLAU,GE (1985)
DC	pKa			
HL 2.51E-016	atm m3/mol	25	EST	MEYLAN,WM & HOWARD,PH (1991)
OH 3.73E-011	cm3/molc sec	25	EST	MEYLAN,WM & HOWARD,PH (1993)

CAS #: 083870-99-9				4-ME-BENZENESULFONYL-L-(N'-PR)GLUTAMINE

Formula: $C_{15}H_{22}N_2O_5S$

Mol Weight: 342.42

MP (deg C): | FP (deg C):

BP (deg C):

BP pressure (mm Hg):

Property/Value	Units	Temp	Data Type	Reference
WS 9.63E+002	mg/L	25	EST	MEYLAN,WM ET AL. (1996)
logP 0.02			EXP	JHA,T ET AL. (1986)
VP 5.42E-012	mm Hg	25	EST	NEELY,WB & BLAU,GE (1985)
DC	pKa			
HL 1.57E-016	atm m3/mol	25	EST	MEYLAN,WM & HOWARD,PH (1991)
OH 3.53E-011	cm3/molc sec	25	EST	MEYLAN,WM & HOWARD,PH (1993)

CAS #: 083871-00-5				4-ME-BENZENESULFONYL-L-(N'-BU)GLUTAMINE

Formula: $C_{16}H_{24}N_2O_5S$

Mol Weight: 356.44

MP (deg C): | FP (deg C):

BP (deg C):

BP pressure (mm Hg):

Property/Value	Units	Temp	Data Type	Reference
WS 3.02E+002	mg/L	25	EST	MEYLAN,WM ET AL. (1996)
logP 0.00			EXP	JHA,T ET AL. (1986)
VP 2.34E-012	mm Hg	25	EST	NEELY,WB & BLAU,GE (1985)
DC	pKa			
HL 2.09E-016	atm m3/mol	25	EST	MEYLAN,WM & HOWARD,PH (1991)
OH 3.67E-011	cm3/molc sec	25	EST	MEYLAN,WM & HOWARD,PH (1993)

CAS #: 083871-02-7				4-ME-BENZENESULFONYL-L-(N'-HEX)GLUTAMINE

Formula: $C_{18}H_{28}N_2O_5S$

Mol Weight: 384.50

MP (deg C): | FP (deg C):

BP (deg C):

BP pressure (mm Hg):

Property/Value	Units	Temp	Data Type	Reference
WS 2.95E+001	mg/L	25	EST	MEYLAN,WM ET AL. (1996)
logP 0.84			EXP	JHA,T ET AL. (1986)
VP 4.31E-013	mm Hg	25	EST	NEELY,WB & BLAU,GE (1985)
DC	pKa			
HL 3.68E-016	atm m3/mol	25	EST	MEYLAN,WM & HOWARD,PH (1991)
OH 3.95E-011	cm3/molc sec	25	EST	MEYLAN,WM & HOWARD,PH (1993)

CAS #: 083871-03-8				4ME-BENZENESULFONYL-L-(N'-CYHX)GLUTAMINE

Formula: $C_{18}H_{26}N_2O_5S$

Mol Weight: 382.48

MP (deg C): | FP (deg C):

BP (deg C):

BP pressure (mm Hg):

Property/Value	Units	Temp	Data Type	Reference
WS 4.37E+001	mg/L	25	EST	MEYLAN,WM ET AL. (1996)
logP 0.10			EXP	JHA,T ET AL. (1986)
VP 2.95E-013	mm Hg	25	EST	NEELY,WB & BLAU,GE (1985)
DC	pKa			
HL 1.62E-016	atm m3/mol	25	EST	MEYLAN,WM & HOWARD,PH (1991)
OH 4.83E-011	cm3/molc sec	25	EST	MEYLAN,WM & HOWARD,PH (1993)

CAS #: 083905-01-5				AZITHROMYCIN

Formula: $C_{38}H_{72}N_2O_{12}$

Mol Weight: 749.00

MP (deg C):   FP (deg C):

BP (deg C):

BP pressure (mm Hg):

Property/Value	Units	Temp	Data Type	Reference
WS 7.09E+000	mg/L	25	EST	MEYLAN,WM ET AL. (1996)
logP 1.61			EXP	MFG DATA SHEET
VP 3.91E-027	mm Hg	25	EST	NEELY,WB & BLAU,GE (1985)
DC	pKa			
HL 5.30E-029	atm m3/mol	25	EST	MEYLAN,WM & HOWARD,PH (1991)
OH 4.24E-010	cm3/molc sec	25	EST	MEYLAN,WM & HOWARD,PH (1993)

CAS #: 084043-25-4				URIDINE, 2'-DEOXY-5-IODO-, 5'-PROPANOATE

Formula: $C_{12}H_{15}IN_2O_6$

Mol Weight: 410.17

MP (deg C):   FP (deg C):

BP (deg C):

BP pressure (mm Hg):

Property/Value	Units	Temp	Data Type	Reference
WS 6.95E+002	mg/L	25	EST	MEYLAN,WM ET AL. (1996)
logP 0.68			EXP	SANGSTER,J (1994)
VP 9.11E-016	mm Hg	25	EST	NEELY,WB & BLAU,GE (1985)
DC	pKa			
HL 8.39E-020	atm m3/mol	25	EST	MEYLAN,WM & HOWARD,PH (1991)
OH 8.10E-011	cm3/molc sec	25	EST	MEYLAN,WM & HOWARD,PH (1993)

CAS #: 084043-26-5				URIDINE, 2'-DEOXY-5-IODO-, 5'-BUTANOATE

Formula: $C_{12}H_{15}IN_2O_6$

Mol Weight: 410.17

MP (deg C):   FP (deg C):

BP (deg C):

BP pressure (mm Hg):

Property/Value	Units	Temp	Data Type	Reference
WS 4.79E+002	mg/L	25	EST	MEYLAN,WM ET AL. (1996)
logP 0.87			EXP	SANGSTER,J (1993)
VP 9.97E-016	mm Hg	25	EST	NEELY,WB & BLAU,GE (1985)
DC	pKa			
HL 8.39E-020	atm m3/mol	25	EST	MEYLAN,WM & HOWARD,PH (1991)
OH 3.55E-011	cm3/molc sec	25	EST	MEYLAN,WM & HOWARD,PH (1993)

CAS #: 084043-27-6				URIDINE, 2'-DEOXY-5-IODO-, 5'-(2-METHYLPROPANOAT

Formula: $C_{12}H_{15}IN_2O_6$

Mol Weight: 410.17

MP (deg C):   FP (deg C):

BP (deg C):

BP pressure (mm Hg):

Property/Value	Units	Temp	Data Type	Reference
WS 5.18E+002	mg/L	25	EST	MEYLAN,WM ET AL. (1996)
logP 0.83			EXP	SANGSTER,J (1993)
VP 1.81E-015	mm Hg	25	EST	NEELY,WB & BLAU,GE (1985)
DC	pKa			
HL 8.39E-020	atm m3/mol	25	EST	MEYLAN,WM & HOWARD,PH (1991)
OH 3.51E-011	cm3/molc sec	25	EST	MEYLAN,WM & HOWARD,PH (1993)

CAS #: 084043-28-7				URIDINE, 2'-DEOXY-5-IODO-, 5'-(2,2-DIMETHYLPROPA

Formula: $C_{14}H_{19}IN_2O_6$

Mol Weight: 438.22

MP (deg C):   FP (deg C):

BP (deg C):

BP pressure (mm Hg):

Property/Value	Units	Temp	Data Type	Reference
WS 1.27E+002	mg/L	25	EST	MEYLAN,WM ET AL. (1996)
logP 1.34			EXP	SANGSTER,J (1994)
VP 3.36E-016	mm Hg	25	EST	NEELY,WB & BLAU,GE (1985)
DC	pKa			
HL 1.48E-019	atm m3/mol	25	EST	MEYLAN,WM & HOWARD,PH (1991)
OH 8.06E-011	cm3/molc sec	25	EST	MEYLAN,WM & HOWARD,PH (1993)

CAS #: 084043-31-2				URIDINE, 2'-DEOXY-5-IODO-, 5'-BENZOATE

Formula: $C_{16}H_{15}IN_2O_6$

Mol Weight: 458.21

MP (deg C):   FP (deg C):

BP (deg C):

BP pressure (mm Hg):

Property/Value	Units	Temp	Data Type	Reference
WS 8.43E+001	mg/L	25	EST	MEYLAN,WM ET AL. (1996)
logP 1.40			EXP	SANGSTER,J (1994)
VP 4.72E-018	mm Hg	25	EST	NEELY,WB & BLAU,GE (1985)
DC	pKa			
HL 1.25E-020	atm m3/mol	25	EST	MEYLAN,WM & HOWARD,PH (1991)
OH 8.08E-011	cm3/molc sec	25	EST	MEYLAN,WM & HOWARD,PH (1993)

CAS #: 084052-69-7				URIDINE, 2'-DEOXY-5-IODO-, 5'-PENTANOATE

Formula: $C_{14}H_{19}IN_2O_6$

Mol Weight: 438.22

MP (deg C):   FP (deg C):

BP (deg C):

BP pressure (mm Hg):

Property/Value	Units	Temp	Data Type	Reference
WS 1.04E+002	mg/L	25	EST	MEYLAN,WM ET AL. (1996)
logP 1.44			EXP	SANGSTER,J (1994)
VP 1.24E-016	mm Hg	25	EST	NEELY,WB & BLAU,GE (1985)
DC	pKa			
HL 1.48E-019	atm m3/mol	25	EST	MEYLAN,WM & HOWARD,PH (1991)
OH 8.37E-011	cm3/molc sec	25	EST	MEYLAN,WM & HOWARD,PH (1993)

CAS #: 084057-95-4				ROPIVACAINE

Formula: $C_{17}H_{26}N_2O$

Mol Weight: 274.41

MP (deg C):   FP (deg C):

BP (deg C):

BP pressure (mm Hg):

Property/Value	Units	Temp	Data Type	Reference
WS 5.76E+001	mg/L	25	EST	MEYLAN,WM ET AL. (1996)
logP 2.90			EXP	HANSCH,C ET AL. (1995)
VP 3.99E-008	mm Hg	25	EST	NEELY,WB & BLAU,GE (1985)
DC	pKa			
HL 1.35E-010	atm m3/mol	25	EST	MEYLAN,WM & HOWARD,PH (1991)
OH 1.32E-010	cm3/molc sec	25	EST	MEYLAN,WM & HOWARD,PH (1993)

CAS #: 084057-96-5	2-PROPANOL, 1-[4-[2-[2-(4-FLUOROPHENYL)ETHOXY]ET

Formula: $C_{22}H_{30}FNO_4$

Mol Weight: 391.49

MP (deg C): | **FP (deg C):**

BP (deg C):

BP pressure (mm Hg):

Property/Value	Units	Temp	Data Type	Reference
WS 2.44E+001	mg/L	25	EST	MEYLAN,WM ET AL. (1996)
logP 3.70			EXP	RECANATINI,M (1992)
VP 3.73E-011	mm Hg	25	EST	NEELY,WB & BLAU,GE (1985)
DC	pKa			
HL 9.40E-016	atm m3/mol	25	EST	MEYLAN,WM & HOWARD,PH (1991)
OH 1.55E-010	cm3/molc sec	25	EST	MEYLAN,WM & HOWARD,PH (1993)

CAS #: 084104-57-4	5-THIA-1-AZABICYCLO[4.2.0]OCT-2-ENE-2-CARBOXYLIC

Formula: $C_{20}H_{15}N_5O_7S_2$

Mol Weight: 501.50

MP (deg C): | **FP (deg C):**

BP (deg C):

BP pressure (mm Hg):

Property/Value	Units	Temp	Data Type	Reference
WS 5.65E+001	mg/L	25	EST	MEYLAN,WM ET AL. (1996)
logP 1.28			EXP	SANGSTER,J (1993)
VP 3.64E-018	mm Hg	25	EST	NEELY,WB & BLAU,GE (1985)
DC	pKa			
HL 2.19E-020	atm m3/mol	25	EST	MEYLAN,WM & HOWARD,PH (1991)
OH 1.08E-010	cm3/molc sec	25	EST	MEYLAN,WM & HOWARD,PH (1993)

CAS #: 084123-05-7	ETHANONE, 1-[1-(2-HYDROXYPROPYL)-5-METHYL-2-NITR

Formula: $C_9H_{13}N_3O_4$

Mol Weight: 227.22

MP (deg C): | **FP (deg C):**

BP (deg C):

BP pressure (mm Hg):

Property/Value	Units	Temp	Data Type	Reference
WS 6.20E+003	mg/L	25	EST	MEYLAN,WM ET AL. (1996)
logP 0.37			EXP	SANGSTER,J (1993)
VP 5.04E-009	mm Hg	25	EST	NEELY,WB & BLAU,GE (1985)
DC	pKa			
HL 4.09E-014	atm m3/mol	25	EST	MEYLAN,WM & HOWARD,PH (1991)
OH 4.60E-011	cm3/molc sec	25	EST	MEYLAN,WM & HOWARD,PH (1993)

CAS #: 084174-20-9	N-NITROSO-N-ME-(4-MEOBENZYL)AMINE

Formula: $C_9H_{12}N_2O_2$

Mol Weight: 180.21

MP (deg C): | **FP (deg C):**

BP (deg C):

BP pressure (mm Hg):

Property/Value	Units	Temp	Data Type	Reference
WS 5.09E+003	mg/L	25	EST	MEYLAN,WM ET AL. (1996)
logP 1.21			EXP	HANSCH,C & LEO,AJ (1985)
VP 3.99E-004	mm Hg	25	EST	NEELY,WB & BLAU,GE (1985)
DC	pKa			
HL 9.83E-009	atm m3/mol	25	EST	MEYLAN,WM & HOWARD,PH (1991)
OH 3.70E-011	cm3/molc sec	25	EST	MEYLAN,WM & HOWARD,PH (1993)

CAS #: 084174-21-0	BENZENEMETHANAMINE, 4-FLUORO-N-METHYL-N-NITROSO-

Formula: $C_8H_9FN_2O$

Mol Weight: 168.17

MP (deg C): | **FP (deg C):**

BP (deg C):

BP pressure (mm Hg):

Property/Value	Units	Temp	Data Type	Reference
WS 2.35E+003	mg/L	25	EST	MEYLAN,WM ET AL. (1996)
logP 1.67			EXP	SINGER,GM ET AL. (1986)
VP 3.78E-003	mm Hg	25	EST	NEELY,WB & BLAU,GE (1985)
DC	pKa			
HL 1.94E-007	atm m3/mol	25	EST	MEYLAN,WM & HOWARD,PH (1991)
OH 1.29E-011	cm3/molc sec	25	EST	MEYLAN,WM & HOWARD,PH (1993)

CAS #: 084174-22-1	BENZENEMETHANAMINE, 4-CHLORO-N-METHYL-N-NITROSO-

Formula: $C_8H_9ClN_2O$

Mol Weight: 184.63

MP (deg C): | **FP (deg C):**

BP (deg C):

BP pressure (mm Hg):

Property/Value	Units	Temp	Data Type	Reference
WS 7.05E+002	mg/L	25	EST	MEYLAN,WM ET AL. (1996)
logP 2.19			EXP	SINGER,GM ET AL. (1986)
VP 5.83E-004	mm Hg	25	EST	NEELY,WB & BLAU,GE (1985)
DC	pKa			
HL 1.23E-007	atm m3/mol	25	EST	MEYLAN,WM & HOWARD,PH (1991)
OH 1.16E-011	cm3/molc sec	25	EST	MEYLAN,WM & HOWARD,PH (1993)

CAS #: 084174-23-2	N-NITROSO-N-ME-(4-CYANOBENZYL)AMINE

Formula: $C_9H_9N_3O$

Mol Weight: 175.19

MP (deg C): | **FP (deg C):**

BP (deg C):

BP pressure (mm Hg):

Property/Value	Units	Temp	Data Type	Reference
WS 4.87E+003	mg/L	25	EST	MEYLAN,WM ET AL. (1996)
logP 0.95			EXP	HANSCH,C & LEO,AJ (1985)
VP 5.57E-005	mm Hg	25	EST	NEELY,WB & BLAU,GE (1985)
DC	pKa			
HL 1.60E-009	atm m3/mol	25	EST	MEYLAN,WM & HOWARD,PH (1991)
OH 1.08E-011	cm3/molc sec	25	EST	MEYLAN,WM & HOWARD,PH (1993)

CAS #: 084174-24-3	BENZENEMETHANAMINE, N-METHYL-4-NITRO-N-NITROSO-

Formula: $C_8H_9N_3O_3$

Mol Weight: 195.18

MP (deg C): | **FP (deg C):**

BP (deg C):

BP pressure (mm Hg):

Property/Value	Units	Temp	Data Type	Reference
WS 3.76E+003	mg/L	25	EST	MEYLAN,WM ET AL. (1996)
logP 0.82			EXP	SINGER,GM ET AL. (1986)
VP 2.68E-005	mm Hg	25	EST	NEELY,WB & BLAU,GE (1985)
DC	pKa			
HL 6.55E-010	atm m3/mol	25	EST	MEYLAN,WM & HOWARD,PH (1991)
OH 1.06E-011	cm3/molc sec	25	EST	MEYLAN,WM & HOWARD,PH (1993)

CAS #: 084199-91-7	2-PYRIDINEPROPANEAMIDE

Formula: $C_8H_{10}N_2O$

Mol Weight: 150.18

MP (deg C):　　FP (deg C):

BP (deg C):

BP pressure (mm Hg):

Property/ Value	Units	Temp	Data Type	Reference
WS 1.00E+006	mg/L	25	EST	MEYLAN,WM ET AL. (1996)
logP -0.27			EXP	HANSCH,C & LEO,AJ (1985)
VP 6.13E-005	mm Hg	25	EST	NEELY,WB & BLAU,GE (1985)
DC	pKa			
HL 1.57E-012	atm m3/mol	25	EST	MEYLAN,WM & HOWARD,PH (1991)
OH 7.42E-012	cm3/molc sec	25	EST	MEYLAN,WM & HOWARD,PH (1993)

CAS #: 084199-97-3	2-PYRIDINEPENTANEAMIDE

Formula: $C_{10}H_{14}N_2O$

Mol Weight: 178.24

MP (deg C):　　FP (deg C):

BP (deg C):

BP pressure (mm Hg):

Property/ Value	Units	Temp	Data Type	Reference
WS 5.31E+005	mg/L	25	EST	MEYLAN,WM ET AL. (1996)
logP 0.38			EXP	HANSCH,C & LEO,AJ (1985)
VP 1.35E-005	mm Hg	25	EST	NEELY,WB & BLAU,GE (1985)
DC	pKa			
HL 2.76E-012	atm m3/mol	25	EST	MEYLAN,WM & HOWARD,PH (1991)
OH 1.08E-011	cm3/molc sec	25	EST	MEYLAN,WM & HOWARD,PH (1993)

CAS #: 084199-99-5	3-PYRIDINEPROPANEAMIDE

Formula: $C_8H_{10}N_2O$

Mol Weight: 150.18

MP (deg C):　　FP (deg C):

BP (deg C):

BP pressure (mm Hg):

Property/ Value	Units	Temp	Data Type	Reference
WS 1.00E+006	mg/L	25	EST	MEYLAN,WM ET AL. (1996)
logP -0.26			EXP	HANSCH,C & LEO,AJ (1985)
VP 6.13E-005	mm Hg	25	EST	NEELY,WB & BLAU,GE (1985)
DC	pKa			
HL 1.57E-012	atm m3/mol	25	EST	MEYLAN,WM & HOWARD,PH (1991)
OH 7.42E-012	cm3/molc sec	25	EST	MEYLAN,WM & HOWARD,PH (1993)

CAS #: 084200-01-1	3-PYRIDINEBUTANAMIDE

Formula: $C_9H_{12}N_2O$

Mol Weight: 164.21

MP (deg C):　　FP (deg C):

BP (deg C):

BP pressure (mm Hg):

Property/ Value	Units	Temp	Data Type	Reference
WS 1.00E+006	mg/L	25	EST	MEYLAN,WM ET AL. (1996)
logP -0.08			EXP	HANSCH,C & LEO,AJ (1985)
VP 3.11E-005	mm Hg	25	EST	NEELY,WB & BLAU,GE (1985)
DC	pKa			
HL 2.08E-012	atm m3/mol	25	EST	MEYLAN,WM & HOWARD,PH (1991)
OH 9.41E-012	cm3/molc sec	25	EST	MEYLAN,WM & HOWARD,PH (1993)

CAS #: 084200-03-3	3-PYRIDINEPENTANOL

Formula: $C_{10}H_{15}NO$

Mol Weight: 165.24

MP (deg C):　　FP (deg C):

BP (deg C):

BP pressure (mm Hg):

Property/ Value	Units	Temp	Data Type	Reference
WS 2.61E+005	mg/L	25	EST	MEYLAN,WM ET AL. (1996)
logP 1.41			EXP	HANSCH,C & LEO,AJ (1985)
VP 1.36E-004	mm Hg	25	EST	NEELY,WB & BLAU,GE (1985)
DC	pKa			
HL 8.84E-010	atm m3/mol	25	EST	MEYLAN,WM & HOWARD,PH (1991)
OH 1.05E-011	cm3/molc sec	25	EST	MEYLAN,WM & HOWARD,PH (1993)

CAS #: 084200-04-4	3-PYRIDINEPENTANEAMIDE

Formula: $C_{10}H_{14}N_2O$

Mol Weight: 178.24

MP (deg C):　　FP (deg C):

BP (deg C):

BP pressure (mm Hg):

Property/ Value	Units	Temp	Data Type	Reference
WS 5.10E+005	mg/L	25	EST	MEYLAN,WM ET AL. (1996)
logP 0.40			EXP	HANSCH,C & LEO,AJ (1985)
VP 1.35E-005	mm Hg	25	EST	NEELY,WB & BLAU,GE (1985)
DC	pKa			
HL 2.76E-012	atm m3/mol	25	EST	MEYLAN,WM & HOWARD,PH (1991)
OH 1.08E-011	cm3/molc sec	25	EST	MEYLAN,WM & HOWARD,PH (1993)

CAS #: 084200-05-5	3-PYRIDINEPENTANEAMINE

Formula: $C_{10}H_{16}N_2$

Mol Weight: 164.25

MP (deg C):　　FP (deg C):

BP (deg C):

BP pressure (mm Hg):

Property/ Value	Units	Temp	Data Type	Reference
WS 8.31E+005	mg/L	25	EST	MEYLAN,WM ET AL. (1996)
logP 1.41			EXP	HANSCH,C & LEO,AJ (1985)
VP 4.71E-003	mm Hg	25	EST	NEELY,WB & BLAU,GE (1985)
DC	pKa			
HL 2.49E-009	atm m3/mol	25	EST	MEYLAN,WM & HOWARD,PH (1991)
OH 3.80E-011	cm3/molc sec	25	EST	MEYLAN,WM & HOWARD,PH (1993)

CAS #: 084200-07-7	4-PYRIDINEPROPANEAMIDE

Formula: $C_8H_{10}N_2O$

Mol Weight: 150.18

MP (deg C):　　FP (deg C):

BP (deg C):

BP pressure (mm Hg):

Property/ Value	Units	Temp	Data Type	Reference
WS 1.00E+006	mg/L	25	EST	MEYLAN,WM ET AL. (1996)
logP -0.25			EXP	HANSCH,C & LEO,AJ (1985)
VP 6.13E-005	mm Hg	25	EST	NEELY,WB & BLAU,GE (1985)
DC	pKa			
HL 1.57E-012	atm m3/mol	25	EST	MEYLAN,WM & HOWARD,PH (1991)
OH 7.42E-012	cm3/molc sec	25	EST	MEYLAN,WM & HOWARD,PH (1993)

CAS #: 084200-11-3 — 4-PYRIDINEPENTANEAMIDE

Formula: $C_{10}H_{14}N_2O$

Mol Weight: 178.24

MP (deg C):   FP (deg C):

BP (deg C):

BP pressure (mm Hg):

Property/Value	Units	Temp	Data Type	Reference
WS 4.91E+005	mg/L	25	EST	MEYLAN,WM ET AL. (1996)
logP 0.42			EXP	HANSCH,C & LEO,AJ (1985)
VP 1.35E-005	mm Hg	25	EST	NEELY,WB & BLAU,GE (1985)
DC	pKa			
HL 2.76E-012	atm m3/mol	25	EST	MEYLAN,WM & HOWARD,PH (1991)
OH 1.08E-011	cm3/molc sec	25	EST	MEYLAN,WM & HOWARD,PH (1993)

CAS #: 084226-14-2 — FLA 797

Formula: $C_{15}H_{21}BrN_2O_3$

Mol Weight: 357.25

MP (deg C):   FP (deg C):

BP (deg C):

BP pressure (mm Hg):

Property/Value	Units	Temp	Data Type	Reference
WS 4.17E+002	mg/L	25	EST	MEYLAN,WM ET AL. (1996)
logP 2.00			EXP	HOEGBERG,T ET AL. (1986)
VP 1.37E-010	mm Hg	25	EST	NEELY,WB & BLAU,GE (1985)
DC	pKa			
HL 1.02E-014	atm m3/mol	25	EST	MEYLAN,WM & HOWARD,PH (1991)
OH 2.15E-010	cm3/molc sec	25	EST	MEYLAN,WM & HOWARD,PH (1993)

CAS #: 084359-13-7 — 2-PYRIDINEBUTANAMINE, MONOHYDROCHLORIDE

Formula: $C_9H_{15}ClN_2$

Mol Weight: 186.69

MP (deg C):   FP (deg C):

BP (deg C):

BP pressure (mm Hg):

Property/Value	Units	Temp	Data Type	Reference
WS 1.00E+006	mg/L	25	EST	MEYLAN,WM ET AL. (1996)
logP -1.60			EXP	SANGSTER,J (1994)
VP 8.21E-008	mm Hg	25	EST	NEELY,WB & BLAU,GE (1985)
DC	pKa			
HL 1.74E-016	atm m3/mol	25	EST	MEYLAN,WM & HOWARD,PH (1991)
OH 1.56E-011	cm3/molc sec	25	EST	MEYLAN,WM & HOWARD,PH (1993)

CAS #: 084359-18-2 — 3-PYRIDINEBUTANAMINE, MONOHYDROCHLORIDE

Formula: $C_9H_{15}ClN_2$

Mol Weight: 186.69

MP (deg C):   FP (deg C):

BP (deg C):

BP pressure (mm Hg):

Property/Value	Units	Temp	Data Type	Reference
WS 1.00E+006	mg/L	25	EST	MEYLAN,WM ET AL. (1996)
logP -1.63			EXP	SANGSTER,J (1994)
VP 8.21E-008	mm Hg	25	EST	NEELY,WB & BLAU,GE (1985)
DC	pKa			
HL 1.74E-016	atm m3/mol	25	EST	MEYLAN,WM & HOWARD,PH (1991)
OH 1.56E-011	cm3/molc sec	25	EST	MEYLAN,WM & HOWARD,PH (1993)

CAS #: 084359-21-7 — 4-PYRIDINEBUTANAMINE, MONOHYDROCHLORIDE

Formula: $C_9H_{15}ClN_2$

Mol Weight: 186.69

MP (deg C):   FP (deg C):

BP (deg C):

BP pressure (mm Hg):

Property/Value	Units	Temp	Data Type	Reference
WS 1.00E+006	mg/L	25	EST	MEYLAN,WM ET AL. (1996)
logP -1.67			EXP	SANGSTER,J (1994)
VP 8.21E-008	mm Hg	25	EST	NEELY,WB & BLAU,GE (1985)
DC	pKa			
HL 1.74E-016	atm m3/mol	25	EST	MEYLAN,WM & HOWARD,PH (1991)
OH 1.56E-011	cm3/molc sec	25	EST	MEYLAN,WM & HOWARD,PH (1993)

CAS #: 084368-16-1 — YCLOPROPANECARBOXYLIC ACID, 3-[(DIHYDRO-2-OXO-3(

Formula: $C_{12}H_{16}O_3S$

Mol Weight: 240.32

MP (deg C):   FP (deg C):

BP (deg C):

BP pressure (mm Hg):

Property/Value	Units	Temp	Data Type	Reference
WS 2.74E+002	mg/L	25	EST	MEYLAN,WM ET AL. (1996)
logP 2.33			EXP	SANGSTER,J (1993)
VP 5.91E-005	mm Hg	25	EST	NEELY,WB & BLAU,GE (1985)
DC	pKa			
HL 1.16E-007	atm m3/mol	25	EST	MEYLAN,WM & HOWARD,PH (1991)
OH 3.37E-011	cm3/molc sec	25	EST	MEYLAN,WM & HOWARD,PH (1993)

CAS #: 084397-27-3 — N7-(6-MEO-PYRID-3-YL)-MITOMYCIN C

Formula: $C_{21}H_{23}N_5O_6$

Mol Weight: 441.45

MP (deg C):   FP (deg C):

BP (deg C):

BP pressure (mm Hg):

Property/Value	Units	Temp	Data Type	Reference
WS 5.39E+003	mg/L	25	EST	MEYLAN,WM ET AL. (1996)
logP 0.59			EXP	HANSCH,C & LEO,AJ (1985)
VP 3.76E-013	mm Hg	25	EST	NEELY,WB & BLAU,GE (1985)
DC	pKa			
HL 4.86E-029	atm m3/mol	25	EST	MEYLAN,WM & HOWARD,PH (1991)
OH 5.55E-011	cm3/molc sec	25	EST	MEYLAN,WM & HOWARD,PH (1993)

CAS #: 084397-36-4 — N7-(5-ME THIAZOL-2-YL)-MITOMYCIN C

Formula: $C_{19}H_{23}N_5O_5S$

Mol Weight: 433.49

MP (deg C):   FP (deg C):

BP (deg C):

BP pressure (mm Hg):

Property/Value	Units	Temp	Data Type	Reference
WS 2.03E+003	mg/L	25	EST	MEYLAN,WM ET AL. (1996)
logP 1.16			EXP	HANSCH,C & LEO,AJ (1985)
VP 5.97E-013	mm Hg	25	EST	NEELY,WB & BLAU,GE (1985)
DC	pKa			
HL 4.93E-028	atm m3/mol	25	EST	MEYLAN,WM & HOWARD,PH (1991)
OH 5.24E-011	cm3/molc sec	25	EST	MEYLAN,WM & HOWARD,PH (1993)

CAS #: 084397-45-5	N7-(3-PYRAZOLYL)-MITOMYCIN C

Formula: $C_{18}H_{20}N_6O_5$

Mol Weight: 400.40

MP (deg C):   FP (deg C):

BP (deg C):

BP pressure (mm Hg):

Property/Value	Units	Temp	Data Type	Reference
WS 1.30E+003	mg/L	25	EST	MEYLAN,WM ET AL. (1996)
logP 0.08			EXP	HANSCH,C & LEO,AJ (1985)
VP 1.97E-013	mm Hg	25	EST	NEELY,WB & BLAU,GE (1985)
DC	pKa			
HL 4.30E-028	atm m3/mol	25	EST	MEYLAN,WM & HOWARD,PH (1991)
OH 2.46E-010	cm3/molc sec	25	EST	MEYLAN,WM & HOWARD,PH (1993)

CAS #: 084408-37-7	ETHANOL, 2-[(2-AMINO-9H-PURIN-9-YL)METHOXY]-

Formula: $C_8H_{11}N_5O_2$

Mol Weight: 209.21

MP (deg C):   FP (deg C):

BP (deg C):

BP pressure (mm Hg):

Property/Value	Units	Temp	Data Type	Reference
WS 1.61E+004	mg/L	25	EST	MEYLAN,WM ET AL. (1996)
logP -1.08			EXP	KRISTL,A ET AL. (1993)
VP 3.07E-010	mm Hg	25	EST	NEELY,WB & BLAU,GE (1985)
DC	pKa			
HL 3.02E-015	atm m3/mol	25	EST	MEYLAN,WM & HOWARD,PH (1991)
OH 2.17E-010	cm3/molc sec	25	EST	MEYLAN,WM & HOWARD,PH (1993)

CAS #: 084472-85-5	URIDINE, 3'-AZIDO-2',3'-DIDEOXY-

Formula: $C_9H_{11}N_5O_4$

Mol Weight: 253.22

MP (deg C):   FP (deg C):

BP (deg C):

BP pressure (mm Hg):

Property/Value	Units	Temp	Data Type	Reference
WS 7.72E+002	mg/L	25	EST	MEYLAN,WM ET AL. (1996)
logP -0.32			EXP	BALZARINI,J ET AL. (1989)
VP 1.16E-019	mm Hg	25	EST	NEELY,WB & BLAU,GE (1985)
DC	pKa			
HL 4.64E-024	atm m3/mol	25	EST	MEYLAN,WM & HOWARD,PH (1991)
OH 7.85E-011	cm3/molc sec	25	EST	MEYLAN,WM & HOWARD,PH (1993)

CAS #: 084472-89-9	CYTIDINE, 3'-AZIDO-2',3'-DIDEOXY-

Formula: $C_9H_{12}N_6O_3$

Mol Weight: 252.23

MP (deg C):   FP (deg C):

BP (deg C):

BP pressure (mm Hg):

Property/Value	Units	Temp	Data Type	Reference
WS 1.65E+003	mg/L	25	EST	MEYLAN,WM ET AL. (1996)
logP -0.70			EXP	SANGSTER,J (1993)
VP 1.17E-015	mm Hg	25	EST	NEELY,WB & BLAU,GE (1985)
DC	pKa			
HL 2.57E-024	atm m3/mol	25	EST	MEYLAN,WM & HOWARD,PH (1991)
OH 1.26E-010	cm3/molc sec	25	EST	MEYLAN,WM & HOWARD,PH (1993)

CAS #: 084496-56-0	CLOMEPROP

Formula: $C_{16}H_{15}Cl_2NO_2$

Mol Weight: 324.21

MP (deg C):   FP (deg C):

BP (deg C):

BP pressure (mm Hg):

Property/Value	Units	Temp	Data Type	Reference
WS 7.04E-001	mg/L	25	EST	MEYLAN,WM ET AL. (1996)
logP 4.80			EXP	HANSCH,C ET AL. (1995)
VP 6.27E-009	mm Hg	25	EST	NEELY,WB & BLAU,GE (1985)
DC	pKa			
HL 1.52E-010	atm m3/mol	25	EST	MEYLAN,WM & HOWARD,PH (1991)
OH 2.67E-011	cm3/molc sec	25	EST	MEYLAN,WM & HOWARD,PH (1993)

CAS #: 084640-21-1	P-MEBENZYL N,N-DIME CARBAMATE

Formula: $C_{11}H_{15}NO_2$

Mol Weight: 193.25

MP (deg C):   FP (deg C):

BP (deg C):

BP pressure (mm Hg):

Property/Value	Units	Temp	Data Type	Reference
WS 2.69E+002	mg/L	25	EST	MEYLAN,WM ET AL. (1996)
logP 2.63			EXP	HANSCH,C & LEO,AJ (1985)
VP 6.16E-003	mm Hg	25	EST	NEELY,WB & BLAU,GE (1985)
DC	pKa			
HL 1.70E-008	atm m3/mol	25	EST	MEYLAN,WM & HOWARD,PH (1991)
OH 2.39E-011	cm3/molc sec	25	EST	MEYLAN,WM & HOWARD,PH (1993)

CAS #: 084640-22-2	P-FLUOROBENZYL N,N-DIME CARBAMATE

Formula: $C_{10}H_{12}FNO_2$

Mol Weight: 197.21

MP (deg C):   FP (deg C):

BP (deg C):

BP pressure (mm Hg):

Property/Value	Units	Temp	Data Type	Reference
WS 4.91E+002	mg/L	25	EST	MEYLAN,WM ET AL. (1996)
logP 2.30			EXP	HANSCH,C & LEO,AJ (1985)
VP 2.33E-002	mm Hg	25	EST	NEELY,WB & BLAU,GE (1985)
DC	pKa			
HL 1.80E-008	atm m3/mol	25	EST	MEYLAN,WM & HOWARD,PH (1991)
OH 2.06E-011	cm3/molc sec	25	EST	MEYLAN,WM & HOWARD,PH (1993)

CAS #: 084640-23-3	M-MEO BENZYL N,N-DIME CARBAMATE

Formula: $C_{11}H_{15}NO_3$

Mol Weight: 209.25

MP (deg C):   FP (deg C):

BP (deg C):

BP pressure (mm Hg):

Property/Value	Units	Temp	Data Type	Reference
WS 6.43E+002	mg/L	25	EST	MEYLAN,WM ET AL. (1996)
logP 2.09			EXP	HANSCH,C & LEO,AJ (1985)
VP 1.81E-003	mm Hg	25	EST	NEELY,WB & BLAU,GE (1985)
DC	pKa			
HL 9.11E-010	atm m3/mol	25	EST	MEYLAN,WM & HOWARD,PH (1991)
OH 7.19E-011	cm3/molc sec	25	EST	MEYLAN,WM & HOWARD,PH (1993)

CAS #: 084640-32-4	P-CYANOBENZYL N,N-DIME CARBAMATE

Formula: $C_{11}H_{12}N_2O_2$

Mol Weight: 204.23

MP (deg C): | FP (deg C):

BP (deg C):

BP pressure (mm Hg):

Property/Value	Units	Temp	Data Type	Reference
WS 8.48E+002	mg/L	25	EST	MEYLAN,WM ET AL. (1996)
logP 1.67			EXP	HANSCH,C & LEO,AJ (1985)
VP 2.32E-004	mm Hg	25	EST	NEELY,WB & BLAU,GE (1985)
DC	pKa			
HL 1.49E-010	atm m3/mol	25	EST	MEYLAN,WM & HOWARD,PH (1991)
OH 1.85E-011	cm3/molc sec	25	EST	MEYLAN,WM & HOWARD,PH (1993)

CAS #: 084640-33-5	P-BROMOBENZYL N,N-DIME CARBAMATE

Formula: $C_{10}H_{12}BrNO_2$

Mol Weight: 258.12

MP (deg C): | FP (deg C):

BP (deg C):

BP pressure (mm Hg):

Property/Value	Units	Temp	Data Type	Reference
WS 5.73E+001	mg/L	25	EST	MEYLAN,WM ET AL. (1996)
logP 3.01			EXP	HANSCH,C & LEO,AJ (1985)
VP 9.13E-004	mm Hg	25	EST	NEELY,WB & BLAU,GE (1985)
DC	pKa			
HL 6.14E-009	atm m3/mol	25	EST	MEYLAN,WM & HOWARD,PH (1991)
OH 1.91E-011	cm3/molc sec	25	EST	MEYLAN,WM & HOWARD,PH (1993)

CAS #: 084640-34-6	P-IODOBENZYL N,N-DIME CARBAMATE

Formula: $C_{10}H_{12}INO_2$

Mol Weight: 305.12

MP (deg C): | FP (deg C):

BP (deg C):

BP pressure (mm Hg):

Property/Value	Units	Temp	Data Type	Reference
WS 1.68E+001	mg/L	25	EST	MEYLAN,WM ET AL. (1996)
logP 3.32			EXP	HANSCH,C & LEO,AJ (1985)
VP 2.03E-004	mm Hg	25	EST	NEELY,WB & BLAU,GE (1985)
DC	pKa			
HL 3.57E-009	atm m3/mol	25	EST	MEYLAN,WM & HOWARD,PH (1991)
OH 1.92E-011	cm3/molc sec	25	EST	MEYLAN,WM & HOWARD,PH (1993)

CAS #: 084640-35-7	P-CF3-BENZYL N,N-DIME CARBAMATE

Formula: $C_{11}H_{12}F_3NO_2$

Mol Weight: 247.22

MP (deg C): | FP (deg C):

BP (deg C):

BP pressure (mm Hg):

Property/Value	Units	Temp	Data Type	Reference
WS 5.74E+001	mg/L	25	EST	MEYLAN,WM ET AL. (1996)
logP 3.08			EXP	HANSCH,C & LEO,AJ (1985)
VP 1.17E-002	mm Hg	25	EST	NEELY,WB & BLAU,GE (1985)
DC	pKa			
HL 1.34E-007	atm m3/mol	25	EST	MEYLAN,WM & HOWARD,PH (1991)
OH 3.58E-012	cm3/molc sec	25	EST	MEYLAN,WM & HOWARD,PH (1993)

CAS #: 084640-36-8	P-PHENO-BENZYL N,N-DIME CARBAMATE

Formula: $C_{16}H_{17}NO_3$

Mol Weight: 271.32

MP (deg C): | FP (deg C):

BP (deg C):

BP pressure (mm Hg):

Property/Value	Units	Temp	Data Type	Reference
WS 1.67E+001	mg/L	25	EST	MEYLAN,WM ET AL. (1996)
logP 3.55			EXP	HANSCH,C ET AL. (1995)
VP 6.07E-006	mm Hg	25	EST	NEELY,WB & BLAU,GE (1985)
DC	pKa			
HL 3.37E-010	atm m3/mol	25	EST	MEYLAN,WM & HOWARD,PH (1991)
OH 2.85E-011	cm3/molc sec	25	EST	MEYLAN,WM & HOWARD,PH (1993)

CAS #: 084640-38-0	P-(OCONME2)BENZYL N,N-DIME CARBAM

Formula: $C_{13}H_{18}N_2O_4$

Mol Weight: 266.30

MP (deg C): | FP (deg C):

BP (deg C):

BP pressure (mm Hg):

Property/Value	Units	Temp	Data Type	Reference
WS 8.41E+002	mg/L	25	EST	MEYLAN,WM ET AL. (1996)
logP 1.59			EXP	HANSCH,C & LEO,AJ (1985)
VP 4.67E-004	mm Hg	25	EST	NEELY,WB & BLAU,GE (1985)
DC	pKa			
HL 3.20E-010	atm m3/mol	25	EST	MEYLAN,WM & HOWARD,PH (1991)
OH 3.73E-011	cm3/molc sec	25	EST	MEYLAN,WM & HOWARD,PH (1993)

CAS #: 084640-39-1	P-METHIOBENZYL N,N-DIME CARBAMATE

Formula: $C_{11}H_{15}NO_2S$

Mol Weight: 225.31

MP (deg C): | FP (deg C):

BP (deg C):

BP pressure (mm Hg):

Property/Value	Units	Temp	Data Type	Reference
WS 1.16E+002	mg/L	25	EST	MEYLAN,WM ET AL. (1996)
logP 2.86			EXP	HANSCH,C ET AL. (1995)
VP 3.14E-004	mm Hg	25	EST	NEELY,WB & BLAU,GE (1985)
DC	pKa			
HL 4.49E-010	atm m3/mol	25	EST	MEYLAN,WM & HOWARD,PH (1991)
OH 3.40E-011	cm3/molc sec	25	EST	MEYLAN,WM & HOWARD,PH (1993)

CAS #: 084652-30-2	L-ASPARAGINAMIDE, N-ACETYL-

Formula: $C_6H_{11}N_3O_3$

Mol Weight: 173.17

MP (deg C): | FP (deg C):

BP (deg C):

BP pressure (mm Hg):

Property/Value	Units	Temp	Data Type	Reference
WS 3.33E+005	mg/L	25	EST	MEYLAN,WM ET AL. (1996)
logP -2.41			EXP	SANGSTER,J (1994)
VP 4.37E-009	mm Hg	25	EST	NEELY,WB & BLAU,GE (1985)
DC	pKa			
HL 2.38E-017	atm m3/mol	25	EST	MEYLAN,WM & HOWARD,PH (1991)
OH 3.56E-011	cm3/molc sec	25	EST	MEYLAN,WM & HOWARD,PH (1993)

CAS #: 084712-77-6	1,3,5-TRIAZINE-2,4-DIAMINE, 6-CHLORO-N-ETHYL-N'-
Formula: $C_{11}H_{18}ClN_5$	
Mol Weight: 255.75	
MP (deg C):	FP (deg C):
BP (deg C):	
BP pressure (mm Hg):	

Property/Value	Units	Temp	Data Type	Reference
WS 1.75E+001	mg/L	25	EST	MEYLAN,WM ET AL. (1996)
logP 3.63			EXP	MITSUTAKE,KI ET AL. (1986)
VP 3.63E-006	mm Hg	25	EST	NEELY,WB & BLAU,GE (1985)
DC	pKa			
HL 4.61E-009	atm m3/mol	25	EST	MEYLAN,WM & HOWARD,PH (1991)
OH 4.33E-011	cm3/molc sec	25	EST	MEYLAN,WM & HOWARD,PH (1993)

CAS #: 084846-21-9	BICYCLO[2.2.1]HEPT-5-ENE-2,3-DICARBOXAMIDE
Formula: $C_{13}H_{20}N_2O_2$	
Mol Weight: 236.32	
MP (deg C):	FP (deg C):
BP (deg C):	
BP pressure (mm Hg):	

Property/Value	Units	Temp	Data Type	Reference
WS 7.68E+003	mg/L	25	EST	MEYLAN,WM ET AL. (1996)
logP 0.66			EXP	SANGSTER,J (1993)
VP 1.28E-008	mm Hg	25	EST	NEELY,WB & BLAU,GE (1985)
DC	pKa			
HL 2.14E-011	atm m3/mol	25	EST	MEYLAN,WM & HOWARD,PH (1991)
OH 9.71E-011	cm3/molc sec	25	EST	MEYLAN,WM & HOWARD,PH (1993)

CAS #: 084852-15-3	N-NONYLPHENOL
Formula: $C_{15}H_{24}O$	
Mol Weight: 220.36	
MP (deg C):	FP (deg C):
BP (deg C): 293-297	
BP pressure (mm Hg):	

Property/Value	Units	Temp	Data Type	Reference
WS 5.00E+003	mg/L	25	EXP	GEYER,H ET AL (1982)
logP 5.92			EST	MEYLAN,WM & HOWARD,PH (1995)
VP 5.39E-004	mm Hg	25	EXT	BOUBLIK,T ET AL. (1984)
DC 11.06	pKa		EXP	LIPNICK,RL ET AL. (1986)
HL 2.40E-009	atm m3/mol	22	EXP	HELLMANN,H (1987)
OH 5.17E-011	cm3/molc sec	25	EST	MEYLAN,WM & HOWARD,PH (1993)

CAS #: 084884-30-0	[3,4'-BIPYRIDINE]-5-CARBONITRILE,6-(DIMETHYLAMIN
Formula: $C_{13}H_{12}N_4$	
Mol Weight: 224.27	
MP (deg C):	FP (deg C):
BP (deg C):	
BP pressure (mm Hg):	

Property/Value	Units	Temp	Data Type	Reference
WS 2.69E+002	mg/L	25	EST	MEYLAN,WM ET AL. (1996)
logP 2.13			EXP	HAGEN,V ET AL. (1989)
VP 2.01E-006	mm Hg	25	EST	NEELY,WB & BLAU,GE (1985)
DC	pKa			
HL 1.09E-013	atm m3/mol	25	EST	MEYLAN,WM & HOWARD,PH (1991)
OH 3.14E-012	cm3/molc sec	25	EST	MEYLAN,WM & HOWARD,PH (1993)

CAS #: 084952-60-3	PHENETHYLAMINE,N-METHYL-A-ETHYL
Formula: $C_{11}H_{17}N$	
Mol Weight: 163.26	
MP (deg C):	FP (deg C):
BP (deg C):	
BP pressure (mm Hg):	

Property/Value	Units	Temp	Data Type	Reference
WS 2.63E+003	mg/L	25	EST	MEYLAN,WM ET AL. (1996)
logP 2.82			EXP	HANSCH,C & LEO,AJ (1985)
VP 6.18E-002	mm Hg	25	EST	NEELY,WB & BLAU,GE (1985)
DC	pKa			
HL 3.14E-006	atm m3/mol	25	EST	MEYLAN,WM & HOWARD,PH (1991)
OH 8.38E-011	cm3/molc sec	25	EST	MEYLAN,WM & HOWARD,PH (1993)

CAS #: 084952-61-4	PHENETHYLAMINE,N-METHYL-A-ISOPROPYL
Formula: $C_{12}H_{19}N$	
Mol Weight: 177.29	
MP (deg C):	FP (deg C):
BP (deg C):	
BP pressure (mm Hg):	

Property/Value	Units	Temp	Data Type	Reference
WS 1.25E+003	mg/L	25	EST	MEYLAN,WM ET AL. (1996)
logP 3.12			EXP	HANSCH,C & LEO,AJ (1985)
VP 4.23E-002	mm Hg	25	EST	NEELY,WB & BLAU,GE (1985)
DC	pKa			
HL 4.17E-006	atm m3/mol	25	EST	MEYLAN,WM & HOWARD,PH (1991)
OH 8.52E-011	cm3/molc sec	25	EST	MEYLAN,WM & HOWARD,PH (1993)

CAS #: 084952-62-5	PHENETHYLAMINE,N-ME-A-ISOBUTYL
Formula: $C_{13}H_{21}N$	
Mol Weight: 191.32	
MP (deg C):	FP (deg C):
BP (deg C):	
BP pressure (mm Hg):	

Property/Value	Units	Temp	Data Type	Reference
WS 3.03E+002	mg/L	25	EST	MEYLAN,WM ET AL. (1996)
logP 3.76			EXP	HANSCH,C & LEO,AJ (1985)
VP 1.66E-002	mm Hg	25	EST	NEELY,WB & BLAU,GE (1985)
DC	pKa			
HL 5.54E-006	atm m3/mol	25	EST	MEYLAN,WM & HOWARD,PH (1991)
OH 8.66E-011	cm3/molc sec	25	EST	MEYLAN,WM & HOWARD,PH (1993)

CAS #: 085033-96-1	PHENYLTHIO-ACETIC ACID
Formula: C_8H_8OS	
Mol Weight: 152.22	
MP (deg C):	FP (deg C):
BP (deg C):	
BP pressure (mm Hg):	

Property/Value	Units	Temp	Data Type	Reference
WS 1.74E+003	mg/L	25	EST	MEYLAN,WM ET AL. (1996)
logP 1.91			EXP	HANSCH,C & LEO,AJ (1985)
VP 1.96E-002	mm Hg	25	EST	NEELY,WB & BLAU,GE (1985)
DC	pKa			
HL 1.68E-005	atm m3/mol	25	EST	MEYLAN,WM & HOWARD,PH (1991)
OH 1.74E-011	cm3/molc sec	25	EST	MEYLAN,WM & HOWARD,PH (1993)

CAS #:	085078-20-2		1,1-DI(P-MEO PHENYL)-2-NITROBUTANE

Formula: $C_{18}H_{21}NO_4$

Mol Weight: 315.37

MP (deg C): FP (deg C):

BP (deg C):

BP pressure (mm Hg):

Property/ Value	Units	Temp	Data Type	Reference
WS 6.44E-001	mg/L	25	EST	MEYLAN,WM ET AL. (1996)
logP 4.45			EXP	HANSCH,C & LEO,AJ (1985)
VP 3.18E-007	mm Hg	25	EST	NEELY,WB & BLAU,GE (1985)
DC	pKa			
HL 2.25E-009	atm m3/mol	25	EST	MEYLAN,WM & HOWARD,PH (1991)
OH 5.41E-011	cm3/molc sec	25	EST	MEYLAN,WM & HOWARD,PH (1993)

CAS #:	085078-27-9		1,1-DI(P-METHOXYPHENYL)-2-NITROETHANE

Formula: $C_{16}H_{17}NO_4$

Mol Weight: 287.32

MP (deg C): FP (deg C):

BP (deg C):

BP pressure (mm Hg):

Property/ Value	Units	Temp	Data Type	Reference
WS 5.30E+000	mg/L	25	EST	MEYLAN,WM ET AL. (1996)
logP 3.57			EXP	HANSCH,C & LEO,AJ (1985)
VP 8.36E-007	mm Hg	25	EST	NEELY,WB & BLAU,GE (1985)
DC	pKa			
HL 1.27E-009	atm m3/mol	25	EST	MEYLAN,WM & HOWARD,PH (1991)
OH 5.38E-011	cm3/molc sec	25	EST	MEYLAN,WM & HOWARD,PH (1993)

CAS #:	085093-33-0		2,4(1H,3H)-PYRIMIDINEDIONE, 1-[(4-CHLOROPHENYL)M

Formula: $C_{11}H_8ClFN_2O_2$

Mol Weight: 254.65

MP (deg C): FP (deg C):

BP (deg C):

BP pressure (mm Hg):

Property/ Value	Units	Temp	Data Type	Reference
WS 8.86E+002	mg/L	25	EST	MEYLAN,WM ET AL. (1996)
logP 1.64			EXP	SANGSTER,J (1994)
VP 1.09E-009	mm Hg	25	EST	NEELY,WB & BLAU,GE (1985)
DC	pKa			
HL 2.18E-011	atm m3/mol	25	EST	MEYLAN,WM & HOWARD,PH (1991)
OH 1.13E-011	cm3/molc sec	25	EST	MEYLAN,WM & HOWARD,PH (1993)

CAS #:	085093-35-2		2,4(1H,3H)-PYRIMIDINEDIONE, 5-FLUORO-1-[(4-METHO

Formula: $C_{12}H_{11}FN_2O_3$

Mol Weight: 250.23

MP (deg C): FP (deg C):

BP (deg C):

BP pressure (mm Hg):

Property/ Value	Units	Temp	Data Type	Reference
WS 3.05E+003	mg/L	25	EST	MEYLAN,WM ET AL. (1996)
logP 1.04			EXP	SANGSTER,J (1994)
VP 7.67E-010	mm Hg	25	EST	NEELY,WB & BLAU,GE (1985)
DC	pKa			
HL 1.74E-012	atm m3/mol	25	EST	MEYLAN,WM & HOWARD,PH (1991)
OH 3.67E-011	cm3/molc sec	25	EST	MEYLAN,WM & HOWARD,PH (1993)

CAS #:	085269-46-1		1,2,3-TRICHLOROOCTANE

Formula: $C_8H_{15}Cl_3$

Mol Weight: 217.57

MP (deg C): FP (deg C):

BP (deg C):

BP pressure (mm Hg):

Property/ Value	Units	Temp	Data Type	Reference
WS 2.39E+000	mg/L	25	EST	MEYLAN,WM ET AL. (1996)
logP 4.89			EST	MEYLAN,WM & HOWARD,PH (1995)
VP 5.01E-002	mm Hg	25	EST	NEELY,WB & BLAU,GE (1985)
DC	pKa			
HL 2.34E-002	atm m3/mol	25	EST	MEYLAN,WM & HOWARD,PH (1991)
OH 5.11E-012	cm3/molc sec	25	EST	MEYLAN,WM & HOWARD,PH (1993)

CAS #:	085326-06-3		GUANOSINE, 2',3'-DIDEOXY-

Formula: $C_{10}H_{13}N_5O_3$

Mol Weight: 251.25

MP (deg C): FP (deg C):

BP (deg C):

BP pressure (mm Hg):

Property/ Value	Units	Temp	Data Type	Reference
WS 8.31E+003	mg/L	25	EST	MEYLAN,WM ET AL. (1996)
logP -1.01			EXP	BALZARINI,J ET AL. (1989)
VP 7.71E-016	mm Hg	25	EST	NEELY,WB & BLAU,GE (1985)
DC	pKa			
HL 3.11E-023	atm m3/mol	25	EST	MEYLAN,WM & HOWARD,PH (1991)
OH 9.49E-011	cm3/molc sec	25	EST	MEYLAN,WM & HOWARD,PH (1993)

CAS #:	085432-35-5		5,5-DIET BARBITURIC ACID,N1-IPR

Formula: $C_{11}H_{18}N_2O_3$

Mol Weight: 226.28

MP (deg C): FP (deg C):

BP (deg C):

BP pressure (mm Hg):

Property/ Value	Units	Temp	Data Type	Reference
WS 3.08E+002	mg/L	25	EST	MEYLAN,WM ET AL. (1996)
logP 2.36			EXP	HANSCH,C & LEO,AJ (1985)
VP 2.24E-010	mm Hg	25	EST	NEELY,WB & BLAU,GE (1985)
DC	pKa			
HL 1.40E-012	atm m3/mol	25	EST	MEYLAN,WM & HOWARD,PH (1991)
OH 1.79E-011	cm3/molc sec	25	EST	MEYLAN,WM & HOWARD,PH (1993)

CAS #:	085432-36-6		BARBITURIC ACID,1-IPR-5-ET-5(PH)

Formula: $C_{15}H_{18}N_2O_3$

Mol Weight: 274.32

MP (deg C): FP (deg C):

BP (deg C):

BP pressure (mm Hg):

Property/ Value	Units	Temp	Data Type	Reference
WS 1.08E+002	mg/L	25	EST	MEYLAN,WM ET AL. (1996)
logP 2.58			EXP	HANSCH,C & LEO,AJ (1985)
VP 2.81E-012	mm Hg	25	EST	NEELY,WB & BLAU,GE (1985)
DC	pKa			
HL 6.39E-014	atm m3/mol	25	EST	MEYLAN,WM & HOWARD,PH (1991)
OH 1.84E-011	cm3/molc sec	25	EST	MEYLAN,WM & HOWARD,PH (1993)

1381

CAS #: 085653-81-2

4H-PYRIDO[1,2-A]PYRIMIDIN-4-ONE, 8-METHYL-

Formula: $C_9H_8N_2O$

Mol Weight: 160.18

MP (deg C): | FP (deg C):

BP (deg C):

BP pressure (mm Hg):

Property/Value	Units	Temp	Data Type	Reference
WS 2.27E+004	mg/L	25	EST	MEYLAN,WM ET AL. (1996)
logP 0.56			EXP	SANGSTER,J (1994)
VP 1.60E-004	mm Hg	25	EST	NEELY,WB & BLAU,GE (1985)
DC	pKa			
HL 2.35E-009	atm m3/mol	25	EST	MEYLAN,WM & HOWARD,PH (1991)
OH 1.27E-010	cm3/molc sec	25	EST	MEYLAN,WM & HOWARD,PH (1993)

CAS #: 085653-87-8

CYCLOPENTA[D]PYRIDO[1,2-A]PYRIMIDIN-10(1H)-ONE,

Formula: $C_{12}H_{16}N_2O$

Mol Weight: 204.27

MP (deg C): | FP (deg C):

BP (deg C):

BP pressure (mm Hg):

Property/Value	Units	Temp	Data Type	Reference
WS 3.93E+003	mg/L	25	EST	MEYLAN,WM ET AL. (1996)
logP 1.20			EXP	SANGSTER,J (1994)
VP 5.41E-006	mm Hg	25	EST	NEELY,WB & BLAU,GE (1985)
DC	pKa			
HL 5.10E-009	atm m3/mol	25	EST	MEYLAN,WM & HOWARD,PH (1991)
OH 2.22E-010	cm3/molc sec	25	EST	MEYLAN,WM & HOWARD,PH (1993)

CAS #: 085653-88-9

CYCLOPENTA[D]PYRIDO[1,2-A]PYRIMIDIN-10(1H)-ONE,

Formula: $C_{12}H_{16}N_2O$

Mol Weight: 204.27

MP (deg C): | FP (deg C):

BP (deg C):

BP pressure (mm Hg):

Property/Value	Units	Temp	Data Type	Reference
WS 3.78E+003	mg/L	25	EST	MEYLAN,WM ET AL. (1996)
logP 1.22			EXP	SANGSTER,J (1994)
VP 5.41E-006	mm Hg	25	EST	NEELY,WB & BLAU,GE (1985)
DC	pKa			
HL 5.10E-009	atm m3/mol	25	EST	MEYLAN,WM & HOWARD,PH (1991)
OH 2.22E-010	cm3/molc sec	25	EST	MEYLAN,WM & HOWARD,PH (1993)

CAS #: 085683-39-2

1,2,4-TRIAZIN-5(4H)-ONE, 6-(4-FLUOROPHENYL)-2,3-

Formula: $C_{10}H_8FN_3OS$

Mol Weight: 237.26

MP (deg C): | FP (deg C):

BP (deg C):

BP pressure (mm Hg):

Property/Value	Units	Temp	Data Type	Reference
WS 2.58E+002	mg/L	25	EST	MEYLAN,WM ET AL. (1996)
logP 2.38			EXP	SANGSTER,J (1994)
VP 6.07E-007	mm Hg	25	EST	NEELY,WB & BLAU,GE (1985)
DC	pKa			
HL 9.58E-011	atm m3/mol	25	EST	MEYLAN,WM & HOWARD,PH (1991)
OH 5.10E-011	cm3/molc sec	25	EST	MEYLAN,WM & HOWARD,PH (1993)

CAS #: 085721-33-1

CIPROFLOXACIN

Formula: $C_{17}H_{18}FN_3O_3$

Mol Weight: 331.35

MP (deg C): | FP (deg C):

BP (deg C):

BP pressure (mm Hg):

Property/Value	Units	Temp	Data Type	Reference
WS 1.15E+004	mg/L	25	EST	MEYLAN,WM ET AL. (1996)
logP 0.28			EXP	TAKACS-NOVAK,K ET AL. (1992)
VP 2.85E-013	mm Hg	25	EST	NEELY,WB & BLAU,GE (1985)
DC	pKa			
HL 5.09E-019	atm m3/mol	25	EST	MEYLAN,WM & HOWARD,PH (1991)
OH 2.79E-010	cm3/molc sec	25	EST	MEYLAN,WM & HOWARD,PH (1993)

CAS #: 085729-23-3

1H-INDOLE-3-CARBOXAMIDE, N-METHYL-

Formula: $C_{10}H_{10}N_2O$

Mol Weight: 174.20

MP (deg C): | FP (deg C):

BP (deg C):

BP pressure (mm Hg):

Property/Value	Units	Temp	Data Type	Reference
WS 5.03E+003	mg/L	25	EST	MEYLAN,WM ET AL. (1996)
logP 1.25			EXP	YAMAGAMI,C ET AL. (1994)
VP 8.66E-007	mm Hg	25	EST	NEELY,WB & BLAU,GE (1985)
DC	pKa			
HL 7.99E-013	atm m3/mol	25	EST	MEYLAN,WM & HOWARD,PH (1991)
OH 1.06E-010	cm3/molc sec	25	EST	MEYLAN,WM & HOWARD,PH (1993)

CAS #: 085785-20-2

ESPROCARB

Formula: $C_{15}H_{23}NOS$

Mol Weight: 265.42

MP (deg C): | FP (deg C):

BP (deg C):

BP pressure (mm Hg):

Property/Value	Units	Temp	Data Type	Reference
WS 2.29E+000	mg/L	25	EST	MEYLAN,WM ET AL. (1996)
logP 4.60			EXP	HANSCH,C ET AL. (1995)
VP 7.05E-006	mm Hg	25	EST	NEELY,WB & BLAU,GE (1985)
DC	pKa			
HL 1.24E-006	atm m3/mol	25	EST	MEYLAN,WM & HOWARD,PH (1991)
OH 3.72E-011	cm3/molc sec	25	EST	MEYLAN,WM & HOWARD,PH (1993)

CAS #: 085819-14-3

O-MEO BENZAMIDE,N-(3-DIMEAMINOPR)

Formula: $C_{13}H_{20}N_2O_2$

Mol Weight: 236.32

MP (deg C): | FP (deg C):

BP (deg C):

BP pressure (mm Hg):

Property/Value	Units	Temp	Data Type	Reference
WS 2.32E+003	mg/L	25	EST	MEYLAN,WM ET AL. (1996)
logP 1.27			EXP	HANSCH,C & LEO,AJ (1985)
VP 1.05E-006	mm Hg	25	EST	NEELY,WB & BLAU,GE (1985)
DC	pKa			
HL 2.51E-013	atm m3/mol	25	EST	MEYLAN,WM & HOWARD,PH (1991)
OH 1.11E-010	cm3/molc sec	25	EST	MEYLAN,WM & HOWARD,PH (1993)

CAS #: 085819-15-4				O-MEO BENZAMIDE,N-(3-DIETAMINOPR)
Formula: $C_{15}H_{24}N_2O_2$				
Mol Weight: 264.37				
MP (deg C):		FP (deg C):		
BP (deg C):				
BP pressure (mm Hg):				

Property/ Value	Units	Temp	Data Type	Reference
WS 6.17E+002	mg/L	25	EST	MEYLAN,WM ET AL. (1996)
logP 1.76			EXP	HANSCH,C & LEO,AJ (1985)
VP 1.93E-007	mm Hg	25	EST	NEELY,WB & BLAU,GE (1985)
DC	pKa			
HL 4.43E-013	atm m3/mol	25	EST	MEYLAN,WM & HOWARD,PH (1991)
OH 1.26E-010	cm3/molc sec	25	EST	MEYLAN,WM & HOWARD,PH (1993)

CAS #: 085856-32-2				BENZENEACETAMIDE, N-(3-AMINOPHENYL)-
Formula: $C_{14}H_{14}N_2O$				
Mol Weight: 226.28				
MP (deg C):		FP (deg C):		
BP (deg C):				
BP pressure (mm Hg):				

Property/ Value	Units	Temp	Data Type	Reference
WS 1.40E+003	mg/L	25	EST	MEYLAN,WM ET AL. (1996)
logP 1.59			EXP	YAMAGAMI,C ET AL. (1984)
VP 3.34E-008	mm Hg	25	EST	NEELY,WB & BLAU,GE (1985)
DC	pKa			
HL 1.76E-013	atm m3/mol	25	EST	MEYLAN,WM & HOWARD,PH (1991)
OH 2.01E-010	cm3/molc sec	25	EST	MEYLAN,WM & HOWARD,PH (1993)

CAS #: 085873-47-8				PROPANAMIDE, 2-METHYL-N-[2-NITRO-4-[(TRIFLUOROME
Formula: $C_{11}H_{11}F_3N_2O_5S$				
Mol Weight: 340.28				
MP (deg C):		FP (deg C):		
BP (deg C):				
BP pressure (mm Hg):				

Property/ Value	Units	Temp	Data Type	Reference
WS 6.51E+000	mg/L	25	EST	MEYLAN,WM ET AL. (1996)
logP 3.10			EXP	MONZANI,A ET AL. (1985)
VP 3.66E-009	mm Hg	25	EST	NEELY,WB & BLAU,GE (1985)
DC	pKa			
HL 8.27E-011	atm m3/mol	25	EST	MEYLAN,WM & HOWARD,PH (1991)
OH 2.52E-012	cm3/molc sec	25	EST	MEYLAN,WM & HOWARD,PH (1993)

CAS #: 085879-18-1				3,4-DICLBENZALDEHYD,O-((MEAM)CO)OXIME
Formula: $C_9H_8Cl_2N_2O_2$				
Mol Weight: 247.08				
MP (deg C):		FP (deg C):		
BP (deg C):				
BP pressure (mm Hg):				

Property/ Value	Units	Temp	Data Type	Reference
WS 8.36E+001	mg/L	25	EST	MEYLAN,WM ET AL. (1996)
logP 2.89			EXP	HANSCH,C & LEO,AJ (1985)
VP 1.44E-004	mm Hg	25	EST	NEELY,WB & BLAU,GE (1985)
DC	pKa			
HL 6.33E-009	atm m3/mol	25	EST	MEYLAN,WM & HOWARD,PH (1991)
OH 8.95E-012	cm3/molc sec	25	EST	MEYLAN,WM & HOWARD,PH (1993)

CAS #: 085879-19-2				3-PHO BENZALDEHYDE, O-((MEAM)CO)OXIME
Formula: $C_{15}H_{14}N_2O_3$				
Mol Weight: 270.29				
MP (deg C):		FP (deg C):		
BP (deg C):				
BP pressure (mm Hg):				

Property/ Value	Units	Temp	Data Type	Reference
WS 3.94E+001	mg/L	25	EST	MEYLAN,WM ET AL. (1996)
logP 3.12			EXP	HANSCH,C & LEO,AJ (1985)
VP 2.09E-006	mm Hg	25	EST	NEELY,WB & BLAU,GE (1985)
DC	pKa			
HL 2.52E-010	atm m3/mol	25	EST	MEYLAN,WM & HOWARD,PH (1991)
OH 2.46E-011	cm3/molc sec	25	EST	MEYLAN,WM & HOWARD,PH (1993)

CAS #: 085879-21-6				3-(METHYLTHIO)PHENYLUREA
Formula: $C_8H_{10}N_2OS$				
Mol Weight: 182.25				
MP (deg C):		FP (deg C):		
BP (deg C):				
BP pressure (mm Hg):				

Property/ Value	Units	Temp	Data Type	Reference
WS 2.45E+003	mg/L	25	EST	MEYLAN,WM ET AL. (1996)
logP 1.57			EXP	HANSCH,C & LEO,AJ (1985)
VP 2.32E-005	mm Hg	25	EST	NEELY,WB & BLAU,GE (1985)
DC	pKa			
HL 5.87E-012	atm m3/mol	25	EST	MEYLAN,WM & HOWARD,PH (1991)
OH 2.02E-010	cm3/molc sec	25	EST	MEYLAN,WM & HOWARD,PH (1993)

CAS #: 085879-22-7				4-(4-BROMOPHENOXY)PHENYLUREA
Formula: $C_{13}H_{11}BrN_2O_2$				
Mol Weight: 307.15				
MP (deg C):		FP (deg C):		
BP (deg C):				
BP pressure (mm Hg):				

Property/ Value	Units	Temp	Data Type	Reference
WS 7.72E+000	mg/L	25	EST	MEYLAN,WM ET AL. (1996)
logP 3.70			EXP	HANSCH,C & LEO,AJ (1985)
VP 5.35E-008	mm Hg	25	EST	NEELY,WB & BLAU,GE (1985)
DC	pKa			
HL 1.75E-012	atm m3/mol	25	EST	MEYLAN,WM & HOWARD,PH (1991)
OH 4.51E-011	cm3/molc sec	25	EST	MEYLAN,WM & HOWARD,PH (1993)

CAS #: 085911-77-9				UREA, N,N-DIMETHYL-N'-(4-PROPYLPHENYL)-
Formula: $C_{12}H_{18}N_2O$				
Mol Weight: 206.29				
MP (deg C):		FP (deg C):		
BP (deg C):				
BP pressure (mm Hg):				

Property/ Value	Units	Temp	Data Type	Reference
WS 2.70E+002	mg/L	25	EST	MEYLAN,WM ET AL. (1996)
logP 2.55			EXP	MITSUTAKE,KI ET AL. (1986)
VP 1.57E-005	mm Hg	25	EST	NEELY,WB & BLAU,GE (1985)
DC	pKa			
HL 1.89E-009	atm m3/mol	25	EST	MEYLAN,WM & HOWARD,PH (1991)
OH 5.66E-011	cm3/molc sec	25	EST	MEYLAN,WM & HOWARD,PH (1993)

085956-22-5 — PRAVASTATIN

Formula: $C_{23}H_{36}O_6$
Mol Weight: 408.54
MP (deg C):   FP (deg C):
BP (deg C):
BP pressure (mm Hg):

Property/Value	Units	Temp	Data Type	Reference
WS 2.39E+001	mg/L	25	EST	MEYLAN,WM ET AL. (1996)
logP 2.42			EXP	HANSCH,C ET AL. (1995)
VP 2.82E-014	mm Hg	25	EST	NEELY,WB & BLAU,GE (1985)
DC	pKa			
HL 5.84E-012	atm m3/mol	25	EST	MEYLAN,WM & HOWARD,PH (1991)
OH 2.37E-010	cm3/molc sec	25	EST	MEYLAN,WM & HOWARD,PH (1993)

086022-72-2 — BENZOBICYCLO(221)HEPTENE,2EN-NH2-7CF3

Formula: $C_{12}H_{12}F_3N$
Mol Weight: 227.23
MP (deg C):   FP (deg C):
BP (deg C):
BP pressure (mm Hg):

Property/Value	Units	Temp	Data Type	Reference
WS 1.18E+003	mg/L	25	EST	MEYLAN,WM ET AL. (1996)
logP 2.85			EXP	HANSCH,C & LEO,AJ (1985)
VP 6.14E-003	mm Hg	25	EST	NEELY,WB & BLAU,GE (1985)
DC	pKa			
HL 2.67E-006	atm m3/mol	25	EST	MEYLAN,WM & HOWARD,PH (1991)
OH 2.42E-011	cm3/molc sec	25	EST	MEYLAN,WM & HOWARD,PH (1993)

086072-07-3 — 2-CHLORO-1-METHYL-5-NITRO-1H-IMIDAZOLE

Formula: $C_4H_4ClN_3O_2$
Mol Weight: 161.55
MP (deg C):   FP (deg C):
BP (deg C):
BP pressure (mm Hg):

Property/Value	Units	Temp	Data Type	Reference
WS 8.11E+003	mg/L	25	EST	MEYLAN,WM ET AL. (1996)
logP 0.62			EXP	SUWINSKI,J ET AL. (1985)
VP 2.83E-004	mm Hg	25	EST	NEELY,WB & BLAU,GE (1985)
DC	pKa			
HL 2.34E-007	atm m3/mol	25	EST	MEYLAN,WM & HOWARD,PH (1991)
OH 1.05E-012	cm3/molc sec	25	EST	MEYLAN,WM & HOWARD,PH (1993)

086073-85-0 — 4-QUINOLINEMETHANOL, 2-(1,1-DIMETHYLETHYL)-à-[2-

Formula: $C_{21}H_{30}N_2O$
Mol Weight: 326.49
MP (deg C):   FP (deg C):
BP (deg C):
BP pressure (mm Hg):

Property/Value	Units	Temp	Data Type	Reference
WS 6.93E+001	mg/L	25	EST	MEYLAN,WM ET AL. (1996)
logP 3.63			EXP	MANNHOLD,R ET AL. (1990)
VP 5.71E-011	mm Hg	25	EST	NEELY,WB & BLAU,GE (1985)
DC	pKa			
HL 3.91E-014	atm m3/mol	25	EST	MEYLAN,WM & HOWARD,PH (1991)
OH 1.42E-010	cm3/molc sec	25	EST	MEYLAN,WM & HOWARD,PH (1993)

086230-36-6 — CARBOFURAN,N(N'-ME, PHCARBAMIDOTHIO)

Formula: $C_{20}H_{22}N_2O_4S$
Mol Weight: 386.47
MP (deg C):   FP (deg C):
BP (deg C):
BP pressure (mm Hg):

Property/Value	Units	Temp	Data Type	Reference
WS 3.81E+000	mg/L	25	EST	MEYLAN,WM ET AL. (1996)
logP 3.50			EXP	HANSCH,C & LEO,AJ (1985)
VP 2.68E-010	mm Hg	25	EST	NEELY,WB & BLAU,GE (1985)
DC	pKa			
HL 9.90E-008	atm m3/mol	25	EST	MEYLAN,WM & HOWARD,PH (1991)
OH 5.13E-011	cm3/molc sec	25	EST	MEYLAN,WM & HOWARD,PH (1993)

086230-37-7 — CARBOFURAN,N(N'-ME, 4-CL-PHCARBAMIDOTHIO)

Formula: $C_{20}H_{21}ClN_2O_4S$
Mol Weight: 420.92
MP (deg C):   FP (deg C):
BP (deg C):
BP pressure (mm Hg):

Property/Value	Units	Temp	Data Type	Reference
WS 8.71E-001	mg/L	25	EST	MEYLAN,WM ET AL. (1996)
logP 4.00			EXP	HANSCH,C & LEO,AJ (1985)
VP 7.07E-011	mm Hg	25	EST	NEELY,WB & BLAU,GE (1985)
DC	pKa			
HL 7.34E-008	atm m3/mol	25	EST	MEYLAN,WM & HOWARD,PH (1991)
OH 4.26E-011	cm3/molc sec	25	EST	MEYLAN,WM & HOWARD,PH (1993)

086230-38-8 — CARBOFURAN,N(N'-ME, 2-CL-PHCARBAMIDOTHIO)

Formula: $C_{20}H_{21}ClN_2O_4S$
Mol Weight: 420.92
MP (deg C):   FP (deg C):
BP (deg C):
BP pressure (mm Hg):

Property/Value	Units	Temp	Data Type	Reference
WS 7.15E-001	mg/L	25	EST	MEYLAN,WM ET AL. (1996)
logP 4.10			EXP	HANSCH,C & LEO,AJ (1985)
VP 7.07E-011	mm Hg	25	EST	NEELY,WB & BLAU,GE (1985)
DC	pKa			
HL 7.34E-008	atm m3/mol	25	EST	MEYLAN,WM & HOWARD,PH (1991)
OH 4.26E-011	cm3/molc sec	25	EST	MEYLAN,WM & HOWARD,PH (1993)

086315-52-8 — ISOMAZOLE

Formula: $C_{14}H_{13}N_3O_2S$
Mol Weight: 287.34
MP (deg C):   FP (deg C):
BP (deg C):
BP pressure (mm Hg):

Property/Value	Units	Temp	Data Type	Reference
WS 1.25E+003	mg/L	25	EST	MEYLAN,WM ET AL. (1996)
logP 1.25			EXP	HANSCH,C ET AL. (1995)
VP 9.87E-012	mm Hg	25	EST	NEELY,WB & BLAU,GE (1985)
DC	pKa			
HL 4.01E-018	atm m3/mol	25	EST	MEYLAN,WM & HOWARD,PH (1991)
OH 1.28E-010	cm3/molc sec	25	EST	MEYLAN,WM & HOWARD,PH (1993)

CAS #: 086393-37-5	AMIFLOXACIN

Formula: $C_{16}H_{19}FN_4O_3$

Mol Weight: 334.35

MP (deg C): 299-301 de

FP (deg C):

BP (deg C):

BP pressure (mm Hg):

Property/Value	Units	Temp	Data Type	Reference
WS 1.22E+004	mg/L	25	EST	MEYLAN,WM ET AL. (1996)
logP 0.23			EXP	HANSCH,C ET AL. (1995)
VP 2.11E-013	mm Hg	25	EST	NEELY,WB & BLAU,GE (1985)
DC	pKa			
HL 3.63E-021	atm m3/mol	25	EST	MEYLAN,WM & HOWARD,PH (1991)
OH 3.89E-010	cm3/molc sec	25	EST	MEYLAN,WM & HOWARD,PH (1993)

CAS #: 086476-06-4	BENZYLOXYCARBONYL MITOMYCIN C

Formula: $C_{23}H_{24}N_4O_7$

Mol Weight: 468.47

MP (deg C):

FP (deg C):

BP (deg C):

BP pressure (mm Hg):

Property/Value	Units	Temp	Data Type	Reference
WS 8.32E+002	mg/L	25	EST	MEYLAN,WM ET AL. (1996)
logP 1.34			EXP	HANSCH,C & LEO,AJ (1985)
VP 5.09E-013	mm Hg	25	EST	NEELY,WB & BLAU,GE (1985)
DC	pKa			
HL 1.67E-024	atm m3/mol	25	EST	MEYLAN,WM & HOWARD,PH (1991)
OH 7.28E-011	cm3/molc sec	25	EST	MEYLAN,WM & HOWARD,PH (1993)

CAS #: 086479-06-3	HEXAFLUMURON

Formula: $C_{16}H_8Cl_2F_6N_2O_3$

Mol Weight: 461.15

MP (deg C):

FP (deg C):

BP (deg C):

BP pressure (mm Hg):

Property/Value	Units	Temp	Data Type	Reference
WS 3.00E-002	mg/L		EXP	YON,D ET AL. (1992)
logP 5.68			EXP	TOMLIN,C (1994)
VP 1.64E-011	mm Hg	25	EST	NEELY,WB & BLAU,GE (1985)
DC	pKa			
HL 1.08E-011	atm m3/mol	25	EST	MEYLAN,WM & HOWARD,PH (1991)
OH 2.10E-011	cm3/molc sec	25	EST	MEYLAN,WM & HOWARD,PH (1993)

CAS #: 086516-51-0	ETHANONE,1-(4-MEOPH)-2-[(4MEPH)SULFONYL]

Formula: $C_{16}H_{16}O_4S$

Mol Weight: 304.37

MP (deg C):

FP (deg C):

BP (deg C):

BP pressure (mm Hg):

Property/Value	Units	Temp	Data Type	Reference
WS 9.00E+001	mg/L	25	EST	MEYLAN,WM ET AL. (1996)
logP 2.47			EXP	WANG,L ET AL. (1993)
VP 3.77E-008	mm Hg	25	EST	NEELY,WB & BLAU,GE (1985)
DC	pKa			
HL 1.08E-011	atm m3/mol	25	EST	MEYLAN,WM & HOWARD,PH (1991)
OH 2.71E-011	cm3/molc sec	25	EST	MEYLAN,WM & HOWARD,PH (1993)

CAS #: 086552-09-2	BENZENAMINE, 4-[(4-AMINOPHENYL)SULFONYL]-N,N-DIM

Formula: $C_{14}H_{16}N_2O_2S$

Mol Weight: 276.36

MP (deg C):

FP (deg C):

BP (deg C):

BP pressure (mm Hg):

Property/Value	Units	Temp	Data Type	Reference
WS 3.36E+002	mg/L	25	EST	MEYLAN,WM ET AL. (1996)
logP 1.99			EXP	ALTOMARE,C ET AL. (1991)
VP 4.47E-008	mm Hg	25	EST	NEELY,WB & BLAU,GE (1985)
DC	pKa			
HL 1.40E-012	atm m3/mol	25	EST	MEYLAN,WM & HOWARD,PH (1991)
OH 1.05E-010	cm3/molc sec	25	EST	MEYLAN,WM & HOWARD,PH (1993)

CAS #: 086569-89-3	N(N"-ME-(DECOXY)CARBAMYLTHIO)METHOMYL

Formula: $C_{17}H_{33}N_3O_4S_2$

Mol Weight: 407.60

MP (deg C):

FP (deg C):

BP (deg C):

BP pressure (mm Hg):

Property/Value	Units	Temp	Data Type	Reference
WS 1.39E-002	mg/L	25	EST	MEYLAN,WM ET AL. (1996)
logP 6.20			EXP	HANSCH,C & LEO,AJ (1985)
VP 2.30E-007	mm Hg	25	EST	NEELY,WB & BLAU,GE (1985)
DC	pKa			
HL 5.38E-008	atm m3/mol	25	EST	MEYLAN,WM & HOWARD,PH (1991)
OH 4.71E-011	cm3/molc sec	25	EST	MEYLAN,WM & HOWARD,PH (1993)

CAS #: 086569-90-6	N(N"-ME(CYHEXO)CARBAMYLTHIO)METHOMYL

Formula: $C_{13}H_{23}N_3O_4S_2$

Mol Weight: 349.47

MP (deg C):

FP (deg C):

BP (deg C):

BP pressure (mm Hg):

Property/Value	Units	Temp	Data Type	Reference
WS 5.26E+000	mg/L	25	EST	MEYLAN,WM ET AL. (1996)
logP 3.60			EXP	HANSCH,C & LEO,AJ (1985)
VP 5.30E-006	mm Hg	25	EST	NEELY,WB & BLAU,GE (1985)
DC	pKa			
HL 7.64E-009	atm m3/mol	25	EST	MEYLAN,WM & HOWARD,PH (1991)
OH 4.58E-011	cm3/molc sec	25	EST	MEYLAN,WM & HOWARD,PH (1993)

CAS #: 086569-91-7	ETHANIMIDOTHIOIC ACID, N-[[[[(ACETYLMETHYLAMINO)

Formula: $C_8H_{15}N_3O_3S_2$

Mol Weight: 265.36

MP (deg C):

FP (deg C):

BP (deg C):

BP pressure (mm Hg):

Property/Value	Units	Temp	Data Type	Reference
WS 1.51E+003	mg/L	25	EST	MEYLAN,WM ET AL. (1996)
logP 1.30			EXP	SANGSTER,J (1993)
VP 5.87E-006	mm Hg	25	EST	NEELY,WB & BLAU,GE (1985)
DC	pKa			
HL 1.19E-009	atm m3/mol	25	EST	MEYLAN,WM & HOWARD,PH (1991)
OH 3.41E-011	cm3/molc sec	25	EST	MEYLAN,WM & HOWARD,PH (1993)

086569-92-8 — N-(N''-ME-ETHYLCARBAMYLTHIO)METHOMYL

Formula: $C_9H_{17}N_3O_3S_2$

Mol Weight: 279.38

MP (deg C):   FP (deg C):

BP (deg C):

BP pressure (mm Hg):

Property/Value	Units	Temp	Data Type	Reference
WS 6.95E+002	mg/L	25	EST	MEYLAN,WM ET AL. (1996)
logP 1.60			EXP	HANSCH,C & LEO,AJ (1985)
VP 2.52E-006	mm Hg	25	EST	NEELY,WB & BLAU,GE (1985)
DC	pKa			
HL 1.58E-009	atm m3/mol	25	EST	MEYLAN,WM & HOWARD,PH (1991)
OH 3.53E-011	cm3/molc sec	25	EST	MEYLAN,WM & HOWARD,PH (1993)

086569-93-9 — N(N''-ME-PR CARBAMYLTHIO)METHOMYL

Formula: $C_{10}H_{19}N_3O_3S_2$

Mol Weight: 293.41

MP (deg C):   FP (deg C):

BP (deg C):

BP pressure (mm Hg):

Property/Value	Units	Temp	Data Type	Reference
WS 1.77E+002	mg/L	25	EST	MEYLAN,WM ET AL. (1996)
logP 2.20			EXP	HANSCH,C & LEO,AJ (1985)
VP 1.07E-006	mm Hg	25	EST	NEELY,WB & BLAU,GE (1985)
DC	pKa			
HL 2.09E-009	atm m3/mol	25	EST	MEYLAN,WM & HOWARD,PH (1991)
OH 3.87E-011	cm3/molc sec	25	EST	MEYLAN,WM & HOWARD,PH (1993)

086569-94-0 — N(N''-ME-IPR CARBAMYLTHIO)METHOMYL

Formula: $C_{10}H_{19}N_3O_3S_2$

Mol Weight: 293.41

MP (deg C):   FP (deg C):

BP (deg C):

BP pressure (mm Hg):

Property/Value	Units	Temp	Data Type	Reference
WS 1.77E+002	mg/L	25	EST	MEYLAN,WM ET AL. (1996)
logP 2.20			EXP	HANSCH,C & LEO,AJ (1985)
VP 1.92E-006	mm Hg	25	EST	NEELY,WB & BLAU,GE (1985)
DC	pKa			
HL 2.09E-009	atm m3/mol	25	EST	MEYLAN,WM & HOWARD,PH (1991)
OH 3.66E-011	cm3/molc sec	25	EST	MEYLAN,WM & HOWARD,PH (1993)

086569-95-1 — N(N''-BU-ME CARBAMYLTHIO)METHOMYL

Formula: $C_{11}H_{21}N_3O_3S_2$

Mol Weight: 307.44

MP (deg C):   FP (deg C):

BP (deg C):

BP pressure (mm Hg):

Property/Value	Units	Temp	Data Type	Reference
WS 9.91E+001	mg/L	25	EST	MEYLAN,WM ET AL. (1996)
logP 2.40			EXP	HANSCH,C & LEO,AJ (1985)
VP 4.55E-007	mm Hg	25	EST	NEELY,WB & BLAU,GE (1985)
DC	pKa			
HL 2.78E-009	atm m3/mol	25	EST	MEYLAN,WM & HOWARD,PH (1991)
OH 4.10E-011	cm3/molc sec	25	EST	MEYLAN,WM & HOWARD,PH (1993)

086569-98-4 — N-THIO-(N'ME-O-TBU-CARBAMYL)CARBOFURAN

Formula: $C_{18}H_{26}N_2O_5S$

Mol Weight: 382.48

MP (deg C):   FP (deg C):

BP (deg C):

BP pressure (mm Hg):

Property/Value	Units	Temp	Data Type	Reference
WS 8.36E-001	mg/L	25	EST	MEYLAN,WM ET AL. (1996)
logP 4.30			EXP	HANSCH,C & LEO,AJ (1985)
VP 1.73E-007	mm Hg	25	EST	NEELY,WB & BLAU,GE (1985)
DC	pKa			
HL 7.92E-009	atm m3/mol	25	EST	MEYLAN,WM & HOWARD,PH (1991)
OH 5.44E-011	cm3/molc sec	25	EST	MEYLAN,WM & HOWARD,PH (1993)

086570-00-5 — CARBOFURAN,N-(N'ME,ET-CARBAMIDOTHIO)

Formula: $C_{16}H_{22}N_2O_4S$

Mol Weight: 338.43

MP (deg C):   FP (deg C):

BP (deg C):

BP pressure (mm Hg):

Property/Value	Units	Temp	Data Type	Reference
WS 1.35E+001	mg/L	25	EST	MEYLAN,WM ET AL. (1996)
logP 3.20			EXP	HANSCH,C & LEO,AJ (1985)
VP 2.04E-008	mm Hg	25	EST	NEELY,WB & BLAU,GE (1985)
DC	pKa			
HL 1.27E-009	atm m3/mol	25	EST	MEYLAN,WM & HOWARD,PH (1991)
OH 5.52E-011	cm3/molc sec	25	EST	MEYLAN,WM & HOWARD,PH (1993)

086570-01-6 — CARBOFURAN,N-(N'-ME,PR CARBAMIDOTHIO)

Formula: $C_{17}H_{24}N_2O_4S$

Mol Weight: 352.46

MP (deg C):   FP (deg C):

BP (deg C):

BP pressure (mm Hg):

Property/Value	Units	Temp	Data Type	Reference
WS 3.40E+000	mg/L	25	EST	MEYLAN,WM ET AL. (1996)
logP 3.80			EXP	HANSCH,C & LEO,AJ (1985)
VP 8.64E-009	mm Hg	25	EST	NEELY,WB & BLAU,GE (1985)
DC	pKa			
HL 1.69E-009	atm m3/mol	25	EST	MEYLAN,WM & HOWARD,PH (1991)
OH 5.86E-011	cm3/molc sec	25	EST	MEYLAN,WM & HOWARD,PH (1993)

086570-02-7 — CARBOFURAN,N-(N'-ME,ISOPR CARBAMIDOTHIO)

Formula: $C_{17}H_{24}N_2O_4S$

Mol Weight: 352.46

MP (deg C):   FP (deg C):

BP (deg C):

BP pressure (mm Hg):

Property/Value	Units	Temp	Data Type	Reference
WS 5.05E+000	mg/L	25	EST	MEYLAN,WM ET AL. (1996)
logP 3.60			EXP	HANSCH,C & LEO,AJ (1985)
VP 1.51E-008	mm Hg	25	EST	NEELY,WB & BLAU,GE (1985)
DC	pKa			
HL 1.69E-009	atm m3/mol	25	EST	MEYLAN,WM & HOWARD,PH (1991)
OH 5.64E-011	cm3/molc sec	25	EST	MEYLAN,WM & HOWARD,PH (1993)

086570-05-0 — CARBOFURAN,N(N'-ME,24CL-PHCARBAMIDOTHIO)

Formula: $C_{20}H_{20}Cl_2N_2O_4S$
Mol Weight: 455.36
MP (deg C):
FP (deg C):
BP (deg C):
BP pressure (mm Hg):

Property/Value	Units	Temp	Data Type	Reference
WS 9.01E-002	mg/L	25	EST	MEYLAN,WM ET AL. (1996)
logP 4.90			EXP	HANSCH,C & LEO,AJ (1985)
VP 1.85E-011	mm Hg	25	EST	NEELY,WB & BLAU,GE (1985)
DC	pKa			
HL 5.44E-008	atm m3/mol	25	EST	MEYLAN,WM & HOWARD,PH (1991)
OH 4.00E-011	cm3/molc sec	25	EST	MEYLAN,WM & HOWARD,PH (1993)

086678-86-6 — 24-CHOLANAMIDE, 3,7,12-TRIHYDROXY-N-(PHENYLMETHY

Formula: $C_{31}H_{47}NO_4$
Mol Weight: 497.72
MP (deg C):
FP (deg C):
BP (deg C):
BP pressure (mm Hg):

Property/Value	Units	Temp	Data Type	Reference
WS 7.02E-001	mg/L	25	EST	MEYLAN,WM ET AL. (1996)
logP 3.54			EXP	SANGSTER,J (1994)
VP 5.52E-020	mm Hg	25	EST	NEELY,WB & BLAU,GE (1985)
DC	pKa			
HL 1.87E-015	atm m3/mol	25	EST	MEYLAN,WM & HOWARD,PH (1991)
OH 7.04E-011	cm3/molc sec	25	EST	MEYLAN,WM & HOWARD,PH (1993)

086811-58-7 — FLUAZURON

Formula: $C_{20}H_{10}Cl_2F_5N_3O_3$
Mol Weight: 506.22
MP (deg C):
FP (deg C):
BP (deg C):
BP pressure (mm Hg):

Property/Value	Units	Temp	Data Type	Reference
WS 2.88E-002	mg/L	25	EST	MEYLAN,WM ET AL. (1996)
logP 5.10			EXP	TOMLIN,C (1994)
VP 3.61E-014	mm Hg	25	EST	NEELY,WB & BLAU,GE (1985)
DC	pKa			
HL 2.00E-014	atm m3/mol	25	EST	MEYLAN,WM & HOWARD,PH (1991)
OH 3.43E-011	cm3/molc sec	25	EST	MEYLAN,WM & HOWARD,PH (1993)

086811-82-7 — 1-(4-BROMOPHENOXY)SILATRANE

Formula: $C_{12}H_{16}BrNO_4Si$
Mol Weight: 346.26
MP (deg C):
FP (deg C):
BP (deg C):
BP pressure (mm Hg):

Property/Value	Units	Temp	Data Type	Reference
WS 2.13E+004	mg/L	25	EST	MEYLAN,WM ET AL. (1996)
logP 0.58			EXP	LUKASIAK,J (1984)
VP 3.34E-006	mm Hg	25	EST	NEELY,WB & BLAU,GE (1985)
DC	pKa			
HL 5.73E-014	atm m3/mol	25	EST	MEYLAN,WM & HOWARD,PH (1991)
OH 1.24E-010	cm3/molc sec	25	EST	MEYLAN,WM & HOWARD,PH (1993)

086825-39-0 — 1-(3-CHLOROPHENOXY)SILATRANE

Formula: $C_{12}H_{16}ClNO_4Si$
Mol Weight: 301.80
MP (deg C):
FP (deg C):
BP (deg C):
BP pressure (mm Hg):

Property/Value	Units	Temp	Data Type	Reference
WS 4.94E+004	mg/L	25	EST	MEYLAN,WM ET AL. (1996)
logP 0.46			EXP	LUKASIAK,J (1984)
VP 8.67E-006	mm Hg	25	EST	NEELY,WB & BLAU,GE (1985)
DC	pKa			
HL 1.07E-013	atm m3/mol	25	EST	MEYLAN,WM & HOWARD,PH (1991)
OH 1.33E-010	cm3/molc sec	25	EST	MEYLAN,WM & HOWARD,PH (1993)

086880-51-5 — EPANOLOL

Formula: $C_{20}H_{23}N_3O_4$
Mol Weight: 369.42
MP (deg C): 118-120
FP (deg C):
BP (deg C):
BP pressure (mm Hg):

Property/Value	Units	Temp	Data Type	Reference
WS 1.60E+003	mg/L	25	EST	MEYLAN,WM ET AL. (1996)
logP 0.92			EXP	SANGSTER,J (1994)
VP 5.60E-016	mm Hg	25	EST	NEELY,WB & BLAU,GE (1985)
DC	pKa			
HL 1.67E-025	atm m3/mol	25	EST	MEYLAN,WM & HOWARD,PH (1991)
OH 1.57E-010	cm3/molc sec	25	EST	MEYLAN,WM & HOWARD,PH (1993)

086886-77-3 — BENZENEPROPANAMIDE, N-(3,4-DICHLOROPHENYL)-

Formula: $C_{15}H_{13}Cl_2NO$
Mol Weight: 294.18
MP (deg C):
FP (deg C):
BP (deg C):
BP pressure (mm Hg):

Property/Value	Units	Temp	Data Type	Reference
WS 9.57E-001	mg/L	25	EST	MEYLAN,WM ET AL. (1996)
logP 4.85			EXP	HANSCH,C ET AL. (1995)
VP 2.14E-008	mm Hg	25	EST	NEELY,WB & BLAU,GE (1985)
DC	pKa			
HL 3.63E-010	atm m3/mol	25	EST	MEYLAN,WM & HOWARD,PH (1991)
OH 1.19E-011	cm3/molc sec	25	EST	MEYLAN,WM & HOWARD,PH (1993)

086943-77-3 — BENZOBICYCLO(221)HEPTENE,2EX-PRAMINO

Formula: $C_{14}H_{19}N$
Mol Weight: 201.31
MP (deg C):
FP (deg C):
BP (deg C):
BP pressure (mm Hg):

Property/Value	Units	Temp	Data Type	Reference
WS 9.32E+002	mg/L	25	EST	MEYLAN,WM ET AL. (1996)
logP 3.30			EXP	HANSCH,C & LEO,AJ (1985)
VP 9.65E-004	mm Hg	25	EST	NEELY,WB & BLAU,GE (1985)
DC	pKa			
HL 1.19E-006	atm m3/mol	25	EST	MEYLAN,WM & HOWARD,PH (1991)
OH 1.04E-010	cm3/molc sec	25	EST	MEYLAN,WM & HOWARD,PH (1993)

086943-78-4 — BENZOBICYCLO(221)HEPTENE,2EX-ETAMINO

Formula: $C_{13}H_{17}N$
Mol Weight: 187.29
MP (deg C): **FP (deg C):**
BP (deg C):
BP pressure (mm Hg):

Property/Value	Units	Temp	Data Type	Reference
WS 2.99E+003	mg/L	25	EST	MEYLAN,WM ET AL. (1996)
logP 2.72			EXP	HANSCH,C & LEO,AJ (1985)
VP 2.73E-003	mm Hg	25	EST	NEELY,WB & BLAU,GE (1985)
DC	pKa			
HL 8.95E-007	atm m3/mol	25	EST	MEYLAN,WM & HOWARD,PH (1991)
OH 1.00E-010	cm3/molc sec	25	EST	MEYLAN,WM & HOWARD,PH (1993)

086943-79-5 — BENZOBICYCLO(221)HEPTENE,2EX-MEAMINO

Formula: $C_{12}H_{15}N$
Mol Weight: 173.26
MP (deg C): **FP (deg C):**
BP (deg C):
BP pressure (mm Hg):

Property/Value	Units	Temp	Data Type	Reference
WS 4.70E+003	mg/L	25	EST	MEYLAN,WM ET AL. (1996)
logP 2.47			EXP	HANSCH,C & LEO,AJ (1985)
VP 6.95E-003	mm Hg	25	EST	NEELY,WB & BLAU,GE (1985)
DC	pKa			
HL 6.74E-007	atm m3/mol	25	EST	MEYLAN,WM & HOWARD,PH (1991)
OH 9.29E-011	cm3/molc sec	25	EST	MEYLAN,WM & HOWARD,PH (1993)

086943-80-8 — 9-OXABENZOBICYCLO(221)HEPTEN,2EX-MEAM

Formula: $C_{11}H_{13}NO$
Mol Weight: 175.23
MP (deg C): **FP (deg C):**
BP (deg C):
BP pressure (mm Hg):

Property/Value	Units	Temp	Data Type	Reference
WS 9.13E+004	mg/L	25	EST	MEYLAN,WM ET AL. (1996)
logP 0.91			EXP	HANSCH,C & LEO,AJ (1985)
VP 5.46E-003	mm Hg	25	EST	NEELY,WB & BLAU,GE (1985)
DC	pKa			
HL 5.64E-010	atm m3/mol	25	EST	MEYLAN,WM & HOWARD,PH (1991)
OH 1.07E-010	cm3/molc sec	25	EST	MEYLAN,WM & HOWARD,PH (1993)

086988-90-1 — FURAZAN, CHLOROMETHYL-, 5-OXIDE

Formula: $C_3H_3ClN_2O_2$
Mol Weight: 134.52
MP (deg C): **FP (deg C):**
BP (deg C):
BP pressure (mm Hg):

Property/Value	Units	Temp	Data Type	Reference
WS 1.19E+004	mg/L	25	EST	MEYLAN,WM ET AL. (1996)
logP 1.02			EXP	CALVINO,R ET AL. (1992)
VP 2.23E-002	mm Hg	25	EST	NEELY,WB & BLAU,GE (1985)
DC	pKa			
HL 1.55E-009	atm m3/mol	25	EST	MEYLAN,WM & HOWARD,PH (1991)
OH 4.14E-012	cm3/molc sec	25	EST	MEYLAN,WM & HOWARD,PH (1993)

086992-67-8 — BENZOBICYCLO(221)HEPTENE,2EN-PRAMINO

Formula: $C_{14}H_{19}N$
Mol Weight: 201.31
MP (deg C): **FP (deg C):**
BP (deg C):
BP pressure (mm Hg):

Property/Value	Units	Temp	Data Type	Reference
WS 9.32E+002	mg/L	25	EST	MEYLAN,WM ET AL. (1996)
logP 3.13			EXP	HANSCH,C & LEO,AJ (1985)
VP 9.65E-004	mm Hg	25	EST	NEELY,WB & BLAU,GE (1985)
DC	pKa			
HL 1.19E-006	atm m3/mol	25	EST	MEYLAN,WM & HOWARD,PH (1991)
OH 1.04E-010	cm3/molc sec	25	EST	MEYLAN,WM & HOWARD,PH (1993)

086992-68-9 — BENZOBICYCLO(221)HEPTENE,2EN-ETAMINO

Formula: $C_{13}H_{17}N$
Mol Weight: 187.29
MP (deg C): **FP (deg C):**
BP (deg C):
BP pressure (mm Hg):

Property/Value	Units	Temp	Data Type	Reference
WS 2.99E+003	mg/L	25	EST	MEYLAN,WM ET AL. (1996)
logP 2.62			EXP	HANSCH,C & LEO,AJ (1985)
VP 2.73E-003	mm Hg	25	EST	NEELY,WB & BLAU,GE (1985)
DC	pKa			
HL 8.95E-007	atm m3/mol	25	EST	MEYLAN,WM & HOWARD,PH (1991)
OH 1.00E-010	cm3/molc sec	25	EST	MEYLAN,WM & HOWARD,PH (1993)

086992-69-0 — BENZOBICYCLO(221)HEPTENE,9EN-MEAMINO

Formula: $C_{12}H_{15}N$
Mol Weight: 173.26
MP (deg C): **FP (deg C):**
BP (deg C):
BP pressure (mm Hg):

Property/Value	Units	Temp	Data Type	Reference
WS 5.72E+003	mg/L	25	EST	MEYLAN,WM ET AL. (1996)
logP 2.37			EXP	HANSCH,C & LEO,AJ (1985)
VP 6.95E-003	mm Hg	25	EST	NEELY,WB & BLAU,GE (1985)
DC	pKa			
HL 6.74E-007	atm m3/mol	25	EST	MEYLAN,WM & HOWARD,PH (1991)
OH 9.29E-011	cm3/molc sec	25	EST	MEYLAN,WM & HOWARD,PH (1993)

086992-70-3 — 9-OXABENZOBICYCLO(221)HEPTEN,2EN-MEAM

Formula: $C_{11}H_{13}NO$
Mol Weight: 175.23
MP (deg C): **FP (deg C):**
BP (deg C):
BP pressure (mm Hg):

Property/Value	Units	Temp	Data Type	Reference
WS 9.13E+004	mg/L	25	EST	MEYLAN,WM ET AL. (1996)
logP 0.95			EXP	HANSCH,C & LEO,AJ (1985)
VP 5.46E-003	mm Hg	25	EST	NEELY,WB & BLAU,GE (1985)
DC	pKa			
HL 5.64E-010	atm m3/mol	25	EST	MEYLAN,WM & HOWARD,PH (1991)
OH 1.07E-010	cm3/molc sec	25	EST	MEYLAN,WM & HOWARD,PH (1993)

CAS #: 087072-64-8				1-ME-1-ACETYL-3(P-ETPH)TRIAZENE

Formula: $C_{11}H_{15}N_3O$

Mol Weight: 205.26

MP (deg C):   **FP (deg C):**

BP (deg C):

BP pressure (mm Hg):

Property/Value	Units	Temp	Data Type	Reference
WS 5.55E+001	mg/L	25	EST	MEYLAN,WM ET AL. (1996)
logP 3.36			EXP	HANSCH,C & LEO,AJ (1985)
VP 1.04E-004	mm Hg	25	EST	NEELY,WB & BLAU,GE (1985)
DC	pKa			
HL 1.16E-009	atm m3/mol	25	EST	MEYLAN,WM & HOWARD,PH (1991)
OH 4.68E-012	cm3/molc sec	25	EST	MEYLAN,WM & HOWARD,PH (1993)

CAS #: 087072-65-9				1-ME-1-ACETYL-3-(P-F PH)TRIAZENE

Formula: $C_9H_{10}FN_3O$

Mol Weight: 195.20

MP (deg C):   **FP (deg C):**

BP (deg C):

BP pressure (mm Hg):

Property/Value	Units	Temp	Data Type	Reference
WS 2.00E+002	mg/L	25	EST	MEYLAN,WM ET AL. (1996)
logP 2.77			EXP	HANSCH,C & LEO,AJ (1985)
VP 8.52E-004	mm Hg	25	EST	NEELY,WB & BLAU,GE (1985)
DC	pKa			
HL 9.25E-010	atm m3/mol	25	EST	MEYLAN,WM & HOWARD,PH (1991)
OH 2.63E-012	cm3/molc sec	25	EST	MEYLAN,WM & HOWARD,PH (1993)

CAS #: 087130-18-5				CARBAMIC ACID, (3,4-DIETHOXYPHENYL)-, METHYL EST

Formula: $C_{12}H_{17}NO_4$

Mol Weight: 239.27

MP (deg C):   **FP (deg C):**

BP (deg C):

BP pressure (mm Hg):

Property/Value	Units	Temp	Data Type	Reference
WS 4.36E+002	mg/L	25	EST	MEYLAN,WM ET AL. (1996)
logP 2.10			EXP	TAKAHASHI,J ET AL. (1988)
VP 8.71E-005	mm Hg	25	EST	NEELY,WB & BLAU,GE (1985)
DC	pKa			
HL 1.35E-010	atm m3/mol	25	EST	MEYLAN,WM & HOWARD,PH (1991)
OH 2.13E-010	cm3/molc sec	25	EST	MEYLAN,WM & HOWARD,PH (1993)

CAS #: 087130-19-6				CARBAMIC ACID, (3,4-DIETHOXYPHENYL)-, ETHYL ESTE

Formula: $C_{13}H_{19}NO_4$

Mol Weight: 253.30

MP (deg C):   **FP (deg C):**

BP (deg C):

BP pressure (mm Hg):

Property/Value	Units	Temp	Data Type	Reference
WS 1.54E+002	mg/L	25	EST	MEYLAN,WM ET AL. (1996)
logP 2.54			EXP	TAKAHASHI,J ET AL. (1988)
VP 3.60E-005	mm Hg	25	EST	NEELY,WB & BLAU,GE (1985)
DC	pKa			
HL 1.79E-010	atm m3/mol	25	EST	MEYLAN,WM & HOWARD,PH (1991)
OH 2.14E-010	cm3/molc sec	25	EST	MEYLAN,WM & HOWARD,PH (1993)

CAS #: 087130-20-9				CARBAMIC ACID, (3,4-DIETHOXYPHENYL)-, 1-METHYLET

Formula: $C_{14}H_{21}NO_4$

Mol Weight: 267.33

MP (deg C):   **FP (deg C):**

BP (deg C):

BP pressure (mm Hg):

Property/Value	Units	Temp	Data Type	Reference
WS 6.19E+001	mg/L	25	EST	MEYLAN,WM ET AL. (1996)
logP 2.91			EXP	TAKAHASHI,J ET AL. (1988)
VP 2.75E-005	mm Hg	25	EST	NEELY,WB & BLAU,GE (1985)
DC	pKa			
HL 2.37E-010	atm m3/mol	25	EST	MEYLAN,WM & HOWARD,PH (1991)
OH 2.16E-010	cm3/molc sec	25	EST	MEYLAN,WM & HOWARD,PH (1993)

CAS #: 087130-27-6				CARBAMIC ACID, (3,4-DIETHOXYPHENYL)-, 2-PROPENYL

Formula: $C_{14}H_{19}NO_4$

Mol Weight: 265.31

MP (deg C):   **FP (deg C):**

BP (deg C):

BP pressure (mm Hg):

Property/Value	Units	Temp	Data Type	Reference
WS 8.71E+001	mg/L	25	EST	MEYLAN,WM ET AL. (1996)
logP 2.75			EXP	TAKAHASHI,J ET AL. (1988)
VP 1.66E-005	mm Hg	25	EST	NEELY,WB & BLAU,GE (1985)
DC	pKa			
HL 1.77E-010	atm m3/mol	25	EST	MEYLAN,WM & HOWARD,PH (1991)
OH 2.40E-010	cm3/molc sec	25	EST	MEYLAN,WM & HOWARD,PH (1993)

CAS #: 087130-34-5				CARBAMIC ACID, (3,4-DIETHOXYPHENYL)-, 2-PROPYNYL

Formula: $C_{14}H_{17}NO_4$

Mol Weight: 263.30

MP (deg C):   **FP (deg C):**

BP (deg C):

BP pressure (mm Hg):

Property/Value	Units	Temp	Data Type	Reference
WS 1.03E+002	mg/L	25	EST	MEYLAN,WM ET AL. (1996)
logP 2.68			EXP	TAKAHASHI,J ET AL. (1988)
VP 1.03E-005	mm Hg	25	EST	NEELY,WB & BLAU,GE (1985)
DC	pKa			
HL 2.94E-011	atm m3/mol	25	EST	MEYLAN,WM & HOWARD,PH (1991)
OH 2.21E-010	cm3/molc sec	25	EST	MEYLAN,WM & HOWARD,PH (1993)

CAS #: 087130-47-0				CARBAMIC ACID, (3,4-DIETHOXYPHENYL)-, 2-BROMO-1-

Formula: $C_{14}H_{19}Br_2NO_4$

Mol Weight: 425.13

MP (deg C):   **FP (deg C):**

BP (deg C):

BP pressure (mm Hg):

Property/Value	Units	Temp	Data Type	Reference
WS 2.19E+000	mg/L	25	EST	MEYLAN,WM ET AL. (1996)
logP 3.50			EXP	TAKAHASHI,J ET AL. (1988)
VP 1.17E-007	mm Hg	25	EST	NEELY,WB & BLAU,GE (1985)
DC	pKa			
HL 3.15E-012	atm m3/mol	25	EST	MEYLAN,WM & HOWARD,PH (1991)
OH 2.14E-010	cm3/molc sec	25	EST	MEYLAN,WM & HOWARD,PH (1993)

087130-55-0

CARBAMIC ACID, (3,4-DIETHOXYPHENYL)-, 2-CHLORO-1

Formula: $C_{16}H_{24}ClNO_5$

Mol Weight: 345.83

MP (deg C): | FP (deg C):

BP (deg C):

BP pressure (mm Hg):

Property/Value	Units	Temp	Data Type	Reference
WS 2.56E+001	mg/L	25	EST	MEYLAN,WM ET AL. (1996)
logP 2.82			EXP	TAKAHASHI,J ET AL. (1988)
VP 2.40E-007	mm Hg	25	EST	NEELY,WB & BLAU,GE (1985)
DC	pKa			
HL 1.30E-012	atm m3/mol	25	EST	MEYLAN,WM & HOWARD,PH (1991)
OH 2.26E-010	cm3/molc sec	25	EST	MEYLAN,WM & HOWARD,PH (1993)

087130-56-1

CARBAMIC ACID, (3,4-DIETHOXYPHENYL)-, 2-CYANOETH

Formula: $C_{14}H_{18}N_2O_4$

Mol Weight: 278.31

MP (deg C): | FP (deg C):

BP (deg C):

BP pressure (mm Hg):

Property/Value	Units	Temp	Data Type	Reference
WS 1.52E+002	mg/L	25	EST	MEYLAN,WM ET AL. (1996)
logP 2.07			EXP	TAKAHASHI,J ET AL. (1988)
VP 5.48E-007	mm Hg	25	EST	NEELY,WB & BLAU,GE (1985)
DC	pKa			
HL 7.42E-014	atm m3/mol	25	EST	MEYLAN,WM & HOWARD,PH (1991)
OH 2.13E-010	cm3/molc sec	25	EST	MEYLAN,WM & HOWARD,PH (1993)

087130-88-9

CARBAMIC ACID, (3,4-DIETHOXYPHENYL)-, PROPYL EST

Formula: $C_{14}H_{21}NO_4$

Mol Weight: 267.33

MP (deg C): | FP (deg C):

BP (deg C):

BP pressure (mm Hg):

Property/Value	Units	Temp	Data Type	Reference
WS 5.95E+001	mg/L	25	EST	MEYLAN,WM ET AL. (1996)
logP 2.93			EXP	TAKAHASHI,J ET AL. (1988)
VP 1.55E-005	mm Hg	25	EST	NEELY,WB & BLAU,GE (1985)
DC	pKa			
HL 2.37E-010	atm m3/mol	25	EST	MEYLAN,WM & HOWARD,PH (1991)
OH 2.16E-010	cm3/molc sec	25	EST	MEYLAN,WM & HOWARD,PH (1993)

087154-30-1

2H-1,3,2-OXAZAPHOSPHORIN-2-AMINE, 3-CHLORO-N,N-BI

Formula: $C_7H_{14}Cl_3N_2O_2P$

Mol Weight: 295.53

MP (deg C): | FP (deg C):

BP (deg C):

BP pressure (mm Hg):

Property/Value	Units	Temp	Data Type	Reference
WS 2.87E+002	mg/L	25	EST	MEYLAN,WM ET AL. (1996)
logP 1.94			EXP	SANGSTER,J (1994)
VP 1.94E-005	mm Hg	25	EST	NEELY,WB & BLAU,GE (1985)
DC	pKa			
HL 2.61E-010	atm m3/mol	25	EST	MEYLAN,WM & HOWARD,PH (1991)
OH 4.28E-011	cm3/molc sec	25	EST	MEYLAN,WM & HOWARD,PH (1993)

087237-48-7

HALOXYFOP-ETOTYL

Formula: $C_{19}H_{19}ClF_3NO_5$

Mol Weight: 433.82

MP (deg C): 56-58 | FP (deg C):

BP (deg C):

BP pressure (mm Hg):

Property/Value	Units	Temp	Data Type	Reference
WS 2.70E+000	mg/L	25	EXP	SHIU,WY ET AL. (1990)
logP 4.33			EXP	TOMLIN,C (1994)
VP 3.15E-008	mm Hg	25	EST	NEELY,WB & BLAU,GE (1985)
DC	pKa			
HL 2.47E-010	atm m3/mol	25	EST	MEYLAN,WM & HOWARD,PH (1991)
OH 4.55E-011	cm3/molc sec	25	EST	MEYLAN,WM & HOWARD,PH (1993)

087269-97-4

CYCLOPENTA[B]PYRROLE-2-CARBOXYLIC ACID, DERIVATI

Formula: $C_{21}H_{28}N_2O_5$

Mol Weight: 388.47

MP (deg C): 109 | FP (deg C):

BP (deg C):

BP pressure (mm Hg):

Property/Value	Units	Temp	Data Type	Reference
WS 9.76E-001	mg/L	25	EST	MEYLAN,WM ET AL. (1996)
logP 0.22			EXP	RANADIVE,SA ET AL. (1992)
VP 3.17E-014	mm Hg	25	EST	NEELY,WB & BLAU,GE (1985)
DC	pKa			
HL 8.10E-019	atm m3/mol	25	EST	MEYLAN,WM & HOWARD,PH (1991)
OH 1.35E-010	cm3/molc sec	25	EST	MEYLAN,WM & HOWARD,PH (1993)

087340-61-2

CARBAMIC ACID, DIMETHYL-, [4-(AMINOCARBONYL)PHEN

Formula: $C_{11}H_{14}N_2O_3$

Mol Weight: 222.25

MP (deg C): | FP (deg C):

BP (deg C):

BP pressure (mm Hg):

Property/Value	Units	Temp	Data Type	Reference
WS 7.37E+003	mg/L	25	EST	MEYLAN,WM ET AL. (1996)
logP 0.77			EXP	SANGSTER,J (1994)
VP 1.38E-006	mm Hg	25	EST	NEELY,WB & BLAU,GE (1985)
DC	pKa			
HL 1.00E-011	atm m3/mol	25	EST	MEYLAN,WM & HOWARD,PH (1991)
OH 2.40E-011	cm3/molc sec	25	EST	MEYLAN,WM & HOWARD,PH (1993)

087359-11-3

1H-IMIDAZO[4,5-C]PYRIDINE, 2-(2,4-DIMETHOXYPHENY

Formula: $C_{14}H_{13}N_3O_2$

Mol Weight: 255.28

MP (deg C): | FP (deg C):

BP (deg C):

BP pressure (mm Hg):

Property/Value	Units	Temp	Data Type	Reference
WS 1.44E+002	mg/L	25	EST	MEYLAN,WM ET AL. (1996)
logP 2.56			EXP	HANSCH,C ET AL. (1995)
VP 8.24E-010	mm Hg	25	EST	NEELY,WB & BLAU,GE (1985)
DC	pKa			
HL 1.29E-013	atm m3/mol	25	EST	MEYLAN,WM & HOWARD,PH (1991)
OH 2.34E-010	cm3/molc sec	25	EST	MEYLAN,WM & HOWARD,PH (1993)

CAS #: 087359-63-5
1H-IMIDAZO[4,5-C]PYRIDINE, 2-[2-(3-CHLOROPROPOXY

Formula: $C_{16}H_{16}ClN_3O_2$

Mol Weight: 317.78

MP (deg C): FP (deg C):

BP (deg C):

BP pressure (mm Hg):

Property/Value	Units	Temp	Data Type	Reference
WS 9.33E+000	mg/L	25	EST	MEYLAN,WM ET AL. (1996)
logP 3.53			EXP	HANSCH,C ET AL. (1995)
VP 1.72E-011	mm Hg	25	EST	NEELY,WB & BLAU,GE (1985)
DC	pKa			
HL 8.00E-014	atm m3/mol	25	EST	MEYLAN,WM & HOWARD,PH (1991)
OH 2.41E-010	cm3/molc sec	25	EST	MEYLAN,WM & HOWARD,PH (1993)

CAS #: 087365-63-7
RH-4663 O-PH N-BUTYLENE CARBAMATE

Formula: $C_{14}H_{13}Cl_2NO_3$

Mol Weight: 314.17

MP (deg C): FP (deg C):

BP (deg C):

BP pressure (mm Hg):

Property/Value	Units	Temp	Data Type	Reference
WS 5.54E+000	mg/L	25	EST	MEYLAN,WM ET AL. (1996)
logP 3.82			EXP	NANDIHALLI,UB ET AL. (1993)
VP 1.50E-006	mm Hg	25	EST	NEELY,WB & BLAU,GE (1985)
DC	pKa			
HL 3.88E-010	atm m3/mol	25	EST	MEYLAN,WM & HOWARD,PH (1991)
OH 3.94E-011	cm3/molc sec	25	EST	MEYLAN,WM & HOWARD,PH (1993)

CAS #: 087374-78-5
RH-6251 O-PH N-PENTYLENE CARBAMATE

Formula: $C_{15}H_{15}Cl_2NO_3$

Mol Weight: 328.20

MP (deg C): FP (deg C):

BP (deg C):

BP pressure (mm Hg):

Property/Value	Units	Temp	Data Type	Reference
WS 1.97E+000	mg/L	25	EST	MEYLAN,WM ET AL. (1996)
logP 4.25			EXP	NANDIHALLI,UB ET AL. (1993)
VP 6.11E-007	mm Hg	25	EST	NEELY,WB & BLAU,GE (1985)
DC	pKa			
HL 5.16E-010	atm m3/mol	25	EST	MEYLAN,WM & HOWARD,PH (1991)
OH 4.32E-011	cm3/molc sec	25	EST	MEYLAN,WM & HOWARD,PH (1993)

CAS #: 087418-35-7
ADENOSINE, 2',3'-DIDEOXY-3'-FLUORO-

Formula: $C_{10}H_{12}FN_5O_2$

Mol Weight: 253.24

MP (deg C): FP (deg C):

BP (deg C):

BP pressure (mm Hg):

Property/Value	Units	Temp	Data Type	Reference
WS 9.50E+002	mg/L	25	EST	MEYLAN,WM ET AL. (1996)
logP 0.08			EXP	BALZARINI,J ET AL. (1989)
VP 4.20E-011	mm Hg	25	EST	NEELY,WB & BLAU,GE (1985)
DC	pKa			
HL 1.66E-019	atm m3/mol	25	EST	MEYLAN,WM & HOWARD,PH (1991)
OH 2.08E-010	cm3/molc sec	25	EST	MEYLAN,WM & HOWARD,PH (1993)

CAS #: 087469-33-8
2H-PURIN-2-ONE, 1,3,6,9-TETRAHYDRO-6-IMINO-1-MET

Formula: $C_{10}H_{13}N_5O$

Mol Weight: 219.25

MP (deg C): FP (deg C):

BP (deg C):

BP pressure (mm Hg):

Property/Value	Units	Temp	Data Type	Reference
WS 3.19E+003	mg/L	25	EST	MEYLAN,WM ET AL. (1996)
logP -0.57			EXP	CARNEY,C & GRAHAM,E (1985)
VP 2.38E-008	mm Hg	25	EST	NEELY,WB & BLAU,GE (1985)
DC	pKa			
HL 2.08E-013	atm m3/mol	25	EST	MEYLAN,WM & HOWARD,PH (1991)
OH 9.95E-011	cm3/molc sec	25	EST	MEYLAN,WM & HOWARD,PH (1993)

CAS #: 087469-35-0
1-METHYLISOGUANINE,9[2(METHOXY)ME]-

Formula: $C_8H_{13}N_5O_2$

Mol Weight: 211.23

MP (deg C): FP (deg C):

BP (deg C):

BP pressure (mm Hg):

Property/Value	Units	Temp	Data Type	Reference
WS 2.12E+004	mg/L	25	EST	MEYLAN,WM ET AL. (1996)
logP -1.22			EXP	CARNEY,C & GRAHAM,E (1985)
VP 2.17E-008	mm Hg	25	EST	NEELY,WB & BLAU,GE (1985)
DC	pKa			
HL 1.86E-015	atm m3/mol	25	EST	MEYLAN,WM & HOWARD,PH (1991)
OH 7.88E-011	cm3/molc sec	25	EST	MEYLAN,WM & HOWARD,PH (1993)

CAS #: 087469-41-8
2H-PURIN-2-ONE, 1,3,6,9-TETRAHYDRO-6-IMINO-9-(2-

Formula: $C_9H_{15}N_5O_2$

Mol Weight: 225.25

MP (deg C): FP (deg C):

BP (deg C):

BP pressure (mm Hg):

Property/Value	Units	Temp	Data Type	Reference
WS 1.07E+004	mg/L	25	EST	MEYLAN,WM ET AL. (1996)
logP -0.96			EXP	CARNEY,C & GRAHAM,E (1985)
VP 9.71E-009	mm Hg	25	EST	NEELY,WB & BLAU,GE (1985)
DC	pKa			
HL 2.47E-015	atm m3/mol	25	EST	MEYLAN,WM & HOWARD,PH (1991)
OH 8.34E-011	cm3/molc sec	25	EST	MEYLAN,WM & HOWARD,PH (1993)

CAS #: 087469-42-9
2H-PURIN-2-ONE, 1,3,6,7-TETRAHYDRO-6-IMINO-7-(2-

Formula: $C_9H_{15}N_5O_2$

Mol Weight: 225.25

MP (deg C): FP (deg C):

BP (deg C):

BP pressure (mm Hg):

Property/Value	Units	Temp	Data Type	Reference
WS 3.03E+004	mg/L	25	EST	MEYLAN,WM ET AL. (1996)
logP -1.49			EXP	CARNEY,C & GRAHAM,E (1985)
VP 9.71E-009	mm Hg	25	EST	NEELY,WB & BLAU,GE (1985)
DC	pKa			
HL 2.47E-015	atm m3/mol	25	EST	MEYLAN,WM & HOWARD,PH (1991)
OH 8.34E-011	cm3/molc sec	25	EST	MEYLAN,WM & HOWARD,PH (1993)

CAS #:	087469-43-0				2H-PURIN-2-ONE, 9-(2-ETHOXYETHYL)-1,3,6,9-TETRAH
Formula: $C_{10}H_{17}N_5O_2$					
Mol Weight: 239.28					
MP (deg C):		FP (deg C):			
BP (deg C):					
BP pressure (mm Hg):					

Property/Value	Units	Temp	Data Type	Reference
WS 2.99E+003	mg/L	25	EST	MEYLAN,WM ET AL. (1996)
logP -0.40			EXP	CARNEY,C & GRAHAM,E (1985)
VP 4.32E-009	mm Hg	25	EST	NEELY,WB & BLAU,GE (1985)
DC	pKa			
HL 3.27E-015	atm m3/mol	25	EST	MEYLAN,WM & HOWARD,PH (1991)
OH 8.88E-011	cm3/molc sec	25	EST	MEYLAN,WM & HOWARD,PH (1993)

CAS #:	087469-44-1				2H-PURIN-2-ONE, 7-(2-ETHOXYETHYL)-1,3,6,7-TETRAH
Formula: $C_{10}H_{17}N_5O_2$					
Mol Weight: 239.28					
MP (deg C):		FP (deg C):			
BP (deg C):					
BP pressure (mm Hg):					

Property/Value	Units	Temp	Data Type	Reference
WS 1.18E+004	mg/L	25	EST	MEYLAN,WM ET AL. (1996)
logP -1.10			EXP	CARNEY,C & GRAHAM,E (1985)
VP 4.32E-009	mm Hg	25	EST	NEELY,WB & BLAU,GE (1985)
DC	pKa			
HL 3.27E-015	atm m3/mol	25	EST	MEYLAN,WM & HOWARD,PH (1991)
OH 8.88E-011	cm3/molc sec	25	EST	MEYLAN,WM & HOWARD,PH (1993)

CAS #:	087469-45-2				2H-PURIN-2-ONE, 9-[2-(ETHYLTHIO)ETHYL]-1,3,6,9-T
Formula: $C_{10}H_{17}N_5OS$					
Mol Weight: 255.34					
MP (deg C):		FP (deg C):			
BP (deg C):					
BP pressure (mm Hg):					

Property/Value	Units	Temp	Data Type	Reference
WS 1.11E+003	mg/L	25	EST	MEYLAN,WM ET AL. (1996)
logP 0.00			EXP	CARNEY,C & GRAHAM,E (1985)
VP 9.73E-010	mm Hg	25	EST	NEELY,WB & BLAU,GE (1985)
DC	pKa			
HL 2.98E-015	atm m3/mol	25	EST	MEYLAN,WM & HOWARD,PH (1991)
OH 9.16E-011	cm3/molc sec	25	EST	MEYLAN,WM & HOWARD,PH (1993)

CAS #:	087469-46-3				2H-PURIN-2-ONE, 7-[2-(ETHYLTHIO)ETHYL]-1,3,6,7-T
Formula: $C_{10}H_{17}N_5OS$					
Mol Weight: 255.34					
MP (deg C):		FP (deg C):			
BP (deg C):					
BP pressure (mm Hg):					

Property/Value	Units	Temp	Data Type	Reference
WS 1.43E+003	mg/L	25	EST	MEYLAN,WM ET AL. (1996)
logP -0.13			EXP	CARNEY,C & GRAHAM,E (1985)
VP 9.73E-010	mm Hg	25	EST	NEELY,WB & BLAU,GE (1985)
DC	pKa			
HL 2.98E-015	atm m3/mol	25	EST	MEYLAN,WM & HOWARD,PH (1991)
OH 9.16E-011	cm3/molc sec	25	EST	MEYLAN,WM & HOWARD,PH (1993)

CAS #:	087469-47-4				1-METHYLISOGUANINE,9-[2-(ETHENYLOXY)ETHYL]-
Formula: $C_{10}H_{15}N_5O_2$					
Mol Weight: 237.26					
MP (deg C):		FP (deg C):			
BP (deg C):					
BP pressure (mm Hg):					

Property/Value	Units	Temp	Data Type	Reference
WS 2.24E+003	mg/L	25	EST	MEYLAN,WM ET AL. (1996)
logP -0.24			EXP	CARNEY,C & GRAHAM,E (1985)
VP 4.59E-009	mm Hg	25	EST	NEELY,WB & BLAU,GE (1985)
DC	pKa			
HL 1.64E-014	atm m3/mol	25	EST	MEYLAN,WM & HOWARD,PH (1991)
OH 1.17E-010	cm3/molc sec	25	EST	MEYLAN,WM & HOWARD,PH (1993)

CAS #:	087469-48-5				1-METHYLISOGUANINE,7-[2-(ETHENYLOXY)ETHYL]-
Formula: $C_{10}H_{15}N_5O_2$					
Mol Weight: 237.26					
MP (deg C):		FP (deg C):			
BP (deg C):					
BP pressure (mm Hg):					

Property/Value	Units	Temp	Data Type	Reference
WS 6.47E+003	mg/L	25	EST	MEYLAN,WM ET AL. (1996)
logP -0.78			EXP	CARNEY,C & GRAHAM,E (1985)
VP 4.59E-009	mm Hg	25	EST	NEELY,WB & BLAU,GE (1985)
DC	pKa			
HL 1.64E-014	atm m3/mol	25	EST	MEYLAN,WM & HOWARD,PH (1991)
OH 1.17E-010	cm3/molc sec	25	EST	MEYLAN,WM & HOWARD,PH (1993)

CAS #:	087469-53-2				1-METHYLISOGUANINE,9[2(4-BRPHENYL)ET]-
Formula: $C_{14}H_{16}BrN_5O_2$					
Mol Weight: 366.22					
MP (deg C):		FP (deg C):			
BP (deg C):					
BP pressure (mm Hg):					

Property/Value	Units	Temp	Data Type	Reference
WS 1.69E+002	mg/L	25	EST	MEYLAN,WM ET AL. (1996)
logP 1.73			EXP	CARNEY,C & GRAHAM,E (1985)
VP 6.34E-012	mm Hg	25	EST	NEELY,WB & BLAU,GE (1985)
DC	pKa			
HL 3.63E-016	atm m3/mol	25	EST	MEYLAN,WM & HOWARD,PH (1991)
OH 8.88E-011	cm3/molc sec	25	EST	MEYLAN,WM & HOWARD,PH (1993)

CAS #:	087469-54-3				1-METHYLISOGUANINE,7[2(4-BRPHENYL)ET]-
Formula: $C_{14}H_{16}BrN_5O_2$					
Mol Weight: 366.22					
MP (deg C):		FP (deg C):			
BP (deg C):					
BP pressure (mm Hg):					

Property/Value	Units	Temp	Data Type	Reference
WS 4.35E+002	mg/L	25	EST	MEYLAN,WM ET AL. (1996)
logP 1.25			EXP	CARNEY,C & GRAHAM,E (1985)
VP 6.34E-012	mm Hg	25	EST	NEELY,WB & BLAU,GE (1985)
DC	pKa			
HL 3.63E-016	atm m3/mol	25	EST	MEYLAN,WM & HOWARD,PH (1991)
OH 8.88E-011	cm3/molc sec	25	EST	MEYLAN,WM & HOWARD,PH (1993)

CAS #: 087469-61-2				1-METHYLISOGUANINE,9[BENZYL]-

Formula: $C_{13}H_{15}N_5O$

Mol Weight: 257.30

MP (deg C):   FP (deg C):

BP (deg C):

BP pressure (mm Hg):

Property/ Value	Units	Temp	Data Type	Reference
WS 6.16E+003	mg/L	25	EST	MEYLAN,WM ET AL. (1996)
logP 0.65			EXP	CARNEY,C & GRAHAM,E (1985)
VP 3.12E-010	mm Hg	25	EST	NEELY,WB & BLAU,GE (1985)
DC	pKa			
HL 1.28E-014	atm m3/mol	25	EST	MEYLAN,WM & HOWARD,PH (1991)
OH 7.81E-011	cm3/molc sec	25	EST	MEYLAN,WM & HOWARD,PH (1993)

CAS #: 087469-62-3				1-METHYLISOGUANINE,7[BENZYL]-

Formula: $C_{13}H_{15}N_5O$

Mol Weight: 257.30

MP (deg C):   FP (deg C):

BP (deg C):

BP pressure (mm Hg):

Property/ Value	Units	Temp	Data Type	Reference
WS 9.44E+002	mg/L	25	EST	MEYLAN,WM ET AL. (1996)
logP 0.07			EXP	CARNEY,C & GRAHAM,E (1985)
VP 3.12E-010	mm Hg	25	EST	NEELY,WB & BLAU,GE (1985)
DC	pKa			
HL 1.28E-014	atm m3/mol	25	EST	MEYLAN,WM & HOWARD,PH (1991)
OH 7.81E-011	cm3/molc sec	25	EST	MEYLAN,WM & HOWARD,PH (1993)

CAS #: 087469-66-7				1-METHYLISOGUANINE,7[ACOET]-

Formula: $C_{10}H_{15}N_5O_3$

Mol Weight: 253.26

MP (deg C):   FP (deg C):

BP (deg C):

BP pressure (mm Hg):

Property/ Value	Units	Temp	Data Type	Reference
WS 1.21E+004	mg/L	25	EST	MEYLAN,WM ET AL. (1996)
logP -1.20			EXP	CARNEY,C & GRAHAM,E (1985)
VP 1.60E-009	mm Hg	25	EST	NEELY,WB & BLAU,GE (1985)
DC	pKa			
HL 5.02E-016	atm m3/mol	25	EST	MEYLAN,WM & HOWARD,PH (1991)
OH 7.54E-011	cm3/molc sec	25	EST	MEYLAN,WM & HOWARD,PH (1993)

CAS #: 087469-68-9				2H-PURIN-2-ONE, 1,3,6,9-TETRAHYDRO-6-IMINO-9-[(2

Formula: $C_{10}H_{17}N_5O_3$

Mol Weight: 255.28

MP (deg C):   FP (deg C):

BP (deg C):

BP pressure (mm Hg):

Property/ Value	Units	Temp	Data Type	Reference
WS 2.21E+004	mg/L	25	EST	MEYLAN,WM ET AL. (1996)
logP -1.52			EXP	CARNEY,C & GRAHAM,E (1985)
VP 1.85E-009	mm Hg	25	EST	NEELY,WB & BLAU,GE (1985)
DC	pKa			
HL 1.45E-016	atm m3/mol	25	EST	MEYLAN,WM & HOWARD,PH (1991)
OH 8.68E-011	cm3/molc sec	25	EST	MEYLAN,WM & HOWARD,PH (1993)

CAS #: 087546-18-7				FLUMICLORAC-PENTYL

Formula: $C_{21}H_{23}ClFNO_5$

Mol Weight: 423.87

MP (deg C):   FP (deg C):

BP (deg C):

BP pressure (mm Hg):

Property/ Value	Units	Temp	Data Type	Reference
WS 1.19E-001	mg/L	25	EST	MEYLAN,WM ET AL. (1996)
logP 4.99			EXP	TOMLIN,C (1994)
VP 1.42E-012	mm Hg	25	EST	NEELY,WB & BLAU,GE (1985)
DC	pKa			
HL 1.67E-010	atm m3/mol	25	EST	MEYLAN,WM & HOWARD,PH (1991)
OH 4.13E-011	cm3/molc sec	25	EST	MEYLAN,WM & HOWARD,PH (1993)

CAS #: 087586-46-7				[4,7]DB33C11-DIBENZO CROWN ETHER

Formula: $C_{30}H_{44}O_{11}$

Mol Weight: 580.68

MP (deg C):   FP (deg C):

BP (deg C):

BP pressure (mm Hg):

Property/ Value	Units	Temp	Data Type	Reference
WS 1.24E+001	mg/L	25	EST	MEYLAN,WM ET AL. (1996)
logP 1.45			EXP	STOLWIJK,TB ET AL. (1989)
VP 3.96E-015	mm Hg	25	EST	NEELY,WB & BLAU,GE (1985)
DC	pKa			
HL 6.18E-022	atm m3/mol	25	EST	MEYLAN,WM & HOWARD,PH (1991)
OH 1.63E-010	cm3/molc sec	25	EST	MEYLAN,WM & HOWARD,PH (1993)

CAS #: 087653-22-3				UREA, N-METHYL-N'-[4-(6-OXO-3-PHENYL-1(6H)-PYRID

Formula: $C_{16}H_{20}N_4O_2$

Mol Weight: 300.36

MP (deg C):   FP (deg C):

BP (deg C):

BP pressure (mm Hg):

Property/ Value	Units	Temp	Data Type	Reference
WS 1.09E+003	mg/L	25	EST	MEYLAN,WM ET AL. (1996)
logP 1.23			EXP	SANGSTER,J (1994)
VP 2.79E-010	mm Hg	25	EST	NEELY,WB & BLAU,GE (1985)
DC	pKa			
HL 8.13E-017	atm m3/mol	25	EST	MEYLAN,WM & HOWARD,PH (1991)
OH 4.54E-011	cm3/molc sec	25	EST	MEYLAN,WM & HOWARD,PH (1993)

CAS #: 087653-23-4				UREA, N-ETHYL-N'-[4-(6-OXO-3-PHENYL-1(6H)-PYRIDA

Formula: $C_{17}H_{22}N_4O_2$

Mol Weight: 314.39

MP (deg C):   FP (deg C):

BP (deg C):

BP pressure (mm Hg):

Property/ Value	Units	Temp	Data Type	Reference
WS 2.77E+002	mg/L	25	EST	MEYLAN,WM ET AL. (1996)
logP 1.83			EXP	SANGSTER,J (1994)
VP 1.22E-010	mm Hg	25	EST	NEELY,WB & BLAU,GE (1985)
DC	pKa			
HL 1.08E-016	atm m3/mol	25	EST	MEYLAN,WM & HOWARD,PH (1991)
OH 4.88E-011	cm3/molc sec	25	EST	MEYLAN,WM & HOWARD,PH (1993)

087653-35-8

UREA, N,N-DIMETHYL-N'-[4-(6-OXO-3-PHENYL-1(6H)-P

Formula: $C_{17}H_{22}N_4O_2$

Mol Weight: 314.39

MP (deg C): FP (deg C):

BP (deg C):

BP pressure (mm Hg):

Property/Value	Units	Temp	Data Type	Reference
WS 4.52E+002	mg/L	25	EST	MEYLAN,WM ET AL. (1996)
logP 1.58			EXP	SANGSTER,J (1994)
VP 1.32E-010	mm Hg	25	EST	NEELY,WB & BLAU,GE (1985)
DC	pKa			
HL 1.79E-016	atm m3/mol	25	EST	MEYLAN,WM & HOWARD,PH (1991)
OH 4.59E-011	cm3/molc sec	25	EST	MEYLAN,WM & HOWARD,PH (1993)

087653-49-4

UREA, [6-(6-OXO-3-PHENYL-1(6H)-PYRIDAZINYL)HEXYL

Formula: $C_{17}H_{22}N_4O_2$

Mol Weight: 314.39

MP (deg C): FP (deg C):

BP (deg C):

BP pressure (mm Hg):

Property/Value	Units	Temp	Data Type	Reference
WS 2.82E+002	mg/L	25	EST	MEYLAN,WM ET AL. (1996)
logP 1.82			EXP	SANGSTER,J (1994)
VP 9.45E-011	mm Hg	25	EST	NEELY,WB & BLAU,GE (1985)
DC	pKa			
HL 6.53E-017	atm m3/mol	25	EST	MEYLAN,WM & HOWARD,PH (1991)
OH 4.76E-011	cm3/molc sec	25	EST	MEYLAN,WM & HOWARD,PH (1993)

087666-25-9

UREA, N-METHYL-N'-[2-(6-OXO-3-PHENYL-1(6H)-PYRID

Formula: $C_{14}H_{16}N_4O_2$

Mol Weight: 272.31

MP (deg C): FP (deg C):

BP (deg C):

BP pressure (mm Hg):

Property/Value	Units	Temp	Data Type	Reference
WS 8.23E+003	mg/L	25	EST	MEYLAN,WM ET AL. (1996)
logP 0.39			EXP	YAMADA,T ET AL. (1983)
VP 1.44E-009	mm Hg	25	EST	NEELY,WB & BLAU,GE (1985)
DC	pKa			
HL 4.61E-017	atm m3/mol	25	EST	MEYLAN,WM & HOWARD,PH (1991)
OH 4.26E-011	cm3/molc sec	25	EST	MEYLAN,WM & HOWARD,PH (1993)

087666-26-0

UREA, [3-(6-OXO-3-PHENYL-1(6H)-PYRIDAZINYL)PROPY

Formula: $C_{14}H_{16}N_4O_2$

Mol Weight: 272.31

MP (deg C): FP (deg C):

BP (deg C):

BP pressure (mm Hg):

Property/Value	Units	Temp	Data Type	Reference
WS 4.56E+003	mg/L	25	EST	MEYLAN,WM ET AL. (1996)
logP 0.69			EXP	YAMADA,T ET AL. (1983)
VP 1.12E-009	mm Hg	25	EST	NEELY,WB & BLAU,GE (1985)
DC	pKa			
HL 2.79E-017	atm m3/mol	25	EST	MEYLAN,WM & HOWARD,PH (1991)
OH 4.34E-011	cm3/molc sec	25	EST	MEYLAN,WM & HOWARD,PH (1993)

087666-27-1

UREA, N-METHYL-N'-[3-(6-OXO-3-PHENYL-1(6H)-PYRID

Formula: $C_{15}H_{18}N_4O_2$

Mol Weight: 286.34

MP (deg C): FP (deg C):

BP (deg C):

BP pressure (mm Hg):

Property/Value	Units	Temp	Data Type	Reference
WS 1.51E+003	mg/L	25	EST	MEYLAN,WM ET AL. (1996)
logP 1.16			EXP	SANGSTER,J (1994)
VP 6.35E-010	mm Hg	25	EST	NEELY,WB & BLAU,GE (1985)
DC	pKa			
HL 6.13E-017	atm m3/mol	25	EST	MEYLAN,WM & HOWARD,PH (1991)
OH 4.40E-011	cm3/molc sec	25	EST	MEYLAN,WM & HOWARD,PH (1993)

087666-28-2

UREA, N-ETHYL-N'-[3-(6-OXO-3-PHENYL-1(6H)-PYRIDA

Formula: $C_{16}H_{20}N_4O_2$

Mol Weight: 300.36

MP (deg C): FP (deg C):

BP (deg C):

BP pressure (mm Hg):

Property/Value	Units	Temp	Data Type	Reference
WS 4.49E+002	mg/L	25	EST	MEYLAN,WM ET AL. (1996)
logP 1.68			EXP	SANGSTER,J (1994)
VP 2.79E-010	mm Hg	25	EST	NEELY,WB & BLAU,GE (1985)
DC	pKa			
HL 8.13E-017	atm m3/mol	25	EST	MEYLAN,WM & HOWARD,PH (1991)
OH 4.74E-011	cm3/molc sec	25	EST	MEYLAN,WM & HOWARD,PH (1993)

087666-30-6

UREA, N,N-DIMETHYL-N'-[3-(6-OXO-3-PHENYL-1(6H)-P

Formula: $C_{16}H_{20}N_4O_2$

Mol Weight: 300.36

MP (deg C): FP (deg C):

BP (deg C):

BP pressure (mm Hg):

Property/Value	Units	Temp	Data Type	Reference
WS 5.15E+002	mg/L	25	EST	MEYLAN,WM ET AL. (1996)
logP 1.61			EXP	SANGSTER,J (1994)
VP 3.01E-010	mm Hg	25	EST	NEELY,WB & BLAU,GE (1985)
DC	pKa			
HL 1.35E-016	atm m3/mol	25	EST	MEYLAN,WM & HOWARD,PH (1991)
OH 4.45E-011	cm3/molc sec	25	EST	MEYLAN,WM & HOWARD,PH (1993)

087666-31-7

UREA, [4-(6-OXO-3-PHENYL-1(6H)-PYRIDAZINYL)BUTYL

Formula: $C_{15}H_{18}N_4O_2$

Mol Weight: 286.34

MP (deg C): FP (deg C):

BP (deg C):

BP pressure (mm Hg):

Property/Value	Units	Temp	Data Type	Reference
WS 1.37E+003	mg/L	25	EST	MEYLAN,WM ET AL. (1996)
logP 1.21			EXP	SANGSTER,J (1994)
VP 4.92E-010	mm Hg	25	EST	NEELY,WB & BLAU,GE (1985)
DC	pKa			
HL 3.70E-017	atm m3/mol	25	EST	MEYLAN,WM & HOWARD,PH (1991)
OH 4.48E-011	cm3/molc sec	25	EST	MEYLAN,WM & HOWARD,PH (1993)

CAS #: 087674-68-8				DIMETHENAMID
Formula: $C_{12}H_{20}ClNO_2S$				
Mol Weight: 277.82				
MP (deg C):		FP (deg C):		
BP (deg C):				
BP pressure (mm Hg):				

Property/Value	Units	Temp	Data Type	Reference
WS 2.47E+002	mg/L	25	EST	MEYLAN,WM ET AL. (1996)
logP 2.15			EXP	TOMLIN,C (1994)
VP 2.15E-006	mm Hg	25	EST	NEELY,WB & BLAU,GE (1985)
DC	pKa			
HL 6.08E-010	atm m3/mol	25	EST	MEYLAN,WM & HOWARD,PH (1991)
OH 5.23E-011	cm3/molc sec	25	EST	MEYLAN,WM & HOWARD,PH (1993)

CAS #: 087691-87-0				1,2-BENZISOTHIAZOLE, 3-(1-PIPERAZINYL)-
Formula: $C_{11}H_{13}N_3S$				
Mol Weight: 219.31				
MP (deg C):		FP (deg C):		
BP (deg C):				
BP pressure (mm Hg):				

Property/Value	Units	Temp	Data Type	Reference
WS 5.26E+003	mg/L	25	EST	MEYLAN,WM ET AL. (1996)
logP 2.14			EXP	CACCIA,S ET AL. (1985)
VP 2.68E-006	mm Hg	25	EST	NEELY,WB & BLAU,GE (1985)
DC	pKa			
HL 8.79E-012	atm m3/mol	25	EST	MEYLAN,WM & HOWARD,PH (1991)
OH 3.06E-010	cm3/molc sec	25	EST	MEYLAN,WM & HOWARD,PH (1993)

CAS #: 087743-55-3				CARBAMIC ACID, (3-ACETYLPHENYL)-, METHYL ESTER
Formula: $C_{10}H_{11}NO_3$				
Mol Weight: 193.20				
MP (deg C):		FP (deg C):		
BP (deg C):				
BP pressure (mm Hg):				

Property/Value	Units	Temp	Data Type	Reference
WS 1.92E+003	mg/L	25	EST	MEYLAN,WM ET AL. (1996)
logP 1.63			EXP	TAKAHASHI,J ET AL. (1988)
VP 6.98E-004	mm Hg	25	EST	NEELY,WB & BLAU,GE (1985)
DC	pKa			
HL 3.98E-011	atm m3/mol	25	EST	MEYLAN,WM & HOWARD,PH (1991)
OH 1.83E-011	cm3/molc sec	25	EST	MEYLAN,WM & HOWARD,PH (1993)

CAS #: 087757-18-4				ISOXAPYRIFOP
Formula: $C_{17}H_{16}Cl_2N_2O_4$				
Mol Weight: 383.23				
MP (deg C):		FP (deg C):		
BP (deg C):				
BP pressure (mm Hg):				

Property/Value	Units	Temp	Data Type	Reference
WS 5.25E+000	mg/L	25	EST	MEYLAN,WM ET AL. (1996)
logP 3.36			EXP	TOMLIN,C (1994)
VP 1.41E-009	mm Hg	25	EST	NEELY,WB & BLAU,GE (1985)
DC	pKa			
HL 5.37E-014	atm m3/mol	25	EST	MEYLAN,WM & HOWARD,PH (1991)
OH 5.12E-011	cm3/molc sec	25	EST	MEYLAN,WM & HOWARD,PH (1993)

CAS #: 087771-40-2				IOVERSOL
Formula: $C_{18}H_{24}I_3N_3O_9$				
Mol Weight: 807.12				
MP (deg C):		FP (deg C):		
BP (deg C):				
BP pressure (mm Hg):				

Property/Value	Units	Temp	Data Type	Reference
WS 1.15E+002	mg/L	25	EST	MEYLAN,WM ET AL. (1996)
logP -2.98			EXP	HANSCH,C ET AL. (1995)
VP 2.98E-031	mm Hg	25	EST	NEELY,WB & BLAU,GE (1985)
DC	pKa			
HL 8.54E-026	atm m3/mol	25	EST	MEYLAN,WM & HOWARD,PH (1991)
OH 6.13E-011	cm3/molc sec	25	EST	MEYLAN,WM & HOWARD,PH (1993)

CAS #: 087818-31-3				CINMETHYLIN
Formula: $C_{18}H_{26}O_2$				
Mol Weight: 274.41				
MP (deg C):		FP (deg C):		
BP (deg C): 313				
BP pressure (mm Hg):				

Property/Value	Units	Temp	Data Type	Reference
WS 6.30E+001	mg/L	20	EXP	SHIU,WY ET AL. (1990)
logP 3.83			EXP	HANSCH,C ET AL. (1995)
VP 7.57E-005	mm Hg	20	EXP	TOMLIN,C (1994)
DC	pKa			
HL 4.34E-007	atm m3/mol	20	EST	VP/WSOL
OH 6.00E-011	cm3/molc sec	25	EST	MEYLAN,WM & HOWARD,PH (1993)

CAS #: 087831-85-4				L-ALANINE, 3-PYRAZINYL-
Formula: $C_7H_9N_3O_2$				
Mol Weight: 167.17				
MP (deg C):		FP (deg C):		
BP (deg C):				
BP pressure (mm Hg):				

Property/Value	Units	Temp	Data Type	Reference
WS 1.30E+005	mg/L	25	EST	MEYLAN,WM ET AL. (1996)
logP -2.79			EXP	SANGSTER,J (1993)
VP 7.19E-009	mm Hg	25	EST	NEELY,WB & BLAU,GE (1985)
DC	pKa			
HL 6.53E-014	atm m3/mol	25	EST	MEYLAN,WM & HOWARD,PH (1991)
OH 3.99E-011	cm3/molc sec	25	EST	MEYLAN,WM & HOWARD,PH (1993)

CAS #: 087871-35-0				4,6-NH2 2,2-DIME1(4-NH2)PH S-TRIAZENE
Formula: $C_{11}H_{16}N_6$				
Mol Weight: 232.29				
MP (deg C):		FP (deg C):		
BP (deg C):				
BP pressure (mm Hg):				

Property/Value	Units	Temp	Data Type	Reference
WS 9.44E+004	mg/L	25	EST	MEYLAN,WM ET AL. (1996)
logP -0.59			EXP	HANSCH,C & LEO,AJ (1985)
VP 2.87E-007	mm Hg	25	EST	NEELY,WB & BLAU,GE (1985)
DC	pKa			
HL 2.57E-016	atm m3/mol	25	EST	MEYLAN,WM & HOWARD,PH (1991)
OH 2.42E-010	cm3/molc sec	25	EST	MEYLAN,WM & HOWARD,PH (1993)

CAS #: 087885-48-1				1,2-PROPANEDIOL, 3-[(3-NITROPYRAZINYL)AMINO]

Formula: $C_7H_{10}N_4O_4$

Mol Weight: 214.18

MP (deg C): | FP (deg C):

BP (deg C):

BP pressure (mm Hg):

Property/Value	Units	Temp	Data Type	Reference
WS 9.68E+004	mg/L	25	EST	MEYLAN,WM ET AL. (1996)
logP -0.49			EXP	HANSCH,C & LEO,AJ (1985)
VP 8.92E-009	mm Hg	25	EST	NEELY,WB & BLAU,GE (1985)
DC	pKa			
HL 4.61E-016	atm m3/mol	25	EST	MEYLAN,WM & HOWARD,PH (1991)
OH 2.59E-011	cm3/molc sec	25	EST	MEYLAN,WM & HOWARD,PH (1993)

CAS #: 087885-53-8				PYRAZINECARBOXYLIC ACID, 6-[(2,3-DIHYDROXYPROPYL

Formula: $C_9H_{12}N_4O_6$

Mol Weight: 272.22

MP (deg C): | FP (deg C):

BP (deg C):

BP pressure (mm Hg):

Property/Value	Units	Temp	Data Type	Reference
WS 2.16E+004	mg/L	25	EST	MEYLAN,WM ET AL. (1996)
logP -0.10			EXP	SANGSTER,J (1993)
VP 1.49E-010	mm Hg	25	EST	NEELY,WB & BLAU,GE (1985)
DC	pKa			
HL 1.36E-019	atm m3/mol	25	EST	MEYLAN,WM & HOWARD,PH (1991)
OH 2.56E-011	cm3/molc sec	25	EST	MEYLAN,WM & HOWARD,PH (1993)

CAS #: 088054-21-1				1H-IMIDAZOLE, 5-NITRO-

Formula: $C_3H_3N_3O_2$

Mol Weight: 113.08

MP (deg C): | FP (deg C):

BP (deg C):

BP pressure (mm Hg):

Property/Value	Units	Temp	Data Type	Reference
WS 5.93E+004	mg/L	25	EST	MEYLAN,WM ET AL. (1996)
logP -0.16			EXP	HANSCH,C ET AL. (1995)
VP 1.75E-004	mm Hg	25	EST	NEELY,WB & BLAU,GE (1985)
DC -0.05	pKa	20	EXP	PERRIN,DD (1965)
HL 1.48E-008	atm m3/mol	25	EST	MEYLAN,WM & HOWARD,PH (1991)
OH 4.50E-012	cm3/molc sec	25	EST	MEYLAN,WM & HOWARD,PH (1993)

CAS #: 088054-22-2				2-METHYL-5-NITROIMIDAZOLE

Formula: $C_4H_5N_3O_2$

Mol Weight: 127.10

MP (deg C): 252-254 | FP (deg C):

BP (deg C):

BP pressure (mm Hg):

Property/Value	Units	Temp	Data Type	Reference
WS 2.01E+004	mg/L	25	EST	MEYLAN,WM ET AL. (1996)
logP 0.33			EXP	HANSCH,C & LEO,AJ (1985)
VP 6.48E-005	mm Hg	25	EST	NEELY,WB & BLAU,GE (1985)
DC	pKa			
HL 1.64E-008	atm m3/mol	25	EST	MEYLAN,WM & HOWARD,PH (1991)
OH 3.99E-012	cm3/molc sec	25	EST	MEYLAN,WM & HOWARD,PH (1993)

CAS #: 088058-85-9				2H-NAPHTH[1,2-B]-1,4-OXAZIN-9-OL,3,4,4A,5,6,10B-

Formula: $C_{15}H_{21}NO_2$

Mol Weight: 247.34

MP (deg C): | FP (deg C):

BP (deg C):

BP pressure (mm Hg):

Property/Value	Units	Temp	Data Type	Reference
WS 7.81E+003	mg/L	25	EST	MEYLAN,WM ET AL. (1996)
logP 2.44			EXP	DIJKSTRA,D ET AL. (1988)
VP 1.71E-006	mm Hg	25	EST	NEELY,WB & BLAU,GE (1985)
DC	pKa			
HL 2.39E-012	atm m3/mol	25	EST	MEYLAN,WM & HOWARD,PH (1991)
OH 2.56E-010	cm3/molc sec	25	EST	MEYLAN,WM & HOWARD,PH (1993)

CAS #: 088058-88-2				NAXAGOLIDE

Formula: $C_{14}H_{19}NO$

Mol Weight: 217.31

MP (deg C): | FP (deg C):

BP (deg C):

BP pressure (mm Hg):

Property/Value	Units	Temp	Data Type	Reference
WS 2.99E+003	mg/L	25	EST	MEYLAN,WM ET AL. (1996)
logP 2.44			EXP	SANGSTER,J (1994)
VP 2.67E-004	mm Hg	25	EST	NEELY,WB & BLAU,GE (1985)
DC	pKa			
HL 3.46E-007	atm m3/mol	25	EST	MEYLAN,WM & HOWARD,PH (1991)
OH 1.62E-010	cm3/molc sec	25	EST	MEYLAN,WM & HOWARD,PH (1993)

CAS #: 088095-60-7				4-PROPOXYPYRAZOLE

Formula: $C_6H_{10}N_2O$

Mol Weight: 126.16

MP (deg C): | FP (deg C):

BP (deg C):

BP pressure (mm Hg):

Property/Value	Units	Temp	Data Type	Reference
WS 5.83E+003	mg/L	25	EST	MEYLAN,WM ET AL. (1996)
logP 1.42			EXP	HANSCH,C & LEO,AJ (1985)
VP 1.37E-002	mm Hg	25	EST	NEELY,WB & BLAU,GE (1985)
DC	pKa			
HL 3.85E-007	atm m3/mol	25	EST	MEYLAN,WM & HOWARD,PH (1991)
OH 2.10E-010	cm3/molc sec	25	EST	MEYLAN,WM & HOWARD,PH (1993)

CAS #: 088107-10-2				ETHANONE, 1-[2-HYDROXY-3-PROPYL-4-[4-(1H-TETRAZO

Formula: $C_{16}H_{22}N_4O_3$

Mol Weight: 318.38

MP (deg C): | FP (deg C):

BP (deg C):

BP pressure (mm Hg):

Property/Value	Units	Temp	Data Type	Reference
WS 1.99E+001	mg/L	25	EST	MEYLAN,WM ET AL. (1996)
logP 3.82			EXP	HANSCH,C ET AL. (1995)
VP 4.95E-011	mm Hg	25	EST	NEELY,WB & BLAU,GE (1985)
DC	pKa			
HL 3.17E-012	atm m3/mol	25	EST	MEYLAN,WM & HOWARD,PH (1991)
OH 2.16E-010	cm3/molc sec	25	EST	MEYLAN,WM & HOWARD,PH (1993)

CAS #: 088116-59-0				IOHEXOL DERIVATIVE

Formula: $C_{19}H_{26}I_3N_3O_9$

Mol Weight: 821.15

MP (deg C):   FP (deg C):

BP (deg C):

BP pressure (mm Hg):

Property/Value	Units	Temp	Data Type	Reference
WS 6.51E+001	mg/L	25	EST	MEYLAN,WM ET AL. (1996)
logP -2.80			EXP	HANSCH,C & LEO,AJ (1985)
VP 6.11E-030	mm Hg	25	EST	NEELY,WB & BLAU,GE (1985)
DC	pKa			
HL 1.04E-026	atm m3/mol	25	EST	MEYLAN,WM & HOWARD,PH (1991)
OH 8.51E-011	cm3/molc sec	25	EST	MEYLAN,WM & HOWARD,PH (1993)

CAS #: 088132-15-4				1-(4-NO2 PHENYL)-3-MEO-3-ME UREA

Formula: $C_9H_{11}N_3O_4$

Mol Weight: 225.21

MP (deg C):   FP (deg C):

BP (deg C):

BP pressure (mm Hg):

Property/Value	Units	Temp	Data Type	Reference
WS 1.06E+003	mg/L	25	EST	MEYLAN,WM ET AL. (1996)
logP 1.74			EXP	HANSCH,C & LEO,AJ (1985)
VP 1.00E-006	mm Hg	25	EST	NEELY,WB & BLAU,GE (1985)
DC	pKa			
HL 8.29E-011	atm m3/mol	25	EST	MEYLAN,WM & HOWARD,PH (1991)
OH 6.91E-012	cm3/molc sec	25	EST	MEYLAN,WM & HOWARD,PH (1993)

CAS #: 088132-16-5				1-MEO-1-ME-3-(P-AMINOPHENYL)UREA

Formula: $C_9H_{13}N_3O_2$

Mol Weight: 195.22

MP (deg C):   FP (deg C):

BP (deg C):

BP pressure (mm Hg):

Property/Value	Units	Temp	Data Type	Reference
WS 3.10E+003	mg/L	25	EST	MEYLAN,WM ET AL. (1996)
logP -0.16			EXP	HANSCH,C & LEO,AJ (1985)
VP 3.30E-006	mm Hg	25	EST	NEELY,WB & BLAU,GE (1985)
DC	pKa			
HL 7.43E-012	atm m3/mol	25	EST	MEYLAN,WM & HOWARD,PH (1991)
OH 1.78E-010	cm3/molc sec	25	EST	MEYLAN,WM & HOWARD,PH (1993)

CAS #: 088132-19-8				1-(3-NO2 PHENYL)-3-MEO-3-ME UREA

Formula: $C_9H_{11}N_3O_4$

Mol Weight: 225.21

MP (deg C):   FP (deg C):

BP (deg C):

BP pressure (mm Hg):

Property/Value	Units	Temp	Data Type	Reference
WS 1.28E+003	mg/L	25	EST	MEYLAN,WM ET AL. (1996)
logP 1.64			EXP	HANSCH,C & LEO,AJ (1985)
VP 1.00E-006	mm Hg	25	EST	NEELY,WB & BLAU,GE (1985)
DC	pKa			
HL 8.29E-011	atm m3/mol	25	EST	MEYLAN,WM & HOWARD,PH (1991)
OH 5.31E-012	cm3/molc sec	25	EST	MEYLAN,WM & HOWARD,PH (1993)

CAS #: 088132-20-1				1(4BB-DICYANOVINYL)PH)-3MEO3ME UREA

Formula: $C_{13}H_{12}N_4O_2$

Mol Weight: 256.27

MP (deg C):   FP (deg C):

BP (deg C):

BP pressure (mm Hg):

Property/Value	Units	Temp	Data Type	Reference
WS 1.60E+002	mg/L	25	EST	MEYLAN,WM ET AL. (1996)
logP 2.19			EXP	HANSCH,C & LEO,AJ (1985)
VP 2.83E-009	mm Hg	25	EST	NEELY,WB & BLAU,GE (1985)
DC	pKa			
HL 5.36E-014	atm m3/mol	25	EST	MEYLAN,WM & HOWARD,PH (1991)
OH 4.39E-011	cm3/molc sec	25	EST	MEYLAN,WM & HOWARD,PH (1993)

CAS #: 088132-23-4				1(4-ACETYLPHENYL)-3-MEO-3-ME UREA

Formula: $C_{11}H_{14}N_2O_3$

Mol Weight: 222.25

MP (deg C):   FP (deg C):

BP (deg C):

BP pressure (mm Hg):

Property/Value	Units	Temp	Data Type	Reference
WS 3.04E+003	mg/L	25	EST	MEYLAN,WM ET AL. (1996)
logP 1.22			EXP	HANSCH,C & LEO,AJ (1985)
VP 2.33E-006	mm Hg	25	EST	NEELY,WB & BLAU,GE (1985)
DC	pKa			
HL 3.83E-011	atm m3/mol	25	EST	MEYLAN,WM & HOWARD,PH (1991)
OH 4.06E-011	cm3/molc sec	25	EST	MEYLAN,WM & HOWARD,PH (1993)

CAS #: 088132-24-5				1-(P-F PHENYL)-3-MEO-3-ME UREA

Formula: $C_9H_{11}FN_2O_2$

Mol Weight: 198.20

MP (deg C):   FP (deg C):

BP (deg C):

BP pressure (mm Hg):

Property/Value	Units	Temp	Data Type	Reference
WS 2.53E+003	mg/L	25	EST	MEYLAN,WM ET AL. (1996)
logP 1.46			EXP	HANSCH,C & LEO,AJ (1985)
VP 9.49E-005	mm Hg	25	EST	NEELY,WB & BLAU,GE (1985)
DC	pKa			
HL 2.45E-008	atm m3/mol	25	EST	MEYLAN,WM & HOWARD,PH (1991)
OH 1.60E-011	cm3/molc sec	25	EST	MEYLAN,WM & HOWARD,PH (1993)

CAS #: 088132-25-6				1(4-ETO-CO PHENYL)-3-MEO-3-ME UREA

Formula: $C_{12}H_{16}N_2O_4$

Mol Weight: 252.27

MP (deg C):   FP (deg C):

BP (deg C):

BP pressure (mm Hg):

Property/Value	Units	Temp	Data Type	Reference
WS 2.81E+002	mg/L	25	EST	MEYLAN,WM ET AL. (1996)
logP 2.24			EXP	HANSCH,C & LEO,AJ (1985)
VP 9.92E-007	mm Hg	25	EST	NEELY,WB & BLAU,GE (1985)
DC	pKa			
HL 1.80E-010	atm m3/mol	25	EST	MEYLAN,WM & HOWARD,PH (1991)
OH 1.70E-011	cm3/molc sec	25	EST	MEYLAN,WM & HOWARD,PH (1993)

CAS #: 088132-26-7 — 1-ME-1-MEO-3-(4-BENZOYLPHENYL)UREA

Formula: $C_{16}H_{16}N_2O_3$

Mol Weight: 284.32

MP (deg C): | FP (deg C):

BP (deg C):

BP pressure (mm Hg):

Property/Value	Units	Temp	Data Type	Reference
WS 5.46E+001	mg/L	25	EST	MEYLAN,WM ET AL. (1996)
logP 2.86			EXP	HANSCH,C & LEO,AJ (1985)
VP 1.28E-008	mm Hg	25	EST	NEELY,WB & BLAU,GE (1985)
DC	pKa			
HL 7.58E-012	atm m3/mol	25	EST	MEYLAN,WM & HOWARD,PH (1991)
OH 4.22E-011	cm3/molc sec	25	EST	MEYLAN,WM & HOWARD,PH (1993)

CAS #: 088132-31-4 — N1-ME-MEO N3-(4PHOBUO)PHENYLUREA

Formula: $C_{19}H_{24}N_2O_4$

Mol Weight: 344.41

MP (deg C): | FP (deg C):

BP (deg C):

BP pressure (mm Hg):

Property/Value	Units	Temp	Data Type	Reference
WS 5.98E+000	mg/L	25	EST	MEYLAN,WM ET AL. (1996)
logP 3.57			EXP	HANSCH,C & LEO,AJ (1985)
VP 9.12E-010	mm Hg	25	EST	NEELY,WB & BLAU,GE (1985)
DC	pKa			
HL 1.26E-011	atm m3/mol	25	EST	MEYLAN,WM & HOWARD,PH (1991)
OH 8.20E-011	cm3/molc sec	25	EST	MEYLAN,WM & HOWARD,PH (1993)

CAS #: 088132-35-8 — 1-FLUORENYL-3-MEO-3-ME UREA

Formula: $C_{16}H_{16}N_2O_2$

Mol Weight: 268.32

MP (deg C): | FP (deg C):

BP (deg C):

BP pressure (mm Hg):

Property/Value	Units	Temp	Data Type	Reference
WS 1.06E+000	mg/L	25	EST	MEYLAN,WM ET AL. (1996)
logP 3.67			EXP	HANSCH,C & LEO,AJ (1985)
VP 1.81E-008	mm Hg	25	EST	NEELY,WB & BLAU,GE (1985)
DC	pKa			
HL 6.53E-010	atm m3/mol	25	EST	MEYLAN,WM & HOWARD,PH (1991)
OH 9.67E-011	cm3/molc sec	25	EST	MEYLAN,WM & HOWARD,PH (1993)

CAS #: 088132-40-5 — 1,1-DIMETHYL-(3-BUTYL)PHENYLUREA

Formula: $C_{13}H_{20}N_2O$

Mol Weight: 220.32

MP (deg C): | FP (deg C):

BP (deg C):

BP pressure (mm Hg):

Property/Value	Units	Temp	Data Type	Reference
WS 5.98E+001	mg/L	25	EST	MEYLAN,WM ET AL. (1996)
logP 3.23			EXP	HANSCH,C & LEO,AJ (1985)
VP 6.82E-006	mm Hg	25	EST	NEELY,WB & BLAU,GE (1985)
DC	pKa			
HL 2.51E-009	atm m3/mol	25	EST	MEYLAN,WM & HOWARD,PH (1991)
OH 1.12E-010	cm3/molc sec	25	EST	MEYLAN,WM & HOWARD,PH (1993)

CAS #: 088132-41-6 — 1,1-DIME-3-(4-CYCLOHXPHENYL)UREA

Formula: $C_{15}H_{22}N_2O$

Mol Weight: 246.36

MP (deg C): | FP (deg C):

BP (deg C):

BP pressure (mm Hg):

Property/Value	Units	Temp	Data Type	Reference
WS 1.33E+001	mg/L	25	EST	MEYLAN,WM ET AL. (1996)
logP 3.83			EXP	HANSCH,C & LEO,AJ (1985)
VP 8.58E-007	mm Hg	25	EST	NEELY,WB & BLAU,GE (1985)
DC	pKa			
HL 1.95E-009	atm m3/mol	25	EST	MEYLAN,WM & HOWARD,PH (1991)
OH 6.34E-011	cm3/molc sec	25	EST	MEYLAN,WM & HOWARD,PH (1993)

CAS #: 088132-43-8 — BENZOIC ACID, 2-CHLORO-4-[[(DIMETHYLAMINO)CARBON

Formula: $C_{19}H_{29}ClN_2O_3$

Mol Weight: 368.91

MP (deg C): | FP (deg C):

BP (deg C):

BP pressure (mm Hg):

Property/Value	Units	Temp	Data Type	Reference
WS 1.90E-001	mg/L	25	EST	MEYLAN,WM ET AL. (1996)
logP 5.15			EXP	HANSCH,C ET AL. (1995)
VP 7.55E-009	mm Hg	25	EST	NEELY,WB & BLAU,GE (1985)
DC	pKa			
HL 4.47E-011	atm m3/mol	25	EST	MEYLAN,WM & HOWARD,PH (1991)
OH 4.24E-011	cm3/molc sec	25	EST	MEYLAN,WM & HOWARD,PH (1993)

CAS #: 088158-44-5 — PREDNISOLONE-21-GLUCOSIDE

Formula: $C_{27}H_{38}O_{10}$

Mol Weight: 522.60

MP (deg C): | FP (deg C):

BP (deg C):

BP pressure (mm Hg):

Property/Value	Units	Temp	Data Type	Reference
WS 4.13E+002	mg/L	25	EST	MEYLAN,WM ET AL. (1996)
logP 0.27			EXP	HANSCH,C ET AL. (1995)
VP 6.64E-022	mm Hg	25	EST	NEELY,WB & BLAU,GE (1985)
DC	pKa			
HL 2.60E-019	atm m3/mol	25	EST	MEYLAN,WM & HOWARD,PH (1991)
OH 1.22E-010	cm3/molc sec	25	EST	MEYLAN,WM & HOWARD,PH (1993)

CAS #: 088203-03-6 — BENZAMIDE,O-DICL ACETYLAMINO-

Formula: $C_9H_8Cl_2N_2O_2$

Mol Weight: 247.08

MP (deg C): | FP (deg C):

BP (deg C):

BP pressure (mm Hg):

Property/Value	Units	Temp	Data Type	Reference
WS 9.02E+002	mg/L	25	EST	MEYLAN,WM ET AL. (1996)
logP 1.68			EXP	HANSCH,C & LEO,AJ (1985)
VP 4.70E-009	mm Hg	25	EST	NEELY,WB & BLAU,GE (1985)
DC	pKa			
HL 4.99E-013	atm m3/mol	25	EST	MEYLAN,WM & HOWARD,PH (1991)
OH 1.35E-011	cm3/molc sec	25	EST	MEYLAN,WM & HOWARD,PH (1993)

CAS #: 088203-04-7				ACETOPHENONE, O-CYANOMETHYLAMINO

Formula: $C_{10}H_{10}N_2O$

Mol Weight: 174.20

MP (deg C):   FP (deg C):

BP (deg C):

BP pressure (mm Hg):

Property/Value	Units	Temp	Data Type	Reference
WS 1.43E+003	mg/L	25	EST	MEYLAN,WM ET AL. (1996)
logP 1.58			EXP	HANSCH,C & LEO,AJ (1985)
VP 1.25E-004	mm Hg	25	EST	NEELY,WB & BLAU,GE (1985)
DC	pKa			
HL 3.16E-012	atm m3/mol	25	EST	MEYLAN,WM & HOWARD,PH (1991)
OH 4.06E-011	cm3/molc sec	25	EST	MEYLAN,WM & HOWARD,PH (1993)

CAS #: 088203-05-8				BENZAMIDE, O-CYANOMETHYLAMINO-

Formula: $C_9H_9N_3O$

Mol Weight: 175.19

MP (deg C):   FP (deg C):

BP (deg C):

BP pressure (mm Hg):

Property/Value	Units	Temp	Data Type	Reference
WS 1.18E+004	mg/L	25	EST	MEYLAN,WM ET AL. (1996)
logP 0.50			EXP	HANSCH,C & LEO,AJ (1985)
VP 7.45E-007	mm Hg	25	EST	NEELY,WB & BLAU,GE (1985)
DC	pKa			
HL 7.13E-016	atm m3/mol	25	EST	MEYLAN,WM & HOWARD,PH (1991)
OH 4.25E-011	cm3/molc sec	25	EST	MEYLAN,WM & HOWARD,PH (1993)

CAS #: 088203-06-9				ANISOLE, O-ACETYLMETHYLAMINO

Formula: $C_{10}H_{13}NO_2$

Mol Weight: 179.22

MP (deg C):   FP (deg C):

BP (deg C):

BP pressure (mm Hg):

Property/Value	Units	Temp	Data Type	Reference
WS 2.97E+003	mg/L	25	EST	MEYLAN,WM ET AL. (1996)
logP 1.49			EXP	HANSCH,C & LEO,AJ (1985)
VP 2.58E-003	mm Hg	25	EST	NEELY,WB & BLAU,GE (1985)
DC	pKa			
HL 3.32E-009	atm m3/mol	25	EST	MEYLAN,WM & HOWARD,PH (1991)
OH 4.43E-011	cm3/molc sec	25	EST	MEYLAN,WM & HOWARD,PH (1993)

CAS #: 088203-07-0				ANISOLE,CARBOXAMIDOMETHYLAMINO-

Formula: $C_9H_{12}N_2O_2$

Mol Weight: 180.21

MP (deg C):   FP (deg C):

BP (deg C):

BP pressure (mm Hg):

Property/Value	Units	Temp	Data Type	Reference
WS 1.19E+004	mg/L	25	EST	MEYLAN,WM ET AL. (1996)
logP 0.78			EXP	HANSCH,C & LEO,AJ (1985)
VP 9.39E-006	mm Hg	25	EST	NEELY,WB & BLAU,GE (1985)
DC	pKa			
HL 7.49E-013	atm m3/mol	25	EST	MEYLAN,WM & HOWARD,PH (1991)
OH 4.62E-011	cm3/molc sec	25	EST	MEYLAN,WM & HOWARD,PH (1993)

CAS #: 088267-61-2				BENZAMIDE, 2-[(2-HYDROXYETHYL)AMINO]-

Formula: $C_9H_{12}N_2O_2$

Mol Weight: 180.21

MP (deg C):   FP (deg C):

BP (deg C):

BP pressure (mm Hg):

Property/Value	Units	Temp	Data Type	Reference
WS 2.97E+003	mg/L	25	EST	MEYLAN,WM ET AL. (1996)
logP -0.05			EXP	LISCIANI,R ET AL. (1986)
VP 2.45E-008	mm Hg	25	EST	NEELY,WB & BLAU,GE (1985)
DC	pKa			
HL 8.34E-017	atm m3/mol	25	EST	MEYLAN,WM & HOWARD,PH (1991)
OH 5.57E-011	cm3/molc sec	25	EST	MEYLAN,WM & HOWARD,PH (1993)

CAS #: 088267-62-3				BENZAMIDE, 2-[(3-HYDROXYPROPYL)AMINO]-

Formula: $C_{10}H_{14}N_2O_2$

Mol Weight: 194.24

MP (deg C):   FP (deg C):

BP (deg C):

BP pressure (mm Hg):

Property/Value	Units	Temp	Data Type	Reference
WS 1.18E+004	mg/L	25	EST	MEYLAN,WM ET AL. (1996)
logP 0.70			EXP	LISCIANI,R ET AL. (1986)
VP 9.20E-009	mm Hg	25	EST	NEELY,WB & BLAU,GE (1985)
DC	pKa			
HL 1.11E-016	atm m3/mol	25	EST	MEYLAN,WM & HOWARD,PH (1991)
OH 5.71E-011	cm3/molc sec	25	EST	MEYLAN,WM & HOWARD,PH (1993)

CAS #: 088280-81-3				4-PYRIMIDINECARBOXYLIC ACID, 1,2,3,6-TETRAHYDRO-

Formula: $C_{13}H_{20}N_2O_4$

Mol Weight: 268.32

MP (deg C):   FP (deg C):

BP (deg C):

BP pressure (mm Hg):

Property/Value	Units	Temp	Data Type	Reference
WS 1.12E+002	mg/L	25	EST	MEYLAN,WM ET AL. (1996)
logP 2.60			EXP	SANGSTER,J (1993)
VP 3.49E-010	mm Hg	25	EST	NEELY,WB & BLAU,GE (1985)
DC	pKa			
HL 1.88E-011	atm m3/mol	25	EST	MEYLAN,WM & HOWARD,PH (1991)
OH 1.86E-011	cm3/molc sec	25	EST	MEYLAN,WM & HOWARD,PH (1993)

CAS #: 088283-41-4				ETHANOL, 1-(2,4-DICHLOROPHENYL)-2-(3-PYRIDINYL)-

Formula: $C_{14}H_{12}Cl_2N_2O$

Mol Weight: 295.17

MP (deg C):   FP (deg C):

BP (deg C):

BP pressure (mm Hg):

Property/Value	Units	Temp	Data Type	Reference
WS 1.81E+002	mg/L	25	EST	MEYLAN,WM ET AL. (1996)
logP 3.70			EXP	TOMLIN,C (1994)
VP 3.49E-006	mm Hg	25	EST	NEELY,WB & BLAU,GE (1985)
DC	pKa			
HL 8.81E-009	atm m3/mol	25	EST	MEYLAN,WM & HOWARD,PH (1991)
OH 3.25E-012	cm3/molc sec	25	EST	MEYLAN,WM & HOWARD,PH (1993)

CAS #: 088437-14-3				4H-PYRIDO[1,2-A]PYRIMIDIN-4-ONE, 6,7,8,9-TETRAHY
Formula: C$_{10}$H$_{14}$N$_2$O				
Mol Weight: 178.24				
MP (deg C):		FP (deg C):		
BP (deg C):				
BP pressure (mm Hg):				

Property/ Value	Units	Temp	Data Type	Reference
WS 1.73E+004	mg/L	25	EST	MEYLAN,WM ET AL. (1996)
logP 0.60			EXP	SANGSTER,J (1993)
VP 7.93E-005	mm Hg	25	EST	NEELY,WB & BLAU,GE (1985)
DC	pKa			
HL 3.29E-009	atm m3/mol	25	EST	MEYLAN,WM & HOWARD,PH (1991)
OH 3.83E-011	cm3/molc sec	25	EST	MEYLAN,WM & HOWARD,PH (1993)

CAS #: 088449-50-7				PHENYLACETIC ACID, 3-MEO-4-IPRO
Formula: C$_{12}$H$_{16}$O$_4$				
Mol Weight: 224.26				
MP (deg C):		FP (deg C):		
BP (deg C):				
BP pressure (mm Hg):				

Property/ Value	Units	Temp	Data Type	Reference
WS 2.60E+003	mg/L	25	EST	MEYLAN,WM ET AL. (1996)
logP 1.75			EXP	HANSCH,C & LEO,AJ (1985)
VP 2.44E-005	mm Hg	25	EST	NEELY,WB & BLAU,GE (1985)
DC	pKa			
HL 2.73E-010	atm m3/mol	25	EST	MEYLAN,WM & HOWARD,PH (1991)
OH 4.51E-011	cm3/molc sec	25	EST	MEYLAN,WM & HOWARD,PH (1993)

CAS #: 088477-43-4				3,4-DIME PYRIDINIUM IODIDE, N-HEXYL
Formula: C$_{13}$H$_{22}$IN				
Mol Weight: 319.23				
MP (deg C):		FP (deg C):		
BP (deg C):				
BP pressure (mm Hg):				

Property/ Value	Units	Temp	Data Type	Reference
WS 1.84E+005	mg/L	25	EST	MEYLAN,WM ET AL. (1996)
logP -1.51			EXP	HANSCH,C & LEO,AJ (1985)
VP 8.97E-005	mm Hg	25	EST	NEELY,WB & BLAU,GE (1985)
DC	pKa			
HL 7.55E-010	atm m3/mol	25	EST	MEYLAN,WM & HOWARD,PH (1991)
OH 1.82E-011	cm3/molc sec	25	EST	MEYLAN,WM & HOWARD,PH (1993)

CAS #: 088477-44-5				3,4-DIME PYRIDINIUM IODIDE, N-DECYL
Formula: C$_{17}$H$_{30}$IN				
Mol Weight: 375.34				
MP (deg C):		FP (deg C):		
BP (deg C):				
BP pressure (mm Hg):				

Property/ Value	Units	Temp	Data Type	Reference
WS 1.19E+003	mg/L	25	EST	MEYLAN,WM ET AL. (1996)
logP 0.66			EXP	HANSCH,C & LEO,AJ (1985)
VP 2.82E-006	mm Hg	25	EST	NEELY,WB & BLAU,GE (1985)
DC	pKa			
HL 2.35E-009	atm m3/mol	25	EST	MEYLAN,WM & HOWARD,PH (1991)
OH 2.38E-011	cm3/molc sec	25	EST	MEYLAN,WM & HOWARD,PH (1993)

CAS #: 088477-45-6				3,4-DIME PYRIDINIUM IODIDE, N-DODECYL
Formula: C$_{19}$H$_{34}$IN				
Mol Weight: 403.39				
MP (deg C):		FP (deg C):		
BP (deg C):				
BP pressure (mm Hg):				

Property/ Value	Units	Temp	Data Type	Reference
WS 9.34E+001	mg/L	25	EST	MEYLAN,WM ET AL. (1996)
logP 1.75			EXP	HANSCH,C & LEO,AJ (1985)
VP 5.96E-007	mm Hg	25	EST	NEELY,WB & BLAU,GE (1985)
DC	pKa			
HL 4.13E-009	atm m3/mol	25	EST	MEYLAN,WM & HOWARD,PH (1991)
OH 2.66E-011	cm3/molc sec	25	EST	MEYLAN,WM & HOWARD,PH (1993)

CAS #: 088477-46-7				3,4-DIME PYRIDINIUM I, N-TETRADECYL
Formula: C$_{21}$H$_{38}$IN				
Mol Weight: 431.45				
MP (deg C):		FP (deg C):		
BP (deg C):				
BP pressure (mm Hg):				

Property/ Value	Units	Temp	Data Type	Reference
WS 6.13E+000	mg/L	25	EST	MEYLAN,WM ET AL. (1996)
logP 2.93			EXP	HANSCH,C & LEO,AJ (1985)
VP 1.06E-007	mm Hg	25	EST	NEELY,WB & BLAU,GE (1985)
DC	pKa			
HL 7.29E-009	atm m3/mol	25	EST	MEYLAN,WM & HOWARD,PH (1991)
OH 2.95E-011	cm3/molc sec	25	EST	MEYLAN,WM & HOWARD,PH (1993)

CAS #: 088477-47-8				3,4-DIME PYRIDINIUM I, N-HEXADECYL
Formula: C$_{23}$H$_{42}$IN				
Mol Weight: 459.50				
MP (deg C):		FP (deg C):		
BP (deg C):				
BP pressure (mm Hg):				

Property/ Value	Units	Temp	Data Type	Reference
WS 2.60E+000	mg/L	25	EST	MEYLAN,WM ET AL. (1996)
logP 3.16			EXP	HANSCH,C & LEO,AJ (1985)
VP 1.84E-008	mm Hg	25	EST	NEELY,WB & BLAU,GE (1985)
DC	pKa			
HL 1.28E-008	atm m3/mol	25	EST	MEYLAN,WM & HOWARD,PH (1991)
OH 3.23E-011	cm3/molc sec	25	EST	MEYLAN,WM & HOWARD,PH (1993)

CAS #: 088477-49-0				2,4-DIME PYRIDINIUM IODIDE, N-OCTYL
Formula: C$_{15}$H$_{26}$IN				
Mol Weight: 347.29				
MP (deg C):		FP (deg C):		
BP (deg C):				
BP pressure (mm Hg):				

Property/ Value	Units	Temp	Data Type	Reference
WS 1.97E+004	mg/L	25	EST	MEYLAN,WM ET AL. (1996)
logP -0.57			EXP	HANSCH,C & LEO,AJ (1985)
VP 1.58E-005	mm Hg	25	EST	NEELY,WB & BLAU,GE (1985)
DC	pKa			
HL 1.33E-009	atm m3/mol	25	EST	MEYLAN,WM & HOWARD,PH (1991)
OH 2.23E-011	cm3/molc sec	25	EST	MEYLAN,WM & HOWARD,PH (1993)

CAS #: 088491-49-0	9H-CYCLOPENTA[D]PYRROLO[1,2-A]PYRIMIDIN-4-ONE, 1
Formula: $C_{11}H_{14}N_2O$	
Mol Weight: 190.25	
MP (deg C):	FP (deg C):
BP (deg C):	
BP pressure (mm Hg):	

Property/Value	Units	Temp	Data Type	Reference
WS 1.83E+003	mg/L	25	EST	MEYLAN,WM ET AL. (1996)
logP 0.22			EXP	SANGSTER,J (1994)
VP 6.17E-005	mm Hg	25	EST	NEELY,WB & BLAU,GE (1985)
DC	pKa			
HL 1.45E-009	atm m3/mol	25	EST	MEYLAN,WM & HOWARD,PH (1991)
OH 3.63E-011	cm3/molc sec	25	EST	MEYLAN,WM & HOWARD,PH (1993)

CAS #: 088491-50-3	PYRROLO[2,1-B]QUINAZOLIN-9(1H)-ONE, 2,3,5,6,7,8-
Formula: $C_{12}H_{16}N_2O$	
Mol Weight: 204.27	
MP (deg C):	FP (deg C):
BP (deg C):	
BP pressure (mm Hg):	

Property/Value	Units	Temp	Data Type	Reference
WS 9.58E+003	mg/L	25	EST	MEYLAN,WM ET AL. (1996)
logP 0.83			EXP	SANGSTER,J (1994)
VP 2.44E-005	mm Hg	25	EST	NEELY,WB & BLAU,GE (1985)
DC	pKa			
HL 1.93E-009	atm m3/mol	25	EST	MEYLAN,WM & HOWARD,PH (1991)
OH 3.92E-011	cm3/molc sec	25	EST	MEYLAN,WM & HOWARD,PH (1993)

CAS #: 088497-75-0	3,5-DIME PYRIDINIUM I,N-HEXADECYL
Formula: $C_{23}H_{42}IN$	
Mol Weight: 459.50	
MP (deg C):	FP (deg C):
BP (deg C):	
BP pressure (mm Hg):	

Property/Value	Units	Temp	Data Type	Reference
WS 1.79E+000	mg/L	25	EST	MEYLAN,WM ET AL. (1996)
logP 3.35			EXP	HANSCH,C & LEO,AJ (1985)
VP 1.84E-008	mm Hg	25	EST	NEELY,WB & BLAU,GE (1985)
DC	pKa			
HL 1.28E-008	atm m3/mol	25	EST	MEYLAN,WM & HOWARD,PH (1991)
OH 3.36E-011	cm3/molc sec	25	EST	MEYLAN,WM & HOWARD,PH (1993)

CAS #: 088497-95-4	3-METHYLSULFONYLPHENYLUREA
Formula: $C_8H_{10}N_2O_3S$	
Mol Weight: 214.24	
MP (deg C):	FP (deg C):
BP (deg C):	
BP pressure (mm Hg):	

Property/Value	Units	Temp	Data Type	Reference
WS 2.29E+003	mg/L	25	EST	MEYLAN,WM ET AL. (1996)
logP -0.12			EXP	HANSCH,C & LEO,AJ (1985)
VP 5.70E-007	mm Hg	25	EST	NEELY,WB & BLAU,GE (1985)
DC	pKa			
HL 4.66E-014	atm m3/mol	25	EST	MEYLAN,WM & HOWARD,PH (1991)
OH 1.12E-011	cm3/molc sec	25	EST	MEYLAN,WM & HOWARD,PH (1993)

CAS #: 088511-47-1	4-PYRIDAZINECARBOXAMIDE
Formula: $C_5H_5N_3O$	
Mol Weight: 123.12	
MP (deg C):	FP (deg C):
BP (deg C):	
BP pressure (mm Hg):	

Property/Value	Units	Temp	Data Type	Reference
WS 3.16E+004	mg/L	25	EST	MEYLAN,WM ET AL. (1996)
logP -0.96			EXP	YAMAGAMI,C ET AL. (1990)
VP 1.49E-005	mm Hg	25	EST	NEELY,WB & BLAU,GE (1985)
DC 1.00	pKa		EXP	PERRIN,DD (1965)
HL 1.18E-012	atm m3/mol	25	EST	MEYLAN,WM & HOWARD,PH (1991)
OH 2.21E-012	cm3/molc sec	25	EST	MEYLAN,WM & HOWARD,PH (1993)

CAS #: 088511-48-2	2-PYRIMIDINECARBOXAMIDE
Formula: $C_5H_5N_3O$	
Mol Weight: 123.12	
MP (deg C):	FP (deg C):
BP (deg C):	
BP pressure (mm Hg):	

Property/Value	Units	Temp	Data Type	Reference
WS 5.07E+004	mg/L	25	EST	MEYLAN,WM ET AL. (1996)
logP -1.20			EXP	YAMAGAMI,C ET AL. (1990)
VP 2.44E-004	mm Hg	25	EST	NEELY,WB & BLAU,GE (1985)
DC	pKa			
HL 1.20E-012	atm m3/mol	25	EST	MEYLAN,WM & HOWARD,PH (1991)
OH 2.21E-012	cm3/molc sec	25	EST	MEYLAN,WM & HOWARD,PH (1993)

CAS #: 088530-27-2	ADAMANTYL-ALANIN-AMIDE,N-ACETYL
Formula: $C_{15}H_{24}N_2O_2$	
Mol Weight: 264.37	
MP (deg C):	FP (deg C):
BP (deg C):	
BP pressure (mm Hg):	

Property/Value	Units	Temp	Data Type	Reference
WS 4.69E+002	mg/L	25	EST	MEYLAN,WM ET AL. (1996)
logP 1.90			EXP	HANSCH,C & LEO,AJ (1985)
VP 1.06E-009	mm Hg	25	EST	NEELY,WB & BLAU,GE (1985)
DC	pKa			
HL 3.77E-010	atm m3/mol	25	EST	MEYLAN,WM & HOWARD,PH (1991)
OH 1.02E-010	cm3/molc sec	25	EST	MEYLAN,WM & HOWARD,PH (1993)

CAS #: 088598-34-9	2,4-DICHLOROBENZYLPENCILLIN
Formula: $C_{16}H_{16}Cl_2N_2O_4S$	
Mol Weight: 403.29	
MP (deg C):	FP (deg C):
BP (deg C):	
BP pressure (mm Hg):	

Property/Value	Units	Temp	Data Type	Reference
WS 4.44E+000	mg/L	25	EST	MEYLAN,WM ET AL. (1996)
logP 3.30			EXP	HANSCH,C & LEO,AJ (1985)
VP 1.66E-013	mm Hg	25	EST	NEELY,WB & BLAU,GE (1985)
DC	pKa			
HL 6.39E-015	atm m3/mol	25	EST	MEYLAN,WM & HOWARD,PH (1991)
OH 8.78E-011	cm3/molc sec	25	EST	MEYLAN,WM & HOWARD,PH (1993)

1402

CAS #: 088644-13-7				A-MESULFONAMIDOBENZYLPENCILLIN
Formula: C17H21N3O6S2				
Mol Weight: 427.50				
MP (deg C):		FP (deg C):		
BP (deg C):				
BP pressure (mm Hg):				

Property/ Value	Units	Temp	Data Type	Reference
WS 3.19E+002	mg/L	25	EST	MEYLAN,WM ET AL. (1996)
logP 0.95			EXP	HANSCH,C & LEO,AJ (1985)
VP 8.53E-016	mm Hg	25	EST	NEELY,WB & BLAU,GE (1985)
DC	pKa			
HL 1.46E-017	atm m3/mol	25	EST	MEYLAN,WM & HOWARD,PH (1991)
OH 1.05E-010	cm3/molc sec	25	EST	MEYLAN,WM & HOWARD,PH (1993)

CAS #: 088695-06-1				1,4-BENZDAZEPIN-2-ON-5(2F-PH)-7,8-DICL
Formula: C15H9Cl2FN2O				
Mol Weight: 323.16				
MP (deg C):		FP (deg C):		
BP (deg C):				
BP pressure (mm Hg):				

Property/ Value	Units	Temp	Data Type	Reference
WS 7.13E+000	mg/L	25	EST	MEYLAN,WM ET AL. (1996)
logP 3.63			EXP	HANSCH,C & LEO,AJ (1985)
VP 2.10E-009	mm Hg	25	EST	NEELY,WB & BLAU,GE (1985)
DC	pKa			
HL 1.54E-010	atm m3/mol	25	EST	MEYLAN,WM & HOWARD,PH (1991)
OH 6.69E-012	cm3/molc sec	25	EST	MEYLAN,WM & HOWARD,PH (1993)

CAS #: 088695-07-2				1,4-BENZDIAZEPIN-2-ONE-5(26F-PH)8CL
Formula: C15H9ClF2N2O				
Mol Weight: 306.70				
MP (deg C):		FP (deg C):		
BP (deg C):				
BP pressure (mm Hg):				

Property/ Value	Units	Temp	Data Type	Reference
WS 3.89E+001	mg/L	25	EST	MEYLAN,WM ET AL. (1996)
logP 2.88			EXP	HANSCH,C & LEO,AJ (1985)
VP 9.35E-009	mm Hg	25	EST	NEELY,WB & BLAU,GE (1985)
DC	pKa			
HL 2.42E-010	atm m3/mol	25	EST	MEYLAN,WM & HOWARD,PH (1991)
OH 1.48E-011	cm3/molc sec	25	EST	MEYLAN,WM & HOWARD,PH (1993)

CAS #: 088695-10-7				1,4-BENZDAZEPIN-2-ON-1-ET-5-PH-7-AMINO
Formula: C17H17N3O				
Mol Weight: 279.34				
MP (deg C):		FP (deg C):		
BP (deg C):				
BP pressure (mm Hg):				

Property/ Value	Units	Temp	Data Type	Reference
WS 4.60E+002	mg/L	25	EST	MEYLAN,WM ET AL. (1996)
logP 1.81			EXP	HANSCH,C & LEO,AJ (1985)
VP 2.34E-009	mm Hg	25	EST	NEELY,WB & BLAU,GE (1985)
DC	pKa			
HL 2.31E-012	atm m3/mol	25	EST	MEYLAN,WM & HOWARD,PH (1991)
OH 1.33E-010	cm3/molc sec	25	EST	MEYLAN,WM & HOWARD,PH (1993)

CAS #: 088695-11-8				14-BENZODIAZPN2-ON,1,3-ME-5(2CLPH)7CL
Formula: C17H14Cl2N2O				
Mol Weight: 333.22				
MP (deg C):		FP (deg C):		
BP (deg C):				
BP pressure (mm Hg):				

Property/ Value	Units	Temp	Data Type	Reference
WS 9.76E+000	mg/L	25	EST	MEYLAN,WM ET AL. (1996)
logP 3.40			EXP	HANSCH,C & LEO,AJ (1985)
VP 4.27E-009	mm Hg	25	EST	NEELY,WB & BLAU,GE (1985)
DC	pKa			
HL 3.58E-009	atm m3/mol	25	EST	MEYLAN,WM & HOWARD,PH (1991)
OH 8.54E-012	cm3/molc sec	25	EST	MEYLAN,WM & HOWARD,PH (1993)

CAS #: 088695-12-9				1,4-BENZDAZEPIN-2-ON-1-MESME-5-PH-7AMINO
Formula: C17H17N3OS				
Mol Weight: 311.41				
MP (deg C):		FP (deg C):		
BP (deg C):				
BP pressure (mm Hg):				

Property/ Value	Units	Temp	Data Type	Reference
WS 2.41E+002	mg/L	25	EST	MEYLAN,WM ET AL. (1996)
logP 1.92			EXP	HANSCH,C & LEO,AJ (1985)
VP 2.23E-010	mm Hg	25	EST	NEELY,WB & BLAU,GE (1985)
DC	pKa			
HL 8.67E-012	atm m3/mol	25	EST	MEYLAN,WM & HOWARD,PH (1991)
OH 1.62E-010	cm3/molc sec	25	EST	MEYLAN,WM & HOWARD,PH (1993)

CAS #: 088695-13-0				1,4-BENZDAZEPN-2-ON-1-MEOME-3-OH-5PH-7AM
Formula: C17H17N3O3				
Mol Weight: 311.34				
MP (deg C):		FP (deg C):		
BP (deg C):				
BP pressure (mm Hg):				

Property/ Value	Units	Temp	Data Type	Reference
WS 2.41E+003	mg/L	25	EST	MEYLAN,WM ET AL. (1996)
logP 0.75			EXP	HANSCH,C & LEO,AJ (1985)
VP 4.14E-013	mm Hg	25	EST	NEELY,WB & BLAU,GE (1985)
DC	pKa			
HL 1.26E-012	atm m3/mol	25	EST	MEYLAN,WM & HOWARD,PH (1991)
OH 1.58E-010	cm3/molc sec	25	EST	MEYLAN,WM & HOWARD,PH (1993)

CAS #: 088793-46-8				1,2-PROPANEDIOL, 3-[(6-AMINO-3-CHLORO-5-NITROPYR
Formula: C7H10ClN5O4				
Mol Weight: 263.64				
MP (deg C):		FP (deg C):		
BP (deg C):				
BP pressure (mm Hg):				

Property/ Value	Units	Temp	Data Type	Reference
WS 3.06E+004	mg/L	25	EST	MEYLAN,WM ET AL. (1996)
logP -0.22			EXP	SANGSTER,J (1994)
VP 5.51E-011	mm Hg	25	EST	NEELY,WB & BLAU,GE (1985)
DC	pKa			
HL 7.61E-018	atm m3/mol	25	EST	MEYLAN,WM & HOWARD,PH (1991)
OH 3.54E-011	cm3/molc sec	25	EST	MEYLAN,WM & HOWARD,PH (1993)

1404

CAS #: 089073-49-4	BENZENESULFONAMIDE, 4-(1,3-DIETHYL-2,3,6,7-TETRA

Formula: $C_{19}H_{26}N_6O_4S$

Mol Weight: 434.52

MP (deg C):   FP (deg C):

BP (deg C):

BP pressure (mm Hg):

Property/ Value	Units	Temp	Data Type	Reference
WS 1.19E+001	mg/L	25	EST	MEYLAN,WM ET AL. (1996)
logP 2.57			EXP	HANSCH,C ET AL. (1995)
VP 5.09E-018	mm Hg	25	EST	NEELY,WB & BLAU,GE (1985)
DC	pKa			
HL 2.57E-020	atm m3/mol	25	EST	MEYLAN,WM & HOWARD,PH (1991)
OH 1.36E-010	cm3/molc sec	25	EST	MEYLAN,WM & HOWARD,PH (1993)

CAS #: 089083-18-1	ACETAMIDE, N-[5-CHLORO-6-[(2,3-DIHYDROXYPROPYL)A

Formula: $C_9H_{12}ClN_5O_5$

Mol Weight: 305.68

MP (deg C):   FP (deg C):

BP (deg C):

BP pressure (mm Hg):

Property/ Value	Units	Temp	Data Type	Reference
WS 5.57E+002	mg/L	25	EST	MEYLAN,WM ET AL. (1996)
logP 0.00			EXP	SANGSTER,J (1993)
VP 7.77E-014	mm Hg	25	EST	NEELY,WB & BLAU,GE (1985)
DC	pKa			
HL 2.47E-020	atm m3/mol	25	EST	MEYLAN,WM & HOWARD,PH (1991)
OH 2.65E-011	cm3/molc sec	25	EST	MEYLAN,WM & HOWARD,PH (1993)

CAS #: 089083-22-7	ACETAMIDE, N-[5-CHLORO-6-(4-METHYL-1-PIPERAZINYL)

Formula: $C_{11}H_{15}ClN_6O_3$

Mol Weight: 314.73

MP (deg C):   FP (deg C):

BP (deg C):

BP pressure (mm Hg):

Property/ Value	Units	Temp	Data Type	Reference
WS 1.53E+003	mg/L	25	EST	MEYLAN,WM ET AL. (1996)
logP 0.50			EXP	SANGSTER,J (1993)
VP 8.11E-010	mm Hg	25	EST	NEELY,WB & BLAU,GE (1985)
DC	pKa			
HL 6.23E-017	atm m3/mol	25	EST	MEYLAN,WM & HOWARD,PH (1991)
OH 1.14E-010	cm3/molc sec	25	EST	MEYLAN,WM & HOWARD,PH (1993)

CAS #: 089145-29-9	CARBAMIC ACID, [5-(3-FLUOROPHENYL)-4,5-DIHYDRO-1

Formula: $C_{13}H_{16}FN_3O_2$

Mol Weight: 265.29

MP (deg C):   FP (deg C):

BP (deg C):

BP pressure (mm Hg):

Property/ Value	Units	Temp	Data Type	Reference
WS 5.21E+002	mg/L	25	EST	MEYLAN,WM ET AL. (1996)
logP 1.84			EXP	SANGSTER,J (1994)
VP 1.30E-005	mm Hg	25	EST	NEELY,WB & BLAU,GE (1985)
DC	pKa			
HL 1.57E-013	atm m3/mol	25	EST	MEYLAN,WM & HOWARD,PH (1991)
OH 9.89E-011	cm3/molc sec	25	EST	MEYLAN,WM & HOWARD,PH (1993)

CAS #: 089145-31-3	CARBAMIC ACID, [5-(4-FLUOROPHENYL)-4,5-DIHYDRO-1

Formula: $C_{13}H_{16}FN_3O_2$

Mol Weight: 265.29

MP (deg C):   FP (deg C):

BP (deg C):

BP pressure (mm Hg):

Property/ Value	Units	Temp	Data Type	Reference
WS 4.63E+002	mg/L	25	EST	MEYLAN,WM ET AL. (1996)
logP 1.90			EXP	SANGSTER,J (1994)
VP 1.30E-005	mm Hg	25	EST	NEELY,WB & BLAU,GE (1985)
DC	pKa			
HL 1.57E-013	atm m3/mol	25	EST	MEYLAN,WM & HOWARD,PH (1991)
OH 9.60E-011	cm3/molc sec	25	EST	MEYLAN,WM & HOWARD,PH (1993)

CAS #: 089145-43-7	CARBAMIC ACID, [4,5-DIHYDRO-5-[4-(HYDROXYMETHYL)

Formula: $C_{14}H_{19}N_3O_3$

Mol Weight: 277.33

MP (deg C):   FP (deg C):

BP (deg C):

BP pressure (mm Hg):

Property/ Value	Units	Temp	Data Type	Reference
WS 3.18E+003	mg/L	25	EST	MEYLAN,WM ET AL. (1996)
logP 0.84			EXP	SANGSTER,J (1994)
VP 3.74E-009	mm Hg	25	EST	NEELY,WB & BLAU,GE (1985)
DC	pKa			
HL 5.41E-018	atm m3/mol	25	EST	MEYLAN,WM & HOWARD,PH (1991)
OH 1.02E-010	cm3/molc sec	25	EST	MEYLAN,WM & HOWARD,PH (1993)

CAS #: 089145-44-8	CARBAMIC ACID, [5-(4-FORMYLPHENYL)-4,5-DIHYDRO-1

Formula: $C_{13}H_{15}N_3O_3$

Mol Weight: 261.28

MP (deg C):   FP (deg C):

BP (deg C):

BP pressure (mm Hg):

Property/ Value	Units	Temp	Data Type	Reference
WS 4.78E+003	mg/L	25	EST	MEYLAN,WM ET AL. (1996)
logP 0.74			EXP	SANGSTER,J (1994)
VP 1.23E-006	mm Hg	25	EST	NEELY,WB & BLAU,GE (1985)
DC	pKa			
HL 2.52E-016	atm m3/mol	25	EST	MEYLAN,WM & HOWARD,PH (1991)
OH 1.10E-010	cm3/molc sec	25	EST	MEYLAN,WM & HOWARD,PH (1993)

CAS #: 089145-54-0	CARBAMIC ACID, [5-(2,5-DIBROMOPHENYL)-4,5-DIHYDR

Formula: $C_{12}H_{13}Br_2N_3O_2$

Mol Weight: 391.07

MP (deg C):   FP (deg C):

BP (deg C):

BP pressure (mm Hg):

Property/ Value	Units	Temp	Data Type	Reference
WS 1.50E+001	mg/L	25	EST	MEYLAN,WM ET AL. (1996)
logP 2.77			EXP	SANGSTER,J (1994)
VP 2.46E-007	mm Hg	25	EST	NEELY,WB & BLAU,GE (1985)
DC	pKa			
HL 1.60E-014	atm m3/mol	25	EST	MEYLAN,WM & HOWARD,PH (1991)
OH 9.25E-011	cm3/molc sec	25	EST	MEYLAN,WM & HOWARD,PH (1993)

CAS #: 089145-58-4 — CARBAMIC ACID, [5-(3-CHLOROPHENYL)-4,5-DIHYDRO-1

Formula: $C_{12}H_{14}ClN_3O_2$
Mol Weight: 267.72
MP (deg C): | FP (deg C):
BP (deg C):
BP pressure (mm Hg):

Property/Value	Units	Temp	Data Type	Reference
WS 3.03E+002	mg/L	25	EST	MEYLAN,WM ET AL. (1996)
logP 2.10			EXP	SANGSTER,J (1994)
VP 6.19E-006	mm Hg	25	EST	NEELY,WB & BLAU,GE (1985)
DC	pKa			
HL 7.49E-014	atm m3/mol	25	EST	MEYLAN,WM & HOWARD,PH (1991)
OH 9.49E-011	cm3/molc sec	25	EST	MEYLAN,WM & HOWARD,PH (1993)

CAS #: 089164-74-9 — 5,12-NAPHTHACENEDIONE PYRANOSYL DERIVATIVE

Formula: $C_{32}H_{34}N_2O_{11}$
Mol Weight: 622.63
MP (deg C): | FP (deg C):
BP (deg C):
BP pressure (mm Hg):

Property/Value	Units	Temp	Data Type	Reference
WS 1.14E+000	mg/L	25	EST	MEYLAN,WM ET AL. (1996)
logP 2.59			EXP	SANGSTER,J (1993)
VP 3.39E-026	mm Hg	25	EST	NEELY,WB & BLAU,GE (1985)
DC	pKa			
HL 1.96E-030	atm m3/mol	25	EST	MEYLAN,WM & HOWARD,PH (1991)
OH 1.78E-010	cm3/molc sec	25	EST	MEYLAN,WM & HOWARD,PH (1993)

CAS #: 089164-79-4 — 2-CYANOMORPHOLINO-DOXORUBICIN,12-IMINO

Formula: $C_{32}H_{35}N_3O_{11}$
Mol Weight: 637.65
MP (deg C): | FP (deg C):
BP (deg C):
BP pressure (mm Hg):

Property/Value	Units	Temp	Data Type	Reference
WS 3.03E-001	mg/L	25	EST	MEYLAN,WM ET AL. (1996)
logP 1.97			EXP	HANSCH,C ET AL. (1995)
VP 8.26E-029	mm Hg	25	EST	NEELY,WB & BLAU,GE (1985)
DC	pKa			
HL 1.80E-030	atm m3/mol	25	EST	MEYLAN,WM & HOWARD,PH (1991)
OH 2.90E-010	cm3/molc sec	25	EST	MEYLAN,WM & HOWARD,PH (1993)

CAS #: 089196-05-4 — MORPHOLINO-DOXORUBICIN,12-IMINO-

Formula: $C_{31}H_{36}N_2O_{11}$
Mol Weight: 612.64
MP (deg C): | FP (deg C):
BP (deg C):
BP pressure (mm Hg):

Property/Value	Units	Temp	Data Type	Reference
WS 1.14E+000	mg/L	25	EST	MEYLAN,WM ET AL. (1996)
logP 1.80			EXP	HANSCH,C ET AL. (1995)
VP 7.84E-027	mm Hg	25	EST	NEELY,WB & BLAU,GE (1985)
DC	pKa			
HL 4.33E-027	atm m3/mol	25	EST	MEYLAN,WM & HOWARD,PH (1991)
OH 3.74E-010	cm3/molc sec	25	EST	MEYLAN,WM & HOWARD,PH (1993)

CAS #: 089196-07-6 — 5,12-NAPHTHACENEDIONE PYRANOSYL DERIVATIVE

Formula: $C_{32}H_{34}N_2O_{12}$
Mol Weight: 638.63
MP (deg C): | FP (deg C):
BP (deg C):
BP pressure (mm Hg):

Property/Value	Units	Temp	Data Type	Reference
WS 2.98E+000	mg/L	25	EST	MEYLAN,WM ET AL. (1996)
logP 1.98			EXP	SANGSTER,J (1993)
VP 1.09E-028	mm Hg	25	EST	NEELY,WB & BLAU,GE (1985)
DC	pKa			
HL 3.05E-028	atm m3/mol	25	EST	MEYLAN,WM & HOWARD,PH (1991)
OH 1.80E-010	cm3/molc sec	25	EST	MEYLAN,WM & HOWARD,PH (1993)

CAS #: 089242-75-1 — 3-PYRIDINEMETHANOL,A-2-PROPYNYL-

Formula: C_9H_9NO
Mol Weight: 147.18
MP (deg C): | FP (deg C):
BP (deg C):
BP pressure (mm Hg):

Property/Value	Units	Temp	Data Type	Reference
WS 4.67E+005	mg/L	25	EST	MEYLAN,WM ET AL. (1996)
logP 1.21			EXP	HANSCH,C ET AL. (1995)
VP 8.45E-004	mm Hg	25	EST	NEELY,WB & BLAU,GE (1985)
DC	pKa			
HL 8.24E-011	atm m3/mol	25	EST	MEYLAN,WM & HOWARD,PH (1991)
OH 1.75E-011	cm3/molc sec	25	EST	MEYLAN,WM & HOWARD,PH (1993)

CAS #: 089242-77-3 — 3-PYRIDINEMETHANOL, à,à-BIS(2-PROPYNYL)-

Formula: $C_{12}H_{11}NO$
Mol Weight: 185.23
MP (deg C): | FP (deg C):
BP (deg C):
BP pressure (mm Hg):

Property/Value	Units	Temp	Data Type	Reference
WS 1.33E+004	mg/L	25	EST	MEYLAN,WM ET AL. (1996)
logP 1.29			EXP	SANGSTER,J (1993)
VP 3.00E-005	mm Hg	25	EST	NEELY,WB & BLAU,GE (1985)
DC	pKa			
HL 2.39E-011	atm m3/mol	25	EST	MEYLAN,WM & HOWARD,PH (1991)
OH 2.21E-011	cm3/molc sec	25	EST	MEYLAN,WM & HOWARD,PH (1993)

CAS #: 089242-78-4 — 3-PYRIDINEMETHANOL, _-(PHENYLMETHYL)-_-(2-PROPEN

Formula: $C_{16}H_{15}NO$
Mol Weight: 237.30
MP (deg C): | FP (deg C):
BP (deg C):
BP pressure (mm Hg):

Property/Value	Units	Temp	Data Type	Reference
WS 2.27E+003	mg/L	25	EST	MEYLAN,WM ET AL. (1996)
logP 1.87			EXP	HANSCH,C ET AL. (1995)
VP 2.22E-007	mm Hg	25	EST	NEELY,WB & BLAU,GE (1985)
DC	pKa			
HL 8.83E-012	atm m3/mol	25	EST	MEYLAN,WM & HOWARD,PH (1991)
OH 1.99E-011	cm3/molc sec	25	EST	MEYLAN,WM & HOWARD,PH (1993)

CAS #: 089242-79-5	3-PYRIDINEMETHANOL, à-CYCLOHEXYL-à-(2-PROPYNYL)-

Formula: $C_{15}H_{19}NO$

Mol Weight: 229.32

MP (deg C):   FP (deg C):

BP (deg C):

BP pressure (mm Hg):

Property/ Value	Units	Temp	Data Type	Reference
WS 1.63E+003	mg/L	25	EST	MEYLAN,WM ET AL. (1996)
logP 2.09			EXP	SANGSTER,J (1993)
VP 1.36E-006	mm Hg	25	EST	NEELY,WB & BLAU,GE (1985)
DC	pKa			
HL 1.99E-010	atm m3/mol	25	EST	MEYLAN,WM & HOWARD,PH (1991)
OH 2.97E-011	cm3/molc sec	25	EST	MEYLAN,WM & HOWARD,PH (1993)

CAS #: 089242-81-9	3-PYRIDINEMETHANOL, à-PHENYL-à-(2-PROPYNYL)-

Formula: $C_{15}H_{13}NO$

Mol Weight: 223.28

MP (deg C):   FP (deg C):

BP (deg C):

BP pressure (mm Hg):

Property/ Value	Units	Temp	Data Type	Reference
WS 4.01E+003	mg/L	25	EST	MEYLAN,WM ET AL. (1996)
logP 1.67			EXP	SANGSTER,J (1993)
VP 5.84E-007	mm Hg	25	EST	NEELY,WB & BLAU,GE (1985)
DC	pKa			
HL 6.65E-012	atm m3/mol	25	EST	MEYLAN,WM & HOWARD,PH (1991)
OH 1.58E-011	cm3/molc sec	25	EST	MEYLAN,WM & HOWARD,PH (1993)

CAS #: 089242-82-0	3-PYRIDINEMETHANOL,A-2-PROPYNYL,A-4-FLUOROPHENYL

Formula: $C_{15}H_{12}FNO$

Mol Weight: 241.27

MP (deg C):   FP (deg C):

BP (deg C):

BP pressure (mm Hg):

Property/ Value	Units	Temp	Data Type	Reference
WS 2.58E+003	mg/L	25	EST	MEYLAN,WM ET AL. (1996)
logP 1.78			EXP	HANSCH,C ET AL. (1995)
VP 6.61E-007	mm Hg	25	EST	NEELY,WB & BLAU,GE (1985)
DC	pKa			
HL 7.76E-012	atm m3/mol	25	EST	MEYLAN,WM & HOWARD,PH (1991)
OH 1.44E-011	cm3/molc sec	25	EST	MEYLAN,WM & HOWARD,PH (1993)

CAS #: 089242-87-5	3-PYRIDINEMETHANOL, _-(4-METHYLPHENYL)-_-(2-PROP

Formula: $C_{16}H_{15}NO$

Mol Weight: 237.30

MP (deg C):   FP (deg C):

BP (deg C):

BP pressure (mm Hg):

Property/ Value	Units	Temp	Data Type	Reference
WS 2.46E+003	mg/L	25	EST	MEYLAN,WM ET AL. (1996)
logP 1.83			EXP	HANSCH,C ET AL. (1995)
VP 2.06E-007	mm Hg	25	EST	NEELY,WB & BLAU,GE (1985)
DC	pKa			
HL 7.34E-012	atm m3/mol	25	EST	MEYLAN,WM & HOWARD,PH (1991)
OH 1.77E-011	cm3/molc sec	25	EST	MEYLAN,WM & HOWARD,PH (1993)

CAS #: 089246-38-8	ACETANILIDE,N-(4-CYANOPHENYL)-

Formula: $C_{15}H_{12}N_2O$

Mol Weight: 236.28

MP (deg C):   FP (deg C):

BP (deg C):

BP pressure (mm Hg):

Property/ Value	Units	Temp	Data Type	Reference
WS 5.97E+001	mg/L	25	EST	MEYLAN,WM ET AL. (1996)
logP 2.82			EXP	HANSCH,C ET AL. (1995)
VP 2.59E-008	mm Hg	25	EST	NEELY,WB & BLAU,GE (1985)
DC	pKa			
HL 4.81E-012	atm m3/mol	25	EST	MEYLAN,WM & HOWARD,PH (1991)
OH 1.40E-011	cm3/molc sec	25	EST	MEYLAN,WM & HOWARD,PH (1993)

CAS #: 089246-39-9	BENZENEACETAMIDE, N-(4-ACETYLPHENYL)-

Formula: $C_{16}H_{15}NO_2$

Mol Weight: 253.30

MP (deg C):   FP (deg C):

BP (deg C):

BP pressure (mm Hg):

Property/ Value	Units	Temp	Data Type	Reference
WS 1.02E+002	mg/L	25	EST	MEYLAN,WM ET AL. (1996)
logP 2.75			EXP	YAMAGAMI,C ET AL. (1984)
VP 2.82E-008	mm Hg	25	EST	NEELY,WB & BLAU,GE (1985)
DC	pKa			
HL 9.07E-013	atm m3/mol	25	EST	MEYLAN,WM & HOWARD,PH (1991)
OH 1.70E-011	cm3/molc sec	25	EST	MEYLAN,WM & HOWARD,PH (1993)

CAS #: 089246-40-2	BENZENEACETAMIDE, N-(3-CYANOPHENYL)-

Formula: $C_{15}H_{12}N_2O$

Mol Weight: 236.28

MP (deg C):   FP (deg C):

BP (deg C):

BP pressure (mm Hg):

Property/ Value	Units	Temp	Data Type	Reference
WS 6.85E+001	mg/L	25	EST	MEYLAN,WM ET AL. (1996)
logP 2.75			EXP	YAMAGAMI,C ET AL. (1984)
VP 2.59E-008	mm Hg	25	EST	NEELY,WB & BLAU,GE (1985)
DC	pKa			
HL 4.81E-012	atm m3/mol	25	EST	MEYLAN,WM & HOWARD,PH (1991)
OH 7.17E-012	cm3/molc sec	25	EST	MEYLAN,WM & HOWARD,PH (1993)

CAS #: 089529-99-7	1-(4-CARBAMOYLPHENYL)-3-METHYL-3-PROPYLTRIAZENE

Formula: $C_{11}H_{16}N_4O$

Mol Weight: 220.28

MP (deg C):   FP (deg C):

BP (deg C):

BP pressure (mm Hg):

Property/ Value	Units	Temp	Data Type	Reference
WS 4.81E+002	mg/L	25	EST	MEYLAN,WM ET AL. (1996)
logP 2.17			EXP	HANSCH,C ET AL. (1995)
VP 1.11E-006	mm Hg	25	EST	NEELY,WB & BLAU,GE (1985)
DC	pKa			
HL 3.89E-013	atm m3/mol	25	EST	MEYLAN,WM & HOWARD,PH (1991)
OH 1.56E-011	cm3/molc sec	25	EST	MEYLAN,WM & HOWARD,PH (1993)

CAS #: 089530-00-7 — 1-(4-CARBAMOYLPHENYL)-3-METHYL-3-HEXYLTRIAZENE

Formula: $C_{14}H_{22}N_4O$

Mol Weight: 262.36

MP (deg C): | FP (deg C):

BP (deg C):

BP pressure (mm Hg):

Property/Value	Units	Temp	Data Type	Reference
WS 1.43E+001	mg/L	25	EST	MEYLAN,WM ET AL. (1996)
logP 3.69			EXP	HANSCH,C ET AL. (1995)
VP 1.04E-007	mm Hg	25	EST	NEELY,WB & BLAU,GE (1985)
DC	pKa			
HL 9.10E-013	atm m3/mol	25	EST	MEYLAN,WM & HOWARD,PH (1991)
OH 1.98E-011	cm3/molc sec	25	EST	MEYLAN,WM & HOWARD,PH (1993)

CAS #: 089530-01-8 — 1-(4-CARBAMOYLPHENYL)-3-METHYL-3-HEPTYLTRIAZENE

Formula: $C_{15}H_{24}N_4O$

Mol Weight: 276.38

MP (deg C): | FP (deg C):

BP (deg C):

BP pressure (mm Hg):

Property/Value	Units	Temp	Data Type	Reference
WS 4.35E+000	mg/L	25	EST	MEYLAN,WM ET AL. (1996)
logP 4.20			EXP	HANSCH,C ET AL. (1995)
VP 4.41E-008	mm Hg	25	EST	NEELY,WB & BLAU,GE (1985)
DC	pKa			
HL 1.21E-012	atm m3/mol	25	EST	MEYLAN,WM & HOWARD,PH (1991)
OH 2.12E-011	cm3/molc sec	25	EST	MEYLAN,WM & HOWARD,PH (1993)

CAS #: 089599-01-9 — BENZENESULFONAMIDE, 3-BROMO-

Formula: $C_6H_6BrNO_2S$

Mol Weight: 236.09

MP (deg C): | FP (deg C):

BP (deg C):

BP pressure (mm Hg):

Property/Value	Units	Temp	Data Type	Reference
WS 1.98E+003	mg/L	25	EST	MEYLAN,WM ET AL. (1996)
logP 1.35			EXP	CAROTTI,A ET AL. (1989)
VP 5.48E-005	mm Hg	25	EST	NEELY,WB & BLAU,GE (1985)
DC	pKa			
HL 1.68E-007	atm m3/mol	25	EST	MEYLAN,WM & HOWARD,PH (1991)
OH 2.62E-013	cm3/molc sec	25	EST	MEYLAN,WM & HOWARD,PH (1993)

CAS #: 089674-85-1 — TETRACHLOROCYCLOHEXENE (345)

Formula: $C_6H_6Cl_4$

Mol Weight: 219.93

MP (deg C): | FP (deg C):

BP (deg C):

BP pressure (mm Hg):

Property/Value	Units	Temp	Data Type	Reference
WS 3.40E+001	mg/L	25	EST	MEYLAN,WM ET AL. (1996)
logP 3.72			EXP	HANSCH,C & LEO,AJ (1985)
VP 4.69E-002	mm Hg	25	EST	NEELY,WB & BLAU,GE (1985)
DC	pKa			
HL 3.70E-003	atm m3/mol	25	EST	MEYLAN,WM & HOWARD,PH (1991)
OH 1.49E-011	cm3/molc sec	25	EST	MEYLAN,WM & HOWARD,PH (1993)

CAS #: 089674-87-3 — TETRACHLOROCYCLOHEXENE (345)

Formula: $C_6H_6Cl_4$

Mol Weight: 219.93

MP (deg C): | FP (deg C):

BP (deg C):

BP pressure (mm Hg):

Property/Value	Units	Temp	Data Type	Reference
WS 3.40E+001	mg/L	25	EST	MEYLAN,WM ET AL. (1996)
logP 3.65			EXP	HANSCH,C & LEO,AJ (1985)
VP 4.69E-002	mm Hg	25	EST	NEELY,WB & BLAU,GE (1985)
DC	pKa			
HL 3.70E-003	atm m3/mol	25	EST	MEYLAN,WM & HOWARD,PH (1991)
OH 1.49E-011	cm3/molc sec	25	EST	MEYLAN,WM & HOWARD,PH (1993)

CAS #: 089674-88-4 — TETRACHLOROCYCLOHEXENE (345)

Formula: $C_6H_6Cl_4$

Mol Weight: 219.93

MP (deg C): | FP (deg C):

BP (deg C):

BP pressure (mm Hg):

Property/Value	Units	Temp	Data Type	Reference
WS 3.40E+001	mg/L	25	EST	MEYLAN,WM ET AL. (1996)
logP 3.52			EXP	HANSCH,C & LEO,AJ (1985)
VP 4.69E-002	mm Hg	25	EST	NEELY,WB & BLAU,GE (1985)
DC	pKa			
HL 3.70E-003	atm m3/mol	25	EST	MEYLAN,WM & HOWARD,PH (1991)
OH 1.49E-011	cm3/molc sec	25	EST	MEYLAN,WM & HOWARD,PH (1993)

CAS #: 089690-74-4 — ACETAMIDE, N-[5-CHLORO-6-[[2,3-DIHYDROXY-1-(HYDR

Formula: $C_{10}H_{14}ClN_5O_6$

Mol Weight: 335.71

MP (deg C): | FP (deg C):

BP (deg C):

BP pressure (mm Hg):

Property/Value	Units	Temp	Data Type	Reference
WS 1.35E+003	mg/L	25	EST	MEYLAN,WM ET AL. (1996)
logP -0.66			EXP	SANGSTER,J (1993)
VP 4.34E-016	mm Hg	25	EST	NEELY,WB & BLAU,GE (1985)
DC	pKa			
HL 2.39E-020	atm m3/mol	25	EST	MEYLAN,WM & HOWARD,PH (1991)
OH 9.51E-011	cm3/molc sec	25	EST	MEYLAN,WM & HOWARD,PH (1993)

CAS #: 089690-76-6 — 1,2,4-BUTANETRIOL, 3-[(5-NITROPYRAZINYL)AMINO]-

Formula: $C_8H_{12}N_4O_5$

Mol Weight: 244.21

MP (deg C): | FP (deg C):

BP (deg C):

BP pressure (mm Hg):

Property/Value	Units	Temp	Data Type	Reference
WS 1.47E+005	mg/L	25	EST	MEYLAN,WM ET AL. (1996)
logP -0.89			EXP	HANSCH,C ET AL. (1995)
VP 8.87E-011	mm Hg	25	EST	NEELY,WB & BLAU,GE (1985)
DC	pKa			
HL 2.04E-017	atm m3/mol	25	EST	MEYLAN,WM & HOWARD,PH (1991)
OH 9.44E-011	cm3/molc sec	25	EST	MEYLAN,WM & HOWARD,PH (1993)

CAS #: 089770-25-2				4-ACETYL-7-CHLOROQUINOLINE

Formula: $C_{11}H_8ClNO$

Mol Weight: 205.65

MP (deg C):   FP (deg C):

BP (deg C):

BP pressure (mm Hg):

Property/ Value	Units	Temp	Data Type	Reference
WS 2.37E+002	mg/L	25	EST	MEYLAN,WM ET AL. (1996)
logP 2.62			EXP	HANSCH,C & LEO,AJ (1985)
VP 8.54E-005	mm Hg	25	EST	NEELY,WB & BLAU,GE (1985)
DC	pKa			
HL 9.29E-010	atm m3/mol	25	EST	MEYLAN,WM & HOWARD,PH (1991)
OH 4.15E-012	cm3/molc sec	25	EST	MEYLAN,WM & HOWARD,PH (1993)

CAS #: 089770-26-3				4-QUINOLINAMINE,8-NITRO

Formula: $C_9H_7N_3O_2$

Mol Weight: 189.18

MP (deg C):   FP (deg C):

BP (deg C):

BP pressure (mm Hg):

Property/ Value	Units	Temp	Data Type	Reference
WS 1.54E+003	mg/L	25	EST	MEYLAN,WM ET AL. (1996)
logP 1.31			EXP	HANSCH,C & LEO,AJ (1985)
VP 5.24E-006	mm Hg	25	EST	NEELY,WB & BLAU,GE (1985)
DC	pKa			
HL 9.60E-013	atm m3/mol	25	EST	MEYLAN,WM & HOWARD,PH (1991)
OH 1.43E-011	cm3/molc sec	25	EST	MEYLAN,WM & HOWARD,PH (1993)

CAS #: 089770-27-4				7-CL-4-PROPIONYLAMINO-QUINOLINE

Formula: $C_{12}H_{11}ClN_2O$

Mol Weight: 234.69

MP (deg C):   FP (deg C):

BP (deg C):

BP pressure (mm Hg):

Property/ Value	Units	Temp	Data Type	Reference
WS 1.34E+002	mg/L	25	EST	MEYLAN,WM ET AL. (1996)
logP 2.73			EXP	HANSCH,C & LEO,AJ (1985)
VP 1.25E-007	mm Hg	25	EST	NEELY,WB & BLAU,GE (1985)
DC	pKa			
HL 7.76E-013	atm m3/mol	25	EST	MEYLAN,WM & HOWARD,PH (1991)
OH 1.31E-011	cm3/molc sec	25	EST	MEYLAN,WM & HOWARD,PH (1993)

CAS #: 089770-28-5				4-MEO-8-NO2 QUINOLINE

Formula: $C_{10}H_8N_2O_3$

Mol Weight: 204.19

MP (deg C):   FP (deg C):

BP (deg C):

BP pressure (mm Hg):

Property/ Value	Units	Temp	Data Type	Reference
WS 3.89E+002	mg/L	25	EST	MEYLAN,WM ET AL. (1996)
logP 1.92			EXP	HANSCH,C & LEO,AJ (1985)
VP 2.38E-005	mm Hg	25	EST	NEELY,WB & BLAU,GE (1985)
DC	pKa			
HL 1.61E-010	atm m3/mol	25	EST	MEYLAN,WM & HOWARD,PH (1991)
OH 5.48E-012	cm3/molc sec	25	EST	MEYLAN,WM & HOWARD,PH (1993)

CAS #: 089770-29-6				6-METHYLSULFONYLQUINOLINE

Formula: $C_{10}H_9NO_2S$

Mol Weight: 207.25

MP (deg C):   FP (deg C):

BP (deg C):

BP pressure (mm Hg):

Property/ Value	Units	Temp	Data Type	Reference
WS 1.01E+004	mg/L	25	EST	MEYLAN,WM ET AL. (1996)
logP 0.70			EXP	HANSCH,C & LEO,AJ (1985)
VP 8.90E-006	mm Hg	25	EST	NEELY,WB & BLAU,GE (1985)
DC	pKa			
HL 1.59E-010	atm m3/mol	25	EST	MEYLAN,WM & HOWARD,PH (1991)
OH 4.14E-012	cm3/molc sec	25	EST	MEYLAN,WM & HOWARD,PH (1993)

CAS #: 089770-30-9				6-PHENYLSULFONYLQUINOLINE

Formula: $C_{15}H_{11}NO_2S$

Mol Weight: 269.32

MP (deg C):   FP (deg C):

BP (deg C):

BP pressure (mm Hg):

Property/ Value	Units	Temp	Data Type	Reference
WS 1.18E+002	mg/L	25	EST	MEYLAN,WM ET AL. (1996)
logP 2.57			EXP	HANSCH,C & LEO,AJ (1985)
VP 5.02E-008	mm Hg	25	EST	NEELY,WB & BLAU,GE (1985)
DC	pKa			
HL 3.19E-011	atm m3/mol	25	EST	MEYLAN,WM & HOWARD,PH (1991)
OH 3.50E-012	cm3/molc sec	25	EST	MEYLAN,WM & HOWARD,PH (1993)

CAS #: 089770-31-0				6-SULFONAMIDOQUINOLINE

Formula: $C_9H_8N_2O_2S$

Mol Weight: 208.24

MP (deg C):   FP (deg C):

BP (deg C):

BP pressure (mm Hg):

Property/ Value	Units	Temp	Data Type	Reference
WS 1.32E+004	mg/L	25	EST	MEYLAN,WM ET AL. (1996)
logP 0.56			EXP	HANSCH,C & LEO,AJ (1985)
VP 2.02E-006	mm Hg	25	EST	NEELY,WB & BLAU,GE (1985)
DC	pKa			
HL 5.39E-011	atm m3/mol	25	EST	MEYLAN,WM & HOWARD,PH (1991)
OH 3.08E-012	cm3/molc sec	25	EST	MEYLAN,WM & HOWARD,PH (1993)

CAS #: 089770-32-1				7-DIMETHYLAMINOQUINOLINE

Formula: $C_{11}H_{12}N_2$

Mol Weight: 172.23

MP (deg C):   FP (deg C):

BP (deg C):

BP pressure (mm Hg):

Property/ Value	Units	Temp	Data Type	Reference
WS 2.91E+002	mg/L	25	EST	MEYLAN,WM ET AL. (1996)
logP 2.71			EXP	HANSCH,C & LEO,AJ (1985)
VP 1.11E-003	mm Hg	25	EST	NEELY,WB & BLAU,GE (1985)
DC	pKa			
HL 1.09E-008	atm m3/mol	25	EST	MEYLAN,WM & HOWARD,PH (1991)
OH 2.03E-010	cm3/molc sec	25	EST	MEYLAN,WM & HOWARD,PH (1993)

CAS #: 089770-33-2	8-PHENYLSULFONYLQUINOLINE
Formula: $C_{15}H_{11}NO_2S$	
Mol Weight: 269.32	
MP (deg C):	FP (deg C):
BP (deg C):	
BP pressure (mm Hg):	

Property/Value	Units	Temp	Data Type	Reference
WS 1.89E+002	mg/L	25	EST	MEYLAN,WM ET AL. (1996)
logP 2.33			EXP	HANSCH,C & LEO,AJ (1985)
VP 5.02E-008	mm Hg	25	EST	NEELY,WB & BLAU,GE (1985)
DC	pKa			
HL 3.19E-011	atm m3/mol	25	EST	MEYLAN,WM & HOWARD,PH (1991)
OH 3.50E-012	cm3/molc sec	25	EST	MEYLAN,WM & HOWARD,PH (1993)

CAS #: 089770-34-3	6-ETHOXYQUINOXALINE
Formula: $C_{10}H_{10}N_2O$	
Mol Weight: 174.20	
MP (deg C):	FP (deg C):
BP (deg C):	
BP pressure (mm Hg):	

Property/Value	Units	Temp	Data Type	Reference
WS 7.46E+002	mg/L	25	EST	MEYLAN,WM ET AL. (1996)
logP 2.22			EXP	HANSCH,C & LEO,AJ (1985)
VP 5.63E-004	mm Hg	25	EST	NEELY,WB & BLAU,GE (1985)
DC	pKa			
HL 2.24E-008	atm m3/mol	25	EST	MEYLAN,WM & HOWARD,PH (1991)
OH 2.82E-011	cm3/molc sec	25	EST	MEYLAN,WM & HOWARD,PH (1993)

CAS #: 089770-35-4	2-METHYLAMINO-4-PHENYLQUINAZOLINE
Formula: $C_{15}H_{13}N_3$	
Mol Weight: 235.29	
MP (deg C):	FP (deg C):
BP (deg C):	
BP pressure (mm Hg):	

Property/Value	Units	Temp	Data Type	Reference
WS 2.50E+001	mg/L	25	EST	MEYLAN,WM ET AL. (1996)
logP 3.58			EXP	HANSCH,C & LEO,AJ (1985)
VP 4.03E-007	mm Hg	25	EST	NEELY,WB & BLAU,GE (1985)
DC	pKa			
HL 6.02E-008	atm m3/mol	25	EST	MEYLAN,WM & HOWARD,PH (1991)
OH 1.80E-011	cm3/molc sec	25	EST	MEYLAN,WM & HOWARD,PH (1993)

CAS #: 089770-36-5	3-BUTOXYPYRIDAZINE
Formula: $C_8H_{12}N_2O$	
Mol Weight: 152.20	
MP (deg C):	FP (deg C):
BP (deg C):	
BP pressure (mm Hg):	

Property/Value	Units	Temp	Data Type	Reference
WS 2.15E+003	mg/L	25	EST	MEYLAN,WM ET AL. (1996)
logP 1.80			EXP	HANSCH,C & LEO,AJ (1985)
VP 1.65E-003	mm Hg	25	EST	NEELY,WB & BLAU,GE (1985)
DC	pKa			
HL 3.62E-006	atm m3/mol	25	EST	MEYLAN,WM & HOWARD,PH (1991)
OH 1.49E-011	cm3/molc sec	25	EST	MEYLAN,WM & HOWARD,PH (1993)

CAS #: 089770-37-6	3-ACETYLCINNOLINE
Formula: $C_{10}H_8N_2O$	
Mol Weight: 172.19	
MP (deg C):	FP (deg C):
BP (deg C):	
BP pressure (mm Hg):	

Property/Value	Units	Temp	Data Type	Reference
WS 3.54E+003	mg/L	25	EST	MEYLAN,WM ET AL. (1996)
logP 1.44			EXP	HANSCH,C & LEO,AJ (1985)
VP 1.40E-005	mm Hg	25	EST	NEELY,WB & BLAU,GE (1985)
DC	pKa			
HL 5.10E-010	atm m3/mol	25	EST	MEYLAN,WM & HOWARD,PH (1991)
OH 1.93E-012	cm3/molc sec	25	EST	MEYLAN,WM & HOWARD,PH (1993)

CAS #: 089770-38-7	7ET-3CINNOLINCARBOXLIC ACID,ET ESTER
Formula: $C_{13}H_{14}N_2O_2$	
Mol Weight: 230.27	
MP (deg C):	FP (deg C):
BP (deg C):	
BP pressure (mm Hg):	

Property/Value	Units	Temp	Data Type	Reference
WS 3.23E+002	mg/L	25	EST	MEYLAN,WM ET AL. (1996)
logP 2.31			EXP	HANSCH,C & LEO,AJ (1985)
VP 8.66E-007	mm Hg	25	EST	NEELY,WB & BLAU,GE (1985)
DC	pKa			
HL 3.51E-009	atm m3/mol	25	EST	MEYLAN,WM & HOWARD,PH (1991)
OH 3.70E-012	cm3/molc sec	25	EST	MEYLAN,WM & HOWARD,PH (1993)

CAS #: 089770-39-8	4-METHYL-6-CHLOROCINNOLINE
Formula: $C_9H_7ClN_2$	
Mol Weight: 178.62	
MP (deg C):	FP (deg C):
BP (deg C):	
BP pressure (mm Hg):	

Property/Value	Units	Temp	Data Type	Reference
WS 1.14E+003	mg/L	25	EST	MEYLAN,WM ET AL. (1996)
logP 1.96			EXP	HANSCH,C & LEO,AJ (1985)
VP 4.41E-005	mm Hg	25	EST	NEELY,WB & BLAU,GE (1985)
DC	pKa			
HL 2.29E-007	atm m3/mol	25	EST	MEYLAN,WM & HOWARD,PH (1991)
OH 1.76E-012	cm3/molc sec	25	EST	MEYLAN,WM & HOWARD,PH (1993)

CAS #: 089770-40-1	4-METHYL-7-CHLOROCINNOLINE
Formula: $C_9H_7ClN_2$	
Mol Weight: 178.62	
MP (deg C):	FP (deg C):
BP (deg C):	
BP pressure (mm Hg):	

Property/Value	Units	Temp	Data Type	Reference
WS 1.14E+003	mg/L	25	EST	MEYLAN,WM ET AL. (1996)
logP 1.98			EXP	HANSCH,C & LEO,AJ (1985)
VP 4.41E-005	mm Hg	25	EST	NEELY,WB & BLAU,GE (1985)
DC	pKa			
HL 2.29E-007	atm m3/mol	25	EST	MEYLAN,WM & HOWARD,PH (1991)
OH 1.76E-012	cm3/molc sec	25	EST	MEYLAN,WM & HOWARD,PH (1993)

CAS #: 089770-41-2	8-METHYL-4-CINNOLINAMINE

Formula: $C_9H_9N_3$
Mol Weight: 159.19
MP (deg C): | FP (deg C):
BP (deg C):
BP pressure (mm Hg):

Property/Value	Units	Temp	Data Type	Reference
WS 3.15E+003	mg/L	25	EST	MEYLAN,WM ET AL. (1996)
logP 1.57			EXP	HANSCH,C & LEO,AJ (1985)
VP 6.53E-006	mm Hg	25	EST	NEELY,WB & BLAU,GE (1985)
DC	pKa			
HL 1.09E-010	atm m3/mol	25	EST	MEYLAN,WM & HOWARD,PH (1991)
OH 3.42E-011	cm3/molc sec	25	EST	MEYLAN,WM & HOWARD,PH (1993)

CAS #: 089789-53-7	PENTANONE,PYRROL-2-YL

Formula: $C_9H_{13}NO$
Mol Weight: 151.21
MP (deg C): | FP (deg C):
BP (deg C):
BP pressure (mm Hg):

Property/Value	Units	Temp	Data Type	Reference
WS 6.43E+002	mg/L	25	EST	MEYLAN,WM ET AL. (1996)
logP 2.42			EXP	HANSCH,C & LEO,AJ (1985)
VP 6.67E-003	mm Hg	25	EST	NEELY,WB & BLAU,GE (1985)
DC	pKa			
HL 3.87E-008	atm m3/mol	25	EST	MEYLAN,WM & HOWARD,PH (1991)
OH 1.07E-010	cm3/molc sec	25	EST	MEYLAN,WM & HOWARD,PH (1993)

CAS #: 089843-47-0	1H-BENZIMIDAZOLE, 6-NITRO-

Formula: $C_7H_5N_3O_2$
Mol Weight: 163.14
MP (deg C): | FP (deg C):
BP (deg C):
BP pressure (mm Hg):

Property/Value	Units	Temp	Data Type	Reference
WS 1.41E+003	mg/L	25	EST	MEYLAN,WM ET AL. (1996)
logP 1.50			EXP	SCHULTZ,TW & APPLEHANS,FM (1985)
VP 9.81E-007	mm Hg	25	EST	NEELY,WB & BLAU,GE (1985)
DC	pKa			
HL 1.45E-009	atm m3/mol	25	EST	MEYLAN,WM & HOWARD,PH (1991)
OH 4.50E-012	cm3/molc sec	25	EST	MEYLAN,WM & HOWARD,PH (1993)

CAS #: 089894-13-3	PHENOXYACETIC ACID, 4-CL-3-NO2

Formula: $C_8H_6ClNO_5$
Mol Weight: 231.59
MP (deg C): | FP (deg C):
BP (deg C):
BP pressure (mm Hg):

Property/Value	Units	Temp	Data Type	Reference
WS 7.94E+002	mg/L	25	EST	MEYLAN,WM ET AL. (1996)
logP 1.85			EXP	HANSCH,C & LEO,AJ (1985)
VP 3.30E-006	mm Hg	25	EST	NEELY,WB & BLAU,GE (1985)
DC 2.96	pKa	25	EXP	KORTUM,G ET AL (1961)
HL 4.91E-011	atm m3/mol	25	EST	MEYLAN,WM & HOWARD,PH (1991)
OH 5.34E-012	cm3/molc sec	25	EST	MEYLAN,WM & HOWARD,PH (1993)

CAS #: 089937-39-3	N(2,5-DIME-PYRROL-1-YL)-6-($-MORPHOL)-3-PYRIDAZ-

Formula: $C_{16}H_{21}N_5O_2$
Mol Weight: 315.38
MP (deg C): | FP (deg C):
BP (deg C):
BP pressure (mm Hg):

Property/Value	Units	Temp	Data Type	Reference
WS 3.97E+002	mg/L	25	EST	MEYLAN,WM ET AL. (1996)
logP 1.64			EXP	HANSCH,C ET AL. (1995)
VP 3.71E-010	mm Hg	25	EST	NEELY,WB & BLAU,GE (1985)
DC	pKa			
HL 1.45E-017	atm m3/mol	25	EST	MEYLAN,WM & HOWARD,PH (1991)
OH 2.75E-010	cm3/molc sec	25	EST	MEYLAN,WM & HOWARD,PH (1993)

CAS #: 090032-54-5	3-FURANMETHANAMINIUM, TETRAHYDRO-N,N,N-TRIMETHYL

Formula: $C_8H_{18}INO$
Mol Weight: 271.14
MP (deg C): | FP (deg C):
BP (deg C):
BP pressure (mm Hg):

Property/Value	Units	Temp	Data Type	Reference
WS 1.00E+006	mg/L	25	EST	MEYLAN,WM ET AL. (1996)
logP -2.65			EXP	PRATESI,P RT AL. (1984)
VP 5.71E-008	mm Hg	25	EST	NEELY,WB & BLAU,GE (1985)
DC	pKa			
HL 4.00E-015	atm m3/mol	25	EST	MEYLAN,WM & HOWARD,PH (1991)
OH 3.71E-011	cm3/molc sec	25	EST	MEYLAN,WM & HOWARD,PH (1993)

CAS #: 090047-59-9	3-ALLYL-6-CHLOROBENZAZEPINE

Formula: $C_{13}H_{16}ClN$
Mol Weight: 221.73
MP (deg C): | FP (deg C):
BP (deg C):
BP pressure (mm Hg):

Property/Value	Units	Temp	Data Type	Reference
WS 5.22E+002	mg/L	25	EST	MEYLAN,WM ET AL. (1996)
logP 3.30			EXP	HANSCH,C ET AL. (1995)
VP 4.78E-004	mm Hg	25	EST	NEELY,WB & BLAU,GE (1985)
DC	pKa			
HL 1.85E-006	atm m3/mol	25	EST	MEYLAN,WM & HOWARD,PH (1991)
OH 1.29E-010	cm3/molc sec	25	EST	MEYLAN,WM & HOWARD,PH (1993)

CAS #: 090110-78-4	2,6-DINO2 ACETANILIDE

Formula: $C_8H_7N_3O_5$
Mol Weight: 225.16
MP (deg C): | FP (deg C):
BP (deg C):
BP pressure (mm Hg):

Property/Value	Units	Temp	Data Type	Reference
WS 4.38E+003	mg/L	25	EST	MEYLAN,WM ET AL. (1996)
logP 0.56			EXP	NAKAGAWA,Y ET AL. (1992)
VP 9.00E-008	mm Hg	25	EST	NEELY,WB & BLAU,GE (1985)
DC	pKa			
HL 5.66E-010	atm m3/mol	25	EST	MEYLAN,WM & HOWARD,PH (1991)
OH 2.96E-013	cm3/molc sec	25	EST	MEYLAN,WM & HOWARD,PH (1993)

CAS #: 090197-63-0				BARBITURIC ACID, 5-T-BU-

Formula: $C_8H_{12}N_2O_3$

Mol Weight: 184.20

MP (deg C): | FP (deg C):

BP (deg C):

BP pressure (mm Hg):

Property/Value	Units	Temp	Data Type	Reference
WS 3.03E+004	mg/L	25	EST	MEYLAN,WM ET AL. (1996)
logP 0.28			EXP	WONG,O & MCKEOWN,RH (1988)
VP 2.35E-010	mm Hg	25	EST	NEELY,WB & BLAU,GE (1985)
DC	pKa			
HL 3.61E-013	atm m3/mol	25	EST	MEYLAN,WM & HOWARD,PH (1991)
OH 3.84E-012	cm3/molc sec	25	EST	MEYLAN,WM & HOWARD,PH (1993)

CAS #: 090355-74-1				2,6-PIPERIDINEDIONE, 3-ETHYL-3-METHYL-

Formula: $C_8H_{13}NO_2$

Mol Weight: 155.20

MP (deg C): | FP (deg C):

BP (deg C):

BP pressure (mm Hg):

Property/Value	Units	Temp	Data Type	Reference
WS 4.58E+004	mg/L	25	EST	MEYLAN,WM ET AL. (1996)
logP 0.23			EXP	HANSCH,C ET AL. (1995)
VP 2.29E-007	mm Hg	25	EST	NEELY,WB & BLAU,GE (1985)
DC	pKa			
HL 7.87E-008	atm m3/mol	25	EST	MEYLAN,WM & HOWARD,PH (1991)
OH 2.49E-011	cm3/molc sec	25	EST	MEYLAN,WM & HOWARD,PH (1993)

CAS #: 090426-03-2				3,5-DICL-2'-ME-4'-NO2 SALICYLANILIDE

Formula: $C_{14}H_{10}Cl_2N_2O_4$

Mol Weight: 341.15

MP (deg C): | FP (deg C):

BP (deg C):

BP pressure (mm Hg):

Property/Value	Units	Temp	Data Type	Reference
WS 1.53E+000	mg/L	25	EST	MEYLAN,WM ET AL. (1996)
logP 4.51			EXP	TERADA,H ET AL. (1988)
VP 9.15E-012	mm Hg	25	EST	NEELY,WB & BLAU,GE (1985)
DC	pKa			
HL 3.83E-013	atm m3/mol	25	EST	MEYLAN,WM & HOWARD,PH (1991)
OH 2.46E-012	cm3/molc sec	25	EST	MEYLAN,WM & HOWARD,PH (1993)

CAS #: 090426-05-4				3,5,-4'-TRICL-4'-NO2 SALICYLANILIDE

Formula: $C_{13}H_7Cl_3N_2O_4$

Mol Weight: 361.57

MP (deg C): | FP (deg C):

BP (deg C):

BP pressure (mm Hg):

Property/Value	Units	Temp	Data Type	Reference
WS 3.75E-001	mg/L	25	EST	MEYLAN,WM ET AL. (1996)
logP 5.08			EXP	TERADA,H ET AL. (1988)
VP 5.32E-012	mm Hg	25	EST	NEELY,WB & BLAU,GE (1985)
DC	pKa			
HL 5.62E-012	atm m3/mol	25	EST	MEYLAN,WM & HOWARD,PH (1991)
OH 8.73E-013	cm3/molc sec	25	EST	MEYLAN,WM & HOWARD,PH (1993)

CAS #: 090467-88-2				DIAZENECARBONITRILE, (1,1-DIMETHYLETHYL)-2-OXIDE

Formula: $C_5H_9N_3O$

Mol Weight: 127.15

MP (deg C): | FP (deg C):

BP (deg C):

BP pressure (mm Hg):

Property/Value	Units	Temp	Data Type	Reference
WS 2.22E+003	mg/L	25	EST	MEYLAN,WM ET AL. (1996)
logP 1.40			EXP	HANSCH,C ET AL. (1995)
VP 2.58E-006	mm Hg	25	EST	NEELY,WB & BLAU,GE (1985)
DC	pKa			
HL 1.92E-010	atm m3/mol	25	EST	MEYLAN,WM & HOWARD,PH (1991)
OH 5.02E-013	cm3/molc sec	25	EST	MEYLAN,WM & HOWARD,PH (1993)

CAS #: 090467-89-3				2-PYRIDYL AZOXYCYANIDE

Formula: $C_6H_4N_4O$

Mol Weight: 148.12

MP (deg C): | FP (deg C):

BP (deg C):

BP pressure (mm Hg):

Property/Value	Units	Temp	Data Type	Reference
WS 1.01E+004	mg/L	25	EST	MEYLAN,WM ET AL. (1996)
logP 0.53			EXP	CALVINO,R R ET AL (1991)
VP 4.31E-008	mm Hg	25	EST	NEELY,WB & BLAU,GE (1985)
DC	pKa			
HL 2.69E-014	atm m3/mol	25	EST	MEYLAN,WM & HOWARD,PH (1991)
OH 7.91E-014	cm3/molc sec	25	EST	MEYLAN,WM & HOWARD,PH (1993)

CAS #: 090507-20-3				1,2,4-OXADIAZOLE-3-PHENYL-5-HYDROXIC ACID

Formula: $C_9H_8N_3O_2$

Mol Weight: 190.18

MP (deg C): | FP (deg C):

BP (deg C):

BP pressure (mm Hg):

Property/Value	Units	Temp	Data Type	Reference
WS 2.36E+002	mg/L	25	EST	MEYLAN,WM ET AL. (1996)
logP 2.72			EXP	HANSCH,C ET AL. (1995)
VP 3.66E-007	mm Hg	25	EST	NEELY,WB & BLAU,GE (1985)
DC	pKa			
HL 1.20E-010	atm m3/mol	25	EST	MEYLAN,WM & HOWARD,PH (1991)
OH 9.33E-012	cm3/molc sec	25	EST	MEYLAN,WM & HOWARD,PH (1993)

CAS #: 090507-21-4				1,2,4-OXADIAZOLE-5-CARBOXALDEHYDE,3-ME,OXIME

Formula: $C_4H_5N_3O_2$

Mol Weight: 127.10

MP (deg C): | FP (deg C):

BP (deg C):

BP pressure (mm Hg):

Property/Value	Units	Temp	Data Type	Reference
WS 3.75E+004	mg/L	25	EST	MEYLAN,WM ET AL. (1996)
logP 0.47			EXP	HANSCH,C & LEO,AJ (1985)
VP 1.61E-003	mm Hg	25	EST	NEELY,WB & BLAU,GE (1985)
DC	pKa			
HL 1.72E-009	atm m3/mol	25	EST	MEYLAN,WM & HOWARD,PH (1991)
OH 6.08E-012	cm3/molc sec	25	EST	MEYLAN,WM & HOWARD,PH (1993)

CAS #: 090507-24-7				1,2,5-THIADIAZOLE-3-CARBOXIMIDOTHIOIC ACID, N-

Formula: $C_9H_{18}N_4OS_2$

Mol Weight: 262.40

MP (deg C): | FP (deg C):

BP (deg C):

BP pressure (mm Hg):

Property/Value	Units	Temp	Data Type	Reference
WS 4.83E+003	mg/L	25	EST	MEYLAN,WM ET AL. (1996)
logP 0.74			EXP	HANSCH,C & LEO,AJ (1985)
VP 1.47E-008	mm Hg	25	EST	NEELY,WB & BLAU,GE (1985)
DC	pKa			
HL 1.63E-013	atm m3/mol	25	EST	MEYLAN,WM & HOWARD,PH (1991)
OH 1.07E-010	cm3/molc sec	25	EST	MEYLAN,WM & HOWARD,PH (1993)

CAS #: 090597-22-1				CPE-C

Formula: $C_{10}H_{13}N_3O_4$

Mol Weight: 239.23

MP (deg C): | FP (deg C):

BP (deg C):

BP pressure (mm Hg):

Property/Value	Units	Temp	Data Type	Reference
WS 1.64E+005	mg/L	25	EST	MEYLAN,WM ET AL. (1996)
logP -2.45			EXP	HANSCH,C ET AL. (1995)
VP 1.55E-011	mm Hg	25	EST	NEELY,WB & BLAU,GE (1985)
DC	pKa			
HL 9.75E-018	atm m3/mol	25	EST	MEYLAN,WM & HOWARD,PH (1991)
OH 1.69E-010	cm3/molc sec	25	EST	MEYLAN,WM & HOWARD,PH (1993)

CAS #: 090656-45-4				BENZOIC ACID, 4-CHLORO-2-[(4-CHLOROPHENYL)AMINO]

Formula: $C_{13}H_9Cl_2NO_2$

Mol Weight: 282.13

MP (deg C): | FP (deg C):

BP (deg C):

BP pressure (mm Hg):

Property/Value	Units	Temp	Data Type	Reference
WS 8.88E-001	mg/L	25	EST	MEYLAN,WM ET AL. (1996)
logP 4.97			EXP	SANGSTER,J (1993)
VP 1.45E-007	mm Hg	25	EST	NEELY,WB & BLAU,GE (1985)
DC	pKa			
HL 1.16E-011	atm m3/mol	25	EST	MEYLAN,WM & HOWARD,PH (1991)
OH 8.15E-011	cm3/molc sec	25	EST	MEYLAN,WM & HOWARD,PH (1993)

CAS #: 090663-30-2				O-MEO BENZAMIDE, N-(4-AMINOBUTYL)

Formula: $C_{12}H_{18}N_2O_2$

Mol Weight: 222.29

MP (deg C): | FP (deg C):

BP (deg C):

BP pressure (mm Hg):

Property/Value	Units	Temp	Data Type	Reference
WS 1.07E+004	mg/L	25	EST	MEYLAN,WM ET AL. (1996)
logP 0.58			EXP	HANSCH,C & LEO,AJ (1985)
VP 5.03E-007	mm Hg	25	EST	NEELY,WB & BLAU,GE (1985)
DC	pKa			
HL 6.92E-014	atm m3/mol	25	EST	MEYLAN,WM & HOWARD,PH (1991)
OH 6.52E-011	cm3/molc sec	25	EST	MEYLAN,WM & HOWARD,PH (1993)

CAS #: 090700-03-1				BENZOIC ACID, 4-CHLORO-2-[(4-IODOPHENYL)AMINO]-

Formula: $C_{13}H_9ClINO_2$

Mol Weight: 373.58

MP (deg C): | FP (deg C):

BP (deg C):

BP pressure (mm Hg):

Property/Value	Units	Temp	Data Type	Reference
WS 1.16E-001	mg/L	25	EST	MEYLAN,WM ET AL. (1996)
logP 5.37			EXP	SANGSTER,J (1993)
VP 1.59E-008	mm Hg	25	EST	NEELY,WB & BLAU,GE (1985)
DC	pKa			
HL 3.63E-012	atm m3/mol	25	EST	MEYLAN,WM & HOWARD,PH (1991)
OH 8.71E-011	cm3/molc sec	25	EST	MEYLAN,WM & HOWARD,PH (1993)

CAS #: 090792-54-4				HYDRAZINECARBOXIMIDAMIDE, 2-[(3-NITROPHENYL)METH

Formula: $C_8H_9N_5O_2$

Mol Weight: 207.19

MP (deg C): | FP (deg C):

BP (deg C):

BP pressure (mm Hg):

Property/Value	Units	Temp	Data Type	Reference
WS 1.61E+004	mg/L	25	EST	MEYLAN,WM ET AL. (1996)
logP 0.01			EXP	SOMAN,G ET AL. (1986)
VP 3.13E-006	mm Hg	25	EST	NEELY,WB & BLAU,GE (1985)
DC	pKa			
HL 1.99E-014	atm m3/mol	25	EST	MEYLAN,WM & HOWARD,PH (1991)
OH 8.63E-011	cm3/molc sec	25	EST	MEYLAN,WM & HOWARD,PH (1993)

CAS #: 090873-90-8				BENZENESULFONAMIDE, 2-[(2-METHYL-1-OXOPROPYL)AMI

Formula: $C_{10}H_{14}N_2O_3S$

Mol Weight: 242.30

MP (deg C): | FP (deg C):

BP (deg C):

BP pressure (mm Hg):

Property/Value	Units	Temp	Data Type	Reference
WS 3.87E+003	mg/L	25	EST	MEYLAN,WM ET AL. (1996)
logP 0.97			EXP	MONZANI,A ET AL. (1985)
VP 2.63E-008	mm Hg	25	EST	NEELY,WB & BLAU,GE (1985)
DC	pKa			
HL 8.52E-013	atm m3/mol	25	EST	MEYLAN,WM & HOWARD,PH (1991)
OH 5.17E-012	cm3/molc sec	25	EST	MEYLAN,WM & HOWARD,PH (1993)

CAS #: 090875-63-1				PROPANAMIDE, N-[2-(AMINOSULFONYL)-5-CHLOROPHENYL

Formula: $C_{10}H_{13}ClN_2O_3S$

Mol Weight: 276.74

MP (deg C): | FP (deg C):

BP (deg C):

BP pressure (mm Hg):

Property/Value	Units	Temp	Data Type	Reference
WS 4.23E+002	mg/L	25	EST	MEYLAN,WM ET AL. (1996)
logP 1.87			EXP	MONZANI,A ET AL. (1985)
VP 7.18E-009	mm Hg	25	EST	NEELY,WB & BLAU,GE (1985)
DC	pKa			
HL 6.32E-013	atm m3/mol	25	EST	MEYLAN,WM & HOWARD,PH (1991)
OH 4.38E-012	cm3/molc sec	25	EST	MEYLAN,WM & HOWARD,PH (1993)

CAS #: 090914-81-1	3-METHYL-4-BROMOACETANILIDE

Formula: $C_9H_{10}BrNO$

Mol Weight: 228.10

MP (deg C):   FP (deg C):

BP (deg C):

BP pressure (mm Hg):

Property/Value	Units	Temp	Data Type	Reference
WS 1.27E+002	mg/L	25	EST	MEYLAN,WM ET AL. (1996)
logP 2.80			EXP	NAKAGAWA,Y ET AL. (1992)
VP 1.36E-005	mm Hg	25	EST	NEELY,WB & BLAU,GE (1985)
DC	pKa			
HL 2.71E-009	atm m3/mol	25	EST	MEYLAN,WM & HOWARD,PH (1991)
OH 9.53E-012	cm3/molc sec	25	EST	MEYLAN,WM & HOWARD,PH (1993)

CAS #: 090936-47-3	1-(3,5-DICHLOROPHENOXY)SILATRANE

Formula: $C_{12}H_{15}Cl_2NO_4Si$

Mol Weight: 336.25

MP (deg C):   FP (deg C):

BP (deg C):

BP pressure (mm Hg):

Property/Value	Units	Temp	Data Type	Reference
WS 6.93E+003	mg/L	25	EST	MEYLAN,WM ET AL. (1996)
logP 1.22			EXP	LUKASIAK,J (1984)
VP 2.07E-006	mm Hg	25	EST	NEELY,WB & BLAU,GE (1985)
DC	pKa			
HL 7.89E-014	atm m3/mol	25	EST	MEYLAN,WM & HOWARD,PH (1991)
OH 1.29E-010	cm3/molc sec	25	EST	MEYLAN,WM & HOWARD,PH (1993)

CAS #: 090936-52-0	1-(3-AMINOPHENOXY)SILATRANE

Formula: $C_{12}H_{18}N_2O_4Si$

Mol Weight: 282.37

MP (deg C):   FP (deg C):

BP (deg C):

BP pressure (mm Hg):

Property/Value	Units	Temp	Data Type	Reference
WS 1.00E+006	mg/L	25	EST	MEYLAN,WM ET AL. (1996)
logP -1.15			EXP	HANSCH,C & LEO,AJ (1985)
VP 1.46E-006	mm Hg	25	EST	NEELY,WB & BLAU,GE (1985)
DC	pKa			
HL 5.08E-017	atm m3/mol	25	EST	MEYLAN,WM & HOWARD,PH (1991)
OH 3.18E-010	cm3/molc sec	25	EST	MEYLAN,WM & HOWARD,PH (1993)

CAS #: 090936-53-1	1-(4-ETHYLPHENOXY)SILATRANE

Formula: $C_{14}H_{21}NO_4Si$

Mol Weight: 295.41

MP (deg C):   FP (deg C):

BP (deg C):

BP pressure (mm Hg):

Property/Value	Units	Temp	Data Type	Reference
WS 4.42E+004	mg/L	25	EST	MEYLAN,WM ET AL. (1996)
logP 0.56			EXP	HANSCH,C & LEO,AJ (1985)
VP 6.67E-006	mm Hg	25	EST	NEELY,WB & BLAU,GE (1985)
DC	pKa			
HL 2.11E-013	atm m3/mol	25	EST	MEYLAN,WM & HOWARD,PH (1991)
OH 1.46E-010	cm3/molc sec	25	EST	MEYLAN,WM & HOWARD,PH (1993)

CAS #: 090955-56-9	1-(3,5-DIMETHYLPHENOXY)SILATRANE

Formula: $C_{14}H_{21}NO_4Si$

Mol Weight: 295.41

MP (deg C):   FP (deg C):

BP (deg C):

BP pressure (mm Hg):

Property/Value	Units	Temp	Data Type	Reference
WS 5.38E+004	mg/L	25	EST	MEYLAN,WM ET AL. (1996)
logP 0.46			EXP	HANSCH,C & LEO,AJ (1985)
VP 6.26E-006	mm Hg	25	EST	NEELY,WB & BLAU,GE (1985)
DC	pKa			
HL 1.75E-013	atm m3/mol	25	EST	MEYLAN,WM & HOWARD,PH (1991)
OH 2.65E-010	cm3/molc sec	25	EST	MEYLAN,WM & HOWARD,PH (1993)

CAS #: 090982-32-4	CHLORIMURON-ETHYL (PH 7)

Formula: $C_{15}H_{15}ClN_4O_6S$

Mol Weight: 414.83

MP (deg C): 198-201   FP (deg C):

BP (deg C):

BP pressure (mm Hg):

Property/Value	Units	Temp	Data Type	Reference
WS 1.10E+001	mg/L	25	EXP	SHIU,WY ET AL. (1990) @ pH=5
logP 2.50			EXP	HAY,JV (1990)
VP 1.50E-005	mm Hg	25	EXP	TOMLIN,C (1994)
DC	pKa			
HL 7.44E-007	atm m3/mol	25	EST	VP/WSOL
OH 4.27E-011	cm3/molc sec	25	EST	MEYLAN,WM & HOWARD,PH (1993)

CAS #: 091027-93-9	5-CHLORO-1,2-DIMETHYL-4-NITRO-1H-IMIDAZOLE

Formula: $C_5H_6ClN_3O_2$

Mol Weight: 175.58

MP (deg C):   FP (deg C):

BP (deg C):

BP pressure (mm Hg):

Property/Value	Units	Temp	Data Type	Reference
WS 9.83E+002	mg/L	25	EST	MEYLAN,WM ET AL. (1996)
logP 1.62			EST	MEYLAN,WM & HOWARD,PH (1995)
VP 1.00E-004	mm Hg	25	EST	NEELY,WB & BLAU,GE (1985)
DC	pKa			
HL 2.58E-007	atm m3/mol	25	EST	MEYLAN,WM & HOWARD,PH (1991)
OH 3.63E-011	cm3/molc sec	25	EST	MEYLAN,WM & HOWARD,PH (1993)

CAS #: 091027-94-0	4-CHLORO-1,2-DIMETHYL-5-NITRO-1H-IMIDAZOLE

Formula: $C_5H_6ClN_3O_2$

Mol Weight: 175.58

MP (deg C):   FP (deg C):

BP (deg C):

BP pressure (mm Hg):

Property/Value	Units	Temp	Data Type	Reference
WS 9.83E+002	mg/L	25	EST	MEYLAN,WM ET AL. (1996)
logP 1.62			EST	MEYLAN,WM & HOWARD,PH (1995)
VP 1.00E-004	mm Hg	25	EST	NEELY,WB & BLAU,GE (1985)
DC	pKa			
HL 2.58E-007	atm m3/mol	25	EST	MEYLAN,WM & HOWARD,PH (1991)
OH 3.63E-011	cm3/molc sec	25	EST	MEYLAN,WM & HOWARD,PH (1993)

CAS #: 091350-72-0	CARBAMIC ACID, 2-(1H-INDOL-3-YL)ETHYL ESTER

Formula: $C_{11}H_{12}N_2O_2$
Mol Weight: 204.23
MP (deg C): | FP (deg C):
BP (deg C):
BP pressure (mm Hg):

Property/Value	Units	Temp	Data Type	Reference
WS 1.50E+003	mg/L	25	EST	MEYLAN,WM ET AL. (1996)
logP 1.69			EXP	TANAKA,M ET AL. (1985)
VP 7.10E-006	mm Hg	25	EST	NEELY,WB & BLAU,GE (1985)
DC	pKa			
HL 6.97E-013	atm m3/mol	25	EST	MEYLAN,WM & HOWARD,PH (1991)
OH 2.05E-010	cm3/molc sec	25	EST	MEYLAN,WM & HOWARD,PH (1993)

CAS #: 091371-14-1	1,6-DIBROMODIBENZO-P-DIOXIN

Formula: $C_{12}H_6Br_2O_2$
Mol Weight: 342.00
MP (deg C): | FP (deg C):
BP (deg C):
BP pressure (mm Hg):

Property/Value	Units	Temp	Data Type	Reference
WS 3.18E-003	mg/L	25	EST	MEYLAN,WM ET AL. (1996)
logP 6.12			EST	MEYLAN,WM & HOWARD,PH (1995)
VP 4.87E-006	mm Hg	25	EST	NEELY,WB & BLAU,GE (1985)
DC	pKa			
HL 1.86E-006	atm m3/mol	25	EST	MEYLAN,WM & HOWARD,PH (1991)
OH 6.19E-012	cm3/molc sec	25	EST	MEYLAN,WM & HOWARD,PH (1993)

CAS #: 091430-64-7	BARBITURIC ACID,5-TBU-5ALLYL

Formula: $C_{11}H_{16}N_2O_3$
Mol Weight: 224.26
MP (deg C): | FP (deg C):
BP (deg C):
BP pressure (mm Hg):

Property/Value	Units	Temp	Data Type	Reference
WS 1.30E+003	mg/L	25	EST	MEYLAN,WM ET AL. (1996)
logP 1.64			EXP	WONG,O & MCKEOWN,RH (1988)
VP 3.21E-011	mm Hg	25	EST	NEELY,WB & BLAU,GE (1985)
DC	pKa			
HL 6.29E-013	atm m3/mol	25	EST	MEYLAN,WM & HOWARD,PH (1991)
OH 3.24E-011	cm3/molc sec	25	EST	MEYLAN,WM & HOWARD,PH (1993)

CAS #: 091431-42-4	1,4-NAPHTHALENEDIOL, 6-CHLORO-2,3-DIMETHOXY-, DI

Formula: $C_{16}H_{15}ClO_6$
Mol Weight: 338.75
MP (deg C): | FP (deg C):
BP (deg C):
BP pressure (mm Hg):

Property/Value	Units	Temp	Data Type	Reference
WS 1.19E+001	mg/L	25	EST	MEYLAN,WM ET AL. (1996)
logP 3.26			EXP	JONES,GH ET AL. (1986)
VP 4.78E-007	mm Hg	25	EST	NEELY,WB & BLAU,GE (1985)
DC	pKa			
HL 1.97E-010	atm m3/mol	25	EST	MEYLAN,WM & HOWARD,PH (1991)
OH 7.26E-011	cm3/molc sec	25	EST	MEYLAN,WM & HOWARD,PH (1993)

CAS #: 091445-76-0	PROPANOIC ACID, 3-(7-BROMO-4H-1,2,4-BENZOTHIADIA

Formula: $C_{10}H_9BrN_2O_4S$
Mol Weight: 333.16
MP (deg C): | FP (deg C):
BP (deg C):
BP pressure (mm Hg):

Property/Value	Units	Temp	Data Type	Reference
WS 3.12E+003	mg/L	25	EST	MEYLAN,WM ET AL. (1996)
logP 0.93			EXP	GAMBERINI,G ET AL. (1989)
VP 1.60E-011	mm Hg	25	EST	NEELY,WB & BLAU,GE (1985)
DC	pKa			
HL 3.64E-013	atm m3/mol	25	EST	MEYLAN,WM & HOWARD,PH (1991)
OH 5.41E-012	cm3/molc sec	25	EST	MEYLAN,WM & HOWARD,PH (1993)

CAS #: 091455-17-3	1,1,1,3,5,7,7,7-OCTAMETHYL-3,5-BIS[2-(TRIMETHYL*

Formula: $C_{18}H_{50}O_3Si_6$
Mol Weight: 483.11
MP (deg C): | FP (deg C):
BP (deg C):
BP pressure (mm Hg):

Property/Value	Units	Temp	Data Type	Reference
WS 1.85E-008	mg/L	25	EST	MEYLAN,WM ET AL. (1996)
logP 10.69			EST	MEYLAN,WM & HOWARD,PH (1995)
VP	mm Hg			
DC	pKa			
HL	atm m3/mol			
OH 7.15E-012	cm3/molc sec	25	EST	MEYLAN,WM & HOWARD,PH (1993)

CAS #: 091465-08-6	LAMBDA-CYHALOTHRIN

Formula: $C_{23}H_{19}ClF_3NO_3$
Mol Weight: 449.86
MP (deg C): 49.2 | FP (deg C):
BP (deg C):
BP pressure (mm Hg):

Property/Value	Units	Temp	Data Type	Reference
WS 8.53E-004	mg/L	25	EST	MEYLAN,WM ET AL. (1996)
logP 7.00			EXP	TOMLIN,C (1994)
VP 1.50E-009	mm Hg	25	EXP	TOMLIN,C (1994)
DC	pKa			
HL 1.35E-005	atm m3/mol	25	EST	MEYLAN,WM & HOWARD,PH (1991)
OH 3.15E-011	cm3/molc sec	25	EST	MEYLAN,WM & HOWARD,PH (1993)

CAS #: 091551-85-8	PROPANOIC ACID, 3-(7-CHLORO-4H-1,2,4-BENZOTHIADI

Formula: $C_{10}H_9ClN_2O_4S$
Mol Weight: 288.71
MP (deg C): | FP (deg C):
BP (deg C):
BP pressure (mm Hg):

Property/Value	Units	Temp	Data Type	Reference
WS 9.13E+003	mg/L	25	EST	MEYLAN,WM ET AL. (1996)
logP 0.69			EXP	GAMBERINI,G ET AL. (1989)
VP 3.79E-011	mm Hg	25	EST	NEELY,WB & BLAU,GE (1985)
DC	pKa			
HL 6.77E-013	atm m3/mol	25	EST	MEYLAN,WM & HOWARD,PH (1991)
OH 5.46E-012	cm3/molc sec	25	EST	MEYLAN,WM & HOWARD,PH (1993)

CAS #: 091564-01-1 — CARBAMIC ACID, DIMETHYL-, 3-PHENYLPROPYL ESTER

Formula: $C_{12}H_{17}NO_2$

Mol Weight: 207.27

MP (deg C): FP (deg C):

BP (deg C):

BP pressure (mm Hg):

Property/Value	Units	Temp	Data Type	Reference
WS 1.26E+002	mg/L	25	EST	MEYLAN,WM ET AL. (1996)
logP 2.93			EXP	TANAKA,M ET AL. (1985)
VP 2.43E-003	mm Hg	25	EST	NEELY,WB & BLAU,GE (1985)
DC	pKa			
HL 2.71E-008	atm m3/mol	25	EST	MEYLAN,WM & HOWARD,PH (1991)
OH 2.54E-011	cm3/molc sec	25	EST	MEYLAN,WM & HOWARD,PH (1993)

CAS #: 091814-09-4 — PROPANOIC ACID, 2,2-DIMETHYL-, 2,3-DIMETHOXY-1,4

Formula: $C_{22}H_{28}O_6$

Mol Weight: 388.46

MP (deg C): FP (deg C):

BP (deg C):

BP pressure (mm Hg):

Property/Value	Units	Temp	Data Type	Reference
WS 4.99E-002	mg/L	25	EST	MEYLAN,WM ET AL. (1996)
logP 5.69			EXP	JONES,GH ET AL. (1986)
VP 5.52E-008	mm Hg	25	EST	NEELY,WB & BLAU,GE (1985)
DC	pKa			
HL 1.46E-009	atm m3/mol	25	EST	MEYLAN,WM & HOWARD,PH (1991)
OH 1.59E-010	cm3/molc sec	25	EST	MEYLAN,WM & HOWARD,PH (1993)

CAS #: 091814-10-7 — 1,4-NAPHTHALENEDIOL, 5-CHLORO-2,3-DIMETHOXY-, DI

Formula: $C_{16}H_{15}ClO_6$

Mol Weight: 338.75

MP (deg C): FP (deg C):

BP (deg C):

BP pressure (mm Hg):

Property/Value	Units	Temp	Data Type	Reference
WS 1.45E+001	mg/L	25	EST	MEYLAN,WM ET AL. (1996)
logP 3.16			EXP	JONES,GH ET AL. (1986)
VP 4.78E-007	mm Hg	25	EST	NEELY,WB & BLAU,GE (1985)
DC	pKa			
HL 1.97E-010	atm m3/mol	25	EST	MEYLAN,WM & HOWARD,PH (1991)
OH 7.26E-011	cm3/molc sec	25	EST	MEYLAN,WM & HOWARD,PH (1993)

CAS #: 091814-12-9 — 1,4-NAPHTHALENEDIOL, 6-FLUORO-2,3-DIMETHOXY-, DI

Formula: $C_{16}H_{15}FO_6$

Mol Weight: 322.29

MP (deg C): FP (deg C):

BP (deg C):

BP pressure (mm Hg):

Property/Value	Units	Temp	Data Type	Reference
WS 4.07E+001	mg/L	25	EST	MEYLAN,WM ET AL. (1996)
logP 2.75			EXP	JONES,GH ET AL. (1986)
VP 2.65E-006	mm Hg	25	EST	NEELY,WB & BLAU,GE (1985)
DC	pKa			
HL 3.11E-010	atm m3/mol	25	EST	MEYLAN,WM & HOWARD,PH (1991)
OH 1.03E-010	cm3/molc sec	25	EST	MEYLAN,WM & HOWARD,PH (1993)

CAS #: 091814-13-0 — 1,4-NAPHTHALENEDIOL, 6-BROMO-2,3-DIMETHOXY-, DIA

Formula: $C_{16}H_{15}BrO_6$

Mol Weight: 383.20

MP (deg C): FP (deg C):

BP (deg C):

BP pressure (mm Hg):

Property/Value	Units	Temp	Data Type	Reference
WS 4.95E+000	mg/L	25	EST	MEYLAN,WM ET AL. (1996)
logP 3.39			EXP	JONES,GH ET AL. (1986)
VP 1.86E-007	mm Hg	25	EST	NEELY,WB & BLAU,GE (1985)
DC	pKa			
HL 1.06E-010	atm m3/mol	25	EST	MEYLAN,WM & HOWARD,PH (1991)
OH 6.82E-011	cm3/molc sec	25	EST	MEYLAN,WM & HOWARD,PH (1993)

CAS #: 091814-24-3 — 1,4-NAPHTHALENEDIOL, 2,3,5-TRIMETHOXY-, DIACETAT

Formula: $C_{17}H_{18}O_7$

Mol Weight: 334.33

MP (deg C): FP (deg C):

BP (deg C):

BP pressure (mm Hg):

Property/Value	Units	Temp	Data Type	Reference
WS 6.47E+001	mg/L	25	EST	MEYLAN,WM ET AL. (1996)
logP 2.43			EXP	JONES,GH ET AL. (1986)
VP 3.33E-007	mm Hg	25	EST	NEELY,WB & BLAU,GE (1985)
DC	pKa			
HL 1.58E-011	atm m3/mol	25	EST	MEYLAN,WM & HOWARD,PH (1991)
OH 2.03E-010	cm3/molc sec	25	EST	MEYLAN,WM & HOWARD,PH (1993)

CAS #: 091814-29-8 — 1-NAPHTHALENECARBONITRILE, 5,8-BIS(ACETYLOXY)-6,

Formula: $C_{17}H_{15}NO_6$

Mol Weight: 329.31

MP (deg C): FP (deg C):

BP (deg C):

BP pressure (mm Hg):

Property/Value	Units	Temp	Data Type	Reference
WS 3.22E+001	mg/L	25	EST	MEYLAN,WM ET AL. (1996)
logP 2.51			EXP	JONES,GH ET AL. (1986)
VP 5.53E-008	mm Hg	25	EST	NEELY,WB & BLAU,GE (1985)
DC	pKa			
HL 2.57E-012	atm m3/mol	25	EST	MEYLAN,WM & HOWARD,PH (1991)
OH 2.56E-011	cm3/molc sec	25	EST	MEYLAN,WM & HOWARD,PH (1993)

CAS #: 091814-30-1 — 2-NAPHTHALENECARBONITRILE, 5,8-BIS(ACETYLOXY)-6,

Formula: $C_{17}H_{15}NO_6$

Mol Weight: 329.31

MP (deg C): FP (deg C):

BP (deg C):

BP pressure (mm Hg):

Property/Value	Units	Temp	Data Type	Reference
WS 5.06E+001	mg/L	25	EST	MEYLAN,WM ET AL. (1996)
logP 2.28			EXP	JONES,GH ET AL. (1986)
VP 5.53E-008	mm Hg	25	EST	NEELY,WB & BLAU,GE (1985)
DC	pKa			
HL 2.57E-012	atm m3/mol	25	EST	MEYLAN,WM & HOWARD,PH (1991)
OH 2.56E-011	cm3/molc sec	25	EST	MEYLAN,WM & HOWARD,PH (1993)

091814-31-2

ACETAMIDE, N-[5,8-BIS(ACETYLOXY)-6,7-DIMETHOXY-1

Formula: $C_{18}H_{19}NO_7$

Mol Weight: 361.35

MP (deg C): FP (deg C):

BP (deg C):

BP pressure (mm Hg):

Property/ Value	Units	Temp	Data Type	Reference
WS 1.04E+002	mg/L	25	EST	MEYLAN,WM ET AL. (1996)
logP 2.00			EXP	SANGSTER,J (1993)
VP 1.80E-010	mm Hg	25	EST	NEELY,WB & BLAU,GE (1985)
DC	pKa			
HL 3.05E-016	atm m3/mol	25	EST	MEYLAN,WM & HOWARD,PH (1991)
OH 2.02E-010	cm3/molc sec	25	EST	MEYLAN,WM & HOWARD,PH (1993)

091814-40-3

1,4-NAPHTHALENEDIOL, 2,3,6-TRIMETHOXY-, DIACETAT

Formula: $C_{17}H_{18}O_7$

Mol Weight: 334.33

MP (deg C): FP (deg C):

BP (deg C):

BP pressure (mm Hg):

Property/ Value	Units	Temp	Data Type	Reference
WS 7.00E+001	mg/L	25	EST	MEYLAN,WM ET AL. (1996)
logP 2.39			EXP	JONES,GH ET AL. (1986)
VP 3.33E-007	mm Hg	25	EST	NEELY,WB & BLAU,GE (1985)
DC	pKa			
HL 1.58E-011	atm m3/mol	25	EST	MEYLAN,WM & HOWARD,PH (1991)
OH 2.03E-010	cm3/molc sec	25	EST	MEYLAN,WM & HOWARD,PH (1993)

091814-42-5

1,4-NAPHTHALENEDIOL, 2,3-DIMETHOXY-6-METHYL-, DI

Formula: $C_{17}H_{18}O_6$

Mol Weight: 318.33

MP (deg C): FP (deg C):

BP (deg C):

BP pressure (mm Hg):

Property/ Value	Units	Temp	Data Type	Reference
WS 3.46E+001	mg/L	25	EST	MEYLAN,WM ET AL. (1996)
logP 2.86			EXP	JONES,GH ET AL. (1986)
VP 9.47E-007	mm Hg	25	EST	NEELY,WB & BLAU,GE (1985)
DC	pKa			
HL 2.94E-010	atm m3/mol	25	EST	MEYLAN,WM & HOWARD,PH (1991)
OH 2.02E-010	cm3/molc sec	25	EST	MEYLAN,WM & HOWARD,PH (1993)

091814-56-1

1,4-NAPHTHALENEDIOL, 2,3-DIMETHOXY-, DIPROPANOAT

Formula: $C_{18}H_{20}O_6$

Mol Weight: 332.36

MP (deg C): FP (deg C):

BP (deg C):

BP pressure (mm Hg):

Property/ Value	Units	Temp	Data Type	Reference
WS 9.13E+000	mg/L	25	EST	MEYLAN,WM ET AL. (1996)
logP 3.44			EXP	SANGSTER,J (1993)
VP 4.61E-007	mm Hg	25	EST	NEELY,WB & BLAU,GE (1985)
DC	pKa			
HL 4.70E-010	atm m3/mol	25	EST	MEYLAN,WM & HOWARD,PH (1991)
OH 1.59E-010	cm3/molc sec	25	EST	MEYLAN,WM & HOWARD,PH (1993)

091814-58-3

1,4-NAPHTHALENEDIOL, 2,3-DIMETHOXY-, BIS(2-METHY

Formula: $C_{20}H_{24}O_6$

Mol Weight: 360.41

MP (deg C): FP (deg C):

BP (deg C):

BP pressure (mm Hg):

Property/ Value	Units	Temp	Data Type	Reference
WS 6.20E-001	mg/L	25	EST	MEYLAN,WM ET AL. (1996)
logP 4.61			EXP	JONES,GH ET AL. (1986)
VP 2.81E-007	mm Hg	25	EST	NEELY,WB & BLAU,GE (1985)
DC	pKa			
HL 8.27E-010	atm m3/mol	25	EST	MEYLAN,WM & HOWARD,PH (1991)
OH 1.61E-010	cm3/molc sec	25	EST	MEYLAN,WM & HOWARD,PH (1993)

091833-49-7

PYRIDO[3,2-F]-1,4-OXAZEPINE-5(2H)-THIONE, 2-[2-(

Formula: $C_{13}H_{19}N_3OS$

Mol Weight: 265.38

MP (deg C): FP (deg C):

BP (deg C):

BP pressure (mm Hg):

Property/ Value	Units	Temp	Data Type	Reference
WS 2.36E+004	mg/L	25	EST	MEYLAN,WM ET AL. (1996)
logP 1.08			EXP	HANSCH,C ET AL. (1995)
VP 3.45E-007	mm Hg	25	EST	NEELY,WB & BLAU,GE (1985)
DC	pKa			
HL 5.14E-013	atm m3/mol	25	EST	MEYLAN,WM & HOWARD,PH (1991)
OH 1.81E-010	cm3/molc sec	25	EST	MEYLAN,WM & HOWARD,PH (1993)

091937-67-6

PROPANEDINITRILE, [(5-CHLORO-2-FURANYL)METHYLENE

Formula: $C_8H_3ClN_2O$

Mol Weight: 178.58

MP (deg C): FP (deg C):

BP (deg C):

BP pressure (mm Hg):

Property/ Value	Units	Temp	Data Type	Reference
WS 4.26E+002	mg/L	25	EST	MEYLAN,WM ET AL. (1996)
logP 2.17			EXP	BALAZ,S ET AL. (1985)
VP 1.34E-004	mm Hg	25	EST	NEELY,WB & BLAU,GE (1985)
DC	pKa			
HL 1.02E-008	atm m3/mol	25	EST	MEYLAN,WM & HOWARD,PH (1991)
OH 2.90E-011	cm3/molc sec	25	EST	MEYLAN,WM & HOWARD,PH (1993)

092245-57-3

2-(4-(PHENYLAZO)PHENOXY)ETHANOL

Formula: $C_{14}H_{14}N_2O_2$

Mol Weight: 242.28

MP (deg C): FP (deg C):

BP (deg C):

BP pressure (mm Hg):

Property/ Value	Units	Temp	Data Type	Reference
WS 1.72E+001	mg/L	25	EST	MEYLAN,WM ET AL. (1996)
logP 3.22			EST	MEYLAN,WM & HOWARD,PH (1995)
VP 2.11E-008	mm Hg	25	EXP	SHIMIZU,T ET AL. (1987)
DC	pKa			
HL 4.22E-011	atm m3/mol	25	EST	MEYLAN,WM & HOWARD,PH (1991)
OH 2.05E-011	cm3/molc sec	25	EST	MEYLAN,WM & HOWARD,PH (1993)

CAS #: 092387-48-9				MONO[1-(4-CHLOROPHENYL)ETHYL] 1,2-BENZENEDICARB*
Formula: $C_{16}H_{13}ClO_4$				
Mol Weight: 304.73				
MP (deg C):			FP (deg C):	
BP (deg C):				
BP pressure (mm Hg):				

Property/ Value	Units	Temp	Data Type	Reference
WS 3.38E+000	mg/L	25	EST	MEYLAN,WM ET AL. (1996)
logP 4.14			EST	MEYLAN,WM & HOWARD,PH (1995)
VP 6.79E-008	mm Hg	25	EST	NEELY,WB & BLAU,GE (1985)
DC	pKa			
HL 5.55E-011	atm m3/mol	25	EST	MEYLAN,WM & HOWARD,PH (1991)
OH 5.93E-012	cm3/molc sec	25	EST	MEYLAN,WM & HOWARD,PH (1993)

CAS #: 092387-49-0				MONI[1-(4-BROMOPHENYL)ETHYL] 1,2-BENZENEDICARBO*
Formula: $C_{16}H_{13}BrO_4$				
Mol Weight: 349.19				
MP (deg C):			FP (deg C):	
BP (deg C):				
BP pressure (mm Hg):				

Property/ Value	Units	Temp	Data Type	Reference
WS 1.13E+000	mg/L	25	EST	MEYLAN,WM ET AL. (1996)
logP 4.38			EST	MEYLAN,WM & HOWARD,PH (1995)
VP 2.53E-008	mm Hg	25	EST	NEELY,WB & BLAU,GE (1985)
DC	pKa			
HL 2.98E-011	atm m3/mol	25	EST	MEYLAN,WM & HOWARD,PH (1991)
OH 5.75E-012	cm3/molc sec	25	EST	MEYLAN,WM & HOWARD,PH (1993)

CAS #: 092562-88-4				GUANOSINE, 2',3'-DIDEOXY-3'-FLUORO-
Formula: $C_{10}H_{12}FN_5O_3$				
Mol Weight: 269.24				
MP (deg C):			FP (deg C):	
BP (deg C):				
BP pressure (mm Hg):				

Property/ Value	Units	Temp	Data Type	Reference
WS 3.31E+003	mg/L	25	EST	MEYLAN,WM ET AL. (1996)
logP -0.66			EXP	BALZARINI,J ET AL. (1989)
VP 9.33E-016	mm Hg	25	EST	NEELY,WB & BLAU,GE (1985)
DC	pKa			
HL 6.18E-023	atm m3/mol	25	EST	MEYLAN,WM & HOWARD,PH (1991)
OH 7.10E-011	cm3/molc sec	25	EST	MEYLAN,WM & HOWARD,PH (1993)

CAS #: 092598-80-6				1H-IMIDAZOLE-5-BUTANOIC ACID, à-ETHYL-à-(HYDROXY
Formula: $C_{14}H_{24}N_2O_3$				
Mol Weight: 268.36				
MP (deg C):			FP (deg C):	
BP (deg C):				
BP pressure (mm Hg):				

Property/ Value	Units	Temp	Data Type	Reference
WS 6.22E+002	mg/L	25	EST	MEYLAN,WM ET AL. (1996)
logP 1.73			EXP	BUNDGAARD,H ET AL. (1986)
VP 9.97E-009	mm Hg	25	EST	NEELY,WB & BLAU,GE (1985)
DC	pKa			
HL 5.59E-011	atm m3/mol	25	EST	MEYLAN,WM & HOWARD,PH (1991)
OH 1.09E-010	cm3/molc sec	25	EST	MEYLAN,WM & HOWARD,PH (1993)

CAS #: 092598-81-7				1H-IMIDAZOLE-5-BUTANOIC ACID, _-ETHYL-beta-(HYDR
Formula: $C_{19}H_{26}N_2O_3$				
Mol Weight: 330.43				
MP (deg C):			FP (deg C):	
BP (deg C):				
BP pressure (mm Hg):				

Property/ Value	Units	Temp	Data Type	Reference
WS 1.01E+002	mg/L	25	EST	MEYLAN,WM ET AL. (1996)
logP 2.23			EXP	SANGSTER,J (1994)
VP 2.87E-011	mm Hg	25	EST	NEELY,WB & BLAU,GE (1985)
DC	pKa			
HL 3.40E-012	atm m3/mol	25	EST	MEYLAN,WM & HOWARD,PH (1991)
OH 1.10E-010	cm3/molc sec	25	EST	MEYLAN,WM & HOWARD,PH (1993)

CAS #: 092598-82-8				1H-IMIDAZOLE-5-BUTANOIC ACID, alpha-ETHYL-beta-(
Formula: $C_{18}H_{24}N_2O_3$				
Mol Weight: 316.40				
MP (deg C):			FP (deg C):	
BP (deg C):				
BP pressure (mm Hg):				

Property/ Value	Units	Temp	Data Type	Reference
WS 2.04E+002	mg/L	25	EST	MEYLAN,WM ET AL. (1996)
logP 1.97			EXP	SANGSTER,J (1993)
VP 4.10E-011	mm Hg	25	EST	NEELY,WB & BLAU,GE (1985)
DC	pKa			
HL 2.56E-012	atm m3/mol	25	EST	MEYLAN,WM & HOWARD,PH (1991)
OH 1.08E-010	cm3/molc sec	25	EST	MEYLAN,WM & HOWARD,PH (1993)

CAS #: 092598-84-0				1H-IMIDAZOLE-5-BUTANOIC ACID, beta-[(BENZOYLOXY)
Formula: $C_{25}H_{28}N_2O_4$				
Mol Weight: 420.51				
MP (deg C):			FP (deg C):	
BP (deg C):				
BP pressure (mm Hg):				

Property/ Value	Units	Temp	Data Type	Reference
WS 4.23E-001	mg/L	25	EST	MEYLAN,WM ET AL. (1996)
logP 4.37			EXP	SANGSTER,J (1993)
VP 4.85E-011	mm Hg	25	EST	NEELY,WB & BLAU,GE (1985)
DC	pKa			
HL 3.30E-011	atm m3/mol	25	EST	MEYLAN,WM & HOWARD,PH (1991)
OH 1.07E-010	cm3/molc sec	25	EST	MEYLAN,WM & HOWARD,PH (1993)

CAS #: 092598-86-2				1H-IMIDAZOLE-5-BUTANOIC ACID, _-ETHYL-beta-(HYDR
Formula: $C_{19}H_{26}N_2O_3$				
Mol Weight: 330.43				
MP (deg C):			FP (deg C):	
BP (deg C):				
BP pressure (mm Hg):				

Property/ Value	Units	Temp	Data Type	Reference
WS 8.65E+001	mg/L	25	EST	MEYLAN,WM ET AL. (1996)
logP 2.31			EXP	SANGSTER,J (1994)
VP 1.49E-011	mm Hg	25	EST	NEELY,WB & BLAU,GE (1985)
DC	pKa			
HL 3.40E-012	atm m3/mol	25	EST	MEYLAN,WM & HOWARD,PH (1991)
OH 1.10E-010	cm3/molc sec	25	EST	MEYLAN,WM & HOWARD,PH (1993)

CAS #: 092598-89-5	1H-IMIDAZOLE-5-BUTANOIC ACID, _-ETHYL-beta-(HYDR

Formula: $C_{18}H_{23}ClN_2O_3$

Mol Weight: 350.85

MP (deg C): FP (deg C):

BP (deg C):

BP pressure (mm Hg):

Property/ Value	Units	Temp	Data Type	Reference
WS 3.09E+001	mg/L	25	EST	MEYLAN,WM ET AL. (1996)
logP 2.69			EXP	SANGSTER,J (1994)
VP 7.02E-012	mm Hg	25	EST	NEELY,WB & BLAU,GE (1985)
DC	pKa			
HL 1.90E-012	atm m3/mol	25	EST	MEYLAN,WM & HOWARD,PH (1991)
OH 1.05E-010	cm3/molc sec	25	EST	MEYLAN,WM & HOWARD,PH (1993)

CAS #: 092598-90-8	1H-IMIDAZOLE-5-BUTANOIC ACID, beta-[(BENZOYLOXY)

Formula: $C_{25}H_{27}ClN_2O_4$

Mol Weight: 454.96

MP (deg C): FP (deg C):

BP (deg C):

BP pressure (mm Hg):

Property/ Value	Units	Temp	Data Type	Reference
WS 9.06E-002	mg/L	25	EST	MEYLAN,WM ET AL. (1996)
logP 4.90			EXP	SANGSTER,J (1993)
VP 1.36E-011	mm Hg	25	EST	NEELY,WB & BLAU,GE (1985)
DC	pKa			
HL 2.45E-011	atm m3/mol	25	EST	MEYLAN,WM & HOWARD,PH (1991)
OH 1.03E-010	cm3/molc sec	25	EST	MEYLAN,WM & HOWARD,PH (1993)

CAS #: 092598-92-0	1H-IMIDAZOLE-5-BUTANOIC ACID, _-ETHYL-beta-(HYDR

Formula: $C_{19}H_{26}N_2O_3$

Mol Weight: 330.43

MP (deg C): FP (deg C):

BP (deg C):

BP pressure (mm Hg):

Property/ Value	Units	Temp	Data Type	Reference
WS 6.96E+001	mg/L	25	EST	MEYLAN,WM ET AL. (1996)
logP 2.42			EXP	SANGSTER,J (1994)
VP 1.39E-011	mm Hg	25	EST	NEELY,WB & BLAU,GE (1985)
DC	pKa			
HL 2.83E-012	atm m3/mol	25	EST	MEYLAN,WM & HOWARD,PH (1991)
OH 1.10E-010	cm3/molc sec	25	EST	MEYLAN,WM & HOWARD,PH (1993)

CAS #: 092598-93-1	1H-IMIDAZOLE-5-BUTANOIC ACID, _-ETHYL-beta-(HYDR

Formula: $C_{19}H_{26}N_2O_3$

Mol Weight: 330.43

MP (deg C): FP (deg C):

BP (deg C):

BP pressure (mm Hg):

Property/ Value	Units	Temp	Data Type	Reference
WS 6.44E+001	mg/L	25	EST	MEYLAN,WM ET AL. (1996)
logP 2.46			EXP	SANGSTER,J (1994)
VP 1.39E-011	mm Hg	25	EST	NEELY,WB & BLAU,GE (1985)
DC	pKa			
HL 2.83E-012	atm m3/mol	25	EST	MEYLAN,WM & HOWARD,PH (1991)
OH 1.10E-010	cm3/molc sec	25	EST	MEYLAN,WM & HOWARD,PH (1993)

CAS #: 092598-94-2	1H-IMIDAZOLE-5-BUTANOIC ACID, beta-[(BENZOYLOXY)

Formula: $C_{26}H_{30}N_2O_4$

Mol Weight: 434.54

MP (deg C): FP (deg C):

BP (deg C):

BP pressure (mm Hg):

Property/ Value	Units	Temp	Data Type	Reference
WS 1.35E-001	mg/L	25	EST	MEYLAN,WM ET AL. (1996)
logP 4.85			EXP	SANGSTER,J (1993)
VP 2.50E-011	mm Hg	25	EST	NEELY,WB & BLAU,GE (1985)
DC	pKa			
HL 3.64E-011	atm m3/mol	25	EST	MEYLAN,WM & HOWARD,PH (1991)
OH 1.08E-010	cm3/molc sec	25	EST	MEYLAN,WM & HOWARD,PH (1993)

CAS #: 092598-95-3	1H-IMIDAZOLE-5-BUTANOIC ACID, alpha-ETHYL-1-METH

Formula: $C_{23}H_{32}N_2O_4$

Mol Weight: 400.52

MP (deg C): FP (deg C):

BP (deg C):

BP pressure (mm Hg):

Property/ Value	Units	Temp	Data Type	Reference
WS 6.10E-001	mg/L	25	EST	MEYLAN,WM ET AL. (1996)
logP 4.33			EXP	SANGSTER,J (1993)
VP 9.70E-010	mm Hg	25	EST	NEELY,WB & BLAU,GE (1985)
DC	pKa			
HL 3.24E-010	atm m3/mol	25	EST	MEYLAN,WM & HOWARD,PH (1991)
OH 1.10E-010	cm3/molc sec	25	EST	MEYLAN,WM & HOWARD,PH (1993)

CAS #: 092598-97-5	1H-IMIDAZOLE-5-BUTANOIC ACID, beta-[(ACETYLOXY)M

Formula: $C_{21}H_{28}N_2O_4$

Mol Weight: 372.47

MP (deg C): FP (deg C):

BP (deg C):

BP pressure (mm Hg):

Property/ Value	Units	Temp	Data Type	Reference
WS 6.74E+000	mg/L	25	EST	MEYLAN,WM ET AL. (1996)
logP 3.31			EXP	SANGSTER,J (1993)
VP 5.98E-009	mm Hg	25	EST	NEELY,WB & BLAU,GE (1985)
DC	pKa			
HL 1.84E-010	atm m3/mol	25	EST	MEYLAN,WM & HOWARD,PH (1991)
OH 1.08E-010	cm3/molc sec	25	EST	MEYLAN,WM & HOWARD,PH (1993)

CAS #: 092598-99-7	1H-IMIDAZOLE-5-BUTANOIC ACID, alpha-ETHYL-beta-(

Formula: $C_{22}H_{32}N_2O_3$

Mol Weight: 372.51

MP (deg C): FP (deg C):

BP (deg C):

BP pressure (mm Hg):

Property/ Value	Units	Temp	Data Type	Reference
WS 3.32E+000	mg/L	25	EST	MEYLAN,WM ET AL. (1996)
logP 3.67			EXP	SANGSTER,J (1993)
VP 1.68E-012	mm Hg	25	EST	NEELY,WB & BLAU,GE (1985)
DC	pKa			
HL 6.61E-012	atm m3/mol	25	EST	MEYLAN,WM & HOWARD,PH (1991)
OH 1.10E-010	cm3/molc sec	25	EST	MEYLAN,WM & HOWARD,PH (1993)

CAS #: 092622-09-8				1H-IMIDAZOLE-5-BUTANOIC ACID, beta-[[(3-CHLOROBE

Formula: $C_{25}H_{27}ClN_2O_4$

Mol Weight: 454.96

MP (deg C):   **FP (deg C):**

BP (deg C):

BP pressure (mm Hg):

Property/Value	Units	Temp	Data Type	Reference
WS 6.36E-002	mg/L	25	EST	MEYLAN,WM ET AL. (1996)
logP 5.08			EXP	SANGSTER,J (1993)
VP 1.36E-011	mm Hg	25	EST	NEELY,WB & BLAU,GE (1985)
DC	pKa			
HL 2.45E-011	atm m3/mol	25	EST	MEYLAN,WM & HOWARD,PH (1991)
OH 1.06E-010	cm3/molc sec	25	EST	MEYLAN,WM & HOWARD,PH (1993)

CAS #: 092627-83-3				3-PYRIDINECARBOXYLIC ACID, 2-[(1-METHYL-1H-IMIDA

Formula: $C_{24}H_{27}N_3O_4$

Mol Weight: 421.50

MP (deg C):   **FP (deg C):**

BP (deg C):

BP pressure (mm Hg):

Property/Value	Units	Temp	Data Type	Reference
WS 5.59E+000	mg/L	25	EST	MEYLAN,WM ET AL. (1996)
logP 3.05			EXP	SANGSTER,J (1993)
VP 3.19E-011	mm Hg	25	EST	NEELY,WB & BLAU,GE (1985)
DC	pKa			
HL 4.32E-014	atm m3/mol	25	EST	MEYLAN,WM & HOWARD,PH (1991)
OH 1.06E-010	cm3/molc sec	25	EST	MEYLAN,WM & HOWARD,PH (1993)

CAS #: 092817-12-4				19-NORPREGNA-1,3,5(10)-TRIEN-20-YNE-3,17-DIOL, 1

Formula: $C_{20}H_{23}FO_2$

Mol Weight: 314.40

MP (deg C):   **FP (deg C):**

BP (deg C):

BP pressure (mm Hg):

Property/Value	Units	Temp	Data Type	Reference
WS 1.89E+002	mg/L	25	EST	MEYLAN,WM ET AL. (1996)
logP 3.30			EXP	SANGSTER,J (1993)
VP 3.19E-009	mm Hg	25	EST	NEELY,WB & BLAU,GE (1985)
DC	pKa			
HL 1.58E-011	atm m3/mol	25	EST	MEYLAN,WM & HOWARD,PH (1991)
OH 1.20E-010	cm3/molc sec	25	EST	MEYLAN,WM & HOWARD,PH (1993)

CAS #: 092854-81-4				DILOXANIDE BENZOATE

Formula: $C_{16}H_{13}Cl_2NO_3$

Mol Weight: 338.19

MP (deg C):   **FP (deg C):**

BP (deg C):

BP pressure (mm Hg):

Property/Value	Units	Temp	Data Type	Reference
WS 4.75E+001	mg/L	25	EST	MEYLAN,WM ET AL. (1996)
logP 2.56			EXP	DUTTA,H ET AL. (1988)
VP 2.68E-008	mm Hg	25	EST	NEELY,WB & BLAU,GE (1985)
DC	pKa			
HL 3.74E-011	atm m3/mol	25	EST	MEYLAN,WM & HOWARD,PH (1991)
OH 1.17E-011	cm3/molc sec	25	EST	MEYLAN,WM & HOWARD,PH (1993)

CAS #: 092897-88-6				[1,3]DIOXEPINO[5,6-D]ISOXAZOLE, 3A,4,8,8A-TETRAH

Formula: $C_{18}H_{17}NO_3$

Mol Weight: 295.34

MP (deg C):   **FP (deg C):**

BP (deg C):

BP pressure (mm Hg):

Property/Value	Units	Temp	Data Type	Reference
WS 1.31E+002	mg/L	25	EST	MEYLAN,WM ET AL. (1996)
logP 3.42			EXP	CAMILLERI,P ET AL. (1989)
VP 2.64E-007	mm Hg	25	EST	NEELY,WB & BLAU,GE (1985)
DC	pKa			
HL 2.17E-009	atm m3/mol	25	EST	MEYLAN,WM & HOWARD,PH (1991)
OH 3.45E-011	cm3/molc sec	25	EST	MEYLAN,WM & HOWARD,PH (1993)

CAS #: 092901-23-0				DEXAMETHASONE-21-GALACTOSIDE

Formula: $C_{28}H_{39}FO_{10}$

Mol Weight: 554.62

MP (deg C):   **FP (deg C):**

BP (deg C):

BP pressure (mm Hg):

Property/Value	Units	Temp	Data Type	Reference
WS 1.21E+002	mg/L	25	EST	MEYLAN,WM ET AL. (1996)
logP 0.49			EXP	HANSCH,C ET AL. (1995)
VP 4.89E-022	mm Hg	25	EST	NEELY,WB & BLAU,GE (1985)
DC	pKa			
HL 6.87E-019	atm m3/mol	25	EST	MEYLAN,WM & HOWARD,PH (1991)
OH 1.21E-010	cm3/molc sec	25	EST	MEYLAN,WM & HOWARD,PH (1993)

CAS #: 092901-24-1				PREGNA-1,4-DIENE-3,20-DIONE, 21-(beta-D-GALACTOP

Formula: $C_{27}H_{38}O_{10}$

Mol Weight: 522.60

MP (deg C):   **FP (deg C):**

BP (deg C):

BP pressure (mm Hg):

Property/Value	Units	Temp	Data Type	Reference
WS 4.13E+002	mg/L	25	EST	MEYLAN,WM ET AL. (1996)
logP 0.11			EXP	SANGSTER,J (1994)
VP 6.64E-022	mm Hg	25	EST	NEELY,WB & BLAU,GE (1985)
DC	pKa			
HL 2.60E-019	atm m3/mol	25	EST	MEYLAN,WM & HOWARD,PH (1991)
OH 1.22E-010	cm3/molc sec	25	EST	MEYLAN,WM & HOWARD,PH (1993)

CAS #: 092901-25-2				PREGN-4-ENE-3,20-DIONE, 21-(á-D-GALACTOPYRANOSYL

Formula: $C_{27}H_{40}O_{10}$

Mol Weight: 524.61

MP (deg C):   **FP (deg C):**

BP (deg C):

BP pressure (mm Hg):

Property/Value	Units	Temp	Data Type	Reference
WS 3.70E+002	mg/L	25	EST	MEYLAN,WM ET AL. (1996)
logP 0.15			EXP	HANSCH,C ET AL. (1995)
VP 7.67E-022	mm Hg	25	EST	NEELY,WB & BLAU,GE (1985)
DC	pKa			
HL 5.55E-019	atm m3/mol	25	EST	MEYLAN,WM & HOWARD,PH (1991)
OH 1.63E-010	cm3/molc sec	25	EST	MEYLAN,WM & HOWARD,PH (1993)

CAS #: 093371-46-1				PROPIONIC ACID,2(4BIPHENYLYL)3OH3(P-MEOPH)
Formula: $C_{22}H_{20}O_4$				
Mol Weight: 348.40				
MP (deg C):		FP (deg C):		
BP (deg C):				
BP pressure (mm Hg):				

Property/Value	Units	Temp	Data Type	Reference
WS 4.37E+001	mg/L	25	EST	MEYLAN,WM ET AL. (1996)
logP 3.59			EXP	HANSCH,C ET AL. (1995)
VP 2.57E-013	mm Hg	25	EST	NEELY,WB & BLAU,GE (1985)
DC	pKa			
HL 7.87E-016	atm m3/mol	25	EST	MEYLAN,WM & HOWARD,PH (1991)
OH 4.50E-011	cm3/molc sec	25	EST	MEYLAN,WM & HOWARD,PH (1993)

CAS #: 093371-50-7				PROPIONIC ACID,2(4-BIPHENYLYL)3OH3(2FPH)
Formula: $C_{21}H_{17}FO_3$				
Mol Weight: 336.37				
MP (deg C):		FP (deg C):		
BP (deg C):				
BP pressure (mm Hg):				

Property/Value	Units	Temp	Data Type	Reference
WS 3.63E+001	mg/L	25	EST	MEYLAN,WM ET AL. (1996)
logP 3.77			EXP	HANSCH,C ET AL. (1995)
VP 2.31E-012	mm Hg	25	EST	NEELY,WB & BLAU,GE (1985)
DC	pKa			
HL 1.55E-014	atm m3/mol	25	EST	MEYLAN,WM & HOWARD,PH (1991)
OH 2.12E-011	cm3/molc sec	25	EST	MEYLAN,WM & HOWARD,PH (1993)

CAS #: 093371-51-8				PROPIONIC ACID,2(4-BIPHENYLYL)-3-OH-3-(3FPH)-
Formula: $C_{21}H_{17}FO_3$				
Mol Weight: 336.37				
MP (deg C):		FP (deg C):		
BP (deg C):				
BP pressure (mm Hg):				

Property/Value	Units	Temp	Data Type	Reference
WS 8.96E+001	mg/L	25	EST	MEYLAN,WM ET AL. (1996)
logP 3.31			EXP	HANSCH,C ET AL. (1995)
VP 2.31E-012	mm Hg	25	EST	NEELY,WB & BLAU,GE (1985)
DC	pKa			
HL 1.55E-014	atm m3/mol	25	EST	MEYLAN,WM & HOWARD,PH (1991)
OH 2.41E-011	cm3/molc sec	25	EST	MEYLAN,WM & HOWARD,PH (1993)

CAS #: 093371-55-2				BUTANOIC ACID, 2-(4-BIPHENYLYL)-3-HYDROXY-3-METHY
Formula: $C_{17}H_{18}O_3$				
Mol Weight: 270.33				
MP (deg C):		FP (deg C):		
BP (deg C):				
BP pressure (mm Hg):				

Property/Value	Units	Temp	Data Type	Reference
WS 2.45E+002	mg/L	25	EST	MEYLAN,WM ET AL. (1996)
logP 3.25			EXP	HANSCH,C ET AL. (1995)
VP 8.90E-010	mm Hg	25	EST	NEELY,WB & BLAU,GE (1985)
DC	pKa			
HL 2.90E-013	atm m3/mol	25	EST	MEYLAN,WM & HOWARD,PH (1991)
OH 1.47E-011	cm3/molc sec	25	EST	MEYLAN,WM & HOWARD,PH (1993)

CAS #: 093371-56-3				3-OH-4-ME-2-(4-BIPHENYLYL)PENTANOIC ACID
Formula: $C_{18}H_{20}O_3$				
Mol Weight: 284.36				
MP (deg C):		FP (deg C):		
BP (deg C):				
BP pressure (mm Hg):				

Property/Value	Units	Temp	Data Type	Reference
WS 6.39E+001	mg/L	25	EST	MEYLAN,WM ET AL. (1996)
logP 3.84			EXP	HANSCH,C ET AL. (1995)
VP 3.05E-010	mm Hg	25	EST	NEELY,WB & BLAU,GE (1985)
DC	pKa			
HL 3.86E-013	atm m3/mol	25	EST	MEYLAN,WM & HOWARD,PH (1991)
OH 2.29E-011	cm3/molc sec	25	EST	MEYLAN,WM & HOWARD,PH (1993)

CAS #: 093414-39-2				2-PROPENOIC ACID, 3-(2-METHOXYPHENYL)-, 2-(4-FOR
Formula: $C_{20}H_{20}O_6$				
Mol Weight: 356.38				
MP (deg C):		FP (deg C):		
BP (deg C):				
BP pressure (mm Hg):				

Property/Value	Units	Temp	Data Type	Reference
WS 4.33E+000	mg/L	25	EST	MEYLAN,WM ET AL. (1996)
logP 3.65			EXP	HANSCH,C ET AL. (1995)
VP 7.43E-009	mm Hg	25	EST	NEELY,WB & BLAU,GE (1985)
DC	pKa			
HL 2.07E-013	atm m3/mol	25	EST	MEYLAN,WM & HOWARD,PH (1991)
OH 7.68E-011	cm3/molc sec	25	EST	MEYLAN,WM & HOWARD,PH (1993)

CAS #: 093414-41-6				BENZOIC ACID, 2-CHLORO-, 2-(4-FORMYL-2-METHOXYPH
Formula: $C_{17}H_{15}ClO_5$				
Mol Weight: 334.76				
MP (deg C):		FP (deg C):		
BP (deg C):				
BP pressure (mm Hg):				

Property/Value	Units	Temp	Data Type	Reference
WS 1.39E+001	mg/L	25	EST	MEYLAN,WM ET AL. (1996)
logP 3.21			EXP	HANSCH,C ET AL. (1995)
VP 7.48E-008	mm Hg	25	EST	NEELY,WB & BLAU,GE (1985)
DC	pKa			
HL 2.18E-011	atm m3/mol	25	EST	MEYLAN,WM & HOWARD,PH (1991)
OH 3.64E-011	cm3/molc sec	25	EST	MEYLAN,WM & HOWARD,PH (1993)

CAS #: 093414-42-7				BENZOIC ACID, 3-CHLORO-, 2-(4-FORMYL-2-METHOXYPH
Formula: $C_{17}H_{15}ClO_5$				
Mol Weight: 334.76				
MP (deg C):		FP (deg C):		
BP (deg C):				
BP pressure (mm Hg):				

Property/Value	Units	Temp	Data Type	Reference
WS 5.62E+000	mg/L	25	EST	MEYLAN,WM ET AL. (1996)
logP 3.67			EXP	HANSCH,C ET AL. (1995)
VP 7.48E-008	mm Hg	25	EST	NEELY,WB & BLAU,GE (1985)
DC	pKa			
HL 2.18E-011	atm m3/mol	25	EST	MEYLAN,WM & HOWARD,PH (1991)
OH 3.63E-011	cm3/molc sec	25	EST	MEYLAN,WM & HOWARD,PH (1993)

093414-43-8

CAS #: 093414-43-8
BENZOIC ACID, 4-CHLORO-, 2-(4-FORMYL-2-METHOXYPH

Formula: $C_{17}H_{15}ClO_5$
Mol Weight: 334.76
MP (deg C):
FP (deg C):
BP (deg C):
BP pressure (mm Hg):

Property/Value	Units	Temp	Data Type	Reference
WS 5.19E+000	mg/L	25	EST	MEYLAN,WM ET AL. (1996)
logP 3.71			EXP	HANSCH,C ET AL. (1995)
VP 7.48E-008	mm Hg	25	EST	NEELY,WB & BLAU,GE (1985)
DC	pKa			
HL 2.18E-011	atm m3/mol	25	EST	MEYLAN,WM & HOWARD,PH (1991)
OH 3.64E-011	cm3/molc sec	25	EST	MEYLAN,WM & HOWARD,PH (1993)

093414-45-0

CAS #: 093414-45-0
BENZOIC ACID, 4-CHLORO-, 2-[4-[(HYDROXYIMINO)MET

Formula: $C_{17}H_{16}ClNO_5$
Mol Weight: 349.77
MP (deg C):
FP (deg C):
BP (deg C):
BP pressure (mm Hg):

Property/Value	Units	Temp	Data Type	Reference
WS 2.79E+000	mg/L	25	EST	MEYLAN,WM ET AL. (1996)
logP 3.92			EXP	HANSCH,C ET AL. (1995)
VP 4.56E-011	mm Hg	25	EST	NEELY,WB & BLAU,GE (1985)
DC	pKa			
HL 5.80E-013	atm m3/mol	25	EST	MEYLAN,WM & HOWARD,PH (1991)
OH 5.62E-011	cm3/molc sec	25	EST	MEYLAN,WM & HOWARD,PH (1993)

093414-47-2

CAS #: 093414-47-2
BENZOIC ACID, 3-CHLORO-, 2-[4-[(HYDROXYIMINO)MET

Formula: $C_{17}H_{16}ClNO_5$
Mol Weight: 349.77
MP (deg C):
FP (deg C):
BP (deg C):
BP pressure (mm Hg):

Property/Value	Units	Temp	Data Type	Reference
WS 2.68E+000	mg/L	25	EST	MEYLAN,WM ET AL. (1996)
logP 3.94			EXP	HANSCH,C ET AL. (1995)
VP 4.56E-011	mm Hg	25	EST	NEELY,WB & BLAU,GE (1985)
DC	pKa			
HL 5.80E-013	atm m3/mol	25	EST	MEYLAN,WM & HOWARD,PH (1991)
OH 5.61E-011	cm3/molc sec	25	EST	MEYLAN,WM & HOWARD,PH (1993)

093414-50-7

CAS #: 093414-50-7
BENZALDEHYDE, 4-(2-HYDROXYETHOXY)-3-METHOXY-, OX

Formula: $C_{10}H_{13}NO_4$
Mol Weight: 211.22
MP (deg C):
FP (deg C):
BP (deg C):
BP pressure (mm Hg):

Property/Value	Units	Temp	Data Type	Reference
WS 7.63E+003	mg/L	25	EST	MEYLAN,WM ET AL. (1996)
logP 0.82			EXP	HANSCH,C ET AL. (1995)
VP 8.37E-008	mm Hg	25	EST	NEELY,WB & BLAU,GE (1985)
DC	pKa			
HL 6.07E-014	atm m3/mol	25	EST	MEYLAN,WM & HOWARD,PH (1991)
OH 5.81E-011	cm3/molc sec	25	EST	MEYLAN,WM & HOWARD,PH (1993)

093414-55-2

CAS #: 093414-55-2
BENZOIC ACID, 3,4,5-TRIMETHOXY-, 2-[4-[[(2-ETHOX

Formula: $C_{24}H_{29}NO_{10}$
Mol Weight: 491.50
MP (deg C):
FP (deg C):
BP (deg C):
BP pressure (mm Hg):

Property/Value	Units	Temp	Data Type	Reference
WS 1.34E+000	mg/L	25	EST	MEYLAN,WM ET AL. (1996)
logP 3.26			EXP	HANSCH,C ET AL. (1995)
VP 3.30E-011	mm Hg	25	EST	NEELY,WB & BLAU,GE (1985)
DC	pKa			
HL 1.16E-015	atm m3/mol	25	EST	MEYLAN,WM & HOWARD,PH (1991)
OH 1.06E-010	cm3/molc sec	25	EST	MEYLAN,WM & HOWARD,PH (1993)

093414-58-5

CAS #: 093414-58-5
1H-BENZOTRIAZOLE-5-CARBOXAMIDE, 6-METHOXY-N-[[1-

Formula: $C_{16}H_{21}N_5O_2$
Mol Weight: 315.38
MP (deg C):
FP (deg C):
BP (deg C):
BP pressure (mm Hg):

Property/Value	Units	Temp	Data Type	Reference
WS 1.54E+003	mg/L	25	EST	MEYLAN,WM ET AL. (1996)
logP 0.95			EXP	VAN DAMME,M ET AL. (1984)
VP 1.17E-011	mm Hg	25	EST	NEELY,WB & BLAU,GE (1985)
DC	pKa			
HL 5.26E-018	atm m3/mol	25	EST	MEYLAN,WM & HOWARD,PH (1991)
OH 1.35E-010	cm3/molc sec	25	EST	MEYLAN,WM & HOWARD,PH (1993)

093414-59-6

CAS #: 093414-59-6
BENZAMIDE, 4-AMINO-N-[2-(DIETHYLAMINO)ETHYL]-5-F

Formula: $C_{14}H_{22}FN_3O_2$
Mol Weight: 283.35
MP (deg C):
FP (deg C):
BP (deg C):
BP pressure (mm Hg):

Property/Value	Units	Temp	Data Type	Reference
WS 1.73E+002	mg/L	25	EST	MEYLAN,WM ET AL. (1996)
logP 2.28			EXP	HANSCH,C ET AL. (1995)
VP 2.04E-008	mm Hg	25	EST	NEELY,WB & BLAU,GE (1985)
DC	pKa			
HL 1.38E-016	atm m3/mol	25	EST	MEYLAN,WM & HOWARD,PH (1991)
OH 3.05E-010	cm3/molc sec	25	EST	MEYLAN,WM & HOWARD,PH (1993)

093414-60-9

CAS #: 093414-60-9
BENZAMIDE, 4-AMINO-N-[2-(DIETHYLAMINO)ETHYL]-5-I

Formula: $C_{14}H_{22}IN_3O_2$
Mol Weight: 391.25
MP (deg C):
FP (deg C):
BP (deg C):
BP pressure (mm Hg):

Property/Value	Units	Temp	Data Type	Reference
WS 7.97E+000	mg/L	25	EST	MEYLAN,WM ET AL. (1996)
logP 3.09			EXP	HANSCH,C ET AL. (1995)
VP 5.48E-010	mm Hg	25	EST	NEELY,WB & BLAU,GE (1985)
DC	pKa			
HL 2.73E-017	atm m3/mol	25	EST	MEYLAN,WM & HOWARD,PH (1991)
OH 3.05E-010	cm3/molc sec	25	EST	MEYLAN,WM & HOWARD,PH (1993)

093414-62-1 — BENZAMIDE, 5-(AMINOSULFONYL)-N-[(1-ETHYL-3-PYRRO

Formula: $C_{15}H_{23}N_3O_4S$
Mol Weight: 341.43
MP (deg C):   **FP (deg C):**
BP (deg C):
BP pressure (mm Hg):

Property/Value	Units	Temp	Data Type	Reference
WS 1.98E+003	mg/L	25	EST	MEYLAN,WM ET AL. (1996)
logP 0.64			EXP	VAN DAMME,M ET AL. (1984)
VP 3.30E-011	mm Hg	25	EST	NEELY,WB & BLAU,GE (1985)
DC	pKa			
HL 1.53E-017	atm m3/mol	25	EST	MEYLAN,WM & HOWARD,PH (1991)
OH 1.11E-010	cm3/molc sec	25	EST	MEYLAN,WM & HOWARD,PH (1993)

093414-63-2 — BENZAMIDE, N-[(1-ETHYL-3-PYRROLIDINYL)METHYL]-5-

Formula: $C_{17}H_{26}N_2O_4S$
Mol Weight: 354.47
MP (deg C):   **FP (deg C):**
BP (deg C):
BP pressure (mm Hg):

Property/Value	Units	Temp	Data Type	Reference
WS 4.43E+002	mg/L	25	EST	MEYLAN,WM ET AL. (1996)
logP 1.31			EXP	VAN DAMME,M ET AL. (1984)
VP 5.67E-011	mm Hg	25	EST	NEELY,WB & BLAU,GE (1985)
DC	pKa			
HL 6.00E-017	atm m3/mol	25	EST	MEYLAN,WM & HOWARD,PH (1991)
OH 1.19E-010	cm3/molc sec	25	EST	MEYLAN,WM & HOWARD,PH (1993)

093414-64-3 — BENZAMIDE, 4-AMINO-5-(AMINOSULFONYL)-N-[(1-ETHYL

Formula: $C_{15}H_{24}N_4O_4S$
Mol Weight: 356.45
MP (deg C):   **FP (deg C):**
BP (deg C):
BP pressure (mm Hg):

Property/Value	Units	Temp	Data Type	Reference
WS 1.55E+003	mg/L	25	EST	MEYLAN,WM ET AL. (1996)
logP 0.66			EXP	VAN DAMME,M ET AL. (1984)
VP 1.52E-012	mm Hg	25	EST	NEELY,WB & BLAU,GE (1985)
DC	pKa			
HL 5.41E-021	atm m3/mol	25	EST	MEYLAN,WM & HOWARD,PH (1991)
OH 3.07E-010	cm3/molc sec	25	EST	MEYLAN,WM & HOWARD,PH (1993)

093414-65-4 — BENZAMIDE, 4-AMINO-N-[2-(DIETHYLAMINO)ETHYL]-5-(

Formula: $C_{16}H_{27}N_3O_4S$
Mol Weight: 357.48
MP (deg C):   **FP (deg C):**
BP (deg C):
BP pressure (mm Hg):

Property/Value	Units	Temp	Data Type	Reference
WS 1.97E+002	mg/L	25	EST	MEYLAN,WM ET AL. (1996)
logP 1.70			EXP	HANSCH,C ET AL. (1995)
VP 9.68E-012	mm Hg	25	EST	NEELY,WB & BLAU,GE (1985)
DC	pKa			
HL 3.62E-020	atm m3/mol	25	EST	MEYLAN,WM & HOWARD,PH (1991)
OH 3.13E-010	cm3/molc sec	25	EST	MEYLAN,WM & HOWARD,PH (1993)

093450-05-6 — CARBAMIC ACID, [2-(PENTYLOXY)PHENYL]-, 2-(1-PIPE

Formula: $C_{20}H_{32}N_2O_3$
Mol Weight: 348.49
MP (deg C):   **FP (deg C):**
BP (deg C):
BP pressure (mm Hg):

Property/Value	Units	Temp	Data Type	Reference
WS 1.08E+001	mg/L	25	EST	MEYLAN,WM ET AL. (1996)
logP 4.42			EXP	HANSCH,C ET AL. (1995)
VP 5.03E-008	mm Hg	25	EST	NEELY,WB & BLAU,GE (1985)
DC	pKa			
HL 5.95E-013	atm m3/mol	25	EST	MEYLAN,WM & HOWARD,PH (1991)
OH 1.76E-010	cm3/molc sec	25	EST	MEYLAN,WM & HOWARD,PH (1993)

093450-23-8 — CARBAMIC ACID, [3-(BUTOXYMETHYL)PHENYL]-, 2-(1-P

Formula: $C_{19}H_{30}N_2O_3$
Mol Weight: 334.46
MP (deg C):   **FP (deg C):**
BP (deg C):
BP pressure (mm Hg):

Property/Value	Units	Temp	Data Type	Reference
WS 5.11E+001	mg/L	25	EST	MEYLAN,WM ET AL. (1996)
logP 3.73			EXP	HANSCH,C ET AL. (1995)
VP 1.18E-007	mm Hg	25	EST	NEELY,WB & BLAU,GE (1985)
DC	pKa			
HL 4.48E-013	atm m3/mol	25	EST	MEYLAN,WM & HOWARD,PH (1991)
OH 2.28E-010	cm3/molc sec	25	EST	MEYLAN,WM & HOWARD,PH (1993)

093450-24-9 — CARBAMIC ACID, [3-[(PENTYLOXY)METHYL]PHENYL]-, 2

Formula: $C_{20}H_{32}N_2O_3$
Mol Weight: 348.49
MP (deg C):   **FP (deg C):**
BP (deg C):
BP pressure (mm Hg):

Property/Value	Units	Temp	Data Type	Reference
WS 1.67E+001	mg/L	25	EST	MEYLAN,WM ET AL. (1996)
logP 4.20			EXP	HANSCH,C ET AL. (1995)
VP 5.03E-008	mm Hg	25	EST	NEELY,WB & BLAU,GE (1985)
DC	pKa			
HL 5.95E-013	atm m3/mol	25	EST	MEYLAN,WM & HOWARD,PH (1991)
OH 2.30E-010	cm3/molc sec	25	EST	MEYLAN,WM & HOWARD,PH (1993)

093450-34-1 — CARBAMIC ACID, [2-(BUTOXYMETHYL)PHENYL]-, 2-(1-P

Formula: $C_{19}H_{30}N_2O_3$
Mol Weight: 334.46
MP (deg C):   **FP (deg C):**
BP (deg C):
BP pressure (mm Hg):

Property/Value	Units	Temp	Data Type	Reference
WS 3.38E+001	mg/L	25	EST	MEYLAN,WM ET AL. (1996)
logP 3.94			EXP	HANSCH,C ET AL. (1995)
VP 1.18E-007	mm Hg	25	EST	NEELY,WB & BLAU,GE (1985)
DC	pKa			
HL 4.48E-013	atm m3/mol	25	EST	MEYLAN,WM & HOWARD,PH (1991)
OH 1.74E-010	cm3/molc sec	25	EST	MEYLAN,WM & HOWARD,PH (1993)

CAS #: 093476-51-8 — N-(M-TOLYL)SUCCINIMIDE

Formula: $C_{11}H_{11}NO_2$

Mol Weight: 189.22

MP (deg C):   FP (deg C):

BP (deg C):

BP pressure (mm Hg):

Property/Value	Units	Temp	Data Type	Reference
WS 1.62E+004	mg/L	25	EST	MEYLAN,WM ET AL. (1996)
logP 0.57			EXP	HANSCH,C & LEO,AJ (1985)
VP 4.03E-007	mm Hg	25	EST	NEELY,WB & BLAU,GE (1985)
DC	pKa			
HL 1.44E-007	atm m3/mol	25	EST	MEYLAN,WM & HOWARD,PH (1991)
OH 3.69E-011	cm3/molc sec	25	EST	MEYLAN,WM & HOWARD,PH (1993)

CAS #: 093484-23-2 — HYDRAZINECARBOXIMIDAMIDE, 2-(1-METHYLHEPTYLIDENE

Formula: $C_9H_{20}N_4$

Mol Weight: 184.29

MP (deg C):   FP (deg C):

BP (deg C):

BP pressure (mm Hg):

Property/Value	Units	Temp	Data Type	Reference
WS 8.78E+002	mg/L	25	EST	MEYLAN,WM ET AL. (1996)
logP 2.08			EXP	SANGSTER,J (1993)
VP 5.26E-004	mm Hg	25	EST	NEELY,WB & BLAU,GE (1985)
DC	pKa			
HL 4.54E-010	atm m3/mol	25	EST	MEYLAN,WM & HOWARD,PH (1991)
OH 9.11E-011	cm3/molc sec	25	EST	MEYLAN,WM & HOWARD,PH (1993)

CAS #: 093677-64-6 — CARBAMIC ACID, DIETHYL-, 2-METHYLPROPYL ESTER

Formula: $C_9H_{19}NO_2$

Mol Weight: 173.26

MP (deg C):   FP (deg C):

BP (deg C):

BP pressure (mm Hg):

Property/Value	Units	Temp	Data Type	Reference
WS 2.32E+002	mg/L	25	EST	MEYLAN,WM ET AL. (1996)
logP 2.82			EXP	TANAKA,M ET AL. (1985)
VP 3.93E-001	mm Hg	25	EST	NEELY,WB & BLAU,GE (1985)
DC	pKa			
HL 7.87E-007	atm m3/mol	25	EST	MEYLAN,WM & HOWARD,PH (1991)
OH 2.76E-011	cm3/molc sec	25	EST	MEYLAN,WM & HOWARD,PH (1993)

CAS #: 094011-82-2 — BAZINAPRINE

Formula: $C_{17}H_{19}N_5O$

Mol Weight: 309.37

MP (deg C):   FP (deg C):

BP (deg C):

BP pressure (mm Hg):

Property/Value	Units	Temp	Data Type	Reference
WS 1.46E+003	mg/L	25	EST	MEYLAN,WM ET AL. (1996)
logP 1.89			EXP	HANSCH,C ET AL. (1995)
VP 1.48E-010	mm Hg	25	EST	NEELY,WB & BLAU,GE (1985)
DC	pKa			
HL 7.45E-018	atm m3/mol	25	EST	MEYLAN,WM & HOWARD,PH (1991)
OH 1.65E-010	cm3/molc sec	25	EST	MEYLAN,WM & HOWARD,PH (1993)

CAS #: 094050-52-9 — FLUCYCLOXURON,(E)-ISOMER

Formula: $C_{25}H_{20}ClF_2N_3O_3$

Mol Weight: 483.91

MP (deg C):   FP (deg C):

BP (deg C):

BP pressure (mm Hg):

Property/Value	Units	Temp	Data Type	Reference
WS 1.16E-003	mg/L	25	EST	MEYLAN,WM ET AL. (1996)
logP 6.97			EXP	TOMLIN,C (1994)
VP 1.18E-015	mm Hg	25	EST	NEELY,WB & BLAU,GE (1985)
DC	pKa			
HL 2.10E-014	atm m3/mol	25	EST	MEYLAN,WM & HOWARD,PH (1991)
OH 5.77E-011	cm3/molc sec	25	EST	MEYLAN,WM & HOWARD,PH (1993)

CAS #: 094050-53-0 — FLUCYCLOXURON,(Z)-ISOMER

Formula: $C_{25}H_{20}ClF_2N_3O_3$

Mol Weight: 483.91

MP (deg C):   FP (deg C):

BP (deg C):

BP pressure (mm Hg):

Property/Value	Units	Temp	Data Type	Reference
WS 1.16E-003	mg/L	25	EST	MEYLAN,WM ET AL. (1996)
logP 6.90			EXP	TOMLIN,C (1994)
VP 1.18E-015	mm Hg	25	EST	NEELY,WB & BLAU,GE (1985)
DC	pKa			
HL 2.10E-014	atm m3/mol	25	EST	MEYLAN,WM & HOWARD,PH (1991)
OH 5.77E-011	cm3/molc sec	25	EST	MEYLAN,WM & HOWARD,PH (1993)

CAS #: 094089-22-2 — PYRAZINE, 2-METHYL-3-(1-METHYLETHOXY)-

Formula: $C_8H_{12}N_2O$

Mol Weight: 152.20

MP (deg C):   FP (deg C):

BP (deg C):

BP pressure (mm Hg):

Property/Value	Units	Temp	Data Type	Reference
WS 9.07E+002	mg/L	25	EST	MEYLAN,WM ET AL. (1996)
logP 2.24			EXP	YAMAGAMI,C ET AL. (1991)
VP 1.13E-001	mm Hg	25	EST	NEELY,WB & BLAU,GE (1985)
DC	pKa			
HL 3.06E-006	atm m3/mol	25	EST	MEYLAN,WM & HOWARD,PH (1991)
OH 1.54E-011	cm3/molc sec	25	EST	MEYLAN,WM & HOWARD,PH (1993)

CAS #: 094452-21-8 — 2,4(1H,3H)-PYRIMIDINEDIONE, 5-FLUORO-3-(1-OXOBUT

Formula: $C_8H_9FN_2O_3$

Mol Weight: 200.17

MP (deg C):   FP (deg C):

BP (deg C):

BP pressure (mm Hg):

Property/Value	Units	Temp	Data Type	Reference
WS 1.17E+004	mg/L	25	EST	MEYLAN,WM ET AL. (1996)
logP 0.67			EXP	SANGSTER,J (1994)
VP 2.17E-008	mm Hg	25	EST	NEELY,WB & BLAU,GE (1985)
DC	pKa			
HL 9.48E-013	atm m3/mol	25	EST	MEYLAN,WM & HOWARD,PH (1991)
OH 1.05E-011	cm3/molc sec	25	EST	MEYLAN,WM & HOWARD,PH (1993)

CAS #: 094497-46-8				ACETAMIDE, N-(3,4-DICHLOROPHENYL)-2-METHOXY-

Formula: $C_9H_9Cl_2NO_2$

Mol Weight: 234.08

MP (deg C):   FP (deg C):

BP (deg C):

BP pressure (mm Hg):

Property/Value	Units	Temp	Data Type	Reference
WS 1.25E+002	mg/L	25	EST	MEYLAN,WM ET AL. (1996)
logP 2.77			EXP	HANSCH,C ET AL. (1995)
VP 4.23E-006	mm Hg	25	EST	NEELY,WB & BLAU,GE (1985)
DC	pKa			
HL 2.81E-010	atm m3/mol	25	EST	MEYLAN,WM & HOWARD,PH (1991)
OH 7.65E-012	cm3/molc sec	25	EST	MEYLAN,WM & HOWARD,PH (1993)

CAS #: 094563-11-8				CARBAMIC ACID, (4-CYANOPHENYL)-, METHYL ESTER

Formula: $C_9H_8N_2O_2$

Mol Weight: 176.18

MP (deg C):   FP (deg C):

BP (deg C):

BP pressure (mm Hg):

Property/Value	Units	Temp	Data Type	Reference
WS 8.71E+002	mg/L	25	EST	MEYLAN,WM ET AL. (1996)
logP 1.82			EXP	TAKAHASHI,J ET AL. (1988)
VP 6.30E-004	mm Hg	25	EST	NEELY,WB & BLAU,GE (1985)
DC	pKa			
HL 2.11E-010	atm m3/mol	25	EST	MEYLAN,WM & HOWARD,PH (1991)
OH 7.75E-012	cm3/molc sec	25	EST	MEYLAN,WM & HOWARD,PH (1993)

CAS #: 094593-91-6				CINOSULFURON

Formula: $C_{15}H_{19}N_5O_7S$

Mol Weight: 413.41

MP (deg C):   FP (deg C):

BP (deg C):

BP pressure (mm Hg):

Property/Value	Units	Temp	Data Type	Reference
WS 4.58E+001	mg/L	25	EST	MEYLAN,WM ET AL. (1996)
logP 2.04			EXP	TOMLIN,C (1994)
VP 6.92E-013	mm Hg	25	EST	NEELY,WB & BLAU,GE (1985)
DC	pKa			
HL 5.24E-015	atm m3/mol	25	EST	MEYLAN,WM & HOWARD,PH (1991)
OH 2.23E-011	cm3/molc sec	25	EST	MEYLAN,WM & HOWARD,PH (1993)

CAS #: 094641-10-8				2-BENZOTHIAZOLESULFONAMIDE, 6-NITRO-

Formula: $C_7H_5N_3O_4S_2$

Mol Weight: 259.26

MP (deg C):   FP (deg C):

BP (deg C):

BP pressure (mm Hg):

Property/Value	Units	Temp	Data Type	Reference
WS 5.90E+002	mg/L	25	EST	MEYLAN,WM ET AL. (1996)
logP 1.36			EXP	ELLER,MG ET AL. (1985)
VP 2.28E-008	mm Hg	25	EST	NEELY,WB & BLAU,GE (1985)
DC	pKa			
HL 1.16E-013	atm m3/mol	25	EST	MEYLAN,WM & HOWARD,PH (1991)
OH 1.56E-013	cm3/molc sec	25	EST	MEYLAN,WM & HOWARD,PH (1993)

CAS #: 094641-11-9				2-BENZOTHIAZOLESULFONAMIDE, 6-AMINO-

Formula: $C_7H_7N_3O_2S_2$

Mol Weight: 229.28

MP (deg C):   FP (deg C):

BP (deg C):

BP pressure (mm Hg):

Property/Value	Units	Temp	Data Type	Reference
WS 1.11E+004	mg/L	25	EST	MEYLAN,WM ET AL. (1996)
logP 0.52			EXP	ELLER,MG ET AL. (1985)
VP 5.64E-008	mm Hg	25	EST	NEELY,WB & BLAU,GE (1985)
DC	pKa			
HL 1.04E-014	atm m3/mol	25	EST	MEYLAN,WM & HOWARD,PH (1991)
OH 1.42E-011	cm3/molc sec	25	EST	MEYLAN,WM & HOWARD,PH (1993)

CAS #: 094779-66-5				BARBITURIC ACID,N-ME,5-ALLYL,5-CYPENTENYL

Formula: $C_{13}H_{16}N_2O_3$

Mol Weight: 248.28

MP (deg C):   FP (deg C):

BP (deg C):

BP pressure (mm Hg):

Property/Value	Units	Temp	Data Type	Reference
WS 5.88E+002	mg/L	25	EST	MEYLAN,WM ET AL. (1996)
logP 1.89			EXP	HANSCH,C & LEO,AJ (1985)
VP 1.55E-011	mm Hg	25	EST	NEELY,WB & BLAU,GE (1985)
DC	pKa			
HL 8.39E-013	atm m3/mol	25	EST	MEYLAN,WM & HOWARD,PH (1991)
OH 1.22E-010	cm3/molc sec	25	EST	MEYLAN,WM & HOWARD,PH (1993)

CAS #: 094789-37-4				2-PYRIMIDINECARBONITRILE, 4-METHOXY-

Formula: $C_6H_5N_3O$

Mol Weight: 135.13

MP (deg C):   FP (deg C):

BP (deg C):

BP pressure (mm Hg):

Property/Value	Units	Temp	Data Type	Reference
WS 8.78E+003	mg/L	25	EST	MEYLAN,WM ET AL. (1996)
logP 0.86			EXP	YAMAGAMI,C ET AL. (1991)
VP 1.35E-002	mm Hg	25	EST	NEELY,WB & BLAU,GE (1985)
DC	pKa			
HL 1.52E-008	atm m3/mol	25	EST	MEYLAN,WM & HOWARD,PH (1991)
OH 1.16E-012	cm3/molc sec	25	EST	MEYLAN,WM & HOWARD,PH (1993)

CAS #: 094921-29-6				2H-NAPHTH[1,2-B]-1,4-OXAZIN-7-OL,3,4,4A,5,6,10B-

Formula: $C_{15}H_{21}NO_2$

Mol Weight: 247.34

MP (deg C):   FP (deg C):

BP (deg C):

BP pressure (mm Hg):

Property/Value	Units	Temp	Data Type	Reference
WS 1.20E+004	mg/L	25	EST	MEYLAN,WM ET AL. (1996)
logP 2.22			EXP	DIJKSTRA,D ET AL. (1988)
VP 1.71E-006	mm Hg	25	EST	NEELY,WB & BLAU,GE (1985)
DC	pKa			
HL 2.39E-012	atm m3/mol	25	EST	MEYLAN,WM & HOWARD,PH (1991)
OH 2.56E-010	cm3/molc sec	25	EST	MEYLAN,WM & HOWARD,PH (1993)

CAS #:	095261-40-8			CARBAMIC ACID, (3,4-DIETHOXYPHENYL)-, 2-METHYLPR

Formula: $C_{15}H_{23}NO_4$

Mol Weight: 281.35

MP (deg C): FP (deg C):

BP (deg C):

BP pressure (mm Hg):

Property/ Value	Units	Temp	Data Type	Reference
WS 2.13E+001	mg/L	25	EST	MEYLAN,WM ET AL. (1996)
logP 3.36			EXP	TAKAHASHI,J ET AL. (1988)
VP 1.19E-005	mm Hg	25	EST	NEELY,WB & BLAU,GE (1985)
DC	pKa			
HL 3.15E-010	atm m3/mol	25	EST	MEYLAN,WM & HOWARD,PH (1991)
OH 2.17E-010	cm3/molc sec	25	EST	MEYLAN,WM & HOWARD,PH (1993)

CAS #:	095261-60-2			CARBAMIC ACID, (3,4-DIETHOXYPHENYL)-, 2-METHOXYE

Formula: $C_{14}H_{21}NO_5$

Mol Weight: 283.33

MP (deg C): FP (deg C):

BP (deg C):

BP pressure (mm Hg):

Property/ Value	Units	Temp	Data Type	Reference
WS 1.77E+002	mg/L	25	EST	MEYLAN,WM ET AL. (1996)
logP 2.27			EXP	TAKAHASHI,J ET AL. (1988)
VP 7.12E-006	mm Hg	25	EST	NEELY,WB & BLAU,GE (1985)
DC	pKa			
HL 2.09E-012	atm m3/mol	25	EST	MEYLAN,WM & HOWARD,PH (1991)
OH 2.22E-010	cm3/molc sec	25	EST	MEYLAN,WM & HOWARD,PH (1993)

CAS #:	095261-65-7			CARBAMIC ACID, (3,4-DIETHOXYPHENYL)-, PHENYLMETH

Formula: $C_{18}H_{21}NO_4$

Mol Weight: 315.37

MP (deg C): FP (deg C):

BP (deg C):

BP pressure (mm Hg):

Property/ Value	Units	Temp	Data Type	Reference
WS 1.55E+001	mg/L	25	EST	MEYLAN,WM ET AL. (1996)
logP 3.29			EXP	TAKAHASHI,J ET AL. (1988)
VP 1.99E-007	mm Hg	25	EST	NEELY,WB & BLAU,GE (1985)
DC	pKa			
HL 1.09E-011	atm m3/mol	25	EST	MEYLAN,WM & HOWARD,PH (1991)
OH 2.14E-010	cm3/molc sec	25	EST	MEYLAN,WM & HOWARD,PH (1993)

CAS #:	095333-49-6			IMIDAZOLE-2-THIONE,3-(3-FLUORO-4-HYDROXYBENZYL)

Formula: $C_{10}H_9FN_2OS$

Mol Weight: 224.26

MP (deg C): FP (deg C):

BP (deg C):

BP pressure (mm Hg):

Property/ Value	Units	Temp	Data Type	Reference
WS 1.40E+005	mg/L	25	EST	MEYLAN,WM ET AL. (1996)
logP 1.12			EXP	HANSCH,C ET AL. (1995)
VP 1.18E-007	mm Hg	25	EST	NEELY,WB & BLAU,GE (1985)
DC	pKa			
HL 1.99E-011	atm m3/mol	25	EST	MEYLAN,WM & HOWARD,PH (1991)
OH 1.72E-010	cm3/molc sec	25	EST	MEYLAN,WM & HOWARD,PH (1993)

CAS #:	095333-60-1			IMIDIAZOLE-2-THIONE,3-(3,5-DIFLUORO-4-HYDROXYBEN

Formula: $C_{10}H_8F_2N_2OS$

Mol Weight: 242.25

MP (deg C): FP (deg C):

BP (deg C):

BP pressure (mm Hg):

Property/ Value	Units	Temp	Data Type	Reference
WS 1.21E+005	mg/L	25	EST	MEYLAN,WM ET AL. (1996)
logP 1.08			EXP	HANSCH,C ET AL. (1995)
VP 1.43E-007	mm Hg	25	EST	NEELY,WB & BLAU,GE (1985)
DC	pKa			
HL 2.32E-011	atm m3/mol	25	EST	MEYLAN,WM & HOWARD,PH (1991)
OH 1.63E-010	cm3/molc sec	25	EST	MEYLAN,WM & HOWARD,PH (1993)

CAS #:	095333-64-5			IMIDAZOLE-2-THIONE,3-(4-HYDROXYBENZYL)

Formula: $C_{10}H_{10}N_2OS$

Mol Weight: 206.27

MP (deg C): FP (deg C):

BP (deg C):

BP pressure (mm Hg):

Property/ Value	Units	Temp	Data Type	Reference
WS 2.12E+005	mg/L	25	EST	MEYLAN,WM ET AL. (1996)
logP 1.02			EXP	HANSCH,C ET AL. (1995)
VP 1.01E-007	mm Hg	25	EST	NEELY,WB & BLAU,GE (1985)
DC	pKa			
HL 1.71E-011	atm m3/mol	25	EST	MEYLAN,WM & HOWARD,PH (1991)
OH 1.99E-010	cm3/molc sec	25	EST	MEYLAN,WM & HOWARD,PH (1993)

CAS #:	095333-80-5			IMIDAZOLE-2-THIONE,3-(3-FLUOROBENZYL)

Formula: $C_{10}H_9FN_2S$

Mol Weight: 208.26

MP (deg C): FP (deg C):

BP (deg C):

BP pressure (mm Hg):

Property/ Value	Units	Temp	Data Type	Reference
WS 1.24E+004	mg/L	25	EST	MEYLAN,WM ET AL. (1996)
logP 1.77			EXP	HANSCH,C ET AL. (1995)
VP 7.93E-006	mm Hg	25	EST	NEELY,WB & BLAU,GE (1985)
DC	pKa			
HL 1.91E-007	atm m3/mol	25	EST	MEYLAN,WM & HOWARD,PH (1991)
OH 1.61E-010	cm3/molc sec	25	EST	MEYLAN,WM & HOWARD,PH (1993)

CAS #:	095333-81-6			IMIDAZOLE-2-THIONE,3-(3,5-DIFLUOROBENZYL)-

Formula: $C_{10}H_8F_2N_2S$

Mol Weight: 226.25

MP (deg C): FP (deg C):

BP (deg C):

BP pressure (mm Hg):

Property/ Value	Units	Temp	Data Type	Reference
WS 6.62E+003	mg/L	25	EST	MEYLAN,WM ET AL. (1996)
logP 1.98			EXP	HANSCH,C ET AL. (1995)
VP 1.07E-005	mm Hg	25	EST	NEELY,WB & BLAU,GE (1985)
DC	pKa			
HL 2.23E-007	atm m3/mol	25	EST	MEYLAN,WM & HOWARD,PH (1991)
OH 1.66E-010	cm3/molc sec	25	EST	MEYLAN,WM & HOWARD,PH (1993)

CAS #: 095384-52-4 — BENZENEACETAMIDE, N-(3-IODOPHENYL)-

Formula: $C_{14}H_{12}INO$

Mol Weight: 337.16

MP (deg C): | FP (deg C):
BP (deg C):
BP pressure (mm Hg):

Property/Value	Units	Temp	Data Type	Reference
WS 3.39E+000	mg/L	25	EST	MEYLAN,WM ET AL. (1996)
logP 3.91			EXP	YAMAGAMI,C ET AL. (1984)
VP 2.46E-008	mm Hg	25	EST	NEELY,WB & BLAU,GE (1985)
DC	pKa			
HL 1.15E-010	atm m3/mol	25	EST	MEYLAN,WM & HOWARD,PH (1991)
OH 1.37E-011	cm3/molc sec	25	EST	MEYLAN,WM & HOWARD,PH (1993)

CAS #: 095384-53-5 — BENZENEACETAMIDE, N-[3-(METHYLAMINO)PHENYL]-

Formula: $C_{15}H_{16}N_2O$

Mol Weight: 240.31

MP (deg C): | FP (deg C):
BP (deg C):
BP pressure (mm Hg):

Property/Value	Units	Temp	Data Type	Reference
WS 2.85E+002	mg/L	25	EST	MEYLAN,WM ET AL. (1996)
logP 2.31			EXP	YAMAGAMI,C ET AL. (1984)
VP 7.03E-008	mm Hg	25	EST	NEELY,WB & BLAU,GE (1985)
DC	pKa			
HL 3.87E-013	atm m3/mol	25	EST	MEYLAN,WM & HOWARD,PH (1991)
OH 2.02E-010	cm3/molc sec	25	EST	MEYLAN,WM & HOWARD,PH (1993)

CAS #: 095384-54-6 — BENZENEACETAMIDE, N-[3-(ETHYLAMINO)PHENYL]-

Formula: $C_{16}H_{18}N_2O$

Mol Weight: 254.33

MP (deg C): | FP (deg C):
BP (deg C):
BP pressure (mm Hg):

Property/Value	Units	Temp	Data Type	Reference
WS 1.37E+002	mg/L	25	EST	MEYLAN,WM ET AL. (1996)
logP 2.59			EXP	YAMAGAMI,C ET AL. (1984)
VP 2.97E-008	mm Hg	25	EST	NEELY,WB & BLAU,GE (1985)
DC	pKa			
HL 5.13E-013	atm m3/mol	25	EST	MEYLAN,WM & HOWARD,PH (1991)
OH 2.10E-010	cm3/molc sec	25	EST	MEYLAN,WM & HOWARD,PH (1993)

CAS #: 095384-55-7 — BENZENEACETAMIDE, N-[3-(ACETYLOXY)PHENYL]-

Formula: $C_{16}H_{15}NO_3$

Mol Weight: 269.30

MP (deg C): | FP (deg C):
BP (deg C):
BP pressure (mm Hg):

Property/Value	Units	Temp	Data Type	Reference
WS 1.75E+002	mg/L	25	EST	MEYLAN,WM ET AL. (1996)
logP 2.37			EXP	YAMAGAMI,C ET AL. (1984)
VP 2.30E-008	mm Hg	25	EST	NEELY,WB & BLAU,GE (1985)
DC	pKa			
HL 5.99E-012	atm m3/mol	25	EST	MEYLAN,WM & HOWARD,PH (1991)
OH 1.85E-011	cm3/molc sec	25	EST	MEYLAN,WM & HOWARD,PH (1993)

CAS #: 095384-56-8 — BENZENEACETAMIDE, N-[3-((METHYLSULFONYL)OXY]PHEN

Formula: $C_{15}H_{15}NO_4S$

Mol Weight: 305.36

MP (deg C): | FP (deg C):
BP (deg C):
BP pressure (mm Hg):

Property/Value	Units	Temp	Data Type	Reference
WS 1.19E+002	mg/L	25	EST	MEYLAN,WM ET AL. (1996)
logP 2.32			EXP	YAMAGAMI,C ET AL. (1984)
VP 3.70E-010	mm Hg	25	EST	NEELY,WB & BLAU,GE (1985)
DC	pKa			
HL 1.38E-013	atm m3/mol	25	EST	MEYLAN,WM & HOWARD,PH (1991)
OH 1.43E-010	cm3/molc sec	25	EST	MEYLAN,WM & HOWARD,PH (1993)

CAS #: 095384-57-9 — BENZENEACETAMIDE, N-(4-HYDROXYPHENYL)-

Formula: $C_{14}H_{13}NO_2$

Mol Weight: 227.27

MP (deg C): | FP (deg C):
BP (deg C):
BP pressure (mm Hg):

Property/Value	Units	Temp	Data Type	Reference
WS 8.79E+002	mg/L	25	EST	MEYLAN,WM ET AL. (1996)
logP 1.82			EXP	YAMAGAMI,C ET AL. (1984)
VP 9.97E-009	mm Hg	25	EST	NEELY,WB & BLAU,GE (1985)
DC	pKa			
HL 5.18E-014	atm m3/mol	25	EST	MEYLAN,WM & HOWARD,PH (1991)
OH 2.31E-011	cm3/molc sec	25	EST	MEYLAN,WM & HOWARD,PH (1993)

CAS #: 095384-58-0 — BENZENEACETAMIDE, N-(2-HYDROXYPHENYL)-

Formula: $C_{14}H_{13}NO_2$

Mol Weight: 227.27

MP (deg C): | FP (deg C):
BP (deg C):
BP pressure (mm Hg):

Property/Value	Units	Temp	Data Type	Reference
WS 2.81E+002	mg/L	25	EST	MEYLAN,WM ET AL. (1996)
logP 2.40			EXP	YAMAGAMI,C ET AL. (1984)
VP 9.97E-009	mm Hg	25	EST	NEELY,WB & BLAU,GE (1985)
DC	pKa			
HL 5.18E-014	atm m3/mol	25	EST	MEYLAN,WM & HOWARD,PH (1991)
OH 2.31E-011	cm3/molc sec	25	EST	MEYLAN,WM & HOWARD,PH (1993)

CAS #: 095384-59-1 — BENZENEACETAMIDE, N-(2-AMINOPHENYL)-

Formula: $C_{14}H_{14}N_2O$

Mol Weight: 226.28

MP (deg C): | FP (deg C):
BP (deg C):
BP pressure (mm Hg):

Property/Value	Units	Temp	Data Type	Reference
WS 1.84E+003	mg/L	25	EST	MEYLAN,WM ET AL. (1996)
logP 1.45			EXP	YAMAGAMI,C ET AL. (1984)
VP 3.34E-008	mm Hg	25	EST	NEELY,WB & BLAU,GE (1985)
DC	pKa			
HL 1.76E-013	atm m3/mol	25	EST	MEYLAN,WM & HOWARD,PH (1991)
OH 6.19E-011	cm3/molc sec	25	EST	MEYLAN,WM & HOWARD,PH (1993)

CAS #: 095384-60-4			BENZENEACETAMIDE, N-(2-METHOXYPHENYL)-

Formula: $C_{15}H_{15}NO_2$

Mol Weight: 241.29

MP (deg C):   FP (deg C):

BP (deg C):

BP pressure (mm Hg):

Property/ Value	Units	Temp	Data Type	Reference
WS 1.36E+002	mg/L	25	EST	MEYLAN,WM ET AL. (1996)
logP 2.68			EXP	YAMAGAMI,C ET AL. (1984)
VP 1.60E-007	mm Hg	25	EST	NEELY,WB & BLAU,GE (1985)
DC	pKa			
HL 2.95E-011	atm m3/mol	25	EST	MEYLAN,WM & HOWARD,PH (1991)
OH 1.76E-011	cm3/molc sec	25	EST	MEYLAN,WM & HOWARD,PH (1993)

CAS #: 095440-71-4			PREDNISONE-17-ACETATE

Formula: $C_{23}H_{28}O_6$

Mol Weight: 400.48

MP (deg C):   FP (deg C):

BP (deg C):

BP pressure (mm Hg):

Property/ Value	Units	Temp	Data Type	Reference
WS 8.28E+001	mg/L	25	EST	MEYLAN,WM ET AL. (1996)
logP 2.43			EXP	HANSCH,C & LEO,AJ (1985)
VP 4.52E-013	mm Hg	25	EST	NEELY,WB & BLAU,GE (1985)
DC	pKa			
HL 3.05E-014	atm m3/mol	25	EST	MEYLAN,WM & HOWARD,PH (1991)
OH 6.60E-011	cm3/molc sec	25	EST	MEYLAN,WM & HOWARD,PH (1993)

CAS #: 095465-99-9			CADUSAFOS

Formula: $C_{10}H_{23}O_2PS_2$

Mol Weight: 270.40

MP (deg C):   FP (deg C):

BP (deg C):

BP pressure (mm Hg):

Property/ Value	Units	Temp	Data Type	Reference
WS 8.49E+000	mg/L	25	EST	MEYLAN,WM ET AL. (1996)
logP 3.90			EXP	TOMLIN,C (1994)
VP 2.84E-004	mm Hg	25	EST	NEELY,WB & BLAU,GE (1985)
DC	pKa			
HL 5.46E-007	atm m3/mol	25	EST	MEYLAN,WM & HOWARD,PH (1991)
OH 1.20E-010	cm3/molc sec	25	EST	MEYLAN,WM & HOWARD,PH (1993)

CAS #: 095549-92-1			PIPERAZINE, 1-(4-AMINO-6,7-DIMETHOXY-2-QUINAZOLI

Formula: $C_{18}H_{25}N_5O_4$

Mol Weight: 375.43

MP (deg C):   FP (deg C):

BP (deg C):

BP pressure (mm Hg):

Property/ Value	Units	Temp	Data Type	Reference
WS 4.61E+002	mg/L	25	EST	MEYLAN,WM ET AL. (1996)
logP 1.14			EXP	SANGSTER,J (1993)
VP 2.67E-011	mm Hg	25	EST	NEELY,WB & BLAU,GE (1985)
DC	pKa			
HL 9.37E-019	atm m3/mol	25	EST	MEYLAN,WM & HOWARD,PH (1991)
OH 2.58E-010	cm3/molc sec	25	EST	MEYLAN,WM & HOWARD,PH (1993)

CAS #: 095579-17-2			O-METHYL THPO

Formula: $C_7H_{11}N_2O_2$

Mol Weight: 155.18

MP (deg C):   FP (deg C):

BP (deg C):

BP pressure (mm Hg):

Property/ Value	Units	Temp	Data Type	Reference
WS 8.01E+005	mg/L	25	EST	MEYLAN,WM ET AL. (1996)
logP -0.04			EXP	HANSCH,C ET AL. (1995)
VP 1.17E-002	mm Hg	25	EST	NEELY,WB & BLAU,GE (1985)
DC	pKa			
HL 4.07E-010	atm m3/mol	25	EST	MEYLAN,WM & HOWARD,PH (1991)
OH 9.34E-011	cm3/molc sec	25	EST	MEYLAN,WM & HOWARD,PH (1993)

CAS #: 095579-22-9			O,N-DIME THPO

Formula: $C_8H_{13}N_2O_2$

Mol Weight: 169.21

MP (deg C):   FP (deg C):

BP (deg C):

BP pressure (mm Hg):

Property/ Value	Units	Temp	Data Type	Reference
WS 2.91E+005	mg/L	25	EST	MEYLAN,WM ET AL. (1996)
logP 0.40			EXP	SAUERBERG,P ET AL. (1986)
VP 2.14E-002	mm Hg	25	EST	NEELY,WB & BLAU,GE (1985)
DC	pKa			
HL 8.93E-010	atm m3/mol	25	EST	MEYLAN,WM & HOWARD,PH (1991)
OH 9.77E-011	cm3/molc sec	25	EST	MEYLAN,WM & HOWARD,PH (1993)

CAS #: 095599-36-3			2-BENZOTHIAZOLESULFONAMIDE, 6-(2-HYDROXYETHOXY)-

Formula: $C_9H_{10}N_2O_4S_2$

Mol Weight: 274.32

MP (deg C):   FP (deg C):

BP (deg C):

BP pressure (mm Hg):

Property/ Value	Units	Temp	Data Type	Reference
WS 5.41E+003	mg/L	25	EST	MEYLAN,WM ET AL. (1996)
logP 0.59			EXP	ELLER,MG ET AL. (1985)
VP 5.02E-011	mm Hg	25	EST	NEELY,WB & BLAU,GE (1985)
DC	pKa			
HL 8.42E-017	atm m3/mol	25	EST	MEYLAN,WM & HOWARD,PH (1991)
OH 1.58E-011	cm3/molc sec	25	EST	MEYLAN,WM & HOWARD,PH (1993)

CAS #: 095635-46-4			2-F 4-BR 5-NO2 ACETANILIDE

Formula: $C_8H_6BrFN_2O_3$

Mol Weight: 277.05

MP (deg C):   FP (deg C):

BP (deg C):

BP pressure (mm Hg):

Property/ Value	Units	Temp	Data Type	Reference
WS 1.07E+002	mg/L	25	EST	MEYLAN,WM ET AL. (1996)
logP 2.11			EXP	NAKAGAWA,Y ET AL. (1992)
VP 5.95E-007	mm Hg	25	EST	NEELY,WB & BLAU,GE (1985)
DC	pKa			
HL 1.13E-011	atm m3/mol	25	EST	MEYLAN,WM & HOWARD,PH (1991)
OH 2.07E-013	cm3/molc sec	25	EST	MEYLAN,WM & HOWARD,PH (1993)

CAS #:	095715-61-0		CYCLOHEXANEACETAMIDE, N,N-DIETHYL-

Formula:	$C_{12}H_{23}NO$

Mol Weight:	197.32

MP (deg C):		FP (deg C):

BP (deg C):

BP pressure (mm Hg):

Property/ Value	Units	Temp	Data Type	Reference
WS 9.78E+001	mg/L	25	EST	MEYLAN,WM ET AL. (1996)
logP 3.12			EXP	SURYANARAYANA,MVS ET AL. (1991)
VP 1.64E+000	mm Hg	25	EXP	SURYANARAYANA,MVS ET AL. (1991)
DC	pKa			
HL 2.29E-007	atm m3/mol	25	EST	MEYLAN,WM & HOWARD,PH (1991)
OH 4.24E-011	cm3/molc sec	25	EST	MEYLAN,WM & HOWARD,PH (1993)

CAS #:	095796-50-2		CLOBETASOL-17-PROPIONATE-DES-F

Formula:	$C_{25}H_{33}ClO_5$

Mol Weight:	448.99

MP (deg C):		FP (deg C):

BP (deg C):

BP pressure (mm Hg):

Property/ Value	Units	Temp	Data Type	Reference
WS 6.87E+000	mg/L	25	EST	MEYLAN,WM ET AL. (1996)
logP 3.34			EXP	HANSCH,C & LEO,AJ (1985)
VP 1.79E-013	mm Hg	25	EST	NEELY,WB & BLAU,GE (1985)
DC	pKa			
HL 1.16E-014	atm m3/mol	25	EST	MEYLAN,WM & HOWARD,PH (1991)
OH 7.21E-011	cm3/molc sec	25	EST	MEYLAN,WM & HOWARD,PH (1993)

CAS #:	095796-51-3		CLOBETASONE-17-PROPIONATE-DES-F

Formula:	$C_{25}H_{31}ClO_5$

Mol Weight:	446.98

MP (deg C):		FP (deg C):

BP (deg C):

BP pressure (mm Hg):

Property/ Value	Units	Temp	Data Type	Reference
WS 2.37E+000	mg/L	25	EST	MEYLAN,WM ET AL. (1996)
logP 3.30			EXP	HANSCH,C & LEO,AJ (1985)
VP 4.79E-011	mm Hg	25	EST	NEELY,WB & BLAU,GE (1985)
DC	pKa			
HL 1.22E-013	atm m3/mol	25	EST	MEYLAN,WM & HOWARD,PH (1991)
OH 6.84E-011	cm3/molc sec	25	EST	MEYLAN,WM & HOWARD,PH (1993)

CAS #:	095853-68-2		4,4,4-TRIFLUORO-3-METHYL-1-BUTANOL

Formula:	$C_5H_9F_3O$

Mol Weight:	142.12

MP (deg C):		FP (deg C):

BP (deg C):

BP pressure (mm Hg):

Property/ Value	Units	Temp	Data Type	Reference
WS 1.80E+004	mg/L	25	EST	MEYLAN,WM ET AL. (1996)
logP 1.37			EXP	MULLER,N (1986)
VP 1.42E+001	mm Hg	25	EST	NEELY,WB & BLAU,GE (1985)
DC	pKa			
HL 6.59E-005	atm m3/mol	25	EST	MEYLAN,WM & HOWARD,PH (1991)
OH 5.91E-012	cm3/molc sec	25	EST	MEYLAN,WM & HOWARD,PH (1993)

CAS #:	095998-69-9		DICHLORO(TRIFLUOROMETHYL)BENZOPHENONE

Formula:	$C_{14}H_7Cl_2F_3O$

Mol Weight:	319.11

MP (deg C):		FP (deg C):

BP (deg C):

BP pressure (mm Hg):

Property/ Value	Units	Temp	Data Type	Reference
WS 2.32E-001	mg/L	25	EST	MEYLAN,WM ET AL. (1996)
logP 5.40			EST	MEYLAN,WM & HOWARD,PH (1995)
VP	mm Hg			
DC	pKa			
HL	atm m3/mol			
OH 1.50E-012	cm3/molc sec	25	EST	MEYLAN,WM & HOWARD,PH (1993)

CAS #:	095998-70-2		ALPHA,ALPHA-DIFLUORODICHLOROTRIFLUOROMETHYL DIP*

Formula:	$C_{14}H_7Cl_2F_5$

Mol Weight:	341.11

MP (deg C):		FP (deg C):

BP (deg C):

BP pressure (mm Hg):

Property/ Value	Units	Temp	Data Type	Reference
WS 2.14E-002	mg/L	25	EST	MEYLAN,WM ET AL. (1996)
logP 6.46			EST	MEYLAN,WM & HOWARD,PH (1995)
VP	mm Hg			
DC	pKa			
HL	atm m3/mol			
OH 1.98E-012	cm3/molc sec	25	EST	MEYLAN,WM & HOWARD,PH (1993)

CAS #:	096000-43-0		1-AC-6-DEME-7-METHOXYMITOSENE

Formula:	$C_{17}H_{20}N_2O_7$

Mol Weight:	364.36

MP (deg C):		FP (deg C):

BP (deg C):

BP pressure (mm Hg):

Property/ Value	Units	Temp	Data Type	Reference
WS 5.02E+002	mg/L	25	EST	MEYLAN,WM ET AL. (1996)
logP 1.19			EXP	HANSCH,C ET AL. (1995)
VP 2.09E-009	mm Hg	25	EST	NEELY,WB & BLAU,GE (1985)
DC	pKa			
HL 9.31E-020	atm m3/mol	25	EST	MEYLAN,WM & HOWARD,PH (1991)
OH 1.50E-010	cm3/molc sec	25	EST	MEYLAN,WM & HOWARD,PH (1993)

CAS #:	096000-44-1		1-ACETYL-6-DEMETHYL-6-CHLORO-7-METHOXYMITOSENE

Formula:	$C_{17}H_{19}ClN_2O_7$

Mol Weight:	398.80

MP (deg C):		FP (deg C):

BP (deg C):

BP pressure (mm Hg):

Property/ Value	Units	Temp	Data Type	Reference
WS 5.37E+001	mg/L	25	EST	MEYLAN,WM ET AL. (1996)
logP 2.08			EXP	HANSCH,C ET AL. (1995)
VP 6.44E-010	mm Hg	25	EST	NEELY,WB & BLAU,GE (1985)
DC	pKa			
HL 6.70E-020	atm m3/mol	25	EST	MEYLAN,WM & HOWARD,PH (1991)
OH 1.39E-010	cm3/molc sec	25	EST	MEYLAN,WM & HOWARD,PH (1993)

CAS #: 096000-47-4 — 1H-PYRROLO[1,2-A]INDOLE-5,8-DIONE DERIVATIVE

Formula: $C_{20}H_{23}N_3O_6$

Mol Weight: 401.42

MP (deg C): **FP (deg C):**

BP (deg C):

BP pressure (mm Hg):

Property/ Value	Units	Temp	Data Type	Reference
WS 1.95E+003	mg/L	25	EST	MEYLAN,WM ET AL. (1996)
logP 1.40			EXP	SANGSTER,J (1993)
VP 9.21E-011	mm Hg	25	EST	NEELY,WB & BLAU,GE (1985)
DC	pKa			
HL 2.28E-021	atm m3/mol	25	EST	MEYLAN,WM & HOWARD,PH (1991)
OH 2.22E-010	cm3/molc sec	25	EST	MEYLAN,WM & HOWARD,PH (1993)

CAS #: 096000-48-5 — 1H-PYRROLO[1,2-A]INDOLE-5,8-DIONE, 1-(ACETYLOXY)

Formula: $C_{16}H_{16}BrN_3O_6$

Mol Weight: 426.23

MP (deg C): **FP (deg C):**

BP (deg C):

BP pressure (mm Hg):

Property/ Value	Units	Temp	Data Type	Reference
WS 2.23E+003	mg/L	25	EST	MEYLAN,WM ET AL. (1996)
logP 1.15			EXP	SANGSTER,J (1993)
VP 9.24E-011	mm Hg	25	EST	NEELY,WB & BLAU,GE (1985)
DC	pKa			
HL 1.21E-022	atm m3/mol	25	EST	MEYLAN,WM & HOWARD,PH (1991)
OH 1.58E-010	cm3/molc sec	25	EST	MEYLAN,WM & HOWARD,PH (1993)

CAS #: 096027-93-9 — BENZOIC ACID, 2-[(TRIFLUOROACETYL)AMINO]ETHYL ES

Formula: $C_{11}H_{10}F_3NO_3$

Mol Weight: 261.20

MP (deg C): **FP (deg C):**

BP (deg C):

BP pressure (mm Hg):

Property/ Value	Units	Temp	Data Type	Reference
WS 2.45E+002	mg/L	25	EST	MEYLAN,WM ET AL. (1996)
logP 2.25			EXP	HANSCH,C ET AL. (1995)
VP 7.53E-006	mm Hg	25	EST	NEELY,WB & BLAU,GE (1985)
DC	pKa			
HL 1.21E-010	atm m3/mol	25	EST	MEYLAN,WM & HOWARD,PH (1991)
OH 7.18E-012	cm3/molc sec	25	EST	MEYLAN,WM & HOWARD,PH (1993)

CAS #: 096237-91-1 — 2,6-DIBROMOBENZAMIDE

Formula: $C_7H_5Br_2NO$

Mol Weight: 278.94

MP (deg C): **FP (deg C):**

BP (deg C):

BP pressure (mm Hg):

Property/ Value	Units	Temp	Data Type	Reference
WS 2.27E+003	mg/L	25	EST	MEYLAN,WM ET AL. (1996)
logP 1.00			EXP	NAKAGAWA,Y ET AL. (1992)
VP 5.86E-006	mm Hg	25	EST	NEELY,WB & BLAU,GE (1985)
DC	pKa			
HL 3.51E-010	atm m3/mol	25	EST	MEYLAN,WM & HOWARD,PH (1991)
OH 2.70E-012	cm3/molc sec	25	EST	MEYLAN,WM & HOWARD,PH (1993)

CAS #: 096258-78-5 — 1H-IMIDAZOLE-1-ETHANOL, 5-IODO-à-(METHOXYMETHYL)

Formula: $C_7H_{10}IN_3O_4$

Mol Weight: 327.08

MP (deg C): **FP (deg C):**

BP (deg C):

BP pressure (mm Hg):

Property/ Value	Units	Temp	Data Type	Reference
WS 1.55E+003	mg/L	25	EST	MEYLAN,WM ET AL. (1996)
logP 0.41			EXP	GUPTA,RP ET AL. (1985)
VP 1.53E-009	mm Hg	25	EST	NEELY,WB & BLAU,GE (1985)
DC	pKa			
HL 5.53E-014	atm m3/mol	25	EST	MEYLAN,WM & HOWARD,PH (1991)
OH 2.09E-011	cm3/molc sec	25	EST	MEYLAN,WM & HOWARD,PH (1993)

CAS #: 096258-79-6 — 1H-IMIDAZOLE-1-ETHANOL, 4-IODO-à-(METHOXYMETHYL)

Formula: $C_7H_{10}IN_3O_4$

Mol Weight: 327.08

MP (deg C): **FP (deg C):**

BP (deg C):

BP pressure (mm Hg):

Property/ Value	Units	Temp	Data Type	Reference
WS 6.02E+002	mg/L	25	EST	MEYLAN,WM ET AL. (1996)
logP 0.89			EXP	GUPTA,RP ET AL. (1985)
VP 1.53E-009	mm Hg	25	EST	NEELY,WB & BLAU,GE (1985)
DC	pKa			
HL 5.53E-014	atm m3/mol	25	EST	MEYLAN,WM & HOWARD,PH (1991)
OH 2.09E-011	cm3/molc sec	25	EST	MEYLAN,WM & HOWARD,PH (1993)

CAS #: 096258-81-0 — 1H-IMIDAZOLE-1-ACETIC ACID, 4-IODO-5-NITRO-, ETH

Formula: $C_7H_8IN_3O_4$

Mol Weight: 325.06

MP (deg C): **FP (deg C):**

BP (deg C):

BP pressure (mm Hg):

Property/ Value	Units	Temp	Data Type	Reference
WS 5.72E+002	mg/L	25	EST	MEYLAN,WM ET AL. (1996)
logP 0.93			EXP	SANGSTER,J (1994)
VP 3.51E-007	mm Hg	25	EST	NEELY,WB & BLAU,GE (1985)
DC	pKa			
HL 2.32E-010	atm m3/mol	25	EST	MEYLAN,WM & HOWARD,PH (1991)
OH 3.84E-012	cm3/molc sec	25	EST	MEYLAN,WM & HOWARD,PH (1993)

CAS #: 096258-83-2 — 1H-IMIDAZOLE-1-ACETAMIDE, 4-IODO-5-NITRO-N-(3-PY

Formula: $C_{11}H_{10}IN_5O_3$

Mol Weight: 387.14

MP (deg C): **FP (deg C):**

BP (deg C):

BP pressure (mm Hg):

Property/ Value	Units	Temp	Data Type	Reference
WS 5.30E+003	mg/L	25	EST	MEYLAN,WM ET AL. (1996)
logP 0.88			EXP	SANGSTER,J (1994)
VP 4.87E-013	mm Hg	25	EST	NEELY,WB & BLAU,GE (1985)
DC	pKa			
HL 1.34E-019	atm m3/mol	25	EST	MEYLAN,WM & HOWARD,PH (1991)
OH 4.69E-011	cm3/molc sec	25	EST	MEYLAN,WM & HOWARD,PH (1993)

CAS #: 096327-32-1 — 1H-PYRROLIZINE-1-CARBOXYLIC ACID, 2,3-DIHYDRO-5-

Formula: $C_{16}H_{15}NO_3$

Mol Weight: 269.30

MP (deg C): FP (deg C):

BP (deg C):

BP pressure (mm Hg):

Property/Value	Units	Temp	Data Type	Reference
WS 7.98E+001	mg/L	25	EST	MEYLAN,WM ET AL. (1996)
logP 3.23			EXP	SANGSTER,J (1993)
VP 4.80E-008	mm Hg	25	EST	NEELY,WB & BLAU,GE (1985)
DC	pKa			
HL 3.70E-013	atm m3/mol	25	EST	MEYLAN,WM & HOWARD,PH (1991)
OH 2.04E-010	cm3/molc sec	25	EST	MEYLAN,WM & HOWARD,PH (1993)

CAS #: 096327-34-3 — 1H-PYRROLIZINE-1-CARBOXYLIC ACID, 5-(4-BROMOBENZ

Formula: $C_{15}H_{12}BrNO_3$

Mol Weight: 334.18

MP (deg C): FP (deg C):

BP (deg C):

BP pressure (mm Hg):

Property/Value	Units	Temp	Data Type	Reference
WS 1.61E+001	mg/L	25	EST	MEYLAN,WM ET AL. (1996)
logP 3.60			EXP	SANGSTER,J (1993)
VP 1.24E-008	mm Hg	25	EST	NEELY,WB & BLAU,GE (1985)
DC	pKa			
HL 6.07E-014	atm m3/mol	25	EST	MEYLAN,WM & HOWARD,PH (1991)
OH 8.89E-011	cm3/molc sec	25	EST	MEYLAN,WM & HOWARD,PH (1993)

CAS #: 096327-38-7 — 1H-PYRROLIZINE-1-CARBOXYLIC ACID, 2,3-DIHYDRO-5-

Formula: $C_{18}H_{19}NO_4$

Mol Weight: 313.36

MP (deg C): FP (deg C):

BP (deg C):

BP pressure (mm Hg):

Property/Value	Units	Temp	Data Type	Reference
WS 3.31E+001	mg/L	25	EST	MEYLAN,WM ET AL. (1996)
logP 3.38			EXP	SANGSTER,J (1993)
VP 6.75E-009	mm Hg	25	EST	NEELY,WB & BLAU,GE (1985)
DC	pKa			
HL 3.49E-014	atm m3/mol	25	EST	MEYLAN,WM & HOWARD,PH (1991)
OH 2.16E-010	cm3/molc sec	25	EST	MEYLAN,WM & HOWARD,PH (1993)

CAS #: 096327-39-8 — 1H-PYRROLIZINE-1-CARBOXYLIC ACID, 2,3-DIHYDRO-5-

Formula: $C_{18}H_{17}NO_4$

Mol Weight: 311.34

MP (deg C): FP (deg C):

BP (deg C):

BP pressure (mm Hg):

Property/Value	Units	Temp	Data Type	Reference
WS 2.79E+001	mg/L	25	EST	MEYLAN,WM ET AL. (1996)
logP 3.48			EXP	SANGSTER,J (1993)
VP 4.40E-009	mm Hg	25	EST	NEELY,WB & BLAU,GE (1985)
DC	pKa			
HL 2.60E-014	atm m3/mol	25	EST	MEYLAN,WM & HOWARD,PH (1991)
OH 2.36E-010	cm3/molc sec	25	EST	MEYLAN,WM & HOWARD,PH (1993)

CAS #: 096327-51-4 — 1H-PYRROLIZINE-1-CARBOXYLIC ACID, 2,3-DIHYDRO-5-

Formula: $C_{16}H_{12}F_3NO_3$

Mol Weight: 323.27

MP (deg C): FP (deg C):

BP (deg C):

BP pressure (mm Hg):

Property/Value	Units	Temp	Data Type	Reference
WS 1.57E+001	mg/L	25	EST	MEYLAN,WM ET AL. (1996)
logP 3.69			EXP	SANGSTER,J (1993)
VP 8.43E-008	mm Hg	25	EST	NEELY,WB & BLAU,GE (1985)
DC	pKa			
HL 2.91E-012	atm m3/mol	25	EST	MEYLAN,WM & HOWARD,PH (1991)
OH 2.04E-010	cm3/molc sec	25	EST	MEYLAN,WM & HOWARD,PH (1993)

CAS #: 096327-52-5 — 1H-PYRROLIZINE-1-CARBOXYLIC ACID, 2,3-DIHYDRO-5-

Formula: $C_{16}H_{12}F_3NO_3$

Mol Weight: 323.27

MP (deg C): FP (deg C):

BP (deg C):

BP pressure (mm Hg):

Property/Value	Units	Temp	Data Type	Reference
WS 1.51E+001	mg/L	25	EST	MEYLAN,WM ET AL. (1996)
logP 3.71			EXP	SANGSTER,J (1993)
VP 8.43E-008	mm Hg	25	EST	NEELY,WB & BLAU,GE (1985)
DC	pKa			
HL 2.91E-012	atm m3/mol	25	EST	MEYLAN,WM & HOWARD,PH (1991)
OH 2.04E-010	cm3/molc sec	25	EST	MEYLAN,WM & HOWARD,PH (1993)

CAS #: 096327-56-9 — 1H-PYRROLIZINE-1-CARBOXYLIC ACID, 5-(2,6-DIFLUOR

Formula: $C_{15}H_{11}F_2NO_3$

Mol Weight: 291.26

MP (deg C): FP (deg C):

BP (deg C):

BP pressure (mm Hg):

Property/Value	Units	Temp	Data Type	Reference
WS 2.66E+002	mg/L	25	EST	MEYLAN,WM ET AL. (1996)
logP 2.47			EXP	SANGSTER,J (1993)
VP 1.54E-007	mm Hg	25	EST	NEELY,WB & BLAU,GE (1985)
DC	pKa			
HL 4.56E-013	atm m3/mol	25	EST	MEYLAN,WM & HOWARD,PH (1991)
OH 2.04E-010	cm3/molc sec	25	EST	MEYLAN,WM & HOWARD,PH (1993)

CAS #: 096327-57-0 — 1H-PYRROLIZINE-1-CARBOXYLIC ACID, 5-(2,4-DIFLUOR

Formula: $C_{15}H_{11}F_2NO_3$

Mol Weight: 291.26

MP (deg C): FP (deg C):

BP (deg C):

BP pressure (mm Hg):

Property/Value	Units	Temp	Data Type	Reference
WS 2.10E+002	mg/L	25	EST	MEYLAN,WM ET AL. (1996)
logP 2.59			EXP	SANGSTER,J (1993)
VP 1.54E-007	mm Hg	25	EST	NEELY,WB & BLAU,GE (1985)
DC	pKa			
HL 4.56E-013	atm m3/mol	25	EST	MEYLAN,WM & HOWARD,PH (1991)
OH 2.04E-010	cm3/molc sec	25	EST	MEYLAN,WM & HOWARD,PH (1993)

CAS #: 096327-58-1				1H-PYRROLIZINE-1-CARBOXYLIC ACID, 5-(2,4-DICHLOR

Formula: $C_{15}H_{11}Cl_2NO_3$

Mol Weight: 324.17

MP (deg C):   FP (deg C):

BP (deg C):

BP pressure (mm Hg):

Property/ Value	Units	Temp	Data Type	Reference
WS 2.00E+001	mg/L	25	EST	MEYLAN,WM ET AL. (1996)
logP 3.56			EXP	SANGSTER,J (1993)
VP 8.02E-009	mm Hg	25	EST	NEELY,WB & BLAU,GE (1985)
DC	pKa			
HL 1.84E-013	atm m3/mol	25	EST	MEYLAN,WM & HOWARD,PH (1991)
OH 2.04E-010	cm3/molc sec	25	EST	MEYLAN,WM & HOWARD,PH (1993)

CAS #: 096327-60-5				1H-PYRROLIZINE-1-CARBOXYLIC ACID, 2,3-DIHYDRO-5-

Formula: $C_{15}H_8F_5NO_3$

Mol Weight: 345.23

MP (deg C):   FP (deg C):

BP (deg C):

BP pressure (mm Hg):

Property/ Value	Units	Temp	Data Type	Reference
WS 2.81E+001	mg/L	25	EST	MEYLAN,WM ET AL. (1996)
logP 3.24			EXP	SANGSTER,J (1994)
VP 2.66E-007	mm Hg	25	EST	NEELY,WB & BLAU,GE (1985)
DC	pKa			
HL 7.26E-013	atm m3/mol	25	EST	MEYLAN,WM & HOWARD,PH (1991)
OH 2.04E-010	cm3/molc sec	25	EST	MEYLAN,WM & HOWARD,PH (1993)

CAS #: 096327-61-6				1H-PYRROLIZINE-1-CARBOXYLIC ACID, 2,3-DIHYDRO-5-

Formula: $C_{21}H_{17}NO_3$

Mol Weight: 331.37

MP (deg C):   FP (deg C):

BP (deg C):

BP pressure (mm Hg):

Property/ Value	Units	Temp	Data Type	Reference
WS 4.49E+000	mg/L	25	EST	MEYLAN,WM ET AL. (1996)
logP 4.27			EXP	SANGSTER,J (1993)
VP 1.10E-010	mm Hg	25	EST	NEELY,WB & BLAU,GE (1985)
DC	pKa			
HL 2.57E-014	atm m3/mol	25	EST	MEYLAN,WM & HOWARD,PH (1991)
OH 2.04E-010	cm3/molc sec	25	EST	MEYLAN,WM & HOWARD,PH (1993)

CAS #: 096327-63-8				1H-PYRROLIZINE-1-CARBOXYLIC ACID, 5-(CYCLOBUTYLC

Formula: $C_{13}H_{15}NO_3$

Mol Weight: 233.27

MP (deg C):   FP (deg C):

BP (deg C):

BP pressure (mm Hg):

Property/ Value	Units	Temp	Data Type	Reference
WS 6.35E+002	mg/L	25	EST	MEYLAN,WM ET AL. (1996)
logP 2.41			EXP	SANGSTER,J (1993)
VP 1.44E-006	mm Hg	25	EST	NEELY,WB & BLAU,GE (1985)
DC	pKa			
HL 1.75E-012	atm m3/mol	25	EST	MEYLAN,WM & HOWARD,PH (1991)
OH 2.19E-010	cm3/molc sec	25	EST	MEYLAN,WM & HOWARD,PH (1993)

CAS #: 096327-65-0				1H-PYRROLIZINE-1-CARBOXYLIC ACID, 2,3-DIHYDRO-5-

Formula: $C_{16}H_{15}NO_3$

Mol Weight: 269.30

MP (deg C):   FP (deg C):

BP (deg C):

BP pressure (mm Hg):

Property/ Value	Units	Temp	Data Type	Reference
WS 3.23E+002	mg/L	25	EST	MEYLAN,WM ET AL. (1996)
logP 2.52			EXP	SANGSTER,J (1993)
VP 4.82E-008	mm Hg	25	EST	NEELY,WB & BLAU,GE (1985)
DC	pKa			
HL 1.37E-013	atm m3/mol	25	EST	MEYLAN,WM & HOWARD,PH (1991)
OH 2.05E-010	cm3/molc sec	25	EST	MEYLAN,WM & HOWARD,PH (1993)

CAS #: 096337-67-6				1-(3-CYANOPHENYL)-3-MEO-3-ME UREA

Formula: $C_{10}H_{11}N_3O_2$

Mol Weight: 205.22

MP (deg C):   FP (deg C):

BP (deg C):

BP pressure (mm Hg):

Property/ Value	Units	Temp	Data Type	Reference
WS 1.64E+003	mg/L	25	EST	MEYLAN,WM ET AL. (1996)
logP 1.33			EXP	HANSCH,C & LEO,AJ (1985)
VP 2.58E-006	mm Hg	25	EST	NEELY,WB & BLAU,GE (1985)
DC	pKa			
HL 2.03E-010	atm m3/mol	25	EST	MEYLAN,WM & HOWARD,PH (1991)
OH 7.17E-012	cm3/molc sec	25	EST	MEYLAN,WM & HOWARD,PH (1993)

CAS #: 096358-98-4				TR-METHYLALLYL-CHRYSANTHEMATE

Formula: $C_{14}H_{22}O_2$

Mol Weight: 222.33

MP (deg C):   FP (deg C):

BP (deg C):

BP pressure (mm Hg):

Property/ Value	Units	Temp	Data Type	Reference
WS 2.23E+000	mg/L	25	EST	MEYLAN,WM ET AL. (1996)
logP 4.89			EXP	HANSCH,C & LEO,AJ (1985)
VP 2.08E-002	mm Hg	25	EST	NEELY,WB & BLAU,GE (1985)
DC	pKa			
HL 1.35E-003	atm m3/mol	25	EST	MEYLAN,WM & HOWARD,PH (1991)
OH 1.17E-010	cm3/molc sec	25	EST	MEYLAN,WM & HOWARD,PH (1993)

CAS #: 096382-70-6				PROPANEDINITRILE, [(3-HYDROXYPHENYL)HYDRAZONO]-

Formula: $C_9H_6N_4O$

Mol Weight: 186.17

MP (deg C):   FP (deg C):

BP (deg C):

BP pressure (mm Hg):

Property/ Value	Units	Temp	Data Type	Reference
WS 2.21E+003	mg/L	25	EST	MEYLAN,WM ET AL. (1996)
logP 2.28			EXP	STURDIK,E ET AL. (1985)
VP 3.18E-007	mm Hg	25	EST	NEELY,WB & BLAU,GE (1985)
DC	pKa			
HL 6.03E-013	atm m3/mol	25	EST	MEYLAN,WM & HOWARD,PH (1991)
OH 2.00E-010	cm3/molc sec	25	EST	MEYLAN,WM & HOWARD,PH (1993)

096448-60-1 — ALLOPURINOL, 1-ACETYL-

Formula:	$C_7H_6N_4O_2$
Mol Weight:	178.15
MP (deg C):	FP (deg C):
BP (deg C):	
BP pressure (mm Hg):	

Property/Value	Units	Temp	Data Type	Reference
WS 5.48E+003	mg/L	25	EST	MEYLAN,WM ET AL. (1996)
logP -0.35			EXP	BUNDGAARD,H & FALCH,E (1985)
VP 3.23E-006	mm Hg	25	EST	NEELY,WB & BLAU,GE (1985)
DC	pKa			
HL 9.14E-015	atm m3/mol	25	EST	MEYLAN,WM & HOWARD,PH (1991)
OH 2.00E-010	cm3/molc sec	25	EST	MEYLAN,WM & HOWARD,PH (1993)

096448-61-2 — ALLOPURINOL, 1-(OXOPROPYL)-

Formula:	$C_8H_8N_4O_2$
Mol Weight:	192.18
MP (deg C):	FP (deg C):
BP (deg C):	
BP pressure (mm Hg):	

Property/Value	Units	Temp	Data Type	Reference
WS 1.30E+003	mg/L	25	EST	MEYLAN,WM ET AL. (1996)
logP 0.30			EXP	BUNDGAARD,H & FALCH,E (1985)
VP 1.41E-006	mm Hg	25	EST	NEELY,WB & BLAU,GE (1985)
DC	pKa			
HL 1.21E-014	atm m3/mol	25	EST	MEYLAN,WM & HOWARD,PH (1991)
OH 2.01E-010	cm3/molc sec	25	EST	MEYLAN,WM & HOWARD,PH (1993)

096448-63-4 — ALLOPURINOL, 1-BENZOYL-

Formula:	$C_{12}H_8N_4O_2$
Mol Weight:	240.22
MP (deg C):	FP (deg C):
BP (deg C):	
BP pressure (mm Hg):	

Property/Value	Units	Temp	Data Type	Reference
WS 2.53E+003	mg/L	25	EST	MEYLAN,WM ET AL. (1996)
logP 1.20			EXP	BUNDGAARD,H & FALCH,E (1985)
VP 5.22E-011	mm Hg	25	EST	NEELY,WB & BLAU,GE (1985)
DC	pKa			
HL 2.23E-014	atm m3/mol	25	EST	MEYLAN,WM & HOWARD,PH (1991)
OH 3.19E-011	cm3/molc sec	25	EST	MEYLAN,WM & HOWARD,PH (1993)

096474-05-4 — ALLOPURINOL, 1-(1-OXOBUTYL)-

Formula:	$C_9H_{10}N_4O_2$
Mol Weight:	206.21
MP (deg C):	FP (deg C):
BP (deg C):	
BP pressure (mm Hg):	

Property/Value	Units	Temp	Data Type	Reference
WS 7.64E+003	mg/L	25	EST	MEYLAN,WM ET AL. (1996)
logP 0.85			EXP	BUNDGAARD,H & FALCH,E (1985)
VP 6.15E-007	mm Hg	25	EST	NEELY,WB & BLAU,GE (1985)
DC	pKa			
HL 1.61E-014	atm m3/mol	25	EST	MEYLAN,WM & HOWARD,PH (1991)
OH 2.05E-010	cm3/molc sec	25	EST	MEYLAN,WM & HOWARD,PH (1993)

096474-17-8 — 2-BENZOTHIAZOLESULFONAMIDE, 4,6-DICHLORO-

Formula:	$C_7H_4Cl_2N_2O_2S_2$
Mol Weight:	283.16
MP (deg C):	FP (deg C):
BP (deg C):	
BP pressure (mm Hg):	

Property/Value	Units	Temp	Data Type	Reference
WS 9.25E+001	mg/L	25	EST	MEYLAN,WM ET AL. (1996)
logP 2.60			EXP	SANGSTER,J (1993)
VP 8.06E-008	mm Hg	25	EST	NEELY,WB & BLAU,GE (1985)
DC	pKa			
HL 1.61E-011	atm m3/mol	25	EST	MEYLAN,WM & HOWARD,PH (1991)
OH 3.07E-013	cm3/molc sec	25	EST	MEYLAN,WM & HOWARD,PH (1993)

096474-18-9 — 2-BENZOTHIAZOLESULFONAMIDE, 6-(PHENYLMETHOXY)-

Formula:	$C_{14}H_{12}N_2O_3S_2$
Mol Weight:	320.39
MP (deg C):	FP (deg C):
BP (deg C):	
BP pressure (mm Hg):	

Property/Value	Units	Temp	Data Type	Reference
WS 1.79E+001	mg/L	25	EST	MEYLAN,WM ET AL. (1996)
logP 3.18			EXP	ELLER,MG ET AL. (1985)
VP 5.92E-010	mm Hg	25	EST	NEELY,WB & BLAU,GE (1985)
DC	pKa			
HL 1.40E-013	atm m3/mol	25	EST	MEYLAN,WM & HOWARD,PH (1991)
OH 1.52E-011	cm3/molc sec	25	EST	MEYLAN,WM & HOWARD,PH (1993)

096474-19-0 — ACETAMIDE, N-[2-(AMINOSULFONYL)-6-BENZOTHIAZOLYL

Formula:	$C_9H_9N_3O_3S_2$
Mol Weight:	271.32
MP (deg C):	FP (deg C):
BP (deg C):	
BP pressure (mm Hg):	

Property/Value	Units	Temp	Data Type	Reference
WS 3.31E+003	mg/L	25	EST	MEYLAN,WM ET AL. (1996)
logP 0.86			EXP	ELLER,MG ET AL. (1985)
VP 2.59E-010	mm Hg	25	EST	NEELY,WB & BLAU,GE (1985)
DC	pKa			
HL 3.36E-017	atm m3/mol	25	EST	MEYLAN,WM & HOWARD,PH (1991)
OH 2.83E-012	cm3/molc sec	25	EST	MEYLAN,WM & HOWARD,PH (1993)

096489-71-3 — PYRIDABEN

Formula:	$C_{19}H_{25}ClN_2OS$
Mol Weight:	364.94
MP (deg C):	FP (deg C):
BP (deg C):	
BP pressure (mm Hg):	

Property/Value	Units	Temp	Data Type	Reference
WS 1.83E-002	mg/L	25	EST	MEYLAN,WM ET AL. (1996)
logP 6.37			EXP	TOMLIN,C (1994)
VP 3.29E-009	mm Hg	25	EST	NEELY,WB & BLAU,GE (1985)
DC	pKa			
HL 2.26E-010	atm m3/mol	25	EST	MEYLAN,WM & HOWARD,PH (1991)
OH 4.08E-011	cm3/molc sec	25	EST	MEYLAN,WM & HOWARD,PH (1993)

CAS #: 096525-23-4 — FLURTAMONE

Formula: $C_{18}H_{14}F_3NO_2$

Mol Weight: 333.31

MP (deg C): FP (deg C):

BP (deg C):

BP pressure (mm Hg):

Property/Value	Units	Temp	Data Type	Reference
WS 1.41E+002	mg/L	25	EST	MEYLAN,WM ET AL. (1996)
logP 3.22			EXP	SANGSTER,J (1994)
VP	mm Hg			
DC	pKa			
HL	atm m3/mol			
OH 9.06E-011	cm3/molc sec	25	EST	MEYLAN,WM & HOWARD,PH (1993)

CAS #: 096662-64-5 — 1-PIPERIDINYLOXY, 4-[[[[BIS(1-AZIRIDINYL)PHOSPHI

Formula: $C_{14}H_{28}N_5O_3P$

Mol Weight: 345.38

MP (deg C): FP (deg C):

BP (deg C):

BP pressure (mm Hg):

Property/Value	Units	Temp	Data Type	Reference
WS 7.49E+002	mg/L	25	EST	MEYLAN,WM ET AL. (1996)
logP -0.42			EXP	SOSNOVSKY,G ET AL. (1986)
VP 2.06E-008	mm Hg	25	EST	NEELY,WB & BLAU,GE (1985)
DC	pKa			
HL 7.45E-026	atm m3/mol	25	EST	MEYLAN,WM & HOWARD,PH (1991)
OH 1.95E-011	cm3/molc sec	25	EST	MEYLAN,WM & HOWARD,PH (1993)

CAS #: 096662-65-6 — TEPA 4-(N-OH-PIPERIDINYL)-UREA DERIVATIV

Formula: $C_{20}H_{22}N_6O_5$

Mol Weight: 426.44

MP (deg C): FP (deg C):

BP (deg C):

BP pressure (mm Hg):

Property/Value	Units	Temp	Data Type	Reference
WS 9.00E+002	mg/L	25	EST	MEYLAN,WM ET AL. (1996)
logP -0.52			EXP	SOSNOVSKY,G ET AL. (1986)
VP 1.78E-010	mm Hg	25	EST	NEELY,WB & BLAU,GE (1985)
DC	pKa			
HL 9.68E-023	atm m3/mol	25	EST	MEYLAN,WM & HOWARD,PH (1991)
OH 8.55E-011	cm3/molc sec	25	EST	MEYLAN,WM & HOWARD,PH (1993)

CAS #: 096662-66-7 — TEPA 4-(PIPERIDINYL)-UREA DERIVATIVE

Formula: $C_{14}H_{28}N_5O_2P$

Mol Weight: 329.39

MP (deg C): FP (deg C):

BP (deg C):

BP pressure (mm Hg):

Property/Value	Units	Temp	Data Type	Reference
WS 1.95E+003	mg/L	25	EST	MEYLAN,WM ET AL. (1996)
logP -0.80			EXP	SOSNOVSKY,G ET AL. (1986)
VP 4.87E-008	mm Hg	25	EST	NEELY,WB & BLAU,GE (1985)
DC	pKa			
HL 4.83E-020	atm m3/mol	25	EST	MEYLAN,WM & HOWARD,PH (1991)
OH 8.25E-011	cm3/molc sec	25	EST	MEYLAN,WM & HOWARD,PH (1993)

CAS #: 096841-03-1 — 2-PROPENOIC ACID, 3-(5-CHLORO-2-FURANYL)-2-CYANO

Formula: $C_9H_6ClNO_3$

Mol Weight: 211.61

MP (deg C): FP (deg C):

BP (deg C):

BP pressure (mm Hg):

Property/Value	Units	Temp	Data Type	Reference
WS 1.85E+002	mg/L	25	EST	MEYLAN,WM ET AL. (1996)
logP 2.40			EXP	BALAZ,S ET AL. (1985)
VP 2.43E-004	mm Hg	25	EST	NEELY,WB & BLAU,GE (1985)
DC	pKa			
HL 6.82E-009	atm m3/mol	25	EST	MEYLAN,WM & HOWARD,PH (1991)
OH 3.19E-011	cm3/molc sec	25	EST	MEYLAN,WM & HOWARD,PH (1993)

CAS #: 096850-25-8 — PHENOL, 2-TRIDECYL-

Formula: $C_{19}H_{32}O$

Mol Weight: 276.47

MP (deg C): FP (deg C):

BP (deg C):

BP pressure (mm Hg):

Property/Value	Units	Temp	Data Type	Reference
WS 2.83E-003	mg/L	25	EST	MEYLAN,WM ET AL. (1996)
logP 8.61			EXP	ITOKAWA,H ET AL. (1989)
VP 9.37E-007	mm Hg	25	EST	NEELY,WB & BLAU,GE (1985)
DC	pKa			
HL 1.85E-005	atm m3/mol	25	EST	MEYLAN,WM & HOWARD,PH (1991)
OH 5.73E-011	cm3/molc sec	25	EST	MEYLAN,WM & HOWARD,PH (1993)

CAS #: 096860-23-0

Formula: $C_{11}H_{16}N_6S_2$

Mol Weight: 296.42

MP (deg C): FP (deg C):

BP (deg C):

BP pressure (mm Hg):

Property/Value	Units	Temp	Data Type	Reference
WS 1.10E+003	mg/L	25	EST	MEYLAN,WM ET AL. (1996)
logP 1.25			EXP	HANSCH,C ET AL. (1995)
VP 3.73E-008	mm Hg	25	EST	NEELY,WB & BLAU,GE (1985)
DC	pKa			
HL 5.46E-014	atm m3/mol	25	EST	MEYLAN,WM & HOWARD,PH (1991)
OH 2.59E-010	cm3/molc sec	25	EST	MEYLAN,WM & HOWARD,PH (1993)

CAS #: 096914-10-2 — 1H-IMIDAZOLE-5-BUTANOIC ACID, à-ETHYL-à-(HYDROXY

Formula: $C_{13}H_{22}N_2O_3$

Mol Weight: 254.33

MP (deg C): FP (deg C):

BP (deg C):

BP pressure (mm Hg):

Property/Value	Units	Temp	Data Type	Reference
WS 5.33E+003	mg/L	25	EST	MEYLAN,WM ET AL. (1996)
logP 0.73			EXP	BUNDGAARD,H ET AL. (1986)
VP 1.69E-008	mm Hg	25	EST	NEELY,WB & BLAU,GE (1985)
DC	pKa			
HL 4.21E-011	atm m3/mol	25	EST	MEYLAN,WM & HOWARD,PH (1991)
OH 1.00E-010	cm3/molc sec	25	EST	MEYLAN,WM & HOWARD,PH (1993)

1435

CAS #: 096914-11-3 — PILOCARPIC ACID, HEXYL ESTER

Formula: $C_{17}H_{30}N_2O_3$

Mol Weight: 310.44

MP (deg C):   FP (deg C):

BP (deg C):

BP pressure (mm Hg):

Property/Value	Units	Temp	Data Type	Reference
WS 5.17E+001	mg/L	25	EST	MEYLAN,WM ET AL. (1996)
logP 2.71			EXP	HANSCH,C ET AL. (1995)
VP 3.72E-010	mm Hg	25	EST	NEELY,WB & BLAU,GE (1985)
DC	pKa			
HL 1.31E-010	atm m3/mol	25	EST	MEYLAN,WM & HOWARD,PH (1991)
OH 1.09E-010	cm3/molc sec	25	EST	MEYLAN,WM & HOWARD,PH (1993)

CAS #: 097055-06-6 — 2-METHYL-6-ETHYLACETANILIDE

Formula: $C_{11}H_{15}NO$

Mol Weight: 177.25

MP (deg C):   FP (deg C):

BP (deg C):

BP pressure (mm Hg):

Property/Value	Units	Temp	Data Type	Reference
WS 3.55E+003	mg/L	25	EST	MEYLAN,WM ET AL. (1996)
logP 1.41			EXP	NAKAGAWA,Y ET AL. (1992)
VP 2.82E-005	mm Hg	25	EST	NEELY,WB & BLAU,GE (1985)
DC	pKa			
HL 9.98E-009	atm m3/mol	25	EST	MEYLAN,WM & HOWARD,PH (1991)
OH 1.99E-011	cm3/molc sec	25	EST	MEYLAN,WM & HOWARD,PH (1993)

CAS #: 097141-29-2 — BENZENESULFONAMIDE, 4-METHYL-2-[(2-METHYL-1-OXOP

Formula: $C_{12}H_{18}N_2O_3S$

Mol Weight: 270.35

MP (deg C):   FP (deg C):

BP (deg C):

BP pressure (mm Hg):

Property/Value	Units	Temp	Data Type	Reference
WS 1.05E+003	mg/L	25	EST	MEYLAN,WM ET AL. (1996)
logP 1.45			EXP	MONZANI,A ET AL. (1985)
VP 5.24E-009	mm Hg	25	EST	NEELY,WB & BLAU,GE (1985)
DC	pKa			
HL 1.25E-012	atm m3/mol	25	EST	MEYLAN,WM & HOWARD,PH (1991)
OH 1.32E-011	cm3/molc sec	25	EST	MEYLAN,WM & HOWARD,PH (1993)

CAS #: 097141-31-6 — BENZENESULFONAMIDE, 4-METHOXY-2-[(2-METHYL-1-OXO

Formula: $C_{11}H_{16}N_2O_4S$

Mol Weight: 272.33

MP (deg C):   FP (deg C):

BP (deg C):

BP pressure (mm Hg):

Property/Value	Units	Temp	Data Type	Reference
WS 1.20E+003	mg/L	25	EST	MEYLAN,WM ET AL. (1996)
logP 1.37			EXP	MONZANI,A ET AL. (1985)
VP 5.08E-009	mm Hg	25	EST	NEELY,WB & BLAU,GE (1985)
DC	pKa			
HL 5.04E-014	atm m3/mol	25	EST	MEYLAN,WM & HOWARD,PH (1991)
OH 3.26E-011	cm3/molc sec	25	EST	MEYLAN,WM & HOWARD,PH (1993)

CAS #: 097141-32-7 — BUTANAMIDE, N-[2-(AMINOSULFONYL)-5-METHYLPHENYL]

Formula: $C_{13}H_{20}N_2O_3S$

Mol Weight: 284.38

MP (deg C):   FP (deg C):

BP (deg C):

BP pressure (mm Hg):

Property/Value	Units	Temp	Data Type	Reference
WS 2.29E+002	mg/L	25	EST	MEYLAN,WM ET AL. (1996)
logP 2.13			EXP	MONZANI,A ET AL. (1985)
VP 2.32E-009	mm Hg	25	EST	NEELY,WB & BLAU,GE (1985)
DC	pKa			
HL 1.66E-012	atm m3/mol	25	EST	MEYLAN,WM & HOWARD,PH (1991)
OH 1.69E-011	cm3/molc sec	25	EST	MEYLAN,WM & HOWARD,PH (1993)

CAS #: 097141-33-8 — BUTANAMIDE, N-[2-(AMINOSULFONYL)-5-(TRIFLUOROMET

Formula: $C_{13}H_{17}F_3N_2O_3S$

Mol Weight: 338.35

MP (deg C):   FP (deg C):

BP (deg C):

BP pressure (mm Hg):

Property/Value	Units	Temp	Data Type	Reference
WS 1.77E+001	mg/L	25	EST	MEYLAN,WM ET AL. (1996)
logP 3.06			EXP	MONZANI,A ET AL. (1985)
VP 4.12E-009	mm Hg	25	EST	NEELY,WB & BLAU,GE (1985)
DC	pKa			
HL 1.31E-011	atm m3/mol	25	EST	MEYLAN,WM & HOWARD,PH (1991)
OH 1.02E-011	cm3/molc sec	25	EST	MEYLAN,WM & HOWARD,PH (1993)

CAS #: 097141-34-9 — BUTANAMIDE, N-[2-(AMINOSULFONYL)-5-METHOXYPHENYL

Formula: $C_{13}H_{20}N_2O_4S$

Mol Weight: 300.38

MP (deg C):   FP (deg C):

BP (deg C):

BP pressure (mm Hg):

Property/Value	Units	Temp	Data Type	Reference
WS 2.35E+002	mg/L	25	EST	MEYLAN,WM ET AL. (1996)
logP 2.01			EXP	MONZANI,A ET AL. (1985)
VP 9.94E-010	mm Hg	25	EST	NEELY,WB & BLAU,GE (1985)
DC	pKa			
HL 8.89E-014	atm m3/mol	25	EST	MEYLAN,WM & HOWARD,PH (1991)
OH 3.99E-011	cm3/molc sec	25	EST	MEYLAN,WM & HOWARD,PH (1993)

CAS #: 097141-35-0 — BUTANAMIDE, N-[2-(AMINOSULFONYL)-5-CHLOROPHENYL]

Formula: $C_{12}H_{17}ClN_2O_3S$

Mol Weight: 304.80

MP (deg C):   FP (deg C):

BP (deg C):

BP pressure (mm Hg):

Property/Value	Units	Temp	Data Type	Reference
WS 7.65E+001	mg/L	25	EST	MEYLAN,WM ET AL. (1996)
logP 2.55			EXP	MONZANI,A ET AL. (1985)
VP 1.41E-009	mm Hg	25	EST	NEELY,WB & BLAU,GE (1985)
DC	pKa			
HL 1.11E-012	atm m3/mol	25	EST	MEYLAN,WM & HOWARD,PH (1991)
OH 1.17E-011	cm3/molc sec	25	EST	MEYLAN,WM & HOWARD,PH (1993)

CAS #: 097183-49-8				HYDRAZINECARBOXIMIDAMIDE, 2-(PHENYL-2-THIENYLMET
Formula: $C_{12}H_{12}N_4S$				
Mol Weight: 244.32				
MP (deg C):		FP (deg C):		
BP (deg C):				
BP pressure (mm Hg):				

Property/Value	Units	Temp	Data Type	Reference
WS 5.39E+002	mg/L	25	EST	MEYLAN,WM ET AL. (1996)
logP 1.96			EXP	JIRA,T ET AL. (1985)
VP 4.23E-007	mm Hg	25	EST	NEELY,WB & BLAU,GE (1985)
DC	pKa			
HL 2.21E-013	atm m3/mol	25	EST	MEYLAN,WM & HOWARD,PH (1991)
OH 1.09E-010	cm3/molc sec	25	EST	MEYLAN,WM & HOWARD,PH (1993)

CAS #: 097183-51-2				HYDRAZINECARBOXIMIDAMIDE, 2-(1-PROPYLBUTYLIDENE)
Formula: $C_8H_{18}N_4$				
Mol Weight: 170.26				
MP (deg C):		FP (deg C):		
BP (deg C):				
BP pressure (mm Hg):				

Property/Value	Units	Temp	Data Type	Reference
WS 2.04E+003	mg/L	25	EST	MEYLAN,WM ET AL. (1996)
logP 1.73			EXP	JIRA,T ET AL. (1985)
VP 1.45E-003	mm Hg	25	EST	NEELY,WB & BLAU,GE (1985)
DC	pKa			
HL 3.42E-010	atm m3/mol	25	EST	MEYLAN,WM & HOWARD,PH (1991)
OH 8.95E-011	cm3/molc sec	25	EST	MEYLAN,WM & HOWARD,PH (1993)

CAS #: 097183-52-3				HYDRAZINECARBOXIMIDAMIDE, 2-(2-THIENYLMETHYLENE)
Formula: $C_6H_8N_4S$				
Mol Weight: 168.22				
MP (deg C):		FP (deg C):		
BP (deg C):				
BP pressure (mm Hg):				

Property/Value	Units	Temp	Data Type	Reference
WS 9.13E+003	mg/L	25	EST	MEYLAN,WM ET AL. (1996)
logP 0.98			EXP	JIRA,T ET AL. (1985)
VP 1.38E-004	mm Hg	25	EST	NEELY,WB & BLAU,GE (1985)
DC	pKa			
HL 2.74E-012	atm m3/mol	25	EST	MEYLAN,WM & HOWARD,PH (1991)
OH 1.09E-010	cm3/molc sec	25	EST	MEYLAN,WM & HOWARD,PH (1993)

CAS #: 097454-00-7				DKA-24
Formula: $C_{10}H_{14}Cl_2N_2O_2$				
Mol Weight: 265.14				
MP (deg C):		FP (deg C):		
BP (deg C):				
BP pressure (mm Hg):				

Property/Value	Units	Temp	Data Type	Reference
WS 2.72E+003	mg/L	25	EST	MEYLAN,WM ET AL. (1996)
logP 1.00			EXP	TOMLIN,C (1994)
VP 1.80E-007	mm Hg	25	EST	NEELY,WB & BLAU,GE (1985)
DC	pKa			
HL 1.22E-010	atm m3/mol	25	EST	MEYLAN,WM & HOWARD,PH (1991)
OH 8.38E-011	cm3/molc sec	25	EST	MEYLAN,WM & HOWARD,PH (1993)

CAS #: 097534-21-9				MERBARONE
Formula: $C_{11}H_9N_3O_3S$				
Mol Weight: 263.28				
MP (deg C):		FP (deg C):		
BP (deg C):				
BP pressure (mm Hg):				

Property/Value	Units	Temp	Data Type	Reference
WS 7.42E+002	mg/L	25	EST	MEYLAN,WM ET AL. (1996)
logP 0.14			EXP	HANSCH,C ET AL. (1995)
VP 2.67E-014	mm Hg	25	EST	NEELY,WB & BLAU,GE (1985)
DC	pKa			
HL 7.32E-017	atm m3/mol	25	EST	MEYLAN,WM & HOWARD,PH (1991)
OH 2.42E-011	cm3/molc sec	25	EST	MEYLAN,WM & HOWARD,PH (1993)

CAS #: 097683-29-9				1-SO3NA-3-ME AZULENE
Formula: $C_{11}H_9NaO_3S$				
Mol Weight: 244.25				
MP (deg C):		FP (deg C):		
BP (deg C):				
BP pressure (mm Hg):				

Property/Value	Units	Temp	Data Type	Reference
WS 4.77E+005	mg/L	25	EST	MEYLAN,WM ET AL. (1996)
logP -1.49			EXP	YANAGISAWA,T ET AL. (1988)
VP 6.82E-013	mm Hg	25	EST	NEELY,WB & BLAU,GE (1985)
DC	pKa			
HL	atm m3/mol			
OH 3.06E-010	cm3/molc sec	25	EST	MEYLAN,WM & HOWARD,PH (1993)

CAS #: 097683-30-2				1-SO3NA-3-ET AZULENE
Formula: $C_{12}H_{11}NaO_3S$				
Mol Weight: 258.27				
MP (deg C):		FP (deg C):		
BP (deg C):				
BP pressure (mm Hg):				

Property/Value	Units	Temp	Data Type	Reference
WS 1.74E+005	mg/L	25	EST	MEYLAN,WM ET AL. (1996)
logP -1.07			EXP	YANAGISAWA,T ET AL. (1988)
VP 2.92E-013	mm Hg	25	EST	NEELY,WB & BLAU,GE (1985)
DC	pKa			
HL	atm m3/mol			
OH 3.07E-010	cm3/molc sec	25	EST	MEYLAN,WM & HOWARD,PH (1993)

CAS #: 097683-31-3				1-SO3NA-3-ET-7-ISOPROPYL AZULENE
Formula: $C_{15}H_{17}NaO_3S$				
Mol Weight: 300.35				
MP (deg C):		FP (deg C):		
BP (deg C):				
BP pressure (mm Hg):				

Property/Value	Units	Temp	Data Type	Reference
WS 5.24E+004	mg/L	25	EST	MEYLAN,WM ET AL. (1996)
logP -0.74			EXP	YANAGISAWA,T ET AL. (1988)
VP 4.06E-014	mm Hg	25	EST	NEELY,WB & BLAU,GE (1985)
DC	pKa			
HL	atm m3/mol			
OH 3.44E-010	cm3/molc sec	25	EST	MEYLAN,WM & HOWARD,PH (1993)

CAS #: 097683-34-6	1-SO3NA-3-BUTYL AZULENE

Formula: $C_{14}H_{15}NaO_3S$

Mol Weight: 286.33

MP (deg C):

FP (deg C):

BP (deg C):

BP pressure (mm Hg):

Property/ Value	Units	Temp	Data Type	Reference
WS 1.63E+004	mg/L	25	EST	MEYLAN,WM ET AL. (1996)
logP -0.05			EXP	YANAGISAWA,T ET AL. (1988)
VP 5.30E-014	mm Hg	25	EST	NEELY,WB & BLAU,GE (1985)
DC	pKa			
HL	atm m3/mol			
OH 3.10E-010	cm3/molc sec	25	EST	MEYLAN,WM & HOWARD,PH (1993)

CAS #: 097713-49-0	BENZOIC ACID, 2-[(3-BROMOPHENYL)AMINO]-4-CHLORO-

Formula: $C_{13}H_9BrClNO_2$

Mol Weight: 326.58

MP (deg C):

FP (deg C):

BP (deg C):

BP pressure (mm Hg):

Property/ Value	Units	Temp	Data Type	Reference
WS 3.36E-001	mg/L	25	EST	MEYLAN,WM ET AL. (1996)
logP 5.16			EXP	SANGSTER,J (1993)
VP 5.61E-008	mm Hg	25	EST	NEELY,WB & BLAU,GE (1985)
DC	pKa			
HL 6.24E-012	atm m3/mol	25	EST	MEYLAN,WM & HOWARD,PH (1991)
OH 1.31E-010	cm3/molc sec	25	EST	MEYLAN,WM & HOWARD,PH (1993)

CAS #: 097871-45-9	TRANS-CINNAMAMIDE,A-BR,N-SEC-BUTYL

Formula: $C_{13}H_{16}BrNO$

Mol Weight: 282.19

MP (deg C):

FP (deg C):

BP (deg C):

BP pressure (mm Hg):

Property/ Value	Units	Temp	Data Type	Reference
WS 8.33E+001	mg/L	25	EST	MEYLAN,WM ET AL. (1996)
logP 3.56			EXP	HANSCH,C & LEO,AJ (1985)
VP 6.69E-007	mm Hg	25	EST	NEELY,WB & BLAU,GE (1985)
DC	pKa			
HL 2.70E-010	atm m3/mol	25	EST	MEYLAN,WM & HOWARD,PH (1991)
OH 2.65E-011	cm3/molc sec	25	EST	MEYLAN,WM & HOWARD,PH (1993)

CAS #: 097871-46-0	CIS-CINNAMAMIDE,A-BR,N-SEC-BUTYL

Formula: $C_{13}H_{16}BrNO$

Mol Weight: 282.19

MP (deg C):

FP (deg C):

BP (deg C):

BP pressure (mm Hg):

Property/ Value	Units	Temp	Data Type	Reference
WS 8.33E+001	mg/L	25	EST	MEYLAN,WM ET AL. (1996)
logP 2.66			EXP	HANSCH,C & LEO,AJ (1985)
VP 6.69E-007	mm Hg	25	EST	NEELY,WB & BLAU,GE (1985)
DC	pKa			
HL 2.70E-010	atm m3/mol	25	EST	MEYLAN,WM & HOWARD,PH (1991)
OH 2.65E-011	cm3/molc sec	25	EST	MEYLAN,WM & HOWARD,PH (1993)

CAS #: 097871-47-1	TRANS-CINNAMAMIDE,A-CL,N-SEC-BUTYL

Formula: $C_{13}H_{16}ClNO$

Mol Weight: 237.73

MP (deg C):

FP (deg C):

BP (deg C):

BP pressure (mm Hg):

Property/ Value	Units	Temp	Data Type	Reference
WS 2.47E+002	mg/L	25	EST	MEYLAN,WM ET AL. (1996)
logP 3.52			EXP	HANSCH,C & LEO,AJ (1985)
VP 3.11E-006	mm Hg	25	EST	NEELY,WB & BLAU,GE (1985)
DC	pKa			
HL 9.76E-010	atm m3/mol	25	EST	MEYLAN,WM & HOWARD,PH (1991)
OH 2.50E-011	cm3/molc sec	25	EST	MEYLAN,WM & HOWARD,PH (1993)

CAS #: 097871-48-2	CIS-CINNAMAMIDE,A-CL,N-SEC-BUTYL

Formula: $C_{13}H_{16}ClNO$

Mol Weight: 237.73

MP (deg C):

FP (deg C):

BP (deg C):

BP pressure (mm Hg):

Property/ Value	Units	Temp	Data Type	Reference
WS 2.47E+002	mg/L	25	EST	MEYLAN,WM ET AL. (1996)
logP 2.40			EXP	HANSCH,C & LEO,AJ (1985)
VP 3.11E-006	mm Hg	25	EST	NEELY,WB & BLAU,GE (1985)
DC	pKa			
HL 9.76E-010	atm m3/mol	25	EST	MEYLAN,WM & HOWARD,PH (1991)
OH 2.50E-011	cm3/molc sec	25	EST	MEYLAN,WM & HOWARD,PH (1993)

CAS #: 097871-49-3	TRANS-CINNAMAMIDE,A-F,N-SEC-BUTYL

Formula: $C_{13}H_{16}FNO$

Mol Weight: 221.28

MP (deg C):

FP (deg C):

BP (deg C):

BP pressure (mm Hg):

Property/ Value	Units	Temp	Data Type	Reference
WS 7.63E+001	mg/L	25	EST	MEYLAN,WM ET AL. (1996)
logP 3.10			EXP	HANSCH,C & LEO,AJ (1985)
VP 9.37E-006	mm Hg	25	EST	NEELY,WB & BLAU,GE (1985)
DC	pKa			
HL 2.60E-009	atm m3/mol	25	EST	MEYLAN,WM & HOWARD,PH (1991)
OH 2.50E-011	cm3/molc sec	25	EST	MEYLAN,WM & HOWARD,PH (1993)

CAS #: 097871-50-6	N-SEC-BUTYLCINNAMAMIDE

Formula: $C_{13}H_{17}NO$

Mol Weight: 203.29

MP (deg C):

FP (deg C):

BP (deg C):

BP pressure (mm Hg):

Property/ Value	Units	Temp	Data Type	Reference
WS 1.55E+002	mg/L	25	EST	MEYLAN,WM ET AL. (1996)
logP 2.85			EXP	HANSCH,C & LEO,AJ (1985)
VP 8.80E-006	mm Hg	25	EST	NEELY,WB & BLAU,GE (1985)
DC	pKa			
HL 1.36E-009	atm m3/mol	25	EST	MEYLAN,WM & HOWARD,PH (1991)
OH 3.83E-011	cm3/molc sec	25	EST	MEYLAN,WM & HOWARD,PH (1993)

097871-51-7 — A-ME-N-SEC-BUTYLCINNAMAMIDE

Formula: $C_{14}H_{19}NO$

Mol Weight: 217.31

MP (deg C): **FP (deg C):**

BP (deg C):

BP pressure (mm Hg):

Property/Value	Units	Temp	Data Type	Reference
WS 1.24E+002	mg/L	25	EST	MEYLAN,WM ET AL. (1996)
logP 2.88			EXP	HANSCH,C & LEO,AJ (1985)
VP 4.36E-006	mm Hg	25	EST	NEELY,WB & BLAU,GE (1985)
DC	pKa			
HL 2.13E-009	atm m3/mol	25	EST	MEYLAN,WM & HOWARD,PH (1991)
OH 4.92E-011	cm3/molc sec	25	EST	MEYLAN,WM & HOWARD,PH (1993)

097871-52-8 — A-MES-N-SEC-BUTYLCINNAMAMIDE

Formula: $C_{14}H_{19}NOS$

Mol Weight: 249.38

MP (deg C): **FP (deg C):**

BP (deg C):

BP pressure (mm Hg):

Property/Value	Units	Temp	Data Type	Reference
WS 6.16E+001	mg/L	25	EST	MEYLAN,WM ET AL. (1996)
logP 3.03			EXP	HANSCH,C & LEO,AJ (1985)
VP 4.50E-007	mm Hg	25	EST	NEELY,WB & BLAU,GE (1985)
DC	pKa			
HL 4.85E-011	atm m3/mol	25	EST	MEYLAN,WM & HOWARD,PH (1991)
OH 5.18E-011	cm3/molc sec	25	EST	MEYLAN,WM & HOWARD,PH (1993)

097871-53-9 — A-MEO-N-SEC-BUTYLCINNAMAMIDE

Formula: $C_{14}H_{19}NO_2$

Mol Weight: 233.31

MP (deg C): **FP (deg C):**

BP (deg C):

BP pressure (mm Hg):

Property/Value	Units	Temp	Data Type	Reference
WS 1.05E+002	mg/L	25	EST	MEYLAN,WM ET AL. (1996)
logP 2.86			EXP	HANSCH,C & LEO,AJ (1985)
VP 2.01E-006	mm Hg	25	EST	NEELY,WB & BLAU,GE (1985)
DC	pKa			
HL 1.26E-010	atm m3/mol	25	EST	MEYLAN,WM & HOWARD,PH (1991)
OH 5.90E-011	cm3/molc sec	25	EST	MEYLAN,WM & HOWARD,PH (1993)

097871-54-0 — A-ACETYLAMINO-N-SEC-BUTYL CINNAMAMIDE

Formula: $C_{15}H_{20}N_2O_2$

Mol Weight: 260.34

MP (deg C): **FP (deg C):**

BP (deg C):

BP pressure (mm Hg):

Property/Value	Units	Temp	Data Type	Reference
WS 2.53E+003	mg/L	25	EST	MEYLAN,WM ET AL. (1996)
logP 1.07			EXP	HANSCH,C & LEO,AJ (1985)
VP 1.43E-009	mm Hg	25	EST	NEELY,WB & BLAU,GE (1985)
DC	pKa			
HL 1.50E-012	atm m3/mol	25	EST	MEYLAN,WM & HOWARD,PH (1991)
OH 4.25E-011	cm3/molc sec	25	EST	MEYLAN,WM & HOWARD,PH (1993)

097871-55-1 — A-ETO-N-SEC-BUTYL CINNAMAMIDE

Formula: $C_{15}H_{21}NO_2$

Mol Weight: 247.34

MP (deg C): **FP (deg C):**

BP (deg C):

BP pressure (mm Hg):

Property/Value	Units	Temp	Data Type	Reference
WS 4.53E+001	mg/L	25	EST	MEYLAN,WM ET AL. (1996)
logP 3.20			EXP	HANSCH,C & LEO,AJ (1985)
VP 8.62E-007	mm Hg	25	EST	NEELY,WB & BLAU,GE (1985)
DC	pKa			
HL 1.67E-010	atm m3/mol	25	EST	MEYLAN,WM & HOWARD,PH (1991)
OH 6.43E-011	cm3/molc sec	25	EST	MEYLAN,WM & HOWARD,PH (1993)

097871-56-2 — A-ETS-N-SEC-BUTYLCINNAMAMIDE

Formula: $C_{15}H_{21}NOS$

Mol Weight: 263.40

MP (deg C): **FP (deg C):**

BP (deg C):

BP pressure (mm Hg):

Property/Value	Units	Temp	Data Type	Reference
WS 2.49E+001	mg/L	25	EST	MEYLAN,WM ET AL. (1996)
logP 3.40			EXP	HANSCH,C & LEO,AJ (1985)
VP 1.92E-007	mm Hg	25	EST	NEELY,WB & BLAU,GE (1985)
DC	pKa			
HL 6.43E-011	atm m3/mol	25	EST	MEYLAN,WM & HOWARD,PH (1991)
OH 5.82E-011	cm3/molc sec	25	EST	MEYLAN,WM & HOWARD,PH (1993)

097871-57-3 — A-BUTYL-N-SEC-BUTYLCINNAMAMIDE

Formula: $C_{17}H_{25}NO$

Mol Weight: 259.39

MP (deg C): **FP (deg C):**

BP (deg C):

BP pressure (mm Hg):

Property/Value	Units	Temp	Data Type	Reference
WS 7.01E+000	mg/L	25	EST	MEYLAN,WM ET AL. (1996)
logP 4.07			EXP	HANSCH,C & LEO,AJ (1985)
VP 4.07E-007	mm Hg	25	EST	NEELY,WB & BLAU,GE (1985)
DC	pKa			
HL 4.97E-009	atm m3/mol	25	EST	MEYLAN,WM & HOWARD,PH (1991)
OH 5.29E-011	cm3/molc sec	25	EST	MEYLAN,WM & HOWARD,PH (1993)

097871-59-5 — 2-PROPENAMIDE, N-(1-METHYLPROPYL)-3-PHENYL-2-(PH

Formula: $C_{19}H_{21}NOS$

Mol Weight: 311.45

MP (deg C): **FP (deg C):**

BP (deg C):

BP pressure (mm Hg):

Property/Value	Units	Temp	Data Type	Reference
WS 1.99E+000	mg/L	25	EST	MEYLAN,WM ET AL. (1996)
logP 4.36			EXP	HANSCH,C ET AL. (1995)
VP 2.19E-009	mm Hg	25	EST	NEELY,WB & BLAU,GE (1985)
DC	pKa			
HL 9.69E-012	atm m3/mol	25	EST	MEYLAN,WM & HOWARD,PH (1991)
OH 6.16E-011	cm3/molc sec	25	EST	MEYLAN,WM & HOWARD,PH (1993)

CAS #: 097871-60-8 — A-BENZOYLAMINO-N-SEC-BU CINNAMAMIDE

Formula: $C_{20}H_{22}N_2O_2$

Mol Weight: 322.41

MP (deg C):
FP (deg C):

BP (deg C):

BP pressure (mm Hg):

Property/Value	Units	Temp	Data Type	Reference
WS 7.05E+001	mg/L	25	EST	MEYLAN,WM ET AL. (1996)
logP 2.47			EXP	HANSCH,C & LEO,AJ (1985)
VP 8.09E-012	mm Hg	25	EST	NEELY,WB & BLAU,GE (1985)
DC	pKa			
HL 2.97E-013	atm m3/mol	25	EST	MEYLAN,WM & HOWARD,PH (1991)
OH 4.41E-011	cm3/molc sec	25	EST	MEYLAN,WM & HOWARD,PH (1993)

CAS #: 097871-63-1 — P-CHLOROCINNAMAMIDE,B-ET,N-SEC-BUTYL

Formula: $C_{15}H_{20}ClNO$

Mol Weight: 265.79

MP (deg C):
FP (deg C):

BP (deg C):

BP pressure (mm Hg):

Property/Value	Units	Temp	Data Type	Reference
WS 3.51E+000	mg/L	25	EST	MEYLAN,WM ET AL. (1996)
logP 4.38			EXP	HANSCH,C & LEO,AJ (1985)
VP 4.85E-007	mm Hg	25	EST	NEELY,WB & BLAU,GE (1985)
DC	pKa			
HL 2.09E-009	atm m3/mol	25	EST	MEYLAN,WM & HOWARD,PH (1991)
OH 4.96E-011	cm3/molc sec	25	EST	MEYLAN,WM & HOWARD,PH (1993)

CAS #: 097871-64-2 — P-CL CINNAMAMIDE,B-PR,N-SEC-BUTYL

Formula: $C_{16}H_{22}ClNO$

Mol Weight: 279.81

MP (deg C):
FP (deg C):

BP (deg C):

BP pressure (mm Hg):

Property/Value	Units	Temp	Data Type	Reference
WS 1.11E+000	mg/L	25	EST	MEYLAN,WM ET AL. (1996)
logP 4.87			EXP	HANSCH,C & LEO,AJ (1985)
VP 2.07E-007	mm Hg	25	EST	NEELY,WB & BLAU,GE (1985)
DC	pKa			
HL 2.78E-009	atm m3/mol	25	EST	MEYLAN,WM & HOWARD,PH (1991)
OH 5.09E-011	cm3/molc sec	25	EST	MEYLAN,WM & HOWARD,PH (1993)

CAS #: 097871-65-3 — P-CL CINNAMAMIDE,B-IPR,N-SEC-BUTYL

Formula: $C_{16}H_{22}ClNO$

Mol Weight: 279.81

MP (deg C):
FP (deg C):

BP (deg C):

BP pressure (mm Hg):

Property/Value	Units	Temp	Data Type	Reference
WS 5.81E+000	mg/L	25	EST	MEYLAN,WM ET AL. (1996)
logP 4.03			EXP	HANSCH,C & LEO,AJ (1985)
VP 3.65E-007	mm Hg	25	EST	NEELY,WB & BLAU,GE (1985)
DC	pKa			
HL 2.78E-009	atm m3/mol	25	EST	MEYLAN,WM & HOWARD,PH (1991)
OH 5.07E-011	cm3/molc sec	25	EST	MEYLAN,WM & HOWARD,PH (1993)

CAS #: 097871-66-4 — P-CL CINNAMAMIDE,B-PHENYL,N-SEC-BU

Formula: $C_{19}H_{20}ClNO$

Mol Weight: 313.83

MP (deg C):
FP (deg C):

BP (deg C):

BP pressure (mm Hg):

Property/Value	Units	Temp	Data Type	Reference
WS 1.58E+000	mg/L	25	EST	MEYLAN,WM ET AL. (1996)
logP 4.46			EXP	HANSCH,C & LEO,AJ (1985)
VP 5.81E-009	mm Hg	25	EST	NEELY,WB & BLAU,GE (1985)
DC	pKa			
HL 4.50E-011	atm m3/mol	25	EST	MEYLAN,WM & HOWARD,PH (1991)
OH 5.03E-011	cm3/molc sec	25	EST	MEYLAN,WM & HOWARD,PH (1993)

CAS #: 097871-67-5 — P-CL-CINNAMAMIDE,B-CL,N-SEC-BUTYL

Formula: $C_{13}H_{15}Cl_2NO$

Mol Weight: 272.18

MP (deg C):
FP (deg C):

BP (deg C):

BP pressure (mm Hg):

Property/Value	Units	Temp	Data Type	Reference
WS 1.47E+001	mg/L	25	EST	MEYLAN,WM ET AL. (1996)
logP 3.61			EXP	HANSCH,C & LEO,AJ (1985)
VP 6.84E-007	mm Hg	25	EST	NEELY,WB & BLAU,GE (1985)
DC	pKa			
HL 7.23E-010	atm m3/mol	25	EST	MEYLAN,WM & HOWARD,PH (1991)
OH 2.44E-011	cm3/molc sec	25	EST	MEYLAN,WM & HOWARD,PH (1993)

CAS #: 097871-68-6 — P-CL-CINNAMAMIDE,B-BR,N-SEC-BUTYL

Formula: $C_{13}H_{15}BrClNO$

Mol Weight: 316.63

MP (deg C):
FP (deg C):

BP (deg C):

BP pressure (mm Hg):

Property/Value	Units	Temp	Data Type	Reference
WS 4.58E+000	mg/L	25	EST	MEYLAN,WM ET AL. (1996)
logP 3.90			EXP	HANSCH,C & LEO,AJ (1985)
VP 1.74E-007	mm Hg	25	EST	NEELY,WB & BLAU,GE (1985)
DC	pKa			
HL 2.00E-010	atm m3/mol	25	EST	MEYLAN,WM & HOWARD,PH (1991)
OH 2.60E-011	cm3/molc sec	25	EST	MEYLAN,WM & HOWARD,PH (1993)

CAS #: 097874-69-6 — 2-PENTENOIC ACID, 2-CYANO-3-[[(4-NITROPHENYL)MET

Formula: $C_{17}H_{21}N_3O_5$

Mol Weight: 347.37

MP (deg C):
FP (deg C):

BP (deg C):

BP pressure (mm Hg):

Property/Value	Units	Temp	Data Type	Reference
WS 8.06E+001	mg/L	25	EST	MEYLAN,WM ET AL. (1996)
logP 2.64			EXP	HANSCH,C ET AL. (1995)
VP 1.90E-009	mm Hg	25	EST	NEELY,WB & BLAU,GE (1985)
DC	pKa			
HL 1.33E-015	atm m3/mol	25	EST	MEYLAN,WM & HOWARD,PH (1991)
OH 9.21E-011	cm3/molc sec	25	EST	MEYLAN,WM & HOWARD,PH (1993)

CAS #: 098006-91-8				PYRAZINECARBONITRILE, 5-METHYL-

Formula: $C_6H_5N_3$

Mol Weight: 119.13

MP (deg C):　　　FP (deg C):

BP (deg C):

BP pressure (mm Hg):

Property/Value	Units	Temp	Data Type	Reference
WS 3.29E+004	mg/L	25	EST	MEYLAN,WM ET AL. (1996)
logP 0.26			EXP	YAMAGAMI,C ET AL. (1991)
VP 4.78E-002	mm Hg	25	EST	NEELY,WB & BLAU,GE (1985)
DC	pKa			
HL 3.11E-008	atm m3/mol	25	EST	MEYLAN,WM & HOWARD,PH (1991)
OH 2.42E-013	cm3/molc sec	25	EST	MEYLAN,WM & HOWARD,PH (1993)

CAS #: 098079-47-1				8-DESFLUORO-LOMEFLOXACIN

Formula: $C_{17}H_{20}FN_3O_3$

Mol Weight: 333.37

MP (deg C):　　　FP (deg C):

BP (deg C):

BP pressure (mm Hg):

Property/Value	Units	Temp	Data Type	Reference
WS 9.92E+003	mg/L	25	EST	MEYLAN,WM ET AL. (1996)
logP 0.34			EXP	TAKACS-NOVAK,K ET AL. (1992)
VP 4.01E-013	mm Hg	25	EST	NEELY,WB & BLAU,GE (1985)
DC	pKa			
HL 1.16E-018	atm m3/mol	25	EST	MEYLAN,WM & HOWARD,PH (1991)
OH 3.33E-010	cm3/molc sec	25	EST	MEYLAN,WM & HOWARD,PH (1993)

CAS #: 098079-51-7				LOMEFLOXACIN

Formula: $C_{17}H_{19}F_2N_3O_3$

Mol Weight: 351.36

MP (deg C): 239-240.5　　FP (deg C):

BP (deg C):

BP pressure (mm Hg):

Property/Value	Units	Temp	Data Type	Reference
WS 2.72E+004	mg/L	25	EST	MEYLAN,WM ET AL. (1996)
logP -0.30			EXP	TAKACS-NOVAK,K ET AL. (1992)
VP 4.75E-013	mm Hg	25	EST	NEELY,WB & BLAU,GE (1985)
DC	pKa			
HL 1.35E-018	atm m3/mol	25	EST	MEYLAN,WM & HOWARD,PH (1991)
OH 1.34E-010	cm3/molc sec	25	EST	MEYLAN,WM & HOWARD,PH (1993)

CAS #: 098087-97-9				2-PYRIDINAMINE, 4-(5-AMINO-1H-1,2,4-TRIAZOL-3-YL

Formula: $C_{16}H_{18}N_6$

Mol Weight: 294.36

MP (deg C):　　　FP (deg C):

BP (deg C):

BP pressure (mm Hg):

Property/Value	Units	Temp	Data Type	Reference
WS 3.29E+001	mg/L	25	EST	MEYLAN,WM ET AL. (1996)
logP 3.05			EXP	LIPINSKI,CA ET AL. (1985)
VP 7.09E-011	mm Hg	25	EST	NEELY,WB & BLAU,GE (1985)
DC	pKa			
HL 5.99E-018	atm m3/mol	25	EST	MEYLAN,WM & HOWARD,PH (1991)
OH 3.23E-011	cm3/molc sec	25	EST	MEYLAN,WM & HOWARD,PH (1993)

CAS #: 098154-93-9				CARBAMIC ACID, DIMETHYL-, 1-METHYL-2-PHENYLETHYL

Formula: $C_{12}H_{17}NO_2$

Mol Weight: 207.27

MP (deg C):　　　FP (deg C):

BP (deg C):

BP pressure (mm Hg):

Property/Value	Units	Temp	Data Type	Reference
WS 2.42E+002	mg/L	25	EST	MEYLAN,WM ET AL. (1996)
logP 2.60			EXP	TANAKA,M ET AL. (1985)
VP 5.02E-003	mm Hg	25	EST	NEELY,WB & BLAU,GE (1985)
DC	pKa			
HL 2.71E-008	atm m3/mol	25	EST	MEYLAN,WM & HOWARD,PH (1991)
OH 2.61E-011	cm3/molc sec	25	EST	MEYLAN,WM & HOWARD,PH (1993)

CAS #: 098154-94-0				CARBAMIC ACID, DIMETHYL-, 2-(1H-INDOL-3-YL)ETHYL

Formula: $C_{13}H_{16}N_2O_2$

Mol Weight: 232.28

MP (deg C):　　　FP (deg C):

BP (deg C):

BP pressure (mm Hg):

Property/Value	Units	Temp	Data Type	Reference
WS 2.00E+002	mg/L	25	EST	MEYLAN,WM ET AL. (1996)
logP 2.54			EXP	TANAKA,M ET AL. (1985)
VP 6.62E-006	mm Hg	25	EST	NEELY,WB & BLAU,GE (1985)
DC	pKa			
HL 3.36E-012	atm m3/mol	25	EST	MEYLAN,WM & HOWARD,PH (1991)
OH 2.19E-010	cm3/molc sec	25	EST	MEYLAN,WM & HOWARD,PH (1993)

CAS #: 098154-95-1				CARBAMIC ACID, DIETHYL-, 2-(1H-INDOL-3-YL)ETHYL

Formula: $C_{15}H_{20}N_2O_2$

Mol Weight: 260.34

MP (deg C):　　　FP (deg C):

BP (deg C):

BP pressure (mm Hg):

Property/Value	Units	Temp	Data Type	Reference
WS 3.15E+001	mg/L	25	EST	MEYLAN,WM ET AL. (1996)
logP 3.30			EXP	TANAKA,M ET AL. (1985)
VP 1.23E-006	mm Hg	25	EST	NEELY,WB & BLAU,GE (1985)
DC	pKa			
HL 5.92E-012	atm m3/mol	25	EST	MEYLAN,WM & HOWARD,PH (1991)
OH 2.26E-010	cm3/molc sec	25	EST	MEYLAN,WM & HOWARD,PH (1993)

CAS #: 098154-96-2				CARBAMIC ACID, 3-(3-PYRIDINYL)PROPYL ESTER

Formula: $C_9H_{12}N_2O_2$

Mol Weight: 180.21

MP (deg C):　　　FP (deg C):

BP (deg C):

BP pressure (mm Hg):

Property/Value	Units	Temp	Data Type	Reference
WS 4.44E+005	mg/L	25	EST	MEYLAN,WM ET AL. (1996)
logP 0.46			EXP	TANAKA,M ET AL. (1985)
VP 1.28E-003	mm Hg	25	EST	NEELY,WB & BLAU,GE (1985)
DC	pKa			
HL 7.37E-012	atm m3/mol	25	EST	MEYLAN,WM & HOWARD,PH (1991)
OH 7.32E-012	cm3/molc sec	25	EST	MEYLAN,WM & HOWARD,PH (1993)

CAS #: 098154-97-3	CARBAMIC ACID, DIMETHYL-, 3-(3-PYRIDINYL)PROPYL

Formula: $C_{11}H_{16}N_2O_2$

Mol Weight: 208.26

MP (deg C): | FP (deg C):

BP (deg C):

BP pressure (mm Hg):

Property/Value	Units	Temp	Data Type	Reference
WS 4.66E+004	mg/L	25	EST	MEYLAN,WM ET AL. (1996)
logP 1.44			EXP	TANAKA,M ET AL. (1985)
VP 9.94E-004	mm Hg	25	EST	NEELY,WB & BLAU,GE (1985)
DC	pKa			
HL 3.55E-011	atm m3/mol	25	EST	MEYLAN,WM & HOWARD,PH (1991)
OH 2.14E-011	cm3/molc sec	25	EST	MEYLAN,WM & HOWARD,PH (1993)

CAS #: 098154-98-4	CARBAMIC ACID, DIMETHYL-, 3-(2-PYRIDINYL)PROPYL

Formula: $C_{13}H_{20}N_2O_2$

Mol Weight: 236.32

MP (deg C): | FP (deg C):

BP (deg C):

BP pressure (mm Hg):

Property/Value	Units	Temp	Data Type	Reference
WS 5.42E+003	mg/L	25	EST	MEYLAN,WM ET AL. (1996)
logP 2.36			EXP	TANAKA,M ET AL. (1985)
VP 1.40E-004	mm Hg	25	EST	NEELY,WB & BLAU,GE (1985)
DC	pKa			
HL 6.26E-011	atm m3/mol	25	EST	MEYLAN,WM & HOWARD,PH (1991)
OH 2.83E-011	cm3/molc sec	25	EST	MEYLAN,WM & HOWARD,PH (1993)

CAS #: 098154-99-5	CARBAMIC ACID, DIMETHYL-, 2-ETHYLBUTYL ESTER

Formula: $C_9H_{19}NO_2$

Mol Weight: 173.26

MP (deg C): | FP (deg C):

BP (deg C):

BP pressure (mm Hg):

Property/Value	Units	Temp	Data Type	Reference
WS 2.10E+002	mg/L	25	EST	MEYLAN,WM ET AL. (1996)
logP 2.87			EXP	TANAKA,M ET AL. (1985)
VP 3.93E-001	mm Hg	25	EST	NEELY,WB & BLAU,GE (1985)
DC	pKa			
HL 7.87E-007	atm m3/mol	25	EST	MEYLAN,WM & HOWARD,PH (1991)
OH 2.42E-011	cm3/molc sec	25	EST	MEYLAN,WM & HOWARD,PH (1993)

CAS #: 098155-00-1	CARBAMIC ACID, DIETHYL-, 2-ETHYLBUTYL ESTER

Formula: $C_{11}H_{23}NO_2$

Mol Weight: 201.31

MP (deg C): | FP (deg C):

BP (deg C):

BP pressure (mm Hg):

Property/Value	Units	Temp	Data Type	Reference
WS 3.63E+001	mg/L	25	EST	MEYLAN,WM ET AL. (1996)
logP 3.60			EXP	TANAKA,M ET AL. (1985)
VP 5.58E-002	mm Hg	25	EST	NEELY,WB & BLAU,GE (1985)
DC	pKa			
HL 1.39E-006	atm m3/mol	25	EST	MEYLAN,WM & HOWARD,PH (1991)
OH 3.11E-011	cm3/molc sec	25	EST	MEYLAN,WM & HOWARD,PH (1993)

CAS #: 098183-16-5	CARBAMIC ACID, DIMETHYL-, 2-PHENYLETHYL ESTER

Formula: $C_{11}H_{15}NO_2$

Mol Weight: 193.25

MP (deg C): | FP (deg C):

BP (deg C):

BP pressure (mm Hg):

Property/Value	Units	Temp	Data Type	Reference
WS 4.22E+002	mg/L	25	EST	MEYLAN,WM ET AL. (1996)
logP 2.40			EXP	TANAKA,M ET AL. (1985)
VP 7.05E-003	mm Hg	25	EST	NEELY,WB & BLAU,GE (1985)
DC	pKa			
HL 2.04E-008	atm m3/mol	25	EST	MEYLAN,WM & HOWARD,PH (1991)
OH 2.39E-011	cm3/molc sec	25	EST	MEYLAN,WM & HOWARD,PH (1993)

CAS #: 098183-17-6	CARBAMIC ACID, DIETHYL-, 2-PHENYLETHYL ESTER

Formula: $C_{13}H_{19}NO_2$

Mol Weight: 221.30

MP (deg C): | FP (deg C):

BP (deg C):

BP pressure (mm Hg):

Property/Value	Units	Temp	Data Type	Reference
WS 6.27E+001	mg/L	25	EST	MEYLAN,WM ET AL. (1996)
logP 3.20			EXP	TANAKA,M ET AL. (1985)
VP 8.73E-004	mm Hg	25	EST	NEELY,WB & BLAU,GE (1985)
DC	pKa			
HL 3.60E-008	atm m3/mol	25	EST	MEYLAN,WM & HOWARD,PH (1991)
OH 3.08E-011	cm3/molc sec	25	EST	MEYLAN,WM & HOWARD,PH (1993)

CAS #: 098204-04-7	ALLOPURINOL,PHENYLGLYCYLOXYMETHYL

Formula: $C_{14}H_{15}N_5O_3$

Mol Weight: 301.31

MP (deg C): | FP (deg C):

BP (deg C):

BP pressure (mm Hg):

Property/Value	Units	Temp	Data Type	Reference
WS 1.67E+004	mg/L	25	EST	MEYLAN,WM ET AL. (1996)
logP -0.15			EXP	HANSCH,C ET AL. (1995)
VP 6.28E-012	mm Hg	25	EST	NEELY,WB & BLAU,GE (1985)
DC	pKa			
HL 2.11E-018	atm m3/mol	25	EST	MEYLAN,WM & HOWARD,PH (1991)
OH 8.34E-011	cm3/molc sec	25	EST	MEYLAN,WM & HOWARD,PH (1993)

CAS #: 098204-05-8	ALLOPURINOL,1-PH-ALANYL-OME-

Formula: $C_{15}H_{17}N_5O_3$

Mol Weight: 315.33

MP (deg C): | FP (deg C):

BP (deg C):

BP pressure (mm Hg):

Property/Value	Units	Temp	Data Type	Reference
WS 4.67E+003	mg/L	25	EST	MEYLAN,WM ET AL. (1996)
logP 0.40			EXP	HANSCH,C ET AL. (1995)
VP 2.71E-012	mm Hg	25	EST	NEELY,WB & BLAU,GE (1985)
DC	pKa			
HL 2.80E-018	atm m3/mol	25	EST	MEYLAN,WM & HOWARD,PH (1991)
OH 8.78E-011	cm3/molc sec	25	EST	MEYLAN,WM & HOWARD,PH (1993)

CAS #: 098204-06-9				ALLOPURINOL,1-LEUCINYLOXYMETHYL-

Formula: $C_{12}H_{17}N_5O_3$

Mol Weight: 279.30

MP (deg C):　　FP (deg C):

BP (deg C):

BP pressure (mm Hg):

Property/Value	Units	Temp	Data Type	Reference
WS 1.11E+004	mg/L	25	EST	MEYLAN,WM ET AL. (1996)
logP 0.19			EXP	HANSCH,C ET AL. (1995)
VP 1.56E-010	mm Hg	25	EST	NEELY,WB & BLAU,GE (1985)
DC	pKa			
HL 8.11E-017	atm m3/mol	25	EST	MEYLAN,WM & HOWARD,PH (1991)
OH 8.60E-011	cm3/molc sec	25	EST	MEYLAN,WM & HOWARD,PH (1993)

CAS #: 098204-08-1				ALLOPURINOL,1-N,N-DIETHYLGYCYCLOMETHYL

Formula: $C_{12}H_{17}N_5O_3$

Mol Weight: 279.30

MP (deg C):　　FP (deg C):

BP (deg C):

BP pressure (mm Hg):

Property/Value	Units	Temp	Data Type	Reference
WS 5.35E+002	mg/L	25	EST	MEYLAN,WM ET AL. (1996)
logP 0.20			EXP	HANSCH,C ET AL. (1995)
VP 5.13E-011	mm Hg	25	EST	NEELY,WB & BLAU,GE (1985)
DC	pKa			
HL 2.22E-016	atm m3/mol	25	EST	MEYLAN,WM & HOWARD,PH (1991)
OH 1.22E-010	cm3/molc sec	25	EST	MEYLAN,WM & HOWARD,PH (1993)

CAS #: 098204-09-2				ALLOPURINOL,1,N,N-DIPROPYLGLYCYLOXYMETHYL-

Formula: $C_{14}H_{23}N_5O_3$

Mol Weight: 309.37

MP (deg C):　　FP (deg C):

BP (deg C):

BP pressure (mm Hg):

Property/Value	Units	Temp	Data Type	Reference
WS 9.15E+002	mg/L	25	EST	MEYLAN,WM ET AL. (1996)
logP 1.27			EXP	HANSCH,C ET AL. (1995)
VP 4.39E-011	mm Hg	25	EST	NEELY,WB & BLAU,GE (1985)
DC	pKa			
HL 3.91E-016	atm m3/mol	25	EST	MEYLAN,WM & HOWARD,PH (1991)
OH 1.41E-010	cm3/molc sec	25	EST	MEYLAN,WM & HOWARD,PH (1993)

CAS #: 098204-10-5				ALLOPURINOL,1,N,N-DIETHYLALANYLOXYMETHYL-

Formula: $C_{13}H_{21}N_5O_3$

Mol Weight: 295.34

MP (deg C):　　FP (deg C):

BP (deg C):

BP pressure (mm Hg):

Property/Value	Units	Temp	Data Type	Reference
WS 3.26E+003	mg/L	25	EST	MEYLAN,WM ET AL. (1996)
logP 0.72			EXP	HANSCH,C ET AL. (1995)
VP 1.66E-010	mm Hg	25	EST	NEELY,WB & BLAU,GE (1985)
DC	pKa			
HL 2.94E-016	atm m3/mol	25	EST	MEYLAN,WM & HOWARD,PH (1991)
OH 1.42E-010	cm3/molc sec	25	EST	MEYLAN,WM & HOWARD,PH (1993)

CAS #: 098270-76-9				(DES-IPR) N,N-DIETAMINOPR-CLOFAZIMINE

Formula: $C_{31}H_{31}Cl_2N_5$

Mol Weight: 544.53

MP (deg C):　　FP (deg C):

BP (deg C):

BP pressure (mm Hg):

Property/Value	Units	Temp	Data Type	Reference
WS 6.59E-002	mg/L	25	EST	MEYLAN,WM ET AL. (1996)
logP 4.39			EXP	QUIGLEY,JM ET AL. (1990)
VP 1.25E-014	mm Hg	25	EST	NEELY,WB & BLAU,GE (1985)
DC	pKa			
HL 3.14E-014	atm m3/mol	25	EST	MEYLAN,WM & HOWARD,PH (1991)
OH 2.31E-010	cm3/molc sec	25	EST	MEYLAN,WM & HOWARD,PH (1993)

CAS #: 098319-26-7				PROSCAR

Formula: $C_{23}H_{36}N_2O_2$

Mol Weight: 372.56

MP (deg C): 252-254　　FP (deg C):

BP (deg C):

BP pressure (mm Hg):

Property/Value	Units	Temp	Data Type	Reference
WS 1.17E+001	mg/L	25	EST	MEYLAN,WM ET AL. (1996)
logP 3.03			EXP	HANSCH,C ET AL. (1995)
VP 7.47E-012	mm Hg	25	EST	NEELY,WB & BLAU,GE (1985)
DC	pKa			
HL 3.77E-011	atm m3/mol	25	EST	MEYLAN,WM & HOWARD,PH (1991)
OH 5.24E-011	cm3/molc sec	25	EST	MEYLAN,WM & HOWARD,PH (1993)

CAS #: 098369-49-4				BENZOIC ACID, 2-[(4-OXO-2-AZETIDINYL)THIO]-, PHE

Formula: $C_{17}H_{15}NO_3S$

Mol Weight: 313.38

MP (deg C):　　FP (deg C):

BP (deg C):

BP pressure (mm Hg):

Property/Value	Units	Temp	Data Type	Reference
WS 5.93E+001	mg/L	25	EST	MEYLAN,WM ET AL. (1996)
logP 2.62			EXP	ARNOLDI,A ET AL. (1988)
VP 3.87E-010	mm Hg	25	EST	NEELY,WB & BLAU,GE (1985)
DC	pKa			
HL 5.62E-012	atm m3/mol	25	EST	MEYLAN,WM & HOWARD,PH (1991)
OH 8.39E-011	cm3/molc sec	25	EST	MEYLAN,WM & HOWARD,PH (1993)

CAS #: 098378-56-4				BENZAMIDE, N-HEXYL-3,4-DIHYDROXY-

Formula: $C_{13}H_{19}NO_3$

Mol Weight: 237.30

MP (deg C):　　FP (deg C):

BP (deg C):

BP pressure (mm Hg):

Property/Value	Units	Temp	Data Type	Reference
WS 5.12E+002	mg/L	25	EST	MEYLAN,WM ET AL. (1996)
logP 2.71			EXP	NAITO,Y ET AL. (1991)
VP 1.14E-009	mm Hg	25	EST	NEELY,WB & BLAU,GE (1985)
DC	pKa			
HL 2.17E-016	atm m3/mol	25	EST	MEYLAN,WM & HOWARD,PH (1991)
OH 3.70E-011	cm3/molc sec	25	EST	MEYLAN,WM & HOWARD,PH (1993)

CAS #: 098486-60-3				ACETAMIDE, N-(1-METHYLETHYL)-N-PROPYL-
Formula: $C_8H_{17}NO$				
Mol Weight: 143.23				
MP (deg C):		FP (deg C):		
BP (deg C):				
BP pressure (mm Hg):				

Property/Value	Units	Temp	Data Type	Reference
WS 4.42E+003	mg/L	25	EST	MEYLAN,WM ET AL. (1996)
logP 1.48			EXP	SANGSTER,J (1993)
VP 2.23E-001	mm Hg		EST	NEELY,WB & BLAU,GE (1985)
DC	pKa			
HL 1.67E-007	atm m3/mol	25	EST	MEYLAN,WM & HOWARD,PH (1991)
OH 2.94E-011	cm3/molc sec	25	EST	MEYLAN,WM & HOWARD,PH (1993)

CAS #: 098730-04-2				BENOXACOR
Formula: $C_{11}H_{11}Cl_2NO_2$				
Mol Weight: 260.12				
MP (deg C):		FP (deg C):		
BP (deg C):				
BP pressure (mm Hg):				

Property/Value	Units	Temp	Data Type	Reference
WS 1.03E+002	mg/L	25	EST	MEYLAN,WM ET AL. (1996)
logP 2.70			EXP	TOMLIN,C (1994)
VP 9.45E-006	mm Hg	25	EST	NEELY,WB & BLAU,GE (1985)
DC	pKa			
HL 5.43E-010	atm m3/mol	25	EST	MEYLAN,WM & HOWARD,PH (1991)
OH 4.65E-011	cm3/molc sec	25	EST	MEYLAN,WM & HOWARD,PH (1993)

CAS #: 098736-45-9				BENZENEMETHANAMINE, 3-METHOXY-N-METHYL-N-NITROSO-
Formula: $C_9H_{12}N_2O_2$				
Mol Weight: 180.21				
MP (deg C):		FP (deg C):		
BP (deg C):				
BP pressure (mm Hg):				

Property/Value	Units	Temp	Data Type	Reference
WS 2.36E+003	mg/L	25	EST	MEYLAN,WM ET AL. (1996)
logP 1.60			EXP	SINGER,GM ET AL. (1986)
VP 3.99E-004	mm Hg	25	EST	NEELY,WB & BLAU,GE (1985)
DC	pKa			
HL 9.83E-009	atm m3/mol	25	EST	MEYLAN,WM & HOWARD,PH (1991)
OH 6.42E-011	cm3/molc sec	25	EST	MEYLAN,WM & HOWARD,PH (1993)

CAS #: 098736-46-0				BENZENEMETHANAMINE, 3-CHLORO-N-METHYL-N-NITROSO-
Formula: $C_8H_9ClN_2O$				
Mol Weight: 184.63				
MP (deg C):		FP (deg C):		
BP (deg C):				
BP pressure (mm Hg):				

Property/Value	Units	Temp	Data Type	Reference
WS 2.96E+003	mg/L	25	EST	MEYLAN,WM ET AL. (1996)
logP 1.46			EXP	SINGER,GM ET AL. (1986)
VP 5.83E-004	mm Hg	25	EST	NEELY,WB & BLAU,GE (1985)
DC	pKa			
HL 1.23E-007	atm m3/mol	25	EST	MEYLAN,WM & HOWARD,PH (1991)
OH 1.34E-011	cm3/molc sec	25	EST	MEYLAN,WM & HOWARD,PH (1993)

CAS #: 098736-47-1				BENZENEMETHANAMINE, 3-BROMO-N-METHYL-N-NITROSO-
Formula: $C_8H_9BrN_2O$				
Mol Weight: 229.08				
MP (deg C):		FP (deg C):		
BP (deg C):				
BP pressure (mm Hg):				

Property/Value	Units	Temp	Data Type	Reference
WS 1.20E+003	mg/L	25	EST	MEYLAN,WM ET AL. (1996)
logP 1.65			EXP	SANGSTER,J (1994)
VP 1.84E-004	mm Hg	25	EST	NEELY,WB & BLAU,GE (1985)
DC	pKa			
HL 6.62E-008	atm m3/mol	25	EST	MEYLAN,WM & HOWARD,PH (1991)
OH 1.30E-011	cm3/molc sec	25	EST	MEYLAN,WM & HOWARD,PH (1993)

CAS #: 098736-48-2				BENZENEMETHANAMINE, N-METHYL-N-NITROSO-3-(TRIFLU
Formula: $C_9H_9F_3N_2O$				
Mol Weight: 218.18				
MP (deg C):		FP (deg C):		
BP (deg C):				
BP pressure (mm Hg):				

Property/Value	Units	Temp	Data Type	Reference
WS 7.03E+002	mg/L	25	EST	MEYLAN,WM ET AL. (1996)
logP 1.99			EXP	SINGER,GM ET AL. (1986)
VP 2.04E-003	mm Hg	25	EST	NEELY,WB & BLAU,GE (1985)
DC	pKa			
HL 1.44E-006	atm m3/mol	25	EST	MEYLAN,WM & HOWARD,PH (1991)
OH 1.07E-011	cm3/molc sec	25	EST	MEYLAN,WM & HOWARD,PH (1993)

CAS #: 098736-50-6				BENZENEMETHANAMINE, 4-BROMO-N-METHYL-N-NITROSO-
Formula: $C_8H_9BrN_2O$				
Mol Weight: 229.08				
MP (deg C):		FP (deg C):		
BP (deg C):				
BP pressure (mm Hg):				

Property/Value	Units	Temp	Data Type	Reference
WS 1.40E+003	mg/L	25	EST	MEYLAN,WM ET AL. (1996)
logP 1.57			EXP	SANGSTER,J (1994)
VP 1.84E-004	mm Hg	25	EST	NEELY,WB & BLAU,GE (1985)
DC	pKa			
HL 6.62E-008	atm m3/mol	25	EST	MEYLAN,WM & HOWARD,PH (1991)
OH 1.14E-011	cm3/molc sec	25	EST	MEYLAN,WM & HOWARD,PH (1993)

CAS #: 098736-51-7				BENZOIC ACID, 4-[(METHYLNITROSOAMINO)METHYL]-, M
Formula: $C_{10}H_{12}N_2O_3$				
Mol Weight: 208.22				
MP (deg C):		FP (deg C):		
BP (deg C):				
BP pressure (mm Hg):				

Property/Value	Units	Temp	Data Type	Reference
WS 3.68E+003	mg/L	25	EST	MEYLAN,WM ET AL. (1996)
logP 1.21			EXP	SANGSTER,J (1993)
VP 5.49E-005	mm Hg	25	EST	NEELY,WB & BLAU,GE (1985)
DC	pKa			
HL 1.07E-009	atm m3/mol	25	EST	MEYLAN,WM & HOWARD,PH (1991)
OH 1.17E-011	cm3/molc sec	25	EST	MEYLAN,WM & HOWARD,PH (1993)

CAS #: 098736-52-8				BENZENEMETHANAMINE, N-METHYL-3-NITRO-N-NITROSO-

Formula: $C_8H_9N_3O_3$

Mol Weight: 195.18

MP (deg C):   FP (deg C):

BP (deg C):

BP pressure (mm Hg):

Property/Value	Units	Temp	Data Type	Reference
WS 1.89E+003	mg/L	25	EST	MEYLAN,WM ET AL. (1996)
logP 1.17			EXP	SINGER,GM ET AL. (1986)
VP 2.68E-005	mm Hg	25	EST	NEELY,WB & BLAU,GE (1985)
DC	pKa			
HL 6.55E-010	atm m3/mol	25	EST	MEYLAN,WM & HOWARD,PH (1991)
OH 1.04E-011	cm3/molc sec	25	EST	MEYLAN,WM & HOWARD,PH (1993)

CAS #: 098827-16-8				ALLOPURINOL,1,5-BIS(PIVALOYLOXYMETHYL)-

Formula: $C_{17}H_{26}N_4O_5$

Mol Weight: 366.42

MP (deg C):   FP (deg C):

BP (deg C):

BP pressure (mm Hg):

Property/Value	Units	Temp	Data Type	Reference
WS 3.71E+001	mg/L	25	EST	MEYLAN,WM ET AL. (1996)
logP 2.50			EXP	BUNDGAARD,H & FALCH,E (1985A)
VP 6.18E-009	mm Hg	25	EST	NEELY,WB & BLAU,GE (1985)
DC	pKa			
HL 7.26E-015	atm m3/mol	25	EST	MEYLAN,WM & HOWARD,PH (1991)
OH 4.82E-011	cm3/molc sec	25	EST	MEYLAN,WM & HOWARD,PH (1993)

CAS #: 098827-17-9				ALLOPURINOL,2,5-BIS(PIVALOYLOXYMETHYL)-

Formula: $C_{17}H_{26}N_4O_5$

Mol Weight: 366.42

MP (deg C):   FP (deg C):

BP (deg C):

BP pressure (mm Hg):

Property/Value	Units	Temp	Data Type	Reference
WS 5.09E+001	mg/L	25	EST	MEYLAN,WM ET AL. (1996)
logP 2.34			EXP	BUNDGAARD,H & FALCH,E (1985A)
VP 6.18E-009	mm Hg	25	EST	NEELY,WB & BLAU,GE (1985)
DC	pKa			
HL 7.26E-015	atm m3/mol	25	EST	MEYLAN,WM & HOWARD,PH (1991)
OH 4.82E-011	cm3/molc sec	25	EST	MEYLAN,WM & HOWARD,PH (1993)

CAS #: 098827-18-0				ALLOPURINOL,1-[PIVALOYLOXYMETHYL]-

Formula: $C_{11}H_{14}N_4O_3$

Mol Weight: 250.26

MP (deg C):   FP (deg C):

BP (deg C):

BP pressure (mm Hg):

Property/Value	Units	Temp	Data Type	Reference
WS 2.88E+003	mg/L	25	EST	MEYLAN,WM ET AL. (1996)
logP 1.07			EXP	BUNDGAARD,H & FALCH,E (1985A)
VP 2.07E-007	mm Hg	25	EST	NEELY,WB & BLAU,GE (1985)
DC	pKa			
HL 2.41E-015	atm m3/mol	25	EST	MEYLAN,WM & HOWARD,PH (1991)
OH 2.02E-010	cm3/molc sec	25	EST	MEYLAN,WM & HOWARD,PH (1993)

CAS #: 098827-19-1				ALLOPURINOL, 1,5-BIS(BUTYRYLOXYMETHYL)-

Formula: $C_{15}H_{22}N_4O_5$

Mol Weight: 338.37

MP (deg C):   FP (deg C):

BP (deg C):

BP pressure (mm Hg):

Property/Value	Units	Temp	Data Type	Reference
WS 2.09E+002	mg/L	25	EST	MEYLAN,WM ET AL. (1996)
logP 1.82			EXP	BUNDGAARD,H & FALCH,E (1985A)
VP 6.18E-009	mm Hg	25	EST	NEELY,WB & BLAU,GE (1985)
DC	pKa			
HL 4.12E-015	atm m3/mol	25	EST	MEYLAN,WM & HOWARD,PH (1991)
OH 5.16E-011	cm3/molc sec	25	EST	MEYLAN,WM & HOWARD,PH (1993)

CAS #: 098827-20-4				ALLOPURINOL, 2,5-BIS(BUTYRYLOXYMETHYL)-

Formula: $C_{15}H_{22}N_4O_5$

Mol Weight: 338.37

MP (deg C):   FP (deg C):

BP (deg C):

BP pressure (mm Hg):

Property/Value	Units	Temp	Data Type	Reference
WS 3.22E+002	mg/L	25	EST	MEYLAN,WM ET AL. (1996)
logP 1.60			EXP	BUNDGAARD,H & FALCH,E (1985A)
VP 6.18E-009	mm Hg	25	EST	NEELY,WB & BLAU,GE (1985)
DC	pKa			
HL 4.12E-015	atm m3/mol	25	EST	MEYLAN,WM & HOWARD,PH (1991)
OH 5.16E-011	cm3/molc sec	25	EST	MEYLAN,WM & HOWARD,PH (1993)

CAS #: 098827-22-6				BUTANOIC ACID, (2,5-DIHYDRO-4-OXO-4H-PYRAZOLO[3,

Formula: $C_{10}H_{12}N_4O_3$

Mol Weight: 236.23

MP (deg C):   FP (deg C):

BP (deg C):

BP pressure (mm Hg):

Property/Value	Units	Temp	Data Type	Reference
WS 7.21E+002	mg/L	25	EST	MEYLAN,WM ET AL. (1996)
logP 0.33			EXP	SANGSTER,J (1994)
VP 5.97E-010	mm Hg	25	EST	NEELY,WB & BLAU,GE (1985)
DC	pKa			
HL 2.24E-014	atm m3/mol	25	EST	MEYLAN,WM & HOWARD,PH (1991)
OH 3.38E-011	cm3/molc sec	25	EST	MEYLAN,WM & HOWARD,PH (1993)

CAS #: 098827-23-7				ALLOPURINOL, 1-[ETO-CO-OXY-ME]-

Formula: $C_9H_{10}N_4O_4$

Mol Weight: 238.20

MP (deg C):   FP (deg C):

BP (deg C):

BP pressure (mm Hg):

Property/Value	Units	Temp	Data Type	Reference
WS 8.90E+002	mg/L	25	EST	MEYLAN,WM ET AL. (1996)
logP 0.21			EXP	BUNDGAARD,H & FALCH,E (1985A)
VP 1.17E-007	mm Hg	25	EST	NEELY,WB & BLAU,GE (1985)
DC	pKa			
HL 4.85E-015	atm m3/mol	25	EST	MEYLAN,WM & HOWARD,PH (1991)
OH 2.03E-010	cm3/molc sec	25	EST	MEYLAN,WM & HOWARD,PH (1993)

CAS #: 098827-27-1 — ALLOPURINOL,1-[N,N-DIETHYLSUCCINAMYL-OXYMETHYL]-

Formula: $C_{14}H_{19}N_5O_4$
Mol Weight: 321.34
MP (deg C): FP (deg C):
BP (deg C):
BP pressure (mm Hg):

Property/Value	Units	Temp	Data Type	Reference
WS 6.94E+002	mg/L	25	EST	MEYLAN,WM ET AL. (1996)
logP -0.22			EXP	BUNDGAARD,H & FALCH,E (1985A)
VP 1.08E-010	mm Hg	25	EST	NEELY,WB & BLAU,GE (1985)
DC	pKa			
HL 1.77E-021	atm m3/mol	25	EST	MEYLAN,WM & HOWARD,PH (1991)
OH 2.28E-010	cm3/molc sec	25	EST	MEYLAN,WM & HOWARD,PH (1993)

CAS #: 098846-64-1 — ALLOPURINOL, 1-[(ACETYLOXY)METHYL]-

Formula: $C_8H_8N_4O_3$
Mol Weight: 208.18
MP (deg C): FP (deg C):
BP (deg C):
BP pressure (mm Hg):

Property/Value	Units	Temp	Data Type	Reference
WS 3.87E+003	mg/L	25	EST	MEYLAN,WM ET AL. (1996)
logP -0.35			EXP	BUNDGAARD,H & FALCH,E (1985A)
VP 1.10E-006	mm Hg	25	EST	NEELY,WB & BLAU,GE (1985)
DC	pKa			
HL 1.03E-015	atm m3/mol	25	EST	MEYLAN,WM & HOWARD,PH (1991)
OH 2.02E-010	cm3/molc sec	25	EST	MEYLAN,WM & HOWARD,PH (1993)

CAS #: 098846-65-2 — ALLOPURINOL, 1-[BENZOYLOXYMETHYL]-

Formula: $C_{13}H_{10}N_4O_3$
Mol Weight: 270.25
MP (deg C): FP (deg C):
BP (deg C):
BP pressure (mm Hg):

Property/Value	Units	Temp	Data Type	Reference
WS 9.54E+002	mg/L	25	EST	MEYLAN,WM ET AL. (1996)
logP 1.50			EXP	BUNDGAARD,H & FALCH,E (1985A)
VP 5.58E-009	mm Hg	25	EST	NEELY,WB & BLAU,GE (1985)
DC	pKa			
HL 2.04E-016	atm m3/mol	25	EST	MEYLAN,WM & HOWARD,PH (1991)
OH 2.02E-010	cm3/molc sec	25	EST	MEYLAN,WM & HOWARD,PH (1993)

CAS #: 098846-66-3 — ALLOPURINOL, 1-[NICOTINOYLOXYMETHYL]-

Formula: $C_{12}H_9N_5O_3$
Mol Weight: 271.24
MP (deg C): FP (deg C):
BP (deg C):
BP pressure (mm Hg):

Property/Value	Units	Temp	Data Type	Reference
WS 5.18E+002	mg/L	25	EST	MEYLAN,WM ET AL. (1996)
logP 0.27			EXP	BUNDGAARD,H & FALCH,E (1985A)
VP 3.73E-009	mm Hg	25	EST	NEELY,WB & BLAU,GE (1985)
DC	pKa			
HL 2.67E-019	atm m3/mol	25	EST	MEYLAN,WM & HOWARD,PH (1991)
OH 2.02E-010	cm3/molc sec	25	EST	MEYLAN,WM & HOWARD,PH (1993)

CAS #: 099027-00-6 — L-VALINAMIDE, N-ACETYL-L-TYROSYL-

Formula: $C_{16}H_{23}N_3O_4$
Mol Weight: 321.38
MP (deg C): FP (deg C):
BP (deg C):
BP pressure (mm Hg):

Property/Value	Units	Temp	Data Type	Reference
WS 2.54E+003	mg/L	25	EST	MEYLAN,WM ET AL. (1996)
logP -0.20			EXP	HANSCH,C ET AL. (1995)
VP 8.03E-015	mm Hg	25	EST	NEELY,WB & BLAU,GE (1985)
DC	pKa			
HL 2.05E-020	atm m3/mol	25	EST	MEYLAN,WM & HOWARD,PH (1991)
OH 8.01E-011	cm3/molc sec	25	EST	MEYLAN,WM & HOWARD,PH (1993)

CAS #: 099035-20-8 — CIMETIDINE,2-(3-NITROPYRIDIN-4-ONE)

Formula: $C_{12}H_{15}N_5O_3S$
Mol Weight: 309.35
MP (deg C): FP (deg C):
BP (deg C):
BP pressure (mm Hg):

Property/Value	Units	Temp	Data Type	Reference
WS 6.66E+004	mg/L	25	EST	MEYLAN,WM ET AL. (1996)
logP -0.20			EXP	HANSCH,C ET AL. (1995)
VP 7.48E-012	mm Hg	25	EST	NEELY,WB & BLAU,GE (1985)
DC	pKa			
HL 7.46E-019	atm m3/mol	25	EST	MEYLAN,WM & HOWARD,PH (1991)
OH 2.18E-010	cm3/molc sec	25	EST	MEYLAN,WM & HOWARD,PH (1993)

CAS #: 099035-21-9 — CIMETIDINE,(METHYLAMINOCYCLOBUTANEDIONE) ANALOG

Formula: $C_{12}H_{16}N_4O_2S$
Mol Weight: 280.35
MP (deg C): FP (deg C):
BP (deg C):
BP pressure (mm Hg):

Property/Value	Units	Temp	Data Type	Reference
WS 1.90E+005	mg/L	25	EST	MEYLAN,WM ET AL. (1996)
logP -0.08			EXP	HANSCH,C ET AL. (1995)
VP 9.39E-011	mm Hg	25	EST	NEELY,WB & BLAU,GE (1985)
DC	pKa			
HL 1.70E-018	atm m3/mol	25	EST	MEYLAN,WM & HOWARD,PH (1991)
OH 2.15E-010	cm3/molc sec	25	EST	MEYLAN,WM & HOWARD,PH (1993)

CAS #: 099069-47-3 — 1H-BENZOTRIAZOLE-1-PROPANAMIDE, á-METHYL-

Formula: $C_{10}H_{12}N_4O$
Mol Weight: 204.23
MP (deg C): FP (deg C):
BP (deg C):
BP pressure (mm Hg):

Property/Value	Units	Temp	Data Type	Reference
WS 9.52E+003	mg/L	25	EST	MEYLAN,WM ET AL. (1996)
logP 0.75			EXP	SPARATORE,F ET AL. (1988)
VP 3.32E-007	mm Hg	25	EST	NEELY,WB & BLAU,GE (1985)
DC	pKa			
HL 8.37E-013	atm m3/mol	25	EST	MEYLAN,WM & HOWARD,PH (1991)
OH 1.16E-011	cm3/molc sec	25	EST	MEYLAN,WM & HOWARD,PH (1993)

099081-88-6 — ALPHA-(P-ISO-PR-PHENYL)-N-PHENYLNITRONE

Formula: $C_{16}H_{17}NO$

Mol Weight: 239.32

MP (deg C):　　FP (deg C):

BP (deg C):

BP pressure (mm Hg):

Property/Value	Units	Temp	Data Type	Reference
WS 3.59E+001	mg/L	25	EST	MEYLAN,WM ET AL. (1996)
logP 3.37			EXP	KIRCHNER,JJ ET AL. (1985)
VP 1.66E-009	mm Hg	25	EST	NEELY,WB & BLAU,GE (1985)
DC	pKa			
HL 1.54E-011	atm m3/mol	25	EST	MEYLAN,WM & HOWARD,PH (1991)
OH 1.02E-011	cm3/molc sec	25	EST	MEYLAN,WM & HOWARD,PH (1993)

099167-62-1 — 4,6(1H,5H)-PYRIMIDINEDIONE, 5,5-DIPROPYL-2-THIOX

Formula: $C_{10}H_{16}N_2O_2S$

Mol Weight: 228.32

MP (deg C):　　FP (deg C):

BP (deg C):

BP pressure (mm Hg):

Property/Value	Units	Temp	Data Type	Reference
WS 1.66E+002	mg/L	25	EST	MEYLAN,WM ET AL. (1996)
logP 2.66			EXP	WONG,O & MCKEOWN,RH (1988)
VP 2.98E-010	mm Hg	25	EST	NEELY,WB & BLAU,GE (1985)
DC	pKa			
HL 4.78E-009	atm m3/mol	25	EST	MEYLAN,WM & HOWARD,PH (1991)
OH 2.26E-011	cm3/molc sec	25	EST	MEYLAN,WM & HOWARD,PH (1993)

099167-69-8 — BARBITURIC ACID,DI-IPR

Formula: $C_{10}H_{16}N_2O_3$

Mol Weight: 212.25

MP (deg C):　　FP (deg C):

BP (deg C):

BP pressure (mm Hg):

Property/Value	Units	Temp	Data Type	Reference
WS 1.76E+003	mg/L	25	EST	MEYLAN,WM ET AL. (1996)
logP 1.56			EXP	WONG,O & MCKEOWN,RH (1988)
VP 8.23E-011	mm Hg	25	EST	NEELY,WB & BLAU,GE (1985)
DC	pKa			
HL 6.36E-013	atm m3/mol	25	EST	MEYLAN,WM & HOWARD,PH (1991)
OH 1.78E-011	cm3/molc sec	25	EST	MEYLAN,WM & HOWARD,PH (1993)

099287-38-4 — 1-SO3NA-3-PROPYL AZULENE

Formula: $C_{13}H_{13}NaO_3S$

Mol Weight: 272.30

MP (deg C):　　FP (deg C):

BP (deg C):

BP pressure (mm Hg):

Property/Value	Units	Temp	Data Type	Reference
WS 1.57E+005	mg/L	25	EST	MEYLAN,WM ET AL. (1996)
logP -1.11			EXP	YANAGISAWA,T ET AL. (1988)
VP 1.25E-013	mm Hg	25	EST	NEELY,WB & BLAU,GE (1985)
DC	pKa			
HL	atm m3/mol			
OH 3.08E-010	cm3/molc sec	25	EST	MEYLAN,WM & HOWARD,PH (1993)

099287-40-8 — 1-SO3NA-3-PENTYL AZULENE

Formula: $C_{15}H_{17}NaO_3S$

Mol Weight: 300.35

MP (deg C):　　FP (deg C):

BP (deg C):

BP pressure (mm Hg):

Property/Value	Units	Temp	Data Type	Reference
WS 3.83E+003	mg/L	25	EST	MEYLAN,WM ET AL. (1996)
logP 0.59			EXP	YANAGISAWA,T ET AL. (1988)
VP 2.25E-014	mm Hg	25	EST	NEELY,WB & BLAU,GE (1985)
DC	pKa			
HL	atm m3/mol			
OH 3.11E-010	cm3/molc sec	25	EST	MEYLAN,WM & HOWARD,PH (1993)

099287-44-2 — 1-SO3NA-3-PROPYL-7-ISOPROPYL AZULENE

Formula: $C_{16}H_{19}NaO_3S$

Mol Weight: 314.38

MP (deg C):　　FP (deg C):

BP (deg C):

BP pressure (mm Hg):

Property/Value	Units	Temp	Data Type	Reference
WS 2.45E+004	mg/L	25	EST	MEYLAN,WM ET AL. (1996)
logP -0.45			EXP	YANAGISAWA,T ET AL. (1988)
VP 1.72E-014	mm Hg	25	EST	NEELY,WB & BLAU,GE (1985)
DC	pKa			
HL	atm m3/mol			
OH 3.45E-010	cm3/molc sec	25	EST	MEYLAN,WM & HOWARD,PH (1993)

099287-46-4 — 1-SO3NA-3-BUTYL-7-ISOPROPYL AZULENE

Formula: $C_{17}H_{21}NaO_3S$

Mol Weight: 328.41

MP (deg C):　　FP (deg C):

BP (deg C):

BP pressure (mm Hg):

Property/Value	Units	Temp	Data Type	Reference
WS 6.86E+003	mg/L	25	EST	MEYLAN,WM ET AL. (1996)
logP 0.10			EXP	YANAGISAWA,T ET AL. (1988)
VP 7.28E-015	mm Hg	25	EST	NEELY,WB & BLAU,GE (1985)
DC	pKa			
HL	atm m3/mol			
OH 3.47E-010	cm3/molc sec	25	EST	MEYLAN,WM & HOWARD,PH (1993)

099287-48-6 — 1-SO3NA-3-PENTYL-7-ISOPROPYL AZULENE

Formula: $C_{18}H_{23}NaO_3S$

Mol Weight: 342.44

MP (deg C):　　FP (deg C):

BP (deg C):

BP pressure (mm Hg):

Property/Value	Units	Temp	Data Type	Reference
WS 1.58E+003	mg/L	25	EST	MEYLAN,WM ET AL. (1996)
logP 0.75			EXP	YANAGISAWA,T ET AL. (1988)
VP 3.07E-015	mm Hg	25	EST	NEELY,WB & BLAU,GE (1985)
DC	pKa			
HL	atm m3/mol			
OH 3.48E-010	cm3/molc sec	25	EST	MEYLAN,WM & HOWARD,PH (1993)

099287-50-0 — 1-SO3NA-3-BENZYL-7-ISOPROPYL AZULENE

Formula: $C_{20}H_{19}NaO_3S$

Mol Weight: 362.43

MP (deg C): FP (deg C):

BP (deg C):

BP pressure (mm Hg):

Property/Value	Units	Temp	Data Type	Reference
WS 3.45E+003	mg/L	25	EST	MEYLAN,WM ET AL. (1996)
logP 0.21			EXP	YANAGISAWA,T ET AL. (1988)
VP 1.82E-016	mm Hg	25	EST	NEELY,WB & BLAU,GE (1985)
DC	pKa			
HL	atm m3/mol			
OH 3.49E-010	cm3/molc sec	25	EST	MEYLAN,WM & HOWARD,PH (1993)

099287-52-2 — 1-SO3NA-3-ME-4-METHOXY AZULENE

Formula: $C_{12}H_{11}NaO_4S$

Mol Weight: 274.27

MP (deg C): FP (deg C):

BP (deg C):

BP pressure (mm Hg):

Property/Value	Units	Temp	Data Type	Reference
WS 2.18E+005	mg/L	25	EST	MEYLAN,WM ET AL. (1996)
logP -1.29			EXP	YANAGISAWA,T ET AL. (1988)
VP 1.30E-013	mm Hg	25	EST	NEELY,WB & BLAU,GE (1985)
DC	pKa			
HL	atm m3/mol			
OH 3.65E-010	cm3/molc sec	25	EST	MEYLAN,WM & HOWARD,PH (1993)

099287-54-4 — 1-SO3NA-3-ET-4-METHOXY AZULENE

Formula: $C_{13}H_{13}NaO_4S$

Mol Weight: 288.30

MP (deg C): FP (deg C):

BP (deg C):

BP pressure (mm Hg):

Property/Value	Units	Temp	Data Type	Reference
WS 9.86E+004	mg/L	25	EST	MEYLAN,WM ET AL. (1996)
logP -0.98			EXP	YANAGISAWA,T ET AL. (1988)
VP 5.52E-014	mm Hg	25	EST	NEELY,WB & BLAU,GE (1985)
DC	pKa			
HL	atm m3/mol			
OH 3.65E-010	cm3/molc sec	25	EST	MEYLAN,WM & HOWARD,PH (1993)

099287-56-6 — 1-SO3NA-3-PROPYL-4-METHOXY AZULENE

Formula: $C_{14}H_{15}NaO_4S$

Mol Weight: 302.33

MP (deg C): FP (deg C):

BP (deg C):

BP pressure (mm Hg):

Property/Value	Units	Temp	Data Type	Reference
WS 4.11E+004	mg/L	25	EST	MEYLAN,WM ET AL. (1996)
logP -0.63			EXP	YANAGISAWA,T ET AL. (1988)
VP 2.34E-014	mm Hg	25	EST	NEELY,WB & BLAU,GE (1985)
DC	pKa			
HL	atm m3/mol			
OH 3.67E-010	cm3/molc sec	25	EST	MEYLAN,WM & HOWARD,PH (1993)

099287-58-8 — 1-SO3NA-3-BUTYL-4-METHOXY AZULENE

Formula: $C_{15}H_{17}NaO_4S$

Mol Weight: 316.35

MP (deg C): FP (deg C):

BP (deg C):

BP pressure (mm Hg):

Property/Value	Units	Temp	Data Type	Reference
WS 1.81E+004	mg/L	25	EST	MEYLAN,WM ET AL. (1996)
logP -0.31			EXP	YANAGISAWA,T ET AL. (1988)
VP 9.91E-015	mm Hg	25	EST	NEELY,WB & BLAU,GE (1985)
DC	pKa			
HL	atm m3/mol			
OH 3.68E-010	cm3/molc sec	25	EST	MEYLAN,WM & HOWARD,PH (1993)

099287-60-2 — 1-SO3NA-3-PENTYL-4-METHOXY AZULENE

Formula: $C_{16}H_{19}NaO_4S$

Mol Weight: 330.38

MP (deg C): FP (deg C):

BP (deg C):

BP pressure (mm Hg):

Property/Value	Units	Temp	Data Type	Reference
WS 8.97E+003	mg/L	25	EST	MEYLAN,WM ET AL. (1996)
logP -0.05			EXP	YANAGISAWA,T ET AL. (1988)
VP 4.19E-015	mm Hg	25	EST	NEELY,WB & BLAU,GE (1985)
DC	pKa			
HL	atm m3/mol			
OH 3.70E-010	cm3/molc sec	25	EST	MEYLAN,WM & HOWARD,PH (1993)

099287-64-6 — 1-SO3NA-3-ME-4-METHOXY-7-IPR AZULENE

Formula: $C_{15}H_{17}NaO_4S$

Mol Weight: 316.35

MP (deg C): FP (deg C):

BP (deg C):

BP pressure (mm Hg):

Property/Value	Units	Temp	Data Type	Reference
WS 4.56E+004	mg/L	25	EST	MEYLAN,WM ET AL. (1996)
logP -0.78			EXP	YANAGISAWA,T ET AL. (1988)
VP 1.79E-014	mm Hg	25	EST	NEELY,WB & BLAU,GE (1985)
DC	pKa			
HL	atm m3/mol			
OH 4.01E-010	cm3/molc sec	25	EST	MEYLAN,WM & HOWARD,PH (1993)

099287-65-7 — 1-SO3NA-3-ET-4-METHOXY-7-IPR AZULENE

Formula: $C_{16}H_{19}NaO_4S$

Mol Weight: 330.38

MP (deg C): FP (deg C):

BP (deg C):

BP pressure (mm Hg):

Property/Value	Units	Temp	Data Type	Reference
WS 1.68E+004	mg/L	25	EST	MEYLAN,WM ET AL. (1996)
logP -0.37			EXP	YANAGISAWA,T ET AL. (1988)
VP 7.58E-015	mm Hg	25	EST	NEELY,WB & BLAU,GE (1985)
DC	pKa			
HL	atm m3/mol			
OH 4.02E-010	cm3/molc sec	25	EST	MEYLAN,WM & HOWARD,PH (1993)

CAS #: 099287-66-8				1-SO3NA-3-PR-4-METHOXY-7-IPR AZULENE

Formula: $C_{17}H_{21}NaO_4S$

Mol Weight: 344.41

MP (deg C):   FP (deg C):

BP (deg C):

BP pressure (mm Hg):

Property/Value	Units	Temp	Data Type	Reference
WS 6.83E+003	mg/L	25	EST	MEYLAN,WM ET AL. (1996)
logP -0.01			EXP	YANAGISAWA,T ET AL. (1988)
VP 3.20E-015	mm Hg	25	EST	NEELY,WB & BLAU,GE (1985)
DC	pKa			
HL	atm m3/mol			
OH 4.04E-010	cm3/molc sec	25	EST	MEYLAN,WM & HOWARD,PH (1993)

CAS #: 099287-70-4				1-SO3NA-3-PE-4-METHOXY-7-IPR AZULENE

Formula: $C_{19}H_{25}NaO_4S$

Mol Weight: 372.46

MP (deg C):   FP (deg C):

BP (deg C):

BP pressure (mm Hg):

Property/Value	Units	Temp	Data Type	Reference
WS 3.05E+003	mg/L	25	EST	MEYLAN,WM ET AL. (1996)
logP 0.20			EXP	YANAGISAWA,T ET AL. (1988)
VP 5.64E-016	mm Hg	25	EST	NEELY,WB & BLAU,GE (1985)
DC	pKa			
HL	atm m3/mol			
OH 4.07E-010	cm3/molc sec	25	EST	MEYLAN,WM & HOWARD,PH (1993)

CAS #: 099414-56-9				BUTANOIC ACID, 4-(3-CHLORO-4-METHOXYBENZOYL)-3-M

Formula: $C_{13}H_{15}ClO_4$

Mol Weight: 270.72

MP (deg C):   FP (deg C):

BP (deg C):

BP pressure (mm Hg):

Property/Value	Units	Temp	Data Type	Reference
WS 2.71E+002	mg/L	25	EST	MEYLAN,WM ET AL. (1996)
logP 2.60			EXP	SANGSTER,J (1993)
VP 1.28E-006	mm Hg	25	EST	NEELY,WB & BLAU,GE (1985)
DC	pKa			
HL 7.48E-012	atm m3/mol	25	EST	MEYLAN,WM & HOWARD,PH (1991)
OH 1.83E-011	cm3/molc sec	25	EST	MEYLAN,WM & HOWARD,PH (1993)

CAS #: 099481-58-0				4-BROMO-S-PROPYL-BENZOYLTHIOFORMOHYDROXIMATE

Formula: $C_{11}H_{12}BrNO_2S$

Mol Weight: 302.20

MP (deg C):   FP (deg C):

BP (deg C):

BP pressure (mm Hg):

Property/Value	Units	Temp	Data Type	Reference
WS 1.55E+001	mg/L	25	EST	MEYLAN,WM ET AL. (1996)
logP 3.38			EXP	HANSCH,C ET AL. (1994)
VP 1.82E-008	mm Hg	25	EST	NEELY,WB & BLAU,GE (1985)
DC	pKa			
HL 8.81E-011	atm m3/mol	25	EST	MEYLAN,WM & HOWARD,PH (1991)
OH 1.31E-011	cm3/molc sec	25	EST	MEYLAN,WM & HOWARD,PH (1993)

CAS #: 099726-76-8				3-QUINOLINECARBOXYLIC ACID, 1-ETHYL-6,8-DIFLUORO

Formula: $C_{16}H_{17}F_2N_3O_3$

Mol Weight: 337.33

MP (deg C):   FP (deg C):

BP (deg C):

BP pressure (mm Hg):

Property/Value	Units	Temp	Data Type	Reference
WS 5.62E+004	mg/L	25	EST	MEYLAN,WM ET AL. (1996)
logP -0.57			EXP	TAKACS-NOVAK,K ET AL. (1992)
VP 8.02E-013	mm Hg	25	EST	NEELY,WB & BLAU,GE (1985)
DC	pKa			
HL 1.02E-018	atm m3/mol	25	EST	MEYLAN,WM & HOWARD,PH (1991)
OH 1.23E-010	cm3/molc sec	25	EST	MEYLAN,WM & HOWARD,PH (1993)

CAS #: 099764-47-3				1,2,4-OXADIAZOL-5-OXIME,3-T-BUTYL-

Formula: $C_7H_{12}N_3O_2$

Mol Weight: 170.19

MP (deg C):   FP (deg C):

BP (deg C):

BP pressure (mm Hg):

Property/Value	Units	Temp	Data Type	Reference
WS 9.05E+002	mg/L	25	EST	MEYLAN,WM ET AL. (1996)
logP 2.15			EXP	HANSCH,C ET AL. (1995)
VP 1.17E-004	mm Hg	25	EST	NEELY,WB & BLAU,GE (1985)
DC	pKa			
HL 4.02E-009	atm m3/mol	25	EST	MEYLAN,WM & HOWARD,PH (1991)
OH 6.44E-012	cm3/molc sec	25	EST	MEYLAN,WM & HOWARD,PH (1993)

CAS #: 099854-42-9				DILOXANIDE ACETATE

Formula: $C_{11}H_{11}Cl_2NO_3$

Mol Weight: 276.12

MP (deg C):   FP (deg C):

BP (deg C):

BP pressure (mm Hg):

Property/Value	Units	Temp	Data Type	Reference
WS 7.85E+002	mg/L	25	EST	MEYLAN,WM ET AL. (1996)
logP 1.56			EXP	DUTTA,H ET AL. (1988)
VP 4.36E-006	mm Hg	25	EST	NEELY,WB & BLAU,GE (1985)
DC	pKa			
HL 1.89E-010	atm m3/mol	25	EST	MEYLAN,WM & HOWARD,PH (1991)
OH 9.99E-012	cm3/molc sec	25	EST	MEYLAN,WM & HOWARD,PH (1993)

CAS #: 099877-03-9				L-TYROSINE, 3-HYDROXY-, 2-HYDROXYPROPYL ESTER

Formula: $C_{12}H_{17}NO_5$

Mol Weight: 255.27

MP (deg C):   FP (deg C):

BP (deg C):

BP pressure (mm Hg):

Property/Value	Units	Temp	Data Type	Reference
WS 1.00E+006	mg/L	25	EST	MEYLAN,WM ET AL. (1996)
logP -0.49			EXP	MARREL,C ET AL. (1985)
VP 3.34E-009	mm Hg	25	EST	NEELY,WB & BLAU,GE (1985)
DC	pKa			
HL 2.69E-020	atm m3/mol	25	EST	MEYLAN,WM & HOWARD,PH (1991)
OH 1.06E-010	cm3/molc sec	25	EST	MEYLAN,WM & HOWARD,PH (1993)

CAS #: 099877-04-0 — L-TYROSINE, 3-HYDROXY-, 2-METHOXY-1-METHYLETHYL

Formula: $C_{13}H_{19}NO_5$

Mol Weight: 269.30

MP (deg C): | FP (deg C):

BP (deg C):

BP pressure (mm Hg):

Property/Value	Units	Temp	Data Type	Reference
WS 6.10E+005	mg/L	25	EST	MEYLAN,WM ET AL. (1996)
logP 0.08			EXP	MARREL,C ET AL. (1985)
VP 3.09E-008	mm Hg	25	EST	NEELY,WB & BLAU,GE (1985)
DC	pKa			
HL 8.62E-018	atm m3/mol	25	EST	MEYLAN,WM & HOWARD,PH (1991)
OH 1.08E-010	cm3/molc sec	25	EST	MEYLAN,WM & HOWARD,PH (1993)

CAS #: 099877-05-1 — L-TYROSINE, 3-HYDROXY-, CYCLOHEXYL ESTER

Formula: $C_{15}H_{21}NO_4$

Mol Weight: 279.34

MP (deg C): | FP (deg C):

BP (deg C):

BP pressure (mm Hg):

Property/Value	Units	Temp	Data Type	Reference
WS 1.85E+004	mg/L	25	EST	MEYLAN,WM ET AL. (1996)
logP 1.79			EXP	MARREL,C ET AL. (1985)
VP 4.61E-009	mm Hg	25	EST	NEELY,WB & BLAU,GE (1985)
DC	pKa			
HL 7.59E-016	atm m3/mol	25	EST	MEYLAN,WM & HOWARD,PH (1991)
OH 1.08E-010	cm3/molc sec	25	EST	MEYLAN,WM & HOWARD,PH (1993)

CAS #: 099877-06-2 — L-TYROSINE, 3-HYDROXY-, (TETRAHYDRO-2H-PYRAN-2-Y

Formula: $C_{15}H_{21}NO_5$

Mol Weight: 295.34

MP (deg C): | FP (deg C):

BP (deg C):

BP pressure (mm Hg):

Property/Value	Units	Temp	Data Type	Reference
WS 2.01E+005	mg/L	25	EST	MEYLAN,WM ET AL. (1996)
logP 0.47		.	EXP	MARREL,C ET AL. (1985)
VP 1.46E-009	mm Hg	25	EST	NEELY,WB & BLAU,GE (1985)
DC	pKa			
HL 8.43E-019	atm m3/mol	25	EST	MEYLAN,WM & HOWARD,PH (1991)
OH 1.29E-010	cm3/molc sec	25	EST	MEYLAN,WM & HOWARD,PH (1993)

CAS #: 099877-08-4 — L-TYROSINE, 3-HYDROXY-, 2-PHENYLETHYL ESTER

Formula: $C_{17}H_{19}NO_4$

Mol Weight: 301.35

MP (deg C): | FP (deg C):

BP (deg C):

BP pressure (mm Hg):

Property/Value	Units	Temp	Data Type	Reference
WS 1.47E+004	mg/L	25	EST	MEYLAN,WM ET AL. (1996)
logP 1.76			EXP	MARREL,C ET AL. (1985)
VP 3.60E-010	mm Hg	25	EST	NEELY,WB & BLAU,GE (1985)
DC	pKa			
HL 4.48E-017	atm m3/mol	25	EST	MEYLAN,WM & HOWARD,PH (1991)
OH 1.04E-010	cm3/molc sec	25	EST	MEYLAN,WM & HOWARD,PH (1993)

CAS #: 099877-09-5 — L-TYROSINE, 3-HYDROXY-, 2-(4-CHLOROPHENYL)ETHYL

Formula: $C_{17}H_{18}ClNO_4$

Mol Weight: 335.79

MP (deg C): | FP (deg C):

BP (deg C):

BP pressure (mm Hg):

Property/Value	Units	Temp	Data Type	Reference
WS 2.99E+003	mg/L	25	EST	MEYLAN,WM ET AL. (1996)
logP 2.33			EXP	MARREL,C ET AL. (1985)
VP 8.21E-011	mm Hg	25	EST	NEELY,WB & BLAU,GE (1985)
DC	pKa			
HL 3.32E-017	atm m3/mol	25	EST	MEYLAN,WM & HOWARD,PH (1991)
OH 1.01E-010	cm3/molc sec	25	EST	MEYLAN,WM & HOWARD,PH (1993)

CAS #: 099877-10-8 — L-TYROSINE, 3-HYDROXY-, 2-(4-METHOXYPHENYL)ETHYL

Formula: $C_{18}H_{21}NO_5$

Mol Weight: 331.37

MP (deg C): | FP (deg C):

BP (deg C):

BP pressure (mm Hg):

Property/Value	Units	Temp	Data Type	Reference
WS 1.12E+004	mg/L	25	EST	MEYLAN,WM ET AL. (1996)
logP 1.69			EXP	SANGSTER,J (1993)
VP 5.54E-011	mm Hg	25	EST	NEELY,WB & BLAU,GE (1985)
DC	pKa			
HL 2.65E-018	atm m3/mol	25	EST	MEYLAN,WM & HOWARD,PH (1991)
OH 1.26E-010	cm3/molc sec	25	EST	MEYLAN,WM & HOWARD,PH (1993)

CAS #: 099877-11-9 — L-TYROSINE, 3-HYDROXY-, 2-PHENOXYETHYL ESTER

Formula: $C_{17}H_{19}NO_5$

Mol Weight: 317.34

MP (deg C): | FP (deg C):

BP (deg C):

BP pressure (mm Hg):

Property/Value	Units	Temp	Data Type	Reference
WS 1.93E+004	mg/L	25	EST	MEYLAN,WM ET AL. (1996)
logP 1.51			EXP	MARREL,C ET AL. (1985)
VP 1.39E-010	mm Hg	25	EST	NEELY,WB & BLAU,GE (1985)
DC	pKa			
HL 2.40E-018	atm m3/mol	25	EST	MEYLAN,WM & HOWARD,PH (1991)
OH 1.26E-010	cm3/molc sec	25	EST	MEYLAN,WM & HOWARD,PH (1993)

CAS #: 100199-27-7 — CYCLOHEPTANOL, 1-(4-CHLOROPHENYL)-2-(1H-1,2,4-TR

Formula: $C_{15}H_{18}ClN_3O$

Mol Weight: 291.78

MP (deg C): | FP (deg C):

BP (deg C):

BP pressure (mm Hg):

Property/Value	Units	Temp	Data Type	Reference
WS 3.61E+001	mg/L	25	EST	MEYLAN,WM ET AL. (1996)
logP 3.54			EXP	SANGSTER,J (1993)
VP 5.08E-009	mm Hg	25	EST	NEELY,WB & BLAU,GE (1985)
DC	pKa			
HL 1.72E-010	atm m3/mol	25	EST	MEYLAN,WM & HOWARD,PH (1991)
OH 2.10E-011	cm3/molc sec	25	EST	MEYLAN,WM & HOWARD,PH (1993)

CAS #: 100199-35-7				CYCLOHEXANOL, 1-(2-CHLOROPHENYL)-2-(1H-1,2,4-TRI
Formula: $C_{14}H_{16}ClN_3O$				
Mol Weight: 277.76				
MP (deg C):		**FP (deg C):**		
BP (deg C):				
BP pressure (mm Hg):				

Property/Value	Units	Temp	Data Type	Reference
WS 8.49E+001	mg/L	25	EST	MEYLAN,WM ET AL. (1996)
logP 2.68			EXP	SANGSTER,J (1993)
VP 1.42E-008	mm Hg	25	EST	NEELY,WB & BLAU,GE (1985)
DC	pKa			
HL 1.30E-010	atm m3/mol	25	EST	MEYLAN,WM & HOWARD,PH (1991)
OH 1.96E-011	cm3/molc sec	25	EST	MEYLAN,WM & HOWARD,PH (1993)

CAS #: 100199-38-0				CYCLOHEXANOL, 1-(2,5-DICHLOROPHENYL)-2-(1H-1,2,4
Formula: $C_{14}H_{15}Cl_2N_3O$				
Mol Weight: 312.20				
MP (deg C):		**FP (deg C):**		
BP (deg C):				
BP pressure (mm Hg):				

Property/Value	Units	Temp	Data Type	Reference
WS 1.78E+001	mg/L	25	EST	MEYLAN,WM ET AL. (1996)
logP 3.24			EXP	SANGSTER,J (1993)
VP 3.14E-009	mm Hg	25	EST	NEELY,WB & BLAU,GE (1985)
DC	pKa			
HL 9.61E-011	atm m3/mol	25	EST	MEYLAN,WM & HOWARD,PH (1991)
OH 1.88E-011	cm3/molc sec	25	EST	MEYLAN,WM & HOWARD,PH (1993)

CAS #: 100515-03-5				2,4-PYRIMIDINEDIAMINE, 5-[(3,5-DIETHOXYPHENYL)ME
Formula: $C_{15}H_{20}N_4O_2$				
Mol Weight: 288.35				
MP (deg C):		**FP (deg C):**		
BP (deg C):				
BP pressure (mm Hg):				

Property/Value	Units	Temp	Data Type	Reference
WS 1.01E+002	mg/L	25	EST	MEYLAN,WM ET AL. (1996)
logP 2.52			EXP	HANSCH,C ET AL. (1995)
VP 1.02E-008	mm Hg	25	EST	NEELY,WB & BLAU,GE (1985)
DC	pKa			
HL 7.11E-013	atm m3/mol	25	EST	MEYLAN,WM & HOWARD,PH (1991)
OH 2.13E-010	cm3/molc sec	25	EST	MEYLAN,WM & HOWARD,PH (1993)

CAS #: 100646-51-3				QUIZALOFOP-P-ETHYL
Formula: $C_{19}H_{17}ClN_2O_4$				
Mol Weight: 372.81				
MP (deg C):		**FP (deg C):**		
BP (deg C):				
BP pressure (mm Hg):				

Property/Value	Units	Temp	Data Type	Reference
WS 9.96E-001	mg/L	25	EST	MEYLAN,WM ET AL. (1996)
logP 4.28			EXP	HANSCH,C ET AL. (1995)
VP 1.95E-009	mm Hg	25	EST	NEELY,WB & BLAU,GE (1985)
DC	pKa			
HL 7.40E-011	atm m3/mol	25	EST	MEYLAN,WM & HOWARD,PH (1991)
OH 3.35E-011	cm3/molc sec	25	EST	MEYLAN,WM & HOWARD,PH (1993)

CAS #: 100668-10-8				2-PROPENAMIDE, 3-(3,4-DIHYDROXYPHENYL)-N-(PHENYL
Formula: $C_{16}H_{15}NO_3$				
Mol Weight: 269.30				
MP (deg C):		**FP (deg C):**		
BP (deg C):				
BP pressure (mm Hg):				

Property/Value	Units	Temp	Data Type	Reference
WS 5.34E+002	mg/L	25	EST	MEYLAN,WM ET AL. (1996)
logP 2.48			EXP	NAITO,Y ET AL. (1991)
VP 1.73E-011	mm Hg	25	EST	NEELY,WB & BLAU,GE (1985)
DC	pKa			
HL 5.07E-019	atm m3/mol	25	EST	MEYLAN,WM & HOWARD,PH (1991)
OH 5.58E-011	cm3/molc sec	25	EST	MEYLAN,WM & HOWARD,PH (1993)

CAS #: 100668-11-9				2-PROPENAMIDE, 3-(3,4-DIHYDROXYPHENYL)-N-HEXYL-
Formula: $C_{15}H_{21}NO_3$				
Mol Weight: 263.34				
MP (deg C):		**FP (deg C):**		
BP (deg C):				
BP pressure (mm Hg):				

Property/Value	Units	Temp	Data Type	Reference
WS 1.02E+002	mg/L	25	EST	MEYLAN,WM ET AL. (1996)
logP 3.37			EXP	NAITO,Y ET AL. (1991)
VP 1.38E-010	mm Hg	25	EST	NEELY,WB & BLAU,GE (1985)
DC	pKa			
HL 2.59E-017	atm m3/mol	25	EST	MEYLAN,WM & HOWARD,PH (1991)
OH 5.74E-011	cm3/molc sec	25	EST	MEYLAN,WM & HOWARD,PH (1993)

CAS #: 100668-21-1				1,2-BENZENEDIOL, 4-(1-HEXENYL)-
Formula: $C_{12}H_{16}O_2$				
Mol Weight: 192.26				
MP (deg C):		**FP (deg C):**		
BP (deg C):				
BP pressure (mm Hg):				

Property/Value	Units	Temp	Data Type	Reference
WS 9.95E+001	mg/L	25	EST	MEYLAN,WM ET AL. (1996)
logP 3.82			EXP	NAITO,Y ET AL. (1991)
VP 6.51E-006	mm Hg	25	EST	NEELY,WB & BLAU,GE (1985)
DC	pKa			
HL 1.10E-010	atm m3/mol	25	EST	MEYLAN,WM & HOWARD,PH (1991)
OH 8.21E-011	cm3/molc sec	25	EST	MEYLAN,WM & HOWARD,PH (1993)

CAS #: 100754-93-6				PHENYLACETIC ACID,2-(2'-CHLORO-6'-FLUORO)ANILINO
Formula: $C_{14}H_{11}ClFNO_2$				
Mol Weight: 279.70				
MP (deg C):		**FP (deg C):**		
BP (deg C):				
BP pressure (mm Hg):				

Property/Value	Units	Temp	Data Type	Reference
WS 2.27E+001	mg/L	25	EST	MEYLAN,WM ET AL. (1996)
logP 3.80			EXP	HANSCH,C ET AL. (1995)
VP 3.10E-007	mm Hg	25	EST	NEELY,WB & BLAU,GE (1985)
DC	pKa			
HL 7.46E-012	atm m3/mol	25	EST	MEYLAN,WM & HOWARD,PH (1991)
OH 1.67E-010	cm3/molc sec	25	EST	MEYLAN,WM & HOWARD,PH (1993)

CAS #: 100866-99-7	ETHANONE, 1-[4-[(4-AMINOPHENYL)SULFONYL]PHENYL]-
Formula: $C_{14}H_{13}NO_3S$	
Mol Weight: 275.33	
MP (deg C):	FP (deg C):
BP (deg C):	
BP pressure (mm Hg):	

Property/Value	Units	Temp	Data Type	Reference
WS 6.39E+002	mg/L	25	EST	MEYLAN,WM ET AL. (1996)
logP 1.67			EXP	ALTOMARE,C ET AL. (1991)
VP 1.85E-008	mm Hg	25	EST	NEELY,WB & BLAU,GE (1985)
DC	pKa			
HL 1.60E-013	atm m3/mol	25	EST	MEYLAN,WM & HOWARD,PH (1991)
OH 2.35E-011	cm3/molc sec	25	EST	MEYLAN,WM & HOWARD,PH (1993)

CAS #: 101054-93-7	4,4,4-TRIFLUORO-2-BUTANOL
Formula: $C_4H_7F_3O$	
Mol Weight: 128.09	
MP (deg C):	FP (deg C):
BP (deg C):	
BP pressure (mm Hg):	

Property/Value	Units	Temp	Data Type	Reference
WS 7.49E+004	mg/L	25	EST	MEYLAN,WM ET AL. (1996)
logP 0.71			EXP	MULLER,N (1986)
VP 7.37E+001	mm Hg	25	EST	NEELY,WB & BLAU,GE (1985)
DC	pKa			
HL 4.96E-005	atm m3/mol	25	EST	MEYLAN,WM & HOWARD,PH (1991)
OH 8.74E-012	cm3/molc sec	25	EST	MEYLAN,WM & HOWARD,PH (1993)

CAS #: 101063-92-7	PROPANOIC ACID, 3-(6-CHLORO-4H-1,2,4-BENZOTHIADI
Formula: $C_{10}H_9ClN_2O_4S$	
Mol Weight: 288.71	
MP (deg C):	FP (deg C):
BP (deg C):	
BP pressure (mm Hg):	

Property/Value	Units	Temp	Data Type	Reference
WS 8.61E+003	mg/L	25	EST	MEYLAN,WM ET AL. (1996)
logP 0.72			EXP	GAMBERINI,G ET AL. (1989)
VP 3.79E-011	mm Hg	25	EST	NEELY,WB & BLAU,GE (1985)
DC	pKa			
HL 6.77E-013	atm m3/mol	25	EST	MEYLAN,WM & HOWARD,PH (1991)
OH 9.21E-012	cm3/molc sec	25	EST	MEYLAN,WM & HOWARD,PH (1993)

CAS #: 101063-93-8	PROPANOIC ACID, 3-(5,7-DICHLORO-4H-1,2,4-BENZOTH
Formula: $C_{10}H_8Cl_2N_2O_4S$	
Mol Weight: 323.16	
MP (deg C):	FP (deg C):
BP (deg C):	
BP pressure (mm Hg):	

Property/Value	Units	Temp	Data Type	Reference
WS 2.88E+003	mg/L	25	EST	MEYLAN,WM ET AL. (1996)
logP 1.04			EXP	GAMBERINI,G ET AL. (1989)
VP 9.89E-012	mm Hg	25	EST	NEELY,WB & BLAU,GE (1985)
DC	pKa			
HL 5.01E-013	atm m3/mol	25	EST	MEYLAN,WM & HOWARD,PH (1991)
OH 3.13E-012	cm3/molc sec	25	EST	MEYLAN,WM & HOWARD,PH (1993)

CAS #: 101063-94-9	PROPANOIC ACID, 3-(6,7-DICHLORO-4H-1,2,4-BENZOTH
Formula: $C_{10}H_8Cl_2N_2O_4S$	
Mol Weight: 323.16	
MP (deg C):	FP (deg C):
BP (deg C):	
BP pressure (mm Hg):	

Property/Value	Units	Temp	Data Type	Reference
WS 1.17E+003	mg/L	25	EST	MEYLAN,WM ET AL. (1996)
logP 1.50			EXP	GAMBERINI,G ET AL. (1989)
VP 9.89E-012	mm Hg	25	EST	NEELY,WB & BLAU,GE (1985)
DC	pKa			
HL 5.01E-013	atm m3/mol	25	EST	MEYLAN,WM & HOWARD,PH (1991)
OH 4.67E-012	cm3/molc sec	25	EST	MEYLAN,WM & HOWARD,PH (1993)

CAS #: 101063-95-0	PROPANOIC ACID, 3-(5,7-DIBROMO-4H-1,2,4-BENZOTHI
Formula: $C_{10}H_8Br_2N_2O_4S$	
Mol Weight: 412.06	
MP (deg C):	FP (deg C):
BP (deg C):	
BP pressure (mm Hg):	

Property/Value	Units	Temp	Data Type	Reference
WS 4.50E+002	mg/L	25	EST	MEYLAN,WM ET AL. (1996)
logP 1.35			EXP	GAMBERINI,G ET AL. (1989)
VP 1.74E-012	mm Hg	25	EST	NEELY,WB & BLAU,GE (1985)
DC	pKa			
HL 1.45E-013	atm m3/mol	25	EST	MEYLAN,WM & HOWARD,PH (1991)
OH 3.07E-012	cm3/molc sec	25	EST	MEYLAN,WM & HOWARD,PH (1993)

CAS #: 101063-96-1	PROPANOIC ACID, 3-(6-METHYL-4H-1,2,4-BENZOTHIADI
Formula: $C_{11}H_{12}N_2O_4S$	
Mol Weight: 268.29	
MP (deg C):	FP (deg C):
BP (deg C):	
BP pressure (mm Hg):	

Property/Value	Units	Temp	Data Type	Reference
WS 3.39E+004	mg/L	25	EST	MEYLAN,WM ET AL. (1996)
logP 0.16			EXP	GAMBERINI,G ET AL. (1989)
VP 6.31E-011	mm Hg	25	EST	NEELY,WB & BLAU,GE (1985)
DC	pKa			
HL 1.01E-012	atm m3/mol	25	EST	MEYLAN,WM & HOWARD,PH (1991)
OH 2.67E-011	cm3/molc sec	25	EST	MEYLAN,WM & HOWARD,PH (1993)

CAS #: 101063-97-2	PROPANOIC ACID, 3-(6-METHOXY-4H-1,2,4-BENZOTHIAD
Formula: $C_{11}H_{12}N_2O_5S$	
Mol Weight: 284.29	
MP (deg C):	FP (deg C):
BP (deg C):	
BP pressure (mm Hg):	

Property/Value	Units	Temp	Data Type	Reference
WS 2.64E+004	mg/L	25	EST	MEYLAN,WM ET AL. (1996)
logP 0.18			EXP	GAMBERINI,G ET AL. (1989)
VP 2.66E-011	mm Hg	25	EST	NEELY,WB & BLAU,GE (1985)
DC	pKa			
HL 5.40E-014	atm m3/mol	25	EST	MEYLAN,WM & HOWARD,PH (1991)
OH 1.04E-010	cm3/molc sec	25	EST	MEYLAN,WM & HOWARD,PH (1993)

1453

CAS #: 101064-06-6 — BUTANOIC ACID, 4-(6-METHOXY-4H-1,2,4-BENZOTHIADI

Formula: $C_{12}H_{14}N_2O_5S$

Mol Weight: 298.32

MP (deg C):   **FP (deg C):**

BP (deg C):

BP pressure (mm Hg):

Property/Value	Units	Temp	Data Type	Reference
WS 2.46E+004	mg/L	25	EST	MEYLAN,WM ET AL. (1996)
logP 0.12			EXP	GAMBERINI,G ET AL. (1989)
VP 1.15E-011	mm Hg	25	EST	NEELY,WB & BLAU,GE (1985)
DC	pKa			
HL 7.17E-014	atm m3/mol	25	EST	MEYLAN,WM & HOWARD,PH (1991)
OH 1.06E-010	cm3/molc sec	25	EST	MEYLAN,WM & HOWARD,PH (1993)

CAS #: 101064-07-7 — BUTANOIC ACID, 4-[6-(TRIFLUOROMETHYL)-4H-1,2,4-B

Formula: $C_{12}H_{11}F_3N_2O_4S$

Mol Weight: 336.29

MP (deg C):   **FP (deg C):**

BP (deg C):

BP pressure (mm Hg):

Property/Value	Units	Temp	Data Type	Reference
WS 1.62E+003	mg/L	25	EST	MEYLAN,WM ET AL. (1996)
logP 1.24			EXP	GAMBERINI,G ET AL. (1989)
VP 4.94E-011	mm Hg	25	EST	NEELY,WB & BLAU,GE (1985)
DC	pKa			
HL 1.05E-011	atm m3/mol	25	EST	MEYLAN,WM & HOWARD,PH (1991)
OH 5.59E-012	cm3/molc sec	25	EST	MEYLAN,WM & HOWARD,PH (1993)

CAS #: 101080-58-4 — 2-CHLORO-6-METHYLBENZAMIDE

Formula: C_8H_8ClNO

Mol Weight: 169.61

MP (deg C):   **FP (deg C):**

BP (deg C):

BP pressure (mm Hg):

Property/Value	Units	Temp	Data Type	Reference
WS 1.07E+004	mg/L	25	EST	MEYLAN,WM ET AL. (1996)
logP 0.89			EXP	NAKAGAWA,Y ET AL. (1992)
VP 6.03E-005	mm Hg	25	EST	NEELY,WB & BLAU,GE (1985)
DC	pKa			
HL 1.81E-009	atm m3/mol	25	EST	MEYLAN,WM & HOWARD,PH (1991)
OH 5.40E-012	cm3/molc sec	25	EST	MEYLAN,WM & HOWARD,PH (1993)

CAS #: 101126-54-9 — 1,4-NAPHTHALENEDIONE, 2,3-DIHYDRO-2-[[[2-[4-HYDR

Formula: $C_{23}H_{27}NO_4$

Mol Weight: 381.48

MP (deg C):   **FP (deg C):**

BP (deg C):

BP pressure (mm Hg):

Property/Value	Units	Temp	Data Type	Reference
WS 3.62E+001	mg/L	25	EST	MEYLAN,WM ET AL. (1996)
logP 4.25			EXP	DALLET,P ET AL. (1985)
VP 4.13E-012	mm Hg	25	EST	NEELY,WB & BLAU,GE (1985)
DC	pKa			
HL 3.03E-018	atm m3/mol	25	EST	MEYLAN,WM & HOWARD,PH (1991)
OH 2.31E-010	cm3/molc sec	25	EST	MEYLAN,WM & HOWARD,PH (1993)

CAS #: 101126-55-0 — ACETIC ACID, 4-[2-(DIMETHYLAMINO)ETHOXY]-2-METHY

Formula: $C_{13}H_{19}NO_3$

Mol Weight: 237.30

MP (deg C):   **FP (deg C):**

BP (deg C):

BP pressure (mm Hg):

Property/Value	Units	Temp	Data Type	Reference
WS 1.13E+004	mg/L	25	EST	MEYLAN,WM ET AL. (1996)
logP 1.64			EXP	DALLET ET AL. (1985)
VP 2.45E-004	mm Hg	25	EST	NEELY,WB & BLAU,GE (1985)
DC	pKa			
HL 2.79E-009	atm m3/mol	25	EST	MEYLAN,WM & HOWARD,PH (1991)
OH 1.47E-010	cm3/molc sec	25	EST	MEYLAN,WM & HOWARD,PH (1993)

CAS #: 101250-97-9 — 1,3-DIMETHYL-2-NITROGUANADINE

Formula: $C_3H_8N_4O_2$

Mol Weight: 132.12

MP (deg C):   **FP (deg C):**

BP (deg C):

BP pressure (mm Hg):

Property/Value	Units	Temp	Data Type	Reference
WS 3.57E+005	mg/L	25	EST	MEYLAN,WM ET AL. (1996)
logP -0.70			EXP	HANSCH,C & LEO,AJ (1985)
VP 9.71E-002	mm Hg	25	EST	NEELY,WB & BLAU,GE (1985)
DC	pKa			
HL 2.17E-011	atm m3/mol	25	EST	MEYLAN,WM & HOWARD,PH (1991)
OH 1.29E-010	cm3/molc sec	25	EST	MEYLAN,WM & HOWARD,PH (1993)

CAS #: 101347-40-4 — PHOSPHINIC ACID, BIS(1-AZIRIDINYL)-, 2-(2-ETHOXY

Formula: $C_{10}H_{21}N_2O_4P$

Mol Weight: 264.26

MP (deg C):   **FP (deg C):**

BP (deg C):

BP pressure (mm Hg):

Property/Value	Units	Temp	Data Type	Reference
WS 1.62E+004	mg/L	25	EST	MEYLAN,WM ET AL. (1996)
logP 0.10			EXP	SANGSTER,J (1994)
VP 5.83E-005	mm Hg	25	EST	NEELY,WB & BLAU,GE (1985)
DC	pKa			
HL 1.45E-014	atm m3/mol	25	EST	MEYLAN,WM & HOWARD,PH (1991)
OH 6.36E-011	cm3/molc sec	25	EST	MEYLAN,WM & HOWARD,PH (1993)

CAS #: 101347-42-6 — PHOSPHINOTHIOIC ACID, BIS(1-AZIRIDINYL)-, O-[2-(

Formula: $C_{10}H_{21}N_2O_3PS$

Mol Weight: 280.33

MP (deg C):   **FP (deg C):**

BP (deg C):

BP pressure (mm Hg):

Property/Value	Units	Temp	Data Type	Reference
WS 7.56E+003	mg/L	25	EST	MEYLAN,WM ET AL. (1996)
logP 0.38			EXP	SANGSTER,J (1994)
VP 7.25E-005	mm Hg	25	EST	NEELY,WB & BLAU,GE (1985)
DC	pKa			
HL 6.72E-012	atm m3/mol	25	EST	MEYLAN,WM & HOWARD,PH (1991)
OH 1.17E-010	cm3/molc sec	25	EST	MEYLAN,WM & HOWARD,PH (1993)

CAS #: 101533-57-7				BENZONITRILE, 4-[(4-AMINOPHENYL)SULFONYL]-

Formula: $C_{13}H_{10}N_2O_2S$

Mol Weight: 258.30

MP (deg C): 　FP (deg C):

BP (deg C):

BP pressure (mm Hg):

Property/ Value	Units	Temp	Data Type	Reference
WS 4.69E+002	mg/L	25	EST	MEYLAN,WM ET AL. (1996)
logP 1.63			EXP	ALTOMARE,C ET AL. (1991)
VP 1.70E-008	mm Hg	25	EST	NEELY,WB & BLAU,GE (1985)
DC	pKa			
HL 8.51E-013	atm m3/mol	25	EST	MEYLAN,WM & HOWARD,PH (1991)
OH 2.31E-011	cm3/molc sec	25	EST	MEYLAN,WM & HOWARD,PH (1993)

CAS #: 101533-58-8				BENZAMIDE, 4-[(4-AMINOPHENYL)SULFONYL]-N,N-DIETH

Formula: $C_{17}H_{20}N_2O_3S$

Mol Weight: 332.42

MP (deg C): 　FP (deg C):

BP (deg C):

BP pressure (mm Hg):

Property/ Value	Units	Temp	Data Type	Reference
WS 4.66E+002	mg/L	25	EST	MEYLAN,WM ET AL. (1996)
logP 1.44			EXP	ALTOMARE,C ET AL. (1991)
VP 1.45E-010	mm Hg	25	EST	NEELY,WB & BLAU,GE (1985)
DC	pKa			
HL 3.08E-016	atm m3/mol	25	EST	MEYLAN,WM & HOWARD,PH (1991)
OH 4.64E-011	cm3/molc sec	25	EST	MEYLAN,WM & HOWARD,PH (1993)

CAS #: 101626-70-4				TALIPEXOLE

Formula: $C_{11}H_{17}N_3S$

Mol Weight: 223.34

MP (deg C): 　FP (deg C):

BP (deg C):

BP pressure (mm Hg):

Property/ Value	Units	Temp	Data Type	Reference
WS 1.59E+004	mg/L	25	EST	MEYLAN,WM ET AL. (1996)
logP 1.48			EXP	HANSCH,C ET AL. (1995)
VP 1.92E-005	mm Hg	25	EST	NEELY,WB & BLAU,GE (1985)
DC	pKa			
HL 6.29E-013	atm m3/mol	25	EST	MEYLAN,WM & HOWARD,PH (1991)
OH 1.26E-010	cm3/molc sec	25	EST	MEYLAN,WM & HOWARD,PH (1993)

CAS #: 101705-35-5				PENTANOIC ACID, 5-(1-NAPHTHALENYLOXY)-

Formula: $C_{15}H_{16}O_3$

Mol Weight: 244.29

MP (deg C): 　FP (deg C):

BP (deg C):

BP pressure (mm Hg):

Property/ Value	Units	Temp	Data Type	Reference
WS 2.42E+001	mg/L	25	EST	MEYLAN,WM ET AL. (1996)
logP 4.00			EXP	SANGSTER,J (1994)
VP 4.99E-007	mm Hg	25	EST	NEELY,WB & BLAU,GE (1985)
DC	pKa			
HL 5.41E-010	atm m3/mol	25	EST	MEYLAN,WM & HOWARD,PH (1991)
OH 2.14E-010	cm3/molc sec	25	EST	MEYLAN,WM & HOWARD,PH (1993)

CAS #: 101705-36-6				HEXANOIC ACID, 6-(1-NAPHTHALENYLOXY)-

Formula: $C_{16}H_{18}O_3$

Mol Weight: 258.32

MP (deg C): 　FP (deg C):

BP (deg C):

BP pressure (mm Hg):

Property/ Value	Units	Temp	Data Type	Reference
WS 7.58E+000	mg/L	25	EST	MEYLAN,WM ET AL. (1996)
logP 4.50			EXP	SANGSTER,J (1994)
VP 2.50E-007	mm Hg	25	EST	NEELY,WB & BLAU,GE (1985)
DC	pKa			
HL 7.19E-010	atm m3/mol	25	EST	MEYLAN,WM & HOWARD,PH (1991)
OH 2.15E-010	cm3/molc sec	25	EST	MEYLAN,WM & HOWARD,PH (1993)

CAS #: 101708-63-8				2-PYRROLIDINONE, 5-(3-PYRIDINYL)-, N-OXIDE, (S)-

Formula: $C_9H_{12}N_2O_2$

Mol Weight: 180.21

MP (deg C): 　FP (deg C):

BP (deg C):

BP pressure (mm Hg):

Property/ Value	Units	Temp	Data Type	Reference
WS 4.09E+005	mg/L	25	EST	MEYLAN,WM ET AL. (1996)
logP -1.01			EXP	LI,NY & GORROD,JW (1992)
VP 3.08E-007	mm Hg	25	EST	NEELY,WB & BLAU,GE (1985)
DC	pKa			
HL 1.52E-017	atm m3/mol	25	EST	MEYLAN,WM & HOWARD,PH (1991)
OH 1.85E-011	cm3/molc sec	25	EST	MEYLAN,WM & HOWARD,PH (1993)

CAS #: 101714-96-9				2-MONOIODODIBENZO-P-DIOXIN

Formula: $C_{12}H_7IO_2$

Mol Weight: 310.09

MP (deg C): 　FP (deg C):

BP (deg C):

BP pressure (mm Hg):

Property/ Value	Units	Temp	Data Type	Reference
WS 1.02E+005	mg/L	25	EST	MEYLAN,WM ET AL. (1996)
logP 5.51			EST	MEYLAN,WM & HOWARD,PH (1995)
VP 1.45E-005	mm Hg	25	EST	NEELY,WB & BLAU,GE (1985)
DC	pKa			
HL 2.71E-006	atm m3/mol	25	EST	MEYLAN,WM & HOWARD,PH (1991)
OH 8.16E-012	cm3/molc sec	25	EST	MEYLAN,WM & HOWARD,PH (1993)

CAS #: 101820-62-6				BENZAMIDE, 2-[(2,3-DIHYDROXYPROPYL)AMINO]-

Formula: $C_{10}H_{14}N_2O_3$

Mol Weight: 210.23

MP (deg C): 　FP (deg C):

BP (deg C):

BP pressure (mm Hg):

Property/ Value	Units	Temp	Data Type	Reference
WS 9.15E+003	mg/L	25	EST	MEYLAN,WM ET AL. (1996)
logP -0.80			EXP	LISCIANI,R ET AL. (1986)
VP 2.17E-010	mm Hg	25	EST	NEELY,WB & BLAU,GE (1985)
DC	pKa			
HL 4.05E-018	atm m3/mol	25	EST	MEYLAN,WM & HOWARD,PH (1991)
OH 6.61E-011	cm3/molc sec	25	EST	MEYLAN,WM & HOWARD,PH (1993)

CAS #:	101820-63-7			BENZAMIDE, 2-[(2-HYDROXYPROPYL)AMINO]-

Formula: $C_{10}H_{14}N_2O_2$

Mol Weight: 194.24

MP (deg C): FP (deg C):

BP (deg C):

BP pressure (mm Hg):

Property/Value	Units	Temp	Data Type	Reference
WS 1.38E+004	mg/L	25	EST	MEYLAN,WM ET AL. (1996)
logP 0.62			EXP	LISCIANI,R ET AL. (1986)
VP 2.33E-008	mm Hg	25	EST	NEELY,WB & BLAU,GE (1985)
DC	pKa			
HL 1.11E-016	atm m3/mol	25	EST	MEYLAN,WM & HOWARD,PH (1991)
OH 6.02E-011	cm3/molc sec	25	EST	MEYLAN,WM & HOWARD,PH (1993)

CAS #:	101820-64-8			BENZAMIDE, 2-[(2-BUTOXYETHYL)AMINO]-

Formula: $C_{13}H_{20}N_2O_2$

Mol Weight: 236.32

MP (deg C): FP (deg C):

BP (deg C):

BP pressure (mm Hg):

Property/Value	Units	Temp	Data Type	Reference
WS 3.24E+002	mg/L	25	EST	MEYLAN,WM ET AL. (1996)
logP 2.27			EXP	LISCIANI,R ET AL. (1986)
VP 3.28E-007	mm Hg	25	EST	NEELY,WB & BLAU,GE (1985)
DC	pKa			
HL 6.25E-014	atm m3/mol	25	EST	MEYLAN,WM & HOWARD,PH (1991)
OH 9.06E-011	cm3/molc sec	25	EST	MEYLAN,WM & HOWARD,PH (1993)

CAS #:	101820-65-9			BENZAMIDE, 2-[(2-BUTOXYETHYL)AMINO]-5-CHLORO-

Formula: $C_{13}H_{19}ClN_2O_2$

Mol Weight: 270.76

MP (deg C): FP (deg C):

BP (deg C):

BP pressure (mm Hg):

Property/Value	Units	Temp	Data Type	Reference
WS 5.06E+001	mg/L	25	EST	MEYLAN,WM ET AL. (1996)
logP 2.99			EXP	LISCIANI,R ET AL. (1986)
VP 7.47E-008	mm Hg	25	EST	NEELY,WB & BLAU,GE (1985)
DC	pKa			
HL 4.63E-014	atm m3/mol	25	EST	MEYLAN,WM & HOWARD,PH (1991)
OH 6.31E-011	cm3/molc sec	25	EST	MEYLAN,WM & HOWARD,PH (1993)

CAS #:	101820-66-0			BENZAMIDE, 2-[[2-(2-METHYLPHENOXY)ETHYL]AMINO]-

Formula: $C_{16}H_{18}N_2O_2$

Mol Weight: 270.33

MP (deg C): FP (deg C):

BP (deg C):

BP pressure (mm Hg):

Property/Value	Units	Temp	Data Type	Reference
WS 5.61E+001	mg/L	25	EST	MEYLAN,WM ET AL. (1996)
logP 2.94			EXP	LISCIANI,R ET AL. (1986)
VP 8.95E-009	mm Hg	25	EST	NEELY,WB & BLAU,GE (1985)
DC	pKa			
HL 1.09E-014	atm m3/mol	25	EST	MEYLAN,WM & HOWARD,PH (1991)
OH 1.05E-010	cm3/molc sec	25	EST	MEYLAN,WM & HOWARD,PH (1993)

CAS #:	101870-22-8			2-PROPOXYPYRIDINE

Formula: $C_8H_{11}NO$

Mol Weight: 137.18

MP (deg C): FP (deg C):

BP (deg C):

BP pressure (mm Hg):

Property/Value	Units	Temp	Data Type	Reference
WS 7.98E+002	mg/L	25	EST	MEYLAN,WM ET AL. (1996)
logP 2.38			EXP	YAMAGAMI,C ET AL. (1990A)
VP 3.44E-001	mm Hg	25	EST	NEELY,WB & BLAU,GE (1985)
DC	pKa			
HL 6.70E-006	atm m3/mol	25	EST	MEYLAN,WM & HOWARD,PH (1991)
OH 1.45E-011	cm3/molc sec	25	EST	MEYLAN,WM & HOWARD,PH (1993)

CAS #:	101881-19-0			1-LAURYL-4-METHOXYCARBONYL-2-PYRROLIDONE

Formula: $C_{18}H_{33}NO_3$

Mol Weight: 311.47

MP (deg C): FP (deg C):

BP (deg C):

BP pressure (mm Hg):

Property/Value	Units	Temp	Data Type	Reference
WS 4.04E+000	mg/L	25	EST	MEYLAN,WM ET AL. (1996)
logP 4.00			EXP	SASAKI,H ET AL. (1988)
VP 1.95E-007	mm Hg	25	EST	NEELY,WB & BLAU,GE (1985)
DC	pKa			
HL 1.70E-009	atm m3/mol	25	EST	MEYLAN,WM & HOWARD,PH (1991)
OH 4.39E-011	cm3/molc sec	25	EST	MEYLAN,WM & HOWARD,PH (1993)

CAS #:	102089-65-6			BUTANAMIDE, 4-AMINO-N-(2,6-DIMETHYLPHENYL)-

Formula: $C_{12}H_{18}N_2O$

Mol Weight: 206.29

MP (deg C): FP (deg C):

BP (deg C):

BP pressure (mm Hg):

Property/Value	Units	Temp	Data Type	Reference
WS 1.01E+004	mg/L	25	EST	MEYLAN,WM ET AL. (1996)
logP 0.71			EXP	SANGSTER,J (1993)
VP 1.20E-006	mm Hg	25	EST	NEELY,WB & BLAU,GE (1985)
DC	pKa			
HL 1.36E-012	atm m3/mol	25	EST	MEYLAN,WM & HOWARD,PH (1991)
OH 5.60E-011	cm3/molc sec	25	EST	MEYLAN,WM & HOWARD,PH (1993)

CAS #:	102089-66-7			PENTANAMIDE, 5-AMINO-N-(2,6-DIMETHYLPHENYL)-

Formula: $C_{13}H_{20}N_2O$

Mol Weight: 220.32

MP (deg C): FP (deg C):

BP (deg C):

BP pressure (mm Hg):

Property/Value	Units	Temp	Data Type	Reference
WS 4.61E+003	mg/L	25	EST	MEYLAN,WM ET AL. (1996)
logP 1.02			EXP	SANGSTER,J (1993)
VP 5.15E-007	mm Hg	25	EST	NEELY,WB & BLAU,GE (1985)
DC	pKa			
HL 1.81E-012	atm m3/mol	25	EST	MEYLAN,WM & HOWARD,PH (1991)
OH 5.74E-011	cm3/molc sec	25	EST	MEYLAN,WM & HOWARD,PH (1993)

CAS #: 102089-67-8				HEXANAMIDE, 6-AMINO-N-(2,6-DIMETHYLPHENYL)-

Formula: $C_{14}H_{22}N_2O$

Mol Weight: 234.34

MP (deg C):   FP (deg C):

BP (deg C):

BP pressure (mm Hg):

Property/Value	Units	Temp	Data Type	Reference
WS 2.28E+003	mg/L	25	EST	MEYLAN,WM ET AL. (1996)
logP 1.29			EXP	SANGSTER,J (1993)
VP 2.20E-007	mm Hg	25	EST	NEELY,WB & BLAU,GE (1985)
DC	pKa			
HL 2.40E-012	atm m3/mol	25	EST	MEYLAN,WM & HOWARD,PH (1991)
OH 5.88E-011	cm3/molc sec	25	EST	MEYLAN,WM & HOWARD,PH (1993)

CAS #: 102089-68-9				PENTANAMIDE, 4-AMINO-N-(2,6-DIMETHYLPHENYL)-

Formula: $C_{13}H_{20}N_2O$

Mol Weight: 220.32

MP (deg C):   FP (deg C):

BP (deg C):

BP pressure (mm Hg):

Property/Value	Units	Temp	Data Type	Reference
WS 4.10E+003	mg/L	25	EST	MEYLAN,WM ET AL. (1996)
logP 1.08			EXP	SANGSTER,J (1993)
VP 8.93E-007	mm Hg	25	EST	NEELY,WB & BLAU,GE (1985)
DC	pKa			
HL 1.81E-012	atm m3/mol	25	EST	MEYLAN,WM & HOWARD,PH (1991)
OH 6.76E-011	cm3/molc sec	25	EST	MEYLAN,WM & HOWARD,PH (1993)

CAS #: 102089-69-0				HEXANAMIDE, 6-(DIETHYLAMINO)-N-(2,6-DIMETHYLPHEN

Formula: $C_{18}H_{30}N_2O$

Mol Weight: 290.45

MP (deg C):   FP (deg C):

BP (deg C):

BP pressure (mm Hg):

Property/Value	Units	Temp	Data Type	Reference
WS 4.22E+001	mg/L	25	EST	MEYLAN,WM ET AL. (1996)
logP 2.95			EXP	SANGSTER,J (1993)
VP 3.57E-008	mm Hg	25	EST	NEELY,WB & BLAU,GE (1985)
DC	pKa			
HL 2.04E-011	atm m3/mol	25	EST	MEYLAN,WM & HOWARD,PH (1991)
OH 1.21E-010	cm3/molc sec	25	EST	MEYLAN,WM & HOWARD,PH (1993)

CAS #: 102151-07-5				2-PROPANOL, 1-(PROPYLAMINO)-3-[4-[5-(2-THIENYL)-

Formula: $C_{19}H_{23}N_3O_2S$

Mol Weight: 357.48

MP (deg C):   FP (deg C):

BP (deg C):

BP pressure (mm Hg):

Property/Value	Units	Temp	Data Type	Reference
WS 2.14E+002	mg/L	25	EST	MEYLAN,WM ET AL. (1996)
logP 2.84			EXP	SANGSTER,J (1993)
VP 3.70E-016	mm Hg	25	EST	NEELY,WB & BLAU,GE (1985)
DC	pKa			
HL 1.83E-017	atm m3/mol	25	EST	MEYLAN,WM & HOWARD,PH (1991)
OH 1.80E-010	cm3/molc sec	25	EST	MEYLAN,WM & HOWARD,PH (1993)

CAS #: 102151-09-7				2-PROPANOL, 1-[(2-ETHOXYETHYL)AMINO]-3-[4-[4-(2-

Formula: $C_{20}H_{25}N_3O_3S$

Mol Weight: 387.50

MP (deg C):   FP (deg C):

BP (deg C):

BP pressure (mm Hg):

Property/Value	Units	Temp	Data Type	Reference
WS 4.46E+002	mg/L	25	EST	MEYLAN,WM ET AL. (1996)
logP 2.25			EXP	SANGSTER,J (1993)
VP 4.82E-017	mm Hg	25	EST	NEELY,WB & BLAU,GE (1985)
DC	pKa			
HL 2.14E-019	atm m3/mol	25	EST	MEYLAN,WM & HOWARD,PH (1991)
OH 2.11E-010	cm3/molc sec	25	EST	MEYLAN,WM & HOWARD,PH (1993)

CAS #: 102151-10-0				2-PROPANOL, 1-[[2-(4-PYRIDINYL)ETHYL]AMINO]-3-[4

Formula: $C_{23}H_{24}N_4O_2S$

Mol Weight: 420.54

MP (deg C):   FP (deg C):

BP (deg C):

BP pressure (mm Hg):

Property/Value	Units	Temp	Data Type	Reference
WS 5.56E+003	mg/L	25	EST	MEYLAN,WM ET AL. (1996)
logP 2.25			EXP	SANGSTER,J (1993)
VP 4.25E-019	mm Hg	25	EST	NEELY,WB & BLAU,GE (1985)
DC	pKa			
HL 1.45E-021	atm m3/mol	25	EST	MEYLAN,WM & HOWARD,PH (1991)
OH 1.80E-010	cm3/molc sec	25	EST	MEYLAN,WM & HOWARD,PH (1993)

CAS #: 102151-13-3				UREA, N-BUTYL-N'-[2-[[3-[4-(5-METHYL-1H-IMIDAZOL

Formula: $C_{20}H_{31}N_5O_3$

Mol Weight: 389.50

MP (deg C):   FP (deg C):

BP (deg C):

BP pressure (mm Hg):

Property/Value	Units	Temp	Data Type	Reference
WS 3.10E+002	mg/L	25	EST	MEYLAN,WM ET AL. (1996)
logP 1.24			EXP	SANGSTER,J (1993)
VP 4.78E-018	mm Hg	25	EST	NEELY,WB & BLAU,GE (1985)
DC	pKa			
HL 2.03E-023	atm m3/mol	25	EST	MEYLAN,WM & HOWARD,PH (1991)
OH 1.99E-010	cm3/molc sec	25	EST	MEYLAN,WM & HOWARD,PH (1993)

CAS #: 102151-23-5				2-PROPANOL, 1-[[2-(3,4-DIMETHOXYPHENYL)ETHYL]AMI

Formula: $C_{24}H_{31}N_3O_5$

Mol Weight: 441.53

MP (deg C):   FP (deg C):

BP (deg C):

BP pressure (mm Hg):

Property/Value	Units	Temp	Data Type	Reference
WS 1.36E+003	mg/L	25	EST	MEYLAN,WM ET AL. (1996)
logP 1.29			EXP	SANGSTER,J (1993)
VP 1.51E-017	mm Hg	25	EST	NEELY,WB & BLAU,GE (1985)
DC	pKa			
HL 1.26E-021	atm m3/mol	25	EST	MEYLAN,WM & HOWARD,PH (1991)
OH 2.78E-010	cm3/molc sec	25	EST	MEYLAN,WM & HOWARD,PH (1993)

CAS #: 102152-14-7 — 2-PROPANOL, 1-[[2-(3,4-DIMETHOXYPHENYL)ETHYL]AMI

Formula: $C_{24}H_{31}N_3O_5$
Mol Weight: 441.53
MP (deg C): | FP (deg C):
BP (deg C):
BP pressure (mm Hg):

Property/Value	Units	Temp	Data Type	Reference
WS 4.17E+002	mg/L	25	EST	MEYLAN,WM ET AL. (1996)
logP 1.89			EXP	SANGSTER,J (1993)
VP 4.43E-018	mm Hg	25	EST	NEELY,WB & BLAU,GE (1985)
DC	pKa			
HL 6.08E-021	atm m3/mol	25	EST	MEYLAN,WM & HOWARD,PH (1991)
OH 2.51E-010	cm3/molc sec	25	EST	MEYLAN,WM & HOWARD,PH (1993)

CAS #: 102152-21-6 — ETHANONE, 1-[2-[4-[3-[[2-(3,4-DIMETHOXYPHENYL)ET

Formula: $C_{25}H_{31}N_3O_5$
Mol Weight: 453.54
MP (deg C): | FP (deg C):
BP (deg C):
BP pressure (mm Hg):

Property/Value	Units	Temp	Data Type	Reference
WS 2.14E+002	mg/L	25	EST	MEYLAN,WM ET AL. (1996)
logP 2.14			EXP	SANGSTER,J (1993)
VP 6.37E-019	mm Hg	25	EST	NEELY,WB & BLAU,GE (1985)
DC	pKa			
HL 1.87E-022	atm m3/mol	25	EST	MEYLAN,WM & HOWARD,PH (1991)
OH 2.02E-010	cm3/molc sec	25	EST	MEYLAN,WM & HOWARD,PH (1993)

CAS #: 102152-24-9 — 2-PROPANOL, 1-[[2-(3,4-DIMETHOXYPHENYL)ETHYL]AMI

Formula: $C_{25}H_{33}N_3O_5$
Mol Weight: 455.56
MP (deg C): | FP (deg C):
BP (deg C):
BP pressure (mm Hg):

Property/Value	Units	Temp	Data Type	Reference
WS 2.44E+002	mg/L	25	EST	MEYLAN,WM ET AL. (1996)
logP 2.06			EXP	SANGSTER,J (1993)
VP 1.61E-018	mm Hg	25	EST	NEELY,WB & BLAU,GE (1985)
DC	pKa			
HL 1.60E-021	atm m3/mol	25	EST	MEYLAN,WM & HOWARD,PH (1991)
OH 2.41E-010	cm3/molc sec	25	EST	MEYLAN,WM & HOWARD,PH (1993)

CAS #: 102152-27-2 — 2-PROPANOL, 1-[4-(5-CHLORO-1H-IMIDAZOL-2-YL)PHEN

Formula: $C_{22}H_{26}ClN_3O_4$
Mol Weight: 431.92
MP (deg C): | FP (deg C):
BP (deg C):
BP pressure (mm Hg):

Property/Value	Units	Temp	Data Type	Reference
WS 1.76E+002	mg/L	25	EST	MEYLAN,WM ET AL. (1996)
logP 2.40			EXP	SANGSTER,J (1993)
VP 1.87E-017	mm Hg	25	EST	NEELY,WB & BLAU,GE (1985)
DC	pKa			
HL 6.90E-020	atm m3/mol	25	EST	MEYLAN,WM & HOWARD,PH (1991)
OH 1.82E-010	cm3/molc sec	25	EST	MEYLAN,WM & HOWARD,PH (1993)

CAS #: 102152-28-3 — 1H-IMIDAZOLE-5-CARBOXYLIC ACID, 2-[4-[3-[[2-(3,4

Formula: $C_{25}H_{31}N_3O_6$
Mol Weight: 469.54
MP (deg C): | FP (deg C):
BP (deg C):
BP pressure (mm Hg):

Property/Value	Units	Temp	Data Type	Reference
WS 1.24E+002	mg/L	25	EST	MEYLAN,WM ET AL. (1996)
logP 2.30			EXP	SANGSTER,J (1993)
VP 4.68E-019	mm Hg	25	EST	NEELY,WB & BLAU,GE (1985)
DC	pKa			
HL 7.96E-022	atm m3/mol	25	EST	MEYLAN,WM & HOWARD,PH (1991)
OH 1.74E-010	cm3/molc sec	25	EST	MEYLAN,WM & HOWARD,PH (1993)

CAS #: 102262-55-5 — DES-ISOPROPYL CLOFAZIMINE

Formula: $C_{24}H_{16}Cl_2N_4$
Mol Weight: 431.33
MP (deg C): | FP (deg C):
BP (deg C):
BP pressure (mm Hg):

Property/Value	Units	Temp	Data Type	Reference
WS 4.07E+000	mg/L	25	EST	MEYLAN,WM ET AL. (1996)
logP 3.14			EXP	QUIGLEY,JM ET AL. (1990)
VP 2.34E-012	mm Hg	25	EST	NEELY,WB & BLAU,GE (1985)
DC	pKa			
HL 9.29E-012	atm m3/mol	25	EST	MEYLAN,WM & HOWARD,PH (1991)
OH 1.34E-010	cm3/molc sec	25	EST	MEYLAN,WM & HOWARD,PH (1993)

CAS #: 102273-19-8 — BENZAMIDE, 2-[(ACETYLOXY)METHOXY]-N-(4-MORPHOLIN

Formula: $C_{15}H_{20}N_2O_5$
Mol Weight: 308.34
MP (deg C): | FP (deg C):
BP (deg C):
BP pressure (mm Hg):

Property/Value	Units	Temp	Data Type	Reference
WS 3.16E+002	mg/L	25	EST	MEYLAN,WM ET AL. (1996)
logP 0.27			EXP	BUNDGAARD,H ET AL. (1986)
VP 3.99E-009	mm Hg	25	EST	NEELY,WB & BLAU,GE (1985)
DC	pKa			
HL 7.36E-016	atm m3/mol	25	EST	MEYLAN,WM & HOWARD,PH (1991)
OH 2.11E-010	cm3/molc sec	25	EST	MEYLAN,WM & HOWARD,PH (1993)

CAS #: 102273-20-1 — BUTANOIC ACID, [2-[[[(4-MORPHOLINYLMETHYL)AMINO]C

Formula: $C_{17}H_{24}N_2O_5$
Mol Weight: 336.39
MP (deg C): | FP (deg C):
BP (deg C):
BP pressure (mm Hg):

Property/Value	Units	Temp	Data Type	Reference
WS 5.37E+002	mg/L	25	EST	MEYLAN,WM ET AL. (1996)
logP 1.34			EXP	BUNDGAARD,H ET AL. (1986)
VP 7.80E-010	mm Hg	25	EST	NEELY,WB & BLAU,GE (1985)
DC	pKa			
HL 1.30E-015	atm m3/mol	25	EST	MEYLAN,WM & HOWARD,PH (1991)
OH 2.13E-010	cm3/molc sec	25	EST	MEYLAN,WM & HOWARD,PH (1993)

CAS #:	102417-86-7			
Formula: $C_{13}H_{17}N$				PYRIDINE, 1,2,3,6-TETRAHYDRO-1-METHYL-4-(2-METHY
Mol Weight: 187.29				
MP (deg C):		FP (deg C):		
BP (deg C):				
BP pressure (mm Hg):				

	Property/Value	Units	Temp	Data Type	Reference
WS	1.16E+003	mg/L	25	EST	MEYLAN,WM ET AL. (1996)
logP	3.10			EXP	ALTOMARE,CA ET AL. (1992)
VP	2.26E-003	mm Hg	25	EST	NEELY,WB & BLAU,GE (1985)
DC		pKa			
HL	1.64E-006	atm m3/mol	25	EST	MEYLAN,WM & HOWARD,PH (1991)
OH	1.80E-010	cm3/molc sec	25	EST	MEYLAN,WM & HOWARD,PH (1993)

CAS #:	102587-48-4			
Formula: $C_{16}H_{18}N_2O$				UREA, N,N-DIMETHYL-N'[3-(PHENYLMETHYL)PHENYL]-
Mol Weight: 254.33				
MP (deg C):		FP (deg C):		
BP (deg C):				
BP pressure (mm Hg):				

	Property/Value	Units	Temp	Data Type	Reference
WS	8.57E+001	mg/L	25	EST	MEYLAN,WM ET AL. (1996)
logP	2.83			EXP	MITSUTAKE,KI ET AL. (1986)
VP	2.04E-007	mm Hg	25	EST	NEELY,WB & BLAU,GE (1985)
DC		pKa			
HL	8.65E-011	atm m3/mol	25	EST	MEYLAN,WM & HOWARD,PH (1991)
OH	1.14E-010	cm3/molc sec	25	EST	MEYLAN,WM & HOWARD,PH (1993)

CAS #:	102587-49-5			
Formula: $C_{17}H_{20}N_2O$				UREA, N,N-DIMETHYL-N'-[3-(2-PHENYLETHYL)PHENYL]-
Mol Weight: 268.36				
MP (deg C):		FP (deg C):		
BP (deg C):				
BP pressure (mm Hg):				

	Property/Value	Units	Temp	Data Type	Reference
WS	3.39E+001	mg/L	25	EST	MEYLAN,WM ET AL. (1996)
logP	3.21			EXP	HANSCH,C ET AL. (1995)
VP	8.70E-008	mm Hg	25	EST	NEELY,WB & BLAU,GE (1985)
DC		pKa			
HL	1.15E-010	atm m3/mol	25	EST	MEYLAN,WM & HOWARD,PH (1991)
OH	1.15E-010	cm3/molc sec	25	EST	MEYLAN,WM & HOWARD,PH (1993)

CAS #:	102587-50-8			
Formula: $C_9H_{14}ClN_5$				1,3,5-TRIAZINE-2,4-DIAMINE, 6-CHLORO-N-CYCLOBUTY
Mol Weight: 227.70				
MP (deg C):		FP (deg C):		
BP (deg C):				
BP pressure (mm Hg):				

	Property/Value	Units	Temp	Data Type	Reference
WS	1.35E+002	mg/L	25	EST	MEYLAN,WM ET AL. (1996)
logP	2.77			EXP	MITSUTAKE,KI ET AL. (1986)
VP	2.15E-005	mm Hg	25	EST	NEELY,WB & BLAU,GE (1985)
DC		pKa			
HL	2.62E-009	atm m3/mol	25	EST	MEYLAN,WM & HOWARD,PH (1991)
OH	1.78E-011	cm3/molc sec	25	EST	MEYLAN,WM & HOWARD,PH (1993)

CAS #:	102587-54-2			
Formula: $C_{12}H_{22}ClN_5$				1,3,5-TRIAZINE-2,4-DIAMINE, 6-CHLORO-N-(1-METHYL
Mol Weight: 271.80				
MP (deg C):		FP (deg C):		
BP (deg C):				
BP pressure (mm Hg):				

	Property/Value	Units	Temp	Data Type	Reference
WS	1.60E+000	mg/L	25	EST	MEYLAN,WM ET AL. (1996)
logP	4.74			EXP	MITSUTAKE,KI ET AL. (1986)
VP	3.99E-006	mm Hg	25	EST	NEELY,WB & BLAU,GE (1985)
DC		pKa			
HL	1.39E-008	atm m3/mol	25	EST	MEYLAN,WM & HOWARD,PH (1991)
OH	3.69E-011	cm3/molc sec	25	EST	MEYLAN,WM & HOWARD,PH (1993)

CAS #:	102626-89-1			
Formula: $C_5H_8N_4O_2$				UREA, N-METHYL-N'-(METHYLFURAZANYL)-
Mol Weight: 156.15				
MP (deg C):		FP (deg C):		
BP (deg C):				
BP pressure (mm Hg):				

	Property/Value	Units	Temp	Data Type	Reference
WS	1.94E+003	mg/L	25	EST	MEYLAN,WM ET AL. (1996)
logP	0.30			EXP	CALVINO,R ET AL. (1992)
VP	2.31E-004	mm Hg	25	EST	NEELY,WB & BLAU,GE (1985)
DC		pKa			
HL	1.71E-011	atm m3/mol	25	EST	MEYLAN,WM & HOWARD,PH (1991)
OH	5.69E-012	cm3/molc sec	25	EST	MEYLAN,WM & HOWARD,PH (1993)

CAS #:	102632-22-4			
Formula: $C_{18}H_{20}O_6$				1,4-NAPHTHALENEDIOL, 2,3-DIMETHOXY-6,7-DIMETHYL-
Mol Weight: 332.36				
MP (deg C):		FP (deg C):		
BP (deg C):				
BP pressure (mm Hg):				

	Property/Value	Units	Temp	Data Type	Reference
WS	1.65E+001	mg/L	25	EST	MEYLAN,WM ET AL. (1996)
logP	3.14			EXP	SANGSTER,J (1993)
VP	3.36E-007	mm Hg	25	EST	NEELY,WB & BLAU,GE (1985)
DC		pKa			
HL	3.24E-010	atm m3/mol	25	EST	MEYLAN,WM & HOWARD,PH (1991)
OH	2.02E-010	cm3/molc sec	25	EST	MEYLAN,WM & HOWARD,PH (1993)

CAS #:	102632-30-4			
Formula: $C_8H_8ClN_5O_2$				HYDRAZINECARBOXIMIDAMIDE, 2-[(4-CHLORO-3-NITROPH
Mol Weight: 241.64				
MP (deg C):		FP (deg C):		
BP (deg C):				
BP pressure (mm Hg):				

	Property/Value	Units	Temp	Data Type	Reference
WS	3.78E+003	mg/L	25	EST	MEYLAN,WM ET AL. (1996)
logP	0.53			EXP	SANGSTER,J (1994)
VP	7.13E-007	mm Hg	25	EST	NEELY,WB & BLAU,GE (1985)
DC		pKa			
HL	1.48E-014	atm m3/mol	25	EST	MEYLAN,WM & HOWARD,PH (1991)
OH	8.61E-011	cm3/molc sec	25	EST	MEYLAN,WM & HOWARD,PH (1993)

CAS #: 102632-31-5 — HYDRAZINECARBOXIMIDAMIDE, 2-[(2-NITROPHENYL)METH

Formula: $C_8H_9N_5O_2$

Mol Weight: 207.19

MP (deg C): FP (deg C):

BP (deg C):

BP pressure (mm Hg):

Property/Value	Units	Temp	Data Type	Reference
WS 5.77E+003	mg/L	25	EST	MEYLAN,WM ET AL. (1996)
logP 0.53			EXP	SOMAN,G ET AL. (1986)
VP 3.13E-006	mm Hg	25	EST	NEELY,WB & BLAU,GE (1985)
DC	pKa			
HL 1.99E-014	atm m3/mol	25	EST	MEYLAN,WM & HOWARD,PH (1991)
OH 8.65E-011	cm3/molc sec	25	EST	MEYLAN,WM & HOWARD,PH (1993)

CAS #: 103066-37-1 — 4,7-METHANO-1H-ISOINDOLE-1,3(2H)-DIONE, 3A,4,7,7

Formula: $C_{21}H_{27}N_5O_2$

Mol Weight: 381.48

MP (deg C): FP (deg C):

BP (deg C):

BP pressure (mm Hg):

Property/Value	Units	Temp	Data Type	Reference
WS 2.75E+002	mg/L	25	EST	MEYLAN,WM ET AL. (1996)
logP 1.36			EXP	SANGSTER,J (1993)
VP 8.35E-013	mm Hg	25	EST	NEELY,WB & BLAU,GE (1985)
DC	pKa			
HL 5.95E-014	atm m3/mol	25	EST	MEYLAN,WM & HOWARD,PH (1991)
OH 2.12E-010	cm3/molc sec	25	EST	MEYLAN,WM & HOWARD,PH (1993)

CAS #: 103188-47-2 — 2-PROPENAMIDE, 3-(3,4-DIHYDROXYPHENYL)-N-(2-PHEN

Formula: $C_{17}H_{17}NO_3$

Mol Weight: 283.33

MP (deg C): FP (deg C):

BP (deg C):

BP pressure (mm Hg):

Property/Value	Units	Temp	Data Type	Reference
WS 2.77E+002	mg/L	25	EST	MEYLAN,WM ET AL. (1996)
logP 2.72			EXP	NAITO,Y ET AL. (1991)
VP 6.81E-012	mm Hg	25	EST	NEELY,WB & BLAU,GE (1985)
DC	pKa			
HL 6.73E-019	atm m3/mol	25	EST	MEYLAN,WM & HOWARD,PH (1991)
OH 5.78E-011	cm3/molc sec	25	EST	MEYLAN,WM & HOWARD,PH (1993)

CAS #: 103188-54-1 — 1,10-DICHLOROPERFLUORO-2,9-DIMETHYLDECANE

Formula: $C_{12}H_6Cl_2F_{18}$

Mol Weight: 563.06

MP (deg C): FP (deg C):

BP (deg C):

BP pressure (mm Hg):

Property/Value	Units	Temp	Data Type	Reference
WS 9.80E-008	mg/L	25	EST	MEYLAN,WM ET AL. (1996)
logP 10.20			EST	MEYLAN,WM & HOWARD,PH (1995)
VP 4.67E+001	mm Hg	25	EST	NEELY,WB & BLAU,GE (1985)
DC	pKa			
HL 4.90E+004	atm m3/mol	25	EST	MEYLAN,WM & HOWARD,PH (1991)
OH 2.72E-013	cm3/molc sec	25	EST	MEYLAN,WM & HOWARD,PH (1993)

CAS #: 103188-55-2 — PERFLUORO-2,9-DIMETHYLDECANE

Formula: $C_{12}H_6F_{20}$

Mol Weight: 530.15

MP (deg C): FP (deg C):

BP (deg C):

BP pressure (mm Hg):

Property/Value	Units	Temp	Data Type	Reference
WS 5.52E-007	mg/L	25	EST	MEYLAN,WM ET AL. (1996)
logP 9.58			EST	MEYLAN,WM & HOWARD,PH (1995)
VP 9.49E+002	mm Hg	25	EST	NEELY,WB & BLAU,GE (1985)
DC	pKa			
HL 1.57E+006	atm m3/mol	25	EST	MEYLAN,WM & HOWARD,PH (1991)
OH 2.72E-013	cm3/molc sec	25	EST	MEYLAN,WM & HOWARD,PH (1993)

CAS #: 103360-35-6 — 24-NH2-5(5-I-2-MEO-BENZYL)PYRIMIDINE

Formula: $C_{12}H_{13}IN_4O$

Mol Weight: 356.17

MP (deg C): FP (deg C):

BP (deg C):

BP pressure (mm Hg):

Property/Value	Units	Temp	Data Type	Reference
WS 2.81E+001	mg/L	25	EST	MEYLAN,WM ET AL. (1996)
logP 2.70			EXP	HANSCH,C & LEO,AJ (1985)
VP 8.81E-009	mm Hg	25	EST	NEELY,WB & BLAU,GE (1985)
DC	pKa			
HL 1.58E-012	atm m3/mol	25	EST	MEYLAN,WM & HOWARD,PH (1991)
OH 1.20E-011	cm3/molc sec	25	EST	MEYLAN,WM & HOWARD,PH (1993)

CAS #: 103499-06-5 — 1,2,4-OXADOAZOL-5-ALDOXIME,5-BENZYL-

Formula: $C_{10}H_{10}N_3O_2$

Mol Weight: 204.21

MP (deg C): FP (deg C):

BP (deg C):

BP pressure (mm Hg):

Property/Value	Units	Temp	Data Type	Reference
WS 4.67E+002	mg/L	25	EST	MEYLAN,WM ET AL. (1996)
logP 2.29			EXP	HANSCH,C ET AL. (1995)
VP 4.13E-007	mm Hg	25	EST	NEELY,WB & BLAU,GE (1985)
DC	pKa			
HL 1.39E-010	atm m3/mol	25	EST	MEYLAN,WM & HOWARD,PH (1991)
OH 1.17E-011	cm3/molc sec	25	EST	MEYLAN,WM & HOWARD,PH (1993)

CAS #: 103499-08-7 — 1,2,4-OXADIAZOL-3-ALDOXIME,5-PHENYL-

Formula: $C_9H_8N_3O_2$

Mol Weight: 190.18

MP (deg C): FP (deg C):

BP (deg C):

BP pressure (mm Hg):

Property/Value	Units	Temp	Data Type	Reference
WS 4.43E+002	mg/L	25	EST	MEYLAN,WM ET AL. (1996)
logP 2.40			EXP	HANSCH,C ET AL. (1995)
VP 3.66E-007	mm Hg	25	EST	NEELY,WB & BLAU,GE (1985)
DC	pKa			
HL 1.20E-010	atm m3/mol	25	EST	MEYLAN,WM & HOWARD,PH (1991)
OH 9.33E-012	cm3/molc sec	25	EST	MEYLAN,WM & HOWARD,PH (1993)

CAS #: 103499-09-8	1,2,4-OXADIAZOLE-5-(DIETHYLAMINOETHYLTHIO)HYDROX
Formula: $C_{13}H_{25}N_4O_2S$	
Mol Weight: 301.43	
MP (deg C):	FP (deg C):
BP (deg C):	
BP pressure (mm Hg):	

Property/Value	Units	Temp	Data Type	Reference
WS 2.30E+002	mg/L	25	EST	MEYLAN,WM ET AL. (1996)
logP 2.02			EXP	HANSCH,C ET AL. (1994)
VP 6.88E-009	mm Hg	25	EST	NEELY,WB & BLAU,GE (1985)
DC	pKa			
HL 4.98E-014	atm m3/mol	25	EST	MEYLAN,WM & HOWARD,PH (1991)
OH 1.10E-010	cm3/molc sec	25	EST	MEYLAN,WM & HOWARD,PH (1993)

CAS #: 103499-11-2	1,2,4-OXADIAZOLE-5-(DIETHYLAMINOETHYL)THIOHYDROX
Formula: $C_{19}H_{22}N_4O_2S$	
Mol Weight: 370.48	
MP (deg C):	FP (deg C):
BP (deg C):	
BP pressure (mm Hg):	

Property/Value	Units	Temp	Data Type	Reference
WS 2.31E+000	mg/L	25	EST	MEYLAN,WM ET AL. (1996)
logP 3.87			EXP	HANSCH,C ET AL. (1995)
VP 8.05E-014	mm Hg	25	EST	NEELY,WB & BLAU,GE (1985)
DC	pKa			
HL 1.44E-016	atm m3/mol	25	EST	MEYLAN,WM & HOWARD,PH (1991)
OH 1.47E-010	cm3/molc sec	25	EST	MEYLAN,WM & HOWARD,PH (1993)

CAS #: 103595-57-9	(N-ME-BENZYL)AMET-(345-TRIMEO)PH SULFONE
Formula: $C_{19}H_{25}NO_5S$	
Mol Weight: 379.48	
MP (deg C):	FP (deg C):
BP (deg C):	
BP pressure (mm Hg):	

Property/Value	Units	Temp	Data Type	Reference
WS 5.74E+002	mg/L	25	EST	MEYLAN,WM ET AL. (1996)
logP 2.18			EXP	HANSCH,C & LEO,AJ (1985)
VP 1.03E-009	mm Hg	25	EST	NEELY,WB & BLAU,GE (1985)
DC	pKa			
HL 1.37E-014	atm m3/mol	25	EST	MEYLAN,WM & HOWARD,PH (1991)
OH 1.47E-010	cm3/molc sec	25	EST	MEYLAN,WM & HOWARD,PH (1993)

CAS #: 103748-25-0	1-METHYL-2-AMINO-5-CHLOROBENZIMIDAZOLE
Formula: $C_8H_8ClN_3$	
Mol Weight: 181.63	
MP (deg C):	FP (deg C):
BP (deg C):	
BP pressure (mm Hg):	

Property/Value	Units	Temp	Data Type	Reference
WS 8.87E+002	mg/L	25	EST	MEYLAN,WM ET AL. (1996)
logP 2.09			EXP	HANSCH,C ET AL. (1995)
VP 4.15E-006	mm Hg	25	EST	NEELY,WB & BLAU,GE (1985)
DC	pKa			
HL 2.05E-009	atm m3/mol	25	EST	MEYLAN,WM & HOWARD,PH (1991)
OH 1.55E-010	cm3/molc sec	25	EST	MEYLAN,WM & HOWARD,PH (1993)

CAS #: 103890-78-4	3,5-PYRIDINEDICARBOXYLIC ACID, 4-[2-[3-(1,1-DIME
Formula: $C_{26}H_{33}NO_6$	
Mol Weight: 455.56	
MP (deg C):	FP (deg C):
BP (deg C):	
BP pressure (mm Hg):	

Property/Value	Units	Temp	Data Type	Reference
WS 5.07E-001	mg/L	25	EST	MEYLAN,WM ET AL. (1996)
logP 5.20			EXP	SANGSTER,J (1994)
VP 2.16E-010	mm Hg	25	EST	NEELY,WB & BLAU,GE (1985)
DC	pKa			
HL 5.87E-014	atm m3/mol	25	EST	MEYLAN,WM & HOWARD,PH (1991)
OH 1.39E-010	cm3/molc sec	25	EST	MEYLAN,WM & HOWARD,PH (1993)

CAS #: 103910-47-0	BENZENEACETIC ACID, à-ETHYL-à-HYDROXY-4-(2-METHY
Formula: $C_{14}H_{20}O_3$	
Mol Weight: 236.31	
MP (deg C):	FP (deg C):
BP (deg C):	
BP pressure (mm Hg):	

Property/Value	Units	Temp	Data Type	Reference
WS 2.99E+002	mg/L	25	EST	MEYLAN,WM ET AL. (1996)
logP 3.37			EXP	KUCHAR,M ET AL. (1985)
VP 1.60E-007	mm Hg	25	EST	NEELY,WB & BLAU,GE (1985)
DC	pKa			
HL 3.14E-008	atm m3/mol	25	EST	MEYLAN,WM & HOWARD,PH (1991)
OH 1.41E-011	cm3/molc sec	25	EST	MEYLAN,WM & HOWARD,PH (1993)

CAS #: 103951-39-9	BUTANOIC ACID, [2-(AMINOCARBONYL)PHENOXY]METHYL
Formula: $C_{12}H_{15}NO_4$	
Mol Weight: 237.26	
MP (deg C):	FP (deg C):
BP (deg C):	
BP pressure (mm Hg):	

Property/Value	Units	Temp	Data Type	Reference
WS 1.06E+003	mg/L	25	EST	MEYLAN,WM ET AL. (1996)
logP 1.66			EXP	BUNDGAARD,H ET AL. (1986)
VP 5.95E-007	mm Hg	25	EST	NEELY,WB & BLAU,GE (1985)
DC	pKa			
HL 5.50E-013	atm m3/mol	25	EST	MEYLAN,WM & HOWARD,PH (1991)
OH 3.29E-011	cm3/molc sec	25	EST	MEYLAN,WM & HOWARD,PH (1993)

CAS #: 103970-46-3	CARBAMIC ACID, (2-IODOPHENYL)-, METHYL ESTER
Formula: $C_8H_8INO_2$	
Mol Weight: 277.06	
MP (deg C):	FP (deg C):
BP (deg C):	
BP pressure (mm Hg):	

Property/Value	Units	Temp	Data Type	Reference
WS 1.37E+002	mg/L	25	EST	MEYLAN,WM ET AL. (1996)
logP 2.44			EXP	TAKAHASHI,J ET AL. (1988)
VP 5.32E-004	mm Hg	25	EST	NEELY,WB & BLAU,GE (1985)
DC	pKa			
HL 5.06E-009	atm m3/mol	25	EST	MEYLAN,WM & HOWARD,PH (1991)
OH 1.43E-011	cm3/molc sec	25	EST	MEYLAN,WM & HOWARD,PH (1993)

CAS #: 103981-92-6				1-PIPERIDINYLOXY, 4-[[[[BIS(2,2-DIMETHYL-1-AZIRI
Formula: $C_{18}H_{36}N_5O_3P$				
Mol Weight: 401.49				
MP (deg C):		FP (deg C):		
BP (deg C):				
BP pressure (mm Hg):				

Property/Value	Units	Temp	Data Type	Reference
WS 1.28E+003	mg/L	25	EST	MEYLAN,WM ET AL. (1996)
logP 0.44			EXP	SOSNOVSKY,G ET AL. (1986)
VP 2.06E-008	mm Hg	25	EST	NEELY,WB & BLAU,GE (1985)
DC	pKa			
HL 2.31E-025	atm m3/mol	25	EST	MEYLAN,WM & HOWARD,PH (1991)
OH 1.61E-011	cm3/molc sec	25	EST	MEYLAN,WM & HOWARD,PH (1993)

CAS #: 103981-93-7				1-PYRROLIDINYLOXY, 3-[[[[BIS(2,2-DIMETHYL-1-AZIR
Formula: $C_{17}H_{34}N_5O_3P$				
Mol Weight: 387.47				
MP (deg C):		FP (deg C):		
BP (deg C):				
BP pressure (mm Hg):				

Property/Value	Units	Temp	Data Type	Reference
WS 1.53E+003	mg/L	25	EST	MEYLAN,WM ET AL. (1996)
logP 0.45			EXP	SOSNOVSKY,G ET AL. (1986)
VP 2.06E-008	mm Hg	25	EST	NEELY,WB & BLAU,GE (1985)
DC	pKa			
HL 1.74E-025	atm m3/mol	25	EST	MEYLAN,WM & HOWARD,PH (1991)
OH 1.60E-011	cm3/molc sec	25	EST	MEYLAN,WM & HOWARD,PH (1993)

CAS #: 103981-94-8				TEPA 4-(N-OH-PIPERIDINYL)-UREA DERIVATIV
Formula: $C_{18}H_{36}N_5O_3P$				
Mol Weight: 401.49				
MP (deg C):		FP (deg C):		
BP (deg C):				
BP pressure (mm Hg):				

Property/Value	Units	Temp	Data Type	Reference
WS 1.17E+003	mg/L	25	EST	MEYLAN,WM ET AL. (1996)
logP 0.48			EXP	SOSNOVSKY,G ET AL. (1986)
VP 1.78E-010	mm Hg	25	EST	NEELY,WB & BLAU,GE (1985)
DC	pKa			
HL 3.01E-022	atm m3/mol	25	EST	MEYLAN,WM & HOWARD,PH (1991)
OH 8.21E-011	cm3/molc sec	25	EST	MEYLAN,WM & HOWARD,PH (1993)

CAS #: 103981-95-9				TEPA 4-(N-OH-PYRROLDINYL)-UREA DERIVATIV
Formula: $C_{13}H_{26}N_5O_3P$				
Mol Weight: 331.36				
MP (deg C):		FP (deg C):		
BP (deg C):				
BP pressure (mm Hg):				

Property/Value	Units	Temp	Data Type	Reference
WS 1.28E+003	mg/L	25	EST	MEYLAN,WM ET AL. (1996)
logP -0.60			EXP	SOSNOVSKY,G ET AL. (1986)
VP 1.78E-010	mm Hg	25	EST	NEELY,WB & BLAU,GE (1985)
DC	pKa			
HL 7.29E-023	atm m3/mol	25	EST	MEYLAN,WM & HOWARD,PH (1991)
OH 8.18E-011	cm3/molc sec	25	EST	MEYLAN,WM & HOWARD,PH (1993)

CAS #: 103981-96-0				TEPA 4-(N-OH-PYRROLDINYL)-UREA DERIVATIV
Formula: $C_{17}H_{34}N_5O_3P$				
Mol Weight: 387.47				
MP (deg C):		FP (deg C):		
BP (deg C):				
BP pressure (mm Hg):				

Property/Value	Units	Temp	Data Type	Reference
WS 1.34E+003	mg/L	25	EST	MEYLAN,WM ET AL. (1996)
logP 0.51			EXP	SOSNOVSKY,G ET AL. (1986)
VP 1.78E-010	mm Hg	25	EST	NEELY,WB & BLAU,GE (1985)
DC	pKa			
HL 2.27E-022	atm m3/mol	25	EST	MEYLAN,WM & HOWARD,PH (1991)
OH 8.20E-011	cm3/molc sec	25	EST	MEYLAN,WM & HOWARD,PH (1993)

CAS #: 103981-97-1				TEPA 4-(PIPERIDINYL)-UREA DERIVATIVE
Formula: $C_{18}H_{36}N_5O_2P$				
Mol Weight: 385.49				
MP (deg C):		FP (deg C):		
BP (deg C):				
BP pressure (mm Hg):				

Property/Value	Units	Temp	Data Type	Reference
WS 1.28E+003	mg/L	25	EST	MEYLAN,WM ET AL. (1996)
logP 0.55			EXP	SOSNOVSKY,G ET AL. (1986)
VP 2.06E-008	mm Hg	25	EST	NEELY,WB & BLAU,GE (1985)
DC	pKa			
HL 1.50E-019	atm m3/mol	25	EST	MEYLAN,WM & HOWARD,PH (1991)
OH 7.91E-011	cm3/molc sec	25	EST	MEYLAN,WM & HOWARD,PH (1993)

CAS #: 103981-98-2				TEPA ADAMANTYL-UREA DERIVATIVE
Formula: $C_{19}H_{33}N_4O_2P$				
Mol Weight: 380.47				
MP (deg C):		FP (deg C):		
BP (deg C):				
BP pressure (mm Hg):				

Property/Value	Units	Temp	Data Type	Reference
WS 8.23E+001	mg/L	25	EST	MEYLAN,WM ET AL. (1996)
logP 1.98			EXP	SOSNOVSKY,G ET AL. (1986)
VP 2.06E-008	mm Hg	25	EST	NEELY,WB & BLAU,GE (1985)
DC	pKa			
HL 2.28E-016	atm m3/mol	25	EST	MEYLAN,WM & HOWARD,PH (1991)
OH 2.44E-011	cm3/molc sec	25	EST	MEYLAN,WM & HOWARD,PH (1993)

CAS #: 103981-99-3				1-PYRROLIDINYLOXY, 3-[[[[BIS(1-AZIRIDINYL)PHOSPH
Formula: $C_{13}H_{26}N_5O_3P$				
Mol Weight: 331.36				
MP (deg C):		FP (deg C):		
BP (deg C):				
BP pressure (mm Hg):				

Property/Value	Units	Temp	Data Type	Reference
WS 1.02E+003	mg/L	25	EST	MEYLAN,WM ET AL. (1996)
logP -0.48			EXP	SOSNOVSKY,G ET AL. (1986)
VP 2.06E-008	mm Hg	25	EST	NEELY,WB & BLAU,GE (1985)
DC	pKa			
HL 5.61E-026	atm m3/mol	25	EST	MEYLAN,WM & HOWARD,PH (1991)
OH 1.58E-011	cm3/molc sec	25	EST	MEYLAN,WM & HOWARD,PH (1993)

CAS #: 104438-02-0	2-THIOPHENESULFONAMIDE, 5-[(3-HYDROXYPROPYL)SULF

Formula: $C_7H_{11}NO_5S_3$

Mol Weight: 285.36

MP (deg C):

FP (deg C):

BP (deg C):

BP pressure (mm Hg):

Property/Value	Units	Temp	Data Type	Reference
WS 2.69E+003	mg/L	25	EST	MEYLAN,WM ET AL. (1996)
logP -0.36			EXP	SHEPARD,KL ET AL. (1991)
VP 3.22E-011	mm Hg	25	EST	NEELY,WB & BLAU,GE (1985)
DC	pKa			
HL 3.42E-015	atm m3/mol	25	EST	MEYLAN,WM & HOWARD,PH (1991)
OH 1.50E-011	cm3/molc sec	25	EST	MEYLAN,WM & HOWARD,PH (1993)

CAS #: 104438-04-2	ACETIC ACID, 3-[[5-(AMINOSULFONY)-2-THIENYL]SUL

Formula: $C_9H_{13}NO_6S_3$

Mol Weight: 327.40

MP (deg C):

FP (deg C):

BP (deg C):

BP pressure (mm Hg):

Property/Value	Units	Temp	Data Type	Reference
WS 5.39E+002	mg/L	25	EST	MEYLAN,WM ET AL. (1996)
logP 0.17			EXP	SHEPARD,KL ET AL. (1991)
VP 2.07E-009	mm Hg	25	EST	NEELY,WB & BLAU,GE (1985)
DC	pKa			
HL 2.23E-013	atm m3/mol	25	EST	MEYLAN,WM & HOWARD,PH (1991)
OH 1.27E-011	cm3/molc sec	25	EST	MEYLAN,WM & HOWARD,PH (1993)

CAS #: 104438-05-3	2-THIOPHENESULFONAMIDE, 5-[[3-(METHOXYACETYL)OXY

Formula: $C_{10}H_{15}NO_7S_3$

Mol Weight: 357.43

MP (deg C):

FP (deg C):

BP (deg C):

BP pressure (mm Hg):

Property/Value	Units	Temp	Data Type	Reference
WS 6.54E+002	mg/L	25	EST	MEYLAN,WM ET AL. (1996)
logP -0.14			EXP	SHEPARD,KL ET AL. (1991)
VP 3.89E-010	mm Hg	25	EST	NEELY,WB & BLAU,GE (1985)
DC	pKa			
HL 1.85E-014	atm m3/mol	25	EST	MEYLAN,WM & HOWARD,PH (1991)
OH 1.77E-011	cm3/molc sec	25	EST	MEYLAN,WM & HOWARD,PH (1993)

CAS #: 104608-36-8	FOSTEDIL,2-OXOPYRROLIDINO ANANLOG

Formula: $C_{20}H_{21}N_2O_3PS$

Mol Weight: 400.44

MP (deg C):

FP (deg C):

BP (deg C):

BP pressure (mm Hg):

Property/Value	Units	Temp	Data Type	Reference
WS 8.34E+000	mg/L	25	EST	MEYLAN,WM ET AL. (1996)
logP 3.00			EXP	HANSCH,C ET AL. (1995)
VP 2.06E-008	mm Hg	25	EST	NEELY,WB & BLAU,GE (1985)
DC	pKa			
HL 1.89E-017	atm m3/mol	25	EST	MEYLAN,WM & HOWARD,PH (1991)
OH 5.96E-011	cm3/molc sec	25	EST	MEYLAN,WM & HOWARD,PH (1993)

CAS #: 104662-93-3	1H-BENZIMIDAZOLE-1-CARBOXYLIC ACID, 5-BENZOYL-2-

Formula: $C_{18}H_{15}N_3O_5$

Mol Weight: 353.34

MP (deg C):

FP (deg C):

BP (deg C):

BP pressure (mm Hg):

Property/Value	Units	Temp	Data Type	Reference
WS 4.60E+001	mg/L	25	EST	MEYLAN,WM ET AL. (1996)
logP 2.47			EXP	SANGSTER,J (1994)
VP 1.04E-010	mm Hg	25	EST	NEELY,WB & BLAU,GE (1985)
DC	pKa			
HL 8.53E-016	atm m3/mol	25	EST	MEYLAN,WM & HOWARD,PH (1991)
OH 1.36E-010	cm3/molc sec	25	EST	MEYLAN,WM & HOWARD,PH (1993)

CAS #: 104668-67-9	PROPANAMIDE, 3,3,3-TRIFLUORO-2-METHYL-N-[4-NITRO

Formula: $C_{11}H_8F_6N_2O_3$

Mol Weight: 330.19

MP (deg C):

FP (deg C):

BP (deg C):

BP pressure (mm Hg):

Property/Value	Units	Temp	Data Type	Reference
WS 2.25E+000	mg/L	25	EST	MEYLAN,WM ET AL. (1996)
logP 3.71			EXP	MORRIS,JJ ET AL. (1991)
VP 1.89E-006	mm Hg	25	EST	NEELY,WB & BLAU,GE (1985)
DC	pKa			
HL 2.94E-009	atm m3/mol	25	EST	MEYLAN,WM & HOWARD,PH (1991)
OH 8.69E-013	cm3/molc sec	25	EST	MEYLAN,WM & HOWARD,PH (1993)

CAS #: 104764-52-5	N-PROPYL-HYDROXYPYRID-4-ONE

Formula: $C_8H_{11}NO_2$

Mol Weight: 153.18

MP (deg C):

FP (deg C):

BP (deg C):

BP pressure (mm Hg):

Property/Value	Units	Temp	Data Type	Reference
WS 1.03E+005	mg/L	25	EST	MEYLAN,WM ET AL. (1996)
logP -0.17			EXP	GYPARAKI,M ET AL. (1986)
VP 4.00E-005	mm Hg	25	EST	NEELY,WB & BLAU,GE (1985)
DC	pKa			
HL 2.16E-008	atm m3/mol	25	EST	MEYLAN,WM & HOWARD,PH (1991)
OH 1.03E-010	cm3/molc sec	25	EST	MEYLAN,WM & HOWARD,PH (1993)

CAS #: 104764-53-6	N-ISOPROPYL-HYDROXYPYRID-4-ONE

Formula: $C_8H_{11}NO_2$

Mol Weight: 153.18

MP (deg C):

FP (deg C):

BP (deg C):

BP pressure (mm Hg):

Property/Value	Units	Temp	Data Type	Reference
WS 7.64E+004	mg/L	25	EST	MEYLAN,WM ET AL. (1996)
logP -0.02			EXP	GYPARAKI,M ET AL. (1986)
VP 8.71E-005	mm Hg	25	EST	NEELY,WB & BLAU,GE (1985)
DC	pKa			
HL 2.16E-008	atm m3/mol	25	EST	MEYLAN,WM & HOWARD,PH (1991)
OH 1.10E-010	cm3/molc sec	25	EST	MEYLAN,WM & HOWARD,PH (1993)

1466

CAS #: 104958-92-1				1,2,4-TRIAZOLE,3-NITRO-1-(3-AZIRIDINYL-2-HYDROXY

Formula: $C_7H_{13}N_5O_3$

Mol Weight: 215.21

MP (deg C):   **FP (deg C):**

BP (deg C):

BP pressure (mm Hg):

Property/ Value	Units	Temp	Data Type	Reference
WS 6.39E+005	mg/L	25	EST	MEYLAN,WM ET AL. (1996)
logP -0.72			EXP	HANSCH,C ET AL. (1995)
VP 2.60E-007	mm Hg	25	EST	NEELY,WB & BLAU,GE (1985)
DC	pKa			
HL 1.81E-015	atm m3/mol	25	EST	MEYLAN,WM & HOWARD,PH (1991)
OH 2.87E-011	cm3/molc sec	25	EST	MEYLAN,WM & HOWARD,PH (1993)

CAS #: 104958-94-3				1,2,4-TRIAZOLE,3-NITRO-1-(3-CHLORO-2-HYDROXY)PRO

Formula: $C_5H_9ClN_4O_3$

Mol Weight: 208.61

MP (deg C):   **FP (deg C):**

BP (deg C):

BP pressure (mm Hg):

Property/ Value	Units	Temp	Data Type	Reference
WS 1.49E+004	mg/L	25	EST	MEYLAN,WM ET AL. (1996)
logP 0.05			EXP	HANSCH,C ET AL. (1995)
VP 7.17E-007	mm Hg	25	EST	NEELY,WB & BLAU,GE (1985)
DC	pKa			
HL 2.92E-012	atm m3/mol	25	EST	MEYLAN,WM & HOWARD,PH (1991)
OH 4.74E-012	cm3/molc sec	25	EST	MEYLAN,WM & HOWARD,PH (1993)

CAS #: 104987-39-5				4-(3-NITRO-1,2,4-TRIAZOL-1-YL)PRIOPIONYLMORPHOLI

Formula: $C_9H_{13}N_5O_4$

Mol Weight: 255.24

MP (deg C):   **FP (deg C):**

BP (deg C):

BP pressure (mm Hg):

Property/ Value	Units	Temp	Data Type	Reference
WS 1.89E+003	mg/L	25	EST	MEYLAN,WM ET AL. (1996)
logP -0.74			EXP	HANSCH,C ET AL. (1995)
VP 2.41E-007	mm Hg	25	EST	NEELY,WB & BLAU,GE (1985)
DC	pKa			
HL 8.57E-016	atm m3/mol	25	EST	MEYLAN,WM & HOWARD,PH (1991)
OH 6.03E-011	cm3/molc sec	25	EST	MEYLAN,WM & HOWARD,PH (1993)

CAS #: 105217-56-9				2-PROPANOL, 1-[(1-METHYLETHYL)AMINO]-3-(2-PHENYL

Formula: $C_{14}H_{23}NO_2$

Mol Weight: 237.34

MP (deg C):   **FP (deg C):**

BP (deg C):

BP pressure (mm Hg):

Property/ Value	Units	Temp	Data Type	Reference
WS 8.88E+003	mg/L	25	EST	MEYLAN,WM ET AL. (1996)
logP 1.76			EXP	MAULEON,D ET AL. (1988)
VP 2.30E-006	mm Hg	25	EST	NEELY,WB & BLAU,GE (1985)
DC	pKa			
HL 2.37E-012	atm m3/mol	25	EST	MEYLAN,WM & HOWARD,PH (1991)
OH 1.25E-010	cm3/molc sec	25	EST	MEYLAN,WM & HOWARD,PH (1993)

CAS #: 105217-57-0				2-PROPANOL, 1-[(1-METHYLETHYL)AMINO]-3-(2-PHENOX

Formula: $C_{14}H_{23}NO_3$

Mol Weight: 253.34

MP (deg C):   **FP (deg C):**

BP (deg C):

BP pressure (mm Hg):

Property/ Value	Units	Temp	Data Type	Reference
WS 9.36E+003	mg/L	25	EST	MEYLAN,WM ET AL. (1996)
logP 1.63			EXP	MAULEON,D ET AL. (1988)
VP 7.22E-007	mm Hg	25	EST	NEELY,WB & BLAU,GE (1985)
DC	pKa			
HL 1.27E-013	atm m3/mol	25	EST	MEYLAN,WM & HOWARD,PH (1991)
OH 1.45E-010	cm3/molc sec	25	EST	MEYLAN,WM & HOWARD,PH (1993)

CAS #: 105217-58-1				2-PROPANOL, 1-[(1-METHYLETHYL)AMINO]-3-[(3-PHENY

Formula: $C_{15}H_{23}NO_2$

Mol Weight: 249.36

MP (deg C):   **FP (deg C):**

BP (deg C):

BP pressure (mm Hg):

Property/ Value	Units	Temp	Data Type	Reference
WS 3.54E+003	mg/L	25	EST	MEYLAN,WM ET AL. (1996)
logP 2.15			EXP	MAULEON,D ET AL. (1988)
VP 6.98E-007	mm Hg	25	EST	NEELY,WB & BLAU,GE (1985)
DC	pKa			
HL 1.30E-012	atm m3/mol	25	EST	MEYLAN,WM & HOWARD,PH (1991)
OH 1.73E-010	cm3/molc sec	25	EST	MEYLAN,WM & HOWARD,PH (1993)

CAS #: 105412-23-5				THIOUREA, N-(2,2,3,3,4,4,4-HEPTAFLUOROBUTYL)-N'-

Formula: $C_6H_7F_7N_2S$

Mol Weight: 272.19

MP (deg C):   **FP (deg C):**

BP (deg C):

BP pressure (mm Hg):

Property/ Value	Units	Temp	Data Type	Reference
WS 3.61E+003	mg/L	25	EST	MEYLAN,WM ET AL. (1996)
logP 1.99			EXP	HANSCH,C ET AL. (1995)
VP 8.88E-001	mm Hg	25	EST	NEELY,WB & BLAU,GE (1985)
DC	pKa			
HL 2.20E-004	atm m3/mol	25	EST	MEYLAN,WM & HOWARD,PH (1991)
OH 1.42E-012	cm3/molc sec	25	EST	MEYLAN,WM & HOWARD,PH (1993)

CAS #: 105412-24-6				BENZENE, 1-[3-(METHYLSULFINYL)PROPOXY]-4-NITRO-

Formula: $C_{10}H_{13}NO_4S$

Mol Weight: 243.28

MP (deg C):   **FP (deg C):**

BP (deg C):

BP pressure (mm Hg):

Property/ Value	Units	Temp	Data Type	Reference
WS 1.69E+003	mg/L	25	EST	MEYLAN,WM ET AL. (1996)
logP 0.93			EXP	LEAHY,DE ET AL. (1989)
VP 2.51E-006	mm Hg	25	EST	NEELY,WB & BLAU,GE (1985)
DC	pKa			
HL 1.49E-012	atm m3/mol	25	EST	MEYLAN,WM & HOWARD,PH (1991)
OH 8.37E-011	cm3/molc sec	25	EST	MEYLAN,WM & HOWARD,PH (1993)

CAS #: 105412-25-7 — P-NO2-(3-ME SULFONYL)PROPOXYBENZENE

Formula: $C_{10}H_{13}NO_5S$

Mol Weight: 259.28

MP (deg C): FP (deg C):

BP (deg C):

BP pressure (mm Hg):

Property/Value	Units	Temp	Data Type	Reference
WS 9.84E+002	mg/L	25	EST	MEYLAN,WM ET AL. (1996)
logP 1.10			EXP	HANSCH,C & LEO,AJ (1985)
VP 1.24E-006	mm Hg	25	EST	NEELY,WB & BLAU,GE (1985)
DC	pKa			
HL 5.13E-010	atm m3/mol	25	EST	MEYLAN,WM & HOWARD,PH (1991)
OH 1.21E-011	cm3/molc sec	25	EST	MEYLAN,WM & HOWARD,PH (1993)

CAS #: 105412-26-8 — 1-PROPANESULFONAMIDE, 3-(4-NITROPHENOXY)-

Formula: $C_9H_{12}N_2O_5S$

Mol Weight: 260.27

MP (deg C): FP (deg C):

BP (deg C):

BP pressure (mm Hg):

Property/Value	Units	Temp	Data Type	Reference
WS 1.25E+003	mg/L	25	EST	MEYLAN,WM ET AL. (1996)
logP 0.97			EXP	HANSCH,C & LEO,AJ (1985)
VP 2.56E-007	mm Hg	25	EST	NEELY,WB & BLAU,GE (1985)
DC	pKa			
HL 6.36E-011	atm m3/mol	25	EST	MEYLAN,WM & HOWARD,PH (1991)
OH 2.27E-011	cm3/molc sec	25	EST	MEYLAN,WM & HOWARD,PH (1993)

CAS #: 105412-27-9 — BENZOIC ACID, 4-CARBOXAMIDOBUTYL ESTER

Formula: $C_{12}H_{15}NO_3$

Mol Weight: 221.26

MP (deg C): FP (deg C):

BP (deg C):

BP pressure (mm Hg):

Property/Value	Units	Temp	Data Type	Reference
WS 2.20E+003	mg/L	25	EST	MEYLAN,WM ET AL. (1996)
logP 1.39			EXP	HANSCH,C & LEO,AJ (1985)
VP 1.61E-006	mm Hg	25	EST	NEELY,WB & BLAU,GE (1985)
DC	pKa			
HL 1.23E-011	atm m3/mol	25	EST	MEYLAN,WM & HOWARD,PH (1991)
OH 1.12E-011	cm3/molc sec	25	EST	MEYLAN,WM & HOWARD,PH (1993)

CAS #: 105456-57-3 — BENZENAMINE, 4-[(2,4-DICHLOROPHENYL)SULFONYL]-

Formula: $C_{12}H_9Cl_2NO_2S$

Mol Weight: 302.18

MP (deg C): FP (deg C):

BP (deg C):

BP pressure (mm Hg):

Property/Value	Units	Temp	Data Type	Reference
WS 2.58E+001	mg/L	25	EST	MEYLAN,WM ET AL. (1996)
logP 3.12			EXP	ALTOMARE,C ET AL. (1991)
VP 3.33E-008	mm Hg	25	EST	NEELY,WB & BLAU,GE (1985)
DC	pKa			
HL 4.84E-011	atm m3/mol	25	EST	MEYLAN,WM & HOWARD,PH (1991)
OH 2.32E-011	cm3/molc sec	25	EST	MEYLAN,WM & HOWARD,PH (1993)

CAS #: 105456-58-4 — BENZENAMINE, 4-[(2,4-DIMETHYLPHENYL)SULFONYL]-

Formula: $C_{14}H_{15}NO_2S$

Mol Weight: 261.35

MP (deg C): FP (deg C):

BP (deg C):

BP pressure (mm Hg):

Property/Value	Units	Temp	Data Type	Reference
WS 9.92E+001	mg/L	25	EST	MEYLAN,WM ET AL. (1996)
logP 2.71			EXP	ALTOMARE,C ET AL. (1991)
VP 1.08E-007	mm Hg	25	EST	NEELY,WB & BLAU,GE (1985)
DC	pKa			
HL 1.07E-010	atm m3/mol	25	EST	MEYLAN,WM & HOWARD,PH (1991)
OH 2.61E-011	cm3/molc sec	25	EST	MEYLAN,WM & HOWARD,PH (1993)

CAS #: 105456-59-5 — BENZENAMINE, 4-[(2,4-DIMETHOXYPHENYL)SULFONYL]-

Formula: $C_{14}H_{15}NO_4S$

Mol Weight: 293.34

MP (deg C): FP (deg C):

BP (deg C):

BP pressure (mm Hg):

Property/Value	Units	Temp	Data Type	Reference
WS 5.34E+002	mg/L	25	EST	MEYLAN,WM ET AL. (1996)
logP 1.64			EXP	ALTOMARE,C ET AL. (1991)
VP 1.61E-008	mm Hg	25	EST	NEELY,WB & BLAU,GE (1985)
DC	pKa			
HL 3.08E-013	atm m3/mol	25	EST	MEYLAN,WM & HOWARD,PH (1991)
OH 7.54E-011	cm3/molc sec	25	EST	MEYLAN,WM & HOWARD,PH (1993)

CAS #: 105456-60-8 — 1,3-BENZENEDIOL, 4-[(4-AMINOPHENYL)SULFONYL]-

Formula: $C_{12}H_{11}NO_4S$

Mol Weight: 265.29

MP (deg C): FP (deg C):

BP (deg C):

BP pressure (mm Hg):

Property/Value	Units	Temp	Data Type	Reference
WS 5.85E+003	mg/L	25	EST	MEYLAN,WM ET AL. (1996)
logP 1.29			EXP	ALTOMARE,C ET AL. (1991)
VP 9.53E-011	mm Hg	25	EST	NEELY,WB & BLAU,GE (1985)
DC	pKa			
HL 9.54E-019	atm m3/mol	25	EST	MEYLAN,WM & HOWARD,PH (1991)
OH 1.45E-010	cm3/molc sec	25	EST	MEYLAN,WM & HOWARD,PH (1993)

CAS #: 105456-65-3 — 1,3,5-BENZENETRIOL, 2-[(4-AMINOPHENYL)SULFONYL]-

Formula: $C_{12}H_{11}NO_5S$

Mol Weight: 281.29

MP (deg C): FP (deg C):

BP (deg C):

BP pressure (mm Hg):

Property/Value	Units	Temp	Data Type	Reference
WS 2.96E+003	mg/L	25	EST	MEYLAN,WM ET AL. (1996)
logP 1.53			EXP	SANGSTER,J (1994)
VP 5.90E-012	mm Hg	25	EST	NEELY,WB & BLAU,GE (1985)
DC	pKa			
HL 9.93E-023	atm m3/mol	25	EST	MEYLAN,WM & HOWARD,PH (1991)
OH 2.00E-010	cm3/molc sec	25	EST	MEYLAN,WM & HOWARD,PH (1993)

<table>
<tr><td colspan="2">CAS #: 105512-06-9</td><td>CLODINAFOP-PROPARGYL</td></tr>
</table>

CAS #: 105512-06-9 — CLODINAFOP-PROPARGYL

Formula: $C_{17}H_{13}ClFNO_4$

Mol Weight: 349.75

MP (deg C): FP (deg C):

BP (deg C):

BP pressure (mm Hg):

Property/ Value	Units	Temp	Data Type	Reference
WS 2.90E+000	mg/L	25	EST	MEYLAN,WM ET AL. (1996)
logP 3.90			EXP	TOMLIN,C (1994)
VP 1.86E-007	mm Hg	25	EST	NEELY,WB & BLAU,GE (1985)
DC	pKa			
HL 3.51E-010	atm m3/mol	25	EST	MEYLAN,WM & HOWARD,PH (1991)
OH 3.92E-011	cm3/molc sec	25	EST	MEYLAN,WM & HOWARD,PH (1993)

CAS #: 105784-83-6 — URIDINE, 2',3'-DIDEOXY-5-IODO-

Formula: $C_9H_{11}IN_2O_4$

Mol Weight: 338.10

MP (deg C): 176 dec FP (deg C):

BP (deg C):

BP pressure (mm Hg):

Property/ Value	Units	Temp	Data Type	Reference
WS 1.19E+003	mg/L	25	EST	MEYLAN,WM ET AL. (1996)
logP -0.61			EXP	BALZARINI,J ET AL. (1989)
VP 8.82E-014	mm Hg	25	EST	NEELY,WB & BLAU,GE (1985)
DC	pKa			
HL 2.66E-017	atm m3/mol	25	EST	MEYLAN,WM & HOWARD,PH (1991)
OH 7.52E-011	cm3/molc sec	25	EST	MEYLAN,WM & HOWARD,PH (1993)

CAS #: 105801-48-7 — 1-PIPERIDINEACETAMIDE, N-(2,3-DIHYDRO-3-BENZOFUR

Formula: $C_{15}H_{20}N_2O_2$

Mol Weight: 260.34

MP (deg C): FP (deg C):

BP (deg C):

BP pressure (mm Hg):

Property/ Value	Units	Temp	Data Type	Reference
WS 2.25E+002	mg/L	25	EST	MEYLAN,WM ET AL. (1996)
logP 2.30			EXP	TURAN-ZITOUNI,G & BERGE,G (1986)
VP 4.96E-008	mm Hg	25	EST	NEELY,WB & BLAU,GE (1985)
DC	pKa			
HL 5.28E-013	atm m3/mol	25	EST	MEYLAN,WM & HOWARD,PH (1991)
OH 1.52E-010	cm3/molc sec	25	EST	MEYLAN,WM & HOWARD,PH (1993)

CAS #: 105801-49-8 — 4-MORPHOLINEACETAMIDE, N-(2,3-DIHYDRO-3-BENZOFUR

Formula: $C_{14}H_{18}N_2O_3$

Mol Weight: 262.31

MP (deg C): FP (deg C):

BP (deg C):

BP pressure (mm Hg):

Property/ Value	Units	Temp	Data Type	Reference
WS 4.11E+003	mg/L	25	EST	MEYLAN,WM ET AL. (1996)
logP 0.81			EXP	TURAN-ZITOUNI,G & BERGE,G (1986)
VP 3.99E-008	mm Hg	25	EST	NEELY,WB & BLAU,GE (1985)
DC	pKa			
HL 3.51E-015	atm m3/mol	25	EST	MEYLAN,WM & HOWARD,PH (1991)
OH 1.92E-010	cm3/molc sec	25	EST	MEYLAN,WM & HOWARD,PH (1993)

CAS #: 105801-50-1 — ACETAMIDE, N-(5-CHLORO-2,3-DIHYDRO-3-BENZOFURANY

Formula: $C_{14}H_{19}ClN_2O_2$

Mol Weight: 282.77

MP (deg C): FP (deg C):

BP (deg C):

BP pressure (mm Hg):

Property/ Value	Units	Temp	Data Type	Reference
WS 5.36E+001	mg/L	25	EST	MEYLAN,WM ET AL. (1996)
logP 2.88			EXP	TURAN-ZITOUNI,G & BERGE,G (1986)
VP 6.52E-008	mm Hg	25	EST	NEELY,WB & BLAU,GE (1985)
DC	pKa			
HL 6.68E-013	atm m3/mol	25	EST	MEYLAN,WM & HOWARD,PH (1991)
OH 1.31E-010	cm3/molc sec	25	EST	MEYLAN,WM & HOWARD,PH (1993)

CAS #: 105801-51-2 — ACETAMIDE, 2-(DIBUTYLAMINO)-N-(5-CHLORO-2,3-DIHY

Formula: $C_{18}H_{27}ClN_2O_2$

Mol Weight: 338.88

MP (deg C): FP (deg C):

BP (deg C):

BP pressure (mm Hg):

Property/ Value	Units	Temp	Data Type	Reference
WS 1.17E+000	mg/L	25	EST	MEYLAN,WM ET AL. (1996)
logP 4.44			EXP	TURAN-ZITOUNI,G & BERGE,G (1986)
VP 2.54E-009	mm Hg	25	EST	NEELY,WB & BLAU,GE (1985)
DC	pKa			
HL 2.08E-012	atm m3/mol	25	EST	MEYLAN,WM & HOWARD,PH (1991)
OH 1.40E-010	cm3/molc sec	25	EST	MEYLAN,WM & HOWARD,PH (1993)

CAS #: 105801-52-3 — 1-PYRROLIDINEACETAMIDE, N-(5-CHLORO-2,3-DIHYDRO-

Formula: $C_{14}H_{17}ClN_2O_2$

Mol Weight: 280.76

MP (deg C): FP (deg C):

BP (deg C):

BP pressure (mm Hg):

Property/ Value	Units	Temp	Data Type	Reference
WS 1.23E+002	mg/L	25	EST	MEYLAN,WM ET AL. (1996)
logP 2.47			EXP	TURAN-ZITOUNI,G & BERGE,G (1986)
VP 3.27E-008	mm Hg	25	EST	NEELY,WB & BLAU,GE (1985)
DC	pKa			
HL 2.95E-013	atm m3/mol	25	EST	MEYLAN,WM & HOWARD,PH (1991)
OH 1.23E-010	cm3/molc sec	25	EST	MEYLAN,WM & HOWARD,PH (1993)

CAS #: 105801-53-4 — 1-PIPERIDINEACETAMIDE, N-(5-CHLORO-2,3-DIHYDRO-3

Formula: $C_{15}H_{19}ClN_2O_2$

Mol Weight: 294.78

MP (deg C): FP (deg C):

BP (deg C):

BP pressure (mm Hg):

Property/ Value	Units	Temp	Data Type	Reference
WS 3.54E+001	mg/L	25	EST	MEYLAN,WM ET AL. (1996)
logP 3.01			EXP	TURAN-ZITOUNI,G & BERGE,G (1986)
VP 1.36E-008	mm Hg	25	EST	NEELY,WB & BLAU,GE (1985)
DC	pKa			
HL 3.91E-013	atm m3/mol	25	EST	MEYLAN,WM & HOWARD,PH (1991)
OH 1.34E-010	cm3/molc sec	25	EST	MEYLAN,WM & HOWARD,PH (1993)

CAS #:	105801-54-5			4-MORPHOLINEACETAMIDE, N-(5-CHLORO-2,3-DIHYDRO-3
Formula:	$C_{14}H_{17}ClN_2O_3$			
Mol Weight:	296.76			
MP (deg C):		FP (deg C):		
BP (deg C):				
BP pressure (mm Hg):				

Property/ Value	Units	Temp	Data Type	Reference
WS 6.09E+002	mg/L	25	EST	MEYLAN,WM ET AL. (1996)
logP 1.55			EXP	TURAN-ZITOUNI,G & BERGE,G (1986)
VP 1.09E-008	mm Hg	25	EST	NEELY,WB & BLAU,GE (1985)
DC	pKa			
HL 2.60E-015	atm m3/mol	25	EST	MEYLAN,WM & HOWARD,PH (1991)
OH 1.73E-010	cm3/molc sec	25	EST	MEYLAN,WM & HOWARD,PH (1993)

CAS #:	105801-55-6			ACETAMIDE, 2-(DIETHYLAMINO)-N-(2,3-DIHYDRO-5-MET
Formula:	$C_{15}H_{22}N_2O_2$			
Mol Weight:	262.35			
MP (deg C):		FP (deg C):		
BP (deg C):				
BP pressure (mm Hg):				

Property/ Value	Units	Temp	Data Type	Reference
WS 1.02E+002	mg/L	25	EST	MEYLAN,WM ET AL. (1996)
logP 2.69			EXP	TURAN-ZITOUNI,G & BERGE,G (1986)
VP 1.17E-007	mm Hg	25	EST	NEELY,WB & BLAU,GE (1985)
DC	pKa			
HL 9.95E-013	atm m3/mol	25	EST	MEYLAN,WM & HOWARD,PH (1991)
OH 1.55E-010	cm3/molc sec	25	EST	MEYLAN,WM & HOWARD,PH (1993)

CAS #:	105801-56-7			ACETAMIDE, 2-(DIBUTYLAMINO)-N-(2,3-DIHYDRO-5-MET
Formula:	$C_{19}H_{30}N_2O_2$			
Mol Weight:	318.46			
MP (deg C):		FP (deg C):		
BP (deg C):				
BP pressure (mm Hg):				

Property/ Value	Units	Temp	Data Type	Reference
WS 1.58E+000	mg/L	25	EST	MEYLAN,WM ET AL. (1996)
logP 4.43			EXP	TURAN-ZITOUNI,G & BERGE,G (1986)
VP 4.18E-009	mm Hg	25	EST	NEELY,WB & BLAU,GE (1985)
DC	pKa			
HL 3.09E-012	atm m3/mol	25	EST	MEYLAN,WM & HOWARD,PH (1991)
OH 1.65E-010	cm3/molc sec	25	EST	MEYLAN,WM & HOWARD,PH (1993)

CAS #:	105801-57-8			1-PYRROLIDINEACETAMIDE, N-(2,3-DIHYDRO-5-METHYL-
Formula:	$C_{15}H_{20}N_2O_2$			
Mol Weight:	260.34			
MP (deg C):		FP (deg C):		
BP (deg C):				
BP pressure (mm Hg):				

Property/ Value	Units	Temp	Data Type	Reference
WS 2.53E+002	mg/L	25	EST	MEYLAN,WM ET AL. (1996)
logP 2.24			EXP	TURAN-ZITOUNI,G & BERGE,G (1986)
VP 5.34E-008	mm Hg	25	EST	NEELY,WB & BLAU,GE (1985)
DC	pKa			
HL 4.39E-013	atm m3/mol	25	EST	MEYLAN,WM & HOWARD,PH (1991)
OH 1.48E-010	cm3/molc sec	25	EST	MEYLAN,WM & HOWARD,PH (1993)

CAS #:	105801-58-9			1-PIPERIDINEACETAMIDE, N-(2,3-DIHYDRO-5-METHYL-3
Formula:	$C_{16}H_{22}N_2O_2$			
Mol Weight:	274.37			
MP (deg C):		FP (deg C):		
BP (deg C):				
BP pressure (mm Hg):				

Property/ Value	Units	Temp	Data Type	Reference
WS 7.29E+001	mg/L	25	EST	MEYLAN,WM ET AL. (1996)
logP 2.78			EXP	TURAN-ZITOUNI,G & BERGE,G (1986)
VP 2.22E-008	mm Hg	25	EST	NEELY,WB & BLAU,GE (1985)
DC	pKa			
HL 5.82E-013	atm m3/mol	25	EST	MEYLAN,WM & HOWARD,PH (1991)
OH 1.58E-010	cm3/molc sec	25	EST	MEYLAN,WM & HOWARD,PH (1993)

CAS #:	105801-59-0			4-MORPHOLINEACETAMIDE, N-(2,3-DIHYDRO-5-METHYL-3
Formula:	$C_{15}H_{20}N_2O_3$			
Mol Weight:	276.34			
MP (deg C):		FP (deg C):		
BP (deg C):				
BP pressure (mm Hg):				

Property/ Value	Units	Temp	Data Type	Reference
WS 1.12E+003	mg/L	25	EST	MEYLAN,WM ET AL. (1996)
logP 1.38			EXP	TURAN-ZITOUNI,G & BERGE,G (1986)
VP 1.79E-008	mm Hg	25	EST	NEELY,WB & BLAU,GE (1985)
DC	pKa			
HL 3.87E-015	atm m3/mol	25	EST	MEYLAN,WM & HOWARD,PH (1991)
OH 1.98E-010	cm3/molc sec	25	EST	MEYLAN,WM & HOWARD,PH (1993)

CAS #:	105801-60-3			ACETAMIDE, 2-(DIBUTYLAMINO)-N-(2,3-DIHYDRO-6-MET
Formula:	$C_{19}H_{30}N_2O_3$			
Mol Weight:	334.46			
MP (deg C):		FP (deg C):		
BP (deg C):				
BP pressure (mm Hg):				

Property/ Value	Units	Temp	Data Type	Reference
WS 1.84E+000	mg/L	25	EST	MEYLAN,WM ET AL. (1996)
logP 4.24			EXP	TURAN-ZITOUNI,G & BERGE,G (1986)
VP 1.79E-009	mm Hg	25	EST	NEELY,WB & BLAU,GE (1985)
DC	pKa			
HL 1.66E-013	atm m3/mol	25	EST	MEYLAN,WM & HOWARD,PH (1991)
OH 3.34E-010	cm3/molc sec	25	EST	MEYLAN,WM & HOWARD,PH (1993)

CAS #:	105801-62-5			1-PIPERIDINEACETAMIDE, N-(2,3-DIHYDRO-6-METHOXY-
Formula:	$C_{16}H_{22}N_2O_3$			
Mol Weight:	290.37			
MP (deg C):		FP (deg C):		
BP (deg C):				
BP pressure (mm Hg):				

Property/ Value	Units	Temp	Data Type	Reference
WS 1.22E+002	mg/L	25	EST	MEYLAN,WM ET AL. (1996)
logP 2.41			EXP	TURAN-ZITOUNI,G & BERGE,G (1986)
VP 9.62E-009	mm Hg	25	EST	NEELY,WB & BLAU,GE (1985)
DC	pKa			
HL 3.12E-014	atm m3/mol	25	EST	MEYLAN,WM & HOWARD,PH (1991)
OH 3.27E-010	cm3/molc sec	25	EST	MEYLAN,WM & HOWARD,PH (1993)

CAS #: 105801-63-6	4-MORPHOLINEACETAMIDE, N-(2,3-DIHYDRO-6-METHOXY-

Formula: $C_{15}H_{20}N_2O_4$
Mol Weight: 292.34
MP (deg C):　FP (deg C):
BP (deg C):
BP pressure (mm Hg):

Property/Value	Units	Temp	Data Type	Reference
WS 2.23E+003	mg/L	25	EST	MEYLAN,WM ET AL. (1996)
logP 0.92			EXP	TURAN-ZITOUNI,G & BERGE,G (1986)
VP 7.73E-009	mm Hg	25	EST	NEELY,WB & BLAU,GE (1985)
DC	pKa			
HL 2.07E-016	atm m3/mol	25	EST	MEYLAN,WM & HOWARD,PH (1991)
OH 3.67E-010	cm3/molc sec	25	EST	MEYLAN,WM & HOWARD,PH (1993)

CAS #: 105801-65-8	ACETAMIDE, N-(2,3-DIHYDRO-3-BENZOFURANYL)-2-(DIE

Formula: $C_{14}H_{20}N_2O_2$
Mol Weight: 248.33
MP (deg C):　FP (deg C):
BP (deg C):
BP pressure (mm Hg):

Property/Value	Units	Temp	Data Type	Reference
WS 3.19E+002	mg/L	25	EST	MEYLAN,WM ET AL. (1996)
logP 2.20			EXP	TURAN-ZITOUNI,G & BERGE,G (1986)
VP 2.93E-007	mm Hg	25	EST	NEELY,WB & BLAU,GE (1985)
DC	pKa			
HL 9.02E-013	atm m3/mol	25	EST	MEYLAN,WM & HOWARD,PH (1991)
OH 1.49E-010	cm3/molc sec	25	EST	MEYLAN,WM & HOWARD,PH (1993)

CAS #: 105801-66-9	ACETAMIDE, 2-(DIBUTYLAMINO)-N-(2,3-DIHYDRO-3-BEN

Formula: $C_{18}H_{28}N_2O_2$
Mol Weight: 304.44
MP (deg C):　FP (deg C):
BP (deg C):
BP pressure (mm Hg):

Property/Value	Units	Temp	Data Type	Reference
WS 3.30E+000	mg/L	25	EST	MEYLAN,WM ET AL. (1996)
logP 4.15			EXP	TURAN-ZITOUNI,G & BERGE,G (1986)
VP 9.42E-009	mm Hg	25	EST	NEELY,WB & BLAU,GE (1985)
DC	pKa			
HL 2.80E-012	atm m3/mol	25	EST	MEYLAN,WM & HOWARD,PH (1991)
OH 1.59E-010	cm3/molc sec	25	EST	MEYLAN,WM & HOWARD,PH (1993)

CAS #: 105801-67-0	1-PYRROLIDINEACETAMIDE, N-(2,3-DIHYDRO-3-BENZOFU

Formula: $C_{14}H_{18}N_2O_2$
Mol Weight: 246.31
MP (deg C):　FP (deg C):
BP (deg C):
BP pressure (mm Hg):

Property/Value	Units	Temp	Data Type	Reference
WS 7.79E+002	mg/L	25	EST	MEYLAN,WM ET AL. (1996)
logP 1.76			EXP	TURAN-ZITOUNI,G & BERGE,G (1986)
VP 1.18E-007	mm Hg	25	EST	NEELY,WB & BLAU,GE (1985)
DC	pKa			
HL 3.98E-013	atm m3/mol	25	EST	MEYLAN,WM & HOWARD,PH (1991)
OH 1.42E-010	cm3/molc sec	25	EST	MEYLAN,WM & HOWARD,PH (1993)

CAS #: 105836-96-2	2,8-DIBROMODIBENZO-P-DIOXIN

Formula: $C_{12}H_6Br_2O_2$
Mol Weight: 342.00
MP (deg C):　FP (deg C):
BP (deg C):
BP pressure (mm Hg):

Property/Value	Units	Temp	Data Type	Reference
WS 3.18E-003	mg/L	25	EST	MEYLAN,WM ET AL. (1996)
logP 6.12			EST	MEYLAN,WM & HOWARD,PH (1995)
VP 4.87E-006	mm Hg	25	EST	NEELY,WB & BLAU,GE (1985)
DC	pKa			
HL 1.86E-006	atm m3/mol	25	EST	MEYLAN,WM & HOWARD,PH (1991)
OH 6.19E-012	cm3/molc sec	25	EST	MEYLAN,WM & HOWARD,PH (1993)

CAS #: 105906-36-3	2-MONOBROMODIBENZO-P-DIOXIN

Formula: $C_{12}H_7BrO_2$
Mol Weight: 263.10
MP (deg C):　FP (deg C):
BP (deg C):
BP pressure (mm Hg):

Property/Value	Units	Temp	Data Type	Reference
WS 5.28E-002	mg/L	25	EST	MEYLAN,WM ET AL. (1996)
logP 5.23			EST	MEYLAN,WM & HOWARD,PH (1995)
VP 4.82E-005	mm Hg	25	EST	NEELY,WB & BLAU,GE (1985)
DC	pKa			
HL 4.66E-006	atm m3/mol	25	EST	MEYLAN,WM & HOWARD,PH (1991)
OH 8.02E-012	cm3/molc sec	25	EST	MEYLAN,WM & HOWARD,PH (1993)

CAS #: 105908-71-2	1-MONOBROMODIBENZO-P-DIOXIN

Formula: $C_{12}H_7BrO_2$
Mol Weight: 263.10
MP (deg C):　FP (deg C):
BP (deg C):
BP pressure (mm Hg):

Property/Value	Units	Temp	Data Type	Reference
WS 5.28E-002	mg/L	25	EST	MEYLAN,WM ET AL. (1996)
logP 5.23			EST	MEYLAN,WM & HOWARD,PH (1995)
VP 4.82E-005	mm Hg	25	EST	NEELY,WB & BLAU,GE (1985)
DC	pKa			
HL 4.66E-006	atm m3/mol	25	EST	MEYLAN,WM & HOWARD,PH (1991)
OH 8.02E-012	cm3/molc sec	25	EST	MEYLAN,WM & HOWARD,PH (1993)

CAS #: 105951-31-3	4H-THIENO[2,3-B]THIOPYRAN-2-SULFONAMIDE, 5,6-DIH

Formula: $C_7H_9NO_3S_3$
Mol Weight: 251.35
MP (deg C):　FP (deg C):
BP (deg C):
BP pressure (mm Hg):

Property/Value	Units	Temp	Data Type	Reference
WS 4.95E+002	mg/L	25	EST	MEYLAN,WM ET AL. (1996)
logP 0.74			EXP	PONTICELLO,GS ET AL. (1987)
VP 1.38E-007	mm Hg	25	EST	NEELY,WB & BLAU,GE (1985)
DC	pKa			
HL 5.37E-012	atm m3/mol	25	EST	MEYLAN,WM & HOWARD,PH (1991)
OH 2.98E-011	cm3/molc sec	25	EST	MEYLAN,WM & HOWARD,PH (1993)

<table>
<tr><td colspan="2">CAS #: 105951-32-4</td><td>5H-THIENO[3,2-B]THIOPYRAN-2-SULFONAMIDE, 6,7-DIH</td></tr>
<tr><td>Formula:</td><td colspan="2">$C_7H_9NO_3S_3$</td></tr>
<tr><td>Mol Weight:</td><td colspan="2">251.35</td></tr>
</table>

Property/Value	Units	Temp	Data Type	Reference
WS 3.55E+002	mg/L	25	EST	MEYLAN,WM ET AL. (1996)
logP 0.91			EXP	PONTICELLO,GS ET AL. (1987)
VP 1.38E-007	mm Hg	25	EST	NEELY,WB & BLAU,GE (1985)
DC	pKa			
HL 5.37E-012	atm m3/mol	25	EST	MEYLAN,WM & HOWARD,PH (1991)
OH 4.12E-011	cm3/molc sec	25	EST	MEYLAN,WM & HOWARD,PH (1993)

<table>
<tr><td colspan="2">CAS #: 105951-35-7</td><td>4H-THIENO[2,3-B]THIOPYRAN-2-SULFONAMIDE, 5,6-DIH</td></tr>
<tr><td>Formula:</td><td colspan="2">$C_7H_9NO_5S_3$</td></tr>
<tr><td>Mol Weight:</td><td colspan="2">283.35</td></tr>
</table>

Property/Value	Units	Temp	Data Type	Reference
WS 1.52E+003	mg/L	25	EST	MEYLAN,WM ET AL. (1996)
logP -0.04			EXP	PONTICELLO,GS ET AL. (1987)
VP 4.57E-009	mm Hg	25	EST	NEELY,WB & BLAU,GE (1985)
DC	pKa			
HL 4.26E-014	atm m3/mol	25	EST	MEYLAN,WM & HOWARD,PH (1991)
OH 2.95E-011	cm3/molc sec	25	EST	MEYLAN,WM & HOWARD,PH (1993)

<table>
<tr><td colspan="2">CAS #: 105951-36-8</td><td>5H-THIENO[3,2-B]THIOPYRAN-2-SULFONAMIDE, 6,7-DIH</td></tr>
<tr><td>Formula:</td><td colspan="2">$C_7H_9NO_4S_3$</td></tr>
<tr><td>Mol Weight:</td><td colspan="2">267.35</td></tr>
</table>

Property/Value	Units	Temp	Data Type	Reference
WS 1.23E+004	mg/L	25	EST	MEYLAN,WM ET AL. (1996)
logP -1.01			EXP	PONTICELLO,GS ET AL. (1987)
VP 7.29E-011	mm Hg	25	EST	NEELY,WB & BLAU,GE (1985)
DC	pKa			
HL 9.97E-018	atm m3/mol	25	EST	MEYLAN,WM & HOWARD,PH (1991)
OH 7.94E-011	cm3/molc sec	25	EST	MEYLAN,WM & HOWARD,PH (1993)

<table>
<tr><td colspan="2">CAS #: 105951-39-1</td><td>5H-THIENO[3,2-B]THIOPYRAN-2-SULFONAMIDE, 6,7-DIH</td></tr>
<tr><td>Formula:</td><td colspan="2">$C_7H_9NO_5S_3$</td></tr>
<tr><td>Mol Weight:</td><td colspan="2">283.35</td></tr>
</table>

Property/Value	Units	Temp	Data Type	Reference
WS 2.08E+003	mg/L	25	EST	MEYLAN,WM ET AL. (1996)
logP -0.20			EXP	PONTICELLO,GS ET AL. (1987)
VP 4.57E-009	mm Hg	25	EST	NEELY,WB & BLAU,GE (1985)
DC	pKa			
HL 4.26E-014	atm m3/mol	25	EST	MEYLAN,WM & HOWARD,PH (1991)
OH 2.97E-011	cm3/molc sec	25	EST	MEYLAN,WM & HOWARD,PH (1993)

<table>
<tr><td colspan="2">CAS #: 105951-71-1</td><td>4H-THIENO[2,3-B]THIOPYRAN-2-SULFONAMIDE, 5,6-DIH</td></tr>
<tr><td>Formula:</td><td colspan="2">$C_7H_{11}NO_4S_3$</td></tr>
<tr><td>Mol Weight:</td><td colspan="2">269.36</td></tr>
</table>

Property/Value	Units	Temp	Data Type	Reference
WS 2.30E+003	mg/L	25	EST	MEYLAN,WM ET AL. (1996)
logP -0.16			EXP	PONTICELLO,GS ET AL. (1987)
VP 4.45E-008	mm Hg	25	EST	NEELY,WB & BLAU,GE (1985)
DC	pKa			
HL 3.43E-011	atm m3/mol	25	EST	MEYLAN,WM & HOWARD,PH (1991)
OH 1.26E-011	cm3/molc sec	25	EST	MEYLAN,WM & HOWARD,PH (1993)

<table>
<tr><td colspan="2">CAS #: 105983-46-8</td><td>2-BROMO-4-METHYL-5-NITRO-1H-IMIDAZOLE</td></tr>
<tr><td>Formula:</td><td colspan="2">$C_4H_4BrN_3O_2$</td></tr>
<tr><td>Mol Weight:</td><td colspan="2">206.00</td></tr>
</table>

Property/Value	Units	Temp	Data Type	Reference
WS 3.06E+003	mg/L	25	EST	MEYLAN,WM ET AL. (1996)
logP 0.86			EXP	SUWINSKI,J ET AL. (1985)
VP 6.50E-006	mm Hg	25	EST	NEELY,WB & BLAU,GE (1985)
DC	pKa			
HL 6.52E-009	atm m3/mol	25	EST	MEYLAN,WM & HOWARD,PH (1991)
OH 3.61E-011	cm3/molc sec	25	EST	MEYLAN,WM & HOWARD,PH (1993)

<table>
<tr><td colspan="2">CAS #: 105983-47-9</td><td>2-BROMO-1,2-DIMETHYL-5-NITRO-1H-IMIDAZOLE</td></tr>
<tr><td>Formula:</td><td colspan="2">$C_5H_6BrN_3O_2$</td></tr>
<tr><td>Mol Weight:</td><td colspan="2">220.03</td></tr>
</table>

Property/Value	Units	Temp	Data Type	Reference
WS 2.80E+003	mg/L	25	EST	MEYLAN,WM ET AL. (1996)
logP 0.82			EXP	SUWINSKI,J ET AL. (1985)
VP 4.35E-005	mm Hg	25	EST	NEELY,WB & BLAU,GE (1985)
DC	pKa			
HL 1.39E-007	atm m3/mol	25	EST	MEYLAN,WM & HOWARD,PH (1991)
OH 3.63E-011	cm3/molc sec	25	EST	MEYLAN,WM & HOWARD,PH (1993)

<table>
<tr><td colspan="2">CAS #: 105994-26-1</td><td>1H-IMIDAZOLE, 2-BROMO-1,5-DIMETHYL-4-NITRO-</td></tr>
<tr><td>Formula:</td><td colspan="2">$C_5H_8BrN_3O_2$</td></tr>
<tr><td>Mol Weight:</td><td colspan="2">222.04</td></tr>
</table>

Property/Value	Units	Temp	Data Type	Reference
WS 3.00E+003	mg/L	25	EST	MEYLAN,WM ET AL. (1996)
logP 0.77			EST	MEYLAN,WM & HOWARD,PH (1995)
VP 1.39E-003	mm Hg	25	EST	NEELY,WB & BLAU,GE (1985)
DC	pKa			
HL 9.28E-011	atm m3/mol	25	EST	MEYLAN,WM & HOWARD,PH (1991)
OH 3.58E-012	cm3/molc sec	25	EST	MEYLAN,WM & HOWARD,PH (1993)

1472

106206-96-6 — CARBONIC ACID, ETHYL (5-FLUORO-3,4-DIHYDRO-2,4-D

Formula: $C_8H_9FN_2O_5$

Mol Weight: 232.17

MP (deg C):   FP (deg C):

BP (deg C):

BP pressure (mm Hg):

Property/Value	Units	Temp	Data Type	Reference
WS 1.57E+003	mg/L	25	EST	MEYLAN,WM ET AL. (1996)
logP -0.04			EXP	SANGSTER,J (1994)
VP 4.71E-009	mm Hg	25	EST	NEELY,WB & BLAU,GE (1985)
DC	pKa			
HL 8.13E-011	atm m3/mol	25	EST	MEYLAN,WM & HOWARD,PH (1991)
OH 1.36E-011	cm3/molc sec	25	EST	MEYLAN,WM & HOWARD,PH (1993)

106206-97-7 — CARBONIC ACID, ETHYL (5-FLUORO-3,6-DIHYDRO-2,6-D

Formula: $C_8H_9FN_2O_5$

Mol Weight: 232.17

MP (deg C):   FP (deg C):

BP (deg C):

BP pressure (mm Hg):

Property/Value	Units	Temp	Data Type	Reference
WS 1.08E+003	mg/L	25	EST	MEYLAN,WM ET AL. (1996)
logP 0.15			EXP	SANGSTER,J (1994)
VP 5.70E-008	mm Hg	25	EST	NEELY,WB & BLAU,GE (1985)
DC	pKa			
HL 8.13E-011	atm m3/mol	25	EST	MEYLAN,WM & HOWARD,PH (1991)
OH 1.36E-011	cm3/molc sec	25	EST	MEYLAN,WM & HOWARD,PH (1993)

106219-34-5 — CARBONIC ACID, (3,6-DIHYDRO-5-FLUORO-2,6-DIOXO-1

Formula: $C_{12}H_9FN_2O_5$

Mol Weight: 280.21

MP (deg C):   FP (deg C):

BP (deg C):

BP pressure (mm Hg):

Property/Value	Units	Temp	Data Type	Reference
WS 1.80E+003	mg/L	25	EST	MEYLAN,WM ET AL. (1996)
logP 1.11			EXP	SANGSTER,J (1994)
VP 5.96E-010	mm Hg	25	EST	NEELY,WB & BLAU,GE (1985)
DC	pKa			
HL 2.26E-011	atm m3/mol	25	EST	MEYLAN,WM & HOWARD,PH (1991)
OH 1.40E-011	cm3/molc sec	25	EST	MEYLAN,WM & HOWARD,PH (1993)

106231-36-1 — 1,3-OXATHIOLAN-4-ONE, 5,5-DIMETHYL-, O-[(METHYLA

Formula: $C_7H_{12}N_2O_3S$

Mol Weight: 204.25

MP (deg C):   FP (deg C):

BP (deg C):

BP pressure (mm Hg):

Property/Value	Units	Temp	Data Type	Reference
WS 8.30E+003	mg/L	25	EST	MEYLAN,WM ET AL. (1996)
logP 0.82			EXP	KURTZ,AP & DURDEN,JA (1987)
VP 1.85E-003	mm Hg	25	EST	NEELY,WB & BLAU,GE (1985)
DC	pKa			
HL 1.39E-011	atm m3/mol	25	EST	MEYLAN,WM & HOWARD,PH (1991)
OH 4.43E-011	cm3/molc sec	25	EST	MEYLAN,WM & HOWARD,PH (1993)

106231-37-2 — 1,4-DITHIAN-2-ONE, O-[(METHYLAMINO)CARBONYL]OXIM

Formula: $C_6H_{10}N_2O_2S_2$

Mol Weight: 206.29

MP (deg C):   FP (deg C):

BP (deg C):

BP pressure (mm Hg):

Property/Value	Units	Temp	Data Type	Reference
WS 1.55E+004	mg/L	25	EST	MEYLAN,WM ET AL. (1996)
logP 0.49			EXP	KURTZ,AP & DURDEN,JA (1987)
VP 4.42E-004	mm Hg	25	EST	NEELY,WB & BLAU,GE (1985)
DC	pKa			
HL 9.52E-012	atm m3/mol	25	EST	MEYLAN,WM & HOWARD,PH (1991)
OH 3.63E-011	cm3/molc sec	25	EST	MEYLAN,WM & HOWARD,PH (1993)

106231-39-4 — 1,4-DITHIAN-2-ONE, 3,3-DIMETHYL-, O-[(METHYLAMIN

Formula: $C_8H_{14}N_2O_2S_2$

Mol Weight: 234.34

MP (deg C):   FP (deg C):

BP (deg C):

BP pressure (mm Hg):

Property/Value	Units	Temp	Data Type	Reference
WS 1.48E+003	mg/L	25	EST	MEYLAN,WM ET AL. (1996)
logP 1.51			EXP	KURTZ,AP & DURDEN,JA (1987)
VP 1.19E-004	mm Hg	25	EST	NEELY,WB & BLAU,GE (1985)
DC	pKa			
HL 1.68E-011	atm m3/mol	25	EST	MEYLAN,WM & HOWARD,PH (1991)
OH 2.77E-011	cm3/molc sec	25	EST	MEYLAN,WM & HOWARD,PH (1993)

106231-45-2 — 1,3-DITHIOLAN-4-ONE, [(METHYLAMINO)CARBONYL]OXIM

Formula: $C_5H_8N_2O_2S_2$

Mol Weight: 192.26

MP (deg C):   FP (deg C):

BP (deg C):

BP pressure (mm Hg):

Property/Value	Units	Temp	Data Type	Reference
WS 1.76E+004	mg/L	25	EST	MEYLAN,WM ET AL. (1996)
logP 0.51			EXP	KURTZ,AP & DURDEN,JA (1987)
VP 1.30E-003	mm Hg	25	EST	NEELY,WB & BLAU,GE (1985)
DC	pKa			
HL 7.17E-012	atm m3/mol	25	EST	MEYLAN,WM & HOWARD,PH (1991)
OH 6.21E-011	cm3/molc sec	25	EST	MEYLAN,WM & HOWARD,PH (1993)

106231-46-3 — 1,3-DITHIOLAN-4-ONE, 5-METHYL-, O-[(METHYLAMINO)

Formula: $C_6H_{10}N_2O_2S_2$

Mol Weight: 206.29

MP (deg C):   FP (deg C):

BP (deg C):

BP pressure (mm Hg):

Property/Value	Units	Temp	Data Type	Reference
WS 6.79E+003	mg/L	25	EST	MEYLAN,WM ET AL. (1996)
logP 0.91			EXP	KURTZ,AP & DURDEN,JA (1987)
VP 6.33E-004	mm Hg	25	EST	NEELY,WB & BLAU,GE (1985)
DC	pKa			
HL 9.52E-012	atm m3/mol	25	EST	MEYLAN,WM & HOWARD,PH (1991)
OH 7.00E-011	cm3/molc sec	25	EST	MEYLAN,WM & HOWARD,PH (1993)

<table>
<tr><td colspan="5">CAS #: 106231-47-4</td><td>1,3-DITHIOLAN-4-ONE, 5,5-DIMETHYL-, O-[(METHYLAM</td></tr>
</table>

Formula: $C_7H_{12}N_2O_2S_2$

Mol Weight: 220.31

MP (deg C):   **FP (deg C):**

BP (deg C):

BP pressure (mm Hg):

Property/Value	Units	Temp	Data Type	Reference
WS 2.61E+003	mg/L	25	EST	MEYLAN,WM ET AL. (1996)
logP 1.31			EXP	KURTZ,AP & DURDEN,JA (1987)
VP 3.31E-004	mm Hg	25	EST	NEELY,WB & BLAU,GE (1985)
DC	pKa			
HL 1.26E-011	atm m3/mol	25	EST	MEYLAN,WM & HOWARD,PH (1991)
OH 5.53E-011	cm3/molc sec	25	EST	MEYLAN,WM & HOWARD,PH (1993)

<table>
<tr><td colspan="5">CAS #: 106231-48-5</td><td>1,3-DITHIOLAN-4-ONE, 2,2-DIMETHYL-, O-[(METHYLAM</td></tr>
</table>

Formula: $C_7H_{12}N_2O_2S_2$

Mol Weight: 220.31

MP (deg C):   **FP (deg C):**

BP (deg C):

BP pressure (mm Hg):

Property/Value	Units	Temp	Data Type	Reference
WS 2.51E+003	mg/L	25	EST	MEYLAN,WM ET AL. (1996)
logP 1.33			EXP	KURTZ,AP & DURDEN,JA (1987)
VP 3.31E-004	mm Hg	25	EST	NEELY,WB & BLAU,GE (1985)
DC	pKa			
HL 1.26E-011	atm m3/mol	25	EST	MEYLAN,WM & HOWARD,PH (1991)
OH 1.70E-011	cm3/molc sec	25	EST	MEYLAN,WM & HOWARD,PH (1993)

<table>
<tr><td colspan="5">CAS #: 106231-50-9</td><td>ACETAMIDE, 2-(BENZOYLOXY)-N-METHYL-</td></tr>
</table>

Formula: $C_{10}H_{11}NO_3$

Mol Weight: 193.20

MP (deg C):   **FP (deg C):**

BP (deg C):

BP pressure (mm Hg):

Property/Value	Units	Temp	Data Type	Reference
WS 6.76E+003	mg/L	25	EST	MEYLAN,WM ET AL. (1996)
logP 0.99			EXP	NIELSEN,LS & BUNDGAARD,H (1988)
VP 1.04E-005	mm Hg	25	EST	NEELY,WB & BLAU,GE (1985)
DC	pKa			
HL 8.19E-011	atm m3/mol	25	EST	MEYLAN,WM & HOWARD,PH (1991)
OH 7.81E-012	cm3/molc sec	25	EST	MEYLAN,WM & HOWARD,PH (1993)

<table>
<tr><td colspan="5">CAS #: 106231-51-0</td><td>ACETAMIDE, 2-(BENZOYLOXY)-N-PROPYL-</td></tr>
</table>

Formula: $C_{12}H_{15}NO_3$

Mol Weight: 221.26

MP (deg C):   **FP (deg C):**

BP (deg C):

BP pressure (mm Hg):

Property/Value	Units	Temp	Data Type	Reference
WS 8.41E+002	mg/L	25	EST	MEYLAN,WM ET AL. (1996)
logP 1.88			EXP	NIELSEN,LS & BUNDGAARD,H (1988)
VP 1.94E-006	mm Hg	25	EST	NEELY,WB & BLAU,GE (1985)
DC	pKa			
HL 1.44E-010	atm m3/mol	25	EST	MEYLAN,WM & HOWARD,PH (1991)
OH 1.33E-011	cm3/molc sec	25	EST	MEYLAN,WM & HOWARD,PH (1993)

<table>
<tr><td colspan="5">CAS #: 106231-52-1</td><td>ACETAMIDE, 2-(BENZOYLOXY)-N-(1,1-DIMETHYLETHYL)-</td></tr>
</table>

Formula: $C_{13}H_{17}NO_3$

Mol Weight: 235.29

MP (deg C):   **FP (deg C):**

BP (deg C):

BP pressure (mm Hg):

Property/Value	Units	Temp	Data Type	Reference
WS 3.35E+002	mg/L	25	EST	MEYLAN,WM ET AL. (1996)
logP 2.26			EXP	NIELSEN,LS & BUNDGAARD,H (1988)
VP 1.78E-006	mm Hg	25	EST	NEELY,WB & BLAU,GE (1985)
DC	pKa			
HL 1.92E-010	atm m3/mol	25	EST	MEYLAN,WM & HOWARD,PH (1991)
OH 7.75E-012	cm3/molc sec	25	EST	MEYLAN,WM & HOWARD,PH (1993)

<table>
<tr><td colspan="5">CAS #: 106231-53-2</td><td>ACETAMIDE, 2-[[(BENZOYLOXY)ACETYL]AMINO]-</td></tr>
</table>

Formula: $C_{11}H_{12}N_2O_4$

Mol Weight: 236.23

MP (deg C):   **FP (deg C):**

BP (deg C):

BP pressure (mm Hg):

Property/Value	Units	Temp	Data Type	Reference
WS 1.16E+003	mg/L	25	EST	MEYLAN,WM ET AL. (1996)
logP 0.09			EXP	NIELSEN,LS & BUNDGAARD,H (1988)
VP 2.90E-009	mm Hg	25	EST	NEELY,WB & BLAU,GE (1985)
DC	pKa			
HL 3.94E-013	atm m3/mol	25	EST	MEYLAN,WM & HOWARD,PH (1991)
OH 1.21E-011	cm3/molc sec	25	EST	MEYLAN,WM & HOWARD,PH (1993)

<table>
<tr><td colspan="5">CAS #: 106231-54-3</td><td>ACETAMIDE, 2-(BENZOYLOXY)-N,N-DIMETHYL-</td></tr>
</table>

Formula: $C_{11}H_{13}NO_3$

Mol Weight: 207.23

MP (deg C):   **FP (deg C):**

BP (deg C):

BP pressure (mm Hg):

Property/Value	Units	Temp	Data Type	Reference
WS 4.90E+003	mg/L	25	EST	MEYLAN,WM ET AL. (1996)
logP 1.07			EXP	NIELSEN,LS & BUNDGAARD,H (1988)
VP 1.14E-004	mm Hg	25	EST	NEELY,WB & BLAU,GE (1985)
DC	pKa			
HL 1.80E-010	atm m3/mol	25	EST	MEYLAN,WM & HOWARD,PH (1991)
OH 1.79E-011	cm3/molc sec	25	EST	MEYLAN,WM & HOWARD,PH (1993)

<table>
<tr><td colspan="5">CAS #: 106231-55-4</td><td>ACETAMIDE, 2-(BENZOYLOXY)-N,N-DIPROPYL-</td></tr>
</table>

Formula: $C_{15}H_{21}NO_3$

Mol Weight: 263.34

MP (deg C):   **FP (deg C):**

BP (deg C):

BP pressure (mm Hg):

Property/Value	Units	Temp	Data Type	Reference
WS 1.09E+002	mg/L	25	EST	MEYLAN,WM ET AL. (1996)
logP 2.65			EXP	NIELSEN,LS & BUNDGAARD,H (1988)
VP 3.74E-006	mm Hg	25	EST	NEELY,WB & BLAU,GE (1985)
DC	pKa			
HL 5.59E-010	atm m3/mol	25	EST	MEYLAN,WM & HOWARD,PH (1991)
OH 2.88E-011	cm3/molc sec	25	EST	MEYLAN,WM & HOWARD,PH (1993)

CAS #: 106231-56-5				ACETAMIDE, 2-(BENZOYLOXY)-N,N-BIS(1-METHYLETHYL)
Formula: $C_{15}H_{21}NO_3$				
Mol Weight: 263.34				
MP (deg C):			FP (deg C):	
BP (deg C):				
BP pressure (mm Hg):				

Property/Value	Units	Temp	Data Type	Reference
WS 1.30E+002	mg/L	25	EST	MEYLAN,WM ET AL. (1996)
logP 2.56			EXP	NIELSEN,LS & BUNDGAARD,H (1988)
VP 1.18E-005	mm Hg	25	EST	NEELY,WB & BLAU,GE (1985)
DC	pKa			
HL 5.59E-010	atm m3/mol	25	EST	MEYLAN,WM & HOWARD,PH (1991)
OH 3.33E-011	cm3/molc sec	25	EST	MEYLAN,WM & HOWARD,PH (1993)

CAS #: 106231-57-6				ACETAMIDE, 2-(BENZOYLOXY)-N,N-DIBUTYL-
Formula: $C_{17}H_{25}NO_3$				
Mol Weight: 291.39				
MP (deg C):			FP (deg C):	
BP (deg C):				
BP pressure (mm Hg):				

Property/Value	Units	Temp	Data Type	Reference
WS 6.31E+000	mg/L	25	EST	MEYLAN,WM ET AL. (1996)
logP 3.91			EXP	NIELSEN,LS & BUNDGAARD,H (1988)
VP 6.78E-007	mm Hg	25	EST	NEELY,WB & BLAU,GE (1985)
DC	pKa			
HL 9.85E-010	atm m3/mol	25	EST	MEYLAN,WM & HOWARD,PH (1991)
OH 3.16E-011	cm3/molc sec	25	EST	MEYLAN,WM & HOWARD,PH (1993)

CAS #: 106231-58-7				ACETAMIDE, 2-(BENZOYLOXY)-N,N-BIS(2-PROPENYL)-
Formula: $C_{15}H_{17}NO_3$				
Mol Weight: 259.31				
MP (deg C):			FP (deg C):	
BP (deg C):				
BP pressure (mm Hg):				

Property/Value	Units	Temp	Data Type	Reference
WS 2.11E+002	mg/L	25	EST	MEYLAN,WM ET AL. (1996)
logP 2.34			EXP	NIELSEN,LS & BUNDGAARD,H (1988)
VP 4.31E-006	mm Hg	25	EST	NEELY,WB & BLAU,GE (1985)
DC	pKa			
HL 3.10E-010	atm m3/mol	25	EST	MEYLAN,WM & HOWARD,PH (1991)
OH 7.70E-011	cm3/molc sec	25	EST	MEYLAN,WM & HOWARD,PH (1993)

CAS #: 106231-59-8				ACETAMIDE, 2-(BENZOYLOXY)-N-(2-HYDROXYETHYL)-N-M
Formula: $C_{12}H_{15}NO_4$				
Mol Weight: 237.26				
MP (deg C):			FP (deg C):	
BP (deg C):				
BP pressure (mm Hg):				

Property/Value	Units	Temp	Data Type	Reference
WS 8.89E+003	mg/L	25	EST	MEYLAN,WM ET AL. (1996)
logP 0.58			EXP	NIELSEN,LS & BUNDGAARD,H (1988)
VP 5.62E-008	mm Hg	25	EST	NEELY,WB & BLAU,GE (1985)
DC	pKa			
HL 8.73E-015	atm m3/mol	25	EST	MEYLAN,WM & HOWARD,PH (1991)
OH 2.62E-011	cm3/molc sec	25	EST	MEYLAN,WM & HOWARD,PH (1993)

CAS #: 106231-60-1				ACETAMIDE, 2-(BENZOYLOXY)-N-ETHYL-N-(2-HYDROXYET
Formula: $C_{13}H_{17}NO_4$				
Mol Weight: 251.28				
MP (deg C):			FP (deg C):	
BP (deg C):				
BP pressure (mm Hg):				

Property/Value	Units	Temp	Data Type	Reference
WS 3.74E+003	mg/L	25	EST	MEYLAN,WM ET AL. (1996)
logP 0.93			EXP	NIELSEN,LS & BUNDGAARD,H (1988)
VP 2.11E-008	mm Hg	25	EST	NEELY,WB & BLAU,GE (1985)
DC	pKa			
HL 1.16E-014	atm m3/mol	25	EST	MEYLAN,WM & HOWARD,PH (1991)
OH 2.96E-011	cm3/molc sec	25	EST	MEYLAN,WM & HOWARD,PH (1993)

CAS #: 106231-61-2				ACETAMIDE, 2-(BENZOYLOXY)-N,N-BIS(2-HYDROXYETHYL
Formula: $C_{13}H_{17}NO_5$				
Mol Weight: 267.28				
MP (deg C):			FP (deg C):	
BP (deg C):				
BP pressure (mm Hg):				

Property/Value	Units	Temp	Data Type	Reference
WS 1.36E+004	mg/L	25	EST	MEYLAN,WM ET AL. (1996)
logP 0.17			EXP	NIELSEN,LS & BUNDGAARD,H (1988)
VP 2.04E-010	mm Hg	25	EST	NEELY,WB & BLAU,GE (1985)
DC	pKa			
HL 4.24E-016	atm m3/mol	25	EST	MEYLAN,WM & HOWARD,PH (1991)
OH 3.45E-011	cm3/molc sec	25	EST	MEYLAN,WM & HOWARD,PH (1993)

CAS #: 106231-62-3				ACETAMIDE, N-(2-AMINO-2-OXOETHYL)-2-(BENZOYLOXY)
Formula: $C_{12}H_{14}N_2O_4$				
Mol Weight: 250.26				
MP (deg C):			FP (deg C):	
BP (deg C):				
BP pressure (mm Hg):				

Property/Value	Units	Temp	Data Type	Reference
WS 9.87E+002	mg/L	25	EST	MEYLAN,WM ET AL. (1996)
logP 0.08			EXP	NIELSEN,LS & BUNDGAARD,H (1988)
VP 2.18E-008	mm Hg	25	EST	NEELY,WB & BLAU,GE (1985)
DC	pKa			
HL 8.64E-013	atm m3/mol	25	EST	MEYLAN,WM & HOWARD,PH (1991)
OH 2.22E-011	cm3/molc sec	25	EST	MEYLAN,WM & HOWARD,PH (1993)

CAS #: 106231-63-4				GLYCINE, N-[(BENZOYLOXY)ACETYL]-N-METHYL-, ETHYL
Formula: $C_{14}H_{17}NO_5$				
Mol Weight: 279.30				
MP (deg C):			FP (deg C):	
BP (deg C):				
BP pressure (mm Hg):				

Property/Value	Units	Temp	Data Type	Reference
WS 7.53E+002	mg/L	25	EST	MEYLAN,WM ET AL. (1996)
logP 1.56			EXP	NIELSEN,LS & BUNDGAARD,H (1988)
VP 1.33E-005	mm Hg	25	EST	NEELY,WB & BLAU,GE (1985)
DC	pKa			
HL 1.13E-011	atm m3/mol	25	EST	MEYLAN,WM & HOWARD,PH (1991)
OH 2.18E-011	cm3/molc sec	25	EST	MEYLAN,WM & HOWARD,PH (1993)

106231-64-5 — GLYCINE, N-[(BENZOYLOXY)ACETYL]-N-METHYL-

Formula: $C_{12}H_{13}NO_5$
Mol Weight: 251.24
MP (deg C): **FP (deg C):**
BP (deg C):
BP pressure (mm Hg):

Property/Value	Units	Temp	Data Type	Reference
WS 6.24E+003	mg/L	25	EST	MEYLAN,WM ET AL. (1996)
logP 0.67			EXP	NIELSEN,LS & BUNDGAARD,H (1988)
VP 2.13E-007	mm Hg	25	EST	NEELY,WB & BLAU,GE (1985)
DC	pKa			
HL 2.67E-014	atm m3/mol	25	EST	MEYLAN,WM & HOWARD,PH (1991)
OH 2.07E-011	cm3/molc sec	25	EST	MEYLAN,WM & HOWARD,PH (1993)

106231-67-8 — PIPERIDINE, 1-[(BENZOYLOXY)ACETYL]-

Formula: $C_{14}H_{17}NO_3$
Mol Weight: 247.30
MP (deg C): **FP (deg C):**
BP (deg C):
BP pressure (mm Hg):

Property/Value	Units	Temp	Data Type	Reference
WS 5.29E+002	mg/L	25	EST	MEYLAN,WM ET AL. (1996)
logP 1.95			EXP	NIELSEN,LS & BUNDGAARD,H (1988)
VP 4.74E-006	mm Hg	25	EST	NEELY,WB & BLAU,GE (1985)
DC	pKa			
HL 1.86E-010	atm m3/mol	25	EST	MEYLAN,WM & HOWARD,PH (1991)
OH 3.04E-011	cm3/molc sec	25	EST	MEYLAN,WM & HOWARD,PH (1993)

106231-68-9 — MORPHOLINE, 4-[(BENZOYLOXY)ACETYL]-

Formula: $C_{13}H_{15}NO_4$
Mol Weight: 249.27
MP (deg C): **FP (deg C):**
BP (deg C):
BP pressure (mm Hg):

Property/Value	Units	Temp	Data Type	Reference
WS 4.07E+003	mg/L	25	EST	MEYLAN,WM ET AL. (1996)
logP 0.90			EXP	NIELSEN,LS & BUNDGAARD,H (1988)
VP 4.06E-006	mm Hg	25	EST	NEELY,WB & BLAU,GE (1985)
DC	pKa			
HL 1.23E-012	atm m3/mol	25	EST	MEYLAN,WM & HOWARD,PH (1991)
OH 5.76E-011	cm3/molc sec	25	EST	MEYLAN,WM & HOWARD,PH (1993)

106231-69-0 — L-PROLINAMIDE,N2-[(BENZOYLOXY)ACETYL]-

Formula: $C_{14}H_{16}N_2O_4$
Mol Weight: 276.29
MP (deg C): **FP (deg C):**
BP (deg C):
BP pressure (mm Hg):

Property/Value	Units	Temp	Data Type	Reference
WS 5.56E+002	mg/L	25	EST	MEYLAN,WM ET AL. (1996)
logP 0.20			EXP	NIELSEN,LS & BUNDGAARD,H (1988)
VP 3.49E-009	mm Hg	25	EST	NEELY,WB & BLAU,GE (1985)
DC	pKa			
HL 6.71E-013	atm m3/mol	25	EST	MEYLAN,WM & HOWARD,PH (1991)
OH 3.31E-011	cm3/molc sec	25	EST	MEYLAN,WM & HOWARD,PH (1993)

106293-84-9 — 1,3-OXATHIOLAN-4-ONE, 2,5-DIMETHYL-, O-[(METHYLA

Formula: $C_7H_{12}N_2O_3S$
Mol Weight: 204.25
MP (deg C): **FP (deg C):**
BP (deg C):
BP pressure (mm Hg):

Property/Value	Units	Temp	Data Type	Reference
WS 6.30E+003	mg/L	25	EST	MEYLAN,WM ET AL. (1996)
logP 0.96			EXP	KURTZ,AP & DURDEN,JA (1987)
VP 1.53E-003	mm Hg	25	EST	NEELY,WB & BLAU,GE (1985)
DC	pKa			
HL 1.39E-011	atm m3/mol	25	EST	MEYLAN,WM & HOWARD,PH (1991)
OH 9.36E-011	cm3/molc sec	25	EST	MEYLAN,WM & HOWARD,PH (1993)

106319-38-4 — 4H-THIENO[2,3-B]THIOPYRAN-2-SULFONAMIDE, 5,6-DIH

Formula: $C_7H_{11}NO_3S_3$
Mol Weight: 253.36
MP (deg C): **FP (deg C):**
BP (deg C):
BP pressure (mm Hg):

Property/Value	Units	Temp	Data Type	Reference
WS 6.11E+003	mg/L	25	EST	MEYLAN,WM ET AL. (1996)
logP 0.68			EXP	PONTICELLO,GS ET AL. (1987)
VP 2.02E-009	mm Hg	25	EST	NEELY,WB & BLAU,GE (1985)
DC	pKa			
HL 1.58E-013	atm m3/mol	25	EST	MEYLAN,WM & HOWARD,PH (1991)
OH 2.18E-011	cm3/molc sec	25	EST	MEYLAN,WM & HOWARD,PH (1993)

106319-42-0 — 4H-THIENO[2,3-B]THIOPYRAN-2-SULFONAMIDE, 5,6-DIH

Formula: $C_7H_{11}NO_5S_3$
Mol Weight: 285.36
MP (deg C): **FP (deg C):**
BP (deg C):
BP pressure (mm Hg):

Property/Value	Units	Temp	Data Type	Reference
WS 3.24E+003	mg/L	25	EST	MEYLAN,WM ET AL. (1996)
logP -0.44			EXP	PONTICELLO,GS ET AL. (1987)
VP 3.73E-011	mm Hg	25	EST	NEELY,WB & BLAU,GE (1985)
DC	pKa			
HL 1.25E-015	atm m3/mol	25	EST	MEYLAN,WM & HOWARD,PH (1991)
OH 2.16E-011	cm3/molc sec	25	EST	MEYLAN,WM & HOWARD,PH (1993)

106319-44-2 — 5H-THIENO[3,2-B]THIOPYRAN-2-SULFONAMIDE, 6,7-DIH

Formula: $C_7H_{11}NO_3S_3$
Mol Weight: 253.36
MP (deg C): **FP (deg C):**
BP (deg C):
BP pressure (mm Hg):

Property/Value	Units	Temp	Data Type	Reference
WS 9.05E+003	mg/L	25	EST	MEYLAN,WM ET AL. (1996)
logP 0.48			EXP	PONTICELLO,GS ET AL. (1987)
VP 2.02E-009	mm Hg	25	EST	NEELY,WB & BLAU,GE (1985)
DC	pKa			
HL 1.58E-013	atm m3/mol	25	EST	MEYLAN,WM & HOWARD,PH (1991)
OH 3.47E-011	cm3/molc sec	25	EST	MEYLAN,WM & HOWARD,PH (1993)

CAS #: 106319-45-3				5H-THIENO[3,2-B]THIOPYRAN-2-SULFONAMIDE, 6,7-DIH

Formula: $C_7H_{11}NO_5S_3$

Mol Weight: 285.36

MP (deg C):   FP (deg C):

BP (deg C):

BP pressure (mm Hg):

Property/Value	Units	Temp	Data Type	Reference
WS 4.35E+003	mg/L	25	EST	MEYLAN,WM ET AL. (1996)
logP -0.59			EXP	PONTICELLO,GS ET AL. (1987)
VP 3.73E-011	mm Hg	25	EST	NEELY,WB & BLAU,GE (1985)
DC	pKa			
HL 1.25E-015	atm m3/mol	25	EST	MEYLAN,WM & HOWARD,PH (1991)
OH 1.94E-011	cm3/molc sec	25	EST	MEYLAN,WM & HOWARD,PH (1993)

CAS #: 106319-46-4				5H-THIENO[3,2-B]THIOPYRAN-2-SULFONAMIDE, 6,7-DIH

Formula: $C_7H_{11}NO_5S_3$

Mol Weight: 285.36

MP (deg C):   FP (deg C):

BP (deg C):

BP pressure (mm Hg):

Property/Value	Units	Temp	Data Type	Reference
WS 5.40E+003	mg/L	25	EST	MEYLAN,WM ET AL. (1996)
logP -0.70			EXP	PONTICELLO,GS ET AL. (1987)
VP 3.73E-011	mm Hg	25	EST	NEELY,WB & BLAU,GE (1985)
DC	pKa			
HL 1.25E-015	atm m3/mol	25	EST	MEYLAN,WM & HOWARD,PH (1991)
OH 2.11E-011	cm3/molc sec	25	EST	MEYLAN,WM & HOWARD,PH (1993)

CAS #: 106325-08-0				BAS 480F

Formula: $C_{17}H_{15}ClFN_3O$

Mol Weight: 331.78

MP (deg C):   FP (deg C):

BP (deg C):

BP pressure (mm Hg):

Property/Value	Units	Temp	Data Type	Reference
WS 9.46E+000	mg/L	25	EST	MEYLAN,WM ET AL. (1996)
logP 3.44			EXP	TOMLIN,C (1994)
VP 1.10E-007	mm Hg	25	EST	NEELY,WB & BLAU,GE (1985)
DC	pKa			
HL 3.98E-010	atm m3/mol	25	EST	MEYLAN,WM & HOWARD,PH (1991)
OH 8.78E-012	cm3/molc sec	25	EST	MEYLAN,WM & HOWARD,PH (1993)

CAS #: 106335-79-9				4H-THIENO[2,3-B]THIOPYRAN-2-SULFONAMIDE, 5,6-DIH

Formula: $C_7H_{11}NO_4S_3$

Mol Weight: 269.36

MP (deg C):   FP (deg C):

BP (deg C):

BP pressure (mm Hg):

Property/Value	Units	Temp	Data Type	Reference
WS 1.03E+004	mg/L	25	EST	MEYLAN,WM ET AL. (1996)
logP -0.92			EXP	PONTICELLO,GS ET AL. (1987)
VP 7.29E-011	mm Hg	25	EST	NEELY,WB & BLAU,GE (1985)
DC	pKa			
HL 9.97E-018	atm m3/mol	25	EST	MEYLAN,WM & HOWARD,PH (1991)
OH 7.99E-011	cm3/molc sec	25	EST	MEYLAN,WM & HOWARD,PH (1993)

CAS #: 106362-29-2				BENZENAMINE, 3-(1,2,3,6-TETRAHYDRO-1-METHYL-4-PY

Formula: $C_{12}H_{16}N_2$

Mol Weight: 188.27

MP (deg C):   FP (deg C):

BP (deg C):

BP pressure (mm Hg):

Property/Value	Units	Temp	Data Type	Reference
WS 2.84E+004	mg/L	25	EST	MEYLAN,WM ET AL. (1996)
logP 1.47			EXP	ALTOMARE,CA ET AL. (1992)
VP 1.37E-004	mm Hg	25	EST	NEELY,WB & BLAU,GE (1985)
DC	pKa			
HL 5.24E-010	atm m3/mol	25	EST	MEYLAN,WM & HOWARD,PH (1991)
OH 2.76E-010	cm3/molc sec	25	EST	MEYLAN,WM & HOWARD,PH (1993)

CAS #: 106362-30-5				BENZENAMINE, 4-(1,2,3,6-TETRAHYDRO-1-METHYL-4-PY

Formula: $C_{12}H_{16}N_2$

Mol Weight: 188.27

MP (deg C):   FP (deg C):

BP (deg C):

BP pressure (mm Hg):

Property/Value	Units	Temp	Data Type	Reference
WS 2.78E+004	mg/L	25	EST	MEYLAN,WM ET AL. (1996)
logP 1.48			EXP	ALTOMARE,CA ET AL. (1992)
VP 1.37E-004	mm Hg	25	EST	NEELY,WB & BLAU,GE (1985)
DC	pKa			
HL 5.24E-010	atm m3/mol	25	EST	MEYLAN,WM & HOWARD,PH (1991)
OH 2.76E-010	cm3/molc sec	25	EST	MEYLAN,WM & HOWARD,PH (1993)

CAS #: 106400-04-8				4H-THIENO[2,3-B]THIOPYRAN-2-SULFONAMIDE, 5,6-DIH

Formula: $C_7H_{11}NO_5S_3$

Mol Weight: 285.36

MP (deg C):   FP (deg C):

BP (deg C):

BP pressure (mm Hg):

Property/Value	Units	Temp	Data Type	Reference
WS 2.71E+003	mg/L	25	EST	MEYLAN,WM ET AL. (1996)
logP -0.35			EXP	PONTICELLO,GS ET AL. (1987)
VP 3.73E-011	mm Hg	25	EST	NEELY,WB & BLAU,GE (1985)
DC	pKa			
HL 1.25E-015	atm m3/mol	25	EST	MEYLAN,WM & HOWARD,PH (1991)
OH 1.99E-011	cm3/molc sec	25	EST	MEYLAN,WM & HOWARD,PH (1993)

CAS #: 106848-84-4				1H-IMIDAZOLE-5-BUTANOIC ACID, alpha-ETHYL-1-METH

Formula: $C_{22}H_{30}N_2O_4$

Mol Weight: 386.50

MP (deg C):   FP (deg C):

BP (deg C):

BP pressure (mm Hg):

Property/Value	Units	Temp	Data Type	Reference
WS 2.20E+000	mg/L	25	EST	MEYLAN,WM ET AL. (1996)
logP 3.78			EXP	SANGSTER,J (1993)
VP 2.85E-009	mm Hg	25	EST	NEELY,WB & BLAU,GE (1985)
DC	pKa			
HL 2.94E-010	atm m3/mol	25	EST	MEYLAN,WM & HOWARD,PH (1991)
OH 1.08E-010	cm3/molc sec	25	EST	MEYLAN,WM & HOWARD,PH (1993)

CAS #: 106848-85-5				1H-IMIDAZOLE-5-BUTANOIC ACID, alpha-ETHYL-1-METH
Formula: $C_{24}H_{34}N_2O_4$				
Mol Weight: 414.55				
MP (deg C):		FP (deg C):		
BP (deg C):				
BP pressure (mm Hg):				

Property/Value	Units	Temp	Data Type	Reference
WS 2.18E-001	mg/L	25	EST	MEYLAN,WM ET AL. (1996)
logP 4.75			EXP	SANGSTER,J (1993)
VP 4.59E-010	mm Hg	25	EST	NEELY,WB & BLAU,GE (1985)
DC	pKa			
HL 5.18E-010	atm m3/mol	25	EST	MEYLAN,WM & HOWARD,PH (1991)
OH 1.11E-010	cm3/molc sec	25	EST	MEYLAN,WM & HOWARD,PH (1993)

CAS #: 106848-86-6				1H-IMIDAZOLE-5-BUTANOIC ACID, beta-[(2,2-DIMETHY
Formula: $C_{24}H_{34}N_2O_4$				
Mol Weight: 414.55				
MP (deg C):		FP (deg C):		
BP (deg C):				
BP pressure (mm Hg):				

Property/Value	Units	Temp	Data Type	Reference
WS 3.24E-001	mg/L	25	EST	MEYLAN,WM ET AL. (1996)
logP 4.55			EXP	SANGSTER,J (1993)
VP 8.07E-010	mm Hg	25	EST	NEELY,WB & BLAU,GE (1985)
DC	pKa			
HL 4.31E-010	atm m3/mol	25	EST	MEYLAN,WM & HOWARD,PH (1991)
OH 1.08E-010	cm3/molc sec	25	EST	MEYLAN,WM & HOWARD,PH (1993)

CAS #: 107036-57-7				CYTIDINE, 5-BROMO-2',3'-DIDEOXY-
Formula: $C_9H_{12}BrN_3O_3$				
Mol Weight: 290.12				
MP (deg C):		FP (deg C):		
BP (deg C):				
BP pressure (mm Hg):				

Property/Value	Units	Temp	Data Type	Reference
WS 1.83E+003	mg/L	25	EST	MEYLAN,WM ET AL. (1996)
logP -0.50			EXP	CHEUNG,AP & KENNEY,D (1990)
VP 2.24E-009	mm Hg	25	EST	NEELY,WB & BLAU,GE (1985)
DC	pKa			
HL 2.33E-017	atm m3/mol	25	EST	MEYLAN,WM & HOWARD,PH (1991)
OH 8.30E-011	cm3/molc sec	25	EST	MEYLAN,WM & HOWARD,PH (1993)

CAS #: 107036-62-4				CYTIDINE, 2',3'-DIDEOXY-5-FLUORO-
Formula: $C_9H_{12}FN_3O_3$				
Mol Weight: 229.21				
MP (deg C):		FP (deg C):		
BP (deg C):				
BP pressure (mm Hg):				

Property/Value	Units	Temp	Data Type	Reference
WS 1.28E+004	mg/L	25	EST	MEYLAN,WM ET AL. (1996)
logP -1.09			EXP	SIDDIQUI,MA ET AL. (1992)
VP 5.39E-008	mm Hg	25	EST	NEELY,WB & BLAU,GE (1985)
DC	pKa			
HL 2.24E-016	atm m3/mol	25	EST	MEYLAN,WM & HOWARD,PH (1991)
OH 8.04E-011	cm3/molc sec	25	EST	MEYLAN,WM & HOWARD,PH (1993)

CAS #: 107085-84-7				HYDROCORTISON-21-HEMIPIMELATE
Formula: $C_{28}H_{40}O_8$				
Mol Weight: 504.63				
MP (deg C):		FP (deg C):		
BP (deg C):				
BP pressure (mm Hg):				

Property/Value	Units	Temp	Data Type	Reference
WS 2.73E+000	mg/L	25	EST	MEYLAN,WM ET AL. (1996)
logP 3.26			EXP	HANSCH,C & LEO,AJ (1985)
VP 3.06E-018	mm Hg	25	EST	NEELY,WB & BLAU,GE (1985)
DC	pKa			
HL 1.44E-016	atm m3/mol	25	EST	MEYLAN,WM & HOWARD,PH (1991)
OH 1.15E-010	cm3/molc sec	25	EST	MEYLAN,WM & HOWARD,PH (1993)

CAS #: 107103-97-9				5,5,5-TRIFLUORO-3-METHYL-1-PENTANOL
Formula: $C_6H_{11}F_3O$				
Mol Weight: 156.15				
MP (deg C):		FP (deg C):		
BP (deg C):				
BP pressure (mm Hg):				

Property/Value	Units	Temp	Data Type	Reference
WS 9.72E+003	mg/L	25	EST	MEYLAN,WM ET AL. (1996)
logP 1.61			EXP	MULLER,N (1986)
VP 4.09E+000	mm Hg	25	EST	NEELY,WB & BLAU,GE (1985)
DC	pKa			
HL 8.75E-005	atm m3/mol	25	EST	MEYLAN,WM & HOWARD,PH (1991)
OH 8.76E-012	cm3/molc sec	25	EST	MEYLAN,WM & HOWARD,PH (1993)

CAS #: 107115-73-1				1H-PYRROLIZINE-1-CARBOXYLIC ACID, 5-BENZOYL-2,3-
Formula: $C_{16}H_{15}NO_3$				
Mol Weight: 269.30				
MP (deg C):		FP (deg C):		
BP (deg C):				
BP pressure (mm Hg):				

Property/Value	Units	Temp	Data Type	Reference
WS 1.69E+002	mg/L	25	EST	MEYLAN,WM ET AL. (1996)
logP 2.85			EXP	SANGSTER,J (1994)
VP 4.80E-008	mm Hg	25	EST	NEELY,WB & BLAU,GE (1985)
DC	pKa			
HL 3.70E-013	atm m3/mol	25	EST	MEYLAN,WM & HOWARD,PH (1991)
OH 2.04E-010	cm3/molc sec	25	EST	MEYLAN,WM & HOWARD,PH (1993)

CAS #: 107115-74-2				1H-PYRROLIZINE-1-CARBOXYLIC ACID, 5-(4-FLUOROBEN
Formula: $C_{16}H_{14}FNO_3$				
Mol Weight: 287.29				
MP (deg C):		FP (deg C):		
BP (deg C):				
BP pressure (mm Hg):				

Property/Value	Units	Temp	Data Type	Reference
WS 1.05E+002	mg/L	25	EST	MEYLAN,WM ET AL. (1996)
logP 2.97			EXP	SANGSTER,J (1994)
VP 5.77E-008	mm Hg	25	EST	NEELY,WB & BLAU,GE (1985)
DC	pKa			
HL 4.32E-013	atm m3/mol	25	EST	MEYLAN,WM & HOWARD,PH (1991)
OH 2.04E-010	cm3/molc sec	25	EST	MEYLAN,WM & HOWARD,PH (1993)

CAS #: 107485-43-8				2-CHLORO-6-METHOXYBENZAMIDE
Formula: $C_8H_8ClNO_2$				
Mol Weight: 185.61				
MP (deg C):		FP (deg C):		
BP (deg C):				
BP pressure (mm Hg):				

Property/Value	Units	Temp	Data Type	Reference
WS 1.62E+004	mg/L	25	EST	MEYLAN,WM ET AL. (1996)
logP 0.59			EXP	NAKAGAWA,Y ET AL. (1992)
VP 2.27E-005	mm Hg	25	EST	NEELY,WB & BLAU,GE (1985)
DC	pKa			
HL 9.71E-011	atm m3/mol	25	EST	MEYLAN,WM & HOWARD,PH (1991)
OH 1.66E-011	cm3/molc sec	25	EST	MEYLAN,WM & HOWARD,PH (1993)

CAS #: 107485-46-1				BENZAMIDE, N-[[(4-CHLOROPHENYL)AMINO]CARBONYL]-2
Formula: $C_{15}H_{12}ClN_3O_4$				
Mol Weight: 333.73				
MP (deg C):		FP (deg C):		
BP (deg C):				
BP pressure (mm Hg):				

Property/Value	Units	Temp	Data Type	Reference
WS 1.98E+000	mg/L	25	EST	MEYLAN,WM ET AL. (1996)
logP 3.75			EXP	SOTOMATSU,T ET AL. (1987)
VP 1.53E-012	mm Hg	25	EST	NEELY,WB & BLAU,GE (1985)
DC	pKa			
HL 3.80E-014	atm m3/mol	25	EST	MEYLAN,WM & HOWARD,PH (1991)
OH 1.40E-011	cm3/molc sec	25	EST	MEYLAN,WM & HOWARD,PH (1993)

CAS #: 107485-58-5				BENZAMIDE, N-[[(4-CHLOROPHENYL)AMINO]CARBONYL]-2
Formula: $C_{15}H_{13}ClN_2O_3$				
Mol Weight: 304.74				
MP (deg C):		FP (deg C):		
BP (deg C):				
BP pressure (mm Hg):				

Property/Value	Units	Temp	Data Type	Reference
WS 2.22E+000	mg/L	25	EST	MEYLAN,WM ET AL. (1996)
logP 4.35			EXP	SOTOMATSU,T ET AL. (1987)
VP 3.59E-011	mm Hg	25	EST	NEELY,WB & BLAU,GE (1985)
DC	pKa			
HL 5.16E-013	atm m3/mol	25	EST	MEYLAN,WM & HOWARD,PH (1991)
OH 3.39E-011	cm3/molc sec	25	EST	MEYLAN,WM & HOWARD,PH (1993)

CAS #: 107485-63-2				2-FLUORO-6-BROMOBENZAMIDE
Formula: C_7H_5BrFNO				
Mol Weight: 218.03				
MP (deg C):		FP (deg C):		
BP (deg C):				
BP pressure (mm Hg):				

Property/Value	Units	Temp	Data Type	Reference
WS 1.02E+004	mg/L	25	EST	MEYLAN,WM ET AL. (1996)
logP 0.63			EXP	NAKAGAWA,Y ET AL. (1992)
VP 6.31E-005	mm Hg	25	EST	NEELY,WB & BLAU,GE (1985)
DC	pKa			
HL 1.03E-009	atm m3/mol	25	EST	MEYLAN,WM & HOWARD,PH (1991)
OH 3.40E-012	cm3/molc sec	25	EST	MEYLAN,WM & HOWARD,PH (1993)

CAS #: 107485-64-3				2-CHLORO-6-NITROBENZAMIDE
Formula: $C_7H_5ClN_2O_3$				
Mol Weight: 200.58				
MP (deg C):		FP (deg C):		
BP (deg C):				
BP pressure (mm Hg):				

Property/Value	Units	Temp	Data Type	Reference
WS 6.85E+002	mg/L	25	EST	MEYLAN,WM ET AL. (1996)
logP 0.12			EXP	NAKAGAWA,Y ET AL. (1992)
VP 1.92E-006	mm Hg	25	EST	NEELY,WB & BLAU,GE (1985)
DC	pKa			
HL 6.47E-012	atm m3/mol	25	EST	MEYLAN,WM & HOWARD,PH (1991)
OH 2.11E-012	cm3/molc sec	25	EST	MEYLAN,WM & HOWARD,PH (1993)

CAS #: 107485-65-4				2-BROMO-6-NITROBENZAMIDE
Formula: $C_7H_5BrN_2O_3$				
Mol Weight: 245.03				
MP (deg C):		FP (deg C):		
BP (deg C):				
BP pressure (mm Hg):				

Property/Value	Units	Temp	Data Type	Reference
WS 2.90E+002	mg/L	25	EST	MEYLAN,WM ET AL. (1996)
logP 0.28			EXP	NAKAGAWA,Y ET AL. (1992)
VP 8.53E-007	mm Hg	25	EST	NEELY,WB & BLAU,GE (1985)
DC	pKa			
HL 3.48E-012	atm m3/mol	25	EST	MEYLAN,WM & HOWARD,PH (1991)
OH 2.10E-012	cm3/molc sec	25	EST	MEYLAN,WM & HOWARD,PH (1993)

CAS #: 107534-96-3				TEBUCONAZOLE
Formula: $C_{16}H_{24}ClN_3O$				
Mol Weight: 309.84				
MP (deg C): 102.4		FP (deg C):		
BP (deg C):				
BP pressure (mm Hg):				

Property/Value	Units	Temp	Data Type	Reference
WS 2.48E+001	mg/L	25	EST	MEYLAN,WM ET AL. (1996)
logP 3.70			EXP	TOMLIN,C (1994)
VP 1.00E-008	mm Hg	25	EST	NEELY,WB & BLAU,GE (1985)
DC	pKa			
HL 5.18E-010	atm m3/mol	25	EST	MEYLAN,WM & HOWARD,PH (1991)
OH 1.15E-011	cm3/molc sec	25	EST	MEYLAN,WM & HOWARD,PH (1993)

CAS #: 107538-70-5				8-(4-CL-PH)-S CYCLIC AMP-O-BENZYL
Formula: $C_{23}H_{23}ClN_5O_6PS$				
Mol Weight: 563.96				
MP (deg C):		FP (deg C):		
BP (deg C):				
BP pressure (mm Hg):				

Property/Value	Units	Temp	Data Type	Reference
WS 3.48E+000	mg/L	25	EST	MEYLAN,WM ET AL. (1996)
logP 2.24			EXP	KORTH,M & ENGELS,J (1987)
VP 1.78E-010	mm Hg	25	EST	NEELY,WB & BLAU,GE (1985)
DC	pKa			
HL 3.26E-029	atm m3/mol	25	EST	MEYLAN,WM & HOWARD,PH (1991)
OH 3.29E-010	cm3/molc sec	25	EST	MEYLAN,WM & HOWARD,PH (1993)

CAS #: 107538-71-6				8-S-T-BUTYL CYCLIC AMP-O-BENZYL

Formula: $C_{21}H_{28}N_5O_6PS$				
Mol Weight: 509.53				
MP (deg C):		FP (deg C):		
BP (deg C):				
BP pressure (mm Hg):				

Property/ Value	Units	Temp	Data Type	Reference
WS 1.14E+001	mg/L	25	EST	MEYLAN,WM ET AL. (1996)
logP 2.05			EXP	KORTH,M & ENGELS,J (1987)
VP 1.78E-010	mm Hg	25	EST	NEELY,WB & BLAU,GE (1985)
DC	pKa			
HL 5.15E-028	atm m3/mol	25	EST	MEYLAN,WM & HOWARD,PH (1991)
OH 3.26E-010	cm3/molc sec	25	EST	MEYLAN,WM & HOWARD,PH (1993)

CAS #: 107538-72-7				8-S-BENZYL CYCLIC AMP-O-BENZYL

Formula: $C_{24}H_{26}N_5O_6PS$				
Mol Weight: 543.54				
MP (deg C):		FP (deg C):		
BP (deg C):				
BP pressure (mm Hg):				

Property/ Value	Units	Temp	Data Type	Reference
WS 8.87E+000	mg/L	25	EST	MEYLAN,WM ET AL. (1996)
logP 1.92			EXP	KORTH,M & ENGELS,J (1987)
VP 1.78E-010	mm Hg	25	EST	NEELY,WB & BLAU,GE (1985)
DC	pKa			
HL 1.78E-029	atm m3/mol	25	EST	MEYLAN,WM & HOWARD,PH (1991)
OH 3.37E-010	cm3/molc sec	25	EST	MEYLAN,WM & HOWARD,PH (1993)

CAS #: 107538-73-8				8-METHYLTHIO CYCLIC AMP-O-BENZYL

Formula: $C_{18}H_{22}N_5O_6PS$				
Mol Weight: 467.44				
MP (deg C):		FP (deg C):		
BP (deg C):				
BP pressure (mm Hg):				

Property/ Value	Units	Temp	Data Type	Reference
WS 4.92E+001	mg/L	25	EST	MEYLAN,WM ET AL. (1996)
logP 1.62			EXP	KORTH,M & ENGELS,J (1987)
VP 1.78E-010	mm Hg	25	EST	NEELY,WB & BLAU,GE (1985)
DC	pKa			
HL 2.20E-028	atm m3/mol	25	EST	MEYLAN,WM & HOWARD,PH (1991)
OH 3.26E-010	cm3/molc sec	25	EST	MEYLAN,WM & HOWARD,PH (1993)

CAS #: 107538-74-9				8-SEBENZYL CYCLIC AMP-O-BENZYL

Formula: $C_{24}H_{26}N_5O_6PSe$				
Mol Weight: 590.44				
MP (deg C):		FP (deg C):		
BP (deg C):				
BP pressure (mm Hg):				

Property/ Value	Units	Temp	Data Type	Reference
WS 7.18E+000	mg/L	25	EST	MEYLAN,WM ET AL. (1996)
logP 1.67			EXP	KORTH,M & ENGELS,J (1987)
VP 1.78E-010	mm Hg	25	EST	NEELY,WB & BLAU,GE (1985)
DC	pKa			
HL 2.45E-029	atm m3/mol	25	EST	MEYLAN,WM & HOWARD,PH (1991)
OH 4.02E-010	cm3/molc sec	25	EST	MEYLAN,WM & HOWARD,PH (1993)

CAS #: 107538-75-0				8-BROMO CYCLIC AMP-O-BENZYL

Formula: $C_{17}H_{19}BrN_5O_6P$				
Mol Weight: 500.25				
MP (deg C):		FP (deg C):		
BP (deg C):				
BP pressure (mm Hg):				

Property/ Value	Units	Temp	Data Type	Reference
WS 3.29E+001	mg/L	25	EST	MEYLAN,WM ET AL. (1996)
logP 1.58			EXP	KORTH,M & ENGELS,J (1987)
VP 1.78E-010	mm Hg	25	EST	NEELY,WB & BLAU,GE (1985)
DC	pKa			
HL 3.01E-027	atm m3/mol	25	EST	MEYLAN,WM & HOWARD,PH (1991)
OH 2.71E-010	cm3/molc sec	25	EST	MEYLAN,WM & HOWARD,PH (1993)

CAS #: 107550-73-2				ADENOSINE, 2-AMINO-2',3'-DIDEOXY-

Formula: $C_{10}H_{14}N_6O_2$				
Mol Weight: 250.26				
MP (deg C):		FP (deg C):		
BP (deg C):				
BP pressure (mm Hg):				

Property/ Value	Units	Temp	Data Type	Reference
WS 2.85E+003	mg/L	25	EST	MEYLAN,WM ET AL. (1996)
logP -0.46			EXP	BALZARINI,J ET AL. (1989)
VP 1.04E-012	mm Hg	25	EST	NEELY,WB & BLAU,GE (1985)
DC	pKa			
HL 1.04E-019	atm m3/mol	25	EST	MEYLAN,WM & HOWARD,PH (1991)
OH 2.32E-010	cm3/molc sec	25	EST	MEYLAN,WM & HOWARD,PH (1993)

CAS #: 108441-45-8				THYMIDINE, 3'-AZIDO-3'-DEOXY-4-THIO-

Formula: $C_{10}H_{13}N_5O_3S$				
Mol Weight: 283.31				
MP (deg C):		FP (deg C):		
BP (deg C):				
BP pressure (mm Hg):				

Property/ Value	Units	Temp	Data Type	Reference
WS 6.92E+002	mg/L	25	EST	MEYLAN,WM ET AL. (1996)
logP 1.07			EXP	PALOMINO,E ET AL. (1990)
VP 7.58E-019	mm Hg	25	EST	NEELY,WB & BLAU,GE (1985)
DC	pKa			
HL 8.61E-023	atm m3/mol	25	EST	MEYLAN,WM & HOWARD,PH (1991)
OH 1.24E-010	cm3/molc sec	25	EST	MEYLAN,WM & HOWARD,PH (1993)

CAS #: 108610-76-0				[3,4'-BIPYRIDINE]-5-CARBONITRILE,6-[(2-HYDROXYET

Formula: $C_{13}H_{12}N_4O$				
Mol Weight: 240.27				
MP (deg C):		FP (deg C):		
BP (deg C):				
BP pressure (mm Hg):				

Property/ Value	Units	Temp	Data Type	Reference
WS 1.46E+003	mg/L	25	EST	MEYLAN,WM ET AL. (1996)
logP 1.17			EXP	HAGEN,V ET AL. (1989)
VP 5.54E-010	mm Hg	25	EST	NEELY,WB & BLAU,GE (1985)
DC	pKa			
HL 2.58E-019	atm m3/mol	25	EST	MEYLAN,WM & HOWARD,PH (1991)
OH 1.55E-011	cm3/molc sec	25	EST	MEYLAN,WM & HOWARD,PH (1993)

CAS #:	108611-26-3	3-PYRIDINECARBONITRILE, 6-METHYL-2-(4-MORPHOLINY

Formula: $C_{17}H_{17}N_3O$

Mol Weight: 279.34

MP (deg C): FP (deg C):

BP (deg C):

BP pressure (mm Hg):

Property/ Value	Units	Temp	Data Type	Reference
WS 6.71E+000	mg/L	25	EST	MEYLAN,WM ET AL. (1996)
logP 3.65			EXP	KLAUSCHENZ,E ET AL. (1989)
VP 3.39E-008	mm Hg	25	EST	NEELY,WB & BLAU,GE (1985)
DC	pKa			
HL 6.29E-013	atm m3/mol	25	EST	MEYLAN,WM & HOWARD,PH (1991)
OH 7.76E-011	cm3/molc sec	25	EST	MEYLAN,WM & HOWARD,PH (1993)

CAS #:	108611-33-2	3-PYRIDINECARBONITRILE, 5-PHENYL-2-(1-PIPERAZINY

Formula: $C_{17}H_{18}N_4$

Mol Weight: 278.36

MP (deg C): FP (deg C):

BP (deg C):

BP pressure (mm Hg):

Property/ Value	Units	Temp	Data Type	Reference
WS 1.85E+002	mg/L	25	EST	MEYLAN,WM ET AL. (1996)
logP 3.15			EXP	HAGEN,V ET AL. (1990)
VP 1.12E-008	mm Hg	25	EST	NEELY,WB & BLAU,GE (1985)
DC	pKa			
HL 1.21E-014	atm m3/mol	25	EST	MEYLAN,WM & HOWARD,PH (1991)
OH 1.09E-010	cm3/molc sec	25	EST	MEYLAN,WM & HOWARD,PH (1993)

CAS #:	108611-54-7	3-PYRIDINECARBONITRILE, 2-CHLORO-5-(3,4-DIMETHOX

Formula: $C_{15}H_{13}ClN_2O_2$

Mol Weight: 288.74

MP (deg C): FP (deg C):

BP (deg C):

BP pressure (mm Hg):

Property/ Value	Units	Temp	Data Type	Reference
WS 2.80E+001	mg/L	25	EST	MEYLAN,WM ET AL. (1996)
logP 2.86			EXP	KLAUSCHENZ,E ET AL. (1989)
VP 1.33E-007	mm Hg	25	EST	NEELY,WB & BLAU,GE (1985)
DC	pKa			
HL 9.45E-010	atm m3/mol	25	EST	MEYLAN,WM & HOWARD,PH (1991)
OH 1.88E-011	cm3/molc sec	25	EST	MEYLAN,WM & HOWARD,PH (1993)

CAS #:	108635-80-9	3-PYRIDINECARBONITRILE, 2-CHLORO-5-(4-METHOXYPHE

Formula: $C_{14}H_{11}ClN_2O$

Mol Weight: 258.71

MP (deg C): FP (deg C):

BP (deg C):

BP pressure (mm Hg):

Property/ Value	Units	Temp	Data Type	Reference
WS 1.49E+001	mg/L	25	EST	MEYLAN,WM ET AL. (1996)
logP 3.38			EXP	KLAUSCHENZ,E ET AL. (1989)
VP 8.43E-007	mm Hg	25	EST	NEELY,WB & BLAU,GE (1985)
DC	pKa			
HL 1.60E-008	atm m3/mol	25	EST	MEYLAN,WM & HOWARD,PH (1991)
OH 1.01E-011	cm3/molc sec	25	EST	MEYLAN,WM & HOWARD,PH (1993)

CAS #:	108966-48-9	BENZENESULFONAMIDE, 4-[(2-HYDROXYETHYL)THIO]-

Formula: $C_8H_{11}NO_3S_2$

Mol Weight: 233.31

MP (deg C): FP (deg C):

BP (deg C):

BP pressure (mm Hg):

Property/ Value	Units	Temp	Data Type	Reference
WS 1.44E+004	mg/L	25	EST	MEYLAN,WM ET AL. (1996)
logP 0.36			EXP	SHEPARD,KL ET AL. (1991)
VP 1.02E-008	mm Hg	25	EST	NEELY,WB & BLAU,GE (1985)
DC	pKa			
HL 5.97E-013	atm m3/mol	25	EST	MEYLAN,WM & HOWARD,PH (1991)
OH 1.58E-011	cm3/molc sec	25	EST	MEYLAN,WM & HOWARD,PH (1993)

CAS #:	108966-49-0	BENZENESULFONAMIDE, 4-[(2-HYDROXYETHYL)SULFONYL]

Formula: $C_8H_{11}NO_5S_2$

Mol Weight: 265.31

MP (deg C): FP (deg C):

BP (deg C):

BP pressure (mm Hg):

Property/ Value	Units	Temp	Data Type	Reference
WS 9.19E+003	mg/L	25	EST	MEYLAN,WM ET AL. (1996)
logP -0.85			EXP	SHEPARD,KL ET AL. (1991)
VP 1.33E-010	mm Hg	25	EST	NEELY,WB & BLAU,GE (1985)
DC	pKa			
HL 4.75E-015	atm m3/mol	25	EST	MEYLAN,WM & HOWARD,PH (1991)
OH 1.32E-011	cm3/molc sec	25	EST	MEYLAN,WM & HOWARD,PH (1993)

CAS #:	108966-51-4	BENZENESULFONAMIDE, 4-[(3-HYDROXYPROPYL)THIO]-

Formula: $C_9H_{13}NO_3S_2$

Mol Weight: 247.34

MP (deg C): FP (deg C):

BP (deg C):

BP pressure (mm Hg):

Property/ Value	Units	Temp	Data Type	Reference
WS 5.08E+003	mg/L	25	EST	MEYLAN,WM ET AL. (1996)
logP 0.80			EXP	SHEPARD,KL ET AL. (1991)
VP 3.81E-009	mm Hg	25	EST	NEELY,WB & BLAU,GE (1985)
DC	pKa			
HL 7.93E-013	atm m3/mol	25	EST	MEYLAN,WM & HOWARD,PH (1991)
OH 1.72E-011	cm3/molc sec	25	EST	MEYLAN,WM & HOWARD,PH (1993)

CAS #:	108966-53-6	BENZENESULFONAMIDE, 4-[3-HYDROXY-3-METHYLBUTYL]T

Formula: $C_{11}H_{17}NO_3S_2$

Mol Weight: 275.39

MP (deg C): FP (deg C):

BP (deg C):

BP pressure (mm Hg):

Property/ Value	Units	Temp	Data Type	Reference
WS 1.09E+003	mg/L	25	EST	MEYLAN,WM ET AL. (1996)
logP 1.40			EXP	SHEPARD,KL ET AL. (1991)
VP 2.75E-009	mm Hg	25	EST	NEELY,WB & BLAU,GE (1985)
DC	pKa			
HL 1.40E-012	atm m3/mol	25	EST	MEYLAN,WM & HOWARD,PH (1991)
OH 1.72E-011	cm3/molc sec	25	EST	MEYLAN,WM & HOWARD,PH (1993)

CAS #: 108966-54-7 — BENZENESULFONAMIDE, 4-[(5-HYDROXYPENTYL)THIO]-

Formula: $C_{11}H_{17}NO_3S_2$

Mol Weight: 275.39

MP (deg C):   FP (deg C):

BP (deg C):

BP pressure (mm Hg):

Property/Value	Units	Temp	Data Type	Reference
WS 2.20E+003	mg/L	25	EST	MEYLAN,WM ET AL. (1996)
logP 1.04			EXP	SHEPARD,KL ET AL. (1991)
VP 5.23E-010	mm Hg	25	EST	NEELY,WB & BLAU,GE (1985)
DC	pKa			
HL 1.40E-012	atm m3/mol	25	EST	MEYLAN,WM & HOWARD,PH (1991)
OH 2.00E-011	cm3/molc sec	25	EST	MEYLAN,WM & HOWARD,PH (1993)

CAS #: 108966-55-8 — BENZENESULFONAMIDE, 4-[(3-HYDROXYPROPYL)SULFONYL

Formula: $C_9H_{13}NO_5S_2$

Mol Weight: 279.34

MP (deg C):   FP (deg C):

BP (deg C):

BP pressure (mm Hg):

Property/Value	Units	Temp	Data Type	Reference
WS 5.06E+003	mg/L	25	EST	MEYLAN,WM ET AL. (1996)
logP -0.64			EXP	SHEPARD,KL ET AL. (1991)
VP 5.15E-011	mm Hg	25	EST	NEELY,WB & BLAU,GE (1985)
DC	pKa			
HL 6.30E-015	atm m3/mol	25	EST	MEYLAN,WM & HOWARD,PH (1991)
OH 1.46E-011	cm3/molc sec	25	EST	MEYLAN,WM & HOWARD,PH (1993)

CAS #: 108966-56-9 — BENZENESULFONAMIDE, 4-[3-HYDROXY-3-METHYLBUTYL]S

Formula: $C_{11}H_{17}NO_5S_2$

Mol Weight: 307.39

MP (deg C):   FP (deg C):

BP (deg C):

BP pressure (mm Hg):

Property/Value	Units	Temp	Data Type	Reference
WS 1.18E+003	mg/L	25	EST	MEYLAN,WM ET AL. (1996)
logP -0.09			EXP	SHEPARD,KL ET AL. (1991)
VP 4.22E-011	mm Hg	25	EST	NEELY,WB & BLAU,GE (1985)
DC	pKa			
HL 1.11E-014	atm m3/mol	25	EST	MEYLAN,WM & HOWARD,PH (1991)
OH 1.46E-011	cm3/molc sec	25	EST	MEYLAN,WM & HOWARD,PH (1993)

CAS #: 108966-58-1 — BENZENESULFONAMIDE, 4-[(2-HYDROXYETHYL)THIO]-3-N

Formula: $C_8H_{10}N_2O_5S_2$

Mol Weight: 278.31

MP (deg C):   FP (deg C):

BP (deg C):

BP pressure (mm Hg):

Property/Value	Units	Temp	Data Type	Reference
WS 3.16E+003	mg/L	25	EST	MEYLAN,WM ET AL. (1996)
logP 0.38			EXP	SHEPARD,KL ET AL. (1991)
VP 7.67E-011	mm Hg	25	EST	NEELY,WB & BLAU,GE (1985)
DC	pKa			
HL 2.36E-015	atm m3/mol	25	EST	MEYLAN,WM & HOWARD,PH (1991)
OH 1.35E-011	cm3/molc sec	25	EST	MEYLAN,WM & HOWARD,PH (1993)

CAS #: 108966-59-2 — BENZENESULFONAMIDE, 4-[(3-HYDROXYPROPYL)THIO]-3-

Formula: $C_9H_{12}N_2O_5S_2$

Mol Weight: 292.33

MP (deg C):   FP (deg C):

BP (deg C):

BP pressure (mm Hg):

Property/Value	Units	Temp	Data Type	Reference
WS 8.39E+002	mg/L	25	EST	MEYLAN,WM ET AL. (1996)
logP 0.96			EXP	SHEPARD,KL ET AL. (1991)
VP 2.96E-011	mm Hg	25	EST	NEELY,WB & BLAU,GE (1985)
DC	pKa			
HL 3.13E-015	atm m3/mol	25	EST	MEYLAN,WM & HOWARD,PH (1991)
OH 1.49E-011	cm3/molc sec	25	EST	MEYLAN,WM & HOWARD,PH (1993)

CAS #: 108966-60-5 — BENZENESULFONAMIDE, 4-[(3-HYDROXYPROPYL)SULFONYL

Formula: $C_9H_{12}N_2O_7S_2$

Mol Weight: 324.33

MP (deg C):   FP (deg C):

BP (deg C):

BP pressure (mm Hg):

Property/Value	Units	Temp	Data Type	Reference
WS 3.53E+002	mg/L	25	EST	MEYLAN,WM ET AL. (1996)
logP -0.05			EXP	SANGSTER,J (1994)
VP 5.14E-013	mm Hg	25	EST	NEELY,WB & BLAU,GE (1985)
DC	pKa			
HL 2.49E-017	atm m3/mol	25	EST	MEYLAN,WM & HOWARD,PH (1991)
OH 1.45E-011	cm3/molc sec	25	EST	MEYLAN,WM & HOWARD,PH (1993)

CAS #: 108966-61-6 — BENZENESULFONAMIDE, 3-AMINO-4-[(3-HYDROXYPROPYL)

Formula: $C_9H_{14}N_2O_5S_2$

Mol Weight: 294.35

MP (deg C):   FP (deg C):

BP (deg C):

BP pressure (mm Hg):

Property/Value	Units	Temp	Data Type	Reference
WS 3.03E+003	mg/L	25	EST	MEYLAN,WM ET AL. (1996)
logP -0.48			EXP	SHEPARD,KL ET AL. (1991)
VP 1.53E-012	mm Hg	25	EST	NEELY,WB & BLAU,GE (1985)
DC	pKa			
HL 2.23E-018	atm m3/mol	25	EST	MEYLAN,WM & HOWARD,PH (1991)
OH 1.95E-011	cm3/molc sec	25	EST	MEYLAN,WM & HOWARD,PH (1993)

CAS #: 108966-66-1 — BENZOIC ACID, 5-(AMINOSULFONYL)-2-[(3-HYDROXYPRO

Formula: $C_{11}H_{15}NO_5S_2$

Mol Weight: 305.37

MP (deg C):   FP (deg C):

BP (deg C):

BP pressure (mm Hg):

Property/Value	Units	Temp	Data Type	Reference
WS 2.67E+003	mg/L	25	EST	MEYLAN,WM ET AL. (1996)
logP 0.74			EXP	SHEPARD,KL ET AL. (1991)
VP 4.97E-011	mm Hg	25	EST	NEELY,WB & BLAU,GE (1985)
DC	pKa			
HL 5.11E-015	atm m3/mol	25	EST	MEYLAN,WM & HOWARD,PH (1991)
OH 1.56E-011	cm3/molc sec	25	EST	MEYLAN,WM & HOWARD,PH (1993)

CAS #: 108966-67-2				BENZOIC ACID, 5-(AMINOSULFONYL)-2-[(3-HYDROXYPRO

Formula: $C_{11}H_{15}NO_7S_2$

Mol Weight: 337.37

MP (deg C): FP (deg C):

BP (deg C):

BP pressure (mm Hg):

Property/ Value	Units	Temp	Data Type	Reference
WS 1.18E+003	mg/L	25	EST	MEYLAN,WM ET AL. (1996)
logP -0.30			EXP	SHEPARD,KL ET AL. (1991)
VP 8.70E-013	mm Hg	25	EST	NEELY,WB & BLAU,GE (1985)
DC	pKa			
HL 4.06E-017	atm m3/mol	25	EST	MEYLAN,WM & HOWARD,PH (1991)
OH 1.48E-011	cm3/molc sec	25	EST	MEYLAN,WM & HOWARD,PH (1993)

CAS #: 108966-70-7				BENZENESULFONAMIDE, 3-FLUORO-4-[(2-HYDROXYETHYL)

Formula: $C_8H_{10}FNO_3S_2$

Mol Weight: 251.30

MP (deg C): FP (deg C):

BP (deg C):

BP pressure (mm Hg):

Property/ Value	Units	Temp	Data Type	Reference
WS 5.65E+003	mg/L	25	EST	MEYLAN,WM ET AL. (1996)
logP 0.72			EXP	SHEPARD,KL ET AL. (1991)
VP 1.18E-008	mm Hg	25	EST	NEELY,WB & BLAU,GE (1985)
DC	pKa			
HL 6.97E-013	atm m3/mol	25	EST	MEYLAN,WM & HOWARD,PH (1991)
OH 1.40E-011	cm3/molc sec	25	EST	MEYLAN,WM & HOWARD,PH (1993)

CAS #: 108966-73-0				BENZENESULFONAMIDE, 3-CHLORO-4-[(3-HYDROXYPROPYL

Formula: $C_9H_{12}ClNO_3S_2$

Mol Weight: 281.78

MP (deg C): FP (deg C):

BP (deg C):

BP pressure (mm Hg):

Property/ Value	Units	Temp	Data Type	Reference
WS 6.87E+002	mg/L	25	EST	MEYLAN,WM ET AL. (1996)
logP 1.59			EXP	SHEPARD,KL ET AL. (1991)
VP 6.78E-010	mm Hg	25	EST	NEELY,WB & BLAU,GE (1985)
DC	pKa			
HL 5.88E-013	atm m3/mol	25	EST	MEYLAN,WM & HOWARD,PH (1991)
OH 1.53E-011	cm3/molc sec	25	EST	MEYLAN,WM & HOWARD,PH (1993)

CAS #: 108966-74-1				BENZENESULFONAMIDE, 3-FLUORO-4-[(3-HYDROXYPROPYL

Formula: $C_9H_{12}FNO_3S_2$

Mol Weight: 265.33

MP (deg C): FP (deg C):

BP (deg C):

BP pressure (mm Hg):

Property/ Value	Units	Temp	Data Type	Reference
WS 2.77E+003	mg/L	25	EST	MEYLAN,WM ET AL. (1996)
logP 0.99			EXP	SHEPARD,KL ET AL. (1991)
VP 4.41E-009	mm Hg	25	EST	NEELY,WB & BLAU,GE (1985)
DC	pKa			
HL 9.26E-013	atm m3/mol	25	EST	MEYLAN,WM & HOWARD,PH (1991)
OH 1.54E-011	cm3/molc sec	25	EST	MEYLAN,WM & HOWARD,PH (1993)

CAS #: 108966-75-2				BENZENESULFONAMIDE, 3-FLUORO-4-[(4-HYDROXYBUTYL)

Formula: $C_{10}H_{14}FNO_3S_2$

Mol Weight: 279.35

MP (deg C): FP (deg C):

BP (deg C):

BP pressure (mm Hg):

Property/ Value	Units	Temp	Data Type	Reference
WS 1.30E+003	mg/L	25	EST	MEYLAN,WM ET AL. (1996)
logP 1.28			EXP	SHEPARD,KL ET AL. (1991)
VP 1.64E-009	mm Hg	25	EST	NEELY,WB & BLAU,GE (1985)
DC	pKa			
HL 1.23E-012	atm m3/mol	25	EST	MEYLAN,WM & HOWARD,PH (1991)
OH 1.69E-011	cm3/molc sec	25	EST	MEYLAN,WM & HOWARD,PH (1993)

CAS #: 108966-76-3				BENZENESULFONAMIDE, 3-FLUORO-4-[(2-HYDROXYETHYL)

Formula: $C_8H_{10}FNO_5S_2$

Mol Weight: 283.30

MP (deg C): FP (deg C):

BP (deg C):

BP pressure (mm Hg):

Property/ Value	Units	Temp	Data Type	Reference
WS 4.44E+003	mg/L	25	EST	MEYLAN,WM ET AL. (1996)
logP -0.60			EXP	SHEPARD,KL ET AL. (1991)
VP 1.65E-010	mm Hg	25	EST	NEELY,WB & BLAU,GE (1985)
DC	pKa			
HL 5.54E-015	atm m3/mol	25	EST	MEYLAN,WM & HOWARD,PH (1991)
OH 1.32E-011	cm3/molc sec	25	EST	MEYLAN,WM & HOWARD,PH (1993)

CAS #: 108966-77-4				BENZENESULFONAMIDE, 3-FLUORO-4-[(3-HYDROXYPROPYL

Formula: $C_9H_{12}FNO_5S_2$

Mol Weight: 297.33

MP (deg C): FP (deg C):

BP (deg C):

BP pressure (mm Hg):

Property/ Value	Units	Temp	Data Type	Reference
WS 2.53E+003	mg/L	25	EST	MEYLAN,WM ET AL. (1996)
logP -0.41			EXP	SHEPARD,KL ET AL. (1991)
VP 6.41E-011	mm Hg	25	EST	NEELY,WB & BLAU,GE (1985)
DC	pKa			
HL 7.35E-015	atm m3/mol	25	EST	MEYLAN,WM & HOWARD,PH (1991)
OH 1.46E-011	cm3/molc sec	25	EST	MEYLAN,WM & HOWARD,PH (1993)

CAS #: 108966-78-5				BENZENESULFONAMIDE, 4-[(4-HYDROXYBUTYL)SULFONYL]

Formula: $C_{10}H_{15}NO_5S_2$

Mol Weight: 293.36

MP (deg C): FP (deg C):

BP (deg C):

BP pressure (mm Hg):

Property/ Value	Units	Temp	Data Type	Reference
WS 3.88E+003	mg/L	25	EST	MEYLAN,WM ET AL. (1996)
logP -0.26			EXP	SHEPARD,KL ET AL. (1991)
VP 1.99E-011	mm Hg	25	EST	NEELY,WB & BLAU,GE (1985)
DC	pKa			
HL 8.36E-015	atm m3/mol	25	EST	MEYLAN,WM & HOWARD,PH (1991)
OH 1.60E-011	cm3/molc sec	25	EST	MEYLAN,WM & HOWARD,PH (1993)

CAS #: 109151-40-8 — (E)-3-NICOTINOYLACRYLIC ACID

Formula: $C_9H_7NO_3$

Mol Weight: 177.16

MP (deg C):   FP (deg C):

BP (deg C):

BP pressure (mm Hg):

Property/Value	Units	Temp	Data Type	Reference
WS 8.46E+004	mg/L	25	EST	MEYLAN,WM ET AL. (1996)
logP 0.26			EXP	DAL POZZO,A ET AL. (1987)
VP 3.67E-005	mm Hg	25	EST	NEELY,WB & BLAU,GE (1985)
DC	pKa			
HL 3.17E-014	atm m3/mol	25	EST	MEYLAN,WM & HOWARD,PH (1991)
OH 7.77E-012	cm3/molc sec	25	EST	MEYLAN,WM & HOWARD,PH (1993)

CAS #: 109777-62-0 — BARBITURIC ACID,5-ME-5(11DIMEET)

Formula: $C_9H_{14}N_2O_3$

Mol Weight: 198.22

MP (deg C):   FP (deg C):

BP (deg C):

BP pressure (mm Hg):

Property/Value	Units	Temp	Data Type	Reference
WS 9.27E+003	mg/L	25	EST	MEYLAN,WM ET AL. (1996)
logP 0.80			EXP	WONG,O & MCKEOWN,RH (1988)
VP 1.59E-010	mm Hg	25	EST	NEELY,WB & BLAU,GE (1985)
DC	pKa			
HL 4.79E-013	atm m3/mol	25	EST	MEYLAN,WM & HOWARD,PH (1991)
OH 3.03E-012	cm3/molc sec	25	EST	MEYLAN,WM & HOWARD,PH (1993)

CAS #: 109871-96-7 — 1H-PYRROL-1-YLOXY, 2,5-DIHYDRO-2,2,5,5-TETRAMETH

Formula: $C_{14}H_{26}N_2O$

Mol Weight: 238.38

MP (deg C):   FP (deg C):

BP (deg C):

BP pressure (mm Hg):

Property/Value	Units	Temp	Data Type	Reference
WS 7.01E+003	mg/L	25	EST	MEYLAN,WM ET AL. (1996)
logP 1.88			EXP	FUCHS,J ET AL. (1990)
VP 2.20E-008	mm Hg	25	EST	NEELY,WB & BLAU,GE (1985)
DC	pKa			
HL 4.36E-014	atm m3/mol	25	EST	MEYLAN,WM & HOWARD,PH (1991)
OH 1.83E-010	cm3/molc sec	25	EST	MEYLAN,WM & HOWARD,PH (1993)

CAS #: 109881-25-6 — ADENOSINE, 2-AMINO-2',3'-DIDEHYDRO-2',3'-DIDEOXY

Formula: $C_{10}H_{12}N_6O_2$

Mol Weight: 248.25

MP (deg C):   FP (deg C):

BP (deg C):

BP pressure (mm Hg):

Property/Value	Units	Temp	Data Type	Reference
WS 3.05E+003	mg/L	25	EST	MEYLAN,WM ET AL. (1996)
logP -0.48			EXP	SANGSTER,J (1994)
VP 9.12E-013	mm Hg	25	EST	NEELY,WB & BLAU,GE (1985)
DC	pKa			
HL 7.29E-019	atm m3/mol	25	EST	MEYLAN,WM & HOWARD,PH (1991)
OH 2.82E-010	cm3/molc sec	25	EST	MEYLAN,WM & HOWARD,PH (1993)

CAS #: 110104-37-5 — ACETAMIDE, N-[6,9-DIHYDRO-9-[(2-HYDROXYETHOXY)ME

Formula: $C_{10}H_{15}N_5O_4$

Mol Weight: 269.26

MP (deg C):   FP (deg C):

BP (deg C):

BP pressure (mm Hg):

Property/Value	Units	Temp	Data Type	Reference
WS 1.20E+004	mg/L	25	EST	MEYLAN,WM ET AL. (1996)
logP -1.30			EXP	KRISTL,A ET AL. (1993)
VP 3.62E-018	mm Hg	25	EST	NEELY,WB & BLAU,GE (1985)
DC	pKa			
HL 1.63E-021	atm m3/mol	25	EST	MEYLAN,WM & HOWARD,PH (1991)
OH 6.40E-011	cm3/molc sec	25	EST	MEYLAN,WM & HOWARD,PH (1993)

CAS #: 110143-05-0 — ADENOSINE, 2',3'-DIDEOXY-2'-FLUORO-

Formula: $C_{10}H_{12}FN_5O_2$

Mol Weight: 253.24

MP (deg C):   FP (deg C):

BP (deg C):

BP pressure (mm Hg):

Property/Value	Units	Temp	Data Type	Reference
WS 1.58E+003	mg/L	25	EST	MEYLAN,WM ET AL. (1996)
logP -0.14			EXP	SANGSTER,J (1993)
VP 4.20E-011	mm Hg	25	EST	NEELY,WB & BLAU,GE (1985)
DC	pKa			
HL 1.66E-019	atm m3/mol	25	EST	MEYLAN,WM & HOWARD,PH (1991)
OH 2.08E-010	cm3/molc sec	25	EST	MEYLAN,WM & HOWARD,PH (1993)

CAS #: 110143-10-7 — 9H-PURIN-6-AMINE, 9-(2,3-DIDEOXY-2-FLUORO-á-D-TH

Formula: $C_{10}H_{12}FN_5O_2$

Mol Weight: 253.24

MP (deg C):   FP (deg C):

BP (deg C):

BP pressure (mm Hg):

Property/Value	Units	Temp	Data Type	Reference
WS 1.58E+003	mg/L	25	EST	MEYLAN,WM ET AL. (1996)
logP -0.18			EXP	FORD,H ET AL. (1991)
VP 4.20E-011	mm Hg	25	EST	NEELY,WB & BLAU,GE (1985)
DC	pKa			
HL 1.66E-019	atm m3/mol	25	EST	MEYLAN,WM & HOWARD,PH (1991)
OH 2.08E-010	cm3/molc sec	25	EST	MEYLAN,WM & HOWARD,PH (1993)

CAS #: 110192-33-1 — 5,8-METHANO-5H-BENZOCYCLOHEPTEN-10-AMINE, 6,7,8,

Formula: $C_{13}H_{14}F_3N$

Mol Weight: 241.26

MP (deg C):   FP (deg C):

BP (deg C):

BP pressure (mm Hg):

Property/Value	Units	Temp	Data Type	Reference
WS 3.30E+002	mg/L	25	EST	MEYLAN,WM ET AL. (1996)
logP 3.65			EXP	SANGSTER,J (1994)
VP 1.95E-003	mm Hg	25	EST	NEELY,WB & BLAU,GE (1985)
DC	pKa			
HL 3.54E-006	atm m3/mol	25	EST	MEYLAN,WM & HOWARD,PH (1991)
OH 3.14E-011	cm3/molc sec	25	EST	MEYLAN,WM & HOWARD,PH (1993)

CAS #:	110192-35-3			5,8-METHANO-5H-BENZOCYCLOHEPTEN-10-AMINE, 6,7,8,	

Formula: $C_{13}H_{14}F_3N$

Mol Weight: 241.26

MP (deg C):   FP (deg C):

BP (deg C):

BP pressure (mm Hg):

Property/Value	Units	Temp	Data Type	Reference
WS 3.05E+002	mg/L	25	EST	MEYLAN,WM ET AL. (1996)
logP 3.59			EXP	SANGSTER,J (1994)
VP 1.95E-003	mm Hg	25	EST	NEELY,WB & BLAU,GE (1985)
DC	pKa			
HL 3.54E-006	atm m3/mol	25	EST	MEYLAN,WM & HOWARD,PH (1991)
OH 3.14E-011	cm3/molc sec	25	EST	MEYLAN,WM & HOWARD,PH (1993)

CAS #:	110235-47-7			MEPANIPYRIM	

Formula: $C_{14}H_{13}N_3$

Mol Weight: 223.28

MP (deg C):   FP (deg C):

BP (deg C):

BP pressure (mm Hg):

Property/Value	Units	Temp	Data Type	Reference
WS 4.05E+001	mg/L	25	EST	MEYLAN,WM ET AL. (1996)
logP 3.41			EXP	TOMLIN,C (1994)
VP 5.00E-006	mm Hg	25	EST	NEELY,WB & BLAU,GE (1985)
DC	pKa			
HL 3.27E-007	atm m3/mol	25	EST	MEYLAN,WM & HOWARD,PH (1991)
OH 1.96E-010	cm3/molc sec	25	EST	MEYLAN,WM & HOWARD,PH (1993)

CAS #:	110267-94-2			5,8-METHANO-5H-BENZOCYCLOHEPTEN-10-AMINE, 6,7,8,	

Formula: $C_{13}H_{14}F_3N$

Mol Weight: 241.26

MP (deg C):   FP (deg C):

BP (deg C):

BP pressure (mm Hg):

Property/Value	Units	Temp	Data Type	Reference
WS 3.30E+002	mg/L	25	EST	MEYLAN,WM ET AL. (1996)
logP 3.41			EXP	SANGSTER,J (1994)
VP 1.95E-003	mm Hg	25	EST	NEELY,WB & BLAU,GE (1985)
DC	pKa			
HL 3.54E-006	atm m3/mol	25	EST	MEYLAN,WM & HOWARD,PH (1991)
OH 3.14E-011	cm3/molc sec	25	EST	MEYLAN,WM & HOWARD,PH (1993)

CAS #:	110267-95-3			5,8-METHANO-5H-BENZOCYCLOHEPTEN-10-AMINE, 6,7,8,	

Formula: $C_{13}H_{14}F_3N$

Mol Weight: 241.26

MP (deg C):   FP (deg C):

BP (deg C):

BP pressure (mm Hg):

Property/Value	Units	Temp	Data Type	Reference
WS 2.55E+002	mg/L	25	EST	MEYLAN,WM ET AL. (1996)
logP 3.54			EXP	SANGSTER,J (1994)
VP 1.95E-003	mm Hg	25	EST	NEELY,WB & BLAU,GE (1985)
DC	pKa			
HL 3.54E-006	atm m3/mol	25	EST	MEYLAN,WM & HOWARD,PH (1991)
OH 3.14E-011	cm3/molc sec	25	EST	MEYLAN,WM & HOWARD,PH (1993)

CAS #:	110267-96-4			5,8-METHANO-5H-BENZOCYCLOHEPTEN-10-AMINE, 6,7,8,	

Formula: $C_{13}H_{14}F_3N$

Mol Weight: 241.26

MP (deg C):   FP (deg C):

BP (deg C):

BP pressure (mm Hg):

Property/Value	Units	Temp	Data Type	Reference
WS 3.05E+002	mg/L	25	EST	MEYLAN,WM ET AL. (1996)
logP 3.45			EXP	SANGSTER,J (1994)
VP 1.95E-003	mm Hg	25	EST	NEELY,WB & BLAU,GE (1985)
DC	pKa			
HL 3.54E-006	atm m3/mol	25	EST	MEYLAN,WM & HOWARD,PH (1991)
OH 3.14E-011	cm3/molc sec	25	EST	MEYLAN,WM & HOWARD,PH (1993)

CAS #:	110301-07-0			L-TYROSINE, 3-HYDROXY-, 1-METHYLETHYL ESTER	

Formula: $C_{12}H_{17}NO_4$

Mol Weight: 239.27

MP (deg C):   FP (deg C):

BP (deg C):

BP pressure (mm Hg):

Property/Value	Units	Temp	Data Type	Reference
WS 3.42E+005	mg/L	25	EST	MEYLAN,WM ET AL. (1996)
logP 0.57			EXP	SANGSTER,J (1993)
VP 2.23E-007	mm Hg	25	EST	NEELY,WB & BLAU,GE (1985)
DC	pKa			
HL 7.36E-016	atm m3/mol	25	EST	MEYLAN,WM & HOWARD,PH (1991)
OH 9.93E-011	cm3/molc sec	25	EST	MEYLAN,WM & HOWARD,PH (1993)

CAS #:	110488-70-5			DIMETHOMORPH	

Formula: $C_{21}H_{22}ClNO_4$

Mol Weight: 387.87

MP (deg C):   FP (deg C):

BP (deg C):

BP pressure (mm Hg):

Property/Value	Units	Temp	Data Type	Reference
WS 1.87E+001	mg/L	25	EST	MEYLAN,WM ET AL. (1996)
logP 2.68			EXP	TOMLIN,C (1994)
VP 3.63E-010	mm Hg	25	EST	NEELY,WB & BLAU,GE (1985)
DC	pKa			
HL 1.01E-015	atm m3/mol	25	EST	MEYLAN,WM & HOWARD,PH (1991)
OH 1.07E-010	cm3/molc sec	25	EST	MEYLAN,WM & HOWARD,PH (1993)

CAS #:	110747-15-4			2-PROPANOL, 1,1'-IMINOBIS[3-(2-METHYLPHENOXY)	

Formula: $C_{20}H_{27}NO_4$

Mol Weight: 345.44

MP (deg C):   FP (deg C):

BP (deg C):

BP pressure (mm Hg):

Property/Value	Units	Temp	Data Type	Reference
WS 4.94E+001	mg/L	25	EST	MEYLAN,WM ET AL. (1996)
logP 3.67			EXP	RECANATINI,M (1992)
VP 6.60E-012	mm Hg	25	EST	NEELY,WB & BLAU,GE (1985)
DC	pKa			
HL 1.58E-015	atm m3/mol	25	EST	MEYLAN,WM & HOWARD,PH (1991)
OH 1.72E-010	cm3/molc sec	25	EST	MEYLAN,WM & HOWARD,PH (1993)

<table>
<tr><td colspan="2">

CAS #: 110933-19-2
1H-IMIDAZOLE-1-ETHANOL, _,_ -BIS(4-FLUOROPHENYL)-

Formula: $C_{17}H_{14}F_2N_2O$

Mol Weight: 300.31

MP (deg C):   FP (deg C):

BP (deg C):

BP pressure (mm Hg):

Property/ Value	Units	Temp	Data Type	Reference
WS 3.70E+001	mg/L	25	EST	MEYLAN,WM ET AL. (1996)
logP 2.95			EXP	SANGSTER,J (1994)
VP 6.60E-010	mm Hg	25	EST	NEELY,WB & BLAU,GE (1985)
DC	pKa			
HL 3.45E-011	atm m3/mol	25	EST	MEYLAN,WM & HOWARD,PH (1991)
OH 4.55E-011	cm3/molc sec	25	EST	MEYLAN,WM & HOWARD,PH (1993)

</td><td colspan="2">

CAS #: 111041-44-2
IMIDODICARBONIC DIAMIDE, N-[2-FLUORO-5-(PHENYLME

Formula: $C_{17}H_{18}FN_3O_2$

Mol Weight: 315.35

MP (deg C):   FP (deg C):

BP (deg C):

BP pressure (mm Hg):

Property/ Value	Units	Temp	Data Type	Reference
WS 4.66E+000	mg/L	25	EST	MEYLAN,WM ET AL. (1996)
logP 3.90			EXP	CAMILLERI,P ET AL. (1989)
VP 8.04E-011	mm Hg	25	EST	NEELY,WB & BLAU,GE (1985)
DC	pKa			
HL 5.60E-014	atm m3/mol	25	EST	MEYLAN,WM & HOWARD,PH (1991)
OH 4.47E-011	cm3/molc sec	25	EST	MEYLAN,WM & HOWARD,PH (1993)

</td></tr>
<tr><td colspan="2">

CAS #: 111041-46-4
IMIDODICARBONIC DIAMIDE. N-(5-ETHYL-2-FLUOROPHEN

Formula: $C_{12}H_{16}FN_3O_2$

Mol Weight: 253.28

MP (deg C):   FP (deg C):

BP (deg C):

BP pressure (mm Hg):

Property/ Value	Units	Temp	Data Type	Reference
WS 3.96E+001	mg/L	25	EST	MEYLAN,WM ET AL. (1996)
logP 3.23			EXP	CAMILLERI,P ET AL. (1989)
VP 1.34E-008	mm Hg	25	EST	NEELY,WB & BLAU,GE (1985)
DC	pKa			
HL 9.21E-013	atm m3/mol	25	EST	MEYLAN,WM & HOWARD,PH (1991)
OH 4.00E-011	cm3/molc sec	25	EST	MEYLAN,WM & HOWARD,PH (1993)

</td><td colspan="2">

CAS #: 111041-47-5
IMIDODICARBONIC DIAMIDE, N-(2,5-DIFLUOROPHENYL)-

Formula: $C_{10}H_{11}F_2N_3O_2$

Mol Weight: 243.21

MP (deg C):   FP (deg C):

BP (deg C):

BP pressure (mm Hg):

Property/ Value	Units	Temp	Data Type	Reference
WS 2.30E+002	mg/L	25	EST	MEYLAN,WM ET AL. (1996)
logP 2.40			EXP	CAMILLERI,P ET AL. (1989)
VP 8.03E-008	mm Hg	25	EST	NEELY,WB & BLAU,GE (1985)
DC	pKa			
HL 7.34E-013	atm m3/mol	25	EST	MEYLAN,WM & HOWARD,PH (1991)
OH 2.12E-011	cm3/molc sec	25	EST	MEYLAN,WM & HOWARD,PH (1993)

</td></tr>
<tr><td colspan="2">

CAS #: 111041-48-6
IMIDODICARBONIC DIAMIDE, N-[2-FLUORO-5-(1-METHYL

Formula: $C_{13}H_{18}FN_3O_2$

Mol Weight: 267.31

MP (deg C):   FP (deg C):

BP (deg C):

BP pressure (mm Hg):

Property/ Value	Units	Temp	Data Type	Reference
WS 1.83E+001	mg/L	25	EST	MEYLAN,WM ET AL. (1996)
logP 3.53			EXP	CAMILLERI,P ET AL. (1989)
VP 9.73E-009	mm Hg	25	EST	NEELY,WB & BLAU,GE (1985)
DC	pKa			
HL 1.22E-012	atm m3/mol	25	EST	MEYLAN,WM & HOWARD,PH (1991)
OH 3.95E-011	cm3/molc sec	25	EST	MEYLAN,WM & HOWARD,PH (1993)

</td><td colspan="2">

CAS #: 111041-55-5
IMIDODICARBONIC DIAMIDE, N-(2-FLUORO-5-IODOPHENY

Formula: $C_{10}H_{11}FIN_3O_2$

Mol Weight: 351.12

MP (deg C):   FP (deg C):

BP (deg C):

BP pressure (mm Hg):

Property/ Value	Units	Temp	Data Type	Reference
WS 5.35E+000	mg/L	25	EST	MEYLAN,WM ET AL. (1996)
logP 3.58			EXP	CAMILLERI,P ET AL. (1989)
VP 2.22E-009	mm Hg	25	EST	NEELY,WB & BLAU,GE (1985)
DC	pKa			
HL 1.46E-013	atm m3/mol	25	EST	MEYLAN,WM & HOWARD,PH (1991)
OH 1.26E-011	cm3/molc sec	25	EST	MEYLAN,WM & HOWARD,PH (1993)

</td></tr>
<tr><td colspan="2">

CAS #: 111041-57-7
IMIDODICARBONIC DIAMIDE, N-[2-FLUORO-5-(1-METHYL

Formula: $C_{13}H_{18}FN_3O_3$

Mol Weight: 283.31

MP (deg C):   FP (deg C):

BP (deg C):

BP pressure (mm Hg):

Property/ Value	Units	Temp	Data Type	Reference
WS 3.59E+001	mg/L	25	EST	MEYLAN,WM ET AL. (1996)
logP 3.08			EXP	CAMILLERI,P ET AL. (1989)
VP 4.19E-009	mm Hg	25	EST	NEELY,WB & BLAU,GE (1985)
DC	pKa			
HL 6.55E-014	atm m3/mol	25	EST	MEYLAN,WM & HOWARD,PH (1991)
OH 1.74E-010	cm3/molc sec	25	EST	MEYLAN,WM & HOWARD,PH (1993)

</td><td colspan="2">

CAS #: 111041-62-4
IMIDODICARBONIC DIAMIDE, N-(5-ACETYL-2-FLUOROPHE

Formula: $C_{12}H_{14}FN_3O_3$

Mol Weight: 267.26

MP (deg C):   FP (deg C):

BP (deg C):

BP pressure (mm Hg):

Property/ Value	Units	Temp	Data Type	Reference
WS 5.28E+002	mg/L	25	EST	MEYLAN,WM ET AL. (1996)
logP 1.82			EXP	CAMILLERI,P ET AL. (1989)
VP 2.75E-009	mm Hg	25	EST	NEELY,WB & BLAU,GE (1985)
DC	pKa			
HL 1.14E-015	atm m3/mol	25	EST	MEYLAN,WM & HOWARD,PH (1991)
OH 9.29E-012	cm3/molc sec	25	EST	MEYLAN,WM & HOWARD,PH (1993)

</td></tr>
</table>

CAS #: 111041-66-8 — IMIDODICARBONIC DIAMIDE, N-(4-FLUORO[1,1'-BIPHEN

Formula: $C_{16}H_{16}FN_3O_2$

Mol Weight: 301.32

MP (deg C): FP (deg C):

BP (deg C):

BP pressure (mm Hg):

Property/Value	Units	Temp	Data Type	Reference
WS 7.13E+000	mg/L	25	EST	MEYLAN,WM ET AL. (1996)
logP 3.78			EXP	CAMILLERI,P ET AL. (1989)
VP 6.74E-011	mm Hg	25	EST	NEELY,WB & BLAU,GE (1985)
DC	pKa			
HL 4.83E-014	atm m3/mol	25	EST	MEYLAN,WM & HOWARD,PH (1991)
OH 3.06E-011	cm3/molc sec	25	EST	MEYLAN,WM & HOWARD,PH (1993)

CAS #: 111041-67-9 — IMIDODICARBONIC DIAMIDE, N-(2-FLUORO-5-METHOXYPH

Formula: $C_{11}H_{14}FN_3O_3$

Mol Weight: 255.25

MP (deg C): FP (deg C):

BP (deg C):

BP pressure (mm Hg):

Property/Value	Units	Temp	Data Type	Reference
WS 3.29E+002	mg/L	25	EST	MEYLAN,WM ET AL. (1996)
logP 2.14			EXP	CAMILLERI,P ET AL. (1989)
VP 1.30E-008	mm Hg	25	EST	NEELY,WB & BLAU,GE (1985)
DC	pKa			
HL 3.72E-014	atm m3/mol	25	EST	MEYLAN,WM & HOWARD,PH (1991)
OH 1.63E-010	cm3/molc sec	25	EST	MEYLAN,WM & HOWARD,PH (1993)

CAS #: 111041-76-0 — IMIDODICARBONIC DIAMIDE, N-[5-(1-ETHYLPROPYL)-2-

Formula: $C_{15}H_{22}FN_3O_2$

Mol Weight: 295.36

MP (deg C): FP (deg C):

BP (deg C):

BP pressure (mm Hg):

Property/Value	Units	Temp	Data Type	Reference
WS 2.03E+000	mg/L	25	EST	MEYLAN,WM ET AL. (1996)
logP 4.46			EXP	CAMILLERI,P ET AL. (1989)
VP 1.92E-009	mm Hg	25	EST	NEELY,WB & BLAU,GE (1985)
DC	pKa			
HL 2.15E-012	atm m3/mol	25	EST	MEYLAN,WM & HOWARD,PH (1991)
OH 4.28E-011	cm3/molc sec	25	EST	MEYLAN,WM & HOWARD,PH (1993)

CAS #: 111041-86-2 — IMIDODICARBONIC DIAMIDE, N-(5-CYCLOHEXYL-2-FLUOR

Formula: $C_{16}H_{22}FN_3O_2$

Mol Weight: 307.37

MP (deg C): FP (deg C):

BP (deg C):

BP pressure (mm Hg):

Property/Value	Units	Temp	Data Type	Reference
WS 1.36E+000	mg/L	25	EST	MEYLAN,WM ET AL. (1996)
logP 4.58			EXP	CAMILLERI,P ET AL. (1989)
VP 3.58E-010	mm Hg	25	EST	NEELY,WB & BLAU,GE (1985)
DC	pKa			
HL 1.26E-012	atm m3/mol	25	EST	MEYLAN,WM & HOWARD,PH (1991)
OH 4.73E-011	cm3/molc sec	25	EST	MEYLAN,WM & HOWARD,PH (1993)

CAS #: 111041-88-4 — IMIDODICARBONIC DIAMIDE, N-[5-(1,1-DIMETHYLETHYL

Formula: $C_{14}H_{20}FN_3O_2$

Mol Weight: 281.33

MP (deg C): FP (deg C):

BP (deg C):

BP pressure (mm Hg):

Property/Value	Units	Temp	Data Type	Reference
WS 6.54E+000	mg/L	25	EST	MEYLAN,WM ET AL. (1996)
logP 3.96			EXP	CAMILLERI,P ET AL. (1989)
VP 5.97E-009	mm Hg	25	EST	NEELY,WB & BLAU,GE (1985)
DC	pKa			
HL 1.62E-012	atm m3/mol	25	EST	MEYLAN,WM & HOWARD,PH (1991)
OH 3.53E-011	cm3/molc sec	25	EST	MEYLAN,WM & HOWARD,PH (1993)

CAS #: 111075-86-6 — 1-ACETYL-4-(3-ME PHENYL) SEMICARBAZIDE

Formula: $C_{10}H_{13}N_3O_2$

Mol Weight: 207.23

MP (deg C): FP (deg C):

BP (deg C):

BP pressure (mm Hg):

Property/Value	Units	Temp	Data Type	Reference
WS 9.19E+003	mg/L	25	EST	MEYLAN,WM ET AL. (1996)
logP 0.75			EXP	STEIN,J ET AL. (1986)
VP 5.16E-008	mm Hg	25	EST	NEELY,WB & BLAU,GE (1985)
DC	pKa			
HL 3.01E-015	atm m3/mol	25	EST	MEYLAN,WM & HOWARD,PH (1991)
OH 1.18E-010	cm3/molc sec	25	EST	MEYLAN,WM & HOWARD,PH (1993)

CAS #: 111075-87-7 — 1-ACETYL-4-(4-T-BU PHENYL) SEMICARBAZIDE

Formula: $C_{13}H_{19}N_3O_2$

Mol Weight: 249.32

MP (deg C): FP (deg C):

BP (deg C):

BP pressure (mm Hg):

Property/Value	Units	Temp	Data Type	Reference
WS 3.99E+002	mg/L	25	EST	MEYLAN,WM ET AL. (1996)
logP 2.08			EXP	STEIN,J ET AL. (1986)
VP 1.03E-008	mm Hg	25	EST	NEELY,WB & BLAU,GE (1985)
DC	pKa			
HL 7.04E-015	atm m3/mol	25	EST	MEYLAN,WM & HOWARD,PH (1991)
OH 5.83E-011	cm3/molc sec	25	EST	MEYLAN,WM & HOWARD,PH (1993)

CAS #: 111075-88-8 — 1-ACETYL-4-(3-F PHENYL) SEMICARBAZIDE

Formula: $C_9H_{10}FN_3O_2$

Mol Weight: 211.20

MP (deg C): FP (deg C):

BP (deg C):

BP pressure (mm Hg):

Property/Value	Units	Temp	Data Type	Reference
WS 9.12E+003	mg/L	25	EST	MEYLAN,WM ET AL. (1996)
logP 0.73			EXP	STEIN,J ET AL. (1986)
VP 1.38E-007	mm Hg	25	EST	NEELY,WB & BLAU,GE (1985)
DC	pKa			
HL 3.18E-015	atm m3/mol	25	EST	MEYLAN,WM & HOWARD,PH (1991)
OH 6.00E-011	cm3/molc sec	25	EST	MEYLAN,WM & HOWARD,PH (1993)

111075-89-9 — 1-ACETYL-4-(4-F PHENYL) SEMICARBAZIDE

CAS #: 111075-89-9

Formula: $C_9H_{10}FN_3O_2$

Mol Weight: 211.20

MP (deg C):

FP (deg C):

BP (deg C):

BP pressure (mm Hg):

Property/Value	Units	Temp	Data Type	Reference
WS 1.58E+004	mg/L	25	EST	MEYLAN,WM ET AL. (1996)
logP 0.45			EXP	STEIN,J ET AL. (1986)
VP 1.38E-007	mm Hg	25	EST	NEELY,WB & BLAU,GE (1985)
DC	pKa			
HL 3.18E-015	atm m3/mol	25	EST	MEYLAN,WM & HOWARD,PH (1991)
OH 2.10E-011	cm3/molc sec	25	EST	MEYLAN,WM & HOWARD,PH (1993)

111075-90-2 — 1-ACETYL-4-(3-CL PHENYL) SEMICARBAZIDE

CAS #: 111075-90-2

Formula: $C_9H_{10}ClN_3O_2$

Mol Weight: 227.65

MP (deg C):

FP (deg C):

BP (deg C):

BP pressure (mm Hg):

Property/Value	Units	Temp	Data Type	Reference
WS 2.29E+003	mg/L	25	EST	MEYLAN,WM ET AL. (1996)
logP 1.33			EXP	STEIN,J ET AL. (1986)
VP 3.16E-008	mm Hg	25	EST	NEELY,WB & BLAU,GE (1985)
DC	pKa			
HL 2.02E-015	atm m3/mol	25	EST	MEYLAN,WM & HOWARD,PH (1991)
OH 3.66E-011	cm3/molc sec	25	EST	MEYLAN,WM & HOWARD,PH (1993)

111075-91-3 — 1-ACETYL-4-(3-BR PHENYL) SEMICARBAZIDE

CAS #: 111075-91-3

Formula: $C_9H_{10}BrN_3O_2$

Mol Weight: 272.11

MP (deg C):

FP (deg C):

BP (deg C):

BP pressure (mm Hg):

Property/Value	Units	Temp	Data Type	Reference
WS 9.68E+002	mg/L	25	EST	MEYLAN,WM ET AL. (1996)
logP 1.48			EXP	STEIN,J ET AL. (1986)
VP 1.37E-008	mm Hg	25	EST	NEELY,WB & BLAU,GE (1985)
DC	pKa			
HL 1.09E-015	atm m3/mol	25	EST	MEYLAN,WM & HOWARD,PH (1991)
OH 3.35E-011	cm3/molc sec	25	EST	MEYLAN,WM & HOWARD,PH (1993)

111075-92-4 — 1-ACETYL-4-(3-IODO PHENYL) SEMICARBAZIDE

CAS #: 111075-92-4

Formula: $C_9H_{10}IN_3O_2$

Mol Weight: 319.10

MP (deg C):

FP (deg C):

BP (deg C):

BP pressure (mm Hg):

Property/Value	Units	Temp	Data Type	Reference
WS 3.29E+002	mg/L	25	EST	MEYLAN,WM ET AL. (1996)
logP 1.71			EXP	STEIN,J ET AL. (1986)
VP 3.85E-009	mm Hg	25	EST	NEELY,WB & BLAU,GE (1985)
DC	pKa			
HL 6.32E-016	atm m3/mol	25	EST	MEYLAN,WM & HOWARD,PH (1991)
OH 3.47E-011	cm3/molc sec	25	EST	MEYLAN,WM & HOWARD,PH (1993)

111075-93-5 — 1-ACETYL-4-(4-IODO PHENYL) SEMICARBAZIDE

CAS #: 111075-93-5

Formula: $C_9H_{10}IN_3O_2$

Mol Weight: 319.10

MP (deg C):

FP (deg C):

BP (deg C):

BP pressure (mm Hg):

Property/Value	Units	Temp	Data Type	Reference
WS 3.70E+002	mg/L	25	EST	MEYLAN,WM ET AL. (1996)
logP 1.65			EXP	STEIN,J ET AL. (1986)
VP 3.85E-009	mm Hg	25	EST	NEELY,WB & BLAU,GE (1985)
DC	pKa			
HL 6.32E-016	atm m3/mol	25	EST	MEYLAN,WM & HOWARD,PH (1991)
OH 2.07E-011	cm3/molc sec	25	EST	MEYLAN,WM & HOWARD,PH (1993)

111075-94-6 — 1-ACETYL-4-(3-CF3 PHENYL) SEMICARBAZIDE

CAS #: 111075-94-6

Formula: $C_{10}H_{10}F_3N_3O_2$

Mol Weight: 261.21

MP (deg C):

FP (deg C):

BP (deg C):

BP pressure (mm Hg):

Property/Value	Units	Temp	Data Type	Reference
WS 8.99E+002	mg/L	25	EST	MEYLAN,WM ET AL. (1996)
logP 1.59			EXP	STEIN,J ET AL. (1986)
VP 9.07E-008	mm Hg	25	EST	NEELY,WB & BLAU,GE (1985)
DC	pKa			
HL 2.37E-014	atm m3/mol	25	EST	MEYLAN,WM & HOWARD,PH (1991)
OH 6.56E-012	cm3/molc sec	25	EST	MEYLAN,WM & HOWARD,PH (1993)

111075-95-7 — 1-ACETYL-4-(3-OME PHENYL) SEMICARBAZIDE

CAS #: 111075-95-7

Formula: $C_{10}H_{13}N_3O_3$

Mol Weight: 223.23

MP (deg C):

FP (deg C):

BP (deg C):

BP pressure (mm Hg):

Property/Value	Units	Temp	Data Type	Reference
WS 8.47E+002	mg/L	25	EST	MEYLAN,WM ET AL. (1996)
logP 0.33			EXP	STEIN,J ET AL. (1986)
VP 2.24E-008	mm Hg	25	EST	NEELY,WB & BLAU,GE (1985)
DC	pKa			
HL 1.61E-016	atm m3/mol	25	EST	MEYLAN,WM & HOWARD,PH (1991)
OH 2.07E-010	cm3/molc sec	25	EST	MEYLAN,WM & HOWARD,PH (1993)

111075-96-8 — 1-ACETYL-4-(4-OET PHENYL) SEMICARBAZIDE

CAS #: 111075-96-8

Formula: $C_{11}H_{15}N_3O_3$

Mol Weight: 237.26

MP (deg C):

FP (deg C):

BP (deg C):

BP pressure (mm Hg):

Property/Value	Units	Temp	Data Type	Reference
WS 1.02E+004	mg/L	25	EST	MEYLAN,WM ET AL. (1996)
logP 0.51			EXP	STEIN,J ET AL. (1986)
VP 1.00E-008	mm Hg	25	EST	NEELY,WB & BLAU,GE (1985)
DC	pKa			
HL 2.14E-016	atm m3/mol	25	EST	MEYLAN,WM & HOWARD,PH (1991)
OH 4.97E-011	cm3/molc sec	25	EST	MEYLAN,WM & HOWARD,PH (1993)

CAS #: 111075-97-9 — 1-ACETYL-4-(4-OPH PHENYL) SEMICARBAZIDE

Formula: $C_{15}H_{15}N_3O_3$
Mol Weight: 285.31
MP (deg C): **FP (deg C):**
BP (deg C):
BP pressure (mm Hg):

Property/Value	Units	Temp	Data Type	Reference
WS 2.87E+002	mg/L	25	EST	MEYLAN,WM ET AL. (1996)
logP 2.01			EXP	STEIN,J ET AL. (1986)
VP 1.36E-010	mm Hg	25	EST	NEELY,WB & BLAU,GE (1985)
DC	pKa			
HL 5.96E-017	atm m3/mol	25	EST	MEYLAN,WM & HOWARD,PH (1991)
OH 5.42E-011	cm3/molc sec	25	EST	MEYLAN,WM & HOWARD,PH (1993)

CAS #: 111075-98-0 — 1-ACETYL-4-(4-DIMEAMINOPH) SEMICARBAZIDE

Formula: $C_{11}H_{16}N_4O_2$
Mol Weight: 236.28
MP (deg C): **FP (deg C):**
BP (deg C):
BP pressure (mm Hg):

Property/Value	Units	Temp	Data Type	Reference
WS 1.11E+003	mg/L	25	EST	MEYLAN,WM ET AL. (1996)
logP 0.11			EXP	STEIN,J ET AL. (1986)
VP 1.06E-008	mm Hg	25	EST	NEELY,WB & BLAU,GE (1985)
DC	pKa			
HL 4.33E-017	atm m3/mol	25	EST	MEYLAN,WM & HOWARD,PH (1991)
OH 2.09E-010	cm3/molc sec	25	EST	MEYLAN,WM & HOWARD,PH (1993)

CAS #: 111075-99-1 — 1-ACETYL-4-(4-OH PHENYL) SEMICARBAZIDE

Formula: $C_9H_{11}N_3O_3$
Mol Weight: 209.21
MP (deg C): **FP (deg C):**
BP (deg C):
BP pressure (mm Hg):

Property/Value	Units	Temp	Data Type	Reference
WS 9.08E+003	mg/L	25	EST	MEYLAN,WM ET AL. (1996)
logP -0.79			EXP	STEIN,J ET AL. (1986)
VP 1.72E-009	mm Hg	25	EST	NEELY,WB & BLAU,GE (1985)
DC	pKa			
HL 2.84E-019	atm m3/mol	25	EST	MEYLAN,WM & HOWARD,PH (1991)
OH 6.13E-011	cm3/molc sec	25	EST	MEYLAN,WM & HOWARD,PH (1993)

CAS #: 111076-00-7 — 1-ACETYL-4-(3-NO2 PHENYL) SEMICARBAZIDE

Formula: $C_9H_{10}N_4O_4$
Mol Weight: 238.20
MP (deg C): **FP (deg C):**
BP (deg C):
BP pressure (mm Hg):

Property/Value	Units	Temp	Data Type	Reference
WS 7.36E+003	mg/L	25	EST	MEYLAN,WM ET AL. (1996)
logP 0.67			EXP	STEIN,J ET AL. (1986)
VP 2.40E-009	mm Hg	25	EST	NEELY,WB & BLAU,GE (1985)
DC	pKa			
HL 1.08E-017	atm m3/mol	25	EST	MEYLAN,WM & HOWARD,PH (1991)
OH 1.03E-011	cm3/molc sec	25	EST	MEYLAN,WM & HOWARD,PH (1993)

CAS #: 111076-01-8 — 1-ACETYL-4-(4-NO2 PHENYL) SEMICARBAZIDE

Formula: $C_9H_{10}N_4O_4$
Mol Weight: 238.20
MP (deg C): **FP (deg C):**
BP (deg C):
BP pressure (mm Hg):

Property/Value	Units	Temp	Data Type	Reference
WS 6.04E+003	mg/L	25	EST	MEYLAN,WM ET AL. (1996)
logP 0.77			EXP	STEIN,J ET AL. (1986)
VP 2.40E-009	mm Hg	25	EST	NEELY,WB & BLAU,GE (1985)
DC	pKa			
HL 1.08E-017	atm m3/mol	25	EST	MEYLAN,WM & HOWARD,PH (1991)
OH 1.19E-011	cm3/molc sec	25	EST	MEYLAN,WM & HOWARD,PH (1993)

CAS #: 111076-02-9 — 1-ACETYL-4-(4-CYANO PHENYL)SEMICARBAZIDE

Formula: $C_{10}H_{10}N_4O_2$
Mol Weight: 218.22
MP (deg C): **FP (deg C):**
BP (deg C):
BP pressure (mm Hg):

Property/Value	Units	Temp	Data Type	Reference
WS 9.05E+003	mg/L	25	EST	MEYLAN,WM ET AL. (1996)
logP 0.38			EXP	STEIN,J ET AL. (1986)
VP 4.48E-009	mm Hg	25	EST	NEELY,WB & BLAU,GE (1985)
DC	pKa			
HL 2.63E-017	atm m3/mol	25	EST	MEYLAN,WM & HOWARD,PH (1991)
OH 1.41E-011	cm3/molc sec	25	EST	MEYLAN,WM & HOWARD,PH (1993)

CAS #: 111079-49-3 — 1H-PURINE-2,6-DIONE, 3,9-DIHYDRO-1,3-DIMETHYL-

Formula: $C_7H_8N_4O_2$
Mol Weight: 180.17
MP (deg C): **FP (deg C):**
BP (deg C):
BP pressure (mm Hg):

Property/Value	Units	Temp	Data Type	Reference
WS 2.91E+003	mg/L	25	EST	MEYLAN,WM ET AL. (1996)
logP -0.04			EXP	SANGSTER,J (1993)
VP 5.12E-009	mm Hg	25	EST	NEELY,WB & BLAU,GE (1985)
DC 8.81	pKa		EXP	KORTUM,G ET AL (1961)
HL 1.68E-012	atm m3/mol	25	EST	MEYLAN,WM & HOWARD,PH (1991)
OH 1.93E-011	cm3/molc sec	25	EST	MEYLAN,WM & HOWARD,PH (1993)

CAS #: 111226-62-1 — 1H-IMIDAZOLE-1-CARBOTHIOIC ACID, O-[1-(PHENOXYME

Formula: $C_{16}H_{20}N_2O_2S$
Mol Weight: 304.41
MP (deg C): **FP (deg C):**
BP (deg C):
BP pressure (mm Hg):

Property/Value	Units	Temp	Data Type	Reference
WS 2.61E+000	mg/L	25	EST	MEYLAN,WM ET AL. (1996)
logP 4.27			EXP	SANGSTER,J (1994)
VP 4.30E-008	mm Hg	25	EST	NEELY,WB & BLAU,GE (1985)
DC	pKa			
HL 1.07E-008	atm m3/mol	25	EST	MEYLAN,WM & HOWARD,PH (1991)
OH 8.29E-011	cm3/molc sec	25	EST	MEYLAN,WM & HOWARD,PH (1993)

111226-63-2

CAS #: **111226-63-2**

1H-IMIDAZOLE-1-CARBOTHIOIC ACID, O-[1-[(4-CHLORO

Formula: $C_{16}H_{19}ClN_2O_2S$

Mol Weight: 338.86

MP (deg C): | FP (deg C):

BP (deg C):

BP pressure (mm Hg):

Property/ Value	Units	Temp	Data Type	Reference
WS 4.20E-001	mg/L	25	EST	MEYLAN,WM ET AL. (1996)
logP 4.96			EXP	SANGSTER,J (1994)
VP 9.40E-009	mm Hg	25	EST	NEELY,WB & BLAU,GE (1985)
DC	pKa			
HL 7.94E-009	atm m3/mol	25	EST	MEYLAN,WM & HOWARD,PH (1991)
OH 6.77E-011	cm3/molc sec	25	EST	MEYLAN,WM & HOWARD,PH (1993)

111226-64-3

CAS #: **111226-64-3**

HYL]-2,2-DIMETHYLPROPYL ESTER

Formula: $C_{16}H_{19}BrN_2O_2S$

Mol Weight: 383.32

MP (deg C): | FP (deg C):

BP (deg C):

BP pressure (mm Hg):

Property/ Value	Units	Temp	Data Type	Reference
WS 1.68E-001	mg/L	25	EST	MEYLAN,WM ET AL. (1996)
logP 5.11			EXP	SANGSTER,J (1994)
VP 3.46E-009	mm Hg	25	EST	NEELY,WB & BLAU,GE (1985)
DC	pKa			
HL 4.27E-009	atm m3/mol	25	EST	MEYLAN,WM & HOWARD,PH (1991)
OH 6.76E-011	cm3/molc sec	25	EST	MEYLAN,WM & HOWARD,PH (1993)

111226-65-4

CAS #: **111226-65-4**

1H-IMIDAZOLE-1-CARBOTHIOIC ACID, O-[1-[(4-METHYL

Formula: $C_{17}H_{22}N_2O_2S$

Mol Weight: 318.44

MP (deg C): | FP (deg C):

BP (deg C):

BP pressure (mm Hg):

Property/ Value	Units	Temp	Data Type	Reference
WS 1.00E+000	mg/L	25	EST	MEYLAN,WM ET AL. (1996)
logP 4.66			EXP	SANGSTER,J (1994)
VP 1.70E-008	mm Hg	25	EST	NEELY,WB & BLAU,GE (1985)
DC	pKa			
HL 1.18E-008	atm m3/mol	25	EST	MEYLAN,WM & HOWARD,PH (1991)
OH 8.79E-011	cm3/molc sec	25	EST	MEYLAN,WM & HOWARD,PH (1993)

111226-67-6

CAS #: **111226-67-6**

1H-IMIDAZOLE-1-CARBOTHIOIC ACID, O-[1-[(4-PROPYL

Formula: $C_{19}H_{26}N_2O_2S$

Mol Weight: 346.50

MP (deg C): | FP (deg C):

BP (deg C):

BP pressure (mm Hg):

Property/ Value	Units	Temp	Data Type	Reference
WS 7.69E-002	mg/L	25	EST	MEYLAN,WM ET AL. (1996)
logP 5.77			EXP	SANGSTER,J (1994)
VP 2.98E-009	mm Hg	25	EST	NEELY,WB & BLAU,GE (1985)
DC	pKa			
HL 2.09E-008	atm m3/mol	25	EST	MEYLAN,WM & HOWARD,PH (1991)
OH 9.01E-011	cm3/molc sec	25	EST	MEYLAN,WM & HOWARD,PH (1993)

111226-68-7

CAS #: **111226-68-7**

1H-IMIDAZOLE-1-CARBOTHIOIC ACID, O-[1-[[4-(1-MET

Formula: $C_{19}H_{26}N_2O_2S$

Mol Weight: 346.50

MP (deg C): | FP (deg C):

BP (deg C):

BP pressure (mm Hg):

Property/ Value	Units	Temp	Data Type	Reference
WS 2.31E-001	mg/L	25	EST	MEYLAN,WM ET AL. (1996)
logP 5.21			EXP	SANGSTER,J (1994)
VP 5.28E-009	mm Hg	25	EST	NEELY,WB & BLAU,GE (1985)
DC	pKa			
HL 2.09E-008	atm m3/mol	25	EST	MEYLAN,WM & HOWARD,PH (1991)
OH 8.96E-011	cm3/molc sec	25	EST	MEYLAN,WM & HOWARD,PH (1993)

111226-70-1

CAS #: **111226-70-1**

1H-IMIDAZOLE-1-CARBOTHIOIC ACID, O-[1-[([1,1'-BI

Formula: $C_{22}H_{24}N_2O_2S$

Mol Weight: 380.51

MP (deg C): | FP (deg C):

BP (deg C):

BP pressure (mm Hg):

Property/ Value	Units	Temp	Data Type	Reference
WS 2.70E-002	mg/L	25	EST	MEYLAN,WM ET AL. (1996)
logP 6.06			EXP	SANGSTER,J (1994)
VP 2.87E-011	mm Hg	25	EST	NEELY,WB & BLAU,GE (1985)
DC	pKa			
HL 8.23E-010	atm m3/mol	25	EST	MEYLAN,WM & HOWARD,PH (1991)
OH 8.02E-011	cm3/molc sec	25	EST	MEYLAN,WM & HOWARD,PH (1993)

111226-71-2

CAS #: **111226-71-2**

1H-IMIDAZOLE-1-CARBOTHIOIC ACID, O-[1-[(4-METHOX

Formula: $C_{17}H_{22}N_2O_3S$

Mol Weight: 334.44

MP (deg C): | FP (deg C):

BP (deg C):

BP pressure (mm Hg):

Property/ Value	Units	Temp	Data Type	Reference
WS 1.60E+000	mg/L	25	EST	MEYLAN,WM ET AL. (1996)
logP 4.31			EXP	SANGSTER,J (1994)
VP 6.52E-009	mm Hg	25	EST	NEELY,WB & BLAU,GE (1985)
DC	pKa			
HL 6.34E-010	atm m3/mol	25	EST	MEYLAN,WM & HOWARD,PH (1991)
OH 8.08E-011	cm3/molc sec	25	EST	MEYLAN,WM & HOWARD,PH (1993)

111226-84-7

CAS #: **111226-84-7**

1H-IMIDAZOLE-1-CARBOTHIOIC ACID, O-[1-[(4-ETHYLP

Formula: $C_{18}H_{24}N_2O_2S$

Mol Weight: 332.47

MP (deg C): | FP (deg C):

BP (deg C):

BP pressure (mm Hg):

Property/ Value	Units	Temp	Data Type	Reference
WS 3.55E-001	mg/L	25	EST	MEYLAN,WM ET AL. (1996)
logP 5.09			EXP	SANGSTER,J (1994)
VP 7.14E-009	mm Hg	25	EST	NEELY,WB & BLAU,GE (1985)
DC	pKa			
HL 1.57E-008	atm m3/mol	25	EST	MEYLAN,WM & HOWARD,PH (1991)
OH 8.87E-011	cm3/molc sec	25	EST	MEYLAN,WM & HOWARD,PH (1993)

CAS #: 111226-99-4 — 1H-IMIDAZOLE-1-CARBOXYLIC ACID, 1-(PHENOXYMETHYL

Formula: $C_{16}H_{20}N_2O_3$

Mol Weight: 288.35

MP (deg C): | FP (deg C):
BP (deg C):
BP pressure (mm Hg):

Property/Value	Units	Temp	Data Type	Reference
WS 1.26E+001	mg/L	25	EST	MEYLAN,WM ET AL. (1996)
logP 3.58			EXP	SANGSTER,J (1994)
VP 2.90E-007	mm Hg	25	EST	NEELY,WB & BLAU,GE (1985)
DC	pKa			
HL 1.07E-007	atm m3/mol	25	EST	MEYLAN,WM & HOWARD,PH (1991)
OH 6.97E-011	cm3/molc sec	25	EST	MEYLAN,WM & HOWARD,PH (1993)

CAS #: 111227-00-0 — 1H-IMIDAZOLE-1-CARBOXYLIC ACID, 1-[(4-FLUOROPHEN

Formula: $C_{16}H_{19}FN_2O_3$

Mol Weight: 306.34

MP (deg C): | FP (deg C):
BP (deg C):
BP pressure (mm Hg):

Property/Value	Units	Temp	Data Type	Reference
WS 8.12E+000	mg/L	25	EST	MEYLAN,WM ET AL. (1996)
logP 3.68			EXP	SANGSTER,J (1994)
VP 3.19E-007	mm Hg	25	EST	NEELY,WB & BLAU,GE (1985)
DC	pKa			
HL 1.25E-007	atm m3/mol	25	EST	MEYLAN,WM & HOWARD,PH (1991)
OH 5.55E-011	cm3/molc sec	25	EST	MEYLAN,WM & HOWARD,PH (1993)

CAS #: 111227-01-1 — 1H-IMIDAZOLE-1-CARBOXYLIC ACID, 1-[(4-CHLOROPHEN

Formula: $C_{16}H_{19}ClN_2O_3$

Mol Weight: 322.79

MP (deg C): | FP (deg C):
BP (deg C):
BP pressure (mm Hg):

Property/Value	Units	Temp	Data Type	Reference
WS 2.38E+000	mg/L	25	EST	MEYLAN,WM ET AL. (1996)
logP 4.19			EXP	SANGSTER,J (1994)
VP 7.31E-008	mm Hg	25	EST	NEELY,WB & BLAU,GE (1985)
DC	pKa			
HL 7.91E-008	atm m3/mol	25	EST	MEYLAN,WM & HOWARD,PH (1991)
OH 5.45E-011	cm3/molc sec	25	EST	MEYLAN,WM & HOWARD,PH (1993)

CAS #: 111227-03-3 — 1H-IMIDAZOLE-1-CARBOXYLIC ACID, 1-[(4-BROMOPHENO

Formula: $C_{16}H_{19}BrN_2O_3$

Mol Weight: 367.25

MP (deg C): | FP (deg C):
BP (deg C):
BP pressure (mm Hg):

Property/Value	Units	Temp	Data Type	Reference
WS 9.03E-001	mg/L	25	EST	MEYLAN,WM ET AL. (1996)
logP 4.37			EXP	SANGSTER,J (1994)
VP 2.73E-008	mm Hg	25	EST	NEELY,WB & BLAU,GE (1985)
DC	pKa			
HL 4.25E-008	atm m3/mol	25	EST	MEYLAN,WM & HOWARD,PH (1991)
OH 5.44E-011	cm3/molc sec	25	EST	MEYLAN,WM & HOWARD,PH (1993)

CAS #: 111227-04-4 — 1H-IMIDAZOLE-1-CARBOXYLIC ACID, 1-[(4-NITROPHENO

Formula: $C_{16}H_{19}N_3O_5$

Mol Weight: 333.35

MP (deg C): | FP (deg C):
BP (deg C):
BP pressure (mm Hg):

Property/Value	Units	Temp	Data Type	Reference
WS 4.38E+000	mg/L	25	EST	MEYLAN,WM ET AL. (1996)
logP 3.35			EXP	SANGSTER,J (1994)
VP 3.51E-009	mm Hg	25	EST	NEELY,WB & BLAU,GE (1985)
DC	pKa			
HL 4.21E-010	atm m3/mol	25	EST	MEYLAN,WM & HOWARD,PH (1991)
OH 5.09E-011	cm3/molc sec	25	EST	MEYLAN,WM & HOWARD,PH (1993)

CAS #: 111227-05-5 — 1H-IMIDAZOLE-1-CARBOXYLIC ACID, 1-[(4-CYANOPHENO

Formula: $C_{17}H_{19}N_3O_3$

Mol Weight: 313.36

MP (deg C): | FP (deg C):
BP (deg C):
BP pressure (mm Hg):

Property/Value	Units	Temp	Data Type	Reference
WS 1.14E+001	mg/L	25	EST	MEYLAN,WM ET AL. (1996)
logP 3.15			EXP	SANGSTER,J (1994)
VP 8.18E-009	mm Hg	25	EST	NEELY,WB & BLAU,GE (1985)
DC	pKa			
HL 1.03E-009	atm m3/mol	25	EST	MEYLAN,WM & HOWARD,PH (1991)
OH 5.20E-011	cm3/molc sec	25	EST	MEYLAN,WM & HOWARD,PH (1993)

CAS #: 111227-07-7 — 1H-IMIDAZOLE-1-CARBOXYLIC ACID, 1-[[4-(1-METHYLE

Formula: $C_{19}H_{26}N_2O_3$

Mol Weight: 330.43

MP (deg C): | FP (deg C):
BP (deg C):
BP pressure (mm Hg):

Property/Value	Units	Temp	Data Type	Reference
WS 6.85E-001	mg/L	25	EST	MEYLAN,WM ET AL. (1996)
logP 4.77			EXP	SANGSTER,J (1994)
VP 4.13E-008	mm Hg	25	EST	NEELY,WB & BLAU,GE (1985)
DC	pKa			
HL 2.08E-007	atm m3/mol	25	EST	MEYLAN,WM & HOWARD,PH (1991)
OH 7.64E-011	cm3/molc sec	25	EST	MEYLAN,WM & HOWARD,PH (1993)

CAS #: 111227-08-8 — 1H-IMIDAZOLE-1-CARBOXYLIC ACID, 1-[(4-METHYLPHEN

Formula: $C_{17}H_{22}N_2O_2$

Mol Weight: 286.38

MP (deg C): | FP (deg C):
BP (deg C):
BP pressure (mm Hg):

Property/Value	Units	Temp	Data Type	Reference
WS 5.02E+000	mg/L	25	EST	MEYLAN,WM ET AL. (1996)
logP 4.06			EXP	SANGSTER,J (1994)
VP 1.60E-007	mm Hg	25	EST	NEELY,WB & BLAU,GE (1985)
DC	pKa			
HL 3.33E-008	atm m3/mol	25	EST	MEYLAN,WM & HOWARD,PH (1991)
OH 9.15E-011	cm3/molc sec	25	EST	MEYLAN,WM & HOWARD,PH (1993)

111227-09-9

CAS #: 111227-09-9 — 1H-IMIDAZOLE-1-CARBOXYLIC ACID, 1-[[[1,1'-BIPHEN

Formula: $C_{22}H_{24}N_2O_3$

Mol Weight: 364.45

MP (deg C):   FP (deg C):

BP (deg C):

BP pressure (mm Hg):

Property/Value	Units	Temp	Data Type	Reference
WS 6.46E-001	mg/L	25	EST	MEYLAN,WM ET AL. (1996)
logP 4.56			EXP	SANGSTER,J (1994)
VP 1.98E-010	mm Hg	25	EST	NEELY,WB & BLAU,GE (1985)
DC	pKa			
HL 8.19E-009	atm m3/mol	25	EST	MEYLAN,WM & HOWARD,PH (1991)
OH 6.70E-011	cm3/molc sec	25	EST	MEYLAN,WM & HOWARD,PH (1993)

111227-10-2

CAS #: 111227-10-2 — 1H-IMIDAZOLE-1-CARBOXYLIC ACID, 1-[[4-(1,1-DIMET

Formula: $C_{20}H_{28}N_2O_3$

Mol Weight: 344.46

MP (deg C):   FP (deg C):

BP (deg C):

BP pressure (mm Hg):

Property/Value	Units	Temp	Data Type	Reference
WS 5.12E-001	mg/L	25	EST	MEYLAN,WM ET AL. (1996)
logP 4.82			EXP	SANGSTER,J (1994)
VP 2.19E-008	mm Hg	25	EST	NEELY,WB & BLAU,GE (1985)
DC	pKa			
HL 2.76E-007	atm m3/mol	25	EST	MEYLAN,WM & HOWARD,PH (1991)
OH 7.45E-011	cm3/molc sec	25	EST	MEYLAN,WM & HOWARD,PH (1993)

111227-13-5

CAS #: 111227-13-5 — 1H-IMIDAZOLE-1-CARBOXYLIC ACID, 1-[(4-ETHYLPHENO

Formula: $C_{18}H_{24}N_2O_3$

Mol Weight: 316.40

MP (deg C):   FP (deg C):

BP (deg C):

BP pressure (mm Hg):

Property/Value	Units	Temp	Data Type	Reference
WS 3.10E+000	mg/L	25	EST	MEYLAN,WM ET AL. (1996)
logP 4.10			EXP	SANGSTER,J (1994)
VP 5.57E-008	mm Hg	25	EST	NEELY,WB & BLAU,GE (1985)
DC	pKa			
HL 1.56E-007	atm m3/mol	25	EST	MEYLAN,WM & HOWARD,PH (1991)
OH 7.55E-011	cm3/molc sec	25	EST	MEYLAN,WM & HOWARD,PH (1993)

111227-14-6

CAS #: 111227-14-6 — 1H-IMIDAZOLE-1-CARBOXYLIC ACID, 1-[(4-PROPYLPHEN

Formula: $C_{19}H_{26}N_2O_3$

Mol Weight: 330.43

MP (deg C):   FP (deg C):

BP (deg C):

BP pressure (mm Hg):

Property/Value	Units	Temp	Data Type	Reference
WS 7.42E-001	mg/L	25	EST	MEYLAN,WM ET AL. (1996)
logP 4.73			EXP	SANGSTER,J (1994)
VP 2.35E-008	mm Hg	25	EST	NEELY,WB & BLAU,GE (1985)
DC	pKa			
HL 2.08E-007	atm m3/mol	25	EST	MEYLAN,WM & HOWARD,PH (1991)
OH 7.69E-011	cm3/molc sec	25	EST	MEYLAN,WM & HOWARD,PH (1993)

111227-15-7

CAS #: 111227-15-7 — 1H-IMIDAZOLE-1-CARBOXYLIC ACID, 1-[(4-BUTYLPHENO

Formula: $C_{20}H_{28}N_2O_3$

Mol Weight: 344.46

MP (deg C):   FP (deg C):

BP (deg C):

BP pressure (mm Hg):

Property/Value	Units	Temp	Data Type	Reference
WS 3.81E-001	mg/L	25	EST	MEYLAN,WM ET AL. (1996)
logP 4.97			EXP	SANGSTER,J (1994)
VP 9.91E-009	mm Hg	25	EST	NEELY,WB & BLAU,GE (1985)
DC	pKa			
HL 2.76E-007	atm m3/mol	25	EST	MEYLAN,WM & HOWARD,PH (1991)
OH 7.83E-011	cm3/molc sec	25	EST	MEYLAN,WM & HOWARD,PH (1993)

111227-17-9

CAS #: 111227-17-9 — 1H-IMIDAZOLE-1-CARBOXYLIC ACID, 1-[(4-METHOXYPHE

Formula: $C_{17}H_{22}N_2O_4$

Mol Weight: 318.38

MP (deg C):   FP (deg C):

BP (deg C):

BP pressure (mm Hg):

Property/Value	Units	Temp	Data Type	Reference
WS 1.06E+001	mg/L	25	EST	MEYLAN,WM ET AL. (1996)
logP 3.46			EXP	SANGSTER,J (1994)
VP 5.10E-008	mm Hg	25	EST	NEELY,WB & BLAU,GE (1985)
DC	pKa			
HL 6.31E-009	atm m3/mol	25	EST	MEYLAN,WM & HOWARD,PH (1991)
OH 6.76E-011	cm3/molc sec	25	EST	MEYLAN,WM & HOWARD,PH (1993)

111227-18-0

CAS #: 111227-18-0 — 1H-IMIDAZOLE-1-CARBOXYLIC ACID, 1-[[4-(METHYLTHI

Formula: $C_{17}H_{22}N_2O_3S$

Mol Weight: 334.44

MP (deg C):   FP (deg C):

BP (deg C):

BP pressure (mm Hg):

Property/Value	Units	Temp	Data Type	Reference
WS 2.78E+000	mg/L	25	EST	MEYLAN,WM ET AL. (1996)
logP 4.03			EXP	SANGSTER,J (1994)
VP 1.10E-008	mm Hg	25	EST	NEELY,WB & BLAU,GE (1985)
DC	pKa			
HL 3.11E-009	atm m3/mol	25	EST	MEYLAN,WM & HOWARD,PH (1991)
OH 6.25E-011	cm3/molc sec	25	EST	MEYLAN,WM & HOWARD,PH (1993)

111227-21-5

CAS #: 111227-21-5 — 1H-IMIDAZOLE-1-CARBOXYLIC ACID, 1-[(4-ETHOXYPHEN

Formula: $C_{18}H_{24}N_2O_4$

Mol Weight: 332.40

MP (deg C):   FP (deg C):

BP (deg C):

BP pressure (mm Hg):

Property/Value	Units	Temp	Data Type	Reference
WS 3.99E+000	mg/L	25	EST	MEYLAN,WM ET AL. (1996)
logP 3.86			EXP	SANGSTER,J (1994)
VP 2.15E-008	mm Hg	25	EST	NEELY,WB & BLAU,GE (1985)
DC	pKa			
HL 8.38E-009	atm m3/mol	25	EST	MEYLAN,WM & HOWARD,PH (1991)
OH 7.30E-011	cm3/molc sec	25	EST	MEYLAN,WM & HOWARD,PH (1993)

CAS #: 111227-26-0				1H-IMIDAZOLE-1-CARBOXYLIC ACID, 1-[(4-PROPOXYPHE

Formula: $C_{19}H_{26}N_2O_4$

Mol Weight: 346.43

MP (deg C):   FP (deg C):

BP (deg C):

BP pressure (mm Hg):

Property/Value	Units	Temp	Data Type	Reference
WS 2.05E+000	mg/L	25	EST	MEYLAN,WM ET AL. (1996)
logP 4.10			EXP	SANGSTER,J (1994)
VP 9.06E-009	mm Hg	25	EST	NEELY,WB & BLAU,GE (1985)
DC	pKa			
HL 1.11E-008	atm m3/mol	25	EST	MEYLAN,WM & HOWARD,PH (1991)
OH 7.73E-011	cm3/molc sec	25	EST	MEYLAN,WM & HOWARD,PH (1993)

CAS #: 111227-41-9				1H-IMIDAZOLE-1-CARBOXYLIC ACID, 1-[[4-(ETHOXYCAR

Formula: $C_{19}H_{24}N_2O_5$

Mol Weight: 360.41

MP (deg C):   FP (deg C):

BP (deg C):

BP pressure (mm Hg):

Property/Value	Units	Temp	Data Type	Reference
WS 1.62E+000	mg/L	25	EST	MEYLAN,WM ET AL. (1996)
logP 4.12			EXP	SANGSTER,J (1994)
VP 1.57E-008	mm Hg	25	EST	NEELY,WB & BLAU,GE (1985)
DC	pKa			
HL 9.13E-010	atm m3/mol	25	EST	MEYLAN,WM & HOWARD,PH (1991)
OH 5.68E-011	cm3/molc sec	25	EST	MEYLAN,WM & HOWARD,PH (1993)

CAS #: 111435-97-3				(DES-IPR) N-PYRROLIDINYL-PR-CLOFAZIMINE

Formula: $C_{31}H_{29}Cl_2N_5$

Mol Weight: 542.52

MP (deg C):   FP (deg C):

BP (deg C):

BP pressure (mm Hg):

Property/Value	Units	Temp	Data Type	Reference
WS 2.54E-001	mg/L	25	EST	MEYLAN,WM ET AL. (1996)
logP 3.72			EXP	QUIGLEY,JM ET AL. (1990)
VP 6.07E-015	mm Hg	25	EST	NEELY,WB & BLAU,GE (1985)
DC	pKa			
HL 1.39E-014	atm m3/mol	25	EST	MEYLAN,WM & HOWARD,PH (1991)
OH 2.33E-010	cm3/molc sec	25	EST	MEYLAN,WM & HOWARD,PH (1993)

CAS #: 111435-98-4				(DES-IPR) N-PIPERIDINYL-PR-CLOFAZIMINE

Formula: $C_{32}H_{31}Cl_2N_5$

Mol Weight: 556.54

MP (deg C):   FP (deg C):

BP (deg C):

BP pressure (mm Hg):

Property/Value	Units	Temp	Data Type	Reference
WS 4.18E-002	mg/L	25	EST	MEYLAN,WM ET AL. (1996)
logP 4.53			EXP	QUIGLEY,JM ET AL. (1990)
VP 2.36E-015	mm Hg	25	EST	NEELY,WB & BLAU,GE (1985)
DC	pKa			
HL 1.84E-014	atm m3/mol	25	EST	MEYLAN,WM & HOWARD,PH (1991)
OH 2.39E-010	cm3/molc sec	25	EST	MEYLAN,WM & HOWARD,PH (1993)

CAS #: 111436-11-4				(DES-IPR)NN-DIETAMINO(1ME)BU-CLOFAZIMINE

Formula: $C_{33}H_{35}Cl_2N_5$

Mol Weight: 572.59

MP (deg C):   FP (deg C):

BP (deg C):

BP pressure (mm Hg):

Property/Value	Units	Temp	Data Type	Reference
WS 4.16E-002	mg/L	25	EST	MEYLAN,WM ET AL. (1996)
logP 4.41			EXP	QUIGLEY,JM ET AL. (1990)
VP 3.75E-015	mm Hg	25	EST	NEELY,WB & BLAU,GE (1985)
DC	pKa			
HL 5.54E-014	atm m3/mol	25	EST	MEYLAN,WM & HOWARD,PH (1991)
OH 2.34E-010	cm3/molc sec	25	EST	MEYLAN,WM & HOWARD,PH (1993)

CAS #: 111436-12-5				(DES-IPR) 4--PIPERIDINYL-PR-CLOFAZIMINE

Formula: $C_{30}H_{27}Cl_2N_5$

Mol Weight: 528.49

MP (deg C):   FP (deg C):

BP (deg C):

BP pressure (mm Hg):

Property/Value	Units	Temp	Data Type	Reference
WS 4.82E-001	mg/L	25	EST	MEYLAN,WM ET AL. (1996)
logP 3.50			EXP	QUIGLEY,JM ET AL. (1990)
VP 4.91E-015	mm Hg	25	EST	NEELY,WB & BLAU,GE (1985)
DC	pKa			
HL 6.31E-015	atm m3/mol	25	EST	MEYLAN,WM & HOWARD,PH (1991)
OH 2.26E-010	cm3/molc sec	25	EST	MEYLAN,WM & HOWARD,PH (1993)

CAS #: 111479-05-1				PROPAQUIZAFOP

Formula: $C_{22}H_{22}ClN_3O_5$

Mol Weight: 443.89

MP (deg C):   FP (deg C):

BP (deg C):

BP pressure (mm Hg):

Property/Value	Units	Temp	Data Type	Reference
WS 1.92E-001	mg/L	25	EST	MEYLAN,WM ET AL. (1996)
logP 4.60			EXP	HANSCH,C ET AL. (1995)
VP 2.65E-011	mm Hg	25	EST	NEELY,WB & BLAU,GE (1985)
DC	pKa			
HL 2.51E-012	atm m3/mol	25	EST	MEYLAN,WM & HOWARD,PH (1991)
OH 3.42E-011	cm3/molc sec	25	EST	MEYLAN,WM & HOWARD,PH (1993)

CAS #: 111802-43-8				PREGNA-1,4-DIENE-16-CARBOXYLIC ACID, 11,21-DIHYD

Formula: $C_{23}H_{30}O_6$

Mol Weight: 402.49

MP (deg C):   FP (deg C):

BP (deg C):

BP pressure (mm Hg):

Property/Value	Units	Temp	Data Type	Reference
WS 1.04E+002	mg/L	25	EST	MEYLAN,WM ET AL. (1996)
logP 1.70			EXP	SANGSTER,J (1994)
VP 8.00E-014	mm Hg	25	EST	NEELY,WB & BLAU,GE (1985)
DC	pKa			
HL 4.13E-013	atm m3/mol	25	EST	MEYLAN,WM & HOWARD,PH (1991)
OH 7.34E-011	cm3/molc sec	25	EST	MEYLAN,WM & HOWARD,PH (1993)

112225-83-9 — 1-(3-CHLORO),2-DIBENZOYL-1-TERT-BUTYLHYDRAZINE

Formula: $C_{18}H_{19}ClN_2O_2$
Mol Weight: 330.82
MP (deg C):
FP (deg C):
BP (deg C):
BP pressure (mm Hg):

Property/Value	Units	Temp	Data Type	Reference
WS 1.28E+001	mg/L	25	EST	MEYLAN,WM ET AL. (1996)
logP 3.28			EXP	OIKAWA,N ET AL. (1994)
VP 2.91E-010	mm Hg	25	EST	NEELY,WB & BLAU,GE (1985)
DC	pKa			
HL 3.58E-011	atm m3/mol	25	EST	MEYLAN,WM & HOWARD,PH (1991)
OH 2.34E-011	cm3/molc sec	25	EST	MEYLAN,WM & HOWARD,PH (1993)

112225-85-1 — 1-(4-FLUORO),2-DIBENZOYL-1-TERT-BUTYLHYDRAZINE

Formula: $C_{18}H_{19}FN_2O_2$
Mol Weight: 314.36
MP (deg C):
FP (deg C):
BP (deg C):
BP pressure (mm Hg):

Property/Value	Units	Temp	Data Type	Reference
WS 3.72E+001	mg/L	25	EST	MEYLAN,WM ET AL. (1996)
logP 2.85			EXP	OIKAWA,N ET AL. (1994)
VP 1.32E-009	mm Hg	25	EST	NEELY,WB & BLAU,GE (1985)
DC	pKa			
HL 5.64E-011	atm m3/mol	25	EST	MEYLAN,WM & HOWARD,PH (1991)
OH 2.50E-011	cm3/molc sec	25	EST	MEYLAN,WM & HOWARD,PH (1993)

112225-86-2 — 1-(2-NITRO),2-DIBENZOYL-1-TERT-BUTYLHYDRAZINE

Formula: $C_{18}H_{19}N_3O_4$
Mol Weight: 341.37
MP (deg C):
FP (deg C):
BP (deg C):
BP pressure (mm Hg):

Property/Value	Units	Temp	Data Type	Reference
WS 3.28E+001	mg/L	25	EST	MEYLAN,WM ET AL. (1996)
logP 2.27			EXP	OIKAWA,N ET AL. (1994)
VP 2.07E-011	mm Hg	25	EST	NEELY,WB & BLAU,GE (1985)
DC	pKa			
HL 1.91E-013	atm m3/mol	25	EST	MEYLAN,WM & HOWARD,PH (1991)
OH 2.29E-011	cm3/molc sec	25	EST	MEYLAN,WM & HOWARD,PH (1993)

112225-87-3 — 1,2-DIBENZOYL-1-TERT-BUTYLHYDRAZINE

Formula: $C_{18}H_{20}N_2O_2$
Mol Weight: 296.37
MP (deg C):
FP (deg C):
BP (deg C):
BP pressure (mm Hg):

Property/Value	Units	Temp	Data Type	Reference
WS 1.04E+002	mg/L	25	EST	MEYLAN,WM ET AL. (1996)
logP 2.45			EXP	OIKAWA,N ET AL. (1994)
VP 1.09E-009	mm Hg	25	EST	NEELY,WB & BLAU,GE (1985)
DC	pKa			
HL 4.83E-011	atm m3/mol	25	EST	MEYLAN,WM & HOWARD,PH (1991)
OH 2.46E-011	cm3/molc sec	25	EST	MEYLAN,WM & HOWARD,PH (1993)

112226-22-9 — 1-(3,4-DICHLORO),2-DIBENZOYL-1-TERT-BUTYLHYDRAZI

Formula: $C_{18}H_{18}Cl_2N_2O_2$
Mol Weight: 365.26
MP (deg C):
FP (deg C):
BP (deg C):
BP pressure (mm Hg):

Property/Value	Units	Temp	Data Type	Reference
WS 1.18E+000	mg/L	25	EST	MEYLAN,WM ET AL. (1996)
logP 4.25			EXP	OIKAWA,N ET AL. (1994)
VP 7.69E-011	mm Hg	25	EST	NEELY,WB & BLAU,GE (1985)
DC	pKa			
HL 2.65E-011	atm m3/mol	25	EST	MEYLAN,WM & HOWARD,PH (1991)
OH 2.31E-011	cm3/molc sec	25	EST	MEYLAN,WM & HOWARD,PH (1993)

112226-54-7 — 1-(4-CHLORO),2-DIBENZOYL-1-TERT-BUTYLHYDRAZINE

Formula: $C_{18}H_{19}ClN_2O_2$
Mol Weight: 330.82
MP (deg C):
FP (deg C):
BP (deg C):
BP pressure (mm Hg):

Property/Value	Units	Temp	Data Type	Reference
WS 9.70E+000	mg/L	25	EST	MEYLAN,WM ET AL. (1996)
logP 3.42			EXP	OIKAWA,N ET AL. (1994)
VP 2.91E-010	mm Hg	25	EST	NEELY,WB & BLAU,GE (1985)
DC	pKa			
HL 3.58E-011	atm m3/mol	25	EST	MEYLAN,WM & HOWARD,PH (1991)
OH 2.40E-011	cm3/molc sec	25	EST	MEYLAN,WM & HOWARD,PH (1993)

112226-56-9 — 1-(4-METHYL),2-DIBENZOYL-1-TERT-BUTYLHYDRAZINE

Formula: $C_{19}H_{22}N_2O_2$
Mol Weight: 310.40
MP (deg C):
FP (deg C):
BP (deg C):
BP pressure (mm Hg):

Property/Value	Units	Temp	Data Type	Reference
WS 2.98E+001	mg/L	25	EST	MEYLAN,WM ET AL. (1996)
logP 2.99			EXP	OIKAWA,N ET AL. (1994)
VP 4.82E-010	mm Hg	25	EST	NEELY,WB & BLAU,GE (1985)
DC	pKa			
HL 5.33E-011	atm m3/mol	25	EST	MEYLAN,WM & HOWARD,PH (1991)
OH 2.76E-011	cm3/molc sec	25	EST	MEYLAN,WM & HOWARD,PH (1993)

112226-58-1 — 1-(2-CHLORO),2-DIBENZOYL-1-TERT-BUTYLHYDRAZINE

Formula: $C_{18}H_{19}ClN_2O_2$
Mol Weight: 330.82
MP (deg C):
FP (deg C):
BP (deg C):
BP pressure (mm Hg):

Property/Value	Units	Temp	Data Type	Reference
WS 4.96E+001	mg/L	25	EST	MEYLAN,WM ET AL. (1996)
logP 2.59			EXP	OIKAWA,N ET AL. (1994)
VP 2.91E-010	mm Hg	25	EST	NEELY,WB & BLAU,GE (1985)
DC	pKa			
HL 3.58E-011	atm m3/mol	25	EST	MEYLAN,WM & HOWARD,PH (1991)
OH 2.40E-011	cm3/molc sec	25	EST	MEYLAN,WM & HOWARD,PH (1993)

CAS #: 112226-76-3				1-(2-FLUORO),2-DIBENZOYL-1-TERT-BUTYLHYDRAZINE
Formula: $C_{18}H_{19}FN_2O_2$				
Mol Weight: 314.36				
MP (deg C):		FP (deg C):		
BP (deg C):				
BP pressure (mm Hg):				

Property/ Value	Units	Temp	Data Type	Reference
WS 9.38E+001	mg/L	25	EST	MEYLAN,WM ET AL. (1996)
logP 2.38			EXP	OIKAWA,N ET AL. (1994)
VP 1.32E-009	mm Hg	25	EST	NEELY,WB & BLAU,GE (1985)
DC	pKa			
HL 5.64E-011	atm m3/mol	25	EST	MEYLAN,WM & HOWARD,PH (1991)
OH 2.50E-011	cm3/molc sec	25	EST	MEYLAN,WM & HOWARD,PH (1993)

CAS #: 112226-77-4				1-(2-METHOXY),2-DIBENZOYL-1-TERT-BUTYLHYDRAZINE
Formula: $C_{19}H_{22}N_2O_3$				
Mol Weight: 326.40				
MP (deg C):		FP (deg C):		
BP (deg C):				
BP pressure (mm Hg):				

Property/ Value	Units	Temp	Data Type	Reference
WS 1.55E+002	mg/L	25	EST	MEYLAN,WM ET AL. (1996)
logP 2.04			EXP	OIKAWA,N ET AL. (1994)
VP 2.05E-010	mm Hg	25	EST	NEELY,WB & BLAU,GE (1985)
DC	pKa			
HL 2.86E-012	atm m3/mol	25	EST	MEYLAN,WM & HOWARD,PH (1991)
OH 4.32E-011	cm3/molc sec	25	EST	MEYLAN,WM & HOWARD,PH (1993)

CAS #: 112226-78-5				1-(2-METHYL),2-DIBENZOYL-1-TERT-BUTYLHYDRAZINE
Formula: $C_{19}H_{22}N_2O_2$				
Mol Weight: 310.40				
MP (deg C):		FP (deg C):		
BP (deg C):				
BP pressure (mm Hg):				

Property/ Value	Units	Temp	Data Type	Reference
WS 4.78E+001	mg/L	25	EST	MEYLAN,WM ET AL. (1996)
logP 2.75			EXP	OIKAWA,N ET AL. (1994)
VP 4.82E-010	mm Hg	25	EST	NEELY,WB & BLAU,GE (1985)
DC	pKa			
HL 5.33E-011	atm m3/mol	25	EST	MEYLAN,WM & HOWARD,PH (1991)
OH 2.76E-011	cm3/molc sec	25	EST	MEYLAN,WM & HOWARD,PH (1993)

CAS #: 112226-82-1				1-(3-NITRO),2-DIBENZOYL-1-TERT-BUTYLHYDRAZINE
Formula: $C_{18}H_{19}N_3O_4$				
Mol Weight: 341.37				
MP (deg C):		FP (deg C):		
BP (deg C):				
BP pressure (mm Hg):				

Property/ Value	Units	Temp	Data Type	Reference
WS 1.33E+001	mg/L	25	EST	MEYLAN,WM ET AL. (1996)
logP 2.73			EXP	OIKAWA,N ET AL. (1994)
VP 2.07E-011	mm Hg	25	EST	NEELY,WB & BLAU,GE (1985)
DC	pKa			
HL 1.91E-013	atm m3/mol	25	EST	MEYLAN,WM & HOWARD,PH (1991)
OH 2.30E-011	cm3/molc sec	25	EST	MEYLAN,WM & HOWARD,PH (1993)

CAS #: 112226-85-4				1-(4-BROMO),2-DIBENZOYL-1-TERT-BUTYLHYDRAZINE
Formula: $C_{18}H_{19}BrN_2O_2$				
Mol Weight: 375.27				
MP (deg C):		FP (deg C):		
BP (deg C):				
BP pressure (mm Hg):				

Property/ Value	Units	Temp	Data Type	Reference
WS 3.26E+000	mg/L	25	EST	MEYLAN,WM ET AL. (1996)
logP 3.66			EXP	OIKAWA,N ET AL. (1994)
VP 1.24E-010	mm Hg	25	EST	NEELY,WB & BLAU,GE (1985)
DC	pKa			
HL 1.93E-011	atm m3/mol	25	EST	MEYLAN,WM & HOWARD,PH (1991)
OH 2.39E-011	cm3/molc sec	25	EST	MEYLAN,WM & HOWARD,PH (1993)

CAS #: 112226-86-5				1-(2,6-DIFLUORO),2-DIBENZOYL-1-TERT-BUTYLHYDRAZI
Formula: $C_{18}H_{18}F_2N_2O_2$				
Mol Weight: 332.35				
MP (deg C):		FP (deg C):		
BP (deg C):				
BP pressure (mm Hg):				

Property/ Value	Units	Temp	Data Type	Reference
WS 1.13E+002	mg/L	25	EST	MEYLAN,WM ET AL. (1996)
logP 2.16			EXP	OIKAWA,N ET AL. (1994)
VP 1.59E-009	mm Hg	25	EST	NEELY,WB & BLAU,GE (1985)
DC	pKa			
HL 6.58E-011	atm m3/mol	25	EST	MEYLAN,WM & HOWARD,PH (1991)
OH 2.56E-011	cm3/molc sec	25	EST	MEYLAN,WM & HOWARD,PH (1993)

CAS #: 112226-92-3				1-(4-METHOXY),2-DIBENZOYL-1-TERT-BUTYLHYDRAZINE
Formula: $C_{19}H_{22}N_2O_3$				
Mol Weight: 326.40				
MP (deg C):		FP (deg C):		
BP (deg C):				
BP pressure (mm Hg):				

Property/ Value	Units	Temp	Data Type	Reference
WS 5.59E+001	mg/L	25	EST	MEYLAN,WM ET AL. (1996)
logP 2.56			EXP	OIKAWA,N ET AL. (1994)
VP 2.05E-010	mm Hg	25	EST	NEELY,WB & BLAU,GE (1985)
DC	pKa			
HL 2.86E-012	atm m3/mol	25	EST	MEYLAN,WM & HOWARD,PH (1991)
OH 4.32E-011	cm3/molc sec	25	EST	MEYLAN,WM & HOWARD,PH (1993)

CAS #: 112226-93-4				1-(2-IODO),2-DIBENZOYL-1-TERT-BUTYLHYDRAZINE
Formula: $C_{18}H_{19}IN_2O_2$				
Mol Weight: 422.27				
MP (deg C):		FP (deg C):		
BP (deg C):				
BP pressure (mm Hg):				

Property/ Value	Units	Temp	Data Type	Reference
WS 8.52E+000	mg/L	25	EST	MEYLAN,WM ET AL. (1996)
logP 2.83			EXP	OIKAWA,N ET AL. (1994)
VP 3.36E-011	mm Hg	25	EST	NEELY,WB & BLAU,GE (1985)
DC	pKa			
HL 1.12E-011	atm m3/mol	25	EST	MEYLAN,WM & HOWARD,PH (1991)
OH 2.40E-011	cm3/molc sec	25	EST	MEYLAN,WM & HOWARD,PH (1993)

CAS #: 112226-95-6 — 1-(2-METHYLTHIO),2-DIBENZOYL-1-TERT-BUTYLHYDRAZI

Formula: $C_{19}H_{22}N_2O_2S$

Mol Weight: 342.46

MP (deg C):   FP (deg C):

BP (deg C):

BP pressure (mm Hg):

Property/ Value	Units	Temp	Data Type	Reference
WS 4.14E+001	mg/L	25	EST	MEYLAN,WM ET AL. (1996)
logP 2.60			EXP	OIKAWA,N ET AL. (1994)
VP 4.50E-011	mm Hg	25	EST	NEELY,WB & BLAU,GE (1985)
DC	pKa			
HL 1.41E-012	atm m3/mol	25	EST	MEYLAN,WM & HOWARD,PH (1991)
OH 3.53E-011	cm3/molc sec	25	EST	MEYLAN,WM & HOWARD,PH (1993)

CAS #: 112226-97-8 — 1-(2,4-DICHLORO),2-DIBENZOYL-1-TERT-BUTYLHYDRAZI

Formula: $C_{18}H_{18}Cl_2N_2O_2$

Mol Weight: 365.26

MP (deg C):   FP (deg C):

BP (deg C):

BP pressure (mm Hg):

Property/ Value	Units	Temp	Data Type	Reference
WS 4.65E+000	mg/L	25	EST	MEYLAN,WM ET AL. (1996)
logP 3.55			EXP	OIKAWA,N ET AL. (1994)
VP 7.69E-011	mm Hg	25	EST	NEELY,WB & BLAU,GE (1985)
DC	pKa			
HL 2.65E-011	atm m3/mol	25	EST	MEYLAN,WM & HOWARD,PH (1991)
OH 2.37E-011	cm3/molc sec	25	EST	MEYLAN,WM & HOWARD,PH (1993)

CAS #: 112227-02-8 — 1-(3-FLUORO),2-DIBENZOYL-1-TERT-BUTYLHYDRAZINE

Formula: $C_{18}H_{19}FN_2O_2$

Mol Weight: 314.36

MP (deg C):   FP (deg C):

BP (deg C):

BP pressure (mm Hg):

Property/ Value	Units	Temp	Data Type	Reference
WS 4.27E+001	mg/L	25	EST	MEYLAN,WM ET AL. (1996)
logP 2.78			EXP	OIKAWA,N ET AL. (1994)
VP 1.32E-009	mm Hg	25	EST	NEELY,WB & BLAU,GE (1985)
DC	pKa			
HL 5.64E-011	atm m3/mol	25	EST	MEYLAN,WM & HOWARD,PH (1991)
OH 2.38E-011	cm3/molc sec	25	EST	MEYLAN,WM & HOWARD,PH (1993)

CAS #: 112249-32-8 — 1-(2-BROMO),2-DIBENZOYL-1-TERT-BUTYLHYDRAZINE

Formula: $C_{18}H_{19}BrN_2O_2$

Mol Weight: 375.27

MP (deg C):   FP (deg C):

BP (deg C):

BP pressure (mm Hg):

Property/ Value	Units	Temp	Data Type	Reference
WS 2.19E+001	mg/L	25	EST	MEYLAN,WM ET AL. (1996)
logP 2.69			EXP	OIKAWA,N ET AL. (1994)
VP 1.24E-010	mm Hg	25	EST	NEELY,WB & BLAU,GE (1985)
DC	pKa			
HL 1.93E-011	atm m3/mol	25	EST	MEYLAN,WM & HOWARD,PH (1991)
OH 2.39E-011	cm3/molc sec	25	EST	MEYLAN,WM & HOWARD,PH (1993)

CAS #: 112410-23-8 — TEBUFENOZIDE

Formula: $C_{22}H_{28}N_2O_2$

Mol Weight: 352.48

MP (deg C):   FP (deg C):

BP (deg C):

BP pressure (mm Hg):

Property/ Value	Units	Temp	Data Type	Reference
WS 1.41E+000	mg/L	25	EST	MEYLAN,WM ET AL. (1996)
logP 4.25			EXP	TOMLIN,C (1994)
VP 4.03E-011	mm Hg	25	EST	NEELY,WB & BLAU,GE (1985)
DC	pKa			
HL 8.62E-011	atm m3/mol	25	EST	MEYLAN,WM & HOWARD,PH (1991)
OH 3.24E-011	cm3/molc sec	25	EST	MEYLAN,WM & HOWARD,PH (1993)

CAS #: 112411-06-0 — BENZOIC ACID, 2,3,4,5,6-PENTAFLUORO-, [1-(1,1-DI

Formula: $C_{18}H_{15}F_5N_2O_2$

Mol Weight: 386.32

MP (deg C):   FP (deg C):

BP (deg C):

BP pressure (mm Hg):

Property/ Value	Units	Temp	Data Type	Reference
WS 6.24E+000	mg/L	25	EST	MEYLAN,WM ET AL. (1996)
logP 3.25			EXP	OIKAWA,N ET AL. (1994)
VP 2.80E-009	mm Hg	25	EST	NEELY,WB & BLAU,GE (1985)
DC	pKa			
HL 1.05E-010	atm m3/mol	25	EST	MEYLAN,WM & HOWARD,PH (1991)
OH 2.32E-011	cm3/molc sec	25	EST	MEYLAN,WM & HOWARD,PH (1993)

CAS #: 112411-46-8 — 1-(3-METHYL),2-DIBENZOYL-1-TERT-BUTYLHYDRAZINE

Formula: $C_{19}H_{22}N_2O_2$

Mol Weight: 310.40

MP (deg C):   FP (deg C):

BP (deg C):

BP pressure (mm Hg):

Property/ Value	Units	Temp	Data Type	Reference
WS 4.42E+001	mg/L	25	EST	MEYLAN,WM ET AL. (1996)
logP 2.79			EXP	OIKAWA,N ET AL. (1994)
VP 4.82E-010	mm Hg	25	EST	NEELY,WB & BLAU,GE (1985)
DC	pKa			
HL 5.33E-011	atm m3/mol	25	EST	MEYLAN,WM & HOWARD,PH (1991)
OH 2.51E-011	cm3/molc sec	25	EST	MEYLAN,WM & HOWARD,PH (1993)

CAS #: 112411-90-2 — 1-(2,3-DIMETHYL),2-DIBENZOYL-1-TERT-BUTYLHYDRAZI

Formula: $C_{20}H_{24}N_2O_2$

Mol Weight: 324.43

MP (deg C):   FP (deg C):

BP (deg C):

BP pressure (mm Hg):

Property/ Value	Units	Temp	Data Type	Reference
WS 1.99E+001	mg/L	25	EST	MEYLAN,WM ET AL. (1996)
logP 3.10			EXP	OIKAWA,N ET AL. (1994)
VP 2.11E-010	mm Hg	25	EST	NEELY,WB & BLAU,GE (1985)
DC	pKa			
HL 5.89E-011	atm m3/mol	25	EST	MEYLAN,WM & HOWARD,PH (1991)
OH 2.87E-011	cm3/molc sec	25	EST	MEYLAN,WM & HOWARD,PH (1993)

CAS #: 112412-44-9				1-(3,4-DIMETHYL),2-DIBENZOYL-1-TERT-BUTYLHYDRAZI
Formula: $C_{20}H_{24}N_2O_2$				
Mol Weight: 324.43				
MP (deg C):		FP (deg C):		
BP (deg C):				
BP pressure (mm Hg):				

Property/Value	Units	Temp	Data Type	Reference
WS 1.24E+001	mg/L	25	EST	MEYLAN,WM ET AL. (1996)
logP 3.34			EXP	OIKAWA,N ET AL. (1994)
VP 2.11E-010	mm Hg	25	EST	NEELY,WB & BLAU,GE (1985)
DC	pKa			
HL 5.89E-011	atm m3/mol	25	EST	MEYLAN,WM & HOWARD,PH (1991)
OH 2.87E-011	cm3/molc sec	25	EST	MEYLAN,WM & HOWARD,PH (1993)

CAS #: 112426-60-5				1-(3-METHOXY),2-DIBENZOYL-1-TERT-BUTYLHYDRAZINE
Formula: $C_{19}H_{22}N_2O_3$				
Mol Weight: 326.40				
MP (deg C):		FP (deg C):		
BP (deg C):				
BP pressure (mm Hg):				

Property/Value	Units	Temp	Data Type	Reference
WS 5.59E+001	mg/L	25	EST	MEYLAN,WM ET AL. (1996)
logP 2.56			EXP	OIKAWA,N ET AL. (1994)
VP 2.05E-010	mm Hg	25	EST	NEELY,WB & BLAU,GE (1985)
DC	pKa			
HL 2.86E-012	atm m3/mol	25	EST	MEYLAN,WM & HOWARD,PH (1991)
OH 3.27E-011	cm3/molc sec	25	EST	MEYLAN,WM & HOWARD,PH (1993)

CAS #: 112426-62-7				1-(4-CYANO),2-DIBENZOYL-1-TERT-BUTYLHYDRAZINE
Formula: $C_{19}H_{19}N_3O_2$				
Mol Weight: 321.38				
MP (deg C):		FP (deg C):		
BP (deg C):				
BP pressure (mm Hg):				

Property/Value	Units	Temp	Data Type	Reference
WS 3.66E+001	mg/L	25	EST	MEYLAN,WM ET AL. (1996)
logP 2.50			EXP	OIKAWA,N ET AL. (1994)
VP 3.92E-011	mm Hg	25	EST	NEELY,WB & BLAU,GE (1985)
DC	pKa			
HL 4.67E-013	atm m3/mol	25	EST	MEYLAN,WM & HOWARD,PH (1991)
OH 2.30E-011	cm3/molc sec	25	EST	MEYLAN,WM & HOWARD,PH (1993)

CAS #: 112426-65-0				1-(4-TRIFLUOROMETHYL),2-DIBENZOYL-1-TERT-BUTYLHY
Formula: $C_{19}H_{19}F_3N_2O_2$				
Mol Weight: 364.37				
MP (deg C):		FP (deg C):		
BP (deg C):				
BP pressure (mm Hg):				

Property/Value	Units	Temp	Data Type	Reference
WS 3.06E+000	mg/L	25	EST	MEYLAN,WM ET AL. (1996)
logP 3.77			EXP	OIKAWA,N ET AL. (1994)
VP 8.59E-010	mm Hg	25	EST	NEELY,WB & BLAU,GE (1985)
DC	pKa			
HL 4.20E-010	atm m3/mol	25	EST	MEYLAN,WM & HOWARD,PH (1991)
OH 2.55E-012	cm3/molc sec	25	EST	MEYLAN,WM & HOWARD,PH (1993)

CAS #: 112426-66-1				1-(3-TRIFLUOROMETHYL),2-DIBENZOYL-1-TERT-BUTYLHY
Formula: $C_{19}H_{19}F_3N_2O_2$				
Mol Weight: 364.37				
MP (deg C):		FP (deg C):		
BP (deg C):				
BP pressure (mm Hg):				

Property/Value	Units	Temp	Data Type	Reference
WS 4.19E+000	mg/L	25	EST	MEYLAN,WM ET AL. (1996)
logP 3.61			EXP	OIKAWA,N ET AL. (1994)
VP 8.59E-010	mm Hg	25	EST	NEELY,WB & BLAU,GE (1985)
DC	pKa			
HL 4.20E-010	atm m3/mol	25	EST	MEYLAN,WM & HOWARD,PH (1991)
OH 2.64E-012	cm3/molc sec	25	EST	MEYLAN,WM & HOWARD,PH (1993)

CAS #: 112426-67-2				1-(2-TRIFLUOROMETHYL),2-DIBENZOYL-1-TERT-BUTYLHY
Formula: $C_{19}H_{19}F_3N_2O_2$				
Mol Weight: 364.37				
MP (deg C):		FP (deg C):		
BP (deg C):				
BP pressure (mm Hg):				

Property/Value	Units	Temp	Data Type	Reference
WS 1.87E+001	mg/L	25	EST	MEYLAN,WM ET AL. (1996)
logP 2.85			EXP	OIKAWA,N ET AL. (1994)
VP 8.59E-010	mm Hg	25	EST	NEELY,WB & BLAU,GE (1985)
DC	pKa			
HL 4.20E-010	atm m3/mol	25	EST	MEYLAN,WM & HOWARD,PH (1991)
OH 2.55E-012	cm3/molc sec	25	EST	MEYLAN,WM & HOWARD,PH (1993)

CAS #: 112426-69-4				1-(3-CYANO),2-DIBENZOYL-1-TERT-BUTYLHYDRAZINE
Formula: $C_{19}H_{19}N_3O_2$				
Mol Weight: 321.38				
MP (deg C):		FP (deg C):		
BP (deg C):				
BP pressure (mm Hg):				

Property/Value	Units	Temp	Data Type	Reference
WS 5.01E+001	mg/L	25	EST	MEYLAN,WM ET AL. (1996)
logP 2.34			EXP	OIKAWA,N ET AL. (1994)
VP 3.92E-011	mm Hg	25	EST	NEELY,WB & BLAU,GE (1985)
DC	pKa			
HL 4.67E-013	atm m3/mol	25	EST	MEYLAN,WM & HOWARD,PH (1991)
OH 2.31E-011	cm3/molc sec	25	EST	MEYLAN,WM & HOWARD,PH (1993)

CAS #: 112426-71-8				1-(2,6-DICHLORO),2-DIBENZOYL-1-TERT-BUTYLHYDRAZI
Formula: $C_{18}H_{18}Cl_2N_2O_2$				
Mol Weight: 365.26				
MP (deg C):		FP (deg C):		
BP (deg C):				
BP pressure (mm Hg):				

Property/Value	Units	Temp	Data Type	Reference
WS 3.26E+001	mg/L	25	EST	MEYLAN,WM ET AL. (1996)
logP 2.56			EXP	OIKAWA,N ET AL. (1994)
VP 7.69E-011	mm Hg	25	EST	NEELY,WB & BLAU,GE (1985)
DC	pKa			
HL 2.65E-011	atm m3/mol	25	EST	MEYLAN,WM & HOWARD,PH (1991)
OH 2.37E-011	cm3/molc sec	25	EST	MEYLAN,WM & HOWARD,PH (1993)

CAS #: 112426-76-3				1-(2,3-DICHLORO),2-DIBENZOYL-1-TERT-BUTYLHYDRAZI

Formula: $C_{18}H_{18}Cl_2N_2O_2$

Mol Weight: 365.26

MP (deg C): FP (deg C):

BP (deg C):

BP pressure (mm Hg):

Property/Value	Units	Temp	Data Type	Reference
WS 6.13E+000	mg/L	25	EST	MEYLAN,WM ET AL. (1996)
logP 3.41			EXP	OIKAWA,N ET AL. (1994)
VP 7.69E-011	mm Hg	25	EST	NEELY,WB & BLAU,GE (1985)
DC	pKa			
HL 2.65E-011	atm m3/mol	25	EST	MEYLAN,WM & HOWARD,PH (1991)
OH 2.31E-011	cm3/molc sec	25	EST	MEYLAN,WM & HOWARD,PH (1993)

CAS #: 112426-79-6				1-(3,5-DIMETHYL),2-DIBENZOYL-1-TERT-BUTYLHYDRAZI

Formula: $C_{20}H_{24}N_2O_2$

Mol Weight: 324.43

MP (deg C): FP (deg C):

BP (deg C):

BP pressure (mm Hg):

Property/Value	Units	Temp	Data Type	Reference
WS 1.12E+001	mg/L	25	EST	MEYLAN,WM ET AL. (1996)
logP 3.39			EXP	OIKAWA,N ET AL. (1994)
VP 2.11E-010	mm Hg	25	EST	NEELY,WB & BLAU,GE (1985)
DC	pKa			
HL 5.89E-011	atm m3/mol	25	EST	MEYLAN,WM & HOWARD,PH (1991)
OH 2.87E-011	cm3/molc sec	25	EST	MEYLAN,WM & HOWARD,PH (1993)

CAS #: 112426-81-0				1-(3-CHLORO-2-METHYL),2-DIBENZOYL-1-TERT-BUTYLHY

Formula: $C_{19}H_{21}ClN_2O_2$

Mol Weight: 344.84

MP (deg C): FP (deg C):

BP (deg C):

BP pressure (mm Hg):

Property/Value	Units	Temp	Data Type	Reference
WS 5.95E+000	mg/L	25	EST	MEYLAN,WM ET AL. (1996)
logP 3.57			EXP	OIKAWA,N ET AL. (1994)
VP 1.28E-010	mm Hg	25	EST	NEELY,WB & BLAU,GE (1985)
DC	pKa			
HL 3.95E-011	atm m3/mol	25	EST	MEYLAN,WM & HOWARD,PH (1991)
OH 2.43E-011	cm3/molc sec	25	EST	MEYLAN,WM & HOWARD,PH (1993)

CAS #: 112427-09-5				1-(3-BROMO),2-DIBENZOYL-1-TERT-BUTYLHYDRAZINE

Formula: $C_{18}H_{19}BrN_2O_2$

Mol Weight: 375.27

MP (deg C): FP (deg C):

BP (deg C):

BP pressure (mm Hg):

Property/Value	Units	Temp	Data Type	Reference
WS 4.55E+000	mg/L	25	EST	MEYLAN,WM ET AL. (1996)
logP 3.49			EXP	OIKAWA,N ET AL. (1994)
VP 1.24E-010	mm Hg	25	EST	NEELY,WB & BLAU,GE (1985)
DC	pKa			
HL 1.93E-011	atm m3/mol	25	EST	MEYLAN,WM & HOWARD,PH (1991)
OH 2.33E-011	cm3/molc sec	25	EST	MEYLAN,WM & HOWARD,PH (1993)

CAS #: 112427-40-4				1-(3-IODO),2-DIBENZOYL-1-TERT-BUTYLHYDRAZINE

Formula: $C_{18}H_{19}IN_2O_2$

Mol Weight: 422.27

MP (deg C): FP (deg C):

BP (deg C):

BP pressure (mm Hg):

Property/Value	Units	Temp	Data Type	Reference
WS 1.48E+000	mg/L	25	EST	MEYLAN,WM ET AL. (1996)
logP 3.72			EXP	OIKAWA,N ET AL. (1994)
VP 3.36E-011	mm Hg	25	EST	NEELY,WB & BLAU,GE (1985)
DC	pKa			
HL 1.12E-011	atm m3/mol	25	EST	MEYLAN,WM & HOWARD,PH (1991)
OH 2.34E-011	cm3/molc sec	25	EST	MEYLAN,WM & HOWARD,PH (1993)

CAS #: 112427-41-5				1-(2-ETHYL),2-DIBENZOYL-1-TERT-BUTYLHYDRAZINE

Formula: $C_{20}H_{24}N_2O_2$

Mol Weight: 324.43

MP (deg C): FP (deg C):

BP (deg C):

BP pressure (mm Hg):

Property/Value	Units	Temp	Data Type	Reference
WS 2.62E+001	mg/L	25	EST	MEYLAN,WM ET AL. (1996)
logP 2.96			EXP	OIKAWA,N ET AL. (1994)
VP 2.11E-010	mm Hg	25	EST	NEELY,WB & BLAU,GE (1985)
DC	pKa			
HL 7.08E-011	atm m3/mol	25	EST	MEYLAN,WM & HOWARD,PH (1991)
OH 2.83E-011	cm3/molc sec	25	EST	MEYLAN,WM & HOWARD,PH (1993)

CAS #: 112427-66-4				1-(3,4-DIMETHOXY),2-DIBENZOYL-1-TERT-BUTYLHYDRAZ

Formula: $C_{20}H_{24}N_2O_4$

Mol Weight: 356.43

MP (deg C): FP (deg C):

BP (deg C):

BP pressure (mm Hg):

Property/Value	Units	Temp	Data Type	Reference
WS 9.30E+001	mg/L	25	EST	MEYLAN,WM ET AL. (1996)
logP 2.09			EXP	OIKAWA,N ET AL. (1994)
VP 3.78E-011	mm Hg	25	EST	NEELY,WB & BLAU,GE (1985)
DC	pKa			
HL 1.69E-013	atm m3/mol	25	EST	MEYLAN,WM & HOWARD,PH (1991)
OH 4.14E-011	cm3/molc sec	25	EST	MEYLAN,WM & HOWARD,PH (1993)

CAS #: 112427-73-3				1-(4-IODO),2-DIBENZOYL-1-TERT-BUTYLHYDRAZINE

Formula: $C_{18}H_{19}IN_2O_2$

Mol Weight: 422.27

MP (deg C): FP (deg C):

BP (deg C):

BP pressure (mm Hg):

Property/Value	Units	Temp	Data Type	Reference
WS 1.32E+000	mg/L	25	EST	MEYLAN,WM ET AL. (1996)
logP 3.78			EXP	OIKAWA,N ET AL. (1994)
VP 3.36E-011	mm Hg	25	EST	NEELY,WB & BLAU,GE (1985)
DC	pKa			
HL 1.12E-011	atm m3/mol	25	EST	MEYLAN,WM & HOWARD,PH (1991)
OH 2.40E-011	cm3/molc sec	25	EST	MEYLAN,WM & HOWARD,PH (1993)

CAS #: 112473-77-5				BUTANOIC ACID, 2-ETHYL-3-METHYL-, PHENYL ESTER
Formula: C$_{13}$H$_{18}$O$_2$				
Mol Weight: 206.29				
MP (deg C):		FP (deg C):		
BP (deg C):				
BP pressure (mm Hg):				

Property/ Value	Units	Temp	Data Type	Reference
WS 1.05E+001	mg/L	25	EST	MEYLAN,WM ET AL. (1996)
logP 4.20			EXP	YANG,HZ ET AL. (1987)
VP 1.17E-002	mm Hg	25	EST	NEELY,WB & BLAU,GE (1985)
DC	pKa			
HL 2.67E-004	atm m3/mol	25	EST	MEYLAN,WM & HOWARD,PH (1991)
OH 8.22E-012	cm3/molc sec	25	EST	MEYLAN,WM & HOWARD,PH (1993)

CAS #: 112473-78-6				BUTANOIC ACID, 2-ETHOXY-3-METHYL-, PHENYLMETHYL
Formula: C$_{14}$H$_{20}$O$_3$				
Mol Weight: 236.31				
MP (deg C):		FP (deg C):		
BP (deg C):				
BP pressure (mm Hg):				

Property/ Value	Units	Temp	Data Type	Reference
WS 4.36E+001	mg/L	25	EST	MEYLAN,WM ET AL. (1996)
logP 3.29			EXP	YANG,HZ ET AL. (1987)
VP 1.36E-003	mm Hg	25	EST	NEELY,WB & BLAU,GE (1985)
DC	pKa			
HL 3.64E-006	atm m3/mol	25	EST	MEYLAN,WM & HOWARD,PH (1991)
OH 2.60E-011	cm3/molc sec	25	EST	MEYLAN,WM & HOWARD,PH (1993)

CAS #: 112740-56-4				2-OXOPROPANAL,1-METHYLSULFOXIDE
Formula: C$_4$H$_7$NO$_3$S				
Mol Weight: 149.17				
MP (deg C):		FP (deg C):		
BP (deg C):				
BP pressure (mm Hg):				

Property/ Value	Units	Temp	Data Type	Reference
WS 2.80E+004	mg/L	25	EST	MEYLAN,WM ET AL. (1996)
logP -0.72			EXP	HANSCH,C ET AL. (1995)
VP 2.57E-005	mm Hg	25	EST	NEELY,WB & BLAU,GE (1985)
DC	pKa			
HL 4.00E-014	atm m3/mol	25	EST	MEYLAN,WM & HOWARD,PH (1991)
OH 6.80E-011	cm3/molc sec	25	EST	MEYLAN,WM & HOWARD,PH (1993)

CAS #: 112740-61-1				2-OXOPROPANAL,1-METHYLSULFIDE
Formula: C$_4$H$_7$NO$_2$S				
Mol Weight: 133.17				
MP (deg C):		FP (deg C):		
BP (deg C):				
BP pressure (mm Hg):				

Property/ Value	Units	Temp	Data Type	Reference
WS 8.27E+003	mg/L	25	EST	MEYLAN,WM ET AL. (1996)
logP 1.21			EXP	HANSCH,C ET AL. (1995)
VP 2.17E-003	mm Hg	25	EST	NEELY,WB & BLAU,GE (1985)
DC	pKa			
HL 6.34E-010	atm m3/mol	25	EST	MEYLAN,WM & HOWARD,PH (1991)
OH 9.68E-012	cm3/molc sec	25	EST	MEYLAN,WM & HOWARD,PH (1993)

CAS #: 112830-95-2				6H-DIBENZO[B,D]PYRAN-9-METHANOL, 3-(1,1-DIMETHYL
Formula: C$_{25}$H$_{38}$O$_3$				
Mol Weight: 386.58				
MP (deg C):		FP (deg C):		
BP (deg C):				
BP pressure (mm Hg):				

Property/ Value	Units	Temp	Data Type	Reference
WS 2.02E-002	mg/L	25	EST	MEYLAN,WM ET AL. (1996)
logP 7.44			EXP	SANGSTER,J (1993)
VP 6.28E-012	mm Hg	25	EST	NEELY,WB & BLAU,GE (1985)
DC	pKa			
HL 2.77E-011	atm m3/mol	25	EST	MEYLAN,WM & HOWARD,PH (1991)
OH 3.14E-010	cm3/molc sec	25	EST	MEYLAN,WM & HOWARD,PH (1993)

CAS #: 112904-74-2				2H-1-BENZOPYRAN-8-OL, 3-(DIPROPYLAMINO)-3,4-DIHY
Formula: C$_{15}$H$_{23}$NO$_2$				
Mol Weight: 249.36				
MP (deg C):		FP (deg C):		
BP (deg C):				
BP pressure (mm Hg):				

Property/ Value	Units	Temp	Data Type	Reference
WS 3.69E+002	mg/L	25	EST	MEYLAN,WM ET AL. (1996)
logP 3.30			EXP	DIJKSTRA,D ET AL. (1988)
VP 2.83E-006	mm Hg	25	EST	NEELY,WB & BLAU,GE (1985)
DC	pKa			
HL 3.30E-011	atm m3/mol	25	EST	MEYLAN,WM & HOWARD,PH (1991)
OH 2.48E-010	cm3/molc sec	25	EST	MEYLAN,WM & HOWARD,PH (1993)

CAS #: 112960-12-0				2H,5H-[1]BENZOPYRANO[4,3-B]-1,4-OXAZIN-9-OL, 3,4
Formula: C$_{14}$H$_{19}$NO$_3$				
Mol Weight: 249.31				
MP (deg C):		FP (deg C):		
BP (deg C):				
BP pressure (mm Hg):				

Property/ Value	Units	Temp	Data Type	Reference
WS 1.92E+004	mg/L	25	EST	MEYLAN,WM ET AL. (1996)
logP 1.97			EXP	SANGSTER,J (1994)
VP 1.27E-006	mm Hg	25	EST	NEELY,WB & BLAU,GE (1985)
DC	pKa			
HL 9.66E-014	atm m3/mol	25	EST	MEYLAN,WM & HOWARD,PH (1991)
OH 2.83E-010	cm3/molc sec	25	EST	MEYLAN,WM & HOWARD,PH (1993)

CAS #: 113024-34-3				1H-1,2,4-TRIAZOLE-3-SULFONAMIDE, N-(2,6-DIFLUORO
Formula: C$_{16}$H$_{14}$F$_2$N$_4$O$_2$S				
Mol Weight: 364.38				
MP (deg C):		FP (deg C):		
BP (deg C):				
BP pressure (mm Hg):				

Property/ Value	Units	Temp	Data Type	Reference
WS 3.51E+000	mg/L	25	EST	MEYLAN,WM ET AL. (1996)
logP 3.70			EXP	SANGSTER,J (1994)
VP 7.94E-010	mm Hg	25	EST	NEELY,WB & BLAU,GE (1985)
DC	pKa			
HL 4.79E-013	atm m3/mol	25	EST	MEYLAN,WM & HOWARD,PH (1991)
OH 9.60E-012	cm3/molc sec	25	EST	MEYLAN,WM & HOWARD,PH (1993)

CAS #: 113024-75-2 — 1H-1,2,4-TRIAZOLE-3-SULFONAMIDE, N-(2,6-DICHLORO

Formula: $C_{16}H_{14}Cl_2N_4O_2S$

Mol Weight: 397.29

MP (deg C):
FP (deg C):
BP (deg C):
BP pressure (mm Hg):

Property/Value	Units	Temp	Data Type	Reference
WS 5.56E-001	mg/L	25	EST	MEYLAN,WM ET AL. (1996)
logP 4.40			EXP	SANGSTER,J (1994)
VP 3.79E-011	mm Hg	25	EST	NEELY,WB & BLAU,GE (1985)
DC	pKa			
HL 1.61E-013	atm m3/mol	25	EST	MEYLAN,WM & HOWARD,PH (1991)
OH 1.33E-011	cm3/molc sec	25	EST	MEYLAN,WM & HOWARD,PH (1993)

CAS #: 113588-95-7 — 2-THIOPHENECARBOXAMIDE, 4-BROMO-5-METHYL-3-(1-ME

Formula: $C_{10}H_{12}BrN_5O_2S$

Mol Weight: 346.21

MP (deg C):
FP (deg C):
BP (deg C):
BP pressure (mm Hg):

Property/Value	Units	Temp	Data Type	Reference
WS 2.22E+002	mg/L	25	EST	MEYLAN,WM ET AL. (1996)
logP 1.72			EXP	MULLICAN,MD ET AL. (1991)
VP 7.94E-011	mm Hg	25	EST	NEELY,WB & BLAU,GE (1985)
DC	pKa			
HL 3.45E-015	atm m3/mol	25	EST	MEYLAN,WM & HOWARD,PH (1991)
OH 1.55E-011	cm3/molc sec	25	EST	MEYLAN,WM & HOWARD,PH (1993)

CAS #: 113588-97-9 — 2-THIOPHENECARBOXAMIDE, 4,5-DIBROMO-3-(1-METHYLE

Formula: $C_9H_9Br_2N_5O_2S$

Mol Weight: 411.09

MP (deg C):
FP (deg C):
BP (deg C):
BP pressure (mm Hg):

Property/Value	Units	Temp	Data Type	Reference
WS 6.88E+001	mg/L	25	EST	MEYLAN,WM ET AL. (1996)
logP 1.85			EXP	SANGSTER,J (1994)
VP 2.02E-011	mm Hg	25	EST	NEELY,WB & BLAU,GE (1985)
DC	pKa			
HL 1.25E-015	atm m3/mol	25	EST	MEYLAN,WM & HOWARD,PH (1991)
OH 1.34E-011	cm3/molc sec	25	EST	MEYLAN,WM & HOWARD,PH (1993)

CAS #: 113589-05-2 — 2-THIOPHENECARBOXAMIDE, 4-BROMO-5-METHOXY-3-(1-M

Formula: $C_{10}H_{12}BrN_5O_3S$

Mol Weight: 362.21

MP (deg C):
FP (deg C):
BP (deg C):
BP pressure (mm Hg):

Property/Value	Units	Temp	Data Type	Reference
WS 3.68E+002	mg/L	25	EST	MEYLAN,WM ET AL. (1996)
logP 1.35			EXP	MULLICAN,MD ET AL. (1991)
VP 3.35E-011	mm Hg	25	EST	NEELY,WB & BLAU,GE (1985)
DC	pKa			
HL 1.85E-016	atm m3/mol	25	EST	MEYLAN,WM & HOWARD,PH (1991)
OH 2.46E-011	cm3/molc sec	25	EST	MEYLAN,WM & HOWARD,PH (1993)

CAS #: 113589-14-3 — 1H-PYRROLE-2-CARBOXAMIDE, 4,5-DIBROMO-3-(1-METHY

Formula: $C_{10}H_{12}Br_2N_6O_2$

Mol Weight: 408.06

MP (deg C):
FP (deg C):
BP (deg C):
BP pressure (mm Hg):

Property/Value	Units	Temp	Data Type	Reference
WS 1.92E+001	mg/L	25	EST	MEYLAN,WM ET AL. (1996)
logP 2.52			EXP	MULLICAN,MD ET AL. (1991)
VP 2.58E-011	mm Hg	25	EST	NEELY,WB & BLAU,GE (1985)
DC	pKa			
HL 8.24E-017	atm m3/mol	25	EST	MEYLAN,WM & HOWARD,PH (1991)
OH 2.02E-011	cm3/molc sec	25	EST	MEYLAN,WM & HOWARD,PH (1993)

CAS #: 113916-37-3 — 3-PYRIDINESULFONAMIDE, N-(AMINOCARBONYL)-

Formula: $C_6H_7N_3O_3S$

Mol Weight: 201.20

MP (deg C):
FP (deg C):
BP (deg C):
BP pressure (mm Hg):

Property/Value	Units	Temp	Data Type	Reference
WS 7.59E+003	mg/L	25	EST	MEYLAN,WM ET AL. (1996)
logP -0.65			EXP	CLOUX,JL ET AL. (1988)
VP 7.83E-007	mm Hg	25	EST	NEELY,WB & BLAU,GE (1985)
DC	pKa			
HL 5.84E-014	atm m3/mol	25	EST	MEYLAN,WM & HOWARD,PH (1991)
OH 2.08E-012	cm3/molc sec	25	EST	MEYLAN,WM & HOWARD,PH (1993)

CAS #: 113932-77-7 — CARBAMIC ACID, (3,4-DIETHOXYPHENYL)-, BUTYL ESTE

Formula: $C_{15}H_{23}NO_4$

Mol Weight: 281.35

MP (deg C):
FP (deg C):
BP (deg C):
BP pressure (mm Hg):

Property/Value	Units	Temp	Data Type	Reference
WS 1.75E+001	mg/L	25	EST	MEYLAN,WM ET AL. (1996)
logP 3.46			EXP	TAKAHASHI,J ET AL. (1988)
VP 6.67E-006	mm Hg	25	EST	NEELY,WB & BLAU,GE (1985)
DC	pKa			
HL 3.15E-010	atm m3/mol	25	EST	MEYLAN,WM & HOWARD,PH (1991)
OH 2.17E-010	cm3/molc sec	25	EST	MEYLAN,WM & HOWARD,PH (1993)

CAS #: 113932-78-8 — CARBAMIC ACID, (3,4-DIETHOXYPHENYL)-, CYCLOPENTY

Formula: $C_{16}H_{23}NO_4$

Mol Weight: 293.37

MP (deg C):
FP (deg C):
BP (deg C):
BP pressure (mm Hg):

Property/Value	Units	Temp	Data Type	Reference
WS 1.92E+001	mg/L	25	EST	MEYLAN,WM ET AL. (1996)
logP 3.33			EXP	HANSCH,C ET AL. (1995)
VP 2.42E-006	mm Hg	25	EST	NEELY,WB & BLAU,GE (1985)
DC	pKa			
HL 1.84E-010	atm m3/mol	25	EST	MEYLAN,WM & HOWARD,PH (1991)
OH 2.21E-010	cm3/molc sec	25	EST	MEYLAN,WM & HOWARD,PH (1993)

CAS #: 113932-79-9 — CARBAMIC ACID, (2-ETHOXYPHENYL)-, METHYL ESTER

Formula: C10H13NO3
Mol Weight: 195.22
MP (deg C):
FP (deg C):
BP (deg C):
BP pressure (mm Hg):

Property/Value	Units	Temp	Data Type	Reference
WS 3.89E+002	mg/L	25	EST	MEYLAN,WM ET AL. (1996)
logP 2.43			EXP	TAKAHASHI,J ET AL. (1988)
VP 1.79E-003	mm Hg	25	EST	NEELY,WB & BLAU,GE (1985)
DC	pKa			
HL 1.71E-009	atm m3/mol	25	EST	MEYLAN,WM & HOWARD,PH (1991)
OH 4.33E-011	cm3/molc sec	25	EST	MEYLAN,WM & HOWARD,PH (1993)

CAS #: 113932-80-2 — CARBAMIC ACID, [2-(TRIFLUOROMETHYL)PHENYL]-, MET

Formula: C9H8F3NO2
Mol Weight: 219.16
MP (deg C):
FP (deg C):
BP (deg C):
BP pressure (mm Hg):

Property/Value	Units	Temp	Data Type	Reference
WS 5.71E+002	mg/L	25	EST	MEYLAN,WM ET AL. (1996)
logP 2.09			EXP	TAKAHASHI,J ET AL. (1988)
VP 3.69E-002	mm Hg	25	EST	NEELY,WB & BLAU,GE (1985)
DC	pKa			
HL 1.90E-007	atm m3/mol	25	EST	MEYLAN,WM & HOWARD,PH (1991)
OH 8.79E-012	cm3/molc sec	25	EST	MEYLAN,WM & HOWARD,PH (1993)

CAS #: 113932-81-3 — CARBAMIC ACID, (3-IODOPHENYL)-, METHYL ESTER

Formula: C8H8INO2
Mol Weight: 277.06
MP (deg C):
FP (deg C):
BP (deg C):
BP pressure (mm Hg):

Property/Value	Units	Temp	Data Type	Reference
WS 4.31E+001	mg/L	25	EST	MEYLAN,WM ET AL. (1996)
logP 3.03			EXP	TAKAHASHI,J ET AL. (1988)
VP 5.32E-004	mm Hg	25	EST	NEELY,WB & BLAU,GE (1985)
DC	pKa			
HL 5.06E-009	atm m3/mol	25	EST	MEYLAN,WM & HOWARD,PH (1991)
OH 2.83E-011	cm3/molc sec	25	EST	MEYLAN,WM & HOWARD,PH (1993)

CAS #: 113932-82-4 — CARBAMIC ACID, (3-ETHYLPHENYL)-, METHYL ESTER

Formula: C10H13NO2
Mol Weight: 179.22
MP (deg C):
FP (deg C):
BP (deg C):
BP pressure (mm Hg):

Property/Value	Units	Temp	Data Type	Reference
WS 2.54E+002	mg/L	25	EST	MEYLAN,WM ET AL. (1996)
logP 2.74			EXP	TAKAHASHI,J ET AL. (1988)
VP 6.17E-003	mm Hg	25	EST	NEELY,WB & BLAU,GE (1985)
DC	pKa			
HL 3.20E-008	atm m3/mol	25	EST	MEYLAN,WM & HOWARD,PH (1991)
OH 1.07E-010	cm3/molc sec	25	EST	MEYLAN,WM & HOWARD,PH (1993)

CAS #: 113932-83-5 — CARBAMIC ACID, (3-ETHOXYPHENYL)-, METHYL ESTER

Formula: C10H13NO3
Mol Weight: 195.22
MP (deg C):
FP (deg C):
BP (deg C):
BP pressure (mm Hg):

Property/Value	Units	Temp	Data Type	Reference
WS 4.93E+002	mg/L	25	EST	MEYLAN,WM ET AL. (1996)
logP 2.31			EXP	TAKAHASHI,J ET AL. (1988)
VP 1.79E-003	mm Hg	25	EST	NEELY,WB & BLAU,GE (1985)
DC	pKa			
HL 1.71E-009	atm m3/mol	25	EST	MEYLAN,WM & HOWARD,PH (1991)
OH 2.06E-010	cm3/molc sec	25	EST	MEYLAN,WM & HOWARD,PH (1993)

CAS #: 113932-84-6 — CARBAMIC ACID, (4-BUTYLPHENYL)-, METHYL ESTER

Formula: C12H17NO2
Mol Weight: 207.27
MP (deg C):
FP (deg C):
BP (deg C):
BP pressure (mm Hg):

Property/Value	Units	Temp	Data Type	Reference
WS 2.83E+001	mg/L	25	EST	MEYLAN,WM ET AL. (1996)
logP 3.69			EXP	TAKAHASHI,J ET AL. (1988)
VP 7.49E-004	mm Hg	25	EST	NEELY,WB & BLAU,GE (1985)
DC	pKa			
HL 5.64E-008	atm m3/mol	25	EST	MEYLAN,WM & HOWARD,PH (1991)
OH 5.61E-011	cm3/molc sec	25	EST	MEYLAN,WM & HOWARD,PH (1993)

CAS #: 114012-04-3 — O,O-DIME O-(2-CN-4-CL)PH PHOSPHORTHIOATE

Formula: C9H9ClNO3PS
Mol Weight: 277.67
MP (deg C):
FP (deg C):
BP (deg C):
BP pressure (mm Hg):

Property/Value	Units	Temp	Data Type	Reference
WS 7.87E+000	mg/L	25	EST	MEYLAN,WM ET AL. (1996)
logP 3.58			EXP	DEBRUIJN,J & HERMENS,J (1991)
VP 4.64E-005	mm Hg	25	EST	NEELY,WB & BLAU,GE (1985)
DC	pKa			
HL 3.05E-007	atm m3/mol	25	EST	MEYLAN,WM & HOWARD,PH (1991)
OH 5.88E-011	cm3/molc sec	25	EST	MEYLAN,WM & HOWARD,PH (1993)

CAS #: 114260-70-7 — ACETIC ACID, [(6-CHLORO-4H-1,2,4-BENZOTHIADIAZIN

Formula: C9H7ClN2O4S2
Mol Weight: 306.75
MP (deg C):
FP (deg C):
BP (deg C):
BP pressure (mm Hg):

Property/Value	Units	Temp	Data Type	Reference
WS 5.55E+003	mg/L	25	EST	MEYLAN,WM ET AL. (1996)
logP 0.82			EXP	GAMBERINI,G ET AL. (1989)
VP 7.98E-012	mm Hg	25	EST	NEELY,WB & BLAU,GE (1985)
DC	pKa			
HL 4.10E-015	atm m3/mol	25	EST	MEYLAN,WM & HOWARD,PH (1991)
OH 1.41E-011	cm3/molc sec	25	EST	MEYLAN,WM & HOWARD,PH (1993)

114260-71-8

ACETIC ACID, [(7-CHLORO-4H-1,2,4-BENZOTHIADIAZIN

Formula: $C_9H_7ClN_2O_4S_2$

Mol Weight: 306.75

MP (deg C):   FP (deg C):

BP (deg C):

BP pressure (mm Hg):

Property/Value	Units	Temp	Data Type	Reference
WS 6.76E+003	mg/L	25	EST	MEYLAN,WM ET AL. (1996)
logP 0.72			EXP	GAMBERINI,G ET AL. (1989)
VP 7.98E-012	mm Hg	25	EST	NEELY,WB & BLAU,GE (1985)
DC	pKa			
HL 4.10E-015	atm m3/mol	25	EST	MEYLAN,WM & HOWARD,PH (1991)
OH 1.03E-011	cm3/molc sec	25	EST	MEYLAN,WM & HOWARD,PH (1993)

114260-72-9

ACETIC ACID, [(5,7-DICHLORO-4H-1,2,4-BENZOTHIADI

Formula: $C_9H_6Cl_2N_2O_4S_2$

Mol Weight: 341.19

MP (deg C):   FP (deg C):

BP (deg C):

BP pressure (mm Hg):

Property/Value	Units	Temp	Data Type	Reference
WS 2.04E+003	mg/L	25	EST	MEYLAN,WM ET AL. (1996)
logP 1.09			EXP	GAMBERINI,G ET AL. (1989)
VP 2.06E-012	mm Hg	25	EST	NEELY,WB & BLAU,GE (1985)
DC	pKa			
HL 3.04E-015	atm m3/mol	25	EST	MEYLAN,WM & HOWARD,PH (1991)
OH 8.02E-012	cm3/molc sec	25	EST	MEYLAN,WM & HOWARD,PH (1993)

114260-73-0

ACETIC ACID, [(6,7-DICHLORO-4H-1,2,4-BENZOTHIADI

Formula: $C_9H_6Cl_2N_2O_4S_2$

Mol Weight: 341.19

MP (deg C):   FP (deg C):

BP (deg C):

BP pressure (mm Hg):

Property/Value	Units	Temp	Data Type	Reference
WS 2.00E+003	mg/L	25	EST	MEYLAN,WM ET AL. (1996)
logP 1.10			EXP	GAMBERINI,G ET AL. (1989)
VP 2.06E-012	mm Hg	25	EST	NEELY,WB & BLAU,GE (1985)
DC	pKa			
HL 3.04E-015	atm m3/mol	25	EST	MEYLAN,WM & HOWARD,PH (1991)
OH 9.56E-012	cm3/molc sec	25	EST	MEYLAN,WM & HOWARD,PH (1993)

114260-74-1

ACETIC ACID, [(7-BROMO-4H-1,2,4-BENZOTHIADIAZIN-

Formula: $C_9H_7BrN_2O_4S_2$

Mol Weight: 351.20

MP (deg C):   FP (deg C):

BP (deg C):

BP pressure (mm Hg):

Property/Value	Units	Temp	Data Type	Reference
WS 2.48E+003	mg/L	25	EST	MEYLAN,WM ET AL. (1996)
logP 0.92			EXP	GAMBERINI,G ET AL. (1989)
VP 3.35E-012	mm Hg	25	EST	NEELY,WB & BLAU,GE (1985)
DC	pKa			
HL 2.20E-015	atm m3/mol	25	EST	MEYLAN,WM & HOWARD,PH (1991)
OH 1.03E-011	cm3/molc sec	25	EST	MEYLAN,WM & HOWARD,PH (1993)

114260-75-2

ACETIC ACID, [(5,7-DIBROMO-4H-1,2,4-BENZOTHIADIA

Formula: $C_9H_6Br_2N_2O_4S_2$

Mol Weight: 430.10

MP (deg C):   FP (deg C):

BP (deg C):

BP pressure (mm Hg):

Property/Value	Units	Temp	Data Type	Reference
WS 5.05E+002	mg/L	25	EST	MEYLAN,WM ET AL. (1996)
logP 1.16			EXP	GAMBERINI,G ET AL. (1989)
VP 3.58E-013	mm Hg	25	EST	NEELY,WB & BLAU,GE (1985)
DC	pKa			
HL 8.77E-016	atm m3/mol	25	EST	MEYLAN,WM & HOWARD,PH (1991)
OH 7.95E-012	cm3/molc sec	25	EST	MEYLAN,WM & HOWARD,PH (1993)

114260-76-3

ACETIC ACID, [(6-METHYL-4H-1,2,4-BENZOTHIADIAZIN

Formula: $C_{10}H_{10}N_2O_4S_2$

Mol Weight: 286.33

MP (deg C):   FP (deg C):

BP (deg C):

BP pressure (mm Hg):

Property/Value	Units	Temp	Data Type	Reference
WS 9.24E+003	mg/L	25	EST	MEYLAN,WM ET AL. (1996)
logP 0.70			EXP	GAMBERINI,G ET AL. (1989)
VP 1.33E-011	mm Hg	25	EST	NEELY,WB & BLAU,GE (1985)
DC	pKa			
HL 6.10E-015	atm m3/mol	25	EST	MEYLAN,WM & HOWARD,PH (1991)
OH 3.16E-011	cm3/molc sec	25	EST	MEYLAN,WM & HOWARD,PH (1993)

114260-77-4

ACETIC ACID, [(6-METHOXY-4H-1,2,4-BENZOTHIADIAZI

Formula: $C_{10}H_{10}N_2O_5S_2$

Mol Weight: 302.33

MP (deg C):   FP (deg C):

BP (deg C):

BP pressure (mm Hg):

Property/Value	Units	Temp	Data Type	Reference
WS 1.43E+004	mg/L	25	EST	MEYLAN,WM ET AL. (1996)
logP 0.37			EXP	GAMBERINI,G ET AL. (1989)
VP 5.58E-012	mm Hg	25	EST	NEELY,WB & BLAU,GE (1985)
DC	pKa			
HL 3.27E-016	atm m3/mol	25	EST	MEYLAN,WM & HOWARD,PH (1991)
OH 1.09E-010	cm3/molc sec	25	EST	MEYLAN,WM & HOWARD,PH (1993)

114260-78-5

ACETIC ACID, [[6-(TRIFLUOROMETHYL)-4H-1,2,4-BENZ

Formula: $C_{10}H_7F_3N_2O_4S_2$

Mol Weight: 340.30

MP (deg C):   FP (deg C):

BP (deg C):

BP pressure (mm Hg):

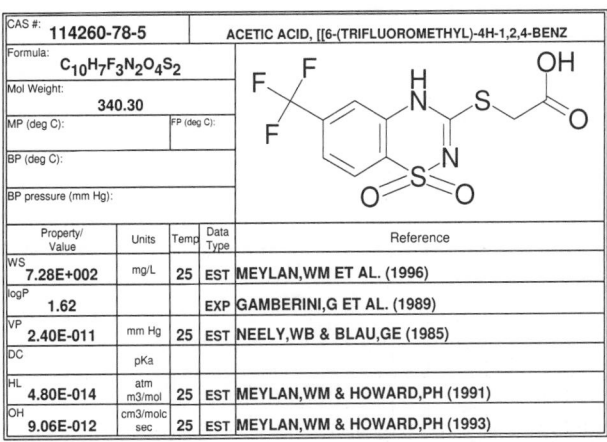

Property/Value	Units	Temp	Data Type	Reference
WS 7.28E+002	mg/L	25	EST	MEYLAN,WM ET AL. (1996)
logP 1.62			EXP	GAMBERINI,G ET AL. (1989)
VP 2.40E-011	mm Hg	25	EST	NEELY,WB & BLAU,GE (1985)
DC	pKa			
HL 4.80E-014	atm m3/mol	25	EST	MEYLAN,WM & HOWARD,PH (1991)
OH 9.06E-012	cm3/molc sec	25	EST	MEYLAN,WM & HOWARD,PH (1993)

CAS #: 114282-93-8 — ACETIC ACID, [(4H-1,2,4-BENZOTHIADIAZIN-3-YL)THI

Formula: $C_9H_8N_2O_4S_2$

Mol Weight: 272.30

MP (deg C): FP (deg C):

BP (deg C):

BP pressure (mm Hg):

Property/Value	Units	Temp	Data Type	Reference
WS 3.09E+004	mg/L	25	EST	MEYLAN,WM ET AL. (1996)
logP 0.18			EXP	GAMBERINI,G ET AL. (1989)
VP 3.07E-011	mm Hg	25	EST	NEELY,WB & BLAU,GE (1985)
DC	pKa			
HL 5.53E-015	atm m3/mol	25	EST	MEYLAN,WM & HOWARD,PH (1991)
OH 1.68E-011	cm3/molc sec	25	EST	MEYLAN,WM & HOWARD,PH (1993)

CAS #: 114298-20-3 — 4,7-ETHENO-1H-CYCLOBUT[F]ISOINDOLE-1,3-(2H)-DION

Formula: $C_{24}H_{29}N_5O_2$

Mol Weight: 419.53

MP (deg C): FP (deg C):

BP (deg C):

BP pressure (mm Hg):

Property/Value	Units	Temp	Data Type	Reference
WS 6.85E+001	mg/L	25	EST	MEYLAN,WM ET AL. (1996)
logP 1.79			EXP	SANGSTER,J (1993)
VP 6.44E-014	mm Hg	25	EST	NEELY,WB & BLAU,GE (1985)
DC	pKa			
HL 5.40E-014	atm m3/mol	25	EST	MEYLAN,WM & HOWARD,PH (1991)
OH 4.98E-010	cm3/molc sec	25	EST	MEYLAN,WM & HOWARD,PH (1993)

CAS #: 114482-97-2 — 1-SO3NA-7-ISOPROPYL AZULENE

Formula: $C_{13}H_{13}NaO_3S$

Mol Weight: 272.30

MP (deg C): FP (deg C):

BP (deg C):

BP pressure (mm Hg):

Property/Value	Units	Temp	Data Type	Reference
WS 1.00E+006	mg/L	25	EST	MEYLAN,WM ET AL. (1996)
logP -2.13			EXP	YANAGISAWA,T ET AL. (1988)
VP 2.08E-013	mm Hg	25	EST	NEELY,WB & BLAU,GE (1985)
DC	pKa			
HL	atm m3/mol			
OH 3.06E-010	cm3/molc sec	25	EST	MEYLAN,WM & HOWARD,PH (1993)

CAS #: 114482-98-3 — 1-SO3NA-3-ME-7-ISOPROPYL AZULENE

Formula: $C_{14}H_{15}NaO_3S$

Mol Weight: 286.33

MP (deg C): FP (deg C):

BP (deg C):

BP pressure (mm Hg):

Property/Value	Units	Temp	Data Type	Reference
WS 8.32E+004	mg/L	25	EST	MEYLAN,WM ET AL. (1996)
logP -0.88			EXP	YANAGISAWA,T ET AL. (1988)
VP 9.55E-014	mm Hg	25	EST	NEELY,WB & BLAU,GE (1985)
DC	pKa			
HL	atm m3/mol			
OH 3.43E-010	cm3/molc sec	25	EST	MEYLAN,WM & HOWARD,PH (1993)

CAS #: 114525-06-3 — 4,4,4-TRIFLUORO-2-METHYL-1-BUTANOL

Formula: $C_5H_9F_3O$

Mol Weight: 142.12

MP (deg C): FP (deg C):

BP (deg C):

BP pressure (mm Hg):

Property/Value	Units	Temp	Data Type	Reference
WS 1.73E+004	mg/L	25	EST	MEYLAN,WM ET AL. (1996)
logP 1.39			EXP	MULLER,N (1986)
VP 1.42E+001	mm Hg	25	EST	NEELY,WB & BLAU,GE (1985)
DC	pKa			
HL 6.59E-005	atm m3/mol	25	EST	MEYLAN,WM & HOWARD,PH (1991)
OH 7.34E-012	cm3/molc sec	25	EST	MEYLAN,WM & HOWARD,PH (1993)

CAS #: 114611-36-8 — HYDROCORTISON-21-PENTYL-CO-O-ME

Formula: $C_{29}H_{42}O_8$

Mol Weight: 518.65

MP (deg C): FP (deg C):

BP (deg C):

BP pressure (mm Hg):

Property/Value	Units	Temp	Data Type	Reference
WS 3.76E-001	mg/L	25	EST	MEYLAN,WM ET AL. (1996)
logP 3.70			EXP	HANSCH,C & LEO,AJ (1985)
VP 2.58E-016	mm Hg	25	EST	NEELY,WB & BLAU,GE (1985)
DC	pKa			
HL 4.61E-014	atm m3/mol	25	EST	MEYLAN,WM & HOWARD,PH (1991)
OH 1.15E-010	cm3/molc sec	25	EST	MEYLAN,WM & HOWARD,PH (1993)

CAS #: 114665-08-6 — BENZOIC ACID, 2-HYDROXY-, 2-(DIMETHYLAMINO)-2-OX

Formula: $C_{11}H_{13}NO_4$

Mol Weight: 223.23

MP (deg C): FP (deg C):

BP (deg C):

BP pressure (mm Hg):

Property/Value	Units	Temp	Data Type	Reference
WS 1.17E+004	mg/L	25	EST	MEYLAN,WM ET AL. (1996)
logP 1.21			EXP	BUNDGAARD,H & NIELSEN,NM (1988)
VP 1.53E-006	mm Hg	25	EST	NEELY,WB & BLAU,GE (1985)
DC	pKa			
HL 2.36E-011	atm m3/mol	25	EST	MEYLAN,WM & HOWARD,PH (1991)
OH 2.81E-011	cm3/molc sec	25	EST	MEYLAN,WM & HOWARD,PH (1993)

CAS #: 114665-09-7 — BENZOIC ACID, 2-HYDROXY-, 2-[(2-HYDROXYETHYL)MET

Formula: $C_{12}H_{15}NO_5$

Mol Weight: 253.26

MP (deg C): FP (deg C):

BP (deg C):

BP pressure (mm Hg):

Property/Value	Units	Temp	Data Type	Reference
WS 1.59E+004	mg/L	25	EST	MEYLAN,WM ET AL. (1996)
logP 0.86			EXP	BUNDGAARD,H & NIELSEN,NM (1988)
VP 2.22E-009	mm Hg	25	EST	NEELY,WB & BLAU,GE (1985)
DC	pKa			
HL 1.14E-015	atm m3/mol	25	EST	MEYLAN,WM & HOWARD,PH (1991)
OH 3.65E-011	cm3/molc sec	25	EST	MEYLAN,WM & HOWARD,PH (1993)

CAS #: 114753-51-4				PHENOL, 5-(1,1-DIMETHYLHEPTYL)-2-(3-HYDROXYCYCLO
Formula: $C_{21}H_{34}O_2$				
Mol Weight: 318.50				
MP (deg C):		FP (deg C):		
BP (deg C):				
BP pressure (mm Hg):				

Property/ Value	Units	Temp	Data Type	Reference
WS 2.32E-001	mg/L	25	EST	MEYLAN,WM ET AL. (1996)
logP 6.68			EXP	SANGSTER,J (1993)
VP 1.31E-009	mm Hg	25	EST	NEELY,WB & BLAU,GE (1985)
DC	pKa			
HL 4.38E-010	atm m3/mol	25	EST	MEYLAN,WM & HOWARD,PH (1991)
OH 1.15E-010	cm3/molc sec	25	EST	MEYLAN,WM & HOWARD,PH (1993)

CAS #: 114753-52-5				ADENOSINE, 2-AMINO-3'-AZIDO-2',3'-DIDEOXY-
Formula: $C_{10}H_{13}N_9O_2$				
Mol Weight: 291.27				
MP (deg C):		FP (deg C):		
BP (deg C):				
BP pressure (mm Hg):				

Property/ Value	Units	Temp	Data Type	Reference
WS 1.56E+002	mg/L	25	EST	MEYLAN,WM ET AL. (1996)
logP 0.24			EXP	BALZARINI,J ET AL. (1989)
VP 1.37E-020	mm Hg	25	EST	NEELY,WB & BLAU,GE (1985)
DC	pKa			
HL 2.29E-027	atm m3/mol	25	EST	MEYLAN,WM & HOWARD,PH (1991)
OH 2.34E-010	cm3/molc sec	25	EST	MEYLAN,WM & HOWARD,PH (1993)

CAS #: 114753-53-6				ADENOSINE, 2-AMINO-2',3'-DIDEOXY-3'-FLUORO-
Formula: $C_{10}H_{13}FN_6O_2$				
Mol Weight: 268.25				
MP (deg C):		FP (deg C):		
BP (deg C):				
BP pressure (mm Hg):				

Property/ Value	Units	Temp	Data Type	Reference
WS 8.30E+002	mg/L	25	EST	MEYLAN,WM ET AL. (1996)
logP 0.05			EXP	BALZARINI,J ET AL. (1989)
VP 1.25E-012	mm Hg	25	EST	NEELY,WB & BLAU,GE (1985)
DC	pKa			
HL 2.08E-019	atm m3/mol	25	EST	MEYLAN,WM & HOWARD,PH (1991)
OH 2.08E-010	cm3/molc sec	25	EST	MEYLAN,WM & HOWARD,PH (1993)

CAS #: 114991-36-5				SPARSOMYCIN,S-PROPYL ANALOG
Formula: $C_{15}H_{23}N_3O_5S_2$				
Mol Weight: 389.50				
MP (deg C):		FP (deg C):		
BP (deg C):				
BP pressure (mm Hg):				

Property/ Value	Units	Temp	Data Type	Reference
WS 6.50E+002	mg/L	25	EST	MEYLAN,WM ET AL. (1996)
logP -0.67			EXP	HANSCH,C ET AL. (1995)
VP 2.68E-021	mm Hg	25	EST	NEELY,WB & BLAU,GE (1985)
DC	pKa	-		
HL 2.94E-029	atm m3/mol	25	EST	MEYLAN,WM & HOWARD,PH (1991)
OH 1.81E-010	cm3/molc sec	25	EST	MEYLAN,WM & HOWARD,PH (1993)

CAS #: 115054-71-2				1H-BENZOTRIAZOLE-1-ACETAMIDE, à-METHYL-
Formula: $C_9H_{10}N_4O$				
Mol Weight: 190.21				
MP (deg C):		FP (deg C):		
BP (deg C):				
BP pressure (mm Hg):				

Property/ Value	Units	Temp	Data Type	Reference
WS 1.21E+003	mg/L	25	EST	MEYLAN,WM ET AL. (1996)
logP 0.35			EXP	SANGSTER,J (1993)
VP 7.30E-007	mm Hg	25	EST	NEELY,WB & BLAU,GE (1985)
DC	pKa			
HL 6.31E-013	atm m3/mol	25	EST	MEYLAN,WM & HOWARD,PH (1991)
OH 4.99E-012	cm3/molc sec	25	EST	MEYLAN,WM & HOWARD,PH (1993)

CAS #: 115054-74-5				1H-BENZOTRIAZOLE-1-ACETIC ACID, à-BUTYL-, ETHYL
Formula: $C_{14}H_{19}N_3O_2$				
Mol Weight: 261.33				
MP (deg C):		FP (deg C):		
BP (deg C):				
BP pressure (mm Hg):				

Property/ Value	Units	Temp	Data Type	Reference
WS 2.99E+001	mg/L	25	EST	MEYLAN,WM ET AL. (1996)
logP 3.32			EXP	SPARATORE,F ET AL. (1988)
VP 3.29E-006	mm Hg	25	EST	NEELY,WB & BLAU,GE (1985)
DC	pKa			
HL 3.07E-008	atm m3/mol	25	EST	MEYLAN,WM & HOWARD,PH (1991)
OH 8.57E-012	cm3/molc sec	25	EST	MEYLAN,WM & HOWARD,PH (1993)

CAS #: 115054-75-6				1H-BENZOTRIAZOLE-1-ACETAMIDE, à-ETHYL-
Formula: $C_{10}H_{12}N_4O$				
Mol Weight: 204.23				
MP (deg C):		FP (deg C):		
BP (deg C):				
BP pressure (mm Hg):				

Property/ Value	Units	Temp	Data Type	Reference
WS 7.38E+003	mg/L	25	EST	MEYLAN,WM ET AL. (1996)
logP 0.88			EXP	SPARATORE,F ET AL. (1988)
VP 3.32E-007	mm Hg	25	EST	NEELY,WB & BLAU,GE (1985)
DC	pKa			
HL 8.37E-013	atm m3/mol	25	EST	MEYLAN,WM & HOWARD,PH (1991)
OH 8.60E-012	cm3/molc sec	25	EST	MEYLAN,WM & HOWARD,PH (1993)

CAS #: 115054-76-7				2H-BENZOTRIAZOLE-2-ACETAMIDE, à-ETHYL-
Formula: $C_{10}H_{12}N_4O$				
Mol Weight: 204.23				
MP (deg C):		FP (deg C):		
BP (deg C):				
BP pressure (mm Hg):				

Property/ Value	Units	Temp	Data Type	Reference
WS 3.56E+003	mg/L	25	EST	MEYLAN,WM ET AL. (1996)
logP 1.25			EXP	SPARATORE,F ET AL. (1988)
VP 3.32E-007	mm Hg	25	EST	NEELY,WB & BLAU,GE (1985)
DC	pKa			
HL 8.37E-013	atm m3/mol	25	EST	MEYLAN,WM & HOWARD,PH (1991)
OH 8.60E-012	cm3/molc sec	25	EST	MEYLAN,WM & HOWARD,PH (1993)

CAS #: 115054-77-8 — 1H-BENZOTRIAZOLE-1-ACETAMIDE, à,à-DIMETHYL-

Formula: $C_{10}H_{12}N_4O$
Mol Weight: 204.23
MP (deg C): | FP (deg C):
BP (deg C):
BP pressure (mm Hg):

Property/Value	Units	Temp	Data Type	Reference
WS 9.34E+003	mg/L	25	EST	MEYLAN,WM ET AL. (1996)
logP 0.76			EXP	SPARATORE,F ET AL. (1988)
VP 4.54E-007	mm Hg	25	EST	NEELY,WB & BLAU,GE (1985)
DC	pKa			
HL 8.37E-013	atm m3/mol	25	EST	MEYLAN,WM & HOWARD,PH (1991)
OH 4.06E-012	cm3/molc sec	25	EST	MEYLAN,WM & HOWARD,PH (1993)

CAS #: 115054-78-9 — 1H-BENZOTRIAZOLE-1-ACETIC ACID, à-ETHYL-, ETHYL

Formula: $C_{12}H_{15}N_3O_2$
Mol Weight: 233.27
MP (deg C): | FP (deg C):
BP (deg C):
BP pressure (mm Hg):

Property/Value	Units	Temp	Data Type	Reference
WS 2.93E+002	mg/L	25	EST	MEYLAN,WM ET AL. (1996)
logP 2.34			EXP	SPARATORE,F ET AL. (1988)
VP 1.43E-005	mm Hg	25	EST	NEELY,WB & BLAU,GE (1985)
DC	pKa			
HL 1.74E-008	atm m3/mol	25	EST	MEYLAN,WM & HOWARD,PH (1991)
OH 5.74E-012	cm3/molc sec	25	EST	MEYLAN,WM & HOWARD,PH (1993)

CAS #: 115054-79-0 — 2H-BENZOTRIAZOLE-2-ACETIC ACID, à,à-DIMETHYL-, E

Formula: $C_{12}H_{17}N_3O_2$
Mol Weight: 235.29
MP (deg C): | FP (deg C):
BP (deg C):
BP pressure (mm Hg):

Property/Value	Units	Temp	Data Type	Reference
WS 8.33E+001	mg/L	25	EST	MEYLAN,WM ET AL. (1996)
logP 2.60			EXP	SPARATORE,F ET AL. (1988)
VP 2.16E-005	mm Hg	25	EST	NEELY,WB & BLAU,GE (1985)
DC	pKa			
HL 1.74E-008	atm m3/mol	25	EST	MEYLAN,WM & HOWARD,PH (1991)
OH 3.00E-012	cm3/molc sec	25	EST	MEYLAN,WM & HOWARD,PH (1993)

CAS #: 115054-80-3 — 2H-BENZOTRIAZOLE-2-ACETIC ACID, à-ETHYL-, ETHYL

Formula: $C_{12}H_{15}N_3O_2$
Mol Weight: 233.27
MP (deg C): | FP (deg C):
BP (deg C):
BP pressure (mm Hg):

Property/Value	Units	Temp	Data Type	Reference
WS 1.01E+002	mg/L	25	EST	MEYLAN,WM ET AL. (1996)
logP 2.88			EXP	SPARATORE,F ET AL. (1988)
VP 1.43E-005	mm Hg	25	EST	NEELY,WB & BLAU,GE (1985)
DC	pKa			
HL 1.74E-008	atm m3/mol	25	EST	MEYLAN,WM & HOWARD,PH (1991)
OH 5.74E-012	cm3/molc sec	25	EST	MEYLAN,WM & HOWARD,PH (1993)

CAS #: 115054-81-4 — 1H-BENZOTRIAZOLE-1-ACETAMIDE, à-BUTYL-

Formula: $C_{12}H_{16}N_4O$
Mol Weight: 232.29
MP (deg C): | FP (deg C):
BP (deg C):
BP pressure (mm Hg):

Property/Value	Units	Temp	Data Type	Reference
WS 7.34E+002	mg/L	25	EST	MEYLAN,WM ET AL. (1996)
logP 1.88			EXP	SPARATORE,F ET AL. (1988)
VP 6.77E-008	mm Hg	25	EST	NEELY,WB & BLAU,GE (1985)
DC	pKa			
HL 1.48E-012	atm m3/mol	25	EST	MEYLAN,WM & HOWARD,PH (1991)
OH 1.20E-011	cm3/molc sec	25	EST	MEYLAN,WM & HOWARD,PH (1993)

CAS #: 115054-82-5 — 2H-BENZOTRIAZOLE-2-ACETIC ACID, à-PROPYL-, ETHYL

Formula: $C_{13}H_{17}N_3O_2$
Mol Weight: 247.30
MP (deg C): | FP (deg C):
BP (deg C):
BP pressure (mm Hg):

Property/Value	Units	Temp	Data Type	Reference
WS 3.18E+001	mg/L	25	EST	MEYLAN,WM ET AL. (1996)
logP 3.38			EXP	SPARATORE,F ET AL. (1988)
VP 7.55E-006	mm Hg	25	EST	NEELY,WB & BLAU,GE (1985)
DC	pKa			
HL 2.31E-008	atm m3/mol	25	EST	MEYLAN,WM & HOWARD,PH (1991)
OH 7.16E-012	cm3/molc sec	25	EST	MEYLAN,WM & HOWARD,PH (1993)

CAS #: 115054-83-6 — 2H-BENZOTRIAZOLE-2-ACETAMIDE, à-PROPYL-

Formula: $C_{11}H_{14}N_4O$
Mol Weight: 218.26
MP (deg C): | FP (deg C):
BP (deg C):
BP pressure (mm Hg):

Property/Value	Units	Temp	Data Type	Reference
WS 1.13E+003	mg/L	25	EST	MEYLAN,WM ET AL. (1996)
logP 1.75			EXP	SPARATORE,F ET AL. (1988)
VP 1.50E-007	mm Hg	25	EST	NEELY,WB & BLAU,GE (1985)
DC	pKa			
HL 1.11E-012	atm m3/mol	25	EST	MEYLAN,WM & HOWARD,PH (1991)
OH 1.06E-011	cm3/molc sec	25	EST	MEYLAN,WM & HOWARD,PH (1993)

CAS #: 115054-84-7 — 2H-BENZOTRIAZOLE-2-ACETIC ACID, à,à-DIMETHYL-, E

Formula: $C_{12}H_{17}N_3O_2$
Mol Weight: 235.29
MP (deg C): | FP (deg C):
BP (deg C):
BP pressure (mm Hg):

Property/Value	Units	Temp	Data Type	Reference
WS 8.33E+001	mg/L	25	EST	MEYLAN,WM ET AL. (1996)
logP 2.98			EXP	SPARATORE,F ET AL. (1988)
VP 2.16E-005	mm Hg	25	EST	NEELY,WB & BLAU,GE (1985)
DC	pKa			
HL 1.74E-008	atm m3/mol	25	EST	MEYLAN,WM & HOWARD,PH (1991)
OH 3.00E-012	cm3/molc sec	25	EST	MEYLAN,WM & HOWARD,PH (1993)

115054-85-8 — 2H-BENZOTRIAZOLE-2-ACETAMIDE, à,à-DIMETHYL-

Formula: $C_{10}H_{12}N_4O$

Mol Weight: 204.23

MP (deg C):   FP (deg C):

BP (deg C):

BP pressure (mm Hg):

Property/Value	Units	Temp	Data Type	Reference
WS 4.25E+003	mg/L	25	EST	MEYLAN,WM ET AL. (1996)
logP 1.16			EXP	SPARATORE,F ET AL. (1988)
VP 4.54E-007	mm Hg	25	EST	NEELY,WB & BLAU,GE (1985)
DC	pKa			
HL 8.37E-013	atm m3/mol	25	EST	MEYLAN,WM & HOWARD,PH (1991)
OH 4.06E-012	cm3/molc sec	25	EST	MEYLAN,WM & HOWARD,PH (1993)

115054-86-9 — 1H-BENZOTRIAZOLE-1-ACETIC ACID, à-PROPYL-, ETHYL

Formula: $C_{13}H_{17}N_3O_2$

Mol Weight: 247.30

MP (deg C):   FP (deg C):

BP (deg C):

BP pressure (mm Hg):

Property/Value	Units	Temp	Data Type	Reference
WS 9.56E+001	mg/L	25	EST	MEYLAN,WM ET AL. (1996)
logP 2.82			EXP	SPARATORE,F ET AL. (1988)
VP 7.55E-006	mm Hg	25	EST	NEELY,WB & BLAU,GE (1985)
DC	pKa			
HL 2.31E-008	atm m3/mol	25	EST	MEYLAN,WM & HOWARD,PH (1991)
OH 7.16E-012	cm3/molc sec	25	EST	MEYLAN,WM & HOWARD,PH (1993)

115054-87-0 — 2H-BENZOTRIAZOLE-2-ACETIC ACID, à-BUTYL-, ETHYL

Formula: $C_{14}H_{19}N_3O_2$

Mol Weight: 261.33

MP (deg C):   FP (deg C):

BP (deg C):

BP pressure (mm Hg):

Property/Value	Units	Temp	Data Type	Reference
WS 9.94E+000	mg/L	25	EST	MEYLAN,WM ET AL. (1996)
logP 3.88			EXP	SPARATORE,F ET AL. (1988)
VP 3.29E-006	mm Hg	25	EST	NEELY,WB & BLAU,GE (1985)
DC	pKa			
HL 3.07E-008	atm m3/mol	25	EST	MEYLAN,WM & HOWARD,PH (1991)
OH 8.57E-012	cm3/molc sec	25	EST	MEYLAN,WM & HOWARD,PH (1993)

115054-88-1 — 2H-BENZOTRIAZOLE-2-ACETAMIDE, à-METHYL-

Formula: $C_9H_{10}N_4O$

Mol Weight: 190.21

MP (deg C):   FP (deg C):

BP (deg C):

BP pressure (mm Hg):

Property/Value	Units	Temp	Data Type	Reference
WS 9.04E+003	mg/L	25	EST	MEYLAN,WM ET AL. (1996)
logP 0.86			EXP	SPARATORE,F ET AL. (1988)
VP 7.30E-007	mm Hg	25	EST	NEELY,WB & BLAU,GE (1985)
DC	pKa			
HL 6.31E-013	atm m3/mol	25	EST	MEYLAN,WM & HOWARD,PH (1991)
OH 4.99E-012	cm3/molc sec	25	EST	MEYLAN,WM & HOWARD,PH (1993)

115054-89-2 — 1H-BENZOTRIAZOLE-1-ACETAMIDE, à-PROPYL-

Formula: $C_{11}H_{14}N_4O$

Mol Weight: 218.26

MP (deg C):   FP (deg C):

BP (deg C):

BP pressure (mm Hg):

Property/Value	Units	Temp	Data Type	Reference
WS 2.33E+003	mg/L	25	EST	MEYLAN,WM ET AL. (1996)
logP 1.38			EXP	SPARATORE,F ET AL. (1988)
VP 1.50E-007	mm Hg	25	EST	NEELY,WB & BLAU,GE (1985)
DC	pKa			
HL 1.11E-012	atm m3/mol	25	EST	MEYLAN,WM & HOWARD,PH (1991)
OH 1.06E-011	cm3/molc sec	25	EST	MEYLAN,WM & HOWARD,PH (1993)

115054-90-5 — 2H-BENZOTRIAZOLE-2-ACETAMIDE, à-BUTYL-

Formula: $C_{12}H_{16}N_4O$

Mol Weight: 232.29

MP (deg C):   FP (deg C):

BP (deg C):

BP pressure (mm Hg):

Property/Value	Units	Temp	Data Type	Reference
WS 3.34E+002	mg/L	25	EST	MEYLAN,WM ET AL. (1996)
logP 2.28			EXP	SPARATORE,F ET AL. (1988)
VP 6.77E-008	mm Hg	25	EST	NEELY,WB & BLAU,GE (1985)
DC	pKa			
HL 1.48E-012	atm m3/mol	25	EST	MEYLAN,WM & HOWARD,PH (1991)
OH 1.20E-011	cm3/molc sec	25	EST	MEYLAN,WM & HOWARD,PH (1993)

115054-91-6 — PENTANOIC ACID, 4-(1H-BENZOTRIAZOL-1-YL)-

Formula: $C_{11}H_{13}N_3O_2$

Mol Weight: 219.25

MP (deg C):   FP (deg C):

BP (deg C):

BP pressure (mm Hg):

Property/Value	Units	Temp	Data Type	Reference
WS 6.42E+004	mg/L	25	EST	MEYLAN,WM ET AL. (1996)
logP 0.15			EXP	SPARATORE,F ET AL. (1988)
VP 1.10E-006	mm Hg	25	EST	NEELY,WB & BLAU,GE (1985)
DC	pKa			
HL 5.44E-011	atm m3/mol	25	EST	MEYLAN,WM & HOWARD,PH (1991)
OH 6.35E-012	cm3/molc sec	25	EST	MEYLAN,WM & HOWARD,PH (1993)

115054-94-9 — 2H-BENZOTRIAZOLE-2-PROPANAMIDE, á-METHYL-

Formula: $C_{10}H_{12}N_4O$

Mol Weight: 204.23

MP (deg C):   FP (deg C):

BP (deg C):

BP pressure (mm Hg):

Property/Value	Units	Temp	Data Type	Reference
WS 5.83E+003	mg/L	25	EST	MEYLAN,WM ET AL. (1996)
logP 1.00			EXP	SPARATORE,F ET AL. (1988)
VP 3.32E-007	mm Hg	25	EST	NEELY,WB & BLAU,GE (1985)
DC	pKa			
HL 8.37E-013	atm m3/mol	25	EST	MEYLAN,WM & HOWARD,PH (1991)
OH 1.16E-011	cm3/molc sec	25	EST	MEYLAN,WM & HOWARD,PH (1993)

CAS #: 115054-96-1				PENTANOIC ACID, 4-(2H-BENZOTRIAZOL-2-YL)-
Formula: $C_{11}H_{13}N_3O_2$				
Mol Weight: 219.25				
MP (deg C):		FP (deg C):		
BP (deg C):				
BP pressure (mm Hg):				

Property/ Value	Units	Temp	Data Type	Reference
WS 2.22E+004	mg/L	25	EST	MEYLAN,WM ET AL. (1996)
logP 0.69			EXP	SPARATORE,F ET AL. (1988)
VP 1.10E-006	mm Hg	25	EST	NEELY,WB & BLAU,GE (1985)
DC	pKa			
HL 5.44E-011	atm m3/mol	25	EST	MEYLAN,WM & HOWARD,PH (1991)
OH 6.35E-012	cm3/molc sec	25	EST	MEYLAN,WM & HOWARD,PH (1993)

CAS #: 115054-97-2				2H-BENZOTRIAZOLE-1-BUTANAMIDE, --METHYL-
Formula: $C_{11}H_{14}N_4O$				
Mol Weight: 218.26				
MP (deg C):		FP (deg C):		
BP (deg C):				
BP pressure (mm Hg):				

Property/ Value	Units	Temp	Data Type	Reference
WS 2.47E+003	mg/L	25	EST	MEYLAN,WM ET AL. (1996)
logP 1.35			EXP	SPARATORE,F ET AL. (1988)
VP 1.50E-007	mm Hg	25	EST	NEELY,WB & BLAU,GE (1985)
DC	pKa			
HL 1.11E-012	atm m3/mol	25	EST	MEYLAN,WM & HOWARD,PH (1991)
OH 1.09E-011	cm3/molc sec	25	EST	MEYLAN,WM & HOWARD,PH (1993)

CAS #: 115174-03-3				FURAZANAMINE, 4-METHYL-, 5-OXIDE
Formula: $C_3H_5N_3O_2$				
Mol Weight: 115.09				
MP (deg C):		FP (deg C):		
BP (deg C):				
BP pressure (mm Hg):				

Property/ Value	Units	Temp	Data Type	Reference
WS 1.46E+005	mg/L	25	EST	MEYLAN,WM ET AL. (1996)
logP -0.17			EXP	CALVINO,R ET AL. (1992)
VP 1.82E-003	mm Hg	25	EST	NEELY,WB & BLAU,GE (1985)
DC	pKa			
HL 7.38E-013	atm m3/mol	25	EST	MEYLAN,WM & HOWARD,PH (1991)
OH 4.14E-012	cm3/molc sec	25	EST	MEYLAN,WM & HOWARD,PH (1993)

CAS #: 115178-63-7				ACETAMIDE, 2-(BENZOYLOXY)-N,N-BIS(2-HYDROXYPROPY
Formula: $C_{15}H_{21}NO_5$				
Mol Weight: 295.34				
MP (deg C):		FP (deg C):		
BP (deg C):				
BP pressure (mm Hg):				

Property/ Value	Units	Temp	Data Type	Reference
WS 3.57E+003	mg/L	25	EST	MEYLAN,WM ET AL. (1996)
logP 0.66			EXP	NIELSEN,LS & BUNDGAARD,H (1988)
VP 1.84E-010	mm Hg	25	EST	NEELY,WB & BLAU,GE (1985)
DC	pKa			
HL 7.47E-016	atm m3/mol	25	EST	MEYLAN,WM & HOWARD,PH (1991)
OH 4.35E-011	cm3/molc sec	25	EST	MEYLAN,WM & HOWARD,PH (1993)

CAS #: 115178-64-8				ACETAMIDE, 2-(BENZOYLOXY)-N,N-BIS(2-METHOXYETHYL
Formula: $C_{15}H_{21}NO_5$				
Mol Weight: 295.34				
MP (deg C):		FP (deg C):		
BP (deg C):				
BP pressure (mm Hg):				

Property/ Value	Units	Temp	Data Type	Reference
WS 1.06E+003	mg/L	25	EST	MEYLAN,WM ET AL. (1996)
logP 1.28			EXP	NIELSEN,LS & BUNDGAARD,H (1988)
VP 6.50E-007	mm Hg	25	EST	NEELY,WB & BLAU,GE (1985)
DC	pKa			
HL 4.34E-014	atm m3/mol	25	EST	MEYLAN,WM & HOWARD,PH (1991)
OH 5.92E-011	cm3/molc sec	25	EST	MEYLAN,WM & HOWARD,PH (1993)

CAS #: 115178-66-0				AZETIDINE, 1-[BENZOYLOXY(ACETYL)]-
Formula: $C_{12}H_{13}NO_3$				
Mol Weight: 219.24				
MP (deg C):		FP (deg C):		
BP (deg C):				
BP pressure (mm Hg):				

Property/ Value	Units	Temp	Data Type	Reference
WS 3.28E+003	mg/L	25	EST	MEYLAN,WM ET AL. (1996)
logP 1.20			EXP	NIELSEN,LS & BUNDGAARD,H (1988)
VP 2.77E-005	mm Hg	25	EST	NEELY,WB & BLAU,GE (1985)
DC	pKa			
HL 1.05E-010	atm m3/mol	25	EST	MEYLAN,WM & HOWARD,PH (1991)
OH 1.98E-011	cm3/molc sec	25	EST	MEYLAN,WM & HOWARD,PH (1993)

CAS #: 115178-68-2				1H-AZEPINE, HEXAHYDRO-1-[(BENZOYLOXY)ACETYL]-
Formula: $C_{15}H_{19}NO_3$				
Mol Weight: 261.32				
MP (deg C):		FP (deg C):		
BP (deg C):				
BP pressure (mm Hg):				

Property/ Value	Units	Temp	Data Type	Reference
WS 2.22E+002	mg/L	25	EST	MEYLAN,WM ET AL. (1996)
logP 2.30			EXP	NIELSEN,LS & BUNDGAARD,H (1988)
VP 1.95E-006	mm Hg	25	EST	NEELY,WB & BLAU,GE (1985)
DC	pKa			
HL 2.46E-010	atm m3/mol	25	EST	MEYLAN,WM & HOWARD,PH (1991)
OH 3.18E-011	cm3/molc sec	25	EST	MEYLAN,WM & HOWARD,PH (1993)

CAS #: 115178-69-3				PIPERIDINE, 1-[(BENZOYLOXY)ACETYL]-2-ETHYL-
Formula: $C_{16}H_{21}NO_3$				
Mol Weight: 275.35				
MP (deg C):		FP (deg C):		
BP (deg C):				
BP pressure (mm Hg):				

Property/ Value	Units	Temp	Data Type	Reference
WS 9.86E+001	mg/L	25	EST	MEYLAN,WM ET AL. (1996)
logP 2.62			EXP	HANSCH,C ET AL. (1995)
VP 1.11E-006	mm Hg	25	EST	NEELY,WB & BLAU,GE (1985)
DC	pKa			
HL 3.27E-010	atm m3/mol	25	EST	MEYLAN,WM & HOWARD,PH (1991)
OH 3.91E-011	cm3/molc sec	25	EST	MEYLAN,WM & HOWARD,PH (1993)

CAS #:	115178-70-6	PIPERIDINE, 1-[(BENZOYLOXY)ACETYL]-2,6-DIMETHYL-

Formula: $C_{16}H_{21}NO_3$

Mol Weight: 275.35

MP (deg C): FP (deg C):

BP (deg C):

BP pressure (mm Hg):

Property/Value	Units	Temp	Data Type	Reference
WS 5.69E+001	mg/L	25	EST	MEYLAN,WM ET AL. (1996)
logP 2.90			EXP	HANSCH,C ET AL. (1995)
VP 1.44E-006	mm Hg	25	EST	NEELY,WB & BLAU,GE (1985)
DC	pKa			
HL 3.27E-010	atm m3/mol	25	EST	MEYLAN,WM & HOWARD,PH (1991)
OH 4.09E-011	cm3/molc sec	25	EST	MEYLAN,WM & HOWARD,PH (1993)

CAS #:	115178-71-7	4-PIPERIDINOL, 1-[(BENZOYLOXY)ACETYL]-

Formula: $C_{14}H_{17}NO_4$

Mol Weight: 263.30

MP (deg C): FP (deg C):

BP (deg C):

BP pressure (mm Hg):

Property/Value	Units	Temp	Data Type	Reference
WS 5.78E+003	mg/L	25	EST	MEYLAN,WM ET AL. (1996)
logP 0.63			EXP	NIELSEN,LS & BUNDGAARD,H (1988)
VP 7.15E-009	mm Hg	25	EST	NEELY,WB & BLAU,GE (1985)
DC	pKa			
HL 6.79E-015	atm m3/mol	25	EST	MEYLAN,WM & HOWARD,PH (1991)
OH 3.94E-011	cm3/molc sec	25	EST	MEYLAN,WM & HOWARD,PH (1993)

CAS #:	115178-74-0	2-AZETIDINECARBOXYLIC ACID, 1-[(BENZOYLOXY)ACETY

Formula: $C_{13}H_{13}NO_5$

Mol Weight: 263.25

MP (deg C): FP (deg C):

BP (deg C):

BP pressure (mm Hg):

Property/Value	Units	Temp	Data Type	Reference
WS 4.22E+003	mg/L	25	EST	MEYLAN,WM ET AL. (1996)
logP 0.79			EXP	NIELSEN,LS & BUNDGAARD,H (1988)
VP 7.22E-008	mm Hg	25	EST	NEELY,WB & BLAU,GE (1985)
DC	pKa			
HL 1.56E-014	atm m3/mol	25	EST	MEYLAN,WM & HOWARD,PH (1991)
OH 2.10E-011	cm3/molc sec	25	EST	MEYLAN,WM & HOWARD,PH (1993)

CAS #:	115178-75-1	L-PROLINE, N-[(BENZOYLOXY)ACETYL]-

Formula: $C_{14}H_{15}NO_5$

Mol Weight: 277.28

MP (deg C): FP (deg C):

BP (deg C):

BP pressure (mm Hg):

Property/Value	Units	Temp	Data Type	Reference
WS 2.24E+003	mg/L	25	EST	MEYLAN,WM ET AL. (1996)
logP 1.02			EXP	NIELSEN,LS & BUNDGAARD,H (1988)
VP 2.89E-008	mm Hg	25	EST	NEELY,WB & BLAU,GE (1985)
DC	pKa			
HL 2.07E-014	atm m3/mol	25	EST	MEYLAN,WM & HOWARD,PH (1991)
OH 2.92E-011	cm3/molc sec	25	EST	MEYLAN,WM & HOWARD,PH (1993)

CAS #:	115178-76-2	L-PROLINE, N-[(BENZOYLOXY)ACETYL]-, METHYL ESTER

Formula: $C_{15}H_{17}NO_5$

Mol Weight: 291.31

MP (deg C): FP (deg C):

BP (deg C):

BP pressure (mm Hg):

Property/Value	Units	Temp	Data Type	Reference
WS 8.45E+002	mg/L	25	EST	MEYLAN,WM ET AL. (1996)
logP 1.42			EXP	NIELSEN,LS & BUNDGAARD,H (1988)
VP 3.24E-006	mm Hg	25	EST	NEELY,WB & BLAU,GE (1985)
DC	pKa			
HL 6.64E-012	atm m3/mol	25	EST	MEYLAN,WM & HOWARD,PH (1991)
OH 2.88E-011	cm3/molc sec	25	EST	MEYLAN,WM & HOWARD,PH (1993)

CAS #:	115178-77-3	PROPANAMIDE, 3-(BENZOYLOXY)-N,N-DIMETHYL-

Formula: $C_{12}H_{15}NO_3$

Mol Weight: 221.26

MP (deg C): FP (deg C):

BP (deg C):

BP pressure (mm Hg):

Property/Value	Units	Temp	Data Type	Reference
WS 2.74E+003	mg/L	25	EST	MEYLAN,WM ET AL. (1996)
logP 1.28			EXP	NIELSEN,LS & BUNDGAARD,H (1988)
VP 4.70E-005	mm Hg	25	EST	NEELY,WB & BLAU,GE (1985)
DC	pKa			
HL 3.37E-011	atm m3/mol	25	EST	MEYLAN,WM & HOWARD,PH (1991)
OH 2.34E-011	cm3/molc sec	25	EST	MEYLAN,WM & HOWARD,PH (1993)

CAS #:	115178-78-4	BUTANAMIDE, 4-(BENZOYLOXY)-N,N-DIMETHYL-

Formula: $C_{13}H_{17}NO_3$

Mol Weight: 235.29

MP (deg C): FP (deg C):

BP (deg C):

BP pressure (mm Hg):

Property/Value	Units	Temp	Data Type	Reference
WS 7.35E+002	mg/L	25	EST	MEYLAN,WM ET AL. (1996)
logP 1.86			EXP	NIELSEN,LS & BUNDGAARD,H (1988)
VP 2.01E-005	mm Hg	25	EST	NEELY,WB & BLAU,GE (1985)
DC	pKa			
HL 4.48E-011	atm m3/mol	25	EST	MEYLAN,WM & HOWARD,PH (1991)
OH 2.39E-011	cm3/molc sec	25	EST	MEYLAN,WM & HOWARD,PH (1993)

CAS #:	115178-79-5	PROPANAMIDE, 2-(BENZOYLOXY)-N,N-DIETHYL-, (ñ)-

Formula: $C_{14}H_{19}NO_3$

Mol Weight: 249.31

MP (deg C): FP (deg C):

BP (deg C):

BP pressure (mm Hg):

Property/Value	Units	Temp	Data Type	Reference
WS 2.59E+002	mg/L	25	EST	MEYLAN,WM ET AL. (1996)
logP 2.30			EXP	NIELSEN,LS & BUNDGAARD,H (1988)
VP 1.54E-005	mm Hg	25	EST	NEELY,WB & BLAU,GE (1985)
DC	pKa			
HL 4.21E-010	atm m3/mol	25	EST	MEYLAN,WM & HOWARD,PH (1991)
OH 2.76E-011	cm3/molc sec	25	EST	MEYLAN,WM & HOWARD,PH (1993)

116383-80-3

CAS #: 116383-80-3

1H-PYRROLE-3,4-DICARBOXYLIC ACID, 2-(5-CHLORO-2-

Formula: $C_{14}H_{13}ClN_2O_4$

Mol Weight: 308.72

MP (deg C):   **FP (deg C):**

BP (deg C):

BP pressure (mm Hg):

Property/Value	Units	Temp	Data Type	Reference
WS 5.40E+001	mg/L	25	EST	MEYLAN,WM ET AL. (1996)
logP 2.70			EXP	SANGSTER,J (1994)
VP 1.75E-007	mm Hg	25	EST	NEELY,WB & BLAU,GE (1985)
DC	pKa			
HL 3.10E-014	atm m3/mol	25	EST	MEYLAN,WM & HOWARD,PH (1991)
OH 3.87E-011	cm3/molc sec	25	EST	MEYLAN,WM & HOWARD,PH (1993)

116383-85-8

CAS #: 116383-85-8

1H-PYRROLE-3,4-DICARBOXYLIC ACID, 2-(4-CHLORO-2-

Formula: $C_{14}H_{13}ClN_2O_4$

Mol Weight: 308.72

MP (deg C):   **FP (deg C):**

BP (deg C):

BP pressure (mm Hg):

Property/Value	Units	Temp	Data Type	Reference
WS 6.20E+001	mg/L	25	EST	MEYLAN,WM ET AL. (1996)
logP 2.63			EXP	SANGSTER,J (1994)
VP 1.75E-007	mm Hg	25	EST	NEELY,WB & BLAU,GE (1985)
DC	pKa			
HL 3.10E-014	atm m3/mol	25	EST	MEYLAN,WM & HOWARD,PH (1991)
OH 3.90E-011	cm3/molc sec	25	EST	MEYLAN,WM & HOWARD,PH (1993)

116482-56-5

CAS #: 116482-56-5

BENZOIC ACID, 2-(ACETYLOXY)-, 2-(DIETHYLAMINO)-2

Formula: $C_{15}H_{19}NO_5$

Mol Weight: 293.32

MP (deg C):   **FP (deg C):**

BP (deg C):

BP pressure (mm Hg):

Property/Value	Units	Temp	Data Type	Reference
WS 1.37E+003	mg/L	25	EST	MEYLAN,WM ET AL. (1996)
logP 1.16			EXP	NIELSEN,LS & BUNDGAARD,H (1989)
VP 4.01E-006	mm Hg	25	EST	NEELY,WB & BLAU,GE (1985)
DC	pKa			
HL 3.81E-012	atm m3/mol	25	EST	MEYLAN,WM & HOWARD,PH (1991)
OH 2.49E-011	cm3/molc sec	25	EST	MEYLAN,WM & HOWARD,PH (1993)

116482-75-8

CAS #: 116482-75-8

BENZOIC ACID, 2-(ACETYLOXY)-, 2-(DIPROPYLAMINO)-

Formula: $C_{17}H_{23}NO_5$

Mol Weight: 321.38

MP (deg C):   **FP (deg C):**

BP (deg C):

BP pressure (mm Hg):

Property/Value	Units	Temp	Data Type	Reference
WS 1.51E+002	mg/L	25	EST	MEYLAN,WM ET AL. (1996)
logP 2.09			EXP	NIELSEN,LS & BUNDGAARD,H (1989)
VP 7.07E-007	mm Hg	25	EST	NEELY,WB & BLAU,GE (1985)
DC	pKa			
HL 6.72E-012	atm m3/mol	25	EST	MEYLAN,WM & HOWARD,PH (1991)
OH 2.89E-011	cm3/molc sec	25	EST	MEYLAN,WM & HOWARD,PH (1993)

116482-76-9

CAS #: 116482-76-9

BENZOIC ACID, 2-(ACETYLOXY)-, 2-[BIS(1-METHYLETH

Formula: $C_{17}H_{23}NO_5$

Mol Weight: 321.38

MP (deg C):   **FP (deg C):**

BP (deg C):

BP pressure (mm Hg):

Property/Value	Units	Temp	Data Type	Reference
WS 1.70E+002	mg/L	25	EST	MEYLAN,WM ET AL. (1996)
logP 2.03			EXP	NIELSEN,LS & BUNDGAARD,H (1989)
VP 2.46E-006	mm Hg	25	EST	NEELY,WB & BLAU,GE (1985)
DC	pKa			
HL 6.72E-012	atm m3/mol	25	EST	MEYLAN,WM & HOWARD,PH (1991)
OH 3.34E-011	cm3/molc sec	25	EST	MEYLAN,WM & HOWARD,PH (1993)

116482-77-0

CAS #: 116482-77-0

BENZOIC ACID, 2-(ACETYLOXY)-, 2-[(2-ETHOXY-2-OXO

Formula: $C_{16}H_{19}NO_7$

Mol Weight: 337.33

MP (deg C):   **FP (deg C):**

BP (deg C):

BP pressure (mm Hg):

Property/Value	Units	Temp	Data Type	Reference
WS 2.77E+002	mg/L	25	EST	MEYLAN,WM ET AL. (1996)
logP 1.67			EXP	NIELSEN,LS & BUNDGAARD,H (1989)
VP 1.89E-007	mm Hg	25	EST	NEELY,WB & BLAU,GE (1985)
DC	pKa			
HL 1.36E-013	atm m3/mol	25	EST	MEYLAN,WM & HOWARD,PH (1991)
OH 2.19E-011	cm3/molc sec	25	EST	MEYLAN,WM & HOWARD,PH (1993)

116482-78-1

CAS #: 116482-78-1

BENZOIC ACID, 2-(ACETYLOXY)-, 2-[(2-AMINO-2-OXOE

Formula: $C_{14}H_{16}N_2O_6$

Mol Weight: 308.29

MP (deg C):   **FP (deg C):**

BP (deg C):

BP pressure (mm Hg):

Property/Value	Units	Temp	Data Type	Reference
WS 1.33E+003	mg/L	25	EST	MEYLAN,WM ET AL. (1996)
logP -0.46			EXP	NIELSEN,LS & BUNDGAARD,H (1989)
VP 2.66E-009	mm Hg	25	EST	NEELY,WB & BLAU,GE (1985)
DC	pKa			
HL 1.04E-014	atm m3/mol	25	EST	MEYLAN,WM & HOWARD,PH (1991)
OH 2.23E-011	cm3/molc sec	25	EST	MEYLAN,WM & HOWARD,PH (1993)

116482-80-5

CAS #: 116482-80-5

BENZOIC ACID, 2-(ACETYLOXY)-, 2-(4-MORPHOLINYL)-

Formula: $C_{15}H_{17}NO_6$

Mol Weight: 307.31

MP (deg C):   **FP (deg C):**

BP (deg C):

BP pressure (mm Hg):

Property/Value	Units	Temp	Data Type	Reference
WS 6.17E+003	mg/L	25	EST	MEYLAN,WM ET AL. (1996)
logP 0.30			EXP	NIELSEN,LS & BUNDGAARD,H (1989)
VP 5.93E-007	mm Hg	25	EST	NEELY,WB & BLAU,GE (1985)
DC	pKa			
HL 1.48E-014	atm m3/mol	25	EST	MEYLAN,WM & HOWARD,PH (1991)
OH 5.77E-011	cm3/molc sec	25	EST	MEYLAN,WM & HOWARD,PH (1993)

CAS #: 116583-45-0				2H-PYRAN, TETRAHYDRO-6-CYCLOPROPYL-3-[1-(ETHOXYI
Formula: $C_{15}H_{23}NO_4$				
Mol Weight: 281.35				
MP (deg C):		FP (deg C):		
BP (deg C):				
BP pressure (mm Hg):				

Property/ Value	Units	Temp	Data Type	Reference
WS 2.49E+001	mg/L	25	EST	MEYLAN,WM ET AL. (1996)
logP 3.28			EXP	SANGSTER,J (1994)
VP 5.69E-007	mm Hg	25	EST	NEELY,WB & BLAU,GE (1985)
DC	pKa			
HL 5.65E-009	atm m3/mol	25	EST	MEYLAN,WM & HOWARD,PH (1991)
OH 1.36E-011	cm3/molc sec	25	EST	MEYLAN,WM & HOWARD,PH (1993)

CAS #: 117121-33-2				UREA, N'-[4-[(1-CYCLOHEXYL-1H-TETRAZOL-5-YL)OXY]
Formula: $C_{16}H_{22}N_6O_2$				
Mol Weight: 330.39				
MP (deg C):		FP (deg C):		
BP (deg C):				
BP pressure (mm Hg):				

Property/ Value	Units	Temp	Data Type	Reference
WS 1.12E+002	mg/L	25	EST	MEYLAN,WM ET AL. (1996)
logP 2.18			EXP	SANGSTER,J (1994)
VP 4.11E-010	mm Hg	25	EST	NEELY,WB & BLAU,GE (1985)
DC	pKa			
HL 9.34E-014	atm m3/mol	25	EST	MEYLAN,WM & HOWARD,PH (1991)
OH 5.50E-011	cm3/molc sec	25	EST	MEYLAN,WM & HOWARD,PH (1993)

CAS #: 117121-34-3				UREA, N,N-DIMETHYL-N'-[4-(1-OCTYL-1H-TETRAZOL-5-
Formula: $C_{18}H_{28}N_6O_2$				
Mol Weight: 360.46				
MP (deg C):		FP (deg C):		
BP (deg C):				
BP pressure (mm Hg):				

Property/ Value	Units	Temp	Data Type	Reference
WS 2.22E+000	mg/L	25	EST	MEYLAN,WM ET AL. (1996)
logP 3.96			EXP	CAMILLERI,P ET AL. (1989)
VP 1.14E-010	mm Hg	25	EST	NEELY,WB & BLAU,GE (1985)
DC	pKa			
HL 3.73E-013	atm m3/mol	25	EST	MEYLAN,WM & HOWARD,PH (1991)
OH 5.46E-011	cm3/molc sec	25	EST	MEYLAN,WM & HOWARD,PH (1993)

CAS #: 117121-35-4				UREA, N,N-DIMETHYL-N'-[4-[[1-(2-PROPENYL)-1H-TET
Formula: $C_{13}H_{18}N_6O_2$				
Mol Weight: 290.33				
MP (deg C):		FP (deg C):		
BP (deg C):				
BP pressure (mm Hg):				

Property/ Value	Units	Temp	Data Type	Reference
WS 2.01E+003	mg/L	25	EST	MEYLAN,WM ET AL. (1996)
logP 1.00			EXP	CAMILLERI,P ET AL. (1989)
VP 7.27E-009	mm Hg	25	EST	NEELY,WB & BLAU,GE (1985)
DC	pKa			
HL 6.75E-014	atm m3/mol	25	EST	MEYLAN,WM & HOWARD,PH (1991)
OH 7.23E-011	cm3/molc sec	25	EST	MEYLAN,WM & HOWARD,PH (1993)

CAS #: 117121-36-5				UREA, N,N-DIMETHYL-N'-[4-[[1-(PHENYLMETHYL)-1H-T
Formula: $C_{17}H_{18}N_6O_2$				
Mol Weight: 338.37				
MP (deg C):		FP (deg C):		
BP (deg C):				
BP pressure (mm Hg):				

Property/ Value	Units	Temp	Data Type	Reference
WS 2.57E+002	mg/L	25	EST	MEYLAN,WM ET AL. (1996)
logP 1.70			EXP	CAMILLERI,P ET AL. (1989)
VP 9.25E-011	mm Hg	25	EST	NEELY,WB & BLAU,GE (1985)
DC	pKa			
HL 4.15E-015	atm m3/mol	25	EST	MEYLAN,WM & HOWARD,PH (1991)
OH 5.08E-011	cm3/molc sec	25	EST	MEYLAN,WM & HOWARD,PH (1993)

CAS #: 117121-37-6				1H-TETRAZOLE-1-ACETIC ACID, 5-[4-[[(DIMETHYLAMIN
Formula: $C_{14}H_{18}N_6O_4$				
Mol Weight: 334.34				
MP (deg C):		FP (deg C):		
BP (deg C):				
BP pressure (mm Hg):				

Property/ Value	Units	Temp	Data Type	Reference
WS 1.94E+003	mg/L	25	EST	MEYLAN,WM ET AL. (1996)
logP 0.70			EXP	CAMILLERI,P ET AL. (1989)
VP 4.80E-010	mm Hg	25	EST	NEELY,WB & BLAU,GE (1985)
DC	pKa			
HL 1.62E-016	atm m3/mol	25	EST	MEYLAN,WM & HOWARD,PH (1991)
OH 4.74E-011	cm3/molc sec	25	EST	MEYLAN,WM & HOWARD,PH (1993)

CAS #: 117121-38-7				UREA, N,N-DIMETHYL-N'-[4-[[1-(2-PROPYNYL)-1H-TET
Formula: $C_{13}H_{14}N_6O_2$				
Mol Weight: 286.30				
MP (deg C):		FP (deg C):		
BP (deg C):				
BP pressure (mm Hg):				

Property/ Value	Units	Temp	Data Type	Reference
WS 2.06E+003	mg/L	25	EST	MEYLAN,WM ET AL. (1996)
logP 1.00			EXP	CAMILLERI,P ET AL. (1989)
VP 5.15E-009	mm Hg	25	EST	NEELY,WB & BLAU,GE (1985)
DC	pKa			
HL 1.12E-014	atm m3/mol	25	EST	MEYLAN,WM & HOWARD,PH (1991)
OH 5.30E-011	cm3/molc sec	25	EST	MEYLAN,WM & HOWARD,PH (1993)

CAS #: 117121-39-8				UREA, N'-[3-CHLORO-4-[(1-ETHYL-1H-TETRAZOL-5-YL)
Formula: $C_{12}H_{15}ClN_6O_2$				
Mol Weight: 310.75				
MP (deg C):		FP (deg C):		
BP (deg C):				
BP pressure (mm Hg):				

Property/ Value	Units	Temp	Data Type	Reference
WS 5.56E+002	mg/L	25	EST	MEYLAN,WM ET AL. (1996)
logP 1.50			EXP	CAMILLERI,P ET AL. (1989)
VP 4.17E-009	mm Hg	25	EST	NEELY,WB & BLAU,GE (1985)
DC	pKa			
HL 5.05E-014	atm m3/mol	25	EST	MEYLAN,WM & HOWARD,PH (1991)
OH 3.35E-011	cm3/molc sec	25	EST	MEYLAN,WM & HOWARD,PH (1993)

117121-41-2

CAS #: 117121-41-2

UREA, N,N-DIMETHYL-N'-[4-[(2-METHYL-2H-TETRAZOL-

Formula: $C_{11}H_{14}N_6O_2$

Mol Weight: 262.27

MP (deg C): **FP (deg C):**

BP (deg C):

BP pressure (mm Hg):

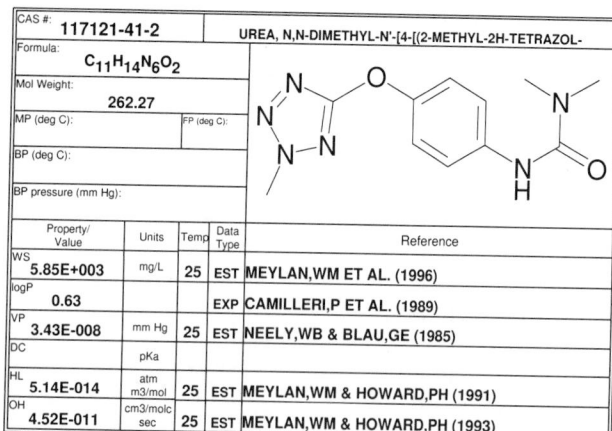

Property/Value	Units	Temp	Data Type	Reference
WS 5.85E+003	mg/L	25	EST	MEYLAN,WM ET AL. (1996)
logP 0.63			EXP	CAMILLERI,P ET AL. (1989)
VP 3.43E-008	mm Hg	25	EST	NEELY,WB & BLAU,GE (1985)
DC	pKa			
HL 5.14E-014	atm m3/mol		EST	MEYLAN,WM & HOWARD,PH (1991)
OH 4.52E-011	cm3/molc sec	25	EST	MEYLAN,WM & HOWARD,PH (1993)

117121-42-3

CAS #: 117121-42-3

UREA, N'-[4-[(2-ETHYL-2H-TETRAZOL-5-YL)OXY]PHENY

Formula: $C_{12}H_{16}N_6O_2$

Mol Weight: 276.30

MP (deg C): **FP (deg C):**

BP (deg C):

BP pressure (mm Hg):

Property/Value	Units	Temp	Data Type	Reference
WS 1.93E+003	mg/L	25	EST	MEYLAN,WM ET AL. (1996)
logP 1.10			EXP	CAMILLERI,P ET AL. (1989)
VP 1.54E-008	mm Hg	25	EST	NEELY,WB & BLAU,GE (1985)
DC	pKa			
HL 6.82E-014	atm m3/mol	25	EST	MEYLAN,WM & HOWARD,PH (1991)
OH 4.61E-011	cm3/molc sec	25	EST	MEYLAN,WM & HOWARD,PH (1993)

117121-43-4

CAS #: 117121-43-4

UREA, N,N-DIMETHYL-N'-[4-[[2-(1-METHYLETHYL)-2H-

Formula: $C_{13}H_{18}N_6O_2$

Mol Weight: 290.33

MP (deg C): **FP (deg C):**

BP (deg C):

BP pressure (mm Hg):

Property/Value	Units	Temp	Data Type	Reference
WS 8.07E+002	mg/L	25	EST	MEYLAN,WM ET AL. (1996)
logP 1.45			EXP	CAMILLERI,P ET AL. (1989)
VP 1.11E-008	mm Hg	25	EST	NEELY,WB & BLAU,GE (1985)
DC	pKa			
HL 9.06E-014	atm m3/mol	25	EST	MEYLAN,WM & HOWARD,PH (1991)
OH 4.73E-011	cm3/molc sec	25	EST	MEYLAN,WM & HOWARD,PH (1993)

117121-44-5

CAS #: 117121-44-5

UREA, N'-[4-[[2-(1,1-DIMETHYLETHYL)-2H-TETRAZOL-

Formula: $C_{14}H_{20}N_6O_2$

Mol Weight: 304.35

MP (deg C): **FP (deg C):**

BP (deg C):

BP pressure (mm Hg):

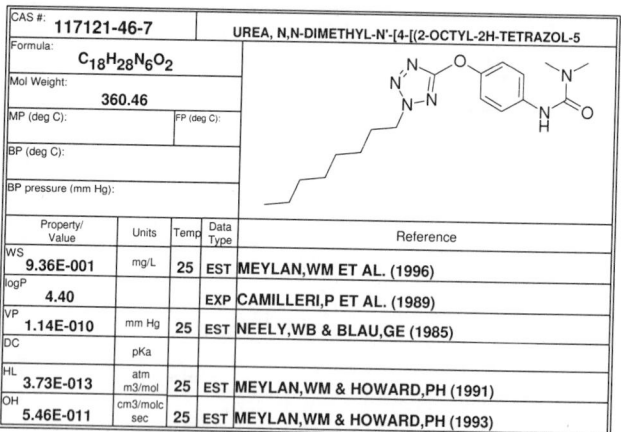

Property/Value	Units	Temp	Data Type	Reference
WS 1.94E+002	mg/L	25	EST	MEYLAN,WM ET AL. (1996)
logP 2.08			EXP	CAMILLERI,P ET AL. (1989)
VP 6.84E-009	mm Hg	25	EST	NEELY,WB & BLAU,GE (1985)
DC	pKa			
HL 1.20E-013	atm m3/mol	25	EST	MEYLAN,WM & HOWARD,PH (1991)
OH 4.55E-011	cm3/molc sec	25	EST	MEYLAN,WM & HOWARD,PH (1993)

117121-45-6

CAS #: 117121-45-6

UREA, N'-[4-[(2-CYCLOHEXYL-2H-TETRAZOL-5-YL)OXY]

Formula: $C_{16}H_{22}N_6O_2$

Mol Weight: 330.39

MP (deg C): **FP (deg C):**

BP (deg C):

BP pressure (mm Hg):

Property/Value	Units	Temp	Data Type	Reference
WS 4.26E+001	mg/L	25	EST	MEYLAN,WM ET AL. (1996)
logP 2.67			EXP	CAMILLERI,P ET AL. (1989)
VP 4.11E-010	mm Hg	25	EST	NEELY,WB & BLAU,GE (1985)
DC	pKa			
HL 9.34E-014	atm m3/mol	25	EST	MEYLAN,WM & HOWARD,PH (1991)
OH 5.50E-011	cm3/molc sec	25	EST	MEYLAN,WM & HOWARD,PH (1993)

117121-46-7

CAS #: 117121-46-7

UREA, N,N-DIMETHYL-N'-[4-[(2-OCTYL-2H-TETRAZOL-5

Formula: $C_{18}H_{28}N_6O_2$

Mol Weight: 360.46

MP (deg C): **FP (deg C):**

BP (deg C):

BP pressure (mm Hg):

Property/Value	Units	Temp	Data Type	Reference
WS 9.36E-001	mg/L	25	EST	MEYLAN,WM ET AL. (1996)
logP 4.40			EXP	CAMILLERI,P ET AL. (1989)
VP 1.14E-010	mm Hg	25	EST	NEELY,WB & BLAU,GE (1985)
DC	pKa			
HL 3.73E-013	atm m3/mol	25	EST	MEYLAN,WM & HOWARD,PH (1991)
OH 5.46E-011	cm3/molc sec	25	EST	MEYLAN,WM & HOWARD,PH (1993)

117121-47-8

CAS #: 117121-47-8

UREA, N,N-DIMETHYL-N'-[4-[[2-(2-PROPENYL)-2H-TET

Formula: $C_{13}H_{16}N_6O_2$

Mol Weight: 288.31

MP (deg C): **FP (deg C):**

BP (deg C):

BP pressure (mm Hg):

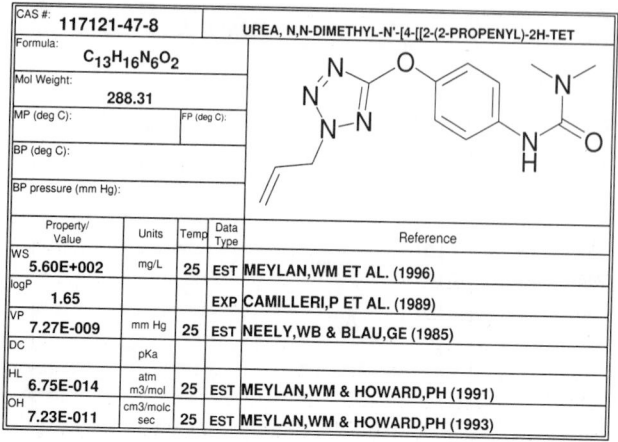

Property/Value	Units	Temp	Data Type	Reference
WS 5.60E+002	mg/L	25	EST	MEYLAN,WM ET AL. (1996)
logP 1.65			EXP	CAMILLERI,P ET AL. (1989)
VP 7.27E-009	mm Hg	25	EST	NEELY,WB & BLAU,GE (1985)
DC	pKa			
HL 6.75E-014	atm m3/mol		EST	MEYLAN,WM & HOWARD,PH (1991)
OH 7.23E-011	cm3/molc sec	25	EST	MEYLAN,WM & HOWARD,PH (1993)

117121-48-9

CAS #: 117121-48-9

UREA, N,N-DIMETHYL-N'-[4-[[2-(PHENYLMETHYL)-2H-T

Formula: $C_{17}H_{18}N_6O_2$

Mol Weight: 338.37

MP (deg C): **FP (deg C):**

BP (deg C):

BP pressure (mm Hg):

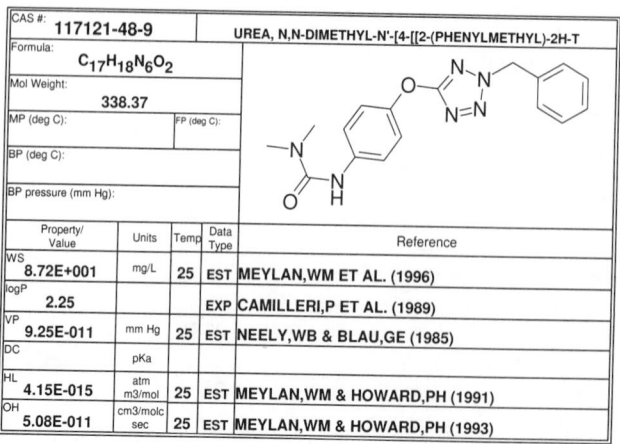

Property/Value	Units	Temp	Data Type	Reference
WS 8.72E+001	mg/L	25	EST	MEYLAN,WM ET AL. (1996)
logP 2.25			EXP	CAMILLERI,P ET AL. (1989)
VP 9.25E-011	mm Hg	25	EST	NEELY,WB & BLAU,GE (1985)
DC	pKa			
HL 4.15E-015	atm m3/mol	25	EST	MEYLAN,WM & HOWARD,PH (1991)
OH 5.08E-011	cm3/molc sec	25	EST	MEYLAN,WM & HOWARD,PH (1993)

CAS #: 117121-49-0	2H-TETRAZOLE-2-ACETIC ACID, 5-[4-[[(DIMETHYLAMIN

Formula: $C_{14}H_{20}N_6O_4$

Mol Weight: 336.35

MP (deg C): 　　　FP (deg C):

BP (deg C):

BP pressure (mm Hg):

Property/ Value	Units	Temp	Data Type	Reference
WS 8.51E+002	mg/L	25	EST	MEYLAN,WM ET AL. (1996)
logP 1.12			EXP	HANSCH,C ET AL. (1995)
VP 4.80E-010	mm Hg	25	EST	NEELY,WB & BLAU,GE (1985)
DC	pKa			
HL 1.62E-016	atm m3/mol	25	EST	MEYLAN,WM & HOWARD,PH (1991)
OH 4.74E-011	cm3/molc sec	25	EST	MEYLAN,WM & HOWARD,PH (1993)

CAS #: 117121-50-3	UREA, N'-[4-[[2-(2-CYANOETHYL)-2H-TETRAZOL-5-YL]

Formula: $C_{13}H_{15}N_7O_2$

Mol Weight: 301.31

MP (deg C): 　　　FP (deg C):

BP (deg C):

BP pressure (mm Hg):

Property/ Value	Units	Temp	Data Type	Reference
WS 1.85E+002	mg/L	25	EST	MEYLAN,WM ET AL. (1996)
logP 0.28			EXP	CAMILLERI,P ET AL. (1989)
VP 2.58E-010	mm Hg	25	EST	NEELY,WB & BLAU,GE (1985)
DC	pKa			
HL 2.83E-017	atm m3/mol	25	EST	MEYLAN,WM & HOWARD,PH (1991)
OH 4.54E-011	cm3/molc sec	25	EST	MEYLAN,WM & HOWARD,PH (1993)

CAS #: 117121-51-4	TETRAZOLE,2-ETHYL-5-(2-CHLORO-4-DIMETHYLUREYL)PH

Formula: $C_{12}H_{15}ClN_6O_2$

Mol Weight: 310.75

MP (deg C): 　　　FP (deg C):

BP (deg C):

BP pressure (mm Hg):

Property/ Value	Units	Temp	Data Type	Reference
WS 3.75E+002	mg/L	25	EST	MEYLAN,WM ET AL. (1996)
logP 1.70			EXP	HANSCH,C ET AL. (1995)
VP 4.17E-009	mm Hg	25	EST	NEELY,WB & BLAU,GE (1985)
DC	pKa			
HL 5.05E-014	atm m3/mol	25	EST	MEYLAN,WM & HOWARD,PH (1991)
OH 3.35E-011	cm3/molc sec	25	EST	MEYLAN,WM & HOWARD,PH (1993)

CAS #: 117144-73-7	UREA, N,N-DIMETHYL-N'-[4-[[2-(2-PROPYNYL)-2H-TET

Formula: $C_{13}H_{14}N_6O_2$

Mol Weight: 286.30

MP (deg C): 　　　FP (deg C):

BP (deg C):

BP pressure (mm Hg):

Property/ Value	Units	Temp	Data Type	Reference
WS 5.75E+002	mg/L	25	EST	MEYLAN,WM ET AL. (1996)
logP 1.65			EXP	CAMILLERI,P ET AL. (1989)
VP 5.15E-009	mm Hg	25	EST	NEELY,WB & BLAU,GE (1985)
DC	pKa			
HL 1.12E-014	atm m3/mol	25	EST	MEYLAN,WM & HOWARD,PH (1991)
OH 5.30E-011	cm3/molc sec	25	EST	MEYLAN,WM & HOWARD,PH (1993)

CAS #: 117491-56-2	1H-PYRROLE-2,5-DIONE, 1-[(ACETYLOXY)METHYL]-3-ME

Formula: $C_8H_9NO_4$

Mol Weight: 183.17

MP (deg C): 　　　FP (deg C):

BP (deg C):

BP pressure (mm Hg):

Property/ Value	Units	Temp	Data Type	Reference
WS 2.19E+004	mg/L	25	EST	MEYLAN,WM ET AL. (1996)
logP 0.45			EXP	NISHIMURA,K ET AL. (1988)
VP 2.04E-006	mm Hg	25	EST	NEELY,WB & BLAU,GE (1985)
DC	pKa			
HL 1.04E-009	atm m3/mol	25	EST	MEYLAN,WM & HOWARD,PH (1991)
OH 3.20E-011	cm3/molc sec	25	EST	MEYLAN,WM & HOWARD,PH (1993)

CAS #: 117491-57-3	1H-PYRROLE-2,5-DIONE, 1-[(ACETYLOXY)METHYL]-3,4-

Formula: $C_9H_{11}NO_4$

Mol Weight: 197.19

MP (deg C): 　　　FP (deg C):

BP (deg C):

BP pressure (mm Hg):

Property/ Value	Units	Temp	Data Type	Reference
WS 1.26E+004	mg/L	25	EST	MEYLAN,WM ET AL. (1996)
logP 0.65			EXP	NISHIMURA,K ET AL. (1988)
VP 8.81E-007	mm Hg	25	EST	NEELY,WB & BLAU,GE (1985)
DC	pKa			
HL 1.62E-009	atm m3/mol	25	EST	MEYLAN,WM & HOWARD,PH (1991)
OH 3.50E-011	cm3/molc sec	25	EST	MEYLAN,WM & HOWARD,PH (1993)

CAS #: 117491-58-4	1H-ISOINDOLE-1-ONE, 2-[(ACETYLOXY)METHYL]-

Formula: $C_{11}H_{11}NO_3$

Mol Weight: 205.22

MP (deg C): 　　　FP (deg C):

BP (deg C):

BP pressure (mm Hg):

Property/ Value	Units	Temp	Data Type	Reference
WS 5.87E+003	mg/L	25	EST	MEYLAN,WM ET AL. (1996)
logP 0.99			EXP	NISHIMURA,K ET AL. (1988)
VP 1.82E-005	mm Hg	25	EST	NEELY,WB & BLAU,GE (1985)
DC	pKa			
HL 1.86E-010	atm m3/mol	25	EST	MEYLAN,WM & HOWARD,PH (1991)
OH 2.87E-011	cm3/molc sec	25	EST	MEYLAN,WM & HOWARD,PH (1993)

CAS #: 117505-21-2	1-NAPHTHALENE AZOXYCYANIDE

Formula: $C_{11}H_7N_3O$

Mol Weight: 197.20

MP (deg C): 　　　FP (deg C):

BP (deg C):

BP pressure (mm Hg):

Property/ Value	Units	Temp	Data Type	Reference
WS 3.28E+001	mg/L	25	EST	MEYLAN,WM ET AL. (1996)
logP 3.17			EXP	CALVINO,R R ET AL (1991)
VP 4.42E-010	mm Hg	25	EST	NEELY,WB & BLAU,GE (1985)
DC	pKa			
HL 2.01E-012	atm m3/mol	25	EST	MEYLAN,WM & HOWARD,PH (1991)
OH 4.62E-012	cm3/molc sec	25	EST	MEYLAN,WM & HOWARD,PH (1993)

CAS #: 117505-22-3		4-QUINOLINE AZOXYCYANIDE

Formula: $C_{10}H_6N_4O$

Mol Weight: 198.19

MP (deg C):   FP (deg C):

BP (deg C):

BP pressure (mm Hg):

Property/ Value	Units	Temp	Data Type	Reference
WS 2.82E+002	mg/L	25	EST	MEYLAN,WM ET AL. (1996)
logP 2.07			EXP	CALVINO,R R ET AL (1991)
VP 3.00E-010	mm Hg	25	EST	NEELY,WB & BLAU,GE (1985)
DC	pKa			
HL 2.63E-015	atm m3/mol	25	EST	MEYLAN,WM & HOWARD,PH (1991)
OH 3.08E-012	cm3/molc sec	25	EST	MEYLAN,WM & HOWARD,PH (1993)

CAS #: 117505-23-4		4-QUINOLINE,1-OXIDE AZOXYCYANIDE

Formula: $C_{10}H_8N_4O_2$

Mol Weight: 216.20

MP (deg C):   FP (deg C):

BP (deg C):

BP pressure (mm Hg):

Property/ Value	Units	Temp	Data Type	Reference
WS 1.51E+003	mg/L	25	EST	MEYLAN,WM ET AL. (1996)
logP 1.12			EXP	CALVINO,R R ET AL (1991)
VP 6.88E-012	mm Hg	25	EST	NEELY,WB & BLAU,GE (1985)
DC	pKa			
HL 2.63E-020	atm m3/mol	25	EST	MEYLAN,WM & HOWARD,PH (1991)
OH 3.08E-012	cm3/molc sec	25	EST	MEYLAN,WM & HOWARD,PH (1993)

CAS #: 117505-24-5		4-PYRIDINE OXIDE AZOXYCYANIDE

Formula: $C_6H_6N_4O_2$

Mol Weight: 166.14

MP (deg C):   FP (deg C):

BP (deg C):

BP pressure (mm Hg):

Property/ Value	Units	Temp	Data Type	Reference
WS 5.43E+004	mg/L	25	EST	MEYLAN,WM ET AL. (1996)
logP -0.41			EXP	CALVINO,R R ET AL (1991)
VP 1.09E-009	mm Hg	25	EST	NEELY,WB & BLAU,GE (1985)
DC	pKa			
HL 2.69E-019	atm m3/mol	25	EST	MEYLAN,WM & HOWARD,PH (1991)
OH 7.91E-014	cm3/molc sec	25	EST	MEYLAN,WM & HOWARD,PH (1993)

CAS #: 117505-25-6		PYRAZINE AZOXYCYANIDE

Formula: $C_5H_3N_5O$

Mol Weight: 149.11

MP (deg C):   FP (deg C):

BP (deg C):

BP pressure (mm Hg):

Property/ Value	Units	Temp	Data Type	Reference
WS 2.33E+004	mg/L	25	EST	MEYLAN,WM ET AL. (1996)
logP 0.10			EXP	CALVINO,R R ET AL (1991)
VP 2.96E-008	mm Hg	25	EST	NEELY,WB & BLAU,GE (1985)
DC	pKa			
HL 1.12E-014	atm m3/mol	25	EST	MEYLAN,WM & HOWARD,PH (1991)
OH 4.92E-014	cm3/molc sec	25	EST	MEYLAN,WM & HOWARD,PH (1993)

CAS #: 117509-89-4		1H-ISOINDOLE-1,3(2H)-DIONE, 2-[(ACETYLOXY)METHYL

Formula: $C_{11}H_{13}NO_4$

Mol Weight: 223.23

MP (deg C):   FP (deg C):

BP (deg C):

BP pressure (mm Hg):

Property/ Value	Units	Temp	Data Type	Reference
WS 4.91E+003	mg/L	25	EST	MEYLAN,WM ET AL. (1996)
logP 0.97			EXP	NISHIMURA,K ET AL. (1988)
VP 1.56E-007	mm Hg	25	EST	NEELY,WB & BLAU,GE (1985)
DC	pKa			
HL 3.18E-009	atm m3/mol	25	EST	MEYLAN,WM & HOWARD,PH (1991)
OH 9.46E-011	cm3/molc sec	25	EST	MEYLAN,WM & HOWARD,PH (1993)

CAS #: 117525-25-4		6H-PURIN-6-ONE, 9-(2,3-DIDEOXY-2-FLUORO-á-D-THRE

Formula: $C_{10}H_{11}FN_4O_3$

Mol Weight: 254.22

MP (deg C):   FP (deg C):

BP (deg C):

BP pressure (mm Hg):

Property/ Value	Units	Temp	Data Type	Reference
WS 1.19E+004	mg/L	25	EST	MEYLAN,WM ET AL. (1996)
logP -1.21			EXP	BARCHI,JJ ET AL. (1991)
VP 7.91E-011	mm Hg	25	EST	NEELY,WB & BLAU,GE (1985)
DC	pKa			
HL 4.88E-020	atm m3/mol	25	EST	MEYLAN,WM & HOWARD,PH (1991)
OH 2.08E-010	cm3/molc sec	25	EST	MEYLAN,WM & HOWARD,PH (1993)

CAS #: 117539-78-3		2-PYRIMIDINE AZOXYCYANIDE

Formula: $C_5H_3N_5O$

Mol Weight: 149.11

MP (deg C):   FP (deg C):

BP (deg C):

BP pressure (mm Hg):

Property/ Value	Units	Temp	Data Type	Reference
WS 3.66E+004	mg/L	25	EST	MEYLAN,WM ET AL. (1996)
logP -0.13			EXP	CALVINO,R R ET AL (1991)
VP 2.96E-008	mm Hg	25	EST	NEELY,WB & BLAU,GE (1985)
DC	pKa			
HL 1.12E-014	atm m3/mol	25	EST	MEYLAN,WM & HOWARD,PH (1991)
OH 4.92E-014	cm3/molc sec	25	EST	MEYLAN,WM & HOWARD,PH (1993)

CAS #: 117574-40-0		2,5-CYCLOHEXADIENE-1,4-DIONE, 3-[(3-PYRIDINYL)ME

Formula: $C_{15}H_{15}NO_2$

Mol Weight: 241.29

MP (deg C):   FP (deg C):

BP (deg C):

BP pressure (mm Hg):

Property/ Value	Units	Temp	Data Type	Reference
WS 3.50E+003	mg/L	25	EST	MEYLAN,WM ET AL. (1996)
logP 2.55			EXP	SANGSTER,J (1994)
VP 1.50E-006	mm Hg	25	EST	NEELY,WB & BLAU,GE (1985)
DC	pKa			
HL 7.77E-013	atm m3/mol	25	EST	MEYLAN,WM & HOWARD,PH (1991)
OH 2.92E-011	cm3/molc sec	25	EST	MEYLAN,WM & HOWARD,PH (1993)

CAS #: 117856-61-8

PYRAZINE, 2,5-DIMETHOXY-

Formula: $C_6H_8N_2O_2$

Mol Weight: 140.14

MP (deg C): FP (deg C):

BP (deg C):

BP pressure (mm Hg):

Property/ Value	Units	Temp	Data Type	Reference
WS 8.89E+003	mg/L	25	EST	MEYLAN,WM ET AL. (1996)
logP 1.14			EXP	YAMAGAMI,C ET AL. (1991)
VP 1.57E-001	mm Hg	25	EST	NEELY,WB & BLAU,GE (1985)
DC	pKa			
HL 8.49E-007	atm m3/mol	25	EST	MEYLAN,WM & HOWARD,PH (1991)
OH 3.85E-012	cm3/molc sec	25	EST	MEYLAN,WM & HOWARD,PH (1993)

CAS #: 118024-67-2

1H-PURINE-2,6-DIONE, 3,9-DIHYDRO-1-METHYL-3-PROP

Formula: $C_9H_{12}N_4O_2$

Mol Weight: 208.22

MP (deg C): FP (deg C):

BP (deg C):

BP pressure (mm Hg):

Property/ Value	Units	Temp	Data Type	Reference
WS 5.34E+003	mg/L	25	EST	MEYLAN,WM ET AL. (1996)
logP 1.02			EXP	SANGSTER,J (1993)
VP 1.00E-009	mm Hg	25	EST	NEELY,WB & BLAU,GE (1985)
DC	pKa			
HL 2.96E-012	atm m3/mol	25	EST	MEYLAN,WM & HOWARD,PH (1991)
OH 2.48E-011	cm3/molc sec	25	EST	MEYLAN,WM & HOWARD,PH (1993)

CAS #: 118198-71-3

BENZENEMETHANOL, 3,4-DIHYDROXY-à-PENTYL-

Formula: $C_{12}H_{18}O_3$

Mol Weight: 210.28

MP (deg C): FP (deg C):

BP (deg C):

BP pressure (mm Hg):

Property/ Value	Units	Temp	Data Type	Reference
WS 4.97E+003	mg/L	25	EST	MEYLAN,WM ET AL. (1996)
logP 2.32			EXP	NAITO,Y ET AL. (1991)
VP 2.42E-007	mm Hg	25	EST	NEELY,WB & BLAU,GE (1985)
DC	pKa			
HL 9.71E-015	atm m3/mol	25	EST	MEYLAN,WM & HOWARD,PH (1991)
OH 6.88E-011	cm3/molc sec	25	EST	MEYLAN,WM & HOWARD,PH (1993)

CAS #: 118198-78-0

7-OCTEN-1-OL, 8-(3,4-DIHYDROXYPHENYL)-6-OXO-

Formula: $C_{14}H_{18}O_4$

Mol Weight: 250.30

MP (deg C): FP (deg C):

BP (deg C):

BP pressure (mm Hg):

Property/ Value	Units	Temp	Data Type	Reference
WS 1.09E+004	mg/L	25	EST	MEYLAN,WM ET AL. (1996)
logP 1.67			EXP	NAITO,Y ET AL. (1991)
VP 1.63E-009	mm Hg	25	EST	NEELY,WB & BLAU,GE (1985)
DC	pKa			
HL 1.44E-018	atm m3/mol	25	EST	MEYLAN,WM & HOWARD,PH (1991)
OH 8.49E-011	cm3/molc sec	25	EST	MEYLAN,WM & HOWARD,PH (1993)

CAS #: 118199-17-0

ETHANONE, 1-(6,7-DIHYDROXY-2-NAPHTHALENYL)-

Formula: $C_{12}H_{10}O_3$

Mol Weight: 202.21

MP (deg C): FP (deg C):

BP (deg C):

BP pressure (mm Hg):

Property/ Value	Units	Temp	Data Type	Reference
WS 1.69E+003	mg/L	25	EST	MEYLAN,WM ET AL. (1996)
logP 2.32			EXP	NAITO,Y ET AL. (1991)
VP 1.63E-007	mm Hg	25	EST	NEELY,WB & BLAU,GE (1985)
DC	pKa			
HL 1.04E-014	atm m3/mol	25	EST	MEYLAN,WM & HOWARD,PH (1991)
OH 1.58E-010	cm3/molc sec	25	EST	MEYLAN,WM & HOWARD,PH (1993)

CAS #: 118247-01-1

BENZOIC ACID, 2-(ACETYLOXY)-, 2-(ETHYLAMINO)-2-O

Formula: $C_{13}H_{15}NO_5$

Mol Weight: 265.27

MP (deg C): FP (deg C):

BP (deg C):

BP pressure (mm Hg):

Property/ Value	Units	Temp	Data Type	Reference
WS 3.31E+003	mg/L	25	EST	MEYLAN,WM ET AL. (1996)
logP 0.90			EXP	NIELSEN,LS & BUNDGAARD,H (1989)
VP 6.43E-007	mm Hg	25	EST	NEELY,WB & BLAU,GE (1985)
DC	pKa			
HL 1.31E-012	atm m3/mol	25	EST	MEYLAN,WM & HOWARD,PH (1991)
OH 1.14E-011	cm3/molc sec	25	EST	MEYLAN,WM & HOWARD,PH (1993)

CAS #: 118247-02-2

BENZOIC ACID, 2-(ACETYLOXY)-, 2-[(2-AMINO-2-OXOE

Formula: $C_{13}H_{14}N_2O_6$

Mol Weight: 294.27

MP (deg C): FP (deg C):

BP (deg C):

BP pressure (mm Hg):

Property/ Value	Units	Temp	Data Type	Reference
WS 1.51E+003	mg/L	25	EST	MEYLAN,WM ET AL. (1996)
logP -0.43			EXP	NIELSEN,LS & BUNDGAARD,H (1989)
VP 2.90E-010	mm Hg	25	EST	NEELY,WB & BLAU,GE (1985)
DC	pKa			
HL 4.73E-015	atm m3/mol	25	EST	MEYLAN,WM & HOWARD,PH (1991)
OH 1.22E-011	cm3/molc sec	25	EST	MEYLAN,WM & HOWARD,PH (1993)

CAS #: 118247-03-3

GLYCINE, N-[[[2-(ACETYLOXY)BENZOYL]OXY]ACETYL]-,

Formula: $C_{15}H_{17}NO_7$

Mol Weight: 323.31

MP (deg C): FP (deg C):

BP (deg C):

BP pressure (mm Hg):

Property/ Value	Units	Temp	Data Type	Reference
WS 1.28E+003	mg/L	25	EST	MEYLAN,WM ET AL. (1996)
logP 0.99			EXP	NIELSEN,LS & BUNDGAARD,H (1989)
VP 1.65E-008	mm Hg	25	EST	NEELY,WB & BLAU,GE (1985)
DC	pKa			
HL 6.22E-014	atm m3/mol	25	EST	MEYLAN,WM & HOWARD,PH (1991)
OH 1.19E-011	cm3/molc sec	25	EST	MEYLAN,WM & HOWARD,PH (1993)

CAS #:	118247-04-4	BENZOIC ACID, 2-(ACETYLOXY)-, 2-(DIMETHYLAMINO)-

Formula: $C_{13}H_{15}NO_5$

Mol Weight: 265.27

MP (deg C): FP (deg C):

BP (deg C):

BP pressure (mm Hg):

Property/ Value	Units	Temp	Data Type	Reference
WS 9.21E+003	mg/L	25	EST	MEYLAN,WM ET AL. (1996)
logP 0.38			EXP	NIELSEN,LS & BUNDGAARD,H (1989)
VP 2.22E-005	mm Hg	25	EST	NEELY,WB & BLAU,GE (1985)
DC	pKa			
HL 2.16E-012	atm m3/mol	25	EST	MEYLAN,WM & HOWARD,PH (1991)
OH 1.80E-011	cm3/molc sec	25	EST	MEYLAN,WM & HOWARD,PH (1993)

CAS #:	118247-05-5	BENZOIC ACID, 2-(ACETYLOXY)-, 2-[(2-HYDROXYETHYL

Formula: $C_{15}H_{19}NO_5$

Mol Weight: 293.32

MP (deg C): FP (deg C):

BP (deg C):

BP pressure (mm Hg):

Property/ Value	Units	Temp	Data Type	Reference
WS 1.51E+004	mg/L	25	EST	MEYLAN,WM ET AL. (1996)
logP -0.06			EXP	NIELSEN,LS & BUNDGAARD,H (1989)
VP 3.87E-010	mm Hg	25	EST	NEELY,WB & BLAU,GE (1985)
DC	pKa			
HL 6.48E-018	atm m3/mol	25	EST	MEYLAN,WM & HOWARD,PH (1991)
OH 2.79E-011	cm3/molc sec	25	EST	MEYLAN,WM & HOWARD,PH (1993)

CAS #:	118247-07-7	BENZOIC ACID, 2-(ACETYLOXY)-, (1-OXOBUTOXY)METHY

Formula: $C_{14}H_{16}O_6$

Mol Weight: 280.28

MP (deg C): FP (deg C):

BP (deg C):

BP pressure (mm Hg):

Property/ Value	Units	Temp	Data Type	Reference
WS 1.17E+002	mg/L	25	EST	MEYLAN,WM ET AL. (1996)
logP 2.50			EXP	NIELSEN,LS & BUNDGAARD,H (1989)
VP 2.16E-004	mm Hg	25	EST	NEELY,WB & BLAU,GE (1985)
DC	pKa			
HL 1.75E-009	atm m3/mol	25	EST	MEYLAN,WM & HOWARD,PH (1991)
OH 5.31E-012	cm3/molc sec	25	EST	MEYLAN,WM & HOWARD,PH (1993)

CAS #:	118247-08-8	BENZOIC ACID, 2-(ACETYLOXY)-, 2-(ACETYLMETHYLAMI

Formula: $C_{14}H_{17}NO_5$

Mol Weight: 279.30

MP (deg C): FP (deg C):

BP (deg C):

BP pressure (mm Hg):

Property/ Value	Units	Temp	Data Type	Reference
WS 9.16E+002	mg/L	25	EST	MEYLAN,WM ET AL. (1996)
logP 1.46			EXP	NIELSEN,LS & BUNDGAARD,H (1989)
VP 9.47E-006	mm Hg	25	EST	NEELY,WB & BLAU,GE (1985)
DC	pKa			
HL 4.06E-013	atm m3/mol	25	EST	MEYLAN,WM & HOWARD,PH (1991)
OH 2.30E-011	cm3/molc sec	25	EST	MEYLAN,WM & HOWARD,PH (1993)

CAS #:	118259-45-3	UREA, N'-[3-CHLORO-4-[[(1,1-DIMETHYLETHYL)-1H-TE

Formula: $C_{14}H_{19}ClN_6O_2$

Mol Weight: 338.80

MP (deg C): FP (deg C):

BP (deg C):

BP pressure (mm Hg):

Property/ Value	Units	Temp	Data Type	Reference
WS 8.67E+001	mg/L	25	EST	MEYLAN,WM ET AL. (1996)
logP 2.25			EXP	CAMILLERI,P ET AL. (1989)
VP 1.85E-009	mm Hg	25	EST	NEELY,WB & BLAU,GE (1985)
DC	pKa			
HL 8.91E-014	atm m3/mol	25	EST	MEYLAN,WM & HOWARD,PH (1991)
OH 3.29E-011	cm3/molc sec	25	EST	MEYLAN,WM & HOWARD,PH (1993)

CAS #:	118259-53-3	UREA, N'-[4-[[1-(1,1-DIMETHYLETHYL)-1H-TETRAZOL-

Formula: $C_{14}H_{20}N_6O_2$

Mol Weight: 304.35

MP (deg C): FP (deg C):

BP (deg C):

BP pressure (mm Hg):

Property/ Value	Units	Temp	Data Type	Reference
WS 4.51E+002	mg/L	25	EST	MEYLAN,WM ET AL. (1996)
logP 1.65			EXP	CAMILLERI,P ET AL. (1989)
VP 6.84E-009	mm Hg	25	EST	NEELY,WB & BLAU,GE (1985)
DC	pKa			
HL 1.20E-013	atm m3/mol	25	EST	MEYLAN,WM & HOWARD,PH (1991)
OH 4.55E-011	cm3/molc sec	25	EST	MEYLAN,WM & HOWARD,PH (1993)

CAS #:	118409-80-6	PHNEYLACETIC ACID,2-(2',6'-DICHLORO-4-METHOXY)AN

Formula: $C_{15}H_{13}Cl_2NO_3$

Mol Weight: 326.18

MP (deg C): FP (deg C):

BP (deg C):

BP pressure (mm Hg):

Property/ Value	Units	Temp	Data Type	Reference
WS 3.45E+000	mg/L	25	EST	MEYLAN,WM ET AL. (1996)
logP 4.44			EXP	HANSCH,C ET AL. (1995)
VP 1.12E-008	mm Hg	25	EST	NEELY,WB & BLAU,GE (1985)
DC	pKa			
HL 2.80E-013	atm m3/mol	25	EST	MEYLAN,WM & HOWARD,PH (1991)
OH 1.71E-010	cm3/molc sec	25	EST	MEYLAN,WM & HOWARD,PH (1993)

CAS #:	118506-53-9	FURAZANAMINE, 4-METHYL-, 2-OXIDE

Formula: $C_3H_5N_3O_2$

Mol Weight: 115.09

MP (deg C): FP (deg C):

BP (deg C):

BP pressure (mm Hg):

Property/ Value	Units	Temp	Data Type	Reference
WS 1.96E+005	mg/L	25	EST	MEYLAN,WM ET AL. (1996)
logP -0.32			EXP	CALVINO,R ET AL. (1992)
VP 1.82E-003	mm Hg	25	EST	NEELY,WB & BLAU,GE (1985)
DC	pKa			
HL 7.38E-013	atm m3/mol	25	EST	MEYLAN,WM & HOWARD,PH (1991)
OH 4.14E-012	cm3/molc sec	25	EST	MEYLAN,WM & HOWARD,PH (1993)

CAS #: 118709-72-1	5-CL-5-T-BU BARBITURIC ACID

Formula: $C_8H_{11}ClN_2O_3$

Mol Weight: 218.64

MP (deg C): 　 FP (deg C):

BP (deg C):

BP pressure (mm Hg):

Property/ Value	Units	Temp	Data Type	Reference
WS 6.98E+003	mg/L	25	EST	MEYLAN,WM ET AL. (1996)
logP 0.82			EXP	WONG,O & MCKEOWN,RH (1988)
VP 9.76E-011	mm Hg	25	EST	NEELY,WB & BLAU,GE (1985)
DC	pKa			
HL 1.27E-013	atm m3/mol	25	EST	MEYLAN,WM & HOWARD,PH (1991)
OH 2.50E-012	cm3/molc sec	25	EST	MEYLAN,WM & HOWARD,PH (1993)

CAS #: 118976-97-9	2-THIOPHENESULFONAMIDE, 4-(4-METHOXYBENZOYL)-

Formula: $C_{12}H_{11}NO_4S_2$

Mol Weight: 297.35

MP (deg C): 　 FP (deg C):

BP (deg C):

BP pressure (mm Hg):

Property/ Value	Units	Temp	Data Type	Reference
WS 3.55E+001	mg/L	25	EST	MEYLAN,WM ET AL. (1996)
logP 1.76			EXP	HARTMAN,GD ET AL. (1992)
VP 1.13E-008	mm Hg	25	EST	NEELY,WB & BLAU,GE (1985)
DC	pKa			
HL 4.89E-012	atm m3/mol	25	EST	MEYLAN,WM & HOWARD,PH (1991)
OH 2.13E-011	cm3/molc sec	25	EST	MEYLAN,WM & HOWARD,PH (1993)

CAS #: 118993-57-0	2-THIOPHENESULFONAMIDE, 4-(4-HYDROXYBENZOYL)-

Formula: $C_{11}H_9NO_4S_2$

Mol Weight: 283.33

MP (deg C): 　 FP (deg C):

BP (deg C):

BP pressure (mm Hg):

Property/ Value	Units	Temp	Data Type	Reference
WS 3.24E+002	mg/L	25	EST	MEYLAN,WM ET AL. (1996)
logP 1.41			EXP	HARTMAN,GD ET AL. (1992)
VP 8.27E-010	mm Hg	25	EST	NEELY,WB & BLAU,GE (1985)
DC	pKa			
HL 8.61E-015	atm m3/mol	25	EST	MEYLAN,WM & HOWARD,PH (1991)
OH 3.14E-011	cm3/molc sec	25	EST	MEYLAN,WM & HOWARD,PH (1993)

CAS #: 118993-61-6	2-FURANSULFONAMIDE, 4-(4-METHOXYBENZOYL)-

Formula: $C_{12}H_{11}NO_5S$

Mol Weight: 281.29

MP (deg C): 　 FP (deg C):

BP (deg C):

BP pressure (mm Hg):

Property/ Value	Units	Temp	Data Type	Reference
WS 7.05E+001	mg/L	25	EST	MEYLAN,WM ET AL. (1996)
logP 1.52			EXP	HARTMAN,GD ET AL. (1992)
VP 6.79E-008	mm Hg	25	EST	NEELY,WB & BLAU,GE (1985)
DC	pKa			
HL 8.98E-012	atm m3/mol	25	EST	MEYLAN,WM & HOWARD,PH (1991)
OH 2.41E-011	cm3/molc sec	25	EST	MEYLAN,WM & HOWARD,PH (1993)

CAS #: 119005-12-8	BENZOIC ACID, 3-[(4-OXO-2-AZETIDINYL)OXY]-, PHEN

Formula: $C_{17}H_{15}NO_4$

Mol Weight: 297.31

MP (deg C): 　 FP (deg C):

BP (deg C):

BP pressure (mm Hg):

Property/ Value	Units	Temp	Data Type	Reference
WS 4.88E+001	mg/L	25	EST	MEYLAN,WM ET AL. (1996)
logP 2.83			EXP	ARNOLDI,A ET AL. (1988)
VP 1.73E-009	mm Hg	25	EST	NEELY,WB & BLAU,GE (1985)
DC	pKa			
HL 4.86E-013	atm m3/mol	25	EST	MEYLAN,WM & HOWARD,PH (1991)
OH 6.99E-011	cm3/molc sec	25	EST	MEYLAN,WM & HOWARD,PH (1993)

CAS #: 119005-13-9	2-AZETIDINONE, 4-(1-NAPHTHALENYLOXY)-

Formula: $C_{13}H_{11}NO_2$

Mol Weight: 213.24

MP (deg C): 　 FP (deg C):

BP (deg C):

BP pressure (mm Hg):

Property/ Value	Units	Temp	Data Type	Reference
WS 2.00E+002	mg/L	25	EST	MEYLAN,WM ET AL. (1996)
logP 2.66			EXP	ARNOLDI,A ET AL. (1988)
VP 1.68E-007	mm Hg	25	EST	NEELY,WB & BLAU,GE (1985)
DC	pKa			
HL 9.13E-011	atm m3/mol	25	EST	MEYLAN,WM & HOWARD,PH (1991)
OH 2.59E-010	cm3/molc sec	25	EST	MEYLAN,WM & HOWARD,PH (1993)

CAS #: 119005-15-1	1,1'-BIPHENYL, 4-[(2-OXOAZETIDIN-4-YL)OXY]-

Formula: $C_{15}H_{13}NO_2$

Mol Weight: 239.28

MP (deg C): 　 FP (deg C):

BP (deg C):

BP pressure (mm Hg):

Property/ Value	Units	Temp	Data Type	Reference
WS 1.60E+002	mg/L	25	EST	MEYLAN,WM ET AL. (1996)
logP 2.61			EXP	ARNOLDI,A ET AL. (1988)
VP 2.34E-008	mm Hg	25	EST	NEELY,WB & BLAU,GE (1985)
DC	pKa			
HL 7.18E-011	atm m3/mol	25	EST	MEYLAN,WM & HOWARD,PH (1991)
OH 7.75E-011	cm3/molc sec	25	EST	MEYLAN,WM & HOWARD,PH (1993)

CAS #: 119005-16-2	2-AZETIDINONE, 4-(9-PHENANTHRENYLOXY)-

Formula: $C_{17}H_{13}NO_2$

Mol Weight: 263.30

MP (deg C): 　 FP (deg C):

BP (deg C):

BP pressure (mm Hg):

Property/ Value	Units	Temp	Data Type	Reference
WS 3.63E+000	mg/L	25	EST	MEYLAN,WM ET AL. (1996)
logP 3.08			EXP	ARNOLDI,A ET AL. (1988)
VP 9.66E-010	mm Hg	25	EST	NEELY,WB & BLAU,GE (1985)
DC	pKa			
HL 8.91E-012	atm m3/mol	25	EST	MEYLAN,WM & HOWARD,PH (1991)
OH 2.59E-010	cm3/molc sec	25	EST	MEYLAN,WM & HOWARD,PH (1993)

CAS #: 119005-17-3 — 2-AZETIDINONE, 4-(4-BENZOYLPHENOXY)-

Formula: $C_{16}H_{13}NO_3$

Mol Weight: 267.29

MP (deg C): FP (deg C):

BP (deg C):

BP pressure (mm Hg):

Property/Value	Units	Temp	Data Type	Reference
WS 2.82E+002	mg/L	25	EST	MEYLAN,WM ET AL. (1996)
logP 2.14			EXP	ARNOLDI,A ET AL. (1988)
VP 4.99E-009	mm Hg	25	EST	NEELY,WB & BLAU,GE (1985)
DC	pKa			
HL 3.37E-013	atm m3/mol	25	EST	MEYLAN,WM & HOWARD,PH (1991)
OH 8.01E-011	cm3/molc sec	25	EST	MEYLAN,WM & HOWARD,PH (1993)

CAS #: 119005-18-4 — 2-AZETIDINONE, 4-[4-(1-OXO-3-PHENYL-2-PROPENYL)P

Formula: $C_{18}H_{15}NO_3$

Mol Weight: 293.33

MP (deg C): FP (deg C):

BP (deg C):

BP pressure (mm Hg):

Property/Value	Units	Temp	Data Type	Reference
WS 8.74E+001	mg/L	25	EST	MEYLAN,WM ET AL. (1996)
logP 2.56			EXP	ARNOLDI,A ET AL. (1988)
VP 7.60E-010	mm Hg	25	EST	NEELY,WB & BLAU,GE (1985)
DC	pKa			
HL 4.02E-014	atm m3/mol	25	EST	MEYLAN,WM & HOWARD,PH (1991)
OH 9.99E-011	cm3/molc sec	25	EST	MEYLAN,WM & HOWARD,PH (1993)

CAS #: 119005-20-8 — 2-AZETIDINONE, 4-[4-(1-OXO-3-PHENYLPROPYL)PHENOX

Formula: $C_{18}H_{17}NO_3$

Mol Weight: 295.34

MP (deg C): FP (deg C):

BP (deg C):

BP pressure (mm Hg):

Property/Value	Units	Temp	Data Type	Reference
WS 9.96E+001	mg/L	25	EST	MEYLAN,WM ET AL. (1996)
logP 2.48			EXP	SANGSTER,J (1993)
VP 9.78E-010	mm Hg	25	EST	NEELY,WB & BLAU,GE (1985)
DC	pKa			
HL 1.82E-013	atm m3/mol	25	EST	MEYLAN,WM & HOWARD,PH (1991)
OH 8.77E-011	cm3/molc sec	25	EST	MEYLAN,WM & HOWARD,PH (1993)

CAS #: 119005-21-9 — 2-AZETIDINONE, 4-(2-BENZOYLPHENOXY)-

Formula: $C_{16}H_{13}NO_3$

Mol Weight: 267.29

MP (deg C): FP (deg C):

BP (deg C):

BP pressure (mm Hg):

Property/Value	Units	Temp	Data Type	Reference
WS 5.18E+002	mg/L	25	EST	MEYLAN,WM ET AL. (1996)
logP 1.83			EXP	ARNOLDI,A ET AL. (1988)
VP 4.99E-009	mm Hg	25	EST	NEELY,WB & BLAU,GE (1985)
DC	pKa			
HL 3.37E-013	atm m3/mol	25	EST	MEYLAN,WM & HOWARD,PH (1991)
OH 8.01E-011	cm3/molc sec	25	EST	MEYLAN,WM & HOWARD,PH (1993)

CAS #: 119005-22-0 — 2-AZETIDINONE, 4-(2-NAPHTHALENYLTHIO)-

Formula: $C_{13}H_{11}NOS$

Mol Weight: 229.30

MP (deg C): FP (deg C):

BP (deg C):

BP pressure (mm Hg):

Property/Value	Units	Temp	Data Type	Reference
WS 1.71E+002	mg/L	25	EST	MEYLAN,WM ET AL. (1996)
logP 2.64			EXP	ARNOLDI,A ET AL. (1988)
VP 3.66E-008	mm Hg	25	EST	NEELY,WB & BLAU,GE (1985)
DC	pKa			
HL 1.05E-009	atm m3/mol	25	EST	MEYLAN,WM & HOWARD,PH (1991)
OH 2.13E-010	cm3/molc sec	25	EST	MEYLAN,WM & HOWARD,PH (1993)

CAS #: 119005-23-1 — 2-AZETIDINONE, 4-(2-PYRIDINYLTHIO)-

Formula: $C_8H_8N_2OS$

Mol Weight: 180.23

MP (deg C): FP (deg C):

BP (deg C):

BP pressure (mm Hg):

Property/Value	Units	Temp	Data Type	Reference
WS 1.66E+004	mg/L	25	EST	MEYLAN,WM ET AL. (1996)
logP 0.61			EXP	HANSCH,C ET AL. (1995)
VP 3.68E-006	mm Hg	25	EST	NEELY,WB & BLAU,GE (1985)
DC	pKa			
HL 5.50E-010	atm m3/mol	25	EST	MEYLAN,WM & HOWARD,PH (1991)
OH 7.59E-011	cm3/molc sec	25	EST	MEYLAN,WM & HOWARD,PH (1993)

CAS #: 119005-26-4 — 1-AZETIDINEACETIC ACID, 2-(2-NAPHTHALENYLOXY)-4-

Formula: $C_{17}H_{17}NO_4$

Mol Weight: 299.33

MP (deg C): FP (deg C):

BP (deg C):

BP pressure (mm Hg):

Property/Value	Units	Temp	Data Type	Reference
WS 5.03E+001	mg/L	25	EST	MEYLAN,WM ET AL. (1996)
logP 2.80			EXP	ARNOLDI,A ET AL. (1988)
VP 1.00E-008	mm Hg	25	EST	NEELY,WB & BLAU,GE (1985)
DC	pKa			
HL 1.26E-011	atm m3/mol	25	EST	MEYLAN,WM & HOWARD,PH (1991)
OH 2.73E-010	cm3/molc sec	25	EST	MEYLAN,WM & HOWARD,PH (1993)

CAS #: 119005-27-5 — 1-AZETIDINEACETIC ACID, 2-(2-NAPHTHALENYLOXY)-4-

Formula: $C_{19}H_{21}NO_4$

Mol Weight: 327.38

MP (deg C): FP (deg C):

BP (deg C):

BP pressure (mm Hg):

Property/Value	Units	Temp	Data Type	Reference
WS 7.13E+000	mg/L	25	EST	MEYLAN,WM ET AL. (1996)
logP 3.60			EXP	ARNOLDI,A ET AL. (1988)
VP 3.90E-009	mm Hg	25	EST	NEELY,WB & BLAU,GE (1985)
DC	pKa			
HL 2.23E-011	atm m3/mol	25	EST	MEYLAN,WM & HOWARD,PH (1991)
OH 2.72E-010	cm3/molc sec	25	EST	MEYLAN,WM & HOWARD,PH (1993)

CAS #: 119005-28-6	1-AZETIDINEACETIC ACID, 2-(4-BENZOYLPHENOXY)-4-O

Formula: $C_{20}H_{19}NO_5$

Mol Weight: 353.38

MP (deg C): FP (deg C):

BP (deg C):

BP pressure (mm Hg):

Property/Value	Units	Temp	Data Type	Reference
WS 1.62E+001	mg/L	25	EST	MEYLAN,WM ET AL. (1996)
logP 3.00			EXP	ARNOLDI,A ET AL. (1988)
VP 2.99E-010	mm Hg	25	EST	NEELY,WB & BLAU,GE (1985)
DC	pKa			
HL 4.67E-014	atm m3/mol	25	EST	MEYLAN,WM & HOWARD,PH (1991)
OH 9.41E-011	cm3/molc sec	25	EST	MEYLAN,WM & HOWARD,PH (1993)

CAS #: 119005-29-7	1-AZETIDINEACETIC ACID, 2-OXO-4-[4-(1-OXO-3-PHEN

Formula: $C_{22}H_{21}NO_5$

Mol Weight: 379.42

MP (deg C): FP (deg C):

BP (deg C):

BP pressure (mm Hg):

Property/Value	Units	Temp	Data Type	Reference
WS 6.23E+000	mg/L	25	EST	MEYLAN,WM ET AL. (1996)
logP 3.30			EXP	ARNOLDI,A ET AL. (1988)
VP 4.43E-011	mm Hg	25	EST	NEELY,WB & BLAU,GE (1985)
DC	pKa			
HL 5.57E-015	atm m3/mol	25	EST	MEYLAN,WM & HOWARD,PH (1991)
OH 1.14E-010	cm3/molc sec	25	EST	MEYLAN,WM & HOWARD,PH (1993)

CAS #: 119005-30-0	1-AZETIDINEACETIC ACID, 2-(2-NAPHTHALENYLTHIO)-4

Formula: $C_{17}H_{17}NO_3S$

Mol Weight: 315.39

MP (deg C): FP (deg C):

BP (deg C):

BP pressure (mm Hg):

Property/Value	Units	Temp	Data Type	Reference
WS 1.25E+001	mg/L	25	EST	MEYLAN,WM ET AL. (1996)
logP 3.40			EXP	ARNOLDI,A ET AL. (1988)
VP 2.13E-009	mm Hg	25	EST	NEELY,WB & BLAU,GE (1985)
DC	pKa			
HL 1.46E-010	atm m3/mol	25	EST	MEYLAN,WM & HOWARD,PH (1991)
OH 2.27E-010	cm3/molc sec	25	EST	MEYLAN,WM & HOWARD,PH (1993)

CAS #: 119018-06-3	2-THIOPHENESULFONAMIDE, 4-(4-METHYLBENZOYL)-

Formula: $C_{12}H_{11}NO_3S_2$

Mol Weight: 281.35

MP (deg C): FP (deg C):

BP (deg C):

BP pressure (mm Hg):

Property/Value	Units	Temp	Data Type	Reference
WS 2.00E+001	mg/L	25	EST	MEYLAN,WM ET AL. (1996)
logP 2.16			EXP	HARTMAN,GD ET AL. (1992)
VP 2.62E-008	mm Hg	25	EST	NEELY,WB & BLAU,GE (1985)
DC	pKa			
HL 9.13E-011	atm m3/mol	25	EST	MEYLAN,WM & HOWARD,PH (1991)
OH 5.64E-012	cm3/molc sec	25	EST	MEYLAN,WM & HOWARD,PH (1993)

CAS #: 119410-37-6	SPARSOMYCIN, S-PENTYL ANALOG

Formula: $C_{17}H_{27}N_3O_5S_2$

Mol Weight: 417.55

MP (deg C): FP (deg C):

BP (deg C):

BP pressure (mm Hg):

Property/Value	Units	Temp	Data Type	Reference
WS 7.32E+002	mg/L	25	EST	MEYLAN,WM ET AL. (1996)
logP 0.60			EXP	HANSCH,C ET AL. (1995)
VP 3.39E-022	mm Hg	25	EST	NEELY,WB & BLAU,GE (1985)
DC	pKa			
HL 5.18E-029	atm m3/mol	25	EST	MEYLAN,WM & HOWARD,PH (1991)
OH 1.84E-010	cm3/molc sec	25	EST	MEYLAN,WM & HOWARD,PH (1993)

CAS #: 119555-47-4	2(1H)-PYRIMIDINONE, 4-AMINO-1-(2,3-DIDEOXY-2-FLU

Formula: $C_9H_{12}FN_3O_3$

Mol Weight: 229.21

MP (deg C): FP (deg C):

BP (deg C):

BP pressure (mm Hg):

Property/Value	Units	Temp	Data Type	Reference
WS 1.53E+004	mg/L	25	EST	MEYLAN,WM ET AL. (1996)
logP -1.18			EXP	SANGSTER,J (1994)
VP 7.07E-008	mm Hg	25	EST	NEELY,WB & BLAU,GE (1985)
DC	pKa			
HL 2.33E-016	atm m3/mol	25	EST	MEYLAN,WM & HOWARD,PH (1991)
OH 6.93E-011	cm3/molc sec	25	EST	MEYLAN,WM & HOWARD,PH (1993)

CAS #: 119644-21-2	ADENOSINE, 2',3'-DIAZIDO-2',3'-DIDEOXY-

Formula: $C_{10}H_{13}N_{11}O_2$

Mol Weight: 319.29

MP (deg C): FP (deg C):

BP (deg C):

BP pressure (mm Hg):

Property/Value	Units	Temp	Data Type	Reference
WS 7.46E+002	mg/L	25	EST	MEYLAN,WM ET AL. (1996)
logP 0.80			EXP	BALZARINI,JM ET AL. (1989)
VP 4.65E-025	mm Hg	25	EST	NEELY,WB & BLAU,GE (1985)
DC	pKa			
HL 1.56E-026	atm m3/mol	25	EST	MEYLAN,WM & HOWARD,PH (1991)
OH 3.01E-011	cm3/molc sec	25	EST	MEYLAN,WM & HOWARD,PH (1993)

CAS #: 119644-22-3	URIDINE, 5-CHLORO-2',3'-DIDEOXY-3'-FLUORO-

Formula: $C_9H_{10}ClFN_2O_4$

Mol Weight: 264.64

MP (deg C): FP (deg C):

BP (deg C):

BP pressure (mm Hg):

Property/Value	Units	Temp	Data Type	Reference
WS 1.34E+003	mg/L	25	EST	MEYLAN,WM ET AL. (1996)
logP -0.17			EXP	BALZARINI,JM ET AL. (1989)
VP 2.38E-012	mm Hg	25	EST	NEELY,WB & BLAU,GE (1985)
DC	pKa			
HL 3.02E-016	atm m3/mol	25	EST	MEYLAN,WM & HOWARD,PH (1991)
OH 1.87E-011	cm3/molc sec	25	EST	MEYLAN,WM & HOWARD,PH (1993)

CAS #: 119644-23-4

URIDINE, 2',3'-DIDEOXY-3'-FLUORO-5-IODO-

Formula: $C_9H_{10}FIN_2O_4$

Mol Weight: 356.09

MP (deg C): FP (deg C):

BP (deg C):

BP pressure (mm Hg):

Property/Value	Units	Temp	Data Type	Reference
WS 1.84E+002	mg/L	25	EST	MEYLAN,WM ET AL. (1996)
logP 0.21			EXP	BALZARINI,JM ET AL. (1989)
VP 1.06E-013	mm Hg	25	EST	NEELY,WB & BLAU,GE (1985)
DC	pKa			
HL 5.28E-017	atm m3/mol	25	EST	MEYLAN,WM & HOWARD,PH (1991)
OH 1.96E-011	cm3/molc sec	25	EST	MEYLAN,WM & HOWARD,PH (1993)

CAS #: 119731-18-9

2-THIOPHENESULFONAMIDE, 4-[(4-METHOXYPHENYL)SULF

Formula: $C_{11}H_{11}NO_5S_3$

Mol Weight: 333.41

MP (deg C): FP (deg C):

BP (deg C):

BP pressure (mm Hg):

Property/Value	Units	Temp	Data Type	Reference
WS 4.78E+001	mg/L	25	EST	MEYLAN,WM ET AL. (1996)
logP 1.36			EXP	HARTMAN,GD ET AL. (1992)
VP 3.91E-010	mm Hg	25	EST	NEELY,WB & BLAU,GE (1985)
DC	pKa			
HL 6.28E-013	atm m3/mol	25	EST	MEYLAN,WM & HOWARD,PH (1991)
OH 5.86E-012	cm3/molc sec	25	EST	MEYLAN,WM & HOWARD,PH (1993)

CAS #: 119738-06-6

QUIZALOFOP-P-TEFURYL

Formula: $C_{22}H_{21}ClN_2O_5$

Mol Weight: 428.88

MP (deg C): FP (deg C):

BP (deg C):

BP pressure (mm Hg):

Property/Value	Units	Temp	Data Type	Reference
WS 4.14E-001	mg/L	25	EST	MEYLAN,WM ET AL. (1996)
logP 4.32			EXP	TOMLIN,C (1994)
VP 4.32E-011	mm Hg	25	EST	NEELY,WB & BLAU,GE (1985)
DC	pKa			
HL 8.47E-014	atm m3/mol	25	EST	MEYLAN,WM & HOWARD,PH (1991)
OH 5.80E-011	cm3/molc sec	25	EST	MEYLAN,WM & HOWARD,PH (1993)

CAS #: 120137-90-8

[3,4'-BIPYRIDINE]-5-CARBONITRILE,6-(ETHYLAMINO)-

Formula: $C_{13}H_{12}N_4$

Mol Weight: 224.27

MP (deg C): FP (deg C):

BP (deg C):

BP pressure (mm Hg):

Property/Value	Units	Temp	Data Type	Reference
WS 1.82E+002	mg/L	25	EST	MEYLAN,WM ET AL. (1996)
logP 2.33			EXP	HAGEN,V ET AL. (1989)
VP 8.95E-007	mm Hg	25	EST	NEELY,WB & BLAU,GE (1985)
DC	pKa			
HL 7.05E-015	atm m3/mol	25	EST	MEYLAN,WM & HOWARD,PH (1991)
OH 9.46E-012	cm3/molc sec	25	EST	MEYLAN,WM & HOWARD,PH (1993)

CAS #: 120137-94-2

[3,4'-BIPYRIDINE]-5-CARBONITRILE,6-(2-PROPENYLAM

Formula: $C_{14}H_{12}N_4$

Mol Weight: 236.28

MP (deg C): FP (deg C):

BP (deg C):

BP pressure (mm Hg):

Property/Value	Units	Temp	Data Type	Reference
WS 2.61E+002	mg/L	25	EST	MEYLAN,WM ET AL. (1996)
logP 2.07			EXP	HAGEN,V ET AL. (1989)
VP 4.11E-007	mm Hg	25	EST	NEELY,WB & BLAU,GE (1985)
DC	pKa			
HL 6.98E-015	atm m3/mol	25	EST	MEYLAN,WM & HOWARD,PH (1991)
OH 3.56E-011	cm3/molc sec	25	EST	MEYLAN,WM & HOWARD,PH (1993)

CAS #: 120137-95-3

[3,4'-BIPYRIDINE]-5-CARBONITRILE,6-(1-PYRROLIDIN

Formula: $C_{15}H_{14}N_4$

Mol Weight: 250.31

MP (deg C): FP (deg C):

BP (deg C):

BP pressure (mm Hg):

Property/Value	Units	Temp	Data Type	Reference
WS 3.87E+001	mg/L	25	EST	MEYLAN,WM ET AL. (1996)
logP 2.95			EXP	HAGEN,V ET AL. (1989)
VP 1.50E-007	mm Hg	25	EST	NEELY,WB & BLAU,GE (1985)
DC	pKa			
HL 8.46E-014	atm m3/mol	25	EST	MEYLAN,WM & HOWARD,PH (1991)
OH 2.00E-011	cm3/molc sec	25	EST	MEYLAN,WM & HOWARD,PH (1993)

CAS #: 120137-97-5

[3,4'-BIPYRIDINE]-5-CARBONITRILE,6-[(3-METHOXYPR

Formula: $C_{15}H_{16}N_4O$

Mol Weight: 268.32

MP (deg C): FP (deg C):

BP (deg C):

BP pressure (mm Hg):

Property/Value	Units	Temp	Data Type	Reference
WS 1.87E+002	mg/L	25	EST	MEYLAN,WM ET AL. (1996)
logP 2.03			EXP	HAGEN,V ET AL. (1990)
VP 6.38E-008	mm Hg	25	EST	NEELY,WB & BLAU,GE (1985)
DC	pKa			
HL 1.10E-016	atm m3/mol	25	EST	MEYLAN,WM & HOWARD,PH (1991)
OH 2.32E-011	cm3/molc sec	25	EST	MEYLAN,WM & HOWARD,PH (1993)

CAS #: 120182-07-2

2-OXAZOLAMINE, 4,5-DIHYDRO-5-[[4-(2-PYRIDINYL)-1

Formula: $C_{13}H_{19}N_5O$

Mol Weight: 261.33

MP (deg C): FP (deg C):

BP (deg C):

BP pressure (mm Hg):

Property/Value	Units	Temp	Data Type	Reference
WS 7.96E+003	mg/L	25	EST	MEYLAN,WM ET AL. (1996)
logP 0.48			EXP	DEMOTES-MAINARD,F ET AL. (1993)
VP 5.97E-007	mm Hg	25	EST	NEELY,WB & BLAU,GE (1985)
DC	pKa			
HL 1.51E-018	atm m3/mol	25	EST	MEYLAN,WM & HOWARD,PH (1991)
OH 1.75E-010	cm3/molc sec	25	EST	MEYLAN,WM & HOWARD,PH (1993)

CAS #: 120182-20-9 — 2-OXAZOLAMINE, 4,5-DIHYDRO-5-[[4-(2-PYRIMIDINYL)

Formula: $C_{12}H_{18}N_6O$

Mol Weight: 262.32

MP (deg C): | FP (deg C):

BP (deg C):

BP pressure (mm Hg):

Property/Value	Units	Temp	Data Type	Reference
WS 1.34E+004	mg/L	25	EST	MEYLAN,WM ET AL. (1996)
logP 0.21			EXP	DEMOTES-MAINARD,F ET AL. (1993)
VP 4.13E-007	mm Hg	25	EST	NEELY,WB & BLAU,GE (1985)
DC	pKa			
HL 2.22E-015	atm m3/mol	25	EST	MEYLAN,WM & HOWARD,PH (1991)
OH 1.64E-010	cm3/molc sec	25	EST	MEYLAN,WM & HOWARD,PH (1993)

CAS #: 120236-83-1 — ISOMAZOLE,PYRIDAZINE ANALOG

Formula: $C_{13}H_{12}N_4O_2S$

Mol Weight: 288.33

MP (deg C): | FP (deg C):

BP (deg C):

BP pressure (mm Hg):

Property/Value	Units	Temp	Data Type	Reference
WS 4.47E+002	mg/L	25	EST	MEYLAN,WM ET AL. (1996)
logP 0.23			EXP	HANSCH,C ET AL. (1995)
VP 4.11E-013	mm Hg	25	EST	NEELY,WB & BLAU,GE (1985)
DC	pKa			
HL 1.63E-018	atm m3/mol	25	EST	MEYLAN,WM & HOWARD,PH (1991)
OH 1.28E-010	cm3/molc sec	25	EST	MEYLAN,WM & HOWARD,PH (1993)

CAS #: 120236-84-2 — ISOMAZOLE,(5,6)PYRIDAZINE ANALOG

Formula: $C_{13}H_{12}N_4O_2S$

Mol Weight: 288.33

MP (deg C): | FP (deg C):

BP (deg C):

BP pressure (mm Hg):

Property/Value	Units	Temp	Data Type	Reference
WS 6.67E+003	mg/L	25	EST	MEYLAN,WM ET AL. (1996)
logP 0.39			EXP	HANSCH,C ET AL. (1995)
VP 4.11E-013	mm Hg	25	EST	NEELY,WB & BLAU,GE (1985)
DC	pKa			
HL 1.63E-018	atm m3/mol	25	EST	MEYLAN,WM & HOWARD,PH (1991)
OH 1.28E-010	cm3/molc sec	25	EST	MEYLAN,WM & HOWARD,PH (1993)

CAS #: 120236-85-3 — ISOMAZOLE,PYRAZINE ANALOG

Formula: $C_{13}H_{12}N_4O_2S$

Mol Weight: 288.33

MP (deg C): | FP (deg C):

BP (deg C):

BP pressure (mm Hg):

Property/Value	Units	Temp	Data Type	Reference
WS 3.92E+003	mg/L	25	EST	MEYLAN,WM ET AL. (1996)
logP 0.66			EXP	HANSCH,C ET AL. (1995)
VP 6.67E-012	mm Hg	25	EST	NEELY,WB & BLAU,GE (1985)
DC	pKa			
HL 1.66E-018	atm m3/mol	25	EST	MEYLAN,WM & HOWARD,PH (1991)
OH 1.28E-010	cm3/molc sec	25	EST	MEYLAN,WM & HOWARD,PH (1993)

CAS #: 120236-88-6 — 1H-PYRROLO[3,2-C]PYRIDINE, 2-(2,4-DIMETHOXYPHENY

Formula: $C_{15}H_{14}N_2O_2$

Mol Weight: 254.29

MP (deg C): | FP (deg C):

BP (deg C):

BP pressure (mm Hg):

Property/Value	Units	Temp	Data Type	Reference
WS 2.30E+001	mg/L	25	EST	MEYLAN,WM ET AL. (1996)
logP 3.50			EXP	HANSCH,C ET AL. (1995)
VP 9.57E-008	mm Hg	25	EST	NEELY,WB & BLAU,GE (1985)
DC	pKa			
HL 3.12E-013	atm m3/mol	25	EST	MEYLAN,WM & HOWARD,PH (1991)
OH 3.62E-010	cm3/molc sec	25	EST	MEYLAN,WM & HOWARD,PH (1993)

CAS #: 120277-90-9 — 7-AZABICYCLO[4.1.0]HEPTANE, 7-[2-HYDROXY-3-(1H-I

Formula: $C_{12}H_{19}N_3O$

Mol Weight: 221.30

MP (deg C): | FP (deg C):

BP (deg C):

BP pressure (mm Hg):

Property/Value	Units	Temp	Data Type	Reference
WS 4.21E+004	mg/L	25	EST	MEYLAN,WM ET AL. (1996)
logP 1.07			EXP	SUTO,MJ ET AL. (1991)
VP 4.93E-008	mm Hg	25	EST	NEELY,WB & BLAU,GE (1985)
DC	pKa			
HL 1.55E-012	atm m3/mol	25	EST	MEYLAN,WM & HOWARD,PH (1991)
OH 1.59E-010	cm3/molc sec	25	EST	MEYLAN,WM & HOWARD,PH (1993)

CAS #: 120277-93-2 — 6-AZABICYCLO[3.1.0]HEXANE, 6-[2-HYDROXY-3-(1H-IM

Formula: $C_{11}H_{17}N_3O$

Mol Weight: 207.28

MP (deg C): | FP (deg C):

BP (deg C):

BP pressure (mm Hg):

Property/Value	Units	Temp	Data Type	Reference
WS 1.98E+005	mg/L	25	EST	MEYLAN,WM ET AL. (1996)
logP 0.37			EXP	SUTO,MJ ET AL. (1991)
VP 1.39E-007	mm Hg	25	EST	NEELY,WB & BLAU,GE (1985)
DC	pKa			
HL 1.16E-012	atm m3/mol	25	EST	MEYLAN,WM & HOWARD,PH (1991)
OH 1.35E-010	cm3/molc sec	25	EST	MEYLAN,WM & HOWARD,PH (1993)

CAS #: 120277-94-3 — 7-AZABICYCLO[4.1.0]HEPTANE, 7-[2-HYDROXY-3-(1H-I

Formula: $C_{13}H_{21}N_3O$

Mol Weight: 235.33

MP (deg C): | FP (deg C):

BP (deg C):

BP pressure (mm Hg):

Property/Value	Units	Temp	Data Type	Reference
WS 1.38E+004	mg/L	25	EST	MEYLAN,WM ET AL. (1996)
logP 1.55			EXP	SANGSTER,J (1994)
VP 2.96E-008	mm Hg	25	EST	NEELY,WB & BLAU,GE (1985)
DC	pKa			
HL 2.05E-012	atm m3/mol	25	EST	MEYLAN,WM & HOWARD,PH (1991)
OH 1.60E-010	cm3/molc sec	25	EST	MEYLAN,WM & HOWARD,PH (1993)

CAS #: 120277-95-4				8-AZABICYCLO[5.1.0]OCTANE, 8-[2-HYDROXY-3-(1H-IM
Formula: $C_{13}H_{21}N_3O$				
Mol Weight: 235.33				
MP (deg C):		FP (deg C):		
BP (deg C):				
BP pressure (mm Hg):				

Property/Value	Units	Temp	Data Type	Reference
WS 1.71E+004	mg/L	25	EST	MEYLAN,WM ET AL. (1996)
logP 1.44			EXP	SANGSTER,J (1994)
VP 2.08E-008	mm Hg	25	EST	NEELY,WB & BLAU,GE (1985)
DC	pKa			
HL 2.05E-012	atm m3/mol	25	EST	MEYLAN,WM & HOWARD,PH (1991)
OH 1.82E-010	cm3/molc sec	25	EST	MEYLAN,WM & HOWARD,PH (1993)

CAS #: 120277-97-6				3-OXA-7-AZABICYCLO[4.1.0]HEPTANE, 7-[2-HYDROXY-3
Formula: $C_{11}H_{17}N_3O_2$				
Mol Weight: 223.28				
MP (deg C):		FP (deg C):		
BP (deg C):				
BP pressure (mm Hg):				

Property/Value	Units	Temp	Data Type	Reference
WS 2.94E+005	mg/L	25	EST	MEYLAN,WM ET AL. (1996)
logP 0.07			EXP	SUTO,MJ ET AL. (1991)
VP 4.25E-008	mm Hg	25	EST	NEELY,WB & BLAU,GE (1985)
DC	pKa			
HL 1.29E-015	atm m3/mol	25	EST	MEYLAN,WM & HOWARD,PH (1991)
OH 1.40E-010	cm3/molc sec	25	EST	MEYLAN,WM & HOWARD,PH (1993)

CAS #: 120282-84-0				BENZENEACETAMIDE,alpha-HYDROXY-N-TRICYCLO[3.3.1.
Formula: $C_{18}H_{23}NO_2$				
Mol Weight: 285.39				
MP (deg C):		FP (deg C):		
BP (deg C):				
BP pressure (mm Hg):				

Property/Value	Units	Temp	Data Type	Reference
WS 3.43E+001	mg/L	25	EST	MEYLAN,WM ET AL. (1996)
logP 3.09			EXP	SANGSTER,J (1993)
VP 9.70E-011	mm Hg	25	EST	NEELY,WB & BLAU,GE (1985)
DC	pKa			
HL 3.39E-010	atm m3/mol	25	EST	MEYLAN,WM & HOWARD,PH (1991)
OH 3.47E-011	cm3/molc sec	25	EST	MEYLAN,WM & HOWARD,PH (1993)

CAS #: 120282-85-1				ADAMANTYL DERIVATIVE
Formula: $C_{21}H_{27}NO_3$				
Mol Weight: 341.45				
MP (deg C):		FP (deg C):		
BP (deg C):				
BP pressure (mm Hg):				

Property/Value	Units	Temp	Data Type	Reference
WS 4.64E+000	mg/L	25	EST	MEYLAN,WM ET AL. (1996)
logP 3.72			EXP	SANGSTER,J (1993)
VP 1.81E-009	mm Hg	25	EST	NEELY,WB & BLAU,GE (1985)
DC	pKa			
HL 4.87E-011	atm m3/mol	25	EST	MEYLAN,WM & HOWARD,PH (1991)
OH 3.26E-011	cm3/molc sec	25	EST	MEYLAN,WM & HOWARD,PH (1993)

CAS #: 120309-36-6				ACETAMIDE, N-(2-BUTOXY-5-NITROPHENYL)-
Formula: $C_{12}H_{16}N_2O_4$				
Mol Weight: 252.27				
MP (deg C):		FP (deg C):		
BP (deg C):				
BP pressure (mm Hg):				

Property/Value	Units	Temp	Data Type	Reference
WS 1.10E+002	mg/L	25	EST	MEYLAN,WM ET AL. (1996)
logP 2.26			EXP	FURST,W & BECHER,M (1990)
VP 7.43E-008	mm Hg	25	EST	NEELY,WB & BLAU,GE (1985)
DC	pKa			
HL 3.37E-012	atm m3/mol	25	EST	MEYLAN,WM & HOWARD,PH (1991)
OH 1.39E-011	cm3/molc sec	25	EST	MEYLAN,WM & HOWARD,PH (1993)

CAS #: 120309-46-8				1H-PYRROLE-2-CARBOXAMIDE, N-PROPYL-
Formula: $C_8H_{12}N_2O$				
Mol Weight: 152.20				
MP (deg C):		FP (deg C):		
BP (deg C):				
BP pressure (mm Hg):				

Property/Value	Units	Temp	Data Type	Reference
WS 6.35E+003	mg/L	25	EST	MEYLAN,WM ET AL. (1996)
logP 1.25			EXP	YAMAGAMI,C ET AL. (1994)
VP 2.51E-005	mm Hg	25	EST	NEELY,WB & BLAU,GE (1985)
DC	pKa			
HL 1.44E-011	atm m3/mol	25	EST	MEYLAN,WM & HOWARD,PH (1991)
OH 1.12E-010	cm3/molc sec	25	EST	MEYLAN,WM & HOWARD,PH (1993)

CAS #: 120399-49-7				L-VALINAMIDE, N-BENZOYLGLYCYL-N-ETHYL-
Formula: $C_{16}H_{23}N_3O_3$				
Mol Weight: 305.38				
MP (deg C):		FP (deg C):		
BP (deg C):				
BP pressure (mm Hg):				

Property/Value	Units	Temp	Data Type	Reference
WS 1.45E+003	mg/L	25	EST	MEYLAN,WM ET AL. (1996)
logP 1.05			EXP	SANGSTER,J (1994)
VP 1.07E-012	mm Hg	25	EST	NEELY,WB & BLAU,GE (1985)
DC	pKa			
HL 1.06E-015	atm m3/mol	25	EST	MEYLAN,WM & HOWARD,PH (1991)
OH 4.04E-011	cm3/molc sec	25	EST	MEYLAN,WM & HOWARD,PH (1993)

CAS #: 120399-50-0				ACETAMIDE, 2-(BENZOYLAMINO)-N-ETHYL-
Formula: $C_{11}H_{14}N_2O_2$				
Mol Weight: 206.25				
MP (deg C):		FP (deg C):		
BP (deg C):				
BP pressure (mm Hg):				

Property/Value	Units	Temp	Data Type	Reference
WS 1.30E+004	mg/L	25	EST	MEYLAN,WM ET AL. (1996)
logP 0.58			EXP	LEAHY,DE ET AL. (1989)
VP 2.31E-008	mm Hg	25	EST	NEELY,WB & BLAU,GE (1985)
DC	pKa			
HL 6.81E-011	atm m3/mol	25	EST	MEYLAN,WM & HOWARD,PH (1991)
OH 1.96E-011	cm3/molc sec	25	EST	MEYLAN,WM & HOWARD,PH (1993)

CAS #: 120503-34-6		2-FURANMETHANOL, 5-(6-CHLORO-9H-PURIN-9-YL)TETRA

Formula: $C_{10}H_{11}ClN_4O_2$

Mol Weight: 254.68

MP (deg C):   FP (deg C):

BP (deg C):

BP pressure (mm Hg):

Property/Value	Units	Temp	Data Type	Reference
WS 6.81E+002	mg/L	25	EST	MEYLAN,WM ET AL. (1996)
logP 0.24			EXP	MURAKAMI,K ET AL. (1991)
VP 2.46E-010	mm Hg	25	EST	NEELY,WB & BLAU,GE (1985)
DC	pKa			
HL 1.10E-014	atm m3/mol	25	EST	MEYLAN,WM & HOWARD,PH (1991)
OH 5.77E-011	cm3/molc sec	25	EST	MEYLAN,WM & HOWARD,PH (1993)

CAS #: 120503-37-9		2-FURANMETHANOL, TETRAHYDRO-5-(6-IODO-9H-PURIN-9

Formula: $C_{10}H_{11}IN_4O_2$

Mol Weight: 346.13

MP (deg C):   FP (deg C):

BP (deg C):

BP pressure (mm Hg):

Property/Value	Units	Temp	Data Type	Reference
WS 2.31E+003	mg/L	25	EST	MEYLAN,WM ET AL. (1996)
logP 0.53			EXP	MURAKAMI,K ET AL. (1991)
VP 2.07E-011	mm Hg	25	EST	NEELY,WB & BLAU,GE (1985)
DC	pKa			
HL 5.46E-017	atm m3/mol	25	EST	MEYLAN,WM & HOWARD,PH (1991)
OH 5.61E-011	cm3/molc sec	25	EST	MEYLAN,WM & HOWARD,PH (1993)

CAS #: 120908-61-4		2-PROPANOL, 1-[[2-HYDROXY-3-(1-NAPHTHALENYLOXY)P

Formula: $C_{22}H_{25}NO_3$

Mol Weight: 351.45

MP (deg C):   FP (deg C):

BP (deg C):

BP pressure (mm Hg):

Property/Value	Units	Temp	Data Type	Reference
WS 3.06E+001	mg/L	25	EST	MEYLAN,WM ET AL. (1996)
logP 3.87			EXP	RECANATINI,M (1992)
VP 2.80E-013	mm Hg	25	EST	NEELY,WB & BLAU,GE (1985)
DC	pKa			
HL 2.36E-015	atm m3/mol	25	EST	MEYLAN,WM & HOWARD,PH (1991)
OH 1.89E-010	cm3/molc sec	25	EST	MEYLAN,WM & HOWARD,PH (1993)

CAS #: 120908-62-5		2-PROPANOL, 1-[[2-HYDROXY-3-(4-NITROPHENOXY)PROP

Formula: $C_{18}H_{22}N_2O_6$

Mol Weight: 362.39

MP (deg C):   FP (deg C):

BP (deg C):

BP pressure (mm Hg):

Property/Value	Units	Temp	Data Type	Reference
WS 1.25E+002	mg/L	25	EST	MEYLAN,WM ET AL. (1996)
logP 2.62			EXP	HANSCH,C ET AL. (1995)
VP 3.28E-013	mm Hg	25	EST	NEELY,WB & BLAU,GE (1985)
DC	pKa			
HL 5.10E-018	atm m3/mol	25	EST	MEYLAN,WM & HOWARD,PH (1991)
OH 1.43E-010	cm3/molc sec	25	EST	MEYLAN,WM & HOWARD,PH (1993)

CAS #: 120908-63-6		ACETAMIDE, N-[4-[2-HYDROXY-3-[(2-HYDROXY-3-PHENO

Formula: $C_{20}H_{26}N_2O_5$

Mol Weight: 374.44

MP (deg C):   FP (deg C):

BP (deg C):

BP pressure (mm Hg):

Property/Value	Units	Temp	Data Type	Reference
WS 2.86E+002	mg/L	25	EST	MEYLAN,WM ET AL. (1996)
logP 1.39			EXP	RECANATINI,M (1992)
VP 1.28E-015	mm Hg	25	EST	NEELY,WB & BLAU,GE (1985)
DC	pKa			
HL 1.48E-021	atm m3/mol	25	EST	MEYLAN,WM & HOWARD,PH (1991)
OH 1.52E-010	cm3/molc sec	25	EST	MEYLAN,WM & HOWARD,PH (1993)

CAS #: 120928-09-8		FENAZAQUIN

Formula: $C_{20}H_{22}N_2O$

Mol Weight: 306.41

MP (deg C):   FP (deg C):

BP (deg C):

BP pressure (mm Hg):

Property/Value	Units	Temp	Data Type	Reference
WS 2.22E-001	mg/L	25	EST	MEYLAN,WM ET AL. (1996)
logP 5.51			EXP	TOMLIN,C (1994)
VP 7.36E-008	mm Hg	25	EST	NEELY,WB & BLAU,GE (1985)
DC	pKa			
HL 4.25E-008	atm m3/mol	25	EST	MEYLAN,WM & HOWARD,PH (1991)
OH 3.86E-011	cm3/molc sec	25	EST	MEYLAN,WM & HOWARD,PH (1993)

CAS #: 121009-77-6		SIMVASTATIN, 5-HYDROXY

Formula: $C_{25}H_{40}O_6$

Mol Weight: 436.59

MP (deg C):   FP (deg C):

BP (deg C):

BP pressure (mm Hg):

Property/Value	Units	Temp	Data Type	Reference
WS 6.84E-001	mg/L	25	EST	MEYLAN,WM ET AL. (1996)
logP 4.47			EXP	HANSCH,C ET AL. (1995)
VP 1.42E-014	mm Hg	25	EST	NEELY,WB & BLAU,GE (1985)
DC	pKa			
HL 9.68E-014	atm m3/mol	25	EST	MEYLAN,WM & HOWARD,PH (1991)
OH 2.36E-010	cm3/molc sec	25	EST	MEYLAN,WM & HOWARD,PH (1993)

CAS #: 121107-43-5		2,2'-4,4'-TETRACHLORO-3-METHYL DIPHENYLMETHANE

Formula: $C_{14}H_{10}Cl_4$

Mol Weight: 320.05

MP (deg C):   FP (deg C):

BP (deg C):

BP pressure (mm Hg):

Property/Value	Units	Temp	Data Type	Reference
WS 6.65E-003	mg/L	25	EST	MEYLAN,WM ET AL. (1996)
logP 7.20			EXP	VAN HAELST,AG ET AL. (1994)
VP 4.38E-006	mm Hg	25	EST	NEELY,WB & BLAU,GE (1985)
DC	pKa			
HL 1.60E-004	atm m3/mol	25	EST	MEYLAN,WM & HOWARD,PH (1991)
OH 3.69E-012	cm3/molc sec	25	EST	MEYLAN,WM & HOWARD,PH (1993)

CAS #: 121107-44-6	2,2'-4,4'-TETRACHLORO-5-METHYL DIPHENYLMETHANE

Formula: $C_{14}H_{10}Cl_4$

Mol Weight: 320.05

MP (deg C):　　FP (deg C):

BP (deg C):

BP pressure (mm Hg):

Property/Value	Units	Temp	Data Type	Reference
WS 4.23E-003	mg/L	25	EST	MEYLAN,WM ET AL. (1996)
logP 7.43			EXP	VAN HAELST,AG ET AL. (1994)
VP 4.38E-006	mm Hg	25	EST	NEELY,WB & BLAU,GE (1985)
DC	pKa			
HL 1.60E-004	atm m3/mol	25	EST	MEYLAN,WM & HOWARD,PH (1991)
OH 3.69E-012	cm3/molc sec	25	EST	MEYLAN,WM & HOWARD,PH (1993)

CAS #: 121107-46-8	2,2'-4,5'-TETRACHLORO-5-METHYL DIPHENYLMETHANE

Formula: $C_{14}H_{10}Cl_4$

Mol Weight: 320.05

MP (deg C):　　FP (deg C):

BP (deg C):

BP pressure (mm Hg):

Property/Value	Units	Temp	Data Type	Reference
WS 3.47E-003	mg/L	25	EST	MEYLAN,WM ET AL. (1996)
logP 7.53			EXP	VAN HAELST,AG ET AL. (1994)
VP 4.38E-006	mm Hg	25	EST	NEELY,WB & BLAU,GE (1985)
DC	pKa			
HL 1.60E-004	atm m3/mol	25	EST	MEYLAN,WM & HOWARD,PH (1991)
OH 3.51E-012	cm3/molc sec	25	EST	MEYLAN,WM & HOWARD,PH (1993)

CAS #: 121107-47-9	2,2'-4,6'-TETRACHLORO-3-METHYL DIPHENYLMETHANE

Formula: $C_{14}H_{10}Cl_4$

Mol Weight: 320.05

MP (deg C):　　FP (deg C):

BP (deg C):

BP pressure (mm Hg):

Property/Value	Units	Temp	Data Type	Reference
WS 6.65E-003	mg/L	25	EST	MEYLAN,WM ET AL. (1996)
logP 7.20			EXP	VAN HAELST,AG ET AL. (1994)
VP 4.38E-006	mm Hg	25	EST	NEELY,WB & BLAU,GE (1985)
DC	pKa			
HL 1.60E-004	atm m3/mol	25	EST	MEYLAN,WM & HOWARD,PH (1991)
OH 3.69E-012	cm3/molc sec	25	EST	MEYLAN,WM & HOWARD,PH (1993)

CAS #: 121107-48-0	2,2'-4,6'-TETRACHLORO-5-METHYL DIPHENYLMETHANE

Formula: $C_{14}H_{10}Cl_4$

Mol Weight: 320.05

MP (deg C):　　FP (deg C):

BP (deg C):

BP pressure (mm Hg):

Property/Value	Units	Temp	Data Type	Reference
WS 1.67E-002	mg/L	25	EST	MEYLAN,WM ET AL. (1996)
logP 6.73			EXP	VAN HAELST,AG ET AL. (1994)
VP 4.38E-006	mm Hg	25	EST	NEELY,WB & BLAU,GE (1985)
DC	pKa			
HL 1.60E-004	atm m3/mol	25	EST	MEYLAN,WM & HOWARD,PH (1991)
OH 3.69E-012	cm3/molc sec	25	EST	MEYLAN,WM & HOWARD,PH (1993)

CAS #: 121107-55-9	2,2'-5,5'-TETRACHLORO-4-METHYL DIPHENYLMETHANE

Formula: $C_{14}H_{10}Cl_4$

Mol Weight: 320.05

MP (deg C):　　FP (deg C):

BP (deg C):

BP pressure (mm Hg):

Property/Value	Units	Temp	Data Type	Reference
WS 3.83E-003	mg/L	25	EST	MEYLAN,WM ET AL. (1996)
logP 7.48			EXP	VAN HAELST,AG ET AL. (1994)
VP 4.38E-006	mm Hg	25	EST	NEELY,WB & BLAU,GE (1985)
DC	pKa			
HL 1.60E-004	atm m3/mol	25	EST	MEYLAN,WM & HOWARD,PH (1991)
OH 3.34E-012	cm3/molc sec	25	EST	MEYLAN,WM & HOWARD,PH (1993)

CAS #: 121107-65-1	2,3'-4,4'-TETRACHLORO-5-METHYL DIPHENYLMETHANE

Formula: $C_{14}H_{10}Cl_4$

Mol Weight: 320.05

MP (deg C):　　FP (deg C):

BP (deg C):

BP pressure (mm Hg):

Property/Value	Units	Temp	Data Type	Reference
WS 5.91E-003	mg/L	25	EST	MEYLAN,WM ET AL. (1996)
logP 7.26			EXP	VAN HAELST,AG ET AL. (1994)
VP 4.38E-006	mm Hg	25	EST	NEELY,WB & BLAU,GE (1985)
DC	pKa			
HL 1.60E-004	atm m3/mol	25	EST	MEYLAN,WM & HOWARD,PH (1991)
OH 3.51E-012	cm3/molc sec	25	EST	MEYLAN,WM & HOWARD,PH (1993)

CAS #: 121107-77-5	2,3'-4,4'-TETRACHLORO-6-METHYL DIPHENYLMETHANE

Formula: $C_{14}H_{10}Cl_4$

Mol Weight: 320.05

MP (deg C):　　FP (deg C):

BP (deg C):

BP pressure (mm Hg):

Property/Value	Units	Temp	Data Type	Reference
WS 4.40E-003	mg/L	25	EST	MEYLAN,WM ET AL. (1996)
logP 7.41			EXP	VAN HAELST,AG ET AL. (1994)
VP 4.38E-006	mm Hg	25	EST	NEELY,WB & BLAU,GE (1985)
DC	pKa			
HL 1.60E-004	atm m3/mol	25	EST	MEYLAN,WM & HOWARD,PH (1991)
OH 3.52E-012	cm3/molc sec	25	EST	MEYLAN,WM & HOWARD,PH (1993)

CAS #: 121107-83-3	2',3-4,6'-TETRACHLORO-6-METHYL DIPHENYLMETHANE

Formula: $C_{14}H_{10}Cl_4$

Mol Weight: 320.05

MP (deg C):　　FP (deg C):

BP (deg C):

BP pressure (mm Hg):

Property/Value	Units	Temp	Data Type	Reference
WS 7.33E-003	mg/L	25	EST	MEYLAN,WM ET AL. (1996)
logP 7.15			EXP	VAN HAELST,AG ET AL. (1994)
VP 4.38E-006	mm Hg	25	EST	NEELY,WB & BLAU,GE (1985)
DC	pKa			
HL 1.60E-004	atm m3/mol	25	EST	MEYLAN,WM & HOWARD,PH (1991)
OH 3.52E-012	cm3/molc sec	25	EST	MEYLAN,WM & HOWARD,PH (1993)

121284-18-2 — 1,4-BIS(2-(2-MEO-ETO)-ETO)BENZENE

Formula: $C_{16}H_{26}O_6$
Mol Weight: 314.38
MP (deg C):
FP (deg C):
BP (deg C):
BP pressure (mm Hg):

Property/Value	Units	Temp	Data Type	Reference
WS 2.27E+003	mg/L	25	EST	MEYLAN,WM ET AL. (1996)
logP 0.76			EXP	STOLWIJK,TB ET AL. (1989)
VP 3.27E-006	mm Hg	25	EST	NEELY,WB & BLAU,GE (1985)
DC	pKa			
HL 1.10E-012	atm m3/mol	25	EST	MEYLAN,WM & HOWARD,PH (1991)
OH 7.63E-011	cm3/molc sec	25	EST	MEYLAN,WM & HOWARD,PH (1993)

121284-19-3 — 1,3-BIS(2-(2-MEO-ETO)-ETO)BENZENE

Formula: $C_{16}H_{26}O_6$
Mol Weight: 314.38
MP (deg C):
FP (deg C):
BP (deg C):
BP pressure (mm Hg):

Property/Value	Units	Temp	Data Type	Reference
WS 1.86E+003	mg/L	25	EST	MEYLAN,WM ET AL. (1996)
logP 0.86			EXP	STOLWIJK,TB ET AL. (1989)
VP 3.27E-006	mm Hg	25	EST	NEELY,WB & BLAU,GE (1985)
DC	pKa			
HL 1.10E-012	atm m3/mol	25	EST	MEYLAN,WM & HOWARD,PH (1991)
OH 2.58E-010	cm3/molc sec	25	EST	MEYLAN,WM & HOWARD,PH (1993)

121284-20-6 — [8,8]DB48C16-DIBENZO CROWN ETHER

Formula: $C_{40}H_{64}O_{16}$
Mol Weight: 800.95
MP (deg C):
FP (deg C):
BP (deg C):
BP pressure (mm Hg):

Property/Value	Units	Temp	Data Type	Reference
WS 2.66E+000	mg/L	25	EST	MEYLAN,WM ET AL. (1996)
logP 0.52			EXP	STOLWIJK,TB ET AL. (1989)
VP 9.85E-021	mm Hg	25	EST	NEELY,WB & BLAU,GE (1985)
DC	pKa			
HL 5.60E-031	atm m3/mol	25	EST	MEYLAN,WM & HOWARD,PH (1991)
OH 2.33E-010	cm3/molc sec	25	EST	MEYLAN,WM & HOWARD,PH (1993)

121325-44-8 — RH-5348 DIPH ETHER

Formula: $C_{15}H_9ClF_3NO_5$
Mol Weight: 375.69
MP (deg C):
FP (deg C):
BP (deg C):
BP pressure (mm Hg):

Property/Value	Units	Temp	Data Type	Reference
WS 4.47E-001	mg/L	25	EST	MEYLAN,WM ET AL. (1996)
logP 4.21			EXP	NANDIHALLI,UB ET AL. (1993)
VP 1.66E-007	mm Hg	25	EST	NEELY,WB & BLAU,GE (1985)
DC	pKa			
HL 1.93E-008	atm m3/mol	25	EST	MEYLAN,WM & HOWARD,PH (1991)
OH 6.29E-013	cm3/molc sec	25	EST	MEYLAN,WM & HOWARD,PH (1993)

121508-87-0 — BENZENEACETAMIDE, N-[2-[[3-(PHENYLMETHOXY)-2-HYD

Formula: $C_{19}H_{24}N_2O_3$
Mol Weight: 328.41
MP (deg C):
FP (deg C):
BP (deg C):
BP pressure (mm Hg):

Property/Value	Units	Temp	Data Type	Reference
WS 1.06E+003	mg/L	25	EST	MEYLAN,WM ET AL. (1996)
logP 1.05			EXP	MAULEON,D ET AL. (1988)
VP 5.61E-013	mm Hg	25	EST	NEELY,WB & BLAU,GE (1985)
DC	pKa			
HL 2.73E-020	atm m3/mol	25	EST	MEYLAN,WM & HOWARD,PH (1991)
OH 1.62E-010	cm3/molc sec	25	EST	MEYLAN,WM & HOWARD,PH (1993)

121508-88-1 — PROPANAMIDE, N-[2-[[2-HYDROXY-3-(PHENYLMETHOXY)P

Formula: $C_{16}H_{26}N_2O_3$
Mol Weight: 294.40
MP (deg C):
FP (deg C):
BP (deg C):
BP pressure (mm Hg):

Property/Value	Units	Temp	Data Type	Reference
WS 3.83E+003	mg/L	25	EST	MEYLAN,WM ET AL. (1996)
logP 0.63			EXP	MAULEON,D ET AL. (1988)
VP 6.15E-011	mm Hg	25	EST	NEELY,WB & BLAU,GE (1985)
DC	pKa			
HL 7.90E-019	atm m3/mol	25	EST	MEYLAN,WM & HOWARD,PH (1991)
OH 1.25E-010	cm3/molc sec	25	EST	MEYLAN,WM & HOWARD,PH (1993)

121508-89-2 — 2-PROPANOL, 1-[(1,1-DIMETHYLETHYL)AMINO]-3-(2-PH

Formula: $C_{15}H_{25}NO_2$
Mol Weight: 251.37
MP (deg C):
FP (deg C):
BP (deg C):
BP pressure (mm Hg):

Property/Value	Units	Temp	Data Type	Reference
WS 1.67E+003	mg/L	25	EST	MEYLAN,WM ET AL. (1996)
logP 2.52			EXP	MAULEON,D ET AL. (1988)
VP 1.17E-006	mm Hg	25	EST	NEELY,WB & BLAU,GE (1985)
DC	pKa			
HL 3.15E-012	atm m3/mol	25	EST	MEYLAN,WM & HOWARD,PH (1991)
OH 1.07E-010	cm3/molc sec	25	EST	MEYLAN,WM & HOWARD,PH (1993)

121508-90-5 — BENZENEACETAMIDE, N-[2-[[2-HYDROXY-3-(2-PHENYLET

Formula: $C_{21}H_{28}N_2O_3$
Mol Weight: 356.47
MP (deg C):
FP (deg C):
BP (deg C):
BP pressure (mm Hg):

Property/Value	Units	Temp	Data Type	Reference
WS 4.94E+002	mg/L	25	EST	MEYLAN,WM ET AL. (1996)
logP 1.24			EXP	MAULEON,D ET AL. (1988)
VP 8.01E-014	mm Hg	25	EST	NEELY,WB & BLAU,GE (1985)
DC	pKa			
HL 4.81E-020	atm m3/mol	25	EST	MEYLAN,WM & HOWARD,PH (1991)
OH 1.33E-010	cm3/molc sec	25	EST	MEYLAN,WM & HOWARD,PH (1993)

121508-91-6

CAS #: 121508-91-6

PROPANAMIDE, N-[2-[[2-HYDROXY-3-(2-PHENYLETHOXY)

Formula: $C_{17}H_{28}N_2O_3$

Mol Weight: 308.42

MP (deg C): **FP (deg C):**

BP (deg C):

BP pressure (mm Hg):

Property/Value	Units	Temp	Data Type	Reference
WS 1.69E+003	mg/L	25	EST	MEYLAN,WM ET AL. (1996)
logP 0.95			EXP	MAULEON,D ET AL. (1988)
VP 2.29E-011	mm Hg	25	EST	NEELY,WB & BLAU,GE (1985)
DC	pKa			
HL 1.05E-018	atm m3/mol	25	EST	MEYLAN,WM & HOWARD,PH (1991)
OH 1.30E-010	cm3/molc sec	25	EST	MEYLAN,WM & HOWARD,PH (1993)

121508-92-7

CAS #: 121508-92-7

BENZENEACETAMIDE, N-[2-[[3-(2-HYDROXY-3-PHENYLPR

Formula: $C_{22}H_{30}N_2O_3$

Mol Weight: 370.50

MP (deg C): **FP (deg C):**

BP (deg C):

BP pressure (mm Hg):

Property/Value	Units	Temp	Data Type	Reference
WS 1.92E+002	mg/L	25	EST	MEYLAN,WM ET AL. (1996)
logP 1.62			EXP	MAULEON,D ET AL. (1988)
VP 3.01E-014	mm Hg	25	EST	NEELY,WB & BLAU,GE (1985)
DC	pKa			
HL 6.38E-020	atm m3/mol	25	EST	MEYLAN,WM & HOWARD,PH (1991)
OH 1.35E-010	cm3/molc sec	25	EST	MEYLAN,WM & HOWARD,PH (1993)

121508-93-8

CAS #: 121508-93-8

PROPANAMIDE, N-[2-[[2-HYDROXY-3-(3-PHENYLPROPOXY

Formula: $C_{18}H_{30}N_2O_3$

Mol Weight: 322.45

MP (deg C): **FP (deg C):**

BP (deg C):

BP pressure (mm Hg):

Property/Value	Units	Temp	Data Type	Reference
WS 7.75E+002	mg/L	25	EST	MEYLAN,WM ET AL. (1996)
logP 1.25			EXP	MAULEON,D ET AL. (1988)
VP 8.79E-012	mm Hg	25	EST	NEELY,WB & BLAU,GE (1985)
DC	pKa			
HL 1.39E-018	atm m3/mol	25	EST	MEYLAN,WM & HOWARD,PH (1991)
OH 1.32E-010	cm3/molc sec	25	EST	MEYLAN,WM & HOWARD,PH (1993)

121508-94-9

CAS #: 121508-94-9

2-PROPANOL, 1-[(1,1-DIMETHYLETHYL)AMINO]-3-(2-PH

Formula: $C_{15}H_{25}NO_3$

Mol Weight: 267.37

MP (deg C): **FP (deg C):**

BP (deg C):

BP pressure (mm Hg):

Property/Value	Units	Temp	Data Type	Reference
WS 1.90E+003	mg/L	25	EST	MEYLAN,WM ET AL. (1996)
logP 2.35			EXP	MAULEON,D ET AL. (1988)
VP 3.92E-007	mm Hg	25	EST	NEELY,WB & BLAU,GE (1985)
DC	pKa			
HL 1.69E-013	atm m3/mol	25	EST	MEYLAN,WM & HOWARD,PH (1991)
OH 1.27E-010	cm3/molc sec	25	EST	MEYLAN,WM & HOWARD,PH (1993)

121508-95-0

CAS #: 121508-95-0

BENZENEACETAMIDE, N-[2-[[2-HYDROXY-3-(2-PHENOXYE

Formula: $C_{21}H_{28}N_2O_4$

Mol Weight: 372.47

MP (deg C): **FP (deg C):**

BP (deg C):

BP pressure (mm Hg):

Property/Value	Units	Temp	Data Type	Reference
WS 1.04E+003	mg/L	25	EST	MEYLAN,WM ET AL. (1996)
logP 0.75			EXP	MAULEON,D ET AL. (1988)
VP 2.90E-014	mm Hg	25	EST	NEELY,WB & BLAU,GE (1985)
DC	pKa			
HL 2.58E-021	atm m3/mol	25	EST	MEYLAN,WM & HOWARD,PH (1991)
OH 1.53E-010	cm3/molc sec	25	EST	MEYLAN,WM & HOWARD,PH (1993)

121508-96-1

CAS #: 121508-96-1

PROPANAMIDE, 2-METHYL-N-[2-[[3-(2-PHENOXYETHOXY)

Formula: $C_{17}H_{28}N_2O_4$

Mol Weight: 324.42

MP (deg C): **FP (deg C):**

BP (deg C):

BP pressure (mm Hg):

Property/Value	Units	Temp	Data Type	Reference
WS 3.43E+003	mg/L	25	EST	MEYLAN,WM ET AL. (1996)
logP 0.48			EXP	MAULEON,D ET AL. (1988)
VP 8.47E-012	mm Hg	25	EST	NEELY,WB & BLAU,GE (1985)
DC	pKa			
HL 5.62E-020	atm m3/mol	25	EST	MEYLAN,WM & HOWARD,PH (1991)
OH 1.50E-010	cm3/molc sec	25	EST	MEYLAN,WM & HOWARD,PH (1993)

121508-97-2

CAS #: 121508-97-2

BENZENEACETAMIDE, N-[2-[[2-HYDROXY-3-[(3-PHENYL-

Formula: $C_{22}H_{28}N_2O_3$

Mol Weight: 368.48

MP (deg C): **FP (deg C):**

BP (deg C):

BP pressure (mm Hg):

Property/Value	Units	Temp	Data Type	Reference
WS 1.50E+002	mg/L	25	EST	MEYLAN,WM ET AL. (1996)
logP 1.76			EXP	MAULEON,D ET AL. (1988)
VP 2.23E-014	mm Hg	25	EST	NEELY,WB & BLAU,GE (1985)
DC	pKa			
HL 2.64E-020	atm m3/mol	25	EST	MEYLAN,WM & HOWARD,PH (1991)
OH 1.81E-010	cm3/molc sec	25	EST	MEYLAN,WM & HOWARD,PH (1993)

121508-98-3

CAS #: 121508-98-3

PROPANAMIDE, 2-METHYL-N-[2-[[3-(3-PHENYL-2-PROPE

Formula: $C_{18}H_{28}N_2O_3$

Mol Weight: 320.44

MP (deg C): **FP (deg C):**

BP (deg C):

BP pressure (mm Hg):

Property/Value	Units	Temp	Data Type	Reference
WS 3.42E+003	mg/L	25	EST	MEYLAN,WM ET AL. (1996)
logP 0.51			EXP	MAULEON,D ET AL. (1988)
VP 6.54E-012	mm Hg	25	EST	NEELY,WB & BLAU,GE (1985)
DC	pKa			
HL 5.76E-019	atm m3/mol	25	EST	MEYLAN,WM & HOWARD,PH (1991)
OH 1.78E-010	cm3/molc sec	25	EST	MEYLAN,WM & HOWARD,PH (1993)

121509-02-2

CAS #: 121509-02-2				BENZENEACETAMIDE, N-[2-[[3-[(1,2,3,4-TETRAHYDRO-
Formula: $C_{23}H_{30}N_2O_3$				
Mol Weight: 382.51				
MP (deg C):		FP (deg C):		
BP (deg C):				
BP pressure (mm Hg):				

Property/ Value	Units	Temp	Data Type	Reference
WS 1.44E+002	mg/L	25	EST	MEYLAN,WM ET AL. (1996)
logP 1.68			EXP	MAULEON,D ET AL. (1988)
VP 8.71E-015	mm Hg	25	EST	NEELY,WB & BLAU,GE (1985)
DC	pKa			
HL 3.10E-020	atm m3/mol	25	EST	MEYLAN,WM & HOWARD,PH (1991)
OH 1.42E-010	cm3/molc sec	25	EST	MEYLAN,WM & HOWARD,PH (1993)

121509-03-3

CAS #: 121509-03-3				PROPANAMIDE, 2-METHYL-N-[2-[[3-[(1,2,3,4-TETRAHY
Formula: $C_{19}H_{30}N_2O_3$				
Mol Weight: 334.46				
MP (deg C):		FP (deg C):		
BP (deg C):				
BP pressure (mm Hg):				

Property/ Value	Units	Temp	Data Type	Reference
WS 1.54E+002	mg/L	25	EST	MEYLAN,WM ET AL. (1996)
logP 1.99			EXP	MAULEON,D ET AL. (1988)
VP 2.62E-012	mm Hg	25	EST	NEELY,WB & BLAU,GE (1985)
DC	pKa			
HL 6.78E-019	atm m3/mol	25	EST	MEYLAN,WM & HOWARD,PH (1991)
OH 1.39E-010	cm3/molc sec	25	EST	MEYLAN,WM & HOWARD,PH (1993)

121552-61-2

CAS #: 121552-61-2				CGA 219417 (CYPRODINIL)
Formula: $C_{14}H_{15}N_3$				
Mol Weight: 225.30				
MP (deg C):		FP (deg C):		
BP (deg C):				
BP pressure (mm Hg):				

Property/ Value	Units	Temp	Data Type	Reference
WS 1.24E+001	mg/L	25	EST	MEYLAN,WM ET AL. (1996)
logP 4.00			EXP	TOMLIN,C (1994)
VP 1.17E-005	mm Hg	25	EST	NEELY,WB & BLAU,GE (1985)
DC	pKa			
HL 1.91E-006	atm m3/mol	25	EST	MEYLAN,WM & HOWARD,PH (1991)
OH 2.00E-010	cm3/molc sec	25	EST	MEYLAN,WM & HOWARD,PH (1993)

121588-32-7

CAS #: 121588-32-7				N(24-NO2-6CF3PH)-2(3CN-BENZO)PYRIDINAMIN
Formula: $C_{17}H_8F_3N_5O_4$				
Mol Weight: 403.28				
MP (deg C):		FP (deg C):		
BP (deg C):				
BP pressure (mm Hg):				

Property/ Value	Units	Temp	Data Type	Reference
WS 7.71E-001	mg/L	25	EST	MEYLAN,WM ET AL. (1996)
logP 3.88			EXP	BRANDT,U ET AL. (1992)
VP 3.21E-011	mm Hg	25	EST	NEELY,WB & BLAU,GE (1985)
DC	pKa			
HL 3.84E-015	atm m3/mol	25	EST	MEYLAN,WM & HOWARD,PH (1991)
OH 2.43E-011	cm3/molc sec	25	EST	MEYLAN,WM & HOWARD,PH (1993)

121788-14-5

CAS #: 121788-14-5				BENZOIC ACID, 4-FLUORO-3-[[[METHYL[(METHYLAMINO)
Formula: $C_{12}H_{14}FN_3O_4$				
Mol Weight: 283.26				
MP (deg C):		FP (deg C):		
BP (deg C):				
BP pressure (mm Hg):				

Property/ Value	Units	Temp	Data Type	Reference
WS 1.73E+002	mg/L	25	EST	MEYLAN,WM ET AL. (1996)
logP 2.28			EXP	CAMILLERI,P ET AL. (1989)
VP 2.15E-009	mm Hg	25	EST	NEELY,WB & BLAU,GE (1985)
DC	pKa			
HL 4.05E-015	atm m3/mol	25	EST	MEYLAN,WM & HOWARD,PH (1991)
OH 6.52E-012	cm3/molc sec	25	EST	MEYLAN,WM & HOWARD,PH (1993)

121788-15-6

CAS #: 121788-15-6				IMIDODICARBONIC DIAMIDE, N-(2-FLUORO-5-NITROPHEN
Formula: $C_{10}H_{11}FN_4O_4$				
Mol Weight: 270.22				
MP (deg C):		FP (deg C):		
BP (deg C):				
BP pressure (mm Hg):				

Property/ Value	Units	Temp	Data Type	Reference
WS 1.98E+002	mg/L	25	EST	MEYLAN,WM ET AL. (1996)
logP 2.30			EXP	CAMILLERI,P ET AL. (1989)
VP 1.38E-009	mm Hg	25	EST	NEELY,WB & BLAU,GE (1985)
DC	pKa			
HL 2.48E-015	atm m3/mol	25	EST	MEYLAN,WM & HOWARD,PH (1991)
OH 4.37E-012	cm3/molc sec	25	EST	MEYLAN,WM & HOWARD,PH (1993)

121788-16-7

CAS #: 121788-16-7				IMIDODICARBONIC DIAMIDE, N-[2-FLUORO-5-[(1-METHY
Formula: $C_{13}H_{18}FN_3O_4S$				
Mol Weight: 331.37				
MP (deg C):		FP (deg C):		
BP (deg C):				
BP pressure (mm Hg):				

Property/ Value	Units	Temp	Data Type	Reference
WS 2.03E+002	mg/L	25	EST	MEYLAN,WM ET AL. (1996)
logP 1.87			EXP	CAMILLERI,P ET AL. (1989)
VP 2.90E-011	mm Hg	25	EST	NEELY,WB & BLAU,GE (1985)
DC	pKa			
HL 2.57E-016	atm m3/mol	25	EST	MEYLAN,WM & HOWARD,PH (1991)
OH 2.17E-011	cm3/molc sec	25	EST	MEYLAN,WM & HOWARD,PH (1993)

121822-75-1

CAS #: 121822-75-1				ALPHA-TOLUENE SULFONYLUREA
Formula: $C_8H_{10}N_2O_3S$				
Mol Weight: 214.24				
MP (deg C):		FP (deg C):		
BP (deg C):				
BP pressure (mm Hg):				

Property/ Value	Units	Temp	Data Type	Reference
WS 1.13E+003	mg/L	25	EST	MEYLAN,WM ET AL. (1996)
logP 0.24			EXP	LIPINSKI,CA ET AL. (1991)
VP 6.08E-007	mm Hg	25	EST	NEELY,WB & BLAU,GE (1985)
DC	pKa			
HL 1.80E-011	atm m3/mol	25	EST	MEYLAN,WM & HOWARD,PH (1991)
OH 1.41E-011	cm3/molc sec	25	EST	MEYLAN,WM & HOWARD,PH (1993)

CAS #: 121822-76-2 — BENZENEMETHANESULFONAMIDE, N-[(METHYLAMINO)CARBO

Formula: $C_9H_{12}N_2O_3S$

Mol Weight: 228.27

MP (deg C): FP (deg C):

BP (deg C):

BP pressure (mm Hg):

Property/Value	Units	Temp	Data Type	Reference
WS 1.08E+004	mg/L	25	EST	MEYLAN,WM ET AL. (1996)
logP 0.54			EXP	CLOUX,JL ET AL. (1988)
VP 3.72E-007	mm Hg	25	EST	NEELY,WB & BLAU,GE (1985)
DC	pKa			
HL 3.96E-011	atm m3/mol	25	EST	MEYLAN,WM & HOWARD,PH (1991)
OH 1.47E-011	cm3/molc sec	25	EST	MEYLAN,WM & HOWARD,PH (1993)

CAS #: 121822-77-3 — BENZENEMETHANESULFONAMIDE, N-[(DIMETHYLAMINO)CAR

Formula: $C_{10}H_{14}N_2O_3S$

Mol Weight: 242.30

MP (deg C): FP (deg C):

BP (deg C):

BP pressure (mm Hg):

Property/Value	Units	Temp	Data Type	Reference
WS 9.02E+003	mg/L	25	EST	MEYLAN,WM ET AL. (1996)
logP 0.54			EXP	CLOUX,JL ET AL. (1988)
VP 1.86E-007	mm Hg	25	EST	NEELY,WB & BLAU,GE (1985)
DC	pKa			
HL 8.69E-011	atm m3/mol	25	EST	MEYLAN,WM & HOWARD,PH (1991)
OH 1.52E-011	cm3/molc sec	25	EST	MEYLAN,WM & HOWARD,PH (1993)

CAS #: 121822-78-4 — BENZENEMETHANESULFONAMIDE, N-METHYL-N-[(METHYLAM

Formula: $C_{10}H_{14}N_2O_3S$

Mol Weight: 242.30

MP (deg C): FP (deg C):

BP (deg C):

BP pressure (mm Hg):

Property/Value	Units	Temp	Data Type	Reference
WS 2.56E+003	mg/L	25	EST	MEYLAN,WM ET AL. (1996)
logP 1.18			EXP	CLOUX,JL ET AL. (1988)
VP 1.86E-007	mm Hg	25	EST	NEELY,WB & BLAU,GE (1985)
DC	pKa			
HL 8.69E-011	atm m3/mol	25	EST	MEYLAN,WM & HOWARD,PH (1991)
OH 1.52E-011	cm3/molc sec	25	EST	MEYLAN,WM & HOWARD,PH (1993)

CAS #: 121822-79-5 — 3-PYRIDINESULFONAMIDE, N-[(METHYLAMINO)CARBONYL]

Formula: $C_7H_9N_3O_3S$

Mol Weight: 215.23

MP (deg C): FP (deg C):

BP (deg C):

BP pressure (mm Hg):

Property/Value	Units	Temp	Data Type	Reference
WS 4.33E+003	mg/L	25	EST	MEYLAN,WM ET AL. (1996)
logP -0.45			EXP	CLOUX,JL ET AL. (1988)
VP 4.55E-007	mm Hg	25	EST	NEELY,WB & BLAU,GE (1985)
DC	pKa			
HL 1.28E-013	atm m3/mol	25	EST	MEYLAN,WM & HOWARD,PH (1991)
OH 2.64E-012	cm3/molc sec	25	EST	MEYLAN,WM & HOWARD,PH (1993)

CAS #: 121822-80-8 — 3-PYRIDINESULFONAMIDE, N-METHYL-N-[(METHYLAMINO)

Formula: $C_8H_{11}N_3O_3S$

Mol Weight: 229.26

MP (deg C): FP (deg C):

BP (deg C):

BP pressure (mm Hg):

Property/Value	Units	Temp	Data Type	Reference
WS 1.35E+004	mg/L	25	EST	MEYLAN,WM ET AL. (1996)
logP 0.42			EXP	CLOUX,JL ET AL. (1988)
VP 2.22E-007	mm Hg	25	EST	NEELY,WB & BLAU,GE (1985)
DC	pKa			
HL 2.82E-013	atm m3/mol	25	EST	MEYLAN,WM & HOWARD,PH (1991)
OH 3.19E-012	cm3/molc sec	25	EST	MEYLAN,WM & HOWARD,PH (1993)

CAS #: 122009-33-0 — 1-PYRAZOLIDINECARBOXALDEHYDE, 5,5-DIMETHYL-3-OXO

Formula: $C_{12}H_{16}N_4O$

Mol Weight: 232.29

MP (deg C): FP (deg C):

BP (deg C):

BP pressure (mm Hg):

Property/Value	Units	Temp	Data Type	Reference
WS 8.69E+002	mg/L	25	EST	MEYLAN,WM ET AL. (1996)
logP 0.26			EXP	SANGSTER,J (1994)
VP 1.45E-007	mm Hg	25	EST	NEELY,WB & BLAU,GE (1985)
DC	pKa			
HL 1.21E-014	atm m3/mol	25	EST	MEYLAN,WM & HOWARD,PH (1991)
OH 6.72E-011	cm3/molc sec	25	EST	MEYLAN,WM & HOWARD,PH (1993)

CAS #: 122224-71-9 — 2-OXAZOLIDINONE DERIVATIVE

Formula: $C_{35}H_{50}N_6O_7$

Mol Weight: 666.82

MP (deg C): FP (deg C):

BP (deg C):

BP pressure (mm Hg):

Property/Value	Units	Temp	Data Type	Reference
WS 7.70E-001	mg/L	25	EST	MEYLAN,WM ET AL. (1996)
logP 2.20			EXP	SANGSTER,J (1993)
VP 3.10E-031	mm Hg	25	EST	NEELY,WB & BLAU,GE (1985)
DC	pKa			
HL 4.32E-033	atm m3/mol	25	EST	MEYLAN,WM & HOWARD,PH (1991)
OH 2.36E-010	cm3/molc sec	25	EST	MEYLAN,WM & HOWARD,PH (1993)

CAS #: 122224-80-0 — 2-OXAZOLIDINONE DERIVATIVE

Formula: $C_{34}H_{50}N_6O_7$

Mol Weight: 654.81

MP (deg C): FP (deg C):

BP (deg C):

BP pressure (mm Hg):

Property/Value	Units	Temp	Data Type	Reference
WS 1.54E+000	mg/L	25	EST	MEYLAN,WM ET AL. (1996)
logP 1.94			EXP	HANSCH,C ET AL. (1995)
VP 9.72E-033	mm Hg	25	EST	NEELY,WB & BLAU,GE (1985)
DC	pKa			
HL 3.06E-032	atm m3/mol	25	EST	MEYLAN,WM & HOWARD,PH (1991)
OH 2.13E-010	cm3/molc sec	25	EST	MEYLAN,WM & HOWARD,PH (1993)

CAS #: 122224-83-3 — 2-OXAZOLIDINONE DERIVATIVE

Formula: $C_{39}H_{58}N_6O_{10}$

Mol Weight: 770.93

MP (deg C): FP (deg C):

BP (deg C):

BP pressure (mm Hg):

Property/ Value	Units	Temp	Data Type	Reference
WS 1.25E-001	mg/L	25	EST	MEYLAN,WM ET AL. (1996)
logP 2.31			EXP	HANSCH,C ET AL. (1995)
VP 1.01E-033	mm Hg	25	EST	NEELY,WB & BLAU,GE (1985)
DC	pKa			
HL 4.61E-038	atm m3/mol	25	EST	MEYLAN,WM & HOWARD,PH (1991)
OH 2.38E-010	cm3/molc sec	25	EST	MEYLAN,WM & HOWARD,PH (1993)

CAS #: 122224-84-4 — 2-OXAZOLIDINONE DERIVATIVE

Formula: $C_{38}H_{58}N_6O_9$

Mol Weight: 742.92

MP (deg C): FP (deg C):

BP (deg C):

BP pressure (mm Hg):

Property/ Value	Units	Temp	Data Type	Reference
WS 2.35E-001	mg/L	25	EST	MEYLAN,WM ET AL. (1996)
logP 2.21			EXP	SANGSTER,J (1993)
VP 6.48E-033	mm Hg	25	EST	NEELY,WB & BLAU,GE (1985)
DC	pKa			
HL 8.94E-036	atm m3/mol	25	EST	MEYLAN,WM & HOWARD,PH (1991)
OH 2.43E-010	cm3/molc sec	25	EST	MEYLAN,WM & HOWARD,PH (1993)

CAS #: 122224-98-0 — 2-OXAZOLIDINONE DERIVATIVE

Formula: $C_{34}H_{47}N_5O_7S$

Mol Weight: 669.85

MP (deg C): FP (deg C):

BP (deg C):

BP pressure (mm Hg):

Property/ Value	Units	Temp	Data Type	Reference
WS 2.13E-001	mg/L	25	EST	MEYLAN,WM ET AL. (1996)
logP 2.83			EXP	SANGSTER,J (1993)
VP 1.29E-028	mm Hg	25	EST	NEELY,WB & BLAU,GE (1985)
DC	pKa			
HL 3.32E-033	atm m3/mol	25	EST	MEYLAN,WM & HOWARD,PH (1991)
OH 1.45E-010	cm3/molc sec	25	EST	MEYLAN,WM & HOWARD,PH (1993)

CAS #: 122224-99-1 — 2-OXAZOLIDINONE DERIVATIVE

Formula: $C_{35}H_{50}N_6O_7$

Mol Weight: 666.82

MP (deg C): FP (deg C):

BP (deg C):

BP pressure (mm Hg):

Property/ Value	Units	Temp	Data Type	Reference
WS 4.36E-001	mg/L	25	EST	MEYLAN,WM ET AL. (1996)
logP 2.49			EXP	SANGSTER,J (1993)
VP 6.36E-030	mm Hg	25	EST	NEELY,WB & BLAU,GE (1985)
DC	pKa			
HL 4.25E-033	atm m3/mol	25	EST	MEYLAN,WM & HOWARD,PH (1991)
OH 2.36E-010	cm3/molc sec	25	EST	MEYLAN,WM & HOWARD,PH (1993)

CAS #: 122225-00-7 — 2-OXAZOLIDINONE DERIVATIVE

Formula: $C_{36}H_{50}N_4O_7S$

Mol Weight: 682.89

MP (deg C): FP (deg C):

BP (deg C):

BP pressure (mm Hg):

Property/ Value	Units	Temp	Data Type	Reference
WS 3.92E-002	mg/L	25	EST	MEYLAN,WM ET AL. (1996)
logP 3.59			EXP	SANGSTER,J (1993)
VP 7.94E-029	mm Hg	25	EST	NEELY,WB & BLAU,GE (1985)
DC	pKa			
HL 3.37E-030	atm m3/mol	25	EST	MEYLAN,WM & HOWARD,PH (1991)
OH 1.70E-010	cm3/molc sec	25	EST	MEYLAN,WM & HOWARD,PH (1993)

CAS #: 122266-90-4 — THIENO[2,3-B]THIOPHENE-2-SULFONAMIDE, 5-[[(2-MET

Formula: $C_{10}H_{16}N_2O_3S_3$

Mol Weight: 308.44

MP (deg C): FP (deg C):

BP (deg C):

BP pressure (mm Hg):

Property/ Value	Units	Temp	Data Type	Reference
WS 4.21E+004	mg/L	25	EST	MEYLAN,WM ET AL. (1996)
logP 0.32			EXP	PRUGH,JD ET AL. (1991)
VP 4.06E-009	mm Hg	25	EST	NEELY,WB & BLAU,GE (1985)
DC	pKa			
HL 4.71E-014	atm m3/mol	25	EST	MEYLAN,WM & HOWARD,PH (1991)
OH 1.18E-010	cm3/molc sec	25	EST	MEYLAN,WM & HOWARD,PH (1993)

CAS #: 122266-93-7 — THIENO[2,3-B]THIOPHENE-2-SULFONAMIDE, 5-[[(2-HYD

Formula: $C_9H_{14}N_2O_3S_3$

Mol Weight: 294.42

MP (deg C): FP (deg C):

BP (deg C):

BP pressure (mm Hg):

Property/ Value	Units	Temp	Data Type	Reference
WS 5.71E+005	mg/L	25	EST	MEYLAN,WM ET AL. (1996)
logP -0.72			EXP	PRUGH,JD ET AL. (1991)
VP 8.56E-012	mm Hg	25	EST	NEELY,WB & BLAU,GE (1985)
DC	pKa			
HL 1.47E-016	atm m3/mol	25	EST	MEYLAN,WM & HOWARD,PH (1991)
OH 9.46E-011	cm3/molc sec	25	EST	MEYLAN,WM & HOWARD,PH (1993)

CAS #: 122266-99-3 — THIENO[2,3-B]THIOPHENE-2-SULFONAMIDE, 5-[[[2-(ME

Formula: $C_{10}H_{15}N_2O_2S_4$

Mol Weight: 324.51

MP (deg C): FP (deg C):

BP (deg C):

BP pressure (mm Hg):

Property/ Value	Units	Temp	Data Type	Reference
WS 9.06E+003	mg/L	25	EST	MEYLAN,WM ET AL. (1996)
logP 1.18			EXP	PRUGH,JD ET AL. (1991)
VP 9.14E-010	mm Hg	25	EST	NEELY,WB & BLAU,GE (1985)
DC	pKa			
HL 4.30E-014	atm m3/mol	25	EST	MEYLAN,WM & HOWARD,PH (1991)
OH 1.02E-010	cm3/molc sec	25	EST	MEYLAN,WM & HOWARD,PH (1993)

CAS #: 122267-01-0	THIENO[2,3-B]THIOPHENE-2-SULFONAMIDE, 5-[[[2-(ME
Formula: C10H15N2O3S4	
Mol Weight: 340.51	

Property/Value	Units	Temp	Data Type	Reference
WS 1.76E+005	mg/L	25	EST	MEYLAN,WM ET AL. (1996)
logP -0.44			EXP	PRUGH,JD ET AL. (1991)
VP 5.04E-011	mm Hg	25	EST	NEELY,WB & BLAU,GE (1985)
DC	pKa			
HL 2.71E-018	atm m3/mol	25	EST	MEYLAN,WM & HOWARD,PH (1991)
OH 1.60E-010	cm3/molc sec	25	EST	MEYLAN,WM & HOWARD,PH (1993)

CAS #: 122321-30-6	THIENO[2,3-B]FURAN-2-SULFONAMIDE, 5-(4-MORPHOLIN
Formula: C11H14N2O4S2	
Mol Weight: 302.37	

Property/Value	Units	Temp	Data Type	Reference
WS 2.15E+004	mg/L	25	EST	MEYLAN,WM ET AL. (1996)
logP 0.88			EXP	HARTMAN,GD ET AL. (1992)
VP 8.87E-009	mm Hg	25	EST	NEELY,WB & BLAU,GE (1985)
DC	pKa			
HL 8.38E-014	atm m3/mol	25	EST	MEYLAN,WM & HOWARD,PH (1991)
OH 1.63E-010	cm3/molc sec	25	EST	MEYLAN,WM & HOWARD,PH (1993)

CAS #: 122321-32-8	THIENO[2,3-B]FURAN-2-SULFONAMIDE, 5-[[(2-FLUOROE
Formula: C9H11FN2O3S2	
Mol Weight: 278.33	

Property/Value	Units	Temp	Data Type	Reference
WS 8.55E+004	mg/L	25	EST	MEYLAN,WM ET AL. (1996)
logP 0.34			EXP	HARTMAN,GD ET AL. (1992)
VP 1.16E-007	mm Hg	25	EST	NEELY,WB & BLAU,GE (1985)
DC	pKa			
HL 1.47E-011	atm m3/mol	25	EST	MEYLAN,WM & HOWARD,PH (1991)
OH 2.80E-011	cm3/molc sec	25	EST	MEYLAN,WM & HOWARD,PH (1993)

CAS #: 122453-73-0	AC 303,630
Formula: C15H13BrClF3N2O	
Mol Weight: 409.64	

Property/Value	Units	Temp	Data Type	Reference
WS 1.12E-001	mg/L	25	EST	MEYLAN,WM ET AL. (1996)
logP 4.83			EXP	TOMLIN,C (1994)
VP 7.36E-008	mm Hg	25	EST	NEELY,WB & BLAU,GE (1985)
DC	pKa			
HL 5.73E-009	atm m3/mol	25	EST	MEYLAN,WM & HOWARD,PH (1991)
OH 1.35E-011	cm3/molc sec	25	EST	MEYLAN,WM & HOWARD,PH (1993)

CAS #: 122568-02-9	URIDINE, 2',3'-DIDEHYDRO-2',3'-DIDEOXY-4-THIO-
Formula: C9H10N2O3S	
Mol Weight: 226.26	

Property/Value	Units	Temp	Data Type	Reference
WS 2.71E+003	mg/L	25	EST	MEYLAN,WM ET AL. (1996)
logP -0.28			EXP	SANGSTER,J (1994)
VP 1.05E-010	mm Hg	25	EST	NEELY,WB & BLAU,GE (1985)
DC	pKa			
HL 1.75E-014	atm m3/mol	25	EST	MEYLAN,WM & HOWARD,PH (1991)
OH 1.47E-010	cm3/molc sec	25	EST	MEYLAN,WM & HOWARD,PH (1993)

CAS #: 122568-03-0	THYMIDINE, 3'-DEOXY-4-THIO-
Formula: C10H14N2O3S	
Mol Weight: 242.30	

Property/Value	Units	Temp	Data Type	Reference
WS 1.10E+004	mg/L	25	EST	MEYLAN,WM ET AL. (1996)
logP 0.44			EXP	PALOMINO,E ET AL. (1990)
VP 4.48E-011	mm Hg	25	EST	NEELY,WB & BLAU,GE (1985)
DC	pKa			
HL 3.92E-015	atm m3/mol	25	EST	MEYLAN,WM & HOWARD,PH (1991)
OH 1.23E-010	cm3/molc sec	25	EST	MEYLAN,WM & HOWARD,PH (1993)

CAS #: 122568-04-1	URIDINE, 2',3'-DIDEOXY-4-THIO-
Formula: C9H12N2O3S	
Mol Weight: 228.27	

Property/Value	Units	Temp	Data Type	Reference
WS 2.30E+003	mg/L	25	EST	MEYLAN,WM ET AL. (1996)
logP -0.21			EXP	SANGSTER,J (1994)
VP 1.19E-010	mm Hg	25	EST	NEELY,WB & BLAU,GE (1985)
DC	pKa			
HL 2.50E-015	atm m3/mol	25	EST	MEYLAN,WM & HOWARD,PH (1991)
OH 1.04E-010	cm3/molc sec	25	EST	MEYLAN,WM & HOWARD,PH (1993)

CAS #: 122970-35-8	9H-PURIN-2-AMINE, 6-CHLORO-9-[TETRAHYDRO-5-(HYDR
Formula: C10H12ClN5O2	
Mol Weight: 269.69	

Property/Value	Units	Temp	Data Type	Reference
WS 5.95E+002	mg/L	25	EST	MEYLAN,WM ET AL. (1996)
logP 0.21			EXP	MURAKAMI,K ET AL. (1991)
VP 7.51E-012	mm Hg	25	EST	NEELY,WB & BLAU,GE (1985)
DC	pKa			
HL 1.38E-014	atm m3/mol	25	EST	MEYLAN,WM & HOWARD,PH (1991)
OH 1.88E-010	cm3/molc sec	25	EST	MEYLAN,WM & HOWARD,PH (1993)

CAS #: 123045-56-7				BENZENESULFONAMIDE, 3-(1-METHYLETHYL)-
Formula: $C_9H_{13}NO_2S$				
Mol Weight: 199.27				
MP (deg C):		FP (deg C):		
BP (deg C):				
BP pressure (mm Hg):				

Property/ Value	Units	Temp	Data Type	Reference
WS 1.26E+003	mg/L	25	EST	MEYLAN,WM ET AL. (1996)
logP 1.81			EXP	SANGSTER,J (1993)
VP 8.66E-005	mm Hg	25	EST	NEELY,WB & BLAU,GE (1985)
DC	pKa			
HL 8.21E-007	atm m3/mol	25	EST	MEYLAN,WM & HOWARD,PH (1991)
OH 3.26E-012	cm3/molc sec	25	EST	MEYLAN,WM & HOWARD,PH (1993)

CAS #: 123045-57-8				BENZENESULFONAMIDE, 3-BUTOXY-
Formula: $C_{10}H_{15}NO_3S$				
Mol Weight: 229.30				
MP (deg C):		FP (deg C):		
BP (deg C):				
BP pressure (mm Hg):				

Property/ Value	Units	Temp	Data Type	Reference
WS 5.34E+002	mg/L	25	EST	MEYLAN,WM ET AL. (1996)
logP 2.06			EXP	CAROTTI,A ET AL. (1989)
VP 8.11E-006	mm Hg	25	EST	NEELY,WB & BLAU,GE (1985)
DC	pKa			
HL 5.84E-008	atm m3/mol	25	EST	MEYLAN,WM & HOWARD,PH (1991)
OH 1.69E-011	cm3/molc sec	25	EST	MEYLAN,WM & HOWARD,PH (1993)

CAS #: 123045-58-9				BENZENESULFONAMIDE, 4-(HEXYLOXY)-3-NITRO-
Formula: $C_{12}H_{18}N_2O_5S$				
Mol Weight: 302.35				
MP (deg C):		FP (deg C):		
BP (deg C):				
BP pressure (mm Hg):				

Property/ Value	Units	Temp	Data Type	Reference
WS 1.36E+001	mg/L	25	EST	MEYLAN,WM ET AL. (1996)
logP 2.99			EXP	CAROTTI,A ET AL. (1989)
VP 2.34E-008	mm Hg	25	EST	NEELY,WB & BLAU,GE (1985)
DC	pKa			
HL 4.06E-010	atm m3/mol	25	EST	MEYLAN,WM & HOWARD,PH (1991)
OH 1.57E-011	cm3/molc sec	25	EST	MEYLAN,WM & HOWARD,PH (1993)

CAS #: 123063-34-3				1,3-BENZENEDIOL, 4-TRIDECYL-
Formula: $C_{19}H_{32}O_2$				
Mol Weight: 292.47				
MP (deg C):		FP (deg C):		
BP (deg C):				
BP pressure (mm Hg):				

Property/ Value	Units	Temp	Data Type	Reference
WS 7.44E-002	mg/L	25	EST	MEYLAN,WM ET AL. (1996)
logP 6.84			EXP	ITOKAWA,H ET AL. (1989)
VP 1.30E-008	mm Hg	25	EST	NEELY,WB & BLAU,GE (1985)
DC	pKa			
HL 1.93E-009	atm m3/mol	25	EST	MEYLAN,WM & HOWARD,PH (1991)
OH 2.17E-010	cm3/molc sec	25	EST	MEYLAN,WM & HOWARD,PH (1993)

CAS #: 123063-36-5				1,2-BENZENEDIOL, 4-UNDECYL-
Formula: $C_{17}H_{28}O_2$				
Mol Weight: 264.41				
MP (deg C):		FP (deg C):		
BP (deg C):				
BP pressure (mm Hg):				

Property/ Value	Units	Temp	Data Type	Reference
WS 8.34E-002	mg/L	25	EST	MEYLAN,WM ET AL. (1996)
logP 6.97			EXP	ITOKAWA,H ET AL. (1989)
VP 8.66E-008	mm Hg	25	EST	NEELY,WB & BLAU,GE (1985)
DC	pKa			
HL 1.09E-009	atm m3/mol	25	EST	MEYLAN,WM & HOWARD,PH (1991)
OH 7.11E-011	cm3/molc sec	25	EST	MEYLAN,WM & HOWARD,PH (1993)

CAS #: 123134-25-8				CYCLOHEXANOL, 2-[(DIMETHYLAMINO)METHYL]-1-(3-MET
Formula: $C_{16}H_{25}NO_2$				
Mol Weight: 263.38				
MP (deg C):		FP (deg C):		
BP (deg C):				
BP pressure (mm Hg):				

Property/ Value	Units	Temp	Data Type	Reference
WS 1.15E+003	mg/L	25	EST	MEYLAN,WM ET AL. (1996)
logP 2.51			EXP	SANGSTER,J (1994)
VP 4.57E-007	mm Hg	25	EST	NEELY,WB & BLAU,GE (1985)
DC	pKa			
HL 1.54E-011	atm m3/mol	25	EST	MEYLAN,WM & HOWARD,PH (1991)
OH 1.47E-010	cm3/molc sec	25	EST	MEYLAN,WM & HOWARD,PH (1993)

CAS #: 123154-38-1				CYCLOHEXANOL, 2-[(DIMETHYLAMINO)METHYL]-1-(3-MET
Formula: $C_{16}H_{25}NO_2$				
Mol Weight: 263.38				
MP (deg C):		FP (deg C):		
BP (deg C):				
BP pressure (mm Hg):				

Property/ Value	Units	Temp	Data Type	Reference
WS 1.15E+003	mg/L	25	EST	MEYLAN,WM ET AL. (1996)
logP 2.63			EXP	SANGSTER,J (1994)
VP 4.57E-007	mm Hg	25	EST	NEELY,WB & BLAU,GE (1985)
DC	pKa			
HL 1.54E-011	atm m3/mol	25	EST	MEYLAN,WM & HOWARD,PH (1991)
OH 1.47E-010	cm3/molc sec	25	EST	MEYLAN,WM & HOWARD,PH (1993)

CAS #: 123297-69-8				CARBAMIC ACID, [[[[(1,3-BENZODIOXOL-5-YLOXY)CARB
Formula: $C_{23}H_{36}N_2O_6S$				
Mol Weight: 468.62				
MP (deg C):		FP (deg C):		
BP (deg C):				
BP pressure (mm Hg):				

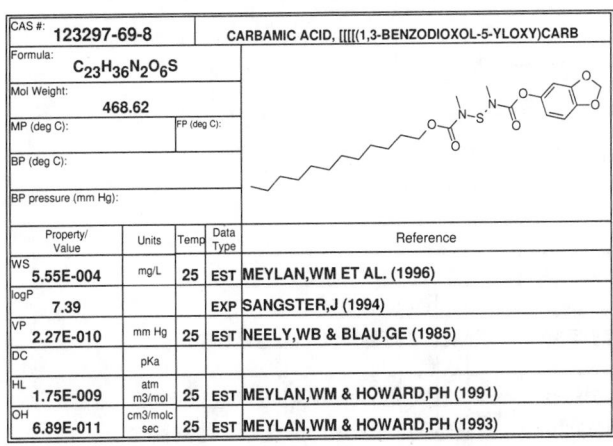

Property/ Value	Units	Temp	Data Type	Reference
WS 5.55E-004	mg/L	25	EST	MEYLAN,WM ET AL. (1996)
logP 7.39			EXP	SANGSTER,J (1994)
VP 2.27E-010	mm Hg	25	EST	NEELY,WB & BLAU,GE (1985)
DC	pKa			
HL 1.75E-009	atm m3/mol	25	EST	MEYLAN,WM & HOWARD,PH (1991)
OH 6.89E-011	cm3/molc sec	25	EST	MEYLAN,WM & HOWARD,PH (1993)

CAS #: 123953-16-2	FURAZANCARBOXALDEHYDE, METHYL-, 5-OXIDE

Formula: $C_4H_4N_2O_3$
Mol Weight: 128.09
MP (deg C): | FP (deg C):
BP (deg C):
BP pressure (mm Hg):

Property/Value	Units	Temp	Data Type	Reference
WS 1.26E+004	mg/L	25	EST	MEYLAN,WM ET AL. (1996)
logP 1.02			EXP	CALVINO,R ET AL. (1992)
VP 2.50E-003	mm Hg	25	EST	NEELY,WB & BLAU,GE (1985)
DC	pKa			
HL 5.20E-012	atm m3/mol	25	EST	MEYLAN,WM & HOWARD,PH (1991)
OH 2.10E-011	cm3/molc sec	25	EST	MEYLAN,WM & HOWARD,PH (1993)

CAS #: 123953-17-3	FURAZANCARBOXALDEHYDE, METHYL-, 2-OXIDE

Formula: $C_4H_4N_2O_3$
Mol Weight: 128.09
MP (deg C): | FP (deg C):
BP (deg C):
BP pressure (mm Hg):

Property/Value	Units	Temp	Data Type	Reference
WS 7.40E+003	mg/L	25	EST	MEYLAN,WM ET AL. (1996)
logP 1.29			EXP	CALVINO,R ET AL. (1992)
VP 2.50E-003	mm Hg	25	EST	NEELY,WB & BLAU,GE (1985)
DC	pKa			
HL 5.20E-012	atm m3/mol	25	EST	MEYLAN,WM & HOWARD,PH (1991)
OH 2.10E-011	cm3/molc sec	25	EST	MEYLAN,WM & HOWARD,PH (1993)

CAS #: 124050-99-3	2H-PYRAN, TETRAHYDRO-3-[1-(ETHOXYIMINO)BUTYL]-2,

Formula: $C_{23}H_{25}NO_4$
Mol Weight: 379.46
MP (deg C): | FP (deg C):
BP (deg C):
BP pressure (mm Hg):

Property/Value	Units	Temp	Data Type	Reference
WS 3.60E-001	mg/L	25	EST	MEYLAN,WM ET AL. (1996)
logP 4.75			EXP	SANGSTER,J (1994)
VP 8.98E-011	mm Hg	25	EST	NEELY,WB & BLAU,GE (1985)
DC	pKa			
HL 2.69E-011	atm m3/mol	25	EST	MEYLAN,WM & HOWARD,PH (1991)
OH 1.38E-011	cm3/molc sec	25	EST	MEYLAN,WM & HOWARD,PH (1993)

CAS #: 124051-00-9	2H-PYRAN, TETRAHYDRO-3-[1-(ETHOXYIMINO)BUTYL]-6-

Formula: $C_{16}H_{25}NO_4$
Mol Weight: 295.38
MP (deg C): | FP (deg C):
BP (deg C):
BP pressure (mm Hg):

Property/Value	Units	Temp	Data Type	Reference
WS 1.63E+001	mg/L	25	EST	MEYLAN,WM ET AL. (1996)
logP 3.40			EXP	HANSCH,C ET AL. (1995)
VP 4.16E-007	mm Hg	25	EST	NEELY,WB & BLAU,GE (1985)
DC	pKa			
HL 1.77E-008	atm m3/mol	25	EST	MEYLAN,WM & HOWARD,PH (1991)
OH 9.26E-011	cm3/molc sec	25	EST	MEYLAN,WM & HOWARD,PH (1993)

CAS #: 124424-25-5	2,4(1H,3H)-PYRIMIDINEDIONE, 1-(2,3-DIDEOXY-2-FLU

Formula: $C_9H_{11}FN_2O_4$
Mol Weight: 230.20
MP (deg C): | FP (deg C):
BP (deg C):
BP pressure (mm Hg):

Property/Value	Units	Temp	Data Type	Reference
WS 8.23E+003	mg/L	25	EST	MEYLAN,WM ET AL. (1996)
logP -0.87			EXP	SIDDIQUI,MA ET AL. (1992)
VP 9.24E-012	mm Hg	25	EST	NEELY,WB & BLAU,GE (1985)
DC	pKa			
HL 4.20E-016	atm m3/mol	25	EST	MEYLAN,WM & HOWARD,PH (1991)
OH 2.23E-011	cm3/molc sec	25	EST	MEYLAN,WM & HOWARD,PH (1993)

CAS #: 124497-72-9	2-OH-2-PH-4-ME MORPHOLINE (HCL)

Formula: $C_{11}H_{15}NO_2$
Mol Weight: 193.25
MP (deg C): | FP (deg C):
BP (deg C):
BP pressure (mm Hg):

Property/Value	Units	Temp	Data Type	Reference
WS 2.33E+004	mg/L	25	EST	MEYLAN,WM ET AL. (1996)
logP 1.54			EXP	REKKA,E & KOUROUNAKIS,P (1989)
VP 3.28E-005	mm Hg	25	EST	NEELY,WB & BLAU,GE (1985)
DC	pKa			
HL 7.39E-013	atm m3/mol	25	EST	MEYLAN,WM & HOWARD,PH (1991)
OH 1.40E-010	cm3/molc sec	25	EST	MEYLAN,WM & HOWARD,PH (1993)

CAS #: 124497-73-0	2-OH-2-PH-4,5-DIME MORPHOLINE HCL

Formula: $C_{12}H_{17}NO_2$
Mol Weight: 207.27
MP (deg C): | FP (deg C):
BP (deg C):
BP pressure (mm Hg):

Property/Value	Units	Temp	Data Type	Reference
WS 1.21E+004	mg/L	25	EST	MEYLAN,WM ET AL. (1996)
logP 1.79			EXP	REKKA,E & KOUROUNAKIS,P (1989)
VP 1.47E-005	mm Hg	25	EST	NEELY,WB & BLAU,GE (1985)
DC	pKa			
HL 9.81E-013	atm m3/mol	25	EST	MEYLAN,WM & HOWARD,PH (1991)
OH 1.28E-010	cm3/molc sec	25	EST	MEYLAN,WM & HOWARD,PH (1993)

CAS #: 124497-74-1	2-OH-2-(P-BR-PH)-4-ME MORPHOLINE HCL

Formula: $C_{11}H_{14}BrNO_2$
Mol Weight: 272.15
MP (deg C): | FP (deg C):
BP (deg C):
BP pressure (mm Hg):

Property/Value	Units	Temp	Data Type	Reference
WS 8.26E+003	mg/L	25	EST	MEYLAN,WM ET AL. (1996)
logP 1.57			EXP	REKKA,E & KOUROUNAKIS,P (1989)
VP 1.45E-006	mm Hg	25	EST	NEELY,WB & BLAU,GE (1985)
DC	pKa			
HL 2.94E-013	atm m3/mol	25	EST	MEYLAN,WM & HOWARD,PH (1991)
OH 1.37E-010	cm3/molc sec	25	EST	MEYLAN,WM & HOWARD,PH (1993)

124497-75-2 — 2-OH-2-PH-3,4-DIME MORPHOLINE HCL

Formula: $C_{12}H_{17}NO_2$

Mol Weight: 207.27

MP (deg C): FP (deg C):

BP (deg C):

BP pressure (mm Hg):

Property/ Value	Units	Temp	Data Type	Reference
WS 9.37E+003	mg/L	25	EST	MEYLAN,WM ET AL. (1996)
logP 1.92			EXP	REKKA,E & KOUROUNAKIS,P (1989)
VP 1.47E-005	mm Hg	25	EST	NEELY,WB & BLAU,GE (1985)
DC	pKa			
HL 9.81E-013	atm m3/mol	25	EST	MEYLAN,WM & HOWARD,PH (1991)
OH 1.72E-010	cm3/molc sec	25	EST	MEYLAN,WM & HOWARD,PH (1993)

124497-76-3 — 2-OH-2,6-DIPH-4,5-DIME MORPHOLINE HCL

Formula: $C_{18}H_{21}NO_2$

Mol Weight: 283.37

MP (deg C): FP (deg C):

BP (deg C):

BP pressure (mm Hg):

Property/ Value	Units	Temp	Data Type	Reference
WS 6.64E+001	mg/L	25	EST	MEYLAN,WM ET AL. (1996)
logP 1.87			EXP	REKKA,E & KOUROUNAKIS,P (1989)
VP 1.58E-008	mm Hg	25	EST	NEELY,WB & BLAU,GE (1985)
DC	pKa			
HL 7.92E-014	atm m3/mol	25	EST	MEYLAN,WM & HOWARD,PH (1991)
OH 1.44E-010	cm3/molc sec	25	EST	MEYLAN,WM & HOWARD,PH (1993)

124497-77-4 — 2-OH-3-PH-4-ME MORPHOLINE HCL

Formula: $C_{11}H_{15}NO_2$

Mol Weight: 193.25

MP (deg C): FP (deg C):

BP (deg C):

BP pressure (mm Hg):

Property/ Value	Units	Temp	Data Type	Reference
WS 6.88E+004	mg/L	25	EST	MEYLAN,WM ET AL. (1996)
logP 0.99			EXP	REKKA,E & KOUROUNAKIS,P (1989)
VP 1.61E-005	mm Hg	25	EST	NEELY,WB & BLAU,GE (1985)
DC	pKa			
HL 7.39E-013	atm m3/mol	25	EST	MEYLAN,WM & HOWARD,PH (1991)
OH 1.83E-010	cm3/molc sec	25	EST	MEYLAN,WM & HOWARD,PH (1993)

124497-78-5 — 2-OH-2-(M-CF3-PH)-4-ME MORPHOLINE HCL

Formula: $C_{12}H_{14}F_3NO_2$

Mol Weight: 261.25

MP (deg C): FP (deg C):

BP (deg C):

BP pressure (mm Hg):

Property/ Value	Units	Temp	Data Type	Reference
WS 1.36E+004	mg/L	25	EST	MEYLAN,WM ET AL. (1996)
logP 1.39			EXP	REKKA,E & KOUROUNAKIS,P (1989)
VP 2.02E-005	mm Hg	25	EST	NEELY,WB & BLAU,GE (1985)
DC	pKa			
HL 6.42E-012	atm m3/mol	25	EST	MEYLAN,WM & HOWARD,PH (1991)
OH 6.99E-011	cm3/molc sec	25	EST	MEYLAN,WM & HOWARD,PH (1993)

124497-80-9 — 2-OPR-2-PH-4-ME MORPHOLINE HCL

Formula: $C_{14}H_{21}NO_2$

Mol Weight: 235.33

MP (deg C): FP (deg C):

BP (deg C):

BP pressure (mm Hg):

Property/ Value	Units	Temp	Data Type	Reference
WS 4.40E+003	mg/L	25	EST	MEYLAN,WM ET AL. (1996)
logP 2.13			EXP	REKKA,E & KOUROUNAKIS,P (1989)
VP 1.86E-004	mm Hg	25	EST	NEELY,WB & BLAU,GE (1985)
DC	pKa			
HL 2.09E-009	atm m3/mol	25	EST	MEYLAN,WM & HOWARD,PH (1991)
OH 1.50E-010	cm3/molc sec	25	EST	MEYLAN,WM & HOWARD,PH (1993)

124497-81-0 — 2-PROPENYLOXY-2-PH-4-ME MORPHOLINE HCL

Formula: $C_{14}H_{19}NO_2$

Mol Weight: 233.31

MP (deg C): FP (deg C):

BP (deg C):

BP pressure (mm Hg):

Property/ Value	Units	Temp	Data Type	Reference
WS 4.88E+003	mg/L	25	EST	MEYLAN,WM ET AL. (1996)
logP 2.09			EXP	REKKA,E & KOUROUNAKIS,P (1989)
VP 2.00E-004	mm Hg	25	EST	NEELY,WB & BLAU,GE (1985)
DC	pKa			
HL 1.56E-009	atm m3/mol	25	EST	MEYLAN,WM & HOWARD,PH (1991)
OH 1.71E-010	cm3/molc sec	25	EST	MEYLAN,WM & HOWARD,PH (1993)

124497-82-1 — 2-PROPYNYLOXY-2-PH-4-ME MORPHOLINE HCL

Formula: $C_{14}H_{17}NO_2$

Mol Weight: 231.30

MP (deg C): FP (deg C):

BP (deg C):

BP pressure (mm Hg):

Property/ Value	Units	Temp	Data Type	Reference
WS 7.72E+003	mg/L	25	EST	MEYLAN,WM ET AL. (1996)
logP 1.87			EXP	REKKA,E & KOUROUNAKIS,P (1989)
VP 1.03E-004	mm Hg	25	EST	NEELY,WB & BLAU,GE (1985)
DC	pKa			
HL 2.59E-010	atm m3/mol	25	EST	MEYLAN,WM & HOWARD,PH (1991)
OH 1.52E-010	cm3/molc sec	25	EST	MEYLAN,WM & HOWARD,PH (1993)

124497-83-2 — 2-ETHOXY-2-PH-3,4-ME MORPHOLINE HCL

Formula: $C_{14}H_{21}NO_2$

Mol Weight: 235.33

MP (deg C): FP (deg C):

BP (deg C):

BP pressure (mm Hg):

Property/ Value	Units	Temp	Data Type	Reference
WS 3.34E+003	mg/L	25	EST	MEYLAN,WM ET AL. (1996)
logP 2.27			EXP	REKKA,E & KOUROUNAKIS,P (1989)
VP 2.46E-004	mm Hg	25	EST	NEELY,WB & BLAU,GE (1985)
DC	pKa			
HL 2.09E-009	atm m3/mol	25	EST	MEYLAN,WM & HOWARD,PH (1991)
OH 1.78E-010	cm3/molc sec	25	EST	MEYLAN,WM & HOWARD,PH (1993)

CAS #: 124497-85-4 — 2-BUTOXY-2-PH-4-ME MORPHOLINE HCL

Formula: $C_{15}H_{23}NO_2$

Mol Weight: 249.36

MP (deg C): FP (deg C):

BP (deg C):

BP pressure (mm Hg):

Property/Value	Units	Temp	Data Type	Reference
WS 3.03E+003	mg/L	25	EST	MEYLAN,WM ET AL. (1996)
logP 2.23			EXP	REKKA,E & KOUROUNAKIS,P (1989)
VP 7.44E-005	mm Hg	25	EST	NEELY,WB & BLAU,GE (1985)
DC	pKa			
HL 2.77E-009	atm m3/mol	25	EST	MEYLAN,WM & HOWARD,PH (1991)
OH 1.52E-010	cm3/molc sec	25	EST	MEYLAN,WM & HOWARD,PH (1993)

CAS #: 124497-86-5 — 2-ISOBUTOXY-2-PH-4-ME MORPHOLINE HCL

Formula: $C_{15}H_{23}NO_2$

Mol Weight: 249.36

MP (deg C): FP (deg C):

BP (deg C):

BP pressure (mm Hg):

Property/Value	Units	Temp	Data Type	Reference
WS 3.41E+003	mg/L	25	EST	MEYLAN,WM ET AL. (1996)
logP 2.17			EXP	REKKA,E & KOUROUNAKIS,P (1989)
VP 1.16E-004	mm Hg	25	EST	NEELY,WB & BLAU,GE (1985)
DC	pKa			
HL 2.77E-009	atm m3/mol	25	EST	MEYLAN,WM & HOWARD,PH (1991)
OH 1.54E-010	cm3/molc sec	25	EST	MEYLAN,WM & HOWARD,PH (1993)

CAS #: 124700-70-5 — 4-PYRIMIDINOL, 2-(METHYLTHIO)- (KETO FORM)

Formula: $C_5H_6N_2OS$

Mol Weight: 142.18

MP (deg C): FP (deg C):

BP (deg C):

BP pressure (mm Hg):

Property/Value	Units	Temp	Data Type	Reference
WS 2.61E+003	mg/L	25	EST	MEYLAN,WM ET AL. (1996)
logP 0.22			EXP	HANSCH,C ET AL. (1995)
VP 6.03E-007	mm Hg	25	EST	NEELY,WB & BLAU,GE (1985)
DC	pKa			
HL 6.32E-012	atm m3/mol	25	EST	MEYLAN,WM & HOWARD,PH (1991)
OH 1.50E-011	cm3/molc sec	25	EST	MEYLAN,WM & HOWARD,PH (1993)

CAS #: 124725-22-0 — 1H-PYRROLE-2-CARBOXAMIDE, N-ETHYL-1-METHYL-

Formula: $C_8H_{12}N_2O$

Mol Weight: 152.20

MP (deg C): FP (deg C):

BP (deg C):

BP pressure (mm Hg):

Property/Value	Units	Temp	Data Type	Reference
WS 8.70E+003	mg/L	25	EST	MEYLAN,WM ET AL. (1996)
logP 1.09			EXP	YAMAGAMI,C ET AL. (1994)
VP 1.25E-004	mm Hg	25	EST	NEELY,WB & BLAU,GE (1985)
DC	pKa			
HL 2.32E-010	atm m3/mol	25	EST	MEYLAN,WM & HOWARD,PH (1991)
OH 1.10E-010	cm3/molc sec	25	EST	MEYLAN,WM & HOWARD,PH (1993)

CAS #: 124802-88-6 — PROPANOIC ACID, 3-[(4H-1,2,4-BENZOTHIADIAZIN-3-Y

Formula: $C_{10}H_{10}N_2O_4S_2$

Mol Weight: 286.33

MP (deg C): FP (deg C):

BP (deg C):

BP pressure (mm Hg):

Property/Value	Units	Temp	Data Type	Reference
WS 1.00E+004	mg/L	25	EST	MEYLAN,WM ET AL. (1996)
logP 0.66			EXP	GAMBERINI,G ET AL. (1989)
VP 1.33E-011	mm Hg	25	EST	NEELY,WB & BLAU,GE (1985)
DC	pKa			
HL 7.34E-015	atm m3/mol	25	EST	MEYLAN,WM & HOWARD,PH (1991)
OH 2.12E-011	cm3/molc sec	25	EST	MEYLAN,WM & HOWARD,PH (1993)

CAS #: 124802-90-0 — PROPANOIC ACID, 3-[(6,7-DICHLORO-4H-1,2,4-BENZOT

Formula: $C_{10}H_8Cl_2N_2O_4S_2$

Mol Weight: 355.22

MP (deg C): FP (deg C):

BP (deg C):

BP pressure (mm Hg):

Property/Value	Units	Temp	Data Type	Reference
WS 4.07E+002	mg/L	25	EST	MEYLAN,WM ET AL. (1996)
logP 1.81			EXP	GAMBERINI,G ET AL. (1989)
VP 8.86E-013	mm Hg	25	EST	NEELY,WB & BLAU,GE (1985)
DC	pKa			
HL 4.03E-015	atm m3/mol	25	EST	MEYLAN,WM & HOWARD,PH (1991)
OH 1.39E-011	cm3/molc sec	25	EST	MEYLAN,WM & HOWARD,PH (1993)

CAS #: 124802-91-1 — PROPANOIC ACID, 3-[(7-BROMO-4H-1,2,4-BENZOTHIADI

Formula: $C_{10}H_9BrN_2O_4S_2$

Mol Weight: 365.23

MP (deg C): FP (deg C):

BP (deg C):

BP pressure (mm Hg):

Property/Value	Units	Temp	Data Type	Reference
WS 3.83E+002	mg/L	25	EST	MEYLAN,WM ET AL. (1996)
logP 1.77			EXP	GAMBERINI,G ET AL. (1989)
VP 1.44E-012	mm Hg	25	EST	NEELY,WB & BLAU,GE (1985)
DC	pKa			
HL 2.92E-015	atm m3/mol	25	EST	MEYLAN,WM & HOWARD,PH (1991)
OH 1.47E-011	cm3/molc sec	25	EST	MEYLAN,WM & HOWARD,PH (1993)

CAS #: 124802-92-2 — PROPANOIC ACID, 3-[(5,7-DIBROMO-4H-1,2,4-BENZOTH

Formula: $C_{10}H_8Br_2N_2O_4S_2$

Mol Weight: 444.12

MP (deg C): FP (deg C):

BP (deg C):

BP pressure (mm Hg):

Property/Value	Units	Temp	Data Type	Reference
WS 7.30E+001	mg/L	25	EST	MEYLAN,WM ET AL. (1996)
logP 2.04			EXP	GAMBERINI,G ET AL. (1989)
VP 1.53E-013	mm Hg	25	EST	NEELY,WB & BLAU,GE (1985)
DC	pKa			
HL 1.16E-015	atm m3/mol	25	EST	MEYLAN,WM & HOWARD,PH (1991)
OH 1.23E-011	cm3/molc sec	25	EST	MEYLAN,WM & HOWARD,PH (1993)

1537

CAS #: 125111-04-8	2,4(1H,3H)-PYRIMIDINEDIONE, 5-FLUORO-1-[(4-METHY

Formula: $C_{12}H_{11}FN_2O_2$

Mol Weight: 234.23

MP (deg C): FP (deg C):

BP (deg C):

BP pressure (mm Hg):

Property/Value	Units	Temp	Data Type	Reference
WS 1.51E+003	mg/L	25	EST	MEYLAN,WM ET AL. (1996)
logP 1.50			EXP	SANGSTER,J (1994)
VP 1.79E-009	mm Hg	25	EST	NEELY,WB & BLAU,GE (1985)
DC	pKa			
HL 3.25E-011	atm m3/mol	25	EST	MEYLAN,WM & HOWARD,PH (1991)
OH 1.60E-011	cm3/molc sec	25	EST	MEYLAN,WM & HOWARD,PH (1993)

CAS #: 125111-05-9	2,4(1H,3H)-PYRIMIDINEDIONE, 1-[(2,4-DICHLOROPHEN

Formula: $C_{10}H_5Cl_2FN_2O_2$

Mol Weight: 275.07

MP (deg C): FP (deg C):

BP (deg C):

BP pressure (mm Hg):

Property/Value	Units	Temp	Data Type	Reference
WS 3.62E+002	mg/L	25	EST	MEYLAN,WM ET AL. (1996)
logP 1.96			EXP	SANGSTER,J (1994)
VP 6.59E-010	mm Hg	25	EST	NEELY,WB & BLAU,GE (1985)
DC	pKa			
HL 4.69E-010	atm m3/mol	25	EST	MEYLAN,WM & HOWARD,PH (1991)
OH 5.89E-012	cm3/molc sec	25	EST	MEYLAN,WM & HOWARD,PH (1993)

CAS #: 125116-23-6	METCONAZOLE

Formula: $C_{17}H_{22}ClN_3O$

Mol Weight: 319.84

MP (deg C): FP (deg C):

BP (deg C):

BP pressure (mm Hg):

Property/Value	Units	Temp	Data Type	Reference
WS 4.13E+000	mg/L	25	EST	MEYLAN,WM ET AL. (1996)
logP 3.93			EXP	TOMLIN,C (1994)
VP 2.01E-009	mm Hg	25	EST	NEELY,WB & BLAU,GE (1985)
DC	pKa			
HL 3.03E-010	atm m3/mol	25	EST	MEYLAN,WM & HOWARD,PH (1991)
OH 1.81E-011	cm3/molc sec	25	EST	MEYLAN,WM & HOWARD,PH (1993)

CAS #: 125225-28-7	IPCONAZOLE

Formula: $C_{18}H_{26}ClN_3O$

Mol Weight: 335.88

MP (deg C): FP (deg C):

BP (deg C):

BP pressure (mm Hg):

Property/Value	Units	Temp	Data Type	Reference
WS 6.36E+000	mg/L	25	EST	MEYLAN,WM ET AL. (1996)
logP 4.21			EXP	TOMLIN,C (1994)
VP 8.46E-010	mm Hg	25	EST	NEELY,WB & BLAU,GE (1985)
DC	pKa			
HL 4.03E-010	atm m3/mol	25	EST	MEYLAN,WM & HOWARD,PH (1991)
OH 2.94E-011	cm3/molc sec	25	EST	MEYLAN,WM & HOWARD,PH (1993)

CAS #: 125476-98-4	1-PROPANONE, 1-(2-METHYLPHENYL)-3-(1-PIPERIDINYL

Formula: $C_{15}H_{21}NO$

Mol Weight: 231.34

MP (deg C): FP (deg C):

BP (deg C):

BP pressure (mm Hg):

Property/Value	Units	Temp	Data Type	Reference
WS 2.89E+002	mg/L	25	EST	MEYLAN,WM ET AL. (1996)
logP 3.54			EXP	SANGSTER,J (1994)
VP 4.55E-005	mm Hg	25	EST	NEELY,WB & BLAU,GE (1985)
DC	pKa			
HL 7.35E-009	atm m3/mol	25	EST	MEYLAN,WM & HOWARD,PH (1991)
OH 1.31E-010	cm3/molc sec	25	EST	MEYLAN,WM & HOWARD,PH (1993)

CAS #: 125573-05-9	1H-PURINE-2,6-DIONE, 3,9-DIHYDRO-3-ETHYL-1-METHY

Formula: $C_8H_{10}N_4O_2$

Mol Weight: 194.19

MP (deg C): FP (deg C):

BP (deg C):

BP pressure (mm Hg):

Property/Value	Units	Temp	Data Type	Reference
WS 1.68E+004	mg/L	25	EST	MEYLAN,WM ET AL. (1996)
logP 0.52			EXP	SANGSTER,J (1993)
VP 2.27E-009	mm Hg	25	EST	NEELY,WB & BLAU,GE (1985)
DC	pKa			
HL 2.23E-012	atm m3/mol	25	EST	MEYLAN,WM & HOWARD,PH (1991)
OH 2.27E-011	cm3/molc sec	25	EST	MEYLAN,WM & HOWARD,PH (1993)

CAS #: 125603-72-7	CYCLOPROPANECARBOXYLIC ACID, 3-[(DIHYDRO*

Formula: $C_{18}H_{20}O_3S$

Mol Weight: 316.42

MP (deg C): FP (deg C):

BP (deg C):

BP pressure (mm Hg):

Property/Value	Units	Temp	Data Type	Reference
WS 3.63E+000	mg/L	25	EST	MEYLAN,WM ET AL. (1996)
logP 4.02			EXP	SANGSTER,J (1993)
VP 1.44E-007	mm Hg	25	EST	NEELY,WB & BLAU,GE (1985)
DC	pKa			
HL 9.38E-009	atm m3/mol	25	EST	MEYLAN,WM & HOWARD,PH (1991)
OH 4.24E-011	cm3/molc sec	25	EST	MEYLAN,WM & HOWARD,PH (1993)

CAS #: 125757-81-5	N-ETHYL-HYDROXYPYRID-4-ONE

Formula: $C_7H_9NO_2$

Mol Weight: 139.16

MP (deg C): FP (deg C):

BP (deg C):

BP pressure (mm Hg):

Property/Value	Units	Temp	Data Type	Reference
WS 1.85E+005	mg/L	25	EST	MEYLAN,WM ET AL. (1996)
logP -0.40			EXP	GYPARAKI,M ET AL. (1986)
VP 1.32E-004	mm Hg	25	EST	NEELY,WB & BLAU,GE (1985)
DC	pKa			
HL 1.63E-008	atm m3/mol	25	EST	MEYLAN,WM & HOWARD,PH (1991)
OH 1.00E-010	cm3/molc sec	25	EST	MEYLAN,WM & HOWARD,PH (1993)

CAS #: 126502-17-8	9H-PURIN-6-AMINE, 9-(2,3-DIDEOXY-2-FLUORO-á-D-TH
Formula: $C_{11}H_{14}FN_5O_2$	
Mol Weight: 267.26	
MP (deg C):	FP (deg C):
BP (deg C):	
BP pressure (mm Hg):	

Property/Value	Units	Temp	Data Type	Reference
WS 5.45E+002	mg/L	25	EST	MEYLAN,WM ET AL. (1996)
logP 0.27			EXP	BARCHI,JJ ET AL. (1991)
VP 8.97E-011	mm Hg	25	EST	NEELY,WB & BLAU,GE (1985)
DC	pKa			
HL 3.64E-019	atm m3/mol	25	EST	MEYLAN,WM & HOWARD,PH (1991)
OH 2.10E-010	cm3/molc sec	25	EST	MEYLAN,WM & HOWARD,PH (1993)

CAS #: 126585-48-6	SELENIUM, BIS[3-[(METHYLSULFONYL)OXY]PROPYL]-
Formula: $C_8H_{18}O_6S_2Se$	
Mol Weight: 353.32	
MP (deg C):	FP (deg C):
BP (deg C):	
BP pressure (mm Hg):	

Property/Value	Units	Temp	Data Type	Reference
WS 2.80E+003	mg/L	25	EST	MEYLAN,WM ET AL. (1996)
logP 0.38			EXP	SANGSTER,J (1994)
VP 4.45E-008	mm Hg	25	EST	NEELY,WB & BLAU,GE (1985)
DC	pKa			
HL 1.19E-011	atm m3/mol	25	EST	MEYLAN,WM & HOWARD,PH (1991)
OH 8.76E-011	cm3/molc sec	25	EST	MEYLAN,WM & HOWARD,PH (1993)

CAS #: 126813-40-9	BENZENESULFONAMIDE, N-[2-(DIMETHYLAMINO)ETHYL]-N
Formula: $C_{14}H_{22}N_4O_6S$	
Mol Weight: 374.42	
MP (deg C):	FP (deg C):
BP (deg C):	
BP pressure (mm Hg):	

Property/Value	Units	Temp	Data Type	Reference
WS 1.12E+003	mg/L	25	EST	MEYLAN,WM ET AL. (1996)
logP 1.42			EXP	HANSCH,C ET AL. (1995)
VP 8.71E-010	mm Hg	25	EST	NEELY,WB & BLAU,GE (1985)
DC	pKa			
HL 5.88E-015	atm m3/mol	25	EST	MEYLAN,WM & HOWARD,PH (1991)
OH 9.14E-011	cm3/molc sec	25	EST	MEYLAN,WM & HOWARD,PH (1993)

CAS #: 127025-29-0	THIENO[3,2-B]THIOPHENE-2-SULFONAMIDE, 5-[[(2-MET
Formula: $C_{10}H_{16}N_2O_3S_3$	
Mol Weight: 308.44	
MP (deg C):	FP (deg C):
BP (deg C):	
BP pressure (mm Hg):	

Property/Value	Units	Temp	Data Type	Reference
WS 4.21E+004	mg/L	25	EST	MEYLAN,WM ET AL. (1996)
logP 0.51			EXP	PRUGH,JD ET AL. (1991)
VP 4.06E-009	mm Hg	25	EST	NEELY,WB & BLAU,GE (1985)
DC	pKa			
HL 4.71E-014	atm m3/mol	25	EST	MEYLAN,WM & HOWARD,PH (1991)
OH 1.18E-010	cm3/molc sec	25	EST	MEYLAN,WM & HOWARD,PH (1993)

CAS #: 127356-05-2	1H-IMIDAZO[4,5-B]PYRIDIN-6-AMINE, 2-[2-METHOXY-4
Formula: $C_{14}H_{14}N_4O_2S$	
Mol Weight: 302.36	
MP (deg C):	FP (deg C):
BP (deg C):	
BP pressure (mm Hg):	

Property/Value	Units	Temp	Data Type	Reference
WS 6.22E+003	mg/L	25	EST	MEYLAN,WM ET AL. (1996)
logP 0.33			EXP	SANGSTER,J (1994)
VP 4.48E-013	mm Hg	25	EST	NEELY,WB & BLAU,GE (1985)
DC	pKa			
HL 1.42E-021	atm m3/mol	25	EST	MEYLAN,WM & HOWARD,PH (1991)
OH 2.65E-010	cm3/molc sec	25	EST	MEYLAN,WM & HOWARD,PH (1993)

CAS #: 127356-43-8	1H-IMIDAZO[4,5-B]PYRIDIN-6-AMINE, 2-(2,4-DIMETHO
Formula: $C_{14}H_{14}N_4O_2$	
Mol Weight: 270.29	
MP (deg C):	FP (deg C):
BP (deg C):	
BP pressure (mm Hg):	

Property/Value	Units	Temp	Data Type	Reference
WS 7.53E+002	mg/L	25	EST	MEYLAN,WM ET AL. (1996)
logP 1.62			EXP	SANGSTER,J (1994)
VP 3.98E-011	mm Hg	25	EST	NEELY,WB & BLAU,GE (1985)
DC	pKa			
HL 4.56E-017	atm m3/mol	25	EST	MEYLAN,WM & HOWARD,PH (1991)
OH 3.71E-010	cm3/molc sec	25	EST	MEYLAN,WM & HOWARD,PH (1993)

CAS #: 127390-68-5	1H-PYRROLE-3,4-DICARBOXYLIC ACID, 2-(6-CHLORO-2-
Formula: $C_{14}H_{13}ClN_2O_4$	
Mol Weight: 308.72	
MP (deg C):	FP (deg C):
BP (deg C):	
BP pressure (mm Hg):	

Property/Value	Units	Temp	Data Type	Reference
WS 5.96E+001	mg/L	25	EST	MEYLAN,WM ET AL. (1996)
logP 2.65			EXP	SANGSTER,J (1994)
VP 1.75E-007	mm Hg	25	EST	NEELY,WB & BLAU,GE (1985)
DC	pKa			
HL 1.95E-012	atm m3/mol	25	EST	MEYLAN,WM & HOWARD,PH (1991)
OH 3.90E-011	cm3/molc sec	25	EST	MEYLAN,WM & HOWARD,PH (1993)

CAS #: 127592-40-9	URIDINE, 5-CHLORO-2',3'-DIDEOXY-
Formula: $C_9H_{11}ClN_2O_4$	
Mol Weight: 246.65	
MP (deg C):	FP (deg C):
BP (deg C):	
BP pressure (mm Hg):	

Property/Value	Units	Temp	Data Type	Reference
WS 1.07E+004	mg/L	25	EST	MEYLAN,WM ET AL. (1996)
logP -1.11			EXP	BALZARINI,J ET AL. (1989)
VP 1.97E-012	mm Hg	25	EST	NEELY,WB & BLAU,GE (1985)
DC	pKa			
HL 1.52E-016	atm m3/mol	25	EST	MEYLAN,WM & HOWARD,PH (1991)
OH 7.43E-011	cm3/molc sec	25	EST	MEYLAN,WM & HOWARD,PH (1993)

CAS #: 127792-23-8				PHENYLACETIC ACID,2-(2'-CHLORO-6'-BROMO)ANILINO

Formula: $C_{14}H_{11}BrClNO_2$

Mol Weight: 340.61

MP (deg C):   FP (deg C):

BP (deg C):

BP pressure (mm Hg):

Property/Value	Units	Temp	Data Type	Reference
WS 1.91E+000	mg/L	25	EST	MEYLAN,WM ET AL. (1996)
logP 4.64			EXP	HANSCH,C ET AL. (1995)
VP 2.52E-008	mm Hg	25	EST	NEELY,WB & BLAU,GE (1985)
DC	pKa			
HL 2.55E-012	atm m3/mol	25	EST	MEYLAN,WM & HOWARD,PH (1991)
OH 1.65E-010	cm3/molc sec	25	EST	MEYLAN,WM & HOWARD,PH (1993)

CAS #: 127792-24-9				PHNEYLACETIC ACID,2-(2'-CHLORO-6'-IODO)ANILINO

Formula: $C_{14}H_{11}ClFNO_2$

Mol Weight: 279.70

MP (deg C):   FP (deg C):

BP (deg C):

BP pressure (mm Hg):

Property/Value	Units	Temp	Data Type	Reference
WS 7.22E-001	mg/L	25	EST	MEYLAN,WM ET AL. (1996)
logP 4.80			EXP	HANSCH,C ET AL. (1995)
VP 7.07E-009	mm Hg	25	EST	NEELY,WB & BLAU,GE (1985)
DC	pKa			
HL 1.48E-012	atm m3/mol	25	EST	MEYLAN,WM & HOWARD,PH (1991)
OH 1.67E-010	cm3/molc sec	25	EST	MEYLAN,WM & HOWARD,PH (1993)

CAS #: 127792-31-8				PHENYLACETIC ACID,2-(2',5'-DICHLORO)ANILINO

Formula: $C_{14}H_{11}Cl_2NO_2$

Mol Weight: 296.16

MP (deg C):   FP (deg C):

BP (deg C):

BP pressure (mm Hg):

Property/Value	Units	Temp	Data Type	Reference
WS 1.94E+000	mg/L	25	EST	MEYLAN,WM ET AL. (1996)
logP 4.94			EXP	HANSCH,C ET AL. (1995)
VP 6.14E-008	mm Hg	25	EST	NEELY,WB & BLAU,GE (1985)
DC	pKa			
HL 4.73E-012	atm m3/mol	25	EST	MEYLAN,WM & HOWARD,PH (1991)
OH 1.82E-010	cm3/molc sec	25	EST	MEYLAN,WM & HOWARD,PH (1993)

CAS #: 127792-33-0				PHENYLACETIC ACID,2-(3',4'-DICHLORO)ANILINO

Formula: $C_{14}H_{11}Cl_2NO_2$

Mol Weight: 296.16

MP (deg C):   FP (deg C):

BP (deg C):

BP pressure (mm Hg):

Property/Value	Units	Temp	Data Type	Reference
WS 2.10E+000	mg/L	25	EST	MEYLAN,WM ET AL. (1996)
logP 4.90			EXP	HANSCH,C ET AL. (1995)
VP 6.14E-008	mm Hg	25	EST	NEELY,WB & BLAU,GE (1985)
DC	pKa			
HL 4.73E-012	atm m3/mol	25	EST	MEYLAN,WM & HOWARD,PH (1991)
OH 1.82E-010	cm3/molc sec	25	EST	MEYLAN,WM & HOWARD,PH (1993)

CAS #: 127836-03-7				2,4-DIF 3,5-DICL ACETANILIDE

Formula: $C_8H_5Cl_2F_2NO$

Mol Weight: 240.04

MP (deg C):   FP (deg C):

BP (deg C):

BP pressure (mm Hg):

Property/Value	Units	Temp	Data Type	Reference
WS 8.96E+001	mg/L	25	EST	MEYLAN,WM ET AL. (1996)
logP 2.90			EXP	NAKAGAWA,Y ET AL. (1992)
VP 3.05E-005	mm Hg	25	EST	NEELY,WB & BLAU,GE (1985)
DC	pKa			
HL 4.61E-009	atm m3/mol	25	EST	MEYLAN,WM & HOWARD,PH (1991)
OH 8.02E-013	cm3/molc sec	25	EST	MEYLAN,WM & HOWARD,PH (1993)

CAS #: 127944-48-3				2-OXAZOLIDINONE DERIVATIVE

Formula: $C_{35}H_{51}N_7O_8$

Mol Weight: 697.84

MP (deg C):   FP (deg C):

BP (deg C):

BP pressure (mm Hg):

Property/Value	Units	Temp	Data Type	Reference
WS 5.72E-001	mg/L	25	EST	MEYLAN,WM ET AL. (1996)
logP 2.11			EXP	SANGSTER,J (1993)
VP 5.22E-033	mm Hg	25	EST	NEELY,WB & BLAU,GE (1985)
DC	pKa			
HL 1.74E-037	atm m3/mol	25	EST	MEYLAN,WM & HOWARD,PH (1991)
OH 2.43E-010	cm3/molc sec	25	EST	MEYLAN,WM & HOWARD,PH (1993)

CAS #: 127944-49-4				2-OXAZOLIDINONE DERIVATIVE

Formula: $C_{35}H_{51}N_7O_9$

Mol Weight: 713.84

MP (deg C):   FP (deg C):

BP (deg C):

BP pressure (mm Hg):

Property/Value	Units	Temp	Data Type	Reference
WS 2.58E+000	mg/L	25	EST	MEYLAN,WM ET AL. (1996)
logP 1.22			EXP	SANGSTER,J (1993)
VP 1.36E-037	mm Hg	25	EST	NEELY,WB & BLAU,GE (1985)
DC	pKa			
HL 2.63E-038	atm m3/mol	25	EST	MEYLAN,WM & HOWARD,PH (1991)
OH 2.27E-010	cm3/molc sec	25	EST	MEYLAN,WM & HOWARD,PH (1993)

CAS #: 127944-50-7				2-OXAZOLIDINONE DERIVATIVE

Formula: $C_{35}H_{53}N_7O_9$

Mol Weight: 715.85

MP (deg C):   FP (deg C):

BP (deg C):

BP pressure (mm Hg):

Property/Value	Units	Temp	Data Type	Reference
WS 1.69E+000	mg/L	25	EST	MEYLAN,WM ET AL. (1996)
logP 1.42			EXP	SANGSTER,J (1993)
VP 1.75E-037	mm Hg	25	EST	NEELY,WB & BLAU,GE (1985)
DC	pKa			
HL 5.97E-038	atm m3/mol	25	EST	MEYLAN,WM & HOWARD,PH (1991)
OH 2.30E-010	cm3/molc sec	25	EST	MEYLAN,WM & HOWARD,PH (1993)

127944-51-8 — 2-OXAZOLIDINONE DERIVATIVE

CAS #: 127944-51-8

Formula: $C_{35}H_{52}N_6O_7$

Mol Weight: 668.84

MP (deg C):
FP (deg C):
BP (deg C):
BP pressure (mm Hg):

Property/Value	Units	Temp	Data Type	Reference
WS 4.57E-001	mg/L	25	EST	MEYLAN,WM ET AL. (1996)
logP 2.45			EXP	SANGSTER,J (1993)
VP 6.91E-031	mm Hg	25	EST	NEELY,WB & BLAU,GE (1985)
DC	pKa			
HL 9.80E-033	atm m3/mol	25	EST	MEYLAN,WM & HOWARD,PH (1991)
OH 2.25E-010	cm3/molc sec	25	EST	MEYLAN,WM & HOWARD,PH (1993)

127944-53-0 — 2-OXAZOLIDINONE DERIVATIVE

CAS #: 127944-53-0

Formula: $C_{36}H_{54}N_6O_8$

Mol Weight: 698.87

MP (deg C):
FP (deg C):
BP (deg C):
BP pressure (mm Hg):

Property/Value	Units	Temp	Data Type	Reference
WS 2.67E-001	mg/L	25	EST	MEYLAN,WM ET AL. (1996)
logP 2.49			EXP	SANGSTER,J (1993)
VP 1.07E-031	mm Hg	25	EST	NEELY,WB & BLAU,GE (1985)
DC	pKa			
HL 5.75E-034	atm m3/mol	25	EST	MEYLAN,WM & HOWARD,PH (1991)
OH 2.29E-010	cm3/molc sec	25	EST	MEYLAN,WM & HOWARD,PH (1993)

127944-54-1 — 2-OXAZOLIDINONE DERIVATIVE

CAS #: 127944-54-1

Formula: $C_{38}H_{59}N_7O_{10}$

Mol Weight: 773.93

MP (deg C):
FP (deg C):
BP (deg C):
BP pressure (mm Hg):

Property/Value	Units	Temp	Data Type	Reference
WS 2.07E-001	mg/L	25	EST	MEYLAN,WM ET AL. (1996)
logP 2.03			EXP	SANGSTER,J (1993)
VP 1.06E-034	mm Hg	25	EST	NEELY,WB & BLAU,GE (1985)
DC	pKa			
HL 3.59E-040	atm m3/mol	25	EST	MEYLAN,WM & HOWARD,PH (1991)
OH 2.50E-010	cm3/molc sec	25	EST	MEYLAN,WM & HOWARD,PH (1993)

127944-55-2 — 2-OXAZOLIDINONE DERIVATIVE

CAS #: 127944-55-2

Formula: $C_{39}H_{59}N_7O_{11}$

Mol Weight: 801.95

MP (deg C):
FP (deg C):
BP (deg C):
BP pressure (mm Hg):

Property/Value	Units	Temp	Data Type	Reference
WS 1.27E-001	mg/L	25	EST	MEYLAN,WM ET AL. (1996)
logP 2.06			EXP	SANGSTER,J (1993)
VP 1.63E-035	mm Hg	25	EST	NEELY,WB & BLAU,GE (1985)
DC	pKa			
HL 1.85E-042	atm m3/mol	25	EST	MEYLAN,WM & HOWARD,PH (1991)
OH 2.45E-010	cm3/molc sec	25	EST	MEYLAN,WM & HOWARD,PH (1993)

127944-56-3 — 2-OXAZOLIDINONE DERIVATIVE

CAS #: 127944-56-3

Formula: $C_{43}H_{67}N_7O_{13}$

Mol Weight: 890.05

MP (deg C):
FP (deg C):
BP (deg C):
BP pressure (mm Hg):

Property/Value	Units	Temp	Data Type	Reference
WS 6.91E-002	mg/L	25	EST	MEYLAN,WM ET AL. (1996)
logP 1.67			EXP	SANGSTER,J (1993)
VP 5.51E-038	mm Hg	25	EST	NEELY,WB & BLAU,GE (1985)
DC	pKa			
HL	atm m3/mol			
OH 2.73E-010	cm3/molc sec	25	EST	MEYLAN,WM & HOWARD,PH (1993)

127944-57-4 — 2-OXAZOLIDINONE DERIVATIVE

CAS #: 127944-57-4

Formula: $C_{43}H_{66}N_6O_{12}$

Mol Weight: 859.04

MP (deg C):
FP (deg C):
BP (deg C):
BP pressure (mm Hg):

Property/Value	Units	Temp	Data Type	Reference
WS 7.57E-002	mg/L	25	EST	MEYLAN,WM ET AL. (1996)
logP 1.87			EXP	HANSCH,C ET AL. (1995)
VP 3.53E-036	mm Hg	25	EST	NEELY,WB & BLAU,GE (1985)
DC	pKa			
HL 1.11E-041	atm m3/mol	25	EST	MEYLAN,WM & HOWARD,PH (1991)
OH 2.66E-010	cm3/molc sec	25	EST	MEYLAN,WM & HOWARD,PH (1993)

128104-25-6 — 1H-IMIDAZOLE-1-ETHANOL, -(4-BROMOPHENYL)- -(4-F

CAS #: 128104-25-6

Formula: $C_{17}H_{14}BrFN_2O$

Mol Weight: 361.22

MP (deg C):
FP (deg C):
BP (deg C):
BP pressure (mm Hg):

Property/Value	Units	Temp	Data Type	Reference
WS 6.36E+000	mg/L	25	EST	MEYLAN,WM ET AL. (1996)
logP 3.42			EXP	SANGSTER,J (1994)
VP 3.24E-011	mm Hg	25	EST	NEELY,WB & BLAU,GE (1985)
DC	pKa			
HL 1.18E-011	atm m3/mol	25	EST	MEYLAN,WM & HOWARD,PH (1991)
OH 4.41E-011	cm3/molc sec	25	EST	MEYLAN,WM & HOWARD,PH (1993)

128104-27-8 — 1-PROPANOL, 1,1-BIS(4-FLUOROPHENYL)-2-(1H-IMIDAZ

CAS #: 128104-27-8

Formula: $C_{18}H_{16}F_2N_2O$

Mol Weight: 314.34

MP (deg C):
FP (deg C):
BP (deg C):
BP pressure (mm Hg):

Property/Value	Units	Temp	Data Type	Reference
WS 2.15E+001	mg/L	25	EST	MEYLAN,WM ET AL. (1996)
logP 3.13			EXP	SANGSTER,J (1994)
VP 4.64E-010	mm Hg	25	EST	NEELY,WB & BLAU,GE (1985)
DC	pKa			
HL 4.58E-011	atm m3/mol	25	EST	MEYLAN,WM & HOWARD,PH (1991)
OH 4.95E-011	cm3/molc sec	25	EST	MEYLAN,WM & HOWARD,PH (1993)

128348-32-3

CAS #: 128348-32-3

2-THIOPHENESULFONAMIDE, 4-[(4-METHYLPHENYL)SULFO

Formula: $C_{11}H_{11}NO_4S_3$

Mol Weight: 317.41

MP (deg C):   **FP (deg C):**

BP (deg C):

BP pressure (mm Hg):

Property/ Value	Units	Temp	Data Type	Reference
WS 3.50E+001	mg/L	25	EST	MEYLAN,WM ET AL. (1996)
logP 1.63			EXP	HARTMAN,GD ET AL. (1992)
VP 9.16E-010	mm Hg	25	EST	NEELY,WB & BLAU,GE (1985)
DC	pKa			
HL 1.17E-011	atm m3/mol	25	EST	MEYLAN,WM & HOWARD,PH (1991)
OH 1.66E-012	cm3/molc sec	25	EST	MEYLAN,WM & HOWARD,PH (1993)

128348-34-5

CAS #: 128348-34-5

2-FURANSULFONAMIDE, 4-[(4-METHYLPHENYL)SULFONYL]

Formula: $C_{11}H_{11}NO_5S_2$

Mol Weight: 301.34

MP (deg C):   **FP (deg C):**

BP (deg C):

BP pressure (mm Hg):

Property/ Value	Units	Temp	Data Type	Reference
WS 5.73E+001	mg/L	25	EST	MEYLAN,WM ET AL. (1996)
logP 1.49			EXP	HARTMAN,GD ET AL. (1992)
VP 5.64E-009	mm Hg	25	EST	NEELY,WB & BLAU,GE (1985)
DC	pKa			
HL 2.15E-011	atm m3/mol	25	EST	MEYLAN,WM & HOWARD,PH (1991)
OH 3.08E-012	cm3/molc sec	25	EST	MEYLAN,WM & HOWARD,PH (1993)

128348-35-6

CAS #: 128348-35-6

2-FURANSULFONAMIDE, 4-[(4-METHOXYPHENYL)SULFONYL

Formula: $C_{11}H_{11}NO_6S_2$

Mol Weight: 317.34

MP (deg C):   **FP (deg C):**

BP (deg C):

BP pressure (mm Hg):

Property/ Value	Units	Temp	Data Type	Reference
WS 8.65E+001	mg/L	25	EST	MEYLAN,WM ET AL. (1996)
logP 1.17			EXP	HARTMAN,GD ET AL. (1992)
VP 2.42E-009	mm Hg	25	EST	NEELY,WB & BLAU,GE (1985)
DC	pKa			
HL 1.15E-012	atm m3/mol	25	EST	MEYLAN,WM & HOWARD,PH (1991)
OH 7.28E-012	cm3/molc sec	25	EST	MEYLAN,WM & HOWARD,PH (1993)

128348-42-5

CAS #: 128348-42-5

2-THIOPHENESULFONAMIDE, 4-[(4-HYDROXYPHENYL)SULF

Formula: $C_{10}H_9NO_5S_3$

Mol Weight: 319.38

MP (deg C):   **FP (deg C):**

BP (deg C):

BP pressure (mm Hg):

Property/ Value	Units	Temp	Data Type	Reference
WS 3.60E+002	mg/L	25	EST	MEYLAN,WM ET AL. (1996)
logP 1.11			EXP	HARTMAN,GD ET AL. (1992)
VP 2.22E-011	mm Hg	25	EST	NEELY,WB & BLAU,GE (1985)
DC	pKa			
HL 1.10E-015	atm m3/mol	25	EST	MEYLAN,WM & HOWARD,PH (1991)
OH 7.70E-012	cm3/molc sec	25	EST	MEYLAN,WM & HOWARD,PH (1993)

128348-43-6

CAS #: 128348-43-6

2-THIOPHENESULFONAMIDE, 4-[(3-HYDROXYPHENYL)SULF

Formula: $C_{10}H_{11}NO_5S_3$

Mol Weight: 321.39

MP (deg C):   **FP (deg C):**

BP (deg C):

BP pressure (mm Hg):

Property/ Value	Units	Temp	Data Type	Reference
WS 2.04E+002	mg/L	25	EST	MEYLAN,WM ET AL. (1996)
logP 1.40			EXP	HARTMAN,GD ET AL. (1992)
VP 2.22E-011	mm Hg	25	EST	NEELY,WB & BLAU,GE (1985)
DC	pKa			
HL 1.10E-015	atm m3/mol	25	EST	MEYLAN,WM & HOWARD,PH (1991)
OH 7.70E-012	cm3/molc sec	25	EST	MEYLAN,WM & HOWARD,PH (1993)

128348-44-7

CAS #: 128348-44-7

2-FURANSULFONAMIDE, 4-[(4-HYDROXYPHENYL)SULFONYL

Formula: $C_{10}H_9NO_6S_2$

Mol Weight: 303.31

MP (deg C):   **FP (deg C):**

BP (deg C):

BP pressure (mm Hg):

Property/ Value	Units	Temp	Data Type	Reference
WS 4.83E+003	mg/L	25	EST	MEYLAN,WM ET AL. (1996)
logP -0.10			EXP	HARTMAN,GD ET AL. (1992)
VP 1.58E-010	mm Hg	25	EST	NEELY,WB & BLAU,GE (1985)
DC	pKa			
HL 2.03E-015	atm m3/mol	25	EST	MEYLAN,WM & HOWARD,PH (1991)
OH 9.12E-012	cm3/molc sec	25	EST	MEYLAN,WM & HOWARD,PH (1993)

128496-09-3

CAS #: 128496-09-3

2(1H)-PYRIMIDINONE, 4-AMINO-1-(2,3-DIDEOXY-2-FLU

Formula: $C_9H_{11}F_2N_3O_3$

Mol Weight: 247.20

MP (deg C):   **FP (deg C):**

BP (deg C):

BP pressure (mm Hg):

Property/ Value	Units	Temp	Data Type	Reference
WS 7.33E+003	mg/L	25	EST	MEYLAN,WM ET AL. (1996)
logP -0.92			EXP	SANGSTER,J (1994)
VP 6.54E-008	mm Hg	25	EST	NEELY,WB & BLAU,GE (1985)
DC	pKa			
HL 4.46E-016	atm m3/mol	25	EST	MEYLAN,WM & HOWARD,PH (1991)
OH 2.54E-011	cm3/molc sec	25	EST	MEYLAN,WM & HOWARD,PH (1993)

128887-24-1

CAS #: 128887-24-1

[3,4'-BIPYRIDINE]-5-CARBONITRILE,6-ETHOXY-

Formula: $C_{13}H_{11}N_3O$

Mol Weight: 225.25

MP (deg C):   **FP (deg C):**

BP (deg C):

BP pressure (mm Hg):

Property/ Value	Units	Temp	Data Type	Reference
WS 1.08E+002	mg/L	25	EST	MEYLAN,WM ET AL. (1996)
logP 2.59			EXP	HAGEN,V ET AL. (1990)
VP 2.01E-006	mm Hg	25	EST	NEELY,WB & BLAU,GE (1985)
DC	pKa			
HL 4.90E-012	atm m3/mol	25	EST	MEYLAN,WM & HOWARD,PH (1991)
OH 6.72E-012	cm3/molc sec	25	EST	MEYLAN,WM & HOWARD,PH (1993)

CAS #: 128887-26-3				[3,4'-BIPYRIDINE]-5-CARBONITRILE,6-METHOXY-2-MET
Formula: $C_{13}H_{11}N_3O$				
Mol Weight: 225.25				
MP (deg C):		FP (deg C):		
BP (deg C):				
BP pressure (mm Hg):				

Property/ Value	Units	Temp	Data Type	Reference
WS 1.53E+002	mg/L	25	EST	MEYLAN,WM ET AL. (1996)
logP 2.41			EXP	HAGEN,V ET AL. (1990)
VP 1.88E-006	mm Hg	25	EST	NEELY,WB & BLAU,GE (1985)
DC	pKa			
HL 4.08E-012	atm m3/mol	25	EST	MEYLAN,WM & HOWARD,PH (1991)
OH 1.53E-012	cm3/molc sec	25	EST	MEYLAN,WM & HOWARD,PH (1993)

CAS #: 129016-38-2				3(2H)-PYRIDAZINONE, 6-HYDROXY-
Formula: $C_4H_4N_2O_2$				
Mol Weight: 112.09				
MP (deg C):		FP (deg C):		
BP (deg C):				
BP pressure (mm Hg):				

Property/ Value	Units	Temp	Data Type	Reference
WS 3.08E+004	mg/L	25	EST	MEYLAN,WM ET AL. (1996)
logP -0.90			EXP	BAKER,EA ET AL. (1992)
VP 1.38E-007	mm Hg	25	EST	NEELY,WB & BLAU,GE (1985)
DC	pKa			
HL 3.00E-013	atm m3/mol	25	EST	MEYLAN,WM & HOWARD,PH (1991)
OH 1.69E-011	cm3/molc sec	25	EST	MEYLAN,WM & HOWARD,PH (1993)

CAS #: 129358-45-8				6-HEPTENOIC ACID, 7-[5-(4-FLUOROPHENYL)-2-(1-MET
Formula: $C_{24}H_{25}FN_3NaO_4$				
Mol Weight: 461.47				
MP (deg C):		FP (deg C):		
BP (deg C):				
BP pressure (mm Hg):				

Property/ Value	Units	Temp	Data Type	Reference
WS 2.94E+003	mg/L	25	EST	MEYLAN,WM ET AL. (1996)
logP -0.43			EXP	SANGSTER,J (1994)
VP 6.25E-027	mm Hg	25	EST	NEELY,WB & BLAU,GE (1985)
DC	pKa			
HL	atm m3/mol			
OH 1.18E-010	cm3/molc sec	25	EST	MEYLAN,WM & HOWARD,PH (1993)

CAS #: 129358-65-2				6-HEPTENOIC ACID,7[4(4-CHLORO-3,5-DIMEPH]-5-(4-F
Formula: $C_{27}H_{29}ClFN_2NaO_4$				
Mol Weight: 522.98				
MP (deg C):		FP (deg C):		
BP (deg C):				
BP pressure (mm Hg):				

Property/ Value	Units	Temp	Data Type	Reference
WS 5.87E+000	mg/L	25	EST	MEYLAN,WM ET AL. (1996)
logP 2.27			EXP	SANGSTER,J (1994)
VP 3.34E-028	mm Hg	25	EST	NEELY,WB & BLAU,GE (1985)
DC	pKa			
HL	atm m3/mol			
OH 1.25E-010	cm3/molc sec	25	EST	MEYLAN,WM & HOWARD,PH (1993)

CAS #: 129358-82-3				6-HEPTENOIC ACID, 7-[4,5-BIS(4-FLUOROPHENYL)-2-(
Formula: $C_{25}H_{25}F_2N_2NaO_4$				
Mol Weight: 478.48				
MP (deg C):		FP (deg C):		
BP (deg C):				
BP pressure (mm Hg):				

Property/ Value	Units	Temp	Data Type	Reference
WS 2.39E+002	mg/L	25	EST	MEYLAN,WM ET AL. (1996)
logP 0.72			EXP	SANGSTER,J (1994)
VP 1.20E-026	mm Hg	25	EST	NEELY,WB & BLAU,GE (1985)
DC	pKa			
HL	atm m3/mol			
OH 1.19E-010	cm3/molc sec	25	EST	MEYLAN,WM & HOWARD,PH (1993)

CAS #: 129358-85-6				6-HEPTENOIC ACID, 7-[4,5-BIS(4-FLUOROPHENYL)-2-(
Formula: $C_{23}H_{18}F_5N_2NaO_4$				
Mol Weight: 504.39				
MP (deg C):		FP (deg C):		
BP (deg C):				
BP pressure (mm Hg):				

Property/ Value	Units	Temp	Data Type	Reference
WS 1.60E+002	mg/L	25	EST	MEYLAN,WM ET AL. (1996)
logP 0.73			EXP	SANGSTER,J (1994)
VP 8.38E-026	mm Hg	25	EST	NEELY,WB & BLAU,GE (1985)
DC	pKa			
HL	atm m3/mol			
OH 1.17E-010	cm3/molc sec	25	EST	MEYLAN,WM & HOWARD,PH (1993)

CAS #: 129586-32-9				SSF-109 FUNGICIDE
Formula: $C_{15}H_{20}ClN_3O$				
Mol Weight: 293.80				
MP (deg C):		FP (deg C):		
BP (deg C):				
BP pressure (mm Hg):				

Property/ Value	Units	Temp	Data Type	Reference
WS 1.17E+002	mg/L	25	EST	MEYLAN,WM ET AL. (1996)
logP 3.02			EXP	TOMLIN,C (1994)
VP 5.08E-009	mm Hg	25	EST	NEELY,WB & BLAU,GE (1985)
DC	pKa			
HL 1.72E-010	atm m3/mol	25	EST	MEYLAN,WM & HOWARD,PH (1991)
OH 2.10E-011	cm3/molc sec	25	EST	MEYLAN,WM & HOWARD,PH (1993)

CAS #: 129625-27-0				BENZOIC ACID, 5-(AMINOSULFONYL)-4-CHLORO-2-[[(2-
Formula: $C_{16}H_{17}ClN_2O_7S$				
Mol Weight: 416.84				
MP (deg C):		FP (deg C):		
BP (deg C):				
BP pressure (mm Hg):				

Property/ Value	Units	Temp	Data Type	Reference
WS 8.03E+000	mg/L	25	EST	MEYLAN,WM ET AL. (1996)
logP 2.90			EXP	SANGSTER,J (1993)
VP 1.46E-010	mm Hg	25	EST	NEELY,WB & BLAU,GE (1985)
DC	pKa			
HL 3.98E-016	atm m3/mol	25	EST	MEYLAN,WM & HOWARD,PH (1991)
OH 1.15E-010	cm3/molc sec	25	EST	MEYLAN,WM & HOWARD,PH (1993)

CAS #: 129865-48-1				1(2H)-PYRIDINEETHANOL, 3,6-DIHYDRO-4-PHENYL-
Formula: $C_{13}H_{17}NO$				
Mol Weight: 203.29				
MP (deg C):		FP (deg C):		
BP (deg C):				
BP pressure (mm Hg):				

Property/ Value	Units	Temp	Data Type	Reference
WS 8.40E+003	mg/L	25	EST	MEYLAN,WM ET AL. (1996)
logP 2.00			EXP	ALTOMARE,CA ET AL. (1992)
VP 2.70E-006	mm Hg	25	EST	NEELY,WB & BLAU,GE (1985)
DC	pKa			
HL 7.20E-011	atm m3/mol	25	EST	MEYLAN,WM & HOWARD,PH (1991)
OH 1.90E-010	cm3/molc sec	25	EST	MEYLAN,WM & HOWARD,PH (1993)

CAS #: 130179-73-6				BENZAMIDE, 4-(1H-IMIDAZO[4,5-C]PYRIDIN-2-YL)-3-M
Formula: $C_{14}H_{12}N_4O_2$				
Mol Weight: 268.28				
MP (deg C):		FP (deg C):		
BP (deg C):				
BP pressure (mm Hg):				

Property/ Value	Units	Temp	Data Type	Reference
WS 1.84E+003	mg/L	25	EST	MEYLAN,WM ET AL. (1996)
logP 1.18			EXP	SANGSTER,J (1994)
VP 1.52E-012	mm Hg	25	EST	NEELY,WB & BLAU,GE (1985)
DC	pKa			
HL 8.96E-019	atm m3/mol	25	EST	MEYLAN,WM & HOWARD,PH (1991)
OH 7.18E-011	cm3/molc sec	25	EST	MEYLAN,WM & HOWARD,PH (1993)

CAS #: 130336-12-8				3-AZAGLUTARAMIDE ANALOG
Formula: $C_{44}H_{75}N_5O_7$				
Mol Weight: 786.12				
MP (deg C):		FP (deg C):		
BP (deg C):				
BP pressure (mm Hg):				

Property/ Value	Units	Temp	Data Type	Reference
WS 1.34E-003	mg/L	25	EST	MEYLAN,WM ET AL. (1996)
logP 4.50			EXP	SANGSTER,J (1994)
VP 2.80E-029	mm Hg	25	EST	NEELY,WB & BLAU,GE (1985)
DC	pKa			
HL 7.74E-031	atm m3/mol	25	EST	MEYLAN,WM & HOWARD,PH (1991)
OH 3.79E-010	cm3/molc sec	25	EST	MEYLAN,WM & HOWARD,PH (1993)

CAS #: 130787-06-3				ACETAMIDE, N-(2-CYCLOPENTYL-2H-BENZOTRIAZOL-5-YL
Formula: $C_{13}H_{16}N_4O$				
Mol Weight: 244.30				
MP (deg C):		FP (deg C):		
BP (deg C):				
BP pressure (mm Hg):				

Property/ Value	Units	Temp	Data Type	Reference
WS 2.06E+002	mg/L	25	EST	MEYLAN,WM ET AL. (1996)
logP 2.45			EXP	CALIENDO,G ET AL. (1990)
VP 1.81E-008	mm Hg	25	EST	NEELY,WB & BLAU,GE (1985)
DC	pKa			
HL 4.91E-012	atm m3/mol	25	EST	MEYLAN,WM & HOWARD,PH (1991)
OH 1.33E-011	cm3/molc sec	25	EST	MEYLAN,WM & HOWARD,PH (1993)

CAS #: 130787-07-4				ACETAMIDE, 2-CHLORO-N-(2-CYCLOPENTYL-2H-BENZOTRI
Formula: $C_{13}H_{15}ClN_4O$				
Mol Weight: 278.74				
MP (deg C):		FP (deg C):		
BP (deg C):				
BP pressure (mm Hg):				

Property/ Value	Units	Temp	Data Type	Reference
WS 6.88E+001	mg/L	25	EST	MEYLAN,WM ET AL. (1996)
logP 2.78			EXP	CALIENDO,G ET AL. (1990)
VP 2.06E-009	mm Hg	25	EST	NEELY,WB & BLAU,GE (1985)
DC	pKa			
HL 1.73E-012	atm m3/mol	25	EST	MEYLAN,WM & HOWARD,PH (1991)
OH 1.35E-011	cm3/molc sec	25	EST	MEYLAN,WM & HOWARD,PH (1993)

CAS #: 130787-08-5				ACETAMIDE, 2,2-DICHLORO-N-(2-CYCLOPENTYL-2H-BENZ
Formula: $C_{13}H_{14}Cl_2N_4O$				
Mol Weight: 313.19				
MP (deg C):		FP (deg C):		
BP (deg C):				
BP pressure (mm Hg):				

Property/ Value	Units	Temp	Data Type	Reference
WS 1.28E+001	mg/L	25	EST	MEYLAN,WM ET AL. (1996)
logP 3.40			EXP	CALIENDO,G ET AL. (1990)
VP 9.24E-010	mm Hg	25	EST	NEELY,WB & BLAU,GE (1985)
DC	pKa			
HL 6.09E-013	atm m3/mol	25	EST	MEYLAN,WM & HOWARD,PH (1991)
OH 1.34E-011	cm3/molc sec	25	EST	MEYLAN,WM & HOWARD,PH (1993)

CAS #: 130787-09-6				ACETAMIDE, 2,2,2-TRICHLORO-N-(2-CYCLOPENTYL-2H-B
Formula: $C_{13}H_{13}Cl_3N_4O$				
Mol Weight: 347.63				
MP (deg C):		FP (deg C):		
BP (deg C):				
BP pressure (mm Hg):				

Property/ Value	Units	Temp	Data Type	Reference
WS 1.94E+000	mg/L	25	EST	MEYLAN,WM ET AL. (1996)
logP 4.12			EXP	CALIENDO,G ET AL. (1990)
VP 8.48E-010	mm Hg	25	EST	NEELY,WB & BLAU,GE (1985)
DC	pKa			
HL 2.15E-013	atm m3/mol	25	EST	MEYLAN,WM & HOWARD,PH (1991)
OH 1.32E-011	cm3/molc sec	25	EST	MEYLAN,WM & HOWARD,PH (1993)

CAS #: 130787-12-1				2H-BENZOTRIAZOLE, 2-CYCLOPENTYL-5-[(ETHOXYCARBON
Formula: $C_{14}H_{18}N_4O_2$				
Mol Weight: 274.33				
MP (deg C):		FP (deg C):		
BP (deg C):				
BP pressure (mm Hg):				

Property/ Value	Units	Temp	Data Type	Reference
WS 2.03E+001	mg/L	25	EST	MEYLAN,WM ET AL. (1996)
logP 3.43			EXP	CALIENDO,G ET AL. (1990)
VP 2.28E-007	mm Hg	25	EST	NEELY,WB & BLAU,GE (1985)
DC	pKa			
HL 2.31E-011	atm m3/mol	25	EST	MEYLAN,WM & HOWARD,PH (1991)
OH 3.04E-011	cm3/molc sec	25	EST	MEYLAN,WM & HOWARD,PH (1993)

CAS #: 130787-13-2 — 2H-BENZOTRIAZOLE, 2-CYCLOPENTYL-5-[[(2-METHYLPRO

Formula: $C_{16}H_{22}N_4O_2$

Mol Weight: 302.38

MP (deg C):
FP (deg C):
BP (deg C):
BP pressure (mm Hg):

Property/Value	Units	Temp	Data Type	Reference
WS 2.63E+000	mg/L	25	EST	MEYLAN,WM ET AL. (1996)
logP 4.28			EXP	CALIENDO,G ET AL. (1990)
VP 7.49E-008	mm Hg	25	EST	NEELY,WB & BLAU,GE (1985)
DC	pKa			
HL 4.06E-011	atm m3/mol	25	EST	MEYLAN,WM & HOWARD,PH (1991)
OH 3.33E-011	cm3/molc sec	25	EST	MEYLAN,WM & HOWARD,PH (1993)

CAS #: 130787-14-3 — BENZAMIDE, N-(2-CYCLOPENTYL-2H-BENZOTRIAZOL-5-YL

Formula: $C_{18}H_{18}N_4O$

Mol Weight: 306.37

MP (deg C):
FP (deg C):
BP (deg C):
BP pressure (mm Hg):

Property/Value	Units	Temp	Data Type	Reference
WS 5.81E+000	mg/L	25	EST	MEYLAN,WM ET AL. (1996)
logP 3.85			EXP	CALIENDO,G ET AL. (1990)
VP 1.09E-010	mm Hg	25	EST	NEELY,WB & BLAU,GE (1985)
DC	pKa			
HL 9.72E-013	atm m3/mol	25	EST	MEYLAN,WM & HOWARD,PH (1991)
OH 1.50E-011	cm3/molc sec	25	EST	MEYLAN,WM & HOWARD,PH (1993)

CAS #: 130787-15-4 — BENZAMIDE, 4-CHLORO-N-(2-CYCLOPENTYL-2H-BENZOTRI

Formula: $C_{18}H_{17}ClN_4O$

Mol Weight: 340.82

MP (deg C):
FP (deg C):
BP (deg C):
BP pressure (mm Hg):

Property/Value	Units	Temp	Data Type	Reference
WS 2.18E+000	mg/L	25	EST	MEYLAN,WM ET AL. (1996)
logP 4.11			EXP	CALIENDO,G ET AT (1990)
VP 2.87E-011	mm Hg	25	EST	NEELY,WB & BLAU,GE (1985)
DC	pKa			
HL 7.20E-013	atm m3/mol	25	EST	MEYLAN,WM & HOWARD,PH (1991)
OH 1.45E-011	cm3/molc sec	25	EST	MEYLAN,WM & HOWARD,PH (1993)

CAS #: 130787-16-5 — BENZAMIDE, N-(2-CYCLOPENTYL-2H-BENZOTRIAZOL-5-YL

Formula: $C_{19}H_{20}N_4O_2$

Mol Weight: 336.40

MP (deg C):
FP (deg C):
BP (deg C):
BP pressure (mm Hg):

Property/Value	Units	Temp	Data Type	Reference
WS 2.71E+000	mg/L	25	EST	MEYLAN,WM ET AL. (1996)
logP 4.03			EXP	CALIENDO,G ET AL. (1992)
VP 2.01E-011	mm Hg	25	EST	NEELY,WB & BLAU,GE (1985)
DC	pKa			
HL 5.75E-014	atm m3/mol	25	EST	MEYLAN,WM & HOWARD,PH (1991)
OH 3.37E-011	cm3/molc sec	25	EST	MEYLAN,WM & HOWARD,PH (1993)

CAS #: 130787-17-6 — BENZAMIDE, N-(2-CYCLOPENTYL-2H-BENZOTRIAZOL-5-YL

Formula: $C_{19}H_{17}F_3N_4O$

Mol Weight: 374.37

MP (deg C):
FP (deg C):
BP (deg C):
BP pressure (mm Hg):

Property/Value	Units	Temp	Data Type	Reference
WS 2.61E-001	mg/L	25	EST	MEYLAN,WM ET AL. (1996)
logP 4.95			EXP	CALIENDO,G ET AL. (1990)
VP 8.58E-011	mm Hg	25	EST	NEELY,WB & BLAU,GE (1985)
DC	pKa			
HL 8.44E-012	atm m3/mol	25	EST	MEYLAN,WM & HOWARD,PH (1991)
OH 1.35E-011	cm3/molc sec	25	EST	MEYLAN,WM & HOWARD,PH (1993)

CAS #: 130787-18-7 — BENZAMIDE, N-(2-CYCLOPENTYL-2H-BENZOTRIAZOL-5-YL

Formula: $C_{18}H_{17}N_5O_3$

Mol Weight: 351.37

MP (deg C):
FP (deg C):
BP (deg C):
BP pressure (mm Hg):

Property/Value	Units	Temp	Data Type	Reference
WS 1.55E+000	mg/L	25	EST	MEYLAN,WM ET AL. (1996)
logP 3.75			EXP	CALIENDO,G ET AL. (1990)
VP 1.98E-012	mm Hg	25	EST	NEELY,WB & BLAU,GE (1985)
DC	pKa			
HL 3.83E-015	atm m3/mol	25	EST	MEYLAN,WM & HOWARD,PH (1991)
OH 1.34E-011	cm3/molc sec	25	EST	MEYLAN,WM & HOWARD,PH (1993)

CAS #: 130787-19-8 — BENZAMIDE, N-(2-CYCLOPENTYL-2H-BENZOTRIAZOL-5-YL

Formula: $C_{18}H_{16}Cl_2N_4O$

Mol Weight: 375.26

MP (deg C):
FP (deg C):
BP (deg C):
BP pressure (mm Hg):

Property/Value	Units	Temp	Data Type	Reference
WS 1.25E-001	mg/L	25	EST	MEYLAN,WM ET AL. (1996)
logP 5.32			EXP	CALIENDO,G ET AL. (1990)
VP 7.47E-012	mm Hg	25	EST	NEELY,WB & BLAU,GE (1985)
DC	pKa			
HL 5.33E-013	atm m3/mol	25	EST	MEYLAN,WM & HOWARD,PH (1991)
OH 1.36E-011	cm3/molc sec	25	EST	MEYLAN,WM & HOWARD,PH (1993)

CAS #: 130787-20-1 — BENZAMIDE, N-(2-CYCLOPENTYL-2H-BENZOTRIAZOL-5-YL

Formula: $C_{21}H_{24}N_4O_4$

Mol Weight: 396.45

MP (deg C):
FP (deg C):
BP (deg C):
BP pressure (mm Hg):

Property/Value	Units	Temp	Data Type	Reference
WS 1.60E+000	mg/L	25	EST	MEYLAN,WM ET AL. (1996)
logP 3.87			EXP	CALIENDO,G ET AL. (1990)
VP 6.52E-013	mm Hg	25	EST	NEELY,WB & BLAU,GE (1985)
DC	pKa			
HL 2.01E-016	atm m3/mol	25	EST	MEYLAN,WM & HOWARD,PH (1991)
OH 1.02E-010	cm3/molc sec	25	EST	MEYLAN,WM & HOWARD,PH (1993)

CAS #: 130787-21-2 — 2H-BENZOTRIAZOLE, 2-CYCLOPENTYL-5-[(METHYLSULFON

Formula: $C_{12}H_{16}N_4O_2S$

Mol Weight: 280.35

MP (deg C): **FP (deg C):**

BP (deg C):

BP pressure (mm Hg):

Property/Value	Units	Temp	Data Type	Reference
WS 1.32E+002	mg/L	25	EST	MEYLAN,WM ET AL. (1996)
logP 2.44			EXP	CALIENDO,G ET AL. (1990)
VP 2.38E-008	mm Hg	25	EST	NEELY,WB & BLAU,GE (1985)
DC	pKa			
HL 9.27E-010	atm m3/mol	25	EST	MEYLAN,WM & HOWARD,PH (1991)
OH 2.98E-011	cm3/molc sec	25	EST	MEYLAN,WM & HOWARD,PH (1993)

CAS #: 130787-22-3 — BENZENESULFONAMIDE, N-(2-CYCLOPENTYL-2H-BENZOTRI

Formula: $C_{18}H_{20}N_4O_2S$

Mol Weight: 356.45

MP (deg C): **FP (deg C):**

BP (deg C):

BP pressure (mm Hg):

Property/Value	Units	Temp	Data Type	Reference
WS 2.59E+000	mg/L	25	EST	MEYLAN,WM ET AL. (1996)
logP 3.91			EXP	CALIENDO,G ET AL. (1990)
VP 6.33E-011	mm Hg	25	EST	NEELY,WB & BLAU,GE (1985)
DC	pKa			
HL 2.04E-010	atm m3/mol	25	EST	MEYLAN,WM & HOWARD,PH (1991)
OH 3.00E-011	cm3/molc sec	25	EST	MEYLAN,WM & HOWARD,PH (1993)

CAS #: 130787-23-4 — 2H-BENZOTRIAZOLE, 2-CYCLOPENTYL-5-[[(4-NITROPHEN

Formula: $C_{17}H_{17}N_5O_4S$

Mol Weight: 387.42

MP (deg C): **FP (deg C):**

BP (deg C):

BP pressure (mm Hg):

Property/Value	Units	Temp	Data Type	Reference
WS 8.16E-001	mg/L	25	EST	MEYLAN,WM ET AL. (1996)
logP 3.82			EXP	CALIENDO,G ET AL. (1990)
VP 2.64E-012	mm Hg	25	EST	NEELY,WB & BLAU,GE (1985)
DC	pKa			
HL 7.31E-013	atm m3/mol	25	EST	MEYLAN,WM & HOWARD,PH (1991)
OH 2.88E-011	cm3/molc sec	25	EST	MEYLAN,WM & HOWARD,PH (1993)

CAS #: 130787-24-5 — 2H-BENZOTRIAZOLE, 5-[[(4-AMINOPHENYL)SULFONYL]AM

Formula: $C_{17}H_{19}N_5O_2S$

Mol Weight: 357.44

MP (deg C): **FP (deg C):**

BP (deg C):

BP pressure (mm Hg):

Property/Value	Units	Temp	Data Type	Reference
WS 4.26E+001	mg/L	25	EST	MEYLAN,WM ET AL. (1996)
logP 2.48			EXP	CALIENDO,G ET AL. (1990)
VP 6.83E-012	mm Hg	25	EST	NEELY,WB & BLAU,GE (1985)
DC	pKa			
HL 6.55E-014	atm m3/mol	25	EST	MEYLAN,WM & HOWARD,PH (1991)
OH 5.18E-011	cm3/molc sec	25	EST	MEYLAN,WM & HOWARD,PH (1993)

CAS #: 130817-92-4 — L-PROLINAMIDE, 5-OXO-L-PROLYL-1-(ETHOXYCARBONYL)

Formula: $C_{19}H_{26}N_6O_6$

Mol Weight: 434.46

MP (deg C): **FP (deg C):**

BP (deg C):

BP pressure (mm Hg):

Property/Value	Units	Temp	Data Type	Reference
WS 1.18E+003	mg/L	25	EST	MEYLAN,WM ET AL. (1996)
logP -1.30			EXP	HANSCH,C ET AL. (1995)
VP 1.14E-019	mm Hg	25	EST	NEELY,WB & BLAU,GE (1985)
DC	pKa			
HL 8.23E-025	atm m3/mol	25	EST	MEYLAN,WM & HOWARD,PH (1991)
OH 1.56E-010	cm3/molc sec	25	EST	MEYLAN,WM & HOWARD,PH (1993)

CAS #: 130817-93-5 — L-PROLINAMIDE, 5-OXO-L-PROLYL-1-[(1-METHYLETHOXY

Formula: $C_{20}H_{28}N_6O_6$

Mol Weight: 448.48

MP (deg C): **FP (deg C):**

BP (deg C):

BP pressure (mm Hg):

Property/Value	Units	Temp	Data Type	Reference
WS 3.60E+002	mg/L	25	EST	MEYLAN,WM ET AL. (1996)
logP -0.80			EXP	HANSCH,C ET AL. (1995)
VP 8.23E-020	mm Hg	25	EST	NEELY,WB & BLAU,GE (1985)
DC	pKa			
HL 1.09E-024	atm m3/mol	25	EST	MEYLAN,WM & HOWARD,PH (1991)
OH 1.58E-010	cm3/molc sec	25	EST	MEYLAN,WM & HOWARD,PH (1993)

CAS #: 130817-94-6 — L-PROLINAMIDE, 5-OXO-L-PROLYL-1-(BUTOXYCARBONYL)

Formula: $C_{21}H_{30}N_6O_6$

Mol Weight: 462.51

MP (deg C): **FP (deg C):**

BP (deg C):

BP pressure (mm Hg):

Property/Value	Units	Temp	Data Type	Reference
WS 1.53E+002	mg/L	25	EST	MEYLAN,WM ET AL. (1996)
logP -0.47			EXP	HANSCH,C ET AL. (1995)
VP 2.55E-020	mm Hg	25	EST	NEELY,WB & BLAU,GE (1985)
DC	pKa			
HL 1.45E-024	atm m3/mol	25	EST	MEYLAN,WM & HOWARD,PH (1991)
OH 1.59E-010	cm3/molc sec	25	EST	MEYLAN,WM & HOWARD,PH (1993)

CAS #: 130817-95-7 — L-PROLINAMIDE, 5-OXO-L-PROLYL-1-[(2-METHYLPROPOX

Formula: $C_{21}H_{30}N_6O_6$

Mol Weight: 462.51

MP (deg C): **FP (deg C):**

BP (deg C):

BP pressure (mm Hg):

Property/Value	Units	Temp	Data Type	Reference
WS 1.45E+002	mg/L	25	EST	MEYLAN,WM ET AL. (1996)
logP -0.44			EXP	HANSCH,C ET AL. (1995)
VP 3.96E-020	mm Hg	25	EST	NEELY,WB & BLAU,GE (1985)
DC	pKa			
HL 1.45E-024	atm m3/mol	25	EST	MEYLAN,WM & HOWARD,PH (1991)
OH 1.59E-010	cm3/molc sec	25	EST	MEYLAN,WM & HOWARD,PH (1993)

CAS #: 130817-96-8 — L-PROLINAMIDE, 5-OXO-L-PROLYL-1-[(HEXYLOXY)CARBO

Formula: $C_{23}H_{34}N_6O_6$

Mol Weight: 490.56

MP (deg C):   **FP (deg C):**

BP (deg C):

BP pressure (mm Hg):

Property/Value	Units	Temp	Data Type	Reference
WS 2.04E+002	mg/L	25	EST	MEYLAN,WM ET AL. (1996)
logP 0.71			EXP	HANSCH,C ET AL. (1995)
VP 5.86E-021	mm Hg	25	EST	NEELY,WB & BLAU,GE (1985)
DC	pKa			
HL 2.56E-024	atm m3/mol	25	EST	MEYLAN,WM & HOWARD,PH (1991)
OH 1.62E-010	cm3/molc sec	25	EST	MEYLAN,WM & HOWARD,PH (1993)

CAS #: 130817-97-9 — L-PROLINAMIDE, 5-OXO-L-PROLYL-1-[(OCTYLOXY)CARBO

Formula: $C_{25}H_{38}N_6O_6$

Mol Weight: 518.62

MP (deg C):   **FP (deg C):**

BP (deg C):

BP pressure (mm Hg):

Property/Value	Units	Temp	Data Type	Reference
WS 1.52E+001	mg/L	25	EST	MEYLAN,WM ET AL. (1996)
logP 1.82			EXP	HANSCH,C ET AL. (1995)
VP 1.34E-021	mm Hg	25	EST	NEELY,WB & BLAU,GE (1985)
DC	pKa			
HL 4.51E-024	atm m3/mol	25	EST	MEYLAN,WM & HOWARD,PH (1991)
OH 1.64E-010	cm3/molc sec	25	EST	MEYLAN,WM & HOWARD,PH (1993)

CAS #: 130817-98-0 — L-PROLINAMIDE, 5-OXO-L-PROLYL-1-[(2-ETHYLHEXYLOX

Formula: $C_{25}H_{38}N_6O_6$

Mol Weight: 518.62

MP (deg C):   **FP (deg C):**

BP (deg C):

BP pressure (mm Hg):

Property/Value	Units	Temp	Data Type	Reference
WS 1.52E+001	mg/L	25	EST	MEYLAN,WM ET AL. (1996)
logP 1.82			EXP	HANSCH,C ET AL. (1995)
VP 2.09E-021	mm Hg	25	EST	NEELY,WB & BLAU,GE (1985)
DC	pKa			
HL 4.51E-024	atm m3/mol	25	EST	MEYLAN,WM & HOWARD,PH (1991)
OH 1.65E-010	cm3/molc sec	25	EST	MEYLAN,WM & HOWARD,PH (1993)

CAS #: 130817-99-1 — L-PROLINAMIDE, 5-OXO-L-PROLYL-1-[(CYCLOHEXYLOXY)

Formula: $C_{23}H_{32}N_6O_6$

Mol Weight: 488.55

MP (deg C):   **FP (deg C):**

BP (deg C):

BP pressure (mm Hg):

Property/Value	Units	Temp	Data Type	Reference
WS 2.61E+002	mg/L	25	EST	MEYLAN,WM ET AL. (1996)
logP 0.60			EXP	HANSCH,C ET AL. (1995)
VP 4.22E-021	mm Hg	25	EST	NEELY,WB & BLAU,GE (1985)
DC	pKa			
HL 1.13E-024	atm m3/mol	25	EST	MEYLAN,WM & HOWARD,PH (1991)
OH 1.62E-010	cm3/molc sec	25	EST	MEYLAN,WM & HOWARD,PH (1993)

CAS #: 130927-06-9 — 5-THIA-1-AZABICYCLO[4.2.0]OCT-2-ENE-2-CARBOXYLIC

Formula: $C_{32}H_{42}N_6O_{13}S_3$

Mol Weight: 814.92

MP (deg C):   **FP (deg C):**

BP (deg C):

BP pressure (mm Hg):

Property/Value	Units	Temp	Data Type	Reference
WS 8.36E-003	mg/L	25	EST	MEYLAN,WM ET AL. (1996)
logP 3.34			EXP	SANGSTER,J (1993)
VP 9.32E-026	mm Hg	25	EST	NEELY,WB & BLAU,GE (1985)
DC	pKa			
HL 1.98E-030	atm m3/mol	25	EST	MEYLAN,WM & HOWARD,PH (1991)
OH 2.13E-010	cm3/molc sec	25	EST	MEYLAN,WM & HOWARD,PH (1993)

CAS #: 131042-60-9 — 2-AZASPIRO[4.4]NONANE-1,3-DIONE, 2-HYDROXY-

Formula: $C_8H_{11}NO_3$

Mol Weight: 169.18

MP (deg C):   **FP (deg C):**

BP (deg C):

BP pressure (mm Hg):

Property/Value	Units	Temp	Data Type	Reference
WS 1.04E+005	mg/L	25	EST	MEYLAN,WM ET AL. (1996)
logP -0.26			EXP	SANGSTER,J (1994)
VP 8.49E-009	mm Hg	25	EST	NEELY,WB & BLAU,GE (1985)
DC	pKa			
HL 6.95E-011	atm m3/mol	25	EST	MEYLAN,WM & HOWARD,PH (1991)
OH 2.43E-011	cm3/molc sec	25	EST	MEYLAN,WM & HOWARD,PH (1993)

CAS #: 131042-61-0 — 2-AZASPIRO[4.4]NONANE-1,3-DIONE, 2-(PHENYLMETHOX

Formula: $C_{15}H_{17}NO_3$

Mol Weight: 259.31

MP (deg C):   **FP (deg C):**

BP (deg C):

BP pressure (mm Hg):

Property/Value	Units	Temp	Data Type	Reference
WS 2.28E+002	mg/L	25	EST	MEYLAN,WM ET AL. (1996)
logP 2.30			EXP	SANGSTER,J (1994)
VP 7.82E-009	mm Hg	25	EST	NEELY,WB & BLAU,GE (1985)
DC	pKa			
HL 1.80E-009	atm m3/mol	25	EST	MEYLAN,WM & HOWARD,PH (1991)
OH 2.93E-011	cm3/molc sec	25	EST	MEYLAN,WM & HOWARD,PH (1993)

CAS #: 131067-26-0 — 2-OXAZOLIDINONE, 5-[[2-BENZYL-3-(4-MORPHOLINYLCA

Formula: $C_{36}H_{52}N_6O_7$

Mol Weight: 680.85

MP (deg C):   **FP (deg C):**

BP (deg C):

BP pressure (mm Hg):

Property/Value	Units	Temp	Data Type	Reference
WS 6.73E-001	mg/L	25	EST	MEYLAN,WM ET AL. (1996)
logP 2.16			EXP	SANGSTER,J (1994)
VP 2.78E-030	mm Hg	25	EST	NEELY,WB & BLAU,GE (1985)
DC	pKa			
HL 9.22E-032	atm m3/mol	25	EST	MEYLAN,WM & HOWARD,PH (1991)
OH 2.36E-010	cm3/molc sec	25	EST	MEYLAN,WM & HOWARD,PH (1993)

CAS #: 131073-11-5 — OXAZOLO[4,5-C]PYRIDINE, 2-(2,4-DIMETHOXYPHENYL)-

Formula: $C_{14}H_{12}N_2O_3$

Mol Weight: 256.26

MP (deg C): FP (deg C):

BP (deg C):

BP pressure (mm Hg):

Property/Value	Units	Temp	Data Type	Reference
WS 1.87E+002	mg/L	25	EST	MEYLAN,WM ET AL. (1996)
logP 2.42			EXP	HANSCH,C ET AL. (1995)
VP 5.45E-007	mm Hg	25	EST	NEELY,WB & BLAU,GE (1985)
DC	pKa			
HL 2.42E-013	atm m3/mol	25	EST	MEYLAN,WM & HOWARD,PH (1991)
OH 1.87E-010	cm3/molc sec	25	EST	MEYLAN,WM & HOWARD,PH (1993)

CAS #: 131229-37-3 — IMIDAZO[5,1-B]QUINAZOLIN-9(2H)-ONE, 1,3-DIHYDRO-

Formula: $C_{12}H_{13}N_3O$

Mol Weight: 215.26

MP (deg C): FP (deg C):

BP (deg C):

BP pressure (mm Hg):

Property/Value	Units	Temp	Data Type	Reference
WS 6.86E+003	mg/L	25	EST	MEYLAN,WM ET AL. (1996)
logP 0.85			EXP	SANGSTER,J (1994)
VP 2.45E-006	mm Hg	25	EST	NEELY,WB & BLAU,GE (1985)
DC	pKa			
HL 7.36E-011	atm m3/mol	25	EST	MEYLAN,WM & HOWARD,PH (1991)
OH 1.21E-010	cm3/molc sec	25	EST	MEYLAN,WM & HOWARD,PH (1993)

CAS #: 131229-38-4 — IMIDAZO[5,1-B]QUINAZOLIN-9(2H)-ONE, 1,3-DIHYDRO-

Formula: $C_{13}H_{15}N_3O$

Mol Weight: 229.28

MP (deg C): FP (deg C):

BP (deg C):

BP pressure (mm Hg):

Property/Value	Units	Temp	Data Type	Reference
WS 4.13E+003	mg/L	25	EST	MEYLAN,WM ET AL. (1996)
logP 1.02			EXP	SANGSTER,J (1994)
VP 1.09E-006	mm Hg	25	EST	NEELY,WB & BLAU,GE (1985)
DC	pKa			
HL 9.77E-011	atm m3/mol	25	EST	MEYLAN,WM & HOWARD,PH (1991)
OH 1.28E-010	cm3/molc sec	25	EST	MEYLAN,WM & HOWARD,PH (1993)

CAS #: 131229-39-5 — IMIDAZO[5,1-B]QUINAZOLIN-9(2H)-ONE, 1,3-DIHYDRO-

Formula: $C_{14}H_{17}N_3O$

Mol Weight: 243.31

MP (deg C): FP (deg C):

BP (deg C):

BP pressure (mm Hg):

Property/Value	Units	Temp	Data Type	Reference
WS 1.64E+003	mg/L	25	EST	MEYLAN,WM ET AL. (1996)
logP 1.40			EXP	SANGSTER,J (1994)
VP 4.95E-007	mm Hg	25	EST	NEELY,WB & BLAU,GE (1985)
DC	pKa			
HL 1.30E-010	atm m3/mol	25	EST	MEYLAN,WM & HOWARD,PH (1991)
OH 1.32E-010	cm3/molc sec	25	EST	MEYLAN,WM & HOWARD,PH (1993)

CAS #: 131229-40-8 — IMIDAZO[5,1-B]QUINAZOLIN-9(2H)-ONE, 2-BUTYL-1,3-

Formula: $C_{15}H_{19}N_3O$

Mol Weight: 257.34

MP (deg C): FP (deg C):

BP (deg C):

BP pressure (mm Hg):

Property/Value	Units	Temp	Data Type	Reference
WS 9.26E+002	mg/L	25	EST	MEYLAN,WM ET AL. (1996)
logP 1.60			EXP	SANGSTER,J (1994)
VP 2.24E-007	mm Hg	25	EST	NEELY,WB & BLAU,GE (1985)
DC	pKa			
HL 1.72E-010	atm m3/mol	25	EST	MEYLAN,WM & HOWARD,PH (1991)
OH 1.33E-010	cm3/molc sec	25	EST	MEYLAN,WM & HOWARD,PH (1993)

CAS #: 131229-41-9 — IMIDAZO[5,1-B]QUINAZOLIN-9(2H)-ONE, 1,3-DIHYDRO-

Formula: $C_{13}H_{15}N_3O_2$

Mol Weight: 245.28

MP (deg C): FP (deg C):

BP (deg C):

BP pressure (mm Hg):

Property/Value	Units	Temp	Data Type	Reference
WS 1.02E+004	mg/L	25	EST	MEYLAN,WM ET AL. (1996)
logP 0.46			EXP	SANGSTER,J (1994)
VP 9.39E-010	mm Hg	25	EST	NEELY,WB & BLAU,GE (1985)
DC	pKa			
HL 3.57E-015	atm m3/mol	25	EST	MEYLAN,WM & HOWARD,PH (1991)
OH 1.34E-010	cm3/molc sec	25	EST	MEYLAN,WM & HOWARD,PH (1993)

CAS #: 131229-42-0 — IMIDAZO[5,1-B]QUINAZOLIN-9(2H)-ONE, 1,3-DIHYDRO-

Formula: $C_{19}H_{19}N_3O$

Mol Weight: 305.38

MP (deg C): FP (deg C):

BP (deg C):

BP pressure (mm Hg):

Property/Value	Units	Temp	Data Type	Reference
WS 3.32E+002	mg/L	25	EST	MEYLAN,WM ET AL. (1996)
logP 1.80			EXP	SANGSTER,J (1993)
VP 3.30E-009	mm Hg	25	EST	NEELY,WB & BLAU,GE (1985)
DC	pKa			
HL 7.89E-012	atm m3/mol	25	EST	MEYLAN,WM & HOWARD,PH (1991)
OH 1.36E-010	cm3/molc sec	25	EST	MEYLAN,WM & HOWARD,PH (1993)

CAS #: 131229-43-1 — IMIDAZO[5,1-B]QUINAZOLIN-9(2H)-ONE, 1,3-DIHYDRO-

Formula: $C_{18}H_{17}N_3O$

Mol Weight: 291.36

MP (deg C): FP (deg C):

BP (deg C):

BP pressure (mm Hg):

Property/Value	Units	Temp	Data Type	Reference
WS 4.68E+002	mg/L	25	EST	MEYLAN,WM ET AL. (1996)
logP 1.72			EXP	SANGSTER,J (1993)
VP 7.44E-009	mm Hg	25	EST	NEELY,WB & BLAU,GE (1985)
DC	pKa			
HL 5.94E-012	atm m3/mol	25	EST	MEYLAN,WM & HOWARD,PH (1991)
OH 1.33E-010	cm3/molc sec	25	EST	MEYLAN,WM & HOWARD,PH (1993)

CAS #: 131229-44-2				IMIDAZO[5,1-B]QUINAZOLIN-9(2H)-ONE, 1,3-DIHYDRO-

Formula: $C_{17}H_{15}N_3O$

Mol Weight: 277.33

MP (deg C):   FP (deg C):

BP (deg C):

BP pressure (mm Hg):

Property/Value	Units	Temp	Data Type	Reference
WS 4.91E+002	mg/L	25	EST	MEYLAN,WM ET AL. (1996)
logP 1.79			EXP	SANGSTER,J (1994)
VP 1.67E-008	mm Hg	25	EST	NEELY,WB & BLAU,GE (1985)
DC	pKa			
HL 1.73E-010	atm m3/mol	25	EST	MEYLAN,WM & HOWARD,PH (1991)
OH 2.00E-010	cm3/molc sec	25	EST	MEYLAN,WM & HOWARD,PH (1993)

CAS #: 131229-45-3				IMIDAZO[5,1-B]QUINAZOLIN-9(2H)-ONE, 1,3-DIHYDRO-

Formula: $C_{18}H_{17}N_3O_2$

Mol Weight: 307.36

MP (deg C):   FP (deg C):

BP (deg C):

BP pressure (mm Hg):

Property/Value	Units	Temp	Data Type	Reference
WS 3.70E+002	mg/L	25	EST	MEYLAN,WM ET AL. (1996)
logP 1.73			EXP	SANGSTER,J (1993)
VP 3.20E-009	mm Hg	25	EST	NEELY,WB & BLAU,GE (1985)
DC	pKa			
HL 1.02E-011	atm m3/mol	25	EST	MEYLAN,WM & HOWARD,PH (1991)
OH 1.81E-010	cm3/molc sec	25	EST	MEYLAN,WM & HOWARD,PH (1993)

CAS #: 131229-46-4				IMIDAZO[5,1-B]QUINAZOLIN-9(2H)-ONE, 1,3-DIHYDRO-

Formula: $C_{18}H_{17}N_3O$

Mol Weight: 291.36

MP (deg C):   FP (deg C):

BP (deg C):

BP pressure (mm Hg):

Property/Value	Units	Temp	Data Type	Reference
WS 1.97E+002	mg/L	25	EST	MEYLAN,WM ET AL. (1996)
logP 2.16			EXP	SANGSTER,J (1993)
VP 7.44E-009	mm Hg	25	EST	NEELY,WB & BLAU,GE (1985)
DC	pKa			
HL 1.91E-010	atm m3/mol	25	EST	MEYLAN,WM & HOWARD,PH (1991)
OH 2.33E-010	cm3/molc sec	25	EST	MEYLAN,WM & HOWARD,PH (1993)

CAS #: 131341-86-1				FLUDIOXONIL

Formula: $C_{12}H_6F_2N_2O_2$

Mol Weight: 248.19

MP (deg C):   FP (deg C):

BP (deg C):

BP pressure (mm Hg):

Property/Value	Units	Temp	Data Type	Reference
WS 3.99E+000	mg/L	25	EST	MEYLAN,WM ET AL. (1996)
logP 4.12			EXP	TOMLIN,C (1994)
VP 1.39E-006	mm Hg	25	EST	NEELY,WB & BLAU,GE (1985)
DC	pKa			
HL 3.10E-011	atm m3/mol	25	EST	MEYLAN,WM & HOWARD,PH (1991)
OH 5.76E-011	cm3/molc sec	25	EST	MEYLAN,WM & HOWARD,PH (1993)

CAS #: 131402-47-6				ALLOPURINOL,N-(3-(MORPHOLINOMETHYL)BENZOYLOXY)ME

Formula: $C_{18}H_{19}N_5O_4$

Mol Weight: 369.38

MP (deg C):   FP (deg C):

BP (deg C):

BP pressure (mm Hg):

Property/Value	Units	Temp	Data Type	Reference
WS 5.12E+002	mg/L	25	EST	MEYLAN,WM ET AL. (1996)
logP 1.13			EXP	HANSCH,C ET AL. (1995)
VP 1.73E-014	mm Hg	25	EST	NEELY,WB & BLAU,GE (1985)
DC	pKa			
HL 9.44E-021	atm m3/mol	25	EST	MEYLAN,WM & HOWARD,PH (1991)
OH 1.82E-010	cm3/molc sec	25	EST	MEYLAN,WM & HOWARD,PH (1993)

CAS #: 131402-48-7				ALLOPURINOL,N-(4-(MORPHOLINOMETHYL)BENZOYLOXY)ME

Formula: $C_{18}H_{19}N_5O_4$

Mol Weight: 369.38

MP (deg C):   FP (deg C):

BP (deg C):

BP pressure (mm Hg):

Property/Value	Units	Temp	Data Type	Reference
WS 5.22E+002	mg/L	25	EST	MEYLAN,WM ET AL. (1996)
logP 1.12			EXP	HANSCH,C ET AL. (1995)
VP 1.73E-014	mm Hg	25	EST	NEELY,WB & BLAU,GE (1985)
DC	pKa			
HL 9.44E-021	atm m3/mol	25	EST	MEYLAN,WM & HOWARD,PH (1991)
OH 1.83E-010	cm3/molc sec	25	EST	MEYLAN,WM & HOWARD,PH (1993)

CAS #: 131402-50-1				ALLOPURINOL,N-(3-(4-METHYLYPIPERIZINYLMETHYL)BEN

Formula: $C_{19}H_{24}N_6O_3$

Mol Weight: 384.44

MP (deg C):   FP (deg C):

BP (deg C):

BP pressure (mm Hg):

Property/Value	Units	Temp	Data Type	Reference
WS 2.10E+002	mg/L	25	EST	MEYLAN,WM ET AL. (1996)
logP 1.49			EXP	HANSCH,C ET AL. (1995)
VP 7.98E-015	mm Hg	25	EST	NEELY,WB & BLAU,GE (1985)
DC	pKa			
HL 4.00E-022	atm m3/mol	25	EST	MEYLAN,WM & HOWARD,PH (1991)
OH 2.17E-010	cm3/molc sec	25	EST	MEYLAN,WM & HOWARD,PH (1993)

CAS #: 131402-51-2				ALLOPURINOL,N-(4-DIMETHYLAMINOMETHYL)BENZOYLOXYM

Formula: $C_{16}H_{19}N_5O_3$

Mol Weight: 329.36

MP (deg C):   FP (deg C):

BP (deg C):

BP pressure (mm Hg):

Property/Value	Units	Temp	Data Type	Reference
WS 1.08E+003	mg/L	25	EST	MEYLAN,WM ET AL. (1996)
logP 1.05			EXP	HANSCH,C ET AL. (1995)
VP 6.24E-013	mm Hg	25	EST	NEELY,WB & BLAU,GE (1985)
DC	pKa			
HL 1.38E-018	atm m3/mol	25	EST	MEYLAN,WM & HOWARD,PH (1991)
OH 1.10E-010	cm3/molc sec	25	EST	MEYLAN,WM & HOWARD,PH (1993)

CAS #: 131402-52-3 — ALLOPURINOL,3-(IMIDAZOL-1-YL)METHYLBENZOYLOXYMET

Formula: $C_{17}H_{16}N_6O_3$

Mol Weight: 352.36

MP (deg C):   FP (deg C):

BP (deg C):

BP pressure (mm Hg):

Property/ Value	Units	Temp	Data Type	Reference
WS 9.16E+002	mg/L	25	EST	MEYLAN,WM ET AL. (1996)
logP 0.97			EXP	HANSCH,C ET AL. (1995)
VP 4.49E-017	mm Hg	25	EST	NEELY,WB & BLAU,GE (1985)
DC	pKa			
HL 1.56E-019	atm m3/mol	25	EST	MEYLAN,WM & HOWARD,PH (1991)
OH 8.90E-011	cm3/molc sec	25	EST	MEYLAN,WM & HOWARD,PH (1993)

CAS #: 131802-71-6 — BENZENEACETAMIDE, à-HYDROXY-à-ETHYL-, (ñ)-

Formula: $C_{10}H_{13}NO_2$

Mol Weight: 179.22

MP (deg C):   FP (deg C):

BP (deg C):

BP pressure (mm Hg):

Property/ Value	Units	Temp	Data Type	Reference
WS 5.25E+003	mg/L	25	EST	MEYLAN,WM ET AL. (1996)
logP 1.20			EXP	SANGSTER,J (1993)
VP 3.59E-007	mm Hg	25	EST	NEELY,WB & BLAU,GE (1985)
DC	pKa			
HL 2.48E-010	atm m3/mol	25	EST	MEYLAN,WM & HOWARD,PH (1991)
OH 1.02E-011	cm3/molc sec	25	EST	MEYLAN,WM & HOWARD,PH (1993)

CAS #: 132133-10-9 — N(24-NO2-6CF3 PH)-2(35CN-6ME)PYRIDINAMIN

Formula: $C_{15}H_7F_3N_6O_4$

Mol Weight: 392.26

MP (deg C):   FP (deg C):

BP (deg C):

BP pressure (mm Hg):

Property/ Value	Units	Temp	Data Type	Reference
WS 7.55E+000	mg/L	25	EST	MEYLAN,WM ET AL. (1996)
logP 2.80			EXP	BRANDT,U ET AL. (1992)
VP 7.85E-011	mm Hg	25	EST	NEELY,WB & BLAU,GE (1985)
DC	pKa			
HL 4.19E-016	atm m3/mol	25	EST	MEYLAN,WM & HOWARD,PH (1991)
OH 1.52E-013	cm3/molc sec	25	EST	MEYLAN,WM & HOWARD,PH (1993)

CAS #: 132169-99-4 — BR-N(2,6-IPRPH)CARBAMOYLME IMINODIACETIC ACID

Formula: $C_{18}H_{25}BrN_2O_5$

Mol Weight: 429.31

MP (deg C):   FP (deg C):

BP (deg C):

BP pressure (mm Hg):

Property/ Value	Units	Temp	Data Type	Reference
WS 5.66E-001	mg/L	25	EST	MEYLAN,WM ET AL. (1996)
logP 0.20			EXP	GODOY,N ET AL. (1989)
VP 6.57E-015	mm Hg	25	EST	NEELY,WB & BLAU,GE (1985)
DC	pKa			
HL 2.03E-018	atm m3/mol	25	EST	MEYLAN,WM & HOWARD,PH (1991)
OH 1.02E-010	cm3/molc sec	25	EST	MEYLAN,WM & HOWARD,PH (1993)

CAS #: 132170-00-4 — BR-N(26-MEPH)CARBAMOYLME IMINODIACETIC ACID

Formula: $C_{14}H_{17}BrN_2O_5$

Mol Weight: 373.21

MP (deg C):   FP (deg C):

BP (deg C):

BP pressure (mm Hg):

Property/ Value	Units	Temp	Data Type	Reference
WS 3.24E+000	mg/L	25	EST	MEYLAN,WM ET AL. (1996)
logP -0.28			EXP	GODOY,N ET AL. (1989)
VP 5.33E-014	mm Hg	25	EST	NEELY,WB & BLAU,GE (1985)
DC	pKa			
HL 6.52E-019	atm m3/mol	25	EST	MEYLAN,WM & HOWARD,PH (1991)
OH 9.86E-011	cm3/molc sec	25	EST	MEYLAN,WM & HOWARD,PH (1993)

CAS #: 132194-21-9 — 2-FURANMETHANOL, 5-(2-AMINO-6-FLUORO-9H-PURIN-9-

Formula: $C_{10}H_{12}FN_5O_2$

Mol Weight: 253.24

MP (deg C):   FP (deg C):

BP (deg C):

BP pressure (mm Hg):

Property/ Value	Units	Temp	Data Type	Reference
WS 1.23E+003	mg/L	25	EST	MEYLAN,WM ET AL. (1996)
logP -0.05			EXP	MURAKAMI,K ET AL. (1991)
VP 4.36E-011	mm Hg	25	EST	NEELY,WB & BLAU,GE (1985)
DC	pKa			
HL 3.45E-016	atm m3/mol	25	EST	MEYLAN,WM & HOWARD,PH (1991)
OH 2.32E-010	cm3/molc sec	25	EST	MEYLAN,WM & HOWARD,PH (1993)

CAS #: 132194-22-0 — 2-FURANMETHANOL, 5-(2-AMINO-6-BROMO-9H-PURIN-9-Y

Formula: $C_{10}H_{11}BrN_4O_2$

Mol Weight: 299.13

MP (deg C):   FP (deg C):

BP (deg C):

BP pressure (mm Hg):

Property/ Value	Units	Temp	Data Type	Reference
WS 3.06E+002	mg/L	25	EST	MEYLAN,WM ET AL. (1996)
logP 0.34			EXP	MURAKAMI,K ET AL. (1991)
VP 9.24E-011	mm Hg	25	EST	NEELY,WB & BLAU,GE (1985)
DC	pKa			
HL 9.38E-017	atm m3/mol	25	EST	MEYLAN,WM & HOWARD,PH (1991)
OH 5.51E-011	cm3/molc sec	25	EST	MEYLAN,WM & HOWARD,PH (1993)

CAS #: 132194-23-1 — 2-FURANMETHANOL, 5-(2-AMINO-6-IODO-9H-PURIN-9-YL

Formula: $C_{10}H_{12}IN_5O_2$

Mol Weight: 361.14

MP (deg C):   FP (deg C):

BP (deg C):

BP pressure (mm Hg):

Property/ Value	Units	Temp	Data Type	Reference
WS 1.91E+003	mg/L	25	EST	MEYLAN,WM ET AL. (1996)
logP 0.52			EXP	MURAKAMI,K ET AL. (1991)
VP 6.09E-013	mm Hg	25	EST	NEELY,WB & BLAU,GE (1985)
DC	pKa			
HL 6.84E-017	atm m3/mol	25	EST	MEYLAN,WM & HOWARD,PH (1991)
OH 1.92E-010	cm3/molc sec	25	EST	MEYLAN,WM & HOWARD,PH (1993)

CAS #: 132194-24-2 — 2-FURANMETHANOL, 5-(6-FLUORO-9H-PURIN-9-YL)TETRA

Formula: $C_{10}H_{11}FN_4O_2$

Mol Weight: 238.22

MP (deg C):   FP (deg C):

BP (deg C):

BP pressure (mm Hg):

Property/Value	Units	Temp	Data Type	Reference
WS 1.35E+003	mg/L	25	EST	MEYLAN,WM ET AL. (1996)
logP 0.00			EXP	MURAKAMI,K ET AL. (1991)
VP 1.38E-009	mm Hg	25	EST	NEELY,WB & BLAU,GE (1985)
DC	pKa			
HL 2.75E-016	atm m3/mol	25	EST	MEYLAN,WM & HOWARD,PH (1991)
OH 7.75E-011	cm3/molc sec	25	EST	MEYLAN,WM & HOWARD,PH (1993)

CAS #: 132194-25-3 — 2-FURANMETHANOL, 5-(6-BROMO-9H-PURIN-9-YL)TETRAH

Formula: $C_{10}H_{11}BrN_4O_2$

Mol Weight: 299.13

MP (deg C):   FP (deg C):

BP (deg C):

BP pressure (mm Hg):

Property/Value	Units	Temp	Data Type	Reference
WS 3.06E+002	mg/L	25	EST	MEYLAN,WM ET AL. (1996)
logP 0.35			EXP	MURAKAMI,K ET AL. (1991)
VP 9.24E-011	mm Hg	25	EST	NEELY,WB & BLAU,GE (1985)
DC	pKa			
HL 9.38E-017	atm m3/mol	25	EST	MEYLAN,WM & HOWARD,PH (1991)
OH 5.51E-011	cm3/molc sec	25	EST	MEYLAN,WM & HOWARD,PH (1993)

CAS #: 132213-87-7 — 1-BUTANAMINE, 4-[(6A,7,10,10A-TETRAHYDRO-6,6,9-T

Formula: $C_{25}H_{39}NO_2$

Mol Weight: 385.60

MP (deg C):   FP (deg C):

BP (deg C):

BP pressure (mm Hg):

Property/Value	Units	Temp	Data Type	Reference
WS 2.67E-003	mg/L	25	EST	MEYLAN,WM ET AL. (1996)
logP 8.38			EXP	SANGSTER,J (1993)
VP 4.16E-009	mm Hg	25	EST	NEELY,WB & BLAU,GE (1985)
DC	pKa			
HL 3.34E-008	atm m3/mol	25	EST	MEYLAN,WM & HOWARD,PH (1991)
OH 3.53E-010	cm3/molc sec	25	EST	MEYLAN,WM & HOWARD,PH (1993)

CAS #: 132213-90-2 — 1,4-ISOQUINOLINEDIONE DERIVATIVE

Formula: $C_{35}H_{45}NO_4$

Mol Weight: 543.75

MP (deg C):   FP (deg C):

BP (deg C):

BP pressure (mm Hg):

Property/Value	Units	Temp	Data Type	Reference
WS 6.72E-006	mg/L	25	EST	MEYLAN,WM ET AL. (1996)
logP 9.07			EXP	SANGSTER,J (1993)
VP 9.70E-016	mm Hg	25	EST	NEELY,WB & BLAU,GE (1985)
DC	pKa			
HL 7.54E-013	atm m3/mol	25	EST	MEYLAN,WM & HOWARD,PH (1991)
OH 3.45E-010	cm3/molc sec	25	EST	MEYLAN,WM & HOWARD,PH (1993)

CAS #: 132213-93-5 — 6H-DIBENZO[B,D]PYRAN, 1-([1,1'-BIPHENYL]-4-YLMET

Formula: $C_{34}H_{40}O_2$

Mol Weight: 480.70

MP (deg C):   FP (deg C):

BP (deg C):

BP pressure (mm Hg):

Property/Value	Units	Temp	Data Type	Reference
WS 2.17E-007	mg/L	25	EST	MEYLAN,WM ET AL. (1996)
logP 11.29			EXP	SANGSTER,J (1993)
VP 1.26E-012	mm Hg	25	EST	NEELY,WB & BLAU,GE (1985)
DC	pKa			
HL 7.29E-007	atm m3/mol	25	EST	MEYLAN,WM & HOWARD,PH (1991)
OH 2.80E-010	cm3/molc sec	25	EST	MEYLAN,WM & HOWARD,PH (1993)

CAS #: 132296-11-8 — PHENOL, 5-(1,1-DIMETHYLOCTYL)-2-(3-HYDROXYCYCLOH

Formula: $C_{22}H_{36}O_2$

Mol Weight: 332.53

MP (deg C):   FP (deg C):

BP (deg C):

BP pressure (mm Hg):

Property/Value	Units	Temp	Data Type	Reference
WS 7.17E-002	mg/L	25	EST	MEYLAN,WM ET AL. (1996)
logP 7.18			EXP	SANGSTER,J (1993)
VP 4.83E-010	mm Hg	25	EST	NEELY,WB & BLAU,GE (1985)
DC	pKa			
HL 5.81E-010	atm m3/mol	25	EST	MEYLAN,WM & HOWARD,PH (1991)
OH 1.16E-010	cm3/molc sec	25	EST	MEYLAN,WM & HOWARD,PH (1993)

CAS #: 132296-12-9 — PHENOL, 5-(1,1-DIMETHYLNONYL)-2-(3-HYDROXYCYCLOH

Formula: $C_{23}H_{38}O_2$

Mol Weight: 346.56

MP (deg C):   FP (deg C):

BP (deg C):

BP pressure (mm Hg):

Property/Value	Units	Temp	Data Type	Reference
WS 8.59E-003	mg/L	25	EST	MEYLAN,WM ET AL. (1996)
logP 8.16			EXP	SANGSTER,J (1993)
VP 1.78E-010	mm Hg	25	EST	NEELY,WB & BLAU,GE (1985)
DC	pKa			
HL 7.72E-010	atm m3/mol	25	EST	MEYLAN,WM & HOWARD,PH (1991)
OH 1.18E-010	cm3/molc sec	25	EST	MEYLAN,WM & HOWARD,PH (1993)

CAS #: 132296-13-0 — PHENOL, 5-(1,1-DIMETHYLUNDECYL)-2-(3-HYDROXYCYCL

Formula: $C_{25}H_{42}O_2$

Mol Weight: 374.61

MP (deg C):   FP (deg C):

BP (deg C):

BP pressure (mm Hg):

Property/Value	Units	Temp	Data Type	Reference
WS 3.48E-003	mg/L	25	EST	MEYLAN,WM ET AL. (1996)
logP 8.42			EXP	SANGSTER,J (1993)
VP 2.36E-011	mm Hg	25	EST	NEELY,WB & BLAU,GE (1985)
DC	pKa			
HL 1.36E-009	atm m3/mol	25	EST	MEYLAN,WM & HOWARD,PH (1991)
OH 1.20E-010	cm3/molc sec	25	EST	MEYLAN,WM & HOWARD,PH (1993)

CAS #: 132296-14-1			PHENOL, 5-(1,1-DIMETHYLHEPTYL)-2-(5-HYDROXY-2-ME	
Formula: $C_{22}H_{36}O_2$				
Mol Weight: 332.53				
MP (deg C):		FP (deg C):		
BP (deg C):				
BP pressure (mm Hg):				

Property/ Value	Units	Temp	Data Type	Reference
WS 1.40E-002	mg/L	25	EST	MEYLAN,WM ET AL. (1996)
logP 7.49			EXP	SANGSTER,J (1993)
VP 6.48E-010	mm Hg	25	EST	NEELY,WB & BLAU,GE (1985)
DC	pKa			
HL 5.81E-010	atm m3/mol	25	EST	MEYLAN,WM & HOWARD,PH (1991)
OH 1.16E-010	cm3/molc sec	25	EST	MEYLAN,WM & HOWARD,PH (1993)

CAS #: 132296-15-2			PHENOL, 5-(1,1-DIMETHYLHEPTYL)-2-[5-HYDROXY-2-(4	
Formula: $C_{25}H_{42}O_3$				
Mol Weight: 390.61				
MP (deg C):		FP (deg C):		
BP (deg C):				
BP pressure (mm Hg):				

Property/ Value	Units	Temp	Data Type	Reference
WS 2.63E-002	mg/L	25	EST	MEYLAN,WM ET AL. (1996)
logP 6.68			EXP	HANSCH,C ET AL. (1995)
VP 2.20E-013	mm Hg	25	EST	NEELY,WB & BLAU,GE (1985)
DC	pKa			
HL 4.97E-011	atm m3/mol	25	EST	MEYLAN,WM & HOWARD,PH (1991)
OH 1.25E-010	cm3/molc sec	25	EST	MEYLAN,WM & HOWARD,PH (1993)

CAS #: 132296-16-3			PHENOL, 5-(1,1-DIMETHYLETHYL)-2-(3-HYDROXYCYCLOH	
Formula: $C_{16}H_{24}O_2$				
Mol Weight: 248.37				
MP (deg C):		FP (deg C):		
BP (deg C):				
BP pressure (mm Hg):				

Property/ Value	Units	Temp	Data Type	Reference
WS 1.03E+002	mg/L	25	EST	MEYLAN,WM ET AL. (1996)
logP 4.05			EXP	SANGSTER,J (1994)
VP 1.51E-007	mm Hg	25	EST	NEELY,WB & BLAU,GE (1985)
DC	pKa			
HL 1.06E-010	atm m3/mol	25	EST	MEYLAN,WM & HOWARD,PH (1991)
OH 1.08E-010	cm3/molc sec	25	EST	MEYLAN,WM & HOWARD,PH (1993)

CAS #: 132296-18-5			PHENOL, 5-(1,1-DIMETHYLBUTYL)-2-(3-HYDROXYCYCLOH	
Formula: $C_{18}H_{28}O_2$				
Mol Weight: 276.42				
MP (deg C):		FP (deg C):		
BP (deg C):				
BP pressure (mm Hg):				

Property/ Value	Units	Temp	Data Type	Reference
WS 1.58E+001	mg/L	25	EST	MEYLAN,WM ET AL. (1996)
logP 4.82			EXP	HANSCH,C ET AL. (1995)
VP 2.14E-008	mm Hg	25	EST	NEELY,WB & BLAU,GE (1985)
DC	pKa			
HL 1.87E-010	atm m3/mol	25	EST	MEYLAN,WM & HOWARD,PH (1991)
OH 1.11E-010	cm3/molc sec	25	EST	MEYLAN,WM & HOWARD,PH (1993)

CAS #: 132296-19-6			PHENOL, 5-(1,1-DIMETHYLPENTYL)-2-(3-HYDROXYCYCLO	
Formula: $C_{19}H_{30}O_2$				
Mol Weight: 290.45				
MP (deg C):		FP (deg C):		
BP (deg C):				
BP pressure (mm Hg):				

Property/ Value	Units	Temp	Data Type	Reference
WS 3.96E+000	mg/L	25	EST	MEYLAN,WM ET AL. (1996)
logP 5.43			EXP	HANSCH,C ET AL. (1995)
VP 8.00E-009	mm Hg	25	EST	NEELY,WB & BLAU,GE (1985)
DC	pKa			
HL 2.48E-010	atm m3/mol	25	EST	MEYLAN,WM & HOWARD,PH (1991)
OH 1.12E-010	cm3/molc sec	25	EST	MEYLAN,WM & HOWARD,PH (1993)

CAS #: 132296-20-9			PHENOL, 5-(1,1-DIMETHYLHEXYL)-2-(3-HYDROXYCYCLOH	
Formula: $C_{20}H_{32}O_2$				
Mol Weight: 304.48				
MP (deg C):		FP (deg C):		
BP (deg C):				
BP pressure (mm Hg):				

Property/ Value	Units	Temp	Data Type	Reference
WS 1.13E+000	mg/L	25	EST	MEYLAN,WM ET AL. (1996)
logP 5.97			EXP	HANSCH,C ET AL. (1995)
VP 3.52E-009	mm Hg	25	EST	NEELY,WB & BLAU,GE (1985)
DC	pKa			
HL 3.30E-010	atm m3/mol	25	EST	MEYLAN,WM & HOWARD,PH (1991)
OH 1.13E-010	cm3/molc sec	25	EST	MEYLAN,WM & HOWARD,PH (1993)

CAS #: 132339-34-5			PHENOL, 5-(1,1-DIMETHYLHEPTYL)-2-(5-HYDROXY-2-ME	
Formula: $C_{22}H_{36}O_2$				
Mol Weight: 332.53				
MP (deg C):		FP (deg C):		
BP (deg C):				
BP pressure (mm Hg):				

Property/ Value	Units	Temp	Data Type	Reference
WS 1.40E-002	mg/L	25	EST	MEYLAN,WM ET AL. (1996)
logP 8.01			EXP	SANGSTER,J (1993)
VP 6.48E-010	mm Hg	25	EST	NEELY,WB & BLAU,GE (1985)
DC	pKa			
HL 5.81E-010	atm m3/mol	25	EST	MEYLAN,WM & HOWARD,PH (1991)
OH 1.16E-010	cm3/molc sec	25	EST	MEYLAN,WM & HOWARD,PH (1993)

CAS #: 132452-02-9			3-PHENANTHRENOL, 1,2,3,4,4A,9,10,10A-OCTAHYDRO-4	
Formula: $C_{15}H_{21}NO$				
Mol Weight: 231.34				
MP (deg C):		FP (deg C):		
BP (deg C):				
BP pressure (mm Hg):				

Property/ Value	Units	Temp	Data Type	Reference
WS 2.89E+003	mg/L	25	EST	MEYLAN,WM ET AL. (1996)
logP 2.37			EXP	SANGSTER,J (1994)
VP 7.39E-007	mm Hg	25	EST	NEELY,WB & BLAU,GE (1985)
DC	pKa			
HL 5.77E-011	atm m3/mol	25	EST	MEYLAN,WM & HOWARD,PH (1991)
OH 9.09E-011	cm3/molc sec	25	EST	MEYLAN,WM & HOWARD,PH (1993)

132453-02-2

CAS #: 132453-02-2		4AH-FLUOREN-4A-AMINE, N-ETHYL-1,2,3,4,9,9A-HEXAH	
Formula: $C_{16}H_{23}N$			
Mol Weight: 229.37			
MP (deg C):	FP (deg C):		
BP (deg C):			
BP pressure (mm Hg):			

Property/Value	Units	Temp	Data Type	Reference
WS 1.78E+002	mg/L	25	EST	MEYLAN,WM ET AL. (1996)
logP 3.80			EXP	HAYS,SJ ET AL. (1993)
VP 1.62E-004	mm Hg	25	EST	NEELY,WB & BLAU,GE (1985)
DC	pKa			
HL 2.09E-006	atm m3/mol	25	EST	MEYLAN,WM & HOWARD,PH (1991)
OH 8.74E-011	cm3/molc sec	25	EST	MEYLAN,WM & HOWARD,PH (1993)

132453-06-6

CAS #: 132453-06-6		4A(2H)-PHENANTHRENAMINE, 1,3,4,9,10,10A-HEXAHYDR	
Formula: $C_{16}H_{23}NO$			
Mol Weight: 245.37			
MP (deg C):	FP (deg C):		
BP (deg C):			
BP pressure (mm Hg):			

Property/Value	Units	Temp	Data Type	Reference
WS 1.60E+002	mg/L	25	EST	MEYLAN,WM ET AL. (1996)
logP 3.75			EXP	SANGSTER,J (1994)
VP 4.59E-005	mm Hg	25	EST	NEELY,WB & BLAU,GE (1985)
DC	pKa			
HL 9.33E-008	atm m3/mol	25	EST	MEYLAN,WM & HOWARD,PH (1991)
OH 1.35E-010	cm3/molc sec	25	EST	MEYLAN,WM & HOWARD,PH (1993)

132453-23-7

CAS #: 132453-23-7		10AH-DIBENZO[B,D]PYRAN-10A-AMINE, 6,6A,7,8,9,10-	
Formula: $C_{14}H_{19}NO$			
Mol Weight: 217.31			
MP (deg C):	FP (deg C):		
BP (deg C):			
BP pressure (mm Hg):			

Property/Value	Units	Temp	Data Type	Reference
WS 5.40E+002	mg/L	25	EST	MEYLAN,WM ET AL. (1996)
logP 3.31			EXP	SANGSTER,J (1994)
VP 2.10E-004	mm Hg	25	EST	NEELY,WB & BLAU,GE (1985)
DC	pKa			
HL 6.37E-008	atm m3/mol	25	EST	MEYLAN,WM & HOWARD,PH (1991)
OH 1.13E-010	cm3/molc sec	25	EST	MEYLAN,WM & HOWARD,PH (1993)

132453-25-9

CAS #: 132453-25-9		1H-FLUORENE-2-METHANAMINE, 2,3,4,4A,9,9A-HEXAHYD	
Formula: $C_{15}H_{22}N_2$			
Mol Weight: 230.36			
MP (deg C):	FP (deg C):		
BP (deg C):			
BP pressure (mm Hg):			

Property/Value	Units	Temp	Data Type	Reference
WS 8.28E+003	mg/L	25	EST	MEYLAN,WM ET AL. (1996)
logP 1.84			EXP	HAYS,SJ ET AL. (1993)
VP 3.06E-005	mm Hg	25	EST	NEELY,WB & BLAU,GE (1985)
DC	pKa			
HL 1.62E-010	atm m3/mol	25	EST	MEYLAN,WM & HOWARD,PH (1991)
OH 1.12E-010	cm3/molc sec	25	EST	MEYLAN,WM & HOWARD,PH (1993)

132453-26-0

CAS #: 132453-26-0		1H-FLUORENE-2-METHANAMINE, 2,3,4,4A,9,9A-HEXAHYD	
Formula: $C_{16}H_{24}N_2$			
Mol Weight: 244.38			
MP (deg C):	FP (deg C):		
BP (deg C):			
BP pressure (mm Hg):			

Property/Value	Units	Temp	Data Type	Reference
WS 1.97E+003	mg/L	25	EST	MEYLAN,WM ET AL. (1996)
logP 2.48			EXP	HAYS,SJ ET AL. (1993)
VP 2.91E-005	mm Hg	25	EST	NEELY,WB & BLAU,GE (1985)
DC	pKa			
HL 3.56E-010	atm m3/mol	25	EST	MEYLAN,WM & HOWARD,PH (1991)
OH 1.55E-010	cm3/molc sec	25	EST	MEYLAN,WM & HOWARD,PH (1993)

132453-38-4

CAS #: 132453-38-4		10AH-DIBENZO[B,D]THIOPYRAN-10A-AMINE, 6,6A,7,8,9	
Formula: $C_{14}H_{19}NS$			
Mol Weight: 233.38			
MP (deg C):	FP (deg C):		
BP (deg C):			
BP pressure (mm Hg):			

Property/Value	Units	Temp	Data Type	Reference
WS 1.44E+002	mg/L	25	EST	MEYLAN,WM ET AL. (1996)
logP 3.88			EXP	SANGSTER,J (1994)
VP 5.09E-005	mm Hg	25	EST	NEELY,WB & BLAU,GE (1985)
DC	pKa			
HL 3.14E-008	atm m3/mol	25	EST	MEYLAN,WM & HOWARD,PH (1991)
OH 9.76E-011	cm3/molc sec	25	EST	MEYLAN,WM & HOWARD,PH (1993)

132453-63-5

CAS #: 132453-63-5		2-CHLORO-6-ACETYLAMINO-PYRAZINE	
Formula: $C_6H_6ClN_3O$			
Mol Weight: 171.59			
MP (deg C):	FP (deg C):		
BP (deg C):			
BP pressure (mm Hg):			

Property/Value	Units	Temp	Data Type	Reference
WS 6.95E+003	mg/L	25	EST	MEYLAN,WM ET AL. (1996)
logP 1.10			EXP	YAMAGAMI,C & TAKAO,N (1991)
VP 2.80E-005	mm Hg	25	EST	NEELY,WB & BLAU,GE (1985)
DC	pKa			
HL 1.56E-010	atm m3/mol	25	EST	MEYLAN,WM & HOWARD,PH (1991)
OH 1.13E-012	cm3/molc sec	25	EST	MEYLAN,WM & HOWARD,PH (1993)

132483-29-5

CAS #: 132483-29-5		BENZENEACETIC ACID, _-METHYL-4-(4-PHENYL-2-THIAZ	
Formula: $C_{18}H_{17}NO_2S$			
Mol Weight: 311.41			
MP (deg C):	FP (deg C):		
BP (deg C):			
BP pressure (mm Hg):			

Property/Value	Units	Temp	Data Type	Reference
WS 1.59E+000	mg/L	25	EST	MEYLAN,WM ET AL. (1996)
logP 4.95			EXP	NAIRO,Y ET AL. (1992)
VP 4.64E-010	mm Hg	25	EST	NEELY,WB & BLAU,GE (1985)
DC	pKa			
HL 2.46E-013	atm m3/mol	25	EST	MEYLAN,WM & HOWARD,PH (1991)
OH 1.13E-011	cm3/molc sec	25	EST	MEYLAN,WM & HOWARD,PH (1993)

132483-32-0 — BENZENEACETIC ACID, à-METHYL-4-(2-THIAZOLYL)-

Formula: $C_{12}H_{11}NO_2S$

Mol Weight: 233.29

MP (deg C): FP (deg C):

BP (deg C):

BP pressure (mm Hg):

Property/Value	Units	Temp	Data Type	Reference
WS 3.01E+002	mg/L	25	EST	MEYLAN,WM ET AL. (1996)
logP 2.79			EXP	NAIRO,Y ET AL. (1992)
VP 5.13E-007	mm Hg	25	EST	NEELY,WB & BLAU,GE (1985)
DC	pKa			
HL 3.21E-012	atm m3/mol	25	EST	MEYLAN,WM & HOWARD,PH (1991)
OH 8.65E-012	cm3/molc sec	25	EST	MEYLAN,WM & HOWARD,PH (1993)

132483-33-1 — BENZENEACETIC ACID, 3-CHLORO-à-METHYL-4-(2-THIAZ

Formula: $C_{12}H_{10}ClNO_2S$

Mol Weight: 267.74

MP (deg C): FP (deg C):

BP (deg C):

BP pressure (mm Hg):

Property/Value	Units	Temp	Data Type	Reference
WS 5.95E+001	mg/L	25	EST	MEYLAN,WM ET AL. (1996)
logP 3.39			EXP	NAIRO,Y ET AL. (1992)
VP 1.19E-007	mm Hg	25	EST	NEELY,WB & BLAU,GE (1985)
DC	pKa			
HL 2.38E-012	atm m3/mol	25	EST	MEYLAN,WM & HOWARD,PH (1991)
OH 6.26E-012	cm3/molc sec	25	EST	MEYLAN,WM & HOWARD,PH (1993)

132483-34-2 — BENZENEACETIC ACID, 3-FLUORO-à-METHYL-4-(2-THIAZ

Formula: $C_{12}H_{10}FNO_2S$

Mol Weight: 251.28

MP (deg C): FP (deg C):

BP (deg C):

BP pressure (mm Hg):

Property/Value	Units	Temp	Data Type	Reference
WS 1.20E+002	mg/L	25	EST	MEYLAN,WM ET AL. (1996)
logP 3.14			EXP	NAIRO,Y ET AL. (1992)
VP 5.74E-007	mm Hg	25	EST	NEELY,WB & BLAU,GE (1985)
DC	pKa			
HL 3.74E-012	atm m3/mol	25	EST	MEYLAN,WM & HOWARD,PH (1991)
OH 8.07E-012	cm3/molc sec	25	EST	MEYLAN,WM & HOWARD,PH (1993)

132483-36-4 — BENZENEACETIC ACID, à,3-DIMETHYL-4-(2-THIAZOLYL)

Formula: $C_{13}H_{13}NO_2S$

Mol Weight: 247.32

MP (deg C): FP (deg C):

BP (deg C):

BP pressure (mm Hg):

Property/Value	Units	Temp	Data Type	Reference
WS 1.48E+002	mg/L	25	EST	MEYLAN,WM ET AL. (1996)
logP 3.06			EXP	NAIRO,Y ET AL. (1992)
VP 2.05E-007	mm Hg	25	EST	NEELY,WB & BLAU,GE (1985)
DC	pKa			
HL 3.54E-012	atm m3/mol	25	EST	MEYLAN,WM & HOWARD,PH (1991)
OH 1.33E-011	cm3/molc sec	25	EST	MEYLAN,WM & HOWARD,PH (1993)

132483-39-7 — BENZENEACETIC ACID, à-METHYL-4-(2-THIAZOLYL)-3-(

Formula: $C_{13}H_{10}F_3NO_2S$

Mol Weight: 301.29

MP (deg C): FP (deg C):

BP (deg C):

BP pressure (mm Hg):

Property/Value	Units	Temp	Data Type	Reference
WS 7.30E+001	mg/L	25	EST	MEYLAN,WM ET AL. (1996)
logP 3.06			EXP	NAIRO,Y ET AL. (1992)
VP 3.51E-007	mm Hg	25	EST	NEELY,WB & BLAU,GE (1985)
DC	pKa			
HL 2.79E-011	atm m3/mol	25	EST	MEYLAN,WM & HOWARD,PH (1991)
OH 5.39E-012	cm3/molc sec	25	EST	MEYLAN,WM & HOWARD,PH (1993)

132483-42-2 — BENZENEACETIC ACID, à-METHYL-4-(4-METHYL-2-THIAZ

Formula: $C_{13}H_{13}NO_2S$

Mol Weight: 247.32

MP (deg C): FP (deg C):

BP (deg C):

BP pressure (mm Hg):

Property/Value	Units	Temp	Data Type	Reference
WS 1.57E+002	mg/L	25	EST	MEYLAN,WM ET AL. (1996)
logP 3.03			EXP	NAIRO,Y ET AL. (1992)
VP 2.05E-007	mm Hg	25	EST	NEELY,WB & BLAU,GE (1985)
DC	pKa			
HL 3.54E-012	atm m3/mol	25	EST	MEYLAN,WM & HOWARD,PH (1991)
OH 8.97E-012	cm3/molc sec	25	EST	MEYLAN,WM & HOWARD,PH (1993)

132483-43-3 — BENZENEACETIC ACID, à-METHYL-4-(5-METHYL-2-THIAZ

Formula: $C_{13}H_{13}NO_2S$

Mol Weight: 247.32

MP (deg C): FP (deg C):

BP (deg C):

BP pressure (mm Hg):

Property/Value	Units	Temp	Data Type	Reference
WS 1.24E+002	mg/L	25	EST	MEYLAN,WM ET AL. (1996)
logP 3.15			EXP	NAIRO,Y ET AL. (1992)
VP 2.05E-007	mm Hg	25	EST	NEELY,WB & BLAU,GE (1985)
DC	pKa			
HL 3.54E-012	atm m3/mol	25	EST	MEYLAN,WM & HOWARD,PH (1991)
OH 8.97E-012	cm3/molc sec	25	EST	MEYLAN,WM & HOWARD,PH (1993)

132483-44-4 — BENZENEACETIC ACID, 4-(4-ETHENYL-2-THIAZOLYL)-à-

Formula: $C_{14}H_{13}NO_2S$

Mol Weight: 259.33

MP (deg C): FP (deg C):

BP (deg C):

BP pressure (mm Hg):

Property/Value	Units	Temp	Data Type	Reference
WS 4.22E+001	mg/L	25	EST	MEYLAN,WM ET AL. (1996)
logP 3.62			EXP	NAIRO,Y ET AL. (1992)
VP 9.36E-008	mm Hg	25	EST	NEELY,WB & BLAU,GE (1985)
DC	pKa			
HL 1.64E-012	atm m3/mol	25	EST	MEYLAN,WM & HOWARD,PH (1991)
OH 3.35E-011	cm3/molc sec	25	EST	MEYLAN,WM & HOWARD,PH (1993)

CAS #: 132483-45-5				BENZENEACETIC ACID, à-METHYL-4-[4-(TRIFLUOROMETH
Formula: C₁₃H₁₀F₃NO₂S				
Mol Weight: 301.29				
MP (deg C):			FP (deg C):	
BP (deg C):				
BP pressure (mm Hg):				

Property/Value	Units	Temp	Data Type	Reference
WS 1.77E+001	mg/L	25	EST	MEYLAN,WM ET AL. (1996)
logP 3.78			EXP	NAIRO,Y ET AL. (1992)
VP 3.51E-007	mm Hg	25	EST	NEELY,WB & BLAU,GE (1985)
DC	pKa			
HL 2.79E-011	atm m3/mol	25	EST	MEYLAN,WM & HOWARD,PH (1991)
OH 6.37E-012	cm3/molc sec	25	EST	MEYLAN,WM & HOWARD,PH (1993)

CAS #: 132483-46-6				BENZENEACETIC ACID, 4-(4,5-DIMETHYL-2-THIAZOLYL)
Formula: C₁₄H₁₅NO₂S				
Mol Weight: 261.35				
MP (deg C):			FP (deg C):	
BP (deg C):				
BP pressure (mm Hg):				

Property/Value	Units	Temp	Data Type	Reference
WS 6.86E+001	mg/L	25	EST	MEYLAN,WM ET AL. (1996)
logP 3.36			EXP	NAIRO,Y ET AL. (1992)
VP 8.73E-008	mm Hg	25	EST	NEELY,WB & BLAU,GE (1985)
DC	pKa			
HL 3.90E-012	atm m3/mol	25	EST	MEYLAN,WM & HOWARD,PH (1991)
OH 7.89E-012	cm3/molc sec	25	EST	MEYLAN,WM & HOWARD,PH (1993)

CAS #: 132483-47-7				BENZENEACETIC ACID, 4-(4-ETHYL-2-THIAZOLYL)-à-ME
Formula: C₁₄H₁₅NO₂S				
Mol Weight: 261.35				
MP (deg C):			FP (deg C):	
BP (deg C):				
BP pressure (mm Hg):				

Property/Value	Units	Temp	Data Type	Reference
WS 3.19E+001	mg/L	25	EST	MEYLAN,WM ET AL. (1996)
logP 3.75			EXP	NAIRO,Y ET AL. (1992)
VP 8.74E-008	mm Hg	25	EST	NEELY,WB & BLAU,GE (1985)
DC	pKa			
HL 4.70E-012	atm m3/mol	25	EST	MEYLAN,WM & HOWARD,PH (1991)
OH 9.81E-012	cm3/molc sec	25	EST	MEYLAN,WM & HOWARD,PH (1993)

CAS #: 132483-50-2				BENZENEACETIC ACID, 4-[(4-(AMINOCARBONYL)-2-THIA
Formula: C₁₃H₁₂N₂O₃S				
Mol Weight: 276.32				
MP (deg C):			FP (deg C):	
BP (deg C):				
BP pressure (mm Hg):				

Property/Value	Units	Temp	Data Type	Reference
WS 2.36E+002	mg/L	25	EST	MEYLAN,WM ET AL. (1996)
logP 2.17			EXP	NAIRO,Y ET AL. (1992)
VP 1.81E-010	mm Hg	25	EST	NEELY,WB & BLAU,GE (1985)
DC	pKa			
HL 1.32E-018	atm m3/mol	25	EST	MEYLAN,WM & HOWARD,PH (1991)
OH 8.64E-012	cm3/molc sec	25	EST	MEYLAN,WM & HOWARD,PH (1993)

CAS #: 132483-51-3				BENZENEACETIC ACID, à-METHYL-4-[4-[(METHYLAMINO)
Formula: C₁₄H₁₄N₂O₃S				
Mol Weight: 290.34				
MP (deg C):			FP (deg C):	
BP (deg C):				
BP pressure (mm Hg):				

Property/Value	Units	Temp	Data Type	Reference
WS 1.13E+002	mg/L	25	EST	MEYLAN,WM ET AL. (1996)
logP 2.45			EXP	NAIRO,Y ET AL. (1992)
VP 1.03E-010	mm Hg	25	EST	NEELY,WB & BLAU,GE (1985)
DC	pKa			
HL 2.89E-018	atm m3/mol	25	EST	MEYLAN,WM & HOWARD,PH (1991)
OH 1.27E-011	cm3/molc sec	25	EST	MEYLAN,WM & HOWARD,PH (1993)

CAS #: 132483-52-4				BENZENEACETIC ACID, 4-[4-[(DIMETHYLAMINO)CARBONY
Formula: C₁₅H₁₆N₂O₃S				
Mol Weight: 304.37				
MP (deg C):			FP (deg C):	
BP (deg C):				
BP pressure (mm Hg):				

Property/Value	Units	Temp	Data Type	Reference
WS 1.69E+002	mg/L	25	EST	MEYLAN,WM ET AL. (1996)
logP 2.15			EXP	NAIRO,Y ET AL. (1992)
VP 7.99E-010	mm Hg	25	EST	NEELY,WB & BLAU,GE (1985)
DC	pKa			
HL 6.35E-018	atm m3/mol	25	EST	MEYLAN,WM & HOWARD,PH (1991)
OH 2.28E-011	cm3/molc sec	25	EST	MEYLAN,WM & HOWARD,PH (1993)

CAS #: 132483-55-7				BENZENEACETIC ACID, 4-(4-BUTYL-2-THIAZOLYL)-à-ME
Formula: C₁₆H₁₉NO₂S				
Mol Weight: 289.40				
MP (deg C):			FP (deg C):	
BP (deg C):				
BP pressure (mm Hg):				

Property/Value	Units	Temp	Data Type	Reference
WS 2.64E+000	mg/L	25	EST	MEYLAN,WM ET AL. (1996)
logP 4.83			EXP	NAIRO,Y ET AL. (1992)
VP 1.76E-008	mm Hg	25	EST	NEELY,WB & BLAU,GE (1985)
DC	pKa			
HL 8.28E-012	atm m3/mol	25	EST	MEYLAN,WM & HOWARD,PH (1991)
OH 1.26E-011	cm3/molc sec	25	EST	MEYLAN,WM & HOWARD,PH (1993)

CAS #: 132483-56-8				BENZENEACETIC ACID, 3-HYDROXY-_-METHYL-4-(2-THIA
Formula: C₁₂H₁₁NO₃S				
Mol Weight: 249.29				
MP (deg C):			FP (deg C):	
BP (deg C):				
BP pressure (mm Hg):				

Property/Value	Units	Temp	Data Type	Reference
WS 4.34E+002	mg/L	25	EST	MEYLAN,WM ET AL. (1996)
logP 3.18			EXP	NAIRO,Y ET AL. (1992)
VP 7.18E-009	mm Hg	25	EST	NEELY,WB & BLAU,GE (1985)
DC	pKa			
HL 3.34E-016	atm m3/mol	25	EST	MEYLAN,WM & HOWARD,PH (1991)
OH 6.12E-011	cm3/molc sec	25	EST	MEYLAN,WM & HOWARD,PH (1993)

CAS #: 132483-57-9				BENZENEACETIC ACID, 3-METHOXY-à-METHYL-4-(2-THIA

Formula: $C_{13}H_{13}NO_3S$

Mol Weight: 263.32

MP (deg C):　　FP (deg C):

BP (deg C):

BP pressure (mm Hg):

Property/Value	Units	Temp	Data Type	Reference
WS 3.85E+002	mg/L	25	EST	MEYLAN,WM ET AL. (1996)
logP 2.47			EXP	NAIRO,Y ET AL. (1992)
VP 8.46E-008	mm Hg	25	EST	NEELY,WB & BLAU,GE (1985)
DC	pKa			
HL 1.90E-013	atm m3/mol	25	EST	MEYLAN,WM & HOWARD,PH (1991)
OH 4.19E-011	cm3/molc sec	25	EST	MEYLAN,WM & HOWARD,PH (1993)

CAS #: 132483-58-0				BENZENEACETIC ACID, à-METHYL-3-(METHYLTHIO)-4-(2

Formula: $C_{13}H_{13}NO_2S_2$

Mol Weight: 279.38

MP (deg C):　　FP (deg C):

BP (deg C):

BP pressure (mm Hg):

Property/Value	Units	Temp	Data Type	Reference
WS 1.06E+002	mg/L	25	EST	MEYLAN,WM ET AL. (1996)
logP 3.02			EXP	NAIRO,Y ET AL. (1992)
VP 1.96E-008	mm Hg	25	EST	NEELY,WB & BLAU,GE (1985)
DC	pKa			
HL 9.34E-014	atm m3/mol	25	EST	MEYLAN,WM & HOWARD,PH (1991)
OH 2.69E-011	cm3/molc sec	25	EST	MEYLAN,WM & HOWARD,PH (1993)

CAS #: 132483-59-1				BENZENEACETIC ACID, à-METHYL-3-NITRO-4-(2-THIAZO

Formula: $C_{12}H_{10}N_2O_4S$

Mol Weight: 278.29

MP (deg C):　　FP (deg C):

BP (deg C):

BP pressure (mm Hg):

Property/Value	Units	Temp	Data Type	Reference
WS 2.89E+002	mg/L	25	EST	MEYLAN,WM ET AL. (1996)
logP 2.06			EXP	NAIRO,Y ET AL. (1992)
VP 9.22E-009	mm Hg	25	EST	NEELY,WB & BLAU,GE (1985)
DC	pKa			
HL 1.26E-014	atm m3/mol	25	EST	MEYLAN,WM & HOWARD,PH (1991)
OH 4.44E-012	cm3/molc sec	25	EST	MEYLAN,WM & HOWARD,PH (1993)

CAS #: 132483-60-4				BENZENEACETIC ACID, 3-AMINO-_-METHYL-4-(2-THIAZO

Formula: $C_{12}H_{12}N_2O_2S$

Mol Weight: 248.31

MP (deg C):　　FP (deg C):

BP (deg C):

BP pressure (mm Hg):

Property/Value	Units	Temp	Data Type	Reference
WS 4.85E+002	mg/L	25	EST	MEYLAN,WM ET AL. (1996)
logP 2.45			EXP	NAIRO,Y ET AL. (1992)
VP 2.29E-008	mm Hg	25	EST	NEELY,WB & BLAU,GE (1985)
DC	pKa			
HL 1.13E-015	atm m3/mol	25	EST	MEYLAN,WM & HOWARD,PH (1991)
OH 1.87E-010	cm3/molc sec	25	EST	MEYLAN,WM & HOWARD,PH (1993)

CAS #: 132483-61-5				BENZENEACETIC ACID, 4-(4-ACETYL-2-THIAZOLYL)-à-M

Formula: $C_{14}H_{13}NO_3S$

Mol Weight: 275.33

MP (deg C):　　FP (deg C):

BP (deg C):

BP pressure (mm Hg):

Property/Value	Units	Temp	Data Type	Reference
WS 2.31E+002	mg/L	25	EST	MEYLAN,WM ET AL. (1996)
logP 2.65			EXP	NAIRO,Y ET AL. (1992)
VP 1.82E-008	mm Hg	25	EST	NEELY,WB & BLAU,GE (1985)
DC	pKa			
HL 5.84E-015	atm m3/mol	25	EST	MEYLAN,WM & HOWARD,PH (1991)
OH 6.74E-012	cm3/molc sec	25	EST	MEYLAN,WM & HOWARD,PH (1993)

CAS #: 132722-90-8				9H-PURINE, 9-(2,3-DIDEOXY-2-FLUORO-á-D-THREO-PEN

Formula: $C_{10}H_{11}FN_4O_2$

Mol Weight: 238.22

MP (deg C):　　FP (deg C):

BP (deg C):

BP pressure (mm Hg):

Property/Value	Units	Temp	Data Type	Reference
WS 2.95E+003	mg/L	25	EST	MEYLAN,WM ET AL. (1996)
logP -0.40			EXP	BARCHI,JJ ET AL. (1991)
VP 1.33E-009	mm Hg	25	EST	NEELY,WB & BLAU,GE (1985)
DC	pKa			
HL 4.69E-016	atm m3/mol	25	EST	MEYLAN,WM & HOWARD,PH (1991)
OH 4.43E-011	cm3/molc sec	25	EST	MEYLAN,WM & HOWARD,PH (1993)

CAS #: 132722-91-9				9H-PURIN-6-AMINE, 9-(2,3-DIDEOXY-2-FLUORO-á-D-TH

Formula: $C_{11}H_{14}FN_5O_2$

Mol Weight: 267.26

MP (deg C):　　FP (deg C):

BP (deg C):

BP pressure (mm Hg):

Property/Value	Units	Temp	Data Type	Reference
WS 7.32E+002	mg/L	25	EST	MEYLAN,WM ET AL. (1996)
logP 0.12			EXP	BARCHI,JJ ET AL. (1991)
VP 1.62E-011	mm Hg	25	EST	NEELY,WB & BLAU,GE (1985)
DC	pKa			
HL 1.83E-019	atm m3/mol	25	EST	MEYLAN,WM & HOWARD,PH (1991)
OH 2.08E-010	cm3/molc sec	25	EST	MEYLAN,WM & HOWARD,PH (1993)

CAS #: 132722-92-0				DIDEOXY-ARA-A,2,N6-DIMETHYL-2'-FLUORO

Formula: $C_{12}H_{16}FN_5O_2$

Mol Weight: 281.29

MP (deg C):　　FP (deg C):

BP (deg C):

BP pressure (mm Hg):

Property/Value	Units	Temp	Data Type	Reference
WS 4.48E+003	mg/L	25	EST	MEYLAN,WM ET AL. (1996)
logP 0.64			EXP	HANSCH,C ET AL. (1995)
VP 3.47E-011	mm Hg	25	EST	NEELY,WB & BLAU,GE (1985)
DC	pKa			
HL 4.01E-019	atm m3/mol	25	EST	MEYLAN,WM & HOWARD,PH (1991)
OH 2.10E-010	cm3/molc sec	25	EST	MEYLAN,WM & HOWARD,PH (1993)

CAS #: 132722-93-1	BENZAMIDE, N-[9-(2,3-DIDEOXY-2-FLUORO-beta-D-THR

Formula: $C_{17}H_{16}FN_5O_3$

Mol Weight: 357.35

MP (deg C): | FP (deg C):

BP (deg C):

BP pressure (mm Hg):

Property/Value	Units	Temp	Data Type	Reference
WS 1.33E+003	mg/L	25	EST	MEYLAN,WM ET AL. (1996)
logP 0.73			EXP	SANGSTER,J (1994)
VP 1.46E-016	mm Hg	25	EST	NEELY,WB & BLAU,GE (1985)
DC	pKa			
HL 1.06E-022	atm m3/mol	25	EST	MEYLAN,WM & HOWARD,PH (1991)
OH 2.10E-010	cm3/molc sec	25	EST	MEYLAN,WM & HOWARD,PH (1993)

CAS #: 132722-94-2	9H-PURIN-6-AMINE, 9-(2,3-DIDEOXY-2-FLUORO-á-D-TH

Formula: $C_{11}H_{14}FN_5O_2$

Mol Weight: 267.26

MP (deg C): | FP (deg C):

BP (deg C):

BP pressure (mm Hg):

Property/Value	Units	Temp	Data Type	Reference
WS 7.62E+002	mg/L	25	EST	MEYLAN,WM ET AL. (1996)
logP 0.10			EXP	BARCHI,JJ ET AL. (1991)
VP 1.62E-011	mm Hg	25	EST	NEELY,WB & BLAU,GE (1985)
DC	pKa			
HL 1.83E-019	atm m3/mol	25	EST	MEYLAN,WM & HOWARD,PH (1991)
OH 2.08E-010	cm3/molc sec	25	EST	MEYLAN,WM & HOWARD,PH (1993)

CAS #: 132722-95-3	9H-PURINE, 6-CHLORO-9-(2,3-DIDEOXY-2-FLUORO-á-D-

Formula: $C_{10}H_{10}ClFN_4O_2$

Mol Weight: 272.67

MP (deg C): | FP (deg C):

BP (deg C):

BP pressure (mm Hg):

Property/Value	Units	Temp	Data Type	Reference
WS 4.61E+002	mg/L	25	EST	MEYLAN,WM ET AL. (1996)
logP 0.32			EXP	BARCHI,JJ ET AL. (1991)
VP 2.94E-010	mm Hg	25	EST	NEELY,WB & BLAU,GE (1985)
DC	pKa			
HL 2.19E-014	atm m3/mol	25	EST	MEYLAN,WM & HOWARD,PH (1991)
OH 3.36E-011	cm3/molc sec	25	EST	MEYLAN,WM & HOWARD,PH (1993)

CAS #: 132723-00-3	9H-PURINE, 9-(2,3-DIDEOXY-2-FLUORO-á-D-THREO-PEN

Formula: $C_{10}H_{13}FN_4O_3$

Mol Weight: 256.24

MP (deg C): | FP (deg C):

BP (deg C):

BP pressure (mm Hg):

Property/Value	Units	Temp	Data Type	Reference
WS 3.38E+005	mg/L	25	EST	MEYLAN,WM ET AL. (1996)
logP -1.38			EXP	BARCHI,JJ ET AL. (1991)
VP 1.83E-011	mm Hg	25	EST	NEELY,WB & BLAU,GE (1985)
DC	pKa			
HL 4.69E-021	atm m3/mol	25	EST	MEYLAN,WM & HOWARD,PH (1991)
OH 4.43E-011	cm3/molc sec	25	EST	MEYLAN,WM & HOWARD,PH (1993)

CAS #: 132765-80-1	L-VALINAMIDE, N-ACETYL-L-ALANYL-

Formula: $C_{10}H_{19}N_3O_3$

Mol Weight: 229.28

MP (deg C): | FP (deg C):

BP (deg C):

BP pressure (mm Hg):

Property/Value	Units	Temp	Data Type	Reference
WS 1.42E+004	mg/L	25	EST	MEYLAN,WM ET AL. (1996)
logP -1.13			EXP	AKAMATSU,M ET AL. (1990)
VP 5.76E-010	mm Hg	25	EST	NEELY,WB & BLAU,GE (1985)
DC	pKa			
HL 2.44E-015	atm m3/mol	25	EST	MEYLAN,WM & HOWARD,PH (1991)
OH 3.48E-011	cm3/molc sec	25	EST	MEYLAN,WM & HOWARD,PH (1993)

CAS #: 132765-81-2	L-VALINAMIDE, N-ACETYL-L-LEUCYL-

Formula: $C_{13}H_{25}N_3O_3$

Mol Weight: 271.36

MP (deg C): | FP (deg C):

BP (deg C):

BP pressure (mm Hg):

Property/Value	Units	Temp	Data Type	Reference
WS 5.27E+002	mg/L	25	EST	MEYLAN,WM ET AL. (1996)
logP 0.26			EXP	HANSCH,C ET AL. (1995)
VP 7.96E-011	mm Hg	25	EST	NEELY,WB & BLAU,GE (1985)
DC	pKa			
HL 5.70E-015	atm m3/mol	25	EST	MEYLAN,WM & HOWARD,PH (1991)
OH 4.29E-011	cm3/molc sec	25	EST	MEYLAN,WM & HOWARD,PH (1993)

CAS #: 132765-82-3	L-VALINAMIDE, N-ACETYL-L-ISOLEUCYL-

Formula: $C_{13}H_{25}N_3O_3$

Mol Weight: 271.36

MP (deg C): | FP (deg C):

BP (deg C):

BP pressure (mm Hg):

Property/Value	Units	Temp	Data Type	Reference
WS 6.42E+002	mg/L	25	EST	MEYLAN,WM ET AL. (1996)
logP 0.16			EXP	HANSCH,C ET AL. (1995)
VP 7.96E-011	mm Hg	25	EST	NEELY,WB & BLAU,GE (1985)
DC	pKa			
HL 5.70E-015	atm m3/mol	25	EST	MEYLAN,WM & HOWARD,PH (1991)
OH 4.65E-011	cm3/molc sec	25	EST	MEYLAN,WM & HOWARD,PH (1993)

CAS #: 132765-83-4	L-VALINAMIDE, N-ACETYL-L-VALYL-

Formula: $C_{12}H_{23}N_3O_3$

Mol Weight: 257.34

MP (deg C): | FP (deg C):

BP (deg C):

BP pressure (mm Hg):

Property/Value	Units	Temp	Data Type	Reference
WS 1.98E+003	mg/L	25	EST	MEYLAN,WM ET AL. (1996)
logP -0.32			EXP	HANSCH,C ET AL. (1995)
VP 1.82E-010	mm Hg	25	EST	NEELY,WB & BLAU,GE (1985)
DC	pKa			
HL 4.30E-015	atm m3/mol	25	EST	MEYLAN,WM & HOWARD,PH (1991)
OH 4.36E-011	cm3/molc sec	25	EST	MEYLAN,WM & HOWARD,PH (1993)

132765-84-5 — L-ISOLEUCINAMIDE, N-ACETYL-L-LEUCYL-

Formula: $C_{14}H_{27}N_3O_3$

Mol Weight: 285.39

MP (deg C):   FP (deg C):

BP (deg C):

BP pressure (mm Hg):

Property/ Value	Units	Temp	Data Type	Reference
WS 3.92E+003	mg/L	25	EST	MEYLAN,WM ET AL. (1996)
logP 0.68			EXP	HANSCH,C ET AL. (1995)
VP 3.47E-011	mm Hg	25	EST	NEELY,WB & BLAU,GE (1985)
DC	pKa			
HL 7.57E-015	atm m3/mol	25	EST	MEYLAN,WM & HOWARD,PH (1991)
OH 4.58E-011	cm3/molc sec	25	EST	MEYLAN,WM & HOWARD,PH (1993)

132765-85-6 — L-ALANINAMIDE, N-ACETYL-L-VALYL-

Formula: $C_{10}H_{19}N_3O_3$

Mol Weight: 229.28

MP (deg C):   FP (deg C):

BP (deg C):

BP pressure (mm Hg):

Property/ Value	Units	Temp	Data Type	Reference
WS 1.42E+004	mg/L	25	EST	MEYLAN,WM ET AL. (1996)
logP -1.14			EXP	AKAMATSU,M ET AL. (1990)
VP 5.76E-010	mm Hg	25	EST	NEELY,WB & BLAU,GE (1985)
DC	pKa			
HL 2.44E-015	atm m3/mol	25	EST	MEYLAN,WM & HOWARD,PH (1991)
OH 3.48E-011	cm3/molc sec	25	EST	MEYLAN,WM & HOWARD,PH (1993)

132765-86-7 — L-VALINAMIDE, N-ACETYL-L-TRYPTOPHYL-

Formula: $C_{18}H_{24}N_4O_3$

Mol Weight: 344.42

MP (deg C):   FP (deg C):

BP (deg C):

BP pressure (mm Hg):

Property/ Value	Units	Temp	Data Type	Reference
WS 1.59E+003	mg/L	25	EST	MEYLAN,WM ET AL. (1996)
logP 0.73			EXP	HANSCH,C ET AL. (1995)
VP 3.07E-015	mm Hg	25	EST	NEELY,WB & BLAU,GE (1985)
DC	pKa			
HL 3.23E-020	atm m3/mol	25	EST	MEYLAN,WM & HOWARD,PH (1991)
OH 2.39E-010	cm3/molc sec	25	EST	MEYLAN,WM & HOWARD,PH (1993)

132765-88-9 — L-PHENYLALANINAMIDE, N-ACETYL-L-SERYL-

Formula: $C_{14}H_{19}N_3O_4$

Mol Weight: 293.33

MP (deg C):   FP (deg C):

BP (deg C):

BP pressure (mm Hg):

Property/ Value	Units	Temp	Data Type	Reference
WS 3.11E+003	mg/L	25	EST	MEYLAN,WM ET AL. (1996)
logP -0.79			EXP	HANSCH,C ET AL. (1995)
VP 3.67E-016	mm Hg	25	EST	NEELY,WB & BLAU,GE (1985)
DC	pKa			
HL 4.08E-021	atm m3/mol	25	EST	MEYLAN,WM & HOWARD,PH (1991)
OH 4.92E-011	cm3/molc sec	25	EST	MEYLAN,WM & HOWARD,PH (1993)

132765-89-0 — L-VALINAMIDE, N-ACETYL-L-THREONYL-

Formula: $C_{11}H_{21}N_3O_4$

Mol Weight: 259.31

MP (deg C):   FP (deg C):

BP (deg C):

BP pressure (mm Hg):

Property/ Value	Units	Temp	Data Type	Reference
WS 1.20E+004	mg/L	25	EST	MEYLAN,WM ET AL. (1996)
logP -1.25			EXP	HANSCH,C ET AL. (1995)
VP 1.10E-013	mm Hg	25	EST	NEELY,WB & BLAU,GE (1985)
DC	pKa			
HL 1.18E-019	atm m3/mol	25	EST	MEYLAN,WM & HOWARD,PH (1991)
OH 6.25E-011	cm3/molc sec	25	EST	MEYLAN,WM & HOWARD,PH (1993)

132765-90-3 — L-ISOLEUCINAMIDE, N-ACETYL-L-THREONYL-

Formula: $C_{12}H_{23}N_3O_4$

Mol Weight: 273.33

MP (deg C):   FP (deg C):

BP (deg C):

BP pressure (mm Hg):

Property/ Value	Units	Temp	Data Type	Reference
WS 4.65E+003	mg/L	25	EST	MEYLAN,WM ET AL. (1996)
logP -0.86			EXP	HANSCH,C ET AL. (1995)
VP 4.13E-014	mm Hg	25	EST	NEELY,WB & BLAU,GE (1985)
DC	pKa			
HL 1.57E-019	atm m3/mol	25	EST	MEYLAN,WM & HOWARD,PH (1991)
OH 6.54E-011	cm3/molc sec	25	EST	MEYLAN,WM & HOWARD,PH (1993)

132765-91-4 — L-VALINAMIDE, N-ACETYL-L-GLUTAMINYL-

Formula: $C_{12}H_{22}N_4O_4$

Mol Weight: 286.33

MP (deg C):   FP (deg C):

BP (deg C):

BP pressure (mm Hg):

Property/ Value	Units	Temp	Data Type	Reference
WS 2.59E+004	mg/L	25	EST	MEYLAN,WM ET AL. (1996)
logP -1.85			EXP	HANSCH,C ET AL. (1995)
VP 7.93E-014	mm Hg	25	EST	NEELY,WB & BLAU,GE (1985)
DC	pKa			
HL 4.91E-022	atm m3/mol	25	EST	MEYLAN,WM & HOWARD,PH (1991)
OH 5.27E-011	cm3/molc sec	25	EST	MEYLAN,WM & HOWARD,PH (1993)

132765-92-5 — L-LEUCINAMIDE, N-ACETYL-L-GLUTAMINYL-

Formula: $C_{13}H_{24}N_4O_4$

Mol Weight: 300.36

MP (deg C):   FP (deg C):

BP (deg C):

BP pressure (mm Hg):

Property/ Value	Units	Temp	Data Type	Reference
WS 8.02E+003	mg/L	25	EST	MEYLAN,WM ET AL. (1996)
logP -1.32			EXP	HANSCH,C ET AL. (1995)
VP 3.37E-014	mm Hg	25	EST	NEELY,WB & BLAU,GE (1985)
DC	pKa			
HL 6.52E-022	atm m3/mol	25	EST	MEYLAN,WM & HOWARD,PH (1991)
OH 5.20E-011	cm3/molc sec	25	EST	MEYLAN,WM & HOWARD,PH (1993)

CAS #: 132765-93-6 — L-PHENYLALANINAMIDE, N2-ACETYL-L-GLUTAMINYL-

Formula: $C_{16}H_{22}N_4O_4$

Mol Weight: 334.38

MP (deg C):　FP (deg C):
BP (deg C):
BP pressure (mm Hg):

Property/Value	Units	Temp	Data Type	Reference
WS 2.86E+003	mg/L	25	EST	MEYLAN,WM ET AL. (1996)
logP -1.14			EXP	HANSCH,C ET AL. (1995)
VP 5.08E-016	mm Hg	25	EST	NEELY,WB & BLAU,GE (1985)
DC	pKa			
HL 2.25E-023	atm m3/mol	25	EST	MEYLAN,WM & HOWARD,PH (1991)
OH 5.33E-011	cm3/molc sec	25	EST	MEYLAN,WM & HOWARD,PH (1993)

CAS #: 132765-94-7 — L-GLUTAMAMIDE, N-ACETYL-L-PHENYLALANYL-

Formula: $C_{16}H_{22}N_4O_4$

Mol Weight: 334.38

MP (deg C):　FP (deg C):
BP (deg C):
BP pressure (mm Hg):

Property/Value	Units	Temp	Data Type	Reference
WS 2.86E+003	mg/L	25	EST	MEYLAN,WM ET AL. (1996)
logP -1.03			EXP	HANSCH,C ET AL. (1995)
VP 5.08E-016	mm Hg	25	EST	NEELY,WB & BLAU,GE (1985)
DC	pKa			
HL 2.25E-023	atm m3/mol	25	EST	MEYLAN,WM & HOWARD,PH (1991)
OH 5.33E-011	cm3/molc sec	25	EST	MEYLAN,WM & HOWARD,PH (1993)

CAS #: 132765-95-8 — L-GLUTINAMIDE, N-ACETYL-L-VALYL-

Formula: $C_{12}H_{22}N_4O_4$

Mol Weight: 286.33

MP (deg C):　FP (deg C):
BP (deg C):
BP pressure (mm Hg):

Property/Value	Units	Temp	Data Type	Reference
WS 2.59E+004	mg/L	25	EST	MEYLAN,WM ET AL. (1996)
logP -1.82			EXP	HANSCH,C ET AL. (1995)
VP 7.93E-014	mm Hg	25	EST	NEELY,WB & BLAU,GE (1985)
DC	pKa			
HL 4.91E-022	atm m3/mol	25	EST	MEYLAN,WM & HOWARD,PH (1991)
OH 5.27E-011	cm3/molc sec	25	EST	MEYLAN,WM & HOWARD,PH (1993)

CAS #: 132765-96-9 — L-ASPARTAMIDE, N-ACETYL-L-ISOLEUCYL-

Formula: $C_{12}H_{22}N_4O_4$

Mol Weight: 286.33

MP (deg C):　FP (deg C):
BP (deg C):
BP pressure (mm Hg):

Property/Value	Units	Temp	Data Type	Reference
WS 1.20E+004	mg/L	25	EST	MEYLAN,WM ET AL. (1996)
logP -1.41			EXP	HANSCH,C ET AL. (1995)
VP 7.93E-014	mm Hg	25	EST	NEELY,WB & BLAU,GE (1985)
DC	pKa			
HL 4.91E-022	atm m3/mol	25	EST	MEYLAN,WM & HOWARD,PH (1991)
OH 5.92E-011	cm3/molc sec	25	EST	MEYLAN,WM & HOWARD,PH (1993)

CAS #: 132765-98-1 — L-ISOLEUCINAMIDE, N-ACETYL-L-ASPARAGINYL-

Formula: $C_{12}H_{22}N_4O_4$

Mol Weight: 286.33

MP (deg C):　FP (deg C):
BP (deg C):
BP pressure (mm Hg):

Property/Value	Units	Temp	Data Type	Reference
WS 1.20E+004	mg/L	25	EST	MEYLAN,WM ET AL. (1996)
logP -1.43			EXP	HANSCH,C ET AL. (1995)
VP 7.93E-014	mm Hg	25	EST	NEELY,WB & BLAU,GE (1985)
DC	pKa			
HL 4.91E-022	atm m3/mol	25	EST	MEYLAN,WM & HOWARD,PH (1991)
OH 5.92E-011	cm3/molc sec	25	EST	MEYLAN,WM & HOWARD,PH (1993)

CAS #: 132765-99-2 — L-PHENYLALANINAMIDE, N-ACETYL-L-ASPARAGINYL-

Formula: $C_{15}H_{20}N_4O_4$

Mol Weight: 320.35

MP (deg C):　FP (deg C):
BP (deg C):
BP pressure (mm Hg):

Property/Value	Units	Temp	Data Type	Reference
WS 4.30E+003	mg/L	25	EST	MEYLAN,WM ET AL. (1996)
logP -1.14			EXP	HANSCH,C ET AL. (1995)
VP 1.21E-015	mm Hg	25	EST	NEELY,WB & BLAU,GE (1985)
DC	pKa			
HL 1.70E-023	atm m3/mol	25	EST	MEYLAN,WM & HOWARD,PH (1991)
OH 5.69E-011	cm3/molc sec	25	EST	MEYLAN,WM & HOWARD,PH (1993)

CAS #: 132766-00-8 — L-ASPARTAMIDE, N-ACETYL-L-LEUCYL-

Formula: $C_{12}H_{22}N_4O_4$

Mol Weight: 286.33

MP (deg C):　FP (deg C):
BP (deg C):
BP pressure (mm Hg):

Property/Value	Units	Temp	Data Type	Reference
WS 9.30E+003	mg/L	25	EST	MEYLAN,WM ET AL. (1996)
logP -1.30			EXP	HANSCH,C ET AL. (1995)
VP 7.93E-014	mm Hg	25	EST	NEELY,WB & BLAU,GE (1985)
DC	pKa			
HL 4.91E-022	atm m3/mol	25	EST	MEYLAN,WM & HOWARD,PH (1991)
OH 5.56E-011	cm3/molc sec	25	EST	MEYLAN,WM & HOWARD,PH (1993)

CAS #: 132766-01-9 — L-ALANINAMIDE, N-ACETYL-L-VALYL-L-ALANYL-

Formula: $C_{13}H_{24}N_4O_4$

Mol Weight: 300.36

MP (deg C):　FP (deg C):
BP (deg C):
BP pressure (mm Hg):

Property/Value	Units	Temp	Data Type	Reference
WS 9.39E+003	mg/L	25	EST	MEYLAN,WM ET AL. (1996)
logP -1.40			EXP	HANSCH,C ET AL. (1995)
VP 7.37E-014	mm Hg	25	EST	NEELY,WB & BLAU,GE (1985)
DC	pKa			
HL 2.15E-020	atm m3/mol	25	EST	MEYLAN,WM & HOWARD,PH (1991)
OH 4.68E-011	cm3/molc sec	25	EST	MEYLAN,WM & HOWARD,PH (1993)

132766-02-0 — L-VALINAMIDE, N-ACETYL-L-VALYL-L-ALANYL-

Formula: $C_{15}H_{28}N_4O_4$

Mol Weight: 328.41

MP (deg C): FP (deg C):

BP (deg C):

BP pressure (mm Hg):

Property/Value	Units	Temp	Data Type	Reference
WS 1.53E+003	mg/L	25	EST	MEYLAN,WM ET AL. (1996)
logP -0.67			EXP	HANSCH,C ET AL. (1995)
VP 2.23E-014	mm Hg	25	EST	NEELY,WB & BLAU,GE (1985)
DC	pKa			
HL 3.79E-020	atm m3/mol	25	EST	MEYLAN,WM & HOWARD,PH (1991)
OH 5.56E-011	cm3/molc sec	25	EST	MEYLAN,WM & HOWARD,PH (1993)

132766-03-1 — GLYCINAMIDE, N-ACETYL-L-VALYL-L-ISOLEUCYL-

Formula: $C_{15}H_{28}N_4O_4$

Mol Weight: 328.41

MP (deg C): FP (deg C):

BP (deg C):

BP pressure (mm Hg):

Property/Value	Units	Temp	Data Type	Reference
WS 9.91E+002	mg/L	25	EST	MEYLAN,WM ET AL. (1996)
logP -0.45			EXP	HANSCH,C ET AL. (1995)
VP 1.33E-014	mm Hg	25	EST	NEELY,WB & BLAU,GE (1985)
DC	pKa			
HL 3.79E-020	atm m3/mol	25	EST	MEYLAN,WM & HOWARD,PH (1991)
OH 5.48E-011	cm3/molc sec	25	EST	MEYLAN,WM & HOWARD,PH (1993)

132766-04-2 — L-VALINAMIDE, N-ACETYL-L-ALANYL-L-LEUCYL-

Formula: $C_{16}H_{30}N_4O_4$

Mol Weight: 342.44

MP (deg C): FP (deg C):

BP (deg C):

BP pressure (mm Hg):

Property/Value	Units	Temp	Data Type	Reference
WS 4.44E+002	mg/L	25	EST	MEYLAN,WM ET AL. (1996)
logP -0.14			EXP	HANSCH,C ET AL. (1995)
VP 9.42E-015	mm Hg	25	EST	NEELY,WB & BLAU,GE (1985)
DC	pKa			
HL 5.04E-020	atm m3/mol	25	EST	MEYLAN,WM & HOWARD,PH (1991)
OH 5.49E-011	cm3/molc sec	25	EST	MEYLAN,WM & HOWARD,PH (1993)

132766-05-3 — L-ALANINAMIDE, N-ACETYL-L-VALYL-L-PHENYLALANYL-

Formula: $C_{19}H_{28}N_4O_4$

Mol Weight: 376.46

MP (deg C): FP (deg C):

BP (deg C):

BP pressure (mm Hg):

Property/Value	Units	Temp	Data Type	Reference
WS 1.86E+002	mg/L	25	EST	MEYLAN,WM ET AL. (1996)
logP 0.06			EXP	HANSCH,C ET AL. (1995)
VP 1.40E-016	mm Hg	25	EST	NEELY,WB & BLAU,GE (1985)
DC	pKa			
HL 1.74E-021	atm m3/mol	25	EST	MEYLAN,WM & HOWARD,PH (1991)
OH 5.62E-011	cm3/molc sec	25	EST	MEYLAN,WM & HOWARD,PH (1993)

132766-06-4 — L-ISOLEUCINAMIDE, N-ACETYL-L-ALANYL-L-VALYL-

Formula: $C_{16}H_{30}N_4O_4$

Mol Weight: 342.44

MP (deg C): FP (deg C):

BP (deg C):

BP pressure (mm Hg):

Property/Value	Units	Temp	Data Type	Reference
WS 5.09E+002	mg/L	25	EST	MEYLAN,WM ET AL. (1996)
logP -0.20			EXP	HANSCH,C ET AL. (1995)
VP 9.42E-015	mm Hg	25	EST	NEELY,WB & BLAU,GE (1985)
DC	pKa			
HL 5.04E-020	atm m3/mol	25	EST	MEYLAN,WM & HOWARD,PH (1991)
OH 5.85E-011	cm3/molc sec	25	EST	MEYLAN,WM & HOWARD,PH (1993)

132766-07-5 — L-ALANINAMIDE, N-ACETYL-L-ISOLEUCYL-L-PHENYLALAN

Formula: $C_{20}H_{30}N_4O_4$

Mol Weight: 390.49

MP (deg C): FP (deg C):

BP (deg C):

BP pressure (mm Hg):

Property/Value	Units	Temp	Data Type	Reference
WS 1.26E+003	mg/L	25	EST	MEYLAN,WM ET AL. (1996)
logP 0.52			EXP	HANSCH,C ET AL. (1995)
VP 5.82E-017	mm Hg	25	EST	NEELY,WB & BLAU,GE (1985)
DC	pKa			
HL 2.31E-021	atm m3/mol	25	EST	MEYLAN,WM & HOWARD,PH (1991)
OH 5.91E-011	cm3/molc sec	25	EST	MEYLAN,WM & HOWARD,PH (1993)

132766-08-6 — L-VALINAMIDE, N-ACETYLGLYCYL-L-ALANYL-

Formula: $C_{12}H_{22}N_4O_4$

Mol Weight: 286.33

MP (deg C): FP (deg C):

BP (deg C):

BP pressure (mm Hg):

Property/Value	Units	Temp	Data Type	Reference
WS 1.55E+004	mg/L	25	EST	MEYLAN,WM ET AL. (1996)
logP -1.56			EXP	HANSCH,C ET AL. (1995)
VP 1.03E-013	mm Hg	25	EST	NEELY,WB & BLAU,GE (1985)
DC	pKa			
HL 1.62E-020	atm m3/mol	25	EST	MEYLAN,WM & HOWARD,PH (1991)
OH 4.32E-011	cm3/molc sec	25	EST	MEYLAN,WM & HOWARD,PH (1993)

132766-09-7 — L-PHENYLALANINAMIDE, N-ACETYL-L-ALANYLGLYCYL-

Formula: $C_{16}H_{22}N_4O_4$

Mol Weight: 334.38

MP (deg C): FP (deg C):

BP (deg C):

BP pressure (mm Hg):

Property/Value	Units	Temp	Data Type	Reference
WS 1.52E+003	mg/L	25	EST	MEYLAN,WM ET AL. (1996)
logP -0.71			EXP	HANSCH,C ET AL. (1995)
VP 6.67E-016	mm Hg	25	EST	NEELY,WB & BLAU,GE (1985)
DC	pKa			
HL 7.43E-022	atm m3/mol	25	EST	MEYLAN,WM & HOWARD,PH (1991)
OH 4.38E-011	cm3/molc sec	25	EST	MEYLAN,WM & HOWARD,PH (1993)

CAS #: 132766-10-0

L-VALINAMIDE, N-ACETYL-L-ISOLEUCYL-L-ALANYL-

Formula: $C_{16}H_{30}N_4O_4$

Mol Weight: 342.44

MP (deg C):　FP (deg C):

BP (deg C):

BP pressure (mm Hg):

	Property/Value	Units	Temp	Data Type	Reference
WS	5.09E+002	mg/L	25	EST	MEYLAN,WM ET AL. (1996)
logP	-0.21			EXP	HANSCH,C ET AL. (1995)
VP	9.42E-015	mm Hg	25	EST	NEELY,WB & BLAU,GE (1985)
DC		pKa			
HL	5.04E-020	atm m3/mol	25	EST	MEYLAN,WM & HOWARD,PH (1991)
OH	5.85E-011	cm3/molc sec	25	EST	MEYLAN,WM & HOWARD,PH (1993)

CAS #: 132766-11-1

GLYCINAMIDE, N-ACETYL-L-PHENYLALANYL-L-ISOLEUCYL

Formula: $C_{19}H_{28}N_4O_4$

Mol Weight: 376.46

MP (deg C):　FP (deg C):

BP (deg C):

BP pressure (mm Hg):

	Property/Value	Units	Temp	Data Type	Reference
WS	1.07E+002	mg/L	25	EST	MEYLAN,WM ET AL. (1996)
logP	0.34			EXP	HANSCH,C ET AL. (1995)
VP	8.24E-017	mm Hg	25	EST	NEELY,WB & BLAU,GE (1985)
DC		pKa			
HL	1.74E-021	atm m3/mol	25	EST	MEYLAN,WM & HOWARD,PH (1991)
OH	5.54E-011	cm3/molc sec	25	EST	MEYLAN,WM & HOWARD,PH (1993)

CAS #: 132766-12-2

L-ISOLEUCINAMIDE, N-ACETYL-L-VALYL-L-VALYL-

Formula: $C_{18}H_{34}N_4O_4$

Mol Weight: 370.50

MP (deg C):　FP (deg C):

BP (deg C):

BP pressure (mm Hg):

	Property/Value	Units	Temp	Data Type	Reference
WS	1.78E+003	mg/L	25	EST	MEYLAN,WM ET AL. (1996)
logP	0.49			EXP	AKAMATSU,M ET AL. (1990)
VP	2.82E-015	mm Hg	25	EST	NEELY,WB & BLAU,GE (1985)
DC		pKa			
HL	8.88E-020	atm m3/mol	25	EST	MEYLAN,WM & HOWARD,PH (1991)
OH	6.72E-011	cm3/molc sec	25	EST	MEYLAN,WM & HOWARD,PH (1993)

CAS #: 132766-13-3

GLYCINAMIDE, N-ACETYLGLYCYL-L-LEUCYL-

Formula: $C_{12}H_{22}N_4O_4$

Mol Weight: 286.33

MP (deg C):　FP (deg C):

BP (deg C):

BP pressure (mm Hg):

	Property/Value	Units	Temp	Data Type	Reference
WS	8.11E+003	mg/L	25	EST	MEYLAN,WM ET AL. (1996)
logP	-1.23			EXP	HANSCH,C ET AL. (1995)
VP	6.18E-014	mm Hg	25	EST	NEELY,WB & BLAU,GE (1985)
DC		pKa			
HL	1.62E-020	atm m3/mol	25	EST	MEYLAN,WM & HOWARD,PH (1991)
OH	3.89E-011	cm3/molc sec	25	EST	MEYLAN,WM & HOWARD,PH (1993)

CAS #: 132766-15-5

GLYCINAMIDE, N-ACETYL-L-TRYPTOPHYL-L-ISOLEUCYL-

Formula: $C_{21}H_{29}N_5O_4$

Mol Weight: 415.50

MP (deg C):　FP (deg C):

BP (deg C):

BP pressure (mm Hg):

	Property/Value	Units	Temp	Data Type	Reference
WS	7.25E+002	mg/L	25	EST	MEYLAN,WM ET AL. (1996)
logP	0.62			EXP	HANSCH,C ET AL. (1995)
VP	1.51E-019	mm Hg	25	EST	NEELY,WB & BLAU,GE (1985)
DC		pKa			
HL	2.86E-025	atm m3/mol	25	EST	MEYLAN,WM & HOWARD,PH (1991)
OH	2.51E-010	cm3/molc sec	25	EST	MEYLAN,WM & HOWARD,PH (1993)

CAS #: 132766-16-6

L-PHENYLALANINAMIDE, N-ACETYL-L-TRYPTOPHYLGLYCYL

Formula: $C_{24}H_{27}N_5O_4$

Mol Weight: 449.51

MP (deg C):　FP (deg C):

BP (deg C):

BP pressure (mm Hg):

	Property/Value	Units	Temp	Data Type	Reference
WS	2.14E+002	mg/L	25	EST	MEYLAN,WM ET AL. (1996)
logP	0.99			EXP	HANSCH,C ET AL. (1995)
VP	3.92E-021	mm Hg	25	EST	NEELY,WB & BLAU,GE (1985)
DC		pKa			
HL	9.86E-027	atm m3/mol	25	EST	MEYLAN,WM & HOWARD,PH (1991)
OH	2.48E-010	cm3/molc sec	25	EST	MEYLAN,WM & HOWARD,PH (1993)

CAS #: 132766-17-7

L-VALINAMIDE, N-ACETYL-L-TRYPTOPHYL-L-ALANYL-

Formula: $C_{21}H_{29}N_5O_4$

Mol Weight: 415.50

MP (deg C):　FP (deg C):

BP (deg C):

BP pressure (mm Hg):

	Property/Value	Units	Temp	Data Type	Reference
WS	1.21E+003	mg/L	25	EST	MEYLAN,WM ET AL. (1996)
logP	0.36			EXP	HANSCH,C ET AL. (1995)
VP	2.58E-019	mm Hg	25	EST	NEELY,WB & BLAU,GE (1985)
DC		pKa			
HL	2.86E-025	atm m3/mol	25	EST	MEYLAN,WM & HOWARD,PH (1991)
OH	2.51E-010	cm3/molc sec	25	EST	MEYLAN,WM & HOWARD,PH (1993)

CAS #: 132766-18-8

L-PHENYLALANINAMIDE, N-ACETYL-L-LEUCYL-L-SERYL-

Formula: $C_{20}H_{30}N_4O_5$

Mol Weight: 406.49

MP (deg C):　FP (deg C):

BP (deg C):

BP pressure (mm Hg):

	Property/Value	Units	Temp	Data Type	Reference
WS	8.70E+001	mg/L	25	EST	MEYLAN,WM ET AL. (1996)
logP	0.23			EXP	HANSCH,C ET AL. (1995)
VP	6.11E-022	mm Hg	25	EST	NEELY,WB & BLAU,GE (1985)
DC		pKa			
HL	8.44E-026	atm m3/mol	25	EST	MEYLAN,WM & HOWARD,PH (1991)
OH	6.92E-011	cm3/molc sec	25	EST	MEYLAN,WM & HOWARD,PH (1993)

132766-19-9

CAS #: 132766-19-9

L-LEUCINAMIDE, N-ACETYL-L-LEUCYL-L-THREONYL-

Formula: $C_{18}H_{34}N_4O_5$

Mol Weight: 386.50

MP (deg C):   **FP (deg C):**

BP (deg C):

BP pressure (mm Hg):

Property/Value	Units	Temp	Data Type	Reference
WS 1.13E+002	mg/L	25	EST	MEYLAN,WM ET AL. (1996)
logP 0.24			EXP	AKAMATSU,M ET AL. (1990)
VP 8.37E-020	mm Hg	25	EST	NEELY,WB & BLAU,GE (1985)
DC	pKa			
HL 3.25E-024	atm m3/mol	25	EST	MEYLAN,WM & HOWARD,PH (1991)
OH 8.18E-011	cm3/molc sec	25	EST	MEYLAN,WM & HOWARD,PH (1993)

132766-20-2

CAS #: 132766-20-2

L-LEUCINAMIDE, N-ACETYL-L-ALANYL-L-TYROSYL-

Formula: $C_{20}H_{30}N_4O_5$

Mol Weight: 406.49

MP (deg C):   **FP (deg C):**

BP (deg C):

BP pressure (mm Hg):

Property/Value	Units	Temp	Data Type	Reference
WS 5.62E+002	mg/L	25	EST	MEYLAN,WM ET AL. (1996)
logP -0.04			EXP	HANSCH,C ET AL. (1995)
VP 1.54E-019	mm Hg	25	EST	NEELY,WB & BLAU,GE (1985)
DC	pKa			
HL 2.40E-025	atm m3/mol	25	EST	MEYLAN,WM & HOWARD,PH (1991)
OH 9.14E-011	cm3/molc sec	25	EST	MEYLAN,WM & HOWARD,PH (1993)

132766-21-3

CAS #: 132766-21-3

L-PHENYLALANINAMIDE, N-ACETYL-L-ALANYL-L-TYROSYL

Formula: $C_{23}H_{28}N_4O_5$

Mol Weight: 440.50

MP (deg C):   **FP (deg C):**

BP (deg C):

BP pressure (mm Hg):

Property/Value	Units	Temp	Data Type	Reference
WS 1.91E+002	mg/L	25	EST	MEYLAN,WM ET AL. (1996)
logP 0.26			EXP	HANSCH,C ET AL. (1995)
VP 1.68E-021	mm Hg	25	EST	NEELY,WB & BLAU,GE (1985)
DC	pKa			
HL 8.28E-027	atm m3/mol	25	EST	MEYLAN,WM & HOWARD,PH (1991)
OH 9.26E-011	cm3/molc sec	25	EST	MEYLAN,WM & HOWARD,PH (1993)

132899-69-5

CAS #: 132899-69-5

PROPANAMIDE, N-(2-ETHOXY-5-NITROPHENYL)-

Formula: $C_{11}H_{14}N_2O_4$

Mol Weight: 238.25

MP (deg C):   **FP (deg C):**

BP (deg C):

BP pressure (mm Hg):

Property/Value	Units	Temp	Data Type	Reference
WS 2.94E+002	mg/L	25	EST	MEYLAN,WM ET AL. (1996)
logP 1.85			EXP	FURST,W & BECHER,M (1990)
VP 1.65E-007	mm Hg	25	EST	NEELY,WB & BLAU,GE (1985)
DC	pKa			
HL 2.54E-012	atm m3/mol	25	EST	MEYLAN,WM & HOWARD,PH (1991)
OH 8.81E-012	cm3/molc sec	25	EST	MEYLAN,WM & HOWARD,PH (1993)

132911-42-3

CAS #: 132911-42-3

1H-PYRROLE-2-CARBOXAMIDE, N-METHYL-

Formula: $C_6H_8N_2O$

Mol Weight: 124.14

MP (deg C):   **FP (deg C):**

BP (deg C):

BP pressure (mm Hg):

Property/Value	Units	Temp	Data Type	Reference
WS 4.24E+004	mg/L	25	EST	MEYLAN,WM ET AL. (1996)
logP 0.42			EXP	YAMAGAMI,C ET AL. (1994)
VP 1.51E-004	mm Hg	25	EST	NEELY,WB & BLAU,GE (1985)
DC	pKa			
HL 8.19E-012	atm m3/mol	25	EST	MEYLAN,WM & HOWARD,PH (1991)
OH 1.06E-010	cm3/molc sec	25	EST	MEYLAN,WM & HOWARD,PH (1993)

133445-74-6

CAS #: 133445-74-6

THIENO[2,3-B]THIOPHENE-2-SULFONAMIDE, 5-[[[2-(ME

Formula: $C_{10}H_{15}N_2O_4S_4$

Mol Weight: 356.51

MP (deg C):   **FP (deg C):**

BP (deg C):

BP pressure (mm Hg):

Property/Value	Units	Temp	Data Type	Reference
WS 9.89E+004	mg/L	25	EST	MEYLAN,WM ET AL. (1996)
logP -0.26			EXP	PRUGH,JD ET AL. (1991)
VP 2.80E-011	mm Hg	25	EST	NEELY,WB & BLAU,GE (1985)
DC	pKa			
HL 3.41E-016	atm m3/mol	25	EST	MEYLAN,WM & HOWARD,PH (1991)
OH 1.00E-010	cm3/molc sec	25	EST	MEYLAN,WM & HOWARD,PH (1993)

133636-94-9

CAS #: 133636-94-9

RH-1965 O-PH N-BUTYLENE CARBAMATE

Formula: $C_{13}H_{13}Cl_2NO_4$

Mol Weight: 318.16

MP (deg C):   **FP (deg C):**

BP (deg C):

BP pressure (mm Hg):

Property/Value	Units	Temp	Data Type	Reference
WS 4.07E+000	mg/L	25	EST	MEYLAN,WM ET AL. (1996)
logP 3.95			EXP	NANDIHALLI,UB ET AL. (1993)
VP 9.38E-006	mm Hg	25	EST	NEELY,WB & BLAU,GE (1985)
DC	pKa			
HL 1.94E-010	atm m3/mol	25	EST	MEYLAN,WM & HOWARD,PH (1991)
OH 2.52E-011	cm3/molc sec	25	EST	MEYLAN,WM & HOWARD,PH (1993)

133636-96-1

CAS #: 133636-96-1

RH-1964 O-PH(DICL) N-BUTYLENE CARBAMATE

Formula: $C_{15}H_{17}Cl_2NO_4$

Mol Weight: 346.21

MP (deg C):   **FP (deg C):**

BP (deg C):

BP pressure (mm Hg):

Property/Value	Units	Temp	Data Type	Reference
WS 6.71E-001	mg/L	25	EST	MEYLAN,WM ET AL. (1996)
logP 4.67			EXP	NANDIHALLI,UB ET AL. (1993)
VP 3.11E-006	mm Hg	25	EST	NEELY,WB & BLAU,GE (1985)
DC	pKa			
HL 3.42E-010	atm m3/mol	25	EST	MEYLAN,WM & HOWARD,PH (1991)
OH 2.89E-011	cm3/molc sec	25	EST	MEYLAN,WM & HOWARD,PH (1993)

CAS #: 133636-98-3

RH-1422 O-PH N-BUTYLENE CARBAMATE

Formula: $C_{15}H_{17}ClFNO_4$

Mol Weight: 329.76

MP (deg C):　　FP (deg C):

BP (deg C):

BP pressure (mm Hg):

Property/ Value	Units	Temp	Data Type	Reference
WS 2.04E+000	mg/L	25	EST	MEYLAN,WM ET AL. (1996)
logP 4.22			EXP	NANDIHALLI,UB ET AL. (1993)
VP 1.68E-005	mm Hg	25	EST	NEELY,WB & BLAU,GE (1985)
DC	pKa			
HL 5.38E-010	atm m3/mol	25	EST	MEYLAN,WM & HOWARD,PH (1991)
OH 2.94E-011	cm3/molc sec	25	EST	MEYLAN,WM & HOWARD,PH (1993)

CAS #: 133959-98-5

1H-PYRROL-1-YLOXY, 3-[[[2-(DIMETHYLAMINO)ETHYL]M

Formula: $C_{14}H_{29}N_3O$

Mol Weight: 255.41

MP (deg C):　　FP (deg C):

BP (deg C):

BP pressure (mm Hg):

Property/ Value	Units	Temp	Data Type	Reference
WS 3.81E+004	mg/L	25	EST	MEYLAN,WM ET AL. (1996)
logP 0.91			EXP	FUCHS,J ET AL. (1990)
VP 2.13E-008	mm Hg	25	EST	NEELY,WB & BLAU,GE (1985)
DC	pKa			
HL 2.78E-017	atm m3/mol	25	EST	MEYLAN,WM & HOWARD,PH (1991)
OH 2.53E-010	cm3/molc sec	25	EST	MEYLAN,WM & HOWARD,PH (1993)

CAS #: 133960-00-6

1(2H)-PYRIDINYLOXY, 3,6-DIHYDRO-2,2,6,6-TETRAMET

Formula: $C_{14}H_{26}N_2O_2$

Mol Weight: 254.38

MP (deg C):　　FP (deg C):

BP (deg C):

BP pressure (mm Hg):

Property/ Value	Units	Temp	Data Type	Reference
WS 2.45E+004	mg/L	25	EST	MEYLAN,WM ET AL. (1996)
logP 1.14			EXP	FUCHS,J ET AL. (1990)
VP 7.32E-009	mm Hg	25	EST	NEELY,WB & BLAU,GE (1985)
DC	pKa			
HL 3.85E-016	atm m3/mol	25	EST	MEYLAN,WM & HOWARD,PH (1991)
OH 2.38E-010	cm3/molc sec	25	EST	MEYLAN,WM & HOWARD,PH (1993)

CAS #: 134419-51-5

2-AZETIDINECARBOXYLIC ACID, 1-[2-HYDROXY-3-(1H-I

Formula: $C_{10}H_{15}N_3O_3$

Mol Weight: 225.25

MP (deg C):　　FP (deg C):

BP (deg C):

BP pressure (mm Hg):

Property/ Value	Units	Temp	Data Type	Reference
WS 1.46E+004	mg/L	25	EST	MEYLAN,WM ET AL. (1996)
logP -2.02			EXP	SUTO,MJ ET AL. (1991)
VP 2.86E-012	mm Hg	25	EST	NEELY,WB & BLAU,GE (1985)
DC	pKa			
HL 2.22E-016	atm m3/mol	25	EST	MEYLAN,WM & HOWARD,PH (1991)
OH 8.18E-011	cm3/molc sec	25	EST	MEYLAN,WM & HOWARD,PH (1993)

CAS #: 134419-52-6

2-AZETIDINECARBOXYLIC ACID, 1-[2-HYDROXY-3-(1H-I

Formula: $C_{11}H_{17}N_3O_3$

Mol Weight: 239.28

MP (deg C):　　FP (deg C):

BP (deg C):

BP pressure (mm Hg):

Property/ Value	Units	Temp	Data Type	Reference
WS 4.88E+005	mg/L	25	EST	MEYLAN,WM ET AL. (1996)
logP -0.29			EXP	SUTO,MJ ET AL. (1991)
VP 2.03E-008	mm Hg	25	EST	NEELY,WB & BLAU,GE (1985)
DC	pKa			
HL 7.12E-014	atm m3/mol	25	EST	MEYLAN,WM & HOWARD,PH (1991)
OH 8.14E-011	cm3/molc sec	25	EST	MEYLAN,WM & HOWARD,PH (1993)

CAS #: 134419-53-7

1H-IMIDAZOLE-1-ETHANOL, à-[[(2-BROMOCYCLOHEXYL)A

Formula: $C_{12}H_{20}BrN_3O$

Mol Weight: 302.22

MP (deg C):　　FP (deg C):

BP (deg C):

BP pressure (mm Hg):

Property/ Value	Units	Temp	Data Type	Reference
WS 1.12E+004	mg/L	25	EST	MEYLAN,WM ET AL. (1996)
logP 1.21			EXP	SUTO,MJ ET AL. (1991)
VP 1.88E-009	mm Hg	25	EST	NEELY,WB & BLAU,GE (1985)
DC	pKa			
HL 2.44E-013	atm m3/mol	25	EST	MEYLAN,WM & HOWARD,PH (1991)
OH 7.40E-011	cm3/molc sec	25	EST	MEYLAN,WM & HOWARD,PH (1993)

CAS #: 134419-55-9

1H-IMIDAZOLE-1-ETHANOL, à-[[(4-BROMO-TETRAHYDRO-

Formula: $C_{11}H_{18}BrN_3O_2$

Mol Weight: 304.19

MP (deg C):　　FP (deg C):

BP (deg C):

BP pressure (mm Hg):

Property/ Value	Units	Temp	Data Type	Reference
WS 1.65E+005	mg/L	25	EST	MEYLAN,WM ET AL. (1996)
logP -0.17			EXP	SUTO,MJ ET AL. (1991)
VP 1.35E-009	mm Hg	25	EST	NEELY,WB & BLAU,GE (1985)
DC	pKa			
HL 2.04E-016	atm m3/mol	25	EST	MEYLAN,WM & HOWARD,PH (1991)
OH 1.04E-010	cm3/molc sec	25	EST	MEYLAN,WM & HOWARD,PH (1993)

CAS #: 134917-52-5

PROPANOIC ACID, 3-[(7-CHLORO-4H-1,2,4-BENZOTHIAD

Formula: $C_{10}H_9ClN_2O_4S_2$

Mol Weight: 320.78

MP (deg C):　　FP (deg C):

BP (deg C):

BP pressure (mm Hg):

Property/ Value	Units	Temp	Data Type	Reference
WS 1.38E+003	mg/L	25	EST	MEYLAN,WM ET AL. (1996)
logP 1.43			EXP	GAMBERINI,G ET AL. (1989)
VP 3.45E-012	mm Hg	25	EST	NEELY,WB & BLAU,GE (1985)
DC	pKa			
HL 5.44E-015	atm m3/mol	25	EST	MEYLAN,WM & HOWARD,PH (1991)
OH 1.47E-011	cm3/molc sec	25	EST	MEYLAN,WM & HOWARD,PH (1993)

CAS #: 134917-53-6 — PROPANOIC ACID, 3-[(5,7-DICHLORO-4H-1,2,4-BENZOT

Formula: $C_{10}H_8Cl_2N_2O_4S_2$

Mol Weight: 355.22

MP (deg C): FP (deg C):

BP (deg C):

BP pressure (mm Hg):

Property/Value	Units	Temp	Data Type	Reference
WS 4.24E+002	mg/L	25	EST	MEYLAN,WM ET AL. (1996)
logP 1.79			EXP	GAMBERINI,G ET AL. (1989)
VP 8.86E-013	mm Hg	25	EST	NEELY,WB & BLAU,GE (1985)
DC	pKa			
HL 4.03E-015	atm m3/mol	25	EST	MEYLAN,WM & HOWARD,PH (1991)
OH 1.24E-011	cm3/molc sec	25	EST	MEYLAN,WM & HOWARD,PH (1993)

CAS #: 135600-74-7 — URIDINE, 2'-DEOXY-5-IODO-, 5'-(4-METHOXYBENZOATE

Formula: $C_{17}H_{17}IN_2O_7$

Mol Weight: 488.24

MP (deg C): FP (deg C):

BP (deg C):

BP pressure (mm Hg):

Property/Value	Units	Temp	Data Type	Reference
WS 5.65E+001	mg/L	25	EST	MEYLAN,WM ET AL. (1996)
logP 1.38			EXP	SANGSTER,J (1994)
VP 5.98E-019	mm Hg	25	EST	NEELY,WB & BLAU,GE (1985)
DC	pKa			
HL 7.40E-022	atm m3/mol	25	EST	MEYLAN,WM & HOWARD,PH (1991)
OH 8.79E-011	cm3/molc sec	25	EST	MEYLAN,WM & HOWARD,PH (1993)

CAS #: 135617-14-0 — URIDINE, 2'-DEOXY-5-IODO-, 5'-(4-NITROBENZOATE)

Formula: $C_{16}H_{14}IN_3O_8$

Mol Weight: 503.21

MP (deg C): FP (deg C):

BP (deg C):

BP pressure (mm Hg):

Property/Value	Units	Temp	Data Type	Reference
WS 2.43E+001	mg/L	25	EST	MEYLAN,WM ET AL. (1996)
logP 1.24			EXP	SANGSTER,J (1994)
VP 3.58E-020	mm Hg	25	EST	NEELY,WB & BLAU,GE (1985)
DC	pKa			
HL 4.94E-023	atm m3/mol	25	EST	MEYLAN,WM & HOWARD,PH (1991)
OH 8.02E-011	cm3/molc sec	25	EST	MEYLAN,WM & HOWARD,PH (1993)

CAS #: 135685-04-0 — BENZENESULFONAMIDE, N-(2-AMINOETHYL)-N-METHYL-4-

Formula: $C_{12}H_{18}N_4O_6S$

Mol Weight: 346.36

MP (deg C): FP (deg C):

BP (deg C):

BP pressure (mm Hg):

Property/Value	Units	Temp	Data Type	Reference
WS 7.53E+003	mg/L	25	EST	MEYLAN,WM ET AL. (1996)
logP 0.65			EXP	SAAN,WS ET AL. (1991)
VP 1.13E-009	mm Hg	25	EST	NEELY,WB & BLAU,GE (1985)
DC	pKa			
HL 1.22E-015	atm m3/mol	25	EST	MEYLAN,WM & HOWARD,PH (1991)
OH 4.39E-011	cm3/molc sec	25	EST	MEYLAN,WM & HOWARD,PH (1993)

CAS #: 135832-36-9 — ACETIC ACID, 2-[[5-(AMINOSULFONYL)-2-THIENYL]SUL

Formula: $C_8H_{11}NO_6S_3$

Mol Weight: 313.37

MP (deg C): FP (deg C):

BP (deg C):

BP pressure (mm Hg):

Property/Value	Units	Temp	Data Type	Reference
WS 7.34E+002	mg/L	25	EST	MEYLAN,WM ET AL. (1996)
logP 0.11			EXP	SHEPARD,KL ET AL. (1991)
VP 4.67E-009	mm Hg	25	EST	NEELY,WB & BLAU,GE (1985)
DC	pKa			
HL 1.68E-013	atm m3/mol	25	EST	MEYLAN,WM & HOWARD,PH (1991)
OH 1.13E-011	cm3/molc sec	25	EST	MEYLAN,WM & HOWARD,PH (1993)

CAS #: 135832-38-1 — 2-THIOPHENESULFONAMIDE, 5-[(4-HYDROXYBUTYL)THIO]

Formula: $C_8H_{13}NO_3S_3$

Mol Weight: 267.39

MP (deg C): FP (deg C):

BP (deg C):

BP pressure (mm Hg):

Property/Value	Units	Temp	Data Type	Reference
WS 2.92E+003	mg/L	25	EST	MEYLAN,WM ET AL. (1996)
logP 0.95			EXP	SHEPARD,KL ET AL. (1991)
VP 6.74E-010	mm Hg	25	EST	NEELY,WB & BLAU,GE (1985)
DC	pKa			
HL 5.72E-013	atm m3/mol	25	EST	MEYLAN,WM & HOWARD,PH (1991)
OH 2.91E-011	cm3/molc sec	25	EST	MEYLAN,WM & HOWARD,PH (1993)

CAS #: 135832-39-2 — 2-THIOPHENESULFONAMIDE, 5-[(4-HYDROXYBUTYL)SULFO

Formula: $C_8H_{13}NO_5S_3$

Mol Weight: 299.39

MP (deg C): FP (deg C):

BP (deg C):

BP pressure (mm Hg):

Property/Value	Units	Temp	Data Type	Reference
WS 2.15E+003	mg/L	25	EST	MEYLAN,WM ET AL. (1996)
logP -0.34			EXP	SHEPARD,KL ET AL. (1991)
VP 1.24E-011	mm Hg	25	EST	NEELY,WB & BLAU,GE (1985)
DC	pKa			
HL 4.54E-015	atm m3/mol	25	EST	MEYLAN,WM & HOWARD,PH (1991)
OH 1.64E-011	cm3/molc sec	25	EST	MEYLAN,WM & HOWARD,PH (1993)

CAS #: 135832-41-6 — BENZENESULFONAMIDE, 4-[(4-HYDROXYBUTYL)THIO]-

Formula: $C_{10}H_{15}NO_3S_2$

Mol Weight: 261.36

MP (deg C): FP (deg C):

BP (deg C):

BP pressure (mm Hg):

Property/Value	Units	Temp	Data Type	Reference
WS 3.28E+003	mg/L	25	EST	MEYLAN,WM ET AL. (1996)
logP 0.93			EXP	SHEPARD,KL ET AL. (1991)
VP 1.42E-009	mm Hg	25	EST	NEELY,WB & BLAU,GE (1985)
DC	pKa			
HL 1.05E-012	atm m3/mol	25	EST	MEYLAN,WM & HOWARD,PH (1991)
OH 1.86E-011	cm3/molc sec	25	EST	MEYLAN,WM & HOWARD,PH (1993)

1566

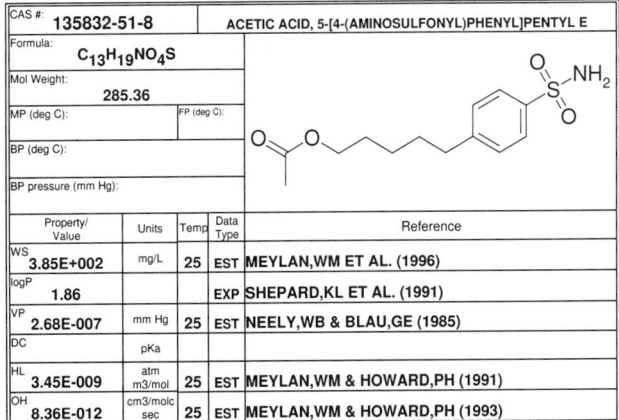

135832-51-8 — ACETIC ACID, 5-[4-(AMINOSULFONYL)PHENYL]PENTYL E

Formula: $C_{13}H_{19}NO_4S$

Mol Weight: 285.36

MP (deg C): | FP (deg C):

BP (deg C):

BP pressure (mm Hg):

Property/Value	Units	Temp	Data Type	Reference
WS 3.85E+002	mg/L	25	EST	MEYLAN,WM ET AL. (1996)
logP 1.86			EXP	SHEPARD,KL ET AL. (1991)
VP 2.68E-007	mm Hg	25	EST	NEELY,WB & BLAU,GE (1985)
DC	pKa			
HL 3.45E-009	atm m3/mol	25	EST	MEYLAN,WM & HOWARD,PH (1991)
OH 8.36E-012	cm3/molc sec	25	EST	MEYLAN,WM & HOWARD,PH (1993)

136309-02-9 — 2-CHLORO-6-ETHOXY-PYRAZINE

Formula: $C_6H_7ClN_2O$

Mol Weight: 158.59

MP (deg C): | FP (deg C):

BP (deg C):

BP pressure (mm Hg):

Property/Value	Units	Temp	Data Type	Reference
WS 8.83E+002	mg/L	25	EST	MEYLAN,WM ET AL. (1996)
logP 2.22			EXP	YAMAGAMI,C & TAKAO,N (1991)
VP 7.86E-002	mm Hg	25	EST	NEELY,WB & BLAU,GE (1985)
DC	pKa			
HL 9.77E-005	atm m3/mol	25	EST	MEYLAN,WM & HOWARD,PH (1991)
OH 7.96E-012	cm3/molc sec	25	EST	MEYLAN,WM & HOWARD,PH (1993)

136309-03-0 — 2-CHLORO-6-PROPOXY-PYRAZINE

Formula: $C_7H_9ClN_2O$

Mol Weight: 172.62

MP (deg C): | FP (deg C):

BP (deg C):

BP pressure (mm Hg):

Property/Value	Units	Temp	Data Type	Reference
WS 2.90E+002	mg/L	25	EST	MEYLAN,WM ET AL. (1996)
logP 2.71			EXP	YAMAGAMI,C & TAKAO,N (1991)
VP 2.44E-002	mm Hg	25	EST	NEELY,WB & BLAU,GE (1985)
DC	pKa			
HL 1.30E-004	atm m3/mol	25	EST	MEYLAN,WM & HOWARD,PH (1991)
OH 1.22E-011	cm3/molc sec	25	EST	MEYLAN,WM & HOWARD,PH (1993)

136309-04-1 — PYRAZINECARBONITRILE, 6-METHYL-

Formula: $C_6H_5N_3$

Mol Weight: 119.13

MP (deg C): | FP (deg C):

BP (deg C):

BP pressure (mm Hg):

Property/Value	Units	Temp	Data Type	Reference
WS 2.31E+004	mg/L	25	EST	MEYLAN,WM ET AL. (1996)
logP 0.44			EXP	YAMAGAMI,C ET AL. (1991)
VP 4.78E-002	mm Hg	25	EST	NEELY,WB & BLAU,GE (1985)
DC	pKa			
HL 3.11E-008	atm m3/mol	25	EST	MEYLAN,WM & HOWARD,PH (1991)
OH 2.15E-013	cm3/molc sec	25	EST	MEYLAN,WM & HOWARD,PH (1993)

136309-05-2 — PYRAZINE, 2-ETHOXY-6-METHOXY-

Formula: $C_7H_{10}N_2O_2$

Mol Weight: 154.17

MP (deg C): | FP (deg C):

BP (deg C):

BP pressure (mm Hg):

Property/Value	Units	Temp	Data Type	Reference
WS 1.48E+003	mg/L	25	EST	MEYLAN,WM ET AL. (1996)
logP 1.98			EXP	YAMAGAMI,C ET AL. (1991)
VP 4.78E-002	mm Hg	25	EST	NEELY,WB & BLAU,GE (1985)
DC	pKa			
HL 1.13E-006	atm m3/mol	25	EST	MEYLAN,WM & HOWARD,PH (1991)
OH 3.50E-011	cm3/molc sec	25	EST	MEYLAN,WM & HOWARD,PH (1993)

136309-06-3 — ACETAMIDE, N-(6-METHOXYPYRAZINYL)-

Formula: $C_7H_9N_3O_2$

Mol Weight: 167.17

MP (deg C): | FP (deg C):

BP (deg C):

BP pressure (mm Hg):

Property/Value	Units	Temp	Data Type	Reference
WS 1.26E+004	mg/L	25	EST	MEYLAN,WM ET AL. (1996)
logP 0.82			EXP	YAMAGAMI,C ET AL. (1991)
VP 1.95E-005	mm Hg	25	EST	NEELY,WB & BLAU,GE (1985)
DC	pKa			
HL 1.80E-012	atm m3/mol	25	EST	MEYLAN,WM & HOWARD,PH (1991)
OH 1.71E-011	cm3/molc sec	25	EST	MEYLAN,WM & HOWARD,PH (1993)

136309-07-4 — PYRAZINECARBONITRILE, 6-METHOXY-

Formula: $C_6H_5N_3O$

Mol Weight: 135.13

MP (deg C): | FP (deg C):

BP (deg C):

BP pressure (mm Hg):

Property/Value	Units	Temp	Data Type	Reference
WS 7.36E+003	mg/L	25	EST	MEYLAN,WM ET AL. (1996)
logP 0.95			EXP	YAMAGAMI,C ET AL. (1991)
VP 1.35E-002	mm Hg	25	EST	NEELY,WB & BLAU,GE (1985)
DC	pKa			
HL 1.52E-008	atm m3/mol	25	EST	MEYLAN,WM & HOWARD,PH (1991)
OH 1.16E-012	cm3/molc sec	25	EST	MEYLAN,WM & HOWARD,PH (1993)

136309-08-5 — PYRAZINECARBOXYLIC ACID, 6-METHOXY-, ETHYL ESTER

Formula: $C_8H_{10}N_2O_3$

Mol Weight: 182.18

MP (deg C): | FP (deg C):

BP (deg C):

BP pressure (mm Hg):

Property/Value	Units	Temp	Data Type	Reference
WS 5.08E+003	mg/L	25	EST	MEYLAN,WM ET AL. (1996)
logP 1.20			EXP	YAMAGAMI,C ET AL. (1991)
VP 3.75E-003	mm Hg	25	EST	NEELY,WB & BLAU,GE (1985)
DC	pKa			
HL 1.35E-008	atm m3/mol	25	EST	MEYLAN,WM & HOWARD,PH (1991)
OH 3.06E-012	cm3/molc sec	25	EST	MEYLAN,WM & HOWARD,PH (1993)

CAS #: 136309-09-6 — PYRAZINAMINE, 6-METHOXY-N,N-DIMETHYL

Formula:	$C_7H_{11}N_3O$		
Mol Weight:	153.19		
MP (deg C):		FP (deg C):	
BP (deg C):			
BP pressure (mm Hg):			

Property/ Value	Units	Temp	Data Type	Reference
WS 1.47E+003	mg/L	25	EST	MEYLAN,WM ET AL. (1996)
logP 1.99			EXP	YAMAGAMI,C ET AL. (1991)
VP 4.87E-002	mm Hg	25	EST	NEELY,WB & BLAU,GE (1985)
DC	pKa			
HL 2.50E-008	atm m3/mol	25	EST	MEYLAN,WM & HOWARD,PH (1991)
OH 2.03E-010	cm3/molc sec	25	EST	MEYLAN,WM & HOWARD,PH (1993)

CAS #: 136309-10-9 — PYRAZINECARBOXYLIC ACID, 6-(DIMETHYLAMINO)-, ETH

Formula:	$C_9H_{13}N_3O_2$		
Mol Weight:	195.22		
MP (deg C):		FP (deg C):	
BP (deg C):			
BP pressure (mm Hg):			

Property/ Value	Units	Temp	Data Type	Reference
WS 4.04E+003	mg/L	25	EST	MEYLAN,WM ET AL. (1996)
logP 1.24			EXP	YAMAGAMI,C ET AL. (1991)
VP 1.30E-003	mm Hg	25	EST	NEELY,WB & BLAU,GE (1985)
DC	pKa			
HL 3.97E-010	atm m3/mol	25	EST	MEYLAN,WM & HOWARD,PH (1991)
OH 1.41E-011	cm3/molc sec	25	EST	MEYLAN,WM & HOWARD,PH (1993)

CAS #: 136309-11-0 — PYRAZINE, 2-CHLORO-5-ETHOXY-

Formula:	$C_6H_7ClN_2O$		
Mol Weight:	158.59		
MP (deg C):		FP (deg C):	
BP (deg C):			
BP pressure (mm Hg):			

Property/ Value	Units	Temp	Data Type	Reference
WS 1.39E+003	mg/L	25	EST	MEYLAN,WM ET AL. (1996)
logP 1.99			EXP	YAMAGAMI,C ET AL. (1991)
VP 7.86E-002	mm Hg	25	EST	NEELY,WB & BLAU,GE (1985)
DC	pKa			
HL 9.77E-005	atm m3/mol	25	EST	MEYLAN,WM & HOWARD,PH (1991)
OH 6.91E-012	cm3/molc sec	25	EST	MEYLAN,WM & HOWARD,PH (1993)

CAS #: 136309-12-1 — ACETAMIDE, N-(5-CHLOROPYRAZINYL)-

Formula:	$C_6H_6ClN_3O$		
Mol Weight:	171.59		
MP (deg C):		FP (deg C):	
BP (deg C):			
BP pressure (mm Hg):			

Property/ Value	Units	Temp	Data Type	Reference
WS 2.01E+004	mg/L	25	EST	MEYLAN,WM ET AL. (1996)
logP 0.56			EXP	YAMAGAMI,C ET AL. (1991)
VP 2.80E-005	mm Hg	25	EST	NEELY,WB & BLAU,GE (1985)
DC	pKa			
HL 1.56E-010	atm m3/mol	25	EST	MEYLAN,WM & HOWARD,PH (1991)
OH 5.30E-013	cm3/molc sec	25	EST	MEYLAN,WM & HOWARD,PH (1993)

CAS #: 136309-13-2 — PYRAZINAMINE, 5-CHLORO-N,N-DIMETHYL-

Formula:	$C_6H_8ClN_3$		
Mol Weight:	157.60		
MP (deg C):		FP (deg C):	
BP (deg C):			
BP pressure (mm Hg):			

Property/ Value	Units	Temp	Data Type	Reference
WS 2.48E+003	mg/L	25	EST	MEYLAN,WM ET AL. (1996)
logP 1.70			EXP	YAMAGAMI,C ET AL. (1991)
VP 8.03E-002	mm Hg	25	EST	NEELY,WB & BLAU,GE (1985)
DC	pKa			
HL 2.17E-006	atm m3/mol	25	EST	MEYLAN,WM & HOWARD,PH (1991)
OH 1.53E-011	cm3/molc sec	25	EST	MEYLAN,WM & HOWARD,PH (1993)

CAS #: 136309-14-3 — PYRAZINAMINE, 5-METHOXY-N,N-DIMETHYL

Formula:	$C_7H_{11}N_3O$		
Mol Weight:	153.19		
MP (deg C):		FP (deg C):	
BP (deg C):			
BP pressure (mm Hg):			

Property/ Value	Units	Temp	Data Type	Reference
WS 2.86E+003	mg/L	25	EST	MEYLAN,WM ET AL. (1996)
logP 1.65			EXP	YAMAGAMI,C ET AL. (1991)
VP 4.87E-002	mm Hg	25	EST	NEELY,WB & BLAU,GE (1985)
DC	pKa			
HL 2.50E-008	atm m3/mol	25	EST	MEYLAN,WM & HOWARD,PH (1991)
OH 4.11E-011	cm3/molc sec	25	EST	MEYLAN,WM & HOWARD,PH (1993)

CAS #: 136389-76-9 — 1H-PYRAZOLE, 3-METHYL-5-[(4-METHYLPHENYL)AMINO]-

Formula:	$C_{17}H_{16}N_4O_2$		
Mol Weight:	308.34		
MP (deg C):		FP (deg C):	
BP (deg C):			
BP pressure (mm Hg):			

Property/ Value	Units	Temp	Data Type	Reference
WS 5.26E+000	mg/L	25	EST	MEYLAN,WM ET AL. (1996)
logP 3.43			EXP	SANGSTER,J (1994)
VP 5.15E-009	mm Hg	25	EST	NEELY,WB & BLAU,GE (1985)
DC	pKa			
HL 1.26E-014	atm m3/mol	25	EST	MEYLAN,WM & HOWARD,PH (1991)
OH 2.00E-010	cm3/molc sec	25	EST	MEYLAN,WM & HOWARD,PH (1993)

CAS #: 136389-77-0 — 1H-PYRAZOLE, 3-METHYL-5-[(3-METHYLPHENYL)AMINO]-

Formula:	$C_{17}H_{16}N_4O_2$		
Mol Weight:	308.34		
MP (deg C):		FP (deg C):	
BP (deg C):			
BP pressure (mm Hg):			

Property/ Value	Units	Temp	Data Type	Reference
WS 7.50E+000	mg/L	25	EST	MEYLAN,WM ET AL. (1996)
logP 3.25			EXP	SANGSTER,J (1994)
VP 5.15E-009	mm Hg	25	EST	NEELY,WB & BLAU,GE (1985)
DC	pKa			
HL 1.26E-014	atm m3/mol	25	EST	MEYLAN,WM & HOWARD,PH (1991)
OH 2.00E-010	cm3/molc sec	25	EST	MEYLAN,WM & HOWARD,PH (1993)

CAS #: 136389-80-5 — 1H-PYRAZOLE, 3-METHYL-4-NITRO-5-[(4-NITROPHENYL)

Formula: $C_{16}H_{13}N_5O_4$

Mol Weight: 339.31

MP (deg C): | FP (deg C):

BP (deg C):

BP pressure (mm Hg):

Property/Value	Units	Temp	Data Type	Reference
WS 3.04E+001	mg/L	25	EST	MEYLAN,WM ET AL. (1996)
logP 2.78			EXP	SANGSTER,J (1994)
VP 2.30E-010	mm Hg	25	EST	NEELY,WB & BLAU,GE (1985)
DC	pKa			
HL 4.50E-017	atm m3/mol	25	EST	MEYLAN,WM & HOWARD,PH (1991)
OH 5.69E-011	cm3/molc sec	25	EST	MEYLAN,WM & HOWARD,PH (1993)

CAS #: 136389-87-2 — 1H-PYRAZOLE, 3-METHYL-4-NITRO-1-PHENYL-5-[[4-(TR

Formula: $C_{17}H_{13}F_3N_4O_2$

Mol Weight: 362.31

MP (deg C): | FP (deg C):

BP (deg C):

BP pressure (mm Hg):

Property/Value	Units	Temp	Data Type	Reference
WS 1.01E+000	mg/L	25	EST	MEYLAN,WM ET AL. (1996)
logP 3.89			EXP	SANGSTER,J (1994)
VP 9.13E-009	mm Hg	25	EST	NEELY,WB & BLAU,GE (1985)
DC	pKa			
HL 9.90E-014	atm m3/mol	25	EST	MEYLAN,WM & HOWARD,PH (1991)
OH 6.81E-011	cm3/molc sec	25	EST	MEYLAN,WM & HOWARD,PH (1993)

CAS #: 136832-80-9 — KETOBEMIDONE, PIVALIC ESTER

Formula: $C_{20}H_{29}NO_3$

Mol Weight: 331.46

MP (deg C): | FP (deg C):

BP (deg C):

BP pressure (mm Hg):

Property/Value	Units	Temp	Data Type	Reference
WS 5.01E+002	mg/L	25	EST	MEYLAN,WM ET AL. (1996)
logP 2.59			EXP	HANSCH,C ET AL. (1995)
VP 4.61E-007	mm Hg	25	EST	NEELY,WB & BLAU,GE (1985)
DC	pKa			
HL 7.65E-011	atm m3/mol	25	EST	MEYLAN,WM & HOWARD,PH (1991)
OH 1.04E-010	cm3/molc sec	25	EST	MEYLAN,WM & HOWARD,PH (1993)

CAS #: 136832-82-1 — KETOBEMIDONE, BENZOYL ESTER

Formula: $C_{22}H_{25}NO_3$

Mol Weight: 351.45

MP (deg C): | FP (deg C):

BP (deg C):

BP pressure (mm Hg):

Property/Value	Units	Temp	Data Type	Reference
WS 1.60E+002	mg/L	25	EST	MEYLAN,WM ET AL. (1996)
logP 3.03			EXP	HANSCH,C ET AL. (1995)
VP 1.21E-008	mm Hg	25	EST	NEELY,WB & BLAU,GE (1985)
DC	pKa			
HL 6.47E-012	atm m3/mol	25	EST	MEYLAN,WM & HOWARD,PH (1991)
OH 1.05E-010	cm3/molc sec	25	EST	MEYLAN,WM & HOWARD,PH (1993)

CAS #: 136944-22-4 — 2-PROPENAMIDE, 3-(3,4-DIHYDROXYPHENYL)-N-(3,4-DI

Formula: $C_{17}H_{17}NO_5$

Mol Weight: 315.33

MP (deg C): | FP (deg C):

BP (deg C):

BP pressure (mm Hg):

Property/Value	Units	Temp	Data Type	Reference
WS 4.12E+002	mg/L	25	EST	MEYLAN,WM ET AL. (1996)
logP 2.30			EXP	NAITO,Y ET AL. (1991)
VP 9.76E-013	mm Hg	25	EST	NEELY,WB & BLAU,GE (1985)
DC	pKa			
HL 5.52E-021	atm m3/mol	25	EST	MEYLAN,WM & HOWARD,PH (1991)
OH 2.22E-010	cm3/molc sec	25	EST	MEYLAN,WM & HOWARD,PH (1993)

CAS #: 136944-23-5 — 2-PROPENAMIDE, 3-(3,4-DIHYDROXYPHENYL)-N-[(3,4-D

Formula: $C_{18}H_{19}NO_5$

Mol Weight: 329.36

MP (deg C): | FP (deg C):

BP (deg C):

BP pressure (mm Hg):

Property/Value	Units	Temp	Data Type	Reference
WS 6.26E+002	mg/L	25	EST	MEYLAN,WM ET AL. (1996)
logP 1.99			EXP	NAITO,Y ET AL. (1991)
VP 3.81E-013	mm Hg	25	EST	NEELY,WB & BLAU,GE (1985)
DC	pKa			
HL 1.77E-021	atm m3/mol	25	EST	MEYLAN,WM & HOWARD,PH (1991)
OH 9.88E-011	cm3/molc sec	25	EST	MEYLAN,WM & HOWARD,PH (1993)

CAS #: 136944-24-6 — 2-PROPENAMIDE, 3-(3,4-DIHYDROXYPHENYL)-N-[2-(3,4

Formula: $C_{19}H_{21}NO_5$

Mol Weight: 343.38

MP (deg C): | FP (deg C):

BP (deg C):

BP pressure (mm Hg):

Property/Value	Units	Temp	Data Type	Reference
WS 4.41E+002	mg/L	25	EST	MEYLAN,WM ET AL. (1996)
logP 2.07			EXP	NAITO,Y ET AL. (1991)
VP 1.48E-013	mm Hg	25	EST	NEELY,WB & BLAU,GE (1985)
DC	pKa			
HL 2.35E-021	atm m3/mol	25	EST	MEYLAN,WM & HOWARD,PH (1991)
OH 1.01E-010	cm3/molc sec	25	EST	MEYLAN,WM & HOWARD,PH (1993)

CAS #: 136944-25-7 — 1-HEXEN-3-ONE, 1-(3,4-DIHYDROXYPHENYL)-

Formula: $C_{12}H_{14}O_3$

Mol Weight: 206.24

MP (deg C): | FP (deg C):

BP (deg C):

BP pressure (mm Hg):

Property/Value	Units	Temp	Data Type	Reference
WS 9.12E+002	mg/L	25	EST	MEYLAN,WM ET AL. (1996)
logP 2.61			EXP	NAITO,Y ET AL. (1991)
VP 8.39E-007	mm Hg	25	EST	NEELY,WB & BLAU,GE (1985)
DC	pKa			
HL 2.23E-014	atm m3/mol	25	EST	MEYLAN,WM & HOWARD,PH (1991)
OH 7.73E-011	cm3/molc sec	25	EST	MEYLAN,WM & HOWARD,PH (1993)

CAS #:	136944-37-1	2-PROPENAMIDE, 3-(3,4-DIHYDROXYPHENOXY)-N-HEXYL-

Formula: $C_{15}H_{21}NO_3$

Mol Weight: 263.34

MP (deg C):   FP (deg C):

BP (deg C):

BP pressure (mm Hg):

Property/Value	Units	Temp	Data Type	Reference
WS 1.02E+002	mg/L	25	EST	MEYLAN,WM ET AL. (1996)
logP 3.36			EXP	HANSCH,C ET AL. (1995)
VP 1.38E-010	mm Hg	25	EST	NEELY,WB & BLAU,GE (1985)
DC	pKa			
HL 2.59E-017	atm m3/mol	25	EST	MEYLAN,WM & HOWARD,PH (1991)
OH 5.74E-011	cm3/molc sec	25	EST	MEYLAN,WM & HOWARD,PH (1993)

CAS #:	136944-43-9	2,3-NAPHTHALENEDIOL, 6-ETHYL-

Formula: $C_{12}H_{12}O_2$

Mol Weight: 188.23

MP (deg C):   FP (deg C):

BP (deg C):

BP pressure (mm Hg):

Property/Value	Units	Temp	Data Type	Reference
WS 3.20E+002	mg/L	25	EST	MEYLAN,WM ET AL. (1996)
logP 3.25			EXP	NAITO,Y ET AL. (1991)
VP 9.54E-007	mm Hg	25	EST	NEELY,WB & BLAU,GE (1985)
DC	pKa			
HL 8.34E-012	atm m3/mol	25	EST	MEYLAN,WM & HOWARD,PH (1991)
OH 2.01E-010	cm3/molc sec	25	EST	MEYLAN,WM & HOWARD,PH (1993)

CAS #:	136944-48-4	2-NAPHTHALENECARBOXAMIDE, N-BUTYL-6,7-DIHYDROXY-

Formula: $C_{15}H_{17}NO_3$

Mol Weight: 259.31

MP (deg C):   FP (deg C):

BP (deg C):

BP pressure (mm Hg):

Property/Value	Units	Temp	Data Type	Reference
WS 3.80E+002	mg/L	25	EST	MEYLAN,WM ET AL. (1996)
logP 2.72			EXP	NAITO,Y ET AL. (1991)
VP 2.62E-011	mm Hg	25	EST	NEELY,WB & BLAU,GE (1985)
DC	pKa			
HL 1.20E-017	atm m3/mol	25	EST	MEYLAN,WM & HOWARD,PH (1991)
OH 1.71E-010	cm3/molc sec	25	EST	MEYLAN,WM & HOWARD,PH (1993)

CAS #:	136944-52-0	2,3-NAPHTHALENEDIOL, 6-ETHOXY-

Formula: $C_{12}H_{12}O_3$

Mol Weight: 204.23

MP (deg C):   FP (deg C):

BP (deg C):

BP pressure (mm Hg):

Property/Value	Units	Temp	Data Type	Reference
WS 7.83E+002	mg/L	25	EST	MEYLAN,WM ET AL. (1996)
logP 2.70			EXP	NAITO,Y ET AL. (1991)
VP 4.19E-007	mm Hg	25	EST	NEELY,WB & BLAU,GE (1985)
DC	pKa			
HL 4.47E-013	atm m3/mol	25	EST	MEYLAN,WM & HOWARD,PH (1991)
OH 2.06E-010	cm3/molc sec	25	EST	MEYLAN,WM & HOWARD,PH (1993)

CAS #:	137057-42-2	9H-PURINE, 6-METHOXY-9-[2,3,5-TRIS-O-(1-OXOPROPY

Formula: $C_{20}H_{26}N_4O_8$

Mol Weight: 450.45

MP (deg C):   FP (deg C):

BP (deg C):

BP pressure (mm Hg):

Property/Value	Units	Temp	Data Type	Reference
WS 6.64E+000	mg/L	25	EST	MEYLAN,WM ET AL. (1996)
logP 2.75			EXP	JONES,LA ET AL. (1992)
VP 3.30E-012	mm Hg	25	EST	NEELY,WB & BLAU,GE (1985)
DC	pKa			
HL 1.10E-019	atm m3/mol	25	EST	MEYLAN,WM & HOWARD,PH (1991)
OH 2.39E-010	cm3/molc sec	25	EST	MEYLAN,WM & HOWARD,PH (1993)

CAS #:	137057-43-3	9H-PURINE, 6-METHOXY-9-[2,3,5-TRIS-O-(1-OXOBUTYL

Formula: $C_{23}H_{32}N_4O_8$

Mol Weight: 492.53

MP (deg C):   FP (deg C):

BP (deg C):

BP pressure (mm Hg):

Property/Value	Units	Temp	Data Type	Reference
WS 2.67E-001	mg/L	25	EST	MEYLAN,WM ET AL. (1996)
logP 4.07			EXP	JONES,GH ET AL. (1986)
VP 2.61E-013	mm Hg	25	EST	NEELY,WB & BLAU,GE (1985)
DC	pKa			
HL 2.57E-019	atm m3/mol	25	EST	MEYLAN,WM & HOWARD,PH (1991)
OH 2.43E-010	cm3/molc sec	25	EST	MEYLAN,WM & HOWARD,PH (1993)

CAS #:	137057-45-5	9H-PURINE, 6-METHOXY-9-[2,3,5-TRIS-O-(2-METHYL-1

Formula: $C_{23}H_{32}N_4O_8$

Mol Weight: 492.53

MP (deg C):   FP (deg C):

BP (deg C):

BP pressure (mm Hg):

Property/Value	Units	Temp	Data Type	Reference
WS 3.32E-001	mg/L	25	EST	MEYLAN,WM ET AL. (1996)
logP 3.96			EXP	JONES,GH ET AL. (1986)
VP 1.21E-012	mm Hg	25	EST	NEELY,WB & BLAU,GE (1985)
DC	pKa			
HL 2.57E-019	atm m3/mol	25	EST	MEYLAN,WM & HOWARD,PH (1991)
OH 2.41E-010	cm3/molc sec	25	EST	MEYLAN,WM & HOWARD,PH (1993)

CAS #:	137057-57-9	9H-PURINE, 9-[2,3-BIS-O-(1-OXOPENTYL)-beta-D-ARA

Formula: $C_{21}H_{30}N_4O_7$

Mol Weight: 450.50

MP (deg C):   FP (deg C):

BP (deg C):

BP pressure (mm Hg):

Property/Value	Units	Temp	Data Type	Reference
WS 7.93E-001	mg/L	25	EST	MEYLAN,WM ET AL. (1996)
logP 3.83			EXP	SANGSTER,J (1994)
VP 9.02E-016	mm Hg	25	EST	NEELY,WB & BLAU,GE (1985)
DC	pKa			
HL 3.94E-021	atm m3/mol	25	EST	MEYLAN,WM & HOWARD,PH (1991)
OH 2.46E-010	cm3/molc sec	25	EST	MEYLAN,WM & HOWARD,PH (1993)

CAS #: 137057-62-6 — 9H-PURINE, 9-[2,3-BIS-O-(2,2-DIMETHYL-1-OXOPROPY

Formula: $C_{21}H_{30}N_4O_7$

Mol Weight: 450.50

MP (deg C):   FP (deg C):

BP (deg C):

BP pressure (mm Hg):

Property/Value	Units	Temp	Data Type	Reference
WS 1.22E+000	mg/L	25	EST	MEYLAN,WM ET AL. (1996)
logP 3.61			EXP	SANGSTER,J (1994)
VP 6.52E-015	mm Hg	25	EST	NEELY,WB & BLAU,GE (1985)
DC	pKa			
HL 3.94E-021	atm m3/mol	25	EST	MEYLAN,WM & HOWARD,PH (1991)
OH 2.39E-010	cm3/molc sec	25	EST	MEYLAN,WM & HOWARD,PH (1993)

CAS #: 137057-67-1 — 9H-PURINE, 9-[2,3-BIS-O-(PHENYLACETYL)-beta-D-AR

Formula: $C_{27}H_{26}N_4O_7$

Mol Weight: 518.53

MP (deg C):   FP (deg C):

BP (deg C):

BP pressure (mm Hg):

Property/Value	Units	Temp	Data Type	Reference
WS 5.81E-001	mg/L	25	EST	MEYLAN,WM ET AL. (1996)
logP 3.48			EXP	SANGSTER,J (1993)
VP 1.72E-019	mm Hg	25	EST	NEELY,WB & BLAU,GE (1985)
DC	pKa			
HL 4.69E-024	atm m3/mol	25	EST	MEYLAN,WM & HOWARD,PH (1991)
OH 2.46E-010	cm3/molc sec	25	EST	MEYLAN,WM & HOWARD,PH (1993)

CAS #: 137057-69-3 — 9H-PURINE, 9-[2,5-BIS-O-(1-OXOPENTYL)-BETA-D-ARA

Formula: $C_{21}H_{30}N_4O_7$

Mol Weight: 450.50

MP (deg C):   FP (deg C):

BP (deg C):

BP pressure (mm Hg):

Property/Value	Units	Temp	Data Type	Reference
WS 1.30E+000	mg/L	25	EST	MEYLAN,WM ET AL. (1996)
logP 3.58			EXP	JONES,LA ET AL. (1992)
VP 1.24E-015	mm Hg	25	EST	NEELY,WB & BLAU,GE (1985)
DC	pKa			
HL 3.94E-021	atm m3/mol	25	EST	MEYLAN,WM & HOWARD,PH (1991)
OH 2.48E-010	cm3/molc sec	25	EST	MEYLAN,WM & HOWARD,PH (1993)

CAS #: 137057-71-7 — 9H-PURINE, 9-[2,5-BIS-O-(2,2-DIMETHYL-1-OXOPROPY

Formula: $C_{21}H_{30}N_4O_7$

Mol Weight: 450.50

MP (deg C):   FP (deg C):

BP (deg C):

BP pressure (mm Hg):

Property/Value	Units	Temp	Data Type	Reference
WS 1.85E+000	mg/L	25	EST	MEYLAN,WM ET AL. (1996)
logP 3.40			EXP	JONES,LA ET AL. (1992)
VP 8.97E-015	mm Hg	25	EST	NEELY,WB & BLAU,GE (1985)
DC	pKa			
HL 3.94E-021	atm m3/mol	25	EST	MEYLAN,WM & HOWARD,PH (1991)
OH 2.42E-010	cm3/molc sec	25	EST	MEYLAN,WM & HOWARD,PH (1993)

CAS #: 137057-72-8 — H-PURINE, 9-(2,5-DI-O-BENZOYL-BETA-D-ARABINOFURA

Formula: $C_{25}H_{22}N_4O_7$

Mol Weight: 490.48

MP (deg C):   FP (deg C):

BP (deg C):

BP pressure (mm Hg):

Property/Value	Units	Temp	Data Type	Reference
WS 1.05E+000	mg/L	25	EST	MEYLAN,WM ET AL. (1996)
logP 3.39			EXP	JONES,LA ET AL. (1992)
VP 1.83E-018	mm Hg	25	EST	NEELY,WB & BLAU,GE (1985)
DC	pKa			
HL 2.82E-023	atm m3/mol	25	EST	MEYLAN,WM & HOWARD,PH (1991)
OH 2.43E-010	cm3/molc sec	25	EST	MEYLAN,WM & HOWARD,PH (1993)

CAS #: 137057-73-9 — 9H-PURINE, 9-[2,5-BIS-O-(4-METHOXYBENZOYL)-beta-

Formula: $C_{26}H_{24}N_4O_8$

Mol Weight: 520.50

MP (deg C):   FP (deg C):

BP (deg C):

BP pressure (mm Hg):

Property/Value	Units	Temp	Data Type	Reference
WS 4.04E-001	mg/L	25	EST	MEYLAN,WM ET AL. (1996)
logP 3.65			EXP	SANGSTER,J (1993)
VP 2.28E-015	mm Hg	25	EST	NEELY,WB & BLAU,GE (1985)
DC	pKa			
HL 2.03E-021	atm m3/mol	25	EST	MEYLAN,WM & HOWARD,PH (1991)
OH 2.41E-010	cm3/molc sec	25	EST	MEYLAN,WM & HOWARD,PH (1993)

CAS #: 137103-53-8 — 1H-IMIDAZOLE-1-ETHANOL, _-(4-FLUOROPHENYL)-_-[4-

Formula: $C_{22}H_{18}FN_3O$

Mol Weight: 359.41

MP (deg C):   FP (deg C):

BP (deg C):

BP pressure (mm Hg):

Property/Value	Units	Temp	Data Type	Reference
WS 4.22E+001	mg/L	25	EST	MEYLAN,WM ET AL. (1996)
logP 2.47			EXP	SANGSTER,J (1994)
VP 7.50E-014	mm Hg	25	EST	NEELY,WB & BLAU,GE (1985)
DC	pKa			
HL 2.97E-015	atm m3/mol	25	EST	MEYLAN,WM & HOWARD,PH (1991)
OH 4.73E-011	cm3/molc sec	25	EST	MEYLAN,WM & HOWARD,PH (1993)

CAS #: 137103-54-9 — 1H-IMIDAZOLE-1-ETHANOL, alpha-(4-FLUOROPHENYL)-a

Formula: $C_{22}H_{18}FN_3O$

Mol Weight: 359.41

MP (deg C):   FP (deg C):

BP (deg C):

BP pressure (mm Hg):

Property/Value	Units	Temp	Data Type	Reference
WS 4.76E+000	mg/L	25	EST	MEYLAN,WM ET AL. (1996)
logP 3.58			EXP	SANGSTER,J (1993)
VP 7.50E-014	mm Hg	25	EST	NEELY,WB & BLAU,GE (1985)
DC	pKa			
HL 2.97E-015	atm m3/mol	25	EST	MEYLAN,WM & HOWARD,PH (1991)
OH 4.73E-011	cm3/molc sec	25	EST	MEYLAN,WM & HOWARD,PH (1993)

CAS #: 137103-55-0 — 1H-IMIDAZOLE-1-ETHANOL, alpha-(2,4-DIFLUOROPHENY

Formula: $C_{22}H_{17}F_2N_3O$

Mol Weight: 377.40

MP (deg C): | FP (deg C):

BP (deg C):

BP pressure (mm Hg):

Property/ Value	Units	Temp	Data Type	Reference
WS 6.29E+000	mg/L	25	EST	MEYLAN,WM ET AL. (1996)
logP 3.31			EXP	SANGSTER,J (1993)
VP 9.39E-014	mm Hg	25	EST	NEELY,WB & BLAU,GE (1985)
DC	pKa			
HL 3.47E-015	atm m3/mol	25	EST	MEYLAN,WM & HOWARD,PH (1991)
OH 4.81E-011	cm3/molc sec	25	EST	MEYLAN,WM & HOWARD,PH (1993)

CAS #: 137103-59-4 — 1H-IMIDAZOLE-1-ETHANOL, alpha-(4-FLUOROPHENYL)-a

Formula: $C_{22}H_{18}FN_3O$

Mol Weight: 359.41

MP (deg C): | FP (deg C):

BP (deg C):

BP pressure (mm Hg):

Property/ Value	Units	Temp	Data Type	Reference
WS 1.78E+001	mg/L	25	EST	MEYLAN,WM ET AL. (1996)
logP 2.91			EXP	SANGSTER,J (1993)
VP 7.50E-014	mm Hg	25	EST	NEELY,WB & BLAU,GE (1985)
DC	pKa			
HL 2.97E-015	atm m3/mol	25	EST	MEYLAN,WM & HOWARD,PH (1991)
OH 5.07E-011	cm3/molc sec	25	EST	MEYLAN,WM & HOWARD,PH (1993)

CAS #: 137103-64-1 — 1H-IMIDAZOLE-1-ETHANOL, alpha-(4-CHLOROPHENYL)-a

Formula: $C_{22}H_{18}ClN_3O$

Mol Weight: 375.86

MP (deg C): | FP (deg C):

BP (deg C):

BP pressure (mm Hg):

Property/ Value	Units	Temp	Data Type	Reference
WS 6.95E+000	mg/L	25	EST	MEYLAN,WM ET AL. (1996)
logP 3.27			EXP	SANGSTER,J (1993)
VP 1.23E-014	mm Hg	25	EST	NEELY,WB & BLAU,GE (1985)
DC	pKa			
HL 1.89E-015	atm m3/mol	25	EST	MEYLAN,WM & HOWARD,PH (1991)
OH 4.60E-011	cm3/molc sec	25	EST	MEYLAN,WM & HOWARD,PH (1993)

CAS #: 137128-33-7 — 1H-IMIDAZOLE-1-ETHANOL, alpha-(2,4-DIFLUOROPHENY

Formula: $C_{22}H_{17}F_2N_3O$

Mol Weight: 377.40

MP (deg C): | FP (deg C):

BP (deg C):

BP pressure (mm Hg):

Property/ Value	Units	Temp	Data Type	Reference
WS 3.22E+000	mg/L	25	EST	MEYLAN,WM ET AL. (1996)
logP 3.65			EXP	SANGSTER,J (1993)
VP 9.39E-014	mm Hg	25	EST	NEELY,WB & BLAU,GE (1985)
DC	pKa			
HL 3.47E-015	atm m3/mol	25	EST	MEYLAN,WM & HOWARD,PH (1991)
OH 4.81E-011	cm3/molc sec	25	EST	MEYLAN,WM & HOWARD,PH (1993)

CAS #: 137302-31-9 — L-THREO-PENTONAMIDE DERIVATIVE

Formula: $C_{35}H_{54}F_2N_6O_8S$

Mol Weight: 756.92

MP (deg C): | FP (deg C):

BP (deg C):

BP pressure (mm Hg):

Property/ Value	Units	Temp	Data Type	Reference
WS 2.87E-002	mg/L	25	EST	MEYLAN,WM ET AL. (1996)
logP 3.17			EXP	SANGSTER,J (1993)
VP 8.48E-025	mm Hg	25	EST	NEELY,WB & BLAU,GE (1985)
DC	pKa			
HL 6.37E-029	atm m3/mol	25	EST	MEYLAN,WM & HOWARD,PH (1991)
OH 2.74E-010	cm3/molc sec	25	EST	MEYLAN,WM & HOWARD,PH (1993)

CAS #: 137302-35-3 — L-NORVALINAMIDE DERIVATIVE

Formula: $C_{35}H_{52}F_2N_6O_8S$

Mol Weight: 754.90

MP (deg C): | FP (deg C):

BP (deg C):

BP pressure (mm Hg):

Property/ Value	Units	Temp	Data Type	Reference
WS 1.92E-002	mg/L	25	EST	MEYLAN,WM ET AL. (1996)
logP 3.39			EXP	SANGSTER,J (1993)
VP 4.25E-025	mm Hg	25	EST	NEELY,WB & BLAU,GE (1985)
DC	pKa			
HL 7.78E-030	atm m3/mol	25	EST	MEYLAN,WM & HOWARD,PH (1991)
OH 2.47E-010	cm3/molc sec	25	EST	MEYLAN,WM & HOWARD,PH (1993)

CAS #: 137302-63-7 — L-NORVALINAMIDE DERIVATIVE

Formula: $C_{36}H_{54}F_2N_6O_9S$

Mol Weight: 784.93

MP (deg C): | FP (deg C):

BP (deg C):

BP pressure (mm Hg):

Property/ Value	Units	Temp	Data Type	Reference
WS 4.42E-002	mg/L	25	EST	MEYLAN,WM ET AL. (1996)
logP 2.73			EXP	SANGSTER,J (1993)
VP 9.09E-026	mm Hg	25	EST	NEELY,WB & BLAU,GE (1985)
DC	pKa			
HL 4.60E-031	atm m3/mol	25	EST	MEYLAN,WM & HOWARD,PH (1991)
OH 2.69E-010	cm3/molc sec	25	EST	MEYLAN,WM & HOWARD,PH (1993)

CAS #: 137429-62-0 — L-NORVALINAMIDE DERIVATIVE

Formula: $C_{36}H_{55}F_2N_7O_7S$

Mol Weight: 767.94

MP (deg C): | FP (deg C):

BP (deg C):

BP pressure (mm Hg):

Property/ Value	Units	Temp	Data Type	Reference
WS 5.75E-002	mg/L	25	EST	MEYLAN,WM ET AL. (1996)
logP 2.73			EXP	SANGSTER,J (1993)
VP 2.16E-025	mm Hg	25	EST	NEELY,WB & BLAU,GE (1985)
DC	pKa			
HL 3.29E-031	atm m3/mol	25	EST	MEYLAN,WM & HOWARD,PH (1991)
OH 2.16E-010	cm3/molc sec	25	EST	MEYLAN,WM & HOWARD,PH (1993)

CAS #: 138568-62-4	BENZENEACETIC ACID, 3-BROMO-à-METHYL-4-(2-THIAZO

Formula: $C_{12}H_{10}BrNO_2S$

Mol Weight: 312.19

MP (deg C):   FP (deg C):

BP (deg C):

BP pressure (mm Hg):

Property/Value	Units	Temp	Data Type	Reference
WS 3.56E+001	mg/L	25	EST	MEYLAN,WM ET AL. (1996)
logP 3.35			EXP	NAIRO,Y ET AL. (1992)
VP 5.21E-008	mm Hg	25	EST	NEELY,WB & BLAU,GE (1985)
DC	pKa			
HL 1.28E-012	atm m3/mol	25	EST	MEYLAN,WM & HOWARD,PH (1991)
OH 6.01E-012	cm3/molc sec	25	EST	MEYLAN,WM & HOWARD,PH (1993)

CAS #: 138568-63-5	BENZENEACETIC ACID, à-METHYL-4-[4-(1-METHYLETHYL

Formula: $C_{15}H_{17}NO_2S$

Mol Weight: 275.37

MP (deg C):   FP (deg C):

BP (deg C):

BP pressure (mm Hg):

Property/Value	Units	Temp	Data Type	Reference
WS 9.17E+000	mg/L	25	EST	MEYLAN,WM ET AL. (1996)
logP 4.29			EXP	NAIRO,Y ET AL. (1992)
VP 6.49E-008	mm Hg	25	EST	NEELY,WB & BLAU,GE (1985)
DC	pKa			
HL 6.23E-012	atm m3/mol	25	EST	MEYLAN,WM & HOWARD,PH (1991)
OH 1.09E-011	cm3/molc sec	25	EST	MEYLAN,WM & HOWARD,PH (1993)

CAS #: 138568-64-6	BENZENEACETIC ACID, à-METHYL-4-[4-(2-PROPENYL)-2

Formula: $C_{15}H_{15}NO_2S$

Mol Weight: 273.36

MP (deg C):   FP (deg C):

BP (deg C):

BP pressure (mm Hg):

Property/Value	Units	Temp	Data Type	Reference
WS 1.34E+001	mg/L	25	EST	MEYLAN,WM ET AL. (1996)
logP 4.11			EXP	NAIRO,Y ET AL. (1992)
VP 4.17E-008	mm Hg	25	EST	NEELY,WB & BLAU,GE (1985)
DC	pKa			
HL 4.64E-012	atm m3/mol	25	EST	MEYLAN,WM & HOWARD,PH (1991)
OH 3.59E-011	cm3/molc sec	25	EST	MEYLAN,WM & HOWARD,PH (1993)

CAS #: 138568-65-7	BENZENEACETIC ACID, 4-[4-(HYDROXYMETHYL)-2-THIAZ

Formula: $C_{13}H_{13}NO_3S$

Mol Weight: 263.32

MP (deg C):   FP (deg C):

BP (deg C):

BP pressure (mm Hg):

Property/Value	Units	Temp	Data Type	Reference
WS 5.93E+002	mg/L	25	EST	MEYLAN,WM ET AL. (1996)
logP 2.25			EXP	NAIRO,Y ET AL. (1992)
VP 1.20E-010	mm Hg	25	EST	NEELY,WB & BLAU,GE (1985)
DC	pKa			
HL 1.29E-016	atm m3/mol	25	EST	MEYLAN,WM & HOWARD,PH (1991)
OH 1.21E-011	cm3/molc sec	25	EST	MEYLAN,WM & HOWARD,PH (1993)

CAS #: 138568-66-8	BENZENEACETIC ACID, 4-[4-(METHOXYMETHYL)-2-THIAZ

Formula: $C_{14}H_{15}NO_3S$

Mol Weight: 277.34

MP (deg C):   FP (deg C):

BP (deg C):

BP pressure (mm Hg):

Property/Value	Units	Temp	Data Type	Reference
WS 2.68E+002	mg/L	25	EST	MEYLAN,WM ET AL. (1996)
logP 2.56			EXP	NAIRO,Y ET AL. (1992)
VP 3.80E-008	mm Hg	25	EST	NEELY,WB & BLAU,GE (1985)
DC	pKa			
HL 4.14E-014	atm m3/mol	25	EST	MEYLAN,WM & HOWARD,PH (1991)
OH 1.52E-011	cm3/molc sec	25	EST	MEYLAN,WM & HOWARD,PH (1993)

CAS #: 138568-67-9	BENZENEACETIC ACID, 4-(5-ETHYL-2-THIAZOLYL)-à-ME

Formula: $C_{14}H_{15}NO_2S$

Mol Weight: 261.35

MP (deg C):   FP (deg C):

BP (deg C):

BP pressure (mm Hg):

Property/Value	Units	Temp	Data Type	Reference
WS 3.38E+001	mg/L	25	EST	MEYLAN,WM ET AL. (1996)
logP 3.72			EXP	NAIRO,Y ET AL. (1992)
VP 8.74E-008	mm Hg	25	EST	NEELY,WB & BLAU,GE (1985)
DC	pKa			
HL 4.70E-012	atm m3/mol	25	EST	MEYLAN,WM & HOWARD,PH (1991)
OH 9.81E-012	cm3/molc sec	25	EST	MEYLAN,WM & HOWARD,PH (1993)

CAS #: 138568-68-0	BENZENEACETIC ACID, 4-(4-ETHYL-5-METHYL-2-THIAZO

Formula: $C_{15}H_{17}NO_2S$

Mol Weight: 275.37

MP (deg C):   FP (deg C):

BP (deg C):

BP pressure (mm Hg):

Property/Value	Units	Temp	Data Type	Reference
WS 1.66E+001	mg/L	25	EST	MEYLAN,WM ET AL. (1996)
logP 3.99			EXP	NAIRO,Y ET AL. (1992)
VP 3.93E-008	mm Hg	25	EST	NEELY,WB & BLAU,GE (1985)
DC	pKa			
HL 5.18E-012	atm m3/mol	25	EST	MEYLAN,WM & HOWARD,PH (1991)
OH 8.86E-012	cm3/molc sec	25	EST	MEYLAN,WM & HOWARD,PH (1993)

CAS #: 138568-69-1	BENZENEACETIC ACID, à-METHYL-4-(4,5,6,7-TETRAHYD

Formula: $C_{16}H_{17}NO_2S$

Mol Weight: 287.38

MP (deg C):   FP (deg C):

BP (deg C):

BP pressure (mm Hg):

Property/Value	Units	Temp	Data Type	Reference
WS 6.18E+000	mg/L	25	EST	MEYLAN,WM ET AL. (1996)
logP 4.41			EXP	NAIRO,Y ET AL. (1992)
VP 1.13E-008	mm Hg	25	EST	NEELY,WB & BLAU,GE (1985)
DC	pKa			
HL 3.03E-012	atm m3/mol	25	EST	MEYLAN,WM & HOWARD,PH (1991)
OH 1.27E-011	cm3/molc sec	25	EST	MEYLAN,WM & HOWARD,PH (1993)

CAS #: 138568-70-4 — BENZENEACETIC ACID, 3-FLUORO-à-METHYL-4-(4-METHY

Formula: $C_{13}H_{12}FNO_2S$

Mol Weight: 265.31

MP (deg C): FP (deg C):

BP (deg C):

BP pressure (mm Hg):

Property/ Value	Units	Temp	Data Type	Reference
WS 4.76E+001	mg/L	25	EST	MEYLAN,WM ET AL. (1996)
logP 3.52			EXP	NAIRO,Y ET AL. (1992)
VP 2.32E-007	mm Hg	25	EST	NEELY,WB & BLAU,GE (1985)
DC	pKa			
HL 4.13E-012	atm m3/mol	25	EST	MEYLAN,WM & HOWARD,PH (1991)
OH 8.72E-012	cm3/molc sec	25	EST	MEYLAN,WM & HOWARD,PH (1993)

CAS #: 138568-71-5 — BENZENEACETIC ACID, 4-(4-ETHYL-2-THIAZOLYL)-3-FL

Formula: $C_{14}H_{14}FNO_2S$

Mol Weight: 279.34

MP (deg C): FP (deg C):

BP (deg C):

BP pressure (mm Hg):

Property/ Value	Units	Temp	Data Type	Reference
WS 1.12E+001	mg/L	25	EST	MEYLAN,WM ET AL. (1996)
logP 4.16			EXP	NAIRO,Y ET AL. (1992)
VP 1.05E-007	mm Hg	25	EST	NEELY,WB & BLAU,GE (1985)
DC	pKa			
HL 5.48E-012	atm m3/mol	25	EST	MEYLAN,WM & HOWARD,PH (1991)
OH 9.57E-012	cm3/molc sec	25	EST	MEYLAN,WM & HOWARD,PH (1993)

CAS #: 138568-72-6 — BENZENEACETIC ACID, 3-FLUORO-à-METHYL-4-[4-(1-ME

Formula: $C_{15}H_{16}FNO_2S$

Mol Weight: 293.36

MP (deg C): FP (deg C):

BP (deg C):

BP pressure (mm Hg):

Property/ Value	Units	Temp	Data Type	Reference
WS 3.78E+000	mg/L	25	EST	MEYLAN,WM ET AL. (1996)
logP 4.62			EXP	NAIRO,Y ET AL. (1992)
VP 7.64E-008	mm Hg	25	EST	NEELY,WB & BLAU,GE (1985)
DC	pKa			
HL 7.28E-012	atm m3/mol	25	EST	MEYLAN,WM & HOWARD,PH (1991)
OH 1.06E-011	cm3/molc sec	25	EST	MEYLAN,WM & HOWARD,PH (1993)

CAS #: 138568-73-7 — BENZENEACETIC ACID, 4-(4-BUTYL-2-THIAZOLYL)-3-FL

Formula: $C_{16}H_{18}FNO_2S$

Mol Weight: 307.39

MP (deg C): FP (deg C):

BP (deg C):

BP pressure (mm Hg):

Property/ Value	Units	Temp	Data Type	Reference
WS 9.25E-001	mg/L	25	EST	MEYLAN,WM ET AL. (1996)
logP 5.24			EXP	NAIRO,Y ET AL. (1992)
VP 2.12E-008	mm Hg	25	EST	NEELY,WB & BLAU,GE (1985)
DC	pKa			
HL 9.66E-012	atm m3/mol	25	EST	MEYLAN,WM & HOWARD,PH (1991)
OH 1.23E-011	cm3/molc sec	25	EST	MEYLAN,WM & HOWARD,PH (1993)

CAS #: 138568-74-8 — BENZENEACETIC ACID, 3-FLUORO-à-METHYL-4-(5-METHY

Formula: $C_{13}H_{12}FNO_2S$

Mol Weight: 265.31

MP (deg C): FP (deg C):

BP (deg C):

BP pressure (mm Hg):

Property/ Value	Units	Temp	Data Type	Reference
WS 4.48E+001	mg/L	25	EST	MEYLAN,WM ET AL. (1996)
logP 3.55			EXP	NAIRO,Y ET AL. (1992)
VP 2.32E-007	mm Hg	25	EST	NEELY,WB & BLAU,GE (1985)
DC	pKa			
HL 4.13E-012	atm m3/mol	25	EST	MEYLAN,WM & HOWARD,PH (1991)
OH 8.72E-012	cm3/molc sec	25	EST	MEYLAN,WM & HOWARD,PH (1993)

CAS #: 138568-75-9 — BENZENEACETIC ACID, 4-(5-ETHYL-2-THIAZOLYL)-3-FL

Formula: $C_{14}H_{14}FNO_2S$

Mol Weight: 279.34

MP (deg C): FP (deg C):

BP (deg C):

BP pressure (mm Hg):

Property/ Value	Units	Temp	Data Type	Reference
WS 1.27E+001	mg/L	25	EST	MEYLAN,WM ET AL. (1996)
logP 4.10			EXP	NAIRO,Y ET AL. (1992)
VP 1.05E-007	mm Hg	25	EST	NEELY,WB & BLAU,GE (1985)
DC	pKa			
HL 5.48E-012	atm m3/mol	25	EST	MEYLAN,WM & HOWARD,PH (1991)
OH 9.57E-012	cm3/molc sec	25	EST	MEYLAN,WM & HOWARD,PH (1993)

CAS #: 138568-76-0 — BENZENEACETIC ACID, 4-(4,5-DIMETHYL-2-THIAZOLYL)

Formula: $C_{14}H_{14}FNO_2S$

Mol Weight: 279.34

MP (deg C): FP (deg C):

BP (deg C):

BP pressure (mm Hg):

Property/ Value	Units	Temp	Data Type	Reference
WS 2.57E+001	mg/L	25	EST	MEYLAN,WM ET AL. (1996)
logP 3.74			EXP	NAIRO,Y ET AL. (1992)
VP 1.05E-007	mm Hg	25	EST	NEELY,WB & BLAU,GE (1985)
DC	pKa			
HL 4.56E-012	atm m3/mol	25	EST	MEYLAN,WM & HOWARD,PH (1991)
OH 7.95E-012	cm3/molc sec	25	EST	MEYLAN,WM & HOWARD,PH (1993)

CAS #: 138568-77-1 — BENZENEACETIC ACID, 4-(4-ETHYL-5-METHYL-2-THIAZO

Formula: $C_{15}H_{16}FNO_2S$

Mol Weight: 293.36

MP (deg C): FP (deg C):

BP (deg C):

BP pressure (mm Hg):

Property/ Value	Units	Temp	Data Type	Reference
WS 5.38E+000	mg/L	25	EST	MEYLAN,WM ET AL. (1996)
logP 4.44			EXP	NAIRO,Y ET AL. (1992)
VP 4.72E-008	mm Hg	25	EST	NEELY,WB & BLAU,GE (1985)
DC	pKa			
HL 6.05E-012	atm m3/mol	25	EST	MEYLAN,WM & HOWARD,PH (1991)
OH 8.92E-012	cm3/molc sec	25	EST	MEYLAN,WM & HOWARD,PH (1993)

CAS #: 138568-78-2	BENZENEACETIC ACID, 3-CHLORO-à-METHYL-4-(4-METHY

Formula: $C_{13}H_{12}ClNO_2S$

Mol Weight: 281.76

MP (deg C):   FP (deg C):

BP (deg C):

BP pressure (mm Hg):

Property/ Value	Units	Temp	Data Type	Reference
WS 2.97E+001	mg/L	25	EST	MEYLAN,WM ET AL. (1996)
logP 3.65			EXP	NAIRO,Y ET AL. (1992)
VP 5.36E-008	mm Hg	25	EST	NEELY,WB & BLAU,GE (1985)
DC	pKa			
HL 2.62E-012	atm m3/mol	25	EST	MEYLAN,WM & HOWARD,PH (1991)
OH 6.90E-012	cm3/molc sec	25	EST	MEYLAN,WM & HOWARD,PH (1993)

CAS #: 138568-79-3	BENZENEACETIC ACID, 3-CHLORO-4-(4-ETHYL-2-THIAZO

Formula: $C_{14}H_{14}ClNO_2S$

Mol Weight: 295.79

MP (deg C):   FP (deg C):

BP (deg C):

BP pressure (mm Hg):

Property/ Value	Units	Temp	Data Type	Reference
WS 7.14E+000	mg/L	25	EST	MEYLAN,WM ET AL. (1996)
logP 4.28			EXP	NAIRO,Y ET AL. (1992)
VP 2.40E-008	mm Hg	25	EST	NEELY,WB & BLAU,GE (1985)
DC	pKa			
HL 3.48E-012	atm m3/mol	25	EST	MEYLAN,WM & HOWARD,PH (1991)
OH 7.76E-012	cm3/molc sec	25	EST	MEYLAN,WM & HOWARD,PH (1993)

CAS #: 138568-80-6	BENZENEACETIC ACID, 3-CHLORO-à-METHYL-4-[4-(1-ME

Formula: $C_{15}H_{16}ClNO_2S$

Mol Weight: 309.82

MP (deg C):   FP (deg C):

BP (deg C):

BP pressure (mm Hg):

Property/ Value	Units	Temp	Data Type	Reference
WS 2.21E+000	mg/L	25	EST	MEYLAN,WM ET AL. (1996)
logP 4.78			EXP	NAIRO,Y ET AL. (1992)
VP 1.75E-008	mm Hg	25	EST	NEELY,WB & BLAU,GE (1985)
DC	pKa			
HL 4.62E-012	atm m3/mol	25	EST	MEYLAN,WM & HOWARD,PH (1991)
OH 8.83E-012	cm3/molc sec	25	EST	MEYLAN,WM & HOWARD,PH (1993)

CAS #: 138568-81-7	BENZENEACETIC ACID, 4-(4-BUTYL-2-THIAZOLYL)-3-CH

Formula: $C_{16}H_{18}ClNO_2S$

Mol Weight: 323.84

MP (deg C):   FP (deg C):

BP (deg C):

BP pressure (mm Hg):

Property/ Value	Units	Temp	Data Type	Reference
WS 5.84E-001	mg/L	25	EST	MEYLAN,WM ET AL. (1996)
logP 5.36			EXP	NAIRO,Y ET AL. (1992)
VP 4.78E-009	mm Hg	25	EST	NEELY,WB & BLAU,GE (1985)
DC	pKa			
HL 6.13E-012	atm m3/mol	25	EST	MEYLAN,WM & HOWARD,PH (1991)
OH 1.05E-011	cm3/molc sec	25	EST	MEYLAN,WM & HOWARD,PH (1993)

CAS #: 138568-82-8	BENZENEACETIC ACID, 3-CHLORO-à-METHYL4-(5-METHYL

Formula: $C_{13}H_{12}ClNO_2S$

Mol Weight: 281.76

MP (deg C):   FP (deg C):

BP (deg C):

BP pressure (mm Hg):

Property/ Value	Units	Temp	Data Type	Reference
WS 2.91E+001	mg/L	25	EST	MEYLAN,WM ET AL. (1996)
logP 3.66			EXP	NAIRO,Y ET AL. (1992)
VP 5.36E-008	mm Hg	25	EST	NEELY,WB & BLAU,GE (1985)
DC	pKa			
HL 2.62E-012	atm m3/mol	25	EST	MEYLAN,WM & HOWARD,PH (1991)
OH 6.90E-012	cm3/molc sec	25	EST	MEYLAN,WM & HOWARD,PH (1993)

CAS #: 138568-83-9	BENZENEACETIC ACID, 3-CHLORO-4-(5-ETHYL-2-THIAZO

Formula: $C_{14}H_{14}ClNO_2S$

Mol Weight: 295.79

MP (deg C):   FP (deg C):

BP (deg C):

BP pressure (mm Hg):

Property/ Value	Units	Temp	Data Type	Reference
WS 5.75E+000	mg/L	25	EST	MEYLAN,WM ET AL. (1996)
logP 4.39			EXP	NAIRO,Y ET AL. (1992)
VP 2.40E-008	mm Hg	25	EST	NEELY,WB & BLAU,GE (1985)
DC	pKa			
HL 3.48E-012	atm m3/mol	25	EST	MEYLAN,WM & HOWARD,PH (1991)
OH 7.76E-012	cm3/molc sec	25	EST	MEYLAN,WM & HOWARD,PH (1993)

CAS #: 138568-84-0	BENZENEACETIC ACID, 3-CHLORO-4-(4,5-DIMETHYL-2-T

Formula: $C_{14}H_{14}ClNO_2S$

Mol Weight: 295.79

MP (deg C):   FP (deg C):

BP (deg C):

BP pressure (mm Hg):

Property/ Value	Units	Temp	Data Type	Reference
WS 1.39E+001	mg/L	25	EST	MEYLAN,WM ET AL. (1996)
logP 3.94			EXP	NAIRO,Y ET AL. (1992)
VP 2.40E-008	mm Hg	25	EST	NEELY,WB & BLAU,GE (1985)
DC	pKa			
HL 2.89E-012	atm m3/mol	25	EST	MEYLAN,WM & HOWARD,PH (1991)
OH 6.14E-012	cm3/molc sec	25	EST	MEYLAN,WM & HOWARD,PH (1993)

CAS #: 138568-85-1	BENZENEACETIC ACID, 3-CHLORO-4-(4-ETHYL-5-METHYL

Formula: $C_{15}H_{16}ClNO_2S$

Mol Weight: 309.82

MP (deg C):   FP (deg C):

BP (deg C):

BP pressure (mm Hg):

Property/ Value	Units	Temp	Data Type	Reference
WS 3.03E+000	mg/L	25	EST	MEYLAN,WM ET AL. (1996)
logP 4.62			EXP	NAIRO,Y ET AL. (1992)
VP 1.07E-008	mm Hg	25	EST	NEELY,WB & BLAU,GE (1985)
DC	pKa			
HL 3.84E-012	atm m3/mol	25	EST	MEYLAN,WM & HOWARD,PH (1991)
OH 7.10E-012	cm3/molc sec	25	EST	MEYLAN,WM & HOWARD,PH (1993)

CAS #: 138686-07-4	2,4(1H,3H)-PYRIMIDINEDIONE, 5-CHLORO-1-(2-FLUORO

Formula: $C_9H_{10}ClFN_2O_4$

Mol Weight: 264.64

MP (deg C): | FP (deg C):

BP (deg C):

BP pressure (mm Hg):

Property/Value	Units	Temp	Data Type	Reference
WS 1.66E+003	mg/L	25	EST	MEYLAN,WM ET AL. (1996)
logP -0.28			EXP	SANGSTER,J (1994)
VP 2.38E-012	mm Hg	25	EST	NEELY,WB & BLAU,GE (1985)
DC	pKa			
HL 3.02E-016	atm m3/mol	25	EST	MEYLAN,WM & HOWARD,PH (1991)
OH 1.93E-011	cm3/molc sec	25	EST	MEYLAN,WM & HOWARD,PH (1993)

CAS #: 138771-70-7	D-PHENYLALANINAMIDE, N-ACETYL-D-PHENYLALANYL-D-P

Formula: $C_{29}H_{32}N_4O_4$

Mol Weight: 500.60

MP (deg C): | FP (deg C):

BP (deg C):

BP pressure (mm Hg):

Property/Value	Units	Temp	Data Type	Reference
WS 7.71E+000	mg/L	25	EST	MEYLAN,WM ET AL. (1996)
logP 2.30			EXP	SANGSTER,J (1994)
VP 2.68E-021	mm Hg	25	EST	NEELY,WB & BLAU,GE (1985)
DC	pKa			
HL 6.43E-024	atm m3/mol	25	EST	MEYLAN,WM & HOWARD,PH (1991)
OH 6.61E-011	cm3/molc sec	25	EST	MEYLAN,WM & HOWARD,PH (1993)

CAS #: 138848-10-9	CYTIDINE, 2',3'-DIDEOXY-N-[(DIMETHYLAMINO)METHYL

Formula: $C_{12}H_{17}FN_4O_3$

Mol Weight: 284.29

MP (deg C): | FP (deg C):

BP (deg C):

BP pressure (mm Hg):

Property/Value	Units	Temp	Data Type	Reference
WS 2.61E+003	mg/L	25	EST	MEYLAN,WM ET AL. (1996)
logP -0.64			EXP	KERR,SG & KALMAN,TI (1992)
VP 6.14E-009	mm Hg	25	EST	NEELY,WB & BLAU,GE (1985)
DC	pKa			
HL 7.40E-017	atm m3/mol	25	EST	MEYLAN,WM & HOWARD,PH (1991)
OH 1.34E-010	cm3/molc sec	25	EST	MEYLAN,WM & HOWARD,PH (1993)

CAS #: 138926-22-4	RH-1224 O-PH N-PENTYLENE(F2)CARBAMATE

Formula: $C_{15}H_{13}Cl_2F_2NO_3$

Mol Weight: 364.18

MP (deg C): | FP (deg C):

BP (deg C):

BP pressure (mm Hg):

Property/Value	Units	Temp	Data Type	Reference
WS 4.54E+000	mg/L	25	EST	MEYLAN,WM ET AL. (1996)
logP 3.57			EXP	NANDIHALLI,UB ET AL. (1993)
VP 8.94E-007	mm Hg	25	EST	NEELY,WB & BLAU,GE (1985)
DC	pKa			
HL 2.04E-009	atm m3/mol	25	EST	MEYLAN,WM & HOWARD,PH (1991)
OH 2.40E-011	cm3/molc sec	25	EST	MEYLAN,WM & HOWARD,PH (1993)

CAS #: 139733-55-4	ACETAMIDE, 2-[4-[[2-(2-HYDROXY-3-PHENOXYPROPYL)A

Formula: $C_{19}H_{24}N_2O_5$

Mol Weight: 360.41

MP (deg C): | FP (deg C):

BP (deg C):

BP pressure (mm Hg):

Property/Value	Units	Temp	Data Type	Reference
WS 6.16E+002	mg/L	25	EST	MEYLAN,WM ET AL. (1996)
logP 1.10			EXP	SANGSTER,J (1994)
VP 5.48E-014	mm Hg	25	EST	NEELY,WB & BLAU,GE (1985)
DC	pKa			
HL 1.69E-021	atm m3/mol	25	EST	MEYLAN,WM & HOWARD,PH (1991)
OH 1.75E-010	cm3/molc sec	25	EST	MEYLAN,WM & HOWARD,PH (1993)

CAS #: 139733-59-8	ACETAMIDE, 2-[[[4-[2-(2-HYDROXY-3-PHENOXY)PROPYL

Formula: $C_{22}H_{30}N_2O_5$

Mol Weight: 402.49

MP (deg C): | FP (deg C):

BP (deg C):

BP pressure (mm Hg):

Property/Value	Units	Temp	Data Type	Reference
WS 2.64E+001	mg/L	25	EST	MEYLAN,WM ET AL. (1996)
logP 2.40			EXP	SANGSTER,J (1993)
VP 3.90E-015	mm Hg	25	EST	NEELY,WB & BLAU,GE (1985)
DC	pKa			
HL 6.56E-021	atm m3/mol	25	EST	MEYLAN,WM & HOWARD,PH (1991)
OH 1.84E-010	cm3/molc sec	25	EST	MEYLAN,WM & HOWARD,PH (1993)

CAS #: 139733-61-2	ACETAMIDE, 2-[[4-[2-(2-HYDROXY-3-PHENOXYPROPYL)A

Formula: $C_{22}H_{30}N_2O_5$

Mol Weight: 402.49

MP (deg C): | FP (deg C):

BP (deg C):

BP pressure (mm Hg):

Property/Value	Units	Temp	Data Type	Reference
WS 2.64E+001	mg/L	25	EST	MEYLAN,WM ET AL. (1996)
logP 2.40			EXP	SANGSTER,J (1993)
VP 7.06E-015	mm Hg	25	EST	NEELY,WB & BLAU,GE (1985)
DC	pKa			
HL 6.56E-021	atm m3/mol	25	EST	MEYLAN,WM & HOWARD,PH (1991)
OH 1.87E-010	cm3/molc sec	25	EST	MEYLAN,WM & HOWARD,PH (1993)

CAS #: 139733-77-0	ACETAMIDE, 2-[4-[2-[(2-HYDROXY-3-PHENOXYPROPYL)A

Formula: $C_{22}H_{30}N_2O_6$

Mol Weight: 418.49

MP (deg C): | FP (deg C):

BP (deg C):

BP pressure (mm Hg):

Property/Value	Units	Temp	Data Type	Reference
WS 8.79E+002	mg/L	25	EST	MEYLAN,WM ET AL. (1996)
logP 0.50			EXP	SANGSTER,J (1993)
VP 2.18E-017	mm Hg	25	EST	NEELY,WB & BLAU,GE (1985)
DC	pKa			
HL 2.40E-022	atm m3/mol	25	EST	MEYLAN,WM & HOWARD,PH (1991)
OH 1.89E-010	cm3/molc sec	25	EST	MEYLAN,WM & HOWARD,PH (1993)

CAS #: 139733-95-2 — ACETAMIDE, 2-[4-[2-(2-HYDROXY-3-PHENOXYPROPYL)AM

Formula: $C_{26}H_{30}N_2O_5$

Mol Weight: 450.54

MP (deg C):   **FP (deg C):**

BP (deg C):

BP pressure (mm Hg):

Property/Value	Units	Temp	Data Type	Reference
WS 2.74E+000	mg/L	25	EST	MEYLAN,WM ET AL. (1996)
logP 3.20			EXP	SANGSTER,J (1993)
VP 2.07E-017	mm Hg	25	EST	NEELY,WB & BLAU,GE (1985)
DC	pKa			
HL 3.00E-022	atm m3/mol	25	EST	MEYLAN,WM & HOWARD,PH (1991)
OH 1.87E-010	cm3/molc sec	25	EST	MEYLAN,WM & HOWARD,PH (1993)

CAS #: 139734-21-7 — 2-PROPANOL, 3-[[2-[4-[2-OXO-2-(2-PIPERIDINYL)ETH

Formula: $C_{24}H_{32}N_2O_5$

Mol Weight: 428.53

MP (deg C):   **FP (deg C):**

BP (deg C):

BP pressure (mm Hg):

Property/Value	Units	Temp	Data Type	Reference
WS 1.22E+001	mg/L	25	EST	MEYLAN,WM ET AL. (1996)
logP 2.60			EXP	SANGSTER,J (1993)
VP 9.00E-015	mm Hg	25	EST	NEELY,WB & BLAU,GE (1985)
DC	pKa			
HL 8.43E-021	atm m3/mol	25	EST	MEYLAN,WM & HOWARD,PH (1991)
OH 2.02E-010	cm3/molc sec	25	EST	MEYLAN,WM & HOWARD,PH (1993)

CAS #: 139734-30-8 — 2-PROPANOL, 3-[[2-[4-[2-OXO-2-(1,2,3,4-TETRAHYDR

Formula: $C_{28}H_{32}N_2O_5$

Mol Weight: 476.58

MP (deg C):   **FP (deg C):**

BP (deg C):

BP pressure (mm Hg):

Property/Value	Units	Temp	Data Type	Reference
WS 8.52E-001	mg/L	25	EST	MEYLAN,WM ET AL. (1996)
logP 3.60			EXP	SANGSTER,J (1993)
VP 5.84E-017	mm Hg	25	EST	NEELY,WB & BLAU,GE (1985)
DC	pKa			
HL 3.21E-022	atm m3/mol	25	EST	MEYLAN,WM & HOWARD,PH (1991)
OH 2.03E-010	cm3/molc sec	25	EST	MEYLAN,WM & HOWARD,PH (1993)

CAS #: 140111-43-9 — 4,4'-BIPYRIDINE, 1,2,3,6-TETRAHYDRO-1-METHYL-

Formula: $C_{11}H_{14}N_2$

Mol Weight: 174.25

MP (deg C):   **FP (deg C):**

BP (deg C):

BP pressure (mm Hg):

Property/Value	Units	Temp	Data Type	Reference
WS 5.12E+004	mg/L	25	EST	MEYLAN,WM ET AL. (1996)
logP 1.25			EXP	ALTOMARE,CA ET AL. (1992)
VP 3.43E-003	mm Hg	25	EST	NEELY,WB & BLAU,GE (1985)
DC	pKa			
HL 1.94E-009	atm m3/mol	25	EST	MEYLAN,WM & HOWARD,PH (1991)
OH 1.75E-010	cm3/molc sec	25	EST	MEYLAN,WM & HOWARD,PH (1993)

CAS #: 141018-14-6 — CYTIDINE, 2',3'-DIDEOXY-N-[(DIETHYLAMINO)METHYLE

Formula: $C_{14}H_{22}N_4O_3$

Mol Weight: 294.36

MP (deg C):   **FP (deg C):**

BP (deg C):

BP pressure (mm Hg):

Property/Value	Units	Temp	Data Type	Reference
WS 7.30E+002	mg/L	25	EST	MEYLAN,WM ET AL. (1996)
logP -0.06			EXP	KERR,SG & KALMAN,TI (1992)
VP 6.94E-010	mm Hg	25	EST	NEELY,WB & BLAU,GE (1985)
DC	pKa			
HL 6.56E-017	atm m3/mol	25	EST	MEYLAN,WM & HOWARD,PH (1991)
OH 2.05E-010	cm3/molc sec	25	EST	MEYLAN,WM & HOWARD,PH (1993)

CAS #: 141018-15-7 — CYTIDINE, 2',3'-DIDEOXY-N-[(DIPROPYLAMINO)METHYL

Formula: $C_{16}H_{26}N_4O_3$

Mol Weight: 322.41

MP (deg C):   **FP (deg C):**

BP (deg C):

BP pressure (mm Hg):

Property/Value	Units	Temp	Data Type	Reference
WS 1.40E+003	mg/L	25	EST	MEYLAN,WM ET AL. (1996)
logP 0.95			EXP	KERR,SG & KALMAN,TI (1992)
VP 9.36E-011	mm Hg	25	EST	NEELY,WB & BLAU,GE (1985)
DC	pKa			
HL 1.16E-016	atm m3/mol	25	EST	MEYLAN,WM & HOWARD,PH (1991)
OH 2.11E-010	cm3/molc sec	25	EST	MEYLAN,WM & HOWARD,PH (1993)

CAS #: 141018-16-8 — CYTIDINE, 2',3'-DIDEOXY-N-[(1-PIPERIDINYL)METHYL

Formula: $C_{15}H_{22}N_4O_3$

Mol Weight: 306.37

MP (deg C):   **FP (deg C):**

BP (deg C):

BP pressure (mm Hg):

Property/Value	Units	Temp	Data Type	Reference
WS 4.91E+002	mg/L	25	EST	MEYLAN,WM ET AL. (1996)
logP 0.06			EXP	KERR,SG & KALMAN,TI (1992)
VP 8.06E-011	mm Hg	25	EST	NEELY,WB & BLAU,GE (1985)
DC	pKa			
HL 3.84E-017	atm m3/mol	25	EST	MEYLAN,WM & HOWARD,PH (1991)
OH 2.08E-010	cm3/molc sec	25	EST	MEYLAN,WM & HOWARD,PH (1993)

CAS #: 141018-17-9 — CYTIDINE, 2',3'-DIDEOXY-N-[(4-MORPHOLINYL)METHYL

Formula: $C_{14}H_{20}N_4O_4$

Mol Weight: 308.34

MP (deg C):   **FP (deg C):**

BP (deg C):

BP pressure (mm Hg):

Property/Value	Units	Temp	Data Type	Reference
WS 3.62E+003	mg/L	25	EST	MEYLAN,WM ET AL. (1996)
logP -0.97			EXP	KERR,SG & KALMAN,TI (1992)
VP 6.21E-011	mm Hg	25	EST	NEELY,WB & BLAU,GE (1985)
DC	pKa			
HL 2.55E-019	atm m3/mol	25	EST	MEYLAN,WM & HOWARD,PH (1991)
OH 2.47E-010	cm3/molc sec	25	EST	MEYLAN,WM & HOWARD,PH (1993)

141018-18-0 — CYTIDINE, 2',3'-DIDEOXY-N-[(1-PYRROLIDINYL)METHY

Formula	$C_{14}H_{20}N_4O_3$
Mol Weight	292.34
MP (deg C)	
FP (deg C)	
BP (deg C)	
BP pressure (mm Hg)	

Property/Value	Units	Temp	Data Type	Reference
WS 1.68E+003	mg/L	25	EST	MEYLAN,WM ET AL. (1996)
logP -0.47			EXP	KERR,SG & KALMAN,TI (1992)
VP 2.36E-010	mm Hg	25	EST	NEELY,WB & BLAU,GE (1985)
DC	pKa			
HL 2.89E-017	atm m3/mol	25	EST	MEYLAN,WM & HOWARD,PH (1991)
OH 1.90E-010	cm3/molc sec	25	EST	MEYLAN,WM & HOWARD,PH (1993)

141018-19-1 — CYTIDINE, 2',3'-DIDEOXY-N-[(DIETHYLAMINO)METHYLE

Formula	$C_{14}H_{21}FN_4O_3$
Mol Weight	312.35
MP (deg C)	
FP (deg C)	
BP (deg C)	
BP pressure (mm Hg)	

Property/Value	Units	Temp	Data Type	Reference
WS 3.24E+002	mg/L	25	EST	MEYLAN,WM ET AL. (1996)
logP 0.23			EXP	KERR,SG & KALMAN,TI (1992)
VP 8.47E-010	mm Hg	25	EST	NEELY,WB & BLAU,GE (1985)
DC	pKa			
HL 1.30E-016	atm m3/mol	25	EST	MEYLAN,WM & HOWARD,PH (1991)
OH 1.49E-010	cm3/molc sec	25	EST	MEYLAN,WM & HOWARD,PH (1993)

141018-20-4 — CYTIDINE, 2',3'-DIDEOXY-3'-FLUORO-N-[(DIPROPYLAM

Formula	$C_{16}H_{25}FN_4O_3$
Mol Weight	340.40
MP (deg C)	
FP (deg C)	
BP (deg C)	
BP pressure (mm Hg)	

Property/Value	Units	Temp	Data Type	Reference
WS 4.98E+002	mg/L	25	EST	MEYLAN,WM ET AL. (1996)
logP 1.35			EXP	KERR,SG & KALMAN,TI (1992)
VP 1.14E-010	mm Hg	25	EST	NEELY,WB & BLAU,GE (1985)
DC	pKa			
HL 2.30E-016	atm m3/mol	25	EST	MEYLAN,WM & HOWARD,PH (1991)
OH 1.56E-010	cm3/molc sec	25	EST	MEYLAN,WM & HOWARD,PH (1993)

141018-21-5 — CYTIDINE, 2',3'-DIDEOXY-3'-FLUORO-N-[[BIS(1-METH

Formula	$C_{16}H_{25}FN_4O_3$
Mol Weight	340.40
MP (deg C)	
FP (deg C)	
BP (deg C)	
BP pressure (mm Hg)	

Property/Value	Units	Temp	Data Type	Reference
WS 7.52E+002	mg/L	25	EST	MEYLAN,WM ET AL. (1996)
logP 1.14			EXP	KERR,SG & KALMAN,TI (1992)
VP 4.18E-010	mm Hg	25	EST	NEELY,WB & BLAU,GE (1985)
DC	pKa			
HL 2.30E-016	atm m3/mol	25	EST	MEYLAN,WM & HOWARD,PH (1991)
OH 1.69E-010	cm3/molc sec	25	EST	MEYLAN,WM & HOWARD,PH (1993)

141018-22-6 — CYTIDINE, 2',3'-DIDEOXY-3'-FLUORO-N-[(1-PIPERIDI

Formula	$C_{15}H_{21}FN_4O_3$
Mol Weight	324.36
MP (deg C)	
FP (deg C)	
BP (deg C)	
BP pressure (mm Hg)	

Property/Value	Units	Temp	Data Type	Reference
WS 4.35E+003	mg/L	25	EST	MEYLAN,WM ET AL. (1996)
logP 0.36			EXP	KERR,SG & KALMAN,TI (1992)
VP 9.86E-011	mm Hg	25	EST	NEELY,WB & BLAU,GE (1985)
DC	pKa			
HL 7.63E-017	atm m3/mol	25	EST	MEYLAN,WM & HOWARD,PH (1991)
OH 1.52E-010	cm3/molc sec	25	EST	MEYLAN,WM & HOWARD,PH (1993)

141018-23-7 — CYTIDINE, 2',3'-DIDEOXY-3'-FLUORO-N-[(4-MORPHOLI

Formula	$C_{14}H_{19}FN_4O_4$
Mol Weight	326.33
MP (deg C)	
FP (deg C)	
BP (deg C)	
BP pressure (mm Hg)	

Property/Value	Units	Temp	Data Type	Reference
WS 1.60E+003	mg/L	25	EST	MEYLAN,WM ET AL. (1996)
logP -0.68			EXP	KERR,SG & KALMAN,TI (1992)
VP 7.45E-011	mm Hg	25	EST	NEELY,WB & BLAU,GE (1985)
DC	pKa			
HL 5.07E-019	atm m3/mol	25	EST	MEYLAN,WM & HOWARD,PH (1991)
OH 1.92E-010	cm3/molc sec	25	EST	MEYLAN,WM & HOWARD,PH (1993)

141018-24-8 — CYTIDINE, 2',3'-DIDEOXY-3'-FLUORO-N-[(1-PYRROLID

Formula	$C_{14}H_{19}FN_4O_3$
Mol Weight	310.33
MP (deg C)	
FP (deg C)	
BP (deg C)	
BP pressure (mm Hg)	

Property/Value	Units	Temp	Data Type	Reference
WS 7.17E+002	mg/L	25	EST	MEYLAN,WM ET AL. (1996)
logP -0.16			EXP	KERR,SG & KALMAN,TI (1992)
VP 2.88E-010	mm Hg	25	EST	NEELY,WB & BLAU,GE (1985)
DC	pKa			
HL 5.75E-017	atm m3/mol	25	EST	MEYLAN,WM & HOWARD,PH (1991)
OH 1.46E-010	cm3/molc sec	25	EST	MEYLAN,WM & HOWARD,PH (1993)

141043-80-3 — CYTIDINE, 2',3'-DIDEOXY-N-[[BIS(1-METHYLETHYL)AM

Formula	$C_{16}H_{26}N_4O_3$
Mol Weight	322.41
MP (deg C)	
FP (deg C)	
BP (deg C)	
BP pressure (mm Hg)	

Property/Value	Units	Temp	Data Type	Reference
WS 2.20E+003	mg/L	25	EST	MEYLAN,WM ET AL. (1996)
logP 0.72			EXP	KERR,SG & KALMAN,TI (1992)
VP 3.42E-010	mm Hg	25	EST	NEELY,WB & BLAU,GE (1985)
DC	pKa			
HL 1.16E-016	atm m3/mol	25	EST	MEYLAN,WM & HOWARD,PH (1991)
OH 2.24E-010	cm3/molc sec	25	EST	MEYLAN,WM & HOWARD,PH (1993)

141249-30-1

CAS #: 141249-30-1 — 2(1H)-PYRIMIDINONE, 4-AMINO-5-CHLORO-1-(2,3-DIDE

Formula: $C_9H_{11}ClFN_3O_3$

Mol Weight: 263.66

MP (deg C):   FP (deg C):

BP (deg C):

BP pressure (mm Hg):

Property/Value	Units	Temp	Data Type	Reference
WS 2.40E+003	mg/L	25	EST	MEYLAN,WM ET AL. (1996)
logP -0.46			EXP	SANGSTER,J (1994)
VP 1.56E-008	mm Hg	25	EST	NEELY,WB & BLAU,GE (1985)
DC	pKa			
HL 1.68E-016	atm m3/mol	25	EST	MEYLAN,WM & HOWARD,PH (1991)
OH 2.54E-011	cm3/molc sec	25	EST	MEYLAN,WM & HOWARD,PH (1993)

141249-32-3

CAS #: 141249-32-3 — 2,4(1H,3H)-PYRIMIDINEDIONE, 1-(2,3-DIDEOXY-2-FLU

Formula: $C_9H_{10}F_2N_2O_4$

Mol Weight: 248.19

MP (deg C):   FP (deg C):

BP (deg C):

BP pressure (mm Hg):

Property/Value	Units	Temp	Data Type	Reference
WS 4.89E+003	mg/L	25	EST	MEYLAN,WM ET AL. (1996)
logP -0.72			EXP	SANGSTER,J (1994)
VP 9.13E-012	mm Hg	25	EST	NEELY,WB & BLAU,GE (1985)
DC	pKa			
HL 8.04E-016	atm m3/mol	25	EST	MEYLAN,WM & HOWARD,PH (1991)
OH 1.83E-011	cm3/molc sec	25	EST	MEYLAN,WM & HOWARD,PH (1993)

141605-11-0

CAS #: 141605-11-0 — 1,2,4-TRIAZIN-5(4H)-ONE, 4-AMINO-3-(METHYLTHIO)-

Formula: $C_9H_{14}N_4O_2S$

Mol Weight: 242.30

MP (deg C):   FP (deg C):

BP (deg C):

BP pressure (mm Hg):

Property/Value	Units	Temp	Data Type	Reference
WS 3.18E+003	mg/L	25	EST	MEYLAN,WM ET AL. (1996)
logP 1.07			EXP	SANGSTER,J (1994)
VP 1.50E-007	mm Hg	25	EST	NEELY,WB & BLAU,GE (1985)
DC	pKa			
HL 1.17E-015	atm m3/mol	25	EST	MEYLAN,WM & HOWARD,PH (1991)
OH 5.13E-011	cm3/molc sec	25	EST	MEYLAN,WM & HOWARD,PH (1993)

141605-15-4

CAS #: 141605-15-4 — 1,2,4-TRIAZIN-5(4H)-ONE, 6-(3-ETHOXYPHENYL)-3-ME

Formula: $C_{13}H_{16}N_4O_2$

Mol Weight: 260.30

MP (deg C):   FP (deg C):

BP (deg C):

BP pressure (mm Hg):

Property/Value	Units	Temp	Data Type	Reference
WS 4.85E+002	mg/L	25	EST	MEYLAN,WM ET AL. (1996)
logP 1.91			EXP	SANGSTER,J (1994)
VP 7.15E-008	mm Hg	25	EST	NEELY,WB & BLAU,GE (1985)
DC	pKa			
HL 1.01E-012	atm m3/mol	25	EST	MEYLAN,WM & HOWARD,PH (1991)
OH 1.33E-010	cm3/molc sec	25	EST	MEYLAN,WM & HOWARD,PH (1993)

141605-16-5

CAS #: 141605-16-5 — 1,2,4-TRIAZIN-5(4H)-ONE, 4-AMINO-6-(3,5-DICHLORO

Formula: $C_{10}H_8Cl_2N_4OS$

Mol Weight: 303.17

MP (deg C):   FP (deg C):

BP (deg C):

BP pressure (mm Hg):

Property/Value	Units	Temp	Data Type	Reference
WS 3.10E+001	mg/L	25	EST	MEYLAN,WM ET AL. (1996)
logP 3.02			EXP	SANGSTER,J (1994)
VP 7.33E-009	mm Hg	25	EST	NEELY,WB & BLAU,GE (1985)
DC	pKa			
HL 2.58E-014	atm m3/mol	25	EST	MEYLAN,WM & HOWARD,PH (1991)
OH 1.99E-011	cm3/molc sec	25	EST	MEYLAN,WM & HOWARD,PH (1993)

141605-18-7

CAS #: 141605-18-7 — 1,2,4-TRIAZIN-5(4H)-ONE, 6-(3,5-DICHLOROPHENYL)-

Formula: $C_{10}H_7Cl_2N_3OS$

Mol Weight: 288.16

MP (deg C):   FP (deg C):

BP (deg C):

BP pressure (mm Hg):

Property/Value	Units	Temp	Data Type	Reference
WS 1.69E+001	mg/L	25	EST	MEYLAN,WM ET AL. (1996)
logP 3.43			EXP	SANGSTER,J (1994)
VP 3.83E-008	mm Hg	25	EST	NEELY,WB & BLAU,GE (1985)
DC	pKa			
HL 4.51E-011	atm m3/mol	25	EST	MEYLAN,WM & HOWARD,PH (1991)
OH 5.02E-011	cm3/molc sec	25	EST	MEYLAN,WM & HOWARD,PH (1993)

141605-19-8

CAS #: 141605-19-8 — 1,2,4-TRIAZIN-5(4H)-ONE, 4-AMINO-6-(3,5-DICHLORO

Formula: $C_9H_6Cl_2N_4OS$

Mol Weight: 289.14

MP (deg C):   FP (deg C):

BP (deg C):

BP pressure (mm Hg):

Property/Value	Units	Temp	Data Type	Reference
WS 7.45E+001	mg/L	25	EST	MEYLAN,WM ET AL. (1996)
logP 2.67			EXP	SANGSTER,J (1994)
VP 1.01E-008	mm Hg	25	EST	NEELY,WB & BLAU,GE (1985)
DC	pKa			
HL 8.58E-014	atm m3/mol	25	EST	MEYLAN,WM & HOWARD,PH (1991)
OH 4.96E-011	cm3/molc sec	25	EST	MEYLAN,WM & HOWARD,PH (1993)

141627-87-4

CAS #: 141627-87-4 — 1,2,4-TRIAZIN-5(4H)-ONE, 4-AMINO-6-(3,5-DICHLORO

Formula: $C_{10}H_9Cl_2N_5O$

Mol Weight: 286.12

MP (deg C):   FP (deg C):

BP (deg C):

BP pressure (mm Hg):

Property/Value	Units	Temp	Data Type	Reference
WS 3.46E+002	mg/L	25	EST	MEYLAN,WM ET AL. (1996)
logP 1.91			EXP	SANGSTER,J (1994)
VP 1.64E-008	mm Hg	25	EST	NEELY,WB & BLAU,GE (1985)
DC	pKa			
HL 5.46E-016	atm m3/mol	25	EST	MEYLAN,WM & HOWARD,PH (1991)
OH 8.14E-011	cm3/molc sec	25	EST	MEYLAN,WM & HOWARD,PH (1993)

CAS #: 141627-88-5	1,2,4-TRIAZIN-5(4H)-ONE, 6-(3-ETHOXYPHENYL)-2,3-

Formula: $C_{12}H_{13}N_3O_2S$

Mol Weight: 263.32

MP (deg C): FP (deg C):

BP (deg C):

BP pressure (mm Hg):

Property/ Value	Units	Temp	Data Type	Reference
WS 1.05E+002	mg/L	25	EST	MEYLAN,WM ET AL. (1996)
logP 2.67			EXP	SANGSTER,J (1994)
VP 4.43E-008	mm Hg	25	EST	NEELY,WB & BLAU,GE (1985)
DC	pKa			
HL 6.45E-012	atm m3/mol	25	EST	MEYLAN,WM & HOWARD,PH (1991)
OH 1.02E-010	cm3/molc sec	25	EST	MEYLAN,WM & HOWARD,PH (1993)

CAS #: 141627-90-9	1,2,4-TRIAZIN-5(4H)-ONE, 6-[3-CHLORO-4-(TRIFLUOR

Formula: $C_{11}H_7ClF_3N_3O_2S$

Mol Weight: 337.71

MP (deg C): FP (deg C):

BP (deg C):

BP pressure (mm Hg):

Property/ Value	Units	Temp	Data Type	Reference
WS 6.07E+000	mg/L	25	EST	MEYLAN,WM ET AL. (1996)
logP 3.61			EXP	SANGSTER,J (1994)
VP 4.78E-008	mm Hg	25	EST	NEELY,WB & BLAU,GE (1985)
DC	pKa			
HL 2.84E-011	atm m3/mol	25	EST	MEYLAN,WM & HOWARD,PH (1991)
OH 4.06E-011	cm3/molc sec	25	EST	MEYLAN,WM & HOWARD,PH (1993)

CAS #: 142657-44-1	1H-IMIDAZOLE-1-ETHANOL, _-(3-BROMOPHENYL)-_-(2,4

Formula: $C_{17}H_{13}BrF_2N_2O$

Mol Weight: 379.21

MP (deg C): FP (deg C):

BP (deg C):

BP pressure (mm Hg):

Property/ Value	Units	Temp	Data Type	Reference
WS 3.47E+000	mg/L	25	EST	MEYLAN,WM ET AL. (1996)
logP 3.60			EXP	SANGSTER,J (1994)
VP 4.03E-011	mm Hg	25	EST	NEELY,WB & BLAU,GE (1985)
DC	pKa			
HL 1.38E-011	atm m3/mol	25	EST	MEYLAN,WM & HOWARD,PH (1991)
OH 4.61E-011	cm3/molc sec	25	EST	MEYLAN,WM & HOWARD,PH (1993)

CAS #: 142657-45-2	1H-IMIDAZOLE-1-ETHANOL, _-(4-BROMOPHENYL)-_-PHEN

Formula: $C_{17}H_{15}BrN_2O$

Mol Weight: 343.23

MP (deg C): FP (deg C):

BP (deg C):

BP pressure (mm Hg):

Property/ Value	Units	Temp	Data Type	Reference
WS 9.38E+000	mg/L	25	EST	MEYLAN,WM ET AL. (1996)
logP 3.35			EXP	SANGSTER,J (1994)
VP 2.74E-011	mm Hg	25	EST	NEELY,WB & BLAU,GE (1985)
DC	pKa			
HL 1.01E-011	atm m3/mol	25	EST	MEYLAN,WM & HOWARD,PH (1991)
OH 4.55E-011	cm3/molc sec	25	EST	MEYLAN,WM & HOWARD,PH (1993)

CAS #: 142657-53-2	1H-IMIDAZOLE-1-ETHANOL, _-(4-HYDROXYPHENYL)-_-[4

Formula: $C_{22}H_{19}N_3O_2$

Mol Weight: 357.42

MP (deg C): FP (deg C):

BP (deg C):

BP pressure (mm Hg):

Property/ Value	Units	Temp	Data Type	Reference
WS 1.59E+002	mg/L	25	EST	MEYLAN,WM ET AL. (1996)
logP 2.49			EXP	SANGSTER,J (1994)
VP 3.13E-015	mm Hg	25	EST	NEELY,WB & BLAU,GE (1985)
DC	pKa			
HL 2.65E-019	atm m3/mol	25	EST	MEYLAN,WM & HOWARD,PH (1991)
OH 8.45E-011	cm3/molc sec	25	EST	MEYLAN,WM & HOWARD,PH (1993)

CAS #: 142978-29-8	N(24-NO2-6CL PH)-2(3CL-5NO2)PYRIDINAMINE

Formula: $C_{11}H_5Cl_2N_5O_6$

Mol Weight: 374.10

MP (deg C): FP (deg C):

BP (deg C):

BP pressure (mm Hg):

Property/ Value	Units	Temp	Data Type	Reference
WS 2.66E+001	mg/L	25	EST	MEYLAN,WM ET AL. (1996)
logP 2.60			EXP	BRANDT,U ET AL. (1992)
VP 2.37E-010	mm Hg	25	EST	NEELY,WB & BLAU,GE (1985)
DC	pKa			
HL 1.01E-015	atm m3/mol	25	EST	MEYLAN,WM & HOWARD,PH (1991)
OH 5.44E-014	cm3/molc sec	25	EST	MEYLAN,WM & HOWARD,PH (1993)

CAS #: 142978-30-1	N(24-NO2-6CL PH)-2(3CL-5CF3)PYRIDINAMINE

Formula: $C_{12}H_5Cl_2F_3N_4O_4$

Mol Weight: 397.10

MP (deg C): FP (deg C):

BP (deg C):

BP pressure (mm Hg):

Property/ Value	Units	Temp	Data Type	Reference
WS 2.04E+000	mg/L	25	EST	MEYLAN,WM ET AL. (1996)
logP 3.74			EXP	BRANDT,U ET AL. (1992)
VP 9.40E-009	mm Hg	25	EST	NEELY,WB & BLAU,GE (1985)
DC	pKa			
HL 2.23E-012	atm m3/mol	25	EST	MEYLAN,WM & HOWARD,PH (1991)
OH 8.17E-014	cm3/molc sec	25	EST	MEYLAN,WM & HOWARD,PH (1993)

CAS #: 142978-32-3	N(24-NO2-6CF3 PH)-2(3CL-5NO2)PYRIDINAMIN

Formula: $C_{12}H_5ClF_3N_5O_6$

Mol Weight: 407.65

MP (deg C): FP (deg C):

BP (deg C):

BP pressure (mm Hg):

Property/ Value	Units	Temp	Data Type	Reference
WS 1.12E+001	mg/L	25	EST	MEYLAN,WM ET AL. (1996)
logP 2.80			EXP	BRANDT,U ET AL. (1992)
VP 7.01E-010	mm Hg	25	EST	NEELY,WB & BLAU,GE (1985)
DC	pKa			
HL 1.19E-014	atm m3/mol	25	EST	MEYLAN,WM & HOWARD,PH (1991)
OH 4.10E-014	cm3/molc sec	25	EST	MEYLAN,WM & HOWARD,PH (1993)

143121-06-6 — RH-1909 O-PH(2-CL) N-BUTYLENE CARBAMATE

Formula: $C_{11}H_{12}ClNO_2$

Mol Weight: 225.68

MP (deg C): FP (deg C):

BP (deg C):

BP pressure (mm Hg):

Property/Value	Units	Temp	Data Type	Reference
WS 6.82E+001	mg/L	25	EST	MEYLAN,WM ET AL. (1996)
logP 3.13			EXP	NANDIHALLI,UB ET AL. (1993)
VP 3.30E-004	mm Hg	25	EST	NEELY,WB & BLAU,GE (1985)
DC	pKa			
HL 4.06E-008	atm m3/mol	25	EST	MEYLAN,WM & HOWARD,PH (1991)
OH 2.58E-011	cm3/molc sec	25	EST	MEYLAN,WM & HOWARD,PH (1993)

143121-07-7 — RH-7160 O-PH(24DICL) N-DIET CARBAMATE

Formula: $C_{14}H_{15}Cl_2NO_3$

Mol Weight: 316.19

MP (deg C): FP (deg C):

BP (deg C):

BP pressure (mm Hg):

Property/Value	Units	Temp	Data Type	Reference
WS 4.18E+000	mg/L	25	EST	MEYLAN,WM ET AL. (1996)
logP 3.95			EXP	NANDIHALLI,UB ET AL. (1993)
VP 3.19E-006	mm Hg	25	EST	NEELY,WB & BLAU,GE (1985)
DC	pKa			
HL 8.81E-010	atm m3/mol	25	EST	MEYLAN,WM & HOWARD,PH (1991)
OH 3.76E-011	cm3/molc sec	25	EST	MEYLAN,WM & HOWARD,PH (1993)

143121-08-8 — RH-9611 O-PH N-BUTYLENEN CARBAMATE

Formula: $C_{11}H_{11}Cl_2NO_3$

Mol Weight: 276.12

MP (deg C): FP (deg C):

BP (deg C):

BP pressure (mm Hg):

Property/Value	Units	Temp	Data Type	Reference
WS 1.83E+002	mg/L	25	EST	MEYLAN,WM ET AL. (1996)
logP 2.98			EXP	NANDIHALLI,UB ET AL. (1993)
VP 1.12E-006	mm Hg	25	EST	NEELY,WB & BLAU,GE (1985)
DC	pKa			
HL 3.13E-012	atm m3/mol	25	EST	MEYLAN,WM & HOWARD,PH (1991)
OH 2.79E-011	cm3/molc sec	25	EST	MEYLAN,WM & HOWARD,PH (1993)

143121-10-2 — RH-0978 O-PH N-BUTYLENE(F2)CARBAMATE

Formula: $C_{14}H_{11}Cl_2F_2NO_3$

Mol Weight: 350.15

MP (deg C): FP (deg C):

BP (deg C):

BP pressure (mm Hg):

Property/Value	Units	Temp	Data Type	Reference
WS 7.42E+000	mg/L	25	EST	MEYLAN,WM ET AL. (1996)
logP 3.42			EXP	NANDIHALLI,UB ET AL. (1993)
VP 2.19E-006	mm Hg	25	EST	NEELY,WB & BLAU,GE (1985)
DC	pKa			
HL 1.54E-009	atm m3/mol	25	EST	MEYLAN,WM & HOWARD,PH (1991)
OH 1.84E-011	cm3/molc sec	25	EST	MEYLAN,WM & HOWARD,PH (1993)

143202-51-1 — 1H-PYRIDO[4,3-B]INDOLE, 5-ACETYL-2,3,4,4A,5,9B-H

Formula: $C_{15}H_{20}N_2O$

Mol Weight: 244.34

MP (deg C): FP (deg C):

BP (deg C):

BP pressure (mm Hg):

Property/Value	Units	Temp	Data Type	Reference
WS 2.18E+003	mg/L	25	EST	MEYLAN,WM ET AL. (1996)
logP 1.25			EXP	SANGSTER,J (1994)
VP 5.26E-006	mm Hg	25	EST	NEELY,WB & BLAU,GE (1985)
DC	pKa			
HL 2.62E-011	atm m3/mol	25	EST	MEYLAN,WM & HOWARD,PH (1991)
OH 1.16E-010	cm3/molc sec	25	EST	MEYLAN,WM & HOWARD,PH (1993)

143502-47-0 — RH-8826 DIPHENYL SULFONE

Formula: $C_{15}H_9ClF_3NO_6S$

Mol Weight: 423.75

MP (deg C): FP (deg C):

BP (deg C):

BP pressure (mm Hg):

Property/Value	Units	Temp	Data Type	Reference
WS 7.93E-001	mg/L	25	EST	MEYLAN,WM ET AL. (1996)
logP 3.57			EXP	NANDIHALLI,UB ET AL. (1993)
VP 1.14E-009	mm Hg	25	EST	NEELY,WB & BLAU,GE (1985)
DC	pKa			
HL 4.08E-011	atm m3/mol	25	EST	MEYLAN,WM & HOWARD,PH (1991)
OH 2.74E-013	cm3/molc sec	25	EST	MEYLAN,WM & HOWARD,PH (1993)

143502-48-1 — RH-8827 DIPHENYL SULFOXIDE

Formula: $C_{15}H_9ClF_3NO_5S$

Mol Weight: 407.75

MP (deg C): FP (deg C):

BP (deg C):

BP pressure (mm Hg):

Property/Value	Units	Temp	Data Type	Reference
WS 8.53E-001	mg/L	25	EST	MEYLAN,WM ET AL. (1996)
logP 3.65			EXP	NANDIHALLI,UB ET AL. (1993)
VP 2.03E-009	mm Hg	25	EST	NEELY,WB & BLAU,GE (1985)
DC	pKa			
HL 3.24E-013	atm m3/mol	25	EST	MEYLAN,WM & HOWARD,PH (1991)
OH 6.03E-011	cm3/molc sec	25	EST	MEYLAN,WM & HOWARD,PH (1993)

144106-11-6 — PROPANAMINE, N-(1-METHYLETHYL)-2-(1-OXOBUTOXY)-3

Formula: $C_{19}H_{29}NO_4$

Mol Weight: 335.45

MP (deg C): FP (deg C):

BP (deg C):

BP pressure (mm Hg):

Property/Value	Units	Temp	Data Type	Reference
WS 3.27E+001	mg/L	25	EST	MEYLAN,WM ET AL. (1996)
logP 3.95			EXP	JORDAN,CGM ET AL. (1992)
VP 8.47E-007	mm Hg	25	EST	NEELY,WB & BLAU,GE (1985)
DC	pKa			
HL 7.29E-011	atm m3/mol	25	EST	MEYLAN,WM & HOWARD,PH (1991)
OH 1.57E-010	cm3/molc sec	25	EST	MEYLAN,WM & HOWARD,PH (1993)

CAS #: 144106-14-9				PROPANAMINE, N-(1-METHYLETHYL)-2-(2-METHYL-1-OXO
Formula: $C_{19}H_{29}NO_4$				
Mol Weight: 335.45				
MP (deg C):		FP (deg C):		
BP (deg C):				
BP pressure (mm Hg):				

Property/Value	Units	Temp	Data Type	Reference
WS 4.66E+001	mg/L	25	EST	MEYLAN,WM ET AL. (1996)
logP 3.77			EXP	JORDAN,CGM ET AL. (1992)
VP 1.52E-006	mm Hg	25	EST	NEELY,WB & BLAU,GE (1985)
DC	pKa			
HL 7.29E-011	atm m3/mol	25	EST	MEYLAN,WM & HOWARD,PH (1991)
OH 1.56E-010	cm3/molc sec	25	EST	MEYLAN,WM & HOWARD,PH (1993)

CAS #: 144106-15-0				CYCLOPROPANECARBOXYLIC ACID, 2-[2-(2-PROPENYLOXY
Formula: $C_{19}H_{27}NO_4$				
Mol Weight: 333.43				
MP (deg C):		FP (deg C):		
BP (deg C):				
BP pressure (mm Hg):				

Property/Value	Units	Temp	Data Type	Reference
WS 2.13E+002	mg/L	25	EST	MEYLAN,WM ET AL. (1996)
logP 3.01			EXP	JORDAN,CGM ET AL. (1992)
VP 6.39E-007	mm Hg	25	EST	NEELY,WB & BLAU,GE (1985)
DC	pKa			
HL 3.21E-011	atm m3/mol	25	EST	MEYLAN,WM & HOWARD,PH (1991)
OH 1.54E-010	cm3/molc sec	25	EST	MEYLAN,WM & HOWARD,PH (1993)

CAS #: 144458-93-5				BENZENEPROPANAMINE, N,N-BIS(2-CARBONYLETHYL)-4-[
Formula: $C_{20}H_{27}NO_2S$				
Mol Weight: 345.51				
MP (deg C):		FP (deg C):		
BP (deg C):				
BP pressure (mm Hg):				

Property/Value	Units	Temp	Data Type	Reference
WS 1.00E+002	mg/L	25	EST	MEYLAN,WM ET AL. (1996)
logP 3.31			EXP	SANGSTER,J (1994)
VP 1.26E-012	mm Hg	25	EST	NEELY,WB & BLAU,GE (1985)
DC	pKa			
HL 2.88E-014	atm m3/mol	25	EST	MEYLAN,WM & HOWARD,PH (1991)
OH 1.35E-010	cm3/molc sec	25	EST	MEYLAN,WM & HOWARD,PH (1993)

CAS #: 144458-95-7				4-PIPERIDINOL, 1-[[3-(PHENOXYMETHYL)PHENYL]METHY
Formula: $C_{19}H_{23}NO_2$				
Mol Weight: 297.40				
MP (deg C):		FP (deg C):		
BP (deg C):				
BP pressure (mm Hg):				

Property/Value	Units	Temp	Data Type	Reference
WS 5.58E+002	mg/L	25	EST	MEYLAN,WM ET AL. (1996)
logP 2.77			EXP	SANGSTER,J (1994)
VP 1.76E-009	mm Hg	25	EST	NEELY,WB & BLAU,GE (1985)
DC	pKa			
HL 5.31E-013	atm m3/mol	25	EST	MEYLAN,WM & HOWARD,PH (1991)
OH 1.49E-010	cm3/molc sec	25	EST	MEYLAN,WM & HOWARD,PH (1993)

CAS #: 144458-99-1				BENZENEMETHANAMINE, N,N-BIS(2-HYDROXYETHYL)-3-(P
Formula: $C_{18}H_{23}NO_3$				
Mol Weight: 301.39				
MP (deg C):		FP (deg C):		
BP (deg C):				
BP pressure (mm Hg):				

Property/Value	Units	Temp	Data Type	Reference
WS 1.26E+003	mg/L	25	EST	MEYLAN,WM ET AL. (1996)
logP 2.33			EXP	SANGSTER,J (1994)
VP 6.56E-011	mm Hg	25	EST	NEELY,WB & BLAU,GE (1985)
DC	pKa			
HL 3.32E-014	atm m3/mol	25	EST	MEYLAN,WM & HOWARD,PH (1991)
OH 1.44E-010	cm3/molc sec	25	EST	MEYLAN,WM & HOWARD,PH (1993)

CAS #: 144459-03-0				MORPHOLINE, 4-[[3-[[(4-METHOXYPHENYL)METHOXY]MET
Formula: $C_{20}H_{25}NO_3$				
Mol Weight: 327.43				
MP (deg C):		FP (deg C):		
BP (deg C):				
BP pressure (mm Hg):				

Property/Value	Units	Temp	Data Type	Reference
WS 3.30E+002	mg/L	25	EST	MEYLAN,WM ET AL. (1996)
logP 2.83			EXP	SANGSTER,J (1994)
VP 8.20E-008	mm Hg	25	EST	NEELY,WB & BLAU,GE (1985)
DC	pKa			
HL 1.25E-012	atm m3/mol	25	EST	MEYLAN,WM & HOWARD,PH (1991)
OH 2.00E-010	cm3/molc sec	25	EST	MEYLAN,WM & HOWARD,PH (1993)

CAS #: 147064-47-9				9AH-FLUOREN-9A-AMINE, 1,4,4A,9-TETRAHYDRO-N-METH
Formula: $C_{14}H_{17}N$				
Mol Weight: 199.30				
MP (deg C):		FP (deg C):		
BP (deg C):				
BP pressure (mm Hg):				

Property/Value	Units	Temp	Data Type	Reference
WS 7.10E+002	mg/L	25	EST	MEYLAN,WM ET AL. (1996)
logP 3.28			EXP	SANGSTER,J (1994)
VP 7.59E-004	mm Hg	25	EST	NEELY,WB & BLAU,GE (1985)
DC	pKa			
HL 1.04E-006	atm m3/mol	25	EST	MEYLAN,WM & HOWARD,PH (1991)
OH 1.32E-010	cm3/molc sec	25	EST	MEYLAN,WM & HOWARD,PH (1993)

CAS #: 147128-62-9				9AH-FLUOREN-9A-AMINE, 1,2,3,4,4A,9-HEXAHYDRO-, C
Formula: $C_{13}H_{17}N$				
Mol Weight: 187.29				
MP (deg C):		FP (deg C):		
BP (deg C):				
BP pressure (mm Hg):				

Property/Value	Units	Temp	Data Type	Reference
WS 1.59E+003	mg/L	25	EST	MEYLAN,WM ET AL. (1996)
logP 2.94			EXP	SANGSTER,J (1994)
VP 1.08E-003	mm Hg	25	EST	NEELY,WB & BLAU,GE (1985)
DC	pKa			
HL 5.41E-007	atm m3/mol	25	EST	MEYLAN,WM & HOWARD,PH (1991)
OH 3.46E-011	cm3/molc sec	25	EST	MEYLAN,WM & HOWARD,PH (1993)

CAS #: 147267-64-9				1(2H)-PYRIDINEACETAMIDE, 3-AMINO-2-OXO-6-PHENYL-
Formula: $C_{19}H_{20}F_3N_3O_3$				
Mol Weight: 395.38				
MP (deg C):		FP (deg C):		
BP (deg C):				
BP pressure (mm Hg):				

Property/Value	Units	Temp	Data Type	Reference
WS 1.07E+002	mg/L	25	EST	MEYLAN,WM ET AL. (1996)
logP 1.74			EXP	SANGSTER,J (1994)
VP 1.03E-011	mm Hg	25	EST	NEELY,WB & BLAU,GE (1985)
DC	pKa			
HL 1.25E-014	atm m3/mol	25	EST	MEYLAN,WM & HOWARD,PH (1991)
OH 4.64E-011	cm3/molc sec	25	EST	MEYLAN,WM & HOWARD,PH (1993)

CAS #: 148727-67-7				4A(4H)-PHENANTHRENAMINE, 1,9,10,10A-TETRAHYDRO-9
Formula: $C_{15}H_{19}N$				
Mol Weight: 213.33				
MP (deg C):		FP (deg C):		
BP (deg C):				
BP pressure (mm Hg):				

Property/Value	Units	Temp	Data Type	Reference
WS 4.75E+002	mg/L	25	EST	MEYLAN,WM ET AL. (1996)
logP 3.40			EXP	SANGSTER,J (1994)
VP 1.67E-004	mm Hg	25	EST	NEELY,WB & BLAU,GE (1985)
DC	pKa			
HL 8.39E-007	atm m3/mol	25	EST	MEYLAN,WM & HOWARD,PH (1991)
OH 9.28E-011	cm3/molc sec	25	EST	MEYLAN,WM & HOWARD,PH (1993)

CAS #: 148727-74-6				9(1H)-PHENANTHRENONE, 2,3,4,4A,10,10A-HEXAHYDRO-
Formula: $C_{15}H_{19}NO$				
Mol Weight: 229.32				
MP (deg C):		FP (deg C):		
BP (deg C):				
BP pressure (mm Hg):				

Property/Value	Units	Temp	Data Type	Reference
WS 2.73E+003	mg/L	25	EST	MEYLAN,WM ET AL. (1996)
logP 2.41			EXP	SANGSTER,J (1994)
VP 2.18E-005	mm Hg	25	EST	NEELY,WB & BLAU,GE (1985)
DC	pKa			
HL 1.96E-009	atm m3/mol	25	EST	MEYLAN,WM & HOWARD,PH (1991)
OH 8.61E-011	cm3/molc sec	25	EST	MEYLAN,WM & HOWARD,PH (1993)

CAS #: 149589-40-2				BENZAMIDE, N-METHYL-3-PHENOXY-N-PROPYL-
Formula: $C_{17}H_{19}NO_2$				
Mol Weight: 269.35				
MP (deg C):		FP (deg C):		
BP (deg C):				
BP pressure (mm Hg):				

Property/Value	Units	Temp	Data Type	Reference
WS 2.13E+001	mg/L	25	EST	MEYLAN,WM ET AL. (1996)
logP 3.44			EXP	SANGSTER,J (1994)
VP 6.85E-007	mm Hg	25	EST	NEELY,WB & BLAU,GE (1985)
DC	pKa			
HL 4.11E-010	atm m3/mol	25	EST	MEYLAN,WM & HOWARD,PH (1991)
OH 2.86E-011	cm3/molc sec	25	EST	MEYLAN,WM & HOWARD,PH (1993)

CAS #: 149589-41-3				BENZAMIDE, N-BUTYL-N-METHYL-3-PHENOXY-
Formula: $C_{18}H_{21}NO_2$				
Mol Weight: 283.37				
MP (deg C):		FP (deg C):		
BP (deg C):				
BP pressure (mm Hg):				

Property/Value	Units	Temp	Data Type	Reference
WS 7.30E+000	mg/L	25	EST	MEYLAN,WM ET AL. (1996)
logP 3.89			EXP	SANGSTER,J (1994)
VP 2.92E-007	mm Hg	25	EST	NEELY,WB & BLAU,GE (1985)
DC	pKa			
HL 5.46E-010	atm m3/mol	25	EST	MEYLAN,WM & HOWARD,PH (1991)
OH 3.00E-011	cm3/molc sec	25	EST	MEYLAN,WM & HOWARD,PH (1993)

CAS #: 149589-42-4				BENZAMIDE, N-METHYL-N-PENTYL-3-PHENOXY-
Formula: $C_{19}H_{23}NO_2$				
Mol Weight: 297.40				
MP (deg C):		FP (deg C):		
BP (deg C):				
BP pressure (mm Hg):				

Property/Value	Units	Temp	Data Type	Reference
WS 2.09E+000	mg/L	25	EST	MEYLAN,WM ET AL. (1996)
logP 4.43			EXP	SANGSTER,J (1994)
VP 1.24E-007	mm Hg	25	EST	NEELY,WB & BLAU,GE (1985)
DC	pKa			
HL 7.25E-010	atm m3/mol	25	EST	MEYLAN,WM & HOWARD,PH (1991)
OH 3.14E-011	cm3/molc sec	25	EST	MEYLAN,WM & HOWARD,PH (1993)

CAS #: 149589-43-5				BENZAMIDE, N-HEXYL-N-METHYL-3-PHENOXY-
Formula: $C_{20}H_{25}NO_2$				
Mol Weight: 311.43				
MP (deg C):		FP (deg C):		
BP (deg C):				
BP pressure (mm Hg):				

Property/Value	Units	Temp	Data Type	Reference
WS 5.54E-001	mg/L	25	EST	MEYLAN,WM ET AL. (1996)
logP 5.01			EXP	SANGSTER,J (1994)
VP 5.23E-008	mm Hg	25	EST	NEELY,WB & BLAU,GE (1985)
DC	pKa			
HL 9.62E-010	atm m3/mol	25	EST	MEYLAN,WM & HOWARD,PH (1991)
OH 3.28E-011	cm3/molc sec	25	EST	MEYLAN,WM & HOWARD,PH (1993)

CAS #: 150423-89-5				TETRADECANAMIDE, 2,2-DIMETHYL-N-METHYL-3-OXO-N-(
Formula: $C_{26}H_{43}NO_5$				
Mol Weight: 449.64				
MP (deg C):		FP (deg C):		
BP (deg C):				
BP pressure (mm Hg):				

Property/Value	Units	Temp	Data Type	Reference
WS 3.02E-003	mg/L	25	EST	MEYLAN,WM ET AL. (1996)
logP 6.67			EXP	SANGSTER,J (1993)
VP 5.99E-011	mm Hg	25	EST	NEELY,WB & BLAU,GE (1985)
DC	pKa			
HL 5.28E-013	atm m3/mol	25	EST	MEYLAN,WM & HOWARD,PH (1991)
OH 2.21E-010	cm3/molc sec	25	EST	MEYLAN,WM & HOWARD,PH (1993)

APPENDIX I

CBr2F2	000075-61-6	DIBROMODIFLUOROMETHANE
CBr4	000558-13-4	CARBON TETRABROMIDE
CBrCl3	000075-62-7	BROMOTRICHLOROMETHANE
CBrClF2	000353-59-3	BROMOCHLORODIFLUOROMETHANE
CBrF3	000075-63-8	BROMOTRIFLUOROMETHANE
CCl2F2	000075-71-8	DICHLORODIFLUOROMETHANE
CCl2O	000075-44-5	PHOSGENE
CCl3F	000075-69-4	TRICHLOROFLUOROMETHANE
CCl3NO2	000076-06-2	TRICHLORONITROMETHANE
CCl4	000056-23-5	CARBON TETRACHLORIDE
CCl4S	000594-42-3	PERCHLOROMETHYL MERCAPTAN
CClF3	000075-72-9	CHLOROTRIFLUOROMETHANE
CF2O	000353-50-4	CARBONIC DIFLUORIDE
CF3NO2	000335-02-4	TRIFLUORONITROMETHANE
CF4	000075-73-0	TETRAFLUOROMETHANE
CHBr2Cl	000124-48-1	DIBROMOCHLOROMETHANE
CHBr3	000075-25-2	BROMOFORM
CHBrCl2	000075-27-4	BROMODICHLOROMETHANE
CHCl2F	000075-43-4	DICHLOROFLUOROMETHANE
CHCl3	000067-66-3	CHLOROFORM
CHClF2	000075-45-6	CHLORODIFLUOROMETHANE
CHF3	000075-46-7	TRIFLUOROMETHANE
CHFO	001493-02-3	FORMYL FLUORIDE
CHI3	000075-47-8	IODOFORM
CH2Br2	000074-95-3	DIBROMOMETHANE
CH2BrCl	000074-97-5	CHLOROBROMOMETHANE
CH2Cl2	000075-09-2	DICHLOROMETHANE
CH2ClF	000593-70-4	CHLOROFLUOROMETHANE
CH2ClI	000593-71-5	CHLOROIODOMETHANE
CH2F2	000075-10-5	DIFLUOROMETHANE
CH2I2	000075-11-6	METHYLENE IODIDE
CH2N2	000334-88-3	DIAZOMETHANE
CH2N4	000288-94-8	1H-TETRAZOLE
CH2O	000050-00-0	FORMALDEHYDE
CH2O	030525-89-4	PARAFORMALDEHYDE
CH2O2	000064-18-6	FORMIC ACID
CH3AsNa2O3	000144-21-8	DSMA
CH3Br	000074-83-9	METHYL BROMIDE
CH3Cl	000074-87-3	METHYL CHLORIDE
CH3Cl3Si	000075-79-6	METHYLTRICHLOROSILANE
CH3ClO2S	000124-63-0	METHYLSULFONYL CHLORIDE
CH3F	000593-53-3	FLUOROMETHANE
CH3FO2S	000558-25-8	FLUOROMETHYL SULFONE
CH3I	000074-88-4	METHYL IODIDE
CH3N5	004418-61-5	5-AMINO-1H-TETRAZOLE
CH3NO	000075-12-7	FORMAMIDE
CH3NO	000075-17-2	FORMAMIDE OXIME
CH3NO2	000075-52-5	NITROMETHANE
CH3NO2	000624-91-9	METHYLNITRITE
CH3NO3	000598-58-3	METHYLNITRATE
CH4	000074-82-8	METHANE
CH4Cl2Si	000075-54-7	DICHLOROMETHYLSILANE
CH4N2O	000057-13-6	UREA
CH4N2O	000624-84-0	N-FORMYLHYDRAZINE
CH4N2O2	000127-07-1	HYDROXYUREA
CH4N2S	000062-56-6	THIOUREA
CH4N4O2	000556-88-7	NITROGUANIDINE

CH4O	000067-56-1	METHANOL
CH4O2	003031-73-0	METHYLHYDROPEROXIDE
CH4O3S	000075-75-2	METHANESULFONIC ACID
CH4O6S2	000503-40-2	METHANEDISULFONIC ACID
CH4S	000074-93-1	METHANETHIOL
CH5N	000074-89-5	METHYLAMINE
CH5N3	000113-00-8	GUANIDINE
CH5N3S	000079-19-6	THIOSEMICARBAZIDE
CH5NO	000067-62-9	O-METHYLHYDROXYLAMINE
CH5NO	000593-77-1	N-METHYLHYDROXYLAMINE
CH5O2P	004206-94-4	METHYLPHOSPHINIC ACID
CH5O3P	000993-13-5	METHYLPHOPHONIC ACID
CH6ClN	000593-51-1	METHANAMINE, HYDROCHLORIDE
CH6ClN3O	000563-41-7	SEMICARBAZIDE
CH6N2	000060-34-4	METHYL HYDRAZINE
C2Br2F4	000124-73-2	1,2-DIBROMOTETRAFLUOROETHANE
C2BrF3	000598-73-2	2-BROMO-1,1,2-TRIFLUOROETHYLENE
C2Cl2F2	000079-35-6	1,1-DICHLORO-2,2-DIFLUOROETHYLENE
C2Cl2F2	000598-88-9	1,2-DICHLORO-1,2-DIFLUOROETHYLENE
C2Cl2F4	000076-14-2	1,2-DICHLOROTETRAFLUOROETHANE
C2Cl2F4	000374-07-2	1,1-DICHLORO-1,2,2,2-TETRAFLUOROETHANE
C2Cl3F3	000076-13-1	1,1,2-TRICHLOROTRIFLUOROETHANE
C2Cl3F3	000354-58-5	1,1,1-TRICHLORO-2,2,2-TRIFLUOROETHANE
C2Cl3N	000545-06-2	TRICHLOROACETONITRILE
C2Cl3NaO2	000650-51-1	SODIUM TRICHLOROACETATE
C2Cl4	000127-18-4	TETRACHLOROETHENE
C2Cl4F2	000076-11-9	1,1,1,2-TETRACHLORO-2,2-DIFLUOROETHANE
C2Cl4F2	000076-12-0	1,1,2,2,-TETRACHLORODIFLUOROETHANE
C2Cl4O	000076-02-8	TRICHLORACETYLCHLORIDE
C2Cl4O	016650-10-5	TETRACHLOROETHYLENE OXIDE
C2Cl6	000067-72-1	HEXACHLOROETHANE
C2ClF3	000079-38-9	CHLOROTRIFLUOROETHYLENE
C2ClF3	000359-29-5	CHLOROTRIFLUOROETHENE
C2ClF5	000076-15-3	CHLOROPENTAFLUOROETHANE
C2F4	000116-14-3	TETRAFLUOROETHYLENE
C2F6	000076-16-4	HEXAFLUOROETHANE
C2HBr2N	003252-43-5	DIBROMOACETONITRILE
C2HBr2N3	007411-23-6	1,2,4-TRIAZOLE,3,5-DIBROMO
C2HBr3	000598-16-3	ETHENE, TRIBROMO-
C2HBr3O	000115-17-3	TRIBROMOACETALDEHYDE
C2HBrClF3	000151-67-7	HALOTHANE
C2HCl2F3	000306-83-2	1,1,1-TRIFLUORO-2,2-DICHLOROETHANE
C2HCl2N	003018-12-0	DICHLOROACETONITRILE
C2HCl3	000079-01-6	TRICHLOROETHENE
C2HCl3O	000075-87-6	TRICHLOROACETALDEHYDE
C2HCl3O	000079-36-7	DICHLOROACETYL CHLORIDE
C2HCl3O	016967-79-6	TRICHLOROOXIRANE
C2HCl3O2	000076-03-9	TRICHLOROACETIC ACID
C2HCl5	000076-01-7	PENTACHLOROETHANE
C2HClF2	000359-10-4	CHLORO-1,1-DIFLUOROETHYLENE
C2HF3O2	000076-05-1	TRIFLUOROACETATE
C2HF5	000354-33-6	1,1,1,2,2-PENTAFLUOROETHANE
C2H2	000074-86-2	ACETYLENE
C2H2AsCl3	000541-25-3	DICHLORO(2-CHLOROVINYL)ARSINE (TRANS)
C2H2Br2	025429-23-6	1,2-DIBROMOETHENE
C2H2Br4	000079-27-6	1,1,2,2-TETRABROMOETHANE
C2H2Cl2	000075-35-4	1,1-DICHLOROETHENE
C2H2Cl2	000156-59-2	1,2-DICHLOROETHENE (CIS)

C2H2Cl2	000156-60-5	1,2-DICHLOROETHENE (TRANS)
C2H2Cl2	000540-59-0	1,2-DICHLOROETHYLENE
C2H2Cl2F2	001649-08-7	1,2-DICHLORO-1,1-DIFLUOROETHANE
C2H2Cl2O	000079-04-9	CHLOROACETYLCHLORIDE
C2H2Cl2O2	000079-43-6	DICHLOROACETIC ACID
C2H2Cl3NO	000594-65-0	TRICHLOROACETAMIDE
C2H2Cl4	000079-34-5	1,1,2,2-TETRACHLOROETHANE
C2H2Cl4	000630-20-6	1,1,1,2-TETRACHLOROETHANE
C2H2ClF3	000075-88-7	1,1,1-TRIFLUORO-2-CHLOROETHANE
C2H2ClN	000107-14-2	CHLOROACETONITRILE
C2H2F2	000075-38-7	1,1-DIFLUOROETHENE
C2H2F3NO	000354-38-1	TRIFLUOROACETAMIDE
C2H2F4	000811-97-2	1,1,1,2-TETRAFLUOROETHANE
C2H2FNaO2	000062-74-8	FLUOROACETIC ACID, SODIUM SALT
C2H2I2	000590-26-1	1,2-DIIODOETHENE (CIS)
C2H2I2	000590-27-2	1,2-DIIODOETHENE (TRANS)
C2H2N2O	000288-37-9	FURAZAN
C2H2O	000463-51-4	KETENE
C2H2O2	000107-22-2	GLYOXAL
C2H2O3	000298-12-4	GLYOXYLIC ACID
C2H2O4	000144-62-7	OXALIC ACID
C2H3Br	000593-60-2	BROMOETHENE
C2H3Br2Cl	027949-36-6	1,1-DIBROMO-2-CHLOROETHANE
C2H3Br3O	000075-80-9	ETHANOL, 2,2,2-TRIBROMO-
C2H3BrO	000506-96-7	ACETYL BROMIDE
C2H3BrO2	000079-08-3	BROMOACETIC ACID
C2H3Cl	000075-01-4	VINYLCHLORIDE
C2H3Cl2F	001717-00-6	1,1-DICHLOROFLUOROETHANE
C2H3Cl2NO	000683-72-7	ACETAMIDE, 2,2-DICHLORO-
C2H3Cl2NO2	000594-72-9	1,1-DICHLORO-1-NITROETHANE
C2H3Cl3	000071-55-6	1,1,1-TRICHLOROETHANE
C2H3Cl3	000079-00-5	1,1,2-TRICHLOROETHANE
C2H3Cl3O	000115-20-8	2,2,2-TRICHLOROETHANOL
C2H3Cl3O2	000302-17-0	CHORAL HYDRATE
C2H3Cl3Si	000075-94-5	VINYLTRICHLOROSILANE
C2H3ClF2	000075-68-3	1-CHLORO-1,1-DIFLUOROETHANE
C2H3ClO	000075-36-5	ACETYLCHLORIDE
C2H3ClO	000107-20-0	CHLOROACETALDEHYDE
C2H3ClO2	000079-11-8	CHLOROACETIC ACID
C2H3ClO2	000079-22-1	METHYL CHLOROFORMATE
C2H3F	000075-02-5	VINYLFLUORIDE
C2H3F3	000420-46-2	1,1,1-TRIFLUOROETHANE
C2H3F3	000430-66-0	1,1,2-TRIFLUOROETHANE
C2H3F3O	000075-89-8	2,2,2-TRIFLUOROETHANOL
C2H3FO2	000144-49-0	FLUOROACETIC ACID
C2H3HgN	002597-97-9	CHIPCOTE
C2H3IO	000507-02-8	ACETYL IODIDE
C2H3N	000075-05-8	ACETONITRILE
C2H3N	000593-75-9	METHYL ISOCYANIDE
C2H3N3	000288-36-8	2H-1,2,3-TRIAZOLE
C2H3N3	000288-88-0	1H-1,2,4-TRIAZOLE
C2H3NaO2	000127-09-3	ACETIC ACID, SODIUM SALT
C2H3NaO3	002836-32-0	GLYCOLIC ACID, SODIUM SALT
C2H3NO	000624-83-9	METHYLISOCYANATE
C2H3NO3	000471-47-6	OXAMIC ACID
C2H3NO5	002278-22-0	PEROXYACETYLNITRATE
C2H3NS	000556-61-6	METHYL ISOTHIOCYANATE
C2H3NS	000556-64-9	METHYL THIOCYANATE

C2H4	000074-85-1	ETHYLENE
C2H4Br2	000106-93-4	1,2-DIBROMOETHANE
C2H4Br2	000557-91-5	1,1-DIBROMOETHANE
C2H4BrCl	000107-04-0	1-BROMO-2-CHLOROETHANE
C2H4BrCl	000593-96-4	1-BROMO-1-CHLOROETHANE
C2H4BrNO	000683-57-8	BROMOACETAMIDE
C2H4Cl2	000075-34-3	1,1-DICHLOROETHANE
C2H4Cl2	000107-06-2	1,2-DICHLOROETHANE
C2H4Cl2O	000542-88-1	BIS(CHLOROMETHYL) ETHER
C2H4Cl2O	000598-38-9	2,2-DICHLOROETHANOL
C2H4Cl3O4P	000306-52-5	TRICHLOROETHYL PHOSPHATE
C2H4ClNO	000079-07-2	CHLOROACETAMIDE
C2H4ClNO2	000598-92-5	ETHANE, 1-CHLORO-1-NITRO-
C2H4F2	000075-37-6	1,1-DIFLUOROETHANE
C2H4F2	000624-72-6	1,2-DIFLUOROETHANE
C2H4F3N	000753-90-2	2,2,2-TRIFLUOROETHYLAMINE
C2H4FNO	000640-19-7	FLUOROACETAMIDE
C2H4I2	000624-73-7	ETHANE, 1,2-DIIODO-
C2H4INO	000144-48-9	IODOACETAMIDE
C2H4N2	000540-61-4	AMINOACETONITRILE
C2H4N2O2	000471-46-5	OXAMIDE
C2H4N2O2	000628-36-4	DIFORMYLHYDRAZINE
C2H4N2O2	009011-05-6	UREA, POLYMER WITH FORMALDEHYDE
C2H4N2O6	000628-96-6	ETHYLENE GLYCOL DINITRATE
C2H4N2S2	000079-40-3	RUBEANIC ACID
C2H4N3O	003775-60-8	2-AMINO-1,3,4-OXADIAZOLE
C2H4N4	000061-82-5	AMITROLE
C2H4N4	000461-58-5	2-CYANOGUANIDINE
C2H4N4O	017220-38-1	FURAZANDIAMINE
C2H4NNaS2	000137-42-8	METHYLDITHIOCARBAMIC ACID, NA SALT
C2H4O	000075-07-0	ACETALDEHYDE
C2H4O	000075-21-8	ETHYLENE OXIDE
C2H4O2	000064-19-7	ACETIC ACID
C2H4O2	000107-31-3	METHYLFORMATE
C2H4O2	000141-46-8	GLYCOLALDEHYDE
C2H4O2S	000068-11-1	MERCAPTOACETIC ACID
C2H4O3	000079-14-1	HYDROXYACETIC ACID
C2H4O3	000079-21-0	PEROXYACETIC ACID
C2H4S	000420-12-2	ETHYLENE SULFIDE
C2H4S3	000289-16-7	1,2,4-TRITHIOLANE
C2H5Br	000074-96-4	BROMOETHANE
C2H5BrO	000540-51-2	2-BROMOETHANOL
C2H5Cl	000075-00-3	CHLOROETHANE
C2H5Cl3Si	000115-21-9	ETHYLTRICHLOROSILANE
C2H5ClHg	000107-27-7	ETHYLMERCURY CHLORIDE
C2H5ClO	000107-07-3	2-CHLOROETHANOL
C2H5ClO	000107-30-2	CHLOROMETHOXYMETHANE
C2H5F	000353-36-6	FLUOROETHANE
C2H5FO	000371-62-0	2-FLUOROETHANOL
C2H5I	000075-03-6	IODOETHANE
C2H5N	000151-56-4	AZIRIDINE
C2H5N3O2	000684-93-5	1-NITROSO-1-METHYLUREA
C2H5N3S	004005-51-0	2-AMINO-1,3,4-THIADIAZOLE
C2H5N5	001455-77-2	1H-1,2,4-TRIAZOLE-3,5-DIAMINE
C2H5N5O3	000070-25-7	N-NITROSO-N-METHYL-N'-NITROGUANIDINE
C2H5NO	000060-35-5	ACETAMIDE
C2H5NO	000107-29-9	ACETALDEHYDE OXIME
C2H5NO	000123-39-7	METHYLFORMAMIDE

C2H5NO2	000056-40-6	GLYCINE
C2H5NO2	000079-24-3	NITROETHANE
C2H5NO2	000109-95-5	ETHYLNITRITE
C2H5NO2	000546-88-3	ACETAMIDE, N-HYDROXY-
C2H5NO2	000598-55-0	O-METHYL CARBAMATE
C2H5NO3	000625-48-9	2-NITROETHANOL
C2H5NO3	000625-58-1	ETHYLNITRATE
C2H5NS	000062-55-5	THIOACETAMIDE
C2H5O5P	004408-78-0	PHOSPHONOACETIC ACID
C2H6	000074-84-0	ETHANE
C2H6Cl2Si	000075-78-5	DICHLORODIMETHYLSILANE
C2H6ClO2PS	002524-03-0	O,O-DIMETHYL PHOSPHOROCHLORIDOTHIOATE
C2H6ClO3P	011672-87-0	ETHEPHON
C2H6ClO3P	016672-87-0	ETHEPHON
C2H6Hg	000593-74-8	DIMETHYL MERCURY
C2H6N2O	000062-75-9	N-NITROSODIMETHYLAMINE
C2H6N2O	000598-50-5	UREA, METHYL-
C2H6N2O	001068-57-1	ACETIC ACID HYDRAZIDE
C2H6N2O2	004164-28-7	DIMETHYLNITRAMINE
C2H6N2O2	007433-43-4	1-METHYLHYDROXYUREA
C2H6N2S	000598-52-7	N-METHYLTHIOUREA
C2H6N4O2S2	014949-00-9	2-NH2-1,3,4-THIADIAZOLE-SO2NH2
C2H6O	000064-17-5	ETHANOL
C2H6O	000115-10-6	DIMETHYL ETHER
C2H6O2	000107-21-1	ETHYLENE GLYCOL
C2H6O2S	000067-71-0	DIMETHYL SULFONE
C2H6O3S	000066-27-3	METHYL METHANE SULFONATE
C2H6O4S	000077-78-1	DIMETHYLSULFATE
C2H6OS	000067-68-5	DIMETHYL SULFOXIDE
C2H6S	000075-08-1	ETHYL MERCAPTAN
C2H6S	000075-18-3	DIMETHYLSULFIDE
C2H6S2	000540-63-6	1,2-ETHANEDITHIOL
C2H6S2	000624-92-0	DIMETHYLDISULFIDE
C2H7AsO2	000075-60-5	CACODYLIC ACID
C2H7HgO4P	002235-25-8	ETHYL MERCURIC PHOSPHATE
C2H7N	000075-04-7	ETHYLAMINE
C2H7N	000124-40-3	DIMETHYLAMINE
C2H7N3	000471-29-4	METHYLGUANIDINE
C2H7N3O	017696-95-6	4-METHYLSEMICARBAZIDE
C2H7NO	000141-43-5	ETHANOLAMINE
C2H7NO	001117-97-1	O,N-DIMETHYLHYDROXYLAMINE
C2H7NO	005725-96-2	METHANAMINE, N-HYDROXY-N-METHYL-
C2H7NO	016645-06-0	N,N-DIMETHYLHYDROXYLAMINE
C2H7NO3S	000107-35-7	TAURINE
C2H7O3P	000868-85-9	DIMETHYL PHOSPHITE
C2H8ClN	000506-59-2	N-METHYLMETHANAMINE, HYDROCHLORIDE
C2H8ClN	000557-66-4	ETHYLAMINE, HYDROCHLORIDE
C2H8N2	000057-14-7	1,1-DIMETHYLHYDRAZINE
C2H8N2	000107-15-3	1,2-DIAMINOETHANE
C2H8N2	000540-73-8	1,2-DIMETHYLHYDRAZINE
C2H8NO2PS	010265-92-6	METHAMIDPHOS
C2H8NO2PS	017321-47-0	PHOSPHORAMIDOTHIOIC ACID, O,O-DIMETHYL ESTER
C3Cl3N3O3	000087-90-1	1,3,5-TRICHLOROISOCYANURIC ACID
C3Cl4	018608-30-5	PERCHLOROALLENE
C3Cl6	001888-71-7	PERCHLOROPROPYLENE
C3Cl6O	000116-16-5	HEXACHLOROACETONE
C3Cl8	000594-90-1	PERCHLOROPROPANE
C3ClF5O	000079-53-8	CHLOROPENTAFLUOROACETONE

C3F6	000116-15-4	HEXAFLUOROPROPENE
C3F6O	000684-16-2	HEXAFLUORO-2-PROPANONE
C3F6O	010057-27-9	HEXAFLUOROACETONE
C3F6O	034202-69-2	HEXAFLUOROACETONE TRIHYDRATE
C3F8	000076-19-7	OCTAFLUOROPROPANE
C3HBr3N2	002034-22-2	2,4,5-TRIBROMOIMIDAZOLE
C3HCl3N2	007682-38-4	2,4,5-TRICHLOROIMIDAZOLE
C3HCl3O2	002257-35-4	2,3,3-TRICHLORO-2-PROPENOIC ACID
C3HI3N2	001746-25-4	2,4,5-TRIIODOIMIDAZOLE
C3H2Br2N2O	010222-01-2	2,2-DIBROMO-2-CYANOACETAMIDE
C3H2BrN3O2	006963-65-1	4-BROMO-5-NITRO-1H-IMIDAZOLE
C3H2BrN3O2	065902-59-2	2-BROMO-4-NITRO-1H-IMIDAZOLE
C3H2Cl2	025523-14-2	3,3-DICHLOROPROPYNE
C3H2Cl2	083682-32-0	1,3-DICHLOROALLENE
C3H2Cl6	003607-78-1	1,1,1,3,3,3-HEXACHLOROPROPANE
C3H2ClF5O	013838-16-9	ETHRANE [HALOETHER]
C3H2ClF5O	026675-46-7	ISOFLURANE
C3H2ClN3O2	006963-66-2	1H-IMIDAZOLE, 5-CHLORO-4-NITRO-
C3H2ClN3O2	057531-37-0	2-CHLORO-4-NITRO-1H-IMIDAZOLE
C3H2ClN3O2	057531-38-1	4-CHLORO-5-NITRO-1H-IMIDAZOLE
C3H2F6O	000920-66-1	1,1,1,3,3,3-HEXAFLUOROPROPAN-2-OL
C3H2IN3O2	076529-48-1	1H-IMIDAZOLE, 4-IODO-5-NITRO-
C3H2N2	000109-77-3	MALONONITRILE
C3H2N4O4	005213-49-0	2,4-DINITRO-1H-IMIDAZOLE
C3H2N4O4	019183-14-3	4,5-DINITROIMIDAZOLE
C3H3Cl	003223-70-9	CHLOROALLENE
C3H3Cl	007747-84-4	1-CHLOROPROPYNE
C3H3Cl2NaO2	000127-20-8	DALAPON (SODIUM SALT)
C3H3Cl3	000096-19-5	1,2,3-TRICHLOROPROPENE
C3H3Cl3	002567-14-8	1,1,3-TRICHLOROPROPENE
C3H3Cl3	021400-25-9	1,1,2-TRICHLOROPROPENE
C3H3Cl3	037077-84-2	2,3,3-TRICHLOROPROPENE
C3H3Cl3O	000918-00-3	1,1,1-TRICHLOROACETONE
C3H3Cl3O2	000598-99-2	METHYL TRICHLOROACETATE
C3H3Cl3O2	003278-46-4	2,2,3-TRICHLOROPROPIONIC ACID
C3H3Cl5	015104-61-7	1,1,2,3,3-PENTACHLOROPROPANE
C3H3Cl5F4N3OP3	059700-57-1	PENTACHLORO(2,2,3,3-TETRAFLUOROPROPOXY)CYCLOTRI*
C3H3ClN2O2	065514-04-7	FURAZAN, CHLOROMETHYL-, 2-OXIDE
C3H3ClN2O2	086988-90-1	FURAZAN, CHLOROMETHYL-, 5-OXIDE
C3H3F3	000677-21-4	3,3,3-TRIFLUORO-1-PROPENE
C3H3F3O	000421-50-1	1,1,1-TRIFLUOROACETONE
C3H3F3O2	000431-47-0	METHYLTRIFLUOROACETATE
C3H3F5	001814-88-6	1,1,1,2,2-PENTAFLUOROPROPANE
C3H3F5O	000422-05-9	PENTAFLUORO-1-PROPANOL
C3H3IN2	003469-69-0	4-IODOPYRAZOLE
C3H3N	000107-13-1	ACRYLONITRILE
C3H3N3	000290-87-9	SYM-TRIAZINE
C3H3N3O2	000071-33-0	5-AZAURACIL
C3H3N3O2	000461-89-2	1,2,4-TRIAZINE-3,5(2H,4H)-DIONE
C3H3N3O2	000527-73-1	2-NITROIMIDAZOLE
C3H3N3O2	002075-46-9	4-NITROPYRAZOLE
C3H3N3O2	003034-38-6	4-NITROIMIDAZOLE
C3H3N3O2	088054-21-1	1H-IMIDAZOLE, 5-NITRO-
C3H3N3O2S	000121-66-4	2-THIAZOLAMINE, 5-NITRO-
C3H3N3O3	000108-80-5	CYANURIC ACID
C3H3N3O3	077666-53-6	FURAZAN, METHYLNITRO
C3H3N3O4	049558-02-3	FURAZAN, METHYLNITRO-, 2-OXIDE
C3H3NO	000288-14-2	ISO-OXAZOLE

C3H3NO	000288-42-6	OXAZOLE
C3H3NO2	000372-09-8	CYANOACETIC ACID
C3H3NS	000288-47-1	THIAZOLE
C3H4	000074-99-7	PROPYNE
C3H4	000463-49-0	1,2-PROPADIENE
C3H4Br2	000513-31-5	2,3-DIBROMO-1-PROPENE
C3H4BrCl	003737-00-6	CHLOROBROMOPROPENE
C3H4Cl2	000078-88-6	2,3-DICHLOROPROPENE
C3H4Cl2	000542-75-6	TRANS-1,3-DICHLOROPROPENE
C3H4Cl2	000563-54-2	1,2-DICHLOROPROPENE
C3H4Cl2	000563-57-5	3,3-DICHLOROPROPENE
C3H4Cl2	000563-58-6	1,1-DICHLOROPROPENE
C3H4Cl2	010061-01-5	1,3-DICHLOROPROPENE, (Z)
C3H4Cl2	010061-02-6	1,3-DICHLOROPROPENE (TRANS)
C3H4Cl2F2O	000076-38-0	METHOXYFLURANE
C3H4Cl2O	000513-88-2	1,1-DICHLOROPROPANONE
C3H4Cl2O	000534-07-6	1,3-DICHLOROACETONE
C3H4Cl2O2	000075-99-0	DALAPON
C3H4Cl2O2	000116-54-1	METHYL DICHLOROACETATE
C3H4Cl2O2	000565-64-0	2,3-DICHLOROPROPIONIC ACID
C3H4Cl2O2	000627-11-2	CHLOROETHYL CHLOROFORMATE
C3H4Cl4	013116-53-5	PROPANE, 1,2,2,3-TETRACHLORO-
C3H4Cl4	018495-30-2	1,1,2,3-TETRACHLOROPROPANE
C3H4ClN	000542-76-7	3-CHLOROPROPIONITRILE
C3H4ClN	001617-17-0	PROPANENITRILE, 2-CHLORO-
C3H4F4O	000076-37-9	2,2,3,3-TETRAFLUOROPROPANOL
C3H4N2	000288-13-1	PYRAZOLE
C3H4N2	000288-32-4	IMIDAZOLE
C3H4N2O2	000461-72-3	HYDANTOIN
C3H4N2S	000096-50-4	2-AMINOTHIAZOLE
C3H4O	000107-02-8	ACROLEIN
C3H4O	000107-19-7	PROPARGYL ALCOHOL
C3H4O	006004-44-0	1-PROPEN-1-ONE(METHYLKETENE)
C3H4O2	000057-57-8	BETA-PROPIOLACTONE
C3H4O2	000078-98-8	METHYL GLYOXAL
C3H4O2	000079-10-7	ACRYLIC ACID
C3H4O2	000692-45-5	VINYL FORMATE
C3H4O2	000765-34-4	GLYCIDYLALDEHYDE
C3H4O3	000096-49-1	1,3-DIOXOLAN-2-ONE
C3H4O3	000127-17-3	PYRUVIC ACID
C3H4O4	000141-82-2	MALONIC ACID
C3H5Br	000106-95-6	3-BROMO-1-PROPENE
C3H5Br2Cl	000096-12-8	1,2-DIBROMO-3-CHLOROPROPANE
C3H5BrO	000598-31-2	1-BROMO-2-PROPANONE
C3H5BrO	003132-64-7	ALPHA-EPIBROMOHYDRIN
C3H5BrO2	000598-72-1	A-BROMOPROPIONIC ACID
C3H5Cl	000107-05-1	3-CHLOROPROPYLENE
C3H5Cl	000557-98-2	2-CHLOROPROPENE
C3H5Cl	000590-21-6	1-CHLOROPROPENE
C3H5Cl2NO2	013698-16-3	ETHYL DICHLOROCARBAMATE
C3H5Cl3	000096-18-4	1,2,3-TRICHLOROPROPANE
C3H5Cl3	000598-77-6	1,1,2-TRICHLOROPROPANE
C3H5Cl3	003175-23-3	1,2,2-TRICHLOROPROPANE
C3H5Cl3	007789-89-1	1,1,1-TRICHLOROPROPANE
C3H5Cl3Si	000107-37-9	ALLYTRICHLOROSILANE
C3H5ClO	000078-95-5	CHLOROACETONE
C3H5ClO	000106-89-8	1-CHLORO-2,3-EPOXYPROPANE
C3H5ClO2	000096-34-4	METHYL CHLOROACETATE

C3H5ClO2	000107-94-8	3-CHLOROPROPIONIC ACID
C3H5ClO2	000541-41-3	ETHYL CHLOROCARBONATE
C3H5ClO2	000598-78-7	2-CHLOROPROPIONIC ACID
C3H5F3O	000374-01-6	1,1,1-TRIFLUORO-2-PROPANOL
C3H5F3O	002240-88-2	3,3,3-TRIFLUORO-1-PROPANOL
C3H5FO	000430-51-3	FLUOROACETONE
C3H5FO	000503-09-3	EPIFLUOROHYDRIN
C3H5IO2	000141-76-4	3-IODOPROPIONIC ACID
C3H5N	000107-12-0	PROPIONITRILE
C3H5N	002450-71-7	2-PROPYNYLAMINE
C3H5N2O	004570-45-0	OXAZOLE-2-AMINE
C3H5N3	028466-26-4	4-AMINOPYRAZOLE
C3H5N3O	017647-70-0	FURAZANAMINE, 4-METHYL-
C3H5N3O2	115174-03-3	FURAZANAMINE, 4-METHYL-, 5-OXIDE
C3H5N3O2	118506-53-9	FURAZANAMINE, 4-METHYL-, 2-OXIDE
C3H5N3O9	000055-63-0	1,2,3-PROPANETRIOL TRINITRATE
C3H5N3OS	061444-94-8	1,2,3-THIADIAZOLE-5-CARBOXALDOXIME
C3H5NO	000078-97-7	2-HYDROXYPROPANENITRILE
C3H5NO	000079-06-1	ACRYLAMIDE
C3H5NO	000109-78-4	ETHYLENE CYANOHYDRIN
C3H5NO2	000306-44-5	HYDROXYIMINOACETONE
C3H5NO2S	002274-42-2	METHYLSULFONYLACETONITRILE
C3H6	000075-19-4	CYCLOPROPANE
C3H6	000115-07-1	PROPENE
C3H6Br2	000078-75-1	1,2-DIBROMOPROPANE
C3H6Br2	000109-64-8	1,3-DIBROMOPROPANE
C3H6Br2O	000096-13-9	2,3-DIBROMO-1-PROPANOL
C3H6BrCl	000109-70-6	1-CHLORO-3-BROMOPROPANE
C3H6BrCl	003017-95-6	2-BROMO-1-CHLOROPROPANE
C3H6Cl2	000078-87-5	1,2-DICHLOROPROPANE
C3H6Cl2	000142-28-9	1,3-DICHLOROPROPANE
C3H6Cl2O	000096-23-1	1,3-DICHLORO-2-PROPANOL
C3H6Cl2O	000616-23-9	2,3-DICHLOROPROPANOL
C3H6Cl2O	026545-73-3	GLYCEROL DICHLOROHYDRIN
C3H6Cl2O	063151-11-1	2,2-DICHLOROPROPANOL
C3H6Cl2O	083682-72-8	3,3-DICHLOROPROPANOL
C3H6ClN3O2	002365-30-2	1-(2-CHLOROETHYL)-1-NITROSOUREA
C3H6ClNO	000079-44-7	DIMETHYLCARBAMYL CHLORIDE
C3H6F2O	000453-13-4	1,3-DIFLUORO-2-PROPANOL
C3H6F3N	002730-67-8	N-ME ETHYLAMINE,2,2,2-TRIFLUORO
C3H6FNO	000367-49-7	A-FLUORO-N-METHYLACETAMIDE
C3H6I2	000627-31-6	PROPANE, 1,3-DIIODO-
C3H6KOS2	000140-89-6	POTASSIUM ETHYLXANTHATE
C3H6N2	001467-79-4	DIMETHYL CYANAMIDE
C3H6N2O	004549-40-0	N-NITROSOMETHYLVINYLAMINE
C3H6N2O3	000462-60-2	N-(AMINOCARBONYL)GLYCINE
C3H6N2O3	013256-22-9	N-NITROSOSARCOSINE
C3H6N2O7	000621-65-8	1,2,3-PROPANETRIOL, 1,2-DINITRATE
C3H6N2O7	000623-87-0	1,2,3-PROPANETRIOL, 1,3-DINITRATE
C3H6N2OS	000591-08-2	1-ACETYL-2-THIOUREA
C3H6N2S	000096-45-7	ETHYLENETHIOUREA
C3H6N3O	052838-39-8	2-AMINO-5-ME-1,3,4-OXADIAZOLE
C3H6N4	001609-07-0	1-METHYL-2-CYANOGUANIDINE
C3H6N4O2	026621-45-4	1,2,4-TRIAZOLE-1-METHYL-3-NITRO-
C3H6N6	000108-78-1	MELAMINE
C3H6N6O6	000121-82-4	1,3,5,-TRINITROHEXAHYDRO-1,3,5-TRIAZINE
C3H6NNaS2	000128-04-1	DIMETHYLDITHIOCARBAMIC ACID, SODIUM SALT
C3H6O	000067-64-1	ACETONE

C3H6O	000075-56-9	PROPYLENE OXIDE
C3H6O	000107-18-6	ALLYL ALCOHOL
C3H6O	000107-25-5	METHYLVINYLETHER
C3H6O	000123-38-6	PROPIONALDEHYDE
C3H6O	000503-30-0	TRIMETHYLENEOXIDE
C3H6O2	000079-09-4	PROPIONIC ACID
C3H6O2	000079-20-9	METHYL ACETATE
C3H6O2	000109-94-4	ETHYL FORMATE
C3H6O2	000116-09-6	HYDROXYACETONE
C3H6O2	000556-52-5	OXIRANEMETHANOL
C3H6O2	000646-06-0	1,3-DIOXALANE
C3H6O2S	000107-96-0	3-MERCAPTOPROPIONIC ACID
C3H6O3	000050-21-5	LACTIC ACID
C3H6O3	000096-26-4	DIHYDROXY ACETONE
C3H6O3	000110-88-3	1,3,5-TRIOXANE
C3H6O3	000598-82-3	A-HYDROXYPROPIONIC ACID
C3H6O3	000625-45-6	METHOXYACETIC ACID
C3H6O3S	001120-71-4	2,2-DIOXIDE-1,2-OXATHIOLANE
C3H6S	000287-27-4	THIACYCLOBUTANE
C3H6S	000870-23-5	ALLYL MERCAPTAN
C3H7Br	000075-26-3	2-BROMOPROPANE
C3H7Br	000106-94-5	1-BROMOPROPANE
C3H7BrHgO	018832-83-2	BROMO(2-HYDROXYPROPYL)MERCURY
C3H7BrO	000627-18-9	3-BROMOPROPANOL
C3H7Cl	000075-29-6	2-CHLOROPROPANE
C3H7Cl	000540-54-5	1-CHLOROPROPANE
C3H7Cl3Si	000141-57-1	PROPYLTRICHLOROSILANE
C3H7ClHgO	000123-88-6	ARETAN
C3H7ClO	000078-89-7	2-CHLORO-1-PROPANOL
C3H7ClO	000127-00-4	1-CHLORO-2-PROPANOL
C3H7ClO	000627-30-5	3-CHLORO-1-PROPANOL
C3H7ClO	003188-13-4	CHLOROMETHYL ETHYL ETHER
C3H7ClO2	000096-24-2	3-CHLORO-1,2-PROPANEDIOL
C3H7F	000460-13-9	1-FLUOROPROPANE
C3H7FO	000462-43-1	3-FLUOROPROPANOL
C3H7I	000075-30-9	2-IODOPROPANE
C3H7I	000107-08-4	1-IODOPROPANE
C3H7N	000075-55-8	2-METHYLAZIRIDINE
C3H7N	000107-11-9	ALLYLAMINE
C3H7N	000503-29-7	AZETIDINE
C3H7N	000765-30-0	CYCLOPROPANAMINE
C3H7N3O2	000759-73-9	1-ETHYL-1-NITROSOUREA
C3H7NO	000068-12-2	N,N'-DIMETHYLFORMAMIDE
C3H7NO	000079-05-0	PROPRIONAMIDE
C3H7NO	000079-16-3	N-METHYLACETAMIDE
C3H7NO	000127-06-0	2-PROPANONE, OXIME
C3H7NO	000627-45-2	N-FORMYLETHYLAMINE
C3H7NO2	000051-79-6	O-ETHYL CARBAMATE (URETHANE)
C3H7NO2	000056-41-7	ALANINE
C3H7NO2	000079-46-9	2-NITROPROPANE
C3H7NO2	000107-95-9	BETA-ALANINE
C3H7NO2	000107-97-1	N-METHYL GLYCINE
C3H7NO2	000108-03-2	1-NITROPROPANE
C3H7NO2	000302-72-7	ALANINE
C3H7NO2	000543-67-9	N-PROPYLNITRITE
C3H7NO2	006642-30-4	N,O-DIMETHYLCARBAMATE
C3H7NO2S	000052-90-4	CYSTEINE
C3H7NO3	000056-45-1	SERINE

C3H7NO3	000627-13-4	1-PROPYLNITRATE
C3H7NO3	001712-64-7	ISOPROPYL NITRATE
C3H7NOS	013749-94-5	1-(MES)-ACETALDEHYDE OXIME
C3H7NS	000542-85-8	ETHYL ISOTHIOCYANATE
C3H8	000074-98-6	PROPANE
C3H8ClN	026245-56-7	N-CHLORO-2-PROPANAMINE
C3H8N2O	000096-31-1	DIMETHYLUREA, SYM
C3H8N2O	000598-94-7	1,1-DIMETHYLUREA
C3H8N2O	000625-52-5	ETHYLUREA
C3H8N2O	010595-95-6	N-METHYL-N-NITROSOETHAMINE
C3H8N2O	024535-11-3	PROPIONIC ACID HYDRAZIDE
C3H8N2O2	005710-11-2	3-ETHYLHYDROXYUREA
C3H8N2O2	007433-42-3	1-ETHYLHYDROXYUREA
C3H8N2O3	000140-95-4	DIMETHYLOL UREA
C3H8N2S	000534-13-4	N,N-DIMETHYLTHIOUREA
C3H8N2S	000625-53-6	N-ETHYLTHIOUREA
C3H8N2S	006972-05-0	THIOUREA, N,N-DIMETHYL-
C3H8N4O2	101250-97-9	1,3-DIMETHYL-2-NITROGUANADINE
C3H8NO5P	001071-83-6	GLYPHOSPHATE
C3H8O	000067-63-0	ISOPROPANOL
C3H8O	000071-23-8	1-PROPANOL
C3H8O	000540-67-0	ETHYL METHYL ETHER
C3H8O2	000057-55-6	1,2-PROPANEDIOL
C3H8O2	000109-86-4	2-METHOXYETHANOL
C3H8O2	000109-87-5	DIMETHOXYMETHANE
C3H8O2	000504-63-2	1,3-PROPANEDIOL
C3H8O3	000056-81-5	GLYCEROL
C3H8O3S	000062-50-0	ETHYL METHANESULFONATE
C3H8S	000075-33-2	2-PROPANETHIOL
C3H8S	000107-03-9	1-PROPANETHIOL
C3H8S	000624-89-5	ETHYL METHYL SULFIDE
C3H9ClSi	000075-77-4	CHLOROTRIMETHYLSILANE
C3H9N	000075-31-0	ISOPROPYLAMINE
C3H9N	000075-50-3	TRIMETHYLAMINE
C3H9N	000107-10-8	PROPYLAMINE
C3H9N	000624-78-2	METHYLETHYLAMINE
C3H9N3	003324-71-8	DIMETHYLGUANIDINE
C3H9N3O	013050-41-4	4-ETHYLSEMICARBAZIDE
C3H9N3O	022718-51-0	4,4-DIMETHYLSEMICARBAZIDE
C3H9N3S	006926-58-5	4,4-DIMETHYLTHIOSEMICARBAZIDE
C3H9NO	000078-96-6	1-AMINO-2-PROPANOL
C3H9NO	000109-83-1	2-(METHYLAMINO)ETHANOL
C3H9NO	000156-87-6	PROPANOLAMINE
C3H9O2P	016391-07-4	Phosphinic acid, methyl-, ethyl ester
C3H9O2PS	018005-40-8	O-ETHYL METHYLPHOSPHONOTHIOATE
C3H9O2PS2	002953-29-9	PHOSPHORODITHIOIC ACID, O,O,S-TRIMETHYL ESTER
C3H9O2PS2	022608-53-3	O,S,S-TRIMETHYL PHOSPHORODITHIOATE
C3H9O3P	000121-45-9	TRIMETHYL PHOSPHITE
C3H9O3P	000756-79-6	Phosphonic acid, methyl-, dimethyl ester
C3H9O3P	001832-53-7	ETHYL METHYL PHOSPHONATE
C3H9O3PS	000152-18-1	O,O,O-TRIMETHYL PHOSPHOROTHIOATE
C3H9O3PS	000152-20-5	PHOSPHOROTHIOIC ACID, O,O,S-TRIMETHYL ESTER
C3H9O4P	000512-56-1	TRIMETHYL PHOSPHATE
C3H10ClN	000556-53-6	PROPYLAMINE HYDROCHLORIDE
C3H10N2	000078-90-0	1,2-DIAMINOPROPANE
C3H10N2	000109-76-2	1,3-PROPANEDIAMINE
C3H10NO2PS	028167-49-9	O,S-DIME-N-ME-PHOSPHORAMIDOTHIOAT
C3H10O3Si	002487-90-3	TRIMETHOXY SILANE

C3H11ClN2	060597-20-8	TRIMETHYL HYDRAZINE, MONOHYDROCHLORIDE
C3H11N2O4P	025954-13-6	FOSAMINE AMMONIUM SALT
C4Cl10	006820-74-2	PERCHLOROISOBUTANE
C4Cl6	000087-68-3	HEXACHLOROBUTADIENE
C4Cl6	056827-79-3	1,2-HEXACHLOROBUTADIENE
C4Cl6	083682-47-7	1-HEXACHLOROBUTYNE
C4Cl6O3	004124-31-6	TRICHLOROACETIC ANHYDRIDE
C4Cl8	002482-68-0	2-PERCHLOROBUTENE
C4F8	000115-25-3	PERFLUOROCYCLOBUTANE
C4F8	000360-89-4	OCTAFLUORO-2-BUTENE
C4F10	000355-25-9	DECAFLUOROBUTANE
C4F6O3	000407-25-0	TRIFLUOROACETIC ANHYDRIDE
C4HCl7	083682-39-7	1-CHLORO-2-TRICHLOROMETHYL-3,3,3-TRICHLOROPROP*
C4HCl9	021483-62-5	1,1,1,2,2,3,3,4,4-NONACHLOROBUTANE
C4H2	000460-12-8	DIACETYLENE
C4H2Br2S	003140-93-0	THIOPHENE, 2,3-DIBROMO-
C4H2Br2S	003141-26-2	3,4-DIBROMOTHIOPHENE
C4H2Br2S	003141-27-3	2,5-DIBROMOTHIOPHENE
C4H2Cl2N2	004774-14-5	2-CHLORO-6-CHLORO-PYRAZINE
C4H2Cl2N2	019745-07-4	PYRAZINE, 2,5-DICHLORO-
C4H2Cl2O2	000627-63-4	FUMARYL CHLORIDE
C4H2Cl2S	003172-52-9	2,5-DICHLOROTHIOPHENE
C4H2Cl4	000921-09-5	1,1,2,3-TETRACHLOROBUTA-1,3-DIENE
C4H2Cl4	001637-31-6	1,2,3,4-TETRACHLOROBUTA-1,3-DIENE
C4H2Cl4	036038-53-6	1,1,4,4-TETRACHLOROBUTA-1,3-DIENE
C4H2Cl6	083682-38-6	2-TRICHLOROMETHYL-3,3,3-TRICHLOROPROPENE
C4H2Cl6	083682-71-7	2,3,4,4,4,4-HEXACHLOROBUT-1-ENE
C4H2Cl8	018791-19-0	1,1,1,2,3,4,4,4-OCTACHLOROBUTANE
C4H2Cl8	020338-26-5	1,1,2,2,3,3,4,4-OCTACHLOROBUTANE
C4H2Cl8	032694-76-1	1,1,1,2,3,3,4,4-OCTACHLOROBUTANE
C4H2ClFN2	033873-10-8	2-CHLORO-6-FLUORO-PYRAZINE
C4H2F2N2	033873-09-5	2,6-DIFLUOROPYRAZINE
C4H2N2	000764-42-1	2-BUTENEDINITRILE, (E)-
C4H2N2O4	000050-71-5	ALLOXAN
C4H2O3	000108-31-6	MALEIC ANHYDRIDE
C4H3BrN2	004595-59-9	5-BROMOPYRIMIDINE
C4H3BrN2	004595-60-2	PYRIMIDINE, 2-BROMO-
C4H3BrN2	056423-63-3	2-BROMOPYRAZINE
C4H3BrN2O2	000051-20-7	5-BROMOURACIL
C4H3BrS	000872-31-1	3-BROMOTHIOPHENE
C4H3BrS	001003-09-4	2-BROMOTHIOPHENE
C4H3Cl3	001573-58-6	1,2,3-TRICHLOROBUTA-1,3-DIENE
C4H3Cl3	002852-07-5	1,1,2-TRICHLORO-1,3-BUTADIENE
C4H3Cl3	034819-62-0	4,4,4-TRICHLORO-1,2-BUTADIENE
C4H3Cl3	039083-26-6	1,2,3-TRICHLOROBUTA-1,3-DIENE(Z)
C4H3Cl3	053978-04-4	1,2,3-TRICHLOROBUTA-1,3-DIENE(E)
C4H3Cl3	058679-08-6	1,1,4-TRICHLORO-1,2-BUTADIENE
C4H3Cl3	083682-43-3	4,4,4-TRICHLORO-1-BUTYNE
C4H3Cl3	083682-46-6	1,1,4-TRICHLOROBUTADIENE
C4H3Cl5	014129-82-9	1,3,3-TRICHLORO-2-(DICHLOROMETHYL)PROPENE
C4H3Cl7	034973-41-6	1,1,2,2,3,4,4-HEPTACHLOROBUTANE
C4H3Cl7	083682-70-6	1,1,1,2,2,3,3-HEPTACHLOROBUTANE
C4H3ClN2	001120-95-2	PYRIDAZINE, 3-CHLORO-
C4H3ClN2	001722-12-9	PYRIMIDINE, 2-CHLORO-
C4H3ClN2	014508-49-7	2-CHLOROPYRAZINE
C4H3ClN2	017180-93-7	PYRIMIDINE, 4-CHLORO-
C4H3ClN2	017180-94-8	PYRIMIDINE, 5-CHLORO-
C4H3ClN2O2	001820-81-1	5-CHLOROURACIL

C4H3ClS	000096-43-5	2-CHLOROTHIOPHENE
C4H3F7O	000375-01-9	2,2,3,3,4,4,-HEPTAFLUOROBUTANOL
C4H3FN2	000675-21-8	PYRIMIDINE, 5-FLUORO-
C4H3FN2	004949-13-7	2-FLUOROPYRAZINE
C4H3FN2	031575-35-6	PYRIMIDINE, 2-FLUORO-
C4H3FN2O2	000051-21-8	5-FLUOROURACIL
C4H3IN2O2	000696-07-1	2,4(1H,3H)-PYRIMIDINEDIONE, 5-IODO-
C4H3IS	003437-95-4	2-IODOTHIOPHENE
C4H3N3	031108-57-3	4-CYANOPYRAZOLE
C4H3N3O2	014080-32-1	PYRIMIDINE, 5-NITRO-
C4H3N5O	002683-90-1	8-AZAHYPOXANTHINE
C4H3NO2	000541-59-3	1H-PYRROLE-2,5-DIONE
C4H3NO2S	000609-40-5	2-NITROTHIOPHENE
C4H4	000689-97-4	1-BUTEN-3-YNE
C4H4BrN3O2	000933-87-9	5-BROMO-1-METHYL-4-NITRO-1H-IMIDAZOLE
C4H4BrN3O2	016681-63-3	2-BROMO-1-METHYL-4-NITRO-1H-IMIDAZOLE
C4H4BrN3O2	018874-52-7	4-BROMO-2-METHYL-5-NITROIMIDAZOLE
C4H4BrN3O2	059177-47-8	4-BROMO-1-METHYL-5-NITRO-1H-IMIDAZOLE
C4H4BrN3O2	105983-46-8	2-BROMO-4-METHYL-5-NITRO-1H-IMIDAZOLE
C4H4BrO	022037-28-1	3-BROMOFURAN
C4H4Cl2	000821-10-3	2-BUTYNE, 1,4-DICHLORO-
C4H4Cl2	001653-19-6	2,3-DICHLOROBUTA-1,3-DIENE
C4H4Cl2	002984-42-1	1,4-DICHLORO-1,3-BUTADIENE
C4H4Cl2	006061-06-9	1,1-DICHLORO-1,3-BUTADIENE
C4H4Cl2	083682-41-1	4,4-DICHLORO-1,2-BUTADIENE
C4H4Cl2	083682-42-2	4,4-DICHLORO-1-BUTYNE
C4H4Cl2	083682-44-4	1,4-DICHLORO-1,2-BUTADIENE
C4H4Cl2	083682-45-5	1,4-DICHLORO-1-BUTYNE
C4H4Cl2O2	000543-20-4	SUCCINYL CHLORIDE
C4H4Cl2O3	000541-88-8	CHLOROACETIC ANHYDRIDE
C4H4Cl6	001573-57-5	1,2,2,3,3,4-HEXACHLOROBUTANE
C4H4Cl6	002431-55-2	1,1,2,2,3,4-HEXACHLOROBUTANE
C4H4Cl6	079458-54-1	1,1,1,4,4,4-HEXACHLOROBUTANE
C4H4Cl6	083682-69-3	1,1,2,2,3,3-HEXACHLOROBUTANE
C4H4ClN3	005469-69-2	3-CL-6-PYRIDAZINAMINE
C4H4ClN3	033332-28-4	2-CHLORO-6-AMINO-PYRAZINE
C4H4ClN3	033332-29-5	PYRAZINAMINE, 5-CHLORO-
C4H4ClN3O2	004897-25-0	5-CHLORO-1-METHYL-4-NITROIMIDAZOLE
C4H4ClN3O2	004897-31-8	4-CHLORO-1-METHYL-5-NITROIMIDAZOLE
C4H4ClN3O2	063634-21-9	2-CHLORO-1-METHYL-4-NITRO-1H-IMIDAZOLE
C4H4ClN3O2	063662-67-9	4-CHLORO-2-METHYL-5-NITRO-1H-IMIDAZOLE
C4H4ClN3O2	086072-07-3	2-CHLORO-1-METHYL-5-NITRO-1H-IMIDAZOLE
C4H4F6N2O	000625-89-8	N-NITROSO-BIS(2,2,2TRIFET)AMINE
C4H4IN3O2	035681-63-1	1H-IMIDAZOLE, 5-IODO-1-METHYL-4-NITRO-
C4H4IN3O2	076529-47-0	1H-IMIDAZOLE, 4-IODO-1-METHYL-5-NITRO-
C4H4N2	000110-61-2	SUCCINONITRILE
C4H4N2	000289-80-5	PYRIDAZINE
C4H4N2	000289-95-2	PYRIMIDINE
C4H4N2	000290-37-9	PYRAZINE
C4H4N2O	000557-01-7	2-PYRIMIDONE
C4H4N2O	004562-27-0	4-PYRIMIDONE
C4H4N2O	006270-63-9	PYRAZINE-2-ONE
C4H4N2O2	000066-22-8	URACIL
C4H4N2O2	000123-33-1	MALEIC HYDRAZIDE
C4H4N2O2	129016-38-2	3(2H)-PYRIDAZINONE, 6-HYDROXY-
C4H4N2O3	000067-52-7	BARBITURIC ACID
C4H4N2O3	123953-16-2	FURAZANCARBOXALDEHYDE, METHYL-, 5-OXIDE
C4H4N2O3	123953-17-3	FURAZANCARBOXALDEHYDE, METHYL-, 2-OXIDE

C4H4N2O4	026537-53-1	SYDNONE, 3-(CARBOXYMETHYL)-
C4H4N2OS	000156-82-1	2-THIOURACIL
C4H4N4O	016339-18-7	N-NITROSO-BIS(CYANOMETHYL)AMINE
C4H4N4O4	005213-50-3	1-METHYL-2,4-DINITRO-1H-IMIDAZOLE
C4H4N4O4	019183-15-4	1-METHYL-4,5-DINITROIMIDAZOLE
C4H4N4O4	019183-16-5	2-METHYL-4,5-DINITROIMIDAZOLE
C4H4N6	001123-54-2	8-AZAADENINE
C4H4N6O	000134-58-7	8-AZAGUANINE
C4H4NO3	000609-39-2	2-NITROFURAN
C4H4O	000110-00-9	FURAN
C4H4O2	000497-23-4	2(5H)-FURANONE
C4H4O2	000674-82-8	DIKETENE
C4H4O2	015506-53-3	1,3-CYCLOBUTANEDIONE
C4H4O3	000108-30-5	SUCCINIC ANHYDRIDE
C4H4O4	000110-16-7	MALEIC ACID
C4H4O4	000110-17-8	FUMARIC ACID
C4H4S	000110-02-1	THIOPHENE
C4H5Cl	000126-99-8	CHLOROPRENE
C4H5Cl	000627-22-5	1-CHLOROPRENE
C4H5Cl	000627-23-6	1-CHLORO-1,2-BUTADIENE
C4H5Cl	021020-24-6	3-CHLORO-1-BUTYNE
C4H5Cl	025790-55-0	4-CHLORO-1,2-BUTADIENE
C4H5Cl	034581-41-4	3-CHLORO-1,2-BUTADIENE
C4H5Cl	051908-64-6	4-CHLORO-1-BUTYNE
C4H5Cl	062981-74-2	1-CHLORO-1-BUTYNE
C4H5Cl3	002431-54-1	1,2,4-TRICHLORO-2-BUTENE
C4H5Cl3	013245-65-3	1,3-DICHLORO-2-CHLOROMETHYLPROPENE
C4H5Cl3	041601-59-6	1,1,4-TRICHLORO-2-BUTENE
C4H5Cl3	057808-36-3	1,1,4-TRICHLOROBUT-2-ENE(E)
C4H5Cl3	060845-51-4	2-CHLOROMETHYL-3,3-DICHLOROPROPENE
C4H5Cl3O2	000515-84-4	ETHYL TRICHLOROACETATE
C4H5Cl5	077753-24-3	1,1,2,3,4-PENTACHLOROBUTANE
C4H5Cl5	083293-82-7	1,2,2,3,3-PENTACHLOROBUTANE
C4H5ClO2	002937-50-0	ALLYL CHLOROCARBONATE
C4H5ClO3	004755-77-5	ETHYL CHLOROGLYOXYLATE
C4H5ClO4	016045-92-4	CHLOROSUCCINIC ACID
C4H5F3O	000406-90-6	FLUROXENE
C4H5F3O2	000383-63-1	TRIFLUOROETHYL ACETATE
C4H5N	000109-75-1	3-BUTENENITRILE
C4H5N	000109-97-7	PYRROLE
C4H5N	000126-98-7	METHACRYLONITRILE
C4H5N	004786-20-3	CROTONONITRILE
C4H5N	005500-21-0	CYCLOPROPANECARBONITRILE
C4H5N3	000109-12-6	2-AMINOPYRIMIDINE
C4H5N3	000591-54-8	4-PYRIMIDINAMINE
C4H5N3	005049-61-6	2-AMINOPYRAZINE
C4H5N3	005469-70-5	2-AMINOPYRIMIDINE
C4H5N3	020744-39-2	4-PYRIDAZINAMINE
C4H5N3	024108-33-6	1,2,4-TRIAZINE, 3-METHYL-
C4H5N3O	000071-30-7	CYTOSINE
C4H5N3O	000108-53-2	2-AMINO-4-PYRIMIDONE
C4H5N3O2	000696-23-1	2-METHYL-4-NITRO-1H-IMIDAZOLE
C4H5N3O2	001671-82-5	1-METHYL-2-NITRO-1H-IMIDAZOLE
C4H5N3O2	003034-41-1	1-METHYL-4-NITRO-IH-IMIDAZOLE
C4H5N3O2	003034-42-2	1-METHYL-5-NITROIMIDAZOLE
C4H5N3O2	005334-39-4	3-ME-4-NO2 PYRAZOLE
C4H5N3O2	014003-66-8	4-METHYL-5-NITRO-1H-IMIDAZOLE
C4H5N3O2	088054-22-2	2-METHYL-5-NITROIMIDAZOLE

C4H5N3O2	090507-21-4	1,2,4-OXADIAZOLE-5-CARBOXALDEHYDE,3-ME,OXIME
C4H5N3O3	037895-44-6	FURAZANCARBOXAMIDE, 4-METHYL-, 5-OXIDE
C4H5N3O3	037895-45-7	FURAZANCARBOXAMIDE, 4-METHYL-, 2-OXIDE
C4H5N3O3	068019-78-3	4-METHOXY-5-NITRO-1H-IMIDAZOLE
C4H5NO2	000105-34-0	METHYL CYANOACETATE
C4H5NO2	000123-56-8	SUCCINIMIDE
C4H5NO2	010004-44-1	3-HYDROXY-5-METHYLISOXAZOLE
C4H5NO2S	000822-84-4	3-NITROTHIOPHENE
C4H5NS	000057-06-7	3-ISOTHIOCYANATO-1-PROPENE
C4H5NS	000693-95-8	4-METHYLTHIAZOLE
C4H6	000106-99-0	1,3-BUTADIENE
C4H6	000107-00-6	1-BUTYNE
C4H6	000503-17-3	2-BUTYNE
C4H6	000590-19-2	1,2-BUTADIENE
C4H6Br4	001529-68-6	1,2,3,4-TETRABROMOBUTANE
C4H6Cl2	000110-57-6	1,4-DICHLORO-2-BUTENE(TRANS)
C4H6Cl2	000760-23-6	3,4-DICHLORO-1-BUTENE
C4H6Cl2	000764-41-0	1,4-DICHLORO-2-BUTENE
C4H6Cl2	000926-57-8	1,3-DICHLORO-2-BUTENE
C4H6Cl2	001476-11-5	1,4-DICHLORO-2-BUTENE(CIS)
C4H6Cl2	001871-57-4	2-(CHLOROMETHYL)-3-CHLORO-1-PROPENE
C4H6Cl2	003375-22-2	1,3-DICHLORO-2-METHYLPROPENE
C4H6Cl2	031423-92-4	1,4-DICHLORO-1-BUTENE
C4H6Cl2N2O	059863-59-1	3,4-DICHLORO-N-NITROSOPYRROLIDINE
C4H6Cl2O2	000535-15-9	ETHYL DICHLOROACETATE
C4H6Cl4	003405-32-1	1,2,3,4-TETRACHLOROBUTANE
C4H6Cl4	013138-51-7	1,2,3,3-TETRACHLOROBUTANE
C4H6Cl4	014499-87-7	2,2,3,3,-TETRACHLOROBUTANE
C4H6Cl4	033455-24-2	1,1,4,4-TETRACHLOROBUTANE
C4H6ClN	000628-20-6	BUTANENITRILE, 4-CHLORO-
C4H6F2O2	000454-31-9	ACETIC ACID, DIFLUORO-, ETHYL ESTER
C4H6HgO4	001600-27-7	MERCURIC ACETATE
C4H6N2	000616-47-7	1H-IMIDAZOLE, 1-METHYL-
C4H6N2	000693-98-1	1H-IMIDAZOLE, 2-METHYL-
C4H6N2	000822-36-6	1H-IMIDAZOLE, 4-METHYL-
C4H6N2	000930-36-9	1-METHYLPYRAZOLE
C4H6N2Na2S4	000142-59-6	NABAM
C4H6N2O	004975-21-7	FURAZAN, DIMETHYL-
C4H6N2O	010552-94-0	N-NITROSO-3-PYRROLINE
C4H6N2O	014884-01-6	4-METHOXYPYRAZOLE
C4H6N2O2	000616-03-5	2,4-IMIDAZOLIDINEDIONE, 5-METHYL-
C4H6N2O2	002518-42-5	FURAZAN, DIMETHYL-, 2-OXIDE
C4H6N2O2	002763-96-4	MUSCIMOL
C4H6N2OS	025602-39-5	FORMAMIDE, N-2-THIAZOLYL
C4H6N2S	000060-56-0	METHIMAZOLE
C4H6N2S	005728-21-2	1,2,5-THIADIAZOLE, 3,4-DIMETHYL-
C4H6N2S	027464-82-0	1,3,4-THIADIAZOLE, 2,5-DIMETHYL-
C4H6N2S4Zn	000142-14-3	ZINEB (ALSO SEE CAS NO. 12122-67-7)
C4H6N2S4Zn	012122-67-7	(ETHYLENEBIS(DITHIOCARBAMATO))ZINC
C4H6N2Se	017505-12-3	1,2,5-SELENADIAZOLE, 3,4-DIMETHYL-
C4H6N4	041536-80-5	2,6-PYRAZINEDIAMINE
C4H6N4O2	000817-99-2	ACETAMIDE, N-(2-AMINO-2-OXOETHYL)-2-DIAZO-
C4H6N4O2	004531-54-8	1-METHYL-4-NITRO-1H-IMIDAZOL-5-AMINE
C4H6N4O3	000097-59-6	ALLANTOIN
C4H6N4O3S2	000059-66-5	ACETAZOLEAMIDE
C4H6N4OS	062036-62-8	3-MES-4-NH2-1,2,4-TRIAZINE-5-ONE
C4H6NO	005765-44-6	ISOXAZOLE, 5-METHYL-
C4H6O	000078-85-3	METHACROLEIN

C4H6O	000078-94-4	METHYLVINYLKETONE
C4H6O	000109-93-3	DIVINYL ETHER
C4H6O	000123-73-9	TRANS-CROTONALDEHYDE
C4H6O	000598-26-5	2-METHYL-1-PROPEN-1-ONE(DIMETHYLKETENE)
C4H6O	001191-95-3	CYCLOBUTANONE
C4H6O	001708-29-8	2,5-DIHYDROFURAN
C4H6O	004170-30-3	CROTONALDEHYDE
C4H6O	020334-52-5	1-BUTEN-1-ONE(ETHYLKETENE)
C4H6O2	000079-41-4	METHACRYLIC ACID
C4H6O2	000096-33-3	METHYL ACRYLATE
C4H6O2	000096-48-0	GAMMA-BUTYROLACTONE
C4H6O2	000107-93-7	CROTONIC ACID
C4H6O2	000108-05-4	VINYL ACETATE
C4H6O2	000110-65-6	2-BUTYNE-1,4-DIOL
C4H6O2	000298-18-0	1,2:3,4-DIEPOXYBUTANE DL
C4H6O2	000431-03-8	2,3-BUTANEDIONE
C4H6O2	000503-64-0	(Z)-2-BUTENOIC ACID (ISOCROTONIC ACID)
C4H6O2	000564-00-1	1,2:3,4-DIEPOXYBUTANE
C4H6O2	001464-53-5	DIEPOXYBUTANE
C4H6O2	001759-53-1	CYCLOPROPANECARBOXYLIC ACID
C4H6O2	003068-88-0	4-METHYL-2-OXETANONE
C4H6O2	003724-65-0	CROTONIC ACID
C4H6O2	030031-64-2	1,2:3,4-DIEPOXYBUTANE (2S,3S)
C4H6O2	030419-67-1	1,2:3,4-DIEPOXYBUTANE (2R,3R)
C4H6O2S	000077-77-0	1,1'-SULFONYLBISETHENE
C4H6O2S	000077-79-2	2,5-DIHYDROTHIOPHENE 1,1-DIOXIDE
C4H6O3	000108-24-7	ACETIC ANHYDRIDE
C4H6O3	000108-32-7	PROPYLENE CARBONATE
C4H6O3	000541-50-4	ACETOACETIC ACID
C4H6O4	000110-15-6	BUTANEDIOC ACID
C4H6O4	000516-05-2	METHYL MALONIC ACID
C4H6O4	000553-90-2	METHYL OXALATE
C4H6O5	006915-15-7	MALIC ACID
C4H6O6	000087-69-4	TARTARIC ACID
C4H6OS	001115-15-7	1,1'-SULFINYLBISETHENE
C4H6OS	039700-44-2	G-THIOBUTYROLACTONE
C4H6S	000627-51-0	1,1'-THIOBISETHENE
C4H7Br	004784-77-4	2-BROMO-2-BUTENE
C4H7Br	005162-44-7	4-BROMO-1-BUTENE
C4H7Br	013294-71-8	2-BROMO-2-BUTENE (CIS)
C4H7Br	031844-98-1	1-BROMO-1-BUTENE
C4H7Br2Cl2O4P	000300-76-5	NALED
C4H7BrO	000814-75-5	3-BROMO-2-BUTANONE
C4H7BrO	000816-40-0	1-BROMO-2-BUTANONE
C4H7BrO2	000080-58-0	A-BROMOBUTYRIC ACID
C4H7BrO2	000105-36-2	BROMOACETIC ACID, ETHYL ESTER
C4H7BrO2	000927-68-4	ETHANOL, 2-BROMO-, ACETATE
C4H7Cl	000513-37-1	1-CHLORO-2-METHYLPROPENE
C4H7Cl	000563-47-3	3-CHLORO-2-METHYLPROPENE
C4H7Cl	000563-52-0	3-CHLORO-1-BUTENE
C4H7Cl	000591-97-9	1-CHLORO-2-BUTENE
C4H7Cl	000927-73-1	4-CHLORO-1-BUTENE
C4H7Cl	004461-42-1	1-CHLORO-1-BUTENE
C4H7Cl2O3P	001185-97-3	OO-DIME DICHLOROVINYLPHOSPHONATE
C4H7Cl2O4P	000062-73-7	DICHLORVOS
C4H7Cl3	001790-22-3	1,2,4-TRICHLOROBUTANE
C4H7Cl3	015187-71-0	1,3,3-TRICHLOROBUTANE
C4H7Cl3O	000057-15-8	B,B,B-TRICHLORO-T-BUTANOL

C4H7Cl3O2	000076-40-4	2,2,3-TRICHLORO-1,1-BUTANEDIOL
C4H7ClO	000110-75-8	2-CHLOROETHYL VINYL ETHER
C4H7ClO	000671-56-7	3-HYDROXY-4-CHLORO-1-BUTENE
C4H7ClO	004091-39-8	3-CHLORO-2-BUTANONE
C4H7ClO	075455-41-3	4-HYDROXY-3-CHLORO-1-BUTENE
C4H7ClO2	000105-39-5	ETHYL CHLOROACETATE
C4H7ClO2	000108-23-6	ISOPROPYL CHLOROFORMATE
C4H7ClOS	040709-82-8	((2-CHLOROETHYL)SULFINYL)ETHENE
C4H7F3O	000461-18-7	4,4,4-TRIFLUORO-1-BUTANOL
C4H7F3O	000507-52-8	2-(TRIFLUORMETHYL)-2-PROPANOL
C4H7F3O	101054-93-7	4,4,4-TRIFLUORO-2-BUTANOL
C4H7IO2	000623-48-3	ETHYL IODOACETATE
C4H7N	000078-82-0	2-METHYLPROPANENITRILE
C4H7N	000109-74-0	BUTYRONITRILE
C4H7N3O	000060-27-5	CREATININE
C4H7N3OS	061444-95-9	2-ME-134-THIADIAZOLE-5-CARBOXALDOXIME
C4H7N3OS	061444-96-0	3-ME-124-THIADIAZOLE-5-CARBOXALDOXIME
C4H7N5	001004-38-2	2,4,6-PYRIMIDINETRIAMINE
C4H7NaO2	000156-54-7	BUTANOIC ACID, SODIUM SALT
C4H7NaOS2	000140-93-2	ISOPROPYLXANTHIC ACID, SODIUM SALT
C4H7NO	000075-86-5	ACETONE CYANOHYDRIN
C4H7NO	000616-45-5	PYRROLIDONE
C4H7NO2	000625-77-4	DIACETAMIDE
C4H7NO2	002517-04-6	2-AZETIDINECARBOXYLIC ACID
C4H7NO2S	112740-61-1	2-OXOPROPANAL,1-METHYLSULFIDE
C4H7NO3S	112740-56-4	2-OXOPROPANAL,1-METHYLSULFOXIDE
C4H7NO4	000056-84-8	ASPARTIC ACID
C4H7NS	002295-35-4	2-AZACYCLOPENTANTHIONE
C4H8	000106-98-9	1-BUTENE
C4H8	000115-11-7	ISOBUTENE
C4H8	000287-23-0	CYCLOBUTANE
C4H8	000590-18-1	CIS-2-BUTENE
C4H8	000594-11-6	METHYLCYCLOPROPANE
C4H8	000624-64-6	TRANS-2-BUTENE
C4H8Br2	000110-52-1	1,4-DIBROMOBUTANE
C4H8Br2	000533-98-2	1,2-DIBROMOBUTANE
C4H8Br2	000594-34-3	1,2-DIBROMO-2-METHYLPROPANE
C4H8Br2	005408-86-6	2,3-DIBROMOBUTANE
C4H8BrNO	005327-00-4	N-ET-BROMOACETAMIDE
C4H8Cl2	000110-56-5	1,4-DICHLOROBUTANE
C4H8Cl2	000541-33-3	1,1-DICHLOROBUTANE
C4H8Cl2	000594-37-6	1,2-DICHLORO-2-METHYLPROPANE
C4H8Cl2	000616-21-7	1,2-DICHLOROBUTANE
C4H8Cl2	007581-97-7	2,3-DICHLOROBUTANE
C4H8Cl2O	000111-44-4	BIS(2-CHLOROETHYL) ETHER
C4H8Cl2O	002419-74-1	1,4-DICHLORO-2-HYDROXYBUTANE
C4H8Cl2O2S	000471-03-4	1,1'-SULFONYLBIS(2-CHLOROETHANE)
C4H8Cl2OS	005819-08-9	1,1'-SULFINYLBIS(2-CHLOROETHANE)
C4H8Cl2S	000505-60-2	DI-2-CHLOROETHYL SULFIDE
C4H8Cl2Si	010138-21-3	ETHYL VINYL DICHLOROSILANE
C4H8Cl3O4P	000052-68-6	TRICHLORFON
C4H8F3N	000819-06-7	2,2,2-TRIF N,N-DIME ETHYLAMINE
C4H8F3N	058171-47-4	ETHANAMINE, N-ETHYL-2,2,2-TRIFLUORO-
C4H8N2	000534-26-9	2-METHYL-2-IMIDAZOLINE
C4H8N2O	000930-55-2	N-NITROSOPYRROLIDINE
C4H8N2O	004549-43-3	N-NITROSOMETHYLALLYLAMINE
C4H8N2O	013256-13-8	N-NITROSO-ETHYLVINYLAMINE
C4H8N2O2	000059-89-2	N-NITROSOMORPHOLINE

C4H8N2O2	000095-45-4	DIMETHYL GLYOXIME
C4H8N2O2	000110-14-5	SUCCINAMIDE
C4H8N2O2	000623-59-6	ACETAMIDE, N- (METHYLAMINO)CARBONYL -
C4H8N2O2	002620-63-5	GLYCIN-AMIDE,N-ACETYL
C4H8N2O3	000070-47-3	ASPARAGINE
C4H8N2O3	000556-50-3	GYCYLGLYCINE
C4H8N2O3	000615-53-2	ETHYL N-METHYLNITROSOCARBAMATE
C4H8N2O3	056856-83-8	N-NITROSO-ACETOXYMETHYLMETHYLAMINE
C4H8N2OS	026541-51-5	N-NITROSOTHIOMORPHOLINE
C4H8N2S	000109-57-9	ALLYLTHIOUREA
C4H8N2S	002055-46-1	N,N'-TRIMETHYLENETHIOUREA
C4H8N2S4	000111-54-6	1,2-ETHANEDIYLBISCARBAMODITHIOIC ACID
C4H8N3O	003775-61-9	2-NH2-5-ET-1,3,4-OXADIAZOLE
C4H8N4	031857-31-5	1,3-DIMETHYL-2-CYANOGUANIDINE
C4H8N4O2	000140-79-4	N,N'-DINITROSOPIPERAZINE
C4H8N8O8	002691-41-0	OCTAHYDRO-1,3,5,7-TETRANITRO-1,3,5,7-TETRAZOC*
C4H8O	000078-84-2	ISOBUTYRALDEHYDE
C4H8O	000078-93-3	2-BUTANONE
C4H8O	000106-88-7	1,2-BUTYLENEOXIDE
C4H8O	000109-92-2	ETHYL VINYL ETHER
C4H8O	000109-99-9	TETRAHYDROFURAN
C4H8O	000116-11-0	ISOPROPENYL METHYL ETHER
C4H8O	000123-72-8	BUTYRALDEHYDE
C4H8O	000558-30-5	2,2-DIMETHYLOXIRANE
C4H8O	004088-60-2	2-BUTEN-1-OL (CIS)
C4H8O2	000079-31-2	ISOBUTYRICACID
C4H8O2	000107-92-6	BUTYRIC ACID
C4H8O2	000110-64-5	2-BUTENE-1,4-DIOL
C4H8O2	000110-74-7	PROPYL FORMATE
C4H8O2	000123-91-1	1,4-DIOXANE
C4H8O2	000141-78-6	ETHYL ACETATE
C4H8O2	000497-06-3	3-BUTENE-1,2-DIOL
C4H8O2	000505-22-6	1,3-DIOXANE
C4H8O2	000513-86-0	ACETOIN
C4H8O2	000554-12-1	METHYL PROPIONATE
C4H8O2	000590-90-9	3-HYDROXY-2-BUTANONE
C4H8O2	000625-55-8	ISOPROPYL FORMATE
C4H8O2	000821-11-4	2-BUTENE-1,4-DIOL (TRANS)
C4H8O2	001338-23-4	PEROXIDE-2-BUTANONE
C4H8O2	005878-19-3	1-METHOXY-2-PROPANONE
C4H8O2	006117-80-2	2-BUTENE-1,4-DIOL (CIS)
C4H8O2	050317-11-8	1,2-DIHYDROXY-2-BUTENE
C4H8O2S	000126-33-0	TETRAHYDROTHIOPHENE-1,1-DIOXIDE
C4H8O3	000547-64-8	2-HYDROXYPROPANOIC ACID, METHYL ESTER
C4H8O3	000594-61-6	A-HYDROXY-I-BUTYRIC ACID
C4H8OS2	000108-25-8	O-ISOPROPYL DITHIOCARBONATE
C4H8S	000110-01-0	TETRAHYDROTHIOPHENE
C4H8S2	000505-29-3	1,4-DITHIANE
C4H9Br	000078-76-2	2-BROMOBUTANE
C4H9Br	000078-77-3	ISOBUTYL BROMIDE
C4H9Br	000109-65-9	1-BROMOBUTANE
C4H9Br	000507-19-7	2-BROMO-2-METHYLPROPANE
C4H9Cl	000078-86-4	2-CHLOROBUTANE
C4H9Cl	000109-69-3	1-CHLOROBUTANE
C4H9Cl	000507-20-0	T-BUTYL CHLORIDE
C4H9Cl	000513-36-0	ISOBUTYL CHLORIDE
C4H9Cl	000918-20-7	2-CHLORO-2-METHYLPROPANE
C4H9ClO	000628-34-2	2-CHLOROETHYLETHER

C4H9ClO	000928-51-8	1-BUTANOL, 4-CHLORO-
C4H9ClOS	000693-30-1	2-((2-CHLOROETHYL)THIO)ETHANOL
C4H9F	002366-52-1	1-FLUOROBUTANE
C4H9I	000513-38-2	1-IODO-2-METHYLPROPANE
C4H9I	000513-48-4	2-IODOBUTANE
C4H9I	000542-69-8	N-BUTYL IODIDE
C4H9N	000123-75-1	PYRROLIDINE
C4H9N	002549-67-9	2-ETHYLAZIRIDINE
C4H9N	002658-24-4	2,2-DIMETHYLETHYLENIMINE
C4H9N3O	005632-47-3	N-NITROSOPIPERAZINE
C4H9N3O	057421-73-5	4-ALLYLSEMICARBAZIDE
C4H9N3O2	000057-00-1	CREATINE
C4H9N3O2	000816-57-9	1-PROPYL-1-NITROSOUREA
C4H9N3O2	003475-63-6	1-NITROSO-TRIMETHYL UREA
C4H9NO	000096-29-7	2-BUTANONE OXIME
C4H9NO	000110-91-8	MORPHOLINE
C4H9NO	000127-19-5	N,N'-DIMETHYLACETAMIDE
C4H9NO	000541-35-5	BUTYRAMIDE
C4H9NO	001187-58-2	N-METHYL PROPIONAMIDE
C4H9NO2	000056-12-2	4-AMINOBUTYRIC ACID
C4H9NO2	000080-60-4	ALPHA-AMINOBUTYRIC ACID
C4H9NO2	000105-40-8	N-METHYLCARBAMIC ACID, ET ESTER
C4H9NO2	000544-16-1	1-BUTYLNITRITE
C4H9NO2	000594-70-7	2-METHYL-2-NITROPROPANE
C4H9NO2	000600-24-8	2-NITROBUTANE
C4H9NO2	000625-74-1	ISONITROBUTANE
C4H9NO2	000627-05-4	1-NITROBUTANE
C4H9NO2	000627-12-3	O-PROPYLCARBAMATE
C4H9NO2	001118-68-9	N,N-DIMETHYLGLYCINE
C4H9NO3	000072-19-5	THREONINE(L)
C4H9NO3	000076-39-1	2-METHYL-2-NITRO-1-PROPANOL
C4H9NO3	000543-29-3	ISOBUTYL NITRATE
C4H9NO3	000609-31-4	2-NITRO-1-BUTANOL
C4H9NO3	000928-45-0	BUTYL NITRATE
C4H9NO4	000077-49-6	2-METHYL-2-NITROPROPANE-1,3-DIOL
C4H9NO5	000126-11-4	2-HYDROXYMETHYL-2-NITRO-1,3-PROPANEDIOL
C4H10	000075-28-5	2-METHYLPROPANE
C4H10	000106-97-8	N-BUTANE
C4H10Cl2Si	001719-53-5	DICHLORODIETHYLSILANE
C4H10FO2P	000107-44-8	SARIN
C4H10N2	000110-85-0	PIPERAZINE
C4H10N2O	000055-18-5	N-NITROSODIETHYLAMINE
C4H10N2O	000924-46-9	N-NITROSOMETHYLPROPYLAMINE
C4H10N2O	003538-65-6	BUTRYIC ACID HYDRAZIDE
C4H10N2O	003619-17-8	I-BUTRYIC ACID HYDRAZIDE
C4H10N2O2	000305-62-4	2,4-DIAMINOBUTYRIC ACID
C4H10N2O2	005710-12-3	3-PROPYLHYDROXYUREA
C4H10N2O2	013147-25-6	(N-NITROSO-N-ETHYL)ETHANOLAMINE
C4H10N2O2	060165-07-3	3-ISOPROPYLHYDROXYUREA
C4H10N2O3	001116-54-7	N-NITROSODIETHANOLAMINE
C4H10NO3PS	030560-19-1	ACEPHATE
C4H10O	000060-29-7	DIETHYL ETHER
C4H10O	000071-36-3	1-BUTANOL
C4H10O	000075-65-0	T-BUTANOL
C4H10O	000078-83-1	ISOBUTYL ALCOHOL
C4H10O	000078-92-2	2-BUTANOL
C4H10O	000557-17-5	METHYL PROPYL ETHER
C4H10O	000598-53-8	METHYL ISOPROPYL ETHER

C4H10O	004712-38-3	1-BUTANOL (D)
C4H10O	004712-39-4	2-BUTANOL (D)
C4H10O2	000075-91-2	TERT-BUTYLHYDROPEROXIDE
C4H10O2	000107-88-0	1,3-BUTANEDIOL
C4H10O2	000107-98-2	1-METHOXY-2-PROPANOL
C4H10O2	000110-63-4	1,4-BUTANEDIOL
C4H10O2	000110-71-4	ETHYLENE GLYCOL DIMETHYL ETHER
C4H10O2	000110-80-5	2-ETHOXYETHANOL
C4H10O2	000513-85-9	2,3-BUTANEDIOL
C4H10O2	000534-15-6	1,1-DIMETHOXYETHANE
C4H10O2	000584-03-2	1,2-BUTANEDIOL
C4H10O2	005341-95-7	2,3-BUTANEDIOL (MESO)
C4H10O2	006982-25-8	DL-2,3-BUTANDIOL
C4H10O2	019132-06-0	2,3-BUTANEDIOL (D)
C4H10O2	024347-58-8	2,3-BUTANEDIOL (L)
C4H10O2S	000111-48-8	2,2'-THIOBISETHANOL
C4H10O2S	000597-35-3	DIETHYL SULFONE
C4H10O3	000111-46-6	DIETHYLENE GLYCOL
C4H10O3	000149-73-5	TRIMETHOXYMETHANE
C4H10O3	004435-50-1	1,2,3-TRIHYDROXYBUTANE
C4H10O4	000149-32-6	ERYTHRITOL
C4H10O4S	000064-67-5	DIETHYL SULFATE
C4H10S	000075-66-1	2-METHYL-2-PROPANETHIOL
C4H10S	000109-79-5	BUTANETHIOL
C4H10S	000352-93-2	DIETHYLSULFIDE
C4H10S	000513-44-0	2-METHYL-1-PROPANETHIOL
C4H10S	000513-53-1	2-BUTANETHIOL
C4H10S	001551-21-9	3-METHYL-2-THIABUTANE
C4H10S	003877-15-4	N-PROPYLMETHYLSULFIDE
C4H10S2	000110-81-6	DIETHYL DISULFIDE
C4H11N	000075-64-9	T-BUTYLAMINE
C4H11N	000078-81-9	ISOBUTYLAMINE
C4H11N	000109-73-9	N-BUTYLAMINE
C4H11N	000109-89-7	DIETHYLAMINE
C4H11N	000513-49-5	SEC-BUTYLAMINE
C4H11N	000598-56-1	ETHYL DIMETHYLAMINE
C4H11N	000627-35-0	N-METHYLPROPYLAMINE
C4H11N	013952-84-6	SEC-BUTYLAMINE
C4H11N3O	057421-72-4	4-PROPYLSEMICARBAZIDE
C4H11N3O	057930-20-8	4-I-PROPYLSEMICARBAZIDE
C4H11NO	000108-01-0	2-DIMETHYLAMINOETHANOL
C4H11NO	000124-68-5	2-AMINO-2-METHYL-1-PROPANOL
C4H11NO	003710-84-7	DIETHYLHYDROXYLAMINE
C4H11NO	005332-73-0	3-METHOXYPROPYLAMINE
C4H11NO2	000111-42-2	DIETHANOLAMINE
C4H11NO2	000115-69-5	2-AMINO-2-METHYLPROPANEDIOL
C4H11NO3	000077-86-1	AMINOTRIS(HYDROXYMETHYL)METHANE
C4H11NO8P2	002439-99-8	GLYPHOSINE
C4H11O3P	000762-04-9	DIETHYL PHOSPHITE
C4H11O3P	001832-54-8	ISOPROPYL METHYLPHOSPHONATE
C4H12BrN	006274-12-0	DIETHYLAMINE, HYDROBROMIDE
C4H12ClN	000660-68-4	N-ETHYLETHANAMINE, HYDROCHLORIDE
C4H12ClN	005041-09-8	ISOBUTYLAMINE HYDROCHLORIDE
C4H12ClO4P	000124-64-1	TETRAKIS(HYDROXYMETHYL)PHOSPHONIUM CHLORIDE
C4H12FN2OP	000115-26-4	DIMEFOX
C4H12IN	019833-78-4	DIETHYLAMINE, HYDROIODIDE
C4H12N2	000110-60-1	1,4-BUTANEDIAMINE
C4H12N2	000110-70-3	1,2-ETHANEDIAMINE, N,N'-DIMETHYL-

C4H12N2	001615-80-1	N,N'-DIETHYLHYDRAZINE
C4H12N2O	000111-41-1	2-(2-AMINOETHYLAMINO)ETHANOL
C4H12NO2PS	016271-10-6	O,S-DIETHYLPHOSPHORAMIDOTHIOATE
C4H12NO2PS	052067-48-8	O=P(OC)(SC)N-ETHYL
C4H12Si	000075-76-3	TETRAMETHYLSILANE
C4H13N3	000111-40-0	DIETHYLENETRIAMINE
C5Cl5N	002176-62-7	2,3,4,5,6-PENTACHLORPYRIDINE
C5Cl6	000077-47-4	HEXACHLOROCYCLOPENTADIENE
C5Cl8	000706-78-5	OCTACHLOROCYCLOPENTENE
C5F12	000678-26-2	PENTANE, DODECAFLUORO-
C5HCl4N	002402-79-1	2,3,5,6-TETRACHLORPYRIDINE
C5H2Cl3N	006515-09-9	2,3,6-TRICHLOROPYRIDINE
C5H2Cl3N	016063-69-7	2,4,6-TRICHLOROPYRIDINE
C5H2Cl3N	016063-70-0	2,3,5-TRICHLOROPYRIDINE
C5H2Cl3NO	001970-40-7	PYRICLOR
C5H2Cl3NO	006515-38-4	3,5,6-TRICHLORO-2-PYRIDINOL
C5H2Cl8	083682-40-0	1,1-DICHLORO-2,3-BIS(TRICHLOROMETHYL)PROPENE
C5H2ClN3	006863-74-7	2-CHLORO-6-CYANO-PYRAZINE
C5H2ClN3	036070-75-4	PYRAZINECARBONITRILE, 5-CHLORO-
C5H3Cl2N	002402-77-9	2,3-DICHLOROPYRIDINE
C5H3Cl2N	002402-78-0	2,6-DICHLOROPYRIDINE
C5H3Cl2N	002457-47-8	3,5-DICHLOROPYRIDINE
C5H3Cl2N	016110-09-1	2,5-DICHLOROPYRIDINE
C5H3F2N	001513-65-1	2,6-DIFLUOROPYRIDINE
C5H3F3N2O2	000054-20-6	TRIFLUOROMETHYLURACIL
C5H3N3	014080-23-0	2-PYRIMIDINECARBONITRILE
C5H3N3	019847-12-2	2-CYANOPYRAZINE
C5H3N3	040805-79-6	5-PYRIMIDINECARBONITRILE
C5H3N3	042839-04-3	4-PYRIMIDINECARBONITRILE
C5H3N3	068776-62-5	4-PYRIDAZINECARBONITRILE
C5H3N3O2	004425-56-3	5-CYANOURACIL
C5H3N5O	117505-25-6	PYRAZINE AZOXYCYANIDE
C5H3N5O	117539-78-3	2-PYRIMIDINE AZOXYCYANIDE
C5H3NO4	000698-63-5	5-NITRO-2-FURALDEHYDE
C5H4BrN	000109-04-6	2-BROMOPYRIDINE
C5H4BrN	000626-55-1	3-BROMOPYRIDINE
C5H4BrN	001120-87-2	4-BROMOPYRIDINE
C5H4ClN	000109-09-1	2-CHLOROPYRIDINE
C5H4ClN	000626-60-8	3-CHLOROPYRIDINE
C5H4ClN	000626-61-9	4-CHLOROPYRIDINE
C5H4ClN3O	036070-79-8	2-CHLORO-6-CONH2-PYRAZINE
C5H4ClN3O	066346-83-6	3-CL-6-PYRIDAZINECARBOXAMIDE
C5H4FN	000372-47-4	3-FLUOROPYRIDINE
C5H4FN	000372-48-5	2-FLUOROPYRIDINE
C5H4IN	001120-90-7	3-IODOPYRIDINE
C5H4N2	004513-94-4	2-CYANOPYRROLE
C5H4N2O2	001122-61-8	4-NITROPYRIDINE
C5H4N2O2	002530-26-9	3-NITROPYRIDINE
C5H4N2O2	015009-91-3	2-NITROPYRIDINE
C5H4N2O3	001195-08-0	5-FORMYLURACIL
C5H4N2O4	000065-86-1	OROTIC ACID
C5H4N4	000120-73-0	1H-PURINE
C5H4N4O	000068-94-0	HYPOXANTHINE
C5H4N4O	000315-30-0	ALLOPURINOL
C5H4N4O2	000069-89-6	XANTHINE
C5H4N4S	000050-44-2	6-PURINETHIOL HYDRATE
C5H4NO	000617-90-3	2-CYANOFURAN
C5H4O2	000098-01-1	FUFURAL

C5H4O2	000498-60-2	FURAN-3-CARBOXALDEHYDE
C5H4O2	000542-28-9	D-VALEROLACTONE
C5H4O2S	000088-13-1	THIOPHENE-3-CARBOXYLIC ACID
C5H4O2S	000527-72-0	THIOPHENE-2-CARBOXYLIC ACID
C5H4O3	000088-14-2	2-FURANCARBOXYLIC ACID
C5H4O3	000488-93-7	FURAN-3-CARBOXYLIC ACID
C5H4OS	000098-03-3	THIOPHENE-2-CARBOXALDEHYDE
C5H4OS	000498-62-4	3-THIOPHENECARBOXALDEHYDE
C5H5BrN2	013534-98-0	4-AMINO-3-BROMOPYRIDINE
C5H5BrN2O3	077317-64-7	3-OH-5-BR-6-ME-PYRIMIDINE-2,4-DIONE
C5H5Cl	041851-50-7	5-CHLORO-1,3-CYCLOPENTADIENE
C5H5Cl3N2OS	002593-15-9	TERRAZOLE
C5H5ClN2	038557-71-0	2-CHLORO-6-METHYL-PYRAZINE
C5H5ClN2	059303-10-5	PYRAZINE, 2-CHLORO-5-METHYL-
C5H5ClN2O	001722-10-7	3-CHLORO-6-METHOXYPYRIDAZINE
C5H5ClN2O	033332-30-8	2-CHLORO-6-METHOXY-PYRAZINE
C5H5ClN2O	033332-31-9	PYRAZINE, 2-CHLORO-5-METHOXY-
C5H5F5O2	000426-65-3	ETHYL PENTAFLUOROPROPIONATE
C5H5N	000110-86-1	PYRIDINE
C5H5N2O4	006236-05-1	NIFUROXIME
C5H5N3O	000098-96-4	2-PYRAZINECARBOXAMIDE
C5H5N3O	005450-54-4	3-PYRIDAZINECARBOXAMIDE
C5H5N3O	028648-86-4	4-PYRIMIDINECARBOXAMIDE
C5H5N3O	040929-49-5	5-PYRIMIDINECARBOXAMIDE
C5H5N3O	088511-47-1	4-PYRIDAZINECARBOXAMIDE
C5H5N3O	088511-48-2	2-PYRIMIDINECARBOXAMIDE
C5H5N3O2	004214-76-0	2-PYRIDINAMINE, 5-NITRO-
C5H5N3O3	039928-74-0	1H-IMIDAZOLE-5-CARBOXALDEHYDE, 1-METHYL-2-NITRO-
C5H5N3S	004537-73-9	PYRIMIDINE-2-THIOCARBOXAMIDE
C5H5N5	000073-24-5	ADENINE
C5H5N5	066224-65-5	9H-PURIN-6-AMINE
C5H5N5O	000073-40-5	GUANINE
C5H5N5S	000154-42-7	THIOGUANINE
C5H5NO	000109-00-2	3-HYDROXYPYRIDINE
C5H5NO	000109-10-4	A-PYRIDONE
C5H5NO	000142-08-5	2-HYDROXYPYRIDINE
C5H5NO	000626-64-2	4-HYDROXYPYRIDINE
C5H5NO	000694-59-7	PYRIDINE, 1-OXIDE
C5H5NO	001003-29-8	PYRROLE-2-CARBOXALDEHYDE
C5H5NO2	000137-05-3	METHYL 2-CYANO-2-PROPENOATE
C5H5NO2	000609-38-1	FURAN-2-CARBOXAMIDE
C5H5NO2	000634-97-9	PYRROLE-2-CARBOXYLIC ACID
C5H5NS	001003-31-2	2-CYANOTHIOPHENE
C5H5NS	002637-34-5	2(1H)-PYRIDINETHIONE
C5H5NS	019829-29-9	4(1H)-PYRIDINETHIONE
C5H6	000542-92-7	CYCLOPENTADIENE
C5H6	004054-38-0	1,3-CYCLOHEPTADIENE
C5H6BrN3O2	021117-52-2	5-BROMO-1,2-DIMETHYL-4-NITROIMIDAZOLE
C5H6BrN3O2	021431-58-3	4-BROMO-1,2-DIMETHYL-5-NITROIMIDAZOLE
C5H6BrN3O2	105983-47-9	2-BROMO-1,2-DIMETHYL-5-NITRO-1H-IMIDAZOLE
C5H6Cl2O2	002873-74-7	GLUTARYL CHLORIDE
C5H6Cl6N2O3	000116-52-9	DICHLORAL UREA
C5H6ClN3O2	091027-93-9	5-CHLORO-1,2-DIMETHYL-4-NITRO-1H-IMIDAZOLE
C5H6ClN3O2	091027-94-0	4-CHLORO-1,2-DIMETHYL-5-NITRO-1H-IMIDAZOLE
C5H6N2	000109-08-0	2-METHYLPYRAZINE
C5H6N2	000462-08-8	3-AMINOPYRIDINE
C5H6N2	000504-24-5	4-AMINOPYRIDINE
C5H6N2	000504-29-0	2-AMINOPYRIDINE

C5H6N2	000544-13-8	PENTANEDINITRILE
C5H6N2	001120-88-3	PYRIDAZINE, 4-METHYL-
C5H6N2	001632-76-4	PYRIDAZINE, 3-METHYL-
C5H6N2	002036-41-1	PYRIMIDINE, 5-METHYL-
C5H6N2	003438-46-8	4-METHYLPYRIMIDINE
C5H6N2	005053-43-0	PYRIMIDINE, 2-METHYL-
C5H6N2	007321-55-3	DIMETHYLMALONONITRILE
C5H6N2O	000931-63-5	PYRIMIDINE, 2-METHOXY-
C5H6N2O	002466-76-4	1H-IMIDAZOLE, 1-ACETYL-
C5H6N2O	003149-28-8	2-METHOXYPYRAZINE
C5H6N2O	004551-72-8	1H-PYRROLE-2-CARBOXAMIDE
C5H6N2O	006104-41-2	PYRIMIDINE, 4-METHOXY-
C5H6N2O	019064-65-4	PYRIDAZINE, 3-METHOXY-
C5H6N2O	020733-11-3	PYRIDAZINE, 4-METHOXY-
C5H6N2O	031458-33-0	PYRIMIDINE, 5-METHOXY-
C5H6N2O2	000065-71-4	5-METHYLURACIL
C5H6N2O2	000615-77-0	1-METHYLURACIL
C5H6N2O2	022390-53-0	3,4-PYRIDAZINEDIOL, 6-METHYL-
C5H6N2O2S	002438-76-8	3-SULFONAMIDOPYRIDINE
C5H6N2O3	001124-33-0	4-NITROPYRIDINE-1-OXIDE
C5H6N2O3	002565-47-1	N-METHYL BARBITURIC ACID
C5H6N2O4	104151-78-2	FURAZANCARBOXYLIC ACID, 4-METHYL-, METHYL ESTER,
C5H6N2O4	104151-90-8	FURAZANCARBOXYLIC ACID, 4-METHYL-, METHYL ESTER,
C5H6N2OS	000056-04-2	METACIL
C5H6N2OS	124700-70-5	4-PYRIMIDINOL, 2-(METHYLTHIO)- (KETO FORM)
C5H6N2S	000823-09-6	PYRIMIDINE, 2-(METHYLTHIO)-
C5H6N2S	021948-70-9	2-METHYLTHIOPYRAZINE
C5H6N4	000274-87-3	TETRAZOLO[1,5-A]PYRIDINE
C5H6N4O3	000069-93-2	URIC ACID
C5H6N4O4	019183-17-6	1,2-DIMETHYL-4,5-DINITROIMIDAZOLE
C5H6NO2	000609-35-8	FURAN-3-CARBOXAMIDE
C5H6NO3	004857-42-5	5-METHYLISOXAZOLE-3-CARBOXYLIC ACID
C5H6O	000534-22-5	2-METHYLFURAN
C5H6O	000930-27-8	3-METHYLFURAN
C5H6O2	000098-00-0	2-HYDROXYMETHYLFURAN
C5H6O2	004412-91-3	3-HYDROXYMETHYLFURAN
C5H6O3	000108-55-4	GLUTARIC ANHYDRIDE
C5H6O3	004100-80-5	PYROTATARIC ANHYDRIDE
C5H6S	000554-14-3	2-METHYLTHIOPHENE
C5H6S	000616-44-4	3-METHYLTHIOPHENE
C5H7N	000096-54-8	N-METHYLPYRROLE
C5H7N3	000108-52-1	2-AMINO-4-METHYLPYRIMIDINE
C5H7N3	003435-28-7	6-METHYL-4-PYRIMIDINAMINE
C5H7N3	005521-56-2	PYRAZINAMINE, 6-METHYL-
C5H7N3	006714-29-0	1H-IMIDAZO[1,2-B]PYRAZOLE, 2,3-DIHYDRO-
C5H7N3	032111-28-7	2-METHYLAMINOPYRAZINE
C5H7N3O	006647-92-3	4-ACETYLAMINOPYRAZOLE
C5H7N3O	006905-47-1	PYRAZINAMINE, 6-METHOXY-
C5H7N3O	018591-86-1	4-PYRIDAZINOL, 3-AMINO-6-METHYL-
C5H7N3O	054013-07-9	PYRAZINAMINE, 5-METHOXY-
C5H7N3O2	000551-92-8	1,2-DIMETHYL-5-NITRO-1H-IMIDAZOLE
C5H7N3O2	007464-68-8	1,5-DIMETHYL-4-NITRO-1H-IMIDAZOLE
C5H7N3O2	013230-04-1	1,2-DIMETHYL-4-NITROIMIDAZOLE
C5H7N3O2	019213-65-1	5-ETHYL-6-AZAURACIL
C5H7N3O3	035687-41-3	5-METHOXY-1-METHYL-3-NITRO-1H-IMIDAZOLE
C5H7N3O3	035687-42-4	4-METHOXY-2-METHYL-5-NITRO-1H-IMIDAZOLE
C5H7N3O3S	000140-40-9	NITHIAMIDE
C5H7N3O4	000115-02-6	AZASERINE

C5H7NO	000617-89-0	2-AMINOMETHYLFURAN
C5H7NO2	000105-56-6	ETHYL CYANOACETATE
C5H7NO2	001121-07-9	N-METHYLSUCCINIMIDE
C5H7NO2	006602-28-4	3-HYDROXYPYRIDINE-N-OXIDE
C5H7NOS	005813-89-8	THIOPHEN-2-CARBOXAMIDE
C5H7NOS	051460-47-0	THIOPHEN-3-CARBOXAMIDE
C5H7NS	000541-58-2	2,4-DIMETHYLTHIAZOLE
C5H7NS2	020300-02-1	THIOPHENE,2-THIOCARBOXAMIDE
C5H7O2	025414-22-6	2-METHOXYFURAN
C5H8	000078-79-5	2-METHYL-1,3-BUTADIENE
C5H8	000142-29-0	CYCLOPENTENE
C5H8	000157-40-4	SPIROPENTANE
C5H8	000185-94-4	BICYCLO(2.1.0)PENTANE
C5H8	000504-60-9	1,3-PENTADIENE
C5H8	000591-93-5	1,4-PENTADIENE
C5H8	000591-95-7	1,2-PENTADIENE
C5H8	000598-25-4	3-METHYL-1,2-BUTADIENE
C5H8	000627-19-0	1-PENTYNE
C5H8	000627-21-4	2-PENTYNE
C5H8	001574-41-0	1,3-PENTADIENE (Z)
C5H8	002004-70-8	1,3-PENTADIENE (E)
C5H8Br2N2O	057541-73-8	3,4-DIBROMO-N-NITROSOPIPERIDINE
C5H8Br4	003229-00-3	PROPANE, 1,3-DIBROMO-2,2-BIS(BROMOMETHYL)-
C5H8BrN3O2	105994-26-1	1H-IMIDAZOLE, 2-BROMO-1,5-DIMETHYL-4-NITRO-
C5H8Cl2N2O	057541-72-7	3,4-DICHLORO-N-NITROSOPIPERIDINE
C5H8Cl3NO3	000541-79-7	CARBAMIC ACID, (2,2,2-TRICHLORO-1-HYDROXYETHYL)-
C5H8ClN5	001007-28-9	DESISOPROPYLATRAZINE
C5H8N2	000067-51-6	1H-PYRAZOLE, 3,5-DIMETHYL-
C5H8N2O	014150-95-9	2-PYRIDINAMINE, 1-OXIDE
C5H8N2O	017647-69-7	FURAZAN, ETHYLMETHYL-
C5H8N2O	055556-92-8	PYRIDINE, 1,2,3,6-TETRAHYDRO-1-NITROSO-
C5H8N2O	081437-10-7	4-ETHOXYPYRAZOLE
C5H8N2O2	000077-71-4	2,4-IMIDAZOLIDINEDIONE, 5,5-DIMETHYL-
C5H8N2O2	050550-65-7	2-PIPERIDINONE, 1-NITROSO-
C5H8N2O2	055556-91-7	N-NITROSO-4-PIPERIDONE
C5H8N2O2S2	106231-45-2	1,3-DITHIOLAN-4-ONE, [(METHYLAMINO)CARBONYL]OXIM
C5H8N2O3	000638-20-0	1,3-DIACETYLUREA
C5H8N2O3S	054266-44-3	1,3-OXATHIOLAN-4-ONE, O-[(METHYLAMINO)CARBONYL]O
C5H8N2OS	002719-23-5	ACETAMIDE, N-2-THIAZOLYL-
C5H8N2S	018453-07-1	2-THIAZOLEETHANAMINE
C5H8N4O12	000078-11-5	PENTAERYTHRITOL TETRANITRATE
C5H8N4O2	021677-57-6	5-AMINO-1,2-DIMETHYL-4-NITROIMIDAZOLE
C5H8N4O2	038726-90-8	ACETAMIDE, 2-[(DIAZOACETYL)AMINO]-N-METHYL-
C5H8N4O2	102626-89-1	UREA, N-METHYL-N'-(METHYLFURAZANYL)-
C5H8N4O3	022903-73-7	2-(2-NO2-1-IMIDAZOLYL)ACETAMIDE
C5H8N4O3S2	000554-57-4	METHAZOLAMIDE
C5H8N4O4	070965-23-0	1-TRIAZOLEACETIC ACID-3-NITRO,METHYL ESTER
C5H8N4OS	018826-96-5	3-MES-4NH2-6-ME-1,2,4-TRIAZIN-5-ON
C5H8N4S	007120-01-6	GUANIDINE, (4-METHYL-2-THIAZOLYL)-
C5H8NNaO4	000142-47-2	MONOSODIUM L-GLUTAMATE
C5H8O	000110-87-2	2H-PYRAN, 3,4-DIHYDRO-
C5H8O	000115-19-5	2-METHYL-3-BUTYN-2-OL
C5H8O	000120-92-3	CYCLOPENTANONE
C5H8O	000625-33-2	3-PENTEN-2-ONE
C5H8O	000765-43-5	CYCLOPROPYL METHYL KETONE
C5H8O2	000080-59-1	TIGLIC ACID
C5H8O2	000080-62-6	METHYL METHACRYLATE
C5H8O2	000105-38-4	VINYL PROPIONATE

C5H8O2	000111-30-8	PENTANE-1,5-DIAL
C5H8O2	000123-54-6	2,4-PENTANEDIONE
C5H8O2	000140-88-5	ETHYL ACRYLATE
C5H8O2	000591-87-7	ALLYL ACETATE
C5H8O2S2	038695-58-8	THIOPHENE, 3-METHYLSULFONYL-
C5H8O3	000105-45-3	METHYL ACETOACETATE
C5H8O3	000123-76-2	LEVULINIC ACID
C5H8O3	000818-61-1	2-HYDROXYETHYL ACRYLATE
C5H8O4	000108-59-8	PROPANEDIOIC ACID, DIMETHYL ESTER
C5H8O4	000110-94-1	PENTANEDIOC ACID
C5H8O4	000498-21-5	METHYLSUCCINIC ACID
C5H8O4	000595-46-0	DIMETHYLMALONIC ACID
C5H8O4	000601-75-2	ETHYLMALONIC ACID
C5H8OS	000636-72-6	2-HYDROXYMETHYLTHIOPHENE
C5H9BrO2	000565-74-2	2-BROMO-3-METHYL-BUTANOIC ACID
C5H9BrO2	002067-33-6	5-BROMOPENTANOIC ACID
C5H9Cl	021450-13-5	1-CHLORO-1-PENTENE
C5H9Cl2N3O2	000154-93-8	1,3-BIS(2-CHLOROET)-1-NITROSOUREA
C5H9ClFN3O2	013908-92-4	3(2-CLET)-1-(2FET)-1-NITROSOUREA
C5H9ClN4O3	104958-94-3	1,2,4-TRIAZOLE,3-NITRO-1-(3-CHLORO-2-HYDROXY)PRO
C5H9ClO2	000105-48-6	ISOPROPYL CHLOROACETATE
C5H9ClO2	000535-13-7	ETHYL ALPHA-CHLOROPROPIONATE
C5H9F3O	000352-61-4	5,5,5-TRIFLUORO-1-PENTANOL
C5H9F3O	095853-68-2	4,4,4-TRIFLUORO-3-METHYL-1-BUTANOL
C5H9F3O	114525-06-3	4,4,4-TRIFLUORO-2-METHYL-1-BUTANOL
C5H9N	000110-59-8	VALERONITRILE
C5H9N	000625-28-5	BUTANENITRILE, 3-METHYL-
C5H9N	000630-18-2	TRIMETHYLACETONITRILE
C5H9N	018936-17-9	2-METHYLBUTYRONITRILE
C5H9N3	000051-45-6	1H-IMIDAZOLE-4-ETHANAMINE
C5H9N3O	090467-88-2	DIAZENECARBONITRILE, (1,1-DIMETHYLETHYL)-2-OXIDE
C5H9N3O3	000936-05-0	1-ME-2-NO2-5-HYDROXYME IMIDAZOLE
C5H9N3O3	005006-67-7	1-(2-HYDROXYET)-2-NITROIMIDAZOLE
C5H9NO	000108-27-0	2-PYRROLIDINONE, 5-METHYL-
C5H9NO	000111-36-4	BUTYL ISOCYANATE
C5H9NO	000675-20-7	DELTA-VALEROLACTAM
C5H9NO	000872-50-4	N-METHYLPYRROLIDONE
C5H9NO	003760-54-1	N-FORMYLPYRROLIDINE
C5H9NO2	000147-85-3	PROLINE
C5H9NO2	004394-85-8	N-FORMYLMORPHOLINE
C5H9NO3	000051-35-4	HYDROXYPROLINE
C5H9NO4	000056-86-0	GLUTAMIC ACID
C5H9NO4	000617-65-2	GLUTAMIC ACID (DL)
C5H9NS	000591-82-2	PROPANE, 1-ISOTHIOCYANATO-2-METHYL-
C5H9NS	000592-82-5	BUTANE, 1-ISOTHIOCYANATO-
C5H9NS	000628-83-1	THIOCYANIC ACID, BUTYL ESTER
C5H9NS	004426-79-3	BUTANE, 2-ISOTHIOCYANATO-
C5H9NS	013070-01-4	2-AZAHEXANTHIONE
C5H10	000109-67-1	1-PENTENE
C5H10	000109-68-2	2-PENTENE
C5H10	000287-92-3	CYCLOPENTANE
C5H10	000513-35-9	2-METHYL-2-BUTENE
C5H10	000563-45-1	3-METHYL-1-BUTENE
C5H10	000563-46-2	2-METHYL-1-BUTENE
C5H10	000598-61-8	METHYLCYCLOBUTANE
C5H10	000627-20-3	CIS-2-PENTENE
C5H10	000646-04-8	TRANS-2-PENTENE
C5H10Br2	019398-53-9	1,2-DIBROMOPENTANE

C5H10Br2O2	003296-90-0	1,3-PROPANEDIOL, 2,2-BIS(BRME)-
C5H10Cl2	000628-76-2	1,5-DICHLOROPENTANE
C5H10Cl2O2	000111-91-1	BIS(2-CHLOROETHOXY)METHANE
C5H10ClN3O3	060784-46-5	HYDROXYETHYL-CLET-NO UREA
C5H10ClNO4	000138-15-8	GLUTAMIC ACID HYDROCHLORIDE
C5H10HgO3	000151-38-2	(METHOXYETHYL)MERCURIC ACETATE
C5H10N2O	000100-75-4	N-NITROSOPIPERIDINE
C5H10N2O2	007606-79-3	GLYCINE,N-ACETYL-N'-MEAMINO AMIDE
C5H10N2O2	015962-47-7	ALANINE-AMIDE,N-ACETYL
C5H10N2O2	023549-54-4	2-METHYLPROPANOYLUREA
C5H10N2O2	055556-85-9	3-OH-N-NITROSOPIPERIDINE
C5H10N2O2	055556-93-9	4-OH-N-NITROSOPIPERIDINE
C5H10N2O2	075881-16-2	2-ME-N-NITROSOMORPHOLINE
C5H10N2O2S	010061-64-0	L-CYSTEINAMIDE, N-ACETYL-
C5H10N2O2S	016752-77-5	METHOMYL
C5H10N2O3	000056-85-9	GLUTAMINE
C5H10N2O3	000614-95-9	ETHYL N-ETHYLNITROSOCARBAMATE
C5H10N2O3	023361-38-8	L-SERINAMIDE, N-ACETYL-
C5H10N2S2	000533-74-4	DMTT
C5H10N3O	065283-97-8	2-AMINO-5-I-PR-1,3,4-OXADIAZOLE
C5H10N3O	069741-89-5	2-AMINO-5-PROPYL-1,3,4-OXADIAZOLE
C5H10N4O2	055556-94-0	2-ME-N,N'-DINITROSOPIPERAZINE
C5H10N4O2	055557-00-1	1H-1,4-DIAZEPINE, HEXAHYDRO-1,4-DINITROSO-
C5H10N6O4	027640-19-3	1,1'-PROPYLEN-BIS(1-NITROSOUREA)
C5H10O	000096-17-3	2-METHYLBUTANAL
C5H10O	000096-22-0	3-PENTANONE
C5H10O	000096-41-3	CYCLOPENTANOL
C5H10O	000096-47-9	2-METHYLTETRAHYDROFURAN
C5H10O	000107-87-9	2-PENTANONE
C5H10O	000110-62-3	1-PENTANAL
C5H10O	000142-68-7	TETRAHYDROPYRAN
C5H10O	000557-31-3	ALLYL ETHYL ETHER
C5H10O	000563-80-4	3-METHYL-2-BUTANONE
C5H10O	000590-86-3	3-METHYL-1-BUTANAL
C5H10O	000616-25-1	1-PENTEN-3-OL
C5H10O	000630-19-3	PIVALALDEHYDE
C5H10O	000763-32-6	3-METHYL-3-BUTEN-1-OL
C5H10O	000821-09-0	4-PENTEN-1-OL
C5H10O	000922-63-4	2-ETHYLPROPENAL
C5H10O	001569-50-2	3-PENTEN-2-OL
C5H10O2	000075-98-9	TRIMETHYLACETIC ACID
C5H10O2	000097-99-4	TETRAHYDROFURFURYL ALCOHOL
C5H10O2	000105-37-3	ETHYL PROPIONATE
C5H10O2	000108-21-4	ISO-PROPYLACETATE
C5H10O2	000109-52-4	PENTANOIC ACID
C5H10O2	000109-60-4	N-PROPYL ACETATE
C5H10O2	000111-45-5	ETHANOL, 2-(2-PROPENYLOXY)-
C5H10O2	000115-22-0	3-HYDROXY-3-METHYL-2-BUTANONE
C5H10O2	000116-53-0	BUTANOIC ACID, 2-METHYL-
C5H10O2	000503-74-2	ISOVALERIC ACID
C5H10O2	000542-55-2	ISOBUTYL FORMATE
C5H10O2	000547-63-7	METHYL ISOBUTYRATE
C5H10O2	000592-84-7	N-BUTYLFORMATE
C5H10O2	000623-42-7	METHYLBUTYRATE
C5H10O2	001120-97-4	4-METHYLDIOXANE
C5H10O2	002916-31-6	1,3-DIOXOLANE, 2,2-DIMETHYL-
C5H10O2	003393-64-4	4-HYDROXY-3-METHYL-2-BUTANONE
C5H10O2S	000872-93-5	3-METHYL SULFOLANE

C5H10O3	000097-64-3	ETHYL LACTATE
C5H10O3	000105-58-8	DIETHYL CARBONATE
C5H10O3	000110-49-6	2-METHOXYETHYL ACETATE
C5H10O3	000617-31-2	PENTANOIC ACID, 2-HYDROXY-
C5H10O3	000925-57-5	METHYL-4-HYDROXYBUTYRATE
C5H10O5	000050-69-1	RIBOSE
C5H10O5	000147-81-9	ARABINOSE
C5H10OS2	000110-50-9	BUTYL XANTHATE
C5H10OS2	000623-80-3	S,S-DIETHYL CARBONODITHIOATE
C5H10S	001613-51-0	TETRAHYDRO-2H-THIOPYRAN
C5H10S	001679-07-8	CYCLOPENTANETHIOL
C5H11Br	000107-81-3	2-BROMOPENTANE
C5H11Br	000107-82-4	1-BROMO-3-METHYLBUTANE
C5H11Br	000110-53-2	1-BROMOPENTANE
C5H11BrO3	019184-65-7	MONOBROMONEOPENTYLTRIOL
C5H11Cl	000543-59-9	1-CHLOROPENTANE
C5H11Cl	000594-36-5	BUTANE, 2-CHLORO-2-METHYL-
C5H11Cl	000616-20-6	3-CHLOROPENTANE
C5H11Cl	000625-29-6	2-CHLOROPENTANE
C5H11Cl2N	000051-75-2	2-CHLORO-N-(2-CHLOROETHYL)-N-METHYLETHANAMINE
C5H11Cl3Si	000107-72-2	TRICHLOROPENTYLSILANE
C5H11ClHgN2O2	000062-37-3	CHLORMERODRIN
C5H11F	000592-50-7	1-FLUOROPENTANE
C5H11I	000628-17-1	1-IODOPENTANE
C5H11I	000637-97-8	2-IODOPENTANE
C5H11N	000110-89-4	PIPERADINE
C5H11N	000120-94-5	1-METHYL-PYRROLIDINE
C5H11N	002424-02-4	ALLYLETHYLAMINE
C5H11N2O2P	000077-81-6	TABUN
C5H11N3O	016339-07-4	4-METHYL-N-NITROSOPIPERAZINE
C5H11N3O2	000760-60-1	UREA, N-(2-METHYLPROPYL)-N-NITROSO-
C5H11N3O2	000869-01-2	1-BUTYL-1-NITROSOUREA
C5H11N3O2	050285-71-7	1-NITROSO-1-ET-3,3-DIMETHYL-UREA
C5H11N5O	055973-02-9	2,4,6-PYRIMIDINETRIAMINE, N -METHYL-1-OXIDE
C5H11NNaS2	000148-18-5	DIETHYLCARBAMODITHIOIC ACID, SODIUM SALT
C5H11NO	000109-02-4	N-METHYLMORPHOLINE
C5H11NO	000617-84-5	N,N-DIETHYLFORMAMIDE
C5H11NO	000758-96-3	PROPANAMIDE, N,N-DIMETHYL-
C5H11NO2	000072-18-4	VALINE
C5H11NO2	000107-43-7	BETAINE
C5H11NO2	000463-04-7	AMYL NITRITE
C5H11NO2	000592-35-8	O-BUTYL CARBAMATE
C5H11NO2	000623-78-9	ETHYL ETHYLCARBAMATE
C5H11NO2	000628-05-7	1-NITROPENTANE
C5H11NO2	000660-88-8	5-AMINOPENTANOIC ACID
C5H11NO2	000687-48-9	N,N-DIMETHYLETHYLCARBAMATE
C5H11NO2	002114-15-0	O-ISOBUTYL CARBAMATE
C5H11NO2	004248-19-5	O-T-BUTYL CARBAMATE
C5H11NO2	006600-40-4	NORVALINE
C5H11NO2	017671-76-0	O-PROPYL-N-METHYLCARBAMATE
C5H11NO2S	000052-67-5	PENCILLAMINE
C5H11NO2S	000063-68-3	METHIONINE
C5H11NO3	000543-87-3	1-BUTANOL, 3-METHYL-, NITRATE
C5H11NO4	000597-09-1	2-ETHYL-2-NITRO-1,3-PROPANEDIOL
C5H11O4P	004202-14-6	PHOSPHONIC ACID, (2-OXOPROPYL)-, DIMETHYL ESTER
C5H12	000078-78-4	ISOPENTANE
C5H12	000109-66-0	N-PENTANE
C5H12	000463-82-1	2,2-DIMETHYLPROPANE

C5H12	064771-72-8	PARAFFINS, C1-C10
C5H12ClN	006091-44-7	PIPERDINE, HYDROCHLOROIC ACID
C5H12ClNO2	000590-46-5	ACIDINE
C5H12N2	002213-43-6	N-AMINOPIPERIDINE
C5H12N2O	000592-31-4	BUTYLUREA
C5H12N2O	000632-22-4	TETRAMETHYLUREA
C5H12N2O	000689-11-2	SEC-BUTYLUREA
C5H12N2O	002504-18-9	N-NITROSOMETHYL-ISOBUTYLAMINE
C5H12N2O	007068-83-9	N-NITROSOMETHYLBUTYLAMINE
C5H12N2O	016339-04-1	N-NITROSO-N-ETHYL-ISOPROPYLAMINE
C5H12N2O	017883-59-9	2,2-DIME PROPIONIC ACID HYDRAZIDE
C5H12N2O	024310-18-7	I-VALERIC ACID HYDRAZIDE
C5H12N2O	038291-82-6	VALERIC ACID HYDRAZIDE
C5H12N2O	050816-31-4	N,N-DIETHYLUREA
C5H12N2O	070195-11-8	2-METHYLBUTYRIC ACID HYDRAZIDE
C5H12N2O2	000070-26-8	ORNITHINE
C5H12N2O2	005681-57-2	3-BUTYLHYDROXYUREA
C5H12N2O2	063491-77-0	3-T-BUTYLHYDROXYUREA
C5H12N2S	000105-55-5	N,N-DIETHYLTHIOUREA
C5H12N2S	002782-91-4	TETRAMETHYLTHIOUREA
C5H12NO3PS2	000060-51-5	DIMETHOATE
C5H12NO4PS	001113-02-6	DIMETHOXON
C5H12O	000071-41-0	1-PENTANOL
C5H12O	000075-84-3	2,2-DIMETHYL-1-PROPANOL
C5H12O	000075-85-4	2-METHYL-2-BUTANOL
C5H12O	000123-51-3	ISOPENTANOL
C5H12O	000137-32-6	2-METHYL-1-BUTANOL
C5H12O	000584-02-1	3-PENTANOL
C5H12O	000598-75-4	3-METHYL-2-BUTANOL
C5H12O	000625-44-5	METHYL ISOBUTYL ETHER
C5H12O	000625-54-7	ETHYL ISOPROPYL ETHER
C5H12O	000628-28-4	METHYL N-BUTYL ETHER
C5H12O	000628-32-0	ETHYL PROPYL ETHER
C5H12O	001634-04-4	METHYL T-BUTYL ETHER
C5H12O	006032-29-7	2-PENTANOL
C5H12O	006795-87-5	SEC-BUTYL METHYL ETHER
C5H12O2	000077-76-9	2,2-DIMETHOXYPROPANE
C5H12O2	000109-59-1	ISOPROPOXYETHANOL
C5H12O2	000111-29-5	1,5-PENTANEDIOL
C5H12O2	000111-32-0	1,3-BUTYLENE GLYCOL METHYL ETHER
C5H12O2	000111-35-3	3-ETHOXY-1-PROPANOL
C5H12O2	000462-95-3	DIETHOXYMETHANE
C5H12O2	002517-43-3	3-METHOXY-1-BUTANOL
C5H12O2	002807-30-9	ETHYLENE GLYCOL MONOPROPYL ETHER
C5H12O2	007778-85-0	1,2-DIMETHOXYPROPANE
C5H12O2	052125-53-8	PROPYLENE GLYCOL MONOETHYL ETHER
C5H12O3	000111-77-3	DIETHYLENE GLYCOL MONOMETHYL ETHER
C5H12O3	001445-45-0	1,1,1-TRIMETHOXYETHANE
C5H12O4	000115-77-5	PENTAERYTHRITOL
C5H12S	000110-66-7	1-PENTANETHIOL
C5H12S	000541-31-1	ISOAMYLTHIOL
C5H12S	000628-29-5	2-THIAHEXANE
C5H12S	001679-09-0	2-METHYL-2-BUTANETHIOL
C5H12S	001878-18-8	2-METHYL-1-BUTANETHIOL
C5H12S	004110-50-3	ETHYLPROPYLSULFIDE
C5H12S	005145-99-3	ETHYL ISOPROPYL SULFIDE
C5H12S2	030453-31-7	ETHYL PROPYL DISULFIDE
C5H13Cl2N	000999-81-5	CHLORMEQUAT CHLORIDE

C5H13N	000110-58-7	N-PENTYLAMINE
C5H13N	000110-68-9	METHYLBUTYLAMINE
C5H13N	019961-27-4	ETHYL ISOPROPYLAMINE
C5H13N3	000080-70-6	GUANIDINE, N,N,N',N'-TETRAMETHYL-
C5H13N3O	020605-19-0	4-BUTYLSEMICARBAZIDE
C5H13N3O	074255-46-2	4-T-BUTYLSEMICARBAZIDE
C5H13N3O	079353-74-5	4-I-BUTYLSEMICARBAZIDE
C5H13NO2	000105-59-9	N-METHYLDIETHANOLAMINE
C5H13O2PS	002511-10-6	Phosphonothioic acid, methyl-, O,S-diethyl ester
C5H13O2PS	006996-81-2	Phosphonothioic acid, methyl-, diethyl ester
C5H13O3P	000683-08-9	DIETHYL METHYLPHOSPHONATE
C5H13O3PS2	002587-90-8	DEMEPHION
C5H14ClN	000541-23-1	ISOAMYLAMINE HYDROCHLORIDE
C5H14IN	000051-93-4	ETHYLTRIMETHYLAMMONIUM IODIDE
C5H14N2	000142-25-6	1,2-ETHANEDIAMINE, N,N,N'-TRIMETHYL-
C5H14NO2PS	016271-16-2	O,S-DIME-N-PR-PHOSPHORAMIDOTHIOAT
C5H14OSi	001825-62-3	ETHOXYTRIMETHYLSILANE
C5H14Si	018143-31-2	SILANE, DIMETHYL PROPYL
C5H15N2OP	002511-17-3	PHOSPHONIC DIAMIDE, PENTAMETHYL-
C6Br6	000087-82-1	HEXABROMOBENZENE
C6BrF5	000344-04-7	BROMOPENTAFLUOROBENZENE
C6Cl14	083682-34-2	PERCHLOROHEXANE
C6Cl3N3O6	002631-68-7	TRICHLORO-2,4,6-TRINITROBENZENE
C6Cl4O2	000118-75-2	CHLORANIL
C6Cl5NaO	000131-52-2	PENTACHLOROPHENOL, NA SALT
C6Cl5NO2	000082-68-8	PENTACHLORONITROBENZENE
C6Cl6	000118-74-1	HEXACHLOROBENZENE
C6Cl8O	004024-81-1	OCH
C6ClF5	000344-07-0	CHLOROPENTAFLUOROBENZENE
C6F12	000355-68-0	PERFLUOROCYCLOHEXANE
C6F14	000355-42-0	PERFLUORO-N-HEXANE
C6F6	000392-56-3	HEXAFLUOROBENZENE
C6HBr5	000608-90-2	PENTABROMOBENZENE
C6HBr5O	000608-71-9	PENTABROMOPHENOL
C6HCl4N3	002338-10-5	4,5,6,7-TETRACHLOROBENZOTRIAZOLE
C6HCl4NO2	000117-18-0	2,3,5,6-TETRACHLORONITROBENZENE
C6HCl4NO2	000879-39-0	2,3,4,5-TETRACHLORONITROBENZENE
C6HCl4NO3	004824-72-0	4-NITRO-2,3,5,6-TETRACHLOROPHENOL
C6HCl5	000608-93-5	PENTACHLOROBENZENE
C6HCl5O	000087-86-5	PENTACHLOROPHENOL
C6HClN4	072111-57-0	2,3-PYRAZINEDICARBONITRILE, 5-CHLORO-
C6HF5	000363-72-4	PENTAFLUOROBENZENE
C6HF5O	000771-61-9	PENTAFLUOROPHENOL
C6H2Br2ClNO	000537-45-1	2,6-DIBROMOQUINONECHLOROIMIDE
C6H2Br4	000634-89-9	1,2,3,5-TETRABROMOBENZENE
C6H2Br4	000636-28-2	1,2,4,5-TETRABROMOBENZENE
C6H2Br4	022311-25-7	1,2,3,4-TETRABROMOBENZENE
C6H2Cl3NO2	000089-69-0	2,4,5-TRICHLORONITROBENZENE
C6H2Cl3NO2	017700-09-3	2,3,4-TRICHLORONITROBENZENE
C6H2Cl3NO2	018708-70-8	BENZENE, 1,3,5-TRICHLORO-2-NITRO-
C6H2Cl3NO3	000082-62-2	3,4,6-TRICHLORO-2-NITROPHENOL
C6H2Cl3NO3	020404-02-8	2,3,6-TRICHLORO-4-NITROPHENOL
C6H2Cl4	000095-94-3	1,2,4,5-TETRACHLOROBENZENE
C6H2Cl4	000634-66-2	1,2,3,4-TETRACHLOROBENZENE
C6H2Cl4	000634-90-2	1,2,3,5-TETRACHLOROBENZENE
C6H2Cl4O	000058-90-2	2,3,4,6-TETRACHLOROPHENOL
C6H2Cl4O	000935-95-5	2,3,5,6-TETRACHLOROPHENOL
C6H2Cl4O	004901-51-3	2,3,4,5-TETRACHLOROPHENOL

C6H2Cl4O2	001198-55-6	TETRACHLORO-1,2-BENZENEDIOL
C6H2Cl5N	000527-20-8	PENTACHLOROANILINE
C6H2Cl5N	001128-16-1	3,5-DICHLORO-2-(TRICHLOROMETHYL)PYRIDINE
C6H2ClN3O6	000088-88-0	2,4,6-TRINITROCHLOROBENZENE
C6H2F4	000327-54-8	1,2,4,5-TETRAFLUOROBENZENE
C6H2F4	000551-62-2	1,2,3,4-TETRAFLUOROBENZENE
C6H2F4	002367-82-0	1,2,3,5-TETRAFLUOROBENZENE
C6H2F5N	000771-60-8	2,3,4,5,6-PENTAFLUOROANILINE
C6H2N4	013481-25-9	2,3-PYRAZINEDICARBONITRILE
C6H3Br2ClO	005324-13-0	2,6-DIBROMO-4-CHLOROPHENOL
C6H3Br2NO3	000099-28-5	2,6-DIBROMO-4-NITROPHENOL
C6H3Br3	000608-21-9	1,2,3-TRIBROMOBENZENE
C6H3Br3	000615-54-3	1,2,4-TRIBROMOBENZENE
C6H3Br3	000626-39-1	1,3,5-TRIBROMOBENZENE
C6H3Br3O	000118-79-6	2,4,6-TRIBROMOPHENOL
C6H3Br3O2	002437-49-2	2,4,6-TRIBROMORESORCINOL
C6H3BrCl2O	003217-15-0	4-BROMO-2,6-DICHLOROPHENOL
C6H3BrClNO2	029682-39-1	1-BROMO-2-CHLORO-4-NITROBENZENE
C6H3BrClNO3	058349-01-2	4-BROMO-2-CHLORO-6-NITROPHENOL
C6H3Cl2NO2	000089-61-2	2,5-DICHLORONITROBENZENE
C6H3Cl2NO2	000099-54-7	3,4-DICHLORONITROBENZENE
C6H3Cl2NO2	000611-06-3	2,4-DICHLORONITROBENZENE
C6H3Cl2NO2	000618-62-2	BENZENE, 1,3-DICHLORO-5-NITRO-
C6H3Cl2NO2	001702-17-6	3,6-DICHLOROPICOLINIC ACID
C6H3Cl2NO2	003209-22-1	2,3-DICHLORONITROBENZENE
C6H3Cl2NO3	000609-89-2	2,4-DICHLORO-6-NITROPHENOL
C6H3Cl2NO3	000618-80-4	2,6-DICHLORO-4-NITROPHENOL
C6H3Cl2NO3	005847-57-4	2,5-DICHLORO-4-NITROPHENOL
C6H3Cl2NO3	039224-65-2	4,5-DICHLORO-2-NITROPHENOL
C6H3Cl3	000087-61-6	1,2,3-TRICHLOROBENZENE
C6H3Cl3	000108-70-3	1,3,5-TRICHLOROBENZENE
C6H3Cl3	000120-82-1	1,2,4-TRICHLOROBENZENE
C6H3Cl3N2O2	001918-02-1	PICLORAM
C6H3Cl3O	000088-06-2	2,4,6-TRICHLOROPHENOL
C6H3Cl3O	000095-95-4	2,4,5-TRICHLOROPHENOL
C6H3Cl3O	000609-19-8	3,4,5-TRICHLOROPHENOL
C6H3Cl3O	000933-75-5	2,3,6-TRICHLOROPHENOL
C6H3Cl3O	000933-78-8	2,3,5-TRICHLOROPHENOL
C6H3Cl3O	015950-66-0	2,3,4-TRICHLOROPHENOL
C6H3Cl3O2	032139-72-3	3,4,6-TRICHLOROCATECHOL
C6H3Cl3O2	056961-20-7	3,4,5-TRICHLOROCATECHOL
C6H3Cl4N	000634-83-3	2,3,4,5-TETRACHLOROANILINE
C6H3Cl4N	001929-82-4	NITRAPYRIN
C6H3Cl4N	003481-20-7	2,3,5,6-TETRACHLOROANILINE
C6H3ClN2O4	000097-00-7	2,4-DINITROCHLOROBENZENE
C6H3ClN2O4	000610-40-2	1,2-DINO2 4-CL BENZENE
C6H3ClN2O5	000088-87-9	4-CHLORO-2,6-DINITROPHENOL
C6H3ClN2O5	000946-31-6	6-CHLORO-2,4-DINITROPHENOL
C6H3ClN2S	016269-66-2	THIENO(2,3-D)-PYRMIDINE,4-CHLORO
C6H3F2NO2	000364-74-9	BENZENE, 1,4-DIFLUORO-2-NITRO-
C6H3F2NO2	002265-94-3	BENZENE, 1,3-DIFLUORO-5-NITRO-
C6H3F3	000367-23-7	BENZENE, 1,2,4-TRIFLUORO-
C6H3F3	000372-38-3	1,3,5-TRIFLUOROBENZENE
C6H3F6N5O2	080616-56-4	ACETAMIDE, N,N'-1H-1,2,4-TRIAZOLE-3,5-DIYLBIS[2,
C6H3FN2O4	000070-34-8	1-FLUORO-2,4-DINITROBENZENE
C6H3I3O	000609-23-4	2,4,6-TRIIODOPHENOL
C6H3KN2O5	014314-69-3	PHENOL, 2,4-DINITRO-, POTASSIUM SALT
C6H3N3O6	000099-35-4	1,3,5-TRINITROBENZENE

C6H3N3O7	000088-89-1	2,4,6-TRINITROPHENOL
C6H3N3O8	000082-71-3	2,4,6-TRINITRORESORCINOL
C6H4Br2	000106-37-6	1,4-DIBROMOBENZENE
C6H4Br2	000108-36-1	M-DIBROMOBENZENE
C6H4Br2	000583-53-9	O-DIBROMOBENZENE
C6H4Br2O	000608-33-3	2,6-DIBROMOPHENOL
C6H4Br2O	000615-56-5	3,4-DIBROMOPHENOL
C6H4Br2O	000615-58-7	2,4-DIBROMOPHENOL
C6H4Br2O	000626-41-5	3,5-DIBROMOPHENOL
C6H4Br2O	028165-52-8	2,5-DIBROMOPHENOL
C6H4Br2O	057383-80-9	2,3-DIBROMOPHENOL
C6H4BrCl	000106-39-8	P-BROMOCHLOROBENZENE
C6H4BrCl	000108-37-2	M-BROMOCHLOROBENZENE
C6H4BrCl	000694-80-4	2-BROMOCHLOROBENZENE
C6H4BrF	000460-00-4	P-BROMOFLUOROBENZENE
C6H4BrF	001073-06-9	M-BROMOFLUOROBENZENE
C6H4BrI	000589-87-7	4-BROMOIODOBENZENE
C6H4BrNO	064693-23-8	N-BROMOBENZOQUINONEMONIMINE
C6H4BrNO2	000577-19-5	O-BROMONITROBENZENE
C6H4BrNO2	000585-79-5	3-BROMO-1-NITROBENZENE
C6H4BrNO2	000586-78-7	4-BROMO-1-NITROBENZENE
C6H4BrNO3	005470-65-5	3-BROMO-4-NITROPHENOL
C6H4BrNO3	005847-59-6	2-BROMO-4-NITROPHENOL
C6H4BrNO3	007693-52-9	4-BROMO-2-NITROPHENOL
C6H4Cl2	000095-50-1	1,2-DICHLOROBENZENE
C6H4Cl2	000106-46-7	1,4-DICHLOROBENZENE
C6H4Cl2	000541-73-1	3-DICHLOROBENZENE
C6H4Cl2N2O2	000099-30-9	2,6-DICHLORO-4-NITROANILINE
C6H4Cl2O	000087-65-0	2,6-DICHLOROPHENOL
C6H4Cl2O	000095-77-2	3,4-DICHLOROPHENOL
C6H4Cl2O	000120-83-2	2,4-DICHLOROPHENOL
C6H4Cl2O	000576-24-9	2,3-DICHLOROPHENOL
C6H4Cl2O	000583-78-8	2,5-DICHLOROPHENOL
C6H4Cl2O	000591-35-5	3,5-DICHLOROPHENOL
C6H4Cl3N	000634-67-3	2,3,4-TRICHLOROANILINE
C6H4Cl3N	000634-91-3	3,4,5-TRICHLOROANILINE
C6H4Cl3N	000634-93-5	2,4,6-TRICHLOROANILINE
C6H4Cl3N	000636-30-6	2,4,5-TRICHLOROANILINE
C6H4Cl6	057722-15-3	HEXACHLOROCYCLOHEXENE
C6H4Cl6	057722-16-4	HEXACHLOROCYCLOHEXENE
C6H4Cl6	057722-17-5	HEXACHLOROCYCLOHEXENE
C6H4ClF	000348-51-6	1-CHLORO-2-FLUOROBENZENE
C6H4ClF	000352-33-0	P-CHLOROFLUOROBENZENE
C6H4ClF	000625-98-9	M-CHLOROFLUOROBENZENE
C6H4ClI	000615-41-8	2-CHLOROIODOBENZENE
C6H4ClI	000625-99-0	3-CHLOROIODOBENZENE
C6H4ClI	000637-87-6	4-CHLOROIODOBENZENE
C6H4ClNO	000637-61-6	QUINONE CHLOROIMIDE
C6H4ClNO2	000088-73-3	2-CHLORO-1-NITROBENZENE
C6H4ClNO2	000100-00-5	P-CHLORONITROBENZENE
C6H4ClNO2	000121-73-3	3-CHLORO-NITROBENZENE
C6H4ClNO2	004684-94-0	6-CHLOROPICOLINIC ACID
C6H4ClNO3	000089-64-5	4-CHLORO-2-NITROPHENOL
C6H4ClNO3	000491-11-2	3-CHLORO-4-NITROPHENOL
C6H4ClNO3	000603-86-1	6-CHLORO-2-NITROPHENOL
C6H4ClNO3	000611-07-4	5-CHLORO-2-NITROPHENOL
C6H4ClNO3	000619-08-9	2-CHLORO-4-NITROPHENOL
C6H4F2	000367-11-3	1,2-DIFLUOROBENZENE

C6H4F2	000372-18-9	M-DIFLUOROBENZENE
C6H4F2	000540-36-3	P-DIFLUOROBENZENE
C6H4F2O	028177-48-2	PHENOL, 2,6-DIFLUORO-
C6H4F7N5O	080616-59-7	BUTANAMIDE, N-(5-AMINO-1H-1,2,4-TRIAZOL-3-YL)-2,
C6H4FNO2	000350-46-9	4-FLUORONITROBENZENE
C6H4FNO2	000402-67-5	3-FLUORONITROBENZENE
C6H4FNO2	001493-27-2	O-FLUORONITROBENZENE
C6H4FNO3	000403-19-0	2-FLUORO-4-NITROPHENOL
C6H4FNO3	000446-36-6	5-FLUORO-2-NITROPHENOL
C6H4I2	000615-42-9	1,2-DIIODOBENZENE
C6H4I2	000624-38-4	1,4-DIIODOBENZENE
C6H4I2	000626-00-6	1,3-DIIODOBENZENE
C6H4INO2	000645-00-1	3-IODONITROBENZENE
C6H4N2	000100-48-1	4-CYANOPYRIDINE
C6H4N2	000100-54-9	3-CYANOPYRIDINE
C6H4N2	000100-70-9	2-CYANOPYRIDINE
C6H4N2O	000273-09-6	BENZOFURAZAN
C6H4N2O2	000480-96-6	BENZOFURAZAN, 1-OXIDE
C6H4N2O4	000099-65-0	1,3-DINITROBENZENE
C6H4N2O4	000100-25-4	P-DINITROBENZENE
C6H4N2O4	000528-29-0	O-DINITROBENZENE
C6H4N2O5	000051-28-5	2,4-DINITROPHENOL
C6H4N2O5	000066-56-8	2,3-DINITROPHENOL
C6H4N2O5	000329-71-5	2,5-DINITROPHENOL
C6H4N2O5	000573-56-8	2,6-DINITROPHENOL
C6H4N2O5	000577-71-9	3,4-DINITROPHENOL
C6H4N2O5	000586-11-8	3,5-DINITROPHENOL
C6H4N2O6	000519-44-8	2,4-DINITRORESORCINOL
C6H4N2Se	000273-15-4	2,1,3-BENZOSELENADIAZOLE
C6H4N4	000091-18-9	PTERIDINE
C6H4N4	000868-54-2	1-PROPENE-1,1,3-TRICARBONITRILE, 2-AMINO-
C6H4N4O	090467-89-3	2-PYRIDYL AZOXYCYANIDE
C6H4NNaO3	014609-74-6	SODIUM P-NITROPHENOXIDE
C6H4NNaO5S	000127-68-4	M-NITROBENZENESULFONIC ACID, SODIUM SALT
C6H4O2	000106-51-4	2,5-CYCLOHEXADIENE-1,4-DIONE
C6H5Br	000108-86-1	BROMOBENZENE
C6H5Br2N	000615-57-6	2,4-DIBROMOANILINE
C6H5BrO	000095-56-7	O-BROMOPHENOL
C6H5BrO	000106-41-2	P-BROMOPHENOL
C6H5BrO	000591-20-8	M-BROMOPHENOL
C6H5Cl	000108-90-7	CHLOROBENZENE
C6H5Cl2N	000095-76-1	3,4-DICHLOROANILINE
C6H5Cl2N	000095-82-9	2,5-DICHLOROANILINE
C6H5Cl2N	000554-00-7	2,4-DICHLOROANILINE
C6H5Cl2N	000608-27-5	2,3-DICHLOROANILINE
C6H5Cl2N	000608-31-1	2,6-DICHLOROANILINE
C6H5Cl2N	000626-43-7	3,5-DICHLOROANILINE
C6H5Cl2NO2S	023815-28-3	BENZENESULFONAMIDE, 3,4-DICHLORO-
C6H5Cl2O2P	000770-12-7	PHENYL DICHLOROPHOSPHATE
C6H5Cl3Si	000098-13-5	PHENYLTRICHLOROSILANE
C6H5Cl5	000319-94-8	PENTACHLOROCYCLOHEXENE
C6H5Cl5	000643-15-2	PENTACHLOROCYCLOHEXENE
C6H5Cl5	051795-30-3	PENTACHLOROCYCLOHEXENE
C6H5Cl5	054083-24-8	PENTACHLOROCYCLOHEXENE
C6H5Cl5	054083-25-9	PENTACHLOROCYCLOHEXENE
C6H5ClFN	000367-21-5	3-CHLORO-4-FLUOROANILINE
C6H5ClHg	000100-56-1	PHENYL MERCURIC CHLORIDE
C6H5ClHgO	000090-03-9	CHLOROMERCURIPHENOL

C6H5ClHgO3S	000554-77-8	P-CHLOROMERCURIPHENYLSULFONIC ACID
C6H5ClN2O2	000089-63-4	4-CHLORO-2-NITROANILINE
C6H5ClN2O2	000121-87-9	2-CHLORO-4-NITROANILINE
C6H5ClN2O2	000635-22-3	BENZENAMINE, 4-CHLORO-3-NITRO-
C6H5ClN2O2	023611-75-8	2-CHLORO-6-CO2ME-PYRAZINE
C6H5ClN2O3	006358-09-4	2-AMINO-6-CHLORO-4-NITROPHENOL
C6H5ClN2O4S	000097-09-6	BENZENESULFONAMIDE, 4-CHLORO-3-NITRO-
C6H5ClO	000095-57-8	2-CHLOROPHENOL
C6H5ClO	000106-48-9	4-CHLOROPHENOL
C6H5ClO	000108-43-0	3-CHLOROPHENOL
C6H5ClO2	000095-88-5	4-CHLORORESORCINOL
C6H5ClO2	000615-67-8	CHLOROHYDROQUINONE
C6H5ClO2S	000098-09-9	BENZENESULFONYL CHLORIDE
C6H5ClS	000106-54-7	4-CHLOROBENZENETHIOL
C6H5ClS	002037-31-2	3-CHLOROBENZENETHIOL
C6H5F	000462-06-6	FLUOROBENZENE
C6H5F2N	000367-25-9	2,4-DIFLUOROANILINE
C6H5F5S	002557-81-5	PHENYLSULFUR PENTAFLUORIDE
C6H5FN2O3	075410-15-0	2,4(1H,3H)-PYRIMIDINEDIONE, 3-ACETYL-5-FLUORO-
C6H5FO	000367-12-4	O-FLUOROPHENOL
C6H5FO	000371-41-5	P-FLUOROPHENOL
C6H5FO	000372-20-3	M-FLUOROPHENOL
C6H5I	000591-50-4	IODOBENZENE
C6H5IO	000533-58-4	2-IODOPHENOL
C6H5IO	000540-38-5	4-IODOPHENOL
C6H5IO	000626-02-8	3-IODOPHENOL
C6H5N3	000095-14-7	1H-BENZOTRIAZOLE
C6H5N3	000272-97-9	3,5-DIAZAINDOLE
C6H5N3	000273-02-9	2H-BENZOTRIAZOLE
C6H5N3	000273-21-2	1H-IMIDAZO[4,5-B]PYRIDINE
C6H5N3	098006-91-8	PYRAZINECARBONITRILE, 5-METHYL-
C6H5N3	136309-04-1	PYRAZINECARBONITRILE, 6-METHYL-
C6H5N3O	094789-37-4	2-PYRIMIDINECARBONITRILE, 4-METHOXY-
C6H5N3O	136309-07-4	PYRAZINECARBONITRILE, 6-METHOXY-
C6H5N3O4	000097-02-9	2,4-DINITROANILINE
C6H5N3O4	000606-22-4	2,6-DINITROANILINE
C6H5N3O4	000618-87-1	3,5-DINITROANILINE
C6H5N3O5	000096-91-3	2-AMINO-4,6-DINITROPHENOL
C6H5N3S	014080-56-9	THIENO[2,3-D]PYRIMIDINE, 4-AMINO-
C6H5NaO	000139-02-6	PHENOL, SODIUM SALT
C6H5NaO2S	000873-55-2	BENZENESULFINIC ACID, SODIUM SALT
C6H5NO	000500-22-1	3-FORMYLPYRIDINE
C6H5NO	000872-85-5	4-FORMYLPYRIDINE
C6H5NO	001121-60-4	2-PYRIDINECARBOXYALDEHYDE
C6H5NO2	000055-22-1	ISONICOTINIC ACID
C6H5NO2	000059-67-6	NICOTINIC ACID
C6H5NO2	000098-95-3	NITROBENZENE
C6H5NO2	000098-98-6	PICOLINIC ACID
C6H5NO2	000104-91-6	P-NITROSOPHENOL
C6H5NO2	000586-96-9	NITROSOBENZENE
C6H5NO2	000637-62-7	P-BENZOQUINONE OXIME
C6H5NO3	000088-75-5	2-NITROPHENOL
C6H5NO3	000100-02-7	4-NITROPHENOL
C6H5NO3	000554-84-7	3-NITROPHENOL
C6H5NO3	002398-81-4	N-OXIDENICOTINIC ACID
C6H5NO3	013602-12-5	ISONICOTINIC ACID N-OXIDE
C6H5NO4	000601-89-8	2-NITRORESORCINOL
C6H5NO4	003163-07-3	1,3-BENZENEDIOL, 4-NITRO-

C6H5NO4	003316-09-4	1,2-DIHYDROXY-4-NITROBENZENE
C6H5NO4	025021-08-3	N-MALEOYLGLYCINE
C6H5NS	020893-30-5	2-CYANOMETHYLTHIOPHENE
C6H6	000071-43-2	BENZENE
C6H6	002809-69-0	2,4-HEXADIYNE
C6H6Br2Cl4	055265-69-5	2,4-DIBR-TETRACHLOROCYCLOHEXANE
C6H6Br2Cl4	055298-45-8	1,2-DIBR-TETRACHLOROCYCLOHEXANE
C6H6Br2Cl4	055332-89-3	2,3-DIBR-TETRACHLOROCYCLOHEXANE
C6H6Br5Cl	000087-84-3	CYCLOHEXANE, 1,2,3,4,5-PENTABROMO-6-CHLORO-
C6H6BrCl4F	055265-51-5	2-F-3-BR-TETRACHLOROCYCLOHEXANE
C6H6BrCl5	036635-03-7	1-BR-PENTACHLOROCYCLOHEXANE
C6H6BrN	000106-40-1	P-BROMOANILINE
C6H6BrN	000591-19-5	M-BROMOANILINE
C6H6BrN	000615-36-1	O-BROMOANILINE
C6H6BrN3O	029849-15-8	2-BROMOISONIAZID
C6H6BrNO2S	000701-34-8	4-BROMOBENZENESULFONAMIDE
C6H6BrNO2S	089599-01-9	BENZENESULFONAMIDE, 3-BROMO-
C6H6BrO3	002527-99-3	2-FURANCARBOXYLIC ACID, 5-BROMO-, METHYL ESTER
C6H6Cl2	069645-07-4	1,6-DICHLOROHEXA-1,3,5-TRIENE
C6H6Cl2	083682-37-5	3,4-DICHLOROHEXA-1,3,5-TRIENE
C6H6Cl2N2	005348-42-5	4,5-DICHLORO-O-PHENYLENEDIAMINE
C6H6Cl4	000319-81-3	TETRACHLOROCYCLOHEXENE
C6H6Cl4	001782-00-9	TETRACHLOROCYCLOHEXENE
C6H6Cl4	028810-38-0	TETRACHLOROCYCLOHEXENE
C6H6Cl4	033875-95-5	TETRACHLOROCYCLOHEXENE
C6H6Cl4	041992-55-6	TETRACHLOROCYCLOHEXENE
C6H6Cl4	089674-85-1	TETRACHLOROCYCLOHEXENE (345)
C6H6Cl4	089674-87-3	TETRACHLOROCYCLOHEXENE (345)
C6H6Cl4	089674-88-4	TETRACHLOROCYCLOHEXENE (345)
C6H6Cl4F8N3O2P3	065601-40-3	2,2,4,4-TETRACHLORO-2,2,4,4,6,6-HEXAHYDRO-6,6-B*
C6H6Cl5F	056086-55-6	2-F-PENTACHLOROCYCLOHEXANE
C6H6Cl5I	033489-27-9	3-I-PENTACHLOROCYCLOHEXANE
C6H6Cl5I	033489-28-0	1-I-PENTACHLOROCYCLOHEXANE
C6H6Cl6	000058-89-9	GAMMA-HEXACHLOROCYCLOHEXANE
C6H6Cl6	000319-84-6	ALPHA-HEXACHLOROCYCLOHEXANE
C6H6Cl6	000319-85-7	BETA-HEXACHLOROCYCLOHEXANE
C6H6Cl6	000319-86-8	DELTA-HEXACHLOROCYCLOHEXANE
C6H6Cl6	000608-73-1	1,2,3,4,5,6-HEXACHLOROCYCLOHEXANE
C6H6Cl6	001725-74-2	1,2,3,4,5,6-HEXACHLORO-3-HEXENE
C6H6Cl6	083682-30-8	1,1,4,4-TETRACHLORO-2,3-DICHLOROMETHYL-2-BUTENE
C6H6Cl6	083682-31-9	2,2,3,4,5,5-HEXACHLORO-3-HEXENE
C6H6ClN	000095-51-2	2-CHLOROANILINE
C6H6ClN	000106-47-8	4-CHLOROANILINE
C6H6ClN	000108-42-9	3-CHLOROANILINE
C6H6ClN3O	058481-04-2	2-CHLOROISONIAZID
C6H6ClN3O	132453-63-5	2-CHLORO-6-ACETYLAMINO-PYRAZINE
C6H6ClN3O	136309-12-1	ACETAMIDE, N-(5-CHLOROPYRAZINYL)-
C6H6ClNO	000095-85-2	PHENOL, 2-AMINO-4-CHLORO-
C6H6ClNO2S	000098-64-6	4-CHLOROBENZENESULFONAMIDE
C6H6ClNO2S	006961-82-6	O-CHLOROBENZENESULFONAMIDE
C6H6ClNO2S	017260-71-8	M-CHLOROBENZENESULFONAMIDE
C6H6ClNO3S	000088-43-7	4-CHLOROANILINE-3-SULFONIC ACID
C6H6ClNO3S	000133-74-4	4-CHLOROANILINE-2-SULFONIC ACID
C6H6FN	000348-54-9	2-FLUOROANILINE
C6H6FN	000371-40-4	4-FLUOROANILINE
C6H6FN	000372-19-0	3-FLUOROANILINE
C6H6FN3O	000369-24-4	2-FLUOROISONIAZID
C6H6FN3O3	056563-18-9	1(2H)-PYRIMIDINECARBOXAMIDE, 5-FLUORO-3,4-DIHYDR

C6H6HgO	000100-57-2	PHENYLMERCURIC HYDROXIDE
C6H6IN	000540-37-4	BENZENAMINE, 4-IODO-
C6H6IN	000615-43-0	2-IODOANILINE
C6H6IN	000626-01-7	BENZENAMINE, 3-IODO-
C6H6IN3O	029247-87-8	4-PYRIDINECARBOXYLIC ACID, 2-IODO-, HYDRAZIDE
C6H6INO2S	000825-86-5	4-IODOBENZENESULFONAMIDE
C6H6INO2S	050702-39-1	BENZENESULFONAMIDE, 3-IODO-
C6H6N2O	000098-92-0	NICOTINAMIDE
C6H6N2O	000696-54-8	4-PYRIDINEALDOXIME
C6H6N2O	000873-69-8	2-PYRIDINEALDOXIME
C6H6N2O	001452-77-3	2-PYRIDINECARBOXAMIDE
C6H6N2O	001453-82-3	I-NICOTINAMIDE
C6H6N2O	014906-59-3	4-CYANOPYRIDINE OXIDE
C6H6N2O	022047-25-2	ACETYLPYRAZINE
C6H6N2O2	000088-74-4	2-NITROANILINE
C6H6N2O2	000099-09-2	3-NITROANILINE
C6H6N2O2	000100-01-6	4-NITROANILINE
C6H6N2O2	000105-11-3	2,5-CYCLOHEXADIENE-1,4-DIONE, DIOXIME
C6H6N2O2	000933-90-4	2-PYRIDINECARBOXAMIDE, 3-HYDROXY-
C6H6N2O2	002450-08-0	4-PYRIMIDINECARBOXYLIC ACID, METHYL ESTER
C6H6N2O2	006164-79-0	2-METHYLPYRAZINE CARBOXYLATE
C6H6N2O2	021203-68-9	2-METHYL-5-NITROPYRIDINE
C6H6N2O2	034231-77-1	4-PYRIDAZINECARBOXYLIC ACID, METHYL ESTER
C6H6N2O2	034253-01-5	5-PYRIMIDINECARBOXYLIC ACID, METHYL ESTER
C6H6N2O2	034253-02-6	3-PYRIDAZINECARBOXYLIC ACID, METHYL ESTER
C6H6N2O2	034253-03-7	2-PYRIMIDINECARBOXYLIC ACID, METHYL ESTER
C6H6N2O2	075773-74-9	2-AMINO-4-NITROSOPHENOL
C6H6N2O2S	001615-06-1	1,3-DIHYDRO-BENZOTHIADIAZOL,22-O2
C6H6N2O3	000099-57-0	2-AMINO-4-NITROPHENOL
C6H6N2O3	000119-34-6	PHENOL, 4-AMINO-2-NITRO-
C6H6N2O3	005446-92-4	2-METHOXY-5-NITROPYRIDINE
C6H6N2O3	006947-77-9	5,7-DIAZASPIRO[2,5]OCTANE-4,6,8-TRIONE
C6H6N2O4	006153-44-2	4-PYRIMIDINECARBOXYLIC ACID, 1,2,3,6-TETRAHYDRO-
C6H6N2O4S	000121-52-8	M-NITROBENZENESULFONAMIDE
C6H6N2O4S	005455-59-4	2-NITROBENZENESULFONAMIDE
C6H6N2O4S	006325-93-5	4-NITROBENZENESULFONAMIDE
C6H6N2S	000273-13-2	2,1,3-BENZOTHIADIAZOLE
C6H6N2S	004621-66-3	THIONICOTINAMIDE
C6H6N2S	005346-38-3	2-PYRIDINETHIOCARBOXAMIDE
C6H6N4O2	001076-22-8	3-METHYLXANTHINE
C6H6N4O2	006136-37-4	1-METHYLXANTHINE
C6H6N4O2	117505-24-5	4-PYRIDINE OXIDE AZOXYCYANIDE
C6H6N4O2S2	040016-42-0	2-IMIDAZOLIDINETHIONE, 1-(5-NITRO-2-THIAZOLYL)-
C6H6N4O3	000605-99-2	3-METHYL URIC ACID
C6H6N4O3	000612-37-3	7-METHYL URIC ACID
C6H6N4O3	000708-79-2	1-METHYL URIC ACID
C6H6N4O3	058481-05-3	2-NITROISONIAZID
C6H6N4O3S	000061-57-4	NIRIDAZOLE
C6H6N4O3S	000831-71-0	HYDRAZINECARBOTHIAMIDE, 2-[(5-NITRO-2-FURANYL)ME
C6H6N4O3S	024240-60-6	OXAZOLIDINE, 2-[(5-NITRO-2-THIAZOLYL)IMINO]-
C6H6N4O4	000059-87-0	NITROFURAZONE
C6H6N4O4	000119-26-6	HYDRAZINE, (2,4-DINITROPHENYL)-
C6H6N4O7	000131-74-8	AMMONIUM PICRATE
C6H6NNaO3S	001126-34-7	BENZENESULFONIC ACID, 3-AMINO-, MONOSODIUM SALT
C6H6NO	002745-25-7	2-CYANOMETHYLFURAN
C6H6NO3	000699-18-3	2-(B-NITROVINYL)FURAN
C6H6NO3	053916-74-8	3-(B-NITROVINYL)FURAN
C6H6O	000108-95-2	PHENOL

C6H6O2	000108-46-3	RESORCINOL
C6H6O2	000120-80-9	CATECHOL
C6H6O2	000123-31-9	HYDROQUINONE
C6H6O2	000620-02-0	2-FURANCARBOXALDEHYDE, 5-METHYL-
C6H6O2	001192-62-7	2-ACETYLFURAN
C6H6O3	000087-66-1	1,2,3-TRIHYDROXYBENZENE
C6H6O3	000108-73-6	1,3,5-TRIHYDROXYBENZENE
C6H6O3	000118-71-8	4H-PYRAN-4-ONE, 3-HYDROXY-2-METHYL-
C6H6O3	000533-73-3	1,2,4-TRIHYDROXYBENZENE (PYROGALLOL)
C6H6O3	000611-13-2	FUROIC ACID, METHYL ESTER
C6H6O3	013129-23-2	3-FURANCARBOXYLIC ACID, METHYL ESTER
C6H6O3S	000098-11-3	BENZENESULFONIC ACID
C6H6O4	000501-30-4	4H-PYRAN-4-ONE, 5-HYDROXY-2-(HYDROXYMETHYL)-
C6H6O4S	000098-67-9	4-HYDROXYPHENYLSULFONIC ACID
C6H6OS	000088-15-3	2-ACETYLTHIOPHENE
C6H6OS	001468-83-3	3-ACETYLTHIOPHENE
C6H6S	000108-98-5	THIOPHENOL
C6H7BrIN	032222-42-7	PYRIDINIUM, 3-BROMO-1-METHYL-, IODIDE
C6H7BrN2	000589-21-9	HYDRAZINE, (4-BROMOPHENYL)-
C6H7BrN2O3	077317-65-8	3-MEO-5-BR-6-ME-PYRIMIDIN-2,4-DIONE
C6H7Cl	083682-35-3	3-CHLOROHEXA-1,2,5-TRIENE
C6H7Cl3	056994-25-3	TRICHLOROCYCLOHEXENE (345)
C6H7Cl5	022138-39-2	1-PENTACHLOROCYCLOHEXANE
C6H7Cl5	056994-23-1	3-PENTACHLOROCYCLOHEXANE
C6H7Cl5O	053861-64-6	1-HYDROXYPENTACHLOROCYCLOHEXANE
C6H7ClN2	000095-83-0	4-CHLORO-1,2-BENZENEDIAMINE
C6H7ClN2	000095-89-6	PYRAZINE, 3-CHLORO-2,5-DIMETHYL-
C6H7ClN2	005131-60-2	1,3-BENZENEDIAMINE, 4-CHLORO-
C6H7ClN2O	136309-02-9	2-CHLORO-6-ETHOXY-PYRAZINE
C6H7ClN2O	136309-11-0	PYRAZINE, 2-CHLORO-5-ETHOXY-
C6H7F3O2	000352-87-4	2,2,2-TRIFLUOROETHYL METHACRYLATE
C6H7F7N2S	105412-23-5	THIOUREA, N-(2,2,3,3,4,4,4-HEPTAFLUOROBUTYL)-N'-
C6H7N	000062-53-3	ANILINE
C6H7N	000108-89-4	4-METHYLPYRIDINE
C6H7N	000108-99-6	3-METHYLPYRIDINE
C6H7N	000109-06-8	2-METHYLPYRIDINE
C6H7N3O	000054-85-3	ISONIAZID
C6H7N3O	005521-57-3	PYRAZINECARBOXAMIDE, 5-METHYL-
C6H7N3O	005521-62-0	PYRAZINECARBOXAMIDE, 6-METHYL-
C6H7N3O	013053-88-8	2-ACETAMIDOPYRIMIDINE
C6H7N3O	013438-65-8	3-PYRIDINECARBOXAMIDE, 2-AMINO-
C6H7N3O	016166-22-6	ACETAMIDE, N-4-PYRIMIDINYL-
C6H7N3O	021352-21-6	2-ACETYLAMINOPYRAZINE
C6H7N3O	045810-14-8	ACETAMIDE, N-5-PYRIMIDINYL-
C6H7N3O2	000099-56-9	2-AMINO-4-NITROANILINE
C6H7N3O2	000100-16-3	HYDRAZINE, (4-NITROPHENYL)-
C6H7N3O2	003694-52-8	3-NITRO-O-PHENYLENEDIAMINE
C6H7N3O2	005307-14-2	1,4-BENZENEDIAMINE, 2-NITRO-
C6H7N3O2	036070-86-7	PYRAZINECARBOXAMIDE, 6-METHOXY-
C6H7N3O3S	113916-37-3	3-PYRIDINESULFONAMIDE, N-(AMINOCARBONYL)-
C6H7N3OS	005419-96-5	HYDRAZINECARBOTHIAMIDE, 2-(2-FURANYLMETHYLENE)-
C6H7N5	000700-00-5	9H-PURIN-6-AMINE, 9-METHYL-
C6H7N5O2S	024240-69-5	2(1,3-IMIDAZOLIDINYLIDEN-2-AMINO)-5NO2-THIAZOLE
C6H7NO	000095-55-6	O-AMINOPHENOL
C6H7NO	000100-55-0	3-PYRIDINEMETHANOL
C6H7NO	000100-65-2	PHENYLHYDROXYLAMINE
C6H7NO	000123-30-8	PHENOL, 4-AMINO-
C6H7NO	000586-95-8	4-PYRIDINEMETHANOL

C6H7NO	000586-98-1	2-PYRIDINEMETHANOL
C6H7NO	000591-27-5	PHENOL, 3-AMINO-
C6H7NO	000620-08-6	4-METHOXYPYRIDINE
C6H7NO	000694-85-9	2(1H)-PYRIDINONE, 1-METHYL-
C6H7NO	000695-19-2	4(1H)-PYRIDINONE, 1-METHYL-
C6H7NO	000931-19-1	2-METHYL-N-OXIDEPYRIDINE
C6H7NO	001003-67-4	4-METHYL-N-OXIDEPYRIDINE
C6H7NO	001003-73-2	3-METHYL-N-OXIDEPYRIDINE
C6H7NO	001072-83-9	2-ACETYLPYRROLE
C6H7NO	001628-89-3	2-METHOXYPYRIDINE
C6H7NO	007295-76-3	3-METHYOXYPYRIDINE
C6H7NO2	001193-62-0	METHYL PYRROLE-2-CARBOXYLATE
C6H7NO2	019365-01-6	2(1H)-PYRIDINONE, 3-HYDROXY-1-METHYL-
C6H7NO2	050700-61-3	N-METHYL-HYDROXYPYRID-4-ONE
C6H7NO2S	000098-10-2	BENZENESULFONAMIDE
C6H7NO2S	000874-84-0	2-(B-NITROVINYL)THIOPHENE
C6H7NO2S	028783-31-5	3-(B-NITROVINYL)THIOPHENE
C6H7NO3S	000088-20-0	O-TOLUENE SULFONIC ACID
C6H7NO3S	000088-21-1	O-ANILINESULFONIC ACID
C6H7NO3S	000121-47-1	3-AMINOBENZENESULFONIC ACID
C6H7NO3S	000121-57-3	BENZENESULFONIC ACID, 4-AMINO-
C6H7NO3S	000599-71-3	BENZENESULFONAMIDE, N-HYDROXY-
C6H7NO3S	001576-43-8	P-HYDROXYBENZENESULFONAMIDE
C6H7NS	000137-07-5	O-AMINOTHIOPHENOL
C6H7NS	001193-02-8	4-AMINOBENZENETHIOL
C6H7NS	013781-53-8	3-CYANOMETHYLTHIOPHENE
C6H7NS	018438-38-5	2-METHYLTHIOPYRIDINE
C6H7NS	022948-02-3	3-AMINOBENZENETHIOL
C6H7O3P	001571-33-1	PHENYLPHOSPHONIC ACID
C6H8	000592-57-4	1,3-CYCLOHEXADIENE
C6H8	000628-41-1	1,4-CYCLOHEXADIENE
C6H8	000821-07-8	TRANS-1,3,5-HEXATRIENE
C6H8	002235-12-3	1,3,5-HEXATRIENE
C6H8	002612-46-6	CIS-1,3,5-HEXATRIENE
C6H8Cl2	067546-51-4	1,6-DICHLOROHEXA-1,5-DIENE
C6H8Cl2	083682-33-1	3,4-DICHLOROHEXA-1,5-DIENE
C6H8Cl6	018585-38-1	1,2,3,4,5,6-HEXACHLOROHEXANE
C6H8Cl6	083682-28-4	2,2,3,4,5,5-HEXACHLOROHEXANE
C6H8Cl6	083682-29-5	1,2,2,5,5,6-HEXACHLOROHEXANE
C6H8ClN	000142-04-1	ANILINE HYDROCHLORIDE
C6H8ClN3	007145-60-0	3-CL-6-PYRIDAZINAMINE,N,N-DIME
C6H8ClN3	061655-72-9	2-CHLORO-6-(N,N-DIMETHYLAMINO)-PYRAZINE
C6H8ClN3	136309-13-2	PYRAZINAMINE, 5-CHLORO-N,N-DIMETHYL-
C6H8IN	000930-73-4	N-METHYLPYRIDINIUM IODIDE
C6H8N2	000095-54-5	1,2-BENZENEDIAMINE
C6H8N2	000100-63-0	PHENYL HYDRAZINE
C6H8N2	000106-50-3	1,4-BENZENEDIAMINE
C6H8N2	000108-45-2	1,3-BENZENEDIAMINE
C6H8N2	000108-50-9	PYRAZINE, 2,6-DIMETHYL-
C6H8N2	000111-69-3	ADIPONITRILE
C6H8N2	000123-32-0	2,5-DIMETHYLPYRAZINE
C6H8N2	001558-17-4	4,6-DIMETHYLPYRIMIDINE
C6H8N2	001603-41-4	2-AMINO-5-METHYLPYRIDINE
C6H8N2	003731-51-9	2-PYRIDINEMETHANEAMINE
C6H8N2	003731-52-0	3-PYRIDINEMETHANEAMINE
C6H8N2	003731-53-1	4-PYRIDINEMETHANEAMINE
C6H8N2	004553-62-2	2-METHYLPENTANEDINITRILE
C6H8N2	004597-87-9	2-METHYLAMINOPYRIDINE

C6H8N2	005910-89-4	PYRAZINE, 2,3-DIMETHYL-
C6H8N2	013925-00-3	2-ETHYL PYRAZINE
C6H8N2O	002847-30-5	PYRAZINE, 2-METHOXY-3-METHYL-
C6H8N2O	002882-21-5	PYRAZINE, 2-METHOXY-6-METHYL-
C6H8N2O	003739-82-0	PYRIMIDINE, 2-ETHOXY-
C6H8N2O	024903-72-8	PYRIDAZINE, 4-ETHOXY-
C6H8N2O	027522-25-4	PYRIMIDINE, 5-ETHOXY-
C6H8N2O	038028-67-0	2-ETHOXYPYRAZINE
C6H8N2O	062567-44-6	PYRIDAZINE, 3-ETHOXY-
C6H8N2O	064230-41-7	1H-PYRROLE-2-CARBOXAMIDE, 1-METHYL-
C6H8N2O	132911-42-3	1H-PYRROLE-2-CARBOXAMIDE, N-METHYL-
C6H8N2O2	000874-14-6	2,4(1H,3H)-PYRIMIDINEDIONE, 1,3-DIMETHYL-
C6H8N2O2	004774-15-6	PYRAZINE, 2,6-DIMETHOXY-
C6H8N2O2	117856-61-8	PYRAZINE, 2,5-DIMETHOXY-
C6H8N2O2S	000063-74-1	SULFANILAMIDE
C6H8N2O2S	000080-17-1	BENZENESULFONIC ACID HYDRAZIDE
C6H8N2O2S	000098-18-0	M-AMINOBENZENESULFONAMIDE
C6H8N2O2S	007729-78-4	4,6(1H,5H)-PYRIMIDINEDIONE, DIHYDRO-5,5-DIMETHYL
C6H8N2O2S	015959-53-2	PHENYLSULFAMIDE
C6H8N2O3	002518-72-1	5-ETHYLBARBITURIC ACID
C6H8N2O3	024448-94-0	2,4,6(1H,3H,5H)-PYRIMIDINETRIONE, 5,5-DIMETHYL-
C6H8N2O4S2	003701-01-7	1,3-BENZENEDISULFONAMIDE
C6H8N2O4S2	016993-45-6	1,4-BENZENEDISULFONAMIDE
C6H8N2O8	000087-33-2	D-GLUCITOL, 1,4:3,6-DIANHYDRO-, DINITRATE
C6H8N2S	002882-20-4	PYRAZINE, 2-METHYL-3-(METHYLTHIO)-
C6H8N2S2	037813-54-0	3,6-DIMETHYLTHIOPYRIDAZINE
C6H8N4O	001116-82-1	N-NITROSO-BIS(2-CYANOET)AMINE
C6H8N4O	005594-16-1	ACETAMIDE, N-(6-METHYLPYRAZINYL)-
C6H8N4O	058481-01-9	2-AMINOISONIAZID
C6H8N4O2	000552-62-5	7-METHYLXANTHINE
C6H8N4O5	021117-51-1	1H-IMIDAZOLE-1-ETHANOL, 2-METHYL-5-NITRO-, NITRA
C6H8N4S	097183-52-3	HYDRAZINECARBOXIMIDAMIDE, 2-(2-THIENYLMETHYLENE)
C6H8NO4	000586-84-5	2-(METHOXYMETHYL)-5-NO2 FURAN
C6H8O	000142-83-6	2,4-HEXADIENAL
C6H8O	000625-86-5	FURAN, 2,5-DIMETHYL-
C6H8O	000930-68-7	2-CYCLOHEXENE-1-ONE
C6H8O	002550-28-9	1-HEXYN-5-ONE
C6H8O	003208-16-0	2-ETHYLFURAN
C6H8O2	000110-44-1	SORBIC ACID
C6H8O2	000820-69-9	3-HEXENE-2,5-DIONE (TRANS)
C6H8O2	017559-81-8	3-HEXENE-2,5-DIONE (CIS)
C6H8O2S	022913-26-4	METHYL THIOPHEN-3-CARBOXYLATE
C6H8O3S	006577-69-1	1,4-OXATHIIN-3-CARBOXYLIC ACID, 5,6-DIHYDRO-2-ME
C6H8O4	000624-48-6	METHYL MALEATE
C6H8O4	000624-49-7	2-BUTENEDIOIC ACID (E)-, DIMETHYL ESTER
C6H8O6	000050-81-7	ASCORBIC ACID
C6H8O7	000077-92-9	CITRIC ACID
C6H8S	000638-02-8	2,5-DIMETHYLTHIOPHENE
C6H8S	000872-55-9	2-ETHYLTHIOPHENE
C6H9F3N4OS	025366-23-8	THIAZAFLURON
C6H9N	000625-84-3	2,5-DIMETHYLPYRROLE
C6H9N3	000111-94-4	PROPANENITRILE, 3,3'-IMINOBIS-
C6H9N3	000461-98-3	2,6-DIMETHYL-4-PYRIMIDINAMINE
C6H9N3	000767-15-7	2-AMINO-4,6-DIMETHYLPYRIMIDINE
C6H9N3	005214-29-9	2-DIMETHYLAMINOPYRAZINE
C6H9N3	005621-02-3	2-PYRIMIDINAMINE, N,N-DIMETHYL-
C6H9N3	017258-31-0	3-PYRIDAZINAMINE,N,N-DIMETHYL
C6H9N3	017258-38-7	4-PYRIDAZINAMINE,N,N-DIMETHYL

C6H9N3	031401-45-3	4-PYRIMIDINAMINE, N,N-DIMETHYL-
C6H9N3	031401-46-4	5-PYRIMIDINAMINE, N,N-DIMETHYL-
C6H9N3O2	000071-00-1	HISTIDINE
C6H9N3O2	000135-20-6	N-HYDROXY-N-NITROSOBENZENAMINE AMMONIUM SALT
C6H9N3O2	013373-32-5	1H-IMIDAZOLE, 2-(1-METHYLETHYL)-4-NITRO-
C6H9N3O2	039070-08-1	1-ME-2-NO2-5-VINYL IMIDAZOLE
C6H9N3O3	000443-48-1	METRONIDAZOLE
C6H9N3O3	000999-29-1	N-DIAZOACETYLGLYCINE,ETHYL ESTER
C6H9N3O3	023571-38-2	1H-IMIDAZOLE-1-ETHANOL, 5-METHYL-2-NITRO-
C6H9N3O3	035687-44-6	5-METHOXY-1,2-DIMETHYL-4-NITRO-1H-IMIDAZOLE
C6H9N3O3	071006-75-2	1-(ACETYLMETHYL)-2-NO2 IMIDAZOLE
C6H9N3O3S	074141-73-4	1H-IMIDAZOLE, 1-[2-(METHYLSULFINYL)ETHYL]-2-NITR
C6H9N3O4	013551-92-3	3-(2-NITROIMIDAZOL-1-YL)-1,2-PROPANEDIOL
C6H9N3O4	022813-31-6	1-(CH2CO2ME)-2-NO2- IMIDAZOLE
C6H9N3O4	040361-79-3	1-ME-2-NO2-5-MEOCARBONYL IMIDAZOLE
C6H9N5O2	060010-81-3	1,2,4-TRIAZOLE,3,5-DIACETAMIDO-
C6H9NO	000088-12-0	N-VINYL-2-PYRROLIDINONE
C6H9NO2	000932-17-2	2-PYRROLIDINONE, 1-ACETYL-
C6H9NO2	023046-86-8	N-FORMYLCYCLOBUTANECARBOXAMIDE
C6H9NO2S	013584-27-5	1,4-OXATHIIN-3-CARBOXAMIDE, 5,6-DIHYDRO-2-METHYL
C6H9NO3	042346-68-9	1-METHYL-4-CARBOXY-2-PYRROLIDONE
C6H9NO3S	001195-16-0	N-(TETRAHYDRO-2-OXO-3-THIENYL)ACETAMIDE
C6H9NO3S3	104437-96-9	2-THIOPHENESULFONAMIDE, 5-[(2-HYDROXYETHYL)THIO]
C6H9NO5S3	104438-00-8	2-THIOPHENESULFONAMIDE, 5-[(2-HYDROXYETHYL)SULFO
C6H9NO6	000139-13-9	NITRILOTRIACETIC ACID
C6H9NO6	016051-77-7	D-GLUCITOL, 1,4:3,6-DIANHYDRO-, 5-NITRATE
C6H9NO6	016106-20-0	D-GLUCITOL, 1,4:3,6-DIANHYDRO-, 2-NITRATE
C6H9NOS	004461-29-4	THIOPHEN-2-YL-ACETAMIDE
C6H9NOS	013053-81-1	2-ACETAMIDOTHIOPHENE
C6H9NOS	013781-66-3	3-ACETAMIDOTHIOPHENE
C6H9NOS	059445-83-9	1-(THIOPHEN-2-YL)ACETALDEHYDE OXIME
C6H10	000110-83-8	CYCLOHEXENE
C6H10	000513-81-5	2,3-DIMETHYL-1,3-BUTADIENE
C6H10	000592-42-7	1,5-HEXADIENE
C6H10	000592-48-3	1,3-HEXADIENE
C6H10	000693-02-7	1-HEXYNE
C6H10	000763-30-4	2-METHYL-1,4-PENTADIENE
C6H10	000926-54-5	2-METHYL-1,3-PENTADIENE
C6H10	000926-56-7	4-METHYL-1,3-PENTADIENE
C6H10	004549-74-0	3-METHYL-1,3-PENTADIENE
C6H10	005194-50-3	2,4-HEXADIENE (CIS,TRANS)
C6H10	005194-51-4	2,4-HEXADIENE (TRANS,TRANS)
C6H10	007319-00-8	TRANS-1,4-HEXADIENE
C6H10	020237-34-7	1,3-HEXADIENE (TRANS)
C6H10Cl2	000822-86-6	1,2-DICHLOROCYCLOHEXANE -TRANS
C6H10Cl2	001121-21-7	1,2-DICHLOROCYCLOHEXANE -CIS
C6H10Cl2Si	003651-23-8	DIALLYLDICHLOROSILANE
C6H10ClN3O3	013551-86-5	1(3-CL-2-OH PR)-2-NO2-IMIDAZOLE
C6H10ClN5	006190-65-4	DESETHYLATRAZINE
C6H10ClNS	000533-45-9	CLOMETHIAZOLE
C6H10FN3O3	013551-89-8	1-(3-F-2-OHPROPYL)-2-NO2 IMIDAZOLE
C6H10N2O	014884-03-8	4-ISOPROPOXYPYRAZOLE
C6H10N2O	016338-97-9	N-NITROSO-DIALLYLAMINE
C6H10N2O	077580-78-0	FURAZAN, METHYLPROPYL-
C6H10N2O	088095-60-7	4-PROPOXYPYRAZOLE
C6H10N2O2	016935-34-5	2,4-IMIDAZOLIDINEDIONE, 5-(1-METHYLETHYL)-
C6H10N2O2S2	106231-37-2	1,4-DITHIAN-2-ONE, O-[(METHYLAMINO)CARBONYL]OXIM
C6H10N2O2S2	106231-46-3	1,3-DITHIOLAN-4-ONE, 5-METHYL-, O-[(METHYLAMINO)

C6H10N2O3S	054266-46-5	1,3-OXATHIOLAN-4-ONE, 5-METHYL-, O-[(METHYLAMINO
C6H10N2O3S	054266-55-6	1,4-OXATHIAN-3-ONE, O-[(METHYLAMINO)CARBONYL]OXI
C6H10N2O4	004033-40-3	L-ASPARAGINE, N -ACETYL-
C6H10N4	000054-95-5	PENTYLENETETRAZOLE
C6H10N4O2	038726-91-9	N-DIAZOACETYLGLYCINE-N'-ETHYLAMIDE
C6H10N4O3	040647-30-1	METHANAMINE, N-[(1-METHYL-2-NITRO-1H-IMIDAZOL-5-
C6H10N4O4	007681-76-7	RONIDAZOLE
C6H10N4OS	021087-59-2	3-SME-4-NH2-6-ET-124-TRIAZIN-5-ONE
C6H10N4OS	022278-77-9	3-SH-4-NH2-6IPR-124TRIAZINE-5-ONE
C6H10N6O	004342-03-4	IMIDAZOL-4-CONH2-5(33DIME-1-TRIAZENO)
C6H10N6O	021466-00-2	3(33DIME-1-TRIAZENO)PYRAZOLE-4-CONH2
C6H10O	000077-75-8	METHYL PENTYNOL
C6H10O	000105-31-7	1-HEXYN-3-OL
C6H10O	000108-94-1	CYCLOHEXANONE
C6H10O	000109-49-9	1-HEXEN-5-ONE
C6H10O	000109-50-2	3-HEXYN-2-OL
C6H10O	000141-79-7	MESITYL OXIDE
C6H10O	000505-57-7	2-HEXENAL(TRANS)
C6H10O	000557-40-4	ALLYL ETHER
C6H10O	003744-02-3	ISOMESITYL OXIDE
C6H10O	006728-26-3	(E)-2-HEXENAL
C6H10O2	000097-63-2	ETHYL METHACRYLATE
C6H10O2	000110-13-4	2,5-HEXANEDIONE
C6H10O2	000502-44-3	E-CAPROLACTONE
C6H10O2	000925-60-0	2-PROPENOIC ACID, PROPYL ESTER
C6H10O2S4	000502-55-6	BIS(ETHYLXANTHOGEN)
C6H10O3	000123-62-6	PROPIONIC ANHYDRIDE
C6H10O3	000141-97-9	ACETOACETIC ESTER
C6H10O3	000624-45-3	PENTANOIC ACID, 4-OXO-, METHYL ESTER
C6H10O3	000868-77-9	2-HYDROXYETHYL METHACRYLATE
C6H10O3	000999-61-1	2-HYDROXYPROPYL ACRYLATE
C6H10O4	000095-92-1	ETHYL OXALATE
C6H10O4	000106-65-0	DIMETHYL SUCCINATE
C6H10O4	000111-55-7	ETHYLENE GLYCOL DIACETATE
C6H10O4	000124-04-9	HEXANEDIOIC ACID ·
C6H10O4	000616-62-6	PROPYLPROPANEDIOIC ACID
C6H10O4	023261-20-3	1,2:5,6-DIANHYDROGALACTITOL
C6H10O4S2	055290-64-7	DIMETHIPIN
C6H10O6	000608-68-4	L-DIMETHYL TARTRATE
C6H10O6	005057-96-5	DIMETHYL TARTRATE (MESO)
C6H10O7	006556-12-3	GLUCURONIC ACID
C6H10S	000592-88-1	DIALLYL SULFIDE
C6H10S	052006-63-0	3-ETHYLTHIOPHENE
C6H11Br	000108-85-0	BROMOCYCLOHEXANE
C6H11BrN2O2	000496-67-3	A-BROMO-I-VALERYLUREA
C6H11BrO2	000584-93-0	2-BROMOPENTANOIC ACID
C6H11BrO2	000600-00-0	ETHYL ALPHA-BROMOISOBUTYRATE
C6H11Cl	000542-18-7	CHLOROCYCLOHEXANE
C6H11Cl4O3PS	054593-83-8	CHLORETHOXYFOS
C6H11ClN2O2	061345-66-2	A-CHLORO-I-VALERYLUREA
C6H11F3O	065611-47-4	6,6,6-TRIFLUORO-1-HEXANOL
C6H11F3O	107103-97-9	5,5,5-TRIFLUORO-3-METHYL-1-PENTANOL
C6H11N	000124-02-7	DIALLYLAMINE
C6H11N	000542-54-1	4-METHYLPENTANITRILE
C6H11N	000628-73-9	HEXANENITRILE
C6H11N2O4PS3	000950-37-8	METHIDATHION
C6H11N3O3	000705-19-1	1-(2-OHET)-2-ME-4-NO2 IMIDAZOLE
C6H11N3O3	013551-91-2	1-(2-OH PROPYL)-2-NO2 IMIDAZOLE

C6H11N3O3	071006-73-0	1-(2-MEO ETHYL)-2-NO2 IMIDAZOLE
C6H11N3O3	084652-30-2	L-ASPARAGINAMIDE, N-ACETYL-
C6H11N3O4	000556-33-2	TRIGLYCINE
C6H11N3O4	006129-11-9	2ME2NO2-PROPANAL-N-MECARBAMOYL OXIME
C6H11N3O4	039070-09-2	1-ME-2-NO2-5-(1,2-DIOHET)IMIDAZOLE
C6H11N3O4S	071006-79-6	ME(2(2-NO2-1-IMIDAZOLYL)ET)SULFONE
C6H11N5O4	104958-85-2	(3-NITRO-1,2,4-TRIAZOL-1-YL)ACETAMIDE,N-(2-HYDRO
C6H11NaO2	000151-33-7	SODIUM HEXANOATE
C6H11NO	000100-64-1	CYCLOHEXANONE OXIME
C6H11NO	000105-60-2	CAPROLACTAM
C6H11NO	002687-91-4	2-PYRROLIDINONE, 1-ETHYL-
C6H11NO	005075-92-3	2-PYRROLIDINONE, 1,5-DIMETHYL-
C6H11NO2	000052-52-8	CYCLOLEUCINE
C6H11NO2	000498-94-2	ISONIPECOTIC ACID
C6H11NO2	000498-95-3	NIPECOTIC ACID
C6H11NO2	000535-75-1	2-PIPERIDINECARBOXYLIC ACID
C6H11NO3	003619-02-1	ALANINE-N-ACETYL,METHYL ESTER
C6H11NO5	074754-55-5	6-NITRATOHEXANOIC ACID
C6H11NS	007203-96-5	2-AZACYCLOHEPTANTHIONE
C6H12	000096-37-7	METHYLCYCLOPENTANE
C6H12	000110-82-7	CYCLOHEXANE
C6H12	000558-37-2	3,3-DIMETHYL-1-BUTENE
C6H12	000563-78-0	2,3-DIMETHYL-1-BUTENE
C6H12	000563-79-1	2,3-DIMETHYL-2-BUTENE
C6H12	000592-41-6	1-HEXENE
C6H12	000592-43-8	2-HEXENE
C6H12	000592-47-2	3-HEXENE
C6H12	000625-27-4	2-METHYL-2-PENTENE
C6H12	000674-76-0	TRANS-4-METHYL-2-PENTENE
C6H12	000691-37-2	4-METHYL-1-PENTENE
C6H12	000691-38-3	4-METHYL-2-PENTENE (CIS)
C6H12	000760-20-3	3-METHYL-1-PENTENE
C6H12	000760-21-4	2-ETHYL-1-BUTENE
C6H12	000763-29-1	2-METHYL-1-PENTENE
C6H12	000922-61-2	3-METHYL-2-PENTENE
C6H12	003638-35-5	ISOPROPYLCYCLOPROPANE
C6H12	004050-45-7	2-HEXENE (TRANS)
C6H12	004461-48-7	4-METHYL-2-PENTENE
C6H12	007688-21-3	2-HEXENE (Z)
C6H12	014850-23-8	4-OCTENE (TRANS)
C6H12Br2O4	000488-41-5	1,6-DIBR-1,6-DIDEOXYMANNITOL
C6H12Br2O4	010318-26-0	MITOLACTOL
C6H12Cl2	002162-92-7	1,2-DICHLOROHEXANE
C6H12Cl2	013275-18-8	2,5-DICHLOROHEXANE
C6H12Cl2O	000108-60-1	DCIP (2,2'-OXYBIS-1-CHLOROPROPANE)
C6H12Cl2O2	000112-26-5	1,2-BIS(2-CHLOROETHOXY)ETHANE
C6H12Cl2O2	000619-33-0	1,1-DICHLORO-2,2-DIETHOXYETHANE
C6H12Cl3N	000555-77-1	TRIS(BETA-CHLOROETHYL)AMINE
C6H12Cl3O4P	000115-96-8	TRI-2-CHLOROETHYL PHOSPHATE
C6H12ClNO	002315-36-8	CDEA
C6H12ClO4P	000311-47-7	2-CHLOROVINYL DIETHYL PHOSPHATE
C6H12F3N	037174-09-7	2,2,2-TRIFLUOROET AMINE,N,N-DIET
C6H12MnN2S4	012427-38-2	MANEB
C6H12N2	000280-57-9	1,4-DIAZABICYCLO(2,2,2)OCTANE (DABCO)
C6H12N2O	000695-94-3	1-(2-OHET)-2-ME IMIDAZOLINE
C6H12N2O	000932-83-2	N-NITROSOHEXAMETHYLENEIMINE
C6H12N2O	007247-89-4	2-METHYL-N-NITROSOPEPERIDINE
C6H12N2O	013603-07-1	3-METHYL-N-NITROSOPIPERIDINE

C6H12N2O	015104-03-7	4-METHYL-N-NITROSOPIPERIDINE
C6H12N2O	055556-86-0	2,5-DIMETHYL-N-NITROSOPYRROLIDINE
C6H12N2O2	001456-28-6	2,6-DIMETHYL-N-NITROSOMORPHOLINE
C6H12N2O2	002274-08-0	3-METHYLBUTANOYLUREA
C6H12N2O2	022715-68-0	ALANINE,N-ACETYL,N'-MEAM-AMIDE
C6H12N2O3	001596-84-5	DAMINOZIDE
C6H12N2O3	060828-33-3	THREONIN-AMIDE, N-ACETYL
C6H12N2O3	066017-91-2	N-NITROSO-ACETOXYMETHYL-N-PROPYLAMINE
C6H12N2O4S2	000056-89-3	CYSTINE
C6H12N2S	014294-09-8	1-PIPERIDINETHIOCARBOXAMIDE
C6H12N2S3	000097-74-5	TETRAMETHYLTHIURAM
C6H12N2S4	000137-26-8	THIRAM
C6H12N2S4Zn	000137-30-4	ZIRAM
C6H12N3O	052838-38-7	2-AMINO-5-BU-1,3,4-OXADIAZOLE
C6H12N3O	069741-90-8	2-AMINO-5-I-BU-1,3,4-OXADIAZOLE
C6H12N3O	069741-91-9	2-AMINO-5-SEC-BUTYL-1,3,4-OXADIAZOLE
C6H12N3O	069741-92-0	2-AMINO-5-T-BU-1,3,4-OXADIAZOLE
C6H12N3OP	000545-55-1	TEPA
C6H12N3PS	000052-24-4	THIOTEPA
C6H12N4	000100-97-0	HEXAMETHYLENETETRAMINE
C6H12N4O2	055380-34-2	2,6-DIME-N,N'-DINITROSOPIPERAZINE
C6H12N4O2	055556-88-2	2,5-DIME-N,N'-DINITROSOPIPERAZINE
C6H12N4O4	104958-86-3	1,2,4-TRIAZOLE,3-NITRO-1-(2-HYDROXY-3-METHOXY)PR
C6H12N5O2PS2	000078-57-9	MENAZON
C6H12NO3PS2	021548-32-3	FOSTHIETAN
C6H12NO4PS2	002540-82-1	FORMOTHION
C6H12O	000066-25-1	HEXANAL
C6H12O	000075-97-8	3,3-DIMETHYL-2-BUTANONE
C6H12O	000097-96-1	2-ETHYLBUTYRALDEHYDE
C6H12O	000108-10-1	4-METHYL-2-PENTANONE
C6H12O	000108-93-0	CYCLOHEXANOL
C6H12O	000109-53-5	ISOBUTYL VINYL ETHER
C6H12O	000111-34-2	BUTYL VINYL ETHER
C6H12O	000123-15-9	2-METHYLPENTALDEHYDE
C6H12O	000544-12-7	3-HEXENE-1-OL
C6H12O	000565-61-7	3-METHYLPENTAN-2-ONE
C6H12O	000565-69-5	2-METHYL-3-PENTANONE
C6H12O	000589-38-8	3-HEXANONE
C6H12O	000591-78-6	2-HEXANONE
C6H12O	000592-90-5	OXEPANE
C6H12O	001436-34-6	OXIRANE, BUTYL-
C6H12O	001471-03-0	ALLYL PROPYL ETHER
C6H12O	002088-07-5	2-METHYL-1-PENTEN-3-OL
C6H12O	002144-41-4	CIS-2,5-DIMETHYLTETRAHYDROFURAN
C6H12O	002390-94-5	TRANS-2,5-DIMETHYLTETRAHYDROFURAN
C6H12O	004798-44-1	1-HEXEN-3-OL
C6H12O	004798-45-2	4-METHYL-1-PENTEN-3-OL
C6H12O	004798-58-7	2-HEXENE-4-OL
C6H12O2	000088-09-5	2-ETHYLBUTYRIC ACID
C6H12O2	000097-61-0	PENTANOIC ACID, 2-METHYL-
C6H12O2	000097-62-1	ETHYL ISOBUTYRATE
C6H12O2	000100-72-1	TETRAHYDROPYRAN-2-METHANOL
C6H12O2	000105-46-4	S-BUTYLACETATE
C6H12O2	000105-54-4	ETHYLBUTYRATE
C6H12O2	000106-36-5	N-PROPYLPROPIONATE
C6H12O2	000110-19-0	SEC-BUTYLACETATE
C6H12O2	000110-45-2	ISOAMYL FORMATE
C6H12O2	000123-42-2	4-HYDROXY-4-METHYL-2-PENTANONE

C6H12O2	000123-86-4	N-BUTYL ACETATE ⌐
C6H12O2	000142-62-1	HEXANOIC ACID
C6H12O2	000540-88-5	T-BUTYL ACETATE ⌐
C6H12O2	000556-24-1	METHYL (3-METHYL)BUTANOATE
C6H12O2	000595-37-9	2,2-DIMETHYL BUTYRIC ACID
C6H12O2	000598-98-1	PROPANOIC ACID, 2,2-DIMETHYL-, METHYL ESTER
C6H12O2	000624-24-8	METHYL VALERATE
C6H12O2	000637-78-5	ISOPROPYL PROPIONATE
C6H12O2	000646-07-1	4-METHYLPENTANOIC ACID ⌐
C6H12O2	000931-17-9	1,2-CYCLOHEXANEDIOL
C6H12O2	001460-57-7	1,2-CYCLOHEXANEDIOL-TRANS
C6H12O2	001792-81-0	1,2-CYCLOHEXANEDIOL-CIS
C6H12O3	000100-79-8	DIOXOLANE
C6H12O3	000108-65-6	PROPYLENE GLYCOL ME ETHER ACETATE
C6H12O3	000111-15-9	ETHOXYETHYLACETATE
C6H12O3	000123-63-7	PARALDEHYDE
C6H12O3	000999-10-0	BUTANOIC ACID, 4-HYDROXY-, ETHYL ESTER
C6H12O3	006064-63-7	HEXANOIC ACID, 2-HYDROXY-
C6H12O6	000050-99-7	GLUCOSE
C6H12O6	000059-23-4	GALACTOSE
C6H12O6	000087-89-8	INOSITOL
C6H12O6	003458-28-4	D-MANNOSE
C6H12O6	007660-25-5	BETA-FRUCTOPYRANOSE (D)
C6H12O7	000526-95-4	D-GLUCONIC ACID
C6H12S	001569-69-3	CYCLOHEXANETHIOL
C6H13Br	000111-25-1	1-BROMOHEXANE
C6H13Br	003377-86-4	2-BROMOHEXANE
C6H13Cl	000544-10-5	1-CHLOROHEXANE
C6H13Cl	000638-28-8	2-CHLOROHEXANE
C6H13Cl3Si	000928-65-4	HEXYLTRICHLOROSILANE
C6H13F	000373-14-8	1-FLUOROHEXANE
C6H13I	000638-45-9	1-IODOHEXANE
C6H13N	000108-91-8	CYCLOHEXANAMINE
C6H13N	000109-05-7	2-METHYLPIPERIDINE
C6H13N	000111-49-9	HEXAMETHYLENEIMINE
C6H13N	000626-67-5	N-METHYLPIPERIDINE
C6H13N	001120-85-0	1-BUTYLAZIRIDINE
C6H13N	005666-21-7	ALLYLPROPYLAMINE
C6H13N3O	067774-31-6	N-NITROSO-3,5-DIME PIPERAZINE
C6H13N3O2	050285-72-8	1-NITROSO-1-METHYL DIETHYLUREA
C6H13N3O3	000372-75-8	L-ORNITHINE, N5-(AMINOCARBONYL)-
C6H13N5O	055921-54-5	24-NH2-6-ETAM PYRIMIDINE-3-OXIDE
C6H13N5O	055921-58-9	24-NH2-6-DIMEAM PYRIMIDINE-3-OXIDE
C6H13NO	000100-74-3	4-ETHYLMORPHOLINE
C6H13NO	000685-91-6	DIETHYLACETAMIDE
C6H13NO2	000060-32-2	6-AMINOHEXANOIC ACID
C6H13NO2	000061-90-5	LEUCINE
C6H13NO2	000073-32-5	ISO-LEUCINE
C6H13NO2	000319-78-8	ISOLEUCINE
C6H13NO2	000327-57-1	A-AMINOCAPROIC ACID
C6H13NO2	000328-39-2	LEUCINE
C6H13NO2	000638-42-6	O-PENTYL CARBAMATE
C6H13NO2	000646-14-0	NITROHEXANE
C6H13NO2	033105-81-6	T-BUTYLGLYCINE
C6H13NO2	038580-89-1	CARBAMIC ACID, DIMETHYL-, 1-METHYLETHYL ESTER
C6H13NO2	039076-02-3	METHYL SEC-BUTYLCARBAMATE
C6H13NO4	019130-96-2	3,4,5-PIPERIDINETRIOL, 2-(HYDROXYMETHYL)-
C6H14	000075-83-2	2,2-DIMETHYLBUTANE

C6H14	000079-29-8	2,3-DIMETHYLBUTANE
C6H14	000096-14-0	3-METHYLPENTANE
C6H14	000107-83-5	2-METHYLPENTANE
C6H14	000110-54-3	N-HEXANE
C6H14ClN	024948-81-0	N-CHLORO-N-(1-METHYETHYL)-2-PROPANAMINE
C6H14INO2	079661-42-0	ETHANAMINIUM, 2-(FORMYLOXY)-N,N,N-TRIMETHYL-, IO
C6H14N2	000106-55-8	2,5-DIMETHYLPIPERAZINE
C6H14N2	000106-58-1	PIPERAZINE, 1,4-DIMETHYL-
C6H14N2	000694-83-7	1,2-CYCLOHEXANEDIAMINE
C6H14N2	007154-73-6	1-PYRROLIDINEETHANAMINE
C6H14N2O	000601-77-4	DI-I-PROPYLNITROSOAMINE
C6H14N2O	000621-64-7	N-NITROSODIPROPYLAMINE
C6H14N2O	002443-62-1	HEXANOIC ACID HYDRAZIDE
C6H14N2O	003398-69-4	N-NITROSO-N-ETHYL-(T-BUTYL)AMINE
C6H14N2O	004549-44-4	N-NITROSOETHYL-N-BUTYLAMINE
C6H14N2O	013256-07-0	N-NITROSO-METHYLAMYLAMINE
C6H14N2O	031820-22-1	N-NITROSO-METHYL-NEOPENTYLAMINE
C6H14N2O2	000056-87-1	LYSINE
C6H14N2O3	067856-65-9	N-NITROSO-BIS(2-MEOET)AMINE
C6H14N4O2	000074-79-3	ARGININE
C6H14NO3PS2	000116-01-8	ETHOATE METHYL
C6H14O	000077-74-7	3-METHYL-3-PENTANOL
C6H14O	000097-95-0	2-ETHYL-1-BUTANOL
C6H14O	000105-30-6	2-METHYL-1-PENTANOL
C6H14O	000108-11-2	4-METHYL-2-PENTANOL
C6H14O	000108-20-3	DI-ISOPROPYL ETHER
C6H14O	000111-27-3	1-HEXANOL
C6H14O	000111-43-3	DI(N-PROPYL) ETHER
C6H14O	000464-07-3	3,3-DIMETHYL-2-BUTANOL
C6H14O	000565-60-6	3-METHYL-2-PENTANOL
C6H14O	000565-67-3	2-METHYL-3-PENTANOL
C6H14O	000590-36-3	2-METHYL-2-PENTANOL
C6H14O	000594-60-5	2,3-DIMETHYL-2-BUTANOL
C6H14O	000623-37-0	3-HEXANOL
C6H14O	000624-95-3	3,3-DIMETHYL-1-BUTANOL
C6H14O	000626-89-1	4-METHYL-1-PENTANOL
C6H14O	000626-93-7	2-HEXANOL
C6H14O	000627-08-7	PROPYL ISOPROPYL ETHER
C6H14O	000628-81-9	BUTYL ETHYL ETHER
C6H14O	000637-92-3	ETHYL T-BUTYL ETHER
C6H14O	000994-05-8	METHYL-T-AMYL ETHER
C6H14O	001185-33-7	2,2-DIMETHYL-1-BUTANOL
C6H14O	001320-98-5	4-METHYLPENTANOL
C6H14O2	000076-09-5	PINACOL
C6H14O2	000105-57-7	1,1-DIETHOXYETHANE
C6H14O2	000107-41-5	2-METHYL-2,4-PENTANEDIOL
C6H14O2	000111-76-2	ETHYLENE GLYCOL N-BUTYL ETHER
C6H14O2	000629-11-8	1,6-HEXANEDIOL
C6H14O2	000629-14-1	1,2-DIETHOXYETHANE
C6H14O2	001569-01-3	1-PROPOXY-2-PROPANOL
C6H14O2	004439-24-1	ETHANOL, 2-(2-METHYLPROPOXY)-
C6H14O2	007580-85-0	2-(T-BUTOXY)ETHANOL
C6H14O2	010215-30-2	2-PROPOXY-1-PROPANOL
C6H14O2S	000598-03-8	PROPYL SULFONE
C6H14O3	000077-99-6	1,1,1-TRIS(HYDROXYMETHYL)PROPANE
C6H14O3	000106-62-7	1-PROPANOL, 2-(2-HYDROXYPROPOXY)-
C6H14O3	000106-69-4	1,2,6-HEXANETRIOL
C6H14O3	000111-90-0	DIETHYLENE GLYCOL MONOETHYL ETHER

C6H14O3	000111-96-6	2-METHOXYETHYLETHER
C6H14O3	014315-97-0	1,1,3-TRIMETHOXYPROPANE
C6H14O4	000112-27-6	TRIETHYLENE GLYCOL
C6H14O6	000050-70-4	SORBITOL
C6H14O6	000069-65-8	MANNITOL
C6H14O6	000133-43-7	DL-MANNITOL
C6H14O6	000608-66-2	GALACTITOL
C6H14O6S2	000055-98-1	MYLERLAN
C6H14S	000111-31-9	1-HEXANETHIOL
C6H14S	000625-80-9	2,4-DIMETHYL-3-THIAPENTANE
C6H14S	000638-46-0	ETHYL BUTYLSULFIDE
C6H14S	001741-83-9	METHYL PENTYL SULFIDE
C6H14S2	000629-19-6	4,5-DITHIAOCTANE
C6H15BrN2O2	028099-10-7	CARBAMIC ACID,O-ETHYL(TRIMEAMM)BR
C6H15ClSi	000994-30-9	CHLOROTRIETHYLSILANE
C6H15FSi	000358-43-0	TRIETHYLFLUOROSILANE
C6H15N	000108-18-9	DIISOPROPYLAMINE
C6H15N	000111-26-2	N-HEXYLAMINE
C6H15N	000121-44-8	TRIETHYLAMINE
C6H15N	000142-84-7	DIPROPYLAMINE
C6H15N	000927-62-8	DIMETHYLBUTYLAMINE
C6H15N3O	050405-18-0	4-PENTYLSEMICARBAZIDE
C6H15N5	000692-13-7	1-BUTYLBIGUANIDE
C6H15NO	000100-37-8	2-(DIETHYLAMINO)-ETHANOL
C6H15NO	004620-70-6	ETHANOL, 2- (1,1-DIMETHYLETHYL)AMINO -
C6H15NO2	000110-97-4	DIISOPROPANOLAMINE
C6H15NO3	000102-71-6	TRIETHANOLAMINE
C6H15O2PS3	000640-15-3	THIOMETON
C6H15O3P	000078-38-6	Phosphonic acid, ethyl-, diethyl ester
C6H15O3P	001809-20-7	PHOSPHONIC ACID, BIS(1-METHYLETHYL) ESTER
C6H15O3PS2	000919-86-8	DEMETON-S-METHYL
C6H15O4P	000078-40-0	TRIETHYL PHOSPHATE
C6H15O4PS2	000301-12-2	OXYDEMETON METHYL
C6H16BrN	000636-70-4	TRIETHYLAMINE, HYDROBROMIDE
C6H16ClN	000554-68-7	TRIETHYLAMINE HYDROCHLORIDE
C6H16FN2OP	000371-86-8	ISOPESTOX
C6H16INO	004238-50-0	2-METHOXYETHYL-TRIME AMMONIUM IODIDE
C6H16N2	000100-36-7	1,2-ETHANEDIAMINE, N,N-DIETHYL-
C6H16N2	000110-18-9	N,N,N',N'-TETRAME ETHYLENEDIAMINE
C6H16N2	000111-74-0	1,2-ETHANEDIAMINE, N,N'-DIETHYL-
C6H16N2	000124-09-4	HEXAMETHYLENEDIAMINE
C6H16N3OP	007778-06-5	1,3,2-DIAZAPHOSPHOLIDIN-2-AMINE, N,N,1,3-TETRAME
C6H16NO2PS	028167-45-5	O,S-DIPROPYLPHOSPHORAMIDOTHIOATE
C6H16NO2PS	052067-44-4	O=P(OC)(SC)N-T-BU
C6H16NO2PS	052067-49-9	O=P(OC)(SC)N-N-BU
C6H16NO2PS	052067-50-2	O=P(OC)(SC)N-S-BU
C6H16NO2PS	052067-51-3	O=P(OC)(SC)N-I-BU
C6H16NO3P	002404-03-7	DIETHYL DIMETHYLPHOSPHORAMIDATE
C6H16O2Si	000078-62-6	DIMETHYL DIETHOXYSILANE
C6H16O5P2	032288-17-8	DIETHYL DIMETHYLDIPHOSPHONATE
C6H16Si	000617-86-7	TRIETHYLSILANE
C6H16Si	001001-52-1	SILANE, BUTYL-DIMETHYL
C6H16Si	003510-70-1	SILANE, TRIMETHYLPROPYL-
C6H17N3	034066-96-1	1,2-ETHANEDIAMINE, N-(2-AMINOETHYL)-N,N'-DIMETHY
C6H18N3OP	000680-31-9	HEXAMETHYLPHOSPHORAMIDE
C6H18N4	000112-24-3	TRIETHYLENETETRAAMINE
C6H18O3Si3	000541-05-9	HEXAMETHYLCYCLOTRISILOXANE
C6H18OSi2	000107-46-0	HEXAMETHYLDISILOXANE

C6H19NSi2	000999-97-3	HEXAMETHYLDISILIZANE
C7F14	000355-02-2	PERFLUOROMETHYLCYCLOHEXANE
C7F16	000335-57-9	PERFLUOROHEPTANE
C7F8	000434-64-0	OCTAFLUOROTOLUENE
C7HCl5N2	007682-34-0	BENZIMIDAZOLE-2,4,5,6,7-PENTACL
C7H2Cl6	003389-71-7	1,2,3,4,7,7-HEXACHLORO-2,5-NORBORNADIENE
C7H2ClF3N2O4	000393-75-9	1,3-DINO2 2-CL 5-CF3 BENZENE
C7H3Br2NO	001689-84-5	BROMOXYNIL
C7H3Br5	000087-83-2	PENTABROMOTOLUENE
C7H3Cl2F3	000320-60-5	2,4-DICHLORO(TRIFLUOROMETHYL)BENZENE
C7H3Cl2F3	000328-84-7	3,4-DICHLOROBENZOTRIFLUORIDE
C7H3Cl2N	001194-65-6	2,6-DICHLOROBENZONITRILE
C7H3Cl2NO	000102-36-3	1,2-DICHLORO-4-ISOCYANATOBENZENE
C7H3Cl2NO4	000088-86-8	DINOBEN
C7H3Cl3O2	000050-31-7	2,3,6-TRICHLOROBENZOIC ACID
C7H3Cl3O2	000050-82-8	2,4,5-TRICHLOROBENZOIC ACID
C7H3Cl4NO3	002438-88-2	2,3,5,6-TETRACHLORO-4-NITROANILINE
C7H3Cl5	000877-11-2	2,3,4,5,6-PENTACHLOROTOLUENE
C7H3Cl5	069911-61-1	PENTACHLOROTOLUENE
C7H3Cl5O	001825-21-4	PENTACHLOROANISOLE
C7H3ClF3NO2	000118-83-2	4-CHLORO-1-NITRO-2(TRIFLUOROMETHYL)BENZENE
C7H3ClN2O6	000118-97-8	4-CHLORO-3,5-DINITROBENZOIC ACID
C7H3ClN4	072126-54-6	2,3-PYRAZINEDICARBONITRILE, 5-CHLORO-6-METHYL-
C7H3F12N3	023757-42-8	1H-IMIDAZOLE-4-AMINE, 2,5-DIHYDRO-2,2,5,5-TETRAK
C7H3F5	000771-56-2	N-PROPYLPENTAFLUOROBENZENE
C7H3F5O2S	000651-85-4	PENTAFLUOROPHENYL METHYL SUFONE
C7H3I2NO	001689-83-4	IOXYNIL
C7H3I3O2	000088-82-4	2,3,5-TRIIODOBENZOIC ACID
C7H3N3	001633-44-9	3,4-PYRIDINEDICARBONITRILE
C7H3N3O4S	001594-56-5	2,4-DINITROPHENYL THIOCYANATE
C7H3N3O8	000129-66-8	2,4,6-TRINITROBENZOIC ACID
C7H4BrN	000623-00-7	4-BROMOBENZONITRILE
C7H4BrN	002042-37-7	2-BROMOBENZONITRILE
C7H4BrN	006952-59-6	3-BROMOBENZONITRILE
C7H4BrN3O	062825-09-6	M-BROMOPHENYL AZOXYCYANIDE
C7H4BrN3O	062825-10-9	P-BROMOPHENYL AZOXYCYANIDE
C7H4BrNS	001985-12-2	P-BROMOPHENYL ISOTHIOCYANATE
C7H4BrNS	002131-59-1	BENZENE, 1-BROMO-3-ISOTHIOCYANATO-
C7H4Cl2F3NO2S	023383-96-2	2,4-DICL CF3-METHANESULFONANILIDE
C7H4Cl2N2O2S2	096474-17-8	2-BENZOTHIAZOLESULFONAMIDE, 4,6-DICHLORO-
C7H4Cl2O	000618-46-2	M-CHLOROBENZOYLCHLORIDE
C7H4Cl2O	006287-38-3	3,4-DICHLOROBENZALDEHYDE
C7H4Cl2O2	000050-30-6	2,6-DICHLOROBENZOIC ACID
C7H4Cl2O2	000050-79-3	2,5-DICHLOROBENZOIC ACID
C7H4Cl2O2	000050-84-0	2,4-DICHLOROBENZOIC ACID
C7H4Cl2O2	000051-36-5	3,5-DICHLOROBENZOIC ACID
C7H4Cl2O2	000051-44-5	3,4-DICHLOROBENZOIC ACID
C7H4Cl3NO3	055335-06-3	TRICLOPYR
C7H4Cl4	001006-31-1	2,3,5,6-TETRACHLOROTOLUENE
C7H4Cl4	001006-32-2	2,3,4,5-TETRACHLOROTOLUENE
C7H4Cl4	002136-89-2	2-CHLOROBENZOTRICHLORIDE
C7H4Cl4	005216-25-1	4-CHLOROBENZOTRICHLORIDE
C7H4Cl4O	000938-22-7	2,3,4,6-TETRACHLOROANISOLE
C7H4Cl4O	000938-86-3	2,3,4,5-TETRACHLOROANISOLE
C7H4Cl4O	006936-40-9	BENZENE, 1,2,4,5-TETRACHLORO-3-METHOXY-
C7H4Cl4O2	002539-17-5	TETRACHLOROGUAIACOL
C7H4ClF3	000088-16-4	CHLORO(TRIFLUOROMETHYL)BENZENE
C7H4ClF3	000098-56-6	4-(TRIFLUOROMETHYL)CHLOROBENZENE

C7H4ClFO2	000434-75-3	BENZOIC ACID, 2-CHLORO-6-FLUORO-
C7H4ClN	000623-03-0	BENZONITRILE, 4-CHLORO-
C7H4ClN3O	054797-22-7	P-CHLOROPHENYL AZOXYCYANIDE
C7H4ClN3O	062825-07-4	M-CHLOROPHENYL AZOXYCYANIDE
C7H4ClNO	000104-12-1	1-CHLORO-4-ISOCYANATOBENZENE
C7H4ClNO	002909-38-8	3-CHLOROPHENYL ISOCYANATE
C7H4ClNO4	000099-60-5	BENZOIC ACID, 2-CHLORO-4-NITRO-
C7H4ClNO4	002516-95-2	2-NITRO-5-CHLOROBENZOIC ACID
C7H4ClNO4	002516-96-3	2-CHLORO-5-NITROBENZOIC ACID
C7H4ClNO4	003970-35-2	2-CHLORO-3-NITROBENZOIC ACID
C7H4ClNO4	005344-49-0	BENZOIC ACID, 2-CHLORO-6-NITRO-
C7H4ClNO4	006280-88-2	2-NITRO-4-CHLOROBENZOIC ACID
C7H4ClNS	002131-55-7	BENZENE, 1-CHLORO-4-ISOTHIOCYANATO-
C7H4ClNS	002392-68-9	1-CHLORO-3-ISOTHIOCYANATO-BENZENE
C7H4F2O2	000385-00-2	BENZOIC ACID, 2,6-DIFLUORO-
C7H4F3N3	013797-63-2	4-PYRIDINE IMIDAZOLE, 2-CF3
C7H4F3N3	019918-36-6	5-PYRIDINE IMADAZOLE, 2-CF3
C7H4F3N3O4	000445-66-9	2,6-NITRO-4-TRIFLUORO-ANILINE
C7H4F3NO2	000098-46-4	M-TRIFLUOROMETHYLNITROBENZENE
C7H4F3NO2	000384-22-5	1-NO2 2-CF3 BENZENE
C7H4F3NO2	000402-54-0	BENZENE, 1-NITRO-4-(TRIFLUOROMETHYL)-
C7H4F3NO3	000088-30-2	3-TRIFLUOROMETHYL-4-NITROPHENOL
C7H4F3NO3	000400-99-7	4-TRIFLUOROMETHYL-2-NITROPHENOL
C7H4F4	000392-85-8	1-FLUORO-2-TRIFLUOROMETHYLBENZENE
C7H4F5NO2S	023384-22-7	2,4-DIF CF3-METHANESULFONANILIDE
C7H4FN	000403-54-3	BENZONITRILE, 3-FLUORO-
C7H4I2O3	000133-91-5	3,5-DIIODOSALICYLIC ACID
C7H4INS	002059-76-9	BENZENE, 1-IODO-4-ISOTHIOCYANATO-
C7H4N2O2	000612-24-8	1-NITRO-2-CYANOBENZENE
C7H4N2O2	000619-24-9	3-CYANO-1-NITROBENZENE
C7H4N2O2	000619-72-7	1-NITRO-4-CYANOBENZENE
C7H4N2O2S	002131-61-5	4-NITROPHENYLISOTHIOCYANATE
C7H4N2O3	039835-28-4	BENZISOXAZOLE, 5-NITRO
C7H4N2O6	000099-34-3	3,5-DINITROBENZOIC ACID
C7H4N2O6	000528-45-0	3,4-DINITROBENZOIC ACID
C7H4N2O6	000610-30-0	2,4-DINITROBENZOIC ACID
C7H4N2O7	000609-99-4	BENZOIC ACID, 2-HYDROXY-3,5-DINITRO-
C7H4N4O3	060142-50-9	P-NITROPHENYL AZOXYCYANIDE
C7H4N4O3	062825-12-1	M-NITROPHENYL AZOXYCYANIDE
C7H4N4O4	031208-76-1	5,7-DINITROBENZPYRAZOLE
C7H4O6	000099-32-1	4-OXO-4H-PYRAN-2,6-DICARBOXYLIC ACID
C7H4O7	000497-59-6	3-HYDROXY-4-OXO-4H-PYRAN-2,6-DICARBOXYLIC ACID
C7H5Br2NO	096237-91-1	2,6-DIBROMOBENZAMIDE
C7H5Br3O	000607-99-8	2,4,6-TRIBROMOANISOLE
C7H5BrClNO	066073-53-8	2-CHLORO-6-BROMOBENZAMIDE
C7H5BrF3NO2S	023384-08-9	M-BROMO CF3-METHANESULFONANILIDE
C7H5BrFNO	107485-63-2	2-FLUORO-6-BROMOBENZAMIDE
C7H5BrN2O3	107485-65-4	2-BROMO-6-NITROBENZAMIDE
C7H5BrO	000618-32-6	BENZOYL BROMIDE
C7H5BrO2	000088-65-3	O-BROMOBENZOIC ACID
C7H5BrO2	000585-76-2	M-BROMOBENZOIC ACID
C7H5BrO2	000586-76-5	P-BROMOBENZOIC ACID
C7H5BrO2	006293-55-6	2-METHYL-6-BROMOBENZOQUINONE
C7H5BrO3	000089-55-4	5-BROMOSALICYLIC ACID
C7H5Cl2N	000622-44-6	PHENYLISOCYANIDE DICHLORIDE
C7H5Cl2NO	002008-58-4	2,6-DICHLOROBENZAMIDE
C7H5Cl2NO2	000133-90-4	3-AMINO-2,5-DICHLOROBENZOIC ACID
C7H5Cl2NS	001918-13-4	CHLORTHIAMID

C7H5Cl3	000094-99-5	2,4-DICHLOROBENZYLCHLORIDE
C7H5Cl3	000098-07-7	BENZOTRICHLORIDE
C7H5Cl3	002077-46-5	2,3,6-TRICHLOROTOLUENE
C7H5Cl3	006639-30-1	2,4,5-TRICHLOROTOLUENE
C7H5Cl3	007359-72-0	2,3,4-TRICHLOROTOLUENE
C7H5Cl3	023749-65-7	2,4,6-TRICHLOROTOLUENE
C7H5Cl3O	000087-40-1	2,4,6-TRICHLOROANISOLE
C7H5Cl3O	000551-76-8	2,4,6-TRICHLORO-3-METHYLPHENOL
C7H5Cl3O	006130-75-2	2,4,5-TRICHLOROANISOLE
C7H5Cl3O	050375-10-5	2,3,6-TRICHLOROANISOLE
C7H5Cl3O	054135-80-7	1,2,3-TRICHLORO-4-METHOXYBENZENE
C7H5Cl3O	054135-81-8	2,3,5-TRICHLOROANISOLE
C7H5Cl3O	054135-82-9	BENZENE, 1,2,3-TRICHLORO-5-METHOXY-
C7H5Cl3O2	002668-24-8	4,5,6-TRICHLOROGUAIACOL
C7H5Cl3O2	057057-83-7	3,4,5-TRICHLORO-2-METHOXYPHENOL
C7H5Cl3S	000701-65-5	TRICHLOROMETHYLTHIOBENZENE
C7H5ClF3N	000121-50-6	BENZENAMINE, 2-CHLORO-5-(TRIFLUOROMETHYL)-
C7H5ClF3NO2S	023384-03-4	M-CHLORO CF3-METHANESULFONANILIDE
C7H5ClF3NO2S	023384-04-5	P-CHLORO CF3-METHANESULFONANILIDE
C7H5ClFNO	066073-54-9	2-FLUORO-6-CHLOROBENZAMIDE
C7H5ClHgO2	000059-85-8	P-(CHLOROMERCURI)BENZOIC ACID
C7H5ClN2O	000061-80-3	ZOXAZOLAMINE
C7H5ClN2O2S	019477-12-4	124-BENZTHIADIAZINE-11-O2-7-CL
C7H5ClN2O2S	019477-31-7	124-BENZTHIADIAZINE-11-O2-6-CL
C7H5ClN2O2S2	088946-20-7	6-CHLOROBENZOTHIAZOLE,2-SULFONAMIDE
C7H5ClN2O3	107485-64-3	2-CHLORO-6-NITROBENZAMIDE
C7H5ClO	000089-98-5	2-CHLOROBENZALDEHYDE
C7H5ClO	000098-88-4	BENZOYL CHLORIDE
C7H5ClO	000104-88-1	4-CHLOROBENZALDEHYDE
C7H5ClO	000587-04-2	M-CHLOROBENZALDEHYDE
C7H5ClO2	000074-11-3	4-CHLOROBENZOIC ACID
C7H5ClO2	000118-91-2	2-CHLOROBENZOIC ACID
C7H5ClO2	000535-80-8	M-CHLOROBENZOIC ACID
C7H5ClO3	000321-14-2	5-CHLOROSALICYLIC ACID
C7H5ClO3	003964-58-7	3-CHLORO-4-HYDROXYBENZOIC ACID
C7H5F2NO	018063-03-1	2,6-DIFLUOROBENZAMIDE
C7H5F3	000098-08-8	BENZOTRIFLUORIDE
C7H5F3INO2S	023384-10-3	P-IODO CF3-METHANESULFONANILIDE
C7H5F3N2O4S	021988-05-6	3-CF3-4-NITROBENZENESULFONAMIDE
C7H5F3O	000098-17-9	M-TRIFLUOROMETHYLPHENOL
C7H5F3O	000402-45-9	P-TRIFLUOROMETHYLPHENOL
C7H5F3O	000444-30-4	O-TRIFLUOROMETHYLPHENOL
C7H5F3O	000456-55-3	TRIFLUOROMETHOXYBENZENE
C7H5F3S	000456-56-4	TRIFLUOROMETHYLTHIOBENZENE
C7H5F4NO2S	023384-00-1	P-FLUORO CF3-METHANESULFONANILIDE
C7H5F4NO2S	023384-01-2	M-FLUORO CF3-METHANESULFONANILIDE
C7H5FO	000459-57-4	BENZALDEHYDE, 4-FLUORO-
C7H5FO2	000445-29-4	2-FLUOROBENZOIC ACID
C7H5FO2	000455-38-9	M-FLUOROBENZOIC ACID
C7H5FO2	000456-22-4	P-FLUOROBENZOIC ACID
C7H5IO2	000088-67-5	2-IODOBENZOIC ACID
C7H5IO2	000618-51-9	3-IODOBENZOIC ACID
C7H5IO2	000619-58-9	4-IODOBENZOIC ACID
C7H5IO3	000119-30-2	BENZOIC ACID, 2-HYDROXY-5-IODO-
C7H5N	000100-47-0	BENZONITRILE
C7H5N3	000322-46-3	PYRIDO(2,3)PYRAZINE
C7H5N3	000622-83-3	DIAZENECARBONITRILE, PHENYL-
C7H5N3O	000090-16-4	1,2,3-BENZOTRIAZIN-4(1H)-ONE

C7H5N3O	054797-20-5	PHENYL AZOXYCYANIDE
C7H5N3O2	089843-47-0	1H-BENZIMIDAZOLE, 6-NITRO-
C7H5N3O4S2	094641-10-8	2-BENZOTHIAZOLESULFONAMIDE, 6-NITRO-
C7H5N3O5	000121-81-3	3,5-DINITROBENZAMIDE
C7H5N3O6	000118-96-7	2,4,6-TRINITROTOLUENE
C7H5N3O7	000606-35-9	2,4,6-TRINITROANISOLE
C7H5N5O8	000479-45-8	TETRYL
C7H5NaO3	000054-21-7	SODIUM SALICYLATE
C7H5NO	000103-71-9	PHENYL ISOCYANATE
C7H5NO	000271-58-9	BENZISOXAZOLE
C7H5NO	000271-95-4	1,2-BENZISOXAZOLE
C7H5NO	000273-53-0	BENZOXAZOLE
C7H5NO	000611-20-1	O-CYANOPHENOL
C7H5NO	000767-00-0	P-CYANOPHENOL
C7H5NO	000873-62-1	M-CYANOPHENOL
C7H5NO2	000059-49-4	O-PHENYLENE CARBAMATE
C7H5NO3	000099-61-6	M-NITROBENZALDEHYDE
C7H5NO3	000552-89-6	O-NITROBENZALDEHYDE
C7H5NO3	000555-16-8	4-NITROBENZALDEHYDE
C7H5NO3	002297-94-1	2,4,6-CYCLOHEPTATRIENE-1-ONE, 2-HYDROXY-5-NITROS
C7H5NO3S	000081-07-2	1,1-DIOX-1,2-BENZISOTHIAZOL-3-ONE
C7H5NO4	000062-23-7	P-NITROBENZOIC ACID
C7H5NO4	000089-00-9	QUINOLINIC ACID
C7H5NO4	000100-26-5	PYRIDINE-2,5-DICARBOXYLIC ACID
C7H5NO4	000121-92-6	M-NITROBENZOIC ACID
C7H5NO4	000490-11-9	CINCHOMERONIC ACID
C7H5NO4	000499-80-9	2,4-PYRIDINEDICARBOXYLIC ACID
C7H5NO4	000499-81-0	DINICOTINIC ACID
C7H5NO4	000499-83-2	PYRIDINE-2,6-DICARBOXYLIC ACID
C7H5NO4	000552-16-9	2-NITROBENZOIC ACID
C7H5NO4	001874-22-2	2-PROPENAL, 3-(5-NITRO-2-FURANYL)-
C7H5NO4	003011-34-5	4-FORMYL-2-NITROPHENOL
C7H5NO5	000085-38-1	P-NITROSALICYLIC ACID
C7H5NO5	000096-97-9	5-NITROSALICYLIC ACID
C7H5NO5	000616-82-0	4-HYDROXY-3-NITROBENZOIC ACID
C7H5NO5	000619-14-7	3-HYDROXY-4-NITROBENZOIC ACID
C7H5NOS	000934-34-9	S-ORTHOPHENYLENETHIOCARBAMATE
C7H5NOS	002131-60-4	4-HYDROXYPHENYLISOTHIOCYANATE
C7H5NOS	003125-63-1	PHENOL, 3-ISOTHIOCYANATO-
C7H5NS	000095-16-9	BENZOTHIAZOLE
C7H5NS	000103-72-0	PHENYLISOTHIOCYANATE
C7H5NS	000271-06-7	6-AZATHIANAPHTHENE
C7H5NS	005285-87-0	PHENYLTHIOCYANATE
C7H5NS2	000149-30-4	2-MERCAPTOBENZOTHIAZOLE
C7H6Br2O2	002183-54-2	BENZENEMETHANOL, 3,5-DIBROMO-2-HYDROXY-
C7H6BrN3O4S2	019367-61-4	BROMOTHIAZIDE
C7H6BrNO	000698-67-9	4-BROMOBENZAMIDE
C7H6BrNO	004001-73-4	2-BROMOBENZAMIDE
C7H6BrNO	022726-00-7	M-BROMOBENZAMIDE
C7H6BrNO2	002486-52-4	2-BR-4-AMINOBENZOIC ACID
C7H6BrNO3	020039-91-2	2-BROMO-4-METHYL-6-NITROPHENOL
C7H6Cl2	000095-73-8	2,4-DICHLOROTOLUENE
C7H6Cl2	000095-75-0	3,4-DICHLOROTOLUENE
C7H6Cl2	000098-87-3	(DICHLOROMETHYL)BENZENE
C7H6Cl2	000104-83-6	BENZENE, 1-CHLORO-4-(CHLOROMETHYL)-
C7H6Cl2	000118-69-4	2,6-DICHLOROTOLUENE
C7H6Cl2	025186-47-4	3,5-DICHLOROTOLUENE
C7H6Cl2N2O	002327-02-8	(3,4-DICHLOROPHENYL)UREA

C7H6Cl2O	001805-32-9	3,4-DICHLOROBENZYL ALCOHOL
C7H6Cl2O	002432-12-4	4-METHYL-2,6-DICHLOROPHENOL
C7H6Cl2O	015258-73-8	BENZENEMETHANOL, 2,6-DICHLORO-
C7H6Cl2O	033719-74-3	3,5-DICHLOROANISOLE
C7H6Cl2O	060211-57-6	BENZENEMETHANOL, 3,5-DICHLORO-
C7H6Cl2O2	002460-49-3	4,5-DICHLORO-2-METHOXYPHENOL
C7H6Cl2O2	006641-02-7	BENZENEMETHANOL, 3,5-DICHLORO-2-HYDROXY-
C7H6ClF	000345-35-7	O-FLUOROBENZYL CHLORIDE
C7H6ClF	000352-11-4	P-FLUOROBENZYL CHLORIDE
C7H6ClF	000456-42-8	M-FLUOROBENZYL CHLORIDE
C7H6ClF2NO2S	001513-31-1	DIFLUOROMETHANESULFONANILIDE,P-CL
C7H6ClN3	005418-93-9	1H-BENZIMIDAZOL-2-AMINE, 5-CHLORO-
C7H6ClN3O2S	037157-99-6	2H-1,2,4-BENZOTHIADIAZIN-3-AMINE,6-CHLORO-, 1,1-
C7H6ClN3O4S2	000058-94-6	CHLOROTHIAZIDE
C7H6ClN3S	040852-07-1	4-PYRIDINE IMIDAZOLE,2-MES-6-CL
C7H6ClNO	000139-71-9	FORMAMIDE, N-(3-CHLOROPHENYL)-
C7H6ClNO	000609-66-5	2-CHLOROBENZAMIDE
C7H6ClNO	000618-48-4	3-CHLOROBENZAMIDE
C7H6ClNO	000619-56-7	4-CHLOROBENZAMIDE
C7H6ClNO	002617-79-0	P-CHLOROFORMANILIDE
C7H6ClNO	003848-36-0	P-CHLOROBENZALDOXIME
C7H6ClNO2	000083-42-1	BENZENE, 1-CHLORO-2-METHYL-3-NITRO-
C7H6ClNO2	000089-59-8	BENZENE, 4-CHLORO-1-METHYL-2-NITRO-
C7H6ClNO2	000121-86-8	2-CHLORO-4-NITROTOLUENE
C7H6ClNO2	000635-21-2	2-AMINO-5-CHLOROBENZOIC ACID
C7H6ClNO2	002457-76-3	2-CHLORO-4-AMINOBENZOIC ACID
C7H6ClNO3	007147-89-9	4-CHLORO-5-METHYL-2-NITROPHENOL
C7H6ClNS	002521-24-6	BENZENECARBOTHIAMIDE, 4-CHLORO-
C7H6F2O	000458-92-4	DIFLUOROMETHOXYBENZENE
C7H6F2O	019064-18-7	BENZENEMETHANOL, 2,6-DIFLUORO-
C7H6F3N	000088-17-5	BENZENAMINE, 2-(TRIFLUOROMETHYL)-
C7H6F3N	000098-16-8	3-TRIFLUOROMETHYLANILINE
C7H6F3N	000455-14-1	P-TRIFLUOROMETHYLANILINE
C7H6F3NO2S	000456-64-4	CF3-METHANESULFONANILIDE
C7H6F3NO3S	023375-12-4	M-OH CF3-METHANESULFONANILIDE
C7H6FN3O4S2	001535-61-1	FLUOROTHIAZIDE
C7H6FNO	000445-28-3	2-FLUOROBENZAMIDE
C7H6FNO	000455-37-8	3-FLUOROBENZAMIDE
C7H6FNO	000459-23-4	P-FLUOROBENZALDOXIME
C7H6FNO	000459-25-6	P-FLUOROFORMANILIDE
C7H6FNO	000824-75-9	P-FLUOROBENZAMIDE
C7H6FNO2	000446-31-1	2-FLUORO-4-AMINOBENZOIC ACID
C7H6FNO2	000455-87-8	BENZOIC ACID, 4-AMINO-3-FLUORO-
C7H6INO	003930-83-4	2-IODOBENZAMIDE
C7H6INO	003956-07-8	4-IODOBENZAMIDE
C7H6INO2	002122-63-6	BENZOIC ACID, 4-AMINO-3-IODO-
C7H6INO2	005326-47-6	2-AMINO-5-IODOBENZOIC ACID
C7H6INO2	073655-51-3	2-IODO-4-AMINOBENZOIC ACID
C7H6N2	000051-17-2	BENZIMIDAZOLE
C7H6N2	000271-29-4	6-AZAINDOLE
C7H6N2	000271-34-1	5-AZAINDOLE
C7H6N2	000271-44-3	INDAZOLE
C7H6N2	000272-49-1	4-AZAINDOLE
C7H6N2	000274-55-5	7-AZAINDOLE
C7H6N2	000873-74-5	P-AMINOBENZONITRILE
C7H6N2	001620-75-3	2-CYANO-6-METHYLPYRIDINE
C7H6N2	002214-53-1	4-CYANO-2-METHYLPYRIDINE
C7H6N2	002237-30-1	M-AMINOBENZONITRILE

C7H6N2	010592-27-5	7-AZAINDOLE
C7H6N2O	000615-16-7	O-PHENYLENE UREA
C7H6N2O	004570-41-6	BENZOXAZOLE, 2-AMINO
C7H6N2O2S	000359-85-3	124-BENZOTHIADIAZINE-1,1-O2
C7H6N2O2S	003118-68-1	BENZENESULFONAMIDE, 3-CYANO-
C7H6N2O2S	003119-02-6	P-CYANOBENZENESULFONAMIDE
C7H6N2O2S	026060-30-0	BENZENECARBOTHIAMIDE, 4-(NITRO)-
C7H6N2O2S2	000433-17-0	BENZOTHIAZOLE-2-SULFONAMIDE
C7H6N2O3	000102-38-5	3-NITROFORMANILIDE
C7H6N2O3	000610-15-1	2-NITROBENZAMIDE
C7H6N2O3	000619-80-7	4-NITROBENZAMIDE
C7H6N2O3	000645-09-0	M-NITROBENZAMIDE
C7H6N2O3	001129-37-9	P-NITROBENZALDOXIME
C7H6N2O3	016135-31-2	P-NITROFORMANILIDE
C7H6N2O3S2	029927-14-8	6-OH BENZOTHIAZOLE-2-SULFONAMIDE
C7H6N2O4	000121-14-2	2,4-DINITROTOLUENE
C7H6N2O4	000602-01-7	2,3-DINITROTOLUENE
C7H6N2O4	000606-20-2	1,3-DINO2 2-METHYL BENZENE
C7H6N2O4	000610-39-9	1,2-DINO2 4-METHYL BENZENE
C7H6N2O4	000619-15-8	2,5-DINITROTOLUENE
C7H6N2O4	000619-17-0	2-AMINO-4-NITROBENZOIC ACID
C7H6N2O4	000710-25-8	2-PROPENAMIDE, 3-(5-NITRO-2-FURANYL)-
C7H6N2O5	000119-27-7	2,4-DINITROANISOLE
C7H6N2O5	000534-52-1	4,6-DINITRO-O-CRESOL
C7H6N2O5	000609-93-8	2,6-DINITRO-P-CRESOL
C7H6N2O5	000616-73-9	2,4-DINITRO-5-METHYLPHENOL
C7H6N2S	000583-39-1	O-PHENYLENETHIOUREA
C7H6N4	005378-52-9	1H-TETRAZOLE, 1-PHENYL-
C7H6N4O2	096448-60-1	ALLOPURINOL, 1-ACETYL-
C7H6N4O4	001081-15-8	FORMALDEHYDE, (2,4-DINITROPHENYL)HYDRAZONE
C7H6N4O6S2	023141-81-3	124-BENZTHIADIAZN,11-O2-7-NH2SO2-6NO2
C7H6NaO2	000532-32-1	BENZOIC ACID, SODIUM SALT
C7H6O	000100-52-7	BENZALDEHYDE
C7H6O2	000065-85-0	BENZOIC ACID
C7H6O2	000090-02-8	O-HYDROXYBENZALDEHYDE
C7H6O2	000100-83-4	3-HYDROXYBENZALDEHYDE
C7H6O2	000123-08-0	4-HYDROXYBENZALDEHYDE
C7H6O2	000274-09-9	1,3-BENZODIOXOLE
C7H6O2	000533-75-5	TROPOLONE
C7H6O2	000553-97-9	2-METHYL-1,4-BENZOQUINONE
C7H6O2	000623-30-3	2-PROPENAL, 3-(2-FURANYL)-
C7H6O2	001864-94-4	PHENYL FORMATE
C7H6O2	003324-76-3	2,4,6-CYCLOHEPTATRIEN-1-ONE, 3-HYDROXY-
C7H6O2S	000147-93-3	2-MERCAPTOBENZOIC ACID
C7H6O3	000069-72-7	SALICYLIC ACID
C7H6O3	000099-06-9	M-HYDROXYBENZOIC ACID
C7H6O3	000099-96-7	P-HYDROXYBENZOIC ACID
C7H6O4	000089-86-1	2,4-DIHYDROXYBENZOIC ACID
C7H6O4	000099-10-5	3,5-DIHYDROXYBENZOIC ACID
C7H6O4	000099-50-3	3,4-DIHYDROXYBENZOIC ACID
C7H6O4	000303-07-1	2,6-DIHYDROXYBENZOIC ACID
C7H6O4	000303-38-8	2,3-DIHYDROXYBENZOIC ACID
C7H6O4	000490-79-9	2,5-DIHYDROXYBENZOIC ACID
C7H6O5	000149-91-7	3,4,5-TRIHYDROXYBENZOIC ACID
C7H6O5	000610-02-6	2,3,4-TRIHYDROXYBENZOIC ACID
C7H7Br	000095-46-5	O-BROMOTOLUENE
C7H7Br	000100-39-0	ALPHA-BROMOTOLUENE
C7H7Br	000106-38-7	P-BROMOTOLUENE

C7H7Br	000591-17-3	M-BROMOTOLUENE
C7H7BrN2O	001967-25-5	P-BROMOPHENYLUREA
C7H7BrN2O	002989-98-2	M-BROMOPHENYLUREA
C7H7BrN2O	005933-32-4	BENZOYLHYDRAZINE, P-BROMO
C7H7BrN2O	029418-67-5	BENZOYLHYDRAZINE, O-BROMO
C7H7BrN2O	039115-96-3	BENZOYLHYDRAZINE, M-BROMO
C7H7BrO	000104-92-7	4-BROMOANISOLE
C7H7BrO	000578-57-4	1-BROMO-2-METHOXYBENZENE
C7H7BrO	002398-37-0	1-BROMO-3-METHOXYBENZENE
C7H7BrO2	002316-64-5	BENZENEMETHANOL, 5-BROMO-2-HYDROXY-
C7H7Cl	000095-49-8	2-CHLOROTOLUENE
C7H7Cl	000100-44-7	ALPHA-CHLOROTOLUENE
C7H7Cl	000106-43-4	P-CHLOROTOLUENE
C7H7Cl	000108-41-8	M-CHLOROTOLUENE
C7H7Cl2N	056462-00-1	N-METHYL-2,6-DICHLOROANILINE
C7H7Cl2N3O	073469-87-1	HYDRAZINECARBOXAMIDE, N-(2,5-DICHLOROPHENYL)-
C7H7Cl2NO	002971-90-6	3,5-DICHLORO-2,6-DIMETHYL-4-PYRIDINOL
C7H7Cl3NO3P	002214-34-8	O=P(OME)(N)O-2,4,5-CLPHENYL
C7H7Cl3NO3PS	005598-13-0	CHLORPYRIFOS METHYL
C7H7ClFNO2S	050585-76-7	FLUOROMETHANESULFONANILIDE,P-CL
C7H7ClN2O	000114-38-5	O-CHLOROPHENYLUREA
C7H7ClN2O	000140-38-5	P-CHLOROPHENYLUREA
C7H7ClN2O	000536-40-3	BENZOYLHYDRAZINE, P-CHLORO
C7H7ClN2O	001673-47-8	BENZOYLHYDRAZINE, M-CHLORO
C7H7ClN2O	001967-27-7	M-CHLOROPHENYLUREA
C7H7ClN2O	005814-05-1	BENZOYLHYDRAZINE, O-CHLORO
C7H7ClN2O2	030085-34-8	3-(P-CHLOROPHENYL)HYDROXYUREA
C7H7ClN2O2	075680-92-1	3-CL-6-ET PYRIDAZINECARBOXYLATE
C7H7ClN2S	005344-82-1	1-(O-CHLOROPHENYL)-2-THIOUREA
C7H7ClN4	014210-25-4	1-PHENYL-5-CHLOROTETRAZOLE
C7H7ClN4O2	000085-18-7	8-CHLOROTHEOPHYLLINE
C7H7ClO	000059-50-7	3-METHYL-4-CHLOROPHENOL
C7H7ClO	000087-64-9	2-METHYL-6-CHLOROPHENOL
C7H7ClO	000615-74-7	PHENOL, 2-CHLORO-5-METHYL-
C7H7ClO	000623-12-1	4-CHLOROANISOLE
C7H7ClO	000766-51-8	O-CHLOROANISOLE
C7H7ClO	000873-63-2	M-CHLOROBENZYL ALCOHOL
C7H7ClO	000873-76-7	P-CHLOROBENZYL ALCOHOL
C7H7ClO	001570-64-5	2-METHYL-4-CHLOROPHENOL
C7H7ClO	002845-89-8	M-CHLOROANISOLE
C7H7ClO	006640-27-3	2-METHYL-4-CHLOROPHENOL
C7H7ClO	017849-38-6	BENZENEMETHANOL, 2-CHLORO-
C7H7ClO	054548-50-4	CHLORO-M-CRESOL
C7H7ClO2	005330-38-1	BENZENEMETHANOL, 5-CHLORO-2-HYDROXY-
C7H7ClO2S	000098-57-7	4-METHYLSULFONYLCHLOROBENZENE
C7H7ClO2S	000098-59-9	P-TOLUENESULFONYL CHLORIDE
C7H7ClO2S	000133-59-5	O-TOLUENESULFONYL CHLORIDE
C7H7ClO2S	007205-98-3	CHLOROMETHYL PHENYL SULFONE
C7H7F	000095-52-3	O-FLUOROTOLUENE
C7H7F	000350-50-5	BENZYL FLUORIDE
C7H7F	000352-32-9	P-FLUOROTOLUENE
C7H7F	000352-70-5	M-FLUOROTOLUENE
C7H7F2NO2S	000658-43-5	DIFLUOROMETHANESULFONANILIDE
C7H7FN2O	000656-31-5	O-FLUOROPHENYLUREA
C7H7FN2O	000659-30-3	P-FLUOROPHENYLUREA
C7H7FN2O	000770-19-4	M-FLUOROPHENYLUREA
C7H7FN2O3	056058-99-2	2,4(1H,3H)-PYRIMIDINEDIONE, 5-FLUORO-1-(1-OXOPRO
C7H7FN2O4	021839-33-8	5-FLUOROURACIL-3-ETHOXYCARBONYL-

C7H7FO	000321-28-8	2-FLUOROANISOLE
C7H7FO	000446-51-5	BENZENEMETHANOL, 2-FLUORO-
C7H7FO	000456-49-5	M-FLUOROANISOLE
C7H7FO	000459-56-3	BENZENEMETHANOL, 4-FLUORO-
C7H7FO	000459-60-9	P-FLUOROANISOLE
C7H7FO2S	000455-16-3	P-FLUOROSULFONYLTOLUENE
C7H7I	000615-37-2	O-IODOTOLUENE
C7H7I	000620-05-3	BENZYL IODIDE
C7H7IN2O	031822-03-4	BENZOYLHYDRAZINE, O-IODO
C7H7IN2O	039115-94-1	BENZOYLHYDRAZINE, M-IODO
C7H7IN2O	039115-95-2	BENZOYLHYDRAZINE, P-IODO
C7H7N	000100-43-6	4-VINYLPYRIDINE
C7H7N	000100-69-6	2-VINYLPYRIDINE
C7H7N2O	036216-80-5	3-AMINO-BENZISOXAZOLE
C7H7N3	000874-05-5	3-AMINOINDAZOLE
C7H7N3	000934-32-7	2-AMINOBENZIMIDAZOLE
C7H7N3	013351-73-0	1-METHYLBENZOTRIAZOLE
C7H7N3	019335-11-6	1H-INDAZOL-5-AMINE
C7H7N3O2	056406-50-9	M-NITROBENZAMIDINE HYDROCHLORIDE
C7H7N3O2	060142-49-6	DIAZENECARBOXAMIDE, PHENYL-, 2-OXIDE
C7H7N3O2S2	094641-11-9	2-BENZOTHIAZOLESULFONAMIDE, 6-AMINO-
C7H7N3O3	000606-26-8	BENZOYLHYDRAZINE, O-NITRO
C7H7N3O3	000618-94-0	3-NITROBENZOYLHYDRAZINE
C7H7N3O3	000636-97-5	4-NITROBENZOYLHYDRAZINE
C7H7N3O4S2	023141-75-5	124-BENZTHIADIAZIN,11-O2-7SULFAMYL
C7H7NO	000055-21-0	BENZAMIDE
C7H7NO	000103-70-8	FORMANILIDE
C7H7NO	000350-03-8	3-ACETYLPYRIDINE
C7H7NO	000622-31-1	BENZALDOXIME,SYN
C7H7NO	000932-90-1	BENZALDOXIME
C7H7NO	001122-54-9	4-ACETYLPYRIDINE
C7H7NO	001122-62-9	2-ACETYLPYRIDINE
C7H7NO2	000065-45-2	O-HYDROXYBENZAMIDE
C7H7NO2	000088-72-2	2-NITROTOLUENE
C7H7NO2	000093-60-7	NICOTINIC ACID, METHYL ESTER
C7H7NO2	000099-05-8	3-AMINOBENZOIC ACID
C7H7NO2	000099-08-1	3-NITROTOLUENE
C7H7NO2	000099-99-0	P-NITROTOLUENE
C7H7NO2	000118-92-3	2-AMINOBENZOIC ACID
C7H7NO2	000150-13-0	4-AMINOBENZOIC ACID
C7H7NO2	000495-18-1	BENZOHYDROXAMIC ACID
C7H7NO2	000618-49-5	3-HYDROXYBENZAMIDE
C7H7NO2	000619-57-8	P-HYDROXYBENZAMIDE
C7H7NO2	000622-42-4	A-NITROTOLUENE
C7H7NO2	000622-46-8	O-PHENYL CARBAMATE
C7H7NO2	002459-07-6	PICOLINIC ACID, METHYL ESTER
C7H7NO2	002459-09-8	I-NICOTINIC ACID, METHYL ESTER
C7H7NO2	006971-38-6	PHENOL, 2-METHYL-4-NITROSO-
C7H7NO2	028356-58-3	4-ACETYLPYRIDINE
C7H7NO3	000065-49-6	P-AMINOSALICYLIC ACID
C7H7NO3	000091-23-6	2-NITROANISOLE
C7H7NO3	000100-17-4	P-NITROANISOLE
C7H7NO3	000119-33-5	4-METHYL-2-NITROPHENOL
C7H7NO3	000555-03-3	M-NITROANISOLE
C7H7NO3	000612-25-9	2-NITROBENZYL ALCOHOL
C7H7NO3	000619-25-0	M-NITROBENZYL ALCOHOL
C7H7NO3	000619-73-8	P-NITROBENZYL ALCOHOL
C7H7NO3	000700-38-9	5-METHYL-2-NITROPHENOL

C7H7NO3	002042-14-0	3-NITRO-P-CRESOL
C7H7NO3	002374-03-0	3-HYDROXY-4-AMINOBENZOIC ACID
C7H7NO3	002581-34-2	3-METHYL-4-NITROPHENOL
C7H7NO3	004920-77-8	3-METHYL-2-NITROPHENOL
C7H7NO3	005428-54-6	PHENOL, 2-METHYL-5-NITRO-
C7H7NO4	001568-70-3	4-METHOXY-2-NITROPHENOL
C7H7NO4	007423-55-4	N-MALEOYL-3-AMINOPROPIONIC ACID
C7H7NO4S	000138-41-0	4-AMINOSULFONYLBENZOIC ACID
C7H7NO4S	002976-32-1	3-METHYLSUFONYLNITROBENZENE
C7H7NS	002227-79-4	THIOBENZAMIDE
C7H8	000108-88-3	TOLUENE
C7H8	000121-46-0	BICYCLO(2.2.1)HEPTA-2,5-DIENE
C7H8	000544-25-2	CYCLOHEPTATRIENE
C7H8BrClN2O3	077317-67-0	3-CLETO-5-BR-6-ME-PYRIMID-2,4-DIONE
C7H8BrN	006933-10-4	3-METHYL-4-BROMOANILINE
C7H8BrN3O	002646-26-6	HYDRAZINECARBOXAMIDE, N-(4-BROMOPHENYL)
C7H8BrN3O	073469-88-2	HYDRAZINECARBOXAMIDE, N-(2-BROMOPHENYL)
C7H8BrN3O4S2	023141-83-5	HYDROBROMOTHIAZIDE
C7H8Cl2Si	000149-74-6	DICHLOROMETHYLPHENYLSILANE
C7H8ClN	000087-60-5	2-METHYL-3-CHLOROANILINE
C7H8ClN	000095-69-2	2-METHYL-4-CHLOROANILINE
C7H8ClN	000095-74-9	3-CHLORO-P-TOLUIDINE
C7H8ClN	000095-79-4	5-CHLORO-O-TOLUIDINE
C7H8ClN3O	035580-76-8	HYDRAZINECARBOXAMIDE, N-(2-CHLOROPHENYL)
C7H8ClN3O	051707-42-7	HYDRAZINECARBOXAMIDE, N-(3-CHLOROPHENYL)
C7H8ClN3O	069194-89-4	HYDRAZINECARBOXAMIDE, N-(4-CHLOROPHENYL)
C7H8ClN3O4S2	000058-93-5	HYDROCHLOROTHIAZIDE
C7H8ClNO	005345-54-0	3-CHLORO-4-METHOXYANILINE
C7H8ClNO2S	004284-51-9	P-CHLORO METHANESULFONANILIDE
C7H8ClNO2S	006333-79-5	BENZENESULFONAMIDE, 4-CHLORO-N-METHYL
C7H8F2Si	000328-57-4	DIFLUOROMETHYLPHENYLSILANE
C7H8FN3O3	058471-47-9	1(2H)-PYRIMIDINECARBOXAMIDE, N-ETHYL-5-FLUORO-3,
C7H8FN3O3	060908-29-4	1(2H)-PYRIMIDINECARBOXAMIDE, 5-FLUORO-3,4-DIHYDR
C7H8FNO2S	002070-61-3	FLUOROMETHANE SULFONANILIDE
C7H8Hg	021392-61-0	PHENYLMETHYL MERCURY
C7H8HgN2O	002279-64-3	AGROX
C7H8IN3O	073469-90-6	HYDRAZINECARBOXAMIDE, N-(4-IODOPHENYL)
C7H8IN3O4	096258-81-0	1H-IMIDAZOLE-1-ACETIC ACID, 4-IODO-5-NITRO-, ETH
C7H8N2	001670-14-0	BENZAMIDINE HCL
C7H8N2O	000064-10-8	PHENYLUREA
C7H8N2O	000088-68-6	BENZAMIDE, 2-AMINO-
C7H8N2O	000114-33-0	N-METHYLNICOTINAMIDE
C7H8N2O	000613-94-5	BENZOYL HYDRAZINE
C7H8N2O	002835-68-9	4-AMINOBENZAMIDE
C7H8N2O	003544-24-9	BENZAMIDE, 3-AMINO-
C7H8N2O	003724-16-1	3-PYRIDINE ACETAMIDE
C7H8N2O	005221-42-1	4-ACETAMIDOPYRIDINE
C7H8N2O	005231-96-9	2-ACETAMIDOPYRIDINE
C7H8N2O	005451-39-8	2-PYRIDINEACETAMIDE
C7H8N2O	005867-45-8	3-ACETAMINOPYRIDINE
C7H8N2O	039640-62-5	4-PYRIDINEACETAMIDE
C7H8N2O2	000089-62-3	2-NITRO-P-TOLUIDINE
C7H8N2O2	000099-52-5	4-NITRO-2-TOLUIDINE
C7H8N2O2	000099-55-8	5-NITRO-2-TOLUIDINE
C7H8N2O2	000100-15-2	P-NITRO-N-METHYLANILINE
C7H8N2O2	000119-32-4	3-NITRO-4-TOLUIDINE
C7H8N2O2	000611-03-0	BENZOIC ACID, 2,4-DIAMINO-
C7H8N2O2	000611-05-2	3-METHYL-4-NITROANILINE

C7H8N2O2	000612-28-2	N-METHYL-O-NITROANILINE
C7H8N2O2	000619-05-6	BENZOIC ACID, 3,4-DIAMINO-
C7H8N2O2	000936-02-7	BENZOYLHYDRAZINE, O-HYDROXY
C7H8N2O2	005351-23-5	BENZOYLHYDRAZINE, P-HYDROXY
C7H8N2O2	005818-06-4	BENZOYLHYDRAZINE, M-HYDROXY
C7H8N2O2	006924-68-1	2-ETHYLPYRAZINE CARBOXYLATE
C7H8N2O2	007335-35-5	3-PHENYLHYDROXYUREA
C7H8N2O2	007409-30-5	P-NITROBENZYLAMINE
C7H8N2O2	039123-39-2	4-PYRIDAZINECARBOXYLIC ACID, ETHYL ESTER
C7H8N2O2	040929-50-8	5-PYRIMIDINECARBOXYLIC ACID, ETHYL ESTER
C7H8N2O2	041110-33-2	PYRAZINECARBOXYLIC ACID, 5-METHYL-, METHYL ESTER
C7H8N2O2	041110-38-7	PYRAZINECARBOXYLIC ACID, 6-METHYL-, METHYL ESTER
C7H8N2O2	042839-08-7	2-PYRIMIDINECARBOXYLIC ACID, ETHYL ESTER
C7H8N2O3	000085-45-0	BENZENAMINE, 2-METHOXY-3-NITRO-
C7H8N2O3	000096-96-8	BENZENAMINE, 4-METHOXY-2-NITRO-
C7H8N2O3	000099-59-2	2-METHOXY-5-NITROANILINE
C7H8N2O3	006128-03-6	6,8-DIAZASPIRO[3.5]NONANE-5,7,9-TRIONE
C7H8N2O3	023813-24-3	PYRAZINECARBOXYLIC ACID, 6-METHOXY-, METHYL ESTE
C7H8N2O3S	001576-42-7	BENZAMIDE, 3-(AMINOSULFONYL)-
C7H8N2O3S	006306-24-7	BENZAMIDE, 4-(AMINOSULFONYL)-
C7H8N2O3S	035207-08-0	BENZENESULFONAMIDE, N-(AMINOCARBONYL)-
C7H8N2O4	001747-53-1	4-PYRIMIDINECARBOXYLIC ACID, 1,2,3,6-TETRAHYDRO-
C7H8N2S	000103-85-5	N-PHENYLTHIOUREA
C7H8N2S	002454-39-9	O-AMINOTHIOBENZAMIDE
C7H8N3O3S	053207-62-8	1(5-NO2-2-FURYL)-2-IMIDAZOLIDINTHIONE
C7H8N3O4	053207-61-7	1(5-NO2-2-FURYL)-2-IMIDAZOLIDINONE
C7H8N4	040262-48-4	PYRAZINECARBONITRILE, 6-(DIMETHYLAMINO)-
C7H8N4O2	000058-55-9	THEOPHYLLINE
C7H8N4O2	000083-67-0	THEOBROMINE
C7H8N4O2	041078-01-7	1H-PURINE-2,6-DIONE, 3,9-DIHYDRO-3-ETHYL-
C7H8N4O2	111079-49-3	1H-PURINE-2,6-DIONE, 3,9-DIHYDRO-1,3-DIMETHYL-
C7H8N4O2S2	037422-15-4	1(5NO2-2-THIAZOLYL)-2MES-2-IMIDAZOLINE
C7H8N4O3	000944-73-0	1,3-DIMETHYL URIC ACID
C7H8N4O3	013087-49-5	3,7-DIMETHYL URIC ACID
C7H8N4O3	017433-92-0	HYDRAZINECARBOXAMIDE, N-(3-NITROPHENYL)
C7H8N4O3	017433-93-1	HYDRAZINECARBOXAMIDE, N-(4-NITROPHENYL)
C7H8N4O3	017433-94-2	HYDRAZINECARBOXAMIDE, N-(2-NITROPHENYL)
C7H8N4O3	033868-03-0	1,7-DIMETHYL URIC ACID
C7H8N4O3S	024240-83-3	2(4-ME-1,3-OXAZOLIDINYLIDEN-2-AMINO)5NO2-THIAZOL
C7H8N4O6S2	023141-88-0	6-NITROHYDROTHIAZIDE
C7H8N6	077314-75-1	2-PYRIDINAMINE, 4-(5-AMINO-1H-1,2,4-TRIAZOL-3-YL
C7H8O	000095-48-7	O-CRESOL
C7H8O	000100-51-6	BENZYL ALCOHOL
C7H8O	000100-66-3	ANISOLE
C7H8O	000106-44-5	P-CRESOL
C7H8O	000108-39-4	M-CRESOL
C7H8O2	000090-01-7	O-HYDROXYBENZYL ALCOHOL
C7H8O2	000090-05-1	2-METHOXYPHENOL
C7H8O2	000095-71-6	1,4-BENZENEDIOL, 2-METHYL-
C7H8O2	000150-19-6	M-METHOXYPHENOL
C7H8O2	000150-76-5	P-METHOXYPHENOL
C7H8O2	000452-86-8	1,2-BENZENEDIOL, 4-METHYL-
C7H8O2	000504-15-4	5-METHYLRESORCINOL
C7H8O2	000620-24-6	M-HYDROXYBENZYL ALCOHOL
C7H8O2	000623-05-2	P-HYDROXYBENZYL ALCOHOL
C7H8O2	001004-36-0	GAMMA-2,6-DIMETHYLPYRONE
C7H8O2S	000426-58-4	SULFONE, PHENYL, METHYL-
C7H8O2S	000536-57-2	P-TOLUENESULFINIC ACID

C7H8O2S	003112-85-4	METHYL PHENYL SULFONE
C7H8O3	000614-99-3	FUROIC ACID, ETHYL ESTER
C7H8O3	000824-46-4	1,4-BENZENEDIOL, 2-METHOXY-
C7H8O3	001335-40-6	ETHYL 2-FUROATE
C7H8O3S	000104-15-4	P-TOLUENESULFONIC ACID
C7H8O3S	016156-59-5	PHENYLMETHANESULFONATE
C7H8OS	001073-72-9	P-METHIOPHENOL
C7H8OS	001193-82-4	METHYL PHENYL SULFOXIDE
C7H8S	000100-68-5	THIOANISOLE
C7H8S	000106-45-6	4-METHYLBENZENETHIOL
C7H8S	000108-40-7	3-METHYLBENZENETHIOL
C7H8S	000137-06-4	2-METHYLBENZENETHIOL
C7H8Se	004346-64-9	PHENYL METHYL SELENIDE
C7H9BrN2O3	077317-66-9	3-ETO-5-BR-6-ME-PYRIMIDIN-2,4-DIONE
C7H9Cl	083682-36-4	1-CHLOROHEPT-1-YNE-5-ENE
C7H9Cl2N	003165-93-3	4-CHLORO-2-METHYLBENZENAMINE HYDROCHLORIDE
C7H9Cl5	056421-44-4	1-METHYLPENTACHLOROCYCLOHEXANE
C7H9Cl5O	056046-08-3	3-METHOXYPENTACHLOROCYCLOHEXANE
C7H9Cl5O	056400-43-2	1-HOME PENTACHLOROCYCLOHEXANE
C7H9Cl5O	056421-31-9	2-METHOXYPENTACHLOROCYCLOHEXANE
C7H9Cl5S	056046-09-4	3-METHIOPENTACHLOROCYCLOHEXANE
C7H9Cl5S	060132-42-5	1-METHIOPENTACHLOROCYCLOHEXANE
C7H9ClN2O	136309-03-0	2-CHLORO-6-PROPOXY-PYRAZINE
C7H9N	000095-53-4	O-TOLUIDINE
C7H9N	000100-46-9	BENZYLAMINE
C7H9N	000100-61-8	N-METHYLANILINE
C7H9N	000100-71-0	2-ETHYLPYRIDINE
C7H9N	000106-49-0	P-TOLUIDINE
C7H9N	000108-44-1	M-TOLUIDINE
C7H9N	000108-47-4	2,4-DIMETHYLPYRIDINE
C7H9N	000108-48-5	2,6-LUTIDINE
C7H9N	000536-75-4	4-ETHYLPYRIDINE
C7H9N	000536-78-7	3-ETHYLPYRIDINE
C7H9N	000583-58-4	3,4-DIMETHYLPYRIDINE
C7H9N	000583-61-9	2,3-DIMETHYLPYRIDINE
C7H9N	000589-93-5	2,5-DIMETHYLPYRIDINE
C7H9N	000591-22-0	3,5-DIMETHYLPYRIDINE
C7H9N3	002002-16-6	2-PHENYLGUANIDINE
C7H9N3	002498-50-2	4-AMINOBENZAMIDINE
C7H9N3	016584-00-2	2-METHYLBENZOTRIAZOLE
C7H9N3O	000103-03-7	HYDRAZINECARBOXAMIDE, 2-PHENYL-
C7H9N3O	000537-47-3	HYDRAZINECARBOXAMIDE, N-PHENYL-
C7H9N3O	001904-58-1	BENZOYLHYDRAZINE, O-AMINO
C7H9N3O	003758-59-6	2-METHYLISONIAZID
C7H9N3O	005351-17-7	4-AMINOBENZOYLHYDRAZINE
C7H9N3O	014062-34-1	3-AMINOBENZOYLHYDRAZINE
C7H9N3O	032743-27-4	N,N-DIMETHYL 2-PYRAZINECARBOXAMIDE
C7H9N3O2	018591-79-2	ACETAMIDE, N-(4-HYDROXY-6-METHYL-3-PYRIDAZINYL)-
C7H9N3O2	019353-97-0	2-METHOXY ISONIAZID
C7H9N3O2	038191-26-3	HYDRAZINECARBOXAMIDE,N-(2-HYDROXYPHENYL)-
C7H9N3O2	054015-45-1	PYRAZINECARBOXAMIDE, 6-ETHOXY-
C7H9N3O2	073469-91-7	HYDRAZINECARBOXAMIDE, N-(3-HYDROXYPHENYL)-
C7H9N3O2	073469-92-8	HYDRAZINECARBOXAMIDE, N-(4-HYDROXYPHENYL)-
C7H9N3O2	074550-87-1	1H-IMIDAZOLE, 2-CYCLOPROPYL-1-METHYL-5-NITRO-
C7H9N3O2	087831-85-4	L-ALANINE, 3-PYRAZINYL-
C7H9N3O2	136309-06-3	ACETAMIDE, N-(6-METHOXYPYRAZINYL)-
C7H9N3O3	003150-53-6	1-(2-OHET)-5-CYANO-2-NO2 PYRROLE
C7H9N3O3S	121822-79-5	3-PYRIDINESULFONAMIDE, N-[(METHYLAMINO)CARBONYL]

C7H9N3O4	001013-51-0	ACETIC ACID,2-METHYL-5-NITROIMIDAZOL-2-YL,METHYL
C7H9N3O4	021787-95-1	2-PROPANONE, 1-METHOXY-2-(2-NITRO-1H-IMIDAZOL-1-
C7H9N3O4S2	023141-82-4	HYDROTHIAZIDE
C7H9N3S	056844-20-3	THIENO(23D)-PYRIMIDINE,4(MEAMINO)
C7H9N5O2S	031052-76-3	2(1ME-13-IMIDAZOLINYLIDEN-2AM)5NO2-THIAZOLE
C7H9N5O2S	037385-10-7	2(4ME-1,3-IMIDAZOLINYLIDEN-2AM)-5NO2-THIAZOLE
C7H9N5O3S	037394-79-9	2(6H-5OH-PYRIMIDINYLIDEN-2AM)-5NO2-THIAZOLE
C7H9NO	000090-04-0	2-METHOXYANILINE
C7H9NO	000095-84-1	PHENOL, 2-AMINO-4-METHYL-
C7H9NO	000103-74-2	2-PYRIDINEETHANOL
C7H9NO	000104-94-9	4-METHOXYANILINE
C7H9NO	000536-90-3	M-METHOXYANILINE
C7H9NO	001073-23-0	2,6-DIMETHYL-N-OXIDEPYRIDINE
C7H9NO	001877-77-6	M-AMINOBENZYL ALCOHOL
C7H9NO	003718-65-8	3,5-DIMETHYLPYRIDINE-N-OXIDE
C7H9NO	005344-27-4	4-PYRIDINEETHANOL
C7H9NO	006293-56-7	3-PYRIDINEETHANOL
C7H9NO	014529-53-4	2-ETHOXYPYRIDINE
C7H9NO	061224-32-6	P-AMINOBENZYLALCOHOL
C7H9NO2	002199-43-1	1H-PYRROLE-2-CARBOXYLIC ACID, ETHYL ESTER
C7H9NO2	006231-18-1	PYRIDINE, 2,6-DIMETHOXY-
C7H9NO2	010586-17-1	ISOPROPYL CYANOACRYLATE
C7H9NO2	037619-24-2	1H-PYRROLE-2-CARBOXYLIC ACID, 1-METHYL-, METHYL
C7H9NO2	040611-76-5	5-METHYLPYRROLE-3-CARBOXYLIC ACID,METHYL ESTER
C7H9NO2	125757-81-5	N-ETHYL-HYDROXYPYRID-4-ONE
C7H9NO2S	000070-55-3	4-METHYLBENZENESULFONAMIDE
C7H9NO2S	000088-19-7	O-METHYLBENZENESULFONAMIDE
C7H9NO2S	001197-22-4	N-PHENYL METHANESULFONAMIDE
C7H9NO2S	001333-07-9	M-TOLUENESULFONAMIDE
C7H9NO2S	001899-94-1	M-METHYLBENZENESULFONAMIDE
C7H9NO2S	004563-33-1	BENZENEMETHANESULFONAMIDE
C7H9NO2S	005183-78-8	N-METHYLBENZENESULFONAMIDE
C7H9NO2S	005470-49-5	4-METHYLSULFONYLANILINE
C7H9NO3S	001129-26-6	P-METHOXYBENZENESULFONAMIDE
C7H9NO3S	058734-57-9	BENZENESULFONAMIDE, 3-METHOXY-
C7H9NO3S3	105951-31-3	4H-THIENO[2,3-B]THIOPYRAN-2-SULFONAMIDE, 5,6-DIH
C7H9NO3S3	105951-32-4	5H-THIENO[3,2-B]THIOPYRAN-2-SULFONAMIDE, 6,7-DIH
C7H9NO4S3	105951-36-8	5H-THIENO[3,2-B]THIOPYRAN-2-SULFONAMIDE, 6,7-DIH
C7H9NO5S3	105951-35-7	4H-THIENO[2,3-B]THIOPYRAN-2-SULFONAMIDE, 5,6-DIH
C7H9NO5S3	105951-39-1	5H-THIENO[3,2-B]THIOPYRAN-2-SULFONAMIDE, 6,7-DIH
C7H9NS	001783-81-9	M-METHIOANILINE
C7H9NS	002987-53-3	O-METHIOANILINE
C7H9O3	000614-98-2	FURAN-3-CARBOXYLIC AC, ET ESTER
C7H9O3	006141-57-7	3-METHYLFUROIC ACID, ME ESTER
C7H10	000187-26-8	TRICYCLO(4.1.0.2,4)HEPTANE
C7H10	000287-13-8	TRICYCLO(4.1.0.02,7)HEPTANE
C7H10	000498-66-8	BICYCLO[2.2.1]-2-HEPTENE
C7H10	016554-83-9	BICYCLO(4.1.0)HEPT-3-ENE
C7H10BrN	001906-79-2	1-ETHYLPYRIDINIUM BROMIDE
C7H10BrN3O4	083200-90-2	1H-IMIDAZOLE-1-ETHANOL, 4-BROMO-α-(METHOXYMETHYL
C7H10Cl4O	056933-31-4	2-METHOXYTETRACHLOROCYCLOHEXANE
C7H10ClN	000636-21-5	2-METHYLBENZENAMINE HYDROCHLORIDE
C7H10ClN	003287-99-8	BENZYLAMINE, HYDROCHLORIDE
C7H10ClN3	000535-89-7	2-CHLORO-4-METHYL-6-(DIMETHYLAMINO)PYRIMIDINE
C7H10ClN5O4	088793-46-8	1,2-PROPANEDIOL, 3-[(6-AMINO-3-CHLORO-5-NITROPYR
C7H10IN3O4	096258-78-5	1H-IMIDAZOLE-1-ETHANOL, 5-IODO-α-(METHOXYMETHYL)
C7H10IN3O4	096258-79-6	1H-IMIDAZOLE-1-ETHANOL, 4-IODO-α-(METHOXYMETHYL)
C7H10N2	000095-70-5	2,5-DIAMINOTOLUENE

C7H10N2	000095-80-7	2,4-TOLUENEDIAMINE
C7H10N2	000108-71-4	5-METHYL-1,3-BENZENEDIAMINE
C7H10N2	000496-72-0	3,4-DIAMINOTOLUENE
C7H10N2	000646-20-8	HEPTANEDINITRILE
C7H10N2	000823-40-5	2,6-DIAMINOTOLUENE
C7H10N2	001122-58-3	4-DIMETHYLAMINOPYRIDINE
C7H10N2	002687-25-4	2,3-DIAMINOTOLUENE
C7H10N2	002706-56-1	3-PYRIDINE ETHANEAMINE
C7H10N2	005683-33-0	2-DIMETHYLAMINOPYRIDINE
C7H10N2	013258-63-4	4-PYRIDINE ETHANEAMINE
C7H10N2	014667-55-1	PYRAZINE, TRIMETHYL-
C7H10N2	015707-23-0	PYRAZINE, 2-ETHYL-3-METHYL-
C7H10N2	020173-04-0	N-METHYL-3-PYRIDYLMETHYLAMINE
C7H10N2	020173-24-4	3-PYRIDINE ETHANAMINE
C7H10N2O	000615-05-4	2,4 DIAMINOANISOLE
C7H10N2O	025680-58-4	2-ETHYL-3-METHOXYPYRAZINE
C7H10N2O	032737-14-7	PYRAZINE, 2-ETHOXY-3-METHYL-
C7H10N2O	070090-28-7	2-PROPOXYPYRAZINE
C7H10N2O2	136309-05-2	PYRAZINE, 2-ETHOXY-6-METHOXY-
C7H10N2O2S	001709-52-0	N1-METHYLSULFANILAMIDE
C7H10N2O2S	016891-79-5	P-METHYLAMINOBENZENESULFONAMIDE
C7H10N2O3	007391-69-7	BARBITURIC ACID, 5-IPR
C7H10N2O3	027653-63-0	BARBITURIC ACID,5-ET-5-ME
C7H10N2O3	083128-50-1	BARBITURIC ACID, N-ME,5-ETHYL
C7H10N2O4S2	004426-90-8	BENZENESULFONAMIDE, 4-[(METHYLSULFONYL)AMINO]-
C7H10N2S	014001-64-0	2-MES-4,6-DIMETHYLPYRIMIDINE
C7H10N4O	040262-55-3	PYRAZINECARBOXAMIDE, 6-(DIMETHYLAMINO)
C7H10N4O2S	000057-67-0	SULFANILYL GUANADINE
C7H10N4O3	007597-60-6	6-NH2-5-(N-FORMYLAMINO)-1,3-DIMETHYLURACIL
C7H10N4O3	033130-54-0	6-NH2-5-(N-METHYLFORMYLAMINO)-1-METHYLURACIL
C7H10N4O3	055782-76-8	6-NH2-5-(N-METHYLFORMYLAMINO)-3-METHYLURACIL
C7H10N4O4	022668-01-5	2(2NO2-IMIDZOL-1YL)N-2ETOH-ACETAMIDE
C7H10N4O4	031478-45-2	CARBAMIC ACID, 2-(2-METHYL-5-NITRO-1H-IMIDAZOL-1
C7H10N4O4	087885-48-1	1,2-PROPANEDIOL, 3-[(3-NITROPYRAZINYL)AMINO]
C7H10N4OS	019494-89-4	5-OHPICOLINALDEHYDE THIOSEMICARBAZONE
C7H10N4S	072189-64-1	HYDRAZINECARBOXIMIDAMIDE, 2-[1-(2-THIENYL)ETHYLI
C7H10NO2	013156-75-7	N,N-DIMETHYL-FURAN-2-CARBOXAMIDE
C7H10NO2	014757-80-3	N,N-DIMETHYL-FURAN-3-CARBOXAMIDE
C7H10NO6P	001883-15-4	3,4-PYRIDINEDIMETHANOL, 6-METHYL-5-(PHOSPHONOOXY
C7H10O	000100-50-5	3-CYCLOHEXENE-1-CARBOXALDEHYDE
C7H10O	000497-38-1	NORCAMPHOR
C7H10O2S	002810-04-0	THIOPHENE, 2-ETHYL CARBOXYLATE
C7H10O2S	019432-68-9	THIOPHEN-2-ACETIC ACID, METHYL ESTER
C7H10O2S	058414-52-1	METHYL THIOPHEN-3-ACETATE
C7H10O4	000869-29-4	ALLYLIDENEDIACETATE
C7H11N2O2	095579-17-2	O-METHYL THPO
C7H11N3	056343-48-7	PYRAZINAMINE, N,N,6-TRIMETHYL-
C7H11N3O	022390-38-1	3-PYRIDAZINAMINE, 4-HYDROXY-N,N,6-TRIMETHYL-
C7H11N3O	027772-24-3	4-PYRIDAZINAMINE, 3-METHOXY-N,N-DIMETHYL-
C7H11N3O	038717-24-7	4-PYRIDAZINEAMINE,3-MEO,N,N-DIME
C7H11N3O	136309-09-6	PYRAZINAMINE, 6-METHOXY-N,N-DIMETHYL
C7H11N3O	136309-14-3	PYRAZINAMINE, 5-METHOXY-N,N-DIMETHYL
C7H11N3O2	014885-29-1	1H-IMIDAZOLE, 1-METHYL-2-(1-METHYLETHYL)-5-NITRO
C7H11N3O2	052236-30-3	6-(TERT-BUTYL)-1,2,4-TIRAZINE-3,5(2H,4H)-DIONE
C7H11N3O3	003366-95-8	1H-IMIDAZOLE-1-ETHANOL, α,2-DIMETHYL-5-NITRO-
C7H11N3O4	013551-87-6	1(2-OH-3-MEO-PR)-2-NITROIMIDAZOLE
C7H11N3S	042270-37-1	PIPERAZINE, 1-(2-THIAZOLYL)-
C7H11N7S	004658-28-0	AZIPROTRYNE

C7H11NO	003173-53-3	ISOCYANATOCYCLOHEXANE
C7H11NO2	000077-67-8	ETHOSUXIMIDE
C7H11NO2	006932-05-4	N-ACETYLCYCLOBUTANECARBOXAMIDE
C7H11NO3	059857-86-2	1-METHYL-4-METHOXYCARBONYL-2-PYRROLIDONE
C7H11NO3S3	104437-99-2	2-THIOPHENESULFONAMIDE, 5-[(3-HYDROXYPROPYL)THIO
C7H11NO3S3	106319-38-4	4H-THIENO[2,3-B]THIOPYRAN-2-SULFONAMIDE, 5,6-DIH
C7H11NO3S3	106319-44-2	5H-THIENO[3,2-B]THIOPYRAN-2-SULFONAMIDE, 6,7-DIH
C7H11NO4S3	105951-71-1	4H-THIENO[2,3-B]THIOPYRAN-2-SULFONAMIDE, 5,6-DIH
C7H11NO4S3	106335-79-9	4H-THIENO[2,3-B]THIOPYRAN-2-SULFONAMIDE, 5,6-DIH
C7H11NO5S3	104438-02-0	2-THIOPHENESULFONAMIDE, 5-[(3-HYDROXYPROPYL)SULF
C7H11NO5S3	106319-42-0	4H-THIENO[2,3-B]THIOPYRAN-2-SULFONAMIDE, 5,6-DIH
C7H11NO5S3	106319-45-3	5H-THIENO[3,2-B]THIOPYRAN-2-SULFONAMIDE, 6,7-DIH
C7H11NO5S3	106319-46-4	5H-THIENO[3,2-B]THIOPYRAN-2-SULFONAMIDE, 6,7-DIH
C7H11NO5S3	106400-04-8	4H-THIENO[2,3-B]THIOPYRAN-2-SULFONAMIDE, 5,6-DIH
C7H11NOS	030717-57-8	N,N-DIME-2-CARBOXAMIDE-THIOPHENE
C7H11NOS	059906-37-5	N,N-DIME THIOPHENE-3-CARBOXAMIDE
C7H12	000279-23-2	BICYCLO(2.2.1.)HEPTANE
C7H12	000286-08-8	BICYCLO(4.1.0)HEPTANE
C7H12	000591-48-0	3-METHYL-1-CYCLOHEXENE
C7H12	000591-49-1	1-METHYLCYCLOHEXENE
C7H12	000628-71-7	1-HEPTYNE
C7H12	000628-92-2	CYCLOHEPTENE
C7H12	001192-37-6	METHYLENE CYCLOHEXANE
C7H12	002396-63-6	1,6-HEPTADIENE
C7H12	004049-81-4	2-METHYL-1,5-HEXADIENE
C7H12ClN3O2S2	033022-04-7	1-NO-1-(2CLET)-3(CYHX-35-DITHIAN)UREA
C7H12ClN3O3	016773-42-5	1(3CL2OHPR)-2-ME-5-NO2 IMIDAZOLE
C7H12ClN3O4	052320-87-3	ALANINE, N-[[(2-CHLOROETHYL)NITROSOAMINO>CARBONY
C7H12ClN3O6S2	033022-05-8	1-NO-1-(2CLET)-3(M-DI-SO2PYRAN)UREA
C7H12ClN5	000122-34-9	SIMAZINE
C7H12ClN5	003004-71-5	NORAZINE
C7H12ClN5	003146-66-5	2-CL TETRAMETHYLMELAMINE
C7H12FN3O2S2	033024-47-4	1-NO-1-(2FLET)-3(CYHX-35-DITHIAN)UREA
C7H12N2O	001801-72-5	DIALLYL UREA
C7H12N2O	035990-32-0	ACETAMIDE, N-(3,6-DIHYDRO-1(2H)-PYRIDINYL)-
C7H12N2O2	016395-58-7	PROLIN-AMIDE, N-ACETYL
C7H12N2O2	058942-03-3	2,4-IMIDAZOLIDINEDIONE, 5-(2-METHYLPROPYL)-
C7H12N2O2S2	106231-47-4	1,3-DITHIOLAN-4-ONE, 5,5-DIMETHYL-, O-[(METHYLAM
C7H12N2O2S2	106231-48-5	1,3-DITHIOLAN-4-ONE, 2,2-DIMETHYL-, O-[(METHYLAM
C7H12N2O3S	106231-36-1	1,3-OXATHIOLAN-4-ONE, 5,5-DIMETHYL-, O-[(METHYLA
C7H12N2O3S	106293-84-9	1,3-OXATHIOLAN-4-ONE, 2,5-DIMETHYL-, O-[(METHYLA
C7H12N2O4	025460-87-1	GLUTAMIC ACID-MONOAMIDE,N-ACETYL
C7H12N3O2	099764-47-3	1,2,4-OXADIAZOL-5-OXIME,3-T-BUTYL-
C7H12N4O2	018826-97-6	3-MEO-4NH2-6-IPR-1,2,4-TRIAZIN-5-ON
C7H12N4O2	052819-97-3	N-DIAZOACETYLGLYCINE-N'-N-PR AMIDE
C7H12N4O2	052819-98-4	N-DIAZOACETYLGLYCINE-N'-I-PR AMIDE
C7H12N4O2	056507-37-0	DIKETOMETRIBUZIN
C7H12N4OS	021087-60-5	3-SME-4-NH2-6-PR-124-TRIAZIN-5-ONE
C7H12N4OS	021087-61-6	3-SME-4-NH2-6-IPR-124-TRIAZIN-5-ONE
C7H12NO	014496-34-5	FURANMETHAMINE, N,N-DIMETHYL
C7H12O	000286-16-8	2-OXABICYCLO[4.1.0]HEPTANE
C7H12O	000502-42-1	CYCLOHEPTANONE
C7H12O	000583-60-8	2-METHYLCYCLOHEXANONE
C7H12O	000589-92-4	4-METHYLCYCLOHEXANONE
C7H12O	000591-24-2	3-METHYLCYCLOHEXANONE
C7H12O	002046-23-3	4-CYCLOPROPYL 2-BUTANONE
C7H12O2	000098-89-5	CYCLOHEXANECARBOXYLIC ACID
C7H12O2	000106-63-8	ISOBUTYL ACRYLATE

C7H12O2	000141-32-2	BUTYL ACRYLATE
C7H12O2	002210-28-8	N-PROPYL METHACRYLATE
C7H12O2	004655-34-9	ISOPROPYL METHACRYLATE
C7H12O2	007424-54-6	3,5-HEPTADIONE
C7H12O3	000539-88-8	ETHYL LEVULINATE
C7H12O3	000923-26-2	2-HYDROXYPROPYL METHACRYLATE
C7H12O4	000105-53-3	DIETHYL MALONATE
C7H12O4	000111-16-0	PIMELIC ACID
C7H12O4	000534-59-8	BUTYL MALONIC ACID
C7H12O4	000626-70-0	2-METHYLHEXANEDIOIC ACID
C7H12O4	000681-57-2	GLUTARIC ACID, 2,2-DIMETHYL-
C7H12O4	001119-40-0	PENTANEDIOIC ACID, DIMETHYL ESTER
C7H13BrN2O2	000077-65-6	CARBROMAL
C7H13ClN2O2	061421-89-4	A-CHLORO-A-ETHYLBUTANOYLUREA
C7H13ClO	002528-61-2	HEPTANOYL CHLORIDE
C7H13N	000100-76-5	QUINUCLIDINE
C7H13N	000629-08-3	HEPTANENITRILE
C7H13N3O2	005401-94-5	1H-INDAZOLE, 5-NITRO-
C7H13N3O2	023571-34-8	1-ME-2-NO2-5-IPR IMIDAZOLE
C7H13N3O3	018839-88-8	GLUTAMINE-AMIDE, N-ACETYL
C7H13N3O3	030802-37-0	L-ALANINAMIDE, N-ACETYL-L-ALANYL-
C7H13N3O3	060628-92-4	1-ME-2-NO2-5-(1-OH-1-MEET)IMIDAZOLE
C7H13N3O3	071006-76-3	1-ETOET-2-NO2 IMIDAZOLE
C7H13N3O3S	023135-22-0	OXAMYL
C7H13N3O4	020417-83-8	3NO2-3ME-2BUTANON(N-ME-CARBMOY)OXIME
C7H13N3O4	062580-80-7	1-(2,3-DIOHPR)-2-ME-5-NO2IMIDAZOLE
C7H13N3O4S	019387-94-1	TINIDAZOLE-METHYL
C7H13N3O4S	068160-70-3	1-(2-ETSO2ET)-2-NO2-IMIDAZOLE
C7H13N5O	021087-57-0	6-IPR-4-NH2-3-MEAM-124-TRIAZ-5-ONE
C7H13N5O3	104958-92-1	1,2,4-TRIAZOLE,3-NITRO-1-(3-AZIRIDINYL-2-HYDROXY
C7H13NO	000673-66-5	2-AZACYCLOOCTANONE
C7H13NO2	006187-24-2	3-NITRO-3-HEPTENE
C7H13NO2	055049-18-8	PIPERIDINOACETIC ACID
C7H13NO3	003619-01-0	BUTANOIC ACID,2-ACETYLAMINO,METHYL ESTER
C7H13NS	022928-63-8	2-AZACYCLOOCTANTHIONE
C7H13NS	026019-17-0	2-(N,N-DIME AMINOMETHYL)THIOPHENE
C7H13O6P	000298-01-1	MEVINPHOS (TRANS)
C7H13O6P	007786-34-7	MEVINPHOS
C7H14	000108-87-2	1-METHYLCYCLOHEXANE
C7H14	000291-64-5	CYCLOHEPTANE
C7H14	000592-76-7	1-HEPTENE
C7H14	000594-56-9	2,3,3-TRIMETHYL-1-BUTENE
C7H14	000690-08-4	4,4-DIMETHYL-2-PENTENE (TRANS)
C7H14	000816-79-5	3-ETHYL-2-PENTENE
C7H14	000822-50-4	1,2-DIMETHYLCYCLOPENTANE (TRANS)
C7H14	001192-18-3	1,2-DIMETHYLCYCLOPENTANE (CIS)
C7H14	001638-26-2	1,1-DIMETHYLCYCLOPENTANE
C7H14	001640-89-7	ETHYLCYCLOPENTANE
C7H14	001759-58-6	1,3-DIMETHYLCYCLOPENTANE (TRANS)
C7H14	002213-32-3	2,4-DIMETHYL-1-PENTENE
C7H14	002452-99-5	1,2-DIMETHYLCYCLOPENTANE
C7H14	002532-58-3	1,3-DIMETHYLCYCLOPENTANE (CIS)
C7H14	002738-19-4	2-METHYL-2-HEXENE
C7H14	003404-61-3	3-METHYL-1-HEXENE
C7H14	003404-65-7	3-METHYL-3-HEXENE
C7H14	003404-72-6	2,3-DIMETHYL-1-PENTENE
C7H14	003404-73-7	3,3-DIMETHYL-1-PENTENE
C7H14	003769-23-1	4-METHYL-1-HEXENE

C7H14	004038-04-4	3-ETHYL-1-PENTENE
C7H14	006094-02-6	2-METHYL-1-HEXENE
C7H14	006443-92-1	2-HEPTENE (CIS)
C7H14	007642-10-6	3-HEPTENE (CIS)
C7H14	010574-37-5	2,3-DIMETHYL-2-PENTENE
C7H14	014686-13-6	2-HEPTENE (TRANS)
C7H14	014686-14-7	3-HEPTENE (TRANS)
C7H14	017618-77-8	3-METHYL-2-HEXENE
C7H14	024910-63-2	3,4-DIMETHYL-2-PENTENE
C7H14	042154-69-8	2-METHYL-3-HEXENE
C7H14BrN3O2	030273-97-3	ETHANAMINIUM, 2-[(DIAZOACETYL)OXY]-N,N,N-TRIMETH
C7H14Cl3N2O2P	087154-30-1	2H-1,3,2-OXAZAPHOSPHORIN-2-AMINE,3-CHLORO-N,N-BI
C7H14N2	000693-13-0	N,N'-METHANETETRAYLBIS-2-PROPANAMINE
C7H14N2O	005432-28-0	N-NITROSO-METHYLCYCLOHEXYLAMINE
C7H14N2O	017721-95-8	2,6-DIMETHYL-N-NITROSOPIPERIDINE
C7H14N2O	020917-49-1	N-NITROSO-HEPTAMETHYLENEIMINE
C7H14N2O	065445-59-2	3,5-DIME-N-NITROSOPIPERIDINE
C7H14N2O2	002274-01-3	3-ETHYLBUTANOYLUREA
C7H14N2O2	037933-88-3	VALINE-AMIDE,N-ACETYL
C7H14N2O2S	000116-06-3	ALDICARB
C7H14N2O2S	023361-37-7	METHIONINE-AMIDE,N-ACETYL
C7H14N2O3	001190-48-3	L-LYSINE, N -FORMYL-
C7H14N2O3	056986-36-8	N-NITROSO-ACETOXYMETHYL-N-BUTYLAMINE
C7H14N2O3S	001646-87-3	ALDICARB SULFOXIDE
C7H14N2O4S	001646-88-4	ALDICARB SULFONE
C7H14N2S	030826-80-3	N,N'-DIET-ETHYLENETHIOUREA
C7H14N3O	069741-93-1	2-AMINO-5-AMYL-1,3,4-OXADIAZOLE
C7H14N4O2	002338-12-7	5-NITROBENZOTRIAZOLE
C7H14N4O4	104958-88-5	1,2,4-TRIAZOLE,3-NITRO-1-(3-ETHOXY-2-HYDROXY)PRO
C7H14N6	002827-47-6	TETRAMETHYLMELAMINE
C7H14NO3PS2	000947-02-4	CYOLANE
C7H14NO5P	006923-22-4	AZODRIN
C7H14O	000106-35-4	3-HEPTANONE
C7H14O	000110-12-3	5-METHYL-2-HEXANONE
C7H14O	000110-43-0	2-HEPTANONE
C7H14O	000111-71-7	HEPTANAL
C7H14O	000123-19-3	DI(N-PROPYL) KETONE
C7H14O	000502-41-0	CYCLOHEPTANOL
C7H14O	000565-80-0	2,4-DIMETHYL-3-PENTANONE
C7H14O	000583-59-5	2-METHYLCYCLOHEXANOL
C7H14O	000589-91-3	4-METHYLCYCLOHEXANOL
C7H14O	000590-67-0	1-METHYLCYCLOHEXANOL
C7H14O	000591-23-1	3-METHYLCYCLOHEXANOL
C7H14O	005454-79-5	3-METHYLCYCLOHEXANOL (CIS)
C7H14O	007443-52-9	2-METHYLCYCLOHEXANOL (TRANS)
C7H14O	007443-55-2	3-METHYLCYCLOHEXANOL (TRANS)
C7H14O	007443-70-1	2-METHYLCYCLOHEXANOL (CIS)
C7H14O	007731-28-4	4-METHYLCYCLOHEXANOL (CIS)
C7H14O	007731-29-5	4-METHYLCYCLOHEXANOL (TRANS)
C7H14O2	000105-66-8	N-PROPYLBUTYRATE
C7H14O2	000106-70-7	METHYL HEXANOATE
C7H14O2	000107-70-0	4-METHOXY-4-METHYL-2-PENTANONE
C7H14O2	000108-64-5	ETHYL ISOVALERATE
C7H14O2	000111-14-8	HEPTANOIC ACID
C7H14O2	000123-92-2	ISOAMYL ACETATE
C7H14O2	000539-82-2	ETHYL VALERATE
C7H14O2	000540-42-1	2-METHYLPROPYL PROPANOATE
C7H14O2	000590-01-2	N-BUTYL PROPIONATE

C7H14O2	000617-50-5	ISOPROPYL ISOBUTYRATE
C7H14O2	000625-16-1	TERT-PENTYLACETATE
C7H14O2	000626-38-0	2-PENTANOL ACETATE
C7H14O2	000628-63-7	N-AMYL ACETATE
C7H14O2	000644-49-5	PROPYL ISOBUTYRATE
C7H14O2	002426-08-6	GLYCIDYL N-BUTYL ETHER
C7H14O2	020225-24-5	2-ETHYLPENTANOIC ACID
C7H14O3	000138-22-7	BUTYL LACTATE
C7H14O3	004435-53-4	3-METHOXYBUTYL ACETATE
C7H14O4	000557-25-5	1-MONOBUTYRIN
C7H14O4	000629-38-9	DIETHYLENE GLYCOL MONOMETHYL ETHER ACETATE
C7H15Br	000629-04-9	1-BROMOHEPTANE
C7H15Br	001974-04-5	2-BROMOHEPTANE
C7H15Cl	000629-06-1	1-CHLOROHEPTANE
C7H15Cl2N2O2P	000050-18-0	CYCLOPHOSPHAMIDE
C7H15Cl2N2O2P	003778-73-2	IFOSFAMIDE
C7H15F	000661-11-0	1-FLUOROHEPTANE
C7H15I	004282-40-0	1-IODOHEPTANE
C7H15N	000766-09-6	PIPERIDINE, 1-ETHYL-
C7H15N	000766-17-6	2,6-DIMETHYLPIPERIDINE (CIS-)
C7H15N3O	052662-76-7	4-CYCLOHEXYLSEMICARBAZIDE
C7H15N3O2	050285-70-6	1-NITROSOTRIETHYLUREA
C7H15NO	003040-44-6	1-PIPERIDINE ETHANOL
C7H15NO2	002114-20-7	O-HEXYL CARBAMATE
C7H15NO2	002594-17-4	O-PENTYL-N-METHYLCARBAMATE
C7H15NO2	024847-58-3	CARBAMIC ACID, 2-ETHYLBUTYL ESTER
C7H15NO2	052113-78-7	CARBAMIC ACID, DIMETHYL-, 2-METHYLPROPYL ESTER
C7H15NO2	052642-50-9	O-T-HEXYMOPYRIMIDINE
C7H15NO2	060577-35-7	3-OXAZOLIDINYLOXY, 2,2,5,5-TETRAMETHYL-
C7H15NO5	069567-10-8	3,4,5-PIPERIDINETRIOL, 2-(HYDROXYMETHYL)-1-METHY
C7H16	000108-08-7	2,4-DIMETHYLPENTANE
C7H16	000142-82-5	N-HEPTANE
C7H16	000464-06-2	2,2,3-TRIMETHYLBUTANE
C7H16	000562-49-2	3,3-DIMETHYLPENTANE
C7H16	000565-59-3	2,3-DIMETHYLPENTANE
C7H16	000589-34-4	3-METHYLHEXANE
C7H16	000590-35-2	2,2-DIMETHYLPENTANE
C7H16	000591-76-4	2-METHYLHEXANE
C7H16	000617-78-7	3-ETHYLPENTANE
C7H16BrNO2	000066-23-9	ACETYLCHOLINE BROMIDE
C7H16ClN3O2S2	015263-52-2	S,S'-(2-(DIMETHYLAMINO)-1,3-PROPANEDIYL) CARBAM*
C7H16FO2P	000096-64-0	SOMAN
C7H16INO2	000541-66-2	1,3-DIOXOLANE-4-METHANAMINIUM, N,N,N-TRIMETHYL-,
C7H16INOS	058326-97-9	1,3-OXATHIOLANE-5-METHANAMINIUM, N,N,N-TRIMETHYL
C7H16N2O	004128-37-4	N,N'-BIS(1-METHYLETHYL)UREA
C7H16N2O	022371-32-0	HEPTANOIC ACID HYDRAZIDE
C7H16N2O	025413-64-3	N-NITROSOPROPYLBUTYLAMINE
C7H16NO4PS2	000919-76-6	AMIDOTHION
C7H16O	000111-70-6	1-HEPTANOL
C7H16O	000543-49-7	2-HEPTANOL
C7H16O	000589-55-9	4-HEPTANOL
C7H16O	000589-82-2	3-HEPTANOL
C7H16O	000594-83-2	2,3,3,-TRIMETHYL-2-BUTANOL
C7H16O	000595-41-5	2,3-DIMETHYL-3-PENTANOL
C7H16O	000597-49-9	3-ETHYL-3-PENTANOL
C7H16O	000597-96-6	3-METHYL-3-HEXANOL
C7H16O	000600-36-2	2,4-DIMETHYL-3-PENTANOL
C7H16O	000617-29-8	2-METHYL-3-HEXANOL

C7H16O	000625-06-9	2,4-DIMETHYL-2-PENTANOL
C7H16O	000625-23-0	2-METHYL-2-HEXANOL
C7H16O	000627-59-8	5-METHYL-2-HEXANOL
C7H16O	000919-94-8	ETHYL(TERT-AMYL) ETHER
C7H16O	002370-12-9	2,2-DIMETHYL-1-PENTANOL
C7H16O	003121-79-7	4,4-DIMETHYL-1-PENTANOL
C7H16O	003970-62-5	2,2-DIMETHYL-3-PENTANOL
C7H16O	004911-70-0	2,3-DIMETHYL-2-PENTANOL
C7H16O	006305-71-1	2,4-DIMETHYL-1-PENTANOL
C7H16O	017348-59-3	ISOPROPYL T-BUTYL ETHER
C7H16O2	000115-76-4	2,2-DIETHYL-1,3-PROPANEDIOL
C7H16O2	000126-84-1	2,2-DIETHOXYPROPANE
C7H16O2	023436-19-3	1-ISOBUTOXY-2-PROPANOL
C7H16O3	000122-51-0	ETHYL ORTHOFORMATE
C7H16O4	000112-35-6	METHOXY TRIETHYLENE GLYCOL
C7H16O4S2	000115-24-2	2,2-BIS(ETHYLSULFONYL)PROPANE
C7H16S	000111-47-7	DI-N-PROPYLSULFIDE
C7H16S	001639-09-4	1-HEPTANETHIOL
C7H17N	000111-68-2	HEPTYLAMINE
C7H17N	007515-80-2	2-PROPANAMINE, 2-METHYL-N-(1-METHYLETHYL)-
C7H17N	020193-21-9	PROPYL-BUTYLAMINE
C7H17N	039190-66-4	PROPYL-ISOBUTYLAMINE
C7H17N	039190-67-5	PROPYL SEC-BUTYLAMINE
C7H17N	044652-67-7	1-HEXANAMINE, 2-METHYL-
C7H17N3O	079353-76-7	4-HEXYLSEMICARBAZIDE
C7H17N5	021306-55-8	IMIDODICARBONIMIDIC DIAMIDE, N-PENTYL
C7H17O2PS	025371-75-9	O,O-DIPROPYL METHYLPHOSPHONOTHIOATE
C7H17O2PS2	000556-75-2	O-ET-S-(ETSET)-ME-PHOSPHONATE
C7H17O2PS3	000298-02-2	PHORATE
C7H17O2PS3	036614-38-7	ISOTHIOATE
C7H17O3P	001445-75-6	PHOSPHONIC ACID, METHYL-, DIISOPROPYL ESTER
C7H17O3PS2	002600-69-3	PHORATE, O-ANALOG
C7H17O4PS2	002588-05-8	PHORATE SULFOXIDE
C7H17O4PS3	002588-04-7	PHORATE SULFONE
C7H17O5PS2	002588-06-9	PHOSPHOROTHIOIC ACID, O,O-DIETHYL S-[(ETHYLSULFO
C7H18IN	000302-57-8	TRIETHYL METHYL AMMONIUM IODIDE
C7H18IN	004153-42-8	ETHANAMINIUM, 1,1-DIMETHYL-N,N,N-TRIMETHYL-, IOD
C7H18IN	007722-19-2	TRIMETHYL BUTYL AMMONIUM IODIDE
C7H18INO	016332-51-7	2-ETHOXYETHYL TRIMETHYL AMMONIUM IODIDE
C7H18Si	001000-49-3	SILANE, BUTYL TRIMETHYL
C8Br4O3	000632-79-1	4,5,6,7-TETRABROMO-1,3-ISOBENZOFURANDIONE
C8Cl4N2	001897-45-6	CHLOROTHANONIL
C8Cl8	029082-74-4	OCTACHLOROSTYRENE
C8HBr4F3N2	002338-30-9	4567-TETRABR 2-CF3 BENZIMIDAZOLE
C8H2Br3F3N2	007682-32-8	BENZIMIDAZOLE-4,5,6-TRIBR-2-CF3
C8H2Cl3F3N2	002338-27-4	1H-BENZIMIDAZOLE, 4,5,6-TRICHLORO-2-(TRIFLUOROME
C8H2Cl3F3N2	003393-59-7	4,5,7-TRICL-2-CF3-BENZIMIDAZOLE
C8H2Cl4O2	027355-22-2	TETRACHLOROPHTHALIDE
C8H3Br2F3N2	006587-21-9	2-CF3-5,6-DIBROMO-BENZIMIDAZOLE
C8H3BrN2O	056656-96-3	PROPANEDINITRILE, [(5-BROMO-2-FURANYL)METHYLENE]
C8H3Cl2F3N2	002338-25-2	2-CF3-5,6-DICHLOROBENZIMIDAZOLE
C8H3Cl2F3N2	003615-21-2	2-CF3-4,5-DICHLOROBENZIMIDAZOLE
C8H3Cl2F3N2	004228-88-0	2-CF3-4,6-DICHLOROBENZIMIDAZOLE
C8H3Cl2F3N2	004228-89-1	2-CF3-4,7-DICHLOROBENZIMIDAZOLE
C8H3ClF3N3O2	006609-40-1	5-CL-6-NO2-2-CF3-BENZIMIDAZOLE
C8H3ClN2O	091937-67-6	PROPANEDINITRILE, [(5-CHLORO-2-FURANYL)METHYLENE
C8H3IN2O	069527-40-8	PROPANEDINITRILE, [(5-IODO-2-FURANYL)METHYLENE]-
C8H4BrF3N2	003671-60-1	2-CF3-5-BROMOBENZIMIDAZOLE

C8H4Cl2F3NO	064694-85-5	ACETAMIDE, N-(3,4-DICHLOROPHENYL)-2,2,2-TRIFLUOR
C8H4Cl2O2	000088-95-9	PHTHALOYL CHLORIDE
C8H4Cl3NaO2	002439-00-1	2,3,6-TRICHLOROPHENYLACETIC ACID, SODIUM SALT
C8H4Cl6	000068-36-0	P-DI(TRICHLOROMETHYL)BENZENE
C8H4Cl8	083682-58-0	2,5-BIS(TRICHLOROMETHYL)-3,4-DICHLORO-1,3,5-HE*
C8H4ClF3N2	000656-49-5	2-CF3-5-CHLOROBENZIMIDAZOLE
C8H4ClF3N2	002338-31-0	2-CF3-4-CHLOROBENZIMIDAZOLE
C8H4ClF3N2O2S	002251-64-1	124-BENZTHIADIZ-1-O2-3-CF3-6-CL
C8H4ClNO2	003481-09-2	N-CHLOROPHTHALIMIDE
C8H4F3N3O2	000327-19-5	2-CF3-5-NO2 BENZIMADAZOLE
C8H4F4O2	000652-32-4	BENZOIC ACID, 2,3,5,6-TETRAFLUORO-4-METHYL-
C8H4F5NO	000653-22-5	2,3,4,5,6-PENTAFLUOROACETANILIDE
C8H4F6	000402-31-3	BENZENE, 1,3-BIS(TRIFLUOROMETHYL)-
C8H4F6	000433-19-2	BENZENE, 1,4-BIS(TRIFLUOROMETHYL)-
C8H4N2	000091-15-6	1,2-DICYANOBENZENE
C8H4N2	000623-26-7	1,4-BENZENEDICARBONITRILE
C8H4N2	000626-17-5	1,3-DICYANOBENZENE
C8H4N2O	003237-22-7	PROPANEDINITRILE, (2-FURANYLMETHYLENE)
C8H4N2O4	000089-40-7	4-NITROPHTHALIMIDE
C8H4N2O4	000603-62-3	3-NITROPHTHALIMIDE
C8H4N2O4	000611-09-6	1H-INDOLE-2,3-DIONE, 5-NITRO-
C8H4N2S	003125-78-8	BENZONITRILE, 3-ISOTHIOCYANATO-
C8H4O3	000085-44-9	PHTHALIC ANHYDRIDE
C8H5Br5	000085-22-3	PENTABROMOETHYLBENZENE
C8H5BrF3NO	024568-11-4	BROMOBENZENE,P-TRIFLUOROACETAMIDO
C8H5BrN2	078593-33-6	CINNOLINE, 3-BROMO-
C8H5Cl2F2NO	127836-03-7	2,4-DIF 3,5-DICL ACETANILIDE
C8H5Cl2NaO3	002702-72-9	2,4-D, SODIUM SALT
C8H5Cl2NO2	018984-21-9	2,4-DICHLORO-B-NITROSTYRENE
C8H5Cl3N2O2S	037157-97-4	124-BENZTHIADIAZN-1-O2-3-DICLME-6-CL
C8H5Cl3O2	000085-34-7	FENAC
C8H5Cl3O3	000093-76-5	2,4,5-TRICHLOROPHENOXYACETIC ACID
C8H5Cl3O3	000575-89-3	2,4,6-TRICHLOROPHENOXYACETIC ACID
C8H5Cl3O3	002307-49-5	TRICAMBA
C8H5ClN2	001448-87-9	2-CHLOROQUINOXALINE
C8H5ClN2	005448-43-1	6-CHLOROQUINOXALINE
C8H5ClN2	007556-99-2	7-CHLOROQUINAZOLINE
C8H5ClN2	017404-90-9	3-CHLOROCINNOLINE
C8H5ClN2	062163-09-1	5-CHLOROQUINOXALINE
C8H5ClN2O	024786-13-8	FURAZAN, CHLOROPHENYL-
C8H5ClN2O2	082775-80-2	FURAZAN, CHLOROPHENYL-, 5-OXIDE
C8H5ClN2O2	082775-81-3	FURAZAN, CHLOROPHENYL-, 2-OXIDE
C8H5ClN2O4	015851-93-1	2-CHLORO-5-NITRO-B-NITROSTYRENE
C8H5ClN4	072111-58-1	2,3-PYRAZINEDICARBONITRILE, 5-CHLORO-6-ETHYL-
C8H5F3N2	000312-73-2	2-CF3 BENZIMIDAZOLE
C8H5F3O	000434-45-7	TRIFLUOROACETOPHENONE
C8H5F3O	000454-89-7	M-(TRIFLUOROMETHYL)BENZALDEHYDE
C8H5F3O2	000454-92-2	M-TRIFLUOROMETHYLBENZOIC ACID
C8H5F3O2	000455-24-3	P-CF3 BENZOIC ACID
C8H5F3O2S	000326-91-0	THENOYLTRIFLUOROMETHYLACETONE
C8H5F5	000309-11-5	PENTAFLUOROETHYLBENZENE
C8H5F6N	000328-74-5	3,5-BIS(TRIFLUOROME)ANILINE
C8H5F6NO2S	023384-11-4	M-CF3 CF3-METHANESULFONANILIDE
C8H5F6NO2S	023384-12-5	P-CF3 CF3-METHANESULFONANILIDE
C8H5N2O4S2	013410-84-9	5-(5-NO2-2-FURFURYLIDENE-2-S-THIAZOLIDIN-4-ONE
C8H5N2O5S	027564-47-2	5(5NO2-2-FURIL)THIAZOLIDIN,24-DION
C8H5N2O5S	052661-48-0	6(5NO2-2-FURYL)-13-THIAZIN-24-DIONE
C8H5N3O2	021905-82-8	3-NITROCINNOLINE

C8H5N3O3	076016-68-7	FURAZAN, NITROPHENYL-
C8H5NO	024964-64-5	M-CYANOBENAZLDEHYDE
C8H5NO2	000085-41-6	PHTHALIMIDE
C8H5NO2	000091-56-5	1H-INDOLE-2,3-DIONE
C8H5NO2	000619-65-8	P-CYANOBENZOIC ACID
C8H5NO2	001877-72-1	M-CYANOBENZOIC ACID
C8H5NO2S	004965-26-8	5-NITROBENZOTHIOPHENE
C8H5NO3	000524-38-9	N-HYDROXYPHTHALIMIDE
C8H5NO6	000603-11-2	3-NITROPHTHALIC ACID
C8H5NOS	002131-62-6	4-CARBOXYPHENYLISOTHIOCYANATE
C8H6	000536-74-3	ETHYNYL BENZENE
C8H6BrClN2O2S	005108-54-3	124-BENZTHIADIAZIN-11SO2-3ME5BR7CL
C8H6BrFN2O3	095635-46-4	2-F 4-BR 5-NO2 ACETANILIDE
C8H6BrN	005798-79-8	ALPHA-BROMOBENZENEACETONITRILE
C8H6BrN	010075-50-0	5-BROMOINDOLE
C8H6Cl2	006607-45-0	(1,2-DICHLOROETHENYL)BENZENE
C8H6Cl2N2O2S	037157-57-6	124-BENZTHIADIAZIN-1-O2-3-CLME-6-CL
C8H6Cl2O2	002905-69-3	METHYL 2,5-DICHLOROBENZOATE
C8H6Cl2O2	014920-87-7	BENZOIC ACID, 2,6-DICHLORO-, METHYL ESTER
C8H6Cl2O2	028165-71-1	ACETIC ACID, 2,6-DICHLOROPHENYL ESTER
C8H6Cl2O3	000094-75-7	2,4-DICHLOROPHENOXYACETIC ACID
C8H6Cl2O3	000587-64-4	3,5-DICHLOROPHENOXYACETIC ACID
C8H6Cl2O3	000588-22-7	3,4-DICHLOROPHENOXYACETIC ACID
C8H6Cl2O3	001918-00-9	DICAMBA
C8H6Cl2O3	002976-74-1	2,3-DICHLOROPHENOXYACETIC ACID
C8H6Cl3NaO5S	003570-61-4	2,4,5-TES
C8H6Cl3NO	020149-84-2	ACETAMIDE, 2-CHLORO-N-(3,4-DICHLOROPHENYL)
C8H6Cl3NO	033715-62-7	3,4,5-TRICHLOROACETANILIDE
C8H6Cl4	000877-10-1	1,2,4,5-TETRACHLORO-3,6-DIMETHYLBENZENE
C8H6Cl6	083682-57-9	2,5-BIS(DICHLOROMETHYL)-3,4-DICHLORO-1,3,5-HEX*
C8H6ClIN2O2S	037148-13-3	124-BENZTHIADIAZIN-11SO2-3ME5I7CL
C8H6ClN	053294-05-6	6-CHLOROINDOLE
C8H6ClN3O4S	031365-75-0	124-BENZTHIADIAZIN-1-O2-3ME-6NO2-7CL
C8H6ClN3O4S	037157-79-2	124-BENZTHIADIAZIN-1-O2-3ME-5NO2-7CL
C8H6ClNO2	000706-07-0	4-CHLORO B-NITROSTYRENE
C8H6ClNO2	003156-34-1	2-CHLORO-B-NITROSTYRENE
C8H6ClNO2	003156-35-2	3-CHLORO-B-NITROSTYRENE
C8H6ClNO3	000099-47-8	ALPHA-CHLORO-M-NITROACETOPHENONE
C8H6ClNO4	042087-80-9	METHYL 4-CHLORO-2-NITROBENZOATE
C8H6ClNO4	080563-87-7	BENZOIC ACID, 2-CHLORO-6-NITRO-, METHYL ESTER
C8H6ClNO5	089894-13-3	PHENOXYACETIC ACID, 4-CL-3-NO2
C8H6F2O2	013671-00-6	BENZOIC ACID, 2,6-DIFLUORO-, METHYL ESTER
C8H6F3N3O4S2	000148-56-1	FLUMETHAZIDE
C8H6F3NO	000360-64-5	2-TRIFLUOROMETHYLBENZAMIDE
C8H6F3NO	000404-24-0	TRIFLUOROACETANILIDE
C8H6F3NO	001891-90-3	P-TRIFLUOROMETHYLBENZAMIDE
C8H6N2	000091-19-0	QUINOXALINE
C8H6N2	000253-52-1	PHTHALAZINE
C8H6N2	000253-66-7	CINNOLINE
C8H6N2	000253-82-7	QUINAZOLINE
C8H6N2O	000119-39-1	1,2(H)-PHTHALAZINONE
C8H6N2O	000491-36-1	QUINAZOLIDIN-4-ONE
C8H6N2O	000825-56-9	1,3,4-OXADIAZOLE, 2-PHENYL-
C8H6N2O	001196-57-2	QUINOXALINE-2-ONE
C8H6N2O	003034-34-2	P-CYANOBENZAMIDE
C8H6N2O	003441-01-8	M-CYANOBENZAMIDE
C8H6N2O	006321-94-4	P-CYANOFORMANILIDE
C8H6N2O	017174-98-0	BENZAMIDE, 2-CYANO-

C8H6N2O	017227-47-3	4-QUINAZOLINOL
C8H6N2O	018514-84-6	CINNOLINE-4-ONE
C8H6N2O	031777-46-5	3-CINNOLONE
C8H6N2O2	000771-39-1	2-NH2-1,3-BENZOXAZIN-4-ONE
C8H6N2O2	001445-69-8	1,4-PHTHALAZINEDIONE, 2,3-DIHYDRO
C8H6N2O2	005319-71-1	4-(3H)-QUINAZOLINONE,3-HYDROXY
C8H6N2O2	006146-52-7	INDOLE, 5-NITRO
C8H6N2O2	006960-42-5	7-NITROINDOLE
C8H6N2O2	015804-19-0	QUINOXALINE-1,4-DIHYDRO-2,3-DIONE
C8H6N2O3	033101-81-4	N-PHENYL-3,5-OXADIAZOLIDIONE
C8H6N2O4	000882-26-8	3-NITRO-B-NITROSTYRENE
C8H6N2O4	003156-39-6	2-NITRO-B-NITROSTYRENE
C8H6N2O4	003156-41-0	4-NITRO-B-NITROSTYRENE
C8H6N2O6	000643-43-6	DINITROPHENYL ACETATE
C8H6N2O6	002702-58-1	BENZOIC ACID, 3,5-DINITRO-, METHYL ESTER
C8H6N3O4S	052661-38-8	2-NH2-5-(5NO2-2-FURFURYL)THIAZOLONE
C8H6N4	063655-40-3	2-AMINO-5-CYANOBENZIMIDAZOLE
C8H6N4O5	000067-20-9	NITROFURANTOIN
C8H6O	000271-89-6	BENZOFURAN
C8H6O2	000087-41-2	PHTHALIDE
C8H6O2S	000825-44-5	BENZO(B)THIOPHENE S,S-DIOXIDE
C8H6O3	000120-57-0	PIPERONAL
C8H6O3	000611-73-4	BENZOYLFORMIC ACID
C8H6O3	000619-21-6	3-FORMYLBENZOIC ACID
C8H6O3	000619-66-9	4-FORMYLBENZOIC ACID
C8H6O3	016859-59-9	1(3H)-ISOBENZOFURANONE-3-HYDROXY
C8H6O4	000088-99-3	O-PHTHALIC ACID
C8H6O4	000100-21-0	TEREPHTHALIC ACID
C8H6O4	000121-91-5	ISOPHTHALIC ACID
C8H6S	000095-15-8	BENZO(B)THIOPHENE
C8H6S	011095-43-5	BENZOTHIOPHENE
C8H6S2	000492-97-7	2,2'-BITHIOPHENE
C8H6Se	000272-30-0	SELANONAPHTHENE
C8H7Br	000103-64-0	BETA-BROMOSTYRENE
C8H7Br	002039-82-9	4-BROMOSTYRENE
C8H7Br2NO	033098-80-5	2,6-DIBROMOACETANILIDE
C8H7Br3O2	023976-66-1	2-(2,4,6-TRIBROMOPHENOXY)ETHANOL
C8H7BrN2O2S	013460-15-6	124-BENZTHIADIAZIN-1-O2-3-ME-7-BR
C8H7BrN2O2S	037148-00-8	124-BENZOTHIADIAZINE-1-O2-3-ME-6-BR
C8H7BrN2O3	077317-71-6	3-C≡CCO-5-BR-6-ME-PYRIMID-2,4-DIONE
C8H7BrO	000070-11-1	ETHANONE, 2-BROMO-1-PHENYL-
C8H7BrO	000099-90-1	P-BROMOACETOPHENONE
C8H7BrO	002142-63-4	M-BROMOACETOPHENONE
C8H7BrO	002142-69-0	ETHANONE, 1-(2-BROMOPHENYL)-
C8H7BrO2	000610-94-6	METHYL 2-BROMOBENZOATE
C8H7BrO2	000618-89-3	METHYL 3-BROMOBENZOATE
C8H7BrO2	000619-42-1	BENZOIC ACID, 4-BROMO-, METHYL ESTER
C8H7BrO2	001829-37-4	2-BROMOPHENYL ACETATE
C8H7BrO2	001878-67-7	M-BROMOPHENYLACETIC ACID
C8H7BrO2	001878-68-8	P-BROMOPHENYLACETIC ACID
C8H7BrO2	006232-88-8	4-(BROMOMETHYL)BENZOIC ACID
C8H7BrO3	001798-99-8	M-BROMOPHENOXYACETIC ACID
C8H7BrO3	001878-91-7	P-BROMOPHENOXYACETIC ACID
C8H7BrO3	001879-56-7	O-BROMOPHENOXYACETIC ACID
C8H7Cl	001073-67-2	P-CHLOROSTYRENE
C8H7Cl	002039-85-2	M-CHLOROSTYRENE
C8H7Cl	002039-87-4	O-CHLOROSTYRENE
C8H7Cl2NO	002150-93-8	N-(3,4-DICHLOROPHENYL)ACETAMIDE

C8H7Cl2NO	002621-62-7	2,5-DICHLOROACETANILIDE
C8H7Cl2NO	006975-29-7	2,4-DICHLOROACETANILIDE
C8H7Cl2NO	017700-54-8	2,6-DICHLOROACETANILIDE
C8H7Cl2NO	023068-36-2	2,3-DICHLOROACETANILIDE
C8H7Cl2NO	031592-84-4	3,5-DICHLOROACETANILIDE
C8H7Cl2NO2	001918-18-9	METHYL (3,4-DICHLOROPHENYL)CARBAMATE
C8H7Cl2NO2	002153-10-8	4-HYDROXY DICHLOROACETANILIDE
C8H7Cl2NO2	007286-84-2	CHLORAMBEN, METHYL ESTER
C8H7Cl2NO2	013538-26-6	N-ME-3,5-DICHLOROPHENYLCARBAMATE
C8H7Cl2NO2	018315-50-9	N-ME-3,4-DICHLOROPHENYLCARBAMATE
C8H7Cl2NO2	018315-62-3	N-ME-2,5-DICHLOROPHENYLCARBAMATE
C8H7Cl2NOS	025052-57-7	CARBANILIC ACID, 3,4-DICHLOROTHIO-, S-METHYL EST
C8H7Cl3O3	002539-26-6	3,4,5-TRICL-2,6-MEO PHENOL
C8H7Cl4N3O4S2	004267-05-4	3-CCL3 HYDROCHLOROTHIAZIDE
C8H7ClF3N3O4S2	001547-10-0	3-CF3 HYDROCHLOROTHIAZIDE
C8H7ClFNO	059280-70-5	2-FLUORO-4-CHLOROACETANILIDE
C8H7ClN2O2S	000364-98-7	124-BENZTHIDIAZN-1,1-O2-3-ME-7-CL
C8H7ClN2O2S	014559-54-7	124-BENZTHIADIAZIN-1-O2-3-ME-6-CL
C8H7ClN2O2S	022680-31-5	124-BENZTHIADIAZN-1-O2-3-ME-8-CL
C8H7ClN2O2S	031363-85-6	124-BENZTHIADIAZIN-1-O2-3-ME-5-CL
C8H7ClN2O3	000881-51-6	2-NITRO-4-CHLOROACETANILIDE
C8H7ClN2S	007692-57-1	BENZIMIDAZOLE-5-CHLORO-2-MES
C8H7ClO	000099-02-5	M-CHLOROACETOPHENONE
C8H7ClO	000099-91-2	1-(4-CHLOROPHENYL)ETHANONE
C8H7ClO	000103-80-0	PHENYLACETYL CHLORIDE
C8H7ClO	000532-27-4	2-CHLORO-1-PHENYLETHANONE
C8H7ClO	002142-68-9	O-CHLOROACETOPHENONE
C8H7ClO2	000501-53-1	BENZYL CHLOROCARBONATE
C8H7ClO2	000610-96-8	METHYL O-CHLOROBENZOATE
C8H7ClO2	001126-46-1	METHYL P-CHLOROBENZOATE
C8H7ClO2	001878-65-5	M-CHLOROPHENYLACETIC ACID
C8H7ClO2	001878-66-6	P-CHLOROPHENYLACETIC ACID
C8H7ClO2	002905-65-9	BENZOIC ACID, 3-CHLORO-, METHYL ESTER
C8H7ClO2	004525-75-1	2-CHLOROPHENYL ACETATE
C8H7ClO2	013031-39-5	3-CHLOROPHENYL ACETATE
C8H7ClO2	021327-86-6	BENZOIC ACID, 2-CHLORO-6-METHYL-
C8H7ClO3	000122-88-3	P-CHLOROPHENOXYACETIC ACID
C8H7ClO3	000588-32-9	M-CHLOROPHENOXYACETIC ACID
C8H7ClO3	000614-61-9	O-CHLOROPHENOXYACETIC ACID
C8H7F2NO	003896-29-5	2,6-DIFLUOROACETANILIDE
C8H7F3N2O	013114-87-9	3-(TRIFLUOROMETHYL)PHENYLUREA
C8H7F3O	000349-75-7	M-(TRIFLUOROME)BENZYL ALCOHOL
C8H7FN2O2S	031365-74-9	124-BENZTHIADIAZIN-1-O2-3-ME-7-F
C8H7FO	000403-42-9	4-FLUOROACETOPHENONE
C8H7FO	000455-36-7	M-FLUOROACETOPHENONE
C8H7FO2	000331-25-9	M-FLUOROPHENYLACETIC ACID
C8H7FO2	000394-35-4	BENZOIC ACID, 2-FLUORO-, METHYL ESTER
C8H7FO2	000403-33-8	BENZOIC ACID, 4-FLUORO-, METHYL ESTER
C8H7FO2	000405-50-5	P-FLUOROPHENYLACETIC ACID
C8H7FO2	000451-82-1	O-FLUOROPHENYLACETIC ACID
C8H7FO2	000701-83-7	3-FLUOROPHENYL ACETATE
C8H7FO2	029650-44-0	2-FLUOROPHENYL ACETATE
C8H7FO3	000348-10-7	O-FLUOROPHENOXYACETIC ACID
C8H7FO3	000404-98-8	M-FLUOROPHENOXYACETIC ACID
C8H7FO3	000405-79-8	P-FLUOROPHENOXYACETIC ACID
C8H7FO5S	034380-54-6	P-SO2F PHENOXYACETIC ACID
C8H7IN2O2S	037148-02-0	124-BENZOTHIADIAZINE,7-I-3-ME,1,1-DIO
C8H7IO2	000610-97-9	BENZOIC ACID, 2-IODO-, METHYL ESTER

C8H7IO2	001798-06-7	P-IODOPHENYLACETIC ACID
C8H7IO2	001878-69-9	M-IODOPHENYLACETIC ACID
C8H7IO2	018698-96-9	BENZENEACETIC ACID, 2-IODO-
C8H7IO2	032865-61-5	2-IODOPHENYLACETATE
C8H7IO3	001878-92-8	O-IODOPHENOXYACETIC ACID
C8H7IO3	001878-93-9	M-IODOPHENOXYACETIC ACID
C8H7IO3	001878-94-0	P-IODOPHENOXYACETIC ACID
C8H7N	000104-85-8	4-METHYLBENZONITRILE
C8H7N	000120-72-9	INDOLE
C8H7N	000140-29-4	PHENYLACETONITRILE
C8H7N	000274-40-8	INDOLIZINE
C8H7N	000529-19-1	2-METHYLBENZONITRILE
C8H7N3	001680-44-0	1H-1,2,3-TRIAZOLE, 4-PHENYL-
C8H7N3	003357-42-4	1H-1,2,4-TRIAZOLE, 3-PHENYL-
C8H7N3	005152-83-0	4-CINNOLINAMINE
C8H7N3	005424-05-5	2-AMINOQUINOXALINE
C8H7N3	006298-37-9	6-AMINOQUINOXALINE
C8H7N3	015018-66-3	4-QUINAZOLINAMINE
C8H7N3	016566-20-4	5-AMINOQUINOXALINE
C8H7N3	017372-79-1	3-CINNOLINAMINE
C8H7N3	019064-69-8	1-PHTHALAZINAMINE
C8H7N3O	010349-14-1	FURAZANAMINE, 4-PHENYL-
C8H7N3O	020198-19-0	2-AMINOQUINAZOLINE-4-ONE
C8H7N3O	022305-44-8	3-METHYLBENZOTRIAZIN-4-ONE
C8H7N3O	054797-21-6	P-METHYLPHENYL AZOXYCYANIDE
C8H7N3O	064462-06-2	M-METHYLPHENYL AZOXYCYANIDE
C8H7N3O2	001792-40-1	1H-BENZIMIDAZOLE, 2-METHYL-5-NITRO-
C8H7N3O2	004144-64-3	1H-BENZOTRIAZOLE-1-ACETIC ACID
C8H7N3O2	006850-23-3	1-METHYL-6-NITROINDAZOLE
C8H7N3O2	029945-54-8	FURAZANAMINE, 4-PHENYL-, 2-OXIDE
C8H7N3O2	030059-86-0	FURAZANAMINE, 4-PHENYL-, 5-OXIDE
C8H7N3O2	062825-14-3	M-METHOXYPHENYL AZOXYCYANIDE
C8H7N3O2	062825-15-4	P-METHOXY PHENYL AZOXYCYANIDE
C8H7N3O4S	037148-07-5	124-BENZOTHIADIAZINE-1-O2-3-ME-6-NO2
C8H7N3O5	000067-45-8	2-OXAZOLIDINONE, 3- (5-NITRO-2-FURANYL)METHYLEN
C8H7N3O5	004551-76-2	3,5-DINITRO-4-METHYLBENZAMIDE
C8H7N3O5	090110-78-4	2,6-DINO2 ACETANILIDE
C8H7N4O3	050832-71-8	3(5NO2-2-FURFURILIDENE)-124-TRIAZOLE
C8H7NO	000059-48-3	OXINDOLE
C8H7NO	000874-90-8	BENZONITRILE, 4-METHOXY-
C8H7NO	001953-54-4	1H-INDOL-5-OL
C8H7NO	014191-95-8	4-HYDROXYBENZYLCYANIDE
C8H7NO2	000102-96-5	BETA-NITROSTYRENE
C8H7NO3	000100-19-6	P-NITROACETOPHENONE
C8H7NO3	000121-89-1	3-NITROACETOPHENONE
C8H7NO3	000577-59-3	2-NITROACETOPHENONE
C8H7NO3	003156-44-3	3-HYDROXY-B-NITROSTYRENE
C8H7NO3	003179-08-6	4-HYDROXY-B-NITROSTYRENE
C8H7NO4	000104-03-0	P-NITROPHENYLACETIC ACID
C8H7NO4	000606-27-9	BENZOIC ACID, 2-NITRO-, METHYL ESTER
C8H7NO4	000610-69-5	O-NITROPHENYL ACETATE
C8H7NO4	000618-95-1	BENZOIC ACID, 3-NITRO-, METHYL ESTER
C8H7NO4	000619-50-1	METHYL P-NITROBENZOATE
C8H7NO4	000830-03-5	P-NITROPHENYLACETATE
C8H7NO4	001523-06-4	M-NITROPHENYL ACETATE
C8H7NO4	001877-73-2	M-NITROPHENYLACETIC ACID
C8H7NO5	001798-11-4	P-NITROPHENOXYACETIC ACID
C8H7NO5	001878-87-1	O-NITROPHENOXYACETIC ACID

C8H7NO5	001878-88-2	M-NITROPHENOXYACETIC ACID
C8H7NOS	002284-20-0	BENZENE, 1-ISOTHIOCYANATO-4-METHOXY
C8H7NOS	002786-62-1	3-METHYLBENZOTHIAZOL-2-ONE
C8H7NOS	003125-64-2	BENZENE, 1-ISOTHIOCYANATO-3-METHOXY
C8H7NS	000120-75-2	2-METHYBENZOTHIAZOLE
C8H7NS	000622-59-3	4-METHYLPHENYLISOTHIOCYANATE
C8H7NS	000622-78-6	BENZYLISOTHIOCYANATE
C8H7NS	003012-37-1	BENZYLTHIOCYANATE
C8H7NS2	000615-22-5	2-METHYLTHIOBENZOTHIAZOLE
C8H8	000100-42-5	STYRENE
C8H8	000629-20-9	1,3,5,7-CYCLOOCTATETRAENE
C8H8Br2	000093-52-7	(1,2-DIBROMOETHYL)BENZENE
C8H8BrCl2O3PS	002104-96-3	BROMOPHOS
C8H8BrNO	000103-88-8	4-BROMOACETANILIDE
C8H8BrNO	000614-76-6	2-BROMOACETANILIDE
C8H8BrNO	000621-38-5	3-BROMOACETANILIDE
C8H8BrNO2	013538-27-7	N-METHYL-2-BROMOPHENYLCARBAMATE
C8H8BrNO2	013538-50-6	N-METHYL-4-BROMOPHENYLCARBAMATE
C8H8BrNO2	013538-60-8	N-METHYL-3-BROMOPHENYLCARBAMATE
C8H8BrNO2	025216-70-0	CARBAMIC ACID, (2-BROMOPHENYL)-, METHYL ESTER
C8H8BrNO2	025216-72-2	CARBAMIC ACID, (3-BROMOPHENYL)-, METHYL ESTER
C8H8BrNO2	040912-73-0	BENZAMIDE, 5-BROMO-2-HYDROXY-3-METHYL
C8H8Cl2	000626-16-4	1,3-(BIS-CHLOROMETHYL)BENZENE
C8H8Cl2IO3PS	018181-70-9	IODOFENPHOS
C8H8Cl2N2O	003567-62-2	3-(3,4-DICHLOROPHENYL)-1-METHYLUREA
C8H8Cl2N4O	021724-58-3	3,4-DICHLOROPHENYLAMIDINOUREA
C8H8Cl2NaO5S	000136-78-7	SESONE
C8H8Cl2O2	000120-67-2	2-(2,4-DICHLOROPHENOXY)ETHANOL
C8H8Cl2O2	002675-77-6	1,4-DICHLORO-2,5-DIMETHOXYBENZENE
C8H8Cl3N3O4S2	000133-67-5	TRICHLORMETHIAZIDE
C8H8Cl3O3PS	000299-84-3	RONNEL
C8H8Cl4	083682-55-7	2,5-BIS(CHLOROMETHYL)-3,4-DICHLORO-1,3,5-HEXAT*
C8H8Cl5F3O	056400-11-4	3(CF3-MEO)PENTACHLOROCYCLOHEXANE
C8H8Cl5F3O	056421-36-4	1(CF3-MEO)PENTACHLOROCYCLOHEXANE
C8H8Cl6	083682-53-5	2,5-BIS(CHLOROMETHYL)-3,3,4,4-TETRACHLORO-1,5-*
C8H8ClN3	103748-25-0	1-METHYL-2-AMINO-5-CHLOROBENZIMIDAZOLE
C8H8ClN3O2S	037148-08-6	124-BENZTHIADIZ-1-O2-3-ME-6-NH2-7-CL
C8H8ClN3O4S2	001025-75-8	CHLOROTHIAZIDE-3-METHYL
C8H8ClN5O2	102632-30-4	HYDRAZINECARBOXIMIDAMIDE, 2-[(4-CHLORO-3-NITROPH
C8H8ClN5O3	046833-92-5	4-NO2-3-CHLOROPHENYLAMIDINOUREA
C8H8ClNO	000533-17-5	2-CHLOROACETANILIDE
C8H8ClNO	000539-03-7	P-CHLOROACETANILIDE
C8H8ClNO	000587-65-5	A-CHLOROACETANILIDE
C8H8ClNO	000588-07-8	3-CHLOROACETANILIDE
C8H8ClNO	101080-58-4	2-CHLORO-6-METHYLBENZAMIDE
C8H8ClNO2	000940-36-3	CARBAMIC ACID, (4-CHLOROPHENYL)-, METHYL ESTER
C8H8ClNO2	002150-88-1	ME-N-(M-CHLOROPHENYL)CARBAMATE
C8H8ClNO2	002620-53-3	N-METHYL-4-CHLOROPHENYLCARBAMATE
C8H8ClNO2	002621-80-9	P-CHLOROBENZYLCARBAMATE
C8H8ClNO2	003942-54-9	N-ME-2-CHLOROPHENYLCARBAMATE
C8H8ClNO2	004090-00-0	N-METHYL-3-CHLOROPHENYLCARBAMATE
C8H8ClNO2	005202-89-1	2-AMINO-5-CHLORO-METHYLBENZOATE
C8H8ClNO2	020668-13-7	CARBAMIC ACID, (2-CHLOROPHENYL)-, METHYL ESTER
C8H8ClNO2	107485-43-8	2-CHLORO-6-METHOXYBENZAMIDE
C8H8F3N3O4S2	000135-09-1	3,4-DIHYDRO-6-(TRIFLUOROMETHYL)-2H(1,2,4)-BENZO*
C8H8F3NO2S	037595-73-6	P-ME CF3-METHANESULFONANILIDE
C8H8F3NO2S2	023375-06-6	P-MES CF3-METHANESULFONANILIDE
C8H8F3NO3S	023384-33-0	M-MEO CF3-METHANESULFONANILIDE

C8H8F3NO3S	023384-34-1	P-MEO CF3-METHANESULFONANILIDE
C8H8F3NO4S2	023375-08-8	M-MESO2 CF3-METHANESULFONANILIDE
C8H8F3NO4S2	023375-10-2	P-MESO2 CF3-METHANESULFONANILIDE
C8H8FN3O2	000777-59-3	1-ME-1-NITROSO-3-(P-F PH)UREA
C8H8FNO	000351-28-0	3-FLUOROACETANILIDE
C8H8FNO	000351-83-7	4-FLUOROACETANILIDE
C8H8FNO	000399-31-5	2-FLUOROACETANILIDE
C8H8FNO2	000704-73-4	N-METHYL-2-FLUOROPHENYLCARBAMATE
C8H8FNO2	000705-48-6	N-METHYL-3-FLUOROPHENYLCARBAMATE
C8H8FNO2	000705-70-4	N-METHYL-4-FLUOROPHENYLCARBAMATE
C8H8FNO2	016664-12-3	CARBAMIC ACID, (2-FLUOROPHENYL)-, METHYL ESTER
C8H8FNO2	016744-99-3	CARBAMIC ACID, (4-FLUOROPHENYL)-, METHYL ESTER
C8H8FNO2	072755-13-6	CARBAMIC ACID, (3-FLUOROPHENYL)-, METHYL ESTER
C8H8FNO3S	000329-20-4	P-ACETAMIDO-BENZENESO2-FLUORIDE
C8H8HgO2	000062-38-4	PHENYLMERCURIC ACETATE
C8H8INO	000622-50-4	4-IODOACETANILIDE
C8H8INO	019230-45-6	ACETAMIDE, N-(3-IODOPHENYL)-
C8H8INO	019591-17-4	2-IODOACETANILIDE
C8H8INO2	013538-28-8	N-METHYL-2-IODOPHENYLCARBAMATE
C8H8INO2	013538-51-7	N-METHYL-4-IODOPHENYLCARBAMATE
C8H8INO2	013941-09-8	N-METHYL-3-IODOPHENYLCARBAMATE
C8H8INO2	103970-46-3	CARBAMIC ACID, (2-IODOPHENYL)-, METHYL ESTER
C8H8INO2	113932-81-3	CARBAMIC ACID, (3-IODOPHENYL)-, METHYL ESTER
C8H8N2	000615-15-6	2-METHYLBENAIMIDAZOLE
C8H8N2	005192-03-0	5-AMINOINDOLE
C8H8N2O	019776-98-8	BENZOXAZOLE, 2-METHYLAMINO-
C8H8N2O	023443-10-9	4H-PYRIDO(1,2-A)PYRIMIDIN-4-ONE
C8H8N2O2	000088-96-0	1,2-BENZENEDICARBOXAMIDE
C8H8N2O2	000524-40-3	RICIDINE
C8H8N2O2	001740-57-4	ISO-PHTHALAMIDE
C8H8N2O2	019727-83-4	1H-INDOLE, 2,3-DIHYDRO-6-NITRO-
C8H8N2O2	032692-19-6	1H-INDOLE, 2,3-DIHYDRO-5-NITRO-
C8H8N2O2S	000360-81-6	124-BENZOTHIADIAZINE-1,1-O2-3-ME
C8H8N2O3	000104-04-1	4-NITROACETANILIDE
C8H8N2O3	000122-28-1	M-NITROACETANILIDE
C8H8N2O3	000552-32-9	2-NITROACETANILIDE
C8H8N2O3	006306-25-8	BENZOIC ACID, 4-[(AMINOCARBONYL)AMINO]-
C8H8N2O3	022026-39-7	1,3-BENZODIOLE-5-CARBOXYLIC ACID, HYDRAZIDE
C8H8N2O4	001943-87-9	CARBAMIC ACID, (4-NITROPHENYL)-, METHYL ESTER
C8H8N2O4	002189-61-9	CARBAMIC ACID, (3-NITROPHENYL)-, METHYL ESTER
C8H8N2O4	005819-21-6	N-METHYL-4-NITROPHENYLCARBAMATE
C8H8N2O4	006132-21-4	N-METHYL-3-NITROPHENYLCARBAMATE
C8H8N2O4	006164-77-8	2,3-PYRAZINEDICARBOXYLIC ACID, DIMETHYL ESTER
C8H8N2O4	007374-06-3	N-METHYL-2-NITROPHENYLCARBAMATE
C8H8N2O4	013725-30-9	CARBAMIC ACID, (2-NITROPHENYL)-, METHYL ESTER
C8H8N2O4	064431-77-2	2-NITRO-6-METHOXYBENZAMIDE
C8H8N2O5	000610-54-8	2,4-DINITROPHENETOLE
C8H8N2OS	119005-23-1	2-AZETIDINONE, 4-(2-PYRIDINYLTHIO)-
C8H8N2S	025445-77-6	4-PHENYL-1,2,3-THIADIAZOLE
C8H8N3O	001455-84-1	5(2-PYRIDYL)-3-ME-1,2,4-OXADIAZOLE
C8H8N3O	010350-68-2	3(2-PYRIDYL)-5-ME-1,2,4-OXADIAZOLE
C8H8N3O	066079-85-4	2(2-PYRIDYL)-5-ME-1,3,4-OXADIAZOLE
C8H8N4	000086-54-4	HYDRALAZINE
C8H8N4	001899-48-5	2,4-DIAMINOQUINAZOLINE
C8H8N4	018489-25-3	1H-TETRAZOLE, 5-(PHENYLMETHYL)-
C8H8N4	041410-39-3	1-PHENYL-3-CYANOGUANIDINE
C8H8N4O	069218-56-0	1-CARBAMYLMETHYLBENZOTRIAZOLE
C8H8N4O2	096448-61-2	ALLOPURINOL, 1-(OXOPROPYL)-

C8H8N4O3	098846-64-1	ALLOPURINOL, 1-[(ACETYLOXY)METHYL]-
C8H8N4O4	021959-57-9	2-(5-NO2-2-FURYLVINYL)5-NH2-3-OXOIMDAZOLINE
C8H8O	000096-09-3	STYRENE OXIDE
C8H8O	000098-86-2	ACETOPHENONE
C8H8O	000104-87-0	P-TOLUALDEHYDE
C8H8O	000122-78-1	PHENYLACETALDEHYDE
C8H8O	000496-14-0	PHTHALAN
C8H8O	000496-16-2	2,3-DIHYDROBENZOFURAN
C8H8O	000529-20-4	2-METHYLBENZALDEHYDE
C8H8O2	000093-58-3	BENZOIC ACID, METHYL ESTER
C8H8O2	000099-04-7	M-TOLUIC ACID
C8H8O2	000099-93-4	P-HYDROXYACETOPHENONE
C8H8O2	000099-94-5	P-TOLUIC ACID
C8H8O2	000103-82-2	PHENYLACETIC ACID
C8H8O2	000118-90-1	O-TOLUIC ACID
C8H8O2	000118-93-4	O-HYDROXYACETOPHENONE
C8H8O2	000121-71-1	M-HYDROXYACETOPHENONE
C8H8O2	000122-79-2	PHENYL ACETATE
C8H8O2	000123-11-5	P-ANISALDEHYDE
C8H8O2	000135-02-4	BENZALDEHYDE, 2-METHOXY-
C8H8O2	000137-18-8	2,5-CYCLOHEXADIENE-1,4-DIONE, 2,5-DIMETHYL-
C8H8O2	000493-09-4	1,4-BENZODIOXAN
C8H8O2	000526-86-3	2,5-CYCLOHEXADIENE-1,4-DIONE, 2,3-DIMETHYL-
C8H8O2	000527-61-7	2,5-CYCLOHEXADIENE-1,4-DIONE, 2,6-DIMETHYL
C8H8O2	000591-31-1	BENZALDEHYDE, 3-METHOXY-
C8H8O2	000623-15-4	3-BUTEN-2-ONE, 4-(2-FURANYL)-
C8H8O2S	000103-04-8	S-PHENYLMERCAPTOACETIC ACID
C8H8O2S	013205-48-6	P-METHIOBENZOIC ACID
C8H8O3	000050-85-1	4-METHYLSALICYLIC ACID
C8H8O3	000083-40-9	3-METHYLSALICYLIC ACID
C8H8O3	000089-56-5	5-METHYLSALICYLIC ACID
C8H8O3	000090-64-2	A-HYDROXYPHENYLACETIC ACID
C8H8O3	000099-76-3	4-HYDROXY METHYL BENZOATE
C8H8O3	000100-09-4	P-METHOXYBENZOIC ACID
C8H8O3	000102-29-4	M-HYDROXYPHENYL ACETATE
C8H8O3	000119-36-8	METHYL SALICYLATE
C8H8O3	000121-33-5	VANILLIN
C8H8O3	000122-59-8	PHENOXYACETIC ACID
C8H8O3	000148-53-8	O-VANILLIN
C8H8O3	000156-38-7	P-HYDROXYPHENYLACETIC ACID
C8H8O3	000495-76-1	1,3-BENZODIOXOLE-5-METHANOL
C8H8O3	000579-75-9	O-METHOXYBENZOIC ACID
C8H8O3	000586-38-9	M-METHOXYBENZOIC ACID
C8H8O3	000614-75-5	O-HYDROXYPHENYLACETIC ACID
C8H8O3	000621-37-4	M-HYDROXYPHENYLACETIC ACID
C8H8O3	000621-59-0	ISOVANILLIN
C8H8O3	019438-10-9	METHYL M-HYDROXYBENZOATE
C8H8O4	000102-32-9	3,4-DIHYDROXYPHENYLACETIC ACID
C8H8O4	000121-34-6	4-HYDROXY-3-METHOXYBENZOIC ACID
C8H8O4	000451-13-8	BENZENEACETIC ACID, 2,5-DIHYDROXY-
C8H8O4	000520-45-6	DEHYDROACETIC ACID
C8H8O4	000530-55-2	2,6-DIMETHOXY-P-BENZOQUINONE
C8H8O4	000877-22-5	3-METHOXYSALICYCLIC ACID
C8H8O4	001878-83-7	M-HYDROXYPHENOXYACETIC ACID
C8H8O4	001878-84-8	P-HYDROXYPHENOXYACETIC ACID
C8H8O4	002150-47-2	METHYL 2,4-DIHYDROXYBENZOATE
C8H8O4	003117-02-0	2,3-DIMETHOXY-P-BENZOQUINONE
C8H8O4	006324-11-4	O-HYDROXYPHENOXYACETIC ACID

C8H8O4S	004052-30-6	P-METHYLSULFONYLBENZOIC ACID
C8H8O5	000099-24-1	METHYL GALLATE
C8H8OS	000934-87-2	THIACETIC ACID, S-PHENYL ESTER
C8H8OS	085033-96-1	PHENYLTHIO-ACETIC ACID
C8H9Br	000103-63-9	(2-BROMOETHYL)BENZENE
C8H9Br	001585-07-5	1-BROMO-4-ETHYLBENZENE
C8H9BrN2O	078508-46-0	(3-METHYL-4-BROMOPHENYL)UREA
C8H9BrN2O	098736-47-1	BENZENEMETHANAMINE, 3-BROMO-N-METHYL-N-NITROSO-
C8H9BrN2O	098736-50-6	BENZENEMETHANAMINE, 4-BROMO-N-METHYL-N-NITROSO-
C8H9BrN2O2	051581-35-2	3-(N,N-DIMECARBAM)-2-BR-PYRIDINE
C8H9BrN2O3	077317-70-5	3-ALLYLO-5-BR-6-ME-PYRIMID-2,4-DION
C8H9BrN4	037873-43-1	HYDRAZINECARBOXIMIDAMIDE, 2-[(4-BROMOPHENYL)METH
C8H9Cl	000104-82-5	P-(CHLOROMETHYL)TOLUENE
C8H9Cl	000615-60-1	3,4-DIMETHYLCHLOROBENZENE
C8H9Cl	000622-24-2	B-PHENYL ETHYL CHLORIDE
C8H9Cl	000672-65-1	(1-CHLOROETHYL)BENZENE
C8H9Cl2N3O4S2	001824-47-1	3-CHLOROMETHYLHYDROCHLOROTHIAZIDE
C8H9Cl2N3O4S2	023141-87-9	3-DICHLOROMETHYLHYDROTHIAZIDE
C8H9Cl2NO3	002307-55-3	2,4-D, AMINE SALT
C8H9Cl3	083682-63-7	1,4-DICHLORO-4-(1-CHLOROETHENYL)CYCLO-1-HEXENE
C8H9Cl3NO3P	002213-70-9	O=P(OME)(NME)O-2,4,5-CLPHENYL
C8H9Cl5O2	060423-87-2	1-ACETOXYPENTACHLOROCYCLOHEXANE
C8H9ClN2O	020940-42-5	N-(3-CHLOROPHENYL)-N'-METHYLUREA
C8H9ClN2O	084174-22-1	BENZENEMETHANAMINE, 4-CHLORO-N-METHYL-N-NITROSO-
C8H9ClN2O	098736-46-0	BENZENEMETHANAMINE, 3-CHLORO-N-METHYL-N-NITROSO-
C8H9ClN2O2	025277-05-8	(3-CHLORO-4-METHOXYPHENYL)UREA
C8H9ClN2O2	051581-34-1	3-(N,N-DIMECARBAM)-2-CL-PYRIDINE
C8H9ClN2O3S	052102-43-9	BENZENESULFONAMIDE, 4-CHLORO-N-[(METHYLAMINO)CAR
C8H9ClN4	013308-88-8	HYDRAZINECARBOXIMIDAMIDE, 2-[(4-CHLOROPHENYL)MET
C8H9ClN4O	058247-24-8	4-CHLOROPHENYLAMIDINO UREA
C8H9ClNO5PS	000500-28-7	3-CHLORO-DIMETHYL PARATHION
C8H9ClNO5PS	002463-84-5	DICAPTHON
C8H9ClNO6P	002255-15-4	3-CHLORO-DIMETHYL PARA-OXON
C8H9ClNO6P	017650-76-9	2-CHLORO-DIMETHYL PARA-OXON
C8H9ClO	000088-04-0	4-CHLORO-3,5-DIMETHYL PHENOL
C8H9ClO2	002100-42-7	1,4-DIMETHOXY-2-CHLOROBENZENE
C8H9ClO2	059365-60-5	1-CHLORO-2-(1,2-ETHANEDIOL)BENZENE
C8H9FN2O	078508-45-9	(4-FLUORO-3-METHYLPHENYL)UREA
C8H9FN2O	084174-21-0	BENZENEMETHANAMINE, 4-FLUORO-N-METHYL-N-NITROSO-
C8H9FN2O3	017902-23-7	TEGAFUR
C8H9FN2O3	094452-21-8	2,4(1H,3H)-PYRIMIDINEDIONE, 5-FLUORO-3-(1-OXOBUT
C8H9FN2O5	106206-96-6	CARBONIC ACID, ETHYL (5-FLUORO-3,4-DIHYDRO-2,4-D
C8H9FN2O5	106206-97-7	CARBONIC ACID, ETHYL (5-FLUORO-3,6-DIHYDRO-2,6-D
C8H9IN2O2	051581-36-3	3-(N,N-DIMECARBAM)-2-I-PYRIDINE
C8H9N	000140-76-1	2-METHYL-5-VINYLPYRIDINE
C8H9N	000622-29-7	METHANAMINE, N-(PHENYLMETHYLENE)-
C8H9N3	001622-57-7	1-METHYL-2-AMINOBENZIMIDAZOLE
C8H9N3	006285-68-3	2-AMINO-5-METHYLBENZIMIDAZOLE
C8H9N3O	001574-10-3	HYDRAZINECARBOXAMIDE, 2-(PHENYLMETHYLENE)-
C8H9N3O	006232-91-3	2-AMINO-5-METHOXYBENZIMIDAZOLE
C8H9N3O2	002845-79-6	1-BENZOYL-2-CARBAMYLHYDRAZINE
C8H9N3O2	004144-68-7	2-CARBOXYMETHYLBENZOTRIAZOLE
C8H9N3O2	006850-22-2	2-METHYL-6-NITROINDAZOLE
C8H9N3O2	017433-31-7	1-ACETYL-2-PICOLINOYL HYDRAZINE
C8H9N3O2	021561-99-9	1-NITROSO-1-METHYL-3-PHENYLUREA
C8H9N3O3	073469-93-9	BENZOIC ACID, 2-[(HYDRAZINECARBONYL)AMINO]-
C8H9N3O3	073469-94-0	BENZOIC ACID, 3-[(HYDRAZINOCARBONYL)AMINO]-
C8H9N3O3	073469-95-1	BENZOIC ACID, 4-[(HYDRAZINOCARBONYL)AMINO]-

C8H9N3O3	084174-24-3	BENZENEMETHANAMINE, N-METHYL-4-NITRO-N-NITROSO-
C8H9N3O3	098736-52-8	BENZENEMETHANAMINE, N-METHYL-3-NITRO-N-NITROSO-
C8H9N3O4	051581-33-0	3-(N,N-DIMECARBAM)-2-NO2-PYRIDINE
C8H9N3O4S2	003256-99-3	METHYLTHIAZIDE
C8H9N3OS	007420-37-3	M-OH BENZALDEHYDETHIOSEMICARBIZONE
C8H9N3S	001627-73-2	BENZALDEHYDE THIOSEMICARBAZONE
C8H9N3S	063893-52-7	THIENO(23D)PYRIMIDINE-4-DEMEAMINO
C8H9N5O2	030068-29-2	HYDRAZINECARBOXIMIDAMIDE, 2-[(4-NITROPHENYL)METH
C8H9N5O2	090792-54-4	HYDRAZINECARBOXIMIDAMIDE, 2-[(3-NITROPHENYL)METH
C8H9N5O2	102632-31-5	HYDRAZINECARBOXIMIDAMIDE, 2-[(2-NITROPHENYL)METH
C8H9N5O3	016018-79-4	3-NITROPHENYLAMIDINO UREA
C8H9NO	000093-61-8	N-METHYLFORMANILIDE
C8H9NO	000099-03-6	M-AMINOACETOPHENONE
C8H9NO	000099-92-3	4-AMINOACETOPHENONE
C8H9NO	000103-81-1	PHENYLACETAMIDE
C8H9NO	000103-84-4	ACETANILIDE
C8H9NO	000527-85-5	2-METHYLBENZAMIDE
C8H9NO	000551-93-9	O-AMINOACETOPHENONE
C8H9NO	000613-93-4	N-METHYLBENZAMIDE
C8H9NO	000618-47-3	3-METHYLBENZAMIDE
C8H9NO	000619-55-6	4-METHYLBENZAMIDE
C8H9NO	001701-69-5	ETHYL-4-PYRIDYL KETONE
C8H9NO	003085-54-9	P-METHYLFORMANILIDE
C8H9NO	003235-02-7	P-METHYLBENZALDOXIME
C8H9NO	003376-32-7	O-METHYLBENZALDOXIME
C8H9NO2	000056-91-7	4-AMINOMETHYLBENZOIC ACID
C8H9NO2	000069-91-0	ALPHA-PHENYLGLYCINE
C8H9NO2	000081-20-9	1,3-DIMETHYL-2-NITROBENZENE
C8H9NO2	000083-41-0	1,2-DIMETHYL-3-NITROBENZENE
C8H9NO2	000089-87-2	4-NITRO-M-XYLENE
C8H9NO2	000099-51-4	1,2-DIMETHYL-4-NITROBENZENE
C8H9NO2	000100-12-9	4-ETHYLNITROBENZENE
C8H9NO2	000103-01-5	N-PHENYLGLYCINE
C8H9NO2	000103-90-2	ACETAMIDE, N-(4-HYDROXYPHENYL)
C8H9NO2	000134-20-3	METHYL ANTHRANILATE
C8H9NO2	000612-22-6	2-ETHYLNITROBENZENE
C8H9NO2	000614-18-6	NICOTINIC ACID, ETHYL ESTER
C8H9NO2	000614-80-2	O-HYDROXYACETANILIDE
C8H9NO2	000619-45-4	METHYL P-AMINOBENZOATE
C8H9NO2	000621-42-1	M-HYDROXYACETANILIDE
C8H9NO2	000621-84-1	O-BENZYL CARBAMATE
C8H9NO2	000621-88-5	PHENOXYACETAMIDE
C8H9NO2	001570-45-2	ISONICOTINIC ACID, ETHYL ESTER
C8H9NO2	001943-79-9	N-METHYLPHENYLCARBAMATE
C8H9NO2	002439-77-2	O-METHOXYBENZAMIDE
C8H9NO2	002486-70-6	4-AMINO-M-TOLUIC ACID
C8H9NO2	002486-75-1	BENZOIC ACID, 4-AMINO-2-METHYL-
C8H9NO2	002524-52-9	PICOLINIC ACID, ETHYL ESTER
C8H9NO2	002603-10-3	METHYL N-PHENYLCARBAMATE
C8H9NO2	002835-06-5	BENZENEACETIC ACID, .ALPHA.-AMINO-, (.+-.)-
C8H9NO2	003424-93-9	P-METHYOXYBENZAMIDE
C8H9NO2	004746-61-6	O-HYDROXY PHENYLACETAMIDE
C8H9NO2	005813-86-5	M-METHOXYBENZAMIDE
C8H9NO2	006125-24-2	B-NITROETHYL BENZENE
C8H9NO2	006282-02-6	BENZAMIDE, N-(HYDROXYMETHYL)-
C8H9NO2	013871-68-6	P-AMINOPHENYL ACETATE
C8H9NO2	014008-60-7	BENZAMIDE, 2-HYDROXY-3-METHYL-
C8H9NO2	027153-17-9	M-METHOXYFORMANILIDE

C8H9NO3	000066-72-8	PYRIDOXAL
C8H9NO3	000100-27-6	2-(P-NITROPHENYL)ETHANOL
C8H9NO3	000100-29-8	P-NITROPHENETOLE
C8H9NO3	002423-71-4	2,6-DIMETHYL-4-NITROPHENOL
C8H9NO3	002486-80-8	BENZOIC ACID, 4-AMINO-2-METHOXY-
C8H9NO3	004837-88-1	1-MEO-2-ME-3-NO2-BENZENE
C8H9NO3S	001565-17-9	P-ACETYLBENZENESULFONAMIDE
C8H9NO3S	035203-88-4	BENZENESULFONAMIDE, 3-ACETYL-
C8H9NO4	001585-79-1	N-MALEOYLGLYCINE, ETHYL ESTER
C8H9NO4	057078-98-5	N-MALEOYL-4-AMINOBUTYRIC ACID
C8H9NO4	117491-56-2	1H-PYRROLE-2,5-DIONE, 1-[(ACETYLOXY)METHYL]-3-ME
C8H9NO4S	022808-73-7	4-SULFAMYLBENZOIC ACID, ME ESTER
C8H9NOS	002362-64-3	BENZENECARBOTHIAMIDE, 4-METHOXY-
C8H9NOS	013509-38-1	CARBANILIC ACID, THIO-, S-METHYL ESTER
C8H9NOS	054705-16-7	BENZAMIDE, 2-(METHYLTHIO)-
C8H9NS	000637-53-6	THIOACETANILIDE
C8H9NS	002362-62-1	BENZENECARBOTHIAMIDE, 4-METHYL-
C8H9NS2	053550-91-7	BENZENECARBOTHIAMIDE, 4-(METHYLTHIO)-
C8H9O3PS	003811-49-2	SALITHION
C8H10	000095-47-6	O-XYLENE
C8H10	000100-41-4	ETHYLBENZENE
C8H10	000106-42-3	P-XYLENE
C8H10	000108-38-3	M-XYLENE
C8H10BrN3O	073469-89-3	HYDRAZINECARBOXAMIDE, N-(3-BROMOPHENYL)
C8H10Cl2	013547-07-4	2-CHLORO-4-(1-CHLOROVINYL)CYCLO-1-HEXENE
C8H10Cl2	014112-00-6	1,2-DICHLORO-1,2-DIVINYLCYCLOBUTANE
C8H10Cl2	065122-21-6	1,4-DICHLORO-4-VINYLCYCLO-1-HEXENE
C8H10Cl2	083682-54-6	2,5-DIMETHYL-1,6-DICHLORO-1,3,5-HEXATRIENE
C8H10Cl2	083682-56-8	2,5-BIS(CHLOROMETHYL)-1,3,5-HEXATRIENE
C8H10Cl2	083682-65-9	1,5-DICHLOROCYCLO-1,4-OCTADIENE
C8H10Cl2NO3P	010363-40-3	O=P(NME)(OME)O-(2,4-DICLPHENYL)
C8H10Cl4	083682-50-2	2,5-DIMETHYL-3,3,4,4-TETRACHLORO-1,5-HEXADIENE
C8H10Cl4	083682-52-4	2,5-BIS(CHLOROMETHYL)-3,4-DICHLORO-1,5-HEXADIE*
C8H10Cl6	083682-64-8	1,2,4-TRICHLORO-4-(1,1,2-TRICHLOROETHYL)CYCLOH*
C8H10ClN3	020241-00-3	1-(O-CL PH)-3,3-DIMETHYLTRIAZENE
C8H10ClN3	020241-05-8	1-(3-CL PH)-3,3-DIMETHYLTRIAZENE
C8H10ClN3O4S2	000890-67-5	3-METHYLHYDROCHLOROTHIAZIDE
C8H10ClNO3	000065-22-5	PYRIDOXAL HYDROCHLORIDE
C8H10FNO3S2	108966-70-7	BENZENESULFONAMIDE, 3-FLUORO-4-[(2-HYDROXYETHYL)
C8H10FNO5S2	108966-76-3	BENZENESULFONAMIDE, 3-FLUORO-4-[(2-HYDROXYETHYL)
C8H10IN3	066974-58-1	1-(O-I PH)-3,3-DIMETHYLTRIAZENE
C8H10N2O	000063-99-0	M-TOLYLUREA
C8H10N2O	000122-80-5	P-AMINOACETANILIDE
C8H10N2O	000138-89-6	P-NITROSO-N,N-DIMETHYLANILINE
C8H10N2O	000538-32-9	BENZYLUREA
C8H10N2O	000937-40-6	N-BENZYL-N-METHYLNITROSOAMINE
C8H10N2O	001007-36-9	1-PHENYL-3-METHYLUREA
C8H10N2O	003619-22-5	BENZOYLHYDRAZINE, P-METHYL
C8H10N2O	004314-66-3	N-ETHYLNICOTINAMIDE
C8H10N2O	004559-87-9	1-METHYL-1-PHENYL UREA
C8H10N2O	006972-69-6	N,N-DIMETHYLNICOTINAMIDE
C8H10N2O	007658-80-2	BENZOYLHYDRAZINE, O-METHYL
C8H10N2O	013050-47-0	BENZOYLHYDRAZINE, M-METHYL
C8H10N2O	014805-91-5	2-PYRIDINECARBOXAMIDE,NN-DIMETHYL
C8H10N2O	065754-05-4	4H-PYRIDO[1,2-A]PYRIMIDIN-4-ONE, 6,7,8,9-TETRAHY
C8H10N2O	084199-91-7	2-PYRIDINEPROPANEAMIDE
C8H10N2O	084199-99-5	3-PYRIDINEPROPANEAMIDE
C8H10N2O	084200-07-7	4-PYRIDINEPROPANEAMIDE

C8H10N2O2	000100-23-2	P-NITRO-N,N-DIMETHYLANILINE
C8H10N2O2	000619-31-8	BENZENAMINE, N,N-DIMETHYL-3-NITRO-
C8H10N2O2	003290-99-1	BENZOYLHYDRAZINE, P-METHOXY
C8H10N2O2	003665-80-3	N-ETHYL-4-NITROBENZENAMINE
C8H10N2O2	005785-06-8	BENZOYLHYDRAZINE, M-METHOXY
C8H10N2O2	006265-73-2	N-HYDROXYETHYLNICOTINAMIDE
C8H10N2O2	007466-54-8	BENZOYLHYDRAZINE, O-METHOXY
C8H10N2O2	010439-77-7	N,2-DIMETHYL-4-NITROBENZENAMINE
C8H10N2O2	016947-63-0	2,6-DIMETHYL-4-NITROBENZENAMINE
C8H10N2O2	034761-82-5	3,5-DIMETHYL-4-NITROBENZENAMINE
C8H10N2O2	041110-34-3	PYRAZINECARBOXYLIC ACID, 5-METHYL-, ETHYL ESTER
C8H10N2O2	041110-39-8	PYRAZINECARBOXYLIC ACID, 6-METHYL-, ETHYL ESTER
C8H10N2O3	056209-30-4	7,9-DIAZASPIRO[4.5]DECANE-6,8,10-TRIONE
C8H10N2O3	136309-08-5	PYRAZINECARBOXYLIC ACID, 6-METHOXY-, ETHYL ESTER
C8H10N2O3S	000080-11-5	BENZENESULFONAMIDE, N,4-DIMETHYL-N-NITROSO-
C8H10N2O3S	000121-61-9	ACETAMIDE, N-[4-(AMINOSULFONYL)PHENYL]
C8H10N2O3S	000144-80-9	SULFANILACETAMIDE
C8H10N2O3S	010518-85-1	4-SULFAMYLBENZAMIDE,N-METHYL
C8H10N2O3S	032094-40-9	BENZYLCARBONYL-NH-SO2-NH2
C8H10N2O3S	052102-38-2	BENZENESULFONAMIDE, N-[(METHYLAMINO)CARBONYL]-
C8H10N2O3S	088497-95-4	3-METHYLSULFONYLPHENYLUREA
C8H10N2O3S	121822-75-1	ALPHA-TOLUENE SULFONYLUREA
C8H10N2O4	004450-03-7	4-PYRIMIDINECARBOXYLIC ACID, 1,2,3,6-TETRAHYDRO-
C8H10N2O4	004450-26-4	TEH-PYRIMIDIN26DIONE,4-IPR CARBOXYLATE
C8H10N2O4S	003337-71-1	ASULAM
C8H10N2O5S2	108966-58-1	BENZENESULFONAMIDE, 4-[(2-HYDROXYETHYL)THIO]-3-N
C8H10N2OS	085879-21-6	3-(METHYLTHIO)PHENYLUREA
C8H10N2S	002724-69-8	1-METHYL-3-PHENYLTHIOUREA
C8H10N2S	004104-75-0	UREA,1-METHYL-1-PHENYL-2-THIO
C8H10N3O3S	053207-66-2	1(5-NO2-2-FURYL)-2-MES IMIDAZOLINE
C8H10N4	003357-37-7	HYDRAZINECARBOXIMIDAMIDE, 2-(PHENYLMETHYLENE)-
C8H10N4O	018697-64-8	3-PYRAZOLIN-5-ONE, 3-METHYL-2-(4-PYRIMIDINYL)-
C8H10N4O	069218-61-7	2-CARBAMYLMETHYLBENZOTRIAZOLE
C8H10N4O2	000058-08-2	CAFFEINE
C8H10N4O2	001135-58-6	UREA, N,N"-1,3-PHENYLENEBIS-
C8H10N4O2	020241-06-9	1-(3-NO2 PH)-3,3-DIMETHYLTRIAZENE
C8H10N4O2	041078-02-8	1H-PURINE-2,6-DIONE, 3,9-DIHYDRO-3-PROPYL-
C8H10N4O2	042548-73-2	1-(4-CARBAMOYLPHENYL)-3-METHYL-3-HYDROXYTRIAZENE
C8H10N4O2	058481-02-0	2-ACETAMIDOISONIAZID
C8H10N4O2	125573-05-9	1H-PURINE-2,6-DIONE, 3,9-DIHYDRO-3-ETHYL-1-METHY
C8H10N4O3	005415-44-1	1,3,7-TRIMETHYL URIC ACID
C8H10N6	077314-76-2	2-PYRIDINAMINE, 4-(5-AMINO-1H-1,2,4-TRIAZOL-3-YL
C8H10N6O	066975-19-7	1H-IMIDAZOLE-4-CARBOXAMIDE, 5-[3-METHYL-3-(2- PR
C8H10NO5PS	000298-00-0	PARATHION METHYL
C8H10NO6P	000054-47-7	PYRIDOXAL-5-PHOSPHATE
C8H10NO6P	000950-35-6	METHYL PARAOXON
C8H10O	000060-12-8	2-PHENYLETHANOL
C8H10O	000090-00-6	O-ETHYLPHENOL
C8H10O	000095-65-8	3,4-DIMETHYLPHENOL
C8H10O	000095-87-4	2,5-DIMETHYLPHENOL
C8H10O	000098-85-1	ALPHA-METHYLBENZYL ALCOHOL
C8H10O	000100-84-5	M-METHYLANISOLE
C8H10O	000103-73-1	ETHOXYBENZENE
C8H10O	000104-93-8	P-METHYLANISOLE
C8H10O	000105-67-9	2,4-DIMETHYLPHENOL
C8H10O	000108-68-9	3,5-DIMETHYLPHENOL
C8H10O	000123-07-9	P-ETHYLPHENOL
C8H10O	000526-75-0	2,3-DIMETHYLPHENOL

C8H10O	000538-86-3	BENZYL METHYL ETHER
C8H10O	000576-26-1	2,6-DIMETHYLPHENOL
C8H10O	000578-58-5	2-METHYLANISOLE
C8H10O	000587-03-1	M-METHYLBENZYL ALCOHOL
C8H10O	000589-18-4	P-METHYL BENZYL ALCOHOL
C8H10O	000620-17-7	M-ETHYLPHENOL
C8H10O	001517-69-7	1-PHENYLETHANOL
C8H10O2	000091-16-7	1,2-DIMETHOXYBENZENE
C8H10O2	000093-51-6	2-METHOXY-4-METHYLPHENOL
C8H10O2	000094-71-3	O-ETHOXYPHENOL
C8H10O2	000105-13-5	P-METHOXYBENZYL ALCOHOL
C8H10O2	000122-99-6	2-PHENOXYETHANOL
C8H10O2	000150-78-7	P-DIMETHOXYBENZENE
C8H10O2	000151-10-0	M-DIMETHOXYBENZENE
C8H10O2	000501-94-0	2-(P-HYDROXYPHENYL)ETHANOL
C8H10O2	000608-43-5	1,4-BENZENEDIOL, 2,3-DIMETHYL-
C8H10O2	000612-16-8	O-METHOXYBENZYL ALCOHOL
C8H10O2	000621-34-1	M-ETHOXYPHENOL
C8H10O2	000622-08-2	ETHYLENE GLYCOL MONOBENZYL ETHER
C8H10O2	000622-62-8	P-ETHOXYPHENOL
C8H10O2	004383-07-7	BENZENEMETHANOL, 2-HYDROXY-5-METHYL-
C8H10O3	000091-10-1	2,6-DIMETHOXYPHENOL
C8H10O3	000500-99-2	3,5-DIMETHOXYPHENOL
C8H10O3	000615-10-1	3-FURANCARBOXYLIC ACID, PROPYL ESTER
C8H10O3S	000515-46-8	ETHYL BENZENESULFONATE
C8H10O3S	004075-58-5	THIOPHEN-2-COCO-O-ETHYL
C8H10O3S	043032-67-3	METHYLSULFONYLANISOLE
C8H10O4	003343-19-9	3,4-DIHYDROXYPHENYLGLYCOL
C8H10O4	052643-52-4	1,4-BENZENEDIOL, 2,3-DIMETHOXY-
C8H10O4S2	019723-86-5	PROPANEDIOIC ACID, 1,3-DITHIOLAN-2-YLIDENE,- DIM
C8H10O5	000145-73-3	ENDOTHAL
C8H10S	000622-38-8	PHENYL ETHYL SULFIDE
C8H11BrN2O2	000314-42-1	ISOCIL
C8H11BrN2O3	077317-68-1	3-PRO-5-BR-6-ME-PYRIMIDIN-2,4-DIONE
C8H11BrN2O3	077317-69-2	3-IPRO-5-BR-6-ME-PYRIMIDI-2,4-DIONE
C8H11Cl2N3O2	000066-75-1	URACIL MUSTARD
C8H11Cl2NO	037764-25-3	ACETAMIDE, 2,2-DICHLORO-N,N-DI-2-PROPENYL-
C8H11Cl3O6	015879-93-3	CHLORALOSE
C8H11Cl5O	056421-32-0	2-ETHOXYPENTACHLOROCYCLOHEXANE
C8H11Cl5O	056421-35-3	3-ETHOXYPENTACHLOROCYCLOHEXANE
C8H11ClN2O3	118709-72-1	5-CL-5-T-BU BARBITURIC ACID
C8H11ClN4O4	013909-02-9	1(2-CLET)-3(2-GLUTARIMIDYL)-1-NO-UREA
C8H11ClNO3P	019608-64-1	O=P(NC)(OC)O-2-CL PHENYL
C8H11ClNO3P	019670-19-0	O=P(NME)(OME)O-(4-CL-PHENYL)
C8H11ClSi	000768-33-2	CHLORODIMETHYLPHENYLSILANE
C8H11FSi	000454-57-9	FLUORODIMETHYLPHENYLSILANE
C8H11N	000064-04-0	2-PHENYLETHYLAMINE
C8H11N	000087-59-2	2,3-DIMETHYLANILINE
C8H11N	000087-62-7	2,6-DIMETHYLANILINE
C8H11N	000095-68-1	2,4-DIMETHYLANILINE
C8H11N	000095-78-3	2,5-DIMETHYLANILINE
C8H11N	000103-67-3	N-METHYLBENZYLAMINE
C8H11N	000103-69-5	N-ETHYLANILINE
C8H11N	000104-84-7	BENZENEMETHANAMINE, 4-METHYL-
C8H11N	000104-90-5	5-ETHYL-2-METHYLPYRIDINE
C8H11N	000108-69-0	3,5-DIMETHYLANILINE
C8H11N	000108-75-8	2,4,6-COLLIDINE
C8H11N	000121-69-7	N,N-DIMETHYLANILINE

C8H11N	000578-54-1	O-ETHYLANILINE
C8H11N	000589-16-2	P-ETHYLANILINE
C8H11N	000611-21-2	N-METHYL-O-TOLUIDINE
C8H11N	000623-08-5	N-METHYL P-TOLUIDINE
C8H11N	000696-44-6	BENZENAMINE, N,3-DIMETHYL-
C8H11N	001122-81-2	4-PROPYLPYRIDINE
C8H11N2O5P	054267-24-2	O=P(NME)(OME)-O-(4-NITROPHENYL)
C8H11N3	007227-91-0	N,N-DIMETHYLTRIAZENOBENZENE
C8H11N3	016584-04-6	2-ETHYLBENZOTRIAZOLE
C8H11N3	016584-05-7	1-ETHYLBENZOTRIAZOLE
C8H11N3NaO3S	000140-56-7	FENAMINOSULF
C8H11N3O	000938-56-7	1-(2-HYDROXYETHYL)BENZOTRIAZOLE
C8H11N3O	000939-72-0	2-(2-HYDROXYETHYL)BENZOTRIAZOLE
C8H11N3O	004608-25-7	2-ETHYLISONIAZID
C8H11N3O	015940-62-2	HYDRAZINECARBOXAMIDE, N-(3-METHYLPHENYL)-
C8H11N3O	015940-63-3	HYDRAZINECARBOXAMIDE, N-(2-METHYLPHENYL)-
C8H11N3O	016956-42-6	4-BENZYLSEMICARBAZIDE
C8H11N3O	059477-92-8	1-PHENYL-3,3-DIMETHYLTRIAZINE OXIDE
C8H11N3O	062774-57-6	HYDRAZINECARBOXAMIDE, N-(4-METHYLPHENYL)-
C8H11N3O2	040262-54-2	PYRAZINECARBOXYLIC ACID, 6-(DIMETHYLAMINO)-, MET
C8H11N3O2	058481-00-8	2-ETHOXYISONIAZID
C8H11N3O2	062774-58-7	HYDRAZINECARBOXAMIDE, N-(2-METHOXYPHENYL)-
C8H11N3O2	062774-59-8	HYDRAZINECARBOXAMIDE, N-(4-METHOXYPHENYL)-
C8H11N3O2	073469-85-9	HYDRAZINECARBOXAMIDE, N-(3-METHOXYPHENYL)-
C8H11N3O3S	121822-80-8	3-PYRIDINESULFONAMIDE, N-METHYL-N-[(METHYLAMINO)
C8H11N3O4	013182-82-6	METRONIDAZOLE ACETATE
C8H11N3O4S2	001824-46-0	6-METHYLHYDROTHIAZIDE
C8H11N3O6	000054-25-1	6-AZAURIDINE
C8H11N3O6	017306-43-3	1-(B-D-RIBOFURANOSYL)-2-NO2IMIDAZOLE
C8H11N3O6	040919-33-3	URICYTIN
C8H11N3S	013207-50-6	4-METHYL-1-PHENYL-3-THIOSEMICARBAZIDE
C8H11N5O2	084408-37-7	ETHANOL, 2-[(2-AMINO-9H-PURIN-9-YL)METHOXY]-
C8H11N5O2S	031052-77-4	2(1ET-13-IMIDAZOLINYLIDEN-2AM)5NO2-THIAZOLE
C8H11N5O2S	031052-78-5	2(14ME-13-IMIDAZOLN-2AM)5NO2-THIAZOLE
C8H11N5O2S	031052-79-6	2(13DIME-13-IMIDAZOLINYLIDEN-2AM)-5NO2-THIAZOLE
C8H11N5O2S	031052-84-3	2(15DIME-13-IMIDAZOLINYLIDEN-2AM)-5NO2-THIAZOLE
C8H11N5O3S	024240-70-8	2(1-OHET-1,3-IMIDAZOLINYLIDEN-2-AMINO)5NO2-THIAZ
C8H11N5O4	104958-84-1	4-[(3-NITRO-1,2,4-TRIAZOL-1-YL)ACETYL]-MORPHOLIN
C8H11N7O	061566-10-7	AMBAZONE [SEMICARBAZONE]
C8H11N7S	000539-21-9	AMBAZONE
C8H11NO	000051-67-2	4-(2-AMINOETHYL)PHENOL
C8H11NO	000094-70-2	2-ETHOXYANILINE
C8H11NO	000099-07-0	PHENOL, 3-(DIMETHYLAMINO)-
C8H11NO	000102-50-1	BENZENAMINE, 4-METHOXY-2-METHYL-
C8H11NO	000104-10-9	P-AMINOPHENYLETHANOL
C8H11NO	000120-71-8	2-METHOXY-5-METHYLBENZENAMINE
C8H11NO	000122-98-5	2-ANILINOETHANOL
C8H11NO	000156-43-4	P-ETHOXYANILINE
C8H11NO	002629-72-3	4-PYRIDINEPROPANOL
C8H11NO	002859-67-8	3-PYRIDINEPROPANOL
C8H11NO	002859-68-9	2-PYRIDINEPROPANOL
C8H11NO	007568-93-6	PHENYLETHANOLAMINE
C8H11NO	101870-22-8	2-PROPOXYPYRIDINE
C8H11NO2	000051-61-6	DOPAMINE
C8H11NO2	000102-56-7	2,5-DIMETHOXYANILINE
C8H11NO2	000104-14-3	NORTRON
C8H11NO2	002199-49-7	PYRROLE-3-CARBOXYLIC ACID-4-METHYL,ETHYL ESTER
C8H11NO2	060708-27-2	2-FURANCARBOXAMIDE, N-PROPYL-

C8H11NO2	104764-52-5	N-PROPYL-HYDROXYPYRID-4-ONE
C8H11NO2	104764-53-6	N-ISOPROPYL-HYDROXYPYRID-4-ONE
C8H11NO2S	000138-38-5	4-ETHYLBENZENESULFONAMIDE
C8H11NO2S	000640-61-9	BENZENESULFONAMIDE, N,4-DIMETHYL-
C8H11NO2S	013229-35-1	METHANESULFONAMIDE, N-METHYL-N-PHENYL-
C8H11NO2S	014417-01-7	N,N-DIMETHYLBENZENESULFONAMIDE
C8H11NO2S2	000741-58-2	BENSULIDE
C8H11NO3	000051-41-2	(-)-NORADRENALINE
C8H11NO3	000065-23-6	PYRIDOXINE
C8H11NO3	030652-21-2	4(1H)-PYRIDINONE, 3-HYDROXY-1-(2-HYDROXYETHYL)-2
C8H11NO3	131042-60-9	2-AZASPIRO[4.4]NONANE-1,3-DIONE, 2-HYDROXY-
C8H11NO3S2	108966-48-9	BENZENESULFONAMIDE, 4-[(2-HYDROXYETHYL)THIO]-
C8H11NO5S2	108966-49-0	BENZENESULFONAMIDE, 4-[(2-HYDROXYETHYL)SULFONYL]
C8H11NO6S3	135832-36-9	ACETIC ACID, 2-[[5-(AMINOSULFONYL)-2-THIENYL]SUL
C8H11NS	004946-22-9	P-(DIMETHYLAMINO)BENZENETHIOL
C8H11O3P	002240-41-7	O,O-DIMETHYLPHENYLPHOSPHONATE
C8H11O3PS	033576-92-0	O,O-DIMETHYL-O-PHENYLPHOSPHOROTHIOATE
C8H11O4P	010113-28-7	O,O-DIMETHYL-O-PHENYLPHOSPHATE
C8H11P	000672-66-2	(DIMETHYL)-PHENYLPHOSPHINE
C8H12	000100-40-3	4-VINYLCYCLOHEXENE
C8H12	000111-78-4	CYCLOOCTA-1,5-DIENE
C8H12	000931-64-6	BICYCLO(2.2.2)OCT-2-ENE
C8H12	002422-85-7	1,2-DIVINYLCYCLOBUTANE
C8H12	004916-63-6	2,5-DIMETHYL-1,3,5-HEXATRIENE
C8H12Br4	003322-93-8	1,2-DIBROMO-4-(1,2-DIBROMOETHYL)CYCLOHEXANE
C8H12Cl2	045803-84-7	4-VINYL-1,2-DICHLOROCYCLOHEXANE
C8H12Cl2	083682-48-8	1,6-DICHLORO-2,5-DIMETHYL-1,5-HEXADIENE
C8H12Cl2	083682-49-9	3,4-DICHLORO-2,5-DIMETHYL-1,5-HEXADIENE
C8H12Cl2	083682-51-3	2,5-BIS(CHLOROMETHYL)-1,5-HEXADIENE
C8H12Cl2O	083682-66-0	BIS(2-CHLORO-3-BUTENYL) ETHER
C8H12Cl4	051962-63-1	4-(1,2-DICHLOROETHYL)-1,2-DICHLOROCYCLOHEXANE
C8H12Cl4	056421-45-5	1,4-DIMETHYLTETRACHLOROCYCLOHEXANE
C8H12Cl4	083682-62-6	1,2-BIS(1,2-DICHLOROETHYL)CYCLOBUTANE
C8H12Cl4O2	056400-12-5	2,3-DIMEO TETRACHLOROCYCLOHEXANE
C8H12Cl4O2	056400-36-3	3,6-DIMEO TETRACHLOROCYCLOHEXANE
C8H12Cl4O2	060132-40-3	1,2-DIMEO TETRACHLOROCYCLOHEXANE
C8H12Cl4OS	060067-85-8	2-MEO 3-MES TETRACHLOROCYCLOHEXANE
C8H12ClN3O6	052320-85-1	CLET-NITROSOUREIDOGLUTARIC ACID
C8H12ClN5	022936-85-2	1,3,5-TRIAZINE-2,4-DIAMINE, 6-CHLORO-N-CYCLOPROP
C8H12ClNO	000093-71-0	2-CHLORO-N,N-DIALLYLACETAMIDE
C8H12ClNO3	000055-27-6	NORADRENALINE (HCL)
C8H12ClNO3	000058-56-0	PYRIDOXINE HYDROCHLORIDE
C8H12F3N3O4	021787-91-7	1(2-OH-6-CF3-ETOPR)-2-NO2-IMIDAZOLE
C8H12MnN4S8Zn	008018-01-7	MANCOZEB
C8H12N2	000099-98-9	N,N-DIMETHYL-P-PHENYLENEDIAMINE
C8H12N2	000629-40-3	OCTANEDINITRILE
C8H12N2	001124-11-4	PYRAZINE, TETRAMETHYL-
C8H12N2	002055-21-2	N,N-DIMETHYL-3-PYRIDYLMETHYLAMINE
C8H12N2	003000-75-7	N-ETHYL-3-PYRIDYLMETHYLAMINE
C8H12N2	005638-76-6	2-PYRIDINEETHANAMINE, N-METHYL-
C8H12N2	015583-16-1	2-PYRIDINEPROPANEAMINE
C8H12N2	015707-24-1	PYRAZINE, 2,3-DIETHYL-
C8H12N2	015986-80-8	PYRAZINE, 2-METHYL-3-PROPYL-
C8H12N2	029460-92-2	2-ISOBUTYLPYRAZINE
C8H12N2	030532-36-6	4-PYRIDINEPROPANEAMINE
C8H12N2	041038-69-1	3-PYRIDINEPROPANEAMINE
C8H12N2O	025773-40-4	2-ISOPROPYL-3-METHOXYPYRAZINE
C8H12N2O	067845-28-7	2-METHYL-6-PROPOXYPYRAZINE

C8H12N2O	089770-36-5	3-BUTOXYPYRIDAZINE
C8H12N2O	094089-22-2	PYRAZINE, 2-METHYL-3-(1-METHYLETHOXY)-
C8H12N2O	120309-46-8	1H-PYRROLE-2-CARBOXAMIDE, N-PROPYL-
C8H12N2O	124725-22-0	1H-PYRROLE-2-CARBOXAMIDE, N-ETHYL-1-METHYL-
C8H12N2O2	000085-87-0	PYRIDOXAMINE
C8H12N2O2	000524-36-7	PYRIDOXAMINE
C8H12N2O2	000822-06-0	HEXAMETHYLENE DIISOCYANATE
C8H12N2O2	033870-85-8	PYRAZINE, 2,6-DIETHOXY-
C8H12N2O2S	000077-32-7	THIOBARBITAL
C8H12N2O2S	000710-15-6	PHENETHYL SULFAMIDE
C8H12N2O2S	001709-59-7	N1-DIMETHYLSULFANILAMIDE
C8H12N2O2S	006162-21-6	N',N'-DIMETHYLSULFANILAMIDE
C8H12N2O3	000057-44-3	BARBITAL
C8H12N2O3	001953-33-9	BARBITURIC ACID, 5-BUTYL
C8H12N2O3	053943-59-2	BARBITURIC ACID,5-ME-5-IPR
C8H12N2O3	077317-85-2	PYRIMIDINE-2,4-DIONE,3-PRO-6-ME
C8H12N2O3	090197-63-0	BARBITURIC ACID, 5-T-BU-
C8H12N2O4S	000154-97-2	PYRIDINIUM, 2-[(HYDROXYIMINO)METHYL]-1-METHYL-,
C8H12N2O4S2	006966-38-7	M-(BIS-METHYLSUFONYLAMIDO)BENZENE
C8H12N2O5S2	135832-45-0	BENZENESULFONAMIDE, 3-AMINO-4-[(2-HYDROXYETHYL)S
C8H12N4	020980-22-7	PYRIMIDINE, 2-(1-PIPERAZINYL)-
C8H12N4O2	054831-57-1	HISTIDINE-AMIDE,N-ACETYL
C8H12N4O2S	055469-64-2	4(33-DIMETRIAZENO)BENZENESULFONAMIDE
C8H12N4O3	033130-55-1	6-NH2-5-(N-METHYLFORMYLAMINO)-1,3-DIMETHYLURACIL
C8H12N4O3	060142-00-9	N-(DIAZOACETYLAMINOACETYL)MORPHOLINE
C8H12N4O3S	042116-76-7	CARNIDAZOLE
C8H12N4O3S2	024240-67-3	2(1-HOET-13THIAZOLIDINYLIDEN-2N)5NO2-THIAZOLE
C8H12N4O4	081892-67-3	1H-IMIDAZOLE-1-ACETAMIDE, N-(2-HYDROXYETHYL)-α-M
C8H12N4O4S	024240-65-1	2-(3-HOET-1,3-OXAZOLIDINYLIDEN-2N)-5-NO2-THIAZOL
C8H12N4O5	065886-71-7	1,3,5-TRIAZIN-2(1H)-ONE, 4-AMINO-1-(beta-D-ARABI
C8H12N4O5	074141-75-6	1H-IMIDAZOLE-1-ACETAMIDE, N-(2,3-DIHYDROXYPROPYL
C8H12N4O5	089690-76-6	1,2,4-BUTANETRIOL, 3-[(5-NITROPYRAZINYL)AMINO]-
C8H12N6O2	059277-86-0	ETHANOL, 2-[(2,6-DIAMINO-9H-PURIN-9-YL)METHOXY]-
C8H12NO5PS2	000115-93-5	CYTHIOATE
C8H12NO6P	000447-05-2	PYRIDOXINE 5'-PHOSPHATE
C8H12NO7P	041468-25-1	PYRIDOXAL-5'-PHOSPHATE MONOHYDRATE
C8H12O	000078-27-3	1-ETHYNYL CYCLOHEXANOL
C8H12O	000106-86-5	7-OXABICYCLO 4.1.0 HEPTANE, 3-ETHENYL-
C8H12O2	000106-87-6	7-OXABICYCLO 4.1.0 HEPTANE, 3-OXIRANYL-
C8H12O2	000126-81-8	5,5-DIMETHYL-1,3-CYCLOHEXANEDIONE
C8H12O4	000141-05-9	DIETHYL MALEATE
C8H12Si	000766-77-8	SILANE, DIMETHYL PHENYL
C8H13ClINO	075523-39-6	2-FURANMETHANAMINIUM, 5-BROMO-N,N,N-TRIMETHYL-,
C8H13ClINO	075523-40-9	2-FURANMETHANAMINIUM, 5-CHLORO-N,N,N-TRIMETHYL-,
C8H13N2O2	095579-22-9	O,N-DIME THPO
C8H13N2O2P	001754-58-1	o-PHENYL N,N'-DIMETHYL PHOSPHORODIAMIDATE
C8H13N2O3PS	000297-97-2	THIONAZIN
C8H13N2O5P	000529-96-4	PYRIDOXAMINE-5'-PHOSPHATE
C8H13N3OS	035045-02-4	DEAMINOMETRIBUZIN
C8H13N5	000707-98-2	9-PROPYL ADENINE
C8H13N5O2	036137-88-9	5(33DIME-1-TRIAZENO)IMDAZL-4-COOET
C8H13N5O2	080616-57-5	PROPANAMIDE, N,N'-1H-1,2,4-TRIAZOLE-3,5-DIYLBIS-
C8H13N5O2	087469-35-0	1-METHYLISOGUANINE,9[2(METHOXY)ME]-
C8H13N5O3	059277-89-3	ACYCLOVIR
C8H13N5O3	104958-89-6	N-BU-3-NO2-124-TRIAZOL-1-ACETAMIDE
C8H13NO2	000063-75-2	ARECOLINE
C8H13NO2	022707-38-6	2-PYRROLIDINONE, 1-(1-OXOBUTYL)-
C8H13NO2	023046-87-9	N-PROPIONYLCYCLOBUTANECARBOXAMIDE

C8H13NO2	090355-74-1	2,6-PIPERIDINEDIONE, 3-ETHYL-3-METHYL-
C8H13NO3S3	135832-38-1	2-THIOPHENESULFONAMIDE, 5-[(4-HYDROXYBUTYL)THIO]
C8H13NO5S3	135832-39-2	2-THIOPHENESULFONAMIDE, 5-[(4-HYDROXYBUTYL)SULFO
C8H14	000280-33-1	BICYCLO[2.2.2]OCTANE
C8H14	000627-58-7	2,5-DIMETHYL-1,5-HEXADIENE
C8H14	000629-05-0	1-OCTYNE
C8H14	000764-13-6	2,5-DIMETHYL-2,4-HEXADIENE
C8H14	001942-45-6	4-OCTYNE
C8H14	002809-67-8	2-OCTYNE
C8H14Cl3O5P	000126-22-7	BUTONATE
C8H14ClN3O2	013909-03-0	3-CYPE-1-(2-CLET)-1-NITROSOUREA
C8H14ClN3O2S	033022-01-4	1-NO-1-(2CLET)-3(4SULFAPYRANYL)UREA
C8H14ClN3O4S	033022-02-5	1-NO-1-(2CLET)-3(4THIOPYRANSO2)UREA
C8H14ClN5	001912-24-9	ATRAZINE
C8H14ClN5O	002904-53-2	135-TRIAZINE,N-ETOH,N'-IPR,4-CL-
C8H14ClNS2	000095-06-7	SULFALLATE
C8H14FN3O2S	032319-90-7	1(2-FLET)-3(4-SULFAPYRANYL)-1-NO UREA
C8H14INO	000541-64-0	2-FURANMETHANAMINIUM, N,N,N-TRIMETHYL-, IODIDE
C8H14INO	022601-05-4	3-FURANMETHANAMINIUM, N,N,N-TRIMETHYL-, IODIDE
C8H14N2	029460-93-3	ISOBUTYLPYRAZINE
C8H14N2	052222-71-6	4-PENTYLPYRAZOLE
C8H14N2O2	024847-46-9	PROLINE,N-ACETYL N'-MEAMINO-AMIDE
C8H14N2O2	068709-03-5	CARBAMIC ACID, (1,2,3,6-TETRAHYDRO-1-PYRIDINYL)-
C8H14N2O2S2	106231-39-4	1,4-DITHIAN-2-ONE, 3,3-DIMETHYL-, O-[(METHYLAMIN
C8H14N2O3S	054266-80-7	1,3-OXATHIOLAN-4-ONE, 5-PROPYL-, O-[(METHYLAMINO
C8H14N2O3S	054266-83-0	1,3-OXATHIOLAN-4-ONE, 5-(1-METHYLETHYL)-, O-[(ME
C8H14N2O4S	054266-84-1	1,3-OXATHIOLAN-4-ONE, 5-(2-METHOXYETHYL)-, O-[(M
C8H14N4	029604-73-7	3,6-PYRIDAZINAMINE-N,N'-TETRAME
C8H14N4	038717-44-1	PYRIDAZINE-3,5-NH2,N,N'-TETRAME
C8H14N4O	033665-20-2	1,2,4-TRIAZIN-5(4H)-ONE, 4-AMINO-6-(1,1-DIMETHYL
C8H14N4O2	060141-98-2	DIAZOACETYLGLYCINE-N'-IBU AMIDE
C8H14N4O3	050846-57-6	1-B-AMINOACETYLET-2ME-5-NO2IMIDAZOLE
C8H14N4OS	021087-62-7	3-SME-4-NH2-6-IBU-124-TRIAZIN-5-ONE
C8H14N4OS	021087-64-9	METRIBUZIN
C8H14N4OS	050917-22-1	3-ETS-4-NH2-6-IPR-1,2,4-TRIAZIN-5-ON
C8H14N4S2	038603-23-5	THIABURIMAMIDE
C8H14N6O	004574-37-2	IMIDAZOL-5-CONH2-4-(3,3-DIET-1-NNN)
C8H14N6O3	021244-66-6	IMIDAZOL-5-CONH2,4(33BIS(2OHET)1-NNN
C8H14NO	020863-54-1	3-ME,2-(N,N-DIMEAMINOME)FURAN
C8H14O	000502-49-8	CYCLOOCTANONE
C8H14O	000645-62-5	2-ETHYL-2-HEXENAL
C8H14O	002363-89-5	2-OCTENAL
C8H14O	003483-39-4	OXIRANE, CYCLOHEXYL-
C8H14O2	000097-86-9	METHACRYLIC ACID, I-BUTYL ESTER
C8H14O2	000097-88-1	METHACRYLIC ACID, N-BUTYL ESTER
C8H14O2	000140-77-2	CYCLOPENTYLPROPRIONIC ACID
C8H14O2	000142-30-3	2,5-DIMETHYL-3-HEXYNE-2,5-DIOL
C8H14O2	000585-07-9	T-BUTYL METHACRYLATE
C8H14O2	005292-21-7	CYCLOHEXYLACETIC ACID
C8H14O2	007307-04-2	5,5-DIMETHYL-2,4-HEXANEDIONE
C8H14O2S2	000062-46-4	1,2-DITHIOLANE-3-PENTANOIC ACID
C8H14O3	000106-31-0	BUTYRIC ANHYDRIDE
C8H14O3	002046-21-1	6-KETO METHYLHEPTANOATE
C8H14O3	050405-44-2	2H-PYRAN-2-ONE, 4-HYDROXY-3,5,6-TRIMETHYL-
C8H14O3	083682-67-1	1,1'-OXYBIS-3-BUTEN-2-OL
C8H14O3	083682-68-2	2,2'-OXYBIS-3-BUTEN-1-OL
C8H14O4	000123-25-1	DIETHYL SUCCINATE
C8H14O4	000505-48-6	SUBERIC ACID

C8H14O4	000616-88-6	PENTYLMALONIC ACID
C8H14O4	000626-86-8	HEXANEDIOIC ACID, MONOETHYL ESTER
C8H14O4	000627-93-0	HEXANEDIOIC ACID, DIMETHYL ESTER
C8H14O4	000828-00-2	2,6-DIMETHYL-1,3-DIOXAN-4-OL ACETATE
C8H14O6	000087-91-2	TARTARIC ACID, DIETHYL ESTER
C8H15Cl3	085269-46-1	1,2,3-TRICHLOROOCTANE
C8H15N	000124-12-9	OCTANONITRILE
C8H15N2O3PS	000108-35-0	PYRAZOTHION
C8H15N2O4P	000108-34-9	DIETHYL 3-METHYL-5-PYRAZOLYL PHOSPHATE
C8H15N3O3S2	086569-91-7	ETHANIMIDOTHIOIC ACID, N-[[[[(ACETYLMETHYLAMINO)
C8H15N3O4	060174-20-1	1-(2-OH-3-MEOPR)-2-ME-5-NO2IMIDAZOLE
C8H15N3O4	065988-94-5	1-(2-OH-3-ETOPR)-2-NO2 IMIDAZOLE
C8H15N3O4	071006-78-5	1-(2-OH-3-MEOPR)-2-ME-4-NO2IMIDAZOLE
C8H15N3O4S	019387-91-8	TINIDAZOLE
C8H15N3O4S	028795-24-6	ME(3(2ME-5NO2-1-IMIDAZ)PROPYL)SULFONE
C8H15N3O4S2	064055-10-3	ETHANIMIDOTHIOIC ACID, N-[[[[(METHOXYCARBONYLMET
C8H15N3O6	052019-05-3	ME-3-DEOXY-3(3ME3NOUREA)XYLOPYRANOSID
C8H15N3O7	018883-66-4	STREPTOZOCIN
C8H15N3O7	037793-01-4	ß-D-GALACTOPYRANOSE, 2-DEOXY-2-[[(METHYLNITROSOA
C8H15N3O7	037793-02-5	ß-D-MANNOPYRANOSE, 2-DEOXY-2-[[(METHYLNITROSOAMI
C8H15N3O7	066395-17-3	ß-D-GLUCOPYRANOSE, 2-DEOXY-2-[[(METHYLNITROSOAMI
C8H15N5O	000673-04-1	SIMETONE
C8H15N5O	002163-68-0	4-ETHYLAMINO-6-ISOPROPYLAMINO-S-TRIAZIN-2-OL
C8H15N5O	003035-45-8	G-32292 (ARATONE)
C8H15N5O	055921-65-8	24-NH2-6-PYRROLIDIN-PYRIMIDINE-3-OXID
C8H15N5O2	055921-64-7	2,4-PYRIMIDINEDIAMINE, 6-(4-MORPHOLINYL)-, 3-OXI
C8H15N5S	001014-69-3	DESMETRYNE
C8H15N5S	001014-70-6	SIMETRYN
C8H15N7O2S3	076824-35-6	FAMOTIDINE
C8H15NO	000120-29-6	ENDO-8-METHYL-8-AZABICYCLO(3.2.1)OCTAN-3-OL
C8H15NO	000935-30-8	2-AZACYCLO-NONANONE
C8H15NO	003829-78-5	ALLYL-ISOPROPYL ACETAMIDE
C8H15NO	019797-08-1	N-ETHYL-EPSILON-CAPROLACTAM
C8H15NO2	002867-47-2	2-(DIMETHYLAMINO)ETHYL 2-METHYL-2-PROPENOATE
C8H15NO2	018381-45-8	BUTANENITRILE, 4,4-DIETHOXY-
C8H15NO3	001522-00-5	N,N-DIETHYL SUCCINAMIC ACID
C8H15NS	021577-72-0	2-AZACYCLO-NONANTHIONE
C8H15NS	026019-18-1	2-CH2CH2NME2 THIOPHENE
C8H16	000107-39-1	2,4,4-TRIMETHYL-1-PENTENE
C8H16	000107-40-4	2,4,4-TRIMETHYL-2-PENTENE
C8H16	000111-66-0	1-OCTENE
C8H16	000111-67-1	2-OCTENE
C8H16	000292-64-8	CYCLOOCTANE
C8H16	000560-23-6	2,3,3-TRIMETHYL-1-PENTENE
C8H16	000565-77-5	2,3,4-TRIMETHYL-2-PENTENE
C8H16	000583-57-3	1,2-DIMETHYLCYCLOHEXANE
C8H16	000589-90-2	1,4-DIMETHYLCYCLOHEXANE
C8H16	000590-66-9	1,1-DIMETHYL CYCLOHEXANE
C8H16	000591-21-9	1,3-DIMETHYLCYCLOHEXANE
C8H16	000592-99-4	4-OCTENE
C8H16	000624-29-3	1,4-DIMETHYLCYCLOHEXANE (CIS)
C8H16	000627-97-4	2-METHYL-2-HEPTENE
C8H16	000638-04-0	1,3-DIMETHYLCYCLOHEXANE (CIS)
C8H16	000930-90-5	1-METHYL-2-ETHYLCYCLOPENTANE (TRANS)
C8H16	001632-16-2	2-ETHYL-1-HEXENE
C8H16	001678-91-7	ETHYLCYCLOHEXANE
C8H16	002040-96-2	PROPYLCYCLOPENTANE
C8H16	002207-01-4	1,2-DIMETHYLCYCLOHEXANE (CIS)

C8H16	002207-03-6	1,3-DIMETHYLCYCLOHEXANE (TRANS)
C8H16	002207-04-7	1,4-DIMETHYLCYCLOHEXANE (TRANS)
C8H16	003404-78-2	2,5-DIMETHYL-2-HEXENE
C8H16	003726-47-4	1-METHYL-3-ETHYLCYCLOPENTANE
C8H16	003875-51-2	ISOPROPYLCYCLOPENTANE
C8H16	004516-69-2	1,1,3-TRIMETHYLCYCLOPENTANE
C8H16	005026-76-6	6-METHYL-1-HEPTENE
C8H16	006876-23-9	1,2-DIMETHYLCYCLOHEXANE (TRANS)
C8H16	007145-20-2	2,3-DIMETHYL-2-HEXENE
C8H16	007642-04-8	2-OCTENE (CIS)
C8H16	013389-42-9	2-OCTENE (TRANS)
C8H16	015870-10-7	2-METHYL-1-HEPTENE
C8H16	015890-40-1	1(TRANS), 2(CIS), 3-TRIMETHYLCYCLOPENTANE
C8H16	015910-22-2	2,5-DIMETHYL-3-HEXENE
C8H16	025167-70-8	2,4,4-TRIMETHYLPENTENE
C8H16N2O	020917-50-4	N-NITROSO-OCTAMETHYLENEIMINE
C8H16N2O2	019701-84-9	VALINE,N-ACETYL,N'-MEAM-AMIDE
C8H16N2O2	028529-34-2	LEUCIN-AMIDE, N-ACETYL
C8H16N2O2	056711-06-9	ISO-LEUCIN-AMIDE, N-ACETYL
C8H16N3O	069741-94-2	2-NH2-5-HEXYL-1,3,4-OXADIAZOLE
C8H16N4O2	023264-57-5	2356-TETRAME-N,N'-DIN=O PIPERAZINE
C8H16N5S8Zn	009006-42-2	METIRAM
C8H16N6	016268-62-5	PENTAMETHYLMELAMINE
C8H16N6O	064124-14-7	1,3,5-TRIAZINE-2,4,6-TRIAMINE, N-HYDROXY-N,N',N'
C8H16N6O4	027640-22-8	1,1'-HEXAMETHYLENE-BIS(1-NITROSOUREA)
C8H16NO3PS2	000950-10-7	PHOSPHORAMIDIC ACID, (4-METHYL-1,3-DITHIOLAN-2-Y
C8H16NO5P	000141-66-2	DICROTOPHOS
C8H16O	000106-68-3	3-OCTANONE
C8H16O	000111-13-7	2-OCTANONE
C8H16O	000123-05-7	2-ETHYLHEXANAL
C8H16O	000124-13-0	OCTANAL
C8H16O	000541-85-5	ETHYL AMYLKETONE
C8H16O	001940-18-7	1-ETHYLCYCLOHEXANOL
C8H16O	003760-20-1	2-ETHYLCYCLOHEXANOL
C8H16O	004442-79-9	CYCLOHEXANEETHANOL
C8H16O	004630-06-2	6-METHYL-5-HEPTEN-2-OL
C8H16O	005337-72-4	2,6-DIMETHYLCYCLOHEXANOL
C8H16O	005441-52-1	3,5-DIMETHYLCYCLOHEXANOL
C8H16O	015045-43-9	2,2,5,5-TETRAMETHYL-TETRAHYDROFURAN
C8H16O	039170-83-7	CYCLOHEXANOL, 2,6-DIMETHYL-, (1α,2α,6ß)-
C8H16O	039170-84-8	CYCLOHEXANOL, 2,6-DIMETHYL-, (1α,2α,6α)-
C8H16O	042846-29-7	CYCLOHEXANOL, 2,6-DIMETHYL-, (1α,2ß,6ß)-
C8H16O2	000097-85-8	ISOBUTYL ISOBUTYRATE
C8H16O2	000099-66-1	VALPROIC ACID
C8H16O2	000106-73-0	METHYL HEPTANOATE
C8H16O2	000108-84-9	4-METHYL-2-PENTYL ACETATE
C8H16O2	000109-21-7	N-BUTYLBUTYRATE
C8H16O2	000123-66-0	ETHYL HEXANOATE
C8H16O2	000124-07-2	OCTANOIC ACID
C8H16O2	000142-92-7	N-HEXYL ACETATE
C8H16O2	000149-57-5	HEXANOIC ACID, 2-ETHYL-
C8H16O2	000557-00-6	PROPYL ISOVALERATE
C8H16O2	000624-54-4	AMYL PROPIONATE
C8H16O3	000112-07-2	2-BUTOXYETHANOL ACETATE
C8H16O3	006382-06-5	AMYL LACTATE
C8H16O4	000112-15-2	CARBITOL ACETATE
C8H16O4	000294-93-9	1,4,7,10-TETRAOXACYCLODODECANE
C8H16O5S	056245-60-4	GALACTOPYRANOSIDE, ETHYL 1-THIO-

C8H16O6	018997-88-1	ß-D-GALACTOPYRANOSIDE, ETHYL
C8H17Br	000111-83-1	1-BROMOOCTANE
C8H17Cl	000111-85-3	1-CHLOROOCTANE
C8H17Cl	000123-04-6	3-(CHLOROMETHYL)HEPTANE
C8H17Cl	000628-61-5	2-CHLOROOCTANE
C8H17F	000463-11-6	1-FLUOROOCTANE
C8H17I	000557-36-8	2-IODOOCTANE
C8H17I	000629-27-6	1-IODOOCTANE
C8H17K2O4P	019045-79-5	DIPOTASSIUM MONOOCTYL ESTER
C8H17N	000104-89-2	2-METHYL-5-ETHYLPIPERIDINE
C8H17N	000458-88-8	2-PROPYLPIPERIDINE
C8H17N	005470-02-0	1-PROPYLPIPERIDINE
C8H17N3O2	019789-60-7	LYSIN-AMIDE, N-ACETYL
C8H17N5O	055921-55-6	2,4-DINH2-6-BUAM-PYRIMIDINE-3-OXIDE
C8H17N5O2	064365-27-1	ARGININ-AMIDE, N-ACETYL
C8H17NO	010315-98-7	N-ISOBUTYLMORPHOLINE
C8H17NO	098486-60-3	ACETAMIDE, N-(1-METHYLETHYL)-N-PROPYL-
C8H17NO2	004248-20-8	O-HEPTYL CARBAMATE
C8H18	000111-65-9	OCTANE
C8H18	000540-84-1	2,2,4-TRIMETHYLPENTANE
C8H18	000560-21-4	2,3,3-TRIMETHYLPENTANE
C8H18	000563-16-6	3,3-DIMETHYL HEXANE
C8H18	000564-02-3	2,2,3-TRIMETHYLPENTANE
C8H18	000565-75-3	2,3,4-TRIMETHYLPENTANE
C8H18	000583-48-2	3,4-DIMETHYLHEXANE
C8H18	000584-94-1	2,3-DIMETHYLHEXANE
C8H18	000589-43-5	2,4-DIMETHYLHEXANE
C8H18	000589-53-7	4-METHYLHEPTANE
C8H18	000589-81-1	3-METHYLHEPTANE
C8H18	000590-73-8	2,2-DIMETHYLHEXANE
C8H18	000592-13-2	2,5-DIMETHYLHEXANE
C8H18	000592-27-8	2-METHYLHEPTANE
C8H18	000594-82-1	2,2,3,3-TETRAMETHYLBUTANE
C8H18	000609-26-7	2-METHYL-3-ETHYLPENTANE
C8H18	000619-99-8	3-ETHYLHEXANE
C8H18	001067-08-9	3-METHYL-3-ETHYLPENTANE
C8H18	026635-64-3	ISOOCTANE
C8H18ClN	000999-33-7	N-BUTYL-N-CHLORO-1-BUTANAMINE
C8H18INO	052303-77-2	2-FURANMETHANAMINIUM, TETRAHYDRO-N,N,N-TRIMETHYL
C8H18INO	090032-54-5	3-FURANMETHANAMINIUM, TETRAHYDRO-N,N,N-TRIMETHYL
C8H18INO2	002209-02-1	2-FURANMETHANAMINIUM, TETRAHYDRO-4-HYDROXY-N,N,N
C8H18INO2	002494-55-5	ETHANAMINIUM, N,N,N-TRIMETHYL-2-(1-OXOPROPOXY)-,
C8H18N2O	000924-16-3	N-NITROSODIBUTYLAMINE
C8H18N2O	000997-95-5	N-NITROSO-DIISOBUTYLAMINE
C8H18N2O	006304-39-8	OCTANOIC ACID HYDRAZIDE
C8H18N2O3	067856-66-0	N-NITROSO-BIS(2-ETOET)AMINE
C8H18N2S	053393-06-9	THIOUREA, N-HEXYL-N'-METHYL-
C8H18N4	097183-51-2	HYDRAZINECARBOXIMIDAMIDE, 2-(1-PROPYLBUTYLIDENE)
C8H18NO4PS2	002275-23-2	VAMIDOTHION
C8H18O	000104-76-7	2-ETHYL-1-HEXANOL
C8H18O	000106-67-2	4-METHYL-2-ETHYL-1-PENTANOL
C8H18O	000111-87-5	1-OCTANOL
C8H18O	000123-44-4	2,2,4-TRIMETHYL-1-PENTANOL
C8H18O	000123-96-6	2-OCTANOL
C8H18O	000142-96-1	DI-N-BUTYL ETHER
C8H18O	000589-62-8	4-OCTANOL
C8H18O	000589-98-0	3-OCTANOL
C8H18O	000598-06-1	3-METHYL-3-HEPTANOL

C8H18O	000625-25-2	2-METHYL-2-HEPTANOL
C8H18O	000628-55-7	DI-ISOBUTYL ETHER
C8H18O	006163-66-2	DI (TERT-BUTYL) ETHER
C8H18O	006863-58-7	DI (SEC-BUTYL) ETHER
C8H18O	007294-05-5	2,2,3-TRIMETHYL-3-PENTANOL
C8H18O	026952-21-6	ISOOCTYL ALCOHOL
C8H18O2	000094-96-2	2-ETHYL-1,3-HEXANDIOL
C8H18O2	000112-25-4	ETHANOL, 2-(HEXYLOXY)-
C8H18O2	000144-19-4	2,2,4-TRIMETHYL-1,3-PENTANEDIOL
C8H18O2	051422-54-9	ETHYL T-BUTYL ETHYL DIETHER
C8H18O3	000112-34-5	DIETHYLENE GLYCOL MONO-N-BUTYL ETHER
C8H18O3	000112-36-7	BIS (2-ETHOXY ETHYL) ETHER
C8H18O3	054446-78-5	1-(2-BUTOXYETHOXY)ETHANOL
C8H18O3Si	000078-08-0	TRIETHOXYVINYLSILANE
C8H18O4	000112-49-2	2,5,8,11-TETRAOXADODECANE
C8H18O4	000112-50-5	ETHOXYTRIETHYLENE GLYCOL
C8H18O4S2	000076-20-0	TRIONAL
C8H18O5	000112-60-7	TETRAETHYLENE GLYCOL
C8H18O6S2Se	126585-48-6	SELENIUM, BIS[3-[(METHYLSULFONYL)OXY]PROPYL]-
C8H18S	000111-88-6	1-OCTANETHIOL
C8H18S	000141-59-3	T-OCTYL MERCAPTAN
C8H18S	000544-40-1	5-THIANONANE
C8H18S	003001-66-9	2-OCTANETHIOL
C8H18S2	000629-45-8	DIBUTYL DISULFIDE
C8H18S2	001518-72-5	DI-TERT-BUTYLDISULFIDE
C8H19N	000104-75-6	2-ETHYLHEXYLAMINE
C8H19N	000110-96-3	DIISOBUTYLAMINE
C8H19N	000111-86-4	N-OCTYLAMINE
C8H19N	000111-92-2	DIBUTYLAMINE
C8H19N	020634-92-8	ETHYL DIPROPYLAMINE
C8H19NO	000096-80-0	2-(BIS(1-METHYLETHYL)AMINO)ETHANOL
C8H19NO6S2	013425-98-4	1-PROPANOL, 3,3'-IMINOBIS-, DIMETHANESULFONATE (
C8H19O2PS2	013194-48-4	O-ETHYL S,S-DIPROPYL PHOSPHORODITHIOATE
C8H19O2PS2	050728-06-8	O=P(OET)(OET-S-(2-THIOET)ET
C8H19O2PS3	000298-04-4	DISULFOTON
C8H19O3P	001809-19-4	DIBUTYL PHOSPHITE
C8H19O3PS3	002497-07-6	DISULFOTON SULFOXIDE
C8H19O4P	000107-66-4	DI-N-BUTYL PHOSPHATE
C8H19O4PS3	002497-06-5	DISULFOTON SULFONE
C8H20BrN	000071-91-0	TETRAETHYL AMMONIUM BROMIDE
C8H20ClN	000056-34-8	TETRAETHYL AMMONIUM CHLORIDE
C8H20IN	000068-05-3	TETRAETHYL AMMONIUM IODIDE
C8H20IN	019109-66-1	TRIMETHYL PENTYL AMMONIUM IODIDE
C8H20INO	005432-44-0	PROPOXYETHYL TRIMETHYL AMMONIUM IODIDE
C8H20INO	021949-06-4	ISOPROPOXYETHYL TRIMETHYL AMMONIUM IODIDE
C8H20N3OP	040725-71-1	PHOSPHONIC DIAMIDE, N,N,N',N'-TETRAMETHYL-P-1-PY
C8H20NO2PS	052067-52-4	O,S-DIME-N-HEXYLPHOSPHOEAMIDITHIOATE
C8H20O4Si	000078-10-4	ETHYL SILICATE
C8H20O5P2S2	003689-24-5	SULFOTEPP
C8H20O7P2	000107-49-3	TETRAETHYL PYROPHOSPHATE
C8H20Pb	000078-00-2	TETRAETHYL LEAD
C8H20Si	000631-36-7	TETRAETHYLSILANE
C8H22N3OP	051754-90-6	PHOSPHORIC TRIAMIDE, N,N-DIETHYL-N',N',N'',N''-TET
C8H23N5	000112-57-2	TETRAETHYLENEPENTAMINE
C8H24N4O3P2	000152-16-9	OCTAMETHYLPYROPHOSPHORAMIDE
C8H24O2Si3	000107-51-7	OCTAMETHYLTRISILOXANE
C8H24O4Si4	000556-67-2	OCTAMETHYLTETRASILOXANE
C9H4Cl3NO2S	000133-07-3	FOLPET

C9H4Cl6O4	000115-28-6	CHLORENDIC ACID
C9H4Cl8O	000297-78-9	ISOBENZAN
C9H4N2S3	000093-75-4	BAYER 30686
C9H4O5	000552-30-7	TRIMELLITIC ANHYDRIDE
C9H5Cl2N	000086-98-6	QUINOLINE, 4,7-DICHLORO-
C9H5Cl3N4	000101-05-3	ANILAZINE
C9H5ClN4	000555-60-2	PROPANEDINITRILE, (3-CHLOROPHENYL)HYDRAZONO -
C9H5ClN4	000946-76-9	PROPANEDINITRILE, [(4-CHLOROPHENYL)HYDRAZONO]-
C9H5N3	023088-24-6	6-CYANOQUINOXALINE
C9H5N5O2	003722-12-1	PROPANEDINITRILE, [(4-NITROPHENYL)HYDRAZONO]-
C9H5N5O2	055653-13-9	PROPANEDINITRILE, [(2-NITROPHENYL)HYDRAZONO]-
C9H6Br2N2O4S2	114260-75-2	ACETIC ACID, [(5,7-DIBROMO-4H-1,2,4-BENZOTHIADIA
C9H6BrN	004965-36-0	7-BROMOQUINOLINE
C9H6BrN	005332-24-1	3-BROMOQUINOLINE
C9H6BrN	005332-25-2	6-BROMOQUINOLINE
C9H6BrNO3	056656-97-4	2-PROPENOIC ACID, 3-(5-BROMO-2-FURANYL)-2-CYANO-
C9H6Cl2N2O	015386-80-8	ACETAMIDE, 2-CYANO-N-(3,4-DICHLOROPHENYL)-
C9H6Cl2N2O3	020354-26-1	METHAZOLE
C9H6Cl2N2O4S2	114260-72-9	ACETIC ACID, [(5,7-DICHLORO-4H-1,2,4-BENZOTHIADI
C9H6Cl2N2O4S2	114260-73-0	ACETIC ACID, [(6,7-DICHLORO-4H-1,2,4-BENZOTHIADI
C9H6Cl2N4OS	141605-19-8	1,2,4-TRIAZIN-5(4H)-ONE, 4-AMINO-6-(3,5-DICHLORO
C9H6Cl2O3	053596-19-3	2-PROPENOIC ACID, 3-(2,4-DICHLOROPHENOXY)-, (E)-
C9H6Cl6	056682-87-2	1,3,5-TRIS(DICHLOROMETHYL)BENZENE
C9H6Cl6O3S	000115-29-7	ENDOSULFAN
C9H6Cl6O3S	000959-98-8	A-ENDOSULFAN
C9H6Cl6O3S	033213-65-9	BETA-ENDOSULFAN
C9H6Cl6O4S	001031-07-8	ENDOSULFAN SULFATE
C9H6ClN	000611-33-6	8-CHLOROQUINOLINE
C9H6ClN	000612-57-7	6-CHLOROQUINOLINE
C9H6ClN	000612-62-4	2-CHLOROQUINOLINE
C9H6ClNO	000130-16-5	8-QUINOLINOL,5-CHLORO-
C9H6ClNO	057334-36-8	4-CHLORO-8-QUINOLINOL
C9H6ClNO3	096841-03-1	2-PROPENOIC ACID, 3-(5-CHLORO-2-FURANYL)-2-CYANO
C9H6ClNO3S	003813-05-6	BENAZOLIN
C9H6F5N3O4S2	023141-80-2	FLUETHIAZIDE
C9H6F6O	000718-64-9	2-PROPANOL,2-PHHENYL-1,1,1,3,3,3-HEXAFLUORO
C9H6FN	000394-68-3	8-FLUOROQUINOLINE
C9H6IN	006560-83-4	2-IODOQUINOLINE
C9H6INO3	069527-41-9	2-PROPENOIC ACID, 2-CYANO-3-(5-IODO-2-FURANYL)-,
C9H6N2	005457-28-3	3-CYANOINDOLE
C9H6N2	015861-24-2	INDOLE-5-CYANO-
C9H6N2	015861-36-6	INDOLE-6-CYANO-
C9H6N2	016136-52-0	INDOLE-4-CYANO-
C9H6N2O	081020-79-3	PROPANEDINITRILE, [(5-METHYL-2-FURANYL)METHYLENE
C9H6N2O2	000091-08-7	TOLUENE-2,6-DIISOCYANATE
C9H6N2O2	000584-84-9	2,4-TOLUENE DIISOCYANATE
C9H6N2O2	000607-34-1	5-NITROQUINOLINE
C9H6N2O2	000607-35-2	8-NITROQUINOLINE
C9H6N2O2	000613-50-3	6-NITROQUINOLINE
C9H6N2O2	000613-51-4	7-NITROQUINOLINE
C9H6N2O2	003741-15-9	4-NITROQUINOLINE
C9H6N2O2	017576-53-3	3-NITROQUINOLINE
C9H6N2O3	000056-57-5	4-NITROQUINOLINE OXIDE
C9H6N2O3	002211-33-8	N-PHENYL-2,4,5-IMIDAZOLITRIONE
C9H6N2O3	004008-48-4	NITROXOLINE
C9H6N2O5S	025603-07-0	2,4-THIAZOLIDINEDIONE, 3-METHYL-5-(5-NITROFURFUR
C9H6N2OS	076542-53-5	PROPANEDINITRILE, [[5-(METHYLTHIO)-2-FURANYL]MET
C9H6N2S	018774-47-5	2-BENZOTHIOPHENAMINE,3-CYANO

I-84

C9H6N2S3	021564-17-0	2-(THIOCYANATEMETHYLTHIO)BENZOTHIAZOLE
C9H6N4	000306-18-3	CARBONYL CYANIDE,PHENYLHYDRAZONE
C9H6N4O	096382-70-6	PROPANEDINITRILE, [(3-HYDROXYPHENYL)HYDRAZONO]-
C9H6O2	000091-64-5	COUMARIN
C9H6O2	000491-38-3	4H-1-BENZOPYRAN-4-ONE
C9H6O2	000606-23-5	1,3-INDANDIONE
C9H6O3	000496-41-3	2-BENZOFURANCARBOXYLIC ACID
C9H6O3	059887-89-7	4H-1-BENZOPYRAN-4-ONE, 7-HYDROXY-
C9H6O4	000485-47-2	1H-INDENE-1,3(2H)-DIONE, 2,2-DIHYDROXY-
C9H6O5	000528-46-1	PHTHALONIC ACID
C9H6O6	000528-44-9	TRIMELLITIC ACID
C9H6O6	000554-95-0	1,3,5-BENZENETRICARBOXYLIC ACID
C9H6O6	000569-51-7	HEMIMELLITIC ACID
C9H7Br2N5	057381-42-7	2',5'-DIBROMOBENZOGUANAMINE
C9H7BrClN5	057381-38-1	2'-BR-5'-CHLOROBENZOGUANAMINE
C9H7BrClN5	057381-45-0	2'-CL-5'-BROMOBENZOGUANAMINE
C9H7BrClN5	057381-50-7	2'-BR-4'-CHLOROBENZOGUANAMINE
C9H7BrFN5	057381-40-5	2'-BR-5'-FLUOROBENZOGUANAMINE
C9H7BrFN5	057381-60-9	5'-BR-2'-FLUOROBENZOGUANAMINE
C9H7BrN2O4S2	114260-74-1	ACETIC ACID, [(7-BROMO-4H-1,2,4-BENZOTHIADIAZIN-
C9H7BrO2	032862-97-8	M-BROMOCINNAMIC ACID
C9H7Cl2N5	057381-26-7	2',5'-DICHLOROBENZOGUANAMINE
C9H7Cl2N5	057381-46-1	2',4'-DICHLOROBENZOGUANAMINE
C9H7Cl2N5	057381-54-1	2',6'-DICHLOROBENZOGUANAMINE
C9H7Cl2NO	001077-74-3	QUINOLINE, 4,7-DICHLORO-, 1-OXIDE
C9H7Cl3O3	000093-72-1	SILVEX
C9H7Cl3O3	001928-37-6	2,4,5-T, METHYL ESTER
C9H7ClF3NO	000344-53-6	2-CF3-4-CHLOROACETANILIDE
C9H7ClF3NO	000348-90-3	3-CF3-4-CHLOROACETANILIDE
C9H7ClFN5	057381-35-8	5'-CL-2'-FLUOROBENZOGUANAMINE
C9H7ClFN5	057381-57-4	2'-CL-5'-FLUOROBENZOGUANAMINE
C9H7ClN2	089770-39-8	4-METHYL-6-CHLOROCINNOLINE
C9H7ClN2	089770-40-1	4-METHYL-7-CHLOROCINNOLINE
C9H7ClN2O3	003741-12-6	6-CHLORO-4-NITROQUINOLINE-1-OXIDE
C9H7ClN2O3	014100-52-8	3-CHLORO-4-NITROQUINOLINE-1-OXIDE
C9H7ClN2O4S2	114260-70-7	ACETIC ACID, [(6-CHLORO-4H-1,2,4-BENZOTHIADIAZIN
C9H7ClN2O4S2	114260-71-8	ACETIC ACID, [(7-CHLORO-4H-1,2,4-BENZOTHIADIAZIN
C9H7ClO3	034385-94-9	5-CL-2,3-DIHYDROBENZOFURAN-2-ACID
C9H7F3N2O2S	000723-57-9	124-BENZTHIADIZ-1-O2-3-ME-6-CF3
C9H7F3N2O2S	020046-63-3	124-BENZTHIADIAZN-1-O2-3-ME-7-CF3
C9H7F3O2	000351-35-9	M-(TRIFLUOROME)PHENYLACETIC ACID
C9H7F3O2	003038-48-0	ACETIC ACID,O-CF3 PHENYL ESTER
C9H7F3O2	032857-62-8	P-TRIFLUOROME PHENYLACETIC ACID
C9H7F3O3	000349-82-6	M-(TRIFLUOROME)PHENOXYACETIC ACID
C9H7F4NO	000349-27-9	2-FLUORO-5-CF3-ACETANILIDE
C9H7FN2O3	019789-69-6	8-FLUORO-4-NITROQUINOLINE-1-OXIDE
C9H7N	000091-22-5	QUINOLINE
C9H7N	000119-65-3	ISOQUINOLINE
C9H7N	001885-38-7	3-PHENYL-2-PROPENENITRILE
C9H7N2O4S2	015913-35-6	3ME5(5NO2-2-FURFURIL)2-S-THIAZOL-4ONE
C9H7N2O4S2	052661-45-7	2-MES-5-(5NO2-2-FURFURILD)THIAZOLONE
C9H7N3	001722-18-5	2-PHENYL-S-TRIAZINE
C9H7N3O	005182-90-1	2-QUINOXALINECARBOXAMIDE
C9H7N3O2	035975-00-9	5-QUINOLINAMINE, 6-NITRO-
C9H7N3O2	089770-26-3	4-QUINOLINAMINE,8-NITRO
C9H7N3O5	001596-52-7	4,6-DINITROQUINOLINE-1-OXIDE
C9H7N3O5	014753-19-6	4,8-DINITROQUINOLINE-1-OXIDE
C9H7N3O5	016238-73-6	4,5-DINITROQUINOLINE-1-OXIDE

C9H7N3S	041814-78-2	TRICYCLAZOLE
C9H7N7O2S	000446-86-6	AZATHIOPRINE
C9H7NO	000059-31-4	2-(1H)-QUINOLINONE
C9H7NO	000148-24-3	8-QUINOLINOL
C9H7NO	000487-89-8	1H-INDOLE-3-CARBOXALDEHYDE
C9H7NO	000491-30-5	1(2H)-ISOQUINOLINONE
C9H7NO	000529-37-3	4(1H)-QUINOLINONE
C9H7NO	000578-67-6	5-QUINOLINOL
C9H7NO	000580-16-5	6-HDROXYQUINOLINE
C9H7NO	000580-18-7	3-QUINOLINOL
C9H7NO	000580-20-1	7-QUINOLINOL
C9H7NO	001443-80-7	P-CYANOACETOPHENONE
C9H7NO	006136-68-1	M-CYANOACETOPHENONE
C9H7NO2	000550-44-7	PHTHALIMIDE,N-METHYL
C9H7NO2	000771-50-6	INDOLE, 3-CARBOXYLIC ACID
C9H7NO2	001477-50-5	INDOLE-2-CARBOXYLIC ACID
C9H7NO2	001878-71-3	M-CYANOPHENYLACETIC ACID
C9H7NO2	002058-74-4	N-METHYLINDOL-2,3-DIONE
C9H7NO2	005715-02-6	2-CYANOPHENYL ACETATE
C9H7NO2	014959-84-3	4,8-DIHYDROXYQUINOLINE
C9H7NO3	001878-82-6	P-CYANOPHENOXYACETIC ACID
C9H7NO3	001879-58-9	M-CYANOPHENOXYACETIC ACID
C9H7NO3	003695-86-1	2-PROPENOIC ACID, 2-CYANO-3-(2-FURANYL)-, METHYL
C9H7NO3	003759-90-8	2,4-OXAZOLIDINEDIONE, 3-PHENYL-
C9H7NO3	005305-00-0	N-PHENYL-2,4-ISOOXAZOLIDIONE
C9H7NO3	005841-63-4	5-PHENYL-2,4-OXAZOLIDIONE
C9H7NO3	109151-40-8	(E)-3-NICOTINOYLACRYLIC ACID
C9H7NO4	000882-06-4	TRANS-P-NITROCINNAMIC ACID
C9H7NS	001826-11-5	THIAZOLE, 2-PHENYL-
C9H8	000095-13-6	INDENE
C9H8Br3N3	038941-33-2	2-(246-TRIBR PH IMINO)IMIDAZOLIDINE
C9H8BrCl2N3	040065-09-6	2-(26-DICL-4-BRPHIMINO)IMIDAZOLIDINE
C9H8BrN5	030101-52-1	3'-BROMOBENZOGUANAMINE
C9H8BrN5	030530-48-4	2'-BROMOBENZOGUANAMINE
C9H8Cl2N2O2	085879-18-1	3,4-DICLBENZALDEHYD,O-((MEAM)CO)OXIME
C9H8Cl2N2O2	088203-03-6	BENZAMIDE,O-DICL ACETYLAMINO-
C9H8Cl2N4O2	065936-26-7	2-(26-DICL-4-NO2-PHIMINO)IMIDAZOLIDN
C9H8Cl2O3	000120-36-5	DICHLORPROP
C9H8Cl2O3	001928-38-7	2,4-D, METHYL ESTER
C9H8Cl3N3	059465-40-6	2-(245-TRICL PH IMINO)IMIDAZOLIDINE
C9H8Cl3N3	059465-51-9	2-(246-TRICL PH IMINO)IMIDAZOLIDINE
C9H8Cl3N3	082780-90-3	2-IMIDAZOLIDINIMINE, N-(2,3,6-TRICHLOROPHENYL)-
C9H8Cl3NO2S	000133-06-2	CAPTAN
C9H8ClN5	000500-42-5	2-AMINO-4-(P-CHLOROANILINO)-S-TRIAZINE
C9H8ClN5	004514-53-8	4'-CHLOROBENZOGUANAMINE
C9H8ClN5	004514-54-9	3'-CHLOROBENZOGUANAMINE
C9H8ClN5	029366-77-6	2'-CHLOROBENZOGUANAMINE
C9H8ClN5O	021702-93-2	4-CYANO-3-CHLOROPHENYLAMIDINOUREA
C9H8ClNO	004637-59-6	4-CHLOROQUINOLINE-1-OXIDE
C9H8ClNOS	068960-60-1	2-AZETIDINONE, 4-[(4-CHLOROPHENYL)THIO]-
C9H8F3NO	000344-62-7	O-CF3 ACETANILIDE
C9H8F3NO	000349-97-3	ACETAMIDE, N-[4-(TRIFLUOROMETHYL)PHENYL]-
C9H8F3NO	000351-36-0	M-(TRIFLUOROME)ACETANILIDE
C9H8F3NO2	001737-06-0	4-OCF3 ACETANILIDE
C9H8F3NO2	014061-27-9	N-METHYL-3-TRIFLUOROMEPHCARBAMATE
C9H8F3NO2	018584-93-5	CARBAMIC ACID, [3-(TRIFLUOROMETHYL)PHENYL]-, MET
C9H8F3NO2	113932-80-2	CARBAMIC ACID, [2-(TRIFLUOROMETHYL)PHENYL]-, MET
C9H8F3NO3S	023375-11-3	M-ACETYL CF3-METHANESULFONANILIDE

C9H8F3NO3S	023383-94-0	P-ACETYL CF3-METHANESULFONANILIDE
C9H8FN5	030530-43-9	3'-FLUOROBENZOGUANAMINE
C9H8FN5	030530-44-0	4'-FLUOROBENZOGUANAMINE
C9H8IN5	072781-91-0	2'-IODOBENZOGUANAMINE
C9H8INO3	000147-58-0	GLYCINE, N-(2-IODOBENZOYL)-
C9H8INO3	052386-94-4	GLYCINE, N-(3-IODOBENZOYL)-
C9H8INO3	055790-22-2	GLYCINE, N-(4-IODOBENZOYL)-
C9H8N2	000578-66-5	8-AMINOQUINOLINE
C9H8N2	000578-68-7	4-AMINOQUINOLINE
C9H8N2	000580-15-4	6-AMINOQUINOLINE
C9H8N2	000580-17-6	3-AMINOQUINOLINE
C9H8N2	000580-22-3	2-AMINOQUINOLINE
C9H8N2	000611-34-7	5-AMINOQUINOLINE
C9H8N2	000670-96-2	1H-IMIDAZOLE, 2-PHENYL-
C9H8N2	001126-00-7	1H-PYRAZOLE, 1-PHENYL-
C9H8N2	001532-72-5	1-ISOQUINOLINAMINE
C9H8N2	001532-84-9	1-ISOQUINOLINAMINE
C9H8N2	007251-61-8	2-METHYLQUINOXALINE
C9H8N2	013708-12-8	5-METHYLQUINOXALINE
C9H8N2O	001670-84-4	1H-INDOLE-2-CARBOXAMIDE
C9H8N2O	001670-86-6	INDOLE-4-CARBOXAMIDE
C9H8N2O	001670-87-7	INDOLE-5-CARBOXAMIDE
C9H8N2O	002436-66-0	4(3H)-QUINAZOLINONE, 3-METHYL-
C9H8N2O	003397-78-2	4-METHOXYCINNOLINE
C9H8N2O	006091-81-2	2-METHYL-2,3-BENZODIAZIN-1-ONE
C9H8N2O	006479-18-1	N-METHYLQUINOXALINE-2-ONE
C9H8N2O	010349-09-4	FURAZAN, METHYLPHENYL-
C9H8N2O	013207-66-4	8-QUINOLINOL, 5-AMINO-
C9H8N2O	016347-95-8	4-METHOXYQUINAZOLINE
C9H8N2O	023443-12-1	4H-PYRIDO[1,2-A]PYRIMIDIN-4-ONE, 9-METHYL-
C9H8N2O	023443-20-1	4H-PYRIDO[1,2-A]PYRIMIDIN-4-ONE, 7-METHYL-
C9H8N2O	025116-00-1	ACETAMIDE, N-(2-CYANOPHENYL)-
C9H8N2O	035704-19-9	ACETAMIDE, N-(4-CYANOPHENYL)-
C9H8N2O	039209-88-6	2-METHOXYQUINOXALINE
C9H8N2O	058202-84-9	ACETAMIDE, N-(3-CYANOPHENYL)-
C9H8N2O	085653-81-2	4H-PYRIDO[1,2-A]PYRIMIDIN-4-ONE, 8-METHYL-
C9H8N2O2	000086-96-4	7-METHYL QUINAZOLINE-2,4-DIONE
C9H8N2O2	000089-24-7	5-PHENYL-2,4-IMIDAZOLIDIONE
C9H8N2O2	000942-79-0	N-METHYL-2-CYANOPHENYLCARBAMATE
C9H8N2O2	000943-49-7	N-METHYL-3-CYANOPHENYLCARBAMATE
C9H8N2O2	002221-13-8	N-PHENYL-2,5-IMIDAZOLIDIONE
C9H8N2O2	006898-86-8	FURAZAN, METHYLPHENYL-, 2-OXIDE
C9H8N2O2	006898-87-9	FURAZAN, METHYLPHENYL-, 5-OXIDE
C9H8N2O2	015414-78-5	N-PHENYL-2,4-IMIDAZOLIDIONE
C9H8N2O2	018315-52-1	N-ME-4-CYANOPHENYLCARBAMATE
C9H8N2O2	028354-19-0	BENZOIC ACID, O-CYANOMETHYLAMINO-
C9H8N2O2	062484-16-6	QUINAZOLINE-2,4-DIONE,6-ME
C9H8N2O2	065448-74-0	CARBAMIC ACID, (3-CYANOPHENYL)-, METHYL ESTER
C9H8N2O2	094563-11-8	CARBAMIC ACID, (4-CYANOPHENYL)-, METHYL ESTER
C9H8N2O2S	016310-37-5	2-METHYLSULFONYLQUINOXALINE
C9H8N2O2S	035203-91-9	8-SULFONAMIDOQUINOLINE
C9H8N2O2S	049739-37-9	FURAZAN, METHYL(PHENYLTHIO)-, 5-OXIDE
C9H8N2O2S	089770-31-0	6-SULFONAMIDOQUINOLINE
C9H8N2O3	007433-86-5	3-NITROQUINOLINE-1-OXIDE
C9H8N2O3	007613-19-6	5-NITROQUINOLINE-1-OXIDE
C9H8N2O3	013675-92-8	6-NITROQUINOLINE-1-OXIDE
C9H8N2O3	014753-17-4	7-NITROQUINOLINE-1-OXIDE
C9H8N2O3	014753-18-5	8-NITROQUINOLINE-1-OXIDE

C9H8N2O3S	049739-42-6	FURAZAN, METHYL(PHENYLSULFONYL)-
C9H8N2O4S	049739-41-5	FURAZAN, METHYL(PHENYLSULFONYL)-, 2-OXIDE
C9H8N2O4S	049739-43-7	FURAZAN, METHYL(PHENYLSULFONYL)-, 5-OXIDE
C9H8N2O4S2	114282-93-8	ACETIC ACID, [(4H-1,2,4-BENZOTHIADIAZIN-3-YL)THI
C9H8N2S	006141-18-0	2-METHYLTHIOQUINAZOLINE
C9H8N2S	021948-73-2	QUINOXALINE, 2-(METHYLTHIO)-
C9H8N3O2	090507-20-3	1,2,4-OXADIAZOLE-3-PHENYL-5-HYDROXIC ACID
C9H8N3O2	103499-08-7	1,2,4-OXADIAZOL-3-ALDOXIME,5-PHENYL-
C9H8N3O3	006756-33-8	2(5-NO2-2-FURFURILIDENE)IMIDAZOLE
C9H8N3O4S	025580-69-2	4-THIAZOLIDINONE, 2-IMINO-3-METHYL-5-[(5-NITRO-2
C9H8N3O4S	025603-06-9	2-MEAM-5-(5-NO2-2-FURFURYL)THIAZOLONE
C9H8N3O4S	052661-53-7	2MEAM-6(5NO2-2FURYL)-13-THIAZIN-4-ONE
C9H8N4O5	021919-05-1	5-(1-AZIRIDINYL)-2,4-NO2BENZAMIDE
C9H8N4OS	051707-55-2	THIDIAZURON
C9H8N6O2	029366-71-0	2'-NITROBENZOGUANAMINE
C9H8N6O2	029366-72-1	3'-NITROBENZOGUANAMINE
C9H8N6O2	029366-73-2	4'-NITROBENZOGUANAMINE
C9H8NO2	000939-05-9	3-OH-5-PHENYLISOXAZOLE
C9H8O	000104-55-2	3-PHENYL-2-PROPENAL
C9H8O	000768-03-6	ACRYLOPHENONE
C9H8O	004265-25-2	2-METHYLBENZOFURAN
C9H8O2	000140-10-3	2-PROPENOIC ACID, 3-PHENYL-, (E)-
C9H8O2	000621-82-9	CINNAMIC ACID
C9H8O2	001075-49-6	BENZOIC ACID, 4-ETHENYL-
C9H8O2	003453-64-3	3-METHYL-1(3H)-ISOBENZOFURANONE
C9H8O2S	015062-75-6	5,7-DIME-2-OXO-1,3-BENZOXATHIOL
C9H8O3	000501-98-4	2-PROPENOIC ACID, 3-(4-HYDROXYPHENYL)-, (E)-
C9H8O3	000577-56-0	O-ACETYLBENZOIC ACID
C9H8O3	000586-89-0	P-ACETYLBENZOIC ACID
C9H8O3	001828-76-8	1(3H)-ISOBENZOFURANONE, 3-HYDROXY-3-METHYL-
C9H8O3	004122-57-0	1(3H)-ISOBENZOFURANONE-3-METHOXYL
C9H8O3	007400-08-0	CINNAMIC ACID, P-HYDROXY
C9H8O3	042103-65-1	2-PROPENOIC ACID, 3-PHENOXY-
C9H8O4	000050-78-2	ACETYLSALICYLIC ACID
C9H8O4	000331-39-5	2-PROPENOIC ACID, 3-(3,4-DIHYDROXYPHENYL)-, (E)-
C9H8O4	000501-89-3	BENZENEACETIC ACID, 4-CARBOXY-
C9H8O4	001679-64-7	METHYL TEREPHTHALATE
C9H8O4	001877-71-0	ISOPHTHALIC ACID, METHYL ESTER
C9H8O4	002084-13-1	M-CARBOXYPHENYLACETIC ACID
C9H8O4	002345-34-8	4-ACETOXYBENZOIC ACID
C9H8O4	004376-18-5	MONOMETHYLPHTHALATE
C9H8O4	022042-71-3	P-FORMYLPHENOXYACETIC ACID
C9H8O4	034385-93-8	1,4-BENZODIOXANE-2-CARBOXYLIC ACID
C9H8O5	001878-61-1	PHENOXYACETIC ACID, M-CARBOXY
C9H9Br2N3	004205-93-0	2-(2,6-DIBR PH IMINO)IMIDAZOLIDINE
C9H9Br2N3	078834-88-5	2-IMIDAZOLIDINIMINE, N-(2,3-DIBROMOPHENYL)-
C9H9Br2N5O2S	113588-97-9	2-THIOPHENECARBOXAMIDE, 4,5-DIBROMO-3-(1-METHYLE
C9H9Br2NO3	000537-24-6	3,5-DIBROMOTYROSINE
C9H9BrClN3	015327-50-1	2-IMIDAZOLIDINIMINE, N-(5-BROMO-2-CHLOROPHENYL)-
C9H9BrClN3	065936-25-6	2-IMIDAZOLIDINIMINE, N-(2-BROMO-6-CHLOROPHENYL)-
C9H9BrClN3	078834-86-3	2-IMIDAZOLIDINIMINE, N-(3-BROMO-2-CHLOROPHENYL)-
C9H9Cl2N3	004205-90-7	2-(2,6-DICL PH IMINO)IMIDAZOLIDINE
C9H9Cl2N3	015327-44-3	2(2,3-DICLPHENYLIMINO)IMIDAZOLINE
C9H9Cl2N3	023830-88-8	2-(2,4-DICL PH IMINO)IMIDAZOLIDINE
C9H9Cl2N3	056514-55-7	2-(2,5-DICL PH IMINO)IMIDAZOLIDINE
C9H9Cl2N3O	057101-48-1	2(2,6-DICL-4-OHPH IMINO)IMIDAZOLIDIN
C9H9Cl2N3O2	013907-78-3	UREA, N-(2-CHLOROETHYL)-N'-(4-CHLOROPHENYL)-N-NI
C9H9Cl2NO	000709-98-8	PROPANIL

C9H9Cl2NO2	000579-38-4	DILOXANIDE
C9H9Cl2NO2	000587-56-4	2-CHLOROETHYL (3-CHLOROPHENYL)CARBAMATE
C9H9Cl2NO2	001966-58-1	3,4-DICHLOROBENZYL METHYLCARBAMATE
C9H9Cl2NO2	002752-68-3	2,4-DICL-5-ME-N-ME PHENYLCARBAMATE
C9H9Cl2NO2	014676-37-0	ANISOLE, O-DICHLOROACETYLAMINO-
C9H9Cl2NO2	094497-46-8	ACETAMIDE, N-(3,4-DICHLOROPHENYL)-2-METHOXY-
C9H9Cl3	017299-97-7	1,3,5-TRI(ALPHA-CHLOROMETHYL)BENZENE
C9H9Cl3F12N3O3P3	065601-41-4	2,2,4-TRICHLORO-2,2,4,4,6,6-HEXAHYDRO-4,6,6-TRI*
C9H9ClFN3	059772-37-1	2-(2-CL-6-F PH IMINO)IMIDAZOLIDINE
C9H9ClFN3	081060-10-8	2-IMIDAZOLIDINIMINE, N-(2-CHLORO-3-FLUOROPHENYL)
C9H9ClFN3	081060-11-9	2-IMIDAZOLIDINIMINE, N-(2-CHLORO-5-FLUOROPHENYL)
C9H9ClN2O2	071059-53-5	4-CL BENZALDEHYDE,O-((MEAM)CO)OXIME
C9H9ClN2O2S	001207-63-2	124-BENZTHIADIZ-1-O2-3-ET-7-CL
C9H9ClN2O2S	014559-55-8	124-BENZTHIADIAZIN-1-O2-3-ET-6-CL
C9H9ClN2O3S	037157-71-4	124-BENZTHIADIAZNE-1-O2-3-MEOME-6-CL
C9H9ClNO3PS	114012-04-3	O,O-DIME O-(2-CN-4-CL)PH PHOSPHORTHIOATE
C9H9ClO2	005406-33-7	P-CHLOROBENZYL ACETATE
C9H9ClO3	000094-74-6	2-METHYL-4-CHLOROPHENOXYACETIC ACID
C9H9ClO3	003307-39-9	2-(P-CHLOROPHENOXY)PROPIONIC ACID
C9H9F2N3	059772-33-7	2-(2,6-DIFL PH IMINO)IMIDAZOLIDINE
C9H9F3N2O	098736-48-2	BENZENEMETHANAMINE, N-METHYL-N-NITROSO-3-(TRIFLU
C9H9I2NO3	000066-02-4	3,5-DIIODOTYROSINE
C9H9N	000083-34-1	3-METHYLINDOLE
C9H9N	000095-20-5	2-METHYLINDOLE
C9H9N	000603-76-9	1-METHYLINDOLE
C9H9N	000614-96-0	5-METHYLINDOLE
C9H9N	000645-59-0	BENZYLACETONITRILE
C9H9N	000933-67-5	7-METHYLINDOLE
C9H9N	002947-61-7	BENZENEACETONITRILE, 4-METHYL-
C9H9N	006921-29-5	TRIPROPYNYL-AMINE
C9H9N2O3S	049561-47-9	2(5NO2-2-FURFURILIDENE)THIAZOLE
C9H9N3	089770-41-2	8-METHYL-4-CINNOLINAMINE
C9H9N3O	084174-23-2	N-NITROSO-N-ME-(4-CYANOBENZYL)AMINE
C9H9N3O	088203-05-8	BENZAMIDE, O-CYANOMETHYLAMINO-
C9H9N3O2	000654-15-9	1-(2-CARBOXYET)BENZOTRIAZOLE
C9H9N3O2	010605-21-7	CARBENDAZIM
C9H9N3O2S2	000072-14-0	SULFATHIOZOLE
C9H9N3O3S	052190-69-9	1,3-BENZODIOXOLE-5-CARBOXYLIC ACID, 2-(AMINOTHIO
C9H9N3O3S2	096474-19-0	ACETAMIDE, N-[2-(AMINOSULFONYL)-6-BENZOTHIAZOLYL
C9H9N5	000091-76-9	BENZOGUANAMINE
C9H9NO	000553-03-7	2(1H)-QUINOLINONE, 3,4-DIHYDRO-
C9H9NO	000621-79-4	CINNAMAMIDE
C9H9NO	001006-94-6	5-METHOXYINDOLE
C9H9NO	001515-85-1	BENZONITRILE, 4-(METHOXYMETHYL)-
C9H9NO	001613-37-2	QUINOLINE-1-OXIDE
C9H9NO	003055-86-5	B-PHENOXYPROPIONITRILE
C9H9NO	019111-74-1	2,6-DIMETHYLBENZONITRILE-N-OXIDE
C9H9NO	066432-03-9	N-FORMYL-STYRYLAMINE
C9H9NO	089242-75-1	3-PYRIDINEMETHANOL,A-2-PROPYNYL-
C9H9NO2	000122-85-0	P-FORMYLACETANILIDE
C9H9NO2	000705-60-2	B-METHYL B-NITROSTYRENE
C9H9NO2	003039-74-5	4-HYDROXYQUINOLINE-1-OXIDE
C9H9NO2	007559-36-6	4-METHYL-B-NITROSTYRENE
C9H9NO2	034222-71-4	2-METHYL-B-NITROSTYRENE
C9H9NO2	041656-75-1	P-ACETYLFORMANILIDE
C9H9NO3	000089-52-1	N-ACETYL O-AMINOBENZOIC ACID
C9H9NO3	000495-69-2	HIPPURIC ACID
C9H9NO3	000556-08-1	P-CARBOXYACETANILIDE

C9H9NO3	000587-48-4	3-ACETYLAMINOBENZOIC ACID
C9H9NO3	003179-09-7	3-METHOXY B-NITROSTYRENE
C9H9NO3	003179-10-0	4-METHOXY-B-NITROSTYRENE
C9H9NO3	051074-95-4	P-ACETOXYBENZAMIDE
C9H9NO3	054335-82-9	N-ME-3-FORMYLPHENYLCARBAMATE
C9H9NO3	054335-83-0	N-ME-4-FORMYLPHENYLCARBAMATE
C9H9NO3	064649-43-0	ACETAMIDE, 2-(BENZOYLOXY)-
C9H9NO3S	027052-09-1	2-PHENYL-1,1-DIOXO-THIAZOL-5-ONE
C9H9NO4	000050-86-2	BENZOIC ACID, 4-(ACETYLAMINO)-2-HYDROXY
C9H9NO4	000099-77-4	ETHYL-P-NITROBENZOATE
C9H9NO4	000487-54-7	GLYCINE, N-(2-HYDROXYBENZOYL)-
C9H9NO4	000618-98-4	ETHYL-M-NITROBENZOATE
C9H9NO4	000619-90-9	ACETIC ACID, NITROBENZYL ESTER
C9H9NO4	005251-93-4	BENZADOX
C9H9NO4	006178-42-3	B-NITRO-4-OH-3-METHOXYSTYRENE
C9H9NO7	000092-55-7	2-NO2-5-DI-(ACETOXYMETHYL)FURANE
C9H9NS	002257-09-2	2-PHENETHYLISOTHIOCYANATE
C9H9NS	004478-92-6	1-PHENETHYLISOTHIOCYANATE
C9H10	000098-83-9	ALPHA-METHYLSTYRENE
C9H10	000100-80-1	M-METHYLSTYRENE
C9H10	000300-57-2	ALLYLBENZENE
C9H10	000496-11-7	INDANE
C9H10	000611-15-4	O-METHYLSTYRENE
C9H10	000622-97-9	P-METHYLSTYRENE
C9H10	000637-50-3	BETA-METHYLSTYRENE
C9H10	000873-49-4	CYCLOPROPYLBENZENE
C9H10	000873-66-5	TRANS-1-PHENYL-1-PROPENE
C9H10BrClN2O2	013360-45-7	CHLORBROMURON
C9H10BrFN2O4	115249-86-0	URIDINE, 5-BROMO-2',3'-DIDEOXY-3'-FLUORO-
C9H10BrN3	063346-73-6	2-(4-BR PHENYLIMINO)IMIDAZOLIDINE
C9H10BrN3O2	002706-22-1	1-ACETYL-4-(4-BR PHENYL) SEMICARBAZIDE
C9H10BrN3O2	111075-91-3	1-ACETYL-4-(3-BR PHENYL) SEMICARBAZIDE
C9H10BrNO	024106-05-6	2-METHYL-4-BROMOACETANILIDE
C9H10BrNO	090914-81-1	3-METHYL-4-BROMOACETANILIDE
C9H10BrNO2	007305-04-6	2-BROMOPHENYLDIMETHYLCARBAMATE
C9H10BrNO2	040912-87-6	BENZAMIDE, 5-BROMO-2-HYDROXY-N,3-DIMETHYL-
C9H10Cl2N2O	000330-54-1	DIURON
C9H10Cl2N2O	010290-38-7	1,1-DIMETHYL-3-(3,5-DICLPH)UREA
C9H10Cl2N2O2	000330-55-2	LINURON
C9H10ClFN2O4	119644-22-3	URIDINE, 5-CHLORO-2',3'-DIDEOXY-3'-FLUORO-
C9H10ClFN2O4	138686-07-4	2,4(1H,3H)-PYRIMIDINEDIONE, 5-CHLORO-1-(2-FLUORO
C9H10ClN3	004749-68-2	2-(2-CHLOROPHENYLIMINO)IMIDAZOLINE
C9H10ClN3O2	013206-67-2	1-NITROSO-1-(2-CLET)-3-PHENYLUREA
C9H10ClN3O2	017075-31-9	1-ACETYL-4-(4-CL PHENYL) SEMICARBAZIDE
C9H10ClN3O2	111075-90-2	1-ACETYL-4-(3-CL PHENYL) SEMICARBAZIDE
C9H10ClN5O2	138261-41-3	IMIDACLOPRID
C9H10ClNO	005900-55-0	5-CHLORO-2-METHYLACETANILIDE
C9H10ClNO	006526-67-6	BENZAMIDE, 2-CHLORO-N,N-DIMETHYL-
C9H10ClNO	007149-79-3	3-CHLORO-4-METHYLACETANILIDE
C9H10ClNO	021352-09-0	2-CHLORO-6-METHYLACETANILIDE
C9H10ClNO	057058-33-0	N-((4-CHLOROPHENYL)METHYL)ACETAMIDE
C9H10ClNO2	001991-78-2	P-CHLOROPHENYLALANINE
C9H10ClNO2	002589-65-3	3-ME-4-CLPHENYL-N-ME CARBAMATE
C9H10ClNO2	007073-42-9	3-CHLORO-4-METHOXYACETANILIDE
C9H10ClNO2	007305-03-5	4-CHLOROPHENYLDIMETHYLCARBAMATE
C9H10ClNS	007738-99-0	THIAZOLIDINE, 2-(P-CL PHENYL)
C9H10ClNS	060980-82-7	THIAZOLIDINE, 2-(M-CHLOROPHENYL)
C9H10F2N2O4	141249-32-3	2,4(1H,3H)-PYRIMIDINEDIONE, 1-(2,3-DIDEOXY-2-FLU

C9H10F3N3	000402-38-0	3,3-DIME-1-(3-CF3 PH)TRIAZENE
C9H10FIN2O4	119644-23-4	URIDINE, 2',3'-DIDEOXY-3'-FLUORO-5-IODO-
C9H10FN3O	087072-65-9	1-ME-1-ACETYL-3-(P-F PH)TRIAZENE
C9H10FN3O2	111075-88-8	1-ACETYL-4-(3-F PHENYL) SEMICARBAZIDE
C9H10FN3O2	111075-89-9	1-ACETYL-4-(4-F PHENYL) SEMICARBAZIDE
C9H10FNO2	000060-17-3	DL-P-FLUOROPHENYLALANINE
C9H10IN3O2	111075-92-4	1-ACETYL-4-(3-IODO PHENYL) SEMICARBAZIDE
C9H10IN3O2	111075-93-5	1-ACETYL-4-(4-IODO PHENYL) SEMICARBAZIDE
C9H10N2	000582-60-5	5,6-DIMETHYLBENZIMIDAZOLE
C9H10N2	001197-19-9	P-CYANO-N,N-DIMETHYLANILINE
C9H10N2	036092-88-3	INDOLE-2-AMINE, N-METHYL
C9H10N2O	000092-43-3	1-PHENYL-3-PYRAZOLIDINONE
C9H10N2O	001693-94-3	4H-PYRIDO(12A)PYRIMIDIN-4-ONE,2ME
C9H10N2O	002508-86-3	4-AMINOQUINOLINE-1-OXIDE
C9H10N2O	003097-21-0	2-BENZIMIDAZOLINONE
C9H10N2O	005980-06-3	2-PYRROLIDINONE, 5-(3-PYRIDINYL)-
C9H10N2O	013858-89-4	BENZOXAZOLE-2-AMINE,N,N-DIMETHYL
C9H10N2O	023443-11-0	4H-PYRIDO(1,2-A)PYRIMIDIN-4-ONE,6-ME
C9H10N2O	028354-25-8	ANISOLE, O-CYANOMETHYLAMINO-
C9H10N2O	039080-57-4	PYRIDO(12A)PYRIMIDIN-4-ONE,3-ME
C9H10N2O2	000063-98-9	PHENACEMIDE
C9H10N2O2	000102-03-4	ACETAMIDE, N-[(PHENYLAMINO)CARBONYL]-
C9H10N2O2	002426-12-2	BENZALDEHYDE,O-((MEAMINO)CO)OXIME
C9H10N2O2	003201-53-4	BENZAMIDE, N-[(METHYLAMINO)CARBONYL]
C9H10N2O2	004637-56-3	4-(HYDROXYAMINO)QUINOLINE 1-OXIDE
C9H10N2O2	033368-89-7	6-METHYLQUINOXALINE-1,4-DIOXIDE
C9H10N2O2	033809-77-7	BENZAMIDE, O-ACETYLAMINO
C9H10N2O2	058202-83-8	P-N-ACETYLAMINOBENZAMIDE
C9H10N2O2S	000831-25-4	THIAZOLIDINE,2-(P-NITROPHENYL)
C9H10N2O2S	031363-88-9	124-BENZTHIADIAZIN-1-O2-3-ME-6-ME
C9H10N2O2S	031363-89-0	124-BENZTHIADIAZIN-1-O2-3-ME-7-ME
C9H10N2O2S	031365-88-5	124-BENZTHIADIAZIN-1-O2-3-ME-5-ME
C9H10N2O2S	075808-93-4	THIAZOLIDINE,2-(M-NITROPHENYL)
C9H10N2O3	000061-78-9	GLYCINE, N-(4-AMINOBENZOYL)-
C9H10N2O3	000612-45-3	2-NITRO-4-METHYLACETANILIDE
C9H10N2O3	007496-53-9	BENZOIC ACID, 2-[(ACETYLAMINO)AMINO]-
C9H10N2O3	056222-10-7	N-((4-NITROPHENYL)METHYL)ACETAMIDE
C9H10N2O3	056751-20-3	DIAZENECARBOXYLIC ACID, PHENYL-, ETHYL ESTER, 2-
C9H10N2O3	059907-22-1	2-METHYL-6-NITROACETANILIDE
C9H10N2O3S	006451-55-4	124-BENZTHIADIZ-1-O2-3-ME-6-MEO
C9H10N2O3S	122568-02-9	URIDINE, 2',3'-DIDEHYDRO-2',3'-DIDEOXY-4-THIO-
C9H10N2O3S2	000452-35-7	ETHOXYZOLAMIDE
C9H10N2O4	000119-81-3	2-NITRO-4-METHOXYACETANILIDE
C9H10N2O4	002922-40-9	P-NITROPHENYLALANINE
C9H10N2O4	003373-86-2	O-NITROPHENYLDIMETHYLCARBAMATE
C9H10N2O4	005974-93-6	2,4(1H,3H)-PYRIMIDINEDIONE, 1-[2,5-DIHYDRO-5-(HY
C9H10N2O4	006169-23-9	PHENOXYACETIC ACID, 3-UREA
C9H10N2O4	007244-70-4	4-NITROPHENYLDIMETHYLCARBAMATE
C9H10N2O4	033721-54-9	ACETAMIDE, N-(2-METHOXY-5-NITROPHENYL)-
C9H10N2O4S2	095599-36-3	2-BENZOTHIAZOLESULFONAMIDE, 6-(2-HYDROXYETHOXY)-
C9H10N2S	002131-64-8	BENZENAMINE, 4-ISOTHIOCYANATO-N,N-DIMETHYL-
C9H10N3O	031803-00-6	2-NH2-5-BENZYL-1,3,4-OXADIAZOLE
C9H10N4	023456-95-3	3,3-DIME-1-(4-CYANOPHENYL)TRIAZENE
C9H10N4O	062825-16-5	P-(N,N-DIMETHYL)PHENYL AZOXYCYANIDE
C9H10N4O	069218-57-1	1-(2-CARBAMYLETHYL)BENZOTRIAZOLE
C9H10N4O	115054-71-2	1H-BENZOTRIAZOLE-1-ACETAMIDE, α-METHYL-
C9H10N4O	115054-88-1	2H-BENZOTRIAZOLE-2-ACETAMIDE, α-METHYL-
C9H10N4O2	023917-23-9	PYRAZOL-3-ON,2(6MEO-4-PYRIMDINYL)5ME

C9H10N4O2	096474-05-4	ALLOPURINOL, 1-(1-OXOBUTYL)-
C9H10N4O2S2	000144-82-1	SULFAMETHIAZOLE
C9H10N4O4	001567-89-1	2-PROPANONE, (2,4-DINITROPHENYL)HYDRAZONE
C9H10N4O4	098827-23-7	ALLOPURINOL, 1-[ETO-CO-OXY-ME]-
C9H10N4O4	111076-00-7	1-ACETYL-4-(3-NO2 PHENYL) SEMICARBAZIDE
C9H10N4O4	111076-01-8	1-ACETYL-4-(4-NO2 PHENYL) SEMICARBAZIDE
C9H10N4S	056610-79-8	1H-TETRAZOLE, 1-METHYL-5-[(PHENYLMETHYL)THIO]-
C9H10NO3PS	002636-26-2	CYANOPHOS
C9H10O	000093-55-0	PROPIOPHENONE
C9H10O	000103-79-7	1-PHENYL-2-PROPANONE
C9H10O	000104-54-1	CINNAMYL ALCOHOL
C9H10O	000122-00-9	P-METHYLACETOPHENONE
C9H10O	000585-74-0	M-METHYLACETOPHENONE
C9H10O	001470-94-6	5-INDANOL
C9H10O	001746-13-0	ALLYLPHENYL ETHER
C9H10O2	000070-70-2	P-HYDROXYPROPIOPHENONE
C9H10O2	000089-71-4	O-TOLUIC ACID, METHYL ESTER
C9H10O2	000093-89-0	ETHYL BENZOATE
C9H10O2	000099-36-5	BENZOIC ACID, 3-METHYL-, METHYL ESTER
C9H10O2	000099-75-2	BENZOIC ACID, 4-METHYL-, METHYL ESTER
C9H10O2	000100-06-1	P-METHOXYACETOPHENONE
C9H10O2	000101-41-7	PHENYLACETIC ACID, METHYL ESTER
C9H10O2	000122-46-3	M-TOLYLACETATE
C9H10O2	000122-60-1	PHENYL GLYDIDYL ETHER
C9H10O2	000140-11-4	BENZYL ACETATE
C9H10O2	000140-39-6	P-TOLYLACETATE
C9H10O2	000492-37-5	ALPHA-PHENYLPROPIONIC ACID
C9H10O2	000501-52-0	BETA-PHENYLPROPIONIC ACID
C9H10O2	000533-18-6	O-TOLYLACETATE
C9H10O2	000586-37-8	M-METHOXYACETOPHENONE
C9H10O2	000610-99-1	O-HYDROXYPROPIOPHENONE
C9H10O2	000619-64-7	4-ETHYLBENZOIC ACID
C9H10O2	000621-36-3	M-METHYLPHENYLACETIC ACID
C9H10O2	000621-87-4	2-PROPANONE, 1-PHENOXY-
C9H10O2	000622-47-9	P-METHYLPHENYLACETIC ACID
C9H10O2	000632-46-2	2,6-DIMETHYLBENZOIC ACID
C9H10O2	000935-92-2	2,5-CYCLOHEXADIENE-1,4-DIONE, 2,3,5-TRIMETHYL-
C9H10O2	001450-72-2	2-ACETYL-4-METHYLPHENOL
C9H10O2	032723-67-4	3-METHYL-4-METHOXYBENZALDEHYDE
C9H10O2S	000103-46-8	S-BENZYLMERCAPTOACETIC ACID
C9H10O3	000094-33-7	1,2-ETHANEDIOL, MONOBENZOATE
C9H10O3	000104-01-8	P-METHOXYPHENYLACETIC ACID
C9H10O3	000118-61-6	ETHYL 2-HYDROXYBENZOATE
C9H10O3	000120-14-9	BENZALDEHYDE, 3,4-DIMETHOXY-
C9H10O3	000120-47-8	P-HYDROXYBENZOIC ACID,ETHYL ESTER
C9H10O3	000121-32-4	BENZALDEHYDE, 3-ETHOXY-4-HYDROXY-
C9H10O3	000121-98-2	P-ANISIC ACID, METHYL ESTER
C9H10O3	000501-97-3	P-HYDROXY-B-PROPIONIC ACID
C9H10O3	000515-30-0	BENZENEACETIC ACID, .ALPHA.-HYDROXY-.ALPHA.-METH
C9H10O3	000529-64-6	TROPIC ACID
C9H10O3	000606-45-1	BENZOIC ACID, 2-METHOXY-, METHYL ESTER
C9H10O3	000613-70-7	O-METHOXYPHENYL ACETATE
C9H10O3	000619-86-3	P-ETHOXYBENZOIC ACID
C9H10O3	000940-64-7	P-METHYLPHENOXYACETIC ACID
C9H10O3	001200-06-2	4-METHOXYPHENYL ACETATE
C9H10O3	001643-15-8	M-METHYLPHENOXYACETIC
C9H10O3	001798-09-0	M-METHOXYPHENYLACETIC ACID
C9H10O3	001878-49-5	O-METHYLPHENOXYACETIC ACID

C9H10O3	002065-23-8	METHYL PHENOXYACETATE
C9H10O3	005451-83-2	ACETIC ACID, 3-METHOXYPHENYL ESTER
C9H10O3	007781-98-8	ETHYL-3-HYDROXYBENZOATE
C9H10O4	000091-52-1	2,4-DIMETHOXYBENZOIC ACID
C9H10O4	000093-07-2	3,4-DIMETHOXYBENZOIC ACID
C9H10O4	000306-08-1	BENZENEACETIC ACID, 4-HYDROXY-3-METHOXY-
C9H10O4	000605-94-7	2,5-CYCLOHEXADIENE-1,4-DIONE, 2,3-DIMETHOXY-5-ME
C9H10O4	001132-21-4	3,5-DIMETHOXYBENZOIC ACID
C9H10O4	001466-76-8	2,6-DIMETHOXYBENZOIC ACID
C9H10O4	001521-38-6	2,3-DIMETHOXYBENZOIC ACID
C9H10O4	001877-75-4	P-METHOXYPHENOXYACETIC ACID
C9H10O4	001878-85-9	O-METHOXY PHENOXYACETIC ACID
C9H10O4	002088-24-6	M-METHOXYPHENOXYACETIC ACID
C9H10O4	010502-44-0	BENZENEACETIC ACID, _-HYDROXY-4-METHOXY-
C9H10O5	000530-57-4	4-OH-3,5-DIMETHOXYBENZIOC ACID
C9H10O5	000831-61-8	ETHYL GALLATE
C9H10O5	003117-05-3	2,3,5-TRIMETHOXY-P-BENZOQUINONE
C9H10O5S	042288-41-5	ACETIC ACID, [4-(METHYLSULFONYL)PHENOXY]-
C9H10S	005296-64-0	ALLYL PHENYL SULFIDE
C9H11Br	000586-61-8	1-ISOPROPYL-4-BROMOBENZENE
C9H11Br	000637-59-2	PROPYLBROMIDE, G-PHENYL
C9H11BrN2O	020940-43-6	1,1-DIMETHYL-3-P-BR-PHENYLUREA
C9H11BrN2O2	003060-89-7	METOBROMURON
C9H11BrN2O4	028616-93-5	URIDINE, 5-BROMO-2',3'-DIDEOXY-
C9H11BrN2O5	000059-14-3	5-BROMODEOXYURIDINE
C9H11Cl	000104-52-9	3-PHENYL-1-CHLOROPROPANE
C9H11Cl2FN2O2S2	001085-98-9	DICHLOFLUANID
C9H11Cl2N3O4S2	000135-07-9	2-METHYL-3-CHLOROMETHYLHYDROCHLOROTHIAZIDE
C9H11Cl2O3PS	057018-04-9	TOLCLOFOS-METHYL
C9H11Cl3NO3P	002213-84-5	O=P(OME)(NET)O-2,4,5-CLPHENYL
C9H11Cl3NO3PS	002921-88-2	CHLORPYRIFOS
C9H11ClFN3O3	141249-30-1	2(1H)-PYRIMIDINONE, 4-AMINO-5-CHLORO-1-(2,3-DIDE
C9H11ClN2O	000150-68-5	MONURON
C9H11ClN2O	000587-34-8	1,1-DIMETHYL-3(M-CHLOROPHENYL)
C9H11ClN2O	022175-22-0	N-(3-CHLORO-4-METHYLPHENYL)-N'-METHYLUREA
C9H11ClN2O2	001746-81-2	MONOLINURON
C9H11ClN2O2	020782-57-4	1-ME-3-(3-CL-4-MEO PHENYL) UREA
C9H11ClN2O3S	024535-70-4	BENZENESULFONAMIDE, 4-CHLORO-N-[(ETHYLAMINO)CARB
C9H11ClN2O4	127592-40-9	URIDINE, 5-CHLORO-2',3'-DIDEOXY-
C9H11ClN4O	056189-70-9	4-CL-3-ME-PHENYLAMIDINO UREA
C9H11F	002038-62-2	PROPYLFLUORIDE, G-PHENYL
C9H11F2N3O3	128496-09-3	2(1H)-PYRIMIDINONE, 4-AMINO-1-(2,3-DIDEOXY-2-FLU
C9H11FN2O	000330-39-2	1,1-DIMETHYL-3-M-FLUOROPHENYLUREA
C9H11FN2O	000332-33-2	1,1-DIMETHYL-3-P-FLUOROPHENYLUREA
C9H11FN2O2	028170-26-5	1-(M-F PHENYL)-3-MEO-3-ME UREA
C9H11FN2O2	088132-24-5	1-(P-F PHENYL)-3-MEO-3-ME UREA
C9H11FN2O3S2	122321-32-8	THIENO[2,3-B]FURAN-2-SULFONAMIDE, 5-[[(2-FLUOROE
C9H11FN2O4	041107-56-6	URIDINE, 2',3'-DIDEOXY-3'-FLUORO-
C9H11FN2O4	124424-25-5	2,4(1H,3H)-PYRIMIDINEDIONE, 1-(2,3-DIDEOXY-2-FLU
C9H11FN2O5	000050-91-9	FLOXURIDINE
C9H11I	004119-41-9	G-PHENYL PROPYLIODIDE
C9H11IN2O4	105784-83-6	URIDINE, 2',3'-DIDEOXY-5-IODO-
C9H11IN2O5	000054-42-2	URIDINE, 2'-DEOXY-5-IODO-
C9H11N	000095-62-5	CYCLOPROPANAMINE, 2-PHENYL-, TRANS
C9H11N	000635-46-1	1,2,3,4-TETRAHYDROQUINOLINE
C9H11N	069684-88-4	CYCLOPROPANAMINE, 2-PHENYL-, CIS-(+)-
C9H11N3	001848-75-5	2-IMIDAZOLIDINIMINE, N-PHENYL-
C9H11N3	004368-68-7	1-BENZYL-1,2,3-TRIAZOLE

C9H11N3	016227-13-7	4-BENZYL-1,2,4-TRIAZOLE
C9H11N3	043170-96-3	IMIDAZO[1,2-A]PYRIDINE-2-ETHANAMINE
C9H11N3	069218-29-7	1-I-PROPYLBENZOTRIAZOLE
C9H11N3O	016543-55-8	3-(1-NITROSO-2-PYRROLIDINYL)PYRIDINE
C9H11N3O	069218-34-4	1-(3-HYDROXYPROPYL)BENZOTRIAZOLE
C9H11N3O2	007203-91-0	BENZOIC ACID, 4-(3,3-DIMETHYL-1-TRIAZENYL)-
C9H11N3O2	016583-99-6	2-(2-CARBOXYET)BENZOTRIAZOLE
C9H11N3O2	020241-07-0	3,3-DIME-1(3-CARBOXYPHENYL)TRIAZENE
C9H11N3O2	054606-23-4	4H-PYRIDO[1,2-A]PYRIMIDINE-3-CARBOXAMIDE, 6,7,8,
C9H11N3O2	054680-35-2	1-ETHYL-1-NITROSO-3-PHENYLUREA
C9H11N3O2	072586-68-6	1,3-DIME-1-NITROSO-3-PHENYLUREA
C9H11N3O3	007159-97-9	1,1-DIMETHYL-3-P-NITROPHENYLUREA
C9H11N3O3	007159-98-0	1,1-DIMETHYL-3-M-NITROPHENYLUREA
C9H11N3O3	007481-88-1	CYTIDINE, 2',3'-DIDEHYDRO-2',3'-DIDEOXY-
C9H11N3O3	111075-99-1	1-ACETYL-4-(4-OH PHENYL) SEMICARBAZIDE
C9H11N3O3S	077317-86-3	3-PRO-5-SCN-6-ME-PYRIMIDIN-2,4-DION
C9H11N3O4	031698-14-3	ANCITABINE
C9H11N3O4	088132-15-4	1-(4-NO2 PHENYL)-3-MEO-3-ME UREA
C9H11N3O4	088132-19-8	1-(3-NO2 PHENYL)-3-MEO-3-ME UREA
C9H11N5O4	084472-85-5	URIDINE, 3'-AZIDO-2',3'-DIDEOXY-
C9H11NO	000100-10-7	P-FORMYL-N,N-DIMETHYLANILINE
C9H11NO	000102-93-2	PROPIONAMIDE, G-PHENYL
C9H11NO	000103-89-9	4-METHYLACETANILIDE
C9H11NO	000120-66-1	ACETAMIDE, N-(2-METHYLPHENYL)-
C9H11NO	000537-92-8	3-METHYLACETANILIDE
C9H11NO	000579-10-2	N-METHYLACETANILIDE
C9H11NO	000588-46-5	N-BENZYLACETAMIDE
C9H11NO	000611-74-5	N,N-DIMETHYLBENZAMIDE
C9H11NO	000620-71-3	PROPIONANILIDE
C9H11NO	001197-50-8	W-PHENYLPROPIONALDEHYDE OXIME
C9H11NO	055321-98-7	2,6-DIMETHYLBENZAMIDE
C9H11NO	067832-97-7	2-ETHYLBENZAMIDE
C9H11NO2	000051-66-1	P-METHOXYACETANILIDE
C9H11NO2	000063-91-2	PHENYLALANINE
C9H11NO2	000087-25-2	O-AMINOBENZOIC ACID, ETHYL ESTER
C9H11NO2	000093-26-5	O-METHOXYACETANILIDE
C9H11NO2	000094-09-7	P-AMINOBENZOIC ACID, ETHYL ESTER
C9H11NO2	000101-99-5	N-PHENYL ETHYLCARBAMATE
C9H11NO2	000150-30-1	DL-PHENYLALANINE
C9H11NO2	000579-58-8	N-METHYL P-HYDROXYACETANILIDE
C9H11NO2	000588-16-9	3-METHOXYACETANILIDE
C9H11NO2	000614-19-7	3-AMINO-3-PHENYLPROPANOIC ACID
C9H11NO2	000619-84-1	4-DIMETHYLAMINO BENZOIC ACID
C9H11NO2	001128-78-5	N-METHYL-O-TOLYLCARBAMATE
C9H11NO2	001129-41-5	N-METHYL-M-TOLYLCARBAMATE
C9H11NO2	001129-48-2	N-METHYL-P-TOLYLCARBAMATE
C9H11NO2	001817-47-6	4-ISOPROPYLNITROBENZENE
C9H11NO2	005602-96-0	CARBAMIC ACID, (4-METHYLPHENYL)-, METHYL ESTER
C9H11NO2	006326-19-8	CARBAMIC ACID, 2-PHENYLETHYL ESTER
C9H11NO2	006526-72-3	1-NITRO-2-ISOPROPYLBENZENE
C9H11NO2	006969-90-0	N,N-DIMETHYLPHENYLCARBAMATE
C9H11NO2	007681-15-4	NICOTINIC ACID, PROPYL ESTER
C9H11NO2	010342-59-3	4-PROPYLNITROBENZENE
C9H11NO2	014983-92-7	CARBAMIC ACID, (2-METHYLPHENYL)-, METHYL ESTER
C9H11NO2	015422-25-0	N-METHYLPHENOXYACETAMIDE
C9H11NO2	016375-90-9	3-METHYL-4-HYDROXYACETANILIDE
C9H11NO2	019340-77-3	N-BENZYLGLYCOLAMIDE
C9H11NO2	030379-59-0	CARBAMIC ACID, METHYL-, PHENYLMETHYL ESTER

C9H11NO2	039076-18-1	CARBAMIC ACID, (3-METHYLPHENYL)-, METHYL ESTER
C9H11NO2	039495-15-3	2-METHYL-4-HYDROXYACETANILIDE
C9H11NO2	042465-53-2	ETHANONE, 1-(2-AMINO-4-METHOXYPHENYL)
C9H11NO2	055836-71-0	P-ETHOXYBENZAMIDE
C9H11NO2S	003938-34-9	N-ME 4-METHYLTHIOPHENYLCARBAMATE
C9H11NO3	000059-07-4	2-ETHOXY-4-AMINOBENZOIC ACID
C9H11NO3	000060-18-4	TYROSINE
C9H11NO3	000556-03-6	DL-TYROSINE
C9H11NO3	003938-24-7	N-ME-2-METHOXYPHENYLCARBAMATE
C9H11NO3	003938-28-1	N-ME-3-METHOXYPHENYLCARBAMATE
C9H11NO3	003938-29-2	N-METHYL-4-METHOXYPHENYLCARBATE
C9H11NO3	014803-72-6	CARBAMIC ACID, (4-METHOXYPHENYL)-, METHYL ESTER
C9H11NO3	014803-73-7	CARBAMIC ACID, (2-METHOXYPHENYL)-, METHYL ESTER
C9H11NO3	021864-67-5	2,6-DIMETHOXYBENZAMIDE
C9H11NO3	025784-00-3	O-(2-HYDROXYET)AMINOBENZOIC ACID
C9H11NO3	026455-31-2	P-NITROPHENYL-I-PROPYL ETHER
C9H11NO3	051422-77-6	CARBAMIC ACID, (3-METHOXYPHENYL)-, METHYL ESTER
C9H11NO3S	022821-80-3	ACETAMIDE, N-[4-(METHYLSULFONYL)PHENYL]
C9H11NO4	000059-92-7	DOPA
C9H11NO4	000063-84-3	DOPA
C9H11NO4	057078-99-6	N-MALEOYL-5-AMINOPENTANOIC ACID
C9H11NO4	057079-05-7	N-MALEOYL-5-NH2PROPIONIC ACID,ET EST
C9H11NO4	117491-57-3	1H-PYRROLE-2,5-DIONE, 1-[(ACETYLOXY)METHYL]-3,4-
C9H11NO4S	005446-77-5	BENZOIC ACID, 4-(AMINOSULFONYL)-, ETHYL ESTER
C9H11NOS	002524-78-9	ACETAMIDE, N-[3-(METHYLTHIO)PHENYL]-
C9H11NOS	006310-41-4	ACETAMIDE, N-[2-(METHYLTHIO)PHENYL]-
C9H11NOS	010352-44-0	ACETAMIDE, N-[4-(METHYLTHIO)PHENYL]
C9H12	000095-63-6	1,2,4-TRIMETHYLBENZENE
C9H12	000098-82-8	CUMENE
C9H12	000103-65-1	N-PROPYLBENZENE
C9H12	000108-67-8	1,3,5-TRIMETHYLBENZENE
C9H12	000526-73-8	1,2,3-TRIMETHYLBENZENE
C9H12	000611-14-3	1-ETHYL-2-METHYLBENZENE
C9H12	000620-14-4	M-ETHYLTOLUENE
C9H12	000622-96-8	P-ETHYLTOLUENE
C9H12	002396-65-8	1,8-NONADIYNE
C9H12	016219-75-3	ETHYLIDENE NORBORNENE
C9H12BrN3O3	107036-57-7	CYTIDINE, 5-BROMO-2',3'-DIDEOXY-
C9H12Cl2O2	059897-94-8	TR-CYPR-CO2ME-22-DIME-3(2DICLETHENYL)
C9H12ClN3O4	031652-78-5	CYTIDINE, 5'-CHLORO-5'-DEOXY-
C9H12ClN3O4S2	001824-58-4	3-ETHYLHYDROCHLOROTHIAZIDE
C9H12ClN5	015468-86-7	1,3,5-TRIAZINE, 6-CHLORO-2,4-BIS[(2-PROPENYL)AMI
C9H12ClN5	060221-92-3	N1-P-CHLOROPHENYL-N5-METHYLBIGUANIDE
C9H12ClN5O5	089083-18-1	ACETAMIDE, N-[5-CHLORO-6-[(2,3-DIHYDROXYPROPYL)A
C9H12ClNO3	019480-39-8	(4-CHLORO-2-METHYLPHENOXY)ACETIC ACID, AMINE SA*
C9H12ClNO3S2	108966-73-0	BENZENESULFONAMIDE, 3-CHLORO-4-[(3-HYDROXYPROPYL
C9H12ClO2PS3	000953-17-3	CARBOPHENOTHION-METHYL
C9H12ClO4P	034783-40-9	HEPTENOPHOS
C9H12FN3O3	051246-79-8	CYTIDINE, 2',3'-DIDEOXY-3'-FLUORO-
C9H12FN3O3	064098-82-4	1(2H)-PYRIMIDINECARBOXAMIDE, N-BUTYL-5-FLUORO-3,
C9H12FN3O3	107036-62-4	CYTIDINE, 2',3'-DIDEOXY-5-FLUORO-
C9H12FN3O3	119555-47-4	2(1H)-PYRIMIDINONE, 4-AMINO-1-(2,3-DIDEOXY-2-FLU
C9H12FNO3S2	108966-74-1	BENZENESULFONAMIDE, 3-FLUORO-4-[(3-HYDROXYPROPYL
C9H12FNO5S2	108966-77-4	BENZENESULFONAMIDE, 3-FLUORO-4-[(3-HYDROXYPROPYL
C9H12HgNS2	032407-99-1	PHENYLMERCURY DIMETHYL DITHIOCARBAMATE
C9H12N2	000494-97-3	NORNICOTINE
C9H12N2O	000101-42-8	FENURON
C9H12N2O	000621-04-5	1-PHENYL-3-ETHYLUREA

C9H12N2O	000938-91-0	1,3-DIMETHYL PHENYL UREA
C9H12N2O	001009-17-2	BENZAMIDE,N-(2-AMINOETHYL)
C9H12N2O	006083-47-2	P-DIMETHYLAMINOBENZAMIDE
C9H12N2O	015582-85-1	2-PYRIDINEBUTANAMIDE
C9H12N2O	018960-16-2	N-ISOPROPYLNICOTINAMIDE
C9H12N2O	032092-29-8	4H-PYRIDO[1,2-A]PYRIMIDIN-4-ONE, 6,7,8,9-TETRAHY
C9H12N2O	033225-17-1	M-DIMETHYLAMINOBENZAMIDE
C9H12N2O	038604-70-5	1-ACETYL-1-METHYL-2-PHENYLHYDRAZINE
C9H12N2O	051055-31-3	N-PROPYLNICOTINAMIDE
C9H12N2O	054606-24-5	4H-PYRIDO[1,2-A]PYRIMIDIN-4-ONE, 6,7,8,9-TETRAHY
C9H12N2O	062783-49-7	BENZENEMETHANAMINE, N,3-DIMETHYL-N-NITROSO-
C9H12N2O	062783-50-0	BENZENEMETHANAMINE, N,4-DIMETHYL-N-NITROSO-
C9H12N2O	071879-58-8	4-PYRIDINEBUTANAMIDE
C9H12N2O	084200-01-1	3-PYRIDINEBUTANAMIDE
C9H12N2O2	001576-17-6	3-PHENYL-1-METHYL-1-METHOXYUREA
C9H12N2O2	002410-24-4	P-AMINOPHENYLALANINE
C9H12N2O2	004849-46-1	1,1-DIME-3-(3-HYDROXYPHENYL)UREA
C9H12N2O2	023138-98-9	UREA, N-(3-METHOXYPHENYL)-N'-METHYL-
C9H12N2O2	025186-43-0	N-(1-METHYLETHYL)-4-NITROBENZENAMINE
C9H12N2O2	084174-20-9	N-NITROSO-N-ME-(4-MEOBENZYL)AMINE
C9H12N2O2	088203-07-0	ANISOLE,CARBOXAMIDOMETHYLAMINO-
C9H12N2O2	088267-61-2	BENZAMIDE, 2-[(2-HYDROXYETHYL)AMINO]-
C9H12N2O2	098154-96-2	CARBAMIC ACID, 3-(3-PYRIDINYL)PROPYL ESTER
C9H12N2O2	098736-45-9	BENZENEMETHANAMINE, 3-METHOXY-N-METHYL-N-NITROSO
C9H12N2O2	101708-63-8	2-PYRROLIDINONE, 5-(3-PYRIDINYL)-, N-OXIDE, (S)-
C9H12N2O3	000052-44-8	2,4-DIAZASPIRO[5.5]UNDECANE-1,3,5-TRIONE
C9H12N2O3	002373-84-4	5-ALLYL-5-ETHYL BARBITURIC ACID
C9H12N2O3	030087-17-3	1-(3-OHPHENYL)-3-MEO-3-ME UREA
C9H12N2O3S	013909-69-8	UREA, N-METHYL-N'-[(4-METHYLPHENYL)SULFONYL]-
C9H12N2O3S	032324-41-7	BENZENESULFONAMIDE, N-[(ETHYLAMINO)CARBONYL]-
C9H12N2O3S	032324-42-8	BENZENESULFONAMIDE, N-[(DIMETHYLAMINO)CARBONYL]-
C9H12N2O3S	059777-62-7	N-ETHYL 4-SULFAMYLBENZAMIDE
C9H12N2O3S	121822-76-2	BENZENEMETHANESULFONAMIDE, N-[(METHYLAMINO)CARBO
C9H12N2O3S	122568-04-1	URIDINE, 2',3'-DIDEOXY-4-THIO-
C9H12N2O4	004598-39-4	4-PYRIMIDINECARBOXYLIC ACID, 1,2,3,6-TETRAHYDRO-
C9H12N2O4	005983-09-5	URIDINE, 2',3'-DIDEOXY-
C9H12N2O4	022754-37-6	4-PYRIMIDINECARBOXYLIC ACID, 1,2,3,6-TETRAHYDRO-
C9H12N2O5	000951-78-0	URIDINE, 2'-DEOXY-
C9H12N2O5S	060084-10-8	4-THIAZOLECARBOXAMIDE, 2-(beta-D-RIBOFURANOSYL)-
C9H12N2O5S	105412-26-8	1-PROPANESULFONAMIDE, 3-(4-NITROPHENOXY)-
C9H12N2O5S2	108966-59-2	BENZENESULFONAMIDE, 4-[(3-HYDROXYPROPYL)THIO]-3-
C9H12N2O6	000058-96-8	URIDINE
C9H12N2O7S2	108966-60-5	BENZENESULFONAMIDE, 4-[(3-HYDROXYPROPYL)SULFONYL
C9H12N2S	000705-62-4	THIOUREA, N,N-DIMETHYL-N'-PHENYL-
C9H12N2S	002740-94-5	1-METHYL-3-BENZYL THIOUREA
C9H12N2S	002741-06-2	1-PHENYL-3-ETHYL THIOUREA
C9H12N4O	013308-82-2	HYDRAZINECARBOXIMIDAMIDE, 2-[(4-METHOXYPHENYL)ME
C9H12N4O	018694-41-2	PYRIMIDINE,4(5MEO-3ME-1-PYRAZOL)
C9H12N4O	033330-89-1	2-(3,3-DIME-1-TRIAZINO)BENZAMIDE
C9H12N4O	033330-91-5	4-(3,3-DIME-1-TRIAZINO)BENZAMIDE
C9H12N4O	069218-62-8	2-(2-CARBAMYLETHYL)BENZOTRIAZOLE
C9H12N4O2	041078-03-9	1H-PURINE-2,6-DIONE, 3,9-DIHYDRO-3-BUTYL-
C9H12N4O2	058247-23-7	4-METHOXYPHENYLAMIDINO UREA
C9H12N4O2	058481-03-1	2-ACETAMIDOMETHYLISONIAZID
C9H12N4O2	066974-76-3	1-(4-CARBAMOYLPHENYL)-3-METHYL-3-METHOXYTRIAZENE
C9H12N4O2	118024-67-2	1H-PURINE-2,6-DIONE, 3,9-DIHYDRO-1-METHYL-3-PROP
C9H12N4O3S	030122-47-5	2(445ME-13OXAZOLID-2AM)5NO2THIAZOLE
C9H12N4O6	087885-53-8	PYRAZINECARBOXYLIC ACID, 6-[(2,3-DIHYDROXYPROPYL

C9H12N6	000051-18-3	2,4,6-TRIS(1-AZIRIDINYL)-1,3,5-TRIAZINE
C9H12N6	077314-77-3	2-PYRIDINAMINE, 4-(5-AMINO-1H-1,2,4-TRIAZOL-3-YL
C9H12N6O	077314-62-6	ETHANOL, 2-[[4-(5-AMINO-1H-1,2,4-TRIAZOL-3-YL)-2
C9H12N6O3	084472-89-9	CYTIDINE, 3'-AZIDO-2',3'-DIDEOXY-
C9H12NO5PS	000122-14-5	FENITROTHION
C9H12NO6P	002255-17-6	3-METHYL-DIMETHYL PARA-OXON
C9H12O	000088-69-7	O-ISOPROPYLPHENOL
C9H12O	000099-89-8	P-ISOPROPYLPHENOL
C9H12O	000122-97-4	3-PHENYLPROPANOL
C9H12O	000496-78-6	2,4,5-TRIMETHYLPHENOL
C9H12O	000527-60-6	2,4,6-TRIMETHYLPHENOL
C9H12O	000539-30-0	BENZYL ETHYL ETHER
C9H12O	000617-94-7	2-PHENYL ISOPROPANOL
C9H12O	000618-45-1	M-ISOPROPYLPHENOL
C9H12O	000622-85-5	PHENYLPROPYL ETHER
C9H12O	000644-35-9	O-PROPYLPHENOL
C9H12O	000645-56-7	4-PROPYLPHENOL
C9H12O	000697-82-5	2,3,5-TRIMETHYLPHENOL
C9H12O	000698-71-5	3-ETHYL-5-METHYLPHENOL
C9H12O	000698-87-3	BENZENEETHANOL, .ALPHA.-METHYL-
C9H12O	001004-66-6	2,6-DIMETHYLANISOLE
C9H12O	002416-94-6	2,3,6-TRIMETHYLPHENOL
C9H12O	014804-32-1	2-ETHYLANISOLE
C9H12O	016308-92-2	2,4-DIMETHYLBENZYL ALCOHOL
C9H12O2	000080-15-9	CUMENE HYDROPEROXIDE
C9H12O2	004169-04-4	PROPYLENE GLYCOL MONOPHENYL ETHER
C9H12O2	004812-20-8	O-ISOPROPOXYPHENOL
C9H12O2	005673-07-4	1,3-DIMETHOXY-2-METHYLBENZENE
C9H12O2	016533-50-9	M-PROPOXYPHENOL
C9H12O2	018979-50-5	PHENOL, 4-PROPOXY-
C9H12O2	038064-90-3	1,3-DIMETHOXY-4-METHYLBENZENE
C9H12O2	041532-81-4	2-METHOXYETHOXYBENZENE
C9H12O3	000538-43-2	PHENYL GLYCEROL
C9H12O3	000634-36-6	BENZENE, 1,2,3-TRIMETHOXY-
C9H12O3	002380-78-1	3-METHOXY-4-HYDROXYPHENYLETHANOL
C9H12O3	032954-58-8	1-PENTANONE, 1-(3-FURANYL)-4-HYDROXY-
C9H12O3S	000080-40-0	ETHYL-P-METHYLBENZENESULFONATE
C9H12O4	000534-82-7	3-METHOXY-4-HYDROXYPHENYLGLYCOL
C9H12O4	003066-90-8	1,4-BENZENEDIOL, 2,3-DIMETHOXY-5-METHYL-
C9H13BrN2O2	000314-40-9	BROMACIL
C9H13BrN2O3	077317-72-7	3-BUO-5-BR-6-ME-PYRIMIDIN-2,4-DIONE
C9H13BrN2O3	077317-73-8	3-SBUO-5-BR-6-ME-PYRIMIDIN-2,4-DION
C9H13BrN2O3	077317-74-9	3-IBUO-5-BR-6-ME-PYRIMIDIN-2,4-DION
C9H13BrN2O3	077317-88-5	3-PRO-5-BR-1,6-ME-PYRIMIDN-2,4-DION
C9H13ClFN3O2	000834-91-3	2,4(1H,3H)-PYRIMIDINEDIONE, 5-[(2-CHLOROETHYL)(2
C9H13ClIN	002373-41-3	BENZENAMINIUM, 3-CHLORO-N,N,N-TRIMETHYL-, IODIDE
C9H13ClN2O2	005902-51-2	TERBACIL
C9H13ClN6	021725-46-2	CYANAZINE
C9H13ClN6O2	042471-28-3	NIMUSTINE
C9H13FNO3PS	039624-86-7	DOWCO 275
C9H13IN2O2	027389-55-5	M-NO2-TRIME-ANILINIUM IODIDE
C9H13N	000060-15-1	3-AMINO-1-PROPYLBENZENE
C9H13N	000099-88-7	P-ISOPROPYLANILINE
C9H13N	000099-97-8	N,N,4-TRIMETHYLANILINE
C9H13N	000103-83-3	N,N-DIMETHYLBENZYLAMINE
C9H13N	000121-72-2	M-TOLUIDINE-N,N-DIMETHYL
C9H13N	000137-17-7	2,4,5-TRIMETHYLANILINE
C9H13N	000156-34-3	BENZENEETHANAMINE, A-METHYL-, (R)-

C9H13N	000300-62-9	ALPHA-METHYLPHENETHYLAMINE
C9H13N	000589-08-2	BENZENEETHANAMINE, N-METHYL-
C9H13N	000609-72-3	N,N-DIMETHYL-O-TOLUIDINE
C9H13N	000622-80-0	N-PROPYLANILINE
C9H13N	000643-28-7	BENZENAMINE, 2-(1-METHYLETHYL)-
C9H13N	002038-57-5	3-PHENYL PROPYLAMINE
C9H13N	002696-84-6	BENZENAMINE, 4-PROPYL-
C9H13N	003261-62-9	BENZENEETHANAMINE, 4-METHYL-
C9H13N	003748-84-3	2,3,5,6-TETRAMETHYLPYRIDINE
C9H13N	003978-81-2	4-(TERT-BUTYL)PYRIDINE
C9H13N	005335-75-1	PYRIDINE, 4-BUTYL-
C9H13N	005944-41-2	2-T-BUTYLPYRIDINE
C9H13N	014321-27-8	BENZENEMETHANAMINE, N-ETHYL-
C9H13N	055755-16-3	BENZENEETHANAMINE, 2-METHYL-
C9H13N3	016584-02-4	1-PROPYLBENZOTRIAZOLE
C9H13N3	016584-03-5	2-PROPYLBENZOTRIAZOLE
C9H13N3	020241-03-6	1-(3-TOLYL)-3,3-DIMETHYLTRIAZENE
C9H13N3	034803-66-2	PIPERAZINE, 1-(2-PYRIDINYL)-
C9H13N3	069218-31-1	2-I-PROPYLBENZOTRIAZOLE
C9H13N3O	000054-92-2	IPRONIAZID
C9H13N3O	000305-33-9	4-PYRIDINECARBOXYLIC ACID, 2-(1-METHYLETHYL)HYDR
C9H13N3O	014339-54-9	2-PROPYLISONIAZID
C9H13N3O	020240-99-7	1-(O-ANISYL)-3,3-DIMETHYLTRIAZENE
C9H13N3O	039938-79-9	1,1-DIME-3-(3-AMINOPHENYL)UREA
C9H13N3O	066974-61-6	1(O-HYDROXYMEPH)-3,3-DIME TRIAZENE
C9H13N3O	069218-38-8	2-(3-HYDROXYPROPYL)BENZOTRIAZOLE
C9H13N3O2	074099-07-3	HYDRAZINECARBOXAMIDE, N-(4-ETHOXYPHENYL)-
C9H13N3O2	088132-16-5	1-MEO-1-ME-3-(P-AMINOPHENYL)UREA
C9H13N3O2	136309-10-9	PYRAZINECARBOXYLIC ACID, 6-(DIMETHYLAMINO)-, ETH
C9H13N3O3	007481-89-2	CYTIDINE, 2',3'-DIDEOXY-
C9H13N3O3S	053207-64-0	N-IPR-3(5NO2-2-THIAZOLYL)ACRYLAMIDE
C9H13N3O4	000951-77-9	CYTIDINE, 2'-DEOXY-
C9H13N3O4	068160-71-4	1H-IMIDAZOLE-1-ETHANOL, 2-NITRO-_-[(1-PROPENYLOX
C9H13N3O4	084123-05-7	ETHANONE, 1-[1-(2-HYDROXYPROPYL)-5-METHYL-2-NITR
C9H13N3O5	000065-46-3	CYTIDINE
C9H13N3O6S	083107-47-5	1(BD-GLUCOTHIOPYRANOSYL)2NO2IMIDAZOLE
C9H13N3O7	083107-46-4	1-(BD-GLUCOPYRANOSYL)-2-NO2IMIDAZOLE
C9H13N3S	052416-13-4	1-(O-THIOANISYL)-33-DIME TRIAZINE
C9H13N3S	052416-14-5	1-(3-THIOANISYL)-33-DIME TRIAZINE
C9H13N5O	066521-48-0	4-(3,3-DIME-1-TRIAZINO)PHENYLUREA
C9H13N5O	066975-13-1	BENZOIC ACID HYDRAZD,O(33DIME)TRIAZIN
C9H13N5O2S	037385-14-1	2(155TRIME-1,3-IMIDAZOLINYLIDEN-2AM)5NO2-THIAZOL
C9H13N5O3S	037385-07-2	5-PYRIMIDINOL, HEXAHYDRO-1,3-DIMETHYL-2-[(5-NITR
C9H13N5O4	082410-32-0	6H-PURIN-6-ONE, 2-AMINO-1,9-DIHYDRO-9-[[2-HYDROX
C9H13N5O4	104987-39-5	4-(3-NITRO-1,2,4-TRIAZOL-1-YL)PRIOPIONYLMORPHOLI
C9H13NO	000104-63-2	ETHANOL, 2- (PHENYLMETHYL)AMINO -
C9H13NO	000700-75-4	N,N-DIMETHYL O-ANISIDINE
C9H13NO	000701-56-4	N,N-DIMETHYL P-ANISIDINE
C9H13NO	005264-15-3	4-PYRIDINEBUTANOL
C9H13NO	014838-15-4	PHENYLPROPANOLAMINE
C9H13NO	017945-79-8	2-PYRIDINEBUTANOL
C9H13NO	031121-11-6	1-PROPANOL, 3-(PHENYLAMINO)-
C9H13NO	036393-56-3	BENZENEMETHANOL, .ALPHA.-(1-AMINOETHYL)-, (R*,R*
C9H13NO	060753-14-2	3-PYRIDINEBUTANOL
C9H13NO	089789-53-7	PENTANONE,PYRROL-2-YL
C9H13NO2	000054-49-9	METARAMINOL
C9H13NO2	000094-07-5	SYNEPHRINE
C9H13NO2	000554-52-9	O-METHYLDOPAMINE

C9H13NO2	002933-75-7	ANISOLE,O-(2-HYDROXYETHYL)AMINO
C9H13NO2	104764-54-7	N-BUTYL-HYDROXYPYRID-4-ONE
C9H13NO2S	001132-18-9	P-PROPYLBENZENESULFONAMIDE
C9H13NO2S	006335-39-3	BENZENESULFONAMIDE, 4-(1-METHYLETHYL)-
C9H13NO2S	123045-56-7	BENZENESULFONAMIDE, 3-(1-METHYLETHYL)-
C9H13NO3	000051-43-4	EPINEPHRINE
C9H13NO3	000097-31-4	NORMETANEPHRINE
C9H13NO3	000633-72-7	3-O-METHYLPYRIDOXOL
C9H13NO3	006539-57-7	A-METHYLNORADRENALINE
C9H13NO3	150900-10-0	2-AZASPIRO[4.4]NONANE-1,3-DIONE, 2-METHOXY-
C9H13NO3S	135832-46-1	BENZENESULFONAMIDE, 4-(3-HYDROXYPROPYL)-
C9H13NO3S2	108966-51-4	BENZENESULFONAMIDE, 4-[(3-HYDROXYPROPYL)THIO]-
C9H13NO5S2	108966-55-8	BENZENESULFONAMIDE, 4-[(3-HYDROXYPROPYL)SULFONYL
C9H13NO6S3	104438-04-2	ACETIC ACID, 3-[[5-(AMINOSULFONYL)-2-THIENYL]SUL
C9H13O6PS	002778-04-3	ENDOTHION
C9H14Br2N2	075523-37-4	2-PYRIDINEMETHANAMINIUM, 6-BROMO-N,N,N-TRIMETHYL
C9H14BrClN2	075523-36-3	2-PYRIDINEMETHANAMINIUM, 6-CHLORO-N,N,N-TRIMETHY
C9H14BrFN2	075523-35-2	2-PYRIDINEMETHANAMINIUM, 6-FLUORO-N,N,N-TRIMETHY
C9H14BrN	000874-80-6	BUTYLPYRIDINIUM BROMIDE
C9H14Cl2	083682-61-5	1,2-DICHLORO-1,5-OCTADIENE
C9H14ClN3O2	033021-94-2	1-NO-1-(2CLET)-3(2CYHEXEN-1-YL)UREA
C9H14ClN3O4	052320-88-4	CYCLOPENTANECARBOXYLIC ACID, 1-[[[(2-CHLOROETHYL
C9H14ClN5	022936-86-3	CYPRAZINE
C9H14ClN5	040533-52-6	1,3,5-TRIAZINE-2,4-DIAMINE, 6-CHLORO-N-(CYCLOPRO
C9H14ClN5	102587-50-8	1,3,5-TRIAZINE-2,4-DIAMINE, 6-CHLORO-N-CYCLOBUTY
C9H14ClNO2	000061-76-7	PHENYLEPHRINE
C9H14ClNO3	000329-63-5	EPINEPHRINE SALT (HCL)
C9H14IN	000098-04-4	PHENYL TRIMETHYL AMMONIUM IODIDE
C9H14INO	002498-27-3	BENZENEMETHANAMINIUM, 3-HYDROXY-N,N,N-TRIMETHYL-
C9H14N2	006021-23-4	3-PYRIDINEBUTANEAMINE
C9H14N2	006304-27-4	2-PYRIDINEETHANAMINE, N,N-DIMETHYL-
C9H14N2	015987-00-5	PYRAZINE, 2-BUTYL-3-METHYL-
C9H14N2	018138-04-0	PYRAZINE, 2,3-DIETHYL-5-METHYL-
C9H14N2	019730-12-2	N-I-PROPYL-3-PYRIDYLMETHYLAMINE
C9H14N2	019730-13-3	N-N-PROPYL-3-PYRIDYLMETHYLAMINE
C9H14N2	019730-15-5	N-ETHYL-2-(3-PYRIDYL)ETHYLAMINE
C9H14N2	020173-26-6	N,N-DIME-2-(3-PYRIDYL)ETHYLAMINE
C9H14N2	034974-00-0	2-PYRIDINEBUTANEAMINE
C9H14N2	062174-83-8	4-PYRIDINEBUTANEAMINE
C9H14N2O	024683-00-9	2-ISOBUTYL-3-METHOXYPYRAZINE
C9H14N2O	080431-10-3	CYCLOPROPANECARBOXAMIDE, N-(3,6-DIHYDRO-1(2H)-PY
C9H14N2O2S	077812-86-3	N,N-DIME-4-MEAM BENZENESULFONAMIDE
C9H14N2O3	000050-11-3	METHARBITAL
C9H14N2O3	000076-76-6	PROBARBITAL
C9H14N2O3	033376-25-9	BARBITURIC ACID, 5-ET,5-PR
C9H14N2O3	034569-18-1	2,4,6(1H,3H,5H)-PYRIMIDINETRIONE, 5-BUTYL-1-METH
C9H14N2O3	109777-62-0	BARBITURIC ACID,5-ME-5(11DIMEET)
C9H14N2O3S3	122266-93-7	THIENO[2,3-B]THIOPHENE-2-SULFONAMIDE, 5-[[(2-HYD
C9H14N2O5S2	108966-61-6	BENZENESULFONAMIDE, 3-AMINO-4-[(3-HYDROXYPROPYL)
C9H14N4O2S	021087-58-1	124-TRIAZIN-5-ONE,4NH2-3MES-6PYRAN2YL
C9H14N4O2S	141605-11-0	1,2,4-TRIAZIN-5(4H)-ONE, 4-AMINO-3-(METHYLTHIO)-
C9H14N4O3	006506-37-2	MORPHOLINE, 4-[2-(5-NITRO-1H-IMIDAZOL-1-YL)ETHYL
C9H14N4O3	036664-18-3	MORPHOLINE, 4-[2-(2-NITRO-1H-IMIDAZOL-1-YL)ETHYL
C9H14N4O3S	051022-76-5	CARBAMOTHIOIC ACID, [2-(2-ETHYL-5-NITRO-1H-IMIDA
C9H14N4O5	074141-74-5	1H-IMIDAZOLE-1-ACETAMIDE, N,N-BIS(2-HYDROXYETHYL
C9H14N4O5	081892-68-4	1H-IMIDAZOLE-1-ACETAMIDE, N-(2,3-DIHYDROXYPROPYL
C9H14N4O6	081892-69-5	1H-IMIDAZOLE-1-ACETAMIDE, N-[2-HYDROXY-1,1-BIS(H
C9H14NO4P	017795-32-3	ME-PHOSPHORAMIDATE,O-ME,O-4-MEOPH

C9H14O	000078-59-1	ISOPHORONE
C9H14O	000504-20-1	PHORONE
C9H14O6	000102-76-1	TRIACETIN
C9H14OSi	013132-25-7	PHENOL, P-(TRIMETHYL SILYL)-
C9H14Si	000768-32-1	SILANE, TRIMETHYL PHENYL
C9H15Br6O4P	000126-72-7	TRIS(2,3-DIBROMOPROPYL) PHOSPHATE
C9H15BrN2	016593-50-3	3-PYRIDYLMETHYL TRIME AMMONIUM BROMIDE
C9H15BrN2	019004-42-3	2-PYRIDYLMETHYL TRIME AMMONIUM BROMIDE
C9H15BrN2	019067-63-1	4-PYRIDYLMETHYL TRIME AMMONIUM BROMIDE
C9H15Cl2N3O2	013909-11-0	1-(2-CLET)-3-(2-CLCYHX)1-N=O UREA
C9H15Cl2N3O2	013909-12-1	UREA, N'-(2-CHLOROCYCLOHEXYL)-N-(2-CHLOROETHYL)-
C9H15Cl3O3	060067-82-5	1,2,3-TRIMEO TRICHLOROCYCLOHEXANE
C9H15Cl6O4P	013674-87-8	TRIS(1,3-DICHLOROISOPROPYL) PHOSPHATE
C9H15Cl6O4P	040120-74-9	TRIS(1,3-DICHLOROPROPYL) PHOSPHATE
C9H15ClINO	066473-27-6	NNN-TRIME-N-(5CLME-2-FURANME) I
C9H15ClN2	084359-13-7	2-PYRIDINEBUTANAMINE, MONOHYDROCHLORIDE
C9H15ClN2	084359-18-2	3-PYRIDINEBUTANAMINE, MONOHYDROCHLORIDE
C9H15ClN2	084359-21-7	4-PYRIDINEBUTANAMINE, MONOHYDROCHLORIDE
C9H15ClN4	005915-41-3	TERBUTHYLAZINE
C9H15N	000102-70-5	2-PROPEN-1-AMINE, N,N-DI-2-PROPENYL-
C9H15N3O2	038507-32-3	2,4(1H,3H)-PYRIMIDINEDIONE, 5-(DIMETHYLAMINO)-1,
C9H15N3O4	021787-89-3	1(2-OH-3-ALLYLOXYPR)2-NO2IMIDAZOLE
C9H15N5	002715-70-0	9-BUTYL ADENINE
C9H15N5O	010521-52-5	ADENINE,9-(1-OHME-PROPYL)
C9H15N5O2	087469-41-8	2H-PURIN-2-ONE, 1,3,6,9-TETRAHYDRO-6-IMINO-9-(2-
C9H15N5O2	087469-42-9	2H-PURIN-2-ONE, 1,3,6,7-TETRAHYDRO-6-IMINO-7-(2-
C9H15NO2	006815-52-7	N-BUTROYLCYCLOBUTANECARBOXAMIDE
C9H15NO2	023046-88-0	N-I-BUTYROYLCYCLOBUTANECARBOXAMIDE
C9H15NO3S	062571-86-2	CAPTOPRIL
C9H15O4P	001623-19-4	TRIALLYLPHOSPHATE
C9H16	003452-09-3	1-NONYNE
C9H16Cl2N4	004080-31-3	N-(3-CHLORALLYL) HEXAMINIUM CHLORIDE
C9H16ClN3O2	013010-47-4	1-N=O-1-(2-CLET)-3-CYCLOHEXYLUREA
C9H16ClN3O3	052049-26-0	1-NO-1(2CLET)-3(4-OH CYHEXYL)UREA
C9H16ClN3O3	056239-24-8	TR-1-NO-1(2CLET)-3(4-OH CYHEXYL)UREA
C9H16ClN3O3	056323-43-4	1-NO-1(CLET)-3(2-OH CYHEXYL)UREA
C9H16ClN3O3	058494-43-2	TR-1-NO-1-(2-CLET)-3(2-OHCYHEX)UREA
C9H16ClN3O4	054749-90-5	CHLOROZOTOCIN
C9H16ClN3O4	058484-09-6	1-NO-1-(2-CLET)-3(2,6-OHCYHEX)UREA
C9H16ClN3O7	070866-07-8	beta-D-GLUCOPYRANOSE, 2-[[[(2-CHLOROETHYL)NITROS
C9H16ClN5	000139-40-2	PROPAZINE
C9H16ClN5	001912-26-1	TRIETAZINE
C9H16INO	001197-60-0	2-FURANMETHANAMINIUM, N,N,N,5-TETRAMETHYL-, IODI
C9H16INO	068724-15-2	3-FURANMETHANAMINIUM, N,N,N,5-TETRAMETHYL-, IODI
C9H16N2O3S	073930-96-8	BUTANOIC ACID, 2-[[(AMINOTHIOXOMETHYL)AMINO]CARB
C9H16N4O3	006497-78-5	MORPHOLINE, 4-[2-(4-NITRO-1H-IMIDAZOL-1-YL)ETHYL
C9H16N4OS	021085-18-7	3-SME-4-NH2-6-IPE-124-TRIAZIN-5-ONE
C9H16N4OS	034014-18-1	TEBUTHIURON
C9H16N4OS	038603-64-4	1-ME-3(2(5ME IMIDAZOL4YL)MESET UREA
C9H16N4OS	050917-23-2	3-PRS-4-NH2-6-IPR-124-TRIAZINE-5-ONE
C9H16N4OS	050917-24-3	3-IPRS-4-NH2-6-IPR-124-TRIAZINE-5-ONE
C9H16N4S	034970-69-9	BURIAMIDE
C9H16N4S2	034839-70-8	METIAMIDE
C9H16N4S2	074633-67-3	THIOUREA, N-METHYL-N'-[2-[[1-METHYL-1H-IMIDAZOL-
C9H16N6O	039980-81-9	IMIDAZOLE-5-CONH2,4(3SBU-3MEN=NN)
C9H16N6O2S	059660-23-0	GUANIDINE, N-METHYL-N'-[2-[[(5-METHYL-1H-IMIDAZO
C9H16NO	049547-83-3	2-(N,N-DIMEAMINOPROPYL)FURAN
C9H16O2	014090-88-1	4,6-NONANEDIONE

C9H16O2	018362-64-6	3,5-HEPTANEDIONE, 2,6-DIMETHYL-
C9H16O4	000123-99-9	NONANEDIOC ACID
C9H16O4	000818-38-2	DIETHYL GLUTARATE
C9H16O4	001732-08-7	HEPTANEDIOIC ACID, DIMETHYL ESTER
C9H17ClN3O3PS	042509-80-8	ISAZOFOS
C9H17N	002243-27-8	NONANONITRILE
C9H17N3O3	034017-18-0	L-VALINAMIDE, N-ACETYLGLYCYL-
C9H17N3O3S2	086569-92-8	N-(N"-ME-ETHYLCARBAMYLTHIO)METHOMYL
C9H17N3O4	071006-72-9	1-(3-IPRO-2-OH-PR)-2-NO2 IMIDAZOLE
C9H17N3O4S	019387-93-0	TINIDAZOLE-PROPYL
C9H17N3O4S	019390-40-0	TINIDAZOLE-ISOPROPYL
C9H17N3O4S2	064028-99-5	N-(N"-ME-ETHOXYCARBAMYLTHIO)METHOMYL
C9H17N3O7	041110-59-2	_-D-GLUCOPYRANOSIDE, METHYL 6-DEOXY-6-[[(METHYLN
C9H17N3O7	052019-10-0	_-D-ALTROPYRANOSIDE, METHYL 3-DEOXY-3-[[(METHYLN
C9H17N3O7	052019-12-2	α-D-GLUCOPYRANOSIDE, METHYL 3-DEOXY-3-[[(METHYLN
C9H17N5O	001610-17-9	ATRATONE
C9H17N5O	007374-53-0	4,6-BIS(ISOPROPYLAMINO)-S-TRIAZIN-2-OL
C9H17N5O	038304-91-5	2,4-NH2-6PIPERIDINO-PYRIMIDINE-3-OXID
C9H17N5S	000834-12-8	AMETRYNE
C9H17NO2	002896-70-0	1-PIPERIDINYLOXY, 2,2,6,6-TETRAMETHYL-4-OXO-
C9H17NO2	060142-96-3	CYCLOHEXANEACETIC ACID, 1-(AMINOMETHYL)-
C9H17NO2S	000112-56-1	2(2-BUTOXYETHOXY)ETHYLTHIOCYANATE
C9H17NO5	000079-83-4	D-PANTOTHENIC ACID
C9H17NOS	002212-67-1	MOLINATE
C9H18	000124-11-8	ISONONENE
C9H18	000696-29-7	ISOPROPYLCYCLOHEXANE
C9H18	001678-92-8	N-PROPYLCYCLOHEXANE
C9H18	003073-66-3	1,1,3-TRIMETHYLCYCLOHEXANE
C9H18Cl3O4P	013674-84-5	2-PROPANOL, 1-CHLORO-, PHOSPHATE (3:1)
C9H18FeN3S6	014484-64-1	FERBAM
C9H18INO2	079661-41-9	ETHANAMINIUM, 2-[(CYCLOPROPYLCARBONYL)OXY]-N,N,N
C9H18N2O	006130-93-4	2,2,6,6-TETME-N-NITROSOPIPERDINE
C9H18N2O	046061-25-0	4-T-BUTYL-N-NITROSOPIPERIDINE
C9H18N2O2	032483-15-1	LEUCINE,N-ACETYL-N'-MEAMINO AMIDE
C9H18N2O2	032483-16-2	ISOLEUCINE,N-ACETYL-N'-MEAM AMIDE
C9H18N2O4	000057-53-4	MEPROBAMATE
C9H18N4O3S2	074385-72-1	N-(N"-DI-ME-N'-ME-URYLTHIO)METHOMYL
C9H18N4OS2	090507-24-7	1,2,5-THIADIAZOLE-3-CARBOXIMIDOTHIOIC ACID, N- H
C9H18N6	000645-05-6	HEXAMETHYL MELAMINE
C9H18O	000108-83-8	2,6-DIMETHYL-4-HEPTANONE
C9H18O	000124-19-6	NONANAL
C9H18O	000502-56-7	5-NONANONE
C9H18O	000815-24-7	3-PENTANONE, 2,2,4,4-TETRAMETHYL-
C9H18O	000821-55-6	2-NONANONE
C9H18O	058654-67-4	2-OCTANONE, 5-METHYL-
C9H18O2	000106-27-4	ISOPENTYL BUTANOATE
C9H18O2	000106-30-9	ETHYL HEPTANOATE
C9H18O2	000111-11-5	METHYL OCTANOATE
C9H18O2	000112-05-0	NONANOIC ACID
C9H18O2	000540-18-1	PENTYL BUTYRATE
C9H18O2	000589-59-3	ISOBUTYL ISOVALERATE
C9H18O2	000626-77-7	PROPYL HEXANOATE
C9H18O2	002311-46-8	ISOPROPYL HEXANOATE
C9H18O2	003274-28-0	HEXANOIC ACID, 2-PROPYL-
C9H18O4	088917-22-0	PROPANOL 1 (OR 2)-2-METHOXYMETHYL ETHOXY, ACETAT
C9H18O5S	000367-93-1	1-ISOPROPYLTHIO-B-GALACTOPYRANSIDE
C9H18O5S	063407-51-2	1-PROPYLTHIO-B-GALACTOPYRANSIDE
C9H18O6	062178-32-9	B-PROPYLGALACTOPYRANOSIDE

C9H19Br	000693-58-3	1-BROMO-N-NONANE
C9H19Cl	002473-01-0	1-CHLORONONANE
C9H19F	000463-18-3	1-FLUORONONANE
C9H19I	004282-42-2	IODONONANE
C9H19N	000768-66-1	PIPERIDINE, 2,2,6,6-TETRAMETHYL-
C9H19N	054704-34-6	CYCLOHEXANEETHANAMINE, α-METHYL-
C9H19NO	000103-00-4	1-CYCLOHEXYLAMINO-2-PROPANOL
C9H19NO2	002029-64-3	O-OCTYL CARBAMATE
C9H19NO2	093677-64-6	CARBAMIC ACID, DIETHYL-, 2-METHYLPROPYL ESTER
C9H19NO2	098154-99-5	CARBAMIC ACID, DIMETHYL-, 2-ETHYLBUTYL ESTER
C9H19NOS	000759-94-4	EPTAM (EPTC)
C9H20	000111-84-2	N-NONANE
C9H20	000921-47-1	2,3,4-TRIMETHYLHEXANE
C9H20	001067-20-5	3,3-DIETHYLPENTANE
C9H20	001068-87-7	2,4-DIMETHYL-3-ETHYLPENTANE
C9H20	001069-53-0	2,3,5-TRIMETHYLHEXANE
C9H20	001070-87-7	2,2,4,4-TETRAMETHYLPENTANE
C9H20	001071-26-7	2,2-DIMETHYLHEPTANE
C9H20	001186-53-4	2,2,3,4-TETRAMETHYLPENTANE
C9H20	002216-33-3	3-METHYLOCTANE
C9H20	002216-34-4	4-METHYLOCTANE
C9H20	003221-61-2	2-METHYLOCTANE
C9H20	003522-94-9	2,2,5-TRIMETHYLHEXANE
C9H20	007154-79-2	2,2,3,3-TETRAMETHYLPENTANE
C9H20	016747-25-4	2,2,3-TRIMETHYLHEXANE
C9H20	016747-26-5	2,2,4-TRIMETHYLHEXANE
C9H20	016747-28-7	2,3,3-TRIMETHYLHEXANE
C9H20	016747-30-1	2,4,4-TRIMETHYLHEXANE
C9H20	016747-31-2	3,3,4-TRIMETHYLHEXANE
C9H20	016747-32-3	2,2-DIMETHYL-3-ETHYLPENTANE
C9H20	016747-38-9	2,3,3,4-TETRAMETHYLPENTANE
C9H20IN	021727-39-9	CYCLOPENTANEMETHANAMINIUM, N,N,N-TRIMETHYL-, IOD
C9H20INO	068724-25-4	3-FURANMETHANAMINIUM, TETRAHYDRO-N,N,N,5-TETRAME
C9H20INO	070135-16-9	2-FURANMETHANAMINIUM, TETRAHYDRO-N,N,N,5-TETRAME
C9H20N2O	000869-79-4	DI-(SEC-BUTYL)UREA
C9H20N2O	014691-88-4	1-PIPERIDINYLOXY, 4-AMINO-2,2,6,6-TETRAMETHYL-
C9H20N2S	000109-46-6	N,N-DIBUTYLTHIOUREA
C9H20N4	093484-23-2	HYDRAZINECARBOXIMIDAMIDE, 2-(1-METHYLHEPTYLIDENE
C9H20NO3PS2	002275-18-5	PROTHOATE
C9H20O	000108-82-7	2,6-DIMETHYL-4-HEPTANOL
C9H20O	000143-08-8	1-NONANOL
C9H20O	000623-93-8	5-NONANOL
C9H20O	000624-51-1	3-NONANOL
C9H20O	000628-99-9	2-NONANOL
C9H20O	002430-22-0	7-METHYL-1-OCTANOL
C9H20O	003452-97-9	ISONONYL ALCOHOL
C9H20O	005932-79-6	4-NONANOL
C9H20O	014202-62-1	2,2-DIETHYL-1-PENTANOL
C9H20O	019549-73-6	2,6-DIMETHYL-3-HEPTANOL
C9H20O	019549-79-2	3,5-DIMETHYL-4-HEPTANOL
C9H20O2	010138-47-3	2-((1-ETHYLPENTYL)OXY)ETHANOL
C9H20S	001455-21-6	1-NONANETHIOL
C9H21N	000102-69-2	TRIPROPYLAMINE
C9H21N	000112-20-9	N-NONYLAMINE
C9H21O2PS3	013071-79-9	TERBUFOS
C9H21O3P	002404-73-1	DIBUTYL METHYLPHOSPHONATE
C9H21O3PS3	005827-05-4	APHIDAN
C9H21O3PS3	010548-10-4	TERBUFOS SULFOXIDE

C9H21O4P	000513-08-6	TRIPROPYL PHOSPHATE
C9H21O4PS2	056165-57-2	PHOSPHOROTHIOIC ACID, S-[[(1,1-DIMETHYLETHYL)SUL
C9H21O4PS3	056070-16-7	TERBUFOS SULFONE
C9H21O5PS2	056070-15-6	PHOSPHOROTHIOIC ACID, S-[[(1,1-DIMETHYLETHYL)SUL
C9H22IN	015066-77-0	TRIMETHYL HEXYL AMMONIUM IODIDE
C9H22NO2PS	073207-98-4	S-(2-(BIS(1-METHYLETHYL)AMINO)ETHYL) METHYL PHO*
C9H22O4P2S4	000563-12-2	ETHION
C10Cl10	002227-17-0	DIENOCHLOR
C10Cl10O	000143-50-0	KEPONE
C10Cl12	002385-85-5	MIREX
C10Cl8	002234-13-1	OCTACHLORONAPHTHALENE
C10F22	003021-63-4	PERFLUORO-2,7-DIMETHYLOCTANE
C10H4Cl2FNO2	041205-21-4	SPARTICIDE
C10H4Cl2O2	000117-80-6	DICHLONE
C10H4Cl4	020020-02-4	1,2,3,4-TETRACHLORONAPHTHALENE
C10H4Cl4	031604-28-1	1,3,5,8-TETRACHLORONAPHTHALENE
C10H4Cl4	053555-63-8	1,2,3,5-TETRACHLORONAPHTHALENE
C10H4Cl4	053555-64-9	1,3,5,7-TETRACHLORONAPHTHALENE
C10H4Cl4	055720-43-9	1,4,6,7-TETRACHLORONAPHTHALENE
C10H5BrO2	002065-37-4	2-BROMO-1,4-NAPHTHOQUINONE
C10H5Cl2FN2O2	125111-05-9	2,4(1H,3H)-PYRIMIDINEDIONE, 1-[(2,4-DICHLOROPHEN
C10H5Cl3	055720-37-1	1,3,7-TRICHLORONAPHTHALENE
C10H5Cl3	055720-40-6	2,3,6-TRICHLORONAPHTHALENE
C10H5Cl7	000076-44-8	HEPTACHLOR
C10H5Cl7O	001024-57-3	HEPTACHLOR EPOXIDE
C10H5Cl9	005103-73-1	CIS-NONACHLOR
C10H5Cl9	039765-80-5	TRANS-NONACHLOR
C10H5ClO2	001010-60-2	2-CHLORO-1,4-NAPHTHOQUINONE
C10H5F3N4O	000370-86-5	PROPANEDINITRILE, [[4-(TRIFLUOROMETHOXY)PHENYL]H
C10H5N3O6	002364-46-7	NAPHTHALENE, 1,3,8-TRINITRO-
C10H6BrNO2	030922-52-2	3-BROMO-2-NITROSO-1-NAPHTHOL
C10H6Cl2	001825-30-5	1,5-DICHLORONAPHTHALENE
C10H6Cl2	001825-31-6	1,4-DICHLORONAPHTHALENE
C10H6Cl2	002050-69-3	1,2-DICHLORONAPHTHALENE
C10H6Cl2	002050-73-9	NAPHTHALENE, 1,7-DICHLORO-
C10H6Cl2	002050-74-0	1,8-DICHLORONAPHTHALENE
C10H6Cl2	002050-75-1	2,3-DICHLORONAPHTHALENE
C10H6Cl2	002065-70-5	2,6-DICHLORONAPHTHALENE
C10H6Cl2	002198-77-8	2,7-DICHLORONAPHTHALENE
C10H6Cl2N2	003740-92-9	FENCLORIM
C10H6Cl2O	002050-76-2	2,4-DICHLORO-1-NAPHTHOL
C10H6Cl3N3O2	036865-56-2	3,4,5-TRICL-C6H2NHN=C(CN)CO-OME
C10H6Cl3N3O2	036905-04-1	ACETIC ACID, CYANO (2,4,5-TRICHLOROPHENYL) HYDRA
C10H6Cl4O4	001861-32-1	DIMETHYL TETRACHLOROTEREPHTHALATE
C10H6Cl6	003734-48-3	CHLORDENE
C10H6Cl8	000057-74-9	CHLORDANE
C10H6Cl8	005103-71-9	ALPHA-CHLORDANE
C10H6Cl8	005103-74-2	CHLORDANE (TRANS)
C10H6Cl8	005566-34-7	GAMMA-CHLORDANE
C10H6Cl8	012789-03-6	CHLORDANE
C10H6ClNO2	002797-51-5	2-AMINO-3-CHLORO-1,4-NAPHTHOQUINONE
C10H6F3N	000317-57-7	8-CF3 QUINOLINE
C10H6F3N	000325-14-4	7-CF3 QUINOLINE
C10H6F3NO	000322-97-4	4-HYDROXY-7-TRIFLUOROMEQUINOLINE
C10H6N2	001436-43-7	2-CYANOQUINOLINE
C10H6N2	002700-22-3	MALONONITRILE, BENZAL
C10H6N2	002973-27-5	4-CYANOQUINOLINE
C10H6N2	034846-64-5	3-CYANOQUINOLINE

I-103

C10H6N2O4	000602-38-0	NAPHTHALENE, 1,8-DINITRO-
C10H6N2O4	000605-71-0	NAPHTHALENE, 1,5-DINITRO-
C10H6N2O4	000606-37-1	NAPHTHALENE, 1,3-DINITRO-
C10H6N2O4S2	027464-57-9	5-(5NO2-FURYLPROPENILIDINE)-2-S-THIAZOLIDEN-2,4-
C10H6N2OS2	002439-01-2	OXYTHIOQUINOX
C10H6N4O	117505-22-3	4-QUINOLINE AZOXYCYANIDE
C10H6O2	000130-15-4	1,4-NAPHTHOQUINONE
C10H6O3	000083-72-7	1,4-NAPHOQUINONE, 2-HYDROXY
C10H6O3	000481-39-0	5-HYDROXY-1,4-NAPHTHOQUINONE
C10H6O4	000475-38-7	1,4-NAPHTHALENEDIONE, 5,8-DIHYDROXY-
C10H6O4	004940-39-0	4H-1-BENZOPYRAN-2-CARBOXYLIC ACID, 4-OXO-
C10H6O8	000089-05-4	PYROMELLITIC ACID
C10H6O8	000476-73-3	MELLOPHANIC ACID
C10H6O8	000479-47-0	1,2,3,5-BENZENETETRACARBOXYLIC ACID
C10H7Br	000090-11-9	1-BROMONAPHTHALENE
C10H7Br	000580-13-2	2-BROMONAPHTHALENE
C10H7Cl	000090-13-1	1-CHLORONAPHTHALENE
C10H7Cl	000091-58-7	2-CHLORONAPHTHALENE
C10H7Cl2N3O	028317-61-5	3,4-DICL-C6H3NHN=C(CN)CO-ME
C10H7Cl2N3O	028317-62-6	3,5-DICL-C6H3NHN=C(CN)CO-ME
C10H7Cl2N3O2	028313-58-8	3,5-DICL-C6H3NHN=C(CN)CO-OME
C10H7Cl2N3OS	141605-18-7	1,2,4-TRIAZIN-5(4H)-ONE, 6-(3,5-DICHLOROPHENYL)-
C10H7Cl2NO2	024096-53-5	OHRIC
C10H7Cl5O2	021757-82-4	ACETIC ACID, 1-(3,4-DICHLOROPHENYL)-2,2,2-TRICHL
C10H7Cl7	014168-01-5	BETA-DIHYDROHEPTACHLOR
C10H7ClF3N5	065052-53-1	2'-CL-5'-CF3-BENZOGUANAMINE
C10H7ClO	000604-44-4	4-CHLORO-1-NAPHTHOL
C10H7ClO4	033607-91-9	6-CL-4-KETOCHROMAN-2-CARBOXLIC ACID
C10H7F	000321-38-0	1-FLUORONAPHTHALENE
C10H7F3N2O4S2	114260-78-5	ACETIC ACID, [[6-(TRIFLUOROMETHYL)-4H-1,2,4-BENZ
C10H7N3O5S	052661-42-4	ACETAMIDE, N-[5-[(5-NITRO-2-FURANYL)METHYLENE]-4
C10H7N3O5S	052661-54-8	ACETAMIDE, N-[6-(5-NITRO-2-FURANYL)-4-OXO-4H-1,3
C10H7N3S	000148-79-8	THIABENDAZOLE
C10H7N5O	037761-96-9	7H-1,2,3-TRIAZOLO[4,5-D]PYRIMIDIN-7-ONE, 1,4-DIH
C10H7N5O5	028317-71-7	BUTANENITRILE, 2-[(2,4-DINITROPHENYL)HYDRAZONO]-
C10H7NO	005470-96-2	2-QUINOLINECARBOXALDEHYDE
C10H7NO2	000086-57-7	1-NITRONAPHTHALENE
C10H7NO2	000131-91-9	1-NITROSO-2-NAPHTHOL
C10H7NO2	000132-53-6	2-NITROSO-1-NAPHTHOL
C10H7NO2	000581-89-5	2-NITRONAPHTHALENE
C10H7NO2	000941-69-5	1H-PYRROLE-2,5-DIONE, 1-PHENYL-
C10H7NO2	002348-81-4	1,4-NAPHTHALENEDIONE, 2-AMINO-
C10H7NO2	004965-30-4	N-OH-1,4-NAPHTHOQUINONEMONIMINE
C10H7NO2	007206-56-6	1-NITROAZULENE
C10H7NO3	006476-18-2	3-PHENYL-2,4,5-PYRROLITRIONE
C10H7NO3	038397-06-7	6-NITRO-1-NAPHTHOL
C10H8	000091-20-3	NAPHTHALENE
C10H8	000275-51-4	AZULENE
C10H8Br2N2O4S	101063-95-0	PROPANOIC ACID, 3-(5,7-DIBROMO-4H-1,2,4-BENZOTHI
C10H8Br2N2O4S2	124802-92-2	PROPANOIC ACID, 3-[(5,7-DIBROMO-4H-1,2,4-BENZOTH
C10H8BrNO2	000574-98-1	N-(2-BROMOETHYL)PHTHALIMIDE
C10H8BrNO2	041167-74-2	P-BROMO-N-PHENYLSUCCINIMIDE
C10H8BrNO2	058714-54-8	M-BROMO-N-PHENYLSUCCINIMIDE
C10H8Cl2N2O4S	101063-93-8	PROPANOIC ACID, 3-(5,7-DICHLORO-4H-1,2,4-BENZOTH
C10H8Cl2N2O4S	101063-94-9	PROPANOIC ACID, 3-(6,7-DICHLORO-4H-1,2,4-BENZOTH
C10H8Cl2N2O4S2	124802-90-0	PROPANOIC ACID, 3-[(6,7-DICHLORO-4H-1,2,4-BENZOT
C10H8Cl2N2O4S2	134917-53-6	PROPANOIC ACID, 3-[(5,7-DICHLORO-4H-1,2,4-BENZOT
C10H8Cl2N4OS	141605-16-5	1,2,4-TRIAZIN-5(4H)-ONE, 4-AMINO-6-(3,5-DICHLORO

C10H8ClN3O	001698-60-8	CHLORIDAZON
C10H8ClN3O	028317-58-0	BUTANENITRILE, 2-[(4-CHLOROPHENYL)HYDRAZONO]-3-O
C10H8ClN3O	028317-59-1	BUTANENITRILE, 2-[(2-CHLOROPHENYL)HYDRAZONO]-3-O
C10H8ClN3O	028317-60-4	BUTANENITRILE, 2-[(3-CHLOROPHENYL)HYDRAZONO]-3-O
C10H8ClN3O2	036874-69-8	ACETIC ACID, ((3-CHLOROPHENYL)HYDRAZONO)CYANO-,
C10H8ClN3O2	051450-97-6	DRAZOXOLON
C10H8ClNO2	006943-00-6	P-CHLORO-N-PHENYLSUCCINIMIDE
C10H8ClNO2	007402-22-4	O-CHLORO-N-PHENYLSUCCINIMIDE
C10H8ClNO2	015386-99-9	M-CHLORO-N-PHENYLSUCCINIMIDE
C10H8ClNOS2	006012-92-6	N-244
C10H8F2N2OS	095333-60-1	IMIDIAZOLE-2-THIONE,3-(3,5-DIFLUORO-4-HYDROXYBEN
C10H8F2N2S	095333-81-6	IMIDAZOLE-2-THIONE,3-(3,5-DIFLUOROBENZYL)-
C10H8F3N5	030508-78-2	3'-TRIFLUOROMETHYLBENZOGUANAMINE
C10H8F3N5S	065052-47-3	3'-(CF3-THIO)BENZOGUANAMINE
C10H8FN3OS	085683-39-2	1,2,4-TRIAZIN-5(4H)-ONE, 6-(4-FLUOROPHENYL)-2,3-
C10H8FNO2	060693-35-8	M-FLUORO-N-PHENYLSUCCINIMIDE
C10H8FNO2	060693-37-0	P-FLUORO-N-PHENYLSUCCINIMIDE
C10H8FNO2	072601-44-6	O-FLUORO-N-PHENYLSUCCINIMIDE
C10H8INO2	072601-45-7	O-IODO-N-PHENYLSUCCINIMIDE
C10H8INO2	072601-46-8	M-IODO-N-PHENYLSUCCINIMIDE
C10H8N2	000366-18-7	2,2-BIPYRIDINE
C10H8N2	000553-26-4	4,4'-DIPYRIDYL
C10H8N2	000581-47-5	(4-PYRIDYL)-2-PYRIDINE
C10H8N2	000581-50-0	2,3'-BIPYRIDINE
C10H8N2	003438-48-0	PYRIMIDINE, 4-PHENYL-
C10H8N2O	000709-79-5	BENZALCYANOACETAMIDE
C10H8N2O	006480-67-7	3-QUINOLINECARBOXAMIDE
C10H8N2O	062937-22-8	8-FORMYLAMIDOQUINOLINE
C10H8N2O	066516-96-9	N-FORMYL-P-CYANOSTYRYLAMINE
C10H8N2O	089770-37-6	3-ACETYLCINNOLINE
C10H8N2O2	000881-07-2	QUINOLINE, 2-METHYL-8-NITRO-
C10H8N2O3	000085-81-4	QUINOLINE, 6-METHOXY-8-NITRO-
C10H8N2O3	022275-34-9	BARBITURIC ACID, 5-PHNEYL-
C10H8N2O3	030345-85-8	N-BENZYL-2,4,5-IMIDAZOLITRIONE
C10H8N2O3	089770-28-5	4-MEO-8-NO2 QUINOLINE
C10H8N2O4	018377-52-1	O-NITRO-N-PHENYLSUCCINIMIDE
C10H8N2O4	031036-66-5	M-NITRO-N-PHENYLSUCCINIMIDE
C10H8N2O4	035488-92-7	P-NITRO-N-PHENYLSUCCINIMIDE
C10H8N3O3	001083-59-6	2(5-NO2-2-FURFURYLIDENE)PYRIMIDINE
C10H8N4	040257-94-1	PROPANEDINITRILE, [(4-METHYLPHENYL)HYDRAZONO]-
C10H8N4O2	117505-23-4	4-QUINOLINE,1-OXIDE AZOXYCYANIDE
C10H8NNaO3S	000130-13-2	1-NAPHTHALENESULFONIC ACID, 4-AMINO-, SODIUM SAL
C10H8O	000090-15-3	1-NAPHTHOL
C10H8O	000135-19-3	2-NAPHTHOL
C10H8O2	000083-56-7	1,5-DIHYDROXYNAPHTHALENE
C10H8O2	000092-44-4	2,3-NAPHTHALENEDIOL
C10H8O2	000132-86-5	NAPHTHALENE-1,3-DIOL
C10H8O2	000571-60-8	1,4-NAPHTHALENEDIOL
C10H8O2	000575-38-2	NAPTHALENE-1,7-DIOL
C10H8O2	000581-43-1	2,6-NAPHTHALENEDIOL
C10H8O2	023251-68-5	2(5H)-FURANONE, 5-(4-HEXEN-2-YNYLIDENE)-
C10H8O3	000090-33-5	7-HYDROXY-4-METHYLCOUMARIN
C10H8O3	001646-27-1	2-BENZOFURANCARBOXYLIC ACID, METHYL ESTER
C10H8O3	017812-07-6	3-BENZOYLACRYLIC ACID (TRANS)
C10H8O3	022609-88-7	5-PHENYL-2,4-FURANDIONE
C10H8O3S	000120-18-3	NAPTHALENE-2-SULFONIC ACID
C10H8O4	000492-94-4	FURAN, 2-COCO-FURYL
C10H8O4	000552-86-3	FURAN, 2-CH(OH)CO-FURYL

C10H8O4	002373-80-0	3,4-(METHYLENEDIOXY)CINNAMIC ACID
C10H8O4	051048-00-1	2H-1-BENZOPYRAN-2-CARBOXYLIC ACID, 3,4-DIHYDRO-4
C10H8O4S	000084-87-7	1-NAPHTHOL-4-SULFONIC ACID
C10H8O4S	000117-59-9	1-NAPHTHALENESULFONIC ACID, 5-HYDROXY-
C10H8O4S	003771-14-0	1-NAPHTHOL-3-SULFONIC ACID
C10H8O6S2	000081-04-9	1,5-NAPHTHALENEDISULFONIC ACID
C10H8OS	001128-05-8	3-ACETYL BENZOTHIOPHENE
C10H8S	000825-55-8	2-PHENYLTHIOPHENE
C10H9BrN2O4S	091445-76-0	PROPANOIC ACID, 3-(7-BROMO-4H-1,2,4-BENZOTHIADIA
C10H9BrN2O4S2	124802-91-1	PROPANOIC ACID, 3-[(7-BROMO-4H-1,2,4-BENZOTHIADI
C10H9Cl2N5O	141627-87-4	1,2,4-TRIAZIN-5(4H)-ONE, 4-AMINO-6-(3,5-DICHLORO
C10H9Cl2NO	002164-09-2	DICRYL
C10H9Cl2NO	002759-71-9	CYPROMIDE
C10H9Cl2NO	010249-33-9	2-BUTENAMIDE, N-(3,4-DICHLOROPHENYL)-
C10H9Cl2NO2	006140-12-1	ACETOPHENONE,O-DICL ACETYLAMINO
C10H9Cl2NO3	003957-74-2	4-ACETOXY DICHLOROACETANILIDE
C10H9Cl3	020057-31-2	A-(2,2,2-TRICHLOROETHYL)STYRENE
C10H9Cl3O3	000093-80-1	2,4,5-TB
C10H9Cl3O3	001928-39-8	2,4,5-T, ETHYL ESTER
C10H9Cl4NO2S	002425-06-1	CAPTAFOL
C10H9Cl4O4P	000961-11-5	STIROFOS
C10H9Cl4O4P	022248-79-9	TETRACHLORVINPHOS
C10H9ClN2	022752-98-3	N-(4-PYRIDYL)PYRIDINIUM CHLORIDE
C10H9ClN2O2S	013460-17-8	124-BENZTHIDIAZIN-1-O2-3-CYPR-6-CL
C10H9ClN2O2S	037157-59-8	124-BENZTHIADIAZN-1-O2-3-ALLYL-6-CL
C10H9ClN2O4S	091551-85-8	PROPANOIC ACID, 3-(7-CHLORO-4H-1,2,4-BENZOTHIADI
C10H9ClN2O4S	101063-92-7	PROPANOIC ACID, 3-(6-CHLORO-4H-1,2,4-BENZOTHIADI
C10H9ClN2O4S2	124850-85-7	PROPANOIC ACID, [(6-CHLORO-4H-1,2,4-BENZOTHIADIA
C10H9ClN2O4S2	134917-52-5	PROPANOIC ACID, 3-[(7-CHLORO-4H-1,2,4-BENZOTHIAD
C10H9ClN4	072111-60-5	2,3-PYRAZINEDICARBONITRILE, 5-BUTYL-6-CHLORO-
C10H9ClO3	040026-24-2	2H-1-BENZOPYRAN-2-CARBOXYLIC ACID, 6-CHLORO-2,3-
C10H9F6N	002924-27-8	BENZENAMINE, N,N-BIS(2,2,2-TRIFLUOROETHYL)-
C10H9FN2O2	015948-56-8	MALONAMIDE,3-FLUOROBENZAL
C10H9FN2OS	095333-49-6	IMIDAZOLE-2-THIONE,3-(3-FLUORO-4-HYDROXYBENZYL)
C10H9FN2S	095333-80-5	IMIDAZOLE-2-THIONE,3-(3-FLUOROBENZYL)
C10H9N	000091-59-8	2-NAPHTHYLAMINE
C10H9N	000091-62-3	6-METHYLQUINOLINE
C10H9N	000091-63-4	2-METHYLQUINOLINE
C10H9N	000134-32-7	1-NAPHTHYLAMINE
C10H9N	000491-35-0	4-METHYLQUINOLINE
C10H9N	000611-32-5	8-METHYLQUINOLINE
C10H9N	000612-58-8	3-METHYLQUINOLINE
C10H9N	000612-60-2	7-METHYLQUINOLINE
C10H9N	001125-80-0	ISOQUINOLONE-3-METHYL
C10H9N	003042-22-6	1H-PYRROLE, 2-PHENYL-
C10H9N2O3S	006448-55-1	4-ME(5-NO2-2-FURFURILIDENE)THIAZOLE
C10H9N2O5S	025603-13-8	3ET-5-(5NO2-FURFURIL)THIAZOLIDIN-4-ON
C10H9N3	001202-34-2	N-2-PYRIDINYL-2-PYRIDINAMINE
C10H9N3O	005382-47-8	6-QUINOLINECARBOXYLIC ACID HYDRAZIDE
C10H9N3O	032451-61-9	8-UREIDO-QUINOLINE
C10H9N3O2	036874-74-5	C6H5NHN=C(CN)CO-OME
C10H9N3O4S	025694-31-9	2(N,N-DIME)-5-(5-NO2-2-FURFURYL)THIAZOLONE
C10H9N3O4S	027472-85-1	2-ETAM-5(5NO2-2-FURFURYL)THIAZOLONE
C10H9N5	066974-82-1	1-(3,4-DICYANOPH)-33-DIME TRIAZENE
C10H9N5O2	053409-75-9	2-PYRIMIDINAMINE, 4-[2-(1-METHYL-5-NITRO-1H-IMID
C10H9NO	000606-43-9	1-METHYL-2(H)-QUINOLINONE
C10H9NO	000826-81-3	2-METHYL 8-QUINOLONOL
C10H9NO	000938-33-0	8-METHOXYQUINOLINE

C10H9NO	003846-73-9	4-METHYL-8-QUINOLINOL
C10H9NO	004964-76-5	7-METHOXYQUINOLINE
C10H9NO	005263-87-6	6-METHOXYQUINOLINE
C10H9NO	005541-67-3	5-METHYL-8-QUINOLINOL
C10H9NO	025314-91-4	3-ACETYL INDOLE
C10H9NO	052986-70-6	6-METHOXYISOQUINOLINE
C10H9NO2	000083-25-0	N-PHENYLSUCCINIMIDE
C10H9NO2	000087-51-4	INDOLE-3-ACETIC ACID
C10H9NO2	000608-08-2	INDOLE-3-OL,ACETATE ESTER
C10H9NO2	000942-24-5	1H-INDOLE-3-CARBOXYLIC ACID, METHYL ESTER
C10H9NO2	001202-04-6	1H-INDOLE-2-CARBOXYLIC ACID, METHYL ESTER
C10H9NO2	002314-79-6	N-(P-TOLYL)SUCCINIMIDE
C10H9NO2	003567-38-2	3-PHENYL-CARBAMOYLOXY-1-PROPYNE
C10H9NO2	019860-27-6	5-PHENYL-2,4-PYRROLIDIONE
C10H9NO2	032387-21-6	1H-INDOLE-3-CARBOXYLIC ACID, 1-METHYL-
C10H9NO2	057334-34-6	4-METHYL-5,8-DIHYDROXYQUINOLINE
C10H9NO2	057334-35-7	5-METHOXY-8-QUINOLINOL
C10H9NO2	064663-59-8	2-BENZOFURANCARBOXAMIDE, N-METHYL-
C10H9NO2S	000606-25-7	1-NAPHTHALENESULFONAMIDE
C10H9NO2S	033321-31-2	5-BENZYL-2,4-THIAZOLIDIONE
C10H9NO2S	089770-29-6	6-METHYLSULFONYLQUINOLINE
C10H9NO3	005813-92-3	3,4-DIOXYMETHYLENECINNAMAMIDE
C10H9NO3	005841-62-3	5-BENZYL-2,4-OXAZOLIDIONE
C10H9NO3	005841-66-7	3-ME-5-PH OXAZOLIDINE-2,4-DIONE
C10H9NO3	073403-31-3	2-PROPENOIC ACID, 2-CYANO-3-(5-METHYL-2-FURANYL)
C10H9NO3S	000081-06-1	1-AMINO-2-NAPHTHALENESULFONIC ACID
C10H9NO3S	000081-16-3	2-AMINO-1-NAPHTHALENESULFONIC ACID
C10H9NO3S	000084-86-6	4-AMINO-1-NAPHTHALENESULFONIC ACID
C10H9NO3S	000119-79-9	1-NAPHTHYLAMINE-6-SULFONIC ACID
C10H9NO3S	076542-54-6	2-PROPENOIC ACID, 2-CYANO-3-[5-(METHYLTHIO)-2-FU
C10H9NO5S3	128348-42-5	2-THIOPHENESULFONAMIDE, 4-[(4-HYDROXYPHENYL)SULF
C10H9NO6S2	000086-65-7	2-NAPHTHYLAMINE-6,8-DISULFONIC ACID
C10H9NO6S2	000118-33-2	2-NAPHTHYLAMINE-5,7-DISULFONIC ACID
C10H9NO6S2	000131-27-1	3-AMINO-1,5-NAPHTHALENE DISULFONIC ACID
C10H9NO6S2	128348-44-7	2-FURANSULFONAMIDE, 4-[(4-HYDROXYPHENYL)SULFONYL
C10H9NOS	054466-88-5	INDOL-3-YL-THIOACETICACID
C10H9NS	040279-26-3	2-METHYLTHIOQUINOLINE
C10H10	000108-57-6	M-DIETHENYLBENZENE
C10H10	000447-53-0	1,2-DIHYDRONAPHTHALENE
C10H10BrN	002516-72-5	N-ME QUINOLINIUM BROMIDE
C10H10BrN5O	065052-55-3	2'-MEO-5'-BR BENZOGUANAMINE
C10H10Cl2O2	014437-17-3	CHLORFENPROP METHYL
C10H10Cl2O3	000094-82-6	2,4-DB
C10H10Cl2O3	000533-23-3	2,4-D, ETHYL ESTER
C10H10Cl3O4P	002274-67-1	PHOSPHORIC ACID,2-CHLORO-(2,4-DICHLOROPHENYL)VIN
C10H10ClFN4O2	132722-95-3	9H-PURINE, 6-CHLORO-9-(2,3-DIDEOXY-2-FLUORO-ß-D-
C10H10ClKO3	001929-86-8	MECOPROP, POTASSIUM SALT
C10H10ClN3O3S	037148-09-7	124BENZOTHIADIAZN-11O2-3ME-6NHAC-7CL
C10H10ClN5O	072459-58-6	TRIAZOXIDE
C10H10F3N3	040065-00-7	2-(2-CF3-PH-IMINO)IMIDAZOLIDINE
C10H10F3N3O2	111075-94-6	1-ACETYL-4-(3-CF3 PHENYL) SEMICARBAZIDE
C10H10F3N3O4	002077-99-8	26NO2-N-PR-AAA-TRIF-P-TOLUIDINE
C10H10N2	000479-27-6	1,8-NAPHTHALENEDIAMINE
C10H10N2	000771-97-1	2,3-NAPHTHALENEDIAMINE
C10H10N2	002243-62-1	1,5-DIAMINONAPHTHALENE
C10H10N2	004238-71-5	1H-IMIDAZOLE, 1-(PHENYLMETHYL)-
C10H10N2	007153-23-3	6,7-DIMETHYLQUINOXALINE
C10H10N2O	016867-29-1	4H-PYRIDO[1,2-A]PYRIMIDIN-4-ONE, 2,3-DIMETHYL-

C10H10N2O	069808-71-5	1H-INDOLE-2-CARBOXAMIDE, N-METHYL-
C10H10N2O	085729-23-3	1H-INDOLE-3-CARBOXAMIDE, N-METHYL-
C10H10N2O	088203-04-7	ACETOPHENONE,O-CYANOMETHYLAMINO
C10H10N2O	089770-34-3	6-ETHOXYQUINOXALINE
C10H10N2O2	002221-12-7	2,4-IMIDAZOLIDINEDIONE, 1-METHYL-3-PHENYL-
C10H10N2O2	002301-40-8	2,4-IMIDAZOLIDINEDIONE, 3-(PHENYLMETHYL)-
C10H10N2O2	002474-50-2	2,3-QUINOXALINEDIONE,1,4-DIHYDRO-6,7-DIMETHYL-
C10H10N2O2	003530-82-3	5-BENZYL-2,4-IMIDAZOLIDIONE
C10H10N2O2	006777-05-5	N-BENZYL-2,4-IMIDAZOLIDIONE
C10H10N2O2	019411-83-7	MALONAMIDE,BENZAL
C10H10N2O2	057315-37-4	1,4-DIMETHOXYPHTHALAZINE
C10H10N2O2	058175-07-8	N,N-DIME-QUINOXALINE-2,3-DIONE
C10H10N2O3	000715-48-0	6-METHYL-4-NITROQUINOLINE-1-OXIDE
C10H10N2O3	004831-62-3	2-METHYL-4-NITROQUINOLINE-1-OXIDE
C10H10N2O3	014073-00-8	3-METHYL-4-NITROQUINOLINE-1-OXIDE
C10H10N2O3	014094-43-0	5-METHYL-4-NITROQUINOLINE-1-OXIDE
C10H10N2O3	014094-45-2	8-METHYL-4-NITROQUINOLINE-1-OXIDE
C10H10N2O3	014753-13-0	7-METHYL-4-NITROQUINOLINE-1-OXIDE
C10H10N2O4S	004826-22-6	PROPANOIC ACID, 3-(4H-1,2,4-BENZOTHIADIAZIN-3-YL
C10H10N2O4S2	114260-76-3	ACETIC ACID, [(6-METHYL-4H-1,2,4-BENZOTHIADIAZIN
C10H10N2O4S2	124802-88-6	PROPANOIC ACID, 3-[(4H-1,2,4-BENZOTHIADIAZIN-3-Y
C10H10N2O5	002644-96-4	P-NITRO-METHYLHIPPURATE
C10H10N2O5S2	114260-77-4	ACETIC ACID, [(6-METHOXY-4H-1,2,4-BENZOTHIADIAZI
C10H10N2OS	028915-24-4	1,3,4-OXADIAZOLE, 2-METHYL-5-[(PHENYLMETHYL)THIO
C10H10N2OS	095333-64-5	IMIDAZOLE-2-THIONE,3-(4-HYDROXYBENZYL)
C10H10N2S2	042755-32-8	1,3,4-THIADIAZOLE, 2-METHYL-5-[(PHENYLMETHYL)THI
C10H10N3O2	103499-06-5	1,2,4-OXADOAZOL-5-ALDOXIME,5-BENZYL-
C10H10N3O4S	025603-08-1	4-THIAZOLIDINONE, 3-ETHYL-2-IMINO-5-[(5-NITRO-2-
C10H10N3O4S	025603-09-2	4-THIAZOLIDINONE, 3-METHYL-2-(MEHTYLIMINO)-5-[(5
C10H10N4O2	111076-02-9	1-ACETYL-4-(4-CYANO PHENYL)SEMICARBAZIDE
C10H10N4O2S	000068-35-9	SULFADIAZINE
C10H10N4O2S	000599-82-6	2-SULFANILAMIDOPYRIMIDINE
C10H10N4OS	021087-63-8	3-SME-4-NH2-6-PH-124-TRIAZIN-5-ONE
C10H10N6O2	062973-76-6	AZANIDAZOLE
C10H10O	000122-57-6	METHYL STYRYL KETONE
C10H10O	000529-34-0	3,4-DIHYDRO-1(2H)-NAPHTHALENONE
C10H10O2	000094-59-7	SAFROLE
C10H10O2	000103-26-4	CINNAMIC ACID, METHYL ESTER
C10H10O2	000120-58-1	1,2-METHYLENEDIOXY-4-PROPENYL BENZENE
C10H10O2	001009-61-6	P-ACETYLACETOPHENONE
C10H10O2	001199-77-5	ALPHA-METHYLCINNAMIC ACID
C10H10O2	001689-09-4	1(3H)ISOBENZOFURANONE-3,3-DIMETHYL
C10H10O2	006781-42-6	M-ACETYLACETOPHENONE
C10H10O3	000830-09-1	CINNAMIC ACID, P-METHOXY-
C10H10O3	000943-89-5	P-METHOXYCINNAMIC ACID (TRANS)
C10H10O3	001011-54-7	2-PROPENOIC ACID, 3-(2-METHOXYPHENYL)-, (E)-
C10H10O3	002051-95-8	BENZENEBUTANOIC ACID, GAMMA -OXO-
C10H10O3	005739-82-2	2-PROPENOIC ACID, 3-PHENOXY-, METHYL ESTER, (E)-
C10H10O3	013031-43-1	P-ACETOXYACETOPHENONE
C10H10O3	014737-91-8	2-PROPENOIC ACID, 3-(2-METHOXYPHENYL)-, (Z)-
C10H10O3	017570-26-2	2-PROPENOIC ACID, 3-(3-METHOXYPHENYL)-, (E)-
C10H10O3	017781-16-7	3-KETOCARBOFURANPHENOL
C10H10O3	051939-71-0	CHROMAN-2-CARBOXYLIC ACID
C10H10O4	000120-61-6	DIMETHYLTEREPHTHALATE
C10H10O4	000131-11-3	DIMETHYL PHTHALATE
C10H10O4	000580-02-9	ACETYLSALICYLIC ACID, METHYL ESTER
C10H10O4	001135-24-6	CINNAMIC ACID, 3-METHOXY-4-HYDROXY
C10H10O4	001878-62-2	O-ACETYLPHENOXYACETIC ACID

C10H10O4	001878-80-4	M-ACETYLPHENOXYACETIC ACID
C10H10O4	001878-81-5	P-ACETYLPHENOXYACETIC ACID
C10H10O4	007500-53-0	1,2-BENZENEDIACETIC ACID
C10H11BrN2O3	000561-86-4	BARBITURIC ACID,5-ALLYL-5(2BR ALLYL)
C10H11BrN4O2	132194-22-0	2-FURANMETHANOL, 5-(2-AMINO-6-BROMO-9H-PURIN-9-Y
C10H11BrN4O2	132194-25-3	2-FURANMETHANOL, 5-(6-BROMO-9H-PURIN-9-YL)TETRAH
C10H11Cl2N3	004201-33-6	2(2,6-DICL-4MEPH IMINO)IMIDAZOLIDINE
C10H11Cl2N3	004201-34-7	2(2,4-DICL-6MEPH IMINO)IMIDAZOLIDINE
C10H11Cl2N3	082801-84-1	2(2,3-DICL-6MEPH IMINO)IMIDAZOLIDINE
C10H11Cl2N3O	065936-24-5	2(2,6-DICL-4MEOPHIMINO)IMIDAZOLIDINE
C10H11Cl2NO	000882-14-4	PROPANAMIDE, N-(3,4-DICHLOROPHENYL)-
C10H11Cl2NO	002150-95-0	BUTANAMIDE, N-(3,4-DICHLOROPHENYL)-
C10H11Cl3O2	001861-44-5	TRITAC
C10H11ClN2O2S	037148-19-9	124-BENZOTHIADIAZINE-1-O2-3-PR-6-CL
C10H11ClN2O2S	037148-20-2	124-BENZOTHIADIAZIN-1-O2-3-IPR-6-CL
C10H11ClN2O2S2	037158-00-2	124-BENZTHIADIAZNE-1-O2-3-ETSME-6-CL
C10H11ClN4O2	120503-34-6	2-FURANMETHANOL, 5-(6-CHLORO-9H-PURIN-9-YL)TETRA
C10H11ClO2	025800-28-6	BENZOIC ACID, 2-CHLORO-, PROPYL ESTER
C10H11ClO3	000093-65-2	MECOPROP
C10H11ClO3	000882-09-7	2-(P-CLPHENOXY)2-ME-PROPIONIC ACID
C10H11ClO3	003547-07-7	4-(4-CHLOROPHENOXY) BUTYRIC ACID
C10H11ClO3	007085-19-0	MCPP
C10H11ClO3	014426-42-7	ETHYL 4-CHLOROPHENOXYACETATE
C10H11F2N3O2	111041-47-5	IMIDODICARBONIC DIAMIDE, N-(2,5-DIFLUOROPHENYL)-
C10H11F3N2O	002164-17-2	FLUOMETURON
C10H11F3N2O2	000838-89-1	1-MEO-1-ME-3-(CF3-PHENYL)UREA
C10H11F3N2O5	000070-00-8	TRIFLURIDINE
C10H11F7O2	017587-22-3	1,1,1,2,2,3,3-HEPTAFLUORO-7,7-DIMETHYL-4,6-OCTA*
C10H11FIN3O2	111041-55-5	IMIDODICARBONIC DIAMIDE, N-(2-FLUORO-5-IODOPHENY
C10H11FN4O2	132194-24-2	2-FURANMETHANOL, 5-(6-FLUORO-9H-PURIN-9-YL)TETRA
C10H11FN4O2	132722-90-8	9H-PURINE, 9-(2,3-DIDEOXY-2-FLUORO-ß-D-THREO-PEN
C10H11FN4O3	117525-25-4	6H-PURIN-6-ONE, 9-(2,3-DIDEOXY-2-FLUORO-ß-D-THRE
C10H11FN4O4	121788-15-6	IMIDODICARBONIC DIAMIDE, N-(2-FLUORO-5-NITROPHEN
C10H11IN4O2	120503-37-9	2-FURANMETHANOL, TETRAHYDRO-5-(6-IODO-9H-PURIN-9
C10H11N	000091-55-4	1H-INDOLE, 2,3-DIMETHYL-
C10H11N	000635-90-5	PYRROLE, 1-PHENYL-
C10H11N	000875-79-6	1,2-DIMETHYLINDOLE
C10H11N	002046-18-6	G-PHENYLPROPYLCYANIDE
C10H11N2O	055752-22-2	1,2,4-OXADIAZOLE,5-METHYL-3-(P-TOLYL)
C10H11N3O	001617-93-2	1,3,4-OXADIAZOL-2-AMINE, 5-(2-PHENYLETHYL)-
C10H11N3O2	000654-13-7	1H-BENZOTRIAZOLE-1-PROPANOIC ACID, ß-METHYL-
C10H11N3O2	000654-19-3	1-(3-CARBOXYPROPYL)BENZOTRIAZOLE
C10H11N3O2	004233-61-8	1H-BENZOTRIAZOLE-1-ACETIC ACID, α,α-DIMETHYL-
C10H11N3O2	004248-18-4	2H-BENZOTRIAZOLE-2-PROPANOIC ACID, ß-METHYL-
C10H11N3O2	016583-98-5	2H-BENZOTRIAZOLE-2-ACETIC ACID, α,α-DIMETHYL-
C10H11N3O2	069218-46-8	1H-BENZOTRIAZOLE-1-ACETIC ACID, ETHYL ESTER
C10H11N3O2	069218-51-5	2H-BENZOTRIAZOLE-2-ACETIC ACID, ETHYL ESTER
C10H11N3O2	096337-67-6	1-(3-CYANOPHENYL)-3-MEO-3-ME UREA
C10H11N3O3S	000723-46-6	SULFAMETHOXAZOLE
C10H11N3O5S	006954-35-4	4-NITROIMIDAZOLE1-ME-5-SO3PH
C10H11N3OS	018740-23-3	THIENO[2,3-D]PYRIMIDINE, 4-MORPHOLINYL-
C10H11N5	019388-12-6	4'-METHYLBENZOGUANAMINE
C10H11N5	030508-25-9	2'-METHYLBENZOGUANAMINE
C10H11N5	077314-61-5	2-PYRIDINAMINE, 4-(5-AMINO-1H-1,2,4-TRIAZOL-3-YL
C10H11N5O	066974-92-3	N-CYANO-2(33DIME-1-TRIAZEN)BENZAMIDE
C10H11N5O	072775-80-5	2'-METHOXYBENZOGUANAMINE
C10H11N5O2	007057-48-9	ADENOSINE, 2',3'-DIDEHYDRO-2',3'-DIDEOXY-
C10H11N5O2	058247-25-9	4-CYANO-3-MEO-PHENYLAMIDINO UREA

C10H11N5O2S	065052-50-8	3'-METHYLSULFONYLBENZOGUANAMINE
C10H11N5O3	053766-80-6	GUANOSINE, 2',3'-DIDEHYDRO-2',3'-DIDEOXY-
C10H11N5S	065052-49-5	3'-METHYLTHIOBENZOGUANAMINE
C10H11NO	002757-10-0	N-METHYLCINNAMAMIDE
C10H11NO	004053-42-3	6-METHYLQUINOLINE, 1-OXIDE
C10H11NO	073159-84-9	9-OXABENZOBICYCLO(221)HEPTENE,2EXNH2
C10H11NO	073208-84-1	9-OXABENZOBICYCLO(221)HEPTENE,2ENNH2
C10H11NO2	000102-01-2	ACETOACETANILIDE
C10H11NO2	001202-32-0	B-ETHYL-B-NITROSTYRENE
C10H11NO2	002719-21-3	ACETAMIDE, N-(4-ACETYLPHENYL)-
C10H11NO2	005234-26-4	ACETOPHENONE, O-ACETYLAMINO
C10H11NO2	007463-31-2	M-ACETYLAMINOACETOPHENONE
C10H11NO2	053643-53-1	TUBERIN
C10H11NO3	001135-43-9	N-METHYL-4-ACETYLPHENYLCARBAMATE
C10H11NO3	001205-08-9	METHYL N-BENZOGLYCINE
C10H11NO3	001457-85-8	OXANILIC ACID,ETHYL ESTER
C10H11NO3	002486-77-3	3-ALLYLOXY-4-AMINOBENZOIC ACID
C10H11NO3	002623-33-8	P-ACETOXYACETANILIDE
C10H11NO3	002719-08-6	BENZOIC ACID, 2-(ACETYLAMINO)-, METHYL ESTER
C10H11NO3	006317-89-1	M-ACETOXYACETANILIDE
C10H11NO3	013414-55-6	7-NITRO-2,2-DIMETHYL-2,3-DIHYDROBENZOFURAN
C10H11NO3	017012-22-5	P-METHOXYCARBONYLACETANILIDE
C10H11NO3	052189-36-3	BENZOIC ACID, 3-(ACETYLAMINO)-, METHYL ESTER
C10H11NO3	078944-67-9	BENZOIC ACID,O-ACETYLMETHYLAMINO
C10H11NO3	087743-55-3	CARBAMIC ACID, (3-ACETYLPHENYL)-, METHYL ESTER
C10H11NO3	106231-50-9	ACETAMIDE, 2-(BENZOYLOXY)-N-METHYL-
C10H11NO3S	020872-53-1	4-METHYLSULFONYLQUINOLINE-1-OXIDE
C10H11NO4	006339-04-4	PHENOXYACETIC ACID, M-ACETAMIDO-
C10H11NO4	021998-12-9	N-ME-4-CARBOMEO PHENYLCARBAMATE
C10H11NO4	102273-25-6	BENZAMIDE, 2-[(ACETYLOXY)METHOXY]-
C10H11NO4Se	034835-05-7	P-NITROBENZYLSELENOPROPIONIC ACID
C10H11NO5S3	128348-43-6	2-THIOPHENESULFONAMIDE, 4-[(3-HYDROXYPHENYL)SULF
C10H12	000077-73-6	DICYCLOPENTADIENE
C10H12	000119-64-2	TETRALIN
C10H12	000768-49-0	(2-METHYL-1-PROPENYL)BENZENE
C10H12	003454-07-7	P-ETHYLSTYRENE
C10H12	007525-62-4	M-ETHYLSTYRENE
C10H12Br2N6O2	113589-14-3	1H-PYRROLE-2-CARBOXAMIDE, 4,5-DIBROMO-3-(1-METHY
C10H12BrCl2O3PS	004824-78-6	BROMOPHOS ETHYL
C10H12BrN3	016822-80-3	2-IMIDAZOLIDINIMINE, N-(5-BROMO-2-METHYLPHENYL)-
C10H12BrN3	016822-94-9	2-IMIDAZOLIDINIMINE, N-(3-BROMO-2-METHYLPHENYL)-
C10H12BrN5O2S	113588-95-7	2-THIOPHENECARBOXAMIDE, 4-BROMO-5-METHYL-3-(1-ME
C10H12BrN5O3S	113589-05-2	2-THIOPHENECARBOXAMIDE, 4-BROMO-5-METHOXY-3-(1-M
C10H12BrNO	052121-41-2	3-ETHYL-4-BROMOACETANILIDE
C10H12BrNO	064835-48-9	ACETAMIDE, N-(4-BROMO-3,5-DIMETHYLPHENYL)-
C10H12BrNO2	040912-88-7	BENZAMIDE, 5-BROMO-N-ETHYL-2-HYDROXY-3-METHYL-
C10H12BrNO2	084640-33-5	P-BROMOBENZYL N,N-DIME CARBAMATE
C10H12BrNO3	017199-23-4	2-BROMO-4-(TERT-BUTYL)-6-NITROPHENOL
C10H12Cl2O	034593-75-4	4-(TERT-BUTYL)-2,6-DICHLOROPHENOL
C10H12Cl3O2PS	000327-98-0	TRICHLORONATE
C10H12ClN	026232-35-9	6-CHLORO-BENZAZEPINE
C10H12ClN3	004201-22-3	2-(2-CL-4-ME-PHIMINO)IMIDAZOLIDINE
C10H12ClN3	004201-24-5	2-(2-CL-6-ME-PHIMINO)IMIDAZOLIDINE
C10H12ClN3	004201-26-7	2-(4-CL-2-ME-PHIMINO)IMIDAZOLIDIN
C10H12ClN3	016822-82-5	BENZENAMINE, 2-CHLORO-N-(2-IMIDAZOLIDINYLIDENE)-
C10H12ClN3	016822-85-8	2(5CL-2ME-PHENYLIMINO)IMIDAZOLIDINE
C10H12ClN3	016822-97-2	2-IMIDAZOLIDINIMINE, N-(2-CHLORO-3-METHYLPHENYL)
C10H12ClN3O3	013909-21-2	1-(2-CLET)-3-(M-MEOPH)-1-N=O UREA

C10H12ClN3O4S2	002854-99-1	3-PROPYLCHLOROTHIAZIDE
C10H12ClN5O2	122970-35-8	9H-PURIN-2-AMINE, 6-CHLORO-9-[TETRAHYDRO-5-(HYDR
C10H12ClNO2	000101-21-3	CHLORPROPHAM
C10H12ClNO2	000671-04-5	BANOL
C10H12ClNO2	001134-47-0	BENZENEPROPANOIC ACID, ß-(AMINOMETHYL)-4-CHLORO-
C10H12ClNO2	058708-52-4	BENZAMIDE, 5-CHLORO-N-ETHYL-2-HYDROXY-3-METHYL-
C10H12ClNO2	062100-41-8	PROPANAMIDE, N-(4-CHLOROPHENYL)-2-HYDROXY-2-METH
C10H12ClNO2	084640-26-6	M-CHLOROBENZYL N,N-DIME CARBAMATE
C10H12ClNO2	084640-27-7	P-CHLOROBENZYL N,N-DIME CARBAMATE
C10H12ClNO3	014593-28-3	4-(TERT-BUTYL)-2-CHLORO-6-NITROPHENOL
C10H12FN3	028125-87-3	2-IMIDAZOLIDINIMINE, N-(5-FLUORO-2-METHYLPHENYL)
C10H12FN3O2	053285-95-3	IMIDODICARBONIC DIAMIDE, N-(2-FLUOROPHENYL)-N',2
C10H12FN5O2	087418-35-7	ADENOSINE, 2',3'-DIDEOXY-3'-FLUORO-
C10H12FN5O2	110143-05-0	ADENOSINE, 2',3'-DIDEOXY-2'-FLUORO-
C10H12FN5O2	110143-10-7	9H-PURIN-6-AMINE, 9-(2,3-DIDEOXY-2-FLUORO-ß-D-TH
C10H12FN5O2	132194-21-9	2-FURANMETHANOL, 5-(2-AMINO-6-FLUORO-9H-PURIN-9-
C10H12FN5O3	092562-88-4	GUANOSINE, 2',3'-DIDEOXY-3'-FLUORO-
C10H12FNO	062008-55-3	2-(M-FLUOROPHENYL)MORPHOLINE
C10H12FNO2	052237-19-1	BENZENEPROPANOIC ACID, ß-(AMINOMETHYL)-4-FLUORO-
C10H12FNO2	084640-22-2	P-FLUOROBENZYL N,N-DIME CARBAMATE
C10H12IN5O2	132194-23-1	2-FURANMETHANOL, 5-(2-AMINO-6-IODO-9H-PURIN-9-YL
C10H12INO2	084640-34-6	P-IODOBENZYL N,N-DIME CARBAMATE
C10H12N2	000059-98-3	1H-IMIDAZOLE, 4,5-DIHYDRO-2-(PHENYLMETHYL)-
C10H12N2	000061-54-1	TRYPTAMINE
C10H12N2	000525-74-6	4,5-DIHYDRONICOTYRINE
C10H12N2Na4O8	000064-02-8	EDTA, SODIUM SALT
C10H12N2O	000050-67-9	5-HYDROXYTRYPTAMINE
C10H12N2O	000486-56-6	2-PYRROLIDINONE, 1-METHYL-5-(3-PYRIDINYL)-, (S)-
C10H12N2O	002403-66-9	2-(G-HYDROXYPROPYL)BENZIMIDAZOLE
C10H12N2O	013140-86-8	N-CYCLOPROPYL-N'-PHENYLUREA
C10H12N2O	016867-28-0	4H-PYRIDO(12A)PYRIMDIN-4-ON,2,6-DIME
C10H12N2O	039080-46-1	PYRIDO(12A)PYRIMIDIN-4-ONE,3,6-DIME
C10H12N2O2	000713-05-3	3-PYRIDINEBUTANAMIDE, N-METHYL-¥-OXO-
C10H12N2O2	002050-85-3	ACETAMIDE, N,N'-(1,2-PHENYLENE)BIS-
C10H12N2O2	002604-08-2	B-NITRO-4-DIMETHYLAMINOSTYRENE
C10H12N2O2	010268-78-7	M-DIACETAMIDOBENZENE
C10H12N2O2	027545-04-6	BENZAMIDE, O-ACETYLMETHYLAMINO
C10H12N2O2	028615-21-6	UREA, N-ETHYL-N'-BENZOYL-
C10H12N2O2	034834-67-8	2-PYRROLIDINONE, 3-HYDROXY-1-METHYL-5-(3-PYRIDIN
C10H12N2O2	037096-14-3	2-PYRROLIDINONE, 3-HYDROXY-1-METHYL-5-(3-PYRIDIN
C10H12N2O2	061192-50-5	2-PYRROLIDINONE, 5-HYDROXY-1-METHYL-5-(3-PYRIDIN
C10H12N2O2	080431-14-7	2-FURANCARBOXAMIDE, N-(3,6-DIHYDRO-1(2H)-PYRIDIN
C10H12N2O2S	037148-03-1	124-BENZOTHIADIAZINE-1-O2-3-ME-6-ET
C10H12N2O3	000052-43-7	ALLOBARBITAL
C10H12N2O3	000606-38-2	2,4-DIME 6-NO2 ACETANILIDE
C10H12N2O3	005259-86-9	P-AMINOHIPPURIC ACID, ME ESTER
C10H12N2O3	080715-99-7	4H-PYRIDO[1,2-A]PYRIMIDIN-3-CARBOXYLIC ACID, 6,7
C10H12N2O3	098736-51-7	BENZOIC ACID, 4-[(METHYLNITROSOAMINO)METHYL]-, M
C10H12N2O3S	005983-08-4	THYMIDINE, 2',3'-DIDEHYDRO-3'-DEOXY-4-THIO-
C10H12N2O3S	025057-89-0	BENTAZONE
C10H12N2O3S2	088946-19-4	BENZOTHIAZOLE,2-N-ME SULFONAMIDO-6ETO
C10H12N2O4	002427-90-9	2,3-PYRAZINEDICARBOXYLIC ACID, DIETHYL ESTER
C10H12N2O4	003056-17-5	THYMIDINE, 2',3'-DIDEHYDRO-3'-DEOXY-
C10H12N2O4	062100-54-3	PROPANAMIDE, 2-HYDROXY-2-METHYL-N-(4-NITROPHENYL
C10H12N2O4	084640-31-3	P-NITROBENZYL N,N-DIME CARBAMATE
C10H12N2O5	000088-85-7	DINOSEB
C10H12N2O5	004097-49-8	4-(TERT-BUTYL)-2,6-DINITROPHENOL
C10H12N2O5S	081792-98-5	BENZENESULFONAMIDE, 2-NITRO-N-(2-METHYL-1-OXOPRO

C10H12N2OS	080431-15-8	2-THIOPHENECARBOXAMIDE, N-(3,6-DIHYDRO-1(2H)-PYR
C10H12N3O3P	067410-39-3	3,5-DIOXA-6-AZA-4-PHOSPAOCT-6-ENE-8-NITRILE, 4-A
C10H12N3O3PS2	000086-50-0	METHYL AZINPHOS
C10H12N3O4PS	000961-22-8	METHYL AZINPHOS O-ANALOG
C10H12N4O	018597-53-0	PYRAZOL-3-ONE,2H1(4,6-ME2PYRIMID)5ME
C10H12N4O	018597-57-4	PYRAZOL-3-ONE,2(4,6-ME4PYRIMID)5ME
C10H12N4O	018697-50-2	PYRAZOL-3-ONE,1(4,6-ME2PYRIMID)5ME
C10H12N4O	069218-58-2	1-(3-CARBAMYLPROPYL)BENZOTRIAZOLE
C10H12N4O	099069-47-3	1H-BENZOTRIAZOLE-1-PROPANAMIDE, ß-METHYL-
C10H12N4O	115054-75-6	1H-BENZOTRIAZOLE-1-ACETAMIDE, α-ETHYL-
C10H12N4O	115054-76-7	2H-BENZOTRIAZOLE-2-ACETAMIDE, α-ETHYL-
C10H12N4O	115054-77-8	1H-BENZOTRIAZOLE-1-ACETAMIDE, α,α-DIMETHYL-
C10H12N4O	115054-85-8	2H-BENZOTRIAZOLE-2-ACETAMIDE, α,α-DIMETHYL-
C10H12N4O	115054-94-9	2H-BENZOTRIAZOLE-2-PROPANAMIDE, ß-METHYL-
C10H12N4O2	018597-55-2	PYRAZOL-3-ONE,2(2ME-6MEO PYRIMID)5ME
C10H12N4O2	023898-90-0	PYRAZOL-3-ON,2(2MEO-6ME-4PYRIMD)5-ME
C10H12N4O2	023906-03-8	PYRAZOL-3-ON,2(6MEO-4-PYRIMDIN)15DIME
C10H12N4O2S2	000094-19-9	ETHAZOLE
C10H12N4O3	023905-85-3	PYRAZOL-3-ON,1(46DIMEO-2PYRIMDIN)5ME
C10H12N4O3	069655-05-6	INOSINE, 2',3'-DIDEOXY-
C10H12N4O3	098827-22-6	BUTANOIC ACID, (2,5-DIHYDRO-4-OXO-4H-PYRAZOLO[3,
C10H12N4O4	000890-38-0	INOSINE, 2'-DEOXY-
C10H12N4O4S2	037427-69-3	2(1-ACETOXYET-13THIAZOLIDINYLIDEN-2AM)5NO2-THIAZ
C10H12N4O5	000058-63-9	INOSINE
C10H12N5O6P	000060-92-4	ADENOSINE, CYCLIC 3',5'-(HYDROGEN PHOSPHATE)
C10H12N6O2	109881-25-6	ADENOSINE, 2-AMINO-2',3'-DIDEHYDRO-2',3'-DIDEOXY
C10H12N8O2	066323-44-2	ADENOSINE, 3'-AZIDO-2',3'-DIDEOXY-
C10H12O	000104-46-1	ANETHOLE
C10H12O	000495-40-9	1-BUTANONE, 1-PHENYL-
C10H12O	000529-33-9	1-NAPHTHALENOL, 1,2,3,4-TETRAHYDRO-
C10H12O	000529-35-1	1-NAPHTHALENOL, 5,6,7,8-TETRAHYDRO-
C10H12O	000611-70-1	1-PROPANONE, 2-METHYL-1-PHENYL-
C10H12O	000936-98-1	TR-2-PHENYLCYCLOPROPYLCARBINOL
C10H12O	000937-30-4	4-ETHYLACETOPHENONE
C10H12O	001125-78-6	2-NAPHTHALENOL, 5,6,7,8-TETRAHYDRO-
C10H12O	004180-23-8	ANETHOLE (TRANS)
C10H12O	005337-93-9	1-(4-METHYLPHENYL)-1-PROPANONE
C10H12O	013037-71-3	4-(2-BUTENYL)PHENOL
C10H12O	020944-88-1	2-(2-METHYLALLYL)PHENOL
C10H12O	022699-70-3	3-ETHYLACETOPHENONE
C10H12O	032783-20-3	BENZENE, [[(1-METHYLETHENYL)OXY]METHYL]-
C10H12O2	000094-58-6	DIHYDROSAFROLE
C10H12O2	000097-53-0	EUGENOL
C10H12O2	000097-54-1	2-METHOXY-4-(1-PROPENYL)PHENOL
C10H12O2	000101-97-3	BENZENEACETIC ACID, ETHYL ESTER
C10H12O2	000103-25-3	METHYL B-PHENYLPROPIONATE
C10H12O2	000103-45-7	B-PHENYLETHYL ACETATE
C10H12O2	000490-91-5	2,5-CYCLOHEXADIENE-1,4-DIONE, 2-METHYL-5-(1-METH
C10H12O2	000527-17-3	DUROQUINONE
C10H12O2	000536-66-3	CUMIC ACID
C10H12O2	000672-76-4	5-I-PROPYLTROPALONE
C10H12O2	000876-98-2	PHENOL, 2,6-DIMETHYL-, ACETATE
C10H12O2	000939-48-0	BENZOIC ACID, 1-METHYLETHYL ESTER
C10H12O2	001563-38-8	CARBOFURAN PHENOL
C10H12O2	001821-12-1	4-PHENYLBUTYRIC ACID
C10H12O2	002216-45-7	P-METHYLBENZYL ACETATE
C10H12O2	002315-68-6	PROPYL BENZOATE
C10H12O2	002438-04-2	O-ISOPROPYLBENZOIC ACID

C10H12O2	002438-05-3	4-PROPYLBENZOIC ACID
C10H12O2	003056-59-5	2-ETHYLPHENYL ACETATE
C10H12O2	003056-60-8	ACETIC ACID, 3-ETHYLPHENYL ESTER
C10H12O2	003245-23-6	4-ETHYLPHENYL ACETATE
C10H12O2	004593-90-2	3-PHENYL-N-BUTYRIC ACID
C10H12O2	014387-10-1	BENZENEACETIC ACID, 4-ETHYL-
C10H12O2	014920-81-1	BENZOIC ACID, 2,6-DIMETHYL-, METHYL ESTER
C10H12O2Se	006926-05-2	PROPANOIC ACID, [(PHENYLMETHYL)SELENO]-
C10H12O3	000094-13-3	P-HYDROXY PROPYL BENZOATE
C10H12O3	000607-90-9	BENZOIC ACID, 2-HYDROXY-, PROPYL ESTER
C10H12O3	000774-40-3	MANDELIC ACID, ETHYL ESTER
C10H12O3	001798-03-4	O-ETHYL PHENOXYACETIC ACID
C10H12O3	001878-51-9	M-ETHYLPHENOXYACETIC ACID
C10H12O3	001929-29-9	P-MEO-B-PHENYLPROPIONIC ACID
C10H12O3	002989-17-5	O-ME PHENOXYACETIC ACID,ME ESTER
C10H12O3	004919-33-9	P-ETHOXYPHENYLACETIC ACID
C10H12O3	005438-19-7	4-PROPOXYBENZOIC ACID
C10H12O3	035480-26-3	M-METHOXYBENZYL ACETATE
C10H12O3	057453-98-2	BENZOIC ACID, 2-METHOXYETHYL ESTER
C10H12O4	000483-54-5	2,3-DIMETHOXY-5,6-DIMETHYLBENZOQUINONE
C10H12O4	000999-21-3	DIALLYL MALEATE
C10H12O4	064673-04-7	BENZALDEHYDE, 4-(2-HYDROXYETHOXY)-3-METHOXY-
C10H12O5	000121-79-9	PROPYL GALLATE
C10H12O5	000573-11-5	2,3,4-TRIMETHOXYBENZOIC ACID
C10H12O5	001138-60-9	ISOPROPYL GALLATE
C10H12O5S	015267-77-3	ACETIC ACID, [3-METHYL-4-(METHYLSULFONYL)PHENOXY
C10H12O6	003117-06-4	2,3,5,6-TETRAMETHOXY-P-BENZOQUINONE
C10H13BrN2O3	000545-93-7	BARBITURIC ACID,5-IPR,5(2BR ALLYL)
C10H13BrN5O6P	039023-66-0	8-BROMO CYCLIC AMP
C10H13Cl2FN2O2S2	000731-27-1	TOLYFLUANIDE
C10H13Cl2N	000553-27-5	N,N-DI-B-CHLOROETHYLANILINE
C10H13Cl2NO3	002008-39-1	2,4-D, DIMETHYLAMINE SALT
C10H13Cl2O3PS	000097-17-6	DICHLOFENTHION
C10H13Cl3NO3P	002213-85-6	O=P(OME)(NPR)O-2,4,5-CLPHENYL
C10H13ClN2	006164-98-3	CHLORDIMEFORM
C10H13ClN2	006640-24-0	PIPERAZINE, 1-(3-CHLOROPHENYL)-
C10H13ClN2	038212-33-8	PIPERAZINE, 1-(4-CHLOROPHENYL)-
C10H13ClN2	039512-50-0	PIPERAZINE, 1-(2-CHLOROPHENYL)-
C10H13ClN2O	015545-48-9	CHLORTOLURON
C10H13ClN2O2	019937-59-8	METOXURON
C10H13ClN2O2	054922-65-5	N'-ET(4-CLPHENOXY)ACETIC A HYDRAZIDE
C10H13ClN2O3S	090875-63-1	PROPANAMIDE, N-[2-(AMINOSULFONYL)-5-CHLOROPHENYL
C10H13ClO	000089-68-9	CHLOROTHYMOL
C10H13ClO	000098-28-2	4-(TERT-BUTYL)-2-CHLOROPHENOL
C10H13FN2	002252-63-3	PIPERAZINE, 1-(4-FLUOROPHENYL)-
C10H13FN2O4	025526-93-6	THYMIDINE, 3'-DEOXY-3'-FLUORO-
C10H13FN4O3	132723-00-3	9H-PURINE, 9-(2,3-DIDEOXY-2-FLUORO-ß-D-THREO-PEN
C10H13FN6O2	114753-53-6	ADENOSINE, 2-AMINO-2',3'-DIDEOXY-3'-FLUORO-
C10H13N	001006-64-0	PYRROLIDINE, 2-PHENYL-
C10H13N	002954-50-9	2-NAPHTHALENAMINE, 1,2,3,4-TETRAHYDRO-
C10H13N	004096-21-3	1-PHENYLPYRROLIDINE
C10H13N	029726-60-1	ISOQUINOLINE, 1,2,3,4-TETRAHYDRO-3-METHYL-
C10H13N11O2	119644-21-2	ADENOSINE, 2',3'-DIAZIDO-2',3'-DIDEOXY-
C10H13N2O4	001951-56-0	N-IPR-3-(5-NO2-2-FURYL)ACRYLAMIDE
C10H13N3	000708-43-0	1-BUTYLBENZOTRIAZOLE
C10H13N3	001131-64-2	2(1H)-ISOQUINOLINECARBOXIMIDAMIDE, 3,4-DIHYDRO-
C10H13N3	036318-56-6	2-IMIDAZOLIDINIMINE, N-(2-METHYLPHENYL)-
C10H13N3	063936-04-9	1-S-BUTYLBENZOTRIAZOLE

C10H13N3O	027843-08-9	W-PHENYLPROPIONALDEHYDE SEMICARBIZONE
C10H13N3O	066975-11-9	P-(3-ME-3-ACETYL-1-TRIAZENO)TOLUENE
C10H13N3O	069218-35-5	1-(4-HYDROXYBUTYL)BENZOTRIAZOLE
C10H13N3O2	004144-70-1	2-(3-CARBOXYPROPYL)BENZOTRIAZOLE
C10H13N3O2	017075-29-5	1-ACETYL-4-(4-ME PHENYL) SEMICARBAZIDE
C10H13N3O2	033484-45-6	4H-PYRIDO[1,2-A]PYRIMIDINE-3-CARBOXAMIDE, 6,7,8,
C10H13N3O2	078326-57-5	1-I-PROPYL-1-NITROSO-3-PHENYLUREA
C10H13N3O2	111075-86-6	1-ACETYL-4-(3-ME PHENYL) SEMICARBAZIDE
C10H13N3O3	017075-30-8	1-ACETYL-4-(4-OME PHENYL) SEMICARBAZIDE
C10H13N3O3	111075-95-7	1-ACETYL-4-(3-OME PHENYL) SEMICARBAZIDE
C10H13N3O4	090597-22-1	CPE-C
C10H13N3O4S2	037157-82-7	124-BENZTHIDIAZIN-1O2-3ME-7SO2N(DIME)
C10H13N5O	087469-33-8	2H-PURIN-2-ONE, 1,3,6,9-TETRAHYDRO-6-IMINO-1-MET
C10H13N5O2	004097-22-7	ADENOSINE, 2',3'-DIDEOXY-
C10H13N5O2	083293-76-9	5(33-DIME-1-TRIAZENO)ISOPHTHALAMIDE
C10H13N5O3	000958-09-8	ADENOSINE, 2'-DEOXY-
C10H13N5O3	085326-06-3	GUANOSINE, 2',3'-DIDEOXY-
C10H13N5O3S	000789-61-7	B-2'-DEOXYTHIOGUANOSINE
C10H13N5O3S	108441-45-8	THYMIDINE, 3'-AZIDO-3'-DEOXY-4-THIO-
C10H13N5O4	000058-61-7	ADENOSINE
C10H13N5O4	005536-17-4	9H-PURIN-6-AMINE, 9-.BETA.-D-ARABINOFURANOSYL-
C10H13N5O5	000118-00-3	GUANOSINE
C10H13N9O2	114753-52-5	ADENOSINE, 2-AMINO-3'-AZIDO-2',3'-DIDEOXY-
C10H13NO	000134-98-5	2,3-DIMETHYLACETANILIDE
C10H13NO	000619-76-1	4-ISOPROPYLBENZAMIDE
C10H13NO	000877-95-2	ACETAMIDE, N-(2-PHENYLETHYL)-
C10H13NO	000940-43-2	BENZENEPROPANAMIDE, N-METHYL-
C10H13NO	001129-50-6	BUTYRANILIDE
C10H13NO	001199-98-0	BUTRAMIDE, 4-PHENYL
C10H13NO	002050-43-3	2,4-DIMETHYLACETANILIDE
C10H13NO	002050-44-4	2,5-DIMETHYLACETANILIDE
C10H13NO	002050-45-5	3,5-DIMETHYLACETANILIDE
C10H13NO	002124-31-4	P-ACETYL-N,N-DIMETHYLANILINE
C10H13NO	002198-53-0	2,6-DIMETHYLACETANILIDE
C10H13NO	002198-54-1	3,4-DIMETHYLACETANILIDE
C10H13NO	003663-34-1	4-ETHYLACETANILIDE
C10H13NO	004406-41-1	I-BUTYRANILIDE
C10H13NO	005465-00-9	BENZENEACETAMIDE, N-ETHYL-
C10H13NO	006935-65-5	BENZAMIDE, N,N,3-TRIMETHYL-
C10H13NO	013575-92-3	1-NAPHTHALENOL, 2-AMINO-1,2,3,4-TETRAHYDRO-, CIS
C10H13NO	013917-17-4	1-NAPHTHALENOL, 2-AMINO-1,2,3,4-TETRAHYDRO-, TRA
C10H13NO	014062-78-3	BENZAMIDE, N,N,4-TRIMETHYL-
C10H13NO	018925-69-4	BENZENEACETAMIDE, N,N-DIMETHYL-
C10H13NO	023972-41-0	2-PHENYLMORPHOLINE
C10H13NO	025079-96-3	N-((4-METHYLPHENYL)METHYL)ACETAMIDE
C10H13NO	026819-07-8	BENZAMIDE, N-ETHYL-3-METHYL-
C10H13NO	026819-08-9	BENZAMIDE, N-ETHYL-4-METHYL-
C10H13NO	033098-65-6	2-ETHYLACETANILIDE
C10H13NO	056177-33-4	2-ISOPROPYLBENZAMIDE
C10H13NO	068298-46-4	7-AMINO-2,2-DIMETHYL-2,3-DIHYDROBENZOFURAN
C10H13NO	069163-87-7	3-PH PROPIONALDEHYDE OXIME,ME ETHER
C10H13NO2	000062-44-2	P-PHENACETIN
C10H13NO2	000094-12-2	PROPYL-P-AMINOBENZOATE
C10H13NO2	000122-42-9	ISOPROPYL PHENYL CARBAMATE
C10H13NO2	000581-08-8	ACETAMIDE, N-(2-ETHOXYPHENYL)-
C10H13NO2	000591-33-3	3-ETHOXYACETANILIDE
C10H13NO2	000673-31-4	CARBAMIC ACID, 3-PHENYLPROPYL ESTER
C10H13NO2	002011-57-6	2,3-DIME-4-HYDROXYACETANILIDE

C10H13NO2	002425-10-7	3,4-DIMEPHENYL N-METHYLCARBAMATE
C10H13NO2	002631-30-3	N-METHYL-4-ETHYLPHENYLCARBAMATE
C10H13NO2	002631-42-7	N-METHYL-2-ETHYLPHENYLCARBAMATE
C10H13NO2	002655-12-1	N-ME-2,3-DIMETHYLPHENYL CARBAMATE
C10H13NO2	002655-14-3	N-ME-3,5-DIMETHYLPHENYL CARBAMATE
C10H13NO2	003382-56-7	4-T-BUTYLNITROBENZENE
C10H13NO2	003971-99-1	N-ME-2,5-DIMETHYLPHENYLCARBAMATE
C10H13NO2	004043-23-6	N-ME-M-ETHYLPHENYLCARBAMATE
C10H13NO2	004764-17-4	3,4-METHYLENEDIOXYAMPHETAMINE
C10H13NO2	005532-90-1	PROPYL N-PHENYLCARBAMATE
C10H13NO2	006337-56-0	2,6-DIME-4-HYDROXYACETANILIDE
C10H13NO2	006938-06-3	NICOTINIC ACID, BUTYL ESTER
C10H13NO2	007290-99-5	M-METHOXY-N,N-DIMETHYLBENZAMIDE
C10H13NO2	007291-00-1	P-METHOXY-N,N-DIMETHYLBENZAMIDE
C10H13NO2	007291-34-1	O-MEO N,N-DIMETHYLBENZAMIDE
C10H13NO2	007305-06-8	O-TOLYLDIMETHYLCARBAMATE
C10H13NO2	007305-07-9	M-TOLYLDIMETHYLCARBAMATE
C10H13NO2	007305-08-0	P-TOLYLDIMETHYLCARBAMATE
C10H13NO2	007403-41-0	BENZAMIDE, N-ETHYL-4-METHOXY-
C10H13NO2	010397-59-8	N,N-DIMETHYLPHENOXYACETAMIDE
C10H13NO2	010507-52-5	BENZYL N,N-DIMETHYLCARBAMATE
C10H13NO2	022900-79-4	3,5-DIMETHYL-4-HYDROXYACETANILIDE
C10H13NO2	028026-77-9	3-ETHYL-4-HYDROXYACETANILIDE
C10H13NO2	028238-55-3	CARBAMIC ACID, (4-ETHYLPHENYL)-, METHYL ESTER
C10H13NO2	035103-34-5	N-((4-METHOXYPHENYL)METHYL)ACETAMIDE
C10H13NO2	035813-38-8	N-METHYL P-METHOXYACETANILIDE
C10H13NO2	040673-68-5	ACETOPHENONE, O-(2-OHET)AMINO
C10H13NO2	050868-74-1	2-METHYL-3-METHOXYACETANILIDE
C10H13NO2	051497-09-7	1,3-BENZODIOXOLE-5-ETHANAMINE, _-METHYL-(±)-
C10H13NO2	069477-71-0	2,5-DIME-4-HYDROXYACETANILIDE
C10H13NO2	088203-06-9	ANISOLE, O-ACETYLMETHYLAMINO
C10H13NO2	113932-82-4	CARBAMIC ACID, (3-ETHYLPHENYL)-, METHYL ESTER
C10H13NO2	131802-71-6	BENZENEACETAMIDE, α-HYDROXY-α-ETHYL-, (±)-
C10H13NO2S	003566-00-5	N-ME CARBAMATE,3-ME-4-MES PHENYL
C10H13NO3	000658-48-0	TYROSINE, A-METHYL-
C10H13NO3	001080-06-4	L-TYROSINE, METHYL ESTER
C10H13NO3	002486-79-5	2-PROPOXY-4-AMINOBENZOIC ACID
C10H13NO3	003279-07-0	4-(TERT-BUTYL)-2-NITROPHENOL
C10H13NO3	003555-18-8	4-(SEC-BUTYL)-2-NITROPHENOL
C10H13NO3	005722-68-9	3-T-BUTYL-4-NITROPHENOL
C10H13NO3	007225-96-9	N-METHYL-3-ETHOXYPHENYLCARBAMATE
C10H13NO3	007244-78-2	P-NITROPHENYL BUTYL ETHER
C10H13NO3	007305-09-1	3-METHOXYPHENYLDIMETHYLCARBAMATE
C10H13NO3	007305-10-4	4-METHOXYPHENYLDIMETHYLCARBAMATE
C10H13NO3	013538-54-0	N-METHYL-4-ETHOXYPHENYLCARBAMATE
C10H13NO3	023409-17-8	N-ME-2-ETHOXYPHENYLCARBAMATE
C10H13NO3	035407-50-2	CARBAMIC ACID, (4-ETHOXYPHENYL)-, METHYL ESTER
C10H13NO3	113932-79-9	CARBAMIC ACID, (2-ETHOXYPHENYL)-, METHYL ESTER
C10H13NO3	113932-83-5	CARBAMIC ACID, (3-ETHOXYPHENYL)-, METHYL ESTER
C10H13NO3S	016697-83-9	ACETAMIDE, N-METHYL-N-[(4-METHYLPHENYL)SULFONYL]
C10H13NO3S	075032-34-7	BENZENE, 1-[3-(METHYLTHIO)PROPOXY]-4-NITRO-
C10H13NO4	000300-48-1	TYROSINE, 3-METHOXY-
C10H13NO4	007101-51-1	L-TYROSINE, 3-HYDROXY-, METHYL ESTER
C10H13NO4	055750-53-3	N-MALEOYL-6-AMINOHEXANOIC ACID
C10H13NO4	093414-50-7	BENZALDEHYDE, 4-(2-HYDROXYETHOXY)-3-METHOXY-, OX
C10H13NO4S	059777-58-1	PROPYL 4-SULFAMYLBENZOATE
C10H13NO4S	105412-24-6	BENZENE, 1-[3-(METHYLSULFINYL)PROPOXY]-4-NITRO-
C10H13NO5S	105412-25-7	P-NO2-(3-ME SULFONYL)PROPOXYBENZENE

C10H13NO7S2	135832-43-8	BENZOIC ACID, 5-(AMINOSULFONYL)-2-[(2-HYDROXYETH
C10H13NS	080353-09-9	PYRIDINE, 1,2,3,6-TETRAHYDRO-1-METHYL-4-(2-THIEN
C10H14	000095-93-2	1,2,4,5-TETRAMETHYLBENZENE
C10H14	000098-06-6	T-BUTYLBENZENE
C10H14	000099-87-6	P-CYMENE
C10H14	000104-51-8	N-BUTYLBENZENE
C10H14	000105-05-5	P-DIETHYLBENZENE
C10H14	000135-01-3	O-DIETHYLBENZENE
C10H14	000135-98-8	SEC-BUTYLBENZENE
C10H14	000141-93-5	M-DIETHYLBENZENE
C10H14	000488-23-3	1,2,3,4-TETRAMETHYLBENZENE
C10H14	000527-53-7	1,2,3,5-TETRAMETHYLBENZENE
C10H14	000527-84-4	1-METHYL-2-ISOPROPYLBENZENE
C10H14	000535-77-3	M-CYMENE
C10H14	000538-93-2	ISOBUTYLBENZENE
C10H14	000874-41-9	1,3-DIMETHYL-4-ETHYLBENZENE
C10H14	000933-98-2	1,2-DIMETHYL-3-ETHYLBENZENE
C10H14	000934-74-7	1,3-DIMETHYL-5-ETHYLBENZENE
C10H14	000934-80-5	4-ETHYL-1,2-DIMETHYLBENZENE
C10H14	001074-17-5	1-METHYL-2-PROPYLBENZENE
C10H14	001074-43-7	1-METHYL-3-PROPYLBENZENE
C10H14	001074-55-1	1-METHYL-4-PROPYLBENZENE
C10H14	001758-88-9	2-ETHYL-1,4-DIMETHYLBENZENE
C10H14	002870-04-4	1,3-DIMETHYL-2-ETHYLBENZENE
C10H14BrNO2	025155-15-1	P-I-PROPYLTOLUENE
C10H14BrNO2	061367-72-4	3,5-DIMEO-4-BROMOPHENETHYLAMINE
C10H14Cl2N2O2	097454-00-7	DKA-24
C10H14Cl2NO2PS	000299-85-4	DMPA
C10H14Cl6N4O2	026644-46-2	TRIFORINE
C10H14ClN	000461-78-9	CHLORPHENTERMINE
C10H14ClN5	060221-93-4	N1-P-CHLOROPHENYL-N5-ETHYLBIGUANIDE
C10H14ClN5O6	089690-74-4	ACETAMIDE, N-[5-CHLORO-6-[[2,3-DIHYDROXY-1-(HYDR
C10H14FNO3S2	108966-75-2	BENZENESULFONAMIDE, 3-FLUORO-4-[(4-HYDROXYBUTYL)
C10H14N2	000054-11-5	NICOTINE
C10H14N2	000092-54-6	1-PHENYLPIPERAZINE
C10H14N2	000370-09-2	3-PYRIDYLMETHYL-N-PYRROLIDINE
C10H14N2	000494-52-0	ANABASINE
C10H14N2	033089-74-6	METHANIMIDAMIDE, N-(2,4-DIMETHYLPHENYL)-N'-METHY
C10H14N2O	000059-26-7	NIKETHAMIDE
C10H14N2O	005363-32-6	BENZAMIDE, O-ISOPROPYLAMINO-
C10H14N2O	006108-74-3	BENZAMIDE,N-(3-AMINOPROPYL)
C10H14N2O	007160-01-2	1,1-DIMETHYL-3-(P-TOLYL)UREA
C10H14N2O	007463-28-7	P-ACETAMIDO-N,N-DIMETHYLANILINE
C10H14N2O	010354-55-9	N-BUTYLNICOTINAMIDE
C10H14N2O	017751-47-2	3-PYRIDYLMETHYL-N-MORPHOLINE
C10H14N2O	018865-38-8	ACETAMIDE, 2-AMINO-N-(2,6-DIMETHYLPHENYL)-
C10H14N2O	023138-95-6	UREA, N-(3-ETHYLPHENYL)-N'-METHYL-
C10H14N2O	039080-49-4	4H-PYRIDO[1,2-A]PYRIMIDIN-4-ONE, 6,7,8,9-TETRAHY
C10H14N2O	052387-57-2	N-(ETHYLAMINOMETHYL)BENZAMIDE
C10H14N2O	054504-62-0	4H-PYRIDO[1,2-A]PYRIMIDIN-4-ONE, 6,7,8,9-TETRAHY
C10H14N2O	059917-58-7	N-(DIMETHYLAMINOMETHYL)BENZAMIDE
C10H14N2O	084199-97-3	2-PYRIDINEPENTANEAMIDE
C10H14N2O	084200-04-4	3-PYRIDINEPENTANEAMIDE
C10H14N2O	084200-11-3	4-PYRIDINEPENTANEAMIDE
C10H14N2O	088437-14-3	4H-PYRIDO[1,2-A]PYRIMIDIN-4-ONE, 6,7,8,9-TETRAHY
C10H14N2O2	002631-39-2	N-ME-3-DIMEAMINOPHENYL CARBAMATE
C10H14N2O2	004138-38-9	N-(1,1-DIMETHYLETHYL)-4-NITROBENZENAMINE
C10H14N2O2	007160-02-3	N'-(4-METHOXYPHENYL)-N,N-DIMETHYLUREA

C10H14N2O2	013171-61-4	2,3,5,6-TETRAMETHYL-4-NITROBENZENAMINE
C10H14N2O2	028170-54-9	N'-(3-METHOXYPHENYL)-N,N-DIMETHYLUREA
C10H14N2O2	052093-77-3	N'-ET-PHENOXYACETIC ACID HYDRAZIDE
C10H14N2O2	053673-10-2	O-MEO BENZAMIDE, N-(2-AMINOETHYL)
C10H14N2O2	084640-25-5	M-AMINOBENZYL N,N-DIME CARBAMATE
C10H14N2O2	088267-62-3	BENZAMIDE, 2-[(3-HYDROXYPROPYL)AMINO]-
C10H14N2O2	101820-63-7	BENZAMIDE, 2-[(2-HYDROXYPROPYL)AMINO]-
C10H14N2O2S	000115-56-0	BARBITURIC,2-THIO-5-ET-5-MEALLYL
C10H14N2O3	000077-02-1	APROBARBITAL
C10H14N2O3	061346-84-7	BARBITURIC ACID,1-ME-5-ALLYL-5-ET
C10H14N2O3	066843-01-4	2,4,6(1H,3H,5H)-PYRIMIDINETRIONE, 5-METHYL-5-(3-
C10H14N2O3	101820-62-6	BENZAMIDE, 2-[(2,3-DIHYDROXYPROPYL)AMINO]-
C10H14N2O3S	001467-23-8	BENZENESULFONAMIDE, N-[(ETHYLAMINOCARBONYL]-4-ME
C10H14N2O3S	003149-01-7	BENZENESULFONAMIDE, N-[[(1-METHYLETHYL)AMINO]CAR
C10H14N2O3S	004932-53-0	BENZENESULFONAMIDE, N-[(PROPYLAMINO)CARBONYL]-
C10H14N2O3S	005462-24-8	4-SULFAMYLBENZAMIDE,N-PROPYL
C10H14N2O3S	077812-87-4	NN-DIME-4-(N-ME-N-FORMYL)BENZENE-SO2N
C10H14N2O3S	081792-73-6	BENZENESULFONAMIDE, 2-AMINO-N-(2-METHYL-1-OXOPRO
C10H14N2O3S	090873-90-8	BENZENESULFONAMIDE, 2-[(2-METHYL-1-OXOPROPYL)AMI
C10H14N2O3S	121822-77-3	BENZENEMETHANESULFONAMIDE, N-[(DIMETHYLAMINO)CAR
C10H14N2O3S	121822-78-4	BENZENEMETHANESULFONAMIDE, N-METHYL-N-[(METHYLAM
C10H14N2O3S	122568-03-0	THYMIDINE, 3'-DEOXY-4-THIO-
C10H14N2O4	002537-29-3	BARBITURIC ACID,5(2HOPR)5-ALLYL
C10H14N2O4	003416-05-5	THYMIDINE, 3'-DEOXY-
C10H14N2O4	004450-00-4	4-PYRIMIDINECARBOXYLIC ACID, 1,2,3,6-TETRAHYDRO-
C10H14N2O4	004450-04-8	4-PYRIMIDINECARBOXYLIC ACID, 1,2,3,6-TETRAHYDRO-
C10H14N2O4	072962-38-0	3-FURANCARBOXAMIDE, N-(AMINOCARBONYL)-TETRAHYDRO
C10H14N2O4	072962-39-1	3-FURANCARBOXAMIDE, N-(AMINOCARBONYL)-TETRAHYDRO
C10H14N2O4S	007236-57-9	THYMIDINE, 4-THIO-
C10H14N2O5	000050-89-5	THYMIDINE
C10H14N2S	002741-14-2	THIOUREA, N'-(PHENYLMETHYL)-N,N-DIMETHYL
C10H14N4O	059708-19-9	1-(4-CARBAMOYLPHENYL)-3-METHYL-3-ETHYLTRIAZENE
C10H14N4O	066974-88-7	N-ME-2-(33-DIME-1-TRIAZINO)BENZAMIDE
C10H14N4O	069218-63-9	2-(3-CARBAMYLPROPYL)BENZOTRIAZOLE
C10H14N4O	082530-96-9	HYDRAZINECARBOXIMIDAMIDE, 2-[(4-ETHOXYPHENYL)MET
C10H14N4O2	018694-42-3	PYRIMIDINE,4(5MEO-3ME-1-PYRAZOL)6MEO
C10H14N4O2	023917-24-0	PYRIMIDINE,2-(3ME-5MEO-1PYRAZOL)4MEO
C10H14N4O2	028822-58-4	1H-PURINE-2,6-DIONE, 3,7-DIHYDRO-1-METHYL-3-(2-M
C10H14N4O2	031542-48-0	1H-PURINE-2,6-DIONE, 3-BUTYL-3,9-DIHYDRO-1-METHY
C10H14N4O2	059708-20-2	1-(4-CARBAMOYLPHENYL)-3-METHYL-3-HYDROXYETHYLTRI
C10H14N4O2	066974-78-5	4-MEO-3-(33-DIME TRIAZENO)BENZAMIDE
C10H14N4O3S	024229-59-2	OXAZOLIDINE, 4,4,5,5-TETRAMETHYL-2-[(5-NITRO-2-T
C10H14N4O4S	004988-64-1	6-MERCAPTOPURINE RIBOSIDE
C10H14N6	077314-58-0	2-PYRIDINAMINE, 4-(5-AMINO-1H-1,2,4-TRIAZOL-3-YL
C10H14N6O	077314-64-8	2-PYRIDINAMINE, 4-(5-AMINO-1H-1,2,4-TRIAZOL-3-YL
C10H14N6O2	107550-73-2	ADENOSINE, 2-AMINO-2',3'-DIDEOXY-
C10H14N6O3	004546-70-7	ADENOSINE, 2-AMINO-2'-DEOXY-
C10H14N6O4	013263-99-5	NSC #128,668
C10H14NNaO3S	005123-63-7	3-(DIETHYLAMINO)BENZENESULFONIC ACID, SODIUM SAL
C10H14NO5PS	000056-38-2	PARATHION
C10H14NO5PS	013074-09-4	3-ETHYL-DIMETHYLPARATHION
C10H14NO5PS	050590-05-1	3,5-DIMETHYL-DIMETHYLPARATHION
C10H14NO6P	000311-45-5	PARAOXON
C10H14NO6P	050590-06-2	3,5-DIMETHYL-DIMETHYLPARA-OXON
C10H14O	000088-18-6	O-T-BUTYLPHENOL
C10H14O	000089-72-5	O-SEC-BUTYLPHENOL
C10H14O	000089-83-8	THYMOL
C10H14O	000098-54-4	P-T-BUTYLPHENOL

C10H14O	000099-49-0	CARVONE
C10H14O	000099-71-8	P-(SEC-BUTYL)PHENOL
C10H14O	000585-34-2	PHENOL, 3-(1,1-DIMETHYLETHYL)-
C10H14O	001006-59-3	PHENOL, 2,6-DIETHYL-
C10H14O	001126-79-0	BUTYL PHENYL ETHER
C10H14O	001638-22-8	P-BUTYLPHENOL
C10H14O	002046-33-5	1-METHOXY-3-PHENYLPROPANE
C10H14O	003180-09-4	2-BUTYLPHENOL
C10H14O	003360-41-6	4-PHENYLBUTANOL
C10H14O	006485-40-1	CARVONE
C10H14O2	000122-94-1	4-BUTOXYPHENOL
C10H14O2	000465-29-2	CAMPHORQUINONE
C10H14O2	000553-84-4	1-PENTANONE, 1-(3-FURANYL)-4-METHYL-
C10H14O2	002525-05-5	1,2-BENZENEDIOL, 4-BUTYL-
C10H14O2	006770-38-3	BENZENE, 1,4-BIS(METHOXYMETHYL)-
C10H14O2	010373-78-1	DL-CAMPHORQUINONE
C10H14O2	018979-72-1	M-BUTOXYPHENOL
C10H14O3	000059-47-2	1,2-PROPANEDIOL-3-(2-TOLYLOXY)
C10H14O3	000707-07-3	TRIMETHOXYMETHYLBENZENE
C10H14O4	000097-90-5	ETHYLENE DIMETHYLACRYLATE
C10H14O4S2	019607-41-1	PROPANEDIOIC ACID, 1,3-DITHIOLAN-2-YLIDENE-, DIE
C10H14O5	004074-88-8	DIETHYLENE GLYCOL DIACRYLATE
C10H14O6	052092-59-8	TETRAMETHOXYHYDROQUINONE
C10H15Br2N	071323-99-4	M-BROMOBENZYL TRIMETHYL AMMONIUM BROMIDE
C10H15BrClN	025251-56-3	M-CL BENZYL TRIMETHYL AMMONIUM BROMIDE
C10H15BrFN	071324-00-0	M-FLUOROBENZYL TRIMETHYL AMMONIUM BROMIDE
C10H15BrIN	071324-02-2	M-IODOBENZYL TRIMETHYL AMMONIUM BROMIDE
C10H15BrN2O2	071323-93-8	M-NITROBENZYL TRIMETHYL AMMONIUM BROMIDE
C10H15BrN2O3	077317-75-0	3-PEO-5-BR-6-ME-PYRIMIDIN-2,4-DIONE
C10H15BrN2O3	077317-87-4	3-PRO-5-BR-6-PR-PYRIMIDIN-2,4-DIONE
C10H15HgNO3	005822-97-9	PHENYLMERCURY MONOETHANOL AMMONIUM ACETATE
C10H15N	000091-66-7	N,N-DIETHYLANILINE
C10H15N	000093-88-9	BENZENEETHANAMINE, N,ß-DIMETHYL-
C10H15N	000102-97-6	BENZENEMETHANAMINE, N-(1-METHYLETHYL)-
C10H15N	000104-13-2	BENZENAMINE, 4-BUTYL-
C10H15N	000122-09-8	PHENTERMINE
C10H15N	000537-46-2	METHAMPHETAMINE
C10H15N	000579-66-8	2,6-DIETHYLANILINE
C10H15N	000769-92-6	BENZENAMINE, 4-(1,1-DIMETHYLETHYL)
C10H15N	001126-78-9	N-BUTYLANILINE
C10H15N	002961-50-4	P-AMYLPYRIDINE
C10H15N	013214-66-9	4-PHENYLBUTYLAMINE
C10H15N	022374-89-6	BENZENEPROPANAMINE, α-METHYL-
C10H15N	053309-89-0	1-BENZYLPROPYLAMINE
C10H15N2O2S4	122266-99-3	THIENO[2,3-B]THIOPHENE-2-SULFONAMIDE, 5-[[[2-(ME
C10H15N2O3S4	122267-01-0	THIENO[2,3-B]THIOPHENE-2-SULFONAMIDE, 5-[[[2-(ME
C10H15N2O4S4	133445-74-6	THIENO[2,3-B]THIOPHENE-2-SULFONAMIDE, 5-[[[2-(ME
C10H15N3	016584-01-3	2-BUTYLBENZOTRIAZOLE
C10H15N3	066382-00-1	2-S-BUTYLBENZOTRIAZOLE
C10H15N3O	058480-99-2	2-I-BUTYLISONIAZID
C10H15N3O	069218-39-9	2-(4-HYDROXYBUTYL)BENZOTRIAZOLE
C10H15N3O3	134419-51-5	2-AZETIDINECARBOXYLIC ACID, 1-[2-HYDROXY-3-(1H-I
C10H15N3O4	025152-20-9	THYMIDINE, 5'-AMINO-5'-DEOXY-
C10H15N3O5	002140-61-6	5-METHYLCYTIDINE
C10H15N3O6	068160-69-0	1(2,3-BIS(ACETYLOXY)PR-2-NO2IMIDAZOLE
C10H15N5	000114-86-3	PHENFORMIN
C10H15N5	038407-85-1	HYDRAZINECARBOXIMIDAMIDE, 2-[[4-(DIMETHYLAMINO)P
C10H15N5O2	087469-47-4	1-METHYLISOGUANINE,9-[2-(ETHENYLOXY)ETHYL]-

C10H15N5O2	087469-48-5	1-METHYLISOGUANINE,7-[2-(ETHENYLOXY)ETHYL]-
C10H15N5O3	087469-66-7	1-METHYLISOGUANINE,7[ACOET]-
C10H15N5O3S	002133-81-5	A-2'-DEOXYTHIOGUANOSINE
C10H15N5O4	000961-07-9	GUANOSINE, 2'-DEOXY-
C10H15N5O4	110104-37-5	ACETAMIDE, N-[6,9-DIHYDRO-9-[(2-HYDROXYETHOXY)ME
C10H15NO	000064-13-1	P-METHOXYAMPHETAMINE
C10H15NO	000090-82-4	BENZENEMETHANOL, .ALPHA.- 1-(METHYLAMINO)ETHYL -
C10H15NO	000092-50-2	ETHANOL, 2-(ETHYLPHENYLAMINO)-
C10H15NO	000103-62-8	N-BUTYL-P-AMINOPHENOL
C10H15NO	000299-42-3	EPHEDRINE
C10H15NO	003217-09-2	N-BENZYL-2-AMINOPROPANOL
C10H15NO	004343-96-8	4-PYRIDINEPENTANOL
C10H15NO	006853-14-1	2-N,N-DIMEAMINO-1-PHENYLETHANOL
C10H15NO	084200-03-3	3-PYRIDINEPENTANOL
C10H15NO2	000120-07-0	PHENYLDIETHANOLAMINE
C10H15NO2	000120-20-7	3,4-DIMETHOXYPHENETHYLAMINE
C10H15NO2S	001135-00-8	P-BUTYLBENZENESULFONAMIDE
C10H15NO3	005001-33-2	METANEPHRINE
C10H15NO3	150900-11-1	2-AZASPIRO[4.4]NONANE-1,3-DIONE, 2-ETHOXY-
C10H15NO3S	001138-58-5	BENZENESULFONAMIDE, 4-BUTOXY-
C10H15NO3S	123045-57-8	BENZENESULFONAMIDE, 3-BUTOXY-
C10H15NO3S	135832-48-3	BENZENESULFONAMIDE, 4-(4-HYDROXYBUTYL)-
C10H15NO3S2	135832-41-6	BENZENESULFONAMIDE, 4-[(4-HYDROXYBUTYL)THIO]-
C10H15NO5S2	108966-78-5	BENZENESULFONAMIDE, 4-[(4-HYDROXYBUTYL)SULFONYL]
C10H15NO5S2	135832-42-7	BENZENESULFONAMIDE, 4-[(4-HYDROXYBUTYL)SULFONYL]
C10H15NO6	039935-49-4	3-DEAZAURIDINE
C10H15NO7S3	104438-05-3	2-THIOPHENESULFONAMIDE, 5-[[3-(METHOXYACETYL)OXY
C10H15O2PS	000944-21-8	FONOPHOS, O-ANALOG
C10H15O3PS	032345-29-2	O,O-DIET-O-PHENYLPHOSPHOROTHIOATE
C10H15O3PS2	000055-38-9	FENTHION
C10H15O4P	002510-86-3	O,O-DIETHYL-O-PHENYLPHOSPHATE
C10H15O4PS2	003761-41-9	FENTHION SULFOXIDE
C10H15OPS2	000944-22-9	FONOPHOS
C10H16	000079-92-5	CAMPHENE
C10H16	000080-56-8	ALPHA-PINENE
C10H16	000099-83-2	α-PHELLANDRENE
C10H16	000099-85-4	GAMMA-TERPINENE
C10H16	000123-35-3	3-METHYLENE-7-METHYL-1,6-OCTADIENE (MYRCENE)
C10H16	000127-91-3	ß-PINENE
C10H16	000138-86-3	LIMONENE
C10H16	000281-23-2	TRICYCLO[3.3.1.1]DECANE
C10H16	000554-61-0	2-CARENE
C10H16	000555-10-2	BETA-PHELLANDRENE
C10H16	000586-62-9	TERPINOLENE
C10H16	005989-27-5	D-LIMONENE
C10H16	005989-54-8	(-)-1-METHYL-4-(1-METHYLETHENYL)CYCLOHEXENE
C10H16	006004-38-2	TRICYCLO(5.2.1.0)DECANE
C10H16	007705-14-8	DIPENTENE (+-)
C10H16	013466-78-9	3-CARENE (DELTA)
C10H16	013877-91-3	3,7-DIMETHYL-1,3,6-OCTATRIENE
C10H16Br2N2O2	000054-91-1	PIPOBROMAN
C10H16BrN	005350-41-4	BENZYLTRIMETHYL AMMONIUM BROMIDE
C10H16BrNO	071323-98-3	M-HYDROXYBENZYL TRIMETHYL AMMONIUM BROMIDE
C10H16Cl3NOS	002303-17-5	DIISOPROPYL CARBAMATE
C10H16ClN	000056-93-9	BENZYL TRIMETHYL AMMONIUM CHLORIDE
C10H16ClN	030684-06-1	BENZENEBUTANAMINE, HYDROCHLORIDE
C10H16ClN3O2	013909-13-2	3(2-NORBORNYL)-1-(2CLET)-1-NO UREA
C10H16ClN3O4	042558-93-0	CIS-1-NO-1(CLET)-3(4COOH CYHEXYL)UREA

C10H16ClN3O4	042558-94-1	TRAN-1NO-1(CLET)-3(4COOH CYHEXYL)UREA
C10H16IN	033046-97-8	M-METHYL-TRIMETHYL ANILINIUM IODIDE
C10H16N2	000093-05-0	N,N-DIETHYL-1,4-BENZENEDIAMINE
C10H16N2	000100-22-1	N,N,N',N'-TETRAMETHYL-P-PHENYLENEDIAMINE
C10H16N2	001871-96-1	DECANEDINITRILE
C10H16N2	002055-14-3	N,N-DIETHYL-3-PYRIDYLMETHYLAMINE
C10H16N2	003000-74-6	4-(N-ME)-3-PYRIDYLBUTYLAMINE
C10H16N2	020173-12-0	N-BUTYL-3-PYRIDYLMETHYLAMINE
C10H16N2	059082-52-9	4-PYRIDINEPENTANEAMINE
C10H16N2	059082-57-4	2-PYRIDINEPENTANEAMINE
C10H16N2	084200-05-5	3-PYRIDINEPENTANEAMINE
C10H16N2O	002820-55-5	PYRIDINE, 3-(1-METHYL-2-PYRROLIDINYL)-, N-OXIDE,
C10H16N2O2S	099167-62-1	4,6(1H,5H)-PYRIMIDINEDIONE, 5,5-DIPROPYL-2-THIOX
C10H16N2O3	000077-28-1	5-BUTYL-5-ETHYLBARBITURIC ACID
C10H16N2O3	000125-40-6	BUTABARBITAL
C10H16N2O3	001135-61-1	BARBITURIC ACID,5ET-5IBU
C10H16N2O3	002217-08-5	BARBITURIC ACID,5,5-DIPR
C10H16N2O3	015379-32-5	BARBITURIC ACID,5ET-5-TBU
C10H16N2O3	056344-90-2	BARBITURIC ACID,5-PR-5-ET-1-ME
C10H16N2O3	099167-69-8	BARBITURIC ACID,DI-IPR
C10H16N2O3S3	122266-90-4	THIENO[2,3-B]THIOPHENE-2-SULFONAMIDE, 5-[[(2-MET
C10H16N2O3S3	127025-29-0	THIENO[3,2-B]THIOPHENE-2-SULFONAMIDE, 5-[[(2-MET
C10H16N2O8	000060-00-4	ETHYLENEDIAMINETETRAACETIC ACID
C10H16N4O	019353-98-1	2-DIETHYLAMINOISONIAZID
C10H16N4O2S	055511-98-3	BUTHIDAZOLE
C10H16N4O3	000603-00-9	PROXYPHYLLINE
C10H16N4O3	000644-64-4	DIMETILAN
C10H16N6S	051481-61-9	CIMETIDINE
C10H16N8S2	069014-14-8	GUANIDINE, N-[2-[[[2-[(AMINOIMINOMETHYL)AMINO]-4
C10H16NO3PS	003735-01-1	PARATHION-AMINO
C10H16NO5PS2	000052-85-7	FAMPHUR
C10H16O	000076-22-2	CAMPHOR
C10H16O	000464-49-3	(1R,4R)-(+)-CAMPHOR
C10H16O	000499-74-1	6-METHYL-3-ISOPROPYL-2-CYCLOHEXEN-1-ONE
C10H16O	000546-80-5	THUJONE
C10H16O	000768-95-6	1-HYDROXYADAMANTANE
C10H16O	001195-79-5	FENCHONE
C10H16O	004832-17-1	DECAHYDRO-2-NAPHTHALENONE
C10H16O	005392-40-5	CITRAL
C10H16O	007764-50-3	2-METHYL-5-(1-METHYLETHENYL)CYCLOHEXANONE
C10H16O2	000101-43-9	2-METHYLCYCLOHEXYL ACRYLATE
C10H16O4	000124-83-4	CAMPHORIC ACID
C10H16O4	005394-83-2	CIS-1,2,2-TRIMETHYL-1,3-CYCLOPENTANEDICARBOXYL*
C10H16O4S	003144-16-9	CAMPHORSULFONIC ACID
C10H16Si	000770-09-2	BENZYLTRIMETHYLSILANE
C10H17BrN2	022337-35-5	2-PYRIDYL-ME-(6NN-TETRA-ME)AMMONIUM BR
C10H17BrN2	071323-96-1	M-AMINOBENZYL TRIMETHYL AMMONIUM BROMIDE
C10H17BrN2	075523-38-5	4-PYRIDYL-ME(6NN-TETRA-ME)AMMONIUM BR
C10H17Cl2NOS	002303-16-4	DIALLATE
C10H17I	000090-14-2	1-IODONAPHTHALENE
C10H17N	000768-94-5	AMANTADINE
C10H17N2O4PS	038260-54-7	ETRIMFOS
C10H17N3O	036067-73-9	4H-OXAZOLO[4,5-D]AZEPIN-2-AMINE, 6-ETHYL-5,6,7,8
C10H17N3O2	000119-38-0	ISOLAN
C10H17N3O3S2	066469-80-5	N(N"-ME-CYPR CARBAMYLTHIO)METHOMYL
C10H17N3O4	082249-23-8	2-PROPANOL, 1-BUTOXY-3-(2-NITRO-1H-IMIDAZOL-1-YL
C10H17N5	002002-36-0	9-PENTYLADENINE
C10H17N5O	055921-61-4	24NH2-6DIALLYLAMINO-PYRIMIDINE-3-OXIDE

C10H17N5O2	050917-19-6	3-MORPHOL-4-NH2-6-IPR-124-TRIAZINE-5-ONE
C10H17N5O2	054387-29-0	PIPERAZINE, 1-METHYL-4-[(1-METHYL-5-NITRO-1H-IMI
C10H17N5O2	074571-56-5	PIPERAZINE, 1-METHYL-4-[2-(5-NITRO-1H-IMIDAZOL-1
C10H17N5O2	080616-58-6	1,2,4-TRIAZOLE-3,5-BUTYRAMIDO-
C10H17N5O2	087469-43-0	2H-PURIN-2-ONE, 9-(2-ETHOXYETHYL)-1,3,6,9-TETRAH
C10H17N5O2	087469-44-1	2H-PURIN-2-ONE, 7-(2-ETHOXYETHYL)-1,3,6,7-TETRAH
C10H17N5O2S	055884-23-6	1,1-ETHENEDIAMINE, N-METHYL-N'-[2-[[(5-METHYL-1H
C10H17N5O2S2	054855-72-0	CIMETIDINE,4-THIOPYRIMIDIN-2-YL-DIOXIDE ANALOG
C10H17N5O3	087469-68-9	2H-PURIN-2-ONE, 1,3,6,9-TETRAHYDRO-6-IMINO-9-[(2
C10H17N5O6	007093-67-6	GLYCINE, N-[N-[N-(N-GLYCYLGLYCYL)GLYCYL]GLYCYL]-
C10H17N5OS	087469-45-2	2H-PURIN-2-ONE, 9-[2-(ETHYLTHIO)ETHYL]-1,3,6,9-T
C10H17N5OS	087469-46-3	2H-PURIN-2-ONE, 7-[2-(ETHYLTHIO)ETHYL]-1,3,6,7-T
C10H17NO2	000125-64-4	DIMERIN
C10H17NO5	083040-20-4	1-PYRROLIDINYLOXY, 3,4-DICARBOXY-2,2,5,5-TETRAME
C10H18	000091-17-8	DECAHYDRONAPHTHALENE
C10H18	000493-01-6	CIS-BICYCLO[4.4.0]DECANE
C10H18	000493-02-7	TRANS-BICYCLO[4.4.0]DECANE
C10H18	000499-94-5	CARVOMETHENE
C10H18Cl2N6O4	060784-43-2	1,1'-TETRAMETHYLENE-BIS-CNU
C10H18ClN3O2	013909-09-6	1-(2-CLET)-3(4-MECYHX)-1-N=O UREA
C10H18ClN5	001912-25-0	IPAZINE
C10H18INO	075523-41-0	2-FURANMETHANAMINIUM, 5-ETHYL-N,N,N-TRIMETHYL-,
C10H18INO2	005822-68-4	PYRIDINIUM, 3-(ETHOXYCARBONYL)-1,2,5,6-TETRAHYDR
C10H18N2O	080431-09-0	1234-H4-PYRIDINE-1-T-BU-CARBONYLAMINO
C10H18N2O2	002917-98-8	PIPERAZINE,1,4-BIS(2,3-EPOXYPROPYL)
C10H18N2O4	073632-76-5	BUTANOIC ACID, 2-[[(AMINOCARBONYL)AMINO]CARBONYL
C10H18N2O5	073632-79-8	BUTANOIC ACID, 2-[[(AMINOCARBONYL)AMINO]CARBONYL
C10H18N3O2	055861-78-4	ISOURON
C10H18N4O2	060141-99-3	N-DIAZOACETYLGLYCINE-N'-HEXYLAMIDE
C10H18N4O4S2	006513-23-1	L-CYSTEINAMIDE, S,S'-BIS[N-ACETYL-
C10H18N4O4S3	059669-26-0	THIODICARB
C10H18N4O5S2	065907-54-2	6-OXA-3-THIA-2,4,7,10-TETRAAZAUNDEC-7-ENOIC ACID
C10H18N4OS	021085-20-1	3-SME-4-NH2-6-HX-124-TRIAZINE-5-ONE
C10H18N4OS	050917-25-4	3-BUS-4-NH2-6-IPR-124-TRIAZINE-5-ONE
C10H18N4S	051264-00-7	THIOUREA, N-METHYL-N'-[4-(5-METHYL-1H-IMIDAZOL-4
C10H18O	000078-70-6	LINALOOL
C10H18O	000089-80-5	5-METHYL-2-(1-METHYLETHYL)-CYCLOHEXANONE (TR*)
C10H18O	000098-55-5	ALPHA-TERPINEOL
C10H18O	000106-23-0	CITRONELLAL
C10H18O	000106-24-1	GERANIOL
C10H18O	000124-76-5	BICYCLO 2.2.1 HEPTAN-2-OL, 1,7,7-TRIMETHYL-, EXO
C10H18O	000138-87-4	BETA-TERPINEOL
C10H18O	000470-67-7	1,4-CINEOL
C10H18O	000470-82-6	1,8-CINEOLE
C10H18O	000507-70-0	BORNEOL
C10H18O	000825-51-4	DECAHYDRO-2-NAPHTHOL
C10H18O	001632-73-1	FENCHYL ALCOHOL
C10H18O	007785-53-7	D-ALPHA-TERPINEOL
C10H18O	008000-41-7	TERPINEOL
C10H18O	010482-56-1	2-(4-METHYL-3-CYLOHEXENYL)ISOPROPANOL
C10H18O	011039-70-6	PLINOL
C10H18O2	000706-14-9	GAMMA-DECALACTONE
C10H18O4	000111-20-6	DECANEDIOIC ACID
C10H18O4	000141-28-6	DIETHYLHEXANEDIOATE
C10H18O4	000925-15-5	BUTANEDIOIC ACID, DIPROPYL ESTER
C10H18O4	002050-60-4	DIBUTYL OXALATE
C10H18O6	000111-21-7	ETHYLENEBIS(OXYETHYLENE)DIACETATE
C10H19ClNO5P	000297-99-4	PHOSPHAMIDON

C10H19ClNO5P	013171-21-6	DIMECRON
C10H19N3O3	132765-80-1	L-VALINAMIDE, N-ACETYL-L-ALANYL-
C10H19N3O3	132765-85-6	L-ALANINAMIDE, N-ACETYL-L-VALYL-
C10H19N3O3S2	086569-93-9	N(N"-ME-PR CARBAMYLTHIO)METHOMYL
C10H19N3O3S2	086569-94-0	N(N"-ME-IPR CARBAMYLTHIO)METHOMYL
C10H19N5O	001610-18-0	PROMETON
C10H19N5O	013532-26-8	2-DIETHYLAMINO-4-ETHYLAMINO-6-METHOXY-S-TRIAZI*
C10H19N5O	026259-45-0	SECBUMETON
C10H19N5O	033665-71-3	3-N-BUAM-4NH2-6IPR-124-TRIAZIN-5-ONE
C10H19N5O	033693-04-8	TERBUMETON
C10H19N5O	055921-57-8	24NH2-6-CYHXAM-PYRIMIDINE-3-OXIDE
C10H19N5S	000886-50-0	TERBUTRYN
C10H19N5S	007287-19-6	PROMETRYNE
C10H19NO	000092-53-5	N-PHENYLMORPHOLINE
C10H19NO	004838-65-7	1-HEXYL-2-PYRROLIDONE
C10H19NO	005830-33-1	CYCLOHEXANEACETAMIDE, N,N-DIMETHYL-
C10H19NO	042327-99-1	2-BUTANONE, 3-METHYL-4-(1-PIPERIDINYL)-
C10H19O6PS2	000121-75-5	MALATHION
C10H20	000099-82-1	P-MENTHANE
C10H20	000293-96-9	CYCLODECANE
C10H20	000872-05-9	1-DECENE
C10H20	001678-93-9	BUTYLCYCLOHEXANE
C10H20	001678-98-4	ISOBUTYLCYCLOHEXANE
C10H20	003178-22-1	TERT-BUTYLCYCLOHEXANE
C10H20	007058-01-7	S-BUTYLCYCLOHEXANE
C10H20Br2	004101-68-2	1,2-DIBROMODECANE
C10H20IN	058944-11-9	NNN-ME-N(1ME-3-CYPENTENE)AMMON I
C10H20IN	074819-88-8	NNN-ME-N(1ME2CYPENTENE)AMMON I
C10H20IN	083283-07-2	NNN-ME-N(1ME-1-CYPENTENE)AMMON I
C10H20N2S	030826-88-1	TRIMETHYLENETHIOUREA,N,N-DIPROPYL
C10H20N2S4	000097-77-8	DISULFIRAM
C10H20N6O	052298-71-2	ETHANOL, 2-[[4,6-BIS(DIMETHYLAMINO)-1,3,5-TRIAZI
C10H20NO4PS	031218-83-4	PROPETAMPHOS
C10H20NO5PS2	002595-54-2	MECARBAM
C10H20O	000089-78-1	MENTHOL
C10H20O	000106-22-9	3,7-DIMETHYL-6-OCTEN-1-OL
C10H20O	000112-31-2	N-DECANAL
C10H20O	000693-54-9	2-DECANONE
C10H20O	000937-05-3	CYCLOHEXANOL, 4-(1,1-DIMETHYLETHYL)-, CIS-
C10H20O	001321-89-7	ISODECYLADEHYDE
C10H20O	002216-51-5	MENTHOL (L)
C10H20O	021862-63-5	CYCLOHEXANOL, 4-(1,1-DIMETHYLETHYL)-, TRANS-
C10H20O	026489-01-0	(+-)-CITRONELLOL
C10H20O2	000103-09-3	2-ETHYLHEXYLACETATE
C10H20O2	000106-32-1	ETHYL OCTANOATE
C10H20O2	000334-48-5	DECANOIC ACID
C10H20O2	000626-82-4	N-BUTYL HEXANOATE
C10H20O2	000659-70-1	ISOAMYL ISOVALERATE
C10H20O2	001731-84-6	METHYL NONANOATE
C10H20O2	003115-28-4	HEXANOIC ACID, 2-BUTYL-
C10H20O2	031080-39-4	2-PROPYLHEPTANOIC ACID
C10H20O2	067952-57-2	6-METHYL-2-HEPTANOL ACETATE
C10H20O4	000124-17-4	DIETHYLENE GLYCOL MONOBUTYL ETHER ACETATE
C10H20O5	033100-27-5	1,4,7,10,13-PENTAOXACYCLOPENTADECANE
C10H20O5S	063407-52-3	1-BUTYLTHIO-B-GALACTOPYRANOSIDE
C10H20O6	039824-09-4	1-BUTYLGALACTOPYRANOSIDE
C10H21Br	000112-29-8	1-BROMODECANE
C10H21Cl	001002-69-3	1-CHLORODECANE

C10H21F	000334-56-5	1-FLUORODECANE
C10H21I	002050-77-3	1-IODODECANE
C10H21N2O2P	014984-65-7	PHOSPHINIC ACID,BIS(DIMEAZ),ET EST
C10H21N2O3PS	101347-42-6	PHOSPHINOTHIOIC ACID, BIS(1-AZIRIDINYL)-, O-[2-(
C10H21N2O4P	101347-40-4	PHOSPHINIC ACID, BIS(1-AZIRIDINYL)-, 2-(2-ETHOXY
C10H21N5O	055921-60-3	24NH2-6DIPROPYLAMINO-PYRIMIDINE-3-OXIDE
C10H21NOS	001114-71-2	PEBULATE
C10H21NOS	001929-77-7	VERNOLATE
C10H22	000124-18-5	N-DECANE
C10H22	000871-83-0	2-METHYLNONANE
C10H22	001071-81-4	2,2,5,5-TETRAMETHYLHEXANE
C10H22	002051-30-1	2,6-DIMETHYLOCTANE
C10H22	013475-81-5	2,2,3,3-TETRAMETHYLHEXANE
C10H22	016747-42-5	2,2,4,5-TETRAMETHYLHEXANE
C10H22IN	041550-04-3	NNN-ME-N(3ME-CYPEME)AMMON I(Z)
C10H22INO	055903-47-4	NNN-ME-N(3OH4MECYPEME)AMMON I
C10H22N2O	042585-33-1	1-PIPERIDINYLOXY, 4-(METHYLAMINO)-2,2,6,6-TETRAM
C10H22N3OP	053439-65-9	PHOSPHINIC AMIDE, N,N-DIMETHYL-P,P-BIS(1-PYRROLI
C10H22O	000112-30-1	1-DECANOL
C10H22O	000544-01-4	DIISOPENTYL ETHER
C10H22O	000693-65-2	DI-N-PENTYLETHER
C10H22O	025339-17-7	ISODECANOL
C10H22O	055505-26-5	8-MEHTYL-1-NONANOL
C10H22O2	000112-48-1	1,2-DIBUTYOXYETHANE
C10H22O3	000112-59-4	ETHANOL, 2- 2-(HEXYLOXY)ETHOXY -
C10H22O3	002451-01-6	TERPIN HYDRATE
C10H22O4	000143-22-6	TRIETHYLENE GLYCOL BUTYL ETHER
C10H22O4	020324-33-8	TRIPROPYLENE GLYCOL METHYL ETHER
C10H22O4	029681-20-7	2-(2-(2-(TERT-BUTOXY)ETHOXY)ETHOXY)ETHANOL
C10H22OS	003547-33-9	ETHANOL, 2-(OCTYLTHIO)-
C10H22S	000143-10-2	1-DECANETHIOL
C10H22S	000544-02-5	DIISOAMYL SULFIDE
C10H22S2	000112-51-6	6,7-DITHIADODECANE
C10H23N	002016-57-1	N-DECYLAMINE
C10H23N	002050-92-2	DIPENTYLAMINE
C10H23NO	029812-79-1	O-DECYLHYDROXYLAMINE
C10H23O2PS2	095465-99-9	CADUSAFOS
C10H23O5P	007598-61-0	PHOSPHINIC ACID, (2,2-DIETHOXYETHYL)-, DIETHYL E
C10H24IN	003531-14-4	NNN-TRIPR-N-ME AMMONIUM IODIDE
C10H24N4	007209-38-3	1,4-PIPERAZINEDIPROPANAMINE
C10H2O6	000089-32-7	PYROMELLITIC DIANHYDRIDE
C10H30O3Si4	000141-62-8	DECAMETHYLTETRASILOXANE
C10H30O3Si4	017928-28-8	METHYLTRIS(TRIMETHYLSILOXY)SILANE
C10H30O5Si5	000541-02-6	DECAMETHYLCYCLOPENTASILOXANE
C11H5Cl2N5O6	142978-29-8	N(24-NO2-6CL PH)-2(3CL-5NO2)PYRIDINAMINE
C11H6Cl2F3N3O2	028313-69-1	2,6-DICL-4-CF3-C6H2NHN=C(CN)COOME
C11H6Cl2N2	074738-17-3	FENPICLONIL
C11H6ClN3	070959-58-9	[3,4'-BIPYRIDINE]-5-CARBONITRILE,6-CHLORO-
C11H6O10	001585-40-6	BENZENEPENTACARBOXYLIC ACID
C11H6O3	000066-97-7	PSORALEN
C11H6O3	000523-50-2	FURO(2,3-H)COUMARIN
C11H7Cl6HgNO2	002597-93-5	50-CS-46
C11H7ClF3N3O	028317-56-8	2-CL-5-CF3-C6H3NHN=C(CN)CO-ME
C11H7ClF3N3O2	028313-77-1	2-CL-5-CF3-C6H3NHN=C(CN)COOME
C11H7ClF3N3O2	036865-72-2	2-CF3-4-CL-C6H3NHN=C(CN)COOME
C11H7ClF3N3O2S	141627-90-9	1,2,4-TRIAZIN-5(4H)-ONE, 6-[3-CHLORO-4-(TRIFLUOR
C11H7F3N4O4	028313-79-3	2-NO2-4-CF3-C6H3NHN=C(CN)COOME
C11H7FN2O3	061251-77-2	3-BENZOYL-5-FLUOROURACIL

C11H7FN2O4	066999-97-1	1-PHENYLOXYCARBONYL-5-FLUOROURACIL
C11H7N3O	117505-21-2	1-NAPHTHALENE AZOXYCYANIDE
C11H7NaO2	017273-79-9	2-NAPHTHOIC ACID, SODIUM SALT
C11H7NaO3	014206-62-3	2-NAPHTHALENECARBOXYLIC ACID,3-HYDROXY-,MONOSODI
C11H7NS	000551-06-4	1-ISOTHIOCYANONAPHTHALENE
C11H7NS	001636-33-5	2-ISOTHIOCYANONAPHTHALENE
C11H8Cl3N3O2	028322-78-3	2,4,5-TRICL-C6H2NHN=C(CN)CO-OET
C11H8Cl3N3O2	036865-51-7	3,4,5-TRICL-C6H2NHN=C(CN)CO-OET
C11H8ClF2N3O2	036874-60-9	3-CHF2-4-CL-C6H3NHN=C(CN)CO-OME
C11H8ClFN2O2	085093-33-0	2,4(1H,3H)-PYRIMIDINEDIONE, 1-[(4-CHLOROPHENYL)M
C11H8ClNO	089770-25-2	4-ACETYL-7-CHLOROQUINOLINE
C11H8F3N3O2	028313-74-8	3-CF3-C6H4NHN=C(CN)CO-OME
C11H8F3N3O2	028313-76-0	4-CF3-C6H4NHN=C(CN)CO-OME
C11H8F3N3O2	028384-50-1	2-CF3-C6H4NHN=C(CN)CO-OME
C11H8F3N3O2S	028313-92-0	4-SCF3-C6H4NHN=C(CN)CO-OME
C11H8F3N3OS	028317-78-4	4-SCF3-C6H4NHN=C(CN)CO-ME
C11H8F3NO2	060050-38-6	M-CF3-N-PHENYLSUCCINIMIDE
C11H8F6N2O3	104668-67-9	PROPANAMIDE, 3,3,3-TRIFLUORO-2-METHYL-N-[4-NITRO
C11H8F6N2O4	072115-11-8	PROPANAMIDE, 3,3,3-TRIFLUORO-2-HYDROXY-2-METHYL-
C11H8FN3O3	056563-17-8	1(2H)-PYRIMIDINECARBOXAMIDE, 5-FLUORO-3,4-DICHLO
C11H8N2	000244-63-3	9H-PYRIDO[3,4-B]INDOLE
C11H8N2	002826-25-7	MALONONITRILE, 4-METHYLBENZAL
C11H8N2	005447-87-0	MALONONITRILE, A-METHYLBENZAL
C11H8N2O	003878-19-1	FUBERIDAZOLE
C11H8N2O2S2	037157-95-2	124-BENZTHIADIAZIN-1-O2-3-(2-THIENYL)
C11H8N2O3S	037157-96-3	124-BENZTHIADIAZIN-1-O2-3-(2-FURYL)
C11H8N4O	055653-16-2	PROPANEDINITRILE, [(4-ACETYLPHENYL)HYDRAZONO]-
C11H8N6O2	017659-57-3	8-(3-NITROPHENYL)-ADENINE
C11H8O	007206-61-3	1-AZULENE CARBOXYALDEHYDE
C11H8O2	000058-27-5	2-METHYL-1,4-NAPHTHOQUINONE
C11H8O2	000086-55-5	1-NAPTHOIC ACID
C11H8O2	000093-09-4	2-NAPHTHOIC ACID
C11H8O2	000605-93-6	6-METHYL-1,4-NAPHTHOQUINONE
C11H8O2	000692-94-4	2-DECENE-4,6,8-TRIYONIC ACID, METHYL ESTER
C11H8O2	001201-25-8	1-AZULENE CARBOXYLIC ACID
C11H8O2	002739-57-3	2-DECENE-4,6,8-TRIYNOIC ACID, METHYL ESTER, (Z)-
C11H8O3	000092-70-6	2-NAPHTHALENECARBOXYLIC ACID, 3-HYDROXY-
C11H8O3	000483-55-6	2-ME-3-OH-1,4-NAPHTHOQUINONE
C11H8O3	002348-82-5	2-METHOXY-1,4-NAPHTHOQUINONE
C11H9Br2N3O2S	030961-41-2	N1(3,5-DIBR-2-PYRIDYL)SULFANILAMI
C11H9Cl2N3O2	028313-59-9	3,5-DICL-C6H3NHN=C(CN)COOET
C11H9Cl2N3O2	036865-77-7	3,4-DICL-C6H3NHN=C(CN)COOET
C11H9Cl2N3O2S	030961-38-7	N1-(3,5-CL-2-PYRIDYL)SULFANILAMID
C11H9Cl2NO2	000101-27-9	BARBAN
C11H9Cl3N4	006981-11-9	2,4-DIAMINOPYRIMIDINE-5(245-CL)BENZYL
C11H9Cl3N4	007520-68-5	2,4-DIAMINOPYRIMIDINE-5(345-CL)BENZYL
C11H9Cl5O3	000136-25-4	ERBON
C11H9ClN4	081865-11-4	PROPANEDINITRILE, [[4-(2-CHLOROETHYL)PHENYL]HYDR
C11H9ClO2S	051527-19-6	TIANAFAC
C11H9F3N2O4S	101063-98-3	PROPANOIC ACID, 3-[6-(TRIFLUOROMETHYL)-4H-1,2,4-
C11H9F3N2O4S2	124850-88-0	PROPANOIC ACID, 3-[[6-(TRIFLUOROMETHYL)-4H-1,2,4
C11H9FN2O2	004871-13-0	2,4(1H,3H)-PYRIMIDINEDIONE, 5-FLUORO-1-(PHENYLME
C11H9N	000939-23-1	4-PHENYLPYRIDINE
C11H9N	001008-88-4	3-PHENYLPYRIDINE
C11H9N	001008-89-5	2-PHENYLPYRIDINE
C11H9N2O5S	025580-68-1	3ME-5-(5NO2-FURPROPENILIDENE)THIAZOLIDN-24-DIONE
C11H9N2O5S	025603-14-9	3AL-5-(5NO2-FURFURIL)THIAZOLIDIN-2,4-ONE
C11H9N3O3S	097534-21-9	MERBARONE

C11H9N3O4S	027472-83-9	2-MEAM-5(5NO2-FURYLPROPENILIDENE)THIAZOLONE
C11H9N3O5S	052661-43-5	ACETAMIDE, N-METHYL-N-[5-[(5-NITRO-2-FURANYL)MET
C11H9N3O5S	052661-66-2	3-ME-2-ACETIMINO-5-(5NO2-2-FURFURILIDENE)THIAZOL
C11H9N5	017720-22-8	8-PHENYLADENINE
C11H9N5O	055121-29-4	ACETAMIDE, N-[4-[(DICYANOMETHYLENE)HYDRAZONO]PHE
C11H9N5O6	028313-64-6	2,4-DINO2-C6H3NHN=C(CN)COOET
C11H9NaO3S	097683-29-9	1-SO3NA-3-ME AZULENE
C11H9NO	002243-81-4	1-NAPHTHALENECARBOXAMIDE
C11H9NO	004783-68-0	2-PHENOXYPYRIDINE
C11H9NO	033021-53-3	3-ACETYLQUINOLINE
C11H9NO	073013-68-0	6-ACETYLQUINOLINE
C11H9NO2	002598-29-0	8-ACETYLOXYQUINOLINE
C11H9NO2	005959-52-4	3-AMINO-2-NAPHTHOIC ACID
C11H9NO2	053951-84-1	METHYL3-QUINOLINECARBOXYLATE
C11H9NO4	005493-24-3	1H-ISOINDOLE-1,3(2H)-DIONE, 2-[(ACETYLOXY)METHYL
C11H9NO4	022509-74-6	N-CARBETHOXYPHTHALIMIDE
C11H9NO4S2	118993-57-0	2-THIOPHENESULFONAMIDE, 4-(4-HYDROXYBENZOYL)-
C11H9NS	028856-77-1	3-PHENYLTHIOPYRIDINE
C11H10	000090-12-0	1-METHYLNAPHTHALENE
C11H10	000091-57-6	2-METHYLNAPHTHALENE
C11H10Br2N2O4S	101064-04-4	BUTANOIC ACID, 4-(5,7-DIBROMO-4H-1,2,4-BENZOTHIA
C11H10BrN3O2S	016805-99-5	N1-(5-BR-2-PYRIDYL)SULFANILAMIDE
C11H10BrN3O2S	030961-39-8	N1-(3-BR-2-PYRIDYL)SULFANILAMIDE
C11H10BrNO2	005460-29-7	N-(3-BROMOPROPYL)PHTHALIMIDE
C11H10Cl2N2O4S	101064-01-1	BUTANOIC ACID, 4-(5,7-DICHLORO-4H-1,2,4-BENZOTHI
C11H10Cl2N2O4S	101064-02-2	BUTANOIC ACID, 4-(6,7-DICHLORO-4H-1,2,4-BENZOTHI
C11H10Cl2N4	007761-45-7	2,4-NH2-5(3,4DICLPH)-6-MEPYRIMIDINE
C11H10ClN3O	001698-62-0	3(2H)-PYRIDAZINONE, 4-CHLORO-5-(METHYLAMINO)-2-P
C11H10ClN3O	028317-64-8	2-ME-4-CL-C6H3NHN=C(CN)CO-ME
C11H10ClN3O2	003994-23-8	2-CL-C6H4NHN=C(CN)COOET
C11H10ClN3O2	036874-67-6	3-CL-C6H4NHN=C(CN)COOET
C11H10ClN3O2S	026807-64-7	N1-(3-CL-2-PYRIDYL)SULFANILAMIDE
C11H10ClN3O2S	030961-36-5	N1-(5-CL-2-PYRIDYL)SULFANILAMIDE
C11H10ClN3O2S	034392-79-5	N1-(2-CL-3-PYRIDYL)SULFANILAMIDE
C11H10ClN3O2S	034392-82-0	N1-(6-CL-3-PYRIDYL)SULFANILAMIDE
C11H10ClNO2	001967-16-4	CHLORBUFAM
C11H10ClNO3S	025059-80-7	BENAZOLIN-ETHYL
C11H10F3NO3	024568-14-7	BENZOIC ACID, 4-[(TRIFLUOROACETYL)AMINO]-,ETHYL
C11H10F3NO3	096027-93-9	BENZOIC ACID, 2-[(TRIFLUOROACETYL)AMINO]ETHYL ES
C11H10FN3O3	004548-15-6	1H-IMIDAZOL-1-ETHANOL, 2-(4-FLUOROPHENYL)-5-NITR
C11H10IN5O3	096258-83-2	1H-IMIDAZOLE-1-ACETAMIDE, 4-IODO-5-NITRO-N-(3-PY
C11H10N2	006631-37-4	2-PYRIDINEAMINE, N-PHENYL
C11H10N2O	000530-53-0	PYRROLO[2,1-B]QUINAZOLIN-9(1H)-ONE, 2,3-DIHYDRO-
C11H10N2O	005417-50-5	3-ACETYLAMINOQUINOLINE
C11H10N2O	022433-76-7	6-ACETAMINOQUINOLINE
C11H10N2O	033757-42-5	8-ACETYLAMINOQUINOLINE
C11H10N2O	036164-42-8	7-ACETYLAMINOQUINOLINE
C11H10N2O	042464-80-2	ACETAMIDE, N-(5-QUINOLINYL)-
C11H10N2O	070026-41-4	CYCLOPENTA[D]PYRIDO[1,2-A]PYRIMIDIN-10(1H)-ONE,
C11H10N2O2	032654-59-4	4-ACETYLOXYAMINOQUINOLINE
C11H10N2O3	000076-94-8	BARBITURIC ACID,5-ME-5-PHENYL-
C11H10N2O5	003688-53-7	2-(2-FURYL)-3-(5-NO2FURYL)ACRYLAMIDE
C11H10N2O5S	027550-11-4	2,4-THIAZOLIDINEDIONE, 3-(1-METHYLETHYL)-5-[(5-N
C11H10N2O5S	052661-71-9	2,4-THIAZOLIDINEDIONE, 5-[(5-NITRO-2-FURANYL)MET
C11H10N2S	000086-88-4	#NAME?
C11H10O	000093-04-9	NAPHTHALENE, 2-METHOXY-
C11H10O	001592-38-7	2-NAPHTHALENEMETHANOL
C11H10O	002216-69-5	NAPHTHALENE, 1-METHOXY-

C11H10O	004780-79-4	1-NAPHTHALENEMETHANOL
C11H10O2	000505-02-2	2,8-DECADIENE-4,6-DIYNOIC ACID, METHYL ESTER
C11H10O2	001552-94-9	2,4-PENTADIENOIC ACID, 5-PHENYL-
C11H10O2	017584-90-6	4H-1-BENZOPYRAN-4-ONE, 2,3-DIMETHYL-
C11H10O2S	054108-51-9	NAPHTHALENE, 1-(METHYLSULFONYL)-
C11H10O3	003160-37-0	PIPERONYL ACETONE
C11H10O3	003199-61-9	2-BENZOFURANCARBOXYLIC ACID, ETHYL ESTER
C11H10O3	003988-76-9	2-PROPEN-1-ONE, 1,3-BIS(2-FURANYL)-
C11H11Br3N2O5	049648-42-2	TRIBROMOAMPHENICOL
C11H11BrN2O4S	101064-03-3	BUTANOIC ACID, 4-(7-BROMO-4H-1,2,4-BENZOTHIADIAZ
C11H11Cl2NO	002790-16-1	CYCLOPROPANECARBOXAMIDE, N-(3,4-DICHLOROPHENYL)-
C11H11Cl2NO	074054-79-8	3-BUTENAMIDE, N-(2,4-DICHLOROPHENYL)-2-METHYL-
C11H11Cl2NO2	040575-34-6	PHENOPYLATE
C11H11Cl2NO2	098730-04-2	BENOXACOR
C11H11Cl2NO3	099854-42-9	DILOXANIDE ACETATE
C11H11Cl2NO3	143121-08-8	RH-9611 O-PH N-BUTYLENEN CARBAMATE
C11H11Cl3N2O5	019934-51-1	TRICHLOROAMPHENICOL
C11H11Cl3O3	063253-20-3	ACETIC ACID, 2,2,2-TRICHLORO-1-(4-METHOXYPHENYL)
C11H11ClN2O2S	037148-24-6	124-BENZTHIADIAZIN-1-O2-3-CYBU-7-CL
C11H11ClN2O4S	101063-99-4	BUTANOIC ACID, 4-(6-CHLORO-4H-1,2,4-BENZOTHIADIA
C11H11ClN2O4S	101064-00-0	BUTANOIC ACID, 4-(7-CHLORO-4H-1,2,4-BENZOTHIADIA
C11H11ClN4	069945-58-0	2,4-PYRIMIDINEDIAMINE, 5-[(3-CHLOROPHENYL)METHYL
C11H11ClO3	022131-79-9	3-CL-4-ALLYLOXYPHENYL ACETIC ACID
C11H11ClO3	074407-83-3	2-PROPENOIC ACID, 3-(2-CHLOROPHENOXY)-, ETHYL ES
C11H11F3N2O3	013311-84-7	PROPANAMIDE, 2-METHYL-N-[4-NITRO-3-(TRIFLUOROMET
C11H11F3N2O4	052806-53-8	PROPANAMIDE, 2-HYDROXY-2-METHYL-N-[4-NITRO-3-(TR
C11H11F3N2O5	042583-67-5	TRIFLUOROAMPHENICOL
C11H11F3N2O5S	085873-47-8	PROPANAMIDE, 2-METHYL-N-[2-NITRO-4-[(TRIFLUOROME
C11H11IN2O	000129-81-7	4-IODOANTIPYRINE
C11H11N	000877-43-0	2,6-DIMETHYLQUINOLINE
C11H11N	001198-37-4	2,4-DIMETHYLQUINOLINE
C11H11N	002216-68-4	1-NAPHTHALENAMINE, N-METHYL-
C11H11N3O	000715-99-1	1-PH-3,5-DIME-4-NITROSOPYRAZOLE
C11H11N3O	074099-08-4	HYDRAZINECARBOXAMIDE, N-(1-NAPHTHALENYL)-
C11H11N3O2S	000144-83-2	SULFAPYRIDINE
C11H11N3O4S	027472-90-8	4(5H)-THIAZOLONE, 5-[(5-NITRO-2-FURANYL)METHYLEN
C11H11N3O4S	027472-92-0	4(5H)-THIAZOLONE, 2-[(1-METHYLETHYL)AMINO]-5-[(5
C11H11N3OS	021085-19-8	3MES-4NH2-6CYHX-124-TRIAZIN-5-ONE
C11H11N5	000094-78-0	PHENAZOPYRIDINE
C11H11N5OS	115787-59-2	4H-PYRROLO[3,2-D]PYRIMIDIN-4-ONE, 2,6-DIAMINO-3,
C11H11N5OS	115812-51-6	4H-PYRROLO[3,2-D]PYRIMIDIN-4-ONE, 2,6-DIAMINO-3,
C11H11NO	015011-28-6	4,5-DIMETHYL-8-QUINOLINOL
C11H11NO	031835-53-7	2-METHYL-4-METHOXYQUINOLINE
C11H11NO	057369-32-1	PYROQUILON
C11H11NO2	000086-34-0	N-METHYL-2-PHENYL SUCCINIMIDE
C11H11NO2	000830-96-6	1H-INDOLE-3-PROPANOIC ACID
C11H11NO2	003770-50-1	1H-INDOLE-2-CARBOXYLIC ACID, ETHYL ESTER
C11H11NO2	006639-06-1	1H-INDOLE-1-PROPANOIC ACID
C11H11NO2	030125-76-9	N-BENZYL-2,4-PYRROLIDIONE
C11H11NO2	057334-38-0	4-ME-5-MEO-8-QUINOLINOL
C11H11NO2	070290-53-8	O-METHYL-N-PHENYLSUCCINIMIDE
C11H11NO2	093476-51-8	N-(M-TOLYL)SUCCINIMIDE
C11H11NO2S	080412-20-0	INDOL-3-YL-2-THIOPROPIONIC ACID
C11H11NO3	002314-80-9	P-METHOXY-N-PHENYLSUCCINIMIDE
C11H11NO3	016141-40-5	M-METHOXY-N-PHENYLSUCCINIMIDE
C11H11NO3	066432-08-4	N-FORMYL-P-ACETOXYSTYRLAMINE
C11H11NO3	117491-58-4	1H-ISOINDOLE-1-ONE, 2-[(ACETYLOXY)METHYL]-
C11H11NO4	068659-48-3	ACETAMIDE, N-ACETYL-2-(BENZOYLOXY)-

C11H11NO4S3	128348-32-3	2-THIOPHENESULFONAMIDE, 4-[(4-METHYLPHENYL)SULFO
C11H11NO5	050785-22-3	BENZOIC ACID, 2-(ACETYLOXY)-, 2-AMINO-2-OXOETHYL
C11H11NO5S2	128348-34-5	2-FURANSULFONAMIDE, 4-[(4-METHYLPHENYL)SULFONYL]
C11H11NO5S3	119731-18-9	2-THIOPHENESULFONAMIDE, 4-[(4-METHOXYPHENYL)SULF
C11H11NO6S2	128348-35-6	2-FURANSULFONAMIDE, 4-[(4-METHOXYPHENYL)SULFONYL
C11H12Br2N2O5	017371-30-1	DIBROMOAMPHENICOL
C11H12BrCl2NO3	023885-59-8	P-BROMO-AMPHENICOL
C11H12BrN	026670-42-8	ETHYLQUINOLINIUM BROMIDE
C11H12BrN3O2	069811-19-4	CARBAMIC ACID, [4-(2-BROMOPHENYL)-4,5-DIHYDRO-1H
C11H12BrN3O2	069811-22-9	CARBAMIC ACID, [4-(3-BROMOPHENYL)-4,5-DIHYDRO-1H
C11H12BrNO2	058708-46-6	BENZAMIDE, 5-BROMO-N-CYCLOPROPYL-2-HYDROXY-3-MET
C11H12BrNO2S	099481-58-0	4-BROMO-S-PROPYL-BENZOYLTHIOFORMOHYDROXIMATE
C11H12Cl2INO3	049648-53-5	P-IODOAMPHENICOL
C11H12Cl2N2O5	000056-75-7	CHLORAMPHENICOL
C11H12Cl2O3	000094-11-1	2,4-D, ISOPROPYL ESTER
C11H12Cl2O3	001928-61-6	2,4-D, PROPYL ESTER
C11H12ClN	002012-81-9	BENZENEACETONITRILE, 4-CHLORO-.ALPHA.-(1-METHYLE
C11H12ClNO2	001759-02-0	RH-1911 O-PH(4-CL) N-BUTYLENE CARBAMATE
C11H12ClNO2	143121-06-6	RH-1909 O-PH(2-CL) N-BUTYLENE CARBAMATE
C11H12F2N2O5	049648-37-5	DIFLUOROAMPHENICOL
C11H12F3NO	031599-68-5	2-(M-CF3 PHENYL)MORPHOLINE
C11H12F3NO2	084640-35-7	P-CF3-BENZYL N,N-DIME CARBAMATE
C11H12FN	001978-59-2	PYRIDINE, 4-(4-FLUOROPHENYL)-1,2,3,6-TETRAHYDRO-
C11H12FN3O2	069811-11-6	CARBAMIC ACID, [4-(3-FLUORPHENYL)-4,5-DIHYDRO-1H
C11H12N2	021154-18-7	2-QUINOLINAMINE, N,N-DIMETHYL-
C11H12N2	029526-42-9	8-DIMETHYLAMINOQUINOLINE
C11H12N2	089770-32-1	7-DIMETHYLAMINOQUINOLINE
C11H12N2O	069808-72-6	1H-INDOLE-2-CARBOXAMIDE, N-ETHYL-
C11H12N2O2	000054-12-6	DL-TRYPTOPHAN
C11H12N2O2	000073-22-3	TRYPTOPHAN
C11H12N2O2	000086-35-1	2,4-IMIDAZOLIDINEDIONE, 3-ETHYL-5-PHENYL-
C11H12N2O2	000631-07-2	2,4-IMIDAZOLIDINEDIONE, 5-ETHYL-5-PHENYL-
C11H12N2O2	015888-02-5	MALONAMIDE, 3-METHYLBENZAL
C11H12N2O2	023484-11-9	4-ACETAMIDOQUINOLINE-1-OXIDE
C11H12N2O2	081310-40-9	BENZENEACETONITRILE, α-(1-METHYLETHYL)-4-NITRO-
C11H12N2O2	084640-32-4	P-CYANOBENZYL N,N-DIME CARBAMATE
C11H12N2O2	091350-72-0	CARBAMIC ACID, 2-(1H-INDOL-3-YL)ETHYL ESTER
C11H12N2O3	015804-68-9	MALONAMIDE, 3-METHOXYBENZAL
C11H12N2O3	032092-18-5	PYRIDO(12A)PYRIMIDIN-4-ONE,3-ETO-CO
C11H12N2O4	106231-53-2	ACETAMIDE, 2-[[(BENZOYLOXY)ACETYL]AMINO]-
C11H12N2O4S	004826-24-8	BUTANOIC ACID, 4-(4H-1,2,4-BENZOTHIADIAZIN-3-YL)
C11H12N2O4S	101063-96-1	PROPANOIC ACID, 3-(6-METHYL-4H-1,2,4-BENZOTHIADI
C11H12N2O4S2	124802-93-3	PROPANOIC ACID, 3-[(6-METHYL-4H-1,2,4-BENZOTHIAD
C11H12N2O5S	101063-97-2	PROPANOIC ACID, 3-(6-METHOXY-4H-1,2,4-BENZOTHIAD
C11H12N2O5S2	124850-87-9	PROPANOIC ACID, 3-[(6-METHOXY-4H-1,2,4-BENZOTHIA
C11H12N2S	061887-92-1	2-THIAZOLEETHANAMINE, 4-PHENYL-
C11H12N4	007319-45-1	2,4-DIAMINO-5-BENZYLPYRIMIDINE
C11H12N4	072112-24-4	2,3-PYRAZINEDICARBONITRILE, 5-BUTYL-6-METHYL-
C11H12N4O2S	000127-79-7	SULFAMERAZINE
C11H12N4O2S	000599-88-2	SULFAPERINE
C11H12N4O3S	000080-35-3	SULFAMETHOXYPYRIDAZINE
C11H12N4O3S	000152-47-6	SULFALENE
C11H12N4O3S	000651-06-9	SULFAMETER
C11H12N4O3S	001220-83-3	BENZENESULFONAMIDE, 4-AMINO-N-(6-METHOXY-4-PYRIM
C11H12N4O5	024570-14-7	5(1-AZIRD)-2,4-NO2 BENZAMIDE,N-ET
C11H12N4O5	027221-03-0	5(1-AZIRD)-2,4-NO2 BENZAMIDE,NN-DIME
C11H12NO4PS2	000732-11-6	PHOSMET
C11H12O	002513-25-9	2,2-DIMETHYLBENZOPYRAN

C11H12O2	000103-36-6	CINNAMIC ACID, ETHYL ESTER
C11H12O2	002495-37-6	BENZYL METHACRYLATE
C11H12O2	064002-57-9	1(1H)-ISOBENZOFURANONE-3-I-PROPYL
C11H12O3	000094-02-0	ETHYL BENZOYLACETATE
C11H12O3	000607-91-0	MYRISTICIN
C11H12O3	001501-05-9	BUTANOIC ACID, 4-BENZOYL-
C11H12O3	001771-65-9	BENZENEBUTANOIC ACID, 2-METHYL-4-OXO-4-
C11H12O3	001878-58-6	5-INDANOXYACETIC ACID
C11H12O3	010031-96-6	EUGENYL FORMATE
C11H12O3	043013-70-3	2-PROPENOIC ACID, 3-PHENOXY-, ETHYL ESTER
C11H12O4	000102-37-4	ETHYL CINNAMATE,3,4-DIHYDROXY
C11H12O4	000581-55-5	BENZAL DIACETATE
C11H12O4	002316-26-9	CINNAMIC ACID,3,4-DIMETHOXY
C11H12O4	003153-44-4	BENZENEBUTANOIC ACID, 4-METHOXY-¥-OXO-
C11H12O4	038353-69-4	BENZOIC ACID, 2-ACETYLOXY ET ESTER
C11H12O4S	076432-30-9	BENZOIC ACID, 2-(ACETYLOXY)-, (METHYLTHIO)METHYL
C11H12O5S	076432-33-2	BENZOIC ACID, 2-(ACETYLOXY)-, (METHYLSULFINYL)ME
C11H12O6S	076432-35-4	BENZOIC ACID, 2-(ACETYLOXY)-, (METHYLSULFONYL)ME
C11H13BrN2O3	064889-77-6	BARBITURIC ACID,N-ME-5ALLYL,5-BR-ALLYL
C11H13BrN2O5	040027-72-3	MONOBROMOAMPHENICOL
C11H13Cl2NO	002150-96-1	PENTANAMIDE, N-(3,4-DICHLOROPHENYL)-
C11H13Cl2NO	007160-22-7	PROPANAMIDE, N-(3,4-DICHLOROPHENYL)-2,2-DIMETHYL
C11H13Cl2NO	007160-25-0	BUTANAMIDE, N-(3,4-DICHLOROPHENYL)-2-METHYL-
C11H13Cl2NO3	025126-19-6	H-(NO NO2) AMPHENICOL
C11H13Cl3N2O3	004482-55-7	FENURON TCA
C11H13ClF3N3O4S3	000346-18-9	POLYTHIAZIDE
C11H13ClN2O2S	013460-16-7	124-BENZTHIADIAZIN-1-O2-3-TBU-7-CL
C11H13ClN2O2S	037148-21-3	124-BENZOTHIADIAZIN-1-O2-3-IBU-6-CL
C11H13ClN2O2S	037148-22-4	124-BENZOTHIADIAZIN-1-O2-3-SBU-7-CL
C11H13ClN2O2S	037157-85-0	124-BENZTHIADIAZINE-1-O2-3-BU-7-CL
C11H13ClN2O5	017278-57-8	MONOCHLOROAMPHENICOL
C11H13ClN3O3P	067410-46-2	3,5-DIOXA-6-AZA-4-PHOSPHAOCT-6-ENE-8-NITRILE, 7-
C11H13ClO3	000094-81-5	MCPB
C11H13ClO3	017413-90-0	2-(P-CLPHENOXY)-2-ET-PROPIONIC AC
C11H13ClO3	025204-89-1	BUTANOIC ACID, 2-(4-CHLORO-2-METHYLPHENOXY)-
C11H13F3N2	015532-75-9	PIPERAZINE, 1-[3-(TRIFLUOROMETHYL)PHENYL]-
C11H13F3N2O3S	053780-34-0	MEFLUIDIDE
C11H13F3N4O4	029091-05-2	DINITRAMINE
C11H13FN2O5	049648-38-6	MONOFLUOROAMPHENICOL
C11H13IN2O5	040027-73-4	MONOIODOAMPHENICOL
C11H13N	010338-69-9	PYRIDINE, 1,2,3,6-TETRAHYDRO-4-PHENYL-
C11H13N	014098-20-5	BENZOBICYCLO(2.2.1)HEPTENE,9EX-NH2
C11H13N	058742-04-4	BENZOBICYCLO(2,2,1)HEPTENE,2-EN-AMINO
C11H13N	062624-26-4	BENZOBICYCLO(2,2,1)HEPTENE,2-EX-AMINO
C11H13N	072597-35-4	BENZOBICYCLO(2,2,1)HEPTENE,9-EN-AMINO
C11H13N3	066480-64-6	1H-BENZOTRIAZOLE, 1-CYCLOPENTYL-
C11H13N3	066536-69-4	2H-BENZOTRIAZOLE, 2-CYCLOPENTYL-
C11H13N3O2	069218-42-4	1H-BENZOTRIAZOLE-1-PENTANOIC ACID
C11H13N3O2	069218-44-6	2H-BENZOTRIAZOLE-2-PENTANOIC ACID
C11H13N3O2	069218-47-9	1H-BENZOTRIAZOLE-1-PROPANOIC ACID, ETHYL ESTER
C11H13N3O2	069218-52-6	2H-BENZOTRIAZOLE-2-PROPANOIC ACID, ETHYL ESTER
C11H13N3O2	075584-51-9	2H-BENZOTRIAZOLE-2-ACETIC ACID, α-METHYL-, ETHYL
C11H13N3O2	075584-60-0	1H-BENZOTRIAZOLE-1-ACETIC ACID, α-METHYL-, ETHYL
C11H13N3O2	115054-91-6	PENTANOIC ACID, 4-(1H-BENZOTRIAZOL-1-YL)-
C11H13N3O2	115054-96-1	PENTANOIC ACID, 4-(2H-BENZOTRIAZOL-2-YL)-
C11H13N3O3	071006-77-4	1-(2-(PHENOXY)ETHYL)2-NO2IMIDAZOLE
C11H13N3O3	075912-69-5	1-(2-(P-NO2PHENOXY)ETHYL)IMIDAZOLE
C11H13N3O3S	000127-69-5	SULFISOXAZOLE

C11H13N3O4S	071006-80-9	PHENYL(2-NO2-1-IMIDAZOLYL)ET)SULFONE
C11H13N3S	087691-87-0	1,2-BENZISOTHIAZOLE, 3-(1-PIPERAZINYL)-
C11H13N5	069945-50-2	2,4-DIAMINOPYRIMIDINE,5(P-NH2BENZYL)
C11H13N5	072775-79-2	2'-ETHYLBENZOGUANAMINE
C11H13N5O	066974-90-1	BENZAMIDE, N-(CYANOMETHYL)-2-(3,3-DIMETHYL-1-THI
C11H13N5O	072775-81-6	2'-ETHOXYLBENZOGUANAMINE
C11H13NO	005291-77-0	N-BENZYL-2-PYRROLIDINONE
C11H13NO	013156-74-6	N,N-DIMETHYLCINNAMAMIDE
C11H13NO	038954-41-5	1-PHENYL-2-ACETAMIDOCYCLOPROPANE
C11H13NO	086943-80-8	9-OXABENZOBICYCLO(221)HEPTEN,2EX-MEAM
C11H13NO	086992-70-3	9-OXABENZOBICYCLO(221)HEPTEN,2EN-MEAM
C11H13NO2	055379-71-0	RH-1908 O-PH N-BUTYLENE CARBAMATE
C11H13NO3	002018-61-3	L-PHENYLALANINE, N-ACETYL-
C11H13NO3	002689-47-6	3-ACETYLPHENYLDIMETHYLCARBAMATE
C11H13NO3	010051-63-5	N-ME-3-PROPIONYLPHENYLCARBAMATE
C11H13NO3	029230-99-7	2-ACETYLPHENYLDIMETHYLCARBAMATE
C11H13NO3	054266-28-3	N-ME-4-PROPIONYLPHENYLCARBAMATE
C11H13NO3	064649-57-6	ACETAMIDE, 2-(BENZOYLOXY)-N-ETHYL-
C11H13NO3	106231-54-3	ACETAMIDE, 2-(BENZOYLOXY)-N,N-DIMETHYL-
C11H13NO4	000537-55-3	N-ACETYL-L-TYROSINE
C11H13NO4	006636-22-2	4-CH3COO-PHENYLALANINE
C11H13NO4	006988-21-2	DIOXACARB
C11H13NO4	022781-23-3	BENDIOCARB
C11H13NO4	114665-08-6	BENZOIC ACID, 2-HYDROXY-, 2-(DIMETHYLAMINO)-2-OX
C11H13NO4	117509-89-4	1H-ISOINDOLE-1,3(2H)-DIONE, 2-[(ACETYLOXY)METHYL
C11H14BrN3	065936-23-4	2-IMIDAZOLIDINIMINE, N-(4-BROMO-2,6-DIMETHYLPHEN
C11H14BrN5	003567-84-8	4,6-NH2-2,2-DIME-1(4BRPH)S-TRIAZINE
C11H14BrNO2	058708-45-5	BENZAMIDE, 5-BROMO-2-HYDROXY-3-METHYL-N-(1-METHY
C11H14BrNO2	124497-74-1	2-OH-2-(P-BR-PH)-4-ME MORPHOLINE HCL
C11H14Cl2N2O	005006-90-6	UREA, N-(3,4-DICHLOROPHENYL)-N'-(2-METHYLPROPYL)
C11H14ClN3	004201-36-9	2(24-DIME-6-CL-PH IMINO)IMIDAZOLINE
C11H14ClN3	004201-38-1	2-IMIDAZOLIDINIMINE, N-(4-CHLORO-2,6-DIMETHYLPHE
C11H14ClN3O2	063346-71-4	2(5-CL-2,4-MEO PH IMINO)IMIDAZOLINE
C11H14ClNO	001918-16-7	PROPACHLOR
C11H14ClNO	010345-79-6	BENZAMIDE, 2-CHLORO-N,N-DIETHYL-
C11H14ClNO2	002164-13-8	BCPC
C11H14FN3O3	111041-67-9	IMIDODICARBONIC DIAMIDE, N-(2-FLUORO-5-METHOXYPH
C11H14FN5	001542-59-2	4,6-NH2-2,2-DIME-1(4FPH)S-TRIAZINE
C11H14FN5O2	126502-17-8	9H-PURIN-6-AMINE, 9-(2,3-DIDEOXY-2-FLUORO-ß-D-TH
C11H14FN5O2	132722-91-9	9H-PURIN-6-AMINE, 9-(2,3-DIDEOXY-2-FLUORO-ß-D-TH
C11H14FN5O2	132722-94-2	9H-PURIN-6-AMINE, 9-(2,3-DIDEOXY-2-FLUORO-ß-D-TH
C11H14IN5	046781-41-3	4,6-NH2-2,2-DIME-1(4IPH)S-TRIAZINE
C11H14N2	140111-43-9	4,4'-BIPYRIDINE, 1,2,3,6-TETRAHYDRO-1-METHYL-
C11H14N2O	000060-80-0	ANTIPYRINE
C11H14N2O	000485-35-8	CYTISINE
C11H14N2O	006652-04-6	4-PHENYL-N-NITROSOPIPERIDINE
C11H14N2O	038326-28-2	PYRIDO(12A)PYRIMIDIN-4-ONE,2ME,6ET
C11H14N2O	057773-19-0	PYRIDO(12A)PYRIMIDINE-4-ONE,3ET,6ME
C11H14N2O	063736-04-9	PYRIDO(12A)PYRIMIDINE-4-ONE,2ET,6ME
C11H14N2O	088491-49-0	9H-CYCLOPENTA[D]PYRROLO[1,2-A]PYRIMIDIN-4-ONE, 1
C11H14N2O2	006574-15-8	N-(P-NITROPHENYL)PIPERIDINE
C11H14N2O2	007376-90-1	PHENYLALANINE-AMIDE,N-ACETYL
C11H14N2O2	034993-08-3	N-NITROSO-PHENMETRAZINE
C11H14N2O2	042865-65-6	1,1-DIME-3-(M-ACETYLPHENYL)UREA
C11H14N2O2	080866-90-6	2-ETHYL-2-PHENYL-MALONAMIDE
C11H14N2O2	120399-50-0	ACETAMIDE, 2-(BENZOYLAMINO)-N-ETHYL-
C11H14N2O3	000718-67-2	BARBITURIC ACID,5ME-5CYHXENE
C11H14N2O3	000780-59-6	N-METHYLALLOBARBITAL

C11H14N2O3	001948-71-6	TYROSINE-AMIDE,N-ACETYL
C11H14N2O3	033484-38-7	4H-PYRIDO[1,2-A]PYRIMIDINE-3-CARBOXYLIC ACID, 6,
C11H14N2O3	038326-36-2	4H-PYRIDO[1,2-A]PYRIMIDINE-3-CARBOXYLIC ACID, 6,
C11H14N2O3	087340-61-2	CARBAMIC ACID, DIMETHYL-, [4-(AMINOCARBONYL)PHEN
C11H14N2O3	088132-23-4	1(4-ACETYLPHENYL)-3-MEO-3-ME UREA
C11H14N2O4	000553-20-8	ACETAMIDE, N-(5-NITRO-2-PROPOXYPHENYL)-
C11H14N2O4	132899-69-5	PROPANAMIDE, N-(2-ETHOXY-5-NITROPHENYL)-
C11H14N2O4S2	122321-30-6	THIENO[2,3-B]FURAN-2-SULFONAMIDE, 5-(4-MORPHOLIN
C11H14N2O5	004097-36-3	DNAP
C11H14N2O5	004423-58-9	AMPHENICOL
C11H14N2O5S	081792-99-6	PROPANAMIDE, 2-METHYL-N-[(4-METHYL-2-NITROPHENYL
C11H14N2O5S	083870-87-5	BENZENESULFONYL-L-GLUTAMINE
C11H14N4O	023898-86-4	PYRAZOL-3-ON,2(26DIME-4PYRIMD)15DIME
C11H14N4O	023898-92-2	PYRAZOL-3-ON,1(46DIME-2PYRIMD)25DIME
C11H14N4O	066974-73-0	BENZAMIDE, 4-[3-METHYL-3-(2-PROPENYL)-1-TRIAZENY
C11H14N4O	069218-59-3	1-(4-CARBAMOYLBUTYL)BENZOTRIAZOLE
C11H14N4O	115054-83-6	2H-BENZOTRIAZOLE-2-ACETAMIDE, α-PROPYL-
C11H14N4O	115054-89-2	1H-BENZOTRIAZOLE-1-ACETAMIDE, α-PROPYL-
C11H14N4O	115054-97-2	2H-BENZOTRIAZOLE-1-BUTANAMIDE, ¥-METHYL-
C11H14N4O2	023898-89-7	PYRAZOL-3-ON,2(2ME6MEO-4PYRIMD)15DIME
C11H14N4O2	023898-95-5	PYRAZOL-3-ON,1(2ME6MEO-4PYRIMD)25DIME
C11H14N4O3	098827-18-0	ALLOPURINOL,1-[PIVALOYLOXYMETHYL]-
C11H14N4O4	030516-87-1	THYMIDINE, 3'-AZIDO-3'-DEOXY-
C11H14N4O4S	000342-69-8	9H-PURINE, 6-(METHYLTHIO)-9-(beta-D-RIBOFURANOSY
C11H14N6O2	017711-74-9	1,3,5-TRIAZINE-2,4-DIAMINE, 1,6-DIHYDRO-6,6-DIME
C11H14N6O2	117121-41-2	UREA, N,N-DIMETHYL-N'-[4-[(2-METHYL-2H-TETRAZOL-
C11H14N6O5	074550-92-8	1H-IMIDAZOLE-1-ETHANOL, 2-METHYL-_-[(2-METHYL-5-
C11H14O	000645-13-6	P-ISOPROPYLACETOPHENONE
C11H14O	000938-16-9	2,2-DIMETHYLPROPIOPHENONE
C11H14O	001518-83-8	4-CYCLOPENTYLPHENOL
C11H14O	002235-83-8	5-PHENYL-2-PENTANONE
C11H14O2	000093-15-2	METHYLEUGENOL
C11H14O2	000098-73-7	BENZOIC ACID, 4-(TERT-BUTYL)-
C11H14O2	000136-60-7	BUTYL BENZOATE
C11H14O2	001608-68-0	2-ISOPROPYLPHENYL ACETATE
C11H14O2	002021-28-5	ETHYL B-PHENYLPROPIONATE
C11H14O2	002046-17-5	METHYL 4-PHENYLBUTYRATE
C11H14O2	002270-20-4	5-PHENYLPENTANOIC ACID
C11H14O2	015806-38-9	BUTANOIC ACID, 3-METHYL-, PHENYL ESTER
C11H14O2	020651-71-2	4-BUTYLBENZOIC ACID
C11H14O2	026114-12-5	BENZENEACETIC ACID, 4-PROPYL-
C11H14O3	000094-26-8	P-HYDROXY BUTYL BENZOATE
C11H14O3	001498-96-0	P-BUTOXYBENZOIC ACID
C11H14O3	001643-16-9	P-I-PROPYLPHENOXYACETIC ACID
C11H14O3	001878-52-0	M-I-PROPYLPHENOXYACETIC ACID
C11H14O3	002052-14-4	BENZOIC ACID, 2-HYDROXY-, BUTYL ESTER
C11H14O3	002084-11-9	M-PROPYLPHENOXYACETIC ACID
C11H14O3	004521-28-2	P-METHOXY-G-PHENYLBUTYRIC ACID
C11H14O3	023676-09-7	ETHYL 4-ETHOXYBENZOATE
C11H14O3	025141-58-6	O-I-PROPYLPHENOXYACETIC ACID
C11H14O3	026118-57-0	BENZENEACETIC ACID, 4-PROPOXY-
C11H14O4	013909-73-4	2,3,4-TRIMETHOXYACETOPHENONE
C11H14O4	020587-61-5	ETHANOL, 2- 2-(BENZOYLOXY)ETHOXY -
C11H14O4	039589-98-5	DIMALONE
C11H14O5	001083-41-6	BUTYL GALLATE
C11H14O5	003856-05-1	ISOBUTYL GALLATE
C11H15BrClO3PS	041198-08-7	PROFENOFOS
C11H15BrF3N	071324-01-1	NNN-ME-N(3-CF3PHME) AMMONIUM BROMIDE

C11H15BrN2	071323-97-2	NNN-ME-N(3(CYANOPHENOL)METHYL) AMMONIUM BROMIDE
C11H15BrN2O	078508-43-7	N'-(4-BROMO-3,5-DIMETHYLPHENYL)-N,N-DIMETHYLU*
C11H15BrN2O3	000125-55-3	BARBITURIC ACID,5-IPR,5-BR-ALLYL,N-ME
C11H15BrN2O3	001142-70-7	BARBITURIC ACID,5-IPR-5-(2BR2ALLYL)
C11H15Cl2NO	000059-61-0	DICHLORISOPROTERENOL
C11H15Cl2O2PS2	034643-46-4	PROTHIOPHOS
C11H15Cl2O2PS3	002275-14-1	PHENKAPTON
C11H15Cl2O4PS	026707-54-0	O=P(OET)(OET)O-2,6-CL-4-SMEPHENYL
C11H15Cl3NO3P	002213-87-8	PHOSPHORAMIDIC ACID, N-BUTYL-, O-METHYL O-(2,4,5
C11H15Cl3NO3P	002213-88-9	PHOSPHORAMIDIC ACID, N-(2-METHYLPROPYL)-, O-METH
C11H15Cl3NO3P	002214-33-7	PHOSPHORAMIDIC ACID, N-(1,1-DIMETHYLETHYL)-, O-
C11H15Cl3NO3P	002388-47-8	PHOSPHORAMIDIC ACID, N-(1-METHYLPROPYL)-, O-METH
C11H15ClN2O2	003544-35-2	N'-IPR-(4-CLPHENOXY)ACETIC HYDRAZIDE
C11H15ClN2O2	051963-48-5	N'-IPR-(2-CLPHENOXY)ACETIC HYDRAZIDE
C11H15ClN2O2	051963-49-6	N'-IPR-(3-CLPHENOXY)ACETIC HYDRAZIDE
C11H15ClN2O3S	000094-20-2	N2-BUTYL-N1-P-CHLOROBENZENESULFONYLUREA
C11H15ClN2O3S	013909-64-3	BENZENESULFONAMIDE, N-[(BUTYLAMINO)CARBONYL]-4-C
C11H15ClN6O3	089083-22-7	ACETAMIDE, N-[5-CHLORO-6-(4-METHYL-1-PIPERAZINYL
C11H15N	000771-99-3	PIPERIDINE, 4-PHENYL-
C11H15N	004096-20-2	N-PHENYLPIPERIDINE
C11H15N	019485-85-9	2-NAPHTHALENAMINE, 1,2,3,4-TETRAHYDRO-N-METHYL-
C11H15N3	004794-83-6	2-(2,4-DIME PHENYLIMINO)IMIDAZOLINE
C11H15N3	004859-06-7	2-(2,6-DIME PHENYLIMINO)IMIDAZOLINE
C11H15N3	067246-30-4	1-S-PENTABENZOTRIAZOLE
C11H15N3	069218-26-4	1-PENTABENZOTRIAZOLE
C11H15N3O	036385-57-6	PIPERAZINE-2-CARBOXANILIDE
C11H15N3O	063346-72-5	2-(2-MEO-4-ME PH IMINO)IMIDAZOLIDINE
C11H15N3O	069218-36-6	1-(5-HYDROXYPENTYL)BENZOTRIAZOLE
C11H15N3O	087072-64-8	1-ME-1-ACETYL-3(P-ETPH)TRIAZENE
C11H15N3O2	022259-30-9	FORMETANATE
C11H15N3O2	050531-51-6	2(25-DIMEO PH IMINO)IMIDAZOLIDINE
C11H15N3O2	064399-29-7	4H-PYRIDO[1,2-A]PYRIMIDINE-3-CARBOXAMIDE, 1,6,7,
C11H15N3O3	111075-96-8	1-ACETYL-4-(4-OET PHENYL) SEMICARBAZIDE
C11H15N3O4	001882-26-4	PYRIDINOL CARBAMATE
C11H15N3O6	013491-47-9	ACETAMIDE, N-(1-beta-D-ARABINOFURANOSYL-1,2-DIHY
C11H15N5	004022-58-6	4,6-NH2 1,2-H 2,2-DIME-1PH TRIAZENE
C11H15N5O2	066974-91-2	BENZAMIDE, N-(2-AMINO-2-OXOETHYL)-2-(3,3-DIMETHY
C11H15N5O2S	074188-69-5	1H-PYRROL-2-AMINE, N-[2-[[(5-METHYL-1H-IMIDAZOL-
C11H15N5OS	054855-60-6	CIMETIDINE,PYRIMIDIN-2-ONE ANALOG
C11H15N5OS	054855-84-4	CIMETIDINE,PYRIMID-4-ONE-2-YL ANALOG
C11H15NO	001467-34-1	3,4,5-TRIMETHYLACETANILIDE
C11H15NO	001696-17-9	N,N-DIETHYLBENZAMIDE
C11H15NO	002364-50-3	VALERANILIDE
C11H15NO	006625-74-7	TERT-VALERANILIDE
C11H15NO	010264-18-3	VALERANILIDE
C11H15NO	010316-00-4	4-BENZYLMORPHOLINE
C11H15NO	019246-04-9	ACETAMIDE, N-[2-(1-METHYLETHYL)PHENYL]-
C11H15NO	020330-99-8	ACETAMIDE, N-(4-PROPYLPHENYL)-
C11H15NO	054394-78-4	SEC-VALERANILIDE
C11H15NO	056108-12-4	4-TERT-BUTYLBENZAMIDE
C11H15NO	062008-56-4	2-(M-TOLYL)MORPHOLINE
C11H15NO	097055-06-6	2-METHYL-6-ETHYLACETANILIDE
C11H15NO2	000064-00-6	N-METHYL O-(3-IPRPHENYL)CARBAMATE
C11H15NO2	000094-25-7	BUTYL-P-AMINOBENZOATE
C11H15NO2	000671-03-4	2,4,5-TRIMEPH N-ME CARBAMATE
C11H15NO2	001538-74-5	BUTYL N-PHENYLCARBAMATE
C11H15NO2	002046-19-7	2-AMINO-5-PHENYLPENTANOIC ACID
C11H15NO2	002631-40-5	N-METHYL-2-ISOPROPYLPHENYLCARBAM

C11H15NO2	002686-99-9	N-ME-3,4,5-TRIMEPHENYL CARBAMATE
C11H15NO2	004089-99-0	N-ME-4-I-PROPYLPHENYLCARBAMATE
C11H15NO2	013780-91-1	3-ISOPROPYL-4-HYDROXYACETANILIDE
C11H15NO2	015482-11-8	N-METHYL-2-PROPYLPHENYLCARBAMATE
C11H15NO2	036592-59-3	2,3,6-TRIMETHYL-4-HYDROXYACETANILIDE
C11H15NO2	069477-72-1	2,3,5-TRIME-4-HYDROXYACETANILIDE
C11H15NO2	073258-87-4	P-BUTOXYBENZAMIDE
C11H15NO2	084640-21-1	P-MEBENZYL N,N-DIME CARBAMATE
C11H15NO2	098183-16-5	CARBAMIC ACID, DIMETHYL-, 2-PHENYLETHYL ESTER
C11H15NO2	124497-72-9	2-OH-2-PH-4-ME MORPHOLINE (HCL)
C11H15NO2	124497-77-4	2-OH-3-PH-4-ME MORPHOLINE HCL
C11H15NO2	138324-57-9	BENZAMIDE, 2-ETHOXY-N-ETHYL-
C11H15NO2S	002032-65-7	BAYER 37344
C11H15NO2S	029973-13-5	CRONETON
C11H15NO2S	084640-39-1	P-METHIOBENZYL N,N-DIME CARBAMATE
C11H15NO3	000114-26-1	PROPOXUR
C11H15NO3	000949-67-7	L-TYROSINE, ETHYL ESTER
C11H15NO3	003938-20-3	N-ME-3-I-PROPOXYPHENYLCARBAMATE
C11H15NO3	084640-23-3	M-MEO BENZYL N,N-DIME CARBAMATE
C11H15NO3	084640-24-4	P-MEO BENZYL N,N-DIME CARBAMATE
C11H15NO4	037178-37-3	L-TYROSINE, 3-HYDROXY-, ETHYL ESTER
C11H15NO4S	059777-59-2	BUTYL 4-SULFAMYLBENZOATE
C11H15NO4S	135832-47-2	ACETIC ACID, 3-[4-(AMINOSULFONYL)PHENYL]PROPYL E
C11H15NO5	000532-03-6	1,2-PROPANEDIOL, 3-(2-METHOXYPHENOXY)-, 1-CARBAM
C11H15NO5S2	108966-66-1	BENZOIC ACID, 5-(AMINOSULFONYL)-2-[(3-HYDROXYPRO
C11H15NO7S2	108966-67-2	BENZOIC ACID, 5-(AMINOSULFONYL)-2-[(3-HYDROXYPRO
C11H16	000098-51-1	P-(T-BUTYL)TOLUENE
C11H16	000538-68-1	PENTYLBENZENE
C11H16	000700-12-9	PENTAMETHYLBENZENE
C11H16	002050-24-0	1,3-DIETHYL-5-METHYLBENZENE
C11H16	029316-05-0	S-AMYLBENZENE
C11H16BrNO2	032156-26-6	2,-5-DIMETHOXY-4-BR AMPHETAMINE
C11H16BrNO2	064778-79-6	3,5-DIMETHOXY-4-BR AMPHETAMINE
C11H16ClN3O2	023422-53-9	FORMETANATE HYDROCHLORIDE
C11H16ClN3O4S2	023141-84-6	3-T-BUTYLHYDROCHLOROTHIAZIDE
C11H16ClN5	000500-92-5	IMIDODICARBONIMIDIC DIAMIDE, N-(4-CHLOROPHENYL)-
C11H16ClN5	049871-96-7	IMIDODICARBONIMIDIC DIAMIDE, N-(4-CHLOROPHENYL)-
C11H16ClO2PS3	000786-19-6	CARBOPHENTHION
C11H16ClO3PS2	021840-66-4	O,O-DIET-O(3-CL-4-MESPH)PHOSPHOROTHIOATE
C11H16ClO3PS2	026512-71-0	O,O-DIET-O(2-CL-4-MESPH)PHOSPHOROTHIOATE
C11H16ClO4PS	024493-78-5	OO-DIET-O(2-CL-4-MESPH)PHOSPHATE
C11H16ClO4PS	026798-03-8	OO-DIET-O-(3-CL-4-MES PH)PHOSPHATE
C11H16FN3O3	061422-45-5	CARMOFUR
C11H16N2	013552-35-7	3-PYRIDYLMETHYL-2-PIPERIDINE
C11H16N2	020173-28-8	3-PYRIDYLETHYL-2-(N-PYRROLIDINE)
C11H16N2	024380-92-5	METHYLANABASINE
C11H16N2	039512-51-1	PIPERAZINE, 1-(2-METHYLPHENYL)-
C11H16N2O	005363-33-7	BENZAMIDE, O-BUTYLAMINO
C11H16N2O	005692-23-9	N-(4-AMINOBUTYL)BENZAMIDE
C11H16N2O	035386-24-4	PIPERAZINE, 1-(2-METHOXYPHENYL)-
C11H16N2O	036627-56-2	1,1-DIME-3-(3,5-DIME PH) UREA
C11H16N2O	041708-72-9	PROPANAMIDE, 2-AMINO-N-(2,6-DIMETHYLPHENYL)-
C11H16N2O	063224-18-0	BENZAMIDE,N-(2-DIMETHYLAMINOETHYL)
C11H16N2O	066675-75-0	PROPANAMIDE, 3-AMINO-N-(2,6-DIMETHYLPHENYL)-
C11H16N2O2	000092-13-7	PILOCARPOL
C11H16N2O2	002032-59-9	AMINOCARB
C11H16N2O2	051963-47-4	N-IPR-1-(PHENOXY)PROPIONIC HYDRAZIDE
C11H16N2O2	052019-60-0	N'-IPR-PHENOXYACETIC ACID HYDRAZIDE

C11H16N2O2	098154-97-3	CARBAMIC ACID, DIMETHYL-, 3-(3-PYRIDINYL)PROPYL
C11H16N2O3	000115-44-6	BARBITURIC ACID, ALLYL,SEC-BUTYL
C11H16N2O3	007413-36-7	1(4-NO2 PH)-2-ISOPROPYLAMINO-ETHANOL
C11H16N2O3	021149-88-2	BARBITURIC ACID,5-ET-5(3ME-2BUTENE)
C11H16N2O3	039080-62-1	4H-PYRIDO[1,2-A]PYRIMIDINE-3-CARBOXYLIC ACID, 1,
C11H16N2O3	091430-64-7	BARBITURIC ACID,5-TBU-5ALLYL
C11H16N2O3S	050618-71-8	BENZENESULFONAMIDE, N-[(DIETHYLAMINO)CARBONYL]-
C11H16N2O3S	052102-41-7	BENZENESULFONAMIDE, N-[[(1,1-DIMETHYLETHYL)AMINO
C11H16N2O3S	059777-63-8	N-BUTYL 4-SULFAMYLBENZAMIDE
C11H16N2O3S	074746-42-2	ACETAMIDE, N-[4-[(DIMETHYLAMINO)SULFONYL]PHENYL]
C11H16N2O3S	081792-74-7	PROPANAMIDE, N-[(2-AMINO-4-METHYLPHENYL)SULFONYL
C11H16N2O4	060035-83-8	N-(TRIOHMEMETHANE)-N'-PHENYL UREA
C11H16N2O4	067460-68-8	2,5-DIMEO-4-NO2-AMPHETAMINE
C11H16N2O4	073632-82-3	4-PENTENOIC ACID, 2-[[(AMINOCARBONYL)AMINO]CARBO
C11H16N2O4S	081792-76-9	BENZENESULFONAMIDE, 2-AMINO-4-METHOXY-N-(2-METHY
C11H16N2O4S	097141-31-6	BENZENESULFONAMIDE, 4-METHOXY-2-[(2-METHYL-1-OXO
C11H16N4O	018694-43-4	PYRIMIDINE,2(5MEO-3ME-1PYRAZ)4,6DIME
C11H16N4O	023903-41-5	PYRIMIDINE,4(3MEO-5ME-1PYRAZOL)26DIME
C11H16N4O	023905-77-3	PYRIMIDINE,2(3MEO-5ME-1PYRAZOL)46DIME
C11H16N4O	059708-23-5	1-(4-CARBAMOYLPHENYL)-3-I-PROPYL-3-METHYLTRIAZEN
C11H16N4O	069218-64-0	2-(4-CARBAMOYLBUTYL)BENZOTRIAZOLE
C11H16N4O	089529-99-7	1-(4-CARBAMOYLPHENYL)-3-METHYL-3-PROPYLTRIAZENE
C11H16N4O2	018694-40-1	PYRIMIDINE,2(5MEO-3ME-1PYRAZ)4MEO6ME
C11H16N4O2	023903-42-6	PYRIMIDINE,6(5MEO3ME-1PYRAZOL)2ME4MEO
C11H16N4O2	111075-98-0	1-ACETYL-4(4-DIMEAMINOPH) SEMICARBAZIDE
C11H16N4O3	023905-98-8	PYRIMIDINE,6(3ME5MEO-1PYRAZOL)24DIMEO
C11H16N4O4	000069-33-0	TUBERCIDIN
C11H16N5O6PS	030630-07-0	8-METHYLTHIO CYCLIC AMP
C11H16N6	077314-79-5	2-PYRIDINAMINE, 4-(5-AMINO-1H-1,2,4-TRIAZOL-3-YL
C11H16N6	087871-35-0	4,6-NH2 2,2-DIME1(4-NH2)PH S-TRIAZENE
C11H16N6S2	096860-23-0	
C11H16NO4PS	024353-61-5	BENZOIC ACID, 2-[(AMINOMETHOXYPHOSPHINOTHIOYL)OX
C11H16NO5PS	001592-82-1	3-ISOPROPYL DIMETHYL PARATHION
C11H16NO6P	013074-11-8	3-ISOPROPYL-DIMETHYLPARA-OXON
C11H16O	000080-46-6	P-TERT-AMYLPHENOL
C11H16O	000098-27-1	4-(T-BUTYL)-2-CRESOL
C11H16O	000877-65-6	4-TERT-BUTYLBENZYL ALCOHOL
C11H16O	002409-55-4	2-(TERT-BUTYL)-4-METHYLPHENOL
C11H16O	014938-35-3	4-PENTYLPHENOL
C11H16O2	002525-11-3	1,2-BENZENEDIOL, 4-PENTYL-
C11H16O2	018979-53-8	PHENOL, 4-(PENTYLOXY)-
C11H16O2	018979-73-2	M-PENTOXYPHENOL
C11H16O4	001188-09-6	2-PROPENOIC ACID, 2-METHYL-, 1,3-PROPANEDIYL EST
C11H16O4S2	050780-76-2	PROPANEDIOIC ACID, 1,3-DITHIOLAN-2-YLIDENE-, ETH
C11H16O5S	071203-60-6	ETHYL-(3,4,5-TRIMEO)PHENYL SULFONE
C11H16O6S	057553-97-6	OHET-(3,4,5-TRIMEO)PHENYL SULFONE
C11H17BrN2O	071323-95-0	M-CARBAMYLBENZYL TRIMETHYL AMMONIUM BROMIDE
C11H17BrN2O3	077317-76-1	3-HXO-5-BR-6-ME-PYRIMIDIN-2,4-DIONE
C11H17ClN2O2	000054-71-7	PILOCARPINE, MONOHYDROCHLORIDE
C11H17N	001199-99-1	N,N-DIME-N-G-PHENYLPROPYLAMINE
C11H17N	002403-22-7	BENZENEMETHANAMINE, N-BUTYL-
C11H17N	013021-15-3	N,N,2,4,6-PENTAMETHYLANILINE
C11H17N	027876-24-0	P-HEXYLPYRIDINE
C11H17N	033228-44-3	BENZENAMINE, 4-PENTYL-
C11H17N	046114-16-3	A-I-PROPYLBENZENE ETHANEAMINE
C11H17N	063951-01-9	A-PROPYL BENZENEETHANEAMINE
C11H17N	084952-60-3	PHENETHYLAMINE,N-METHYL-A-ETHYL
C11H17N3	069218-27-5	2-PENTABENZOTRIAZOLE

C11H17N3	069218-32-2	2-S-PENTABENZOTRIAZOLE
C11H17N3O	069218-40-2	2-(5-HYDROXYPENTYL)BENZOTRIAZOLE
C11H17N3O	120277-93-2	6-AZABICYCLO[3.1.0]HEXANE, 6-[2-HYDROXY-3-(1H-IM
C11H17N3O2	120277-97-6	3-OXA-7-AZABICYCLO[4.1.0]HEPTANE, 7-[2-HYDROXY-3
C11H17N3O3	134419-52-6	2-AZETIDINECARBOXYLIC ACID, 1-[2-HYDROXY-3-(1H-I
C11H17N3O3S	000339-43-5	BENZENESULFONAMIDE, 4-AMINO-N- (BUTYLAMINO)CARBO
C11H17N3S	101626-70-4	TALIPEXOLE
C11H17N5O6	038099-11-5	URIDINE, 5-(3,3-DIMETHYL-1-TRIAZENYL)-
C11H17NO	001201-56-5	BENZENEMETHANOL, _-[1-(DIMETHYLAMINO)ETHYL]-, (R
C11H17NO	002455-15-4	ETHANAMINE, N,N-DIMETHYL-2-(3-METHYLPHENOXY)-
C11H17NO	031828-71-4	2-PROPANAMINE, 1-(2,6-DIMETHYLPHENOXY)-
C11H17NO	051344-14-0	ETHANAMINE, N,N-DIMETHYL-2-(4-METHYLPHENOXY)-
C11H17NO	065570-20-9	ANISOLE, O-BUTYLAMINO-
C11H17NO	074974-49-5	P-DIETHYLAMINOBENZYL ALCOHOL
C11H17NO	092936-20-4	BENZENEETHANAMINE, ß-METHOXY-N,α-DIMETHYL-, (R*,
C11H17NO	092936-21-5	BENZENEETHANAMINE, ß-METHOXY-N,α-DIMETHYL-, (R*,
C11H17NO2	000120-26-3	BENZENEETHANAMINE, 3,4-DIMETHOXY-A-METHYL
C11H17NO2	002801-68-5	2,5-DIMETHOXYAMPHETAMINE
C11H17NO2	003077-12-1	ETHANOL, 2,2'- (4-METHYLPHENYL)IMINO BIS-
C11H17NO2	015402-81-0	BENZENEETHANAMINE, 2,3-DIMETHOXY-α-METHYL-
C11H17NO2	023690-13-3	2,4-DIMETHOXYAMPHETAMINE
C11H17NO2	051344-12-8	ETHANAMINE, 2-(4-METHOXYPHENOXY)-N,N-DIMETHYL-
C11H17NO2S	061638-09-3	2,5-DIMETHOXY-4-MES-PHENETHYLAMINE
C11H17NO3	000054-04-6	MESCALINE
C11H17NO3	000122-15-6	DIMETAN
C11H17NO3S	135832-50-7	BENZENESULFONAMIDE, 4-(5-HYDROXYPENTYL)-
C11H17NO3S2	108966-53-6	BENZENESULFONAMIDE, 4-[3-HYDROXY-3-METHYLBUTYL)T
C11H17NO3S2	108966-54-7	BENZENESULFONAMIDE, 4-[(5-HYDROXYPENTYL)THIO]-
C11H17NO5S2	108966-56-9	BENZENESULFONAMIDE, 4-[3-HYDROXY-3-METHYLBUTYL)S
C11H17O3PS2	003070-15-3	FENSULFOTHION SULFIDE
C11H17O4PS	003070-13-1	O,O-DIET-O-(4-METHIOPH)PHOSPHATE
C11H17O4PS2	000115-90-2	FENSULFOTHION
C11H17O5PS2	014255-72-2	FENSULFOTHION SULFONE
C11H18BrN	021949-11-1	M-MEBENZYL-TRIMETHYL-AMMONIUM BR
C11H18BrN	074440-81-6	HEXYLPYRIDINIUM BROMIDE
C11H18BrN3O2	134419-55-9	1H-IMIDAZOLE-1-ETHANOL, α-[[(4-BROMO-TETRAHYDRO-
C11H18BrNO	071323-94-9	M-METHOXYBENZYL TRIMETHYL AMMONIUM BROMIDE
C11H18ClN	053429-15-5	BENZENEPENTANAMINE, HYDROCHLORIDE
C11H18ClN3O4	059414-37-8	CIS-1-NO-1(2-CLET)3(AC OXYCYHEX)UREA
C11H18ClN3O4	059414-38-9	1-NO-1-(2-CLET)-3(ACETYLOXYCYHEX)UREA
C11H18ClN5	084712-77-6	1,3,5-TRIAZINE-2,4-DIAMINE, 6-CHLORO-N-ETHYL-N'-
C11H18ClNO3	000051-30-9	1,2-BENZENEDIOL, 4-[1-HYDROXY-2-[(1-METHYLETHYL)
C11H18IN	001007-67-6	DIET-METHYL-PHENYL-AMMONIUM IODIDE
C11H18IN	046122-12-7	N-HEXYLPYRIDINIUM IODIDE
C11H18N2	001441-44-7	4-(N,N-DIME)-3-PYRIDYLBUTYLAMINE
C11H18N2	020173-34-6	N,N-DIETHYL-3-PYRIDYLETHYLAMINE
C11H18N2O2S	000076-75-5	THIOPENTAL
C11H18N2O2S	004388-79-8	BARBITURIC ACID,5-ET-5-IPE-2-THIO
C11H18N2O3	000057-43-2	AMOBARBITAL
C11H18N2O3	000076-74-4	PENTOBARBITAL
C11H18N2O3	000115-58-2	BARBITURIC ACID, 5-ET-5-AMYL
C11H18N2O3	085432-35-5	5,5-DIET BARBITURIC ACID,N1-IPR
C11H18N2O3	085432-38-8	2,4(3H,5H)-PYRIMIDINEDIONE, 5,5-DIETHYL-6-(1-MET
C11H18N2O3	085445-03-0	4,6(1H,5H)-PYRIMIDINEDIONE, 5,5-DIETHYL-2-(1-MET
C11H18N2O4	004241-40-1	BARBITURIC ACID,5-ET,5-(3-OH-1-ME)BUTYL
C11H18N2O6	073632-80-1	BUTANOIC ACID, 2-[[(AMINOCARBONYL)AMINO]CARBONYL
C11H18N4O2	023103-98-2	PIRIMICARB
C11H18N4O3	070132-50-2	PIMONIDAZOLE

C11H18N4O4	063677-95-2	PENTOSTATIN
C11H18O2	024141-52-4	TRANS-METHYL CHYRSANTHEMATE
C11H19N	033103-93-4	1-AMINOADAMATANE, 3-METHYL
C11H19NO3	116167-27-2	1-HEXYL-4-CARBOXY-2-PYRROLIDONE
C11H19NOS	026530-20-1	3(2H)-ISOTHIAZOLONE, 2-OCTYL-
C11H20ClN5	000580-48-3	CHLORAZINE
C11H20ClN5O	001824-09-5	G 34698
C11H20N2O4	073632-77-6	BUTANOIC ACID, 2-[[(AMINOCARBONYL)AMINO]CARBONYL
C11H20N3O3PS	029232-93-7	PIRIMIPHOS-METHYL
C11H20N4O3	074550-94-0	1H-IMIDAZOLE-1-ETHANOL, α-[(DIETHYLAMINO)METHYL]
C11H20N4O3S2	074385-86-7	ETHANIMIDOTHIOIC ACID, N-[(4-CYCLOPROPYL-2,6-DIM
C11H20O	000878-13-7	CYCLOUNDECANONE
C11H20O	002371-42-8	2-METHYLISOBORNEOL
C11H20O2	000103-11-7	2-ETHYLHEXYL ACRYLATE
C11H20O2	000112-38-9	UNDECYLENIC ACID
C11H20O2	001942-48-9	5,7-UNDECANEDIONE
C11H20O2	002499-59-4	OCTYL ACRYLATE
C11H20O4	000760-55-4	OCTYLMALONIC ACID
C11H20O4	002050-20-6	DIETHYL PIMELATE
C11H20OSi2	014920-92-4	PENTAMETHYLPHENYLDISILOXANE
C11H21N3O3	078233-72-4	L-LEUCINAMIDE, N-ACETYL-L-ALANYL-
C11H21N3O3S2	086569-95-1	N(N"-BU-ME CARBAMYLTHIO)METHOMYL
C11H21N3O4	132765-89-0	L-VALINAMIDE, N-ACETYL-L-THREONYL-
C11H21N3O4S2	064029-01-2	N(N"-ME-TBUO CARBAMYLTHIO)METHOMYL
C11H21N3O4S2	066381-55-3	N(N"-BU-MEO CARBAMYLTHIO)METHOMYL
C11H21N5O	003004-70-4	IPATONE
C11H21N5OS	000841-06-5	METHOPROPTRYNE
C11H21N5S	004147-51-7	DIPROPETRYNE
C11H21N5S	022936-75-0	DIMETHAMRTRYNE
C11H21NOS	001134-23-2	S-ETHYL CYCLOHEXYLETHYLCARBAMOTHIOATE
C11H22	000821-95-4	1-UNDECENE
C11H22	004292-92-6	PENTYLCYCLOHEXANE
C11H22N2O	002163-69-1	CYCLURON
C11H22N2O2	014691-89-5	4-NHAC-2266-TETRAME-PIPERIDIN-N-OXIDE
C11H22N6	016268-75-0	N6,N6-DIET-TETRAME-MELAMINE
C11H22O	000112-12-9	UNDECANONE
C11H22O	000927-49-1	6-UNDECANONE
C11H22O2	000110-42-9	METHYL DECANOATE
C11H22O2	000112-37-8	UNDECANOIC ACID
C11H22O2	000123-29-5	ETHYL NONANOATE
C11H22O2	005454-28-4	BUTYL HEPTANOATE
C11H22O2	005458-59-3	ISOPROPYL OCTANOATE
C11H22O6	039824-10-7	B-PENTYLGALACTOPYRANOSIDE
C11H23Br	000693-67-4	1-BROMOUNDECANE
C11H23F	000506-05-8	1-FLUOROUNDECANE
C11H23I	004282-44-4	1-IODOUNDECANE
C11H23N	082700-02-5	CYCLOOCTANEETHANAMINE, α-METHYL-
C11H23NO2	098155-00-1	CARBAMIC ACID, DIETHYL-, 2-ETHYLBUTYL ESTER
C11H23NOS	002008-41-5	BUTYLATE
C11H24	001120-21-4	N-UNDECANE
C11H24	006975-98-0	2-METHYLDECANE
C11H24N2O	071335-68-7	1-PIPERIDINYLOXY, 4-(DIMETHYLAMINO)-2,2,6,6-TETR
C11H24O	000112-42-5	1-UNDECANOL
C11H24O	001653-30-1	2-UNDECANOL
C11H25O3P	001000-36-8	Phosphonic acid, methyl- dipentyl ester
C11H26IN	014251-76-4	OCTYL-TRIMETHYL-AMMONIUM IODIDE
C11H26NO2PS	050782-69-9	VX
C11H28O4Si4	035331-58-9	HEXAMETHYL(SILACYCLOHEXYL)CYCLOTETRASILOXANE

C12Br10	013654-09-6	DECABROMOBIPHENYL
C12Br10O	001163-19-5	DECABROMODIPHENYL ETHER
C12Br6Cl2O2	002170-44-7	1,2,4,6,7,9-HEXABROMO-3,8-DICHLORODIBENZO-P-DIO*
C12Br8O2	002170-45-8	OCTABROMODIBENZO-P-DIOXIN
C12Cl10	002051-24-3	DECACHLOROBIPHENYL
C12Cl8O	039001-02-0	OCTACHLORODIBENZOFURAN
C12Cl8O2	003268-87-9	1,2,3,4,6,7,8,9-OCTACHLORODIBENZO-P-DIOXIN
C12F27N	000311-89-7	HEPTACOSAFLUOROTRIBUTYLAMINE
C12HCl7O	067562-39-4	DIBENZOFURAN, 1,2,3,4,6,7,8-HEPTACHLORO-
C12HCl7O2	035822-46-9	1,2,3,4,6,7,8-HEPTACHLORODIBENZO-P-DIOXIN
C12HCl7O2	037871-00-4	HEPTACHLORODIBENZO-P-DIOXIN
C12HCl9	040186-72-9	2,2',3,3',4,4',5,5',6-NONACHLOROBIPHENYL
C12HCl9	052663-77-1	2,3,4,5,6,2',3',5',6'-PCB
C12HCl9	053742-07-7	NONACHLORO-1,1'-BIPHENYL
C12H2Cl6O2	019408-74-3	1,2,3,7,8,9-HEXACHLORODIBENZO-P-DIOXIN
C12H2Cl6O2	039227-28-6	1,2,3,4,7,8-HEXACHLORODIBENZO-P-DIOXIN
C12H2Cl6O2	039227-62-8	1,2,4,6,7,9-HEXACHLORODIBENZO-P-DIOXIN
C12H2Cl6O2	057653-85-7	1,2,3,6,7,8-HEXACHLORODIBENZO-P-DIOXIN
C12H2Cl8	002136-99-4	2,2',3,3',5,5',6,6'-OCTACHLOROBIPHENYL
C12H2Cl8	035694-08-7	2,2',3,3',4,4',5,5'-OCTACHLOROBIPHENYL
C12H3Cl5O	057117-31-4	2,3,4,7,8-PENTACHLORODIBENZOFURAN
C12H3Cl5O	057117-41-6	DIBENZOFURAN, 1,2,3,7,8-PENTACHLORO-
C12H3Cl5O	057117-42-7	DIBENZOFURAN, 1,2,3,6,7-PENTACHLORO-
C12H3Cl5O	067481-22-5	DIBENZOFURAN, 2,3,4,6,8-PENTACHLORO-
C12H3Cl5O	067517-48-0	DIBENZOFURAN, 1,2,3,4,8-PENTACHLORO-
C12H3Cl5O	070648-24-7	DIBENZOFURAN, 1,2,4,6,9-PENTACHLORO-
C12H3Cl5O	070872-82-1	DIBENZOFURAN, 1,2,6,7,9-PENTACHLORO-
C12H3Cl5O2	039227-61-7	1,2,3,4,7-PENTACHLORODIBENZO-P-DIOXIN
C12H3Cl5O2	040321-76-4	1,2,3,7,8-PENTACHLORODIBENZO-P-DIOXIN
C12H3Cl5O2	058802-08-7	1,2,4,7,8-PENTACHLORODIBENZO-P-DIOXIN
C12H3Cl7	011096-82-5	AROCLOR 1260
C12H3Cl7	035065-29-3	2,2',3,4,4',5,5'-HEPTACHLOROBIPHENYL
C12H3Cl7	035065-30-6	2,2',3,3',4,4',5-HEPTACHLOROBIPHENYL
C12H3Cl7	052663-64-6	2,2',3,3',5,6,6'-HEPTACHLOROBIPHENYL
C12H3Cl7	052663-68-0	2,2',3,4',5,5',6-HEPTACHLOROBIPHENYL
C12H3Cl7	052663-71-5	2,2',3,3',4,4',6-HEPTACHLOROBIPHENYL
C12H3Cl7	052712-05-7	2,2',3,4,5,5',6-HEPTACHLORO-1,1'-BIPHENYL
C12H4Br2Cl2O2	050585-40-5	2,3-DIBROMO-7,8-DICHLORODIBENZO-P-DIOXIN
C12H4Br2F2O2	050585-43-8	2,3-DIBROMO-7,8-DIFLUORODIBENZO-P-DIOXIN
C12H4Br4O2	050585-41-6	2,3,7,8-TETRABROMODIBENZO-P-DIOXIN
C12H4Cl2F2O2	050585-42-7	2,3-DICHLORO-7,8-DIFLUORODIBENZO-P-DIOXIN
C12H4Cl4	007090-41-7	BIPHENYLENE, 2,3,6,7-TETRACHLORO-
C12H4Cl4O	024478-72-6	DIBENZOFURAN, 1,2,3,4-TETRACHLORO-
C12H4Cl4O	051207-31-9	2,3,7,8-TETRACHLORODIBENZOFURAN
C12H4Cl4O	057117-37-0	DIBENZOFURAN, 2,3,6,8-TETRACHLORO-
C12H4Cl4O	057117-38-1	DIBENZOFURAN, 2,4,6,7-TETRACHLORO-
C12H4Cl4O	057117-39-2	DIBENZOFURAN, 2,3,6,7-TETRACHLORO-
C12H4Cl4O	058802-20-3	DIBENZOFURAN, 1,2,7,8-TETRACHLORO-
C12H4Cl4O	064560-17-4	DIBENZOFURAN, 1,3,7,9-TETRACHLORO-
C12H4Cl4O	070648-16-7	DIBENZOFURAN, 1,3,4,7-TETRACHLORO-
C12H4Cl4O	071998-72-6	DIBENZOFURAN, 1,3,6,8-TETRACHLORO-
C12H4Cl4O	083704-27-2	DIBENZOFURAN, 1,3,4,6-TETRACHLORO-
C12H4Cl4O	083704-30-7	DIBENZOFURAN, 2,3,4,6-TETRACHLORO-
C12H4Cl4O2	001746-01-6	2,3,7,8-TETRACHLORODIBENZO-P-DIOXIN
C12H4Cl4O2	030746-58-8	1,2,3,4-TETRACHLORODIBENZO-P-DIOXIN
C12H4Cl4O2	033423-92-6	1,3,6,8-TETRACHLORODIBENZO-P-DIOXIN
C12H4Cl4O2	050585-46-1	1,3,7,8-TETRACHLORODIBENZO-P-DIOXIN
C12H4Cl4O2	062470-53-5	1,3,7,9-TETRACHLORODIBENZO-P-DIOXIN

C12H4Cl4O2	067028-18-6	1,2,3,7-TETRACHLORODIBENZODIOXIN
C12H4Cl6	032774-16-6	3,3',4,4',5,5'-PCB
C12H4Cl6	033979-03-2	2,2',4,4',6,6'-PCB
C12H4Cl6	035065-27-1	2,4,5,2',4',5'-PCB
C12H4Cl6	035065-28-2	2,2',3,4,4',5'-HEXACHLOROBIPHENYL
C12H4Cl6	035694-06-5	2,2',3,4,4'-PCB
C12H4Cl6	038380-04-0	2,2',3,4',5',6-PCB
C12H4Cl6	038380-07-3	2,2',3,3',4,4'-PCB
C12H4Cl6	038380-08-4	2,3,3',4,4',5-HEXACHLOROBIPHENYL
C12H4Cl6	038411-22-2	2,2',3,3',6,6'-HEXACHLOROBIPHENYL
C12H4Cl6	051908-16-8	2,2',3,4',5,5'-HEXACHLOROBIPHENYL
C12H4Cl6	052663-63-5	2,2',3,5,5',6-HEXACHLOROBIPHENYL
C12H4Cl6	052663-66-8	2,2',3,3',4,5-HEXACHLOROBIPHENYL
C12H4Cl6	052663-72-6	2,3',4,4',5,5'-HEXACHLOROBIPHENYL
C12H4Cl6	052704-70-8	2,2'3,3',5,6-HEXACHLOROBIPHENYL
C12H4Cl6	052712-04-6	2,2',3,4,5,5'-HEXACHLOROBIPHENYL
C12H4Cl6	052744-13-5	2,2',3,3',5,6'-HEXACHLOROBIPHENYL
C12H4Cl6	055215-18-4	2,2',3,3',4,5-HEXACHLOROBIPHENYL
C12H4Cl6	074472-46-1	2,3,3',5,5',6-HEXACHLOROBIPHENYL
C12H5Br5	067888-96-4	2,2',4,5,5'-PENTABROMOBIPHENYL
C12H5Cl2F3N4O4	078879-25-1	N(2,4-NO2-6-CF3 PH)-2(3,5CL)PYRIDINAMINE
C12H5Cl2F3N4O4	142978-30-1	N(24-NO2-6CL PH)-2(3CL-5CF3)PYRIDINAMINE
C12H5Cl3O2	033857-28-2	2,3,7-TRICHLORODIBENZO-P-DIOXIN
C12H5Cl3O2	039227-58-2	1,2,4-TRICHLORODIBENZO-P-DIOXIN
C12H5Cl4N	058910-96-6	9H-CARBAZOLE, 1,3,6,8-TETRACHLORO-
C12H5Cl5	011097-69-1	AROCLOR 1254
C12H5Cl5	018259-05-7	2,3,4,5,6-PCB
C12H5Cl5	031508-00-6	2',3,4,4',5'-PCB
C12H5Cl5	032598-14-4	2,3,3',4,4'-PENTACHLORO-1,1'-BIPHENYL
C12H5Cl5	037680-73-2	2,4,5,2',5'-PCB
C12H5Cl5	038379-99-6	2,2',3,5',6-PCB
C12H5Cl5	038380-01-7	2,2',4,4',5-PENTACHLOROBIPHENYL
C12H5Cl5	038380-02-8	2,2',3,4,5'-PENTACHLOROBIPHENYL
C12H5Cl5	038380-03-9	2,3,3',4',6-PENTACHLOROBIPHENYL
C12H5Cl5	039485-83-1	2,2',4,4',6-PENTACHLOROBIPHENYL
C12H5Cl5	041464-51-1	2,2',3',4,5-PENTACHLOROBIPHENYL
C12H5Cl5	052663-60-2	2,2',3,3',6-PENTACHLOROBIPHENYL
C12H5Cl5	052663-61-3	2,2',3,5,5'-PENTACHLOROBIPHENYL
C12H5Cl5	052663-62-4	2,2',3,4,5-PENTACHLOROBIPHENYL
C12H5Cl5	055215-17-3	2,2',3,4,6-PENTACHLOROBIPHENYL
C12H5Cl5	055312-69-1	2,2',3,4,5-PENTACHLOROBIPHENYL
C12H5Cl5	065510-45-4	2,2',3,4,4'-PENTACHLOROBIPHENYL
C12H5Cl5	068194-05-8	2,2',3,4',6-PENTACHLOROBIPHENYL
C12H5Cl5	068194-12-7	2,3',4,5,5'-PENTACHLOROBIPHENYL
C12H5ClF3N5O6	142978-32-3	N(24-NO2-6CF3 PH)-2(3CL-5NO2)PYRIDINAMIN
C12H5N7O12	000131-73-7	2,2',4,4',6,6'-HEXANITRODIPHENYLAMINE
C12H6Br2O2	039073-07-9	2,7-DIBROMODIBENZO-P-DIOXIN
C12H6Br2O2	050585-37-0	2,3-DIBROMODIBENZO-P-DIOXIN
C12H6Br2O2	091371-14-1	1,6-DIBROMODIBENZO-P-DIOXIN
C12H6Br2O2	105836-96-2	2,8-DIBROMODIBENZO-P-DIOXIN
C12H6Cl2F18	103188-54-1	1,10-DICHLOROPERFLUORO-2,9-DIMETHYLDECANE
C12H6Cl2O	005409-83-6	2,8-DICHLORODIBENZOFURAN
C12H6Cl2O2	029446-15-9	2,3-DICHLORODIBENZO-P-DIOXIN
C12H6Cl2O2	033857-26-0	2,7-DICHLORODIBENZO-P-DIOXIN
C12H6Cl2O2	038178-38-0	1,6-DICHLORODIBENZO-P-DIOXIN
C12H6Cl2O2	038964-22-6	2,8-DICHLORODIBENZO-P-DIOXIN
C12H6Cl2O2	050585-39-2	1,3-DICHLORODIBENZO-P-DIOXIN
C12H6Cl3NO3	001836-77-7	CHLORNITROFEN

C12H6Cl4	002437-79-8	2,2',4,4'-TETRACHLOROBIPHENYL
C12H6Cl4	015968-05-5	2,2',6,6'-PCB
C12H6Cl4	032598-10-0	2,3',4,4'-TETRACHLOROBIPHENYL
C12H6Cl4	032598-11-1	2,3',4',5-PCB
C12H6Cl4	032598-13-3	3,3',4,4'-PCB
C12H6Cl4	032690-93-0	2,4,4',5-PCB
C12H6Cl4	033025-41-1	2,3,4,4'-PCB
C12H6Cl4	033284-52-5	3,3',5,5'-TETRACHLOROBIPHENYL
C12H6Cl4	033284-53-6	2,3,4,5-PCB
C12H6Cl4	035693-99-3	2,2',5,5'-PCB
C12H6Cl4	038444-93-8	2,2',3,3'-TETRACHLOROBIPHENYL
C12H6Cl4	041464-39-5	2,2',3,5'-TETRACHLOROBIPHENYL
C12H6Cl4	041464-40-8	2,2',4,5'-TETRACHLOROBIPHENYL
C12H6Cl4	041464-41-9	2,2',5,6'-TETRACHLOROBIPHENYL
C12H6Cl4	052663-59-9	2,2',3,4-TETRACHLOROBIPHENYL
C12H6Cl4	062796-65-0	2,2',4,6-TETRACHLOROBIPHENYL
C12H6Cl4	070362-47-9	2,2'4,5-TETRACHLOROBIPHENYL
C12H6Cl4N2	014047-09-7	3,3',4,4'-TETRACHLOROAZOBENZENE
C12H6Cl4O	028076-73-5	2,2',4,4'-TETRACHLORODIPHENYL ETHER
C12H6Cl4O2S	000116-29-0	TETRADIFON
C12H6Cl4S	002227-13-6	TETRASUL
C12H6F20	103188-55-2	PERFLUORO-2,9-DIMETHYLDECANE
C12H6F2N2O2	131341-86-1	FLUDIOXONIL
C12H6F2O2	050585-38-1	2,3-DIFLUORODIBENZO-P-DIOXIN
C12H6O12	000517-60-2	MELLITIC ACID
C12H7BrO2	105906-36-3	2-MONOBROMODIBENZO-P-DIOXIN
C12H7BrO2	105908-71-2	1-MONOBROMODIBENZO-P-DIOXIN
C12H7Cl2N2NaO3S	062959-41-5	4-((2,4-DICHLOROPHENYL)AZO)BENZENESULFONIC ACID*
C12H7Cl2NO3	001836-75-5	NITROFEN
C12H7Cl3	007012-37-5	2,4,4'-PCB
C12H7Cl3	015862-07-4	2,4,5-TRICHLOROBIPHENYL
C12H7Cl3	016606-02-3	2,4',5-TRICHLOROBIPHENYL
C12H7Cl3	035693-92-6	2,4,6-TRICHLOROBIPHENYL
C12H7Cl3	037680-65-2	2,2',5-TRICHLOROBIPHENYL
C12H7Cl3	037680-66-3	2,2',4-TRICHLOROBIPHENYL
C12H7Cl3	038444-73-4	2,2',6-PCB
C12H7Cl3	038444-77-8	2,4',6-TRICHLOROBIPHENYL
C12H7Cl3	038444-78-9	2,2',3-PCB
C12H7Cl3	038444-81-4	2,3',5-PCB
C12H7Cl3	038444-84-7	2,3,3'-PCB
C12H7Cl3	038444-85-8	2,3,4'-PCB
C12H7Cl3	038444-86-9	2',3,4-TRICHLOROBIPHENYL
C12H7Cl3	038444-87-0	3,3',5-TRICHLOROBIPHENYL
C12H7Cl3	038444-90-5	3,4,4'-TRICHLOROBIPHENYL
C12H7Cl3	055702-45-9	2,3,6-TRICHLOROBIPHENYL
C12H7Cl3	055702-46-0	2,3,4-TRICHLORO-1,1'-BIPHENYL
C12H7Cl3O	052322-80-2	BENZENE, 1,2,4-TRICHLORO-5-PHENOXY-
C12H7Cl3O2	003380-34-5	5-CHLORO-2-(2,4-DICHLOROPHENOXY)PHENOL
C12H7ClO2	039227-53-7	1-CHLORODIBENZO-P-DIOXIN
C12H7ClO2	039227-54-8	2-CHLORODIBENZO-P-DIOXIN
C12H7F4N3O3	036865-60-8	4,5-(-OCF2CF2O-)C6H3NHN=C(CN)CO-ME
C12H7F4N3O4	036865-53-9	4,5-(-OCF2CF2O-)C6H3NHN=C(CN)CO-OME
C12H7F6N3O2	036865-54-0	3,5-DI-CF3-C6H3NHN=C(CN)COOME-CIS
C12H7F6N3O2	036905-00-7	3,5-DI-CF3-C6H3NHN=C(CN)COOME-TRANS
C12H7IO2	101714-96-9	2-MONOIODODIBENZO-P-DIOXIN
C12H7N5O2	036874-72-3	2,5-DICYANO-C6H3NHN=C(CN)COOME
C12H8	000208-96-8	ACENAPHTHYLENE
C12H8Br2	000092-86-4	4,4'-DIBROMOBIPHENYL

C12H8Cl2	002050-67-1	3,3'-DICHLOROBIPHENYL
C12H8Cl2	002050-68-2	4,4'-DICHLOROBIPHENYL
C12H8Cl2	002974-90-5	BIPHENYL, 3,4'-DICHLORO-
C12H8Cl2	002974-92-7	3,4-DICHLOROBIPHENYL
C12H8Cl2	013029-08-8	2,2'-DICHLOROBIPHENYL
C12H8Cl2	016605-91-7	2,3-DICHLOROBIPHENYL
C12H8Cl2	025569-80-6	2,3'-PCB
C12H8Cl2	033146-45-1	2,6-DICHLOROBIPHENYL
C12H8Cl2	033284-50-3	2,4-DICHLOROBIPHENYL
C12H8Cl2	034883-39-1	2,5-DICHLOROBIPHENYL
C12H8Cl2	034883-41-5	3,5-DICHLORO-1,1'-BIPHENYL
C12H8Cl2	034883-43-7	2,4'-DICHLOROBIPHENYL
C12H8Cl2O	005335-24-0	2,4-DICHLORO-6-PHENYLPHENOL
C12H8Cl2O3S	000080-33-1	CHLORFENSON
C12H8Cl2O3S	000097-16-5	GENITE
C12H8Cl6	000309-00-2	ALDRIN
C12H8Cl6	000465-73-6	ISODRIN
C12H8Cl6O	000060-57-1	DIELDRIN
C12H8Cl6O	000072-20-8	ENDRIN
C12H8Cl6O	007421-93-4	ENDRIN ALDEHYDE
C12H8ClN2NaO2S	002777-05-1	CHLOROAZOBENZENE-4'-SULFONIC ACID, NA SALT
C12H8ClN3O4	016220-58-9	N-(3-CHLOROPHENYL)-2,4-DINITROBENZENAMINE
C12H8ClNO	006318-51-0	4-(4-CHLOROBENZOYL)PYRIDINE
C12H8N2	000066-71-7	O-PHENANTHROLINE
C12H8N2	000092-82-0	PHENAZINE
C12H8N2	000230-07-9	4,7-PHENANTHROLINE
C12H8N2	000230-17-1	5,6-DIAZAPHENANTHRENE
C12H8N2	000230-46-6	1,7-PHENANTHROLINE
C12H8N2O	000578-96-1	11H-PYRIDO[2,1-B]QUINAZOLIN-11-ONE
C12H8N2O3	027693-38-5	4-(4-NITROBENZOYL)-PYRIDINE
C12H8N2O4	001528-74-1	4,4'-DINITROBIPHENYL
C12H8N2O4S2	000537-91-7	DISULFIDE, BIS(3-NITROPHENYL)
C12H8N2O4S2	001155-00-6	DISULFIDE, BIS(2-NITROPHENYL)
C12H8N2O5	000731-92-0	2,4-DINITRO-6-PHENYLPHENOL
C12H8N4O2	096448-63-4	ALLOPURINOL, 1-BENZOYL-
C12H8N4O6	000970-76-3	2,4-DINITRO-N-(4-NITROPHENYL)BENZENEAMINE
C12H8N4O6	000970-91-2	2,4-DINITRO-N-(3-NITROPHENYL)BENZENAMINE
C12H8O	000132-64-9	DIBENZOFURAN
C12H8O2	000082-86-0	1,2-ACENAPHTHYLENEDIONE
C12H8O2	000262-12-4	DIBENZO-P-DIOXIN
C12H8O2	000363-03-1	2,5-CYCLOHEXADIENE-1,4-DIONE, 2-PHENYL-
C12H8O3	006457-92-7	2H-FURO[2,3-H]-1-BENZOPYRAN-2-ONE, 4-METHYL-
C12H8O3	015798-77-3	2H-FURO[2,3-H]-1-BENZOPYRAN-2-ONE, 8-METHYL-
C12H8O3	073459-03-7	2H-FURO(2,3-H)-1-BENZOPYRAN-2-ONE, 5-METHYL-
C12H8O3	078982-40-8	2H-FURO[2,3-H]-1-BENZOPYRAN-2-ONE, 9-METHYL-
C12H8OS	000262-20-4	PHENOTHIOXIN
C12H8S	000132-65-0	DIBENZOTHIOPHENE
C12H8S2	000092-85-3	THIANTHRENE
C12H9Br	000092-66-0	4-BROMOBIPHENYL
C12H9Br	002052-07-5	O-BROMOBIPHENYL
C12H9Br	002113-57-7	3-BROMOBIPHENYL
C12H9BrO	000101-55-3	4-BROMOPHENYL PHENYL ETHER
C12H9Cl	002051-60-7	2-CHLOROBIPHENYL
C12H9Cl	002051-61-8	3-CHLOROBIPHENYL
C12H9Cl	002051-62-9	4-CHLOROBIPHENYL
C12H9Cl	011141-16-5	AROCLOR 1232
C12H9Cl2N	015979-79-0	3-CHLORO-N-(4-CHLOROPHENYL)BENZENAMINE
C12H9Cl2NO2S	105456-57-3	BENZENAMINE, 4-[(2,4-DICHLOROPHENYL)SULFONYL]-

C12H9Cl2NO3	050471-44-8	VINCLOZOLIN
C12H9Cl9	083682-60-4	1,3,5-TRIS(2,2,2-TRICHLOROETHYL)BENZENE
C12H9ClF3N3O	027314-13-2	NORFLURAZON
C12H9ClF3N3O2	028313-52-2	2-CL-5-CF3-C6H3NHN=C(CN)COOET
C12H9ClF3N3O2	028313-57-7	4-CL-3-CF3-C6H3NHN=C(CN)COOET
C12H9ClN2	004340-77-6	4-CHLOROAZOBENZENE
C12H9ClN2O	006657-05-2	2-CHLORO-4-PHENYLAZOPHENOL
C12H9ClN2O2	015979-85-8	3-CHLORO-N-(4-NITROPHENYL)BENZENAMINE
C12H9ClN2O3	037107-20-3	ACETAMIDE, N-[1-(4-CHLOROPHENYL)-2,5-DIOXO-3-PYR
C12H9ClN2O3	074070-46-5	ACLONIFEN
C12H9ClO	000607-12-5	4-CHLORO-2-PHENYLPHENOL
C12H9ClO	007005-72-3	4-CHLOROPHENYL PHENYL ETHER
C12H9ClO2S	000080-00-2	SULPHENONE
C12H9ClO3S	000080-38-6	4-CHLOROPHENYL BENZENESULFONATE
C12H9F	000324-74-3	4-FLUOROBIPHENYL
C12H9FN2O4	066999-98-2	URACIL,1-BENZYLOXYCARBONYL-5-FLUORO-
C12H9FN2O5	106219-34-5	CARBONIC ACID, (3,6-DIHYDRO-5-FLUORO-2,6-DIOXO-1
C12H9N	000086-74-8	CARBAZOLE
C12H9N2NaO3S	042975-18-8	4-(PHENYLAZO)BENZENESULFONIC ACID, SODIUM SALT
C12H9N2NaO4S	002623-36-1	4-((4-HYDROXYPHENYL)AZO)BENZENESULFONIC ACID,MO*
C12H9N2NaO5S	000547-57-9	ACID ORANGE 6
C12H9N3O	070959-61-4	[3,4'-BIPYRIDINE]-5-CARBONITRILE,6-METHOXY-
C12H9N3O2	002491-52-3	(4-NITROPHENYL)PHENYLDIAZENE
C12H9N3O3	001435-60-5	4-(4-NITROPHENYLAZO)PHENOL
C12H9N3O3	055936-40-8	2-NITRO-4-PHENYLAZOPHENOL
C12H9N3O4	000961-68-2	2,4-DINITRO-N-PHENYLBENZENAMINE
C12H9N3O4	015979-87-0	3-NITRO-N-(4-NITROPHENYL)BENZENAMINE
C12H9N3O5	000119-15-3	P-(2,4-DINITROANILINE)PHENOL
C12H9N3O6S	075333-79-8	BENZENAMINE, 4-[(2,4-DINITROPHENYL)SULFONYL]-
C12H9N5O3	098846-66-3	ALLOPURINOL, 1-[NICOTINOYLOXYMETHYL]-
C12H9NaO	000132-27-4	2-PHENYLPHENOL, SODIUM SALT
C12H9NaO	003645-61-2	SODIUM P-PHENYLPHENOXIDE
C12H9NaO2	056892-91-2	CYCLOPROPANECARBOXYLIC ACID, 2-(PHENYLETHYNYL)-,
C12H9NaO3	082609-03-8	ACETIC ACID, (1-NAPHTHALENYLOXY)-, SODIUM SALT
C12H9NaO3S	002217-82-5	P-BIPHENYLSULFONIC ACID, SODIUM SALT
C12H9NO	000091-02-1	PHENYL-A-PYRIDYLKETONE
C12H9NO	000135-67-1	PHENOXAZINE
C12H9NO	005424-19-1	PHENYL-B-PYRIDYL KETONE
C12H9NO	014548-46-0	4-BENZOYLPYRIDINE
C12H9NO2	000092-93-3	P-NITROBIPHENYL
C12H9NO2	000602-87-9	ACENAPHTHYLENE, 1,2-DIHYDRO-5-NITRO-
C12H9NO2	002113-58-8	3-NITROBIPHENYL
C12H9NO2	022526-29-0	4-(2-HYDROXYBENZOYL)-PYRIDINE
C12H9NO2	051246-77-6	4-(4-HYDROXYBENZOYL)-PYRIDINE
C12H9NO3	000620-88-2	4-NITRO DIPHENYL ETHER
C12H9NO3	000885-82-5	4-PHENYL-2-NITROPHENOL
C12H9NO3	002348-74-5	1,4-NAPHTHOQUINONE,2-ACETAMIDO
C12H9NS	000092-84-2	PHENOTHIAZINE
C12H9NS	017112-82-2	1-NAPHTHYLMETHYLISOTHIOCYANATE
C12H9NS	019495-05-7	B-NAPHTHYLMETHYLISOTHIOCYANATE
C12H10	000083-32-9	ACENAPHTHENE
C12H10	000092-52-4	BIPHENYL
C12H10BrNO2S	006626-22-8	BENZENAMINE, 4-[(4-BROMOPHENYL)SULFONYL]-
C12H10BrNO2S	138568-62-4	BENZENEACETIC ACID, 3-BROMO-α-METHYL-4-(2-THIAZO
C12H10Cl2F3NO	061213-25-0	FLUROCHLORIDONE
C12H10Cl2N2	000091-94-1	3,3'-DICHLOROBENZIDINE
C12H10Cl2Si	000080-10-4	DICHLORODIPHENYLSILANE
C12H10ClN	000101-17-7	3-CHLORODIPHENYL ETHER

C12H10ClN	001205-71-6	4-CHLORO-N-PHENYLBENZENAMINE
C12H10ClNO2S	007146-68-1	BENZENAMINE, 4-[(4-CHLOROPHENYL)SULFONYL]-
C12H10ClNO2S	132483-33-1	BENZENEACETIC ACID, 3-CHLORO-α-METHYL-4-(2-THIAZ
C12H10FNO2S	000312-35-6	BENZENAMINE, 4-[(4-FLUOROPHENYL)SULFONYL]-
C12H10FNO2S	132483-34-2	BENZENEACETIC ACID, 3-FLUORO-α-METHYL-4-(2-THIAZ
C12H10N2	000103-33-3	AZOBENZENE
C12H10N2	001437-15-6	1,2-(DI-(A-PYRIDYL)ETHYLENE
C12H10N2O	000086-30-6	DIPHENYLNITROSAMINE
C12H10N2O	000156-10-5	4-NITROSO-N-PHENYLBENZENAMINE
C12H10N2O	000525-57-5	HARMALOL
C12H10N2O	001689-82-3	P-PHENYLAZOPHENOL
C12H10N2O	001752-96-1	NICOTINANILIDE
C12H10N2O	002362-57-4	2-(PHENYLAZO)PHENOL
C12H10N2O	013256-06-9	N-NITROSO-DIPHENYLAMINE
C12H10N2O2	000119-75-5	2-NITRO-N-PHENYLBENZENAMINE
C12H10N2O2	000836-30-6	4-NITRO-N-PHENYLBENZENAMINE
C12H10N2O2	001083-48-3	PYRIDINE, 4- (4-NITROPHENYL)METHYL -
C12H10N2O2	004531-79-7	3-NITRO-N-PHENYLBENZENAMINE
C12H10N2O3	007090-25-7	N-ME-N-NITROSO-1-NAPHTHYLCARBAMAT
C12H10N2O3S	051246-76-5	4-(BENZOYL SULFONAMIDE)-PYRIDINE
C12H10N2O4	013411-16-0	2-PYRIDINEMETHANOL, 6-[2-(5-NITRO-2-FURANYL)ETHE
C12H10N2O4S	001948-92-1	BENZENAMINE, 4-[(4-NITROPHENYL)SULFONYL]-
C12H10N2O4S	132483-59-1	BENZENEACETIC ACID, α-METHYL-3-NITRO-4-(2-THIAZO
C12H10N3NaO3S	002491-71-6	4-((4-AMINOPHENYL)AZO)BENZENESULFONIC ACID,MONO*
C12H10N4	000655-86-7	2,3-DIAMINOPHENAZINE
C12H10N4O2	000730-40-5	4-((4-NITROPHENYL)AZO)BENZENAMINE
C12H10N4O4	006373-73-5	N-(2,4-DINITROPHENYL)-1,4-BENZENEDIAMINE
C12H10O	000090-43-7	2-PHENYLPHENOL
C12H10O	000092-69-3	P-PHENYLPHENOL
C12H10O	000093-08-3	1-(2-NAPHTHALENYL)ETHANONE
C12H10O	000101-84-8	DIPHENYL ETHER
C12H10O	000580-51-8	3-PHENYLPHENOL
C12H10O	000941-98-0	ALPHA-NAPHTHYL METHYL KETONE
C12H10O	007206-57-7	1-ACETYLAZULENE
C12H10O2	000086-87-3	NAPHTHALENEACETIC ACID
C12H10O2	000092-88-6	4,4'-BIPHENOL
C12H10O2	000581-96-4	2-NAPHTHALENEACETIC ACID
C12H10O2	000713-68-8	3-PHENOXYPHENOL
C12H10O2	000830-81-9	1-ACETOXYNAPHTHALENE
C12H10O2	000831-82-3	PHENOL, 4-PHENOXY-
C12H10O2	001806-29-7	2,2'-BIPHENOL
C12H10O2	002202-79-1	1,4-NAPHTHOQUINONE,6,7-DIMETHYL
C12H10O2S	000127-63-9	DIPHENYL SULFONE
C12H10O2S	002664-63-3	PHENOL, 4,4'-THIOBIS-
C12H10O3	000120-23-0	ACETIC ACID, 2-NAPHTHYLOXY-
C12H10O3	002976-75-2	ACETIC ACID, (1-NAPHTHALENYLOXY)-
C12H10O3	005416-18-2	1,4-NAPHTHOQUINONE,2-ME-3-MEO
C12H10O3	118199-17-0	ETHANONE, 1-(6,7-DIHYDROXY-2-NAPHTHALENYL)-
C12H10O3S	004358-63-8	PHENYLBENZENESULFONATE
C12H10O4	000106-34-3	2,5-CYCLOHEXADIENE-1,4-DIONE, W/1,4-BENZENEDIOL
C12H10OS	000945-51-7	DIPHENYL SULFOXIDE
C12H10S	000139-66-2	DIPHENYLSULFIDE
C12H10S2	000882-33-7	DIPHENYL DISULFIDE
C12H11BrFNO2S	033802-91-4	N-(4-SO2F-BENZYL)-PYRIDINIUM BROMIDE
C12H11BrN4O2	083157-97-5	2,4-PYRIMIDINEDIAMINE, 5-[(6-BROMO-1,3-BENZODIOX
C12H11BrO2	015795-19-4	ACETYLACETONE,4-BROMOBENZAL
C12H11Cl2NO	023950-58-5	PRONAMIDE
C12H11ClF3NO3	088485-37-4	FLUXOFENIM

C12H11ClN2O	089770-27-4	7-CL-4-PROPIONYLAMINO-QUINOLINE
C12H11ClN2O2S	016803-92-2	N1-(4-CHLOROPHENYL)SULFANILAMIDE
C12H11ClN2O2S	019837-85-5	N1-(2-CHLOROPHENYL)SULFANILAMIDE
C12H11ClN2O5S	000054-31-9	FUROSEMIDE
C12H11ClO2	015725-14-1	2-CHLOROBENZAL ACETYLACETONE
C12H11ClO2	015725-15-2	3-CHLOROBENZAL ACETYLACETONE
C12H11ClO2	019411-75-7	ACETYLACETONE,4-CHLOROBENZAL
C12H11F3N2O4S	101064-07-7	BUTANOIC ACID, 4-[6-(TRIFLUOROMETHYL)-4H-1,2,4-B
C12H11FN2O2	125111-04-8	2,4(1H,3H)-PYRIMIDINEDIONE, 5-FLUORO-1-[(4-METHY
C12H11FN2O3	085093-35-2	2,4(1H,3H)-PYRIMIDINEDIONE, 5-FLUORO-1-[(4-METHO
C12H11FO2	015851-94-2	4-FLUOROBENZAL ACETYLACETONE
C12H11N	000090-41-5	2-AMINOBIPHENYL
C12H11N	000092-67-1	4-AMINOBIPHENYL
C12H11N	000101-82-6	2-BENZYLPYRIDINE
C12H11N	000122-39-4	DIPHENYLAMINE
C12H11N	002116-65-6	4-BENZYLPYRIDINE
C12H11N3	000060-09-3	P-PHENYLAZOANILINE
C12H11N3	000136-35-6	1,3-DIPHENYL-1-TRIAZENE
C12H11N3O	000103-18-4	4-HYDROXY-4'-AMINOAZOBENZENE
C12H11N3O	058481-06-4	2-PHENYLISONIAZID
C12H11N3O2	006149-34-4	N-(4-NITROPHENYL)-1,4-BENZENDIAMINE
C12H11N3O4S	006829-82-9	N1-(4-NO2-PHENYL)SULFANILAMIDE
C12H11N3O4S	019837-88-8	BENZENESULFONAMIDE, 4-AMINO-N-(2-NITROPHENYL)-
C12H11N4NaO3S	010190-66-6	C.I. DIRECT BROWN 191
C12H11N5	001214-39-7	1H-PURIN-6-AMINE, N-(PHENYLMETHYL)-
C12H11N5	004261-14-7	9H-PURIN-6-AMINE, 9-(PHENYLMETHYL)-
C12H11N7	000396-01-0	2,4,7-TRIAMINO-6-PH PTERIDINE
C12H11N7O	001226-52-4	2,4,7-PTERIDINETRIAMINE, 6-(4-HYDROXYPHENYL)-
C12H11NaO3S	097683-30-2	1-SO3NA-3-ET AZULENE
C12H11NaO4S	099287-52-2	1-SO3NA-3-ME-4-METHOXY AZULENE
C12H11NO	000122-37-2	PHENOL, 4-(PHENYLAMINO)-
C12H11NO	000139-59-3	P-PHENOXYANILINE
C12H11NO	002688-84-8	O-PHENOXYANILINE
C12H11NO	014159-57-0	PHENYL-A-PYRIDYLCARBINOL
C12H11NO	019434-42-5	4-AMINO-2-PHENYLPHENOL
C12H11NO	023702-21-8	1-ACETAMIDOAZULENE
C12H11NO	089242-77-3	3-PYRIDINEMETHANOL, α,α-BIS(2-PROPYNYL)-
C12H11NO2	000063-25-2	1-NAPHTHYL-N-METHYLCARBAMATE
C12H11NO2	002025-40-3	2-CYANO-3-PHENYL ET-2-PROPENOATE
C12H11NO2	004089-04-7	N-METHYL-2-NAPHTHYLCARBAMATE
C12H11NO2	073987-38-9	6-QUINOLINECARBOXYLIC ACID,ET ESTER
C12H11NO2S	001678-25-7	BENZENESULFANILIDE
C12H11NO2S	004371-23-7	[1,1'-BIPHENYL]-4-SULFONAMIDE
C12H11NO2S	007019-01-4	BENZENAMINE, 4-(PHENYLSULFONYL)-
C12H11NO2S	132483-32-0	BENZENEACETIC ACID, α-METHYL-4-(2-THIAZOLYL)-
C12H11NO3	042322-29-2	QUINOLINE,8-ETHYLCARBONATE
C12H11NO3S	025963-47-7	PHENOL, 4- (4-AMINOPHENYL)SULFONYL -
C12H11NO3S	132483-56-8	BENZENEACETIC ACID, 3-HYDROXY-_-METHYL-4-(2-THIA
C12H11NO3S2	119018-06-3	2-THIOPHENESULFONAMIDE, 4-(4-METHYLBENZOYL)-
C12H11NO4	072601-52-6	M-METHOXYCARBONYL-N-PHENYLSUCCINIMIDE
C12H11NO4S	105456-60-8	1,3-BENZENEDIOL, 4-[(4-AMINOPHENYL)SULFONYL]-
C12H11NO4S2	002618-96-4	BENZENESULFONAMIDE, N-(PHENYLSULFONYL)-
C12H11NO4S2	118976-97-9	2-THIOPHENESULFONAMIDE, 4-(4-METHOXYBENZOYL)-
C12H11NO5S	105456-65-3	1,3,5-BENZENETRIOL, 2-[(4-AMINOPHENYL)SULFONYL]-
C12H11NO5S	118993-61-6	2-FURANSULFONAMIDE, 4-(4-METHOXYBENZOYL)-
C12H11O4P	000838-85-7	DIPHENYL PHOSPHATE
C12H12	000569-41-5	1,8-DIMETHYLNAPHTHALENE
C12H12	000571-58-4	1,4-DIMETHYLNAPHTHALENE

C12H12	000571-61-9	1,5-DIMETHYLNAPHTHALENE
C12H12	000573-98-8	1,2-DIMETHYLNAPHTHALENE
C12H12	000575-37-1	1,7-DIMETHYLNAPHTHALENE
C12H12	000575-41-7	1,3-DIMETHYLNAPTHALENE
C12H12	000575-43-9	1,6-DIMETHYLNAPHTHALENE
C12H12	000581-40-8	2,3-DIMETHYLNAPHTHALENE
C12H12	000581-42-0	2,6-DIMETHYLNAPHTHALENE
C12H12	000582-16-1	2,7-DIMETHYLNAPHTHALENE
C12H12	000939-27-5	2-ETHYLNAPHTHALENE
C12H12	001127-76-0	1-ETHYLNAPHTHALENE
C12H12Br2N2	000085-00-7	DIQUAT DIBROMIDE
C12H12Br2N4O	085544-53-2	24NH2-5(24-BR-5-MEO-BENZYL)PYRIMIDINE
C12H12BrN	002589-31-3	BENZYLPYRIDINIUM BROMIDE
C12H12BrN3O2S	030961-42-3	N1(5-BR-3-ME-2-PYRIDYL)SULFANILAMIDE
C12H12BrN3O2S	030961-43-4	N1(3-BR-5-ME-2-PYRIDYL)SULFANILAMIDE
C12H12BrNO2	003236-48-4	N-(4-BROMOBUTYL)PHTHALIMIDE
C12H12Cl2F16N3O4P3	065601-42-5	
2,2-DICHLORO-2,2,4,4,6,6-HEXAHYDRO-4,4,6,6-TETR*		
C12H12Cl2N2	004032-26-2	DIQUAT DICHLORIDE
C12H12Cl2N2O3	023885-61-2	P-CYANO-AMPHENICOL
C12H12Cl2N4	018588-57-3	2,4-PYRIMIDINEDIAMINE, 5-(3,4-DICHLOROPHENYL)-6-
C12H12Cl2N4O	083158-04-7	2,4-NH2PYRIMIDINE,5(35-CL-4MEO)BENZYL
C12H12Cl6	083682-59-1	1,3,5-TRIS(2,2-DICHLOROETHYL)BENZENE
C12H12ClN3O	003707-98-0	3(2H)-PYRIDAZINONE, 4-CHLORO-5-(DIMETHYLAMINO)-2
C12H12ClN5O4S	064902-72-3	CHLORSULFURON (PH 7)
C12H12F3N	083118-48-3	BENZOBICYCLO(221)HEPTENE,2EN-NH2-6CF3
C12H12F3N	083118-50-7	BENZOBICYCLO(221)HEPTENE,2EX-NH2-6CF3
C12H12F3N	083118-51-8	BENZOBICYCLO(221)HEPTENE,2EX-NH2-7CF3
C12H12F3N	086022-72-2	BENZOBICYCLO(221)HEPTENE,2EN-NH2-7CF3
C12H12IN	036913-39-0	PYRIDINIUM, 1-METHYL-4-PHENYL-, IODIDE
C12H12N2	000092-87-5	BENZIDINE
C12H12N2	000101-54-2	4-AMINODIPHENYLAMINE
C12H12N2	000122-66-7	HYDRAZOBENZENE
C12H12N2	000530-50-7	HYDRAZINE, 1,1-DIPHENYL-
C12H12N2	001454-80-4	[1,1'-BIPHENYL]-2,2'-DIAMINE
C12H12N2O	000101-80-4	4,4'-DIAMINODIPHENYL ETHER
C12H12N2O	002446-62-0	11H-PYRIDO[2,1-B]QUINAZOLIN-11-ONE, 6,7,8,9-TETR
C12H12N2O	025283-63-0	3-QUINOLINECARBOXAMIDE,N,N-DIMETHYL
C12H12N2O	070026-51-6	CYCLOPENTA[D]PYRIDO[1,2-A]PYRIMIDIN-10(1H)-ONE,
C12H12N2O	070934-02-0	PYRROLO[2,1-B]QUINAZOLIN-9(1H)-ONE, 2,3-DIHYDRO-
C12H12N2O2S	000080-08-0	DI(P-AMINOPHENYL)SULFONE
C12H12N2O2S	000127-77-5	N1-PHENYLSULFANILIDE
C12H12N2O2S	027147-69-9	BENZENAMINE, 2-[(4-AMINOPHENYL)SULFONYL]-
C12H12N2O2S	034262-32-3	BENZENAMINE, 3-[(4-AMINOPHENYL)SULFONYL]-
C12H12N2O2S	037157-89-4	2H-1,2,4-BENZOTHIADIAZINE, 3-(1-CYCLOPENTEN-1-YL
C12H12N2O2S	132483-60-4	BENZENEACETIC ACID, 3-AMINO-_-METHYL-4-(2-THIAZO
C12H12N2O3	000050-06-6	PHENOBARBITAL
C12H12N2O3	000389-08-2	NALIDIXIC ACID
C12H12N2O3	016867-53-1	PYRIDO(12A)PYRIMIDIN-4-ON,3ETO-CO-6ME
C12H12N2O3	030820-34-9	N-PHENYL-3-ACETYLAMINOSUCCINIMIDE
C12H12N2O3	074782-23-3	OXABETRINIL
C12H12N2O3S	032695-27-5	BENZENAMINE, 4-[(4-AMINOPHENYL)SULFONYL]-N-HYDRO
C12H12N2S	000139-65-1	4,4'-THIODIANILINE
C12H12N2S2	001141-88-4	BENZENAMINE, 2,2'-DITHIOBIS-
C12H12N4	000495-54-5	2,4-DIAMINOAZOBENZENE
C12H12N4	000538-41-0	4,4'-DIAMINOAZOBENZENE
C12H12N4O2	013932-40-6	2,4-NH2PYRIMIDINE,5-(3,4OCH2O)BENZYL
C12H12N4O3	022994-85-0	1H-IMIDAZOLE-1-ACETAMIDE, 2-NITRO-N-(PHENYLMETHY

C12H12N4O5	024570-16-9	5(1-AZIRD)-2,4-NO2 BENZAMIDE,N-CYPR
C12H12N4S	097183-49-8	HYDRAZINECARBOXIMIDAMIDE, 2-(PHENYL-2-THIENYLMET
C12H12NO2	010354-48-0	N-BENZYL-2-FURAMIDE
C12H12O	040295-80-5	2-NAPHTHALENEMETHANOL, α-METHYL-, (\)-
C12H12O2	000711-82-0	ETHANOL, 2-(1-NAPHTHALENYLOXY)-
C12H12O2	004335-90-4	BENZAL ACETYLACETONE
C12H12O2	073090-06-9	1-ISOBENZOFURANONE-3-SPIROPENTYL
C12H12O2	136944-43-9	2,3-NAPHTHALENEDIOL, 6-ETHYL-
C12H12O3	136944-52-0	2,3-NAPHTHALENEDIOL, 6-ETHOXY-
C12H12O4	006626-84-2	DIMETHYLMALONATE, BENZAL
C12H12O6	032620-68-1	BENZOIC ACID, 2-(ACETYLOXY)-, (ACETYLOXY)METHYL
C12H12S2	001134-61-8	2,2'-BITHIOPHENE, 5-(3-BUTEN-1-YNYL)-
C12H13Br2N3O2	089145-54-0	CARBAMIC ACID, [5-(2,5-DIBROMOPHENYL)-4,5-DIHYDR
C12H13BrN2O6S	026973-80-8	CEPHALOSPORIC ACID,7-BR-ACAMINO
C12H13BrN4O	056183-32-5	24NH2-5(3-BR-4-MEO-BENZYL)PYRIMIDINE
C12H13BrN4O	085544-51-0	24NH2-5(5-BR-2-MEO-BENZYL)PYRIMIDINE
C12H13BrN4O	085544-52-1	24NH2-5(2-BR-5-MEO-BENZYL)PYRIMIDINE
C12H13Cl2N3O2	060207-31-0	AZACONAZOL
C12H13Cl2NO	015907-82-1	CYCLOPENTANECARBOXAMIDE, N-(3,4-DICHLOROPHENYL)-
C12H13Cl2NO3	051114-25-1	ET-N-CL-ACETYL-N-(2CLPH)GLYCINATE
C12H13Cl3O3	000093-79-8	2,4,5-T, N-BUTYL ESTER
C12H13ClF3N3O4	033245-39-5	FLUCHLORALIN
C12H13ClN2O	003766-60-7	BUTURON
C12H13ClN4	000058-14-0	PYRIMETHAMINE
C12H13ClN4O	073264-25-2	24-NH2-5(4-CL-3-MEO-BENZYL)PYRIMIDINE
C12H13ClN4O	073275-70-4	24-NH2-5(3-CL-4-MEO-BENZYL)PYRIMIDINE
C12H13ClO4	071354-31-9	BUTANOIC ACID, 4-(3-CHLORO-4-METHOXYBENZOYL)-
C12H13IN4O	085544-43-0	24-NH2-5(3-I-4-MEO-BENZYL)PYRIMIDINE
C12H13IN4O	103360-35-6	24-NH2-5(5-I-2-MEO-BENZYL)PYRIMIDINE
C12H13N	000086-56-6	1-NAPHTHALENAMINE, N,N-DIMETHYL-
C12H13N	000118-44-5	1-NAPHTHALENAMINE, N-ETHYL-
C12H13N2O3P	033862-44-1	PHOSPHOROHYDRAZIDIC ACID, DIPHENYL ESTER
C12H13N3O	131229-37-3	IMIDAZO[5,1-B]QUINAZOLIN-9(2H)-ONE, 1,3-DIHYDRO-
C12H13N3O2	000068-76-8	TRIAZIQUONE
C12H13N3O2S	035880-91-2	1,3-BENZENEDIAMINE, 4-[(4-AMINOPHENYL)SULFONYL]-
C12H13N3O2S	051543-29-4	N1-METHYL-N1-(2-PYRIDYL)SULFANILAMIDE
C12H13N3O2S	141627-88-5	1,2,4-TRIAZIN-5(4H)-ONE, 6-(3-ETHOXYPHENYL)-2,3-
C12H13N3O3S	059729-37-2	1H-IMIDAZOLE, 1-METHYL-2-[[4-(METHYLTHIO)PHENOXY
C12H13N3O4S2	000547-52-4	N1-(4-SO2NH2-PH)SULFANILAMIDE
C12H13N3O5	023885-71-4	CYANAMPHENICOL
C12H13N5	006364-34-7	4-((4-AMINOPHENYL)AZO)1,3-BENZENEDIAMINE
C12H13NO2	000077-41-8	METHSUXIMIDE
C12H13NO2	000133-32-4	3-INDOLEBUTYRIC ACID
C12H13NO2	072601-47-9	M-ETHYL-N-PHENYLSUCCINIMIDE
C12H13NO2	072601-54-8	P-ETHYL-N-PHENYLSUCCINIMIDE
C12H13NO2	077298-35-2	BENZOIC ACID, 4-CYANOBUTYL-
C12H13NO2S	005234-68-4	CARBOXIN
C12H13NO3	115178-66-0	AZETIDINE, 1-[BENZOYLOXY(ACETYL)]-
C12H13NO3S	017757-70-9	1,4-OXATHIIN-3-CARBOXAMIDE, 5,6-DIHYDRO-2-METHYL
C12H13NO4	115178-80-8	ACETAMIDE, N-ACETYL-2-(BENZOYLOXY)-N-METHYL-
C12H13NO4S	005259-88-1	OXYCARBOXIN
C12H13NO4S2	000842-00-2	ET-4-SULFONAMIDONAPHTHYLSULFONE
C12H13NO5	106231-64-5	GLYCINE, N-[(BENZOYLOXY)ACETYL]-N-METHYL-
C12H13NOS	003915-60-4	2-THIOBENZYLPYRIDINE OXIDE
C12H14	031017-40-0	1-PHENYLCYCLOHEXENE
C12H14BrN	006294-92-4	N-PROPYLQUINOLINIUM BROMIDE
C12H14Cl2N2	000531-85-1	BENZIDINE DIHYDROCHLORIDE
C12H14Cl2N2	004685-14-7	PARAQUAT

C12H14Cl2N2O4	039960-99-1	P-AMIDO-AMPHENICOL
C12H14Cl2O3	000094-80-4	2,4-D, BUTYL ESTER
C12H14Cl3O4P	000470-90-6	CHLORFENVINPHOS
C12H14ClN	005048-08-8	PYRIDINE, 4-(4-CHLOROPHENYL)-1,2,3,6-TETRAHYDRO-
C12H14ClN3O2	066308-27-8	CARBAMIC ACID, [5-(2-CHLOROPHENYL)-4,5-DIHYDRO-1
C12H14ClN3O2	066308-28-9	CARBAMIC ACID, [5-(4-CHLOROPHENYL)-4,5-DIHYDRO-1
C12H14ClN3O2	089145-58-4	CARBAMIC ACID, [5-(3-CHLOROPHENYL)-4,5-DIHYDRO-1
C12H14ClNO	022342-21-8	PIPERIDINE, 1-(2-CHLOROBENZOYL)-
C12H14ClNO2	081777-89-1	CLOMAZONE
C12H14ClNO3	032562-52-0	P(N-2CLET-N-ACETYL)AMINOPHACETIC ACID
C12H14ClNO3	051114-26-2	ET-N-CL-ACETYL-N-(PHENYL)GLYCINATE
C12H14F3N5	047071-11-4	4,6-NH2 2,2-DIME1(4-CF3)PH S-TRIAZENE
C12H14F3NO2	124497-78-5	2-OH-2-(M-CF3-PH)-4-ME MORPHOLINE HCL
C12H14FN3O2	066308-26-7	CARBAMIC ACID, [5-(2-FLUOROPHENYL)-4,5-DIHYDRO-1
C12H14FN3O3	111041-62-4	IMIDODICARBONIC DIAMIDE, N-(5-ACETYL-2-FLUOROPHE
C12H14FN3O4	121788-14-5	BENZOIC ACID, 4-FLUORO-3-[[[METHYL[(METHYLAMINO)
C12H14N2	002764-72-9	DIQUAT
C12H14N2O	001016-47-3	ACETAMIDE, N-[2-(1H-INDOL-3-YL)ETHYL]-
C12H14N2O	069397-85-9	1H-INDOLE-3-PROPANAMIDE, N-METHYL-
C12H14N2O	070381-44-1	4H-PYRIDO[1,2-A]PYRIMIDIN-4-ONE, 3-ETHYL-2,6-DIM
C12H14N2O	070381-53-2	4H-PYRIDO[1,2-A]PYRIMIDIN-4-ONE, 2-METHYL-3-PROP
C12H14N2O	070381-71-4	4H-PYRIDO[1,2-A]PYRIMIDIN-4-ONE, 2-ETHYL-3,6-DIM
C12H14N2O2	000125-33-7	5-ETHYLDIHYDRO-5-PHENYL-4,6(1H,5H)-PYRINIDINEDI*
C12H14N2O2S	001431-39-6	1-DIME AMINO-5-SO2NH2-NAPHTHALENE
C12H14N2O2S	020434-64-4	124-BENZTHIADIAZN-1-O2-3-CYCLOPE
C12H14N2O3	000076-68-6	BARBITURIC ACID,5-ALLYL,5-(2CYPENTEN1YL)
C12H14N2O3	005435-82-5	PYRIDO(12A)PYRIMIDIN-4-ON,3ETO-CO-7ME
C12H14N2O3	034667-64-6	PYRIDO(12A)PYRIMIDIN-4-ON,3ETO-CO-8ME
C12H14N2O4	106231-62-3	ACETAMIDE, N-(2-AMINO-2-OXOETHYL)-2-(BENZOYLOXY)
C12H14N2O4	115193-30-1	PROPANAMIDE, 2-[[(BENZOYLOXY)ACETYL]AMINO]-
C12H14N2O4S	101064-05-5	BUTANOIC ACID, 4-(6-METHYL-4H-1,2,4-BENZOTHIADIA
C12H14N2O5	000131-89-5	2-CYCLOHEXYL-4,6-DINITROPHENOL
C12H14N2O5	004097-58-9	4-CYCLOHEXYL-2,6-DINITROPHENOL
C12H14N2O5S	101064-06-6	BUTANOIC ACID, 4-(6-METHOXY-4H-1,2,4-BENZOTHIADI
C12H14N2O6	002813-95-8	DINOSEB ACETATE
C12H14N3O2	000059-63-2	ISOCARBOXAZID
C12H14N4	102408-26-4	2-METHYLAMINO-3-METHYLIMIDAZO[45F]QUINOLINE
C12H14N4O	020285-70-5	24NH2-5(4-MEO-BENZYL)PYRIMIDINE
C12H14N4O	059481-28-6	2,4-DIAMINO-5(3-MEOBENZYL)PYRIMIDINE
C12H14N4O2	073356-40-8	2,4-NH2PYRIMIDINE,5(4-OH-3MEO)BENZYL
C12H14N4O2	083166-76-1	2,4-NH2PYRIMIDINE,5(3-OH-4MEO)BENZYL
C12H14N4O2S	000057-68-1	SULFAMETHAZINE
C12H14N4O2S	000515-64-0	SULFISOMIDINE
C12H14N4O2S	003271-01-0	2-SULFANILAMIDO-5-ETHYLPYRIMIDINE
C12H14N4O2S	007411-79-2	BENZENESULFONAMIDE, 4-AMINO-N-(2,4-DIMETHYL-5-PY
C12H14N4O3	071525-06-9	2,4-NH2PYRIMIDINE,5(34-OH-5MEO)BENZYL
C12H14N4O3	083158-00-3	2,4-NH2PYRIMIDINE,5(35-OH-4MEO)BENZYL
C12H14N4O3S	003772-76-7	SULFAMETHOMIDINE
C12H14N4O4S	000122-11-2	SULFADIMETHOXINE
C12H14N4O4S	002447-57-6	BENZENESULFONAMIDE, 4-AMINO-N-(5,6-DIMETHOXY-4-P
C12H14N4O4S2	023564-05-8	THIOPHANATE-METHYL
C12H14N6O	039942-91-1	1H-IMIDAZOLE-4-CARBOXAMIDE, 5-[3-METHYL-3-(PHENY
C12H14NO4PS	005131-24-8	DITALIMFOS (LAPTRAN)
C12H14O	004894-75-1	CYCLOHEXANONE, 4-PHENYL-
C12H14O2	001134-87-8	3-BENZYL-2,4-PENTANEDIONE
C12H14O2	004770-31-4	1(3H)-ISOBENZOFURANONE,3,3-DIETHYL
C12H14O2	006066-49-5	1(3H)-ISOBENZOFURANONE, 3-BUTYL-
C12H14O2	017598-02-6	2,2-DIMETHYL-7-METHOXYBENZOPYRAN

C12H14O2	093133-67-6	1(3H)-ISOBENZOFURANONE, 3-BUTYL-,(±)-
C12H14O3	000093-28-7	EUGENYL ACETATE
C12H14O3	001878-59-7	2(5678-TETRAHYDRONAPHTHYLOXY)ACETIC ACID
C12H14O3	136944-25-7	1-HEXEN-3-ONE, 1-(3,4-DIHYDROXYPHENYL)-
C12H14O4	000084-66-2	DIETHYL PHTHALATE
C12H14O4	004609-10-3	BUTANOIC ACID, 4-(4-METHOXYBENZOYL)-
C12H14O6	000959-26-2	BIS(2-HYDROXYETHYL)TEREPHTHALATE
C12H14S3	001081-34-1	2,5-THIOPHENYL-THIOPHENE
C12H15Cl2N3O4	039961-07-4	P-UREA-AMPHENICOL
C12H15Cl2NO	002533-89-3	KARSIL
C12H15Cl2NO4Si	090936-47-3	1-(3,5-DICHLOROPHENOXY)SILATRANE
C12H15Cl2NO5S	000847-25-6	P-METHYLSULFONYLAMPHENICOL
C12H15Cl3	018226-46-5	1,3,5-TRIS(CHLOROETHYL)BENZENE
C12H15ClN2O2S	037157-54-3	124-BENZTHIADIAZN-1-O2-3-NEOPE-7-CL
C12H15ClN2O5	049648-35-3	MONOCHLOROMETHYLAMPHENICOL
C12H15ClN6O2	117121-39-8	UREA, N'-[3-CHLORO-4-[(1-ETHYL-1H-TETRAZOL-5-YL)
C12H15ClN6O2	117121-51-4	TETRAZOLE,2-ETHYL-5-(2-CHLORO-4-DIMETHYLUREYL)PH
C12H15ClNO4PS2	002310-17-0	PHOSALONE
C12H15ClO6	004756-30-3	4-CHLOROPHENYL GLUCOPYRANOSIDE
C12H15IN2O6	084043-25-4	URIDINE, 2'-DEOXY-5-IODO-, 5'-PROPANOATE
C12H15IN2O6	084043-26-5	URIDINE, 2'-DEOXY-5-IODO-, 5'-BUTANOATE
C12H15IN2O6	084043-27-6	URIDINE, 2'-DEOXY-5-IODO-, 5'-(2-METHYLPROPANOAT
C12H15IO6	007234-29-9	GLUCOPYRANOSIDE,2-IODOPHENYL
C12H15IO6	020838-40-8	4-IODOPHENYL GLUCOPYRANOSIDE
C12H15N	014342-36-0	2-EN-NH2 BENZOBICYCLO(222)OCTENE
C12H15N	015537-20-9	2-EX-NH2 BENZOBICYCLO(222)OCTENE
C12H15N	028289-54-5	MPTP
C12H15N	058742-05-5	1,4-METHANONAPHTHALEN-2-AMINE, 1,2,3,4-TETRAHYDR
C12H15N	062624-27-5	BENZOBICYCLO(2,2,1)HEPTENE,2-EX-MEAM
C12H15N	086943-79-5	BENZOBICYCLO(221)HEPTENE,2EX-MEAMINO
C12H15N	086992-69-0	BENZOBICYCLO(221)HEPTENE,9EN-MEAMINO
C12H15N2O3PS	013593-03-8	QUINALPHOS
C12H15N2O3PS	014816-18-3	PHOXIM
C12H15N3	066480-65-7	2H-BENZOTRIAZOLE, 2-CYCLOHEXYL-
C12H15N3	066536-70-7	1H-BENZOTRIAZOLE, 1-CYCLOHEXYL-
C12H15N3O	080431-02-3	3-PYRIDINEACETAMIDE, N-(3,6-DIHYDRO-1(2H)-PYRIDI
C12H15N3O2	001143-78-8	1H-BENZOTRIAZOLE-1-PROPANOIC ACID, ß-METHYL-, ET
C12H15N3O2	001612-15-3	2H-BENZOTRIAZOLE-2-PROPANOIC ACID, ß-METHYL-, ET
C12H15N3O2	066308-18-7	CARBAMIC ACID, (4,5-DIHYDRO-1-METHYL-5-PHENYL-1H
C12H15N3O2	069218-43-5	1-(5-CARBOXYPENTYL)BENZOTRIAZOLE
C12H15N3O2	069218-48-0	1-(3-CARBETHOXYPROPYL)BENZOTRIAZOLE
C12H15N3O2	115054-78-9	1H-BENZOTRIAZOLE-1-ACETIC ACID, α-ETHYL-, ETHYL
C12H15N3O2	115054-80-3	2H-BENZOTRIAZOLE-2-ACETIC ACID, α-ETHYL-, ETHYL
C12H15N3O3	071006-74-1	1-(3-PHENOXYPROPYL)-2-NO2 -IMIDAZOLE
C12H15N3O4	021788-11-4	1-(3-PHO-2-OH-PROPYL)-2-NO2-IMIDAZOLE
C12H15N3O4S	013153-62-3	9H-PURINE, 9-(ß-D-ARABINOFURANOSYL)-6-(METHYLTHI
C12H15N4O5P	067410-48-4	3,5-DIOXA-6-AZA-4-PHOSPHAOCT-6-ENE-8-NITRILE, 4-
C12H15N5O	085544-44-1	24-NH2-5(3-NH2-4-MEO-BENZYL)PYRIMIDINE
C12H15N5O	085544-45-2	24-NH2-5(4-NH2-3-MEO-BENZYL)PYRIMIDINE
C12H15N5O2	065873-69-0	2,4NH2PYRIMIDIN,5(26MEOPYRIDIN3YL)CH2
C12H15N5O2	083157-99-7	2,4NH2PYRIMIDIN,5(26MEOPYRIDIN4YL)CH2
C12H15N5O3S	099035-20-8	CIMETIDINE,2-(3-NITROPYRIDIN-4-ONE)
C12H15N5O6S2	079277-27-3	HARMONY
C12H15NO	000776-75-0	PIPERIDINE, 1-BENZOYL-
C12H15NO	053207-52-6	1,3,3-TRIME-2,3-DIHYDRO-4-QUINOLONE
C12H15NO3	001563-66-2	CARBOFURAN
C12H15NO3	018495-00-6	N-PHENOXYACETYLMORPHOLINE
C12H15NO3	021156-62-7	PHENYLALANINE-N-ACETYL,METHYL ESTER

C12H15NO3	057440-16-1	ACETAMIDE, N-[2-(BENZOYLOXY)ETHYL]-N-METHYL-
C12H15NO3	105412-27-9	BENZOIC ACID,4-CARBOXAMIDOBUTYL ESTER
C12H15NO3	106231-51-0	ACETAMIDE, 2-(BENZOYLOXY)-N-PROPYL-
C12H15NO3	115178-77-3	PROPANAMIDE, 3-(BENZOYLOXY)-N,N-DIMETHYL-
C12H15NO3	115193-27-6	ACETAMIDE, 2-(BENZOYLOXY)-N-(1-METHYLETHYL)-
C12H15NO3	115193-32-3	ACETAMIDE, 2-(BENZOYLOXY)-N-ETHYL-N-METHYL-
C12H15NO4	002440-79-1	TYROSINE-N-ACETYL,METHYL ESTER
C12H15NO4	103951-39-9	BUTANOIC ACID, [2-(AMINOCARBONYL)PHENOXY]METHYL
C12H15NO4	106231-59-8	ACETAMIDE, 2-(BENZOYLOXY)-N-(2-HYDROXYETHYL)-N-M
C12H15NO5	114665-09-7	BENZOIC ACID, 2-HYDROXY-, 2-[(2-HYDROXYETHYL)MET
C12H15NO8	002492-87-7	BETA-D-GLUCOPYRANOSIDE, 4-NITROPHENYL
C12H15NO8	002816-24-2	beta-D-GLUCOPYRANOSIDE, 2-NITROPHENYL
C12H15NO8	003150-24-1	GALACTOPYRANOSIDE,2-NITROPHENYL
C12H15NO8	003767-28-0	α-D-GLUCOPYRANOSIDE, 4-NITROPHENYL
C12H15NO8	010357-27-4	_-D-MANNOPYRANOSIDE, 4-NITROPHENYL
C12H15NO8	020838-44-2	3-NITROPHENYL GLUCOPYRANOSIDE
C12H16	000827-52-1	PHENYL CYCLOHEXANE
C12H16BrNO2	040912-89-8	BENZAMIDE, 5-BROMO-N-BUTYL-2-HYDROXY-3-METHYL-
C12H16BrNO4Si	083801-60-9	1-(3-BROMOPHENOXY)SILATRANE
C12H16BrNO4Si	086811-82-7	1-(4-BROMOPHENOXY)SILATRANE
C12H16Cl2N2O	000555-37-3	NEBURON
C12H16ClN3O4S2	002854-98-0	CHLOROTHIAZIDE-3-AMYL
C12H16ClNO4Si	051466-76-3	1-(4-CHLOROPHENOXY)SILATRANE
C12H16ClNO4Si	086825-39-0	1-(3-CHLOROPHENOXY)SILATRANE
C12H16ClNOS	028249-77-6	THIOBENCARB
C12H16ClNOS	034622-58-7	ORBENCARB
C12H16F3N	000458-24-2	FENFLURAMINE
C12H16FN3O2	111041-46-4	IMIDODICARBONIC DIAMIDE. N-(5-ETHYL-2-FLUOROPHEN
C12H16FN5O2	132722-92-0	DIDEOXY-ARA-A,2,N6-DIMETHYL-2'-FLUORO
C12H16N2	000061-50-7	N,N-DIMETHYLTRYPTAMINE
C12H16N2	001885-87-6	N-BENZALAMINOPIPERIDINE
C12H16N2	106362-29-2	BENZENAMINE, 3-(1,2,3,6-TETRAHYDRO-1-METHYL-4-PY
C12H16N2	106362-30-5	BENZENAMINE, 4-(1,2,3,6-TETRAHYDRO-1-METHYL-4-PY
C12H16N2O	013140-89-1	N-CYCLOPENTYL-N'-PHENYLUREA
C12H16N2O	070026-79-8	CYCLOPENTA[D]PYRIDO[1,2-A]PYRIMIDIN-10(1H)-ONE,
C12H16N2O	075349-23-4	1-METHYL-4-BENZOYLPIPERAZINE
C12H16N2O	085653-87-8	CYCLOPENTA[D]PYRIDO[1,2-A]PYRIMIDIN-10(1H)-ONE,
C12H16N2O	085653-88-9	CYCLOPENTA[D]PYRIDO[1,2-A]PYRIMIDIN-10(1H)-ONE,
C12H16N2O	088491-50-3	PYRROLO[2,1-B]QUINAZOLIN-9(1H)-ONE, 2,3,5,6,7,8-
C12H16N2O2	017186-60-6	PH-ALANINE,N-ACETYL,N'-MEAM-AMIDE
C12H16N2O2S	007752-09-2	2H-1,2,4-BENZOTHIADIAZINE, 3-(1-ETHYLPROPYL)-, 1
C12H16N2O3	000052-31-3	CYCLOBARBITAL
C12H16N2O3	000056-29-1	HEXABARITAL
C12H16N2O3	003663-21-6	2-NITRO-4-BUTYLACETANILIDE
C12H16N2O3	006367-14-2	TYROSINE,N-ACETYL,N'-MEAM-AMIDE
C12H16N2O3	017357-04-9	BENZAMIDE, 2-HYDROXY-N-(4-MORPHOLINYLMETHYL)-
C12H16N2O3	032092-14-1	4H-PYRIDO[1,2-A]PYRIMIDINE-3-CARBOXYLIC ACID, 6,
C12H16N2O3	064405-35-2	4H-PYRIDO[1,2-A]PYRIMIDINE-3-CARBOXYLIC ACID, 6,
C12H16N2O3	070999-47-2	4H-PYRIDO[1,2-A]PYRIMIDINE-3-CARBOXYLIC ACID, 6,
C12H16N2O4	004450-01-5	4-PYRIMIDINECARBOXYLIC ACID, 1,2,3,6-TETRAHYDRO-
C12H16N2O4	065300-53-0	N-(2-(P-NO2-PHO)ETHYL)MORPHOLINE
C12H16N2O4	088132-25-6	1(4-ETO-CO PHENYL)-3-MEO-3-ME UREA
C12H16N2O4	120309-36-6	ACETAMIDE, N-(2-BUTOXY-5-NITROPHENYL)-
C12H16N2O4S	000542-16-5	ANILINE SULFATE
C12H16N2O5	004099-65-4	2,4-DINITRO-6-HEXYLPHENOL
C12H16N2O5S	081793-03-5	BUTANAMIDE, 2-ETHYL-N-[(2-NITROPHENYL)SULFONYL]-
C12H16N3O3P	067410-45-1	3,5-DIOXA-6-AZA-4-PHOSPHAOCT-6-ENE-8-NITRILE, 4-
C12H16N3O3PS2	002642-71-9	AZINPHOS ETHYL

C12H16N3O4PS	039923-25-6	AZINPHOS-ETHYL, O-ANALOG
C12H16N4O	069218-60-6	1-(5-CARBAMYLPENTYL)BENZOTRIAZOLE
C12H16N4O	115054-81-4	1H-BENZOTRIAZOLE-1-ACETAMIDE, α-BUTYL-
C12H16N4O	115054-90-5	2H-BENZOTRIAZOLE-2-ACETAMIDE, α-BUTYL-
C12H16N4O	122009-33-0	1-PYRAZOLIDINECARBOXALDEHYDE, 5,5-DIMETHYL-3-OXO
C12H16N4O2S	099035-21-9	CIMETIDINE,(METHYLAMINOCYCLOBUTANEDIONE) ANALOG
C12H16N4O2S	130787-21-2	2H-BENZOTRIAZOLE, 2-CYCLOPENTYL-5-[(METHYLSULFON
C12H16N4OS	054855-69-5	CIMETIDINE,2(PYRIDIN-4-ONE) ANALOG
C12H16N6O2	117121-42-3	UREA, N'-[4-[(2-ETHYL-2H-TETRAZOL-5-YL)OXY]PHENY
C12H16O	001131-60-8	P-CYCLOHEXYLPHENOL
C12H16O2	000094-46-2	1-BUTANOL, 3-METHYL-, BENZOATE
C12H16O2	000103-38-8	BUTANOIC ACID, 3-METHYL-, PHENYLMETHYL ESTER
C12H16O2	001553-60-2	BENZENEACETIC ACID, 4-(2-METHYLPROPYL)-
C12H16O2	005581-75-9	6-PHENYLCAPROIC ACID
C12H16O2	014377-19-6	BENZENEACETIC ACID, 4-BUTYL-
C12H16O2	032857-63-9	BENZENEACETIC ACID, 4-(1,1-DIMETHYLETHYL)-
C12H16O2	072615-27-1	BENZENEACETIC ACID, α-(1-METHYLETHYL)-, METHYL E
C12H16O2	100668-21-1	1,2-BENZENEDIOL, 4-(1-HEXENYL)-
C12H16O3	001878-53-1	M-BUTYLPHENOXYACETIC ACID
C12H16O3	001878-55-3	M-T-BUTYLPHENOXYACETIC ACID
C12H16O3	004917-89-9	P-SEC-BUTYLPHENOXYACETIC ACID
C12H16O3	019271-90-0	O-T-BUTYLPHENOXYACETIC ACID
C12H16O3	076343-98-1	O-SEC-BUTYLPHENOXYACETIC ACID
C12H16O3S	084368-16-1	YCLOPROPANECARBOXYLIC ACID, 3-[(DIHYDRO-2-OXO-3(
C12H16O4	088449-50-7	PHENYLACETIC ACID, 3-MEO-4-IPRO
C12H16O4S	068505-69-1	BENFURESATE
C12H16O4S2	059937-28-9	MALOTILATE
C12H16O5	002486-02-4	ISOAMYL GALLATE
C12H16O5	004568-93-8	AMYL GALLATE
C12H16O5S	016758-34-2	1-PHENYLTHIO-B-GALACTOPYRANOSIDE
C12H16O5S	057553-99-8	ALLYL-(3,4,5-MEO)PHENYL SULFONE
C12H16O6	001464-44-4	BETA-D-GLUCOPYRANOSIDE, PHENYL
C12H16O6	002818-58-8	BETA-D-GALACTOPYRANOSIDE, PHENYL
C12H16O7	000497-76-7	ARBUTIN
C12H17BrN2O3	001216-40-6	BARBITURIC ACID,5(1-MEBU),5-B-BROMALLYL
C12H17BrN2O3	069519-61-5	BARBITURIC ACID,N-ME,5-S-BU,5-BROMALLYL
C12H17BrN2O3S	075852-13-0	NN-DIME-4(N-ME-N(2BRPROPIONYL)BZSO2N
C12H17ClN2O2	051963-50-9	N-IPR1(2CL-PHENOXY)PROPIONIC HYDRAZID
C12H17ClN2O2	051963-51-0	N-IPR1(3CL-PHENOXY)PROPIONIC HYDRAZID
C12H17ClN2O2	051963-52-1	N-IPR1(4CL-PHENOXY)PROPIONIC HYDRAZID
C12H17ClN2O3S	075852-11-8	NN-DIME-4(N-ME-N(2CLPROPIONYL)BZSO2N
C12H17ClN2O3S	081792-82-7	BUTANAMIDE, N-[(2-AMINO-4-CHLOROPHENYL)SULFONYL]
C12H17ClN2O3S	097141-35-0	BUTANAMIDE, N-[2-(AMINOSULFONYL)-5-CHLOROPHENYL]
C12H17FN4O3	138848-10-9	CYTIDINE, 2',3'-DIDEOXY-N-[(DIMETHYLAMINO)METHYL
C12H17IN2O3S	075852-16-3	NN-DIME-4(N-ME-N(2I PROPIONYL)BZSO2N
C12H17N	006373-50-8	BENZENAMINE, 4-CYCLOHEXYL-
C12H17N3	000883-57-8	1-HEXYLBENZOTRIAZOLE
C12H17N3	004201-40-5	2-(246-TRIME PH IMINO)IMIDAZOLIDINE
C12H17N3	069218-30-0	1-S-HEXYLBENZOTRIAZOLE
C12H17N3O	036385-59-8	PIPERAZINE-2-CARBOXANILIDE,2'-ME
C12H17N3O	069218-37-7	1-(6-HYDROXYHEXYL)BENZOTRIAZOLE
C12H17N3O2	069218-45-7	2-(5-CARBOXYPENTYL)BENZOTRIAZOLE
C12H17N3O2	069218-53-7	2-(3-CARBETHOXYPROPYL)BENZOTRIAZOLE
C12H17N3O2	115054-79-0	2H-BENZOTRIAZOLE-2-ACETIC ACID, α,α-DIMETHYL-, E
C12H17N3O2	115054-84-7	2H-BENZOTRIAZOLE-2-ACETIC ACID, α,α-DIMETHYL-, E
C12H17N3O3S	001220-55-9	UREA, N-(1-PYRROLIDINYL)-N'-[(4-METHYLPHENYL)SUL
C12H17N3O5S	075852-04-9	NN-DIME-4(N-ME-N(2NO2PROPIONYL)BZSO2N
C12H17N5	015233-37-1	4,6-NH2 2,2-DIME-1(4-MEPH)S-TRIAZINE

C12H17N5	075842-06-7	3,6-PYRIDAZINEDIAMINE, N'-(2,5-DIMETHYL-1H-PYRRO
C12H17N5O	021316-30-3	4,6-NH2 2,2-DIME-1(4-MEOPH)S-TRIAZINE
C12H17N5O3	098204-06-9	ALLOPURINOL,1-LEUCINYLOXYMETHYL-
C12H17N5O3	098204-08-1	ALLOPURINOL,1-N,N-DIETHYLGYCYCLOMETHYL
C12H17NO	000134-62-3	DEET [N,N,-DIET-3-ME BENZAMIDE]
C12H17NO	000621-15-8	CAPRYLANILIDE
C12H17NO	002431-96-1	BENZENEACETAMIDE, N,N-DIETHYL-
C12H17NO	002728-04-3	BENZAMIDE, N,N-DIETHYL-2-METHYL-
C12H17NO	002728-05-4	BENZAMIDE, N,N-DIETHYL-4-METHYL-
C12H17NO	003663-20-5	4-BUTYLACETANILIDE
C12H17NO	005407-61-4	B-PH-B-HYDROXY-N-ETHYLPYRROLIDINE
C12H17NO	016665-89-7	2,6-DIETHYLACETANILIDE
C12H17NO	020330-45-4	ACETAMIDE, N-[4-(1,1-DIMETHYLETHYL)PHENYL]-
C12H17NO	023702-98-9	A-(2-PIPERIDYL)PHENYLCARBINOL
C12H17NO2	000780-11-0	M-T-BUTYLPHENYL N-METHYLCARBAMATE
C12H17NO2	002626-81-5	N-ME-2-T-BUTYLPHENYLCARBAMATE
C12H17NO2	002626-83-7	N-ME-4-T-BUTYLPHENYLCARBAMATE
C12H17NO2	002631-37-0	3-I-PR-5-MEPHENYL-N-ME CARBAMATE
C12H17NO2	003766-81-2	N-METHYL O-SEC-BUTYL PHENYL CARBAMATE
C12H17NO2	003766-82-3	N-ME-3-ME-4-IPRPHENYL CARBAMATE
C12H17NO2	003942-51-6	N-ME-4-SEC-BUTYLPHENYLCARBAMATE
C12H17NO2	004151-47-7	3-T-BUTYL-4-HYDROXYACETANILIDE
C12H17NO2	013110-37-7	BENZOIC ACID, 4-AMINO-,PENTY ESTER
C12H17NO2	018659-24-0	N-ME-3-ME-6-I-PR PHENYLCARBAMATE
C12H17NO2	023597-82-2	N-HEXYL NICOTINOATE
C12H17NO2	025007-30-1	2-I-PR-PHENYLDIMETHYLCARBAMATE
C12H17NO2	051170-56-0	CARBAMIC ACID, DIETHYL-, PHENYLMETHYL ESTER
C12H17NO2	055205-89-5	3,5-DIETHYL-4-HYDROXYACETANILIDE
C12H17NO2	063075-06-9	N-PENTYL N-PHENYLCARBAMATE
C12H17NO2	091564-01-1	CARBAMIC ACID, DIMETHYL-, 3-PHENYLPROPYL ESTER
C12H17NO2	098154-93-9	CARBAMIC ACID, DIMETHYL-, 1-METHYL-2-PHENYLETHYL
C12H17NO2	113932-84-6	CARBAMIC ACID, (4-BUTYLPHENYL)-, METHYL ESTER
C12H17NO2	124497-73-0	2-OH-2-PH-4,5-DIME MORPHOLINE HCL
C12H17NO2	124497-75-2	2-OH-2-PH-3,4-DIME MORPHOLINE HCL
C12H17NO3	002438-72-4	BUFEXAMIC ACID
C12H17NO3	003978-68-5	N-ME-3-BUTOXYPHENYLCARBAMATE
C12H17NO3	003978-69-6	N-ME-4-BUTOXYPHENYLCARBAMATE
C12H17NO3	061566-63-0	2-PENTOXY-4-AMINOBENZOIC ACID
C12H17NO3	061566-64-1	2-I-PENTOXY-4-AMINOBENZOIC ACID
C12H17NO3	069148-74-9	2-IPRO PHENYLDIMETHYLCARBAMATE
C12H17NO4	039638-51-2	L-TYROSINE, 3-HYDROXY-, PROPYL ESTER
C12H17NO4	057079-00-2	N-MALEOYL-8-AMINOOCTANOIC ACID
C12H17NO4	057079-07-9	N-MALEOYL-6-NH2HEXANOIC ACID,ET ESTER
C12H17NO4	087130-18-5	CARBAMIC ACID, (3,4-DIETHOXYPHENYL)-, METHYL EST
C12H17NO4	110301-07-0	L-TYROSINE, 3-HYDROXY-, 1-METHYLETHYL ESTER
C12H17NO4S	059777-60-5	PENTYL 4-SULFAMYLBENZOATE
C12H17NO4Si	051466-74-1	1-PHENOXYSILATRANE
C12H17NO5	099877-03-9	L-TYROSINE, 3-HYDROXY-, 2-HYDROXYPROPYL ESTER
C12H17NO6	007265-01-2	GLUCOPYRANOSIDE,2-AMINOPHENYL
C12H17NO6	020818-25-1	GLUCOPYRANOSIDE,4-AMINOPHENYL
C12H17O4PS2	002597-03-7	FENTHOATE
C12H18	000087-85-4	HEXAMETHYLBENZENE
C12H18	000099-62-7	1,3-DIISOPROPYLBENZENE
C12H18	000100-18-5	1,4-DIISOPROPYLBENZENE
C12H18	000102-25-0	1,3,5-TRIETHYLBENZENE
C12H18	000577-55-9	1,2-DIISOPROPYLBENZENE
C12H18	001077-16-3	N-HEXYLBENZENE
C12H18	004904-61-4	1,5,9-CYCLODODECATRIENE

C12H18Br6	003194-55-6	1,2,5,6,9,10-HEXABROMOCYCLODODECANE
C12H18N2	013450-66-3	3-PYRIDYLETHYL-2-N-PIPERIDINE
C12H18N2O	007728-40-7	ACETAMIDE, N-(2,6-DIMETHYLPHENYL)-2-(ETHYLAMINO)
C12H18N2O	021236-54-4	2(DIMEAMINO)-N-(2,6-DIMEPH)ACETAMIDE
C12H18N2O	034123-59-6	ISOPROTURON
C12H18N2O	040948-30-9	BENZAMIDE,N-(3-DIME AMINOPROPYL)
C12H18N2O	052387-58-3	N-(ISO-BUTYLAMINOMETHYL)BENZAMIDE
C12H18N2O	059359-46-5	BUTANAMIDE, 2-AMINO-N-(2,6-DIMETHYLPHENYL)-
C12H18N2O	070381-46-3	4H-PYRIDO[1,2-A]PYRIMIDIN-4-ONE, 3-ETHYL-2,6-DIM
C12H18N2O	085911-77-9	UREA, N,N-DIMETHYL-N'-(4-PROPYLPHENYL)-
C12H18N2O	102089-65-6	BUTANAMIDE, 4-AMINO-N-(2,6-DIMETHYLPHENYL)-
C12H18N2O2	000315-18-4	MEXACARBATE
C12H18N2O2	034861-40-0	1-(4-I-PRPHENYL)-3-MEO-3-ME UREA
C12H18N2O2	054922-60-0	N'-IPR(3-TOLYLOXY)ACETIC AC HYDRAZID
C12H18N2O2	054922-61-1	N'-IPR(4-TOLYLOXY)ACETIC AC HYDRAZID
C12H18N2O2	062251-73-4	N'-IPR(2-TOLYLOXY)ACETIC AC HYDRAZID
C12H18N2O2	078798-46-6	O-MEO BENZAMIDE,N-(2-DIMEAMINOET)
C12H18N2O2	084640-28-8	M-DIMEAMINOBZY N,N-DIME CARBAMATE
C12H18N2O2	090663-30-2	O-MEO BENZAMIDE,N-(4-AMINOBUTYL)
C12H18N2O2S	000077-27-0	THIAMYLAL
C12H18N2O3	000076-73-3	SECOBARBITAL
C12H18N2O3	000561-83-1	BARBITURIC ACID,5-ALLYL-5-NEOPENTYL
C12H18N2O3	054922-62-2	N-IPR-1(4-MEO PHENOXY)ACETIC HYDRAZID
C12H18N2O3	062251-74-5	N-IPR-1(2-MEO PHENOXY)ACETIC HYDRAZID
C12H18N2O3	062251-75-6	N-IPR-1(3-MEO PHENOXY)ACETIC HYDRAZID
C12H18N2O3S	000064-77-7	TOLBUTAMIDE
C12H18N2O3S	007752-27-4	BUTANAMIDE, N-[2-(AMINOSULFONYL)PHENYL]-2-ETHYL-
C12H18N2O3S	059777-64-9	N-PENTYL 4-SULFAMYLBENZAMIDE
C12H18N2O3S	064674-86-8	ACETIC ACID, [4-OXO-5-(1-PIPERIDINYL)-2-THIAZOLI
C12H18N2O3S	074746-43-3	NN-DIME-4(N-ME-N(PROPIONYL)BZSO2N
C12H18N2O3S	081792-78-1	BUTANAMIDE, N-[(2-AMINOPHENYL)SULFONYL]-2-ETHYL-
C12H18N2O3S	097141-29-2	BENZENESULFONAMIDE, 4-METHYL-2-[(2-METHYL-1-OXOP
C12H18N2O4	124979-37-9	4-PYRIMIDINECARBOXYLIC ACID, 1,2,3,6-TETRAHYDRO-
C12H18N2O4S	075851-99-9	PROPANAMIDE, N-[4-[(DIMETHYLAMINO)SULFONYL]PHENY
C12H18N2O4Si	090936-52-0	1-(3-AMINOPHENOXY)SILATRANE
C12H18N2O5S	123045-58-9	BENZENESULFONAMIDE, 4-(HEXYLOXY)-3-NITRO-
C12H18N4O	059708-21-3	4-(3-ME-3-BU-1-TRIAZENO)BENZAMIDE
C12H18N4O	059708-25-7	4-(3-ME-3-T-BU-1-TRIAZENO)BENZAMIDE
C12H18N4O	069218-65-1	2-(5-CARBAMYLPENTYL)BENZOTRIAZOLE
C12H18N4O6S	019044-88-3	ORYZALIN
C12H18N4O6S	135685-04-0	BENZENESULFONAMIDE, N-(2-AMINOETHYL)-N-METHYL-4-
C12H18N5O4	024632-47-1	NIFURPIPONE
C12H18N6O	120182-20-9	2-OXAZOLAMINE, 4,5-DIHYDRO-5-[[4-(2-PYRIMIDINYL)
C12H18NO5P	013538-14-2	T-BU PHOSPHONIC ACID,ET,PNO2PH ESTR
C12H18O	000096-70-8	2-(T-BUTYL)-4-ETHYLPHENOL
C12H18O	001502-22-3	CYCLOHEXANONE, 2-(1-CYCLOHEXEN-1-YL)-
C12H18O	001879-09-0	2-(1,1-DIMETHYLETHYL)-4,6-DIMETHYLPHENOL
C12H18O	002078-54-8	PHENOL, 2,6-BIS(1-METHYLETHYL)-
C12H18O2	000136-77-6	4-HEXYLRESORCINOL
C12H18O2	018979-55-0	4-HEXYLOXYPHENOL
C12H18O3	118198-71-3	BENZENEMETHANOL, 3,4-DIHYDROXY-α-PENTYL-
C12H18O4	000532-34-3	INDALONE
C12H18O4	002082-81-7	1,4-BUTANEDIYL BIS(2-METHYL-2-PROPENOATE)
C12H18O4	026074-74-8	CYCLOPROPANECARBOXYLIC ACID, 3-(3-METHOXY-2-METH
C12H18O4	033332-53-5	1,2-BIS(2-MEO-ETO)BENZENE
C12H18O4S2	050512-35-1	IPT (ISOPROTHIOLANE)
C12H18O4S2	050780-68-2	PROPANEDIOIC ACID, 1,3-DITHIOLAN-2-YLIDENE-, DIP
C12H18O5	002358-84-1	DIETHYLENE GLYCOL DIMETHACRYLATE

C12H18O5S	071203-62-8	PROPYL-(3,4,5-MEO)PHENYL-SULFONE
C12H18O5S2	052303-69-2	PROPANEDIOIC ACID, 1,3-DITHIOLAN-2-YLIDENE-, BIS
C12H18O6	001680-21-3	2-PROPENOIC ACID, 1,2-ETHANEDIYLBIS(OXY-2,1-ETHA
C12H18O7	018467-77-1	.ALPHA.-L-XYLO-2-HEXULOFURANOSONIC ACID, 2,3:4,6
C12H19ClNO3P	000299-86-5	CRUFOMATE
C12H19IN2O	065481-90-5	3-CARBAMOYLPYRIDINIUM IODIDE,N-HEXYL
C12H19N	040089-90-5	PYRIDINE, 4-HEPTYL-
C12H19N	067309-36-8	A-BUTYL BENZENEETHANEAMINE
C12H19N	067309-37-9	A-T-BUTYL BENZENEETHANEAMINE
C12H19N	067309-38-0	A-I-BUTYL BENZENEETHANEAMINE
C12H19N	084952-61-4	PHENETHYLAMINE,N-METHYL-A-ISOPROPYL
C12H19N3	069218-28-6	2-HEXYLBENZOTRIAZOLE
C12H19N3	069218-33-3	2-S-HEXYLBENZOTRIAZOLE
C12H19N3O	000671-16-9	PROCARBAZINE
C12H19N3O	069218-41-3	2-(6-HYDROXYHEXYL)BENZOTRIAZOLE
C12H19N3O	120277-90-9	7-AZABICYCLO[4.1.0]HEPTANE, 7-[2-HYDROXY-3-(1H-I
C12H19N3O3S	056046-62-9	N-(2-ETHYL(3-METHYL-4-NITROPHENYL)AMINO)ETHYL)-M
C12H19NO	004249-64-3	2-N,N-DIETAMINO-1-PHENYLETHANOL
C12H19NO2	007695-63-8	2-OH-I-PROPYLAMINO PHENYL ETHER
C12H19NO2	015588-95-1	2,5-DIMETHOXY-4-METHYLAMPHETAMINE
C12H19NO2S	061638-07-1	2,5-DIMETHOXY-4-MES AMPHETAMINE
C12H19NO3	001082-23-1	2,3,4-TRIMEO AMPHETAMINE
C12H19NO3	001082-88-8	3,4,5-TRIMEO AMPHETAMINE
C12H19NO3	001083-09-6	2,4,5-TRIMEO AMPHETAMINE
C12H19NO3	015402-79-6	2,4,6-TRIMETHOXYAMPHETAMINE
C12H19NO3	039201-82-6	3,5-DIMEO-4-ETO-PHENETHYLAMINE
C12H19NO3S	000959-24-0	SOTALOL
C12H19NO3S	001145-46-6	BENZENESULFONAMIDE, 4-(HEXYLOXY)-
C12H19O2PS3	035400-43-2	SULPROFOS
C12H19O3PS2	002670-77-1	OO-DIET O-2-METHYLTHIOPH P=S
C12H19O4PS	004799-59-1	O,O-DIET-O-(3-ME-4-SMEPH)P=O
C12H19O4PS	026512-63-0	O=P O,O-DIET-O-(2-ME-4-SMEPHENYL)
C12H20BrN	071324-04-4	M-ETHYLBENZYL TRIMETHYL AMMONIUM BROMIDE
C12H20BrN3O	134419-53-7	1H-IMIDAZOLE-1-ETHANOL, α-[[(2-BROMOCYCLOHEXYL)A
C12H20Cl2N6O4	013907-57-8	UREA, N,N"-1,4-CYCLOHEXANEDIYLBIS[N'-(2-CHLOROET
C12H20ClN3O4	061137-49-3	UREA, N'-[4-[(ACETYLOXY)METHYL]CYCLOHEXYL]-N-(2-
C12H20ClNO2S	087674-68-8	DIMETHENAMID
C12H20F2N6O4	013908-98-0	CYCLOHEXANE,1,4-DI(1-NO-1ETF UREA)
C12H20IN	002125-48-6	PHENYLPROPYL-TRIME-AMMONIUM IODIDE
C12H20IN	014402-24-5	2-METHYLPYRIDINIUM IODIDE,N-HEXYL
C12H20N2	020173-18-6	N,N-DI-I-PR-3-PYRIDYLMETHYLAMINE
C12H20N2O3	000077-30-5	5-ETHYL-5-HEXYLBARBITURIC ACID
C12H20N2O3	017013-41-1	BARBITURIC ACID,5,5-DIBUTYL-
C12H20N2O3	036380-48-0	BARBITURIC ACID,5-ET-5(2,3-DIMEBU)
C12H20N2O3	057562-99-9	BARBITURIC ACID,5(1MEBU)-5-ET-1-ME
C12H20N2O3S	003930-20-9	SOTALOL
C12H20N2O4S2	135832-44-9	BENZENESULFONAMIDE, 4-[[2-[(2-METHYLPROPYL)AMINO
C12H20N2S	025444-83-1	THIOUREA, N-METHYL-N'-TRICYCLO[3.3.1.1,3,7]DEC-1
C12H20N4O2	051235-04-2	HEXAZINONE
C12H20NO3P	019590-04-6	O=P(NME)(OME)O-(4-T-BU PHENYL)
C12H20NO3P	019590-05-7	O=P(NC)(OC)O-3-T-BU PHENYL
C12H20O2	000076-49-3	BORNYL ACETATE
C12H20O2	000105-87-3	GERANIOL ACETATE
C12H20O2	000115-95-7	LINALYL ACETATE
C12H20O3Si	000780-69-8	SILANE, TRIETHOXYPHENYL-
C12H20O4	000105-76-0	DIBUTYL MALEATE
C12H20O6	000139-45-7	TRIPROPIONIN
C12H20O7	000077-93-0	TRIETHYL CITRATE

C12H21N	019982-08-2	1-AMINOADAMANTANE,3,5-DIMETHYL
C12H21N	041100-45-2	1-AMINOADAMANTANE,3-ETHYL
C12H21N2O3PS	000333-41-5	DIAZINON
C12H21N2O3PS	005826-91-5	G-24622
C12H21N2O4P	000962-58-3	DIAZINON, O-ANALOG
C12H21NO3	071548-53-3	1-HEXYL-4-METHOXYCARBONYL-2-PYRROLIDONE
C12H22	000092-51-3	DICYCLOHEXYL
C12H22ClN5	102587-54-2	1,3,5-TRIAZINE-2,4-DIAMINE, 6-CHLORO-N-(1-METHYL
C12H22N2O2	000050-12-4	MEPHENYTOIN
C12H22N4O3S	051461-71-3	VEL 3510
C12H22N4O4	132765-91-4	L-VALINAMIDE, N -ACETYL-L-GLUTAMINYL-
C12H22N4O4	132765-95-8	L-GLUTINAMIDE, N-ACETYL-L-VALYL-
C12H22N4O4	132765-96-9	L-ASPARTAMIDE, N-ACETYL-L-ISOLEUCYL-
C12H22N4O4	132765-98-1	L-ISOLEUCINAMIDE,N-ACETYL-L-ASPARAGINYL-
C12H22N4O4	132766-00-8	L-ASPARTAMIDE, N-ACETYL-L-LEUCYL-
C12H22N4O4	132766-08-6	L-VALINAMIDE, N-ACETYLGLYCYL-L-ALANYL-
C12H22N4O4	132766-13-3	GLYCINAMIDE, N-ACETYLGLYCYL-L-LEUCYL-
C12H22O	000286-99-7	CYCLODODECANE EPOXIDE
C12H22O	019700-21-1	GEOSMIN
C12H22O11	000057-50-1	SUCROSE
C12H22O11	000063-42-3	LACTOSE
C12H22O11	000069-79-4	MALTOSE
C12H22O2	000688-84-6	2-ETHYLHEXYL 2-METHYL-2-PROPENOATE
C12H22O2	016409-45-3	MENTHYL ACETATE
C12H22O2	038421-90-8	5-DECEN-1-OL ACETATE (E)
C12H22O2	067446-07-5	5-DECEN-1-OL ACETATE (Z)
C12H22O2	067452-27-1	4-DECENYL ACETATE (Z)
C12H22O2	068760-70-3	6-DECENYL ACETATE (Z)
C12H22O2	081634-99-3	3-DECENYL ACETATE (Z)
C12H22O4	002050-23-9	DIETHYL SUBERATE
C12H22O6	005284-99-1	GLUCOPYRANOSIDE,CYCLOHEXYL
C12H23KO2	010124-65-9	LAURIC ACID, POTASSIUM SALT
C12H23N	002437-25-4	LAURONITRILE
C12H23N3O3	132765-83-4	L-VALINAMIDE, N-ACETYL-L-VALYL-
C12H23N3O4	132765-90-3	L-ISOLEUCINAMIDE, N-ACETYL-L-THREONYL-
C12H23N5O	015438-85-4	2,4-BIS(DIETHYLAMINO)-6-METHOXY-S-TRIAZINE
C12H23NO	000947-04-6	AZACYCLOTRIDECAN-2-ONE
C12H23NO	095715-61-0	CYCLOHEXANEACETAMIDE, N,N-DIETHYL-
C12H24	000112-41-4	N-DODECENE
C12H24	000294-62-2	CYCLODODECANE
C12H24	000830-13-7	CYCLODODECANONE
C12H24	001124-14-7	1-DODECENE
C12H24	004292-75-5	1-CYCLOHEXYLHEXANE
C12H24N3OP	006415-07-2	PHOSPHINE OXIDE, TRIS-(1-PYRROLIDINYL)-
C12H24N6	064124-20-5	N2,N4,N6-TRIME-N,N,N-TRIME-MELAMINE
C12H24O	000112-54-9	DODECANAL
C12H24O	001724-39-6	CYCLODODECANOL
C12H24O	006175-49-1	METHYL DECYL KETONE
C12H24O	020056-92-2	7-DODECEN-1-OL (Z)
C12H24O2	000110-38-3	ETHYL DECANOATE
C12H24O2	000112-17-4	DECYL ACETATE
C12H24O2	000143-07-7	DODECANOIC ACID
C12H24O2	000589-75-3	BUTYL OCTANOATE
C12H24O3	002388-12-7	DODECANEPEROXOIC ACID
C12H24O5S	051575-56-5	1-HEXYLTHIO-B-GALACTOPYRANOSIDE
C12H24O6	039824-11-8	B-HEXYLGALACTOPYRANOSIDE
C12H25Br	000143-15-7	1-BROMO-N-DODECANE
C12H25Cl	000112-52-7	1-CHLORODODECANE

C12H25F	000334-68-9	1-FLUORODODECANE
C12H25I	004292-19-7	1-IODODODECANE
C12H25N5O	055921-62-5	24NH2-6DIBUTYLAMINO-PYRIMIDINE-3-OXIDE
C12H25NaO3S	002386-53-0	DODECYL SULFONATE, SODIUM SALT
C12H25NaO4S	000151-21-3	DODECYL SULFATE, SODIUM SALT
C12H26	000112-40-3	DODECANE
C12H26	064771-71-7	PARAFFINS (C>10)
C12H26N2O2	042013-74-1	1-PIPERIDINYLOXY, 4-[(2-HYDROXYETHYL)METHYLAMINO
C12H26O	000112-53-8	DODECANOL
C12H26O	000112-58-3	DI-N-HEXYL ETHER
C12H26O3	000112-73-2	DIETHYLENE GLYCOL DIBUTYL ETHER
C12H26O6	017455-13-9	1,4,7,10,13,16-HEXAOXACYCLOOCTADECANE
C12H26O6P2S4	000078-34-2	DIOXATHION
C12H26O7	002615-15-8	HEXAETHYLENE GLYCOL
C12H26S	000112-55-0	DODECYL MERCAPTAN
C12H27N	000102-82-9	TRI N-BUTYLAMINE
C12H27O4P	000126-73-8	TRIBUTYLPHOSPHATE
C12H27OPS3	000078-48-8	DEF
C12H27PS3	000150-50-5	MERPHOS
C12H28ClN	000929-73-7	1-DODECANAMINE, HYDROCHLORIDE
C12H28IN	000631-40-3	TETRAPROPYL AMMONIUM IODIDE
C12H28N3OP	069981-38-0	PHOSPHONIC DIAMIDE, N,N,N',N'-TETRAETHYL-P-(1-PY
C12H28O5P2S2	003244-90-4	TETRAPROPYL THIOPYROPHOSPHORATE
C12H30N3OP	002622-07-3	PHOSPHORIC TRIAMIDE, HEXAETHYL-
C12H30O13P4	000757-58-4	HEXAETHYL TETRAPHOSPHATE
C12H34O2Si4	018077-53-7	1,1,1,3,5,5,5-HEPTAMETHYL-3-(2-(TRIMETHYLSILY*))
C12H36O4Si5	000141-63-9	DODECAMETHYLPENTASILOXANE
C12H36O4Si5	003555-47-3	TETRAKIS(TRIMETHYLSILOXY)SILANE
C12H36O6Si6	000540-97-6	DODECAMETHYLCYCLOHEXASILOXANE
C13H4Cl2F6N4O4	079622-59-6	FLUAZINAM
C13H4Cl6O	038178-99-3	1,2,4,5,7,8-HEXACHLORO-9H-XANTHENE
C13H4N4O9	000746-53-2	9H-FLUOREN-9-ONE, 2,4,5,7-TETRANITRO-
C13H5N3O7	000129-79-3	9H-FLUOREN-9-ONE, 2,4,7-TRINITRO-
C13H6Cl6O2	000070-30-4	PHENOL,2,2'-METHYLENEBIS 3,4,6-CL
C13H6ClF5N4O4	078879-38-6	N(24-NO2-6CF3 PH)-2(3CL-5CF3)PYRIDINAMIN
C13H6N2O5	031551-45-8	9H-FLUOREN-9-ONE, 2,7-DINITRO-
C13H7Br2ClN2O4	072699-09-3	3,5-DIBR-2'-NO2-4'-CL SALICYLANILIDE
C13H7Br2F2NO2	079402-07-6	3,5-DIBR-2',4'-DIFLUORO SALICYLANILIDE
C13H7Br2N3O6	013181-17-4	BROMOFENOXIM
C13H7Cl2F2NO2	080033-99-4	3,5-DICL-2',4'-DIFLUORO SALICYLANILIDE
C13H7Cl3N2O4	090426-05-4	3,5,-4'-TRICL-4'-NO2 SALICYLANILIDE
C13H7ClF3NO3	042874-01-1	NITROFLUORFEN
C13H7F3N2O5	015457-05-3	FLUORODIFEN
C13H7NO3	042135-22-8	9H-FLUOREN-9-ONE, 3-NITRO-
C13H8	004300-27-0	1-PHENYL-1,3,5-HEPTATRIYNE
C13H8Cl2FNO2	054850-02-1	3,5-DICL-4'-F SALICYLANILIDE
C13H8Cl2N2O4	000050-65-7	NICLOSAMIDE
C13H8Cl2O3	036417-16-0	2-OH NAPHTHOQUINONE,3(33DICLALLYL)
C13H8Cl3NO2	001151-51-5	3,5,4'-TRICL SALICYLANILIDE
C13H8ClFO2	062433-26-5	METHANONE, (4-CHLOROPHENYL)(5-FLUORO-2-HYDROXYPH
C13H8F2O3	022494-42-4	DIFLUNISAL
C13H8F3N3O4	001869-67-6	2,4-DINITRO-N-(3-(TRIFLUOROMETHYL)PHENYL)BENZE*
C13H8F3N3O4	013744-79-1	2,4-DINITRO-4'-(TRIFLUOROMETHYL)DIPHENYLAMINE
C13H8N2O4	005405-53-8	9H-FLUORENE, 2,7-DINITRO-
C13H8O	000486-25-9	FLUORENONE
C13H8O	020252-42-0	7-PHENYLHEPT-2-ENE-4,6-DIYNAL
C13H8OS	000492-22-8	9H-THIOXANTHEN-9-ONE
C13H9Br	001133-80-8	2-BROMO-9H-FLUORENE

C13H9BrClNO2	003679-64-9	4'-CL-5-BR SALICYLANILIDE
C13H9BrClNO2	097713-49-0	BENZOIC ACID, 2-[(3-BROMOPHENYL)AMINO]-4-CHLORO-
C13H9Cl	002523-44-6	2-CHLOROFLOURENE
C13H9Cl2NO	010286-75-6	N-(3,4-DICHLOROPHENYL)BENZAMIDE
C13H9Cl2NO2	004214-48-6	3,5-DICL SALICYLANILIDE
C13H9Cl2NO2	017870-85-8	BENZOIC ACID, 4-CHLORO-2-[(3-CHLOROPHENYL)AMINO]
C13H9Cl2NO2	037183-28-1	2',4'-DICL SALICYLANILIDE
C13H9Cl2NO2	090656-45-4	BENZOIC ACID, 4-CHLORO-2-[(4-CHLOROPHENYL)AMINO]
C13H9Cl2NO4	032861-85-1	CHLOMETHOXYNIL
C13H9ClINO2	090700-03-1	BENZOIC ACID, 4-CHLORO-2-[(4-IODOPHENYL)AMINO]-
C13H9ClN2	010176-63-3	3-PYRIDINECARBONITRILE, 2-CHLORO-6-METHYL-5-PHEN
C13H9ClN2O4	037399-40-9	2'-NO2-4'-CL SALICYLANILIDE
C13H9F3N2O2	000369-90-4	N-(4-NITROPHENYL)-3-(TRIFLUOROMETHYL)BENZENAMINE
C13H9N	000085-02-9	BENZO(F)QUINOLINE
C13H9N	000229-87-8	PHENANTHRIDINE
C13H9N	000230-27-3	BENZO(H)QUINOLINE
C13H9N	000260-94-6	ACRIDINE
C13H9N3S	007612-96-6	4-ISOTHIOCYANOAZOBENZENE
C13H9NaO2	017264-58-3	P-BIPHENYLCARBOXYLIC ACID, NA SALT
C13H9NO	000833-50-1	BENZOXAZOLE, 2-PHENYL-
C13H9NO2	000607-57-8	2-NITROFLUORENE
C13H9NO2	000835-64-3	2-(2-HYDROXYPHENYL)BENZOXAZOLE
C13H9NOS	003411-95-8	2-(2-HYDROXYPHENYL)BENZOXTHIAZOLE
C13H9NS	000883-93-2	2-PHENYLBENZTHIAZOLE
C13H9NS	001510-24-3	4-BIPHENYLISOTHIOCYANATE
C13H9O2	000090-47-1	9H-XANTHEN-9-ONE
C13H10	000086-73-7	9H-FLUORENE
C13H10	000203-80-5	1H-PHENALENE
C13H10	000244-36-0	1H-FLUORENE
C13H10	013678-98-3	HEPT-1,3-DIYN-5-ENYLBENZENE
C13H10BrCl2O2PS	021609-90-5	LEPTOPHOS
C13H10BrCl2O3P	025006-32-0	LEPTOPHOS-O-ANALOG
C13H10BrN	006638-60-4	9H-FLUOREN-2-AMINE, 7-BROMO-
C13H10BrNO2	002627-77-2	4'-BR SALICYLANILIDE
C13H10BrNO3	056356-31-1	2-BROMOPHENYL 4-AMINOSALICYLATE
C13H10BrNO3	056356-32-2	3-BROMOPHENYL 4-AMINOSALICYLATE
C13H10BrNO3	056356-33-3	4-BROMOPHENYL 4-AMINOSALICYLATE
C13H10Cl2N2O	002008-73-3	1-(3,4-DICLPHENYL)-3-PHENYLUREA
C13H10Cl2O2	000097-23-4	PHENOL,2,2'-METHYLENEBIS 4-CHLORO-
C13H10Cl2O2	000555-89-5	NEOTRAN
C13H10Cl2S	000103-17-3	CHLORBENSIDE
C13H10ClFS	000405-30-1	FLUORBENSIDE
C13H10ClN	017784-47-3	ACRIDINE, HYDROCHLORIDE
C13H10ClN3O	030195-30-3	2H-1,4-BENZODIAZEPIN-2-ONE, 7-CHLORO-1,3-DIHYDRO
C13H10ClN3O4S2	001163-51-5	CHLOROTHIAZIDE-3-PHENYL
C13H10ClNO2	001697-18-3	2'-CL SALICYLANILIDE
C13H10ClNO2	003679-63-8	4'-CL SALICYLANILIDE
C13H10ClNO2	004638-48-6	5-CHLOROSALICYLANILIDE
C13H10ClNO2	013278-36-9	BENZOIC ACID, 2-[(3-CHLOROPHENYL)AMINO]-
C13H10ClNO2	019218-88-3	BENZOIC ACID, 4-CHLORO-2-(PHENYLAMINO)-
C13H10ClNO3	056356-28-6	2-CHLOROPHENYL 4-AMINOSALICYLATE
C13H10ClNO3	056356-29-7	3-CHLOROPHENYL 4-AMINOSALICYLATE
C13H10ClNO3	056356-30-0	4-CHLOROPHENYL 4-AMINOSALICYLATE
C13H10F3N	000101-23-5	N-PHENYL-3-(TRIFLUOROMETHYL)BENZENAMINE
C13H10F3NO2S	050585-77-8	CF3-SULFONANILIDE, P-PHENYL
C13H10F3NO2S	132483-39-7	BENZENEACETIC ACID, α-METHYL-4-(2-THIAZOLYL)-3-(
C13H10F3NO2S	132483-45-5	BENZENEACETIC ACID, α-METHYL-4-[4-(TRIFLUOROMETH
C13H10FNO3	056356-25-3	2-FLUOROPHENYL 4-AMINOSALICYLATE

C13H10FNO3	056356-26-4	3-FLUOROPHENYL 4-AMINOSALICYLATE
C13H10FNO3	056356-27-5	4-FLUOROPHENYL 4-AMINOSALICYLATE
C13H10N2	000090-45-9	9-AMINOACRIDINE
C13H10N2	000578-06-3	1-AMINOACRIDINE
C13H10N2	000578-07-4	4-AMINOACRIDINE
C13H10N2	000581-28-2	2-AMINOACRIDINE
C13H10N2	000581-29-3	3-AMINOACRIDINE
C13H10N2	000716-79-0	2-PHENYLBENZIMIDAZOLE
C13H10N2O	002963-66-8	2-(2-HYDROXYPHENYL)BENZOIMIDAZOLE
C13H10N2O2	001214-32-0	9H-FLUOREN-2-AMINE, 7-NITRO-
C13H10N2O2S	018818-44-5	124-BENZTHIADIAZIN-1-O2-3-PHENYL
C13H10N2O2S	101533-57-7	BENZONITRILE, 4-[(4-AMINOPHENYL)SULFONYL]-
C13H10N2O4	000050-35-1	THALIDOMIDE
C13H10N2O4	002389-37-9	5-NO2 SALICYLANILIDE
C13H10N2O4	027693-70-5	N-PHENYLANTHRANILIC ACID,3'-NITRO
C13H10N2O4	037183-26-9	2'-NO2 SALICYLANILIDE
C13H10N2S	023246-36-8	4-ISOTHIOCYANODIPHENYL AMINE
C13H10N4O3	098846-65-2	ALLOPURINOL, 1-[BENZOYLOXYMETHYL]-
C13H10NNaO2	006232-32-2	N-PHENYLANTHRANILIC ACID, SODIUM SALT
C13H10O	000092-83-1	9H-XANTHENE
C13H10O	000119-61-9	BENZOPHENONE
C13H10O	013641-62-8	7-PHENYL-4,6-DIYN-HEPT-2-EN-1-OL
C13H10O2	000092-92-2	1,1'-BIPHENYL -4-CARBOXYLIC ACID
C13H10O2	000093-99-2	PHENYL BENZOATE
C13H10O2	000117-99-7	O-HYDROXYBENZOPHENONE
C13H10O2	000579-44-2	DL-BENZOIN
C13H10O2	000947-84-2	[1,1'-BIPHENYL]-2-CARBOXYLIC ACID
C13H10O2	001137-42-4	P-HYDROXYBENZOPHENONE
C13H10O2	039515-51-0	3-PHENOXYBENZALDEHYDE
C13H10O3	000102-09-0	CARBONIC ACID, DIPHENYL ESTER
C13H10O3	000118-55-8	PHENYL 2-HYDROXYBENZOATE
C13H10O3	002215-77-2	P-PHENOXYBENZOIC ACID
C13H10O3	002243-42-7	O-PHENOXYBENZOIC ACID
C13H10O3	003739-38-6	M-PHENOXYBENZOIC ACID
C13H11BrN2O2	085879-22-7	4-(4-BROMOPHENOXY)PHENYLUREA
C13H11Cl2NO2	032809-16-8	PROCYMIDONE
C13H11Cl2NO5	072391-46-9	CHLOZOLINATE
C13H11ClN2O3S	025270-44-4	BENZENESULFONAMIDE, 4-CHLORO-N-[(PHENYLAMINO)CAR
C13H11ClO	000119-56-2	BENZENEMETHANOL, 4-CHLORO-.ALPHA.-PHENYL-
C13H11ClO	000120-32-1	5-CHLORO-2-HYDROXYDIPHENYLMETHANE
C13H11ClO	002060-58-4	3,11-TRIDECADIENE-5,7,9-TRIYN-1-OL, 2-CHLORO-
C13H11N	000153-78-6	2-AMINOFLUORENE
C13H11N	000538-51-2	BENZALANILINE
C13H11N	006344-63-4	9H-FLUOREN-1-AMINE
C13H11N2NaO3S	062959-39-1	4-((4-METHYLPHENYL)AZO)BENZENESULFONIC ACID,SOD*
C13H11N3	000092-62-6	3,6-DIAMINOACRIDINE
C13H11N3O	002440-22-4	2-(2-HYDROXY-5-METHYLPHENYL)BENZOTRIAZOLE
C13H11N3O	128887-24-1	[3,4'-BIPYRIDINE]-5-CARBONITRILE,6-ETHOXY-
C13H11N3O	128887-26-3	[3,4'-BIPYRIDINE]-5-CARBONITRILE,6-METHOXY-2-MET
C13H11N3O2	000495-73-8	BAYER 15080
C13H11N3O2	007030-18-4	2-METHYL-4'-NITROAZOBENZENE
C13H11N3O2	029418-57-3	2-METHYL-4-NITROAZOBENZENE
C13H11N3O2	029418-58-4	(4-METHYLPHENYL)(4-NITROPHENYL)DIAZENE
C13H11N3O2S	025612-07-1	N1-(4-CYANOPHENYL)SULFANILAMIDE
C13H11N3O4	000964-79-4	N-(2,4-DINITROPHENYL)-M-TOLUIDINE
C13H11N3O4S2	029152-10-1	SUDOXICAM
C13H11N3OS	057237-97-5	1H-BENZIMIDAZOLE, 2-[(2-PYRIDINYLMETHYL)SULFINYL
C13H11NO	000093-98-1	BENZANILIDE

C13H11NO	000574-66-3	METHANONE, DIPHENYL-, OXIME
C13H11NO	000779-84-0	2-HYDROXYBENZALANILINE
C13H11NO	001017-24-9	BENZYL-4-PYRIDYL KETONE
C13H11NO	001137-41-3	4-BENZOYLANILINE
C13H11NO	001137-96-8	ALPHA-PHENYL-N-PHENYLNITRONE
C13H11NO	014548-30-2	4-(4-METHYLBENZOYL)-PYRIDINE
C13H11NO	058889-08-0	PROPANENITRILE, 3-(1-NAPHTHALENYLOXY)-
C13H11NO2	000087-17-2	SALICYLANILIDE
C13H11NO2	000091-40-7	N-PHENYL O-AMINOBENZOIC ACID
C13H11NO2	000094-44-0	3-PYRIDINECARBOXYLIC ACID, PHENYLMETHYL ESTER
C13H11NO2	004930-03-4	N,O-DIPHENYLCARBAMATE
C13H11NO2	014548-47-1	4-(4-METHOXYBENZOYL)PYRIDINE
C13H11NO2	031898-94-9	2-AZETIDINONE, 4-(2-NAPHTHALENYLOXY)-
C13H11NO2	051246-73-2	4-(2-METHOXYBENZOYL)-PYRIDINE
C13H11NO2	119005-13-9	2-AZETIDINONE, 4-(1-NAPHTHALENYLOXY)-
C13H11NO3	000133-11-9	PHENYL-4-AMINOSALICYLATE
C13H11NO3	002303-25-5	3-ME-4'-NO2-DIPHENYL ETHER
C13H11NO3	021003-78-1	3'-OH-N-PHENYL AMINOBENZOIC ACID
C13H11NO5	014698-29-4	OXOLINIC ACID
C13H11NOS	119005-22-0	2-AZETIDINONE, 4-(2-NAPHTHALENYLTHIO)-
C13H11NS	000636-04-4	BENZENECARBOTHIOAMIDE, N-PHENYL-
C13H12	000101-81-5	BENZENE, 1,1'-METHYLENEBIS-
C13H12	000643-58-3	2-METHYLBIPHENYL
C13H12	000643-93-6	3-METHYLBIPHENYL
C13H12	000644-08-6	4-METHYL-1,1'-BIPHENYL
C13H12Cl2N2	000101-14-4	4,4'-METHYLENEBIS(2-CHLOROANILINE)
C13H12Cl2O3	015725-29-8	ETHYLACETOACETATE,2,4-DICLBENZAL
C13H12Cl2O3	015725-30-1	ETHYLACETOACETATE,2,6-DICLBENZAL
C13H12Cl2O3	015725-31-2	ETHYLACETOACETATE,3,4-DICLBENZAL
C13H12Cl3N3O	028343-28-4	2,4,5-TRICL-C6H2NHN=C(CN)CO-T-BU
C13H12ClNO2S	138568-78-2	BENZENEACETIC ACID, 3-CHLORO-α-METHYL-4-(4-METHY
C13H12ClNO2S	138568-82-8	BENZENEACETIC ACID, 3-CHLORO-α-METHYL4-(5-METHYL
C13H12FNO2S	138568-70-4	BENZENEACETIC ACID, 3-FLUORO-α-METHYL-4-(4-METHY
C13H12FNO2S	138568-74-8	BENZENEACETIC ACID, 3-FLUORO-α-METHYL-4-(5-METHY
C13H12N2	000949-87-1	(4-METHYLPHENYL)PHENYLDIAZENE
C13H12N2	005350-57-2	METHANONE, DIPHENYL-, HYDRAZONE
C13H12N2	006676-90-0	(2-METHYLPHENYL)PHENYLDIAZENE
C13H12N2O	000102-07-8	DIPHENYLUREA, SYM
C13H12N2O	000442-51-3	9H-PYRIDO[3,4-B]INDOLE-, 7-METHOXY-1-METHYL-
C13H12N2O	000603-54-3	DIPHENYL UREA, UNSYM
C13H12N2O	000612-98-6	N-BENZYL-N-NITROSOANILINE
C13H12N2O	002396-60-3	(4-METHOXYPHENYL)PHENYLDIAZENE
C13H12N2O	006319-21-7	(2-METHOXYPHENYL)PHENYLDIAZENE
C13H12N2O	029418-37-9	O-(O-TOLYLAZO)PHENOL
C13H12N2O	029418-38-0	2-(PHENYLAZO)-M-CRESOL
C13H12N2O2	015979-82-5	3-METHYL-N-(4-NITROPHENYL)BENZENAMINE
C13H12N2O2	078508-44-8	(4-PHENOXYPHENYL)UREA
C13H12N2O3	000115-43-5	PHENALLYMAL
C13H12N2O3	056356-23-1	3-AMINOPHENYL 4-AMINOSALICYLATE
C13H12N2O3	056356-24-2	4-AMINOPHENYL 4-AMINOSALICYLATE
C13H12N2O3S	024454-46-4	BENZAMIDE, 4-[(4-AMINOPHENYL)SULFONYL]-
C13H12N2O3S	132483-50-2	BENZENEACETIC ACID, 4-[(4-(AMINOCARBONYL)-2-THIA
C13H12N2O5	062100-83-8	ACETAMIDE, N-(2,3-DIMETHYL-8-NITRO-4-OXO-4H-1-BE
C13H12N2O5S	051803-78-2	METHANESULFONAMIDE, N-(4-NITRO-2-PHENOXYPHENYL)-
C13H12N2O5S	074604-76-5	N[4(14BENZDIOXAN6YL)2THIAZOLYL]OXAMIC ACID
C13H12N2S	000102-08-9	THIOCARBANILIDE
C13H12N4	084884-30-0	[3,4'-BIPYRIDINE]-5-CARBONITRILE,6-(DIMETHYLAMIN
C13H12N4	120137-90-8	[3,4'-BIPYRIDINE]-5-CARBONITRILE,6-(ETHYLAMINO)-

C13H12N4O	108610-76-0	[3,4'-BIPYRIDINE]-5-CARBONITRILE,6-[(2-HYDROXYET
C13H12N4O2	062308-10-5	3-METHYL-4-((4-NITROPHENYL)AZO)BENZENAMINE
C13H12N4O2	088132-20-1	1(4BB-DICYANOVINYL)PH)-3MEO3ME UREA
C13H12N4O2S	077456-53-2	ISOMAZOLE,PYRIMIDYL ANALOG
C13H12N4O2S	120236-83-1	ISOMAZOLE,PYRIDAZINE ANALOG
C13H12N4O2S	120236-84-2	ISOMAZOLE,(5,6)PYRIDAZINE ANALOG
C13H12N4O2S	120236-85-3	ISOMAZOLE,PYRAZINE ANALOG
C13H12O	000091-01-0	BENZHYDROL
C13H12O	000101-53-1	4-HYDROXYDIPHENYLMETHANE
C13H12O	000613-37-6	4-METHOXYBIPHENYL
C13H12O	000946-80-5	BENZYL PHENYL ETHER
C13H12O2	000103-16-2	P-(BENZYLOXY) PHENOL
C13H12O2	000620-92-8	4,4'-DIHYDROXYDIPHENYLMETHANE
C13H12O2	001206-88-8	1-ETHYL AZULENE CARBOXYLATE
C13H12O2	001695-04-1	O-PHENOXYANISOLE
C13H12O2S	003112-88-7	BENZYL PHENYL SULFONE
C13H12O3	016563-41-0	PROPANOIC ACID, 3-(1-NAPHTHALENYLOXY)-
C13H13BrFNO2S	033803-13-3	4-ME-N-(4-SO2F-BENZYL)PYRIDINIUM BR
C13H13Cl2N3O3	036734-19-7	ROVRAL
C13H13Cl2NO4	133636-94-9	RH-1965 O-PH N-BUTYLENE CARBAMATE
C13H13Cl3N4O	130787-09-6	ACETAMIDE, 2,2,2-TRICHLORO-N-(2-CYCLOPENTYL-2H-B
C13H13ClO2	019411-80-4	ETHYLACETOACETATE,4-CHLOROBENZAL
C13H13ClO3	015725-22-1	ETHYLACETOACETATE,2-CHLOROBENZAL
C13H13ClO3	015725-23-2	ETHYLACETOACETATE,3-CHLOROBENZAL
C13H13FO3	015725-21-0	ETHYLACETOACETATE,3-FLUOROBENZAL
C13H13N	000103-32-2	N-BENZYLANILINE
C13H13N	000552-82-9	DIPHENYL METHYLAMINE
C13H13N	000620-84-8	4-METHYL-N-PHENYLBENZAMINE
C13H13N	001205-64-7	3-METHYL DIPHENYL AMINE
C13H13N3	000102-06-7	N,N'-DIPHENYLGUANIDINE
C13H13N3O	058481-07-5	2-BENZYLISONIAZID
C13H13N3O5S	052661-68-4	ACETAMIDE, N-[5-[(5-NITRO-2-FURANYL)METHYLENE]-4
C13H13N3O6S	010206-21-0	CEPHACETRILE
C13H13N5O6S	079510-48-8	METSULFURON-METHYL (PH5-7)
C13H13NaO3S	099287-38-4	1-SO3NA-3-PROPYL AZULENE
C13H13NaO3S	114482-97-2	1-SO3NA-7-ISOPROPYL AZULENE
C13H13NaO4S	099287-54-4	1-SO3NA-3-ET-4-METHOXY AZULENE
C13H13NO	000101-16-6	3-METHOXY-N-PHENYLBENZENAMINE
C13H13NO	001208-86-2	4-METHOXY-N-PHENYLBENZENAMINE
C13H13NO2S	004094-38-6	BENZENAMINE, 4-[(4-METHYLPHENYL)SULFONYL]-
C13H13NO2S	015979-81-4	4-METHYLSULFONYL-N-PHENYLBENZENAMINE
C13H13NO2S	132483-36-4	BENZENEACETIC ACID, α,3-DIMETHYL-4-(2-THIAZOLYL)
C13H13NO2S	132483-42-2	BENZENEACETIC ACID, α-METHYL-4-(4-METHYL-2-THIAZ
C13H13NO2S	132483-43-3	BENZENEACETIC ACID, α-METHYL-4-(5-METHYL-2-THIAZ
C13H13NO2S2	132483-58-0	BENZENEACETIC ACID, α-METHYL-3-(METHYLTHIO)-4-(2
C13H13NO3S	017078-72-7	BENZENAMINE, 4-[(4-METHOXYPHENYL)SULFONYL]-
C13H13NO3S	132483-57-9	BENZENEACETIC ACID, 3-METHOXY-α-METHYL-4-(2-THIA
C13H13NO3S	138568-65-7	BENZENEACETIC ACID, 4-[4-(HYDROXYMETHYL)-2-THIAZ
C13H13NO5	115178-74-0	2-AZETIDINECARBOXYLIC ACID, 1-[(BENZOYLOXY)ACETY
C13H13O4P	000115-89-9	DIPHENYLMETHYL PHOSPHATE
C13H14	000829-26-5	2,3,6-TRIMETHYLNAPHTHALENE
C13H14	002027-17-0	2-ISOPROPYLNAPHTHALENE
C13H14	002131-41-1	1,4,5-TRIMETHYLNAPHTHALENE
C13H14	002245-38-7	1,6,7-TRIMETHYLNAPHTHALENE
C13H14	006158-45-8	1-ISOPROPYLNAPHTHALENE
C13H14BrCl2N3O	116255-48-2	BROMUCONAZOLE
C13H14BrN	006324-18-1	B-PHENYLETHYLPYRIDINIUM BROMIDE
C13H14Cl2N4O	130787-08-5	ACETAMIDE, 2,2-DICHLORO-N-(2-CYCLOPENTYL-2H-BENZ

C13H14Cl2N4O2	083157-96-4	2,4-PYRIMIDINEDIAMINE, 5-[(2,6-DICHLORO-3,5-DIME
C13H14ClN3O	028317-91-1	3-CL-C6H4NHN=C(CN)CO-T-BU
C13H14F3N	110192-33-1	5,8-METHANO-5H-BENZOCYCLOHEPTEN-10-AMINE, 6,7,8,
C13H14F3N	110192-35-3	5,8-METHANO-5H-BENZOCYCLOHEPTEN-10-AMINE, 6,7,8,
C13H14F3N	110267-94-2	5,8-METHANO-5H-BENZOCYCLOHEPTEN-10-AMINE, 6,7,8,
C13H14F3N	110267-95-3	5,8-METHANO-5H-BENZOCYCLOHEPTEN-10-AMINE, 6,7,8,
C13H14F3N	110267-96-4	5,8-METHANO-5H-BENZOCYCLOHEPTEN-10-AMINE, 6,7,8,
C13H14F3N3O4	055283-68-6	BENZENAMINE, N-ETHYL-N-(2-METHYL-2-PROPENYL)-2,6
C13H14IN	026863-17-2	PYRIDINIUM, 1-METHYL-4-(4-METHYLPHENYL)-, IODIDE
C13H14N2	000101-77-9	DI-(P-AMINOPHENYL)METHANE
C13H14N2	000321-64-2	TACRINE
C13H14N2O	004425-23-4	AZEPINO[2,1-B]QUINAZOLIN-12(6H)-ONE, 7,8,9,10-TE
C13H14N2O2	089770-38-7	7ET-3CINNOLINCARBOXLIC ACID,ET ESTER
C13H14N2O2S	016803-95-5	N1-(4-METHYLPHENYL)SULFANILAMIDE
C13H14N2O2S	016803-96-6	N1-(2-METHYLPHENYL)SULFANILAMIDE
C13H14N2O2S	037157-91-8	124-BENZTHIADIAZIN-1-O2-3CYHEXEN-3-YL
C13H14N2O3	000115-38-8	MEPHOBARBITAL
C13H14N2O3S	019837-74-2	N1-(4-METHOXYPHENYL)SULFANILAMIDE
C13H14N2O3S	019837-84-4	N1-(2-METHOXYPHENYL)SULFANILAMIDE
C13H14N2O4S2	000067-99-2	GLIOTOXIN
C13H14N2O6	118247-02-2	BENZOIC ACID, 2-(ACETYLOXY)-, 2-[(2-AMINO-2-OXOE
C13H14N4O2	075889-44-0	24-NH2PYRIMIDINE,5(2ME-45OCH2O)BENZYL
C13H14N6O2	117121-38-7	UREA, N,N-DIMETHYL-N'-[4-[[1-(2-PROPYNYL)-1H-TET
C13H14N6O2	117144-73-7	UREA, N,N-DIMETHYL-N'-[4-[[2-(2-PROPYNYL)-2H-TET
C13H14O2	015818-09-4	4-METHYLBENZAL ACETYLACETONE
C13H14O2	019411-71-3	ACETYLACETONE,2-METHYLBENZAL
C13H14O3	000620-80-4	ETHYLACETOACETATE, BENZAL
C13H14O3	015725-16-3	2-METHOXYBENZAL ACETYLACETONE
C13H14O3	015725-17-4	4-METHOXYBENZAL ACETYLACETONE
C13H14O3	015818-10-7	3-METHOXYBENZAL ACETYLACETONE
C13H14O6	050785-24-5	BENZOIC ACID, 2-(ACETYLOXY)-, 2-ETHOXY-2-OXOETHY
C13H15BrClNO	097871-68-6	P-CL-CINNAMAMIDE,B-BR,N-SEC-BUTYL
C13H15BrN4O2	024849-83-0	2,4-PYRIMIDINEDIAMINE, 5-[(2-BROMO-4,5-DIMETHOXY
C13H15BrN4O2	056518-41-3	24-NH2PYRIMIDINE,5(3,5MEO-4-BR)BENZYL
C13H15Cl2N3O3S2	082774-16-1	N(N"ME(24DICLPH)CARBAMYLTHIO)METHOMYL
C13H15Cl2N3O3S2	082784-14-3	N(N"ME(26DICLPH)CARBAMYLTHIO)METHOMYL
C13H15Cl2NO	015907-85-4	CYCLOHEXANECARBOXAMIDE, N-(3,4-DICHLOROPHENYL)-
C13H15Cl2NO	097871-67-5	P-CL-CINNAMAMIDE,B-CL,N-SEC-BUTYL
C13H15Cl2NO4	014433-36-4	P-ACETYL AMPHENICOL
C13H15ClN4O	130787-07-4	ACETAMIDE, 2-CHLORO-N-(2-CYCLOPENTYL-2H-BENZOTRI
C13H15ClN4O2	020285-74-9	24-NH2PYRIMIDINE,5(4,5MEO-3-CL)BENZYL
C13H15ClN4O2	056518-45-7	24-NH2PYRIMIDINE,5(3,5MEO-4-CL)BENZYL
C13H15ClN4O2	083166-74-9	24-NH2PYRIMIDINE,5(3,5MEO-2-CL)BENZYL
C13H15ClO4	099414-56-9	BUTANOIC ACID, 4-(3-CHLORO-4-METHOXYBENZOYL)-3-M
C13H15F3N2O5S	081792-70-3	BUTANAMIDE, 2-ETHYL-N-[[3-NITRO-5-(TRIFLUOROMETH
C13H15F3O6	020772-25-2	3-CF3 PHENYL GLUCOPYRANOSIDE
C13H15IN4O2	056518-58-2	2,4-NH2PYRIMIDINE,5(3,5MEO-4-I)BENZYL
C13H15IN4O2	083166-73-8	2,4-NH2PYRIMIDINE,5(3,4MEO-5-I)BENZYL
C13H15N3	004774-24-7	QUINOLINE, 2-(1-PIPERAZINYL)-
C13H15N3	041229-10-1	4-N-PIPERIDINYLQUINAZOLINE
C13H15N3O	131229-38-4	IMIDAZO[5,1-B]QUINAZOLIN-9(2H)-ONE, 1,3-DIHYDRO-
C13H15N3O2	000087-47-8	PYROLAN
C13H15N3O2	010346-41-5	TRYPTOPHAN-AMIDE,N-ACETYL
C13H15N3O2	031992-01-5	2,4(1H,3H)-PYRIMIDINEDIONE, 5-(DIMETHYLAMINO)-6-
C13H15N3O2	131229-41-9	IMIDAZO[5,1-B]QUINAZOLIN-9(2H)-ONE, 1,3-DIHYDRO-
C13H15N3O3	081334-34-1	IMAZAPYR
C13H15N3O3	089145-44-8	CARBAMIC ACID, [5-(4-FORMYLPHENYL)-4,5-DIHYDRO-1
C13H15N5O	087469-61-2	1-METHYLISOGUANINE,9[BENZYL]-

C13H15N5O	087469-62-3	1-METHYLISOGUANINE,7[BENZYL]-
C13H15N5O2	080407-58-5	46-NH2-(2-OCH2-CON)BENZYLPYRIMIDINE
C13H15N7O2	117121-50-3	UREA, N'-[4-[[2-(2-CYANOETHYL)-2H-TETRAZOL-5-YL]
C13H15NO	002094-99-7	1-(1-ISOCYANATO-1-METHYLETHYL)-3-(1-METHYLETHE*)
C13H15NO	002889-58-9	1-(1-ISOCYANATO-1-METHYLETHYL)-4-(1-METHYLETHE*)
C13H15NO2	000077-21-4	2-ETHYL-2-PHENYLGLUTERIMIDE
C13H15NO2	024691-76-7	PYRACARBOLID
C13H15NO3	096327-63-8	1H-PYRROLIZINE-1-CARBOXYLIC ACID, 5-(CYCLOBUTYLC
C13H15NO4	106231-68-9	MORPHOLINE, 4-[(BENZOYLOXY)ACETYL]-
C13H15NO5	118247-01-1	BENZOIC ACID, 2-(ACETYLOXY)-, 2-(ETHYLAMINO)-2-O
C13H15NO5	118247-04-4	BENZOIC ACID, 2-(ACETYLOXY)-, 2-(DIMETHYLAMINO)-
C13H16BrNO	097871-45-9	TRANS-CINNAMAMIDE,A-BR,N-SEC-BUTYL
C13H16BrNO	097871-46-0	CIS-CINNAMAMIDE,A-BR,N-SEC-BUTYL
C13H16Cl2N2O	022010-25-9	ACETAMIDE, N-(3,4-DICHLOROPHENYL)-2-(1-PIPERIDIN
C13H16Cl3NO4	064700-56-7	TRICLOPYR ESTER
C13H16ClN	090047-59-9	3-ALLYL-6-CHLOROBENZAZEPINE
C13H16ClN3O3S2	082774-06-9	N(N"-ME(4-CLPH)CARBAMYLTHIO)METHOMYL
C13H16ClN3O3S2	082774-07-0	N(N"-ME(2-CLPH)CARBAMYLTHIO)METHOMYL
C13H16ClN3O3S2	082774-14-9	N(N"-ME(3-CLPH)CARBAMYLTHIO)METHOMYL
C13H16ClNO	097871-47-1	TRANS-CINNAMAMIDE,A-CL,N-SEC-BUTYL
C13H16ClNO	097871-48-2	CIS-CINNAMAMIDE,A-CL,N-SEC-BUTYL
C13H16ClNO4	060145-78-0	ET-N-CHLOROACETYL-N-(2MEOPH)GLYCINATE
C13H16F3N3O4	001582-09-8	TRIFLURALIN
C13H16F3N3O4	001861-40-1	BENEFIN
C13H16FN3O2	089145-29-9	CARBAMIC ACID, [5-(3-FLUOROPHENYL)-4,5-DIHYDRO-1
C13H16FN3O2	089145-31-3	CARBAMIC ACID, [5-(4-FLUOROPHENYL)-4,5-DIHYDRO-1
C13H16FNO	097871-49-3	TRANS-CINNAMAMIDE,A-F,N-SEC-BUTYL
C13H16N2O2	098154-94-0	CARBAMIC ACID, DIMETHYL-, 2-(1H-INDOL-3-YL)ETHYL
C13H16N2O2S	037157-88-3	124-BENZTHIADIAZIN-1-O2-3-CYHEXYL
C13H16N2O3	094779-66-5	BARBITURIC ACID,N-ME,5-ALLYL,5-CYPENTENYL
C13H16N2O4	073632-81-2	BENZENEACETIC ACID, α-[[(AMINOCARBONYL)AMINO]CAR
C13H16N4	102408-27-5	IMIDAZO[45F]QUINOLINE,3-METHYL-2-DIEMTHYLAMINO
C13H16N4O	073264-18-3	2,4-DINH2PYRIMIDINE,5(4MEO-3ME)BENZYL
C13H16N4O	073264-19-4	2,4-DINH2PYRIMIDINE,5(3MEO-4ME)BENZYL
C13H16N4O	130787-06-3	ACETAMIDE, N-(2-CYCLOPENTYL-2H-BENZOTRIAZOL-5-YL
C13H16N4O2	005355-16-8	2,4-NH2-5(3,4-DIMEOBZ)PYRIMIDINE
C13H16N4O2	020344-69-8	2,4-NH2PYRIMIDINE,5(3MEO-5-MEO)BENZYL
C13H16N4O2	141605-15-4	1,2,4-TRIAZIN-5(4H)-ONE, 6-(3-ETHOXYPHENYL)-3-ME
C13H16N4O2S	073498-01-8	N1-(5-IPR-2-PYRIMIDYL)SULFANILIDE
C13H16N4O3	021253-58-7	2,4-NH2PYRIMIDINE,5(35MEO-4-OH)BENZYL
C13H16N4O3S	007756-44-7	5ET-6MEO-4SULFANILAMIDOPYRIMIDINE
C13H16N4O4S	005018-56-4	5MEO-6ETO-4-SULFANILAMIDOPYRIMIDIN
C13H16N4O4S	005532-46-7	5-ETO-6-MEO-4-SULFANILAMIDOPYRIMIDIN
C13H16N4OS	068885-48-3	2,4-NH2PYRIMIDINE,5(3MEO-4-MES)BENZYL
C13H16N6O2	117121-47-8	UREA, N,N-DIMETHYL-N'-[4-[[2-(2-PROPENYL)-2H-TET
C13H16O2	025209-50-1	BENZENEACETIC ACID, 4-CYCLOPENTYL-
C13H16O3	000620-79-1	BENZYLACETOACETIC ACID, ETHYL ESTER
C13H16O4	016926-87-7	2-PROPENOIC ACID, 2-METHYL-, 2-HYDROXY-3-PHENOXY
C13H16O4	022020-28-6	2-PROPENOIC ACID, 3-(3,4-DIHYDROXYPHENYL)-, BUTY
C13H16O5	051264-74-5	PROPANOIC ACID, 2-(4-FORMYL-2-METHOXYPHENOXY)ETH
C13H17BrN2O	053316-92-0	N(M-BRPH)-3-N'-PIPERIDINOACETAMIDE
C13H17BrN2O	058479-86-0	N(P-BRPH)3-N'-PIPERIDINOACETAMIDE
C13H17BrN2O	058479-89-3	N(O-BRPH)3-N'-PIPERIDINOACETAMIDE
C13H17Cl2NO	001867-66-9	KETAMINE
C13H17ClN2O	027471-82-5	N-(P-CL PH)-3-N'-PIPERIDINOACETAMIDE
C13H17ClN2O	038367-19-0	N(M-CLPH)-3-N'-PIPERIDINOACETAMIDE
C13H17ClN2O	038367-23-6	N(O-CLPH)-3-N'-PIPERIDINOACETAMIDE
C13H17ClN2O3S	000963-03-1	BENZENESULFONAMIDE, 4-CHLORO-N-[(CYCLOHEXYLAMINO

C13H17ClO3	010443-70-6	MCPB-ETHYL
C13H17F3N2O3S	097141-33-8	BUTANAMIDE, N-[2-(AMINOSULFONYL)-5-(TRIFLUOROMET
C13H17FN2O	037163-41-0	N(P-FPH)-3-N'-PIPERIDINOACETAMIDE
C13H17FN2O	058479-88-2	N-(O-FPH)3-N'-PIPERIDINOACETAMIDE
C13H17FN2O	058479-90-6	N-(M-FPH)3-N'-PIPERIDINOACETAMIDE
C13H17IN2O	058479-87-1	N-(P-IPH)3-N'-PIPERIDINOACETAMIDE
C13H17IN2O	058479-91-7	N-(M-IPH)3-N'-PIPERIDINOACETAMIDE
C13H17N	013314-63-1	PYRIDINE, 1-ETHYL-1,2,3,6-TETRAHYDRO-4-PHENYL-
C13H17N	018883-05-1	BENZOBICYCLO(222)OCTENE,2-EXO-MEAMINO
C13H17N	018883-06-2	BENZOBICYCLO(222)OCTENE,2ENDO-MEAMINO
C13H17N	086943-78-4	BENZOBICYCLO(221)HEPTENE,2EX-ETAMINO
C13H17N	086992-68-9	BENZOBICYCLO(221)HEPTENE,2EN-ETAMINO
C13H17N	102417-86-7	PYRIDINE, 1,2,3,6-TETRAHYDRO-1-METHYL-4-(2-METHY
C13H17N	147128-62-9	9AH-FLUOREN-9A-AMINE, 1,2,3,4,4A,9-HEXAHYDRO-, C
C13H17N2O3PS	024017-47-8	TRIAZOPHOS
C13H17N3O	000058-15-1	AMINOPYRINE
C13H17N3O2	069218-49-1	1-(4-CARBETHOXYBUTYL)BENZOTRIAZOLE
C13H17N3O2	115054-82-5	2H-BENZOTRIAZOLE-2-ACETIC ACID, α-PROPYL-, ETHYL
C13H17N3O2	115054-86-9	1H-BENZOTRIAZOLE-1-ACETIC ACID, α-PROPYL-, ETHYL
C13H17N3O3	029701-43-7	GLYCINAMIDE, N-ACETYL-L-PHENYLALANYL-
C13H17N3O3	034017-16-8	L-PHENYLALANINAMIDE, N-ACETYLGLYCYL-
C13H17N3O3	035204-11-6	N-(O-NO2PH)-3-N'-PIPERIDINOACETAMIDE
C13H17N3O3	035763-43-0	N-(M-NO2PH)-3-N'-PIPERIDINOACETAMIDE
C13H17N3O3	038367-22-5	N(P-NO2PH)-3-N'-PIPERIDINOACETAMIDE
C13H17N3O3S	075852-01-6	PROPANAMIDE, 3-CYANO-N-[4-[(DIMETHYLAMINO)SULFON
C13H17N3O3S2	075852-08-3	PROPANAMIDE, N-[4-[(DIMETHYLAMINO)SULFONYL]PHENY
C13H17N3O3S2	082774-01-4	N(N''-ME(PHENYL)CARBAMYLTHIO)METHOMYL
C13H17N5	069945-51-3	2,4-NH2PYRIMIDIN,5(4-DIMEAMINO)BENZYL
C13H17N5O	070579-34-9	4,6-NH2 2,2-DIME-1(3-ACPH)S-TRIAZENE
C13H17N5O2	056066-19-4	2,4-NH2PYRIMIDIN,5(3,5MEO-4NH2)BENZYL
C13H17NO	003626-62-8	ETHANONE, 2-PHENYL-1-(1-PIPERIDINYL)-
C13H17NO	010342-85-5	4'-PIPERIDINOACETOPHENONE
C13H17NO	013290-48-7	PIPERIDINE, 1-(3-METHYLBENZOYL)-
C13H17NO	013707-23-8	PIPERIDINE, 1-(4-METHYLBENZOYL)-
C13H17NO	097871-50-6	N-SEC-BUTYLCINNAMAMIDE
C13H17NO	129865-48-1	1(2H)-PYRIDINEETHANOL, 3,6-DIHYDRO-4-PHENYL-
C13H17NO2	036405-75-1	N-PHENOXYACETYLPIPERIDINE
C13H17NO2	057700-94-4	PIPERIDINE, 1-(4-METHOXYBENZOYL)-
C13H17NO3	023742-02-1	A,A-DIET-PHENYLACETAMIDE,3,4-DIOXYME
C13H17NO3	064649-63-4	BENZOIC ACID, 2-(DIETHYLAMINO)-2-OXOETHYL ESTER
C13H17NO3	106231-52-1	ACETAMIDE, 2-(BENZOYLOXY)-N-(1,1-DIMETHYLETHYL)-
C13H17NO3	115178-78-4	BUTANAMIDE, 4-(BENZOYLOXY)-N,N-DIMETHYL-
C13H17NO3	115193-28-7	ACETAMIDE, 2-(BENZOYLOXY)-N-BUTYL-
C13H17NO4	000136-45-8	DIPROPYL PYRIDINE-2,5-DICARBOXYLATE
C13H17NO4	106231-60-1	ACETAMIDE, 2-(BENZOYLOXY)-N-ETHYL-N-(2-HYDROXYET
C13H17NO4PS	018854-01-8	ISOXATHION
C13H17NO5	106231-61-2	ACETAMIDE, 2-(BENZOYLOXY)-N,N-BIS(2-HYDROXYETHYL
C13H17NO6	115193-31-2	ACETAMIDE, 2-(BENZOYLOXY)-N-[2-HYDROXY-1,1-BIS(H
C13H18Cl2N2O	071225-61-1	BENZAMIDE,N(DIETAMINOET),3,5-DICL
C13H18Cl2N2O2	000148-82-3	4-(BIS(2-CHLOROETHYL)AMINO)-L-PHENYLALANINE
C13H18ClN	023007-86-5	2,6-METHANO-3-BENZAZOCINE, 1,2,3,4,5,6-HEXAHYDRO
C13H18ClN3O2	033021-93-1	1-NO-1-(2CLET)-3-(2(2BENZYL)PR)UREA
C13H18ClN3O9	054138-85-1	UREA, N-(2-CHLOROETHYL)-N-NITROSO-N'-(2,3,4-TRI-
C13H18ClNO	002307-68-8	SOLAN
C13H18ClNO	007287-36-7	MONALIDE (POTABLAN)
C13H18ClNO	061151-66-4	2-(M-CL-PHENYL)-4-IPR-MORPHOLINE
C13H18ClNO	070657-63-5	BENZAMIDE, 2-CHLORO-N,N-BIS(1-METHYLETHYL)-
C13H18F3N3O4S2	001766-91-2	2H-1,2,4-BENZOTHIADIAZINE-7-SULFONAMIDE, 3,4-DIH

C13H18FN3O2	111041-48-6	IMIDODICARBONIC DIAMIDE, N-[2-FLUORO-5-(1-METHYL
C13H18FN3O3	111041-57-7	IMIDODICARBONIC DIAMIDE, N-[2-FLUORO-5-(1-METHYL
C13H18FN3O4S	121788-16-7	IMIDODICARBONIC DIAMIDE, N-[2-FLUORO-5-[(1-METHY
C13H18N2	069380-11-6	1-(1-IMIDAZOLYL)ADAMANTANE
C13H18N2O	000886-59-9	3-PHENYL-1-CYCLOHEXYLUREA
C13H18N2O	004671-97-0	N-PHENYL-3-N'-PIPERIDINOACETAMIDE
C13H18N2O	042175-93-9	1-PYRROLIDINEACETAMIDE, N-(PHENYLMETHYL)-
C13H18N2O2	002164-08-1	LENACIL
C13H18N2O2	072336-20-0	4-MORPHOLINEACETAMIDE, N-(PHENYLMETHYL)-
C13H18N2O3	000509-86-4	HEPTABARBITAL
C13H18N2O3	000726-78-3	N-METHYLCYCLOBARBITAL
C13H18N2O4	084640-38-0	P-(OCONME2)BENZYL N,N-DIME CARBAM
C13H18N2O5	018048-95-8	DIMETHYLAMPHENICOL
C13H18N2O5	023885-69-0	ETHYLAMPHENICOL
C13H18N2O5S	081793-04-6	BUTANAMIDE, 2-ETHYL-N-[(4-METHYL-2-NITROPHENYL)S
C13H18N2O5S	083870-89-7	BENZENESULFONYL-L-(N'-ET)GLUTAMINE
C13H18N3O4P	067410-49-5	3,5-DIOXA-6-AZA-4-PHOSPHAOCT-6-ENE-8-NITRILE, 4-
C13H18N6	061544-27-2	2,4-NH2PYRIMIDINE,5(3,5-NHME)BENZYL
C13H18N6O2	117121-35-4	UREA, N,N-DIMETHYL-N'-[4-[[1-(2-PROPENYL)-1H-TET
C13H18N6O2	117121-43-4	UREA, N,N-DIMETHYL-N'-[4-[[2-(1-METHYLETHYL)-2H-
C13H18O2	014377-21-0	BENZENEACETIC ACID, 4-PENTYL-
C13H18O2	015687-27-1	IBUPROFEN
C13H18O2	026114-14-7	BENZENEACETIC ACID, 4-(1,1-DIMETHYLPROPYL)-
C13H18O2	112473-77-5	BUTANOIC ACID, 2-ETHYL-3-METHYL-, PHENYL ESTER
C13H18O3	001083-27-8	HEXYL P-HYDROXYBENZOATE
C13H18O3	082412-82-6	BENZENEACETIC ACID, 4-BUTYL-α-HYDROXY-α-METHYL-
C13H18O5	001087-26-9	HEXYL GALLATE
C13H18O5	023022-51-7	ETHANOL, 2-[2-[2-(BENZOYLOXY)ETHOXY]ETHOXY]-
C13H18O5S	026225-79-6	ETHOFUMESATE
C13H18O5S	063407-53-4	1-BENZYLTHIO-B-GALACTOPYRANOSIDE
C13H18O6	004304-12-5	BENZYL GLUCOPYRANOSIDE
C13H18O6	006092-25-7	GLUCOPYRANOSIDE,3-METHYLPHENYL
C13H18O6	007234-31-3	GLUCOPYRANOSIDE,2-METHYLPHENYL
C13H18O6	020274-94-6	GLUCOPYRANOSIDE,4-METHYLPHENYL
C13H18O7	000138-52-3	SALICIN
C13H18O7	006032-32-2	4-METHOXYPHENYL GLUCOPYRANOSIDE
C13H18O7	006092-24-6	2-MEO PHENYL GLUCOPYRANOSIDE
C13H18O7	014062-61-4	3-MEO PHENYL GLUCOPYRANOSIDE
C13H19ClN2O2	000133-16-4	CHLOROPROCAINE
C13H19ClN2O2	054922-63-3	N'-IPR-2-ME-2(4CL-PHO)PROPIONICACIDHYDRAZIDE
C13H19ClN2O2	101820-65-9	BENZAMIDE, 2-[(2-BUTOXYETHYL)AMINO]-5-CHLORO-
C13H19ClNO3PS2	064249-01-0	ANILOFOS
C13H19N3	004751-48-8	2-(2,6-DIETPHENYLIMINO)IMIDAZOLIDINE
C13H19N3O	036371-18-3	PIPERAZINE-2-CARBOXANILIDE,2',6'-DIME
C13H19N3O2	069218-54-8	2-(4-CARBETHOXYBUTYL)BENZOTRIAZOLE
C13H19N3O2	111075-87-7	1-ACETYL-4-(4-T-BU PHENYL) SEMICARBAZIDE
C13H19N3O3S	001443-94-3	UREA, N-(1-PIPERIDINYL)-N'-[(4-METHYLPHENYL)SULF
C13H19N3O4	001918-08-7	DIPROPALIN
C13H19N3O4	040487-42-1	PENDIMETHALIN
C13H19N3O4S	075851-97-7	BUTANEDIAMIDE, N-[4-[(DIMETHYLAMINO)SULFONYL]PHE
C13H19N3O4S2	023141-86-8	3-CYCLOPENTYLMETHYLHYDROTHIAZIDE
C13H19N3O5S2	001404-64-4	SPARSOMYCIN
C13H19N3O6S	004726-14-1	NITRALIN
C13H19N3OS	091833-49-7	PYRIDO[3,2-F]-1,4-OXAZEPINE-5(2H)-THIONE, 2-[2-(
C13H19N5O	120182-07-2	2-OXAZOLAMINE, 4,5-DIHYDRO-5-[[4-(2-PYRIDINYL)-1
C13H19NO	006577-49-7	B-PHENYL-B-MEO-N-ET PYRROLIDINE
C13H19NO	020383-28-2	BENZAMIDE, N,N-BIS(1-METHYLETHYL)-
C13H19NO	023222-62-0	2-PHENYL-4-I-PROPYLMORPHOLINE

C13H19NO2	002798-54-1	N-ME-3-ME-6-T-BU PHENYLCARBAMATE
C13H19NO2	003688-82-2	BENZAMIDE, 2-ETHOXY-N,N-DIETHYL-
C13H19NO2	013476-55-6	BENZOIC ACID, 4-AMINO-, HEXYL ESTER
C13H19NO2	018659-33-1	N-ME-3-ME-5-T-BU PHENYLCARBAMATE
C13H19NO2	055379-70-9	2-S-BUTYLPHENYLDIMETHYLCARBAMATE
C13H19NO2	098183-17-6	CARBAMIC ACID, DIETHYL-, 2-PHENYLETHYL ESTER
C13H19NO2S	000115-31-1	THANITE
C13H19NO2S	062850-32-2	FENOTHIOCARB
C13H19NO3	006292-90-6	L-TYROSINE,BUTYL ESTER
C13H19NO3	024397-14-6	2-HEXYLOXY-4-AMINOBENZOIC ACID
C13H19NO3	084640-29-9	M-I-PRO BENZYL N,N-DIME CARBAMATE
C13H19NO3	098378-56-4	BENZAMIDE, N-HEXYL-3,4-DIHYDROXY-
C13H19NO3	101126-55-0	ACETIC ACID, 4-[2-(DIMETHYLAMINO)ETHOXY]-2-METHY
C13H19NO4	087130-19-6	CARBAMIC ACID, (3,4-DIETHOXYPHENYL)-, ETHYL ESTE
C13H19NO4S	000057-66-9	PROBENECID
C13H19NO4S	135832-51-8	ACETIC ACID, 5-[4-(AMINOSULFONYL)PHENYL]PENTYL E
C13H19NO4Si	051466-75-2	1-(4-METHYLPHENOXY)SILATRANE
C13H19NO4Si	071357-27-2	1-(3-METHYLPHENOXY)SILATRANE
C13H19NO5	099877-04-0	L-TYROSINE, 3-HYDROXY-, 2-METHOXY-1-METHYLETHYL
C13H20	001078-71-3	1-PHENYLHEPTANE
C13H20ClN3O	000891-60-1	BENZAMIDE, 4-AMINO-3-CHLORO-N-[2-(DIETHYLAMINO)E
C13H20ClN3O2	014039-10-2	3(ADAMANTYL)-1(2CLET)-1-NO UREA
C13H20FN3O3	064098-85-7	1-PYRIMIDINECARBOXAMIDE, 5-FLUORO-1,3-DIHYDRO-N-
C13H20N2O	000721-50-6	PRILOCAINE
C13H20N2O	003690-53-7	N-(2-DIETHYLAMINOETHYL)-BENZAMIDE
C13H20N2O	019216-39-8	PROPANAMIDE, 2-(DIMETHYLAMINO)-N-(2,6-DIMETHYLPH
C13H20N2O	032745-69-0	1,1-DIME-3-(P-T-BUTYLPHENYL)UREA
C13H20N2O	072336-16-4	ACETAMIDE, 2-(DIETHYLAMINO)-N-(PHENYLMETHYL)-
C13H20N2O	088132-40-5	1,1-DIMETHYL-(3-BUTYL)PHENYLUREA
C13H20N2O	102089-66-7	PENTANAMIDE, 5-AMINO-N-(2,6-DIMETHYLPHENYL)-
C13H20N2O	102089-68-9	PENTANAMIDE, 4-AMINO-N-(2,6-DIMETHYLPHENYL)-
C13H20N2O2	000059-46-1	PROCAINE
C13H20N2O2	051963-53-2	N'-IPR-2ME-PHENOXYPROPIONIC HYDRAZIDE
C13H20N2O2	070477-14-4	1-(3-T-BUPH)-3-MEO-3-METHYL UREA
C13H20N2O2	084846-21-9	BICYCLO[2.2.1]HEPT-5-ENE-2,3-DICARBOXAMIDE
C13H20N2O2	085819-14-3	O-MEO BENZAMIDE,N-(3-DIMEAMINOPR)
C13H20N2O2	098154-98-4	CARBAMIC ACID, DIMETHYL-, 3-(2-PYRIDINYL)PROPYL
C13H20N2O2	101820-64-8	BENZAMIDE, 2-[(2-BUTOXYETHYL)AMINO]-
C13H20N2O3S	059777-65-0	4-SULFAMYLBENZAMIDE,N-HEXYL
C13H20N2O3S	075852-14-1	NN-DIME-4(N-ME-N-BUTROYL)BZ SULFONAMIDE
C13H20N2O3S	081792-79-2	BUTANAMIDE, N-[(2-AMINO-4-METHYLPHENYL)SULFONYL]
C13H20N2O3S	097141-32-7	BUTANAMIDE, N-[2-(AMINOSULFONYL)-5-METHYLPHENYL]
C13H20N2O3S2	075852-12-9	NN-DIME-4(N-ME-N(2MESPROPIONYL)BZSO2N
C13H20N2O4	073632-84-5	PROPANAMIDE, 2-(1-CYCLOHEXENYL)-N-[(METHYLAMINO)
C13H20N2O4	088280-81-3	4-PYRIMIDINECARBOXYLIC ACID, 1,2,3,6-TETRAHYDRO-
C13H20N2O4S	075852-05-0	NN-DIME-4(N-ME-N(2MEOPROPIONYL)BZSO2N
C13H20N2O4S	081792-81-6	BUTANAMIDE, N-[(2-AMINO-4-METHOXYPHENYL)SULFONYL
C13H20N2O4S	097141-34-9	BUTANAMIDE, N-[2-(AMINOSULFONYL)-5-METHOXYPHENYL
C13H20N2O4S2	075851-96-6	NN-DIME4(N-ME-N(2MESOPROPIONYL)BZSO2N
C13H20N4O	059708-22-4	1-(4-CARBAMOYLPHENYL)-3-METHYL-3-PENTYLTRIAZENE
C13H20N6	077314-59-1	2-PYRIDINAMINE, 4-(5-AMINO-1H-1,2,4-BENZOTRIAZOL
C13H20O	000127-41-3	ALPHA-IONONE
C13H20O	007597-97-9	2-(TERT-BUTYL)-4-ISOPROPYLPHENOL
C13H20O2	013037-86-0	4-(HEPTYLOXY)PHENOL
C13H20O4	001985-51-9	NEOPENTYLGLYCOL DIMETHACRYLATE
C13H21BrN2O3	077317-78-3	3-OCO-5-BR-6-ME-PYRIMIDIN-2,4-DIONE
C13H21ClN2	002519-75-7	N-DIETHYLAMINOETANILINE,3-CL-4-ME
C13H21ClN2O2	000051-05-8	PROCAINE HYDROCHLORIDE

C13H21N	000585-48-8	2,6-DI-T-BUTYLPYRIDINE
C13H21N	084952-62-5	PHENETHYLAMINE,N-ME-A-ISOBUTYL
C13H21N2O4PS	036335-67-8	BUTAMIFOS
C13H21N3O	000051-06-9	PROCAINAMIDE
C13H21N3O	120277-94-3	7-AZABICYCLO[4.1.0]HEPTANE, 7-[2-HYDROXY-3-(1H-I
C13H21N3O	120277-95-4	8-AZABICYCLO[5.1.0]OCTANE, 8-[2-HYDROXY-3-(1H-IM
C13H21N5O3	098204-10-5	ALLOPURINOL,1,N,N-DIETHYLALANYLOXYMETHYL-
C13H21NO2	002933-44-0	DOBEROL
C13H21NO2	019343-24-9	2-PROPANOL, 1-[(1-METHYLETHYL)AMINO]-3-(PHENYLME
C13H21NO2	022004-32-6	2,5-DIMETHOXY-4-ETHYLAMPHETAMINE
C13H21NO2	064980-40-1	2-PROPANOL, 1-[(1,1-DIMETHYLETHYL)AMINO]-3-PHENO
C13H21NO2	070873-69-7	ADAMANTYLALANINE
C13H21NO2	072732-50-4	MOXISYLYTE,N-DESMETHYL,DESACETYL
C13H21NO3	005741-22-0	2-PROPANOL, 1-(2-METHOXYPHENOXY)-3-[(1-METHYLETH
C13H21NO3	039201-78-0	3,5-DIMEO-4-PROPOXYPHENETHYLAMINE
C13H21NO3	061367-70-2	3,5-DIMEO-4-I-PROPOXYPHENETHYLAMINE
C13H21O3PS	026087-47-8	IBP (IPROBENFOS)
C13H21O4PS	007292-16-2	PROPAPHOS
C13H22BrN	002534-66-9	OCTYLPYRIDINIUM BROMIDE
C13H22IN	010291-06-2	N-OCTYLPYRIDINIUM IODIDE
C13H22IN	088477-43-4	3,4-DIME PYRIDINIUM IODIDE,N-HEXYL
C13H22IN	088477-54-7	3,5-DIME PYRIDINIUM IODIDE,N-HEXYL
C13H22N2O	002163-79-3	3-(HEXAHYDRO-4,7-METHANOINDAN-5-YL)-1,1-DIMETHY*
C13H22N2O	018530-56-8	NOREA
C13H22N2O3	060784-70-5	BARBITURIC ACID,5-ET,5-HEPTYL
C13H22N2O3	096914-10-2	1H-IMIDAZOLE-5-BUTANOIC ACID, α-ETHYL-ß-(HYDROXY
C13H22N4O2	001028-33-7	1-HEXYL-3,7-DIMETHYL XANTHINE
C13H22N4O3S	066357-35-5	ZANTAC (RANITIDINE HYDROCHLORIDE)
C13H22NO3PS	022224-92-6	FENAMIPHOS
C13H23N	015210-60-3	1-AMINOADAMANTANE,3,5,7-TRIMETHYL
C13H23N	080121-61-5	1-AMINOADAMANTANE,3-PROPYL
C13H23N3O4S2	086569-90-6	N(N"-ME(CYHEXO)CARBAMYLTHIO)METHOMYL
C13H23NO	072299-07-1	PIPERIDINE, 1-(CYCLOHEXYLACETYL)-
C13H24N2S	001212-29-9	THIOUREA, N,N'-DICYCLOHEXYL-
C13H24N3O3PS	023505-41-1	PIRIMIPHOS ETHYL
C13H24N4O3S	026839-75-8	TIMOLOL
C13H24N4O3S	041483-43-6	BUPIRIMATE
C13H24N4O4	132765-92-5	L-LEUCINAMIDE, N -ACETYL-L-GLUTAMINYL-
C13H24N4O4	132766-01-9	L-ALANINAMIDE, N-ACETYL-L-VALYL-L-ALANYL-
C13H24N6O	056010-86-7	1H-IMIDAZOLE-4-CARBOXAMIDE, 5-(3-METHYL-3-OCTYL-
C13H24O2	001330-61-6	ISODECYL ACRYLATE
C13H24O4	000624-17-9	DIETHYL AZELATE
C13H24O4	004372-29-6	DECYLPROPANEDIOIC ACID
C13H25N	007560-83-0	CYCLOHEXANAMINE, N-CYCLOHEXYL-N-METHYL-
C13H25N3O3	132765-81-2	L-VALINAMIDE, N-ACETYL-L-LEUCYL-
C13H25N3O3	132765-82-3	L-VALINAMIDE, N-ACETYL-L-ISOLEUCYL-
C13H25N3O4S2	066381-49-5	N-(N"-ME-HXOCARBAMYLTHIO)METHOMYL
C13H25N4O2S	103499-09-8	1,2,4-OXADIAZOLE-5-(DIETHYLAMINOETHYLTHIO)HYDROX
C13H26	002437-56-1	1-TRIDECENE
C13H26	005617-41-4	HEPTYLCYCLOHEXANE
C13H26N2O2	021270-93-9	4-BUTYRAMID-2266-TEME-PIPERDIN-N-OXID
C13H26N3O2PS	051526-59-1	1-PIPERIDINYLOXY, 4-[[BIS(1-AZIRIDINYL)PHOSPHINO
C13H26N4O3S2	074385-89-0	N-(N"-DI-IPR-N'-ME-URYLTHIO)METHOMYL
C13H26N5O3P	103981-95-9	TEPA 4-(N-OH-PYRROLDINYL)-UREA DERIVATIV
C13H26N5O3P	103981-99-3	1-PYRROLIDINYLOXY, 3-[[[[BIS(1-AZIRIDINYL)PHOSPH
C13H26O	000462-18-0	TRIDECAN-7-ONE
C13H26O	000593-08-8	2-TRIDECANONE
C13H26O2	000111-82-0	METHYL LAURATE

C13H26O2	000627-90-7	ETHYL UNDECANOATE
C13H26O2	000638-53-9	TRIDECANOIC ACID
C13H26O2	001731-81-3	UNDECYL ALCOHOL ACETATE
C13H26O2	050623-57-9	BUTYL NONANOATE
C13H26O2Si3	018407-16-4	1,1,1,3,3,5,5-HEPTAMETHYL-5-PHENYLTRISILOXANE
C13H26O4Si4	010448-09-6	HEPTAMETHYLPHENYLCYCLOTETRASILOXANE
C13H26O5S	050615-69-5	beta-D-GALACTOPYRANOSIDE, HEPTYL 1-THIO-
C13H26O6	078617-12-6	ß-D-GALACTOPYRANOSIDE, HEPTYL
C13H27Br	000765-09-3	1-BROMO-N-TRIDECANE
C13H27N4OPS	033683-34-0	1-PIPERIDINYLOXY, 4-[[BIS(1-AZIRIDINYL)PHOSPHINO
C13H28	000629-50-5	N-TRIDECANE
C13H28O	000112-70-9	1-TRIDECANOL
C13H28S	019484-26-5	1-TRIDECANETHIOL
C13H29O3P	007040-58-6	BIS(1,2,2-TRIMETHYLPROPYL) METHYLPHOSPHONATE
C13H30IN	003085-79-8	1-BUTANAMINIUM, N,N-DIBUTYL-N-METHYL-, IODIDE
C13H30IN	007447-24-7	1-DECANAMINIUM, N,N,N-TRIMETHYL-, IODIDE
C14H4N2O2S2	003347-22-6	DITHIANONE
C14H6Cl2F4N2O2	083121-18-0	TEFLUBENZURON
C14H6Cl2N2O4	066121-41-3	DIS. A. 18
C14H6N2O6	057875-61-3	1,2-DINITROANTHRAQUINONE
C14H7Br2NO2	000081-49-2	1-AMINO-2,4-DIBROMOANTHRAQUINONE
C14H7Cl2F3O	095998-69-9	DICHLORO(TRIFLUOROMETHYL)BENZOPHENONE
C14H7Cl2F5	095998-70-2	ALPHA,ALPHA-DIFLUORODICHLOROTRIFLUOROMETHYL DIP*
C14H7ClF3NO5	050594-66-6	BENZOIC ACID, 5- 2-CHLORO-4-(TRIFLUOROMETHYL)PHE
C14H7ClO2	000082-44-0	1-CHLOROANTHRAQUINONE
C14H7F3NNaO4	066287-52-3	BENZOIC ACID, 3-NITRO-2-[[3-(TRIFLUOROMETHYL)PHE
C14H7NO4	000082-34-8	1-NITROANTHRAQUINONE
C14H8BrNO3	000116-82-5	1-AMINO-2-BROMO-4-HYDROXY-9,10-ANTHRACENEDIONE
C14H8Cl2N4	074115-24-5	CLOFENTEZINE
C14H8Cl4	000072-55-9	P,P'-DDE
C14H8Cl4	003424-82-6	O,P'-DDE
C14H8ClF3NNaO2	066287-53-4	BENZOIC ACID, 4-CHLORO-2-[[3-(TRIFLUOROMETHYL)PH
C14H8N2S2	018705-45-8	4,4'-DIISOTHIOCYANATEBIPHENYL
C14H8N2S4	000120-78-5	2,2'-DITHIOBISBENZOTHIAZOLE
C14H8NaO7S	000130-22-3	9,10-DIHYDRO-3,4-DIHYDROXY-9,10-DIOXO-2-ANTHRA*
C14H8O2	000084-11-7	9,10-PHENANTHRENEDIONE
C14H8O2	000084-65-1	ANTHRAQUINONE
C14H8O3	000129-43-1	1-HYDROXYANTHRAQUINONE
C14H8O3	000605-32-3	2-HYDROXYANTHRAQUINONE
C14H8O4	000072-48-0	ALIZARINE
C14H8O4	000081-64-1	1,4-DIHYDROXY-9,10-ANTHRACENEDIONE
C14H8O4	000084-60-6	2,6-DIHYDROXY-ANTHRAQUINONE
C14H8O4	000117-12-4	1,5-DIHYDROXYANTHRAQUINONE
C14H8O4	040274-67-7	XANTHONE-2-CARBOXYLIC ACID
C14H8O5	000081-54-9	PURPURIN
C14H8O6	000081-61-8	QUINALIZARIN
C14H8O8S2	000082-48-4	1,8-ANTHRAQUINONEDISULFONIC ACID
C14H8O8S2	000117-14-6	1,5-ANTHRAQUINONEDISULFONIC ACID
C14H9BrN2O	033443-53-7	4-PHENYL-6-BROMOQUINAZOLIN-2-ONE
C14H9Cl	024423-11-8	PHENANTHRENE, 2-CHLORO-
C14H9Cl2NO5	042576-02-3	BIFENOX
C14H9Cl3	001022-22-6	1,1-BIS(P-CHLOROPHENYL)-2-CHLOROETHYLENE
C14H9Cl3F2	000475-26-3	DFDT
C14H9Cl5	000050-29-3	P,P'-DDT
C14H9Cl5	000789-02-6	O,P'-DDT
C14H9Cl5O	000115-32-2	DICOFOL
C14H9ClF2N2O2	035367-38-5	DIFLUBENZURON
C14H9F3N2O2	035367-40-9	BENZAMIDE, 2,6-DIFLUORO-N-[[(4-FLUOROPHENYL)AMIN

C14H9F3NNaO2	001977-00-0	BENZOIC ACID, 2-[[3-(TRIFLUOROMETHYL)PHENYL]AMIN
C14H9N	002523-48-0	2-CYANO-9H-FLUORENE
C14H9N3O4	000082-33-7	C.I. DISPERSE VIOLET 8
C14H9NO2	000082-45-1	1-AMINOANTHRAQUINONE
C14H9NO2	000117-79-3	2-AMINOANTHRAQUINONE
C14H9NO2	000520-03-6	1H-ISOINDOLE-1,3(2H)-DIONE, 2-PHENYL-
C14H9NO2	000602-60-8	ANTHRACENE, 9-NITRO-
C14H9NO2	002021-26-3	1H-ISOINDOLE-1,3(2H)-DIONE, 5-PHENYL-
C14H9NO2S	049540-85-4	4-ISOTHIOCYANOPHENYL BENZOATE
C14H9NO3	000116-85-8	1-AMINO-4-HYDROXYANTHRAQUINONE
C14H9NOS	026328-59-6	4-ISOTHIOCYANOBENZOPHENONE
C14H10	000085-01-8	PHENANTHRENE
C14H10	000120-12-7	ANTHRACENE
C14H10	000501-65-5	DIPHENYLACETYLENE
C14H10BrN3O	001812-30-2	BROMAZEPAM
C14H10Cl2	000951-86-0	1,2-DICHLORO-1,2-DIPHENYLETHENE (TRANS)
C14H10Cl2KNO2	015307-81-0	BENZENEACETIC ACID, 2-[(2,6-DICHLOROPHENYL)AMINO
C14H10Cl2N2O4	090426-03-2	3,5-DICL-2'-ME-4'-NO2 SALICYLANILIDE
C14H10Cl2NNaO2	015307-79-6	BENZENEACETIC ACID, 2-[(2,6-DICHLOROPHENYL)AMINO
C14H10Cl2O3	034645-84-6	FENCLOFENAC
C14H10Cl2O4	051338-10-4	(4-(2,4-DICHLOROPHENOXY)PHENOXY)ACETIC ACID
C14H10Cl4	000053-19-0	O,P'-DDD
C14H10Cl4	000072-54-8	P,P'-DDD
C14H10Cl4	013700-81-7	1,2-DIPHENYL TETRACHLOROETHANE
C14H10Cl4	121107-43-5	2,2'-4,4'-TETRACHLORO-3-METHYL DIPHENYLMETHANE
C14H10Cl4	121107-44-6	2,2'-4,4'-TETRACHLORO-5-METHYL DIPHENYLMETHANE
C14H10Cl4	121107-46-8	2,2'-4,5'-TETRACHLORO-5-METHYL DIPHENYLMETHANE
C14H10Cl4	121107-47-9	2,2'-4,6'-TETRACHLORO-3-METHYL DIPHENYLMETHANE
C14H10Cl4	121107-48-0	2,2'-4,6'-TETRACHLORO-5-METHYL DIPHENYLMETHANE
C14H10Cl4	121107-55-9	2,2'-5,5'-TETRACHLORO-4-METHYL DIPHENYLMETHANE
C14H10Cl4	121107-65-1	2,3'-4,4'-TETRACHLORO-5-METHYL DIPHENYLMETHANE
C14H10Cl4	121107-77-5	2,3'-4,4'-TETRACHLORO-6-METHYL DIPHENYLMETHANE
C14H10Cl4	121107-83-3	2',3-4,6'-TETRACHLORO-6-METHYL DIPHENYLMETHANE
C14H10F3NO2	000530-78-9	FLUFENAMIC ACID
C14H10F3NO3	056356-34-4	3-CF3 PHENYL 4-AMINOSALICYLATE
C14H10F3NO3S	022731-28-8	CF3-SULFONANILIDE,P-BENZOYL
C14H10F4	000425-32-1	BENZENE, 1,1'-(1,1,2,2-TETRAFLUORO-1,2-ETHANEDIY
C14H10HgO4	000583-15-3	BENZOIC ACID, MERCURIC SALT
C14H10N2	005021-43-2	QUINOXALINE, 2-PHENYL-
C14H10N2	017629-01-5	4-PHENYLQUINAZOLINE
C14H10N2	025855-20-3	QUINAZOLINE, 2-PHENYL-
C14H10N2	058814-67-8	ISOQUINOLINE, 3-(2-PYRIDINYL)-
C14H10N2O	001022-45-3	4(1H)-QUINAZOLINONE, 2-PHENYL-
C14H10N2O	005004-45-5	1(2H)-PHTHALAZINONE, 4-PHENYL-
C14H10N2O	019768-02-6	FURAZAN, DIPHENYL-
C14H10N2O	023441-75-0	2(1H)-QUINAZOLINONE, 4-PHENYL-
C14H10N2O2	000128-95-0	DISPERSE VIOLET 1
C14H10N2O2	005585-14-8	FURAZAN, DIPHENYL-, 2-OXIDE
C14H10N4O5	007261-97-4	DANTROLENE
C14H10O	000090-44-8	9(10H)-ANTHRACENONE
C14H10O2	000134-81-6	BENZIL
C14H10O2	001989-33-9	9-CARBOXYFLUORENE
C14H10O2	005398-11-8	1(3H)-ISOBENZOFURANONE-3-PHENYL
C14H10O2S	017394-14-8	9-CARBOXYTHIOXANTHENE
C14H10O3	000082-07-5	XANTHENE-9-CARBOXYLIC ACID
C14H10O3	000093-97-0	BENZOIC ACID, ANHYDRIDE
C14H10O4	000094-36-0	BENZOYL PEROXIDE
C14H10O4	000482-05-3	1,1'-BIPHENYL -2,2'-DICARBOXYLIC ACID

C14H10O5	000552-94-3	2-CARBOXYPHENYL 2-HYDROXYBENZOATE
C14H11BrClNO2	127792-23-8	PHENYLACETIC ACID,2-(2'-CHLORO-6'-BROMO)ANILINO
C14H11BrFNO2	077068-04-3	5-F-2'-ME-4'-BR SALICYANILIDE
C14H11BrO	013140-73-3	ACETAMIDE, N-(3-BROMOPHENYL)-2-PHENYL-
C14H11Cl	001460-06-6	1-CHLORO-1,2-DIPHENYLETHENE
C14H11Cl2F2NO3	143121-10-2	RH-0978 O-PH N-BUTYLENE(F2)CARBAMATE
C14H11Cl2NO	027816-82-6	BENZENEACETAMIDE, N-(3,4-DICHLOROPHENYL)-
C14H11Cl2NO2	015307-86-5	DICLOFENAC
C14H11Cl2NO2	070172-32-6	PHENYLACETIC ACID,2-(2',3'-DICHLORO)ANILINO
C14H11Cl2NO2	127792-31-8	PHENYLACETIC ACID,2-(2',5'-DICHLORO)ANILINO
C14H11Cl2NO2	127792-33-0	PHENYLACETIC ACID,2-(3',4'-DICHLORO)ANILINO
C14H11Cl2NO3	064118-84-9	PHENYLACETIC ACID,2-(2',6'-DICHLORO-4'-HYDROXY)A
C14H11Cl2NO4	003736-81-0	DILOXANIDE FUROATE
C14H11Cl3	002971-22-4	2,2-DIPHENYL-1,1,1-TRICL ETHANE
C14H11ClFNO2	077068-02-1	5-F-2'-ME-4'-CL SALICYANILIDE
C14H11ClFNO2	100754-93-6	PHENYLACETIC ACID,2-(2'-CHLORO-6'-FLUORO)ANILINO
C14H11ClFNO2	127792-24-9	PHNEYLACETIC ACID,2-(2'-CHLORO-6'-IODO)ANILINO
C14H11ClN2O	108635-80-9	3-PYRIDINECARBONITRILE, 2-CHLORO-5-(4-METHOXYPHE
C14H11ClO3S	038488-19-6	ETHANONE, 1-(4-CHLOROPHENYL)-2-(PHENYLSULFONYL)-
C14H11F2NO3S	022736-85-2	DIFLUMIDONE
C14H11N	000086-29-3	DIPHENYLACETONITRILE
C14H11N	000256-96-2	5-AZADIBENZO(A,E)CYCLOHEPTATRIENE
C14H11N	000610-49-1	1-ANTHRACENAMINE
C14H11N	000613-13-8	2-AMINOANTHRACENE
C14H11N	000947-73-9	9-PHENANTHRENAMINE
C14H11N3	014005-50-6	2-AMINO-4-PHENYLQUINAZOLINE
C14H11N3	037989-04-1	1-ISOQUINOLINAMINE, 3-(2-PYRIDINYL)-
C14H11NO	003456-79-9	2H-INDOL-2-ONE, 1,3-DIHYDRO-3-PHENYL-
C14H11NO3	005663-74-1	N-BENZOYLSALICYLAMIDE
C14H11NO4	001972-71-0	DISALICYLIMIDE
C14H11NS	003550-21-8	BENZENE, 1,1'-(ISOTHIOCYANATOMETHYLENE)BIS-
C14H12	000103-30-0	STILBENE
C14H12	000530-48-3	1,1-DIPHENYLETHYLENE
C14H12	000588-59-0	STILBENE
C14H12	000613-31-0	9,10-DIHYDROANTHRACENE
C14H12	000645-49-8	1,1'-(1,2-ETHENEDIYL)BISBENZENE (CIS)
C14H12	000776-35-2	9,10-DIHYDROPHENANTHRENE
C14H12	001730-37-6	1-METHYLFLUORENE
C14H12	002523-37-7	9-METHYL-9H-FLUORENE
C14H12BrN	007496-72-2	2-(4-BROMOPHENYL)INDOLIZINE
C14H12BrN	080488-98-8	2-(3-BROMOPHENYL)INDOLIZINE
C14H12BrNO	007495-11-6	ACETAMIDE, N-(4-BROMOPHENYL)-2-PHENYL-
C14H12Cl2	002387-16-8	1,1-DICHLORO-2,2-DIPHENYLETHANE
C14H12Cl2	005963-49-5	1,2-DICHLORO-1,2-DIPHENYLETHANE
C14H12Cl2N2O	088283-41-4	ETHANOL, 1-(2,4-DICHLOROPHENYL)-2-(3-PYRIDINYL)-
C14H12Cl2O	000080-06-8	DIMITE
C14H12ClN	007496-73-3	2-(4-CHLOROPHENYL)INDOLIZINE
C14H12ClN	082682-70-0	2-(3-CHLOROPHENYL)INDOLIZINE
C14H12ClN3O4S2	003211-40-3	CHLOROTHIAZIDE-3-BENZYL
C14H12ClNO	002990-06-9	ACETAMIDE, N-(4-CHLOROPHENYL)-2-PHENYL-
C14H12ClNO	018109-43-8	BENZENEACETAMIDE, N-(3-CHLOROPHENYL)-
C14H12ClNO2	013710-19-5	TOLFENAMIC ACID
C14H12ClNO2	025933-30-6	5-CL-2'-ME SALICYLANILIDE
C14H12ClNO2	077067-91-5	2'-ME-4'-CL SALICYLANILIDE
C14H12F3NO4S2	037924-13-3	PERFLUIDONE
C14H12FNO	005215-26-9	BENZENEACETAMIDE, N-(2-FLUOROPHENYL)-
C14H12FNO	005215-27-0	ACETAMIDE, N-(3-FLUOROPHENYL)-2-PHENYL-
C14H12FNO	005215-28-1	BENZENEACETAMIDE, N-(4-FLUOROPHENYL)-

C14H12INO	095384-52-4	BENZENEACETAMIDE, N-(3-IODOPHENYL)-
C14H12N2O	063020-20-2	2-ACETAMIDOCARBAZOLE
C14H12N2O2	001821-33-6	BENZAMIDE, N-[(PHENYLAMINO)CARBONYL]-
C14H12N2O2	007703-36-8	3-FURANCARBONITRILE, 5-[(DIMETHYLAMINO)METHYLENE
C14H12N2O2S	020434-66-6	124-BENZTHIADIAZN-1-O2-3-BENZYL
C14H12N2O3	001504-26-3	BENZENEACETAMIDE, N-(2-NITROPHENYL)-
C14H12N2O3	013140-76-6	BENZENEACETAMIDE, N-(3-NITROPHENYL)-
C14H12N2O3	013140-77-7	BENZENEACETAMIDE, N-(4-NITROPHENYL)-
C14H12N2O3	024367-68-8	O-(P-NITROBENZYL)BENZAMIDE
C14H12N2O3	131073-11-5	OXAZOLO[4,5-C]PYRIDINE, 2-(2,4-DIMETHOXYPHENYL)-
C14H12N2O3S2	096474-18-9	2-BENZOTHIAZOLESULFONAMIDE, 6-(PHENYLMETHOXY)-
C14H12N2O4	082568-68-1	BENZAMIDE, 2-[(4-NITROPHENYL)METHOXY]-
C14H12N4	120137-94-2	[3,4'-BIPYRIDINE]-5-CARBONITRILE,6-(2-PROPENYLAM
C14H12N4O2	002475-45-8	1,4,5,8-TETRAAMINOANTHRAQUINONE
C14H12N4O2	102361-56-8	BENZAMIDE, 4-(2-IMIDAZO[1,2-A]PYRIMIDINYL)-3-MET
C14H12N4O2	130179-73-6	BENZAMIDE, 4-(1H-IMIDAZO[4,5-C]PYRIDIN-2-YL)-3-M
C14H12N4O2S	000059-40-5	BENZENESULFONAMIDE, 4-AMINO-N-2-QUINOXALINYL-
C14H12O	000092-91-1	4-ACETYLBIPHENYL
C14H12O	000451-40-1	2-PHENYLACETOPHENONE
C14H12O	002523-46-8	2-METHOXY-9H-FLUORENE
C14H12O	024324-17-2	9H-FLUORENE-9-METHANOL
C14H12O2	000117-34-0	DIPHENYLACETIC ACID
C14H12O2	000119-53-9	BENZOIN
C14H12O2	000120-51-4	BENZYLBENZOATE
C14H12O2	004796-68-3	BICYCLO[2.2.1]HEPTA-2,5-DIEN-7-OL, BENZOATE
C14H12O3	000076-93-7	BENZILIC ACID
C14H12O3	000131-57-7	METHANONE, (2-HYDROXY-4-METHOXYPHENYL)PHENYL-
C14H12O3	001878-57-5	M-PHENYLPHENOXYACETIC ACID
C14H12O3	005348-75-4	O-PHENYLPHENOXYACETIC ACID
C14H12O4	005527-84-4	4H-NAPHTHO[2,3-B]PYRAN-2-CARBOXYLIC ACID, 6,7,8,
C14H13Cl2NO3	087365-63-7	RH-4663 O-PH N-BUTYLENE CARBAMATE
C14H13ClN2O4	116383-80-3	1H-PYRROLE-3,4-DICARBOXYLIC ACID, 2-(5-CHLORO-2-
C14H13ClN2O4	116383-85-8	1H-PYRROLE-3,4-DICARBOXYLIC ACID, 2-(4-CHLORO-2-
C14H13ClN2O4	127390-68-5	1H-PYRROLE-3,4-DICARBOXYLIC ACID, 2-(6-CHLORO-2-
C14H13N	000086-28-2	N-ETHYLCARBAZOL
C14H13N	022627-00-5	O-METHYLBENZOPHENONEIMINE
C14H13N	025379-20-8	2-PHENYLINDOLIZINE
C14H13N2NaO3S	062959-40-4	4-((2,4-DIMETHYLPHENYL)AZO)BENZENESULFONIC ACID*
C14H13N3	110235-47-7	MEPANIPYRIM
C14H13N3O	004128-71-6	N-(4-(PHENYLAZO)PHENYL)ACETAMIDE
C14H13N3O2	005302-39-6	N-(4-((4-HYDROXYPHENYL)AZO)PHENYL)ACETAMIDE
C14H13N3O2	077303-19-6	1H-IMIDAZO[4,5-B]PYRIDINE, 2-(2,4-DIMETHOXYPHENY
C14H13N3O2	087359-11-3	1H-IMIDAZO[4,5-C]PYRIDINE, 2-(2,4-DIMETHOXYPHENY
C14H13N3O2	108611-04-7	[3,4'-BIPYRIDINE]-5-CARBONITRILE,6-(2-HYDROXYPRO
C14H13N3O2S	073384-60-8	1H-IMIDAZO[4,5-B]PYRIDINE, 2-[2-METHOXY-4-(METHY
C14H13N3O2S	086315-52-8	ISOMAZOLE
C14H13N3O2S	093276-61-0	IMIDAZO[1,2-A]PYRIMIDINE, 2-[2-METHOXY-4-(METHYL
C14H13N3O2S	102362-14-1	IMIDAZO[1,2-A]PYRAZINE, 2-[2-METHOXY-4-(METHYLSU
C14H13N3O3	029418-61-9	4'-METHOXY-2-METHYL-4-NITROAZOBENZENE
C14H13N3O5	074550-86-0	1H-IMIDAZOLE-1-ETHANOL, 2-[2-(1,3-BENZODIOXOL-5-
C14H13NO	000519-87-9	N,N-DIPHENYLACETAMIDE
C14H13NO	000621-06-7	BENZENEACETAMIDE, N-PHENYL-
C14H13NO	001934-92-5	N-METHYLBENZANILIDE
C14H13NO	002113-47-5	2-PHENYLACETANILIDE
C14H13NO	016350-99-5	N-BENZYL-N-FORMYLANILINE
C14H13NO	019865-55-5	ALPHA-(P-ME-PHENYL)-N-PHENYLNITRONE
C14H13NO2	000094-18-8	BENZYL-4-AMINOBENZOATE
C14H13NO2	003585-93-1	ALPHA-(P-MEO-PHENYL)-N-PHENYLNITRONE

C14H13NO2	003743-11-1	2-BENZAMIDO-5-METHYLPHENOL
C14H13NO2	006312-87-4	ACETAMIDE, N-(4-PHENOXYPHENYL)-
C14H13NO2	014680-18-3	N-(O-HYDROXYPHENYL)METHYLBENZAMIDE
C14H13NO2	016524-22-4	BENZOIC ACID, 2-[(3-METHYLPHENYL)AMINO]-
C14H13NO2	018705-01-6	PHENOXYACETANILIDE
C14H13NO2	023478-26-4	BENZENEACETAMIDE, N-(3-HYDROXYPHENYL)-
C14H13NO2	041859-85-2	N-(P-HYDROXYPHENYL)METHYLBENZAMIDE
C14H13NO2	050789-46-3	BENZAMIDE, N-METHYL-3-PHENOXY-
C14H13NO2	070172-33-7	2-ANILINOPHENYLACETIC ACID
C14H13NO2	095384-57-9	BENZENEACETAMIDE, N-(4-HYDROXYPHENYL)-
C14H13NO2	095384-58-0	BENZENEACETAMIDE, N-(2-HYDROXYPHENYL)-
C14H13NO2S	132483-44-4	BENZENEACETIC ACID, 4-(4-ETHENYL-2-THIAZOLYL)-α-
C14H13NO3	027693-73-8	BENZOIC ACID, 2-[(3-METHOXYPHENYL)AMINO]-
C14H13NO3	056356-13-9	2-TOLYL 4-AMINOSALICYLATE
C14H13NO3	056356-14-0	3-TOLYL 4-AMINOSALICYLATE
C14H13NO3	056356-15-1	4-TOLYL 4-AMINOSALICYLATE
C14H13NO3S	100866-99-7	ETHANONE, 1-[4-[(4-AMINOPHENYL)SULFONYL]PHENYL]-
C14H13NO3S	132483-61-5	BENZENEACETIC ACID, 4-(4-ACETYL-2-THIAZOLYL)-α-M
C14H13NO4	004465-61-6	4-MEO PHENYL 4-AMINOSALICYLATE
C14H13NO4	056356-17-3	2-MEO PHENYL 4-AMINOSALICYLATE
C14H13NO4	056356-18-4	3-MEO PHENYL 4-AMINOSALICYLATE
C14H13NO4S	034037-45-1	BENZOIC ACID, 4-[(4-AMINOPHENYL)SULFONYL]-
C14H13NO6	032620-72-7	BENZOIC ACID, 2-(ACETYLOXY)-, (2,5-DIOXO-1-PYRRO
C14H13NS	002628-58-2	N-METHYLTHIOBENZANILIDE
C14H14	000103-29-7	BIBENZYL
C14H14	000612-00-0	1,1-DIPHENYLETHANE
C14H14	000612-75-9	3,3'-DIMETHYLBIPHENYL
C14H14	000613-33-2	P-TOLYLTOLUENE
C14H14ClN3	000086-40-8	3,6-DIAMINO-10-METHYL-ACRIDINIUM CHLORIDE
C14H14ClN3O4S2	001824-50-6	3-BENZYLHYDROCHLOROTHIAZIDE
C14H14ClNO2S	138568-79-3	BENZENEACETIC ACID, 3-CHLORO-4-(4-ETHYL-2-THIAZO
C14H14ClNO2S	138568-83-9	BENZENEACETIC ACID, 3-CHLORO-4-(5-ETHYL-2-THIAZO
C14H14ClNO2S	138568-84-0	BENZENEACETIC ACID, 3-CHLORO-4-(4,5-DIMETHYL-2-T
C14H14FNO2S	138568-71-5	BENZENEACETIC ACID, 4-(4-ETHYL-2-THIAZOLYL)-3-FL
C14H14FNO2S	138568-75-9	BENZENEACETIC ACID, 4-(5-ETHYL-2-THIAZOLYL)-3-FL
C14H14FNO2S	138568-76-0	BENZENEACETIC ACID, 4-(4,5-DIMETHYL-2-THIAZOLYL)
C14H14N2	000501-60-0	BIS(4-METHYLPHENYL)DIAZENE
C14H14N2	000584-90-7	BIS(2-METHYLPHENYL)DIAZENE
C14H14N2	000588-04-5	BIS(3-METHYLPHENYL)DIAZENE
C14H14N2	000621-09-0	N,N'-DIPHENYLACETAMIDINE
C14H14N2	000835-31-4	NAPHAZOLINE
C14H14N2	017590-87-3	(2,6-DIMETHYLPHENYL)PHENYLDIAZENE
C14H14N2	029418-21-1	2,4-DIMETHYLAZOBENZENE
C14H14N2	029418-22-2	O,P'-AZOTOLUENE
C14H14N2O	029268-77-7	2-METHOXY-6-METHYLAZOBENZENE
C14H14N2O	029268-78-8	2-METHOXY-2'-METHYLAZOBENZENE
C14H14N2O	029418-41-5	2-METHOXY-4'-METHYLAZOBENZENE
C14H14N2O	029418-42-6	4-METHOXY-2-METHYLAZOBENZENE
C14H14N2O	029418-43-7	(4-METHOXYPHENYL)(2-METHYLPHENYL)DIAZENE
C14H14N2O	029418-44-8	(4-METHOXYPHENYL)(4-METHYLPHENYL)DIAZENE
C14H14N2O	085856-32-2	BENZENEACETAMIDE, N-(3-AMINOPHENYL)-
C14H14N2O	095384-59-1	BENZENEACETAMIDE, N-(2-AMINOPHENYL)-
C14H14N2O2	000501-58-6	BIS(4-METHOXYPHENOL)DIAZENE
C14H14N2O2	000613-55-8	BIS(2-METHOXYPHENYL)DIAZENE
C14H14N2O2	006319-23-9	3,3'-DIMETHOXYAZOBENZENE
C14H14N2O2	029418-46-0	2,4-DIMETHOXYAZOBENZENE
C14H14N2O2	029418-47-1	2,4'-DIMETHOXYAZOBENZENE
C14H14N2O2	029418-48-2	2,6-DIMETHOXYAZOBENZENE

C14H14N2O2	092245-57-3	2-(4-(PHENYLAZO)PHENOXY)ETHANOL
C14H14N2O2S	037157-92-9	124-BENZTHIADIAZIN-1-O2-3(5NORBEN2YL)
C14H14N2O3	001562-94-3	4,4'-DIMETHOXYAZOXYBENZENE
C14H14N2O3S	013909-63-2	BENZENESULFONAMIDE, 4-METHYL-N-[(PHENYLAMINO)CAR
C14H14N2O3S	132483-51-3	BENZENEACETIC ACID, α-METHYL-4-[4-[(METHYLAMINO)
C14H14N3NaO3S	000547-58-0	METHYL ORANGE
C14H14N4O	108610-79-3	[3,4'-BIPYRIDINE]-5-CARBONITRILE,6-[(2-HYDROXYPR
C14H14N4O	108610-82-8	[3,4'-BIPYRIDINE]-5-CARBONITRILE,6-[(3-HYDROXYPR
C14H14N4O2	002491-74-9	4-NITRO-((4-(N-DIMETHYL)AMINOPHENYL)AZO)BENZENE
C14H14N4O2	003010-38-6	DIS. A. 15
C14H14N4O2	127356-43-8	1H-IMIDAZO[4,5-B]PYRIDIN-6-AMINE, 2-(2,4-DIMETHO
C14H14N4O2S	127356-05-2	1H-IMIDAZO[4,5-B]PYRIDIN-6-AMINE, 2-[2-METHOXY-4
C14H14N4O5S	074223-56-6	SULFOMETURON
C14H14N8O4S3	025953-19-9	CEFAZOLIN
C14H14NO4PS	002104-64-5	EPN
C14H14O	000103-50-4	DIBENZYL ETHER
C14H14O	001517-63-1	PHNEYL-P-TOLYLCARBINOL
C14H14O	005472-13-9	PHENYL-O-TOLYLCARBINOL
C14H14O2	000104-66-5	1,2-DIPHENOXYETHANE
C14H14O2	000579-43-1	MESO-HYDROBENZOIN
C14H14O2	000655-48-1	1,2-ETHANEDIOL, 1,2-DIPHENYL-, (R*,R*)-(±)-
C14H14O2S	059094-49-4	1,4-NAPHTHOQUINONE,2-BUTYLTHIO
C14H14O3	000083-26-1	PINDONE
C14H14O3	000083-28-3	VALONE
C14H14O3	016563-45-4	BUTANOIC ACID, 4-(1-NAPHTHALENYLOXY)-
C14H14O3	022204-53-1	NAPROSYN
C14H14O3	036429-48-8	3-(2-PHENOXYETHOXY)PHENOL
C14H14O3	041643-81-6	ACETIC ACID, (1-NAPHTHALENYLOXY)-, ETHYL ESTER
C14H14O4	000131-17-9	DIALLYLPHTHALATE
C14H15BrO4	022399-01-5	DIETHYLMALONATE,4-BR BENZAL
C14H15Cl2N	000494-03-1	CHLORONAPHAZINE
C14H15Cl2N3O	100199-38-0	CYCLOHEXANOL, 1-(2,5-DICHLOROPHENYL)-2-(1H-1,2,4
C14H15Cl2NO3	143121-07-7	RH-7160 O-PH(24DICL) N-DIET CARBAMATE
C14H15ClO4	006768-20-3	DIETHYLMALONATE,2-CL BENZAL
C14H15ClO4	006768-21-4	DIETHYLMALONATE,3-CL BENZAL
C14H15ClO4	006827-40-3	DIETHYLMALONATE,4-CL BENZAL
C14H15FO4	000790-53-4	DIETHYLMALONATE,4-F BENZAL
C14H15FO4	002262-52-4	DIETHYLMALONATE,2-F BENZAL
C14H15N	000103-49-1	DIBENZYL AMINE
C14H15N	000614-30-2	N-METHYL-N-BENZYLANILINE
C14H15N3	000060-11-7	4-(N,N-DIMETHYLAMINO)AZOBENZENE
C14H15N3	000097-56-3	2-METHYL-4-((2-METHYLPHENYL)AZO)BENZENAMINE
C14H15N3	121552-61-2	CGA 219417 (CYPRODINIL)
C14H15N5	108611-20-7	[3,4'-BIPYRIDINE]-5-CARBONITRILE, 6-[(2-AMINOETH
C14H15N5O	018371-12-5	N-(4-((2,4-DIAMINOPHENYL)AZO)PHENYL)ACETAMIDE
C14H15N5O2	055687-58-6	2,4-NH2PYRIMIDIN,5(3,5MEO-4C≡N)BENZYL
C14H15N5O3	098204-04-7	ALLOPURINOL,PHENYLGLYCYLOXYMETHYL
C14H15N5O3S	104795-66-6	BENZO[B]THIOPHENE-2-CARBOXAMIDE, 5-METHOXY-3-(1-
C14H15N5O6S	074223-64-6	METSULFURON-METHYL
C14H15NaO3S	097683-34-6	1-SO3NA-3-BUTYL AZULENE
C14H15NaO3S	114482-98-3	1-SO3NA-3-ME-7-ISOPROPYL AZULENE
C14H15NaO4S	099287-56-6	1-SO3NA-3-PROPYL-4-METHOXY AZULENE
C14H15NO2	059094-50-7	1,4-NAPHTHOQUINONE,2-BUTYLAMINO-
C14H15NO2S	105456-58-4	BENZENAMINE, 4-[(2,4-DIMETHYLPHENYL)SULFONYL]-
C14H15NO2S	132483-46-6	BENZENEACETIC ACID, 4-(4,5-DIMETHYL-2-THIAZOLYL)
C14H15NO2S	132483-47-7	BENZENEACETIC ACID, 4-(4-ETHYL-2-THIAZOLYL)-α-ME
C14H15NO2S	138568-67-9	BENZENEACETIC ACID, 4-(5-ETHYL-2-THIAZOLYL)-α-ME
C14H15NO3	150900-12-2	2-AZASPIRO[4.4]NONANE-1,3-DIONE, 2-PHENOXY-

C14H15NO3S	138568-66-8	BENZENEACETIC ACID, 4-[4-(METHOXYMETHYL)-2-THIAZ
C14H15NO4S	105456-59-5	BENZENAMINE, 4-[(2,4-DIMETHOXYPHENYL)SULFONYL]-
C14H15NO5	115178-75-1	L-PROLINE, N-[(BENZOYLOXY)ACETYL]-
C14H15NO6	006331-45-9	DIETHYLMALONATE,3-NO2 BENZAL
C14H15NO6	017422-56-9	DIETHYLMALONATE,2-NO2 BENZAL
C14H15NO6	022399-00-4	DIETHYLMALONATE,4-NO2 BENZAL
C14H15NOS	079467-22-4	BENZENEMETHANOL, 2-[[2-(AMINOMETHYL)PHENYL]THIO]
C14H15O2PS2	017109-49-8	EDIFENPHOS
C14H15O3	039856-64-9	3-FURANEMETHANOL-5-BENZYL,ACETATE
C14H16	000187-78-0	CYCLOPENT[FG]ACENAPHTHYLENE
C14H16BrN	053394-58-4	G-PHENYLPROPYLPYRIDINIUM BROMIDE
C14H16BrN5O2	087469-53-2	1-METHYLISOGUANINE,9[2(4-BRPHENYL)ET]-
C14H16BrN5O2	087469-54-3	1-METHYLISOGUANINE,7[2(4-BRPHENYL)ET]-
C14H16Cl2N2O	035554-44-0	IMAZALIL BASE
C14H16ClN	020455-68-9	DIBENZYLAMMONIUM CHLORIDE
C14H16ClN3O	100199-35-7	CYCLOHEXANOL, 1-(2-CHLOROPHENYL)-2-(1H-1,2,4-TRI
C14H16ClN3O2	043121-43-3	TRIADIMEFON
C14H16ClN3O4S2	002259-96-3	CYCLOTHIAZIDE
C14H16ClO5PS	000056-72-4	COUMAPHOS
C14H16F3N3O3S2	082774-15-0	N(N"-ME(3-CF3PH)CARBAMYLTHIO)METHOMYL
C14H16F3N3O4	026399-36-0	PROFLURALIN
C14H16N2	000119-93-7	BIANISIDINE
C14H16N2O2	000119-90-4	3,3'-DIMETHOXYBENZIDINE
C14H16N2O2	002778-41-8	1,4-BIS(1-ISOCYANATO-1-METHYLETHYL)BENZENE
C14H16N2O2	002778-42-9	1,3-BIS(1-ISOCYANATO-1-METHYLETHYL)BENZENE
C14H16N2O2	033125-97-2	1H-IMIDAZOLE-5-CARBOXYLIC ACID, 1-(1-PHENYLETHYL
C14H16N2O2S	003572-34-7	BENZENAMINE, 4-[(4-AMINOPHENYL)SULFONYL]-N-ETHYL
C14H16N2O2S	086552-09-2	BENZENAMINE, 4-[(4-AMINOPHENYL)SULFONYL]-N,N-DIM
C14H16N2O4	106231-69-0	L-PROLINAMIDE,N2-[(BENZOYLOXY)ACETYL]-
C14H16N2O4S2	034691-02-6	CEPHALOTHIN ANALOG (3-CH3)
C14H16N2O6	116482-78-1	BENZOIC ACID, 2-(ACETYLOXY)-, 2-[(2-AMINO-2-OXOE
C14H16N4	000539-17-3	C.I. DISPERSE BLACK 3
C14H16N4O2	087666-25-9	UREA, N-METHYL-N'-[2-(6-OXO-3-PHENYL-1(6H)-PYRID
C14H16N4O2	087666-26-0	UREA, [3-(6-OXO-3-PHENYL-1(6H)-PYRIDAZINYL)PROPY
C14H16O4	005292-53-5	DIETHYLMALONATE, BENZAL
C14H16O4	015725-18-5	ACETYLACETONE,2,4-DIMEO BENZAL
C14H16O4	015725-24-3	ETHYLACETOACETATE,2-MEO BENZAL
C14H16O4	015725-25-4	ETHYLACETOACETATE,3-MEO BENZAL
C14H16O4	015725-26-5	ETHYLACETOACETATE,4-MEO BENZAL
C14H16O6	000084-72-0	ETHYL CARBETHOXYMETHYL PHTHALATE
C14H16O6	118247-07-7	BENZOIC ACID, 2-(ACETYLOXY)-, (1-OXOBUTOXY)METHY
C14H17BrN2O5	132170-00-4	BR-N(26-MEPH)CARBAMOYLME IMINODIACETIC ACID
C14H17Cl2N3O2	060207-93-4	1-[[2-(2,4-DICHLOROPHENYL)-4-ETHYL-1,3-DIOXOLAN-
C14H17Cl3O4	002545-59-7	2,4,5-T, BUTOXYETHYL ESTER
C14H17ClN2O2	105801-52-3	1-PYRROLIDINEACETAMIDE, N-(5-CHLORO-2,3-DIHYDRO-
C14H17ClN2O3	105801-54-5	4-MORPHOLINEACETAMIDE, N-(5-CHLORO-2,3-DIHYDRO-3
C14H17ClNO4PS2	010311-84-9	DIALIFOR
C14H17IN4O3	085544-57-6	2,4-PYRIMIDINEDIAMINE, 5-[(2-IODO-3,4,5-TRIMETHO
C14H17N	147064-47-9	9AH-FLUOREN-9A-AMINE, 1,4,4A,9-TETRAHYDRO-N-METH
C14H17N2O4PS	000119-12-0	PYRIDAPHENTHION
C14H17N3O	131229-39-5	IMIDAZO[5,1-B]QUINAZOLIN-9(2H)-ONE, 1,3-DIHYDRO-
C14H17N3O2S	034392-61-5	N1-(4-DIMEAMINOPH)SULFANILAMIDE
C14H17N3O3	104098-48-8	AC 263,222 HERBICIDE
C14H17N3O9	002169-64-4	6-AZAURIDINETRIACETATE
C14H17N5O3	051940-44-4	PIPEMIDIC ACID
C14H17N5O3	055687-57-5	24-NH2PYRIMIDIN,5(35MEO-4CONH2)BENZYL
C14H17NO	005422-81-1	N,N-PENTAMETHYLENE CINNAMAMIDE
C14H17NO2	124497-82-1	2-PROPYNYLOXY-2-PH-4-ME MORPHOLINE HCL

C14H17NO3	106231-67-8	PIPERIDINE, 1-[(BENZOYLOXY)ACETYL]-
C14H17NO4	087130-34-5	CARBAMIC ACID, (3,4-DIETHOXYPHENYL)-, 2-PROPYNYL
C14H17NO4	115178-71-7	4-PIPERIDINOL, 1-[(BENZOYLOXY)ACETYL]-
C14H17NO5	106231-63-4	GLYCINE, N-[(BENZOYLOXY)ACETYL]-N-METHYL-, ETHYL
C14H17NO5	118247-08-8	BENZOIC ACID, 2-(ACETYLOXY)-, 2-(ACETYLMETHYLAMI
C14H17O5PS	000299-45-6	E-838
C14H18Cl2O4	001929-73-3	2,4-D, BUTOXYETHYL ESTER
C14H18ClNO3	060145-76-8	ET-N-CL-ACETYL-N-(2ETPH)GLYCINATE
C14H18N2O	000479-92-5	3H-PYRAZOL-3-ONE, 1,2-DIHYDRO-1,5-DIMETHYL-4-(1-
C14H18N2O2	105801-67-0	1-PYRROLIDINEACETAMIDE, N-(2,3-DIHYDRO-3-BENZOFU
C14H18N2O3	003625-25-0	REPOSAL
C14H18N2O3	105801-49-8	4-MORPHOLINEACETAMIDE, N-(2,3-DIHYDRO-3-BENZOFUR
C14H18N2O3S	077812-88-5	4-PENTYNAMIDE, N-[4-[(DIMETHYLAMINO)SULFONYL]PHE
C14H18N2O4	087130-56-1	CARBAMIC ACID, (3,4-DIETHOXYPHENYL)-, 2-CYANOETH
C14H18N2O5	022839-47-0	ASPARTAME
C14H18N2O7	000973-21-7	DINOBUTON
C14H18N3Na5O10	000140-01-2	DIETHYLENETRIAMINEPENTAACETIC ACID, PENTA SODIU*
C14H18N4O	073576-30-4	2,4-NH2PYRIMIDIN,5(35-ME-4-MEO)BENZYL
C14H18N4O2	006981-18-6	2,4-NH2PYRIMIDIN,5(45-MEO-2-ME)BENZYL
C14H18N4O2	024798-19-4	2,4-NH2PYRIMIDIN,5(34-MEO-5-ME)BENZYL
C14H18N4O2	049845-48-9	2,4-NH2PYRIMIDIN,5(35-MEO-4-ME)BENZYL
C14H18N4O2	073046-15-8	2,4-PYRIMIDINEDIAMINE, 5-[(3-ETHOXY-4-METHOXYPHE
C14H18N4O2	085544-41-8	2,4-PYRIMIDINEDIAMINE, 5-[(4-ETHOXY-3-METHOXYPHE
C14H18N4O2	130787-12-1	2H-BENZOTRIAZOLE, 2-CYCLOPENTYL-5-[(ETHOXYCARBON
C14H18N4O2S	068902-57-8	24-NH2PYRIMIDIN,5(35-MEO-4-MES)BENZYL
C14H18N4O3	000738-70-5	TRIMETHOPRIM
C14H18N4O3	006981-01-7	2,4-NH2PYRIMIDINE,5(2,4,5-MEO)BENZYL
C14H18N4O3	006981-04-0	2,4-PYRIMIDINEDIAMINE, 5-[(2,3,4-TRIMETHOXYPHENY
C14H18N4O3	017804-35-2	BENOMYL
C14H18N4O3	055687-49-5	24-NH2PYRIMIDN,5(35MEO-4-CH2OH)BENZYL
C14H18N4O4S	001164-13-2	2,6-ETO-4-SULFANILAMIDOPYRIMIDINE
C14H18N4O4S	005018-16-6	BENZENESULFONAMIDE, 4-AMINO-N-[6-METHOXY-5-(1-ME
C14H18N6O4	117121-37-6	1H-TETRAZOLE-1-ACETIC ACID, 5-[4-[[(DIMETHYLAMIN
C14H18O2	035889-00-0	BENZENEACETIC ACID, 4-CYCLOHEXYL-
C14H18O2	053446-96-1	9,10-PHENANTHRENEDIOL, 1,2,3,4,4A,9,10,10A-OCTAH
C14H18O2	053446-97-2	9,10-PHENANTHRENEDIOL, 1,2,3,4,4A,9,10,10A-OCTAH
C14H18O2	053446-99-4	9,10-PHENANTHRENEDIOL, 1,2,3,4,4A,9,10,10A-OCTAH
C14H18O3	001878-56-4	PHENOXYACETIC ACID,4-CYCLOHEXYL
C14H18O4	000131-16-8	DIPROPYL PHTHALATE
C14H18O4	000605-45-8	DI-I-PROPYL PHTHALATE
C14H18O4	000607-81-8	BENZYLMALONIC ACID, DIETHYL ESTER
C14H18O4	118198-78-0	7-OCTEN-1-OL, 8-(3,4-DIHYDROXYPHENYL)-6-OXO-
C14H18O6	000117-82-8	BIS(METHOXYETHYL)PHTHALATE
C14H19Br2NO4	087130-47-0	CARBAMIC ACID, (3,4-DIETHOXYPHENYL)-, 2-BROMO-1-
C14H19Cl2NO2	000305-03-3	4-(BIS(2-CHLOROETHYL)AMINO)BENZENEBUTANOIC AC*
C14H19ClN2O2	105801-50-1	ACETAMIDE, N-(5-CHLORO-2,3-DIHYDRO-3-BENZOFURANY
C14H19ClN6O2	118259-45-3	UREA, N'-[3-CHLORO-4-[[(1,1-DIMETHYLETHYL)-1H-TE
C14H19FN4O3	141018-24-8	CYTIDINE, 2',3'-DIDEOXY-3'-FLUORO-N-[(1-PYRROLID
C14H19FN4O4	141018-23-7	CYTIDINE, 2',3'-DIDEOXY-3'-FLUORO-N-[(4-MORPHOLI
C14H19IN2O6	084043-28-7	URIDINE, 2'-DEOXY-5-IODO-, 5'-(2,2-DIMETHYLPROPA
C14H19IN2O6	084052-69-7	URIDINE, 2'-DEOXY-5-IODO-, 5'-PENTANOATE
C14H19N	086943-77-3	BENZOBICYCLO(221)HEPTENE,2EX-PRAMINO
C14H19N	086992-67-8	BENZOBICYCLO(221)HEPTENE,2EN-PRAMINO
C14H19N3O2	069218-50-4	1(5-CARBETHOXYPENTYL)BENZOTRIAZOLE
C14H19N3O2	069810-98-6	CARBAMIC ACID, [4,5-DIHYDRO-4-[4-(1-METHYLETHYL)
C14H19N3O2	115054-74-5	1H-BENZOTRIAZOLE-1-ACETIC ACID, α-BUTYL-, ETHYL
C14H19N3O2	115054-87-0	2H-BENZOTRIAZOLE-2-ACETIC ACID, α-BUTYL-, ETHYL
C14H19N3O3	089145-43-7	CARBAMIC ACID, [4,5-DIHYDRO-5-[4-(HYDROXYMETHYL)

C14H19N3O3S2	082774-00-3	N(N"-ME-(4ME-PH)CARBAMYLTHIO)METHOMYL
C14H19N3O4	132765-88-9	L-PHENYLALANINAMIDE, N-ACETYL-L-SERYL-
C14H19N3S	000091-80-5	METHAPYRILENE
C14H19N5O	075841-82-6	MOPIDRALAZINE
C14H19N5O	085544-46-3	2,4-PYRIMIDINEDIAMINE, 5-[[3-(DIMETHYLAMINO)-4-M
C14H19N5O	085544-47-4	2,4-PYRIMIDINEDIAMINE, 5-[[4-(DIMETHYLAMINO)-3-M
C14H19N5O3S	022181-94-8	GLYCINE, N-[1-OXO-5-(1H-PURIN-6-YLTHIO)PENTYL]-,
C14H19N5O4	098827-27-1	ALLOPURINOL,1-[N,N-DIETHYLSUCCINAMYL-OXYMETHYL]-
C14H19NO	000091-53-2	ETHOXYQUIN
C14H19NO	088058-88-2	NAXAGOLIDE
C14H19NO	097871-51-7	A-ME-N-SEC-BUTYLCINNAMAMIDE
C14H19NO	132453-23-7	10AH-DIBENZO[B,D]PYRAN-10A-AMINE, 6,6A,7,8,9,10-
C14H19NO2	020308-67-2	PIPERIDINE, 1-(2-ETHOXYBENZOYL)-
C14H19NO2	097871-53-9	A-MEO-N-SEC-BUTYLCINNAMAMIDE
C14H19NO2	124497-81-0	2-PROPENYLOXY-2-PH-4-ME MORPHOLINE HCL
C14H19NO3	112960-12-0	2H,5H-[1]BENZOPYRANO[4,3-B]-1,4-OXAZIN-9-OL, 3,4
C14H19NO3	115178-79-5	PROPANAMIDE, 2-(BENZOYLOXY)-N,N-DIETHYL-, (±)-
C14H19NO3	116005-01-7	2H,5H-[1]BENZOPYRANO[4,3-B]-1,4-OXAZIN-7-OL, 3,4
C14H19NO4	087130-27-6	CARBAMIC ACID, (3,4-DIETHOXYPHENYL)-, 2-PROPENYL
C14H19NOS	097871-52-8	A-MES-N-SEC-BUTYLCINNAMAMIDE
C14H19NS	132453-38-4	10AH-DIBENZO[B,D]THIOPYRAN-10A-AMINE, 6,6A,7,8,9
C14H19O6P	007700-17-6	CROTOXYPHOS
C14H20BrNO3	017199-22-3	2-BROMO-6-NITRO-4-(1,1,3,3-TETRAMETHYLBUTYL)-PH*
C14H20Cl2O	073986-52-4	2,6-DICHLORO-4-OCTYLPHENOL
C14H20ClFN2O2	004925-19-3	BENZAMIDE,N(DIETAMET),2OME,3CL,5F
C14H20ClN3O2	055219-65-3	TRIADIMENOL
C14H20ClN3O9	058845-59-3	1NO1(2CLET)3(TRIACETYLXYLOSYL)UREA
C14H20ClNO	070732-30-8	BENZENEMETHANOL, _-[(CYCLOHEXYLAMINO)METHYL]-, H
C14H20ClNO2	015972-60-8	ALACHLOR
C14H20ClNO2	034256-82-1	ACETOCHLOR
C14H20ClNO3	017199-21-2	2-CHLORO-6-NITRO-4-(1,1,3,3-TETRAMETHYLBUTYL)-P*
C14H20FN3O2	111041-88-4	IMIDODICARBONIC DIAMIDE, N-[5-(1,1-DIMETHYLETHYL
C14H20N2O	001611-63-8	UREA, N-(1-METHYLCYCLOHEXYL)-N'-PHENYL-
C14H20N2O	001982-49-6	SIDURON
C14H20N2O	005429-42-5	N-(P-TOLYL)-3-N'-PIPERIDINOACETAMIDE
C14H20N2O	013993-02-7	N-(O-TOLYL)-3-N'-PIPERIDINOACETAMIDE
C14H20N2O	019095-79-5	N-CYCLOHEPTYL-N'-PHENYLUREA
C14H20N2O	038367-20-3	N-(M-TOLYL)-3-N'-PIPERIDINOACETAMIDE
C14H20N2O	072336-19-7	1-PIPERIDINEACETAMIDE, N-(PHENYLMETHYL)-
C14H20N2O2	013523-86-9	PINDOLOL
C14H20N2O2	034915-68-9	BUNITROLOL
C14H20N2O2	040297-47-0	1-PIPERIDINEACETAMIDE, N-(3-METHOXYPHENYL)-
C14H20N2O2	058479-93-9	1-PIPERIDINEACETAMIDE, N-(4-METHOXYPHENYL)-
C14H20N2O2	105801-65-8	ACETAMIDE, N-(2,3-DIHYDRO-3-BENZOFURANYL)-2-(DIE
C14H20N2O3	065329-13-7	BARBITURIC ACID,5-(1-CYCLOHEPTEN-1-YL)-5-ETHYL-1
C14H20N2O3S	000664-95-9	TOLCYCLAMIDE
C14H20N2O5	049648-50-2	TRIMETHYLAMPHENICOL
C14H20N2O5S	075852-03-8	PROPANAMIDE, 3-(ACETYLOXY)-N-[4-[(DIMETHYLAMINO)
C14H20N2O5S	083870-90-0	BENZENESULFONYL-L-(N'-PR)GLUTAMINE
C14H20N3O5PS	013457-18-6	PYRAZOPHOS
C14H20N4O3	141018-18-0	CYTIDINE, 2',3'-DIDEOXY-N-[(1-PYRROLIDINYL)METHY
C14H20N4O4	141018-17-9	CYTIDINE, 2',3'-DIDEOXY-N-[(4-MORPHOLINYL)METHYL
C14H20N6O2	117121-44-5	UREA, N'-[4-[[2-(1,1-DIMETHYLETHYL)-2H-TETRAZOL-
C14H20N6O2	118259-53-3	UREA, N'-[4-[[1-(1,1-DIMETHYLETHYL)-1H-TETRAZOL-
C14H20N6O4	117121-49-0	2H-TETRAZOLE-2-ACETIC ACID, 5-[4-[[(DIMETHYLAMIN
C14H20O2	000719-22-2	2,6-BIS(T-BUTYL)-2,5-CYCLOHEXADIENE-1,4-DIONE
C14H20O2	014377-22-1	BENZENEACETIC ACID, 4-HEXYL-
C14H20O3	001085-12-7	HEPTYL P-HYDROXYBENZOATE

C14H20O3	060003-46-5	BENZENEACETIC ACID, 4-(HEXYLOXY)-
C14H20O3	103910-47-0	BENZENEACETIC ACID, α-ETHYL-α-HYDROXY-4-(2-METHY
C14H20O3	112473-78-6	BUTANOIC ACID, 2-ETHOXY-3-METHYL-, PHENYLMETHYL
C14H20O5	014098-44-3	BENZO-15-CROWN-5-ETHER
C14H20O5S	063407-54-5	1(2-PHENETS)-B-GALACTOPYRANOSIDE
C14H20O6	020772-21-8	GLUCOPYRANOSIDE,3,5-DIMETHYLPHENYL
C14H20O6	020838-34-0	GLUCOPYRANOSIDE, 3-ETHYLPHENYL
C14H20O8	006174-95-4	ETHENE TETRACARBOXYLIC ACID,TETRAETHYL ESTER
C14H21BrN2O2	071225-60-0	BENZAMIDE,N(DIETAMET),2-MEO,5-BR
C14H21ClN2O2	040256-75-5	BENZAMIDE,N(DIETAMET),2MEO-5CL
C14H21ClO	017199-24-5	2-CHLORO-4-(1,1,3,3-TETRAMETHYLBUTYL)PHENOL
C14H21FN2O2	055236-14-1	BENZAMIDE,N-(DIETAMET),2-MEO,5-F
C14H21FN4O3	141018-19-1	CYTIDINE, 2',3'-DIDEOXY-N-[(DIETHYLAMINO)METHYLE
C14H21N3O2	069218-55-9	2(5-CARBETHOXYPENTYL)BENZOTRIAZOLE
C14H21N3O3	004849-32-5	KASUGAMYCIN
C14H21N3O3	021738-42-1	6-QUINOLINEMETHANOL, 1,2,3,4-TETRAHYDRO-2-[[(1-M
C14H21N3O3S	000565-33-3	1-CYCLOHEXYL-3-(4-METHYLMETANILYL)UREA
C14H21N3O3S	001156-19-0	TOLAZAMIDE
C14H21N3O4	033629-47-9	BUTRALIN
C14H21N3O4S	075851-98-8	PROPANAMIDE, 3-(ACETYLAMINO)-N-[4-[(DIMETHYLAMIN
C14H21NO	005448-36-2	BENZAMIDE, 3-METHYL-N,N-BIS(1-METHYLETHYL)-
C14H21NO	006937-52-6	BENZAMIDE, 4-METHYL-N,N-BIS(1-METHYLETHYL)-
C14H21NO	016637-13-1	2,6-DIISOPROPYLACETANILIDE
C14H21NO	034251-46-2	BENZENEACETAMIDE, N,N-BIS(1-METHYLETHYL)-
C14H21NO2	000330-64-3	3,5-DI(IPR)-N-ME-PHENYLCARBAMATE
C14H21NO2	000644-26-8	1-(DIMETHYLAMINO)-2-METHYL-2-BUTANOL BENZOATE
C14H21NO2	014309-40-1	BENZOIC ACID, 4-AMINO-, HEPTYL ESTER
C14H21NO2	079606-43-2	BENZAMIDE, 4-METHOXY-N,N-BIS(1-METHYLETHYL)-
C14H21NO2	124497-80-9	2-OPR-2-PH-4-ME MORPHOLINE HCL
C14H21NO2	124497-83-2	2-ETHOXY-2-PH-3,4-ME MORPHOLINE HCL
C14H21NO4	087130-20-9	CARBAMIC ACID, (3,4-DIETHOXYPHENYL)-, 1-METHYLET
C14H21NO4	087130-88-9	CARBAMIC ACID, (3,4-DIETHOXYPHENYL)-, PROPYL EST
C14H21NO4Si	090936-53-1	1-(4-ETHYLPHENOXY)SILATRANE
C14H21NO4Si	090955-56-9	1-(3,5-DIMETHYLPHENOXY)SILATRANE
C14H21NO5	095261-60-2	CARBAMIC ACID, (3,4-DIETHOXYPHENYL)-, 2-METHOXYE
C14H21NOS	052888-80-9	PROSULFOCARB
C14H22	002189-60-8	OCTYLBENZENE
C14H22BrN3O2	004093-35-0	BROMOPRIDE
C14H22ClN3O2	000364-62-5	METOCLOPRAMIDE
C14H22ClNO2	014556-46-8	BUPRANOLOL
C14H22FN3O2	093414-59-6	BENZAMIDE, 4-AMINO-N-[2-(DIETHYLAMINO)ETHYL]-5-F
C14H22IN3O2	093414-60-9	BENZAMIDE, 4-AMINO-N-[2-(DIETHYLAMINO)ETHYL]-5-I
C14H22N2O	000137-58-6	LIDOCAINE
C14H22N2O	066999-80-2	BENZAMIDE,N-(3-DIETHYLAMINOPROPYL)
C14H22N2O	077470-88-3	HEXANAMIDE, 5-AMINO-N-(2,6-DIMETHYLPHENYL)-
C14H22N2O	102089-67-8	HEXANAMIDE, 6-AMINO-N-(2,6-DIMETHYLPHENYL)-
C14H22N2O2	005180-59-6	2,6-BIS(1,1-DIMETHYLETHYL)-4-NITROBENZENAMINE
C14H22N2O2	065016-34-4	O-MEO BENZAMIDE,N-(DIETAMINOET)
C14H22N2O3	006673-35-4	PRACTOLOL
C14H22N2O3	006673-38-7	ACETAMIDE, N-[3-[2-HYDROXY-3-[(1-METHYLETHYL)AMI
C14H22N2O3	029121-29-7	BENZENEACETAMIDE, 3-[2-HYDROXY-3-[(1-METHYLETHYL
C14H22N2O3	029122-68-7	ATENOLOL
C14H22N2O3S	075852-17-4	PENTANAMIDE, N-[[4-(DIMETHYLAMINO)SULFONYL]PHENY
C14H22N2O4S	004493-18-9	6-(N-HEXANOYLAMINO)PENCILLANIC ACID
C14H22N2O5	073632-86-7	1-CYCLOHEXENE-1-ACETIC ACID, _-METHYL-_-[[[(METH
C14H22N4O	089530-00-7	1-(4-CARBAMOYLPHENYL)-3-METHYL-3-HEXYLTRIAZENE
C14H22N4O3	141018-14-6	CYTIDINE, 2',3'-DIDEOXY-N-[(DIETHYLAMINO)METHYLE
C14H22N4O4	004093-42-9	2MEO-4NH2-5NO2 BENZAMIDE,N(DIETAMET)

C14H22N4O6S	126813-40-9	BENZENESULFONAMIDE, N-[2-(DIMETHYLAMINO)ETHYL]-N
C14H22N5O6PS	081835-01-0	8-S-T-BUTYL CYCLIC AMP
C14H22O	000096-76-4	2,4-DI-T-BUTYLPHENOL
C14H22O	000128-39-2	2,6-DI-T-BUTYLPHENOL
C14H22O	000140-66-9	P-(1,1,3,3-TETRAMETHYLBUTYL)PHENOL
C14H22O	005510-99-6	2,6-DI-SEC-BUTYLPHENOL
C14H22O2	083123-89-1	TR-CROTYL-CHRYSANTHEMATE
C14H22O2	096358-98-4	TR-METHYLALLYL-CHRYSANTHEMATE
C14H22O4S2	050780-71-7	PROPANEDIOIC ACID, 1,3-DITHIOLAN-2-YLIDENE-, BIS
C14H22O4S2	050780-72-8	PROPANEDIOIC ACID, 1,3-DITHIOLAN-2-YLIDENE-, BIS
C14H22O5S	071203-61-7	I-PENTYL-(345-MEO)PHENYL-SULFONE
C14H22O6	000109-16-0	TRIETHYLENEGLYCOL DIMETHACRYLATE
C14H22O7	017831-71-9	2-PROPENOIC ACID, OXYBIS(2,1-ETHANEDIYLOXY-2,1-E
C14H23IN2O	035041-48-6	3-CARBAMOYLPYRIDINIUM IODIDE,N-OCTYL
C14H23N3O2	003761-48-6	BENZAMIDE, 4-AMINO-N-[2-(DIETHYLAMINO)ETHYL]-2-M
C14H23N5O3	098204-09-2	ALLOPURINOL,1,N,N-DIPROPYLGLYCYLOXYMETHYL-
C14H23NO	000091-46-3	ETHANAMINE, N,N-DIMETHYL-2-[5-METHYL-2-(1-METHYL
C14H23NO2	035231-36-8	PHENOL, 4-[2-(DIMETHYLAMINO)ETHOXY]-2-METHYL-5-(
C14H23NO2	063474-05-5	2-PROPANOL, 1-[(1,1-DIMETHYLETHYL)AMINO]-3-(PHEN
C14H23NO2	064638-08-0	2,5-DIMEO-4-PROPYLAMPHETAMINE
C14H23NO2	105217-56-9	2-PROPANOL, 1-[(1-METHYLETHYL)AMINO]-3-(2-PHENYL
C14H23NO3	064778-75-2	3,5-DIMEO-4-BUO-PHENETHYLAMINE
C14H23NO3	105217-57-0	2-PROPANOL, 1-[(1-METHYLETHYL)AMINO]-3-(2-PHENOX
C14H23O4P	002528-36-1	DIBUTYL PHENYL PHOSPHATE
C14H24IN	013515-66-7	2-METHYLPYRIDINIUM IODIDE,N-OCTYL
C14H24N2O3	064810-90-8	BARBITURIC ACID,5-ET-5-OCTYL
C14H24N2O3	092598-80-6	1H-IMIDAZOLE-5-BUTANOIC ACID, α-ETHYL-ß-(HYDROXY
C14H25N	080121-60-4	1-AMINOADAMANTANE,3,5-DIETHYL
C14H25NO2	033602-03-8	2-PYRROLIDINONE, 1-(1-OXODECYL)-
C14H26N2O	109871-96-7	1H-PYRROL-1-YLOXY, 2,5-DIHYDRO-2,2,5,5-TETRAMETH
C14H26N2O2	133960-00-6	1(2H)-PYRIDINYLOXY, 3,6-DIHYDRO-2,2,6,6-TETRAMET
C14H26N4O4	137605-59-5	L-ALANINAMIDE, N2-(N-ACETYL-L-ALANYLGLYCYL)-N1-(
C14H26O2	014959-86-5	7-DODECEN-1-OL ACETATE (Z)
C14H26O2	016974-11-1	9-DODECEN-1-OL ACETATE (Z)
C14H26O2	029964-84-9	ISODECYL METHACRYLATE
C14H26O2	035148-19-7	9-DODECENYL ACETATE (TRANS)
C14H26O4	000110-40-7	DIETHYL SEBACATE
C14H27N3O3	132765-84-5	L-ISOLEUCINAMIDE, N-ACETYL-L-LEUCYL-
C14H27N3O4S2	066381-51-9	N-(N''-BU-BUO-CARBAMYLTHIO)METHOMYL
C14H28	001120-36-1	1-TETRADECENE
C14H28	001795-15-9	1-CYCLOHEXYLOCTANE
C14H28N4O3S2	074399-87-4	N-(N''-DI-ME-N'-HEX-URYLTHIO)METHOMYL
C14H28N5O2P	096662-66-7	TEPA 4-(PIPERIDINYL)-UREA DERIVATIVE
C14H28N5O3P	096662-64-5	1-PIPERIDINYLOXY, 4-[[[[BIS(1-AZIRIDINYL)PHOSPHI
C14H28NO3PS2	024151-93-7	PIPEROPHOS
C14H28O	034010-15-6	11-TETRADECEN-1-OL (Z)
C14H28O	035153-15-2	9-TETRADECEN-1-OL (Z)
C14H28O2	000106-33-2	ETHYL LAURATE
C14H28O2	000112-66-3	DODECYL ACETATE
C14H28O2	000544-63-8	TETRADECANOIC ACID
C14H28O2	001731-88-0	METHYL TRIDECANOATE
C14H28O2	030673-36-0	BUTYL DECANOATE
C14H28O6	040427-75-6	beta-D-GALACTOPYRANOSIDE, OCTYL
C14H29Br	000112-71-0	1-BROMO-N-TETRADECANE
C14H29Cl	002425-54-9	1-CHLOROTETRADECANE
C14H29I	019218-94-1	1-IODOTETRADECANE
C14H29N3O	133959-98-5	1H-PYRROL-1-YLOXY, 3-[[[2-(DIMETHYLAMINO)ETHYL]M
C14H29N5O	055921-56-7	24-NH2-6-DECYLAMINOPYRIMIDINE-3-OXIDE

C14H29NO2	000142-78-9	2-AMINOETHANE DODECANOATE
C14H30	000629-59-4	TETRADECANE
C14H30O	000112-72-1	1-TETRADECANOL
C14H31N	002016-42-4	1-TETRADECANAMINE
C14H31NO	001643-20-5	N,N-DIMETHYLDODECYLAMINE OXIDE
C14H42O5Si6	000107-52-8	TETRADECAMETHYLHEXASILOXANE
C14H42O7Si7	000107-50-6	TETRADECAMETHYLCYCLOHEPTASILOXANE
C15H7F3N6O4	132133-10-9	N(24-NO2-6CF3 PH)-2(35CN-6ME)PYRIDINAMIN
C15H8F5NO3	096327-60-5	1H-PYRROLIZINE-1-CARBOXYLIC ACID, 2,3-DIHYDRO-5-
C15H9BrO2	001146-98-1	1H-INDENE-1,3(2H)-DIONE, 2-(4-BROMOPHENYL)-
C15H9Cl2F3O3	057025-76-0	MC-15608 [2,4'-DICL-4-CF3-3'-CO2ME-DIPH ETHER]
C15H9Cl2FN2O	088695-06-1	1,4-BENZDAZEPIN-2-ON-5(2F-PH)-7,8-DICL
C15H9ClF2N2O	028910-86-3	1,4-BENZDIAZEPIN-2-ONE-5(26F-PH)7CL
C15H9ClF2N2O	088695-07-2	1,4-BENZDIAZEPIN-2-ONE-5(26F-PH)8CL
C15H9ClF3NO4S	061405-48-9	RH-1460 DIPH THIOETHER
C15H9ClF3NO5	050594-67-7	ACIFLUORFEN-ME
C15H9ClF3NO5	121325-44-8	RH-5348 DIPH ETHER
C15H9ClF3NO5S	143502-48-1	RH-8827 DIPHENYL SULFOXIDE
C15H9ClF3NO6S	143502-47-0	RH-8826 DIPHENYL SULFONE
C15H9ClO2	000129-35-1	1-CHLORO-2-METHYLANTHRAQUINONE
C15H9ClO2	001146-99-2	1H-INDENE-1,3(2H)-DIONE, 2-(4-CHLOROPHENYL)-
C15H9ClO2	001470-44-6	1H-INDENE-1,3(2H)-DIONE, 2-(3-CHLOROPHENYL)-
C15H9N	001210-12-4	9-ANTHRACENECARBONITRILE
C15H9NO4	000082-24-6	1-AMINOANTHRAQUINONE-2-CARBOXYLIC ACID
C15H9NO4	004649-27-8	4-CARBOXY-N-PHENYLPHATHALIMIDE
C15H9NS	007613-10-7	2-ISOTHIOCYANO-ANTHRACENE
C15H10	000203-64-5	4,5-METHYLENEPHENANTHRENE
C15H10Br2N2O	065247-10-1	14-BENZDIAZEPIN-2-ONE,5-(2-BRPH)7-BR
C15H10Br2N2O	065247-11-2	14-BENZDIAZEPIN-2-ONE,5-(3-BRPH)7-BR
C15H10Br2N2O	065247-12-3	14-BENZDIAZEPIN-2-ONE,5-(4-BRPH)7-BR
C15H10BrClN2O	051753-57-2	1,4-BENZDAZEPIN-2-ONE-5-(2CLPH)-7-BR
C15H10BrClN2O	063574-83-4	14-BENZDIAZEPIN-2-ONE,5-(2-BRPH)7-CL
C15H10BrClN2O	065247-13-4	14-BENZDIAZEPIN-2-ONE,5-(3-CLPH)7-BR
C15H10BrClN2O	065247-14-5	14-BENZDIAZEPIN-2-ONE,5-(4-CLPH)7-BR
C15H10Cl2N2O	002894-67-9	1,4-BENZDIAZEPIN-2-ONE-5(2-CLPH)-7-CL
C15H10Cl2N2O2	000846-49-1	LORAZEPAM
C15H10ClF3N2O6S	072178-02-0	FOMESAFEN
C15H10ClFN2O	001492-96-2	1,4-BENZODIAZEPIN-2-ONE,5(4FPH..
C15H10ClFN2O	002886-65-9	1,4-BENZDIAZPIN-2-ONE,5(2FPH)7CL
C15H10ClN3O3	001622-61-3	CLONAZEPAM
C15H10FN3O3	002558-30-7	1,4-BENZODIAZPIN-2-ONE,5-FPH,7NO2
C15H10N2O2	000101-68-8	DIPHENYL METHANE DIISOCYANATE
C15H10N4O5	004980-73-8	1,4-BENZDIAZP2-ON,5(2NO2PH)7NO2
C15H10O2	000083-12-5	2-PHENYL-1H-INDENE-1,3(2H)-DIONE
C15H10O2	000525-82-6	FLAVONE
C15H10O2	000723-62-6	ANTHRACENE-9-CARBOXYLIC ACID
C15H10O3	000082-39-3	1-METHOXY-9,10-ANTHRAQUINONE
C15H10O3	000491-78-1	5-HYDROXYFLAVONE
C15H10O3	003274-20-2	2-METHOXY-9,10-ANTHRACENEDIONE
C15H10O3	004143-63-9	2H-1-BENZOPYRAN-2-ONE, 2-(4-HYDROXYPHENYL)-
C15H10O3	006665-86-7	4H-1-BENZOPYRAN-4-ONE, 7-HYDROXY-2-PHENYL-
C15H10O4	000480-40-0	5,7-HYDROXYFLAVONE
C15H10O5	000520-36-5	APIGENIN
C15H10O5	019852-25-6	5,3',4'-TRIHYDROXYFLAVONE
C15H10O6	000491-70-3	LUTEOLIN
C15H10O7	000117-39-5	QUERCETIN
C15H10O7	000480-16-0	MORIN
C15H11BrN2O	002894-61-3	2H-1,4-BENZODIAZEPIN-2-ONE, 7-BROMO-1,3-DIHYDRO-

C15H11Cl2NO3	096327-58-1	1H-PYRROLIZINE-1-CARBOXYLIC ACID, 5-(2,4-DICHLOR
C15H11ClF3NO4	042874-03-3	OXYFLUORFEN
C15H11ClN2O	001088-11-5	1,4-BENZDIAZEPIN-2-ONE-5-PH-7-CL
C15H11ClN2O2	000604-75-1	OXAZEPAM
C15H11ClN2O2	000963-39-3	2H-1,4-BENZODIAZEPIN-2-ONE, 7-CHLORO-1,3-DIHYDRO
C15H11ClO2S	001152-72-3	3-BUTYN-1-OL, 2-CHLORO-4[5-(1,3-PENTADIYNYL)-2-
C15H11ClO3	002536-31-4	CHLORFLURECOL METHYL
C15H11F2NO3	096327-56-9	1H-PYRROLIZINE-1-CARBOXYLIC ACID, 5-(2,6-DIFLUOR
C15H11F2NO3	096327-57-0	1H-PYRROLIZINE-1-CARBOXYLIC ACID, 5-(2,4-DIFLUOR
C15H11FN2O	002648-00-2	1,4-BENZODIAZPIN-2-ONE,5-PH,7-F
C15H11FN2O	002648-01-3	1,4-BENZODIAZPIN-2-ONE,5-(2-F PH)
C15H11I4NO4	000051-48-9	THYROXINE
C15H11N	000612-96-4	2-PHENYLQUINOLINE
C15H11N3	001148-79-4	2,2':6',2"-TERPYRIDINE
C15H11N3O3	000146-22-5	NITRAZEPAM
C15H11NO	038035-81-3	2(1H)-QUINOLINONE, 3-PHENYL-
C15H11NO2	000082-28-0	1-AMINO-2-METHYL-9,10-ANTHRACENEDIONE
C15H11NO2	000082-38-2	DISPERSE RED 9
C15H11NO2S	089770-30-9	6-PHENYLSULFONYLQUINOLINE
C15H11NO2S	089770-33-2	8-PHENYLSULFONYLQUINOLINE
C15H11NO4	010245-51-9	1H-BENZ[DE]ISOQUINOLINE-1,3(2H)-DIONE, 2-[(ACETY
C15H11NO4S	074134-16-0	ACETAMIDE, N-(9-OXO-9H-THIOXANTHEN-3-YL)-, S,S-D
C15H11NS	010319-35-4	4-ISOTHIOCYANO-STILBENE
C15H12	000610-48-0	1-METHYLANTHRACENE
C15H12	000613-12-7	2-METHYLANTHRACENE
C15H12	000779-02-2	9-METHYLANTHRACENE
C15H12	000832-69-9	1-METHYLPHENANTHRENE
C15H12	000832-71-3	3-METHYLPHENANTHRENE
C15H12	002531-84-2	2-METHYLPHENANTHRENE
C15H12Br4O2	000079-94-7	2,2-BIS(4-HYDROXY-3,5-DIBROMOPHENYL)PROPANE
C15H12BrClN4S	057801-81-7	BROTIZOLAM
C15H12BrNO3	096327-34-3	1H-PYRROLIZINE-1-CARBOXYLIC ACID, 5-(4-BROMOBENZ
C15H12Cl2	003141-42-2	BENZENE, 1,1'-(2,2-DICHLOROCYCLOPROPYLIDENE)BIS-
C15H12Cl2O3	000094-83-7	2,4-DEB
C15H12Cl2O4	040843-25-2	DICLOFOP
C15H12Cl4	017925-97-2	CHLOROMETHYLCHLOR
C15H12ClN3O	007722-15-8	3H-1,4-BENZODIAZEPIN-2-AMINE, 7-CHLORO-5-PHENYL-
C15H12ClN3O4	107485-46-1	BENZAMIDE, N-[[(4-CHLOROPHENYL)AMINO]CARBONYL]-2
C15H12ClN5O4	070528-90-4	AZO DYE N2
C15H12FNO	089242-82-0	3-PYRIDINEMETHANOL,A-2-PROPYNYL,A-4-FLUOROPHENYL
C15H12FNO3	029736-22-9	P-FLUOROPHENYLHIPPURATE
C15H12FNO3	066635-90-3	1H-PYRROLIZINE-1-CARBOXYLIC ACID, 5-(4-FLUOROBEN
C15H12I3NO4	006893-02-3	O-(4-HYDROXY-3-IODOPHENYL)-3,5-DIIODO-L-TYROSINE
C15H12N2	061453-53-0	2-ME-4-PHENYLQUINOZALINE
C15H12N2O	000298-46-4	CARBAMAZEPINE
C15H12N2O	002898-08-0	1,4-BENZDIAZEPIN-2-ONE-5-PHENYL
C15H12N2O	013961-64-3	2(1H)-QUINAZOLINONE, 6-METHYL-4-PHENYL-
C15H12N2O	017629-04-8	QUINAZOLIN-2-ONE,1-ME-4-PHENYL
C15H12N2O	017629-09-3	2-METHOXY-4-PHENYLQUINAZOLINE
C15H12N2O	089246-38-8	ACETANILIDE,N-(4-CYANOPHENYL)-
C15H12N2O	089246-40-2	BENZENEACETAMIDE, N-(3-CYANOPHENYL)-
C15H12N2O2	000057-41-0	PHENYTOIN
C15H12N2O2	001220-94-6	C.I. DISPERSE VIOLET 4
C15H12N2O2	002652-77-9	2,5-PYRAZOLIDINDIONE,1,2-DIPHENYL
C15H12N2O3	002784-27-2	HYDANTOIN,5-PH-5-(P-HYDROXY)PH
C15H12N2O3	002872-48-2	DISPERSE RED 11
C15H12N2O3	072570-99-1	ACETAMIDE, N-(7-NITRO-9H-FLUOREN-2-YL)-
C15H12N2O5	002979-53-5	M-NITROPHENYLHIPPURATE

C15H12N4O5	027091-58-3	5(AZIRIDINYL)-2,4-DINO2-N-PH BENZAMIDE
C15H12NO	000092-71-7	OXAZOLE, 2,5-DIPHENYL-
C15H12O	000094-41-7	BENZALACETOPHENONE
C15H12O2	000487-26-3	FLAVANONE
C15H12O2	007011-98-5	1(3H)-ISOBENZOFURANONE-3-BENZYL
C15H12O2	020252-43-1	ACETIC ACID, 2-HEPTEN-4,6-DIYNYL ESTER
C15H12O3	000491-59-8	CHRYSAROBIN
C15H12O3	000606-28-0	BENZOIC ACID, 2-BENZOYL-, METHYL ESTER
C15H12O4	000134-55-4	BENZOIC ACID, 2-(ACETYLOXY)-, PHENYL ESTER
C15H12O5	000480-41-1	NARINGENINE
C15H12O6	000552-58-9	ERIODICTYOL
C15H12O6	020725-03-5	FUSTIN
C15H12O7	000480-18-2	3,5,7,3',4'-PENTAHYDROXYFLAVANONE
C15H13BrClF3N2O	122453-73-0	AC 303,630
C15H13BrO3	005884-48-0	PHENYLACETIC ACID,3-BR-4-PHMEO
C15H13Cl2F2NO3	138926-22-4	RH-1224 O-PH N-PENTYLENE(F2)CARBAMATE
C15H13Cl2NO	086886-77-3	BENZENEPROPANAMIDE, N-(3,4-DICHLOROPHENYL)-
C15H13Cl2NO2	000117-27-1	PROLAN
C15H13Cl2NO2	015307-71-8	PHENYLACETIC ACID,2-(2',6'-DICHLORO-3-METHYL)ANI
C15H13Cl2NO3	118409-80-6	PHNEYLACETIC ACID,2-(2',6'-DICHLORO-4-METHOXY)AN
C15H13ClN2O2	108611-54-7	3-PYRIDINECARBONITRILE, 2-CHLORO-5-(3,4-DIMETHOX
C15H13ClN2O3	107485-58-5	BENZAMIDE, N-[[(4-CHLOROPHENYL)AMINO]CARBONYL]-2
C15H13ClO3	060736-83-6	PHENYLACETIC ACID,3-CL-4-PHMEO
C15H13ClO3S	061820-94-8	ETHANONE,1-(4-CHLOROPHENYL)-2-[(4-METHYLPHENYL)SULF
C15H13FO2	005104-49-4	FLURBIPROFEN
C15H13N3	089770-35-4	2-METHYLAMINO-4-PHENYLQUINAZOLINE
C15H13N3O	004928-02-3	1,4-BENZDIAZEPIN-2-ONE-5-PH-7-NH2
C15H13N3O4S	036322-90-4	PIROXICAM
C15H13NO	000053-96-3	N-2-FLUORENYLACETAMIDE
C15H13NO	001022-66-8	2(1H)-QUINOLINONE, 3,4-DIHYDRO-3-PHENYL-
C15H13NO	003056-73-3	CINNAMANILIDE
C15H13NO	004888-33-9	2(1H)-QUINOLINONE, 3,4-DIHYDRO-4-PHENYL
C15H13NO	089242-81-9	3-PYRIDINEMETHANOL, α-PHENYL-α-(2-PROPYNYL)-
C15H13NO2	032557-55-4	2-BENZOYL-N-METHYLBENZAMIDE
C15H13NO2	119005-15-1	1,1'-BIPHENYL, 4-[(2-OXOAZETIDIN-4-YL)OXY]-
C15H13NO3	002979-54-6	PHENYLHIPPURATE
C15H13NO3	027696-28-2	BENZOIC ACID, 2-[(3-ACETYLPHENYL)AMINO]-
C15H13NO4	001878-89-3	PHENOXYACETIC ACID, M-BENZAMIDO
C15H14	002294-82-8	9-ETHYL-9H-FLUORENE
C15H14Cl2N4O3	006232-56-0	C.I. DISPERSE ORGANGE 5
C15H14ClN3O2	065542-14-5	BENZOIC ACID, 4-[3-[(4-CHLOROPHENYL)METHYL]-3-ME
C15H14ClNO2	023189-28-8	PHENYLACETIC ACID,2-(2'-CL-6-ME)ANILINO
C15H14ClNO2	037984-36-4	PHENYLACETIC ACID,2-(2'-ME-3'-CL)ANILINO
C15H14F3N3O4S2	000073-48-3	BENDROFLUMETHIAZIDE
C15H14FN3O3	078755-81-4	FLUMAZENIL
C15H14N2O2	120236-88-6	1H-PYRROLO[3,2-C]PYRIDINE, 2-(2,4-DIMETHOXYPHENY
C15H14N2O3	030022-13-0	P-AMINOPHENYLHIPPURATE
C15H14N2O3	085879-19-2	3-PHO BENZALDEHYDE,O-((MEAM)CO)OXIME
C15H14N4	065542-21-4	P(3-BENZYL-3ME-1TRIAZENO)BENZONITRILE
C15H14N4	120137-95-3	[3,4'-BIPYRIDINE]-5-CARBONITRILE,6-(1-PYRROLIDIN
C15H14N4O2S	000526-08-9	SULFAPHENAZOLE
C15H14N4O4	065542-15-6	P(3-PNO2BENZYL-3ME-1TRIAZN)BENZOIC AC
C15H14NO2PS	013067-93-1	CYANOFENPHOS
C15H14O	000102-04-5	DIBENZYL KETONE
C15H14O2	000094-47-3	BENZOIC ACID, 2-PHENYLETHYL ESTER
C15H14O2	000118-58-1	BENZYL SALICYLATE
C15H14O2	005558-66-7	A,A-DIPHENYLPROPIONIC ACID
C15H14O3	006547-53-1	PHENYLACETIC ACID,P-PHENYLMETHOXY

C15H14O3S	031378-03-7	PHENACYL P-TOLYL SULFONE
C15H14O3S	038488-14-1	ETHANONE, 1-(4-METHYLPHENYL)-2-(PHENYLSULFONYL)-
C15H14O4S	027918-37-2	PHENYLSULFONYL(P-METHOXYBENZOYL)METHANE
C15H14O6	000154-23-4	CIANIDANOL
C15H15Cl2NO3	087374-78-5	RH-6251 O-PH N-PENTYLENE CARBAMATE
C15H15ClN2O2	001982-47-4	CHLOROXURON
C15H15ClN2O2	054922-68-8	N'-BENZYL-(4-CLPHO)ACETICACIDHYDRAZIDE
C15H15ClN2S	002095-17-2	10H-PHENOTHIAZINE-10-PROPANAMINE,2-CHLORO-
C15H15ClN4O6S	090982-32-4	CHLORIMURON-ETHYL (PH 7)
C15H15ClO	032669-06-0	2-CHLOROETHYLBENZHYDRYL ETHER
C15H15F3N2O2	056425-91-3	FLURPRIMIDOL
C15H15N	007496-81-3	2-(4-TOLYL)INDOLIZINE
C15H15N	016620-75-0	P,P'-DIMETHYLBENZOPHENONEIMINE
C15H15N	022627-01-6	O,O'-DIMETHYLBENZOPHENONEIMINE
C15H15N	022627-02-7	2,6-DIMETHYLBENZOPHENONEIMINE
C15H15N	082682-69-7	2-(3-TOLYL)INDOLIZINE
C15H15N3O2	000493-52-7	2-((4-(DIMETHYLAMINO)PHENYL)AZO)BENZOIC ACID
C15H15N3O2	002832-40-8	4'-((6-HYDROXY-M-TOLYL)AZO)ACETANILIDE
C15H15N3O2	006268-49-1	4-((4-(DIMETHYLAMINO)PHENYL)AZO)BENZOIC ACID
C15H15N3O2	020691-84-3	3-((4-(DIMETHYLAMINO)PHENYL)AZO)BENZOIC ACID
C15H15N3O2	065587-38-4	BENZOIC ACID, 4-[3-METHYL-3-(PHENYLMETHYL)-1-TRI
C15H15N3O3	111075-97-9	1-ACETYL-4-(4-OPH PHENYL) SEMICARBAZIDE
C15H15N3O3S	102361-46-6	IMIDAZO[1,2-A]PYRAZINE, 8-METHOXY-2-[2-METHOXY-4
C15H15N5O4	013425-39-3	ETOFYLLINE NICOTINATE
C15H15NO	006876-65-9	ACETAMIDE, N-PHENYL-N-(4-METHYLPHENYL)-
C15H15NO	007496-82-4	2-(4-METHOXYPHENYL)INDOLIZINE
C15H15NO	050916-16-0	BENZENEACETAMIDE, N-(3-METHYLPHENYL)-
C15H15NO	082682-68-6	2-(3-METHOXYPHENYL)INDOLIZINE
C15H15NO2	000061-68-7	N-PH-ANTHRANILIC ACID,2',3'-DIMETHYL
C15H15NO2	018859-21-7	N-METHYLPHENOXYACETANILIDE
C15H15NO2	050916-19-3	BENZENEACETAMIDE, N-(3-METHOXYPHENYL)-
C15H15NO2	050916-21-7	BENZENEACETAMIDE, N-(4-METHOXYPHENYL)-
C15H15NO2	065261-13-4	BENZAMIDE, N,N-DIMETHYL-3-PHENOXY-
C15H15NO2	088837-70-1	1,1-DIPHENYL-2-NITROPROPANE
C15H15NO2	095384-60-4	BENZENEACETAMIDE, N-(2-METHOXYPHENYL)-
C15H15NO2	117574-40-0	2,5-CYCLOHEXADIENE-1,4-DIONE, 3-[(3-PYRIDINYL)ME
C15H15NO2S	138568-64-6	BENZENEACETIC ACID, α-METHYL-4-[4-(2-PROPENYL)-2
C15H15NO3	026171-23-3	TOLMETIN
C15H15NO3	056356-16-2	BENZOIC ACID, 4-AMINO-2-HYDROXY-, 2,6-DIMETHYLPH
C15H15NO3	082568-69-2	BENZAMIDE, 2-[(4-METHOXYPHENYL)METHOXY]-
C15H15NO3S	010533-83-2	BENZAMIDE, N-METHYL-N-[(4-METHYLPHENYL)SULFONYL]
C15H15NO4S	095384-56-8	BENZENEACETAMIDE, N-[3-[(METHYLSULFONYL)OXY]PHEN
C15H16	013540-50-6	PHENYL XYLYLMETHANE
C15H16	020282-30-8	3-ISOPROPYLBIPHENYL
C15H16	025640-78-2	ISOPROPYLBIPHENYL
C15H16ClNO2S	138568-80-6	BENZENEACETIC ACID, 3-CHLORO-α-METHYL-4-[4-(1-ME
C15H16ClNO2S	138568-85-1	BENZENEACETIC ACID, 3-CHLORO-4-(4-ETHYL-5-METHYL
C15H16FNO2S	138568-72-6	BENZENEACETIC ACID, 3-FLUORO-α-METHYL-4-[4-(1-ME
C15H16FNO2S	138568-77-1	BENZENEACETIC ACID, 4-(4-ETHYL-5-METHYL-2-THIAZO
C15H16HgNO2S	000517-16-8	CERESAN M
C15H16N2	006319-26-2	(2,6-DIMETHYLPHENYL)(2-METHYLPHENYL)DIAZENE
C15H16N2	029418-23-3	2,2',4-TRIMETHYLAZOBENZENE
C15H16N2	029418-24-4	2,4,4'-TRIMETHYLAZOBENZENE
C15H16N2	029418-26-6	2,4,6-TRIMETHYLAZOBENZENE
C15H16N2	029418-27-7	2,4',6-TRIMETHYLAZOBENZENE
C15H16N2O	002990-01-4	1,1-DIPHENYL-3,3-DIMETHYLUREA
C15H16N2O	029418-45-9	4-METHOXY-O,O'-AZOTOLUENE
C15H16N2O	095384-53-5	BENZENEACETAMIDE, N-[3-(METHYLAMINO)PHENYL]-

C15H16N2O2	012771-68-5	ANCYMIDOL
C15H16N2O2	029418-49-3	2,4-DIMETHOXY-2'-METHYLAZOBENZENE
C15H16N2O2	029418-50-6	2,4-DIMETHOXY-4'-METHYLAZOBENZENE
C15H16N2O2	029418-51-7	2,6-DIMETHOXY-2'-METHYLAZOBENZENE
C15H16N2O2	029418-52-8	2,6-DIMETHOXY-4'-METHYLAZOBENZENE
C15H16N2O2	029418-53-9	4,4'-DIMETHOXY-2-METHYLAZOBENZENE
C15H16N2O2	052093-78-4	N-BENZYL-PHENOXY ACETIC HYDRAZIDE
C15H16N2O3S	020037-50-7	BENZAMIDE, N-[(4-AMINOPHENYL)SULFONYL]-2,5-DIMET
C15H16N2O3S	132483-52-4	BENZENEACETIC ACID, 4-[4-[(DIMETHYLAMINO)CARBONY
C15H16N2O4S	000061-33-6	BENZYLPENICILLIN
C15H16N2O5	124897-29-6	1,8-NAPHTHYRIDINE-3-CARBOXYLIC ACID, 1-ETHYL-7-M
C15H16N4O	059708-24-6	P-(3-BENZYL-3ME-1-TRIAZENO)BENZAMIDE
C15H16N4O	108610-80-6	[3,4'-BIPYRIDINE]-5-CARBONITRILE,6-[[1-(HYDROXYM
C15H16N4O	108610-85-1	[3,4'-BIPYRIDINE]-5-CARBONITRILE,6-[(4-HYDROXYBU
C15H16N4O	120137-97-5	[3,4'-BIPYRIDINE]-5-CARBONITRILE,6-[(3-METHOXYPR
C15H16N4O2	004313-13-7	DIS. A. 16
C15H16N4O2	004313-14-8	DIS. A. 17
C15H16N4O2	075922-48-4	1,3-DIETHYL-8-PHENYLXANTHINE
C15H16N4O5S	074222-97-2	SULFOMETURON (pH 5-7)
C15H16N6	077314-69-3	2-PYRIDINAMINE, 4-(5-AMINO-1H-1,2,4-TRIAZOL-3-YL
C15H16N6	077314-73-9	2-PYRIDINAMINE, 4-(5-AMINO-1H-1,2,4-TRIAZOL-3-YL
C15H16O2	000080-05-7	DIPHENYLOLPROPANE
C15H16O3	101705-35-5	PENTANOIC ACID, 5-(1-NAPHTHALENYLOXY)-
C15H16O6	010247-71-9	5H-3,6-METHENOFURO[3,2-G]OXIRENO[D]OXACYCLOUNDEC
C15H17Br2NO2	001689-99-2	BROMOXYNIL OCTANOATE
C15H17BrN4O	014337-53-2	2(5BR-2-PYRIDYLAZO)-5(DIETAM)PHENOL
C15H17Cl2NO4	133636-96-1	RH-1964 O-PH(DICL) N-BUTYLENE CARBAMATE
C15H17ClF3N3O	068694-11-1	TRIFLUMIZOLE
C15H17ClFNO4	133636-98-3	RH-1422 O-PH N-BUTYLENE CARBAMATE
C15H17FN4O3	074011-58-8	ENOXACIN
C15H17N	004275-43-8	A-BENYZL BENZENEETHANEAMINE
C15H17N3	000055-80-1	4-DIMETHYLAMINO-3'-METHYLAZOBENZENE
C15H17N5O3	083166-77-2	24NH2-PYRIMIDIN,5(3,5-MEO-4OCH2CN)BENZYL
C15H17N5O3	098204-05-8	ALLOPURINOL,1-PH-ALANYL-OME-
C15H17N7O5S3	056796-20-4	5-THIA-1-AZABICYCLO[4.2.0]OCT-2-ENE-2-CARBOXYLIC
C15H17NaO3S	097683-31-3	1-SO3NA-3-ET-7-ISOPROPYL AZULENE
C15H17NaO3S	099287-40-8	1-SO3NA-3-PENTYL AZULENE
C15H17NaO4S	099287-58-8	1-SO3NA-3-BUTYL-4-METHOXY AZULENE
C15H17NaO4S	099287-64-6	1-SO3NA-3-ME-4-METHOXY-7-IPR AZULENE
C15H17NO2S	033597-78-3	BENZENAMINE, 4-[(2,4,6-TRIMETHYLPHENYL)SULFONYL]
C15H17NO2S	138568-63-5	BENZENEACETIC ACID, α-METHYL-4-[4-(1-METHYLETHYL
C15H17NO2S	138568-68-0	BENZENEACETIC ACID, 4-(4-ETHYL-5-METHYL-2-THIAZO
C15H17NO3	023434-86-8	1H-INDENE-1,3(2H)-DIONE, 5-[3-OXO-3-(1-PIPERIDIN
C15H17NO3	106231-58-7	ACETAMIDE, 2-(BENZOYLOXY)-N,N-BIS(2-PROPENYL)-
C15H17NO3	131042-61-0	2-AZASPIRO[4.4]NONANE-1,3-DIONE, 2-(PHENYLMETHOX
C15H17NO3	136944-48-4	2-NAPHTHALENECARBOXAMIDE, N-BUTYL-6,7-DIHYDROXY-
C15H17NO5	115178-76-2	L-PROLINE, N-[(BENZOYLOXY)ACETYL]-, METHYL ESTER
C15H17NO6	116482-80-5	BENZOIC ACID, 2-(ACETYLOXY)-, 2-(4-MORPHOLINYL)-
C15H17NO7	118247-03-3	GLYCINE, N-[[[2-(ACETYLOXY)BENZOYL]OXY]ACETYL]-,
C15H18Cl2N2O3	019666-30-9	OXADIAZON
C15H18Cl2N2O4	039961-02-9	CYCLOPROPANECARBOXAMIDE, N-[4-[2-[(DICHLOROACETY
C15H18Cl3N3O2	067747-09-5	PROCHLORAZ
C15H18ClN3O	100199-27-7	CYCLOHEPTANOL, 1-(4-CHLOROPHENYL)-2-(1H-1,2,4-TR
C15H18ClN3O3S	032527-55-2	TIARAMIDE
C15H18N2O3	061346-87-0	BARBITURIC ACID,1-ME-5-ET-5(PHET)
C15H18N2O3	085432-36-6	BARBITURIC ACID,1-IPR-5-ET-5(PH)
C15H18N2O3	085432-37-7	4,6(1H,5H)-PYRIMIDINEDIONE, 5-ETHYL-2-(1-METHYLE
C15H18N2O3	085432-39-9	2,4(3H,5H)-PYRIMIDINEDIONE, 5-ETHYL-6-(1-METHYLE

C15H18N2O4S3	026722-85-0	CEPHALOTHIN ANALOG (3-CH2-S-CH3)
C15H18N2O6	000485-31-4	BINAPACRYL
C15H18N4O2	087666-27-1	UREA, N-METHYL-N'-[3-(6-OXO-3-PHENYL-1(6H)-PYRID
C15H18N4O2	087666-31-7	UREA, [4-(6-OXO-3-PHENYL-1(6H)-PYRIDAZINYL)BUTYL
C15H18N4O3	055687-76-8	2,4-NH2PYRIMIDN,5(3,5MEO-4-AC)BENZYL
C15H18N4O4	055687-40-6	24-NH2PYRIMIDN,5(35MEO-4-CO2ME)BENZYL
C15H18N4O5	000050-07-7	MITOMYCIN C
C15H18O3	000481-06-1	SANTONIN
C15H18O3	000509-93-3	AMBROSIN
C15H18O3	005945-42-6	AROMATICIN
C15H18O4	000508-59-8	PARTHENIN
C15H18O4	005945-41-5	MEXICANIN-1
C15H18O4	006754-13-8	HELENALIN
C15H18O4	015725-27-6	ETHYLACETOACETATE, 3-ETO BENZAL
C15H18O4	015725-33-4	DIETHYLMALONATE,3-ME BENZAL
C15H18O4	024331-75-7	DIETHYLMALONATE,2-ME BENZAL
C15H18O5	006768-22-5	DIETHYLMALONATE,2-MEO BENZAL
C15H18O5	006768-23-6	DIETHYLMALONATE,4-MEO BENZAL
C15H18O5	006771-54-6	DIETHYLMALONATE,3-MEO BENZAL
C15H18O5	015725-32-3	ETHYLACETOACETATE,2,4-DIMEO BENZAL
C15H18O5	020451-62-1	ETHYLACETOACETATE,3,4-DIMEO BENZAL
C15H18O6	024331-83-7	DIETHYLMALONATE,3-MEO 4-OH BENZAL
C15H19BrN2O5	078266-06-5	BR-N(246-MEPH)CARBAMOYLME IMINODIACETIC ACID
C15H19BrN4O2	004891-98-9	2,4-PYRIMIDINEDIAMINE, 5-[(2-BROMO-5-METHOXY-4-P
C15H19Cl2FN2O3	124840-52-4	BENZAMIDE, 3,5-DICHLORO-N-[[1-(2-FLUOROETHYL)-2-
C15H19Cl2N3O	083657-24-3	DINICONAZOLE
C15H19Cl2N3O2	060207-90-1	PROPICONAZOLE
C15H19ClN2O2	105801-53-4	1-PIPERIDINEACETAMIDE, N-(5-CHLORO-2,3-DIHYDRO-3
C15H19ClN4O3	034205-21-5	DIMEFURON
C15H19N	148727-67-7	4A(4H)-PHENANTHRENAMINE, 1,9,10,10A-TETRAHYDRO-9
C15H19N3O	131229-40-8	IMIDAZO[5,1-B]QUINAZOLIN-9(2H)-ONE, 2-BUTYL-1,3-
C15H19N3O5	024279-91-2	2,5-CYCLOHEXADIENE-1,4-DIONE, 2-[2-[(AMINOCARBON
C15H19N3O8	006742-07-0	2(1H)-PYRIMIDINONE, 4-AMINO-1-(2,3,5-TRI-O-ACETY
C15H19N5O7S	094593-91-6	CINOSULFURON
C15H19NO	000054-80-8	PRONETHALOL
C15H19NO	089242-79-5	3-PYRIDINEMETHANOL, α-CYCLOHEXYL-α-(2-PROPYNYL)-
C15H19NO	148727-74-6	9(1H)-PHENANTHRENONE, 2,3,4,4A,10,10A-HEXAHYDRO-
C15H19NO2	000537-26-8	TROPACOCAINE
C15H19NO3	115178-68-2	1H-AZEPINE, HEXAHYDRO-1-[(BENZOYLOXY)ACETYL]-
C15H19NO5	116482-56-5	BENZOIC ACID, 2-(ACETYLOXY)-, 2-(DIETHYLAMINO)-2
C15H19NO5	118247-05-5	BENZOIC ACID, 2-(ACETYLOXY)-, 2-[(2-HYDROXYETHYL
C15H20ClN3O	076738-62-0	PACLOBUTRAZOL
C15H20ClN3O	129586-32-9	SSF-109 FUNGICIDE
C15H20ClNO	097871-63-1	P-CHLOROCINNAMAMIDE,B-ET,N-SEC-BUTYL
C15H20ClNO4	060145-79-1	GLYCINE, N-(CHLOROACETYL)-N-[2-(1-METHYLETHOXY)P
C15H20FNO2	072836-48-7	P-F CINNAMIC ACID,DIETHYLAMINOETHYLESTER
C15H20N2O	143202-51-1	1H-PYRIDO[4,3-B]INDOLE, 5-ACETYL-2,3,4,4A,5,9B-H
C15H20N2O2	097871-54-0	A-ACETYLAMINO-N-SEC-BUTYL CINNAMAMIDE
C15H20N2O2	098154-95-1	CARBAMIC ACID, DIETHYL-, 2-(1H-INDOL-3-YL)ETHYL
C15H20N2O2	105801-48-7	1-PIPERIDINEACETAMIDE, N-(2,3-DIHYDRO-3-BENZOFUR
C15H20N2O2	105801-57-8	1-PYRROLIDINEACETAMIDE, N-(2,3-DIHYDRO-5-METHYL-
C15H20N2O3	105801-59-0	4-MORPHOLINEACETAMIDE, N-(2,3-DIHYDRO-5-METHYL-3
C15H20N2O4	073632-78-7	BUTANOIC ACID, 2-[[(AMINOCARBONYL)AMINO]CARBONYL
C15H20N2O4	105801-63-6	4-MORPHOLINEACETAMIDE, N-(2,3-DIHYDRO-6-METHOXY-
C15H20N2O4S	000968-81-0	DIMELIN
C15H20N2O4S	068840-05-1	CARBOFURAN,N-(N'-ME,MECARBAMIDOTHIO)
C15H20N2O5	102273-19-8	BENZAMIDE, 2-[(ACETYLOXY)METHOXY]-N-(4-MORPHOLIN
C15H20N2O5S	065907-26-8	N(THIO-N,O-DIME CARBAMYL)CARBOFURAN

C15H20N4O	073264-20-7	2,4-PYRIMIDINEDIAMINE, 5-[[4-METHOXY-3-(1-METHYL
C15H20N4O	073264-21-8	2,4-PYRIMIDINEDIAMINE, 5-[[3-METHOXY-4-(1-METHYL
C15H20N4O2	078026-01-4	24NH2-PYRIMIDINE,5(3,5-MEO-4-ET)BENZYL
C15H20N4O2	080267-16-9	2,4-PYRIMIDINEDIAMINE, 5-[(3,4-DIETHOXYPHENYL)ME
C15H20N4O2	080267-17-0	2,4-PYRIMIDINEDIAMINE, 5-[(3-METHOXY-4-PROPOXYPH
C15H20N4O2	083166-75-0	24NH2PYRIMIDIN,5(3-MEO-4-C(OH)ME2)BENZYL
C15H20N4O2	085555-96-0	2,4-PYRIMIDINEDIAMINE, 5-[(4-METHOXY-3-PROPOXYPH
C15H20N4O2	100515-03-5	2,4-PYRIMIDINEDIAMINE, 5-[(3,5-DIETHOXYPHENYL)ME
C15H20N4O2S	071125-73-0	24NH2-PYRIMIDINE,5(3,5-MEO-4-ETS)BENZYL
C15H20N4O3	055687-86-0	24NH2PYRIMIDIN,5(35-MEO-4-MEO-CH2)BENZYL
C15H20N4O3	078025-68-0	24NH2-PYRIMIDINE,5(3,5-MEO-4-ETO)BENZYL
C15H20N4O4	132765-99-2	L-PHENYLALANINAMIDE, N -ACETYL-L-ASPARAGINYL-
C15H20O2	000470-17-7	NAPHTHO[2,3-B]FURAN-2(3H)-ONE, DECAHYDRO-8A-METH
C15H20O2	000546-43-0	ALANTOLACTONE
C15H20O4	002571-81-5	AZULENO[4,5-B]FURAN-2,9-DIONE, DECAHYDRO-6A-HYDR
C15H20O4	017066-68-1	AZULENO[4,5-B]FURAN-2,9-DIONE, DECAHYDRO-4-HYDRO
C15H21BrN2O3	084226-14-2	FLA 797
C15H21Cl2N3O	075736-33-3	DICLOBUTRAZOL ISOMERS
C15H21FN4O3	141018-22-6	CYTIDINE, 2',3'-DIDEOXY-3'-FLUORO-N-[(1-PIPERIDI
C15H21N	001209-98-9	FENCAMFAMINE
C15H21N3O2	000057-47-6	PHYSOSTIGMINE
C15H21N3O2	066308-35-8	CARBAMIC ACID, [5-(2,3-DIMETHYLPHENYL)-4,5-DIHYD
C15H21N5O2	056066-43-4	24NH2PYRIMIDINE,5(3,5-ETO-4-NH2)BENZYL
C15H21N5O2	056066-63-8	2,4-PYRIMIDINEDIAMINE, 5-[[4-(DIMETHYLAMINO)-3,5
C15H21N5O3S	023374-51-8	ß-D-ALANINE, N-[1-OXO-5-(1H-PURIN-6-YLTHIO)PENTY
C15H21N5O3S	028610-09-5	GLYCINE, N-[1-OXO-5-(1H-PURIN-6-YLTHIO)PENTYL]-,
C15H21NO	013552-46-0	1-PROPANONE, 1-(4-METHYLPHENYL)-3-(1-PIPERIDINYL
C15H21NO	022385-99-5	1-PROPANONE, 2-METHYL-1-PHENYL-3-(1-PIPERIDINYL)
C15H21NO	125476-98-4	1-PROPANONE, 1-(2-METHYLPHENYL)-3-(1-PIPERIDINYL
C15H21NO	132452-02-9	3-PHENANTHRENOL, 1,2,3,4,4A,9,10,10A-OCTAHYDRO-4
C15H21NO2	000057-42-1	DEMEROL
C15H21NO2	000500-34-5	BETA-EUCAINE
C15H21NO2	010369-88-7	2-PROPENOIC ACID, 3-PHENYL-, 2-(DIETHYLAMINO)ETH
C15H21NO2	088058-85-9	2H-NAPHTH[1,2-B]-1,4-OXAZIN-9-OL,3,4,4A,5,6,10B-
C15H21NO2	094921-29-6	2H-NAPHTH[1,2-B]-1,4-OXAZIN-7-OL,3,4,4A,5,6,10B-
C15H21NO2	097871-55-1	A-ETO-N-SEC-BUTYL CINNAMAMIDE
C15H21NO3	100668-11-9	2-PROPENAMIDE, 3-(3,4-DIHYDROXYPHENYL)-N-HEXYL-
C15H21NO3	106231-55-4	ACETAMIDE, 2-(BENZOYLOXY)-N,N-DIPROPYL-
C15H21NO3	106231-56-5	ACETAMIDE, 2-(BENZOYLOXY)-N,N-BIS(1-METHYLETHYL)
C15H21NO3	115193-29-8	ACETAMIDE, 2-(BENZOYLOXY)-N-HEXYL-
C15H21NO3	136944-37-1	2-PROPENAMIDE, 3-(3,4-DIHYDROXYPHENOXY)-N-HEXYL-
C15H21NO4	057837-19-1	METALAXYL
C15H21NO4	099877-05-1	L-TYROSINE, 3-HYDROXY-, CYCLOHEXYL ESTER
C15H21NO5	099877-06-2	L-TYROSINE, 3-HYDROXY-, (TETRAHYDRO-2H-PYRAN-2-Y
C15H21NO5	115178-63-7	ACETAMIDE, 2-(BENZOYLOXY)-N,N-BIS(2-HYDROXYPROPY
C15H21NO5	115178-64-8	ACETAMIDE, 2-(BENZOYLOXY)-N,N-BIS(2-METHOXYETHYL
C15H21NOS	097871-56-2	A-ETS-N-SEC-BUTYLCINNAMAMIDE
C15H22ClN3O9	074729-54-7	α-D-GLUCOPYRANOSE, 2-[[[(2-CHLOROETHYL)NITROSOAM
C15H22ClNO2	000050-13-5	MEPERIDINE
C15H22ClNO2	051218-45-2	METOLACHLOR
C15H22FN3O2	111041-76-0	IMIDODICARBONIC DIAMIDE, N-[5-(1-ETHYLPROPYL)-2-
C15H22N2	132453-25-9	1H-FLUORENE-2-METHANAMINE, 2,3,4,4A,9,9A-HEXAHYD
C15H22N2O	000096-88-8	2-PIPERIDINECARBOXAMIDE, N-(2,6-DIMETHYLPHENYL)-
C15H22N2O	006514-54-1	1-(N,N-DIETHYLAMINO-ACETYL-AMINO)INDANE
C15H22N2O	065446-98-2	1-PIPERIDINEACETAMIDE, N-(4-ETHYLPHENYL)-
C15H22N2O	088132-41-6	1,1-DIME-3-(4-CYCLOHXPHENYL)UREA
C15H22N2O2	023694-81-7	2-PROPANOL, 1-[(1-METHYLETHYL)AMINO]-3-[2-METHY
C15H22N2O2	058479-94-0	1-PIPERIDINEACETAMIDE, N-(4-ETHOXYPHENYL)-

C15H22N2O2	065570-49-2	1-PIPERIDINEACETAMIDE, N-(3-ETHOXYPHENYL)-
C15H22N2O2	105801-55-6	ACETAMIDE, 2-(DIETHYLAMINO)-N-(2,3-DIHYDRO-5-MET
C15H22N2O5	049648-49-9	DIETHYLAMPHENICOL
C15H22N2O5S	075852-09-4	BUTANOIC ACID, 4-[[4-[(DIMETHYLAMINO)SULFONYL]PH
C15H22N2O5S	083870-99-9	4-ME-BENZENESULFONYL-L-(N'-PR)GLUTAMINE
C15H22N4	031935-08-7	2,4-DIAMINO-5-(1-ADAMANTYL)-6-METHYLPYRIMIDINE
C15H22N4O2	065016-46-8	BENZAMIDE,N(DIETAMET),2MEO-4NH2-5C≡N
C15H22N4O3	141018-16-8	CYTIDINE, 2',3'-DIDEOXY-N-[(1-PIPERIDINYL)METHYL
C15H22N4O5	098827-19-1	ALLOPURINOL, 1,5-BIS(BUTYRYLOXYMETHYL)-
C15H22N4O5	098827-20-4	ALLOPURINOL, 2,5-BIS(BUTYRYLOXYMETHYL)-
C15H22N6	061573-39-5	24-NH2PYRIMIDN,5(3,5-DIMEAMINO)BENZYL
C15H22N6	075841-91-7	3-PYRIDAZINAMINE, N-(2,5-DIMETHYL-1H-PYRROL-1-YL
C15H22N6O	061573-40-8	24NH2PYRIMDN,5(3MEAM4MEO5DIMEAM)BNZYL
C15H22O5	001034-01-1	OCTYL GALLATE
C15H22O5S	050615-70-8	1(3-PHENPROPYL)-B-GALACTOPYRANOSIDE
C15H22O6	000051-14-9	SESAMEX
C15H23ClO	060044-33-9	2-CHLORO-4-NONYLPHENOL
C15H23ClO4S	000140-57-8	ARAMITE
C15H23N3	063346-74-7	2-IMIDAZOLIDINIMINE, N-[2,6-BIS(1-METHYLETHYL)PH
C15H23N3O4	033820-53-0	ISOPROPALIN
C15H23N3O4S	003485-14-1	CYCLACILLIN
C15H23N3O4S	015676-16-1	SULPIRIDE
C15H23N3O4S	093414-62-1	BENZAMIDE, 5-(AMINOSULFONYL)-N-[(1-ETHYL-3-PYRRO
C15H23N3O5S2	114991-36-5	SPARSOMYCIN,S-PROPYL ANALOG
C15H23N5	004653-73-0	1,3,5-TRIAZINE-2,4-DIAMINE, 1-(4-BUTYLPHENYL)-1,
C15H23N5O	070606-63-2	4,6-NH2 2,2-DIME-1-(3-BUO PH)S-TRIAZINE
C15H23N5O2	075841-95-1	2-PROPANOL, 1-[[6-[(2,5-DIMETHYL-1H-PYRROL-1-YL)
C15H23NO2	013655-52-2	ALPRENOLOL
C15H23NO2	014309-41-2	BENZOIC ACID, 4-AMINO-, OCTYL ESTER
C15H23NO2	027325-36-6	2-PROPANOL, 1-(2-CYCLOPROPYLPHENOXY)-3-[(1-METHY
C15H23NO2	105217-58-1	2-PROPANOL, 1-[(1-METHYLETHYL)AMINO]-3-[(3-PHENY
C15H23NO2	112904-74-2	2H-1-BENZOPYRAN-8-OL, 3-(DIPROPYLAMINO)-3,4-DIHY
C15H23NO2	116005-04-0	2H-1-BENZOPYRAN-6-OL, 3-(DIPROPYLAMINO)-3,4-DIHY
C15H23NO2	124497-85-4	2-BUTOXY-2-PH-4-ME MORPHOLINE HCL
C15H23NO2	124497-86-5	2-ISOBUTOXY-2-PH-4-ME MORPHOLINE HCL
C15H23NO2	138324-58-0	BENZAMIDE, 2-ETHOXY-N,N-BIS(1-METHYLETHYL)-
C15H23NO3	006452-71-7	OXPRENALOL
C15H23NO4	000066-81-9	CYCLOHEXIMIDE
C15H23NO4	095261-40-8	CARBAMIC ACID, (3,4-DIETHOXYPHENYL)-, 2-METHYLPR
C15H23NO4	113932-77-7	CARBAMIC ACID, (3,4-DIETHOXYPHENYL)-, BUTYL ESTE
C15H23NO4	116583-45-0	2H-PYRAN, TETRAHYDRO-6-CYCLOPROPYL-3-[1-(ETHOXYI
C15H23NO6S	057554-05-9	(N-MORPHOLINYL)ET-(345-MEO)PH SULFONE
C15H23NOS	085785-20-2	ESPROCARB
C15H24	000717-74-8	1,3,5-TRIISOPROPYLBENZENE
C15H24	001081-77-2	N-NONYLBENZENE
C15H24N2O	000616-68-2	TRIMECAIN
C15H24N2O	020682-52-4	FORMAMIDE, N-[2-(DIETHYLAMINO)ETHYL]-N-(2,6-DIME
C15H24N2O	021236-52-2	PROPANAMIDE, 3-(DIETHYLAMINO)-N-(2,6-DIMETHYLPHE
C15H24N2O	039000-84-5	PROPANAMIDE, 2-(DIETHYLAMINO)-N-(2,6-DIMETHYLPHE
C15H24N2O	072336-17-5	ACETAMIDE, 2-(DIPROPYLAMINO)-N-(PHENYLMETHYL)-
C15H24N2O2	000094-24-6	TETRACAINE
C15H24N2O2	085819-15-4	O-MEO BENZAMIDE,N-(3-DIETAMINOPR)
C15H24N2O2	088530-27-2	ADAMANTYL-ALANIN-AMIDE,N-ACETYL
C15H24N2O3	034190-59-5	ACETAMIDE, N-[[3-[2-HYDROXY-3-[(1-METHYLETHYL)AM
C15H24N2O3	039617-16-8	ACETAMIDE, N-[2-[3-[(1,1-DIMETHYLETHYL)AMINO]-2-
C15H24N2O3	083029-57-6	PROPANAMIDE, N-[2-[(2-HYDROXY-3-PHENOXYPROPYL)AM
C15H24N2O3S	075852-19-6	HEXANAMIDE, N-[4-(DIMETHYLAMINO)SULFONYL]PHENYL]
C15H24N2O4	034918-50-8	ACETAMIDE, 2-[[3-[2-HYDROXY-3-(1-METHYLETHYL)AMI

C15H24N2O4	034939-39-4	ACETAMIDE, 2-[4-[2-HYDROXY-3-[(1-METHYLETHYL)AMI
C15H24N2O4S	051012-32-9	BENZAMIDE,N(DIETAMET),2-MEO,5-SO2ME
C15H24N2O4S	075852-15-2	PROPANAMIDE, N-[4-(DIMETHYLAMINO)SULFONYL]PHENYL
C15H24N4O	089530-01-8	1-(4-CARBAMOYLPHENYL)-3-METHYL-3-HEPTYLTRIAZENE
C15H24N4O4S	093414-64-3	BENZAMIDE, 4-AMINO-5-(AMINOSULFONYL)-N-[(1-ETHYL
C15H24NO4PS	025311-71-1	ISOFENPHOS
C15H24O	000104-40-5	P-NONYLPHENOL
C15H24O	000128-37-0	2,6-DI-T-BUTYL-4-METHYLPHENOL (BHT)
C15H24O	000136-83-4	PHENOL, 2-NONYL-
C15H24O	000139-84-4	PHENOL, 3-NONYL-
C15H24O	084852-15-3	N-NONYLPHENOL
C15H25BrN2O3	077317-79-4	3-DEO-5-BR-6-ME-PYRIMIDIN-2,4-DIONE
C15H25N4O2P	065101-39-5	TEPA ADAMANTYL-UREA DERIVATIVE
C15H25NO2	041538-42-5	2,5-DIMEO-4-T-BU-AMPHETAMINE
C15H25NO2	064638-09-1	BENZENEETHANAMINE, 4-BUTYL-2,5-DIMETHOXY-α-METHY
C15H25NO2	080617-74-9	2-PROPANOL, 1-[(1-METHYLETHYL)AMINO]-3-(3-PHENYL
C15H25NO2	121508-89-2	2-PROPANOL, 1-[(1,1-DIMETHYLETHYL)AMINO]-3-(2-PH
C15H25NO3	037350-58-6	METOPROLOL
C15H25NO3	121508-94-9	2-PROPANOL, 1-[(1,1-DIMETHYLETHYL)AMINO]-3-(2-PH
C15H26BrN	002534-65-8	DECYLPYRIDINIUM BROMIDE
C15H26IN	007295-91-2	N-DECYLPYRIDINIUM IODIDE
C15H26IN	053242-40-3	3,4-DIMEPYRIDINIUMIODIDE,N-OCTYL
C15H26IN	088477-49-0	2,4-DIME PYRIDINIUM IODIDE,N-OCTYL
C15H26N2	000090-39-1	L-SPARTEINE
C15H26N2O3	064810-91-9	BARBITURIC ACID,5-ET-5-NONYL
C15H26O5	081496-81-3	3,12-EPOXY-12H-PYRANO[4,3-J]-1,2-BENZODIOXEPIN-1
C15H26O6	000060-01-5	TRIBUTYRIN
C15H27NO2S	000301-11-1	LETHANE 60
C15H28ClN3O2	013909-14-3	1(2CLET)-1-(NO)-3-CYCLODODECYLUREA
C15H28N4O4	132766-02-0	L-VALINAMIDE, N-ACETYL-L-VALYL-L-ALANYL-
C15H28N4O4	132766-03-1	GLYCINAMIDE, N-ACETYL-L-VALYL-L-ISOLEUCYL-
C15H28N4O4	137605-60-8	L-ALANINAMIDE, N2-(N-L-ALANYL-L-ALANYL)-N1-(1,1-
C15H28O	000502-72-7	CYCLOPENTADECANONE
C15H28O4	001472-87-3	DIMETHYL TRIDECANEDIOATE
C15H30	002883-02-5	N-NONYLCYCLOHEXANE
C15H30	013360-61-7	1-PENTADECENE
C15H30O	000818-23-5	8-PENTADECANONE
C15H30O2	000124-10-7	METHYL MYRISTATE
C15H30O2	001002-84-2	PENTADECANOIC ACID
C15H30O2	001072-33-9	TRIDECYL ACETATE
C15H30O2	010233-13-3	ISOPROPYL LAURATE
C15H30O2	010580-24-2	UNDECANOIC ACID, BUTYL ESTER
C15H31Br	000629-72-1	1-BROMOPENTADECANE
C15H31NO4	000120-40-1	N,N-DI(2-HYDROXYETHYL)LAURAMIDE
C15H32	000629-62-9	N-PENTADECANE
C15H32O	000629-76-5	1-PENTADECANOL
C15H32O3Si4	017906-09-1	1,1,1,3,3,5,7,7,7-NONAMETHYL-5-PHENYLTETRASILO*
C15H33N3O2	002439-10-3	DODINE
C16H6Br4N2O2	002475-31-2	C.I. VAT BLUE 5
C16H8Cl2F6N2O3	086479-06-3	HEXAFLUMURON
C16H9BrN2O2	006492-73-5	C.I. VAT BLUE 3
C16H9ClF3NO7	077501-60-1	BENZOFLUORFEN
C16H9ClN2Na2O8S2	003624-68-8	C.I. MORDANT BLUE 9, DISODIUM SALT
C16H9ClN2Na2O9S2	001058-92-0	MORDANT BLUE 13
C16H9F3O2	006723-40-6	1H-INDENE-1,3(2H)-DIONE, 2-[4-(TRIFLUOROMETHYL)P
C16H9F3O2	019055-70-0	1H-INDENE-1,3(2H)-DIONE, 2-[3-(TRIFLUOROMETHYL)P
C16H10	000129-00-0	PYRENE
C16H10	000206-44-0	FLUORANTHENE

C16H10ClF3N2O	003864-49-1	1,4-BENZODIAZPN2-ON,5(2CF3PH)7CL
C16H10ClNO2	001090-16-0	1,4-NAPHTHOQUINONE,3-ANILINO-2-CHLORO
C16H10F3N3O3	001427-45-8	1,4-BENZODIAZPN2-ON,5(2CF3PH)7NO2
C16H10KNO5S	018690-90-9	14NAPHTHOQUINONE,2ANILINO-3SULFONATE K SALT
C16H10N2	010394-96-4	MALONONITRILE, ALPHA-PHENYLBENZAL
C16H10N2Na2O7S2	001936-15-8	C.I. 16230
C16H10N2O2	000482-89-3	2-(1,3-DIHYDRO-3-OXO-2H-INDOL-2-YLIDENE)-1,2-DI*
C16H10O	000243-42-5	BENZO[B]NAPHTHO[2,3-D]FURAN
C16H10S	000239-35-0	BENZO(B)NAPHTHO(2,1-D)THIOPHENE
C16H10S	000243-46-9	BENZO[B]NAPHTHO[2,3-D]THIOPHENE
C16H11Cl3N2O	030144-88-8	1,4-BENZODIAZPN2-ON,1-ME-5(26CLPH)7CL
C16H11F3N2O	002285-16-7	1,4-BENZODIAZEPIN-2-ONE,5(CF3)PH
C16H11F3N2O	002730-05-4	1,4-BENZODIAZPIN-2-ONE,5-(2CF3PH)
C16H11F3N2O	003894-63-1	1,4-BENZODIAZPIN2-ONE,5(4CF3PH)
C16H11N	002693-46-1	3-FLUORANTHENAMINE
C16H11N3O	017562-53-7	1,4-BENZDIAZEPIN-2-ONE-5-PH-7CN
C16H11NO2	000132-60-5	2-PHENYLCINCHONINIC ACID
C16H11NO2	006628-97-3	1,4-NAPHTHOQUINONE, 2-ANILINO
C16H12ClFN2O	001959-97-3	1,4-BENZODIAZPN2-ON,1-ME5(4-CLPH)7F
C16H12ClFN2O	003900-31-0	FLUDIAZEPAM
C16H12ClN3O3	067027-56-9	14-BENZDIAZPIN-2-ONE,3ME5-(2CLPH)7NO2
C16H12F2N2O	002024-34-2	1,4-BENZODIAZEPIN-2-ONE,7-FLUORO-5-(2-FLUOROPHEN
C16H12F3NO3	096327-51-4	1H-PYRROLIZINE-1-CARBOXYLIC ACID, 2,3-DIHYDRO-5-
C16H12F3NO3	096327-52-5	1H-PYRROLIZINE-1-CARBOXYLIC ACID, 2,3-DIHYDRO-5-
C16H12FIN2O	034932-78-0	1,4-BENZDAZEPIN-2-ON-1ME-5(2FPH)-7I
C16H12FN3O3	001622-62-4	FLUNITRAZEPAM
C16H12N2	004238-66-8	6H-PYRIDO[4,3-B]CARBAZOLE, 5-METHYL-
C16H12N2O	000842-07-9	1-(PHENYLAZO)-2-NAPHTHALENOL
C16H12N2O	015804-61-2	BENZALCYANOACETANILIDE
C16H12N2O3	021914-07-8	BARBITURIC ACID,5,5-DIPHENYL-
C16H12N2O3	081592-05-4	P-CYANO PHENYLHIPPURATE
C16H12N2O3	081592-12-3	M-CYANO PHENYLHIPPURATE
C16H12O2	000084-51-5	2-ETHYLANTHRAQUINONE
C16H12O2	004070-75-1	1,2-DIBENZOYLETHYLENE
C16H12O2	015432-98-1	1H-INDENE-1,3(2H)-DIONE, 2-(3-METHYLPHENYL)-
C16H12O2	071972-66-2	4H-1-BENZOPYRAN-4-ONE, 3-METHYL-2-PHENYL-
C16H12O3	000117-37-3	1H-INDENE-1,3(2H)-DIONE, 2-(4-METHOXYPHENYL)-
C16H12O3	006149-23-1	1H-INDENE-1,3(2H)-DIONE, 2-(3-METHOXYPHENYL)-
C16H12O3S	055863-26-8	DIBENZO[B,E]THIEPIN-2-ACETIC ACID, 6,11-DIHYDRO-
C16H12O5	002035-15-6	6H-[1,3]DIOXOLO[5,6]BENZOFURO[3,2-C][1]BENZOPYRA
C16H12O6	000520-34-3	DIOSMETINE
C16H13BrO4	092387-49-0	MONI[1-(4-BROMOPHENYL)ETHYL] 1,2-BENZENEDICARBO*
C16H13Cl2NO3	092854-81-4	DILOXANIDE BENZOATE
C16H13ClF3NO4	069806-40-2	PROPANOIC ACID, 2-[4-[[3-CHLORO-5-(TRIFLUOROMETH
C16H13ClFNO3	058667-63-3	FLAMPROP
C16H13ClN2O	000439-14-5	DIAZEPAM
C16H13ClN2O	002894-68-0	2H-1,4-BENZODIAZEPIN-2-ONE, 7-CHLORO-5-(2-CHLORO
C16H13ClN2O	004699-82-5	2H-1,4-BENZODIAZEPIN-2-ONE, 7-CHLORO-1,3-DIHYDRO
C16H13ClN2O	005358-35-0	1,4-BENZODIAZP2-ON,5(OMEPH)7CL
C16H13ClN2O2	000846-50-4	TEMAZEPAM
C16H13ClN2O2	003023-44-7	1,4-BENZODIAZP-2-ON,5(2MEOPH)7CL
C16H13ClN2O2	005358-92-9	1,4-BENZODIAZP2-ON,5(3MEOPH)7CL
C16H13ClN2OS	000846-53-7	1,4-BENZODIAZPN2-ON,5(2-MES PH)7CL
C16H13ClNO3	051234-28-7	ORAFLEX (UNIPROFEN) (OPREN) (COXIGON)
C16H13ClO4	092387-48-9	MONO[1-(4-CHLOROPHENYL)ETHYL] 1,2-BENZENEDICARB*
C16H13F2N3O	076674-21-0	FLUTRIAFOL (PP450)
C16H13N	000090-30-2	1-NAPHTHALENAMINE, N-PHENYL-
C16H13N	000135-88-6	N-PHENYL-2-NAPHTHYLAMINE

C16H13N	001606-67-3	1-PYRENAMINE
C16H13N3	000131-22-6	4-(PHENYLAZO)-1-NAPHTHALENAMINE
C16H13N3O3	002011-67-8	NIMETAZEPAM
C16H13N3O3	004941-45-1	1,4-BENZODIAZEPIN-2-ONE,1,3-DIHYDRO-9-METHYL-7-N
C16H13N3O3	031431-39-7	MEBENDAZOLE
C16H13N5O4	136389-80-5	1H-PYRAZOLE, 3-METHYL-4-NITRO-5-[(4-NITROPHENYL)
C16H13NO	000093-45-8	4-(2-NAPHTHYLAMINO)PHENOL
C16H13NO3	004465-58-1	C.I. DISPERSE RED 3
C16H13NO3	119005-17-3	2-AZETIDINONE, 4-(4-BENZOYLPHENOXY)-
C16H13NO3	119005-21-9	2-AZETIDINONE, 4-(2-BENZOYLPHENOXY)-
C16H13NO4S	074134-17-1	ACETAMIDE, N-(7-METHYL-9-OXO-9H-THIOXANTHEN-3-YL
C16H14	000613-26-3	2,6-DIMETHYLANTHRACENE
C16H14	000781-43-1	9,10-DIMETHYLANTHRACENE
C16H14	000886-65-7	(1,4-DIPHENYL)-1,3-BUTADIENE
C16H14	001576-67-6	3,6-DIMETHYLPHENANTHRENE
C16H14BrN	026323-01-3	N-BENZYLQUINOLINIUM BROMIDE
C16H14Cl2N2O2	074350-90-6	ACETAMIDE, 2,2-DICHLORO-N-[2-OXO-2-(PHENYLAMINO)
C16H14Cl2N4O2S	113024-75-2	1H-1,2,4-TRIAZOLE-3-SULFONAMIDE, N-(2,6-DICHLORO
C16H14Cl2O	014088-71-2	PROCLONOL
C16H14Cl2O3	000510-15-6	CHLOROBENZILATE
C16H14Cl2O4	051338-27-3	DICLOFOP-METHYL
C16H14ClN3O	000058-25-3	LIBRIUM
C16H14ClNO3	083073-36-3	1H-PYRROLIZINE-1-CARBOXYLIC ACID, 5-(4-CHLOROBEN
C16H14F2N4O2S	113024-34-3	1H-1,2,4-TRIAZOLE-3-SULFONAMIDE, N-(2,6-DIFLUORO
C16H14FN3O	034084-50-9	14-BENZODIAZPN2-ON,1-ME-5(2-F PH)7NH2
C16H14FNO3	107115-74-2	1H-PYRROLIZINE-1-CARBOXYLIC ACID, 5-(4-FLUOROBEN
C16H14IN3O8	135617-14-0	URIDINE, 2'-DEOXY-5-IODO-, 5'-(4-NITROBENZOATE)
C16H14N2O	000072-44-6	METHAQUALONE
C16H14N2O	005571-63-1	2H-1,4-BENZODIAZEPIN-2-ONE, 1,3-DIHYDRO-7-METHYL
C16H14N2O2	002475-44-7	1,4-BIS(METHYLAMINO)ANTHRAQUINONE
C16H14N2O2	004224-00-4	2,4-IMIDAZOLIDINEDIONE, 5,5-DIPHENYL-3-METHYL-
C16H14N2O2	005358-96-3	1,4-BENZODIAZP2-ON,5(PH)7MEO
C16H14N2O2	075389-00-3	FURAZAN, BIS(PHENYLMETHYL)-, 2-OXIDE
C16H14N2O2S	073250-68-7	MEFENACET
C16H14N2O4	003860-63-7	C.I. DISPERSE BLUE 26
C16H14N2O4	081592-11-2	M-CONH2 PHENYLHIPPURATE
C16H14N2OS	002891-12-5	1,4-BENZODIAZPIN-2-ONE,5(PH)7MES
C16H14N4	006054-48-4	C.I..DISPERSE BLACK 1
C16H14N4O2	065542-16-7	P-(3-(P-CYANOBENZYL)-3-METHYL-1-TRIAZENO)-
C16H14O2	000103-41-3	BENZYL CINNAMATE
C16H14O3	022071-15-4	BENZENEACETIC ACID, 3-BENZOYL-.ALPHA.-METHYL-
C16H14O3	036330-85-5	FENBUFEN
C16H14O4	033533-53-8	MONO(ALPHA-METHYLBENZYL) PHTHALATE
C16H14O6	000520-33-2	HESPERETINE
C16H15BrO6	091814-13-0	1,4-NAPHTHALENEDIOL, 6-BROMO-2,3-DIMETHOXY-, DIA
C16H15Cl2NO2	000117-26-0	BULAN
C16H15Cl2NO2	084496-56-0	CLOMEPROP
C16H15Cl3	004413-31-4	1,1'-(2,2,2-TRICHLOROETHYLIDENE)BIS(4-METHYLB*)
C16H15Cl3O2	000072-43-5	METHOXYCHLOR
C16H15Cl3OS	034197-16-5	METHOXYMETHIOCHLOR
C16H15Cl3S2	019679-38-0	1,1'-(2,2,2-TRICHLOROETHYLIDENE)BIS(4-METHYLT*)
C16H15ClN2	002898-12-6	MEDAZEPAM
C16H15ClN2O2	074339-86-9	ACETAMIDE, 2-CHLORO-N-[2-OXO-1-(PHENYLAMINO)ETHY
C16H15ClN2OS	033671-46-4	2H-THIENO[2,3-E]-1,4-DIAZEPIN-2-ONE, 5-(2-CHLORO
C16H15ClO3	077269-58-0	PHENYLACETIC ACID,3-CL-4-PHENYLETO
C16H15ClO6	091431-42-4	1,4-NAPHTHALENEDIOL, 6-CHLORO-2,3-DIMETHOXY-, DI
C16H15ClO6	091814-10-7	1,4-NAPHTHALENEDIOL, 5-CHLORO-2,3-DIMETHOXY-, DI
C16H15F3O2	000384-97-4	111F3-22-DI(P-MEOPHENYL)ETHANE

I-185

C16H15FO6	091814-12-9	1,4-NAPHTHALENEDIOL, 6-FLUORO-2,3-DIMETHOXY-, DI
C16H15IN2O6	084043-31-2	URIDINE, 2'-DEOXY-5-IODO-, 5'-BENZOATE
C16H15N3	023099-85-6	4-PH-1-PHTHALAZINAMINE,N,N-DIME
C16H15N3O	004959-16-4	1,4-BENZODIAZPN2-ONE,1-ME-5(PH)7NH2
C16H15NO	089242-78-4	3-PYRIDINEMETHANOL, _-(PHENYLMETHYL)-_-(2-PROPEN
C16H15NO	089242-87-5	3-PYRIDINEMETHANOL, _-(4-METHYLPHENYL)-_-(2-PROP
C16H15NO2	072116-69-9	BENZENEACETAMIDE, N-(3-ACETYLPHENYL)-
C16H15NO2	089246-39-9	BENZENEACETAMIDE, N-(4-ACETYLPHENYL)-
C16H15NO2S	013004-52-9	2(4,4'-DIHYDROXYDIPHENLMETHYL)THIAZOLE
C16H15NO3	029736-20-7	P-METHYLPHENYLHIPPURATE
C16H15NO3	066635-84-5	1H-PYRROLIZINE-1-CARBOXYLIC ACID, 2,3-DIHYDRO-5-
C16H15NO3	066895-70-3	M-METHYLPHENYLHIPPURATE
C16H15NO3	095384-55-7	BENZENEACETAMIDE, N-[3-(ACETYLOXY)PHENYL]-
C16H15NO3	096327-32-1	1H-PYRROLIZINE-1-CARBOXYLIC ACID, 2,3-DIHYDRO-5-
C16H15NO3	096327-65-0	1H-PYRROLIZINE-1-CARBOXYLIC ACID, 2,3-DIHYDRO-5-
C16H15NO3	100668-10-8	2-PROPENAMIDE, 3-(3,4-DIHYDROXYPHENYL)-N-(PHENYL
C16H15NO3	107115-73-1	1H-PYRROLIZINE-1-CARBOXYLIC ACID, 5-BENZOYL-2,3-
C16H15NO4	029736-21-8	P-METHOXYPHENYLHIPPURATE
C16H16	003299-99-8	9-(1-METHYLETHYL)-9H-FLUORENE
C16H16BrN3O6	096000-48-5	1H-PYRROLO[1,2-A]INDOLE-5,8-DIONE, 1-(ACETYLOXY)
C16H16Cl2N2O4S	088598-34-9	2,4-DICHLOROBENZYLPENCILLIN
C16H16Cl2O2	006012-83-5	ETOXINOL
C16H16Cl2O2	007388-31-0	METHOXYCHLOR-DDD
C16H16ClN3O2	087359-63-5	1H-IMIDAZO[4,5-C]PYRIDINE, 2-[2-(3-CHLOROPROPOXY
C16H16FN3O2	111041-66-8	IMIDODICARBONIC DIAMIDE, N-(4-FLUORO[1,1'-BIPHEN
C16H16N2O2	059040-30-1	NAFAZATROM
C16H16N2O2	061706-06-7	1,1-DIME-3-(4-BENZOYLPHENYL)UREA
C16H16N2O2	074339-88-1	ACETAMIDE, 2-(ACETYLPHENYLAMINO)-N-PHENYL-
C16H16N2O2	088132-35-8	1-FLUORENYL-3-MEO-3-ME UREA
C16H16N2O3	088132-26-7	1-ME-1-MEO-3-(4-BENZOYLPHENYL)UREA
C16H16N2O4	013684-56-5	DESMEDIPHAM
C16H16N2O4	013684-63-4	PHENMEDIPHAM
C16H16N2O6S2	000153-61-7	CEPHALOTHIN
C16H16N4O2	015446-39-6	N,N'-(AZODI-4,1-PHENYLENE)BISACETAMIDE
C16H16N4O8S	055268-75-2	CEFUROXIM
C16H16O3	050463-48-4	BENZENEPROPANOIC ACID, 4-(PHENYLMETHOXY)-
C16H16O3	064360-42-5	PHENYLACETIC ACID,3-ME-4-PHMEO
C16H16O3	093371-36-9	BUTANOIC ACID,3-OH-2-(4-BIPHENYLYL)-
C16H16O3S	061820-95-9	ETHANONE,1-(4-METHYLPHENYL)-2-[(4-METHYLPHENYL)SULF
C16H16O4	029973-91-9	PHENYLACETIC ACID, 3-MEO,4-PHMEO
C16H16O4S	086516-51-0	ETHANONE,1-(4-METHOXYPHENYL)-2-[(4-METHYLPHENYL)SULF
C16H16O5	060404-04-8	3-FURANCARBOXYLIC ACID, 4,5-DIHYDRO-5-[(4-METHOX
C16H16O6	061601-23-8	1,4-NAPHTHALENEDIOL, 2,3-DIMETHOXY-, DIACETATE
C16H17ClN2O	010379-14-3	TETRAZEPAM
C16H17ClN2O7S	129625-27-0	BENZOIC ACID, 5-(AMINOSULFONYL)-4-CHLORO-2-[[(2-
C16H17ClN2S	001225-64-5	10H-PHENOTHIAZINE-10-PROPANAMINE, 2-CHLORO-N-MET
C16H17ClN4O3	003180-81-2	C.I. 11115
C16H17ClN4O4	004540-00-5	C.I. DISPERSE RED 7
C16H17ClN5O6PS	041941-66-6	8-(4-CL-PH)-S CYCLIC AMP
C16H17F2N3O3	099726-76-8	3-QUINOLINECARBOXYLIC ACID, 1-ETHYL-6,8-DIFLUORO
C16H17N3O	108611-24-1	3-PYRIDINECARBONITRILE, 2-[(2-HYDROXYPROPYL)AMIN
C16H17N3O	108611-25-2	3-PYRIDINECARBONITRILE, 2-[(3-HYDROXYPROPYL)AMIN
C16H17N3O2	065542-17-8	P(3(P-MEBENZYL)3ME-1TRIAZENO)BENZOICACID
C16H17N3O3	065542-18-9	P(3(PMEBENZYL)3MEO-1TRIAZENO)BENZOICACID
C16H17N3O4	053525-76-1	P(4-NO2-PHO-PRO)BENZAMIDINE HCL
C16H17N3O4	053525-79-4	P(3-NO2-PHO-PRO)BENZAMIDINE HCL
C16H17N3O4S	015686-71-2	CEPHALEXIN
C16H17N3O5	004836-13-9	CYTIDINE, N-BENZOYL-2'-DEOXY-

C16H17N3O6	013089-48-0	CYTIDINE, N-BENZOYL-
C16H17N3O7S2	035607-66-0	CEFOXITIN
C16H17N5O5	062570-20-1	DIS. A. 12
C16H17N5O6	060129-67-1	DIS. A. 14
C16H17NO	000957-51-7	DIPHENAMID
C16H17NO	099081-88-6	ALPHA-(P-ISO-PR-PHENYL)-N-PHENYLNITRONE
C16H17NO2	016053-34-2	BENZENEACETAMIDE, N-(3-ETHOXYPHENYL)-
C16H17NO2	065262-21-7	BENZAMIDE, 3-(2,5-DIMETHYLPHENOXY)-N-METHYL-
C16H17NO2S	138568-69-1	BENZENEACETIC ACID, α-METHYL-4-(4,5,6,7-TETRAHYD
C16H17NO3	056356-19-5	4-AMINOSALICYLIC ACID,4-IPR ESTER
C16H17NO3	056356-20-8	4-AMINOSALICYLIC ACID,2-IPR ESTER
C16H17NO3	084640-36-8	P-PHENO-BENZYL N,N-DIME CARBAMATE
C16H17NO4	055720-47-3	L-TYROSINE, 3-HYDROXY-, PHENYLMETHYL ESTER
C16H17NO4	085078-27-9	1,1-DI(P-METHOXYPHENYL)-2-NITROETHANE
C16H17NO4S	052980-49-1	BENZAMIDE, N,N-DIMETHYL-2-[[(4-METHYLPHENYL)SULF
C16H18	007116-95-2	4-ISOPROPYLBIPHENYL
C16H18ClN3S	000061-73-4	METHYLENE BLUE
C16H18ClNO2S	138568-81-7	BENZENEACETIC ACID, 4-(4-BUTYL-2-THIAZOLYL)-3-CH
C16H18FN3O3	070458-96-7	NORFLOXACIN
C16H18FNO2S	138568-73-7	BENZENEACETIC ACID, 4-(4-BUTYL-2-THIAZOLYL)-3-FL
C16H18N2	006311-44-0	2,2',5,5'-TETRAMETHYLAZOBENZENE
C16H18N2	029418-25-5	BIS(2,4-DIMETHYLPHENYL)DIAZENE
C16H18N2	029418-29-9	2,4,4',6-TETRAMETHYLAZOBENZENE
C16H18N2	029418-30-2	2,2',4,6-TETRAMETHYLAZOBENZENE
C16H18N2	029418-31-3	BIS(2,6-DIMETHYLPHENYL)DIAZENE
C16H18N2	029418-34-6	2,2',3,3'-TETRAMETHYLAZOBENZENE
C16H18N2	029418-36-8	BIS(3,4-DIMETHYLPHENYL)DIAZENE
C16H18N2O	095384-54-6	BENZENEACETAMIDE, N-[3-(ETHYLAMINO)PHENYL]-
C16H18N2O	102587-48-4	UREA, N,N-DIMETHYL-N'[3-(PHENYLMETHYL)PHENYL]-
C16H18N2O2	029418-54-0	4,4'-DIMETHOXY-O,O'-AZOTOLUENE
C16H18N2O2	052093-79-5	N'(1-ME-BENZYL)PHO-ACETIC ACID HYDRAZIDE
C16H18N2O2	101820-66-0	BENZAMIDE, 2-[[2-(2-METHYLPHENOXY)ETHYL]AMINO]-
C16H18N2O3	014214-32-5	DIFENOXURON
C16H18N2O3	025998-87-2	1-(4-BENZYLOXYPH)-3-MEO-3-ME UREA
C16H18N2O3	074109-81-2	1(3-BENZYLOXYPHENYL)-3-(ME-MEO) UREA
C16H18N2O4	029418-55-1	BIS(2,4-DIMETHOXYPHENYL)DIAZENE
C16H18N2O5	124897-30-9	1,8-NAPHTHYRIDINE-3-CARBOXYLIC ACID, 1-ETHYL-1,4
C16H18N2O5S	000087-08-1	PHENOXYMETHYLPENICILLIN
C16H18N2O5S	004759-24-4	ALPHA-HYDROXYBENZYLPENCILLIN
C16H18N2O6	014985-34-3	URIDINE, 3-(PHENYLMETHYL)-
C16H18N2O6S2	000058-71-9	CEPHALOTHIN
C16H18N2O7S2	041744-40-5	SULBENCILLIN
C16H18N2S	000522-24-7	FENETHAZINE
C16H18N4O2	000051-12-7	NIALAMIDE
C16H18N4O2	003025-52-3	DIS. A. 5
C16H18N4O3	002872-52-8	DISPERSE RED 1
C16H18N4O4	002734-52-3	C.I. DISPERSE RED 19
C16H18N4O7S	083055-99-6	BENSULFURON-METHYL (PH7)
C16H18N6	098087-97-9	2-PYRIDINAMINE, 4-(5-AMINO-1H-1,2,4-TRIAZOL-3-YL
C16H18O2	010543-21-2	1,1-DI-(P-METHOXYPHENYL)ETHANE
C16H18O3	101705-36-6	HEXANOIC ACID, 6-(1-NAPHTHALENYLOXY)-
C16H18O6	006044-30-0	GLUCOPYRANOSIDE,2-NAPHTHYL
C16H19BrN2O2S	111226-64-3	HYL]-2,2-DIMETHYLPROPYL ESTER
C16H19BrN2O3	111227-03-3	1H-IMIDAZOLE-1-CARBOXYLIC ACID, 1-[(4-BROMOPHENO
C16H19ClN2	000132-22-9	CHLORPHENIRAMINE
C16H19ClN2O2S	111226-63-2	1H-IMIDAZOLE-1-CARBOTHIOIC ACID, O-[1-[(4-CHLORO
C16H19ClN2O3	111227-01-1	1H-IMIDAZOLE-1-CARBOXYLIC ACID, 1-[(4-CHLOROPHEN
C16H19FN2O3	111227-00-0	1H-IMIDAZOLE-1-CARBOXYLIC ACID, 1-[(4-FLUOROPHEN

C16H19FN4O3	086393-37-5	AMIFLOXACIN
C16H19N	053660-20-1	BENZENEETHANAMINE, N-METHYL-α-(PHENYLMETHYL)-
C16H19N3	002481-94-9	C.I. SOLVENT YELLOW 56
C16H19N3O2	002452-84-8	C.I. SOLVENT YELLOW 58
C16H19N3O2	032150-60-0	2,4(1H,3H)-PYRIMIDINEDIONE, 5-(DIMETHYLAMINO)-6-
C16H19N3O2	032150-73-5	2,4(1H,3H)-PYRIMIDINEDIONE, 1,6-DIMETHYL-3-PHENY
C16H19N3O2	032150-76-8	2,4(1H,3H)-PYRIMIDINEDIONE, 5-(DIMETHYLAMINO)-6-
C16H19N3O3	032150-74-6	2,4(1H,3H)-PYRIMIDINEDIONE, 1,6-DIMETHYL-5-(4-MO
C16H19N3O4S	000069-53-4	AMPICILLIN
C16H19N3O4S	000800-79-3	AMPICILLIN
C16H19N3O5	111227-04-4	1H-IMIDAZOLE-1-CARBOXYLIC ACID, 1-[(4-NITROPHENO
C16H19N3O5S	026787-78-0	AMOXICILLIN
C16H19N3O6	004055-39-4	MITOMYCIN A
C16H19N3O6	004055-40-7	MITOMYCIN B
C16H19N5O3	131402-51-2	ALLOPURINOL,N-(4-DIMETHYLAMINOMETHYL)BENZOYLOXYM
C16H19NaO3S	099287-44-2	1-SO3NA-3-PROPYL-7-ISOPROPYL AZULENE
C16H19NaO4S	099287-60-2	1-SO3NA-3-PENTYL-4-METHOXY AZULENE
C16H19NaO4S	099287-65-7	1-SO3NA-3-ET-4-METHOXY-7-IPR AZULENE
C16H19NO2	000525-66-6	PROPRANOL
C16H19NO2S	132483-55-7	BENZENEACETIC ACID, 4-(4-BUTYL-2-THIAZOLYL)-α-ME
C16H19NO4	010078-46-3	2-PROPEN-1-ONE, 3-(2,5-DIHYDRO-1H-PYRROL-1-YL)-1
C16H19NO4	150900-05-3	2-AZASPIRO[4.4]NONANE-1,3-DIONE, 2-[(4-METHOXYPH
C16H19NO7	116482-77-0	BENZOIC ACID, 2-(ACETYLOXY)-, 2-[(2-ETHOXY-2-OXO
C16H20	030176-62-6	1-PHENYLADAMANTANE
C16H20N2	000366-29-0	N,N,N',N'-TETRAMETHYLBENZIDINE
C16H20N2O2S	051688-32-5	BENZENAMINE, 4-[(4-AMINOPHENYL)SULFONYL]-N,N-DIE
C16H20N2O2S	111226-62-1	1H-IMIDAZOLE-1-CARBOTHIOIC ACID, O-[1-(PHENOXYME
C16H20N2O3	111226-99-4	1H-IMIDAZOLE-1-CARBOXYLIC ACID, 1-(PHENOXYMETHYL
C16H20N2O4	053447-04-4	ARBONATE, (4A_,9_,10_,10Abeta)-
C16H20N2O4	053447-05-5	ARBAMATE, (4A_,9beta,10_,10Abeta)-
C16H20N2O4	053447-06-6	ARBONATE, (4A_,9_,10beta,10Abeta)-
C16H20N4O2	069194-91-8	24-NH2PYRIMIDINE,5(35MEO-4-PRENYL)BENZYL
C16H20N4O2	087653-22-3	UREA, N-METHYL-N'-[4-(6-OXO-3-PHENYL-1(6H)-PYRID
C16H20N4O2	087666-28-2	UREA, N-ETHYL-N'-[3-(6-OXO-3-PHENYL-1(6H)-PYRIDA
C16H20N4O2	087666-30-6	UREA, N,N-DIMETHYL-N'-[3-(6-OXO-3-PHENYL-1(6H)-P
C16H20N4O4	055687-48-4	24-NH2PYRIMIDN,5(35-MEO-4CO2ET)BENZYL
C16H20N4O5	083166-78-3	ACETIC ACID, [4-[(2,4-DIAMINO-5-PYRIMIDINYL)METH
C16H20N4O5S	055687-74-6	24NH2PYRIMIDN,5(35MEO-4COCH2SO2ME)BENZYL
C16H20N4O6	057998-68-2	DIAZIQUONE
C16H20N6O4S3	032924-66-6	CEPHALOTHIN ANALOG (S-TETRAZOLE)
C16H20O2Si	002553-19-7	DIPHENYLDIETHOXYSILANE
C16H20O4	015725-28-7	ETHYLACETOACETATE,3-PRO BENZAL
C16H20O5	015725-37-8	DIETHYLMALONATE, 3-ETHOXYBENZAL
C16H20O6	015818-13-0	PROPANEDIOIC ACID, [(3,4-DIMETHOXYPHENYL)METHYLE
C16H20O6P2S3	003383-96-8	ABATE
C16H21Cl3O3	001928-47-8	2,4,5-T, 2-ETHYLHEXYL ESTER
C16H21Cl3O3	025168-15-4	2,4,5-T, ISOOCTYL ESTER
C16H21N3	061019-78-1	FENAPANIL
C16H21N3O2	031991-98-7	2,4(1H,3H)-PYRIMIDINEDIONE, 5-(BUTYLAMINO)-1,6-D
C16H21N3O2	031991-99-8	2,4(1H,3H)-PYRIMIDINEDIONE, 5-(DIETHYLAMINO)-1,6
C16H21N3O2	032150-38-2	2,4(1H,3H)-PYRIMIDINEDIONE, 5-(DIETHYLAMINO)-3,6
C16H21N3O2	072775-85-0	2,4(1H,3H)-PYRIMIDINEDIONE, 1,6-DIMETHYL-5-[(2-M
C16H21N5O2	059338-93-1	1H-BENZOTRIAZOLE-5-CARBOXAMIDE, 6-METHOXY-N-[[1-
C16H21N5O2	089937-39-3	N(2,5-DIME-PYRROL-1-YL)-6-($-MORPHOL)-3-PYRIDAZ-
C16H21N5O2	093414-58-5	1H-BENZOTRIAZOLE-5-CARBOXAMIDE, 6-METHOXY-N-[[1-
C16H21N5O3	074731-64-9	2,4-NH2PYRIMD,5(35-MEO-4CONME2)BENZYL
C16H21NO3	000087-00-3	HOMATROPIN
C16H21NO3	115178-69-3	PIPERIDINE, 1-[(BENZOYLOXY)ACETYL]-2-ETHYL-

C16H21NO3	115178-70-6	PIPERIDINE, 1-[(BENZOYLOXY)ACETYL]-2,6-DIMETHYL-
C16H21NO4	039552-01-7	ETHANONE, 1-[7-[2-HYDROXY-3-[(1-METHYLETHYL)AMIN
C16H22BrNO3	000051-56-9	HOMATROPINE HYDROBROMIDE
C16H22Cl2O3	001928-44-5	2,4-D, OCTYL ESTER
C16H22Cl2O3	025168-26-7	2,4-D, ISOOCTYL ESTER
C16H22ClNO	097871-64-2	P-CL CINNAMAMIDE,B-PR,N-SEC-BUTYL
C16H22ClNO	097871-65-3	P-CL CINNAMAMIDE,B-IPR,N-SEC-BUTYL
C16H22ClNO3	038727-55-8	ANTOR
C16H22FN3O2	111041-86-2	IMIDODICARBONIC DIAMIDE, N-(5-CYCLOHEXYL-2-FLUOR
C16H22N2O2	105801-58-9	1-PIPERIDINEACETAMIDE, N-(2,3-DIHYDRO-5-METHYL-3
C16H22N2O3	105801-62-5	1-PIPERIDINEACETAMIDE, N-(2,3-DIHYDRO-6-METHOXY-
C16H22N2O4S	086570-00-5	CARBOFURAN,N-(N'ME,ET-CARBAMIDOTHIO)
C16H22N2O5S	065907-27-9	CARBOFURAN,N-(N'ME-O-ET-CARBAMYLTHIO)
C16H22N4O2	074731-66-1	2,4-NH2PYRIMD,5(35-MEO-4-IPR)BENZYL
C16H22N4O2	130787-13-2	2H-BENZOTRIAZOLE, 2-CYCLOPENTYL-5-[[(2-METHYLPRO
C16H22N4O3	055687-91-7	24NH2PYRIMIDINE,5(35-MEO-4-COHME2)BENZYL
C16H22N4O3	088107-10-2	ETHANONE, 1-[2-HYDROXY-3-PROPYL-4-[4-(1H-TETRAZO
C16H22N4O4	053808-87-0	24NH2PYRIMIDINE,5(35-MEO-4-OETOME)BENZYL
C16H22N4O4	132765-93-6	L-PHENYLALANINAMIDE, N2-ACETYL-L-GLUTAMINYL-
C16H22N4O4	132765-94-7	L-GLUTAMAMIDE, N-ACETYL-L-PHENYLALANYL-
C16H22N4O4	132766-09-7	L-PHENYLALANINAMIDE, N-ACETYL-L-ALANYLGLYCYL-
C16H22N6O	061544-28-3	24NH2PYRIMDN,5(3NHME-4ME-5NMECOME)BENZYL
C16H22N6O2	117121-33-2	UREA, N'-[4-[(1-CYCLOHEXYL-1H-TETRAZOL-5-YL)OXY]
C16H22N6O2	117121-45-6	UREA, N'-[4-[(2-CYCLOHEXYL-2H-TETRAZOL-5-YL)OXY]
C16H22N6O4	024305-27-9	L-PROLINAMIDE, 5-OXO-L-PROLYL-L-HISTIDYL-
C16H22N6O4S	022181-95-9	GLYCINE, N-[N-[1-OXO-5-(7H-PURIN-6-YLTHIO)PENTYL
C16H22O11	000604-68-2	.ALPHA.-D-GLUCOPYRANOSE, PENTAACETATE
C16H22O11	003891-59-6	D-GLUCOSE, 2,3,4,5,6-PENTAACETATE
C16H22O4	000084-69-5	DI-ISOBUTYLPHTHALATE
C16H22O4	000084-74-2	DIBUTYL PHTHALATE
C16H22O4	001962-75-0	P-DIBUTYLPHTHALATE
C16H22O4	003126-90-7	1,3-BENZENEDICARBOXYLIC ACID, DIBUTYL ESTER
C16H22O4	075221-43-1	PHENYLACETIC ACID,3-MEO-4-CYHXMEO
C16H22OSi2	000056-33-7	1,1,3,3-TETRAMETHYL-1,3-DIPHENYLDISILOXANE
C16H23BrN2O3	080125-14-0	REMOXIPRIDE
C16H23N	132453-02-2	4AH-FLUOREN-4A-AMINE, N-ETHYL-1,2,3,4,9,9A-HEXAH
C16H23N3O3	065118-48-1	L-VALINAMIDE, N-ACETYL-L-PHENYLALANYL-
C16H23N3O3	120399-49-7	L-VALINAMIDE, N-BENZOYLGLYCYL-N-ETHYL-
C16H23N3O3S2	082774-17-2	N(N''-IPR(4ME-PH)CARBAMYLTHIO)METHOMYL
C16H23N3O3S2	082774-18-3	N(N''-BU-(PHENYL)CARBAMYLTHIO)METHOMYL
C16H23N3O4	099027-00-6	L-VALINAMIDE, N-ACETYL-L-TYROSYL-
C16H23NO	000728-88-1	1-PROPANONE, 2-METHYL-1-(4-METHYLPHENYL)-3-(1-PI
C16H23NO	132453-06-6	4A(2H)-PHENANTHRENAMINE, 1,3,4,9,10,10A-HEXAHYDR
C16H23NO2	000077-20-3	ALPHAPRODINE
C16H23NO2	054340-62-4	BUFURALOL
C16H23NO3	047082-97-3	2-PROPANOL, 1-[(1,1-DIMETHYLETHYL)AMINO]-3-[2-(2
C16H23NO3	072836-46-5	P-MEO-CINNAMIC ACID,DIET-AMINO-ET ESTER
C16H23NO4	113932-78-8	CARBAMIC ACID, (3,4-DIETHOXYPHENYL)-, CYCLOPENTY
C16H24ClN3O	107534-96-3	TEBUCONAZOLE
C16H24ClN3O10	074751-37-4	BETA-D-GLUCOPYRANOSIDE, METHYL 3-[[[(2-CHLOROETH
C16H24ClNO5	087130-55-0	CARBAMIC ACID, (3,4-DIETHOXYPHENYL)-, 2-CHLORO-1
C16H24N2	132453-26-0	1H-FLUORENE-2-METHANAMINE, 2,3,4,4A,9,9A-HEXAHYD
C16H24N2O4	028197-69-5	ACETAMIDE, N-[3-ACETYL-4-[2-HYDROXY-3-[(1-METHYL
C16H24N2O5S	083870-92-2	BENZENESULFONYL-L-(N'-PENTYL)GLUTAMINE
C16H24N2O5S	083871-00-5	4-ME-BENZENESULFONYL-L-(N'-BU)GLUTAMINE
C16H24N4O4	059886-46-3	2,5-CYCLOHEXADIENE-1,4-DIONE, 2,5-BIS(1-AZIRIDIN
C16H24N4O6	059886-55-4	2,5-CYCLOHEXADIENE-1,4-DIONE, 2,5-BIS(1-AZIRIDIN
C16H24N6	061544-20-5	2,4-NH2PYRIMD,5(35-NME2-4-ME)BENZYL

C16H24N6O	061544-22-7	2,4-NH2PYRIMD,5(35-NME2-4-MEO)BENZYL
C16H24O2	132296-16-3	PHENOL, 5-(1,1-DIMETHYLETHYL)-2-(3-HYDROXYCYCLOH
C16H24O6	014098-24-9	B18C6-BENZO CROWN ETHER
C16H24O6	020772-23-0	GLUCOPYRANOSIDE,2-IPR-5-ME PHENYL
C16H24O6	020838-36-2	GLUCOPYRANOSIDE, 3-T-BUTYL PHENYL
C16H24O6	029074-04-2	4-T-BUTYLPHENYL GLUCOPYRANOSIDE
C16H25FN4O3	141018-20-4	CYTIDINE, 2',3'-DIDEOXY-3'-FLUORO-N-[(DIPROPYLAM
C16H25FN4O3	141018-21-5	CYTIDINE, 2',3'-DIDEOXY-3'-FLUORO-N-[[BIS(1-METH
C16H25N5O2	075841-97-3	2-PROPANOL, 1,1'-[[6-[(2,5-DIMETHYL-1H-PYRROL-1-
C16H25N5O2	075842-02-3	3,6-PYRIDAZINEDIAMINE, N'-(2,5-DIMETHYL-1H-PYRRO
C16H25NaO3S	073602-65-0	(1-METHYLNONYL)BENZENESULFONIC ACID, SODIUM SALT
C16H25NaO3S	073602-67-2	(1-BUTYLHEXYL)BENZENESULFONIC ACID, SODIUM SALT
C16H25NO2	002655-19-8	3,5-BIS(TERT-BUTYL)PHENOL METHYLCARBAMATE
C16H25NO2	080617-78-3	2-PROPANOL, 1-[(1,1-DIMETHYLETHYL)AMINO]-3-[(3-P
C16H25NO2	123134-25-8	CYCLOHEXANOL, 2-[(DIMETHYLAMINO)METHYL]-1-(3-MET
C16H25NO2	123154-38-1	CYCLOHEXANOL, 2-[(DIMETHYLAMINO)METHYL]-1-(3-MET
C16H25NO3	000054-32-0	MOXISYLYTE
C16H25NO4	124051-00-9	2H-PYRAN, TETRAHYDRO-3-[1-(ETHOXYIMINO)BUTYL]-6-
C16H25NO5S	057554-04-8	(N-PIPERIDINYL)ET-(345-MEO)PH SULFONE
C16H25NOS	036756-79-3	TIOCARBAZIL
C16H26	000104-72-3	1-PHENYLDECANE
C16H26	004537-11-5	BENZENE, (1-BUTYLHEXYL)-
C16H26	004537-12-6	BENZENE, (1-PROPYLHEPTYL)-
C16H26	004537-13-7	BENZENE, (1-METHYLNONYL)-
C16H26	004621-36-7	BENZENE, (1-ETHYLOCTYL)-
C16H26N2O	021236-53-3	BUTANAMIDE, 4-(DIETHYLAMINO)-N-(2,6-DIMETHYLPHEN
C16H26N2O	077471-00-2	BUTANAMIDE, 3-(DIETHYLAMINO)-N-(2,6-DIMETHYLPHEN
C16H26N2O3	121508-88-1	PROPANAMIDE, N-[2-[[2-HYDROXY-3-(PHENYLMETHOXY)P
C16H26N2O3S	074746-44-4	HEPTANAMIDE, N-[4-(DIMETHYLAMINO)SULFONYL]PHENYL
C16H26N2O4	034919-98-7	CETAMOLOL
C16H26N2O4S	000525-97-3	6-(OCTANOYLAMINO)PENICILLANIC ACID
C16H26N4O	066521-49-1	1-(4-CARBAMOYLPHENYL)-3-METHYL-3-OCTYLTRIAZENE
C16H26N4O3	141018-15-7	CYTIDINE, 2',3'-DIDEOXY-N-[(DIPROPYLAMINO)METHYL
C16H26N4O3	141043-80-3	CYTIDINE, 2',3'-DIDEOXY-N-[[BIS(1-METHYLETHYL)AM
C16H26O	004130-92-1	2,6-DI(TERT-BUTYL)-4-ETHYLPHENOL
C16H26O6	057721-93-4	1,2-BIS(2-(2-MEO-ETO)-ETO)BENZENE
C16H26O6	121284-18-2	1,4-BIS(2-(2-MEO-ETO)-ETO)BENZENE
C16H26O6	121284-19-3	1,3-BIS(2-(2-MEO-ETO)-ETO)BENZENE
C16H26O7	000109-17-1	TETRAETHYLENEGLYCOL DIMETHACRYLATE
C16H27IN2O	035041-49-7	3-CARBAMOYLPYRIDINIUM IODIDE, N-DECYL
C16H27N3O4S	093414-65-4	BENZAMIDE, 4-AMINO-N-[2-(DIETHYLAMINO)ETHYL]-5-(
C16H27NO	083593-63-9	2-N,N-DIBUAMINO-1-PHENYLETHANOL
C16H27NO2	064638-10-4	2,5-DIMETHOXY-4-PENTYLAMPHETAMINE
C16H27NO2	080617-76-1	2-PROPANOL, 1-[(1,1-DIMETHYLETHYL)AMINO]-3-(3-PH
C16H27NO5S	071203-64-0	(N-BU-ME)AMET-(345-MEO)PHENYL-SULFONE
C16H28IN	014402-23-4	2-METHYLPYRIDINIUM IODIDE, N-DECYL
C16H28N4O6	137605-65-3	L-ALANINAMIDE, N2-(N-ACETYL-L-ALANYL-L-ASPARAGYL
C16H29N5O	055921-63-6	24NH2-6-DICYCLOHXAMPYRIMIDIN-3-OXIDE
C16H30N4O4	132766-04-2	L-VALINAMIDE, N-ACETYL-L-ALANYL-L-LEUCYL-
C16H30N4O4	132766-06-4	L-ISOLEUCINAMIDE, N-ACETYL-L-ALANYL-L-VALYL-
C16H30N4O4	132766-10-0	L-VALINAMIDE, N-ACETYL-L-ISOLEUCYL-L-ALANYL-
C16H30O2	000142-90-5	DODECYL 2-METHYL-2-PROPENOATE
C16H30O2	016725-53-4	9-TETRADECEN-1-OL ACETATE (Z)
C16H30O2	017004-62-5	8-HEXADECENOIC ACID
C16H31NO	002687-96-9	1-LAURYL-2-PYRROLIDONE
C16H32	000629-73-2	1-HEXADECENE
C16H32	001795-16-0	N-DECYLCYCLOHEXANE
C16H32NaO2	000408-35-5	HEXADECANOIC ACID, SODIUM SALT

C16H32O	056683-54-6	11-HEXADECEN-1-OL (Z)
C16H32O2	000057-10-3	PALMITIC ACID
C16H32O2	000124-06-1	ETHYLMYRISTATE
C16H32O2	000638-59-5	1-TETRADECANOL ACETATE
C16H32O2	007132-64-1	METHYL PENTADECANOATE
C16H32O4Si4	000177-49-1	TETRACYCLOTETRAMETHYLENECYCLOTETRASILOXANE
C16H33Br	000112-82-3	1-BROMOHEXADECANE
C16H33Cl	004860-03-1	1-CHLOROHEXADECANE
C16H33F9O5Si5	022474-57-3	2,2,4,4,6,8,10-HEPTAMETHYL-6,8,10-TRIS(3,3,3-TR*
C16H33I	000544-77-4	1-IODOHEXADECANE
C16H34	000544-76-3	N-HEXADECANE
C16H34NaO6S	003088-31-1	DIETHYLENE GLYCOL MOMOLAURYL ETHER SODIUM SULFAT
C16H34O	036653-82-4	1-HEXADECANOL
C16H34S	002917-26-2	1-HEXADECANETHIOL
C16H35N	000143-27-1	N-HEXADECYLAMINE
C16H35O4P	000298-07-7	BIS(2-ETHYLHEXYL)PHOSPHATE
C16H35O4P	027215-10-7	DIISOCTYL PHOSPHATE
C16H36N2S2	065332-44-7	N,N'-(DITHIODI-2,1-ETHANEDIYL)BIS(N-(1-METHYL*))
C16H38O6P2S4	008000-97-3	DEMETON
C16H48O6Si7	000541-01-5	HEXADECAMETHYLHEPTASILOXANE
C16H48O8Si8	000556-68-3	HEXADECAMETHYLCYCLOOCTASILOXANE
C17H8F3N5O4	121588-32-7	N(24-NO2-6CF3PH)-2(3CN-BENZO)PYRIDINAMIN
C17H10O	000082-05-3	BENZANTHRONE
C17H11ClF4N2O	049606-44-2	2H-1,4-BENZODIAZEPIN-2-ONE, 7-CHLORO-5-(2-FLUORO
C17H11ClF4N2S	036735-22-5	2H-1,4-BENZODIAZEPINE-2-THIONE, 7-CHLORO-5-(2-FL
C17H11N	000225-11-6	BENZ(A)ACRIDINE
C17H11N	000225-51-4	BENZ(C)ACRIDINE
C17H12	000205-12-9	7H-BENZO(C)FLUORENE
C17H12	000238-84-6	11H-BENZO(A)FLUORENE
C17H12	000243-17-4	11H-BENZO(B)FLUORENE
C17H12	019561-31-0	4H-CYCLOHEPTA(DEF)PHENANTHRENE
C17H12	030777-18-5	BENZO(A)FLUORENE
C17H12	030777-19-6	BENZO(B)FLUORENE
C17H12Cl2N2O	060168-88-9	FENARIMOL
C17H12ClF3N2O	023092-17-3	2H-1,4-BENZODIAZEPIN-2-ONE, 7-CHLORO-1,3-DIHYDRO
C17H12ClFN2O	063284-71-9	5-PYRIMIDINEMETHANOL, .ALPHA.-(2-CHLOROPHENYL)-.
C17H12I2O3	000068-90-6	BENZIODARONE
C17H13BrF2N2O	142657-44-1	1H-IMIDAZOLE-1-ETHANOL, _-(3-BROMOPHENYL)-_-(2,4
C17H13ClFNO4	105512-06-9	CLODINAFOP-PROPARGYL
C17H13ClO3	058182-63-1	ITANOXONE
C17H13F3N4O2	136389-87-2	1H-PYRAZOLE, 3-METHYL-4-NITRO-1-PHENYL-5-[[4-(TR
C17H13FN2O2	036093-54-6	14-BENZODIAZPN2-ON,5(2-F PH)7ACETYL
C17H13N	003558-69-8	2,6-DIPHENYLPYRIDINE
C17H13N2NaO4S	005850-86-2	C.I. ACID ORANGE 8, SODIUM SALT
C17H13N2NaO5S	005858-39-9	C.I. ACID RED 4, MONOSODIUM SALT
C17H13N3O	003489-59-6	2H-1,4-BENZODIAZEPIN-2-ONE,1,3-DIHYDRO-1-METHYL-
C17H13NO2	119005-16-2	2-AZETIDINONE, 4-(9-PHENANTHRENYLOXY)-
C17H14BrFN2O	128104-25-6	1H-IMIDAZOLE-1-ETHANOL, _-(4-BROMOPHENYL)-_-(4-F
C17H14Cl2N2O	088695-11-8	14-BENZODIAZPN2-ON,1,3-ME-5(2CLPH)7CL
C17H14Cl2N2O2	029442-60-2	14-BENZODIAZP2-ON,1-MEOME-5(2CLPH)7CL
C17H14ClF7O2	079538-32-2	TETRAFLUTHRIN
C17H14ClNO2S	018046-21-4	FENTIAZAC
C17H14F2N2O	110933-19-2	1H-IMIDAZOLE-1-ETHANOL, _,_-BIS(4-FLUOROPHENYL)-
C17H14N2	000519-23-3	ELLIPTICINE
C17H14N2O2	001229-55-6	SOLVENT RED 1
C17H14N4O4	027016-91-7	1H-1,4-BENZODIAZEPINE-1-CARBOXAMIDE, 2,3-DIHYDRO
C17H14O2	006549-64-0	1H-INDENE-1,3(2H)-DIONE, 2-(3,5-DIMETHYLPHENYL)-
C17H14O2	007561-67-3	1H-INDENE-1,3(2H)-DIONE, 2-(4-ETHYLPHENYL)-

I-191

C17H14O4	019055-50-6	1H-INDENE-1,3(2H)-DIONE, 2-(3,5-DIMETHOXYPHENYL)
C17H14O5	000117-52-2	COUMAFURYL
C17H14O6	000469-01-2	6H-[1,3]DIOXOLO[5,6]BENZOFURO[3,2-C][1]BENZOPYRA
C17H15BrN2O	142657-45-2	1H-IMIDAZOLE-1-ETHANOL, _-(4-BROMOPHENYL)-_-PHEN
C17H15ClFN3O	106325-08-0	BAS 480F
C17H15ClFNO3	052756-25-9	FLAMPROP-METHYL
C17H15ClN2O	005358-34-9	1,4-BENZDAZEPIN-2-ON-1-ME-5-(2ME-PH)7-CL
C17H15ClN2O	005571-65-3	1,4-BENZODIAZP2-ON,1-ET-5-PH-7CL
C17H15ClN2O2	029442-59-9	14-BENZODIAZP2-ON,1-MEOME-5(PH)-7CL
C17H15ClN2O2	072430-63-8	14-BENZDIAZPIN-2-ONE,1ME5-(4MEOPH)7CL
C17H15ClN2O3	056875-80-0	2H-1,4-BENZODIAZEPIN-2-ONE, 7-CHLORO-1,3-DIHYDRO
C17H15ClN2OS	029442-82-8	1,4-BENZDAZEPIN-2-ON-1-MESME-5-PH-7-CL
C17H15ClO5	093414-41-6	BENZOIC ACID, 2-CHLORO-, 2-(4-FORMYL-2-METHOXYPH
C17H15ClO5	093414-42-7	BENZOIC ACID, 3-CHLORO-, 2-(4-FORMYL-2-METHOXYPH
C17H15ClO5	093414-43-8	BENZOIC ACID, 4-CHLORO-, 2-(4-FORMYL-2-METHOXYPH
C17H15FN2O	051307-86-9	1,4-BENZDAZEPIN-2-ONE-5-(2FPH)-7-ET
C17H15N3O	131229-44-2	IMIDAZO[5,1-B]QUINAZOLIN-9(2H)-ONE, 1,3-DIHYDRO-
C17H15N3O4	029442-58-8	1,4-BENZDAZEPIN-2-ON-1MEOME-5PH-7NO2
C17H15NO3	031842-01-0	BENZENEACETIC ACID, 4-(1,4-DIHYDRO-1-OXO-2H-ISOI
C17H15NO3S	098369-49-4	BENZOIC ACID, 2-[(4-OXO-2-AZETIDINYL)THIO]-, PHE
C17H15NO4	119005-12-8	BENZOIC ACID, 3-[(4-OXO-2-AZETIDINYL)OXY]-, PHEN
C17H15NO4S	074134-18-2	ACETAMIDE, N-(7-ETHYL-9-OXO-9H-THIOXANTHEN-3-YL)
C17H15NO5	005003-48-5	BENZOIC ACID, 2-(ACETYLOXY)-, 4-(ACETYLAMINO)PHE
C17H15NO6	091814-29-8	1-NAPHTHALENECARBONITRILE, 5,8-BIS(ACETYLOXY)-6,
C17H15NO6	091814-30-1	2-NAPHTHALENECARBONITRILE, 5,8-BIS(ACETYLOXY)-6,
C17H16Br2O3	018181-80-1	ISOPROPYL 4,4'DIBROMOBENZILATE
C17H16Cl2N2O4	087757-18-4	ISOXAPYRIFOP
C17H16Cl2O3	005836-10-2	ISOPROPYL-4,4'-DICHLOROBENZILATE
C17H16ClFN2O2	062666-20-0	PROGABIDE
C17H16ClN5O3	006657-33-6	PROPANENITRILE, 3- 4- (2-CHLORO-4-NITROPHENYL)A
C17H16ClNO4	054870-28-9	MEGLITINIDE
C17H16ClNO5	093414-45-0	BENZOIC ACID, 4-CHLORO-, 2-[4-[(HYDROXYIMINO)MET
C17H16ClNO5	093414-47-2	BENZOIC ACID, 3-CHLORO-, 2-[4-[(HYDROXYIMINO)MET
C17H16FN3O2	040837-34-1	14-BENZODIAZP2-ON,1-MEOME-5(2FPH)-7AM
C17H16FN5O3	132722-93-1	BENZAMIDE, N-[9-(2,3-DIDEOXY-2-FLUORO-beta-D-THR
C17H16FNO3	083073-37-4	1H-PYRROLIZINE-1-CARBOXYLIC ACID, 6-ETHYL-5-(4-F
C17H16N2O	005571-62-0	1,4-BENZODIAZEPIN-2-ONE,1,3-DUHYDRO-7,9-DIMETHYL
C17H16N2O2	002221-11-6	2,4-IMIDAZOLIDINEDIONE, 3,5-BIS(PHENYLMETHYL)-
C17H16N2O2	023186-94-9	2,4-IMIDAZOLIDINEDIONE, 5,5-BIS(PHENYLMETHYL)-
C17H16N2O3	002475-46-9	C.I. DISPERSE BLUE 3
C17H16N2O4	081592-09-8	M-ACETAMIDOPHENYLHIPPURATE
C17H16N2O4S	021413-25-2	2,4-IMIDAZOLIDINEDIONE, 5-ETHYL-5-PHENYL-1-(PHEN
C17H16N2OS	023193-98-8	1,4BENZODIAZEPIN2ONE,1ME-5PH-7MES
C17H16N4O2	136389-76-9	1H-PYRAZOLE, 3-METHYL-5-[(4-METHYLPHENYL)AMINO]-
C17H16N4O2	136389-77-0	1H-PYRAZOLE, 3-METHYL-5-[(3-METHYLPHENYL)AMINO]-
C17H16N6O3	131402-52-3	ALLOPURINOL,3-(IMIDAZOL-1-YL)METHYLBENZOYLOXYMET
C17H16O3	051994-35-5	PENTANOIC ACID, 5-[(1,1'-BIPHENYL)-4-YL]-5-OXO-
C17H16O3	063472-04-8	BUTANOIC ACID, 4-[(1,1'-BIPHENYL)-4-YL]-2-METHYL
C17H16O4	023005-56-3	MONO(P,ALPHA-DIMETHYLBENZYL) PHTHALATE
C17H16O5	033533-57-2	MONO(P-METHOXY-ALPHA-METHYLBENZYL) PHTHALATE
C17H17Cl2NO3	023885-56-5	P-PHENYL-AMPHENICOL
C17H17Cl3O	034197-05-2	METHYLETHOXYCHLOR
C17H17ClN2O2	074339-89-2	PROPANAMIDE, 2-CHLORO-N-[2-OXO-2-(PHENYLAMINO)ET
C17H17ClO3	077269-59-1	PHENYLACETIC ACID,3-CL-4-PHPRO
C17H17ClO6	000126-07-8	GRISEOFULVIN
C17H17IN2O7	135600-74-7	URIDINE, 2'-DEOXY-5-IODO-, 5'-(4-METHOXYBENZOATE
C17H17N3O	030144-56-0	1,4-BENZDAZEPIN-2-ON-5PH-7-DIMEAMINO
C17H17N3O	088695-10-7	1,4-BENZDAZEPIN-2-ON-1-ET-5-PH-7-AMINO

C17H17N3O	108611-26-3	3-PYRIDINECARBONITRILE, 6-METHYL-2-(4-MORPHOLINY
C17H17N3O2	040837-24-9	1,4-BENZDAZEPIN-2-ON-1-MEOME-5-PH-7AMINO
C17H17N3O3	088695-13-0	1,4-BENZDAZEPN-2-ON-1-MEOME-3-OH-5PH-7AM
C17H17N3O3S	078090-11-6	1H-BENZIMIDAZOLE-5-CARBOXYLIC ACID, 6-METHYL-2-[
C17H17N3O6S2	021593-23-7	CEPHAPIRIN
C17H17N3OS	088695-12-9	1,4-BENZDAZEPIN-2-ON-1-MESME-5-PH-7AMINO
C17H17N5O4	004546-72-9	ADENOSINE, N-BENZOYL-2'-DEOXY-
C17H17N5O4S	130787-23-4	2H-BENZOTRIAZOLE, 2-CYCLOPENTYL-5-[[(4-NITROPHEN
C17H17N5O5	004546-55-8	ADENOSINE, N-BENZOYL-
C17H17NO2	000058-00-4	6A-BETA-APORPHINE-10,11-DIOL
C17H17NO3	081592-15-6	3,5-DIMETHYLPHENYLHIPPURATE
C17H17NO3	103188-47-2	2-PROPENAMIDE, 3-(3,4-DIHYDROXYPHENYL)-N-(2-PHEN
C17H17NO3S	119005-30-0	1-AZETIDINEACETIC ACID, 2-(2-NAPHTHALENYLTHIO)-4
C17H17NO4	119005-26-4	1-AZETIDINEACETIC ACID, 2-(2-NAPHTHALENYLOXY)-4-
C17H17NO5	081592-16-7	3,5-DIMETHOXYPHENYLHIPPURATE
C17H17NO5	136944-22-4	2-PROPENAMIDE, 3-(3,4-DIHYDROXYPHENYL)-N-(3,4-DI
C17H18	017114-78-2	9-(1,1-DIMETHYLETHYL)-9H-FLUORENE
C17H18Cl2N4O4	058528-60-2	DIS. A. 10
C17H18ClN3O	001159-93-9	CLOBENZEPAM
C17H18ClN5O6	065125-87-3	DIS. A. 8
C17H18ClNO2	000314-19-2	6,ALPHA,BETA-APORPHINE-10,11-DIOL, HYDROCHLOR*
C17H18ClNO4	099877-09-5	L-TYROSINE, 3-HYDROXY-, 2-(4-CHLOROPHENYL)ETHYL
C17H18F3N3O3	079660-72-3	FLEROXACIN
C17H18FN3O2	111041-44-2	IMIDODICARBONIC DIAMIDE, N-[2-FLUORO-5-(PHENYLME
C17H18FN3O3	085721-33-1	CIPROFLOXACIN
C17H18N2O5	049648-47-7	PHENYLAMPHENICOL
C17H18N2O6	021829-25-4	NIFEDIPINE
C17H18N2O6S	004697-36-3	CARBENICILLIN
C17H18N4	108611-33-2	3-PYRIDINECARBONITRILE, 5-PHENYL-2-(1-PIPERAZINY
C17H18N4O3	065542-19-0	BENZOIC ACID, 4-[3-[[4-(ACETYLAMINO)PHENYL]METHY
C17H18N4O3S	072301-79-2	1H-BENZIMIDAZOL-2-AMINE, 6-[(HYDROXYIMINO)PHENYL
C17H18N6O2	117121-36-5	UREA, N,N-DIMETHYL-N'-[4-[[1-(PHENYLMETHYL)-1H-T
C17H18N6O2	117121-48-9	UREA, N,N-DIMETHYL-N'-[4-[[2-(PHENYLMETHYL)-2H-T
C17H18O3	093371-55-2	BUTANOIC ACID,2-(4-BIPHENYLYL)-3-HYDROXY-3-METHY
C17H18O6	091814-42-5	1,4-NAPHTHALENEDIOL, 2,3-DIMETHOXY-6-METHYL-, DI
C17H18O7	091814-24-3	1,4-NAPHTHALENEDIOL, 2,3,5-TRIMETHOXY-, DIACETAT
C17H18O7	091814-40-3	1,4-NAPHTHALENEDIOL, 2,3,6-TRIMETHOXY-, DIACETAT
C17H19BrN5O6P	107538-75-0	8-BROMO CYCLIC AMP-O-BENZYL
C17H19ClN2O7	096000-44-1	1-ACETYL-6-DEMETHYL-6-CHLORO-7-METHOXYMITOSENE
C17H19ClN2OS	000969-99-3	OPROMAZINE
C17H19ClN2OS	003926-67-8	10H-PHENOTHIAZIN-2-OL, 8-CHLORO-10-[3-(DIMETHYLA
C17H19ClN2S	000050-53-3	CHLORPROMAZINE
C17H19ClN4O4	003769-57-1	DISPERSE RED 5
C17H19F2N3O3	075338-41-9	8-FLUORO-PEFLOXACIN
C17H19F2N3O3	098079-51-7	LOMEFLOXACIN
C17H19N3O3	111227-05-5	1H-IMIDAZOLE-1-CARBOXYLIC ACID, 1-[(4-CYANOPHENO
C17H19N3O3S	073590-58-6	OMEPRAZOLE
C17H19N5O	094011-82-2	BAZINAPRINE
C17H19N5O2	072732-56-0	PIRITREXIM
C17H19N5O2S	130787-24-5	2H-BENZOTRIAZOLE, 5-[[(4-AMINOPHENYL)SULFONYL]AM
C17H19N5O6	041541-13-3	DIS. A. 1
C17H19NO2	055814-41-0	MEPRONIL
C17H19NO2	149589-40-2	BENZAMIDE, N-METHYL-3-PHENOXY-N-PROPYL-
C17H19NO3	000057-27-2	MORPHINE
C17H19NO3	000094-62-2	1-PIPEROYL-(E,E)-PIPERIDINE
C17H19NO3	084640-30-2	N,N-DIME CARBAMATE,P-BENZYLOXYBENZYL
C17H19NO4	000076-41-5	OXYMORPHONE
C17H19NO4	034197-26-7	1,1-BIS(P-MEO-PH)-2-NITROPROPANE

C17H19NO4	057646-30-7	N(26DIMEPH)-N-(2FURANOYL)ALANINE,ME
C17H19NO4	072490-01-8	FENOXYCARB
C17H19NO4	079127-80-3	FENOXYCARB
C17H19NO4	099877-08-4	L-TYROSINE, 3-HYDROXY-, 2-PHENYLETHYL ESTER
C17H19NO5	099877-11-9	L-TYROSINE, 3-HYDROXY-, 2-PHENOXYETHYL ESTER
C17H20F6N2O3	054143-55-4	FLECANIDE
C17H20FN3O3	070458-92-3	PEFLOXACIN
C17H20FN3O3	098079-47-1	8-DESFLUORO-LOMEFLOXACIN
C17H20N2O	000085-98-3	DIETHYL DIPHENYL UREA
C17H20N2O	000090-94-8	4,4-BIS(DIMETHYLAMINO)BENZOPHENONE
C17H20N2O	102587-49-5	UREA, N,N-DIMETHYL-N'-[3-(2-PHENYLETHYL)PHENYL]-
C17H20N2O2	070859-35-7	1,1-DIME-3-(3-PHENYLETOPH) UREA
C17H20N2O3	069922-40-3	N1(ME-MEO)N3(P-PHETO)PHENYLUREA
C17H20N2O3S	101533-58-8	BENZAMIDE, 4-[(4-AMINOPHENYL)SULFONYL]-N,N-DIETH
C17H20N2O5S	000147-55-7	PHENETHICILLIN
C17H20N2O5S	001752-26-7	PENICILLIN, A-PHENOXYETHYL
C17H20N2O6S	000061-32-5	METHICILLIN
C17H20N2O7	096000-43-0	1-AC-6-DEME-7-METHOXYMITOSENE
C17H20N2S	000058-40-2	PROMAZINE
C17H20N2S	000060-87-7	PROMETHAZINE
C17H20N4O4	003179-89-3	C.I. DISPERSE RED 17
C17H20N4O4S4	026970-95-6	CEPHALOTHIN ANALOG
C17H20N4O5	041541-14-4	DIS. A. 13
C17H20N4O6	000083-88-5	RIBOFLAVIN
C17H20N5O6PS	032487-38-0	8-S-BENZYL CYCLIC AMP
C17H20N5O6PSe	052109-40-7	8-SEBENZYL CYCLIC AMP
C17H20O6S	071203-63-9	2-OH-2-PH-ET-(3,4,5-MEO)PHENYLSUFONE
C17H21ClN2S	000053-60-1	PROMAZINE HYDROCHLORIDE
C17H21N3	000492-80-8	AURAMINE
C17H21N3O10S	083116-90-9	1H-IMIDAZOLE, 2-NITRO-1-(2,3,4,6-TETRA-O-ACETYL-
C17H21N3O11	067774-11-2	1H-IMIDAZOLE, 2-NITRO-1-(2,3,4,6-TETRA-O-ACETYL-
C17H21N3O2	003771-38-8	DIS. A. 2
C17H21N3O5	097874-69-6	2-PENTENOIC ACID, 2-CYANO-3-[[(4-NITROPHENYL)MET
C17H21N3O6S2	088644-13-7	A-MESULFONAMIDOBENZYLPENCILLIN
C17H21N3O9	015981-93-8	O-TRIACETYL-N-ACETYL-ARA-C
C17H21N5O2	071125-75-2	24NH2PYRIMIDIN,5(35MEO-4(1PYRRYL))BENZYL
C17H21NaO3S	099287-46-4	1-SO3NA-3-BUTYL-7-ISOPROPYL AZULENE
C17H21NaO4S	099287-66-8	1-SO3NA-3-PR-4-METHOXY-7-IPR AZULENE
C17H21NO	000058-73-1	DIPHENHYDRAMINE
C17H21NO2	015299-99-7	N,N-DIETHYL-2-(1-NAPHTHALENYLOXY)PROPANAMIDE
C17H21NO3	002176-16-1	BENZENEETHANAMINE, 3,5-DIMETHOXY-4-(PHENYLMETHOX
C17H21NO3S	074772-77-3	CIGLITAZONE
C17H21NO4	000050-36-2	COCAINE
C17H21NO4	000051-34-3	SCOPOLAMINE
C17H21NO4	082488-02-6	M-NITROBENZYL-TR-CHRYSANTHEMATE
C17H21NO4S4	017606-31-4	BENSULTAP
C17H22BrNO4	000114-49-8	SCOPOLAMINE BROMIDE
C17H22ClN3	002465-27-2	BASIC YELLOW 2
C17H22ClN3O	125116-23-6	METCONAZOLE
C17H22ClNO4	000053-21-4	COCAINE HYDROCHLORIDE
C17H22I3N3O7	076350-28-2	M-BENZAMIDE,NN'BIS(23DIOHPR)5ACE..
C17H22N2	000101-61-1	BIS(4-DIMETHYLAMINOPHENYL)METHANE
C17H22N2O2	111227-08-8	1H-IMIDAZOLE-1-CARBOXYLIC ACID, 1-[(4-METHYLPHEN
C17H22N2O2S	111226-65-4	1H-IMIDAZOLE-1-CARBOTHIOIC ACID, O-[1-[(4-METHYL
C17H22N2O3S	111226-71-2	1H-IMIDAZOLE-1-CARBOTHIOIC ACID, O-[1-[(4-METHOX
C17H22N2O3S	111227-18-0	1H-IMIDAZOLE-1-CARBOXYLIC ACID, 1-[[4-(METHYLTHI
C17H22N2O4	111227-17-9	1H-IMIDAZOLE-1-CARBOXYLIC ACID, 1-[(4-METHOXYPHE
C17H22N4O	025905-77-5	MINAPRINE

C17H22N4O2	085544-42-9	24NH2PYRIMIDN,5(3MEO-4PRENYL-5ETO)BENZYL
C17H22N4O2	087653-23-4	UREA, N-ETHYL-N'-[4-(6-OXO-3-PHENYL-1(6H)-PYRIDA
C17H22N4O2	087653-35-8	UREA, N,N-DIMETHYL-N'-[4-(6-OXO-3-PHENYL-1(6H)-P
C17H22N4O2	087653-49-4	UREA, [6-(6-OXO-3-PHENYL-1(6H)-PYRIDAZINYL)HEXYL
C17H22N4O3	055687-73-5	24NH2PYRIMIDN,5(35ETO-4-ACETYL)BENZYL
C17H22N4O4	055687-46-2	24NH2PYRIMIDN,5(35MEO-4CO2IPR)BENZYL
C17H22O2	064312-78-3	BENZYL-T-CHYRSANTHEMATE
C17H22O5	015725-38-9	DIETHYLMALONATE,3-PRO BENZAL
C17H22O5	019202-92-7	TENULIN
C17H23FN2O7	003343-22-4	URIDINE, 2'-DEOXY-5-FLUORO-, 3',5'-DIBUTANOATE
C17H23N3O	000091-84-9	PYRILAMINE
C17H23N3O	002531-04-6	PIPERYLONE
C17H23NO	000077-07-6	LEVORPHANOL
C17H23NO	000125-73-5	MORPHINAN-3-OL, 17-METHYL-, (9_,13_,14_)-
C17H23NO3	000051-55-8	ATROPINE
C17H23NO5	116482-75-8	BENZOIC ACID, 2-(ACETYLOXY)-, 2-(DIPROPYLAMINO)-
C17H23NO5	116482-76-9	BENZOIC ACID, 2-(ACETYLOXY)-, 2-[BIS(1-METHYLETH
C17H24ClN3O11	058484-10-9	UREA, N-(2-CHLOROETHYL)-N-NITROSO-N'-(2,3,4,6-TE
C17H24ClN3S	061477-97-2	DAZOLICINE
C17H24N2O4S	086570-01-6	CARBOFURAN,N-(N'-ME,PR CARBAMIDOTHIO)
C17H24N2O4S	086570-02-7	CARBOFURAN,N-(N'-ME,ISOPR CARBAMIDOTHIO)
C17H24N2O5	102273-20-1	BUTANOIC ACID, [2-[[(4-MORPHOLINYLMETHYL)AMINO]C
C17H24N4O2	083158-03-6	2,4-NH2PYRIMD,5(2-ME-4MEO-5BUO)BENZYL
C17H24N4O3	077453-35-1	24NH2PYRIMIDIN,5(3ETO4COHME2-5MEO)BENZYL
C17H24N6O2	061544-21-6	24NH2PYRMDN,5(3NME2-4MEO-5NMECOME)BENZYL
C17H24N6O3	050574-87-3	MORPHOLINE, 4-[[4-(4,6-DIAMINO-2,2-DIMETHYL-1,3,
C17H24O8	060835-75-8	BENZO-18-CROWN-6-ETHER,M-CARBOXY-
C17H25N3O4	065356-76-5	L-LEUCINAMIDE, N-ACETYL-L-TYROSYL-
C17H25N3O4S2	083130-01-2	ALANYCARB
C17H25N3O5S	066644-81-3	VERALIPRIDE
C17H25N5O3S	023401-43-6	DL-VALINE, N-[1-OXO-5-(1H-PURIN-6-YLTHIO)PENTYL]
C17H25NO	097871-57-3	A-BUTYL-N-SEC-BUTYLCINNAMAMIDE
C17H25NO2	000064-39-1	PROMEDOL
C17H25NO2	000113-48-4	MGK 264
C17H25NO2	000468-50-8	4-PIPERIDINOL, 3-ETHYL-1-METHYL-4-PHENYL-, PROPI
C17H25NO3	047141-42-4	LEVOBUNOLOL
C17H25NO3	106231-57-6	ACETAMIDE, 2-(BENZOYLOXY)-N,N-DIBUTYL-
C17H25NO3	115193-33-4	ACETAMIDE, 2-(BENZOYLOXY)-N,N-BIS(2-METHYLPROPYL
C17H26ClNO2	023184-66-9	BUTACHLOR
C17H26ClNO2	051218-49-6	PRETILCHLOR
C17H26N2O	084057-95-4	ROPIVACAINE
C17H26N2O3	080171-69-3	CARBAMIC ACID, [2-(ETHOXYMETHYL)PHENYL]-, 2-(1-P
C17H26N2O4S	053583-79-2	SULTOPRIDE
C17H26N2O4S	093414-63-2	BENZAMIDE, N-[(1-ETHYL-3-PYRROLIDINYL)METHYL]-5-
C17H26N4O5	098827-16-8	ALLOPURINOL,1,5-BIS(PIVALOYLOXYMETHYL)-
C17H26N4O5	098827-17-9	ALLOPURINOL,2,5-BIS(PIVALOYLOXYMETHYL)-
C17H26O3	005519-23-3	BENZOIC ACID, 4-(DECYLOXY)-
C17H26O4Si4	013093-12-4	HEXAMETHYL(SILAACENAPHTHENYL)CYCLOTETRASILOXAN*
C17H27N3O4S	071675-85-9	AMISULPRIDE
C17H27N3O5S2	119410-37-6	SPARSOMYCIN,S-PENTYL ANALOG
C17H27NO2	001918-11-2	2,6-DI-T-BUTYL-P-TOLYL METHYLCARBAMATE
C17H27NO2	080617-86-3	2-PROPANOL, 1-[(1,1-DIMETHYLETHYL)AMINO]-3-[(1,2
C17H27NO4	022664-55-7	METIPRANOLOL
C17H27NO4	042200-33-9	NADOLOL
C17H28	004536-86-1	BENZENE, (1-PROPYLOCTYL)-
C17H28	004536-87-2	BENZENE, (1-ETHYLNONYL)-
C17H28	004536-88-3	BENZENE, (1-METHYLDECYL)-
C17H28	004537-14-8	BENZENE, (1-PENTYLHEXYL)-

C17H28	004537-15-9	BENZENE, (1-BUTYLHEPTYL)-
C17H28	006742-54-7	1-PHENYLUNDECANE
C17H28N2O	036637-18-0	ETIDOCAINE (DURANEST)
C17H28N2O3	121508-91-6	PROPANAMIDE, N-[2-[[2-HYDROXY-3-(2-PHENYLETHOXY)
C17H28N2O4	121508-96-1	PROPANAMIDE, 2-METHYL-N-[2-[[3-(2-PHENOXYETHOXY)
C17H28N6	077314-60-4	2-PYRIDINAMINE, 4-(5-AMINO-1H-1,2,4-TRIAZOL-3-YL
C17H28O	020056-71-7	PHENOL, 2-UNDECYL-
C17H28O	020056-72-8	PHENOL, 3-UNDECYL-
C17H28O	020056-73-9	PHENOL, 4-UNDECYL-
C17H28O2	021704-31-4	1,2-BENZENEDIOL, 3-UNDECYL-
C17H28O2	123063-36-5	1,2-BENZENEDIOL, 4-UNDECYL-
C17H30BrN	000104-73-4	DODECYLPYRIDINIUM BROMIDE
C17H30IN	088477-44-5	3,4-DIME PYRIDINIUM IODIDE,N-DECYL
C17H30IN	088477-50-3	2,4-DIME PYRIDINIUM IODIDE,N-DECYL
C17H30IN	088477-56-9	3,5-DIME PYRIDINIUM IODIDE,N-DECYL
C17H30N2O3	096914-11-3	PILOCARPIC ACID, HEXYL ESTER
C17H31NO3	010054-21-4	1-LAURYL-4-CARBOXY-2-PYRROLIDONE
C17H32N2O6S	002256-16-8	D-ERYTHRO-A-D-GALACTO-OCTOPYRANOSIDE DERIVATIVE
C17H33N3O4S2	086569-89-3	N(N"-ME-(DECOXY)CARBAMYLTHIO)METHOMYL
C17H34	006765-39-5	1-HEPTADECENE
C17H34N5O3P	103981-93-7	1-PYRROLIDINYLOXY, 3-[[[[BIS(2,2-DIMETHYL-1-AZIR
C17H34N5O3P	103981-96-0	TEPA 4-(N-OH-PYRROLDINYL)-UREA DERIVATIV
C17H34O2	000110-27-0	ISOPROPYL MYRISTATE
C17H34O2	000112-39-0	METHYL HEXADECANOATE
C17H34O2	000506-12-7	HEPTADECANOIC ACID
C17H35Br	003508-00-7	1-BROMOHEPTADECANE
C17H35Cl	062016-75-5	1-CHLOROHEPTADECANE
C17H36	000629-78-7	HEPTADECANE
C17H36O	001454-85-9	HEPTADECYL ALCOHOL
C17H37O3P	001832-68-4	DIOCTYL METHYLPHOSPHONATE
C17H39N2O2P	057856-12-9	BIS(2-DIISOPROPYLAMINO)ETHYL METHYLPHOSPHONITE
C18H10Cl2O2S2	002379-74-0	C.I. VAT RED 1
C18H10Cl2O2S2	005462-29-3	C.I. VIOLET 2
C18H10N2O2S	002866-43-5	DISPERSE BRIGHTENER
C18H10O2	002498-66-0	BENZ A ANTHRACENE-7,12-DIONE
C18H11NO2	000083-08-9	QUINOPHALONE
C18H11NO2	008003-22-3	SOLVENT YELLOW 33
C18H11NO3	007576-65-0	DISPERSE YELLOW 54
C18H12	000056-55-3	BENZ(A)ANTHRACENE
C18H12	000092-24-0	TETRACENE
C18H12	000195-19-7	BENZO(C)PHENANTHRENE
C18H12	000217-59-4	TRIPHENYLENE
C18H12	000218-01-9	CHRYSENE
C18H12N2	000119-91-5	2,2'-BIQUINOLINE
C18H13N	002642-98-0	6-AMINOCHRYSENE
C18H13NNaO3	000132-67-2	N-1-NAPHTHYLPHTHALMIC ACID
C18H13NO3	000132-66-1	NPA
C18H14	000084-15-1	O-TERPHENYL
C18H14	000092-06-8	m-TERPHENYL
C18H14	000092-94-4	1,1':4',1"-TERPHENYL
C18H14F3NO2	096525-23-4	FLURTAMONE
C18H14N2O	031785-60-1	4(1H)-QUINAZOLINONE, 2,3-DIHYDRO-2-(1-NAPHTHALEN
C18H14N2O3	058907-81-6	1,4-NAPHTHOQUINONE,2-ACETAMIDO,3-ANILINO
C18H14N4O	006250-23-3	C.I. DISPERSE YELLOW 23
C18H14N4O2	002581-69-3	C.I. DISPERSE ORANGE 1
C18H15F5N2O2	112411-06-0	BENZOIC ACID, 2,3,4,5,6-PENTAFLUORO-, [1-(1,1-DI
C18H15N	000603-34-9	TRIPHENYLAMINE
C18H15N3O4S	005124-25-4	C.I. DISPERSE YELLOW 42

C18H15N3O5	104662-93-3	1H-BENZIMIDAZOLE-1-CARBOXYLIC ACID, 5-BENZOYL-2-
C18H15NO3	119005-18-4	2-AZETIDINONE, 4-[4-(1-OXO-3-PHENYL-2-PROPENYL)P
C18H15O4P	000115-86-6	TRIPHENYLPHOSPHATE
C18H15OP	000791-28-6	TRIPHENYLPHOSPHINE OXIDE
C18H15P	000603-35-0	TRIPHENYLPHOSPHINE
C18H16Cl2N2O6S2	059521-86-7	CEPHALOTHIN ANALOG(7-DICLPH-S-ACAM
C18H16Cl2N4O	130787-19-8	BENZAMIDE, N-(2-CYCLOPENTYL-2H-BENZOTRIAZOL-5-YL
C18H16ClFN2O3	052829-30-8	14-BENZDIAZPN-2-ON,1(23DIOHPR)5(2FPH)7CL
C18H16ClNO5	082110-72-3	FENOXAPROP-ETHYL
C18H16F2N2O	128104-27-8	1-PROPANOL, 1,1-BIS-(4-FLUOROPHENYL)-2-(1H-IMIDAZ
C18H16FNO3	123692-64-8	[1,3]DIOXEPINO[5,6-D]ISOXAZOLE, 6-(4-FLUOROPHENY
C18H16FNO3	123808-64-0	[1,3]DIOXEPINO[5,6-D]ISOXAZOLE, 6-(4-FLUOROPHENY
C18H16N2	000074-31-7	N,N'-DIPHENYL-P-BENZENEDIAMINE
C18H16N2O	003118-97-6	C.I. SOLVENT ORANGE 7
C18H16N2O6S	000134-31-6	OXYQUINOLINESULPHATE
C18H16O2	000122-69-0	3-PHENYLPROP-2-ENYL CINNAMATE
C18H16O2	013935-94-9	1H-INDENE-1,3(2H)-DIONE, 2-[3-(1-METHYLETHYL)PHE
C18H16O2	055994-21-3	1H-INDENE-1,3(2H)-DIONE, 2-[4-(1-METHYLETHYL)PHE
C18H16O3	000435-97-2	PHENPROCOUMON
C18H16O7	000125-46-2	1,3(2H,9BH)-DIBENZOFURANDIONE, 2,6-DIACETYL-7,9-
C18H16OSn	000076-87-9	TRIPHENYLTIN HYDROXIDE
C18H17Cl2NO3	022212-55-1	BENZOYLPROP ETHYL
C18H17ClN4O	130787-15-4	BENZAMIDE, 4-CHLORO-N-(2-CYCLOPENTYL-2H-BENZOTRI
C18H17N3O	131229-43-1	IMIDAZO[5,1-B]QUINAZOLIN-9(2H)-ONE, 1,3-DIHYDRO-
C18H17N3O	131229-46-4	IMIDAZO[5,1-B]QUINAZOLIN-9(2H)-ONE, 1,3-DIHYDRO-
C18H17N3O2	131229-45-3	IMIDAZO[5,1-B]QUINAZOLIN-9(2H)-ONE, 1,3-DIHYDRO-
C18H17N3O5	049648-48-8	CYANO-PHENYL-AMPHENICOL
C18H17N3O5	061554-12-9	14-BENZDIAZPN-2-ON,1(23DIOHPR)5PH7NO2
C18H17N5O3	130787-18-7	BENZAMIDE, N-(2-CYCLOPENTYL-2H-BENZOTRIAZOL-5-YL
C18H17NO2S	006577-34-0	1,4-OXATHIIN-3-CARBOXAMIDE, N-[1,1'-BIPHENYL]-2-
C18H17NO2S	132483-29-5	BENZENEACETIC ACID, _-METHYL-4-(4-PHENYL-2-THIAZ
C18H17NO3	092897-88-6	[1,3]DIOXEPINO[5,6-D]ISOXAZOLE, 3A,4,8,8A-TETRAH
C18H17NO3	092937-45-6	[1,3]DIOXEPINO[5,6-D]ISOXAZOLE, 3A,4,8,8A-TETRAH
C18H17NO3	119005-20-8	2-AZETIDINONE, 4-[4-(1-OXO-3-PHENYLPROPYL)PHENOX
C18H17NO4	096327-39-8	1H-PYRROLIZINE-1-CARBOXYLIC ACID, 2,3-DIHYDRO-5-
C18H17NO4S	074134-19-3	ACETAMIDE, N-(9-OXO-7-PROPYL-9H-THIOXANTHEN-3-YL
C18H18Cl2N2O2	112226-22-9	1-(3,4-DICHLORO),2-DIBENZOYL-1-TERT-BUTYLHYDRAZI
C18H18Cl2N2O2	112226-97-8	1-(2,4-DICHLORO),2-DIBENZOYL-1-TERT-BUTYLHYDRAZI
C18H18Cl2N2O2	112426-71-8	1-(2,6-DICHLORO),2-DIBENZOYL-1-TERT-BUTYLHYDRAZI
C18H18Cl2N2O2	112426-76-3	1-(2,3-DICHLORO),2-DIBENZOYL-1-TERT-BUTYLHYDRAZI
C18H18ClN3O	030144-75-3	1,4-DIAZEPIN-2-ONE,1,3-DIHYDRO-1-ME-7-DIMETHYLAM
C18H18ClNO3	060145-77-9	GLYCINE, N-(CHLOROACETYL)-N-[1,1'-BIPHENYL]-2-YL
C18H18ClNO4	030544-61-7	CLANOBUTIN
C18H18ClNO5	029104-30-1	BENZOXIMATE
C18H18ClNS	000113-59-7	CHLORPROTHIXENE
C18H18F2N2O2	112226-86-5	1-(2,6-DIFLUORO),2-DIBENZOYL-1-TERT-BUTYLHYDRAZI
C18H18N2O4	004471-41-4	C.I. DISPERSE BLUE 23
C18H18N2O6S	075994-55-7	CEPHALOTHIN ANALOG(7PHOACETYLAMINO
C18H18N2O7S	010390-44-0	PHENOXYMETHYLCEPHALOSPRIN
C18H18N2O7S	027910-26-5	CEPHALOSPORANIC ACID,7-MANDELAMIDO
C18H18N4O	130787-14-3	BENZAMIDE, N-(2-CYCLOPENTYL-2H-BENZOTRIAZOL-5-YL
C18H18O3	007320-97-0	PENTANOIC ACID, 5-[(1,1'-BIPHENYL)-4-YL]-3-METHY
C18H18O5	000120-55-8	DIETHYLENE GLYCOL DIBENZOATE
C18H19BrN2O2	112226-85-4	1-(4-BROMO),2-DIBENZOYL-1-TERT-BUTYLHYDRAZINE
C18H19BrN2O2	112249-32-8	1-(2-BROMO),2-DIBENZOYL-1-TERT-BUTYLHYDRAZINE
C18H19BrN2O2	112427-09-5	1-(3-BROMO),2-DIBENZOYL-1-TERT-BUTYLHYDRAZINE
C18H19Cl3O2	004329-03-7	ETHOXYCHLOR
C18H19ClN2O2	112225-83-9	1-(3-CHLORO),2-DIBENZOYL-1-TERT-BUTYLHYDRAZINE

I-197

C18H19ClN2O2	112226-54-7	1-(4-CHLORO),2-DIBENZOYL-1-TERT-BUTYLHYDRAZINE
C18H19ClN2O2	112226-58-1	1-(2-CHLORO),2-DIBENZOYL-1-TERT-BUTYLHYDRAZINE
C18H19ClN4	005786-21-0	CLOZAPINE
C18H19F3N2S	000146-54-3	TRIFLUPROMAZINE
C18H19FN2O2	112225-85-1	1-(4-FLUORO),2-DIBENZOYL-1-TERT-BUTYLHYDRAZINE
C18H19FN2O2	112226-76-3	1-(2-FLUORO),2-DIBENZOYL-1-TERT-BUTYLHYDRAZINE
C18H19FN2O2	112227-02-8	1-(3-FLUORO),2-DIBENZOYL-1-TERT-BUTYLHYDRAZINE
C18H19IN2O2	112226-93-4	1-(2-IODO),2-DIBENZOYL-1-TERT-BUTYLHYDRAZINE
C18H19IN2O2	112427-40-4	1-(3-IODO),2-DIBENZOYL-1-TERT-BUTYLHYDRAZINE
C18H19IN2O2	112427-73-3	1-(4-IODO),2-DIBENZOYL-1-TERT-BUTYLHYDRAZINE
C18H19N3O	002891-09-0	1,4-BENZDIAZ-2-ON,1ME5(PH)7DIMEAM
C18H19N3O4	067523-85-7	1-(4-NITRO),2-DIBENZOYL-1-TERT-BUTYLHYDRAZINE
C18H19N3O4	112225-86-2	1-(2-NITRO),2-DIBENZOYL-1-TERT-BUTYLHYDRAZINE
C18H19N3O4	112226-82-1	1-(3-NITRO),2-DIBENZOYL-1-TERT-BUTYLHYDRAZINE
C18H19N5O3	006054-58-6	C.I. DISPERSE DYE
C18H19N5O4	131402-47-6	ALLOPURINOL,N-(3-(MORPHOLINOMETHYL)BENZOYLOXY)ME
C18H19N5O4	131402-48-7	ALLOPURINOL,N-(4-(MORPHOLINOMETHYL)BENZOYLOXY)ME
C18H19N5O5S	078327-27-2	N7-(2-THIAZOLYL)-MITOMYCIN C
C18H19NO4	096327-38-7	1H-PYRROLIZINE-1-CARBOXYLIC ACID, 2,3-DIHYDRO-5-
C18H19NO5	136944-23-5	2-PROPENAMIDE, 3-(3,4-DIHYDROXYPHENYL)-N-[(3,4-D
C18H19NO7	091814-31-2	ACETAMIDE, N-[5,8-BIS(ACETYLOXY)-6,7-DIMETHOXY-1
C18H20Cl2	000072-56-0	1,1-DICHLORO-2,2-BIS(ETHYLPHENYL)ETHANE
C18H20Cl2O2	007388-32-1	1,1-DICHLORO-2,2-BIS(P-ETHOXYPHENYL)ETHANE
C18H20FN3O4	082419-36-1	OFLOXACIN
C18H20N2O2	112225-87-3	1,2-DIBENZOYL-1-TERT-BUTYLHYDRAZINE
C18H20N2S	001982-37-2	METHDILAZINE
C18H20N4O2S	130787-22-3	BENZENESULFONAMIDE, N-(2-CYCLOPENTYL-2H-BENZOTRI
C18H20N6O5	084397-45-5	N7-(3-PYRAZOLYL)-MITOMYCIN C
C18H20N6O7S4	069712-34-1	CEPHALOSPORIN DERIVATIVE
C18H20NO3PS	075889-62-2	PHOSPHONIC ACID, [[4-(2-BENZOTHIAZOLYL)PHENYL]ME
C18H20O2	000056-53-1	DES
C18H20O3	093371-38-1	3-OH-2-(4-BIPHENYLYL)HEXANOIC ACID
C18H20O3	093371-56-3	3-OH-4-ME-2(4-BIPHENYLYL)PENTANOIC ACID
C18H20O3S	125603-72-7	CYCLOPROPANECARBOXYLIC ACID, 3-DIHYDRO-2-OXO
C18H20O4	063837-33-2	C1CCCCC1OC2CCC(OCC3OC(CC)OC3)CC2
C18H20O4	080916-23-0	2-PROPENOIC ACID, 2-METHYL-, 2-[(1-NAPHTHALENYLO
C18H20O6	091814-56-1	1,4-NAPHTHALENEDIOL, 2,3-DIMETHOXY-, DIPROPANOAT
C18H20O6	102632-22-4	1,4-NAPHTHALENEDIOL, 2,3-DIMETHOXY-6,7-DIMETHYL-
C18H21ClN2S	000060-91-3	10H-PHENOTHIAZINE-10-ETHANAMINE, N,N-DIETHYL-
C18H21ClO	056265-27-1	1-(2-CHLORO-1-(4-ETHOXYPHENYL)PROPYL)-4-METHBE*
C18H21N3O2	015893-46-6	PHENYLALANYLPHENYLALANAMIDE
C18H21N3O5	038087-61-5	4-THIA-1-AZABICYCLO[3.2.0]HEPTANE DERIVATIVE
C18H21N5O	079508-78-4	46-NH2 22-DIME-1(3-PHOMEPH)S-TRIAZINE
C18H21NO2	065261-51-0	BENZAMIDE, N-(1-METHYLETHYL)-3-(2,5-DIMETHYLPHEN
C18H21NO2	082535-32-8	P-CYANOBENZYL-TR-CHRYSANTHEMATE
C18H21NO2	124497-76-3	2-OH-2,6-DIPH-4,5-DIME MORPHOLINE HCL
C18H21NO2	149589-41-3	BENZAMIDE, N-BUTYL-N-METHYL-3-PHENOXY-
C18H21NO3	000076-57-3	CODEINE
C18H21NO3	000125-29-1	DIHYDROCODEINONE
C18H21NO3	054010-81-0	1-ETHOXY-4-(1(4-METHYLPHENYL)-2-NITROPROPYL)-B*
C18H21NO4	085078-20-2	1,1-DI(P-MEO PHENYL)-2-NITROBUTANE
C18H21NO4	095261-65-7	CARBAMIC ACID, (3,4-DIETHOXYPHENYL)-, PHENYLMETH
C18H21NO5	099877-10-8	L-TYROSINE, 3-HYDROXY-, 2-(4-METHOXYPHENYL)ETHYL
C18H22ClNO3	001422-07-7	CODEINE HYDROCHLORIDE
C18H22I3N3O8	031112-62-6	METRIZAMIDE
C18H22N2	000050-47-5	DESIPRAMINE
C18H22N2	005692-66-0	BIS(2,4,6-TRIMETHYLPHENYL)DIAZENE
C18H22N2O2S	003689-50-7	OXOMEMAZINE

C18H22N2O3	068358-79-2	3(ME-MEO)1(2(4MEPH)ETO) PH UREA
C18H22N2O5S	000551-27-9	PROPICILLIN
C18H22N2O5S	004780-24-9	ISOPROPICILLIN
C18H22N2O6	120908-62-5	2-PROPANOL, 1-[[2-HYDROXY-3-(4-NITROPHENOXY)PROP
C18H22N2OS	000061-01-8	METHOXYPROMAZINE
C18H22N2S	000084-96-8	TRIMEPRAZINE
C18H22N4O5	041541-11-1	DIS. A. 9
C18H22N5O6PS	107538-73-8	8-METHYLTHIO CYCLIC AMP-O-BENZYL
C18H22O2	000053-16-7	ESTRONE
C18H22O2	000080-43-3	PEROXIDE, BIS(1-METHYL-1-PHENYLETHYL)
C18H23ClN2O3	092598-89-5	1H-IMIDAZOLE-5-BUTANOIC ACID, _-ETHYL-beta-(HYDR
C18H23NaO3S	099287-48-6	1-SO3NA-3-PENTYL-7-ISOPROPYL AZULENE
C18H23NO	000083-98-7	ORPHENADRINE
C18H23NO2	120282-84-0	BENZENEACETAMIDE,alpha-HYDROXY-N-TRICYCLO[3.3.1.
C18H23NO3	004290-62-4	2-PROPANOL, 1-[(1-METHYLETHYL)AMINO]-3-(1-NAPHTH
C18H23NO3	144458-99-1	BENZENEMETHANAMINE, N,N-BIS(2-HYDROXYETHYL)-3-(P
C18H23NO4	024316-19-6	CEPHALOTAXINE
C18H24I2O12	060331-52-4	4,6-DIIDORESORCYL-1,3-DIGLUCOSIDE
C18H24I2O12	060406-79-3	2,4-DIIDORESORCYL-1,3-DIGLUCOSIDE
C18H24I3N3O7	031122-84-6	1,3-BENZENEDICARBOXAMIDE, 5-(ACETYLAMINO)-N,N'-B
C18H24I3N3O8	066108-93-8	M-BENZAMIDE,NNBIS(23DIOHPR),5N....
C18H24I3N3O8	073334-07-3	IOPROMIDE
C18H24I3N3O9	087771-40-2	IOVERSOL
C18H24N2O2S	111226-84-7	1H-IMIDAZOLE-1-CARBOTHIOIC ACID, O-[1-[(4-ETHYLP
C18H24N2O3	092598-82-8	1H-IMIDAZOLE-5-BUTANOIC ACID, alpha-ETHYL-beta-(
C18H24N2O3	111227-13-5	1H-IMIDAZOLE-1-CARBOXYLIC ACID, 1-[(4-ETHYLPHENO
C18H24N2O4	111227-21-5	1H-IMIDAZOLE-1-CARBOXYLIC ACID, 1-[(4-ETHOXYPHEN
C18H24N2O5	076420-72-9	L-PROLINE, 1-[N-(1-CARBOXY-3-PHENYLPROPYL)-L-ALA
C18H24N2O6	000131-72-6	KARATHANE
C18H24N2O6	039300-45-3	DINOCAP
C18H24N4O3	132765-86-7	L-VALINAMIDE, N-ACETYL-L-TRYPTOPHYL-
C18H24N4O4	055687-53-1	24NH2PYRIMIDINE,5(35-ETO-4CO2ET)BENZYL
C18H24N4O5S	055687-72-4	24NH2PYRIMD,5(35ETO4COSO2ME)BENZYL
C18H24NO7P	000052-28-8	CODEINE PHOSPHATE
C18H24O2	000050-28-2	ESTRADIOL
C18H24O3	000050-27-1	ESTRIOL
C18H24O4	000084-64-0	BUTYLCYCLOHEXYL PHTHALATE
C18H24O5	015725-39-0	DIETHYLMALONATE,3-BUTOXYBENZAL
C18H24O6	000085-70-1	BUTYLGLYCOLYL BUTYL PHTHALATE
C18H25BrN2O5	132169-99-4	BR-N(2,6-IPRPH)CARBAMOYLME IMINODIACETIC ACID
C18H25N3O5	066203-00-7	UREA, N-[4,7-DIMETHOXY-6-[2-(1-PYRROLIDINYL)ETHO
C18H25N5O4	095549-92-1	PIPERAZINE, 1-(4-AMINO-6,7-DIMETHOXY-2-QUINAZOLI
C18H25N5O5S	023404-73-1	L-ASPARTIC ACID, N-[1-OXO-5-(7H-PURIN-6-YLTHIO)P
C18H25N7O5S	023374-45-0	GLYCINE, N-[N-[N-[1-OXO-5-(7H-PURIN-6-YLTHIO)PEN
C18H25NO13	056846-39-0	ß-D-GLUCOPYRANOSIDE, 4-NITROPHENYL 4-O-_-D-GLUCO
C18H25NO3	023795-32-6	N,N-DI-I-BU-3,4-(-OCH2O-)CINNAMAMIDE
C18H26BrNO3	002870-71-5	ATROPINE METHYL BROMIDE
C18H26ClN3	000054-05-7	CHLOROQUINE
C18H26ClN3O	125225-28-7	IPCONAZOLE
C18H26N2O5	102273-21-2	PROPANIOIC ACID, 2,2-DIMETHYL-, [2-[[(4-MORPHOLI
C18H26N2O5S	065907-30-4	N-THIO-(N'-ME-O-BU-CARBAMYL)CARBOFURAN
C18H26N2O5S	068672-91-3	N-THIO-(N'-BU-O-ME-CARBAMYL)CARBOFURAN
C18H26N2O5S	083871-03-8	4ME-BENZENESULFONYL-L-(N'-CYHX)GLUTAMINE
C18H26N2O5S	086569-98-4	N-THIO-(N'ME-O-TBU-CARBAMYL)CARBOFURAN
C18H26N4O3	055687-56-4	24NH2PYRIMD,5(35ETO4COSO2ME)BENZYL
C18H26O2	000434-22-0	NANDROLONE
C18H26O2	087818-31-3	CINMETHYLIN
C18H26O4	000131-18-0	DI-N-AMYL PHTHALATE

C18H26O5Si4	040169-27-5	4,4,6,6,8,8-HEXAMETHYLSPIRO(CYCLOTETRASILOXANE*)
C18H26O6	003290-92-4	TRIMETHYLOLPROPANE TRIMETHACRYLATE
C18H27ClN2O2	105801-51-2	ACETAMIDE, 2-(DIBUTYLAMINO)-N-(5-CHLORO-2,3-DIHY
C18H27N	055040-03-4	PIPERIDINE, 1-(4-METHYL-1-PHENYLCYCLOHEXYL)-, TR
C18H27N	058283-42-4	PIPERIDINE, 1-(3-METHYL-1-PHENYLCYCLOHEXYL)-, CI
C18H27NO5	072836-47-6	3,4,5-TRIMEO-CINNAMIC ACID,DIETAMET ESTER
C18H28N2O	002180-92-9	BUPIVACAINE
C18H28N2O2	105801-66-9	ACETAMIDE, 2-(DIBUTYLAMINO)-N-(2,3-DIHYDRO-3-BEN
C18H28N2O3	121508-98-3	PROPANAMIDE, 2-METHYL-N-[2-[[3-(3-PHENYL-2-PROPE
C18H28N2O4	037517-30-9	ACEBUTOLOL
C18H28N2O5S	083871-02-7	4-ME-BENZENESULFONYL-L-(N'-HEX)GLUTAMINE
C18H28N6O2	117121-34-3	UREA, N,N-DIMETHYL-N'-[4-(1-OCTYL-1H-TETRAZOL-5-
C18H28N6O2	117121-46-7	UREA, N,N-DIMETHYL-N'-[4-[(2-OCTYL-2H-TETRAZOL-5
C18H28O2	132296-18-5	PHENOL, 5-(1,1-DIMETHYLBUTYL)-2-(3-HYDROXYCYCLOH
C18H28O3S	000120-62-7	SULFOXIDE
C18H28O4Si4	004657-20-9	2,2,4,6,6,8-HEXAMETHYL-4,8-DIPHENYLCYCLOTETRAS*
C18H28O4Si4	018604-02-9	2,2,4,4,6,8-HEXAMETHYL-6,8-DIPHENYLCYCLOTETRAS*
C18H28O7	067950-78-1	B21C7-BENZO CROWN ETHER
C18H29NaO3S	002211-99-6	4-(1-METHYLUNDECYL)BENZENESULFONIC ACID, SODIUM*
C18H29NaO3S	002212-52-4	4-(1-PENTYLHEPTYL)BENZENESULFONIC ACID, SODIUM *
C18H29NO2	038363-40-5	PENBUTOLOL
C18H29NO3	063659-18-7	2-PROPANOL, 1-[4-[2-(CYCLOPROPYLMETHOXY)ETHYL]PH
C18H30	000123-01-3	DODECYLBENZENE
C18H30	002400-00-2	BENZENE, (1-ETHYLDECYL)-
C18H30	002719-61-1	BENZENE, (1-METHYLUNDECYL)-
C18H30	002719-62-2	BENZENE, (1-PENTYLHEPTYL)-
C18H30	002719-63-3	BENZENE, (1-BUTYLOCTYL)-
C18H30	002719-64-4	(1-PROPYLNONYL)BENZENE
C18H30	068648-87-3	ALKYLBENZENE (C10-C15)
C18H30N2O	077470-99-6	HEXANAMIDE, 5-(DIETHYLAMINO)-N-(2,6-DIMETHYLPHEN
C18H30N2O	102089-69-0	HEXANAMIDE, 6-(DIETHYLAMINO)-N-(2,6-DIMETHYLPHEN
C18H30N2O3	121508-93-8	PROPANAMIDE, N-[2-[[2-HYDROXY-3-(3-PHENYLPROPOXY
C18H30O	000104-43-8	PHENOL, 4-DODECYL-
C18H30O	000732-26-3	2,4,6-TRI(TERT-BUTYL)PHENOL
C18H30O2	000463-40-1	9,12,15-OCTADECATRIENOIC ACID, (Z,Z,Z)-
C18H30O3S	001886-81-3	DODECYL BENZENESULFONATE
C18H31IN2O	035096-55-0	3-CARBAMOYLPYRIDINIUM IODIDE,DODECYL
C18H31N	000104-42-7	P-DODECYLANILINE
C18H31NO4	066722-44-9	2-PROPANOL, 1-[(1-METHYLETHYL)AMINO]-3-[4-[[2-(1
C18H32IN	014402-22-3	2-METHYLPYRIDINIUM IODIDE,N-DODECYL
C18H32O16	000512-69-6	RAFFINOSE
C18H32O2	000060-33-3	LINOLEIC ACID
C18H33ClN2O5S	018323-44-9	CLINDAMYCIN
C18H33NO3	101881-19-0	1-LAURYL-4-METHOXYCARBONYL-2-PYRROLIDONE
C18H34N2O6S	000154-21-2	LINCOMYCIN
C18H34N4O4	132766-12-2	L-ISOLEUCINAMIDE, N-ACETYL-L-VALYL-L-VALYL-
C18H34N4O5	132766-19-9	L-LEUCINAMIDE, N-ACETYL-L-LEUCYL-L-THREONYL-
C18H34NaO2	000143-19-1	OLEIC ACID, SODIUM SALT
C18H34O2	000112-80-1	OLEIC ACID
C18H34O2	002027-47-6	9-OCTADECENOIC ACID
C18H34O3	000141-22-0	12-HYDROXY-9-OCTADECENOIC ACID (CIS)
C18H34O4	000109-43-3	DIBUTYL SEBACATE
C18H34O4	000110-33-8	DIHEXYL ADIPATE
C18H34O6	000095-08-9	TRIETHYLENE GLYCOL BIS(2-ETHYLBUTYRATE)
C18H36	000112-88-9	1-OCTADECENE
C18H36	000143-28-2	OLEYL ALCOHOL (CIS)
C18H36	001795-17-1	DODECYLCYCLOHEXANE
C18H36N5O2P	103981-97-1	TEPA 4-(PIPERIDINYL)-UREA DERIVATIVE

C18H36N5O3P	103981-92-6	1-PIPERIDINYLOXY, 4-[[[[BIS(2,2-DIMETHYL-1-AZIRI
C18H36N5O3P	103981-94-8	TEPA 4-(N-OH-PIPERIDINYL)-UREA DERIVATIV
C18H36O2	000057-11-4	STEARIC ACID
C18H36O2	000628-97-7	ETHYL PALMITATE
C18H37Br	000112-89-0	1-BROMOOCTADECANE
C18H37Cl	003386-33-2	1-CHLOROOCTADECANE
C18H37Cl3Si	000112-04-9	OCTADECYLTRICHLOROSILANE
C18H37I	000629-93-6	1-IODOOCTADECANE
C18H38	000593-45-3	OCTADECANE
C18H38O	000112-92-5	1-OCTADECANOL
C18H38O	002456-27-1	DINONYL ETHER
C18H38O10	003386-18-3	NONAETHYLENE GLYCOL
C18H38S	002885-00-9	1-OCTADECANETHIOL
C18H39O7P	000078-51-3	TRI-2-BUTOXYETHYL PHOSPHATE
C18H50O3Si6	091455-17-3	1,1,1,3,5,7,7,7-OCTAMETHYL-3,5-BIS[2-(TRIMETHYL*
C18H54O7Si8	000556-69-4	OCTADECAMETHYLOCTASILOXANE
C18H54O9Si9	000556-71-8	OCTADECAMETHYLCYCLONONASILOXANE
C19H11F5N2O2	083164-33-4	3-PYRIDINECARBOXAMIDE, N-(2,4-DIFLUOROPHENYL)-2-
C19H12	042126-84-1	1H-BENZO(CD)FLUOANTHENE
C19H12O6	000066-76-2	DICOUMAROL
C19H13Cl	021846-07-1	9-(4-CHLOROPHENYL)-9H-FLUORENE
C19H13Cl	032377-11-0	9-(3-CHLOROPHENYL)-9H-FLUORENE
C19H14	000789-24-2	9-PHENYL-9H-FLUORENE
C19H14	001705-85-7	6-METHYLCHRYSENE
C19H14	002381-15-9	10-METHYL-BENZ(A)ANTHRACENE
C19H14	002422-79-9	9-METHYLBENZ(A)ANTHRACENE
C19H14	002498-77-3	1-METHYLBENZ(A)ANTHRACENE
C19H14	002541-69-7	BENZ(A)ANTHRACENE-7-METHYL
C19H14	003697-24-3	5-METHYLCHRYSENE
C19H14F3NO	059756-60-4	FLURIDONE
C19H14O3	000603-45-2	ROSOLIC ACID
C19H14O5S	000143-74-8	PHENOLSULPHONEPHTHALEIN
C19H15Cl	000076-83-5	BENZENE, 1,1',1"-(CHLOROMETHYLIDYNE)TRIS-
C19H15ClF3NO7	077501-63-4	LACTOFEN
C19H15ClO4	000081-82-3	COUMACHLOR
C19H15N	000963-89-3	7,9-DIMETHYLBENZ(C)ACRIDINE
C19H15NO6	000152-72-7	ACENOCOUMARIN
C19H16	000519-73-3	TRIPHENYLMETHANE
C19H16ClNO4	000053-86-1	INDOMETHACIN
C19H16N2O	005663-04-7	TRIPHENYLUREA
C19H16N4O	006300-37-4	C.I. DISPERSE YELLOW 7
C19H16O	000076-84-6	BENZENEMETHANOL, .ALPHA.,.ALPHA.-DIPHENYL-
C19H16O2	004081-02-1	4,4'-DIHYDROXYTRIPHENYLMETHANE
C19H16O2	004412-35-5	4H-1-BENZOPYRAN-4-ONE, 3-METHYL-2-PHENYL-8-(2-PR
C19H16O4	000081-81-2	WARFARIN
C19H17Cl2N3O5S	003116-76-5	DICLOXACILLIN
C19H17Cl2N3O6S	052248-39-2	4-THIA-1-AZABICYCLO[3.2.0]HEPTANE-2-COOH DERIVAT
C19H17ClN2O	002955-38-6	PRAZEPAM
C19H17ClN2O4	076578-14-8	QUIZALOFOP-ETHYL
C19H17ClN2O4	100646-51-3	QUIZALOFOP-P-ETHYL
C19H17F3N4O	130787-17-6	BENZAMIDE, N-(2-CYCLOPENTYL-2H-BENZOTRIAZOL-5-YL
C19H17N3	000569-61-9	PARAROSANILINE
C19H17NO7	069049-73-6	NEDOCROMIL
C19H17O4P	026444-49-5	CRESYL DIPHENYL PHOSPHATE
C19H18Cl2N2O4S	058011-68-0	PYRAZOLATE
C19H18ClFN3O5S	005250-39-5	FLOXACILLIN
C19H18ClN3O6S	055390-39-1	4-THIA-1-AZABICYCLO[3.2.0]HEPTANE-2-CARBOXYLIC A
C19H18O2	013935-95-0	1H-INDENE-1,3(2H)-DIONE, 2-[3-(1,1-DIMETHYLETHYL

C19H18O2	013935-96-1	1H-INDENE-1,3(2H)-DIONE, 2-[4-(1,1-DIMETHYLETHYL
C19H18O2	055994-23-5	1H-INDENE-1,3(2H)-DIONE, 2-(3,5-DIETHYLPHENYL)-
C19H19ClF3NO5	087237-48-7	HALOXYFOP-ETOTYL
C19H19ClFNO3	052756-22-6	FLAMPROP-ISOPROPYL
C19H19ClN3O5S	000061-72-3	CLOXACILLIN
C19H19F3N2O2	112426-65-0	1-(4-TRIFLUOROMETHYL),2-DIBENZOYL-1-TERT-BUTYLHY
C19H19F3N2O2	112426-66-1	1-(3-TRIFLUOROMETHYL),2-DIBENZOYL-1-TERT-BUTYLHY
C19H19F3N2O2	112426-67-2	1-(2-TRIFLUOROMETHYL),2-DIBENZOYL-1-TERT-BUTYLHY
C19H19N3O	131229-42-0	IMIDAZO[5,1-B]QUINAZOLIN-9(2H)-ONE, 1,3-DIHYDRO-
C19H19N3O2	112426-62-7	1-(4-CYANO),2-DIBENZOYL-1-TERT-BUTYLHYDRAZINE
C19H19N3O2	112426-69-4	1-(3-CYANO),2-DIBENZOYL-1-TERT-BUTYLHYDRAZINE
C19H19N7O6	000059-30-3	FOLIC ACID
C19H19NO3	123750-56-1	[1,3]DIOXEPINO[5,6-D]ISOXAZOLE, 3A,4,8,8A-TETRAH
C19H20Br4O4	004162-45-2	ETHOXYLATED TETRABROMOBISPHENOL A
C19H20ClNO	097871-66-4	P-CL CINNAMAMIDE,B-PHENYL,N-SEC-BU
C19H20F3N3O3	147267-64-9	1(2H)-PYRIDINEACETAMIDE, 3-AMINO-2-OXO-6-PHENYL-
C19H20N2O	073931-96-1	DENZIMOL
C19H20N2O2	000050-33-9	PHENYLBUTAZONE
C19H20N2O3	000129-20-4	OXYPHENBUTAZONE
C19H20N3O5S	000066-79-5	OXACILLIN
C19H20N3O6S	004914-62-9	5-HYDROXY-OXACILLIN
C19H20N4O2	078233-99-5	24-NH2PYRIMIDINE,5(4MEO-3PHMEO)BENZYL
C19H20N4O2	083158-06-9	24-NH2PYRIMIDINE,5(3MEO-4PHMEO)BENZYL
C19H20N4O2	130787-16-5	BENZAMIDE, N-(2-CYCLOPENTYL-2H-BENZOTRIAZOL-5-YL
C19H20O4	000085-68-7	BUTYL BENZYL PHTHALATE
C19H20O6	035413-85-5	2-PROPENOIC ACID, 2-METHYL-, 5A-ETHENYLDECAHYDRO
C19H20O7	013017-11-3	PROPENOIC ACID, ESTER DERIVATIVE
C19H21ClN2O	066063-05-6	PENCYCURON
C19H21ClN2O2	112426-81-0	1-(3-CHLORO-2-METHYL),2-DIBENZOYL-1-TERT-BUTYLHY
C19H21N5O2	069472-19-1	DIS. A. 6
C19H21NO3	000062-67-9	NALORPHINE
C19H21NO3	000115-37-7	PARAMORPHINE
C19H21NO4	000465-65-6	NALOXONE
C19H21NO4	119005-27-5	1-AZETIDINEACETIC ACID, 2-(2-NAPHTHALENYLOXY)-4-
C19H21NO5	136944-24-6	2-PROPENAMIDE, 3-(3,4-DIHYDROXYPHENYL)-N-[2-(3,4
C19H21NOS	097871-59-5	2-PROPENAMIDE, N-(1-METHYLPROPYL)-3-PHENYL-2-(PH
C19H21NS	000113-53-1	DOLSULEPINE
C19H22Cl2O2	056265-23-7	1,1-BIS(P-ETHOXYPHENYL)-2,2-DICHLOROPROPANE
C19H22FN3O	001649-18-9	AZAPERONE
C19H22N2	000486-12-4	TRIPOLIDINE
C19H22N2O	000118-10-5	CINCHONINE
C19H22N2O	000485-71-2	CINCHONIDINE
C19H22N2O	072336-12-0	9((DIETAMINO)ACETYLAMINO) FLUORENE
C19H22N2O2	112226-56-9	1-(4-METHYL),2-DIBENZOYL-1-TERT-BUTYLHYDRAZINE
C19H22N2O2	112226-78-5	1-(2-METHYL),2-DIBENZOYL-1-TERT-BUTYLHYDRAZINE
C19H22N2O2	112411-46-8	1-(3-METHYL),2-DIBENZOYL-1-TERT-BUTYLHYDRAZINE
C19H22N2O2S	112226-95-6	1-(2-METHYLTHIO),2-DIBENZOYL-1-TERT-BUTYLHYDRAZI
C19H22N2O3	112226-77-4	1-(2-METHOXY),2-DIBENZOYL-1-TERT-BUTYLHYDRAZINE
C19H22N2O3	112226-92-3	1-(4-METHOXY),2-DIBENZOYL-1-TERT-BUTYLHYDRAZINE
C19H22N2O3	112426-60-5	1-(3-METHOXY),2-DIBENZOYL-1-TERT-BUTYLHYDRAZINE
C19H22N4O2S	103499-11-2	1,2,4-OXADIAZOLE-5-(DIETHYLAMINOETHYL)THIOHYDROX
C19H22O6	000077-06-5	GIBBERELLIC ACID
C19H23ClN2	000303-49-1	CHLORIMIPRAMINE
C19H23ClN2S	000084-01-5	CHLORPROETHAZINE
C19H23ClO2	056265-22-6	1,1-BIS(P-ETHOXYPHENYL)-2-CHLOROPROPANE
C19H23N3	033089-61-1	AMITRAZ
C19H23N3O2S	102151-07-5	2-PROPANOL, 1-(PROPYLAMINO)-3-[4-[5-(2-THIENYL)-
C19H23N5O3	052128-35-5	24NH2-5ME-6(345MEOPHNME)QUINAZOLINE

C19H23N5O5S	084397-36-4	N7-(5-ME THIAZOL-2-YL)-MITOMYCIN C
C19H23N5O5S	088854-64-2	N7-(5-ME ISOTHIAZOL-3-YL)-MITIMYCIN C
C19H23N7O2	070579-43-0	46NH2-22ME-1(3URIDOPHO)MEPH S-TRIAZINE
C19H23NO2	065261-99-6	BENZAMIDE, N-(1,1-DIMETHYLETHYL)-3-(2,5-DIMETHYL
C19H23NO2	144458-95-7	4-PIPERIDINOL, 1-[[3-(PHENOXYMETHYL)PHENYL]METHY
C19H23NO2	149589-42-4	BENZAMIDE, N-METHYL-N-PENTYL-3-PHENOXY-
C19H23NO3	000076-58-4	DIONINE
C19H23NO4	026258-70-8	1,1-BIS(4-ETHOXYPHENYL)-2-NITROPROPANE
C19H23NO5	000596-15-6	MORPHINE ACETATE
C19H24N2	000050-49-7	IMIPRAMINE
C19H24N2O3	036894-69-6	LABETALOL
C19H24N2O3	053670-99-8	BENZENEACETAMIDE, N-[2-[(3-PHENOXY-2-HYDROXYPROP
C19H24N2O3	121508-87-0	BENZENEACETAMIDE, N-[2-[[3-(PHENYLMETHOXY)-2-HYD
C19H24N2O4	038103-61-6	BENZAMIDE, 4-[2-[[2-HYDROXY-3-(2-METHYLPHENOXY)P
C19H24N2O4	088132-31-4	N1-ME-MEO N3-(4PHOBUO)PHENYLUREA
C19H24N2O5	111227-41-9	1H-IMIDAZOLE-1-CARBOXYLIC ACID, 1-[[4-(ETHOXYCAR
C19H24N2O5	139733-55-4	ACETAMIDE, 2-[4-[[2-(2-HYDROXY-3-PHENOXYPROPYL)A
C19H24N2OS	000060-99-1	METHOTRIMEPRAZINE
C19H24N2S	000522-00-9	ETHOPROPAZINE
C19H24N6O3	131402-50-1	ALLOPURINOL,N-(3-(4-METHYLYPIPERIZINYLMETHYL)BEN
C19H24O	056265-26-0	1-ETHOXY-4-(2-METHYL-1-(4-METHYLPHENYL)PROPYL)*
C19H24O2	004741-74-6	1,1-BIS(P-METHOXYPHENYL)-2,2-DIMETHYLPROPANE
C19H24O3	000119-89-1	PIPERONYL CYCLONENE
C19H24O3	023031-36-9	PRALLETHRIN
C19H25ClN2OS	096489-71-3	PYRIDABEN
C19H25N3S	000058-37-7	AMINOPROMAZINE
C19H25N5O2	073090-70-7	24NH2PYRIMIDINE,5(35ETO4(1PYRRYL))BENZYL
C19H25NaO4S	099287-70-4	1-SO3NA-3-PE-4-METHOXY-7-IPR AZULENE
C19H25NO	000126-04-5	PROPOXYPHENE CARBINOL
C19H25NO	000152-02-3	LEVALLORPHAN
C19H25NO4	007696-12-0	PHTHALTHRIN
C19H25NO5S	103595-57-9	(N-ME-BENZYL)AMET-(345-TRIMEO)PH SULFONE
C19H26ClNO3	000125-30-4	CODETHYLINE HYDROCHLORIDE
C19H26I3N3O9	066108-95-0	IOHEXOL
C19H26I3N3O9	088116-59-0	IOHEXOL DERIVATIVE
C19H26N2O2S	111226-67-6	1H-IMIDAZOLE-1-CARBOTHIOIC ACID, O-[1-[(4-PROPYL
C19H26N2O2S	111226-68-7	1H-IMIDAZOLE-1-CARBOTHIOIC ACID, O-[1-[[4-(1-MET
C19H26N2O3	092598-81-7	1H-IMIDAZOLE-5-BUTANOIC ACID, _-ETHYL-beta-(HYDR
C19H26N2O3	092598-86-2	1H-IMIDAZOLE-5-BUTANOIC ACID, _-ETHYL-beta-(HYDR
C19H26N2O3	092598-92-0	1H-IMIDAZOLE-5-BUTANOIC ACID, _-ETHYL-beta-(HYDR
C19H26N2O3	092598-93-1	1H-IMIDAZOLE-5-BUTANOIC ACID, _-ETHYL-beta-(HYDR
C19H26N2O3	111227-07-7	1H-IMIDAZOLE-1-CARBOXYLIC ACID, 1-[[4-(1-METHYLE
C19H26N2O3	111227-14-6	1H-IMIDAZOLE-1-CARBOXYLIC ACID, 1-[(4-PROPYLPHEN
C19H26N2O4	111227-26-0	1H-IMIDAZOLE-1-CARBOXYLIC ACID, 1-[(4-PROPOXYPHE
C19H26N6O4S	089073-49-4	BENZENESULFONAMIDE, 4-(1,3-DIETHYL-2,3,6,7-TETRA
C19H26N6O6	130817-92-4	L-PROLINAMIDE, 5-OXO-L-PROLYL-1-(ETHOXYCARBONYL)
C19H26O12	000490-67-5	MONOTROPITOSIDE
C19H26O2	000063-05-8	ANDROSTENEDIONE
C19H26O2	000070-38-2	DIMETHRIN
C19H26O3	000584-79-2	ALLETHRIN
C19H26O3	021507-14-2	ESTRA-1,3,5(10)-TRIENE-3,17-DIOL, 11-METHOXY-, (
C19H26O4S	002312-35-8	PROPARGITE
C19H27FN2O7	007207-62-7	URIDINE, 2'-DEOXY-5-FLUORO-, 3',5'-DIPENTANOATE
C19H27NO4	144106-15-0	CYCLOPROPANECARBOXYLIC ACID, 2-[2-(2-PROPENYLOXY
C19H28N4O4	065118-51-6	L-LEUCINAMIDE, N-ACETYL-L-PHENYLALANYLGLYCYL-
C19H28N4O4	132766-05-3	L-ALANINAMIDE, N-ACETYL-L-VALYL-L-PHENYLALANYL-
C19H28N4O4	132766-11-1	GLYCINAMIDE, N-ACETYL-L-PHENYLALANYL-L-ISOLEUCYL
C19H28O2	000053-43-0	PRASTERONE

C19H28O2	000058-22-0	TESTOSTERONE
C19H29ClN2O3	088132-43-8	BENZOIC ACID, 2-CHLORO-4-[[(DIMETHYLAMINO)CARBON
C19H29N	072165-35-6	PIPERIDINE, 1-(4,4-DIMETHYL-1-PHENYLCYCLOHEXYL)-
C19H29NO2	066451-06-7	2-PROPANOL, 1-(2-BICYCLO[2.2.1]HEPT-2-YLPHENOXY)
C19H29NO4	144106-11-6	PROPANAMINE, N-(1-METHYLETHYL)-2-(1-OXOBUTOXY)-3
C19H29NO4	144106-14-9	PROPANAMINE, N-(1-METHYLETHYL)-2-(2-METHYL-1-OXO
C19H30N2O2	105801-56-7	ACETAMIDE, 2-(DIBUTYLAMINO)-N-(2,3-DIHYDRO-5-MET
C19H30N2O3	093450-23-8	CARBAMIC ACID, [3-(BUTOXYMETHYL)PHENYL]-, 2-(1-P
C19H30N2O3	093450-34-1	CARBAMIC ACID, [2-(BUTOXYMETHYL)PHENYL]-, 2-(1-P
C19H30N2O3	105801-60-3	ACETAMIDE, 2-(DIBUTYLAMINO)-N-(2,3-DIHYDRO-6-MET
C19H30N2O3	121509-03-3	PROPANAMIDE, 2-METHYL-N-[2-[[3-[(1,2,3,4-TETRAHY
C19H30O2	000053-41-8	ANDROSTERONE
C19H30O2	000521-18-6	ANDROSTAN-3-ONE, 17-HYDROXY-, (5ALPHA,17BETA)-
C19H30O2	132296-19-6	PHENOL, 5-(1,1-DIMETHYLPENTYL)-2-(3-HYDROXYCYCLO
C19H30O5	000051-03-6	PIPERONYL BUTOXIDE
C19H32	000123-02-4	1-PHENYLTRIDECANE
C19H32	004534-49-0	BENZENE, (1-PENTYLOCTYL)-
C19H32	004534-50-3	BENZENE, (1-BUTYLNONYL)-
C19H32	004534-51-4	BENZENE, (1-PROPYLDECYL)-
C19H32	004534-52-5	BENZENE, (1-ETHYLUNDECYL)-
C19H32	004534-53-6	BENZENE, (1-METHYLDODECYL)-
C19H32N2O	077470-84-9	HEPTANAMIDE, 7-(DIETHYLAMINO)-N-(2,6-DIMETHYLPHE
C19H32O	052780-43-5	PHENOL, 4-TRIDECYL-
C19H32O	072424-02-3	PHENOL, 3-TRIDECYL-
C19H32O	096850-25-8	PHENOL, 2-TRIDECYL-
C19H32O2	028165-91-5	1,2-BENZENEDIOL, 3-TRIDECYL-
C19H32O2	123063-34-3	1,3-BENZENEDIOL, 4-TRIDECYL-
C19H33N4O2P	103981-98-2	TEPA ADAMANTYL-UREA DERIVATIVE
C19H33NO5S	071203-66-2	N-DIBUAMET 3,4,5-TRIMEO PH SULFONE
C19H34BrN	001155-74-4	TETRADECYLPYRIDINIUM BROMIDE
C19H34IN	088477-45-6	3,4-DIME PYRIDINIUM IODIDE,N-DODECYL
C19H34IN	088477-51-4	2,4-DIME PYRIDINIUM IODIDE,N-DODECYL
C19H34IN	088477-57-0	3,5-DIME PYRIDINIUM IODIDE,N-DODECYL
C19H34O2	000112-63-0	METHYL LINOLEATE
C19H34O3	040596-69-8	ALTOSID
C19H35ClN2O5S	013441-66-2	L-THREO-ALPHA-D-GALACTO-OCTOPYRANOSIDE, DERIVATI
C19H35ClN2O5S	019096-41-4	L-THREO-ALPHA-D-GALACTO-OCTOPYRANOSIDE
C19H36N2O6S	002256-17-9	D-ERYTHRO-ALPHA-D-GALACTO-OCTOPYRANOSIDE DERIVAT
C19H36O2	000112-62-9	METHYL OLEATE
C19H38	006006-33-3	N-TRIDECYLCYCLOHEXANE
C19H38	018435-45-5	1-NONADECENE
C19H38O2	000112-61-8	METHYL STEARATE
C19H38O2	000142-91-6	ISOPROPYL PALMITATE
C19H39Br	004434-66-6	1-BROMONONADECANE
C19H40	000629-92-5	NONADECANE
C19H40O	052783-43-4	NONADECANOL
C19H42ClN	000112-02-7	CETRIMONIUM CHLORIDE
C20H9Cl3F5N3O3	071422-67-8	CHLORFLUAZURON
C20H10Cl2F5N3O3	086811-58-7	FLUAZURON
C20H11N2Na3O10S3	000915-67-3	AMARANTH
C20H11N2Na3O10S3	002611-82-7	ACID RED 18
C20H12	000050-32-8	BENZO(A)PYRENE
C20H12	000192-97-2	BENZO(E)PYRENE
C20H12	000198-55-0	PERYLENE
C20H12	000205-82-3	BENZO(J)FLUORANTHENE
C20H12	000205-99-2	BENZO(B)FLUORANTHENE
C20H12	000207-08-9	BENZO(K)FLUORANTHENE
C20H12O5	000518-45-6	FLUORESCEIN

C20H13F3	032377-12-1	9-(3-(TRIFLUOROMETHYL)PHENYL)-9H-FLUORENE
C20H13N	000194-59-2	7H-DIBENZO(C,G)CARBAZOLE
C20H13N	032377-09-6	9-PHENYL-9H-FLUORENE-4-CARBONITRILE
C20H13N2NaO5S	002538-85-4	CALCON
C20H13NO4	017418-58-5	C.I. DISPERSE RED 60
C20H14	000479-23-2	CHOLANTHRENE
C20H14	000602-55-1	9-PHENYLANTHRACENE
C20H14N2O2	004395-65-7	C.I. DISPERSE BLUE 19
C20H14O4	000077-09-8	PHENOLPHTHALEIN
C20H14O4	000084-62-8	DIPHENYL PHTHALATE
C20H15N5O7S2	084104-57-4	5-THIA-1-AZABICYCLO[4.2.0]OCT-2-ENE-2-CARBOXYLIC
C20H15NS	001726-94-9	TRIPHENYLMETHYLISOTHIOCYANATE
C20H16	000057-97-6	7,12-DIME-BENZ(A)ANTHRACENE
C20H16	001572-46-9	9-(PHENYLMETHYL)-9H-FLUORENE
C20H16	003697-27-6	5,6-DIMETHYLCHRYSENE
C20H16	003697-30-1	7-ETHYLBENZ(A)ANTHRACENE
C20H16	018153-42-9	9-(3-METHYLPHENYL)-9H-FLUORENE
C20H16	018153-43-0	9-(4-METHYLPHENYL)-9H-FLUORENE
C20H16	018429-70-4	BENZ(A)ANTHRACENE-4,5-DIMETHYLENE
C20H16	035187-28-1	7,11-DIMETHYLBENZ(A)ANTHRACENE
C20H16N2O2	007154-31-6	1,4-BENZENEDICARBOXAMIDE, N,N'-DIPHENYL-
C20H16O	021846-08-2	9-(4-METHOXYPHENYL)-9H-FLUORENE
C20H16O	032377-13-2	9-(3-METHOXYPHENYL)-9H-FLUORENE
C20H16O4S2	003263-31-8	C.I. VAT ORANGE 5
C20H17FO3S	038194-50-2	SULINDAC
C20H18ClNO6	000303-47-9	OCHRATOXIN A
C20H18N2O6S2	017942-66-4	5-THIA-1-AZABICYCLO[4.2.0]OCT-2-ENE-2-CARBOXYLIC
C20H18O3	000081-92-5	BENZENEMETHANOL, 2-[BIS(4-HYDROXYPHENYL)METHYL]-
C20H19NaO3S	099287-50-0	1-SO3NA-3-BENZYL-7-ISOPROPYL AZULENE
C20H19NO5	119005-28-6	1-AZETIDINEACETIC ACID, 2-(4-BENZOYLPHENOXY)-4-O
C20H20Cl2N2O4S	086570-05-0	CARBOFURAN,N(N'-ME,24CL-PHCARBAMIDOTHIO)
C20H20Cl2N8O5	000528-74-5	DICHLOROMETHOTREXATE
C20H20ClN3	000632-99-5	ROSANILINE
C20H20N6O4S	068133-69-7	AZO DYE RA
C20H20O6	093414-39-2	2-PROPENOIC ACID, 3-(2-METHOXYPHENYL)-, 2-(4-FOR
C20H21ClFNO2	104182-04-9	1-PROPANONE, 3-[4-(4-CHLOROPHENYL)-4-HYDROXY-1-P
C20H21ClN2O4S	086230-37-7	CARBOFURAN,N(N'-ME,4-CL-PHCARBAMIDOTHIO)
C20H21ClN2O4S	086230-38-8	CARBOFURAN,N(N'-ME,2-CL-PHCARBAMIDOTHIO)
C20H21ClOSi2	053634-34-7	1-CHLORO-1,3-DIMETHYL-1,3,3-TRIPHENYLDISILOXANE
C20H21N	000239-64-5	13H-DIBENZO(A,I)CARBAZOLE
C20H21N2O3PS	104608-36-8	FOSTEDIL,2-OXOPYRROLIDINO ANANLOG
C20H21N5O5	078142-97-9	N7-(3-PYRIDYL)-MITOMYCIN C
C20H21NO3	001165-48-6	DIMEFLINE
C20H21NO4	000058-74-2	PAPAVERINE
C20H22ClN	052845-72-4	1-CHLOROAMITRIPTYLINE
C20H22N2O	120928-09-8	FENAZAQUIN
C20H22N2O2	097871-60-8	A-BENZOYLAMINO-N-SEC-BU CINNAMAMIDE
C20H22N2O4S	086230-36-6	CARBOFURAN,N(N'-ME,PHCARBAMIDOTHIO)
C20H22N4O10S	064544-07-6	CEFUROXIME AXETIL
C20H22N4O2	083158-05-8	24NH2-PYRIMD,5(2ME-4MEO-5PHMEO)BENZYL
C20H22N4O3	021822-34-4	24NH2-PYRIMIDINE,5(35MEO-4PHMEO)BENZYL
C20H22N4O3	027810-94-2	24NH2-PYRIMIDINE,5(34MEO-5PHMEO)BENZYL
C20H22N6O5	088854-59-5	N7-(6-AMINO-PYRID-3-YL)-MITOMYCIN C
C20H22N6O5	096662-65-6	TEPA 4-(N-OH-PIPERIDINYL)-UREA DERIVATIV
C20H22N8O5	000059-05-2	METHOTREXATE
C20H22O4	033533-56-1	MONO(P-(TERT-BUTYL)ALPHA-METHYLBENZYL) PHTHALATE
C20H22O5	027138-31-4	DIPROPYLENE GLYCOL, DIBENZOATE
C20H22O7	021899-50-3	2-BUTENOIC ACID, ESTER DERIVATIVE

C20H23FO2	092817-12-4	19-NORPREGNA-1,3,5(10)-TRIEN-20-YNE-3,17-DIOL, 1
C20H23N	000050-48-6	AMITRYPTYLINE
C20H23N3O3	024809-26-5	D-PHENYLALANINAMIDE, N-ACETYL-D-PHENYLALANYL-
C20H23N3O4	052329-50-7	L-PHENYLALANINAMIDE, N-ACETYL-L-TYROSYL-
C20H23N3O4	086880-51-5	EPANOLOL
C20H23N3O6	096000-47-4	1H-PYRROLO[1,2-A]INDOLE-5,8-DIONE DERIVATIVE
C20H23NO3	071626-11-4	BENALAXYL
C20H23NO5	003476-50-4	N-DESACETYL-COLCHICINE
C20H24FNO2S	034933-71-6	4-PIPERIDINOL, 4-(4-CHLOROPHENYL)-1-[3-[(4-FLUOR
C20H24N2O	074517-78-5	9H-FLUORENE-9-CARBOXAMIDE, 9-[3-[(1-METHYLETHYL)
C20H24N2O	074602-43-0	BENZAMIDE, N-(2,6-DIMETHYLPHENYL)-3-(1-PYRROLIDI
C20H24N2O2	000056-54-2	QUINIDINE
C20H24N2O2	000130-95-0	QUININE
C20H24N2O2	074602-48-5	BENZAMIDE, N-(2,6-DIMETHYLPHENYL)-3-(4-MORPHOLIN
C20H24N2O2	112411-90-2	1-(2,3-DIMETHYL),2-DIBENZOYL-1-TERT-BUTYLHYDRAZI
C20H24N2O2	112412-44-9	1-(3,4-DIMETHYL),2-DIBENZOYL-1-TERT-BUTYLHYDRAZI
C20H24N2O2	112426-79-6	1-(3,5-DIMETHYL),2-DIBENZOYL-1-TERT-BUTYLHYDRAZI
C20H24N2O2	112427-41-5	1-(2-ETHYL),2-DIBENZOYL-1-TERT-BUTYLHYDRAZINE
C20H24N2O3	070116-00-6	CINCHONINE OXALATE
C20H24N2O4	112427-66-4	1-(3,4-DIMETHOXY),2-DIBENZOYL-1-TERT-BUTYLHYDRAZ
C20H24N6O2	079508-86-4	ACETAMIDE, N-[3-[[3-(4,6-DIAMINO-2,2-DIMETHYL-1,
C20H24O2	000057-63-6	19-NORPREGNA-1,3,5(10)-TRIEN-20-YNE-3,17-DIOL,
C20H24O6	014187-32-7	[3,3]DB18C6-DIBENZO CROWN ETHER
C20H24O6	091814-58-3	1,4-NAPHTHALENEDIOL, 2,3-DIMETHOXY-, BIS(2-METHY
C20H24O7	020071-53-8	2-BUTENOIC ACID, ESTER DERIVATIVE
C20H25BrN2O2	000549-49-5	QUININE HYDROBROMIDE
C20H25ClN2O2	000130-89-2	QUININE, MONOHYDROCHLORIDE
C20H25ClN2O2	001668-99-1	QUINIDINE HYDROCHLORIDE
C20H25ClO2	056265-24-8	1,1'-(2-CHLORO-2-METHYLPROPYLIDENE)BIS-4-ETHOX*
C20H25N3O	000050-37-3	LSD
C20H25N3O2	055179-31-2	BITERTANOL
C20H25N3O2S2	014759-06-9	SULFORIDAZINE
C20H25N3O3S	102151-09-7	2-PROPANOL, 1-[(2-ETHOXYETHYL)AMINO]-3-[4-[4-(2-
C20H25N3S	000084-97-9	PERAZINE
C20H25N5O5S	088854-60-8	N7-(4,5-DIME THIAZOL-2-YL)-MITOMYCIN C
C20H25NO2	149589-43-5	BENZAMIDE, N-HEXYL-N-METHYL-3-PHENOXY-
C20H25NO3	144459-03-0	MORPHOLINE, 4-[[3-[[(4-METHOXYPHENYL)METHOXY]MET
C20H26N2O	074433-34-4	BENZAMIDE, 4-[(DIETHYLAMINO)METHYL]-N-(2,6-DIMET
C20H26N2O2	000522-66-7	HYDROQUININE
C20H26N2O5	120908-63-6	ACETAMIDE, N-[4-[2-HYDROXY-3-[(2-HYDROXY-3-PHENO
C20H26N4O8	137057-42-2	9H-PURINE, 6-METHOXY-9-[2,3,5-TRIS-O-(1-OXOPROPY
C20H26O2	056265-21-5	1,1'-(2-METHYLPROPYLIDENE)BIS(4-ETHOXYBENZENE)
C20H26O4	000084-61-7	DICYCLOHEXYL PHTHALATE
C20H26O6	000083-59-0	PROPYL ISOME
C20H27NO2S	144458-93-5	BENZENEPROPANAMINE, N,N-BIS(2-CARBONYLETHYL)-4-[
C20H27NO4	059170-23-9	BEVANTOLOL
C20H27NO4	110747-15-4	2-PROPANOL, 1,1'-IMINOBIS[3-(2-METHYLPHENOXY)
C20H27O4P	000115-88-8	OCTYLDIPHENYL PHOSPHATE
C20H27O4P	001241-94-7	2-ETHYLHEXYL DIPHENYL PHOSPHATE
C20H28I3N3O8	031122-82-4	1,3-BENZENEDICARBOXAMIDE, TRIIODO DERIVATIVE
C20H28N2O2	004360-12-7	AJMALINE
C20H28N2O3	111227-10-2	1H-IMIDAZOLE-1-CARBOXYLIC ACID, 1-[[4-(1,1-DIMET
C20H28N2O3	111227-15-7	1H-IMIDAZOLE-1-CARBOXYLIC ACID, 1-[(4-BUTYLPHENO
C20H28N6O6	130817-93-5	L-PROLINAMIDE, 5-OXO-L-PROLYL-1-[(1-METHYLETHOXY
C20H28O2	000302-79-4	RETINOIC ACID
C20H29N3O2	000085-79-0	DIBUCAINE
C20H29N5O3	034661-75-1	2,4(1H,3H)-PYRIMIDINEDIONE, 6-[[3-[4-(2-METHOXYP
C20H29NO2	071990-00-6	2,6-METHANO-3-BENZAZOCIN-8-OL, 6-ETHYL-1,2,3,4,5

C20H29NO3	136832-80-9	KETOBEMIDONE,PIVALIC ESTER
C20H29NO4	016590-41-3	NALTREXONE
C20H30N2O5S	082560-54-1	BENFURACARB
C20H30N4O4	132766-07-5	L-ALANINAMIDE, N-ACETYL-L-ISOLEUCYL-L-PHENYLALAN
C20H30N4O5	132766-18-8	L-PHENYLALANINAMIDE, N-ACETYL-L-LEUCYL-L-SERYL-
C20H30N4O5	132766-20-2	L-LEUCINAMIDE, N-ACETYL-L-ALANYL-L-TYROSYL-
C20H30O2	000058-18-4	17-A-METHYL TESTOSTERONE
C20H30O2	003704-09-4	ESTR-4-EN-3-ONE, 17-HYDROXY-7,17-DIMETHYL-, (7AL
C20H30O4	000084-75-3	DIHEXYL PHTHALATE
C20H30O4	000085-69-8	BUTYL (2-ETHYLHEXYL) PHTHALATE
C20H30O4	000146-50-9	DIISOHEXYL PHTHALATE
C20H31FO3	000076-43-7	ANDROST-4-EN-3-ONE, 9-FLUORO-11,17-DIHYDROXY-17-
C20H31N	072242-01-4	PIPERIDINE, 1-(3,3,5-TRIMETHYL-1-PHENYLCYCLOHEXY
C20H31N5O3	102151-13-3	UREA, N-BUTYL-N'-[2-[[3-[4-(5-METHYL-1H-IMIDAZOL
C20H31NO	000144-11-6	TRIHEXYLPHENEDYL
C20H32N2O3	093450-05-6	CARBAMIC ACID, [2-(PENTYLOXY)PHENYL]-, 2-(1-PIPE
C20H32N2O3	093450-24-9	CARBAMIC ACID, [3-[(PENTYLOXY)METHYL]PHENYL]-, 2
C20H32O2	000058-19-5	DROMOSTANOLONE
C20H32O2	000506-32-1	5,8,11,14-EICOSATETRAENOIC ACID
C20H32O2	132296-20-9	PHENOL, 5-(1,1-DIMETHYLHEXYL)-2-(3-HYDROXYCYCLOH
C20H32O5	000363-24-6	PROSTAGLANDIN E2
C20H32O6	034213-80-4	3,5-DI-T-BU PHENYL GLUCOPYRANOSIDE
C20H32O8	034206-60-5	GRAYANOTOXANE-5,6,7,9,10,14,16-HEPTOL, 2,3-EPOXY
C20H32O8	072216-45-6	B24C8-BENZO CROWN ETHER
C20H33N3O4	056980-93-9	UREA, N'-[3-ACETYL-4-[3-[(1,1-DIMETHYLETHYL)AMIN
C20H33NaO3S	013419-31-3	4-(1-METHYLTRIDECYL)BENZENESULFONIC ACID, SODIU*
C20H34	001459-10-5	1-PHENYLTETRADECANE
C20H34	004534-55-8	BENZENE, (1-PENTYLNONYL)-
C20H34	004534-56-9	BENZENE, (1-BUTYLDECYL)-
C20H34	004534-57-0	BENZENE, (1-PROPYLUNDECYL)-
C20H34	004534-58-1	BENZENE, (1-ETHYLDODECYL)-
C20H34	004534-59-2	BENZENE, (1-METHYLTRIDECYL)-
C20H34O3Si4	013270-97-8	1,1,1,3,5,7,7,7-OCTAMETHYL-3,5-DIPHENYLTETRASIL*
C20H34O5	000551-11-1	PROSTAGLANDIN
C20H34O5	038776-76-0	A-DIHYDROGRAYANOTOXIN
C20H34O6	004678-45-9	GRAYANOTOXIN III
C20H34O8	057721-95-6	1,2-BIS(2-(2-MEO-ETO)-ETO-ETO)BENZENE
C20H35IN2O	035129-56-7	3-CARBAMOYLPYRIDINIUM IODIDE,N-C14
C20H36IN	074639-30-8	2-ME PYRIDINIUM IODIDE,N-TETRADECYL
C20H37NaO7S	000577-11-7	BIS(2-ETHYLHEXYL) SODIUM SULFOSUCCINATE
C20H37NaO7S	001639-66-3	DI-N-OCTYL SODIUM SULFOSUCCINATE
C20H38N2O6S	017057-68-0	D-ERYTHRO-ALPHA-D-GALACTO-OCTOPYRANOSIDE, METHYL
C20H38N2O6S	021085-65-4	D-ERYTHRO-ALPHA-D-GALACTO-OCTOPYRANOSIDE, ETHYL
C20H40	001795-18-2	N-TETRADECYLCYCLOHEXANE
C20H40	003452-07-1	1-EICOSENE
C20H40O2	000111-06-8	BUTYL PALMITATE
C20H40O2	000111-61-5	ETHYL STEARATE
C20H40O2	000506-30-9	EICOSANOIC ACID
C20H41Br	004276-49-7	1-BROMOEICOSANE
C20H41Cl	042217-02-7	1-CHLOROEICOSANE
C20H42	000112-95-8	EICOSANE
C20H42O	000629-96-9	1-EICOSANOL
C20H44IN	002498-20-6	TETRAPENTYLAMMONIUM IODIDE
C20H60O8Si9	002652-13-3	NONASILOXANE, EICOSAMETHYL-
C21H13N	000224-42-0	DIBENZ[a,j]ACRIDINE
C21H13N	000226-36-8	DIBENZ(A,H)ACRIDINE
C21H14	002381-39-7	6-METHYLBENZO(A)PYRENE
C21H14O2	002156-14-1	1H-INDENE-1,3(2H)-DIONE, 2-[1,1'-BIPHENYL]-4-YL-

C21H16	000056-49-5	3-METHYLCHOLANTHRENE
C21H17FO3	093371-50-7	PROPIONIC ACID,2(4-BIPHENYLYL)3OH3(2FPH)
C21H17FO3	093371-51-8	PROPIONIC ACID,2(4-BIPHENYLYL)-3-OH-3-(3FPH)-
C21H17NO3	096327-61-6	1H-PYRROLIZINE-1-CARBOXYLIC ACID, 2,3-DIHYDRO-5-
C21H18O9	024385-10-2	5,12-NAPHTHACENEDIONE, 7,8,9,10-TETRAHYDRO-6,8,1
C21H19ClN2O5	056875-82-2	2H-1-BENZOPYRAN-2-ONE, 3-(ACETYLOXY)-1-[2-(ACETY
C21H19N	032377-15-4	4-(9H-FLUOREN-9-YL)-N,N-DIMETHYLBENZENAMINE
C21H20Br4O2	025327-89-3	TETRABROMOBISPHENOL A BIS(ALLYL ETHER)
C21H20BrN7O6	068877-63-4	AZO DYE N1
C21H20Cl2O3	052645-53-1	PERMETHRIN
C21H20Cl2O3	061949-76-6	CIS-PERMETHRIN
C21H20Cl2O3	061949-77-7	TRANS-PERMETHRIN
C21H20F3NO5S	123643-51-6	COLCHICINE,3-OH-7-TRIFLUOROACETAMIDO-10-METHIO A
C21H20O11	000522-12-3	QUERCITRIN
C21H21N	000129-03-3	CYPROHEPTADINE
C21H21NO6	000118-08-1	HYDRASTINE
C21H21O4P	000078-32-0	TRI-P-CRESYL PHOSPHATE
C21H21O4P	001330-78-5	TRICRESYL PHOSPHATE
C21H21O4P	028108-99-8	ISOPROPYL PHENYL DIPHENYL PHOSPHATE
C21H22ClN	007673-07-6	TRIBENZYLAMINE HYDROCHLORIDE
C21H22ClNO4	110488-70-5	DIMETHOMORPH
C21H22N2O2	000057-24-9	STRYCHNINE
C21H22N2O3	000465-62-3	PSEUDO-STRYCHNINE
C21H22N4O5	014896-01-6	N7-PHENYL-MITOMYCIN C
C21H22O2	055994-24-6	1H-INDENE-1,3(2H)-DIONE, 2-[3,5-BIS(1-METHYLETHY
C21H22O9	001415-73-2	ALOIN
C21H23ClFNO2	000052-86-8	HALOPERIDOL
C21H23ClFNO5	087546-18-7	FLUMICLORAC-PENTYL
C21H23IN2O2	052748-69-3	STRYCHNINE HYDRIODIDE
C21H23N3O2	003215-24-5	2-AMINOSTRYCHNINE
C21H23N3O4	029854-52-2	STRYCHNINE, 2-NITRO-
C21H23N3O5	000066-32-0	STRYCHNINE NITRATE
C21H23N3OS	002622-26-6	PERICYAZINE
C21H23N5O6	084397-27-3	N7-(6-MEO-PYRID-3-YL)-MITOMYCIN C
C21H23NO5	000561-27-3	HEROIN
C21H23NO6	000477-27-0	COLCHICEINE
C21H24ClN3OS	000084-04-8	PIPAMAZINE
C21H24ClNO5	001502-95-0	DIACETYLMORPHINE, HYDROCHLORIDE
C21H24F3N3S	000117-89-5	TRIFLUORPERAZINE
C21H24N2O2	015006-14-1	21,22-DIHYDROSTRYCHNINE
C21H24N4O4	130787-20-1	BENZAMIDE, N-(2-CYCLOPENTYL-2H-BENZOTRIAZOL-5-YL
C21H24O3Si3	003424-57-5	2,4,6-TRIMETHYL-2,4,6-TRIPHENYLCYCLOTRISILOXANE
C21H24O3Si3	006138-53-0	2,4,6-TRIMETHYL-2,4,6-TRIPHENYLCYCLOTRISILOXANE
C21H24OSi2	014920-93-5	1,1,3-TRIMETHYL-1,3,3-TRIPHENYL-DISILOXANE
C21H25N5O2	071351-79-6	4(1H)-PYRIMIDINONE, 2-[[4-(3-METHOXY-2-PYRIDINYL
C21H25NO3	055096-26-9	NALMEFENE
C21H26ClN3OS	000058-39-9	PERPHENAZINE
C21H26N2O	072336-22-2	ACETAMIDE, 2-(DIPROPYLAMINO)-N-(9H-FLUOREN-9-YL)
C21H26N2O	074602-46-3	BENZAMIDE, N-(2,6-DIMETHYLPHENYL)-4-(1-PIPERIDIN
C21H26N2O3	000146-48-5	APHRODINE
C21H26N2S2	000050-52-2	THIORIDAZINE
C21H26O3	034816-55-2	19-NORPREGNA-1,3,5(10)-TRIEN-20-YNE-3,17-DIOL, 1
C21H26O3	055079-83-9	2,4,6,8-NONATETRAENOIC ACID, 9-(4-METHOXY-2,3,6-
C21H26O4	017080-02-3	FURETHRIN
C21H26O5	000053-03-2	PREDNISONE
C21H27ClN2O3	000065-19-0	YOHIMBINE HYDROCHLORIDE
C21H27FO5	000079-60-7	CORTISONE-9A-FLUORO
C21H27FO5	000338-95-4	PREDNISOLONE-9A-FLUORO

C21H27FO6	000124-94-7	TRIAMCINOLONE
C21H27N3O3	076252-06-7	QUINOLINE, 1,2,3,4-TETRAHYDRO-8-[2-HYDROXY-3-[(1
C21H27N3O6S	060903-21-1	4-THIA-1-AZABICYCLO[3.2.0]HEPTANE-2-CARBOXYLIC A
C21H27N5O2	103066-37-1	4,7-METHANO-1H-ISOINDOLE-1,3(2H)-DIONE, 3A,4,7,7
C21H27N5O4S	029094-61-9	GLIPIZIDE
C21H27NO	000076-99-3	METHADONE
C21H27NO3	120282-85-1	ADAMANTYL DERIVATIVE
C21H28N2O3	121508-90-5	BENZENEACETAMIDE, N-[2-[[2-HYDROXY-3-(2-PHENYLET
C21H28N2O4	092598-97-5	1H-IMIDAZOLE-5-BUTANOIC ACID, beta-[(ACETYLOXY)M
C21H28N2O4	121508-95-0	BENZENEACETAMIDE, N-[2-[[2-HYDROXY-3-(2-PHENOXYE
C21H28N2O5	000138-56-7	TRIMETHOBENZAMIDE (TIGAN)
C21H28N2O5	087269-97-4	CYCLOPENTA[B]PYRROLE-2-CARBOXYLIC ACID, DERIVATI
C21H28N2S2	051308-54-4	BUTHIOBATE
C21H28N5O6PS	107538-71-6	8-S-T-BUTYL CYCLIC AMP-O-BENZYL
C21H28O2	027955-87-9	1,1-BIS(P-ETHOXYPHENYL)-2,2-DIMETHYLPROPANE
C21H28O3	000121-21-1	PYRETHRIN I
C21H28O5	000050-24-8	PREDNISOLONE
C21H28O5	000052-39-1	ALDOSTERONE
C21H28O5	000053-06-5	CORTISONE
C21H29FO5	000127-31-1	HYDROCORTISONE-9A-FLUORO
C21H29N	031790-44-0	PIPERIDINE, 1-[4-(1,1-DIMETHYLETHYL)-1-PHENYLCYC
C21H29N	070785-03-4	BENZENEPROPANAMINE, 4-(1,1-DIMETHYLETHYL)-N,N-DI
C21H29N3O	003737-09-5	DISOPYRAMIDE
C21H29N5O4	132766-15-5	GLYCINAMIDE, N-ACETYL-L-TRYPTOPHYL-L-ISOLEUCYL-
C21H29N5O4	132766-17-7	L-VALINAMIDE, N-ACETYL-L-TRYPTOPHYL-L-ALANYL-
C21H29NO	000514-65-8	BIPERIDEN
C21H29NO	017199-55-2	BENZENEETHANOL, beta-[2-(DIMETHYLAMINO)PROPYL]-_
C21H30FN3O2	001893-33-0	PIPAMPERONE
C21H30N2	070785-07-8	N-PH-N(4-TBUBNZYL)-N(NNDIME)ET AMINE
C21H30N2O	086073-85-0	4-QUINOLINEMETHANOL, 2-(1,1-DIMETHYLETHYL)-α-[2-
C21H30N4O7	137057-57-9	9H-PURINE, 9-[2,3-BIS-O-(1-OXOPENTYL)-beta-D-ARA
C21H30N4O7	137057-62-6	9H-PURINE, 9-[2,3-BIS-O-(2,2-DIMETHYL-1-OXOPROPY
C21H30N4O7	137057-69-3	9H-PURINE, 9-[2,5-BIS-O-(1-OXOPENTYL)-BETA-D-ARA
C21H30N4O7	137057-71-7	9H-PURINE, 9-[2,5-BIS-O-(2,2-DIMETHYL-1-OXOPROPY
C21H30N6O6	130817-94-6	L-PROLINAMIDE, 5-OXO-L-PROLYL-1-(BUTOXYCARBONYL)
C21H30N6O6	130817-95-7	L-PROLINAMIDE, 5-OXO-L-PROLYL-1-[(2-METHYLPROPOX
C21H30O2	000057-83-0	PROGESTERONE
C21H30O2	005957-75-5	6H-DIBENZO[B,D]PYRAN-1-OL, 6A,7,10,10A-TETRAHYDR
C21H30O2	033086-25-8	6H-DIBENZO[B,D]PYRAN-1-OL, 6A,7,8,10A-TETRAHYDRO
C21H30O3	000064-85-7	DEOXYCORTICOSTERONE
C21H30O3	000068-96-2	17-ALPHA-HYDROXYPROGESTERONE
C21H30O3	000312-90-3	11-A-HYDROXYPROGESTERONE
C21H30O3	036557-05-8	6H-DIBENZO[B,D]PYRAN-9-METHANOL, 6A,7,8,10A-TETR
C21H30O4	000050-22-6	CORTICOSTERONE
C21H30O4	000152-58-9	CORTEXOLONE
C21H30O5	000050-23-7	HYDROCORTISONE
C21H31N3O5	076547-98-3	L-PROLINE, 1-[N -(1-CARBOXY-3-PHENYLPROPYL)-L-LY
C21H32N4O4	137605-63-1	AC-ALA-PHE-ALA-NTBU
C21H32N6O3	071195-58-9	ALFENTANIL
C21H32O2	000145-13-1	PREGNENOLONE
C21H33NO7	000303-34-4	LASIOCARPINE
C21H34O2	114753-51-4	PHENOL, 5-(1,1-DIMETHYLHEPTYL)-2-(3-HYDROXYCYCLO
C21H34O9	063144-76-3	B27C9-BENZO CROWN ETHER
C21H35ClO	006964-19-8	4-CHLORO-3-PENTADECYLPHENOL
C21H36	002131-18-2	1-PHENYLPENTADECANE
C21H36O2	016825-54-0	1,3-BENZENEDIOL, 4-PENTADECYL-
C21H38IN	088477-46-7	3,4-DIME PYRIDINIUM I,N-TETRADECYL
C21H38IN	088477-52-5	2,4-DIME PYRIDINIUM I,N-TETRADECYL

C21H38IN	088477-58-1	3,5-DIME PYRIDINIUM I,N-TETRADECYL
C21H38O6	000621-70-5	TRICAPROIN
C21H39N7O12	000057-92-1	STREPTOMYCIN
C21H42	006006-95-7	N-PENTADECYLCYCLOHEXANE
C21H42O2	000112-10-7	ISOPROPYL STEARATE
C21H42O2	018281-04-4	ETHYL NONADECANOATE
C21H44	000629-94-7	HENEICOSANE
C22H12	000191-24-2	BENZO(GHI)PERYLENE
C22H12	000191-26-4	DIBENZO(DEF,MNO)CHRYSENE
C22H12	000193-39-5	INDENO(1,2,3-CD)PYRENE
C22H14	000053-70-3	DIBENZ(A,H)ANTHRACENE
C22H14	000135-48-8	PENTACENE
C22H14	000194-69-4	BENZO[C]CHRYSENE
C22H14	000213-46-7	PICENE
C22H14	000214-17-5	BENZO[B]CHRYSENE
C22H14	000215-58-7	1,2,3,4-DIBENZANTHRACENE
C22H14	000224-41-9	DIBENZ(A,J)ANTHRACENE
C22H14	000226-88-0	BENZO[A]NAPHTHACENE
C22H15NO4S	018690-91-0	1,4-NAPHTHALENEDIONE, 2-(PHENYLAMINO)-3-(PHENYLS
C22H16N3NaO6S	006424-85-7	ALIZARIN AZUROL A 2G
C22H16N4O	006253-10-7	C.I. DISPERSE ORANGE 13
C22H17F2N3O	137103-55-0	1H-IMIDAZOLE-1-ETHANOL, alpha-(2,4-DIFLUOROPHENY
C22H17F2N3O	137128-33-7	1H-IMIDAZOLE-1-ETHANOL, alpha-(2,4-DIFLUOROPHENY
C22H18Cl2FNO3	068359-37-5	CYFLUTHRIN
C22H18ClN3O	137103-64-1	1H-IMIDAZOLE-1-ETHANOL, alpha-(4-CHLOROPHENYL)-a
C22H18FN3O	137103-53-8	1H-IMIDAZOLE-1-ETHANOL, _-(4-FLUOROPHENYL)-_-[4-
C22H18FN3O	137103-54-9	1H-IMIDAZOLE-1-ETHANOL, alpha-(4-FLUOROPHENYL)-a
C22H18FN3O	137103-59-4	1H-IMIDAZOLE-1-ETHANOL, alpha-(4-FLUOROPHENYL)-a
C22H19Br2NO3	052918-63-5	DELTAMETHRIN
C22H19Cl2NO3	052315-07-8	CYPERMETHRIN
C22H19Cl2NO3	065731-84-2	BETA-CYPERMETHRIN ISOMER
C22H19Cl2NO3	065732-07-2	BETA-CYPERMETHRIN ISOMER
C22H19N3O2	142657-53-2	1H-IMIDAZOLE-1-ETHANOL, _-(4-HYDROXYPHENYL)-_-[4
C22H20	093037-15-1	12-BUTYL-BENZ[A]ANTHRACENE
C22H20O13	001260-17-9	CARMINIC ACID
C22H20O4	093371-46-1	PROPIONIC ACID,2(4BIPHENYLYL)3OH3(P-MEOPH)
C22H21ClN2O5	056875-83-3	2H-1,4-BENZODIAZEPIN-2-ONE, 3-(ACETYLOXY)-1-[3-(
C22H21ClN2O5	119738-06-6	QUIZALOFOP-P-TEFURYL
C22H21NO2S	002799-07-7	L-CYSTEINE, S-(TRIPHENYLMETHYL)-
C22H21NO5	119005-29-7	1-AZETIDINEACETIC ACID, 2-OXO-4-[4-(1-OXO-3-PHEN
C22H22ClN3O5	111479-05-1	PROPAQUIZAFOP
C22H22FN3O2	000548-73-2	DROPERIDOL
C22H22O8	000518-28-5	PODOPHYLLOTOXIN
C22H23ClN2O2	079794-75-5	LORATADINE
C22H23ClN2O8	000057-62-5	AUREOMYCIN
C22H23NO3	039515-41-8	FENPROPATHRIN
C22H23NO4S2	081872-10-8	L-PROLINE, 1-[3-(BENZOYLTHIO)-2-METHYL-1-OXOPROP
C22H23NO7	006035-40-1	NARCOTINE
C22H23O4P	056803-37-3	T-BUTYLPHENYL DIPHENYL PHOSPHATE
C22H24ClN5O2	057808-66-9	DOMPERIDONE
C22H24FN3O2	002062-84-2	BENPERIDOL
C22H24N2O2S	111226-70-1	1H-IMIDAZOLE-1-CARBOTHIOIC ACID, O-[1-[((1,1'-BI
C22H24N2O3	000509-36-4	2-METHOXYSTRYCHNINE
C22H24N2O3	111227-09-9	1H-IMIDAZOLE-1-CARBOXYLIC ACID, 1-[((1,1'-BIPHEN
C22H24N2O4	000091-96-3	C.I. AZOIC COUPLING COMPONENT 5
C22H24N2O8	000060-54-8	TETRACYCLINE
C22H24N2O8	000564-25-0	VIBRAMYCIN
C22H24N2O9	000079-57-2	OXYTETRACYLCINE

C22H25N3O4S	031883-05-3	MORICIZINE
C22H25NO3	120908-61-4	2-PROPANOL, 1-[[2-HYDROXY-3-(1-NAPHTHALENYLOXY)P
C22H25NO3	136832-82-1	KETOBEMIDONE,BENZOYL ESTER
C22H25NO6	000064-86-8	COLCHICINE
C22H26ClFO4	054063-32-0	CLOBETASONE
C22H26ClN3O4	102152-27-2	2-PROPANOL, 1-[4-(5-CHLORO-1H-IMIDAZOL-2-YL)PHEN
C22H26F3N3OS	000069-23-8	FLUPHENAZINE
C22H26N2O4S	033286-22-5	DILTIAZEM
C22H26N2O5	055565-48-5	BENZYLOXYCARBONYL-PHALANINYLVALINE
C22H26N2O6	000862-26-0	L-TYROSINE, N-(N-CARBOXY-L-VALYL)-, N-BENZYL EST
C22H26N2O8S2	068661-74-5	5-THIA-1-AZABICYCLO[4.2.0]OCT-2-ENE-2-CARBOXYLIC
C22H26N8O5	034378-65-9	METHOTREXATE DIMETHYL ESTER
C22H26O3	010453-86-8	RESMETHRIN
C22H26O8	010215-89-1	SPIRO[AZULENO[4,5-B]FURAN-6(2H),2'-OXIRAN]-2-ONE
C22H26O8	034175-79-6	2-BUTENOIC ACID, 2-[(ACETYLOXY)METHYL]- DERIVATI
C22H27ClF2O5	050629-82-8	2-CHLOROFLUMETHASONE
C22H27ClN2O	059729-31-6	LORCAINIDE
C22H27ClO8	020501-52-4	2-BUTENOIC ACID, 2-METHYL-, 7-(ACETYLOXY)-6-(CHL
C22H27N3O5	000057-64-7	PHYSOSTIGMINE, MONOSALICYLATE
C22H27N3OS	104076-38-2	2-BENZOTHIAZOLAMINE, N-[3-[3-(1-PIPERIDINYLMETHY
C22H27N5O9S2	065243-53-0	5-THIA-1-AZABICYCLO[4.2.0]OCT-2-ENE-2-CARBOXYLIC
C22H27N9O7S3	065243-35-8	CEFMENOXIME,PIVALOYL-OME ESTER
C22H27NO2	000090-69-7	LOBELINE
C22H28ClFO4	025122-41-2	CLOBETASOL
C22H28F2O5	002135-17-3	FLUMETHASONE
C22H28N2O	000437-38-7	FENTANYL
C22H28N2O2	112410-23-8	TEBUFENOZIDE
C22H28N2O3	121508-97-2	BENZENEACETAMIDE, N-[2-[[2-HYDROXY-3-[(3-PHENYL-
C22H28O2	039845-47-1	19-NORPREGNA-1,3,5(10)-TRIEN-20-YNE-3,17-DIOL, 1
C22H28O3	000976-71-6	CANRENONE
C22H28O4	050394-27-9	PREGN-4ENE 11A-EPOXYLACTONE DERIV
C22H28O6	091814-09-4	PROPANOIC ACID, 2,2-DIMETHYL-, 2,3-DIMETHOXY-1,4
C22H28O7	033854-15-8	2-BUTENOIC ACID, 2-METHYL-, 9-(ACETYLOXY)- ESTER
C22H28O7	075887-63-7	CARBONIC ACID, DECAHYDRO-3,6,9-TRIMETHYL-3,12-EP
C22H29FO4	000382-67-2	BETAMETHASONE-21-DESOXY
C22H29FO4	000426-13-1	FLUOROMETHALONE
C22H29FO4	001582-75-8	PREGN-4-ENE-21-CARBOXYLIC ACID, 9-FLUORO-11,17-D
C22H29FO5	000050-02-2	DEXAMETHASONE
C22H29FO5	000378-44-9	BETAMETHASONE
C22H29N3O	025973-55-1	2-(2H-BENZOTRIAZOL-2-YL)-4,6-DI-(TERT-PENTYL)PH*
C22H29N3S2	001420-55-9	ETHYLTHIOPERAZINE
C22H29N7O5	000053-79-2	PUROMYCIN
C22H29NO2	000469-62-5	PROPOXYPHENE
C22H30Cl2N10	000055-56-1	CHLORHEXIDINE
C22H30FNO4	084057-96-5	2-PROPANOL, 1-[4-[2-[2-(4-FLUOROPHENYL)ETHOXY]ET
C22H30N2	037640-71-4	APRINDINE
C22H30N2O2S	056030-54-7	SUFENTANIL
C22H30N2O3	121508-92-7	BENZENEACETAMIDE, N-[2-[[3-(2-HYDROXY-3-PHENYLPR
C22H30N2O4	106848-84-4	1H-IMIDAZOLE-5-BUTANOIC ACID, alpha-ETHYL-1-METH
C22H30N2O5	139733-59-8	ACETAMIDE, 2-[[[4-[2-(2-HYDROXY-3-PHENOXY)PROPYL
C22H30N2O5	139733-61-2	ACETAMIDE, 2-[[4-[2-(2-HYDROXY-3-PHENOXYPROPYL)A
C22H30N2O6	115656-32-1	ACETAMIDE, 2-[4-[2-[(2-HYDROXY-3-PHENOXYPROPYL)A
C22H30N2O6	139733-77-0	ACETAMIDE, 2-[4-[2-[(2-HYDROXY-3-PHENOXYPROPYL)A
C22H30O3	000976-70-5	6,7-DIHYDROCANRENONE
C22H30O3	003642-85-1	PREGN-4-ENE-3,11,20-TRIONE, 6-METHYL-, (6_)-
C22H31N	070785-06-7	3(4-T-BU BENZHYDRYLPROPYL)AMINE,NN-DIME
C22H31O4P	029761-21-5	ISODECYL DIPHENYL PHOSPHATE
C22H32N2O3	092598-99-7	1H-IMIDAZOLE-5-BUTANOIC ACID, alpha-ETHYL-beta-(

C22H32O2	001451-20-3	6H-DIBENZO[B,D]PYRAN, 6A,7,10,10A-TETRAHYDRO-1-M
C22H32O2	027179-29-9	6H-DIBENZO[B,D]PYRAN, 6A,7,8,9,10,10A-HEXAHYDRO-
C22H32O5	000121-29-9	PYRETHRIN II
C22H34N6O4	059886-40-7	2,5-CYCLOHEXADIENE-1,4-DIONE, 2,5-BIS(1-AZIRIDIN
C22H34O4	003648-21-3	DIHEPTYL PHTHALATE
C22H35NO2	013358-11-7	N(2-ETHEXYL)-1-ISOPROPYL-4-METHYLBICYCLO[2.2.2]-
C22H36O2	132296-11-8	PHENOL, 5-(1,1-DIMETHYLOCTYL)-2-(3-HYDROXYCYCLOH
C22H36O2	132296-14-1	PHENOL, 5-(1,1-DIMETHYLHEPTYL)-2-(5-HYDROXY-2-ME
C22H36O2	132339-34-5	PHENOL, 5-(1,1-DIMETHYLHEPTYL)-2-(5-HYDROXY-2-ME
C22H36O7	004720-09-6	ANDROMEDOTOXIN
C22H38	001459-09-2	N-HEXADECYLBENZENE
C22H38O5	059122-46-2	MISOPROSTOL
C22H39O3P	001754-47-8	DIOCTYL PHENYL PHOSPHONATE
C22H40O4Si5	020252-66-8	1,1,1,3,5,5,7,9,9,9-DECAMETHYL-3,7-DIPHENYLPEN*
C22H42O2	000142-77-8	BUTYL OLEATE
C22H42O4	000103-23-1	DI-2-ETHYLHEXYL ADIPATE
C22H42O4	000123-79-5	DIOCTYL ADIPATE
C22H43NO	000112-84-5	13-DECOSENAMIDE (CIS)
C22H44O2	000112-85-6	DOCOSANOIC ACID
C22H44O2	000123-95-5	N-BUTYL STEARATE
C22H44O2	018281-05-5	ETHYL EICOSANOATE
C22H46	000629-97-0	DOCOSANE
C22H46N2O	000095-19-2	2-HEPTADECYL-4,5-DIHYDRO-2-IMIDAZOLINE-1-ETHANOL
C23H16	032377-10-9	7-PHENYL-7H-BENZO(C)FLUORENE
C23H16O11	016110-51-3	CROMOGYLYCIC ACID
C23H16O3	000082-66-6	DIPHACINONE
C23H18F5N2NaO4	129358-85-6	6-HEPTENOIC ACID, 7-[4,5-BIS(4-FLUOROPHENYL)-2-(
C23H18N2O2	005102-79-4	2-DIPHACETYL-1,3INDANDIONE-1HYDRAZONE
C23H19ClF3NO3	068085-85-8	CYHALOTHRIN
C23H19ClF3NO3	091465-08-6	LAMBDA-CYHALOTHRIN
C23H20N2O3S	000057-96-5	SULFOXYPHENYL PYRAZOLIDINE
C23H22N2O6S	038212-89-4	CARBENCILLIN, PHENYL
C23H22O6	000083-79-4	ROTENONE
C23H23ClN5O6PS	107538-70-5	8-(4-CL-PH)-S CYCLIC AMP-O-BENZYL
C23H23NO	001420-06-0	TRIFENMORPH
C23H24N4O2S	102151-10-0	2-PROPANOL, 1-[[2-(4-PYRIDINYL)ETHYL]AMINO]-3-[4
C23H24N4O7	086476-06-4	BENZYLOXYCARBONYL MITOMYCIN C
C23H24N6O4	072828-64-9	AZO DYE N5
C23H25BrN6O10	003956-55-6	DISPERSE BLUE 79
C23H25ClN2	000569-64-2	MALACHITE GREEN
C23H25F3N2OS	002709-56-0	FLUPENTIXOL
C23H25F3N2OS	053772-82-0	1-PIPERAZINEETHANOL, 4-[3-[2-(TRIFLUOROMETHYL)-9
C23H25F3N2OS	053772-85-3	1-PIPERAZINEETHANOL, 4-[3-[2-(TRIFLUOROMETHYL)-9
C23H25N5O3	072593-17-0	BENZAMIDE, N-[2-(3,4,6,7,12,12A-HEXAHYDROPYRAZIN
C23H25NO4	124050-99-3	2H-PYRAN, TETRAHYDRO-3-[1-(ETHOXYIMINO)BUTYL]-2,
C23H26FN3O2	000749-02-0	SPIROPERIDOL
C23H26N2O2	021888-98-2	DEXETIMIDE
C23H26N2O4	000357-57-3	BRUCINE
C23H26N4O	·072593-21-6	BENZAMIDE, N-[2-(3,4,6,7,12,12A-HEXAHYDROPYRAZIN
C23H26O2	013936-01-1	1H-INDENE-1,3(2H)-DIONE, 2-[3,5-BIS(1,1-DIMETHYL
C23H26O2	055994-22-4	1H-INDENE-1,3(2H)-DIONE, 2-(4-OCTYLPHENYL)-
C23H26O3	026002-80-2	PHENOTHRIN
C23H27N3O7	010118-89-5	2-NAPHTHACENECARBOXAMIDE DERIVATIVE
C23H27N3O7	010118-90-8	MINOCYCLINE
C23H27N3O8S2	080929-58-4	5-THIA-1-AZABICYCLO[4.2.0]OCT-2-ENE-2-CARBOXYLIC
C23H27NO4	101126-54-9	1,4-NAPHTHALENEDIONE, 2,3-DIHYDRO-2-[[[2-[4-HYDR
C23H27NO8	000131-28-2	NARCEINE
C23H28N2O4	028879-93-8	21,22-DIHYDROBRUCINE

C23H28N4O5	132766-21-3	L-PHENYLALANINAMIDE, N-ACETYL-L-ALANYL-L-TYROSYL
C23H28O6	095440-71-4	PREDNISONE-17-ACETATE
C23H29ClFN3O4	081098-60-4	CISAPRIDE
C23H29N3O2S2	022189-31-7	THIOTHIXENE
C23H29NO3	000562-26-5	4-PIPERIDINECARBOXYLIC ACID, 1-(3-HYDROXY-3-PHEN
C23H30N2O3	121509-02-2	BENZENEACETAMIDE, N-[2-[[3-[(1,2,3,4-TETRAHYDRO-
C23H30N4O14	083107-50-0	D-GLYCEROL-A-D-GALACTO DERIVATIVE
C23H30O4Si4	010448-10-9	2,2,4,6,8-PENTAMETHYL-4,6,8-TRIPHENYLCYCLOTETR*
C23H30O6	000050-04-4	CORTISONE ACETATE
C23H30O6	000052-21-1	PREDNISOLONE ACETATE
C23H30O6	039791-38-3	PREGN-4-ENE-3,11,20-TRIONE, 17-(ACETYLOXY)-21-HY
C23H30O6	111802-43-8	PREGNA-1,4-DIENE-16-CARBOXYLIC ACID, 11,21-DIHYD
C23H31NO2	000509-74-0	ACETYLMETHADOL
C23H32N2O	018109-55-2	BENZENEACETAMIDE, N-[3-(DIETHYLAMINO)PROPYL]-N-(
C23H32N2O4	092598-95-3	1H-IMIDAZOLE-5-BUTANOIC ACID, alpha-ETHYL-1-METH
C23H32N4O8	137057-43-3	9H-PURINE, 6-METHOXY-9-[2,3,5-TRIS-O-(1-OXOBUTYL
C23H32N4O8	137057-45-5	9H-PURINE, 6-METHOXY-9-[2,3,5-TRIS-O-(2-METHYL-1
C23H32N6O6	130817-99-1	L-PROLINAMIDE, 5-OXO-L-PROLYL-1-[(CYCLOHEXYLOXY)
C23H32O2	000119-47-1	BIS (2-HYDROXY-3TERT-BUTYL-5-METHYLPHENYL) METHA
C23H32O4	000056-47-3	DEOXYCORTICOSTERONE ACETATE
C23H32O6	000050-03-3	HYDROCORTISONE ACETATE
C23H32O6	000066-28-4	STROPHANTHIDIN
C23H32O6	016463-74-4	PREGN-4-ENE-3,20-DIONE, 17-(ACETYLOXY)-11,21-DIH
C23H33N5O4	137605-64-2	AC-ALA-TRP-ALA-NTBU
C23H34N6O6	130817-96-8	L-PROLINAMIDE, 5-OXO-L-PROLYL-1-[(HEXYLOXY)CARBO
C23H34O4	000143-62-4	DIGITOXIGENIN
C23H34O5	000545-26-6	GITOXIGENIN
C23H34O5	001672-46-4	DIGOXIGENIN
C23H34O7S	002488-78-0	DIGITOXIGENIN-3-A-SULFATE
C23H34O7S	002488-80-4	DIGITOXIGENIN-3-B-SULFATE
C23H34O8	000508-52-1	OUABAGENIN
C23H36N2O2	098319-26-7	PROSCAR
C23H36N2O6S	123297-69-8	CARBAMIC ACID, [[[[(1,3-BENZODIOXOL-5-YLOXY)CARB
C23H36O5	073573-88-3	MEVASTATIN
C23H36O6	085956-22-5	PRAVASTATIN
C23H38ClN3O	068284-69-5	1-PIPERIDINEBUTANAMIDE, _-[2-[BIS(1-METHYLETHYL)
C23H38O2	132296-12-9	PHENOL, 5-(1,1-DIMETHYLNONYL)-2-(3-HYDROXYCYCLOH
C23H42IN	007206-40-8	N-OCTADECYLPYRIDINIUM IODIDE
C23H42IN	078191-89-6	2,4-DIME PYRIDINIUM I,N-HEXADECYL
C23H42IN	088477-47-8	3,4-DIME PYRIDINIUM I,N-HEXADECYL
C23H42IN	088497-75-0	3,5-DIME PYRIDINIUM I,N-HEXADECYL
C23H48	000638-67-5	TRICOSANE
C24H11Cl9	012672-29-6	AROCLOR 1248
C24H12	000191-07-1	CORONENE
C24H12O2	000128-66-5	DIBENZO(B,DEF)CHRYSENE-7,14-DIONE
C24H13Cl7	012674-11-2	AROCLOR 1016
C24H14	000189-55-9	DIBENZO(A,I)PYRENE
C24H14	000189-64-0	DIBENZO[A,H]PYRENE
C24H14	000191-30-0	DIBENZO[a,l]PYRENE
C24H14	000192-65-4	DIBENZO[a,e]PYRENE
C24H14	000203-20-3	BENZ(A,J)ACEANTHRYLENE
C24H16Cl2N4	102262-55-5	DES-ISOPROPYL CLOFAZIMINE
C24H16N2O2	001806-34-4	2,2'-(1,4-PHENYLENE)BIS(5-PHENYLOXAZOLE)
C24H16N3NaO3S	037571-27-0	BENZENESULFONIC ACID, 2-[2-[4-(2H-NAPHTHO[1,2-D]
C24H18	000612-71-5	5'-PHENYL-1,1':3',1"-TERPHENYL
C24H21Cl6O6P	000094-84-8	2,4-DEP
C24H25FN3NaO4	129358-45-8	6-HEPTENOIC ACID, 7-[5-(4-FLUOROPHENYL)-2-(1-MET
C24H26N5O6PS	107538-72-7	8-S-BENZYL CYCLIC AMP-O-BENZYL

C24H26N5O6PSe	107538-74-9	8-SEBENZYL CYCLIC AMP-O-BENZYL
C24H27BrN6O10	012239-34-8	DISPERSE BLUE 79
C24H27N3O4	092627-83-3	3-PYRIDINECARBOXYLIC ACID, 2-[(1-METHYL-1H-IMIDA
C24H27N5O4	132766-16-6	L-PHENYLALANINAMIDE, N-ACETYL-L-TRYPTOPHYLGLYCYL
C24H27O4P	025653-16-1	TRI(3,5-XYLENYL) PHOSPHATE
C24H28N6O7S2	058333-58-7	5-THIA-1-AZABICYCLO[4.2.0]OCT-2-ENE-2-CARBOXYLIC
C24H29ClF2O6	065466-50-4	PREGNA-1,4-DIENE-3,20-DIONE, 2-CHLORO-6,9-DIFLUO
C24H29N	051772-35-1	N-((1,1,3,3-TETRAMETHYLBUTYL)PHENYL)-1-NAPHTHA*
C24H29N5O2	114298-20-3	4,7-ETHENO-1H-CYCLOBUT[F]ISOINDOLE-1,3-(2H)-DION
C24H29NO10	093414-55-2	BENZOIC ACID, 3,4,5-TRIMETHOXY-, 2-[4-[[(2-ETHOX
C24H30F2O6	000067-73-2	FLUOCINOLONE ACETONIDE
C24H30F3N3O2S	003093-23-0	ETHANOL, 2-[2-[4-[3-[2-(TRIFLUOROMETHYL)PHENOTHI
C24H30N8O5	043170-88-3	L-GLUTAMIC ACID, N-[4-[[(2,4-DIAMINO-6-PTERIDINY
C24H31FO5	003801-06-7	FLUOROMETHOLONE ACETATE
C24H31FO6	000076-25-5	TRIAMCINOLONE ACETONIDE
C24H31FO6	000987-24-6	BETAMETHASONE-17-ACETATE
C24H31FO6	001177-87-3	DEXAMETHASONE-17-ACETATE
C24H31N3O5	102151-23-5	2-PROPANOL, 1-[[2-(3,4-DIMETHOXYPHENYL)ETHYL]AMI
C24H31N3O5	102152-14-7	2-PROPANOL, 1-[[2-(3,4-DIMETHOXYPHENYL)ETHYL]AMI
C24H31N3O5	102281-28-7	2-PROPANOL, 1-[[2-(3,4-DIMETHOXYPHENYL)ETHYL]AMI
C24H32N2O	067023-19-2	BENZENEACETAMIDE, N-(2,6-DIMETHYLPHENYL)-N-[3-(1
C24H32N2O5	139734-21-7	2-PROPANOL, 3-[[2-[4-[2-OXO-2-(2-PIPERIDINYL)ETH
C24H32O3	006987-59-3	SEARLE, SC-17127
C24H32O4S	000052-01-7	SPIRONOLACTONE
C24H32O4S	033784-05-3	SPIRONOLACTONE,7-BETA ANALOG
C24H32O5	041020-65-9	MEXRENONE
C24H32O8	014174-09-5	[4,4]DB24C6-DIBENZO CROWN ETHER
C24H34N2O4	106848-85-5	1H-IMIDAZOLE-5-BUTANOIC ACID, alpha-ETHYL-1-METH
C24H34N2O4	106848-86-6	1H-IMIDAZOLE-5-BUTANOIC ACID, beta-[(2,2-DIMETHY
C24H34O6	006677-98-1	HYDROCORTISONE-21-PROPIONATE
C24H34O6	065980-97-4	HYDROCORTISONE-17-PROPIONATE
C24H38O4	000117-81-7	BIS(2-ETHYLHEXYL)PHTHALATE
C24H38O4	000117-84-0	DIOCTYL PHTHALATE
C24H38O4	006422-86-2	BIS(2-ETHYLHEXYL) TEREPHTHALATE
C24H38O4	027554-26-3	DIISOOCTYL PHTHALATE
C24H38O5	075330-75-5	LOVASTATIN
C24H40O10	077963-50-9	B30C10-BENZO CROWN ETHER
C24H40O3	083002-04-4	PHENOL, 5-(1,1-DIMETHYLHEPTYL)-2-[5-HYDROXY-2-(3
C24H40O4	000083-44-3	DEOXYCHOLIC ACID
C24H40O4	000083-49-8	HYODEOXYCHOLIC ACID
C24H40O4	000128-13-2	URSODEOXYCHOLIC ACID
C24H40O4	000474-25-9	CHENODEOXYCHOLIC ACID
C24H40O5	000081-25-4	CHOLIC ACID
C24H40O5	000547-75-1	HYOCHOLIC ACID
C24H40O8	005281-13-0	TROPITAL
C24H41NO2	000103-99-1	N-STEAROYL-P-AMINOPHENOL
C24H48O2	005908-87-2	ETHYL DOCOSANOATE
C24H50	000646-31-1	TETRACOSANE
C24H51N	001116-76-3	TRI(N-OCTYL)AMINE
C24H51O4P	000078-42-2	TRIS(2-ETHYLHEXYL) PHOSPHATE
C25H18	017165-86-5	9-(1,1'-BIPHENYL)-4-YL-9H-FLUORENE
C25H20ClF2N3O3	094050-52-9	FLUCYCLOXURON,(E)-ISOMER
C25H20ClF2N3O3	094050-53-0	FLUCYCLOXURON,(Z)-ISOMER
C25H20N2O2	020222-29-1	4-NITRO-N-(TRIPHENYLMETHYL)BENZENAMINE
C25H21ClN2O6	018035-92-2	BENZOIC ACID, 3,4,5-TRIMETHOXY-, 7-CHLORO-2,3-DI
C25H22ClNO3	051630-58-1	FENVALERATE
C25H22ClNO3	066230-04-4	ESFENVALERATE
C25H22N4O7	137057-72-8	H-PURINE, 9-(2,5-DI-O-BENZOYL-BETA-D-ARABINOFURA

C25H22N6O7S	123822-90-2	5-THIA-1-AZABICYCLO[4.2.0]OCT-2-ENE-2-CARBOXYLIC
C25H23N7O9S2	123822-92-4	5-THIA-1-AZABICYCLO[4.2.0]OCT-2-ENE-2-CARBOXYLIC
C25H24N4O10	005967-77-1	THEBAINE PICRATE
C25H25F2N2NaO4	129358-82-3	6-HEPTENOIC ACID, 7-[4,5-BIS(4-FLUOROPHENYL)-2-(
C25H27ClN2O4	092598-90-8	1H-IMIDAZOLE-5-BUTANOIC ACID, beta-[(BENZOYLOXY)
C25H27ClN2O4	092622-09-8	1H-IMIDAZOLE-5-BUTANOIC ACID, beta-[[(3-CHLOROBE
C25H28N2O4	092598-84-0	1H-IMIDAZOLE-5-BUTANOIC ACID, beta-[(BENZOYLOXY)
C25H28O3	080844-07-1	ETOFENPROX
C25H29N9O8S2	062893-19-0	CEFOPERAZONE
C25H30BrFO5	025122-43-4	BROBETASONE-17-PROPIONATE
C25H30ClFO5	025122-56-9	CLOBETASONE-17-PROPIONATE
C25H30ClN3	000548-62-9	CRYSTAL VIOLET (BASIC VIOLET 3)
C25H31ClF2O6	064272-18-0	2-CHLOROFLUMETHASONE-17-PROPIONATE
C25H31ClO5	095796-51-3	CLOBETASONE-17-PROPIONATE-DES-F
C25H31FO5	004351-59-1	CLOBETASONE-17-PROPIONATE-DES-CL
C25H31FO8	000067-78-7	TRIAMCINOLONE DIACETATE
C25H31N3O5	102152-21-6	ETHANONE, 1-[2-[4-[3-[[2-(3,4-DIMETHOXYPHENYL)ET
C25H31N3O6	102152-28-3	1H-IMIDAZOLE-5-CARBOXYLIC ACID, 2-[4-[3-[[2-(3,4
C25H32Cl2O5	025122-49-0	CLOBETASOL-17-PROPIONATE-9-CL
C25H32ClFO5	025122-46-7	CLOBETASOL-17-PROPIONATE
C25H32N2O2	000357-56-2	DEXTROMORAMIDE
C25H32N2O3	061380-40-3	LOFENTANIL
C25H33ClO5	095796-50-2	CLOBETASOL-17-PROPIONATE-DES-F
C25H33FO5	004351-48-8	BETAMETHASONE-17-PROPIONATE-21-DESOXY
C25H33FO6	075883-07-7	BETAMETHASONE-21-PROPIONATE
C25H33N3O5	102152-24-9	2-PROPANOL, 1-[[2-(3,4-DIMETHOXYPHENYL)ETHYL]AMI
C25H33NO4	014521-96-1	ETORPHINE
C25H34O5	041020-77-3	ETHYL MEXRENONE
C25H36O3	123331-83-9	6H-DIBENZO[B,D]PYRAN-9-CARBOXALDEHYDE, 3-(1,1-DI
C25H36O6	000465-15-6	CARD-20(22)-ENOLIDE, 16-(ACETYLOXY)-3,14-DIHYDRO
C25H36O6	006677-99-2	PREGN-4-ENE-3,20-DIONE, 11,17-DIHYDROXY-21-(1-OX
C25H36O6	013609-67-1	HYDROCORTISONE-17-BUTYRATE
C25H38N6O6	130817-97-9	L-PROLINAMIDE, 5-OXO-L-PROLYL-1-[(OCTYLOXY)CARBO
C25H38N6O6	130817-98-0	L-PROLINAMIDE, 5-OXO-L-PROLYL-1-[(2-ETHYLHEXYLOX
C25H38O2	061597-27-1	6H-DIBENZO[B,D]PYRAN-1-OL, 3-(1,1-DIMETHYLHEPTYL
C25H38O3	112830-95-2	6H-DIBENZO[B,D]PYRAN-9-METHANOL, 3-(1,1-DIMETHYL
C25H38O3	123331-84-0	6H-DIBENZO[B,D]PYRAN-9-METHANOL, 3-(1,1-DIMETHYL
C25H39NO2	132213-87-7	1-BUTANAMINE, 4-[(6A,7,10,10A-TETRAHYDRO-6,6,9-T
C25H40O5	079902-63-9	SIMVASTATIN
C25H40O6	121009-77-6	SIMVASTATIN, 5-HYDROXY
C25H42O2	132296-13-0	PHENOL, 5-(1,1-DIMETHYLUNDECYL)-2-(3-HYDROXYCYCL
C25H42O3	132296-15-2	PHENOL, 5-(1,1-DIMETHYLHEPTYL)-2-[5-HYDROXY-2-(4
C25H43N3O6	031088-06-9	2(1H)-PYRIMIDINONE, 4-AMINO-1-[5-O-(1-OXOHEXADEC
C25H50O2	018281-07-7	ETHYL TRICOSANOATE
C25H52	000629-99-2	PENTACOSANE
C26H16	000191-68-4	DIBENZO[G,P]CHRYSENE
C26H16N4Na2O8S2	005850-16-8	C.I. ACID BROWN 14, DISODIUM SALT
C26H18	001499-10-1	9,10-DIPHENYLANTHRACENE
C26H22ClF3N2O3	069409-94-5	FLUVALINATE
C26H23F2NO4	070124-77-5	FLUCYTHRINATE
C26H24N4O8	137057-73-9	9H-PURINE, 9-[2,5-BIS-O-(4-METHOXYBENZOYL)-beta-
C26H25N2NaO6S	026605-69-6	4-THIA-1-AZABICYCLO[3.2.0]HEPTANE-2-CARBOXYLIC A
C26H26N2O6S	035531-88-5	CARBENICILLIN, INDANYL
C26H26N4O3	083158-01-4	24NH2-PYRIMIDIN,5(3MEO-45PHCO)BENZYL
C26H26N4O3	083158-02-5	24NH2-PYRIMIDIN,5(4MEO-35PHCO)BENZYL
C26H26OSi2	000807-28-3	1,3-DIMETHYL-1,1,3,3-TETRAPHENYLDISILOXANE
C26H28Cl2N4O4	065277-42-1	PIPERAZINE, 1-ACETYL-4-[4-[[2-(2,4-DICHLOROPHENY
C26H29N3O6	055985-32-5	3,5-PYRIDINEDICARBOXYLIC ACID, DERIVATIVE

C26H30N2O4	092598-94-2	1H-IMIDAZOLE-5-BUTANOIC ACID, beta-[(BENZOYLOXY)
C26H30N2O5	139733-95-2	ACETAMIDE, 2-[4-[2-(2-HYDROXY-3-PHENOXYPROPYL)AM
C26H32ClFO5	025122-57-0	CLOBETASONE-17-BUTYRATE
C26H32F2O7	000356-12-7	FLUOCINONIDE
C26H33NO6	103890-78-4	3,5-PYRIDINEDICARBOXYLIC ACID, 4-[2-[3-(1,1-DIME
C26H34ClFO5	025122-47-8	CLOBETASOL-17-BUTYRATE
C26H34O6	038196-45-1	PREDISONE-17-VALERATE
C26H35FO6	056933-60-9	BETAMETHASONE-21-BUTYRATE
C26H36O5	041020-79-5	DICIRENONE
C26H36O9	061260-08-0	[2,7]DB27C9-DIBENZO CROWN ETHER
C26H38O6	006678-00-8	HYDROCORTISONE-21-VALERATE
C26H38O6	057524-89-7	HYDROCORTISONE-17-VALERATE
C26H40O3	000315-37-7	TESTOSTERONE ENANTYHATE
C26H42O4	000119-07-3	N-OCTYL N-DECYL PHTHALATE
C26H42O4	028553-12-0	DIISONONYL PHTHALATE
C26H43NO5	000360-65-6	GLYCODEOXYCHOLIC ACID
C26H43NO5	000640-79-9	GLYCOCHENODEOXYCHOLIC ACID
C26H43NO5	013042-33-6	GLYCOHYODEOXYCHOLIC ACID
C26H43NO5	064480-66-6	GLYCOURSODEOXYCHOLIC ACID
C26H43NO5	150423-89-5	TETRADECANAMIDE, 2,2-DIMETHYL-N-METHYL-3-OXO-N-(
C26H43NO6	000475-31-0	GLYCOCHOLIC ACID
C26H43NO6	032747-08-3	GLYCOHYOCHOLIC ACID
C26H44O11	104946-62-5	B33C11-BENZO CROWN ETHER
C26H54	000630-01-3	HEXACOSANE
C27H22Cl2N4	002030-63-9	CLOFAZIMINE
C27H26N4O7	137057-67-1	9H-PURINE, 9-[2,3-BIS-O-(PHENYLACETYL)-beta-D-AR
C27H29ClFN2NaO4	129358-65-2	6-HEPTENOIC ACID,7[4(4-CHLORO-3,5-DIMEPH)-5-(4-F
C27H29NO10	020830-81-3	DAUNOMYCIN
C27H29NO11	023214-92-8	ADRIAMYCIN
C27H32O14	010236-47-2	NARINGINE
C27H33O4P	064532-97-4	4-NONYLPHENYL DIPHENYL PHOSPATE
C27H34N4O	000302-41-0	PIRITRAMIDE
C27H35Cl2FO6	064272-13-5	PREGNA-1,4-DIENE-3,20-DIONE, 2,9-DICHLORO-21-(2,
C27H35NO4	071048-87-8	1,9-PHENANTHRIDINEDIOL, 5,6,6A,7,8,9,10,10A-OCTA
C27H35NO4	074430-97-0	1,9-PHENANTHRIDINEDIOL, 5,6,6A,7,8,9,10,10A-OCTA
C27H36F2O4	015845-96-2	PREGNA-1,4-DIENE-3,20-DIONE, 21-(2,2-DIMETHYL-1-
C27H36F2O6	002002-29-1	FLUMETHASONE-21-PIVALATE
C27H37FO6	002152-44-5	BETAMETHASONE-17-VALERATE
C27H37FO6	002240-28-0	BETAMETHASONE-21-VALERATE
C27H38N2O4	000052-53-9	VERAPAMIL
C27H38O10	088158-44-5	PREDNISOLONE-21-GLUCOSIDE
C27H38O10	092901-24-1	PREGNA-1,4-DIENE-3,20-DIONE, 21-(beta-D-GALACTOP
C27H40O10	092901-25-2	PREGN-4-ENE-3,20-DIONE, 21-(ß-D-GALACTOPYRANOSYL
C27H46O	000057-88-5	CHOLESTEROL
C27H50O6	000538-23-8	TRIOCTONOIN
C27H56	000593-49-7	HEPTACOSANE
C28H12Cl2N2O4	000130-20-1	C.I. VAT BLUE 6
C28H14N2O4	000081-77-6	C.I. VAT BLUE 4
C28H28ClF2N3O	053179-12-7	2H-BENZIMIDAZOL-2-ONE, 1-[1-[4,4-BIS(4-FLUOROPHE
C28H29F2N3O	002062-78-4	PIMOZIDE
C28H32ClN2O3	000081-88-9	RHODAMINE B
C28H32N2O5	139734-30-8	2-PROPANOL, 3-[[2-[4-[2-OXO-2-(1,2,3,4-TETRAHYDR
C28H32O15	000520-27-4	DIOSMINE
C28H32O2Si3	003982-82-9	1,3,3,5-TETRAMETHYL-1,1,5,5-TETRAPHENYLTRISILO*
C28H32O4Si4	000077-63-4	2,4,6,8-TETRAMETHYL-2,4,6,8-TETRAPHENYLCYCLOTE*
C28H36Cl2N2O4	005853-29-2	CEPHAELINE DIHYDROCHLORIDE
C28H36O11	041451-75-6	BRUCEANTIN
C28H37FO7	005593-20-4	BETAMETHASONE 17,21-DIPROPIONATE

C28H37NO4	078513-74-3	1,9-PHENANTHRIDINEDIOL, 5,6,6A,7,8,9,10,10A-OCTA
C28H39FO10	092901-23-0	DEXAMETHASONE-21-GALACTOSIDE
C28H40O10	017455-25-3	[5,5]DB30C10-DIBENZO CROWN ETHER
C28H40O8	107085-84-7	HYDROCORTISON-21-HEMIPIMELATE
C28H46O4	000084-77-5	DIDECYL PHTHALATE
C28H46O4	026761-40-0	DIISODECYL PHTHALATE
C28H58	000630-02-4	OCTACOSANE
C29H19N5Na2O8S2	002610-11-9	DIRECT RED 81
C29H32N4O4	138771-70-7	D-PHENYLALANINAMIDE, N-ACETYL-D-PHENYLALANYL-D-P
C29H32O13	033419-42-0	ETOPSIDE
C29H33F2N3O	003416-26-0	LIDOFLAZINE
C29H40N2O4	000483-18-1	6',7',10,11-TETRAMETHOXYEMETAN
C29H42Cl2N2O4	000316-42-7	EMETINE DIHYDROCHLORIDE
C29H42O10	000508-75-8	CONVALLATOXIN
C29H42O8	114611-36-8	HYDROCORTISON-21-PENTYL-CO-O-ME
C29H42O9	000630-64-8	HELVETICOSIDE
C29H44O10	003253-62-1	CONVALLOTOXOL
C29H44O12	000630-60-4	OUABAIN
C29H44O7	031087-87-3	CARD-20(22)-ENOLIDE, 3-[(2,6-DIDEOXY-3-O-METHYL-
C29H44O8	000508-93-0	EVOMONOSIDE
C29H44O8	000545-27-7	CARD-20(22)-ENOLIDE, 3-[(2,6-DIDEOXY-BETA-D-RIBO
C29H44O8	005352-63-6	CARD-20(22)-ENOLIDE, 3-[(2,6-DIDEOXY-BETA-D-RIBO
C29H44O9	018695-02-8	HELVETICOSOL
C29H60	000630-03-5	NONACOSANE
C30H23BrO4	028772-56-7	BROMADIOLONE
C30H27Cl2N5	111436-12-5	(DES-IPR) 4--PIPERIDINYL-PR-CLOFAZIMINE
C30H29Cl2N5	078182-93-1	(DES-IPR) N,N-DIETAMINOET-CLOFAZIMINE
C30H42O8	000466-06-8	PROSCILLARIDIN
C30H44O11	087586-46-7	[4,7]DB33C11-DIBENZO CROWN ETHER
C30H44O9	000508-77-0	CYMARIN
C30H44O9	001182-87-2	PERUVOSIDE
C30H46O8	000466-07-9	NERIIFOLIN
C30H46O8	018404-43-8	CARD-20(22)-ENOLIDE, 3-[(2,6-DIDEOXY-beta-D-RIBO
C30H46O8	089016-31-9	CARD-20(22)-ENOLIDE, 3-[(2,6-DIDEOXY-3-O-METHYL-
C30H46O9	000465-84-9	CYMAROL
C30H46O9	000595-21-1	STROSPESIDE
C30H50O4	003648-20-2	DIUNDECYL PHTHALATE
C30H52O	000473-03-0	AMBREIN
C30H53NO11	000104-31-4	BENZONATATE
C30H62	000638-68-6	TRIACONTANE
C31H15NO3	003271-76-9	C.I.VAT GREEN 3
C31H29Cl2N5	111435-97-3	(DES-IPR) N-PYRROLIDINYL-PR-CLOFAZIMINE
C31H31Cl2N5	098270-76-9	(DES-IPR) N,N-DIETAMINOPR-CLOFAZIMINE
C31H32N4O2	015301-48-1	BEZITRAMIDE
C31H35NO11	079867-78-0	5,12-NAPHTHACENEDIONE PYRANOSYL DERIVATIVE
C31H35NO12	080790-68-7	5,12-NAPHTHACENEDIONE PYRANOSYL DERIVATIVE
C31H36N2O11	089196-05-4	MORPHOLINO-DOXORUBICIN,12-IMINO-
C31H37NO11	079951-58-9	MORPHOLINO-DAUNORUBICIN,13-DIHYDRO
C31H47NO4	086678-86-6	24-CHOLANAMIDE, 3,7,12-TRIHYDROXY-N-(PHENYLMETHY
C31H64	000630-04-6	HENTRIACONTANE
C32H21N5Na2O6S2	003351-05-1	ACID BLUE 113
C32H22N6Na2O6S2	000573-58-0	CONGO RED
C32H31Cl2N5	111435-98-4	(DES-IPR) N-PIPERIDINYL-PR-CLOFAZIMINE
C32H34N2O11	089164-74-9	5,12-NAPHTHACENEDIONE PYRANOSYL DERIVATIVE
C32H34N2O12	089196-07-6	5,12-NAPHTHACENEDIONE PYRANOSYL DERIVATIVE
C32H34O13S	029767-20-2	TENIPOSIDE
C32H35N3O11	089164-79-4	2-CYANOMORPHOLINO-DOXORUBICIN,12-IMINO
C32H42N6O13S3	130927-06-9	5-THIA-1-AZABICYCLO[4.2.0]OCT-2-ENE-2-CARBOXYLIC

C32H45FO7	038196-44-0	BETAMETHASONE 17,21-DIVALERATE
C32H46O9	055147-64-3	4'-DEHYDROOLEANDRIN
C32H48O9	000465-16-7	OLEANDRIN
C32H48O9	000508-22-5	4'-EPIOLEANDRIN
C32H49NO9	000062-59-9	CEVADINE
C32H54O4	002432-90-8	DIDODECYLPHTHALATE
C32H66	000544-85-4	DOTRIACONTANE
C33H35Cl2N5	111436-11-4	(DES-IPR)NN-DIETAMINO(1ME)BU-CLOFAZIMINE
C33H40N2O9	000050-55-5	RESERPINE
C33H54O6	003319-31-1	TRIS(2-ETHYLHEXYL) TRIMELLITATE
C33H60O7Si8	060617-40-5	1,1,1,3,3,5,7,7,9,11,11,13,15,15,15-PENTADECAM*
C33H68	000630-05-7	TRITRIACONTANE
C34H40O2	132213-93-5	6H-DIBENZO[B,D]PYRAN, 1-([1,1'-BIPHENYL]-4-YLMET
C34H46ClN3O10	035846-53-8	MAYTANSINE
C34H47N5O7S	122224-98-0	2-OXAZOLIDINONE DERIVATIVE
C34H47NO11	000302-27-2	ACONITINE
C34H50N6O7	122224-80-0	2-OXAZOLIDINONE DERIVATIVE
C34H50O10	051008-91-4	ACETYLOLEANDRIN
C34H50O5Si6	013271-58-4	1,1,1,3,5,7,9,11,11,11-DECAMETHYL-3,5,7,9-TETRA*
C34H58O4	000119-06-2	DITRIDECYL PHTHALATE
C34H70	014167-59-0	N-TETRATRIACONTANE
C35H27N5Na2O9S2	006409-90-1	DIRECT YELLOW 62, DISODIUM SALT
C35H45NO4	132213-90-2	1,4-ISOQUINOLINEDIONE DERIVATIVE
C35H50N6O7	122224-71-9	2-OXAZOLIDINONE DERIVATIVE
C35H50N6O7	122224-99-1	2-OXAZOLIDINONE DERIVATIVE
C35H51N7O8	127944-48-3	2-OXAZOLIDINONE DERIVATIVE
C35H51N7O9	127944-49-4	2-OXAZOLIDINONE DERIVATIVE
C35H52F2N6O8S	137302-35-3	L-NORVALINAMIDE DERIVATIVE
C35H52N6O7	127944-51-8	2-OXAZOLIDINONE DERIVATIVE
C35H53N7O9	127944-50-7	2-OXAZOLIDINONE DERIVATIVE
C35H54F2N6O8S	137302-31-9	L-THREO-PENTONAMIDE DERIVATIVE
C35H54O10	016479-50-8	CARD-20(22)-ENOLIDE, 3-[[2,6-DIDEOXY-4-O-(2,6-DI
C35H54O11	005297-05-2	CARD-20(22)-ENOLIDE DERIVATIVE
C35H61N3O14	025423-22-7	ERYTHROMYCIN, 3'-AZIDO-3'-DE(DIMETHYLAMINO)-4'-H
C35H62O3	002082-79-3	OCTADECYL 3,5-BIS(TERT-BUTYL)-4-HYDROXYBENZENEP*
C35H72	000630-07-9	N-PENTATRIACONTANE
C36H27N7Na2O7S2	001937-37-7	C.I. DIRECT BLACK 38
C36H28N2O6	000091-92-9	NAPHTHOL AS-BR
C36H50N4O7S	122225-00-7	2-OXAZOLIDINONE DERIVATIVE
C36H52N6O7	131067-26-0	2-OXAZOLIDINONE, 5-[[2-BENZYL-3-(4-MORPHOLINYLCA
C36H54F2N6O9S	137302-63-7	L-NORVALINAMIDE DERIVATIVE
C36H54N6O8	127944-53-0	2-OXAZOLIDINONE DERIVATIVE
C36H55F2N7O7S	137429-62-0	L-NORVALINAMIDE DERIVATIVE
C36H65NO13	000992-62-1	ERYTHROMYCIN, N-DEMETHYL-
C36H65NO13	001675-02-1	ERYTHROMYCIN, 3"-O-DEMETHYL-
C36H74	000630-06-8	HEXATRIACONTANE
C37H67NO12	000527-75-3	ERYTHROMYCIN, 12-DEOXY-
C37H67NO13	000114-07-8	ERYTHROMYCIN
C37H67NO14	025405-65-6	ERYTHROMYCIN, 4'-HYDROXY-
C37H76	007194-84-5	N-HEPTATRIACONTANE
C38H58N6O9	122224-84-4	2-OXAZOLIDINONE DERIVATIVE
C38H59N7O10	127944-54-1	2-OXAZOLIDINONE DERIVATIVE
C38H66NO13	031357-44-5	ERYTHROMYCIN, 12-DEOXY-, 11-ACETATE
C38H67NO13	031357-41-2	ERYTHROMYCIN, 12-DEOXY-, 4"-FORMATE
C38H67NO14	031357-16-1	ERYTHROMYCIN, 4"-FORMATE
C38H72N2O12	083905-01-5	AZITHROMYCIN
C38H78	007194-85-6	N-OCTATRIACONTANE
C39H58N6O10	122224-83-3	2-OXAZOLIDINONE DERIVATIVE

C39H59N7O11	127944-55-2	2-OXAZOLIDINONE DERIVATIVE
C39H69NO13	031357-18-3	ERYTHROMYCIN, 12-DEOXY-, 4"-ACETATE
C39H69NO14	023893-07-4	ERYTHROMYCIN,4"-ACETATE
C40H42O3Si4	038421-40-8	1,3,5,7-TETRAMETHYL-1,1,3,5,7,7-HEXAPHENYLTETR*
C40H64O16	121284-20-6	[8,8]DB48C16-DIBENZO CROWN ETHER
C40H82	004181-95-7	TETRACONTANE
C41H64O13	000071-63-6	DIGITOXIN
C41H64O14	004562-36-1	GITOXIN
C41H64O14	020830-75-5	DIGOXIN
C41H66O13	003786-76-3	DIGITOXIN, DIHYDRO-
C41H66O14	005297-10-9	DIGOXIN (NON-OLEFINIC BOND)
C42H62O16	001405-86-3	GLYCYRRHIZIC ACID
C42H66O14	030685-43-9	CARD-20(22)-ENOLIDE DERIVATIVE
C42H66O14	031962-94-4	DIGOXIN, ALPHA-METHYL
C42H68O13	055692-91-6	DIGITOXIN, BETA-METHYL
C42H69NO15	056689-45-3	JOSAMYCIN
C42H69NO15	074014-51-0	ROKITAMYCIN
C43H66N6O12	127944-57-4	2-OXAZOLIDINONE DERIVATIVE
C43H66O15	005511-98-8	DIGOXIN ESTER DERIVATIVE
C43H66O15	020991-71-3	CARD-20(22)-ENOLIDE DERIVATIVE
C43H67N7O13	127944-56-3	2-OXAZOLIDINONE DERIVATIVE
C43H68O14	051740-66-0	DIGOXIN,ALPHA-BETA-DIMETHYL-
C44H68O15	051740-69-3	DIGOXIN,12-ACETYL-BETA-METHYL-
C44H75N5O7	130336-12-8	3-AZAGLUTARAMIDE ANALOG
C44H90	007098-22-8	TETRATETRACONTANE
C45H68O16	040246-14-8	ALPHA-BETA-DIACETYLDIGOXIN
C45H73NO15	020562-02-1	ALPHA-SOLANINE
C60H122	007667-80-3	N-HEXACONTANE
CN4O8	000509-14-8	TETRANITROMETHANE
CO2	000124-38-9	CARBON DIOXIDE
COS	000463-58-1	CARBONYL SULFIDE
CS2	000075-15-0	CARBON DISULFIDE
	008001-35-2	TOXAPHENE
	008032-32-4	LIGROIN

APPENDIX II

5O-CS-46	002597-93-5
ABATE	003383-96-8
AC 263,222 HERBICIDE	104098-48-8
AC 303,630	122453-73-0
AC-ALA-PHE-ALA-NTBU	137605-63-1
AC-ALA-TRP-ALA-NTBU	137605-64-2
1-AC-6-DEME-7-METHOXYMITOSENE	096000-43-0
ACEBUTOLOL	037517-30-9
ACENAPHTHENE	000083-32-9
ACENAPHTHYLENE	000208-96-8
ACENAPHTHYLENE, 1,2-DIHYDRO-5-NITRO-	000602-87-9
1,2-ACENAPHTHYLENEDIONE	000082-86-0
ACENOCOUMARIN	000152-72-7
ACEPHATE	030560-19-1
ACETALDEHYDE	000075-07-0
ACETALDEHYDE OXIME	000107-29-9
1-(MES)-ACETALDEHYDE OXIME	013749-94-5
ACETAMIDE, 2-(BENZOYLOXY)-N,N-BIS(2-METHOXYETHYL	115178-64-8
ACETAMIDE, 2-(BENZOYLOXY)-N,N-BIS(2-HYDROXYETHYL	106231-61-2
ACETAMIDE	000060-35-5
ACETAMIDE, 2-(BENZOYLOXY)-N,N-BIS(2-METHYLPROPYL	115193-33-4
ACETAMIDE, 2-(BENZOYLOXY)-N,N-BIS(1-METHYLETHYL)	106231-56-5
ACETAMIDE, N-ACETYL-2-(BENZOYLOXY)-	068659-48-3
ACETAMIDE, N-ACETYL-2-(BENZOYLOXY)-N-METHYL-	115178-80-8
ACETAMIDE, N-[3-ACETYL-4-[2-HYDROXY-3-[(1-METHYL	028197-69-5
ACETAMIDE, N-[5,8-BIS(ACETYLOXY)-6,7-DIMETHOXY-1	091814-31-2
ACETAMIDE, 2-(ACETYLPHENYLAMINO)-N-PHENYL-	074339-88-1
ACETAMIDE, N-(4-ACETYLPHENYL)-	002719-21-3
ACETAMIDE, 2-AMINO-N-(2,6-DIMETHYLPHENYL)-	018865-38-8
ACETAMIDE, N-(2-AMINO-2-OXOETHYL)-2-(BENZOYLOXY)	106231-62-3
ACETAMIDE, N-(2-AMINO-2-OXOETHYL)-2-DIAZO-	000817-99-2
ACETAMIDE, N-[4-(AMINOSULFONYL)PHENYL]	000121-61-9
ACETAMIDE, N-[2-(AMINOSULFONYL)-6-BENZOTHIAZOLYL	096474-19-0
ACETAMIDE, 2-(BENZOYLAMINO)-N-ETHYL-	120399-50-0
ACETAMIDE, 2-(BENZOYLOXY)-N,N-BIS(2-HYDROXYPROPY	115178-63-7
ACETAMIDE, 2-(BENZOYLOXY)-N-ETHYL-N-(2-HYDROXYET	106231-60-1
ACETAMIDE, 2-(BENZOYLOXY)-N-(1-METHYLETHYL)-	115193-27-6
ACETAMIDE, 2-(BENZOYLOXY)-N-METHYL-	106231-50-9
ACETAMIDE, 2-(BENZOYLOXY)-	064649-43-0
ACETAMIDE, 2-[[(BENZOYLOXY)ACETYL]AMINO]-	106231-53-2
ACETAMIDE, 2-(BENZOYLOXY)-N,N-BIS(2-PROPENYL)-	106231-58-7
ACETAMIDE, 2-(BENZOYLOXY)-N,N-DIBUTYL-	106231-57-6
ACETAMIDE, 2-(BENZOYLOXY)-N,N-DIMETHYL-	106231-54-3
ACETAMIDE, 2-(BENZOYLOXY)-N-(1,1-DIMETHYLETHYL)-	106231-52-1
ACETAMIDE, 2-(BENZOYLOXY)-N,N-DIPROPYL-	106231-55-4
ACETAMIDE, 2-(BENZOYLOXY)-N-PROPYL-	106231-51-0
ACETAMIDE, 2-(BENZOYLOXY)-N-[2-HYDROXY-1,1-BIS(H	115193-31-2
ACETAMIDE, 2-(BENZOYLOXY)-N-(2-HYDROXYETHYL)-N-M	106231-59-8
ACETAMIDE, 2-(BENZOYLOXY)-N-HEXYL-	115193-29-8
ACETAMIDE, 2-(BENZOYLOXY)-N-ETHYL-	064649-57-6
ACETAMIDE, 2-(BENZOYLOXY)-N-ETHYL-N-METHYL-	115193-32-3
ACETAMIDE, 2-(BENZOYLOXY)-N-BUTYL-	115193-28-7
ACETAMIDE, N-[2-(BENZOYLOXY)ETHYL]-N-METHYL-	057440-16-1
ACETAMIDE, N-(4-BROMO-3,5-DIMETHYLPHENYL)-	064835-48-9
ACETAMIDE, N-(3-BROMOPHENYL)-2-PHENYL-	013140-73-3
ACETAMIDE, N-(4-BROMOPHENYL)-2-PHENYL-	007495-11-6
ACETAMIDE, N-(2-BUTOXY-5-NITROPHENYL)-	120309-36-6
ACETAMIDE, 2-CHLORO-N-(2-CYCLOPENTYL-2H-BENZOTRI	130787-07-4

ACETAMIDE, 2-CHLORO-N-(3,4-DICHLOROPHENYL)   020149-84-2
ACETAMIDE, N-(5-CHLORO-2,3-DIHYDRO-3-BENZOFURANY   105801-50-1
ACETAMIDE, N-[5-CHLORO-6-[(2,3-DIHYDROXYPROPYL)A   089083-18-1
ACETAMIDE, N-[5-CHLORO-6-(4-METHYL-1-PIPERAZINYL   089083-22-7
ACETAMIDE, 2-CHLORO-N-[2-OXO-1-(PHENYLAMINO)ETHY   074339-86-9
ACETAMIDE, N-[1-(4-CHLOROPHENYL)-2,5-DIOXO-3-PYR   037107-20-3
ACETAMIDE, N-(4-CHLOROPHENYL)-2-PHENYL-   002990-06-9
ACETAMIDE, N-(5-CHLOROPYRAZINYL)-   136309-12-1
ACETAMIDE, 2-CYANO-N-(3,4-DICHLOROPHENYL)-   015386-80-8
ACETAMIDE, N-(3-CYANOPHENYL)-   058202-84-9
ACETAMIDE, N-(4-CYANOPHENYL)-   035704-19-9
ACETAMIDE, N-(2-CYANOPHENYL)-   025116-00-1
ACETAMIDE, N-(2-CYCLOPENTYL-2H-BENZOTRIAZOL-5-YL   130787-06-3
ACETAMIDE, N-[3-[[3-(4,6-DIAMINO-2,2-DIMETHYL-1,   079508-86-4
ACETAMIDE, 2-[(DIAZOACETYL)AMINO]-N-METHYL-   038726-90-8
ACETAMIDE, 2-(DIBUTYLAMINO)-N-(2,3-DIHYDRO-3-BEN   105801-66-9
ACETAMIDE, 2-(DIBUTYLAMINO)-N-(2,3-DIHYDRO-6-MET   105801-60-3
ACETAMIDE, 2-(DIBUTYLAMINO)-N-(2,3-DIHYDRO-5-MET   105801-56-7
ACETAMIDE, 2-(DIBUTYLAMINO)-N-(5-CHLORO-2,3-DIHY   105801-51-2
ACETAMIDE, 2,2-DICHLORO-   000683-72-7
ACETAMIDE, 2,2-DICHLORO-N-(2-CYCLOPENTYL-2H-BENZ   130787-08-5
ACETAMIDE, 2,2-DICHLORO-N,N-DI-2-PROPENYL-   037764-25-3
ACETAMIDE, 2,2-DICHLORO-N-[2-OXO-2-(PHENYLAMINO)   074350-90-6
ACETAMIDE, N-(3,4-DICHLOROPHENYL)-2-(1-PIPERIDIN   022010-25-9
ACETAMIDE, N-(3,4-DICHLOROPHENYL)-2,2,2-TRIFLUOR   064694-85-5
ACETAMIDE, N-(3,4-DICHLOROPHENYL)-2-METHOXY-   094497-46-8
ACETAMIDE, N-[4-[(DICYANOMETHYLENE)HYDRAZONO]PHE   055121-29-4
ACETAMIDE, 2-(DIETHYLAMINO)-N-(2,3-DIHYDRO-5-MET   105801-55-6
ACETAMIDE, 2-(DIETHYLAMINO)-N-(PHENYLMETHYL)-   072336-16-4
ACETAMIDE, N-(2,3-DIHYDRO-3-BENZOFURANYL)-2-(DIE   105801-65-8
ACETAMIDE, N-[6,9-DIHYDRO-9-[(2-HYDROXYETHOXY)ME   110104-37-5
ACETAMIDE, N-(3,6-DIHYDRO-1(2H)-PYRIDINYL)-   035990-32-0
ACETAMIDE, N-[5-CHLORO-6-[[2,3-DIHYDROXY-1-(HYDR   089690-74-4
ACETAMIDE, N-[4-[(DIMETHYLAMINO)SULFONYL]PHENYL]   074746-42-2
ACETAMIDE, N-[2-[3-[(1,1-DIMETHYLETHYL)AMINO]-2-   039617-16-8
ACETAMIDE, N-[4-(1,1-DIMETHYLETHYL)PHENYL]-   020330-45-4
ACETAMIDE, N-(2,3-DIMETHYL-8-NITRO-4-OXO-4H-1-BE   062100-83-8
ACETAMIDE, N-(2,6-DIMETHYLPHENYL)-2-(ETHYLAMINO)   007728-40-7
ACETAMIDE, 2-(DIPROPYLAMINO)-N-(9H-FLUOREN-9-YL)   072336-22-2
ACETAMIDE, 2-(DIPROPYLAMINO)-N-(PHENYLMETHYL)-   072336-17-5
ACETAMIDE, N-(2-ETHOXYPHENYL)-   000581-08-8
ACETAMIDE, N-(7-ETHYL-9-OXO-9H-THIOXANTHEN-3-YL)   074134-18-2
ACETAMIDE, N-(3-FLUOROPHENYL)-2-PHENYL-   005215-27-0
ACETAMIDE, N-HYDROXY-   000546-88-3
ACETAMIDE, N-[4-[2-HYDROXY-3-[(2-HYDROXY-3-PHENO   120908-63-6
ACETAMIDE, N-[[3-[2-HYDROXY-3-[(1-METHYLETHYL)AM   034190-59-5
ACETAMIDE, N-[3-[2-HYDROXY-3-[(1-METHYLETHYL)AMI   006673-38-7
ACETAMIDE, N-(4-HYDROXY-6-METHYL-3-PYRIDAZINYL)-   018591-79-2
ACETAMIDE, 2-[[3-[2-HYDROXY-3-(1-METHYLETHYL)AMI   034918-50-8
ACETAMIDE, 2-[4-[2-HYDROXY-3-[(1-METHYLETHYL)AMI   034939-39-4
ACETAMIDE, 2-[4-[2-[(2-HYDROXY-3-PHENOXYPROPYL)A   115656-32-1
ACETAMIDE, 2-[4-[2-[(2-HYDROXY-3-PHENOXYPROPYL)A   139733-77-0
ACETAMIDE, N-(4-HYDROXYPHENYL)   000103-90-2
ACETAMIDE, 2-[[[4-[2-(2-HYDROXY-3-PHENOXY)PROPYL   139733-59-8
ACETAMIDE, 2-[4-[2-(2-HYDROXY-3-PHENOXYPROPYL)AM   139733-95-2
ACETAMIDE, 2-[4-[[2-(2-HYDROXY-3-PHENOXYPROPYL)A   139733-55-4
ACETAMIDE, 2-[[4-[2-(2-HYDROXY-3-PHENOXYPROPYL)A   139733-61-2
ACETAMIDE, N-[2-(1H-INDOL-3-YL)ETHYL]-   001016-47-3

ACETAMIDE, N-(3-IODOPHENYL)-	019230-45-6
ACETAMIDE, N-(2-METHOXY-5-NITROPHENYL)-	033721-54-9
ACETAMIDE, N-(6-METHOXYPYRAZINYL)-	136309-06-3
ACETAMIDE, N- (METHYLAMINO)CARBONYL -	000623-59-6
ACETAMIDE, N-[2-(1-METHYLETHYL)PHENYL]-	019246-04-9
ACETAMIDE, N-(1-METHYLETHYL)-N-PROPYL-	098486-60-3
ACETAMIDE, N-METHYL-N-[(4-METHYLPHENYL)SULFONYL]	016697-83-9
ACETAMIDE, N-METHYL-N-[5-[(5-NITRO-2-FURANYL)MET	052661-43-5
ACETAMIDE, N-(7-METHYL-9-OXO-9H-THIOXANTHEN-3-YL	074134-17-1
ACETAMIDE, N-(2-METHYLPHENYL)-	000120-66-1
ACETAMIDE, N-(6-METHYLPYRAZINYL)-	005594-16-1
ACETAMIDE, N-[4-(METHYLSULFONYL)PHENYL]	022821-80-3
ACETAMIDE, N-[4-(METHYLTHIO)PHENYL]	010352-44-0
ACETAMIDE, N-[3-(METHYLTHIO)PHENYL]-	002524-78-9
ACETAMIDE, N-[2-(METHYLTHIO)PHENYL]-	006310-41-4
ACETAMIDE, N-(7-NITRO-9H-FLUOREN-2-YL)-	072570-99-1
ACETAMIDE, N-[6-(5-NITRO-2-FURANYL)-4-OXO-4H-1,3	052661-54-8
ACETAMIDE, N-[5-[(5-NITRO-2-FURANYL)METHYLENE]-4	052661-42-4
ACETAMIDE, N-[5-[(5-NITRO-2-FURANYL)METHYLENE]-4	052661-68-4
ACETAMIDE, N-(5-NITRO-2-PROPOXYPHENYL)-	000553-20-8
ACETAMIDE, N-(9-OXO-7-PROPYL-9H-THIOXANTHEN-3-YL	074134-19-3
ACETAMIDE, N-(9-OXO-9H-THIOXANTHEN-3-YL)-, S,S-D	074134-16-0
ACETAMIDE, N-(4-PHENOXYPHENYL)-	006312-87-4
ACETAMIDE, N-[(PHENYLAMINO)CARBONYL]-	000102-03-4
ACETAMIDE, N,N'-(1,2-PHENYLENE)BIS-	002050-85-3
ACETAMIDE, N-(2-PHENYLETHYL)-	000877-95-2
ACETAMIDE, N-PHENYL-N-(4-METHYLPHENYL)-	006876-65-9
ACETAMIDE, N-(4-PROPYLPHENYL)-	020330-99-8
ACETAMIDE, N-5-PYRIMIDINYL-	045810-14-8
ACETAMIDE, N-4-PYRIMIDINYL-	016166-22-6
ACETAMIDE, N-(5-QUINOLINYL)-	042464-80-2
ACETAMIDE, N-(1-beta-D-ARABINOFURANOSYL-1,2-DIHY	013491-47-9
ACETAMIDE, N-2-THIAZOLYL-	002719-23-5
ACETAMIDE, THIOPHEN-2-YL	004461-29-4
ACETAMIDE, N,N'-1H-1,2,4-TRIAZOLE-3,5-DIYLBIS[2,	080616-56-4
ACETAMIDE, 2,2,2-TRICHLORO-N-(2-CYCLOPENTYL-2H-B	130787-09-6
ACETAMIDE, N-[4-(TRIFLUOROMETHYL)PHENYL]-	000349-97-3
1-ACETAMIDOAZULENE	023702-21-8
2-ACETAMIDOCARBAZOLE	063020-20-2
P-ACETAMIDO-N,N-DIMETHYLANILINE	007463-28-7
P-ACETAMIDO-BENZENESO2-FLUORIDE	000329-20-4
2-ACETAMIDOISONIAZID	058481-02-0
2-ACETAMIDOMETHYLISONIAZID	058481-03-1
M-ACETAMIDOPHENYLHIPPURATE	081592-09-8
4-ACETAMIDOPYRIDINE	005221-42-1
2-ACETAMIDOPYRIDINE	005231-96-9
2-ACETAMIDOPYRIMIDINE	013053-88-8
4-ACETAMIDOQUINOLINE-1-OXIDE	023484-11-9
2-ACETAMIDOTHIOPHENE	013053-81-1
3-ACETAMIDOTHIOPHENE	013781-66-3
3-ACETAMINOPYRIDINE	005867-45-8
6-ACETAMINOQUINOLINE	022433-76-7
ACETANILIDE	000103-84-4
ACETANILIDE,N-(4-CYANOPHENYL)-	089246-38-8
4"-ACETATE ERYTHROMYCIN	023893-07-4
ACETAZOLEAMIDE	000059-66-5
ACETIC ACID	000064-19-7
ACETIC ACID, 2-[[5-(AMINOSULFONYL)-2-THIENYL]SUL	135832-36-9

ACETIC ACID, 3-[[5-(AMINOSULFONYL)-2-THIENYL]SUL	104438-04-2
ACETIC ACID, 3-[4-(AMINOSULFONYL)PHENYL]PROPYL E	135832-47-2
ACETIC ACID, 5-[4-(AMINOSULFONYL)PHENYL]PENTYL E	135832-51-8
ACETIC ACID, [(4H-1,2,4-BENZOTHIADIAZIN-3-YL)THI	114282-93-8
ACETIC ACID, [(7-BROMO-4H-1,2,4-BENZOTHIADIAZIN-	114260-74-1
ACETIC ACID,O-CF3 PHENYL ESTER	003038-48-0
ACETIC ACID, [(6-CHLORO-4H-1,2,4-BENZOTHIADIAZIN	114260-70-7
ACETIC ACID, [(7-CHLORO-4H-1,2,4-BENZOTHIADIAZIN	114260-71-8
ACETIC ACID, ((3-CHLOROPHENYL)HYDRAZONO)CYANO-,	036874-69-8
ACETIC ACID, CYANO (2,4,5-TRICHLOROPHENYL) HYDRA	036905-04-1
ACETIC ACID, [4-[(2,4-DIAMINO-5-PYRIMIDINYL)METH	083166-78-3
ACETIC ACID, [(5,7-DIBROMO-4H-1,2,4-BENZOTHIADIA	114260-75-2
ACETIC ACID, [(6,7-DICHLORO-4H-1,2,4-BENZOTHIADI	114260-73-0
ACETIC ACID, [(5,7-DICHLORO-4H-1,2,4-BENZOTHIADI	114260-72-9
ACETIC ACID, 2,6-DICHLOROPHENYL ESTER	028165-71-1
ACETIC ACID, 1-(3,4-DICHLOROPHENYL)-2,2,2-TRICHL	021757-82-4
ACETIC ACID, DIFLUORO-, ETHYL ESTER	000454-31-9
ACETIC ACID, 4-[2-(DIMETHYLAMINO)ETHOXY]-2-METHY	101126-55-0
ACETIC ACID, 3-ETHYLPHENYL ESTER	003056-60-8
ACETIC ACID, 2-HEPTEN-4,6-DIYNYL ESTER	020252-43-1
ACETIC ACID HYDRAZIDE	001068-57-1
ACETIC ACID, (1-NAPHTHALENYLOXY)-	002976-75-2
ACETIC ACID, [(6-METHOXY-4H-1,2,4-BENZOTHIADIAZI	114260-77-4
ACETIC ACID, 3-METHOXYPHENYL ESTER	005451-83-2
ACETIC ACID, [(6-METHYL-4H-1,2,4-BENZOTHIADIAZIN	114260-76-3
ACETIC ACID, [3-METHYL-4-(METHYLSULFONYL)PHENOXY	015267-77-3
ACETIC ACID,2-METHYL-5-NITROIMIDAZOL-2-YL,METHYL	001013-51-0
ACETIC ACID, [4-(METHYLSULFONYL)PHENOXY]-	042288-41-5
ACETIC ACID, (1-NAPHTHALENYLOXY)-, ETHYL ESTER	041643-81-6
ACETIC ACID, (1-NAPHTHALENYLOXY)-, SODIUM SALT	082609-03-8
ACETIC ACID, 2-NAPHTHYLOXY-	000120-23-0
ACETIC ACID, NITROBENZYL ESTER	000619-90-9
ACETIC ACID, [4-OXO-5-(1-PIPERIDINYL)-2-THIAZOLI	064674-86-8
ACETIC ACID, PHENYLY THIO	085033-96-1
ACETIC ACID, SODIUM SALT	000127-09-3
ACETIC ACID, 2,2,2-TRICHLORO-1-(4-METHOXYPHENYL)	063253-20-3
ACETIC ACID, [[6-(TRIFLUOROMETHYL)-4H-1,2,4-BENZ	114260-78-5
ACETIC ANHYDRIDE	000108-24-7
ACETOACETANILIDE	000102-01-2
ACETOACETIC ACID	000541-50-4
ACETOACETIC ESTER	000141-97-9
ACETOCHLOR	034256-82-1
ACETOIN	000513-86-0
ACETONE	000067-64-1
ACETONE CYANOHYDRIN	000075-86-5
ACETONITRILE	000075-05-8
ACETOPHENONE	000098-86-2
ACETOPHENONE, O-ACETYLAMINO	005234-26-4
ACETOPHENONE,O-CYANOMETHYLAMINO	088203-04-7
ACETOPHENONE,O-DICL ACETYLAMINO	006140-12-1
ACETOPHENONE, O-(2-OHET)AMINO	040673-68-5
P-ACETOXYACETANILIDE	002623-33-8
M-ACETOXYACETANILIDE	006317-89-1
P-ACETOXYACETOPHENONE	013031-43-1
P-ACETOXYBENZAMIDE	051074-95-4
4-ACETOXYBENZOIC ACID	002345-34-8
4-ACETOXY DICHLOROACETANILIDE	003957-74-2
2(1-ACETOXYET-13THIAZOLIDINYLIDEN-2AM)5NO2-THIAZ	037427-69-3

2-NO2-5-DI-(ACETOXYMETHYL)FURANE	000092-55-7
1-ACETOXYNAPHTHALENE	000830-81-9
1-ACETOXYPENTACHLOROCYCLOHEXANE	060423-87-2
ACETYLACETONE,4-BROMOBENZAL	015795-19-4
ACETYLACETONE,4-CHLOROBENZAL	019411-75-7
ACETYLACETONE,2,4-DIMEO BENZAL	015725-18-5
ACETYLACETONE,2-METHYLBENZAL	019411-71-3
M-ACETYLACETOPHENONE	006781-42-6
P-ACETYLACETOPHENONE	001009-61-6
L-ALANINAMIDE, N2-(N-ACETYL-L-ALANYL-L-ASPARAGYL	137605-65-3
M-ACETYLAMINOACETOPHENONE	007463-31-2
P-N-ACETYLAMINOBENZAMIDE	058202-83-8
N-ACETYL O-AMINOBENZOIC ACID	000089-52-1
3-ACETYLAMINOBENZOIC ACID	000587-48-4
A-ACETYLAMINO-N-SEC-BUTYL CINNAMAMIDE	097871-54-0
2-ACETYLAMINOPYRAZINE	021352-21-6
4-ACETYLAMINOPYRAZOLE	006647-92-3
3-ACETYLAMINOQUINOLINE	005417-50-5
8-ACETYLAMINOQUINOLINE	033757-42-5
7-ACETYLAMINOQUINOLINE	036164-42-8
P-ACETYL AMPHENICOL	014433-36-4
1-ACETYLAZULENE	007206-57-7
P-ACETYLBENZENESULFONAMIDE	001565-17-9
O-ACETYLBENZOIC ACID	000577-56-0
P-ACETYLBENZOIC ACID	000586-89-0
3-ACETYL BENZOTHIOPHENE	001128-05-8
4-ACETYLBIPHENYL	000092-91-1
ACETYL BROMIDE	000506-96-7
1-ACETYL-4-(3-BR PHENYL) SEMICARBAZIDE	111075-91-3
1-ACETYL-4-(4-BR PHENYL) SEMICARBAZIDE	002706-22-1
1-ACETYL-4-(4-T-BU PHENYL) SEMICARBAZIDE	111075-87-7
M-ACETYL CF3-METHANESULFONANILIDE	023375-11-3
P-ACETYL CF3-METHANESULFONANILIDE	023383-94-0
ACETYL-4-(3-CF3 PHENYL) SEMICARBAZIDE	111075-94-6
ACETYLCHLORIDE	000075-36-5
4-ACETYL-7-CHLOROQUINOLINE	089770-25-2
ACETYLCHOLINE BROMIDE	000066-23-9
3-ACETYLCINNOLINE	089770-37-6
1-ACETYL-4-(4-CL PHENYL) SEMICARBAZIDE	017075-31-9
1-ACETYL-4-(3-CL PHENYL) SEMICARBAZIDE	111075-90-2
1-ACETYL-4-(4-CYANO PHENYL)SEMICARBAZIDE	111076-02-9
N-ACETYLCYCLOBUTANECARBOXAMIDE	006932-05-4
1-ACETYL-4(4-DIMEAMINOPH) SEMICARBAZIDE	111075-98-0
1-ACETYL-6-DEMETHYL-6-CHLORO-7-METHOXYMITOSENE	096000-44-1
P-ACETYL-N,N-DIMETHYLANILINE	002124-31-4
ACETYLENE	000074-86-2
P-ACETYLFORMANILIDE	041656-75-1
1-ACETYL-4-(3-F PHENYL) SEMICARBAZIDE	111075-88-8
1-ACETYL-4-(4-F PHENYL) SEMICARBAZIDE	111075-89-9
2-ACETYLFURAN	001192-62-7
3-ACETYL INDOLE	025314-91-4
ACETYL IODIDE	000507-02-8
1-ACETYL-4-(4-IODO PHENYL) SEMICARBAZIDE	111075-93-5
1-ACETYL-4-(3-IODO PHENYL) SEMICARBAZIDE	111075-92-4
1-ACETYL-4-(3-ME PHENYL) SEMICARBAZIDE	111075-86-6
1-ACETYL-4-(4-ME PHENYL) SEMICARBAZIDE	017075-29-5
ACETYLMETHADOL	000509-74-0
1-(ACETYLMETHYL)-2-NO2 IMIDAZOLE	071006-75-2

2-ACETYL-4-METHYLPHENOL	001450-72-2
1-ACETYL-1-METHYL-2-PHENYLHYDRAZINE	038604-70-5
1-ACETYL-4-(4-NO2 PHENYL) SEMICARBAZIDE	111076-01-8
1-ACETYL-4-(3-NO2 PHENYL) SEMICARBAZIDE	111076-00-7
1-ACETYL-4-(4-OET PHENYL) SEMICARBAZIDE	111075-96-8
1-ACETYL-4-(4-OH PHENYL) SEMICARBAZIDE	111075-99-1
ACETYLOLEANDRIN	051008-91-4
1-ACETYL-4-(4-OME PHENYL) SEMICARBAZIDE	017075-30-8
1-ACETYL-4-(3-OME PHENYL) SEMICARBAZIDE	111075-95-7
1-ACETYL-4-(4-OPH PHENYL) SEMICARBAZIDE	111075-97-9
4-ACETYLOXYAMINOQUINOLINE	032654-59-4
8-ACETYLOXYQUINOLINE	002598-29-0
M-ACETYLPHENOXYACETIC ACID	001878-80-4
P-ACETYLPHENOXYACETIC ACID	001878-81-5
O-ACETYLPHENOXYACETIC ACID	001878-62-2
2-ACETYLPHENYLDIMETHYLCARBAMATE	029230-99-7
3-ACETYLPHENYLDIMETHYLCARBAMATE	002689-47-6
1(4-ACETYLPHENYL)-3-MEO-3-ME UREA	088132-23-4
1-ACETYL-2-PICOLINOYL HYDRAZINE	017433-31-7
ACETYLPYRAZINE	022047-25-2
2-ACETYLPYRIDINE	001122-62-9
4-ACETYLPYRIDINE	028356-58-3
4-ACETYLPYRIDINE	001122-54-9
3-ACETYLPYRIDINE	000350-03-8
2-ACETYLPYRROLE	001072-83-9
6-ACETYLQUINOLINE	073013-68-0
3-ACETYLQUINOLINE	033021-53-3
ACETYLSALICYLIC ACID	000050-78-2
ACETYLSALICYLIC ACID, METHYL ESTER	000580-02-9
3-ACETYLTHIOPHENE	001468-83-3
2-ACETYLTHIOPHENE	000088-15-3
1-ACETYL-2-THIOUREA	000591-08-2
N-ACETYL-L-TYROSINE	000537-55-3
ACID BLUE 113	003351-05-1
ACIDINE	000590-46-5
ACID ORANGE 6	000547-57-9
ACID RED 18	002611-82-7
ACIFLUORFEN-ME	050594-67-7
ACLONIFEN	074070-46-5
ACONITINE	000302-27-2
ACRIDINE	000260-94-6
ACRIDINE, HYDROCHLORIDE	017784-47-3
ACROLEIN	000107-02-8
ACRYLAMIDE	000079-06-1
ACRYLIC ACID	000079-10-7
ACRYLONITRILE	000107-13-1
ACRYLOPHENONE	000768-03-6
ACYCLOVIR	059277-89-3
ADAMANTYL-ALANIN-AMIDE,N-ACETYL	088530-27-2
ADAMANTYLALANINE	070873-69-7
3(ADAMANTYL)-1(2CLET)-1-NO UREA	014039-10-2
ADAMANTYL DERIVATIVE	120282-85-1
ADENINE	000073-24-5
ADENINE,9-(1-OHME-PROPYL)	010521-52-5
ADENOSINE	000058-61-7
ADENOSINE, 2-AMINO-3'-AZIDO-2',3'-DIDEOXY-	114753-52-5
ADENOSINE, 2-AMINO-2'-DEOXY-	004546-70-7
ADENOSINE, 2-AMINO-2',3'-DIDEHYDRO-2',3'-DIDEOXY	109881-25-6

ADENOSINE, 2-AMINO-2',3'-DIDEOXY-   107550-73-2
ADENOSINE, 2-AMINO-2',3'-DIDEOXY-3'-FLUORO-   114753-53-6
ADENOSINE, 3'-AZIDO-2',3'-DIDEOXY-   066323-44-2
ADENOSINE, N-BENZOYL-   004546-55-8
ADENOSINE, N-BENZOYL-2'-DEOXY-   004546-72-9
ADENOSINE, CYCLIC 3',5'-(HYDROGEN PHOSPHATE)   000060-92-4
ADENOSINE, 2'-DEOXY-   000958-09-8
ADENOSINE, 2',3'-DIAZIDO-2',3'-DIDEOXY-   119644-21-2
ADENOSINE, 2',3'-DIDEHYDRO-2',3'-DIDEOXY-   007057-48-9
ADENOSINE, 2',3'-DIDEOXY-   004097-22-7
ADENOSINE, 2',3'-DIDEOXY-2'-FLUORO-   110143-05-0
ADENOSINE, 2',3'-DIDEOXY-3'-FLUORO-   087418-35-7
ADIPONITRILE   000111-69-3
ADRIAMYCIN   023214-92-8
AGROX   002279-64-3
AJMALINE   004360-12-7
ALACHLOR   015972-60-8
L-ALANINAMIDE, N-ACETYL-L-ALANYL-   030802-37-0
L-ALANINAMIDE, N2-(N-ACETYL-L-ALANYLGLYCYL)-N1-(137605-59-5
L-ALANINAMIDE, N-ACETYL-L-ISOLEUCYL-L-PHENYLALAN   132766-07-5
L-ALANINAMIDE, N-ACETYL-L-VALYL-L-PHENYLALANYL-   132766-05-3
L-ALANINAMIDE, N-ACETYL-L-VALYL-   132765-85-6
L-ALANINAMIDE, N-ACETYL-L-VALYL-L-ALANYL-   132766-01-9
L-ALANINAMIDE, N2-(N-L-ALANYL-L-ALANYL)-N1-(1,1-   137605-60-8
ALANINE, N-[[(2-CHLOROETHYL)NITROSOAMINO>CARBONY   052320-87-3
BETA-ALANINE   000107-95-9
ALANINE   000302-72-7
ALANINE   000056-41-7
ALANINE,N-ACETYL,N'-MEAM-AMIDE   022715-68-0
ALANINE-N-ACETYL,METHYL ESTER   003619-02-1
ALANINE-AMIDE,N-ACETYL   015962-47-7
ß-D-ALANINE, N-[1-OXO-5-(1H-PURIN-6-YLTHIO)PENTY   023374-51-8
L-ALANINE, 3-PYRAZINYL-   087831-85-4
ALANTOLACTONE   000546-43-0
ALANYCARB   083130-01-2
ALDICARB   000116-06-3
ALDICARB SULFONE   001646-88-4
ALDICARB SULFOXIDE   001646-87-3
ALDOSTERONE   000052-39-1
ALDRIN   000309-00-2
ALFENTANIL   071195-58-9
ALIZARIN AZUROL A 2G   006424-85-7
ALIZARINE   000072-48-0
ALKYLBENZENE (C10-C15)   068648-87-3
ALLANTOIN   000097-59-6
ALLETHRIN   000584-79-2
ALLOBARBITAL   000052-43-7
ALLOPURINOL,2,5-BIS(PIVALOYLOXYMETHYL)-   098827-17-9
ALLOPURINOL   000315-30-0
ALLOPURINOL,PHENYLGLYCYLOXYMETHYL   098204-04-7
ALLOPURINOL, 1-ACETYL-   096448-60-1
ALLOPURINOL, 1-[(ACETYLOXY)METHYL]-   098846-64-1
ALLOPURINOL, 1-BENZOYL-   096448-63-4
ALLOPURINOL, 1-[BENZOYLOXYMETHYL]-   098846-65-2
ALLOPURINOL, 2,5-BIS(BUTYRYLOXYMETHYL)-   098827-20-4
ALLOPURINOL, 1,5-BIS(BUTYRYLOXYMETHYL)-   098827-19-1
ALLOPURINOL,1,5-BIS(PIVALOYLOXYMETHYL)-   098827-16-8
ALLOPURINOL,1,N,N-DIETHYLALANYLOXYMETHYL-   098204-10-5

ALLOPURINOL,1-N,N-DIETHYLGYCYCLOMETHYL	098204-08-1
ALLOPURINOL,1-[N,N-DIETHYLSUCCINAMYL-OXYMETHYL]-	098827-27-1
ALLOPURINOL,N-(4-DIMETHYLAMINOMETHYL)BENZOYLOXYM	131402-51-2
ALLOPURINOL,1,N,N-DIPROPYLGLYCYLOXYMETHYL-	098204-09-2
ALLOPURINOL, 1-[ETO-CO-OXY-ME]-	098827-23-7
ALLOPURINOL,3-(IMIDAZOL-1-YL)METHYLBENZOYLOXYMET	131402-52-3
ALLOPURINOL,1-LEUCINYLOXYMETHYL-	098204-06-9
ALLOPURINOL,N-(4-(MORPHOLINOMETHYL)BENZOYLOXY)ME	131402-48-7
ALLOPURINOL,N-(3-(MORPHOLINOMETHYL)BENZOYLOXY)ME	131402-47-6
ALLOPURINOL, 1-[NICOTINOYLOXYMETHYL]-	098846-66-3
ALLOPURINOL,N-(3-(4-METHYLPIPERIZINYLMETHYL)BEN	131402-50-1
ALLOPURINOL, 1-(1-OXOBUTYL)-	096474-05-4
ALLOPURINOL, 1-(OXOPROPYL)-	096448-61-2
ALLOPURINOL,1-PH-ALANYL-OME-	098204-05-8
ALLOPURINOL,1-[PIVALOYLOXYMETHYL]-	098827-18-0
ALLOXANE	000050-71-5
ALLYL ACETATE	000591-87-7
ALLYL ALCOHOL	000107-18-6
ALLYLAMINE	000107-11-9
ALLYLBENZENE	000300-57-2
3-ALLYL-6-CHLOROBENZAZEPINE	090047-59-9
ALLYL CHLOROCARBONATE	002937-50-0
ALLYL ETHER	000557-40-4
ALLYLETHYLAMINE	002424-02-4
5-ALLYL-5-ETHYL BARBITURIC ACID	002373-84-4
ALLYL ETHYL ETHER	000557-31-3
ALLYLIDENEDIACETATE	000869-29-4
ALLYL-ISOPROPYL ACETAMIDE	003829-78-5
ALLYL-(3,4,5-MEO)PHENYL SULFONE	057553-99-8
ALLYL MERCAPTAN	000870-23-5
3-ALLYLO-5-BR-6-ME-PYRIMID-2,4-DION	077317-70-5
3-ALLYLOXY-4-AMINOBENZOIC ACID	002486-77-3
ALLYLPHENYL ETHER	001746-13-0
ALLYL PHENYL SULFIDE	005296-64-0
ALLYLPROPYLAMINE	005666-21-7
ALLYL PROPYL ETHER	001471-03-0
4-ALLYLSEMICARBAZIDE	057421-73-5
ALLYLTHIOUREA	000109-57-9
ALLYTRICHLOROSILANE	000107-37-9
3AL-5-(5NO2-FURFURIL)THIAZOLIDIN-2,4-ONE	025603-14-9
ALOIN	001415-73-2
ALPHAPRODINE	000077-20-3
ALPRENOLOL	013655-52-2
ALTOSID	040596-69-8
D-ALTROPYRANOSIDE, METHYL 3-DEOXY-3-[[(METHYLN	052019-10-0
AMANTADINE	000768-94-5
AMARANTH	000915-67-3
AMBAZONE	000539-21-9
AMBAZONE [SEMICARBAZONE]	061566-10-7
AMBREIN	000473-03-0
AMBROSIN	000509-93-3
AMETRYNE	000834-12-8
P-AMIDO-AMPHENICOL	039960-99-1
AMIDOTHION	000919-76-6
AMIFLOXACIN	086393-37-5
P-AMINOACETANILIDE	000122-80-5
AMINOACETONITRILE	000540-61-4
M-AMINOACETOPHENONE	000099-03-6

O-AMINOACETOPHENONE	000551-93-9
4-AMINOACETOPHENONE	000099-92-3
1-B-AMINOACETYLET-2ME-5-NO2IMIDAZOLE	050846-57-6
4-AMINOACRIDINE	000578-07-4
1-AMINOACRIDINE	000578-06-3
9-AMINOACRIDINE	000090-45-9
2-AMINOACRIDINE	000581-28-2
3-AMINOACRIDINE	000581-29-3
1-AMINOADAMANTANE,3,5-DIETHYL	080121-60-4
1-AMINOADAMANTANE,3,5-DIMETHYL	019982-08-2
1-AMINOADAMANTANE,3-ETHYL	041100-45-2
1-AMINOADAMANTANE,3-PROPYL	080121-61-5
1-AMINOADAMANTANE,3,5,7-TRIMETHYL	015210-60-3
1-AMINOADAMATANE, 3-METHYL	033103-93-4
2-AMINO-5-AMYL-1,3,4-OXADIAZOLE	069741-93-1
2-AMINOANTHRACENE	000613-13-8
1-AMINOANTHRAQUINONE	000082-45-1
2-AMINOANTHRAQUINONE	000117-79-3
1-AMINOANTHRAQUINONE-2-CARBOXYLIC ACID	000082-24-6
4-AMINOBENZAMIDE	002835-68-9
4-AMINOBENZAMIDINE	002498-50-2
M-AMINOBENZENESULFONAMIDE	000098-18-0
3-AMINOBENZENESULFONIC ACID	000121-47-1
4-AMINOBENZENETHIOL	001193-02-8
3-AMINOBENZENETHIOL	022948-02-3
2-AMINOBENZIMIDAZOLE	000934-32-7
3-AMINO-BENZISOXAZOLE	036216-80-5
4-AMINOBENZOIC ACID	000150-13-0
3-AMINOBENZOIC ACID	000099-05-8
2-AMINOBENZOIC ACID	000118-92-3
O-AMINOBENZOIC ACID, ETHYL ESTER	000087-25-2
P-AMINOBENZOIC ACID, ETHYL ESTER	000094-09-7
M-AMINOBENZONITRILE	002237-30-1
P-AMINOBENZONITRILE	000873-74-5
4-AMINOBENZOYLHYDRAZINE	005351-17-7
3-AMINOBENZOYLHYDRAZINE	014062-34-1
P-AMINOBENZYLALCOHOL	061224-32-6
M-AMINOBENZYL ALCOHOL	001877-77-6
M-AMINOBENZYL N,N-DIME CARBAMATE	084640-25-5
M-AMINOBENZYL TRIMETHYL	071323-96-1
4-AMINOBIPHENYL	000092-67-1
2-AMINOBIPHENYL	000090-41-5
1-AMINO-2-BROMO-4-HYDROXY-9,10-ANTHRACENEDIONE	000116-82-5
2-FURANMETHANOL, 5-(2-AMINO-6-BROMO-9H-PURIN-9-Y	132194-22-0
4-AMINO-3-BROMOPYRIDINE	013534-98-0
2-AMINO-5-BU-1,3,4-OXADIAZOLE	052838-38-7
N-(4-AMINOBUTYL)BENZAMIDE	005692-23-9
2-AMINO-5-SEC-BUTYL-1,3,4-OXADIAZOLE	069741-91-9
4-AMINOBUTYRIC ACID	000056-12-2
ALPHA-AMINOBUTYRIC ACID	000080-60-4
A-AMINOCAPROIC ACID	000327-57-1
AMINOCARB	002032-59-9
N-(AMINOCARBONYL)GLYCINE	000462-60-2
2-AMINO-4-(P-CHLOROANILINO)-S-TRIAZINE	000500-42-5
2-AMINO-5-CHLOROBENZOIC ACID	000635-21-2
2(1H)-PYRIMIDINONE, 4-AMINO-5-CHLORO-1-(2,3-DIDE	141249-30-1
2-AMINO-5-CHLORO-METHYLBENZOATE	005202-89-1
2-AMINO-3-CHLORO-1,4-NAPHTHOQUINONE	002797-51-5

2-AMINO-6-CHLORO-4-NITROPHENOL	006358-09-4
6-AMINOCHRYSENE	002642-98-0
2-AMINO-5-CYANOBENZIMIDAZOLE	063655-40-3
1-AMINO-2,4-DIBROMOANTHRAQUINONE	000081-49-2
3-AMINO-2,5-DICHLOROBENZOIC ACID	000133-90-4
1,2,4-TRIAZIN-5(4H)-ONE, 4-AMINO-6-(1,1-DIMETHYL	033665-20-2
7-AMINO-2,2-DIMETHYL-2,3-DIHYDROBENZOFURAN	068298-46-4
5-AMINO-1,2-DIMETHYL-4-NITROIMIDAZOLE	021677-57-6
2-AMINO-4,6-DIMETHYLPYRIMIDINE	000767-15-7
2-AMINO-4,6-DINITROPHENOL	000096-91-3
4-AMINODIPHENYLAMINE	000101-54-2
2-AMINOETHANE DODECANOATE	000142-78-9
2-(2-AMINOETHYLAMINO)ETHANOL	000111-41-1
4-(2-AMINOETHYL)PHENOL	000051-67-2
2-AMINOFLUORENE	000153-78-6
6-AMINOHEXANOIC ACID	000060-32-2
P-AMINOHIPPURIC ACID, ME ESTER	005259-86-9
1-AMINO-4-HYDROXYANTHRAQUINONE	000116-85-8
3-AMINOINDAZOLE	000874-05-5
5-AMINOINDOLE	005192-03-0
2-AMINO-5-IODOBENZOIC ACID	005326-47-6
2-AMINO-5-I-PR-1,3,4-OXADIAZOLE	065283-97-8
2-AMINOISONIAZID	058481-01-9
2-AMINO-5-ME-1,3,4-OXADIAZOLE	052838-39-8
2-AMINO-5-METHOXYBENZIMIDAZOLE	006232-91-3
1-AMINO-2-METHYL-9,10-ANTHRACENEDIONE	000082-28-0
2-AMINO-5-METHYLBENZIMIDAZOLE	006285-68-3
4-AMINOMETHYLBENZOIC ACID	000056-91-7
2-AMINOMETHYLFURAN	000617-89-0
2-AMINO-2-METHYLPROPANEDIOL	000115-69-5
2-AMINO-2-METHYL-1-PROPANOL	000124-68-5
2-AMINO-5-METHYLPYRIDINE	001603-41-4
2-AMINO-4-METHYLPYRIMIDINE	000108-52-1
3-AMINO-1,5-NAPHTHALENE DISULFONIC ACID	000131-27-1
2-AMINO-1-NAPHTHALENESULFONIC ACID	000081-16-3
1-AMINO-2-NAPHTHALENESULFONIC ACID	000081-06-1
4-AMINO-1-NAPHTHALENESULFONIC ACID	000084-86-6
3-AMINO-2-NAPHTHOIC ACID	005959-52-4
2-AMINO-4-NITROANILINE	000099-56-9
2-AMINO-4-NITROBENZOIC ACID	000619-17-0
2-AMINO-4-NITROPHENOL	000099-57-0
2-AMINO-4-NITROSOPHENOL	075773-74-9
2-AMINO-1,3,4-OXADIAZOLE	003775-60-8
5-AMINOPENTANOIC ACID	000660-88-8
O-AMINOPHENOL	000095-55-6
1-(3-AMINOPHENOXY)SILATRANE	090936-52-0
P-AMINOPHENYL ACETATE	013871-68-6
P-AMINOPHENYLALANINE	002410-24-4
3-AMINOPHENYL 4-AMINOSALICYLATE	056356-23-1
4-AMINOPHENYL 4-AMINOSALICYLATE	056356-24-2
-((4-AMINOPHENYL)AZO)BENZENESULFONIC ACID,MONO*	002491-71-6
P-AMINOPHENYLETHANOL	000104-10-9
P-AMINOPHENYLHIPPURATE	030022-13-0
2-AMINO-5-PHENYLPENTANOIC ACID	002046-19-7
4-AMINO-2-PHENYLPHENOL	019434-42-5
3-AMINO-3-PHENYLPROPANOIC ACID	000614-19-7
2-AMINO-4-PHENYLQUINAZOLINE	014005-50-6
N-AMINOPIPERIDINE	002213-43-6

AMINOPROMAZINE	000058-37-7
1-AMINO-2-PROPANOL	000078-96-6
3-AMINO-1-PROPYLBENZENE	000060-15-1
2-AMINO-5-PROPYL-1,3,4-OXADIAZOLE	069741-89-5
2-AMINOPYRAZINE	005049-61-6
4-AMINOPYRAZOLE	028466-26-4
4-AMINOPYRIDINE	000504-24-5
2-AMINOPYRIDINE	000504-29-0
3-AMINOPYRIDINE	000462-08-8
N7-(6-AMINO-PYRID-3-YL)-MITOMYCIN C	088854-59-5
2-AMINOPYRIMIDINE	005469-70-5
2-AMINOPYRIMIDINE	000109-12-6
2-AMINO-4-PYRIMIDONE	000108-53-2
AMINOPYRINE	000058-15-1
2-AMINOQUINAZOLINE-4-ONE	020198-19-0
4-AMINOQUINOLINE	000578-68-7
3-AMINOQUINOLINE	000580-17-6
6-AMINOQUINOLINE	000580-15-4
5-AMINOQUINOLINE	000611-34-7
2-AMINOQUINOLINE	000580-22-3
8-AMINOQUINOLINE	000578-66-5
4-AMINOQUINOLINE-1-OXIDE	002508-86-3
5-AMINOQUINOXALINE	016566-20-4
6-AMINOQUINOXALINE	006298-37-9
2-AMINOQUINOXALINE	005424-05-5
P-AMINOSALICYLIC ACID	000065-49-6
4-AMINOSALICYLIC ACID,2-IPR ESTER	056356-20-8
4-AMINOSALICYLIC ACID,4-IPR ESTER	056356-19-5
2-AMINOSTRYCHNINE	003215-24-5
4-AMINOSULFONYLBENZOIC ACID	000138-41-0
5-AMINO-1H-TETRAZOLE	004418-61-5
2-AMINO-1,3,4-THIADIAZOLE	004005-51-0
2-AMINOTHIAZOLE	000096-50-4
O-AMINOTHIOBENZAMIDE	002454-39-9
O-AMINOTHIOPHENOL	000137-07-5
4-AMINO-M-TOLUIC ACID	002486-70-6
AMINOTRIS(HYDROXYMETHYL)METHANE	000077-86-1
AMISULPRIDE	071675-85-9
AMITRAZ	033089-61-1
AMITROLE	000061-82-5
AMITRYPTYLINE	000050-48-6
AMMONIUM PICRATE	000131-74-8
AMOBARBITAL	000057-43-2
AMOXICILLIN	026787-78-0
AMPHENICOL	004423-58-9
AMPICILLIN	000800-79-3
AMPICILLIN	000069-53-4
N-AMYL ACETATE	000628-63-7
S-AMYLBENZENE	029316-05-0
AMYL GALLATE	004568-93-8
AMYL LACTATE	006382-06-5
AMYL NITRITE	000463-04-7
P-TERT-AMYLPHENOL	000080-46-6
AMYL PROPIONATE	000624-54-4
P-AMYLPYRIDINE	002961-50-4
ANABASINE	000494-52-0
ANCITABINE	031698-14-3
ANCYMIDOL	012771-68-5

ANCYMIDOL	012771-68-5
ANDROMEDOTOXIN	004720-09-6
ANDROSTAN-3-ONE, 17-HYDROXY-, (5ALPHA,17BETA)-	000521-18-6
ANDROSTENEDIONE	000063-05-8
ANDROST-4-EN-3-ONE, 9-FLUORO-11,17-DIHYDROXY-17-	000076-43-7
ANDROSTERONE	000053-41-8
ANETHOLE	000104-46-1
ANETHOLE (TRANS)	004180-23-8
ANILAZINE	000101-05-3
ANILINE	000062-53-3
ANILINE HYDROCHLORIDE	000142-04-1
ANILINE SULFATE	000542-16-5
O-ANILINESULFONIC ACID	000088-21-1
2-ANILINOETHANOL	000122-98-5
2-ANILINOPHENYLACETIC ACID	070172-33-7
ANILOFOS	064249-01-0
P-ANISALDEHYDE	000123-11-5
P-ANISIC ACID, METHYL ESTER	000121-98-2
ANISOLE	000100-66-3
ANISOLE, O-ACETYLMETHYLAMINO	088203-06-9
ANISOLE, O-BUTYLAMINO-	065570-20-9
ANISOLE,CARBOXAMIDOMETHYLAMINO-	088203-07-0
ANISOLE, O-CYANOMETHYLAMINO-	028354-25-8
ANISOLE, O-DICHLOROACETYLAMINO-	014676-37-0
ANISOLE,O-(2-HYDROXYETHYL)AMINO	002933-75-7
1-(O-ANISYL)-3,3-DIMETHYLTRIAZENE	020240-99-7
1-ANTHRACENAMINE	000610-49-1
ANTHRACENE	000120-12-7
9-ANTHRACENECARBONITRILE	001210-12-4
ANTHRACENE-9-CARBOXYLIC ACID	000723-62-6
ANTHRACENE, 9-NITRO-	000602-60-8
9(10H)-ANTHRACENONE	000090-44-8
ANTHRAQUINONE	000084-65-1
1,5-ANTHRAQUINONEDISULFONIC ACID	000117-14-6
1,8-ANTHRAQUINONEDISULFONIC ACID	000082-48-4
ANTIPYRINE	000060-80-0
ANTOR	038727-55-8
APHIDAN	005827-05-4
APHRODINE	000146-48-5
APIGENIN	000520-36-5
6A-BETA-APORPHINE-10,11-DIOL	000058-00-4
6,ALPHA,BETA-APORPHINE-10,11-DIOL, HYDROCHLOR*	000314-19-2
CISAPRIDE	081098-60-4
APRINDINE	037640-71-4
APROBARBITAL	000077-02-1
ARABINOSE	000147-81-9
ARAMITE	000140-57-8
G-32292 (ARATONE)	003035-45-8
ARBAMATE, (4A_,9beta,10_,10Abeta)-	053447-05-5
ARBONATE, (4A_,9_,10_,10Abeta)-	053447-04-4
ARBONATE, (4A_,9_,10beta,10Abeta)-	053447-06-6
ARBUTIN	000497-76-7
ARECOLINE	000063-75-2
ARETAN	000123-88-6
ARGININ-AMIDE, N-ACETYL	064365-27-1
ARGININE	000074-79-3
AROCHLOR 1221	011104-28-2
AROCLOR 1260	011096-82-5

AROCLOR 1254	011097-69-1
AROCLOR 1016	012674-11-2
AROCLOR 1232	011141-16-5
AROCLOR 1242	053469-21-9
AROCLOR 1248	012672-29-6
AROMATICIN	005945-42-6
ASCORBIC ACID	000050-81-7
L-ASPARAGINAMIDE, N-ACETYL-	084652-30-2
ASPARAGINE	000070-47-3
L-ASPARAGINE, N -ACETYL-	004033-40-3
ASPARTAME	022839-47-0
L-ASPARTAMIDE, N-ACETYL-L-ISOLEUCYL-	132765-96-9
L-ASPARTAMIDE, N-ACETYL-L-LEUCYL-	132766-00-8
ASPARTIC ACID	000056-84-8
L-ASPARTIC ACID, N-[1-OXO-5-(7H-PURIN-6-YLTHIO)P	023404-73-1
ASULAM	003337-71-1
ATENOLOL	029122-68-7
ATRATONE	001610-17-9
ATRAZINE	001912-24-9
ATROPINE	000051-55-8
ATROPINE METHYL BROMIDE	002870-71-5
AURAMINE	000492-80-8
AUREOMYCIN	000057-62-5
8-AZAADENINE	001123-54-2
7-AZABICYCLO[4.1.0]HEPTANE, 7-[2-HYDROXY-3-(1H-I	120277-94-3
7-AZABICYCLO[4.1.0]HEPTANE, 7-[2-HYDROXY-3-(1H-I	120277-90-9
6-AZABICYCLO[3.1.0]HEXANE, 6-[2-HYDROXY-3-(1H-IM	120277-93-2
8-AZABICYCLO[5.1.0]OCTANE, 8-[2-HYDROXY-3-(1H-IM	120277-95-4
AZACONAZOL	060207-31-0
2-AZACYCLOHEPTANTHIONE	007203-96-5
2-AZACYCLO-NONANONE	000935-30-8
2-AZACYCLOOCTANONE	000673-66-5
2-AZACYCLOOCTANTHIONE	022928-63-8
2-AZACYCLOPENTANTHIONE	002295-35-4
AZACYCLOTRIDECAN-2-ONE	000947-04-6
-AZADIBENZO(A,E)CYCLOHEPTATRIENE	000256-96-2
3-AZAGLUTARAMIDE ANALOG	130336-12-8
8-AZAGUANINE	000134-58-7
2-AZAHEXANTHIONE	013070-01-4
8-AZAHYPOXANTHINE	002683-90-1
4-AZAINDOLE	000272-49-1
6-AZAINDOLE	000271-29-4
7-AZAINDOLE	000274-55-5
7-AZAINDOLE	010592-27-5
5-AZAINDOLE	000271-34-1
AZANIDAZOLE	062973-76-6
AZAPERONE	001649-18-9
AZASERINE	000115-02-6
2-AZASPIRO[4.4]NONANE-1,3-DIONE, 2-[(4-METHOXYPH	150900-05-3
2-AZASPIRO[4.4]NONANE-1,3-DIONE, 2-HYDROXY-	131042-60-9
2-AZASPIRO[4.4]NONANE-1,3-DIONE, 2-ETHOXY-	150900-11-1
2-AZASPIRO[4.4]NONANE-1,3-DIONE, 2-PHENOXY-	150900-12-2
2-AZASPIRO[4.4]NONANE-1,3-DIONE, 2-METHOXY-	150900-10-0
2-AZASPIRO[4.4]NONANE-1,3-DIONE, 2-(PHENYLMETHOX	131042-61-0
6-AZATHIANAPHTHENE	000271-06-7
AZATHIOPRINE	000446-86-6
5-AZAURACIL	000071-33-0
6-AZAURIDINE	000054-25-1

6-AZAURIDINETRIACETATE	002169-64-4
2-AZACYCLO-NONANTHIONE	021577-72-0
1H-AZEPINE, HEXAHYDRO-1-[(BENZOYLOXY)ACETYL]-	115178-68-2
AZEPINO[2,1-B]QUINAZOLIN-12(6H)-ONE, 7,8,9,10-TE	004425-23-4
AZETIDINE	000503-29-7
1-AZETIDINEACETIC ACID, 2-(4-BENZOYLPHENOXY)-4-O	119005-28-6
1-AZETIDINEACETIC ACID, 2-(2-NAPHTHALENYLOXY)-4-	119005-26-4
1-AZETIDINEACETIC ACID, 2-(2-NAPHTHALENYLTHIO)-4	119005-30-0
1-AZETIDINEACETIC ACID, 2-(2-NAPHTHALENYLOXY)-4-	119005-27-5
1-AZETIDINEACETIC ACID, 2-OXO-4-[4-(1-OXO-3-PHEN	119005-29-7
AZETIDINE, 1-[BENZOYLOXY(ACETYL)]-	115178-66-0
2-AZETIDINECARBOXYLIC ACID, 1-[(BENZOYLOXY)ACETY	115178-74-0
2-AZETIDINECARBOXYLIC ACID	002517-04-6
-AZETIDINECARBOXYLIC ACID, 1-[2-HYDROXY-3-(1H-I	134419-52-6
-AZETIDINECARBOXYLIC ACID, 1-[2-HYDROXY-3-(1H-I	134419-51-5
2-AZETIDINONE, 4-(2-BENZOYLPHENOXY)-	119005-21-9
2-AZETIDINONE, 4-(4-BENZOYLPHENOXY)-	119005-17-3
2-AZETIDINONE, 4-[(4-CHLOROPHENYL)THIO]-	068960-60-1
2-AZETIDINONE, 4-(1-NAPHTHALENYLOXY)-	119005-13-9
2-AZETIDINONE, 4-(2-NAPHTHALENYLTHIO)-	119005-22-0
2-AZETIDINONE, 4-(2-NAPHTHALENYLOXY)-	031898-94-9
2-AZETIDINONE, 4-[4-(1-OXO-3-PHENYLPROPYL)PHENOX	119005-20-8
2-AZETIDINONE, 4-[4-(1-OXO-3-PHENYL-2-PROPENYL)P	119005-18-4
2-AZETIDINONE, 4-(9-PHENANTHRENYLOXY)-	119005-16-2
2-AZETIDINONE, 4-(2-PYRIDINYLTHIO)-	119005-23-1
AZINPHOS ETHYL	002642-71-9
AZINPHOS-ETHYL, O-ANALOG	039923-25-6
AZIPROTRYNE	004658-28-0
5(1-AZIRD)-2,4-NO2 BENZAMIDE,N-ET	024570-14-7
5(1-AZIRD)-2,4-NO2 BENZAMIDE,N-CYPR	024570-16-9
5(1-AZIRD)-2,4-NO2 BENZAMIDE,NN-DIME	027221-03-0
AZIRIDINE	000151-56-4
5(AZIRIDINYL)-2,4-DINO2-N-PH BENZAMIDE	027091-58-3
5-(1-AZIRIDINYL)-2,4-NO2BENZAMIDE	021919-05-1
AZITHROMYCIN	083905-01-5
AZOBENZENE	000103-33-3
N,N'-(AZODI-4,1-PHENYLENE)BISACETAMIDE	015446-39-6
AZODRIN	006923-22-4
AZO DYE N5	072828-64-9
AZO DYE N2	070528-90-4
AZO DYE N1	068877-63-4
AZO DYE RA	068133-69-7
O,P'-AZOTOLUENE	029418-22-2
AZULENE	000275-51-4
1-AZULENE CARBOXYALDEHYDE	007206-61-3
1-AZULENE CARBOXYLIC ACID	001201-25-8
AZULENO[4,5-B]FURAN-2,9-DIONE, DECAHYDRO-6A-HYDR	002571-81-5
AZULENO[4,5-B]FURAN-2,9-DIONE, DECAHYDRO-4-HYDRO	017066-68-1
BANOL	000671-04-5
BARBAN	000101-27-9
BARBITAL	000057-44-3
BARBITURIC ACID	000067-52-7
BARBITURIC ACID, ALLYL,SEC-BUTYL	000115-44-6
BARBITURIC ACID,5-ALLYL-5(2BR ALLYL)	000561-86-4
BARBITURIC ACID,5-ALLYL,5-(2CYPENTEN1YL)	000076-68-6
BARBITURIC ACID,5-ALLYL-5-NEOPENTYL	000561-83-1
BARBITURIC ACID, 5-T-BU-	090197-63-0
BARBITURIC ACID,5-TBU-5ALLYL	091430-64-7

BARBITURIC ACID, 5-BUTYL	001953-33-9
BARBITURIC ACID,5-(1-CYCLOHEPTEN-1-YL)-5-ETHYL-1	065329-13-7
BARBITURIC ACID,5ME-5CYHXENE	000718-67-2
BARBITURIC ACID,5,5-DIBUTYL-	017013-41-1
BARBITURIC ACID,DI-IPR	099167-69-8
BARBITURIC ACID,5,5-DIPHENYL-	021914-07-8
BARBITURIC ACID,5,5-DIPR	002217-08-5
BARBITURIC ACID, 5-ET-5-AMYL	000115-58-2
BARBITURIC ACID,5ET-5-TBU	015379-32-5
BARBITURIC ACID,5-ET-5(2,3-DIMEBU)	036380-48-0
BARBITURIC ACID,5-ET,5-HEPTYL	060784-70-5
BARBITURIC ACID,5ET-5IBU	001135-61-1
BARBITURIC ACID,5-ET-5-IPE-2-THIO	004388-79-8
BARBITURIC ACID,5-ET-5(3ME-2BUTENE)	021149-88-2
BARBITURIC ACID,5-ET-5-NONYL	064810-91-9
BARBITURIC ACID,5-ET-5-OCTYL	064810-90-8
BARBITURIC ACID,5-ET,5-(3-OH-1-ME)BUTYL	004241-40-1
BARBITURIC ACID, 5-ET,5-PR	033376-25-9
BARBITURIC ACID,5(2HOPR)5-ALLYL	002537-29-3
BARBITURIC ACID, 5-IPR	007391-69-7
BARBITURIC ACID,5-IPR,5(2BR ALLYL)	000545-93-7
BARBITURIC ACID,5-IPR,5-BR-ALLYL,N-ME	000125-55-3
BARBITURIC ACID,5-IPR-5(2BR2ALLYL)	001142-70-7
BARBITURIC ACID,1-IPR-5-ET-5(PH)	085432-36-6
BARBITURIC ACID,N-ME,5-ALLYL,5-CYPENTENYL	094779-66-5
BARBITURIC ACID,1-ME-5-ALLYL-5-ET	061346-84-7
BARBITURIC ACID,N-ME-5ALLYL,5-BR-ALLYL	064889-77-6
BARBITURIC ACID,5(1-MEBU),5-B-BROMALLYL	001216-40-6
BARBITURIC ACID,N-ME,5-S-BU,5-BROMALLYL	069519-61-5
BARBITURIC ACID,5(1MEBU)-5-ET-1-ME	057562-99-9
BARBITURIC ACID,5-ME-5(11DIMEET)	109777-62-0
BARBITURIC ACID, N-ME,5-ETHYL	083128-50-1
BARBITURIC ACID,5-ME-5-IPR	053943-59-2
BARBITURIC ACID,5-ME-5-PHENYL-	000076-94-8
BARBITURIC ACID,1-ME-5-ET-5(PHET)	061346-87-0
BARBITURIC ACID, 5-PHNEYL-	022275-34-9
BARBITURIC ACID,5-PR-5-ET-1-ME	056344-90-2
BARBITURIC ACID,5-ET-5-ME	027653-63-0
BARBITURIC,2-THIO-5-ET-5-MEALLYL	000115-56-0
BAS 480F	106325-08-0
BASIC YELLOW 2	002465-27-2
BAYER 30686	000093-75-4
BAYER 15080	000495-73-8
BAYER 37344	002032-65-7
BAZINAPRINE	094011-82-2
BCPC	002164-13-8
BR-N(246-MEPH)CARBAMOYLME IMINODIACETIC ACID	078266-06-5
BENALAXYL	071626-11-4
BENAZOLIN	003813-05-6
BENAZOLIN-ETHYL	025059-80-7
BENDIOCARB	022781-23-3
BENDROFLUMETHIAZIDE	000073-48-3
BENEFIN	001861-40-1
BENFURACARB	082560-54-1
BENFURESATE	068505-69-1
BENOMYL	017804-35-2
BENOXACOR	098730-04-2
BENPERIDOL	002062-84-2

BENSULFURON-METHYL (PH7)	083055-99-6
BENSULIDE	000741-58-2
BENSULTAP	017606-31-4
BENTAZONE	025057-89-0
A-BENYZL BENZENEETHANEAMINE	004275-43-8
BENZ(A)ACRIDINE	000225-11-6
BENZ(A)ANTHRACENE	000056-55-3
BENZ(A)ANTHRACENE-4,5-DIMETHYLENE	018429-70-4
BENZ A ANTHRACENE-7,12-DIONE	002498-66-0
BENZ(A)ANTHRACENE-7-METHYL	002541-69-7
BENZ(C)ACRIDINE	000225-51-4
BENZADOX	005251-93-4
BENZ(A,J)ACEANTHRYLENE	000203-20-3
BENZALACETOPHENONE	000094-41-7
BENZAL ACETYLACETONE	004335-90-4
N-BENZALAMINOPIPERIDINE	001885-87-6
BENZALANILINE	000538-51-2
BENZALCYANOACETAMIDE	000709-79-5
BENZALCYANOACETANILIDE	015804-61-2
BENZALDEHYDE	000100-52-7
BENZALDEHYDE, 3,4-DIMETHOXY-	000120-14-9
BENZALDEHYDE, 3-ETHOXY-4-HYDROXY-	000121-32-4
BENZALDEHYDE, 4-FLUORO-	000459-57-4
BENZALDEHYDE, 4-(2-HYDROXYETHOXY)-3-METHOXY-, OX	093414-50-7
BENZALDEHYDE, 4-(2-HYDROXYETHOXY)-3-METHOXY-	064673-04-7
BENZALDEHYDE,O-((MEAMINO)CO)OXIME	002426-12-2
BENZALDEHYDE, 2-METHOXY-	000135-02-4
BENZALDEHYDE, 3-METHOXY-	000591-31-1
BENZALDEHYDE THIOSEMICARBAZONE	001627-73-2
BENZAL DIACETATE	000581-55-5
BENZALDOXIME	000932-90-1
BENZALDOXIME,SYN	000622-31-1
BENZAMIDE	000055-21-0
BENZAMIDE, O-ACETYLAMINO	033809-77-7
BENZAMIDE, O-ACETYLMETHYLAMINO	027545-04-6
BENZAMIDE, 2-[(ACETYLOXY)METHOXY]-N-(4-MORPHOLIN	102273-19-8
BENZAMIDE, 2-[(ACETYLOXY)METHOXY]-	102273-25-6
BENZAMIDE, 2-AMINO-	000088-68-6
BENZAMIDE, 3-AMINO-	003544-24-9
BENZAMIDE, 4-AMINO-5-(AMINOSULFONYL)-N-[(1-ETHYL	093414-64-3
BENZAMIDE, 4-AMINO-3-CHLORO-N-[2-(DIETHYLAMINO)E	000891-60-1
BENZAMIDE, 4-AMINO-N-[2-(DIETHYLAMINO)ETHYL]-5-I	093414-60-9
BENZAMIDE, 4-AMINO-N-[2-(DIETHYLAMINO)ETHYL]-5-(	093414-65-4
BENZAMIDE, 4-AMINO-N-[2-(DIETHYLAMINO)ETHYL]-5-F	093414-59-6
BENZAMIDE, 4-AMINO-N-[2-(DIETHYLAMINO)ETHYL]-2-M	003761-48-6
BENZAMIDE,N-(2-AMINOETHYL)	001009-17-2
BENZAMIDE, N-(2-AMINO-2-OXOETHYL)-2-(3,3-DIMETHY	066974-91-2
BENZAMIDE, 4-[(4-AMINOPHENYL)SULFONYL]-	024454-46-4
BENZAMIDE, N-[(4-AMINOPHENYL)SULFONYL]-2,5-DIMET	020037-50-7
BENZAMIDE, 4-[(4-AMINOPHENYL)SULFONYL]-N,N-DIETH	101533-58-8
BENZAMIDE,N-(3-AMINOPROPYL)	006108-74-3
BENZAMIDE, 4-(AMINOSULFONYL)-	006306-24-7
BENZAMIDE, 3-(AMINOSULFONYL)-	001576-42-7
BENZAMIDE, 5-(AMINOSULFONYL)-N-[(1-ETHYL-3-PYRRO	093414-62-1
M-BENZAMIDE,NNBIS(23DIOHPR),5N....	066108-93-8
M-BENZAMIDE,NN'BIS(23DIOHPR)5ACE..	076350-28-2
BENZAMIDE, N,N-BIS(1-METHYLETHYL)-	020383-28-2
BENZAMIDE, 5-BROMO-N-BUTYL-2-HYDROXY-3-METHYL-	040912-89-8

BENZAMIDE, 5-BROMO-N-CYCLOPROPYL-2-HYDROXY-3-MET   058708-46-6
BENZAMIDE, 5-BROMO-N-ETHYL-2-HYDROXY-3-METHYL-   040912-88-7
BENZAMIDE, 5-BROMO-2-HYDROXY-3-METHYL-N-(1-METHY   058708-45-5
BENZAMIDE, 5-BROMO-2-HYDROXY-3-METHYL   040912-73-0
BENZAMIDE, 5-BROMO-2-HYDROXY-N,3-DIMETHYL-   040912-87-6
BENZAMIDE, 2-[(2-BUTOXYETHYL)AMINO]-5-CHLORO-   101820-65-9
BENZAMIDE, 2-[(2-BUTOXYETHYL)AMINO]-   101820-64-8
BENZAMIDE, O-BUTYLAMINO   005363-33-7
BENZAMIDE, N-BUTYL-N-METHYL-3-PHENOXY-   149589-41-3
BENZAMIDE, 2-CHLORO-N,N-BIS(1-METHYLETHYL)-   070657-63-5
BENZAMIDE, 4-CHLORO-N-(2-CYCLOPENTYL-2H-BENZOTRI   130787-15-4
BENZAMIDE, 2-CHLORO-N,N-DIETHYL-   010345-79-6
BENZAMIDE, 2-CHLORO-N,N-DIMETHYL-   006526-67-6
BENZAMIDE, 5-CHLORO-N-ETHYL-2-HYDROXY-3-METHYL-   058708-52-4
BENZAMIDE, N-[[(4-CHLOROPHENYL)AMINO]CARBONYL]-2   107485-58-5
BENZAMIDE, N-[[(4-CHLOROPHENYL)AMINO]CARBONYL]-2   107485-46-1
BENZAMIDE, 2-CYANO-   017174-98-0
BENZAMIDE, N-(CYANOMETHYL)-2-(3,3-DIMETHYL-1-THI   066974-90-1
BENZAMIDE, O-CYANOMETHYLAMINO-   088203-05-8
BENZAMIDE, N-(2-CYCLOPENTYL-2H-BENZOTRIAZOL-5-YL   130787-14-3
BENZAMIDE, N-(2-CYCLOPENTYL-2H-BENZOTRIAZOL-5-YL   130787-18-7
BENZAMIDE, N-(2-CYCLOPENTYL-2H-BENZOTRIAZOL-5-YL   130787-20-1
BENZAMIDE, N-(2-CYCLOPENTYL-2H-BENZOTRIAZOL-5-YL   130787-16-5
BENZAMIDE, N-(2-CYCLOPENTYL-2H-BENZOTRIAZOL-5-YL   130787-17-6
BENZAMIDE, N-(2-CYCLOPENTYL-2H-BENZOTRIAZOL-5-YL   130787-19-8
BENZAMIDE, 3,5-DICHLORO-N-[[1-(2-FLUOROETHYL)-2-   124840-52-4
BENZAMIDE,O-DICL ACETYLAMINO-   088203-03-6
BENZAMIDE, N-[9-(2,3-DIDEOXY-2-FLUORO-beta-D-THR   132722-93-1
BENZAMIDE,N-(DIETAMET),2-MEO,5-F   055236-14-1
BENZAMIDE,N(DIETAMET),2MEO-5CL   040256-75-5
BENZAMIDE,N(DIETAMET),2-MEO,5-SO2ME   051012-32-9
BENZAMIDE,N(DIETAMET),2MEO-4NH2-5C≡N   065016-46-8
BENZAMIDE,N(DIETAMET),2-MEO,5-BR   071225-60-0
BENZAMIDE,N(DIETAMET),2OME,3CL,5F   004925-19-3
BENZAMIDE, 4-[(DIETHYLAMINO)METHYL]-N-(2,6-DIMET   074433-34-4
BENZAMIDE,N-(3-DIETHYLAMINOPROPYL)   066999-80-2
BENZAMIDE, N,N-DIETHYL-2-METHYL-   002728-04-3
BENZAMIDE, N,N-DIETHYL-4-METHYL-   002728-05-4
BENZAMIDE, 2,6-DIFLUORO-N-[[(4-FLUOROPHENYL)AMIN   035367-40-9
BENZAMIDE, 2-[(2,3-DIHYDROXYPROPYL)AMINO]-   101820-62-6
BENZAMIDE,N-(3-DIME AMINOPROPYL)   040948-30-9
BENZAMIDE,N-(2-DIMETHYLAMINOETHYL)   063224-18-0
BENZAMIDE, N-(1,1-DIMETHYLETHYL)-3-(2,5-DIMETHYL   065261-99-6
BENZAMIDE, N,N-DIMETHYL-2-[[(4-METHYLPHENYL)SULF   052980-49-1
BENZAMIDE, N-(2,6-DIMETHYLPHENYL)-4-(1-PIPERIDIN   074602-46-3
BENZAMIDE, N-(2,6-DIMETHYLPHENYL)-3-(1-PYRROLIDI   074602-43-0
BENZAMIDE, N,N-DIMETHYL-3-PHENOXY-   065261-13-4
BENZAMIDE, 3-(2,5-DIMETHYLPHENOXY)-N-METHYL-   065262-21-7
BENZAMIDE, N-(2,6-DIMETHYLPHENYL)-3-(4-MORPHOLIN   074602-48-5
BENZAMIDE, 2-ETHOXY-N,N-BIS(1-METHYLETHYL)-   138324-58-0
BENZAMIDE, 2-ETHOXY-N,N-DIETHYL-   003688-82-2
BENZAMIDE, 2-ETHOXY-N-ETHYL-   138324-57-9
BENZAMIDE, N-ETHYL-4-METHOXY-   007403-41-0
BENZAMIDE, N-ETHYL-4-METHYL-   026819-08-9
BENZAMIDE, N-ETHYL-3-METHYL-   026819-07-8
BENZAMIDE, N-METHYL-3-PHENOXY-N-PROPYL-   149589-40-2
BENZAMIDE, N-[(1-ETHYL-3-PYRROLIDINYL)METHYL]-5-   093414-63-2
BENZAMIDE, N-HEXYL-3,4-DIHYDROXY-   098378-56-4

BENZAMIDE, N-HEXYL-N-METHYL-3-PHENOXY-   149589-43-5
BENZAMIDE, 2-[(2-HYDROXYETHYL)AMINO]-   088267-61-2
BENZAMIDE, N-(HYDROXYMETHYL)-   006282-02-6
BENZAMIDE, 2-HYDROXY-3-METHYL-   014008-60-7
BENZAMIDE, 4-[2-[[2-HYDROXY-3-(2-METHYLPHENOXY)P   038103-61-6
BENZAMIDE, 2-HYDROXY-N-(4-MORPHOLINYLMETHYL)-   017357-04-9
BENZAMIDE, 2-[(3-HYDROXYPROPYL)AMINO]-   088267-62-3
BENZAMIDE, 2-[(2-HYDROXYPROPYL)AMINO]-   101820-63-7
BENZAMIDE, 4-(2-IMIDAZO[1,2-A]PYRIMIDINYL)-3-MET   102361-56-8
BENZAMIDE, 4-(1H-IMIDAZO[4,5-C]PYRIDIN-2-YL)-3-M   130179-73-6
BENZAMIDE, O-ISOPROPYLAMINO-   005363-32-6
BENZAMIDE, 4-METHOXY-N,N-BIS(1-METHYLETHYL)-   079606-43-2
BENZAMIDE, 2-[(4-METHOXYPHENYL)METHOXY]-   082568-69-2
BENZAMIDE, N-[(METHYLAMINO)CARBONYL]   003201-53-4
BENZAMIDE, 3-METHYL-N,N-BIS(1-METHYLETHYL)-   005448-36-2
BENZAMIDE, 4-METHYL-N,N-BIS(1-METHYLETHYL)-   006937-52-6
N-BENZAMIDE, N-(1-METHYLETHYL)-3-(2,5-DIMETHYLPHEN   065261-51-0
BENZAMIDE, N-METHYL-N-[(4-METHYLPHENYL)SULFONYL]   010533-83-2
BENZAMIDE, N-METHYL-N-PENTYL-3-PHENOXY-   149589-42-4
BENZAMIDE, N-METHYL-3-PHENOXY-   050789-46-3
BENZAMIDE, 2-[[2-(2-METHYLPHENOXY)ETHYL]AMINO]-   101820-66-0
BENZAMIDE, 4-[3-METHYL-3-(2-PROPENYL)-1-TRIAZENY   066974-73-0
BENZAMIDE, 2-(METHYLTHIO)-   054705-16-7
BENZAMIDE,N(DIETAMINOET),3,5-DICL   071225-61-1
BENZAMIDE, N-[2-(3,4,6,7,12,12A-HEXAHYDROPYRAZIN   072593-21-6
BENZAMIDE, N-[2-(3,4,6,7,12,12A-HEXAHYDROPYRAZIN   072593-17-0
BENZAMIDE, 2-[(4-NITROPHENYL)METHOXY]-   082568-68-1
BENZAMIDE, N-[(PHENYLAMINO)CARBONYL]-   001821-33-6
BENZAMIDE, N,N,4-TRIMETHYL-   014062-78-3
BENZAMIDE, N,N,3-TRIMETHYL-   006935-65-5
BENZAMIDINE HCL   001670-14-0
2-BENZAMIDO-5-METHYLPHENOL   003743-11-1
BENZANILIDE   000093-98-1
BENZANTHRONE   000082-05-3
1,4-BENZDAZEPIN-2-ONE-5-(2FPH)-7-ET   051307-86-9
1,4-BENZDAZEPIN-2-ONE-5-(2CLPH)-7BR   051753-57-2
1,4-BENZDAZEPIN-2-ON-1MEOME-5PH-7NO2   029442-58-8
1,4-BENZDAZEPIN-2-ON-1-ET-5-PH-7-AMINO   088695-10-7
1,4-BENZDAZEPIN-2-ON-1ME-5(2FPH)-7I   034932-78-0
1,4-BENZDAZEPIN-2-ON-1-ME-5-(2ME-PH)7-CL   005358-34-9
1,4-BENZDAZEPIN-2-ON-1-MEOME-5-PH-7AMINO   040837-24-9
1,4-BENZDAZEPIN-2-ON-1-MESME-5-PH-7-CL   029442-82-8
1,4-BENZDAZEPIN-2-ON-1-MESME-5-PH-7AMINO   088695-12-9
1,4-BENZDAZEPN-2-ON-1-MEOME-3-OH-5PH-7AM   088695-13-0
14-BENZDIAZEPIN-2-ONE,5-(4-BRPH)7-BR   065247-12-3
14-BENZDIAZEPIN-2-ONE,5-(3-BRPH)7-BR   065247-11-2
14-BENZDIAZEPIN-2-ONE,5-(2-BRPH)7-BR   065247-10-1
14-BENZDIAZEPIN-2-ONE,5-(2-BRPH)7-CL   063574-83-4
1,4-BENZDIAZEPIN-2-ONE-5(2-CLPH)-7-CL   002894-67-9
14-BENZDIAZEPIN-2-ONE,5-(4-CLPH)7-BR   065247-14-5
14-BENZDIAZEPIN-2-ONE,5-(3-CLPH)7-BR   065247-13-4
1,4-BENZDIAZEPIN-2-ONE-5(26F-PH)8CL   088695-07-2
1,4-BENZDIAZEPIN-2-ONE-5(26F-PH)7CL   028910-86-3
1,4-BENZDIAZEPIN-2-ONE-5-PH-7-CL   001088-11-5
1,4-BENZDIAZEPIN-2-ONE-5-PH-7CN   017562-53-7
1,4-BENZDIAZEPIN-2-ONE-5-PHENYL   002898-08-0
1,4-BENZDIAZEPIN-2-ONE-5-PH-7-NH2   004928-02-3
1,4-BENZDAZEPIN-2-ON-5-(2F-PH)-7,8-DICL   088695-06-1

1,4-BENZDAZEPIN-2-ON-5PH-7-DIMEAMINO	030144-56-0
1,4-BENZDIAZ-2-ON,1ME5(PH)7DIMEAM	002891-09-0
14-BENZDIAZPIN-2-ONE,3ME5-(2CLPH)7NO2	067027-56-9
14-BENZDIAZPIN-2-ONE,1ME5-(4MEOPH)7CL	072430-63-8
14-BENZDIAZPN-2-ON,1(23DIOHPR)5(2FPH)7CL	052829-30-8
14-BENZDIAZPN-2-ON,1(23DIOHPR)5PH7NO2	061554-12-9
N[4(14BENZDIOXAN6YL)2THIAZOLYL]OXAMIC ACID	074604-76-5
BENZENAMINE, 2-[(4-AMINOPHENYL)SULFONYL]-	027147-69-9
BENZENAMINE, 4-[(4-AMINOPHENYL)SULFONYL]-N-ETHYL	003572-34-7
BENZENAMINE, 4-[(4-AMINOPHENYL)SULFONYL]-N-HYDRO	032695-27-5
BENZENAMINE, 4-[(4-AMINOPHENYL)SULFONYL]-N,N-DIM	086552-09-2
BENZENAMINE, 3-[(4-AMINOPHENYL)SULFONYL]-	034262-32-3
BENZENAMINE, 4-[(4-AMINOPHENYL)SULFONYL]-N,N-DIE	051688-32-5
BENZENAMINE, N,N-BIS(2,2,2-TRIFLUOROETHYL)-	002924-27-8
BENZENAMINE, 4-[(4-BROMOPHENYL)SULFONYL]-	006626-22-8
BENZENAMINE, 4-BUTYL-	000104-13-2
BENZENAMINE, 2-CHLORO-N-(2-IMIDAZOLIDINYLIDENE)-	016822-82-5
BENZENAMINE, 4-CHLORO-3-NITRO-	000635-22-3
BENZENAMINE, 4-[(4-CHLOROPHENYL)SULFONYL]-	007146-68-1
BENZENAMINE, 2-CHLORO-5-(TRIFLUOROMETHYL)-	000121-50-6
BENZENAMINE, 4-CYCLOHEXYL-	006373-50-8
BENZENAMINE, 4-[(2,4-DICHLOROPHENYL)SULFONYL]-	105456-57-3
BENZENAMINE, 4-[(2,4-DIMETHOXYPHENYL)SULFONYL]-	105456-59-5
BENZENAMINE, 4-(1,1-DIMETHYLETHYL)	000769-92-6
BENZENAMINE, 4-[(2,4-DIMETHYLPHENYL)SULFONYL]-	105456-58-4
BENZENAMINE, N,3-DIMETHYL-	000696-44-6
BENZENAMINE, N,N-DIMETHYL-3-NITRO-	000619-31-8
BENZENAMINE, 4-[(2,4-DINITROPHENYL)SULFONYL]-	075333-79-8
BENZENAMINE, 2,2'-DITHIOBIS-	001141-88-4
BENZENAMINE, N-ETHYL-N-(2-METHYL-2-PROPENYL)-2,6	055283-68-6
BENZENAMINE, 4-[(4-FLUOROPHENYL)SULFONYL]-	000312-35-6
BENZENAMINE, 4-IODO-	000540-37-4
BENZENAMINE, 3-IODO-	000626-01-7
BENZENAMINE, 4-ISOTHIOCYANATO-N,N-DIMETHYL-	002131-64-8
BENZENAMINE, 4-METHOXY-2-METHYL-	000102-50-1
BENZENAMINE, 4-METHOXY-2-NITRO-	000096-96-8
BENZENAMINE, 2-METHOXY-3-NITRO-	000085-45-0
BENZENAMINE, 4-[(4-METHOXYPHENYL)SULFONYL]-	017078-72-7
BENZENAMINE, 2-(1-METHYLETHYL)-	000643-28-7
BENZENAMINE, 4-[(4-METHYLPHENYL)SULFONYL]-	004094-38-6
BENZENAMINE, 4-[(4-NITROPHENYL)SULFONYL]-	001948-92-1
BENZENAMINE, 4-PENTYL-	033228-44-3
BENZENAMINE, 4-(PHENYLSULFONYL)-	007019-01-4
BENZENAMINE, 4-PROPYL-	002696-84-6
BENZENAMINE, 4-(1,2,3,6-TETRAHYDRO-1-METHYL-4-PY	106362-30-5
BENZENAMINE, 3-(1,2,3,6-TETRAHYDRO-1-METHYL-4-PY	106362-29-2
BENZENAMINE, 2-(TRIFLUOROMETHYL)-	000088-17-5
BENZENAMINE, 4-[(2,4,6-TRIMETHYLPHENYL)SULFONYL]	033597-78-3
BENZENAMINIUM, 3-CHLORO-N,N,N-TRIMETHYL-, IODIDE	002373-41-3
BENZENE, 1-ISOTHIOCYANATO-4-METHOXY	002284-20-0
BENZENE	000071-43-2
BENZENEACETAMIDE, N-(3-ACETYLPHENYL)-	072116-69-9
BENZENEACETAMIDE, N-(4-ACETYLPHENYL)-	089246-39-9
BENZENEACETAMIDE, N-[3-(ACETYLOXY)PHENYL]-	095384-55-7
BENZENEACETAMIDE, N-(3-AMINOPHENYL)-	085856-32-2
BENZENEACETAMIDE, N-(2-AMINOPHENYL)-	095384-59-1
BENZENEACETAMIDE, N,N-BIS(1-METHYLETHYL)-	034251-46-2
BENZENEACETAMIDE, N-(3-CHLOROPHENYL)-	018109-43-8

BENZENEACETAMIDE, N-(3-CYANOPHENYL)-	089246-40-2
BENZENEACETAMIDE, N-(3,4-DICHLOROPHENYL)-	027816-82-6
BENZENEACETAMIDE, N,N-DIETHYL-	002431-96-1
BENZENEACETAMIDE, N-[3-(DIETHYLAMINO)PROPYL]-N-(	018109-55-2
BENZENEACETAMIDE, N-(2,6-DIMETHYLPHENYL)-N-[3-(1	067023-19-2
BENZENEACETAMIDE, N,N-DIMETHYL-	018925-69-4
BENZENEACETAMIDE, N-[3-(ETHYLAMINO)PHENYL]-	095384-54-6
BENZENEACETAMIDE, N-ETHYL-	005465-00-9
BENZENEACETAMIDE, N-(3-ETHOXYPHENYL)-	016053-34-2
BENZENEACETAMIDE, N-(4-FLUOROPHENYL)-	005215-28-1
BENZENEACETAMIDE, N-(2-FLUOROPHENYL)-	005215-26-9
BENZENEACETAMIDE, N-[2-[[3-(2-HYDROXY-3-PHENYLPR	121508-92-7
BENZENEACETAMIDE, N-[2-[[2-HYDROXY-3-(2-PHENOXYE	121508-95-0
BENZENEACETAMIDE, N-(3-HYDROXYPHENYL)-	023478-26-4
BENZENEACETAMIDE, N-(4-HYDROXYPHENYL)-	095384-57-9
BENZENEACETAMIDE, 3-[2-HYDROXY-3-[(1-METHYLETHYL	029121-29-7
BENZENEACETAMIDE, N-(2-HYDROXYPHENYL)-	095384-58-0
BENZENEACETAMIDE, N-[2-[[2-HYDROXY-3-[(3-PHENYL-	121508-97-2
BENZENEACETAMIDE, α-HYDROXY-α-ETHYL-, (±)-	131802-71-6
BENZENEACETAMIDE,alpha-HYDROXY-N-TRICYCLO[3.3.1.	120282-84-0
BENZENEACETAMIDE, N-[2-[[2-HYDROXY-3-(2-PHENYLET	121508-90-5
BENZENEACETAMIDE, N-(3-IODOPHENYL)-	095384-52-4
BENZENEACETAMIDE, N-[3-[(METHYLSULFONYL)OXY]PHEN	095384-56-8
BENZENEACETAMIDE, N-(2-METHOXYPHENYL)-	095384-60-4
BENZENEACETAMIDE, N-(3-METHYLPHENYL)-	050916-16-0
BENZENEACETAMIDE, N-[3-(METHYLAMINO)PHENYL]-	095384-53-5
BENZENEACETAMIDE, N-(3-METHOXYPHENYL)-	050916-19-3
BENZENEACETAMIDE, N-(4-METHOXYPHENYL)-	050916-21-7
BENZENEACETAMIDE, N-(3-NITROPHENYL)-	013140-76-6
BENZENEACETAMIDE, N-(4-NITROPHENYL)-	013140-77-7
BENZENEACETAMIDE, N-(2-NITROPHENYL)-	001504-26-3
BENZENEACETAMIDE, N-PHENYL-	000621-06-7
BENZENEACETAMIDE, N-[2-[[3-(PHENYLMETHOXY)-2-HYD	121508-87-0
BENZENEACETAMIDE, N-[2-[(3-PHENOXY-2-HYDROXYPROP	053670-99-8
BENZENEACETIC ACID, 3-CHLORO-α-METHYL4-(5-METHYL	138568-82-8
BENZENEACETIC ACID, 4-(4-ACETYL-2-THIAZOLYL)-α-M	132483-61-5
BENZENEACETIC ACID, 3-AMINO-_-METHYL-4-(2-THIAZO	132483-60-4
BENZENEACETIC ACID, 4-[(4-(AMINOCARBONYL)-2-THIA	132483-50-2
BENZENEACETIC ACID, α-[[(AMINOCARBONYL)AMINO]CAR	073632-81-2
BENZENEACETIC ACID, .ALPHA.-AMINO-, (.+-.)-	002835-06-5
BENZENEACETIC ACID, 3-BENZOYL-.ALPHA.-METHYL-	022071-15-4
BENZENEACETIC ACID, 3-BROMO-α-METHYL-4-(2-THIAZO	138568-62-4
BENZENEACETIC ACID, 4-(4-BUTYL-2-THIAZOLYL)-α-ME	132483-55-7
BENZENEACETIC ACID, 4-(4-BUTYL-2-THIAZOLYL)-3-CH	138568-81-7
BENZENEACETIC ACID, 4-BUTYL-	014377-19-6
BENZENEACETIC ACID, 4-BUTYL-α-HYDROXY-α-METHYL-	082412-82-6
BENZENEACETIC ACID, 4-(4-BUTYL-2-THIAZOLYL)-3-FL	138568-73-7
BENZENEACETIC ACID, 4-CARBOXY-	000501-89-3
BENZENEACETIC ACID, 3-CHLORO-α-METHYL-4-[4-(1-ME	138568-80-6
BENZENEACETIC ACID, 3-CHLORO-4-(5-ETHYL-2-THIAZO	138568-83-9
BENZENEACETIC ACID, 3-CHLORO-4-(4-ETHYL-2-THIAZO	138568-79-3
BENZENEACETIC ACID, 3-CHLORO-α-METHYL-4-(2-THIAZ	132483-33-1
BENZENEACETIC ACID, 3-CHLORO-4-(4-ETHYL-5-METHYL	138568-85-1
BENZENEACETIC ACID, 3-CHLORO-α-METHYL-4-(4-METHY	138568-78-2
BENZENEACETIC ACID, 3-CHLORO-4-(4,5-DIMETHYL-2-T	138568-84-0
BENZENEACETIC ACID, 4-CYCLOPENTYL-	025209-50-1
BENZENEACETIC ACID, 4-CYCLOHEXYL-	035889-00-0
BENZENEACETIC ACID, 2-[(2,6-DICHLOROPHENYL)AMINO	015307-81-0

BENZENEACETIC ACID, 2-[(2,6-DICHLOROPHENYL)AMINO   015307-79-6
BENZENEACETIC ACID, 4-(1,1-DIMETHYLPROPYL)-   026114-14-7
BENZENEACETIC ACID, 4-(1,1-DIMETHYLETHYL)-   032857-63-9
BENZENEACETIC ACID, α,3-DIMETHYL-4-(2-THIAZOLYL)   132483-36-4
BENZENEACETIC ACID, 4-[4-[(DIMETHYLAMINO)CARBONY   132483-52-4
BENZENEACETIC ACID, 4-(1,4-DIHYDRO-1-OXO-2H-ISOI   031842-01-0
BENZENEACETIC ACID, 4-(4,5-DIMETHYL-2-THIAZOLYL)   132483-46-6
BENZENEACETIC ACID, 4-(4,5-DIMETHYL-2-THIAZOLYL)   138568-76-0
BENZENEACETIC ACID, 2,5-DIHYDROXY-   000451-13-8
BENZENEACETIC ACID, 4-(4-ETHYL-2-THIAZOLYL)-3-FL   138568-71-5
BENZENEACETIC ACID, 4-(4-ETHYL-2-THIAZOLYL)-α-ME   132483-47-7
BENZENEACETIC ACID, 4-(4-ETHYL-5-METHYL-2-THIAZO   138568-68-0
BENZENEACETIC ACID, 4-(4-ETHYL-5-METHYL-2-THIAZO   138568-77-1
BENZENEACETIC ACID, 4-(4-ETHENYL-2-THIAZOLYL)-α-   132483-44-4
BENZENEACETIC ACID, 4-ETHYL-   014387-10-1
BENZENEACETIC ACID, α-ETHYL-α-HYDROXY-4-(2-METHY   103910-47-0
BENZENEACETIC ACID, 4-(5-ETHYL-2-THIAZOLYL)-α-ME   138568-67-9
BENZENEACETIC ACID, 4-(5-ETHYL-2-THIAZOLYL)-3-FL   138568-75-9
BENZENEACETIC ACID, 3-FLUORO-α-METHYL-4-[4-(1-ME   138568-72-6
BENZENEACETIC ACID, 3-FLUORO-α-METHYL-4-(2-THIAZ   132483-34-2
BENZENEACETIC ACID, 3-FLUORO-α-METHYL-4-(4-METHY   138568-70-4
BENZENEACETIC ACID, 3-FLUORO-α-METHYL-4-(5-METHY   138568-74-8
BENZENEACETIC ACID, 4-(HEXYLOXY)-   060003-46-5
BENZENEACETIC ACID, 4-HEXYL-   014377-22-1
BENZENEACETIC ACID, .ALPHA.-HYDROXY-.ALPHA.-METH   000515-30-0
BENZENEACETIC ACID, 4-HYDROXY-3-METHOXY-   000306-08-1
BENZENEACETIC ACID, 4-[4-(HYDROXYMETHYL)-2-THIAZ   138568-65-7
BENZENEACETIC ACID, _-HYDROXY-4-METHOXY-   010502-44-0
BENZENEACETIC ACID, 3-HYDROXY-_-METHYL-4-(2-THIA   132483-56-8
BENZENEACETIC ACID, 2-IODO-   018698-96-9
BENZENEACETIC ACID, α-METHYL-4-[4-(2-PROPENYL)-2   138568-64-6
BENZENEACETIC ACID, α-METHYL-4-(2-THIAZOLYL)-3-(132483-39-7
BENZENEACETIC ACID, _-METHYL-4-(4-PHENYL-2-THIAZ   132483-29-5
BENZENEACETIC ACID, α-METHYL-4-(4-METHYL-2-THIAZ   132483-42-2
BENZENEACETIC ACID, α-METHYL-4-(5-METHYL-2-THIAZ   132483-43-3
BENZENEACETIC ACID, 4-(2-METHYLPROPYL)-   001553-60-2
BENZENEACETIC ACID, α-METHYL-3-NITRO-4-(2-THIAZO   132483-59-1
BENZENEACETIC ACID, 3-METHOXY-α-METHYL-4-(2-THIA   132483-57-9
BENZENEACETIC ACID, 4-[4-(METHOXYMETHYL)-2-THIAZ   138568-66-8
BENZENEACETIC ACID, α-METHYL-4-(4,5,6,7-TETRAHYD   138568-69-1
BENZENEACETIC ACID, α-METHYL-4-[4-(TRIFLUOROMETH   132483-45-5
BENZENEACETIC ACID, α-METHYL-4-[4-[(METHYLAMINO)   132483-51-3
BENZENEACETIC ACID, α-METHYL-4-(2-THIAZOLYL)-   132483-32-0
BENZENEACETIC ACID, α-(1-METHYLETHYL)-, METHYL E   072615-27-1
BENZENEACETIC ACID, α-METHYL-3-(METHYLTHIO)-4-(2   132483-58-0
BENZENEACETIC ACID, α-METHYL-4-[4-(1-METHYLETHYL   138568-63-5
BENZENEACETIC ACID, 4-PENTYL-   014377-21-0
BENZENEACETIC ACID, 4-PROPYL-   026114-12-5
BENZENEACETIC ACID, 4-PROPOXY-   026118-57-0
BENZENEACETIC ACID, ETHYL ESTER   000101-97-3
BENZENEACETONITRILE, 4-CHLORO-.ALPHA.-(1-METHYLE   002012-81-9
BENZENEACETONITRILE, 4-METHYL-   002947-61-7
BENZENEACETONITRILE, α-(1-METHYLETHYL)-4-NITRO-   081310-40-9
BENZENE, 1,4-BIS(METHOXYMETHYL)-   006770-38-3
BENZENE, 1,3-BIS(TRIFLUOROMETHYL)-   000402-31-3
BENZENE, 1,4-BIS(TRIFLUOROMETHYL)-   000433-19-2
BENZENE, 1-BROMO-3-ISOTHIOCYANATO-   002131-59-1
BENZENEBUTANAMINE, HYDROCHLORIDE   030684-06-1

BENZENEBUTANOIC ACID, 2-METHYL-4-OXO-4-	001771-65-9
BENZENEBUTANOIC ACID, 4-METHOXY-¥-OXO-	003153-44-4
BENZENEBUTANOIC ACID, GAMMA -OXO-	002051-95-8
BENZENE, (1-BUTYLDECYL)-	004534-56-9
BENZENE, (1-BUTYLHEPTYL)-	004537-15-9
BENZENE, (1-BUTYLHEXYL)-	004537-11-5
BENZENE, (1-BUTYLNONYL)-	004534-50-3
BENZENE, (1-BUTYLOCTYL)-	002719-63-3
BENZENECARBOTHIAMIDE, 4-METHYL-	002362-62-1
BENZENECARBOTHIAMIDE, 4-(NITRO)-	026060-30-0
BENZENECARBOTHIAMIDE, 4-(METHYLTHIO)-	053550-91-7
BENZENECARBOTHIAMIDE, 4-METHOXY-	002362-64-3
BENZENECARBOTHIAMIDE, 4-CHLORO-	002521-24-6
BENZENECARBOTHIOAMIDE, N-PHENYL-	000636-04-4
BENZENE, 1-CHLORO-4-(CHLOROMETHYL)-	000104-83-6
BENZENE, 1-CHLORO-4-ISOTHIOCYANATO-	002131-55-7
BENZENE, 4-CHLORO-1-METHYL-2-NITRO-	000089-59-8
BENZENE, 1,1',1"-(CHLOROMETHYLIDYNE)TRIS-	000076-83-5
BENZENE, 1-CHLORO-2-METHYL-3-NITRO-	000083-42-1
1,2-BENZENEDIACETIC ACID	007500-53-0
1,4-BENZENEDIAMINE	000106-50-3
1,3-BENZENEDIAMINE	000108-45-2
4-((4-AMINOPHENYL)AZO)1,3-BENZENEDIAMINE	006364-34-7
1,2-BENZENEDIAMINE	000095-54-5
1,3-BENZENEDIAMINE, 4-[(4-AMINOPHENYL)SULFONYL]-	035880-91-2
1,3-BENZENEDIAMINE, 4-CHLORO-	005131-60-2
1,4-BENZENEDIAMINE, 2-NITRO-	005307-14-2
1,4-BENZENEDICARBONITRILE	000623-26-7
1,3-BENZENEDICARBOXAMIDE, TRIIODO DERIVATIVE	031122-82-4
1,3-BENZENEDICARBOXAMIDE, 5-(ACETYLAMINO)-N,N'-B	031122-84-6
1,4-BENZENEDICARBOXAMIDE, N,N'-DIPHENYL-	007154-31-6
1,2-BENZENEDICARBOXAMIDE	000088-96-0
1,3-BENZENEDICARBOXYLIC ACID, DIBUTYL ESTER	003126-90-7
BENZENE, 1,1'-(2,2-DICHLOROCYCLOPROPYLIDENE)BIS-	003141-42-2
BENZENE, 1,3-DICHLORO-5-NITRO-	000618-62-2
BENZENE, 1,3-DIFLUORO-5-NITRO-	002265-94-3
BENZENE, 1,4-DIFLUORO-2-NITRO-	000364-74-9
1,2-BENZENEDIOL, 4-[1-HYDROXY-2-[(1-METHYLETHYL)	000051-30-9
1,3-BENZENEDIOL, 4-[(4-AMINOPHENYL)SULFONYL]-	105456-60-8
1,2-BENZENEDIOL, 4-BUTYL-	002525-05-5
1,4-BENZENEDIOL, 2,3-DIMETHOXY-5-METHYL-	003066-90-8
1,4-BENZENEDIOL, 2,3-DIMETHOXY-	052643-52-4
1,4-BENZENEDIOL, 2,3-DIMETHYL-	000608-43-5
1,2-BENZENEDIOL, 4-(1-HEXENYL)-	100668-21-1
1,4-BENZENEDIOL, 2-METHOXY-	000824-46-4
1,4-BENZENEDIOL, 2-METHYL-	000095-71-6
1,2-BENZENEDIOL, 4-METHYL-	000452-86-8
1,3-BENZENEDIOL, 4-NITRO-	003163-07-3
1,3-BENZENEDIOL, 4-PENTADECYL-	016825-54-0
1,2-BENZENEDIOL, 4-PENTYL-	002525-11-3
1,3-BENZENEDIOL, 4-TRIDECYL-	123063-34-3
1,2-BENZENEDIOL, 3-TRIDECYL-	028165-91-5
1,2-BENZENEDIOL, 3-UNDECYL-	021704-31-4
1,2-BENZENEDIOL, 4-UNDECYL-	123063-36-5
1,4-BENZENEDISULFONAMIDE	016993-45-6
1,3-BENZENEDISULFONAMIDE	003701-01-7
BENZENEETHANAMINE, 4-BUTYL-2,5-DIMETHOXY-α-METHY	064638-09-1
BENZENEETHANAMINE, 3,5-DIMETHOXY-4-(PHENYLMETHOX	002176-16-1

BENZENEETHANAMINE, 3,4-DIMETHOXY-A-METHYL	000120-26-3
BENZENEETHANAMINE, 2,3-DIMETHOXY-α-METHYL-	015402-81-0
BENZENEETHANAMINE, N,ß-DIMETHYL-	000093-88-9
BENZENEETHANAMINE, A-METHYL-, (R)-	000156-34-3
BENZENEETHANAMINE, 2-METHYL-	055755-16-3
BENZENEETHANAMINE, 4-METHYL-	003261-62-9
BENZENEETHANAMINE, ß-METHOXY-N,α-DIMETHYL-, (R*,	092936-21-5
BENZENEETHANAMINE, ß-METHOXY-N,α-DIMETHYL-, (R*,	092936-20-4
BENZENEETHANAMINE, N-METHYL-	000589-08-2
BENZENEETHANAMINE, N-METHYL-α-(PHENYLMETHYL)-	053660-20-1
BENZENEETHANOL, beta-[2-(DIMETHYLAMINO)PROPYL]-_	017199-55-2
BENZENEETHANOL, .ALPHA.-METHYL-	000698-87-3
BENZENE, (1-ETHYLDODECYL)-	004534-58-1
BENZENE, (1-ETHYLNONYL)-	004536-87-2
BENZENE, (1-ETHYLOCTYL)-	004621-36-7
BENZENE, (1-ETHYLUNDECYL)-	004534-52-5
BENZENE, 1-IODO-4-ISOTHIOCYANATO-	002059-76-9
BENZENE, 1,1'-(ISOTHIOCYANATOMETHYLENE)BIS-	003550-21-8
BENZENE, 1-ISOTHIOCYANATO-3-METHOXY	003125-64-2
BENZENEMETHANAMINE, N,N-BIS(2-HYDROXYETHYL)-3-(P	144458-99-1
BENZENEMETHANAMINE, N-BUTYL-	002403-22-7
BENZENEMETHANAMINE, 3-BROMO-N-METHYL-N-NITROSO-	098736-47-1
BENZENEMETHANAMINE, 4-BROMO-N-METHYL-N-NITROSO-	098736-50-6
BENZENEMETHANAMINE, 3-CHLORO-N-METHYL-N-NITROSO-	098736-46-0
BENZENEMETHANAMINE, 4-CHLORO-N-METHYL-N-NITROSO-	084174-22-1
BENZENEMETHANAMINE, N,3-DIMETHYL-N-NITROSO-	062783-49-7
BENZENEMETHANAMINE, N,4-DIMETHYL-N-NITROSO-	062783-50-0
BENZENEMETHANAMINE, N-ETHYL-	014321-27-8
BENZENEMETHANAMINE, 4-FLUORO-N-METHYL-N-NITROSO-	084174-21-0
BENZENEMETHANAMINE, N-METHYL-N-NITROSO-3-(TRIFLU	098736-48-2
BENZENEMETHANAMINE, N-(1-METHYLETHYL)-	000102-97-6
BENZENEMETHANAMINE, 3-METHOXY-N-METHYL-N-NITROSO	098736-45-9
BENZENEMETHANAMINE, 4-METHYL-	000104-84-7
BENZENEMETHANAMINE, N-METHYL-3-NITRO-N-NITROSO-	098736-52-8
BENZENEMETHANAMINE, N-METHYL-4-NITRO-N-NITROSO-	084174-24-3
BENZENEMETHANAMINIUM, 3-HYDROXY-N,N,N-TRIMETHYL-	002498-27-3
BENZENEMETHANESULFONAMIDE, N-[(METHYLAMINO)CARBO	121822-76-2
BENZENEMETHANESULFONAMIDE, N-METHYL-N-[(METHYLAM	121822-78-4
BENZENEMETHANESULFONAMIDE	004563-33-1
BENZENEMETHANESULFONAMIDE, N-[(DIMETHYLAMINO)CAR	121822-77-3
BENZENEMETHANOL, .ALPHA.-(1-AMINOETHYL)-, (R*,R*	036393-56-3
BENZENEMETHANOL, 2-[[2-(AMINOMETHYL)PHENYL]THIO]	079467-22-4
BENZENEMETHANOL, 2-[BIS(4-HYDROXYPHENYL)METHYL]-	000081-92-5
BENZENEMETHANOL, 5-BROMO-2-HYDROXY-	002316-64-5
BENZENEMETHANOL, 4-CHLORO-.ALPHA.-PHENYL-	000119-56-2
BENZENEMETHANOL, 2-CHLORO-	017849-38-6
BENZENEMETHANOL, 5-CHLORO-2-HYDROXY-	005330-38-1
BENZENEMETHANOL, _-[(CYCLOHEXYLAMINO)METHYL]-, H	070732-30-8
BENZENEMETHANOL, 3,5-DIBROMO-2-HYDROXY-	002183-54-2
BENZENEMETHANOL, 3,5-DICHLORO-2-HYDROXY-	006641-02-7
BENZENEMETHANOL, 2,6-DICHLORO-	015258-73-8
BENZENEMETHANOL, 3,5-DICHLORO-	060211-57-6
BENZENEMETHANOL, 2,6-DIFLUORO-	019064-18-7
BENZENEMETHANOL, 3,4-DIHYDROXY-α-PENTYL-	118198-71-3
BENZENEMETHANOL, _-[1-(DIMETHYLAMINO)ETHYL]-, (R	001201-56-5
BENZENEMETHANOL, .ALPHA.,.ALPHA.-DIPHENYL-	000076-84-6
BENZENEMETHANOL, 4-FLUORO-	000459-56-3
BENZENEMETHANOL, 2-FLUORO-	000446-51-5

BENZENEMETHANOL, 2-HYDROXY-5-METHYL-	004383-07-7
BENZENEMETHANOL, .ALPHA.- 1-(METHYLAMINO)ETHYL -	000090-82-4
BENZENE, (1-METHYLDECYL)-	004536-88-3
BENZENE, (1-METHYLDODECYL)-	004534-53-6
BENZENE, 1,1'-METHYLENEBIS-	000101-81-5
BENZENE, [[(1-METHYLETHENYL)OXY]METHYL]-	032783-20-3
BENZENE, (1-METHYLNONYL)-	004537-13-7
BENZENE, 1-[3-(METHYLSULFINYL)PROPOXY]-4-NITRO-	105412-24-6
BENZENE, 1-[3-(METHYLTHIO)PROPOXY]-4-NITRO-	075032-34-7
BENZENE, (1-METHYLTRIDECYL)-	004534-59-2
BENZENE, (1-METHYLUNDECYL)-	002719-61-1
BENZENE, 1-NITRO-4-(TRIFLUOROMETHYL)-	000402-54-0
BENZENEPENTACARBOXYLIC ACID	001585-40-6
BENZENEPENTANAMINE, HYDROCHLORIDE	053429-15-5
BENZENE, (1-PENTYLHEPTYL)-	002719-62-2
BENZENE, (1-PENTYLHEXYL)-	004537-14-8
BENZENE, (1-PENTYLNONYL)-	004534-55-8
BENZENE, (1-PENTYLOCTYL)-	004534-49-0
BENZENEPROPANAMIDE, N-(3,4-DICHLOROPHENYL)-	086886-77-3
BENZENEPROPANAMIDE, N-METHYL-	000940-43-2
BENZENEPROPANAMINE, N,N-BIS(2-CARBONYLETHYL)-4-[	144458-93-5
BENZENEPROPANAMINE, 4-(1,1-DIMETHYLETHYL)-N,N-DI	070785-03-4
BENZENEPROPANAMINE, α-METHYL-	022374-89-6
BENZENEPROPANOIC ACID, 4-(PHENYLMETHOXY)-	050463-48-4
BENZENEPROPANOIC ACID, ß-(AMINOMETHYL)-4-FLUORO-	052237-19-1
BENZENEPROPANOIC ACID, ß-(AMINOMETHYL)-4-CHLORO-	001134-47-0
BENZENE, (1-PROPYLDECYL)-	004534-51-4
BENZENE, (1-PROPYLHEPTYL)-	004537-12-6
BENZENE, (1-PROPYLOCTYL)-	004536-86-1
BENZENE, (1-PROPYLUNDECYL)-	004534-57-0
BENZENESULFANILIDE	001678-25-7
BENZENESULFINIC ACID, SODIUM SALT	000873-55-2
BENZENESULFONAMIDE, 4-CHLORO-N-[(CYCLOHEXYLAMINO	000963-03-1
BENZENESULFONAMIDE, 3-CHLORO-4-[(3-HYDROXYPROPYL	108966-73-0
BENZENESULFONAMIDE	000098-10-2
BENZENESULFONAMIDE, 4-AMINO-N-(2-NITROPHENYL)-	019837-88-8
BENZENESULFONAMIDE, 4-AMINO-N-[6-METHOXY-5-(1-ME	005018-16-6
BENZENESULFONAMIDE, 4-AMINO-N-(5,6-DIMETHOXY-4-P	002447-57-6
BENZENESULFONAMIDE, 4-AMINO-N-(2,4-DIMETHYL-5-PY	007411-79-2
BENZENESULFONAMIDE, 3-AMINO-4-[(2-HYDROXYETHYL)S	135832-45-0
BENZENESULFONAMIDE, 2-AMINO-4-METHOXY-N-(2-METHY	081792-76-9
BENZENESULFONAMIDE, 4-AMINO-N-(6-METHOXY-4-PYRIM	001220-83-3
BENZENESULFONAMIDE, 2-AMINO-N-(2-METHYL-1-OXOPRO	081792-73-6
BENZENESULFONAMIDE, 3-AMINO-4-[(3-HYDROXYPROPYL)	108966-61-6
BENZENESULFONAMIDE, 4-AMINO-N- (BUTYLAMINO)CARBO	000339-43-5
BENZENESULFONAMIDE, N-(AMINOCARBONYL)-	035207-08-0
BENZENESULFONAMIDE, 4-AMINO-N-2-QUINOXALINYL-	000059-40-5
BENZENESULFONAMIDE, 3-ACETYL-	035203-88-4
BENZENESULFONAMIDE, N-(2-AMINOETHYL)-N-METHYL-4-	135685-04-0
BENZENESULFONAMIDE, 4-BUTOXY-	001138-58-5
BENZENESULFONAMIDE, 3-BROMO-	089599-01-9
BENZENESULFONAMIDE, 3-BUTOXY-	123045-57-8
BENZENESULFONAMIDE, N-[(BUTYLAMINO)CARBONYL]-4-C	013909-64-3
BENZENESULFONAMIDE, N-(2-CYCLOPENTYL-2H-BENZOTRI	130787-22-3
BENZENESULFONAMIDE, 4-CHLORO-N-[(ETHYLAMINO)CARB	024535-70-4
BENZENESULFONAMIDE, 4-CHLORO-3-NITRO-	000097-09-6
BENZENESULFONAMIDE, 3-CYANO-	003118-68-1
BENZENESULFONAMIDE, 4-CHLORO-N-METHYL	006333-79-5

BENZENESULFONAMIDE, 4-CHLORO-N-[(PHENYLAMINO)CAR	025270-44-4
BENZENESULFONAMIDE, 4-CHLORO-N-[(METHYLAMINO)CAR	052102-43-9
BENZENESULFONAMIDE, N-[[(1,1-DIMETHYLETHYL)AMINO	052102-41-7
BENZENESULFONAMIDE, 4-(1,3-DIETHYL-2,3,6,7-TETRA	089073-49-4
BENZENESULFONAMIDE, N-[(DIETHYLAMINO)CARBONYL]-	050618-71-8
BENZENESULFONAMIDE, N-[(DIMETHYLAMINO)CARBONYL]-	032324-42-8
BENZENESULFONAMIDE, N,4-DIMETHYL-N-NITROSO-	000080-11-5
BENZENESULFONAMIDE, N,4-DIMETHYL-	000640-61-9
BENZENESULFONAMIDE, 3,4-DICHLORO-	023815-28-3
BENZENESULFONAMIDE, N-[2-(DIMETHYLAMINO)ETHYL]-N	126813-40-9
BENZENESULFONAMIDE, N-[(ETHYLAMINOCARBONYL]-4-ME	001467-23-8
BENZENESULFONAMIDE, N-[(ETHYLAMINO)CARBONYL]-	032324-41-7
BENZENESULFONAMIDE, 3-FLUORO-4-[(3-HYDROXYPROPYL	108966-77-4
BENZENESULFONAMIDE, 3-FLUORO-4-[(3-HYDROXYPROPYL	108966-74-1
BENZENESULFONAMIDE, 3-FLUORO-4-[(4-HYDROXYBUTYL)	108966-75-2
BENZENESULFONAMIDE, 3-FLUORO-4-[(2-HYDROXYETHYL)	108966-76-3
BENZENESULFONAMIDE, 3-FLUORO-4-[(2-HYDROXYETHYL)	108966-70-7
BENZENESULFONAMIDE, 4-[(4-HYDROXYBUTYL)SULFONYL]	135832-42-7
BENZENESULFONAMIDE, 4-[(4-HYDROXYBUTYL)SULFONYL]	108966-78-5
BENZENESULFONAMIDE, 4-[(2-HYDROXYETHYL)THIO]-	108966-48-9
BENZENESULFONAMIDE, 4-(4-HYDROXYBUTYL)-	135832-48-3
BENZENESULFONAMIDE, 4-[(5-HYDROXYPENTYL)THIO]-	108966-54-7
BENZENESULFONAMIDE, 4-(5-HYDROXYPENTYL)-	135832-50-7
BENZENESULFONAMIDE, 4-[(4-HYDROXYBUTYL)THIO]-	135832-41-6
BENZENESULFONAMIDE, 4-[(3-HYDROXYPROPYL)THIO]-3-	108966-59-2
BENZENESULFONAMIDE, 4-[(2-HYDROXYETHYL)THIO]-3-N	108966-58-1
BENZENESULFONAMIDE, 4-[(3-HYDROXYPROPYL)THIO]-	108966-51-4
BENZENESULFONAMIDE, 4-[3-HYDROXY-3-METHYLBUTYL)T	108966-53-6
BENZENESULFONAMIDE, 4-(HEXYLOXY)-3-NITRO-	123045-58-9
BENZENESULFONAMIDE, 4-(HEXYLOXY)-	001145-46-6
BENZENESULFONAMIDE, 4-(3-HYDROXYPROPYL)-	135832-46-1
BENZENESULFONAMIDE, 4-[3-HYDROXY-3-METHYLBUTYL)S	108966-56-9
BENZENESULFONAMIDE, N-HYDROXY-	000599-71-3
BENZENESULFONAMIDE, 4-[(3-HYDROXYPROPYL)SULFONYL	108966-60-5
BENZENESULFONAMIDE, 4-[(3-HYDROXYPROPYL)SULFONYL	108966-55-8
BENZENESULFONAMIDE, 4-[(2-HYDROXYETHYL)SULFONYL]	108966-49-0
BENZENESULFONAMIDE, 3-IODO-	050702-39-1
BENZENESULFONAMIDE, 2-[(2-METHYL-1-OXOPROPYL)AMI	090873-90-8
BENZENESULFONAMIDE, 4-METHYL-N-[(PHENYLAMINO)CAR	013909-63-2
BENZENESULFONAMIDE, 3-(1-METHYLETHYL)-	123045-56-7
BENZENESULFONAMIDE, 4-[[2-[(2-METHYLPROPYL)AMINO	135832-44-9
BENZENESULFONAMIDE, 3-METHOXY-	058734-57-9
BENZENESULFONAMIDE, 4-[(METHYLSULFONYL)AMINO]-	004426-90-8
BENZENESULFONAMIDE, 4-(1-METHYLETHYL)-	006335-39-3
BENZENESULFONAMIDE, N-[(METHYLAMINO)CARBONYL]-	052102-38-2
BENZENESULFONAMIDE, 4-METHOXY-2-[(2-METHYL-1-OXO	097141-31-6
BENZENESULFONAMIDE, N-[[(1-METHYLETHYL)AMINO]CAR	003149-01-7
BENZENESULFONAMIDE, 4-METHYL-2-[(2-METHYL-1-OXOP	097141-29-2
BENZENESULFONAMIDE, 2-NITRO-N-(2-METHYL-1-OXOPRO	081792-98-5
BENZENESULFONAMIDE, N-(PHENYLSULFONYL)-	002618-96-4
BENZENESULFONAMIDE, N-[(PROPYLAMINO)CARBONYL]-	004932-53-0
BENZENESULFONIC ACID, 2-[2-[4-(2H-NAPHTHO[1,2-D]	037571-27-0
BENZENESULFONIC ACID HYDRAZIDE	000080-17-1
BENZENESULFONIC ACID, 3-AMINO-, MONOSODIUM SALT	001126-34-7
BENZENESULFONIC ACID, 4-AMINO-	000121-57-3
BENZENESULFONIC ACID	000098-11-3
4-ME-BENZENESULFONYL-L-(N'-BU)GLUTAMINE	083871-00-5
BENZENESULFONYL CHLORIDE	000098-09-9

4ME-BENZENESULFONYL-L-(N'-CYHX)GLUTAMINE	083871-03-8
BENZENESULFONYL-L-(N'-ET)GLUTAMINE	083870-89-7
BENZENESULFONYL-L-GLUTAMINE	083870-87-5
4-ME-BENZENESULFONYL-L-(N'-HEX)GLUTAMINE	083871-02-7
BENZENESULFONYL-L-(N'-PENTYL)GLUTAMINE	083870-92-2
4-ME-BENZENESULFONYL-L-(N'-PR)GLUTAMINE	083870-99-9
BENZENESULFONYL-L-(N'-PR)GLUTAMINE	083870-90-0
1,2,3,5-BENZENETETRACARBOXYLIC ACID	000479-47-0
BENZENE, 1,2,4,5-TETRACHLORO-3-METHOXY-	006936-40-9
BENZENE, 1,1'-(1,1,2,2-TETRAFLUORO-1,2-ETHANEDIY	000425-32-1
BENZENE, (1-ETHYLDECYL)-	002400-00-2
1,3,5-BENZENETRICARBOXYLIC ACID	000554-95-0
BENZENE, 1,2,3-TRICHLORO-5-METHOXY-	054135-82-9
BENZENE, 1,3,5-TRICHLORO-2-NITRO-	018708-70-8
BENZENE, 1,2,4-TRICHLORO-5-PHENOXY-	052322-80-2
BENZENE, 1,2,4-TRIFLUORO-	000367-23-7
BENZENE, 1,2,3-TRIMETHOXY-	000634-36-6
1,3,5-BENZENETRIOL, 2-[(4-AMINOPHENYL)SULFONYL]-	105456-65-3
BENZHYDROL	000091-01-0
BENZIDINE	000092-87-5
BENZIDINE DIHYDROCHLORIDE	000531-85-1
BENZIL	000134-81-6
BENZILIC ACID	000076-93-7
1H-BENZIMIDAZOL-2-AMINE, 5-CHLORO-	005418-93-9
1H-BENZIMIDAZOL-2-AMINE, 6-[(HYDROXYIMINO)PHENYL	072301-79-2
4,5,7-TRICL-2-CF3-BENZIMIDAZOLE	003393-59-7
2-CF3-5,6-DIBROMO-BENZIMIDAZOLE	006587-21-9
BENZIMIDAZOLE	000051-17-2
5-CL-6-NO2-2-CF3-BENZIMIDAZOLE	006609-40-1
1H-BENZIMIDAZOLE-5-CARBOXYLIC ACID, 6-METHYL-2-[	078090-11-6
1H-BENZIMIDAZOLE-1-CARBOXYLIC ACID, 5-BENZOYL-2-	104662-93-3
BENZIMIDAZOLE-5-CHLORO-2-MES	007692-57-1
1H-BENZIMIDAZOLE, 2-METHYL-5-NITRO-	001792-40-1
1H-BENZIMIDAZOLE, 6-NITRO-	089843-47-0
BENZIMIDAZOLE-2,4,5,6,7-PENTACL	007682-34-0
1H-BENZIMIDAZOLE, 2-[(2-PYRIDINYLMETHYL)SULFINYL	057237-97-5
BENZIMIDAZOLE-4,5,6-TRIBR-2-CF3	007682-32-8
1H-BENZIMIDAZOLE, 4,5,6-TRICHLORO-2-(TRIFLUOROME	002338-27-4
2-BENZIMIDAZOLINONE	003097-21-0
2H-BENZIMIDAZOL-2-ONE, 1-[1-[4,4-BIS(4-FLUOROPHE	053179-12-7
BENZIODARONE	000068-90-6
1H-BENZ[DE]ISOQUINOLINE-1,3(2H)-DIONE, 2-[(ACETY	010245-51-9
1,2-BENZISOTHIAZOLE, 3-(1-PIPERAZINYL)-	087691-87-0
BENZISOXAZOLE	000271-58-9
1,2-BENZISOXAZOLE	000271-95-4
BENZISOXAZOLE, 5-NITRO	039835-28-4
BENZOBICYCLO(2.2.1)HEPTENE,9EX-NH2	014098-20-5
BENZOBICYCLO(221)HEPTENE,2EX-MEAMINO	086943-79-5
BENZOBICYCLO(221)HEPTENE,2EX-NH2-6CF3	083118-50-7
BENZOBICYCLO(221)HEPTENE,2EN-NH2-6CF3	083118-48-3
BENZOBICYCLO(2,2,1)HEPTENE,9-EN-AMINO	072597-35-4
BENZOBICYCLO(221)HEPTENE,2EN-NH2-7CF3	086022-72-2
BENZOBICYCLO(221)HEPTENE,2EX-NH2-7CF3	083118-51-8
BENZOBICYCLO(2,2,1)HEPTENE,2-EX-AMINO	062624-26-4
BENZOBICYCLO(2,2,1)HEPTENE,2-EX-MEAM	062624-27-5
BENZOBICYCLO(2,2,1)HEPTENE,2-EN-AMINO	058742-04-4
BENZOBICYCLO(221)HEPTENE,2EN-ETAMINO	086992-68-9
BENZOBICYCLO(221)HEPTENE,2EX-ETAMINO	086943-78-4

BENZOBICYCLO(221)HEPTENE,9EN-MEAMINO	086992-69-0
BENZOBICYCLO(221)HEPTENE,2EN-PRAMINO	086992-67-8
BENZOBICYCLO(221)HEPTENE,2EX-PRAMINO	086943-77-3
BENZOBICYCLO(222)OCTENE,2ENDO-MEAMINO	018883-06-2
BENZOBICYCLO(222)OCTENE,2-EXO-MEAMINO	018883-05-1
BENZO[B]CHRYSENE	000214-17-5
BENZO[C]CHRYSENE	000194-69-4
B33C11-BENZO CROWN ETHER	104946-62-5
B27C9-BENZO CROWN ETHER	063144-76-3
B30C10-BENZO CROWN ETHER	077963-50-9
BENZO-15-CROWN-5-ETHER	014098-44-3
B18C6-BENZO CROWN ETHER	014098-24-9
B21C7-BENZO CROWN ETHER	067950-78-1
B24C8-BENZO CROWN ETHER	072216-45-6
BENZO-18-CROWN-6-ETHER,M-CARBOXY-	060835-75-8
H-1,4-BENZODIAZEPIN-2-AMINE, 7-CHLORO-5-PHENYL-	007722-15-8
H-1,4-BENZODIAZEPINE-1-CARBOXAMIDE, 2,3-DIHYDRO	027016-91-7
2H-1,4-BENZODIAZEPIN-2-ONE, 7-CHLORO-5-(2-FLUORO	049606-44-2
2H-1,4-BENZODIAZEPINE-2-THIONE, 7-CHLORO-5-(2-FL	036735-22-5
2H-1,4-BENZODIAZEPIN-2-ONE, 3-(ACETYLOXY)-1-[3-(	056875-83-3
H-1,4-BENZODIAZEPIN-2-ONE, 7-BROMO-1,3-DIHYDRO-	002894-61-3
1,4-BENZODIAZEPIN-2-ONE,5(CF3)PH	002285-16-7
1,4-BENZODIAZEPIN-2-ONE,5(4FPH..	001492-96-2
2H-1,4-BENZODIAZEPIN-2-ONE, 7-CHLORO-5-(2-CHLORO	002894-68-0
H-1,4-BENZODIAZEPIN-2-ONE, 7-CHLORO-1,3-DIHYDRO	004699-82-5
H-1,4-BENZODIAZEPIN-2-ONE, 7-CHLORO-1,3-DIHYDRO	000963-39-3
H-1,4-BENZODIAZEPIN-2-ONE, 7-CHLORO-1,3-DIHYDRO	030195-30-3
H-1,4-BENZODIAZEPIN-2-ONE, 7-CHLORO-1,3-DIHYDRO	023092-17-3
H-1,4-BENZODIAZEPIN-2-ONE, 7-CHLORO-1,3-DIHYDRO	056875-80-0
2H-1,4-BENZODIAZEPIN-2-ONE,1,3-DIHYDRO-1-METHYL-	003489-59-6
1,4-BENZODIAZEPIN-2-ONE,1,3-DIHYDRO-9-METHYL-7-N	004941-45-1
H-1,4-BENZODIAZEPIN-2-ONE, 1,3-DIHYDRO-7-METHYL	005571-63-1
1,4-BENZODIAZEPIN-2-ONE,1,3-DUHYDRO-7,9-DIMETHYL	005571-62-0
1,4-BENZODIAZEPIN-2-ONE,7-FLUORO-5-(2-FLUOROPHEN	002024-34-2
1,4BENZODIAZEPIN2ONE,1ME-5PH-7MES	023193-98-8
1,4-BENZODIAZPIN-2-ONE,5-(2CF3PH)	002730-05-4
1,4-BENZODIAZPIN2ONE,5(4CF3PH)	003894-63-1
1,4-BENZODIAZPIN-2-ONE,5-(2-F PH)	002648-01-3
1,4-BENZODIAZPIN-2-ONE,5(2FPH)7CL	002886-65-9
1,4-BENZODIAZPIN-2-ONE,5-FPH,7NO2	002558-30-7
1,4-BENZODIAZPIN-2-ONE,5-PH,7-F	002648-00-2
1,4-BENZODIAZPIN-2-ONE,5(PH)7MES	002891-12-5
14-BENZODIAZP2-ON,1-MEOME-5(2FPH)-7AM	040837-34-1
14-BENZODIAZP2-ON,1-MEOME-5(PH)-7CL	029442-59-9
1,4-BENZODIAZPN2-ON,5(2-MES PH)7CL	000846-53-7
1,4-BENZODIAZPN2-ON,5(2CF3PH)7CL	003864-49-1
1,4-BENZODIAZPN2-ON,5(2CF3PH)7NO2	001427-45-8
1,4-BENZODIAZPN2-ONE,1-ME-5(PH)7NH2	004959-16-4
14-BENZODIAZPN2-ON,5(2-F PH)7ACETYL	036093-54-6
14-BENZODIAZPN2-ON,1,3-ME-5(2CLPH)7CL	088695-11-8
1,4-BENZODIAZPN2-ON,1-ME-5(26CLPH)7CL	030144-88-8
14-BENZODIAZPN2-ON,1-ME-5(2-F PH)7NH2	034084-50-9
1,4-BENZODIAZP2-ON,1-ET-5-PH-7CL	005571-65-3
1,4-BENZODIAZPN2-ON,1-ME5(4-CLPH)7F	001959-97-3
14-BENZODIAZP2-ON,1-MEOME-5(2CLPH)7CL	029442-60-2
1,4-BENZODIAZP-2-ON,5(2MEOPH)7CL	003023-44-7
1,4-BENZODIAZP2-ON,5(3MEOPH)7CL	005358-92-9
1,4-BENZODIAZP2-ON,5(2NO2PH)7NO2	004980-73-8

1,4-BENZODIAZP2-ON,5(OMEPH)7CL	005358-35-0
1,4-BENZODIAZP2-ON,5(PH)7MEO	005358-96-3
1,3-BENZODIOLE-5-CARBOXYLIC ACID, HYDRAZIDE	022026-39-7
1,4-BENZODIOXAN	000493-09-4
1,4-BENZODIOXANE-2-CARBOXYLIC ACID	034385-93-8
1,3-BENZODIOXOLE	000274-09-9
1,3-BENZODIOXOLE-5-CARBOXYLIC ACID, 2-(AMINOTHIO	052190-69-9
1,3-BENZODIOXOLE-5-ETHANAMINE, _-METHYL-(±)-	051497-09-7
1,3-BENZODIOXOLE-5-METHANOL	000495-76-1
1H-BENZO(CD)FLUOANTHENE	042126-84-1
BENZO(B)FLUORANTHENE	000205-99-2
BENZO(J)FLUORANTHENE	000205-82-3
BENZO(K)FLUORANTHENE	000207-08-9
BENZO(A)FLUORENE	030777-18-5
7H-BENZO(C)FLUORENE	000205-12-9
BENZO(B)FLUORENE	030777-19-6
11H-BENZO(A)FLUORENE	000238-84-6
11H-BENZO(B)FLUORENE	000243-17-4
BENZOFLUORFEN	077501-60-1
BENZOFURAN	000271-89-6
2-BENZOFURANCARBOXAMIDE, N-METHYL-	064663-59-8
2-BENZOFURANCARBOXYLIC ACID, ETHYL ESTER	003199-61-9
2-BENZOFURANCARBOXYLIC ACID	000496-41-3
2-BENZOFURANCARBOXYLIC ACID, METHYL ESTER	001646-27-1
BENZOFURAZAN	000273-09-6
BENZOFURAZAN, 1-OXIDE	000480-96-6
BENZOGUANAMINE	000091-76-9
2'-CL-5'-CF3-BENZOGUANAMINE	065052-53-1
BENZO(H)QUINOLINE	000230-27-3
BENZOHYDROXAMIC ACID	000495-18-1
BENZOIC ACID,O-ACETYLMETHYLAMINO	078944-67-9
BENZOIC ACID, 1-METHYLETHYL ESTER	000939-48-0
BENZOIC ACID, 5-(AMINOSULFONYL)-2-[(3-HYDROXYPRO	108966-66-1
BENZOIC ACID, METHYL ESTER	000093-58-3
BENZOIC ACID	000065-85-0
BENZOIC ACID, 2-CHLORO-4-[[(DIMETHYLAMINO)CARBON	088132-43-8
BENZOIC ACID, 4-CHLORO-2-[(4-CHLOROPHENYL)AMINO]	090656-45-4
BENZOIC ACID, 4-CHLORO-2-[(3-CHLOROPHENYL)AMINO]	017870-85-8
BENZOIC ACID, 2-[(ACETYLAMINO)AMINO]-	007496-53-9
BENZOIC ACID, 4-(ACETYLAMINO)-2-HYDROXY	000050-86-2
BENZOIC ACID, 3-(ACETYLAMINO)-, METHYL ESTER	052189-36-3
BENZOIC ACID, 4-[3-[[4-(ACETYLAMINO)PHENYL]METHY	065542-19-0
BENZOIC ACID, 2-(ACETYLAMINO)-, METHYL ESTER	002719-08-6
BENZOIC ACID, 2-(ACETYLOXY)-, 2-[(2-AMINO-2-OXOE	116482-78-1
BENZOIC ACID, 2-(ACETYLOXY)-, 2-(ACETYLMETHYLAMI	118247-08-8
BENZOIC ACID, 2-(ACETYLOXY)-, 2-[(2-ETHOXY-2-OXO	116482-77-0
BENZOIC ACID, 2-(ACETYLOXY)-, 2-[(2-HYDROXYETHYL	118247-05-5
BENZOIC ACID, 2-(ACETYLOXY)-, 2-[BIS(1-METHYLETH	116482-76-9
BENZOIC ACID, 2-(ACETYLOXY)-, 2-[(2-AMINO-2-OXOE	118247-02-2
BENZOIC ACID, 2-ACETYLOXY ET ESTER	038353-69-4
BENZOIC ACID, 2-(ACETYLOXY)-, (2,5-DIOXO-1-PYRRO	032620-72-7
BENZOIC ACID, 2-(ACETYLOXY)-, 2-AMINO-2-OXOETHYL	050785-22-3
BENZOIC ACID, 2-(ACETYLOXY)-, 2-(ETHYLAMINO)-2-O	118247-01-1
BENZOIC ACID, 2-(ACETYLOXY)-, 2-(DIETHYLAMINO)-2	116482-56-5
BENZOIC ACID, 2-(ACETYLOXY)-, (METHYLSULFONYL)ME	076432-35-4
BENZOIC ACID, 2-(ACETYLOXY)-, (ACETYLOXY)METHYL	032620-68-1
BENZOIC ACID, 2-(ACETYLOXY)-, 2-(DIMETHYLAMINO)-	118247-04-4
BENZOIC ACID, 2-(ACETYLOXY)-, PHENYL ESTER	000134-55-4

BENZOIC ACID, 2-(ACETYLOXY)-, 2-(DIPROPYLAMINO)-	116482-75-8
BENZOIC ACID, 2-(ACETYLOXY)-, (1-OXOBUTOXY)METHY	118247-07-7
BENZOIC ACID, 2-(ACETYLOXY)-, 2-(4-MORPHOLINYL)-	116482-80-5
BENZOIC ACID, 2-(ACETYLOXY)-, (METHYLTHIO)METHYL	076432-30-9
BENZOIC ACID, 2-(ACETYLOXY)-, 4-(ACETYLAMINO)PHE	005003-48-5
BENZOIC ACID, 2-(ACETYLOXY)-, (METHYLSULFINYL)ME	076432-33-2
BENZOIC ACID, 2-[(3-ACETYLPHENYL)AMINO]-	027696-28-2
BENZOIC ACID, 4-[(AMINOCARBONYL)AMINO]-	006306-25-8
BENZOIC ACID, 4-AMINO-3-FLUORO-	000455-87-8
BENZOIC ACID, 4-AMINO-, HEPTYL ESTER	014309-40-1
BENZOIC ACID, 4-AMINO-, HEXYL ESTER	013476-55-6
BENZOIC ACID, 4-AMINO-2-HYDROXY-, 2,6-DIMETHYLPH	056356-16-2
BENZOIC ACID, 4-AMINO-3-IODO-	002122-63-6
BENZOIC ACID, 2-[(AMINOMETHOXYPHOSPHINOTHIOYL)OX	024353-61-5
BENZOIC ACID, 4-AMINO-2-METHOXY-	002486-80-8
BENZOIC ACID, 4-AMINO-2-METHYL-	002486-75-1
BENZOIC ACID, 4-AMINO-, OCTYL ESTER	014309-41-2
BENZOIC ACID, 4-AMINO-,PENTY ESTER	013110-37-7
BENZOIC ACID, 4-[(4-AMINOPHENYL)SULFONYL]-	034037-45-1
BENZOIC ACID, 5-(AMINOSULFONYL)-4-CHLORO-2-[[(2-	129625-27-0
BENZOIC ACID, 5-(AMINOSULFONYL)-2-[(3-HYDROXYPRO	108966-67-2
BENZOIC ACID, 5-(AMINOSULFONYL)-2-[(2-HYDROXYETH	135832-43-8
BENZOIC ACID, 4-(AMINOSULFONYL)-, ETHYL ESTER	005446-77-5
BENZOIC ACID, ANHYDRIDE	000093-97-0
BENZOIC ACID, 2-BENZOYL-, METHYL ESTER	000606-28-0
BENZOIC ACID, 4-BROMO-, METHYL ESTER	000619-42-1
BENZOIC ACID, 2-[(3-BROMOPHENYL)AMINO]-4-CHLORO-	097713-49-0
BENZOIC ACID, 4-(TERT-BUTYL)-	000098-73-7
BENZOIC ACID,4-CARBOXAMIDOBUTYL ESTER	105412-27-9
BENZOIC ACID, 2-CHLORO-6-FLUORO-	000434-75-3
BENZOIC ACID, 2-CHLORO-, 2-(4-FORMYL-2-METHOXYPH	093414-41-6
BENZOIC ACID, 4-CHLORO-, 2-(4-FORMYL-2-METHOXYPH	093414-43-8
BENZOIC ACID, 3-CHLORO-, 2-(4-FORMYL-2-METHOXYPH	093414-42-7
BENZOIC ACID, 3-CHLORO-, 2-[4-[(HYDROXYIMINO)MET	093414-47-2
BENZOIC ACID, 4-CHLORO-, 2-[4-[(HYDROXYIMINO)MET	093414-45-0
BENZOIC ACID, 4-CHLORO-2-[(4-IODOPHENYL)AMINO]-	090700-03-1
BENZOIC ACID, 2-CHLORO-6-METHYL-	021327-86-6
BENZOIC ACID, 3-CHLORO-, METHYL ESTER	002905-65-9
BENZOIC ACID, 2-CHLORO-6-NITRO-, METHYL ESTER	080563-87-7
BENZOIC ACID, 2-CHLORO-6-NITRO-	005344-49-0
BENZOIC ACID, 2-CHLORO-4-NITRO-	000099-60-5
BENZOIC ACID, 4-CHLORO-2-(PHENYLAMINO)-	019218-88-3
BENZOIC ACID, 2-[(3-CHLOROPHENYL)AMINO]-	013278-36-9
BENZOIC ACID, 4-[3-[(4-CHLOROPHENYL)METHYL]-3-ME	065542-14-5
BENZOIC ACID, 2-CHLORO-, PROPYL ESTER	025800-28-6
BENZOIC ACID, 4-CHLORO-2-[[3-(TRIFLUOROMETHYL)PH	066287-53-4
BENZOIC ACID, 5- 2-CHLORO-4-(TRIFLUOROMETHYL)PHE	050594-66-6
BENZOIC ACID, 4-CYANOBUTYL-	077298-35-2
BENZOIC ACID, O-CYANOMETHYLAMINO-	028354-19-0
BENZOIC ACID, 4-(DECYLOXY)-	005519-23-3
BENZOIC ACID, 3,4-DIAMINO-	000619-05-6
BENZOIC ACID, 2,4-DIAMINO-	000611-03-0
BENZOIC ACID, 2,6-DICHLORO-, METHYL ESTER	014920-87-7
BENZOIC ACID, 2-(DIETHYLAMINO)-2-OXOETHYL ESTER	064649-63-4
BENZOIC ACID, 2,6-DIFLUORO-, METHYL ESTER	013671-00-6
BENZOIC ACID, 2,6-DIFLUORO-	000385-00-2
BENZOIC ACID, 4-(3,3-DIMETHYL-1-TRIAZENYL)-	007203-91-0
BENZOIC ACID, 2,6-DIMETHYL-, METHYL ESTER	014920-81-1

BENZOIC ACID, 3,5-DINITRO-, METHYL ESTER 002702-58-1
BENZOIC ACID, 2-(ACETYLOXY)-, 2-ETHOXY-2-OXOETHY 050785-24-5
BENZOIC ACID, 4-ETHENYL- 001075-49-6
BENZOIC ACID, 4-FLUORO-, METHYL ESTER 000403-33-8
BENZOIC ACID, 2-FLUORO-, METHYL ESTER 000394-35-4
BENZOIC ACID, 4-FLUORO-3-[[[METHYL[(METHYLAMINO) 121788-14-5
BENZOIC ACID HYDRAZD,O(33DIME)TRIAZIN 066975-13-1
BENZOIC ACID, 3-[(HYDRAZINOCARBONYL)AMINO]- 073469-94-0
BENZOIC ACID, 4-[(HYDRAZINOCARBONYL)AMINO]- 073469-95-1
BENZOIC ACID, 2-[(HYDRAZINECARBONYL)AMINO]- 073469-93-9
BENZOIC ACID, 2-HYDROXY-, BUTYL ESTER 002052-14-4
BENZOIC ACID, 2-HYDROXY-, 2-(DIMETHYLAMINO)-2-OX 114665-08-6
BENZOIC ACID, 2-HYDROXY-3,5-DINITRO- 000609-99-4
BENZOIC ACID, 2-HYDROXY-, 2-[(2-HYDROXYETHYL)MET 114665-09-7
BENZOIC ACID, 2-HYDROXY-5-IODO- 000119-30-2
BENZOIC ACID, 2-HYDROXY-, PROPYL ESTER 000607-90-9
BENZOIC ACID, 2-IODO-, METHYL ESTER 000610-97-9
BENZOIC ACID, MERCURIC SALT 000583-15-3
BENZOIC ACID, 2-METHOXYETHYL ESTER 057453-98-2
BENZOIC ACID, 2-METHOXY-, METHYL ESTER 000606-45-1
BENZOIC ACID, 2-[(3-METHOXYPHENYL)AMINO]- 027693-73-8
BENZOIC ACID, 4-METHYL-, METHYL ESTER 000099-75-2
BENZOIC ACID, 3-METHYL-, METHYL ESTER 000099-36-5
BENZOIC ACID, 4-[(METHYLNITROSOAMINO)METHYL]-, M 098736-51-7
BENZOIC ACID, 2-[(3-METHYLPHENYL)AMINO]- 016524-22-4
BENZOIC ACID, 4-[3-METHYL-3-(PHENYLMETHYL)-1-TRI 065587-38-4
BENZOIC ACID, 2-NITRO-, METHYL ESTER 000606-27-9
BENZOIC ACID, 3-NITRO-, METHYL ESTER 000618-95-1
BENZOIC ACID, 3-NITRO-2-[[3-(TRIFLUOROMETHYL)PHE 066287-52-3
BENZOIC ACID, 2-[(4-OXO-2-AZETIDINYL)THIO]-, PHE 098369-49-4
BENZOIC ACID, 3-[(4-OXO-2-AZETIDINYL)OXY]-, PHEN 119005-12-8
BENZOIC ACID, 2,3,4,5,6-PENTAFLUORO-, [1-(1,1-DI 112411-06-0
BENZOIC ACID, 2-PHENYLETHYL ESTER 000094-47-3
BENZOIC ACID, SODIUM SALT 000532-32-1
BENZOIC ACID, 2,3,5,6-TETRAFLUORO-4-METHYL- 000652-32-4
BENZOIC ACID, 4-[(TRIFLUOROACETYL)AMINO]-,ETHYL 024568-14-7
BENZOIC ACID, 2-[(TRIFLUOROACETYL)AMINO]ETHYL ES 096027-93-9
BENZOIC ACID, 2-[[3-(TRIFLUOROMETHYL)PHENYL]AMIN 001977-00-0
BENZOIC ACID, 3,4,5-TRIMETHOXY-, 2-[4-[[(2-ETHOX 093414-55-2
BENZOIC ACID, 3,4,5-TRIMETHOXY-, 7-CHLORO-2,3-DI 018035-92-2
DL-BENZOIN 000579-44-2
BENZOIN 000119-53-9
BENZO[A]NAPHTHACENE 000226-88-0
BENZO[B]NAPHTHO[2,3-D]FURAN 000243-42-5
BENZO(B)NAPHTHO(2,1-D)THIOPHENE 000239-35-0
BENZO[B]NAPHTHO[2,3-D]THIOPHENE 000243-46-9
BENZONATATE 000104-31-4
BENZONITRILE 000100-47-0
BENZONITRILE, 4-[(4-AMINOPHENYL)SULFONYL]- 101533-57-7
BENZONITRILE, 4-CHLORO- 000623-03-0
BENZONITRILE, 3-FLUORO- 000403-54-3
BENZONITRILE, 3-ISOTHIOCYANATO- 003125-78-8
BENZONITRILE, 4-METHOXY- 000874-90-8
BENZONITRILE, 4-(METHOXYMETHYL)- 001515-85-1
BENZO(GHI)PERYLENE 000191-24-2
BENZO(C)PHENANTHRENE 000195-19-7
BENZOPHENONE 000119-61-9
H-1-BENZOPYRAN-2-CARBOXYLIC ACID, 3,4-DIHYDRO-4 051048-00-1

H-1-BENZOPYRAN-2-CARBOXYLIC ACID, 4-OXO-	004940-39-0
H-1-BENZOPYRAN-2-CARBOXYLIC ACID, 6-CHLORO-2,3-	040026-24-2
2H,5H-[1]BENZOPYRANO[4,3-B]-1,4-OXAZIN-7-OL, 3,4	116005-01-7
2H,5H-[1]BENZOPYRANO[4,3-B]-1,4-OXAZIN-9-OL, 3,4	112960-12-0
H-1-BENZOPYRAN-8-OL, 3-(DIPROPYLAMINO)-3,4-DIHY	112904-74-2
H-1-BENZOPYRAN-6-OL, 3-(DIPROPYLAMINO)-3,4-DIHY	116005-04-0
H-1-BENZOPYRAN-4-ONE	000491-38-3
H-1-BENZOPYRAN-2-ONE, 3-(ACETYLOXY)-1-[2-(ACETY	056875-82-2
H-1-BENZOPYRAN-4-ONE, 2,3-DIMETHYL-	017584-90-6
H-1-BENZOPYRAN-2-ONE, 2-(4-HYDROXYPHENYL)-	004143-63-9
H-1-BENZOPYRAN-4-ONE, 7-HYDROXY-2-PHENYL-	006665-86-7
H-1-BENZOPYRAN-4-ONE, 7-HYDROXY-	059887-89-7
H-1-BENZOPYRAN-4-ONE, 3-METHYL-2-PHENYL-	071972-66-2
4H-1-BENZOPYRAN-4-ONE, 3-METHYL-2-PHENYL-8-(2-PR	004412-35-5
BENZO(E)PYRENE	000192-97-2
BENZO(A)PYRENE	000050-32-8
BENZO(F)QUINOLINE	000085-02-9
P-BENZOQUINONE OXIME	000637-62-7
2,1,3-BENZOSELENADIAZOLE	000273-15-4
H-1,2,4-BENZOTHIADIAZIN-3-AMINE,6-CHLORO-, 1,1-	037157-99-6
2H-1,2,4-BENZOTHIADIAZINE, 3-(1-CYCLOPENTEN-1-YL	037157-89-4
H-1,2,4-BENZOTHIADIAZINE, 3-(1-ETHYLPROPYL)-, 1	007752-09-2
124-BENZOTHIADIAZINE,7-I-3-ME,1,1-DIO	037148-02-0
124-BENZOTHIADIAZINE-1,1-O2	000359-85-3
124-BENZOTHIADIAZINE-1-O2-3-ME-6-BR	037148-00-8
124-BENZOTHIADIAZINE-1,1-O2-3-ME	000360-81-6
124-BENZOTHIADIAZINE-1-O2-3-ME-6-NO2	037148-07-5
124-BENZOTHIADIAZINE[1-O2-3-ME-6-ET	037148-03-1
124-BENZOTHIADIAZINE[1-O2-3-PR-6-CL	037148-19-9
H-1,2,4-BENZOTHIADIAZINE-7-SULFONAMIDE, 3,4-DIH	001766-91-2
124-BENZOTHIADIAZIN-1-O2-3-SBU-7-CL	037148-22-4
124-BENZOTHIADIAZIN-1-O2-3-IBU-6-CL	037148-21-3
124-BENZOTHIADIAZIN-1-O2-3-IPR-6-CL	037148-20-2
124BENZOTHIADIAZN-11O2-3ME-6NHAC-7CL	037148-09-7
2,1,3-BENZOTHIADIAZOLE	000273-13-2
2-BENZOTHIAZOLAMINE, N-[3-[3-(1-PIPERIDINYLMETHY	104076-38-2
BENZOTHIAZOLE	000095-16-9
BENZOTHIAZOLE,2-N-ME SULFONAMIDO-6ETO	088946-19-4
2-BENZOTHIAZOLESULFONAMIDE, 6-(2-HYDROXYETHOXY)-	095599-36-3
BENZOTHIAZOLE-2-SULFONAMIDE	000433-17-0
2-BENZOTHIAZOLESULFONAMIDE, 4,6-DICHLORO-	096474-17-8
2-BENZOTHIAZOLESULFONAMIDE, 6-NITRO-	094641-10-8
2-BENZOTHIAZOLESULFONAMIDE, 6-(PHENYLMETHOXY)-	096474-18-9
2-BENZOTHIAZOLESULFONAMIDE, 6-AMINO-	094641-11-9
2-BENZOTHIOPHENAMINE,3-CYANO	018774-47-5
BENZOTHIOPHENE	011095-43-5
BENZO(B)THIOPHENE	000095-15-8
BENZO[B]THIOPHENE-2-CARBOXAMIDE, 5-METHOXY-3-(1-	104795-66-6
BENZO(B)THIOPHENE S,S-DIOXIDE	000825-44-5
1,2,3-BENZOTRIAZIN-4(1H)-ONE	000090-16-4
1H-BENZOTRIAZOLE	000095-14-7
2H-BENZOTRIAZOLE	000273-02-9
2H-BENZOTRIAZOLE-2-ACETAMIDE, α-ETHYL-	115054-76-7
1H-BENZOTRIAZOLE-1-ACETAMIDE, α-PROPYL-	115054-89-2
1H-BENZOTRIAZOLE-1-ACETAMIDE, α-METHYL-	115054-71-2
2H-BENZOTRIAZOLE-2-ACETAMIDE, α,α-DIMETHYL-	115054-85-8
1H-BENZOTRIAZOLE-1-ACETAMIDE, α-ETHYL-	115054-75-6
1H-BENZOTRIAZOLE-1-ACETAMIDE, α-BUTYL-	115054-81-4

2H-BENZOTRIAZOLE-2-ACETAMIDE, α-METHYL-	115054-88-1
2H-BENZOTRIAZOLE-2-ACETAMIDE, α-PROPYL-	115054-83-6
2H-BENZOTRIAZOLE-2-ACETAMIDE, α-BUTYL-	115054-90-5
1H-BENZOTRIAZOLE-1-ACETAMIDE, α,α-DIMETHYL-	115054-77-8
1H-BENZOTRIAZOLE-1-ACETIC ACID, α-ETHYL-, ETHYL	115054-78-9
2H-BENZOTRIAZOLE-2-ACETIC ACID, α,α-DIMETHYL-, E	115054-84-7
2H-BENZOTRIAZOLE-2-ACETIC ACID, α,α-DIMETHYL-, E	115054-79-0
2H-BENZOTRIAZOLE-2-ACETIC ACID, α-ETHYL-, ETHYL	115054-80-3
2H-BENZOTRIAZOLE-2-ACETIC ACID, α,α-DIMETHYL-	016583-98-5
1H-BENZOTRIAZOLE-1-ACETIC ACID, α,α-DIMETHYL-	004233-61-8
2H-BENZOTRIAZOLE-2-ACETIC ACID, α-BUTYL-, ETHYL	115054-87-0
1H-BENZOTRIAZOLE-1-ACETIC ACID	004144-64-3
1H-BENZOTRIAZOLE-1-ACETIC ACID, ETHYL ESTER	069218-46-8
2H-BENZOTRIAZOLE-2-ACETIC ACID, ETHYL ESTER	069218-51-5
1H-BENZOTRIAZOLE-1-ACETIC ACID, α-METHYL-, ETHYL	075584-60-0
1H-BENZOTRIAZOLE-1-ACETIC ACID, α-PROPYL-, ETHYL	115054-86-9
2H-BENZOTRIAZOLE-2-ACETIC ACID, α-METHYL-, ETHYL	075584-51-9
2H-BENZOTRIAZOLE-2-ACETIC ACID, α-PROPYL-, ETHYL	115054-82-5
1H-BENZOTRIAZOLE-1-ACETIC ACID, α-BUTYL-, ETHYL	115054-74-5
2H-BENZOTRIAZOLE, 5-[[(4-AMINOPHENYL)SULFONYL]AM	130787-24-5
2H-BENZOTRIAZOLE-1-BUTANAMIDE, ¥-METHYL-	115054-97-2
1H-BENZOTRIAZOLE-5-CARBOXAMIDE, 6-METHOXY-N-[[1-	093414-58-5
1H-BENZOTRIAZOLE-5-CARBOXAMIDE, 6-METHOXY-N-[[1-	059338-93-1
2H-BENZOTRIAZOLE, 2-CYCLOHEXYL-	066480-65-7
1H-BENZOTRIAZOLE, 1-CYCLOHEXYL-	066536-70-7
1H-BENZOTRIAZOLE, 1-CYCLOPENTYL-	066480-64-6
2H-BENZOTRIAZOLE, 2-CYCLOPENTYL-	066536-69-4
2H-BENZOTRIAZOLE, 2-CYCLOPENTYL-5-[(METHYLSULFON	130787-21-2
2H-BENZOTRIAZOLE, 2-CYCLOPENTYL-5-[[(4-NITROPHEN	130787-23-4
2H-BENZOTRIAZOLE, 2-CYCLOPENTYL-5-[[(2-METHYLPRO	130787-13-2
2H-BENZOTRIAZOLE, 2-CYCLOPENTYL-5-[(ETHOXYCARBON	130787-12-1
1H-BENZOTRIAZOLE-1-PENTANOIC ACID	069218-42-4
2H-BENZOTRIAZOLE-2-PENTANOIC ACID	069218-44-6
2H-BENZOTRIAZOLE-2-PROPANOIC ACID, ETHYL ESTER	069218-52-6
1H-BENZOTRIAZOLE-1-PROPANOIC ACID, ß-METHYL-, ET	001143-78-8
1H-BENZOTRIAZOLE-1-PROPANAMIDE, ß-METHYL-	099069-47-3
2H-BENZOTRIAZOLE-2-PROPANAMIDE, ß-METHYL-	115054-94-9
2H-BENZOTRIAZOLE-2-PROPANOIC ACID, ß-METHYL-, ET	001612-15-3
1H-BENZOTRIAZOLE-1-PROPANOIC ACID, ETHYL ESTER	069218-47-9
2H-BENZOTRIAZOLE-2-PROPANOIC ACID, ß-METHYL-	004248-18-4
1H-BENZOTRIAZOLE-1-PROPANOIC ACID, ß-METHYL-	000654-13-7
2-(2H-BENZOTRIAZOL-2-YL)-4,6-DI-(TERT-PENTYL)PH*	025973-55-1
BENZOTRICHLORIDE	000098-07-7
BENZOTRIFLUORIDE	000098-08-8
BENZOXAZOLE	000273-53-0
BENZOXAZOLE-2-AMINE,N,N-DIMETHYL	013858-89-4
BENZOXAZOLE, 2-AMINO	004570-41-6
BENZOXAZOLE, 2-METHYLAMINO-	019776-98-8
BENZOXAZOLE, 2-PHENYL-	000833-50-1
BENZOXIMATE	029104-30-1
3-BENZOYLACRYLIC ACID (TRANS)	017812-07-6
A-BENZOYLAMINO-N-SEC-BU CINNAMAMIDE	097871-60-8
4-BENZOYLANILINE	001137-41-3
BENZOYL BROMIDE	000618-32-6
1-BENZOYL-2-CARBAMYLHYDRAZINE	002845-79-6
BENZOYL CHLORIDE	000098-88-4
3-BENZOYL-5-FLUOROURACIL	061251-77-2
BENZOYLFORMIC ACID	000611-73-4

BENZOYL HYDRAZINE	000613-94-5
BENZOYLHYDRAZINE, O-AMINO	001904-58-1
BENZOYLHYDRAZINE, O-BROMO	029418-67-5
BENZOYLHYDRAZINE, P-BROMO	005933-32-4
BENZOYLHYDRAZINE, M-BROMO	039115-96-3
BENZOYLHYDRAZINE, O-CHLORO	005814-05-1
BENZOYLHYDRAZINE, P-CHLORO	000536-40-3
BENZOYLHYDRAZINE, M-CHLORO	001673-47-8
BENZOYLHYDRAZINE, M-HYDROXY	005818-06-4
BENZOYLHYDRAZINE, O-HYDROXY	000936-02-7
BENZOYLHYDRAZINE, P-HYDROXY	005351-23-5
BENZOYLHYDRAZINE, P-IODO	039115-95-2
BENZOYLHYDRAZINE, M-IODO	039115-94-1
BENZOYLHYDRAZINE, O-IODO	031822-03-4
BENZOYLHYDRAZINE, O-METHOXY	007466-54-8
BENZOYLHYDRAZINE, P-METHYL	003619-22-5
BENZOYLHYDRAZINE, O-METHYL	007658-80-2
BENZOYLHYDRAZINE, M-METHOXY	005785-06-8
BENZOYLHYDRAZINE, M-METHYL	013050-47-0
BENZOYLHYDRAZINE, P-METHOXY	003290-99-1
BENZOYLHYDRAZINE, O-NITRO	000606-26-8
2-BENZOYL-N-METHYLBENZAMIDE	032557-55-4
ETHANOL, 2- 2-(BENZOYLOXY)ETHOXY -	020587-61-5
BENZOYL PEROXIDE	000094-36-0
BENZOYLPROP ETHYL	022212-55-1
4-BENZOYLPYRIDINE	014548-46-0
N-BENZOYLSALICYLAMIDE	005663-74-1
4-(BENZOYL SULFONAMIDE)-PYRIDINE	051246-76-5
124-BENZTHIADIAZIN-1-O2-3CYHEXEN-3-YL	037157-91-8
124-BENZTHIADIAZINE-1-O2-3-BU-7-CL	037157-85-0
124-BENZTHIADIAZINE-11-O2-6-CL	019477-31-7
124-BENZTHIADIAZINE-11-O2-7-CL	019477-12-4
124-BENZTHIADIAZIN-11SO2-3ME5BR7CL	005108-54-3
124-BENZTHIADIAZIN-11SO2-3ME5I7CL	037148-13-3
124-BENZTHIADIAZIN-1-O2-3ME-6NO2-7CL	031365-75-0
124-BENZTHIADIAZIN-1-O2-3ME-5NO2-7CL	037157-79-2
124-BENZTHIADIAZIN-1-O2-3(5NORBEN2YL)	037157-92-9
124-BENZTHIADIAZIN-1-O2-3-CLME-6-CL	037157-57-6
124-BENZTHIADIAZIN-1-O2-3-CYBU-7-CL	037148-24-6
124-BENZTHIADIAZIN-1-O2-3-CYHEXYL	037157-88-3
124-BENZTHIADIAZIN-1-O2-3-ET-6-CL	014559-55-8
124-BENZTHIADIAZIN-1-O2-3-(2-FURYL)	037157-96-3
124-BENZTHIADIAZIN-1-O2-3-ME-7-BR	013460-15-6
124-BENZTHIADIAZIN-1-O2-3-ME-6-CL	014559-54-7
124-BENZTHIADIAZIN-1-O2-3-ME-5-CL	031363-85-6
124-BENZTHIADIAZIN-1-O2-3-ME-7-F	031365-74-9
124-BENZTHIADIAZIN-1-O2-3-ME-5-ME	031365-88-5
124-BENZTHIADIAZIN-1-O2-3-ME-6-ME	031363-88-9
124-BENZTHIADIAZIN-1-O2-3-ME-7-ME	031363-89-0
124-BENZTHIADIAZIN-1-O2-3-PHENYL	018818-44-5
124-BENZTHIADIAZIN-1-O2-3-TBU-7-CL	013460-16-7
124-BENZTHIADIAZIN-1-O2-3-(2-THIENYL)	037157-95-2
124-BENZTHIADIAZIN,11-O2-7SULFAMYL	023141-75-5
124-BENZTHIADIAZNE-1-O2-3-ETSME-6-CL	037158-00-2
124-BENZTHIADIAZNE-1-O2-3-MEOME-6-CL	037157-71-4
124-BENZTHIADIAZN-1-O2-3-ALLYL-6-CL	037157-59-8
124-BENZTHIADIAZN-1-O2-3-BENZYL	020434-66-6
124-BENZTHIADIAZN-1-O2-3-CYCLOPE	020434-64-4

124-BENZTHIADIAZN-1-O2-3-DICLME-6-CL	037157-97-4
124-BENZTHIADIAZN-1-O2-3-ME-7-CF3	020046-63-3
124-BENZTHIADIAZN-1-O2-3-ME-8-CL	022680-31-5
124-BENZTHIADIAZN-1-O2-3-NEOPE-7-CL	037157-54-3
124-BENZTHIADIAZN,11-O2-7-NH2SO2-6NO2	023141-81-3
124-BENZTHIADIZ-1-O2-3-CF3-6-CL	002251-64-1
124-BENZTHIADIZ-1-O2-3-ET-7-CL	001207-63-2
124-BENZTHIADIZ-1-O2-3-ME-6-CF3	000723-57-9
124-BENZTHIADIZ-1-O2-3-ME-6-MEO	006451-55-4
124-BENZTHIADIZ-1-O2-3-ME-6-NH2-7-CL	037148-08-6
124-BENZTHIDIAZIN-1O2-3ME-7SO2N(DIME)	037157-82-7
124-BENZTHIDIAZIN-1-O2-3-CYPR-6-CL	013460-17-8
124-BENZTHIDIAZN-1,1-O2-3-ME-7-CL	000364-98-7
N-BENZYLACETAMIDE	000588-46-5
BENZYL ACETATE	000140-11-4
BENZYLACETOACETIC ACID, ETHYL ESTER	000620-79-1
BENZYLACETONITRILE	000645-59-0
BENZYL ALCOHOL	000100-51-6
BENZYLAMINE	000100-46-9
BENZYLAMINE, HYDROCHLORIDE	003287-99-8
BENZYL-4-AMINOBENZOATE	000094-18-8
N-BENZYL-2-AMINOPROPANOL	003217-09-2
N-BENZYLANILINE	000103-32-2
BENZYLBENZOATE	000120-51-4
O-BENZYL CARBAMATE	000621-84-1
BENZYLCARBONYL-NH-SO2-NH2	032094-40-9
BENZYL CHLOROCARBONATE	000501-53-1
BENZYL-T-CHYRSANTHEMATE	064312-78-3
BENZYL CINNAMATE	000103-41-3
N'-BENZYL-(4-CLPHO)ACETICACIDHYDRAZIDE	054922-68-8
8-S-BENZYL CYCLIC AMP	032487-38-0
8-S-BENZYL CYCLIC AMP-O-BENZYL	107538-72-7
BENZYL N,N-DIMETHYLCARBAMATE	010507-52-5
BENZYL ETHYL ETHER	000539-30-0
BENZYL FLUORIDE	000350-50-5
N-BENZYL-N-FORMYLANILINE	016350-99-5
N-BENZYL-2-FURAMIDE	010354-48-0
BENZYL GLUCOPYRANOSIDE	004304-12-5
N-BENZYLGLYCOLAMIDE	019340-77-3
3-BENZYLHYDROCHLOROTHIAZIDE	001824-50-6
5-BENZYL-2,4-IMIDAZOLIDIONE	003530-82-3
N-BENZYL-2,4-IMIDAZOLIDIONE	006777-05-5
N-BENZYL-2,4,5-IMIDAZOLITRIONE	030345-85-8
BENZYL IODIDE	000620-05-3
2-BENZYLISONIAZID	058481-07-5
BENZYLISOTHIOCYANATE	000622-78-6
BENZYLMALONIC ACID, DIETHYL ESTER	000607-81-8
S-BENZYLMERCAPTOACETIC ACID	000103-46-8
BENZYL METHACRYLATE	002495-37-6
BENZYL METHYL ETHER	000538-86-3
N-BENZYL-N-METHYLNITROSOAMINE	000937-40-6
P-(3-BENZYL-3ME-1-TRIAZENO)BENZAMIDE	059708-24-6
P(3-BENZYL-3ME-1TRIAZENO)BENZONITRILE	065542-21-4
4-BENZYLMORPHOLINE	010316-00-4
N-BENZYL-N-NITROSOANILINE	000612-98-6
5-BENZYL-2,4-OXAZOLIDIONE	005841-62-3
BENZYLOXYCARBONYL MITOMYCIN C	086476-06-4
BENZYLOXYCARBONYL-PHALANINYLVALINE	055565-48-5

P-(BENZYLOXY) PHENOL	000103-16-2
1(3-BENZYLOXYPHENYL)-3-(ME-MEO) UREA	074109-81-2
1-(4-BENZYLOXYPH)-3-MEO-3-ME UREA	025998-87-2
BENZYLPENICILLIN	000061-33-6
3-BENZYL-2,4-PENTANEDIONE	001134-87-8
N-BENZYL-PHENOXY ACETIC HYDRAZIDE	052093-78-4
BENZYL PHENYL ETHER	000946-80-5
BENZYL PHENYL SULFONE	003112-88-7
1-BENZYLPROPYLAMINE	053309-89-0
2-BENZYLPYRIDINE	000101-82-6
4-BENZYLPYRIDINE	002116-65-6
BENZYLPYRIDINIUM BROMIDE	002589-31-3
BENZYL-4-PYRIDYL KETONE	001017-24-9
24-NH2-5(3-CL-4-MEO-BENZYL)PYRIMIDINE	073275-70-4
24-NH2-5(5-I-2-MEO-BENZYL)PYRIMIDINE	103360-35-6
24-NH2-5(4-NH2-3-MEO-BENZYL)PYRIMIDINE	085544-45-2
24-NH2-5(3-I-4-MEO-BENZYL)PYRIMIDINE	085544-43-0
24-NH2-5(3-NH2-4-MEO-BENZYL)PYRIMIDINE	085544-44-1
24NH2-5(24-BR-5-MEO-BENZYL)PYRIMIDINE	085544-53-2
24NH2-5(5-BR-2-MEO-BENZYL)PYRIMIDINE	085544-51-0
24NH2-5(3-BR-4-MEO-BENZYL)PYRIMIDINE	056183-32-5
24NH2-5(2-BR-5-MEO-BENZYL)PYRIMIDINE	085544-52-1
24NH2-5(4-MEO-BENZYL)PYRIMIDINE	020285-70-5
24-NH2-5(4-CL-3-MEO-BENZYL)PYRIMIDINE	073264-25-2
N-BENZYL-2-PYRROLIDINONE	005291-77-0
N-BENZYL-2,4-PYRROLIDIONE	030125-76-9
N-BENZYLQUINOLINIUM BROMIDE	026323-01-3
BENZYL SALICYLATE	000118-58-1
4-BENZYLSEMICARBAZIDE	016956-42-6
5-BENZYL-2,4-THIAZOLIDIONE	033321-31-2
BENZYLTHIOCYANATE	003012-37-1
1-BENZYLTHIO-B-GALACTOPYRANOSIDE	063407-53-4
1-BENZYL-1,2,3-TRIAZOLE	004368-68-7
4-BENZYL-1,2,4-TRIAZOLE	016227-13-7
BENZYLTRIMETHYL AMMONIUM BROMIDE	005350-41-4
BENZYL TRIMETHYL AMMONIUM CHLORIDE	000056-93-9
BENZYLTRIMETHYLSILANE	000770-09-2
BENZYLUREA	000538-32-9
BETAINE	000107-43-7
BEVANTOLOL	059170-23-9
BEZITRAMIDE	015301-48-1
BIANISIDINE	000119-93-7
BIBENZYL	000103-29-7
TRANS-BICYCLO[4.4.0]DECANE	000493-02-7
CIS-BICYCLO[4.4.0]DECANE	000493-01-6
BICYCLO(2.2.1)HEPTA-2,5-DIENE	000121-46-0
BICYCLO[2.2.1]HEPTA-2,5-DIEN-7-OL, BENZOATE	004796-68-3
BICYCLO(2.2.1.)HEPTANE	000279-23-2
BICYCLO(4.1.0)HEPTANE	000286-08-8
BICYCLO 2.2.1 HEPTAN-2-OL, 1,7,7-TRIMETHYL-, EXO	000124-76-5
BICYCLO(4.1.0)HEPT-3-ENE	016554-83-9
BICYCLO[2.2.1]-2-HEPTENE	000498-66-8
BICYCLO[2.2.1]HEPT-5-ENE-2,3-DICARBOXAMIDE	084846-21-9
BICYCLO[2.2.2]OCTANE	000280-33-1
BICYCLO(2.2.2)OCT-2-ENE	000931-64-6
BICYCLO(2.1.0)PENTANE	000185-94-4
BIFENOX	042576-02-3
BINAPACRYL	000485-31-4

BIPERIDEN	000514-65-8
4,4'-BIPHENOL	000092-88-6
2,2'-BIPHENOL	001806-29-7
BIPHENYL	000092-52-4
1,1'-BIPHENYL, 4-[(2-OXOAZETIDIN-4-YL)OXY]-	119005-15-1
1,1'-BIPHENYL -4-CARBOXYLIC ACID	000092-92-2
P-BIPHENYLCARBOXYLIC ACID, NA SALT	017264-58-3
[1,1'-BIPHENYL]-2-CARBOXYLIC ACID	000947-84-2
[1,1'-BIPHENYL]-2,2'-DIAMINE	001454-80-4
1,1'-BIPHENYL -2,2'-DICARBOXYLIC ACID	000482-05-3
BIPHENYL, 3,4'-DICHLORO-	002974-90-5
BIPHENYLENE, 2,3,6,7-TETRACHLORO-	007090-41-7
4-BIPHENYLISOTHIOCYANATE	001510-24-3
[1,1'-BIPHENYL]-4-SULFONAMIDE	004371-23-7
P-BIPHENYLSULFONIC ACID, SODIUM SALT	002217-82-5
9-(1,1'-BIPHENYL)-4-YL-9H-FLUORENE	017165-86-5
2,2-BIPYRIDINE	000366-18-7
2,3'-BIPYRIDINE	000581-50-0
[3,4'-BIPYRIDINE]-5-CARBONITRILE, 6-[(2-AMINOETH	108611-20-7
[3,4'-BIPYRIDINE]-5-CARBONITRILE,6-(2-PROPENYLAM	120137-94-2
[3,4'-BIPYRIDINE]-5-CARBONITRILE,6-[(2-HYDROXYET	108610-76-0
[3,4'-BIPYRIDINE]-5-CARBONITRILE,6-[[1-(HYDROXYM	108610-80-6
[3,4'-BIPYRIDINE]-5-CARBONITRILE,6-[(3-HYDROXYPR	108610-82-8
[3,4'-BIPYRIDINE]-5-CARBONITRILE,6-METHOXY-	070959-61-4
[3,4'-BIPYRIDINE]-5-CARBONITRILE,6-[(2-HYDROXYPR	108610-79-3
[3,4'-BIPYRIDINE]-5-CARBONITRILE,6-ETHOXY-	128887-24-1
[3,4'-BIPYRIDINE]-5-CARBONITRILE,6-CHLORO-	070959-58-9
[3,4'-BIPYRIDINE]-5-CARBONITRILE,6-(2-HYDROXYPRO	108611-04-7
[3,4'-BIPYRIDINE]-5-CARBONITRILE,6-[(4-HYDROXYBU	108610-85-1
[3,4'-BIPYRIDINE]-5-CARBONITRILE,6-(1-PYRROLIDIN	120137-95-3
[3,4'-BIPYRIDINE]-5-CARBONITRILE,6-[(3-METHOXYPR	120137-97-5
[3,4'-BIPYRIDINE]-5-CARBONITRILE,6-(DIMETHYLAMIN	084884-30-0
[3,4'-BIPYRIDINE]-5-CARBONITRILE,6-(ETHYLAMINO)-	120137-90-8
[3,4'-BIPYRIDINE]-5-CARBONITRILE,6-METHOXY-2-MET	128887-26-3
4,4'-BIPYRIDINE, 1,2,3,6-TETRAHYDRO-1-METHYL-	140111-43-9
2,2'-BIQUINOLINE	000119-91-5
1(2,3-BIS(ACETYLOXY)PR-2-NO2IMIDAZOLE	068160-69-0
2,6-BIS(T-BUTYL)-2,5-CYCLOHEXADIENE-1,4-DIONE	000719-22-2
3,5-BIS(TERT-BUTYL)PHENOL METHYLCARBAMATE	002655-19-8
BIS(2-CHLORO-3-BUTENYL) ETHER	083682-66-0
1,2-BIS(2-CHLOROETHOXY)ETHANE	000112-26-5
BIS(2-CHLOROETHOXY)METHANE	000111-91-1
4-(BIS(2-CHLOROETHYL)AMINO)BENZENEBUTANOIC AC*	000305-03-3
4-(BIS(2-CHLOROETHYL)AMINO)-L-PHENYLALANINE	000148-82-3
BIS(2-CHLOROETHYL) ETHER	000111-44-4
1,3-BIS(2-CHLOROET)-1-NITROSOUREA	000154-93-8
1,3-(BIS-CHLOROMETHYL)BENZENE	000626-16-4
BIS(CHLOROMETHYL) ETHER	000542-88-1
2,5-BIS(CHLOROMETHYL)-1,5-HEXADIENE	083682-51-3
2,5-BIS(CHLOROMETHYL)-1,3,5-HEXATRIENE	083682-56-8
1,1-BIS(P-CHLOROPHENYL)-2-CHLOROETHYLENE	001022-22-6
1,2-BIS(1,2-DICHLOROETHYL)CYCLOBUTANE	083682-62-6
2,4-BIS(DIETHYLAMINO)-6-METHOXY-S-TRIAZINE	015438-85-4
BIS(2-DIISOPROPYLAMINO)ETHYL METHYLPHOSPHONITE	057856-12-9
BIS(2,4-DIMETHOXYPHENYL)DIAZENE	029418-55-1
4,4-BIS(DIMETHYLAMINO)BENZOPHENONE	000090-94-8
BIS(4-DIMETHYLAMINOPHENYL)METHANE	000101-61-1
2,6-BIS(1,1-DIMETHYLETHYL)-4-NITROBENZENAMINE	005180-59-6

BIS(3,4-DIMETHYLPHENYL)DIAZENE	029418-36-8
BIS(2,6-DIMETHYLPHENYL)DIAZENE	029418-31-3
BIS(2,4-DIMETHYLPHENYL)DIAZENE	029418-25-5
BIS (2-ETHOXY ETHYL)	ETHER
	000112-36-7
1,1-BIS(P-ETHOXYPHENYL)-2-CHLOROPROPANE	056265-22-6
1,1-BIS(P-ETHOXYPHENYL)-2,2-DICHLOROPROPANE	056265-23-7
1,1-BIS(P-ETHOXYPHENYL)-2,2-DIMETHYLPROPANE	027955-87-9
1,1-BIS(4-ETHOXYPHENYL)-2-NITROPROPANE	026258-70-8
BIS(2-ETHYLHEXYL)PHOSPHATE	000298-07-7
BIS(2-ETHYLHEXYL)PHTHALATE	000117-81-7
BIS(2-ETHYLHEXYL) SODIUM SULFOSUCCINATE	000577-11-7
BIS(2-ETHYLHEXYL) TEREPHTHALATE	006422-86-2
2,2-BIS(ETHYLSULFONYL)PROPANE	000115-24-2
BIS(ETHYLXANTHOGEN)	000502-55-6
1-PROPANOL, 1,1-BIS(4-FLUOROPHENYL)-2-(1H-IMIDAZ	128104-27-8
2,5-BIS(CHLOROMETHYL)-3,4-DICHLORO-1,3,5-HEXAT*	083682-55-7
2,5-BIS(CHLOROMETHYL)-3,4-DICHLORO-1,5-HEXADIE*	083682-52-4
2,5-BIS(CHLOROMETHYL)-3,3,4,4-TETRACHLORO-1,5-*	083682-53-5
BIS (2-HYDROXY-3TERT-BUTYL-5-METHYLPHENYL) METHA	000119-47-1
2,2-BIS(4-HYDROXY-3,5-DIBROMOPHENYL)PROPANE	000079-94-7
BIS(2-HYDROXYETHYL)TEREPHTHALATE	000959-26-2
2,5-BIS(DICHLOROMETHYL)-3,4-DICHLORO-1,3,5-HEX*	083682-57-9
1,4-BIS(1-ISOCYANATO-1-METHYLETHYL)BENZENE	002778-41-8
1,3-BIS(1-ISOCYANATO-1-METHYLETHYL)BENZENE	002778-42-9
4,6-BIS(ISOPROPYLAMINO)-S-TRIAZIN-2-OL	007374-53-0
1,2-BIS(2-MEO-ETO)BENZENE	033332-53-5
1,3-BIS(2-(2-MEO-ETO)-ETO)BENZENE	121284-19-3
1,2-BIS(2-(2-MEO-ETO)-ETO)BENZENE	057721-93-4
1,4-BIS(2-(2-MEO-ETO)-ETO)BENZENE	121284-18-2
1,2-BIS(2-(2-MEO-ETO)-ETO-ETO)BENZENE	057721-95-6
BIS(METHOXYETHYL)PHTHALATE	000117-82-8
BIS(4-METHOXYPHENOL)DIAZENE	000501-58-6
BIS(2-METHOXYPHENYL)DIAZENE	000613-55-8
1,1-BIS(P-METHOXYPHENYL)-2,2-DIMETHYLPROPANE	004741-74-6
1,4-BIS(METHYLAMINO)ANTHRAQUINONE	002475-44-7
S-(2-(BIS(1-METHYLETHYL)AMINO)ETHYL) METHYL PHO*	073207-98-4
2-(BIS(1-METHYLETHYL)AMINO)ETHANOL	000096-80-0
N,N'-BIS(1-METHYLETHYL)UREA	004128-37-4
BIS(4-METHYLPHENYL)DIAZENE	000501-60-0
BIS(3-METHYLPHENYL)DIAZENE	000588-04-5
BIS(2-METHYLPHENYL)DIAZENE	000584-90-7
M-(BIS-METHYLSUFONYLAMIDO)BENZENE	006966-38-7
1,1-BIS(P-MEO-PH)-2-NITROPROPANE	034197-26-7
2,5-BIS(TRICHLOROMETHYL)-3,4-DICHLORO-1,3,5-HE*	083682-58-0
N-NITROSO-BIS(2,2,2TRIFET)AMINE	000625-89-8
3,5-BIS(TRIFLUOROME)ANILINE	000328-74-5
BIS(2,4,6-TRIMETHYLPHENYL)DIAZENE	005692-66-0
BIS(1,2,2-TRIMETHYLPROPYL) METHYLPHOSPHONATE	007040-58-6
BITERTANOL	055179-31-2
2,2'-BITHIOPHENE	000492-97-7
2,2'-BITHIOPHENE, 5-(3-BUTEN-1-YNYL)-	001134-61-8
BORNEOL	000507-70-0
BORNYL ACETATE	000076-49-3
BR-N(2,6-IPRPH)CARBAMOYLME IMINODIACETIC ACID	132169-99-4
BR-N(26-MEPH)CARBAMOYLME IMINODIACETIC ACID	132170-00-4
N1(5-BR-3-ME-2-PYRIDYL)SULFANILAMIDE	030961-42-3
N1(3-BR-5-ME-2-PYRIDYL)SULFANILAMIDE	030961-43-4

BROBETASONE-17-PROPIONATE	025122-43-4
BROMACIL	000314-40-9
BROMADIOLONE	028772-56-7
BROMAZEPAM	001812-30-2
BROMOACETAMIDE	000683-57-8
2-BROMOACETANILIDE	000614-76-6
4-BROMOACETANILIDE	000103-88-8
3-BROMOACETANILIDE	000621-38-5
BROMOACETIC ACID	000079-08-3
ROMOACETIC ACID, ETHYL ESTER	000105-36-2
P-BROMOACETOPHENONE	000099-90-1
M-BROMOACETOPHENONE	002142-63-4
2-BROMO-4-AMINOBENZOIC ACID	002486-52-4
P-BROMO-AMPHENICOL	023885-59-8
M-BROMOANILINE	000591-19-5
O-BROMOANILINE	000615-36-1
P-BROMOANILINE	000106-40-1
4-BROMOANISOLE	000104-92-7
2-BROMOBENZAMIDE	004001-73-4
4-BROMOBENZAMIDE	000698-67-9
M-BROMOBENZAMIDE	022726-00-7
BROMOBENZENE	000108-86-1
ALPHA-BROMOBENZENEACETONITRILE	005798-79-8
4-BROMOBENZENESULFONAMIDE	000701-34-8
BROMOBENZENE,P-TRIFLUOROACETAMIDO	024568-11-4
3'-BROMOBENZOGUANAMINE	030101-52-1
2'-BROMOBENZOGUANAMINE	030530-48-4
M-BROMOBENZOIC ACID	000585-76-2
P-BROMOBENZOIC ACID	000586-76-5
O-BROMOBENZOIC ACID	000088-65-3
4-BROMOBENZONITRILE	000623-00-7
2-BROMOBENZONITRILE	002042-37-7
3-BROMOBENZONITRILE	006952-59-6
N-BROMOBENZOQUINONEMONIMINE	064693-23-8
P-BROMOBENZYL N,N-DIME CARBAMATE	084640-33-5
M-BROMOBENZYL TRIMETHYL AMMONIUM BROMIDE	071323-99-4
4-BROMOBIPHENYL	000092-66-0
O-BROMOBIPHENYL	002052-07-5
3-BROMOBIPHENYL	002113-57-7
2-BROMOBUTANE	000078-76-2
1-BROMOBUTANE	000109-65-9
3-BROMO-2-BUTANONE	000814-75-5
1-BROMO-2-BUTANONE	000816-40-0
4-BROMO-1-BUTENE	005162-44-7
1-BROMO-1-BUTENE	031844-98-1
2-BROMO-2-BUTENE	004784-77-4
2-BROMO-2-BUTENE (CIS)	013294-71-8
2-BROMO-4-(TERT-BUTYL)-6-NITROPHENOL	017199-23-4
N-(4-BROMOBUTYL)PHTHALIMIDE	003236-48-4
A-BROMOBUTYRIC ACID	000080-58-0
2-BROMOCHLOROBENZENE	000694-80-4
M-BROMOCHLOROBENZENE	000108-37-2
P-BROMOCHLOROBENZENE	000106-39-8
2'-BR-5'-CHLOROBENZOGUANAMINE	057381-38-1
2'-BR-4'-CHLOROBENZOGUANAMINE	057381-50-7
BROMOCHLORODIFLUOROMETHANE	000353-59-3
1-BROMO-1-CHLOROETHANE	000593-96-4

1-BROMO-2-CHLOROETHANE	000107-04-0
1-BROMO-2-CHLORO-4-NITROBENZENE	029682-39-1
4-BROMO-2-CHLORO-6-NITROPHENOL	058349-01-2
2-BROMO-1-CHLOROPROPANE	003017-95-6
M-BROMOCINNAMIC ACID	032862-97-8
8-BROMO CYCLIC AMP	039023-66-0
8-BROMO CYCLIC AMP-O-BENZYL	107538-75-0
BROMOCYCLOHEXANE	000108-85-0
1-BROMODECANE	000112-29-8
5-BROMODEOXYURIDINE	000059-14-3
1-(2-BROMO),2-DIBENZOYL-1-TERT-BUTYLHYDRAZINE	112249-32-8
1-(4-BROMO),2-DIBENZOYL-1-TERT-BUTYLHYDRAZINE	112226-85-4
1-(3-BROMO),2-DIBENZOYL-1-TERT-BUTYLHYDRAZINE	112427-09-5
BROMODICHLOROMETHANE	000075-27-4
4-BROMO-2,6-DICHLOROPHENOL	003217-15-0
2-BROMO-1,2-DIMETHYL-5-NITRO-1H-IMIDAZOLE	105983-47-9
5-BROMO-1,2-DIMETHYL-4-NITROIMIDAZOLE	021117-52-2
4-BROMO-1,2-DIMETHYL-5-NITROIMIDAZOLE	021431-58-3
2-IMIDAZOLIDINIMINE, N-(4-BROMO-2,6-DIMETHYLPHEN	065936-23-4
N'-(4-BROMO-3,5-DIMETHYLPHENYL)-N,N-DIMETHYLU*	078508-43-7
1-BROMO-N-DODECANE	000143-15-7
1-BROMOEICOSANE	004276-49-7
BROMOETHANE	000074-96-4
2-BROMOETHANOL	000540-51-2
BROMOETHENE	000593-60-2
1-BROMO-4-ETHYLBENZENE	001585-07-5
(2-BROMOETHYL)BENZENE	000103-63-9
N-(2-BROMOETHYL)PHTHALIMIDE	000574-98-1
BROMOFENOXIM	013181-17-4
2-BROMO-9H-FLUORENE	001133-80-8
P-BROMOFLUOROBENZENE	000460-00-4
M-BROMOFLUOROBENZENE	001073-06-9
2'-BR-5'-FLUOROBENZOGUANAMINE	057381-40-5
5'-BR-2'-FLUOROBENZOGUANAMINE	057381-60-9
BROMOFORM	000075-25-2
3-BROMOFURAN	022037-28-1
1-BROMOHEPTADECANE	003508-00-7
2-BROMOHEPTANE	001974-04-5
1-BROMOHEPTANE	000629-04-9
1-BROMOHEXADECANE	000112-82-3
1-BROMOHEXANE	000111-25-1
2-BROMOHEXANE	003377-86-4
BROMO(2-HYDROXYPROPYL)MERCURY	018832-83-2
5-BROMOINDOLE	010075-50-0
4-BROMOIODOBENZENE	000589-87-7
2-BROMOISONIAZID	029849-15-8
M-BROMO CF3-METHANESULFONANILIDE	023384-08-9
1-BROMO-2-METHOXYBENZENE	000578-57-4
1-BROMO-3-METHOXYBENZENE	002398-37-0
4-(BROMOMETHYL)BENZOIC ACID	006232-88-8
1-BROMO-3-METHYLBUTANE	000107-82-4
4-BROMO-2-METHYL-5-NITROIMIDAZOLE	018874-52-7
5-BROMO-1-METHYL-4-NITRO-1H-IMIDAZOLE	000933-87-9
4-BROMO-1-METHYL-5-NITRO-1H-IMIDAZOLE	059177-47-8
2-BROMO-4-METHYL-5-NITRO-1H-IMIDAZOLE	105983-46-8
2-BROMO-1-METHYL-4-NITRO-1H-IMIDAZOLE	016681-63-3
2-BROMO-4-METHYL-6-NITROPHENOL	020039-91-2
2-BROMO-2-METHYLPROPANE	000507-19-7

2-BROMONAPHTHALENE	000580-13-2
1-BROMONAPHTHALENE	000090-11-9
2-BROMO-1,4-NAPHTHOQUINONE	002065-37-4
2-BROMO-6-NITROBENZAMIDE	107485-65-4
4-BROMO-1-NITROBENZENE	000586-78-7
O-BROMONITROBENZENE	000577-19-5
3-BROMO-1-NITROBENZENE	000585-79-5
4-BROMO-5-NITRO-1H-IMIDAZOLE	006963-65-1
2-BROMO-4-NITRO-1H-IMIDAZOLE	065902-59-2
3-BROMO-4-NITROPHENOL	005470-65-5
2-BROMO-4-NITROPHENOL	005847-59-6
4-BROMO-2-NITROPHENOL	007693-52-9
3-BROMO-2-NITROSO-1-NAPHTHOL	030922-52-2
2-BROMO-6-NITRO-4-(1,1,3,3-TETRAMETHYLBUTYL)-PH*	017199-22-3
1-BROMONONADECANE	004434-66-6
1-BROMO-N-NONANE	000693-58-3
1-BROMOOCTADECANE	000112-89-0
1-BROMOOCTANE	000111-83-1
1-BROMOPENTADECANE	000629-72-1
BROMOPENTAFLUOROBENZENE	000344-04-7
1-BROMOPENTANE	000110-53-2
2-BROMOPENTANE	000107-81-3
2-BROMOPENTANOIC ACID	000584-93-0
5-BROMOPENTANOIC ACID	002067-33-6
O-BROMOPHENOL	000095-56-7
M-BROMOPHENOL	000591-20-8
P-BROMOPHENOL	000106-41-2
P-BROMOPHENOXYACETIC ACID	001878-91-7
M-BROMOPHENOXYACETIC ACID	001798-99-8
O-BROMOPHENOXYACETIC ACID	001879-56-7
4-(4-BROMOPHENOXY)PHENYLUREA	085879-22-7
1-(4-BROMOPHENOXY)SILATRANE	086811-82-7
1-(3-BROMOPHENOXY)SILATRANE	083801-60-9
2-BROMOPHENYL ACETATE	001829-37-4
P-BROMOPHENYLACETIC ACID	001878-68-8
M-BROMOPHENYLACETIC ACID	001878-67-7
4-BROMOPHENYL 4-AMINOSALICYLATE	056356-33-3
2-BROMOPHENYL 4-AMINOSALICYLATE	056356-31-1
3-BROMOPHENYL 4-AMINOSALICYLATE	056356-32-2
P-BROMOPHENYL AZOXYCYANIDE	062825-10-9
M-BROMOPHENYL AZOXYCYANIDE	062825-09-6
2-BROMOPHENYLDIMETHYLCARBAMATE	007305-04-6
2-(4-BROMO PHENYLIMINO)IMIDAZOLIDINE	063346-73-6
2-(4-BROMOPHENYL)INDOLIZINE	007496-72-2
2-(3-BROMOPHENYL)INDOLIZINE	080488-98-8
P-BROMOPHENYL ISOTHIOCYANATE	001985-12-2
4-BROMOPHENYL PHENYL ETHER	000101-55-3
P-BROMO-N-PHENYLSUCCINIMIDE	041167-74-2
M-BROMO-N-PHENYLSUCCINIMIDE	058714-54-8
P-BROMOPHENYLUREA	001967-25-5
M-BROMOPHENYLUREA	002989-98-2
BROMOPHOS	002104-96-3
BROMOPHOS ETHYL	004824-78-6
BROMOPRIDE	004093-35-0
2-BROMOPROPANE	000075-26-3
1-BROMOPROPANE	000106-94-5
3-BROMOPROPANOL	000627-18-9
1-BROMO-2-PROPANONE	000598-31-2

3-BROMO-1-PROPENE	000106-95-6
A-BROMOPROPIONIC ACID	000598-72-1
4-BROMO-S-PROPYL-BENZOYLTHIOFORMOHYDROXIMATE	099481-58-0
N-(3-BROMOPROPYL)PHTHALIMIDE	005460-29-7
2-BROMOPYRAZINE	056423-63-3
3-BROMOPYRIDINE	000626-55-1
4-BROMOPYRIDINE	001120-87-2
2-BROMOPYRIDINE	000109-04-6
N1-(5-BROMO-2-PYRIDYL)SULFANILAMIDE	016805-99-5
N1-(3-BROMO-2-PYRIDYL)SULFANILAMIDE	030961-39-8
5-BROMOPYRIMIDINE	004595-59-9
3-BROMOQUINOLINE	005332-24-1
6-BROMOQUINOLINE	005332-25-2
7-BROMOQUINOLINE	004965-36-0
4'-BROMOSALICYLANILIDE	002627-77-2
5-BROMOSALICYLIC ACID	000089-55-4
4-BROMOSTYRENE	002039-82-9
BETA-BROMOSTYRENE	000103-64-0
1-BROMO-N-TETRADECANE	000112-71-0
BROMOTHIAZIDE	019367-61-4
2-BROMOTHIOPHENE	001003-09-4
3-BROMOTHIOPHENE	000872-31-1
M-BROMOTOLUENE	000591-17-3
ALPHA-BROMOTOLUENE	000100-39-0
P-BROMOTOLUENE	000106-38-7
O-BROMOTOLUENE	000095-46-5
BROMOTRICHLOROMETHANE	000075-62-7
1-BROMO-N-TRIDECANE	000765-09-3
2-BROMO-1,1,2-TRIFLUOROETHYLENE	000598-73-2
BROMOTRIFLUOROMETHANE	000075-63-8
1-BROMOUNDECANE	000693-67-4
5-BROMOURACIL	000051-20-7
A-BROMO-I-VALERYLUREA	000496-67-3
BROMOXYNIL	001689-84-5
BROMOXYNIL OCTANOATE	001689-99-2
BROMUCONAZOLE	116255-48-2
BROTIZOLAM	057801-81-7
N(P-BRPH)3-N'-PIPERIDINOACETAMIDE	058479-86-0
N(O-BRPH)3-N'-PIPERIDINOACETAMIDE	058479-89-3
BRUCEANTIN	041451-75-6
BRUCINE	000357-57-3
3-N-BUAM-4NH2-6IPR-124-TRIAZIN-5-ONE	033665-71-3
	BUBENZHYDRYLPROPYLAMI 3
NEDIME	3(4-T-BU
	070785-06-7
BENZHYDRYLPROPYL)AMINE,NN-DIME	066381-51-9
N-(N"-BU-BUO-CARBAMYLTHIO)METHOMYL	002438-72-4
BUFEXAMIC ACID	054340-62-4
BUFURALOL	000117-26-0
BULAN	071203-64-0
(N-BU-ME)AMET-(345-MEO)PHENYL-SULFONE	066381-55-3
N(N"-BU-MEO CARBAMYLTHIO)METHOMYL	034915-68-9
BUNITROLOL	104958-89-6
N-BU-3-NO2-124-TRIAZOL-1-ACETAMIDE	077317-72-7
3-BUO-5-BR-6-ME-PYRIMIDIN-2,4-DIONE	077317-73-8
3-SBUO-5-BR-6-ME-PYRIMIDIN-2,4-DION	069741-92-0
2-AMINO-5-T-BU-1,3,4-OXADIAZOLE	069741-90-8
2-AMINO-5-I-BU-1,3,4-OXADIAZOLE	

O=P(NME)(OME)O-(4-T-BU PHENYL)	019590-04-6
O=P(NC)(OC)O-3-T-BU PHENYL	019590-05-7
N(N''-BU-(PHENYL)CARBAMYLTHIO)METHOMYL	082774-18-3
1-(3-T-BUPH)-3-MEO-3-METHYL UREA	070477-14-4
T-BU PHOSPHONIC ACID,ET,PNO2PH ESTR	013538-14-2
BUPIRIMATE	041483-43-6
BUPIVACAINE	002180-92-9
BUPRANOLOL	014556-46-8
BURIAMIDE	034970-69-9
3-BUS-4-NH2-6-IPR-124-TRIAZINE-5-ONE	050917-25-4
BUTABARBITAL	000125-40-6
BUTACHLOR	023184-66-9
1,2-BUTADIENE	000590-19-2
1,3-BUTADIENE	000106-99-0
BUTAMIFOS	036335-67-8
BUTANAMIDE, N-[(2-AMINO-4-CHLOROPHENYL)SULFONYL]	081792-82-7
BUTANAMIDE, 2-AMINO-N-(2,6-DIMETHYLPHENYL)-	059359-46-5
BUTANAMIDE, 4-AMINO-N-(2,6-DIMETHYLPHENYL)-	102089-65-6
BUTANAMIDE, N-[(2-AMINO-4-METHOXYPHENYL)SULFONYL	081792-81-6
BUTANAMIDE, N-[(2-AMINO-4-METHYLPHENYL)SULFONYL]	081792-79-2
BUTANAMIDE, N-[(2-AMINOPHENYL)SULFONYL]-2-ETHYL-	081792-78-1
BUTANAMIDE, N-[2-(AMINOSULFONYL)-5-CHLOROPHENYL]	097141-35-0
BUTANAMIDE, N-[2-(AMINOSULFONYL)-5-METHYLPHENYL]	097141-32-7
BUTANAMIDE, N-[2-(AMINOSULFONYL)PHENYL]-2-ETHYL-	007752-27-4
BUTANAMIDE, N-[2-(AMINOSULFONYL)-5-METHOXYPHENYL	097141-34-9
BUTANAMIDE, N-[2-(AMINOSULFONYL)-5-(TRIFLUOROMET	097141-33-8
BUTANAMIDE, N-(5-AMINO-1H-1,2,4-TRIAZOL-3-YL)-2,	080616-59-7
BUTANAMIDE, 4-(BENZOYLOXY)-N,N-DIMETHYL-	115178-78-4
BUTANAMIDE, N-(3,4-DICHLOROPHENYL)-	002150-95-0
BUTANAMIDE, N-(3,4-DICHLOROPHENYL)-2-METHYL-	007160-25-0
BUTANAMIDE, 3-(DIETHYLAMINO)-N-(2,6-DIMETHYLPHEN	077471-00-2
BUTANAMIDE, 4-(DIETHYLAMINO)-N-(2,6-DIMETHYLPHEN	021236-53-3
BUTANAMIDE, 2-ETHYL-N-[(4-METHYL-2-NITROPHENYL)S	081793-04-6
BUTANAMIDE, 2-ETHYL-N-[[3-NITRO-5-(TRIFLUOROMETH	081792-70-3
BUTANAMIDE, 2-ETHYL-N-[(2-NITROPHENYL)SULFONYL]-	081793-03-5
1-BUTANAMINE, 4-[(6A,7,10,10A-TETRAHYDRO-6,6,9-T	132213-87-7
1-BUTANAMINIUM, N,N-DIBUTYL-N-METHYL-, IODIDE	003085-79-8
N-BUTANE	000106-97-8
BUTANE, 2-CHLORO-2-METHYL-	000594-36-5
BUTANEDIAMIDE, N-[4-[(DIMETHYLAMINO)SULFONYL]PHE	075851-97-7
1,4-BUTANEDIAMINE	000110-60-1
BUTANEDIOC ACID	000110-15-6
BUTANEDIOIC ACID, DIPROPYL ESTER	000925-15-5
1,2-BUTANEDIOL	000584-03-2
1,4-BUTANEDIOL	000110-63-4
1,3-BUTANEDIOL	000107-88-0
2,3-BUTANEDIOL	000513-85-9
2,3-BUTANEDIOL (D)	019132-06-0
2,3-BUTANEDIOL (L)	024347-58-8
2,3-BUTANEDIOL (MESO)	005341-95-7
2,3-BUTANEDIONE	000431-03-8
1,4-BUTANEDIYL BIS(2-METHYL-2-PROPENOATE)	002082-81-7
BUTANE, 2-ISOTHIOCYANATO-	004426-79-3
BUTANE, 1-ISOTHIOCYANATO-	000592-82-5
BUTANENITRILE, 4-CHLORO-	000628-20-6
BUTANENITRILE, 2-[(2-CHLOROPHENYL)HYDRAZONO]-3-O	028317-59-1
BUTANENITRILE, 2-[(4-CHLOROPHENYL)HYDRAZONO]-3-O	028317-58-0
BUTANENITRILE, 2-[(3-CHLOROPHENYL)HYDRAZONO]-3-O	028317-60-4

BUTANENITRILE, 4,4-DIETHOXY-	018381-45-8
BUTANENITRILE, 2-[(2,4-DINITROPHENYL)HYDRAZONO]-	028317-71-7
BUTANENITRILE, 3-METHYL-	000625-28-5
BUTANETHIOL	000109-79-5
2-BUTANETHIOL	000513-53-1
1,2,4-BUTANETRIOL, 3-[(5-NITROPYRAZINYL)AMINO]-	089690-76-6
BUTANOIC ACID,2-ACETYLAMINO,METHYL ESTER	003619-01-0
BUTANOIC ACID, 2-[[(AMINOCARBONYL)AMINO]CARBONYL	073632-79-8
BUTANOIC ACID, 2-[[(AMINOCARBONYL)AMINO]CARBONYL	073632-76-5
BUTANOIC ACID, 2-[[(AMINOCARBONYL)AMINO]CARBONYL	073632-80-1
BUTANOIC ACID, [2-(AMINOCARBONYL)PHENOXY]METHYL	103951-39-9
BUTANOIC ACID, 2-[[(AMINOCARBONYL)AMINO]CARBONYL	073632-77-6
BUTANOIC ACID, 2-[[(AMINOCARBONYL)AMINO]CARBONYL	073632-78-7
BUTANOIC ACID, 2-[[(AMINOTHIOXOMETHYL)AMINO]CARB	073930-96-8
BUTANOIC ACID, 4-(4H-1,2,4-BENZOTHIADIAZIN-3-YL)	004826-24-8
BUTANOIC ACID, 4-BENZOYL-	001501-05-9
BUTANOIC ACID, 4-[(1,1'-BIPHENYL)-4-YL]-2-METHYL	063472-04-8
BUTANOIC ACID, 4-(7-BROMO-4H-1,2,4-BENZOTHIADIAZ	101064-03-3
2-BROMO-3-METHYL-BUTANOIC ACID	000565-74-2
BUTANOIC ACID, 4-(6-CHLORO-4H-1,2,4-BENZOTHIADIA	101063-99-4
BUTANOIC ACID, 4-(7-CHLORO-4H-1,2,4-BENZOTHIADIA	101064-00-0
BUTANOIC ACID, 2-(4-CHLORO-2-METHYLPHENOXY)-	025204-89-1
BUTANOIC ACID, 4-(3-CHLORO-4-METHOXYBENZOYL)-	071354-31-9
BUTANOIC ACID, 4-(3-CHLORO-4-METHOXYBENZOYL)-3-M	099414-56-9
BUTANOIC ACID, 4-(5,7-DIBROMO-4H-1,2,4-BENZOTHIA	101064-04-4
BUTANOIC ACID, 4-(5,7-DICHLORO-4H-1,2,4-BENZOTHI	101064-01-1
BUTANOIC ACID, 4-(6,7-DICHLORO-4H-1,2,4-BENZOTHI	101064-02-2
BUTANOIC ACID, (2,5-DIHYDRO-4-OXO-4H-PYRAZOLO[3,	098827-22-6
BUTANOIC ACID, 4-[[4-[(DIMETHYLAMINO)SULFONYL]PH	075852-09-4
BUTANOIC ACID, 2-ETHOXY-3-METHYL-, PHENYLMETHYL	112473-78-6
BUTANOIC ACID, 2-ETHYL-3-METHYL-, PHENYL ESTER	112473-77-5
BUTANOIC ACID, 4-HYDROXY-, ETHYL ESTER	000999-10-0
BUTANOIC ACID, 4-(4-METHOXYBENZOYL)-	004609-10-3
BUTANOIC ACID, 4-(6-METHOXY-4H-1,2,4-BENZOTHIADI	101064-06-6
BUTANOIC ACID, 2-METHYL-	000116-53-0
BUTANOIC ACID, 4-(6-METHYL-4H-1,2,4-BENZOTHIADIA	101064-05-5
BUTANOIC ACID, 3-METHYL-, PHENYLMETHYL ESTER	000103-38-8
BUTANOIC ACID, 3-METHYL-, PHENYL ESTER	015806-38-9
BUTANOIC ACID, 4-(1-NAPHTHALENYLOXY)-	016563-45-4
BUTANOIC ACID,3-OH-2-(4-BIPHENYLYL)-	093371-36-9
BUTANOIC ACID, SODIUM SALT	000156-54-7
BUTANOIC ACID, 4-[6-(TRIFLUOROMETHYL)-4H-1,2,4-B	101064-07-7
BUTANOIC ACID,2-(4-BIPHENYLYL)-3-HYDROXY-3-METHY	093371-55-2
T-BUTANOL	000075-65-0
1-BUTANOL	000071-36-3
2-BUTANOL	000078-92-2
1-BUTANOL, 4-CHLORO-	000928-51-8
1-BUTANOL (D)	004712-38-3
2-BUTANOL (D)	004712-39-4
1-BUTANOL, 3-METHYL-, BENZOATE	000094-46-2
1-BUTANOL, 3-METHYL-, NITRATE	000543-87-3
2-BUTANONE	000078-93-3
2-BUTANONE, 3-METHYL-4-(1-PIPERIDINYL)-	042327-99-1
2-BUTANONE OXIME	000096-29-7
1-BUTANONE, 1-PHENYL-	000495-40-9
3-BUTENAMIDE, N-(2,4-DICHLOROPHENYL)-2-METHYL-	074054-79-8
2-BUTENAMIDE, N-(3,4-DICHLOROPHENYL)-	010249-33-9
1-BUTENE	000106-98-9

CIS-2-BUTENE	000590-18-1
TRANS-2-BUTENE	000624-64-6
2-BUTENEDINITRILE, (E)-	000764-42-1
2-BUTENEDIOIC ACID (E)-, DIMETHYL ESTER	000624-49-7
2-BUTENE-1,4-DIOL (TRANS)	000821-11-4
3-BUTENE-1,2-DIOL	000497-06-3
2-BUTENE-1,4-DIOL	000110-64-5
2-BUTENE-1,4-DIOL (CIS)	006117-80-2
3-BUTENENITRILE	000109-75-1
2-BUTENOIC ACID, 2-[(ACETYLOXY)METHYL]- DERIVATI	034175-79-6
2-BUTENOIC ACID, ESTER DERIVATIVE	021899-50-3
2-BUTENOIC ACID, ESTER DERIVATIVE	020071-53-8
(Z)-2-BUTENOIC ACID (ISOCROTONIC ACID)	000503-64-0
2-BUTENOIC ACID, 2-METHYL-, 9-(ACETYLOXY)- ESTER	033854-15-8
2-BUTENOIC ACID, 2-METHYL-, 7-(ACETYLOXY)-6-(CHL	020501-52-4
2-BUTEN-1-OL (CIS)	004088-60-2
1-BUTEN-1-ONE(ETHYLKETENE)	020334-52-5
3-BUTEN-2-ONE, 4-(2-FURANYL)-	000623-15-4
4-(2-BUTENYL)PHENOL	013037-71-3
1-BUTEN-3-YNE	000689-97-4
BUTHIDAZOLE	055511-98-3
BUTHIOBATE	051308-54-4
BUTONATE	000126-22-7
P-BUTOXYBENZAMIDE	073258-87-4
P-BUTOXYBENZOIC ACID	001498-96-0
2-(T-BUTOXY)ETHANOL	007580-85-0
2-BUTOXYETHANOL ACETATE	000112-07-2
1-(2-BUTOXYETHOXY)ETHANOL	054446-78-5
2-(2-(2-(TERT-BUTOXY)ETHOXY)ETHOXY)ETHANOL	029681-20-7
2(2-BUTOXYETHOXY)ETHYLTHIOCYANATE	000112-56-1
4-BUTOXYPHENOL	000122-94-1
M-BUTOXYPHENOL	018979-72-1
2-BUTOXY-2-PH-4-ME MORPHOLINE HCL	124497-85-4
3-BUTOXYPYRIDAZINE	089770-36-5
BUTRALIN	033629-47-9
BUTRAMIDE, 4-PHENYL	001199-98-0
4-(3-ME-3-T-BU-1-TRIAZENO)BENZAMIDE	059708-25-7
N-BUTROYLCYCLOBUTANECARBOXAMIDE	006815-52-7
BUTRYIC ACID HYDRAZIDE	003538-65-6
I-BUTRYIC ACID HYDRAZIDE	003619-17-8
BUTURON	003766-60-7
4-BUTYLACETANILIDE	003663-20-5
SEC-BUTYLACETATE	000110-19-0
N-BUTYL ACETATE	000123-86-4
S-BUTYLACETATE	000105-46-4
T-BUTYL ACETATE	000540-88-5
BUTYL ACRYLATE	000141-32-2
9-BUTYL ADENINE	002715-70-0
SEC-BUTYLAMINE	000513-49-5
N-BUTYLAMINE	000109-73-9
SEC-BUTYLAMINE	013952-84-6
T-BUTYLAMINE	000075-64-9
BUTYL-P-AMINOBENZOATE	000094-25-7
N-(ISO-BUTYLAMINOMETHYL)BENZAMIDE	052387-58-3
N-BUTYL-P-AMINOPHENOL	000103-62-8
N-BUTYLANILINE	001126-78-9
BUTYLATE	002008-41-5
1-BUTYLAZIRIDINE	001120-85-0

12-BUTYL-BENZ[A]ANTHRACENE	093037-15-1
4-TERT-BUTYLBENZAMIDE	056108-12-4
SEC-BUTYLBENZENE	000135-98-8
T-BUTYLBENZENE	000098-06-6
N-BUTYLBENZENE	000104-51-8
A-BUTYL BENZENEETHANEAMINE	067309-36-8
A-T-BUTYL BENZENEETHANEAMINE	067309-37-9
A-I-BUTYL BENZENEETHANEAMINE	067309-38-0
P-BUTYLBENZENESULFONAMIDE	001135-00-8
BUTYL BENZOATE	000136-60-7
4-BUTYLBENZOIC ACID	020651-71-2
1-BUTYLBENZOTRIAZOLE	000708-43-0
2-S-BUTYLBENZOTRIAZOLE	066382-00-1
2-BUTYLBENZOTRIAZOLE	016584-01-3
1-S-BUTYLBENZOTRIAZOLE	063936-04-9
4-TERT-BUTYLBENZYL ALCOHOL	000877-65-6
BUTYL BENZYL PHTHALATE	000085-68-7
1-BUTYLBIGUANIDE	000692-13-7
A-BUTYL-N-SEC-BUTYLCINNAMAMIDE	097871-57-3
N-BUTYLBUTYRATE	000109-21-7
O-BUTYL CARBAMATE	000592-35-8
O-T-BUTYL CARBAMATE	004248-19-5
T-BUTYL CHLORIDE	000507-20-0
N2-BUTYL-N1-P-CHLOROBENZENESULFONYLUREA	000094-20-2
N-BUTYL-N-CHLORO-1-BUTANAMINE	000999-33-7
4-(TERT-BUTYL)-2-CHLORO-6-NITROPHENOL	014593-28-3
4-(TERT-BUTYL)-2-CHLOROPHENOL	000098-28-2
A-ETS-N-SEC-BUTYLCINNAMAMIDE	097871-56-2
A-ETO-N-SEC-BUTYL CINNAMAMIDE	097871-55-1
A-MES-N-SEC-BUTYLCINNAMAMIDE	097871-52-8
N-SEC-BUTYLCINNAMAMIDE	097871-50-6
4-(T-BUTYL)-2-CRESOL	000098-27-1
8-S-T-BUTYL CYCLIC AMP	081835-01-0
8-S-T-BUTYL CYCLIC AMP-O-BENZYL	107538-71-6
TERT-BUTYLCYCLOHEXANE	003178-22-1
BUTYLCYCLOHEXANE	001678-93-9
S-BUTYLCYCLOHEXANE	007058-01-7
BUTYLCYCLOHEXYL PHTHALATE	000084-64-0
BUTYL DECANOATE	030673-36-0
4-(TERT-BUTYL)-2,6-DICHLOROPHENOL	034593-75-4
4-(TERT-BUTYL)-2,6-DINITROPHENOL	004097-49-8
1,3-BUTYLENE GLYCOL METHYL ETHER	000111-32-0
1,2-BUTYLENEOXIDE	000106-88-7
5-BUTYL-5-ETHYLBARBITURIC ACID	000077-28-1
BUTYL ETHYL ETHER	000628-81-9
BUTYL (2-ETHYLHEXYL) PHTHALATE	000085-69-8
2-(T-BUTYL)-4-ETHYLPHENOL	000096-70-8
N-BUTYLFORMATE	000592-84-7
1-BUTYLGALACTOPYRANOSIDE	039824-09-4
BUTYL GALLATE	001083-41-6
T-BUTYLGLYCINE	033105-81-6
BUTYLGLYCOLYL BUTYL PHTHALATE	000085-70-1
BUTYL HEPTANOATE	005454-28-4
N-BUTYL HEXANOATE	000626-82-4
(1-BUTYLHEXYL)BENZENESULFONIC ACID, SODIUM SALT	073602-67-2
1,2-DIBENZOYL-1-TERT-BUTYLHYDRAZINE	112225-87-3
3-T-BUTYLHYDROCHLOROTHIAZIDE	023141-84-6
TERT-BUTYLHYDROPEROXIDE	000075-91-2

3-T-BUTYL-4-HYDROXYACETANILIDE	004151-47-7
N-BUTYL-HYDROXYPYRID-4-ONE	104764-54-7
3-BUTYLHYDROXYUREA	005681-57-2
3-T-BUTYLHYDROXYUREA	063491-77-0
N-BUTYL IODIDE	000542-69-8
BUTYL ISOCYANATE	000111-36-4
2-I-BUTYLISONIAZID	058480-99-2
2-(TERT-BUTYL)-4-ISOPROPYLPHENOL	007597-97-9
BUTYL LACTATE	000138-22-7
BUTYL MALONIC ACID	000534-59-8
T-BUTYL METHACRYLATE	000585-07-9
SEC-BUTYL METHYL ETHER	006795-87-5
2-(TERT-BUTYL)-4-METHYLPHENOL	002409-55-4
N-BUTYLNICOTINAMIDE	010354-55-9
BUTYL NITRATE	000928-45-0
1-BUTYLNITRITE	000544-16-1
4-T-BUTYLNITROBENZENE	003382-56-7
4-(SEC-BUTYL)-2-NITROPHENOL	003555-18-8
3-T-BUTYL-4-NITROPHENOL	005722-68-9
4-(TERT-BUTYL)-2-NITROPHENOL	003279-07-0
4-T-BUTYL-N-NITROSOPIPERIDINE	046061-25-0
1-BUTYL-1-NITROSOUREA	000869-01-2
BUTYL NONANOATE	050623-57-9
BUTYL OCTANOATE	000589-75-3
BUTYL OLEATE	000142-77-8
BUTYL PALMITATE	000111-06-8
P-BUTYLPHENOL	001638-22-8
P-(SEC-BUTYL)PHENOL	000099-71-8
2-BUTYLPHENOL	003180-09-4
O-SEC-BUTYLPHENOL	000089-72-5
O-T-BUTYLPHENOL	000088-18-6
M-BUTYLPHENOXYACETIC ACID	001878-53-1
O-T-BUTYLPHENOXYACETIC ACID	019271-90-0
O-SEC-BUTYLPHENOXYACETIC ACID	076343-98-1
P-SEC-BUTYLPHENOXYACETIC ACID	004917-89-9
BUTYL N-PHENYLCARBAMATE	001538-74-5
2-S-BUTYLPHENYLDIMETHYLCARBAMATE	055379-70-9
T-BUTYLPHENYL DIPHENYL PHOSPHATE	056803-37-3
BUTYL PHENYL ETHER	001126-79-0
4-T-BUTYLPHENYL GLUCOPYRANOSIDE	029074-04-2
N-BUTYL PROPIONATE	000590-01-2
2-T-BUTYLPYRIDINE	005944-41-2
4-(TERT-BUTYL)PYRIDINE	003978-81-2
BUTYLPYRIDINIUM BROMIDE	000874-80-6
N-BUTYL-3-PYRIDYLMETHYLAMINE	020173-12-0
4-BUTYLSEMICARBAZIDE	020605-19-0
4-T-BUTYLSEMICARBAZIDE	074255-46-2
4-I-BUTYLSEMICARBAZIDE	079353-74-5
N-BUTYL STEARATE	000123-95-5
N-BUTYL 4-SULFAMYLBENZAMIDE	059777-63-8
BUTYL 4-SULFAMYLBENZOATE	059777-59-2
1-BUTYLTHIO-B-GALACTOPYRANOSIDE	063407-52-3
6-(TERT-BUTYL)-1,2,4-TIRAZINE-3,5(2H,4H)-DIONE	052236-30-3
P-(T-BUTYL)TOLUENE	000098-51-1
SEC-BUTYLUREA	000689-11-2
BUTYLUREA	000592-31-4
DI-(SEC-BUTYL)UREA	000869-79-4
BUTYL VINYL ETHER	000111-34-2

BUTYL XANTHATE	000110-50-9
P-T-BUTYLPHENOL	000098-54-4
1-BUTYNE	000107-00-6
2-BUTYNE	000503-17-3
2-BUTYNE, 1,4-DICHLORO-	000821-10-3
2-BUTYNE-1,4-DIOL	000110-65-6
3-BUTYN-1-OL, 2-CHLORO-4[5-(1,3-PENTADIYNYL)-2-	001152-72-3
BUTYRALDEHYDE	000123-72-8
BUTYRAMIDE	000541-35-5
4-BUTYRAMID-2266-TEME-PIPERDIN-N-OXID	021270-93-9
I-BUTYRANILIDE	004406-41-1
BUTYRANILIDE	001129-50-6
BUTYRIC ACID	000107-92-6
BUTYRIC ANHYDRIDE	000106-31-0
GAMMA-BUTYROLACTONE	000096-48-0
BUTYRONITRILE	000109-74-0
N-I-BUTYROYLCYCLOBUTANECARBOXAMIDE	023046-88-0
C1CCCCC1OC2CCC(OCC3OC(CC)OC3)CC2	063837-33-2
CACODYLIC ACID	000075-60-5
CADUSAFOS	095465-99-9
CAFFEINE	000058-08-2
CALCON	002538-85-4
CAMPHENE	000079-92-5
CAMPHOR—	000076-22-2 —
(1R,4R)-(+)-CAMPHOR	000464-49-3
CAMPHORIC ACID	000124-83-4
DL-CAMPHORQUINONE	010373-78-1
CAMPHORQUINONE	000465-29-2
CAMPHORSULFONIC ACID	003144-16-9
CANRENONE	000976-71-6
CAPROLACTAM	000105-60-2
E-CAPROLACTONE	000502-44-3
CAPRYLANILIDE	000621-15-8
CAPTAFOL	002425-06-1
CAPTAN	000133-06-2
CAPTOPRIL	062571-86-2
CARBAMAZEPINE	000298-46-4
CARBAMIC ACID, (3-ACETYLPHENYL)-, METHYL ESTER	087743-55-3
CARBAMIC ACID, [[[[(1,3-BENZODIOXOL-5-YLOXY)CARB	123297-69-8
CARBAMIC ACID, [4-(2-BROMOPHENYL)-4,5-DIHYDRO-1H	069811-19-4
CARBAMIC ACID, [4-(3-BROMOPHENYL)-4,5-DIHYDRO-1H	069811-22-9
CARBAMIC ACID, (3-BROMOPHENYL)-, METHYL ESTER	025216-72-2
CARBAMIC ACID, (2-BROMOPHENYL)-, METHYL ESTER	025216-70-0
CARBAMIC ACID, [2-(BUTOXYMETHYL)PHENYL]-, 2-(1-P	093450-34-1
CARBAMIC ACID, [3-(BUTOXYMETHYL)PHENYL]-, 2-(1-P	093450-23-8
CARBAMIC ACID, (4-BUTYLPHENYL)-, METHYL ESTER	113932-84-6
CARBAMIC ACID, [5-(4-CHLOROPHENYL)-4,5-DIHYDRO-1	066308-28-9
CARBAMIC ACID, [5-(3-CHLOROPHENYL)-4,5-DIHYDRO-1	089145-58-4
CARBAMIC ACID, [5-(2-CHLOROPHENYL)-4,5-DIHYDRO-1	066308-27-8
CARBAMIC ACID, (4-CHLOROPHENYL)-, METHYL ESTER	000940-36-3
CARBAMIC ACID, (2-CHLOROPHENYL)-, METHYL ESTER	020668-13-7
CARBAMIC ACID, (4-CYANOPHENYL)-, METHYL ESTER	094563-11-8
CARBAMIC ACID, (3-CYANOPHENYL)-, METHYL ESTER	065448-74-0
CARBAMIC ACID, [5-(2,5-DIBROMOPHENYL)-4,5-DIHYDR	089145-54-0
CARBAMIC ACID, (3,4-DIETHOXYPHENYL)-, 2-CHLORO-1	087130-55-0
CARBAMIC ACID, (3,4-DIETHOXYPHENYL)-, 2-BROMO-1-	087130-47-0
CARBAMIC ACID, (3,4-DIETHOXYPHENYL)-, BUTYL ESTE	113932-77-7
CARBAMIC ACID, (3,4-DIETHOXYPHENYL)-, 2-CYANOETH	087130-56-1

CARBAMIC ACID, (3,4-DIETHOXYPHENYL)-, CYCLOPENTY   113932-78-8
CARBAMIC ACID, (3,4-DIETHOXYPHENYL)-, ETHYL ESTE   087130-19-6
CARBAMIC ACID, (3,4-DIETHOXYPHENYL)-, 2-METHOXYE   095261-60-2
CARBAMIC ACID, (3,4-DIETHOXYPHENYL)-, 2-METHYLPR   095261-40-8
CARBAMIC ACID, (3,4-DIETHOXYPHENYL)-, 1-METHYLET   087130-20-9
CARBAMIC ACID, (3,4-DIETHOXYPHENYL)-, METHYL EST   087130-18-5
CARBAMIC ACID, (3,4-DIETHOXYPHENYL)-, PHENYLMETH   095261-65-7
CARBAMIC ACID, (3,4-DIETHOXYPHENYL)-, 2-PROPENYL   087130-27-6
CARBAMIC ACID, (3,4-DIETHOXYPHENYL)-, PROPYL EST   087130-88-9
CARBAMIC ACID, (3,4-DIETHOXYPHENYL)-, 2-PROPYNYL   087130-34-5
CARBAMIC ACID, DIETHYL-, 2-ETHYLBUTYL ESTER   098155-00-1
CARBAMIC ACID, DIETHYL-, 2-(1H-INDOL-3-YL)ETHYL   098154-95-1
CARBAMIC ACID, DIETHYL-, 2-METHYLPROPYL ESTER   093677-64-6
CARBAMIC ACID, DIETHYL-, PHENYLMETHYL ESTER   051170-56-0
CARBAMIC ACID, DIETHYL-, 2-PHENYLETHYL ESTER   098183-17-6
CARBAMIC ACID, [4,5-DIHYDRO-5-[4-(HYDROXYMETHYL)   089145-43-7
CARBAMIC ACID, (4,5-DIHYDRO-1-METHYL-5-PHENYL-1H   066308-18-7
CARBAMIC ACID, [4,5-DIHYDRO-4-[4-(1-METHYLETHYL)   069810-98-6
CARBAMIC ACID, DIMETHYL-, [4-(AMINOCARBONYL)PHEN   087340-61-2
CARBAMIC ACID, [5-(2,3-DIMETHYLPHENYL)-4,5-DIHYD   066308-35-8
CARBAMIC ACID, DIMETHYL-, 2-ETHYLBUTYL ESTER   098154-99-5
CARBAMIC ACID, DIMETHYL-, 2-(1H-INDOL-3-YL)ETHYL   098154-94-0
CARBAMIC ACID, DIMETHYL-, 1-METHYL-2-PHENYLETHYL   098154-93-9
CARBAMIC ACID, DIMETHYL-, 2-METHYLPROPYL ESTER   052113-78-7
CARBAMIC ACID, DIMETHYL-, 1-METHYLETHYL ESTER   038580-89-1
CARBAMIC ACID, DIMETHYL-, 2-PHENYLETHYL ESTER   098183-16-5
CARBAMIC ACID, DIMETHYL-, 3-PHENYLPROPYL ESTER   091564-01-1
CARBAMIC ACID, DIMETHYL-, 3-(3-PYRIDINYL)PROPYL   098154-97-3
CARBAMIC ACID, DIMETHYL-, 3-(2-PYRIDINYL)PROPYL   098154-98-4
CARBAMIC ACID, [2-(ETHOXYMETHYL)PHENYL]-, 2-(1-P   080171-69-3
CARBAMIC ACID, (2-ETHOXYPHENYL)-, METHYL ESTER   113932-79-9
CARBAMIC ACID, (4-ETHOXYPHENYL)-, METHYL ESTER   035407-50-2
CARBAMIC ACID, (3-ETHOXYPHENYL)-, METHYL ESTER   113932-83-5
CARBAMIC ACID, 2-ETHYLBUTYL ESTER   024847-58-3
CARBAMIC ACID, (3-ETHYLPHENYL)-, METHYL ESTER   113932-82-4
CARBAMIC ACID, (4-ETHYLPHENYL)-, METHYL ESTER   028238-55-3
CARBAMIC ACID, [5-(3-FLUOROPHENYL)-4,5-DIHYDRO-1   089145-29-9
CARBAMIC ACID, [4-(3-FLUORPHENYL)-4,5-DIHYDRO-1H   069811-11-6
CARBAMIC ACID, [5-(4-FLUOROPHENYL)-4,5-DIHYDRO-1   089145-31-3
CARBAMIC ACID, [5-(2-FLUOROPHENYL)-4,5-DIHYDRO-1   066308-26-7
CARBAMIC ACID, (4-FLUOROPHENYL)-, METHYL ESTER   016744-99-3
CARBAMIC ACID, (3-FLUOROPHENYL)-, METHYL ESTER   072755-13-6
CARBAMIC ACID, (2-FLUOROPHENYL)-, METHYL ESTER   016664-12-3
CARBAMIC ACID, [5-(4-FORMYLPHENYL)-4,5-DIHYDRO-1   089145-44-8
CARBAMIC ACID, 2-(1H-INDOL-3-YL)ETHYL ESTER   091350-72-0
CARBAMIC ACID, (2-IODOPHENYL)-, METHYL ESTER   103970-46-3
CARBAMIC ACID, (3-IODOPHENYL)-, METHYL ESTER   113932-81-3
CARBAMIC ACID, (3-METHOXYPHENYL)-, METHYL ESTER   051422-77-6
CARBAMIC ACID, (4-METHOXYPHENYL)-, METHYL ESTER   014803-72-6
CARBAMIC ACID, (2-METHOXYPHENYL)-, METHYL ESTER   014803-73-7
CARBAMIC ACID, 2-(2-METHYL-5-NITRO-1H-IMIDAZOL-1   031478-45-2
CARBAMIC ACID, (4-METHYLPHENYL)-, METHYL ESTER   005602-96-0
CARBAMIC ACID, (2-METHYLPHENYL)-, METHYL ESTER   014983-92-7
CARBAMIC ACID, (3-METHYLPHENYL)-, METHYL ESTER   039076-18-1
CARBAMIC ACID, METHYL-, PHENYLMETHYL ESTER   030379-59-0
CARBAMIC ACID, (4-NITROPHENYL)-, METHYL ESTER   001943-87-9
CARBAMIC ACID, (2-NITROPHENYL)-, METHYL ESTER   013725-30-9
CARBAMIC ACID, (3-NITROPHENYL)-, METHYL ESTER   002189-61-9

CARBAMIC ACID, [3-[(PENTYLOXY)METHYL]PHENYL]-, 2	093450-24-9
CARBAMIC ACID, [2-(PENTYLOXY)PHENYL]-, 2-(1-PIPE	093450-05-6
CARBAMIC ACID, 2-PHENYLETHYL ESTER	006326-19-8
CARBAMIC ACID, 3-PHENYLPROPYL ESTER	000673-31-4
CARBAMIC ACID, 3-(3-PYRIDINYL)PROPYL ESTER	098154-96-2
CARBAMIC ACID, (1,2,3,6-TETRAHYDRO-1-PYRIDINYL)-	068709-03-5
CARBAMIC ACID, (2,2,2-TRICHLORO-1-HYDROXYETHYL)-	000541-79-7
CARBAMIC ACID, [2-(TRIFLUOROMETHYL)PHENYL]-, MET	113932-80-2
CARBAMIC ACID, [3-(TRIFLUOROMETHYL)PHENYL]-, MET	018584-93-5
CARBAMIC ACID,O-ETHYL(TRIMEAMM)BR	028099-10-7
CARBAMOTHIOIC ACID, [2-(2-ETHYL-5-NITRO-1H-IMIDA	051022-76-5
2-(4-CARBAMOYLBUTYL)BENZOTRIAZOLE	069218-64-0
1-(4-CARBAMOYLBUTYL)BENZOTRIAZOLE	069218-59-3
1-(4-CARBAMOYLPHENYL)-3-I-PROPYL-3-METHYLTRIAZEN	059708-23-5
1-(4-CARBAMOYLPHENYL)-3-METHYL-3-ETHYLTRIAZENE	059708-19-9
1-(4-CARBAMOYLPHENYL)-3-METHYL-3-METHOXYTRIAZENE	066974-76-3
1-(4-CARBAMOYLPHENYL)-3-METHYL-3-OCTYLTRIAZENE	066521-49-1
1-(4-CARBAMOYLPHENYL)-3-METHYL-3-HEPTYLTRIAZENE	089530-01-8
1-(4-CARBAMOYLPHENYL)-3-METHYL-3-PENTYLTRIAZENE	059708-22-4
1-(4-CARBAMOYLPHENYL)-3-METHYL-3-HEXYLTRIAZENE	089530-00-7
1-(4-CARBAMOYLPHENYL)-3-METHYL-3-HYDROXYTRIAZENE	042548-73-2
1-(4-CARBAMOYLPHENYL)-3-METHYL-3-HYDROXYETHYLTRI	059708-20-2
1-(4-CARBAMOYLPHENYL)-3-METHYL-3-PROPYLTRIAZENE	089529-99-7
3-CARBAMOYLPYRIDINIUM IODIDE,DODECYL	035096-55-0
3-CARBAMOYLPYRIDINIUM IODIDE,N-HEXYL	065481-90-5
3-CARBAMOYLPYRIDINIUM IODIDE,N-C14	035129-56-7
3-CARBAMOYLPYRIDINIUM IODIDE,N-OCTYL	035041-48-6
3-CARBAMOYLPYRIDINIUM IODIDE, N-DECYL	035041-49-7
M-CARBAMYLBENZYL TRIMETHYL AMMONIUM BROMIDE	071323-95-0
2-(2-CARBAMYLETHYL)BENZOTRIAZOLE	069218-62-8
1-(2-CARBAMYLETHYL)BENZOTRIAZOLE	069218-57-1
2-CARBAMYLMETHYLBENZOTRIAZOLE	069218-61-7
1-CARBAMYLMETHYLBENZOTRIAZOLE	069218-56-0
1-(5-CARBAMYLPENTYL)BENZOTRIAZOLE	069218-60-6
2-(5-CARBAMYLPENTYL)BENZOTRIAZOLE	069218-65-1
2-(3-CARBAMYLPROPYL)BENZOTRIAZOLE	069218-63-9
1-(3-CARBAMYLPROPYL)BENZOTRIAZOLE	069218-58-2
CARBANILIC ACID, 3,4-DICHLOROTHIO-, S-METHYL EST	025052-57-7
CARBANILIC ACID, THIO-, S-METHYL ESTER	013509-38-1
CARBAZOLE	000086-74-8
9H-CARBAZOLE, 1,3,6,8-TETRACHLORO-	058910-96-6
CARBENCILLIN, PHENYL	038212-89-4
CARBENDAZIM	010605-21-7
CARBENICILLIN	004697-36-3
CARBENICILLIN, INDANYL	035531-88-5
1-(4-CARBETHOXYBUTYL)BENZOTRIAZOLE	069218-49-1
2-(4-CARBETHOXYBUTYL)BENZOTRIAZOLE	069218-54-8
2(5-CARBETHOXYPENTYL)BENZOTRIAZOLE	069218-55-9
1(5-CARBETHOXYPENTYL)BENZOTRIAZOLE	069218-50-4
N-CARBETHOXYPHTHALIMIDE	022509-74-6
2-(3-CARBETHOXYPROPYL)BENZOTRIAZOLE	069218-53-7
1-(3-CARBETHOXYPROPYL)BENZOTRIAZOLE	069218-48-0
CARBITOL ACETATE	000112-15-2
CARBOFURAN	001563-66-2
CARBOFURAN,N(N'-ME,4-CL-PHCARBAMIDOTHIO)	086230-37-7
CARBOFURAN,N(N'-ME,2-CL-PHCARBAMIDOTHIO)	086230-38-8
CARBOFURAN,N(N'-ME,24CL-PHCARBAMIDOTHIO)	086570-05-0
CARBOFURAN,N-(N'ME,ET-CARBAMIDOTHIO)	086570-00-5

CARBOFURAN,N-(N'-ME,ISOPR CARBAMIDOTHIO)	086570-02-7
CARBOFURAN,N-(N'-ME,MECARBAMIDOTHIO)	068840-05-1
CARBOFURAN,N-(N'ME-O-ET-CARBAMYLTHIO)	065907-27-9
CARBOFURAN,N(N'-ME,PHCARBAMIDOTHIO)	086230-36-6
CARBOFURAN,N-(N'-ME,PR CARBAMIDOTHIO)	086570-01-6
CARBOFURAN PHENOL	001563-38-8
CARBON DIOXIDE	000124-38-9
CARBON DISULFIDE	000075-15-0
CARBONIC ACID, DECAHYDRO-3,6,9-TRIMETHYL-3,12-EP	075887-63-7
CARBONIC ACID, (3,6-DIHYDRO-5-FLUORO-2,6-DIOXO-1	106219-34-5
CARBONIC ACID, DIPHENYL ESTER	000102-09-0
CARBONIC ACID, ETHYL (5-FLUORO-3,6-DIHYDRO-2,6-D	106206-97-7
CARBONIC ACID, ETHYL (5-FLUORO-3,4-DIHYDRO-2,4-D	106206-96-6
CARBONIC DIFLUORIDE	000353-50-4
CARBON TETRABROMIDE	000558-13-4
CARBON TETRACHLORIDE	000056-23-5
CARBONYL CYANIDE,PHENYLHYDRAZONE	000306-18-3
CARBONYL SULFIDE	000463-58-1
CARBOPHENOTHION-METHYL	000953-17-3
CARBOPHENTHION	000786-19-6
CARBOXIN	005234-68-4
P-CARBOXYACETANILIDE	000556-08-1
1-(2-CARBOXYET)BENZOTRIAZOLE	000654-15-9
2-(2-CARBOXYET)BENZOTRIAZOLE	016583-99-6
9-CARBOXYFLUORENE	001989-33-9
2-CARBOXYMETHYLBENZOTRIAZOLE	004144-68-7
2-(5-CARBOXYPENTYL)BENZOTRIAZOLE	069218-45-7
1-(5-CARBOXYPENTYL)BENZOTRIAZOLE	069218-43-5
M-CARBOXYPHENYLACETIC ACID	002084-13-1
2-CARBOXYPHENYL 2-HYDROXYBENZOATE	000552-94-3
4-CARBOXYPHENYLISOTHIOCYANATE	002131-62-6
4-CARBOXY-N-PHENYLPHTHALIMIDE	004649-27-8
3,3-DIME-1-(3-CARBOXYPHENYL)TRIAZENE	020241-07-0
1-(3-CARBOXYPROPYL)BENZOTRIAZOLE	000654-19-3
2-(3-CARBOXYPROPYL)BENZOTRIAZOLE	004144-70-1
9-CARBOXYTHIOXANTHENE	017394-14-8
CARBROMAL	000077-65-6
CARD-20(22)-ENOLIDE, 16-(ACETYLOXY)-3,14-DIHYDRO	000465-15-6
CARD-20(22)-ENOLIDE DERIVATIVE	005297-05-2
CARD-20(22)-ENOLIDE DERIVATIVE	030685-43-9
CARD-20(22)-ENOLIDE DERIVATIVE	020991-71-3
CARD-20(22)-ENOLIDE, 3-[(2,6-DIDEOXY-BETA-D-RIBO	005352-63-6
CARD-20(22)-ENOLIDE, 3-[(2,6-DIDEOXY-BETA-D-RIBO	000545-27-7
CARD-20(22)-ENOLIDE, 3-[(2,6-DIDEOXY-beta-D-RIBO	018404-43-8
CARD-20(22)-ENOLIDE, 3-[(2,6-DIDEOXY-3-O-METHYL-	031087-87-3
CARD-20(22)-ENOLIDE, 3-[(2,6-DIDEOXY-3-O-METHYL-	089016-31-9
CARD-20(22)-ENOLIDE, 3-[[2,6-DIDEOXY-4-O-(2,6-DI	016479-50-8
2-CARENE	000554-61-0
3-CARENE (DELTA)	013466-78-9
CARMINIC ACID	001260-17-9
CARMOFUR	061422-45-5
CARNIDAZOLE	042116-76-7
CARVOMETHENE	000499-94-5
CARVONE	006485-40-1
CARVONE	000099-49-0
CATECHOL	000120-80-9
3-C≡CCO-5-BR-6-ME-PYRIMID-2,4-DIONE	077317-71-6
3-CCL3 HYDROCHLOROTHIAZIDE	004267-05-4

CDEA	002315-36-8
CEFAZOLIN	025953-19-9
CEFMENOXIME,PIVALOYL-OME ESTER	065243-35-8
CEFOPERAZONE	062893-19-0
CEFOXITIN	035607-66-0
CEFUROXIM	055268-75-2
CEFUROXIME AXETIL	064544-07-6
CEPHACETRILE	010206-21-0
CEPHAELINE DIHYDROCHLORIDE	005853-29-2
CEPHALEXIN	015686-71-2
CEPHALOSPORANIC ACID,7-MANDELAMIDO	027910-26-5
CEPHALOSPORIC ACID,7-BR-ACAMINO	026973-80-8
CEPHALOSPORIN DERIVATIVE	069712-34-1
CEPHALOTAXINE	024316-19-6
CEPHALOTHIN	000153-61-7
CEPHALOTHIN	000058-71-9
CEPHALOTHIN ANALOG	026970-95-6
CEPHALOTHIN ANALOG (3-CH2-S-CH3)	026722-85-0
CEPHALOTHIN ANALOG (3-CH3)	034691-02-6
CEPHALOTHIN ANALOG(7-DICLPH-S-ACAM	059521-86-7
CEPHALOTHIN ANALOG(7PHOACETYLAMINO	075994-55-7
CEPHALOTHIN ANALOG (S-TETRAZOLE)	032924-66-6
CEPHAPIRIN	021593-23-7
CERESAN M	000517-16-8
CETAMOLOL	034919-98-7
CETRIMONIUM CHLORIDE	000112-02-7
CEVADINE	000062-59-9
O-CF3 ACETANILIDE	000344-62-7
2-CF3 BENZIMIDAZOLE	000312-73-2
P-CF3 BENZOIC ACID	000455-24-3
P-CF3-BENZYL N,N-DIME CARBAMATE	084640-35-7
2-CF3-5-BROMOBENZIMIDAZOLE	003671-60-1
2-CF3-4-CHLOROACETANILIDE	000344-53-6
3-CF3-4-CHLOROACETANILIDE	000348-90-3
2-CF3-5-CHLOROBENZIMIDAZOLE	000656-49-5
2-CF3-4-CHLOROBENZIMIDAZOLE	002338-31-0
4-SCF3-C6H4NHN=C(CN)CO-ME	028317-78-4
4-CF3-C6H4NHN=C(CN)CO-OME	028313-76-0
2-CF3-C6H4NHN=C(CN)CO-OME	028384-50-1
3-CF3-C6H4NHN=C(CN)CO-OME	028313-74-8
4-SCF3-C6H4NHN=C(CN)CO-OME	028313-92-0
2-CF3-4-CL-C6H3NHN=C(CN)COOME	036865-72-2
2-CF3-4,6-DICHLOROBENZIMIDAZOLE	004228-88-0
2-CF3-5,6-DICHLOROBENZIMIDAZOLE	002338-25-2
2-CF3-4,5-DICHLOROBENZIMIDAZOLE	003615-21-2
2-CF3-4,7-DICHLOROBENZIMIDAZOLE	004228-89-1
3-CF3 HYDROCHLOROTHIAZIDE	001547-10-0
2-(2-CF3-PH-IMINO)IMIDAZOLIDINE	040065-00-7
1(CF3-MEO)PENTACHLOROCYCLOHEXANE	056421-36-4
3(CF3-MEO)PENTACHLOROCYCLOHEXANE	056400-11-4
P-CF3 CF3-METHANESULFONANILIDE	023384-12-5
M-CF3 CF3-METHANESULFONANILIDE	023384-11-4
CF3-METHANESULFONANILIDE	000456-64-4
3-CF3-4-NITROBENZENESULFONAMIDE	021988-05-6
2-CF3-5-NO2 BENZIMADAZOLE	000327-19-5
3-CF3 PHENYL 4-AMINOSALICYLATE	056356-34-4
3-CF3 PHENYL GLUCOPYRANOSIDE	020772-25-2
2-(M-CF3 PHENYL)MORPHOLINE	031599-68-5

M-CF3-N-PHENYLSUCCINIMIDE	060050-38-6
7-CF3 QUINOLINE	000325-14-4
8-CF3 QUINOLINE	000317-57-7
CF3-SULFONANILIDE,P-BENZOYL	022731-28-8
3'-(CF3-THIO)BENZOGUANAMINE	065052-47-3
2-CHLORO-6-(N,N-DIMETHYLAMINO)-PYRAZINE	061655-72-9
2-CH2CH2NME2 THIOPHENE	026019-18-1
1-(CH2CO2ME)-2-NO2- IMIDAZOLE	022813-31-6
CHENODEOXYCHOLIC ACID	000474-25-9
3-CHF2-4-CL-C6H3NHN=C(CN)CO-OME	036874-60-9
CHIPCOTE	002597-97-9
CHLOMETHOXYNIL	032861-85-1
N-(3-CHLORALLYL) HEXAMINIUM CHLORIDE	004080-31-3
CHLORALOSE	015879-93-3
CHLORAMBEN, METHYL ESTER	007286-84-2
CHLORAMPHENICOL	000056-75-7
CHLORANIL	000118-75-2
CHLORAZINE	000580-48-3
CHLORBENSIDE	000103-17-3
CHLORBROMURON	013360-45-7
CHLORBUFAM	001967-16-4
CHLORDANE	000057-74-9
CHLORDANE	012789-03-6
ALPHA-CHLORDANE	005103-71-9
GAMMA-CHLORDANE	005566-34-7
CHLORDANE (TRANS)	005103-74-2
CHLORDENE	003734-48-3
CHLORDIMEFORM	006164-98-3
CHLORENDIC ACID	000115-28-6
CHLORETHOXYFOS	054593-83-8
CHLORFENPROP METHYL	014437-17-3
CHLORFENSON	000080-33-1
CHLORFENVINPHOS	000470-90-6
CHLORFLUAZURON	071422-67-8
CHLORFLURECOL METHYL	002536-31-4
CHLORHEXIDINE	000055-56-1
CHLORIDAZON	001698-60-8
CHLORIMIPRAMINE	000303-49-1
CHLORIMURON-ETHYL (PH 7)	090982-32-4
CHLORMEQUAT CHLORIDE	000999-81-5
CHLORMERODRIN	000062-37-3
CHLORNITROFEN	001836-77-7
CHLOROACETALDEHYDE	000107-20-0
CHLOROACETAMIDE	000079-07-2
3-CHLOROACETANILIDE	000588-07-8
2-CHLOROACETANILIDE	000533-17-5
A-CHLOROACETANILIDE	000587-65-5
P-CHLOROACETANILIDE	000539-03-7
CHLOROACETIC ACID	000079-11-8
CHLOROACETIC ANHYDRIDE	000541-88-8
CHLOROACETONE	000078-95-5
CHLOROACETONITRILE	000107-14-2
O-CHLOROACETOPHENONE	002142-68-9
M-CHLOROACETOPHENONE	000099-02-5
2-CHLORO-6-ACETYLAMINO-PYRAZINE	132453-63-5
CHLOROACETYLCHLORIDE	000079-04-9
CHLOROALLENE	003223-70-9
2-CHLORO-4-AMINOBENZOIC ACID	002457-76-3

2-CHLORO-6-AMINO-PYRAZINE	033332-28-4
1-CHLOROAMITRIPTYLINE	052845-72-4
4-CHLOROANILINE	000106-47-8
3-CHLOROANILINE	000108-42-9
2-CHLOROANILINE	000095-51-2
4-CHLOROANILINE-2-SULFONIC ACID	000133-74-4
4-CHLOROANILINE-3-SULFONIC ACID	000088-43-7
M-CHLOROANISOLE	002845-89-8
O-CHLOROANISOLE	000766-51-8
4-CHLOROANISOLE	000623-12-1
1-CHLOROANTHRAQUINONE	000082-44-0
4-CHLOROAZOBENZENE	004340-77-6
CHLOROAZOBENZENE-4'-SULFONIC ACID, NA SALT	002777-05-1
3-CHLOROBENZAL ACETYLACETONE	015725-15-2
2-CHLOROBENZAL ACETYLACETONE	015725-14-1
M-CHLOROBENZALDEHYDE	000587-04-2
4-CHLOROBENZALDEHYDE	000104-88-1
2-CHLOROBENZALDEHYDE	000089-98-5
P-CHLOROBENZALDOXIME	003848-36-0
3-CHLOROBENZAMIDE	000618-48-4
4-CHLOROBENZAMIDE	000619-56-7
2-CHLOROBENZAMIDE	000609-66-5
6-CHLORO-BENZAZEPINE	026232-35-9
CHLOROBENZENE	000108-90-7
4-CHLORO-1,2-BENZENEDIAMINE	000095-83-0
4-CHLOROBENZENESULFONAMIDE	000098-64-6
O-CHLOROBENZENESULFONAMIDE	006961-82-6
M-CHLOROBENZENESULFONAMIDE	017260-71-8
3-CHLOROBENZENETHIOL	002037-31-2
4-CHLOROBENZENETHIOL	000106-54-7
CHLOROBENZILATE	000510-15-6
3'-CHLOROBENZOGUANAMINE	004514-54-9
2'-CHLOROBENZOGUANAMINE	029366-77-6
4'-CHLOROBENZOGUANAMINE	004514-53-8
4-CHLOROBENZOIC ACID	000074-11-3
M-CHLOROBENZOIC ACID	000535-80-8
2-CHLOROBENZOIC ACID	000118-91-2
6-CHLOROBENZOTHIAZOLE,2-SULFONAMIDE	088946-20-7
4-CHLOROBENZOTRICHLORIDE	005216-25-1
2-CHLOROBENZOTRICHLORIDE	002136-89-2
M-CHLOROBENZOYLCHLORIDE	000618-46-2
4-(4-CHLOROBENZOYL)PYRIDINE	006318-51-0
P-CHLOROBENZYL ACETATE	005406-33-7
P-CHLOROBENZYL ALCOHOL	000873-76-7
M-CHLOROBENZYL ALCOHOL	000873-63-2
P-CHLOROBENZYLCARBAMATE	002621-80-9
M-CHLOROBENZYL N,N-DIME CARBAMATE	084640-26-6
P-CHLOROBENZYL N,N-DIME CARBAMATE	084640-27-7
2-CHLOROBIPHENYL	002051-60-7
4-CHLOROBIPHENYL	002051-62-9
3-CHLOROBIPHENYL	002051-61-8
2-CHLORO-6-BROMOBENZAMIDE	066073-53-8
CHLOROBROMOMETHANE	000074-97-5
1-CHLORO-3-BROMOPROPANE	000109-70-6
CHLOROBROMOPROPENE	003737-00-6
3-CHLORO-1,2-BUTADIENE	034581-41-4
4-CHLORO-1,2-BUTADIENE	025790-55-0
1-CHLORO-1,2-BUTADIENE	000627-23-6

2-CHLOROBUTANE	000078-86-4
1-CHLOROBUTANE	000109-69-3
3-CHLORO-2-BUTANONE	004091-39-8
1-CHLORO-2-BUTENE	000591-97-9
3-CHLORO-1-BUTENE	000563-52-0
4-CHLORO-1-BUTENE	000927-73-1
1-CHLORO-1-BUTENE	004461-42-1
3-CHLORO-1-BUTYNE	021020-24-6
1-CHLORO-1-BUTYNE	062981-74-2
4-CHLORO-1-BUTYNE	051908-64-6
P-CHLORO CF3-METHANESULFONANILIDE	023384-04-5
2-CHLORO-N-(2-CHLOROETHYL)-N-METHYLETHANAMINE	000051-75-2
3-CHLORO-N-(4-CHLOROPHENYL)BENZENAMINE	015979-79-0
2-CHLORO-6-CHLORO-PYRAZINE	004774-14-5
2-CHLORO-4-(1-CHLOROVINYL)CYCLO-1-HEXENE	013547-07-4
P-CHLOROCINNAMAMIDE,B-ET,N-SEC-BUTYL	097871-63-1
3-CHLOROCINNOLINE	017404-90-9
2-CHLORO-6-CONH2-PYRAZINE	036070-79-8
CHLORO-M-CRESOL	054548-50-4
2-CHLORO-6-CYANO-PYRAZINE	006863-74-7
CHLOROCYCLOHEXANE	000542-18-7
5-CHLORO-1,3-CYCLOPENTADIENE	041851-50-7
1-CHLORODECANE	001002-69-3
2-CHLORO-N,N-DIALLYLACETAMIDE	000093-71-0
1-CHLORODIBENZO-P-DIOXIN	039227-53-7
2-CHLORODIBENZO-P-DIOXIN	039227-54-8
1-(4-CHLORO),2-DIBENZOYL-1-TERT-BUTYLHYDRAZINE	112226-54-7
1-(2-CHLORO),2-DIBENZOYL-1-TERT-BUTYLHYDRAZINE	112226-58-1
1-(3-CHLORO),2-DIBENZOYL-1-TERT-BUTYLHYDRAZINE	112225-83-9
5-CHLORO-2-(2,4-DICHLOROPHENOXY)PHENOL	003380-34-5
1-CHLORO-1,1-DIFLUOROETHANE	000075-68-3
CHLORO-1,1-DIFLUOROETHYLENE	000359-10-4
CHLORODIFLUOROMETHANE	000075-45-6
1-PYRROLIDINEACETAMIDE, N-(5-CHLORO-2,3-DIHYDRO-	105801-52-3
4-CHLORO-1,2-DIMETHYL-5-NITRO-1H-IMIDAZOLE	091027-94-0
5-CHLORO-1,2-DIMETHYL-4-NITRO-1H-IMIDAZOLE	091027-93-9
2-CHLORO-DIMETHYL PARA-OXON	017650-76-9
3-CHLORO-DIMETHYL PARA-OXON	002255-15-4
4-CHLORO-3,5-DIMETHYL PHENOL	000088-04-0
CHLORODIMETHYLPHENYLSILANE	000768-33-2
4-CHLORO-3,5-DINITROBENZOIC ACID	000118-97-8
4-CHLORO-2,6-DINITROPHENOL	000088-87-9
6-CHLORO-2,4-DINITROPHENOL	000946-31-6
1-CHLORO-1,2-DIPHENYLETHENE	001460-06-6
3-CHLORODIPHENYL ETHER	000101-17-7
1-CHLORODODECANE	000112-52-7
1-CHLOROEICOSANE	042217-02-7
1-CHLORO-2,3-EPOXYPROPANE	000106-89-8
CHLOROETHANE	000075-00-3
1-CHLORO-2-(1,2-ETHANEDIOL)BENZENE	059365-60-5
2-CHLOROETHANOL	000107-07-3
1-(2-CHLORO-1-(4-ETHOXYPHENYL)PROPYL)-4-METHBE*	056265-27-1
2-CHLORO-6-ETHOXY-PYRAZINE	136309-02-9
(1-CHLOROETHYL)BENZENE	000672-65-1
2-CHLOROETHYLBENZHYDRYL ETHER	032669-06-0
A-CHLORO-A-ETHYLBUTANOYLUREA	061421-89-4
CHLOROETHYL CHLOROFORMATE	000627-11-2
2-CHLOROETHYL (3-CHLOROPHENYL)CARBAMATE	000587-56-4

2-CHLOROETHYLETHER	000628-34-2
1-(2-CHLOROETHYL)-1-NITROSOUREA	002365-30-2
((2-CHLOROETHYL)SULFINYL)ETHENE	040709-82-8
2-((2-CHLOROETHYL)THIO)ETHANOL	000693-30-1
2-CHLOROETHYL VINYL ETHER	000110-75-8
2-CHLOROFLOURENE	002523-44-6
2-CHLOROFLUMETHASONE	050629-82-8
2-CHLOROFLUMETHASONE-17-PROPIONATE	064272-18-0
3-CHLORO-4-FLUOROANILINE	000367-21-5
M-CHLOROFLUOROBENZENE	000625-98-9
P-CHLOROFLUOROBENZENE	000352-33-0
1-CHLORO-2-FLUOROBENZENE	000348-51-6
CHLOROFLUOROMETHANE	000593-70-4
2-CHLORO-6-FLUORO-PYRAZINE	033873-10-8
CHLOROFORM	000067-66-3
P-CHLOROFORMANILIDE	002617-79-0
1-CHLOROHEPTADECANE	062016-75-5
1-CHLOROHEPTANE	000629-06-1
1-CHLOROHEPT-1-YNE-5-ENE	083682-36-4
1-CHLOROHEXADECANE	004860-03-1
2-CHLOROHEXANE	000638-28-8
1-CHLOROHEXANE	000544-10-5
3-CHLOROHEXA-1,2,5-TRIENE	083682-35-3
CHLOROHYDROQUINONE	000615-67-8
3-CHLORO-4-HYDROXYBENZOIC ACID	003964-58-7
5-CHLORO-2-HYDROXYDIPHENYLMETHANE	000120-32-1
1-CHLORO-1,3-DIMETHYL-1,3,3-TRIPHENYLDISILOXANE	053634-34-7
6-CHLOROINDOLE	053294-05-6
2-CHLOROIODOBENZENE	000615-41-8
4-CHLOROIODOBENZENE	000637-87-6
3-CHLOROIODOBENZENE	000625-99-0
CHLOROIODOMETHANE	000593-71-5
1-CHLORO-4-ISOCYANATOBENZENE	000104-12-1
2-CHLOROISONIAZID	058481-04-2
1-CHLORO-3-ISOTHIOCYANATO-BENZENE	002392-68-9
P-(CHLOROMERCURI)BENZOIC ACID	000059-85-8
CHLOROMERCURIPHENOL	000090-03-9
P-CHLOROMERCURIPHENYLSULFONIC ACID	000554-77-8
P-CHLORO METHANESULFONANILIDE	004284-51-9
M-CHLORO CF3-METHANESULFONANILIDE	023384-03-4
3-CHLORO-4-METHOXYACETANILIDE	007073-42-9
3-CHLORO-4-METHOXYANILINE	005345-54-0
2-CHLORO-6-METHOXYBENZAMIDE	107485-43-8
CHLOROMETHOXYMETHANE	000107-30-2
3-PYRIDINECARBONITRILE, 2-CHLORO-5-(4-METHOXYPHE	108635-80-9
(3-CHLORO-4-METHOXYPHENYL)UREA	025277-05-8
2-CHLORO-6-METHOXY-PYRAZINE	033332-30-8
3-CHLORO-6-METHOXYPYRIDAZINE	001722-10-7
N-CHLORO-N-(1-METHYETHYL)-2-PROPANAMINE	024948-81-0
3-CHLORO-4-METHYLACETANILIDE	007149-79-3
5-CHLORO-2-METHYLACETANILIDE	005900-55-0
2-CHLORO-6-METHYLACETANILIDE	021352-09-0
1-CHLORO-2-METHYLANTHRAQUINONE	000129-35-1
2-CHLORO-6-METHYLBENZAMIDE	101080-58-4
4-CHLORO-2-METHYLBENZENAMINE HYDROCHLORIDE	003165-93-3
CHLOROMETHYLCHLOR	017925-97-2
2-(CHLOROMETHYL)-3-CHLORO-1-PROPENE	001871-57-4
1-(3-CHLORO-2-METHYL),2-DIBENZOYL-1-TERT-BUTYLHY	112426-81-0

2-CHLOROMETHYL-3,3-DICHLOROPROPENE	060845-51-4
2-CHLORO-4-METHYL-6-(DIMETHYLAMINO)PYRIMIDINE	000535-89-7
CHLOROMETHYL ETHYL ETHER	003188-13-4
3-(CHLOROMETHYL)HEPTANE	000123-04-6
3-CHLOROMETHYLHYDROCHLOROTHIAZIDE	001824-47-1
4-CHLORO-2-METHYL-5-NITRO-1H-IMIDAZOLE	063662-67-9
4-CHLORO-1-METHYL-5-NITROIMIDAZOLE	004897-31-8
2-CHLORO-1-METHYL-4-NITRO-1H-IMIDAZOLE	063634-21-9
2-CHLORO-1-METHYL-5-NITRO-1H-IMIDAZOLE	086072-07-3
5-CHLORO-1-METHYL-4-NITROIMIDAZOLE	004897-25-0
4-CHLORO-5-METHYL-2-NITROPHENOL	007147-89-9
(4-CHLORO-2-METHYLPHENOXY)ACETIC ACID, AMINE SA*	019480-39-8
N-(3-CHLORO-4-METHYLPHENYL)-N'-METHYLUREA	022175-22-0
CHLOROMETHYL PHENYL SULFONE	007205-98-3
2-CHLORO-2-METHYLPROPANE	000918-20-7
1-CHLORO-2-METHYLPROPENE	000513-37-1
3-CHLORO-2-METHYLPROPENE	000563-47-3
1,1'-(2-CHLORO-2-METHYLPROPYLIDENE)BIS-4-ETHOX*	056265-24-8
2-CHLORO-6-METHYL-PYRAZINE	038557-71-0
P-(CHLOROMETHYL)TOLUENE	000104-82-5
CHLORONAPHAZINE	000494-03-1
1-CHLORONAPHTHALENE	000090-13-1
2-CHLORONAPHTHALENE	000091-58-7
4-CHLORO-1-NAPHTHOL	000604-44-4
2-CHLORO-1,4-NAPHTHOQUINONE	001010-60-2
ALPHA-CHLORO-M-NITROACETOPHENONE	000099-47-8
2-CHLORO-4-NITROANILINE	000121-87-9
4-CHLORO-2-NITROANILINE	000089-63-4
2-CHLORO-6-NITROBENZAMIDE	107485-64-3
P-CHLORONITROBENZENE	000100-00-5
2-CHLORO-1-NITROBENZENE	000088-73-3
3-CHLORO-NITROBENZENE	000121-73-3
2-CHLORO-3-NITROBENZOIC ACID	003970-35-2
2-CHLORO-5-NITROBENZOIC ACID	002516-96-3
2-CHLORO-4-NITRO-1H-IMIDAZOLE	057531-37-0
4-CHLORO-5-NITRO-1H-IMIDAZOLE	057531-38-1
2-CHLORO-5-NITRO-B-NITROSTYRENE	015851-93-1
5-CHLORO-2-NITROPHENOL	000611-07-4
2-CHLORO-4-NITROPHENOL	000619-08-9
3-CHLORO-4-NITROPHENOL	000491-11-2
4-CHLORO-2-NITROPHENOL	000089-64-5
6-CHLORO-2-NITROPHENOL	000603-86-1
3-CHLORO-N-(4-NITROPHENYL)BENZENAMINE	015979-85-8
3-CHLORO-4-NITROQUINOLINE-1-OXIDE	014100-52-8
6-CHLORO-4-NITROQUINOLINE-1-OXIDE	003741-12-6
3-CHLORO-B-NITROSTYRENE	003156-35-2
2-CHLORO-B-NITROSTYRENE	003156-34-1
4-CHLORO B-NITROSTYRENE	000706-07-0
2-CHLORO-6-NITRO-4-(1,1,3,3-TETRAMETHYLBUTYL)-P*	017199-21-2
2-CHLORO-4-NITROTOLUENE	000121-86-8
4-CHLORO-1-NITRO-2(TRIFLUOROMETHYL)BENZENE	000118-83-2
1-CHLORONONANE	002473-01-0
2-CHLORO-4-NONYLPHENOL	060044-33-9
1-CHLOROOCTADECANE	003386-33-2
1-CHLOROOCTANE	000111-85-3
2-CHLOROOCTANE	000628-61-5
4-CHLORO-3-PENTADECYLPHENOL	006964-19-8
CHLOROPENTAFLUOROACETONE	000079-53-8

CHLOROPENTAFLUOROBENZENE	000344-07-0
CHLOROPENTAFLUOROETHANE	000076-15-3
1-CHLOROPENTANE	000543-59-9
3-CHLOROPENTANE	000616-20-6
2-CHLOROPENTANE	000625-29-6
1-CHLORO-1-PENTENE	021450-13-5
3-CHLOROPHENOL	000108-43-0
4-CHLOROPHENOL	000106-48-9
2-CHLOROPHENOL	000095-57-8
O-CHLOROPHENOXYACETIC ACID	000614-61-9
M-CHLOROPHENOXYACETIC ACID	000588-32-9
P-CHLOROPHENOXYACETIC ACID	000122-88-3
4-(4-CHLOROPHENOXY) BUTYRIC ACID	003547-07-7
2-(P-CHLOROPHENOXY)PROPIONIC ACID	003307-39-9
1-(3-CHLOROPHENOXY)SILATRANE	086825-39-0
1-(4-CHLOROPHENOXY)SILATRANE	051466-76-3
1,1-DIMETHYL-3-(M-CHLOROPHENYL)	000587-34-8
3-CHLOROPHENYL ACETATE	013031-39-5
2-CHLOROPHENYL ACETATE	004525-75-1
M-CHLOROPHENYLACETIC ACID	001878-65-5
P-CHLOROPHENYLACETIC ACID	001878-66-6
P-CHLOROPHENYLALANINE	001991-78-2
4-CHLOROPHENYLAMIDINO UREA	058247-24-8
3-CHLOROPHENYL 4-AMINOSALICYLATE	056356-29-7
4-CHLOROPHENYL 4-AMINOSALICYLATE	056356-30-0
2-CHLOROPHENYL 4-AMINOSALICYLATE	056356-28-6
2-CHLORO-4-PHENYLAZOPHENOL	006657-05-2
P-CHLOROPHENYL AZOXYCYANIDE	054797-22-7
M-CHLOROPHENYL AZOXYCYANIDE	062825-07-4
4-CHLORO-N-PHENYLBENZENAMINE	001205-71-6
4-CHLOROPHENYL BENZENESULFONATE	000080-38-6
4-CHLOROPHENYLDIMETHYLCARBAMATE	007305-03-5
N-(3-CHLOROPHENYL)-2,4-DINITROBENZENAMINE	016220-58-9
2-CHLORO-1-PHENYLETHANONE	000532-27-4
1-(4-CHLOROPHENYL)ETHANONE	000099-91-2
N1-P-CHLOROPHENYL-N5-ETHYLBIGUANIDE	060221-93-4
9-(3-CHLOROPHENYL)-9H-FLUORENE	032377-11-0
9-(4-CHLOROPHENYL)-9H-FLUORENE	021846-07-1
4-CHLOROPHENYL GLUCOPYRANOSIDE	004756-30-3
1-PROPANONE, 3-[4-(4-CHLOROPHENYL)-4-HYDROXY-1-P	104182-04-9
3-(P-CHLOROPHENYL)HYDROXYUREA	030085-34-8
2-(2-CHLOROPHENYLIMINO)IMIDAZOLINE	004749-68-2
2-(4-CHLOROPHENYL)INDOLIZINE	007496-73-3
2-(3-CHLOROPHENYL)INDOLIZINE	082682-70-0
3-CHLOROPHENYL ISOCYANATE	002909-38-8
2,4-PYRIMIDINEDIAMINE, 5-[(3-CHLOROPHENYL)METHYL	069945-58-0
N-((4-CHLOROPHENYL)METHYL)ACETAMIDE	057058-33-0
N1-P-CHLOROPHENYL-N5-METHYLBIGUANIDE	060221-92-3
N-(3-CHLOROPHENYL)-N'-METHYLUREA	020940-42-5
4-CHLORO-2-PHENYLPHENOL	000607-12-5
4-CHLOROPHENYL PHENYL ETHER	007005-72-3
M-CHLORO-N-PHENYLSUCCINIMIDE	015386-99-9
P-CHLORO-N-PHENYLSUCCINIMIDE	006943-00-6
O-CHLORO-N-PHENYLSUCCINIMIDE	007402-22-4
N1-(2-CHLOROPHENYL)SULFANILAMIDE	019837-85-5
N1-(4-CHLOROPHENYL)SULFANILAMIDE	016803-92-2
1-(O-CHLOROPHENYL)-2-THIOUREA	005344-82-1
M-CHLOROPHENYLUREA	001967-27-7

O-CHLOROPHENYLUREA	000114-38-5
P-CHLOROPHENYLUREA	000140-38-5
N-CHLOROPHTHALIMIDE	003481-09-2
6-CHLOROPICOLINIC ACID	004684-94-0
1-CHLOROPRENE	000627-22-5
CHLOROPRENE	000126-99-8
CHLOROPROCAINE	000133-16-4
N-CHLORO-2-PROPANAMINE	026245-56-7
1-CHLOROPROPANE	000540-54-5
2-CHLOROPROPANE	000075-29-6
3-CHLORO-1,2-PROPANEDIOL	000096-24-2
3-CHLORO-1-PROPANOL	000627-30-5
2-CHLORO-1-PROPANOL	000078-89-7
1-CHLORO-2-PROPANOL	000127-00-4
1-CHLOROPROPENE	000590-21-6
2-CHLOROPROPENE	000557-98-2
2-CHLOROPROPIONIC ACID	000598-78-7
3-CHLOROPROPIONIC ACID	000107-94-8
3-CHLOROPROPIONITRILE	000542-76-7
2-CHLORO-6-PROPOXY-PYRAZINE	136309-03-0
3-CHLOROPROPYLENE	000107-05-1
1-CHLOROPROPYNE	007747-84-4
2-CHLOROPYRAZINE	014508-49-7
2-CHLOROPYRIDINE	000109-09-1
3-CHLOROPYRIDINE	000626-60-8
4-CHLOROPYRIDINE	000626-61-9
7-CHLOROQUINAZOLINE	007556-99-2
CHLOROQUINE	000054-05-7
2-CHLOROQUINOLINE	000612-62-4
8-CHLOROQUINOLINE	000611-33-6
6-CHLOROQUINOLINE	000612-57-7
4-CHLOROQUINOLINE-1-OXIDE	004637-59-6
4-CHLORO-8-QUINOLINOL	057334-36-8
2-CHLOROQUINOXALINE	001448-87-9
5-CHLOROQUINOXALINE	062163-09-1
6-CHLOROQUINOXALINE	005448-43-1
4-CHLORORESORCINOL	000095-88-5
5-CHLOROSALICYLANILIDE	004638-48-6
5-CHLOROSALICYLIC ACID	000321-14-2
M-CHLOROSTYRENE	002039-85-2
O-CHLOROSTYRENE	002039-87-4
P-CHLOROSTYRENE	001073-67-2
CHLOROSUCCINIC ACID	016045-92-4
1-CHLOROTETRADECANE	002425-54-9
2-CHLORO-4-(1,1,3,3-TETRAMETHYLBUTYL)PHENOL	017199-24-5
CHLOROTHANONIL	001897-45-6
8-CHLOROTHEOPHYLLINE	000085-18-7
CHLOROTHIAZIDE	000058-94-6
CHLOROTHIAZIDE-3-AMYL	002854-98-0
CHLOROTHIAZIDE-3-BENZYL	003211-40-3
CHLOROTHIAZIDE-3-METHYL	001025-75-8
CHLOROTHIAZIDE-3-PHENYL	001163-51-5
2-CHLOROTHIOPHENE	000096-43-5
CHLOROTHYMOL	000089-68-9
2-CHLOROTOLUENE	000095-49-8
ALPHA-CHLOROTOLUENE	000100-44-7
P-CHLOROTOLUENE	000106-43-4
M-CHLOROTOLUENE	000108-41-8

3-CHLORO-P-TOLUIDINE	000095-74-9
5-CHLORO-O-TOLUIDINE	000095-79-4
1-CHLORO-2-TRICHLOROMETHYL-3,3,3-TRICHLOROPROP*	083682-39-7
CHLOROTRIETHYLSILANE	000994-30-9
CHLOROTRIFLUOROETHENE	000359-29-5
CHLOROTRIFLUOROETHYLENE	000079-38-9
CHLORO(TRIFLUOROMETHYL)BENZENE	000088-16-4
CHLOROTRIFLUOROMETHANE	000075-72-9
CHLOROTRIMETHYLSILANE	000075-77-4
5-CHLOROURACIL	001820-81-1
A-CHLORO-I-VALERYLUREA	061345-66-2
2-CHLOROVINYL DIETHYL PHOSPHATE	000311-47-7
CHLOROXURON	001982-47-4
CHLOROZOTOCIN	054749-90-5
CHLORPHENIRAMINE	000132-22-9
CHLORPHENTERMINE	000461-78-9
CHLORPROETHAZINE	000084-01-5
CHLORPROMAZINE	000050-53-3
CHLORPROPHAM	000101-21-3
CHLORPROTHIXENE	000113-59-7
CHLORPYRIFOS	002921-88-2
CHLORPYRIFOS METHYL	005598-13-0
CHLORSULFURON (PH 7)	064902-72-3
CHLORTHIAMID	001918-13-4
CHLORTOLURON	015545-48-9
CHLOZOLINATE	072391-46-9
2-CHLORO-6-CO2ME-PYRAZINE	023611-75-8
C6H5NHN=C(CN)CO-OME	036874-74-5
24-CHOLANAMIDE, 3,7,12-TRIHYDROXY-N-(PHENYLMETHY	086678-86-6
CHOLANTHRENE	000479-23-2
CHOLESTEROL	000057-88-5
CHOLIC ACID	000081-25-4
CHORAL HYDRATE	000302-17-0
CHROMAN-2-CARBOXYLIC ACID	051939-71-0
CHRYSAROBIN	000491-59-8
CHRYSENE	000218-01-9
C.I. 11115	003180-81-2
C.I. 16230	001936-15-8
C.I. ACID BROWN 14, DISODIUM SALT	005850-16-8
C.I. ACID ORANGE 8, SODIUM SALT	005850-86-2
C.I. ACID RED 4, MONOSODIUM SALT	005858-39-9
CIANIDANOL	000154-23-4
C.I. AZOIC COUPLING COMPONENT 5	000091-96-3
C.I. DIRECT BLACK 38	001937-37-7
C.I. DIRECT BROWN 191	010190-66-6
C.I..DISPERSE BLACK 1	006054-48-4
C.I. DISPERSE BLACK 3	000539-17-3
C.I. DISPERSE BLUE 26	003860-63-7
C.I. DISPERSE BLUE 3	002475-46-9
C.I. DISPERSE BLUE 23	004471-41-4
C.I. DISPERSE BLUE 19	004395-65-7
C.I. DISPERSE DYE	006054-58-6
C.I. DISPERSE ORANGE 13	006253-10-7
C.I. DISPERSE ORANGE 1	002581-69-3
C.I. DISPERSE ORGANGE 5	006232-56-0
C.I. DISPERSE RED 7	004540-00-5
C.I. DISPERSE RED 60	017418-58-5
C.I. DISPERSE RED 17	003179-89-3

C.I. DISPERSE RED 3	004465-58-1
C.I.DISPERSE RED 19	002734-52-3
C.I. DISPERSE VIOLET 8	000082-33-7
C.I. DISPERSE VIOLET 4	001220-94-6
C.I. DISPERSE YELLOW 23	006250-23-3
C.I. DISPERSE YELLOW 7	006300-37-4
C.I. DISPERSE YELLOW 42	005124-25-4
CIGLITAZONE	074772-77-3
CIMETIDINE	051481-61-9
CIMETIDINE,4-THIOPYRIMIDIN-2-YL-DIOXIDE ANALOG	054855-72-0
CIMETIDINE,(METHYLAMINOCYCLOBUTANEDIONE) ANALOG	099035-21-9
CIMETIDINE,2-(3-NITROPYRIDIN-4-ONE)	099035-20-8
CIMETIDINE,2(PYRIDIN-4-ONE) ANALOG	054855-69-5
CIMETIDINE,PYRIMIDIN-2-ONE ANALOG	054855-60-6
CIMETIDINE,PYRIMID-4-ONE-2-YL ANALOG	054855-84-4
C.I. MORDANT BLUE 9, DISODIUM SALT	003624-68-8
CINCHOMERONIC ACID	000490-11-9
CINCHONIDINE	000485-71-2
CINCHONINE	000118-10-5
CINCHONINE OXALATE	070116-00-6
1,4-CINEOL	000470-67-7
1,8-CINEOLE	000470-82-6
CINMETHYLIN	087818-31-3
CINNAMAMIDE	000621-79-4
CIS-CINNAMAMIDE,A-BR,N-SEC-BUTYL	097871-46-0
TRANS-CINNAMAMIDE,A-BR,N-SEC-BUTYL	097871-45-9
CIS-CINNAMAMIDE,A-CL,N-SEC-BUTYL	097871-48-2
TRANS-CINNAMAMIDE,A-CL,N-SEC-BUTYL	097871-47-1
TRANS-CINNAMAMIDE,A-F,N-SEC-BUTYL	097871-49-3
CINNAMANILIDE	003056-73-3
CINNAMIC ACID	000621-82-9
CINNAMIC ACID, ETHYL ESTER	000103-36-6
P-F CINNAMIC ACID,DIETHYLAMINOETHYLESTER	072836-48-7
CINNAMIC ACID, P-HYDROXY	007400-08-0
CINNAMIC ACID, P-METHOXY-	000830-09-1
CINNAMIC ACID, 3-METHOXY-4-HYDROXY	001135-24-6
CINNAMIC ACID, METHYL ESTER	000103-26-4
CINNAMIC ACID,3,4-DIMETHOXY	002316-26-9
CINNAMYL ALCOHOL	000104-54-1
4-CINNOLINAMINE	005152-83-0
3-CINNOLINAMINE	017372-79-1
7ET-3CINNOLINCARBOXLIC ACID,ET ESTER	089770-38-7
CINNOLINE	000253-66-7
CINNOLINE, 3-BROMO-	078593-33-6
CINNOLINE-4-ONE	018514-84-6
3-CINNOLONE	031777-46-5
CINOSULFURON	094593-91-6
CIPROFLOXACIN	085721-33-1
C.I. SOLVENT ORANGE 7	003118-97-6
C.I. SOLVENT YELLOW 56	002481-94-9
C.I. SOLVENT YELLOW 58	002452-84-8
CITRAL	005392-40-5
CITRIC ACID	000077-92-9
CITRONELLAL	000106-23-0
(+-)-CITRONELLOL	026489-01-0
C.I. VAT BLUE 5	002475-31-2
C.I. VAT BLUE 6	000130-20-1
C.I. VAT BLUE 3	006492-73-5

C.I. VAT BLUE 4	000081-77-6
C.I.VAT GREEN 3	003271-76-9
C.I. VAT ORANGE 5	003263-31-8
C.I. VAT RED 1	002379-74-0
C.I. VIOLET 2	005462-29-3
3-CL-4-ALLYLOXYPHENYL ACETIC ACID	022131-79-9
CLANOBUTIN	030544-61-7
4-CL BENZALDEHYDE,O-((MEAM)CO)OXIME	071059-53-5
M-CL BENZYL TRIMETHYL AMMONIUM BROMIDE	025251-56-3
2'-CL-5'-BROMOBENZOGUANAMINE	057381-45-0
4'-CL-5-BR SALICYLANILIDE	003679-64-9
5-CL-5-T-BU BARBITURIC ACID	118709-72-1
2-CL-5-CF3-C6H3NHN=C(CN)CO-ME	028317-56-8
2-CL-5-CF3-C6H3NHN=C(CN)COOET	028313-52-2
2-CL-5-CF3-C6H3NHN=C(CN)COOME	028313-77-1
3-CL-C6H4NHN=C(CN)CO-T-BU	028317-91-1
3-CL-C6H4NHN=C(CN)COOET	036874-67-6
2-CL-C6H4NHN=C(CN)COOET	003994-23-8
P-CL-CINNAMAMIDE,B-BR,N-SEC-BUTYL	097871-68-6
P-CL-CINNAMAMIDE,B-CL,N-SEC-BUTYL	097871-67-5
P-CL CINNAMAMIDE,B-IPR,N-SEC-BUTYL	097871-65-3
P-CL CINNAMAMIDE,B-PHENYL,N-SEC-BU	097871-66-4
P-CL CINNAMAMIDE,B-PR,N-SEC-BUTYL	097871-64-2
5-CL-2,3-DIHYDROBENZOFURAN-2-ACID	034385-94-9
P(N-2CLET-N-ACETYL)AMINOPHACETIC ACID	032562-52-0
1-(2-CLET)-3-(2-CLCYHX)1-N=O UREA	013909-11-0
3(2-CLET)-1-(2FET)-1-NITROSOUREA	013908-92-4
1(2-CLET)-3(2-GLUTARIMIDYL)-1-NO-UREA	013909-02-9
1-(2-CLET)-3-(4-MECYHX)-1-N=O UREA	013909-09-6
CLET-NITROSOUREIDOGLUTARIC ACID	052320-85-1
1(2CLET)-1-(NO)-3-CYCLODODECYLUREA	013909-14-3
1-(2-CLET)-3-(M-MEOPH)-1-N=O UREA	013909-21-2
3-CLETO-5-BR-6-ME-PYRIMID-2,4-DIONE	077317-67-0
3-CL-6-ET PYRIDAZINECARBOXYLATE	075680-92-1
5'-CL-2'-FLUOROBENZOGUANAMINE	057381-35-8
2'-CL-5'-FLUOROBENZOGUANAMINE	057381-57-4
2-(2-CL-6-F PH IMINO)IMIDAZOLIDINE	059772-37-1
CLINDAMYCIN	018323-44-9
6-CL-4-KETOCHROMAN-2-CARBOXLIC ACID	033607-91-9
2(5-CL-2,4-MEO PH IMINO)IMIDAZOLINE	063346-71-4
4-CL-3-ME-PHENYLAMIDINO UREA	056189-70-9
2(5CL-2ME-PHENYLIMINO)IMIDAZOLIDINE	016822-85-8
2-(2-CL-6-ME-PHIMINO)IMIDAZOLIDINE	004201-24-5
2-(2-CL-4-ME-PHIMINO)IMIDAZOLIDINE	004201-22-3
2-(4-CL-2-ME-PHIMINO)IMIDAZOLIDIN	004201-26-7
5-CL-2'-ME SALICYLANILIDE	025933-30-6
CLOBENZEPAM	001159-93-9
CLOBETASOL	025122-41-2
CLOBETASOL-17-BUTYRATE	025122-47-8
CLOBETASOL-17-PROPIONATE	025122-46-7
CLOBETASOL-17-PROPIONATE-9-CL	025122-49-0
CLOBETASOL-17-PROPIONATE-DES-F	095796-50-2
CLOBETASONE	054063-32-0
CLOBETASONE-17-BUTYRATE	025122-57-0
CLOBETASONE-17-PROPIONATE-DES-F	095796-51-3
CLOBETASONE-17-PROPIONATE-DES-CL	004351-59-1
CLOBETASONE-17-PROPIONATE	025122-56-9
CLODINAFOP-PROPARGYL	105512-06-9

CLOFAZIMINE	002030-63-9
CLOFENTEZINE	074115-24-5
1(3-CL-2-OH PR)-2-NO2-IMIDAZOLE	013551-86-5
CLOMAZONE	081777-89-1
CLOMEPROP	084496-56-0
CLOMETHIAZOLE	000533-45-9
CLONAZEPAM	001622-61-3
1(3CL2OHPR)-2-ME-5-NO2 IMIDAZOLE	016773-42-5
CLOXACILLIN	000061-72-3
CLOZAPINE	005786-21-0
1-(3-CL PH)-3,3-DIMETHYLTRIAZENE	020241-05-8
1-(O-CL PH)-3,3-DIMETHYLTRIAZENE	020241-00-3
2-(P-CLPHENOXY)2-ME-PROPIONIC ACID	000882-09-7
2-(P-CLPHENOXY)-2-ET-PROPIONIC AC	017413-90-0
2-(M-CL-PHENYL)-4-IPR-MORPHOLINE	061151-66-4
N-(P-CL PH)-3-N'-PIPERIDINOACETAMIDE	027471-82-5
7-CL-4-PROPIONYLAMINO-QUINOLINE	089770-27-4
3-CL-6-PYRIDAZINAMINE	005469-69-2
3-CL-6-PYRIDAZINAMINE,N,N-DIME	007145-60-0
3-CL-6-PYRIDAZINECARBOXAMIDE	066346-83-6
N1-(3,5-CL-2-PYRIDYL)SULFANILAMID	030961-38-7
N1-(5-CL-2-PYRIDYL)SULFANILAMIDE	030961-36-5
N1-(6-CL-3-PYRIDYL)SULFANILAMIDE	034392-82-0
N1-(3-CL-2-PYRIDYL)SULFANILAMIDE	026807-64-7
N1-(2-CL-3-PYRIDYL)SULFANILAMIDE	034392-79-5
2'-CL SALICYLANILIDE	001697-18-3
4'-CL SALICYLANILIDE	003679-63-8
8-(4-CL-PH)-S CYCLIC AMP	041941-66-6
8-(4-CL-PH)-S CYCLIC AMP-O-BENZYL	107538-70-5
2-CL TETRAMETHYLMELAMINE	003146-66-5
COCAINE	000050-36-2
COCAINE HYDROCHLORIDE	000053-21-4
CODEINE	000076-57-3
CODEINE HYDROCHLORIDE	001422-07-7
CODEINE PHOSPHATE	000052-28-8
CODETHYLINE HYDROCHLORIDE	000125-30-4
COLCHICEINE	000477-27-0
COLCHICINE	000064-86-8
COLCHICINE,3-OH-7-TRIFLUOROACETAMIDO-10-METHIO A	123643-51-6
2,4,6-COLLIDINE	000108-75-8
CONGO RED	000573-58-0
M-CONH2 PHENYLHIPPURATE	081592-11-2
CONVALLATOXIN	000508-75-8
CONVALLOTOXOL	003253-62-1
CORONENE	000191-07-1
CORTEXOLONE	000152-58-9
CORTICOSTERONE	000050-22-6
CORTISONE	000053-06-5
CORTISONE ACETATE	000050-04-4
CORTISONE-9A-FLUORO	000079-60-7
COUMACHLOR	000081-82-3
COUMAFURYL	000117-52-2
COUMAPHOS	000056-72-4
COUMARIN	000091-64-5
CPE-C	090597-22-1
CREATINE	000057-00-1
CREATININE	000060-27-5
P-CRESOL	000106-44-5

O-CRESOL	000095-48-7
M-CRESOL	000108-39-4
CRESYL DIPHENYL PHOSPHATE	026444-49-5
CROMOGYLYCIC ACID	016110-51-3
CRONETON	029973-13-5
CROTONALDEHYDE	004170-30-3
TRANS-CROTONALDEHYDE	000123-73-9
CROTONIC ACID	003724-65-0
CROTONIC ACID	000107-93-7
CROTONONITRILE	004786-20-3
CROTOXYPHOS	007700-17-6
TR-CROTYL-CHRYSANTHEMATE	083123-89-1
CRUFOMATE	000299-86-5
CRYSTAL VIOLET (BASIC VIOLET 3)	000548-62-9
CUMENE	000098-82-8
CUMENE HYDROPEROXIDE	000080-15-9
CUMIC ACID	000536-66-3
CYANAMPHENICOL	023885-71-4
CYANAZINE	021725-46-2
CYANOACETIC ACID	000372-09-8
P-CYANOACETOPHENONE	001443-80-7
M-CYANOACETOPHENONE	006136-68-1
CYANO-PHENYL-AMPHENICOL	049648-48-8
P-CYANO-AMPHENICOL	023885-61-2
M-CYANOBENAZLDEHYDE	024964-64-5
P-CYANOBENZAMIDE	003034-34-2
M-CYANOBENZAMIDE	003441-01-8
P-CYANOBENZENESULFONAMIDE	003119-02-6
P-CYANOBENZOIC ACID	000619-65-8
M-CYANOBENZOIC ACID	001877-72-1
P-CYANOBENZYL-TR-CHRYSANTHEMATE	082535-32-8
P-CYANOBENZYL N,N-DIME CARBAMATE	084640-32-4
P-(3-(P-CYANOBENZYL)-3-METHYL-1-TRIAZENO)-	065542-16-7
4-CYANO-3-CHLOROPHENYLAMIDINOUREA	021702-93-2
1-(4-CYANO),2-DIBENZOYL-1-TERT-BUTYLHYDRAZINE	112426-62-7
1-(3-CYANO),2-DIBENZOYL-1-TERT-BUTYLHYDRAZINE	112426-69-4
P-CYANO-N,N-DIMETHYLANILINE	001197-19-9
N-CYANO-2(33DIME-1-TRIAZEN)BENZAMIDE	066974-92-3
CYANOFENPHOS	013067-93-1
2-CYANO-9H-FLUORENE	002523-48-0
P-CYANOFORMANILIDE	006321-94-4
2-CYANOFURAN	000617-90-3
CYANOGEN	000460-19-5
2-CYANOGUANIDINE	000461-58-5
3-CYANOINDOLE	005457-28-3
4-CYANO-3-MEO-PHENYLAMIDINO UREA	058247-25-9
2-CYANOMETHYLFURAN	002745-25-7
2-CYANO-6-METHYLPYRIDINE	001620-75-3
4-CYANO-2-METHYLPYRIDINE	002214-53-1
3-CYANOMETHYLTHIOPHENE	013781-53-8
2-CYANOMETHYLTHIOPHENE	020893-30-5
2-CYANOMORPHOLINO-DOXORUBICIN,12-IMINO	089164-79-4
3-CYANO-1-NITROBENZENE	000619-24-9
M-CYANOPHENOL	000873-62-1
O-CYANOPHENOL	000611-20-1
P-CYANOPHENOL	000767-00-0
P-CYANOPHENOXYACETIC ACID	001878-82-6
M-CYANOPHENOXYACETIC ACID	001879-58-9

2-CYANOPHENYL ACETATE	005715-02-6
M-CYANOPHENYLACETIC ACID	001878-71-3
2-CYANO-3-PHENYL ET-2-PROPENOATE	002025-40-3
P-CYANO PHENYLHIPPURATE	081592-05-4
M-CYANO PHENYLHIPPURATE	081592-12-3
1-(3-CYANOPHENYL)-3-MEO-3-ME UREA	096337-67-6
N1-(4-CYANOPHENYL)SULFANILAMIDE	025612-07-1
CYANOPHOS	002636-26-2
2-CYANOPYRAZINE	019847-12-2
4-CYANOPYRAZOLE	031108-57-3
4-CYANOPYRIDINE	000100-48-1
2-CYANOPYRIDINE	000100-70-9
3-CYANOPYRIDINE	000100-54-9
4-CYANOPYRIDINE OXIDE	014906-59-3
2-CYANOPYRROLE	004513-94-4
2-CYANOQUINOLINE	001436-43-7
4-CYANOQUINOLINE	002973-27-5
3-CYANOQUINOLINE	034846-64-5
6-CYANOQUINOXALINE	023088-24-6
2-CYANOTHIOPHENE	001003-31-2
5-CYANOURACIL	004425-56-3
CYANURIC ACID	000108-80-5
CYCLACILLIN	003485-14-1
CYCLOBARBITAL	000052-31-3
CYCLOBUTANE	000287-23-0
1,3-CYCLOBUTANEDIONE	015506-53-3
CYCLOBUTANONE	001191-95-3
CYCLODECANE	000293-96-9
CYCLODODECANE	000294-62-2
CYCLODODECANE EPOXIDE	000286-99-7
CYCLODODECANOL	001724-39-6
CYCLODODECANONE	000830-13-7
1,5,9-CYCLODODECATRIENE	004904-61-4
1,3-CYCLOHEPTADIENE	004054-38-0
CYCLOHEPTANE	000291-64-5
CYCLOHEPTANOL	000502-41-0
CYCLOHEPTANOL, 1-(4-CHLOROPHENYL)-2-(1H-1,2,4-TR	100199-27-7
CYCLOHEPTANONE	000502-42-1
4H-CYCLOHEPTA(DEF)PHENANTHRENE	019561-31-0
CYCLOHEPTATRIENE	000544-25-2
2,4,6-CYCLOHEPTATRIENE-1-ONE, 2-HYDROXY-5-NITROS	002297-94-1
2,4,6-CYCLOHEPTATRIEN-1-ONE, 3-HYDROXY-	003324-76-3
CYCLOHEPTENE	000628-92-2
N-CYCLOHEPTYL-N'-PHENYLUREA	019095-79-5
1,3-CYCLOHEXADIENE	000592-57-4
1,4-CYCLOHEXADIENE	000628-41-1
2,5-CYCLOHEXADIENE-1,4-DIONE	000106-51-4
2,5-CYCLOHEXADIENE-1,4-DIONE, 2,6-DIMETHYL	000527-61-7
2,5-CYCLOHEXADIENE-1,4-DIONE, DIOXIME	000105-11-3
2,5-CYCLOHEXADIENE-1,4-DIONE, 2,5-DIMETHYL-	000137-18-8
2,5-CYCLOHEXADIENE-1,4-DIONE, 2,3-DIMETHYL-	000526-86-3
2,5-CYCLOHEXADIENE-1,4-DIONE, 2-PHENYL-	000363-03-1
2,5-CYCLOHEXADIENE-1,4-DIONE, 2-[2-[(AMINOCARBON	024279-91-2
2,5-CYCLOHEXADIENE-1,4-DIONE, 2,3,5-TRIMETHYL-	000935-92-2
2,5-CYCLOHEXADIENE-1,4-DIONE, 2,3-DIMETHOXY-5-ME	000605-94-7
2,5-CYCLOHEXADIENE-1,4-DIONE, W/1,4-BENZENEDIOL	000106-34-3
2,5-CYCLOHEXADIENE-1,4-DIONE, 2,5-BIS(1-AZIRIDIN	059886-46-3
2,5-CYCLOHEXADIENE-1,4-DIONE, 2,5-BIS(1-AZIRIDIN	059886-55-4

2,5-CYCLOHEXADIENE-1,4-DIONE, 2,5-BIS(1-AZIRIDIN	059886-40-7
2,5-CYCLOHEXADIENE-1,4-DIONE, 2-METHYL-5-(1-METH	000490-91-5
2,5-CYCLOHEXADIENE-1,4-DIONE, 3-[(3-PYRIDINYL)ME	117574-40-0
CYCLOHEXANAMINE	000108-91-8
CYCLOHEXANAMINE, N-CYCLOHEXYL-N-METHYL-	007560-83-0
CYCLOHEXANE	000110-82-7
CYCLOHEXANEACETAMIDE, N,N-DIMETHYL-	005830-33-1
CYCLOHEXANEACETAMIDE, N,N-DIETHYL-	095715-61-0
CYCLOHEXANEACETIC ACID, 1-(AMINOMETHYL)-	060142-96-3
CYCLOHEXANECARBOXAMIDE, N-(3,4-DICHLOROPHENYL)-	015907-85-4
CYCLOHEXANECARBOXYLIC ACID	000098-89-5
1,2-CYCLOHEXANEDIAMINE	000694-83-7
CYCLOHEXANE,1,4-DI(1-NO-1ETF UREA)	013908-98-0
1,2-CYCLOHEXANEDIOL	000931-17-9
1,2-CYCLOHEXANEDIOL-CIS	001792-81-0
1,2-CYCLOHEXANEDIOL-TRANS	001460-57-7
CYCLOHEXANEETHANAMINE, α-METHYL-	054704-34-6
CYCLOHEXANEETHANOL	004442-79-9
CYCLOHEXANE, 1,2,3,4,5-PENTABROMO-6-CHLORO-	000087-84-3
CYCLOHEXANETHIOL	001569-69-3
CYCLOHEXANOL	000108-93-0
CYCLOHEXANOL, 1-(2-CHLOROPHENYL)-2-(1H-1,2,4-TRI	100199-35-7
CYCLOHEXANOL, 1-(2,5-DICHLOROPHENYL)-2-(1H-1,2,4	100199-38-0
CYCLOHEXANOL, 2-[(DIMETHYLAMINO)METHYL]-1-(3-MET	123134-25-8
CYCLOHEXANOL, 2-[(DIMETHYLAMINO)METHYL]-1-(3-MET	123154-38-1
CYCLOHEXANOL, 2,6-DIMETHYL-, (1α,2α,6α)-	039170-84-8
CYCLOHEXANOL, 2,6-DIMETHYL-, (1α,2α,6ß)-	039170-83-7
CYCLOHEXANOL, 2,6-DIMETHYL-, (1α,2ß,6ß)-	042846-29-7
CYCLOHEXANOL, 4-(1,1-DIMETHYLETHYL)-, TRANS-	021862-63-5
CYCLOHEXANOL, 4-(1,1-DIMETHYLETHYL)-, CIS-	000937-05-3
CYCLOHEXANONE	000108-94-1
CYCLOHEXANONE, 2-(1-CYCLOHEXEN-1-YL)-	001502-22-3
CYCLOHEXANONE OXIME	000100-64-1
CYCLOHEXANONE, 4-PHENYL-	004894-75-1
CYCLOHEXENE	000110-83-8
1-CYCLOHEXENE-1-ACETIC ACID, _-METHYL-_-[[[(METH	073632-86-7
3-CYCLOHEXENE-1-CARBOXALDEHYDE	000100-50-5
2-CYCLOHEXENE-1-ONE	000930-68-7
CYCLOHEXIMIDE	000066-81-9
CYCLOHEXYLACETIC ACID	005292-21-7
1-CYCLOHEXYLAMINO-2-PROPANOL	000103-00-4
2-CYCLOHEXYL-4,6-DINITROPHENOL	000131-89-5
4-CYCLOHEXYL-2,6-DINITROPHENOL	004097-58-9
1-CYCLOHEXYLHEXANE	004292-75-5
1-CYCLOHEXYL-3-(4-METHYLMETANILYL)UREA	000565-33-3
1-CYCLOHEXYLOCTANE	001795-15-9
P-CYCLOHEXYLPHENOL	001131-60-8
4-CYCLOHEXYLSEMICARBAZIDE	052662-76-7
CYCLOLEUCINE	000052-52-8
CYCLOOCTA-1,5-DIENE	000111-78-4
CYCLOOCTANE	000292-64-8
CYCLOOCTANEETHANAMINE, α-METHYL-	082700-02-5
CYCLOOCTANONE	000502-49-8
1,3,5,7-CYCLOOCTATETRAENE	000629-20-9
CYCLOPENT[FG]ACENAPHTHYLENE	000187-78-0
CYCLOPENTADECANONE	000502-72-7
CYCLOPENTADIENE	000542-92-7
9H-CYCLOPENTA[D]PYRROLO[1,2-A]PYRIMIDIN-4-ONE, 1	088491-49-0

CYCLOPENTANE	000287-92-3
CYCLOPENTANECARBOXAMIDE, N-(3,4-DICHLOROPHENYL)-	015907-82-1
CYCLOPENTANECARBOXYLIC ACID, 1-[[[(2-CHLOROETHYL	052320-88-4
CYCLOPENTANEMETHANAMINIUM, N,N,N-TRIMETHYL-, IOD	021727-39-9
CYCLOPENTANETHIOL	001679-07-8
CYCLOPENTANOL	000096-41-3
CYCLOPENTANONE	000120-92-3
CYCLOPENTA[D]PYRIDO[1,2-A]PYRIMIDIN-10(1H)-ONE,	085653-88-9
CYCLOPENTA[D]PYRIDO[1,2-A]PYRIMIDIN-10(1H)-ONE,	085653-87-8
CYCLOPENTA[D]PYRIDO[1,2-A]PYRIMIDIN-10(1H)-ONE,	070026-79-8
CYCLOPENTA[D]PYRIDO[1,2-A]PYRIMIDIN-10(1H)-ONE,	070026-51-6
CYCLOPENTA[D]PYRIDO[1,2-A]PYRIMIDIN-10(1H)-ONE,	070026-41-4
CYCLOPENTA[B]PYRROLE-2-CARBOXYLIC ACID, DERIVATI	087269-97-4
CYCLOPENTENE	000142-29-0
3-CYCLOPENTYLMETHYLHYDROTHIAZIDE	023141-86-8
4-CYCLOPENTYLPHENOL	001518-83-8
N-CYCLOPENTYL-N'-PHENYLUREA	013140-89-1
CYCLOPENTYLPROPRIONIC ACID	000140-77-2
CYCLOPHOSPHAMIDE	000050-18-0
CYCLOPROPANAMINE	000765-30-0
CYCLOPROPANAMINE, 2-PHENYL-, TRANS	000095-62-5
CYCLOPROPANAMINE, 2-PHENYL-, CIS-(+)-	069684-88-4
CYCLOPROPANE	000075-19-4
CYCLOPROPANECARBONITRILE	005500-21-0
CYCLOPROPANECARBOXAMIDE, N-(3,4-DICHLOROPHENYL)-	002790-16-1
CYCLOPROPANECARBOXAMIDE, N-(3,6-DIHYDRO-1(2H)-PY	080431-10-3
CYCLOPROPANECARBOXAMIDE, N-[4-[2-[(DICHLOROACETY	039961-02-9
CYCLOPROPANECARBOXYLIC ACID	001759-53-1
CYCLOPROPANECARBOXYLIC ACID, 3-[(DIHYDRO*	125603-72-7
CYCLOPROPANECARBOXYLIC ACID, 3-[(DIHYDRO-2-OXO-3(	084368-16-1
CYCLOPROPANECARBOXYLIC ACID, 3-(3-METHOXY-2-METH	026074-74-8
CYCLOPROPANECARBOXYLIC ACID, 2-(PHENYLETHYNYL)-,	056892-91-2
CYCLOPROPANECARBOXYLIC ACID, 2-[2-(2-PROPENYLOXY	144106-15-0
CYCLOPROPYLBENZENE	000873-49-4
4-CYCLOPROPYL 2-BUTANONE	002046-23-3
CYCLOPROPYL METHYL KETONE	000765-43-5
N-CYCLOPROPYL-N'-PHENYLUREA	013140-86-8
CYCLOTHIAZIDE	002259-96-3
CYCLOUNDECANONE	000878-13-7
CYCLURON	002163-69-1
CYFLUTHRIN	068359-37-5
CYHALOTHRIN	068085-85-8
LAMBDA-CYHALOTHRIN	091465-08-6
CYMARIN	000508-77-0
CYMAROL	000465-84-9
P-CYMENE	000099-87-6
M-CYMENE	000535-77-3
CYOLANE	000947-02-4
3-CYPE-1-(2-CLET)-1-NITROSOUREA	013909-03-0
CYPERMETHRIN	052315-07-8
BETA-CYPERMETHRIN ISOMER	065732-07-2
BETA-CYPERMETHRIN ISOMER	065731-84-2
CYPRAZINE	022936-86-3
TR-CYPR-CO2ME-22-DIME-3(2DICLETHENYL)	059897-94-8
CGA 219417 (CYPRODINIL)	121552-61-2
CYPROHEPTADINE	000129-03-3
CYPROMIDE	002759-71-9
L-CYSTEINAMIDE, N-ACETYL-	010061-64-0

L-CYSTEINAMIDE, S,S'-BIS[N-ACETYL-	006513-23-1
CYSTEINE	000052-90-4
L-CYSTEINE, S-(TRIPHENYLMETHYL)-	002799-07-7
CYSTINE	000056-89-3
CYTHIOATE	000115-93-5
CYTIDINE, 2',3'-DIDEOXY-N-[(1-PYRROLIDINYL)METHY	141018-18-0
CYTIDINE	000065-46-3
CYTIDINE, 2',3'-DIDEOXY-N-[(DIETHYLAMINO)METHYLE	141018-14-6
CYTIDINE, 2',3'-DIDEOXY-N-[(DIETHYLAMINO)METHYLE	141018-19-1
CYTIDINE, 2',3'-DIDEOXY-N-[(DIMETHYLAMINO)METHYL	138848-10-9
CYTIDINE, 2',3'-DIDEOXY-N-[(DIPROPYLAMINO)METHYL	141018-15-7
CYTIDINE, 2',3'-DIDEOXY-3'-FLUORO-N-[[BIS(1-METH	141018-21-5
CYTIDINE, 2',3'-DIDEOXY-3'-FLUORO-N-[(DIPROPYLAM	141018-20-4
CYTIDINE, 2',3'-DIDEOXY-3'-FLUORO-N-[(4-MORPHOLI	141018-23-7
CYTIDINE, 2',3'-DIDEOXY-3'-FLUORO-N-[(1-PIPERIDI	141018-22-6
CYTIDINE, 2',3'-DIDEOXY-3'-FLUORO-N-[(1-PYRROLID	141018-24-8
CYTIDINE, 2',3'-DIDEOXY-N-[(4-MORPHOLINYL)METHYL	141018-17-9
CYTIDINE, 2',3'-DIDEOXY-N-[(1-PIPERIDINYL)METHYL	141018-16-8
CYTIDINE, 3'-AZIDO-2',3'-DIDEOXY-	084472-89-9
CYTIDINE, N-BENZOYL-	013089-48-0
CYTIDINE, N-BENZOYL-2'-DEOXY-	004836-13-9
CYTIDINE, 5-BROMO-2',3'-DIDEOXY-	107036-57-7
CYTIDINE, 5'-CHLORO-5'-DEOXY-	031652-78-5
CYTIDINE, 2'-DEOXY-	000951-77-9
CYTIDINE, 2',3'-DIDEOXY-	007481-89-2
CYTIDINE, 2',3'-DIDEOXY-N-[[BIS(1-METHYLETHYL)AM	141043-80-3
CYTIDINE, 2',3'-DIDEOXY-5-FLUORO-	107036-62-4
CYTIDINE, 2',3'-DIDEOXY-3'-FLUORO-	051246-79-8
CYTIDINE, 2',3'-DIDEHYDRO-2',3'-DIDEOXY-	007481-88-1
CYTISINE	000485-35-8
CYTOSINE	000071-30-7
D-GLYCEROL-A-D-GALACTO DERIVATIVE	083107-50-0
DALAPON	000075-99-0
DALAPON (SODIUM SALT)	000127-20-8
2,4-D, AMINE SALT	002307-55-3
DAMINOZIDE	001596-84-5
DANTROLENE	007261-97-4
DAUNOMYCIN	020830-81-3
DAZOLICINE	061477-97-2
2,4-DB	000094-82-6
2,4-D, BUTOXYETHYL ESTER	001929-73-3
2,4-D, BUTYL ESTER	000094-80-4
DCIP(2,2'-OXYBIS-1-CHLOROPROPANE)	000108-60-1
P,P'-DDD	000072-54-8
O,P'-DDD	000053-19-0
O,P'-DDE	003424-82-6
P,P'-DDE	000072-55-9
2,4-D, DIMETHYLAMINE SALT	002008-39-1
O,P'-DDT	000789-02-6
P,P'-DDT	000050-29-3
DEAMINOMETRIBUZIN	035045-02-4
3-DEAZAURIDINE	039935-49-4
2,4-DEB	000094-83-7
DECABROMOBIPHENYL	013654-09-6
DECABROMODIPHENYL ETHER	001163-19-5
DECACHLOROBIPHENYL	002051-24-3
2,8-DECADIENE-4,6-DIYNOIC ACID, METHYL ESTER	000505-02-2
DECAFLUOROBUTANE	000355-25-9

DECAHYDRONAPHTHALENE	000091-17-8
DECAHYDRO-2-NAPHTHALENONE	004832-17-1
DECAHYDRO-2-NAPHTHOL	000825-51-4
GAMMA-DECALACTONE	000706-14-9
1,1,1,3,5,5,7,9,9,9-DECAMETHYL-3,7-DIPHENYLPEN*	020252-66-8
DECAMETHYLCYCLOPENTASILOXANE	000541-02-6
1,1,1,3,5,7,9,11,11,11-DECAMETHYL-3,5,7,9-TETRA*	013271-58-4
DECAMETHYLTETRASILOXANE	000141-62-8
N-DECANAL	000112-31-2
1-DECANAMINIUM, N,N,N-TRIMETHYL-, IODIDE	007447-24-7
N-DECANE	000124-18-5
DECANEDINITRILE	001871-96-1
DECANEDIOIC ACID	000111-20-6
1-DECANETHIOL	000143-10-2
DECANOIC ACID	000334-48-5
1-DECANOL	000112-30-1
2-DECANONE	000693-54-9
1-DECENE	000872-05-9
2-DECENE-4,6,8-TRIYNOIC ACID, METHYL ESTER, (Z)-	002739-57-3
2-DECENE-4,6,8-TRIYONIC ACID, METHYL ESTER	000692-94-4
5-DECEN-1-OL ACETATE (E)	038421-90-8
5-DECEN-1-OL ACETATE (Z)	067446-07-5
3-DECENYL ACETATE (Z)	081634-99-3
6-DECENYL ACETATE (Z)	068760-70-3
4-DECENYL ACETATE (Z)	067452-27-1
13-DECOSENAMIDE (CIS)	000112-84-5
DECYL ACETATE	000112-17-4
N-DECYLAMINE	002016-57-1
N-DECYLCYCLOHEXANE	001795-16-0
O-DECYLHYDROXYLAMINE	029812-79-1
DECYLPROPANEDIOIC ACID	004372-29-6
DECYLPYRIDINIUM BROMIDE	002534-65-8
N-DECYLPYRIDINIUM IODIDE	007295-91-2
DEET [N,N,-DIET-3-ME BENZAMIDE]	000134-62-3
DEF	000078-48-8
DEHYDROACETIC ACID	000520-45-6
4'-DEHYDROOLEANDRIN	055147-64-3
DELTAMETHRIN	052918-63-5
DEMEPHION	002587-90-8
DEMEROL	000057-42-1
DEMETON	008000-97-3
DEMETON-S-METHYL	000919-86-8
DENZIMOL	073931-96-1
3-DEO-5-BR-6-ME-PYRIMIDIN-2,4-DIONE	077317-79-4
DEOXYCHOLIC ACID	000083-44-3
DEOXYCORTICOSTERONE	000064-85-7
DEOXYCORTICOSTERONE ACETATE	000056-47-3
B-2'-DEOXYTHIOGUANOSINE	000789-61-7
A-2'-DEOXYTHIOGUANOSINE	002133-81-5
2,4-DEP	000094-84-8
DES	000056-53-1
N-DESACETYL-COLCHICINE	003476-50-4
DESETHYLATRAZINE	006190-65-4
8-DESFLUORO-LOMEFLOXACIN	098079-47-1
DESIPRAMINE	000050-47-5
(DES-IPR)NN-DIETAMINO(1ME)BU-CLOFAZIMINE	111436-11-4
(DES-IPR) N,N-DIETAMINOPR-CLOFAZIMINE	098270-76-9
(DES-IPR) N,N-DIETAMINOET-CLOFAZIMINE	078182-93-1

DESISOPROPYLATRAZINE	001007-28-9
DESMEDIPHAM	013684-56-5
DESMETRYNE	001014-69-3
2,4-D, ETHYL ESTER	000533-23-3
DEXAMETHASONE	000050-02-2
DEXAMETHASONE-17-ACETATE	001177-87-3
DEXAMETHASONE-21-GALACTOSIDE	092901-23-0
DEXETIMIDE	021888-98-2
DEXTROMORAMIDE	000357-56-2
DFDT	000475-26-3
ß-D-GALACTOPYRANOSE, 2-DEOXY-2-[[(METHYLNITROSOA	037793-01-4
D-GLUCITOL, 1,4:3,6-DIANHYDRO-, 5-NITRATE	016051-77-7
D-GLUCITOL, 1,4:3,6-DIANHYDRO-, 2-NITRATE	016106-20-0
beta-D-GLUCOPYRANOSE, 2-[[[(2-CHLOROETHYL)NITROS	070866-07-8
α-D-GLUCOPYRANOSE, 2-[[[(2-CHLOROETHYL)NITROSOAM	074729-54-7
ß-D-GLUCOPYRANOSE, 2-DEOXY-2-[[(METHYLNITROSOAMI	066395-17-3
_-D-GLUCOPYRANOSIDE, METHYL 6-DEOXY-6-[[(METHYLN	041110-59-2
α-D-GLUCOPYRANOSIDE, METHYL 3-DEOXY-3-[[(METHYLN	052019-12-2
DIACETAMIDE	000625-77-4
M-DIACETAMIDOBENZENE	010268-78-7
ALPHA-BETA-DIACETYLDIGOXIN	040246-14-8
DIACETYLENE	000460-12-8
DIACETYLMORPHINE, HYDROCHLORIDE	001502-95-0
1,3-DIACETYLUREA	000638-20-0
DIALIFOR	010311-84-9
DIALLATE	002303-16-4
DIALLYLAMINE	000124-02-7
DIALLYLDICHLOROSILANE	003651-23-8
DIALLYL MALEATE	000999-21-3
DIALLYLPHTHALATE	000131-17-9
DIALLYL SULFIDE	000592-88-1
DIALLYL UREA	001801-72-5
3,6-DIAMINOACRIDINE	000092-62-6
2,4-DIAMINO-5-(1-ADAMANTYL)-6-METHYLPYRIMIDINE	031935-08-7
2,4 DIAMINOANISOLE	000615-05-4
4,4'-DIAMINOAZOBENZENE	000538-41-0
2,4-DIAMINOAZOBENZENE	000495-54-5
2,4-DIAMINO-5-BENZYLPYRIMIDINE	007319-45-1
2,4-DIAMINOBUTYRIC ACID	000305-62-4
4,4'-DIAMINODIPHENYL ETHER	000101-80-4
1,2-DIAMINOETHANE	000107-15-3
2,4-DIAMINO-5-(3-MEOBENZYL)PYRIMIDINE	059481-28-6
3,6-DIAMINO-10-METHYL-ACRIDINIUM CHLORIDE	000086-40-8
1,5-DIAMINONAPHTHALENE	002243-62-1
2,3-DIAMINOPHENAZINE	000655-86-7
N-(4-((2,4-DIAMINOPHENYL)AZO)PHENYL)ACETAMIDE	018371-12-5
DI-(P-AMINOPHENYL)METHANE	000101-77-9
DI(P-AMINOPHENYL)SULFONE	000080-08-0
1,2-DIAMINOPROPANE	000078-90-0
2,4-DIAMINOPYRIMIDINE,5(P-NH2BENZYL)	069945-50-2
2,4-DIAMINOPYRIMIDINE-5(245-CL)BENZYL	006981-11-9
2,4-DIAMINOPYRIMIDINE-5(345-CL)BENZYL	007520-68-5
2,4-DIAMINOQUINAZOLINE	001899-48-5
2,5-DIAMINOTOLUENE	000095-70-5
2,3-DIAMINOTOLUENE	002687-25-4
3,4-DIAMINOTOLUENE	000496-72-0
2,6-DIAMINOTOLUENE	000823-40-5
DI-N-AMYL PHTHALATE	000131-18-0

1,2:5,6-DIANHYDROGALACTITOL	023261-20-3
1,2-(DI-(A-PYRIDYL)ETHYLENE	001437-15-6
1,4-DIAZABICYCLO(2,2,2)OCTANE (DABCO)	000280-57-9
3,5-DIAZAINDOLE	000272-97-9
5,6-DIAZAPHENANTHRENE	000230-17-1
1,3,2-DIAZAPHOSPHOLIDIN-2-AMINE, N,N,1,3-TETRAME	007778-06-5
7,9-DIAZASPIRO[4.5]DECANE-6,8,10-TRIONE	056209-30-4
6,8-DIAZASPIRO[3.5]NONANE-5,7,9-TRIONE	006128-03-6
5,7-DIAZASPIRO[2,5]OCTANE-4,6,8-TRIONE	006947-77-9
2,4-DIAZASPIRO[5.5]UNDECANE-1,3,5-TRIONE	000052-44-8
DIAZENECARBONITRILE, (1,1-DIMETHYLETHYL)-2-OXIDE	090467-88-2
DIAZENECARBONITRILE, PHENYL-	000622-83-3
DIAZENECARBOXAMIDE, PHENYL-, 2-OXIDE	060142-49-6
DIAZENECARBOXYLIC ACID, PHENYL-, ETHYL ESTER, 2-	056751-20-3
DIAZEPAM	000439-14-5
1H-1,4-DIAZEPINE, HEXAHYDRO-1,4-DINITROSO-	055557-00-1
1,4-DIAZEPIN-2-ONE,1,3-DIHYDRO-1-ME-7-DIMETHYLAM	030144-75-3
DIAZINON	000333-41-5
DIAZINON, O-ANALOG	000962-58-3
DIAZIQUONE	057998-68-2
N-(DIAZOACETYLAMINOACETYL)MORPHOLINE	060142-00-9
N-DIAZOACETYLGLYCINE-N'-ETHYLAMIDE	038726-91-9
N-DIAZOACETYLGLYCINE,ETHYL ESTER	000999-29-1
N-DIAZOACETYLGLYCINE-N'-HEXYLAMIDE	060141-99-3
DIAZOACETYLGLYCINE-N'-IBU AMIDE	060141-98-2
N-DIAZOACETYLGLYCINE-N'-I-PR AMIDE	052819-98-4
N-DIAZOACETYLGLYCINE-N'-N-PR AMIDE	052819-97-3
DIAZOMETHANE	000334-88-3
DIBENZ(A,H)ACRIDINE	000226-36-8
DIBENZ[a,j]ACRIDINE	000224-42-0
DIBENZ(A,J)ANTHRACENE	000224-41-9
DIBENZ(A,H)ANTHRACENE	000053-70-3
1,2,3,4-DIBENZANTHRACENE	000215-58-7
7H-DIBENZO(C,G)CARBAZOLE	000194-59-2
13H-DIBENZO(A,I)CARBAZOLE	000239-64-5
DIBENZO(DEF,MNO)CHRYSENE	000191-26-4
DIBENZO(B,DEF)CHRYSENE-7,14-DIONE	000128-66-5
[2,7]DB27C9-DIBENZO CROWN ETHER	061260-08-0
[3,3]DB18C6-DIBENZO CROWN ETHER	014187-32-7
[4,4]DB24C6-DIBENZO CROWN ETHER	014174-09-5
[8,8]DB48C16-DIBENZO CROWN ETHER	121284-20-6
[5,5]DB30C10-DIBENZO CROWN ETHER	017455-25-3
[4,7]DB33C11-DIBENZO CROWN ETHER	087586-46-7
DIBENZO-P-DIOXIN	000262-12-4
DIBENZOFURAN	000132-64-9
1,3(2H,9BH)-DIBENZOFURANDIONE, 2,6-DIACETYL-7,9-	000125-46-2
DIBENZOFURAN, 1,2,3,4,6,7,8-HEPTACHLORO-	067562-39-4
DIBENZOFURAN, 1,2,6,7,9-PENTACHLORO-	070872-82-1
DIBENZOFURAN, 1,2,4,6,9-PENTACHLORO-	070648-24-7
DIBENZOFURAN, 2,3,4,6,8-PENTACHLORO-	067481-22-5
DIBENZOFURAN, 1,2,3,7,8-PENTACHLORO-	057117-41-6
DIBENZOFURAN, 1,2,3,4,8-PENTACHLORO-	067517-48-0
DIBENZOFURAN, 1,2,3,6,7-PENTACHLORO-	057117-42-7
DIBENZOFURAN, 1,3,4,6-TETRACHLORO-	083704-27-2
DIBENZOFURAN, 2,3,6,8-TETRACHLORO-	057117-37-0
DIBENZOFURAN, 2,3,4,6-TETRACHLORO-	083704-30-7
DIBENZOFURAN, 1,3,6,8-TETRACHLORO-	071998-72-6
DIBENZOFURAN, 1,3,4,7-TETRACHLORO-	070648-16-7

DIBENZOFURAN, 2,3,6,7-TETRACHLORO-	057117-39-2
DIBENZOFURAN, 1,2,3,4-TETRACHLORO-	024478-72-6
DIBENZOFURAN, 1,2,7,8-TETRACHLORO-	058802-20-3
DIBENZOFURAN, 1,3,7,9-TETRACHLORO-	064560-17-4
DIBENZOFURAN, 2,4,6,7-TETRACHLORO-	057117-38-1
DIBENZO[G,P]CHRYSENE	000191-68-4
6H-DIBENZO[B,D]PYRAN-1-OL, 3-(1,1-DIMETHYLHEPTYL	061597-27-1
10AH-DIBENZO[B,D]PYRAN-10A-AMINE, 6,6A,7,8,9,10-	132453-23-7
6H-DIBENZO[B,D]PYRAN, 1-([1,1'-BIPHENYL]-4-YLMET	132213-93-5
6H-DIBENZO[B,D]PYRAN-9-CARBOXALDEHYDE, 3-(1,1-DI	123331-83-9
6H-DIBENZO[B,D]PYRAN, 6A,7,8,9,10,10A-HEXAHYDRO-	027179-29-9
6H-DIBENZO[B,D]PYRAN-9-METHANOL, 6A,7,8,10A-TETR	036557-05-8
6H-DIBENZO[B,D]PYRAN-9-METHANOL, 3-(1,1-DIMETHYL	112830-95-2
6H-DIBENZO[B,D]PYRAN-9-METHANOL, 3-(1,1-DIMETHYL	123331-84-0
6H-DIBENZO[B,D]PYRAN-1-OL, 6A,7,8,10A-TETRAHYDRO	033086-25-8
6H-DIBENZO[B,D]PYRAN-1-OL, 6A,7,10,10A-TETRAHYDR	005957-75-5
6H-DIBENZO[B,D]PYRAN, 6A,7,10,10A-TETRAHYDRO-1-M	001451-20-3
DIBENZO[a,l]PYRENE	000191-30-0
DIBENZO[a,e]PYRENE	000192-65-4
DIBENZO(A,I)PYRENE	000189-55-9
DIBENZO[A,H]PYRENE	000189-64-0
DIBENZO[B,E]THIEPIN-2-ACETIC ACID, 6,11-DIHYDRO-	055863-26-8
DIBENZOTHIOPHENE	000132-65-0
10AH-DIBENZO[B,D]THIOPYRAN-10A-AMINE, 6,6A,7,8,9	132453-38-4
1,2-DIBENZOYLETHYLENE	004070-75-1
DIBENZYL AMINE	000103-49-1
DIBENZYLAMMONIUM CHLORIDE	020455-68-9
DIBENZYL ETHER	000103-50-4
DIBENZYL KETONE	000102-04-5
1,6-DIBR-1,6-DIDEOXYMANNITOL	000488-41-5
3,5-DIBR-2',4'-DIFLUORO SALICYLANILIDE	079402-07-6
3,5-DIBR-2'-NO2-4'-CL SALICYLANILIDE	072699-09-3
2,6-DIBROMOACETANILIDE	033098-80-5
DIBROMOACETONITRILE	003252-43-5
DIBROMOAMPHENICOL	017371-30-1
2,4-DIBROMOANILINE	000615-57-6
2,6-DIBROMOBENZAMIDE	096237-91-1
M-DIBROMOBENZENE	000108-36-1
O-DIBROMOBENZENE	000583-53-9
1,4-DIBROMOBENZENE	000106-37-6
2',5'-DIBROMOBENZOGUANAMINE	057381-42-7
4,4'-DIBROMOBIPHENYL	000092-86-4
2,3-DIBROMOBUTANE	005408-86-6
1,2-DIBROMOBUTANE	000533-98-2
1,4-DIBROMOBUTANE	000110-52-1
1,1-DIBROMO-2-CHLOROETHANE	027949-36-6
DIBROMOCHLOROMETHANE	000124-48-1
2,6-DIBROMO-4-CHLOROPHENOL	005324-13-0
1,2-DIBROMO-3-CHLOROPROPANE	000096-12-8
2,2-DIBROMO-2-CYANOACETAMIDE	010222-01-2
1,2-DIBROMODECANE	004101-68-2
2,8-DIBROMODIBENZO-P-DIOXIN	105836-96-2
2,3-DIBROMODIBENZO-P-DIOXIN	050585-37-0
2,7-DIBROMODIBENZO-P-DIOXIN	039073-07-9
1,6-DIBROMODIBENZO-P-DIOXIN	091371-14-1
1,2-DIBROMO-4-(1,2-DIBROMOETHYL)CYCLOHEXANE	003322-93-8
2,3-DIBROMO-7,8-DICHLORODIBENZO-P-DIOXIN	050585-40-5
2,3-DIBROMO-7,8-DIFLUORODIBENZO-P-DIOXIN	050585-43-8

DIBROMODIFLUOROMETHANE	000075-61-6
1,2-DIBROMOETHANE	000106-93-4
1,1-DIBROMOETHANE	000557-91-5
1,2-DIBROMOETHENE	025429-23-6
(1,2-DIBROMOETHYL)BENZENE	000093-52-7
DIBROMOMETHANE	000074-95-3
1,2-DIBROMO-2-METHYLPROPANE	000594-34-3
2,6-DIBROMO-4-NITROPHENOL	000099-28-5
3,4-DIBROMO-N-NITROSOPIPERIDINE	057541-73-8
1,2-DIBROMOPENTANE	019398-53-9
2,3-DIBROMOPHENOL	057383-80-9
2,4-DIBROMOPHENOL	000615-58-7
3,5-DIBROMOPHENOL	000626-41-5
2,5-DIBROMOPHENOL	028165-52-8
3,4-DIBROMOPHENOL	000615-56-5
2,6-DIBROMOPHENOL	000608-33-3
1,2-DIBROMOPROPANE	000078-75-1
1,3-DIBROMOPROPANE	000109-64-8
2,3-DIBROMO-1-PROPANOL	000096-13-9
2,3-DIBROMO-1-PROPENE	000513-31-5
2,6-DIBROMOQUINONECHLOROIMIDE	000537-45-1
1,2-DIBROMOTETRAFLUOROETHANE	000124-73-2
2,5-DIBROMOTHIOPHENE	003141-27-3
3,4-DIBROMOTHIOPHENE	003141-26-2
1,2,4-TRIAZOLE,3,5-DIBROMO	007411-23-6
3,5-DIBROMOTYROSINE	000537-24-6
2-(2,6-DIBR PH IMINO)IMIDAZOLIDINE	004205-93-0
N1(3,5-DIBR-2-PYRIDYL)SULFANILAMI	030961-41-2
1,2-DIBR-TETRACHLOROCYCLOHEXANE	055298-45-8
2,3-DIBR-TETRACHLOROCYCLOHEXANE	055332-89-3
2,4-DIBR-TETRACHLOROCYCLOHEXANE	055265-69-5
N-DIBUAMET 3,4,5-TRIMEO PH SULFONE	071203-66-2
2-N,N-DIBUAMINO-1-PHENYLETHANOL	083593-63-9
DIBUCAINE	000085-79-0
N,N-DI-I-BU-3,4-(-OCH2O-)CINNAMAMIDE	023795-32-6
3,5-DI-T-BU PHENYL GLUCOPYRANOSIDE	034213-80-4
DIBUTYLAMINE	000111-92-2
DIBUTYL DISULFIDE	000629-45-8
DI-TERT-BUTYLDISULFIDE	001518-72-5
DI (SEC-BUTYL) ETHER	006863-58-7
DI (TERT-BUTYL) ETHER	006163-66-2
DI-N-BUTYL ETHER	000142-96-1
2,6-DI(TERT-BUTYL)-4-ETHYLPHENOL	004130-92-1
DIBUTYL MALEATE	000105-76-0
2,6-DI-T-BUTYL-4-METHYLPHENOL (BHT)	000128-37-0
DIBUTYL METHYLPHOSPHONATE	002404-73-1
DIBUTYL OXALATE	002050-60-4
2,4-DI-T-BUTYLPHENOL	000096-76-4
2,6-DI-SEC-BUTYLPHENOL	005510-99-6
2,6-DI-T-BUTYLPHENOL	000128-39-2
DIBUTYL PHENYL PHOSPHATE	002528-36-1
DI-N-BUTYL PHOSPHATE	000107-66-4
DIBUTYL PHOSPHITE	001809-19-4
DIBUTYL PHTHALATE	000084-74-2
P-DIBUTYLPHTHALATE	001962-75-0
2,6-DI-T-BUTYLPYRIDINE	000585-48-8
DIBUTYL SEBACATE	000109-43-3
N,N-DIBUTYLTHIOUREA	000109-46-6

2,6-DI-T-BUTYL-P-TOLYL METHYLCARBAMATE	001918-11-2
1,2-DIBUTYOXYETHANE	000112-48-1
DICAMBA	001918-00-9
DICAPTHON	002463-84-5
MC-15608 [2,4'-DICL-4-CF3-3'-CO2ME-DIPH ETHER]	057025-76-0
3,5-DI-CF3-C6H3NHN=C(CN)COOME-CIS	036865-54-0
3,5-DI-CF3-C6H3NHN=C(CN)COOME-TRANS	036905-00-7
DICHLOFENTHION	000097-17-6
DICHLOFLUANID	001085-98-9
DICHLONE	000117-80-6
DICHLORAL UREA	000116-52-9
DICHLORISOPROTERENOL	000059-61-0
2,3-DICHLOROACETANILIDE	023068-36-2
3,5-DICHLOROACETANILIDE	031592-84-4
2,6-DICHLOROACETANILIDE	017700-54-8
2,5-DICHLOROACETANILIDE	002621-62-7
2,4-DICHLOROACETANILIDE	006975-29-7
DICHLOROACETIC ACID	000079-43-6
1,3-DICHLOROACETONE	000534-07-6
DICHLOROACETONITRILE	003018-12-0
DICHLOROACETYL CHLORIDE	000079-36-7
1,3-DICHLOROALLENE	083682-32-0
3,4-DICHLOROANILINE	000095-76-1
2,6-DICHLOROANILINE	000608-31-1
3,5-DICHLOROANILINE	000626-43-7
2,5-DICHLOROANILINE	000095-82-9
2,3-DICHLOROANILINE	000608-27-5
2,4-DICHLOROANILINE	000554-00-7
3,5-DICHLOROANISOLE	033719-74-3
3,4-DICHLOROBENZALDEHYDE	006287-38-3
2,6-DICHLOROBENZAMIDE	002008-58-4
3-DICHLOROBENZENE	000541-73-1
1,2-DICHLOROBENZENE	000095-50-1
1,4-DICHLOROBENZENE	000106-46-7
3,3'-DICHLOROBENZIDINE	000091-94-1
2',6'-DICHLOROBENZOGUANAMINE	057381-54-1
2',4'-DICHLOROBENZOGUANAMINE	057381-46-1
2',5'-DICHLOROBENZOGUANAMINE	057381-26-7
2,6-DICHLOROBENZOIC ACID	000050-30-6
3,4-DICHLOROBENZOIC ACID	000051-44-5
3,5-DICHLOROBENZOIC ACID	000051-36-5
2,5-DICHLOROBENZOIC ACID	000050-79-3
2,4-DICHLOROBENZOIC ACID	000050-84-0
2,6-DICHLOROBENZONITRILE	001194-65-6
3,4-DICHLOROBENZOTRIFLUORIDE	000328-84-7
3,4-DICHLOROBENZYL ALCOHOL	001805-32-9
2,4-DICHLOROBENZYLCHLORIDE	000094-99-5
3,4-DICHLOROBENZYL METHYLCARBAMATE	001966-58-1
2,4-DICHLOROBENZYLPENCILLIN	088598-34-9
3,3'-DICHLOROBIPHENYL	002050-67-1
2,4'-DICHLOROBIPHENYL	034883-43-7
3,4-DICHLOROBIPHENYL	002974-92-7
2,2'-DICHLOROBIPHENYL	013029-08-8
3,5-DICHLORO-1,1'-BIPHENYL	034883-41-5
2,3-DICHLOROBIPHENYL	016605-91-7
4,4'-DICHLOROBIPHENYL	002050-68-2
2,4-DICHLOROBIPHENYL	033284-50-3
2,5-DICHLOROBIPHENYL	034883-39-1

2,6-DICHLOROBIPHENYL	033146-45-1
1,1-DICHLORO-2,2-BIS(P-ETHOXYPHENYL)ETHANE	007388-32-1
1,1-DICHLORO-2,2-BIS(ETHYLPHENYL)ETHANE	000072-56-0
1,1-DICHLORO-2,3-BIS(TRICHLOROMETHYL)PROPENE	083682-40-0
4,4-DICHLORO-1,2-BUTADIENE	083682-41-1
1,1-DICHLORO-1,3-BUTADIENE	006061-06-9
1,4-DICHLORO-1,2-BUTADIENE	083682-44-4
1,4-DICHLORO-1,3-BUTADIENE	002984-42-1
2,3-DICHLOROBUTA-1,3-DIENE	001653-19-6
2,3-DICHLOROBUTANE	007581-97-7
1,4-DICHLOROBUTANE	000110-56-5
1,2-DICHLOROBUTANE	000616-21-7
1,1-DICHLOROBUTANE	000541-33-3
1,4-DICHLORO-2-BUTENE(TRANS)	000110-57-6
1,3-DICHLORO-2-BUTENE	000926-57-8
3,4-DICHLORO-1-BUTENE	000760-23-6
1,4-DICHLORO-1-BUTENE	031423-92-4
1,4-DICHLORO-2-BUTENE	000764-41-0
1,4-DICHLORO-2-BUTENE(CIS)	001476-11-5
1,4-DICHLORO-1-BUTYNE	083682-45-5
4,4-DICHLORO-1-BUTYNE	083682-42-2
1,4-DICHLORO-4-(1-CHLOROETHENYL)CYCLO-1-HEXENE	083682-63-7
1,3-DICHLORO-2-CHLOROMETHYLPROPENE	013245-65-3
DICHLORO(2-CHLOROVINYL)ARSINE (TRANS)	000541-25-3
1,2-DICHLOROCYCLOHEXANE -TRANS	000822-86-6
1,2-DICHLOROCYCLOHEXANE -CIS	001121-21-7
1,5-DICHLOROCYCLO-1,4-OCTADIENE	083682-65-9
1,6-DICHLORODIBENZO-P-DIOXIN	038178-38-0
2,8-DICHLORODIBENZO-P-DIOXIN	038964-22-6
1,3-DICHLORODIBENZO-P-DIOXIN	050585-39-2
2,7-DICHLORODIBENZO-P-DIOXIN	033857-26-0
2,3-DICHLORODIBENZO-P-DIOXIN	029446-15-9
2,8-DICHLORODIBENZOFURAN	005409-83-6
1-(2,6-DICHLORO),2-DIBENZOYL-1-TERT-BUTYLHYDRAZI	112426-71-8
1-(3,4-DICHLORO),2-DIBENZOYL-1-TERT-BUTYLHYDRAZI	112226-22-9
1-(2,3-DICHLORO),2-DIBENZOYL-1-TERT-BUTYLHYDRAZI	112426-76-3
1-(2,4-DICHLORO),2-DIBENZOYL-1-TERT-BUTYLHYDRAZI	112226-97-8
1,1-DICHLORO-2,2-DIETHOXYETHANE	000619-33-0
DICHLORODIETHYLSILANE	001719-53-5
2,3-DICHLORO-7,8-DIFLUORODIBENZO-P-DIOXIN	050585-42-7
1,2-DICHLORO-1,1-DIFLUOROETHANE	001649-08-7
1,2-DICHLORO-1,2-DIFLUOROETHYLENE	000598-88-9
1,1-DICHLORO-2,2-DIFLUOROETHYLENE	000079-35-6
DICHLORODIFLUOROMETHANE	000075-71-8
2,4-PYRIMIDINEDIAMINE, 5-[(2,6-DICHLORO-3,5-DIME	083157-96-4
1,4-DICHLORO-2,5-DIMETHOXYBENZENE	002675-77-6
1,6-DICHLORO-2,5-DIMETHYL-1,5-HEXADIENE	083682-48-8
3,4-DICHLORO-2,5-DIMETHYL-1,5-HEXADIENE	083682-49-9
3,5-DICHLORO-2,6-DIMETHYL-4-PYRIDINOL	002971-90-6
DICHLORODIMETHYLSILANE	000075-78-5
1,1-DICHLORO-2,2-DIPHENYLETHANE	002387-16-8
1,2-DICHLORO-1,2-DIPHENYLETHENE (TRANS)	000951-86-0
1,2-DICHLORO-1,2-DIPHENYLETHANE	005963-49-5
DICHLORODIPHENYLSILANE	000080-10-4
1,2-DICHLORO-1,2-DIVINYLCYCLOBUTANE	014112-00-6
1,1-DICHLOROETHANE	000075-34-3
1,2-DICHLOROETHANE	000107-06-2
2,2-DICHLOROETHANOL	000598-38-9

1,1-DICHLOROETHENE	000075-35-4
1,2-DICHLOROETHENE (CIS)	000156-59-2
1,2-DICHLOROETHENE (TRANS)	000156-60-5
(1,2-DICHLOROETHENYL)BENZENE	006607-45-0
N,N-DI-B-CHLOROETHYLANILINE	000553-27-5
4-(1,2-DICHLOROETHYL)-1,2-DICHLOROCYCLOHEXANE	051962-63-1
1,2-DICHLOROETHYLENE	000540-59-0
DI-2-CHLOROETHYL SULFIDE	000505-60-2
1,1-DICHLOROFLUOROETHANE	001717-00-6
DICHLOROFLUOROMETHANE	000075-43-4
3,4-DICHLOROHEXA-1,5-DIENE	083682-33-1
1,6-DICHLOROHEXA-1,5-DIENE	067546-51-4
2,2-DICHLORO-2,2,4,4,6,6-HEXAHYDRO-4,4,6,6-TETR*	065601-42-5
1,2-DICHLOROHEXANE	002162-92-7
2,5-DICHLOROHEXANE	013275-18-8
3,4-DICHLOROHEXA-1,3,5-TRIENE	083682-37-5
1,6-DICHLOROHEXA-1,3,5-TRIENE	069645-07-4
1,4-DICHLORO-2-HYDROXYBUTANE	002419-74-1
1,2-DICHLORO-4-ISOCYANATOBENZENE	000102-36-3
DICHLOROMETHANE	000075-09-2
DICHLOROMETHOTREXATE	000528-74-5
4,5-DICHLORO-2-METHOXYPHENOL	002460-49-3
(DICHLOROMETHYL)BENZENE	000098-87-3
3-DICHLOROMETHYLHYDROTHIAZIDE	023141-87-9
DICHLOROMETHYLPHENYLSILANE	000149-74-6
1,2-DICHLORO-2-METHYLPROPANE	000594-37-6
1,3-DICHLORO-2-METHYLPROPENE	003375-22-2
DICHLOROMETHYLSILANE	000075-54-7
1,8-DICHLORONAPHTHALENE	002050-74-0
2,3-DICHLORONAPHTHALENE	002050-75-1
2,7-DICHLORONAPHTHALENE	002198-77-8
2,6-DICHLORONAPHTHALENE	002065-70-5
1,2-DICHLORONAPHTHALENE	002050-69-3
1,5-DICHLORONAPHTHALENE	001825-30-5
1,4-DICHLORONAPHTHALENE	001825-31-6
2,4-DICHLORO-1-NAPHTHOL	002050-76-2
2,6-DICHLORO-4-NITROANILINE	000099-30-9
2,4-DICHLORONITROBENZENE	000611-06-3
3,4-DICHLORONITROBENZENE	000099-54-7
2,5-DICHLORONITROBENZENE	000089-61-2
2,3-DICHLORONITROBENZENE	003209-22-1
1,1-DICHLORO-1-NITROETHANE	000594-72-9
2,4-DICHLORO-6-NITROPHENOL	000609-89-2
2,6-DICHLORO-4-NITROPHENOL	000618-80-4
4,5-DICHLORO-2-NITROPHENOL	039224-65-2
2,5-DICHLORO-4-NITROPHENOL	005847-57-4
3,4-DICHLORO-N-NITROSOPIPERIDINE	057541-72-7
3,4-DICHLORO-N-NITROSOPYRROLIDINE	059863-59-1
2,4-DICHLORO-B-NITROSTYRENE	018984-21-9
1,2-DICHLORO-1,5-OCTADIENE	083682-61-5
2,6-DICHLORO-4-OCTYLPHENOL	073986-52-4
1,5-DICHLOROPENTANE	000628-76-2
1,10-DICHLOROPERFLUORO-2,9-DIMETHYLDECANE	103188-54-1
2,3-DICHLOROPHENOL	000576-24-9
3,4-DICHLOROPHENOL	000095-77-2
2,4-DICHLOROPHENOL	000120-83-2
2,5-DICHLOROPHENOL	000583-78-8
2,6-DICHLOROPHENOL	000087-65-0

3,5-DICHLOROPHENOL	000591-35-5
3,4-DICHLOROPHENOXYACETIC ACID	000588-22-7
2,3-DICHLOROPHENOXYACETIC ACID	002976-74-1
2,4-DICHLOROPHENOXYACETIC ACID	000094-75-7
3,5-DICHLOROPHENOXYACETIC ACID	000587-64-4
2-(2,4-DICHLOROPHENOXY)ETHANOL	000120-67-2
(4-(2,4-DICHLOROPHENOXY)PHENOXY)ACETIC ACID	051338-10-4
1-(3,5-DICHLOROPHENOXY)SILATRANE	090936-47-3
N-(3,4-DICHLOROPHENYL)ACETAMIDE	002150-93-8
3,4-DICHLOROPHENYLAMIDINOUREA	021724-58-3
4-((2,4-DICHLOROPHENYL)AZO)BENZENESULFONIC ACID*	062959-41-5
N-(3,4-DICHLOROPHENYL)BENZAMIDE	010286-75-6
4,5-DICHLORO-O-PHENYLENEDIAMINE	005348-42-5
1-[[2-(2,4-DICHLOROPHENYL)-4-ETHYL-1,3-DIOXOLAN-	060207-93-4
3-(3,4-DICHLOROPHENYL)-1-METHYLUREA	003567-62-2
2,4-DICHLORO-6-PHENYLPHENOL	005335-24-0
(3,4-DICHLOROPHENYL)UREA	002327-02-8
3,6-DICHLOROPICOLINIC ACID	001702-17-6
1,3-DICHLOROPROPANE	000142-28-9
1,2-DICHLOROPROPANE	000078-87-5
2,2-DICHLOROPROPANOL	063151-11-1
3,3-DICHLOROPROPANOL	083682-72-8
1,3-DICHLORO-2-PROPANOL	000096-23-1
2,3-DICHLOROPROPANOL	000616-23-9
1,1-DICHLOROPROPANONE	000513-88-2
1,1-DICHLOROPROPENE	000563-58-6
2,3-DICHLOROPROPENE	000078-88-6
1,3-DICHLOROPROPENE, (Z)	010061-01-5
3,3-DICHLOROPROPENE	000563-57-5
1,3-DICHLOROPROPENE (TRANS)	010061-02-6
1,2-DICHLOROPROPENE	000563-54-2
TRANS-1,3-DICHLOROPROPENE	000542-75-6
2,3-DICHLOROPROPIONIC ACID	000565-64-0
3,3-DICHLOROPROPYNE	025523-14-2
3,5-DICHLOROPYRIDINE	002457-47-8
2,3-DICHLOROPYRIDINE	002402-77-9
2,5-DICHLOROPYRIDINE	016110-09-1
2,6-DICHLOROPYRIDINE	002402-78-0
1,2-DICHLOROTETRAFLUOROETHANE	000076-14-2
1,1-DICHLORO-1,2,2,2-TETRAFLUOROETHANE	000374-07-2
2,5-DICHLOROTHIOPHENE	003172-52-9
3,4-DICHLOROTOLUENE	000095-75-0
3,5-DICHLOROTOLUENE	025186-47-4
2,4-DICHLOROTOLUENE	000095-73-8
2,6-DICHLOROTOLUENE	000118-69-4
3,5-DICHLORO-2-(TRICHLOROMETHYL)PYRIDINE	001128-16-1
2,4-DICHLORO(TRIFLUOROMETHYL)BENZENE	000320-60-5
DICHLORO(TRIFLUOROMETHYL)BENZOPHENONE	095998-69-9
1,4-DICHLORO-4-VINYLCYCLO-1-HEXENE	065122-21-6
DICHLORPROP	000120-36-5
DICHLORVOS	000062-73-7
DICIRENONE	041020-79-5
3,4-DICLBENZALDEHYD,O-((MEAM)CO)OXIME	085879-18-1
2-(26-DICL-4-BRPHIMINO)IMIDAZOLIDINE	040065-09-6
4-CL-3-CF3-C6H3NHN=C(CN)COOET	028313-57-7
2,6-DICL-4-CF3-C6H2NHN=C(CN)COOME	028313-69-1
2,4-DICL CF3-METHANESULFONANILIDE	023383-96-2
3,5-DICL-C6H3NHN=C(CN)COOET	028313-59-9

3,4-DICL-C6H3NHN=C(CN)COOET	036865-77-7
3,5-DICL-C6H3NHN=C(CN)CO-OME	028313-58-8
3,4-DICL-C6H3NHN=C(CN)CO-ME	028317-61-5
3,5-DICL-C6H3NHN=C(CN)CO-ME	028317-62-6
3,5-DICL-2',4'-DIFLUORO SALICYLANILIDE	080033-99-4
3,5-DICL-4'-F SALICYLANILIDE	054850-02-1
2,4-DICL-5-ME-N-ME PHENYLCARBAMATE	002752-68-3
3,5-DICL-2'-ME-4'-NO2 SALICYLANILIDE	090426-03-2
2(2,6-DICL-4MEOPHIMINO)IMIDAZOLIDINE	065936-24-5
2(2,6-DICL-4MEPH IMINO)IMIDAZOLIDINE	004201-33-6
2(2,3-DICL-6MEPH IMINO)IMIDAZOLIDINE	082801-84-1
2(2,4-DICL-6MEPH IMINO)IMIDAZOLIDINE	004201-34-7
2-(26-DICL-4-NO2-PHIMINO)IMIDAZOLIDN	065936-26-7
DICLOBUTRAZOL ISOMERS	075736-33-3
DICLOFENAC	015307-86-5
DICLOFOP	040843-25-2
DICLOFOP-METHYL	051338-27-3
2(2,6-DICL-4-OHPH IMINO)IMIDAZOLIDIN	057101-48-1
DICLOXACILLIN	003116-76-5
2(2,3-DICLPHENYLIMINO)IMIDAZOLINE	015327-44-3
1-(3,4-DICLPHENYL)-3-PHENYLUREA	002008-73-3
2-(2,6-DICL PH IMINO)IMIDAZOLIDINE	004205-90-7
2-(2,5-DICL PH IMINO)IMIDAZOLIDINE	056514-55-7
2-(2,4-DICL PH IMINO)IMIDAZOLIDINE	023830-88-8
2',4'-DICL SALICYLANILIDE	037183-28-1
3,5-DICL SALICYLANILIDE	004214-48-6
DICOFOL	000115-32-2
DICOUMAROL	000066-76-2
DICROTOPHOS	000141-66-2
DICRYL	002164-09-2
1,3-DICYANOBENZENE	000626-17-5
1,2-DICYANOBENZENE	000091-15-6
2,5-DICYANO-C6H3NHN=C(CN)COOME	036874-72-3
1-(3,4-DICYANOPH)-33-DIME TRIAZENE	066974-82-1
1(4BB-DICYANOVINYL)PH)-3MEO3ME UREA	088132-20-1
DICYCLOHEXYL	000092-51-3
DICYCLOHEXYL PHTHALATE	000084-61-7
DICYCLOPENTADIENE	000077-73-6
DIDECYL PHTHALATE	000084-77-5
DIDEOXY-ARA-A,2,N6-DIMETHYL-2'-FLUORO	132722-92-0
DIDODECYLPHTHALATE	002432-90-8
DIELDRIN	000060-57-1
NN-DIME-4-(N-ME-N-FORMYL)BENZENE-SO2N	077812-87-4
DIENOCHLOR	002227-17-0
DIEPOXYBUTANE	001464-53-5
1,2:3,4-DIEPOXYBUTANE	000564-00-1
1,2:3,4-DIEPOXYBUTANE (2R,3R)	030419-67-1
1,2:3,4-DIEPOXYBUTANE (2S,3S)	030031-64-2
1,2:3,4-DIEPOXYBUTANE DL	000298-18-0
9((DIETAMINO)ACETYLAMINO) FLUORENE	072336-12-0
2-N,N-DIETAMINO-1-PHENYLETHANOL	004249-64-3
2(5BR-2-PYRIDYLAZO)-5(DIETAM)PHENOL	014337-53-2
5,5-DIET BARBITURIC ACID,N1-IPR	085432-35-5
OO-DIET-O-(3-CL-4-MES PH)PHOSPHATE	026798-03-8
N,N'-DIET-ETHYLENETHIOUREA	030826-80-3
DIETHANOLAMINE	000111-42-2
M-DIETHENYLBENZENE	000108-57-6
1,1-DIETHOXYETHANE	000105-57-7

1,2-DIETHOXYETHANE	000629-14-1
DIETHOXYMETHANE	000462-95-3
2,2-DIETHOXYPROPANE	000126-84-1
DIETHYLACETAMIDE	000685-91-6
2,6-DIETHYLACETANILIDE	016665-89-7
DIETHYLAMINE	000109-89-7
DIETHYLAMINE, HYDROBROMIDE	006274-12-0
DIETHYLAMINE, HYDROIODIDE	019833-78-4
1-(N,N-DIETHYLAMINO-ACETYL-AMINO)INDANE	006514-54-1
3-(DIETHYLAMINO)BENZENESULFONIC ACID, SODIUM SAL	005123-63-7
P-DIETHYLAMINOBENZYL ALCOHOL	074974-49-5
N-DIETHYLAMINOETANILINE,3-CL-4-ME	002519-75-7
2-DIETHYLAMINO-4-ETHYLAMINO-6-METHOXY-S-TRIAZI*	013532-26-8
N-(2-DIETHYLAMINOETHYL)-BENZAMIDE	003690-53-7
2-DIETHYLAMINOISONIAZID	019353-98-1
DIETHYLAMPHENICOL	049648-49-9
2,6-DIETHYLANILINE	000579-66-8
N,N-DIETHYLANILINE	000091-66-7
DIETHYL AZELATE	000624-17-9
N,N-DIETHYLBENZAMIDE	001696-17-9
P-DIETHYLBENZENE	000105-05-5
O-DIETHYLBENZENE	000135-01-3
M-DIETHYLBENZENE	000141-93-5
N,N-DIETHYL-1,4-BENZENEDIAMINE	000093-05-0
DIETHYLCARBAMODITHIOIC ACID, SODIUM SALT	000148-18-5
DIETHYL CARBONATE	000105-58-8
S,S-DIETHYL CARBONODITHIOATE	000623-80-3
DIETHYL DIMETHYLDIPHOSPHONATE	032288-17-8
DIETHYL DIMETHYLPHOSPHORAMIDATE	002404-03-7
DIETHYL DIPHENYL UREA	000085-98-3
DIETHYL DISULFIDE	000110-81-6
DIETHYLENE GLYCOL	000111-46-6
DIETHYLENE GLYCOL DIACRYLATE	004074-88-8
DIETHYLENE GLYCOL DIBENZOATE	000120-55-8
DIETHYLENE GLYCOL DIBUTYL ETHER	000112-73-2
DIETHYLENE GLYCOL DIMETHACRYLATE	002358-84-1
DIETHYLENE GLYCOL MOMOLAURYL ETHER SODIUM SULFAT	003088-31-1
DIETHYLENE GLYCOL MONOBUTYL ETHER ACETATE	000124-17-4
DIETHYLENE GLYCOL MONOETHYL ETHER	000111-90-0
DIETHYLENE GLYCOL MONO-N-BUTYL ETHER	000112-34-5
DIETHYLENE GLYCOL MONOMETHYL ETHER ACETATE	000629-38-9
DIETHYLENE GLYCOL MONOMETHYL ETHER	000111-77-3
DIETHYLENETRIAMINE	000111-40-0
DIETHYLENETRIAMINEPENTAACETIC ACID, PENTA SODIU*	000140-01-2
DIETHYL ETHER	· 000060-29-7
N,N-DIETHYLFORMAMIDE	000617-84-5
DIETHYL GLUTARATE	000818-38-2
DIETHYLHEXANEDIOATE	000141-28-6
DI-2-ETHYLHEXYL ADIPATE	000103-23-1
N,N'-DIETHYLHYDRAZINE	001615-80-1
3,5-DIETHYL-4-HYDROXYACETANILIDE	055205-89-5
DIETHYLHYDROXYLAMINE	003710-84-7
DIETHYL MALEATE	000141-05-9
DIETHYL MALONATE	000105-53-3
DIETHYLMALONATE, BENZAL	005292-53-5
DIETHYLMALONATE,4-BR BENZAL	022399-01-5
DIETHYLMALONATE,3-BUTOXYBENZAL	015725-39-0
DIETHYLMALONATE,3-CL BENZAL	006768-21-4

DIETHYLMALONATE,2-CL BENZAL	006768-20-3
DIETHYLMALONATE,4-CL BENZAL	006827-40-3
DIETHYLMALONATE, 3-ETHOXYBENZAL	015725-37-8
DIETHYLMALONATE,4-F BENZAL	000790-53-4
DIETHYLMALONATE,2-F BENZAL	002262-52-4
DIETHYLMALONATE,3-ME BENZAL	015725-33-4
DIETHYLMALONATE,2-ME BENZAL	024331-75-7
DIETHYLMALONATE,4-MEO BENZAL	006768-23-6
DIETHYLMALONATE,3-MEO BENZAL	006771-54-6
DIETHYLMALONATE,2-MEO BENZAL	006768-22-5
DIETHYLMALONATE,3-MEO 4-OH BENZAL	024331-83-7
DIETHYLMALONATE,3-NO2 BENZAL	006331-45-9
DIETHYLMALONATE,2-NO2 BENZAL	017422-56-9
DIETHYLMALONATE,4-NO2 BENZAL	022399-00-4
DIETHYLMALONATE,3-PRO BENZAL	015725-38-9
1,3-DIETHYL-5-METHYLBENZENE	002050-24-0
DIETHYL METHYLPHOSPHONATE	000683-08-9
DIETHYL 3-METHYL-5-PYRAZOLYL PHOSPHATE	000108-34-9
N,N-DIETHYL-2-(1-NAPHTHALENYLOXY)PROPANAMIDE	015299-99-7
3,3-DIETHYLPENTANE	001067-20-5
2,2-DIETHYL-1-PENTANOL	014202-62-1
O,O-DIETHYL-O-PHENYLPHOSPHATE	002510-86-3
1,3-DIETHYL-8-PHENYLXANTHINE	075922-48-4
DIETHYL PHOSPHITE	000762-04-9
O,S-DIETHYLPHOSPHORAMIDOTHIOATE	016271-10-6
DIETHYL PHTHALATE	000084-66-2
DIETHYL PIMELATE	002050-20-6
2,2-DIETHYL-1,3-PROPANEDIOL	000115-76-4
N,N-DIETHYL-3-PYRIDYLETHYLAMINE	020173-34-6
N,N-DIETHYL-3-PYRIDYLMETHYLAMINE	002055-14-3
DIETHYL SEBACATE	000110-40-7
DIETHYL SUBERATE	002050-23-9
N,N-DIETHYL SUCCINAMIC ACID	001522-00-5
DIETHYL SUCCINATE	000123-25-1
DIETHYL SULFATE	000064-67-5
DIETHYLSULFIDE	000352-93-2
DIETHYL SULFONE	000597-35-3
N,N-DIETHYLTHIOUREA	000105-55-5
N,N-DIETHYLUREA	050816-31-4
N6,N6-DIET-TETRAME-MELAMINE	016268-75-0
O,O-DIET-O-(3-ME-4-SMEPH)P=O	004799-59-1
O,O-DIET-O-(4-METHIOPH)PHOSPHATE	003070-13-1
DIET-METHYL-PHENYL-AMMONIUM IODIDE	001007-67-6
OO-DIET O-2-METHYLTHIOPH P=S	002670-77-1
A,A-DIET-PHENYLACETAMIDE,3,4-DIOXYME	023742-02-1
2-(2,6-DIETPHENYLIMINO)IMIDAZOLIDINE	004751-48-8
O,O-DIET-O-PHENYLPHOSPHOROTHIOATE	032345-29-2
2,4-DIF 3,5-DICL ACETANILIDE	127836-03-7
DIFENOXURON	014214-32-5
2-(2,6-DIFL PH IMINO)IMIDAZOLIDINE	059772-33-7
DIFLUBENZURON	035367-38-5
DIFLUMIDONE	022736-85-2
DIFLUNISAL	022494-42-4
2,6-DIFLUOROACETANILIDE	003896-29-5
DIFLUOROAMPHENICOL	049648-37-5
2,4-DIFLUOROANILINE	000367-25-9
2,6-DIFLUOROBENZAMIDE	018063-03-1
1,2-DIFLUOROBENZENE	000367-11-3

M-DIFLUOROBENZENE	000372-18-9
P-DIFLUOROBENZENE	000540-36-3
2,3-DIFLUORODIBENZO-P-DIOXIN	050585-38-1
1-(2,6-DIFLUORO),2-DIBENZOYL-1-TERT-BUTYLHYDRAZI	112226-86-5
ALPHA,ALPHA-DIFLUORODICHLOROTRIFLUOROMETHYL DIP*	095998-70-2
1,2-DIFLUOROETHANE	000624-72-6
1,1-DIFLUOROETHANE	000075-37-6
1,1-DIFLUOROETHENE	000075-38-7
DIFLUOROMETHANE	000075-10-5
DIFLUOROMETHANESULFONANILIDE	000658-43-5
DIFLUOROMETHANESULFONANILIDE,P-CL	001513-31-1
DIFLUOROMETHOXYBENZENE	000458-92-4
DIFLUOROMETHYLPHENYLSILANE	000328-57-4
1,3-DIFLUORO-2-PROPANOL	000453-13-4
2,6-DIFLUOROPYRAZINE	033873-09-5
2,6-DIFLUOROPYRIDINE	001513-65-1
2,4-DIF CF3-METHANESULFONANILIDE	023384-22-7
DIFORMYLHYDRAZINE	000628-36-4
DIGITOXIGENIN	000143-62-4
DIGITOXIGENIN-3-A-SULFATE	002488-78-0
DIGITOXIGENIN-3-B-SULFATE	002488-80-4
DIGITOXIN	000071-63-6
DIGITOXIN, DIHYDRO-	003786-76-3
DIGITOXIN, BETA-METHYL	055692-91-6
DIGOXIGENIN	001672-46-4
DIGOXIN	020830-75-5
DIGOXIN,12-ACETYL-BETA-METHYL-	051740-69-3
DIGOXIN,ALPHA-BETA-DIMETHYL-	051740-66-0
DIGOXIN ESTER DERIVATIVE	005511-98-8
DIGOXIN, ALPHA-METHYL	031962-94-4
DIGOXIN (NON-OLEFINIC BOND)	005297-10-9
DIHEPTYL PHTHALATE	003648-21-3
DIHEXYL ADIPATE	000110-33-8
DI-N-HEXYL ETHER	000112-58-3
DIHEXYL PHTHALATE	000084-75-3
9,10-DIHYDROANTHRACENE	000613-31-0
2,3-DIHYDROBENZOFURAN	000496-16-2
1,3-DIHYDRO-BENZOTHIADIAZOL,22-O2	001615-06-1
21,22-DIHYDROBRUCINE	028879-93-8
6,7-DIHYDROCANRENONE	000976-70-5
DIHYDROCODEINONE	000125-29-1
9,10-DIHYDRO-3,4-DIHYDROXY-9,10-DIOXO-2-ANTHRA*	000130-22-3
2,5-DIHYDROFURAN	001708-29-8
A-DIHYDROGRAYANOTOXIN	038776-76-0
BETA-DIHYDROHEPTACHLOR	014168-01-5
1,2-DIHYDRONAPHTHALENE	000447-53-0
3,4-DIHYDRO-1(2H)-NAPHTHALENONE	000529-34-0
4,5-DIHYDRONICOTYRINE	000525-74-6
2-(1,3-DIHYDRO-3-OXO-2H-INDOL-2-YLIDENE)-1,2-DI*	000482-89-3
9,10-DIHYDROPHENANTHRENE	000776-35-2
DIHYDROSAFROLE	000094-58-6
21,22-DIHYDROSTRYCHNINE	015006-14-1
2,5-DIHYDROTHIOPHENE 1,1-DIOXIDE	000077-79-2
3,4-DIHYDRO-6-(TRIFLUOROMETHYL)-2H(1,2,4)-BENZO*	000135-09-1
DIHYDROXY ACETONE	000096-26-4
1,4-DIHYDROXY-9,10-ANTHRACENEDIONE	000081-64-1
1,5-DIHYDROXYANTHRAQUINONE	000117-12-4
2,6-DIHYDROXY-ANTHRAQUINONE	000084-60-6

3,5-DIHYDROXYBENZOIC ACID	000099-10-5
2,6-DIHYDROXYBENZOIC ACID	000303-07-1
2,5-DIHYDROXYBENZOIC ACID	000490-79-9
2,4-DIHYDROXYBENZOIC ACID	000089-86-1
2,3-DIHYDROXYBENZOIC ACID	000303-38-8
3,4-DIHYDROXYBENZOIC ACID	000099-50-3
1,2-DIHYDROXY-2-BUTENE	050317-11-8
2(4,4'-DIHYDROXYDIPHENLMETHYL)THIAZOLE	013004-52-9
4,4'-DIHYDROXYDIPHENYLMETHANE	000620-92-8
N,N-DI(2-HYDROXYETHYL)LAURAMIDE	000120-40-1
1,5-DIHYDROXYNAPHTHALENE	000083-56-7
1,2-DIHYDROXY-4-NITROBENZENE	003316-09-4
3,4-DIHYDROXYPHENYLACETIC ACID	000102-32-9
3,4-DIHYDROXYPHENYLGLYCOL	003343-19-9
4,8-DIHYDROXYQUINOLINE	014959-84-3
4,4'-DIHYDROXYTRIPHENYLMETHANE	004081-02-1
2,4-DIIDORESORCYL-1,3-DIGLUCOSIDE	060406-79-3
4,6-DIIDORESORCYL-1,3-DIGLUCOSIDE	060331-52-4
1,2-DIIODOBENZENE	000615-42-9
1,3-DIIODOBENZENE	000626-00-6
1,4-DIIODOBENZENE	000624-38-4
1,2-DIIODOETHENE (CIS)	000590-26-1
1,2-DIIODOETHENE (TRANS)	000590-27-2
3,5-DIIODOSALICYLIC ACID	000133-91-5
3,5-DIIODOTYROSINE	000066-02-4
3,5-DI(IPR)-N-ME-PHENYLCARBAMATE	000330-64-3
N-(N"-DI-IPR-N'-ME-URYLTHIO)METHOMYL	074385-89-0
DIISOAMYL SULFIDE	000544-02-5
DIISOBUTYLAMINE	000110-96-3
DI-ISOBUTYL ETHER	000628-55-7
DI-ISOBUTYLPHTHALATE	000084-69-5
DIISOCTYL PHOSPHATE	027215-10-7
DIISODECYL PHTHALATE	026761-40-0
DIISOHEXYL PHTHALATE	000146-50-9
DIISONONYL PHTHALATE	028553-12-0
DIISOOCTYL PHTHALATE	027554-26-3
DIISOPENTYL ETHER	000544-01-4
DIISOPROPANOLAMINE	000110-97-4
2,6-DIISOPROPYLACETANILIDE	016637-13-1
DIISOPROPYLAMINE	000108-18-9
1,3-DIISOPROPYLBENZENE	000099-62-7
1,2-DIISOPROPYLBENZENE	000577-55-9
1,4-DIISOPROPYLBENZENE	000100-18-5
DIISOPROPYL CARBAMATE	002303-17-5
4,4'-DIISOTHIOCYANATEBIPHENYL	018705-45-8
DIKETENE	000674-82-8
DIKETOMETRIBUZIN	056507-37-0
DILOXANIDE	000579-38-4
DILOXANIDE ACETATE	099854-42-9
DILOXANIDE BENZOATE	092854-81-4
DILOXANIDE FUROATE	003736-81-0
DILTIAZEM	033286-22-5
DIMALONE	039589-98-5
1,1-DIME-3-(M-ACETYLPHENYL)UREA	042865-65-6
M-DIMEAMINOBZY N,N-DIME CARBAMATE	084640-28-8
2(DIMEAMINO)-N-(2,6-DIMEPH)ACETAMIDE	021236-54-4
3-ME,2-(N,N-DIMEAMINOME)FURAN	020863-54-1
2-(N,N-DIME AMINOMETHYL)THIOPHENE	026019-17-0

1-DIME AMINO-5-SO2NH2-NAPHTHALENE	001431-39-6
2-N,N-DIMEAMINO-1-PHENYLETHANOL	006853-14-1
1,1-DIME-3-(3-AMINOPHENYL)UREA	039938-79-9
N1-(4-DIMEAMINOPH)SULFANILAMIDE	034392-61-5
2-(N,N-DIMEAMINOPROPYL)FURAN	049547-83-3
7,12-DIME-BENZ(A)ANTHRACENE	000057-97-6
1,1-DIME-3-(4-BENZOYLPHENYL)UREA	061706-06-7
N,N-DIME CARBAMATE,P-BENZYLOXYBENZYL	084640-30-2
3-(N,N-DIMECARBAM)-2-BR-PYRIDINE	051581-35-2
3-(N,N-DIMECARBAM)-2-CL-PYRIDINE	051581-34-1
3-(N,N-DIMECARBAM)-2-NO2-PYRIDINE	051581-33-0
3-(N,N-DIMECARBAM)-2-I-PYRIDINE	051581-36-3
N,N-DIME-2-CARBOXAMIDE-THIOPHENE	030717-57-8
3,3-DIME-1-(3-CF3 PH)TRIAZENE	000402-38-0
2(24-DIME-6-CL-PH IMINO)IMIDAZOLINE	004201-36-9
O,O-DIME O-(2-CN-4-CL)PH PHOSPHORTHIOATE	114012-04-3
DIMECRON	013171-21-6
3,3-DIME-1-(4-CYANOPHENYL)TRIAZENE	023456-95-3
1,1-DIME-3-(4-CYCLOHXPHENYL)UREA	088132-41-6
OO-DIME DICHLOROVINYLPHOSPHONATE	001185-97-3
1,1-DIME-3-(3,5-DIME PH) UREA	036627-56-2
2,6-DIME-N,N'-DINITROSOPIPERAZINE	055380-34-2
2,5-DIME-N,N'-DINITROSOPIPERAZINE	055556-88-2
DIMEFLINE	001165-48-6
DIMEFOX	000115-26-4
DIMEFURON	034205-21-5
N-(N''-DI-ME-N'-HEX-URYLTHIO)METHOMYL	074399-87-4
O,S-DIME-N-HEXYLPHOSPHOEAMIDITHIOATE	052067-52-4
2,6-DIME-4-HYDROXYACETANILIDE	006337-56-0
2,3-DIME-4-HYDROXYACETANILIDE	002011-57-6
2,5-DIME-4-HYDROXYACETANILIDE	069477-71-0
1,1-DIME-3-(3-HYDROXYPHENYL)UREA	004849-46-1
2(15DIME-13-IMIDAZOLINYLIDEN-2AM)-5NO2-THIAZOLE	031052-84-3
2(13DIME-13-IMIDAZOLINYLIDEN-2AM)-5NO2-THIAZOLE	031052-79-6
DIMELIN	000968-81-0
N,N-DIME-4-MEAM BENZENESULFONAMIDE	077812-86-3
NN-DIME-4(N-ME-N(2BRPROPIONYL)BZSO2N	075852-13-0
NN-DIME-4(N-ME-N-BUTROYL)BZ SULFONAMIDE	075852-14-1
NN-DIME-4(N-ME-N(2CLPROPIONYL)BZSO2N	075852-11-8
NN-DIME-4(N-ME-N(2I PROPIONYL)BZSO2N	075852-16-3
NN-DIME-4(N-ME-N(2MEOPROPIONYL)BZSO2N	075852-05-0
NN-DIME-4(N-ME-N(2MESPROPIONYL)BZSO2N	075852-12-9
NN-DIME-4(N-ME-N(2NO2PROPIONYL)BZSO2N	075852-04-9
NN-DIME-4(N-ME-N(PROPIONYL)BZSO2N	074746-43-3
N-(N''-DI-ME-N'-ME-URYLTHIO)METHOMYL	074385-72-1
1,3-DIME-1-NITROSO-3-PHENYLUREA	072586-68-6
3,5-DIME-N-NITROSOPIPERIDINE	065445-59-2
NN-DIME4(N-ME-N(2MESOPROPIONYL)BZSO2N	075851-96-6
2(N,N-DIME)-5(5-NO2-2-FURFURYL)THIAZOLONE	025694-31-9
2,4-DIME 6-NO2 ACETANILIDE	000606-38-2
3,5-DIMEO-4-BROMOPHENETHYLAMINE	061367-72-4
2,5-DIMEO-4-NO2-AMPHETAMINE	067460-68-8
1,1-DI(P-MEO PHENYL)-2-NITROBUTANE	085078-20-2
2(25-DIMEO PH IMINO)IMIDAZOLIDINE	050531-51-6
3,5-DIMEO-4-PROPOXYPHENETHYLAMINE	039201-78-0
3,5-DIMEO-4-I-PROPOXYPHENETHYLAMINE	061367-70-2
2,5-DIMEO-4-PROPYLAMPHETAMINE	064638-08-0
2,5-DIMEO-4-T-BU-AMPHETAMINE	041538-42-5

1,2-DIMEO TETRACHLOROCYCLOHEXANE	060132-40-3
2,3-DIMEO TETRACHLOROCYCLOHEXANE	056400-12-5
3,6-DIMEO TETRACHLOROCYCLOHEXANE	056400-36-3
5,7-DIME-2-OXO-1,3-BENZOXATHIOL	015062-75-6
1,1-DIME-3-(P-T-BUTYLPHENYL)UREA	032745-69-0
3,5-DIMEO-4-BUO-PHENETHYLAMINE	064778-75-2
3,5-DIMEO-4-ETO-PHENETHYLAMINE	039201-82-6
1,1-DIME-3-(3-PHENYLETOPH) UREA	070859-35-7
2-(2,4-DIME PHENYLIMINO)IMIDAZOLINE	004794-83-6
2-(2,6-DIME PHENYLIMINO)IMIDAZOLINE	004859-06-7
3,4-DIMEPHENYL N-METHYLCARBAMATE	002425-10-7
N,N-DIME-N-G-PHENYLPROPYLAMINE	001199-99-1
N(26DIMEPH)-N-(2FURANOYL)ALANINE,ME	057646-30-7
2,2-DIME PROPIONIC ACID HYDRAZIDE	017883-59-9
3,4-DIME PYRIDINIUM I,N-HEXADECYL	088477-47-8
3,5-DIME PYRIDINIUM I,N-HEXADECYL	088497-75-0
2,4-DIME PYRIDINIUM I,N-HEXADECYL	078191-89-6
2,4-DIME PYRIDINIUM IODIDE,N-DECYL	088477-50-3
3,5-DIME PYRIDINIUM IODIDE,N-DECYL	088477-56-9
3,4-DIME PYRIDINIUM IODIDE,N-DECYL	088477-44-5
3,5-DIME PYRIDINIUM IODIDE,N-DODECYL	088477-57-0
2,4-DIME PYRIDINIUM IODIDE,N-DODECYL	088477-51-4
3,4-DIME PYRIDINIUM IODIDE,N-DODECYL	088477-45-6
3,4-DIME PYRIDINIUM IODIDE,N-HEXYL	088477-43-4
3,5-DIME PYRIDINIUM IODIDE,N-HEXYL	088477-54-7
2,4-DIME PYRIDINIUM IODIDE,N-OCTYL	088477-49-0
3,4-DIME PYRIDINIUMIODIDE,N-OCTYL	053242-40-3
2,4-DIME PYRIDINIUM I,N-TETRADECYL	088477-52-5
3,5-DIME PYRIDINIUM I,N-TETRADECYL	088477-58-1
3,4-DIME PYRIDINIUM I,N-TETRADECYL	088477-46-7
4-(N,N-DIME)-3-PYRIDYLBUTYLAMINE	001441-44-7
N,N-DIME-2-(3-PYRIDYL)ETHYLAMINE	020173-26-6
N(2,5-DIME-PYRROL-1-YL)-6-($-MORPHOL)-3-PYRIDAZ-	089937-39-3
N,N-DIME-QUINOXALINE-2,3-DIONE	058175-07-8
DIMERIN	000125-64-4
DIMETAN	000122-15-6
DIMETHAMRTRYNE	022936-75-0
DIMETHENAMID	087674-68-8
N7-(4,5-DIME THIAZOL-2-YL)-MITOMYCIN C	088854-60-8
N,N-DIME THIOPHENE-3-CARBOXAMIDE	059906-37-5
DIMETHIPIN	055290-64-7
DIMETHOATE	000060-51-5
DIMETHOMORPH	110488-70-5
DIMETHOXON	001113-02-6
2,4-DIMETHOXYAMPHETAMINE	023690-13-3
2,5-DIMETHOXYAMPHETAMINE	002801-68-5
2,5-DIMETHOXYANILINE	000102-56-7
3,3'-DIMETHOXYAZOBENZENE	006319-23-9
2,4-DIMETHOXYAZOBENZENE	029418-46-0
2,6-DIMETHOXYAZOBENZENE	029418-48-2
2,4'-DIMETHOXYAZOBENZENE	029418-47-1
4,4'-DIMETHOXY-O,O'-AZOTOLUENE	029418-54-0
4,4'-DIMETHOXYAZOXYBENZENE	001562-94-3
2,6-DIMETHOXYBENZAMIDE	021864-67-5
P-DIMETHOXYBENZENE	000150-78-7
1,2-DIMETHOXYBENZENE	000091-16-7
M-DIMETHOXYBENZENE	000151-10-0
3,3'-DIMETHOXYBENZIDINE	000119-90-4

3,5-DIMETHOXYBENZOIC ACID	001132-21-4
3,4-DIMETHOXYBENZOIC ACID	000093-07-2
2,6-DIMETHOXYBENZOIC ACID	001466-76-8
2,3-DIMETHOXYBENZOIC ACID	001521-38-6
2,4-DIMETHOXYBENZOIC ACID	000091-52-1
2,6-DIMETHOXY-P-BENZOQUINONE	000530-55-2
2,3-DIMETHOXY-P-BENZOQUINONE	003117-02-0
3,5-DIMETHOXY-4-BR AMPHETAMINE	064778-79-6
2,-5-DIMETHOXY-4-BR AMPHETAMINE	032156-26-6
1,4-DIMETHOXY-2-CHLOROBENZENE	002100-42-7
1-(3,4-DIMETHOXY),2-DIBENZOYL-1-TERT-BUTYLHYDRAZ	112427-66-4
2,3-DIMETHOXY-5,6-DIMETHYLBENZOQUINONE	000483-54-5
1,1-DIMETHOXYETHANE	000534-15-6
2,5-DIMETHOXY-4-ETHYLAMPHETAMINE	022004-32-6
2,5-DIMETHOXY-4-MES AMPHETAMINE	061638-07-1
DIMETHOXYMETHANE	000109-87-5
2,5-DIMETHOXY-4-METHYLAMPHETAMINE	015588-95-1
2,6-DIMETHOXY-2'-METHYLAZOBENZENE	029418-51-7
2,6-DIMETHOXY-4'-METHYLAZOBENZENE	029418-52-8
4,4'-DIMETHOXY-2-METHYLAZOBENZENE	029418-53-9
2,4-DIMETHOXY-2'-METHYLAZOBENZENE	029418-49-3
2,4-DIMETHOXY-4'-METHYLAZOBENZENE	029418-50-6
1,3-DIMETHOXY-4-METHYLBENZENE	038064-90-3
1,3-DIMETHOXY-2-METHYLBENZENE	005673-07-4
2,5-DIMETHOXY-4-PENTYLAMPHETAMINE	064638-10-4
3,4-DIMETHOXYPHENETHYLAMINE	000120-20-7
3,5-DIMETHOXYPHENOL	000500-99-2
2,6-DIMETHOXYPHENOL	000091-10-1
3,5-DIMETHOXYPHENYLHIPPURATE	081592-16-7
1,1-DI(P-METHOXYPHENYL)-2-NITROETHANE	085078-27-9
1,4-DIMETHOXYPHTHALAZINE	057315-37-4
1,2-DIMETHOXYPROPANE	007778-85-0
2,2-DIMETHOXYPROPANE	000077-76-9
O,N-DIME THPO	095579-22-9
DIMETHRIN	000070-38-2
N,N'-DIMETHYLACETAMIDE	000127-19-5
2,5-DIMETHYLACETANILIDE	002050-44-4
2,6-DIMETHYLACETANILIDE	002198-53-0
3,4-DIMETHYLACETANILIDE	002198-54-1
2,4-DIMETHYLACETANILIDE	002050-43-3
2,3-DIMETHYLACETANILIDE	000134-98-5
3,5-DIMETHYLACETANILIDE	002050-45-5
N,N-DIMETHYL,2-ACETYLPRYIDINE-THIOCARBONOHYDRAZO	096860-23-0
DIMETHYLAMINE	000124-40-3
4-(N,N-DIMETHYLAMINO)AZOBENZENE	000060-11-7
M-DIMETHYLAMINOBENZAMIDE	033225-17-1
P-DIMETHYLAMINOBENZAMIDE	006083-47-2
P-(DIMETHYLAMINO)BENZENETHIOL	004946-22-9
4-DIMETHYLAMINO BENZOIC ACID	000619-84-1
2-DIMETHYLAMINOETHANOL	000108-01-0
2-(DIMETHYLAMINO)ETHYL 2-METHYL-2-PROPENOATE	002867-47-2
4-DIMETHYLAMINO-3'-METHYLAZOBENZENE	000055-80-1
1-(DIMETHYLAMINO)-2-METHYL-2-BUTANOL BENZOATE	000644-26-8
N-(DIMETHYLAMINOMETHYL)BENZAMIDE	059917-58-7
2-((4-(DIMETHYLAMINO)PHENYL)AZO)BENZOIC ACID	000493-52-7
3-((4-(DIMETHYLAMINO)PHENYL)AZO)BENZOIC ACID	020691-84-3
4-((4-(DIMETHYLAMINO)PHENYL)AZO)BENZOIC ACID	006268-49-1
S,S'-(2-(DIMETHYLAMINO)-1,3-PROPANEDIYL) CARBAM*	015263-52-2

2-DIMETHYLAMINOPYRAZINE	005214-29-9
2-DIMETHYLAMINOPYRIDINE	005683-33-0
4-DIMETHYLAMINOPYRIDINE	001122-58-3
7-DIMETHYLAMINOQUINOLINE	089770-32-1
8-DIMETHYLAMINOQUINOLINE	029526-42-9
DIMETHYLAMPHENICOL	018048-95-8
2,6-DIMETHYLANILINE	000087-62-7
2,5-DIMETHYLANILINE	000095-78-3
2,4-DIMETHYLANILINE	000095-68-1
2,3-DIMETHYLANILINE	000087-59-2
3,5-DIMETHYLANILINE	000108-69-0
N,N-DIMETHYLANILINE	000121-69-7
N,N-DIMETHYL O-ANISIDINE	000700-75-4
N,N-DIMETHYL P-ANISIDINE	000701-56-4
2,6-DIMETHYLANISOLE	001004-66-6
9,10-DIMETHYLANTHRACENE	000781-43-1
2,6-DIMETHYLANTHRACENE	000613-26-3
2,4-DIMETHYLAZOBENZENE	029418-21-1
7,11-DIMETHYLBENZ(A)ANTHRACENE	035187-28-1
N,N-DIMETHYLBENZAMIDE	000611-74-5
2,6-DIMETHYLBENZAMIDE	055321-98-7
7,9-DIMETHYLBENZ(C)ACRIDINE	000963-89-3
N,N-DIMETHYLBENZENESULFONAMIDE	014417-01-7
5,6-DIMETHYLBENZIMIDAZOLE	000582-60-5
2,6-DIMETHYLBENZOIC ACID	000632-46-2
2,6-DIMETHYLBENZONITRILE-N-OXIDE	019111-74-1
P,P'-DIMETHYLBENZOPHENONEIMINE	016620-75-0
2,6-DIMETHYLBENZOPHENONEIMINE	022627-02-7
O,O'-DIMETHYLBENZOPHENONEIMINE	022627-01-6
2,2-DIMETHYLBENZOPYRAN	002513-25-9
2,4-DIMETHYLBENZYL ALCOHOL	016308-92-2
N,N-DIMETHYLBENZYLAMINE	000103-83-3
3,3'-DIMETHYLBIPHENYL	000612-75-9
2,3-DIMETHYL-1,3-BUTADIENE	000513-81-5
2,3-DIMETHYLBUTANE	000079-29-8
2,2-DIMETHYLBUTANE	000075-83-2
2,3-DIMETHYL-2-BUTANOL	000594-60-5
2,2-DIMETHYL-1-BUTANOL	001185-33-7
3,3-DIMETHYL-2-BUTANOL	000464-07-3
3,3-DIMETHYL-1-BUTANOL	000624-95-3
3,3-DIMETHYL-2-BUTANONE	000075-97-8
2,3-DIMETHYL-1-BUTENE	000563-78-0
3,3-DIMETHYL-1-BUTENE	000558-37-2
2,3-DIMETHYL-2-BUTENE	000563-79-1
DIMETHYLBUTYLAMINE	000927-62-8
1,1-DIMETHYL-(3-BUTYL)PHENYLUREA	088132-40-5
2,2-DIMETHYL BUTYRIC ACID	000595-37-9
N,O-DIMETHYLCARBAMATE	006642-30-4
DIMETHYLCARBAMYL CHLORIDE	000079-44-7
3,4-DIMETHYLCHLOROBENZENE	000615-60-1
5,6-DIMETHYLCHRYSENE	003697-27-6
N,N-DIMETHYLCINNAMAMIDE	013156-74-6
DIMETHYL CYANAMIDE	001467-79-4
1,3-DIMETHYL-2-CYANOGUANIDINE	031857-31-5
1,3-DIMETHYLCYCLOHEXANE	000591-21-9
1,3-DIMETHYLCYCLOHEXANE (CIS)	000638-04-0
1,4-DIMETHYLCYCLOHEXANE (CIS)	000624-29-3
5,5-DIMETHYL-1,3-CYCLOHEXANEDIONE	000126-81-8

1,2-DIMETHYLCYCLOHEXANE (TRANS)	006876-23-9
1,3-DIMETHYLCYCLOHEXANE (TRANS)	002207-03-6
1,4-DIMETHYLCYCLOHEXANE (TRANS)	002207-04-7
1,1-DIMETHYL CYCLOHEXANE	000590-66-9
1,4-DIMETHYLCYCLOHEXANE	000589-90-2
1,2-DIMETHYLCYCLOHEXANE	000583-57-3
1,2-DIMETHYLCYCLOHEXANE (CIS)	002207-01-4
3,5-DIMETHYLCYCLOHEXANOL	005441-52-1
2,6-DIMETHYLCYCLOHEXANOL	005337-72-4
1,2-DIMETHYLCYCLOPENTANE	002452-99-5
1,3-DIMETHYLCYCLOPENTANE (TRANS)	001759-58-6
1,1-DIMETHYLCYCLOPENTANE	001638-26-2
1,2-DIMETHYLCYCLOPENTANE (TRANS)	000822-50-4
1,3-DIMETHYLCYCLOPENTANE (CIS)	002532-58-3
1,2-DIMETHYLCYCLOPENTANE (CIS)	001192-18-3
1-(3,5-DIMETHYL),2-DIBENZOYL-1-TERT-BUTYLHYDRAZI	112426-79-6
1-(3,4-DIMETHYL),2-DIBENZOYL-1-TERT-BUTYLHYDRAZI	112412-44-9
1-(2,3-DIMETHYL),2-DIBENZOYL-1-TERT-BUTYLHYDRAZI	112411-90-2
2,5-DIMETHYL-1,6-DICHLORO-1,3,5-HEXATRIENE	083682-54-6
1,1-DIMETHYL-3-(3,5-DICLPH)UREA	010290-38-7
DIMETHYL DIETHOXYSILANE	000078-62-6
1,2-DIMETHYL-4,5-DINITROIMIDAZOLE	019183-17-6
2,6-DIMETHYL-1,3-DIOXAN-4-OL ACETATE	000828-00-2
DIMETHYLDISULFIDE	000624-92-0
DIMETHYLDITHIOCARBAMIC ACID, SODIUM SALT	000128-04-1
N,N-DIMETHYLDODECYLAMINE OXIDE	001643-20-5
DIMETHYL ETHER	000115-10-6
1,3-DIMETHYL-5-ETHYLBENZENE	000934-74-7
1,3-DIMETHYL-4-ETHYLBENZENE	000874-41-9
1,2-DIMETHYL-3-ETHYLBENZENE	000933-98-2
1,3-DIMETHYL-2-ETHYLBENZENE	002870-04-4
N,N-DIMETHYLETHYLCARBAMATE	000687-48-9
2-(1,1-DIMETHYLETHYL)-4,6-DIMETHYLPHENOL	001879-09-0
2,2-DIMETHYLETHYLENIMINE	002658-24-4
9-(1,1-DIMETHYLETHYL)-9H-FLUORENE	017114-78-2
N-(1,1-DIMETHYLETHYL)-4-NITROBENZENAMINE	004138-38-9
2,4-DIMETHYL-3-ETHYLPENTANE	001068-87-7
2,2-DIMETHYL-3-ETHYLPENTANE	016747-32-3
1,1-DIMETHYL-3-P-FLUOROPHENYLUREA	000332-33-2
1,1-DIMETHYL-3-M-FLUOROPHENYLUREA	000330-39-2
N,N'-DIMETHYLFORMAMIDE	000068-12-2
N,N-DIMETHYL-FURAN-3-CARBOXAMIDE	014757-80-3
N,N-DIMETHYL-FURAN-2-CARBOXAMIDE	013156-75-7
N,N-DIMETHYLGLYCINE	001118-68-9
DIMETHYL GLYOXIME	000095-45-4
DIMETHYLGUANIDINE	003324-71-8
2,2-DIMETHYLHEPTANE	001071-26-7
2,6-DIMETHYL-4-HEPTANOL	000108-82-7
2,6-DIMETHYL-3-HEPTANOL	019549-73-6
3,5-DIMETHYL-4-HEPTANOL	019549-79-2
2,6-DIMETHYL-4-HEPTANONE	000108-83-8
2,5-DIMETHYL-2,4-HEXADIENE	000764-13-6
2,5-DIMETHYL-1,5-HEXADIENE	000627-58-7
2,4-DIMETHYLHEXANE	000589-43-5
2,3-DIMETHYLHEXANE	000584-94-1
2,2-DIMETHYLHEXANE	000590-73-8
3,3-DIMETHYL HEXANE	000563-16-6
3,4-DIMETHYLHEXANE	000583-48-2

2,5-DIMETHYLHEXANE	000592-13-2
5,5-DIMETHYL-2,4-HEXANEDIONE	007307-04-2
2,5-DIMETHYL-1,3,5-HEXATRIENE	004916-63-6
2,5-DIMETHYL-3-HEXENE	015910-22-2
2,3-DIMETHYL-2-HEXENE	007145-20-2
2,5-DIMETHYL-2-HEXENE	003404-78-2
2,5-DIMETHYL-3-HEXYNE-2,5-DIOL	000142-30-3
1,1-DIMETHYLHYDRAZINE	000057-14-7
1,2-DIMETHYLHYDRAZINE	000540-73-8
3,5-DIMETHYL-4-HYDROXYACETANILIDE	022900-79-4
N,N-DIMETHYLHYDROXYLAMINE	016645-06-0
O,N-DIMETHYLHYDROXYLAMINE	001117-97-1
1,2-DIMETHYLINDOLE	000875-79-6
DIMETHYLMALONATE, BENZAL	006626-84-2
DIMETHYLMALONIC ACID	000595-46-0
DIMETHYLMALONONITRILE	007321-55-3
DIMETHYL MERCURY	000593-74-8
2,2-DIMETHYL-7-METHOXYBENZOPYRAN	017598-02-6
1,6-DIMETHYLNAPHTHALENE	000575-43-9
2,3-DIMETHYLNAPHTHALENE	000581-40-8
2,6-DIMETHYLNAPHTHALENE	000581-42-0
2,7-DIMETHYLNAPHTHALENE	000582-16-1
1,4-DIMETHYLNAPHTHALENE	000571-58-4
1,7-DIMETHYLNAPHTHALENE	000575-37-1
1,5-DIMETHYLNAPHTHALENE	000571-61-9
1,2-DIMETHYLNAPHTHALENE	000573-98-8
1,8-DIMETHYLNAPHTHALENE	000569-41-5
1,3-DIMETHYLNAPTHALENE	000575-41-7
N,N-DIMETHYLNICOTINAMIDE	006972-69-6
DIMETHYLNITRAMINE	004164-28-7
2,6-DIMETHYL-4-NITROBENZENAMINE	016947-63-0
1,2-DIMETHYL-3-NITROBENZENE	000083-41-0
1,3-DIMETHYL-2-NITROBENZENE	000081-20-9
3,5-DIMETHYL-4-NITROBENZENAMINE	034761-82-5
1,2-DIMETHYL-4-NITROBENZENE	000099-51-4
N,2-DIMETHYL-4-NITROBENZENAMINE	010439-77-7
1,3-DIMETHYL-2-NITROGUANADINE	101250-97-9
1,5-DIMETHYL-4-NITRO-1H-IMIDAZOLE	007464-68-8
1,2-DIMETHYL-4-NITROIMIDAZOLE	013230-04-1
1,2-DIMETHYL-5-NITRO-1H-IMIDAZOLE	000551-92-8
2,6-DIMETHYL-4-NITROPHENOL	002423-71-4
1,1-DIMETHYL-3-M-NITROPHENYLUREA	007159-98-0
1,1-DIMETHYL-3-P-NITROPHENYLUREA	007159-97-9
2,6-DIMETHYL-N-NITROSOMORPHOLINE	001456-28-6
2,6-DIMETHYL-N-NITROSOPIPERIDINE	017721-95-8
2,5-DIMETHYL-N-NITROSOPYRROLIDINE	055556-86-0
2,6-DIMETHYLOCTANE	002051-30-1
3,7-DIMETHYL-1,3,6-OCTATRIENE	013877-91-3
3,7-DIMETHYL-6-OCTEN-1-OL	000106-22-9
DIMETHYLOL UREA	000140-95-4
2,6-DIMETHYL-N-OXIDEPYRIDINE	001073-23-0
2,2-DIMETHYLOXIRANE	000558-30-5
3,5-DIMETHYL-DIMETHYLPARA-OXON	050590-06-2
3-ETHYL-DIMETHYLPARATHION	013074-09-4
3-CHLORO-DIMETHYL PARATHION	000500-28-7
3,5-DIMETHYL-DIMETHYLPARATHION	050590-05-1
2,4-DIMETHYLPENTANE	000108-08-7
2,2-DIMETHYLPENTANE	000590-35-2

2,3-DIMETHYLPENTANE	000565-59-3
3,3-DIMETHYLPENTANE	000562-49-2
2,3-DIMETHYL-2-PENTANOL	004911-70-0
2,2-DIMETHYL-3-PENTANOL	003970-62-5
2,4-DIMETHYL-3-PENTANOL	000600-36-2
2,3-DIMETHYL-3-PENTANOL	000595-41-5
2,2-DIMETHYL-1-PENTANOL	002370-12-9
2,4-DIMETHYL-2-PENTANOL	000625-06-9
2,4-DIMETHYL-1-PENTANOL	006305-71-1
4,4-DIMETHYL-1-PENTANOL	003121-79-7
2,4-DIMETHYL-3-PENTANONE	000565-80-0
3,4-DIMETHYL-2-PENTENE	024910-63-2
4,4-DIMETHYL-2-PENTENE (TRANS)	000690-08-4
2,3-DIMETHYL-2-PENTENE	010574-37-5
2,3-DIMETHYL-1-PENTENE	003404-72-6
2,4-DIMETHYL-1-PENTENE	002213-32-3
3,3-DIMETHYL-1-PENTENE	003404-73-7
3,6-DIMETHYLPHENANTHRENE	001576-67-6
3,5-DIMETHYLPHENOL	000108-68-9
2,3-DIMETHYLPHENOL	000526-75-0
2,4-DIMETHYLPHENOL	000105-67-9
2,5-DIMETHYLPHENOL	000095-87-4
2,6-DIMETHYLPHENOL	000576-26-1
3,4-DIMETHYLPHENOL	000095-65-8
N,N-DIMETHYLPHENOXYACETAMIDE	010397-59-8
1-(3,5-DIMETHYLPHENOXY)SILATRANE	090955-56-9
4-((2,4-DIMETHYLPHENYL)AZO)BENZENESULFONIC ACID*	062959-40-4
P-(N,N-DIMETHYL)PHENYL AZOXYCYANIDE	062825-16-5
N,N-DIMETHYLPHENYLCARBAMATE	006969-90-0
N,N-DIMETHYL-P-PHENYLENEDIAMINE	000099-98-9
3,5-DIMETHYLPHENYLHIPPURATE	081592-15-6
(2,6-DIMETHYLPHENYL)(2-METHYLPHENYL)DIAZENE	006319-26-2
(2,6-DIMETHYLPHENYL)PHENYLDIAZENE	017590-87-3
O,O-DIMETHYLPHENYLPHOSPHONATE	002240-41-7
O,O-DIMETHYL-O-PHENYLPHOSPHOROTHIOATE	033576-92-0
O,O-DIMETHYL-O-PHENYLPHOSPHATE	010113-28-7
(DIMETHYL)-PHENYLPHOSPHINE	000672-66-2
1,3-DIMETHYL PHENYL UREA	000938-91-0
DIMETHYL PHOSPHITE	000868-85-9
O,O-DIMETHYL PHOSPHOROCHLORIDOTHIOATE	002524-03-0
DIMETHYL PHTHALATE	000131-11-3
2,5-DIMETHYLPIPERAZINE	000106-55-8
2,6-DIMETHYLPIPERIDINE (CIS-)	000766-17-6
1,1-DIMETHYL-3-P-BR-PHENYLUREA	020940-43-6
2,2-DIMETHYLPROPANE	000463-82-1
2,2-DIMETHYL-1-PROPANOL	000075-84-3
2,2-DIMETHYLPROPIOPHENONE	000938-16-9
2,2-DIMETHYLPROPYL ESTER	111226-64-3
2,5-DIMETHYLPYRAZINE	000123-32-0
N,N-DIMETHYL 2-PYRAZINECARBOXAMIDE	032743-27-4
2,5-DIMETHYLPYRIDINE	000589-93-5
2,4-DIMETHYLPYRIDINE	000108-47-4
2,3-DIMETHYLPYRIDINE	000583-61-9
3,4-DIMETHYLPYRIDINE	000583-58-4
3,5-DIMETHYLPYRIDINE	000591-22-0
3,5-DIMETHYLPYRIDINE-N-OXIDE	003718-65-8
N,N-DIMETHYL-3-PYRIDYLMETHYLAMINE	002055-21-2
2,6-DIMETHYL-4-PYRIMIDINAMINE	000461-98-3

4,6-DIMETHYLPYRIMIDINE	001558-17-4
GAMMA-2,6-DIMETHYLPYRONE	001004-36-0
2,5-DIMETHYLPYRROLE	000625-84-3
2,6-DIMETHYLQUINOLINE	000877-43-0
2,4-DIMETHYLQUINOLINE	001198-37-4
4,5-DIMETHYL-8-QUINOLINOL	015011-28-6
6,7-DIMETHYLQUINOXALINE	007153-23-3
4,4-DIMETHYLSEMICARBAZIDE	022718-51-0
DIMETHYL SUCCINATE	000106-65-0
N1-DIMETHYLSULFANILAMIDE	001709-59-7
N',N'-DIMETHYLSULFANILAMIDE	006162-21-6
DIMETHYLSULFATE	000077-78-1
DIMETHYLSULFIDE	000075-18-3
DIMETHYL SULFONE	000067-71-0
DIMETHYL SULFOXIDE	000067-68-5
L-DIMETHYL TARTRATE	000608-68-4
DIMETHYL TARTRATE (MESO)	005057-96-5
DIMETHYLTEREPHTHALATE	000120-61-6
DIMETHYL TETRACHLOROTEREPHTHALATE	001861-32-1
1,4-DIMETHYLTETRACHLOROCYCLOHEXANE	056421-45-5
2,5-DIMETHYL-3,3,4,4-TETRACHLORO-1,5-HEXADIENE	083682-50-2
CIS-2,5-DIMETHYLTETRAHYDROFURAN	002144-41-4
TRANS-2,5-DIMETHYLTETRAHYDROFURAN	002390-94-5
1,3-DIMETHYL-1,1,3,3-TETRAPHENYLDISILOXANE	000807-28-3
2,4-DIMETHYL-3-THIAPENTANE	000625-80-9
2,4-DIMETHYLTHIAZOLE	000541-58-2
2,5-DIMETHYLTHIOPHENE	000638-02-8
3,6-DIMETHYLTHIOPYRIDAZINE	037813-54-0
4,4-DIMETHYLTHIOSEMICARBAZIDE	006926-58-5
N,N-DIMETHYLTHIOUREA	000534-13-4
N,N-DIMETHYL-O-TOLUIDINE	000609-72-3
1,1-DIMETHYL-3-(P-TOLYL)UREA	007160-01-2
N,N-DIMETHYLTRIAZENOBENZENE	007227-91-0
DIMETHYL TRIDECANEDIOATE	001472-87-3
N,N-DIMETHYLTRYPTAMINE	000061-50-7
DIMETHYLUREA, SYM	000096-31-1
1,1-DIMETHYLUREA	000598-94-7
3,7-DIMETHYL URIC ACID	013087-49-5
1,7-DIMETHYL URIC ACID	033868-03-0
1,3-DIMETHYL URIC ACID	000944-73-0
DIMETILAN	000644-64-4
2,5-DIMETHOXY-4-MES-PHENETHYLAMINE	061638-09-3
4(33-DIMETRIAZENO)BENZENESULFONAMIDE	055469-64-2
5(33DIME-1-TRIAZENO)IMDAZL-4-COOET	036137-88-9
5(33-DIME-1-TRIAZENO)ISOPHTHALAMIDE	083293-76-9
3(33DIME-1-TRIAZENO)PYRAZOLE-4-CONH2	021466-00-2
2-(3,3-DIME-1-TRIAZINO)BENZAMIDE	033330-89-1
4-(3,3-DIME-1-TRIAZINO)BENZAMIDE	033330-91-5
4-(3,3-DIME-1-TRIAZINO)PHENYLUREA	066521-48-0
DIMITE	000080-06-8
2,4-DINH2-6-BUAM-PYRIMIDINE-3-OXIDE	055921-55-6
2,4-DINH2PYRIMIDINE,5(4MEO-3ME)BENZYL	073264-18-3
2,4-DINH2PYRIMIDINE,5(3MEO-4ME)BENZYL	073264-19-4
DINICONAZOLE	083657-24-3
DINICOTINIC ACID	000499-81-0
2,4-DINITRO-N-(3-(TRIFLUOROMETHYL)PHENYL)BENZE*	001869-67-6
DINITRAMINE	029091-05-2
2,6-DINITROANILINE	000606-22-4

3,5-DINITROANILINE	000618-87-1
2,4-DINITROANILINE	000097-02-9
P-(2,4-DINITROANILINE)PHENOL	000119-15-3
2,4-DINITROANISOLE	000119-27-7
1,2-DINITROANTHRAQUINONE	057875-61-3
3,5-DINITROBENZAMIDE	000121-81-3
1,3-DINITROBENZENE	000099-65-0
P-DINITROBENZENE	000100-25-4
O-DINITROBENZENE	000528-29-0
3,4-DINITROBENZOIC ACID	000528-45-0
3,5-DINITROBENZOIC ACID	000099-34-3
2,4-DINITROBENZOIC ACID	000610-30-0
5,7-DINITROBENZPYRAZOLE	031208-76-1
4,4'-DINITROBIPHENYL	001528-74-1
2,4-DINITROCHLOROBENZENE	000097-00-7
4,6-DINITRO-O-CRESOL	000534-52-1
2,6-DINITRO-P-CRESOL	000609-93-8
2,4-DINITRO-6-HEXYLPHENOL	004099-65-4
4,5-DINITROIMIDAZOLE	019183-14-3
2,4-DINITRO-1H-IMIDAZOLE	005213-49-0
3,5-DINITRO-4-METHYLBENZAMIDE	004551-76-2
2,4-DINITRO-5-METHYLPHENOL	000616-73-9
2,4-DINITRO-N-(3-NITROPHENYL)BENZENAMINE	000970-91-2
2,4-DINITRO-N-(4-NITROPHENYL)BENZENEAMINE	000970-76-3
2,4-DINITROPHENETOLE	000610-54-8
2,3-DINITROPHENOL	000066-56-8
3,5-DINITROPHENOL	000586-11-8
3,4-DINITROPHENOL	000577-71-9
2,5-DINITROPHENOL	000329-71-5
2,6-DINITROPHENOL	000573-56-8
2,4-DINITROPHENOL	000051-28-5
DINITROPHENYL ACETATE	000643-43-6
2,4-DINITRO-N-PHENYLBENZENAMINE	000961-68-2
N-(2,4-DINITROPHENYL)-1,4-BENZENEDIAMINE	006373-73-5
2,4-DINITRO-6-PHENYLPHENOL	000731-92-0
2,4-DINITROPHENYL THIOCYANATE	001594-56-5
N-(2,4-DINITROPHENYL)-M-TOLUIDINE	000964-79-4
4,8-DINITROQUINOLINE-1-OXIDE	014753-19-6
4,6-DINITROQUINOLINE-1-OXIDE	001596-52-7
4,5-DINITROQUINOLINE-1-OXIDE	016238-73-6
2,4-DINITRORESORCINOL	000519-44-8
N,N'-DINITROSOPIPERAZINE	000140-79-4
2,4-DINITROTOLUENE	000121-14-2
2,3-DINITROTOLUENE	000602-01-7
2,5-DINITROTOLUENE	000619-15-8
2,4-DINITRO-4'-(TRIFLUOROMETHYL)DIPHENYLAMINE	013744-79-1
2,6-DINO2 ACETANILIDE	090110-78-4
DINOBEN	000088-86-8
DINOBUTON	000973-21-7
DINOCAP	039300-45-3
2,4-DINO2-C6H3NHN=C(CN)COOET	028313-64-6
1,2-DINO2 4-CL BENZENE	000610-40-2
1,3-DINO2 2-CL 5-CF3 BENZENE	000393-75-9
1,2-DINO2 4-METHYL BENZENE	000610-39-9
1,3-DINO2 2-METHYL BENZENE	000606-20-2
DINONYL ETHER	002456-27-1
DINOSEB	000088-85-7
DINOSEB ACETATE	002813-95-8

DIOCTYL ADIPATE	000123-79-5
DIOCTYL METHYLPHOSPHONATE	001832-68-4
DIOCTYL PHENYL PHOSPHONATE	001754-47-8
DIOCTYL PHTHALATE	000117-84-0
DI-N-OCTYL SODIUM SULFOSUCCINATE	001639-66-3
1-(2,3-DIOHPR)-2-ME-5-NO2IMIDAZOLE	062580-80-7
DIONINE	000076-58-4
DIOSMETINE	000520-34-3
DIOSMINE	000520-27-4
3,5-DIOXA-6-AZA-4-PHOSPAOCT-6-ENE-8-NITRILE, 4-A	067410-39-3
3,5-DIOXA-6-AZA-4-PHOSPHAOCT-6-ENE-8-NITRILE, 4-	067410-48-4
3,5-DIOXA-6-AZA-4-PHOSPHAOCT-6-ENE-8-NITRILE, 4-	067410-45-1
3,5-DIOXA-6-AZA-4-PHOSPHAOCT-6-ENE-8-NITRILE, 7-	067410-46-2
3,5-DIOXA-6-AZA-4-PHOSPHAOCT-6-ENE-8-NITRILE, 4-	067410-49-5
DIOXACARB	006988-21-2
1,3-DIOXALANE	000646-06-0
1,3-DIOXANE	000505-22-6
1,4-DIOXANE	000123-91-1
DIOXATHION	000078-34-2
1,1-DIOX-1,2-BENZISOTHIAZOL-3-ONE	000081-07-2
[1,3]DIOXEPINO[5,6-D]ISOXAZOLE, 3A,4,8,8A-TETRAH	092897-88-6
[1,3]DIOXEPINO[5,6-D]ISOXAZOLE, 3A,4,8,8A-TETRAH	092937-45-6
[1,3]DIOXEPINO[5,6-D]ISOXAZOLE, 3A,4,8,8A-TETRAH	123750-56-1
[1,3]DIOXEPINO[5,6-D]ISOXAZOLE, 6-(4-FLUOROPHENY	123808-64-0
[1,3]DIOXEPINO[5,6-D]ISOXAZOLE, 6-(4-FLUOROPHENY	123692-64-8
2,2-DIOXIDE-1,2-OXATHIOLANE	001120-71-4
DIOXOLANE	000100-79-8
1,3-DIOXOLANE, 2,2-DIMETHYL-	002916-31-6
1,3-DIOXOLANE-4-METHANAMINIUM, N,N,N-TRIMETHYL-,	000541-66-2
1,3-DIOXOLAN-2-ONE	000096-49-1
6H-[1,3]DIOXOLO[5,6]BENZOFURO[3,2-C][1]BENZOPYRA	002035-15-6
6H-[1,3]DIOXOLO[5,6]BENZOFURO[3,2-C][1]BENZOPYRA	000469-01-2
3,4-DIOXYMETHYLENECINNAMAMIDE	005813-92-3
DIPENTENE (+-)	007705-14-8
DIPENTYLAMINE	002050-92-2
DI-N-PENTYLETHER	000693-65-2
2-DIPHACETYL-1,3INDANDIONE-1HYDRAZONE	005102-79-4
DIPHACINONE	000082-66-6
DIPHENAMID	000957-51-7
DIPHENHYDRAMINE	000058-73-1
1,2-DIPHENOXYETHANE	000104-66-5
N,N-DIPHENYLACETAMIDE	000519-87-9
N,N'-DIPHENYLACETAMIDINE	000621-09-0
DIPHENYLACETIC ACID	000117-34-0
DIPHENYLACETONITRILE	000086-29-3
DIPHENYLACETYLENE	000501-65-5
DIPHENYLAMINE	000122-39-4
9,10-DIPHENYLANTHRACENE	001499-10-1
N,N'-DIPHENYL-P-BENZENEDIAMINE	000074-31-7
(1,4-DIPHENYL)-1,3-BUTADIENE	000886-65-7
N,O-DIPHENYLCARBAMATE	004930-03-4
DIPHENYLDIETHOXYSILANE	002553-19-7
1,1-DIPHENYL-3,3-DIMETHYLUREA	002990-01-4
DIPHENYL DISULFIDE	000882-33-7
1,1-DIPHENYLETHANE	000612-00-0
DIPHENYL ETHER	000101-84-8
3-ME-4'-NO2-DIPHENYL ETHER	002303-25-5
1,1-DIPHENYLETHYLENE	000530-48-3

N,N'-DIPHENYLGUANIDINE	000102-06-7
DIPHENYL METHANE DIISOCYANATE	000101-68-8
DIPHENYL METHYLAMINE	000552-82-9
DIPHENYLMETHYL PHOSPHATE	000115-89-9
1,1-DIPHENYL-2-NITROPROPANE	088837-70-1
DIPHENYLNITROSAMINE	000086-30-6
DIPHENYLOLPROPANE	000080-05-7
DIPHENYL PHOSPHATE	000838-85-7
DIPHENYL PHTHALATE	000084-62-8
A,A-DIPHENYLPROPIONIC ACID	005558-66-7
2,6-DIPHENYLPYRIDINE	003558-69-8
DIPHENYLSULFIDE	000139-66-2
DIPHENYL SULFONE	000127-63-9
RH-8826 DIPHENYL SULFONE	143502-47-0
DIPHENYL SULFOXIDE	000945-51-7
RH-8827 DIPHENYL SULFOXIDE	143502-48-1
1,2-DIPHENYL TETRACHLOROETHANE	013700-81-7
1,3-DIPHENYL-1-TRIAZENE	000136-35-6
2,2-DIPHENYL-1,1,1-TRICL ETHANE	002971-22-4
IPHENYLUREA, SYM	000102-07-8
IPHENYL UREA, UNSYM	000603-54-3
RH-5348 DIPH ETHER	121325-44-8
RH-1460 DIPH THIOETHER	061405-48-9
DIPOTASSIUM MONOOCTYL ESTER	019045-79-5
DIPROPALIN	001918-08-7
DIPROPETRYNE	004147-51-7
DIPROPYLAMINE	000142-84-7
DIPROPYLENE GLYCOL, DIBENZOATE	027138-31-4
DI(N-PROPYL) ETHER	000111-43-3
DI(N-PROPYL) KETONE	000123-19-3
O,O-DIPROPYL METHYLPHOSPHONOTHIOATE	025371-75-9
DI-I-PROPYLNITROSOAMINE	000601-77-4
O,S-DIPROPYLPHOSPHORAMIDOTHIOATE	028167-45-5
DIPROPYL PHTHALATE	000131-16-8
DI-I-PROPYL PHTHALATE	000605-45-8
DIPROPYL PYRIDINE-2,5-DICARBOXYLATE	000136-45-8
DI-N-PROPYLSULFIDE	000111-47-7
N,N-DI-I-PR-3-PYRIDYLMETHYLAMINE	020173-18-6
4,4'-DIPYRIDYL	000553-26-4
DIQUAT	002764-72-9
DIQUAT DIBROMIDE	000085-00-7
DIQUAT DICHLORIDE	004032-26-2
DIRECT RED 81	002610-11-9
DIRECT YELLOW 62, DISODIUM SALT	006409-90-1
DIS. A. 1	041541-13-3
DIS. A. 10	058528-60-2
DIS. A. 12	062570-20-1
DIS. A. 13	041541-14-4
DIS. A. 14	060129-67-1
DIS. A. 15	003010-38-6
DIS. A. 16	004313-13-7
DIS. A. 17	004313-14-8
DIS. A. 18	066121-41-3
DIS. A. 2	003771-38-8
DIS. A. 5	003025-52-3
DIS. A. 6	069472-19-1
DIS. A. 8	065125-87-3
DIS. A. 9	041541-11-1

DISALICYLIMIDE	001972-71-0
2,4-D, ISOOCTYL ESTER	025168-26-7
2,4-D, ISOPROPYL ESTER	000094-11-1
DISOPYRAMIDE	003737-09-5
DISPERSE BLUE 79	012239-34-8
DISPERSE BLUE 79	003956-55-6
DISPERSE BRIGHTENER	002866-43-5
DISPERSE RED 1	002872-52-8
DISPERSE RED 5	003769-57-1
DISPERSE RED 9	000082-38-2
DISPERSE RED 11	002872-48-2
DISPERSE VIOLET 1	000128-95-0
DISPERSE YELLOW 54	007576-65-0
DISULFIDE, BIS(2-NITROPHENYL)	001155-00-6
DISULFIDE, BIS(3-NITROPHENYL)	000537-91-7
DISULFIRAM	000097-77-8
DISULFOTON	000298-04-4
DISULFOTON SULFONE	002497-06-5
DISULFOTON SULFOXIDE	002497-07-6
DITALIMFOS (LAPTRAN)	005131-24-8
6,7-DITHIADODECANE	000112-51-6
1,4-DITHIANE	000505-29-3
DITHIANONE	003347-22-6
1,4-DITHIAN-2-ONE, 3,3-DIMETHYL-, O-[(METHYLAMIN	106231-39-4
1,4-DITHIAN-2-ONE, O-[(METHYLAMINO)CARBONYL]OXIM	106231-37-2
4,5-DITHIAOCTANE	000629-19-6
2,2'-DITHIOBISBENZOTHIAZOLE	000120-78-5
N,N'-(DITHIODI-2,1-ETHANEDIYL)BIS(N-(1-METHYL*))	065332-44-7
1,2-DITHIOLANE-3-PENTANOIC ACID	000062-46-4
1,3-DITHIOLAN-4-ONE, [(METHYLAMINO)CARBONYL]OXIM	106231-45-2
1,3-DITHIOLAN-4-ONE, 2,2-DIMETHYL-, O-[(METHYLAM	106231-48-5
1,3-DITHIOLAN-4-ONE, 5,5-DIMETHYL-, O-[(METHYLAM	106231-47-4
1,3-DITHIOLAN-4-ONE, 5-METHYL-, O-[(METHYLAMINO)	106231-46-3
P-DI(TRICHLOROMETHYL)BENZENE	000068-36-0
DITRIDECYL PHTHALATE	000119-06-2
DIUNDECYL PHTHALATE	003648-20-2
DIURON	000330-54-1
1,2-DIVINYLCYCLOBUTANE	002422-85-7
DIVINYL ETHER	000109-93-3
DKA-24	097454-00-7
DL-2,3-BUTANDIOL	006982-25-8
DL-P-FLUOROPHENYLALANINE	000060-17-3
DL-VALINE, N-[1-OXO-5-(1H-PURIN-6-YLTHIO)PENTYL]	023401-43-6
ß-D-MANNOPYRANOSE, 2-DEOXY-2-[[(METHYLNITROSOAMI	037793-02-5
2,4-D, METHYL ESTER	001928-38-7
DMPA	000299-85-4
DMTT	000533-74-4
DNAP	004097-36-3
DOBEROL	002933-44-0
DOCOSANE	000629-97-0
DOCOSANOIC ACID	000112-85-6
2,4-D, OCTYL ESTER	001928-44-5
DODECAMETHYLCYCLOHEXASILOXANE	000540-97-6
DODECAMETHYLPENTASILOXANE	000141-63-9
DODECANAL	000112-54-9
1-DODECANAMINE, HYDROCHLORIDE	000929-73-7
DODECANE	000112-40-3
DODECANEPEROXOIC ACID	002388-12-7

DODECANOIC ACID	000143-07-7
DODECANOL	000112-53-8
N-DODECENE	000112-41-4
1-DODECENE	001124-14-7
7-DODECEN-1-OL (Z)	020056-92-2
7-DODECEN-1-OL ACETATE (Z)	014959-86-5
9-DODECEN-1-OL ACETATE (Z)	016974-11-1
9-DODECENYL ACETATE (TRANS)	035148-19-7
DODECYL ACETATE	000112-66-3
P-DODECYLANILINE	000104-42-7
DODECYLBENZENE	000123-01-3
DODECYL BENZENESULFONATE	001886-81-3
DODECYLCYCLOHEXANE	001795-17-1
DODECYL MERCAPTAN	000112-55-0
DODECYL 2-METHYL-2-PROPENOATE	000142-90-5
DODECYLPYRIDINIUM BROMIDE	000104-73-4
DODECYL SULFATE, SODIUM SALT	000151-21-3
DODECYL SULFONATE, SODIUM SALT	002386-53-0
DODINE	002439-10-3
DOLSULEPINE	000113-53-1
DOMPERIDONE	057808-66-9
DOPA	000063-84-3
DOPA	000059-92-7
DOPAMINE	000051-61-6
DOTRIACONTANE	000544-85-4
DOWCO 275	039624-86-7
2,4-D, PROPYL ESTER	001928-61-6
DRAZOXOLON	051450-97-6
DROMOSTANOLONE	000058-19-5
DROPERIDOL	000548-73-2
DSMA	000144-21-8
2,4-D, SODIUM SALT	002702-72-9
D-ALPHA-TERPINEOL	007785-53-7
DUROQUINONE	000527-17-3
E-838	000299-45-6
EDIFENPHOS	017109-49-8
EDTA, SODIUM SALT	000064-02-8
EICOSANE	000112-95-8
EICOSANOIC ACID	000506-30-9
1-EICOSANOL	000629-96-9
5,8,11,14-EICOSATETRAENOIC ACID	000506-32-1
1-EICOSENE	003452-07-1
ELLIPTICINE	000519-23-3
EMETINE DIHYDROCHLORIDE	000316-42-7
ENDO-8-METHYL-8-AZABICYCLO(3.2.1)OCTAN-3-OL	000120-29-6
ENDOSULFAN	000115-29-7
A-ENDOSULFAN	000959-98-8
BETA-ENDOSULFAN	033213-65-9
ENDOSULFAN SULFATE	001031-07-8
ENDOTHAL	000145-73-3
ENDOTHION	002778-04-3
ENDRIN	000072-20-8
ENDRIN ALDEHYDE	007421-93-4
ENOXACIN	074011-58-8
EPANOLOL	086880-51-5
EPHEDRINE	000299-42-3
ALPHA-EPIBROMOHYDRIN	003132-64-7
EPIFLUOROHYDRIN	000503-09-3

EPINEPHRINE	000051-43-4
EPINEPHRINE SALT (HCL)	000329-63-5
4'-EPIOLEANDRIN	000508-22-5
EPN	002104-64-5
3,12-EPOXY-12H-PYRANO[4,3-J]-1,2-BENZODIOXEPIN-1	081496-81-3
EPTAM (EPTC)	000759-94-4
ERBON	000136-25-4
ERIODICTYOL	000552-58-9
ERYTHRITOL	000149-32-6
D-ERYTHRO-ALPHA-D-GALACTO-OCTOPYRANOSIDE, ETHYL	021085-65-4
D-ERYTHRO-ALPHA-D-GALACTO-OCTOPYRANOSIDE, METHYL	017057-68-0
D-ERYTHRO-ALPHA-D-GALACTO-OCTOPYRANOSIDE DERIVAT	002256-17-9
D-ERYTHRO-A-D-GALACTO-OCTOPYRANOSIDE DERIVATIVE	002256-16-8
ERYTHROMYCIN	000114-07-8
ERYTHROMYCIN, 3'-AZIDO-3'-DE(DIMETHYLAMINO)-4'-H	025423-22-7
ERYTHROMYCIN, 3"-O-DEMETHYL-	001675-02-1
ERYTHROMYCIN, 12-DEOXY-	000527-75-3
ERYTHROMYCIN, 12-DEOXY-, 4"-ACETATE	031357-18-3
ERYTHROMYCIN, 12-DEOXY-, 11-ACETATE	031357-44-5
ERYTHROMYCIN, 12-DEOXY-, 4"-FORMATE	031357-41-2
ERYTHROMYCIN, N-DEMETHYL-	000992-62-1
ERYTHROMYCIN, 4"-FORMATE	031357-16-1
ERYTHROMYCIN, 4'-HYDROXY-	025405-65-6
ESFENVALERATE	066230-04-4
ESPROCARB	085785-20-2
ESTRADIOL	000050-28-2
ESTRA-1,3,5(10)-TRIENE-3,17-DIOL, 11-METHOXY-, (	021507-14-2
ESTR-4-EN-3-ONE, 17-HYDROXY-7,17-DIMETHYL-, (7AL	003704-09-4
ESTRIOL	000050-27-1
ESTRONE	000053-16-7
2-ETAM-5(5NO2-2-FURFURYL)THIAZOLONE	027472-85-1
N-ET-BROMOACETAMIDE	005327-00-4
ET-N-CHLOROACETYL-N-(2MEOPH)GLYCINATE	060145-78-0
ET-N-CL-ACETYL-N-(2CLPH)GLYCINATE	051114-25-1
ET-N-CL-ACETYL-N-(2ETPH)GLYCINATE	060145-76-8
ET-N-CL-ACETYL-N-(PHENYL)GLYCINATE	051114-26-2
N'-ET(4-CLPHENOXY)ACETIC A HYDRAZIDE	054922-65-5
O-ET-S-(ETSET)-ME-PHOSPHONATE	000556-75-2
ETHANAMINE, N,N-DIMETHYL-2-(3-METHYLPHENOXY)-	002455-15-4
ETHANAMINE, N,N-DIMETHYL-2-(4-METHYLPHENOXY)-	051344-14-0
ETHANAMINE, N,N-DIMETHYL-2-[5-METHYL-2-(1-METHYL	000091-46-3
ETHANAMINE, N-ETHYL-2,2,2-TRIFLUORO-	058171-47-4
ETHANAMINE, 2-(4-METHOXYPHENOXY)-N,N-DIMETHYL-	051344-12-8
ETHANAMINIUM, 2-[(CYCLOPROPYLCARBONYL)OXY]-N,N,N	079661-41-9
ETHANAMINIUM, 2-[(DIAZOACETYL)OXY]-N,N,N-TRIMETH	030273-97-3
ETHANAMINIUM, 1,1-DIMETHYL-N,N,N-TRIMETHYL-, IOD	004153-42-8
ETHANAMINIUM, 2-(FORMYLOXY)-N,N,N-TRIMETHYL-, IO	079661-42-0
ETHANAMINIUM, N,N,N-TRIMETHYL-2-(1-OXOPROPOXY)-,	002494-55-5
ETHANE	000074-84-0
ETHANE, 1-CHLORO-1-NITRO-	000598-92-5
1,2-ETHANEDIAMINE, N-(2-AMINOETHYL)-N,N'-DIMETHY	034066-96-1
1,2-ETHANEDIAMINE, N,N'-DIETHYL-	000111-74-0
1,2-ETHANEDIAMINE, N,N-DIETHYL-	000100-36-7
1,2-ETHANEDIAMINE, N,N'-DIMETHYL-	000110-70-3
1,2-ETHANEDIAMINE, N,N,N'-TRIMETHYL-	000142-25-6
ETHANE, 1,2-DIIODO-	000624-73-7
1,2-ETHANEDIOL, 1,2-DIPHENYL-, (R*,R*)-(±)-	000655-48-1
1,2-ETHANEDIOL, MONOBENZOATE	000094-33-7

1,2-ETHANEDITHIOL	000540-63-6
1,2-ETHANEDIYLBISCARBAMODITHIOIC ACID	000111-54-6
ETHANIMIDOTHIOIC ACID, N-[(4-CYCLOPROPYL-2,6-DIM	074385-86-7
ETHANIMIDOTHIOIC ACID, N-[[[[(ACETYLMETHYLAMINO)	086569-91-7
ETHANIMIDOTHIOIC ACID, N-[[[[(METHOXYCARBONYLMET	064055-10-3
ETHANOL	000064-17-5
ETHANOLAMINE	000141-43-5
ETHANOL, 2-[(2-AMINO-9H-PURIN-9-YL)METHOXY]-	084408-37-7
ETHANOL, 2-[[4-(5-AMINO-1H-1,2,4-TRIAZOL-3-YL)-2	077314-62-6
ETHANOL, 2-[2-[2-(BENZOYLOXY)ETHOXY]ETHOXY]-	023022-51-7
ETHANOL, 2-[[4,6-BIS(DIMETHYLAMINO)-1,3,5-TRIAZI	052298-71-2
ETHANOL, 2-BROMO-, ACETATE	000927-68-4
ETHANOL, 2-[(2,6-DIAMINO-9H-PURIN-9-YL)METHOXY]-	059277-86-0
ETHANOL, 1-(2,4-DICHLOROPHENYL)-2-(3-PYRIDINYL)-	088283-41-4
2-(DIETHYLAMINO)-ETHANOL	000100-37-8
ETHANOL, 2- (1,1-DIMETHYLETHYL)AMINO -	004620-70-6
ETHANOL, 2-(ETHYLPHENYLAMINO)-	000092-50-2
ETHANOL, 2-(HEXYLOXY)-	000112-25-4
ETHANOL, 2- 2-(HEXYLOXY)ETHOXY -	000112-59-4
ETHANOL, 2,2'- (4-METHYLPHENYL)IMINO BIS-	003077-12-1
ETHANOL, 2-(2-METHYLPROPOXY)-	004439-24-1
ETHANOL, 2-(1-NAPHTHALENYLOXY)-	000711-82-0
1(4-NO2 PH)-2-ISOPROPYLAMINO-ETHANOL	007413-36-7
ETHANOL, 2-(OCTYLTHIO)-	003547-33-9
ETHANOL, 2- (PHENYLMETHYL)AMINO -	000104-63-2
ETHANOL, 2-(2-PROPENYLOXY)-	000111-45-5
ETHANOL, 2,2,2-TRIBROMO-	000075-80-9
ETHANOL, 2-[2-[4-[3-[2-(TRIFLUOROMETHYL)PHENOTHI	003093-23-0
ETHANONE, 1-(2-AMINO-4-METHOXYPHENYL)	042465-53-2
ETHANONE, 1-[4-[(4-AMINOPHENYL)SULFONYL]PHENYL]-	100866-99-7
ETHANONE, 1-(2-BROMOPHENYL)-	002142-69-0
ETHANONE, 2-BROMO-1-PHENYL-	000070-11-1
ETHANONE, 1-(4-CHLOROPHENYL)-2-(PHENYLSULFONYL)-	038488-19-6
ETHANONE, 1-(6,7-DIHYDROXY-2-NAPHTHALENYL)-	118199-17-0
ETHANONE, 1-[2-[4-[3-[[2-(3,4-DIMETHOXYPHENYL)ET	102152-21-6
ETHANONE, 1-(4-CHLOROPHENYL)-2-[(4-METHYLPHENYL)SULFONYL-	061820-94-8
ETHANONE, 1-[7-[2-HYDROXY-3-[(1-METHYLETHYL)AMIN	039552-01-7
ETHANONE, 1-[1-(2-HYDROXYPROPYL)-5-METHYL-2-NITR	084123-05-7
ETHANONE, 1-[2-HYDROXY-3-PROPYL-4-[4-(1H-TETRAZO	088107-10-2
ETHANONE, 1-(4-METHOXYPHENYL)-2-[(4-METHYLPHENYL)SULFONYL-	086516-51-0
ETHANONE, 1-(4-METHYLPHENYL)-2-(PHENYLSULFONYL)-	038488-14-1
ETHANONE, 1-(4-METHYLPHENYL)-2-[(4-METHYLPHENYL)SULFONYL-	061820-95-9
ETHANONE, 2-PHENYL-1-(1-PIPERIDINYL)-	003626-62-8
ETHAZOLE	000094-19-9
1,1-ETHENEDIAMINE, N-METHYL-N'-[2-[[(5-METHYL-1H	055884-23-6
1,1'-(1,2-ETHENEDIYL)BISBENZENE (CIS)	000645-49-8
ETHENE TETRACARBOXYLIC ACID,TETRAETHYL ESTER	006174-95-4
ETHENE, TRIBROMO-	000598-16-3
4,7-ETHENO-1H-CYCLOBUT[F]ISOINDOLE-1,3-(2H)-DION	114298-20-3
ETHEPHON	011672-87-0
ETHEPHON	016672-87-0
N(2-ETHEXYL)-1-ISOPROPYL-4-METHYLBICYCLO[2.2.2]-	013358-11-7
ETHION	000563-12-2
ETHOATE METHYL	000116-01-8
ETHOFUMESATE	026225-79-6
ETHOPROPAZINE	000522-00-9
ETHOSUXIMIDE	000077-67-8
3-ETHOXYACETANILIDE	000591-33-3

2-ETHOXY-4-AMINOBENZOIC ACID	000059-07-4
P-ETHOXYANILINE	000156-43-4
2-ETHOXYANILINE	000094-70-2
P-ETHOXYBENZAMIDE	055836-71-0
ETHOXYBENZENE	000103-73-1
P-ETHOXYBENZOIC ACID	000619-86-3
ETHOXYCHLOR	004329-03-7
2-ETHOXYETHANOL	000110-80-5
ETHOXYETHYLACETATE	000111-15-9
2-ETHOXYETHYL TRIMETHYL AMMONIUM IODIDE	016332-51-7
2-ETHOXYISONIAZID	058481-00-8
O=P(OC)(SC)N-ETHYL	052067-48-8
ETHOXYLATED TETRABROMOBISPHENOL A	004162-45-2
2'-ETHOXYLBENZOGUANAMINE	072775-81-6
1-ETHOXY-4-(2-METHYL-1-(4-METHYLPHENYL)PROPYL)*	056265-26-0
1-ETHOXY-4-(1(4-METHYLPHENYL)-2-NITROPROPYL)-B*-	054010-81-0
2-ETHOXYPENTACHLOROCYCLOHEXANE	056421-32-0
3-ETHOXYPENTACHLOROCYCLOHEXANE	056421-35-3
P-ETHOXYPHENOL	000622-62-8
M-ETHOXYPHENOL	000621-34-1
O-ETHOXYPHENOL	000094-71-3
P-ETHOXYPHENYLACETIC ACID	004919-33-9
2-ETHOXY-2-PH-3,4-ME MORPHOLINE HCL	124497-83-2
3-ETHOXY-1-PROPANOL	000111-35-3
2-ETHOXYPYRAZINE	038028-67-0
4-ETHOXYPYRAZOLE	081437-10-7
2-ETHOXYPYRIDINE	014529-53-4
ETHOXYQUIN	000091-53-2
6-ETHOXYQUINOXALINE	089770-34-3
ETHOXYTRIETHYLENE GLYCOL	000112-50-5
ETHOXYTRIMETHYLSILANE	001825-62-3
ETHOXYZOLAMIDE	000452-35-7
ETHRANE [HALOETHER]	013838-16-9
2-ETHYLACETANILIDE	033098-65-6
4-ETHYLACETANILIDE	003663-34-1
ETHYL ACETATE	000141-78-6
ETHYLACETOACETATE, BENZAL	000620-80-4
ETHYLACETOACETATE,3-CHLOROBENZAL	015725-23-2
ETHYLACETOACETATE,2-CHLOROBENZAL	015725-22-1
ETHYLACETOACETATE,4-CHLOROBENZAL	019411-80-4
ETHYLACETOACETATE,2,4-DICLBENZAL	015725-29-8
ETHYLACETOACETATE,3,4-DICLBENZAL	015725-31-2
ETHYLACETOACETATE,3,4-DIMEO BENZAL	020451-62-1
ETHYLACETOACETATE,2,4-DIMEO BENZAL	015725-32-3
ETHYLACETOACETATE,2,6-DICLBENZAL	015725-30-1
ETHYLACETOACETATE, 3-ETO BENZAL	015725-27-6
ETHYLACETOACETATE,3-FLUOROBENZAL	015725-21-0
ETHYLACETOACETATE,2-MEO BENZAL	015725-24-3
ETHYLACETOACETATE,4-MEO BENZAL	015725-26-5
ETHYLACETOACETATE,3-MEO BENZAL	015725-25-4
ETHYLACETOACETATE,3-PRO BENZAL	015725-28-7
4-ETHYLACETOPHENONE	000937-30-4
3-ETHYLACETOPHENONE	022699-70-3
ETHYL ACRYLATE	000140-88-5
ETHYLAMINE	000075-04-7
ETHYLAMINE, HYDROCHLORIDE	000557-66-4
4-ETHYLAMINO-6-ISOPROPYLAMINO-S-TRIAZIN-2-OL	002163-68-0
N-(ETHYLAMINOMETHYL)BENZAMIDE	052387-57-2

ETHYLAMPHENICOL	023885-69-0
ETHYL(TERT-AMYL) ETHER	000919-94-8
ETHYL AMYLKETONE	000541-85-5
O-ETHYLANILINE	000578-54-1
P-ETHYLANILINE	000589-16-2
N-ETHYLANILINE	000103-69-5
2-ETHYLANISOLE	014804-32-1
2-ETHYLANTHRAQUINONE	000084-51-5
5-ETHYL-6-AZAURACIL	019213-65-1
2-ETHYLAZIRIDINE	002549-67-9
1-ETHYL AZULENE CARBOXYLATE	001206-88-8
5-ETHYLBARBITURIC ACID	002518-72-1
7-ETHYLBENZ(A)ANTHRACENE	003697-30-1
2-ETHYLBENZAMIDE	067832-97-7
ETHYLBENZENE	000100-41-4
4-ETHYLBENZENESULFONAMIDE	000138-38-5
ETHYL BENZENESULFONATE	000515-46-8
ETHYL BENZOATE	000093-89-0
2'-ETHYLBENZOGUANAMINE	072775-79-2
4-ETHYLBENZOIC ACID	000619-64-7
1-ETHYLBENZOTRIAZOLE	016584-05-7
2-ETHYLBENZOTRIAZOLE	016584-04-6
ETHYL BENZOYLACETATE	000094-02-0
M-ETHYLBENZYL TRIMETHYL AMMONIUM BROMIDE	071324-04-4
3-ETHYL-4-BROMOACETANILIDE	052121-41-2
ETHYL ALPHA-BROMOISOBUTYRATE	000600-00-0
2-ETHYL-1-BUTANOL	000097-95-0
3-ETHYLBUTANOYLUREA	002274-01-3
2-ETHYL-1-BUTENE	000760-21-4
ETHYL T-BUTYL ETHER	000637-92-3
ETHYL T-BUTYL ETHYL DIETHER	051422-54-9
ETHYL BUTYLSULFIDE	000638-46-0
2-ETHYLBUTYRALDEHYDE	000097-96-1
ETHYLBUTYRATE	000105-54-4
2-ETHYLBUTYRIC ACID	000088-09-5
N-ETHYL-EPSILON-CAPROLACTAM	019797-08-1
O-ETHYL CARBAMATE (URETHANE)	000051-79-6
N-ETHYLCARBAZOL	000086-28-2
ETHYL CARBETHOXYMETHYL PHTHALATE	000084-72-0
ETHYL CHLOROACETATE	000105-39-5
ETHYL CHLOROCARBONATE	000541-41-3
ETHYL CHLOROGLYOXYLATE	004755-77-5
ETHYL 4-CHLOROPHENOXYACETATE	014426-42-7
ETHYL ALPHA-CHLOROPROPIONATE	000535-13-7
ETHYL CINNAMATE,3,4-DIHYDROXY	000102-37-4
ETHYL CYANOACETATE	000105-56-6
ETHYLCYCLOHEXANE	001678-91-7
1-ETHYLCYCLOHEXANOL	001940-18-7
2-ETHYLCYCLOHEXANOL	003760-20-1
S-ETHYL CYCLOHEXYLETHYLCARBAMOTHIOATE	001134-23-2
ETHYLCYCLOPENTANE	001640-89-7
ETHYL DECANOATE	000110-38-3
1-(2-ETHYL),2-DIBENZOYL-1-TERT-BUTYLHYDRAZINE	112427-41-5
ETHYL DICHLOROACETATE	000535-15-9
ETHYL DICHLOROCARBAMATE	013698-16-3
5-ETHYLDIHYDRO-5-PHENYL-4,6(1H,5H)-PYRINIDINEDI*	000125-33-7
ETHYL DIMETHYLAMINE	000598-56-1
4-ETHYL-1,2-DIMETHYLBENZENE	000934-80-5

2-ETHYL-1,4-DIMETHYLBENZENE	001758-88-9
ETHYL DIPROPYLAMINE	020634-92-8
O-ETHYL S,S-DIPROPYL PHOSPHORODITHIOATE	013194-48-4
ETHYL DOCOSANOATE	005908-87-2
ETHYL EICOSANOATE	018281-05-5
ETHYLENE	000074-85-1
(ETHYLENEBIS(DITHIOCARBAMATO))ZINC	012122-67-7
ETHYLENEBIS(OXYETHYLENE)DIACETATE	000111-21-7
ETHYLENE CYANOHYDRIN	000109-78-4
ETHYLENEDIAMINETETRAACETIC ACID	000060-00-4
ETHYLENE DIMETHYLACRYLATE	000097-90-5
ETHYLENE GLYCOL	000107-21-1
ETHYLENE GLYCOL N-BUTYL ETHER	000111-76-2
ETHYLENE GLYCOL DIACETATE	000111-55-7
ETHYLENE GLYCOL DIMETHYL ETHER	000110-71-4
ETHYLENE GLYCOL DINITRATE	000628-96-6
ETHYLENE GLYCOL MONOBENZYL ETHER	000622-08-2
ETHYLENE GLYCOL MONOPROPYL ETHER	002807-30-9
ETHYLENE OXIDE	000075-21-8
ETHYLENE SULFIDE	000420-12-2
ETHYLENETHIOUREA	000096-45-7
N-ETHYLETHANAMINE, HYDROCHLORIDE	000660-68-4
ETHYL 4-ETHOXYBENZOATE	023676-09-7
ETHYL ETHYLCARBAMATE	000623-78-9
ETHYL N-ETHYLNITROSOCARBAMATE	000614-95-9
9-ETHYL-9H-FLUORENE	002294-82-8
ETHYL FORMATE	000109-94-4
2-ETHYLFURAN	003208-16-0
ETHYL 2-FUROATE	001335-40-6
ETHYL GALLATE	000831-61-8
ETHYL HEPTANOATE	000106-30-9
2-ETHYLHEXANAL	000123-05-7
2-ETHYL-1,3-HEXANDIOL	000094-96-2
3-ETHYLHEXANE	000619-99-8
ETHYL HEXANOATE	000123-66-0
2-ETHYL-1-HEXANOL	000104-76-7
2-ETHYL-2-HEXENAL	000645-62-5
2-ETHYL-1-HEXENE	001632-16-2
2-ETHYLHEXYLACETATE	000103-09-3
2-ETHYLHEXYL ACRYLATE	000103-11-7
2-ETHYLHEXYLAMINE	000104-75-6
5-ETHYL-5-HEXYLBARBITURIC ACID	000077-30-5
2-ETHYLHEXYL DIPHENYL PHOSPHATE	001241-94-7
2-ETHYLHEXYL 2-METHYL-2-PROPENOATE	000688-84-6
3-ETHYLHYDROCHLOROTHIAZIDE	001824-58-4
3-ETHYL-4-HYDROXYACETANILIDE	028026-77-9
ETHYL 2-HYDROXYBENZOATE	000118-61-6
ETHYL-3-HYDROXYBENZOATE	007781-98-8
N-ETHYL-HYDROXYPYRID-4-ONE	125757-81-5
1-ETHYLHYDROXYUREA	007433-42-3
3-ETHYLHYDROXYUREA	005710-11-2
ETHYLIDENE NORBORNENE	016219-75-3
ETHYL IODOACETATE	000623-48-3
ETHYL ISOBUTYRATE	000097-62-1
2-ETHYLISONIAZID	004608-25-7
ETHYL ISOPROPYLAMINE	019961-27-4
ETHYL ISOPROPYL ETHER	000625-54-7
ETHYL ISOPROPYL SULFIDE	005145-99-3

ETHYL ISOTHIOCYANATE	000542-85-8
ETHYL ISOVALERATE	000108-64-5
ETHYL LACTATE	000097-64-3
ETHYL LAURATE	000106-33-2
ETHYL LEVULINATE	000539-88-8
ETHYLMALONIC ACID	000601-75-2
ETHYL MERCAPTAN	000075-08-1
ETHYL MERCURIC PHOSPHATE	002235-25-8
ETHYLMERCURY CHLORIDE	000107-27-7
ETHYL METHACRYLATE	000097-63-2
ETHYL METHANESULFONATE	000062-50-0
2-ETHYL-3-METHOXYPYRAZINE	025680-58-4
1-ETHYL-2-METHYLBENZENE	000611-14-3
ETHYL-P-METHYLBENZENESULFONATE	000080-40-0
ETHYL METHYL ETHER	000540-67-0
N-(2-ETHYL(3-METHYL-4-NITROPHENYL)AMINO)ETHYL)-M	056046-62-9
ETHYL N-METHYLNITROSOCARBAMATE	000615-53-2
3-ETHYL-5-METHYLPHENOL	000698-71-5
O-ETHYL METHYLPHOSPHONOTHIOATE	018005-40-8
ETHYL METHYL PHOSPHONATE	001832-53-7
5-ETHYL-2-METHYLPYRIDINE	000104-90-5
ETHYL METHYL SULFIDE	000624-89-5
ETHYL MEXRENONE	041020-77-3
4-ETHYLMORPHOLINE	000100-74-3
ETHYLMYRISTATE	000124-06-1
2-ETHYLNAPHTHALENE	000939-27-5
1-ETHYLNAPHTHALENE	001127-76-0
N-ETHYLNICOTINAMIDE	004314-66-3
ETHYLNITRATE	000625-58-1
ETHYLNITRITE	000109-95-5
N-ETHYL-4-NITROBENZENAMINE	003665-80-3
4-ETHYLNITROBENZENE	000100-12-9
2-ETHYLNITROBENZENE	000612-22-6
ETHYL-P-NITROBENZOATE	000099-77-4
ETHYL-M-NITROBENZOATE	000618-98-4
2-ETHYL-2-NITRO-1,3-PROPANEDIOL	000597-09-1
1-ETHYL-1-NITROSO-3-PHENYLUREA	054680-35-2
1-ETHYL-1-NITROSOUREA	000759-73-9
B-ETHYL-B-NITROSTYRENE	001202-32-0
ETHYL NONADECANOATE	018281-04-4
ETHYL NONANOATE	000123-29-5
ETHYL OCTANOATE	000106-32-1
ETHYL ORTHOFORMATE	000122-51-0
ETHYL OXALATE	000095-92-1
ETHYL PALMITATE	000628-97-7
ETHYL PENTAFLUOROPROPIONATE	000426-65-3
3-ETHYLPENTANE	000617-78-7
2-ETHYLPENTANOIC ACID	020225-24-5
3-ETHYL-3-PENTANOL	000597-49-9
3-ETHYL-1-PENTENE	004038-04-4
3-ETHYL-2-PENTENE	000816-79-5
2-((1-ETHYLPENTYL)OXY)ETHANOL	010138-47-3
O-ETHYLPHENOL	000090-00-6
M-ETHYLPHENOL	000620-17-7
P-ETHYLPHENOL	000123-07-9
O-ETHYL PHENOXYACETIC ACID	001798-03-4
M-ETHYLPHENOXYACETIC ACID	001878-51-9
1-(4-ETHYLPHENOXY)SILATRANE	090936-53-1

4-ETHYLPHENYL ACETATE	003245-23-6
2-ETHYLPHENYL ACETATE	003056-59-5
2-ETHYL-2-PHENYLGLUTERIMIDE	000077-21-4
2-ETHYL-2-PHENYL-MALONAMIDE	080866-90-6
ETHYL B-PHENYLPROPIONATE	002021-28-5
P-ETHYL-N-PHENYLSUCCINIMIDE	072601-54-8
M-ETHYL-N-PHENYLSUCCINIMIDE	072601-47-9
2-ETHYLPROPENAL	000922-63-4
ETHYL PROPIONATE	000105-37-3
ETHYL PROPYL DISULFIDE	030453-31-7
ETHYL PROPYL ETHER	000628-32-0
ETHYLPROPYLSULFIDE	004110-50-3
2-ETHYL PYRAZINE	013925-00-3
2-ETHYLPYRAZINE CARBOXYLATE	006924-68-1
4-ETHYLPYRIDINE	000536-75-4
3-ETHYLPYRIDINE	000536-78-7
2-ETHYLPYRIDINE	000100-71-0
1-ETHYLPYRIDINIUM BROMIDE	001906-79-2
N-ETHYL-2-(3-PYRIDYL)ETHYLAMINE	019730-15-5
ETHYL-4-PYRIDYL KETONE	001701-69-5
N-ETHYL-3-PYRIDYLMETHYLAMINE	003000-75-7
ETHYLQUINOLINIUM BROMIDE	026670-42-8
4-ETHYLSEMICARBAZIDE	013050-41-4
ETHYL SILICATE	000078-10-4
ETHYL STEARATE	000111-61-5
M-ETHYLSTYRENE	007525-62-4
P-ETHYLSTYRENE	003454-07-7
N-ETHYL 4-SULFAMYLBENZAMIDE	059777-62-7
ETHYLTHIOPERAZINE	001420-55-9
2-ETHYLTHIOPHENE	000872-55-9
3-ETHYLTHIOPHENE	052006-63-0
N-ETHYLTHIOUREA	000625-53-6
P-ETHYLTOLUENE	000622-96-8
M-ETHYLTOLUENE	000620-14-4
ETHYL TRICHLOROACETATE	000515-84-4
ETHYLTRICHLOROSILANE	000115-21-9
ETHYL TRICOSANOATE	018281-07-7
ETHYL-(3,4,5-TRIMEO)PHENYL SULFONE	071203-60-6
ETHYLTRIMETHYLAMMONIUM IODIDE	000051-93-4
ETHYL UNDECANOATE	000627-90-7
ETHYLUREA	000625-52-5
ETHYL VALERATE	000539-82-2
ETHYL VINYL DICHLOROSILANE	010138-21-3
ETHYL VINYL ETHER	000109-92-2
ETHYNYL BENZENE	000536-74-3
1-ETHYNYL CYCLOHEXANOL	000078-27-3
ETIDOCAINE (DURANEST)	036637-18-0
2(1ET-13-IMIDAZOLINYLIDEN-2AM)5NO2-THIAZOLE	031052-77-4
5ET-6MEO-4SULFANILAMIDOPYRIMIDINE	007756-44-7
3ET-5-(5NO2-FURFURIL)THIAZOLIDIN-4-ON	025603-13-8
3-ETO-5-BR-6-ME-PYRIMIDIN-2,4-DIONE	077317-66-9
1(4-ETO-CO PHENYL)-3-MEO-3-ME UREA	088132-25-6
1-ETOET-2-NO2 IMIDAZOLE	071006-76-3
ETOFENPROX	080844-07-1
ETOFYLLINE NICOTINATE	013425-39-3
5-ETO-6-MEO-4-SULFANILAMIDOPYRIMIDIN	005532-46-7
ETOPSIDE	033419-42-0
ETORPHINE	014521-96-1

2,6-ETO-4-SULFANILAMIDOPYRIMIDINE	001164-13-2
2-(3-HOET-1,3-OXAZOLIDINYLIDEN-2N)-5-NO2-THIAZOL	024240-65-1
ETOXINOL	006012-83-5
N'-ET-PHENOXYACETIC ACID HYDRAZIDE	052093-77-3
ETRIMFOS	038260-54-7
3-ETS-4-NH2-6-IPR-1,2,4-TRIAZIN-5-ON	050917-22-1
1-(2-ETSO2ET)-2-NO2-IMIDAZOLE	068160-70-3
ET-4-SULFONAMIDONAPHTHYLSULFONE	000842-00-2
BETA-EUCAINE	000500-34-5
EUGENOL	000097-53-0
EUGENYL ACETATE	000093-28-7
EUGENYL FORMATE	010031-96-6
EVOMONOSIDE	000508-93-0
111F3-22-DI(P-MEOPHENYL)ETHANE	000384-97-4
FAMOTIDINE	076824-35-6
FAMPHUR	000052-85-7
2-F 4-BR 5-NO2 ACETANILIDE	095635-46-4
FENAC	000085-34-7
FENAMINOSULF	000140-56-7
FENAMIPHOS	022224-92-6
FENAPANIL	061019-78-1
FENARIMOL	060168-88-9
FENAZAQUIN	120928-09-8
FENBUFEN	036330-85-5
FENCAMFAMINE	001209-98-9
FENCHONE	001195-79-5
FENCHYL ALCOHOL	001632-73-1
FENCLOFENAC	034645-84-6
FENCLORIM	003740-92-9
FENETHAZINE	000522-24-7
FENFLURAMINE	000458-24-2
FENITROTHION	000122-14-5
FENOTHIOCARB	062850-32-2
FENOXAPROP-ETHYL	082110-72-3
FENOXYCARB	079127-80-3
FENOXYCARB	072490-01-8
FENPICLONIL	074738-17-3
FENPROPATHRIN	039515-41-8
FENSULFOTHION	000115-90-2
FENSULFOTHION SULFIDE	003070-15-3
FENSULFOTHION SULFONE	014255-72-2
FENTANYL	000437-38-7
FENTHION	000055-38-9
FENTHION SULFOXIDE	003761-41-9
FENTHOATE	002597-03-7
FENTIAZAC	018046-21-4
FENURON	000101-42-8
FENURON TCA	004482-55-7
FENVALERATE	051630-58-1
FERBAM	014484-64-1
FLA 797	084226-14-2
FLAMPROP	058667-63-3
FLAMPROP-ISOPROPYL	052756-22-6
FLAMPROP-METHYL	052756-25-9
FLAVANONE	000487-26-3
FLAVONE	000525-82-6
FLECANIDE	054143-55-4
FLEROXACIN	079660-72-3

1(2-FLET)-3(4-SULFAPYRANYL)-1-NO UREA	032319-90-7
FLOXACILLIN	005250-39-5
FLOXURIDINE	000050-91-9
FLUAZINAM	079622-59-6
FLUAZURON	086811-58-7
FLUCHLORALIN	033245-39-5
FLUCYCLOXURON,(E)-ISOMER	094050-52-9
FLUCYCLOXURON,(Z)-ISOMER	094050-53-0
FLUCYTHRINATE	070124-77-5
FLUDIAZEPAM	003900-31-0
FLUDIOXONIL	131341-86-1
FLUETHIAZIDE	023141-80-2
FLUFENAMIC ACID	000530-78-9
FLUMAZENIL	078755-81-4
FLUMETHASONE	002135-17-3
FLUMETHASONE-21-PIVALATE	002002-29-1
FLUMETHAZIDE	000148-56-1
FLUMICLORAC-PENTYL	087546-18-7
FLUNITRAZEPAM	001622-62-4
FLUOCINOLONE ACETONIDE	000067-73-2
FLUOCINONIDE	000356-12-7
FLUOMETURON	002164-17-2
3-FLUORANTHENAMINE	002693-46-1
FLUORANTHENE	000206-44-0
FLUORBENSIDE	000405-30-1
9H-FLUOREN-1-AMINE	006344-63-4
9H-FLUOREN-2-AMINE, 7-BROMO-	006638-60-4
4AH-FLUOREN-4A-AMINE, N-ETHYL-1,2,3,4,9,9A-HEXAH	132453-02-2
9AH-FLUOREN-9A-AMINE, 1,2,3,4,4A,9-HEXAHYDRO-, C	147128-62-9
9H-FLUOREN-2-AMINE, 7-NITRO-	001214-32-0
9AH-FLUOREN-9A-AMINE, 1,4,4A,9-TETRAHYDRO-N-METH	147064-47-9
1H-FLUORENE	000244-36-0
9H-FLUORENE	000086-73-7
9H-FLUORENE-9-CARBOXAMIDE, 9-[3-[(1-METHYLETHYL)	074517-78-5
9H-FLUORENE, 2,7-DINITRO-	005405-53-8
1H-FLUORENE-2-METHANAMINE, 2,3,4,4A,9,9A-HEXAHYD	132453-26-0
1H-FLUORENE-2-METHANAMINE, 2,3,4,4A,9,9A-HEXAHYD	132453-25-9
9H-FLUORENE-9-METHANOL	024324-17-2
FLUORENONE	000486-25-9
9H-FLUOREN-9-ONE, 2,7-DINITRO-	031551-45-8
9H-FLUOREN-9-ONE, 3-NITRO-	042135-22-8
9H-FLUOREN-9-ONE, 2,4,5,7-TETRANITRO-	000746-53-2
9H-FLUOREN-9-ONE, 2,4,7-TRINITRO-	000129-79-3
N-2-FLUORENYLACETAMIDE	000053-96-3
4-(9H-FLUOREN-9-YL)-N,N-DIMETHYLBENZENAMINE	032377-15-4
1-FLUORENYL-3-MEO-3-ME UREA	088132-35-8
FLUORESCEIN	000518-45-6
FLUOROACETAMIDE	000640-19-7
4-FLUOROACETANILIDE	000351-83-7
3-FLUOROACETANILIDE	000351-28-0
2-FLUOROACETANILIDE	000399-31-5
2-FLUORO-5-CF3-ACETANILIDE	000349-27-9
FLUOROACETIC ACID	000144-49-0
FLUOROACETIC ACID, SODIUM SALT	000062-74-8
FLUOROACETONE	000430-51-3
M-FLUOROACETOPHENONE	000455-36-7
4-FLUOROACETOPHENONE	000403-42-9
2-FLUORO-4-AMINOBENZOIC ACID	000446-31-1

4-FLUOROANILINE	000371-40-4
2-FLUOROANILINE	000348-54-9
3-FLUOROANILINE	000372-19-0
2-FLUOROANISOLE	000321-28-8
P-FLUOROANISOLE	000459-60-9
M-FLUOROANISOLE	000456-49-5
4-FLUOROBENZAL ACETYLACETONE	015851-94-2
P-FLUOROBENZALDOXIME	000459-23-4
P-FLUOROBENZAMIDE	000824-75-9
3-FLUOROBENZAMIDE	000455-37-8
2-FLUOROBENZAMIDE	000445-28-3
FLUOROBENZENE	000462-06-6
4'-FLUOROBENZOGUANAMINE	030530-44-0
3'-FLUOROBENZOGUANAMINE	030530-43-9
2-FLUOROBENZOIC ACID	000445-29-4
M-FLUOROBENZOIC ACID	000455-38-9
P-FLUOROBENZOIC ACID	000456-22-4
P-FLUOROBENZYL CHLORIDE	000352-11-4
M-FLUOROBENZYL CHLORIDE	000456-42-8
O-FLUOROBENZYL CHLORIDE	000345-35-7
P-FLUOROBENZYL N,N-DIME CARBAMATE	084640-22-2
M-FLUOROBENZYL TRIMETHYL AMMONIUM BROMIDE	071324-00-0
4-FLUOROBIPHENYL	000324-74-3
2-FLUORO-6-BROMOBENZAMIDE	107485-63-2
1-FLUOROBUTANE	002366-52-1
2-FLUORO-4-CHLOROACETANILIDE	059280-70-5
2-FLUORO-6-CHLOROBENZAMIDE	066073-54-9
1-FLUORODECANE	000334-56-5
1-(3-FLUORO),2-DIBENZOYL-1-TERT-BUTYLHYDRAZINE	112227-02-8
1-(4-FLUORO),2-DIBENZOYL-1-TERT-BUTYLHYDRAZINE	112225-85-1
1-(2-FLUORO),2-DIBENZOYL-1-TERT-BUTYLHYDRAZINE	112226-76-3
FLUORODIFEN	015457-05-3
FLUORODIMETHYLPHENYLSILANE	000454-57-9
1-FLUORO-2,4-DINITROBENZENE	000070-34-8
1-FLUORODODECANE	000334-68-9
FLUOROETHANE	000353-36-6
2-FLUOROETHANOL	000371-62-0
P-FLUOROFORMANILIDE	000459-25-6
1-FLUOROHEPTANE	000661-11-0
1-FLUOROHEXANE	000373-14-8
2-FLUOROISONIAZID	000369-24-4
FLUOROMETHALONE	000426-13-1
FLUOROMETHANE	000593-53-3
FLUOROMETHANESULFONANILIDE,P-CL	050585-76-7
P-FLUORO CF3-METHANESULFONANILIDE	023384-00-1
M-FLUORO CF3-METHANESULFONANILIDE	023384-01-2
FLUOROMETHANE SULFONANILIDE	002070-61-3
FLUOROMETHOLONE ACETATE	003801-06-7
A-FLUORO-N-METHYLACETAMIDE	000367-49-7
(4-FLUORO-3-METHYLPHENYL)UREA	078508-45-9
FLUOROMETHYL SULFONE	000558-25-8
1-FLUORONAPHTHALENE	000321-38-0
4-FLUORONITROBENZENE	000350-46-9
3-FLUORONITROBENZENE	000402-67-5
O-FLUORONITROBENZENE	001493-27-2
5-FLUORO-2-NITROPHENOL	000446-36-6
2-FLUORO-4-NITROPHENOL	000403-19-0
8-FLUORO-4-NITROQUINOLINE-1-OXIDE	019789-69-6

1-FLUORONONANE	000463-18-3
1-FLUOROOCTANE	000463-11-6
8-FLUORO-PEFLOXACIN	075338-41-9
1-FLUOROPENTANE	000592-50-7
P-FLUOROPHENOL	000371-41-5
M-FLUOROPHENOL	000372-20-3
O-FLUOROPHENOL	000367-12-4
P-FLUOROPHENOXYACETIC ACID	000405-79-8
M-FLUOROPHENOXYACETIC ACID	000404-98-8
O-FLUOROPHENOXYACETIC ACID	000348-10-7
3-FLUOROPHENYL ACETATE	000701-83-7
2-FLUOROPHENYL ACETATE	029650-44-0
O-FLUOROPHENYLACETIC ACID	000451-82-1
P-FLUOROPHENYLACETIC ACID	000405-50-5
M-FLUOROPHENYLACETIC ACID	000331-25-9
3-FLUOROPHENYL 4-AMINOSALICYLATE	056356-26-4
2-FLUOROPHENYL 4-AMINOSALICYLATE	056356-25-3
4-FLUOROPHENYL 4-AMINOSALICYLATE	056356-27-5
P-FLUOROPHENYLHIPPURATE	029736-22-9
2-(M-FLUOROPHENYL)MORPHOLINE	062008-55-3
M-FLUORO-N-PHENYLSUCCINIMIDE	060693-35-8
O-FLUORO-N-PHENYLSUCCINIMIDE	072601-44-6
P-FLUORO-N-PHENYLSUCCINIMIDE	060693-37-0
P-FLUOROPHENYLUREA	000659-30-3
O-FLUOROPHENYLUREA	000656-31-5
M-FLUOROPHENYLUREA	000770-19-4
1-FLUOROPROPANE	000460-13-9
3-FLUOROPROPANOL	000462-43-1
2-FLUOROPYRAZINE	004949-13-7
2-FLUOROPYRIDINE	000372-48-5
3-FLUOROPYRIDINE	000372-47-4
8-FLUOROQUINOLINE	000394-68-3
P-FLUOROSULFONYLTOLUENE	000455-16-3
FLUOROTHIAZIDE	001535-61-1
O-FLUOROTOLUENE	000095-52-3
P-FLUOROTOLUENE	000352-32-9
M-FLUOROTOLUENE	000352-70-5
1-FLUORO-2-TRIFLUOROMETHYLBENZENE	000392-85-8
1-FLUOROUNDECANE	000506-05-8
5-FLUOROURACIL	000051-21-8
5-FLUOROURACIL-3-ETHOXYCARBONYL-	021839-33-8
FLUPENTIXOL	002709-56-0
FLUPHENAZINE	000069-23-8
FLURBIPROFEN	005104-49-4
FLURIDONE	059756-60-4
FLUROCHLORIDONE	061213-25-0
FLUROXENE	000406-90-6
FLURPRIMIDOL	056425-91-3
FLURTAMONE	096525-23-4
FLUTRIAFOL (PP450)	076674-21-0
FLUVALINATE	069409-94-5
FLUXOFENIM	088485-37-4
5-F-2'-ME-4'-BR SALICYANILIDE	077068-04-3
5-F-2'-ME-4'-CL SALICYANILIDE	077068-02-1
1-(3-F-2-OHPROPYL)-2-NO2 IMIDAZOLE	013551-89-8
FOLIC ACID	000059-30-3
FOLPET	000133-07-3
FOMESAFEN	072178-02-0

FONOPHOS	000944-22-9
FONOPHOS, O-ANALOG	000944-21-8
FORMALDEHYDE	000050-00-0
FORMALDEHYDE, (2,4-DINITROPHENYL)HYDRAZONE	001081-15-8
FORMAMIDE	000075-12-7
FORMAMIDE, N-(3-CHLOROPHENYL)-	000139-71-9
FORMAMIDE, N-[2-(DIETHYLAMINO)ETHYL]-N-(2,6-DIME	020682-52-4
FORMAMIDE, N-2-THIAZOLYL	025602-39-5
FORMAMIDE OXIME	000075-17-2
FORMANILIDE	000103-70-8
FORMETANATE	022259-30-9
FORMETANATE HYDROCHLORIDE	023422-53-9
FORMIC ACID	000064-18-6
FORMOTHION	002540-82-1
P-FORMYLACETANILIDE	000122-85-0
N-FORMYL-P-ACETOXYSTYRLAMINE	066432-08-4
8-FORMYLAMIDOQUINOLINE	062937-22-8
4-FORMYLBENZOIC ACID	000619-66-9
3-FORMYLBENZOIC ACID	000619-21-6
N-FORMYL-P-CYANOSTYRYLAMINE	066516-96-9
N-FORMYLCYCLOBUTANECARBOXAMIDE	023046-86-8
P-FORMYL-N,N-DIMETHYLANILINE	000100-10-7
N-FORMYLETHYLAMINE	000627-45-2
FORMYL FLUORIDE	001493-02-3
N-FORMYLHYDRAZINE	000624-84-0
N-FORMYLMORPHOLINE	004394-85-8
4-FORMYL-2-NITROPHENOL	003011-34-5
P-FORMYLPHENOXYACETIC ACID	022042-71-3
4-FORMYLPYRIDINE	000872-85-5
3-FORMYLPYRIDINE	000500-22-1
N-FORMYLPYRROLIDINE	003760-54-1
5-FORMYLURACIL	001195-08-0
FOSAMINE AMMONIUM SALT	025954-13-6
FOSTEDIL,2-OXOPYRROLIDINO ANANLOG	104608-36-8
FOSTHIETAN	021548-32-3
1-(M-F PHENYL)-3-MEO-3-ME UREA	028170-26-5
1-(P-F PHENYL)-3-MEO-3-ME UREA	088132-24-5
N-(O-FPH)3-N'-PIPERIDINOACETAMIDE	058479-88-2
N-(M-FPH)3-N'-PIPERIDINOACETAMIDE	058479-90-6
N(P-FPH)-3-N'-PIPERIDINOACETAMIDE	037163-41-0
N-FORMYL-STYRYLAMINE	066432-03-9
BETA-FRUCTOPYRANOSE (D)	007660-25-5
FUBERIDAZOLE	003878-19-1
FUFURAL	000098-01-1
FUMARIC ACID	000110-17-8
FUMARYL CHLORIDE	000627-63-4
FURAN	000110-00-9
3-FURANCARBONITRILE, 5-[(DIMETHYLAMINO)METHYLENE	007703-36-8
2-FURANCARBOXALDEHYDE, 5-METHYL-	000620-02-0
FURAN-3-CARBOXALDEHYDE	000498-60-2
FURAN-3-CARBOXAMIDE	000609-35-8
FURAN-2-CARBOXAMIDE	000609-38-1
3-FURANCARBOXAMIDE, N-(AMINOCARBONYL)-TETRAHYDRO	072962-38-0
3-FURANCARBOXAMIDE, N-(AMINOCARBONYL)-TETRAHYDRO	072962-39-1
2-FURANCARBOXAMIDE, N-(3,6-DIHYDRO-1(2H)-PYRIDIN	080431-14-7
2-FURANCARBOXAMIDE, N-PROPYL-	060708-27-2
FURAN-3-CARBOXYLIC AC, ET ESTER	000614-98-2
3-FURANCARBOXYLIC ACID, 4,5-DIHYDRO-5-[(4-METHOX	060404-04-8

2-FURANCARBOXYLIC ACID, 5-BROMO-, METHYL ESTER	002527-99-3
3-FURANCARBOXYLIC ACID, PROPYL ESTER	000615-10-1
3-FURANCARBOXYLIC ACID, METHYL ESTER	013129-23-2
2-FURANCARBOXYLIC ACID	000088-14-2
FURAN-3-CARBOXYLIC ACID	000488-93-7
FURAN, 2-CH(OH)CO-FURYL	000552-86-3
FURAN, 2-COCO-FURYL	000492-94-4
FURAN, 2,5-DIMETHYL-	000625-86-5
3-FURANEMETHANOL-5-BENZYL,ACETATE	039856-64-9
FURANMETHAMINE, N,N-DIMETHYL	014496-34-5
2-FURANMETHANAMINIUM, 5-CHLORO-N,N,N-TRIMETHYL-,	075523-40-9
2-FURANMETHANAMINIUM, 5-ETHYL-N,N,N-TRIMETHYL-,	075523-41-0
2-FURANMETHANAMINIUM, N,N,N,5-TETRAMETHYL-, IODI	001197-60-0
3-FURANMETHANAMINIUM, N,N,N,5-TETRAMETHYL-, IODI	068724-15-2
2-FURANMETHANAMINIUM, 5-BROMO-N,N,N-TRIMETHYL-,	075523-39-6
3-FURANMETHANAMINIUM, TETRAHYDRO-N,N,N,5-TETRAME	068724-25-4
2-FURANMETHANAMINIUM, TETRAHYDRO-N,N,N-TRIMETHYL	052303-77-2
2-FURANMETHANAMINIUM, TETRAHYDRO-4-HYDROXY-N,N,N	002209-02-1
3-FURANMETHANAMINIUM, N,N,N-TRIMETHYL-, IODIDE	022601-05-4
2-FURANMETHANAMINIUM, N,N,N-TRIMETHYL-, IODIDE	000541-64-0
2-FURANMETHANAMINIUM, TETRAHYDRO-N,N,N,5-TETRAME	070135-16-9
3-FURANMETHANAMINIUM, TETRAHYDRO-N,N,N-TRIMETHYL	090032-54-5
2-FURANMETHANOL, 5-(2-AMINO-6-FLUORO-9H-PURIN-9-	132194-21-9
2-FURANMETHANOL, 5-(2-AMINO-6-IODO-9H-PURIN-9-YL	132194-23-1
2-FURANMETHANOL, 5-(6-BROMO-9H-PURIN-9-YL)TETRAH	132194-25-3
2-FURANMETHANOL, 5-(6-CHLORO-9H-PURIN-9-YL)TETRA	120503-34-6
2-FURANMETHANOL, 5-(6-FLUORO-9H-PURIN-9-YL)TETRA	132194-24-2
2-FURANMETHANOL, TETRAHYDRO-5-(6-IODO-9H-PURIN-9	120503-37-9
2(5H)-FURANONE	000497-23-4
2(5H)-FURANONE, 5-(4-HEXEN-2-YNYLIDENE)-	023251-68-5
2-FURANSULFONAMIDE, 4-[(4-HYDROXYPHENYL)SULFONYL	128348-44-7
2-FURANSULFONAMIDE, 4-(4-METHOXYBENZOYL)-	118993-61-6
2-FURANSULFONAMIDE, 4-[(4-METHYLPHENYL)SULFONYL]	128348-34-5
2-FURANSULFONAMIDE, 4-[(4-METHOXYPHENYL)SULFONYL	128348-35-6
FURAZAN	000288-37-9
FURAZANAMINE, 4-METHYL-	017647-70-0
FURAZANAMINE, 4-METHYL-, 5-OXIDE	115174-03-3
FURAZANAMINE, 4-METHYL-, 2-OXIDE	118506-53-9
FURAZANAMINE, 4-PHENYL-	010349-14-1
FURAZANAMINE, 4-PHENYL-, 2-OXIDE	029945-54-8
FURAZANAMINE, 4-PHENYL-, 5-OXIDE	030059-86-0
FURAZAN, BIS(PHENYLMETHYL)-, 2-OXIDE	075389-00-3
FURAZANCARBOXALDEHYDE, METHYL-, 2-OXIDE	123953-17-3
FURAZANCARBOXALDEHYDE, METHYL-, 5-OXIDE	123953-16-2
FURAZANCARBOXAMIDE, 4-METHYL-, 5-OXIDE	037895-44-6
FURAZANCARBOXAMIDE, 4-METHYL-, 2-OXIDE	037895-45-7
FURAZANCARBOXYLIC ACID, 4-METHYL-, METHYL ESTER,	104151-90-8
FURAZANCARBOXYLIC ACID, 4-METHYL-, METHYL ESTER,	104151-78-2
FURAZAN, CHLOROMETHYL-, 5-OXIDE	086988-90-1
FURAZAN, CHLOROMETHYL-, 2-OXIDE	065514-04-7
FURAZAN, CHLOROPHENYL-	024786-13-8
FURAZAN, CHLOROPHENYL-, 2-OXIDE	082775-81-3
FURAZAN, CHLOROPHENYL-, 5-OXIDE	082775-80-2
FURAZANDIAMINE	017220-38-1
FURAZAN, DIMETHYL-	004975-21-7
FURAZAN, DIMETHYL-, 2-OXIDE	002518-42-5
FURAZAN, DIPHENYL-	019768-02-6
FURAZAN, DIPHENYL-, 2-OXIDE	005585-14-8

FURAZAN, ETHYLMETHYL-	017647-69-7
FURAZAN, METHYLNITRO	077666-53-6
FURAZAN, METHYLNITRO-, 2-OXIDE	049558-02-3
FURAZAN, METHYLPHENYL-	010349-09-4
FURAZAN, METHYL(PHENYLSULFONYL)-	049739-42-6
FURAZAN, METHYL(PHENYLTHIO)-, 5-OXIDE	049739-37-9
FURAZAN, METHYL(PHENYLSULFONYL)-, 5-OXIDE	049739-43-7
FURAZAN, METHYLPHENYL-, 2-OXIDE	006898-86-8
FURAZAN, METHYL(PHENYLSULFONYL)-, 2-OXIDE	049739-41-5
FURAZAN, METHYLPHENYL-, 5-OXIDE	006898-87-9
FURAZAN, METHYLPROPYL-	077580-78-0
FURAZAN, NITROPHENYL-	076016-68-7
FURETHRIN	017080-02-3
2H-FURO[2,3-H]-1-BENZOPYRAN-2-ONE, 9-METHYL-	078982-40-8
2H-FURO[2,3-H]-1-BENZOPYRAN-2-ONE, 8-METHYL-	015798-77-3
2H-FURO(2,3-H)-1-BENZOPYRAN-2-ONE, 5-METHYL-	073459-03-7
2H-FURO[2,3-H]-1-BENZOPYRAN-2-ONE, 4-METHYL-	006457-92-7
FURO(2,3-H)COUMARIN	000523-50-2
FUROIC ACID, ETHYL ESTER	000614-99-3
FUROIC ACID, METHYL ESTER	000611-13-2
FUROSEMIDE	000054-31-9
2-(2-FURYL)-3-(5-NO2FURYL)ACRYLAMIDE	003688-53-7
FUSTIN	020725-03-5
G 34698	001824-09-5
G-24622	005826-91-5
GALACTITOL	000608-66-2
GALACTOPYRANOSIDE, ETHYL 1-THIO-	056245-60-4
ß-D-GALACTOPYRANOSIDE, ETHYL	018997-88-1
beta-D-GALACTOPYRANOSIDE, HEPTYL 1-THIO-	050615-69-5
ß-D-GALACTOPYRANOSIDE, HEPTYL	078617-12-6
GALACTOPYRANOSIDE,2-NITROPHENYL	003150-24-1
beta-D-GALACTOPYRANOSIDE, OCTYL	040427-75-6
BETA-D-GALACTOPYRANOSIDE, PHENYL	002818-58-8
GALACTOSE	000059-23-4
GENITE	000097-16-5
GEOSMIN	019700-21-1
GERANIOL	000106-24-1
GERANIOL ACETATE	000105-87-3
GIBBERELLIC ACID	000077-06-5
GITOXIGENIN	000545-26-6
GITOXIN	004562-36-1
GLIOTOXIN	000067-99-2
GLIPIZIDE	029094-61-9
D-GLUCITOL, 1,4:3,6-DIANHYDRO-, DINITRATE	000087-33-2
D-GLUCONIC ACID	000526-95-4
.ALPHA.-D-GLUCOPYRANOSE, PENTAACETATE	000604-68-2
GLUCOPYRANOSIDE,4-AMINOPHENYL	020818-25-1
GLUCOPYRANOSIDE,2-AMINOPHENYL	007265-01-2
GLUCOPYRANOSIDE, 3-T-BUTYL PHENYL	020838-36-2
GLUCOPYRANOSIDE,CYCLOHEXYL	005284-99-1
GLUCOPYRANOSIDE,3,5-DIMETHYLPHENYL	020772-21-8
GLUCOPYRANOSIDE, 3-ETHYLPHENYL	020838-34-0
GLUCOPYRANOSIDE,2-IODOPHENYL	007234-29-9
GLUCOPYRANOSIDE,2-IPR-5-ME PHENYL	020772-23-0
BETA-D-GLUCOPYRANOSIDE, METHYL 3-[[[(2-CHLOROETH	074751-37-4
GLUCOPYRANOSIDE,3-METHYLPHENYL	006092-25-7
GLUCOPYRANOSIDE,2-METHYLPHENYL	007234-31-3
GLUCOPYRANOSIDE,4-METHYLPHENYL	020274-94-6

GLUCOPYRANOSIDE,2-NAPHTHYL	006044-30-0
α-D-GLUCOPYRANOSIDE, 4-NITROPHENYL	003767-28-0
BETA-D-GLUCOPYRANOSIDE, 4-NITROPHENYL	002492-87-7
beta-D-GLUCOPYRANOSIDE, 2-NITROPHENYL	002816-24-2
ß-D-GLUCOPYRANOSIDE, 4-NITROPHENYL 4-O-_-D-GLUCO	056846-39-0
BETA-D-GLUCOPYRANOSIDE, PHENYL	001464-44-4
1-(BD-GLUCOPYRANOSYL)-2-NO2IMIDAZOLE	083107-46-4
GLUCOSE	000050-99-7
D-GLUCOSE, 2,3,4,5,6-PENTAACETATE	003891-59-6
1(BD-GLUCOTHIOPYRANOSYL)2NO2IMIDAZOLE	083107-47-5
GLUCURONIC ACID	006556-12-3
L-GLUTAMAMIDE, N-ACETYL-L-PHENYLALANYL-	132765-94-7
GLUTAMIC ACID (DL)	000617-65-2
GLUTAMIC ACID	000056-86-0
L-GLUTAMIC ACID, N-[4-[[(2,4-DIAMINO-6-PTERIDINY	043170-88-3
GLUTAMIC ACID HYDROCHLORIDE	000138-15-8
GLUTAMIC ACID-MONOAMIDE,N-ACETYL	025460-87-1
GLUTAMINE	000056-85-9
GLUTAMINE-AMIDE, N-ACETYL	018839-88-8
GLUTARIC ACID, 2,2-DIMETHYL-	000681-57-2
GLUTARIC ANHYDRIDE	000108-55-4
GLUTARYL CHLORIDE	002873-74-7
L-GLUTINAMIDE, N-ACETYL-L-VALYL-	132765-95-8
GLYCEROL	000056-81-5
GLYCEROL DICHLOROHYDRIN	026545-73-3
GLYCIDYLALDEHYDE	000765-34-4
GLYCIDYL N-BUTYL ETHER	002426-08-6
GLYCIN-AMIDE,N-ACETYL	002620-63-5
GLYCINAMIDE, N-ACETYLGLYCYL-L-LEUCYL-	132766-13-3
GLYCINAMIDE, N-ACETYL-L-PHENYLALANYL-L-ISOLEUCYL	132766-11-1
GLYCINAMIDE, N-ACETYL-L-PHENYLALANYL-	029701-43-7
GLYCINAMIDE, N-ACETYL-L-TRYPTOPHYL-L-ISOLEUCYL-	132766-15-5
GLYCINAMIDE, N-ACETYL-L-VALYL-L-ISOLEUCYL-	132766-03-1
GLYCINE, N-[N-[N-(N-GLYCYLGLYCYL)GLYCYL]GLYCYL]-	007093-67-6
GLYCINE	000056-40-6
GLYCINE,N-ACETYL-N'-MEAMINO AMIDE	007606-79-3
GLYCINE, N-[[[2-(ACETYLOXY)BENZOYL]OXY]ACETYL]-,	118247-03-3
GLYCINE, N-(4-AMINOBENZOYL)-	000061-78-9
GLYCINE, N-[(BENZOYLOXY)ACETYL]-N-METHYL-, ETHYL	106231-63-4
GLYCINE, N-[(BENZOYLOXY)ACETYL]-N-METHYL-	106231-64-5
GLYCINE, N-(CHLOROACETYL)-N-[2-(1-METHYLETHOXY)P	060145-79-1
GLYCINE, N-(CHLOROACETYL)-N-[1,1'-BIPHENYL]-2-YL	060145-77-9
GLYCINE, N-(2-HYDROXYBENZOYL)-	000487-54-7
GLYCINE, N-(3-IODOBENZOYL)-	052386-94-4
GLYCINE, N-(4-IODOBENZOYL)-	055790-22-2
GLYCINE, N-(2-IODOBENZOYL)-	000147-58-0
GLYCINE, N-[N-[1-OXO-5-(7H-PURIN-6-YLTHIO)PENTYL	022181-95-9
GLYCINE, N-[1-OXO-5-(1H-PURIN-6-YLTHIO)PENTYL]-,	028610-09-5
GLYCINE, N-[1-OXO-5-(1H-PURIN-6-YLTHIO)PENTYL]-,	022181-94-8
GLYCINE, N-[N-[N-[1-OXO-5-(7H-PURIN-6-YLTHIO)PEN	023374-45-0
GLYCOCHENODEOXYCHOLIC ACID	000640-79-9
GLYCOCHOLIC ACID	000475-31-0
GLYCODEOXYCHOLIC ACID	000360-65-6
GLYCOHYOCHOLIC ACID	032747-08-3
GLYCOHYODEOXYCHOLIC ACID	013042-33-6
GLYCOLALDEHYDE	000141-46-8
GLYCOLIC ACID, SODIUM SALT	002836-32-0
GLYCOURSODEOXYCHOLIC ACID	064480-66-6

GLYCYRRHIZIC ACID	001405-86-3
GLYOXAL	000107-22-2
GLYOXYLIC ACID	000298-12-4
GLYPHOSINE	002439-99-8
GLYPHOSPHATE	001071-83-6
GRAYANOTOXANE-5,6,7,9,10,14,16-HEPTOL, 2,3-EPOXY	034206-60-5
GRAYANOTOXIN III	004678-45-9
GRISEOFULVIN	000126-07-8
GUANIDINE	000113-00-8
GUANIDINE, N-[2-[[[2-[(AMINOIMINOMETHYL)AMINO]-4	069014-14-8
GUANIDINE, N-METHYL-N'-[2-[[(5-METHYL-1H-IMIDAZO	059660-23-0
GUANIDINE, (4-METHYL-2-THIAZOLYL)-	007120-01-6
GUANIDINE, N,N,N',N'-TETRAMETHYL-	000080-70-6
GUANINE	000073-40-5
GUANOSINE	000118-00-3
GUANOSINE, 2'-DEOXY-	000961-07-9
GUANOSINE, 2',3'-DIDEHYDRO-2',3'-DIDEOXY-	053766-80-6
GUANOSINE, 2',3'-DIDEOXY-	085326-06-3
GUANOSINE, 2',3'-DIDEOXY-3'-FLUORO-	092562-88-4
GYCYLGLYCINE	000556-50-3
HALOPERIDOL	000052-86-8
HALOTHANE	000151-67-7
HALOXYFOP-ETOTYL	087237-48-7
HARMALOL	000525-57-5
HARMONY	079277-27-3
HELENALIN	006754-13-8
HELVETICOSIDE	000630-64-8
HELVETICOSOL	018695-02-8
HEMIMELLITIC ACID	000569-51-7
HENEICOSANE	000629-94-7
HENTRIACONTANE	000630-04-6
HEPTABARBITAL	000509-86-4
HEPTACHLOR	000076-44-8
HEPTACHLOR EPOXIDE	001024-57-3
2,2',3,4,4',5,5'-HEPTACHLOROBIPHENYL	035065-29-3
2,2',3,4',5,5',6-HEPTACHLOROBIPHENYL	052663-68-0
2,2',3,3',5,6,6'-HEPTACHLOROBIPHENYL	052663-64-6
2,2',3,3',4,4',6-HEPTACHLOROBIPHENYL	052663-71-5
2,2',3,3',4,4',5-HEPTACHLOROBIPHENYL	035065-30-6
2,2',3,4,5,5',6-HEPTACHLORO-1,1'-BIPHENYL	052712-05-7
1,1,1,2,2,3,3-HEPTACHLOROBUTANE	083682-70-6
1,1,2,2,3,4,4-HEPTACHLOROBUTANE	034973-41-6
HEPTACHLORODIBENZO-P-DIOXIN	037871-00-4
1,2,3,4,6,7,8-HEPTACHLORODIBENZO-P-DIOXIN	035822-46-9
HEPTACOSAFLUOROTRIBUTYLAMINE	000311-89-7
HEPTACOSANE	000593-49-7
HEPTADECANE	000629-78-7
HEPTADECANOIC ACID	000506-12-7
1-HEPTADECENE	006765-39-5
HEPTADECYL ALCOHOL	001454-85-9
2-HEPTADECYL-4,5-DIHYDRO-2-IMIDAZOLINE-1-ETHANOL	000095-19-2
1,6-HEPTADIENE	002396-63-6
3,5-HEPTADIONE	007424-54-6
2,2,3,3,4,4,-HEPTAFLUOROBUTANOL	000375-01-9
1,1,1,2,2,3,3-HEPTAFLUORO-7,7-DIMETHYL-4,6-OCTA*	017587-22-3
HEPTAMETHYLPHENYLCYCLOTETRASILOXANE	010448-09-6
1,1,1,3,3,5,5-HEPTAMETHYL-5-PHENYLTRISILOXANE	018407-16-4
1,1,1,3,5,5,5-HEPTAMETHYL-3-(2-(TRIMETHYLSILY*))	018077-53-7

2,2,4,4,6,8,10-HEPTAMETHYL-6,8,10-TRIS(3,3,3-TR*	022474-57-3
HEPTANAL	000111-71-7
HEPTANAMIDE, 7-(DIETHYLAMINO)-N-(2,6-DIMETHYLPHE	077470-84-9
HEPTANAMIDE, N-[4-(DIMETHYLAMINO)SULFONYL]PHENYL	074746-44-4
N-HEPTANE	000142-82-5
HEPTANEDINITRILE	000646-20-8
HEPTANEDIOIC ACID, DIMETHYL ESTER	001732-08-7
3,5-HEPTANEDIONE, 2,6-DIMETHYL-	018362-64-6
HEPTANENITRILE	000629-08-3
1-HEPTANETHIOL	001639-09-4
HEPTANOIC ACID	000111-14-8
HEPTANOIC ACID HYDRAZIDE	022371-32-0
3-HEPTANOL	000589-82-2
2-HEPTANOL	000543-49-7
1-HEPTANOL	000111-70-6
4-HEPTANOL	000589-55-9
3-HEPTANONE	000106-35-4
2-HEPTANONE	000110-43-0
HEPTANOYL CHLORIDE	002528-61-2
N-HEPTATRIACONTANE	007194-84-5
HEPT-1,3-DIYN-5-ENYLBENZENE	013678-98-3
2-HEPTENE (TRANS)	014686-13-6
3-HEPTENE (TRANS)	014686-14-7
1-HEPTENE	000592-76-7
2-HEPTENE (CIS)	006443-92-1
3-HEPTENE (CIS)	007642-10-6
6-HEPTENOIC ACID, 7-[4,5-BIS(4-FLUOROPHENYL)-2-(	129358-82-3
6-HEPTENOIC ACID, 7-[4,5-BIS(4-FLUOROPHENYL)-2-(	129358-85-6
6-HEPTENOIC ACID,7[4(4-CHLORO-3,5-DIMEPH)-5-(4-F	129358-65-2
6-HEPTENOIC ACID, 7-[5-(4-FLUOROPHENYL)-2-(1-MET	129358-45-8
HEPTENOPHOS	034783-40-9
HEPTYLAMINE	000111-68-2
O-HEPTYL CARBAMATE	004248-20-8
HEPTYLCYCLOHEXANE	005617-41-4
HEPTYL P-HYDROXYBENZOATE	001085-12-7
4-(HEPTYLOXY)PHENOL	013037-86-0
1-HEPTYNE	000628-71-7
HEROIN	000561-27-3
HESPERETINE	000520-33-2
HEXABARITAL	000056-29-1
HEXABROMOBENZENE	000087-82-1
1,2,5,6,9,10-HEXABROMOCYCLODODECANE	003194-55-6
1,2,4,6,7,9-HEXABROMO-3,8-DICHLORODIBENZO-P-DIO*	002170-44-7
HEXACHLOROACETONE	000116-16-5
HEXACHLOROBENZENE	000118-74-1
2,2',3,4',5,5'-HEXACHLOROBIPHENYL	051908-16-8
2,2',3,5,5',6-HEXACHLOROBIPHENYL	052663-63-5
2,2',3,4,5,5'-HEXACHLOROBIPHENYL	052712-04-6
2,2',3,3',4,5-HEXACHLOROBIPHENYL	055215-18-4
2,2'3,3',5,6-HEXACHLOROBIPHENYL	052704-70-8
2,3,3',4,4',5-HEXACHLOROBIPHENYL	038380-08-4
2,3,3',5,5',6-HEXACHLOROBIPHENYL	074472-46-1
2,2',3,3',4,5'-HEXACHLOROBIPHENYL	052663-66-8
2,2',3,3',5,6'-HEXACHLOROBIPHENYL	052744-13-5
2,3',4,4',5,5'-HEXACHLOROBIPHENYL	052663-72-6
2,2',3,4,4',5'-HEXACHLOROBIPHENYL	035065-28-2
2,2',3,3',6,6'-HEXACHLOROBIPHENYL	038411-22-2
1,2-HEXACHLOROBUTADIENE	056827-79-3

HEXACHLOROBUTADIENE	000087-68-3
1,1,2,2,3,3-HEXACHLOROBUTANE	083682-69-3
1,1,2,2,3,4-HEXACHLOROBUTANE	002431-55-2
1,2,2,3,3,4-HEXACHLOROBUTANE	001573-57-5
1,1,1,4,4,4-HEXACHLOROBUTANE	079458-54-1
2,3,3,4,4,4-HEXACHLOROBUT-1-ENE	083682-71-7
1-HEXACHLOROBUTYNE	083682-47-7
1,2,3,4,5,6-HEXACHLOROCYCLOHEXANE	000608-73-1
GAMMA-HEXACHLOROCYCLOHEXANE	000058-89-9
DELTA-HEXACHLOROCYCLOHEXANE	000319-86-8
ALPHA-HEXACHLOROCYCLOHEXANE	000319-84-6
BETA-HEXACHLOROCYCLOHEXANE	000319-85-7
HEXACHLOROCYCLOHEXENE	057722-17-5
HEXACHLOROCYCLOHEXENE	057722-15-3
HEXACHLOROCYCLOHEXENE	057722-16-4
HEXACHLOROCYCLOPENTADIENE	000077-47-4
1,2,3,7,8,9-HEXACHLORODIBENZO-P-DIOXIN	019408-74-3
1,2,4,6,7,9-HEXACHLORODIBENZO-P-DIOXIN	039227-62-8
1,2,3,4,7,8-HEXACHLORODIBENZO-P-DIOXIN	039227-28-6
1,2,3,6,7,8-HEXACHLORODIBENZO-P-DIOXIN	057653-85-7
HEXACHLOROETHANE	000067-72-1
1,2,2,5,5,6-HEXACHLOROHEXANE	083682-29-5
1,2,3,4,5,6-HEXACHLOROHEXANE	018585-38-1
2,2,3,4,5,5-HEXACHLOROHEXANE	083682-28-4
2,2,3,4,5,5-HEXACHLORO-3-HEXENE	083682-31-9
1,2,3,4,5,6-HEXACHLORO-3-HEXENE	001725-74-2
1,2,3,4,7,7-HEXACHLORO-2,5-NORBORNADIENE	003389-71-7
1,1,1,3,3,3-HEXACHLOROPROPANE	003607-78-1
1,2,4,5,7,8-HEXACHLORO-9H-XANTHENE	038178-99-3
N-HEXACONTANE	007667-80-3
HEXACOSANE	000630-01-3
HEXADECAMETHYLCYCLOOCTASILOXANE	000556-68-3
HEXADECAMETHYLHEPTASILOXANE	000541-01-5
N-HEXADECANE	000544-76-3
1-HEXADECANETHIOL	002917-26-2
HEXADECANOIC ACID, SODIUM SALT	000408-35-5
1-HEXADECANOL	036653-82-4
1-HEXADECENE	000629-73-2
8-HEXADECENOIC ACID	017004-62-5
11-HEXADECEN-1-OL (Z)	056683-54-6
N-HEXADECYLAMINE	000143-27-1
N-HEXADECYLBENZENE	001459-09-2
2,4-HEXADIENAL	000142-83-6
1,3-HEXADIENE	000592-48-3
1,5-HEXADIENE	000592-42-7
TRANS-1,4-HEXADIENE	007319-00-8
2,4-HEXADIENE (CIS,TRANS)	005194-50-3
1,3-HEXADIENE (TRANS)	020237-34-7
2,4-HEXADIENE (TRANS,TRANS)	005194-51-4
2,4-HEXADIYNE	002809-69-0
HEXAETHYLENE GLYCOL	002615-15-8
HEXAETHYL TETRAPHOSPHATE	000757-58-4
HEXAFLUMURON	086479-06-3
HEXAFLUOROACETONE	010057-27-9
HEXAFLUOROACETONE TRIHYDRATE	034202-69-2
HEXAFLUOROBENZENE	000392-56-3
HEXAFLUOROETHANE	000076-16-4
1,1,1,3,3,3-HEXAFLUOROPROPAN-2-OL	000920-66-1

HEXAFLUORO-2-PROPANONE	000684-16-2
HEXAFLUOROPROPENE	000116-15-4
3-(HEXAHYDRO-4,7-METHANOINDAN-5-YL)-1,1-DIMETHY*	002163-79-3
HEXAMETHYLBENZENE	000087-85-4
HEXAMETHYLCYCLOTRISILOXANE	000541-05-9
2,2,4,6,6,8-HEXAMETHYL-4,8-DIPHENYLCYCLOTETRAS*	004657-20-9
2,2,4,4,6,8-HEXAMETHYL-6,8-DIPHENYLCYCLOTETRAS*	018604-02-9
HEXAMETHYLDISILIZANE	000999-97-3
HEXAMETHYLDISILOXANE	000107-46-0
1,1'-HEXAMETHYLENE-BIS(1-NITROSOUREA)	027640-22-8
HEXAMETHYLENEDIAMINE	000124-09-4
HEXAMETHYLENE DIISOCYANATE	000822-06-0
HEXAMETHYLENEIMINE	000111-49-9
HEXAMETHYLENETETRAMINE	000100-97-0
HEXAMETHYL(SILAACENAPHTHENYL)CYCLOTETRASILOXAN*	013093-12-4
HEXAMETHYL MELAMINE	000645-05-6
HEXAMETHYLPHOSPHORAMIDE	000680-31-9
HEXAMETHYL(SILACYCLOHEXYL)CYCLOTETRASILOXANE	035331-58-9
4,4,6,6,8,8-HEXAMETHYLSPIRO(CYCLOTETRASILOXANE*)	040169-27-5
HEXANAL	000066-25-1
HEXANAMIDE, 6-AMINO-N-(2,6-DIMETHYLPHENYL)-	102089-67-8
HEXANAMIDE, 5-AMINO-N-(2,6-DIMETHYLPHENYL)-	077470-88-3
HEXANAMIDE, 6-(DIETHYLAMINO)-N-(2,6-DIMETHYLPHEN	102089-69-0
HEXANAMIDE, 5-(DIETHYLAMINO)-N-(2,6-DIMETHYLPHEN	077470-99-6
HEXANAMIDE, N-[4-(DIMETHYLAMINO)SULFONYL]PHENYL]	075852-19-6
1-HEXANAMINE, 2-METHYL-	044652-67-7
N-HEXANE	000110-54-3
HEXANEDIOIC ACID	000124-04-9
HEXANEDIOIC ACID, DIMETHYL ESTER	000627-93-0
HEXANEDIOIC ACID, MONOETHYL ESTER	000626-86-8
1,6-HEXANEDIOL	000629-11-8
2,5-HEXANEDIONE	000110-13-4
HEXANENITRILE	000628-73-9
1-HEXANETHIOL	000111-31-9
1,2,6-HEXANETRIOL	000106-69-4
2,2',4,4',6,6'-HEXANITRODIPHENYLAMINE	000131-73-7
HEXANOIC ACID	000142-62-1
HEXANOIC ACID, 2-BUTYL-	003115-28-4
HEXANOIC ACID, 2-ETHYL-	000149-57-5
HEXANOIC ACID HYDRAZIDE	002443-62-1
HEXANOIC ACID, 2-HYDROXY-	006064-63-7
HEXANOIC ACID, 6-(1-NAPHTHALENYLOXY)-	101705-36-6
HEXANOIC ACID, 2-PROPYL-	003274-28-0
2-HEXANOL	000626-93-7
1-HEXANOL	000111-27-3
3-HEXANOL	000623-37-0
3-HEXANONE	000589-38-8
2-HEXANONE	000591-78-6
6-(N-HEXANOYLAMINO)PENCILLANIC ACID	004493-18-9
1,4,7,10,13,16-HEXAOXACYCLOOCTADECANE	017455-13-9
HEXATRIACONTANE	000630-06-8
1,3,5-HEXATRIENE	002235-12-3
CIS-1,3,5-HEXATRIENE	002612-46-6
TRANS-1,3,5-HEXATRIENE	000821-07-8
HEXAZINONE	051235-04-2
(E)-2-HEXENAL	006728-26-3
2-HEXENAL(TRANS)	000505-57-7
3-HEXENE	000592-47-2

1-HEXENE	000592-41-6
2-HEXENE	000592-43-8
3-HEXENE-2,5-DIONE (CIS)	017559-81-8
3-HEXENE-2,5-DIONE (TRANS)	000820-69-9
2-HEXENE-4-OL	004798-58-7
3-HEXENE-1-OL	000544-12-7
2-HEXENE (TRANS)	004050-45-7
2-HEXENE (Z)	007688-21-3
1-HEXEN-3-OL	004798-44-1
1-HEXEN-5-ONE	000109-49-9
1-HEXEN-3-ONE, 1-(3,4-DIHYDROXYPHENYL)-	136944-25-7
N-HEXYL ACETATE	000142-92-7
N-HEXYLAMINE	000111-26-2
N-HEXYLBENZENE	001077-16-3
2-HEXYLBENZOTRIAZOLE	069218-28-6
1-HEXYLBENZOTRIAZOLE	000883-57-8
1-S-HEXYLBENZOTRIAZOLE	069218-30-0
2-S-HEXYLBENZOTRIAZOLE	069218-33-3
O-HEXYL CARBAMATE	002114-20-7
1-HEXYL-4-CARBOXY-2-PYRROLIDONE	116167-27-2
1-HEXYL-3,7-DIMETHYL XANTHINE	001028-33-7
B-HEXYLGALACTOPYRANOSIDE	039824-11-8
HEXYL GALLATE	001087-26-9
HEXYL P-HYDROXYBENZOATE	001083-27-8
1-HEXYL-4-METHOXYCARBONYL-2-PYRROLIDONE	071548-53-3
N-HEXYL NICOTINOATE	023597-82-2
2-HEXYLOXY-4-AMINOBENZOIC ACID	024397-14-6
4-HEXYLOXYPHENOL	018979-55-0
P-HEXYLPYRIDINE	027876-24-0
HEXYLPYRIDINIUM BROMIDE	074440-81-6
N-HEXYLPYRIDINIUM IODIDE	046122-12-7
1-HEXYL-2-PYRROLIDONE	004838-65-7
4-HEXYLRESORCINOL	000136-77-6
4-HEXYLSEMICARBAZIDE	079353-76-7
1-HEXYLTHIO-B-GALACTOPYRANOSIDE	051575-56-5
HEXYLTRICHLOROSILANE	000928-65-4
1-HEXYNE	000693-02-7
1-HEXYN-3-OL	000105-31-7
3-HEXYN-2-OL	000109-50-2
1-HEXYN-5-ONE	002550-28-9
HIPPURIC ACID	000495-69-2
HISTIDINE	000071-00-1
HISTIDINE-AMIDE,N-ACETYL	054831-57-1
2(1-HOET-13THIAZOLIDINYLIDEN-2N)5NO2-THIAZOLE	024240-67-3
HOMATROPIN	000087-00-3
HOMATROPINE HYDROBROMIDE	000051-56-9
1-HOME PENTACHLOROCYCLOHEXANE	056400-43-2
3-HXO-5-BR-6-ME-PYRIMIDIN-2,4-DIONE	077317-76-1
HYDANTOIN	000461-72-3
HYDANTOIN,5-PH-5-(P-HYDROXY)PH	002784-27-2
HYDRALAZINE	000086-54-4
HYDRASTINE	000118-08-1
HYDRAZINE, (4-BROMOPHENYL)-	000589-21-9
HYDRAZINECARBOTHIAMIDE, 2-(2-FURANYLMETHYLENE)-	005419-96-5
HYDRAZINECARBOTHIAMIDE, 2-[(5-NITRO-2-FURANYL)ME	000831-71-0
HYDRAZINECARBOXAMIDE, N-(3-CHLOROPHENYL)	051707-42-7
HYDRAZINECARBOXAMIDE, N-(4-METHOXYPHENYL)-	062774-59-8
HYDRAZINECARBOXAMIDE, N-(1-NAPHTHALENYL)-	074099-08-4

HYDRAZINECARBOXAMIDE, N-(4-CHLOROPHENYL)	069194-89-4
HYDRAZINECARBOXAMIDE, N-(3-METHYLPHENYL)-	015940-62-2
HYDRAZINECARBOXAMIDE, 2-(PHENYLMETHYLENE)-	001574-10-3
HYDRAZINECARBOXAMIDE, N-(3-HYDROXYPHENYL)-	073469-91-7
HYDRAZINECARBOXAMIDE, N-(2,5-DICHLOROPHENYL)-	073469-87-1
HYDRAZINECARBOXAMIDE, N-(2-METHOXYPHENYL)-	062774-58-7
HYDRAZINECARBOXAMIDE, N-(2-METHYLPHENYL)-	015940-63-3
HYDRAZINECARBOXAMIDE,N-(2-HYDROXYPHENYL)-	038191-26-3
HYDRAZINECARBOXAMIDE, 2-PHENYL-	000103-03-7
HYDRAZINECARBOXAMIDE, N-(4-IODOPHENYL)	073469-90-6
HYDRAZINECARBOXAMIDE, N-(4-HYDROXYPHENYL)-	073469-92-8
HYDRAZINECARBOXAMIDE, N-PHENYL-	000537-47-3
HYDRAZINECARBOXAMIDE, N-(3-NITROPHENYL)	017433-92-0
HYDRAZINECARBOXAMIDE, N-(2-NITROPHENYL)	017433-94-2
HYDRAZINECARBOXAMIDE, N-(3-METHOXYPHENYL)-	073469-85-9
HYDRAZINECARBOXAMIDE, N-(4-NITROPHENYL)	017433-93-1
HYDRAZINECARBOXAMIDE, N-(2-CHLOROPHENYL)	035580-76-8
HYDRAZINECARBOXAMIDE, N-(2-BROMOPHENYL)	073469-88-2
HYDRAZINECARBOXAMIDE, N-(4-METHYLPHENYL)-	062774-57-6
HYDRAZINECARBOXAMIDE, N-(4-BROMOPHENYL)	002646-26-6
HYDRAZINECARBOXAMIDE, N-(3-BROMOPHENYL)	073469-89-3
HYDRAZINECARBOXAMIDE, N-(4-ETHOXYPHENYL)-	074099-07-3
HYDRAZINECARBOXIMIDAMIDE, 2-[(4-METHOXYPHENYL)ME	013308-82-2
HYDRAZINECARBOXIMIDAMIDE, 2-[(4-CHLOROPHENYL)MET	013308-88-8
HYDRAZINECARBOXIMIDAMIDE, 2-[[4-(DIMETHYLAMINO)P	038407-85-1
HYDRAZINECARBOXIMIDAMIDE, 2-[(4-ETHOXYPHENYL)MET	082530-96-9
HYDRAZINECARBOXIMIDAMIDE, 2-[(4-CHLORO-3-NITROPH	102632-30-4
HYDRAZINECARBOXIMIDAMIDE, 2-(PHENYLMETHYLENE)-	003357-37-7
HYDRAZINECARBOXIMIDAMIDE, 2-[(4-BROMOPHENYL)METH	037873-43-1
HYDRAZINECARBOXIMIDAMIDE, 2-(1-METHYLHEPTYLIDENE	093484-23-2
HYDRAZINECARBOXIMIDAMIDE, 2-[(3-NITROPHENYL)METH	090792-54-4
HYDRAZINECARBOXIMIDAMIDE, 2-[(4-NITROPHENYL)METH	030068-29-2
HYDRAZINECARBOXIMIDAMIDE, 2-[(2-NITROPHENYL)METH	102632-31-5
HYDRAZINECARBOXIMIDAMIDE, 2-(PHENYL-2-THIENYLMET	097183-49-8
HYDRAZINECARBOXIMIDAMIDE, 2-(1-PROPYLBUTYLIDENE)	097183-51-2
HYDRAZINECARBOXIMIDAMIDE, 2-[1-(2-THIENYL)ETHYLI	072189-64-1
HYDRAZINECARBOXIMIDAMIDE, 2-(2-THIENYLMETHYLENE)	097183-52-3
HYDRAZINE, (2,4-DINITROPHENYL)-	000119-26-6
HYDRAZINE, 1,1-DIPHENYL-	000530-50-7
HYDRAZINE, (4-NITROPHENYL)-	000100-16-3
HYDRAZOBENZENE	000122-66-7
MESO-HYDROBENZOIN	000579-43-1
HYDROBROMOTHIAZIDE	023141-83-5
HYDROCHLOROTHIAZIDE	000058-93-5
HYDROCORTISONE	000050-23-7
HYDROCORTISONE ACETATE	000050-03-3
HYDROCORTISONE-17-BUTYRATE	013609-67-1
HYDROCORTISONE-9A-FLUORO	000127-31-1
HYDROCORTISONE-17-PROPIONATE	065980-97-4
HYDROCORTISONE-21-PROPIONATE	006677-98-1
HYDROCORTISONE-21-VALERATE	006678-00-8
HYDROCORTISONE-17-VALERATE	057524-89-7
HYDROCORTISON-21-HEMIPIMELATE	107085-84-7
HYDROCORTISON-21-PENTYL-CO-O-ME	114611-36-8
HYDROQUININE	000522-66-7
HYDROQUINONE	000123-31-9
HYDROTHIAZIDE	023141-82-4
M-HYDROXYACETANILIDE	000621-42-1

O-HYDROXYACETANILIDE	000614-80-2
HYDROXYACETIC ACID	000079-14-1
HYDROXYACETONE	000116-09-6
M-HYDROXYACETOPHENONE	000121-71-1
P-HYDROXYACETOPHENONE	000099-93-4
O-HYDROXYACETOPHENONE	000118-93-4
1-HYDROXYADAMANTANE	000768-95-6
4-HYDROXY-4'-AMINOAZOBENZENE	000103-18-4
3-HYDROXY-4-AMINOBENZOIC ACID	002374-03-0
4-(HYDROXYAMINO)QUINOLINE 1-OXIDE	004637-56-3
1-HYDROXYANTHRAQUINONE	000129-43-1
2-HYDROXYANTHRAQUINONE	000605-32-3
2-HYDROXYBENZALANILINE	000779-84-0
4-HYDROXYBENZALDEHYDE	000123-08-0
O-HYDROXYBENZALDEHYDE	000090-02-8
3-HYDROXYBENZALDEHYDE	000100-83-4
3-HYDROXYBENZAMIDE	000618-49-5
O-HYDROXYBENZAMIDE	000065-45-2
P-HYDROXYBENZAMIDE	000619-57-8
P-HYDROXYBENZENESULFONAMIDE	001576-43-8
M-HYDROXYBENZOIC ACID	000099-06-9
P-HYDROXYBENZOIC ACID	000099-96-7
P-HYDROXYBENZOIC ACID,ETHYL ESTER	000120-47-8
O-HYDROXYBENZOPHENONE	000117-99-7
P-HYDROXYBENZOPHENONE	001137-42-4
4-(4-HYDROXYBENZOYL)-PYRIDINE	051246-77-6
4-(2-METHOXYBENZOYL)-PYRIDINE	051246-73-2
4-(2-HYDROXYBENZOYL)-PYRIDINE	022526-29-0
M-HYDROXYBENZYL ALCOHOL	000620-24-6
O-HYDROXYBENZYL ALCOHOL	000090-01-7
P-HYDROXYBENZYL ALCOHOL	000623-05-2
4-HYDROXYBENZYLCYANIDE	014191-95-8
ALPHA-HYDROXYBENZYLPENCILLIN	004759-24-4
M-HYDROXYBENZYL TRIMETHYL AMMONIUM BROMIDE	071323-98-3
3-HYDROXY-2-BUTANONE	000590-90-9
P-HYDROXY BUTYL BENZOATE	000094-26-8
2-(4-HYDROXYBUTYL)BENZOTRIAZOLE	069218-39-9
1-(4-HYDROXYBUTYL)BENZOTRIAZOLE	069218-35-5
2-THIOPHENESULFONAMIDE, 5-[(4-HYDROXYBUTYL)SULFO	135832-39-2
A-HYDROXY-I-BUTYRIC ACID	000594-61-6
4-HYDROXY-3-CHLORO-1-BUTENE	075455-41-3
3-HYDROXY-4-CHLORO-1-BUTENE	000671-56-7
4-HYDROXY DICHLOROACETANILIDE	002153-10-8
4-HYDROXYDIPHENYLMETHANE	000101-53-1
O-(2-HYDROXYET)AMINOBENZOIC ACID	025784-00-3
2-HYDROXYETHYL ACRYLATE	000818-61-1
1-(2-HYDROXYETHYL)BENZOTRIAZOLE	000938-56-7
2-(2-HYDROXYETHYL)BENZOTRIAZOLE	000939-72-0
2-HYDROXYETHYL METHACRYLATE	000868-77-9
N-HYDROXYETHYLNICOTINAMIDE	006265-73-2
HYDROXYETHYL-CLET-NO UREA	060784-46-5
1-(2-HYDROXYET)-2-NITROIMIDAZOLE	005006-67-7
5-HYDROXYFLAVONE	000491-78-1
5,7-HYDROXYFLAVONE	000480-40-0
1-(6-HYDROXYHEXYL)BENZOTRIAZOLE	069218-37-7
2-(6-HYDROXYHEXYL)BENZOTRIAZOLE	069218-41-3
HYDROXYIMINOACETONE	000306-44-5
O-(4-HYDROXY-3-IODOPHENYL)-3,5-DIIODO-L-TYROSINE	006893-02-3

1(O-HYDROXYMEPH)-3,3-DIME TRIAZENE	066974-61-6
4-HYDROXY-3-METHOXYBENZOIC ACID	000121-34-6
4-HYDROXY METHYL BENZOATE	000099-76-3
3-HYDROXY-3-METHYL-2-BUTANONE	000115-22-0
4-HYDROXY-3-METHYL-2-BUTANONE	003393-64-4
7-HYDROXY-4-METHYLCOUMARIN	000090-33-5
3-HYDROXYMETHYLFURAN	004412-91-3
2-HYDROXYMETHYLFURAN	000098-00-0
3-HYDROXY-5-METHYLISOXAZOLE	010004-44-1
2-HYDROXYMETHYL-2-NITRO-1,3-PROPANEDIOL	000126-11-4
4-HYDROXY-4-METHYL-2-PENTANONE	000123-42-2
2-(2-HYDROXY-5-METHYLPHENYL)BENZOTRIAZOLE	002440-22-4
2-HYDROXYMETHYLTHIOPHENE	000636-72-6
5-HYDROXY-1,4-NAPHTHOQUINONE	000481-39-0
3-HYDROXY-4-NITROBENZOIC ACID	000619-14-7
4-HYDROXY-3-NITROBENZOIC ACID	000616-82-0
N-HYDROXY-N-NITROSOBENZENAMINE AMMONIUM SALT	000135-20-6
4-HYDROXY-B-NITROSTYRENE	003179-08-6
3-HYDROXY-B-NITROSTYRENE	003156-44-3
12-HYDROXY-9-OCTADECENOIC ACID (CIS)	000141-22-0
5-HYDROXY-OXACILLIN	004914-62-9
3-HYDROXY-4-OXO-4H-PYRAN-2,6-DICARBOXYLIC ACID	000497-59-6
1-HYDROXYPENTACHLOROCYCLOHEXANE	053861-64-6
1-(5-HYDROXYPENTYL)BENZOTRIAZOLE	069218-36-6
2-(5-HYDROXYPENTYL)BENZOTRIAZOLE	069218-40-2
O-HYDROXYPHENOXYACETIC ACID	006324-11-4
M-HYDROXYPHENOXYACETIC ACID	001878-83-7
P-HYDROXYPHENOXYACETIC ACID	001878-84-8
O-HYDROXY PHENYLACETAMIDE	004746-61-6
M-HYDROXYPHENYL ACETATE	000102-29-4
O-HYDROXYPHENYLACETIC ACID	000614-75-5
A-HYDROXYPHENYLACETIC ACID	000090-64-2
M-HYDROXYPHENYLACETIC ACID	000621-37-4
P-HYDROXYPHENYLACETIC ACID	000156-38-7
4-((4-HYDROXYPHENYL)AZO)BENZENESULFONIC ACID,MO*	002623-36-1
N-(4-((4-HYDROXYPHENYL)AZO)PHENYL)ACETAMIDE	005302-39-6
2-(2-HYDROXYPHENYL)BENZOIMIDAZOLE	002963-66-8
2-(2-HYDROXYPHENYL)BENZOXAZOLE	000835-64-3
2-(2-HYDROXYPHENYL)BENZOXTHIAZOLE	003411-95-8
2-(P-HYDROXYPHENYL)ETHANOL	000501-94-0
4-HYDROXYPHENYLISOTHIOCYANATE	002131-60-4
N-(O-HYDROXYPHENYL)METHYLBENZAMIDE	014680-18-3
N-(P-HYDROXYPHENYL)METHYLBENZAMIDE	041859-85-2
4-HYDROXYPHENYLSULFONIC ACID	000098-67-9
N-HYDROXYPHTHALIMIDE	000524-38-9
17-ALPHA-HYDROXYPROGESTERONE	000068-96-2
11-A-HYDROXYPROGESTERONE	000312-90-3
HYDROXYPROLINE	000051-35-4
2-HYDROXYPROPANENITRILE	000078-97-7
2-HYDROXYPROPANOIC ACID, METHYL ESTER	000547-64-8
A-HYDROXYPROPIONIC ACID	000598-82-3
P-HYDROXY-B-PROPIONIC ACID	000501-97-3
P-HYDROXYPROPIOPHENONE	000070-70-2
O-HYDROXYPROPIOPHENONE	000610-99-1
2-HYDROXYPROPYL ACRYLATE	000999-61-1
2-(G-HYDROXYPROPYL)BENZIMIDAZOLE	002403-66-9
P-HYDROXY PROPYL BENZOATE	000094-13-3
2-(3-HYDROXYPROPYL)BENZOTRIAZOLE	069218-38-8

1-(3-HYDROXYPROPYL)BENZOTRIAZOLE	069218-34-4
2-HYDROXYPROPYL METHACRYLATE	000923-26-2
2-THIOPHENESULFONAMIDE, 5-[(3-HYDROXYPROPYL)SULF	104438-02-0
3-HYDROXYPYRIDINE	000109-00-2
4-HYDROXYPYRIDINE	000626-64-2
2-HYDROXYPYRIDINE	000142-08-5
3-HYDROXYPYRIDINE-N-OXIDE	006602-28-4
6-HYDROXYQUINOLINE	000580-16-5
4-HYDROXYQUINOLINE-1-OXIDE	003039-74-5
4'-((6-HYDROXY-M-TOLYL)AZO)ACETANILIDE	002832-40-8
4-HYDROXY-7-TRIFLUOROMEQUINOLINE	000322-97-4
5-HYDROXYTRYPTAMINE	000050-67-9
HYDROXYUREA	000127-07-1
HYOCHOLIC ACID	000547-75-1
HYODEOXYCHOLIC ACID	000083-49-8
HYPOXANTHINE	000068-94-0
IBP (IPROBENFOS)	026087-47-8
3-IBUO-5-BR-6-ME-PYRIMIDIN-2,4-DION	077317-74-9
IBUPROFEN	015687-27-1
IFOSFAMIDE	003778-73-2
IMAZALIL BASE	035554-44-0
IMAZAPYR	081334-34-1
IMIDACLOPRID	138261-41-3
IMIDAZO[1,2-A]PYRAZINE, 2-[2-METHOXY-4-(METHYLSU	102362-14-1
IMIDAZO[1,2-A]PYRIDINE-2-ETHANAMINE	043170-96-3
IMIDAZO[1,2-A]PYRIMIDINE, 2-[2-METHOXY-4-(METHYL	093276-61-0
1H-IMIDAZO[1,2-B]PYRAZOLE, 2,3-DIHYDRO-	006714-29-0
1H-IMIDAZO[4,5-B]PYRIDINE	000273-21-2
1H-IMIDAZO[4,5-B]PYRIDINE, 2-[2-METHOXY-4-(METHY	073384-60-8
IMIDAZO[5,1-B]QUINAZOLIN-9(2H)-ONE, 1,3-DIHYDRO-	131229-44-2
IMIDAZO[5,1-B]QUINAZOLIN-9(2H)-ONE, 1,3-DIHYDRO-	131229-46-4
IMIDAZO[5,1-B]QUINAZOLIN-9(2H)-ONE, 1,3-DIHYDRO-	131229-43-1
IMIDAZO[5,1-B]QUINAZOLIN-9(2H)-ONE, 1,3-DIHYDRO-	131229-37-3
IMIDAZO[5,1-B]QUINAZOLIN-9(2H)-ONE, 2-BUTYL-1,3-	131229-40-8
IMIDAZO[5,1-B]QUINAZOLIN-9(2H)-ONE, 1,3-DIHYDRO-	131229-39-5
IMIDAZO[5,1-B]QUINAZOLIN-9(2H)-ONE, 1,3-DIHYDRO-	131229-42-0
IMIDAZO[5,1-B]QUINAZOLIN-9(2H)-ONE, 1,3-DIHYDRO-	131229-41-9
IMIDAZO[5,1-B]QUINAZOLIN-9(2H)-ONE, 1,3-DIHYDRO-	131229-45-3
IMIDAZO[5,1-B]QUINAZOLIN-9(2H)-ONE, 1,3-DIHYDRO-	131229-38-4
1H-IMIDAZO[4,5-C]PYRIDINE, 2-[2-(3-CHLOROPROPOXY	087359-63-5
IMIDAZOL-5-CONH2-4-(3,3-DIET-1-NNN)	004574-37-2
IMIDAZOL-4-CONH2-5(33DIME-1-TRIAZENO)	004342-03-4
IMIDAZOL-5-CONH2,4(33BIS(2OHET)1-NNN	021244-66-6
IMIDAZOLE	000288-32-4
1H-IMIDAZOLE-1-ACETAMIDE, N,N-BIS(2-HYDROXYETHYL	074141-74-5
1H-IMIDAZOLE-1-ACETAMIDE, N-(2,3-DIHYDROXYPROPYL	074141-75-6
1H-IMIDAZOLE-1-ACETAMIDE, N-(2,3-DIHYDROXYPROPYL	081892-68-4
1H-IMIDAZOLE-1-ACETAMIDE, N-(2-HYDROXYETHYL)-α-M	081892-67-3
1H-IMIDAZOLE-1-ACETAMIDE, 4-IODO-5-NITRO-N-(3-PY	096258-83-2
1H-IMIDAZOLE-1-ACETAMIDE, 2-NITRO-N-(PHENYLMETHY	022994-85-0
1H-IMIDAZOLE-1-ACETAMIDE, N-[2-HYDROXY-1,1-BIS(H	081892-69-5
1H-IMIDAZOLE-1-ACETIC ACID, 4-IODO-5-NITRO-, ETH	096258-81-0
1H-IMIDAZOLE, 1-ACETYL-	002466-76-4
1H-IMIDAZOLE-4-AMINE, 2,5-DIHYDRO-2,2,5,5-TETRAK	023757-42-8
1H-IMIDAZOLE, 2-BROMO-1,5-DIMETHYL-4-NITRO-	105994-26-1
1H-IMIDAZOLE-5-BUTANOIC ACID, beta-[(BENZOYLOXY)	092598-94-2
1H-IMIDAZOLE-5-BUTANOIC ACID, beta-[(2,2-DIMETHY	106848-86-6
1H-IMIDAZOLE-5-BUTANOIC ACID, beta-[[(3-CHLOROBE	092622-09-8

1H-IMIDAZOLE-5-BUTANOIC ACID, alpha-ETHYL-1-METH	106848-85-5
1H-IMIDAZOLE-5-BUTANOIC ACID, _-ETHYL-beta-(HYDR	092598-93-1
1H-IMIDAZOLE-5-BUTANOIC ACID, alpha-ETHYL-beta-(	092598-82-8
1H-IMIDAZOLE-5-BUTANOIC ACID, _-ETHYL-beta-(HYDR	092598-89-5
1H-IMIDAZOLE-5-BUTANOIC ACID, _-ETHYL-beta-(HYDR	092598-86-2
1H-IMIDAZOLE-5-BUTANOIC ACID, α-ETHYL-ß-(HYDROXY	096914-10-2
1H-IMIDAZOLE-5-BUTANOIC ACID, alpha-ETHYL-1-METH	106848-84-4
1H-IMIDAZOLE-5-BUTANOIC ACID, _-ETHYL-beta-(HYDR	092598-92-0
1H-IMIDAZOLE-5-BUTANOIC ACID, _-ETHYL-beta-(HYDR	092598-81-7
1H-IMIDAZOLE-5-BUTANOIC ACID, _-ETHYL-beta-(HYDR	092598-95-3
1H-IMIDAZOLE-5-BUTANOIC ACID, alpha-ETHYL-1-METH	092598-99-7
1H-IMIDAZOLE-5-BUTANOIC ACID, alpha-ETHYL-beta-(	092598-90-8
1H-IMIDAZOLE-5-BUTANOIC ACID, beta-[(BENZOYLOXY)	092598-97-5
1H-IMIDAZOLE-5-BUTANOIC ACID, beta-[(ACETYLOXY)M	092598-84-0
1H-IMIDAZOLE-5-BUTANOIC ACID, beta-[(BENZOYLOXY)	092598-80-6
1H-IMIDAZOLE-5-BUTANOIC ACID, α-ETHYL-ß-(HYDROXY	111226-84-7
1H-IMIDAZOLE-1-CARBOTHIOIC ACID, O-[1-[(4-ETHYLP	111226-68-7
1H-IMIDAZOLE-1-CARBOTHIOIC ACID, O-[1-[[4-(1-MET	111226-62-1
1H-IMIDAZOLE-1-CARBOTHIOIC ACID, O-[1-(PHENOXYME	111226-70-1
1H-IMIDAZOLE-1-CARBOTHIOIC ACID, O-[1-[([1,1'-BI	111226-63-2
1H-IMIDAZOLE-1-CARBOTHIOIC ACID, O-[1-[(4-CHLORO	111226-67-6
1H-IMIDAZOLE-1-CARBOTHIOIC ACID, O-[1-[(4-PROPYL	111226-65-4
1H-IMIDAZOLE-1-CARBOTHIOIC ACID, O-[1-[(4-METHYL	111226-71-2
1H-IMIDAZOLE-1-CARBOTHIOIC ACID, O-[1-[(4-METHOX	039928-74-0
1H-IMIDAZOLE-5-CARBOXALDEHYDE, 1-METHYL-2-NITRO-	056010-86-7
1H-IMIDAZOLE-4-CARBOXAMIDE, 5-(3-METHYL-3-OCTYL-	066975-19-7
1H-IMIDAZOLE-4-CARBOXAMIDE, 5-[3-METHYL-3-(2- PR	039942-91-1
1H-IMIDAZOLE-4-CARBOXAMIDE, 5-[3-METHYL-3-(PHENY	111227-14-6
1H-IMIDAZOLE-1-CARBOXYLIC ACID, 1-[(4-PROPYLPHEN	102152-28-3
1H-IMIDAZOLE-5-CARBOXYLIC ACID, 2-[4-[3-[[2-(3,4	111227-18-0
1H-IMIDAZOLE-1-CARBOXYLIC ACID, 1-[[4-(METHYLTHI	111227-09-9
1H-IMIDAZOLE-1-CARBOXYLIC ACID, 1-[([1,1'-BIPHEN	111226-99-4
1H-IMIDAZOLE-1-CARBOXYLIC ACID, 1-(PHENOXYMETHYL	111227-10-2
1H-IMIDAZOLE-1-CARBOXYLIC ACID, 1-[[4-(1,1-DIMET	111227-07-7
1H-IMIDAZOLE-1-CARBOXYLIC ACID, 1-[[4-(1-METHYLE	111227-00-0
1H-IMIDAZOLE-1-CARBOXYLIC ACID, 1-[(4-FLUOROPHEN	111227-04-4
1H-IMIDAZOLE-1-CARBOXYLIC ACID, 1-[(4-NITROPHENO	111227-08-8
1H-IMIDAZOLE-1-CARBOXYLIC ACID, 1-[(4-METHYLPHEN	111227-26-0
1H-IMIDAZOLE-1-CARBOXYLIC ACID, 1-[(4-PROPOXYPHE	033125-97-2
1H-IMIDAZOLE-5-CARBOXYLIC ACID, 1-(1-PHENYLETHYL	111227-17-9
1H-IMIDAZOLE-1-CARBOXYLIC ACID, 1-[(4-METHOXYPHE	111227-05-5
1H-IMIDAZOLE-1-CARBOXYLIC ACID, 1-[(4-CYANOPHENO	111227-41-9
1H-IMIDAZOLE-1-CARBOXYLIC ACID, 1-[[4-(ETHOXYCAR	111227-01-1
1H-IMIDAZOLE-1-CARBOXYLIC ACID, 1-[(4-CHLOROPHEN	111227-15-7
1H-IMIDAZOLE-1-CARBOXYLIC ACID, 1-[(4-BUTYLPHENO	111227-13-5
1H-IMIDAZOLE-1-CARBOXYLIC ACID, 1-[(4-ETHYLPHENO	111227-03-3
1H-IMIDAZOLE-1-CARBOXYLIC ACID, 1-[(4-BROMOPHENO	111227-21-5
1H-IMIDAZOLE-1-CARBOXYLIC ACID, 1-[(4-ETHOXYPHEN	006963-66-2
1H-IMIDAZOLE, 5-CHLORO-4-NITRO-	039980-81-9
IMIDAZOLE-5-CONH2,4(3SBU-3MEN=NN)	074550-87-1
1H-IMIDAZOLE, 2-CYCLOPROPYL-1-METHYL-5-NITRO-	000059-98-3
1H-IMIDAZOLE, 4,5-DIHYDRO-2-(PHENYLMETHYL)-	000051-45-6
1H-IMIDAZOLE-4-ETHANAMINE	074550-86-0
1H-IMIDAZOLE-1-ETHANOL, 2-[2-(1,3-BENZODIOXOL-5-	110933-19-2
1H-IMIDAZOLE-1-ETHANOL, _,_-BIS(4-FLUOROPHENYL)-	142657-45-2
1H-IMIDAZOLE-1-ETHANOL, _-(4-BROMOPHENYL)-_-PHEN	083200-90-2
1H-IMIDAZOLE-1-ETHANOL, 4-BROMO-α-(METHOXYMETHYL	142657-44-1
1H-IMIDAZOLE-1-ETHANOL, _-(3-BROMOPHENYL)-_-(2,4	128104-25-6
1H-IMIDAZOLE-1-ETHANOL, _-(4-BROMOPHENYL)-_-(4-F	

1H-IMIDAZOLE-1-ETHANOL, α-[[(4-BROMO-TETRAHYDRO-	134419-55-9
1H-IMIDAZOLE-1-ETHANOL, α-[[(2-BROMOCYCLOHEXYL)A	134419-53-7
1H-IMIDAZOLE-1-ETHANOL, alpha-(4-CHLOROPHENYL)-a	137103-64-1
1H-IMIDAZOLE-1-ETHANOL, α-[(DIETHYLAMINO)METHYL]	074550-94-0
1H-IMIDAZOLE-1-ETHANOL, alpha-(2,4-DIFLUOROPHENY	137103-55-0
1H-IMIDAZOLE-1-ETHANOL, alpha-(2,4-DIFLUOROPHENY	137128-33-7
1H-IMIDAZOLE-1-ETHANOL, α,2-DIMETHYL-5-NITRO-	003366-95-8
1H-IMIDAZOLE-1-ETHANOL, _-(4-FLUOROPHENYL)-_-[4-	137103-53-8
1H-IMIDAZOLE-1-ETHANOL, alpha-(4-FLUOROPHENYL)-a	137103-59-4
1H-IMIDAZOLE-1-ETHANOL, alpha-(4-FLUOROPHENYL)-a	137103-54-9
1H-IMIDAZOLE-1-ETHANOL, _-(4-HYDROXYPHENYL)-_-[4	142657-53-2
1H-IMIDAZOLE-1-ETHANOL, 4-IODO-α-(METHOXYMETHYL)	096258-79-6
1H-IMIDAZOLE-1-ETHANOL, 5-IODO-α-(METHOXYMETHYL)	096258-78-5
1H-IMIDAZOLE-1-ETHANOL, 2-METHYL-_-[(2-METHYL-5-	074550-92-8
1H-IMIDAZOLE-1-ETHANOL, 2-METHYL-5-NITRO-, NITRA	021117-51-1
1H-IMIDAZOLE-1-ETHANOL, 5-METHYL-2-NITRO-	023571-38-2
1H-IMIDAZOLE-1-ETHANOL, 2-NITRO-_-[(1-PROPENYLOX	068160-71-4
1H-IMIDAZOLE, 4-IODO-1-METHYL-5-NITRO-	076529-47-0
1H-IMIDAZOLE, 5-IODO-1-METHYL-4-NITRO-	035681-63-1
1H-IMIDAZOLE, 4-IODO-5-NITRO-	076529-48-1
1H-IMIDAZOLE, 2-METHYL-	000693-98-1
1H-IMIDAZOLE, 1-METHYL-	000616-47-7
1H-IMIDAZOLE, 4-METHYL-	000822-36-6
1H-IMIDAZOLE, 2-(1-METHYLETHYL)-4-NITRO-	013373-32-5
1H-IMIDAZOLE, 1-METHYL-2-[[4-(METHYLTHIO)PHENOXY	059729-37-2
1H-IMIDAZOLE, 1-METHYL-2-(1-METHYLETHYL)-5-NITRO	014885-29-1
1H-IMIDAZOLE, 1-[2-(METHYLSULFINYL)ETHYL]-2-NITR	074141-73-4
1H-IMIDAZOLE, 5-NITRO-	088054-21-1
1H-IMIDAZOLE, 2-NITRO-1-(2,3,4,6-TETRA-O-ACETYL-	083116-90-9
1H-IMIDAZOLE, 2-NITRO-1-(2,3,4,6-TETRA-O-ACETYL-	067774-11-2
1H-IMIDAZOLE, 2-PHENYL-	000670-96-2
1H-IMIDAZOLE, 1-(PHENYLMETHYL)-	004238-71-5
1H-IMIDAZOL-1-ETHANOL, 2-(4-FLUOROPHENYL)-5-NITR	004548-15-6
IMIDAZOLE-2-THIONE,3-(3,5-DIFLUOROBENZYL)-	095333-81-6
IMIDAZOLE-2-THIONE,3-(3-FLUOROBENZYL)	095333-80-5
IMIDAZOLE-2-THIONE,3-(3-FLUORO-4-HYDROXYBENZYL)	095333-49-6
IMIDAZOLE-2-THIONE,3-(4-HYDROXYBENZYL)	095333-64-5
2,4-IMIDAZOLIDINEDIONE, 5,5-BIS(PHENYLMETHYL)-	023186-94-9
2,4-IMIDAZOLIDINEDIONE, 3,5-BIS(PHENYLMETHYL)-	002221-11-6
2,4-IMIDAZOLIDINEDIONE, 5,5-DIMETHYL-	000077-71-4
2,4-IMIDAZOLIDINEDIONE, 5,5-DIPHENYL-3-METHYL-	004224-00-4
2,4-IMIDAZOLIDINEDIONE, 5-ETHYL-5-PHENYL-	000631-07-2
2,4-IMIDAZOLIDINEDIONE, 5-ETHYL-5-PHENYL-1-(PHEN	021413-25-2
2,4-IMIDAZOLIDINEDIONE, 3-ETHYL-5-PHENYL-	000086-35-1
2,4-IMIDAZOLIDINEDIONE, 5-METHYL-	000616-03-5
2,4-IMIDAZOLIDINEDIONE, 1-METHYL-3-PHENYL-	002221-12-7
2,4-IMIDAZOLIDINEDIONE, 5-(2-METHYLPROPYL)-	058942-03-3
2,4-IMIDAZOLIDINEDIONE, 5-(1-METHYLETHYL)-	016935-34-5
2,4-IMIDAZOLIDINEDIONE, 3-(PHENYLMETHYL)-	002301-40-8
2-IMIDAZOLIDINETHIONE, 1-(5-NITRO-2-THIAZOLYL)-	040016-42-0
2-IMIDAZOLIDINIMINE, N-[2,6-BIS(1-METHYLETHYL)PH	063346-74-7
2-IMIDAZOLIDINIMINE, N-(3-BROMO-2-CHLOROPHENYL)-	078834-86-3
2-IMIDAZOLIDINIMINE, N-(3-BROMO-2-METHYLPHENYL)-	016822-94-9
2-IMIDAZOLIDINIMINE, N-(5-BROMO-2-METHYLPHENYL)-	016822-80-3
2-IMIDAZOLIDINIMINE, N-(5-BROMO-2-CHLOROPHENYL)-	015327-50-1
2-IMIDAZOLIDINIMINE, N-(2-BROMO-6-CHLOROPHENYL)-	065936-25-6
2-IMIDAZOLIDINIMINE, N-(2-CHLORO-3-METHYLPHENYL)	016822-97-2
2-IMIDAZOLIDINIMINE, N-(4-CHLORO-2,6-DIMETHYLPHE	004201-38-1

2-IMIDAZOLIDINIMINE, N-(2-CHLORO-5-FLUOROPHENYL)	081060-11-9
2-IMIDAZOLIDINIMINE, N-(2-CHLORO-3-FLUOROPHENYL)	081060-10-8
2-IMIDAZOLIDINIMINE, N-(2,3-DIBROMOPHENYL)-	078834-88-5
2-IMIDAZOLIDINIMINE, N-(5-FLUORO-2-METHYLPHENYL)	028125-87-3
2-IMIDAZOLIDINIMINE, N-(2-METHYLPHENYL)-	036318-56-6
2-IMIDAZOLIDINIMINE, N-PHENYL-	001848-75-5
2-IMIDAZOLIDINIMINE, N-(2,3,6-TRICHLOROPHENYL)-	082780-90-3
2(1,3-IMIDAZOLIDINYLIDEN-2-AMINO)-5NO2-THIAZOLE	024240-69-5
1-(1-IMIDAZOLYL)ADAMANTANE	069380-11-6
IMIDAZO[1,2-A]PYRAZINE, 8-METHOXY-2-[2-METHOXY-4	102361-46-6
1H-IMIDAZO[4,5-B]PYRIDIN-6-AMINE, 2-(2,4-DIMETHO	127356-43-8
1H-IMIDAZO[4,5-B]PYRIDIN-6-AMINE, 2-[2-METHOXY-4	127356-05-2
1H-IMIDAZO[4,5-C]PYRIDINE, 2-(2,4-DIMETHOXYPHENY	087359-11-3
1H-IMIDAZO[4,5-B]PYRIDINE, 2-(2,4-DIMETHOXYPHENY	077303-19-6
IMIDAZO[45F]QUINOLINE,3-METHYL-2-DIEMTHYLAMINO	102408-27-5
IMIDIAZOLE-2-THIONE,3-(3,5-DIFLUORO-4-HYDROXYBEN	095333-60-1
IMIDODICARBONIC DIAMIDE, N-(2,5-DIFLUOROPHENYL)-	111041-47-5
IMIDODICARBONIC DIAMIDE, N-(2-FLUOROPHENYL)-N',2	053285-95-3
IMIDODICARBONIC DIAMIDE, N-(2-FLUORO-5-METHOXYPH	111041-67-9
IMIDODICARBONIC DIAMIDE, N-(2-FLUORO-5-IODOPHENY	111041-55-5
IMIDODICARBONIC DIAMIDE. N-(5-ETHYL-2-FLUOROPHEN	111041-46-4
IMIDODICARBONIC DIAMIDE, N-[5-(1-ETHYLPROPYL)-2-	111041-76-0
IMIDODICARBONIC DIAMIDE, N-(2-FLUORO-5-NITROPHEN	121788-15-6
IMIDODICARBONIC DIAMIDE, N-(5-CYCLOHEXYL-2-FLUOR	111041-86-2
IMIDODICARBONIC DIAMIDE, N-[2-FLUORO-5-(PHENYLME	111041-44-2
IMIDODICARBONIC DIAMIDE, N-(5-ACETYL-2-FLUOROPHE	111041-62-4
IMIDODICARBONIC DIAMIDE, N-[5-(1,1-DIMETHYLETHYL	111041-88-4
IMIDODICARBONIC DIAMIDE, N-(4-FLUORO[1,1'-BIPHEN	111041-66-8
IMIDODICARBONIC DIAMIDE, N-[2-FLUORO-5-[(1-METHY	121788-16-7
IMIDODICARBONIC DIAMIDE, N-[2-FLUORO-5-(1-METHYL	111041-57-7
IMIDODICARBONIC DIAMIDE, N-[2-FLUORO-5-(1-METHYL	111041-48-6
IMIDODICARBONIMIDIC DIAMIDE, N-(4-CHLOROPHENYL)-	049871-96-7
IMIDODICARBONIMIDIC DIAMIDE, N-(4-CHLOROPHENYL)-	000500-92-5
IMIDODICARBONIMIDIC DIAMIDE, N-PENTYL	021306-55-8
2-PROPANOL, 1,1'-IMINOBIS[3-(2-METHYLPHENOXY)	110747-15-4
IMIPRAMINE	000050-49-7
INDALONE	000532-34-3
1,3-INDANDIONE	000606-23-5
INDANE	000496-11-7
5-INDANOL	001470-94-6
5-INDANOXYACETIC ACID	001878-58-6
1H-INDAZOL-5-AMINE	019335-11-6
INDAZOLE	000271-44-3
1H-INDAZOLE, 5-NITRO-	005401-94-5
INDENE	000095-13-6
1H-INDENE-1,3(2H)-DIONE, 2-[1,1'-BIPHENYL]-4-YL-	002156-14-1
1H-INDENE-1,3(2H)-DIONE, 2-[3,5-BIS(1,1-DIMETHYL	013936-01-1
1H-INDENE-1,3(2H)-DIONE, 2-[3,5-BIS(1-METHYLETHY	055994-24-6
1H-INDENE-1,3(2H)-DIONE, 2-(4-BROMOPHENYL)-	001146-98-1
1H-INDENE-1,3(2H)-DIONE, 2-(3-CHLOROPHENYL)-	001470-44-6
1H-INDENE-1,3(2H)-DIONE, 2-(4-CHLOROPHENYL)-	001146-99-2
1H-INDENE-1,3(2H)-DIONE, 2-(3,5-DIETHYLPHENYL)-	055994-23-5
1H-INDENE-1,3(2H)-DIONE, 2,2-DIHYDROXY-	000485-47-2
1H-INDENE-1,3(2H)-DIONE, 2-(3,5-DIMETHOXYPHENYL)	019055-50-6
1H-INDENE-1,3(2H)-DIONE, 2-(3,5-DIMETHYLPHENYL)-	006549-64-0
1H-INDENE-1,3(2H)-DIONE, 2-[4-(1,1-DIMETHYLETHYL	013935-96-1
1H-INDENE-1,3(2H)-DIONE, 2-[3-(1,1-DIMETHYLETHYL	013935-95-0
1H-INDENE-1,3(2H)-DIONE, 2-(4-ETHYLPHENYL)-	007561-67-3

1H-INDENE-1,3(2H)-DIONE, 2-(3-METHOXYPHENYL)-	006149-23-1
1H-INDENE-1,3(2H)-DIONE, 2-(4-METHOXYPHENYL)-	000117-37-3
1H-INDENE-1,3(2H)-DIONE, 2-[3-(1-METHYLETHYL)PHE	013935-94-9
1H-INDENE-1,3(2H)-DIONE, 2-[4-(1-METHYLETHYL)PHE	055994-21-3
1H-INDENE-1,3(2H)-DIONE, 2-(3-METHYLPHENYL)-	015432-98-1
1H-INDENE-1,3(2H)-DIONE, 2-(4-OCTYLPHENYL)-	055994-22-4
1H-INDENE-1,3(2H)-DIONE, 5-[3-OXO-3-(1-PIPERIDIN	023434-86-8
1H-INDENE-1,3(2H)-DIONE, 2-[3-(TRIFLUOROMETHYL)P	019055-70-0
1H-INDENE-1,3(2H)-DIONE, 2-[4-(TRIFLUOROMETHYL)P	006723-40-6
INDENO(1,2,3-CD)PYRENE	000193-39-5
INDOLE	000120-72-9
INDOLE-3-ACETIC ACID	000087-51-4
INDOLE-2-AMINE, N-METHYL	036092-88-3
3-INDOLEBUTYRIC ACID	000133-32-4
1H-INDOLE-3-CARBOXALDEHYDE	000487-89-8
INDOLE-5-CARBOXAMIDE	001670-87-7
1H-INDOLE-2-CARBOXAMIDE	001670-84-4
INDOLE-4-CARBOXAMIDE	001670-86-6
1H-INDOLE-2-CARBOXAMIDE, N-ETHYL-	069808-72-6
1H-INDOLE-3-CARBOXAMIDE, N-METHYL-	085729-23-3
1H-INDOLE-2-CARBOXAMIDE, N-METHYL-	069808-71-5
INDOLE, 3-CARBOXYLIC ACID	000771-50-6
1H-INDOLE-3-CARBOXYLIC ACID, METHYL ESTER	000942-24-5
1H-INDOLE-2-CARBOXYLIC ACID, METHYL ESTER	001202-04-6
1H-INDOLE-2-CARBOXYLIC ACID, ETHYL ESTER	003770-50-1
INDOLE-2-CARBOXYLIC ACID	001477-50-5
1H-INDOLE-3-CARBOXYLIC ACID, 1-METHYL-	032387-21-6
INDOLE-4-CYANO-	016136-52-0
INDOLE-6-CYANO-	015861-36-6
INDOLE-5-CYANO-	015861-24-2
1H-INDOLE, 2,3-DIHYDRO-5-NITRO-	032692-19-6
1H-INDOLE, 2,3-DIHYDRO-6-NITRO-	019727-83-4
1H-INDOLE, 2,3-DIMETHYL-	000091-55-4
1H-INDOLE-2,3-DIONE	000091-56-5
1H-INDOLE-2,3-DIONE, 5-NITRO-	000611-09-6
INDOLE, 5-NITRO	006146-52-7
INDOLE-3-OL,ACETATE ESTER	000608-08-2
1H-INDOLE-3-PROPANAMIDE, N-METHYL-	069397-85-9
1H-INDOLE-3-PROPANOIC ACID	000830-96-6
1H-INDOLE-1-PROPANOIC ACID	006639-06-1
INDOLIZINE	000274-40-8
1H-INDOL-5-OL	001953-54-4
2H-INDOL-2-ONE, 1,3-DIHYDRO-3-PHENYL-	003456-79-9
INDOL-3-YL-THIOACETICACID	054466-88-5
INDOL-3-YL-2-THIOPROPIONIC ACID	080412-20-0
INDOMETHACIN	000053-86-1
INOSINE	000058-63-9
INOSINE, 2'-DEOXY-	000890-38-0
INOSINE, 2',3'-DIDEOXY-	069655-05-6
INOSITOL	000087-89-8
IODOACETAMIDE	000144-48-9
4-IODOACETANILIDE	000622-50-4
2-IODOACETANILIDE	019591-17-4
2-IODO-4-AMINOBENZOIC ACID	073655-51-3
P-IODOAMPHENICOL	049648-53-5
2-IODOANILINE	000615-43-0
4-IODOANTIPYRINE	000129-81-7
4-IODOBENZAMIDE	003956-07-8

2-IODOBENZAMIDE	003930-83-4
IODOBENZENE	000591-50-4
4-IODOBENZENESULFONAMIDE	000825-86-5
2'-IODOBENZOGUANAMINE	072781-91-0
3-IODOBENZOIC ACID	000618-51-9
4-IODOBENZOIC ACID	000619-58-9
2-IODOBENZOIC ACID	000088-67-5
P-IODOBENZYL N,N-DIME CARBAMATE	084640-34-6
M-IODOBENZYL TRIMETHYL AMMONIUM BROMIDE	071324-02-2
2-IODOBUTANE	000513-48-4
P-IODO CF3-METHANESULFONANILIDE	023384-10-3
1-IODODECANE	002050-77-3
1-(2-IODO),2-DIBENZOYL-1-TERT-BUTYLHYDRAZINE	112226-93-4
1-(3-IODO),2-DIBENZOYL-1-TERT-BUTYLHYDRAZINE	112427-40-4
1-(4-IODO),2-DIBENZOYL-1-TERT-BUTYLHYDRAZINE	112427-73-3
1-IODODODECANE	004292-19-7
IODOETHANE	000075-03-6
IODOFENPHOS	018181-70-9
IODOFORM	000075-47-8
1-IODOHEPTANE	004282-40-0
1-IODOHEXADECANE	000544-77-4
1-IODOHEXANE	000638-45-9
1-IODO-2-METHYLPROPANE	000513-38-2
1-IODONAPHTHALENE	000090-14-2
3-IODONITROBENZENE	000645-00-1
IODONONANE	004282-42-2
1-IODOOCTADECANE	000629-93-6
2-IODOOCTANE	000557-36-8
1-IODOOCTANE	000629-27-6
2-IODOPENTANE	000637-97-8
1-IODOPENTANE	000628-17-1
3-IODOPHENOL	000626-02-8
2-IODOPHENOL	000533-58-4
4-IODOPHENOL	000540-38-5
M-IODOPHENOXYACETIC ACID	001878-93-9
O-IODOPHENOXYACETIC ACID	001878-92-8
P-IODOPHENOXYACETIC ACID	001878-94-0
2-IODOPHENYLACETATE	032865-61-5
M-IODOPHENYLACETIC ACID	001878-69-9
P-IODOPHENYLACETIC ACID	001798-06-7
4-IODOPHENYL GLUCOPYRANOSIDE	020838-40-8
M-IODO-N-PHENYLSUCCINIMIDE	072601-46-8
O-IODO-N-PHENYLSUCCINIMIDE	072601-45-7
2-IODOPROPANE	000075-30-9
1-IODOPROPANE	000107-08-4
3-IODOPROPIONIC ACID	000141-76-4
4-IODOPYRAZOLE	003469-69-0
3-IODOPYRIDINE	001120-90-7
2-IODOQUINOLINE	006560-83-4
1-IODOTETRADECANE	019218-94-1
2-IODOTHIOPHENE	003437-95-4
O-IODOTOLUENE	000615-37-2
1-IODOUNDECANE	004282-44-4
IOHEXOL	066108-95-0
IOHEXOL DERIVATIVE	088116-59-0
ALPHA-IONONE	000127-41-3
IOPROMIDE	073334-07-3
IOVERSOL	087771-40-2

IOXYNIL	001689-83-4
IPATONE	003004-70-4
IPAZINE	001912-25-0
IPCONAZOLE	125225-28-7
1-(O-I PH)-3,3-DIMETHYLTRIAZENE	066974-58-1
2-I-PR-PHENYLDIMETHYLCARBAMATE	025007-30-1
N-(M-IPH)3-N'-PIPERIDINOACETAMIDE	058479-91-7
N-(P-IPH)3-N'-PIPERIDINOACETAMIDE	058479-87-1
N'-IPR-(4-CLPHENOXY)ACETIC HYDRAZIDE	003544-35-2
N'-IPR-(2-CLPHENOXY)ACETIC HYDRAZIDE	051963-48-5
N'-IPR-(3-CLPHENOXY)ACETIC HYDRAZIDE	051963-49-6
N'-IPR-2-ME-2(4CL-PHO)PROPIONICACIDHYDRAZIDE	054922-63-3
N-IPR-1(2-MEO PHENOXY)ACETIC HYDRAZID	062251-74-5
N-IPR-1(4-MEO PHENOXY)ACETIC HYDRAZID	054922-62-2
N-IPR-1(3-MEO PHENOXY)ACETIC HYDRAZID	062251-75-6
N(N"-IPR(4ME-PH)CARBAMYLTHIO)METHOMYL	082774-17-2
3-I-PR-5-MEPHENYL-N-ME CARBAMATE	002631-37-0
6-IPR-4-NH2-3-MEAM-124-TRIAZ-5-ONE	021087-57-0
N-IPR-3-(5-NO2-2-FURYL)ACRYLAMIDE	001951-56-0
N-IPR-3(5NO2-2-THIAZOLYL)ACRYLAMIDE	053207-64-0
3-IPRO-5-BR-6-ME-PYRIMIDI-2,4-DIONE	077317-69-2
IPRONIAZID	000054-92-2
1-(3-IPRO-2-OH-PR)-2-NO2 IMIDAZOLE	071006-72-9
2-IPRO PHENYLDIMETHYLCARBAMATE	069148-74-9
N'-IPR-PHENOXYACETIC ACID HYDRAZIDE	052019-60-0
N-IPR1(4CL-PHENOXY)PROPIONIC HYDRAZID	051963-52-1
N-IPR1(3CL-PHENOXY)PROPIONIC HYDRAZID	051963-51-0
N-IPR1(2CL-PHENOXY)PROPIONIC HYDRAZID	051963-50-9
N-IPR-1-(PHENOXY)PROPIONIC HYDRAZIDE	051963-47-4
N'-IPR-2ME-PHENOXYPROPIONIC HYDRAZIDE	051963-53-2
1-(4-I-PRPHENYL)-3-MEO-3-ME UREA	034861-40-0
(DES-IPR) 4--PIPERIDINYL-PR-CLOFAZIMINE	111436-12-5
(DES-IPR) N-PIPERIDINYL-PR-CLOFAZIMINE	111435-98-4
N1-(5-IPR-2-PYRIMIDYL)SULFANILIDE	073498-01-8
(DES-IPR) N-PYRROLIDINYL-PR-CLOFAZIMINE	111435-97-3
3-IPRS-4-NH2-6-IPR-124-TRIAZINE-5-ONE	050917-24-3
N'-IPR(2-TOLYLOXY)ACETIC AC HYDRAZID	062251-73-4
N'-IPR(4-TOLYLOXY)ACETIC AC HYDRAZID	054922-61-1
N'-IPR(3-TOLYLOXY)ACETIC AC HYDRAZID	054922-60-0
IPT (ISOPROTHIOLANE)	050512-35-1
ISAZOFOS	042509-80-8
ISOAMYL ACETATE	000123-92-2
ISOAMYLAMINE HYDROCHLORIDE	000541-23-1
ISOAMYL FORMATE	000110-45-2
ISOAMYL GALLATE	002486-02-4
ISOAMYL ISOVALERATE	000659-70-1
ISOAMYLTHIOL	000541-31-1
ISOBENZAN	000297-78-9
1(3H)-ISOBENZOFURANONE-3-BENZYL	007011-98-5
1(3H)-ISOBENZOFURANONE, 3-BUTYL-	006066-49-5
1(3H)-ISOBENZOFURANONE, 3-BUTYL-,(±)-	093133-67-6
1(3H)-ISOBENZOFURANONE,3,3-DIETHYL	004770-31-4
1(3H)ISOBENZOFURANONE-3,3-DIMETHYL	001689-09-4
1(3H)-ISOBENZOFURANONE, 3-HYDROXY-3-METHYL-	001828-76-8
1(3H)-ISOBENZOFURANONE-3-HYDROXY	016859-59-9
1(3H)-ISOBENZOFURANONE-3-METHOXYL	004122-57-0
1(3H)-ISOBENZOFURANONE-3-PHENYL	005398-11-8
1(1H)-ISOBENZOFURANONE-3-I-PROPYL	064002-57-9

1-ISOBENZOFURANONE-3-SPIROPENTYL	073090-06-9
ISOBUTENE	000115-11-7
2-ISOBUTOXY-2-PH-4-ME MORPHOLINE HCL	124497-86-5
1-ISOBUTOXY-2-PROPANOL	023436-19-3
ISOBUTYL ACRYLATE	000106-63-8
ISOBUTYL ALCOHOL	000078-83-1
ISOBUTYLAMINE	000078-81-9
ISOBUTYLAMINE HYDROCHLORIDE	005041-09-8
ISOBUTYLBENZENE	000538-93-2
ISOBUTYL BROMIDE	000078-77-3
O-ISOBUTYL CARBAMATE	002114-15-0
ISOBUTYL CHLORIDE	000513-36-0
ISOBUTYLCYCLOHEXANE	001678-98-4
ISOBUTYL FORMATE	000542-55-2
ISOBUTYL GALLATE	003856-05-1
ISOBUTYL ISOBUTYRATE	000097-85-8
ISOBUTYL ISOVALERATE	000589-59-3
2-ISOBUTYL-3-METHOXYPYRAZINE	024683-00-9
N-ISOBUTYLMORPHOLINE	010315-98-7
ISOBUTYL NITRATE	000543-29-3
ISOBUTYLPYRAZINE	029460-93-3
2-ISOBUTYLPYRAZINE	029460-92-2
ISOBUTYL VINYL ETHER	000109-53-5
ISOBUTYRALDEHYDE	000078-84-2
ISOBUTYRICACID	000079-31-2
ISOCARBOXAZID	000059-63-2
ISOCIL	000314-42-1
ISOCYANATOCYCLOHEXANE	003173-53-3
1-(1-ISOCYANATO-1-METHYLETHYL)-4-(1-METHYLETHE*)	002889-58-9
1-(1-ISOCYANATO-1-METHYLETHYL)-3-(1-METHYLETHE*)	002094-99-7
ISODECANOL	025339-17-7
ISODECYL ACRYLATE	001330-61-6
ISODECYLADEHYDE	001321-89-7
ISODECYL DIPHENYL PHOSPHATE	029761-21-5
ISODECYL METHACRYLATE	029964-84-9
ISODRIN	000465-73-6
ISOFENPHOS	025311-71-1
ISOFLURANE	026675-46-7
1H-ISOINDOLE-1,3(2H)-DIONE, 2-[(ACETYLOXY)METHYL	005493-24-3
1H-ISOINDOLE-1,3(2H)-DIONE, 2-[(ACETYLOXY)METHYL	117509-89-4
1H-ISOINDOLE-1,3(2H)-DIONE, 2-PHENYL-	000520-03-6
1H-ISOINDOLE-1,3(2H)-DIONE, 5-PHENYL-	002021-26-3
1H-ISOINDOLE-1-ONE, 2-[(ACETYLOXY)METHYL]-	117491-58-4
ISOLAN	000119-38-0
ISO-LEUCIN-AMIDE, N-ACETYL	056711-06-9
L-ISOLEUCINAMIDE,N-ACETYL-L-ASPARAGINYL-	132765-98-1
L-ISOLEUCINAMIDE, N-ACETYL-L-VALYL-L-VALYL-	132766-12-2
L-ISOLEUCINAMIDE, N-ACETYL-L-LEUCYL-	132765-84-5
L-ISOLEUCINAMIDE, N-ACETYL-L-THREONYL-	132765-90-3
L-ISOLEUCINAMIDE, N-ACETYL-L-ALANYL-L-VALYL-	132766-06-4
ISOLEUCINE	000319-78-8
ISOLEUCINE,N-ACETYL-N'-MEAM AMIDE	032483-16-2
ISOMAZOLE	086315-52-8
ISOMAZOLE,PYRAZINE ANALOG	120236-85-3
ISOMAZOLE,PYRIDAZINE ANALOG	120236-83-1
ISOMAZOLE,(5,6)PYRIDAZINE ANALOG	120236-84-2
ISOMAZOLE,PYRIMIDYL ANALOG	077456-53-2
ISOMESITYL OXIDE	003744-02-3

ISONIAZID	000054-85-3
ISONICOTINIC ACID	000055-22-1
ISONICOTINIC ACID, ETHYL ESTER	001570-45-2
ISONICOTINIC ACID N-OXIDE	013602-12-5
ISONIPECOTIC ACID	000498-94-2
ISONITROBUTANE	000625-74-1
ISONONENE	000124-11-8
ISONONYL ALCOHOL	003452-97-9
ISOOCTANE	026635-64-3
ISOOCTYL ALCOHOL	026952-21-6
ISOPENTANE	000078-78-4
ISOPENTANOL	000123-51-3
ISOPENTYL BUTANOATE	000106-27-4
ISOPESTOX	000371-86-8
ALPHA-(P-ISO-PR-PHENYL)-N-PHENYLNITRONE	099081-88-6
ISOPHORONE	000078-59-1
ISO-PHTHALAMIDE	001740-57-4
ISOPHTHALIC ACID	000121-91-5
ISOPHTHALIC ACID, METHYL ESTER	001877-71-0
ISOPROPALIN	033820-53-0
ISOPROPANOL	000067-63-0
ISOPROPENYL METHYL ETHER	000116-11-0
ISOPROPICILLIN	004780-24-9
ISOPROPOXYETHANOL	000109-59-1
ISOPROPOXYETHYL TRIMETHYL AMMONIUM IODIDE	021949-06-4
O-ISOPROPOXYPHENOL	004812-20-8
4-ISOPROPOXYPYRAZOLE	014884-03-8
P-ISOPROPYLACETOPHENONE	000645-13-6
ISOPROPYLAMINE	000075-31-0
P-ISOPROPYLANILINE	000099-88-7
2-ISOPROPYLBENZAMIDE	056177-33-4
4-ISOPROPYLBENZAMIDE	000619-76-1
O-ISOPROPYLBENZOIC ACID	002438-04-2
ISOPROPYLBIPHENYL	025640-78-2
3-ISOPROPYLBIPHENYL	020282-30-8
4-ISOPROPYLBIPHENYL	007116-95-2
1-ISOPROPYL-4-BROMOBENZENE	000586-61-8
ISOPROPYL T-BUTYL ETHER	017348-59-3
ISOPROPYL CHLOROACETATE	000105-48-6
ISOPROPYL CHLOROFORMATE	000108-23-6
DES-ISOPROPYL CLOFAZIMINE	102262-55-5
ISOPROPYL CYANOACRYLATE	010586-17-1
ISOPROPYLCYCLOHEXANE	000696-29-7
ISOPROPYLCYCLOPENTANE	003875-51-2
ISOPROPYLCYCLOPROPANE	003638-35-5
ISOPROPYL 4,4'DIBROMOBENZILATE	018181-80-1
ISOPROPYL-4,4'-DICHLOROBENZILATE	005836-10-2
3-ISOPROPYL DIMETHYL PARATHION	001592-82-1
O-ISOPROPYL DITHIOCARBONATE	000108-25-8
DI-ISOPROPYL ETHER	000108-20-3
ISOPROPYL FORMATE	000625-55-8
ISOPROPYL GALLATE	001138-60-9
ISOPROPYL HEXANOATE	002311-46-8
3-ISOPROPYL-4-HYDROXYACETANILIDE	013780-91-1
N-ISOPROPYL-HYDROXYPYRID-4-ONE	104764-53-6
3-ISOPROPYLHYDROXYUREA	060165-07-3
ISOPROPYL ISOBUTYRATE	000617-50-5
ISOPROPYL LAURATE	010233-13-3

ISOPROPYL METHACRYLATE	004655-34-9
2-ISOPROPYL-3-METHOXYPYRAZINE	025773-40-4
ISOPROPYL METHYLPHOSPHONATE	001832-54-8
ISOPROPYL MYRISTATE	000110-27-0
2-ISOPROPYLNAPHTHALENE	002027-17-0
1-ISOPROPYLNAPHTHALENE	006158-45-8
N-ISOPROPYLNICOTINAMIDE	018960-16-2
ISOPROPYL NITRATE	001712-64-7
4-ISOPROPYLNITROBENZENE	001817-47-6
ISOPROPYL OCTANOATE	005458-59-3
3-ISOPROPYL-DIMETHYLPARA-OXON	013074-11-8
ISOPROPYL PALMITATE	000142-91-6
P-ISOPROPYLPHENOL	000099-89-8
M-ISOPROPYLPHENOL	000618-45-1
O-ISOPROPYLPHENOL	000088-69-7
2-ISOPROPYLPHENYL ACETATE	001608-68-0
ISOPROPYL PHENYL CARBAMATE	000122-42-9
ISOPROPYL PHENYL DIPHENYL PHOSPHATE	028108-99-8
ISOPROPYL PROPIONATE	000637-78-5
ISOPROPYL STEARATE	000112-10-7
1-ISOPROPYLTHIO-B-GALACTOPYRANSIDE	000367-93-1
ISOPROPYLXANTHIC ACID, SODIUM SALT	000140-93-2
ISOPROTURON	034123-59-6
1-ISOQUINOLINAMINE	001532-84-9
1-ISOQUINOLINAMINE	001532-72-5
1-ISOQUINOLINAMINE, 3-(2-PYRIDINYL)-	037989-04-1
ISOQUINOLINE	000119-65-3
2(1H)-ISOQUINOLINECARBOXIMIDAMIDE, 3,4-DIHYDRO-	001131-64-2
1,4-ISOQUINOLINEDIONE DERIVATIVE	132213-90-2
ISOQUINOLINE, 3-(2-PYRIDINYL)-	058814-67-8
ISOQUINOLINE, 1,2,3,4-TETRAHYDRO-3-METHYL-	029726-60-1
1(2H)-ISOQUINOLINONE	000491-30-5
ISOQUINOLONE-3-METHYL	001125-80-0
3(2H)-ISOTHIAZOLONE, 2-OCTYL-	026530-20-1
ISOTHIOATE	036614-38-7
3-ISOTHIOCYANATO-1-PROPENE	000057-06-7
2-ISOTHIOCYANO-ANTHRACENE	007613-10-7
4-ISOTHIOCYANOAZOBENZENE	007612-96-6
4-ISOTHIOCYANOBENZOPHENONE	026328-59-6
4-ISOTHIOCYANODIPHENYL AMINE	023246-36-8
2-ISOTHIOCYANONAPHTHALENE	001636-33-5
1-ISOTHIOCYANONAPHTHALENE	000551-06-4
4-ISOTHIOCYANOPHENYL BENZOATE	049540-85-4
4-ISOTHIOCYANO-STILBENE	010319-35-4
ISOURON	055861-78-4
ISOVALERIC ACID	000503-74-2
ISOVANILLIN	000621-59-0
ISOXAPYRIFOP	087757-18-4
ISOXATHION	018854-01-8
ISOXAZOLE, 5-METHYL-	005765-44-6
ITANOXONE	058182-63-1
JOSAMYCIN	056689-45-3
KARATHANE	000131-72-6
KARSIL	002533-89-3
KASUGAMYCIN	004849-32-5
KEPONE	000143-50-0
KETAMINE	001867-66-9
KETENE	000463-51-4

KETOBEMIDONE,BENZOYL ESTER	136832-82-1
KETOBEMIDONE,PIVALIC ESTER	136832-80-9
3-KETOCARBOFURANPHENOL	017781-16-7
6-KETO METHYLHEPTANOATE	002046-21-1
LABETALOL	036894-69-6
LACTIC ACID	000050-21-5
LACTOFEN	077501-63-4
LACTOSE	000063-42-3
LASIOCARPINE	000303-34-4
LAURIC ACID, POTASSIUM SALT	010124-65-9
LAURONITRILE	002437-25-4
1-LAURYL-4-CARBOXY-2-PYRROLIDONE	010054-21-4
1-LAURYL-4-METHOXYCARBONYL-2-PYRROLIDONE	101881-19-0
1-LAURYL-2-PYRROLIDONE	002687-96-9
LENACIL	002164-08-1
LEPTOPHOS	021609-90-5
LEPTOPHOS-O-ANALOG	025006-32-0
LETHANE 60	000301-11-1
LEUCIN-AMIDE, N-ACETYL	028529-34-2
L-LEUCINAMIDE, N-ACETYL-L-ALANYL-L-TYROSYL-	132766-20-2
L-LEUCINAMIDE, N-ACETYL-L-ALANYL-	078233-72-4
L-LEUCINAMIDE, N -ACETYL-L-GLUTAMINYL-	132765-92-5
L-LEUCINAMIDE, N-ACETYL-L-LEUCYL-L-THREONYL-	132766-19-9
L-LEUCINAMIDE, N-ACETYL-L-PHENYLALANYLGLYCYL-	065118-51-6
L-LEUCINAMIDE, N-ACETYL-L-TYROSYL-	065356-76-5
LEUCINE	000061-90-5
LEUCINE	000328-39-2
LEUCINE,N-ACETYL-N'-MEAMINO AMIDE	032483-15-1
ISO-LEUCINE	000073-32-5
LEVALLORPHAN	000152-02-3
LEVOBUNOLOL	047141-42-4
LEVORPHANOL	000077-07-6
LEVULINIC ACID	000123-76-2
LIBRIUM	000058-25-3
LIDOCAINE	000137-58-6
LIDOFLAZINE	003416-26-0
LIGROIN	008032-32-4
D-LIMONENE	005989-27-5
LIMONENE	000138-86-3
LINALOOL	000078-70-6
LINALYL ACETATE	000115-95-7
LINCOMYCIN	000154-21-2
LINOLEIC ACID	000060-33-3
LINURON	000330-55-2
LOBELINE	000090-69-7
LOFENTANIL	061380-40-3
LOMEFLOXACIN	098079-51-7
LORATADINE	079794-75-5
LORAZEPAM	000846-49-1
LORCAINIDE	059729-31-6
LOVASTATIN	075330-75-5
LSD	000050-37-3
LUTEOLIN	000491-70-3
2,6-LUTIDINE	000108-48-5
LYSIN-AMIDE, N-ACETYL	019789-60-7
L-LYSINE, N -FORMYL-	001190-48-3
LYSINE	000056-87-1
MALACHITE GREEN	000569-64-2

MALATHION	000121-75-5
MALEIC ACID	000110-16-7
MALEIC ANHYDRIDE	000108-31-6
MALEIC HYDRAZIDE	000123-33-1
N-MALEOYL-4-AMINOBUTYRIC ACID	057078-98-5
N-MALEOYL-6-AMINOHEXANOIC ACID	055750-53-3
N-MALEOYL-8-AMINOOCTANOIC ACID	057079-00-2
N-MALEOYL-5-AMINOPENTANOIC ACID	057078-99-6
N-MALEOYL-3-AMINOPROPIONIC ACID	007423-55-4
N-MALEOYLGLYCINE	025021-08-3
N-MALEOYLGLYCINE, ETHYL ESTER	001585-79-1
N-MALEOYL-6-NH2HEXANOIC ACID,ET ESTER	057079-07-9
N-MALEOYL-5-NH2PROPIONIC ACID,ET EST	057079-05-7
MALIC ACID	006915-15-7
MALONAMIDE,BENZAL	019411-83-7
MALONAMIDE,3-FLUOROBENZAL	015948-56-8
MALONAMIDE, 3-METHOXYBENZAL	015804-68-9
MALONAMIDE, 3-METHYLBENZAL	015888-02-5
MALONIC ACID	000141-82-2
MALONONITRILE	000109-77-3
MALONONITRILE, BENZAL	002700-22-3
MALONONITRILE, A-METHYLBENZAL	005447-87-0
MALONONITRILE, 4-METHYLBENZAL	002826-25-7
MALONONITRILE, ALPHA-PHENYLBENZAL	010394-96-4
MALOTILATE	059937-28-9
MALTOSE	000069-79-4
MANCOZEB	008018-01-7
MANDELIC ACID, ETHYL ESTER	000774-40-3
MANEB	012427-38-2
DL-MANNITOL	000133-43-7
MANNITOL	000069-65-8
-D-MANNOPYRANOSIDE, 4-NITROPHENYL	010357-27-4
D-MANNOSE	003458-28-4
MAYTANSINE	035846-53-8
N(M-BRPH)-3-N'-PIPERIDINOACETAMIDE	053316-92-0
M-T-BUTYLPHENOXYACETIC ACID	001878-55-3
M-T-BUTYLPHENYL N-METHYLCARBAMATE	000780-11-0
N(M-CLPH)-3-N'-PIPERIDINOACETAMIDE	038367-19-0
MCPB	000094-81-5
MCPB-ETHYL	010443-70-6
MCPP	007085-19-0
M-MEO BENZYL N,N-DIME CARBAMATE	084640-23-3
N(N"ME(24DICLPH)CARBAMYLTHIO)METHOMYL	082774-16-1
N(N"ME(26DICLPH)CARBAMYLTHIO)METHOMYL	082784-14-3
3-ME-2-ACETIMINO-5-(5NO2-2-FURFURILIDENE)THIAZOL	052661-66-2
1-ME-1-ACETYL-3(P-ETPH)TRIAZENE	087072-64-8
1-ME-1-ACETYL-3-(P-F PH)TRIAZENE	087072-65-9
P-(3-ME-3-ACETYL-1-TRIAZENO)TOLUENE	066975-11-9
2-MEAM-5-(5-NO2-2-FURFURYL)THIAZOLONE	025603-06-9
2-MEAM-5(5NO2-FURYLPROPENILIDENE)THIAZOLONE	027472-83-9
2MEAM-6(5NO2-2FURYL)-13-THIAZIN-4-ONE	052661-53-7
MEBENDAZOLE	031431-39-7
(N-ME-BENZYL)AMET-(345-TRIMEO)PH SULFONE	103595-57-9
M-MEBENZYL-TRIMETHYL-AMMONIUM BR	021949-11-1
P-MEBENZYL N,N-DIME CARBAMATE	084640-21-1
P(3(PMEBENZYL)3MEO-1TRIAZENO)BENZOICACID	065542-18-9
N'(1-ME-BENZYL)PHO-ACETIC ACID HYDRAZIDE	052093-79-5
N(N"-BU-ME CARBAMYLTHIO)METHOMYL	086569-95-1

N(N"-ME-TBUO CARBAMYLTHIO)METHOMYL	064029-01-2
N-ME-3-BUTOXYPHENYLCARBAMATE	003978-68-5
N-ME-4-BUTOXYPHENYLCARBAMATE	003978-69-6
4-(3-ME-3-BU-1-TRIAZENO)BENZAMIDE	059708-21-3
A-ME-N-SEC-BUTYLCINNAMAMIDE	097871-51-7
N-ME-4-SEC-BUTYLPHENYLCARBAMATE	003942-51-6
N-ME-2-T-BUTYLPHENYLCARBAMATE	002626-81-5
N-ME-4-T-BUTYLPHENYLCARBAMATE	002626-83-7
MECARBAM	002595-54-2
N-ME CARBAMATE,3-ME-4-MES PHENYL	003566-00-5
N-ME-4-CARBOMEO PHENYLCARBAMATE	021998-12-9
P-ME CF3-METHANESULFONANILIDE	037595-73-6
N(N"-ME(3-CF3PH)CARBAMYLTHIO)METHOMYL	082774-15-0
NNN-ME-N(3-CF3PHME) AMMONIUM BROMIDE	071324-01-1
N-ME-2-CHLOROPHENYLCARBAMATE	003942-54-9
ME-N-(M-CHLOROPHENYL)CARBAMATE	002150-88-1
2-ME-4-CL-C6H3NHN=C(CN)CO-ME	028317-64-8
1-ME-3-(3-CL-4-MEO PHENYL) UREA	020782-57-4
N(N"-ME(4-CLPH)CARBAMYLTHIO)METHOMYL	082774-06-9
N(N"-ME(2-CLPH)CARBAMYLTHIO)METHOMYL	082774-07-0
N(N"-ME(3-CLPH)CARBAMYLTHIO)METHOMYL	082774-14-9
3-ME-4-CLPHENYL-N-ME CARBAMATE	002589-65-3
2'-ME-4'-CL SALICYLANILIDE	077067-91-5
MECOPROP	000093-65-2
MECOPROP, POTASSIUM SALT	001929-86-8
NNN-ME-N(3(CYANOPHENOL)METHYL) AMMONIUM BROMIDE	071323-97-2
N-ME-4-CYANOPHENYLCARBAMATE	018315-52-1
N(N"-ME(CYHEXO)CARBAMYLTHIO)METHOMYL	086569-90-6
N(N"-ME-CYPR CARBAMYLTHIO)METHOMYL	066469-80-5
MEDAZEPAM	002898-12-6
N(N"-ME-(DECOXY)CARBAMYLTHIO)METHOMYL	086569-89-3
ME-3-DEOXY-3(3ME3NOUREA)XYLOPYRANOSID	052019-05-3
N-ME-3,5-DICHLOROPHENYLCARBAMATE	013538-26-6
N-ME-3,4-DICHLOROPHENYLCARBAMATE	018315-50-9
N-ME-2,5-DICHLOROPHENYLCARBAMATE	018315-62-3
N-ME-3-DIMEAMINOPHENYL CARBAMATE	002631-39-2
N-ME-2,3-DIMETHYLPHENYL CARBAMATE	002655-12-1
N-ME-3,5-DIMETHYLPHENYL CARBAMATE	002655-14-3
N-ME-2,5-DIMETHYLPHENYLCARBAMATE	003971-99-1
N-ME-2-(33-DIME-1-TRIAZINO)BENZAMIDE	066974-88-7
2-ME-N,N'-DINITROSOPIPERAZINE	055556-94-0
N-(N"-ME-ETHOXYCARBAMYLTHIO)METHOMYL	064028-99-5
N-ME-2-ETHOXYPHENYLCARBAMATE	023409-17-8
N-ME ETHYLAMINE,2,2,2-TRIFLUORO	002730-67-8
N-(N"-ME-ETHYLCARBAMYLTHIO)METHOMYL	086569-92-8
N-ME-M-ETHYLPHENYLCARBAMATE	004043-23-6
MEFENACET	073250-68-7
MEFLUIDIDE	053780-34-0
N-ME-4-FORMYLPHENYLCARBAMATE	054335-83-0
N-ME-3-FORMYLPHENYLCARBAMATE	054335-82-9
MEGLITINIDE	054870-28-9
8-MEHTYL-1-NONANOL	055505-26-5
N-(N"-ME-HXOCARBAMYLTHIO)METHOMYL	066381-49-5
2(4ME-1,3-IMIDAZOLINYLIDEN-2AM)-5NO2-THIAZOLE	037385-10-7
2(1ME-13-IMIDAZOLINYLIDEN-2AM)5NO2-THIAZOLE	031052-76-3
2(14ME-13-IMIDAZOLN-2AM)5NO2-THIAZOLE	031052-78-5
N(N"-ME-IPR CARBAMYLTHIO)METHOMYL	086569-94-0
N-ME-3-I-PROPOXYPHENYLCARBAMATE	003938-20-3

N7-(5-ME ISOTHIAZOL-3-YL)-MITIMYCIN C	088854-64-2
MELAMINE	000108-78-1
MELLITIC ACID	000517-60-2
MELLOPHANIC ACID	000476-73-3
N-ME-3-ME-5-T-BU PHENYLCARBAMATE	018659-33-1
N-ME-3-ME-6-T-BU PHENYLCARBAMATE	002798-54-1
NNN-ME-N(3ME-CYPEME)AMMON I(Z)	041550-04-3
NNN-ME-N(1ME-1-CYPENTENE)AMMON I	083283-07-2
NNN-ME-N(1ME-3-CYPENTENE)AMMON I	058944-11-9
NNN-ME-N(1ME2CYPENTENE)AMMON I	074819-88-8
1-ME-3(2(5ME IMIDAZOL4YL)MESET UREA	038603-64-4
N-ME-3-ME-4-IPRPHENYL CARBAMATE	003766-82-3
N-ME-3-ME-6-I-PR PHENYLCARBAMATE	018659-24-0
1-ME-1-MEO-3-(4-BENZOYLPHENYL)UREA	088132-26-7
3(ME-MEO)1(2(4MEPH)ETO) PH UREA	068358-79-2
N1(ME-MEO)N3(P-PHETO)PHENYLUREA	069922-40-3
N1-ME-MEO N3-(4PHOBUO)PHENYLUREA	088132-31-4
4-ME-5-MEO-8-QUINOLINOL	057334-38-0
N(N"-ME-(4ME-PH)CARBAMYLTHIO)METHOMYL	082774-00-3
N-ME-2-METHOXYPHENYLCARBAMATE	003938-24-7
N-ME-3-METHOXYPHENYLCARBAMATE	003938-28-1
N-ME 4-METHYLTHIOPHENYLCARBAMATE	003938-34-9
MENAZON	000078-57-9
1-ME-1-NITROSO-3-(P-F PH)UREA	000777-59-3
2-ME-N-NITROSOMORPHOLINE	075881-16-2
N-ME-N-NITROSO-1-NAPHTHYLCARBAMAT	007090-25-7
2ME2NO2-PROPANAL-N-MECARBAMOYL OXIME	006129-11-9
1-ME-2-NO2-5-(1,2-DIOHET)IMIDAZOLE	039070-09-2
4-ME(5-NO2-2-FURFURILIDENE)THIAZOLE	006448-55-1
3ME5(5NO2-2-FURFURIL)2-S-THIAZOL-4ONE	015913-35-6
3ME-5-(5NO2-FURPROPENILIDENE)THIAZOLIDN-24-DIONE	025580-68-1
1-ME-2-NO2-5-HYDROXYME IMIDAZOLE	000936-05-0
ME(2(2-NO2-1-IMIDAZOLYL)ET)SULFONE	071006-79-6
ME(3(2ME-5NO2-1-IMIDAZ)PROPYL)SULFONE	028795-24-6
1-ME-2-NO2-5-IPR IMIDAZOLE	023571-34-8
1-ME-2-NO2-5-MEOCARBONYL IMIDAZOLE	040361-79-3
1-ME-2-NO2-5-(1-OH-1-MEET)IMIDAZOLE	060628-92-4
3-ME-4-NO2 PYRAZOLE	005334-39-4
1-ME-2-NO2-5-VINYL IMIDAZOLE	039070-08-1
P-MENTHANE	000099-82-1
MENTHOL	000089-78-1
MENTHOL (L)	002216-51-5
MENTHYL ACETATE	016409-45-3
O-MEO BENZAMIDE,N-(4-AMINOBUTYL)	090663-30-2
O-MEO BENZAMIDE, N-(2-AMINOETHYL)	053673-10-2
O-MEO BENZAMIDE,N-(3-DIETAMINOPR)	085819-15-4
O-MEO BENZAMIDE,N-(DIETAMINOET)	065016-34-4
O-MEO BENZAMIDE,N-(3-DIMEAMINOPR)	085819-14-3
O-MEO BENZAMIDE,N-(2-DIMEAMINOET)	078798-46-6
P-MEO BENZYL N,N-DIME CARBAMATE	084640-24-4
2'-MEO-5'-BR BENZOGUANAMINE	065052-55-3
3-MEO-5-BR-6-ME-PYRIMIDIN-2,4-DIONE	077317-65-8
A-MEO-N-SEC-BUTYLCINNAMAMIDE	097871-53-9
P-MEO-CINNAMIC ACID,DIET-AMINO-ET ESTER	072836-46-5
O-MEO N,N-DIMETHYLBENZAMIDE	007291-34-1
4-MEO-3-(33-DIME TRIAZENO)BENZAMIDE	066974-78-5
1-(2-MEO ETHYL)-2-NO2 IMIDAZOLE	071006-73-0
5MEO-6ETO-4-SULFANILAMIDOPYRIMIDIN	005018-56-4

NNN-ME-N(3OH4MECYPEME)AMMON I	055903-47-4
2-ME-3-OH-1,4-NAPHTHOQUINONE	000483-55-6
1-MEO-1-ME-3-(P-AMINOPHENYL)UREA	088132-16-5
O=P(NME)(OME)O-(2,4-DICLPHENYL)	010363-40-3
1-MEO-2-ME-3-NO2-BENZENE	004837-88-1
1-MEO-1-ME-3-(CF3-PHENYL)UREA	000838-89-1
2-(2-MEO-4-ME PH IMINO)IMIDAZOLIDINE	063346-72-5
2-MEO 3-MES TETRACHLOROCYCLOHEXANE	060067-85-8
M-MEO CF3-METHANESULFONANILIDE	023384-33-0
P-MEO CF3-METHANESULFONANILIDE	023384-34-1
3-MEO-4NH2-6-IPR-1,2,4-TRIAZIN-5-ON	018826-97-6
4-MEO-8-NO2 QUINOLINE	089770-28-5
2MEO-4NH2-5NO2 BENZAMIDE,N(DIETAMET)	004093-42-9
2-MEO PHENYL 4-AMINOSALICYLATE	056356-17-3
3-MEO PHENYL 4-AMINOSALICYLATE	056356-18-4
4-MEO PHENYL 4-AMINOSALICYLATE	004465-61-6
2-MEO PHENYL GLUCOPYRANOSIDE	006092-24-6
3-MEO PHENYL GLUCOPYRANOSIDE	014062-61-4
P-MEO-B-PHENYLPROPIONIC ACID	001929-29-9
N7-(6-MEO-PYRID-3-YL)-MITOMYCIN C	084397-27-3
2(445ME-13OXAZOLID-2AM)5NO2THIAZOLE	030122-47-5
2(4-ME-1,3-OXAZOLIDINYLIDEN-2-AMINO)5NO2-THIAZOL	024240-83-3
MEPANIPYRIM	110235-47-7
MEPERIDINE	000050-13-5
O-ME PHENOXYACETIC ACID,ME ESTER	002989-17-5
N(N"-ME(PHENYL)CARBAMYLTHIO)METHOMYL	082774-01-4
2-ME-4-PHENYLQUINOZALINE	061453-53-0
MEPHENYTOIN	000050-12-4
MEPHOBARBITAL	000115-38-8
ME-PHOSPHORAMIDATE,O-ME,O-4-MEOPH	017795-32-3
3-ME-5-PH OXAZOLIDINE-2,4-DIONE	005841-66-7
N(N"-ME-PR CARBAMYLTHIO)METHOMYL	086569-93-9
MEPROBAMATE	000057-53-4
MEPRONIL	055814-41-0
N-ME-3-PROPIONYLPHENYLCARBAMATE	010051-63-5
N-ME-4-PROPIONYLPHENYLCARBAMATE	054266-28-3
N-ME-4-I-PROPYLPHENYLCARBAMATE	004089-99-0
4-(N-ME)-3-PYRIDYLBUTYLAMINE	003000-74-6
N-ME QUINOLINIUM BROMIDE	002516-72-5
MERBARONE	097534-21-9
MERCAPTOACETIC ACID	000068-11-1
2-MERCAPTOBENZOIC ACID	000147-93-3
2-MERCAPTOBENZOTHIAZOLE	000149-30-4
3-MERCAPTOPROPIONIC ACID	000107-96-0
6-MERCAPTOPURINE RIBOSIDE	004988-64-1
MERCURIC ACETATE	001600-27-7
MERPHOS	000150-50-5
3MES-4NH2-6CYHX-124-TRIAZIN-5-ONE	021085-19-8
MESCALINE	000054-04-6
P-MES CF3-METHANESULFONANILIDE	023375-06-6
2-MES-4,6-DIMETHYLPYRIMIDINE	014001-64-0
MESITYL OXIDE	000141-79-7
3-MES-4NH2-6-ME-1,2,4-TRIAZIN-5-ON	018826-96-5
3-MES-4-NH2-1,2,4-TRIAZINE-5-ONE	062036-62-8
2-MES-5-(5NO2-2-FURFURILD)THIAZOLONE	052661-45-7
M-MESO2 CF3-METHANESULFONANILIDE	023375-08-8
P-MESO2 CF3-METHANESULFONANILIDE	023375-10-2
4-ME-N-(4-SO2F-BENZYL)PYRIDINIUM BR	033803-13-3

A-MESULFONAMIDOBENZYLPENCILLIN	088644-13-7
METACIL	000056-04-2
METALAXYL	057837-19-1
METANEPHRINE	005001-33-2
METARAMINOL	000054-49-9
METCONAZOLE	125116-23-6
METHACROLEIN	000078-85-3
METHACRYLIC ACID	000079-41-4
METHACRYLIC ACID, N-BUTYL ESTER	000097-88-1
METHACRYLIC ACID, I-BUTYL ESTER	000097-86-9
METHACRYLONITRILE	000126-98-7
METHADONE	000076-99-3
METHAMIDPHOS	010265-92-6
METHAMPHETAMINE	000537-46-2
METHANAMINE, HYDROCHLORIDE	000593-51-1
METHANAMINE, N-HYDROXY-N-METHYL-	005725-96-2
METHANAMINE, N-[(1-METHYL-2-NITRO-1H-IMIDAZOL-5-	040647-30-1
METHANAMINE, N-(PHENYLMETHYLENE)-	000622-29-7
METHANE	000074-82-8
METHANEDISULFONIC ACID	000503-40-2
METHANESULFONAMIDE, N-(4-NITRO-2-PHENOXYPHENYL)-	051803-78-2
METHANESULFONAMIDE, N-METHYL-N-PHENYL-	013229-35-1
METHANESULFONIC ACID	000075-75-2
N,N'-METHANETETRAYLBIS-2-PROPANAMINE	000693-13-0
METHANETHIOL	000074-93-1
METHANIMIDAMIDE, N-(2,4-DIMETHYLPHENYL)-N'-METHY	033089-74-6
2,6-METHANO-3-BENZAZOCINE, 1,2,3,4,5,6-HEXAHYDRO	023007-86-5
2,6-METHANO-3-BENZAZOCIN-8-OL, 6-ETHYL-1,2,3,4,5	071990-00-6
5,8-METHANO-5H-BENZOCYCLOHEPTEN-10-AMINE, 6,7,8,	110192-33-1
5,8-METHANO-5H-BENZOCYCLOHEPTEN-10-AMINE, 6,7,8,	110267-94-2
5,8-METHANO-5H-BENZOCYCLOHEPTEN-10-AMINE, 6,7,8,	110267-96-4
5,8-METHANO-5H-BENZOCYCLOHEPTEN-10-AMINE, 6,7,8,	110267-95-3
5,8-METHANO-5H-BENZOCYCLOHEPTEN-10-AMINE, 6,7,8,	110192-35-3
4,7-METHANO-1H-ISOINDOLE-1,3(2H)-DIONE, 3A,4,7,7	103066-37-1
METHANOL	000067-56-1
1,4-METHANONAPHTHALEN-2-AMINE, 1,2,3,4-TETRAHYDR	058742-05-5
METHANONE, (4-CHLOROPHENYL)(5-FLUORO-2-HYDROXYPH	062433-26-5
METHANONE, DIPHENYL-, HYDRAZONE	005350-57-2
METHANONE, DIPHENYL-, OXIME	000574-66-3
METHANONE, (2-HYDROXY-4-METHOXYPHENYL)PHENYL-	000131-57-7
METHAPYRILENE	000091-80-5
METHAQUALONE	000072-44-6
METHARBITAL	000050-11-3
BETAMETHASONE	000378-44-9
BETAMETHASONE-17-ACETATE	000987-24-6
BETAMETHASONE-21-BUTYRATE	056933-60-9
BETAMETHASONE-21-DESOXY	000382-67-2
BETAMETHASONE 17,21-DIPROPIONATE	005593-20-4
BETAMETHASONE 17,21-DIVALERATE	038196-44-0
BETAMETHASONE-21-PROPIONATE	075883-07-7
BETAMETHASONE-17-PROPIONATE-21-DESOXY	004351-48-8
BETAMETHASONE-17-VALERATE	002152-44-5
BETAMETHASONE-21-VALERATE	002240-28-0
METHAZOLAMIDE	000554-57-4
METHAZOLE	020354-26-1
METHDILAZINE	001982-37-2
5H-3,6-METHENOFURO[3,2-G]OXIRENO[D]OXACYCLOUNDEC	010247-71-9
2-ME-134-THIADIAZOLE-5-CARBOXALDOXIME	061444-95-9

3-ME-124-THIADIAZOLE-5-CARBOXALDOXIME	061444-96-0
N7-(5-ME THIAZOL-2-YL)-MITOMYCIN C	084397-36-4
METHICILLIN	000061-32-5
METHIDATHION	000950-37-8
METHIMAZOLE	000060-56-0
O-METHIOANILINE	002987-53-3
M-METHIOANILINE	001783-81-9
P-METHIOBENZOIC ACID	013205-48-6
P-METHIOBENZYL N,N-DIME CARBAMATE	084640-39-1
METHIONINE	000063-68-3
METHIONINE-AMIDE,N-ACETYL	023361-37-7
3-METHIOPENTACHLOROCYCLOHEXANE	056046-09-4
1-METHIOPENTACHLOROCYCLOHEXANE	060132-42-5
P-METHIOPHENOL	001073-72-9
METHOMYL	016752-77-5
METHOPROPTRYNE	000841-06-5
METHOTREXATE	000059-05-2
METHOTREXATE DIMETHYL ESTER	034378-65-9
METHOTRIMEPRAZINE	000060-99-1
3-METHOXYACETANILIDE	000588-16-9
P-METHOXYACETANILIDE	000051-66-1
O-METHOXYACETANILIDE	000093-26-5
METHOXYACETIC ACID	000625-45-6
P-METHOXYACETOPHENONE	000100-06-1
M-METHOXYACETOPHENONE	000586-37-8
P-METHOXYAMPHETAMINE	000064-13-1
M-METHOXYANILINE	000536-90-3
2-METHOXYANILINE	000090-04-0
4-METHOXYANILINE	000104-94-9
2-METHOXY-9,10-ANTHRACENEDIONE	003274-20-2
1-METHOXY-9,10-ANTHRAQUINONE	000082-39-3
4-METHOXY-O,O'-AZOTOLUENE	029418-45-9
3-METHOXYBENZAL ACETYLACETONE	015818-10-7
2-METHOXYBENZAL ACETYLACETONE	015725-16-3
4-METHOXYBENZAL ACETYLACETONE	015725-17-4
M-METHOXYBENZAMIDE	005813-86-5
O-METHOXYBENZAMIDE	002439-77-2
P-METHOXYBENZENESULFONAMIDE	001129-26-6
2'-METHOXYBENZOGUANAMINE	072775-80-5
M-METHOXYBENZOIC ACID	000586-38-9
O-METHOXYBENZOIC ACID	000579-75-9
P-METHOXYBENZOIC ACID	000100-09-4
4-(4-METHOXYBENZOYL)PYRIDINE	014548-47-1
M-METHOXYBENZYL ACETATE	035480-26-3
O-METHOXYBENZYL ALCOHOL	000612-16-8
P-METHOXYBENZYL ALCOHOL	000105-13-5
M-METHOXYBENZYL TRIMETHYL AMMONIUM BROMIDE	071323-94-9
4-METHOXYBIPHENYL	000613-37-6
3-METHOXY-1-BUTANOL	002517-43-3
3-METHOXYBUTYL ACETATE	004435-53-4
P-METHOXYCARBONYLACETANILIDE	017012-22-5
M-METHOXYCARBONYL-N-PHENYLSUCCINIMIDE	072601-52-6
METHOXYCHLOR	000072-43-5
METHOXYCHLOR-DDD	007388-31-0
P-METHOXYCINNAMIC ACID (TRANS)	000943-89-5
4-METHOXYCINNOLINE	003397-78-2
1-(3-METHOXY),2-DIBENZOYL-1-TERT-BUTYLHYDRAZINE	112426-60-5
1-(2-METHOXY),2-DIBENZOYL-1-TERT-BUTYLHYDRAZINE	112226-77-4

1-(4-METHOXY),2-DIBENZOYL-1-TERT-BUTYLHYDRAZINE	112226-92-3
P-METHOXY-N,N-DIMETHYLBENZAMIDE	007291-00-1
M-METHOXY-N,N-DIMETHYLBENZAMIDE	007290-99-5
5-METHOXY-1,2-DIMETHYL-4-NITRO-1H-IMIDAZOLE	035687-44-6
2-METHOXYETHANOL	000109-86-4
2-METHOXYETHOXYBENZENE	041532-81-4
2-METHOXYETHYL ACETATE	000110-49-6
2-METHOXYETHYLETHER	000111-96-6
(METHOXYETHYL)MERCURIC ACETATE	000151-38-2
2-METHOXYETHYL-TRIME AMMONIUM IODIDE	004238-50-0
2-METHOXY-9H-FLUORENE	002523-46-8
METHOXYFLURANE	000076-38-0
M-METHOXYFORMANILIDE	027153-17-9
2-METHOXYFURAN	025414-22-6
3-METHOXY-4-HYDROXYPHENYLETHANOL	002380-78-1
3-METHOXY-4-HYDROXYPHENYLGLYCOL	000534-82-7
5-METHOXYINDOLE	001006-94-6
2-METHOXY ISONIAZID	019353-97-0
6-METHOXYISOQUINOLINE	052986-70-6
METHOXYMETHIOCHLOR	034197-16-5
2-METHOXY-4'-METHYLAZOBENZENE	029418-41-5
2-METHOXY-2'-METHYLAZOBENZENE	029268-78-8
4-METHOXY-2-METHYLAZOBENZENE	029418-42-6
2-METHOXY-6-METHYLAZOBENZENE	029268-77-7
2-METHOXY-5-METHYLBENZENAMINE	000120-71-8
4'-METHOXY-2-METHYL-4-NITROAZOBENZENE	029418-61-9
5-METHOXY-1-METHYL-3-NITRO-1H-IMIDAZOLE	035687-41-3
4-METHOXY-2-METHYL-5-NITRO-1H-IMIDAZOLE	035687-42-4
2-(METHOXYMETHYL)-5-NO2 FURAN	000586-84-5
4-METHOXY-4-METHYL-2-PENTANONE	000107-70-0
2-METHOXY-4-METHYLPHENOL	000093-51-6
2-METHOXY-1,4-NAPHTHOQUINONE	002348-82-5
2-METHOXY-5-NITROANILINE	000099-59-2
4-METHOXY-5-NITRO-1H-IMIDAZOLE	068019-78-3
4-METHOXY-2-NITROPHENOL	001568-70-3
2-METHOXY-5-NITROPYRIDINE	005446-92-4
4-METHOXY-B-NITROSTYRENE	003179-10-0
3-METHOXY B-NITROSTYRENE	003179-09-7
3-METHOXYPENTACHLOROCYCLOHEXANE	056046-08-3
2-METHOXYPENTACHLOROCYCLOHEXANE	056421-31-9
M-METHOXYPHENOL	000150-19-6
2-METHOXYPHENOL	000090-05-1
P-METHOXYPHENOL	000150-76-5
O-METHOXY PHENOXYACETIC ACID	001878-85-9
M-METHOXYPHENOXYACETIC ACID	002088-24-6
P-METHOXYPHENOXYACETIC ACID	001877-75-4
4-METHOXYPHENYL ACETATE	001200-06-2
O-METHOXYPHENYL ACETATE	000613-70-7
M-METHOXYPHENYLACETIC ACID	001798-09-0
P-METHOXYPHENYLACETIC ACID	000104-01-8
4-METHOXYPHENYLAMIDINO UREA	058247-23-7
P-METHOXY PHENYL AZOXYCYANIDE	062825-15-4
M-METHOXYPHENYL AZOXYCYANIDE	062825-14-3
3-METHOXY-N-PHENYLBENZENAMINE	000101-16-6
4-METHOXY-N-PHENYLBENZENAMINE	001208-86-2
P-METHOXY-G-PHENYLBUTYRIC ACID	004521-28-2
N'-(3-METHOXYPHENYL)-N,N-DIMETHYLUREA	028170-54-9
3-METHOXYPHENYLDIMETHYLCARBAMATE	007305-09-1

N'-(4-METHOXYPHENYL)-N,N-DIMETHYLUREA	007160-02-3
4-METHOXYPHENYLDIMETHYLCARBAMATE	007305-10-4
1,1-DI-(P-METHOXYPHENYL)ETHANE	010543-21-2
9-(3-METHOXYPHENYL)-9H-FLUORENE	032377-13-2
9-(4-METHOXYPHENYL)-9H-FLUORENE	021846-08-2
4-METHOXYPHENYL GLUCOPYRANOSIDE	006032-32-2
P-METHOXYPHENYLHIPPURATE	029736-21-8
2-(4-METHOXYPHENYL)INDOLIZINE	007496-82-4
2-(3-METHOXYPHENYL)INDOLIZINE	082682-68-6
(4-METHOXYPHENYL)(4-METHYLPHENYL)DIAZENE	029418-44-8
(4-METHOXYPHENYL)(2-METHYLPHENYL)DIAZENE	029418-43-7
N-((4-METHOXYPHENYL)METHYL)ACETAMIDE	035103-34-5
(4-METHOXYPHENYL)PHENYLDIAZENE	002396-60-3
(2-METHOXYPHENYL)PHENYLDIAZENE	006319-21-7
1-METHOXY-3-PHENYLPROPANE	002046-33-5
2-METHOXY-4-PHENYLQUINAZOLINE	017629-09-3
P-METHOXY-N-PHENYLSUCCINIMIDE	002314-80-9
M-METHOXY-N-PHENYLSUCCINIMIDE	016141-40-5
N1-(4-METHOXYPHENYL)SULFANILAMIDE	019837-74-2
N1-(2-METHOXYPHENYL)SULFANILAMIDE	019837-84-4
METHOXYPROMAZINE	000061-01-8
1-METHOXY-2-PROPANOL	000107-98-2
1-METHOXY-2-PROPANONE	005878-19-3
2-METHOXY-4-(1-PROPENYL)PHENOL	000097-54-1
3-METHOXYPROPYLAMINE	005332-73-0
2-METHOXYPYRAZINE	003149-28-8
4-METHOXYPYRAZOLE	014884-01-6
2-METHOXYPYRIDINE	001628-89-3
4-METHOXYPYRIDINE	000620-08-6
4-METHOXYQUINAZOLINE	016347-95-8
6-METHOXYQUINOLINE	005263-87-6
8-METHOXYQUINOLINE	000938-33-0
7-METHOXYQUINOLINE	004964-76-5
5-METHOXY-8-QUINOLINOL	057334-35-7
2-METHOXYQUINOXALINE	039209-88-6
3-METHOXYSALICYCLIC ACID	000877-22-5
2-METHOXYSTRYCHNINE	000509-36-4
2-METHOXYTETRACHLOROCYCLOHEXANE	056933-31-4
METHOXY TRIETHYLENE GLYCOL	000112-35-6
METHSUXIMIDE	000077-41-8
2-METHYBENZOTHIAZOLE	000120-75-2
4-SULFAMYLBENZAMIDE,N-METHYL	010518-85-1
N-METHYLACETAMIDE	000079-16-3
3-METHYLACETANILIDE	000537-92-8
4-METHYLACETANILIDE	000103-89-9
N-METHYLACETANILIDE	000579-10-2
METHYL ACETATE	000079-20-9
METHYL ACETOACETATE	000105-45-3
M-METHYLACETOPHENONE	000585-74-0
P-METHYLACETOPHENONE	000122-00-9
N-METHYL-4-ACETYLPHENYLCARBAMATE	001135-43-9
METHYL ACRYLATE	000096-33-3
N-METHYLALLOBARBITAL	000780-59-6
TR-METHYLALLYL-CHRYSANTHEMATE	096358-98-4
2-(2-METHYLALLYL)PHENOL	020944-88-1
METHYLAMINE	000074-89-5
P-METHYLAMINOBENZENESULFONAMIDE	016891-79-5
1-METHYL-2-AMINOBENZIMIDAZOLE	001622-57-7

METHYL P-AMINOBENZOATE	000619-45-4
1-METHYL-2-AMINO-5-CHLOROBENZIMIDAZOLE	103748-25-0
2-(METHYLAMINO)ETHANOL	000109-83-1
2-METHYLAMINO-3-METHYLIMIDAZO[45F]QUINOLINE	102408-26-4
2-METHYLAMINO-4-PHENYLQUINAZOLINE	089770-35-4
2-METHYLAMINOPYRAZINE	032111-28-7
2-METHYLAMINOPYRIDINE	004597-87-9
METHYL-T-AMYL ETHER	000994-05-8
METHYLANABASINE	024380-92-5
N-METHYLANILINE	000100-61-8
P-METHYLANISOLE	000104-93-8
M-METHYLANISOLE	000100-84-5
2-METHYLANISOLE	000578-58-5
9-METHYLANTHRACENE	000779-02-2
2-METHYLANTHRACENE	000613-12-7
1-METHYLANTHRACENE	000610-48-0
METHYL ANTHRANILATE	000134-20-3
METHYL AZINPHOS	000086-50-0
METHYL AZINPHOS O-ANALOG	000961-22-8
2-METHYLAZIRIDINE	000075-55-8
N-METHYL BARBITURIC ACID	002565-47-1
2-METHYLBENAIMIDAZOLE	000615-15-6
9-METHYLBENZ(A)ANTHRACENE	002422-79-9
1-METHYLBENZ(A)ANTHRACENE	002498-77-3
10-METHYL-BENZ(A)ANTHRACENE	002381-15-9
4-METHYLBENZAL ACETYLACETONE	015818-09-4
2-METHYLBENZALDEHYDE	000529-20-4
P-METHYLBENZALDOXIME	003235-02-7
O-METHYLBENZALDOXIME	003376-32-7
4-METHYLBENZAMIDE	000619-55-6
3-METHYLBENZAMIDE	000618-47-3
2-METHYLBENZAMIDE	000527-85-5
N-METHYLBENZAMIDE	000613-93-4
N-METHYLBENZANILIDE	001934-92-5
2-METHYLBENZENAMINE HYDROCHLORIDE	000636-21-5
5-METHYL-1,3-BENZENEDIAMINE	000108-71-4
M-METHYLBENZENESULFONAMIDE	001899-94-1
N-METHYLBENZENESULFONAMIDE	005183-78-8
4-METHYLBENZENESULFONAMIDE	000070-55-3
O-METHYLBENZENESULFONAMIDE	000088-19-7
4-METHYLBENZENETHIOL	000106-45-6
2-METHYLBENZENETHIOL	000137-06-4
3-METHYLBENZENETHIOL	000108-40-7
6-METHYLBENZO(A)PYRENE	002381-39-7
2-METHYL-2,3-BENZODIAZIN-1-ONE	006091-81-2
2-METHYLBENZOFURAN	004265-25-2
METHYL N-BENZOGLYCINE	001205-08-9
2'-METHYLBENZOGUANAMINE	030508-25-9
4'-METHYLBENZOGUANAMINE	019388-12-6
4-METHYLBENZONITRILE	000104-85-8
2-METHYLBENZONITRILE	000529-19-1
O-METHYLBENZOPHENONEIMINE	022627-00-5
2-METHYL-1,4-BENZOQUINONE	000553-97-9
3-METHYLBENZOTHIAZOL-2-ONE	002786-62-1
3-METHYLBENZOTRIAZIN-4-ONE	022305-44-8
1-METHYLBENZOTRIAZOLE	013351-73-0
2-METHYLBENZOTRIAZOLE	016584-00-2
1-METHYL-4-BENZOYLPIPERAZINE	075349-23-4

4-(4-METHYLBENZOYL)-PYRIDINE	014548-30-2
P-METHYLBENZYL ACETATE	002216-45-7
ALPHA-METHYLBENZYL ALCOHOL	000098-85-1
M-METHYLBENZYL ALCOHOL	000587-03-1
P-METHYL BENZYL ALCOHOL	000589-18-4
N-METHYLBENZYLAMINE	000103-67-3
N-METHYL-N-BENZYLANILINE	000614-30-2
1-METHYL-3-BENZYL THIOUREA	002740-94-5
4-METHYL-1,1'-BIPHENYL	000644-08-6
2-METHYLBIPHENYL	000643-58-3
3-METHYLBIPHENYL	000643-93-6
METHYL BROMIDE	000074-83-9
3-METHYL-4-BROMOACETANILIDE	090914-81-1
2-METHYL-4-BROMOACETANILIDE	024106-05-6
3-METHYL-4-BROMOANILINE	006933-10-4
METHYL 3-BROMOBENZOATE	000618-89-3
METHYL 2-BROMOBENZOATE	000610-94-6
2-METHYL-6-BROMOBENZOQUINONE	006293-55-6
N-METHYL-4-BROMOPHENYLCARBAMATE	013538-50-6
N-METHYL-2-BROMOPHENYLCARBAMATE	013538-27-7
N-METHYL-3-BROMOPHENYLCARBAMATE	013538-60-8
(3-METHYL-4-BROMOPHENYL)UREA	078508-46-0
3-METHYL-1,2-BUTADIENE	000598-25-4
2-METHYL-1,3-BUTADIENE	000078-79-5
3-METHYL-1-BUTANAL	000590-86-3
2-METHYLBUTANAL	000096-17-3
2-METHYL-2-BUTANETHIOL	001679-09-0
2-METHYL-1-BUTANETHIOL	001878-18-8
3-METHYL-2-BUTANOL	000598-75-4
2-METHYL-2-BUTANOL	000075-85-4
2-METHYL-1-BUTANOL	000137-32-6
3-METHYL-2-BUTANONE	000563-80-4
3-METHYLBUTANOYLUREA	002274-08-0
3-METHYL-1-BUTENE	000563-45-1
2-METHYL-2-BUTENE	000513-35-9
2-METHYL-1-BUTENE	000563-46-2
3-METHYL-3-BUTEN-1-OL	000763-32-6
METHYLBUTYLAMINE	000110-68-9
METHYL SEC-BUTYLCARBAMATE	039076-02-3
METHYL N-BUTYL ETHER	000628-28-4
METHYL T-BUTYL ETHER	001634-04-4
2-METHYL-3-BUTYN-2-OL	000115-19-5
METHYLBUTYRATE	000623-42-7
2-METHYLBUTYRIC ACID HYDRAZIDE	070195-11-8
2-METHYLBUTYRONITRILE	018936-17-9
O-METHYL CARBAMATE	000598-55-0
N-METHYLCARBAMIC ACID, ET ESTER	000105-40-8
1-METHYL-4-CARBOXY-2-PYRROLIDONE	042346-68-9
METHYL CHLORIDE	000074-87-3
METHYL CHLOROACETATE	000096-34-4
2-METHYL-4-CHLOROANILINE	000095-69-2
2-METHYL-3-CHLOROANILINE	000087-60-5
METHYL P-CHLOROBENZOATE	001126-46-1
METHYL O-CHLOROBENZOATE	000610-96-8
4-METHYL-7-CHLOROCINNOLINE	089770-40-1
4-METHYL-6-CHLOROCINNOLINE	089770-39-8
METHYL CHLOROFORMATE	000079-22-1
2-METHYL-3-CHLOROMETHYLHYDROCHLOROTHIAZIDE	000135-07-9

METHYL 4-CHLORO-2-NITROBENZOATE	042087-80-9
3-METHYL-4-CHLOROPHENOL	000059-50-7
2-METHYL-6-CHLOROPHENOL	000087-64-9
2-METHYL-4-CHLOROPHENOL	006640-27-3
2-METHYL-4-CHLOROPHENOL	001570-64-5
2-METHYL-4-CHLOROPHENOXYACETIC ACID	000094-74-6
N-METHYL-3-CHLOROPHENYLCARBAMATE	004090-00-0
N-METHYL-4-CHLOROPHENYLCARBAMATE	002620-53-3
3-METHYLCHOLANTHRENE	000056-49-5
6-METHYLCHRYSENE	001705-85-7
5-METHYLCHRYSENE	003697-24-3
TRANS-METHYL CHYRSANTHEMATE	024141-52-4
N-METHYLCINNAMAMIDE	002757-10-0
ALPHA-METHYLCINNAMIC ACID	001199-77-5
8-METHYL-4-CINNOLINAMINE	089770-41-2
METHYL CYANOACETATE	000105-34-0
1-METHYL-2-CYANOGUANIDINE	001609-07-0
N-METHYL-3-CYANOPHENYLCARBAMATE	000943-49-7
N-METHYL-2-CYANOPHENYLCARBAMATE	000942-79-0
METHYL 2-CYANO-2-PROPENOATE	000137-05-3
N-METHYLCYCLOBARBITAL	000726-78-3
METHYLCYCLOBUTANE	000598-61-8
1-METHYLCYCLOHEXANE	000108-87-2
2-METHYLCYCLOHEXANOL	000583-59-5
4-METHYLCYCLOHEXANOL	000589-91-3
3-METHYLCYCLOHEXANOL	000591-23-1
1-METHYLCYCLOHEXANOL	000590-67-0
3-METHYLCYCLOHEXANOL (CIS)	005454-79-5
4-METHYLCYCLOHEXANOL (CIS)	007731-28-4
2-METHYLCYCLOHEXANOL (CIS)	007443-70-1
3-METHYLCYCLOHEXANOL (TRANS)	007443-55-2
4-METHYLCYCLOHEXANOL (TRANS)	007731-29-5
2-METHYLCYCLOHEXANOL (TRANS)	007443-52-9
2-METHYLCYCLOHEXANONE	000583-60-8
3-METHYLCYCLOHEXANONE	000591-24-2
4-METHYLCYCLOHEXANONE	000589-92-4
1-METHYLCYCLOHEXENE	000591-49-1
3-METHYL-1-CYCLOHEXENE	000591-48-0
2-METHYLCYCLOHEXYL ACRYLATE	000101-43-9
METHYLCYCLOPENTANE	000096-37-7
METHYLCYCLOPROPANE	000594-11-6
2-(4-METHYL-3-CYLOHEXENYL)ISOPROPANOL	010482-56-1
5-METHYLCYTIDINE	002140-61-6
2-METHYLDECANE	006975-98-0
METHYL DECANOATE	000110-42-9
METHYL DECYL KETONE	006175-49-1
1-(2-METHYL),2-DIBENZOYL-1-TERT-BUTYLHYDRAZINE	112226-78-5
1-(3-METHYL),2-DIBENZOYL-1-TERT-BUTYLHYDRAZINE	112411-46-8
1-(4-METHYL),2-DIBENZOYL-1-TERT-BUTYLHYDRAZINE	112226-56-9
METHYL DICHLOROACETATE	000116-54-1
N-METHYL-2,6-DICHLOROANILINE	056462-00-1
METHYL 2,5-DICHLOROBENZOATE	002905-69-3
4-METHYL-2,6-DICHLOROPHENOL	002432-12-4
METHYL (3,4-DICHLOROPHENYL)CARBAMATE	001918-18-9
N-METHYLDIETHANOLAMINE	000105-59-9
METHYL 2,4-DIHYDROXYBENZOATE	002150-47-2
4-METHYL-5,8-DIHYDROXYQUINOLINE	057334-34-6
3-METHYL-DIMETHYL PARA-OXON	002255-17-6

1-METHYL-2,4-DINITRO-1H-IMIDAZOLE	005213-50-3
1-METHYL-4,5-DINITROIMIDAZOLE	019183-15-4
2-METHYL-4,5-DINITROIMIDAZOLE	019183-16-5
4-METHYLDIOXANE	001120-97-4
3-METHYL DIPHENYL AMINE	001205-64-7
METHYLDITHIOCARBAMIC ACID, NA SALT	000137-42-8
O-METHYLDOPAMINE	000554-52-9
4,4'-METHYLENEBIS(2-CHLOROANILINE)	000101-14-4
METHYLENE BLUE	000061-73-4
METHYLENE CYCLOHEXANE	001192-37-6
3,4-METHYLENEDIOXYAMPHETAMINE	004764-17-4
3,4-(METHYLENEDIOXY)CINNAMIC ACID	002373-80-0
1,2-METHYLENEDIOXY-4-PROPENYL BENZENE	000120-58-1
METHYLENE IODIDE	000075-11-6
3-METHYLENE-7-METHYL-1,6-OCTADIENE (MYRCENE)	000123-35-3
4,5-METHYLENEPHENANTHRENE	000203-64-5
METHYLETHOXYCHLOR	034197-05-2
N-METHYL-3-ETHOXYPHENYLCARBAMATE	007225-96-9
N-METHYL-4-ETHOXYPHENYLCARBAMATE	013538-54-0
2-METHYL-6-ETHYLACETANILIDE	097055-06-6
METHYLETHYLAMINE	000624-78-2
1-METHYL-3-ETHYLCYCLOPENTANE	003726-47-4
1-METHYL-2-ETHYLCYCLOPENTANE (TRANS)	000930-90-5
9-(1-METHYLETHYL)-9H-FLUORENE	003299-99-8
N-(1-METHYLETHYL)-4-NITROBENZENAMINE	025186-43-0
2-METHYL-3-ETHYLPENTANE	000609-26-7
3-METHYL-3-ETHYLPENTANE	001067-08-9
4-METHYL-2-ETHYL-1-PENTANOL	000106-67-2
N-METHYL-2-ETHYLPHENYLCARBAMATE	002631-42-7
N-METHYL-4-ETHYLPHENYLCARBAMATE	002631-30-3
2-METHYL-5-ETHYLPIPERIDINE	000104-89-2
METHYLEUGENOL	000093-15-2
9-METHYL-9H-FLUORENE	002523-37-7
1-METHYLFLUORENE	001730-37-6
N-METHYL-2-FLUOROPHENYLCARBAMATE	000704-73-4
N-METHYL-3-FLUOROPHENYLCARBAMATE	000705-48-6
N-METHYL-4-FLUOROPHENYLCARBAMATE	000705-70-4
METHYLFORMAMIDE	000123-39-7
P-METHYLFORMANILIDE	003085-54-9
N-METHYLFORMANILIDE	000093-61-8
METHYLFORMATE	000107-31-3
3-METHYLFURAN	000930-27-8
2-METHYLFURAN	000534-22-5
3-METHYLFUROIC ACID, ME ESTER	006141-57-7
METHYL GALLATE	000099-24-1
N-METHYL GLYCINE	000107-97-1
METHYL GLYOXAL	000078-98-8
METHYLGUANIDINE	000471-29-4
2-METHYLHEPTANE	000592-27-8
4-METHYLHEPTANE	000589-53-7
3-METHYLHEPTANE	000589-81-1
METHYL HEPTANOATE	000106-73-0
3-METHYL-3-HEPTANOL	000598-06-1
2-METHYL-2-HEPTANOL	000625-25-2
6-METHYL-2-HEPTANOL ACETATE	067952-57-2
2-METHYL-2-HEPTENE	000627-97-4
6-METHYL-1-HEPTENE	005026-76-6
2-METHYL-1-HEPTENE	015870-10-7

6-METHYL-5-HEPTEN-2-OL	004630-06-2
METHYL HEXADECANOATE	000112-39-0
2-METHYL-1,5-HEXADIENE	004049-81-4
2-METHYLHEXANE	000591-76-4
3-METHYLHEXANE	000589-34-4
2-METHYLHEXANEDIOIC ACID	000626-70-0
METHYL HEXANOATE	000106-70-7
2-METHYL-2-HEXANOL	000625-23-0
2-METHYL-3-HEXANOL	000617-29-8
5-METHYL-2-HEXANOL	000627-59-8
3-METHYL-3-HEXANOL	000597-96-6
5-METHYL-2-HEXANONE	000110-12-3
4-METHYL-1-HEXENE	003769-23-1
3-METHYL-1-HEXENE	003404-61-3
3-METHYL-3-HEXENE	003404-65-7
2-METHYL-1-HEXENE	006094-02-6
2-METHYL-2-HEXENE	002738-19-4
3-METHYL-2-HEXENE	017618-77-8
2-METHYL-3-HEXENE	042154-69-8
METHYL HYDRAZINE	000060-34-4
3-METHYLHYDROCHLOROTHIAZIDE	000890-67-5
METHYLHYDROPEROXIDE	003031-73-0
6-METHYLHYDROTHIAZIDE	001824-46-0
3-METHYL-4-HYDROXYACETANILIDE	016375-90-9
2-METHYL-4-HYDROXYACETANILIDE	039495-15-3
N-METHYL P-HYDROXYACETANILIDE	000579-58-8
METHYL M-HYDROXYBENZOATE	019438-10-9
METHYL-4-HYDROXYBUTYRATE	000925-57-5
N-METHYLHYDROXYLAMINE	000593-77-1
O-METHYLHYDROXYLAMINE	000067-62-9
N-METHYL-HYDROXYPYRID-4-ONE	050700-61-3
1-METHYLHYDROXYUREA	007433-43-4
2-METHYL-2-IMIDAZOLINE	000534-26-9
N-METHYLINDOL-2,3-DIONE	002058-74-4
1-METHYLINDOLE	000603-76-9
7-METHYLINDOLE	000933-67-5
5-METHYLINDOLE	000614-96-0
3-METHYLINDOLE	000083-34-1
2-METHYLINDOLE	000095-20-5
METHYL IODIDE	000074-88-4
N-METHYL-2-IODOPHENYLCARBAMATE	013538-28-8
N-METHYL-4-IODOPHENYLCARBAMATE	013538-51-7
N-METHYL-3-IODOPHENYLCARBAMATE	013941-09-8
N-METHYL O-(3-IPRPHENYL)CARBAMATE	000064-00-6
3-METHYL-1(3H)-ISOBENZOFURANONE	003453-64-3
2-METHYLISOBORNEOL	002371-42-8
METHYL ISOBUTYL ETHER	000625-44-5
METHYL ISOBUTYRATE	000547-63-7
METHYLISOCYANATE	000624-83-9
METHYL ISOCYANIDE	000593-75-9
1-METHYLISOGUANINE,7[ACOET]-	087469-66-7
1-METHYLISOGUANINE,7[BENZYL]-	087469-62-3
1-METHYLISOGUANINE,9[BENZYL]-	087469-61-2
METHYLISOGUANINE,9[2(4-BRPHENYL)ET]-	087469-53-2
1-METHYLISOGUANINE,9[2(METHOXY)ME]-	087469-35-0
-METHYLISOGUANINE,7[2(4-BRPHENYL)ET]-	087469-54-3
1-METHYLISOGUANINE,9-[2-(ETHENYLOXY)ETHYL]-	087469-47-4
1-METHYLISOGUANINE,7-[2-(ETHENYLOXY)ETHYL]-	087469-48-5

2-METHYLISONIAZID	003758-59-6
1-METHYL-2-ISOPROPYLBENZENE	000527-84-4
6-METHYL-3-ISOPROPYL-2-CYCLOHEXEN-1-ONE	000499-74-1
METHYL ISOPROPYL ETHER	000598-53-8
N-METHYL-2-ISOPROPYLPHENYLCARBAM	002631-40-5
METHYL ISOTHIOCYANATE	000556-61-6
5-METHYLISOXAZOLE-3-CARBOXYLIC ACID	004857-42-5
METHYL LAURATE	000111-82-0
METHYL LINOLEATE	000112-63-0
METHYL MALEATE	000624-48-6
METHYL MALONIC ACID	000516-05-2
METHYL METHACRYLATE	000080-62-6
N-METHYLMETHANAMINE, HYDROCHLORIDE	000506-59-2
METHYL METHANE SULFONATE	000066-27-3
2-METHYL-3-METHOXYACETANILIDE	050868-74-1
N-METHYL P-METHOXYACETANILIDE	035813-38-8
3-METHYL-4-METHOXYBENZALDEHYDE	032723-67-4
1-METHYL-4-METHOXYCARBONYL-2-PYRROLIDONE	059857-86-2
N-METHYL-4-METHOXYPHENYLCARBATE	003938-29-2
2-METHYL-4-METHOXYQUINOLINE	031835-53-7
METHYL (3-METHYL)BUTANOATE	000556-24-1
2-METHYL-5-(1-METHYLETHENYL)CYCLOHEXANONE	007764-50-3
(-)-1-METHYL-4-(1-METHYLETHENYL)CYCLOHEXENE	005989-54-8
5-METHYL-2-(1-METHYLETHYL)-CYCLOHEXANONE (TR*)	000089-80-5
2-METHYL-4-((2-METHYLPHENYL)AZO)BENZENAMINE	000097-56-3
N-METHYLMORPHOLINE	000109-02-4
METHYL MYRISTATE	000124-10-7
2-METHYLNAPHTHALENE	000091-57-6
1-METHYLNAPHTHALENE	000090-12-0
2-METHYL-1,4-NAPHTHOQUINONE	000058-27-5
6-METHYL-1,4-NAPHTHOQUINONE	000605-93-6
N-METHYL-2-NAPHTHYLCARBAMATE	004089-04-7
N-METHYLNICOTINAMIDE	000114-33-0
METHYLNITRATE	000598-58-3
METHYLNITRITE	000624-91-9
2-METHYL-6-NITROACETANILIDE	059907-22-1
3-METHYL-4-NITROANILINE	000611-05-2
N-METHYL-O-NITROANILINE	000612-28-2
2-METHYL-4'-NITROAZOBENZENE	007030-18-4
2-METHYL-4-NITROAZOBENZENE	029418-57-3
METHYL P-NITROBENZOATE	000619-50-1
1-METHYL-5-NITROIMIDAZOLE	003034-42-2
1-METHYL-4-NITRO-IH-IMIDAZOLE	003034-41-1
4-METHYL-5-NITRO-1H-IMIDAZOLE	014003-66-8
2-METHYL-4-NITRO-1H-IMIDAZOLE	000696-23-1
2-METHYL-5-NITROIMIDAZOLE	088054-22-2
1-METHYL-2-NITRO-1H-IMIDAZOLE	001671-82-5
1-METHYL-4-NITRO-1H-IMIDAZOL-5-AMINE	004531-54-8
1-METHYL-6-NITROINDAZOLE	006850-23-3
2-METHYL-6-NITROINDAZOLE	006850-22-2
4-METHYL-2-NITROPHENOL	000119-33-5
3-METHYL-2-NITROPHENOL	004920-77-8
3-METHYL-4-NITROPHENOL	002581-34-2
5-METHYL-2-NITROPHENOL	000700-38-9
3-METHYL-4-((4-NITROPHENYL)AZO)BENZENAMINE	062308-10-5
3-METHYL-N-(4-NITROPHENYL)BENZENAMINE	015979-82-5
N-METHYL-4-NITROPHENYLCARBAMATE	005819-21-6
N-METHYL-3-NITROPHENYLCARBAMATE	006132-21-4

N-METHYL-2-NITROPHENYLCARBAMATE	007374-06-3
2-METHYL-2-NITROPROPANE	000594-70-7
2-METHYL-2-NITROPROPANE-1,3-DIOL	000077-49-6
2-METHYL-2-NITRO-1-PROPANOL	000076-39-1
2-METHYL-5-NITROPYRIDINE	021203-68-9
3-METHYL-4-NITROQUINOLINE-1-OXIDE	014073-00-8
6-METHYL-4-NITROQUINOLINE-1-OXIDE	000715-48-0
2-METHYL-4-NITROQUINOLINE-1-OXIDE	004831-62-3
8-METHYL-4-NITROQUINOLINE-1-OXIDE	014094-45-2
5-METHYL-4-NITROQUINOLINE-1-OXIDE	014094-43-0
7-METHYL-4-NITROQUINOLINE-1-OXIDE	014753-13-0
N-METHYL-N-NITROSOETHAMINE	010595-95-6
2-METHYL-N-NITROSOPEPERIDINE	007247-89-4
4-METHYL-N-NITROSOPIPERAZINE	016339-07-4
4-METHYL-N-NITROSOPIPERIDINE	015104-03-7
3-METHYL-N-NITROSOPIPERIDINE	013603-07-1
4-METHYL-B-NITROSTYRENE	007559-36-6
2-METHYL-B-NITROSTYRENE	034222-71-4
B-METHYL B-NITROSTYRENE	000705-60-2
2-METHYLNONANE	000871-83-0
METHYL NONANOATE	001731-84-6
(1-METHYLNONYL)BENZENESULFONIC ACID, SODIUM SALT	073602-65-0
A-METHYLNORADRENALINE	006539-57-7
N-METHYL O-SEC-BUTYL PHENYL CARBAMATE	003766-81-2
3-METHYLOCTANE	002216-33-3
4-METHYLOCTANE	002216-34-4
2-METHYLOCTANE	003221-61-2
METHYL OCTANOATE	000111-11-5
7-METHYL-1-OCTANOL	002430-22-0
METHYL OLEATE	000112-62-9
METHYL ORANGE	000547-58-0
METHYL OXALATE	000553-90-2
4-METHYL-2-OXETANONE	003068-88-0
4-METHYL-N-OXIDEPYRIDINE	001003-67-4
2-METHYL-N-OXIDEPYRIDINE	000931-19-1
3-METHYL-N-OXIDEPYRIDINE	001003-73-2
METHYL PARAOXON	000950-35-6
1-METHYLPENTACHLOROCYCLOHEXANE	056421-44-4
METHYL PENTADECANOATE	007132-64-1
2-METHYL-1,4-PENTADIENE	000763-30-4
4-METHYL-1,3-PENTADIENE	000926-56-7
3-METHYL-1,3-PENTADIENE	004549-74-0
2-METHYL-1,3-PENTADIENE	000926-54-5
2-METHYLPENTALDEHYDE	000123-15-9
3-METHYLPENTANE	000096-14-0
2-METHYLPENTANE	000107-83-5
2-METHYLPENTANEDINITRILE	004553-62-2
2-METHYL-2,4-PENTANEDIOL	000107-41-5
4-METHYLPENTANITRILE	000542-54-1
4-METHYLPENTANOIC ACID	000646-07-1
3-METHYL-3-PENTANOL	000077-74-7
3-METHYL-2-PENTANOL	000565-60-6
4-METHYL-1-PENTANOL	000626-89-1
2-METHYL-2-PENTANOL	000590-36-3
4-METHYLPENTANOL	001320-98-5
4-METHYL-2-PENTANOL	000108-11-2
2-METHYL-3-PENTANOL	000565-67-3
2-METHYL-1-PENTANOL	000105-30-6

3-METHYLPENTAN-2-ONE	000565-61-7
4-METHYL-2-PENTANONE	000108-10-1
2-METHYL-3-PENTANONE	000565-69-5
3-METHYL-2-PENTENE	000922-61-2
3-METHYL-1-PENTENE	000760-20-3
4-METHYL-2-PENTENE	004461-48-7
2-METHYL-1-PENTENE	000763-29-1
4-METHYL-1-PENTENE	000691-37-2
2-METHYL-2-PENTENE	000625-27-4
TRANS-4-METHYL-2-PENTENE	000674-76-0
4-METHYL-2-PENTENE (CIS)	000691-38-3
2-METHYL-1-PENTEN-3-OL	002088-07-5
4-METHYL-1-PENTEN-3-OL	004798-45-2
4-METHYL-2-PENTYL ACETATE	000108-84-9
METHYL PENTYL SULFIDE	001741-83-9
METHYL PENTYNOL	000077-75-8
1-METHYLPHENANTHRENE	000832-69-9
3-METHYLPHENANTHRENE	000832-71-3
2-METHYLPHENANTHRENE	002531-84-2
ALPHA-METHYLPHENETHYLAMINE	000300-62-9
N-METHYLPHENOXYACETAMIDE	015422-25-0
N-METHYLPHENOXYACETANILIDE	018859-21-7
METHYL PHENOXYACETATE	002065-23-8
O-METHYLPHENOXYACETIC ACID	001878-49-5
M-METHYLPHENOXYACETIC	001643-15-8
P-METHYLPHENOXYACETIC ACID	000940-64-7
1-(3-METHYLPHENOXY)SILATRANE	071357-27-2
1-(4-METHYLPHENOXY)SILATRANE	051466-75-2
P-METHYLPHENYLACETIC ACID	000622-47-9
M-METHYLPHENYLACETIC ACID	000621-36-3
4-((4-METHYLPHENYL)AZO)BENZENESULFONIC ACID,SOD*	062959-39-1
M-METHYLPHENYL AZOXYCYANIDE	064462-06-2
P-METHYLPHENYL AZOXYCYANIDE	054797-21-6
4-METHYL-N-PHENYLBENZAMINE	000620-84-8
METHYL 4-PHENYLBUTYRATE	002046-17-5
METHYL N-PHENYLCARBAMATE	002603-10-3
N-METHYLPHENYLCARBAMATE	001943-79-9
9-(3-METHYLPHENYL)-9H-FLUORENE	018153-42-9
9-(4-METHYLPHENYL)-9H-FLUORENE	018153-43-0
M-METHYLPHENYLHIPPURATE	066895-70-3
P-METHYLPHENYLHIPPURATE	029736-20-7
4-METHYLPHENYLISOTHIOCYANATE	000622-59-3
N-((4-METHYLPHENYL)METHYL)ACETAMIDE	025079-96-3
(4-METHYLPHENYL)(4-NITROPHENYL)DIAZENE	029418-58-4
(2-METHYLPHENYL)PHENYLDIAZENE	006676-90-0
(4-METHYLPHENYL)PHENYLDIAZENE	000949-87-1
1-(4-METHYLPHENYL)-1-PROPANONE	005337-93-9
METHYL B-PHENYLPROPIONATE	000103-25-3
O-METHYL-N-PHENYLSUCCINIMIDE	070290-53-8
N-METHYL-2-PHENYL SUCCINIMIDE	000086-34-0
2-THIOPHENESULFONAMIDE, 4-[(4-METHYLPHENYL)SULFO	128348-32-3
N1-(2-METHYLPHENYL)SULFANILAMIDE	016803-96-6
N1-(4-METHYLPHENYL)SULFANILAMIDE	016803-95-5
METHYL PHENYL SULFONE	003112-85-4
METHYL PHENYL SULFOXIDE	001193-82-4
4-METHYL-1-PHENYL-3-THIOSEMICARBAZIDE	013207-50-6
1-METHYL-3-PHENYLTHIOUREA	002724-69-8
1-METHYL-1-PHENYL UREA	004559-87-9

METHYLPHOPHONIC ACID	000993-13-5
METHYLPHOSPHINIC ACID	004206-94-4
N-METHYLPIPERIDINE	000626-67-5
2-METHYLPIPERIDINE	000109-05-7
2-METHYLPROPANE	000075-28-5
2-METHYLPROPANENITRILE	000078-82-0
2-METHYL-1-PROPANETHIOL	000513-44-0
2-METHYL-2-PROPANETHIOL	000075-66-1
2-METHYLPROPANOYLUREA	023549-54-4
2-METHYL-1-PROPEN-1-ONE(DIMETHYLKETENE)	000598-26-5
(2-METHYL-1-PROPENYL)BENZENE	000768-49-0
N-METHYL PROPIONAMIDE	001187-58-2
METHYL PROPIONATE	000554-12-1
2-METHYL-6-PROPOXYPYRAZINE	067845-28-7
N-METHYLPROPYLAMINE	000627-35-0
1-METHYL-4-PROPYLBENZENE	001074-55-1
1-METHYL-3-PROPYLBENZENE	001074-43-7
1-METHYL-2-PROPYLBENZENE	001074-17-5
METHYL PROPYL ETHER	000557-17-5
1,1'-(2-METHYLPROPYLIDENE)BIS(4-ETHOXYBENZENE)	056265-21-5
N-METHYL-2-PROPYLPHENYLCARBAMATE	015482-11-8
2-METHYLPROPYL PROPANOATE	000540-42-1
2-METHYLPYRAZINE	000109-08-0
2-METHYLPYRAZINE CARBOXYLATE	006164-79-0
1-METHYLPYRAZOLE	000930-36-9
2-METHYLPYRIDINE	000109-06-8
3-METHYLPYRIDINE	000108-99-6
4-METHYLPYRIDINE	000108-89-4
2-METHYLPYRIDINIUM IODIDE,N-DODECYL	014402-22-3
2-METHYLPYRIDINIUM IODIDE,N-HEXYL	014402-24-5
N-METHYLPYRIDINIUM IODIDE	000930-73-4
2-METHYLPYRIDINIUM IODIDE, N-DECYL	014402-23-4
2-METHYLPYRIDINIUM IODIDE,N-OCTYL	013515-66-7
3-O-METHYLPYRIDOXOL	000633-72-7
N-METHYL-3-PYRIDYLMETHYLAMINE	020173-04-0
N1-METHYL-N1-(2-PYRIDYL)SULFANILAMIDE	051543-29-4
6-METHYL-4-PYRIMIDINAMINE	003435-28-7
4-METHYLPYRIMIDINE	003438-46-8
N-METHYLPYRROLE	000096-54-8
5-METHYLPYRROLE-3-CARBOXYLIC ACID,METHYL ESTER	040611-76-5
METHYL PYRROLE-2-CARBOXYLATE	001193-62-0
1-METHYL-PYRROLIDINE	000120-94-5
N-METHYLPYRROLIDONE	000872-50-4
7-METHYL QUINAZOLINE-2,4-DIONE	000086-96-4
2-METHYLQUINOLINE	000091-63-4
4-METHYLQUINOLINE	000491-35-0
6-METHYLQUINOLINE	000091-62-3
7-METHYLQUINOLINE	000612-60-2
3-METHYLQUINOLINE	000612-58-8
8-METHYLQUINOLINE	000611-32-5
METHYL3-QUINOLINECARBOXYLATE	053951-84-1
6-METHYLQUINOLINE, 1-OXIDE	004053-42-3
4-METHYL-8-QUINOLINOL	003846-73-9
5-METHYL-8-QUINOLINOL	005541-67-3
1-METHYL-2(H)-QUINOLINONE	000606-43-9
2-METHYL 8-QUINOLONOL	000826-81-3
2-METHYLQUINOXALINE	007251-61-8
5-METHYLQUINOXALINE	013708-12-8

6-METHYLQUINOXALINE-1,4-DIOXIDE	033368-89-7
N-METHYLQUINOXALINE-2-ONE	006479-18-1
5-METHYLRESORCINOL	000504-15-4
METHYL SALICYLATE	000119-36-8
4-METHYLSALICYLIC ACID	000050-85-1
3-METHYLSALICYLIC ACID	000083-40-9
5-METHYLSALICYLIC ACID	000089-56-5
4-METHYLSEMICARBAZIDE	017696-95-6
METHYL STEARATE	000112-61-8
ALPHA-METHYLSTYRENE	000098-83-9
BETA-METHYLSTYRENE	000637-50-3
M-METHYLSTYRENE	000100-80-1
O-METHYLSTYRENE	000611-15-4
P-METHYLSTYRENE	000622-97-9
METHYL STYRYL KETONE	000122-57-6
METHYLSUCCINIC ACID	000498-21-5
N-METHYLSUCCINIMIDE	001121-07-9
3-METHYLSUFONYLNITROBENZENE	002976-32-1
N1-METHYLSULFANILAMIDE	001709-52-0
3-METHYL SULFOLANE	000872-93-5
METHYLSULFONYLACETONITRILE	002274-42-2
P-METHYLSULFONYLAMPHENICOL	000847-25-6
4-METHYLSULFONYLANILINE	005470-49-5
METHYLSULFONYLANISOLE	043032-67-3
P-METHYLSULFONYLBENZOIC ACID	004052-30-6
3'-METHYLSULFONYLBENZOGUANAMINE	065052-50-8
4-METHYLSULFONYLCHLOROBENZENE	000098-57-7
METHYLSULFONYL CHLORIDE	000124-63-0
4-METHYLSULFONYL-N-PHENYLBENZENAMINE	015979-81-4
3-METHYLSULFONYLPHENYLUREA	088497-95-4
6-METHYLSULFONYLQUINOLINE	089770-29-6
4-METHYLSULFONYLQUINOLINE-1-OXIDE	020872-53-1
2-METHYLSULFONYLQUINOXALINE	016310-37-5
METHYL TEREPHTHALATE	001679-64-7
17-A-METHYL TESTOSTERONE	000058-18-4
2-METHYLTETRAHYDROFURAN	000096-47-9
3-METHYL-2-THIABUTANE	001551-21-9
METHYLTHIAZIDE	003256-99-3
4-METHYLTHIAZOLE	000693-95-8
N-METHYLTHIOBENZANILIDE	002628-58-2
3'-METHYLTHIOBENZOGUANAMINE	065052-49-5
2-METHYLTHIOBENZOTHIAZOLE	000615-22-5
METHYL THIOCYANATE	000556-64-9
8-METHYLTHIO CYCLIC AMP-O-BENZYL	107538-73-8
8-METHYLTHIO CYCLIC AMP	030630-07-0
1-(2-METHYLTHIO),2-DIBENZOYL-1-TERT-BUTYLHYDRAZI	112226-95-6
METHYL THIOPHEN-3-ACETATE	058414-52-1
METHYL THIOPHEN-3-CARBOXYLATE	022913-26-4
3-METHYLTHIOPHENE	000616-44-4
2-METHYLTHIOPHENE	000554-14-3
3-(METHYLTHIO)PHENYLUREA	085879-21-6
2-METHYLTHIOPYRAZINE	021948-70-9
2-METHYLTHIOPYRIDINE	018438-38-5
2-METHYLTHIOQUINAZOLINE	006141-18-0
2-METHYLTHIOQUINOLINE	040279-26-3
N-METHYLTHIOUREA	000598-52-7
O-METHYL THPO	095579-17-2
N-METHYL P-TOLUIDINE	000623-08-5

N-METHYL-O-TOLUIDINE	000611-21-2
N-METHYL-O-TOLYLCARBAMATE	001128-78-5
N-METHYL-P-TOLYLCARBAMATE	001129-48-2
N-METHYL-M-TOLYLCARBAMATE	001129-41-5
METHYL TRICHLOROACETATE	000598-99-2
METHYLTRICHLOROSILANE	000075-79-6
METHYL TRIDECANOATE	001731-88-0
4-(1-METHYLTRIDECYL)BENZENESULFONIC ACID, SODIU*	013419-31-3
METHYLTRIFLUOROACETATE	000431-47-0
N-METHYL-3-TRIFLUOROMEPHCARBAMATE	014061-27-9
M-METHYL-TRIMETHYL ANILINIUM IODIDE	033046-97-8
METHYLTRIS(TRIMETHYLSILOXY)SILANE	017928-28-8
4-(1-METHYLUNDECYL)BENZENESULFONIC ACID, SODIUM*	002211-99-6
5-METHYLURACIL	000065-71-4
1-METHYLURACIL	000615-77-0
7-METHYL URIC ACID	000612-37-3
3-METHYL URIC ACID	000605-99-2
1-METHYL URIC ACID	000708-79-2
METHYL VALERATE	000624-24-8
METHYLVINYLETHER	000107-25-5
METHYLVINYLKETONE	000078-94-4
2-METHYL-5-VINYLPYRIDINE	000140-76-1
7-METHYLXANTHINE	000552-62-5
1-METHYLXANTHINE	006136-37-4
3-METHYLXANTHINE	001076-22-8
P-METHYOXYBENZAMIDE	003424-93-9
3-METHYOXYPYRIDINE	007295-76-3
METIAMIDE	034839-70-8
METIPRANOLOL	022664-55-7
METIRAM	009006-42-2
METOBROMURON	003060-89-7
METOCLOPRAMIDE	000364-62-5
METOLACHLOR	051218-45-2
METOPROLOL	037350-58-6
METOXURON	019937-59-8
METRIBUZIN	021087-64-9
N-ME-3,4,5-TRIMEPHENYL CARBAMATE	002686-99-9
METRIZAMIDE	031112-62-6
METRONIDAZOLE	000443-48-1
METRONIDAZOLE ACETATE	013182-82-6
METSULFURON-METHYL (PH5-7)	079510-48-8
METSULFURON-METHYL	074223-64-6
MEVASTATIN	073573-88-3
MEVINPHOS	007786-34-7
MEVINPHOS (TRANS)	000298-01-1
MEXACARBATE	000315-18-4
MEXICANIN-1	005945-41-5
MEXRENONE	041020-65-9
MGK 264	000113-48-4
MINAPRINE	025905-77-5
MINOCYCLINE	010118-90-8
M-I-PRO BENZYL N,N-DIME CARBAMATE	084640-29-9
MIREX	002385-85-5
MISOPROSTOL	059122-46-2
MITOLACTOL	010318-26-0
MITOMYCIN A	004055-39-4
MITOMYCIN B	004055-40-7
MITOMYCIN C	000050-07-7

MOLINATE	002212-67-1
MONALIDE (POTABLAN)	007287-36-7
MONI[1-(4-BROMOPHENYL)ETHYL] 1,2-BENZENEDICARBO*	092387-49-0
MONOBROMOAMPHENICOL	040027-72-3
1-MONOBROMODIBENZO-P-DIOXIN	105908-71-2
2-MONOBROMODIBENZO-P-DIOXIN	105906-36-3
MONOBROMONEOPENTYLTRIOL	019184-65-7
1-MONOBUTYRIN	000557-25-5
MONOCHLOROAMPHENICOL	017278-57-8
MONOCHLOROMETHYLAMPHENICOL	049648-35-3
MONO[1-(4-CHLOROPHENYL)ETHYL] 1,2-BENZENEDICARB*	092387-48-9
MONO(P,ALPHA-DIMETHYLBENZYL) PHTHALATE	023005-56-3
MONOFLUOROAMPHENICOL	049648-38-6
MONOIODOAMPHENICOL	040027-73-4
2-MONOIODODIBENZO-P-DIOXIN	101714-96-9
MONOLINURON	001746-81-2
MONO(P-METHOXY-ALPHA-METHYLBENZYL) PHTHALATE	033533-57-2
MONO(ALPHA-METHYLBENZYL) PHTHALATE	033533-53-8
MONOMETHYLPHTHALATE	004376-18-5
MONO(P-(TERT-BUTYL)ALPHA-METHYLBENZYL) PHTHALATE	033533-56-1
MONOSODIUM L-GLUTAMATE	000142-47-2
MONOTROPITOSIDE	000490-67-5
MONURON	000150-68-5
MOPIDRALAZINE	075841-82-6
MORDANT BLUE 13	001058-92-0
MORICIZINE	031883-05-3
MORIN	000480-16-0
MORPHINAN-3-OL, 17-METHYL-, (9_,13_,14_)-	000125-73-5
MORPHINE	000057-27-2
MORPHINE ACETATE	000596-15-6
MORPHOLINE	000110-91-8
-MORPHOLINEACETAMIDE, N-(2,3-DIHYDRO-5-METHYL-3	105801-59-0
4-MORPHOLINEACETAMIDE, N-(2,3-DIHYDRO-3-BENZOFUR	105801-49-8
-MORPHOLINEACETAMIDE, N-(5-CHLORO-2,3-DIHYDRO-3	105801-54-5
4-MORPHOLINEACETAMIDE, N-(2,3-DIHYDRO-6-METHOXY-	105801-63-6
4-MORPHOLINEACETAMIDE, N-(PHENYLMETHYL)-	072336-20-0
MORPHOLINE, 4-[(BENZOYLOXY)ACETYL]-	106231-68-9
MORPHOLINE, 4-[[4-(4,6-DIAMINO-2,2-DIMETHYL-1,3,	050574-87-3
MORPHOLINE, 4-[[3-[[(4-METHOXYPHENYL)METHOXY]MET	144459-03-0
MORPHOLINE, 4-[2-(4-NITRO-1H-IMIDAZOL-1-YL)ETHYL	006497-78-5
MORPHOLINE, 4-[2-(5-NITRO-1H-IMIDAZOL-1-YL)ETHYL	006506-37-2
MORPHOLINE, 4-[2-(2-NITRO-1H-IMIDAZOL-1-YL)ETHYL	036664-18-3
MORPHOLINO-DAUNORUBICIN,13-DIHYDRO	079951-58-9
MORPHOLINO-DOXORUBICIN,12-IMINO-	089196-05-4
(N-MORPHOLINYL)ET-(345-MEO)PH SULFONE	057554-05-9
BUTANOIC ACID, [2-[[(4-MORPHOLINYLMETHYL)AMINO]C	102273-20-1
3-MORPHOL-4-NH2-6-IPR-124-TRIAZINE-5-ONE	050917-19-6
MOXISYLYTE	000054-32-0
MOXISYLYTE,N-DESMETHYL,DESACETYL	072732-50-4
M-I-PROPYLPHENOXYACETIC ACID	001878-52-0
MPTP	028289-54-5
MUSCIMOL	002763-96-4
MYLERLAN	000055-98-1
MYRISTICIN	000607-91-0
N-244	006012-92-6
NABAM	000142-59-6
NADOLOL	042200-33-9
NAFAZATROM	059040-30-1

NALED	000300-76-5
NALIDIXIC ACID	000389-08-2
NALMEFENE	055096-26-9
NALORPHINE	000062-67-9
NALOXONE	000465-65-6
NALTREXONE	016590-41-3
NANDROLONE	000434-22-0
NAPHAZOLINE	000835-31-4
1,4-NAPHOQUINONE, 2-HYDROXY	000083-72-7
2-NAPHTHACENECARBOXAMIDE DERIVATIVE	010118-89-5
5,12-NAPHTHACENEDIONE PYRANOSYL DERIVATIVE	089196-07-6
5,12-NAPHTHACENEDIONE PYRANOSYL DERIVATIVE	089164-74-9
5,12-NAPHTHACENEDIONE PYRANOSYL DERIVATIVE	080790-68-7
5,12-NAPHTHACENEDIONE PYRANOSYL DERIVATIVE	079867-78-0
5,12-NAPHTHACENEDIONE, 7,8,9,10-TETRAHYDRO-6,8,1	024385-10-2
1-NAPHTHALENAMINE, N,N-DIMETHYL-	000086-56-6
1-NAPHTHALENAMINE, N-ETHYL-	000118-44-5
1-NAPHTHALENAMINE, N-METHYL-	002216-68-4
1-NAPHTHALENAMINE, N-PHENYL-	000090-30-2
2-NAPHTHALENAMINE, 1,2,3,4-TETRAHYDRO-	002954-50-9
2-NAPHTHALENAMINE, 1,2,3,4-TETRAHYDRO-N-METHYL-	019485-85-9
NAPHTHALENE	· 000091-20-3
2-NAPHTHALENEACETIC ACID	000581-96-4
NAPHTHALENEACETIC ACID	000086-87-3
1-NAPHTHALENE AZOXYCYANIDE	117505-21-2
2-NAPHTHALENECARBONITRILE, 5,8-BIS(ACETYLOXY)-6,	091814-30-1
1-NAPHTHALENECARBONITRILE, 5,8-BIS(ACETYLOXY)-6,	091814-29-8
2-NAPHTHALENECARBOXAMIDE, N-BUTYL-6,7-DIHYDROXY-	136944-48-4
1-NAPHTHALENECARBOXAMIDE	002243-81-4
2-NAPHTHALENECARBOXYLIC ACID,3-HYDROXY-,MONOSODI	014206-62-3
2-NAPHTHALENECARBOXYLIC ACID, 3-HYDROXY-	000092-70-6
1,8-NAPHTHALENEDIAMINE	000479-27-6
2,3-NAPHTHALENEDIAMINE	000771-97-1
NAPHTHALENE, 1,7-DICHLORO-	002050-73-9
NAPHTHALENE, 1,8-DINITRO-	000602-38-0
NAPHTHALENE, 1,3-DINITRO-	000606-37-1
NAPHTHALENE, 1,5-DINITRO-	000605-71-0
2,3-NAPHTHALENEDIOL	000092-44-4
1,4-NAPHTHALENEDIOL	000571-60-8
NAPHTHALENE-1,3-DIOL	000132-86-5
2,6-NAPHTHALENEDIOL	000581-43-1
1,4-NAPHTHALENEDIOL, 6-BROMO-2,3-DIMETHOXY-, DIA	091814-13-0
1,4-NAPHTHALENEDIOL, 6-CHLORO-2,3-DIMETHOXY-, DI	091431-42-4
1,4-NAPHTHALENEDIOL, 5-CHLORO-2,3-DIMETHOXY-, DI	091814-10-7
1,4-NAPHTHALENEDIOL, 2,3-DIMETHOXY-, DIPROPANOAT	091814-56-1
1,4-NAPHTHALENEDIOL, 2,3-DIMETHOXY-6-METHYL-, DI	091814-42-5
1,4-NAPHTHALENEDIOL, 2,3-DIMETHOXY-6,7-DIMETHYL-	102632-22-4
1,4-NAPHTHALENEDIOL, 2,3-DIMETHOXY-, BIS(2-METHY	091814-58-3
1,4-NAPHTHALENEDIOL, 2,3-DIMETHOXY-, DIACETATE	061601-23-8
2,3-NAPHTHALENEDIOL, 6-ETHOXY-	136944-52-0
2,3-NAPHTHALENEDIOL, 6-ETHYL-	136944-43-9
1,4-NAPHTHALENEDIOL, 6-FLUORO-2,3-DIMETHOXY-, DI	091814-12-9
1,4-NAPHTHALENEDIOL, 2,3,5-TRIMETHOXY-, DIACETAT	091814-24-3
1,4-NAPHTHALENEDIOL, 2,3,6-TRIMETHOXY-, DIACETAT	091814-40-3
1,4-NAPHTHALENEDIONE, 2-AMINO-	002348-81-4
1,4-NAPHTHALENEDIONE, 2,3-DIHYDRO-2-[[[2-[4-HYDR	101126-54-9
1,4-NAPHTHALENEDIONE, 5,8-DIHYDROXY-	000475-38-7
1,4-NAPHTHALENEDIONE, 2-(PHENYLAMINO)-3-(PHENYLS	018690-91-0

1,5-NAPHTHALENEDISULFONIC ACID	000081-04-9
2-NAPHTHALENEMETHANOL	001592-38-7
1-NAPHTHALENEMETHANOL	004780-79-4
2-NAPHTHALENEMETHANOL, α-METHYL-, (\)-	040295-80-5
NAPHTHALENE, 1-METHOXY-	002216-69-5
NAPHTHALENE, 2-METHOXY-	000093-04-9
NAPHTHALENE, 1-(METHYLSULFONYL)-	054108-51-9
1-NAPHTHALENESULFONAMIDE	000606-25-7
1-NAPHTHALENESULFONIC ACID, 4-AMINO-, SODIUM SAL	000130-13-2
1-NAPHTHALENESULFONIC ACID, 5-HYDROXY-	000117-59-9
NAPHTHALENE, 1,3,8-TRINITRO-	002364-46-7
1-NAPHTHALENOL, 2-AMINO-1,2,3,4-TETRAHYDRO-, TRA	013917-17-4
1-NAPHTHALENOL, 2-AMINO-1,2,3,4-TETRAHYDRO-, CIS	013575-92-3
2-NAPHTHALENOL, 5,6,7,8-TETRAHYDRO-	001125-78-6
1-NAPHTHALENOL, 5,6,7,8-TETRAHYDRO-	000529-35-1
1-NAPHTHALENOL, 1,2,3,4-TETRAHYDRO-	000529-33-9
1-(2-NAPHTHALENYL)ETHANONE	000093-08-3
2H-NAPHTH[1,2-B]-1,4-OXAZIN-7-OL,3,4,4A,5,6,10B-	094921-29-6
2H-NAPHTH[1,2-B]-1,4-OXAZIN-9-OL,3,4,4A,5,6,10B-	088058-85-9
NAPHTHO[2,3-B]FURAN-2(3H)-ONE, DECAHYDRO-8A-METH	000470-17-7
4H-NAPHTHO[2,3-B]PYRAN-2-CARBOXYLIC ACID, 6,7,8,	005527-84-4
2-NAPHTHOIC ACID	000093-09-4
2-NAPHTHOIC ACID, SODIUM SALT	017273-79-9
1-NAPHTHOL	000090-15-3
2-NAPHTHOL	000135-19-3
NAPHTHOL AS-BR	000091-92-9
1-NAPHTHOL-3-SULFONIC ACID	003771-14-0
1-NAPHTHOL-4-SULFONIC ACID	000084-87-7
1,4-NAPHTHOQUINONE	000130-15-4
1,4-NAPHTHOQUINONE,2-ACETAMIDO	002348-74-5
1,4-NAPHTHOQUINONE,2-ACETAMIDO,3-ANILINO	058907-81-6
1,4-NAPHTHOQUINONE,3-ANILINO-2-CHLORO	001090-16-0
1,4-NAPHTHOQUINONE, 2-ANILINO	006628-97-3
14NAPHTHOQUINONE,2ANILINO-3SULFONATE K SALT	018690-90-9
1,4-NAPHTHOQUINONE,2-BUTYLTHIO	059094-49-4
1,4-NAPHTHOQUINONE,2-BUTYLAMINO-	059094-50-7
1,4-NAPHTHOQUINONE,6,7-DIMETHYL	002202-79-1
1,4-NAPHTHOQUINONE,2-ME-3-MEO	005416-18-2
2-NAPHTHYLAMINE	000091-59-8
1-NAPHTHYLAMINE	000134-32-7
2-NAPHTHYLAMINE-6,8-DISULFONIC ACID	000086-65-7
2-NAPHTHYLAMINE-5,7-DISULFONIC ACID	000118-33-2
1-NAPHTHYLAMINE-6-SULFONIC ACID	000119-79-9
4-(2-NAPHTHYLAMINO)PHENOL	000093-45-8
1-NAPHTHYL-N-METHYLCARBAMATE	000063-25-2
1-NAPHTHYLMETHYLISOTHIOCYANATE	017112-82-2
B-NAPHTHYLMETHYLISOTHIOCYANATE	019495-05-7
ALPHA-NAPHTHYL METHYL KETONE	000941-98-0
N-1-NAPHTHYLPHTHALMIC ACID	000132-67-2
1-NAPHTHYLTHIOUREA	000086-88-4
1,8-NAPHTHYRIDINE-3-CARBOXYLIC ACID, 1-ETHYL-1,4	124897-30-9
1,8-NAPHTHYRIDINE-3-CARBOXYLIC ACID, 1-ETHYL-7-M	124897-29-6
NAPROSYN	022204-53-1
NAPTHALENE-1,7-DIOL	000575-38-2
NAPTHALENE-2-SULFONIC ACID	000120-18-3
1-NAPTHOIC ACID	000086-55-5
NARCEINE	000131-28-2
NARCOTINE	006035-40-1

NARINGENINE	000480-41-1
NARINGINE	010236-47-2
NAXAGOLIDE	088058-88-2
NEBURON	000555-37-3
NEDOCROMIL	069049-73-6
NEOPENTYLGLYCOL DIMETHACRYLATE	001985-51-9
NEOTRAN	000555-89-5
NERIIFOLIN	000466-07-9
2-NH2-5-(5NO2-2-FURFURYL)THIAZOLONE	052661-38-8
2-EX-NH2 BENZOBICYCLO(222)OCTENE	015537-20-9
2-EN-NH2 BENZOBICYCLO(222)OCTENE	014342-36-0
2-NH2-1,3-BENZOXAZIN-4-ONE	000771-39-1
2-NH2-5-BENZYL-1,3,4-OXADIAZOLE	031803-00-6
24NH2-6-CYHXAM-PYRIMIDINE-3-OXIDE	055921-57-8
24-NH2-6-DECYLAMINOPYRIMIDINE-3-OXIDE	055921-56-7
24NH2-6DIALLYLAMINO-PYRIMIDINE-3-OXIDE	055921-61-4
24NH2-6DIBUTYLAMINO-PYRIMIDINE-3-OXIDE	055921-62-5
24NH2-6-DICYCLOHXAMPYRIMIDIN-3-OXIDE	055921-63-6
4,6-NH2 2,2-DIME-1(3-ACPH)S-TRIAZENE	070579-34-9
24-NH2-6-DIMEAM PYRIMIDINE-3-OXIDE	055921-58-9
4,6-NH2-2,2-DIME-1(4BRPH)S-TRIAZINE	003567-84-8
4,6-NH2 2,2-DIME-1-(3-BUO PH)S-TRIAZINE	070606-63-2
4,6-NH2 2,2-DIME1(4-CF3)PH S-TRIAZENE	047071-11-4
4,6-NH2-2,2-DIME-1(4FPH)S-TRIAZINE	001542-59-2
4,6-NH2-2,2-DIME-1(4IPH)S-TRIAZINE	046781-41-3
4,6-NH2 2,2-DIME-1(4-MEOPH)S-TRIAZINE	021316-30-3
4,6-NH2 2,2-DIME-1(4-MEPH)S-TRIAZINE	015233-37-1
4,6-NH2 2,2-DIME1(4-NH2)PH S-TRIAZENE	087871-35-0
46-NH2 22-DIME-1(3-PHOMEPH)S-TRIAZINE	079508-78-4
4,6-NH2 1,2-H 2,2-DIME-1PH TRIAZENE	004022-58-6
24NH2-6DIPROPYLAMINO-PYRIMIDINE-3-OXIDE	055921-60-3
24-NH2-6-ETAM PYRIMIDINE-3-OXIDE	055921-54-5
2-NH2-5-ET-1,3,4-OXADIAZOLE	003775-61-9
6-NH2-5-(N-FORMYLAMINO)-1,3-DIMETHYLURACIL	007597-60-6
2-NH2-5-HEXYL-1,3,4-OXADIAZOLE	069741-94-2
24NH2-5ME-6(345MEOPHNME)QUINAZOLINE	052128-35-5
6-NH2-5-(N-METHYLFORMYLAMINO)-3-METHYLURACIL	055782-76-8
6-NH2-5-(N-METHYLFORMYLAMINO)-1-METHYLURACIL	033130-54-0
6-NH2-5-(N-METHYLFORMYLAMINO)-1,3-DIMETHYLURACIL	033130-55-1
46-NH2-(2-OCH2-CON)BENZYLPYRIMIDINE	080407-58-5
2,4-NH2-6PIPERIDINO-PYRIMIDINE-3-OXID	038304-91-5
2,4-NH2PYRIMD,5(35-MEO-4CONME2)BENZYL	074731-64-9
24NH2PYRIMD,5(35ETO4COSO2ME)BENZYL	055687-72-4
24NH2PYRIMD,5(35ETO4COSO2ME)BENZYL	055687-56-4
2,4-NH2PYRIMD,5(35-NME2-4-MEO)BENZYL	061544-22-7
2,4-NH2PYRIMD,5(2-ME-4MEO-5BUO)BENZYL	083158-03-6
2,4-NH2PYRIMD,5(35-NME2-4-ME)BENZYL	061544-20-5
24NH2-PYRIMD,5(2ME-4MEO-5PHMEO)BENZYL	083158-05-8
2,4-NH2PYRIMD,5(35-MEO-4-IPR)BENZYL	074731-66-1
24NH2PYRIMDN,5(3MEAM4MEO5DIMEAM)BNZYL	061573-40-8
24NH2PYRIMDN,5(3NHME-4ME-5NMECOME)BENZYL	061544-28-3
2,4-NH2PYRIMIDIN,5(4-DIMEAMINO)BENZYL	069945-51-3
2,4-NH2PYRIMIDINE,5(35-CL-4MEO)BENZYL	083158-04-7
24NH2PYRIMIDINE,5(3,5-ETO-4-NH2)BENZYL	056066-43-4
24NH2PYRIMIDINE,5(35ETO4(1PYRRYL))BENZYL	073090-70-7
24NH2PYRIMIDINE,5(35-ETO-4CO2ET)BENZYL	055687-53-1
2,4-NH2PYRIMIDINE,5(2,4,5-MEO)BENZYL	006981-01-7
24-NH2PYRIMIDINE,5(3,5MEO-4-BR)BENZYL	056518-41-3

24-NH2PYRIMIDINE,5(2ME-45OCH2O)BENZYL	075889-44-0
24-NH2PYRIMIDINE,5(4,5MEO-3-CL)BENZYL	020285-74-9
24-NH2PYRIMIDINE,5(3,5MEO-2-CL)BENZYL	083166-74-9
24-NH2PYRIMIDINE,5(3,5MEO-4-CL)BENZYL	056518-45-7
24NH2PYRIMIDINE,5(35-MEO-4-COHME2)BENZYL	055687-91-7
24NH2-PYRIMIDINE,5(3,5-MEO-4-ET)BENZYL	078026-01-4
24NH2-PYRIMIDINE,5(3,5-MEO-4-ETO)BENZYL	078025-68-0
24NH2-PYRIMIDINE,5(3,5-MEO-4-ETS)BENZYL	071125-73-0
2,4-NH2PYRIMIDINE,5(3,4MEO-5-I)BENZYL	083166-73-8
2,4-NH2PYRIMIDINE,5(3,5MEO-4-I)BENZYL	056518-58-2
2,4-NH2PYRIMIDINE,5(3MEO-5-MEO)BENZYL	020344-69-8
2,4-NH2PYRIMIDINE,5(3MEO-4-MES)BENZYL	068885-48-3
24NH2PYRIMIDINE,5(35-MEO-4-OETOME)BENZYL	053808-87-0
2,4-NH2PYRIMIDINE,5(35MEO-4-OH)BENZYL	021253-58-7
24-NH2PYRIMIDINE,5(3MEO-4PHMEO)BENZYL	083158-06-9
24-NH2PYRIMIDINE,5(4MEO-3PHMEO)BENZYL	078233-99-5
24NH2-PYRIMIDINE,5(35MEO-4PHMEO)BENZYL	021822-34-4
24NH2-PYRIMIDINE,5(34MEO-5PHMEO)BENZYL	027810-94-2
24-NH2PYRIMIDINE,5(35MEO-4-PRENYL)BENZYL	069194-91-8
2,4-NH2PYRIMIDINE,5(3,5-NHME)BENZYL	061544-27-2
2,4-NH2PYRIMIDINE,5-(3,4OCH2O)BENZYL	013932-40-6
2,4-NH2PYRIMIDINE,5(35-OH-4MEO)BENZYL	083158-00-3
2,4-NH2PYRIMIDINE,5(4-OH-3MEO)BENZYL	073356-40-8
2,4-NH2PYRIMIDINE,5(34-OH-5MEO)BENZYL	071525-06-9
2,4-NH2PYRIMIDINE,5(3-OH-4MEO)BENZYL	083166-76-1
24NH2PYRIMIDIN,5(3ETO4COHME2-5MEO)BENZYL	077453-35-1
2,4-NH2PYRIMIDIN,5(35-ME-4-MEO)BENZYL	073576-30-4
2,4-NH2PYRIMIDIN,5(3,5MEO-4C≡N)BENZYL	055687-58-6
24NH2PYRIMIDIN,5(3-MEO-4-C(OH)ME2)BENZYL	083166-75-0
24-NH2PYRIMIDIN,5(35MEO-4CONH2)BENZYL	055687-57-5
2,4-NH2PYRIMIDIN,5(34-MEO-5-ME)BENZYL	024798-19-4
2,4-NH2PYRIMIDIN,5(35-MEO-4-ME)BENZYL	049845-48-9
2,4-NH2PYRIMIDIN,5(45-MEO-2-ME)BENZYL	006981-18-6
24NH2PYRIMIDIN,5(35-MEO-4-MEO-CH2)BENZYL	055687-86-0
24-NH2PYRIMIDIN,5(35-MEO-4-MES)BENZYL	068902-57-8
24NH2-PYRIMIDIN,5(3,5-MEO-4OCH2CN)BENZYL	083166-77-2
24NH2-PYRIMIDIN,5(3MEO-45PHCO)BENZYL	083158-01-4
24NH2-PYRIMIDIN,5(4MEO-35PHCO)BENZYL	083158-02-5
2,4NH2PYRIMIDIN,5(26MEOPYRIDIN4YL)CH2	083157-99-7
2,4NH2PYRIMIDIN,5(26MEOPYRIDIN3YL)CH2	065873-69-0
24NH2PYRIMIDIN,5(35MEO-4(1PYRRYL))BENZYL	071125-75-2
2,4-NH2PYRIMIDIN,5(3,5MEO-4NH2)BENZYL	056066-19-4
24-NH2PYRIMIDN,5(3,5-DIMEAMINO)BENZYL	061573-39-5
24NH2PYRIMIDN,5(35ETO-4-ACETYL)BENZYL	055687-73-5
24NH2PYRIMIDN,5(3MEO-4PRENYL-5ETO)BENZYL	085544-42-9
2,4-NH2PYRIMIDN,5(3,5MEO-4-AC)BENZYL	055687-76-8
24-NH2PYRIMIDN,5(35MEO-4-CH2OH)BENZYL	055687-49-5
24NH2PYRIMIDN,5(35MEO-4COCH2SO2ME)BENZYL	055687-74-6
24-NH2PYRIMIDN,5(35-MEO-4CO2ET)BENZYL	055687-48-4
24NH2PYRIMIDN,5(35MEO-4CO2IPR)BENZYL	055687-46-2
24-NH2PYRIMIDN,5(35MEO-4-CO2ME)BENZYL	055687-40-6
24NH2PYRMDN,5(3NME2-4MEO-5NMECOME)BENZYL	061544-21-6
24-NH2-6-PYRROLIDIN-PYRIMIDINE-3-OXID	055921-65-8
2-NH2-1,3,4-THIADIAZOLE-SO2NH2	014949-00-9
46NH2-22ME-1(3URIDOPHO)MEPH S-TRIAZINE	070579-43-0
4-NHAC-2266-TETRAME-PIPERIDIN-N-OXIDE	014691-89-5
2,4-NH2-5(3,4DICLPH)-6-MEPYRIMIDINE	007761-45-7
2,4-NH2-5(3,4-DIMEOBZ)PYRIMIDINE	005355-16-8

NIALAMIDE	000051-12-7
NICLOSAMIDE	000050-65-7
I-NICOTINAMIDE	001453-82-3
NICOTINAMIDE	000098-92-0
NICOTINANILIDE	001752-96-1
NICOTINE	000054-11-5
NICOTINIC ACID	000059-67-6
NICOTINIC ACID, BUTYL ESTER	006938-06-3
NICOTINIC ACID, ETHYL ESTER	000614-18-6
NICOTINIC ACID, METHYL ESTER	000093-60-7
I-NICOTINIC ACID, METHYL ESTER	002459-09-8
NICOTINIC ACID, PROPYL ESTER	007681-15-4
(E)-3-NICOTINOYLACRYLIC ACID	109151-40-8
NIFEDIPINE	021829-25-4
NIFUROXIME	006236-05-1
NIFURPIPONE	024632-47-1
NIKETHAMIDE	000059-26-7
NIMETAZEPAM	002011-67-8
NIMUSTINE	042471-28-3
NIPECOTIC ACID	000498-95-3
NIRIDAZOLE	000061-57-4
NITHIAMIDE	000140-40-9
N-NITROSO-METHYLAMYLAMINE	013256-07-0
N-NITROSO-METHYLCYCLOHEXYLAMINE	005432-28-0
8-(3-NITROPHENYL)-ADENINE	017659-57-3
NITRALIN	004726-14-1
NITRAPYRIN	001929-82-4
6-NITRATOHEXANOIC ACID	074754-55-5
NITRAZEPAM	000146-22-5
NITRILOTRIACETIC ACID	000139-13-9
2-NITROACETANILIDE	000552-32-9
4-NITROACETANILIDE	000104-04-1
M-NITROACETANILIDE	000122-28-1
2-NITROACETOPHENONE	000577-59-3
3-NITROACETOPHENONE	000121-89-1
P-NITROACETOPHENONE	000100-19-6
3-NITROANILINE	000099-09-2
2-NITROANILINE	000088-74-4
4-NITROANILINE	000100-01-6
2-NITROANISOLE	000091-23-6
M-NITROANISOLE	000555-03-3
P-NITROANISOLE	000100-17-4
1-NITROANTHRAQUINONE	000082-34-8
1-NITROAZULENE	007206-56-6
M-NITROBENZALDEHYDE	000099-61-6
4-NITROBENZALDEHYDE	000555-16-8
O-NITROBENZALDEHYDE	000552-89-6
P-NITROBENZALDOXIME	001129-37-9
2-NITROBENZAMIDE	000610-15-1
4-NITROBENZAMIDE	000619-80-7
M-NITROBENZAMIDE	000645-09-0
M-NITROBENZAMIDINE HYDROCHLORIDE	056406-50-9
NITROBENZENE	000098-95-3
M-NITROBENZENESULFONAMIDE	000121-52-8
4-NITROBENZENESULFONAMIDE	006325-93-5
2-NITROBENZENESULFONAMIDE	005455-59-4
M-NITROBENZENESULFONIC ACID, SODIUM SALT	000127-68-4
4'-NITROBENZOGUANAMINE	029366-73-2

2'-NITROBENZOGUANAMINE	029366-71-0
3'-NITROBENZOGUANAMINE	029366-72-1
2-NITROBENZOIC ACID	000552-16-9
M-NITROBENZOIC ACID	000121-92-6
P-NITROBENZOIC ACID	000062-23-7
5-NITROBENZOTHIOPHENE	004965-26-8
5-NITROBENZOTRIAZOLE	002338-12-7
4-NITROBENZOYLHYDRAZINE	000636-97-5
3-NITROBENZOYLHYDRAZINE	000618-94-0
4-(4-NITROBENZOYL)-PYRIDINE	027693-38-5
M-NITROBENZYL ALCOHOL	000619-25-0
2-NITROBENZYL ALCOHOL	000612-25-9
P-NITROBENZYL ALCOHOL	000619-73-8
P-NITROBENZYLAMINE	007409-30-5
O-(P-NITROBENZYL)BENZAMIDE	024367-68-8
M-NITROBENZYL-TR-CHRYSANTHEMATE	082488-02-6
P-NITROBENZYL N,N-DIME CARBAMATE	084640-31-3
P-NITROBENZYLSELENOPROPIONIC ACID	034835-05-7
M-NITROBENZYL TRIMETHYL AMMONIUM BROMIDE	071323-93-8
P-NITROBIPHENYL	000092-93-3
3-NITROBIPHENYL	002113-58-8
1-NITROBUTANE	000627-05-4
2-NITROBUTANE	000600-24-8
2-NITRO-1-BUTANOL	000609-31-4
2-NITRO-4-BUTYLACETANILIDE	003663-21-6
2-NITRO-4-CHLOROACETANILIDE	000881-51-6
2-NITRO-4-CHLOROBENZOIC ACID	006280-88-2
2-NITRO-5-CHLOROBENZOIC ACID	002516-95-2
TRANS-P-NITROCINNAMIC ACID	000882-06-4
3-NITROCINNOLINE	021905-82-8
3-NITRO-P-CRESOL	002042-14-0
1-NITRO-4-CYANOBENZENE	000619-72-7
1-NITRO-2-CYANOBENZENE	000612-24-8
1-(2-NITRO),2-DIBENZOYL-1-TERT-BUTYLHYDRAZINE	112225-86-2
1-(4-NITRO),2-DIBENZOYL-1-TERT-BUTYLHYDRAZINE	067523-85-7
1-(3-NITRO),2-DIBENZOYL-1-TERT-BUTYLHYDRAZINE	112226-82-1
4-NITRO-((4-(N-DIMETHYL)AMINOPHENYL)AZO)BENZENE	002491-74-9
B-NITRO-4-DIMETHYLAMINOSTYRENE	002604-08-2
P-NITRO-N,N-DIMETHYLANILINE	000100-23-2
7-NITRO-2,2-DIMETHYL-2,3-DIHYDROBENZOFURAN	013414-55-6
4-NITRO DIPHENYL ETHER	000620-88-2
NITROETHANE	000079-24-3
2-NITROETHANOL	000625-48-9
B-NITROETHYL BENZENE	006125-24-2
NITROFEN	001836-75-5
2-NITROFLUORENE	000607-57-8
NITROFLUORFEN	042874-01-1
3-NITROFORMANILIDE	000102-38-5
P-NITROFORMANILIDE	016135-31-2
5-NITRO-2-FURALDEHYDE	000698-63-5
2-NITROFURAN	000609-39-2
NITROFURANTOIN	000067-20-9
2-PYRIDINEMETHANOL, 6-[2-(5-NITRO-2-FURANYL)ETHE	013411-16-0
NITROFURAZONE	000059-87-0
NITROGUANIDINE	000556-88-7
3-NITRO-3-HEPTENE	006187-24-2
NITROHEXANE	000646-14-0
6-NITROHYDROTHIAZIDE	023141-88-0

4-NITROIMIDAZOLE	003034-38-6
2-NITROIMIDAZOLE	000527-73-1
4-NITROIMIDAZOLE1-ME-5-SO3PH	006954-35-4
3-(2-NITROIMIDAZOL-1-YL)-1,2-PROPANEDIOL	013551-92-3
7-NITROINDOLE	006960-42-5
2-NITROISONIAZID	058481-05-3
1-NITRO-2-ISOPROPYLBENZENE	006526-72-3
NITROMETHANE	000075-52-5
2-NITRO-4-METHOXYACETANILIDE	000119-81-3
2-NITRO-6-METHOXYBENZAMIDE	064431-77-2
2-NITRO-4-METHYLACETANILIDE	000612-45-3
P-NITRO-N-METHYLANILINE	000100-15-2
P-NITRO-METHYLHIPPURATE	002644-96-4
2-NITRONAPHTHALENE	000581-89-5
1-NITRONAPHTHALENE	000086-57-7
6-NITRO-1-NAPHTHOL	038397-06-7
3-NITRO-N-(4-NITROPHENYL)BENZENAMINE	015979-87-0
4-NITRO-B-NITROSTYRENE	003156-41-0
2-NITRO-B-NITROSTYRENE	003156-39-6
3-NITRO-B-NITROSTYRENE	000882-26-8
B-NITRO-4-OH-3-METHOXYSTYRENE	006178-42-3
1-NITROPENTANE	000628-05-7
P-NITROPHENETOLE	000100-29-8
3-NITROPHENOL	000554-84-7
4-NITROPHENOL	000100-02-7
2-NITROPHENOL	000088-75-5
P-NITROPHENOXYACETIC ACID	001798-11-4
O-NITROPHENOXYACETIC ACID	001878-87-1
M-NITROPHENOXYACETIC ACID	001878-88-2
O-NITROPHENYL ACETATE	000610-69-5
M-NITROPHENYL ACETATE	001523-06-4
P-NITROPHENYLACETATE	000830-03-5
M-NITROPHENYLACETIC ACID	001877-73-2
P-NITROPHENYLACETIC ACID	000104-03-0
P-NITROPHENYLALANINE	002922-40-9
3-NITROPHENYLAMIDINO UREA	016018-79-4
4-((4-NITROPHENYL)AZO)BENZENAMINE	000730-40-5
2-NITRO-4-PHENYLAZOPHENOL	055936-40-8
4-(4-NITROPHENYLAZO)PHENOL	001435-60-5
M-NITROPHENYL AZOXYCYANIDE	062825-12-1
P-NITROPHENYL AZOXYCYANIDE	060142-50-9
2-NITRO-N-PHENYLBENZENAMINE	000119-75-5
4-NITRO-N-PHENYLBENZENAMINE	000836-30-6
3-NITRO-N-PHENYLBENZENAMINE	004531-79-7
N-(4-NITROPHENYL)-1,4-BENZENDIAMINE	006149-34-4
P-NITROPHENYL BUTYL ETHER	007244-78-2
4-NITROPHENYLDIMETHYLCARBAMATE	007244-70-4
O-NITROPHENYLDIMETHYLCARBAMATE	003373-86-2
3-NITRO-O-PHENYLENEDIAMINE	003694-52-8
2-(P-NITROPHENYL)ETHANOL	000100-27-6
3-NITROPHENYL GLUCOPYRANOSIDE	020838-44-2
M-NITROPHENYLHIPPURATE	002979-53-5
4-NITROPHENYLISOTHIOCYANATE	002131-61-5
N-((4-NITROPHENYL)METHYL)ACETAMIDE	056222-10-7
(4-NITROPHENYL)PHENYLDIAZENE	002491-52-3
N-(P-NITROPHENYL)PIPERIDINE	006574-15-8
P-NITROPHENYL-I-PROPYL ETHER	026455-31-2
O-NITRO-N-PHENYLSUCCINIMIDE	018377-52-1

M-NITRO-N-PHENYLSUCCINIMIDE	031036-66-5
P-NITRO-N-PHENYLSUCCINIMIDE	035488-92-7
N-(4-NITROPHENYL)-3-(TRIFLUOROMETHYL)BENZENAMINE	000369-90-4
3-NITROPHTHALIC ACID	000603-11-2
4-NITROPHTHALIMIDE	000089-40-7
3-NITROPHTHALIMIDE	000603-62-3
1-NITROPROPANE	000108-03-2
2-NITROPROPANE	000079-46-9
4-NITROPYRAZOLE	002075-46-9
3-NITROPYRIDINE	002530-26-9
2-NITROPYRIDINE	015009-91-3
4-NITROPYRIDINE	001122-61-8
4-NITROPYRIDINE-1-OXIDE	001124-33-0
8-NITROQUINOLINE	000607-35-2
3-NITROQUINOLINE	017576-53-3
7-NITROQUINOLINE	000613-51-4
6-NITROQUINOLINE	000613-50-3
5-NITROQUINOLINE	000607-34-1
4-NITROQUINOLINE	003741-15-9
5-NITROQUINOLINE-1-OXIDE	007613-19-6
8-NITROQUINOLINE-1-OXIDE	014753-18-5
7-NITROQUINOLINE-1-OXIDE	014753-17-4
3-NITROQUINOLINE-1-OXIDE	007433-86-5
4-NITROQUINOLINE OXIDE	000056-57-5
6-NITROQUINOLINE-1-OXIDE	013675-92-8
2-NITRORESORCINOL	000601-89-8
5-NITROSALICYLIC ACID	000096-97-9
P-NITROSALICYLIC ACID	000085-38-1
N-NITROSO-ACETOXYMETHYL-N-PROPYLAMINE	066017-91-2
N-NITROSO-ACETOXYMETHYLMETHYLAMINE	056856-83-8
N-NITROSO-ACETOXYMETHYL-N-BUTYLAMINE	056986-36-8
NITROSOBENZENE	000586-96-9
N-NITROSO-BIS(2-CYANOET)AMINE	001116-82-1
N-NITROSO-BIS(CYANOMETHYL)AMINE	016339-18-7
N-NITROSO-BIS(2-ETOET)AMINE	067856-66-0
N-NITROSO-BIS(2-MEOET)AMINE	067856-65-9
1-NITROSO-1-(2-CLET)-3-PHENYLUREA	013206-67-2
N-NITROSO-DIALLYLAMINE	016338-97-9
N-NITROSODIBUTYLAMINE	000924-16-3
N-NITROSODIETHANOLAMINE	001116-54-7
N-NITROSODIETHYLAMINE	000055-18-5
N-NITROSO-DIISOBUTYLAMINE	000997-95-5
N-NITROSO-3,5-DIME PIPERAZINE	067774-31-6
N-NITROSODIMETHYLAMINE	000062-75-9
P-NITROSO-N,N-DIMETHYLANILINE	000138-89-6
N-NITROSO-DIPHENYLAMINE	013256-06-9
N-NITROSODIPROPYLAMINE	000621-64-7
1-NITROSO-1-ET-3,3-DIMETHYL-UREA	050285-71-7
N-NITROSOETHYL-N-BUTYLAMINE	004549-44-4
N-NITROSO-N-ETHYL-(T-BUTYL)AMINE	003398-69-4
(N-NITROSO-N-ETHYL)ETHANOLAMINE	013147-25-6
N-NITROSO-N-ETHYL-ISOPROPYLAMINE	016339-04-1
N-NITROSO-ETHYLVINYLAMINE	013256-13-8
N-NITROSO-HEPTAMETHYLENEIMINE	020917-49-1
N-NITROSOHEXAMETHYLENEIMINE	000932-83-2
N-NITROSO-N-ME-(4-CYANOBENZYL)AMINE	084174-23-2
N-NITROSO-N-ME-(4-MEOBENZYL)AMINE	084174-20-9
N-NITROSOMETHYLALLYLAMINE	004549-43-3

N-NITROSOMETHYLBUTYLAMINE	007068-83-9
1-NITROSO-1-METHYL DIETHYLUREA	050285-72-8
N-NITROSOMETHYL-ISOBUTYLAMINE	002504-18-9
N-NITROSO-N-METHYL-N'-NITROGUANIDINE	000070-25-7
1-NITROSO-1-METHYL-3-PHENYLUREA	021561-99-9
N-NITROSOMETHYLPROPYLAMINE	000924-46-9
1-NITROSO-1-METHYLUREA	000684-93-5
N-NITROSOMETHYLVINYLAMINE	004549-40-0
N-NITROSOMORPHOLINE	000059-89-2
1-NITROSO-2-NAPHTHOL	000131-91-9
2-NITROSO-1-NAPHTHOL	000132-53-6
N-NITROSO-METHYL-NEOPENTYLAMINE	031820-22-1
N-NITROSO-OCTAMETHYLENEIMINE	020917-50-4
N-NITROSO-PHENMETRAZINE	034993-08-3
P-NITROSOPHENOL	000104-91-6
4-NITROSO-N-PHENYLBENZENAMINE	000156-10-5
N-NITROSOPIPERAZINE	005632-47-3
N-NITROSOPIPERIDINE	000100-75-4
N-NITROSO-4-PIPERIDONE	055556-91-7
N-NITROSOPROPYLBUTYLAMINE	025413-64-3
N-NITROSOPYRROLIDINE	000930-55-2
3-(1-NITROSO-2-PYRROLIDINYL)PYRIDINE	016543-55-8
N-NITROSO-3-PYRROLINE	010552-94-0
N-NITROSOSARCOSINE	013256-22-9
N-NITROSOTHIOMORPHOLINE	026541-51-5
1-NITROSOTRIETHYLUREA	050285-70-6
1-NITROSO-TRIMETHYL UREA	003475-63-6
BETA-NITROSTYRENE	000102-96-5
4-NITRO-2,3,5,6-TETRACHLOROPHENOL	004824-72-0
2-NITROTHIOPHENE	000609-40-5
3-NITROTHIOPHENE	000822-84-4
3-NITROTOLUENE	000099-08-1
P-NITROTOLUENE	000099-99-0
A-NITROTOLUENE	000622-42-4
2-NITROTOLUENE	000088-72-2
4-NITRO-2-TOLUIDINE	000099-52-5
3-NITRO-4-TOLUIDINE	000119-32-4
5-NITRO-2-TOLUIDINE	000099-55-8
2-NITRO-P-TOLUIDINE	000089-62-3
(3-NITRO-1,2,4-TRIAZOL-1-YL)ACETAMIDE,N-(2-HYDRO	104958-85-2
4-[(3-NITRO-1,2,4-TRIAZOL-1-YL)ACETYL]-MORPHOLIN	104958-84-1
4-(3-NITRO-1,2,4-TRIAZOL-1-YL)PRIOPIONYLMORPHOLI	104987-39-5
2,6-NITRO-4-TRIFLUORO-ANILINE	000445-66-9
4-NITRO-N-(TRIPHENYLMETHYL)BENZENAMINE	020222-29-1
2-(B-NITROVINYL)FURAN	000699-18-3
3-(B-NITROVINYL)FURAN	053916-74-8
2-(B-NITROVINYL)THIOPHENE	000874-84-0
3-(B-NITROVINYL)THIOPHENE	028783-31-5
NITROXOLINE	004008-48-4
4-NITRO-M-XYLENE	000089-87-2
P(3-PNO2BENZYL-3ME-1TRIAZN)BENZOIC AC	065542-15-6
1-NO2 2-CF3 BENZENE	000384-22-5
2-NO2-4-CF3-C6H3NHN=C(CN)COOME	028313-79-3
N(24-NO2-6CF3PH)-2(3CN-BENZO)PYRIDINAMIN	121588-32-7
N(24-NO2-6CF3 PH)-2(3CL-5CF3)PYRIDINAMIN	078879-38-6
N(24-NO2-6CF3 PH)-2(3CL-5NO2)PYRIDINAMIN	142978-32-3
N(24-NO2-6CL PH)-2(3CL-5CF3)PYRIDINAMINE	142978-30-1
N(24-NO2-6CF3 PH)-2(35CN-6ME)PYRIDINAMIN	132133-10-9

4-NO2-3-CHLOROPHENYLAMIDINOUREA	046833-92-5
1-NO-1-(2CLET)-3-(2(2BENZYL)PR)UREA	033021-93-1
2'-NO2-4'-CL SALICYLANILIDE	037399-40-9
2(5-NO2-2-FURFURILIDENE)IMIDAZOLE	006756-33-8
2(5NO2-2-FURFURILIDENE)THIAZOLE	049561-47-9
3(5NO2-2-FURFURILIDENE)-124-TRIAZOLE	050832-71-8
5-(5-NO2-2-FURFURYLIDENE-2-S-THIAZOLIDIN-4-ONE	013410-84-9
2(5-NO2-2-FURFURYLIDENE)PYRIMIDINE	001083-59-6
5(5NO2-2-FURIL)THIAZOLIDIN,24-DION	027564-47-2
1(5-NO2-2-FURYL)-2-IMIDAZOLIDINONE	053207-61-7
1(5-NO2-2-FURYL)-2-IMIDAZOLIDINTHIONE	053207-62-8
1(5-NO2-2-FURYL)-2-MES IMIDAZOLINE	053207-66-2
5-(5NO2-FURYLPROPENILIDINE)-2-S-THIAZOLIDEN-2,4-	027464-57-9
6(5NO2-2-FURYL)-13-THIAZIN-24-DIONE	052661-48-0
2-(2-NO2-1-IMIDAZOLYL)ACETAMIDE	022903-73-7
2(2NO2-IMIDZOL-1YL)N-2ETOH-ACETAMIDE	022668-01-5
3NO2-3ME-2BUTANON(N-ME-CARBMOY)OXIME	020417-83-8
P-NO2-(3-ME SULFONYL)PROPOXYBENZENE	105412-25-7
N(24-NO2-6CL PH)-2(3CL-5NO2)PYRIDINAMINE	142978-29-8
1-(3-NO2 PH)-3,3-DIMETHYLTRIAZENE	020241-06-9
1-(2-(P-NO2PHENOXY)ETHYL)IMIDAZOLE	075912-69-5
1-(3-NO2 PHENYL)-3-MEO-3-ME UREA	088132-19-8
1-(4-NO2 PHENYL)-3-MEO-3-ME UREA	088132-15-4
N1-(4-NO2-PHENYL)SULFANILAMIDE	006829-82-9
N-(2-(P-NO2-PHO)ETHYL)MORPHOLINE	065300-53-0
P(3-NO2-PHO-PRO)BENZAMIDINE HCL	053525-79-4
P(4-NO2-PHO-PRO)BENZAMIDINE HCL	053525-76-1
N-(O-NO2PH)-3-N'-PIPERIDINOACETAMIDE	035204-11-6
N(P-NO2PH)-3-N'-PIPERIDINOACETAMIDE	038367-22-5
26NO2-N-PR-AAA-TRIF-P-TOLUIDINE	002077-99-8
2'-NO2 SALICYLANILIDE	037183-26-9
5-NO2 SALICYLANILIDE	002389-37-9
1(5NO2-2-THIAZOLYL)-2MES-2-IMIDAZOLINE	037422-15-4
M-NO2-TRIME-ANILINIUM IODIDE	027389-55-5
N(2,4-NO2-6-CF3 PH)-2(3,5CL)PYRIDINAMINE	078879-25-1
CIS-1-NO-1(2-CLET)3(AC OXYCYHEX)UREA	059414-37-8
1NO1(2CLET)3(TRIACETYLXYLOSYL)UREA	058845-59-3
1-NO-1-(2-CLET)-3(ACETYLOXYCYHEX)UREA	059414-38-9
CIS-1-NO-1(CLET)-3(4COOH CYHEXYL)UREA	042558-93-0
TRAN-1NO-1(CLET)-3(4COOH CYHEXYL)UREA	042558-94-1
1-N=O-1-(2-CLET)-3-CYCLOHEXYLUREA	013010-47-4
1-NO-1-(2CLET)-3(2CYHEXEN-1-YL)UREA	033021-94-2
1-NO-1-(2FLET)-3(CYHX-35-DITHIAN)UREA	033024-47-4
1-NO-1-(2CLET)-3(CYHX-35-DITHIAN)UREA	033022-04-7
1-NO-1-(2CLET)-3(M-DI-SO2PYRAN)UREA	033022-05-8
1-NO-1-(2-CLET)-3(2,6-OHCYHEX)UREA	058484-09-6
1-NO-1-(2CLET)-3(4-OH CYHEXYL)UREA	052049-26-0
1-NO-1-(CLET)-3(2-OH CYHEXYL)UREA	056323-43-4
1-NO-1-(2CLET)-3(4SULFAPYRANYL)UREA	033022-01-4
1-NO-1-(2CLET)-3(4THIOPYRANSO2)UREA	033022-02-5
2-(5-NO2-2-FURYLVINYL)5-NH2-3-OXOIMDAZOLINE	021959-57-9
CIS-NONACHLOR	005103-73-1
TRANS-NONACHLOR	039765-80-5
2,2',3,3',4,4',5,5',6-NONACHLOROBIPHENYL	040186-72-9
NONACHLORO-1,1'-BIPHENYL	053742-07-7
1,1,1,2,2,3,3,4,4-NONACHLOROBUTANE	021483-62-5
NONACOSANE	000630-03-5
NONADECANE	000629-92-5

NONADECANOL	052783-43-4
1-NONADECENE	018435-45-5
1,8-NONADIYNE	002396-65-8
NONAETHYLENE GLYCOL	003386-18-3
1,1,1,3,3,5,7,7,7-NONAMETHYL-5-PHENYLTETRASILO*	017906-09-1
NONANAL	000124-19-6
N-NONANE	000111-84-2
NONANEDIOC ACID	000123-99-9
4,6-NONANEDIONE	014090-88-1
1-NONANETHIOL	001455-21-6
NONANOIC ACID	000112-05-0
5-NONANOL	000623-93-8
4-NONANOL	005932-79-6
1-NONANOL	000143-08-8
3-NONANOL	000624-51-1
2-NONANOL	000628-99-9
2-NONANONE	000821-55-6
5-NONANONE	000502-56-7
NONANONITRILE	002243-27-8
NONASILOXANE, EICOSAMETHYL-	002652-13-3
4,6,8-NONATETRAENOIC ACID, 9-(4-METHOXY-2,3,6-	055079-83-9
H-(NO NO2) AMPHENICOL	025126-19-6
N-NONYLAMINE	000112-20-9
N-NONYLBENZENE	001081-77-2
N-NONYLCYCLOHEXANE	002883-02-5
P-NONYLPHENOL	000104-40-5
N-NONYLPHENOL	084852-15-3
4-NONYLPHENYL DIPHENYL PHOSPATE	064532-97-4
1-NONYNE	003452-09-3
N-(M-NO2PH)-3-N'-PIPERIDINOACETAMIDE	035763-43-0
(-)-NORADRENALINE	000051-41-2
NORADRENALINE (HCL)	000055-27-6
NORAZINE	003004-71-5
3(2-NORBORNYL)-1-(2CLET)-1-NO UREA	013909-13-2
NORCAMPHOR	000497-38-1
NOREA	018530-56-8
NORFLOXACIN	070458-96-7
NORFLURAZON	027314-13-2
NORMETANEPHRINE	000097-31-4
NORNICOTINE	000494-97-3
19-NORPREGNA-1,3,5(10)-TRIEN-20-YNE-3,17-DIOL,	000057-63-6
19-NORPREGNA-1,3,5(10)-TRIEN-20-YNE-3,17-DIOL, 1	039845-47-1
19-NORPREGNA-1,3,5(10)-TRIEN-20-YNE-3,17-DIOL, 1	034816-55-2
19-NORPREGNA-1,3,5(10)-TRIEN-20-YNE-3,17-DIOL, 1	092817-12-4
NORTRON	000104-14-3
L-NORVALINAMIDE DERIVATIVE	137429-62-0
L-NORVALINAMIDE DERIVATIVE	137302-63-7
L-NORVALINAMIDE DERIVATIVE	137302-35-3
NORVALINE	006600-40-4
NPA	000132-66-1
NSC #128,668	013263-99-5
O,O-DIET-O(3-CL-4-MESPH)PHOSPHOROTHIOATE	021840-66-4
O,S-DIME-N-ME-PHOSPHORAMIDOTHIOAT	028167-49-9
O,O-DIET-O(2-CL-4-MESPH)PHOSPHOROTHIOATE	026512-71-0
O,S-DIME-N-PR-PHOSPHORAMIDOTHIOAT	016271-16-2
O=P O,O-DIET-O-(2-ME-4-SMEPHENYL)	026512-63-0
O=P(NME)(OME)O-(4-CL-PHENYL)	019670-19-0
O=P(NC)(OC)O-2-CL PHENYL	019608-64-1

O=P(NME)(OME)-O-(4-NITROPHENYL)	054267-24-2
O=P(OC)(SC)N-I-BU	052067-51-3
O=P(OC)(SC)N-T-BU	052067-44-4
O=P(OC)(SC)N-N-BU	052067-49-9
O=P(OC)(SC)N-S-BU	052067-50-2
O=P(OET)(OET-S-(2-THIOET)ET	050728-06-8
O=P(OET)(OET)O-2,6-CL-4-SMEPHENYL	026707-54-0
O=P(OME)(NET)O-2,4,5-CLPHENYL	002213-84-5
O=P(OME)(NME)O-2,4,5-CLPHENYL	002213-70-9
O=P(OME)(N)O-2,4,5-CLPHENYL	002214-34-8
O=P(OME)(NPR)O-2,4,5-CLPHENYL	002213-85-6
4-OCF3 ACETANILIDE	001737-06-0
4,5-(-OCF2CF2O-)C6H3NHN=C(CN)CO-OME	036865-53-9
4,5-(-OCF2CF2O-)C6H3NHN=C(CN)CO-ME	036865-60-8
OCH	004024-81-1
OCHRATOXIN A	000303-47-9
N(O-CLPH)-3-N'-PIPERIDINOACETAMIDE	038367-23-6
3-OCO-5-BR-6-ME-PYRIMIDIN-2,4-DIONE	077317-78-3
P-(OCONME2)BENZYL N,N-DIME CARBAM	084640-38-0
OCTABROMODIBENZO-P-DIOXIN	002170-45-8
2,2',3,3',5,5',6,6'-OCTACHLOROBIPHENYL	002136-99-4
2,2',3,3',4,4',5,5'-OCTACHLOROBIPHENYL	035694-08-7
1,1,1,2,3,3,4,4-OCTACHLOROBUTANE	032694-76-1
1,1,1,2,3,4,4,4-OCTACHLOROBUTANE	018791-19-0
1,1,2,2,3,3,4,4-OCTACHLOROBUTANE	020338-26-5
OCTACHLOROCYCLOPENTENE	000706-78-5
1,2,3,4,6,7,8,9-OCTACHLORODIBENZO-P-DIOXIN	003268-87-9
OCTACHLORODIBENZOFURAN	039001-02-0
OCTACHLORONAPHTHALENE	002234-13-1
OCTACHLOROSTYRENE	029082-74-4
OCTACOSANE	000630-02-4
OCTADECAMETHYLCYCLONONASILOXANE	000556-71-8
OCTADECAMETHYLOCTASILOXANE	000556-69-4
OCTADECANE	000593-45-3
1-OCTADECANETHIOL	002885-00-9
1-OCTADECANOL	000112-92-5
9,12,15-OCTADECATRIENOIC ACID, (Z,Z,Z)-	000463-40-1
1-OCTADECENE	000112-88-9
9-OCTADECENOIC ACID	002027-47-6
OCTADECYL 3,5-BIS(TERT-BUTYL)-4-HYDROXYBENZENEP*	002082-79-3
N-OCTADECYLPYRIDINIUM IODIDE	007206-40-8
OCTADECYLTRICHLOROSILANE	000112-04-9
OCTAFLUORO-2-BUTENE	000360-89-4
OCTAFLUOROPROPANE	000076-19-7
OCTAFLUOROTOLUENE	000434-64-0
OCTAHYDRO-1,3,5,7-TETRANITRO-1,3,5,7-TETRAZOC*	002691-41-0
1,1,1,3,5,7,7,7-OCTAMETHYL-3,5-BIS[2-(TRIMETHYL*	091455-17-3
1,1,1,3,5,7,7,7-OCTAMETHYL-3,5-DIPHENYLTETRASIL*	013270-97-8
OCTAMETHYLPYROPHOSPHORAMIDE	000152-16-9
OCTAMETHYLTETRASILOXANE	000556-67-2
OCTAMETHYLTRISILOXANE	000107-51-7
OCTANAL	000124-13-0
OCTANE	000111-65-9
OCTANEDINITRILE	000629-40-3
1-OCTANETHIOL	000111-88-6
2-OCTANETHIOL	003001-66-9
OCTANOIC ACID	000124-07-2
OCTANOIC ACID HYDRAZIDE	006304-39-8

1-OCTANOL	000111-87-5
4-OCTANOL	000589-62-8
2-OCTANOL	000123-96-6
3-OCTANOL	000589-98-0
2-OCTANONE	000111-13-7
3-OCTANONE	000106-68-3
2-OCTANONE, 5-METHYL-	058654-67-4
OCTANONITRILE	000124-12-9
6-(OCTANOYLAMINO)PENICILLANIC ACID	000525-97-3
N-OCTATRIACONTANE	007194-85-6
2-OCTENAL	002363-89-5
2-OCTENE	000111-67-1
1-OCTENE	000111-66-0
4-OCTENE	000592-99-4
2-OCTENE (CIS)	007642-04-8
2-OCTENE (TRANS)	013389-42-9
4-OCTENE (TRANS)	014850-23-8
7-OCTEN-1-OL, 8-(3,4-DIHYDROXYPHENYL)-6-OXO-	118198-78-0
OCTYL ACRYLATE	002499-59-4
N-OCTYLAMINE	000111-86-4
OCTYLBENZENE	002189-60-8
O-OCTYL CARBAMATE	002029-64-3
N-OCTYL N-DECYL PHTHALATE	000119-07-3
OCTYLDIPHENYL PHOSPHATE	000115-88-8
OCTYL GALLATE	001034-01-1
OCTYLMALONIC ACID	000760-55-4
T-OCTYL MERCAPTAN	000141-59-3
OCTYLPYRIDINIUM BROMIDE	002534-66-9
N-OCTYLPYRIDINIUM IODIDE	010291-06-2
OCTYL-TRIMETHYL-AMMONIUM IODIDE	014251-76-4
1-OCTYNE	000629-05-0
2-OCTYNE	002809-67-8
4-OCTYNE	001942-45-6
OFLOXACIN	082419-36-1
1(2-OH-3-ALLYLOXYPR)2-NO2IMIDAZOLE	021787-89-3
M-OH BENZALDEHYDETHIOSEMICARBIZONE	007420-37-3
6-OH BENZOTHIAZOLE-2-SULFONAMIDE	029927-14-8
3-OH-2-(4-BIPHENYLYL)HEXANOIC ACID	093371-38-1
3-OH-5-BR-6-ME-PYRIMIDINE-2,4-DIONE	077317-64-7
1(2-OH-6-CF3-ETOPR)-2-NO2-IMIDAZOLE	021787-91-7
M-OH CF3-METHANESULFONANILIDE	023375-12-4
2-OH-2-(M-CF3-PH)-4-ME MORPHOLINE HCL	124497-78-5
4-OH-3,5-DIMETHOXYBENZIOC ACID	000530-57-4
2-OH-2,6-DIPH-4,5-DIME MORPHOLINE HCL	124497-76-3
1-(2-OHET)-5-CYANO-2-NO2 PYRROLE	003150-53-6
2(1-OHET-1,3-IMIDAZOLINYLIDEN-2-AMINO)5NO2-THIAZ	024240-70-8
1-(2-OHET)-2-ME IMIDAZOLINE	000695-94-3
1-(2-OHET)-2-ME-4-NO2 IMIDAZOLE	000705-19-1
1-(2-OH-3-ETOPR)-2-NO2 IMIDAZOLE	065988-94-5
OHET-(3,4,5-TRIMEO)PHENYL SULFONE	057553-97-6
O-T-HEXYMOPYRIMIDINE	052642-50-9
2-OH-I-PROPYLAMINO PHENYL ETHER	007695-63-8
3-OH-4-ME-2(4-BIPHENYLYL)PENTANOIC ACID	093371-56-3
1-(2-OH-3-MEOPR)-2-ME-4-NO2IMIDAZOLE	071006-78-5
1-(2-OH-3-MEOPR)-2-ME-5-NO2IMIDAZOLE	060174-20-1
1(2-OH-3-MEO-PR)-2-NITROIMIDAZOLE	013551-87-6
2-OH NAPHTHOQUINONE,3(33DICLALLYL)	036417-16-0
N-OH-1,4-NAPHTHOQUINONEMONIMINE	004965-30-4

4-OH-N-NITROSOPIPERIDINE	055556-93-9
3-OH-N-NITROSOPIPERIDINE	055556-85-9
2-OH-2-(P-BR-PH)-4-ME MORPHOLINE HCL	124497-74-1
2-OH-2-PH-3,4-DIME MORPHOLINE HCL	124497-75-2
2-OH-2-PH-4,5-DIME MORPHOLINE HCL	124497-73-0
3'-OH-N-PHENYL AMINOBENZOIC ACID	021003-78-1
3-OH-5-PHENYLISOXAZOLE	000939-05-9
1-(3-OHPHENYL)-3-MEO-3-ME UREA	030087-17-3
2-OH-3-PH-4-ME MORPHOLINE HCL	124497-77-4
2-OH-2-PH-4-ME MORPHOLINE (HCL)	124497-72-9
2-OH-2-PH-ET-(3,4,5-MEO)PHENYLSUFONE	071203-63-9
5-OHPICOLINALDEHYDE THIOSEMICARBAZONE	019494-89-4
1-(2-OH PROPYL)-2-NO2 IMIDAZOLE	013551-91-2
2(6H-5OH-PYRIMIDINYLIDEN-2AM)-5NO2-THIAZOLE	037394-79-9
OHRIC	024096-53-5
OLEANDRIN	000465-16-7
ALPHA-OLEFINS (PETROLEUM)	064743-02-8
OLEIC ACID	000112-80-1
OLEIC ACID, SODIUM SALT	000143-19-1
OLEYL ALCOHOL (CIS)	000143-28-2
OMEPRAZOLE	073590-58-6
OO-DIET-O(2-CL-4-MESPH)PHOSPHATE	024493-78-5
2-OPR-2-PH-4-ME MORPHOLINE HCL	124497-80-9
ORAFLEX (UNIPROFEN) (OPREN) (COXIGON)	051234-28-7
ORBENCARB	034622-58-7
ORNITHINE	000070-26-8
L-ORNITHINE, N5-(AMINOCARBONYL)-	000372-75-8
OROTIC ACID	000065-86-1
ORPHENADRINE	000083-98-7
S-ORTHOPHENYLENETHIOCARBAMATE	000934-34-9
ORYZALIN	019044-88-3
OUABAGENIN	000508-52-1
OUABAIN	000630-60-4
3-OXA-7-AZABICYCLO[4.1.0]HEPTANE, 7-[2-HYDROXY-3	120277-97-6
9-OXABENZOBICYCLO(221)HEPTENE,2ENNH2	073208-84-1
9-OXABENZOBICYCLO(221)HEPTEN,2EN-MEAM	086992-70-3
9-OXABENZOBICYCLO(221)HEPTEN,2EX-MEAM	086943-80-8
9-OXABENZOBICYCLO(221)HEPTENE,2EXNH2	073159-84-9
OXABETRINIL	074782-23-3
2-OXABICYCLO[4.1.0]HEPTANE	000286-16-8
7-OXABICYCLO 4.1.0 HEPTANE, 3-ETHENYL-	000106-86-5
7-OXABICYCLO 4.1.0 HEPTANE, 3-OXIRANYL-	000106-87-6
OXACILLIN	000066-79-5
1,2,4-OXADIAZOL-3-ALDOXIME,5-PHENYL-	103499-08-7
1,3,4-OXADIAZOL-2-AMINE, 5-(2-PHENYLETHYL)-	001617-93-2
1,2,4-OXADIAZOLE-5-CARBOXALDEHYDE,3-ME,OXIME	090507-21-4
1,2,4-OXADIAZOLE-5-(DIETHYLAMINOETHYL)THIOHYDROX	103499-11-2
1,2,4-OXADIAZOLE-5-(DIETHYLAMINOETHYLTHIO)HYDROX	103499-09-8
1,3,4-OXADIAZOLE, 2-METHYL-5-[(PHENYLMETHYL)THIO	028915-24-4
1,2,4-OXADIAZOLE,5-METHYL-3-(P-TOLYL)	055752-22-2
1,3,4-OXADIAZOLE, 2-PHENYL-	000825-56-9
1,2,4-OXADIAZOLE-3-PHENYL-5-HYDROXIC ACID	090507-20-3
1,2,4-OXADIAZOL-5-OXIME,3-T-BUTYL-	099764-47-3
OXADIAZON	019666-30-9
1,2,4-OXADOAZOL-5-ALDOXIME,5-BENZYL-	103499-06-5
OXALIC ACID	000144-62-7
OXAMIC ACID	000471-47-6
OXAMIDE	000471-46-5

OXAMYL	023135-22-0
OXANILIC ACID,ETHYL ESTER	001457-85-8
1,4-OXATHIAN-3-ONE, O-[(METHYLAMINO)CARBONYL]OXI	054266-55-6
6-OXA-3-THIA-2,4,7,10-TETRAAZAUNDEC-7-ENOIC ACID	065907-54-2
1,4-OXATHIIN-3-CARBOXAMIDE, 5,6-DIHYDRO-2-METHYL	017757-70-9
1,4-OXATHIIN-3-CARBOXAMIDE, N-[1,1'-BIPHENYL]-2-	006577-34-0
1,4-OXATHIIN-3-CARBOXAMIDE, 5,6-DIHYDRO-2-METHYL	013584-27-5
1,4-OXATHIIN-3-CARBOXYLIC ACID, 5,6-DIHYDRO-2-ME	006577-69-1
1,3-OXATHIOLANE-5-METHANAMINIUM, N,N,N-TRIMETHYL	058326-97-9
1,3-OXATHIOLAN-4-ONE, 5,5-DIMETHYL-, O-[(METHYLA	106231-36-1
1,3-OXATHIOLAN-4-ONE, 2,5-DIMETHYL-, O-[(METHYLA	106293-84-9
1,3-OXATHIOLAN-4-ONE, 5-(2-METHOXYETHYL)-, O-[(M	054266-84-1
1,3-OXATHIOLAN-4-ONE, O-[(METHYLAMINO)CARBONYL]O	054266-44-3
1,3-OXATHIOLAN-4-ONE, 5-METHYL-, O-[(METHYLAMINO	054266-46-5
1,3-OXATHIOLAN-4-ONE, 5-(1-METHYLETHYL)-, O-[(ME	054266-83-0
1,3-OXATHIOLAN-4-ONE, 5-PROPYL-, O-[(METHYLAMINO	054266-80-7
2H-1,3,2-OXAZAPHOSPHORIN-2-AMINE,3-CHLORO-N,N-BI	087154-30-1
OXAZEPAM	000604-75-1
2-OXAZOLAMINE, 4,5-DIHYDRO-5-[[4-(2-PYRIMIDINYL)	120182-20-9
2-OXAZOLAMINE, 4,5-DIHYDRO-5-[[4-(2-PYRIDINYL)-1	120182-07-2
ISO-OXAZOLE	000288-14-2
OXAZOLE	000288-42-6
OXAZOLE-2-AMINE	004570-45-0
OXAZOLE, 2,5-DIPHENYL-	000092-71-7
2,4-OXAZOLIDINEDIONE, 3-PHENYL-	003759-90-8
OXAZOLIDINE, 2-[(5-NITRO-2-THIAZOLYL)IMINO]-	024240-60-6
OXAZOLIDINE, 4,4,5,5-TETRAMETHYL-2-[(5-NITRO-2-T	024229-59-2
2-OXAZOLIDINONE, 5-[[2-BENZYL-3-(4-MORPHOLINYLCA	131067-26-0
2-OXAZOLIDINONE DERIVATIVE	122225-00-7
2-OXAZOLIDINONE DERIVATIVE	127944-56-3
2-OXAZOLIDINONE DERIVATIVE	122224-80-0
2-OXAZOLIDINONE DERIVATIVE	122224-84-4
2-OXAZOLIDINONE DERIVATIVE	127944-53-0
2-OXAZOLIDINONE DERIVATIVE	127944-48-3
2-OXAZOLIDINONE DERIVATIVE	122224-83-3
2-OXAZOLIDINONE DERIVATIVE	127944-49-4
2-OXAZOLIDINONE DERIVATIVE	127944-51-8
2-OXAZOLIDINONE DERIVATIVE	127944-57-4
2-OXAZOLIDINONE DERIVATIVE	127944-54-1
2-OXAZOLIDINONE DERIVATIVE	122224-98-0
2-OXAZOLIDINONE DERIVATIVE	122224-71-9
2-OXAZOLIDINONE DERIVATIVE	122224-99-1
2-OXAZOLIDINONE DERIVATIVE	127944-55-2
2-OXAZOLIDINONE DERIVATIVE	127944-50-7
2-OXAZOLIDINONE, 3- (5-NITRO-2-FURANYL)METHYLEN	000067-45-8
3-OXAZOLIDINYLOXY, 2,2,5,5-TETRAMETHYL-	060577-35-7
OXAZOLO[4,5-C]PYRIDINE, 2-(2,4-DIMETHOXYPHENYL)-	131073-11-5
4H-OXAZOLO[4,5-D]AZEPIN-2-AMINE, 6-ETHYL-5,6,7,8	036067-73-9
OXEPANE	000592-90-5
N-OXIDENICOTINIC ACID	002398-81-4
OXINDOLE	000059-48-3
OXIRANE, BUTYL-	001436-34-6
OXIRANE, CYCLOHEXYL-	003483-39-4
OXIRANEMETHANOL	000556-52-5
OXOLINIC ACID	014698-29-4
OXOMEMAZINE	003689-50-7
2-OXOPROPANAL,1-METHYLSULFOXIDE	112740-56-4
2-OXOPROPANAL,1-METHYLSULFIDE	112740-61-1

4-OXO-4H-PYRAN-2,6-DICARBOXYLIC ACID	000099-32-1
OXPRENALOL	006452-71-7
2,2'-OXYBIS-3-BUTEN-1-OL	083682-68-2
1,1'-OXYBIS-3-BUTEN-2-OL	083682-67-1
OXYCARBOXIN	005259-88-1
OXYDEMETON METHYL	000301-12-2
OXYFLUORFEN	042874-03-3
OXYMORPHONE	000076-41-5
OXYPHENBUTAZONE	000129-20-4
OXYQUINOLINESULPHATE	000134-31-6
OXYTETRACYLCINE	000079-57-2
OXYTHIOQUINOX	002439-01-2
PACLOBUTRAZOL	076738-62-0
PALMITIC ACID	000057-10-3
D-PANTOTHENIC ACID	000079-83-4
PAPAVERINE	000058-74-2
PARAFFINS (C>10)	064771-71-7
PARAFFINS, C1-C10	064771-72-8
PARAFORMALDEHYDE	030525-89-4
PARALDEHYDE	000123-63-7
PARAMORPHINE	000115-37-7
PARAOXON	000311-45-5
PARAQUAT	004685-14-7
PARAROSANILINE	000569-61-9
PARATHION	000056-38-2
PARATHION-AMINO	003735-01-1
PARATHION METHYL	000298-00-0
PARTHENIN	000508-59-8
2,3'-PCB	025569-80-6
3,3',4,4'-PCB	032598-13-3
2,2',3'-PCB	038444-78-9
2,2',3,4,4'-PCB	035694-06-5
2,2',3,5',6-PCB	038379-99-6
2,2',6-PCB	038444-73-4
2,3',4',5-PCB	032598-11-1
2,4,4'-PCB	007012-37-5
2,3,4,4'-PCB	033025-41-1
2,3,4'-PCB	038444-85-8
2,3,3'-PCB	038444-84-7
2,3',5-PCB	038444-81-4
2,2',6,6'-PCB	015968-05-5
2,4,4',5-PCB	032690-93-0
2,2',5,5'-PCB	035693-99-3
2,2',3,3',4,4'-PCB	038380-07-3
2,2',3,4',5',6-PCB	038380-04-0
2,4,5,2',5'-PCB	037680-73-2
2,2',4,4',6,6'-PCB	033979-03-2
2,4,5,2',4',5'-PCB	035065-27-1
3,3',4,4',5,5'-PCB	032774-16-6
2,3,4,5,6-PCB	018259-05-7
2,3,4,5,6,2',3',5',6'-PCB	052663-77-1
2',3,4,4',5'-PCB	031508-00-6
2,3,4,5-PCB	033284-53-6
PEBULATE	001114-71-2
PEFLOXACIN	070458-92-3
PENBUTOLOL	038363-40-5
PENCILLAMINE	000052-67-5
PENCYCURON	066063-05-6

PENDIMETHALIN	040487-42-1
PENICILLIN, A-PHENOXYETHYL	001752-26-7
1-S-PENTABENZOTRIAZOLE	067246-30-4
2-PENTABENZOTRIAZOLE	069218-27-5
2-S-PENTABENZOTRIAZOLE	069218-32-2
1-PENTABENZOTRIAZOLE	069218-26-4
PENTABROMOBENZENE	000608-90-2
2,2',4,5,5'-PENTABROMOBIPHENYL	067888-96-4
PENTABROMOETHYLBENZENE	000085-22-3
PENTABROMOPHENOL	000608-71-9
PENTABROMOTOLUENE	000087-83-2
PENTACENE	000135-48-8
PENTACHLOROANILINE	000527-20-8
PENTACHLOROANISOLE	001825-21-4
PENTACHLOROBENZENE	000608-93-5
2,2',3,5,5'-PENTACHLOROBIPHENYL	052663-61-3
2,2',3,4,5'-PENTACHLOROBIPHENYL	038380-02-8
2,2',3,4,4'-PENTACHLOROBIPHENYL	065510-45-4
2,2',3,4',6-PENTACHLOROBIPHENYL	068194-05-8
2,2',3,3',6-PENTACHLOROBIPHENYL	052663-60-2
2,2',3',4,5-PENTACHLOROBIPHENYL	041464-51-1
2,2',4,4',5-PENTACHLOROBIPHENYL	038380-01-7
2,2',4,4',6-PENTACHLOROBIPHENYL	039485-83-1
2,3',4,5,5'-PENTACHLOROBIPHENYL	068194-12-7
2,3,3',4',6-PENTACHLOROBIPHENYL	038380-03-9
2,3,3',4,4'-PENTACHLORO-1,1'-BIPHENYL	032598-14-4
2,2',3,4,6-PENTACHLOROBIPHENYL	055215-17-3
2,2',3,4,5-PENTACHLOROBIPHENYL	052663-62-4
2,2',3,4,5-PENTACHLOROBIPHENYL	055312-69-1
1,1,2,3,4-PENTACHLOROBUTANE	077753-24-3
1,2,2,3,3-PENTACHLOROBUTANE	083293-82-7
1-I-PENTACHLOROCYCLOHEXANE	033489-28-0
PENTACHLOROCYCLOHEXENE	000643-15-2
PENTACHLOROCYCLOHEXENE	000319-94-8
3-PENTACHLOROCYCLOHEXANE	056994-23-1
1-PENTACHLOROCYCLOHEXANE	022138-39-2
PENTACHLOROCYCLOHEXENE	054083-25-9
PENTACHLOROCYCLOHEXENE	051795-30-3
PENTACHLOROCYCLOHEXENE	054083-24-8
2-F-PENTACHLOROCYCLOHEXANE	056086-55-6
1-BR-PENTACHLOROCYCLOHEXANE	036635-03-7
3-I-PENTACHLOROCYCLOHEXANE	033489-27-9
1,2,4,7,8-PENTACHLORODIBENZO-P-DIOXIN	058802-08-7
1,2,3,7,8-PENTACHLORODIBENZO-P-DIOXIN	040321-76-4
1,2,3,4,7-PENTACHLORODIBENZO-P-DIOXIN	039227-61-7
2,3,4,7,8-PENTACHLORODIBENZOFURAN	057117-31-4
PENTACHLOROETHANE	000076-01-7
PENTACHLORONITROBENZENE	000082-68-8
PENTACHLOROPHENOL	000087-86-5
PENTACHLOROPHENOL, NA SALT	000131-52-2
1,1,2,3,3-PENTACHLOROPROPANE	015104-61-7
PENTACHLORO(2,2,3,3-TETRAFLUOROPROPOXY)CYCLOTRI*	059700-57-1
2,3,4,5,6-PENTACHLOROTOLUENE	000877-11-2
PENTACHLOROTOLUENE	069911-61-1
2,3,4,5,6-PENTACHLORPYRIDINE	002176-62-7
PENTACOSANE	000629-99-2
1,1,1,3,3,5,7,7,9,11,11,13,15,15,15-PENTADECAM*	060617-40-5
N-PENTADECANE	000629-62-9

PENTADECANOIC ACID	001002-84-2
1-PENTADECANOL	000629-76-5
8-PENTADECANONE	000818-23-5
1-PENTADECENE	013360-61-7
N-PENTADECYLCYCLOHEXANE	006006-95-7
1,2-PENTADIENE	000591-95-7
1,3-PENTADIENE	000504-60-9
1,4-PENTADIENE	000591-93-5
1,3-PENTADIENE (E)	002004-70-8
1,3-PENTADIENE (Z)	001574-41-0
2,4-PENTADIENOIC ACID, 5-PHENYL-	001552-94-9
PENTAERYTHRITOL	000115-77-5
PENTAERYTHRITOL TETRANITRATE	000078-11-5
2,3,4,5,6-PENTAFLUOROACETANILIDE	000653-22-5
2,3,4,5,6-PENTAFLUOROANILINE	000771-60-8
PENTAFLUOROBENZENE	000363-72-4
1,1,1,2,2-PENTAFLUOROETHANE	000354-33-6
PENTAFLUOROETHYLBENZENE	000309-11-5
PENTAFLUOROPHENOL	000771-61-9
PENTAFLUOROPHENYL METHYL SUFONE	000651-85-4
1,1,1,2,2-PENTAFLUOROPROPANE	001814-88-6
PENTAFLUORO-1-PROPANOL	000422-05-9
3,5,7,3',4'-PENTAHYDROXYFLAVANONE	000480-18-2
N,N,2,4,6-PENTAMETHYLANILINE	013021-15-3
PENTAMETHYLBENZENE	000700-12-9
N,N-PENTAMETHYLENE CINNAMAMIDE	005422-81-1
PENTAMETHYLMELAMINE	016268-62-5
PENTAMETHYLPHENYLDISILOXANE	014920-92-4
2,2,4,6,8-PENTAMETHYL-4,6,8-TRIPHENYLCYCLOTETR*	010448-10-9
1-PENTANAL	000110-62-3
PENTANAMIDE, 4-AMINO-N-(2,6-DIMETHYLPHENYL)-	102089-68-9
PENTANAMIDE, 5-AMINO-N-(2,6-DIMETHYLPHENYL)-	102089-66-7
PENTANAMIDE, N-(3,4-DICHLOROPHENYL)-	002150-96-1
PENTANAMIDE, N-[[4-(DIMETHYLAMINO)SULFONYL]PHENY	075852-17-4
N-PENTANE	000109-66-0
PENTANE-1,5-DIAL	000111-30-8
PENTANEDINITRILE	000544-13-8
PENTANEDIOC ACID	000110-94-1
PENTANEDIOIC ACID, DIMETHYL ESTER	001119-40-0
1,5-PENTANEDIOL	000111-29-5
2,4-PENTANEDIONE	000123-54-6
PENTANE, DODECAFLUORO-	000678-26-2
1-PENTANETHIOL	000110-66-7
PENTANOIC ACID	000109-52-4
PENTANOIC ACID, 4-(2H-BENZOTRIAZOL-2-YL)-	115054-96-1
PENTANOIC ACID, 4-(1H-BENZOTRIAZOL-1-YL)-	115054-91-6
PENTANOIC ACID, 5-[(1,1'-BIPHENYL)-4-YL]-3-METHY	007320-97-0
PENTANOIC ACID, 5-[(1,1'-BIPHENYL)-4-YL]-5-OXO-	051994-35-5
PENTANOIC ACID, 2-HYDROXY-	000617-31-2
PENTANOIC ACID, 2-METHYL-	000097-61-0
PENTANOIC ACID, 5-(1-NAPHTHALENYLOXY)-	101705-35-5
PENTANOIC ACID, 4-OXO-, METHYL ESTER	000624-45-3
2-PENTANOL	006032-29-7
1-PENTANOL	000071-41-0
3-PENTANOL	000584-02-1
2-PENTANOL ACETATE	000626-38-0
2-PENTANONE	000107-87-9
3-PENTANONE	000096-22-0

1-PENTANONE, 1-(3-FURANYL)-4-HYDROXY-	032954-58-8
1-PENTANONE, 1-(3-FURANYL)-4-METHYL-	000553-84-4
PENTANONE,PYRROL-2-YL	089789-53-7
3-PENTANONE, 2,2,4,4-TETRAMETHYL-	000815-24-7
1,4,7,10,13-PENTAOXACYCLOPENTADECANE	033100-27-5
N-PENTATRIACONTANE	000630-07-9
2-PENTENE	000109-68-2
1-PENTENE	000109-67-1
CIS-2-PENTENE	000627-20-3
TRANS-2-PENTENE	000646-04-8
4-PENTENOIC ACID, 2-[[(AMINOCARBONYL)AMINO]CARBO	073632-82-3
2-PENTENOIC ACID, 2-CYANO-3-[[(4-NITROPHENYL)MET	097874-69-6
3-PENTEN-2-OL	001569-50-2
1-PENTEN-3-OL	000616-25-1
4-PENTEN-1-OL	000821-09-0
3-PENTEN-2-ONE	000625-33-2
PENTOBARBITAL	000076-74-4
PENTOSTATIN	063677-95-2
2-I-PENTOXY-4-AMINOBENZOIC ACID	061566-64-1
2-PENTOXY-4-AMINOBENZOIC ACID	061566-63-0
M-PENTOXYPHENOL	018979-73-2
TERT-PENTYLACETATE	000625-16-1
9-PENTYLADENINE	002002-36-0
N-PENTYLAMINE	000110-58-7
PENTYLBENZENE	000538-68-1
PENTYL BUTYRATE	000540-18-1
O-PENTYL CARBAMATE	000638-42-6
PENTYLCYCLOHEXANE	004292-92-6
PENTYLENETETRAZOLE	000054-95-5
B-PENTYLGALACTOPYRANOSIDE	039824-10-7
4-(1-PENTYLHEPTYL)BENZENESULFONIC ACID, SODIUM *	002212-52-4
PENTYLMALONIC ACID	000616-88-6
I-PENTYL-(345-MEO)PHENYL-SULFONE	071203-61-7
O-PENTYL-N-METHYLCARBAMATE	002594-17-4
4-PENTYLPHENOL	014938-35-3
N-PENTYL N-PHENYLCARBAMATE	063075-06-9
4-PENTYLPYRAZOLE	052222-71-6
4-PENTYLSEMICARBAZIDE	050405-18-0
N-PENTYL 4-SULFAMYLBENZAMIDE	059777-64-9
PENTYL 4-SULFAMYLBENZOATE	059777-60-5
4-PENTYNAMIDE, N-[4-[(DIMETHYLAMINO)SULFONYL]PHE	077812-88-5
1-PENTYNE	000627-19-0
2-PENTYNE	000627-21-4
3-PEO-5-BR-6-ME-PYRIMIDIN-2,4-DIONE	077317-75-0
PERAZINE	000084-97-9
PERCHLOROALLENE	018608-30-5
2-PERCHLOROBUTENE	002482-68-0
PERCHLOROHEXANE	083682-34-2
PERCHLOROISOBUTANE	006820-74-2
PERCHLOROMETHYL MERCAPTAN	000594-42-3
PERCHLOROPROPANE	000594-90-1
PERCHLOROPROPYLENE	001888-71-7
PERFLUIDONE	037924-13-3
PERFLUOROCYCLOBUTANE	000115-25-3
PERFLUOROCYCLOHEXANE	000355-68-0
PERFLUORO-2,9-DIMETHYLDECANE	103188-55-2
PERFLUORO-2,7-DIMETHYLOCTANE	003021-63-4
PERFLUOROHEPTANE	000335-57-9

PERFLUORO-N-HEXANE	000355-42-0
PERFLUOROMETHYLCYCLOHEXANE	000355-02-2
PERICYAZINE	002622-26-6
CIS-PERMETHRIN	061949-76-6
PERMETHRIN	052645-53-1
TRANS-PERMETHRIN	061949-77-7
PEROXIDE, BIS(1-METHYL-1-PHENYLETHYL)	000080-43-3
PEROXIDE-2-BUTANONE	001338-23-4
PEROXYACETIC ACID	000079-21-0
PEROXYACETYLNITRATE	002278-22-0
PERPHENAZINE	000058-39-9
PERUVOSIDE	001182-87-2
PERYLENE	000198-55-0
PH-ALANINE,N-ACETYL,N'-MEAM-AMIDE	017186-60-6
N-PH-ANTHRANILIC ACID,2',3'-DIMETHYL	000061-68-7
N-PH-N(4-TBUBNZYL)-N(NNDIME)ET AMINE	070785-07-8
1-PH-3,5-DIME-4-NITROSOPYRAZOLE	000715-99-1
BETA-PHELLANDRENE	000555-10-2
α-PHELLANDRENE	000099-83-2
PHENACEMIDE	000063-98-9
P-PHENACETIN	000062-44-2
1H-PHENALENE	000203-80-5
PHENALLYMAL	000115-43-5
9-PHENANTHRENAMINE	000947-73-9
4A(2H)-PHENANTHRENAMINE, 1,3,4,9,10,10A-HEXAHYDR	132453-06-6
A(4H)-PHENANTHRENAMINE, 1,9,10,10A-TETRAHYDRO-9	148727-67-7
PHENANTHRENE	000085-01-8
PHENANTHRENE, 2-CHLORO-	024423-11-8
9,10-PHENANTHRENEDIOL, 1,2,3,4,4A,9,10,10A-OCTAH	053446-99-4
9,10-PHENANTHRENEDIOL, 1,2,3,4,4A,9,10,10A-OCTAH	053446-97-2
9,10-PHENANTHRENEDIOL, 1,2,3,4,4A,9,10,10A-OCTAH	053446-96-1
9,10-PHENANTHRENEDIONE	000084-11-7
3-PHENANTHRENOL, 1,2,3,4,4A,9,10,10A-OCTAHYDRO-4	132452-02-9
9(1H)-PHENANTHRENONE, 2,3,4,4A,10,10A-HEXAHYDRO-	148727-74-6
PHENANTHRIDINE	000229-87-8
1,9-PHENANTHRIDINEDIOL, 5,6,6A,7,8,9,10,10A-OCTA	078513-74-3
1,9-PHENANTHRIDINEDIOL, 5,6,6A,7,8,9,10,10A-OCTA	074430-97-0
1,9-PHENANTHRIDINEDIOL, 5,6,6A,7,8,9,10,10A-OCTA	071048-87-8
1,7-PHENANTHROLINE	000230-46-6
O-PHENANTHROLINE	000066-71-7
4,7-PHENANTHROLINE	000230-07-9
PHENAZINE	000092-82-0
PHENAZOPYRIDINE	000094-78-0
PHENETHICILLIN	000147-55-7
PHENETHYLAMINE,N-ME-A-ISOBUTYL	084952-62-5
PHENETHYLAMINE,N-METHYL-A-ETHYL	084952-60-3
PHENETHYLAMINE,N-METHYL-A-ISOPROPYL	084952-61-4
1-PHENETHYLISOTHIOCYANATE	004478-92-6
2-PHENETHYLISOTHIOCYANATE	002257-09-2
PHENETHYL SULFAMIDE	000710-15-6
1(2-PHENETS)-B-GALACTOPYRANOSIDE	063407-54-5
PHENFORMIN	000114-86-3
PHENKAPTON	002275-14-1
PHENMEDIPHAM	013684-63-4
PHENOBARBITAL	000050-06-6
P-PHENO-BENZYL N,N-DIME CARBAMATE	084640-36-8
PHENOL	000108-95-2
PHENOL, 3-AMINO-	000591-27-5

PHENOL, 4-AMINO-	000123-30-8
PHENOL, 2-AMINO-4-CHLORO-	000095-85-2
PHENOL, 2-AMINO-4-METHYL-	000095-84-1
PHENOL, 4-AMINO-2-NITRO-	000119-34-6
PHENOL, 4- (4-AMINOPHENYL)SULFONYL -	025963-47-7
PHENOL, 2,6-BIS(1-METHYLETHYL)-	002078-54-8
PHENOL, 2-CHLORO-5-METHYL-	000615-74-7
PHENOL, 2,6-DIETHYL-	001006-59-3
PHENOL, 2,6-DIMETHYL-, ACETATE	000876-98-2
PHENOL, 4-[2-(DIMETHYLAMINO)ETHOXY]-2-METHYL-5-(	035231-36-8
PHENOL, 3-(DIMETHYLAMINO)-	000099-07-0
PHENOL, 5-(1,1-DIMETHYLBUTYL)-2-(3-HYDROXYCYCLOH	132296-18-5
PHENOL, 5-(1,1-DIMETHYLETHYL)-2-(3-HYDROXYCYCLOH	132296-16-3
PHENOL, 3-(1,1-DIMETHYLETHYL)-	000585-34-2
PHENOL, 5-(1,1-DIMETHYLHEPTYL)-2-(3-HYDROXYCYCLO	114753-51-4
PHENOL, 5-(1,1-DIMETHYLHEPTYL)-2-(5-HYDROXY-2-ME	132339-34-5
PHENOL, 5-(1,1-DIMETHYLHEPTYL)-2-(5-HYDROXY-2-ME	132296-14-1
PHENOL, 5-(1,1-DIMETHYLHEPTYL)-2-[5-HYDROXY-2-(4	132296-15-2
PHENOL, 5-(1,1-DIMETHYLHEPTYL)-2-[5-HYDROXY-2-(3	083002-04-4
PHENOL, 5-(1,1-DIMETHYLHEXYL)-2-(3-HYDROXYCYCLOH	132296-20-9
PHENOL, 5-(1,1-DIMETHYLNONYL)-2-(3-HYDROXYCYCLOH	132296-12-9
PHENOL, 5-(1,1-DIMETHYLOCTYL)-2-(3-HYDROXYCYCLOH	132296-11-8
PHENOL, 5-(1,1-DIMETHYLPENTYL)-2-(3-HYDROXYCYCLO	132296-19-6
PHENOL, 5-(1,1-DIMETHYLUNDECYL)-2-(3-HYDROXYCYCL	132296-13-0
PHENOL, 2,4-DINITRO-, POTASSIUM SALT	014314-69-3
PHENOL, 4-DODECYL-	000104-43-8
PHENOL, 2,6-DIFLUORO-	028177-48-2
PHENOL, 3-ISOTHIOCYANATO-	003125-63-1
PHENOL,2,2'-METHYLENEBIS 3,4,6-CL	000070-30-4
PHENOL,2,2'-METHYLENEBIS 4-CHLORO-	000097-23-4
PHENOL, 2-METHYL-5-NITRO-	005428-54-6
PHENOL, 2-METHYL-4-NITROSO-	006971-38-6
PHENOL, 3-NONYL-	000139-84-4
PHENOL, 2-NONYL-	000136-83-4
PHENOL, 4-(PENTYLOXY)-	018979-53-8
PHENOL, 4-PHENOXY-	000831-82-3
PHENOL, 4-(PHENYLAMINO)-	000122-37-2
PHENOLPHTHALEIN	000077-09-8
PHENOL, 4-PROPOXY-	018979-50-5
PHENOL, SODIUM SALT	000139-02-6
PHENOLSULPHONEPHTHALEIN	000143-74-8
PHENOL, 4,4'-THIOBIS-	002664-63-3
PHENOL, 3-TRIDECYL-	072424-02-3
PHENOL, 2-TRIDECYL-	096850-25-8
PHENOL, 4-TRIDECYL-	052780-43-5
PHENOL, P-(TRIMETHYL SILYL)-	013132-25-7
PHENOL, 2-UNDECYL-	020056-71-7
PHENOL, 4-UNDECYL-	020056-73-9
PHENOL, 3-UNDECYL-	020056-72-8
PHENOPYLATE	040575-34-6
PHENOTHIAZINE	000092-84-2
10H-PHENOTHIAZINE-10-ETHANAMINE, N,N-DIETHYL-	000060-91-3
10H-PHENOTHIAZINE-10-PROPANAMINE,2-CHLORO-	002095-17-2
10H-PHENOTHIAZINE-10-PROPANAMINE, 2-CHLORO-N-MET	001225-64-5
10H-PHENOTHIAZIN-2-OL, 8-CHLORO-10-[3-(DIMETHYLA	003926-67-8
PHENOTHIOXIN	000262-20-4
PHENOTHRIN	026002-80-2
PHENOXAZINE	000135-67-1

PHENOXYACETAMIDE	000621-88-5
PHENOXYACETANILIDE	018705-01-6
PHENOXYACETIC ACID	000122-59-8
PHENOXYACETIC ACID, M-ACETAMIDO-	006339-04-4
PHENOXYACETIC ACID, M-BENZAMIDO	001878-89-3
PHENOXYACETIC ACID, M-CARBOXY	001878-61-1
PHENOXYACETIC ACID, 4-CL-3-NO2	089894-13-3
PHENOXYACETIC ACID, 3-UREA	006169-23-9
PHENOXYACETIC ACID,4-CYCLOHEXYL	001878-56-4
N-PHENOXYACETYLMORPHOLINE	018495-00-6
N-PHENOXYACETYLPIPERIDINE	036405-75-1
O-PHENOXYANILINE	002688-84-8
P-PHENOXYANILINE	000139-59-3
O-PHENOXYANISOLE	001695-04-1
3-PHENOXYBENZALDEHYDE	039515-51-0
O-PHENOXYBENZOIC ACID	002243-42-7
M-PHENOXYBENZOIC ACID	003739-38-6
P-PHENOXYBENZOIC ACID	002215-77-2
2-PHENOXYETHANOL	000122-99-6
3-(2-PHENOXYETHOXY)PHENOL	036429-48-8
1-(2-(PHENOXY)ETHYL)2-NO2IMIDAZOLE	071006-77-4
PHENOXYMETHYLCEPHALOSPRIN	010390-44-0
PHENOXYMETHYLPENICILLIN	000087-08-1
3-PHENOXYPHENOL	000713-68-8
(4-PHENOXYPHENYL)UREA	078508-44-8
B-PHENOXYPROPIONITRILE	003055-86-5
1-(3-PHENOXYPROPYL)-2-NO2 -IMIDAZOLE	071006-74-1
2-PHENOXYPYRIDINE	004783-68-0
1-PHENOXYSILATRANE	051466-74-1
PHENPROCOUMON	000435-97-2
1(3-PHENPROPYL)-B-GALACTOPYRANOSIDE	050615-70-8
PHENTERMINE	000122-09-8
PHENYLACETALDEHYDE	000122-78-1
PHENYLACETAMIDE	000103-81-1
1-PHENYL-2-ACETAMIDOCYCLOPROPANE	038954-41-5
2-PHENYLACETANILIDE	002113-47-5
PHENYL ACETATE	000122-79-2
PHENYLACETIC ACID, METHYL ESTER	000101-41-7
PHENYLACETIC ACID	000103-82-2
PHENYLACETIC ACID, 3-MEO,4-PHMEO	029973-91-9
PHENYLACETIC ACID, 3-MEO-4-IPRO	088449-50-7
PHENYLACETIC ACID,3-BR-4-PHMEO	005884-48-0
PHENYLACETIC ACID,2-(2'-CHLORO-6'-BROMO)ANILINO	127792-23-8
PHENYLACETIC ACID,2-(2'-CHLORO-6'-FLUORO)ANILINO	100754-93-6
PHENYLACETIC ACID,2-(2'-CL-6-ME)ANILINO	023189-28-8
PHENYLACETIC ACID,3-CL-4-PHENYLETO	077269-58-0
PHENYLACETIC ACID,3-CL-4-PHMEO	060736-83-6
PHENYLACETIC ACID,3-CL-4-PHPRO	077269-59-1
PHENYLACETIC ACID,2-(2',5'-DICHLORO)ANILINO	127792-31-8
PHENYLACETIC ACID,2-(2',6'-DICHLORO-3-METHYL)ANI	015307-71-8
PHENYLACETIC ACID,2-(2',3'-DICHLORO)ANILINO	070172-32-6
PHENYLACETIC ACID,2-(3',4'-DICHLORO)ANILINO	127792-33-0
PHENYLACETIC ACID,2-(2',6'-DICHLORO-4'-HYDROXY)A	064118-84-9
PHENYLACETIC ACID,2-(2'-ME-3'-CL)ANILINO	037984-36-4
PHENYLACETIC ACID,3-MEO-4-CYHXMEO	075221-43-1
PHENYLACETIC ACID,3-ME-4-PHMEO	064360-42-5
PHENYLACETIC ACID,P-PHENYLMETHOXY	006547-53-1
PHENYLACETONITRILE	000140-29-4

2-PHENYLACETOPHENONE	000451-40-1
N-PHENYL-3-ACETYLAMINOSUCCINIMIDE	030820-34-9
PHENYLACETYL CHLORIDE	000103-80-0
PHENYLACYL P-TOLYL SULFONE	031378-03-7
1-PHENYLADAMANTANE	030176-62-6
8-PHENYLADENINE	017720-22-8
L-PHENYLALANINAMIDE, N-ACETYLGLYCYL-	034017-16-8
L-PHENYLALANINAMIDE, N-ACETYL-L-ALANYL-L-TYROSYL	132766-21-3
D-PHENYLALANINAMIDE, N-ACETYL-D-PHENYLALANYL-D-P	138771-70-7
L-PHENYLALANINAMIDE, N-ACETYL-L-TRYPTOPHYLGLYCYL	132766-16-6
L-PHENYLALANINAMIDE, N-ACETYL-L-LEUCYL-L-SERYL-	132766-18-8
L-PHENYLALANINAMIDE, N-ACETYL-L-ALANYLGLYCYL-	132766-09-7
L-PHENYLALANINAMIDE, N-ACETYL-L-TYROSYL-	052329-50-7
L-PHENYLALANINAMIDE, N-ACETYL-L-SERYL-	132765-88-9
L-PHENYLALANINAMIDE, N -ACETYL-L-ASPARAGINYL-	132765-99-2
D-PHENYLALANINAMIDE, N-ACETYL-D-PHENYLALANYL-	024809-26-5
L-PHENYLALANINAMIDE, N2-ACETYL-L-GLUTAMINYL-	132765-93-6
DL-PHENYLALANINE	000150-30-1
4-CH3COO-PHENYLALANINE	006636-22-2
PHENYLALANINE	000063-91-2
PHENYLALANINE-N-ACETYL,METHYL ESTER	021156-62-7
L-PHENYLALANINE, N-ACETYL-	002018-61-3
PHENYLALANINE-AMIDE,N-ACETYL	007376-90-1
PHENYLALANYLPHENYLALANAMIDE	015893-46-6
N-PHENYL O-AMINOBENZOIC ACID	000091-40-7
PHENYL-4-AMINOSALICYLATE	000133-11-9
P-PHENYL-AMPHENICOL	023885-56-5
PHENYLAMPHENICOL	049648-47-7
9-PHENYLANTHRACENE	000602-55-1
N-PHENYLANTHRANILIC ACID, SODIUM SALT	006232-32-2
N-PHENYLANTHRANILIC ACID,3'-NITRO	027693-70-5
P-PHENYLAZOANILINE	000060-09-3
4-(PHENYLAZO)BENZENESULFONIC ACID, SODIUM SALT	042975-18-8
2-(PHENYLAZO)-M-CRESOL	029418-38-0
4-(PHENYLAZO)-1-NAPHTHALENAMINE	000131-22-6
1-(PHENYLAZO)-2-NAPHTHALENOL	000842-07-9
2-(PHENYLAZO)PHENOL	002362-57-4
P-PHENYLAZOPHENOL	001689-82-3
2-(4-(PHENYLAZO)PHENOXY)ETHANOL	092245-57-3
N-(4-(PHENYLAZO)PHENYL)ACETAMIDE	004128-71-6
PHENYL AZOXYCYANIDE	054797-20-5
PHENYLBENZENESULFONATE	004358-63-8
2-PHENYLBENZIMIDAZOLE	000716-79-0
PHENYL BENZOATE	000093-99-2
7-PHENYL-7H-BENZO(C)FLUORENE	032377-10-9
2-PHENYLBENZTHIAZOLE	000883-93-2
4-PHENYL-6-BROMOQUINAZOLIN-2-ONE	033443-53-7
4-PHENYLBUTANOL	003360-41-6
PHENYLBUTAZONE	000050-33-9
4-PHENYLBUTYLAMINE	013214-66-9
4-PHENYLBUTYRIC ACID	001821-12-1
3-PHENYL-N-BUTYRIC ACID	004593-90-2
6-PHENYLCAPROIC ACID	005581-75-9
O-PHENYL CARBAMATE	000622-46-8
3-PHENYL-CARBAMOYLOXY-1-PROPYNE	003567-38-2
3-PHENYL-1-CHLOROPROPANE	000104-52-9
1-PHENYL-5-CHLOROTETRAZOLE	014210-25-4
2-PHENYLCINCHONINIC ACID	000132-60-5

1-PHENYL-3-CYANOGUANIDINE	041410-39-3
PHENYL CYCLOHEXANE	000827-52-1
1-PHENYLCYCLOHEXENE	031017-40-0
3-PHENYL-1-CYCLOHEXYLUREA	000886-59-9
1-PHENYLDECANE	000104-72-3
PHENYL DICHLOROPHOSPHATE	000770-12-7
PHENYLDIETHANOLAMINE	000120-07-0
o-PHENYL N,N'-DIMETHYL PHOSPHORODIAMIDATE	001754-58-1
1-PHENYL-3,3-DIMETHYLTRIAZINE OXIDE	059477-92-8
2-PHENYL-1,1-DIOXO-THIAZOL-5-ONE	027052-09-1
7-PHENYL-4,6-DIYN-HEPT-2-EN-1-OL	013641-62-8
2,2'-(1,4-PHENYLENE)BIS(5-PHENYLOXAZOLE)	001806-34-4
O-PHENYLENE CARBAMATE	000059-49-4
O-PHENYLENETHIOUREA	000583-39-1
O-PHENYLENE UREA	000615-16-7
PHENYLEPHRINE	000061-76-7
1-PHENYLETHANOL	001517-69-7
2-PHENYLETHANOL	000060-12-8
PHENYLETHANOLAMINE	007568-93-6
B-PHENYLETHYL ACETATE	000103-45-7
2-PHENYLETHYLAMINE	000064-04-0
N-PHENYL ETHYLCARBAMATE	000101-99-5
B-PHENYL ETHYL CHLORIDE	000622-24-2
B-PHENYLETHYLPYRIDINIUM BROMIDE	006324-18-1
PHENYL ETHYL SULFIDE	000622-38-8
1-PHENYL-3-ETHYL THIOUREA	002741-06-2
1-PHENYL-3-ETHYLUREA	000621-04-5
9-PHENYL-9H-FLUORENE	000789-24-2
9-PHENYL-9H-FLUORENE-4-CARBONITRILE	032377-09-6
PHENYL FORMATE	001864-94-4
5-PHENYL-2,4-FURANDIONE	022609-88-7
PHENYL GLYCEROL	000538-43-2
N-PHENYLGLYCINE	000103-01-5
ALPHA-PHENYLGLYCINE	000069-91-0
PHENYL GLYDIDYL ETHER	000122-60-1
2-PHENYLGUANIDINE	002002-16-6
1-PHENYLHEPTANE	001078-71-3
1-PHENYL-1,3,5-HEPTATRIYNE	004300-27-0
7-PHENYLHEPT-2-ENE-4,6-DIYNAL	020252-42-0
PHENYLHIPPURATE	002979-54-6
PHENYL HYDRAZINE	000100-63-0
PHENYL 2-HYDROXYBENZOATE	000118-55-8
PHENYLHYDROXYLAMINE	000100-65-2
3-PHENYLHYDROXYUREA	007335-35-5
N-PHENYL-2,5-IMIDAZOLIDIONE	002221-13-8
N-PHENYL-2,4-IMIDAZOLIDIONE	015414-78-5
5-PHENYL-2,4-IMIDAZOLIDIONE	000089-24-7
N-PHENYL-2,4,5-IMIDAZOLITRIONE	002211-33-8
2-PHENYL-1H-INDENE-1,3(2H)-DIONE	000083-12-5
2-PHENYLINDOLIZINE	025379-20-8
2-PHENYL-4-I-PROPYLMORPHOLINE	023222-62-0
PHENYL ISOCYANATE	000103-71-9
PHENYLISOCYANIDE DICHLORIDE	000622-44-6
2-PHENYLISONIAZID	058481-06-4
N-PHENYL-2,4-ISOOXAZOLIDIONE	005305-00-0
2-PHENYL ISOPROPANOL	000617-94-7
PHENYLISOTHIOCYANATE	000103-72-0
B-PHENYL-B-MEO-N-ET PYRROLIDINE	006577-49-7

S-PHENYLMERCAPTOACETIC ACID	000103-04-8
PHENYLMERCURIC ACETATE	000062-38-4
PHENYL MERCURIC CHLORIDE	000100-56-1
PHENYLMERCURIC HYDROXIDE	000100-57-2
PHENYLMERCURY DIMETHYL DITHIOCARBAMATE	032407-99-1
PHENYLMERCURY MONOETHANOL AMMONIUM ACETATE	005822-97-9
PHENYLMETHANESULFONATE	016156-59-5
N-PHENYL METHANESULFONAMIDE	001197-22-4
9-(PHENYLMETHYL)-9H-FLUORENE	001572-46-9
PHENYLMETHYL MERCURY	021392-61-0
3-PHENYL-1-METHYL-1-METHOXYUREA	001576-17-6
PHENYL METHYL SELENIDE	004346-64-9
1-PHENYL-3-METHYLUREA	001007-36-9
N7-PHENYL-MITOMYCIN C	014896-01-6
N-PHENYLMORPHOLINE	000092-53-5
2-PHENYLMORPHOLINE	023972-41-0
N-PHENYL-2-NAPHTHYLAMINE	000135-88-6
4-PHENYL-2-NITROPHENOL	000885-82-5
4-PHENYL-N-NITROSOPIPERIDINE	006652-04-6
PHENYL(2-NO2-1-IMIDAZOLYL)ET)SULFONE	071006-80-9
N-PHENYL-3,5-OXADIAZOLIDIONE	033101-81-4
5-PHENYL-2,4-OXAZOLIDIONE	005841-63-4
1-PHENYLPENTADECANE	002131-18-2
5-PHENYLPENTANOIC ACID	002270-20-4
5-PHENYL-2-PENTANONE	002235-83-8
2-PHENYLPHENOL	000090-43-7
P-PHENYLPHENOL	000092-69-3
3-PHENYLPHENOL	000580-51-8
2-PHENYLPHENOL, SODIUM SALT	000132-27-4
O-PHENYLPHENOXYACETIC ACID	005348-75-4
M-PHENYLPHENOXYACETIC ACID	001878-57-5
ALPHA-PHENYL-N-PHENYLNITRONE	001137-96-8
PHENYLPHOSPHONIC ACID	001571-33-1
3-PYRIDINECARBONITRILE, 5-PHENYL-2-(1-PIPERAZINY	108611-33-2
1-PHENYLPIPERAZINE	000092-54-6
N-PHENYLPIPERIDINE	004096-20-2
N-PHENYL-3-N'-PIPERIDINOACETAMIDE	004671-97-0
3-PHENYLPROPANOL	000122-97-4
PHENYLPROPANOLAMINE	014838-15-4
1-PHENYL-2-PROPANONE	000103-79-7
3-PHENYL-2-PROPENAL	000104-55-2
TRANS-1-PHENYL-1-PROPENE	000873-66-5
3-PHENYL-2-PROPENENITRILE	001885-38-7
3-PHENYLPROP-2-ENYL CINNAMATE	000122-69-0
W-PHENYLPROPIONALDEHYDE SEMICARBIZONE	027843-08-9
W-PHENYLPROPIONALDEHYDE OXIME	001197-50-8
BETA-PHENYLPROPIONIC ACID	000501-52-0
ALPHA-PHENYLPROPIONIC ACID	000492-37-5
3-PHENYL PROPYLAMINE	002038-57-5
G-PHENYLPROPYLCYANIDE	002046-18-6
PHENYLPROPYL ETHER	000622-85-5
G-PHENYL PROPYLIODINE	004119-41-9
G-PHENYLPROPYLPYRIDINIUM BROMIDE	053394-58-4
PHENYLPROPYL-TRIME-AMMONIUM IODIDE	002125-48-6
1-PHENYL-3-PYRAZOLIDINONE	000092-43-3
4-PHENYLPYRIDINE	000939-23-1
2-PHENYLPYRIDINE	001008-89-5
3-PHENYLPYRIDINE	001008-88-4

PHENYL-A-PYRIDYLCARBINOL	014159-57-0
PHENYL-B-PYRIDYL KETONE	005424-19-1
PHENYL-A-PYRIDYLKETONE	000091-02-1
1-PHENYLPYRROLIDINE	004096-21-3
5-PHENYL-2,4-PYRROLIDIONE	019860-27-6
3-PHENYL-2,4,5-PYRROLITRIONE	006476-18-2
4-PHENYLQUINAZOLINE	017629-01-5
2-PHENYLQUINOLINE	000612-96-4
N-PHENYLSUCCINIMIDE	000083-25-0
PHENYLSULFAMIDE	015959-53-2
N1-PHENYLSULFANILIDE	000127-77-5
PHENYLSULFONYL(P-METHOXYBENZOYL)METHANE	027918-37-2
8-PHENYLSULFONYLQUINOLINE	089770-33-2
6-PHENYLSULFONYLQUINOLINE	089770-30-9
PHENYLSULFUR PENTAFLUORIDE	002557-81-5
5'-PHENYL-1,1':3',1"-TERPHENYL	000612-71-5
1-PHENYLTETRADECANE	001459-10-5
4-PHENYL-1,2,3-THIADIAZOLE	025445-77-6
PHENYLTHIOCYANATE	005285-87-0
1-PHENYLTHIO-B-GALACTOPYRANOSIDE	016758-34-2
2-PHENYLTHIOPHENE	000825-55-8
3-PHENYLTHIOPYRIDINE	028856-77-1
N-PHENYLTHIOUREA	000103-85-5
PHENYL-O-TOLYLCARBINOL	005472-13-9
2-PHENYL-S-TRIAZINE	001722-18-5
PHENYLTRICHLOROSILANE	000098-13-5
1-PHENYLTRIDECANE	000123-02-4
N-PHENYL-3-(TRIFLUOROMETHYL)BENZENAMINE	000101-23-5
PHENYL TRIMETHYL AMMONIUM IODIDE	000098-04-4
1-PHENYLUNDECANE	006742-54-7
PHENYLUREA	000064-10-8
PHENYL XYLYLMETHANE	013540-50-6
PHENYTOIN	000057-41-0
B-PH-B-HYDROXY-N-ETHYLPYRROLIDINE	005407-61-4
PHNEYLACETIC ACID,2-(2'-CHLORO-6'-IODO)ANILINO	127792-24-9
PHNEYLACETIC ACID,2-(2',6'-DICHLORO-4-METHOXY)AN	118409-80-6
PHNEYL-P-TOLYLCARBINOL	001517-63-1
3-PHO BENZALDEHYDE,O-((MEAM)CO)OXIME	085879-19-2
1-(3-PHO-2-OH-PROPYL)-2-NO2-IMIDAZOLE	021788-11-4
PHORATE	000298-02-2
PHORATE, O-ANALOG	002600-69-3
PHORATE SULFONE	002588-04-7
PHORATE SULFOXIDE	002588-05-8
PHORONE	000504-20-1
PHOSALONE	002310-17-0
PHOSGENE	000075-44-5
PHOSMET	000732-11-6
PHOSPHAMIDON	000297-99-4
PHOSPHINE OXIDE, TRIS-(1-PYRROLIDINYL)-	006415-07-2
PHOSPHINIC ACID, BIS(1-AZIRIDINYL)-, 2-(2-ETHOXY	101347-40-4
PHOSPHINIC ACID,BIS(DIMEAZ),ET EST	014984-65-7
PHOSPHINIC ACID, (2,2-DIETHOXYETHYL)-, DIETHYL E	007598-61-0
PHOSPHINIC ACID, METHYL-, ETHYL ESTER	016391-07-4
PHOSPHINIC AMIDE, N,N-DIMETHYL-P,P-BIS(1-PYRROLI	053439-65-9
PHOSPHINOTHIOIC ACID, BIS(1-AZIRIDINYL)-, O-[2-(	101347-42-6
PHOSPHONIC ACID, METHYL-, DIISOPROPYL ESTER	001445-75-6
PHOSPHONIC ACID, [[4-(2-BENZOTHIAZOLYL)PHENYL]ME	075889-62-2
PHOSPHONIC ACID, BIS(1-METHYLETHYL) ESTER	001809-20-7

PHOSPHONIC ACID, METHYL-, DIMETHYL ESTER	000756-79-6
PHOSPHONIC ACID, METHYL- DIPENTYL ESTER	001000-36-8
PHOSPHONIC ACID, (2-OXOPROPYL)-, DIMETHYL ESTER	004202-14-6
PHOSPHONIC DIAMIDE, PENTAMETHYL-	002511-17-3
PHOSPHONIC DIAMIDE, N,N,N',N'-TETRAETHYL-P-(1-PY	069981-38-0
PHOSPHONIC DIAMIDE, N,N,N',N'-TETRAMETHYL-P-1-PY	040725-71-1
PHOSPHONIC ACID, ETHYL-, DIETHYL ESTER	000078-38-6
PHOSPHONOACETIC ACID	004408-78-0
PHOSPHONOTHIOIC ACID, METHYL-, DIETHYL ESTER	006996-81-2
PHOSPHONOTHIOIC ACID, METHYL-, O,S-DIETHYL ESTER	002511-10-6
PHOSPHORAMIDIC ACID, N-BUTYL-, O-METHYL O-(2,4,5	002213-87-8
PHOSPHORAMIDIC ACID, N-(1,1-DIMETHYLETHYL)-, O-	002214-33-7
PHOSPHORAMIDIC ACID, N-(2-METHYLPROPYL)-, O-METH	002213-88-9
PHOSPHORAMIDIC ACID, N-(1-METHYLPROPYL)-, O-METH	002388-47-8
PHOSPHORAMIDIC ACID, (4-METHYL-1,3-DITHIOLAN-2-Y	000950-10-7
PHOSPHORAMIDOTHIOIC ACID, O,O-DIMETHYL ESTER	017321-47-0
PHOSPHORIC ACID,2-CHLORO-(2,4-DICHLOROPHENYL)VIN	002274-67-1
PHOSPHORIC TRIAMIDE, N,N-DIETHYL-N',N',N'',N''-TET	051754-90-6
PHOSPHORIC TRIAMIDE, HEXAETHYL-	002622-07-3
PHOSPHORODITHIOIC ACID, O,O,S-TRIMETHYL ESTER	002953-29-9
PHOSPHOROHYDRAZIDIC ACID, DIPHENYL ESTER	033862-44-1
PHOSPHOROTHIOIC ACID, S-[[(1,1-DIMETHYLETHYL)SUL	056070-15-6
PHOSPHOROTHIOIC ACID, O,O-DIETHYL S-[(ETHYLSULFO	002588-06-9
PHOSPHOROTHIOIC ACID, S-[[(1,1-DIMETHYLETHYL)SUL	056165-57-2
PHOSPHOROTHIOIC ACID, O,O,S-TRIMETHYL ESTER	000152-20-5
PHOXIM	014816-18-3
4-PH-1-PHTHALAZINAMINE,N,N-DIME	023099-85-6
3-PH PROPIONALDEHYDE OXIME,ME ETHER	069163-87-7
PHTHALAN	000496-14-0
1-PHTHALAZINAMINE	019064-69-8
PHTHALAZINE	000253-52-1
1,4-PHTHALAZINEDIONE, 2,3-DIHYDRO	001445-69-8
1,2(H)-PHTHALAZINONE	000119-39-1
1(2H)-PHTHALAZINONE, 4-PHENYL-	005004-45-5
O-PHTHALIC ACID	000088-99-3
PHTHALIC ANHYDRIDE	000085-44-9
PHTHALIDE	000087-41-2
PHTHALIMIDE	000085-41-6
PHTHALIMIDE,N-METHYL	000550-44-7
PHTHALONIC ACID	000528-46-1
PHTHALOYL CHLORIDE	000088-95-9
PHTHALTHRIN	007696-12-0
PHYSOSTIGMINE	000057-47-6
PHYSOSTIGMINE, MONOSALICYLATE	000057-64-7
PICENE	000213-46-7
PICLORAM	001918-02-1
PICOLINIC ACID	000098-98-6
PICOLINIC ACID, ETHYL ESTER	002524-52-9
PICOLINIC ACID, METHYL ESTER	002459-07-6
PILOCARPIC ACID, HEXYL ESTER	096914-11-3
PILOCARPINE, MONOHYDROCHLORIDE	000054-71-7
PILOCARPOL	000092-13-7
PIMELIC ACID	000111-16-0
PIMONIDAZOLE	070132-50-2
PIMOZIDE	002062-78-4
PINACOL	000076-09-5
PINDOLOL	013523-86-9
PINDONE	000083-26-1

ALPHA-PINENE	000080-56-8
ß-PINENE	000127-91-3
PIPAMAZINE	000084-04-8
PIPAMPERONE	001893-33-0
PIPEMIDIC ACID	051940-44-4
PIPERADINE	000110-89-4
PIPERAZINE	000110-85-0
PIPERAZINE, 1-ACETYL-4-[4-[[2-(2,4-DICHLOROPHENY	065277-42-1
PIPERAZINE, 1-(4-AMINO-6,7-DIMETHOXY-2-QUINAZOLI	095549-92-1
PIPERAZINE,1,4-BIS(2,3-EPOXYPROPYL)	002917-98-8
PIPERAZINE-2-CARBOXANILIDE	036385-57-6
PIPERAZINE-2-CARBOXANILIDE,2'-ME	036385-59-8
PIPERAZINE-2-CARBOXANILIDE,2',6'-DIME	036371-18-3
PIPERAZINE, 1-(3-CHLOROPHENYL)-	006640-24-0
PIPERAZINE, 1-(2-CHLOROPHENYL)-	039512-50-0
PIPERAZINE, 1-(4-CHLOROPHENYL)-	038212-33-8
PIPERAZINE, 1,4-DIMETHYL-	000106-58-1
1,4-PIPERAZINEDIPROPANAMINE	007209-38-3
1-PIPERAZINEETHANOL, 4-[3-[2-(TRIFLUOROMETHYL)-9	053772-82-0
1-PIPERAZINEETHANOL, 4-[3-[2-(TRIFLUOROMETHYL)-9	053772-85-3
PIPERAZINE, 1-(4-FLUOROPHENYL)-	002252-63-3
PIPERAZINE, 1-(2-METHOXYPHENYL)-	035386-24-4
PIPERAZINE, 1-METHYL-4-[(1-METHYL-5-NITRO-1H-IMI	054387-29-0
PIPERAZINE, 1-METHYL-4-[2-(5-NITRO-1H-IMIDAZOL-1	074571-56-5
PIPERAZINE, 1-(2-METHYLPHENYL)-	039512-51-1
PIPERAZINE, 1-(2-PYRIDINYL)-	034803-66-2
PIPERAZINE, 1-(2-THIAZOLYL)-	042270-37-1
PIPERAZINE, 1-[3-(TRIFLUOROMETHYL)PHENYL]-	015532-75-9
PIPERDINE, HYDROCHLOROIC ACID	006091-44-7
1-PIPERIDINEACETAMIDE, N-(PHENYLMETHYL)-	072336-19-7
1-PIPERIDINEACETAMIDE, N-(4-ETHYLPHENYL)-	065446-98-2
1-PIPERIDINEACETAMIDE, N-(3-ETHOXYPHENYL)-	065570-49-2
1-PIPERIDINEACETAMIDE, N-(5-CHLORO-2,3-DIHYDRO-3	105801-53-4
1-PIPERIDINEACETAMIDE, N-(2,3-DIHYDRO-6-METHOXY-	105801-62-5
1-PIPERIDINEACETAMIDE, N-(2,3-DIHYDRO-3-BENZOFUR	105801-48-7
1-PIPERIDINEACETAMIDE, N-(4-METHOXYPHENYL)-	058479-93-9
1-PIPERIDINEACETAMIDE, N-(4-ETHOXYPHENYL)-	058479-94-0
1-PIPERIDINEACETAMIDE, N-(2,3-DIHYDRO-5-METHYL-3	105801-58-9
1-PIPERIDINEACETAMIDE, N-(3-METHOXYPHENYL)-	040297-47-0
PIPERIDINE, 1-BENZOYL-	000776-75-0
PIPERIDINE, 1-[(BENZOYLOXY)ACETYL]-	106231-67-8
PIPERIDINE, 1-[(BENZOYLOXY)ACETYL]-2-ETHYL-	115178-69-3
PIPERIDINE, 1-[(BENZOYLOXY)ACETYL]-2,6-DIMETHYL-	115178-70-6
1-PIPERIDINEBUTANAMIDE, _-[2-[BIS(1-METHYLETHYL)	068284-69-5
2-PIPERIDINECARBOXAMIDE, N-(2,6-DIMETHYLPHENYL)-	000096-88-8
4-PIPERIDINECARBOXYLIC ACID, 1-(3-HYDROXY-3-PHEN	000562-26-5
2-PIPERIDINECARBOXYLIC ACID	000535-75-1
PIPERIDINE, 1-(2-CHLOROBENZOYL)-	022342-21-8
PIPERIDINE, 1-(CYCLOHEXYLACETYL)-	072299-07-1
PIPERIDINE, 1-[4-(1,1-DIMETHYLETHYL)-1-PHENYLCYC	031790-44-0
PIPERIDINE, 1-(4,4-DIMETHYL-1-PHENYLCYCLOHEXYL)-	072165-35-6
2,6-PIPERIDINEDIONE, 3-ETHYL-3-METHYL-	090355-74-1
1-PIPERIDINE ETHANOL	003040-44-6
PIPERIDINE, 1-(2-ETHOXYBENZOYL)-	020308-67-2
PIPERIDINE, 1-ETHYL-	000766-09-6
PIPERIDINE, 1-(3-METHYLBENZOYL)-	013290-48-7
PIPERIDINE, 1-(4-METHOXYBENZOYL)-	057700-94-4
PIPERIDINE, 1-(4-METHYLBENZOYL)-	013707-23-8

PIPERIDINE, 1-(4-METHYL-1-PHENYLCYCLOHEXYL)-, TR	055040-03-4
PIPERIDINE, 1-(3-METHYL-1-PHENYLCYCLOHEXYL)-, CI	058283-42-4
PIPERIDINE, 4-PHENYL-	000771-99-3
PIPERIDINE, 2,2,6,6-TETRAMETHYL-	000768-66-1
1-PIPERIDINETHIOCARBOXAMIDE	014294-09-8
PIPERIDINE, 1-(3,3,5-TRIMETHYL-1-PHENYLCYCLOHEXY	072242-01-4
3,4,5-PIPERIDINETRIOL, 2-(HYDROXYMETHYL)-	019130-96-2
3,4,5-PIPERIDINETRIOL, 2-(HYDROXYMETHYL)-1-METHY	069567-10-8
PIPERIDINOACETIC ACID	055049-18-8
4'-PIPERIDINOACETOPHENONE	010342-85-5
4-PIPERIDINOL, 1-[(BENZOYLOXY)ACETYL]-	115178-71-7
4-PIPERIDINOL, 4-(4-CHLOROPHENYL)-1-[3-[(4-FLUOR	034933-71-6
4-PIPERIDINOL, 3-ETHYL-1-METHYL-4-PHENYL-, PROPI	000468-50-8
4-PIPERIDINOL, 1-[[3-(PHENOXYMETHYL)PHENYL]METHY	144458-95-7
2-PIPERIDINONE, 1-NITROSO-	050550-65-7
(N-PIPERIDINYL)ET-(345-MEO)PH SULFONE	057554-04-8
1-PIPERIDINYLOXY, 4-AMINO-2,2,6,6-TETRAMETHYL-	014691-88-4
1-PIPERIDINYLOXY, 4-[[BIS(1-AZIRIDINYL)PHOSPHINO	051526-59-1
1-PIPERIDINYLOXY, 4-[[BIS(1-AZIRIDINYL)PHOSPHINO	033683-34-0
1-PIPERIDINYLOXY, 4-(DIMETHYLAMINO)-2,2,6,6-TETR	071335-68-7
1-PIPERIDINYLOXY, 4-[(2-HYDROXYETHYL)METHYLAMINO	042013-74-1
1-PIPERIDINYLOXY, 4-(METHYLAMINO)-2,2,6,6-TETRAM	042585-33-1
1-PIPERIDINYLOXY, 2,2,6,6-TETRAMETHYL-4-OXO-	002896-70-0
4-N-PIPERIDINYLQUINAZOLINE	041229-10-1
A-(2-PIPERIDYL)PHENYLCARBINOL	023702-98-9
PIPERONAL	000120-57-0
PIPERONYL ACETONE	003160-37-0
PIPERONYL BUTOXIDE	000051-03-6
PIPERONYL CYCLONENE	000119-89-1
PIPEROPHOS	024151-93-7
1-PIPEROYL-(E,E)-PIPERIDINE	000094-62-2
PIPERYLONE	002531-04-6
PIPOBROMAN	000054-91-1
PIRIMICARB	023103-98-2
PIRIMIPHOS ETHYL	023505-41-1
PIRIMIPHOS-METHYL	029232-93-7
PIRITRAMIDE	000302-41-0
PIRITREXIM	072732-56-0
PIROXICAM	036322-90-4
PIVALALDEHYDE	000630-19-3
PLINOL	011039-70-6
PODOPHYLLOTOXIN	000518-28-5
POLYTHIAZIDE	000346-18-9
POTASSIUM ETHYLXANTHATE	000140-89-6
P(3(P-MEBENZYL)3ME-1TRIAZENO)BENZOICACID	065542-17-8
ALPHA-(P-MEO-PHENYL)-N-PHENYLNITRONE	003585-93-1
ALPHA-(P-ME-PHENYL)-N-PHENYLNITRONE	019865-55-5
P-I-PROPYLPHENOXYACETIC ACID	001643-16-9
P-I-PROPYLTOLUENE	025155-15-1
PRACTOLOL	006673-35-4
PRALLETHRIN	023031-36-9
PRASTERONE	000053-43-0
PRAVASTATIN	085956-22-5
PRAZEPAM	002955-38-6
PREDISONE-17-VALERATE	038196-45-1
PREDNISOLONE	000050-24-8
PREDNISOLONE ACETATE	000052-21-1
PREDNISOLONE-9A-FLUORO	000338-95-4

PREDNISOLONE-21-GLUCOSIDE	088158-44-5
PREDNISONE	000053-03-2
PREDNISONE-17-ACETATE	095440-71-4
PREGNA-1,4-DIENE-16-CARBOXYLIC ACID, 11,21-DIHYD	111802-43-8
PREGNA-1,4-DIENE-3,20-DIONE, 2-CHLORO-6,9-DIFLUO	065466-50-4
PREGNA-1,4-DIENE-3,20-DIONE, 2,9-DICHLORO-21-(2,	064272-13-5
PREGNA-1,4-DIENE-3,20-DIONE, 21-(2,2-DIMETHYL-1-	015845-96-2
PREGNA-1,4-DIENE-3,20-DIONE, 21-(beta-D-GALACTOP	092901-24-1
PREGN-4-ENE-21-CARBOXYLIC ACID, 9-FLUORO-11,17-D	001582-75-8
PREGN-4-ENE-3,20-DIONE, 17-(ACETYLOXY)-11,21-DIH	016463-74-4
PREGN-4-ENE-3,20-DIONE, 11,17-DIHYDROXY-21-(1-OX	006677-99-2
PREGN-4-ENE-3,20-DIONE, 21-(ß-D-GALACTOPYRANOSYL	092901-25-2
PREGN-4ENE 11A-EPOXYLACTONE DERIV	050394-27-9
PREGN-4-ENE-3,11,20-TRIONE, 17-(ACETYLOXY)-21-HY	039791-38-3
PREGN-4-ENE-3,11,20-TRIONE, 6-METHYL-, (6_)-	003642-85-1
PREGNENOLONE	000145-13-1
PRETILCHLOR	051218-49-6
PRILOCAINE	000721-50-6
PROBARBITAL	000076-76-6
PROBENECID	000057-66-9
3-PRO-5-BR-6-ME-PYRIMIDIN-2,4-DIONE	077317-68-1
3-PRO-5-BR-1,6-ME-PYRIMIDN-2,4-DION	077317-88-5
3-PRO-5-BR-6-PR-PYRIMIDIN-2,4-DIONE	077317-87-4
PROCAINAMIDE	000051-06-9
PROCAINE	000059-46-1
PROCAINE HYDROCHLORIDE	000051-05-8
PROCARBAZINE	000671-16-9
PROCHLORAZ	067747-09-5
PROCLONOL	014088-71-2
3-PRO-5-SCN-6-ME-PYRIMIDIN-2,4-DION	077317-86-3
PROCYMIDONE	032809-16-8
PROFENOFOS	041198-08-7
PROFLURALIN	026399-36-0
PROGABIDE	062666-20-0
PROGESTERONE	000057-83-0
PROLAN	000117-27-1
PROLIN-AMIDE, N-ACETYL	016395-58-7
L-PROLINAMIDE,N2-[(BENZOYLOXY)ACETYL]-	106231-69-0
L-PROLINAMIDE, 5-OXO-L-PROLYL-1-[(OCTYLOXY)CARBO	130817-97-9
L-PROLINAMIDE, 5-OXO-L-PROLYL-1-[(HEXYLOXY)CARBO	130817-96-8
L-PROLINAMIDE, 5-OXO-L-PROLYL-1-[(2-METHYLPROPOX	130817-95-7
L-PROLINAMIDE, 5-OXO-L-PROLYL-1-(BUTOXYCARBONYL)	130817-94-6
L-PROLINAMIDE, 5-OXO-L-PROLYL-1-(ETHOXYCARBONYL)	130817-92-4
L-PROLINAMIDE, 5-OXO-L-PROLYL-1-[(2-ETHYLHEXYLOX	130817-98-0
L-PROLINAMIDE, 5-OXO-L-PROLYL-1-[(1-METHYLETHOXY	130817-93-5
L-PROLINAMIDE, 5-OXO-L-PROLYL-L-HISTIDYL-	024305-27-9
L-PROLINAMIDE, 5-OXO-L-PROLYL-1-[(CYCLOHEXYLOXY)	130817-99-1
PROLINE	000147-85-3
PROLINE,N-ACETYL N'-MEAMINO-AMIDE	024847-46-9
L-PROLINE, N-[(BENZOYLOXY)ACETYL]-	115178-75-1
L-PROLINE, N-[(BENZOYLOXY)ACETYL]-, METHYL ESTER	115178-76-2
L-PROLINE, 1-[3-(BENZOYLTHIO)-2-METHYL-1-OXOPROP	081872-10-8
L-PROLINE, 1-[N-(1-CARBOXY-3-PHENYLPROPYL)-L-ALA	076420-72-9
L-PROLINE, 1-[N -(1-CARBOXY-3-PHENYLPROPYL)-L-LY	076547-98-3
PROMAZINE	000058-40-2
OPROMAZINE	000969-99-3
PROMAZINE HYDROCHLORIDE	000053-60-1
PROMEDOL	000064-39-1

PROMETHAZINE	000060-87-7
PROMETON	001610-18-0
PROMETRYNE	007287-19-6
PRONAMIDE	023950-58-5
PRONETHALOL	000054-80-8
PROPACHLOR	001918-16-7
1,2-PROPADIENE	000463-49-0
PROPANAMIDE, 3-(ACETYLAMINO)-N-[4-[(DIMETHYLAMIN	075851-98-8
PROPANAMIDE, 3-(ACETYLOXY)-N-[4-[(DIMETHYLAMINO)	075852-03-8
PROPANAMIDE, 2-AMINO-N-(2,6-DIMETHYLPHENYL)-	041708-72-9
PROPANAMIDE, 3-AMINO-N-(2,6-DIMETHYLPHENYL)-	066675-75-0
PROPANAMIDE, N-[(2-AMINO-4-METHYLPHENYL)SULFONYL	081792-74-7
PROPANAMIDE, N-[2-(AMINOSULFONYL)-5-CHLOROPHENYL	090875-63-1
PROPANAMIDE, 2-(BENZOYLOXY)-N,N-DIETHYL-, (±)-	115178-79-5
PROPANAMIDE, 3-(BENZOYLOXY)-N,N-DIMETHYL-	115178-77-3
PROPANAMIDE, 2-[[(BENZOYLOXY)ACETYL]AMINO]-	115193-30-1
PROPANAMIDE, 2-CHLORO-N-[2-OXO-2-(PHENYLAMINO)ET	074339-89-2
PROPANAMIDE, N-(4-CHLOROPHENYL)-2-HYDROXY-2-METH	062100-41-8
PROPANAMIDE, 3-CYANO-N-[4-[(DIMETHYLAMINO)SULFON	075852-01-6
PROPANAMIDE, 2-(1-CYCLOHEXENYL)-N-[(METHYLAMINO)	073632-84-5
PROPANAMIDE, N-(3,4-DICHLOROPHENYL)-2,2-DIMETHYL	007160-22-7
PROPANAMIDE, N-(3,4-DICHLOROPHENYL)-	000882-14-4
PROPANAMIDE, 3-(DIETHYLAMINO)-N-(2,6-DIMETHYLPHE	021236-52-2
PROPANAMIDE, 2-(DIETHYLAMINO)-N-(2,6-DIMETHYLPHE	039000-84-5
PROPANAMIDE, N-[4-[(DIMETHYLAMINO)SULFONYL]PHENY	075851-99-9
PROPANAMIDE, 2-(DIMETHYLAMINO)-N-(2,6-DIMETHYLPH	019216-39-8
PROPANAMIDE, N,N-DIMETHYL-	000758-96-3
PROPANAMIDE, N-[4-[(DIMETHYLAMINO)SULFONYL]PHENY	075852-08-3
PROPANAMIDE, N-[4-(DIMETHYLAMINO)SULFONYL]PHENYL	075852-15-2
PROPANAMIDE, N-(2-ETHOXY-5-NITROPHENYL)-	132899-69-5
PROPANAMIDE, 2-HYDROXY-2-METHYL-N-[4-NITRO-3-(TR	052806-53-8
PROPANAMIDE, 2-HYDROXY-2-METHYL-N-(4-NITROPHENYL	062100-54-3
PROPANAMIDE, N-[2-[[2-HYDROXY-3-(3-PHENYLPROPOXY	121508-93-8
PROPANAMIDE, N-[2-[(2-HYDROXY-3-PHENOXYPROPYL)AM	083029-57-6
PROPANAMIDE, N-[2-[[2-HYDROXY-3-(PHENYLMETHOXY)P	121508-88-1
PROPANAMIDE, N-[2-[[2-HYDROXY-3-(2-PHENYLETHOXY)	121508-91-6
PROPANAMIDE, 2-METHYL-N-[(4-METHYL-2-NITROPHENYL	081792-99-6
PROPANAMIDE, 2-METHYL-N-[2-NITRO-4-[(TRIFLUOROME	085873-47-8
PROPANAMIDE, 2-METHYL-N-[2-[[3-(2-PHENOXYETHOXY)	121508-96-1
PROPANAMIDE, 2-METHYL-N-[2-[[3-(3-PHENYL-2-PROPE	121508-98-3
PROPANAMIDE, 2-METHYL-N-[2-[[3-[(1,2,3,4-TETRAHY	121509-03-3
PROPANAMIDE, 2-METHYL-N-[4-NITRO-3-(TRIFLUOROMET	013311-84-7
PROPANAMIDE, N,N'-1H-1,2,4-TRIAZOLE-3,5-DIYLBIS-	080616-57-5
PROPANAMIDE, 3,3,3-TRIFLUORO-2-HYDROXY-2-METHYL-	072115-11-8
PROPANAMIDE, 3,3,3-TRIFLUORO-2-METHYL-N-[4-NITRO	104668-67-9
2-PROPANAMINE, 1-(2,6-DIMETHYLPHENOXY)-	031828-71-4
PROPANAMINE, N-(1-METHYLETHYL)-2-(1-OXOBUTOXY)-3	144106-11-6
PROPANAMINE, N-(1-METHYLETHYL)-2-(2-METHYL-1-OXO	144106-14-9
2-PROPANAMINE, 2-METHYL-N-(1-METHYLETHYL)-	007515-80-2
PROPANE	000074-98-6
PROPANE, 1,3-DIBROMO-2,2-BIS(BROMOMETHYL)-	003229-00-3
1,3-PROPANEDIAMINE	000109-76-2
PROPANEDINITRILE, [(4-ACETYLPHENYL)HYDRAZONO]-	055653-16-2
PROPANEDINITRILE, [(5-BROMO-2-FURANYL)METHYLENE]	056656-96-3
PROPANEDINITRILE, [(4-CHLOROPHENYL)HYDRAZONO]-	000946-76-9
PROPANEDINITRILE, [(5-CHLORO-2-FURANYL)METHYLENE	091937-67-6
PROPANEDINITRILE, [[4-(2-CHLOROETHYL)PHENYL]HYDR	081865-11-4
PROPANEDINITRILE, (3-CHLOROPHENYL)HYDRAZONO -	000555-60-2

PROPANEDINITRILE, (2-FURANYLMETHYLENE)	003237-22-7
PROPANEDINITRILE, [(3-HYDROXYPHENYL)HYDRAZONO]-	096382-70-6
PROPANEDINITRILE, [(5-IODO-2-FURANYL)METHYLENE]-	069527-40-8
PROPANEDINITRILE, [[5-(METHYLTHIO)-2-FURANYL]MET	076542-53-5
PROPANEDINITRILE, [(4-METHYLPHENYL)HYDRAZONO]-	040257-94-1
PROPANEDINITRILE, [(5-METHYL-2-FURANYL)METHYLENE	081020-79-3
PROPANEDINITRILE, [(2-NITROPHENYL)HYDRAZONO]-	055653-13-9
PROPANEDINITRILE, [(4-NITROPHENYL)HYDRAZONO]-	003722-12-1
PROPANEDINITRILE, [[4-(TRIFLUOROMETHOXY)PHENYL]H	000370-86-5
PROPANE, 1,3-DIIODO-	000627-31-6
PROPANEDIOIC ACID, [(3,4-DIMETHOXYPHENYL)METHYLE	015818-13-0
PROPANEDIOIC ACID, DIMETHYL ESTER	000108-59-8
PROPANEDIOIC ACID, 1,3-DITHIOLAN-2-YLIDENE-, BIS	050780-71-7
PROPANEDIOIC ACID, 1,3-DITHIOLAN-2-YLIDENE-, DIE	019607-41-1
PROPANEDIOIC ACID, 1,3-DITHIOLAN-2-YLIDENE-, DIP	050780-68-2
PROPANEDIOIC ACID, 1,3-DITHIOLAN-2-YLIDENE-, BIS	050780-72-8
PROPANEDIOIC ACID, 1,3-DITHIOLAN-2-YLIDENE-, ETH	050780-76-2
PROPANEDIOIC ACID, 1,3-DITHIOLAN-2-YLIDENE,- DIM	019723-86-5
PROPANEDIOIC ACID, 1,3-DITHIOLAN-2-YLIDENE-, BIS	052303-69-2
1,2-PROPANEDIOL	000057-55-6
1,3-PROPANEDIOL	000504-63-2
1,2-PROPANEDIOL, 3-[(6-AMINO-3-CHLORO-5-NITROPYR	088793-46-8
1,3-PROPANEDIOL, 2,2-BIS(BRME)-	003296-90-0
1,2-PROPANEDIOL, 3-(2-METHOXYPHENOXY)-, 1-CARBAM	000532-03-6
1,2-PROPANEDIOL, 3-[(3-NITROPYRAZINYL)AMINO]	087885-48-1
1,2-PROPANEDIOL-3-(2-TOLYLOXY)	000059-47-2
PROPANE, 1-ISOTHIOCYANATO-2-METHYL-	000591-82-2
PROPANENITRILE, 3- 4- (2-CHLORO-4-NITROPHENYL)A	006657-33-6
PROPANENITRILE, 2-CHLORO-	001617-17-0
PROPANENITRILE, 3,3'-IMINOBIS-	000111-94-4
PROPANENITRILE, 3-(1-NAPHTHALENYLOXY)-	058889-08-0
1-PROPANESULFONAMIDE, 3-(4-NITROPHENOXY)-	105412-26-8
PROPANE, 1,2,2,3-TETRACHLORO-	013116-53-5
2-PROPANETHIOL	000075-33-2
1-PROPANETHIOL	000107-03-9
1,2,3-PROPANETRIOL, 1,2-DINITRATE	000621-65-8
1,2,3-PROPANETRIOL, 1,3-DINITRATE	000623-87-0
1,2,3-PROPANETRIOL TRINITRATE	000055-63-0
PROPANIL	000709-98-8
PROPANIOIC ACID, 2,2-DIMETHYL-, [2-[[(4-MORPHOLI	102273-21-2
PROPANOIC ACID, 3-(4H-1,2,4-BENZOTHIADIAZIN-3-YL	004826-22-6
PROPANOIC ACID, 3-[(4H-1,2,4-BENZOTHIADIAZIN-3-Y	124802-88-6
PROPANOIC ACID, 3-[(7-BROMO-4H-1,2,4-BENZOTHIADI	124802-91-1
PROPANOIC ACID, 3-(7-BROMO-4H-1,2,4-BENZOTHIADIA	091445-76-0
PROPANOIC ACID, 2-[4-[[3-CHLORO-5-(TRIFLUOROMETH	069806-40-2
PROPANOIC ACID, 3-(7-CHLORO-4H-1,2,4-BENZOTHIADI	091551-85-8
PROPANOIC ACID, [(6-CHLORO-4H-1,2,4-BENZOTHIADIA	124850-85-7
PROPANOIC ACID, 3-(6-CHLORO-4H-1,2,4-BENZOTHIADI	101063-92-7
PROPANOIC ACID, 3-[(7-CHLORO-4H-1,2,4-BENZOTHIAD	134917-52-5
PROPANOIC ACID, 3-[(5,7-DIBROMO-4H-1,2,4-BENZOTH	124802-92-2
PROPANOIC ACID, 3-(5,7-DIBROMO-4H-1,2,4-BENZOTHI	101063-95-0
PROPANOIC ACID, 3-(6,7-DICHLORO-4H-1,2,4-BENZOTH	101063-94-9
PROPANOIC ACID, 3-[(6,7-DICHLORO-4H-1,2,4-BENZOT	124802-90-0
PROPANOIC ACID, 3-[(5,7-DICHLORO-4H-1,2,4-BENZOT	134917-53-6
PROPANOIC ACID, 3-(5,7-DICHLORO-4H-1,2,4-BENZOTH	101063-93-8
PROPANOIC ACID, 2,2-DIMETHYL-, METHYL ESTER	000598-98-1
PROPANOIC ACID, 2,2-DIMETHYL-, 2,3-DIMETHOXY-1,4	091814-09-4
PROPANOIC ACID, 2-(4-FORMYL-2-METHOXYPHENOXY)ETH	051264-74-5

PROPANOIC ACID, 3-(6-METHOXY-4H-1,2,4-BENZOTHIAD	101063-97-2
PROPANOIC ACID, 3-[(6-METHOXY-4H-1,2,4-BENZOTHIA	124850-87-9
PROPANOIC ACID, 3-[(6-METHYL-4H-1,2,4-BENZOTHIAD	124802-93-3
PROPANOIC ACID, 3-(6-METHYL-4H-1,2,4-BENZOTHIADI	101063-96-1
PROPANOIC ACID, 3-(1-NAPHTHALENYLOXY)-	016563-41-0
PROPANOIC ACID, [(PHENYLMETHYL)SELENO]-	006926-05-2
PROPANOIC ACID, 3-[[6-(TRIFLUOROMETHYL)-4H-1,2,4	124850-88-0
PROPANOIC ACID, 3-[6-(TRIFLUOROMETHYL)-4H-1,2,4-	101063-98-3
1-PROPANOL	000071-23-8
PROPANOLAMINE	000156-87-6
2-PROPANOL, 1-(2-BICYCLO[2.2.1]HEPT-2-YLPHENOXY)	066451-06-7
2-PROPANOL, 1-BUTOXY-3-(2-NITRO-1H-IMIDAZOL-1-YL	082249-23-8
2-PROPANOL, 1-[4-(5-CHLORO-1H-IMIDAZOL-2-YL)PHEN	102152-27-2
2-PROPANOL, 1-CHLORO-, PHOSPHATE (3:1)	013674-84-5
2-PROPANOL, 1-[4-[2-(CYCLOPROPYLMETHOXY)ETHYL]PH	063659-18-7
2-PROPANOL, 1-[[2-(3,4-DIMETHOXYPHENYL)ETHYL]AMI	102281-28-7
2-PROPANOL, 1-[[2-(3,4-DIMETHOXYPHENYL)ETHYL]AMI	102152-14-7
2-PROPANOL, 1-[[2-(3,4-DIMETHOXYPHENYL)ETHYL]AMI	102151-23-5
2-PROPANOL, 1-[[2-(3,4-DIMETHOXYPHENYL)ETHYL]AMI	102152-24-9
2-PROPANOL, 1-[(1,1-DIMETHYLETHYL)AMINO]-3-(3-PH	080617-76-1
2-PROPANOL, 1-[(1,1-DIMETHYLETHYL)AMINO]-3-(PHEN	063474-05-5
2-PROPANOL, 1-[(1,1-DIMETHYLETHYL)AMINO]-3-[2-(2	047082-97-3
2-PROPANOL, 1-[(1,1-DIMETHYLETHYL)AMINO]-3-[(3-P	080617-78-3
2-PROPANOL, 1-[(1,1-DIMETHYLETHYL)AMINO]-3-(2-PH	121508-89-2
2-PROPANOL, 1-[(1,1-DIMETHYLETHYL)AMINO]-3-(2-PH	121508-94-9
2-PROPANOL, 1-[(1,1-DIMETHYLETHYL)AMINO]-3-[(1,2	080617-86-3
2-PROPANOL, 1-[(1,1-DIMETHYLETHYL)AMINO]-3-PHENO	064980-40-1
2-PROPANOL, 1,1'-[[6-[(2,5-DIMETHYL-1H-PYRROL-1-	075841-97-3
2-PROPANOL, 1-[[6-[(2,5-DIMETHYL-1H-PYRROL-1-YL)	075841-95-1
2-PROPANOL, 1-[(2-ETHOXYETHYL)AMINO]-3-[4-[4-(2-	102151-09-7
2-PROPANOL, 1-[4-[2-[2-(4-FLUOROPHENYL)ETHOXY]ET	084057-96-5
2-PROPANOL, 1-[[2-HYDROXY-3-(1-NAPHTHALENYLOXY)P	120908-61-4
2-PROPANOL, 1-[[2-HYDROXY-3-(4-NITROPHENOXY)PROP	120908-62-5
1-PROPANOL, 2-(2-HYDROXYPROPOXY)-	000106-62-7
1-PROPANOL, 3,3'-IMINOBIS-, DIMETHANESULFONATE (	013425-98-4
PROPANOL 1 (OR 2)-2-METHOXYMETHYL ETHOXY, ACETAT	088917-22-0
2-PROPANOL, 1-(2-METHOXYPHENOXY)-3-[(1-METHYLETH	005741-22-0
2-PROPANOL, 1-[(1-METHYLETHYL)AMINO]-3-(PHENYLME	019343-24-9
2-PROPANOL, 1-[(1-METHYLETHYL)AMINO]-3-(2-PHENYL	105217-56-9
2-PROPANOL, 1-[(1-METHYLETHYL)AMINO]-3-[(2-METHY	023694-81-7
2-PROPANOL, 1-[(1-METHYLETHYL)AMINO]-3-(1-NAPHTH	004290-62-4
2-PROPANOL, 1-[(1-METHYLETHYL)AMINO]-3-[(3-PHENY	105217-58-1
2-PROPANOL, 1-[(1-METHYLETHYL)AMINO]-3-(2-PHENOX	105217-57-0
2-PROPANOL, 1-[(1-METHYLETHYL)AMINO]-3-[4-[[2-(1	066722-44-9
2-PROPANOL, 1-[(1-METHYLETHYL)AMINO]-3-(3-PHENYL	080617-74-9
2-PROPANOL, 3-[[2-[4-[2-OXO-2-(2-PIPERIDINYL)ETH	139734-21-7
2-PROPANOL, 3-[[2-[4-[2-OXO-2-(1,2,3,4-TETRAHYDR	139734-30-8
1-PROPANOL, 3-(PHENYLAMINO)-	031121-11-6
2-PROPANOL,2-PHENYL-1,1,1,3,3,3-HEXAFLUORO	000718-64-9
2-PROPANOL, 1-(PROPYLAMINO)-3-[4-[5-(2-THIENYL)-	102151-07-5
2-PROPANOL, 1-[[2-(4-PYRIDINYL)ETHYL]AMINO]-3-[4	102151-10-0
2-PROPANOL, 1-(2-CYCLOPROPYLPHENOXY)-3-[(1-METHY	027325-36-6
2-PROPANONE, (2,4-DINITROPHENYL)HYDRAZONE	001567-89-1
2-PROPANONE, 1-METHOXY-2-(2-NITRO-1H-IMIDAZOL-1-	021787-95-1
1-PROPANONE, 2-METHYL-1-(4-METHYLPHENYL)-3-(1-PI	000728-88-1
1-PROPANONE, 2-METHYL-1-PHENYL-3-(1-PIPERIDINYL)	022385-99-5
1-PROPANONE, 1-(2-METHYLPHENYL)-3-(1-PIPERIDINYL	125476-98-4
1-PROPANONE, 1-(4-METHYLPHENYL)-3-(1-PIPERIDINYL	013552-46-0

1-PROPANONE, 2-METHYL-1-PHENYL-	000611-70-1
2-PROPANONE, OXIME	000127-06-0
2-PROPANONE, 1-PHENOXY-	000621-87-4
PROPAPHOS	007292-16-2
PROPAQUIZAFOP	111479-05-1
PROPARGITE	002312-35-8
PROPARGYL ALCOHOL	000107-19-7
PROPAZINE	000139-40-2
2-PROPENAL, 3-(2-FURANYL)-	000623-30-3
2-PROPENAL, 3-(5-NITRO-2-FURANYL)-	001874-22-2
2-PROPENAMIDE, 3-(3,4-DIHYDROXYPHENOXY)-N-HEXYL-	136944-37-1
2-PROPENAMIDE, 3-(3,4-DIHYDROXYPHENYL)-N-(3,4-DI	136944-22-4
2-PROPENAMIDE, 3-(3,4-DIHYDROXYPHENYL)-N-[(3,4-D	136944-23-5
2-PROPENAMIDE, 3-(3,4-DIHYDROXYPHENYL)-N-(PHENYL	100668-10-8
2-PROPENAMIDE, 3-(3,4-DIHYDROXYPHENYL)-N-(2-PHEN	103188-47-2
2-PROPENAMIDE, 3-(3,4-DIHYDROXYPHENYL)-N-[2-(3,4	136944-24-6
2-PROPENAMIDE, 3-(3,4-DIHYDROXYPHENYL)-N-HEXYL-	100668-11-9
2-PROPENAMIDE, N-(1-METHYLPROPYL)-3-PHENYL-2-(PH	097871-59-5
2-PROPENAMIDE, 3-(5-NITRO-2-FURANYL)-	000710-25-8
2-PROPEN-1-AMINE, N,N-DI-2-PROPENYL-	000102-70-5
PROPENE	000115-07-1
1-PROPENE-1,1,3-TRICARBONITRILE, 2-AMINO-	000868-54-2
2-PROPENOIC ACID, 3-(2-METHOXYPHENYL)-, (E)-	001011-54-7
2-PROPENOIC ACID, 3-(5-BROMO-2-FURANYL)-2-CYANO-	056656-97-4
2-PROPENOIC ACID, 3-(5-CHLORO-2-FURANYL)-2-CYANO	096841-03-1
2-PROPENOIC ACID, 3-(2-CHLOROPHENOXY)-, ETHYL ES	074407-83-3
2-PROPENOIC ACID, 2-CYANO-3-(2-FURANYL)-, METHYL	003695-86-1
2-PROPENOIC ACID, 2-CYANO-3-(5-IODO-2-FURANYL)-,	069527-41-9
2-PROPENOIC ACID, 2-CYANO-3-[5-(METHYLTHIO)-2-FU	076542-54-6
2-PROPENOIC ACID, 2-CYANO-3-(5-METHYL-2-FURANYL)	073403-31-3
2-PROPENOIC ACID, 3-(2,4-DICHLOROPHENOXY)-, (E)-	053596-19-3
2-PROPENOIC ACID, 3-(3,4-DIHYDROXYPHENYL)-, BUTY	022020-28-6
2-PROPENOIC ACID, 3-(3,4-DIHYDROXYPHENYL)-, (E)-	000331-39-5
PROPENOIC ACID, ESTER DERIVATIVE	013017-11-3
2-PROPENOIC ACID, 1,2-ETHANEDIYLBIS(OXY-2,1-ETHA	001680-21-3
2-PROPENOIC ACID, 3-(4-HYDROXYPHENYL)-, (E)-	000501-98-4
2-PROPENOIC ACID, 2-METHYL-, 2-HYDROXY-3-PHENOXY	016926-87-7
2-PROPENOIC ACID, 3-(3-METHOXYPHENYL)-, (E)-	017570-26-2
2-PROPENOIC ACID, 3-(2-METHOXYPHENYL)-, 2-(4-FOR	093414-39-2
2-PROPENOIC ACID, 3-(2-METHOXYPHENYL)-, (Z)-	014737-91-8
2-PROPENOIC ACID, 2-METHYL-, 2-[(1-NAPHTHALENYLO	080916-23-0
2-PROPENOIC ACID, 2-METHYL-, 5A-ETHENYLDECAHYDRO	035413-85-5
2-PROPENOIC ACID, 2-METHYL-, 1,3-PROPANEDIYL EST	001188-09-6
2-PROPENOIC ACID, OXYBIS(2,1-ETHANEDIYLOXY-2,1-E	017831-71-9
2-PROPENOIC ACID, 3-PHENOXY-, METHYL ESTER, (E)-	005739-82-2
2-PROPENOIC ACID, 3-PHENOXY-	042103-65-1
2-PROPENOIC ACID, 3-PHENOXY-, ETHYL ESTER	043013-70-3
2-PROPENOIC ACID, 3-PHENYL-, (E)-	000140-10-3
2-PROPENOIC ACID, 3-PHENYL-, 2-(DIETHYLAMINO)ETH	010369-88-7
2-PROPENOIC ACID, PROPYL ESTER	000925-60-0
2-PROPEN-1-ONE, 3-(2,5-DIHYDRO-1H-PYRROL-1-YL)-1	010078-46-3
2-PROPEN-1-ONE, 1,3-BIS(2-FURANYL)-	003988-76-9
1-PROPEN-1-ONE(METHYLKETENE)	006004-44-0
2-PROPENYLOXY-2-PH-4-ME MORPHOLINE HCL	124497-81-0
PROPETAMPHOS	031218-83-4
PROPICILLIN	000551-27-9
PROPICONAZOLE	060207-90-1
BETA-PROPIOLACTONE	000057-57-8

PROPIONALDEHYDE	000123-38-6
PROPIONAMIDE, G-PHENYL	000102-93-2
PROPIONANILIDE	000620-71-3
PROPIONIC ACID	000079-09-4
PROPIONIC ACID,2(4-BIPHENYLYL)3OH3(2FPH)	093371-50-7
PROPIONIC ACID,2(4BIPHENYLYL)3OH3(P-MEOPH)	093371-46-1
PROPIONIC ACID HYDRAZIDE	024535-11-3
PROPIONIC ANHYDRIDE	000123-62-6
PROPIONIC ACID,2(4-BIPHENYLYL)-3-OH-3-(3FPH)-	093371-51-8
PROPIONITRILE	000107-12-0
N-PROPIONYLCYCLOBUTANECARBOXAMIDE	023046-87-9
PROPIOPHENONE	000093-55-0
PROPOXUR	000114-26-1
2-PROPOXY-4-AMINOBENZOIC ACID	002486-79-5
4-PROPOXYBENZOIC ACID	005438-19-7
PROPOXYETHYL TRIMETHYL AMMONIUM IODIDE	005432-44-0
PROPOXYPHENE	000469-62-5
PROPOXYPHENE CARBINOL	000126-04-5
M-PROPOXYPHENOL	016533-50-9
2-PROPOXY-1-PROPANOL	010215-30-2
1-PROPOXY-2-PROPANOL	001569-01-3
2-PROPOXYPYRAZINE	070090-28-7
4-PROPOXYPYRAZOLE	088095-60-7
2-PROPOXYPYRIDINE	101870-22-8
PROPRANOL	000525-66-6
PROPRIONAMIDE	000079-05-0
N-PROPYL ACETATE	000109-60-4
ISO-PROPYLACETATE	000108-21-4
9-PROPYL ADENINE	000707-98-2
PROPYLAMINE	000107-10-8
PROPYLAMINE HYDROCHLORIDE	000556-53-6
PROPYL-P-AMINOBENZOATE	000094-12-2
N-PROPYLANILINE	000622-80-0
N-PROPYLBENZENE	000103-65-1
A-PROPYL BENZENEETHANEAMINE	063951-01-9
A-I-PROPYLBENZENE ETHANEAMINE	046114-16-3
P-PROPYLBENZENESULFONAMIDE	001132-18-9
PROPYL BENZOATE	002315-68-6
4-PROPYLBENZOIC ACID	002438-05-3
2-I-PROPYLBENZOTRIAZOLE	069218-31-1
1-PROPYLBENZOTRIAZOLE	016584-02-4
1-I-PROPYLBENZOTRIAZOLE	069218-29-7
2-PROPYLBENZOTRIAZOLE	016584-03-5
PROPYLBROMIDE, G-PHENYL	000637-59-2
PROPYL SEC-BUTYLAMINE	039190-67-5
PROPYL-BUTYLAMINE	020193-21-9
N-PROPYLBUTYRATE	000105-66-8
O-PROPYLCARBAMATE	000627-12-3
3-PROPYLCHLOROTHIAZIDE	002854-99-1
N-PROPYLCYCLOHEXANE	001678-92-8
PROPYLCYCLOPENTANE	002040-96-2
1,1'-PROPYLENE-BIS(1-NITROSOUREA)	027640-19-3
PROPYLENE CARBONATE	000108-32-7
PROPYLENE GLYCOL ME ETHER ACETATE	000108-65-6
PROPYLENE GLYCOL MONOETHYL ETHER	052125-53-8
PROPYLENE GLYCOL MONOPHENYL ETHER	004169-04-4
PROPYLENE OXIDE	000075-56-9
PROPYLFLUORIDE, G-PHENYL	002038-62-2

PROPYL FORMATE	000110-74-7
B-PROPYLGALACTOPYRANOSIDE	062178-32-9
PROPYL GALLATE	000121-79-9
2-PROPYLHEPTANOIC ACID	031080-39-4
PROPYL HEXANOATE	000626-77-7
N-PROPYL-HYDROXYPYRID-4-ONE	104764-52-5
3-PROPYLHYDROXYUREA	005710-12-3
PROPYL-ISOBUTYLAMINE	039190-66-4
PROPYL ISOBUTYRATE	000644-49-5
PROPYL ISOME	000083-59-0
2-PROPYLISONIAZID	014339-54-9
PROPYL ISOPROPYL ETHER	000627-08-7
PROPYL ISOVALERATE	000557-00-6
PROPYL-(3,4,5-MEO)PHENYL-SULFONE	071203-62-8
N-PROPYL METHACRYLATE	002210-28-8
O-PROPYL-N-METHYLCARBAMATE	017671-76-0
N-PROPYLMETHYLSULFIDE	003877-15-4
N-PROPYLNICOTINAMIDE	051055-31-3
1-PROPYLNITRATE	000627-13-4
N-PROPYLNITRITE	000543-67-9
4-PROPYLNITROBENZENE	010342-59-3
1-I-PROPYL-1-NITROSO-3-PHENYLUREA	078326-57-5
1-PROPYL-1-NITROSOUREA	000816-57-9
(1-PROPYLNONYL)BENZENE	002719-64-4
N-PROPYLPENTAFLUOROBENZENE	000771-56-2
4-PROPYLPHENOL	000645-56-7
O-PROPYLPHENOL	000644-35-9
M-PROPYLPHENOXYACETIC ACID	002084-11-9
O-I-PROPYLPHENOXYACETIC ACID	025141-58-6
PROPYL N-PHENYLCARBAMATE	005532-90-1
1-PROPYLPIPERIDINE	005470-02-0
2-PROPYLPIPERIDINE	000458-88-8
PROPYLPROPANEDIOIC ACID	000616-62-6
N-PROPYLPROPIONATE	000106-36-5
4-PROPYLPYRIDINE	001122-81-2
N-N-PROPYL-3-PYRIDYLMETHYLAMINE	019730-13-3
N-I-PROPYL-3-PYRIDYLMETHYLAMINE	019730-12-2
N-PROPYLQUINOLINIUM BROMIDE	006294-92-4
4-PROPYLSEMICARBAZIDE	057421-72-4
4-I-PROPYLSEMICARBAZIDE	057930-20-8
PROPYL 4-SULFAMYLBENZOATE	059777-58-1
PROPYL SULFONE	000598-03-8
1-PROPYLTHIO-B-GALACTOPYRANSIDE	063407-51-2
PROPYLTRICHLOROSILANE	000141-57-1
5-I-PROPYLTROPALONE	000672-76-4
PROPYNE	000074-99-7
2-PROPYNYLAMINE	002450-71-7
2-PROPYNYLOXY-2-PH-4-ME MORPHOLINE HCL	124497-82-1
PROSCAR	098319-26-7
PROSCILLARIDIN	000466-06-8
PROSTAGLANDIN	000551-11-1
PROSTAGLANDIN E2	000363-24-6
PROSULFOCARB	052888-80-9
PROTHIOPHOS	034643-46-4
PROTHOATE	002275-18-5
PROXYPHYLLINE	000603-00-9
3-PRS-4-NH2-6-IPR-124-TRIAZINE-5-ONE	050917-23-2
PSEUDO-STRYCHNINE	000465-62-3

PSORALEN	000066-97-7
PTERIDINE	000091-18-9
2,4,7-PTERIDINETRIAMINE, 6-(4-HYDROXYPHENYL)-	001226-52-4
9H-PURIN-6-AMINE	066224-65-5
9H-PURIN-6-AMINE, 9-.BETA.-D-ARABINOFURANOSYL-	005536-17-4
9H-PURIN-2-AMINE, 6-CHLORO-9-[TETRAHYDRO-5-(HYDR	122970-35-8
9H-PURIN-6-AMINE, 9-(2,3-DIDEOXY-2-FLUORO-ß-D-TH	110143-10-7
9H-PURIN-6-AMINE, 9-(2,3-DIDEOXY-2-FLUORO-ß-D-TH	132722-91-9
9H-PURIN-6-AMINE, 9-(2,3-DIDEOXY-2-FLUORO-ß-D-TH	132722-94-2
9H-PURIN-6-AMINE, 9-(2,3-DIDEOXY-2-FLUORO-ß-D-TH	126502-17-8
9H-PURIN-6-AMINE, 9-METHYL-	000700-00-5
1H-PURIN-6-AMINE, N-(PHENYLMETHYL)-	001214-39-7
9H-PURIN-6-AMINE, 9-(PHENYLMETHYL)-	004261-14-7
1H-PURINE	000120-73-0
9H-PURINE, 9-(ß-D-ARABINOFURANOSYL)-6-(METHYLTHI	013153-62-3
9H-PURINE, 9-[2,5-BIS-O-(2,2-DIMETHYL-1-OXOPROPY	137057-71-7
9H-PURINE, 9-[2,3-BIS-O-(2,2-DIMETHYL-1-OXOPROPY	137057-62-6
9H-PURINE, 9-[2,3-BIS-O-(1-OXOPENTYL)-beta-D-ARA	137057-57-9
9H-PURINE, 9-[2,5-BIS-O-(1-OXOPENTYL)-BETA-D-ARA	137057-69-3
9H-PURINE, 9-[2,3-BIS-O-(PHENYLACETYL)-beta-D-AR	137057-67-1
9H-PURINE, 6-CHLORO-9-(2,3-DIDEOXY-2-FLUORO-ß-D-	132722-95-3
H-PURINE, 9-(2,5-DI-O-BENZOYL-BETA-D-ARABINOFURA	137057-72-8
9H-PURINE, 9-(2,3-DIDEOXY-2-FLUORO-ß-D-THREO-PEN	132723-00-3
9H-PURINE, 9-(2,3-DIDEOXY-2-FLUORO-ß-D-THREO-PEN	132722-90-8
1H-PURINE-2,6-DIONE, 3-BUTYL-3,9-DIHYDRO-1-METHY	031542-48-0
1H-PURINE-2,6-DIONE, 3,9-DIHYDRO-3-BUTYL-	041078-03-9
1H-PURINE-2,6-DIONE, 3,9-DIHYDRO-1,3-DIMETHYL-	111079-49-3
1H-PURINE-2,6-DIONE, 3,9-DIHYDRO-3-ETHYL-1-METHY	125573-05-9
1H-PURINE-2,6-DIONE, 3,9-DIHYDRO-3-ETHYL-	041078-01-7
1H-PURINE-2,6-DIONE, 3,7-DIHYDRO-1-METHYL-3-(2-M	028822-58-4
1H-PURINE-2,6-DIONE, 3,9-DIHYDRO-1-METHYL-3-PROP	118024-67-2
1H-PURINE-2,6-DIONE, 3,9-DIHYDRO-3-PROPYL-	041078-02-8
9H-PURINE, 9-[2,5-BIS-O-(4-METHOXYBENZOYL)-beta-	137057-73-9
9H-PURINE, 6-METHOXY-9-[2,3,5-TRIS-O-(2-METHYL-1	137057-45-5
9H-PURINE, 6-METHOXY-9-[2,3,5-TRIS-O-(1-OXOBUTYL	137057-43-3
9H-PURINE, 6-METHOXY-9-[2,3,5-TRIS-O-(1-OXOPROPY	137057-42-2
9H-PURINE, 6-(METHYLTHIO)-9-(beta-D-RIBOFURANOSY	000342-69-8
6-PURINETHIOL HYDRATE	000050-44-2
6H-PURIN-6-ONE, 2-AMINO-1,9-DIHYDRO-9-[[2-HYDROX	082410-32-0
6H-PURIN-6-ONE, 9-(2,3-DIDEOXY-2-FLUORO-ß-D-THRE	117525-25-4
2H-PURIN-2-ONE, 7-(2-ETHOXYETHYL)-1,3,6,7-TETRAH	087469-44-1
2H-PURIN-2-ONE, 9-(2-ETHOXYETHYL)-1,3,6,9-TETRAH	087469-43-0
2H-PURIN-2-ONE, 7-[2-(ETHYLTHIO)ETHYL]-1,3,6,7-T	087469-46-3
2H-PURIN-2-ONE, 9-[2-(ETHYLTHIO)ETHYL]-1,3,6,9-T	087469-45-2
2H-PURIN-2-ONE, 1,3,6,9-TETRAHYDRO-6-IMINO-1-MET	087469-33-8
2H-PURIN-2-ONE, 1,3,6,9-TETRAHYDRO-6-IMINO-9-[(2	087469-68-9
2H-PURIN-2-ONE, 1,3,6,9-TETRAHYDRO-6-IMINO-9-(2-	087469-41-8
2H-PURIN-2-ONE, 1,3,6,7-TETRAHYDRO-6-IMINO-7-(2-	087469-42-9
PUROMYCIN	000053-79-2
PURPURIN	000081-54-9
PYRACARBOLID	024691-76-7
2H-PYRAN, TETRAHYDRO-6-CYCLOPROPYL-3-[1-(ETHOXYI	116583-45-0
2H-PYRAN, 3,4-DIHYDRO-	000110-87-2
4H-PYRAN-4-ONE, 5-HYDROXY-2-(HYDROXYMETHYL)-	000501-30-4
4H-PYRAN-4-ONE, 3-HYDROXY-2-METHYL-	000118-71-8
2H-PYRAN-2-ONE, 4-HYDROXY-3,5,6-TRIMETHYL-	050405-44-2
2H-PYRAN, TETRAHYDRO-3-[1-(ETHOXYIMINO)BUTYL]-6-	124051-00-9
2H-PYRAN, TETRAHYDRO-3-[1-(ETHOXYIMINO)BUTYL]-2,	124050-99-3

PYRAZINAMINE, 5-CHLORO-	033332-29-5
PYRAZINAMINE, 5-CHLORO-N,N-DIMETHYL-	136309-13-2
PYRAZINAMINE, 6-METHOXY-	006905-47-1
PYRAZINAMINE, 5-METHOXY-N,N-DIMETHYL	136309-14-3
PYRAZINAMINE, 6-METHOXY-N,N-DIMETHYL	136309-09-6
PYRAZINAMINE, 5-METHOXY-	054013-07-9
PYRAZINAMINE, 6-METHYL-	005521-56-2
PYRAZINAMINE, N,N,6-TRIMETHYL-	056343-48-7
PYRAZINE	000290-37-9
PYRAZINE AZOXYCYANIDE	117505-25-6
PYRAZINE, 2-BUTYL-3-METHYL-	015987-00-5
PYRAZINECARBONITRILE, 6-METHYL-	136309-04-1
PYRAZINECARBONITRILE, 5-METHYL-	098006-91-8
PYRAZINECARBONITRILE, 6-METHOXY-	136309-07-4
PYRAZINECARBONITRILE, 5-CHLORO-	036070-75-4
PYRAZINECARBONITRILE, 6-(DIMETHYLAMINO)-	040262-48-4
PYRAZINECARBOXAMIDE, 5-METHYL-	005521-57-3
PYRAZINECARBOXAMIDE, 6-(DIMETHYLAMINO)	040262-55-3
PYRAZINECARBOXAMIDE, 6-METHYL-	005521-62-0
PYRAZINECARBOXAMIDE, 6-METHOXY-	036070-86-7
2-PYRAZINECARBOXAMIDE	000098-96-4
PYRAZINECARBOXAMIDE, 6-ETHOXY-	054015-45-1
PYRAZINECARBOXYLIC ACID, 5-METHYL-, ETHYL ESTER	041110-34-3
PYRAZINECARBOXYLIC ACID, 6-(DIMETHYLAMINO)-, MET	040262-54-2
PYRAZINECARBOXYLIC ACID, 6-[(2,3-DIHYDROXYPROPYL	087885-53-8
PYRAZINECARBOXYLIC ACID, 6-METHYL-, ETHYL ESTER	041110-39-8
PYRAZINECARBOXYLIC ACID, 6-(DIMETHYLAMINO)-, ETH	136309-10-9
PYRAZINECARBOXYLIC ACID, 6-METHYL-, METHYL ESTER	041110-38-7
PYRAZINECARBOXYLIC ACID, 6-METHOXY-, METHYL ESTE	023813-24-3
PYRAZINECARBOXYLIC ACID, 5-METHYL-, METHYL ESTER	041110-33-2
PYRAZINECARBOXYLIC ACID, 6-METHOXY-, ETHYL ESTER	136309-08-5
PYRAZINE, 3-CHLORO-2,5-DIMETHYL-	000095-89-6
PYRAZINE, 2-CHLORO-5-ETHOXY-	136309-11-0
PYRAZINE, 2-CHLORO-5-METHOXY-	033332-31-9
PYRAZINE, 2-CHLORO-5-METHYL-	059303-10-5
2,6-PYRAZINEDIAMINE	041536-80-5
2,3-PYRAZINEDICARBONITRILE, 5-CHLORO-6-ETHYL-	072111-58-1
2,3-PYRAZINEDICARBONITRILE	013481-25-9
2,3-PYRAZINEDICARBONITRILE, 5-CHLORO-6-METHYL-	072126-54-6
2,3-PYRAZINEDICARBONITRILE, 5-CHLORO-	072111-57-0
2,3-PYRAZINEDICARBONITRILE, 5-BUTYL-6-METHYL-	072112-24-4
2,3-PYRAZINEDICARBONITRILE, 5-BUTYL-6-CHLORO-	072111-60-5
2,3-PYRAZINEDICARBOXYLIC ACID, DIMETHYL ESTER	006164-77-8
2,3-PYRAZINEDICARBOXYLIC ACID, DIETHYL ESTER	002427-90-9
PYRAZINE, 2,5-DICHLORO-	019745-07-4
PYRAZINE, 2,6-DIETHOXY-	033870-85-8
PYRAZINE, 2,3-DIETHYL-	015707-24-1
PYRAZINE, 2,3-DIETHYL-5-METHYL-	018138-04-0
PYRAZINE, 2,5-DIMETHOXY-	117856-61-8
PYRAZINE, 2,6-DIMETHOXY-	004774-15-6
PYRAZINE, 2,3-DIMETHYL-	005910-89-4
PYRAZINE, 2,6-DIMETHYL-	000108-50-9
PYRAZINE, 2-ETHOXY-6-METHOXY-	136309-05-2
PYRAZINE, 2-ETHOXY-3-METHYL-	032737-14-7
PYRAZINE, 2-ETHYL-3-METHYL-	015707-23-0
PYRAZINE, 2-METHOXY-6-METHYL-	002882-21-5
PYRAZINE, 2-METHOXY-3-METHYL-	002847-30-5
PYRAZINE, 2-METHYL-3-(1-METHYLETHOXY)-	094089-22-2

PYRAZINE, 2-METHYL-3-(METHYLTHIO)-	002882-20-4
PYRAZINE, 2-METHYL-3-PROPYL-	015986-80-8
PYRAZINE-2-ONE	006270-63-9
PYRAZINE, TETRAMETHYL-	001124-11-4
PYRAZINE, TRIMETHYL-	014667-55-1
PYRAZOLATE	058011-68-0
PYRAZOLE	000288-13-1
1H-PYRAZOLE, 3,5-DIMETHYL-	000067-51-6
1H-PYRAZOLE, 3-METHYL-5-[(4-METHYLPHENYL)AMINO]-	136389-76-9
1H-PYRAZOLE, 3-METHYL-5-[(3-METHYLPHENYL)AMINO]-	136389-77-0
1H-PYRAZOLE, 3-METHYL-4-NITRO-5-[(4-NITROPHENYL)	136389-80-5
1H-PYRAZOLE, 3-METHYL-4-NITRO-1-PHENYL-5-[[4-(TR	136389-87-2
1H-PYRAZOLE, 1-PHENYL-	001126-00-7
2,5-PYRAZOLIDINDIONE,1,2-DIPHENYL	002652-77-9
1-PYRAZOLIDINECARBOXALDEHYDE, 5,5-DIMETHYL-3-OXO	122009-33-0
3-PYRAZOLIN-5-ONE, 3-METHYL-2-(4-PYRIMIDINYL)-	018697-64-8
PYRAZOL-3-ON,1(46DIMEO-2PYRIMDIN)5ME	023905-85-3
PYRAZOL-3-ON,2(26DIME-4PYRIMD)15DIME	023898-86-4
PYRAZOL-3-ON,1(46DIME-2PYRIMD)25DIME	023898-92-2
3H-PYRAZOL-3-ONE, 1,2-DIHYDRO-1,5-DIMETHYL-4-(1-	000479-92-5
PYRAZOL-3-ONE,2(2ME-6MEO PYRIMID)5ME	018597-55-2
PYRAZOL-3-ONE,1(4,6-ME2PYRIMID)5ME	018697-50-2
PYRAZOL-3-ONE,2H1(4,6-ME2PYRIMID)5ME	018597-53-0
PYRAZOL-3-ONE,2(4,6-ME4PYRIMID)5ME	018597-57-4
PYRAZOL-3-ON,2(2ME6MEO-4PYRIMD)15DIME	023898-89-7
PYRAZOL-3-ON,1(2ME6MEO-4PYRIMD)25DIME	023898-95-5
PYRAZOL-3-ON,2(2MEO-6ME-4PYRIMD)5-ME	023898-90-0
PYRAZOL-3-ON,2(6MEO-4-PYRIMDINYL)5ME	023917-23-9
PYRAZOL-3-ON,2(6MEO-4-PYRIMDIN)15DIME	023906-03-8
N7-(3-PYRAZOLYL)-MITOMYCIN C	084397-45-5
PYRAZOPHOS	013457-18-6
PYRAZOTHION	000108-35-0
1-PYRENAMINE	001606-67-3
PYRENE	000129-00-0
PYRETHRIN I	000121-21-1
PYRETHRIN II	000121-29-9
PYRICLOR	001970-40-7
PYRIDABEN	096489-71-3
PYRIDAPHENTHION	000119-12-0
4-PYRIDAZINAMINE	020744-39-2
3-PYRIDAZINAMINE,N,N-DIMETHYL	017258-31-0
4-PYRIDAZINAMINE,N,N-DIMETHYL	017258-38-7
3-PYRIDAZINAMINE, 4-HYDROXY-N,N,6-TRIMETHYL-	022390-38-1
4-PYRIDAZINAMINE, 3-METHOXY-N,N-DIMETHYL-	027772-24-3
3,6-PYRIDAZINAMINE-N,N'-TETRAME	029604-73-7
PYRIDAZINE	000289-80-5
4-PYRIDAZINEAMINE,3-MEO,N,N-DIME	038717-24-7
4-PYRIDAZINECARBONITRILE	068776-62-5
4-PYRIDAZINECARBOXAMIDE	088511-47-1
3-PYRIDAZINECARBOXAMIDE	005450-54-4
3-PYRIDAZINECARBOXYLIC ACID, METHYL ESTER	034253-02-6
4-PYRIDAZINECARBOXYLIC ACID, METHYL ESTER	034231-77-1
4-PYRIDAZINECARBOXYLIC ACID, ETHYL ESTER	039123-39-2
PYRIDAZINE, 3-CHLORO-	001120-95-2
3,6-PYRIDAZINEDIAMINE, N'-(2,5-DIMETHYL-1H-PYRRO	075842-06-7
3-PYRIDAZINAMINE, N-(2,5-DIMETHYL-1H-PYRROL-1-YL	075841-91-7
3,6-PYRIDAZINEDIAMINE, N'-(2,5-DIMETHYL-1H-PYRRO	075842-02-3
3,4-PYRIDAZINEDIOL, 6-METHYL-	022390-53-0

PYRIDAZINE, 4-ETHOXY-	024903-72-8
PYRIDAZINE, 3-ETHOXY-	062567-44-6
PYRIDAZINE, 4-METHOXY-	020733-11-3
PYRIDAZINE, 3-METHOXY-	019064-65-4
PYRIDAZINE, 3-METHYL-	001632-76-4
PYRIDAZINE, 4-METHYL-	001120-88-3
PYRIDAZINE-3,5-NH2,N,N'-TETRAME	038717-44-1
4-PYRIDAZINOL, 3-AMINO-6-METHYL-	018591-86-1
3(2H)-PYRIDAZINONE, 4-CHLORO-5-(DIMETHYLAMINO)-2	003707-98-0
3(2H)-PYRIDAZINONE, 4-CHLORO-5-(METHYLAMINO)-2-P	001698-62-0
3(2H)-PYRIDAZINONE, 6-HYDROXY-	129016-38-2
3-PYRIDINEMETHANOL,A-2-PROPYNYL-	089242-75-1
2-PYRIDINAMINE, 4-(5-AMINO-1H-1,2,4-BENZOTRIAZOL	077314-59-1
2-PYRIDINAMINE, 4-(5-AMINO-1H-1,2,4-TRIAZOL-3-YL	077314-75-1
2-PYRIDINAMINE, 4-(5-AMINO-1H-1,2,4-TRIAZOL-3-YL	077314-77-3
2-PYRIDINAMINE, 4-(5-AMINO-1H-1,2,4-TRIAZOL-3-YL	077314-73-9
2-PYRIDINAMINE, 4-(5-AMINO-1H-1,2,4-TRIAZOL-3-YL	077314-60-4
2-PYRIDINAMINE, 4-(5-AMINO-1H-1,2,4-TRIAZOL-3-YL	077314-69-3
2-PYRIDINAMINE, 4-(5-AMINO-1H-1,2,4-TRIAZOL-3-YL	098087-97-9
2-PYRIDINAMINE, 4-(5-AMINO-1H-1,2,4-TRIAZOL-3-YL	077314-76-2
2-PYRIDINAMINE, 4-(5-AMINO-1H-1,2,4-TRIAZOL-3-YL	077314-61-5
2-PYRIDINAMINE, 4-(5-AMINO-1H-1,2,4-TRIAZOL-3-YL	077314-79-5
2-PYRIDINAMINE, 4-(5-AMINO-1H-1,2,4-TRIAZOL-3-YL	077314-58-0
2-PYRIDINAMINE, 4-(5-AMINO-1H-1,2,4-TRIAZOL-3-YL	077314-64-8
2-PYRIDINAMINE, 5-NITRO-	004214-76-0
2-PYRIDINAMINE, 1-OXIDE	014150-95-9
PYRIDINE	000110-86-1
3-PYRIDINE ACETAMIDE	003724-16-1
2-PYRIDINEACETAMIDE	005451-39-8
4-PYRIDINEACETAMIDE	039640-62-5
1(2H)-PYRIDINEACETAMIDE, 3-AMINO-2-OXO-6-PHENYL-	147267-64-9
3-PYRIDINEACETAMIDE, N-(3,6-DIHYDRO-1(2H)-PYRIDI	080431-02-3
2-PYRIDINEALDOXIME	000873-69-8
4-PYRIDINEALDOXIME	000696-54-8
2-PYRIDINEAMINE, N-PHENYL	006631-37-4
1234-H4-PYRIDINE-1-T-BU-CARBONYLAMINO	080431-09-0
4-PYRIDINEBUTANAMIDE	071879-58-8
3-PYRIDINEBUTANAMIDE	084200-01-1
2-PYRIDINEBUTANAMIDE	015582-85-1
3-PYRIDINEBUTANAMIDE, N-METHYL-¥-OXO-	000713-05-3
4-PYRIDINEBUTANAMINE, MONOHYDROCHLORIDE	084359-21-7
2-PYRIDINEBUTANAMINE, MONOHYDROCHLORIDE	084359-13-7
3-PYRIDINEBUTANAMINE, MONOHYDROCHLORIDE	084359-18-2
3-PYRIDINEBUTANEAMINE	006021-23-4
2-PYRIDINEBUTANEAMINE	034974-00-0
4-PYRIDINEBUTANEAMINE	062174-83-8
3-PYRIDINEBUTANOL	060753-14-2
2-PYRIDINEBUTANOL	017945-79-8
4-PYRIDINEBUTANOL	005264-15-3
PYRIDINE, 4-BUTYL-	005335-75-1
3-PYRIDINECARBONITRILE, 2-[(2-HYDROXYPROPYL)AMIN	108611-24-1
3-PYRIDINECARBONITRILE, 2-[(3-HYDROXYPROPYL)AMIN	108611-25-2
3-PYRIDINECARBONITRILE, 2-CHLORO-5-(3,4-DIMETHOX	108611-54-7
3-PYRIDINECARBONITRILE, 6-METHYL-2-(4-MORPHOLINY	108611-26-3
3-PYRIDINECARBONITRILE, 2-CHLORO-6-METHYL-5-PHEN	010176-63-3
3-PYRIDINECARBOXAMIDE, N-(2,4-DIFLUOROPHENYL)-2-	083164-33-4
2-PYRIDINECARBOXAMIDE,NN-DIMETHYL	014805-91-5
3-PYRIDINECARBOXAMIDE, 2-AMINO-	013438-65-8

2-PYRIDINECARBOXAMIDE	001452-77-3
2-PYRIDINECARBOXAMIDE, 3-HYDROXY-	000933-90-4
2-PYRIDINECARBOXYALDEHYDE	001121-60-4
3-PYRIDINECARBOXYLIC ACID, PHENYLMETHYL ESTER	000094-44-0
4-PYRIDINECARBOXYLIC ACID, 2-IODO-, HYDRAZIDE	029247-87-8
3-PYRIDINECARBOXYLIC ACID, 2-[(1-METHYL-1H-IMIDA	092627-83-3
4-PYRIDINECARBOXYLIC ACID, 2-(1-METHYLETHYL)HYDR	000305-33-9
PYRIDINE, 4-(4-CHLOROPHENYL)-1,2,3,6-TETRAHYDRO-	005048-08-8
3,4-PYRIDINEDICARBONITRILE	001633-44-9
3,5-PYRIDINEDICARBOXYLIC ACID, DERIVATIVE	055985-32-5
PYRIDINE-2,6-DICARBOXYLIC ACID	000499-83-2
3,5-PYRIDINEDICARBOXYLIC ACID, 4-[2-[3-(1,1-DIME	103890-78-4
2,4-PYRIDINEDICARBOXYLIC ACID	000499-80-9
PYRIDINE-2,5-DICARBOXYLIC ACID	000100-26-5
3,4-PYRIDINEDIMETHANOL, 6-METHYL-5-(PHOSPHONOOXY	001883-15-4
PYRIDINE, 2,6-DIMETHOXY-	006231-18-1
3-PYRIDINE ETHANAMINE	020173-24-4
2-PYRIDINEETHANAMINE, N,N-DIMETHYL-	006304-27-4
2-PYRIDINEETHANAMINE, N-METHYL-	005638-76-6
4-PYRIDINE ETHANEAMINE	013258-63-4
3-PYRIDINE ETHANEAMINE	002706-56-1
4-PYRIDINEETHANOL	005344-27-4
3-PYRIDINEETHANOL	006293-56-7
2-PYRIDINEETHANOL	000103-74-2
1(2H)-PYRIDINEETHANOL, 3,6-DIHYDRO-4-PHENYL-	129865-48-1
PYRIDINE, 1-ETHYL-1,2,3,6-TETRAHYDRO-4-PHENYL-	013314-63-1
PYRIDINE, 4-(4-FLUOROPHENYL)-1,2,3,6-TETRAHYDRO-	001978-59-2
PYRIDINE, 4-HEPTYL-	040089-90-5
5-PYRIDINE IMADAZOLE, 2-CF3	019918-36-6
4-PYRIDINE IMIDAZOLE, 2-CF3	013797-63-2
4-PYRIDINE IMIDAZOLE,2-MES-6-CL	040852-07-1
2-PYRIDINEMETHANAMINIUM, 6-BROMO-N,N,N-TRIMETHYL	075523-37-4
2-PYRIDINEMETHANAMINIUM, 6-CHLORO-N,N,N-TRIMETHY	075523-36-3
2-PYRIDINEMETHANAMINIUM, 6-FLUORO-N,N,N-TRIMETHY	075523-35-2
2-PYRIDINEMETHANEAMINE	003731-51-9
4-PYRIDINEMETHANEAMINE	003731-53-1
3-PYRIDINEMETHANEAMINE	003731-52-0
4-PYRIDINEMETHANOL	000586-95-8
2-PYRIDINEMETHANOL	000586-98-1
3-PYRIDINEMETHANOL	000100-55-0
3-PYRIDINEMETHANOL, α,α-BIS(2-PROPYNYL)-	089242-77-3
3-PYRIDINEMETHANOL, α-CYCLOHEXYL-α-(2-PROPYNYL)-	089242-79-5
3-PYRIDINEMETHANOL, _-(4-METHYLPHENYL)-_-(2-PROP	089242-87-5
3-PYRIDINEMETHANOL, _-(PHENYLMETHYL)-_-(2-PROPEN	089242-78-4
3-PYRIDINEMETHANOL, α-PHENYL-α-(2-PROPYNYL)-	089242-81-9
3-PYRIDINEMETHANOL,A-2-PROPYNYL,A-4-FLUOROPHENYL	089242-82-0
PYRIDINE, 3-(1-METHYL-2-PYRROLIDINYL)-, N-OXIDE,	002820-55-5
PYRIDINE, 4- (4-NITROPHENYL)METHYL -	001083-48-3
PYRIDINE, 1-OXIDE	000694-59-7
4-PYRIDINE OXIDE AZOXYCYANIDE	117505-24-5
2-PYRIDINEPENTANEAMIDE	084199-97-3
4-PYRIDINEPENTANEAMIDE	084200-11-3
3-PYRIDINEPENTANEAMIDE	084200-04-4
2-PYRIDINEPENTANEAMINE	059082-57-4
4-PYRIDINEPENTANEAMINE	059082-52-9
3-PYRIDINEPENTANEAMINE	084200-05-5
3-PYRIDINEPENTANOL	084200-03-3
4-PYRIDINEPENTANOL	004343-96-8

2-PYRIDINEPROPANEAMIDE	084199-91-7
4-PYRIDINEPROPANEAMIDE	084200-07-7
3-PYRIDINEPROPANEAMIDE	084199-99-5
3-PYRIDINEPROPANEAMINE	041038-69-1
4-PYRIDINEPROPANEAMINE	030532-36-6
2-PYRIDINEPROPANEAMINE	015583-16-1
2-PYRIDINEPROPANOL	002859-68-9
3-PYRIDINEPROPANOL	002859-67-8
4-PYRIDINEPROPANOL	002629-72-3
3-PYRIDINESULFONAMIDE, N-METHYL-N-[(METHYLAMINO)	121822-80-8
3-PYRIDINESULFONAMIDE, N-[(METHYLAMINO)CARBONYL]	121822-79-5
3-PYRIDINESULFONAMIDE, N-(AMINOCARBONYL)-	113916-37-3
PYRIDINE, 1,2,3,6-TETRAHYDRO-1-METHYL-4-(2-THIEN	080353-09-9
PYRIDINE, 1,2,3,6-TETRAHYDRO-1-METHYL-4-(2-METHY	102417-86-7
PYRIDINE, 1,2,3,6-TETRAHYDRO-1-NITROSO-	055556-92-8
PYRIDINE, 1,2,3,6-TETRAHYDRO-4-PHENYL-	010338-69-9
2-PYRIDINETHIOCARBOXAMIDE	005346-38-3
2(1H)-PYRIDINETHIONE	002637-34-5
4(1H)-PYRIDINETHIONE	019829-29-9
PYRIDINIUM, 3-BROMO-1-METHYL-, IODIDE	032222-42-7
PYRIDINIUM, 3-(ETHOXYCARBONYL)-1,2,5,6-TETRAHYDR	005822-68-4
PYRIDINIUM, 2-[(HYDROXYIMINO)METHYL]-1-METHYL-,	000154-97-2
PYRIDINIUM, 1-METHYL-4-(4-METHYLPHENYL)-, IODIDE	026863-17-2
PYRIDINIUM, 1-METHYL-4-PHENYL-, IODIDE	036913-39-0
PYRIDINOL CARBAMATE	001882-26-4
4(1H)-PYRIDINONE, 3-HYDROXY-1-(2-HYDROXYETHYL)-2	030652-21-2
2(1H)-PYRIDINONE, 3-HYDROXY-1-METHYL-	019365-01-6
2(1H)-PYRIDINONE, 1-METHYL-	000694-85-9
4(1H)-PYRIDINONE, 1-METHYL-	000695-19-2
1(2H)-PYRIDINYLOXY, 3,6-DIHYDRO-2,2,6,6-TETRAMET	133960-00-6
N-2-PYRIDINYL-2-PYRIDINAMINE	001202-34-2
6H-PYRIDO[4,3-B]CARBAZOLE, 5-METHYL-	004238-66-8
9H-PYRIDO[3,4-B]INDOLE	000244-63-3
1H-PYRIDO[4,3-B]INDOLE, 5-ACETYL-2,3,4,4A,5,9B-H	143202-51-1
9H-PYRIDO[3,4-B]INDOLE-, 7-METHOXY-1-METHYL-	000442-51-3
A-PYRIDONE	000109-10-4
PYRIDO[3,2-F]-1,4-OXAZEPINE-5(2H)-THIONE, 2-[2-(	091833-49-7
PYRIDO(2,3)PYRAZINE	000322-46-3
4H-PYRIDO[1,2-A]PYRIMIDIN-3-CARBOXYLIC ACID, 6,7	080715-99-7
4H-PYRIDO[1,2-A]PYRIMIDINE-3-CARBOXYLIC ACID, 6,	070999-47-2
4H-PYRIDO[1,2-A]PYRIMIDINE-3-CARBOXAMIDE, 6,7,8,	054606-23-4
4H-PYRIDO[1,2-A]PYRIMIDINE-3-CARBOXYLIC ACID, 6,	064405-35-2
4H-PYRIDO[1,2-A]PYRIMIDINE-3-CARBOXYLIC ACID, 6,	032092-14-1
4H-PYRIDO[1,2-A]PYRIMIDINE-3-CARBOXAMIDE, 1,6,7,	064399-29-7
4H-PYRIDO[1,2-A]PYRIMIDINE-3-CARBOXAMIDE, 6,7,8,	033484-45-6
4H-PYRIDO[1,2-A]PYRIMIDINE-3-CARBOXYLIC ACID, 1,	039080-62-1
4H-PYRIDO[1,2-A]PYRIMIDINE-3-CARBOXYLIC ACID, 6,	033484-38-7
4H-PYRIDO[1,2-A]PYRIMIDINE-3-CARBOXYLIC ACID, 6,	038326-36-2
PYRIDO(12A)PYRIMIDINE-4-ONE,2ET,6ME	063736-04-9
PYRIDO(12A)PYRIMIDINE-4-ONE,3ET,6ME	057773-19-0
4H-PYRIDO(12A)PYRIMDIN-4-ON,2,6-DIME	016867-28-0
4H-PYRIDO(1,2-A)PYRIMIDIN-4-ONE	023443-10-9
PYRIDO(12A)PYRIMIDIN-4-ONE,3-ETO-CO	032092-18-5
PYRIDO(12A)PYRIMIDIN-4-ONE,3,6-DIME	039080-46-1
4H-PYRIDO[1,2-A]PYRIMIDIN-4-ONE, 2,3-DIMETHYL-	016867-29-1
4H-PYRIDO[1,2-A]PYRIMIDIN-4-ONE, 3-ETHYL-2,6-DIM	070381-44-1
4H-PYRIDO[1,2-A]PYRIMIDIN-4-ONE, 3-ETHYL-2,6-DIM	070381-46-3
4H-PYRIDO[1,2-A]PYRIMIDIN-4-ONE, 2-ETHYL-3,6-DIM	070381-71-4

PYRIDO(12A)PYRIMIDIN-4-ONE,2ME,6ET	038326-28-2
PYRIDO(12A)PYRIMIDIN-4-ONE,3-ME	039080-57-4
4H-PYRIDO(12A)PYRIMIDIN-4-ONE,2ME	001693-94-3
4H-PYRIDO(1,2-A)PYRIMIDIN-4-ONE,6-ME	023443-11-0
4H-PYRIDO[1,2-A]PYRIMIDIN-4-ONE, 7-METHYL-	023443-20-1
4H-PYRIDO[1,2-A]PYRIMIDIN-4-ONE, 2-METHYL-3-PROP	070381-53-2
4H-PYRIDO[1,2-A]PYRIMIDIN-4-ONE, 8-METHYL-	085653-81-2
4H-PYRIDO[1,2-A]PYRIMIDIN-4-ONE, 9-METHYL-	023443-12-1
4H-PYRIDO[1,2-A]PYRIMIDIN-4-ONE, 6,7,8,9-TETRAHY	032092-29-8
PYRIDO(12A)PYRIMIDIN-4-ON,3ETO-CO-8ME	034667-64-6
4H-PYRIDO[1,2-A]PYRIMIDIN-4-ONE, 6,7,8,9-TETRAHY	039080-49-4
4H-PYRIDO[1,2-A]PYRIMIDIN-4-ONE, 6,7,8,9-TETRAHY	065754-05-4
4H-PYRIDO[1,2-A]PYRIMIDIN-4-ONE, 6,7,8,9-TETRAHY	054504-62-0
4H-PYRIDO[1,2-A]PYRIMIDIN-4-ONE, 6,7,8,9-TETRAHY	054606-24-5
4H-PYRIDO[1,2-A]PYRIMIDIN-4-ONE, 6,7,8,9-TETRAHY	088437-14-3
PYRIDO(12A)PYRIMIDIN-4-ON,3ETO-CO-7ME	005435-82-5
PYRIDO(12A)PYRIMIDIN-4-ON,3ETO-CO-6ME	016867-53-1
11H-PYRIDO[2,1-B]QUINAZOLIN-11-ONE, 6,7,8,9-TETR	002446-62-0
11H-PYRIDO[2,1-B]QUINAZOLIN-11-ONE	000578-96-1
PYRIDOXAL	000066-72-8
PYRIDOXAL HYDROCHLORIDE	000065-22-5
PYRIDOXAL-5-PHOSPHATE	000054-47-7
PYRIDOXAL-5'-PHOSPHATE MONOHYDRATE	041468-25-1
PYRIDOXAMINE	000085-87-0
PYRIDOXAMINE	000524-36-7
PYRIDOXAMINE-5'-PHOSPHATE	000529-96-4
PYRIDOXINE	000065-23-6
PYRIDOXINE HYDROCHLORIDE	000058-56-0
PYRIDOXINE 5'-PHOSPHATE	000447-05-2
2-PYRIDYL-ME-(6NN-TETRA-ME)AMMONIUM BR	022337-35-5
2-PYRIDYL AZOXYCYANIDE	090467-89-3
3-PYRIDYLETHYL-2-N-PIPERIDINE	013450-66-3
3-PYRIDYLETHYL-2-(N-PYRROLIDINE)	020173-28-8
2(2-PYRIDYL)-5-ME-1,3,4-OXADIAZOLE	066079-85-4
5(2-PYRIDYL)-3-ME-1,2,4-OXADIAZOLE	001455-84-1
3(2-PYRIDYL)-5-ME-1,2,4-OXADIAZOLE	010350-68-2
4-PYRIDYL-ME(6NN-TETRA-ME)AMMONIUM BR	075523-38-5
3-PYRIDYLMETHYL-N-MORPHOLINE	017751-47-2
3-PYRIDYLMETHYL-2-PIPERIDINE	013552-35-7
3-PYRIDYLMETHYL-N-PYRROLIDINE	000370-09-2
2-PYRIDYLMETHYL TRIME AMMONIUM BROMIDE	019004-42-3
4-PYRIDYLMETHYL TRIME AMMONIUM BROMIDE	019067-63-1
3-PYRIDYLMETHYL TRIME AMMONIUM BROMIDE	016593-50-3
N7-(3-PYRIDYL)-MITOMYCIN C	078142-97-9
(4-PYRIDYL)-2-PYRIDINE	000581-47-5
N-(4-PYRIDYL)PYRIDINIUM CHLORIDE	022752-98-3
PYRILAMINE	000091-84-9
PYRIMETHAMINE	000058-14-0
4-PYRIMIDINAMINE	000591-54-8
2-PYRIMIDINAMINE, 4-[2-(1-METHYL-5-NITRO-1H-IMID	053409-75-9
5-PYRIMIDINAMINE, N,N-DIMETHYL-	031401-46-4
4-PYRIMIDINAMINE, N,N-DIMETHYL-	031401-45-3
2-PYRIMIDINAMINE, N,N-DIMETHYL-	005621-02-3
PYRIMIDINE, 5-CHLORO-	017180-94-8
PYRIMIDINE, 5-METHOXY-	031458-33-0
PYRIMIDINE	000289-95-2
PYRIMIDINE, 5-ETHOXY-	027522-25-4
PYRIMIDINE, 5-FLUORO-	000675-21-8

PYRIMIDINE, 5-NITRO-   014080-32-1
2-PYRIMIDINE AZOXYCYANIDE   117539-78-3
PYRIMIDINE, 2-BROMO-   004595-60-2
2-PYRIMIDINECARBONITRILE, 4-METHOXY-   094789-37-4
4-PYRIMIDINECARBONITRILE   042839-04-3
5-PYRIMIDINECARBONITRILE   040805-79-6
2-PYRIMIDINECARBONITRILE   014080-23-0
1-PYRIMIDINECARBOXAMIDE, 5-FLUORO-1,3-DIHYDRO-N-   064098-85-7
1(2H)-PYRIMIDINECARBOXAMIDE, N-ETHYL-5-FLUORO-3,   058471-47-9
1(2H)-PYRIMIDINECARBOXAMIDE, N-BUTYL-5-FLUORO-3,   064098-82-4
5-PYRIMIDINECARBOXAMIDE   040929-49-5
4-PYRIMIDINECARBOXAMIDE   028648-86-4
1(2H)-PYRIMIDINECARBOXAMIDE, 5-FLUORO-3,4-DIHYDR   056563-18-9
1(2H)-PYRIMIDINECARBOXAMIDE, 5-FLUORO-3,4-DIHYDR   060908-29-4
1(2H)-PYRIMIDINECARBOXAMIDE, 5-FLUORO-3,4-DICHLO   056563-17-8
2-PYRIMIDINECARBOXAMIDE   088511-48-2
4-PYRIMIDINECARBOXYLIC ACID, 1,2,3,6-TETRAHYDRO-   004450-00-4
4-PYRIMIDINECARBOXYLIC ACID, 1,2,3,6-TETRAHYDRO-   088280-81-3
4-PYRIMIDINECARBOXYLIC ACID, 1,2,3,6-TETRAHYDRO-   001747-53-1
4-PYRIMIDINECARBOXYLIC ACID, 1,2,3,6-TETRAHYDRO-   004598-39-4
4-PYRIMIDINECARBOXYLIC ACID, 1,2,3,6-TETRAHYDRO-   004450-04-8
4-PYRIMIDINECARBOXYLIC ACID, 1,2,3,6-TETRAHYDRO-   004450-01-5
4-PYRIMIDINECARBOXYLIC ACID, 1,2,3,6-TETRAHYDRO-   006153-44-2
4-PYRIMIDINECARBOXYLIC ACID, 1,2,3,6-TETRAHYDRO-   022754-37-6
4-PYRIMIDINECARBOXYLIC ACID, 1,2,3,6-TETRAHYDRO-   004450-03-7
4-PYRIMIDINECARBOXYLIC ACID, 1,2,3,6-TETRAHYDRO-   124979-37-9
2-PYRIMIDINECARBOXYLIC ACID, METHYL ESTER   034253-03-7
5-PYRIMIDINECARBOXYLIC ACID, ETHYL ESTER   040929-50-8
2-PYRIMIDINECARBOXYLIC ACID, ETHYL ESTER   042839-08-7
4-PYRIMIDINECARBOXYLIC ACID, METHYL ESTER   002450-08-0
5-PYRIMIDINECARBOXYLIC ACID, METHYL ESTER   034253-01-5
PYRIMIDINE, 4-CHLORO-   017180-93-7
PYRIMIDINE, 2-CHLORO-   001722-12-9
2,4-PYRIMIDINEDIAMINE, 5-[(2-BROMO-4,5-DIMETHOXY   024849-83-0
2,4-PYRIMIDINEDIAMINE, 5-[(2-BROMO-5-METHOXY-4-P   004891-98-9
2,4-PYRIMIDINEDIAMINE, 5-[(6-BROMO-1,3-BENZODIOX   083157-97-5
2,4-PYRIMIDINEDIAMINE, 5-[(3,4-DIETHOXYPHENYL)ME   080267-16-9
2,4-PYRIMIDINEDIAMINE, 5-[[4-(DIMETHYLAMINO)-3-M   085544-47-4
2,4-PYRIMIDINEDIAMINE, 5-[(3,5-DIETHOXYPHENYL)ME   100515-03-5
2,4-PYRIMIDINEDIAMINE, 5-(3,4-DICHLOROPHENYL)-6-   018588-57-3
2,4-PYRIMIDINEDIAMINE, 5-[[3-(DIMETHYLAMINO)-4-M   085544-46-3
2,4-PYRIMIDINEDIAMINE, 5-[[4-(DIMETHYLAMINO)-3,5   056066-63-8
2,4-PYRIMIDINEDIAMINE, 5-[(4-ETHOXY-3-METHOXYPHE   085544-41-8
2,4-PYRIMIDINEDIAMINE, 5-[(3-ETHOXY-4-METHOXYPHE   073046-15-8
2,4-PYRIMIDINEDIAMINE, 5-[(2-IODO-3,4,5-TRIMETHO   085544-57-6
2,4-PYRIMIDINEDIAMINE, 5-[[4-METHOXY-3-(1-METHYL   073264-20-7
2,4-PYRIMIDINEDIAMINE, 5-[[3-METHOXY-4-(1-METHYL   073264-21-8
2,4-PYRIMIDINEDIAMINE, 5-[(4-METHOXY-3-PROPOXYPH   085555-96-0
2,4-PYRIMIDINEDIAMINE, 5-[(3-METHOXY-4-PROPOXYPH   080267-17-0
2,4-PYRIMIDINEDIAMINE, 6-(4-MORPHOLINYL)-, 3-OXI   055921-64-7
2,4-PYRIMIDINEDIAMINE, 5-[(2,3,4-TRIMETHOXYPHENY   006981-04-0
2,4(1H,3H)-PYRIMIDINEDIONE, 3-ACETYL-5-FLUORO-   075410-15-0
2,4(1H,3H)-PYRIMIDINEDIONE, 5-(BUTYLAMINO)-1,6-D   031991-98-7
2,4(1H,3H)-PYRIMIDINEDIONE, 5-CHLORO-1-(2-FLUORO   138686-07-4
2,4(1H,3H)-PYRIMIDINEDIONE, 1-[(4-CHLOROPHENYL)M   085093-33-0
2,4(1H,3H)-PYRIMIDINEDIONE, 5-[(2-CHLOROETHYL)(2   000834-91-3
2,4(1H,3H)-PYRIMIDINEDIONE, 1-[(2,4-DICHLOROPHEN   125111-05-9
2,4(1H,3H)-PYRIMIDINEDIONE, 1-(2,3-DIDEOXY-2-FLU   141249-32-3

2,4(1H,3H)-PYRIMIDINEDIONE, 1-(2,3-DIDEOXY-2-FLU	124424-25-5
2,4(1H,3H)-PYRIMIDINEDIONE, 5-(DIETHYLAMINO)-3,6	032150-38-2
2,4(3H,5H)-PYRIMIDINEDIONE, 5,5-DIETHYL-6-(1-MET	085432-38-8
4,6(1H,5H)-PYRIMIDINEDIONE, 5,5-DIETHYL-2-(1-MET	085445-03-0
2,4(1H,3H)-PYRIMIDINEDIONE, 1-[2,5-DIHYDRO-5-(HY	005974-93-6
4,6(1H,5H)-PYRIMIDINEDIONE, DIHYDRO-5,5-DIMETHYL	007729-78-4
2,4(1H,3H)-PYRIMIDINEDIONE, 5-(DIMETHYLAMINO)-6-	032150-76-8
2,4(1H,3H)-PYRIMIDINEDIONE, 1,6-DIMETHYL-5-(4-MO	032150-74-6
2,4(1H,3H)-PYRIMIDINEDIONE, 5-(DIMETHYLAMINO)-6-	031992-01-5
2,4(1H,3H)-PYRIMIDINEDIONE, 5-(DIMETHYLAMINO)-6-	032150-60-0
2,4(1H,3H)-PYRIMIDINEDIONE, 5-(DIETHYLAMINO)-1,6	031991-99-8
2,4(1H,3H)-PYRIMIDINEDIONE, 1,3-DIMETHYL-	000874-14-6
2,4(1H,3H)-PYRIMIDINEDIONE, 5-(DIMETHYLAMINO)-1,	038507-32-3
2,4(1H,3H)-PYRIMIDINEDIONE, 1,6-DIMETHYL-5-[(2-M	072775-85-0
2,4(1H,3H)-PYRIMIDINEDIONE, 1,6-DIMETHYL-3-PHENY	032150-73-5
4,6(1H,5H)-PYRIMIDINEDIONE, 5,5-DIPROPYL-2-THIOX	099167-62-1
2,4(3H,5H)-PYRIMIDINEDIONE, 5-ETHYL-6-(1-METHYLE	085432-39-9
4,6(1H,5H)-PYRIMIDINEDIONE, 5-ETHYL-2-(1-METHYLE	085432-37-7
2,4(1H,3H)-PYRIMIDINEDIONE, 5-FLUORO-1-[(4-METHY	125111-04-8
2,4(1H,3H)-PYRIMIDINEDIONE, 5-FLUORO-1-(PHENYLME	004871-13-0
2,4(1H,3H)-PYRIMIDINEDIONE, 5-FLUORO-3-(1-OXOBUT	094452-21-8
2,4(1H,3H)-PYRIMIDINEDIONE, 5-FLUORO-1-(1-OXOPRO	056058-99-2
2,4(1H,3H)-PYRIMIDINEDIONE, 5-FLUORO-1-[(4-METHO	085093-35-2
2,4(1H,3H)-PYRIMIDINEDIONE, 5-IODO-	000696-07-1
2,4(1H,3H)-PYRIMIDINEDIONE, 6-[[3-[4-(2-METHOXYP	034661-75-1
2,4,6(1H,3H,5H)-PYRIMIDINETRIONE, 5-METHYL-5-(3-	066843-01-4
PYRIMIDINE-2,4-DIONE,3-PRO-6-ME	077317-85-2
PYRIMIDINE, 2-ETHOXY-	003739-82-0
PYRIMIDINE, 2-FLUORO-	031575-35-6
PYRIMIDINE,2-(3ME-5MEO-1PYRAZOL)4MEO	023917-24-0
PYRIMIDINE,6(3ME5MEO-1PYRAZOL)24DIMEO	023905-98-8
PYRIMIDINE,2(5MEO-3ME-1PYRAZ)4MEO6ME	018694-40-1
PYRIMIDINE,4(5MEO-3ME-1-PYRAZOL)	018694-41-2
PYRIMIDINE,4(3MEO-5ME-1PYRAZOL)26DIME	023903-41-5
PYRIMIDINE,4(5MEO-3ME-1-PYRAZOL)6MEO	018694-42-3
PYRIMIDINE,2(3MEO-5ME-1PYRAZOL)46DIME	023905-77-3
PYRIMIDINE,6(5MEO3ME-1PYRAZOL)2ME4MEO	023903-42-6
PYRIMIDINE,2(5MEO-3ME-1PYRAZ)4,6DIME	018694-43-4
5-PYRIMIDINEMETHANOL, .ALPHA.-(2-CHLOROPHENYL)-.	063284-71-9
PYRIMIDINE, 2-METHOXY-	000931-63-5
PYRIMIDINE, 4-METHOXY-	006104-41-2
PYRIMIDINE, 2-METHYL-	005053-43-0
PYRIMIDINE, 5-METHYL-	002036-41-1
PYRIMIDINE, 2-(METHYLTHIO)-	000823-09-6
PYRIMIDINE, 4-PHENYL-	003438-48-0
PYRIMIDINE, 2-(1-PIPERAZINYL)-	020980-22-7
PYRIMIDINE-2-THIOCARBOXAMIDE	004537-73-9
2,4,6-PYRIMIDINETRIAMINE	001004-38-2
2,4,6-PYRIMIDINETRIAMINE, N -METHYL-1-OXIDE	055973-02-9
2,4,6(1H,3H,5H)-PYRIMIDINETRIONE, 5-BUTYL-1-METH	034569-18-1
2,4,6(1H,3H,5H)-PYRIMIDINETRIONE, 5,5-DIMETHYL-	024448-94-0
5-PYRIMIDINOL, HEXAHYDRO-1,3-DIMETHYL-2-[(5-NITR	037385-07-2
4-PYRIMIDINOL, 2-(METHYLTHIO)- (KETO FORM)	124700-70-5
2(1H)-PYRIMIDINONE, 4-AMINO-1-(2,3-DIDEOXY-2-FLU	119555-47-4
2(1H)-PYRIMIDINONE, 4-AMINO-1-(2,3-DIDEOXY-2-FLU	128496-09-3
2(1H)-PYRIMIDINONE, 4-AMINO-1-[5-O-(1-OXOHEXADEC	031088-06-9
2(1H)-PYRIMIDINONE, 4-AMINO-1-(2,3,5-TRI-O-ACETY	006742-07-0
4(1H)-PYRIMIDINONE, 2-[[4-(3-METHOXY-2-PYRIDINYL	071351-79-6

2-PYRIMIDONE	000557-01-7
4-PYRIMIDONE	004562-27-0
PYROLAN	000087-47-8
PYROMELLITIC ACID	000089-05-4
PYROMELLITIC DIANHYDRIDE	000089-32-7
PYROQUILON	057369-32-1
PYROTATARIC ANHYDRIDE	004100-80-5
1H-PYRROL-2-AMINE, N-[2-[[(5-METHYL-1H-IMIDAZOL-	074188-69-5
PYRROLE	000109-97-7
PYRROLE-2-CARBOXALDEHYDE	001003-29-8
1H-PYRROLE-2-CARBOXAMIDE	004551-72-8
1H-PYRROLE-2-CARBOXAMIDE, 4,5-DIBROMO-3-(1-METHY	113589-14-3
1H-PYRROLE-2-CARBOXAMIDE, N-ETHYL-1-METHYL-	124725-22-0
1H-PYRROLE-2-CARBOXAMIDE, 1-METHYL-	064230-41-7
1H-PYRROLE-2-CARBOXAMIDE, N-METHYL-	132911-42-3
1H-PYRROLE-2-CARBOXAMIDE, N-PROPYL-	120309-46-8
1H-PYRROLE-2-CARBOXYLIC ACID, 1-METHYL-, METHYL	037619-24-2
PYRROLE-2-CARBOXYLIC ACID	000634-97-9
1H-PYRROLE-2-CARBOXYLIC ACID, ETHYL ESTER	002199-43-1
PYRROLE-3-CARBOXYLIC ACID-4-METHYL,ETHYL ESTER	002199-49-7
1H-PYRROLE-3,4-DICARBOXYLIC ACID, 2-(5-CHLORO-2-	116383-80-3
1H-PYRROLE-3,4-DICARBOXYLIC ACID, 2-(4-CHLORO-2-	116383-85-8
1H-PYRROLE-3,4-DICARBOXYLIC ACID, 2-(6-CHLORO-2-	127390-68-5
1H-PYRROLE-2,5-DIONE	000541-59-3
1H-PYRROLE-2,5-DIONE, 1-[(ACETYLOXY)METHYL]-3-ME	117491-56-2
1H-PYRROLE-2,5-DIONE, 1-[(ACETYLOXY)METHYL]-3,4-	117491-57-3
1H-PYRROLE-2,5-DIONE, 1-PHENYL-	000941-69-5
1H-PYRROLE, 2-PHENYL-	003042-22-6
PYRROLE, 1-PHENYL-	000635-90-5
PYRROLIDINE	000123-75-1
1-PYRROLIDINEACETAMIDE, N-(2,3-DIHYDRO-5-METHYL-	105801-57-8
1-PYRROLIDINEACETAMIDE, N-(2,3-DIHYDRO-3-BENZOFU	105801-67-0
1-PYRROLIDINEACETAMIDE, N-(PHENYLMETHYL)-	042175-93-9
1-PYRROLIDINEETHANAMINE	007154-73-6
PYRROLIDINE, 2-PHENYL-	001006-64-0
2-PYRROLIDINONE, 1-ACETYL-	000932-17-2
2-PYRROLIDINONE, 1,5-DIMETHYL-	005075-92-3
2-PYRROLIDINONE, 1-ETHYL-	002687-91-4
2-PYRROLIDINONE, 5-HYDROXY-1-METHYL-5-(3-PYRIDIN	061192-50-5
2-PYRROLIDINONE, 3-HYDROXY-1-METHYL-5-(3-PYRIDIN	034834-67-8
2-PYRROLIDINONE, 3-HYDROXY-1-METHYL-5-(3-PYRIDIN	037096-14-3
2-PYRROLIDINONE, 5-METHYL-	000108-27-0
2-PYRROLIDINONE, 1-METHYL-5-(3-PYRIDINYL)-, (S)-	000486-56-6
2-PYRROLIDINONE, 1-(1-OXOBUTYL)-	022707-38-6
2-PYRROLIDINONE, 1-(1-OXODECYL)-	033602-03-8
2-PYRROLIDINONE, 5-(3-PYRIDINYL)-, N-OXIDE, (S)-	101708-63-8
2-PYRROLIDINONE, 5-(3-PYRIDINYL)-	005980-06-3
1-PYRROLIDINYLOXY, 3-[[[[BIS(1-AZIRIDINYL)PHOSPH	103981-99-3
1-PIPERIDINYLOXY, 4-[[[[BIS(1-AZIRIDINYL)PHOSPHI	096662-64-5
1-PIPERIDINYLOXY, 4-[[[[BIS(2,2-DIMETHYL-1-AZIRI	103981-92-6
1-PYRROLIDINYLOXY, 3-[[[[BIS(2,2-DIMETHYL-1-AZIR	103981-93-7
1-PYRROLIDINYLOXY, 3,4-DICARBOXY-2,2,5,5-TETRAME	083040-20-4
PYRROLIDONE	000616-45-5
1H-PYRROLIZINE-1-CARBOXYLIC ACID, 2,3-DIHYDRO-5-	096327-61-6
1H-PYRROLIZINE-1-CARBOXYLIC ACID, 5-BENZOYL-2,3-	107115-73-1
1H-PYRROLIZINE-1-CARBOXYLIC ACID, 2,3-DIHYDRO-5-	096327-65-0
1H-PYRROLIZINE-1-CARBOXYLIC ACID, 6-ETHYL-5-(4-F	083073-37-4
1H-PYRROLIZINE-1-CARBOXYLIC ACID, 5-(CYCLOBUTYLC	096327-63-8

1H-PYRROLIZINE-1-CARBOXYLIC ACID, 5-(4-CHLOROBEN	083073-36-3
1H-PYRROLIZINE-1-CARBOXYLIC ACID, 5-(4-FLUOROBEN	066635-90-3
1H-PYRROLIZINE-1-CARBOXYLIC ACID, 5-(4-BROMOBENZ	096327-34-3
1H-PYRROLIZINE-1-CARBOXYLIC ACID, 5-(2,4-DICHLOR	096327-58-1
1H-PYRROLIZINE-1-CARBOXYLIC ACID, 5-(2,6-DIFLUOR	096327-56-9
1H-PYRROLIZINE-1-CARBOXYLIC ACID, 5-(4-FLUOROBEN	107115-74-2
1H-PYRROLIZINE-1-CARBOXYLIC ACID, 5-(2,4-DIFLUOR	096327-57-0
1H-PYRROLIZINE-1-CARBOXYLIC ACID, 2,3-DIHYDRO-5-	096327-38-7
1H-PYRROLIZINE-1-CARBOXYLIC ACID, 2,3-DIHYDRO-5-	096327-32-1
1H-PYRROLIZINE-1-CARBOXYLIC ACID, 2,3-DIHYDRO-5-	096327-51-4
1H-PYRROLIZINE-1-CARBOXYLIC ACID, 2,3-DIHYDRO-5-	066635-84-5
1H-PYRROLIZINE-1-CARBOXYLIC ACID, 2,3-DIHYDRO-5-	096327-52-5
1H-PYRROLIZINE-1-CARBOXYLIC ACID, 2,3-DIHYDRO-5-	096327-39-8
1H-PYRROLIZINE-1-CARBOXYLIC ACID, 2,3-DIHYDRO-5-	096327-60-5
1H-PYRROLO[1,2-A]INDOLE-5,8-DIONE, 1-(ACETYLOXY)	096000-48-5
1H-PYRROLO[1,2-A]INDOLE-5,8-DIONE DERIVATIVE	096000-47-4
1H-PYRROLO[3,2-C]PYRIDINE, 2-(2,4-DIMETHOXYPHENY	120236-88-6
4H-PYRROLO[3,2-D]PYRIMIDIN-4-ONE, 2,6-DIAMINO-3,	115812-51-6
4H-PYRROLO[3,2-D]PYRIMIDIN-4-ONE, 2,6-DIAMINO-3,	115787-59-2
PYRROLO[2,1-B]QUINAZOLIN-9(1H)-ONE, 2,3-DIHYDRO-	000530-53-0
PYRROLO[2,1-B]QUINAZOLIN-9(1H)-ONE, 2,3-DIHYDRO-	070934-02-0
PYRROLO[2,1-B]QUINAZOLIN-9(1H)-ONE, 2,3,5,6,7,8-	088491-50-3
1H-PYRROL-1-YLOXY, 2,5-DIHYDRO-2,2,5,5-TETRAMETH	109871-96-7
1H-PYRROL-1-YLOXY, 3-[[[2-(DIMETHYLAMINO)ETHYL]M	133959-98-5
PYRUVIC ACID	000127-17-3
QUERCETIN	000117-39-5
QUERCITRIN	000522-12-3
QUINALIZARIN	000081-61-8
QUINALPHOS	013593-03-8
QUINAZOLIDIN-4-ONE	000491-36-1
4-QUINAZOLINAMINE	015018-66-3
QUINAZOLINE	000253-82-7
QUINAZOLINE-2,4-DIONE,6-ME	062484-16-6
QUINAZOLINE, 2-PHENYL-	025855-20-3
4-QUINAZOLINOL	017227-47-3
4(1H)-QUINAZOLINONE, 2,3-DIHYDRO-2-(1-NAPHTHALEN	031785-60-1
4-(3H)-QUINAZOLINONE,3-HYDROXY	005319-71-1
QUINAZOLIN-2-ONE,1-ME-4-PHENYL	017629-04-8
2(1H)-QUINAZOLINONE, 6-METHYL-4-PHENYL-	013961-64-3
4(3H)-QUINAZOLINONE, 3-METHYL-	002436-66-0
2(1H)-QUINAZOLINONE, 4-PHENYL-	023441-75-0
4(1H)-QUINAZOLINONE, 2-PHENYL-	001022-45-3
QUINIDINE	000056-54-2
QUINIDINE HYDROCHLORIDE	001668-99-1
QUININE	000130-95-0
QUININE HYDROBROMIDE	000549-49-5
QUININE, MONOHYDROCHLORIDE	000130-89-2
2-QUINOLINAMINE, N,N-DIMETHYL-	021154-18-7
4-QUINOLINAMINE,8-NITRO	089770-26-3
5-QUINOLINAMINE, 6-NITRO-	035975-00-9
QUINOLINE	000091-22-5
4-QUINOLINE AZOXYCYANIDE	117505-22-3
2-QUINOLINECARBOXALDEHYDE	005470-96-2
3-QUINOLINECARBOXAMIDE,N,N-DIMETHYL	025283-63-0
3-QUINOLINECARBOXAMIDE	006480-67-7
6-QUINOLINECARBOXYLIC ACID HYDRAZIDE	005382-47-8
3-QUINOLINECARBOXYLIC ACID, 1-ETHYL-6,8-DIFLUORO	099726-76-8
6-QUINOLINECARBOXYLIC ACID,ET ESTER	073987-38-9

QUINOLINE, 4,7-DICHLORO-	000086-98-6
QUINOLINE, 4,7-DICHLORO-, 1-OXIDE	001077-74-3
QUINOLINE,8-ETHYLCARBONATE	042322-29-2
4-QUINOLINEMETHANOL, 2-(1,1-DIMETHYLETHYL)-α-[2-	086073-85-0
6-QUINOLINEMETHANOL, 1,2,3,4-TETRAHYDRO-2-[[(1-M	021738-42-1
QUINOLINE, 6-METHOXY-8-NITRO-	000085-81-4
QUINOLINE, 2-METHYL-8-NITRO-	000881-07-2
QUINOLINE-1-OXIDE	001613-37-2
4-QUINOLINE,1-OXIDE AZOXYCYANIDE	117505-23-4
QUINOLINE, 2-(1-PIPERAZINYL)-	004774-24-7
QUINOLINE, 1,2,3,4-TETRAHYDRO-8-[2-HYDROXY-3-[(1	076252-06-7
QUINOLINIC ACID	000089-00-9
3-QUINOLINOL	000580-18-7
7-QUINOLINOL	000580-20-1
5-QUINOLINOL	000578-67-6
8-QUINOLINOL	000148-24-3
8-QUINOLINOL, 5-AMINO-	013207-66-4
8-QUINOLINOL,5-CHLORO-	000130-16-5
2-(1H)-QUINOLINONE	000059-31-4
4(1H)-QUINOLINONE	000529-37-3
2(1H)-QUINOLINONE, 3,4-DIHYDRO-	000553-03-7
2(1H)-QUINOLINONE, 3,4-DIHYDRO-3-PHENYL-	001022-66-8
2(1H)-QUINOLINONE, 3,4-DIHYDRO-4-PHENYL	004888-33-9
2(1H)-QUINOLINONE, 3-PHENYL-	038035-81-3
QUINONE CHLOROIMIDE	000637-61-6
QUINOPHALONE	000083-08-9
QUINOXALINE	000091-19-0
2-QUINOXALINECARBOXAMIDE	005182-90-1
QUINOXALINE-1,4-DIHYDRO-2,3-DIONE	015804-19-0
2,3-QUINOXALINEDIONE,1,4-DIHYDRO-6,7-DIMETHYL-	002474-50-2
QUINOXALINE, 2-(METHYLTHIO)-	021948-73-2
QUINOXALINE-2-ONE	001196-57-2
QUINOXALINE, 2-PHENYL-	005021-43-2
QUINUCLIDINE	000100-76-5
QUIZALOFOP-P-ETHYL	100646-51-3
QUIZALOFOP-ETHYL	076578-14-8
QUIZALOFOP-P-TEFURYL	119738-06-6
RAFFINOSE	000512-69-6
REMOXIPRIDE	080125-14-0
REPOSAL	003625-25-0
RESERPINE	000050-55-5
RESMETHRIN	010453-86-8
RESORCINOL	000108-46-3
RETINOIC ACID	000302-79-4
RH-1909 O-PH(2-CL) N-BUTYLENE CARBAMATE	143121-06-6
RH-1911 O-PH(4-CL) N-BUTYLENE CARBAMATE	001759-02-0
RH-1422 O-PH N-BUTYLENE CARBAMATE	133636-98-3
RH-1964 O-PH(DICL) N-BUTYLENE CARBAMATE	133636-96-1
RH-4663 O-PH N-BUTYLENE CARBAMATE	087365-63-7
RH-1965 O-PH N-BUTYLENE CARBAMATE	133636-94-9
RH-1908 O-PH N-BUTYLENE CARBAMATE	055379-71-0
RH-0978 O-PH N-BUTYLENE(F2)CARBAMATE	143121-10-2
RH-9611 O-PH N-BUTYLENEN CARBAMATE	143121-08-8
RHODAMINE B	000081-88-9
RH-7160 O-PH(24DICL) N-DIET CARBAMATE	143121-07-7
RH-6251 O-PH N-PENTYLENE CARBAMATE	087374-78-5
RH-1224 O-PH N-PENTYLENE(F2)CARBAMATE	138926-22-4
RIBOFLAVIN	000083-88-5

1-(B-D-RIBOFURANOSYL)-2-NO2IMIDAZOLE	017306-43-3
RIBOSE	000050-69-1
RICIDINE	000524-40-3
ROKITAMYCIN	074014-51-0
RONIDAZOLE	007681-76-7
RONNEL	000299-84-3
ROPIVACAINE	084057-95-4
ROSANILINE	000632-99-5
ROSOLIC ACID	000603-45-2
ROTENONE	000083-79-4
ROVRAL	036734-19-7
RUBEANIC ACID	000079-40-3
SAFROLE	000094-59-7
SALICIN	000138-52-3
SALICYLANILIDE	000087-17-2
SALICYLIC ACID	000069-72-7
SALITHION	003811-49-2
SANTONIN	000481-06-1
SARIN	000107-44-8
SCOPOLAMINE	000051-34-3
SCOPOLAMINE BROMIDE	000114-49-8
SEARLE, SC-17127	006987-59-3
8-SEBENZYL CYCLIC AMP	052109-40-7
8-SEBENZYL CYCLIC AMP-O-BENZYL	107538-74-9
SECBUMETON	026259-45-0
SECOBARBITAL	000076-73-3
SELANONAPHTHENE	000272-30-0
1,2,5-SELENADIAZOLE, 3,4-DIMETHYL-	017505-12-3
SELENIUM, BIS[3-[(METHYLSULFONYL)OXY]PROPYL]-	126585-48-6
SEMICARBAZIDE	000563-41-7
L-SERINAMIDE, N-ACETYL-	023361-38-8
SERINE	000056-45-1
SESAMEX	000051-14-9
SESONE	000136-78-7
3-SH-4-NH2-6IPR-124TRIAZINE-5-ONE	022278-77-9
SIDURON	001982-49-6
SILANE, BUTYL-DIMETHYL	001001-52-1
SILANE, BUTYL TRIMETHYL	001000-49-3
SILANE, DIMETHYL PHENYL	000766-77-8
SILANE, DIMETHYL PROPYL	018143-31-2
SILANE, TRIETHOXYPHENYL-	000780-69-8
SILANE, TRIMETHYL PHENYL	000768-32-1
SILANE, TRIMETHYLPROPYL-	003510-70-1
SILVEX	000093-72-1
SIMAZINE	000122-34-9
SIMETONE	000673-04-1
SIMETRYN	001014-70-6
SIMVASTATIN	079902-63-9
SIMVASTATIN, 5-HYDROXY	121009-77-6
3-SME-4-NH2-6-ET-124-TRIAZIN-5-ONE	021087-59-2
3-SME-4-NH2-6-HX-124-TRIAZINE-5-ONE	021085-20-1
3-SME-4-NH2-6-IBU-124-TRIAZIN-5-ONE	021087-62-7
3-SME-4-NH2-6-IPE-124-TRIAZIN-5-ONE	021085-18-7
3-SME-4-NH2-6-IPR-124-TRIAZIN-5-ONE	021087-61-6
3-SME-4-NH2-6-PH-124-TRIAZIN-5-ONE	021087-63-8
3-SME-4-NH2-6-PR-124-TRIAZIN-5-ONE	021087-60-5
P-SO2F PHENOXYACETIC ACID	034380-54-6
N-(4-SO2F-BENZYL)-PYRIDINIUM BROMIDE	033802-91-4

N1-(4-SO2NH2-PH)SULFANILAMIDE	000547-52-4
1-SO3NA-3-BENZYL-7-ISOPROPYL AZULENE	099287-50-0
1-SO3NA-3-BUTYL AZULENE	097683-34-6
1-SO3NA-3-BUTYL-7-ISOPROPYL AZULENE	099287-46-4
1-SO3NA-3-BUTYL-4-METHOXY AZULENE	099287-58-8
1-SO3NA-3-ET AZULENE	097683-30-2
1-SO3NA-3-ET-7-ISOPROPYL AZULENE	097683-31-3
1-SO3NA-3-ET-4-METHOXY AZULENE	099287-54-4
1-SO3NA-3-ET-4-METHOXY-7-IPR AZULENE	099287-65-7
1-SO3NA-7-ISOPROPYL AZULENE	114482-97-2
1-SO3NA-3-ME AZULENE	097683-29-9
1-SO3NA-3-ME-7-ISOPROPYL AZULENE	114482-98-3
1-SO3NA-3-ME-4-METHOXY AZULENE	099287-52-2
1-SO3NA-3-ME-4-METHOXY-7-IPR AZULENE	099287-64-6
1-SO3NA-3-PE-4-METHOXY-7-IPR AZULENE	099287-70-4
1-SO3NA-3-PENTYL AZULENE	099287-40-8
1-SO3NA-3-PENTYL-7-ISOPROPYL AZULENE	099287-48-6
1-SO3NA-3-PENTYL-4-METHOXY AZULENE	099287-60-2
1-SO3NA-3-PR-4-METHOXY-7-IPR AZULENE	099287-66-8
1-SO3NA-3-PROPYL AZULENE	099287-38-4
1-SO3NA-3-PROPYL-7-ISOPROPYL AZULENE	099287-44-2
1-SO3NA-3-PROPYL-4-METHOXY AZULENE	099287-56-6
SODIUM HEXANOATE	000151-33-7
SODIUM P-NITROPHENOXIDE	014609-74-6
SODIUM P-PHENYLPHENOXIDE	003645-61-2
SODIUM SALICYLATE	000054-21-7
SODIUM TRICHLOROACETATE	000650-51-1
SOLAN	002307-68-8
ALPHA-SOLANINE	020562-02-1
SOLVENT RED 1	001229-55-6
SOLVENT YELLOW 33	008003-22-3
SOMAN	000096-64-0
SORBIC ACID	000110-44-1
SORBITOL	000050-70-4
SOTALOL	000959-24-0
SOTALOL	003930-20-9
SPARSOMYCIN	001404-64-4
SPARSOMYCIN,S-PENTYL ANALOG	119410-37-6
SPARSOMYCIN,S-PROPYL ANALOG	114991-36-5
L-SPARTEINE	000090-39-1
SPARTICIDE	041205-21-4
SPIRO[AZULENO[4,5-B]FURAN-6(2H),2'-OXIRAN]-2-ONE	010215-89-1
SPIRONOLACTONE	000052-01-7
SPIRONOLACTONE,7-BETA ANALOG	033784-05-3
SPIROPENTANE	000157-40-4
SPIROPERIDOL	000749-02-0
SSF-109 FUNGICIDE	129586-32-9
STEARIC ACID	000057-11-4
N-STEAROYL-P-AMINOPHENOL	000103-99-1
STILBENE	000103-30-0
STILBENE	000588-59-0
STIROFOS	000961-11-5
STREPTOMYCIN	000057-92-1
STREPTOZOCIN	018883-66-4
STROPHANTHIDIN	000066-28-4
STROSPESIDE	000595-21-1
STRYCHNINE	000057-24-9
STRYCHNINE HYDRIODIDE	052748-69-3

STRYCHNINE NITRATE	000066-32-0
STRYCHNINE, 2-NITRO-	029854-52-2
STYRENE	000100-42-5
STYRENE OXIDE	000096-09-3
SUBERIC ACID	000505-48-6
SUCCINAMIDE	000110-14-5
SUCCINIC ANHYDRIDE	000108-30-5
SUCCINIMIDE	000123-56-8
SUCCINONITRILE	000110-61-2
SUCCINYL CHLORIDE	000543-20-4
SUCROSE	000057-50-1
SUDOXICAM	029152-10-1
SUFENTANIL	056030-54-7
SULBENCILLIN	041744-40-5
SULFADIAZINE	000068-35-9
SULFADIMETHOXINE	000122-11-2
SULFALENE	000152-47-6
SULFALLATE	000095-06-7
SULFAMERAZINE	000127-79-7
SULFAMETER	000651-06-9
SULFAMETHAZINE	000057-68-1
SULFAMETHIAZOLE	000144-82-1
SULFAMETHOMIDINE	003772-76-7
SULFAMETHOXAZOLE	000723-46-6
SULFAMETHOXYPYRIDAZINE	000080-35-3
4-SULFAMYLBENZAMIDE,N-HEXYL	059777-65-0
4-SULFAMYLBENZAMIDE,N-PROPYL	005462-24-8
4-SULFAMYLBENZOIC ACID, ME ESTER	022808-73-7
SULFANILACETAMIDE	000144-80-9
SULFANILAMIDE	000063-74-1
2-SULFANILAMIDO-5-ETHYLPYRIMIDINE	003271-01-0
2-SULFANILAMIDOPYRIMIDINE	000599-82-6
SULFANILYL GUANADINE	000057-67-0
SULFAPERINE	000599-88-2
SULFAPHENAZOLE	000526-08-9
SULFAPYRIDINE	000144-83-2
SULFATHIOZOLE	000072-14-0
1,1'-SULFINYLBIS(2-CHLOROETHANE)	005819-08-9
1,1'-SULFINYLBISETHENE	001115-15-7
SULFISOMIDINE	000515-64-0
SULFISOXAZOLE	000127-69-5
SULFOMETURON (pH 5-7)	074222-97-2
SULFOMETURON	074223-56-6
3-SULFONAMIDOPYRIDINE	002438-76-8
6-SULFONAMIDOQUINOLINE	089770-31-0
8-SULFONAMIDOQUINOLINE	035203-91-9
CF3-SULFONANILIDE, P-PHENYL	050585-77-8
SULFONE, PHENYL, METHYL-	000426-58-4
1,1'-SULFONYLBIS(2-CHLOROETHANE)	000471-03-4
1,1'-SULFONYLBISETHENE	000077-77-0
SULFORIDAZINE	014759-06-9
SULFOTEPP	003689-24-5
SULFOXIDE	000120-62-7
SULFOXYPHENYL PYRAZOLIDINE	000057-96-5
SULINDAC	038194-50-2
SULPHENONE	000080-00-2
SULPIRIDE	015676-16-1
SULPROFOS	035400-43-2

SULTOPRIDE	053583-79-2
SYDNONE, 3-(CARBOXYMETHYL)-	026537-53-1
SYNEPHRINE	000094-07-5
TABUN	000077-81-6
TACRINE	000321-64-2
TALIPEXOLE	101626-70-4
TARTARIC ACID, DIETHYL ESTER	000087-91-2
TARTARIC ACID	000087-69-4
TAURINE	000107-35-7
2,4,5-TB	000093-80-1
2,4,5-T, BUTOXYETHYL ESTER	002545-59-7
2,4,5-T, N-BUTYL ESTER	000093-79-8
TEBUCONAZOLE	107534-96-3
TEBUFENOZIDE	112410-23-8
TEBUTHIURON	034014-18-1
TEFLUBENZURON	083121-18-0
TEGAFUR	017902-23-7
TEH-PYRIMIDIN26DIONE,4-IPR CARBOXYLATE	004450-26-4
TEMAZEPAM	000846-50-4
TENIPOSIDE	029767-20-2
TENULIN	019202-92-7
TEPA	000545-55-1
TEPA ADAMANTYL-UREA DERIVATIVE	103981-98-2
TEPA ADAMANTYL-UREA DERIVATIVE	065101-39-5
TEPA 4-(N-OH-PIPERIDINYL)-UREA DERIVATIV	096662-65-6
TEPA 4-(N-OH-PIPERIDINYL)-UREA DERIVATIV	103981-94-8
TEPA 4-(N-OH-PYRROLDINYL)-UREA DERIVATIV	103981-96-0
TEPA 4-(N-OH-PYRROLDINYL)-UREA DERIVATIV	103981-95-9
TEPA 4-(PIPERIDINYL)-UREA DERIVATIVE	096662-66-7
TEPA 4-(PIPERIDINYL)-UREA DERIVATIVE	103981-97-1
1,1,4,4-TETRACHLORO-2,3-DICHLOROMETHYL-2-BUTENE	083682-30-8
TERBACIL	005902-51-2
TERBUFOS	013071-79-9
TERBUFOS SULFONE	056070-16-7
TERBUFOS SULFOXIDE	010548-10-4
TERBUMETON	033693-04-8
TERBUTHYLAZINE	005915-41-3
TERBUTRYN	000886-50-0
TEREPHTHALIC ACID	000100-21-0
O-TERPHENYL	000084-15-1
1,1':4',1"-TERPHENYL	000092-94-4
m-TERPHENYL	000092-06-8
GAMMA-TERPINENE	000099-85-4
TERPINEOL	008000-41-7
BETA-TERPINEOL	000138-87-4
ALPHA-TERPINEOL	000098-55-5
TERPIN HYDRATE	002451-01-6
TERPINOLENE	000586-62-9
2,2':6',2"-TERPYRIDINE	001148-79-4
TERRAZOLE	002593-15-9
2,4,5-TES	003570-61-4
TESTOSTERONE	000058-22-0
TESTOSTERONE ENANTYHATE	000315-37-7
2,4,5-T, ETHYL ESTER	001928-39-8
2,4,5-T, 2-ETHYLHEXYL ESTER	001928-47-8
2,2,6,6-TETME-N-NITROSOPIPERDINE	006130-93-4
1,4,5,8-TETRAAMINOANTHRAQUINONE	002475-45-8
4567-TETRABR 2-CF3 BENZIMIDAZOLE	002338-30-9

1,2,4,5-TETRABROMOBENZENE	000636-28-2
1,2,3,4-TETRABROMOBENZENE	022311-25-7
1,2,3,5-TETRABROMOBENZENE	000634-89-9
TETRABROMOBISPHENOL A BIS(ALLYL ETHER)	025327-89-3
1,2,3,4-TETRABROMOBUTANE	001529-68-6
2,3,7,8-TETRABROMODIBENZO-P-DIOXIN	050585-41-6
1,1,2,2-TETRABROMOETHANE	000079-27-6
4,5,6,7-TETRABROMO-1,3-ISOBENZOFURANDIONE	000632-79-1
TETRACAINE	000094-24-6
TETRACENE	000092-24-0
2,3,5,6-TETRACHLOROANILINE	003481-20-7
2,3,4,5-TETRACHLOROANILINE	000634-83-3
2,3,4,5-TETRACHLOROANISOLE	000938-86-3
2,3,4,6-TETRACHLOROANISOLE	000938-22-7
3,3',4,4'-TETRACHLOROAZOBENZENE	014047-09-7
1,2,3,5-TETRACHLOROBENZENE	000634-90-2
1,2,3,4-TETRACHLOROBENZENE	000634-66-2
1,2,4,5-TETRACHLOROBENZENE	000095-94-3
TETRACHLORO-1,2-BENZENEDIOL	001198-55-6
4,5,6,7-TETRACHLOROBENZOTRIAZOLE	002338-10-5
2,2',5,6'-TETRACHLOROBIPHENYL	041464-41-9
2,2',4,5'-TETRACHLOROBIPHENYL	041464-40-8
2,3',4,4'-TETRACHLOROBIPHENYL	032598-10-0
2,2',4,6-TETRACHLOROBIPHENYL	062796-65-0
2,2'4,5-TETRACHLOROBIPHENYL	070362-47-9
2,2',3,4-TETRACHLOROBIPHENYL	052663-59-9
3,3',5,5'-TETRACHLOROBIPHENYL	033284-52-5
2,2',3,5'-TETRACHLOROBIPHENYL	041464-39-5
2,2',3,3'-TETRACHLOROBIPHENYL	038444-93-8
2,2',4,4'-TETRACHLOROBIPHENYL	002437-79-8
1,2,3,4-TETRACHLOROBUTA-1,3-DIENE	001637-31-6
1,1,4,4-TETRACHLOROBUTA-1,3-DIENE	036038-53-6
1,1,2,3-TETRACHLOROBUTA-1,3-DIENE	000921-09-5
1,2,3,3-TETRACHLOROBUTANE	013138-51-7
2,2,3,3,-TETRACHLOROBUTANE	014499-87-7
1,1,4,4-TETRACHLOROBUTANE	033455-24-2
1,2,3,4-TETRACHLOROBUTANE	003405-32-1
TETRACHLOROCYCLOHEXENE	033875-95-5
TETRACHLOROCYCLOHEXENE	000319-81-3
TETRACHLOROCYCLOHEXENE	028810-38-0
TETRACHLOROCYCLOHEXENE	041992-55-6
TETRACHLOROCYCLOHEXENE	001782-00-9
2-F-3-BR-TETRACHLOROCYCLOHEXANE	055265-51-5
TETRACHLOROCYCLOHEXENE (345)	089674-88-4
TETRACHLOROCYCLOHEXENE (345)	089674-85-1
TETRACHLOROCYCLOHEXENE (345)	089674-87-3
1,3,7,8-TETRACHLORODIBENZO-P-DIOXIN	050585-46-1
1,2,3,4-TETRACHLORODIBENZO-P-DIOXIN	030746-58-8
1,3,6,8-TETRACHLORODIBENZO-P-DIOXIN	033423-92-6
1,2,3,7-TETRACHLORODIBENZODIOXIN	067028-18-6
2,3,7,8-TETRACHLORODIBENZO-P-DIOXIN	001746-01-6
1,3,7,9-TETRACHLORODIBENZO-P-DIOXIN	062470-53-5
2,3,7,8-TETRACHLORODIBENZOFURAN	051207-31-9
1,1,1,2-TETRACHLORO-2,2-DIFLUOROETHANE	000076-11-9
1,1,2,2,-TETRACHLORODIFLUOROETHANE	000076-12-0
1,2,4,5-TETRACHLORO-3,6-DIMETHYLBENZENE	000877-10-1
2,2',4,4'-TETRACHLORODIPHENYL ETHER	028076-73-5
1,1,1,2-TETRACHLOROETHANE	000630-20-6

1,1,2,2-TETRACHLOROETHANE	000079-34-5
TETRACHLOROETHENE	000127-18-4
TETRACHLOROETHYLENE OXIDE	016650-10-5
TETRACHLOROGUAIACOL	002539-17-5
2,2,4,4-TETRACHLORO-2,2,4,4,6,6-HEXAHYDRO-6,6-B*	065601-40-3
2,2'-4,6'-TETRACHLORO-3-METHYL DIPHENYLMETHANE	121107-47-9
2,2'-4,4'-TETRACHLORO-3-METHYL DIPHENYLMETHANE	121107-43-5
2,2'-4,5'-TETRACHLORO-5-METHYL DIPHENYLMETHANE	121107-46-8
2,3'-4,4'-TETRACHLORO-5-METHYL DIPHENYLMETHANE	121107-65-1
2,3'-4,4'-TETRACHLORO-6-METHYL DIPHENYLMETHANE	121107-77-5
2',3-4,6'-TETRACHLORO-6-METHYL DIPHENYLMETHANE	121107-83-3
2,2'-4,4'-TETRACHLORO-5-METHYL DIPHENYLMETHANE	121107-44-6
2,2'-4,6'-TETRACHLORO-5-METHYL DIPHENYLMETHANE	121107-48-0
2,2'-5,5'-TETRACHLORO-4-METHYL DIPHENYLMETHANE	121107-55-9
1,2,3,5-TETRACHLORONAPHTHALENE	053555-63-8
1,3,5,7-TETRACHLORONAPHTHALENE	053555-64-9
1,4,6,7-TETRACHLORONAPHTHALENE	055720-43-9
1,2,3,4-TETRACHLORONAPHTHALENE	020020-02-4
1,3,5,8-TETRACHLORONAPHTHALENE	031604-28-1
2,3,5,6-TETRACHLORO-4-NITROANILINE	002438-88-2
2,3,5,6-TETRACHLORONITROBENZENE	000117-18-0
2,3,4,5-TETRACHLORONITROBENZENE	000879-39-0
2,3,5,6-TETRACHLOROPHENOL	000935-95-5
2,3,4,5-TETRACHLOROPHENOL	004901-51-3
2,3,4,6-TETRACHLOROPHENOL	000058-90-2
TETRACHLOROPHTHALIDE	027355-22-2
1,1,2,3-TETRACHLOROPROPANE	018495-30-2
2,3,5,6-TETRACHLOROTOLUENE	001006-31-1
2,3,4,5-TETRACHLOROTOLUENE	001006-32-2
2,3,5,6-TETRACHLORPYRIDINE	002402-79-1
TETRACHLORVINPHOS	022248-79-9
TETRACONTANE	004181-95-7
TETRACOSANE	000646-31-1
TETRACYCLINE	000060-54-8
TETRACYCLOTETRAMETHYLENECYCLOTETRASILOXANE	000177-49-1
TETRADECAMETHYLCYCLOHEPTASILOXANE	000107-50-6
TETRADECAMETHYLHEXASILOXANE	000107-52-8
TETRADECANAMIDE, 2,2-DIMETHYL-N-METHYL-3-OXO-N-(	150423-89-5
1-TETRADECANAMINE	002016-42-4
TETRADECANE	000629-59-4
TETRADECANOIC ACID	000544-63-8
1-TETRADECANOL	000112-72-1
1-TETRADECANOL ACETATE	000638-59-5
1-TETRADECENE	001120-36-1
9-TETRADECEN-1-OL ACETATE (Z)	016725-53-4
11-TETRADECEN-1-OL (Z)	034010-15-6
9-TETRADECEN-1-OL (Z)	035153-15-2
2-ME PYRIDINIUM IODIDE,N-TETRADECYL	074639-30-8
N-TETRADECYLCYCLOHEXANE	001795-18-2
TETRADECYLPYRIDINIUM BROMIDE	001155-74-4
TETRADIFON	000116-29-0
TETRAETHYL AMMONIUM BROMIDE	000071-91-0
TETRAETHYL AMMONIUM CHLORIDE	000056-34-8
TETRAETHYL AMMONIUM IODIDE	000068-05-3
TETRAETHYLENEGLYCOL DIMETHACRYLATE	000109-17-1
TETRAETHYLENE GLYCOL	000112-60-7
TETRAETHYLENEPENTAMINE	000112-57-2
TETRAETHYL LEAD	000078-00-2

TETRAETHYL PYROPHOSPHATE	000107-49-3
TETRAETHYLSILANE	000631-36-7
1,2,3,4-TETRAFLUOROBENZENE	000551-62-2
1,2,3,5-TETRAFLUOROBENZENE	002367-82-0
1,2,4,5-TETRAFLUOROBENZENE	000327-54-8
1,1,1,2-TETRAFLUOROETHANE	000811-97-2
TETRAFLUOROETHYLENE	000116-14-3
TETRAFLUOROMETHANE	000075-73-0
2,2,3,3-TETRAFLUOROPROPANOL	000076-37-9
TETRAFLUTHRIN	079538-32-2
BENZENEACETAMIDE, N-[2-[[3-[(1,2,3,4-TETRAHYDRO-	121509-02-2
TETRAHYDROFURAN	000109-99-9
2,2,5,5-TETRAMETHYL-TETRAHYDROFURAN	015045-43-9
TETRAHYDROFURFURYL ALCOHOL	000097-99-4
2(5678-TETRAHYDRONAPHTHYLOXY)ACETIC ACID	001878-59-7
N-(TETRAHYDRO-2-OXO-3-THIENYL)ACETAMIDE	001195-16-0
TETRAHYDROPYRAN	000142-68-7
TETRAHYDROPYRAN-2-METHANOL	000100-72-1
1,2,3,4-TETRAHYDROQUINOLINE	000635-46-1
TETRAHYDROTHIOPHENE-1,1-DIOXIDE	000126-33-0
TETRAHYDROTHIOPHENE	000110-01-0
TETRAHYDRO-2H-THIOPYRAN	001613-51-0
TETRAKIS(HYDROXYMETHYL)PHOSPHONIUM CHLORIDE	000124-64-1
TETRAKIS(TRIMETHYLSILOXY)SILANE	003555-47-3
TETRALIN	000119-64-2
2356-TETRAME-N,N'-DIN=O PIPERAZINE	023264-57-5
N,N,N',N'-TETRAME ETHYLENEDIAMINE	000110-18-9
2,3,5,6-TETRAMETHOXY-P-BENZOQUINONE	003117-06-4
6',7',10,11-TETRAMETHOXYEMETAN	000483-18-1
TETRAMETHOXYHYDROQUINONE	052092-59-8
1,3,5,7-TETRAMETHYL-1,1,3,5,7,7-HEXAPHENYLTETR*	038421-40-8
2,2',5,5'-TETRAMETHYLAZOBENZENE	006311-44-0
2,2',4,6-TETRAMETHYLAZOBENZENE	029418-30-2
2,4,4',6-TETRAMETHYLAZOBENZENE	029418-29-9
2,2',3,3'-TETRAMETHYLAZOBENZENE	029418-34-6
1,2,3,4-TETRAMETHYLBENZENE	000488-23-3
1,2,4,5-TETRAMETHYLBENZENE	000095-93-2
1,2,3,5-TETRAMETHYLBENZENE	000527-53-7
N,N,N',N'-TETRAMETHYLBENZIDINE	000366-29-0
2,2,3,3-TETRAMETHYLBUTANE	000594-82-1
N-((1,1,3,3-TETRAMETHYLBUTYL)PHENYL)-1-NAPHTHA*	051772-35-1
P-(1,1,3,3-TETRAMETHYLBUTYL)PHENOL	000140-66-9
1,1,3,3-TETRAMETHYL-1,3-DIPHENYLDISILOXANE	000056-33-7
1,1'-TETRAMETHYLENE-BIS-CNU	060784-43-2
2,2,5,5-TETRAMETHYLHEXANE	001071-81-4
2,2,3,3-TETRAMETHYLHEXANE	013475-81-5
2,2,4,5-TETRAMETHYLHEXANE	016747-42-5
TETRAMETHYLMELAMINE	002827-47-6
2,3,5,6-TETRAMETHYL-4-NITROBENZENAMINE	013171-61-4
2,2,4,4-TETRAMETHYLPENTANE	001070-87-7
2,2,3,4-TETRAMETHYLPENTANE	001186-53-4
2,2,3,3-TETRAMETHYLPENTANE	007154-79-2
2,3,3,4-TETRAMETHYLPENTANE	016747-38-9
N,N,N',N'-TETRAMETHYL-P-PHENYLENEDIAMINE	000100-22-1
2,3,5,6-TETRAMETHYLPYRIDINE	003748-84-3
TETRAMETHYLSILANE	000075-76-3
1,3,3,5-TETRAMETHYL-1,1,5,5-TETRAPHENYLTRISILO*	003982-82-9
TETRAMETHYLTHIOUREA	002782-91-4

TETRAMETHYLTHIURAM	000097-74-5
2,4,6,8-TETRAMETHYL-2,4,6,8-TETRAPHENYLCYCLOTE*	000077-63-4
TETRAMETHYLUREA	000632-22-4
TETRANITROMETHANE	000509-14-8
1,4,7,10-TETRAOXACYCLODODECANE	000294-93-9
2,5,8,11-TETRAOXADODECANE	000112-49-2
TETRAPENTYLAMMONIUM IODIDE	002498-20-6
TETRAPROPYL AMMONIUM IODIDE	000631-40-3
TETRAPROPYL THIOPYROPHOSPHORATE	003244-90-4
TETRASUL	002227-13-6
TETRATETRACONTANE	007098-22-8
N-TETRATRIACONTANE	014167-59-0
TETRAZEPAM	010379-14-3
1H-TETRAZOLE	000288-94-8
2H-TETRAZOLE-2-ACETIC ACID, 5-[4-[[(DIMETHYLAMIN	117121-49-0
1H-TETRAZOLE-1-ACETIC ACID, 5-[4-[[(DIMETHYLAMIN	117121-37-6
TETRAZOLE,2-ETHYL-5-(2-CHLORO-4-DIMETHYLUREYL)PH	117121-51-4
1H-TETRAZOLE, 1-METHYL-5-[(PHENYLMETHYL)THIO]-	056610-79-8
1H-TETRAZOLE, 1-PHENYL-	005378-52-9
1H-TETRAZOLE, 5-(PHENYLMETHYL)-	018489-25-3
TETRAZOLO[1,5-A]PYRIDINE	000274-87-3
TETRYL	000479-45-8
THALIDOMIDE	000050-35-1
THANITE	000115-31-1
THEBAINE PICRATE	005967-77-1
THENOYLTRIFLUOROMETHYLACETONE	000326-91-0
THEOBROMINE	000083-67-0
THEOPHYLLINE	000058-55-9
4-THIA-1-AZABICYCLO[3.2.0]HEPTANE-2-CARBOXYLIC A	026605-69-6
4-THIA-1-AZABICYCLO[3.2.0]HEPTANE DERIVATIVE	038087-61-5
4-THIA-1-AZABICYCLO[3.2.0]HEPTANE-2-COOH DERIVAT	052248-39-2
4-THIA-1-AZABICYCLO[3.2.0]HEPTANE-2-CARBOXYLIC A	055390-39-1
4-THIA-1-AZABICYCLO[3.2.0]HEPTANE-2-CARBOXYLIC A	060903-21-1
5-THIA-1-AZABICYCLO[4.2.0]OCT-2-ENE-2-CARBOXYLIC	017942-66-4
5-THIA-1-AZABICYCLO[4.2.0]OCT-2-ENE-2-CARBOXYLIC	068661-74-5
5-THIA-1-AZABICYCLO[4.2.0]OCT-2-ENE-2-CARBOXYLIC	080929-58-4
5-THIA-1-AZABICYCLO[4.2.0]OCT-2-ENE-2-CARBOXYLIC	130927-06-9
5-THIA-1-AZABICYCLO[4.2.0]OCT-2-ENE-2-CARBOXYLIC	065243-53-0
5-THIA-1-AZABICYCLO[4.2.0]OCT-2-ENE-2-CARBOXYLIC	058333-58-7
5-THIA-1-AZABICYCLO[4.2.0]OCT-2-ENE-2-CARBOXYLIC	056796-20-4
5-THIA-1-AZABICYCLO[4.2.0]OCT-2-ENE-2-CARBOXYLIC	084104-57-4
5-THIA-1-AZABICYCLO[4.2.0]OCT-2-ENE-2-CARBOXYLIC	123822-90-2
5-THIA-1-AZABICYCLO[4.2.0]OCT-2-ENE-2-CARBOXYLIC	123822-92-4
THIABENDAZOLE	000148-79-8
THIABURIMAMIDE	038603-23-5
THIACETIC ACID, S-PHENYL ESTER	000934-87-2
THIACYCLOBUTANE	000287-27-4
1,2,3-THIADIAZOLE-5-CARBOXALDOXIME	061444-94-8
1,2,5-THIADIAZOLE-3-CARBOXIMIDOTHIOIC ACID, N- H	090507-24-7
1,3,4-THIADIAZOLE, 2,5-DIMETHYL-	027464-82-0
1,2,5-THIADIAZOLE, 3,4-DIMETHYL-	005728-21-2
1,3,4-THIADIAZOLE, 2-METHYL-5-[(PHENYLMETHYL)THI	042755-32-8
2-THIAHEXANE	000628-29-5
THIAMYLAL	000077-27-0
5-THIANONANE	000544-40-1
THIANTHRENE	000092-85-3
THIAZAFLURON	025366-23-8
2-THIAZOLAMINE, 5-NITRO-	000121-66-4

THIAZOLE	000288-47-1
4-THIAZOLECARBOXAMIDE, 2-(beta-D-RIBOFURANOSYL)-	060084-10-8
2-THIAZOLEETHANAMINE	018453-07-1
2-THIAZOLEETHANAMINE, 4-PHENYL-	061887-92-1
THIAZOLE, 2-PHENYL-	001826-11-5
THIAZOLIDINE, 2-(M-CHLOROPHENYL)	060980-82-7
THIAZOLIDINE, 2-(P-CL PHENYL)	007738-99-0
2,4-THIAZOLIDINEDIONE, 3-(1-METHYLETHYL)-5-[(5-N	027550-11-4
2,4-THIAZOLIDINEDIONE, 3-METHYL-5-(5-NITROFURFUR	025603-07-0
2,4-THIAZOLIDINEDIONE, 5-[(5-NITRO-2-FURANYL)MET	052661-71-9
THIAZOLIDINE,2-(P-NITROPHENYL)	000831-25-4
THIAZOLIDINE,2-(M-NITROPHENYL)	075808-93-4
4-THIAZOLIDINONE, 3-ETHYL-2-IMINO-5-[(5-NITRO-2-	025603-08-1
4-THIAZOLIDINONE, 2-IMINO-3-METHYL-5-[(5-NITRO-2	025580-69-2
4-THIAZOLIDINONE, 3-METHYL-2-(MEHTYLIMINO)-5-[(5	025603-09-2
4(5H)-THIAZOLONE, 2-[(1-METHYLETHYL)AMINO]-5-[(5	027472-92-0
4(5H)-THIAZOLONE, 5-[(5-NITRO-2-FURANYL)METHYLEN	027472-90-8
N7-(2-THIAZOLYL)-MITOMYCIN C	078327-27-2
THIDIAZURON	051707-55-2
2H-THIENO[2,3-E]-1,4-DIAZEPIN-2-ONE, 5-(2-CHLORO	033671-46-4
THIENO[2,3-B]FURAN-2-SULFONAMIDE, 5-[[(2-FLUOROE	122321-32-8
THIENO[2,3-B]FURAN-2-SULFONAMIDE, 5-(4-MORPHOLIN	122321-30-6
THIENO[2,3-D]PYRIMIDINE, 4-AMINO-	014080-56-9
THIENO(2,3-D)-PYRMIDINE,4-CHLORO	016269-66-2
THIENO(23D)PYRIMIDINE-4-DEMEAMINO	063893-52-7
THIENO(23D)-PYRIMIDINE,4(MEAMINO)	056844-20-3
THIENO[2,3-D]PYRIMIDINE, 4-MORPHOLINYL-	018740-23-3
THIENO[2,3-B]THIOPHENE-2-SULFONAMIDE, 5-[[[2-(ME	133445-74-6
THIENO[2,3-B]THIOPHENE-2-SULFONAMIDE, 5-[[[2-(ME	122267-01-0
THIENO[2,3-B]THIOPHENE-2-SULFONAMIDE, 5-[[[2-(ME	122266-99-3
THIENO[2,3-B]THIOPHENE-2-SULFONAMIDE, 5-[[(2-MET	122266-90-4
THIENO[3,2-B]THIOPHENE-2-SULFONAMIDE, 5-[[(2-MET	127025-29-0
THIENO[2,3-B]THIOPHENE-2-SULFONAMIDE, 5-[[(2-HYD	122266-93-7
4H-THIENO[2,3-B]THIOPYRAN-2-SULFONAMIDE, 5,6-DIH	105951-31-3
4H-THIENO[2,3-B]THIOPYRAN-2-SULFONAMIDE, 5,6-DIH	106335-79-9
4H-THIENO[2,3-B]THIOPYRAN-2-SULFONAMIDE, 5,6-DIH	105951-35-7
4H-THIENO[2,3-B]THIOPYRAN-2-SULFONAMIDE, 5,6-DIH	106400-04-8
4H-THIENO[2,3-B]THIOPYRAN-2-SULFONAMIDE, 5,6-DIH	105951-71-1
4H-THIENO[2,3-B]THIOPYRAN-2-SULFONAMIDE, 5,6-DIH	106319-38-4
5H-THIENO[3,2-B]THIOPYRAN-2-SULFONAMIDE, 6,7-DIH	106319-44-2
5H-THIENO[3,2-B]THIOPYRAN-2-SULFONAMIDE, 6,7-DIH	105951-36-8
4H-THIENO[2,3-B]THIOPYRAN-2-SULFONAMIDE, 5,6-DIH	106319-42-0
5H-THIENO[3,2-B]THIOPYRAN-2-SULFONAMIDE, 6,7-DIH	105951-32-4
5H-THIENO[3,2-B]THIOPYRAN-2-SULFONAMIDE, 6,7-DIH	106319-46-4
5H-THIENO[3,2-B]THIOPYRAN-2-SULFONAMIDE, 6,7-DIH	106319-45-3
5H-THIENO[3,2-B]THIOPYRAN-2-SULFONAMIDE, 6,7-DIH	105951-39-1
THIOACETAMIDE	000062-55-5
THIOACETANILIDE	000637-53-6
THIOANISOLE	000100-68-5
1-(O-THIOANISYL)-33-DIME TRIAZINE	052416-13-4
1-(3-THIOANISYL)-33-DIME TRIAZINE	052416-14-5
THIOBARBITAL	000077-32-7
THIOBENCARB	028249-77-6
THIOBENZAMIDE	002227-79-4
2-THIOBENZYLPYRIDINE OXIDE	003915-60-4
2,2'-THIOBISETHANOL	000111-48-8
1,1'-THIOBISETHENE	000627-51-0
N-THIO-(N'-BU-O-ME-CARBAMYL)CARBOFURAN	068672-91-3

G-THIOBUTYROLACTONE	039700-44-2
THIOCARBANILIDE	000102-08-9
2-(THIOCYANATEMETHYLTHIO)BENZOTHIAZOLE	021564-17-0
THIOCYANIC ACID, BUTYL ESTER	000628-83-1
4,4'-THIODIANILINE	000139-65-1
THIODICARB	059669-26-0
N(THIO-N,O-DIME CARBAMYL)CARBOFURAN	065907-26-8
THIOGUANINE	000154-42-7
N-THIO-(N'ME-O-TBU-CARBAMYL)CARBOFURAN	086569-98-4
N-THIO-(N'-ME-O-BU-CARBAMYL)CARBOFURAN	065907-30-4
THIOMETON	000640-15-3
THIONAZIN	000297-97-2
THIONICOTINAMIDE	004621-66-3
THIOPENTAL	000076-75-5
THIOPHANATE-METHYL	023564-05-8
THIOPHEN-2-ACETIC ACID, METHYL ESTER	019432-68-9
THIOPHEN-2-CARBOXAMIDE	005813-89-8
THIOPHEN-3-CARBOXAMIDE	051460-47-0
THIOPHEN-2-COCO-O-ETHYL	004075-58-5
THIOPHENE	000110-02-1
2,5-THIOPHENYL-THIOPHENE	001081-34-1
THIOPHENE-2-CARBOXALDEHYDE	000098-03-3
3-THIOPHENECARBOXALDEHYDE	000498-62-4
2-THIOPHENECARBOXAMIDE, 4-BROMO-5-METHOXY-3-(1-M	113589-05-2
2-THIOPHENECARBOXAMIDE, N-(3,6-DIHYDRO-1(2H)-PYR	080431-15-8
2-THIOPHENECARBOXAMIDE, 4,5-DIBROMO-3-(1-METHYLE	113588-97-9
2-THIOPHENECARBOXAMIDE, 4-BROMO-5-METHYL-3-(1-ME	113588-95-7
THIOPHENE-2-CARBOXYLIC ACID	000527-72-0
THIOPHENE-3-CARBOXYLIC ACID	000088-13-1
THIOPHENE, 2,3-DIBROMO-	003140-93-0
THIOPHENE, 2-ETHYL CARBOXYLATE	002810-04-0
THIOPHENE, 3-METHYLSULFONYL-	038695-58-8
2-THIOPHENESULFONAMIDE, 4-(4-HYDROXYBENZOYL)-	118993-57-0
2-THIOPHENESULFONAMIDE, 5-[(2-HYDROXYETHYL)THIO]	104437-96-9
2-THIOPHENESULFONAMIDE, 5-[(2-HYDROXYETHYL)SULFO	104438-00-8
2-THIOPHENESULFONAMIDE, 4-(4-METHYLBENZOYL)-	119018-06-3
2-THIOPHENESULFONAMIDE, 4-(4-METHOXYBENZOYL)-	118976-97-9
2-THIOPHENESULFONAMIDE, 5-[[3-(METHOXYACETYL)OXY	104438-05-3
2-THIOPHENESULFONAMIDE, 4-[(4-METHOXYPHENYL)SULF	119731-18-9
2-THIOPHENESULFONAMIDE, 5-[(4-HYDROXYBUTYL)THIO]	135832-38-1
2-THIOPHENESULFONAMIDE, 5-[(3-HYDROXYPROPYL)THIO	104437-99-2
2-THIOPHENESULFONAMIDE, 4-[(3-HYDROXYPHENYL)SULF	128348-43-6
2-THIOPHENESULFONAMIDE, 4-[(4-HYDROXYPHENYL)SULF	128348-42-5
THIOPHENE,2-THIOCARBOXAMIDE	020300-02-1
THIOPHENOL	000108-98-5
1-(THIOPHEN-2-YL)ACETALDEHYDE OXIME	059445-83-9
THIORIDAZINE	000050-52-2
THIOSEMICARBAZIDE	000079-19-6
THIOTEPA	000052-24-4
THIOTHIXENE	022189-31-7
2-THIOURACIL	000156-82-1
THIOUREA	000062-56-6
THIOUREA, N,N'-DICYCLOHEXYL-	001212-29-9
THIOUREA, N,N-DIMETHYL-	006972-05-0
THIOUREA, N,N-DIMETHYL-N'-PHENYL-	000705-62-4
THIOUREA, N-(2,2,3,3,4,4,4-HEPTAFLUOROBUTYL)-N'-	105412-23-5
THIOUREA, N-HEXYL-N'-METHYL-	053393-06-9
THIOUREA, N-METHYL-N'-[2-[[1-METHYL-1H-IMIDAZOL-	074633-67-3

THIOUREA, N-METHYL-N'-[4-(5-METHYL-1H-IMIDAZOL-4	051264-00-7
THIOUREA, N-METHYL-N'-TRICYCLO[3.3.1.1,3,7]DEC-1	025444-83-1
THIOUREA, N'-(PHENYLMETHYL)-N,N-DIMETHYL	002741-14-2
9H-THIOXANTHEN-9-ONE	000492-22-8
THIRAM	000137-26-8
L-THREO-ALPHA-D-GALACTO-OCTOPYRANOSIDE, DERIVATI	013441-66-2
L-THREO-ALPHA-D-GALACTO-OCTOPYRANOSIDE	019096-41-4
THREONIN-AMIDE, N-ACETYL	060828-33-3
THREONINE(L)	000072-19-5
L-THREO-PENTONAMIDE DERIVATIVE	137302-31-9
THUJONE	000546-80-5
THYMIDINE	000050-89-5
THYMIDINE, 5'-AMINO-5'-DEOXY-	025152-20-9
THYMIDINE, 3'-AZIDO-3'-DEOXY-	030516-87-1
THYMIDINE, 3'-AZIDO-3'-DEOXY-4-THIO-	108441-45-8
THYMIDINE, 3'-DEOXY-	003416-05-5
THYMIDINE, 3'-DEOXY-3'-FLUORO-	025526-93-6
THYMIDINE, 3'-DEOXY-4-THIO-	122568-03-0
THYMIDINE, 2',3'-DIDEHYDRO-3'-DEOXY-4-THIO-	005983-08-4
THYMIDINE, 2',3'-DIDEHYDRO-3'-DEOXY-	003056-17-5
THYMIDINE, 4-THIO-	007236-57-9
THYMOL	000089-83-8
THYROXINE	000051-48-9
TIANAFAC	051527-19-6
TIARAMIDE	032527-55-2
TIGLIC ACID	000080-59-1
TIMOLOL	026839-75-8
TINIDAZOLE	019387-91-8
TINIDAZOLE-METHYL	019387-94-1
TINIDAZOLE-PROPYL	019387-93-0
TINIDAZOLE-ISOPROPYL	019390-40-0
TIOCARBAZIL	036756-79-3
2,4,5-T, ISOOCTYL ESTER	025168-15-4
2,4,5-T, METHYL ESTER	001928-37-6
TOLAZAMIDE	001156-19-0
TOLBUTAMIDE	000064-77-7
TOLCLOFOS-METHYL	057018-04-9
TOLCYCLAMIDE	000664-95-9
TOLFENAMIC ACID	013710-19-5
TOLMETIN	026171-23-3
P-TOLUALDEHYDE	000104-87-0
TOLUENE	000108-88-3
2,4-TOLUENEDIAMINE	000095-80-7
2,4-TOLUENE DIISOCYANATE	000584-84-9
TOLUENE-2,6-DIISOCYANATE	000091-08-7
P-TOLUENESULFINIC ACID	000536-57-2
M-TOLUENESULFONAMIDE	001333-07-9
P-TOLUENESULFONIC ACID	000104-15-4
O-TOLUENE SULFONIC ACID	000088-20-0
O-TOLUENESULFONYL CHLORIDE	000133-59-5
P-TOLUENESULFONYL CHLORIDE	000098-59-9
ALPHA-TOLUENE SULFONYLUREA	121822-75-1
P-TOLUIC ACID	000099-94-5
M-TOLUIC ACID	000099-04-7
O-TOLUIC ACID	000118-90-1
O-TOLUIC ACID, METHYL ESTER	000089-71-4
P-TOLUIDINE	000106-49-0
M-TOLUIDINE	000108-44-1

O-TOLUIDINE	000095-53-4
M-TOLUIDINE-N,N-DIMETHYL	000121-72-2
TOLYFLUANIDE	000731-27-1
O-TOLYLACETATE	000533-18-6
M-TOLYLACETATE	000122-46-3
P-TOLYLACETATE	000140-39-6
2-TOLYL 4-AMINOSALICYLATE	056356-13-9
4-TOLYL 4-AMINOSALICYLATE	056356-15-1
3-TOLYL 4-AMINOSALICYLATE	056356-14-0
O-(O-TOLYLAZO)PHENOL	029418-37-9
M-TOLYLDIMETHYLCARBAMATE	007305-07-9
O-TOLYLDIMETHYLCARBAMATE	007305-06-8
P-TOLYLDIMETHYLCARBAMATE	007305-08-0
1-(3-TOLYL)-3,3-DIMETHYLTRIAZENE	020241-03-6
2-(3-TOLYL)INDOLIZINE	082682-69-7
2-(4-TOLYL)INDOLIZINE	007496-81-3
2-(M-TOLYL)MORPHOLINE	062008-56-4
N-(M-TOLYL)-3-N'-PIPERIDINOACETAMIDE	038367-20-3
N-(O-TOLYL)-3-N'-PIPERIDINOACETAMIDE	013993-02-7
N-(P-TOLYL)-3-N'-PIPERIDINOACETAMIDE	005429-42-5
N-(P-TOLYL)SUCCINIMIDE	002314-79-6
N-(M-TOLYL)SUCCINIMIDE	093476-51-8
P-TOLYLTOLUENE	000613-33-2
M-TOLYLUREA	000063-99-0
TOXAPHENE	008001-35-2
TRIACETIN	000102-76-1
O-TRIACETYL-N-ACETYL-ARA-C	015981-93-8
TRIACONTANE	000638-68-6
TRIADIMEFON	043121-43-3
TRIADIMENOL	055219-65-3
TRIALLYLPHOSPHATE	001623-19-4
TRIAMCINOLONE	000124-94-7
TRIAMCINOLONE ACETONIDE	000076-25-5
TRIAMCINOLONE DIACETATE	000067-78-7
2,4,7-TRIAMINO-6-PH PTERIDINE	000396-01-0
SYM-TRIAZINE	000290-87-9
1,3,5-TRIAZINE, 6-CHLORO-2,4-BIS[(2-PROPENYL)AMI	015468-86-7
1,3,5-TRIAZINE-2,4-DIAMINE, 1-(4-BUTYLPHENYL)-1,	004653-73-0
1,3,5-TRIAZINE-2,4-DIAMINE, 6-CHLORO-N-CYCLOBUTY	102587-50-8
1,3,5-TRIAZINE-2,4-DIAMINE, 6-CHLORO-N-CYCLOPROP	022936-85-2
1,3,5-TRIAZINE-2,4-DIAMINE, 6-CHLORO-N-ETHYL-N'-	084712-77-6
1,3,5-TRIAZINE-2,4-DIAMINE, 6-CHLORO-N-(1-METHYL	102587-54-2
1,3,5-TRIAZINE-2,4-DIAMINE, 6-CHLORO-N-(CYCLOPRO	040533-52-6
1,3,5-TRIAZINE-2,4-DIAMINE, 1,6-DIHYDRO-6,6-DIME	017711-74-9
1,2,4-TRIAZINE-3,5(2H,4H)-DIONE	000461-89-2
135-TRIAZINE,N-ETOH,N'-IPR,4-CL-	002904-53-2
1,2,4-TRIAZINE, 3-METHYL-	024108-33-6
1,3,5-TRIAZINE-2,4,6-TRIAMINE, N-HYDROXY-N,N',N'	064124-14-7
1,3,5-TRIAZIN-2(1H)-ONE, 4-AMINO-1-(beta-D-ARABI	065886-71-7
1,2,4-TRIAZIN-5(4H)-ONE, 4-AMINO-6-(3,5-DICHLORO	141605-19-8
1,2,4-TRIAZIN-5(4H)-ONE, 4-AMINO-6-(3,5-DICHLORO	141627-87-4
1,2,4-TRIAZIN-5(4H)-ONE, 4-AMINO-6-(3,5-DICHLORO	141605-16-5
1,2,4-TRIAZIN-5(4H)-ONE, 4-AMINO-3-(METHYLTHIO)-	141605-11-0
1,2,4-TRIAZIN-5(4H)-ONE, 6-[3-CHLORO-4-(TRIFLUOR	141627-90-9
1,2,4-TRIAZIN-5(4H)-ONE, 6-(3,5-DICHLOROPHENYL)-	141605-18-7
1,2,4-TRIAZIN-5(4H)-ONE, 6-(3-ETHOXYPHENYL)-3-ME	141605-15-4
1,2,4-TRIAZIN-5(4H)-ONE, 6-(3-ETHOXYPHENYL)-2,3-	141627-88-5
1,2,4-TRIAZIN-5(4H)-ONE, 6-(4-FLUOROPHENYL)-2,3-	085683-39-2

124-TRIAZIN-5-ONE,4NH2-3MES-6PYRAN2YL	021087-58-1
TRIAZIQUONE	000068-76-8
1,2,4-TRIAZOLE,3-NITRO-1-(3-ETHOXY-2-HYDROXY)PRO	104958-88-5
1H-1,2,4-TRIAZOLE	000288-88-0
2H-1,2,3-TRIAZOLE	000288-36-8
1-TRIAZOLEACETIC ACID-3-NITRO,METHYL ESTER	070965-23-0
1,2,4-TRIAZOLE-3,5-BUTYRAMIDO-	080616-58-6
1,2,4-TRIAZOLE,3,5-DIACETAMIDO-	060010-81-3
1H-1,2,4-TRIAZOLE-3,5-DIAMINE	001455-77-2
1,2,4-TRIAZOLE-1-METHYL-3-NITRO-	026621-45-4
1,2,4-TRIAZOLE,3-NITRO-1-(3-AZIRIDINYL-2-HYDROXY	104958-92-1
1,2,4-TRIAZOLE,3-NITRO-1-(3-CHLORO-2-HYDROXY)PRO	104958-94-3
1,2,4-TRIAZOLE,3-NITRO-1-(2-HYDROXY-3-METHOXY)PR	104958-86-3
1H-1,2,4-TRIAZOLE, 3-PHENYL-	003357-42-4
1H-1,2,3-TRIAZOLE, 4-PHENYL-	001680-44-0
1H-1,2,4-TRIAZOLE-3-SULFONAMIDE, N-(2,6-DIFLUORO	113024-34-3
1H-1,2,4-TRIAZOLE-3-SULFONAMIDE, N-(2,6-DICHLORO	113024-75-2
7H-1,2,3-TRIAZOLO[4,5-D]PYRIMIDIN-7-ONE, 1,4-DIH	037761-96-9
TRIAZOPHOS	024017-47-8
TRIAZOXIDE	072459-58-6
TRIBENZYLAMINE HYDROCHLORIDE	007673-07-6
TRIBROMOACETALDEHYDE	000115-17-3
TRIBROMOAMPHENICOL	049648-42-2
2,4,6-TRIBROMOANISOLE	000607-99-8
1,3,5-TRIBROMOBENZENE	000626-39-1
1,2,3-TRIBROMOBENZENE	000608-21-9
1,2,4-TRIBROMOBENZENE	000615-54-3
2,4,5-TRIBROMOIMIDAZOLE	002034-22-2
2,4,6-TRIBROMOPHENOL	000118-79-6
2-(2,4,6-TRIBROMOPHENOXY)ETHANOL	023976-66-1
2,4,6-TRIBROMORESORCINOL	002437-49-2
2-(246-TRIBR PH IMINO)IMIDAZOLIDINE	038941-33-2
TRI-2-BUTOXYETHYL PHOSPHATE	000078-51-3
TRI N-BUTYLAMINE	000102-82-9
2,4,6-TRI(TERT-BUTYL)PHENOL	000732-26-3
TRIBUTYLPHOSPHATE	000126-73-8
TRIBUTYRIN	000060-01-5
TRICAMBA	002307-49-5
TRICAPROIN	000621-70-5
TRICHLORACETYLCHLORIDE	000076-02-8
TRICHLORFON	000052-68-6
TRICHLORMETHIAZIDE	000133-67-5
TRICHLOROACETALDEHYDE	000075-87-6
TRICHLOROACETAMIDE	000594-65-0
3,4,5-TRICHLOROACETANILIDE	033715-62-7
TRICHLOROACETIC ACID	000076-03-9
TRICHLOROACETIC ANHYDRIDE	004124-31-6
1,1,1-TRICHLOROACETONE	000918-00-3
TRICHLOROACETONITRILE	000545-06-2
TRICHLOROAMPHENICOL	019934-51-1
2,4,6-TRICHLOROANILINE	000634-93-5
2,4,5-TRICHLOROANILINE	000636-30-6
2,3,4-TRICHLOROANILINE	000634-67-3
3,4,5-TRICHLOROANILINE	000634-91-3
2,3,5-TRICHLOROANISOLE	054135-81-8
2,3,6-TRICHLOROANISOLE	050375-10-5
2,4,5-TRICHLOROANISOLE	006130-75-2
2,4,6-TRICHLOROANISOLE	000087-40-1

1,2,4-TRICHLOROBENZENE	000120-82-1
1,3,5-TRICHLOROBENZENE	000108-70-3
1,2,3-TRICHLOROBENZENE	000087-61-6
2,3,6-TRICHLOROBENZOIC ACID	000050-31-7
2,4,5-TRICHLOROBENZOIC ACID	000050-82-8
2,4',6-TRICHLOROBIPHENYL	038444-77-8
3,4,4'-TRICHLOROBIPHENYL	038444-90-5
2,3,4-TRICHLORO-1,1'-BIPHENYL	055702-46-0
2,3,6-TRICHLOROBIPHENYL	055702-45-9
2,4,5-TRICHLOROBIPHENYL	015862-07-4
2,4,6-TRICHLOROBIPHENYL	035693-92-6
2',3,4-TRICHLOROBIPHENYL	038444-86-9
3,3',5-TRICHLOROBIPHENYL	038444-87-0
2,2',4-TRICHLOROBIPHENYL	037680-66-3
2,2',5-TRICHLOROBIPHENYL	037680-65-2
2,4',5-TRICHLOROBIPHENYL	016606-02-3
1,2,3-TRICHLOROBUTA-1,3-DIENE	001573-58-6
1,1,4-TRICHLORO-1,2-BUTADIENE	058679-08-6
4,4,4-TRICHLORO-1,2-BUTADIENE	034819-62-0
1,1,2-TRICHLORO-1,3-BUTADIENE	002852-07-5
1,1,4-TRICHLOROBUTADIENE	083682-46-6
1,2,3-TRICHLOROBUTA-1,3-DIENE(E)	053978-04-4
1,2,3-TRICHLOROBUTA-1,3-DIENE(Z)	039083-26-6
1,3,3-TRICHLOROBUTANE	015187-71-0
1,2,4-TRICHLOROBUTANE	001790-22-3
2,2,3-TRICHLORO-1,1-BUTANEDIOL	000076-40-4
B,B,B-TRICHLORO-T-BUTANOL	000057-15-8
1,1,4-TRICHLORO-2-BUTENE	041601-59-6
1,2,4-TRICHLORO-2-BUTENE	002431-54-1
1,1,4-TRICHLOROBUT-2-ENE(E)	057808-36-3
4,4,4-TRICHLORO-1-BUTYNE	083682-43-3
3,4,5-TRICHLOROCATECHOL	056961-20-7
3,4,6-TRICHLOROCATECHOL	032139-72-3
TRICHLOROCYCLOHEXENE (345)	056994-25-3
1,2,4-TRICHLORODIBENZO-P-DIOXIN	039227-58-2
2,3,7-TRICHLORODIBENZO-P-DIOXIN	033857-28-2
1,3,3-TRICHLORO-2-(DICHLOROMETHYL)PROPENE	014129-82-9
1,1,1-TRICHLOROETHANE	000071-55-6
1,1,2-TRICHLOROETHANE	000079-00-5
2,2,2-TRICHLOROETHANOL	000115-20-8
TRICHLOROETHENE	000079-01-6
1,1'-(2,2,2-TRICHLOROETHYLIDENE)BIS(4-METHYLT*)	019679-38-0
1,1'-(2,2,2-TRICHLOROETHYLIDENE)BIS(4-METHYLB*)	004413-31-4
TRI-2-CHLOROETHYL PHOSPHATE	000115-96-8
TRICHLOROETHYL PHOSPHATE	000306-52-5
A-(2,2,2-TRICHLOROETHYL)STYRENE	020057-31-2
TRICHLOROFLUOROMETHANE	000075-69-4
4,5,6-TRICHLOROGUAIACOL	002668-24-8
2,2,4-TRICHLORO-2,2,4,4,6,6-HEXAHYDRO-4,6,6-TRI*	065601-41-4
2,4,5-TRICHLOROIMIDAZOLE	007682-38-4
1,3,5-TRICHLOROISOCYANURIC ACID	000087-90-1
1,2,3-TRICHLORO-4-METHOXYBENZENE	054135-80-7
3,4,5-TRICHLORO-2-METHOXYPHENOL	057057-83-7
1,3,5-TRI(ALPHA-CHLOROMETHYL)BENZENE	017299-97-7
2,4,6-TRICHLORO-3-METHYLPHENOL	000551-76-8
TRICHLOROMETHYLTHIOBENZENE	000701-65-5
2-TRICHLOROMETHYL-3,3,3-TRICHLOROPROPENE	083682-38-6
1,3,7-TRICHLORONAPHTHALENE	055720-37-1

2,3,6-TRICHLORONAPHTHALENE	055720-40-6
TRICHLORONATE	000327-98-0
2,3,4-TRICHLORONITROBENZENE	017700-09-3
2,4,5-TRICHLORONITROBENZENE	000089-69-0
TRICHLORONITROMETHANE	000076-06-2
2,3,6-TRICHLORO-4-NITROPHENOL	020404-02-8
3,4,6-TRICHLORO-2-NITROPHENOL	000082-62-2
1,2,3-TRICHLOROOCTANE	085269-46-1
TRICHLOROOXIRANE	016967-79-6
TRICHLOROPENTYLSILANE	000107-72-2
2,4,5-TRICHLOROPHENOL	000095-95-4
3,4,5-TRICHLOROPHENOL	000609-19-8
2,4,6-TRICHLOROPHENOL	000088-06-2
2,3,4-TRICHLOROPHENOL	015950-66-0
2,3,5-TRICHLOROPHENOL	000933-78-8
2,3,6-TRICHLOROPHENOL	000933-75-5
2,4,5-TRICHLOROPHENOXYACETIC ACID	000093-76-5
2,4,6-TRICHLOROPHENOXYACETIC ACID	000575-89-3
2,3,6-TRICHLOROPHENYLACETIC ACID, SODIUM SALT	002439-00-1
1,1,2-TRICHLOROPROPANE	000598-77-6
1,2,3-TRICHLOROPROPANE	000096-18-4
1,2,2-TRICHLOROPROPANE	003175-23-3
1,1,1-TRICHLOROPROPANE	007789-89-1
1,1,3-TRICHLOROPROPENE	002567-14-8
2,3,3-TRICHLOROPROPENE	037077-84-2
1,2,3-TRICHLOROPROPENE	000096-19-5
1,1,2-TRICHLOROPROPENE	021400-25-9
2,3,3-TRICHLORO-2-PROPENOIC ACID	002257-35-4
2,2,3-TRICHLOROPROPIONIC ACID	003278-46-4
2,3,5-TRICHLOROPYRIDINE	016063-70-0
2,4,6-TRICHLOROPYRIDINE	016063-69-7
2,3,6-TRICHLOROPYRIDINE	006515-09-9
3,5,6-TRICHLORO-2-PYRIDINOL	006515-38-4
2,3,6-TRICHLOROTOLUENE	002077-46-5
2,4,6-TRICHLOROTOLUENE	023749-65-7
2,4,5-TRICHLOROTOLUENE	006639-30-1
2,3,4-TRICHLOROTOLUENE	007359-72-0
1,2,4-TRICHLORO-4-(1,1,2-TRICHLOROETHYL)CYCLOH*	083682-64-8
1,1,2-TRICHLOROTRIFLUOROETHANE	000076-13-1
1,1,1-TRICHLORO-2,2,2-TRIFLUOROETHANE	000354-58-5
TRICHLORO-2,4,6-TRINITROBENZENE	002631-68-7
2,4,5-TRICL-C6H2NHN=C(CN)CO-T-BU	028343-28-4
2,4,5-TRICL-C6H2NHN=C(CN)CO-OET	028322-78-3
3,4,5-TRICL-C6H2NHN=C(CN)CO-OET	036865-51-7
3,4,5-TRICL-C6H2NHN=C(CN)CO-OME	036865-56-2
3,4,5-TRICL-2,6-MEO PHENOL	002539-26-6
3,5,-4'-TRICL-4'-NO2 SALICYLANILIDE	090426-05-4
TRICLOPYR	055335-06-3
TRICLOPYR ESTER	064700-56-7
2-(245-TRICL PH IMINO)IMIDAZOLIDINE	059465-40-6
2-(246-TRICL PH IMINO)IMIDAZOLIDINE	059465-51-9
3,5,4'-TRICL SALICYLANILIDE	001151-51-5
TRICOSANE	000638-67-5
TRI-P-CRESYL PHOSPHATE	000078-32-0
TRICRESYL PHOSPHATE	001330-78-5
TRICYCLAZOLE	041814-78-2
TRICYCLO[3.3.1.1]DECANE	000281-23-2
TRICYCLO(5.2.1.0)DECANE	006004-38-2

TRICYCLO(4.1.0.2,4)HEPTANE	000187-26-8
TRICYCLO(4.1.0.02,7)HEPTANE	000287-13-8
3,11-TRIDECADIENE-5,7,9-TRIYN-1-OL, 2-CHLORO-	002060-58-4
N-TRIDECANE	000629-50-5
1-TRIDECANETHIOL	019484-26-5
TRIDECANOIC ACID	000638-53-9
1-TRIDECANOL	000112-70-9
2-TRIDECANONE	000593-08-8
TRIDECAN-7-ONE	000462-18-0
1-TRIDECENE	002437-56-1
TRIDECYL ACETATE	001072-33-9
N-TRIDECYLCYCLOHEXANE	006006-33-3
TRIETAZINE	001912-26-1
TRIETHANOLAMINE	000102-71-6
TRIETHOXYVINYLSILANE	000078-08-0
TRIETHYLAMINE	000121-44-8
TRIETHYLAMINE, HYDROBROMIDE	000636-70-4
TRIETHYLAMINE HYDROCHLORIDE	000554-68-7
1,3,5-TRIETHYLBENZENE	000102-25-0
TRIETHYL CITRATE	000077-93-0
TRIETHYLENE GLYCOL	000112-27-6
TRIETHYLENE GLYCOL BIS(2-ETHYLBUTYRATE)	000095-08-9
TRIETHYLENE GLYCOL BUTYL ETHER	000143-22-6
TRIETHYLENEGLYCOL DIMETHACRYLATE	000109-16-0
TRIETHYLENETETRAAMINE	000112-24-3
TRIETHYLFLUOROSILANE	000358-43-0
TRIETHYL METHYL AMMONIUM IODIDE	000302-57-8
TRIETHYL PHOSPHATE	000078-40-0
TRIETHYLSILANE	000617-86-7
2,2,2-TRIF N,N-DIME ETHYLAMINE	000819-06-7
TRIFENMORPH	001420-06-0
TRIFLUMIZOLE	068694-11-1
2-(TRIFLUORMETHYL)-2-PROPANOL	000507-52-8
TRIFLUOROACETAMIDE	000354-38-1
TRIFLUOROACETANILIDE	000404-24-0
TRIFLUOROACETATE	000076-05-1
TRIFLUOROACETIC ANHYDRIDE	000407-25-0
1,1,1-TRIFLUOROACETONE	000421-50-1
TRIFLUOROACETOPHENONE	000434-45-7
TRIFLUOROAMPHENICOL	042583-67-5
1,3,5-TRIFLUOROBENZENE	000372-38-3
4,4,4-TRIFLUORO-2-BUTANOL	101054-93-7
4,4,4-TRIFLUORO-1-BUTANOL	000461-18-7
1,1,1-TRIFLUORO-2-CHLOROETHANE	000075-88-7
1,1,1-TRIFLUORO-2,2-DICHLOROETHANE	000306-83-2
2,2,2-TRIFLUOROET AMINE,N-DIET	037174-09-7
1,1,1-TRIFLUOROETHANE	000420-46-2
1,1,2-TRIFLUOROETHANE	000430-66-0
2,2,2-TRIFLUOROETHANOL	000075-89-8
TRIFLUOROETHYL ACETATE	000383-63-1
2,2,2-TRIFLUOROETHYLAMINE	000753-90-2
2,2,2-TRIFLUOROETHYL METHACRYLATE	000352-87-4
6,6,6-TRIFLUORO-1-HEXANOL	065611-47-4
M-(TRIFLUOROME)ACETANILIDE	000351-36-0
M-(TRIFLUOROME)BENZYL ALCOHOL	000349-75-7
M-(TRIFLUOROME)PHENOXYACETIC ACID	000349-82-6
P-TRIFLUOROME PHENYLACETIC ACID	032857-62-8
M-(TRIFLUOROME)PHENYLACETIC ACID	000351-35-9

TRIFLUOROMETHANE	000075-46-7
TRIFLUOROMETHOXYBENZENE	000456-55-3
P-TRIFLUOROMETHYLANILINE	000455-14-1
3-TRIFLUOROMETHYLANILINE	000098-16-8
2-TRIFLUOROMETHYLBENZAMIDE	000360-64-5
P-TRIFLUOROMETHYLBENZAMIDE	001891-90-3
3'-TRIFLUOROMETHYLBENZOGUANAMINE	030508-78-2
M-(TRIFLUOROMETHYL)BENZALDEHYDE	000454-89-7
M-TRIFLUOROMETHYLBENZOIC ACID	000454-92-2
4,4,4-TRIFLUORO-3-METHYL-1-BUTANOL	095853-68-2
4,4,4-TRIFLUORO-2-METHYL-1-BUTANOL	114525-06-3
4-(TRIFLUOROMETHYL)CHLOROBENZENE	000098-56-6
1-(4-TRIFLUOROMETHYL),2-DIBENZOYL-1-TERT-BUTYLHY	112426-65-0
1-(3-TRIFLUOROMETHYL),2-DIBENZOYL-1-TERT-BUTYLHY	112426-66-1
1-(2-TRIFLUOROMETHYL),2-DIBENZOYL-1-TERT-BUTYLHY	112426-67-2
M-TRIFLUOROMETHYLNITROBENZENE	000098-46-4
3-TRIFLUOROMETHYL-4-NITROPHENOL	000088-30-2
4-TRIFLUOROMETHYL-2-NITROPHENOL	000400-99-7
5,5,5-TRIFLUORO-3-METHYL-1-PENTANOL	107103-97-9
P-TRIFLUOROMETHYLPHENOL	000402-45-9
3-(TRIFLUOROMETHYL)PHENYLUREA	013114-87-9
9-(3-(TRIFLUOROMETHYL)PHENYL)-9H-FLUORENE	032377-12-1
M-TRIFLUOROMETHYLPHENOL	000098-17-9
O-TRIFLUOROMETHYLPHENOL	000444-30-4
TRIFLUOROMETHYLTHIOBENZENE	000456-56-4
TRIFLUOROMETHYLURACIL	000054-20-6
TRIFLUORONITROMETHANE	000335-02-4
5,5,5-TRIFLUORO-1-PENTANOL	000352-61-4
3,3,3-TRIFLUORO-1-PROPANOL	002240-88-2
1,1,1-TRIFLUORO-2-PROPANOL	000374-01-6
3,3,3-TRIFLUORO-1-PROPENE	000677-21-4
TRIFLUORPERAZINE	000117-89-5
TRIFLUPROMAZINE	000146-54-3
TRIFLURALIN	001582-09-8
TRIFLURIDINE	000070-00-8
TRIFORINE	026644-46-2
TRIGLYCINE	000556-33-2
TRIHEXYLPHENEDYL	000144-11-6
1,2,3-TRIHYDROXYBENZENE	000087-66-1
1,3,5-TRIHYDROXYBENZENE	000108-73-6
1,2,4-TRIHYDROXYBENZENE (PYROGALLOL)	000533-73-3
2,3,4-TRIHYDROXYBENZOIC ACID	000610-02-6
3,4,5-TRIHYDROXYBENZOIC ACID	000149-91-7
1,2,3-TRIHYDROXYBUTANE	004435-50-1
5,3',4'-TRIHYDROXYFLAVONE	019852-25-6
2,3,5-TRIIODOBENZOIC ACID	000088-82-4
2,4,5-TRIIODOIMIDAZOLE	001746-25-4
2,4,6-TRIIODOPHENOL	000609-23-4
1,3,5-TRIISOPROPYLBENZENE	000717-74-8
TRIMECAIN	000616-68-2
NNN-TRIME-N-(5CLME-2-FURANME) I	066473-27-6
1,3,3-TRIME-2,3-DIHYDRO-4-QUINOLONE	053207-52-6
2,3,5-TRIME-4-HYDROXYACETANILIDE	069477-72-1
2(155TRIME-1,3-IMIDAZOLINYLIDEN-2AM)5NO2-THIAZOL	037385-14-1
TRIMELLITIC ACID	000528-44-9
TRIMELLITIC ANHYDRIDE	000552-30-7
2,3,4-TRIMEO AMPHETAMINE	001082-23-1
3,4,5-TRIMEO AMPHETAMINE	001082-88-8

2,4,5-TRIMEO AMPHETAMINE	001083-09-6
3,4,5-TRIMEO-CINNAMIC ACID,DIETAMET ESTER	072836-47-6
1,2,3-TRIMEO TRICHLOROCYCLOHEXANE	060067-82-5
2-(246-TRIME PH IMINO)IMIDAZOLIDINE	004201-40-5
2,4,5-TRIMEPH N-ME CARBAMATE	000671-03-4
TRIMEPRAZINE	000084-96-8
TRIMETHOBENZAMIDE (TIGAN)	000138-56-7
TRIMETHOPRIM	000738-70-5
2,3,4-TRIMETHOXYACETOPHENONE	013909-73-4
2,4,6-TRIMETHOXYAMPHETAMINE	015402-79-6
2,3,4-TRIMETHOXYBENZOIC ACID	000573-11-5
2,3,5-TRIMETHOXY-P-BENZOQUINONE	003117-05-3
1,1,1-TRIMETHOXYETHANE	001445-45-0
TRIMETHOXYMETHANE	000149-73-5
TRIMETHOXYMETHYLBENZENE	000707-07-3
1,1,3-TRIMETHOXYPROPANE	014315-97-0
TRIMETHOXY SILANE	002487-90-3
3,4,5-TRIMETHYLACETANILIDE	001467-34-1
TRIMETHYLACETIC ACID	000075-98-9
TRIMETHYLACETONITRILE	000630-18-2
TRIMETHYLAMINE	000075-50-3
TRIMETHYLAMPHENICOL	049648-50-2
N,N,4-TRIMETHYLANILINE	000099-97-8
2,4,5-TRIMETHYLANILINE	000137-17-7
2,4,4'-TRIMETHYLAZOBENZENE	029418-24-4
2,4',6-TRIMETHYLAZOBENZENE	029418-27-7
2,4,6-TRIMETHYLAZOBENZENE	029418-26-6
2,2',4-TRIMETHYLAZOBENZENE	029418-23-3
1,3,5-TRIMETHYLBENZENE	000108-67-8
1,2,4-TRIMETHYLBENZENE	000095-63-6
1,2,3-TRIMETHYLBENZENE	000526-73-8
2,2,3-TRIMETHYLBUTANE	000464-06-2
2,3,3,-TRIMETHYL-2-BUTANOL	000594-83-2
2,3,3-TRIMETHYL-1-BUTENE	000594-56-9
TRIMETHYL BUTYL AMMONIUM IODIDE	007722-19-2
1,1,3-TRIMETHYLCYCLOHEXANE	003073-66-3
CIS-1,2,2-TRIMETHYL-1,3-CYCLOPENTANEDICARBOXYL*	005394-83-2
1,1,3-TRIMETHYLCYCLOPENTANE	004516-69-2
1(TRANS), 2(CIS), 3-TRIMETHYLCYCLOPENTANE	015890-40-1
TRIMETHYLENEOXIDE	000503-30-0
N,N'-TRIMETHYLENETHIOUREA	002055-46-1
TRIMETHYLENETHIOUREA,N,N-DIPROPYL	030826-88-1
3,3,4-TRIMETHYLHEXANE	016747-31-2
2,3,5-TRIMETHYLHEXANE	001069-53-0
2,3,4-TRIMETHYLHEXANE	000921-47-1
2,3,3-TRIMETHYLHEXANE	016747-28-7
2,2,5-TRIMETHYLHEXANE	003522-94-9
2,4,4-TRIMETHYLHEXANE	016747-30-1
2,2,4-TRIMETHYLHEXANE	016747-26-5
2,2,3-TRIMETHYLHEXANE	016747-25-4
TRIMETHYL HEXYL AMMONIUM IODIDE	015066-77-0
TRIMETHYL HYDRAZINE, MONOHYDROCHLORIDE	060597-20-8
2,3,6-TRIMETHYL-4-HYDROXYACETANILIDE	036592-59-3
2,3,6-TRIMETHYLNAPHTHALENE	000829-26-5
1,4,5-TRIMETHYLNAPHTHALENE	002131-41-1
1,6,7-TRIMETHYLNAPHTHALENE	002245-38-7
TRIMETHYLOLPROPANE TRIMETHACRYLATE	003290-92-4
2,3,3-TRIMETHYLPENTANE	000560-21-4

2,2,3-TRIMETHYLPENTANE	000564-02-3
2,3,4-TRIMETHYLPENTANE	000565-75-3
2,2,4-TRIMETHYLPENTANE	000540-84-1
2,2,4-TRIMETHYL-1,3-PENTANEDIOL	000144-19-4
2,2,4-TRIMETHYL-1-PENTANOL	000123-44-4
2,2,3-TRIMETHYL-3-PENTANOL	007294-05-5
2,4,4-TRIMETHYL-1-PENTENE	000107-39-1
2,3,4-TRIMETHYL-2-PENTENE	000565-77-5
2,3,3-TRIMETHYL-1-PENTENE	000560-23-6
2,4,4-TRIMETHYL-2-PENTENE	000107-40-4
2,4,4-TRIMETHYLPENTENE	025167-70-8
TRIMETHYL PENTYL AMMONIUM IODIDE	019109-66-1
2,4,5-TRIMETHYLPHENOL	000496-78-6
2,3,6-TRIMETHYLPHENOL	002416-94-6
2,3,5-TRIMETHYLPHENOL	000697-82-5
2,4,6-TRIMETHYLPHENOL	000527-60-6
TRIMETHYL PHOSPHATE	000512-56-1
TRIMETHYL PHOSPHITE	000121-45-9
O,S,S-TRIMETHYL PHOSPHORODITHIOATE	022608-53-3
O,O,O-TRIMETHYL PHOSPHOROTHIOATE	000152-18-1
1,1,3-TRIMETHYL-1,3,3-TRIPHENYL-DISILOXANE	014920-93-5
2,4,6-TRIMETHYL-2,4,6-TRIPHENYLCYCLOTRISILOXANE	006138-53-0
2,4,6-TRIMETHYL-2,4,6-TRIPHENYLCYCLOTRISILOXANE	003424-57-5
1,3,7-TRIMETHYL URIC ACID	005415-44-1
N2,N4,N6-TRIME-N,N,N-TRIME-MELAMINE	064124-20-5
2,4,6-TRINITROANISOLE	000606-35-9
1,3,5-TRINITROBENZENE	000099-35-4
2,4,6-TRINITROBENZOIC ACID	000129-66-8
2,4,6-TRINITROCHLOROBENZENE	000088-88-0
1,3,5,-TRINITROHEXAHYDRO-1,3,5-TRIAZINE	000121-82-4
2,4,6-TRINITROPHENOL	000088-89-1
2,4,6-TRINITRORESORCINOL	000082-71-3
2,4,6-TRINITROTOLUENE	000118-96-7
TRIOCTONOIN	000538-23-8
TRI(N-OCTYL)AMINE	001116-76-3
N-(TRIOHMEMETHANE)-N'-PHENYL UREA	060035-83-8
TRIONAL	000076-20-0
1,3,5-TRIOXANE	000110-88-3
TRIPHENYLAMINE	000603-34-9
TRIPHENYLENE	000217-59-4
TRIPHENYLMETHANE	000519-73-3
TRIPHENYLMETHYLISOTHIOCYANATE	001726-94-9
TRIPHENYLPHOSPHATE	000115-86-6
TRIPHENYLPHOSPHINE	000603-35-0
TRIPHENYLPHOSPHINE OXIDE	000791-28-6
TRIPHENYLTIN HYDROXIDE	000076-87-9
TRIPHENYLUREA	005663-04-7
TRIPOLIDINE	000486-12-4
NNN-TRIPR-N-ME AMMONIUM IODIDE	003531-14-4
TRIPROPIONIN	000139-45-7
TRIPROPYLAMINE	000102-69-2
TRIPROPYLENE GLYCOL METHYL ETHER	020324-33-8
TRIPROPYL PHOSPHATE	000513-08-6
TRIPROPYNYL-AMINE	006921-29-5
2,4,6-TRIS(1-AZIRIDINYL)-1,3,5-TRIAZINE	000051-18-3
TRIS(BETA-CHLOROETHYL)AMINE	000555-77-1
1,3,5-TRIS(CHLOROETHYL)BENZENE	018226-46-5
TRIS(2,3-DIBROMOPROPYL) PHOSPHATE	000126-72-7

1,3,5-TRIS(2,2-DICHLOROETHYL)BENZENE	083682-59-1
TRIS(1,3-DICHLOROISOPROPYL) PHOSPHATE	013674-87-8
1,3,5-TRIS(DICHLOROMETHYL)BENZENE	056682-87-2
TRIS(1,3-DICHLOROPROPYL) PHOSPHATE	040120-74-9
TRIS(2-ETHYLHEXYL) PHOSPHATE	000078-42-2
TRIS(2-ETHYLHEXYL) TRIMELLITATE	003319-31-1
1,1,1-TRIS(HYDROXYMETHYL)PROPANE	000077-99-6
1,3,5-TRIS(2,2,2-TRICHLOROETHYL)BENZENE	083682-60-4
TRITAC	001861-44-5
1,2,4-TRITHIOLANE	000289-16-7
TRITRIACONTANE	000630-05-7
TRI(3,5-XYLENYL) PHOSPHATE	025653-16-1
TR-1-NO-1-(2-CLET)-3(2-OHCYHEX)UREA	058494-43-2
TR-1-NO-1(2CLET)-3(4-OH CYHEXYL)UREA	056239-24-8
TROPACOCAINE	000537-26-8
TROPIC ACID	000529-64-6
TROPITAL	005281-13-0
TROPOLONE	000533-75-5
TR-2-PHENYLCYCLOPROPYLCARBINOL	000936-98-1
TRYPTAMINE	000061-54-1
DL-TRYPTOPHAN	000054-12-6
TRYPTOPHAN	000073-22-3
TRYPTOPHAN-AMIDE,N-ACETYL	010346-41-5
TUBERCIDIN	000069-33-0
TUBERIN	053643-53-1
DL-TYROSINE	000556-03-6
TYROSINE	000060-18-4
TYROSINE,N-ACETYL,N'-MEAM-AMIDE	006367-14-2
TYROSINE-N-ACETYL,METHYL ESTER	002440-79-1
TYROSINE-AMIDE,N-ACETYL	001948-71-6
L-TYROSINE,BUTYL ESTER	006292-90-6
L-, N-(N-CARBOXY-L-VALYL)-, N-BENZYL EST	000862-26-0
L-TYROSINE, ETHYL ESTER	000949-67-7
L-TYROSINE, 3-HYDROXY-, 2-(4-CHLOROPHENYL)ETHYL	099877-09-5
L-TYROSINE, 3-HYDROXY-, CYCLOHEXYL ESTER	099877-05-1
L-TYROSINE, 3-HYDROXY-, ETHYL ESTER	037178-37-3
L-TYROSINE, 3-HYDROXY-, 2-HYDROXYPROPYL ESTER	099877-03-9
L-TYROSINE, 3-HYDROXY-, 2-METHOXY-1-METHYLETHYL	099877-04-0
L-TYROSINE, 3-HYDROXY-, 2-(4-METHOXYPHENYL)ETHYL	099877-10-8
L-TYROSINE, 3-HYDROXY-, 1-METHYLETHYL ESTER	110301-07-0
L-TYROSINE, 3-HYDROXY-, METHYL ESTER	007101-51-1
L-TYROSINE, 3-HYDROXY-, PHENYLMETHYL ESTER	055720-47-3
L-TYROSINE, 3-HYDROXY-, 2-PHENOXYETHYL ESTER	099877-11-9
L-TYROSINE, 3-HYDROXY-, 2-PHENYLETHYL ESTER	099877-08-4
L-TYROSINE, 3-HYDROXY-, PROPYL ESTER	039638-51-2
L-TYROSINE, 3-HYDROXY-, (TETRAHYDRO-2H-PYRAN-2-Y	099877-06-2
TYROSINE, 3-METHOXY-	000300-48-1
TYROSINE, A-METHYL-	000658-48-0
L-TYROSINE, METHYL ESTER	001080-06-4
N-UNDECANE	001120-21-4
5,7-UNDECANEDIONE	001942-48-9
UNDECANOIC ACID	000112-37-8
UNDECANOIC ACID, BUTYL ESTER	010580-24-2
2-UNDECANOL	001653-30-1
1-UNDECANOL	000112-42-5
6-UNDECANONE	000927-49-1
UNDECANONE	000112-12-9
1-UNDECENE	000821-95-4

UNDECYL ALCOHOL ACETATE	001731-81-3
UNDECYLENIC ACID	000112-38-9
URACIL	000066-22-8
URACIL,1-BENZYLOXYCARBONYL-5-FLUORO-	066999-98-2
URACIL MUSTARD	000066-75-1
UREA	000057-13-6
UREA, N'-[4-[(ACETYLOXY)METHYL]CYCLOHEXYL]-N-(2-	061137-49-3
P-UREA-AMPHENICOL	039961-07-4
UREA, N-BUTYL-N'-[2-[[3-[4-(5-METHYL-1H-IMIDAZOL	102151-13-3
UREA, N'-[3-ACETYL-4-[3-[(1,1-DIMETHYLETHYL)AMIN	056980-93-9
UREA, N'-(2-CHLOROCYCLOHEXYL)-N-(2-CHLOROETHYL)-	013909-12-1
UREA, N'-[3-CHLORO-4-[[(1,1-DIMETHYLETHYL)-1H-TE	118259-45-3
UREA, N-(2-CHLOROETHYL)-N'-(4-CHLOROPHENYL)-N-NI	013907-78-3
UREA, N-(2-CHLOROETHYL)-N-NITROSO-N'-(2,3,4,6-TE	058484-10-9
UREA, N-(2-CHLOROETHYL)-N-NITROSO-N'-(2,3,4-TRI-	054138-85-1
UREA, N'-[3-CHLORO-4-[(1-ETHYL-1H-TETRAZOL-5-YL)	117121-39-8
UREA, N'-[4-[[2-(2-CYANOETHYL)-2H-TETRAZOL-5-YL]	117121-50-3
UREA, N,N''-1,4-CYCLOHEXANEDIYLBIS[N'-(2-CHLOROET	013907-57-8
UREA, N'-[4-[(2-CYCLOHEXYL-2H-TETRAZOL-5-YL)OXY]	117121-45-6
UREA, N'-[4-[(1-CYCLOHEXYL-1H-TETRAZOL-5-YL)OXY]	117121-33-2
UREA, N-(3,4-DICHLOROPHENYL)-N'-(2-METHYLPROPYL)	005006-90-6
UREA, N-[4,7-DIMETHOXY-6-[2-(1-PYRROLIDINYL)ETHO	066203-00-7
UREA, N'-[4-[[2-(1,1-DIMETHYLETHYL)-2H-TETRAZOL-	117121-44-5
UREA, N'-[4-[[1-(1,1-DIMETHYLETHYL)-1H-TETRAZOL-	118259-53-3
UREA, N,N-DIMETHYL-N'-[4-[[2-(1-METHYLETHYL)-2H-	117121-43-4
UREA, N,N-DIMETHYL-N'-[4-[(2-METHYL-2H-TETRAZOL-	117121-41-2
UREA, N,N-DIMETHYL-N'-[4-(1-OCTYL-1H-TETRAZOL-5-	117121-34-3
UREA, N,N-DIMETHYL-N'-[4-[(2-OCTYL-2H-TETRAZOL-5	117121-46-7
UREA, N,N-DIMETHYL-N'-[3-(6-OXO-3-PHENYL-1(6H)-P	087666-30-6
UREA, N,N-DIMETHYL-N'-[4-(6-OXO-3-PHENYL-1(6H)-P	087653-35-8
UREA, N,N-DIMETHYL-N'-[3-(2-PHENYLETHYL)PHENYL]-	102587-49-5
UREA, N,N-DIMETHYL-N'[3-(PHENYLMETHYL)PHENYL]-	102587-48-4
UREA, N,N-DIMETHYL-N'-[4-[[2-(PHENYLMETHYL)-2H-T	117121-48-9
UREA, N,N-DIMETHYL-N'-[4-[[1-(PHENYLMETHYL)-1H-T	117121-36-5
UREA, N,N-DIMETHYL-N'-[4-[[2-(2-PROPENYL)-2H-TET	117121-47-8
UREA, N,N-DIMETHYL-N'-[4-[[1-(2-PROPENYL)-1H-TET	117121-35-4
UREA, N,N-DIMETHYL-N'-(4-PROPYLPHENYL)-	085911-77-9
UREA, N,N-DIMETHYL-N'-[4-[[1-(2-PROPYNYL)-1H-TET	117121-38-7
UREA, N,N-DIMETHYL-N'-[4-[[2-(2-PROPYNYL)-2H-TET	117144-73-7
UREA, N-ETHYL-N'-BENZOYL-	028615-21-6
UREA, N-ETHYL-N'-[4-(6-OXO-3-PHENYL-1(6H)-PYRIDA	087653-23-4
UREA, N-ETHYL-N'-[3-(6-OXO-3-PHENYL-1(6H)-PYRIDA	087666-28-2
UREA, N-(3-ETHYLPHENYL)-N'-METHYL-	023138-95-6
UREA, N'-[4-[(2-ETHYL-2H-TETRAZOL-5-YL)OXY]PHENY	117121-42-3
UREA, N-(3-METHOXYPHENYL)-N'-METHYL-	023138-98-9
UREA, METHYL-	000598-50-5
UREA, N-(1-METHYLCYCLOHEXYL)-N'-PHENYL-	001611-63-8
UREA, N-METHYL-N'-(METHYLFURAZANYL)-	102626-89-1
UREA, N-METHYL-N'-[(4-METHYLPHENYL)SULFONYL]-	013909-69-8
UREA, N-METHYL-N'-[4-(6-OXO-3-PHENYL-1(6H)-PYRID	087653-22-3
UREA, N-METHYL-N'-[2-(6-OXO-3-PHENYL-1(6H)-PYRID	087666-25-9
UREA, N-METHYL-N'-[3-(6-OXO-3-PHENYL-1(6H)-PYRID	087666-27-1
UREA,1-METHYL-1-PHENYL-2-THIO	004104-75-0
UREA, N-(2-METHYLPROPYL)-N-NITROSO-	000760-60-1
UREA, [4-(6-OXO-3-PHENYL-1(6H)-PYRIDAZINYL)BUTYL	087666-31-7
UREA, [3-(6-OXO-3-PHENYL-1(6H)-PYRIDAZINYL)PROPY	087666-26-0
UREA, [6-(6-OXO-3-PHENYL-1(6H)-PYRIDAZINYL)HEXYL	087653-49-4
UREA, N,N''-1,3-PHENYLENEBIS-	001135-58-6

UREA, N-(1-PIPERIDINYL)-N'-[(4-METHYLPHENYL)SULF	001443-94-3
UREA, POLYMER WITH FORMALDEHYDE	009011-05-6
UREA, N-(1-PYRROLIDINYL)-N'-[(4-METHYLPHENYL)SUL	001220-55-9
8-UREIDO-QUINOLINE	032451-61-9
O-ETHYL CARBAMATE (URETHANE)	000051-79-6
URIC ACID	000069-93-2
URICYTIN	040919-33-3
URIDINE	000058-96-8
URIDINE, 3'-AZIDO-2',3'-DIDEOXY-	084472-85-5
URIDINE, 5-BROMO-2',3'-DIDEOXY-3'-FLUORO-	115249-86-0
URIDINE, 5-CHLORO-2',3'-DIDEOXY-3'-FLUORO-	119644-22-3
URIDINE, 5-CHLORO-2',3'-DIDEOXY-	127592-40-9
URIDINE, 2'-DEOXY-	000951-78-0
URIDINE, 2'-DEOXY-5-FLUORO-, 3',5'-DIPENTANOATE	007207-62-7
URIDINE, 2'-DEOXY-5-FLUORO-, 3',5'-DIBUTANOATE	003343-22-4
URIDINE, 2'-DEOXY-5-IODO-	000054-42-2
URIDINE, 2'-DEOXY-5-IODO-, 5'-BENZOATE	084043-31-2
URIDINE, 2'-DEOXY-5-IODO-, 5'-BUTANOATE	084043-26-5
URIDINE, 2'-DEOXY-5-IODO-, 5'-(2,2-DIMETHYLPROPA	084043-28-7
URIDINE, 2'-DEOXY-5-IODO-, 5'-(2-METHYLPROPANOAT	084043-27-6
URIDINE, 2'-DEOXY-5-IODO-, 5'-(4-METHOXYBENZOATE	135600-74-7
URIDINE, 2'-DEOXY-5-IODO-, 5'-(4-NITROBENZOATE)	135617-14-0
URIDINE, 2'-DEOXY-5-IODO-, 5'-PENTANOATE	084052-69-7
URIDINE, 2'-DEOXY-5-IODO-, 5'-PROPANOATE	084043-25-4
URIDINE, 2',3'-DIDEHYDRO-2',3'-DIDEOXY-4-THIO-	122568-02-9
URIDINE, 2',3'-DIDEOXY-	005983-09-5
URIDINE, 2',3'-DIDEOXY-3'-FLUORO-	041107-56-6
URIDINE, 2',3'-DIDEOXY-3'-FLUORO-5-IODO-	119644-23-4
URIDINE, 2',3'-DIDEOXY-5-IODO-	105784-83-6
URIDINE, 2',3'-DIDEOXY-4-THIO-	122568-04-1
URIDINE, 5-(3,3-DIMETHYL-1-TRIAZENYL)-	038099-11-5
URIDINE, 5-BROMO-2',3'-DIDEOXY-	028616-93-5
URIDINE, 3-(PHENYLMETHYL)-	014985-34-3
URSODEOXYCHOLIC ACID	000128-13-2
VALERANILIDE	002364-50-3
SEC-VALERANILIDE	054394-78-4
VALERANILIDE	010264-18-3
TERT-VALERANILIDE	006625-74-7
VALERIC ACID HYDRAZIDE	038291-82-6
I-VALERIC ACID HYDRAZIDE	024310-18-7
DELTA-VALEROLACTAM	000675-20-7
D-VALEROLACTONE	000542-28-9
VALERONITRILE	000110-59-8
L-VALINAMIDE, N-ACETYL-L-ALANYL-L-LEUCYL-	132766-04-2
L-VALINAMIDE, N-ACETYL-L-ALANYL-	132765-80-1
L-VALINAMIDE, N -ACETYL-L-GLUTAMINYL-	132765-91-4
L-VALINAMIDE, N-ACETYLGLYCYL-L-ALANYL-	132766-08-6
L-VALINAMIDE, N-ACETYLGLYCYL-	034017-18-0
L-VALINAMIDE, N-ACETYL-L-ISOLEUCYL-L-ALANYL-	132766-10-0
L-VALINAMIDE, N-ACETYL-L-ISOLEUCYL-	132765-82-3
L-VALINAMIDE, N-ACETYL-L-LEUCYL-	132765-81-2
L-VALINAMIDE, N-ACETYL-L-PHENYLALANYL-	065118-48-1
L-VALINAMIDE, N-ACETYL-L-THREONYL-	132765-89-0
L-VALINAMIDE, N-ACETYL-L-TRYPTOPHYL-L-ALANYL-	132766-17-7
L-VALINAMIDE, N-ACETYL-L-TRYPTOPHYL-	132765-86-7
L-VALINAMIDE, N-ACETYL-L-TYROSYL-	099027-00-6
L-VALINAMIDE, N-ACETYL-L-VALYL-	132765-83-4
L-VALINAMIDE, N-ACETYL-L-VALYL-L-ALANYL-	132766-02-0

L-VALINAMIDE, N-BENZOYLGLYCYL-N-ETHYL-	120399-49-7
VALINE	000072-18-4
VALINE,N-ACETYL,N'-MEAM-AMIDE	019701-84-9
VALINE-AMIDE,N-ACETYL	037933-88-3
VALONE	000083-28-3
VALPROIC ACID	000099-66-1
VAMIDOTHION	002275-23-2
O-VANILLIN	000148-53-8
VANILLIN	000121-33-5
VEL 3510	051461-71-3
VERALIPRIDE	066644-81-3
VERAPAMIL	000052-53-9
VERNOLATE	001929-77-7
VIBRAMYCIN	000564-25-0
VINCLOZOLIN	050471-44-8
VINYL ACETATE	000108-05-4
VINYLCHLORIDE	000075-01-4
4-VINYLCYCLOHEXENE	000100-40-3
4-VINYL-1,2-DICHLOROCYCLOHEXANE	045803-84-7
VINYLFLUORIDE	000075-02-5
VINYL FORMATE	000692-45-5
VINYL PROPIONATE	000105-38-4
2-VINYLPYRIDINE	000100-69-6
4-VINYLPYRIDINE	000100-43-6
N-VINYL-2-PYRROLIDINONE	000088-12-0
VINYLTRICHLOROSILANE	000075-94-5
VX	050782-69-9
WARFARIN	000081-81-2
9H-XANTHENE	000092-83-1
XANTHENE-9-CARBOXYLIC ACID	000082-07-5
9H-XANTHEN-9-ONE	000090-47-1
XANTHINE	000069-89-6
XANTHONE-2-CARBOXYLIC ACID	040274-67-7
O-XYLENE	000095-47-6
P-XYLENE	000106-42-3
M-XYLENE	000108-38-3
.ALPHA.-L-XYLO-2-HEXULOFURANOSONIC ACID, 2,3:4,6	018467-77-1
YOHIMBINE HYDROCHLORIDE	000065-19-0
ZANTAC (RANITIDINE HYDROCHLORIDE)	066357-35-5
ZINEB (ALSO SEE CAS NO. 12122-67-7)	000142-14-3
ZIRAM	000137-30-4
ZOXAZOLAMINE	000061-80-3

APPENDIX III

ABRAHAM,M.H.; CHADHA,H.S.; WHITING,G.S.; MITCHELL,R.C.; HYDROGREN BONDING. 32. AN ANALYSIS OF WATER-OCTANOL AND WATER-ALKANE PARTITIONING AND THE LOG P PARAMETER OF SEILER.; J. PHARM. SCI.; 83:1085-100; 1994

ACHORN,P.J.; HASELTINE,W.G.; MILLER,J.K.; PHYSICOCHEMICAL PROPERTIES OF MONO- AND DIISOCYANATES.; J. CHEM. ENG. DATA; 31:385-7; 1986

ADDISON,R.F.; PATERSON,S.; MACKAY,D.; THE PREDICTED ENVIRONMENTAL DISTRIBUTIONS OF SOME PCB REPLACEMENTS.; CHEMOSPHERE.; 12:827-34.; 1983

AINSWORTH,C.C.; ZACHARA,J.M.; SMITH,S.C.; CARBAZOLE SORPTION BY SURFACE AND SUBSURFACE MATERIALS: INFLUENCE OF SORBENT AND SOLVENT PROPERTIES.; SOIL. SCI. SOC. AM. J.; 53:1391-1401; 1989

AKAMATSU, M., S.-I. OKUTANI, K. NAKAO, N.J. HONG AND T. FUJITA, "HYDROPHOBICITY OF N-ACETYL-DI- AND TRIPEPTIDE AMIDES HAVING UN-IONIZABLE SIDE CHAINS AND CORRELATION WITH SUBSTITUENT AND STRUCTURAL PARAMETERS", QUANT. STRUCT.-ACT. RELAT., 9(3), 189-194 (1990).

AKERBLOM, E.B., "SYNTHESIS AND STRUCTURE-ACTIVITY RELATIONSHIPS OF A SERIES OF ANTIBACTERIALLY ACTIVE 5-(5-NITRO-2-FURFURYLIDENE)THIAZOLONES, 5-(5-NITRO-2-FURYLPROPENYLIDENE) THIAZOLONES AND 6-(5-NITRO-2-FURYL)-4H-1,3-THIAZINONES", J. MED. CHEM., 17(6), 609-615 (1974).

ALCORN, C.J., R.J. SIMPSON, D.E. LEAHY AND T.J.PETERS, "PARTITION AND DISTRIBUTION COEFFICIENTS OF SOLUTES AND DRUGS IN BRUSH BORDER MEMBRANE VESICLES", BIOCHEM. PHARMACOL., 45(9), 1775-1782 (1993).

ALBANESE,V.; MILANO,J.C.; VERNET,J.L.; A STUDY OF THE EVAPORATION OF TRACE CONCENTRATIONS OF LOW MOLECULAR WEIGHT HALOGENATED HYDROCARBONS DISSOLVED IN WATER.; ENVIRON. TECHNOL. LETT.; 8:657-68; 1987

ALBERT,A; GOLDACRE,R.; PHILLIPS,J.; STRENGTH OF HETEROCYCLIC BASES.; J. CHEM. SOC.; 1948:2240-9.; 1948

ALDRICH.; CATALOG HANDBOOK OF FINE CHEMICALS 1988-1989; MILWAUKEE, WI: ALDRICH CHEM CO.; 1988

ALDRICH.; CATALOG HANDBOOK OF FINE CHEMICALS 1988-1989 (and other annual editions); MILWAUKEE, WI: ALDRICH CHEM CO.

ALHAIDER, A.A., C.D. SELASSIE, S.-O. CHUA AND E.J. LIEN, "MEASUREMENTS OF IONIZATION CONSTANTS AND PARTITION COEFFICIENTS OF GUANAZOLE PRODRUGS", J. PHARM. SCI., 71(1), 89-94 (1982).

ALTOMARE, C., R.-S. TSAI, N. EL TAYAR., B. TESTA. A. CAROTTI, S. CELLAMARE AND P.G. DE BENEDETTI, "DETERMINATION OF LIPOPHILICITY AND HYDROGEN-BOND DONOR ACIDITY OF BIOACTIVE SULPHONYL-CONTAINING COMPOUNDS BY REVERSED-PHASE HPLC AND CENTRIFUGAL PARTITION CHROMATOGRAPHY AND THEIR APPLICATION TO STRUCTURE-ACTIVITY RELATIONS", J. PHARM. PHARMACOL., 43(3), 191-197 (1991).

ALTOMARE, C.A., P.A. CARRUPT, P. GAILLARD, N. EL TAYAR, B. TESTA AND A. CAROTTI, "QUANTITATIVE STRUCTURE-METABOLISM RELATIONSHIP ANALYSES OF MAO-MEDIATED TOXICATION OF 1-METHYL-4-PHENYL-1,2,3,6-TETRAHYDROPYRIDINE AND ANALOGS", CHEM. RES. TOXICOL., 5(3), 366-375 (1992).

AL-SAADI, D., W.E. SNEADER, A.J. BAILLIE AND N.G. WATON, "INHIBITION OF MOBILITY OF TETRAHYMENA PYRIFORMIS BY CERTAIN NEW LOCAL ANAESTHETICS: AN ALTERNATIVE IN VITRO APPROACH TO DRUG SCREENING", PHARMAZIE, 48(8), 627-628 (1993).

AMBROSE,D.; ELLENDER,J.H.; LEES,E.B.; SPRAKE,C.H.S.; TOWNSEND,R.; THERMODYNAMIC PROPERTIES OF ORGANIC OXYGEN COMPOUNDS. XXXVIII. VAPOUR PRESSURES OF SOME ALIPHATIC KETONES.; J. CHEM. THERM.; 7:453-72.; 1975

AMBROSE,D.; CONNETT,J.E.; GREEN,J.H.S.; HALES,J.L.; HEAD,A.J.; MARTIN,J.F.; THERMODYNAMIC PROPERTIES OF ORGANIC OXYGEN COMPOUNDS. 42. PHYSICAL AND THERMODYNAMIC PROPERTIES OF BENZALDEHYDE.; J. CHEM. THERM.; 7:1143-57.; 1975A

AMBROSE,D.; ELLENDER,J.H.; SPRAKE,C.H.S.; TOWNSEND,R.; THERMODYNAMIC PROPERTIES OF FLOURINE COMPOUNDS. VAPOR PRESSURE OF THE THREE TETRAFLOUROBENZENES AND 1,3,5,-TRICHLORO-2,4,6-TRIFLUOROBENZENE.; J. CHEM. SOC. FARADAY TRANS. 1.; 71:35-41.; 1975C

AMBROSE,D.; ELLENDER,J.H.; SPRAKE, C.H.S.; TOWNSEND, R.; THERMODYNAMIC PROPERTIES OF ORGANIC OXYGEN COMPOUNDS. XLIII. VAPOR PRESSURES OF SOME ETHERS.; J. CHEM. THERMODYN.; 8:165-78.; 1976

AMBROSE,D.; ELLENDER,J.H.; GUNDRY,H.A.; LEE,D.A.; TOWNSEND,R.; THERMODYNAMIC PROPERTIES OF ORGANIC OXYGEN COMPOUNDS. LI. THE VAPOUR PRESSURES OF SOME ESTERS AND FATTY ACIDS.; J. CHEM. THERM.; 13:795-802.; 1981

AMIDON,G.L.; YALKOWSKY,S.H.; LEUNG,S.; SOLUBILITY OF NONELECTROLYTES IN POLAR SOLVENTS. SOLUBILITY OF ALIPHATIC ALCOHOLS IN WATER.; J. PHARM. SCI.; 63:1858-66.; 1974

AMOORE,J.E.; HAUTALA,E.; ODOR AS AN AID TO CHEMICAL SAFETY: ODOR THRESHOLDS COMPARED WITH THRESHOLD LIMIT VALUES AND VOLATILITIES FOR 214 INDUSTRIAL CHEMICALS IN AIR AND WATER DILUTION; J. APPL. TOXICOL.; 3:272-290; 1983

ANDON,R.J.L.; COX,J.D.; HERINGTON,E.F.G.; PHASE RELATIONSHIP IN THE PYRIDINE SERIES. PART V. THE THERMODYNAMIC PROPERTIES OF DILUTE SOLUTIONS OF PYRIDINE BASES IN WATER AT 25 DEGREES AND 40 DEGREES.; J. AMER. CHEM. SOC.; 76:3188-96; 1954

ARCHER,W.L.; OTHER CHLOROETHANES.; IN: KIRK-OTHMER ENCYCL. CHEM. TECH. 3RD ED. 5:722-42.; 1979

ARNOLDI, A., S. GRASSO, G. MEINARDI AND L. MERLINI, "SYNTHESIS AND ANTIFUNGAL ACTIVITY OF SIMPLE BETA-LACTAMS", EUR. J. MED. CHEM., 23(2), 149-154 (1988).

ARTHUR D. LITTLE, INC.; DEVELOPMENT OF CANDIDATE CHEMICAL LIST: THE EVALUATION OF CANDIDATE CHEMICAL SIMULANTS WHICH MAY BE USED IN CHEMICALLY HAZARDOUS OPERATIONS.; AIR FORCE AERO. MED. RES. LAB., WRIGHT-PATTERSON AFB, OH, AFAMRL-TR-82-87. NTIS AD-B070947. 144 PP.; 1982

ARTIOLA-FORTUNY,J.; FULLER,W.H.; ADSORPTION OF SOME MONOHYDROXYBENZENE DERIVATIVES BY SOILS.; SOIL SCI.; 133:18-26.; 1982

ASHFORD,RD. DICTIONARY OF INDUSTRIAL CHEMICALS; LONDON, ENGLAND: WAVELENGTH PUBLICATIONS LTD. (1994).

ASHCROFT,S.J.; VAPOR PRESSURES AND ENTHALPIES OF VAPORIZATION OF BENZYL HALIDES.; J. CHEM. ENGIN. DATA; 21:397-8.; 1976

ATKINSON,R.; KINETICS AND MECHANISMS OF THE GAS-PHASE REACTIONS OF HYDROXYL RADICAL WITH ORGANIC COMPOUNDS UNDER ATMOSPHERIC CONDITIONS.; CHEM. REV.; 85:69-201; 1985

ATKINSON,R.; A STRUCTURE-ACTIVITY RELATIONSHIP FOR THE ESTIMATION OF RATE CONSTANTS FOR THE GAS-PHASE REACTIONS OF OH RADICALS WITH ORGANIC COMPOUNDS.; INT. J. CHEM. KINETICS; 19:799-828; 1987A

ATKINSON,R.; ESTIMATION OF GAS-PHASE HYDROXYL RADICAL RATE CONSTANTS FOR ORGANIC CHEMICALS; ENVIRON TOXICOL CHEM; 7:435-62; 1988

ATKINSON,R.; KINETICS AND MECHANISMS OF THE GAS-PHASE REACTIONS OF THE HYDROXYL RADICAL WITH ORGANIC COMPOUNDS.; JOURNAL OF PHYSICAL AND CHEMICAL REFERENCE DATA. MONOGRAPH NO. 1.; 1989

ATKINSON,R.; GAS-PHASE TROPOSPHERIC CHEMISTRY OF ORGANIC COMPOUNDS: A REVIEW.; ATMOS. ENVIRON.; 24A:1-41; 1990

ATKINSON,R.; LIFETIMES AND FATES OF TOXIC AIR CONTAMINANTS IN CALIFORNIA'S ATMOSPHERE.; STATEWIDE AIR POLLUT. RES. CENT., UNIV. CALIFORNIA, RIVERSIDE, CA USA. REPORT 1990, ARB-R-90/441; PB90-253204, 290 PP.; 1990A

ATKINSON,R.; KINETICS OF THE GAS-PHASE REACTIONS OF A SERIES OF ORGANOSILICON COMPOUNDS WITH OH AND NO3 AT 297 PLUS OR MINUS 2K.; ENVIRON. SCI. TECHNOL.; 25:863-6; 1991

ATKINSON,R.; ATMOSPHERIC LIFETIMES OF DIBENZO-P-DIOXINS AND DIBENZOFURANS.; SCI. TOTAL ENVIRON.; 104:17-33; 1991A

ATKINSON,R.; ASCMANN,S.M.; GOODMAN,M.A.; WINER,A.M.; KINETICS OF THE GAS-PHASE REACTIONS OF THE OH RADICAL WITH (ETO)3PO AND (ME)2P(S)CL AT 296 +/- 2 DEG K.; INT. J. CHEM. KINET.; 20:273-81; 1988

ATKINSON,R.; BAULCH,D.L.; COX,R.A.; HAMPSON,R.F. JR.; KERR,J.A.; TROE,J.; EVALUATED KINETIC AND PHOTOCHEMICAL DATA FOR ATMOSPHERIC CHEMISTRY, SUPPLEMENT IV.; J. PHYS. CHEM. REF. DATA; 21:1125-1569; 1992C

ATLAS,E.; VELASCO,A.; SULLIVAN,K.; GIAM,C.S.; A RADIOTRACER STUDY OF AIR-WATER EXCHANGE OF SYNTHETIC ORGANIC COMPOUNDS.; CHEMOSPHERE.; 12:1251-8.; 1983

AUBRY,M.; MAYORAL,M.N.; VILLARDRY,P.; DETERMINATION OF THE VAPOR PRESSURE OF LOW VOLATILITY COMPOUNDS BY GAS CHROMATOGRAPHY. APPLICATION TO DIBENZOTHIOPHENE.; BULL. SOC. CHIM. FR. 1975. (3-4, PT.1). PP.500-2.; 1975

AUGUSTIN, J., S. BALAZ, J. HANES AND E. STURDIK, "PARTITIONING BEHAVIOR OF ISOTHIOCYANATES IN THE TWO-PHASE SYSTEM 1-OCTANOL/WATER", CHEM. PAP., 41(3), 401-405 (1987).

AUGUSTIJN-BECKERS,P.W.M.; HORNSBY,A.G.; WAUCHOPE,R.D.; THE SCS/ARS/CES PESTICIDE PROPERTIES DATABASE FOR ENVIRONMENTAL DECISION-MAKING. II. ADDITIONAL COMPOUNDS.; REV. ENVIRON. CONTAM. TOXICOL.; 137:1-82; 1994

AVDEEF,A.; PH-METRIC LOG P. II. REFINEMENT OF PARTITION COEFFICIENTS AND IONIZAITION CONSTANTS OF MULTIPROTIC SUBSTANCES.; J. PHARM. SCI.; 82:183-90; 1993

BACCANARI,D.P.; NOVINSKI,J.A.; PAN,Y.C.; YEVITZ,M.M.; SWAIN,H.A. JR.; HEATS OF SUBLIMATION AND VAPORIZATION AT 25 DEGREES OF LONG-CHAIN FATTY ACIDS AND METHYL ESTERS.; TRANS. FARADAY SOC.; 64:1201-5.; 1968

BACCI,E.; CALAMARI,D.; GAGGI,C.; VIGHI,M.; BIOCONCENTRATION OF ORGANIC CHEMICAL VAPORS IN PLANT LEAVES: EXPERIMENTAL MEASUREMENTS AND CORRELATION.; ENVIRON. SCI. TECHNOL.; 24:885-9; 1990

BAGLAY,A.K.; GURARLY,L.L.; KULESHOV,G.G.; PHYSICAL PROPERTIES OF COMPOUNDS USED IN VITAMIN SYNTHESIS.; J. CHEM. ENG. DATA.; 33:512-8 .; 1988

BAILEY,G.W.; WHITE,J.L.; HERBICIDES: A COMPILATION OF THEIR PHYSICAL, CHEMICAL, AND BIOLOGICAL PROPERTIES.; RES. REV.; 10:97-123.; 1965

BAILEY,G.W.; WHITE,J.L.; ROTHBERG,T.; ADSORPTION OF ORGANIC HERBICIDES BY MONTMORILLONITE: ROLE OF PH AND CHEMICAL CHARACTER OF ADSORBATE.; SOIL SCI. SOC. OF AMER. PROC.; 32:222-34.; 1968

BAILEY,G.W.; WHITE,J.L.; ROTHBERG,T.; ADSORPTION OF ORGANIC HERBICIDES BY MONTMORILLONITE: ROLE OF PH AND CHEMICAL CHARACTER OF ADSORBATE.; SOIL SCI. SOC. OF AMER. PROC.; 32:222-34.; 1968

BAKER,E.A.; HAYES,A.L.; BUTLER,R.C.; PHYSICOCHEMICAL PROPERTIES OF AGROCHEMICALS: THEIR EFFECTS ON FOLIAR PENETRATION.; PESTIC. SCI.; 34:167-82; 1992

BAKER,E.A.; HAYES,A.L.; BUTLER,R.C.; PHYSICOCHEMICAL PROPERTIES OF AGROCHEMICALS: THEIR EFFECTS ON FOLIAR PENETRATION.; PESTIC. SCI.; 34:167-82; 1992

BAKER,P.J. JR.; BOLLMEIER,A.F. JR.; NITROPARAFFINS.; IN: KIRK-OTHMER ENCYCL. CHEM. TECH. 15:969-87.; 1981

BALAZ, S., D. ILAVSKY, E. STURDIK AND J. KOVAC, "ANTIMICROBIAL ACTIVITY OF METHYL ESTERS AND NITRILES OF 2-CYANO-3-(5'-R-2'-FURYL)PROPENOIC ACID", FOLIA MICROBIOL. (PRAGUE), 30, 34-41 (1985).

BALAZ, S., A. KUCHAR, E. STURDIK, M. ROSENBERG, L. STIBRANYI AND D. ILAVSKY, "INTERPHASE DISTRIBUTIONS OF 2-FURYLETHYLENES", COLL. CZECH. CHEM. COMMUN., 50(8), 1642-1647 (1985).

BALKE, N.E. AND T.P. PRICE, "RELATIONSHIP OF LIPOPHILICITY TO INFLUX AND EFFLUX OF TRIAZINE HERBICIDES IN OAT ROOTS", PESTIC. BIOCHEM. PHYSIOL., 30(3), 228-237 (1988).

BALLSCHMITER,K.; WITTLINGER,R.; INTERHEMISPHERE EXCHANGE OF HEXACHLOROCYCLOHEXANE, HEXACHLOROBENZENE, POLYCHLOROBIPHENYLS, AND 1,1,1-TRICHLORO-2,2-BIS(P-CHLOROPHENYL)ETHANE IN THE LOWER TROPOSPHERE.; ENVIRON. SCI. TECHNOL.; 25:1103-11; 1991

BALZARINI,J, M. COOLS AND E. DE CLERQ, "ESTIMATION OF THE LIPOPHILICITY OF ANTI-HIV NUCLEOSIDE ANALOGUES BY DETERMINATION OF THE PARTITION COEFFICIENT AND RETENTION TIME ON A LICHROSPHER 60 RP-8 HPLC COLUMN", BIOCHEM. BIOPHYS. RES. COMMUN., 158(2), 413-422 (1989).

BANERJEE,S.; SOLUBILITY OF ORGANIC MIXTURES IN WATER.; ENVIRON. SCI. TECHNOL.; 18:587-91.; 1984

BANERJEE,S.; YALKOWSKY,S.H.; VALVANI,S.C.; WATER SOLUBILITY AND OCTANOL/WATER PARTITION COEFFICIENTS OF ORGANIC LIMITATIONS OF THE SOLUBILITY PARTITION COEFFICIENT CORRELATION.; ENVIRON. SCI. TECHNOL.; 14:1227-9.; 1980

BANWART,W.L.; HASSETT,J.J.; WOOD,S.G.; MEANS,J.C.; SORPTION OF NITROGEN-HETEROCYCLIC COMPOUNDS BY SOILS AND SEDIMENTS.; SOIL SCI.; 133:42-7.; 1982

BARBER,L.B. II.; THURMAN,E.M.; SCHROEDER,M.P.; LEBLANC,D.R.; LONG-TERM FATE OF ORGANIC MICROPOLLUTANTS IN SEWAGE-CONTAMINATED GROUND WATER.; ENVIRON. SCI. TECHNOL.; 22:205-11; 1988

BARCHI, J.J., V.E. MARQUEZ, J.S. DRISCOLL, H. FORD, H. MITSUYA, T. SHIRASAKA, S. AOKI AND J.A. KELLEY, "POTENTIAL ANTI-AIDS DRUGS. LIPOPHILIC ADENOSINE DEAMINASE-ACTIVATED PRODRUGS", J. MED. CHEM., 34(5), 1647-1655 (1991).

BARNALES,C.A.; BUSHMAN,S.E.; KRALJIC,J.; OXALIC ACID.; IN: KIRK-OTHMER ENCYCL. CHEM. TECHNOL. 3RD ED. NEW YORK, NEW YORK 16:618-36. JOHN WILEY & SONS.; 1981

BARRACLOUGH,P ET AL.; "INOTROPIC 2-ARYLIMIDAZO[1,2-A]PYRIMIDINES", EUR. J. MED. CHEM., 27 (3), 207-217, (1992)

BARTON,A.F.M.; ALCOHOLS WITH WATER.; INTERNATIONAL UNION OF PURE AND APPLIED CHEMISTRY. SOLUBILITY DATA SERIES. VOL. 15, 438 PP.; 1984

BAUGHMAN,G.L.; PERENICH,T.A.; FATE OF DYES IN AQUATIC SYSTEMS: I. SOLUBILITY AND PARTITIONING OF SOME HYDROPHOBIC DYES AND RELATED COMPOUNDS.; ENVIR. TOXIC. CHEM.; 7:183-99; 1988

BAUGHMAN,G.L.; PERENICH,T.A.; INVESTIGATING THE FATE OF DYES IN THE ENVIRONMENT; AM DYEST REP; 77:19-20; 1988A

BAUGHMAN,G.L.; WEBER,E.J.; ESTIMATION OF WATER SOLUBILITY AND OCTANOL/WATER PARTITION COEFFICIENT OF HYDROPHOBIC DYES. PART I. RELATIONSHIP BETWEEN SOLUBILITY AND PARTITION COEFFICIENT.; DYES PIGM.; 16:261-71; 1991

BAUGHMAN,G.L.; BANERJEE,S.; PERENICH,T.A.; DYE SOLUBILITIES.; IN: ADVANCES IN COLOR CHEMISTRY, FREEMAN,M & PETERS,MT EDS., ELSEVIER, IN PRESS.; 1993

BAXTER,WFJR CROTONALDEHYDE; IN "KIRK-OTHMER ENCYCLOPEDIA OF CHEMICAL TECHNOLOGY, WILEY-INTERSCIENCE, NEW YORK, 3RD EDITION, VOL.7 PP.207-218, 1979."; 1979

BAZACO, J.F. AND C.M. COCA, "RELATION BETWEEN THE PARTITION COEFFICIENT (LOG POW) AND CAPACITY FACTOR (LOG K') DERIVED FROM REVERSED-PHASE HPLC OF A SERIES OF BENZALDEHYDE DERIVATIVES", CIENC. IND. FARM., 8(10-12), 223-226 (1989).

BEILSTEIN; ORGANISCHEN CHEMIE.; BERLIN: SPRINGER-VERLAG. 4TH ED.; NA--

BEILSTEIN ONLINE (1991)

BEMIS,A.G.; DINDORF,J.A.; HORWOOD,B.; SAMANS,C.; PHTHALIC ACIDS, AND OTHER BENZENEPOLYCARBOXYLIC ACIDS.; IN: KIRK-OTHMER ENCYCL. CHEM. TECH. 3RD ED. 17:732-77.; 1982

BEMIS,A.G.; DINDORF,J.A.; HORWOOD,B.; SAMANS,C.; PHTHALIC ACIDS, AND OTHER BENZENEPOLYCARBOXYLIC ACIDS.; IN: KIRK-OTHMER ENCYCL. CHEM. TECH. 3RD ED. 17:732-77.; 1982

BENNETT,G.M.; PHILIP, W.G.; CCLII. THE INFLUENCE OF STRUCTURE ON THE SOLUBILITIES OF ETHERS. PART I. ALIPHATIC ETHERS.; J. CHEM. SOC.; PP. 1930-7; 1928

BENNETT,S.R.; BANE,J.M.; BENFORD,P.J.; PYATT,R.L.; ENVIRONMENTAL HAZARDS OF CHEMICAL AGENT SIMULANTS; CRDC-TR-84055, ABERDEEN PROVING GROUND, MD; 1984

BENSCHOP,H.P.; WESSELMAN,H.C.; PHARMAKINETICS OF SOMAN SIMULANT 1,2,2-TRIMETHYLPROPYLDIMETHYLPHOSPHINATE (PDP) IN RATS; ARCH. TOXICOL.; 63:238-43; 1989

BENSON,F.R.; ALCOHOLS, POLYHYDRIC.; IN: KIRK-OTHMER ENCYCL. CHEM. TECH. 3RD ED. 1:754-78.; 1978

BEYNON,K.I.; HUTSON,D.H.; WRIGHT,A.N.; THE METABOLISM AND DEGRADATION OF VINYL PHOSPHATE INSECTICIDES.; RES. REV.; 47:55-142.; 1973

BETTERTON,E.A.; HOFFMANN,M.R.; HENRY'S LAW CONSTANTS OF SOME ENVIRONMENTALLY IMPORTANT ALDEHYDES; ENVIRON SCI TECHNOL; 22:1415-8; 1988

BEWICK,D.W.; HILL,I.R.; PLUCKROSE,J.; STEVENS,J.E.B.; WEISSLER,M.S.; THE ROLE OF LABORATORY AND FIELD STUDIES, USING RADIOLABELED MATERIALS, IN THE INVESTIGATION OF THE DEGRADATION AND MOBILITY OF TEFLUTHRIN IN SOIL.; PROC. BR. CROP. CONF., PEST DIS. 2:159-68; 1986

BEYER,E.M. JR.; DUFFY,M.J.; HAY,J.V.; SCHLUETER,D.D.; SULFONYLUREAS.; IN: HERBICIDES: CHEMISTRY, DEGRADATION, AND MODE OF ACTION. KEARNEY, PC & KAUFMAN, DD EDS. MARCEL DEKKER, INC: NEW YORK AND BASEL. PP.117-89; 117-89; NA--

BEYELER, S. ET AL. 1988. FLAVONVOIDS AS INHIBITORS OF RAT MONOOXYGENASE ACTIVITIES. BIOCHEM. PHARMACOL. 37: 1971-9.

BHATTACHARYYA,D.; HAMRIN,C.E.J.R.; NORTHEY,R.P.; OXIDATION OF HAZARDOUS ORGANICS IN TWO-PHASE FLUOROCARBON-WATER SYSTEM.; HAZ. WASTE HAZ. MATER.; 3:405-29; 1986

BIAGI,GL ET AL. "STUDY OF THE LIPOPHILIC CHARACTER OF A SERIES OF BETA-CARBOLINES", J. CHROMATOGR., 469, 121-126 (1989).

BIDDISCOMBE,D.P.; HANDLEY,R.; HARROP,D.; HEAD,A.J.; LEWIS,G.B.; MARTIN,J.F.; SPRAKE,C.H.S.; THERMODYNAMIC PROPERTIES OF ORGANIC OXYGEN COMPOUNDS. XIII. PREPARATION AND PHYSICAL PROPERTIES OF PURE ETHYLPHENOLS.; J. CHEM. SOC. PP.5764-8.; 1963

BIDLEMAN,T.F.; ESTIMATION OF VAPOR PRESSURES FOR NONPOLAR ORGANIC COMPOUNDS BY CAPILLARY GAS CHROMATOGRAPHY.; ANAL.CHEM.; 56:2490-6.; 1984

BIDLEMAN,T.F.; RENBERG,L.; DETERMINATION OF VAPOR PRESSURES FOR CHLOROGUAIACOLS, CHLOROVERATROLES, AND NONYLPHENOL BY GAS CHROMATOGRAPHY.; CHEMOSPHERE; 14:1475-81; 1985

BIDLEMAN,T.F.; FOREMAN,W.T.; VAPOR-PARTICLE PARTITIONING OF SEMIVOLATILE ORGANIC COMPOUNDS; ADV. CHEM. SER.; 216:27-56; 1987

BIDLEMAN,TF; ENVIRON SCI TECHNOL 22: 361-367 (1988)

BIGGAR,J.W.; RIGGS,R.I.; APPARENT SOLUBILITY OF ORGANOCHLORINE INSECTICIDES IN
WATER AT VARIOUS TEMPERATURES; HILGARDIA; 42:383-91; 1974

BIKALES,N.M.; KOLODNY,E.R.; ACRYLAMIDE.; IN: KIRK-OTHMER ENCYCL. CHEM. TECH.
2ND ED. 1:274-84.; 1963

BINTEIN,S.; DEVILLERS,J.; QSAR FOR ORGANIC CHEMICAL SORPTION IN SOILS AND
SEDIMENTS.; CHEMOSPHERE; 28:1171-88; 1994

BINTEIN,S.; DEVILLERS,J.; KARCHER,W.; NONLINEAR DEPENDENCE OF FISH
BIOCONCENTRATION ON N-OCTANOL/WATER PARTITION COEFFICIENT.; SAR QSAR ENVIRON.
RES.; 1:29-39; 1993

BOCEK,K.; RELATIONS AMONG ACTIVITY COEFFICIENTS, PARTITION COEFFICIENTS AND
SOLUBILITIES.; EXPERIENTIA, SUPPL.; 23:231-40.; 1976

BODOR, N. AND M. HUANG. 1992. AN EXTENDED VERSION OF A NOVEL METHOD FOR THE ESTIMATION OF PARTITION
COEFFICIENTS. J. PHARM. SCI. 81: 272-81.

BOETHLING,R.S.; COOPER,J.C.; ENVIRONMENTAL FATE AND EFFECTS OF TRIARYL AND
TRI-ALKYL/ARYL PHOSPHATE ESTERS.; RES. REV.; 94:49-99.; 1985

BOETHLING,R.S.; ENVIRONMENTAL ASPECTS OF CATIONIC SURFACTANTS.; IN: CATIONIC
SURFACTANTS: ANALYTICAL AND BIOLOGICAL EVALUATION. CROSS,J & SINGER,EJ EDS.,
MARCEL DEKKER INC.: NEW YORK, NY. CHAP 4 PP. 95-135; 1994

BOGYO,D.A.; LANDE,S.S.; MEYLEN,W.M.; HOWARD,P.H.; SANTODONATO,J.; INVESTIGATION
OF SELECTED ENVIRONMENTAL CONTAMINANTS: EPOXIDES.; EPA-560/11-80-005.
WASHINGTON,D.C.:US EPA. PP.202.; 1980

BOLLMEIER,AF; ALKANOAMINES; IN: KIRK-OTHMER ENCYCL. CHEM. TECH. 4TH ED. VOL 2: 3.; 1991

BONOHORST,C.W.; ALTHOUSE,P.M.; TRIEBOLD,H.O.; ESTERS OF NATURALLY OCCURRING
FATTY ACIDS.; IND. ENG. CHEM.; 40:2379-84.; 1948

BOSCHE,H AND SCHNEIDER,K; "BUTANEDIOLS, BUTENEDIOL, AND BUTYNEDIOL" IN: ULLMANN'S ENCYCLOPEDIA OF
INDUSTRIAL CHEMICALS. 5TH ED., VOL. A4: 461.; 1985

BOSCH,S.J.; DRINKING WATER CRITERIA DOCUMENT FOR 2,4-D.; SRC TR-83-707.
SYRACUSE RESEARCH CORPORATION, SYRACUSE, NY.; 1983

BOUBLIK,T.; FRIED,V.; HALA,E.; THE VAPOR PRESSURES OF PURE SUBSTANCES: SELECTED
VALUES OF THE TEMPERATURE DEPENDENCE OF THE VAPOUR PRESSURES OF SOME PURE
SUBSTANCES IN THE NORMAL AND LOW PRESSURE REGION. VOL. 17.; AMSTERDAM,
NETHERLANDS: ELSEVIER SCI. PUBL.; 1984

BOULIN,C.H.; SIMON,L.J.; ACTION DE L'EAU SUR LE SULFATE DIMETHYLIQUE(FR.);
COMPT. REND.; 170:392-4; 1920

BOWMAN,B.T.; SANS,W.W.; FURTHER WATER SOLUBILITY DETERMINATION OF INSECTICIDAL
COMPOUNDS.; J. ENVIRON. SCI. HEALTH.; B18:221-7.; 1983

BOWMAN,B.T.; SANS,W.W.; DETERMINATION OF OCTANOL-WATER PARTIONING COEFFICIENTS
(KOW) OF 61 ORGANOPHOSPHORUS AND CARBAMATE INSECTICIDES AND THEIR RELATIONSHIP
TO RESPECTIVE WATER SOLUBILITY (S) VALUES.; J. ENVIRON. SCI. HEALTH.;
B18:667-83.; 1983A

BOWMAN,B.T.; SANS,W.W.; PARTITIONING BEHAVIOR OF INSECTICIDES IN SOIL-WATER
SYSTEMS: I. ADSORBENT CONCENTRATION EFFECTS.; J. ENVIRON. QUAL.; 14:265-9.; 1985

BOYD,R.H.; CHRISTENSEN,R.L.; PUA,R.; THE HEATS OF COMBUSTION OF ACENAPHTHENE,
ACENAPHTHYLENE, AND FLUORANTHENE. STRAIN AND DELOCALIZATION IN BRIDGED
NAPHTHALENES.; J. AM. CHEM. SOC.; 87:3554-9.; 1965

BOYD,R.H.; SANWAL,S.N.; SHARY-TEHRANY,S.; MCNALLY,D.; THERMOCHEMISTRY,

THERMODYNAMIC FUNCTIONS, AND MOLECULAR STRUCTURES OF SOME CYCLIC HYDROCARBONS.;
J. PHYS. CHEM.; 75:1264-71.; 1971

BRADLEY,R.S.; COTSON,S.; COX,E.G.; THE DETERMINATION OF VAPOR PRESSURE OF THE
ORDER OF 1 MM OF MERCURY BY A MODIFICATION OF SMITH AND MENZIES'S METHOD. THE
VAPOR PRESSURE OF DIMETHYLNITRAMINE.; J. CHEM. SOC.; PP. 740-1; 1952

BRANDT, U. ET AL. 1992. UNCOUPLING ACTIVITY AND PHYSICOCHEMICAL PROPERTIES OF DERIVATIVES OF
FLUAZINAM. BIOCHIM. BIOPHYS. ACTA 1101: 41-7.

BRAUMANN, T. AND L.H. GRIMME, "DETERMINATION OF HYDROPHOBIC PARAMETERS FOR PYRIDAZINONE HERBICIDES
BY LIQUID-LIQUID PARTITION AND REVERSED-PHASE HIGH-PERFORMANCE LIQUID CHROMATOGRAPHY", J.
CHROMATOGR., 206(1), 7-15 (1981).

BRANSON,D.R.; A NEW CAPACITOR FLUID-A CASE STUDY IN PRODUCT STEWARDSHIP.; ASTM
STP 634. IN: AQUATIC TOXICOLOGY AND HAZARD EVALUATION.; PHILADELPHIA, PA:
AMERICAN SOCIETY FOR TESTING AND MATERIALS. PP. 44-61.; 1977

BRIGGS,G.G.; THEORETICAL AND EXPERIMENTAL RELATIONSHIPS BETWEEN SOIL ADSORPTION,
OCTANOL-WATER PARTITION COEFFICIENTS, WATER SOLUBILITIES, BIOCONCENTRATION
FACTORS AND THE PARACHOR.; J. AGRIC. FOOD CHEM.; 29:1050-9.; 1981

BRITTON,K.B.; LOW TEMPERATURE EFFECTS ON SORPTION, HYDROLYSIS AND PHOTOLYSIS OF
ORGANOPHOSPHONATES: A LITERATURE REVIEW; REPORT; AD-A178349/7/GAR; 1986

BRONAUGH, R.L. AND E.R. CONGDON, "PERCUTANEOUS ABSORPTION OF HAIR DYES: CORRELATION WITH PARTITION
COEFFICIENTS", J. INVEST. DERMATOL., 83(2), 124-127 (1984).

BROOKE,D.; NIELSEN,I.; DE BRUIJN,J.; HERMENS,J.; AN INTERLABORATORY EVALAUATION
OF THE STIR-FLASK METHOD FOR THE DETERMINATION OF OCTANOL-WATER PARTITION
COEFFOCIENTS (LOG POW); CHEMOSPHERE; 21:119-33; 1990

BROWN,A.C.; CANOSA-MAS,C.E.; PARR,A.D.; WAYNE,R.P.; LABORATORY STUDIES OF SOME
HALOGENATED ETHANES AND ETHERS: MEASUREMENTS OF RATES OF REACTION WITH OH AND
INFRARED ABSORPTION CROSS-SECTIONS.; ATMOS. ENVIRON.; 24A:2499-511; 1990A

BROWN,D.S.; FLAGG,E.W.; EMPIRICAL PREDICTION OF ORGANIC POLLUTANT SORPTION IN
NATURAL SEDIMENTS.; J. ENVIRON. QUAL.; 10:382-6.; 1981

BROWN,E.S.; HAUSER,C.F.; REAM,B.C.; BERTHOLD,R.V.; ETHYLENE GLYCOL AND PROPYLENE
GLYCOL.; IN: KIRK-OTHMER ENCYCL. CHEM. TECH. 3RD ED. 11:933-56.; 1980

BROWN,S.L.; CHAN,F.Y.; JONES,J.L.; LIU,D.H.; MCCALEB,K.E.; RESEARCH PROGRAM ON
HAZARD PRIORITY RANKING OF MANUFACTURED CHEMICALS (CHEMICALS 1-20).; NTIS
PB-263 161. MENLO PARK,CA: STANFORD RES. INST. PP.191.; 1975A

BROWN,S.L.; CHAN,F.Y.; JONES,J.L.; LIU,D.H.; MCCALEB,K.E.; MILL,T.; SAPIOS,K.N.;
SCHENDEL,D.E.; RESEARCH PROGRAM ON HAZARD PRIORITY RANKING OF MANUFACTURED
CHEMICALS (CHEMICALS 41-60).; NTIS PB-263 163. STANFORD RES. INST., MENLO PARK,
CA. 191 PP.; 1975B

BROWN,S.L.; CHAN,F.Y.; JONES,J.L.; LIU,D.H.; MCCALEB,K.E.; RESEARCH PROGRAM ON
HAZARD PRIORITY RANKING OF MANUFACTURED CHEMICALS(CHEMICALS 61-79).; NTIS
PB-263 164. MENLO PARK,CA: STANFORD RES. INST.; 1975C

BROWNLEE,B.G.; CAREY,J.H.; MACINNIS,G.A.; AQUATIC ENVIRONMENTAL CHEMISTRY OF
Z-(THIOCYANOMETHYLTHIO)BENZOTHIAZOLE AND RELATED BENZOTHIAZOLES.; ENVIRON.
TOXICOL. CHEM.; 11:1153-68; 1992

BRUCKMANN,P.W.; WILLNER,H.; INFRARED SPECTROSCOPIC STUDY OF PEROXYACETYLNITRATE
(PAN) AND ITS DECOMPOSITION PRODUCTS.; ENVIRON. SCI TECHNOL.; 17:352-7.; 1983

BRUGGEMAN,W.A.; WEBER-FUNG,D.; OPPERHUIZEN,A.; VANDERSTEEN,J.; WIJBERGER,A.;
HUTZINGER,O.; ADSORPTION AND RETENTION OF POLYDIMETHYLSILOXANES (SILICONES) IN
FISH: PRELIMINARY EXPERIMENTS.; TOX. ENVIRON. CHEM.; 7:287-96.; 1984

BUDAVARI,S; MERCK INDEX.; THE MERCK INDEX. AN ENCYCLOPEDIA OF CHEMICALS AND DRUGS 11TH ED.; WINDHOLZ,M ED. RAHWAY, N.J.: MERCK AND CO., INC.; 1989

BUDVARI-BARANY, Z., G. SZASZ, K. TAKACS-NOVAK, J. VAMOS, J. KOKOSI AND M. JOZAN, "INVESTIGATION OF STRUCTURE-PROPERTY RELATIONSHIPS FOR IMIDAZOQUINOLONE DERIVATIVES. I. RELATION BETWEEN PARTITION COEFFICIENT AND CHROMATOGRAPHIC RETENTION", J. LIQ. CHROMATOGR., 13(8), 1485-1497 (1990).

BUIKEMA,A.L.J.R.; MCGINNISS,M.J.; CAIRNS,J.J.R.; PHENOLICS IN AQUATIC ECOSYSTEMS: A SELECTED REVIEW OF RECENT LITERATURE.; MAR. ENVIRON. RES.; 2:87-181; 1979

BULL,HB ET AL.; SOLUBILITIES OF AMINO ACIDS, BIPMA, 17, 1091-1100, (1978)

BUNDGAARD, H. AND N.M. NIELSEN, "GLYCOLAMIDE ESTERS AS A NOVEL BIOLABILE PRODRUGTYPE FOR NON-STEROIDAL ANTI-INFLAMMATORY CARBOXYLIC ACID DRUGS", INT. J. PHARM., 43(1-2), 101-110 (1988).

BUNDGAARD, H. AND E. FALCH, "ALLOPURINAL PRODRUGS. I. SYNTHESIS, STABILITY AND PHYSICOCHEMICAL PROPERTIES OF VARIOUS N(1)-ACYL ALLOPURINOL DERIVATIVES", INT. J. PHARM., 23, 223-237 (1985).

BUNDGAARD, H., U. KLIXBULL AND E. FALCH, "PRODRUGS AS DRUG DELIVERY SYSTEMS. 44. O-ACYLOXYMETHYL, O-ACYL AND N-ACYL SALICYLAMIDE DERIVATIVES AS POSSIBLE PRODRUGS FOR SALICYLAMIDE", INT. J. PHARM., 30(2-3), 111-121 (1986).

BUNDGAARD, H. AND E. FALCH, "ALLOPURINOL PRODRUGS. II. SYNTHESIS, HYDROLYSIS, KINETICS AND PHYSICOCHEMICAL PROPERTIES OF VARIOUS N-ACYLOXYMETHYL ALLOPURINOL DERIVATIVES", INT. J. PHARM., 24, 307-325 (1985).

BURAKEVICH,J.V.; CYANURIC AND ISOCYANURIC ACIDS.; IN: KIRK-OTHMER ENCYCL. CHEM. TECH. 3RD ED. 7:397-410.; 1979

BURGOT, G. ET AL. 1990. THERMODYNAMICS OF PARTITIONONG IN THE N-OCTANOL/WATER SYSTEM OF SOME BETA-BLOCKERS. INT. J. PHARM. 63: 73-6.

BURKHARD,L.P.; KUEHL,D.W.; N-OCTANOL/WATER PARTITION COEFFICIENTS BY REVERSE PHASE LIQUID CHROMATOGRAPHY/MASS SPECTROMETRY FOR EIGHT TETRACHLORINATED PLANAR MOLECULES.; CHEMOSPHERE; 15:163-7; 1986

BURKHARD,L.P.; KUEHL,D.W.; VEITH,G.D.; EVALUATION OF REVERSE PHASE LIQUID CHROMATOGRAPH/MASS SPECTROMETRY FOR ESTIMATION OF N-OCTANOL/WATER PARTITION COEFFECIENTS FOR ORGANIC CHEMICALS.; CHEMOSPHERE; 14:1551-60; 1985B

BURKHARD,L.P.; ARMSTRONG,D.E.; ANDREN,A.W.; VAPOR PRESSURES FOR BIPHENYL, 4-CHLOROBIPHENYL, 2,2',3,3',5,5',6,6'-OCTACHLOROBIPHENYL, AND DECACHLOROBIPHENYL.; J. CHEM. ENG. DATA.; 29:248-50.; 1984

BURKHARD,L.P.; ANDREN,A.W.; ARMSTRONG,D.E.; ESTIMATION OF VAPOR PRESSURES FOR POLYCHLORINATED BIPHENYLS: A COMPARISON OF ELEVEN PREDICTIVE METHODS.; ENVIRON. SCI. TECHNOL.; 19:500-7.; 1985

BURKHARD,L.P.; ARMSTRONG,D.E.; ANDREN,A.W.; HENRY'S LAWS CONSTANTS FOR THE POLYCHLORINATED BIPHENYLS.; ENVIRON. SCI. TECH.; 19:590-6.; 1985A

BURKHARD,L.P.; KUEHL,D.W.; VEITH,G.D.; EVALUATION OF REVERSE PHASE LIQUID CHROMATOGRAPH/MASS SPECTROMETRY FOR ESTIMATION OF N-OCTANOL/WATER PARTITION COEFFECIENTS FOR ORGANIC CHEMICALS.; CHEMOSPHERE; 14:1551-60; 1985B

BUTLER,J.A.V.; RAMCHANDANI,C.N.; THE SOLUBILITY OF NON-ELECTROLYTES. PART II. THE INFLUENCE OF THE POLAR GROUP ON THE FREE ENERGY OF HYDRATION OF ALIPHATE COMPOUNDS.; J. CHEM. SOC.; PP. 952-5; 1935

BUTLER,J.A.V.; RAMCHANDANI,C.N.; THOMPSON,D.W.; THE SOLUBILITY OF NON-ELECTROLYTES. PART I. THE FREE ENERGY OF HYDRATION OF SOME ALIPHATIC ALCOHOLS.; J. CHEM. SOC.; 280-285; 1935

BUTTERY,R.G.; GUADAGNI,D.G.; OKANO,S.; AIR-WATER PARTITION COEFFICIENTS OF SOME ALDEHYDES.; J. SCI. FOOD AGRIC.; 16:691-2; 1965

BUTTERY,R.G.; LING,L.C.; GUADAGNI,D.G.; FOOD VOLATILES: VOLATILIES OF ALDEHYDES, KETONES, AND ESTERS IN DILUTE WATER SOLUTION.; J. AGRIC. FOOD CHEM.; 17:385-9; 1969

BUTTERY,R.G.; BOMBEN,J.L.; GUADAGNI,D.G.; LING,L.C.; SOME CONSIDERATIONS OF THE VOLATILITIES OF ORGANIC FLAVOR COMPOUNDS IN FOODS.; J. AGRIC. FOOD CHEM.; 19:1045-8; 1971

BUTZ,R.G.; ATALLAH,Y.H.; YU,C.C.; CALO,C.J.; ENVIRONMENTAL SAFETY ASSESSMENT OF DIPROPYLENE GLYCOL DIBENZOATE.; ENVIRON. TOXICOL. CHEM.; 1:337-46.; 1982A

BUUR, A. AND H. BUNDGAARD, "PRODRUGS OF 5-FLUOROURACIL. III. HYDROLYSIS KINETICS IN AQUEOUS SOLUTION AND BIOLOGICAL MEDIA, LIPOPHILICITY AND SOLUBILITY OF VARIOUS 1-CARBAMYL DERIVATIVES OF 5-FLUOROURACIL", INT. J. PHARM., 23, 209-222 (1985).

BUXTON,G.V.; GREENSTOCK,C.L.; HELMAN,W.P.; ROSS,A.B.; CRITICAL REVIEW OF RATE CONSTANTS FOR REACTIONS OF HYDRATED ELECTRONS, HYDROGEN ATOMS AND HYDROXYL RADICALS (-OH/O-) IN AQUEOUS SOLUTION.; J. PHYS. CHEM. REF. DATA; 17:513-882; 1988

BYKOVA,L.N.; PETROV,S.I.; BLAGODATSKAVA,Z.G.; RELATIVE ACIDITY OF PHENOL AND ITS DERIVATIVES IN A MEDIUM OF NONAQUEOUS SOLVENTS.; ZH. OBSHCH. KHIM.; 40:2295-300.; 1970

CABANI,S.; CONTI,G.; LEPORI,L.; THERMODYNAMIC STUDY ON AQUEOUS DILUTE SOLUTIONS OF ORGANIC COMPOUNDS. PART I. CYCLIC AMINES.; TRANS. FARADAY. SOC.; 67: 1933-42; 1971

CABANI,S.; CONTI,G.; LEPORI,L.; THERMODYNAMIC STUDY ON AQUEOUS DILUTE SOLUTIONS OF ORGANIC COMPOUNDS. PART 2 - CYCLIC ETHERS.; TRANS. FARADAY. SOC.; 67:1943-50; 1971A

CACCIA, S., M.H. FONG AND R. URSO, "IONIZATION CONSTANTS AND PARTITION COEFFICIENTS OF 1-ARYLPIPERAZINE DERIVATIVES", J. PHARM. PHARMACOL., 37(8), 567-570 (1985).

CALIENDO, G., E. NOVELLINO, G. SAGLIOCCO, V. SANTAGADA, C. SILIPO AND A. VITTORIA, "SYNTHESIS AND ANTIMICROBIAL ACTIVITIES OF 5-AMIDOBENZOTRIAZOLE DERIVATIVES", EUR. J. MED. CHEM., 25(4), 343-350 (1990).

CALIENDO, G., E. NOVELLINO, G. SAGLIOCCO, V. SANTAGADA, C. SILIPO AND A. VITTORIA, "SYNTHESIS, ANTIMICROBIAL DATA AND CORRELATIONAL ANALYSIS IN A SET OF 2-ALKYL-5-AMIDOBENZOTRIAZOLES", EUR. J. MED. CHEM., 27(2), 161-166 (1992).

CALL,F.; DETERMINATION OF THE VAPOUR PRESSURE OF ETHYLENE DIBROMIDE.; J. SCI. FOOD. AGRIC.; 8:81-5.; 1957

CALLAHAN,M.A.; SLIMAK,M.W.; GABEL,N.W.; MAY,I.P.; FOWLER,C.F.; FREED,J.R.; JENNINGS,P.; DURFEE,R.L.; WHITMORE,F.C.; MAESTRI,B. ET AL.; WATER-RELATED ENVIRONMENTAL FATE OF 129 PRIORITY POLLUTANTS.; EPA-440/4-79-029A. WASHINGTON,DC: U.S.EPA.; 1979

CALLAHAN,M.A.; SLIMAK,M.W.; GABEL,N.W.; MAY,I.P.; FOWLER,C.F.; FREED,J.R.; JENNINGS,P.; DURFEE,R.L.; WHITMORE,F.C.; MAESTRI,B. ET AL.; WATER-RELATED ENVIRONMENTAL FATE OF 129 PRIORITY POLLUTANTS-VOLUME II.; EPA-440/4-79-029B. WASHINGTON,DC: U.S. EPA.; 1979A

CALVINO, R., A. GASCO AND A. LEO, "AN ANALYSIS OF THE LIPOPHILICITY OF FURAZAN AND FUROXAN DERIVATIVES USING THE CLOGP ALGORITHM", J. CHEM. SOC., PERKIN TRANS. II, (9), 1643-1646 (1992).

CALVINO, R. ET AL. 1991. A REVERSED-PHASE HIGH-PERFORMANCE LIQUID CHROMATOGRAPHIC STUDY OF THE LIPOPHILICITY OF A SERIES OF ANALOGUES OF THE ANTIBIOTIC CALVATIC ACID. J. CHROMATOGR 547: 167-73.

CAMILLERI, P., M.D. BARKER, M.W. KERR, M.K. WHITEHOUSE, J.R. BOWYER AND R.J. LEWIS, "STRUCTURE-ACTIVITY STUDIES ON THE INHIBITION OF PHOTOSYSTEM II ELECTRON TRANSPORT BY PHENYL BIURETS", J. AGRIC. FOOD CHEM., 37(6), 1509-1513 (1989).

CAMPANELLI,A.R.; FERRO,D.; PAVEL,N.V.; STUDY OF THE 4:1 INCLUSION COMPOUND BETWEEN DEOXYCHOLIC ACID AND (E)-P-DIMETHYLAMINOAZOBENZENE BY VAPOR PRESSURE MEASUREMENTS.; THERMOCHIM. ACTA; 87:231-8; 1985

CANTELLI-FORTI, G., M.C. GUERRA, A.M. BARBARO, P. HRELIA, G.L. BIAGI AND P.A. BOREA, "RELATIONSHIP BETWEEN LIPOPHILIC CHARACTER AND URINARY EXCRETION OF NITROIMIDAZOLES AND NITROTHIAZOLES IN RATS", J. MED. CHEM., 29(4), 555-561 (1986).

CANTON, J.H AND R.C.C. WEGMAN, "STUDIES ON THE TOXICITY OF TRIBROMOETHENE, CYCLOHEXENE AND BROMOCYCLOHEXANE TO DIFFERENT FRESHWATER ORGANISMS", WATER RES., 17(7), 743-747 (1983).

CARNEY, C. AND E. GRAHAM, "CORRELATION BETWEEN OCTANOL SOLVENT DISTRIBUTION RATIO AND LIQUID CHROMATOGRAPHIC RETENTION FOR SOME 7- AND 9-SUBSTITUTED 1-METHYLISOQUANINES", ARZNEIM.-FORSCH., 35(1A), 228-233 (1985).

CAROTTI, A., C. RAGUSEO, F. CAMPAGNA, R. LANGRIDGE AND T.E. KLEIN, "INHIBITION OF CARBONIC ANYDRASE BY SUBSTITUTED BENZENESULFONAMIDES. A RE-INVESTIGATION BY QSAR AND MOLECULAR GRAPHICS ANALYSIS", QUANT. STRUCT.-ACT. RELAT., 8(1), 1-10 (1989).

CARRINGER,R.D.; WEBER,J.B.; MONACO,T.J.; ADSORPTION-DESORPTION OF SELECTED PESTICIDES BY ORGANIC MATTER AND MONTMORILLONITE.; J. AGR. FOOD CHEM.; 23:568-72.; 1975

CARSEL,R.F.; MULKEY,L.A.; LORBER,M.N.; BASKIN,L.B.; THE PESTICIDE ROOT ZONE MODEL (PRZM): A PROCEDURE FOR EVALUATING PESTICIDE LEACHING THREATS TO GROUNDWATER; ECOLOGICAL MODELLING; 30:49-69; 1985

CATZ, P. AND D.R. FRIEND, "ALKYL ESTERS AS SKIN PERMEATION ENHANCERS FOR INDOMETHACIN", INT. J. PHARM., 55(1), 17-23 (1989)

CESSNA,A.J.; GROVER,R.; SPECTROSCOPIC DETERMINATION OF DISSOCIATION CONSTANTS OF SELECTED ACIDIC HERBICIDES.; J. AGRIC. FOOD CHEM.; 26:289-92.; 1978

CHAMBERLAIN, K., D.N. BUTCHER AND J.C. WHITE, "RELATIONSHIPS BETWEEN CHEMICAL STRUCTURE AND PHLOEM MOBILITY IN RICINUS COMMUNIS, VAR. GIBSONII WITH REFERENCE TO A SERIES OF OMEGA-(1-NAPHTHOXY) ALKANOIC ACIDS", PESTIC. SCI., 17(1), 48-52 (1986).

CHAO,J.; LIN,C.T.; CHUNG,T.H.; VAPOR PRESSURE OF COAL CHEMICALS.; J. PHYS. CHEM. REF. DATA.; 12:1033-63.; 1983

CHEM INSPECT TEST INST. 1992. BIODEGRADATION AND BIOACCUMULATION DATA OF EXISTING CHEMICALS BASED ON THE CSCL JAPAN. COMPILED UNDER THE SUPERVISION OF CHEMICAL PRODUCTS SAFETY DIV, BASIC INDUST BUREAU, MINISTRY OF INTERN TADE & INDUSTRY JAPAN (EDITED BY CHEMICALS INSPECTION & TESTING INSTITUTE JAPAN). PUBLISHED BY JAPAN CHEMICAL INDUSTRY ECOLOGY-TOXICOLOGY & INFORMATION CENTER. ISBN 4-89074-101-1.

CHESSELLS,M.; HAWKER,D.W.; CONNELL,D.W.; INFLUENCE OF SOLUBILITY IN LIPID ON BIOCONCENTRATION OF HYDROPHOBIC COMPOUNDS.; ECOTOXICOL. ENVIRON. SAFETY; 23:260-73; 1992

CHESTERS,G.; SIMSIMAN,G.V.; LEVY,J.; ALHAJJAR,B.J.; FATHULLA,R.N.; HARKIN,J.M.; ENVIRONMENTAL FATE OF ALACHLOR AND METOLACHLOR.; IN: REVIEWS OF ENVIRON. CONTAM. TOXICOL. VOL. 110, PP. 1-74.; 1989

CHEUNG, A.P. AND D. KENNEY, "PARTITION COEFFICIENTS AND CAPACITY FACTORS OF SOME NUCLEOSIDE ANALOGUES", J. CHROMATOGR., 506, 119-131 (1990).

CHIOU,C.T.; FREED,V.H.; SCHMEDDING,D.W.; KOHNERT,R.L.; PARTITION COEFFICIENT AND BIOACCUMULATION OF SELECTED ORGANIC CHEMICALS.; ENVIRON. SCI. TECHNOL.; 11:475-8.; 1977

CHIOU,C.T.; FREED,V.H.; PETERS,L.J.; KOHNERT,R.L.; EVAPORATION OF SOLUTES FROM WATER.; ENVIRON. INTER.; 3:231-6.; 1980

CHIOU,C.T.; SCHMEDDING,D.W.; MAINES,M.; PARTITIONING OF ORGANIC COMPOUNDS IN OCTANOL-WATER SYSTEMS.; ENVIRON. SCI. TECHNOL.; 16:4-10.; 1982

CHIOU,C.T.; PORTER,P.E.; SCHMEDDING,D.W.; PARTITION EQUILIBRIA OF NONIONIC ORGANIC COMPOUNDS BETWEEN SOIL ORGANIC MATTER AND WATER.; ENVIRON. SCI. TECHNOL.; 17:227-31.; 1983

CHIOU,C.T.; MALCOLM,R.L.; BRINTON,T.I.; KILE,D.E.; WATER SOLUBILITY ENHANCEMENT OF SOME ORGANIC POLLUTANTS AND PESTICIDES BY DISSOLVED HUMIC AND FULVIC ACIDS.; ENVIRON. SCI. TECHNOL.; 20:502-8; 1986

CHIRICO,R.D.; GAMMON,B.E.; KNIPMEYER,S.E.; NGUYEN,A.; STRUBE,M.M.; TSONOPOULOS,C.; STEELE,W.V.; THE THERMODYNAMIC PROPERTIES OF DIBENZOFURAN.; J. CHEM. THERMODYN.; 22:1075-96; 1990

CHMELIK, J., J. HUDECEK, K. PUTYERA, J. MAKOVICKA, V. KALOUS AND J. CHMELIKOVA, "CHARACTERIZATION OF THE HYDROPHOBIC PROPERTIES OF AMINO ACIDS ON THE BASIS OF THEIR PARTITION AND DISTRIBUTION COEFFICIENTS IN THE 1-OCTANOL/WATER SYSTEM", COLL. CZECH. CHEM. COMMUN., 56(10), 2030-2041 (1991).

CHREMOS,G.; ZIMMERMAN,H.K. JR.; AMINO ALCOHOLS AS BUFFERING AGENTS. II. PHOTOLYSIS EQUILIBRIUMS OF DIETHANOLAMINE.; TEXAS J. SCI.; 11:467-70.; 1959

CHRISTIE,A.O.; CRISP,D.J.; ACTIVITY COEFFICIENTS OF THE N-PRIMARY, SECONDARY AND TERTIARY ALIPHATIC AMINES IN AQUEOUS SOLUTION.; J. APPL. CHEM.; 17:11-4; 1967

CHUNG,R.H.; ANTHRAQUINONE.; IN: KIRK-OTHMER ENCYCL. CHEM. TECH. 3RD ED. 2:700-7.; 1978

CLAYTON,GD AND CLAYTON,FE; PATTY'S INDUSTRIAL HYGEINE AND TOXICOLOGY, 3RD ED.; VOL I (1977)

CLAYTON,GD AND CLAYTON,FE; PATTY'S INDUSTRIAL HYGEINE AND TOXICOLOGY, 3RD ED.; VOL IIB (1981)

CLAYTON,GD AND CLAYTON,FE; PATTY'S INDUSTRIAL HYGEINE AND TOXICOLOGY, 3RD ED.; VOL IIC (1982)

CLAYTON,GD AND CLAYTON,FE; PATTY'S INDUSTRIAL HYGEINE AND TOXICOLOGY, 3RD ED.; (1985)

CLOGP; GEMS-GRAPHIC EXPOSURE MODELING SYSTEM CLOGP USEPA (1986)

CLOUX, J.L., J. CROMMEN, J. DELARGE, M.L. PIRARD AND L. THUNUS, "ATTEMPTS AT CALCULATING HYDROPHOBIC CONSTANT OF THE SULFONYLUREA GROUP", J. PHARM. BELG., 43(3), 141-151 (1988).

COATES,M.; CONNELL D.W.; BARRON D.W.; AQUEOUS SOLUBILITY AND OCTAN-1-OL-WATER PARITITION COEFFICIENTS OF ALIPHATIC HYDROCARBONS.; ENVIRON. SCI. TECHNOL.; 19:628-32.; 1985

COCA,J.; DIAZ,R.; EXTRACTION OF FURFURAL FROM AQUEOUS SOLUTIONS WITH CHLORINATED HYDROCARBONS; J CHEM ENG DATA; 25:80-3; 1980

COFFEN,D.L.; VITAMINS (B6); IN: KIRK OTHMER ENCYCL. CHEM. TECH. 3RD NEW YORK,NY: WILEY-INTERSCIENCE. 24:94-107; 1978

COLOMINA,M.; JIMENEZ,P.; ROUX,M.V.; TURRION,C.; THERMOCHEMICAL PROPERTIES OF BENZOIC ACID DERIVATIVES. XII. VAPOR PRESSURES AND ENTHALPIES OF SUBLIMATION AND FORMATION OF FIVE DIMETHOXYBENZOIC ACIDS.; J. CHEM. THERMODYN.; 17:1091-6; 1985

COLOMINA,M.; JIMENEZ,P.; ROUX,M.V.; TURRION,C.; THERMOCHEMICAL PROPERTIES OF BENZOIC ACID DERIVATIVES.; AN. QUIM.; 82:126-30; 1986

COLOMINA,M.; JIMENEZ,P.; PEREZ-OSSORIO,R.; ROUX,M.V.; TURRION,C.; THERMOCHEMICAL PROPERTIES OF BENZOIC ACID DERIVATIVES. XIV. ENTHALPIES OF COMBUSTION, VAPOR PRESSURES, ENTHALPIES OF SUBLIMATION AND ENTHALPIES OF FORMATION OF O-, M-, AND P-ISOPROPYLBENZOIC ACIDS.; J. CHEM. THERMODYN.; 19:155-62; 1987

COMPTON,J.A.F.; MILITARY CHEMICAL AND BIOLOGICAL AGENTS. CHEMICAL AND TOXICOLOGICAL PROPERTIES.; TELFORD PRESS: CALDWELL, NJ 457 PP.; 1987

CONWAY,R.A.; WAGGY,G.T.; SPIEGEL,M.H.; BERGLUND,R.L.; ENVIRONMENTAL FATE AND

EFFECTS OF ETHYLENE OXIDE.; ENVIRON. SCI. TECHNOL.; 17:107-12.; 1983

COOK,A.G.; MASON,G.W.; BASICITIES OF (X) (Y) PO (CH3) WITH VARIATION OF X AND Y SUBSTITUENTS.; J. INORG. NUCL. CHEM.; 35:2090-3; 1973

COOVER,M.P.; SIMS,R.C.; THE EFFECTS OF TEMPERATURE ON POLYCYCLIC AROMATIC HYDROCARBON PERSISTENCE IN AN UNACCLIMATED AGRICULTURAL SOIL; HAZ. WASTE HAZ. MAT.; 4:69-82; 1987

CORCHNOY,S.B.; ATKINSON,R.; KINETICS OF THE GAS-PHASE REACTIONS OF OH AND NO3 RADICALS WITH 2-CARENE, 1,8-CINEOLE, P-CYMENE, AND TERPINOLENE.; ENVIRON. SCI. TECHNOL.; 24:1497-1502; 1990

CRABTREE,E.V.; SARVER,E.W.; EASP: REVIEW OF ANALYTICAL PROCEDURES FOR GB, VX AND THEIR DEGRADATION PRODUCTS; IN EASP:76021; DEPT OF ARMY EDGEWOOD ARSENAL; ADBO16557; 1977

CRAIG,RG; "DENTAL MATERIALS" IN: ULLMANN'S ENCYCLOPEDIA OF INDUSTRIAL CHEMICALS. 5TH ED., VOL. A8: 251-259.; 1987

CROSSLAND,N.O.; PREDICTING THE HAZARDS OF CHEMICALS TO AQUATIC ENVIRONMENTS.; CHEM. IND.; 21:740-4; 1986

DA, Z.A. ET AL. 1992. ENERGY ASPECTS OF OIL/WATER PARTITION LEADING TO THE NOVEL HYDROPHOBIC PARAMETERS FOR THE ANALYSIS OF QUANTITATIVE STRUCTURE-ACTIVITY RELATIONSHIPS. J. MED. CHEM. 35: 3382-7.

DAL POZZO, A., G. DONZELLI, L. RODRIQUEZ AND A. TAJANA, "IN VITRO MODEL FOR THE EVALUATION OF DRUG DISTRIBUTION AND PLASMA PROTEIN-BINDING RELATIONSHIPS", INT. J. PHARM., 50(2), 97-101 (1989).

DAL POZZO, A. ET AL. 1987. CARBONYLACRYLIC DERIVATIVES AS POTENTIAL ANTIMICROBIAL GENTS. J. MED. CHEM. 30: 1674-7.

DALLET, P., J.-P. DUBOST, J.-C. COLLETER, E. AUDRY AND M.-H. CREUZET, "QSAR IN A SERIES OF ALPHA-ADRENOLITIC MOLECULES: OPTIMIZATION OF PHYSICOCHEMICAL PARAMETERS", EUR. J. MED. CHEM., 20(6), 551-557 (1985).

DAMLE,S.B.; CARBONIC AND CARBONOCHLORIDIC ESTERS.; IN: KIRK-OTHMER ENCYCL. CHEM. TECH., 4TH ED., VOL. 5, PP.77-97; 1993

DANIELS,S.L.; HOERGER,F.D.; MOOLENAAR,R.J.; ENVIRONMENTAL EXPOSURE ASSESMENT: EXPERIENCE UNDER THE TOXIC SUBSTANCE CONTROL ACT.; ENVIRON. TOXICOL. CHEM.; 4:107-17.; 1985

DANIELS,S.L.; HOERGER,F.D.; MOOLENAAR,R.J.; ENVIRONMENTAL EXPOSURE ASSESMENT: EXPERIENCE UNDER THE TOXIC SUBSTANCE CONTROL ACT.; ENVIRON. TOXICOL. CHEM.; 4:107-17.; 1985

DANLY,D.E.; CAMPBELL,C.R.; ADIPIC ACID.; IN: KIRK-OTHMER ENCYCL. CHEM. TECH. 3RD ED. 1:510-31.; 1978

DAUBERT,T.E.; DANNER,R.P.; DATA COMPILATION TABLES OF PROPERTIES OF PURE COMPOUNDS.; AMERICAN INSTITUTE OF CHEMICAL ENGINEERS. PP.450; 1985

DAUBERT,T.E.; DANNER, R.P.; PHYSICAL AND THERMODYNAMIC PROPERTIES OF PURE CHEMICALS: DATA COMPILATION.; DESIGN INSTITUTE FOR PHYSICAL PROPERTY DATA, AMERICAN INSTITUTE OF CHEMICAL ENGINEERS. HEMISPHERE PUB. CORP., NEW YORK, NY., 4 VOL.; 1989

DAUBERT,T.E.; DANNER,R.P.; PHYSICAL AND THERMODYNAMIC PROPERTIES OF PURE CHEMICALS: DATA COMPILATION. SUPPLEMENT 1.; DESIGN INSTITUTE FOR PHYSICAL PROPERTY DATA, AMERICAN INSTITUTE OF CHEMICAL ENGINEERS, HEMISPHERE PUB. CORP., NEW YORK, NY.; 1991

DAUBERT,T.E.; DANNER,R.P.; PHYSICAL AND THERMODYNAMIC PROPERTIES OF PURE CHEMICALS: DATA COMPILATION. SUPPLEMENT 1.; DESIGN INSTITUTE FOR PHYSICAL PROPERTY DATA, AMERICAN INSTITUTE OF CHEMICAL ENGINEERS, HEMISPHERE PUB. CORP., NEW YORK, NY.; 1991

DAUBERT,T.E.; DANNER,R.P.; PHYSICAL AND THERMODYNAMIC PROPERTIES OF PURE
CHEMICALS: DATA COMPILATION.; DESIGN INSTITUTE FOR PHYSICAL PROPERTY DATA,
AMERICAN INSTITUTE OF CHEMICAL ENGINEERS. TAYLOR & FRANCIS, WASHINGTON, DC.;
1993

DAUBERT,T.E.; JALOWKA,J.W.; GOREN,V.; VAPOR PRESSURE OF 22 PURE INDUSTRIAL
CHEMICALS.; AICHE SYMP. SER.; 83:128-56; 1987

DAUBERT,T.E.; JALOWKA,J.W.; GOREN,V.; VAPOR PRESSURE OF 22 PURE INDUSTRIAL CHEMICALS.; AICHE SYMP. SER.;
83:128-56; 1987

DAUBERT,T.E.; DANNER, R.P.; PHYSICAL AND THERMODYNAMIC PROPERTIES OF PURE CHEMICALS: DATA COMPILATION.;
DESIGN INSTITUTE FOR PHYSICAL PROPERTY DATA, AMERICAN INSTITUTE OF CHEMICAL ENGINEERS. HEMISPHERE PUB.
CORP., NEW YORK, NY., 4 VOL.; 1989.

DAUBERT,T.E.; DANNER,R.P.; DATA COMPILATION TABLES OF PROPERTIES OF PURE COMPOUNDS.; AMERICAN INSTITUTE
OF CHEMICAL ENGINEERS. PP.450; 1985, 1986, 1987, 1989, 1992, 1994.

DAVIDSON,J.M.; OU,L.T.; RAO,P.S.C.; ADSORPTION, MOVEMENT, AND BIOLOGICAL
DEGRADATION OF HIGH CONCENTRATIONS OF SELECTED PESTICIDES IN SOILS.; IN:
EPA-600/9-78-016. LAND DISPOSAL OF HAZARDOUS WASTES. 4TH ANNUAL RES. SYMP. PROC.
SAN ANTONIO,TX MARCH 6-8, 1978 SHULTZ,D.W. ED. CINCINNATI,OH: U.S EPA,
PP.233-44.; 1978

DAVIS,L.N.; SANTODONATO,J.; HOWARD,P.H.; SAXENA,J.; INVESTIGATION OF SELECTED
POTENTIAL ENVIRONMENTAL CONTAMINANTS: BENZOTRIAZOLES.; EPA-560/2-77-001.
WASHINGTON,D.C.: U.S. EPA.; 1977

DE, A.U., C. SENGUPTA, D. PAL, A. MANDAL AND J. CHATTERJEE, "PARTITION COEFFICIENTS OF SOME METRONIDAZOLE
AND FURACIN ANALOGS IN RELATION TO BIOLOGICAL ACTIVITY", INDIAN J. PHARM. SCI., 45(3), 123-125 (1983).

DEAN,J.A.; LANGE'S HANDBOOK OF CHEMISTRY 13TH ED.; NEW YORK, NY: MCGRAW-HILL BOOK CO.; 1985

DEAN,J.A.; HANDBOOK OF ORGANIC CHEMISTRY; NEW YORK,NY: MCGRAW-HILL BOOK CO.; 1987

DEBNATH,A.K.; HANSCH,C.; STRUCTURE-ACTIVITY RELATIONSHIP OF GENOTOXIC POLYCYCLIC
AROMATIC NITRO COMPOUNDS: FURTHER EVIDENCE FOR THE IMPORTANCE OF HYDROPHOBICITY
AND MOLECULAR ORBITAL ENERGIES IN GENETIC TOXICITY.; ENVIRON. MOL. MUTAGEN.;
20:140-44; 1992

DEBNATH, A.K., R.L. LOPEZ DE COMPADRE, G. DEBNATH, A.J. SHUSTERMAN AND C. HANSCH, "STRUCTURE-ACTIVITY
RELATIONSHIP OF MUTAGENIC AROMATIC AND HETEROAROMATIC NITRO COMPOUNDS. CORRELATION WITH
MOLECULAR ORBITAL ENERGIES AND HYDROPHOBICITY", J. MED. CHEM., 34(2), 786-797 (1991).

DEBNATH, A.K., G. DEBNATH, A.J. SHUSTERMAN AND C. HANSCH, "A QSAR INVESTIGATION OF THE ROLE OF
HYDROPHOBICITY IN REGULATING MUTAGENICITY IN THE AMES TEST.1.", ENVIRON. MOL. MUTAGEN., 19(1), 37-52 (1992).

DEBORD, J. AND M. LABADIE, "REVERSIBLE INHIBITION OF BUTYLCHOLINESTERASE BY SOME ALIPHATIC
PHOSPHORAMIDES", PHOSPHORUS SULFUR, 22(1), 121-130 (1985).

DEBRUIJN,J.; HERMENS,J.; UPTAKE AND ELIMINATION KINETICS OF ORGANOPHOSPHORUS
PESTICIDES IN THE GUPPY (POECILIA RETICULATA): CORRELATIONS WITH THE
OCTANOL/WATER PARTITION COEFFICIENT.; ENVIRON. TOXICOL. CHEM.; 10:791-804; 1991

DEBRUIJN,J.; BUSSER,F.; SEINEN,W.; HERMENS,J.; DETERMINATION OF OCTANOL/WATER
PARTITION COEFFICIENTS FOR HYDROPHOBIC ORGANIC CHEMICALS WITH THE "SLOW
STIRRING" METHOD.; ENVIRON. TOXICOL. CHEM.; 8:499-512; 1989

DEHN,W.M.; COMPARATIVE SOLUBILITIES IN WATER, IN PYRIDINE AND IN AQUEOUS
PYRIDINE.; J. AM. CHEM. SOC.; 39:1399-404; 1917

DE JONGE, A., P.B.M.W.M. TIMMERMANS AND P.A. VAN ZWIETEN, "QUANTITATIVE DESCRIPTION OF ALPHA2-ADRENERGIC
POTENCYIN TERMS OF RECEPTOR AFFINITY AND INTRINSIC ACTIVITY", QUANT. STRUCT.-ACT. RELAT. PHARMACOL.,
CHEM., BIOL., 3(4), 138-143 (1984).

DELASSUS,P.T.; SCHMIDT,D.D.; SOLUBILITIES OF VINYL CHLORIDE AND VINYLIDINE
CHLORIDE IN WATER.; J. CHEM. ENG. DATA; 26:274-6.; 1981

DEMEK,M.M.; DAVIS,G.T.; DENNIS,W.H.; HILL,A.L.; FARRAND,R.L.; MUSSELMAN,N.P.;
EDGEWOOD ARSENAL TECHNICAL REPORT, EATR 4417: BEHAVIOUR OF CHEMICAL AGENTS IN
SEAWATER; IN: EATR 4417: TASK # 1B662706409501; DEPT ARMY EDGEWOOD ARSENAL, MD;
1970

DEMOTES-MAINARD, F., J. THOMAS, J.J. BOSC, G. DEVAUX AND C. JARRY, "RP-HPLC OF NEW ANTIDEPRESSANT
2-AMINO-OXAZOLINES: A COMPARATIVE STUDY OF THEIR LIPOPHILICITY", J. LIQ. CHROMATOGR., 16(3), 767-776 (1993).

DENEER,J.W.; SINNIGE,T.L.; SEINEN,W.; HERMENS,J.L.M.; QUANTITATIVE
STRUCTURE-ACTIVITY FOR THE TOXICITY AND BIOCONCENTRATION FACTOR OF NITROBENZENE
DERIVATIVES TOWARDS THE GUPPY (POECILIA RETICULATA).; AQUAT. TOXICOL.;
10:115-29; 1987

DENEER, J.W., T.L. SINNIGE, W. SEINEN AND J.L.M. HERMENS, "A QUANTITATIVE ATRUCTURE-ACTIVITY RELATIONSHIP
FOR THE ACUTE TOXICITY OF SOME EPOXY COMPOUNDS TO THE GUPPY", AQUATIC TOXICOL., 13(3), 195-204 (1988).

DE VOOGT,P.; MUIR, D.C.G.; WEBSTER, G.R.B.; GOVERS,H.; QUANTITATIVE
STRUCTURE-ACTIVITY RELATIONSHIPS FOR THE BIOCONCENTRATION IN FISH OF SEVEN
POLYCHLORINATED DIBENZODIOXINS.; CHEMOSPHERE; 21:1385-96; 1990

DEWEY,R.H.; BOLLMEIER,A.F. JR.; NITRO ALCOHOLS.; IN: KIRK OTHMER. ENCY. OF
CHEM. TECHNOL. 3RD. ED. NEW YORK,NY: WILEY-INTERSCIENCE 15:910-6; 1978

DIERBERG,F.E.; GIVEN,C.J.; ALDICARB STUDIES IN GROUND WATERS FROM FLORIDA CITRUS
GROVES AND THEIR RELATION TO GROUND-WATER PROTECTION.; GROUND WATER; 24:16-22;
1986

DIETRICH,A.M.; CHESNUTT,S.A.; STONE,L.A.; GALLAGHER,D.L.; DETERMINATION OF
ORGANIC POLLUTANTS IN LAND APPLIED MUNICIPAL WASTEWATER SLUDGES BY TOXICITY
CHARACTERISTIC LEACHING PROCEDURE (TCLP) AND EXTRACTION PROCEDURE TOXICITY TEST
(EP).; WATER ENVIRON. RES.; 65:612-19; 1993

DIJKSTRA, D., T.B.A. MULDER, H. ROLLEMA, P.G. TEPPER, J. VAN DER WEIDE AND A.S. HORN, "SYNTHESIS AND
PHARMACOLOGY OF TRANS-4-N-PROPYL-3,4,4A,10B-TETRAHYDRO-2H,5H-1-BENZOPYRANO[4,3-B]-1,4-OXAZIN-7- AND
-9-OLS: THE SIGNIFICANCE OF NITROGEN PKA VALUES FOR CENTRAL DOPAMINE RECEPTOR ACTIVATION", J. MED. CHEM.,
31(11), 2178-2182 (1988).

DILLING,W.L.; INTERPHASE TRANSFER PROCESSES. II. EVAPORATION RATES OF
CHLOROMETHANES, ETHANES, ETHYLENES, PROPANES, AND PROPYLENES FROM DILUTE AQUEOUS
SOLUTIONS. COMPARISONS WITH THEORETICAL PREDICTIONS.; ENVIRON. SCI. TECHNOL.;
11:405-9.; 1977

DILLING,W.L.; ATMOSPHERIC ENVIRONMENT. CHAPTER 5.; IN: ENVIRONMENTAL RISK
ANALYSIS FOR CHEMICALS. CONWAY,R.A. ED.; NEW YORK, NY: VAN NOSTRAND REINHOLD CO.
PP. 154-97.; 1982

DITSENT,V.E.; SKOROKHODOV,I.I.; TERENTEVA,N.A.; ZOLOTAREVA,M.N.; LOTAREV,M.B.;
SATURATED VAPOR PRESSURE OF POLYDIORGANOSILOXANES. X. CYCLOTETRASILOXANES
CONTAINING SILACYCLOHEXANE, SILAACENAPHTHALENE, AND SILACYCLOPENTENE RADICAL.;
ZH. FIZ. KHIM.; 48:2152.; 1974A

DOBBIN, P.S., R.C. HIDER, A.D. HALL, P.D. TAYLOR, P. SARPONG, J.B. PORTER, G. XIAO AND D. VAN DER HELM, "SYNTHESIS,
PHYSICOCHEMICAL PROPERTIES AND BIOLOGICAL EVALUATION OF N-SUBSTITUTED
2-ALKYL-3-HYDROXY-4(1H)-PYRIDINONES: ORALLY ACTIVE IRON CHELATORS WITH CLINICAL POTENTIAL", J. MED.
CHEM., 36(17), 2448-2458 (1993).

DOBBS,A.J.; THE DETERMINATION OF VAPOUR PRESSURES FROM RELATIVE VOLATILIZATION
RATES.; CHEMOSPHERE.; 13:687-92.; 1984

DOBRY,A.; KELLER,R.; VAPOR PRESSURES OF SOME PHOSPHATE AND PHOSPHONATE ESTERS.;
J. PHYS. CHEM.; 61:1448-9.; 1957

DOLFING,J.; HARRISON,B.K.; GIBBS FREE ENERGY OF FORMATION OF HALOGENATED

AROMATIC COMPOUNDS AND THEIR POTENTIAL ROLE AS ELECTION ACCEPTORS IN ANAEROBIC ENVIRONMENTS.; ENVIRON. SCI. TECHNOL.; 26:2213-18; 1992

DONAGHY,T.; SHANAHAN,I.; HANDE,M.; FITZPATRICK,S.; RATE CONSTANTS AND ATMOSPHERIC LIFETIMES FOR THE REACTIONS OF HYDROXYL RADICALS AND CHLORINE ATOMS WITH HALOALKANES.; INT. J. CHEM. KINET.; 25:273-84; 1993

DORIGAN,J.; FULLER,B.; DUFFY,R.; PRELIMINARY SCORING OF SELECTED ORGANIC AIR POLLUTANTS. APPENDIX III-CHEMISTRY, PRODUCTION, AND TOXICITY OF CHEMICALS F THROUGH N.; EPA-450/3-77-008D. RESEARCH TRIANGLE PARK,NC: U.S. EPA. PP.303.; 1976

DORIGAN,J.; FULLER,B.; DUFFY,R.; PRELIMINARY SCORING OF SELECTED ORGANIC AIR POLLUTANTS. APPENDIX II-CHEMISTRY, PRODUCTION, AND TOXICITY OF CHEMICALS D THROUGH E.; EPA-450/3-77-008C. RESEARCH TRIANGLE PARK,NC: US EPA PP.327.; 1976A

DORN,P.B.; CHOU,C.S,; GENTEMPO,J.J.; DEGRADATION OF BISPHENOL A IN NATURAL WATERS.; CHEMOSPHERE; 16:1501-7; 1987

DOUCETTE,W.J.; ANDREN,A.W.; CORRELATION OF OCTANOL/WATER PARTITION COEFFICIENTS AND TOTAL MOLECULAR SURFACE AREA FOR HIGHLY HYDROPHOBIC AROMATIC COMPOUNDS.; ENVIRON. SCI. TECHNOL.; 21:821-4; 1987

DOUCETTE,W.J.; ANDREN,A.W.; AQUEOUS SOLUBILITY OF SELECTED BIPHENYL, FURAN AND DIOXIN CONGENERS; CHEMOSPHERE; 17:243-52; 1988

DOW CHEMICAL COMPANY.; THE ALKANOLAMINES HANDBOOK.; MIDLAND, MI: DOW CHEMICAL CO.; 1980

DOW CHEMICAL COMPANY.; THE GLYCOL ETHERS HANDBOOK.; MIDLAND, MI: DOW CHEMICAL CO.; 1981

DOW CHEMICAL COMPANY; THE GLYCOL ETHERS HANDBOOK.; THE DOW CHEMICAL COMPANY, MIDLAND, MICHIGAN 97 PP.; 1990

DRAGUN,J.; HELLING,C.S.; PHYSICOCHEMICAL AND STRUCTURAL RELATIONSHIPS OF ORGANIC CHEMICALS UNDERGOING SOIL- AND CLAY-CATALYZED FREE-RADICAL OXIDATION.; SOIL SCI.; 139:100-11; 1985

DRAHONOVSKY,J.; VACEK,Z.; DISSOCIATION CONSTANTS AND COLUMN CHROMATOGRAPHY OF CHLORINATED PHENOLS.; COLLECT. CZECH. CHEM. COMMUN.; 36:3431-40.; 1971

DREISBACH,R.R.; PHYSICAL PROPERTIES OF CHEMICAL COMPOUNDS. I.; ADVANCES IN CHEMISTRY SERIES NO. 15.; WASHINGTON, DC: AMERICAN CHEMICAL SOCIETY.; 1955

DREISBACH,R.R.; PHYSICAL PROPERTIES OF CHEMICAL COMPOUNDS. III.; ADVANCES IN CHEMISTRY. SERIES NO. 29.; WASHINGTON, DC: AMERICAN CHEMICAL SOCIETY.; 1961

DUFFEL, M.W., I.S. ING, T.M. SEGARRA, J.A. DIXSON, C.F. BARFKNECHT AND R.D. SCHOENWALD, "N-SUBSTITUTED SULFONAMIDE CARBONIC ANHYDRASE INHIBITORS WITH TOPICAL EFFECTS ON INTRAOCULAR PRESSURE", J. MED. CHEM., 29(8), 1488-1494 (1986).

DUNCAN, J.D., G. HALLSTROM, U. PAULSEN-SORMAN, B. LINDEKE AND A.K. CHAO, "EFFECTS OF ALPHA-CARBON SUBSTITUENTS ON THE N-DEMETHYLATION OF N-METHYL-2-PHENYLETHYLAMINES BY RAT LIVER MICROSOMES", DRUG METAB. DISPOSITION: BIOL. FATE CHEM., 11(1), 15-20 (1983).

DUNN, W.J. III ET AL. 1983. NONADDITIVITY OF 1-OCTANOL/WATER PARTITION COEFFICIENTS OF DISUBSTITUTED BENZENES: AN EXPLANATION AND ITS CONSIDERATION OF LOG P ESTIMATION. QUANT. STRUCT-ACT. RELAT. 2: 156-63.

DUNN,W.J.II; JOHANSSON,E.; WOLDS,S.; NONADDITIVITY OF 1-OCTANOL/WATER PARTITION COEFFICIENTS OF DISUBSTITUTED BENZENES: AN EXPLANATION AND ITS CONSIDERATION IN LOG P ESTIMATION.; QUANT. STRUCT. -ACT. RELAT. PHARMACOL. CHEM. BIOL.; 2 2:156-63.; 1983

DUPONT DE NEMOURS CO.; FREON PRODUCTS INFORMATION B-2.; A98825. 12/80.; 1980

DUTTA, H. ET AL. 1988. DILOXANIDE FUROATE: STUDIES ON ITS INTERACTION WITH BLOOD PHOSPHOLIPID IN RELATION

TO PARTITION COEFFICIENTS OF RELATED ANALOGS. INDIAN J. PHARM. SCI. 50: 328-31.

DYKYJ,J.; REPAS,M.; SATURATED VAPOR PRESSURE OF ORGANIC COMPOUNDS; PETROCHEMIA;
18:179-98; 1973

DYKYJ,J.; VANKO,A.; VAPOR PRESSURES OF HALOGENATED HYDROCARBONS C7 AND HIGHER;
PETROCHEMIA; 3-22; 1970

DYKYJ,J.; SATURATED VAPOR PRESSURES OF C1-4 COMPOUNDS; PETROCHEMICAL;
10:187-206; 1970

DYKYJ,J.; SEPRAKOVA,M.; PAULECH,J.; THE VAPOR OF TWO C8 ALDEHYDES.; CHEM.
ZVESTI.; 15:465-8.; 1961

EBERLING,C.L.; FORMAMIDE; IN: KIRK-OTHMER ENCYCL. CHEM. TECH. 3RD ED.
11:258-63.; 1980

EDNEY,E.O.; CORSE,E.W.; HYDROXYL RADICLE RATE CONSTANT INTERCOMPARISON STUDY.;
EPA-600/S3-86/056. NTIS PB87-111 142. RESEARCH TRIANGLE PARK, NC: USEPA ATMOS.
SCI. RES. LAB. PROJECT SUMMARY.; 1987

EISENREICH, SJ; ENVIRON SCI TECHNOL 15: 30-38; (1981)

ELLER, M.G., R.D. SCHOENWALD, J.A. DIXSON, T. SEGARRA AND C.F. BARFKNECHT, "TOPICAL CARBONIC ANHYDRASE
INHIBITORS. III. OPTIMIZATION MODEL FOR CORNEAL PENETRATION OF ETHOXAZOLAMIDE", J. PHARM. SCI., 74(2), 155-160
(1985).

ELLGEHAUSEN,H.; DHONDT,C.; FUERER,R.; REVERSED-PHASE CHROMATOGRAPHY AS A GENERAL
METHOD FOR DETERMINING OCTAN-1-OL/WATER PARITION COEFFICIENTS; PESTIC. SCI.;
12:219-27; 1981

ELLINGTON,J.J.; STANCIL,F.E.; OCTANOL/WATER PARTITION COEFFICIENTS FOR
EVALUATION OF HAZARDOUS WASTE LAND DISPOSAL: SELECTED CHEMICALS.; REPORT 1988,
EPA/600/M-88/010; PB89-120760, 6 PP.; 1988

ELLIOTT,S.; THE SOLUBILTY OF CARBON DISULFIDE VAPOR IN NATURAL AQUEOUS SYSTEMS.;
ATMOS. ENVIRON.; 23:1977-80; 1989

EL TAYAR, N. ET AL. 1985. LIPOPHILICITY MEASUREMENTS OF PROTONATED BASIC COMPOUNDS BY REVERSE-PHASE
HIGH-PERFORMANCE LIQUID CHROMATOGRAPHY. J. CHROMATOGR 320: 305-12.

EL TAYAR, N., R.-S. TSAI, P. VALLAT, C. ALTOMARE AND B. TESTA, "MEASUREMENT OF PARTITION COEFFICIENTS BY
VARIOUS CENTRIFUGAL PARTITION CHROMATOGRAPHIC TECHNIQUES. A COMPARATIVE EVALUATION", J.
CHROMATOGR., 556, 181-194 (1991).

ENFIELD,C.G.; WALTERS,D.M.; WILSON,J.T.; PIWONI,M.D.; BEHAVIOR OF ORGANIC
POLLUTANTS DURING RAPID-INFILTRATION OF WASTEWATER INTO SOIL: III. MATHEMATICAL
DESCRIPTION OF TRANSPORT AND TRANSFORMATION.; HAZ. WASTE HAZ. MATER.; 3:57-76;
1986

ENGINEERING SCIENCES DATA UNIT.; VAPOUR PRESSURES AND CRITICAL POINTS OF PURE
SUBSTANCES. VI. C3 TO C15 ALIPHATIC KETONES.; ENG. SCI. DATA ITEM. 75025.
PP.29.; 1975

ENGINEERING SCIENCES DATA UNIT.; VAPOR PRESSURES AND CRITICAL POINTS OF LIQUIDS.
VII. HALOGENATED ETHANES AND ETHYLENES.; ENG. SCI. DATA ITEM. 76004. PP.43.;
1976

ENVIRONMENTAL RESEARCH LABORATORY; ACEPHATE, ALDICARB, CARBOPHENOTHION, DEF,
EPN, ETHOPROP, METHYL PARATHION AND PHORATE; THEIR ACUTE AND CHRONIC TOXICITY,
BIOCONCENTRATION POTENTIAL AND PERSISTENCE AS RELATED TO MARINE ENVIRONMENTS.;
EPA 600/4-81-041 (NTIS PB81-244477). WASHINGTON,DC: USEPA PP.275; 1981

EPSTEIN,J.; PROPERTIES OF GB IN WATER; WAT TECH/QUALITY; 31-7; 1974

ETZWEILER,F.; SENN,E.; SCHMIDT,H.W.H.; DETERMING SOLUBILITY.; ANAL. CHEM.;

67:655-8; 1995

EVANS,T.W.; EDLUND,K.R.; TERTIARY ALKYL ETHERS PREPARATION AND PROPERTIES; IND.
ENG. CHEM.; 28:1186-8; 1936

EVANS,T.W.; EDLUND,K.R.; TERTIARY ALKYL ETHERS PREPARATION AND PROPERTIES; IND.
ENG. CHEM.; 28:1186-8; 1936

FALBE,J.; HENKEL,K.G.; DUSSELDORF,P.L.; WEBER,J.; ALDEHYDES, ALPIHATIC AND
ARALIPHATIC.; IN: ULLMANN'S ENCYCLOPEDIA OF INDUSTRIAL CHEMISTRY. WEIHEIM,
GERMANY. VCH VERLAGSGESELLSCHAFT. VOL. A1:321-52.; 1985

FARMER,W.J.; YANG,M.; LETEY,J.; SPENCER,W.F.; LAND DISPOSAL OF ORGANIC HAZARDOUS
WASTES CONTAINING HCB.; IN: EPA/EQS/IT NATL. CONFERENCE ON DISPOSAL OF RESIDUES
ON LAND, PROC. PP.83-6; 1976

FARRAJ, N.F., S.S. DAVIS, G.D. PARR AND H.N.E. STEVENS, "DISSOCIATION AND PARTITIONING OF PROGABIDE AND ITS
DEGRADATION PRODUCTS", INT. J. PHARM., 46(3), 231-239 (1988).

FAZEKAS,G.B.; TAKACS,G.A.; PHOTOCHEMICAL STABILITY OF ATMOSPHERIC
FLUORODICHLORONITROMETHANE (CFCL2NO2), DIFLUOROCHLORONITROMETHANE (CF2CLNO2) AND
TRIFLUORONITROMETHANE (CF3NO2).; J. PHOTOCHEM.; 21:9-18.; 1983

FEDERAL REGISTER; CARBOFURAN INTERMEDIATES: RESPONSE TO THE INTERAGENCY TESTING
COMMITTEE.; FEDERAL REGISTER; 50:29761-4; 1985

FELDER,J.D.; ADAMS,W.J.; SAEGEER,V.W.; ASSESSMENT OF THE SAFETY OF DIOCTYL
ADIPATE IN FRESHWATER ENVIRONMENTS.; ENVION. TOXICOL. CHEM.; 5:777-84; 1986

FELDHAKE,C.J.; STEVENS,C.D.; THE SOLUBILITY OF TETRAETHYL LEAD IN WATER.; J.
CHEM. ENG. DATA.; 8:196-7.; 1963

FELSOT,A.; DAHM,P.A.; SORPTION OF ORGANOPHOSPHORUS AND CARBAMATE INSECTICIDES BY
SOIL.; J. AGRIC. FOOD CHEM.; 27:557-63.; 1979

FENDINGER,N.J.; GLOTFELTY,D.E.; A LABORATORY METHOD FOR THE EXPERIMENTAL
DETERMINATION OF AIR-WATER HENRY'S LAW CONSTANTS FOR SEVERAL PESTICIDES; ENVIRON
SCI TECHNOL; 22:1298-93; 1988

FENDINGER,N.J.; GLOTFELTY,D.E.; HENRY'S LAW CONSTANTS FOR SELECTED PESTICIDES,
PAHS AND PCBS.; ENVIRON. TOXICOL. CHEM.; 9:731-5; 1990

FERREIRA, G.A.L.; SEIBER,J.N.; VOLATILIZATION AND EXUDATION LOSSES OF
3-N-METHYLCARBAMATE INSECTICIDES APPLIED SYSTEMICALLY TO RICE; J AGRIC FOOD
CHEM; 29:93-9; 1981

FERRO,D.; PIACENTE,V.; HEAT OF VAPORIZATION OF 0-, M-, P-NITROANILINE.;
THERMOCHIM. ACTA; 90:387-9; 1985

FINIZIO,A.; DIGUARDO,A.; ARNOLDI,A.; VIGHI,M.; FANELLI,R.; DIFFERENT APPROACHES
FOR THE EVALUATION OF KOW FOR S-TRIAZINE HERBICIDES.; CHEMOSPHERE; 23:801-12;
1991

FINLEY,KT; "BENZOQUINONE" IN: ULLMANN'S ENCYCLOPEDIA OF INDUSTRIAL CHEMICALS. 5TH ED., VOL. A3: 571-574.;
1985

FIX, J.A., J. ALEXANDER, M. CORTESE, K. ENGLE, P. LEPPERT AND A.J. REPTA, "SHORT-CHAIN ALKYL ESTERS OF L-DOPA AS
PRODRUGS FOR RECTAL ABSORPTION", PHARM. RES., 6(6), 501-505 (1989).

FLANINGAM,O.L.; VAPOR PRESSURES OF POLY(DIMETHYLSILOXANE) OLIGOMERS.; J. CHEM.
ENG. DATA; 31:266-72; 1986

FLETCHER,C.L.; MCKAY,W.A.; POLYCHORINATED DIBENZO-P-DIOXINS (PCDDS) AND
DIBENZOFURANS (PCDFS) IN THE AQUATIC ENVIRONMENT. A LITERATURE REVIEW.;
CHEMOSPHERE; 26:1041-69; 1993

FLICK,E.W.; INDUSTRIAL SOLVENTS HANDBOOK. 4TH EDITION.; NOYES DATA CORP., PARK
RIDGE, NJ.; 1991

FORD, H., C.L. MERSKI AND J.A. KELLEY, "A RAPID MICROSCALE METHOD FOR THE DETERMINATION OF PARTITION
COEFFICIENTS BY HPLC", J. LIQ. CHROMATOGR., 14(18), 3356-3386 (1991).

FORWARD,M.V.; BOWDEN,S.T.; JONES,W.J.; PHYSICAL PROPERTIES OF TRIPHENYL
COMPOUNDS OF GROUP VB ELMENTS.; J. CHEM. SOC. S121-6.; 1949

FORZIATI,A.F.; NORRIS,W.R.; ROSSINI,F.D.; VAPOR PRESSURE AND BOILING POINTS OF
SIXTY API-NBS HYDROCARBONS.; J. RES. NATL. BUREAU STANDARDS; 43:555-63; 1949

FOY,C.L.; THE CHLORINATED ALIPHATIC ACIDS.; IN: DEGRADATION OF PESTICIDES.
KEARNEY,P.C. & KAUFMAN,D.D. EDS. NEW YORK,NY: MARCEL DEKKER. PP.207-53.; 1969

FREAR,D.S.; THE BENZOIC ACID HERBICIDES.; HERBIC.: CHEMISTRY DEGRADATION AND
MODE ACTION, 2ND ED. KEARNEY,P.C. & KAUFMAN,D.D. EDS.; 2:541-607.; 1976

FREED,V.H.; HAGUE,R.; VERNETTI,J.; THERMODYNAMIC PROPERTIES OF SOME CARBAMATES
AND THIOLCARBAMATES IN AQUEOUS SOLUTION.; J. AGRIC. FOOD CHEM.; 15:1121-3.; 1967

FREED,V.H.; CHIOU,C.T.; HAGUE,R.; CHEMODYNAMICS: TRANSPORT AND BEHAVIOR OF
CHEMICALS IN THE ENVIRONMENT-A PROBLEM IN ENVIRONMENTAL HEALTH.; ENVIRON. HEALTH
PERSPECT.; 20:55-70.; 1977

FREED,V.H.; SCHMEDDING,D.; KOHNERT,R.; HAGUE,R.; PHYSICAL CHEMICAL PROPERTIES OF
SEVERAL ORGANOPHOSPHATE: SOME IMPLICATION IN ENVIRONMENTAL AND BIOLOGICAL
BEHAVIOR.; PESTIC. BIOCHEM. PHYSIOL.; 10:203-11.; 1979A

FREITER,E.R.; HALOGENATED DERIVATIVES.; IN: KIRK-OTHMER ENCYCL. CHEM. TECH. 3RD
ED. 1:171-8.; 1978

FREITER,E.R.; CHLOROPHENOLS.; IN: KIRK-OTHMER ENCYCL. CHEM. TECH. 3RD NEW
YORK,NY; WILEY INTERSCIENCE. 5:864-72; 1978A

FRIESEN,K.J.; VILK,J.; MUIR,D.C.G.; AQUEOUS SOLUBILITIES OF SELECTED
2,3,7,8-SUBSTITUTED POLYCHLORINATED DIBENZOFURANS (PCDFS).; CHEMOSPHERE;
20:27-32; 1990B

FUCHS, J., W.H. NITSCHMANN, L. PACKER, O.H. HANKOVSZKY AND K. HIDEG, "PKA VALUES AND PARTITION COEFFICIENTS
OF NITROXIDE SPIN PROBES FOR MEMBRANE BIOENERGETICS MEASUREMENTS", FREE RADICAL RES. COMMUN., 10(6),
315-323 (1990).

FUNASAKI,N.; HADA,S.; SUZUKI,K.; THE DISSOLUTION STATE OF A TRIGLYCERIDE
MOLECULE IN WATER AND ITS ORIENTATION STATE AT THE AIR-WATER INTERFACE.; CHEM.
PHARM. BULL.; 24:731-5; 1976

FUNASAKI, N., S. HADA AND S. NEYA, "PREDICTION OF RETENTION TIMES IN REVERSED-PHASE HIGH-PERFORMANCE
LIQUID CHROMATOGRAPHY FROM THE CHEMICAL STRUCTURE", J. CHROMATOGR., 361, 33-45 (1986).

FUNASAKI, N. ET AL. 1985. PARTITION COEFFICIENTS OF ALIPHATIC ETHERS-MOLECULAR SURFACE AREA APPROACH. J.
PHYS. CHEM. 84: 3046-9.

FUNASAKI, N., S. HADA, S. NEYA AND K. MACHIDA, "INTRAMOLECULAR HYDROPHOBIC ASSOCIATION OF TWO CHAINS OF
OLIGOETHYLENEGLYCOL DIETHERS AND DIESTERS IN WATER", J. PHYS. CHEM., 88(24), 5786-5790 (1984).

FURER,R.; GEIGER,M.; A SIMPLE METHOD OF DETERMINING THE AQUEOUS SOLUBILITY OF
ORGANIC SUBSTANCES.; PESTICIDE SCI.; 8:337; 1977

FURST, W. AND M. BECHER, "HOMOLOGUES OF 2-ACETAMIDO-4-NITROPROPOXYBENZENE (FALIMINT). I. PREPARATION,
PHYSICOCHEMICAL AND ORGANOLEPTIC PROPERTIES", PHARMAZIE, 45(12), 934-935 (1990).

FURST, W., R. NEUBERT, T. JURKSCHAT AND L. LUCKE, "PRODRUG APPROACH OF OROTIC ACID USING AN ABSORPTION
MODEL", INT. J. PHARM., 61(1-2), 43-49 (1990).

GAFFNEY,J.S.; STREIT,G.E.; SPALL,W.D.; HALL,J.H.; BEYOND ACID RAIN - DO SOLUBLE

OXIDANTS AND ORGANIC TOXINS INTERACT WITH SO2 AND NOX TO INCREASE ECOSYSTEM
EFFECTS.; ENVIRON. SCI. TECHNOL.; 21:519-23; 1987

GAMBERINI, G., C. PARENTI, L. COSTANTINO, M. DI BELLA, L. RAFFA AND T. ROSSI, "HETEROARYLALKANOIC ACIDS WITH
POSSIBLE ANTI-INFLAMMATORIC ACTIVITIES. 6. 3-(4H-1,2,4-BENZOTHIADIAZINE-1,1-DIOXIDE-3-YLTHIO)PROPANOIC ACID
DERIVATIVES", PHARMAZIE, 44(9), 601-604 (1989).

GANG,Y.; XIAOBAI,X.U.; INVESTIGATION OF AQUEOUS SOLUBILITIES OF NITRO-PAH BY DYNAMIC COUPLE-COLUMN
HPLC.; CHEMOSPHERE; 24:1699-1705; 1992

GARTEN,C.T. JR.; TRABALKA,J.R.; EVALUATION OF MODELS FOR PREDICTING TERRESTRIAL
FOOD CHAIN BEHAVIOR OF XENOBIOTICS.; ENVIRON. SCI. TECHNOL.; 17:590-5.; 1983

GASPARI, F. AND M. BONATI, "CORRELATION BETWEEN N-OCTANOL/WATER PARTITION COEFFICIENTS AND LIQUID
CHRMATOGRAPHIC RETENTION TIME FOR CAFFEINE AND ITS METABOLITES, AND SOME STRUCTURE-PHARMACOKINETIC
CONSIDERATIONS", J. PHARM. PHARMACOL., 39(4), 252-260 (1987).

GEHRING,P.J.; TORKELSON,T.R.; OYEN,F.; A COMPARISON OF THE LETHALITY OF
CHLORINATED PYRIDINES AND A STUDY OF THE ACUTE TOXICITY OF 2-CHLOROPYRIDINE.;
TOX. APPL. PHARMC.; 11:361-71.; 1967

GERSTL,Z.; HELLING,C.S.; EVALUATION OF MOLECULAR CONNECTIVITY AS A PREDICTIVE
METHOD FOR THE ADSORPTION OF PESTICIDES BY SOILS.; J. ENVIRON. SCI. HEALTH B;
22:55-69; 1987

GEYER,H.; SHEEHAN,P.; KOTZIAS,D.; FREITAG,D.; KORTE,F.; PREDICTION OF
ECOTOXICOLOGICAL BEHAVIOR OF CHEMICALS: RELATIONSHIP BETWEEN PHYSICOCHEMICAL
PROPERTIES AND BIOACCUMULATION OF ORGANIC CHEMICALS IN THE MUSSEL MYTILUS
EDULIS.; CHEMOSPHERE.; 11:1121-34.; 1982

GEYER,H.; VISWANATHAN,R.; FREITAG,D.; FORTE,F.; RELATIONSHIP BETWEEN WATER
SOLUBILITY OF ORGANIC CHEMICALS AND THEIR BIOACCUMULATION BY THE ALGA
CHLORELLA.; CHEMOSPHERE.; 10:1307-13.; 1981

GHERINI,S.A.; SUMMERS,K.V.; MUNSON,R.K.; MILLS,W.B.; CHEMICAL DATA FOR
PREDICTING THE FATE OF ORGANIC COMPOUNDS IN WATER: VOLUME 1, TECHNICAL BASIS:
FINAL REPORT.; W.B. ELECTR. POWER RES. INST., PALO ALTO, CA. REPORT 1989,
EPRI-EA-5818-VOL.1, 187 PP.; 1989

GIAM,C.S.; ATLAS,E.; POWERS,M.A.J.R.; LEONARD,J.E.; PHTHALIC ACID ESTERS; IN:
THE HANDBOOK OF ENVIRONMENTAL CHEMISTRY; ANTHROPOGENIC SUBSTANCES. GERMANY:
SPRINGER-VERLAG BERLIN. 3:67-142; 1984A

GILE,J.D.; GILLETT,J.W.; FATE OF SELECTED FUNGICIDES IN A TERRESTRIAL LABORATORY
ECOSYSTEM.; J. AGRIC. FOOD CHEM.; 27:1159-64.; 1979

GO, M.L. AND T.L. NGIAM, "HYDROPHOBICITY CHANGE ON N-OXIDATION OF SOME 4-AMINOQUINOLINES", CHEM. PHARM.
BULL., 36(4), 1393-1398 (1988).

GODOY, N. ET AL; 1989. SYNTHESIS, LABELING AND BIODISTRIBUTION OF TECHNETIUM-M99-LABELED BROMO
IMINODIACETIC ACIDS. NUCLEOTECNICA 9: 25-32.

GOE,G.L.; PYRIDINE AND PYRIDINE DERIVATIVES.; KIRK-OTHMER ENCYCL. CHEM. TECH.
3RD ED. NY,NY: WILEY INTERSCIENCE, 19:454-83.; 1982

GLOTFELTY,D.E.; TAYLOR,A.W.; TURNER,B.C.; ZOLLER,W.H.; VOLATILIZATION OF
SURFACE-APPLIED PESTICIDES FROM FALLOW SOIL.; J. AGRIC. FOOD CHEM.; 32:638-44.;
1984A

GOICOLEA,A.; ARRANZ,J.F.; BARRIO,R.J.; POLAROGRAPHIC DETERMINATION OF
DISSOCIATION CONSTANTS OF SELECTED HERBICIDES.; AN. QUIM.; 87:163-6; 1991

GOLDBERG,N.A.; KUCHERYAVYI,V.I.; SOME PHYSICOCHEMICAL PROPERTIES OF HEXAMETHYLENE DIISOCYNATE.; PRIKLAD. KHIM.; 33:1912-13.; 1960

GOLDMAN,M.; DACRE,J.C.; LEWISITE: ITS CHEMISTRY, TOXICOLOGY, AND BIOLOGICAL EFFECTS.; IN: ENVIRON. CONTAM. TOXICOL., VOL. 110, PP. 75-115.; 1989

GOSSETT,J.M.; MEASUREMENT OF HENRY'S LAW CONSTANT FOR C1 AND C2 CHLORINATED HYDROCARBONS; ENVIRON SCI TECH; 21:202-6; 1987

GOVERS,H.; RUEPERT,C.; STEVENS,T.; VANLEEUWEN,C.J.; EXPERIMENTAL DETERMINATION AND PREDICTION PARTITION COEFFICIENTS OF THIOUREAS AND THEIR TOXICITY TO PHOTOBACTERIUM PHOSPHOREUM.; CHEMOSPHERE; 15:383-93; 1986

GRAYSON,B.T.; FOSBRAEY,L.A.; DETERMINATION OF THE VAPOUR PRESSURE OF PESTICIDES.; PESTIC. SCI.; 13:269-78.; 1982

GRAYSON,B.T.; LANGNER,E.; WELLS,D.; COMPARISON OF TWO GAS SATURATION METHODS FOR THE DETERMINATION OF THE VAPOR PRESSURE OF CYPERMETHRIN.; PESTIC. SCI.; 13:522-6; 1982

GREEN,F.J.; THE SIGMA-ALDRICH HANDBOOK OF STAINS, DYES AND INDICATORS.; ALDRICH CHEMICAL COMPANY, INC.: MILWAUKEE, WISCONSIN, 776 PP.; 1990

GREEN, P.G., R.S. HINZ, C. CULLANDER, G. YAMANE AND R.H. GUY, "IONTOPHORETIC DELIVERY OF AMINO ACIDS AND AMINO ACID DERIVATIVES ACROSS THE SKIN IN VITRO", PHARM. RES., 8(9), 1113-1120 (1991).

GREENWALD,I.; ON THE SOLUBILITY OF SOME PICRATES AND THE DETERMINATION OF GUANIDINES IN URINE.; BIOCHEM. J.; 20:666; 1926

GROSS,P.M.; SAYLOR,J.H.; THE SOLUBILITIES OF CERTAIN SLIGHTLY SOLUBLE ORGANIC COMPOUNDS IN WATER.; J. AMER. CHEM. SOC.; 53:1744-51.; 1931

GROSS,P.M.; J. AMER. CHEM. SOC.; 51:2362. (IN YALKOWSKY'S DATA BASE).; 1929

GROVER,R.; KERR,L.A.; BOWREN,K.E.; KHAN,S.U.; AIRBORNE RESIDUES OF TRIALLATE AND TRIFLURALIN IN SASKATCHEWAN; BULL ENVIRON CONTAM TOXICOL; 40:683-8; 1988

GUERRA, M.C., A.M. BARBARA, G.C. FORTI, M.T. FOFFANI, G.L. BIAGI, P.A. BOREA AND A. FINI, "RM AND LOG P OF 5-NITROIMIDAZOLES", J. CHROMATOGR., 216, 93-102 (1981).

GUNTHER,F.A.; WESTLAKE,W.E.; JAGLAN,P.S.; REPORTED SOLUBILITIES OF 738 PESTICIDE CHEMICALS IN WATER.; RES. REV.; 20:1-148.; 1968

GUPTA, R.P., C.A. LARROQUETTE, K.C. AGRAWAL, J. GRODKOWSKI AND P. NETA, "POTENTIAL RADIOSENSITIZING AGENTS. 7. 4(5)-IODO-5(4)-NITROIMIDAZOLE DERIVATIVES", J. MED. CHEM., 28(8), 987-991 (1985).

GUY, R.H., E.M. CARLSTROM, D.A.W. BUCKS, R.S. HINZ AND H.I. MAIBACH, "PERCUTANEOUS PENETRATION OF NICOTINATES: IN VIVO AND IN VITRO MEASUREMENTS", J. PHARM. SCI., 75(10), 968-972 (1986).

GYPARAKI, M. ET AL. 1986. EVALUATION OF IN VIVO HYDROXYPYRID-4-ONE IRON CHELATORS INTENDED FOR THE TREATMENT OR IRON OVERLOAD BY THE ORAL ROUTE. BIOCHEM. SOC. TRANS. 14: 1181.

HAAG,W.R.; MILL,T.; DIRECT AND INDIRECT PHOTOLYSIS OF WATER SOLUBLE AZODYES: KINETIC MEASUREMENTS AND STRUCTURE-ACTIVITY RELATIONSHIPS.; ENVIRON. TOXICOL. CHEM.; 6:359-69; 1987

HAAS,J.M.; EARHART,H.W.; TODD,A.S.; ENVIRONMENTAL GUIDE TO DYE CARRIER SELECTION.; AMERICAN DYESTUFF REPORTER; 64:34,36,49; 1975

HAGEN, V., E. KLAUSCHENZ, A. RUMLER, A. HAGEN, S. HEER, R. MITZNER, H. NIEDRICH AND D. LOHMANN, "POTENTIAL CARDIOTONIC AGENTS. 6. SYNTHESIS AND CARDIOVASCULAR PROPERTIES OF 5-(4-PYRIDINYL)-6-METHYL-5-(4-PYRIDINYL)- AND 6-METHYL-5-PHENYL-SUBSTITUTED 3-CYANO-2-OXAALKOXY PYRIDINES", PHARMAZIE, 45(3), 189-190 (1990).

HAGEN, V., A. HAGEN, S. HEER, R. MITZNER AND H. NIEDRICH, "POTENTIAL CARDIOTONIC AGENTS. I. SYNTHESIS AND

PHARMACOLOGICAL PROPERTIES OF 3-CYANO-2-OXAALKYLAMINO-5-(4-PYRIDINYL) PYRIMIDINES", PHARMAZIE, 44(1), 20-23 (1989).

HALBACH,S.; THE OCTANOL/WATER DISTRIBUTION OF MERCURY COMPOUNDS.; ARCH. TOXICOL.; 57:139-41; 1985

HALL, I.H., C.B. OSWALD, S.D. WYRICK, J.H. MAGUIRE AND R.P. SHREWSBURY, "DISPOSITION OF THE HYPOLIPIDAEMIC AGENT 2,3-DIHYDROPHTHALAZINE-1,4-DIONE IN SPRAGUE DAWLEY RATS", J. PHARM. PHARMACOL., 41, 394-397 (1989).

HAM,GE IMINES, CYCLIC.; KIRK-OTHMER ENCYCL CHEM TECH 3RD ED. NY,NY: WILEY-INTERSCIENCE. 13:142-66; 1981

HAMAKER,J.W.; KERLINGER,H.O.; VAPOR PRESSURES OF PESTICIDES.; ADV. CHEM. SER.; 86:39-54; 1969

HAMILTON,D.J.; GAS CHROMATOGRAPHIC MEASUREMENT OF VOLATILITY OF HERBICIDE ESTERS.; J. CHROMATOGR.; 195:75-83.; 1980

HANN,R.W. JR.; JENSEN,P.A.; WATER QUALITY CHARACTERISTICS OF HAZARDOUS MATERIALS.; NTIS-PB-285946. TEXAS A&M UNIV., COLLEGE STATION ENVIRONMENTAL ENGINEERING DIV. 1751 PP.; 1977

HANSCH,C.; QUINLAN,J.E.; LAWRENCE,G.L.; LINEAR FREE ENERGY RELATION BETWEEN PARTITION COEFFECIENTS AND THE AQUEOUS SOLUBILITY OF ORGANIC LIQUIDS.; J. ORG. CHEM.; 33:347.; 1968

HANSCH,C.; LEO,A.J.; MEDCHEM PROJECT. ISSUE NO. 19; CLAREMONT, CA: POMONA COLLEGE.; 1981

HANSCH,C.; LEO,A.J.; MEDCHEM PROJECT.; CLAREMONT, CA: POMONA COLLEGE. ISSUE # 26.; 1985

HANSCH,C.; LEO,A.J. POMONA. 1987. POMONA COLLEGE MEDICINAL CHEMISTRY PROJECT, CLAREMONT, CA 91711, LOG P DATABASE, JULY 1987 EDITION.

HANSCH. C., A. LEO AND D. HOEKMAN. 1995. EXPLORING QSAR. HYDROPHOBIC, ELECTRONIC, AND STERIC CONSTANTS. ACS PROFESSIONAL REFERENCE BOOK. WASHINGTON, DC: AMERICAN CHEMICAL SOCIETY.

HANSCH. C., A. LEO AND D. HOEKMAN. 1995. EXPLORING QSAR. HYDROPHOBIC, ELECTRONIC, AND STERIC CONSTANTS. ACS PROFESSIONAL REFERENCE BOOK. WASHINGTON, DC: AMERICAN CHEMICAL SOCIETY.

HAQUE,R.; SCHMEDDING,D.; STUDIES ON THE ADSORPTION OF SELECTED POLYCHLORINATED BIPHENYL ISOMERS ON SEVERAL SURFACES.; J. ENVIRON. SCI. HEALTH.; B11:129-37.; 1976

HARNER,T.; MACKAY,D.; MEASUREMENT OF OCTANOL-AIR PARTITION COEFFICIENTS FOR CHLOROBENZENES, PCBS AND DDT.; ENVIRON. SCI. TECHNOL.; 29:1599-1606; 1995

HARTLEY,D.; KIDD,H.; THE AGROCHEMICALS HANDBOOK; OLD WOKING SURREY, ENGLAND: UNWIN BROS. LTD.; 1983

HARTLEY,D.; KIDD,H.; THE AGROCHEMICALS HANDBOOK; OLD WOKING SURREY, ENGLAND: UNWIN BROS. LTD; 1985

HARTLEY,D.; KIDD,H.; THE AGROCHEMICALS HANDBOOK; OLD WOKING SURREY, ENGLAND: UNWIN BROS. LTD; 1987

HARTLEY,D.; KIDD,H.; THE AGROCHEMICALS HANDBOOK; OLD WOKING SURREY, ENGLAND: UNWIN BROS. LTD; 1991

HARTMAN, G.D., W. HALCZENKO, R.L. SMITH, M.F. SUGRUE, P.J. MALLORGA, S.R. MICHELSON, W.C. RANDALL, H. SCHWAM AND J.M. SONDEY, "4-SUBSTITUTED THIOPHENE- AND FURAN-2-SULFONAMIDES AS TOPICAL CARBONIC ANHYDRASE INHIBITORS", J. MED. CHEM., 35(21), 3822-3831 (1992).

HARTMAN, G.D., W. HALCZENKO, J.D. PRUGH, R.L. SMITH, M.F. SUGRUE, P. MALLORGA, S.R. MICHELSON, W.C. RANDALL, H. SCHWAM AND J.M. SONDEY, "THIENO[2,3-B]FURAN-2-SULFONAMIDES AS TOPICAL CARBONIC ANHYDRASE INHIBITORS", J. MED. CHEM., 35(16), 3027-3033 (1992).

HARVEY, J JR. ET AL. 1985. PROPERTIES OF SULFOMETURON METHYL AFFECTING ITS ENVIRONMENTAL FATE: AQUEOUS HYDROLYSIS AND PHOTOLYSIS, MOBILITY, AND ADSORPTION ON SOILS, AND BIOACCUMULATION POTENTIAL. J. AGRIC. FOOD CHEM. 33: 590-6.

HASSETT,J.J.; MEANS,J.C.; BANWART,W.L.; WOOD,S.G.; SORPTION PROPERTIES OF SEDIMENTS AND ENERGY-RELATED POLLUTANTS.; EPA-600/3-80-041. ATHENS,GA: U.S. EPA. PP.133.; 1980

HASSETT,J.J.; MEANS,J.C.; BANWART,W.L.; WOOD,S.G.; ALI,S.; KHAN,A.; SORPTION OF DIBENZOTHIOPHENE BY SOIL AND SEDIMENTS.; J. ENVIRON. QUAL.; 9:184-6.; 1980A

HAWLEY,G.G.; THE CONDENSED CHEMICAL DISTIONARY. 10TH ED.; NEW YORK,NY: VAN NOSTRAND REINHOLD CO. PP.1135.; 1981

HAWTHORNE,S.B.; SIEVERS,R.E.; BARKLEY,R.M.; ORGANIC EMISSIONS FROM SHALE OIL WASTEWATERS AND THEIR IMPLICATIONS FOR AIR QUALITY.; ENVIRON. SCI. TECH.; 19:922-7.; 1985

HAY,J.V.; CHEMISTRY OF SULFONYLUREA HERBICIDES.; PESTIC. SCI.; 29:247-61; 1990

HAYS, S.J., P.M. NOVAK, D.F. ORTWINE, C.F. BIGGE, N.L. COLBRY, G. JOHNSON, L.J. LESCOSKY, T.C. MALONE, A. MICHAEL, M.D. REILY, L.L. COUGHENOUR, L.J. BRAHCE, J.L. SHILLIS AND A. PROBERT, "SYNTHESIS AND PHARMACOLOGICAL EVALUATION OF HEXAHYDROFLUORENAMINES AS NONCOMPETITIVE ANTAGONISTS AT THE N-METHYL-D-ASPARTATE RECEPTOR", J. MED. CHEM., 36(6), 654-670 (1993).

HEATH,J.; AHMAD,A.; LEAHEY,J.P.; A SIMPLE PROCEDURE TO MEASURE THE VOLATILITY OF AGROCHEMICALS FROM SOIL AND LEAF SURFACES.; IN: BRIGHTON CROP PROT. CONF. -- PESTS DIS. 2:835-40.; 1992

HEATH,R.R.; TUMLINSON,J.H.; CORRELATION OF RETENTION TIMES ON LIQUID CRYSTAL CAPILLARY COLUMN WITH REPORTED VAPOR PRESSURES AND HALF-LIVES OF COMPOUNDS USED IN PHEROMONE FORMULATIONS.; J. CHEM. ECOL.; 12:2081-8; 1986

HEFTER,G.T.; SQLUBILITY DATA. 2-BUTANOL-WATER SYSTEM.; IN: SOLUB. DATA SER., 15:94-119; 1984A

HEIKES,D.L.; PURGE AND TRAP METHOD FOR DETERMINATION OF VOLATILE HYDROCARBONS AND CARBON DISULFIDE IN TABLE-READY FOODS.; J. ASSOC. OFF. ANAL. CHEM.; 70:215-77; 1987

HEIMLICH,F.; NOLTE,J.; DETERMINATION OF THE PK VALUES OF 2,4-DINITROPHENOL HERBICIDES USING UV SPECTROSCOPY.; SCI. TOTAL. ENVIRON.; 132:125-31; 1993

HEISLER,S.L.; FRIEDLANDER,S.K.; GAS-TO-PARTICLE CONVERSION IN PHOTOCHEMICAL SMOG: AEROSOL GROWTH LOWS AND MECHANISMS FOR ORGANICS.; ATMOS. ENVIRON.; 11:157-68; 1977

HELLMANN,H.; MODEL TESTS ON VOLATILIZATION OF ORGANIC TRACE SUBSTANCES IN SURFACE WATERS.; FRESENIUS' Z. ANAL. CHEM.; 328:475-9; 1987

HEMPHILL,L.; SWANSON,W.S.; SORPTION OF ORGANIC ACIDS BY PURE CLAY MINERALS IN AQUEOUS SOLUTION.; PROC. OF THE 18TH INDUSTRIAL WASTE CONF., ENG. BULL. PURDUE U., LAFAYETTE, IN.; 18:204-17.; 1964

HERMANN,B.W.; ATMOSPHERE RESIDUES OF THE DEFOLIANTS DEF., FOLEX, AND THEIR SULFUR-CONTAINING BY-PRODUCTS NEAR TREATED COTTON FIELDS; DISS ABSTR INT B; 41:4104; 1980

HILAL,SH; CARREIRA,LA; KARICKOFF,SW; ESTIMATION OF CHEMICAL REACTIVITY PARAMETERS AND PHYSICAL PROPERTIES OF ORGANIC MOLECULES USING SPARC. IN: QUANTITATIVE TREATMENTS OF SOLUTE/ SOLVENT INTERACTIONS: THEORECTICAL AND COMPUTATIONALY CHEMISTRY VOL 1. 291-??; 1994

HINCKLEY,D.A.; BIDLEMAN,T.F.; FOREMAN,W.T.; TUSCHALL,J.R.; DETERMINATION OF VAPOR PRESSURES OF NONPOLAR AND SEMIPOLAR ORGANIC COMPOUNDS FROM GAS CHROMATOGRAPHIC RETENTION DATA.; J. CHEM. ENG. DATA.; 35:232-7; 1990

HINE,J.; MOOKERJEE,P.K.; THE INTRINSIC HYDROPHILIC CHARACTER OF ORGANIC COMPOUNDS. CORRELATIONS IN TERMS OF STRUCTURAL CONTRIBUTIONS.; J. ORG. CHEM.; 40:292-8.; 1975

HIRT,R.C.; STEGER,J.E.; SIMARD,G.L.; VAPOR PRESSURE OF 2,4,6-TRIAMINO-S-TRIAZINE (MELAMINE).; J. POLYMER SCI.; 43:319-23; 1960

HOEGBERG, T. ET AL. 1986. SOLID STATE CONFORMATIONS AND ANTIDOPAMINERGIC EFFECTS OF REMOXIPRIDE HYDROCHLORIDE AND A CLOSELY RELATED SALICYCLAMIDE, FLA 797, IN RELATION TO DOPAMINE RECEPTOR MODELS. MOL. PHARMACOL. 30: 345-51.

HOEFNAGEL,M.A.; VAN VEEN,A.; WEBSTER,B.M.; PROTONATION OF AZO-COMPOUNDS. PART II. THE STRUCTURE OF THE CONJUGATE ACID OF TRANS-AZOBENZENE.; REC. TRAV. CHIM.; 88:562.; 1969

HOFF,M.C.; IM,U.K.; HAUSCHILDT,W.F.; PUSKAS,I.; BUTYLENES.; IN: KIRK-OTHMER ENCYCLO. CHEM. TECH. 3RD ED. 4:346-75.; 1978

HOLLIFIELD,H.C.; RAPID NEPHELOMETRIC ESTIMATE OF WATER SOLUBILITY OF HIGHLY INSOLUBLE ORGANIC CHEMICALS OF ENVIRONMENTAL INTEREST.; BULL. ENVIRON. CONTAM. TOXICOL.; 23:579-86.; 1979

HORN,A.S.; NEUROREGULATORS; IN: KIRK-OTHMER ENCYCLOPEDIA OF CHEMICAL TECHNOLOGY, NEW YORK,NY: WILEY INTERSCIENCE. 15:754-86; 1981

HORT,E.V.; ACETYLENE-DERIVED CHEMICALS.; IN: KIRK-OTHMER ENCYCL.CHEM. TECH. 3RD ED. 1:244-76.; 1978

HORVATH,A.L.; SIMILARITY-SYMETRY-ANALOGY-PRINCIPLE. SOLUBILITY OF HALOGENATED HYDROCARBONS IN WATER.; CHEM.-ING.-TECH.; 47:815; 1975

HORVATH,A.L.; HALOGENATED HYDROCARBONS: SOLUBILITY-MISCIBILITY WITH WATER.; NEW YORK,NY: MARCEL DEKKER, INC. PP.889.; 1982

HOU,M.; BAUGHMAN,G.L.; PERENICH,T.A.; ESTIMATION OF WATER SOLUBILITY AND OCTANOL/WATER PARTITION COEFFICIENT OF HYDROPHOBIC DYES. PART II. REVERSE-PHASE HIGH-PERFORMANCE LIQUID CHROMATOGRAPHY.; DYES PIGM.; 16:291-7; 1991

HOWARD,C.J.; RATE CONSTANTS FOR THE GAS-PHASE REACTIONS OF HYDROXYL RADICALS WITH ETHYLENE AND HALOGENATED ETHYLENE COMPOUNDS.; J. CHEM. PHYS.; 65:4771-7.; 1976

HOWARD,P.H.; SANTODONATO,J.; SAXENA,J.; MALLING,J.; GRENINGER,D.; INVESTIGATION OF SELECTED POTENTIAL ENVIRONMENTAL CONTAMINANTS: NITROAROMATICS (DRAFT).; EPA-560/2-76-010. RESEARCH TRIANGLE PARK,NC: U.S. EPA. PP.600.; 1976

HOWARD,P.H.; BANERJEE,S.; ROBRILLARD,K.H.; MEASUREMENT OF WATER SOLUBILITIES, OCTANOL-WATER PARTITION COEFFICIENTS AND VAPOR PRESSURES OF COMMERCIAL PHTHALATE ESTERS.; ENVIRON. TOX. CHEM.; 4:653-61.; 1985

HOWARD,P.H. AND NEAL,M; DICTIONARY OR CHEMICAL NAMES AND SYNONYMS; CHELSEA, MI: LEWIS PUBLISHERS, 1992.

HOYER,H.; PEPERLE,W.; DETERMINATION OF THE VAPOR PRESSURE AND HEAT OF SUBLIMATION OF ORGANIC COMPOUNDS.; Z. ELEKTROCHEM.; 62:61-6.; 1958

HU, J AND X.-F. LENG, "DETERMINATION OF PARTITION COEFFICIENTS FOR SOME PESTICIDES BY USING REVERSED-PHASE HIGH PERFORMANCE LIQUID CHROMATOGRAPHY (HPLC)", SEPU, 10(6), 344-346 (1992).

HUANG, C.H. ET AL. 1985. MECHANISM OF NASAL ABSORPTION OF DRUGS. II. J. PHARM. SCI. 74: 1298-301.

HUGHES,D.W.; MALONIC ACID AND DERIVATIVES.; KIRK-OTHMER ENCYCL. CHEM. TECH. 3RD ED. NY,NY : WILEY-INTERSCIENCE. 14:794-810.; 1981

HUININK,J.; VANMILTENBURG,J.C.; OONK,H.A.; SCHUIJFF,A.; VAPOR-PRESSURE MEASUREMENTS AND THERMODYNAMIC PROPERTIES. 1,3,5-TRIBROMOBENZENE.; RECL. TRAV.

CHIM. PAYS-BAS.; 107:273-7; 1988

HUNTER-SMITH,R.J.; BALLS,P.W.; LISS,P.S.; HENRY'S LAW CONSTANTS AND THE AIR-SEA EXCHANGE OF VARIOUS LOW-MOLECULAR WEIGHT HALOCARBON GASES; TELLUS SER B; 35B:170-6; 1983

HUTCHINSON,T.C.; HELLEBUST,J.A.; MACKAY,D.; TAM,D.; KAUSS,P.; RELATIONSHIP OF HYDROCARBON SOLUBILITY TO TOXICITY IN ALGAE AND CELLULAR MEMBRANE EFFECTS.; IN: PUBL. 4308, AM. PET. INST. ISS PROC. PP.541-7; 1979

HUYSKENS,P.; MULLENS,J.; GOMEZ,A.; TACK,J.; SOLUBILITY OF ALCOHOLS, PHENOLS AND ANILINES IN WATER.; BULL. SOC. CHIM. BELG.; 84:253-62.; 1975

IARC.; SOME AZIRIDINES, N-, S- AND O- MUSTARDS AND SELENIUM; IN: INTERNAT AGNCY FOR RES ON CANCER; LYON, FRANCE; 9:135; 1975

IARC.; GLYCIDYLALDEHYDE; LYON, FRANCE: INTER., AGENCY RES CANCER ;11:175; 1976

IARC; MONOGRAPHS ON THE EVALUATION OF THE CARCINOGENIC RISK TO HUMANS. VOL.17. SOME N-NITROSO COMPOUNDS.; LYON,FRANCE: INTERNATIONAL AGENCY FOR RESEARCH ON CANCER.; 1978

IARC; MONOGRAPH ON THE EVALUATION OF THE CARCINOGENIC RISK OF CHEMICALS TO HUMANS. VOL. 20. SOME HALOGENATED HYDROCARBONS.; LYON,FRANCE:INTERNATIONAL AGENCY FOR RESEARCH ON CANCER.; 1979

IARC.; MONOGRAPH OF THE EVALUATION OF THE CARCINOGENIC RISK OF CHEMICALS TO HUMANS. VOL 30. MISCELLANEOUS PESTICIDES.; LYON,FRANCE: INTERNATIONAL AGENCY FOR REASEARCH ON CANCER.; 1983

IARC; MONOGRAPH ON THE EVALUATION OF THE CARCINOGENIC RISK OF CHEMICALS TO HUMANS. VOL. 41. SOME HALOGENATED HYDROCARBONS AND PESTICIDE EXPOSURES.; LYON,FRANCE:INTERNATIONAL AGENCY FOR RESEARCH ON CANCER.; 1986

IARC; MONOGRAPH ON THE EVALUATION OF THE CARCINOGENIC RISK OF CHEMICALS TO HUMANS. VOL. 53. OCCUPATIONAL EXPOSURES IN INSECTICIDE APPLICATION AND SOME PESTICIDES.; LYON,FRANCE:INTERNATIONAL AGENCY FOR RESEARCH ON CANCER.; 1991

ILCHMANN, A., G. WIENKE, T. MEYER AND J. GMEHLING, "COUNTERCURRENT LIQUID/LIQUID CHROMATOGRAPHY - A RELIABLE METHOD FOR THE DETERMINATION OF PARTITION COEFFICIENTS", CHEM.-ING.-TECH., 65(1), 72-75 (1993).

INST NATL DE RECHERCHE ET DE SECURITE; TOXICOLOGY CARD. 190. METHANETHIOL (CH3-SH), ETHANETHIOL (C2H5-SH) AND 1-BUTANETHIOL (C4-H9-SH); CAH NOTES DOC; 113:597-600; 1983

INUI,H.; ITOH,K.; MATSUO,M.; MIYAMOTO,J.; STUDIES ON BIODEGRADATION OF 2,6-DI-TERT-BUTYL-4-METHYLPHENOL (BHT) IN THE ENVIRONMENT: PART-III: BIODEGRADABILITY OF BHT WITH ACTIVATED SLUDGE.; CHEMOSPHERE.; 8:383-91.; 1979A

ISHOW.; INFORMATION SYSTEM FOR HAZARDOUS ORGANICS IN WATER.; DULUTH, MN: DEPT. OF CHEM., UNIV. OF MINNESOTA.; NA--

ISLAM, M.S. AND M.M. NARURKAR, "SOLUBILITY, STABILITY AND IONIZATION BEHAVIOUR OF FAMOTIDINE", J. PHARM. PHARMACOL., 45(8) , 682-686 (1993).

ISNARD,P.;LAMBERT,S.; AQUEOUS SOLUBILITY AND N-OCTANOL/WATER PARTITION COEFFICIENT CORRELATIONS.; CHEMOSPHERE; 18:1837-53; 1989

ITOKAWA, H., N. TOTSUKA, K. NAKAHARA, M. MAEZURU, K. TAKEYA, M. KONDO, M. INAMATSU AND H. MORITA, "A QUANTITATIVE STRUCTURE-ACTIVITY RELATIONSHIP FOR ANTITUMOR ACTIVITY OF LONG-CHAIN PHENOLS FROM GINKGO BILOBA L.", CHEM. PHARM. BULL., 37(6), 1619-1621 (1989).

JABER,H.M.; MABEY,W.R.; LIU,A.T.; CHOU,T.W.; JOHNSON,H.L.; DATA ACQUISITION FOR ENVIRONMENTAL TRANSPORT AND FATE SCREENING; EPA-600/6-84/009. (NTIS

PB84-243906, PB84-243955). MENLO PARK,CA.: SRI INTER. PP.312.; 1984

JABER,H.M.; MABEY,W.R.; LIU,A.T.; CHOU,T.W.; JOHNSON,H.L.; DATA AQUISITION FOR ENVIRONMENTAL TRANSPORT AND FATE SCREENING FOR COMPOUNDS OF INTEREST TO THE OFFICE OF EMERGENCY AND REMEDIAL RESPONCE.; EPA-600/6-84-011. (NTIS PB84-245281). MENLO PARK,CA: SRI INTERNATIONAL. PP.156.; 1984B

JACK, D.B., J.L. HAWKER, L. ROONEY, M. BEERAHEE, J. LOBOAND P. PATEL, "MEASUREMENT OF THE DISTRIBUTION COEFFICIENTS OF SEVERAL CLASSES OF DRUG USING REVERSED-PHASE THIN-LAYER CHROMATOGRAPHY", J. CHROMATOGR., 452, 257-264 (1988).

JAFVERT,C.T.; WESTALL,J.C.; GRIEDER,E.; SCHWARZENBACH,R.P.; DISTRIBUTION OF HYDROPHOBIC IONOGENIC ORGANIC COMPOUNDS BETWEEN OCTANOL AND WATER: ORGANIC ACIDS.; ENVIRON. SCI. TECHNOL.; 24:1795-1803; 1990

JANSEN, A.C.A., H.W. HILBERS, X.R. NI, S.P. VAN HELDEN AND L.H.M. JANSSEN, "SOME PHYSICO-CHEMICAL PROPERTIES OF DANTROLENE AND TWO OF ITS ANALOGUES", INT. J. PHARM., 75(2-3), 193-199 (1991).

JAPAR,S.M.; WALLINGTON,T.J.; ANDINO,J.M.; BALL,J.C.; ATMOSPHERIC CHEMISTRY OF GASEOUS DIETHYL SULFATE.; ENVIRON. SCI. TECHNOL.; 24:894-7; 1990A

JHA, T. ET AL. 1986. HANSCH ANALYSIS OF ANTINEOPLASTIC AGENTS, 5-N-SUBSTITUTED-2-P-SUBSTITUTED-BENZENESULFONYL-L-GLUTAMINES. INDIAN J. CHEM., SECT. B 25B: 169-74.

JIRA, T., B. FLEISCHMANN, G. BURGHARDT, T. BEYRICH, E. MARTIN AND H. KUHMSTEDT, "HPLC STUDIES ON THE DISTRIBUTION BEHAVIOR OF GUANYLIC AND N-PHENYLGUANYLIC HYDRAZONE DERIVATIVES", PHARMAZIE, 40(1), 34-36 (1985).

JONES,A.H.; SUBLIMATION-PRESSURE DATA FOR ORGANIC COMPOUNDS.; J. CHEM. ENG. DATA.; 5:196-200.; 1960

JONES,C.J.; HUDSON,B.C.; MCGUGAN,P.J.; SMITH,A.J.; THE LEACHING OF SOME HALOGENATED ORGANIC COMPOUNDS FROM DOMESTIC WASTE; J HAZARDOUS MATERIALS; 2:227-33; 1978

JONES, G.H., M.C. VENUTI, J.M. YOUNG, D.V. KRISHNAMURTHY, B.E. LEE, R.A. SIMPSON, A.H. BERKS, D.A. SPIRES, P.J. MALONEY, M. KRUSEMAN, S. ROUHAFZA, K.C. KAPPAS, C.C. BEARD, S.H. UNGER AND P.S. CHEUNG, "TOPICAL NONSTEROIDAL ANTISPORIATIC AGENTS. 1. 1,2,3,4-TETRAOXYGENATED NAPHTHALENE DERIVATIVES", J. MED. CHEM., 29(8), 1504-1511 (1986).
JONES, L.A., A.R. MOORMAN, S.D. CHAMBERLAIN, P. DE MIRANDA, D.J. REYNOLDS, C.L. BURNS AND T.A. KRENITSKY, "DI- AND TRIESTER PRODRUGS OF THE VARICELLA-ZOSTER ANTIVIRAL AGENT 6-METHOXYPURINE ARABINOSIDE", J. MED. CHEM., 35(1), 56-63 (1992).

JORDAN, C.G.M., J.M. QUIGLEY AND R.F. TIMONEY, "SYNTHESIS, HYDROLYSIS KINETICS AND LIPOPHILICITY OF O-ACYL ESTERS OF OXPRENOLOL", INT. J. PHARM., 84(2), 175-189 (1992).

JORDAN,L.S.; FARMER,W.J.; GOODIN,J.R.; DAY,B.E.; NONBIOLOGICAL DETOXICATION OF S-TRIAZINE HERBICIDES.; RES. REV.; 32:267-86.; 1970

JORDAN,T.E.; VAPOR PRESSURE OF ORGANIC COMPOUNDS.; INTERSCIENCE PUBLISHERS, INC., 250 FIFTH AVENUE, NEW YORK, NY. PP. 1-256; 1954

JURY,W.A.; SPENCER,W.F.; FARMER,W.J.; USE OF MODELS FOR ASSESSING RELATIVE VOLATILITY, MOBILITY, AND PERSISTENCE OF PESTICIDES AND OTHER TRACE ORGANICS IN SOIL SYSTEMS.; IN: HAZARD ASSESSMENT OF CHEMICALS J., SAXENA ED., NEW YORK, N.; .: ACADEMIC PRESS. 2:1-43.; 1983

KANAZAWA,J.; PREDICTION OF BIOLOGICAL CONCENTRATION POTENTIAL OF PESTICIDES IN AQUATIC ORGANISMS.; REV. PLANT PROT. RES.; 13:27-36; 1980

KANAZAWA,J.; A METHOD OF PREDICTING THE BIOCONCENTRATION POTENTIAL OF PESTICIDES BY USING FISH.; JARQ.; 17:173-9.; 1983

KARYAKIN,N.V.; RABINOVICH,I.B.; PAKHOMOV,L.G.; HEATS OF SUBLIMATION OF NAPHTHALENE AND ITS MONODERIVATIVES.; ZH. FIZ. KHIM.; 42:1814-6.; 1968

KATAYAMA,H.; VAPOR PRESSURES OF METHYL, ETHYL, N-PROPYL, ISOBUTYL AND N-BUTYL BENZOATES AT REDUCED PRESSURES; J CHEM ENG DATA; 33:75-7; 1988

KAWAMOTO,K.;URANO,K.; PARAMETERS FOR PREDICTING FATE OF ORGANOCHLORINE PESTICIDES IN THE ENVIRONMENT. (I) OCTANOL-WATER AND AIR-WATER PARTITION COEFFICIENTS.; CHEMOSPHERE; 18:1987-96; 1989

KEARNEY,P.C.; KAUFMAN,D.D.; HERBICIDES: CHEMISTRY, DEGRADATION AND MODE OF ACTION VOL 1; MARCEL DEKKER, INC: NY,NY 2CD ED; 1975

KEARNEY,P.C.; KAUFMAN,D.D.; HERBICIDES: CHEMISTRY, DEGRADATION AND MODE OF ACTION VOL 2; MARCEL DEKKER, INC: NY,NY 2CD ED; 1975A

KEMP,M.D.; GOLDHAGEN,S.; ZIHLMAN,F.A.; VAPOR PRESSURES AND CRYOSCOPIC DATA FOR SOME ALIPHATIC DINITROXY AND TRNITROXY COMPOUNDS.; J. PHYS. CHEM.; 61:240-2.; 1957

KENAGA,E.E.; PREDICTED BIOCONCENTRATION FACTORS AND SOIL SORPTION COEFFICIENTS OF PESTICIDES AND OTHER CHEMICALS.; ECOTOXICOL. ENVIRON. SAFETY.; 4:26-38.; 1980

KERTES,A.S.; HYDROCARBONS WITH WATER AND SEAWATER PART II. HYDROCARBONS C8 TO C31; IN: SOLUBILITY DATA SERIES VOL 38; SHAW,PC EDS; PERGAMON PRESS,UK:553PP; 1989

KERR, S.G. AND T.I. KALMAN, "HIGHLY WATER-SOLUBLE LIPOPHILIC PRODRUGS OF THE ANTI-HIV NUCLEOSIDE ANALOGUE 2',3'-DIDEOXYCYTIDINE AND ITS 3'-FLUORO DERIVATIVES", J. MED. CHEM., 35(11), 1996-2001 (1992).

KIECKBUSCH,T.G.; KING,C.J.; AN IMPROVED METHOD OF DETERMINING VAPOR-LIQUID EQUILIBRIA FOR DILUTE ORGANICS IN AQUEOUS SOLUTION.; J. CHROMAT. SCI.; 17:273-6; 1979

KIKKOJI, T., M. GUMBLETON, N. HIGO, R.H. GUY AND L.Z. BENET, "PERCUTANEOUS PENETRATION KINETICS OF NITROGLYCERIN AND ITS DINITRATE METABOLITES ACROSS HAIRLESS MOUSE SKIN IN VITRO", PHARM. RES., 8(10), 1231-1237 (1991).

KILZER,L.; SCHEUNERT,I.; GEYER,H.; KLEIN,W.; KORTE,F.; LABORATORY SCREENING OF THE VOLATILIZATION RATES OF ORGANIC CHEMICALS FROM WATER AND SOIL.; CHEMOSPHERE.; 8:751-61.; 1979

KIM, C.K., M.-S. HONG, Y.-B. KIM AND S.-K. HAN, "EFFECT OF PENETRATION ENHANCERS (PYRROLIDONE DERIVATIVES) ON MULTILAMELLAR LIPOSOMES OF STRATUM CORNEUM LIPID: A STUDY BY UV SPECTROSCOPY AND DIFFERENTIAL SCANNING CALORIMETRY", INT. J. PHARM., 95(1-3), 43-50 (1993).

KIMURA, R. ET AL. 1991. RELATIONSHIP BETWEEN NASAL ABSORPTION AND PHYSICOCHEMICAL PROPERTIES OF QUATERNARY AMMONIUM COMPOUNDS. ARCH. INT. PHARMACODYN. THER. 310: 13-21

KIM,Y.H.; WOODROW,J.E.; SEIBER,J.N.; EVALUATION OF A GAS CHROMATOGRAPHIC METHOD FOR CALCULATING VAPOR PRESSURES WITH ORGANOPHOSPHORUS PESTICIDES.; J. CHROMATOGR.; 314:37-53.; 1984

KIRCHNER, J.J. ET AL. 1985. OCTANOL-WATER PARTITION COEFFICIENTS OF SUBSTITUTED A,N-DIPHENYLNITRONES AND BENZONITRILE N-OXIDES. J. PHARM. SCI. 74: 1129-30.

KLAUSCHENZ, E., A. RUMLER, V. HAGEN, A. HAGEN, S. HEER, R. MITZNER AND H. NIEDRICH, "POTENTIAL CARDIOTONIC AGENTS. 2. SYNTHESIS AND PHARMACOLOGICAL PROPERTIES OF 5-(4-PYRIDINYL)- AND 5-PHENYL SUBSTITUTED 3-CYANO-6-METHYL-2-OXAALKYLAMINOPYRIDINES", PHARMAZIE, 44(1), 23-25 (1989).

KLEEMANN,A ET AL.. IN: ULLMANN'S ENCYCLOPEDIA OF INDUSTRIAL CHEMICALS. 5TH ED., VOL. A2:63.; 1985

KLEIN,R.G.; CALCULATIONS AND MEASUREMENTS ON THE VOLATILITY OF N-NITROSAMINES AND THEIR AQUEOUS SOLUTIONS.; TOXICOLOGY; 23:135-47; 1982

KOLSET,K.; HEIBERG,A.; EVALUATION OF THE 'FUGACITY' (FEQUM) AND THE 'EXAMS' CHEMICAL FATE AND TRANSPORT MODELS: A CASE STUDY ON THE POLLUTION OF THE NORRSUNDET BAY (SWEDEN); WATER SCI TECHNOL; 20:1-12; 1988

KONTOGHIORGHES, G.J., "STRUCTURE/RED BLOOD CELL PERMEABILITY. ACTIVITY OF IRON(III) CHELATOR COMPLEXES", INORG. CHIM. ACTA, 151(2), 101-106 (1988).

KORENMAN,Y.I.; POLUMESTNAYA,E.I.; TISHCHENKO,E.M.; ISOMERISM AND SOLUBILITY OF HYDROXYNAPHTHALENES.; ZH. FIZ. KHIM.; 54:1232-4.; 1980

KORENMAN, Y.I. AND V.N. DANILOV, "EXTRACTION OF ALKYL SALICYLATES FROM AQUEOUS MEDIA", ZH. PRIKL. KHIM. (LENINGRAD), 63(1), 125-129 (1990).

KORTH, M. AND J. ENGELS. 1987. INOTROPIC AND ELECTROPHYSICAL EFFECTS OF 8-SUBSTITUTED CYCLIC AMP ANALOGS ON GUINEA PIG PAPILLARY MUSCLE. NAUNYN-SCHMIEDEBERG'S ARCH. PHARMACOL. 335: 77-85.

KORTUM,G ET AL.; PURE AND APPLIED CHEMISTRY, VOL. 1, NO. 2-3, 1961.

KOSANOVIC, D., D. DUMANOVIC AND J. JOVANOVIC, "SOLUBILITY IN OCTANOL AND THE OCTANOL-WATER PARTITION COEFFICIENTS OF SOME 4(5)- AND 5-NITROIMIDAZOLES", J. SERB. CHEM. SOC., 53(10), 559-563 (1988).

KOSOLAPOFF,G.M.; VAPOUR PRESSURE AND DENSITIES OF SOME LOWER ALKYLPHOSPHONATES.; J. CHEM. SOC.; P. 2964-5; 1955

KRAMARCZYK, K. 1987. ABOUT THE DISTRIBUTION BEHAVIOUR OF THE ANTINEOPLASTIC COMPOUND AMBAZONE AND CONGENERIC DERIVATIVES IN AN N-OCTANOL-WATER SYSTEM. STUD. BIOPHYS. 117: 49-53.

KRAMER, C.R. AND L. BECK, "DISTRIBUTION PROPERTIES OF AROMATICS IN THE SYSTEM N-OCTANOL/WATER. 2. DISTRIBUTION OF 4-ARYLSEMICARBAZIDES", Z. CHEM., 21(6), 228-229 (1981).

KRAMER, C.R. AND U. HENZE, "PARTITIONING PROPERTIES OF BENZENE DERIVATIVES. I. TEMPERATURE DEPENDENCE OF THE PARTITIONING OF MONOSUBSTITUTED BENZENES AND NITROBENZENES IN THE N-OCTANOL/WATER SYSTEM", Z. PHYS. CHEM. (LEIPZIG), 271(3), 503-513 (1990).

KRAWCHUK,B.P.; WEBSTER,G.R.B.; MOVEMENT OF PESTICIDES TO GROUND WATER IN AN IRRIGATED SOIL; WATER POLLUT RES J CAN; 22:129-46; 1987

KRIKORIAN,S.E.; CHORN,T.A.; KING,J.W.; DETERMINATION OF O/W PARTITION COEFFICIENTS OF CERTAIN ORGANOPHOSPHORUS COMPOUNDS USING HIGH PERFORMANCE LIQUID CHROMATOGRAPHY.; QUAT. STRUCT. ACT. RELAT.; 6:65-70; 1987

KRIL, M.B. AND H.-L. FUNG, "INFLUENCE OF HYDROPHOBICITY ON THE ION EXCHANGE SELECTIVITY COEFFICIENTS OF AROMATIC AMINES", J. PHARM. SCI., 79(5), 440-443 (1990).

KRISTL, A., G. VESNAVER, A. MRHAR AND F. KOZJEK, "EVALUATION OF PARTITIONING AND SOLUBILITY DATA FOR SOME GUANINE DERIVATIVES IN TERMS OF MUTUAL MISCIBILITY OF OCTANOL AND WATER PHASES", PHARMAZIE, 48(8), 608-610 (1993).

KUCHAR, M., E. KRAUS, M. JELINKOVA, V. REJHOLEC AND V. MILLER, "THIN-LAYER AND HIGH-PERFORMANCE LIQUID CHROMATOGRAPHY IN THE EVALUATION OF THE LIPOPHILICITY OF ARYLOXO ALKANOIC AND ARYLHYDROXY ALKANOIC ACIDS", J. CHROMATOGR., 347, 335-342 (1985).

KUDCHADKER,A.P.; KUDCHADKER,S.A.; SHUKLA,R.P.; PATNAIK,P.R.; VAPOR PRESSURES AND BOILING POINTS OD SELECTED HALOMETHANES.; J. PHYS. CHEM. REF. DATA.; 8:499-517.; 1979

KUHNE,R.; EBERT,R.U.; KLEINT,F.; SCHMIDT,G.; SCHUURMANN,G.; GROUP CONTRIBUTION METHODS TO ESTIMATE WATER SOLUBIITY OF ORGANIC CHEMICALS.; CHEMOSPHERE; 30:2061-77; 1995

KULEVSKY,N.; WANG,C.T.; STENBERG,V.I.; PHOTOCHEMICAL OXIDATIONS. II. RATE AND PRODUCT FORMATION STUDIES ON THE PHOTOCHEMICAL OXIDATION OF ETHERS.; J. ORG. CHEM.; 34:1345-8.; 1969

KUNDEL,H.; LILLE,V.; KAIDAS,N.; DETERMINATION OF THE SATURATED VAPOR PRESSURE FOR ALKYL RESORCINOLS AND THEIR METHYL ETHERS.; TR. TALLIN. POLITEKH. INST.; 390:107-15.; 1975

KUO,H.W.; WANG,J.D.; LIN,J.M.; DETERMINATION OF VAPOR PRESSURE FOR
4,4'-BIPYRIDINE BY GAS SATURATION AND GAS CHROMATOGRAPHIC METHODS.; CHEMOSPHERE;
24:1679-86; 1992

KUROIWA,S.; OGASAWARA,S.; DISPERSED STATE OF DYES AND THEIR DYEING PROPERTIES.
VIII. SOLUBILITIES OF DISPERSE DYES IN WATER.; NIPPON KAGAKU KAISHI.
PP.1738-43.; 1973

KURTZ, A.P. AND J.A. DURDEN, "QUANTITATIVE STRUCTURE-ACTIVITY RELATIONSHIPS IN INSECTICIDAL OXATHIOLANE
AND DITHIOLANE OXIME CARBAMATES AND RELATED COMPOUNDS", J. AGRIC. FOOD CHEM., 35(1), 115-121 (1987).

KWOK,E.S.C.; AREY,J.; ATKINSON,R.; GAS-PHASE ATMOSPHERIC CHEMISTRY OF
DIBENZO-P-DIOXIN AND DIBENZOFURAN.; ENVIRON. SCI. TECHNOL.; 28:528-33; 1994A

LAHANN, T.R., L.J. DEKREY AND B.D. TARR, "CAPSAICIN ANALGESIA: PREDICTIONS BASED ON PHYSICOCHEMICAL
PARAMETERS", PROC. WEST. PHARMACOL. SOC., 32, 201-204 (1989).

LAM, S.P., D.J. BARLOW AND J.W. GORROD, "CONFORMATIONAL ANALYSIS OF 9-SUBSTITUTED ADENINES IN RELATION TO
THEIR MICROSOMAL N1-OXIDATION", J. PHARM. PHARMACOL., 41(6), 373-378 (1989).

LANDE,S.S.; DURKIN,P.R.; CHRISTOPHER,D.H.; HOWARD,P.H.; SAXENA,J.; INVESTIGATION
OF SELECTED POTENTIAL ENVIRONMENTAL CONTAMINANTS: KETONIC SOLVENTS.;
EPA-560/2-76-003. WASHINGTON,DC: U.S. EPA, OFFICE OF TOXIC SUBSTANCES. PP.331.;
1976

LANDE,S.S.; BOGYO,D.A.; HOWARD,P.H.; SANTODONATO,J.; MEYLAN,W.M.; INVESTIGATION
OF SELECTED POTENTIAL ENVIRONMENTAL CONTAMINANTS: HALOALCOHOLS.;
EPA-560/11-80-004. WASHINGTON,DC: U.S.EPA. PP.163.; 1980

LANDE,S.S.; HAGEN.D.F.; SEAVER,A.E.; COMPUTATION OF TOTAL MOLECULAR SURFACE
AREAS FROM GAS PHASE ION MOBILITY DATA AND IT'S CORRELATION WITH AQUEOUS
SOLUBILITIES OF HYDROCARBONS.; ENVIRON. TOXICOL. CHEM.; 4:325-34.; 1985

LANDE,S.S.; ELNABARAWY,M.T.; REINER.E.A.; WELTER,A.N.; ROBIDEAU,R.R.; LABORATORY
ASSESSMENT OF ENVIRONMENTAL IMPACT OF PHTHALAZINE.; BULL. ENVIRON. CONTAM.
TOXICOL.; 38:332-6; 1987

LAPKIN,M.; EPOXIDES.; IN: KIRK-OTHMER ENCYCL. CHEM TECH. 2ND ED. 8:263-93.;
1965

LARSEN, J.D., H. BUNDGAARD AND V.H.L. LEE, "PRODRUG FORMS FOR THE SULFONAMIDE GROUP. II. WATER-SOLUBLE
AMINOACID DERIVATIVES OF N-METHYLSULFONAMIDES AS POSSIBLE PRODRUGS", INT. J. PHARM., 47(1-3), 103-110 (1988).

LAWRENCE,A.H.; ELIAS,L.; AUTHIER-MARTIN,M.; DETERMINATION OF AMPHETAMINE,
COCAINE, AND HEROIN VAPOR PRESSURES USING A DYNAMIC GAS BLENDING SYSTEM AND GAS
CHROMATOGRAPHIC ANALYSIS.; CAN. J. CHEM.; 62:1886-8.; 1984

LAZNICEK, M., J. KVETINA, J. MAZAK AND V. KRCH, "PLASMA PROTEIN BINDING-LIPOPHILICITY RELATIONSHIPS:
INTERSPECIES COMPARISON OF SOME ORGANIC ACIDS", J. PHARM. PHARMACOL., 39(2),79-83 (1987).

LAZNICEK, M. AND J. KVETINA, "THE EFFECT OF MOLECULAR STRUCTURE ON THE DISTRIBUTION OF SOME ORGANIC
ACIDS IN RATS", QUANT. STRUCT.-ACT. RELAT., 7(4), 234-239 (1988).

LEAHY, D.E., A.L.J. DE MEERE, A.R. WAIT, P.J. TAYLOR AND J.A. TOMENSON, "A GENERAL DESCRIPTION OF WATER-OIL
PARTITIONING RATIOS USING THE ROTATING DIFFUSION CELL", INT. J. PHARM., 50(2) , 117-132 (1989).

LEGGETT,D.C.; PERSISTENCE OF CHEMICAL AGENTS ON THE WINTER BATTLEFIELD PART 1.
LITERATURE REVIEW AND THEORETICAL EVALUATION; IN: DEFENSE TECHNICAL INFO CENTER
US ARMY COLD REGIONS RES ENG LAB; CRREL REPORT # 87-12; 1987

LEIFER,A.; DETERMINATION OF RATES OF REACTION IN THE GAS-PHASE IN THE
TROPOSHPERE. THEORY AND PRACTICE. 5.; EPA-744/R-93-001. WASHINGTON,DC: USEPA.
OFF. POLLUT. PREVENT. TOXICS. PP. 141.; 1993

LEHNER, S.J., B.W. MULLER AND J.K. SEYDEL, "INTERACTIONS BETWEEN P-HYDROXYBENZOIC ACID ESTERS AND

HYDROXYPROPYL-BETA-CYCLODEXTRIN AND THEIR ANTIMICROBIAL EFFECT AGAINST CANDIDA ALBICANS", INT. J. PHARM., 93(1-3), 201-208 (1993).

LEIGHTON,D.T.JR.; CALO,J.M.; DISTRIBUTION COEFFICIENTS OF CHLORINATED HYDROCARBONS IN DILUTE AIR-WATER SYSTEMS FOR GROUNDWATER CONTAMINATION APPLICATIONS; J CHEM ENG; 26:382-5; 1981

LEISTRA,M.; THE DISTRIBUTION OF 1,3-DICHLOROPROPENE OVER THE PHASES IN SOIL.; J. AGR. FOOD CHEM.; 18:1124-6; 1970

LEUENBERGER,C.; LIGOCKI,M.P.; PANKOW,J.F.; TRACE ORGANIC COMPOUNDS IN RAIN. 4. IDENTITIES, CONCENTRATIONS, AND SCAVENGING MECHANISMS FOR PHENOLS IN URBAN AIR AND RAIN.; ENVIRON. SCI. TECHNOL.; 19:1053-8; 1985A

LEUNG,H.W.; PAUSTENBACH,D.J.; ORGANIC ACIDS AND BASES: REVIEW OF TOXICOLOGICAL STUDIES.; AM. J. IND. MED.; 18:717-35; 1990

LEVANOVA,S.V.; GARTER,T.E.; ANDREEVSKI,D.N.;; VAPOR PRESSURES AND APPROXIMATE VALUES OF THERMODYNAMIC FUNCTIONS OF LIQUID ALKYL HALIDES.; NEFTEKHIMIYA.; 7:286-92.; 1967

LEWIS, RJ. HAWLEY'S CONDESNED CHEMICAL DICTIONARY. 12TH ED. NEW YORK, NY: VAN NOSTRAND REINHOLD COMPANY. (1993).

LEYDER,F.; BOULANGER,P.; ULTRAVIOLET ABSORPTION, AQUEOUS SOLUBILITY, AND OCTANOL-WATER PARTITION FOR SEVERAL PHTHALATES.; BULL. ENVIRON. CONTAM. TOXICOL.; 30:152-7.; 1983

LI, J. AND E.M. PERDUE. 1995. PHYSICOCHEMICAL PROPERTIES OF SELECTED MONOTERPENES. PRE-PRINT EXTENDED ABSTRACT, PRESENTED BEFORE THE DIVISION OF ENVIRONMENTAL CHEMISTRY, AMER. CHEM. SOC., ANAHEIM, CA, APRIL 2-7, 1995.

LI,J.; PERDUE,E.M.; PHYSICOCHEMICAL PROPERTIES OF SELECTED MONOTERPENES.; PREPRINTS OF PAPERS PRESENTED AT THE 209TH ACS NATIONAL MEETING; ANAHEIM, CA; APRIL 2-7, 1995; 35(1):134-7.; 1995

LI,J.C.M.; ROSSINI,F.D.; VAPOR PRESSURES AND BOILING POINTS OF THE 1-FLUOROALKANES, 1-CHLOROALKANES, 1-BROMOALKANES, AND 1-IODOALKANES, C1 TO C20.; J. CHEM. ENG. DATA; 6:268-70; 1953

LI, N.Y. AND J.W. GORROD, "DETERMINATION OF PARTITION COEFFICIENTS AND IONISATION CONSTANTS OF (S)(-)-NICOTINE AND CERTAIN METABOLITES", MED. SCI. RES., 20(23), 901-902 (1992).

LIANG, W.Q. AND W. LIN, "THE EFFECT OF MOLECULAR VOLUME AND PARTITION COEFFICIENT ON PERCUTANEOUS ABSORPTION", YAOXUE XUEBAO, 27(9), 684-689 (1992).

LIDE, D.R.; CRC HANDBOOK OF CHEMISTRY AND PHYSICS 75TH ED; CRC PRESS; BOCA RATON, FLA; 1994

LIN, T.S., "SYNTHESIS AND IN VITRO ANTIVIRAL ACTIVITY OF 3'-O-ACYLDERIVATIVES OF 5'-AMINO-5'-DEOXYTHYMIDINE: POTENTIAL PRODRUGS FOR TOPICAL APPLICATION", J. PHARM. SCI., 73(11), 1568-1571 (1984).

LIND,J.A.; KOK,G.L.; HENRY'S LAW DETERMINATION FOR AQUEOUS SOLUTIONS OF HYDROGEN PEROXIDE, METHYLHYDROPEROXIDE, AND PEROXYACETIC ACID.; J. GEOPHYS. RES.; 91:7889-95; 1986

LINDGAARD-JORGENSEN,P AND JACOBSEN,BN (1986); A DATABASE ON BEHAVIOR AND EFFECTS OF ORGANIC POLLUTANTS IN WASTE WATER TREATMENT PROCESSES; IN: COMM. EUR. COMMUNITIES, EUR 10388. ORG. MICROPULLUT. AQUAT. ENVIRON. PP. 429-439.

LION,L.W.; GARBARINI,D.; PARTITIONING EQUILIBRIA OF VOLATILE POLLUTANTS IN THREE-PHASE SYSTEMS.; AFESC/ESL-TR-83-51. (NTIS PBAD-A137 207/7). ST. CLOUD,FL: SOUTHEASTERN CENTER FOR ELECTRICAL ENGINEERING EDUCATION, INC. PP.34.; 1983

LIPNICK,R.L.; BICKINGS,C.K.; JOHNSON,D.E.; EASTMOND,D.A.; COMPARISON OF QSAR PREDICTIONS WITH FISH TOXICITY SCREENING DATA FOR 110 PHENOLS.; IN: ASTM SPEC.

PUBL. 891(AQUAT. TOXICOL. HAZ. ASSESS. 8TH SYMP.):153-176; 1986

LIPINSKI,C.A.; FIESE,E.F.; KORST,R.J.; PKA, LOG P AND MEDCHEM CLOGP FRAGMENT VALUES OF ACIDIC HETEROCYLIC POTENTIAL BIOISOSTERES.; QUANT. STRUCT. ACT. RELAT.; 10:109-17; 1991

LIPINSKI, C.A., J.L. LAMATTINA AND L.A. HOHNKE, "PSEUDOSYMMETRY AND BIOISOSTERISM IN BIARYL PYRIDYL COMPETITIVE HISTAMINE H2-RECEPTOR ANTAGONISTS", J. MED. CHEM., 28(11), 1628-1636 (1985).

LISCIANI, R., S. LEMBO, S. COZZOLINO, C. SILIPO AND A. VITTORIA, "STRUCTURE-ANALGESIC ACTIVITY RELATIONSHIPS IN A SET OF 2-AMINOBENZAMIDE DERIVATIVES", FARMACO, ED. SCI., 41(2), 89-102 (1986).

LITTLEWOOD,R.; VAPOR PRESSURES OF SOME SOLID ORGANIC COMPOUNDS.; J. CHEM. SOC. PP.2419-20.; 1957

LONG,L.H.; CATTANACH,J.; ANTOINE VAPOR-PRESSURES EQUATINS AND HEATS OF VAPORIZATION FOR THE DIMETHYLS OF ZINC, CADMIUM, AND MERCURY.; J. INORG. NUCLEAR. CHEM.; 20:340-2.; 1961

LOVELL,R.J.; KEY,J.A.; STANDIFER,R.L.; KALCEVIC,V.; LAWSON,J.F.; DYLEWSKI,S.W.; SCHOMER,T.L.; ORGANIC CHEMICAL MANUFACTURING. VOLUME 9. SELECTED PROCESSES.; EPA-450/3-80-028D. USEPA OFF. AIR QUAL. PLANN. STAND., WASHINGTON,DC: PP.545; 1980

LUKASIAK J. 1984. STUDIES ON 1-ARYLOXYSILATRANES, III. ACTA CHIM. HUNG. 117: 271-86.

LYMAN,W.J.; REEHE,W.F.; ROSENBLATT,D.H.; HANDBOOK OF CHEMICAL PROPERTY ESTIMATION METHODS. ENVIRONMENTAL BEHAVIOR OF ORGANIC COMPOUNDS.; NEW YORK,NY: MCGRAW-HILL BOOK CO. PP.960.; 1982

LYMAN,W.J.; REEHL,W.F.; ROSENBLATT,D.H.; HANDBOOK OF CHEMICAL PROPERTY ESTIMATION METHODS.; AMERICAN CHEMICAL SOCIETY. WASHINGTON, DC. 2ND PRINTING.; 1990

MABEY,W.R.; SMITH,J.H.; PODOLL,R.T.; JOHNSON,H.L.; MILL,T.; CHOU,T.W.; GATES,J.; PARTRIDGE,I.W.; VANDENBERG,D.; AQUATIC FATE PROCESS DATA FOR ORGANIC PRIORITY POLLUTANTS.; EPA-440/4-81-014. WASHINGTON,DC: U.S. EPA. PP.434.; 1981

MACALUSO,P.; CARBON SULFIDES AND THEIR INORGANIC DERIVATIVES; IN: KIRK-OTHMER ENCYCL CHEM TECH 2ND; NY,NY: WILEY-INTRSCI 19:371-424; 1969

MACINTYRE,W.G.; DEFUR,P.O.; THE EFFECT OF HYDROCARBON MIXTURES ON ADSORPTION OF SUSTITUTED NAPHTHALENES BY CLAY AND SEDIMENT FROM WATER.; CHEMOSPHERE.; 14:103-11.; 1985

MACKAY,D.; SHIU,W.Y.; AQUEOUS SOLUBILITY OF POLYNUCLEAR AROMATIC HYDROCARBONS.; J. CHEM. ENG. DATA.; 22:399-402.; 1977

MACKAY,D.; SHIU,W.Y.; A CRITICAL REVIEW OF HENRY'S LAW CONSTANTS FOR CHEMICALS OF ENVIRONMENTAL INTEREST.; J. PHYS. CHEM. REF. DATA.; 19:1175-99.; 1981

MACKAY,D.; SHIU,W.Y.; SUTHERLAND,R.P.; DETERMINATION OF AIR-WATER HENRY'S LAW CONSTANTS FOR HYDROPHOBIC POLLUTANTS.; ENVIRON. SCI. TECHNOL.; 13:333-6.; 1979

MACKAY,D.; BOBRA,A.; SHIU,W.Y.; YALKOWSKY,S.H.; RELATIONSHIPS BETWEEN AQUEOUS SOLUBILITY AND OCTANOL-WATER PARTITION COEFFICIENTS.; CHEMOSPHERE.; 9:701-11.; 1980

MACKAY,D.; BOBRA,A.; CHAN,D.W.; SHIU,W.Y.; VAPOR PRESSURE CORRELATIONS FOR LOW-VOLATILITY ENVIRONMENTAL CHEMICALS.; ENVIRON. SCI. TECHNOL.; 16:645-9.; 1982

MACKAY,D.; SHIU,W.Y.; BOBRA,A.; BILLINGTON,J.; CHAU,E.; YEUN,A.; NG,C; SZETO,F.; VOLATILIZATION OF ORGANIC POLLUTANTS FROM WATER.; EPA-600/53-82-019. (NTIS PB 82-230 939). ATHENS,GA: US EPA.; 1982A

MACKAY,D.; PATERSON,S.; SCHROEDER,W.H.; MODEL DESCRIBING THE RATES OF TRANSFER PROCESSES OF ORGANIC CHEMICALS BETWEEN ATMOSPHERE AND WATER.; ENVIRON. SCI. TECHNOL.; 20:810-6; 1986

MACKNICK,A.B.; PRAUSNITZ,J.M.; VAPOR PRESSURES OF HIGH-MONECULAR-WEIGHT HYDROCARBONS.; J. CHEM. ENG. DATA.; 24:175-8.; 1979

MAJURY,T.G.; VAPOR PRESSURES AND HEATS OF SUBLIMATION OF P-NITROANILINE, N,N-DIMETHYL-P-NITROANILINE, P-AMINOAZOBENZENE AND N,N-DIMETHYL-P-AMINOAZOBENZENE.; CHEMISTRY INDUSTRY; PP. 349-50; 1956

MAKSIMOV,Y.Y.; VAPOR PRESSURES OF AROMATIC NITRO COMPOUNDS AT VARIOUS TEMPERATURES.; ZH. FIZ. KHIM.; 42:2921-5.; 1968

MALASPINA,L.; GIGLI,R.; BARDI,G.; MICROCALORIMETRIC DETERMINATION OF THE ENTHALPY OF SUBLIMINATION OF BENZOIC ACID AND ANTHRACENE.; J. CHEM. PHYS.; 59:387-94.; 1973

MALASPINA,L.; BARDI,G.; GIGLI,R.; SIMULTANEOUS MICROCALORIMETRIC-KNUDSEN EFFUSION DETERMINATION OF THE SUBLIMATION ENTHALPY AND VAPOR PRESSURE OF 1,2- AND 1,4-DIHYDROXYANTHRAQUINONES.; J. CHEM. THERMODYN.; 5:845-50.; 1973A

MALASPINA,L.; GIGLI,R.; BARDI,G.; DEMARIA,G.; SIMULTANEOUS DETERMINATION BY KNUDSEN EFFUSION MICROCALORIMETRY OF THE VAPOR PRESSURE AND THE ENTHALPY OF SUBLIMATION OF P- AND M-NITROANILINE.; J. CHEM. THERMODYN.; 5:699-706.; 1973B

MANNHOLD, R., K.P. DROSS AND R.F. REKKER, "DRUG LIPOPHILICITY IN QSAR PRACTICE. I. A COMPARISON OF EXPERIMENTAL WITH CALCULATIVE APPROACHES", QUANT. STRUCT.-ACT. RELAT., 9(1), 21-28 (1990).

MARKERT, F. AND NIELSEN, O.J. 1992. CHEM. PHYS. LETTERS 189: 171-4.

MARPLE,L.; BRUNCK,R.; THROOP,L.; WATER SOLUBILITY OF 2,3,7,8-TETRACHLORODIBENZO-O-DIOXIN.; ENVIRON. SCI. TECHNOL.; 20:180-2; 1986

MARREL, C., G. BOSS, H. VAN DE WATERBEEMD, B. TESTA, D. COOPER, P. JENNER AND C.D. MARSDEN, "L-DOPA ESTERS AS POTENTIAL PRODRUGS. I. PHYSICOCHEMICAL PROPERTIES", EUR. J. MED. CHEM., 20(5), 459-465 (1985).

MARTIN,H.; WORTHING,C.R.; PESTICIDE MANUAL. FIFTH EDITION, BRITISH CROP PROTECTION COUNCIL. 563 PP.; 1977

MARTIN-VILLODRE, A., J.M. PLA-DELFINA, J. MORENO, M.D. PEREZ-BUENDIA, J. MIRALLES, E.F. COLLADO, E. SANCHEZ-MOYANO AND A. DEL POZO, "STUDIES ON THE RELIABILITY OF A BIHYPERBOLIC FUNCTIONAL ABSORPTION MODEL. I. RING-SUBSTITUTED ANILINES", J. PHARMACOKINET. BIOPHARM., 14(6), 615-633 (1986).

MARTISKA, A. AND V. BEKAREK. 1990. APPLICATION OF THE EFFECTIVE BORN'S RELATIVE PERMITTIVITY FUNCTIONS FOR THE EVALUATION OF POLARITY EFFECTS OF SOLUTES ON THE OCTANOL-WATER PARTITION COEFFICIENTS. ACTA UNIV. PALACKI OLOMUC., FAC. RERUM NAT. 97(CHEM. 29): 63-7.

MARUTHAMUTHU,P.; HUIE,R.E.; FERRIC ION ASSISTED PHOTOOXIDATION OF HALOACETATES.; CHEMOSPHERE; 30:2199-207; 1995

MARUYAMA, T., M.A. FURUIE, S. HIBINO AND M. OTAGIRI, "COMPARATIVE STUDY OF INTERACTION MODE OF DIAZEPINES WITH HUMAN SERUM ALBUMIN AND ALPHA1-ACID GLYCOPROTEIN", J. PHARM. SCI., 81(1), 16-20 (1992).

MASSALDI,H.A.; KING,C.J.; SIMPLE TECHNIQUE TO DETERMINE SOLUBILITIES OF SPARINGLY SOLUBLE ORGANICS. SOLUBILITY AND ACTIVITY COEFFICIENTS OF D-LIMINONE, BUTYLBENZENE, AND N-HEXYL ACETATE IN WATER AND SUCROSE SOLUTIONS.; J. CHEM. ENG. DATA.; 18:393-7.; 1973

MATSUGUMA,H.J.; NITROPHENOLS.; IN: KIRK-OTHMER ENCYCL. CHEM. TECH. 2ND. ED. 13:888-94.; 1963

MATSUDA, H., T. FURUYA, T. ICHIMURA AND Y. KOBAYASHI, "DETERMINATION OF PARTITION COEFFICIENTS OF A SERIES OF CEPHALOSPORINS BY HIGH-PERFORMANCE LIQUID CHROMATOGRAPHY", YAMANOUCHI SEIYAKU KENKYU HOKOKU, 4, 107-111 (1980).

MAULEON, D., M. DOLORS PUJOL AND G. ROSELL, "BETA-ADRENERGIC ANTAGONISTS: N-ALKYL AND N-AMIDOETHYL(ARYLOXY) PROPANOLAMINES RELATED TO PROPANOLOL", EUR. J. MED. CHEM., 23(5), 421-426 (1988).

MAY,W.E.; WASIK,S.P.; FREEMAN,D.H.; DETERMINATION OF THE SOLUBILITY BEHAVIOR OF SOME POLYCYCLIC AROMATIC HYDROCARBONS IN WATER.; ANAL. CHEM.; 50:997-1000.; 1978

MAY,W.E.; WASIK,D.P.; MILLER,M.M.; TEWARI,Y.B.; BROWN-THIMA,J.M.; GOLDBERG,R.N.; SOLUTION THERMODYNAMICS OF SOME SLIGHTLY SOLUBLE HYDROCARBONS IN WATER.; J. CHEM. REF. DATA.; 28:197-200.; 1983

MAYER,F.L.; ADAMS,W.J.; FINLEY,M.T.; MICHAEL,P.R.; MEHRLE,P.M.; SAEGER,V.W.; PHOSPHATE ESTER HYDRAULIC FLUIDS: AN AQUATIC ENVIRONMENTAL ASSESSMENT OF PYDRAULS 50E AND 115E.; IN: ASTM STP 737. AQUATIC TOXICOLOGY AND HAZARD ASSESSMENT. 4TH CONF. BRANSON,D.R. & DICKSON,K.L. EDS. PHILADELPHIA, PA. P.103-23.; 1981

MAZUR,R.; SWEETENERS.; IN: KIRK-OTHMER ENCYCL. CHEM. TECH., THIRD ED., GRAYSON,M. ED., JOHN WILEY & SONS, NY, VOL 22 PP. 453-8; 1983

MCAULIFFE,C.; SOLUBILITY IN WATER OF PARAFFIN, CYCLOPARAFFIN, OLEFIN, ACETYLENE, CYCLOOLEFIN AND AROMATIC HYDROCARBON.; J. PHYS. CHEM.; 70:1267-75.; 1966

MCCORMACK,W.B.; LAWES,B.C.; SULFURIC AND SULFUROUS ESTERS; IN: KIRK-OTHMER ENCYCL CHEM TECH 3RD., NY,NY: WILEY INTRSCI PUBL:22:233-54; 1983

MCEACHERN,D.M.; INIGUEZ,J.C.; ORNELAS,H.C.; ENTHALPIES OF COMBUSTION AND SUBLIMATION AND VAPOR PRESSURES OF THREE BENZOQUINOLINES.; J. CHEM. ENG. DATA.; 20:226-8.; 1975

MCKILLIP,W.J.; SHERMAN,E.; FURAN DERIVATIVES.; IN: KIRK-OTHMER ENCYCL. CHEM. TECH. 3RD ED. 11:499-527.; 1978

MCLACHLAN, D. ET AL. 1986. STRUCTURE-FUNCTION RELATIONSHIPS IN THE PHOTOTOXICITY OF ACETYLENES FROM THE ASTERACEAE. BIOCHEM. SYST. ECOL. 14: 17-23.

MCLACHLAN,MS ET AL.; ENVIRON SCI TECHNOL, 29: 1998-20004, 1995.

MEANS,J.C.; WOOD,S.G.; HASSETT,J.J.; BANWART,W.L.; SORPTION OF AMINO- AND CARBOXY-SUBSTITUTED POLYNUCLEAR AROMATIC HYDROCARBONS BY SEDIMENTS AND SOILS.; ENVIRON. SCI. TECHNOL.; 16:93-8.; 1982

MEEKS,A.C.; GOLDFARB,I.J.; VAPOR PRESSURES OF FLUOROALCOHOLS.; J. CHEM. ENG. DATA.; 12:196.; 1967

MELNIKOV,N.N.; CHEMISTRY OF PESTICIDES; RES. REV.; 36:1-11; 1971

MERCK INDEX; THE MERCK INDEX. AN ENCYCLOPEDIA OF CHEMICALS AND DRUGS. 9TH ED.; RAHWAY, NJ:MERCK AND CO., INC.; 1976

MERCK INDEX.; THE MERCK INDEX. AN ENCYCLOPEDIA OF CHEMICALS AND DRUGS 10TH ED.; WINDHOLZ,M ED. RAHWAY, N.J.: MERCK AND CO., INC.; 1983

MERCK INDEX.; THE MERCK INDEX. AN ENCYCLOPEDIA OF CHEMICALS AND DRUGS 10TH ED.; RAHWAY, N.J.: MERCK AND CO., INC.; 1983

MERCK INDEX; THE MERCK INDEX. AN ENCYCLOPEDIA OF CHEMICALS AND DRUGS. 11TH ED.; RAHWAY, NJ:MERCK AND CO., INC.; 1989

MERCK INDEX; THE MERCK INDEX. AN ENCYCLOPEDIA OF CHEMICALS AND DRUGS. 12TH ED.; RAHWAY, NJ:MERCK AND CO., INC.; 1996

METCALF,R.L.; BIOLOGICAL FATE AND TRANSFORMATION OF POLLUTANTS IN WATER; ADV. ENVIRON. SCI. TECHNOL.; 8(FATE POLLUT AIR WATER ENVIRON):195-221; 1977

METCALF,R.L.; INSECT CONTROL TECHNOLOGY.; IN: KIRK-OTHMER ENCYCY. CHEM. TECHNOL. 3RD. ED. NEW YORK,NY: WILEY-INTERSCIENCE 13:467; 1978

METCALF,RL; "INSECT CONTROL" IN: ULLMANN'S ENCYCLOPEDIA OF INDUSTRIAL CHEMICALS. 5TH ED., VOL. A5: 263-320; 1989

METCALFE,C.D.; MCLEESE,D.W.; ZITKO,V.; RATE OF VOLATILIZATION OF FENITROTHION FROM FRESH WATER.; CHEMOSPHERE; 9:151-5; 1980

MEYLAN,W.M.; HOWARD,P.H.; BOND CONTRIBUTION METHOD FOR ESTIMATING HENRY'S LAW CONSTANTS.; ENVIRON. TOXICOL. CHEM.; 10:1283-93; 1991

MEYLAN,W.M.; HOWARD,P.H.; COMPUTER ESTIMATION OF THE ATMOSPHERIC GAS-PHASE REACTION RATE OF ORGANIC COMPOUNDS WITH HYDROXYL RADICALS AND OZONE.; CHEMOSPHERE; 26:2293-99; 1993

MEYLAN, W.M. AND P.H. HOWARD. 1995. ATOM/FRAGMENT CONTRIBUTION METHOD FOR ESTIMATING OCTANOL-WATER PARTITION COEFFICIENTS. J. PHARM. SCI. 84: 83-92.

MEYLAN, W.M. ET AL.. 1996. IMPROVED METHOD FOR ESTIMATING WATER SOLUBILITY FROM OCTANOL/WATER PARTITION COEFFICIENT. ENVIRONMENTAL TOXICOLOGY AND CHEMISTRY 15: 100-106.

MIDWEST RESEARCH INSTITUTE.; SUBSTITUTE CHEMICAL PROGRAM. INITIAL SCIENTIFIC AND MINIECONOMIC REVIEW OF CAPTAN.; EPA/540/1-75-012. WASHINGTON,DC: U.S. EPA. PP.184.; 1975B

MIDWEST RESEARCH INSTITUTE.; ASSESSMENT OF THE NEED FOR THE CHARACTER OF AND THE IMPACT RESULTING FROM LIMITATIONS ON ARYL PHOSPHATES.; MRI 4309-L. KANSAS CITY, MO. MIDWEST RES. INST. 279 P.; 1977

MILL,T.; MABEY,W.; PHOTOCHEMICAL TRANSFORMATIONS.; IN: ENVIRONMENTAL EXPOSURE FROM CHEMICALS. VOL. 1. NEELY,WB & BLAU,GE, EDS. BOCA RATON,FL: CRC PRESS. PP.175-216.; 1985

MILLER,M.M.; GHODBANE,S.; WASIK,S.P.; TEWARI,Y.B.; MARTINE,D.E.; AQUEOUS SOLUBILITIES, OCTANOL/WATER PARTITION COEFFICIENTS, AND ENTROPIES OF MELTING OF CHLORINATED BENZENES AND BIPHENYLS.; J. CHEM. ENG. DATA.; 29:184-90.; 1984

MILLER,M.M.; WASIK,S.P.; HUANG,G.; SHIU,W.; MACKAY,D.; RELATIONSHIPS BETWEEN OCTANOL-WATER PARTITION COEFFICIENTS AND AQUEOUS SOLUBILITY.; ENVIRON. SCI. TECHNOL.; 19:522-9.; 1985

MILLIGAN,B.; GILBERT,K.E.; DIAMINOTOLUENES; IN:KIRK-OTHMER ENCYCLO CHEM TECH 3RD ED NY,NY: WILEY INTERSCI 2:321-9; 1978

MILLS,M.S.; THURMAN,E.M.; REDUCTION OF NONPOINT SOURCE CONTAMINATION OF SURFACE WATER AND GROUNDWATER BY STARCH ENCAPSULATION OF HERBICIDES.; ENVIRON. SCI. TECHNOL.; 28:73-9; 1994A

MIRVISH,S.S.; ISSENBERG,P.; SORNSON,H.C.; AIR-WATER AND ETHER-WATER DISTRIBUTION OF N-NITROSO COMPOUNDS: IMPLICATIONS FOR LABORATORY SAFETY, ANALYTIC METHODOLOGY AND CARCINOGENICITY FOR THE RAT ESOPHAGUS, NOSE AND LIVER; J NATL CANCER INST; 56:1125-9; 1976

MITSUTAKE, K.I., H. IWAMURA, R. SHIMIZU AND T. FUJITA, "QUANTITATIVE STRUCTURE-ACTIVITY RELATIONSHIP OF PHOTOSYSTEM II INHIBITORS IN CHLOROPLASTS AND ITS LINK TO HERBICIDAL ACTION", J. AGRIC. FOOD CHEM., 34(4), 725-732 (1986).

MIYAKE, F., F. KITAURA, N. MIZUNO AND H. TERADA, "DETERMINATION OF PARTITION COEFFICIENT AND ACID DISSOCIATION CONSTANT BY HIGH-PERFORMANCE LIQUID CHROMATOGRAPHY ON POROUS POLYMER GEL AS A STATIONARY PHASE", CHEM. PHARM. BULL., 35(1), 377-388 (1987).

MIYAKE, K., N. MIZUNO AND H. TERADA, "METHOD FOR DETERMINATION OF PARTITION COEFFICIENTS BY HIGH-PERFORMANCE LIQUID CHROMATOGRAPHY ON AN OCTADECYLSILANE COLUMN. EXAMINATION OF ITS APPLICABILITY", CHEM. PHARM. BULL., 34(11), 4787-4796 (1986).

MIYATA,H.;TAKAYAMA,K.; MIMURA,M.;KASHIMOTO,T.;FUKUSHIMA,S.; SPECIFIC CONGENER PROFILES OF POLYCHLORINATED DIBENZO-P-DIOXINS AND DIBENZOFURANS IN BLUE MUSSEL

IN OSAKA BAY IN JAPAN: AQUEOUS SOLUBILITIES OF PCDDS AND PCDFS.; BULL. ENVIRON. CONTAM. TOXICOL.; 43:342-9; 1989A

MJOS,K.; AMINES (CYCLIC).; IN: KIRK-OTHMER ENCYCL. CHEM. TECHNOL. 3RD ED. 2:295-308.; 1978

MOODY,R.P.; CARROLL,J.M.; KRESTA,A.M.E.; AUTOMATED HIGH PERFORMANCE LIQUID CHROMATOGRAPHY AND LIQUID SCINTILLATION COUNTING DETERMINATION OF PESTICIDE MIXTURE OCTANOL/WATER PARTITION RATES.; TOXICOL. IND. HEALTH; 3:479-90; 1987

MOORE,R.M.; GEEN,C.E.; TAIT,V.K.; DETERMINATION OF HENRY'S LAW CONSTANTS FOR A SUITE OF NATURALLY OCCURRING HALOGENATED METHANES IN SEAWATER.; CHEMOSPHERE.; 30:1183-91.; 1995

MOORE,W.M.; METHYLENEDIANILINE; IN: KIRK-OTHMER ENCYCLOPEDIA OF CHEMICAL TECHNOLOGY. 3RD ED. NEW YORK: WILEY & SONS 2:338-48; 1978

MONSANTO MATERIAL SAFETY DATA SHEET;   NA--

MORRIS, J.J., L.R. HUGHES, A.T. GLEN AND P.J. TAYLOR, "NON-STEROIDAL ANTI-ANDROGENS. DESIGN OF NOVEL COMPOUNDS BASED ON AN INFRARED STUDY OF THE DOMINANT CONFORMATION AND HYDROGEN-BONDING PROPERTIES OF A SERIES OF ANILIDE ANTI-ANDROGENS", J. MED. CHEM., 34(1), 447-455 (1991).

MUELLER,J.A.; TORO,D.M.D.; MAIELLO,J.A.; FATE OF OCTAMETHYLCYCLOTETRASILOXANE (OMCTS) IN THE ATMOSPHERE AND IN SEWAGE TREATMENT PLANTS AS AN ESTIMATION OF AQUATIC EXPOSURE.; ENVIRON. TOXICOL. CHEM.; 14:1657-66; 1995

MUELLER,M.; KLEIN,W.; COMPARATIVE EVALUATION OF METHODS PREDICTING WATER SOLUBILITY FOR ORGANIC COMPOUNDS.; CHEMOSPHERE; 25:769-82; 1992

MULLER, N. 1986. WHEN IS A TRIFLUOROMETHYL GROUP MORE LIPOPHILIC THAN A METHYL GROUP? PARTITION COEFFICIENTS AND SELECTED CHEMICAL SHIFTS OF ALIPHATIC ALCOHOLS AND TRIFLUOROALCOHOLS. J. PHARM. SCI. 75: 987-91.

MUIR,D.C.G.; PHOSPHATE ESTERS; IN: THE HANDBOOK OF ENVIRONMENTAL CHEMISTRY; ANTHROPOGENIC SUBSTANCES. GERMANY: SPRINGER-BERLAG BERLIN, 3:41-66; 1984

MUIR,D.C.G.; LINT,D.; GRIFT,N.P.; FATE OF THREE PHOSPHATE ESTER FLAME RETARDENTS IN SMALL PONDS.; ENVIRON. TOXICOL. CHEM.; 4:663-75.; 1985C

MULLICAN, M.D., R.J. SORENSON, D.T. CONNOR, D.O. THUESON, J.A. KENNEDY AND M.C. CONROY, "NOVEL THIOPHENE-, PYRROLE-, FURAN- AND BENZENECARBOXAMIDOTETRAZOLES AS POTENTIAL ANTIALLERGY AGENTS", J. MED. CHEM., 34(7), 2186-2194 (1991).

MULLINS,R.M.; ALKANOLAMINES FROM OLEFIN OXIDES AND AMMONIA.; IN: KIRK-OTHMER ENCYCL. CHEM. TECH. 3RD ED. 1:944-60.; 1978

MUNNECKE,D.E.; VANGUNDY,S.D.; MOVEMENT OF FUMIGANTS IN SOIL, DOSAGE, RESPONSES, AND DIFFERENTIAL EFFECTS.; ANNU. REV. PHYTOPATHOL.; 17:405-29; 1979

MUNZ,C.; ROBERTS,P.V.; AIR-WATER PHASE EQUILIBRIA OF VOLATILE ORGANIC SOLUTES.; J. - AM. WATER WORKS ASSOC.; 79:62-9; 1987

MURAKAMI, K., T. SHIRASAKA, H. YOSHIOKA, E. KOJIMA, S. AOKI, H. FORD, J.S. DRISCOLL, J. A. KELLEY AND H. MITSUYA, "ESCHERICHIA COLI MEDIATED BIOSYNTHESIS AND IN VITRO ANTI-HIV ACTIVITY OF LIPOPHILIC 6-HALO-2',3'-DIDEOXYPURINE NUCLEOSIDES" J. MED. CHEM., 34(5), 1606-1612 (1991).

MURPHY,T.J.; MULLIN,M.D.; MEYER,J.A.; EQUILIBRATION OF POLYCHLORINATED BIPHENYLS AND TOXAPHENE WITH AIR AND WATER.; ENVIRON. SCI. TECHNOL.; 21:155-62; 1987

MURRAY,J.J.; POTTIE,R.F.; PUPP,C.; VAPOR PRESSURES AND ENTHALPIES OF SUBLIMATION OF FIVE POLYCYCLIC AROMATIC HYDROCARBONS.; CAN. J. CHEM.; 52:557-63.; 1974

NADAIS,M.H.; BERNARDO-GIL,M.G.; VAPOR-LIQUID EQUILIBRIA OF ALPHA-PINENE + LIMONENE AT REDUCED PRESSURES.; FLUID PHASE EQUILIB.; 91:321-30; 1993

NAHUM, A. AND C. HORVATH, "EVALUATION OF OCTANOL-WATER PARTITION COEFFICIENTS BY USING HIGH-PRESSURE LIQUID CHROMATOGRAPHY", J. CHROMATOGR., 192(2), 315-322 (1980).

NAIRO, Y., Y. YAMAURA, Y. INOUE, C. FUKAYA, K. YOKOYAMA, Y. NAKAGAWA AND T. FUJITA, "QUANTITATIVE STRUCTURE-ACTIVITY RELATIONSHIPS OF 2-[4-(THIAZOL-2-YL)PHENYL]PROPIONIC ACID DERIVATIVES INHIBITING CYCLOOXYGENASE", EUR. J. MED. CHEM., 27(7), 645-654 (1992).

NAITO, Y., M. SUGIURA, Y. YAMAURA, C. FUKAYA, K. YOKOYAMA, Y. NAKAGAWA, T. IKEDA, M. SENDA AND T. FUJITA, "QUANTITATIVE STRUCTURE-ACTIVITY RELATIONSHIP OF CATECHOL DERIVATIVES INHIBITING 5-LIPOXYGENASE", CHEM. PHARM. BULL., 39(7), 1736-1745 (1991).

NAKAGAWA, Y., K. IZUMI, N. OIKAWA, T. SOTOMATSU, M. SHIGEMURA AND T. FUJITA. 1992. ANALYSIS AND PREDICTION OF HYDROPHOBICITY PARAMETERS OF SUBSTITUTED ACETANILIDES, BENZAMIDES AND RELATED AROMATIC COMPOUNDS. ENVIRON. TOXICOL. CHEM. 11: 901-16.

NANDIHALLI, U.B., M.V. DUKE AND S.O. DUKE, "QUANTITATIVE STRUCTURE-ACTIVITY RELATIONSHIPS OF PROTOPORPHYRINOGEN OXIDASE-INHIBITING DIPHENYL ETHER HERBICIDES", PESTIC. BIOCHEM. PHYSIOL., 43(3), 193-211 (1992).

NANDIHALLI, U.B. ET AL. 1993. PREDICTION OF RP-HPLC LOG P FROM SEMIEMPIRICAL MOLECULAR PROPERTIES OF DIPHENYL ETHER AND PHENOPYLATE HERBICIDES. J. AGRIC. FOOD. CHEM. 41: 582-7.

NARURKAR, M.M. AND A.K. MITRA, "SYNTHESIS, PHYSICOCHEMICAL PROPERTIES AND CYTOTOXICITY OF A SERIES OF 5'-ESTER PRODRUGS OF 5-IODO-2'-DEOXYURIDINE", PHARM. RES., 5(11), 734-737 (1988).

NASH,R.G.; COMPARATIVE VOLATILIZATION AND DISSIPATION RATES OF SEVERAL PESTICIDES FROM SOIL.; J. AGRIC. FOOD. CHEM.; 31:210-7.; 1983A

NASH,R.G.; DETERMINING ENVIRONMENTAL FATE OF PESTICIDES WITH MICROAGROECOSYSTEMS.; RES. REV.; 85:199-216; 1983B

NEARY,D.G.; PARSHALL,B.B.; MICHAELS,J.L.; FATE, DISSIPATION AND ENVIRONMENTAL EFFECTS OF PESTICIDES IN A SOUTHERN FORESTS: A REVIEW OF A DECADE OF RESEARCH PROGRESS.; ENVIRON. TOXICOL. CHEM.; 12:411-28; 1993

NEELY,W.B.; BLAU,G.E.; ENVIRONMENTAL EXPOSURE FROM CHEMICALS. VOLUME 1; BACO RATON, FLA: CRC PRESS 245PP; 1985

NEELY,W.B.; PREDICTING THE FLUX OF ORGANICS ACROSS THE AIR/WATER INTERFACE.; CONTROL HAZARD MATER. SPILLS, PROC. NAT. CONF. 3RD. PP 197-200.; 1976

NEELY,W.B.; AN ANALYSIS OF AQUATIC TOXICITY DATA: WATER SOLUBILITY AND ACUTE LC50 FISH DATA.; CHEMOSPHERE.; 13:813-19.; 1984

NELSON,N.H.; FAUST,S.D.; ACIDIC DISSOCIATION CONSTANTS OF SELECTED AQUATIC HERBICIDES.; ENVIRON. SCI. TECH.; 3:1186-8.; 1969

NEMEC,J.W.; BAUER,W. JR.; ACRYLIC ACID AND DERIVATIVES.; IN: KIRK-OTHMER ENCYCL. CHEM. TECHNOL. 3RD. ED. NEW YORK,NY: WILEY-INTERSCIENCE 1:330-54; 1978

NEMEC,J.W.; KIRSCH,L.S.; METHACRYLIC ACID AND DERIVATIVES.; IN: KIRK-OTHMER ENCYCL. CHEM. TECHNOL. 3RD. ED. 15:346-76.; 1981

NICOLINI,E.R.; THE VAPORIZATION OF PURE ORGANIC LIQUIDS.; ANN. CHIM.; 6:582-629; 1951

NIELSEN,O.J.; O'FARRELL,D.J.; TREACY,J.J.; SIDEBOTTOM,H.W.; RATE CONSTANTS FOR THE GAS-PHASE REACTIONS OF HYDROXYL RADICALS WITH TETRAMETHYLLEAD AND TETRAETHYLLEAD.; ENVIRON. SCI. TECHNOL.; 25:1098-103; 1991

NIELSEN, L.S., H. BUNDGAARD AND E. FALCH, "PRODRUGS OF THIABENDAZOLE WITH INCREASED WATER-SOLUBILITY", ACTA PHARM. NORD., 4(1), 43-49 (1992).

NIELSEN, N.M. AND H. BUNDGAARD, "EVALUATION OF GLYCOLAMIDE ESTERS AND VARIOUS OTHER ESTERS OF ASPIRIN

AS TRUE ASPIRIN PRODRUGS", J. MED. CHEM., 32(3), 727-734 (1989).

NIELSEN, N.M. AND H. BUNDGAARD, "GLYCOLAMIDE ESTERS AS BIOLABILE PRODRUGS OF CARBOXYLIC ACID AGENTS: SYNTHESIS, BIOCONVERSION AND PHYSICOCHEMICAL PROPERTIES", J. PHARM. SCI., 77(4), 285-298 (1988).

NISHIMURA, K., T. KITAHARA, Y. IKEMOTO AND T. FUJITA, "QUANTITATIVE STRUCTURE-ACTIVITY STUDIES OF PYRETHROIDS.14.", PESTIC. BIOCHEM. PHYSIOL., 31(2), 155-165 (1988).

NISHIMURA, K., K. HIRAYAMA, T. KOBAYASHI, T. FUJITA AND G. HOLAN, "QUANTITATIVE STRUCTURE-ACTIVITY RELATIONSHIPS OF INSECTICIDAL DIPHENYLDICHLOROCYCLOPROPANES", PESTIC. BIOCHEM. PHYSIOL., 25(2), 153-162 (1986).

NISHIMURA, K., M. OHOKA AND T. FUJITA, "QUANTITATIVE STRUCTURE-ACTIVITY STUDIES OF PYRETHROIDS.10.", PESTIC. BIOCHEM. PHYSIOL., 28(2), 257-270 (1987).

NIIMI,A.J.; LEE,H.B.; KISSOON,G.P.; OCTANOL/WATER PARTITION COEFFICIENTS AND BIOCONCENTRATION FACTORS OF CHLORONITROBENZENES IN RAINBOW TROUT (SALMO GAIRDNERI); ENVIRON. TOXICOL. CHEM.; 8:817-23; 1989

NIKOLAEV,A.V.; GRISHIN,G.M.; KOLESNIKOV,A.A.; POGADAEV,G.I.; TRI-N-OCTYLAMINE SOLUBILITY IN AQUEOUS SOLUTIONS OF SOME ACIDS AND SALTS.; IZV. SIB. OTD. AKAD. NAUK SSSR, SER. KHIM. NAUK. 1969, 145-8.; 1969

NIRMALAKHANDAN,N.N.; SPEECE,R.E.; PREDICTION OF AQUEOUS SOLUBILTY OF ORGANIC CHEMICALS BASED ON MOLECULAR STRUCTURE.; ENVIRON. SCI. TECHNOL.; 22:328-38; 1988

NIYAZOV,A.N.; AMANOV,K.B.; TRAPEZNIKOVA,V.F.; CHIRKOVA,B.V.; SOLUBILITY AND DISSOCIATION CONSTANTS OF SOME ALICYCLIC ACIDS.; IZV. AKAD. NAUK TURKM. SSR, SER. FIZ.-TEKH., KHIM. GEOL. NAUK.; 4:121-3.; 1975

NORRIS,J.M.; EHRMANTRAUT,J.W.; GIBBONS,C.L.; KOCIBA,R.J.; SCHWETZ,B.A.; ROSE,J.Q.; HUMISTON,C.G.; JEWETT,G.L.; CRUMMETT,W.B.; GEHRING,P.J.; TIRSELL,J. TOXICOLOGICAL AND ENVIRONMENTAL FACTORS INVOLVED IN THE SELECTION OF DECABROMODIPHENYL OXIDE AS A FIRE RETARDANT CHEMICAL.; APPL. POLYMER SYMP.; 22:195-219; 1973

NORTON,L.C.; SCHERZINGER,R.A.; PENT-OXONES AND PENT-OXOL.; PAINT VARNISH PROD.; 51:25-31.; 1961

NOYES DATA CORP.; PESTICIDE FACT HANDBBOOK; NOYES DATA CORPORATION, MILL ROAD, PARK RIDGE, NEW JERSEY 07656.; 1988

OBENAUS,F.; DROSTE,W.; NEWMEISTER,J.; MARL,H.A.G.; BUTENES.; ULLMAN'S ENCYCL. IND. CHEM. VCH VERLAGSGESELLSCHAFT. WEINHEIM, GERMANY. A4:483-94.; 1985

OGATA,M.; FUJISAWA,K.; OGINO,Y.; MANO,E.; PARTITION COEFFICIENTS AS A MEASURE OF BIOCONCENTRATION POTENTIAL OF CRUDE OIL COMPOUNDS IN FISH AND SHELLFISH.; BULL. ENVIRON. CONTAM. TOXICOL.; 33:561-7.; 1984

OHE,S.; COMPUTER AIDED DATA BOOK OF VAPOR PRESSURE.; DATA BOOK PUBL. CO., TOKYO, JAPAN.; 1976

OHE,S.; VAPOR-LIQUID EQUILIBRIUM DATA.; KODANSHA LTD, TOKYO AND ELSEVIER SCIENCE PUBLISHERS, AMSTERDAM. 742PP.; 1989

OHNISHI,R.; TANABE,K.; A NEW METHOD OF SOLUBILITY DETERMINATION OF HYDROLYZING SOLUTE-SOLUBILITY OF BENZYL CHLORIDE IN WATER.; BULL. CHEM. SOC. JAPAN; 41:2647-9.; 1971

OIKAWA, N. ET AL. 1994. QUANTITATIVE STRUCTURE-ACTIVITY ANALYSIS OF LARVICIDAL 1-(SUBSTITUTED BENZOYL)-2-BENZOYL-1-TERT-BUTYLHYDRAZINES AGAINST CHILO SUPPRESSALIS. PESTIC. SCI. 41: 139-48.

OKAMOTO, H., M. HASHIDA AND H. SEZAKI, "EFFECT OF 1-ALKYL- OR 1-ALKENYLAZACYCLOALKANONE DERIVATIVES ON THE PENETRATION OF DRUGS WITH DIFFERENT LIPOPHILICITIES THROUGH GUINEA PIG SKIN", J. PHARM. SCI., 80(1), 39-45 (1991).

OKOUCHI,S.; SAEGUSA,H.; NOJIMA,O.; PREDICTION OF ENVIRONMENTAL PARAMETERS BY ADSORBABILITY INDEX: WATER SOLUBILITIES OF HYDROPHOBIC ORGANIC POLLUTANTS.; ENVIRON. INT.; 18:249-61; 1992

OLIVER,B.G.; DESORPTION OF CHLORINATED HYDROCARBONS FROM SPIKED AND ANTHROPOGENICALLY CONTAMINATED SEDIMENTS.; CHEMOSPHERE; 14:1087-106; 1985

OMAR,M.M.; VAPOR PRESSURES, HEATS OF SUBLIMATION, AND HEATS OF VAPORIZATION FOR STRAIGHT-CHAIN ETHYL ESTERS.; J. CHEM. SOC. C. PP.2038-40.; 1967

OPPERHUIZEN,A.; VOORS,P.I.; UPTAKE AND ELIMINATION OF POLYCHLORINATED AROMATIC ETHERS BY FISH: CHLOROANISOLES.; CHEMOSPHERE; 16:953-62; 1987

OPPERHUIZEN,A.; BIOCONCENTRATION OF HYDROPHOBIC CHEMICALS IN FISH.; IN: ASTM SPEC. TECH. PUBL. 921(AQUAT. TOXICOL. ENVIRON. FATE) 9:304-15; 1986

OPPERHUIZEN,A.; RELATIONSHIPS BETWEEN OCTAN-1-OL WATER PARTITION COEFFICIENTS. AQUEOUS ACTIVITY COEFFICIENTS AND REVERSED PHASE HPLC CAPACITY FACTORS OF ALKYL BENZENES, CHLOROBENZENES, CHLORONAPHTHALENE AND CHLOROBIPHENYLS.; OX. ENVIRON. CHEM.; 15:249-654; 1987

OPPERHUIZEN,A.; VELDE,E.W.; GOBAS,F.A.P.C; LIEM,D.A.K.; STEEN,J.M.D.; HUTZINGER,O.; RELATIONSHIP BETWEEN BIOCONCENTRATION IN FISH AND STERIC FACTORS OF HYDROPHOBIC CHEMICALS.; CHEMOSPHERE; 14:1871-96; 1985

OSBORN,A.G.; SCOTT,D.W.; VAPOR PRESSURE OF 17 MISCELLANEOUS ORGANIC COMPOUNDS; J CHEM THERMODYNAMIC; 12:429-38; 1980

OZRETICH,R.J.; SMITH,L.M.; ROBERTS,F.A.; REVERSED-PHASE SEPARATION OF ESTUARINE INTERSTITIAL WATER FRACTIONS AND THE CONSEQUENCES OF C18 RETENTION OF ORGANIC MATTER.; ENVIRON. TOXICOL. CHEM.; 14:1261-72; 1995

PAGE,F.M.; PURNELL,J.H.; PHYSICAL PROPERTIES OF THE LOWER DIALKYL HYDROGEN PHOSPHITES.; J. CHEM. SOC. 1958. PP.621-3.; 1958

PAGOU, M. AND A. KOUTSELINIS, "LIPOPHILICITY AND CHROMATOGRAPHIC BEHAVIOUR OF BENZOIC ACID DERIVATIVES", CHEM. PHARM. BULL., 41(2), 319-324 (1993).

PALOMINO, E., B.R. MELTSNER, D. KESSEL AND J.P. HORWITZ, "SYNTHESIS AND IN VITRO EVALUATION OF SOME MODIFIED 4-THIOPYRIMIDINE NUCLEOSIDES FOR PREVENTION OR REVERSAL OF AIDS-ASSOCIATED NEUROLOGICAL DISORDERS", J. MED. CHEM., 33(1), 258-263 (1990).

PANDIT, N.K. ET AL. 1989. THE EFFECT OF SALTS ON THE DISTRIBUTION AND SOLUBILITY OD AN ACIDIC DRUG. INT. J. PHARM. 50: 7-13.

PAPA,A.J.; SHERMAN,P.D. JR.; KETONES.; IN: KIRK-OTHMER ENCYCL. CHEM. TECH. 3RD ED. 13:894-941.; 1981

PARIS,D.F.; WOLFE,N.L.; STEEN,W.C.; MICROBIAL TRANSFORMATION OF ESTERS OF CHLORINATED CARBOXYLIC ACIDS.; APPL. ENVIRON. MICROBIOL.; 47:7-11.; 1984

PARK,J.H.; HUSSAM,A.; CONASNON,P.; FRITZ,D.; CARC,P.W.; EXPERIMENTAL REEXAMINATION OF SELECTED PARTITION COEFFICIENTS FROM ROHRSCHNEIDER'S DATA SET.; ANAL. CHEM.; 59:1970-6; 1987

PARK,T.; RETTICH,T.R.; BATTINO,R.; PETERSON,D.; WILHEIM,E.; SOLUBILITY OF GASES IN LIQUIDS. 14. BUSEN COEFFICIENTS FOR SEVERAL FLUORINE-CONTAINING GASES (FREONS) DISSOLVED IN WATER AT 298.15 K.; J. CHEM. ENG. DATA; 27:324-6; 1982

PARKIN, J.E., "GUIDE TO THE SELECTION OF DETECTING AGENTS FOR THE INDIRECT PHOTOMETRIC DETECTION OF ALIPHATIC ALCOHOLS AND OTHER COMPOUNDS BY REVERSED-PHASE HIGH-PERFORMANCE LIQUID CHROMATOGRAPHY", J. CHROMATOGR., 351(3), 532-540 (1986).

PARMEGGIANI,L; ENCYCLOPEDIA OF OCCUPATIONAL HEALTH AND SAFETY; GENEVA, SWITZERLAND: INTERNATIONAL LABOUR ORGANISATION; (1983)

PARRISH,C.F.; SOLVENTS, INDUSTRIAL; IN:KIRK-OTHMER ENCYCL CHEM TECH 3RD; NY,NY: WILEY-INTRSCI 21:377-401; 1983

PARSONS,G.H.; ROCHESTER,C.H.; WOOD,C.E.C.; EFFECT OF 4-SUBSTITUTION ON THE THERMODYNAMICS OF HYDRATION OF PHENOL AND THE PHENOXIDE ANION.; J. CHEM. SOC., B.; 533-36; 1971

PARSONS,G.H.; ROCHESTER,C.H.; ROSTRON,A.; SYKES,P.C.; THE THERMODYNAMICS OF HYDRATION OF PHENOLS.; J. CHEM. SOC., PERKIN TRANS.; 2:136-38; 1972

PATIL,S.G.; NICHOLLS,P.H.; CHAMBERLAIN,K.; BRIGGS,G.G.; BROMILOW,R.H.; DEGRADATION RATES IN SOIL OF 1-BENZYLTRIAZOLES AND TWO TRIAZOLE FUNGICIDES; PEST SCI; 22:333-42; 1988

PATTE,F.; ETCHETO,M.; LAFFORT,P.; SOLUBILITY FACTORS FOR 240 SOLUTES AND 207 STATIONARY PHASES IN GAS-LIQUID CHROMATOGRAPHY.; ANAL. CHEM.; 54:2239-47.; 1982

PEAKALL,D.B.; PHTHALATE ESTERS: OCCURRENCE AND BIOLOGICAL EFFECTS.; RES. REV.; 54:1-41.; 1975

PEARCE,P.J.; SIMKINS,R.J.J.; ACID STRENGTHS OF SOME SUBSTITUTED PICRIC ACIDS.; CAN. J. CHEM.; 46:241-8.; 1968

PEARLMAN,R.S.; YALKOWSKI,S.H.; BANERJEE,S.; WATER SOLUBILITIES OF POLYNUCLEAR AROMATIC AND HETEROAROMATIC COMPOUNDS.; J. CHEM. REF. DATA.; 13:555-62.; 1984

PERRISSOUD, D. AND B. TESTA. 1986. INHIBITING OR POTENTIATING EFFECTS OF FLAVONVOIDS ON CARBON TETRACHLORIDE-INDUCED TOXICITY IN ISOLATED RAT HEPATOCYTES. ARZNEIM.-FORSCH. 36: 1249-53.

PCCHEM; GEMS-GRAPHIC EXPOSURE MODELING SYSTEM PCCHEM USEPA (1987)

PEDDLE,C.J.; TURNER,W.E.S.; SOLUBILITIES OF SALTS OF AMMONIUM BASES IN WATER AND IN CHLOROFORM. I. SOLUBILITY AS A CONSTITUTIVE PROPERTY.; J. CHEM. SOC.; 103:1202-9; 1913

PELLA,P.A.; MEASUREMENT OF THE VAPOR PRESSURES OF TNT, 2,4-DNT, AND EGDN.; J. CHEM. THERMODYN.; 9:301-5.; 1977

PERRIN,D.D.; THE EFFECT OF TEMPERATURE ON PK VALUES OF ORGANIC BASES.; AUST. J. CHEM.; 17:484-8.; 1964

PERRIN,D.D.; DISSOCIATION CONSTANTS OF ORGANIC BASES IN AQUEOUS SOLUTION.; IUPAC CHEMICAL DATA SERIES, BUTTERSWORTH, LONDON.; 1965

PERRIN,D.D.; DISSOCIATION CONSTANTS OF ORGANIC BASES IN AQUEOUS SOLUTION.; IUPAC CHEMICAL DATA SERIES: SUPPLEMENT 1972. BUTTERSWORTH, LONDON.; 1972

PERRY,R.H.; GREEN,D.; PERRY'S CHEMICAL HANDBOOK. PHYSICAL AND CHEMICAL DATA.; NEW YORK,NY: MCGRAW HILL. 6TH ED.; 1984

PERRY,R.H.; GREEN,D.W.; MALONEY,J.O.; PERRY'S CHEMICAL ENGINEERS' HANDBOOK; MCGRAW HILL; CRAWFORD & ECKES EDS.; 1984

PERWAK,J.; BYRNE,M.; COONS,S.; GOYER,M.; HARRIS,J.; CRUSE,P.; DEROSIER,R.; MOSS,K.; WENDT,S.; EXPOSURE AND RISK ASSESSMENT FOR BENZO(A)PYRENE AND OTHER POLYCYCLIC AROMATIC HYDROCARBONS. VOL 4.; EPA-440/4-85-020-V4. (NTIS PB 85 222-586). WASHINGTON, DC: USEPA. 215 PP.; 1982B

PIACENTE,V.; SCARDALA,P.; FERRO,C.; GIGLI,R.; VAPORIZATION STUDY OF O-, M-, AND P-CHLOROANILENE BY TORSION-WEIGHING EFFUSION VAPOR PRESSURE MEASUREMENTS.; J. CHEM. ENG. DATA; 30:372-6; 1985

PICKRELL,J.A.; MOKLER,B.V.; GRIFFIS,L.C.; HOBBS,C.H.; BATHIJA,A.; FORMALDEHYDE RELEASE RATE COEFFICIENTS FROM SELECTED CONSUMER PRODUCTS.; ENVIRON. SCI. TECHNOL.; 17:753-7; 1983

PLATFORD,R.F.; THE OCTANOL-WATER PARTITIONING OF SOME HYDROPHOBIC AND HYDROPHILIC COMPOUNDS.; CHEMOSPHERE.; 12:1107-12.; 1983

PLETCHER,T.; CORDES,E.H.; THE BASICITY AND COUPLING CONSTANTS OF CERTAIN ACETALS, KETALS, AND ORTHO ESTERS.; J. ORG. CHEM.; 32:2294-7.; 1967

PLIMMER,J.R.; VOLATILITY.; IN: HERBICIDES: CHEMISTRY, DEGRADATION, AND MODE OF ACTION. 2ND.ED. KEARNEY,P.C. & KAUFMAN,D.D.,EDS. NEW YORK,NY: MARCEL DEKKER, INC. 2:891-934.; 1976

POLASEK, M., D. KOHOUTKOVA AND K. WAISSER, "KINETIC SPECTROPHOTOMETRIC DETERMINATION OF THIOBENZAMIDES AND THEIR PARTITION COEFFICIENTS IN WATER/1-OCTANOL BY USING AN IODINE/AZIDE INDICATOR REACTION", ANAL. CHIM. ACTA, 212(1-2), 279-284 (1988).

PONTICELLO, G.S., M.B. FREEDMAN, C.N. HABECKER, P.A. LYLE, H. SCHWAM. S.L. VARGA, M.E. CHRISTY, W.C. RANDALL AND J.J. BALDWIN, "THIENOTHIOPYRAN-2-SULFONAMIDES: A NOVEL CLASS OF WATER-SOLUBLE CARBONIC ANHYDRASE INHIBITORS", J. MED. CHEM., 30(4), 591-597 (1987).

PRANKERD, R.J. AND R.H. MCKEOWN, "PHYSICO-CHEMICAL PROPERTIES OF BARBITURIC ACID DERIVATIVES. II. PARTITION COEFFICIENTS OF 5,5-DISUBSTITUTED BARBITURIC ACIDS AT 25 C", INT. J. PHARM., 83(1-3), 25-37 (1992).

PRANKERD, R,J. AND R.H. MCKEOWN, "PHYSICOCHEMICAL PROPERTIES OF BARBITURIC ACID DERIVATIVES. III. PARTITION COEFFICIENTS OF CYCLOHEXANE-1',5-SPIROBARBITURIC ACIDS AT 25 C", INT. J. PHARM., 83(1-3), 39-45 (1992).

PRATESI, P, E. GRANA, M.G.S. BARBONE, M.I. LAROTONDA, C. SILIPO AND A. VITTORIA, "QSAR IN A SERIES OF MUSCARINIC AGENTS. VI. FIVE-MEMBERED CYCLIC COMPOUNDS", FARMACO, ED. SCI., 41(5), 335-345 (1986).

PRATESI, P., L. VILLA, V. FERRI, C. DE MICHELI, M. DE AMICI, E. GRANA, M.G.S. BARBONE, C. SILIPO, A. VITTORIA AND B. CAPPELLO, "QSAR IN A SERIES OF MUSCARINIC AGENTS. V. NEW SETS OF RIGID AND FLEXIBLE LIGANDS", FARMACO, ED. SCI., 39(3), 171-188 (1984).

PROBST,G.W.; GOLAB,T.; WRIGHT,W.L.; DINITROANILINES.; IN: HERBICIDES: CHEMISTRY, DEGRADATION, AND MODE OF ACTION. VOL 1, CHPT 9, PP. 453-500. KERANEY,PC & KAUFMAN,DD, EDS. NEW YORK,NY: MARCEL DEKKER.; 1975

PRUGH, J.D., G.D. HARTMAN, P.J. MALLORGA, B.M. MCKEEVER, S.R. MICHELSON, M.A. MURCKO, H. SCHWAM, R.L. SMITH, J.M. SONDEY, J.P. SPRINGER AND M.F. SUGRUE, "NEW ISOMERIC CLASSES OF TOPICALLY ACTIVE OCULAR HYPOTENSIVE CARBONIC ANHYDRASE INHIBITORS: 5-SUBSTITUTED THIENO[2,3-B]THIOPHENE-2-SULFONAMIDES AND 5-SUBSTITUTED THIENO[3,2-B]THIOPHENE-2-SULFONAMIDES", J. MED. CHEM., 34(6), 1805-1818 (1991).

PUPP,C.; LAO,R.C.; MURRAY,J.J.; POTTIE,R.F.; EQUILIBRIUM VAPOR CONCENTRATIONS OF SOME POLYCYCLIC AROMATIC HYDROCARBONS, AS4O6 AND SEO2 AND THE COLLECTION EFFICIENCIES OF THESE AIR POLLUTANTS.; ATMOS. ENVIRON.; 8:915-25; 1974

QUIGLEY, J.M. ET AL. 1990. TEMPERATURE DEPENDENCE AND THEROMDYNAMICS OF PARTITIONING OF CLOFAZIMINE ANALOGS IN THE OCTANOL/WATER SYSTEM. INT. J. PHARM. 58: 107-13.

RACKE,K.D.; DEGRADATION OF ORGANOPHOSPHORUS INSECTICIDES IN ENVIRONMENTAL MATRICES.; IN: ORGANOPHOSPHATES: CHEMISTRY, FATE, AND EFFECTS. CHAMBERS,JE & LEVI,PE, EDS., ACADEMIC PRESS, INC: SAN DIEGO, CALIF. PP. 47-78; 1992

RADZICKA, A. AND R. WOLFENDEN, "COMPARING THE POLARITIES OF THE AMINOACIDS: SIDE-CHAIN DISTRIBUTION COEFFICIENTS BETWEEN THE VAPOR PHASE, CYCLOHEXANE, 1-OCTANOL AND NEUTRAL AQUEOUS SOLUTION", BIOCHEMISTRY, 27(5), 1664-1670 (1988).

RAHMAN,A.; MATTHEWS,L.J.; EFFECT OF SOIL ORGANIC MATTER ON THE PHYTOTOXICITY OF THIRTEEN S-TRIAZINE HERBICIDES.; WEED SCI.; 27:158-61; 1979

RANADIVE, S.A., A.X. CHEN AND A.T.M. SERAJUDDIN, "RELATIVE LIPOPHILICITIES AND STRUCTURAL-PHARMOCOLOGICAL CONSIDERATIONS OF VARIOUS ANGIOTENSIN-CONVERTING ENZYME (ACE) INHIBITORS", PHARM. RES., 9(11), 1480-1486 (1992).

RAO,P.S.C.; DAVIDSON,J.M.; RETENTION AND TRANSFORMATION OF SELECTED PESTICIDES AND PHOSPHORUS IN SOIL-WATER SYSTEMS: A CRITICAL REVIEW.; EPA-600/S3-82-060. ATHENS,GA: U.S. EPA.; 1982

RAO,P.S.C.; HORNSBY,A.G.; JESSUP,R.E.; INDICES FOR RANKING THE POTENTIAL FOR
PESTICIDE CONTAMINATION OF GROUNDWATER.; SOIL CROP SCI. SOC. FL. PROC.; 44:1-8;
1985A

RAPAPORT,R.A.; EISENREICH,S.J.; CHROMATOGRAPHIC DETERMINATION OF OCTANOL-WATER
PARTITION COEFFICIENTS (KOW'S) FOR 58 POLYCHLORINATED BIPHENYL CONGENERS.;
ENVIRON. SCI. TECHNOL.; 18:163-70.; 1984

RECANATINI, M., "PARTITION AND DISTRIBUTION COEFFICIENTS OF ARYLOXYPROPANOLAMINE BETA-ADRENOCEPTOR
ANTAGONISTS", J. PHARM. PHARMACOL., 44(1), 68-70 (1992).
REDDY,K.N.; LOCKE,M.A.; BRYSON,C.T.; FOLIAR WASHOFF AND RUNOFF LOSSES OF
LACTOFEN, NORFLURAZON, AND FLUOMETURON UNDER SIMULATED RAINFALL.; J. AGRIC. FOOD
CHEM.; 42:2338-43; 1994

REINBOLD,K.A.; HASSETT,J.J.; MEANS,J.C.; BANWART,W.L.; ADSORPTION OF
ENERGY-RELATED ORGANIC POLLUTANTS: A LITERATURE REVIEW.; EPA-600/3-79-086.
ATHENS,GA: U.S.EPA. PP.180.; 1979

REINERT, K.H. AND J.H. RODGERS, JR. 1984. INFLUENCE OF SEDIMENT TYPES ON THE SORPTION OF ENDOTHALL. BULL.
ENVIRON. CONTAM. TOXICOL. 32: 557-64.

REKKA, E. AND P. KOUROUNAKIS. 1989. SYNTHESIS, PHYSICOCHEMICAL PROPERTIES AND BIOLOGICAL STUDIES OF
SOME SUBSTITUTED 2-ALKOXY-4-METHYLMORPHOLINES. EUR. J. MED. CHEM. 24: 179-84.

RIBO,J.M.; THE OCTANOL/WATER PARTITION COEFFICIENT OF THE HERBICIDE
CHLORSULFURON AS A FUNCTION OF PH; CHEMOSPHERE; 17:709-715; 1988

RIBEIRODASILVA,M.A.V.; MATOS,A.R.; MONTE,M.J.S.; ALVES,M.C.B.; VIEIRA,J.M.A.P.;
ENTHALPIES OF COMBUSTION, VAPOR PRESSURES, AND ETHALPIES OF SUBLIMATION OF
3-,5-, 6-, AND 8-AMINOQUINOLINE.; J. CHEM. THERMODYN.; 25:579-90.; 1993
RICCO,G.; LOPEZ,A.; CIANNARELLE,R.; SORPTION OF SUBSTITUTED PHENOLS BY
CLAY-BASED SUBSOILS.; QUAD. -IST. RIC. ACQUE; 96:7.1-7.9; 1994

RICE,C.P.; CHERNYAK,S.M.; DETERMINATION OF HENRY'S LAW CONSTANTS OF HALOGENATED
CURRENT-USE PESTICIDES AS A FUNCTION OF ENVIRONMENTAL CONDITIONS.; IN:
ORGANOHALOGEN COMPD., VOL. 24 (DIOXIN 95, 15TH INTER. SYMP. ON; 1995

RICH, P.R., "PARTITION COEFFICIENTS OF QUINONES AND THEIR RELATION TO BIOCHEMICAL REACTIVITY", HIGHLIGHTS
IN UBIQUINONE RESEARCH, PROC. INT. SYMPOSIUM "BIOCHEM., BIOENER. CLIN. APPL. UBIQUINONE", 136-141 (1990).

RICHARDSON,L.T.; MILLER,D.M.; FUNGITOXICITY OF CHLORINATED HYDROCARBON
INSECTICIDES IN RELATION TO WATER SOLUBILITY AND VAPOR PRESURE.; CAN. J.
BOTANY.; 38:163-75.; 1960

RIDDICK,J.A.; BUNGER,W.B.; SAKANO,T.K.; ORGANIC SOLVENTS: PHYSICAL PROPERTIES
AND METHODS OF PURIFICATION.; TECHNIQUES OF CHEMISTRY. 4TH ED. NEW YORK,NY:
WILEY-INTERSCIENCE. 2:PP.1325; 1986

RIEDERER,M.; ESTIMATING PARTITIONING AND TRANSPORT OF ORGANIC CHEMICALS IN THE
FOLIAGE/ATMOSPHERE SYSTEM: DISCUSSION OF A FUGACITY-BASED MODEL.; ENVIRON. SCI.
TECHNOL.; 24:829-37; 1990

RIEMENSCHNEIDER,W.; CARBOXYLIC ACIDS, ALIPHATIC.; IN: ULLMANN'S ENCYCL. OF
INDUST. CHEM., NY:VCH PUBL., VOL. A5:235-48; 1986

RINGK,W.; THEIMER,E.T.; BENZYL ALCOHOL.; IN: KIRK-OTHMER ENCYCL. CHEM. TECH.
3RD ED. 3:793-802.; 1978

RINKENBACK,W.H.; EXPLOSIVES.; IN: KIRK-OTHMER ENCYCL. CHEM. TECH. 2ND ED.
8:581-658.; 1965

RITZ,J.; FUCHS,H.; CAPROLACTAM.; IN: ULLMAN'S ENCYCL. IND. CHEM. 5TH ED., VOL.
A5:31-50.; 1986

ROBERTS,M.S.; ANDERSON,R.A.; SWARBRICK,J.; PERMEABILITY OF HUMAN EPIDERMS TO
PHENOLIC COMPOUNDS.; J. PHARM. PHARMACOL.; 29:677-83.; 1977

ROBINSON,R.A.; THE DISSOCIATION CONSTANTS OF SOME DISUBSTITUTED ANILINES AND PHENOLS IN AQUEOUS SOLUTION AT 25 DEGC.; J. RES. NAT. BUR. STAND.; 71:213-18; 1967

ROCHESTER,C.H.; SYMONDS,J.R.; THERMODYNAMIC STUDIES OF FLUORO ALCOHOLS. 1. VAPOR PRESSURES AND ENTHALPIES OF VAPORAZATION.; J. CHEM. SOC. FARADAY TRANS. 1.; 69:1267-73.; 1973

RODA, A. ET AL. 1990. BILE ACID STRUCTURE-ACTIVITY RELATIONSHIPS: J. LIPID RES. 31: 1433-43.

ROGERS,R.D.; MCFARLANE,J.C.; SORPTION OF CARBON TETRACHLORIDE, ETHYLENE DIBROMIDE AND TRICHLOROETHYLENE IN SOIL AND CLAY.; ENVIRON. MONIT. ASSESS.; 1:155-8.; 1981

RORDORF,B.F.; PREDICTION OF VAPOR PRESSURES, BOILING POINTS AND ENTHALPIES OF FUSION FOR TWENTY-NINE HALOGENATED DIBENZO-P-DIOXINS.; THERMOCHIM. ACTA.; 112:117-22; 1987

RORDORF,B.F.; PREDICTION OF VAPOR PRESSURES, BOILING POINTS AND ENTHALPIES OF FUSION FOR TWENTY-NINE HALOGENATED DIBENZO-P-DIOXINS AND FIFTY-FIVE DIBENZOFURANS BY VAPOR PRESSURE CORRELATION METHOD; CHEMOSPHERE; 18:783-88; 1989

ROSENBLATT,D.H.; MILLER,T.A.; DACRE,J.C.; MUUL,I.; COGLEY,D.R.; PROBLEM DEFINITION STUDIES ON POTENTIAL ENVIRONMENTAL POLLUTANTS II. PHYSICAL, CHEMICAL, TOXICOLOGICAL, AND BIOLOGICAL PROPERTIES OF 16 SUBSTANCES.; TR-7509. FORT DETRICK,MD: ARMY MED. BIOENG. RES. DEVELOP. LAB. PP.290.; 1975

ROSS, D.L., S.K. ELKINTON AND C.M. RILEY, "PHYSICOCHEMICAL PROPERTIES OF THE FLUOROQUINOLONE ANTIMICROBIALS. III. 1-OCTANOL/WATER PARTITION COEFFICIENTS AND THEIR RELATIONSHIPS TO STRUCTURE", INT. J. PHARM., 88(1-3), 379-389 (1992).

ROY,S.D.; FLYNN,G.L.; SOLUBILITY AND RELATED PHYSICOCHEMICAL PROPERTIES OF NARCOTIC ANALGESICS.; PHARMACEUT. RES.; 5:580-86; 1988

ROY,W.R.; GRIFFIN,R.A.; MOBILITY OF ORGANIC SOLVENTS IN WATER-SATURATED SOIL MATERIALS.; ENVIRON. GEOL. WATER SCI.; 7:241-7; 1985

RUZICKA,V.; ZABRANSKY,M.; RUZICKA,K.; MAJER,V.; VAPOR PRESSURE FOR A GROUP OF HIGH-BOILING ALKYLBENZENES UNDER ENVIRONMENTAL CONDITIONS.; THERMOCHIM. ACTA; 245:121-44; 1994

RYON,M.G.; PAL,B.C.; TALMAGE,S.S.; ROSS,R.H.; DATABASE ASSESSMENT OF THE HEALTH AND ENVIRONMENTAL EFFECTS OF MUNITION PRODUCTION WASTE PRODUCTS. FINAL REPORT.; ORNL-6018. (NTIS DE84-016512). OAK RIDGE,TN: OAK RIDGE NATL. LAB PP.217; 1984

SAAN, W.S., J.E. SCHWERING, P.A. LYLE AND E.L. ENGELHARDT, "SYNTHESIS AND EVALUATION OF SOME NITROBENZENESULFONAMIDES CONTAINING NITROISOPROPYL AND (UREIDOOXY)METHYL GROUPS AS NOVEL HYPOXIC CELL SELECTIVE CYTOTOXIC AGENTS", J. MED. CHEM., 34(10), 3132-3138 (1991).

SAEGER,V.W.; HICKS,O.; KALEY,R.G.; MICHAEL,P.R.; MIEURE,J.P.; TUCKER,S.E.; ENVIRONMENTAL FATE OF SELECTED PHOSPHATE ESTERS.; ENVIRON. SCI. TECHNOL.; 13:840-4.; 1979

SAGEBIEL,J.C.; SEIBER,J.N.; WOODROW,J.E.; COMPARISON ON HEADSPACE AND GAS-STRIPPING METHODS FOR DETERMIMINIG THE HENRY'S LAW CONSTANT (H) FOR ORGANIC COMPOUNDS OF LOW TO INTERMEDIATE H.; CHEMOSPHERE; 25:1763-68; 1992

SAITO,H.; KOYASU,J.; SHIGEOKA,T.; CYTOTOXICITY OF ANILINES AND ALDEHYDES TO GOLDFISH GFS CELLS AND RELATIONSHIPS WITH 1-OCTANOL/WATER PARTITION COEFFICIENTS.; CHEMOSPHERE; 27:1553-60; 1993

SANBORN,J.R.; METCALF,R.L.; BRUCE,W.N.; LU,P.Y.; THE FATE OF CHLORDANE AND TOXAPHENE IN A TERRESTRIAL-AQUATIC MODEL ECOSYSTEM.; ENTOMOL.; 5:533-8.; 1976

SANBORN,J.R.; FRANCIS,B.M.; METCALF,R.L.; THE DEGRADATION OF SELECTED PESTICIDES IN SOIL: A REVIEW OF PUBLISHED LITERATURE.; EPA-600/9-77-022. CINCINNATI,OH:

U.S.EPA. PP.616.; 1977

SANEMASA,I.; ARAKI,M.; DEGUCHI,T.; NAGAI,H; SOLUBILITY MEASUREMENTS OF BENZENE AND THE ALKYLBENZENES IN WATER BY MAKING USE OF SOLUTE VAPOR.; BULL. CHEM. SOC. JPN.; 55:1054-62; 1982

SANGSTER,J.; LOGKOW DATABANK.; SANGSTER RES. LAB., MONTREAL, QUEBEC, CANADA.; 1993

SANGSTER, J. 1994. LOGKOW DATABANK. A DATABANK OF EVALUATED OCTANOL-WATER PARTITION COEFFICIENTS (LOG P) ON MICROCOMPUTER DISKETTE. MONTREAL, QUEBEC, CANADA: SANGSTER RESEARCH LABORATORIES.

SASAKI, H. ET AL. 1988. ENHANCING EFFECT OF PYRROLIDONE DERIVATIVES ON TRANSDERMAL DRUG DELIVERY. INT. J. PHARM. 44: 15-24.

SASAKI, H., T. TAKAHASHI, Y. MORI, J. NAKAMURA AND J. SHIBASAKI, "TRANSDERMAL DELIVERY OF 5-FLUOROURACIL AND ITS ALKYLCARBAMOYL DERIVATIVES", INT. J. PHARM., 60(1), 1-9 (1990).

SASAKI, H., Y. MORI, J. NAKAMURA AND J. SHIBASAKI, "SYNTHESIS AND ANTICONVULSANT ACTIVITY OF 1-ACYL-2-PYRROLIDINONE DERIVATIVES", J. MED. CHEM., 34(2), 628-633 (1991).
SANTODONATO,J.; DAVIS,L.N.; HOWARD,P.H.; SAXENA,J.; INVESTIGATION OF SELECTED POTENTIAL ENVIRONMENTAL CONTAMINANTS: MERCAPTOBENZOTHIAZOLES.; EPA-560/2-76-006. WASHINGTON,DC: U.S. EPA. PP.160.; 1976

SAUERBERG, P. ET AL. 1986. PHARMACOLOGICAL PROFILE OF A NOVEL CLASS OF MUSCARINIC ACETYLCHOLINE RECEPTOR AGONISTS. EUR. J. PHARMACOL. 130: 125-31.
SAX,N.I.; LEWIS,R.J.S.R.; HAWLEY'S CONDENSED CHEMICAL DICTIONARY 11TH ED; VANNOSTRAND REINHOLD CO.; N.Y.,N.Y.:P496; 1987

SCALA,A.J.; BANERJEE,S.; VAPOR PRESSURE INTERLABORATORY REPORT.; SYRACUSE,NY: SYRACUSE RESEARCH CORPORATION. FINAL REPORT FOR NATIONAL BUREAU OF STANDARDS. 8 PP. AND APPENDICES.; 1982

SCHELLENBERG,K.; LEUENBERGER,C.; SCHWARZENBACH,R.P.; SORPTION OF CHLORINATED PHENOLS BY NATURAL SEDIMENTS AND AQUIFER MATERIALS.; ENVIRON. SCI. TECHNOL.; 18:652-7.; 1984

SCHIESSL,H.W. HYDRAZINE AND ITS DERIVATIVES.; KIRK-OTHMER ENCYCL. CHEM. TECH. 3RD ED. NY,NY: WILEY-INTERSCIENCE. 12:734-71; 1980

SCHMIDT-BLEEK,F.; HABERLAND,W.; KLEIN,A.W.; CAROLI,S.; STEPS TOWARDS ENVIRONMENTAL HAZARD ASSESSMENT OF NEW CHEMICALS (INCLUDING A HAZARD RANKING SCHEME, BASED UPON DIRECTIVE 79/831/EEC).; CHEMOSPERE.; 11:383-415.; 1982

SCHOENE,K.; BOEHMER,W.; STEINHANSES,J.; DETERMINATION OF VAPOR PRESSURES DOWN TO 0.01 PA BY HEADSPACE GAS-CHROMATOGRAPHY.; FRESENIUS' Z. ANAL. CHEM.; 319:903-6.; 1984

SCHUEUERMAN,G.; DO HAMMETT CONSTANTS MODEL ELECTRONIC PROPERTIES IN QSARS?; SCI. TOTAL ENVIRON.; 109/110:221-35; 1991

SCHULTZ,T.W.; MOULTON,B.A.; STRUCTUR-ACTIVITY RELATIONSHIPS OF SELECTED PYRIDINES 1. SUBSTITUENT CONSTANT ANALYSIS; ECOTOX ENVIRON SAFT; 10:97-111; 1985

SCHULTZ,T.W.; RELATIVE TOXICITY OF PARA-SUBSTITUTED PHENOLS: LOG KOW AND PKA - DEPENDENT STRUCTURE - ACTIVITY RELATIONSHIPS.; BULL. ENVIRON. CONTAM. TOXICOL.; 38:994-9; 1987

SCHULTZ,T.W.; THE USE OF THE IONIZATION CONSTANT (PKA) IN SELECTING MODELS OF TOXICITY IN PHENOLS.; ECOTOX. ENVIRON. SAF.; 14:178-83; 1987A

SCHULTZ, T.W. AND F.M. APPLEHANS, "CORRELATIONS FOR THE ACUTE TOXICITY OF MULTIPLE NITROGEN-SUBSTITUTED AROMATIC MOLECULES", ECOTOXICOL. ENVIRON. SAF., 10(1), 75-85 (1985).

SCHULTZE,H.C.; ETHYLENE OXIDE; IN: KIRK-OTHMER ENCYCL. CHEM. TECHNOL. 2ND ED. NEW YORK, NY: WILEY-INTERSCIENCE. 8:523-58; 1965

SCHWABE,K.; LEGLER,C.; VAPOR PRESSURES, HEATS, AND ENTROPIES OF SUBLIMATION OF THE HEXACHLOROCYCLOHEXANES.; Z. ELEKTROCHEM.; 64:902-5.; 1960
SCHWABE,K.; LEGLER,C.; VAPOR PRESSURES, HEATS, AND ENTROPIES OF SUBLIMATION OF THE HEXACHLOROCYCLOHEXANES.; Z. ELEKTROCHEM.; 64:902-5.; 1960

SCHWARZ, K., L.A. PORTER AND A. FREDGA, "SOME REGULARITIES IN THE STRUCTURE-ACTIVITY FUNCTION RELATIONSHIP OF ORGANO-SELENIUM COMPOUNDS EFFECTIVE AGAINST DIETARY LIVER NECROSIS", ANN. N. Y. ACAD. SCI., 192, 200-214 (1972).

SCHWARZ,F.P.; WASIK,S.P.; A FLUORESCENCE METHOD FOR THE MEASUREMENT OF THE PARTITION COEFFICIENT OF NAPHTHALENE, 1-METHYL NAPHTHALENE AND 1-ETHYL NAPHTHALENE IN WATER; J CHEM ENGINEER DATA; 22:270-3; 1977

SCHWARZ,F.P.; DETERMINATION OF TEMPERATURE DEPENDENCE OF SOLUBILITIES OF POLYCYCLIC AROMATIC HYDROCARBONS IN AQUEOUS SOLUTION BY A FLUORESCENCE METHOD.; J. CHEM. ENG. DATA.; 22:273-7.; 1977

SCHWARZENBACH,R.P.; STIERLI,R.; FOLSOM,B.R.; ZEYER,J.; COMPOUND PROPERTIES RELEVANT FOR ASSESSING THE ENVIRONMENTAL PARTITIONING OF NITROPHENOLS.; ENVIRON. SCI. TECHNOL.; 22:83-92; 1988

SCHWEIZER,A.E.; FOWLKES,R.L.; MCMAKIN,J.H.; WHYTE,T.E. JR.; AMINES (LOWER ALIPHATIC).; IN: KIRK-OTHMER ENCYCL. CHEM. TECHNOL. 3RD ED. 2:272-83.; 1978

SCOW,K.; GOYER,M.; PAYNE,E.; PERWACK,J.; THOMAS,R.; WALLACE, .; WALKER,P.; WOOD,M.; AN EXPOSURE AND RISK ASSESSMENT FOR PENTACHLOROPHENOL.; EPA-440/4-81-021. (NTIS PB85-211944). CAMBRIDGE,MA: ARTHUR D. LITTLE INC. PP.195.; 1980

SEARS,G.W.; HOPKE,E.R.; VAPOR PRESSURE OF THE ISOMERIC TRICHLOROBENZENES IN THE LOW-PRESSURE REGION.; J. AM. CHEM. SOC.; 71:2575-6; 1949A

SEIBER,J.N.; WOODROW,J.E.; AIRBORNE RESIDUES AND HUMAN EXPOSURE.; IN: STUDIES IN ENVIRONMENTAL SCIENCE 24(DETERMINATION AND ASSESSMENT OF PESTICIDE EXPOSURE). SIEWIERSKI,M. EDS. PP.133-46; 1984

SEIDELL,A.; SOLUBITIES OF ORGANIC COMPOUNDS.; NEW YORK,NY: D. VAN NORSTRAND CO. INC.; 1941

SERJEANT,E.P.; DEMPSEY,B.; IONISATION CONSTANTS OF ORGANIC ACIDS IN AQUEOUS SOLUTION.; IUPAC CHEMICAL DATA SERIES NO.23. NEW YORK,NY: PERGAMON PRESS. PP.989.; 1979

SERRENTINO, R., L. CITTI, P.G. GERVASI AND G. TURCHI, "THE TYPE I BINDING OF SOME ALIPHATIC EPOXIDES TO CYTOCHROME P-450 AND THEIR INHIBITION OF MONO-OXYGENASE ACTIVITY", ITAL. J. BIOCHEM., 32(2), 111-123 (1983).

SERVOS,M.R.; MUIR,D.C.G.; WEBSTER,G.R.B.; ENVIRONMENTAL FATE OF POLYCHLORINATED DIBENZO-P-DIOXINS IN LAKE ENCLOSURES.; CAN. J. FISH AQUAT. SCI.; 49:722-34; 1992

SHAFER, W.E., "SORPTION OF THE CYTOKININ N6-BENZYLADENINE BY LEAF CUTICLES: PREDICTION FROM N-OCTANOL:WATER PARTITION COEFFICIENTS", PHYSIOL. PLANT., 78(1), 43-50 (1990).

SHAROM,M.S.; MILES,J.R.W.; HARRIS,C.R.; MCEWEN,F.L.; BEHAVIOR OF 12 INSECTICIDES IN SOIL AND AQUEOUS SUSPENSIONS OF SOIL AND SEDIMENT.; WATER RES.; 14:1095-100.; 1980A

SHAW,D.G.; HYDROCARBONS WITH WATER AND SEAWATER. PART II: HYDROCARBONS C8 TO C36.; INTERNATIONAL UNION OF PURE AND APPLIED CHEMISTRY. SOLUBILITY DATA SERIES. VOL 38, 561 PP.; 1989

SHEEHAN,R.J.; LANGER,S.H.; VAPOR PRESSURES OF FLUORINE- AND SILICON-CONTAINING DERIVATIVES OF SOME HYDROXYLIC COMPOUNDS.; J. CHEM. ENG. DATA.; 14:248-50.; 1969

SHEPPARD,C.S.; MAGELI,O.L.; PEROXIDES AND PEROXY COMPOUNDS, ORGANIC; IN: KIRK-OTHMER ENCYCLO CHEM TECHNOL 3RD NY,NY: WILEY INTERSCI 17:27-90; 1982
SHERMAN,P.D.; ALDEHYDES.; IN: KIRK-OTHMER ENCYCL. CHEM. TECH. 3RD. ED. 1:790-8.; 1978

SHEPPARD,C.S.; MAGELI,O.L.; PEROXIDES AND PEROXY COMPOUNDS, ORGANIC; IN:
KIRK-OTHMER ENCYCLO CHEM TECHNOL 3RD NY,NY: WILEY INTERSCI 17:27-90; 1982

SHEPARD, K.L., S.L. GRAHAM, R.J. HUDCOSKY, S.R. MICHELSON, T.H. SCHOLZ, H. SCHWAM, A.M. SMITH, J.M. SONDEY, K.M.
STROHMAIER, R.L. SMITH AND M.F. SUGRUE, "TOPICALLY ACTIVE CARBONIC ANHYDRASE INHIBITORS. 4.
[(HYDROXYALKYL)SULFONYL] BENZENE- AND [(HYDROXYALKYLSULFONYL]THIOPHENESULFONAMIDES", J. MED. CHEM.,
34(10), 3098-3105 (1991).

SHERBLOM, P.M., P.M. GSCHWEND AND R.P. EGANHOUSE, "AQUEOUS SOLUBILITIES, VAPOR PRESSURES AND
1-OCTANOL/WATER PARTITION COEFFICIENTS FOR C9-C14 LINEAR ALKYLBENZENES", J. CHEM. ENG. DATA, 37(4), 394-399
(1992).

SHERBLOM, P.M. AND R.P. EGANHOUSE, "CORRELATIONS BETWEEN OCTANOL-WATER PARTITION COEFFICIENTS AND
REVERSED-PHASE HIGH-PERFORMANCE LIQUID CHROMATOGRAPHY CAPACITY FACTORS", J. CHROMATOGR., 454, 37-50
(1988).

SHERMAN,P.D.; ALDEHYDES.; IN: KIRK-OTHMER ENCYCL. CHEM. TECH. 3RD. ED.
1:790-8.; 1978

SHIMIZU,T.; OHKUBO,S.; KIMURA,M.; TABATA,I.; HORI,T.; THE VAPOR PRESSURES AND
HEATS OF SUBLIMATION OF MODEL DISPERSE DYES.; J. SOC. DYERS. COLOUR.; 103:132-7;
1987

SHIU,W.Y.; MACKAY,D.; A CRITICAL REVIEW OF AQUEOUS SOLUBILITIES, VAPOR
PRESSURES, HENRY'S LAW CONSTANTS, AND OCTANOL-WATER PARTITION COEFFICIENTS OF
THE POLYCHLORINATED BIPHENYLS.; J. PHYS. CHEM. REF. DATA; 15:911-29; 1986

SHIU,W.Y.; DOUCETTE,W.; GOBAS,F.A.P.C.; ANDREN,A.; MACKAY,D.; PHYSICAL-CHEMICAL
PROPERTIES OF CHLORINATED DIBENZO-P-DIOXINS; ENVIRON SCI TECHNOL; 22:651-8; 1988

SHIU,W.Y.; MA,A.; MACKAY,D.; SEIBER,J.N.; WAUCHOPE,R.D.; SOLUBILITIES OF
PESTICIDE CHEMICALS IN WATER - PART II: DATA COMPILATION.; REV. ENVIRON. CONTAM.
TOXICOL.; 116:15-187; 1990

SHIU,W.Y.; MA,K.C.; VARHANICKOVA,D.; MACKAY,D.; CHLOROPHENOLS AND ALKYLPHENOLS:
A REVIEW AND CORRELATION OF ENVIRONMENTALLY RELEVANT PROPERTIES AND FATE IN AN
EVALUATIVE ENVIRONMENT.; CHEMOSPHERE.; 29:1155-224.; 1994
SIDDIQUI, M.A., J.S. DRISCOLL, V.E. MARQUEZ, J.S. ROTH, T. SHIRASAKA, H. MITSUYA, J.J. BARCHI AND J.A. KELLEY,
"CHEMISTRY AND ANTI-HIV PROPERTIES OF 2'-FLUORO-2',3'-DIDEOXYARABINOFURANOSYLPYRIMIDINES", J. MED. CHEM.,
35(12), 2195-2201 (1992).

SIJM,D.T.H.M.; WEVER,H.; OPPERHUIZEN,A.; INFLUENCE OF BIOTRANSFORMATION ON THE
ACCUMULATION OF PCDDS FROM FLY-ASH IN FISH.; CHEMOSPHERE; 19:475-80; 1989

SINGER, G.M., A.W. ANDREWS AND S.-M. GUO, "QUANTITATIVE STRUCTURE-ACTIVITY RELATIONSHIP OF THE
MUTAGENICITY OF SUBSTITUTED N-NITROSO-N-BENZYL-METHYLAMINES: POSSIBLE IMPLICATIONS FOR
CARCINOGENICITY", J. MED. CHEM., 29(1), 40-44 (1986).

SINGER,M.M.; TJEERDEMA,R.S.; FATE AND EFFECTS OF THE SURFACTANT SODIUM DODECYL
SULFATE.; IN: REVIEWS OF ENVIR. CONTAM. TOXICOL. SPRINGER-VERLAG,NY: VOL.133,
PP.95-149.; 1993

SMALL,M.J.; SOIL DETECTION LIMITS FOR POTENTIAL CHEMICAL WARFARE-RELATED
CONTAMINANTS AT FORT MCCLELLAN, ALABAMA; NTIS-AD-B077 091; TECH REPORT 8208; US
ARMY TOXIC HAZ MAT AGENCY; 1983

SMART,B.E.; FLUORINATED ALIPHATIC COMPOUNDS.; IN: KIRK-OTHMER ENCYCL. CHEM.
TECH. 3RD ED. 10:856-70.; 1980

SMILEY,R.A.; NITRILES.; IN: KIRK-OTHMER ENCYCL. CHEM. TECH. 3RD ED.
15:888-909.; 1981

SMITH,A.E.; DEGRADATION, FATE, AND PERSISTENCE OF PHENOXYALKANOIC ACID
HERBICIDES IN SOIL.; REV. WEED SCI.; 4:1-24; 1989A

SMITH,J.H.; MABEY,W.R.; BOHONUS,N.; HOLT,B.R.; LEE,S.S.; CHOU,T.W.;

VENBERGER,D.C.; MILL,T.; ENVIRONMENTAL PATHWAYS OF SELECTED CHEMICALS IN FRESHWATER SYSTEMS. PART II. LABORATORY STUDIES.; EPA-600/7-78-074. ATHENS,GA: U.S.EPA.; 1978

SMITH,M.A.; APPLEGATE,V.C.; JOHNSON,B.G.H.; PHYSICAL PROPERTIES OF SOME HALO-NITROPHENOLS.; J. CHEM. ENG. DATA.; 6:607-8.; 1961

SNIDER,J.R.; DAWSON,G.A.; TROPOSPHERIC LIGHT ALCOHOLS, CARBONYLS, AND ACETONITRILE CONCENTRATIONS IN THE SOUTHWESTERN UNITED STATES AND HENRY'S LAW DATA.; J. GEOPHYS. RES., D: ATMOS.; 90:3797-805; 1985

SNOW,AW;BARGER,WR; A CHEMICAL COMPARISON OF METHANESULFONYL FLUORIDE WITH ORGANOFLUOROPHOSPHORUS ESTER ANTICHOLINESTERASE COMPOUNDS; CHEM. RES. TOXICOL.; 1(6),379-84; 1988

SOKOLOV,V.B.; KARAPETYANTS,M.K.; STUNZHAS,P.A.; SOKOLOV,L.A.; TEMPERATURE DEPENDENCE OF THE SATURATED VAPOR PRESSURE OF SOME ORGANOHALOSILANES.; TR. KHIM. KHIM. TEKHNOL. PP.80-2.; 1975

SOMAN, G., J. NARAYANAN, B.L. MARTIN AND D.J. GRAVES, "USE OF SUBSTITUTED BENZYLIDINAMINOGUANIDINES IN THE STUDY OF GUANIDINO GROUP SPECIFIC ADP-RIBOSYLTRANSFERASE", BIOCHEMISTRY, 25(14), 4113-4119 (1986).

SOMMERIADE, R. ET AL.   1993.   ENVIRON. SCI. TECHNOL. 27: 2435-40.

SOSNOVSKY,G. ET AL.   1986.   IN THE SEARCH FOR NEW ANTICANCER DRUGS. 19. A PREDICTIVE DESIGN.   J. MED. CHEM. 29: 2225-30.

SOSNOVSKY,G AND N.U.M. RAO.   1985.   IN THE SEARCH FOR NEW ANTICANCER DRUGS. XV. VARIOUS ALIPHATIC AND AROMATIC HYDRAZONES.   CANCER LETTERS 29: 309-22.

SOSNOVSKY, G. AND N.U.M. RAO, "IN THE SEARCH FOR NEW ANTICANCER DRUGS. XXIII. EXPLORATION OF A PREDICTIVE DESIGN FOR ANTICANCER DRUGS OF CARBOHYDRATES CONTAINING N-NITROSOMETHYL AND N-NITROSOAMINOXYL COMPONENTS", J. PHARM. SCI., 80(7), 693-699 (1991).

SOTOMATSU, T., Y. NAKAGAWA AND T. FUJITA, "QUANTITATIVE STRUCTURE-ACTIVITY STUDIES OF BENZOYLPHENYLUREA LARVICIDES. IV. BENZOYL ORTHO SUBSTITUENT EFFECTS AND MOLECULAR CONFORMATION", PESTIC. BIOCHEM. PHYSIOL., 27(2), 156-164 (1987).

SOTOMATSU, T., M. SHIGEMURA, Y. MURATA AND T. FUJITA, "OCTANOL/WATER PARTITION COEFFICIENT OF ORTHO-SUBSTITUTED AROMATIC SOLUTES", J. PHARM. SCI., 82(8), 776-781 (1993).

SPARATORE, F., M.I. LAROTONDA, G. PAGLIETTI, E. RAMUNDO, C. SILIPOAND A. VITTORIA, "ACTIVITY OF BENZOTRIAZOLE DERIVATIVES IN PLANT GROWTH. PREPARATION, CHARACTERISTICS AND CORRELATION BETWEEN CHEMICAL PHYSICAL PROPERTIES AND STRUCTURE", FARMACO, ED. SCI., 33(12), 901-923 (1978).

STEIN, J ET AL.   1986.   SUBSTITUENT EFFECTS ON THE PARTITION OF 1-ACETYL-4-ARYL-SEMICARBAZIDES IN A 1-OCTANOL/WATER SYSTEM. Z. CHEM. 26: 258.

STOLWIJK, T.B. ET AL.   1989.   PARTITION COEFFICIENTS OF CROWN ETHERS.   RECL. TRAV. CHIM. PAYS-BAS 108: 103-8.

STURDIK, E., S. BALAZ, M. ANTALIK AND P. SULO, "LIPOPHILIC-HYDROPHILIC PROPERTIES AND RETENTION OF PHENYLHYDRAZONO PROPANENITRILES BY BIOLOGICAL SYSTEMS", COLL. CZECH. CHEM. COMMUN., 50(2), 538-550 (1985).

SUTO, M.J., M.A. STIER, L.M. WERBEL, C.M. ARUNDEL-SUTO, W.R. LEOPOLD, W.E. ELLIOTT AND J.S. SEBOLT-LEOPOLD, "A NEW CLASS OF ANALOGUES OF THE BIFUNCTIONAL RADIOSENSITIZER ALPHA-(1-AZIRIDINYLMETHYL)-2-NITRO-1H-IMIDAZOLE-1-ETHANOL (RSU 1069): THE CYCLOALKYLAZIRIDINES", J. MED. CHEM., 34(8), 2484-2488 (1991).
SOULIE,M.A.; BARES,D.; METZGER,J.; VAPOR PRESSURES OF SOME AZAAROMATIC COMPOUNDS.; C.R.HEBD. SEANCES ACAD. SCI., SER. C.; 281:341-2.; 1975

SOUTHWORTH,G.R.; KELLER,J.L.; HYDROPHOBIC SORPTION OF POLAR ORGANICS BY LOW ORGANIC CARBON SOILS.; WATER AIR SOIL POLL.; 28:239-48; 1986

SOUTHWORTH,G.R.; THE ROLE OF VOLATILIZATION IN REMOVING POLYCYCLIC AROMATIC HYDROCARBONS FROM AQUATIC ENVIRONMENTS.; BULL. ENVIRON. CONTAM. TOXICOL.; 21:507-14.; 1979A

SPANGGORD,R.J.; MILL,T.; CHOU,T.W.; MABEY,W.R.; SMITH,J.H.; LEE,S.; ENVIRONMENTAL FATE STUDIES ON CERTAIN MUNITIONS WASTEWATER CONSTITUENTS. FINAL REPORT, PHASE I - LITERATURE REVIEW.; SRI PROJECT NO. LSU-7934. CONTRACT NO. DAMD 17-78-C-8081. FORT DETRICK,MD: U.S. ARMY MEDICAL RES. & DEVELOP. COMMAND.; 1980

SPENCER,W.F.; CLIATH,M.M.; FACTORS AFFECTING VAPOR LOSS OF TRIFLURALIN FROM SOIL.; J. AGRIC. FOOD. CHEM.; 22:987-91.; 1974

SPENCER,W.F.; CLIATH,M.M.; MEASUREMENT OF PESTICIDE VAPOR PRESSURES.; RES. REV.; 85:57-71.; 1983

SPENCER,W.F.; SHOUP,T.D.; CLEATH,M.M.; FARMER,W.J.; HAQUE,R.; VAPOR PRESSURE AND RELATIVE VOLATILITY OF ETHYL AND METHYL PARATHION.; J. AGRIC. FOOD CHEM.; 27:273-8.; 1979

STEPHEN,H.; STEPHEN,T.; SOLUBILITIES OF INORGANIC AND ORGANIC COMPOUNDS.; IN: BINARY SYSTEMS. STEPHEN, H ET AL. EDS., NY, NEW YORK: 1:1-79,1604-43; 1963

STONE,A.T.; TORRENTS,A.; SMOLEN,J.; VASUDEVAN,D.; HADLEY,J.; ADSORPTION OF ORGANIC COMPOUNDS POSSESSING LIGAND DONOR GROUPS AT THE OXIDE/WATER INTERFACE.; ENVIRON. SCI. TECHNOL.; 27:895-909; 1993

STRANSKI,I.N.; KLIPPING,G.; BOGENSCHUETZ,A.F.; HEINRICH,H.J.; MAENNIG,H.; THERMAL DECOMPOSITION OF HEXAMETHYLENETETRAMINE.; ADVANCES IN CATALYSIS.; 9:406-14.; 1957

SUNTIO,L.R.; SHIU,W.Y.; MACKAY,D.; CRITICAL REVIEW OF HENRY'S LAW CONSTANTS FOR PESTICIDES; REV ENVIRON CONTAM TOX; 103:1-59; 1988

SUNTIO,L.R.; SHIU,W.Y.; MACKAY,D.; A REVIEW OF THE NATURE AND PROPERTIES OF CHEMICALS PRESENT IN PULP MILL EFFLUENTS; CHEMOSPHERE; 17:1249-90; 1988A

SURYANARAYANA,M.V.S.; PANDEY,K.S.; PRAKASH,S.; RAGHUVEERAN,C.D.; DANGI,R.S.; SWAMY,R.V.; RAO,K.M.; STRUCTURE-ACTIVITY RELATIONSHIP STUDIES WITH MOSQUITO REPELLENT AMIDES.; J. PHARM. SCI.; 80:1055-7.; 1991

SUTTON,C.; CALDER,J.A.; SOLUBILITY OF HIGHER-MOLECULAR-WEIGHT N-PARAFFINS IN DISTILLED WATER AND SEAWATER.; ENVIRON. SCI. TECHNOL.; 8:654-7.; 1974

SUWINSKI,J.; SALWINSKA.E.; WATRAS,J.; WIDEL,M.; NITROIMIDAZOLES. VI. PARTITION COEFFICIENTS AND TAUTOMERISM OF SIMPLE NITROIMIDAZOLES.; ACTA POL. PHARM.; 42:352-8; 1985

SUZUKI,T.; DEVELOPMENT OF AN AUTOMATIC ESTIMATION SYSTEM FOR BOTH THE PARTITION COEFFICIENT AND AQUEOUS SOLUBILITY.; J. COMPUTER-AIDED MOLECULAR DESIGN; 5:149-66; 1991

SWAIN,H.A. JR.; KWAN,C.Y.; SUNG,H.; MEASUREMENT OF VAPOR PRESSURES FROM 20 TO 30 DEGREE C OF LONG-CHAIN PEROXY ACIDS.; J. PHYS. CHEM.; 84:1347-9.; 1980

SWANN,R.L.; LASKOWSKI,D.A.; MCCALL,P.J.; VANDERKUY,K.; DISHBURGER,H.J. ; A RAPID METHOD FOR THE ESTIMATION OF THE ENVIRONMENTAL PARAMETERS OCTANOL, WATER PARTITION COEFFICIENT, SOIL SORPTION CONSTANT, WATER TO AIR RATIO, AND WATER SOLUBILITY.; RES. REV.; 85:17-28.; 1983

SYRACUSE RESEARCH CORP.; INFORMATION PROFILES ON POTENTIAL OCCUPATIONAL HAZARDS. VOLUME I. SINGLE CHEMICALS.; CONTRACT NO. 210-78-0019. PREPARED FOR NIOSH. SYRACUSE,NY: SYRACUSE RESEARCH CORP.; 1979

TAFT,R.W.; ABRAHAM,M.H.; DOUGHERTY,R.M.; KAMLET,M.J.; THE MOLECULAR PROPERTIES GOVERNING SOLUBILITIES ON NONELECTROLYTES IN WATER.; NATURE.; 313:384-6.; 1985

TAKACS-NOVAK, K., M. JOZAN, I. HERMECZ AND G. SZASZ, "LIPOPHILICITY ANTIBACTERIAL FLUOROQUINOLONES", INT. J. PHARM., 79(2-3) 89-96 (1992).

TAKAGI, T., A. KIMURA, Y.-Z. DA, H. NAKAI AND H. FUJIWARA, "PARTITION COEFFICIENTS IN A LIPOSOME/WATER SYSTEM. METHODS FOR DETERMINATION AND ERROR LIMITS", ANAL. SCI., 8(6), 761-765 (1992).

TAKAGISHI,T.; KATAYAMA,A.; KUROKI,N.; SOLUBILITIES OF AZOBENZENE DERIVATIVES IN WATER.; KOLLOID Z. Z. POLYM.; 232:693-9.; 1969

TAKAHASHI, J., O. KIRINO, C. TAKAYAMA AND K. KAMOSHITA, "STUDIES ON FUNGICIDAL ACTIVITY OF N-PHENYLCARBAMATES. IV. DETERMINATION OF THE HYDROPHOBICITY OF N-PHENYLCARBAMATES BY HIGH-PERFORMANCE LIQUID CHROMATOGRAPHY", J. CHROMATOGR., 436(2), 316-322 (1988).

TALIAN,S.F.; AMOS,C.K.JR.; KNIGHT,F.J.; REMEDIATING TOXIC CONTAMINATION OF A WATER TREATMENT PLANT.; IN: PROC. -AWWA WATER QUAL. TECHNOL. CONF. HARRISBURG,PA: GANNETT FLEMING WATER RESOURC. ENG. INC. PP.525-42; 1986

TANAKA, M., K. HORISAKA, C. YAMAGAMI, N. TAKAO AND T. FUJITA, "QUANTITATIVE STRUCTURE-ACTIVITY RELATIONSHIPS OF ANTICONVULSANT ARALKYL AND ALKYLCARBAMATES", CHEM. PHARM. BULL., 33(6), 2403-2410 (1985).

TANII, H. AND K. HASHIMOTO, "STRUCTURE-ACUTE TOXICITY RELATIONSHIP OF DINITRILES IN MICE", ARCH. TOXICOL., 57(2), 88-93 (1985).

TANII, H., H. TSUJI AND K. HASHIMOTO, "STRUCTURE-TOXICITY RELATIONSHIP ON MONOKETONES", TOXICOL. LETT., 30(1), 13-17 (1986).

TANII, H., S. SAITO AND K. HASHIMOTO, "STRUCTURE-ACTIVITY RELATIONSHIP OF ETHYLENEGLYCOL ETHERS", ARCH. TOXICOL., 66(5), 368-371 (1992).

TANII, H. AND K. HASHIMOTO, "STUDIES ON THE MECHANISM OF ACUTE TOXICITY OF NITRILES IN MICE", ARCH. TOXICOL., 55(1), 47-54 (1984).

TERADA, H., S. MURAOKA AND T. FUJITA, "STRUCTURE-ACTIVITY RELATIONSHIPS OF FENAMIC ACIDS", J. MED. CHEM., 17(3), 330-334 (1974).

TERADA, H., Y. KOSUGE, W. MURAYAMA, N. NAKAYA, Y. NUNOGAKI AND K.-I. NUNOGAKI, "CORRELATION OF HYDROPHOBIC PARAMETERS OF ORGANIC COMPOUNDS DETERMINED BY CENTRIFUGAL PARTITION CHROMATOGRAPHY WITH PARTITION COEFFICIENTS BETWEEN OCTANOL AND WATER", J. CHROMATOGR., 400, 343-351 (1987).

TERADA, H. ET AL. 1988. STRUCTURAL REQUIREMENTS OF SALICYLANIIDES FOR UNCOUPLING ACTIVITY IN MITOCHONDRIA: BIOCHIM. BIOPHYS. ACTA 936: 504-12.

TEWARI, Y.B., M.M. MILLER, S.P. WASIK AND D.E. MARTIRE, "AQUEOUS SOLUBILITY AND OCTANOL-WATER PARTITION COEFFICIENTS OF ORGANIC COMPOUNDS AT 298 K", J. CHEM. ENG. DATA, 27(4), 451-454 (1982).

TEWARI,Y.B.; MILLER,M.M.; WASIK,S.P.; CALCULATION OF AQUEOUS SOLUBILITY OF ORGANIC COMPOUNDS.; J. RES. NBS.; 87:155-8.; 1982

TEWARI,Y.B.; MILLER,M.M.; WASIK,S.P.; MARTINE,D.E.; AQUEOUS SOLUBILITY AND OCTANOL/WATER PARTITION COEFFICIENT OF ORGANIC COMPOUNDS AT 25.0 C.; J. CHEM. ENG. DATA.; 27:451-4.; 1982A

THINGVOLD,D.A.; LEE,G.F.; PERSISTENCE OF 3-(TRIFLUOROMETHYL)-4-NITROPHENOL IN AQUATIC ENVIRONMENTS.; ENVIRON. SCI. TECHNOL.; 15:1335-40.; 1981

TOMLIN, C. 1994. THE PESTICIDE MANUAL, TENTH EDITION, CROP PROTECTION PUBLICATIONS; BRITISH CROP PROTECTION COUNCIL, 49 DOWNING ST, FARNHAM, SURVEY GU9 7PH, UNITED KINGDOM.

TOMLINSON,E.; HAFKENSCHEID,T.L.; AQUEOUS SOLUBILITY AND PARTITION COEFFICIENT ESTIMATION FROM HPLC DATA.; IN: PARTITION COEFFICIENT; DETERMINATION AND ESTIMATION. DUNN,JIII ET AL. ED. ELMSFORD,NY: PERGAMON PRESS. PP.101-41; 1986

TOXICOLOGY DATA BASE.; NA--

TSAI, R.S. ET AL. 1991. STRUCTURE-LIPOPHILICITY RELATIONSHIPS OF ZWITTERIONIC AMINO ACIDS. J. CHEM. SOC., PERKIN TRANS. 2: 1797-802.

TSAI, R.S. ET AL. 1991. TOROIDAL COIL CENTRIFUGAL PARTITION CHROMATOGRAPHY, A METHOD FOR MEASURING

PARTITION COEFFICIENTS. J. CHROMATOGR. 538: 119-23.

TSCATS. TOXIC SUBSTANCES CONTROL ACT TEST SUBMISSIONS DATABASE. MANUFACTURER SUBMISSIONS (4, 8D, 8E, FYI) TO THE USEPA.

TURAN-ZITOUNI, G. AND G. BERGE, "DETERMINATION OF PARTITION COEFFICIENTS OF AMINOACETYL-2,3-DIHYDRO-3-BENZOFURANAMINE DERIVATIVES BY EXPERIMENTAL AND THEORETICAL METHODS", DOGA: TIP ECZACILIK, 10(2), 214-221 (1986).

TOWLE,P.H.; BALDWIN,R.H.; MEYER,D.H.; PHTHALIC ACIDS.; IN: KIRK-OTHMER ENCYCL. CHEM. TECH. 2ND ED. 15:444.; 1968

ULFVARSON,U.; ORGANIC MERCURIES.; IN: FUNGICIDES VOL. 2. TORGESON,DC ED. NEW YORK,NY ACADEMIC PRESS: PP.303-29; 1969

UNION CARBIDE; ALDEHYDES.; NEW YORK,NY: UNION CARBIDE CORP. PP.14; 1974

UNION CARBIDE; 1979-1980 CHEMICALS AND PLASTICS PHYSICAL PROPERTIES.; UNION CARBIDE CORP., NEW YORK. P. 17.; 1979

U.S. EPA; TREATABILITY MANUAL I. TREATABILITY DATA.; EPA-600/2-82-001A. WASHINGTON,DC: U.S. EPA.; 1981

US EPA; COMPILATION AND SPECIATION OF NATIONAL EMISSIONS FACTORS FOR CONSUMER/COMMERCIAL SOLVENT USE. INFORMATION COMPILED TO SUPPORT URBAN AIR TOXICS ASSESSMENT STUDIES.; U.S. EPA, OFF. AIR RADIAT., OFF. AIR QUAL. PLAN. STAND., RESEARCH TRIANGLE PARK, NC. EPA/450/2-89/008; PB89-207203; 1989A

US EPA; DRINKING WATER CRITERIA DOCUMENT FOR PHTHALIC ACID ESTERS (PAES).; US ENVIRONMENTAL PROTECTION AGENCY, ENVIRON. CRITER. ASSESS. OFF., CINC., OH. REPORT 1991, ECAO-CIN-D009; PB92-173442, 321 PP.; 1991S

VALSARAJ,K.T.; ON THE PHYSICO-CHEMICAL ASPECTS OF PARTITIONING OF NON-POLAR HYDROPHOBIC ORGANICS AT THE AIR-WATER INTERFACE; CHEMOSPHERE; 17:875-87; 1988

VALVANI,S.C.; YALKOWSKY,S.H.; ROSEMAN,T.J.; SOLUBILITY AND PARTITIONING. IV. AQUEOUS SOLUBILITY AND OCTANOL-WATER PARTITION COEFFICIENTS OF LIQUID NONELECTROLYTES.; J. PHARM. SCI.; 70:502-7.; 1981

VAN DAMME, M., M. HANOCQ AND L. MOLLE, "SPECIFIC ROLE OF SOLVENTS ON THE PARTITION OF SOME SUBSTITUTED BENZAMIDES BETWEEN OCTAN-1-OL AND WATER", PHARM. ACTA HELV., 59(8), 235-240 (1984).

VAN HAELST, A.G. ET AL. 1994. DETERMINATION OF N-OCTANOL/WATER PARTITION COEFFICIENTS OF TETRACHLOROBENZYLTOLUENES INDIVIDUALLY AND IN MIXTURE BY THE SLOW STIRRING METHOD. CHEMOSPHERE 29: 1651-60.

VARUSHCHENKO,R.M.; DRUZHININA,A.I.; GALCHENKO,G.L.; ENTHALPIES OF VAPORIZATION OF SOME BI- AND TRICYCLIC HYDROCARBONS WITH THREE-MEMBERED RINGS.; TR. KHIM. KHIM. TEKHNOL. PP.93-9.; 1974

VEITH,G.D.; DELORE,D.L.; BERGSTEDT,B.V.; MEASURING AND ESTIMATING THE BIOCONCENTRATION FACTOR OF CHEMICALS IN FISH.; J. FISH RES. BOARD CAN.; 36:1040-8.; 1979

VEITH,G.D.; MACEK,K.J.; PETROCELLI,S.R.;CARROLL,J.; AN EVALUATION OF USING PARTITION COEFFICIENTS AND WATER SOLUBILITY TO ESTIMATE BIOCONCENTRATION FACTORS FOR ORGANIC CHEMICALS IN FISH.; ASTM STP 707. AQUATIC TOXICOLOGY. EASTON,J.G. ET AL EDS. AMER. SOC . TEST. MATER. PP.116-29; 1980

VERA,A.; MONTES,M.; USERO,J.L.; CASADO,J.; QUANTITATIVE STRUCTURE-ACTIVITY RELATIONSHIP STUDY OF THE BIOPHYSICOCHEMICAL BEHAVIOR OF NITROSAMINE.; J. PHARM. SCI.; 81:791-96; 1992

VERSCHUEREN,K.; HANDBOOK OF ENVIRONMENTAL DATA ON ORGANIC CHEMICALS.; NEW YORK, NY: VAN NOSTRAND REINHOLD COMPANY.; 1977

VERSCHUEREN,K.; HANDBOOK OF ENVIRONMENTAL DATA ON ORGANIC CHEMICALS. 2ND ED.; NEW YORK,NY: VAN NOSTRAND REINHOLD CO. INC.; 1983

VITENBERG,A.G.; IOFFE,B.V.; ST. DIMITROVA,Z.; BUFAEVA,I.L.; DETERMINATION OF GAS-LIQUID PARTITION COEFFICIENTS BY MEANS OF GAS CHROMATOGRAPHIC ANALYSIS.; J. CHROMAT.; 112:319-27.; 1975

VLASOV,O.N.; GRANZHAN,V.A.; LIQUID-VAPOR EQUILIBRIUM IN THE SYSTEM DIMETHYL ESTER OF ADIPIC ACID-MONOMETHYL ESTER OF ADIPIC ACID.; ZH. PRIKL. KHIM.; 38:1167-9; 1965

VOITKEVICH,S.A.; SURFACE TENSION AND VAPOR PRESSURE OF LIQUID ORGANIC COMPOUNDS. I. HYDROCARBONS, ETHERS, AND ESTERS.; RUSS. J. PHYS. CHEM.; 37:540-4.; 1963

VOORMAN,R.; PENNER,D.; FATE OF MBOCA (4,4'-METHYLENE-BIS(2-CHLOROANILINE)) IN SOIL.; ARCH. ENVIRON. CONTAM. TOXICOL.; 15:595-602; 1986A

VP/WSOL; ESTIMATION METHOD FROM THE AVAILABLE VAPOR PRESSURE AND WATER SOLUBILITY VALUES

VP/HENRY'S LAW CONSTANT; ESTIMATION METHOD FROM THE AVAILABLE VAPOR PRESSURE AND HENRY'S LAW CONSTANT VALUES

WAKITA,K.; YOSHIMOTO,M.; MIYAMOTO,S.; WATANABE,H.; A METHOD FOR CALCULATIONS OF THE AQUEOUS SOLUBILITIES OF ORGANIC CONSTANTS; CHEM PHARM BULL; 34:4463-81; 1986

WALLINGTON,T.J.; KURYLO,M.J.; FLASH PHOTOLYSIS RESONANCE FLUORESCENCE INVESTIGATION OF THE GAS-PHASE REACTIONS OF HYDROXYL RADICALS WITH A SERIES OF ALIPHATIC KETONES OVER THE TEMPERATURE RANGE 240-440 K.; J. PHYS. CHEM.; 91:5050-4; 1987

WALSH,P.N.; SMITH,N.O.; SUBLIMATION PRESSURE OF ALPHA-P-DICHLORO-, BETA-P-DICHLORO-, P-DIBROMO-, AND P-BROMOCHLOROBENZENE.; CHEM. ENG. DATA.; 6:33-5.; 1961

WALTERS,R.W.; LUTHY,R.G.; LIQUID/SUSPENDED SOLID PHASE PARTITIONING OF POLYCYCLIC AROMATIC HYDROCARBON IN COAL COKING WASTE WATERS.; WATER RES.; 18:795-809; 1984A

WALTON,J.; VAPOR PRESSURES AND CRITICAL POINTS OF LIQUIDS. XI. HETEROCYCLIC NITROGEN COMPOUNDS.; ENG. SCI. DATA. ITEM 77019. PP.29.; 1977

WANG,L.; GOVIND,R.; DOBBS,R.A.; SORPTION OF TOXIC ORGANIC COMPOUNDS ON WASTEWATER SOLIDS: MECHANISM AND MODELING.; ENVIRON. SCI. TECHNOL.; 27:152-64; 1993

WANG, L., X. WANG, O. XU ANDL TIAN, "DETERMINATION OF THE N-OCTANOL/WATER PARTITION COEFFICIENTS OF POLYCYCLIC AROMATIC HYDROCARBONS AND ESTIMATION OF THEIR AQUEOUS SOLUBILITIES", HUANJING KEXUE XUEBAO, 6(4), 491-497 (1986).

WANG, W., L. WANG, L. TIAN, Z. ZHANG AND J. QIU, "PARTITION COEFFICIENTS OF THE INTERMEDIATE OF PYRETHROID IN N-OCTANOL/WATER SYSTEM", HUAZHONG GONGXUEYUAN XUEBAO, 15(1), 135-137 (1987).

WARNER, V.D., D.M. LYNCH AND R.S. AJEMIAN, "SYNTHESIS, PHYSICOCHEMICAL PARAMETERS AND IN VITRO EVALUATION OF N-P-CHLOROPHENYL-N-ALKYLBIGUANIDES", J. PHARM. SCI., 65(7), 1070-1072 (1976).

WARD,T.M.; WEBER,J.B.; AQUEOUS SOLUBILITY OF ALKYLAMINO-S-TRIAZINE AS A FUNCTION OF PH AND MOLECULAR STRUCTURE.; J. AGR. FOOD CHEM.; 16:959-61; 1968

WARNER,H.P.; COHEN,J.M.; IRELAND,J.C.; DETERMINATION OF HENRY'S LAW CONSTANTS OF SELECTED PRIORITY POLLUTANTS; EPA/600/D-87/229; NTIS PB87-212684; 1987

WARNER,M.J.; WEISS,R.F.; SOLUBILITIES OF CHLOROFLUOROCARBONS 11 AND 12 IN WATER AND SEAWATER.; DEEP-SEA RES., PART A; 32:1485-97; 1985

WASIK,S.P.; TSANG,W.; GAS CHROMATOGRAPHIC DETERMINATION OF PARTITION COEFFICIENTS OF SOME UNSATURATED HYDROCARBONS AND THEIR DEUTERATED ISOMERS IN

AQUEOUS SILVER NITRATE SOLUTIONS.; J. PHYS. CHEM.; 74:2970-6; 1970

WASIK,S.P.; TEWARI,Y.B.; MILLER,M.M.; MARTIRE,D.E.; OCTANOL/WATER PARTITION COEFFICIENTS AND AQUEOUS SOLUBILITIES OF ORGANIC COMPOUNDS.; NBSIR81-2406. WASHINGTON,DC: US DEPT. COMM. NATL. BUR. STD. PP.66.; 1981

WATANABE,T.; RELATIONSHIP BETWEEN VOLATILIZATION RATES AND PHYSICOCHEMICAL PROPERTIES OF SOME PESTICIDES.; J. PESTIC. SCI. (INT ED); 18:201-9.; 1993

WAUCHOPE,R.D.; GETZEN,F.W.; TEMPERATURE DEPENDENCE OF SOLUBILITIES IN WATER AND HEATS OF FUSION OF SOLID AROMATIC HYDROCARBONS.; J. CHEM. ENG. DATA.; 17:38-41.; 1972

WAUCHOPE,R.D.; YOUNG,J.R.; CHALFANT,R.B.; MARTI,L.R.; SUMNER,H.R.; DEPOSITION, MOBILITY AND PERSISTENCE OF SPRINKLER-IRRIGATION-APPLIED CHLORPYRIFOS ON CORN FOLIAGE AND IN SOIL.; PESTIC. SCI.; 32:235-243; 1991

WAUCHOPE, R.D., T.M. BUTTLER, A.G. HORNSBY, P.W.M. AUGUSTIN-BECKERS AND J.P. BURT. 1991. THE SCS/ARS/CES PESTICIDE PROPERTIES DATABASE FOR ENVIRONMENTAL DECISION-MAKING. REV. ENVIRON. CONTAM. TOXICOL. 123: 1-35.

WAUCHOPE,R.D.; BUTTLER,T.M.; HORNSBY,A.G.; AUGUSTIJN-BECKERS,P.W.M.; BURT,J.P.; THE SCS/ARS/CES PESTICIDE PROPERTIES DATABASE FOR ENVIRONMENTAL DECISION-MAKING; REV. ENVIRON. CONTAM. TOXICOL.; 123:1-36; 1991A

WAYAKU,M.; EQUATIONS BETWEEN VAPOR PRESSURE AND TEMPERATURE FOR CHEMICAL SUBSTANCES.IV.; AROMATIK.; 34:219-31.; 1982

WEAST,R.C.; HANDBOOK OF CHEMISTRY AND PHYSICS, 53RD ED.; CLEVELAND, OH: THE CHEMICAL RUBBER COMPANY.; 1972

WEAST,R.C.; ASTLE,M.J.; BEYER,W.H.; CRC HANDBOOK OF CHEMISTRY AND PHYSICS 66TH ED; CRC PRESS; BOCA RATON, FLA; 1985

WONG, O. AND R.H. MCKEOWN, "SUBSTITUENT EFFECTS ON PARTITION COEFFICIENTS OF BARBITURIC ACIDS", J. PHARM. SCI., 77(11), 926-932 (1988).

WSSA.; HERBICIDE HANDBOOK 6TH ED; WEED SCI SOC OF AM: CHAMPAIGNE, IL; 1989

WEBER,J.; DALEY,J.; OTHER POLYHYDRIC ALCOHOLS.; IN: KIRK-OTHMER ENCYCL. CHEM. TECH. 3RD ED. 1:778-89.; 1978

WEBER,J.B.; SWAIN,L.R.; SORPTION OF DINICONAZOLE AND METOLACHLOR BY FOUR SOILS, CALCIUM-ORGANIC MATTER, AND CALCIUM-MONTMORILLONITE.; SOIL SCI.; 156:171-7; 1993

WEBER,J.B.; WHITACRE,D.M.; MOBILITY OF HERBICIDES IN SOIL COLUMNS UNDER SATURATED- AND UNSATURATED-FLOW CONDITIONS.; WEED SCI.; 30:579-84; 1982

WEBER,J.B.; MECHANISMS OF ADSORPTION OF S-TRIAZINES BY CLAY COLLOIDS AND FACTORS AFFECTING PLANT AVAILABILITY.; RES. REV.; 32:93-130.; 1970

WEBER,J.B.; INTERACTION OF ORGANIC PESTICIDES WITH PARTICULATE MATTER IN AQUATIC AND SOIL SYSTEMS.; ADVAN. CHEM. SER.; 111:55-120.; 1972

WEBER,J.B.; ADSORPTION OF BUTHIDAZOLE, VEL, 3510, TEBUTHIURON, AND FLURIDONE BY ORGANIC MATTER, MONTMORILLONITE CLAY, EXCHANGE RESINS, AND A SANDY LOAM SOIL.; WEED SCI.; 28:478-83; 1980

WEBER,J.B.; BEHAVIOR OF DINTROANILINE HERBICIDES IN SOILS.; WEED TECHNOL.; 4:394-406; 1990

WEBER,J.B.; PROPERTIES AND BEHAVIOR OF PESTICIDES IN SOIL.; IN: MECHANISMS OF PESTICIDE MOVEMENT INTO GROUND WATER. HONEYCUTT,RC & SCHADEBACKER,DJ, EDS. ANN ARBOR,MI: LEWIS PUBL., CRC PRESS, INC. PP. 15-41.; 1994

WEBER,R.C.; PARKER,P.A.; BOWSER,M.; VAPOR PRESSURE DISTRIBUTION OF SELECTED
ORGANIC CHEMICALS.; EPA-600/2-81-021. CINCINNATI,OH: U.S.EPA. PP.39.; 1981

WEBSTER,G.R.B.; FRIESEN,K.L.; SARNA,L.P.; MUIR,D.C.G.; ENVIRONMENTAL FATE
MODELLING OF CHLORODIOXINS: DETERMINATION OF PHYSICAL CONSTANTS.; CHEMOSPHERE;
14:609-22; 1985

WEIL,L.; DURE,G.; QUENTIN,K.E.; SOLUBILITY IN WATER OF INSECTICIDE CHLORINATED
HYDROCARBONS AND POLYCHLORINATED BIPHENYLS IN VIEW OF WATER POLLUTION.; Z.
WASSER ABWASSER FORSCH.; 7:169-75.; 1974

WERNER,A.C.; VAPOR PRESSURES OF PHTHALATE ESTERS.; IND. ENG. CHEM.; 44:2736-40.;
1952

WERNER,A.F.; TAULLI,T.A.; MICHAEL,P.R.; WILLIAMS,M.A.; ESTIMATION AND
VERIFICATION OF THE ENVIRONMENTAL FATE OF O-BENZYL-P-CHLOROPHENOL.; ARCH.
ENVIRON. CONTAM. TOXICOL.; 12:569-575.; 1983

WHELPTON, R., "IONIZATION CONSTANTS, OCTANOL PARTITION COEFFICIENTS AND CHOLINESTERASE INHIBITOR
CONSTANTS FOR CHLORPROMAZINE AND ITS METABOLITES", J. PHARM. PHARMACOL., 41(12) , 856-858 (1989).

WIEDEMANN,H.G.; APPLICATIONS OF THERMOGRAVIMETRY FOR VAPOR PRESSURE
DETERMINATION.; THERMOCHIMICA ACTA.; 3:355-66.; 1972

WILHOIT,R.C.; ZWOLINSKI,B.J.; PHYSICAL AND THERMODYNAMIC PROPERTIES OF ALIPHATIC
ALCOHOL.; J. PHYS. CHEM. REF. DATA.; 2(SUPPL. 1):1-420.; 1973

WILLIS,G.H.; MCDOWELL,L.L.; PESTICIDES IN AGRICULTURAL RUNOFF AND THEIR EFFECTS
ON DOWNSTREAM WATER QUALITY.; ENVIRON. TOX. CHEM.; 1:267-79.; 1982

WINDHOLZ,R ED. MERCK INDEX.; THE MERCK INDEX. AN ENCYCLOPEDIA OF CHEMICALS AND DRUGS 10TH ED.; RAHWAY,
N.J.: MERCK AND CO., INC.; 1983

WITTEKIND, H.H., B. TESTA AND J. ESTREICHER, "THE REVERSIBLE PROXIBARBAL-VALOFAN ISOMERIZATION. II. KINETIC
STUDIES IN A BIPHASIC OCTANOL/WATER SYSTEM", HELV. CHIM. ACTA, 71(5), 1228-1234 (1988).

WOLF,K.; GILBERT,P.A.; EDTA-ETHYLENE DIAMINETETRAACETIC ACID.; IN: THE HANDBOOK
OF ENVIRONMENTAL CHEMISTRY, VOL. 3, PART F, HUTZINGER,O ED., SPRINGER-VERLAG:
HEIDELBERG, BERLIN, PP. 243-59.; 1992

WOLFE,N.L.; ORGANOPHOSPHATE AND ORGANOPHOSPHOROTHIONATE ESTERS: APPLICATON OF
LINEAR FREE ENERGY RELATIONSHIPS TO ESTIMATE HYDROLYSIS RATE CONSTANTS FOR USE
IN ENVIRONMENTAL FATE ASSESSMENT.; CHEMOSPHERE.; 9:571-9.; 1980

WOLFE,N.L.; STEEN,W.C.; BURNS,L.A.; PHTHALATE ESTER HYDROLYSIS: LINEAR FREE
ENERGY RELATIONSHIPS.; CHEMOSPHERE.; 9:403-8.; 1980B

WOLFE,N.L.; ZEPP,R.G.; SCHLOTZHAUER,P.; SINK,M.; TRANSFORMATION PATHWAYS OF
HEXACHLOROCYCLOPENTADIENE IN THE AQUATIC ENVIRONMENT.; CHEMOSPHERE.; 11:91-101.;
1982

WOLFENDEN,R.; WILLIAMS,R.; AFFINITIES OF PHOSPHORIC ACIDS, ESTERS, AND AMIDES
FOR SOLVENT WATER.; J. AMER. CHEM. SOC.; 105:1028-31.; 1983

WONG,O AND RH MCKEOWN, "SUBSTITUENTEFFECTS ON PARTITION COEFFICIENTS OF BARBITURIC ACIDS", J. PHARM.
SCI., 77 (11), 926-932, 1988.

WOODROW,J.E.; MCCHESNEY,M.M.; SEIBER,J.N.; MODELING THE VOLATILIZATION OF
PESTICIDES AND THEIR DISTRIBUTION IN THE ATMOSPHERE.; IN: LONG RANGE TRANSP.
PESTIC. PP. 61-81.; 1990

WORTHING,C.R.; WALKER,S.B.; PESTICIDE MANUAL 7TH ED; BRITISH CROP PROTECTION
COUNCIL. LAVENHAM SUFFOLK, ENGLAND: LAVENHAM PRESS LTD; 1983

WORTHING,C.R.; WALKER,S.B.; THE PESTICIDE MANUAL: A WORLD COMPENDIUM - 8TH ED.; THE BRITISH CROP PROTECTION COUNCIL, 20 BRIDPORT ROAD, THORNTON HEATH CR4 7QG, UK.; 1987

WRIGHT,C.G.; LEIDY,R.B.; DUPREE,H.E.,JR.; INSECTICIDES IN THE AMBIENT AIR OF ROOMS FOLLOWING THEIR APPLICATION FOR CONTROL OF PESTS.; BULL. ENVIRON. CONTAM. TOXICOL.; 26:548-53.; 1981

WSSA.; HERBICIDE HANDBOOK 5TH ED; WEED SCI SOC OF AM: CHAMPAIGNE, IL; 1983

WSSA.; HERBICIDE HANDBOOK 6TH ED; WEED SCI SOC OF AM: CHAMPAIGNE, IL; 1989

XIE,T.M.; HULTHE,B.; FOLESTAD,S.; DETERMINATION OF PARTITION COEFFICIENTS OF CHLORINATED PHENOLS, GUAIACOLS AND CATECHOLS BY SHAKE-FLASK GC AND HPLC.; CHEMOSPHERE; 13:445-59; 1984

YALKOWSKY,S.H.; DANNENFELSER,R.M.; AQUASOL DATABASE OF AQUEOUS SOLUBILITY. VERSION 5.; COLLEGE OF PHARMACY, UNIVERSITY OF ARIZONA - TUCSON,AZ. PC VERSION.; 1992

YALKOWSKY,S.H.; VALVANI,S.C.; SOLUBILITY AND PARTITIONING I: SOLUBILITY OF NONELECTROLYTES IN WATER.; J. PHARM. SCI.; 69:912-22.; 1980

YALKOWSKY,S.H.; VALVANI,S.C.L.; MACKAY,D.; ESTIMATION OF THE AQUEOUS SOLUBILITY OF SOME AROMATIC COMPOUNDS.; RESIDUE REVIEWS.; 85:43-55.; 1983

YAMADA, T., Y. TSUKAMOTO, H. SHIMAMURA, S. BANNO AND M. SATO, "PYRIDAZINONES. IV. SYNTHESIS, ANTISECRETORY AND ANTIULCER ACTIVITIES OF UREA DERIVATIVES", EUR. J. MED. CHEM., 18(3), 209-214 (1983).

YAMAGAMI, C., H. TAKAMI, K. YAMAMOTO, K. MIYOSHI AND N. TAKAO, "HYDROPHOBIC PROPERTIES OF ANTICONVULSANT PHENYLACETANILIDES. RELATIONSHIP BETWEEN OCTANOL-WATER PARTITION COEFFICIENT AND CAPACITY FACTOR DETRMINED BY REVERSED-PHASE LIQUID CHROMATOGRAPHY", CHEM. PHARM. BULL., 32(12), 4994-5002 (1984).

YAMAGAMI, C., T. OGURA AND N. TAKAO. 1990. HYDROPHOBICITY PARAMETERS DETERMINED BY REVERSED-PHASE LIQUID CHROMATOGRAPHY. I. RELATIONSHIPS BETWEEN CAPACITY FACTORS AND OCTANOL-WATER PARTITION COEFFICIENTS FOR MONOSUBSTITUTED PYRAZINES AND THE RELATED PYRIDINES. J. CHROMAT. 514: 123-36.

YAMAGAMI, C., N. TAKAO AND T. FUJITA, "HYDROPHOBICITY PARAMETER OF DIAZINES. 1. ANALYSIS AND PREDICTION OF PARTITION COEFFICIENTS OF MONOSUBSTITUTED DIAZINES", QUANT. STRUCT.-ACT. RELAT., 9(4), 313-320 (1990).

YAMAGAMI,C.; OGURA,T.; TAKAO,N.; HYDROPHOBICITY PARAMETERS DETERMINED BY REVERSED-PHASE LIQUID CHROMATOGRAPHY. I. RELATIONSHIP BETWEEN CAPACITY FACTORS AND OCTANOL-WATER PARTITION COEFFICIENTS FOR MONOSUBSTITUTED PYRAZINES AND THE RELATED PYRIDINES.; J. CHROMATOGR.; 514:123-36; 1990A

YAMAGAMI, C. ET AL. 1991. HYDROPHOBICITY PARAMETERS DETERMINED BY REVERSED-PHASE LIQUID CHROMATOGRAPHY. III. CHEM. PHARM. BULL. 39: 2924-9.

YAMAGAMI, C., N. TAKAO AND T. FUJITA, "HYDROPHOBICITY PARAMETER OF DIAZINES. II. ANALYSIS AND PREDICTION OF PARTITION COEFFICIENTS OF DISUBSTITUTED PYRAZINES", J. PHARM. SCI., 80(8), 772-777 (1991).

YAMAGAMI, C. AND N. TAKAO, "HYDROPHOBICITY PARAMETERS DETERMINED BY REVERSED-PHASE LIQUID CHROMATOGRAPHY. V. RELATIONSHIP BETWEEN THE CAPACITY FACTOR AND THE OCTANOL-WATER PARTITION COEFFICIENT FOR SIMPLE HETEROAROMATIC COMPOUNDS AND THEIR ESTER DERIVATIVES", CHEM. PHARM. BULL., 40(4), 925-929 (1992).

YAMAGAMI, C., M. YOKOTA AND N. TAKAO, "HYDROPHOBICITY PARAMETERS DETERMINED BY REVERSED-PHASE LIQUID CHROMATOGRAPHY", J. CHROMATOGR. A, 662(1), 49-60 (1994).

YAMAMOTO, I., T. KIMURA, Y. TATEOKA, K. WATANABE AND I.K. HO, "N-SUBSTITUTED OXOPYRIMIDINES AND NUCLEOSIDES; STRUCTURE-ACTIVITY RELATIONSHIP FOR HYPNOTIC ACTIVITY AS CENTRAL NERVOUS SYSTEM DEPRESSANT", J. MED. CHEM., 30(12), 2227-2231 (1987).

YAMAZAKI, M., T. SUZUKA, Y. ITO, S. ITOH, M. KITAMURA, K. OHASHI, Y. TAKEDA, A. KAMADA, Y. ORITAAND H. NAKAHAMA, "BIOPHARMACEUTICAL STUDIES OF THIAZIDE DIURETICS. I. DETERMINATION OF PKA VALUES AND

PARTITION COEFFICIENTS OF THIAZIDE DIURETICS", CHEM. PHARM. BULL., 32(6), 2380-2386 (1984).

YANAGISAWA, T. ET AL. 1988. SYNTHESIS AND ANTIULCER ACTIVITIES OF SODIUM ALKYLAZULENESULFONATES. CHEM. PHARM. BULL. 36: 641-7.

YANG, D.J., E.P. LAHODA, P.I. BROWN AND G.O. RANKIN, "STRUCTURE-NEPHROTOXICITY RELATIONSHIPS FOR P-SUBSTITUTED N-PHENYLSUCCINIMIDES IN SPRAGUE-DAWLEY AND FISCHER 344 RATS", TOXICOLOGY, 36(1), 23-35 (1985).

YANG, H.Z., M. NISHIMURA, K. NISHIMURA, S. KURODA AND T. FUJITA, "QUANTITATIVE STRUCTURE-ACTIVITY STUDIES OF PYRETHROIDS. 12. PHYSICOCHEMICAL SUBSTITUENT EFFECTS OF META-PHENOXYBENZYL ALPHA,ALPHA-DISUBSTITUTED ACETATE ON INSECTICIDAL ACTIVITY", PESTIC. BIOCHEM. PHYSIOL., 29(3), 217-232 (1987).

YANG,R.S.H.; PROPYLENE CHLOROHYDRINS: TOXICOLOGY, METABOLISM AND ENVIRONMENTAL FATE.; REV. ENVIRON. CONTAM. TOXICOL.; 99:46-81; 1987

YAWS,C.; YANG,H.C.; PAN,X.; HENRY'S LAW CONSTANT FOR 362 ORGANIC COMPOUNDS IN WATER.; CHEM. ENGIN.; NOVEMBER 1991, PP. 179-85; 1991

YAWS,C.L.; YANG,H.C.; WATER SOLUBILITY DATA FOR ORGANIC COMPOUNDS.; POLLUT. ENG.; 22:70-5; 1990

YAWS,C.L.; IN: HANDBOOK OF VAPOR PRESSURE. VOLUME 1 - C1 TO C4 COMPOUNDS. GULF PUBLISHING CO.: HOUSTON, TX, 347 PP.; 1994

YAWS,C.L.; HANDBOOK OF VAPOR PRESSURE. VOLUME 2 - C5 TO C7 COMPOUNDS. GULF PUBLISHING CO.: HOUSTON, TX, 391 PP.; 1994A

YAWS,C.L.; HANDBOOK OF VAPOR PRESSURE. VOLUME 3 - C8 TO C28 COMPOUNDS. GULF PUBLISHING CO.: HOUSTON, TX, 387 PP.; 1994B

YAWS,C.L.; YANG,H.C.; HOPPIER,J.R.; HANSEN,K.C.; ORGANIC CHEMICALS: WATER SOLUBILITY DATA.; CHEM. ENGN.; 97:115-8; 1990A

YEN,C.; PERENICH,T.A.; BAUGHMAN,G.L.; FATE OF DYES IN AQUATIC SYSTEMS. II. SOLUBILITY AND OCTANOL/WATER PARTITION COEFFICIENTS OF DISPERSE DYES.; ENVIRON. TOXICOL. CHEM.; 8:981-6; 1989

YEN,C.P.C.; PERENICH,T.A.; BAUGHMAN,G.L.; FATES OF DYES IN AQUATIC SYSTEMS II. SOLUBILITY AND OCTANOL/WATER PARTITION COEFFICIENTS OF DISPERSE DYES.; ENVIRON. TOXICOL. CHEM.; 8:981-6; 1989

YIH, T.D. AND J.M. VAN ROSSUM, "KS VALUES OF SOME HOMOLOGOUS SERIES BARBITURATES AND THE RELATIONSHIP WITH THE LIPOPHILICITY AND METABOLIC CLEARANCE", BIOCHEM. PHARMACOL., 26(22), 2117-2120 (1977).

YIN,C.; HASSETT,J.P.; GAS-PARTITIONING APPROACH FOR LABORATORY AND FIELD STUDIES OF MIREX FUGACITY IN WATER.; ENVIRON. SCI. TECHNOL.; 20:1213-7; 1986

YON,D.; OSBORNE, K.; MCGIBBON, A; BALOCH, R.; LACEY, R.; THE ENVIRONMENTAL DISTRIBUTION OF HEXAFLUMURON.; IN: BRIGHTON CROP PROT. CONF. - PESTS DIS. 1992, (2). PP. 907-12.; 1992

YOSHIDA,T.; L-MONSODUIM GLUTAMATE (MSG).; IN: KIRK-OTHMER ENCYCL. CHEM. TECH. 3RD ED. 2:410-21.; 1978

YOUNG, R.C., C.R. GANELLIN, R. GRIFFITHS, R.C. MITCHELL, M.E. PARSONS, D. SAUNDERS AND N.E. SORE, "AN APPROACH TO THE DESIGN OF BRAIN-PENETRATING HISTAMINERGIC AGONISTS", EUR. J. MED. CHEM., 28(3), 201-211 (1993).

ZWOLINSKI,B.J.; WILHOIT,R.C.; HANDBOOK OF VAPOR PRESSURES AND HEATS OF VAPORIZATON OF HYDROCARBONS AND RELATED COMPOUNDS.; API44-TRC101. COLLEGE STATION,TX: THERMODYNAMCS RESEARCH CENTER.; 1971